BERGEY'S MANUAL OF
Systematic Bacteriology
Volume 1

BERGEY'S MANUAL OF
Systematic
Bacteriology

Volume 1

NOEL R. KRIEG
EDITOR, VOLUME 1

JOHN G. HOLT
EDITOR-IN-CHIEF

EDITORIAL BOARD

R. G. E. Murray, Chairman, **Don J. Brenner**
Marvin P. Bryant, John G. Holt, Noel R. Krieg
James W. Moulder, Norbert Pfennig
Peter H. A. Sneath, James T. Staley

WITH CONTRIBUTIONS FROM
124 COLLEAGUES

WILLIAMS & WILKINS
Baltimore/London

Editor: Barbara Tansill
Associate Editor: Carol-Lynn Brown
Copy Editor: Caral S. Nolley
Design: Joanne Janowiak
Illustration Planning: Lorraine Wrozsek
Production: Raymond E. Reter

Library of Congress Cataloging in Publication Data

Main entry under title:

Bergey's manual of systematic bacteriology.

  Based on: Bergey's manual of determinative bacteriology.
  Includes index.
  1. Bacteriology—Classification. I. Bergey, D. H. (David Hendricks), 1860-1937. II.
Holt, John G. III. Krieg, Noel R. IV. Bergey's manual of determinative bacteriology.
[DNLM: 1. Bacteriology—Terminology. 2. Bacteria—Classification. QW 4 B832m]
QR81.B46 1983   589.9′0012   82-21760
ISBN 0-683-04108-8 (v. 1)

Composed and printed in the United States of America

Contributors

Linda Baumann
Department of Bacteriology, University of California,
Davis, California 95616 USA

Paul Baumann
Department of Bacteriology, University of California,
Davis, California 95616 USA

Jan-Hendrik Becking
Institute for Atomic Sciences in Agriculture,
6 Keyenbergseweg, Postbus 48, Wageningen,
The Netherlands

Herve Bercovier
Department of Clinical Microbiology, The Hebrew University,
Hadassah Medical School, P.O. Box 1172, Jerusalem, Israel

Ernst L. Biberstein
Department of Veterinary Medicine, University of California,
Davis, California 95616 USA

Reneé Borrall
Department of Microbiology, Louisiana State University,
Baton Rouge, Louisiana 70803 USA

Kjell Bøvre
Kaptein W. Wilhelmsen og Frues Bakteriologiske Institutt,
University of Oslo, Rikshospitalet, Universitetet I Oslo, Oslo
1, Norway

J. F. Bradbury
Commonwealth Mycological Institute, Ferry Lane, Kew,
Surrey TW9 3AF, England

Don J. Brenner
Molecular Biology Laboratory, Biotechnology Branch,
Division of Bacterial Diseases, Center for Infectious Diseases,
Centers for Disease Control, 1600 Clifton Road, N.E.,
Atlanta, Georgia 30333 USA

John A. Breznak
Department of Microbiology and Public Health, Michigan
State University, East Lansing, Michigan 48824 USA

W. J. Brinley-Morgan
Central Veterinary Laboratory, New Haw, Weybridge, Surrey
KT15 3NB, England

Thomas D. Brock
Department of Bacteriology, 1550 Linden Drive, University of
Wisconsin, Madison, Wisconsin 53706 USA

George H. Brownell
Department of Cell and Molecular Biology, Medical College of
Georgia, Augusta, Georgia 30902 USA

Marvin P. Bryant
Department of Dairy Science, University of Illinois,
Urbana, Illinois 61801 USA

Jeffrey C. Burnham
Department of Microbiology, Medical College of Ohio, C.S.
#10008, Toledo, Ohio 43699 USA

E. Canale-Parola
Department of Microbiology, University of Massachusetts,
Amherst, Massachusetts 01003 USA

G. R. Carter
Division of Pathobiology and Public Practice, Virginia-
Maryland Regional College of Veterinary Medicine, Virginia
Polytechnic Institute and State University, Blacksburg,
Virginia 24062 USA

Kwang-Poo Chang
The Rockefeller University, 1230 York Avenue, New York,
New York 10021 USA

Samuel F. Conti
A217 Graduate Research Center, University of
Massachusetts, Amherst, Massachusetts 01003 USA

M. J. Corbel
Central Veterinary Laboratory, New Haw, Weybridge, Surrey
KT15 3NB, England

Gregory A. Dasch
Rickettsial Diseases Branch, Mail Stop 32, Naval Medical
Research Institute, Bethesda, Maryland 20814 USA

Jozef De Ley
Laboratorium voor Microbiologie, K. L. Ledeganckstraat 35,
9000 Gent, Belgium

Robert S. Dickey
Department of Plant Pathology, Cornell University, Ithaca,
New York 14853 USA

Robert B. Dienst
Department of Cell and Molecular Biology, Medical College of
Georgia, Augusta, Georgia 30902 USA

Johanna Döbereiner
EMBRAPA-SNLCS-PFN, Km 47, 23460 Seropédica,
Rio de Janeiro, Brazil

Henry T. Eigelsbach
Western Maryland College, Westminster, Maryland 21157
USA

S. Faine
Department of Microbiology, Monash University, Clayton,
Victoria 3168, Australia

John J. Farmer, III
Enteric Bacteriology Section, Center for Infectious Diseases,
Centers for Disease Control, Atlanta, Georgia 30333 USA

James C. Feeley
Field Investigations Laboratory Section, Respiratory and
Special Pathogens Branch, Centers for Disease Control,
Atlanta, Georgia 30333 USA

E. A. Freundt
FAO/WHO Collaborating Centre for Animal Mycoplasmas,
Institute of Medical Microbiology, Bartholin Building,
University of Aarhus, DK-8000 Aarhus C., Denmark

A. L. Furniss
Public Health Laboratory, Preston Hall Hospital, Maidstone, Kent ME20 7NH, England

Michel J. Gauthier
Institut National de la Santé et de la Recherche Médicale, Unite n°40, 1, Avenue Jean Lorrain, 06300 Nice, France

Monique Gillis
Laboratorium voor Microbiologie, K.L. Ledeganckstraat 35, 9000 Gent, Belgium

Y. Goodman
Department of Medical Bacteriology, Medical Sciences Building, University of Alberta, Edmonton, Alberta, Canada

Francis Gosselé
Laboratorium voor Microbiologie, K.L. Ledeganckstraat 35, 9000 Gent, Belgium

Rainer Gothe
Institut fur Parasitologie, Rudolf-Buchheim-Strasse 2, D-6300 Giessen, Federal Republic of Germany

R. Nigel Gourlay
ARC Institute for Research on Animal Diseases, Compton, near Newbury, Berkshire RG16 0NN, England

J. R. Greenwood
Orange County Public Health Laboratory, 1729 W. 17th St., Santa Ana, California 92706, USA

Francine Grimont
Entérobactéries, Institut Pasteur, 28, Rue du Docteur Roux, 75724 Paris Cedex 15, France

Patrick A. D. Grimont
Entérobactéries, Institut Pasteur, 28, Rue du Docteur Roux, 75724 Paris Cedex 15, France

Roger J. Gross
Division of Enteric Pathogens, Central Public Health Laboratory, Colindale Avenue, London NW9 5HT, England

Thomas P. Hatch
Department of Microbiology & Immunology, University of Tennessee, Center for Health Sciences, Memphis, Tennessee 38163 USA

Robert B. Hespell
Department of Dairy Science, University of Illinois, Urbana, Illinois 61801 USA

Peter Hirsch
Institut für Allgemeine Mikrobiologie, Olshausenstrasse 40/60 Biologiezentrum, 2300 Kiel 1, Federal Republic of Germany

Tor Hofstad
Department of Microbiology and Immunology, The Gade Institute, University of Bergen, N-5000 Bergen, Norway

Lillian V. Holdeman
Department of Anaerobic Microbiology, Virginia Polytechnic Institute and State University, Blacksburg, Virginia 24061 USA

Barry Holmes
National Collection of Type Cultures, Central Public Health Laboratory, Colindale Avenue, London NW9 5HT, England

John G. Holt
Department of Microbiology, 205 Sciences I, Iowa State University, Ames, Iowa 50011 USA

David L. Huxsoll
Military Disease Hazards, Department of the Army, U.S. Army Medical Research and Development Command, Ft. Detrick, Frederick, Maryland 21701 USA

Roar L. Irgens
Department of Biology, Southwest Missouri State University, Springfield, Missouri 65802 USA

F. L. Jackson
Department of Medical Bacteriology, Medical Sciences Building, University of Alberta, Edmonton, Alberta, Canada

John L. Johnson
Department of Anaerobic Microbiology, Virginia Polytechnic Institute and State University, Blacksburg, Virginia 24061 USA

Russell C. Johnson
Department of Microbiology, University of Minnesota Medical School, 1060 Mayo Memorial, Box 196, Minneapolis, Minnesota 55455 USA

Dorothy Jones
Department of Microbiology, School of Medicine and School of Biological Sciences, University of Leicester, University Road, Leicester LE1 7RH, England

D. Carlyle Jordan
Department of Microbiology, College of Biological Science, University of Guelph, Guelph, Ontario N1G 2W1, Canada

Elliot Juni
Department of Microbiology and Immunology, University of Michigan Medical School, Ann Arbor, Michigan 48109 USA

Roger W. Kelley
Department of Anaerobic Microbiology, Virginia Polytechnic Institute and State University, Blacksburg, Virginia 24061 USA

Richard T. Kelly
Department of Pathology, Baptist Memorial Hospital, Memphis, Tennessee 38146 USA

Karel Kersters
Laboratorium voor Microbiologie, K. L. Ledeganckstraat 35, 9000 Gent, Belgium

Mogens Kilian
Department of Oral Biology, The Royal Dental College, Vennelyst Boulevard, DK-8000 Aarhus C., Denmark

Dieter H. Knösel
Institut für Angewandte Botanik, Universität Hamburg, Marseiller Str. 7, D-2000 Hamburg, 36 Federal Republic of Germany

Miloslav Kocur
Czechoslovak Collection of Microorganisms, J. E. Purkyně University, 662 43 Brno, Czechoslovakia

A. E. Konopka
Department of Biological Science, Lilly Hall of Life Sciences, Purdue University, West Lafayette, Indiana 47906 USA

Julius P. Kreier
Department of Microbiology, Ohio State University, 484 West 12th Avenue, Columbus, Ohio 43210 USA

Noel R. Krieg
Department of Biology, Virginia Polytechnic Institute and State University, Blacksburg, Virginia 24061 USA

Cho-chou Kuo
Department of Pathobiology SC-38, School of Public Health and Community Medicine, University of Washington, Seattle, Washington 98195 USA

Thomas A. Langworthy
Department of Microbiology, School of Medicine, University of South Dakota, Vermillion, South Dakota 57069 USA

Stephen Lapage
27 Salisbury Road, Fordingsbridge Hants SP6 1EH, England

John M. Larkin
Department of Microbiology, Louisiana State University, Baton Rouge, Louisiana 70803 USA

Helge Larsen
Department of Biochemistry, Norwegian Institute of Technology, University of Trondheim, N-7034 Trondheim-NTH, Norway

John V. Lee
Environmental Microbiology and Safety Reference Laboratory, CAMR, PHLS, Porton Down, Salisbury, Wiltshire SP4 0JG, England

Sun Y. Lee
Research and Development Department, Adolph Coors Company, Golden, Colorado 80401 USA

R. A. Lelliott
Ministry of Agriculture, Fisheries and Food, ADAS Harpenden Laboratory, Hatching Green, Harpenden, Herts AL5 2BD, England

L. Le Minor
Entérobactéries, Institut Pasteur, 28, Rue du Docteur Roux, 75724 Paris Cedex 15, France

Walter Mannheim
Department of Bacteriology, Hygiene-Institut und Medizinal-Untersuchungensamt der Universität Marburg, 3550 Marburg/Lahn, Marburg, Federal Republic of Germany

Randolph E. McCoy
University of Florida, Institute of Food and Agricultural Sciences, Agricultural Research and Education Center, 3205 S. W. College Avenue, Fort Lauderdale, Florida 33314 USA

Virginia G. McGann
U.S. Army Medical Research Institute of Infectious Diseases, Frederick, Maryland 21701 USA

Thomas A. McMeekin
Department of Agricultural Science, University of Tasmania, Box 252C, G.P.O., Hobart, Tasmania 7001, Australia

Alma C. McWhorter
Enteric Bacteriology Section, Center for Infectious Diseases, Centers for Disease Control, Atlanta, Georgia 30333 USA

Henri H. Mollaret
Institut Pasteur, 28 Rue du Docteur Roux, 75724 Paris Cedex 15, France

W. E. C. Moore
Department of Anaerobic Microbiology, Virginia Polytechnic Institute and State University, Blacksburg, Virginia 24061 USA

James W. Moulder
Department of Microbiology, Cummings Life Science Center, University of Chicago, 920 East 58th Street, Chicago, Illinois 60637

R. G. E. Murray
Department of Microbiology and Immunology, University of Western Ontario, London, Ontario N6A 5C1, Canada

Reinier Mutters
Department of Bacteriology, Medizinisches Zentrum für Hygiene und Med. Mikrobiologie, Pilgrimstein 2, D-3550 Marburg, Federal Republic of Germany

Peter B. New
Department of Microbiology, University of Sydney, New South Wales 2006, Australia

Frits Ørskov
International Escherichia and Klebsiella Centre, Statens Seruminstitut, Amager Boulevard 80, DK-2300 Copenhagen S, Denmark

Ida Ørskov
International Escherichia and Klebsiella Centre, Statens Seruminstitut, Amager Boulevard 80, DK-2300 Copenhagen S, Denmark

Robert J. Owen
National Collection of Type Cultures, Central Public Health Laboratory, Colindale Avenue, London NW9, 5HT, England

Norberto J. Palleroni
Chemical Research Department, Hoffman-LaRoche, Nutley, New Jersey 07110 USA

G. B. Patel
Division of Biological Sciences, National Research Council of Canada, Ottawa, Ontario K1A 0R6, Canada

John L. Penner
Department of Medical Microbiology, University of Toronto, Banting Institute, 100 College Street, Toronto, Ontario M5G 1L5, Canada

Jerome J. Perry
Department of Microbiology, North Carolina State University, Raleigh, North Carolina 27650 USA

Norbert Pfennig
Faculty of Biology, University of Konstanz, P.O.B. 5560, D-775 Konstanz, Federal Republic of Germany

J. E. Phillips
Department of Veterinary Pathology, Royal (Dick) School of Veterinary Studies, Edinburgh EH9 1QH, Scotland

M. J. Pickett
Department of Microbiology, University of California, Los Angeles, California 90024 USA

Margaret Pittman
Guest Worker, Office of Biologics, National Center for Drugs and Biologics, Food and Drug Administration, 8800 Rockville Pike, Bethesda, Maryland 20205 USA

Michel Popoff
Entérobactéries, Institut Pasteur, 28, Rue du Docteur Roux, 75724 Paris Cedex 15, France

John R. Postgate
ARC Unit of Nitrogen Fixation, University of Sussex, Brighton BN1 9RQ, England

John R. Preer, Jr.
Department of Biology, Indiana University, Bloomington, Indiana 47405 USA

Louise B. Preer
Department of Biology, Indiana University, Bloomington, Indiana 47405 USA

Rudolf A. Prins
Research Institute for Nature Management, P.O. Box 46, 3956 ZR Leersum, The Netherlands

Shmuel Razin
Department of Membrane and Ultrastructure Research, The Hebrew University—Hadassah Medical School, P.O.B. 1172, Jerusalem 91010, Israel

C. Richard
Entérobactéries, Institut Pasteur, 28, Rue du Docteur Roux, 75724 Paris Cedex 15, France

Miodrag Ristic
College of Veterinary Medicine, University of Illinois, Urbana, Illinois 61801 USA

Isadore M. Robinson
National Animal Disease Center, Agricultural Research Service, U.S. Department of Agriculture, Ames, Iowa 50010 USA

Morrison Rogosa
National Institutes of Health, Building 31, Room 3B04, Bethesda, Maryland 20014 USA

Bernard Rowe
Division of Enteric Pathogens, Central Public Health Laboratory, Colindale Avenue, London NW9 5HT, England

Riichi Sakazaki
Enterobacteriology Laboratories, National Institute of Health, 10-35 Kamiosaki-2-chome, Shinagawa-ku, Tokyo, Japan

Norman Savage
Department of Biology, Pan American University, Edinburg, Texas 78539 USA

Julius Schachter
Department of Laboratory Medicine, University of California, San Francisco, San Francisco, California 94143 USA

Hans G. Schlegel
Institut fur Mikrobiologie der Universität Göttingen, 3400 Göttingen, Grisebachstrasse 8, Federal Republic of Germany

Ralph H. W. Schubert
Zentrum der Hygiene, Abt. für Allgemeine und Umwelthygiene, 6000 Frankfurt/Main, den, Paul-Ehrlich Strasse 40, Federal Republic of Germany

Robert M. Smibert
Department of Anaerobic Microbiology, Virginia Polytechnic Institute and State University, Blacksburg, Virginia 24061 USA

Paul F. Smith
Department of Microbiology, School of Medicine, University of South Dakota, Vermillion, South Dakota 57069 USA

Peter H. A. Sneath
Department of Microbiology, School of Medicine and School of Biological Sciences, University of Leicester, University Road, Leicester LE1 7RH, England

Jeremy J. S. Snell
Division of Microbiological Reagents and Quality Control, Central Public Health Laboratory, 175 Colindale Avenue, London NW9 5HT, England

Sigmund S. Socransky
Forsyth Dental Center, 140 The Fenway, Boston, Massachusetts 02115 USA

James T. Staley
Department of Microbiology and Immunology, University of Washington, Seattle, Washington 98195 USA

Johannes Storz
Department of Veterinary Microbiology and Parasitology, School of Veterinary Medicine, Louisiana State University, Baton Rouge, Louisiana 70803 USA

Jean Swings
Laboratorium voor Microbiologie, K.L. Ledeganckstraat 35, 9000 Gent, Belgium

Anne C. R. Tanner
Forsyth Dental Center, 140 The Fenway, Boston, Massachusetts 02115 USA

David Taylor-Robinson
Clinical Research Centre, Watford Road, Harrow, Middlesex HA1 3UJ, England

Yao-Tseng Tchan
Department of Microbiology, University of Sydney, New South Wales 2006, Australia

Joseph G. Tully
Mycoplasma Section, Laboratory of Molecular Microbiology, National Institute of Allergy and Infectious Diseases, Building 550, Frederick Cancer Research Facility, Frederick, Maryland 21701 USA

Richard F. Unz
Department of Civil Engineering, 212 Sackett Building, The Pennsylvania State University, University Park, Pennsylvania 16802 USA

Neylan A. Vedros
Department of Biomedical and Environmental Health Sciences, School of Public Health, University of California, Berkeley, California 94720 USA

Russell H. Vreeland
Department of Biological Sciences, University of New Orleans, Lakefront, New Orleans, Louisiana 70148

Robert E. Weaver
Respiratory and Special Pathogens Branch, Division of Bacterial Diseases, Center for Infectious Diseases, Centers for Disease Control, Atlanta, Georgia 30333 USA

Emilio Weiss
Naval Medical Research Institute, Bethesda, Maryland 20014 USA

Robert F. Whitcomb
Building 465, U.S. Department of Agriculture, BARCE, Beltsville, Maryland 20705 USA

Roger Whittenbury
Department of Biological Sciences, University of Warwick, Coventry CV4 7AL, England

Friedrich Widdel
Faculty of Biology, University of Konstanz, P.O. Box 5560, D-775 Konstanz, Federal Republic of Germany

Jürgen K. W. Wiegel
Department of Microbiology, University of Georgia, Athens, Georgia 30602

Advisory Committee Members

Preface to First Edition of Bergey's Manual of Systematic Bacteriology

Many microbiologists advised the Trust that a new edition of the *Manual* was urgently needed. Of great concern to us was the steadily increasing time interval between editions; this interval reached a maximum of 17 years between the seventh and eighth editions. To be useful the *Manual* must reflect relatively recent information; a new edition is soon dated or obsolete in parts because of the nearly exponential rate at which new information accumulates. A new approach to publication was needed, and from this conviction came our plan to publish the *Manual* as a sequence of four subvolumes concerned with systematic bacteriology as it applies to taxonomy. The four subvolumes are divided roughly as follows: (a) the Gram-negatives of general, medical or industrial importance; (b) the Gram-positives other than actinomycetes; (c) the archaeobacteria, cyanobacteria and remaining Gram-negatives; and (d) the actinomycetes. The Trust believed that more attention and care could be given to preparation of the various descriptions within each subvolume, and also that each subvolume could be prepared, published, and revised as the area demanded, more rapidly than could be the case if the *Manual* were to remain as a single, comprehensive volume as in the past. Moreover, microbiologists would have the option of purchasing only that particular subvolume containing the organisms in which they were interested.

The Trust also believed that the scope of the *Manual* needed to be expanded to include more information of importance for systematic bacteriology and bring together information dealing with ecology, enrichment and isolation, descriptions of species and their determinative characters, maintenance and preservation, all focused on the illumination of bacterial taxonomy. To reflect this change in scope, the title of the *Manual* was changed and the primary publication becomes *Bergey's Manual of Systematic Bacteriology*. This contains not only determinative material such as diagnostic keys and tables useful for identification, but also all of the detailed descriptive information and taxonomic comments. Upon completion of each subvolume, the purely determinative information will be assembled for eventual incorporation into a much smaller publication which will continue the original name of the *Manual, Bergey's Manual of Determinative Bacteriology*, which will be a similar but improved version of the present *Shorter Bergey's Manual*.

So, in the end there will be two publications, one systematic and one determinative in character.

An important task of the Trust was to decide which genera should be covered in the first and subsequent subvolumes. We were assisted in this decision by the recommendations of our Advisory Committees, composed of prominent taxonomic authorities to whom we are most grateful. Authors were chosen on the basis of constant surveillance of the literature of bacterial systematics and by recommendations from our Advisory Committees.

The activation of the 1976 Code had introduced some novel problems. We decided to include not only those genera that had been published in the Approved Lists of Bacterial Names in January 1980 or that had been subsequently validly published, but also certain genera whose names had no current standing in nomenclature. We also decided to include descriptions of certain organisms which had no formal taxonomic nomenclature, such as the endosymbionts of insects. Our goal was to omit no important group of cultivated bacteria and also to stimulate taxonomic research on "neglected" groups and on some groups of undoubted bacteria that have not yet been cultivated and subjected to conventional studies.

The invited authors were provided with instructions and exemplary chapters in June 1980 and, although the intended deadline for receipt of manuscripts was March 1981, all contributions were assembled in January 1982 for the final preparations. The *Manual* was forwarded to the publisher in June 1982.

Some readers will note the consistent use of the stem -var instead of -type in words such as biovar, serovar and pathovar. This is in keeping with the recommendations of the Bacteriological Code and was done against the wishes of some of the authors.

We have deleted much of the synonymy of scientific names which was contained in past editions. The adoption of the new starting date of January 1, 1980 and publication of the Approved Lists of Bacterial Names has made mention of past synonymy obsolete. We have included synonyms of a name only if they have been published since the new starting date, or if they were also on the Approved Lists and, in rare cases with certain pathogens, if the mention of an old name would help readers associate the organism with a clinical problem.

If the reader is interested in tracing the history of a name we suggest he or she consult past editions of the *Manual* or the *Index Bergeyana* and its *Supplement*. In citations of names we have used the abbreviation *AL* to denote the inclusion of the name on the Approved Lists of Bacterial Names and *VP* to show the name has been validly published.

In the matter of citation of the *Manual* in the scientific literature we again stress the fact that the *Manual* is a collection of authored chapters and the citation should refer to the author, the chapter title and its inclusive pages, not the Editor.

To all contributors, the sincere thanks of the Trust is due; the Editor is especially grateful for the good grace with which the authors accepted comments, criticisms and editing of their manuscripts. It is only because of the voluntary and dedicated efforts of these authors that the *Manual* can continue to serve the science of bacteriology on an international basis.

A number of institutions and individuals deserve special acknowledgment from the Trust for their help in bringing about the publication of this volume. We are grateful to the Department of Biology of the Virginia Polytechnic Institute and State University for providing space, facilities and, above all, tolerance for the diverted time taken by the Editor during the preparation of the book. The Department of Microbiology at Iowa State University of Science and Technology continues to provide a welcome home for the main editorial offices and archives of the Trust and we acknowledge their continued support. A grant (LM-03707) from the National Library of Medicine, National Institutes of Health to assist in the preparation of this and the next volume of the *Manual* is gratefully acknowledged.

A number of individuals deserve special mention and thanks for their help. Professor Thomas O. McAdoo of the Department of Foreign Languages and Literatures at the Virginia Polytechnic Institute and State University has given invaluable advice on the etymology and correctness of scientific names. Those assisting the Editor in the Blacksburg office were R. Martin Roop II, Don D. Lee, Eileen C. Falk and Michael W. Friedman and their help is sincerely appreciated. In the Ames office we were ably assisted by Gretchen Colletti and Diane Triggs during the early period of preparation and by Cynthia Pease during the major portion of the editing process. Mrs. Pease has been responsible for the construction of the List of References and her willingness to handle the cumbersome details of text editing on a big computer is gratefully acknowledged.

Comments on this edition of the *Manual* will be welcomed and should be addressed to the Bergey's Manual Trust, c/o The Williams & Wilkins Co., 428 E. Preston St., Baltimore, Md. 21202, U.S.A.

Preface to First Edition of Bergey's Manual of Determinative Bacteriology

The elaborate system of classification of the bacteria into families, tribes and genera by a Committee on Characterization and Classification of the Society of American Bacteriologists (1917, 1920) has made it very desirable to be able to place in the hands of students a more detailed key for the identification of species than any that is available at present. The valuable book on "Determinative Bacteriology" by Professor F. D. Chester, published in 1901, is now of very little assistance to the student, and all previous classifications are of still less value, especially as earlier systems of classification were based entirely on morphologic characters.

It is hoped that this manual will serve to stimulate efforts to perfect the classification of bacteria, especially by emphasizing the valuable features as well as the weaker points in the new system which the Committee of the Society of American Bacteriologists has promulgated. The Committee does not regard the classification of species offered here as in any sense final, but merely a progress report leading to more satisfactory classification in the future.

The Committee desires to express its appreciation and thanks to those members of the society who gave valuable aid in the compilation of material and the classification of certain species...

The assistance of all bacteriologists is earnestly solicited in the correction of possible errors in the text; in the collection of descriptions of all bacteria that may have been omitted from the text; in supplying more detailed descriptions of such organisms as are described incompletely; and in furnishing complete descriptions of new organisms that may be discovered, or in directing the attention of the Committee to publications of such newly described bacteria.

DAVID H. BERGEY, *Chairman*
FRANCIS C. HARRISON
ROBERT S. BREED
BERNARD W. HAMMER
FRANK M. HUNTOON
Committee on Manual.

August, 1923.

DAVID HENDRICKS BERGEY
1860–1937
Bergey set up the Trust on January 2, 1936

History of the Manual

The first edition of *Bergey's Manual of Determinative Bacteriology* was initiated by action of the Society of American Bacteriologists (now called the American Society for Microbiology) by appointment of an Editorial Board consisting of David H. Bergey, Chairman, Francis C. Harrison, Robert S. Breed, Bernard W. Hammer, and Frank M. Huntoon. This Board, under auspices of the Society of American Bacteriologists who, then as now, published the *Journal of Bacteriology* as a service to science, brought the first edition of the *Manual* into print in 1923. The Board, with some changes in membership and Dr. David Bergey as Chairman, published a second edition of the *Manual* in 1925 and a third edition in 1930.

In 1934, during preparation of the fourth edition, Dr. Bergey requested that the Society of American Bacteriologists make available the royalties paid to the Treasurer of the Society from the sale of the earlier editions to defray the expense of preparing the fourth edition for publication. The Society made such provision, but the use of the Society's fiscal machinery proved cumbersome, both to the Society and the Editorial Board. Subsequently, it was agreed by the Society and Dr. Bergey that the Society would transfer to Dr. Bergey all of its rights, title, and interest in the *Manual* and that Dr. Bergey would, in turn, create an educational trust to which all rights would be transferred.

Dr. Bergey was then the nominal owner of the *Manual* and he executed a Trust Indenture on January 2, 1936 designating David H. Bergey, Robert S. Breed, and E. G. D. Murray as the initial trustees, and transferring to the Trustees and their successors the ownership of the *Manual*, its copyrights, and the right to receive the income arising from its publication. The Trust is a nonprofit organization and its income is used solely for the purpose of preparing, editing, and publishing revisions and successive editions of the *Manual* and any supplementary publications, as well as providing for any research that may be necessary or desirable in such activities.

Since the creation of the Trust, the Trustees have published, successively, the fourth, fifth, sixth, seventh, and eighth editions of the *Manual* (dated 1934, 1939, 1948, 1957, and 1974, respectively). In 1977 the Trust published an abbreviated version of the eighth edition, called *The Shorter Bergey's Manual of Determinative Bacteriology*; this contained the outline classification of the bacteria, the descriptions of all genera and higher taxa, all of the keys and tables for the diagnosis of species, all of the illustrations, and two of the introductory chapters; however, it did not contain the detailed species descriptions, most of the taxonomic comments, the etymology of names, and references to authors.

Other ventures in producing books to assist those engaged in bacteriology and bacterial taxonomy in particular include the *Index Bergeyana* (1966), a *Supplement to Index Bergeyana* (1981), and a planned future volume bringing the lists of published names up to date. The Trust is presently publishing the first edition of *Bergey's Manual of Systematic Bacteriology*, which has a much broader scope than the previous publications and is intended to act as the amplified source for revision of the determinative *Manual*.

Through the years the *Manual* has become a widely used international reference work for bacterial taxonomy. Similarly, the Bergey's Manual Trust has become international in its composition, in the location of its meetings and in the breadth of its consultations. In addition to its publication activities, the Trust attempts to foster and support various aspects of taxonomic research. One of the ways in which it does this is by recognizing those individuals who have made outstanding contributions to bacterial taxonomy, through its periodic presentation of the Bergey Award, an effort jointly supported by funds from the Trust and The Williams & Wilkins Company who have been involved in the production of the *Manual* from its beginning.

The following individuals have served as members of the Editorial Board and Board of Trustees.

On Using the Manual

Noel R. Krieg

ARRANGEMENT OF THE MANUAL

One important goal of the *Manual* is to assist in the identification of bacteria, but another goal, equally important, is to indicate the relationships that exist between the various kinds of bacteria. The methods of molecular biology have now made it possible to envision the eventual development of a comprehensive classification of bacteria based on their relatedness to one another. Such a general classification scheme would lead to more unifying concepts of bacterial taxa, to greater stability and predictability, to the development of more reliable identification schemes, and to an understanding of how bacteria have evolved.

Such a general scheme, however, cannot yet be perceived fully. The relatedness within and between some bacterial groups has been intensively studied, but for other groups very little work has been done. Moreover, the relatedness studies that have been done often have involved the use of one or another method without confirmation by other methods. Studies have been done at differing levels of resolution, and the interpretation of the data may not yet be entirely clear. Still

another major difficulty is the conflict between "practical" classification vs. strange groupings that may be indicated by molecular biology methods. This is because some of the phenotypic characteristics traditionally used in bacterial classification (e.g. cell shape, flagellar arrangement, fermentative vs. respiratory types of metabolism, etc.) do not always correlate well with groups established on the basis of relatedness. This conflict will eventually be relieved by the finding of nontraditional, easily determined, phenotypic characteristics that *do* correlate well with relatedness groups, but much work needs to be done in this regard.

Such considerations have forced the present edition of the *Manual* to adhere largely to traditional characteristics in arranging bacterial taxa. It should be understood, however, that reassessments of these groupings will soon need to be made on a broad, comprehensive scale. The present classification, although of considerable practical value, must be regarded as only an interim arrangement.

THE SECTIONS

The *Manual* is presented as various "sections" based on a few readily determined criteria. Each section bears a vernacular name. All accepted genera have been placed in what seems the most appropriate section, although allocation of certain genera has presented difficulties, as indicated by the following examples:

(a) The genus *Gardnerella*. The organisms of this genus have had a checkered taxonomic history and it is still not entirely clear whether they should be placed in Volume 1 with Gram-negative bacteria or in Volume 2 with Gram-positive bacteria.

(b) The genus *Butyrivibrio*. Although the cells stain Gram-negative the ultrastructure of the cell wall is of the Gram-positive type. It is not clear whether the genus should be placed in Volume 1 or Volume 2.

(c) The genus *Xanthobacter*. The cells stain Gram-positive or Gram-variable, yet the cell wall structure and composition, as well as nucleic acid hybridization data, indicate that the organisms are of the Gram-negative type.

(d) The genus *Chromobacterium*. Although 80% of the strains attack glucose fermentatively and grow well anaerobically, the remainder attack glucose oxidatively and grow slowly under anaerobic conditions. It is consequently difficult to assign the organisms definitively to either Section 5 (Facultatively Anaerobic Gram-

Negative Rods) or Section 4 (Gram-Negative Aerobic Rods and Cocci). Nucleic acid hybridization studies indicate a relationship to certain genera of aerobic rods.

(e) The genus *Zymomonas*. Although the organisms are facultatively anaerobic (a few obligately anaerobic), they are related genetically, phenotypically and ecologically to the acetic acid bacteria, which are aerobic. Moreover, the occurrence of the Entner-Doudoroff pathway is typical of aerobic bacteria.

(f) The genus *Thermoplasma*. The lack of a cell wall makes this genus compatible with Section 10 (The Mycoplasmas); however, studies of the ribosomal RNA, as well as various phenotypic characteristics, indicate that the genus is related to the archaeobacteria, covered in Volume 3 of the *Manual*.

(g) The genera *Halobacterium* and *Halococcus*. Although these extreme halophiles are compatible with Section 4 (Gram-Negative Aerobic Rods and Cocci), nucleic acid studies and certain phenotypic characteristics indicate the genus is related to the archaeobacteria, covered in Volume 3.

As an interim solution to some of these problems, some taxa are described not only in Volume 1 but in an appropriate subsequent volume as well.

SECTIONS VS. TAXONOMIC NAMES

Each section bears a vernacular name, but sometimes it also bears the name of a taxon. For example, Section 10 (The Mycoplasmas) is

the Division Tenericutes, Class *Mollicutes*; Section 1 (The Spirochetes) is the order *Spirochetales*; and Section 8 (Anaerobic Gram-Negative

Cocci) is the family *Veillonellaceae.* Some sections may contain more than one order (e.g. Section 9) or family (e.g. Section 5), and some may contain no taxa whatever above the level of genus (e.g. Section 7). As indicated previously, no attempt has been made to provide a complete formal hierarchy of higher taxa throughout the *Manual,* and the vernacular names of the sections form the primary basis for the organization of the *Manual;* however, a suggested hierarchy for higher taxa has been proposed in one of the introductory articles (see The Higher Taxa, or a Place for Everything?).

Some families recognized in the *Manual* represent groups of related genera (e.g. the family *Enterobacteriaceae*). Others, however, are merely families based on practical convenience rather than any known degree of relatedness (e.g. the family *Methylococcaceae*).

In sections containing one or more families, there may be an appendix entitled "Other Organisms." While these genera belong to a particular section, they have not been accepted into any of the recognized families and cannot themselves be grouped into families on the information presently available. For example, Section 5 (Facultatively Anaerobic Gram-Negative Rods) consists of the families *Enterobacteriaceae, Vibrionaceae,* and *Pasteurellaceae* and concludes with an appended list of seven additional genera that do not belong to any family (*Zymomonas, Chromobacterium, Cardiobacterium, Calymmatobacterium, Gardnerella, Eikenella,* and *Streptobacillus*).

Certain sections of the *Manual* may conclude with descriptions of organisms which, for various reasons, have not yet been assigned to a genus. For example, Section 1 concludes with an article on Hindgut Spirochetes of Termites and *Cryptocercus punctulatus,* i.e. a group of spirochetes which have not been cultured. Section 11 (Endosymbionts) deals mainly with unclassified endosymbionts of insects and other organisms. The purpose of including such unclassified organisms in the *Manual* is to stimulate research on their taxonomy.

ARTICLES

Each article dealing with a bacterial genus is presented wherever possible in a definite sequence as follows.

(a) *Name of the Genus.* Accepted names are in **boldface,** followed by the authority for the name, the year of the original description, and the page on which the taxon was named and described. The superscript *AL* indicates that the name was included on the Approved Lists of Bacterial Names, published in January 1980. The superscript *VP* indicates that the name, although not on the Approved Lists of Bacterial Names, was subsequently validly published in the International Journal of Systematic Bacteriology. Names given within quotation marks have no standing in nomenclature; as of the date of preparation of the *Manual* they had not been validly published in the International Journal of Systematic Bacteriology, although they had been "effectively published" elsewhere. Names followed by the term "gen. nov." are newly proposed but will not be validly published until they appear in the International Journal of Systematic Bacteriology; their proposal in the *Manual* constitutes only "effective publication," not valid publication.

(b) *Name of Author(s).* The person or persons who prepared the Bergey article are indicated. The address of each author can be found in the list of Contributors at the beginning of the *Manual.*

(c) *Synonyms.* In some instances a list is given of synonyms which have been used in the past for the same genus. The synonymy may not always be complete, and usually is not given at all, as the Editorial Board believes that the earlier synonyms have been covered adequately in the *Index Bergeyana* or the *Supplement to the Index Bergeyana.*

(d) *Etymology of the Genus Name.* Etymologies are provided as in previous editions, and many (but undoubtedly not all) errors have been corrected. It is often difficult, however, to determine why a particular name was chosen, or the nuance intended, if the details were not provided in the original publication. Those authors who propose new names are urged to consult a Greek and Latin authority before publishing, in order to ensure grammatical correctness and also to ensure that the name means what it is intended to mean. An excellent authority to communicate with in this regard is Dr. Thomas O. MacAdoo, Department of Foreign Languages, Virginia Polytechnic Institute and State University, Blacksburg, Virginia U.S.A. 24061.

(e) *Capsule Description.* This is a brief resume of the salient features of the genus. The most important characteristics are given in **boldface.** The name of the type species of the genus is also indicated.

(f) *Further Descriptive Information.* This portion elaborates on the various features of the genus, particularly those features having significance for systematic bacteriology. The treatment serves to acquaint the reader with the overall biology of the organisms but is not meant to be a comprehensive review. The information is presented in a definite sequence, as follows:
Morphological characteristics

Colonial morphology and pigmentation
Growth conditions and nutrition
Physiology and metabolism
Genetics, plasmids, and bacteriophages
Antigenic structure
Pathogenicity
Ecology

(g) *Enrichment and Isolation.* A few selected methods are presented, together with the pertinent media formulations.

(h) *Maintenance Procedures.* Methods used for maintenance of stock cultures and preservation of strains are given.

(i) *Procedures for Testing Special Characters.* This portion provides methodology for testing for unusual characteristics or performing tests of special importance.

(j) *Differentiation of the Genus from Other Genera.* Those characteristics that are especially useful for distinguishing the genus from similar or related organisms are indicated here, usually in a tabular form.

(k) *Taxonomic Comments.* This summarizes the available information about the taxonomic placement of the genus and indicates the justification for considering the genus to be a distinct taxon. Particular emphasis is given to the methods of molecular biology for estimating the relatedness to other taxa, where such information is available. Taxonomic information regarding the arrangement and status of the various species within the genus follows. Where taxonomic controversy exists, the problems are delineated and the various alternative viewpoints are discussed.

(l) *Further Reading.* A list of selected references, usually of a general nature, is given to enable the reader to gain access to additional sources of information about the genus.

(m) *Differentiation of the Species of the Genus.* Those characteristics that are important for distinguishing from one another the various species within the genus are presented, usually with reference to a table summarizing the information.

(n) *List of the Species of the Genus.* The citation of each species is given, followed in some instances by a brief list of objective synonyms. The etymology of the specific epithet is indicated. Descriptive information for the species is usually presented in tabular form, but special information may be given in the text. Because of the emphasis on tabular data the species descriptions are usually brief. The type strain of each species is indicated, together with the collection in which it can be be found. (Addresses of the various culture collections are given in the chapter List of Culture Collections.)

(o) *Species Incertae Sedis.* The List of Species may be followed in some instances by a listing of additional species under the heading "Species Incertae Sedis." The taxonomic placement or status of such species is questionable and the reasons for the uncertainty are presented.

(p) *Literature Cited.* All references given in the article are listed alphabetically at the end of the volume rather than at the end of each article.

TABLES

In each article dealing with a genus, there are generally three kinds of tables: (a) those that differentiate the genus from similar or related genera, (b) those that differentiate the species within the genus, and (c) those that provide additional information about the species, such information not being particularly useful for differentiation. Unless otherwise indicated, the meanings of symbols are as follows:

+ 90% or more of the strains are positive.

d 11–89% of the strains are positive.

– 90% or more of the strains are negative.

D different reactions occur in different taxa (species of a genus or genera of a family).

v strain instability (NOT equivalent to "d").

Exceptions to the use of these symbols, as well as the meaning of additional symbols, are clearly indicated in footnotes to the tables.

USE OF THE MANUAL FOR DETERMINATIVE PURPOSES

Entry into the *Manual* is best achieved by studying the titles of the various sections, as listed in the Contents. These titles provide an elementary, but by no means perfect, key to the various kinds of bacteria. Each section has keys or tables for differentiation of the various taxa contained therein. Suggestions on identification may be found in the article Identification of Bacteria. For identification of species, it is important to read both the generic and species descriptions because characteristics listed in the generic descriptions are not usually repeated in the species descriptions.

The index is useful in locating the names of unfamiliar taxa or in discovering what has been done with a particular taxon. Every bacterial name mentioned in the *Manual* is listed in the index.

ERRORS, COMMENTS, SUGGESTIONS

As indicated in the Preface to the first edition of *Bergey's Manual of Determinative Bacteriology*, the assistance of all bacteriologists is earnestly solicited in the correction of possible errors in the text. Comments on the presentation will also be welcomed, as well as suggestions for future editions. Correspondence should be addressed to the Bergey's Manual Board of Trustees c/o The Williams & Wilkins Co., 428 East Preston St., Baltimore, Md. 21202, U.S.A.

Contents

Classification of Procaryotic Organisms: An Overview

James T. Staley and Noel R. Krieg

CLASSIFICATION, NOMENCLATURE AND IDENTIFICATION

Classification, nomenclature, and identification are the three separate, but interrelated, areas of taxonomy. **Classification** is the arranging of organisms into taxonomic groups (taxa) on the basis of similarities or relationships. **Nomenclature** is the assignment of names to the taxonomic groups according to international rules. **Identification** is the process of determining that a new isolate belongs to one of the established, named taxa.

There are numerous procaryotic organisms and great diversity in their types. In any endeavor aimed at an understanding of large numbers of entities it is convenient to arrange, or classify, the objects into groups based upon their similarities. Thus, classification has been used to organize the bewildering and seemingly chaotic array of individual bacteria into an orderly framework.

Classification of organisms requires knowledge of their characteristics. For procaryotes, this knowledge is obtained by experimental as well as observational techniques, because biochemical, physiological and genetic characteristics are often necessary, in addition to morphological features, for an adequate description of a taxon.

The process of classification may be applied to existing, named, taxa or to newly described organisms. If the taxa have already been described, named, and classified, either new characteristics about the organisms or a reinterpretation of existing knowledge of characteristics is used to formulate a new classification. However, if the organisms are new, i.e. cannot be identified as existing taxa, they are named according to the rules of nomenclature and placed in an appropriate position in an existing classification.

Taxonomic Ranks

Several levels or ranks are used in bacterial classification. All procaryotic organisms are placed in the Kingdom *Procaryotae*. Divisions, classes, orders, families, genera and species are successively smaller, nonoverlapping subsets of the Kingdom, and the names of these subsets are given formal recognition (have "standing in nomenclature"). An example is given in Table I.1.

In addition to these formal, hierarchical taxonomic categories, informal or vernacular groups that are defined by common descriptive names are often used; the names of such groups have no official standing in nomenclature. Examples of such groups are: the procaryotes, the spirochetes, dissimilatory sulfate- and sulfur-reducing bacteria, the methane-oxidizing bacteria, etc.

Species

The basic taxonomic group in bacterial systematics is the species. The concept of a bacterial species is less definitive than for higher organisms. This difference should not seem surprising, because bacteria, being procaryotic organisms, differ markedly from higher organisms. Sexuality, for example, is not used in bacterial species definitions because relatively few bacteria undergo conjugation. Likewise, morphological features alone are usually of little classificatory significance; this is because most procaryotic organisms are too simple morphologically to provide much useful taxonomic information. Consequently, morphological features are relegated to a less important role in bacterial taxonomy in comparison with the taxonomy of higher organisms.

A bacterial species may be regarded as a collection of strains that share many features in common and differ considerably from other strains. (A strain is made up of the descendants of a single isolation in pure culture, and usually is made up of a succession of cultures ultimately derived from an initial single colony.) One strain of a species is designated as the **type strain**; this strain serves as the name-bearer strain of the species and is the permanent example of the species, i.e. the *reference specimen for the name*. (See the chapter on *Nomenclature* for more detailed information about nomenclatural types.) The type strain has great importance for classification at the species level, because *a species consists of the type strain and all other strains that are considered to be sufficiently similar to it as to warrant inclusion with it in the species*. This concept of a species obviously involves making subjective judgments, and it is not surprising that some bacterial species have greater phenotypic and genetic diversity than others. A more uniform and rigorous species definition would be desirable. For example, the level of DNA homology exhibited among a group of strains might be used as a basis for defining a species, i.e. definition on the basis of a particular degree of genetic relatedness. The advantage of adopting this or a similarly restrictive species definition must be weighed against its potential impact on well established and accepted bacterial groups. For practical reasons, classifications and nomenclature should remain stable because changes create confusion, particularly at the genus and species levels and result in costly modifications of identification schemes and texts. However, classifications have *never* remained static and probably never will, because new information bearing on the taxonomy of bacteria is continually being generated by researchers.

Though classification schemes based on genetic relatedness are rather recent, they promise to be quite reliable and stable. This view may have to be reassessed, however, when we more fully understand the impact that transposable elements might have upon the stability of the procaryotic genome. Genetic studies have already resolved many instances of confusion concerning which strains belong to a given

Table I.1.
Taxonomic ranks

Formal Rank	Example
Kingdom	*Procaryotae*
Division	*Gracilicutes*
Class	*Scotobacteria*
Order	*Spirochaetales*
Family	*Leptospiraceae*
Genus	*Leptospira*
Species	*Leptospira interrogans*

Table I.2.
Infrasubspecific ranks

Preferred Name	Synonym	Applied to Strains Having:
Biovar	Biotype	Special biochemical or physiological properties
Serovar	Serotype	Distinctive antigenic properties
Pathovar	Pathotype	Pathogenic properties for certain hosts
Phagovar	Phagotype	Ability to be lysed by certain bacteriophages
Morphovar	Morphotype	Special morphological features

species, and DNA homology is increasingly being used for establishing new species and for resolving taxonomic problems at the species level.

Subspecies

A species may be divided into two or more subspecies based on minor but consistent phenotypic variations within the species or on genetically determined clusters of strains within the species. It is the lowest taxonomic rank that has official standing in nomenclature.

Infrasubspecific Ranks

Ranks below subspecies, such as biovars, serovars, phagovars, are often used to indicate groups of strains that can be distinguished by some special character, such as antigenic makeup, reactions to bacteriophage, or the like. Such ranks have no official standing in nomenclature but often have great practical usefulness. A list of some common infrasubspecific categories is given in Table I.2.

Genus

All species are assigned to a genus (although not always with a high degree of certainty as to which genus is the best choice). In this regard, bacteriologists conform to the binomial system of nomenclature of Linnaeus in which the organism is designated by its combined genus and species name. The bacterial genus is usually a well-defined group that is clearly separated from other genera, and the thorough descriptions of genera in this edition of *Bergey's Manual* exemplify the depth to which this taxonomic group is usually known. However, there is so far no general agreement on the *definition* of a genus in bacterial taxonomy, and considerable subjectivity is involved at the genus level. Indeed, what is perceived to be a genus by one person may be perceived as being merely a species by another systematist. The use of genetic relatedness (e.g. ribosomal RNA (rRNA) homology or rRNA oligonucleotide cataloging) offers hope for greater objectivity and has already been useful in several instances.

Higher Taxa

Classificatory relationships at the familial and ordinal levels are even less certain than those at the genus and species levels. Frequently there is little basis for ascription of taxa at these higher levels, except in a few cases (e.g. the family *Enterobacteriaceae*) where there is evidence for genetic relatedness. Thus, rather than formalize families and orders upon uncertain relationships, many systematists frequently adopt a provisional, ad hoc ranking in which purely descriptive and vernacular names for groups are applied (e.g. in this edition of *Bergey's Manual* see Section 7 on "Dissimilatory Sulfate- or Sulfur-Reducing Bacteria"). As more is learned about the similarities among these bacteria, familial and ordinal placements will likely ensue. A recent example that illustrates the effect that increased knowledge has on the taxonomy of groups concerns the methane-producing bacteria. In the eighth edition of the *Manual,* the methanogens were treated as a single family of bacteria, with three genera. Authorities for this group now propose that three *orders* are required for the circumscription of these organisms (Balch et al., 1979).

In this edition of the *Manual* the procaryotes have been classified into four divisions, these being subdivided into classes (see the chapter by Murray on "The Higher Taxa"). There is no general agreement about this or any other arrangement of divisions and classes, however, and even at the kingdom level of classification controversy exists. Recent information based on rRNA oligonucleotide catalogues and biochemical features has led some authorities to propose that not all bacteria are procaryotes, and that some represent a kingdom of life distinct from both procaryotes and eucaryotes (i.e. the so-called "archaebacteria") (see Fox et al., 1980, and Woese, 1981, for summaries). That this group possesses a number of unique features is beyond question, and there is strong evidence that it has taken an evolutionary path distinct from that of other bacteria, but so far there is no general agreement as to what level of classification is applicable to the group.

MAJOR DEVELOPMENTS IN BACTERIAL CLASSIFICATION

A century elapsed between Antony van Leeuwenhoek's discovery of bacteria and Müller's initial acknowledgement of bacteria in a classification scheme (Müller, 1773). Another century passed before techniques and procedures had advanced sufficiently to permit a fairly inclusive and meaningful classification of these organisms. For a comprehensive review of the early development of bacterial classification, readers should consult the introductory sections of the first, second, and third editions of *Bergey's Manual.* A less detailed treatment of early classifications can be found in the sixth edition of the *Manual* in which post-1923 developments were emphasized.

Two primary difficulties beset early bacterial classification systems. First, they relied heavily upon morphological criteria. For example, cell shape was often considered to be an extremely important feature. Thus, the cocci were often classified together in one group (family or order). In contrast, contemporary schemes rely much more strongly on physi-

ological characteristics. For example, the fermentative cocci are now separated from the photosynthetic cocci, which are separated from the methanogenic cocci, which are in turn separated from the nitrifying cocci, and so forth. Secondly, the pure culture technique which revolutionized microbiology was not developed until the latter half of the 19th century. In addition to dispelling the concept of "polymorphism," this technical development of Robert Koch's laboratory had great impact on the development of modern procedures in bacterial systematics. Pure cultures are analogous to herbarium specimens in botany. However, pure cultures are much more useful because they can be (a) maintained in a viable state, (b) subcultured, (c) subjected indefinitely to experimental tests, and (d) shipped from one laboratory to another. A natural outgrowth of the pure culture technique was the establishment of *type strains* of species which are deposited in repositories referred to as "culture collections" (a more suitable term would be "strain collections"). These type strains can be obtained from culture collections and used as reference strains for direct comparison with new isolates.

Before the development of computer-assisted numerical taxonomy and subsequent taxonomic methods based on molecular biology, the traditional method of classifying bacteria was to characterize them as thoroughly as possible and then to arrange them according to the intuitive judgment of the systematist. Although the subjective aspects of this method resulted in classifications that were often drastically revised by other systematists who were likely to make different intuitive judgments, many of the arrangements have survived to the present day, even under scrutiny by modern methods. One explanation for this is that the systematists usually *knew their organisms thoroughly*, and their intuitive judgments were based on a wealth of information. Their data, while not computer processed, were at least processed by an active mind to give fairly accurate impressions of the relationships existing between organisms. Moreover, some of the characteristics that were given great weight in classification were, in fact, highly correlated with many other characteristics. This principle of *correlation of characteristics* appears to have started with the Winslows (1908), who noted that parasitic cocci tended to grow poorly on ordinary nutrient media, were strongly Gram-positive, and formed acid from sugars, in contrast to saprophytic cocci which grew abundantly on ordinary media, were generally only weakly Gram-positive and formed no acid. This division of the cocci that were studied by the Winslows (equivalent to the present genus *Micrococcus* (the saprophytes) and the genera *Staphylococcus* and *Streptococcus* (the parasites)) has held up reasonably well even to the present day.

Other classifications have not been so fortunate. A classic example of one which was not is that of the genus "*Paracolobactrum*." This genus was proposed in 1944 and is described in the seventh edition of *Bergey's Manual* in 1957. It was created to contain certain lactose-negative members of the family *Enterobacteriaceae*. Because of the importance of a lactose-negative reaction in *identification* of enteric pathogens (i.e. *Salmonella* and *Shigella*), the reaction was mistakenly given great taxonomic weight in *classification* as well. However, for the organisms placed in "*Paracolobactrum*," the lactose reaction was not highly correlated with other characteristics. In fact, the organisms were merely lactose-negative variants of other lactose-positive species; for example, "*Paracolobactrum coliforme*" resembled *Escherichia coli* in every way except in being lactose-negative. Absurd arrangements such as this eventually led to the development of more objective methods of classification, i.e. numerical taxonomy, in order to avoid giving great weight to any single characteristic.

Phylogenetic Classifications

Classification systems for many higher organisms are based to a large extent upon evolutionary evidence obtained from the fossil record and appropriate sedimentary dating procedures. Such classifications are termed "natural" or "phylogenetic," and are distinguished from "practical" or "artificial" classifications, which are based entirely on phenotypic characteristics. Until about 20 years ago, however, there was no convincing evidence of fossil microorganisms. Now, micropaleontological evidence indicates that microorganisms existed during the Precambrian period. Indeed, many scientists believe that bacteria existed at least 3.5 billion years ago on an earth that is 4.5 billion years old. Of course, the discovery of fossil microorganisms in early sedimentary rocks tells very little about the phylogeny of procaryotic groups. Micropaleontologists are far from reconstructing an evolutionary scheme based upon the presently available fossil record.

Despite the absence of a complete fossil record, proposals have been made since the early part of this century regarding the evolution of bacteria. Until recently, these proposals have been entirely speculative in nature. Orla-Jensen (1909) proposed that autotrophic bacteria were the most primitive group, and he devised an extensive phylogenetic scheme based on this premise. Today, most microbiologists would agree that the premise is probably incorrect, but Orla-Jensen's classification

did provide a coherent framework for thinking about the relationships among bacteria. Another notable phylogenetic scheme was that devised by Kluyver and van Niel (1936); in contrast to Orla-Jensen's scheme, which had been based almost entirely on physiological characteristics, Kluyver and van Niel's scheme was based on morphology. The basic premise was that the simplest morphological form, the coccus, was also the most primitive, and from this form developed more complex forms such as spirilla, rods, and branching filaments.

As recently as the seventh edition of *Bergey's Manual* (i.e. 1957), before convincing evidence of Precambrian microbes had been discovered, the view was expressed that bacteria were a primitive group of organisms, and the classification scheme presented in that edition of the *Manual* claimed to be a natural scheme in which the photosynthetic bacteria were treated first, because they were regarded as the most primitive bacterial group. However, because of the lack of objective evidence for this (or any other) phylogenetic scheme, the eighth edition of the *Manual* abandoned all attempts at a phylogenetic approach to bacterial classification and concentrated instead on providing groupings of organisms under vernacular headings for purposes of recognition and identification; i.e. it was a purely practical and admittedly artificial classification.

Phylogenetic information has increased since the eighth edition, however, largely through the increasing use of methods for measuring genetic relatedness (i.e. DNA/DNA hybridization, DNA/rRNA hybridization, rRNA oligonucleotide cataloging, and protein sequencing). A record of bacterial evolution appears to exist in the amino acid sequences of bacterial proteins and in the nucleotide sequences of bacterial DNA and RNA. Unfortunately, the phylogenetic information is still in a fragmentary form, and it seems probable that the interpretation of the data is still not entirely clear. Not all of the bacterial groups have been surveyed, and it is likely that surprises and strange associations will continue to come from further work. Available phylogenetic information is presented throughout this *Manual* in the *Taxonomic Comments* sections of the various chapters, and some preliminary rearrangements of taxa have already been made based upon phylogenetic information.

Official Classifications

Some microbiologists seem to have the impression that the classification presented in *Bergey's Manual* is the "official classification" to be used in microbiology. It seems important to correct that impression. **There is no "official" classification of bacteria.** (This is in contrast to bacterial *nomenclature*, where each taxon has one and only one valid name, according to internationally agreed-upon rules, and judicial decisions are rendered in instances of controversy about the validity of a name.) The closest approximation to an "official" classification of bacteria would be one that is widely accepted by the community of microbiologists. A classification that is of little use to microbiologists, no matter how fine a scheme or who devised it, will soon be ignored or significantly modified.

It also seems worthwhile to emphasize something that has often been said before, viz. **bacterial classifications are devised for microbiologists, not for the entities being classified.** Bacteria show little interest in the matter of their classification. For the systematist, this is sometimes a very sobering thought!

Further Reading

Cowan, S.T. 1971. Sense and nonsense in bacterial taxonomy. J. Gen. Microbiol. 67: 1–8.
    An incisive, personal view of bacterial taxonomy, with some "heretical" suggestions.
Cowan, S.T. 1974. *Cowan and Steel's manual for the identification of medical bacteria.* Cambridge University Press, Cambridge, England.
    Chapters 1 and 9 of this work provide a concise statement of many principles of bacterial taxonomy.
Gerhardt, P., R.G.E. Murray, R.N. Costilow, E.W. Nester, W.A. Wood, N.R. Krieg and G.B. Phillips (Editors). 1981. Manual of methods for general bacteriology. American Society for Microbiology, Washington, D. C.
    Section V of this book gives a brief introduction to phenotypic characteriza-

tion, numerical taxonomy, genetic characterization, and classification of bacteria.

Johnson, J.L. 1973. Use of nucleic acid homologies in the taxonomy of anaerobic bacteria. Int. J. Syst. Bacteriol. *23:* 308–315.

This paper proposes a unifying concept of a bacterial species and stresses the importance of correlating nucleic acid homology with phenotypic tests to allow differentiation among species.

Margulis, L. 1968. Evolutionary criteria in Thallophytes: a radical alternative. Science (Washington) *161:* 1020–1022.

This paper presents the hypothesis that eucaryotic organisms evolved from procaryotic organisms through endosymbioses.

Schopf, J.W. 1978. The evolution of the earliest cells. Sci. Amer. *239:* 110–138.

A micropaleontologist's view of microbial evolution.

Schwartz, R.M. and M.O. Dayhoff. 1978. Origins of procaryotes, eucaryotes, mitochondria, and chloroplasts. Science (Washington) *199:* 395–403.

A discussion of results obtained from the analysis of protein and nucleic acid sequence data as they pertain to the phylogeny of organisms.

Sneath, P.H.A. 1978. Classification of microorganisms. *In* Norris and Richmond (Editors), *Essays in Microbiology.* John Wiley , Chichester, UK, pp. 9/1–9/31.

An excellent general introduction to bacterial classification.

Trüper, H.G. and J. Krämer. 1981. Principles of characterization and identification of prokaryotes. *In* Starr, Stolp, Trüper, Balows and Schlegel (Editors), *The Prokaryotes: a Handbook on Habitats, Isolation and Identification of Bacteria.* Springer-Verlag, Berlin, pp. 176–193.

A brief overview of systematic bacteriology, including developments and trends in taxonomy.

Woese, C.R. and G.E. Fox. 1977. Phylogenetic structure of the procaryotic domain: the primary kingdoms. Proc. Nat. Acad. Sci. USA *74:* 5088–5090.

The authors recognize three distinct groups: the eubacteria, the archaebacteria, and the urcaryotes (cytoplasmic components of eucaryotes).

BACTERIAL CLASSIFICATION II

Numerical Taxonomy

Peter H. A. Sneath

Numerical taxonomy (sometimes called **taxometrics**) developed in the late 1950s as part of multivariate analyses and in parallel with the development of computers. Its aim was to devise a consistent set of methods for classification of organisms. Much of the impetus in bacteriology came from the problem of handling the tables of data that result from examination of their physiological, biochemical and other properties. Such tables of results are not readily analyzed by eye, in contrast to the elaborate morphological detail that is usually available from examination of higher plants and animals. There was thus a need for an objective method of taxonomic analyses, whose first aim was to sort individual strains of bacteria into homogeneous groups (conventionally species), and which would also assist in the arrangement of species into genera and higher groupings. Such numerical methods also promised to improve the exactitude in measuring taxonomic, phylogenetic, serological, and other forms of relationship, together with other benefits that can accrue from quantitation (such as improved methods for bacterial identification; see the discussion by Sneath of numerical identification on p. 26 of this *Manual*).

Numerical taxonomy has been broadly successful in most of these aims, particularly in defining homogeneous **clusters** of strains, and in integrating data of different kinds (morphological, physiological, antigenic). There are still problems in constructing satisfactory groups at high taxonomic levels, e.g. families and orders, although this may be due to inadequacies in the available data rather than any fundamental weakness in the numerical methods themselves.

The application of the concepts of numerical taxonomy was made possible only through the use of computers, because of the heavy load of routine calculations. However, the principles can be easily illustrated in hand-worked examples. In addition, two problems had to be solved: the first was to decide how to weight different variables or characters; the second was to analyze similarities so as to reveal the **taxonomic structure** of groups, species, or clusters. A full description of numerical taxonomic methods may be found in Sneath (1972) and Sneath and Sokal (1973). Briefer descriptions and illustrations in bacteriology are given by Skerman (1967), Lockhart and Liston (1970), and Sneath (1978a). A thorough review of applications to bacteria is that of Colwell (1973).

It is important to bear in mind certain definitions. Relationships between organisms can be of several kinds. Two broad classes are as follows.

Similarity on Observed Properties. Similarity, or **resemblance,** refers to the attributes that an organism possesses today, without reference to how those attributes arose. It is expressed as proportions of similarities and differences, for example, in existing attributes, and is called **phenetic relationship**. This includes both similarities in phenotype (e.g. motility) and in genotype (e.g. DNA pairing).

Relationship by Ancestry, or Evolutionary Relationship. This refers to the **phylogeny** of organisms, and not necessarily to their present attributes. It is expressed as the time to a common ancestor, or the amount of change that has occurred in an evolutionary lineage. It is not expressed as a proportion of similar attributes, or as the amount of DNA pairing and the like, although evolutionary relationship may sometimes be *deduced* from phenetics *on the assumption* that evolution has indeed proceeded in some orderly and defined way. To give an analogy, individuals from different nations may occasionally look more similar than brothers or sisters of one family: their phenetic resemblance (in the properties observed) may be high though their evolutionary relationship is distant.

Numerical taxonomy is concerned primarily with phenetic relationships. It has in recent years been extended to phylogenetic work, by using rather different techniques: these seek to build up on the assumed regularities of evolution so as to give, from *phenetic data,* the *most probable phylogenetic reconstructions.* Relatively little has been done so far in bacteriology, but a review of the area is given by Sneath (1974).

The basic taxonomic category is the species. It is noted in the chapter on "Nomenclature" that it is useful to distinguish a **taxospecies** (a cluster of strains of high mutual phenetic similarity) from a **genospecies** (a group of strains capable of gene exchange), and both of these from a **nomenspecies** (a group bearing a binominal name whatever its status in other respects). Numerical taxonomy attempts to define taxospecies. Whether these are justified as genospecies or nomenspecies turns on other criteria. It should be emphasized that groups with high genomic similarity are not necessarily genospecies: genomic resemblance is included in phenetic resemblance; genospecies are defined by gene exchange.

Groups can be of two important types. In the first, the possession of certain invariant properties defines the group without permitting any exception. All triangles, for example, have three sides, not four. Such groupings are termed **monothetic**. Taxonomic groups are, however, not of this kind. Exceptions to the most invariant characters are always possible. Instead, taxa are **polythetic**, that is they consist of assemblages whose members share a high proportion of common attributes, but not necessary any invariable set. Numerical taxonomy produces polythetic groups and thus permits the occasional exception on any character.

LOGICAL STEPS IN CLASSIFICATION

The steps in the process of classification are as follows:

1. Collection of data. The **bacterial strains** that are to be classified have to be chosen, and they must be examined for a number of relevant properties (**taxonomic characters**).
2. The data must be coded and scaled in an appropriate fashion.
3. The **similarity** or **resemblance** between the strains is calculated. This yields a table of similarities (**similarity matrix**) based on the chosen set of characters.
4. The similarities are analyzed for **taxonomic structure**, to yield the groups or clusters that are present, and the strains are arranged into **phenons** (phenetic groups), which are broadly equated with taxonomic groups (**taxa**).

5. The properties of the phenons can be tabulated for publication or further study, and the most appropriate characters (**diagnostic characters**) can be chosen on which to set up **identification systems** that will allow the best identification of additional strains.

It may be noted that those steps must be carried out in the above order. One cannot, for example, find diagnostic characters before finding the groups of which they are diagnostic. Furthermore, it is important to obtain complete data, determined under well standardized conditions.

Data for numerical taxonomy

The data needed for numerical taxonomy must be adequate in quantity and quality. It is a common experience that data from the literature are inadequate on both counts: most often it is necessary to examine bacterial strains afresh by an appropriate set of tests.

Organisms

Most taxonomic work with bacteria consists of examining individual strains of bacteria. However, the entities that can be classified may be of various forms,—strains, species, genera,— for which no common term is available. These entities, t in number, are therefore called **operational taxonomic units (OTUs)**. In most studies OTUs will be strains. A numerical taxonomic study, therefore should contain a good selection of strains of the groups under study, together with type strains of the taxa and of related taxa. Where possible, recently isolated strains, and strains from different parts of the world, should be included.

Characters

A **character** is defined as any property that can vary between OTUs. The values it can assume are **character states**. Thus, "length of spore" is a character and "1.5 μm" is one of its states. It is obviously important to compare the same character in different organisms, and the recognition that characters are the same is called the **determination of homology**. This may sometimes pose problems, but in bacteriology these are seldom serious. A single character treated as independent of others is called a **unit character**. Sets of characters that are related in some way are called **character complexes**.

There are many kinds of characters that can be used in taxonomy. The descriptions in the *Manual* give many examples. For numerical taxonomy, the characters should cover a broad range of properties: morphological, physiological, biochemical. It should be noted that certain data are not characters in the above sense. Thus the degree of serological cross-reaction or the percent pairing of DNA are analogous, not to character states, but to similarity measures.

Numbers of Characters

Although it is well to include a number of strains of each known species, numerical taxonomies are not greatly affected by having only a few strains of a species. This is not so, however, for characters. The similarity values should be thought of as estimates of values that would be obtained if one could include a very large number of phenotypic features. The accuracy of such estimates depends critically on having a reasonably large number of characters. The number, n, should be 50 or more. Several hundred are desirable, though the taxonomic gain falls off with very large numbers.

Quality of Data

The quality of the characters is also important. Microbiological data are prone to more experimental error than is commonly realized. The average difference in replicate tests on the same strain is commonly about 5%. Efforts should be made to keep this figure low, particularly by rigorous standardization of test methods. It is very difficult to obtain reasonably reproducible results with some tests, and they should be excluded from the analysis. As a check on the quality of the data, it is useful to reduplicate a few of the strains and carry them through as separate OTUs: the average test error is about half the percentage discrepancy in similarity of such replicates (e.g. 90% similarity implies about 5% of experimental variation).

Coding of the Results

The test reactions and character states now need coding for numerical analysis. There are several satisfactory ways of doing this, but for the present purposes of illustration only one common scheme will be described. This is the familiar process of coding the reactions or states into positive and negative form. The resulting table, therefore, contains entries + and − (or 1 and 0, which are more convenient for computation), for t OTUs scored for n characters. Naturally, there should be as few gaps as possible.

The question arises as to what weight should be given to each character relative to the rest. The usual practice in numerical taxonomy is to give each character equal weight. More specifically, it may be argued that unit characters should have unit weight, and if character complexes are broken into a number of unit characters (each carrying one unit of taxonomic information) it is logical to accord unit weight to each unit character. The difficulties of deciding what weight should be given *before* making a classification (and hence in a fashion that does not prejudge the taxonomy) are considerable. This philosophy derives from the opinions of the 18th century botanist Adanson, and therefore numerical taxonomies are sometimes referred to as Adansonian.

Similarity

The $n \times t$ table can then be analyzed to yield similarities between OTUs. The simplest way is to count, for any pair of OTUs, the number of characters in which they are identical (i.e. both are positive or both are negative). These **matches** can be expressed as a percentage or a proportion, symbolized as S_{SM} (for simple matching coefficient). This is the commonest coefficient in bacteriology. Other coefficients are sometimes used because of particular advantages. Thus the Gower coefficient S_G accommodates both presence-absence characters and quantitative ones, the Jacquard coefficient S_J discounts matches between two negative results, and the Pattern coefficient S_P corrects for apparent differences that are caused solely by differences between strains in growth rate and hence metabolic vigor. These coefficients emphasize different aspects of the phenotype (as is quite legitimate in taxonomy) so one cannot regard one or other as necessarily the correct coefficient, but fortunately this makes little practical difference in most studies.

The similarity values between all pairs of OTUs yields a checkerboard of entries, a square table of similarities known as a **similarity matrix** or **S matrix**. The entries are percentages, with 100% indicating identity and 0% indicating complete dissimilarity between OTUs. Such a table is symmetrical (the similarity of a to b is the same as that of b to a), so that usually only one half, the left lower triangle, is filled in.

These similarities can also be expressed in a complementary form, as *dissimilarities*. Dissimilarities can be treated as analogues of distances when preparing "taxonomic maps" of the OTUs, and it is a convenient property that the quantity $d = \sqrt{(1 - S_{SM})}$ is equivalent geometrically to a *distance* between points representing the OTUs in a space of many dimensions (a **phenetic hyperspace**).

Taxonomic structure

A table of similarities does not of itself make evident the **taxonomic structure** of the OTUs. The strains will be in an arbitrary order which will not reflect the species or other groups. These similarities therefore require further manipulation. It will be seen that a table of serological cross-reactions, if complete and expressed in quantitative terms, is analogous to a table of percentage similarities, and the same is true of a table of DNA pairing values. Such tables can be analyzed by the methods described below, though in serological and nucleic studies there are some particular difficulties on which further work is needed.

There are two main types of analyses to reveal the taxonomic structure, **cluster analysis**, and **ordination**. The result of the former is a tree-like diagram or **dendrogram** (more precisely a **phenogram**, because it expresses phenetic relationships), in which the tightest bunches of twigs represent clusters of very similar OTUs. The result of the latter is an **ordination diagram** or **taxonomic map**, in which closely similar OTUs are placed close together. The mathematical methods can be elaborate, so only a nontechnical account is given here.

In cluster analysis, the principle is to search the table of similarities for high values that indicate the most similar pairs of OTUs. These form the nuclei of the clusters and the computer searches for the next highest similarity values and adds the corresponding OTUs onto these cluster nuclei. Ultimately all OTUs fuse into one group, represented by the basal stem of the dendrogram. Lines drawn across the dendrogram at descending similarity levels define, in turn, phenons that correspond to a reasonable approximation to species, genera, etc. The commonest cluster methods are the **unweighted pair group method with averages (UPGMA)** and **single linkage.**

In ordination, the similarities (or their mathematical equivalents) are analyzed so that the phenetic hyperspace is summarized in a space of only a few dimensions. In two dimensions this is a scattergram of the positions of OTUs from which one can recognize clusters by eye. Three-dimensional perspective drawings can also be made. The commonest ordination methods are **principal components analysis** and **principal coordinates analysis.**

A number of other representations are also used. One example is a similarity matrix in which the OTUs have first been rearranged into the order given by a clustering method and then the cells of the matrix have been shaded, with the highest similarities shown in the darkest tone. In these "shaded S matrices," clusters are shown by dark triangles. Another representation is a table of the mean similarities between OTUs of the same cluster and of different clusters (**inter-** and **intra-group similarity table**): if based on S_{SM} with UPGMA clustering, this table expresses the positions and radii of clusters (Sneath, 1979a) and consequently the distance between them and their probable overlap—properties of importance in numerical identification as discussed later.

For general purposes a dendrogram is the most useful representation, but the others can be very instructive, since each method emphasizes somewhat different aspects of the taxonomy.

The analysis for taxonomic structure should lead logically to the establishment or revision of taxonomic groups. We lack, at present, objective criteria for different taxonomic ranks, that is, one cannot automatically equate a phenon with a taxon. It is however commonly found that phenetic groups formed at about 80% S are equivalent to bacterial species. Similarly, we lack good tests for the statistical significance of clusters and for determining how much they overlap, though some progress is being made here (Sneath, 1977, 1979b). The fidelity with which the dendrogram summarizes the S matrix can be assessed by the **cophenetic correlation coefficient**, and similar statistics can be used to compare the **congruence** between two taxonomies if they are in quantitative form (e.g. phenetic and serological taxonomies). Good scientific judgement in the light of other knowledge is indispensible for interpreting the results of numerical taxonomy.

Descriptions of the groups can now be made by referring back to the original table of strain data. The better diagnostic characters can be chosen—those whose states are very constant within groups but vary between groups. It is better to give percentages or proportions than to use symbols such as +, (+), v, d, or − for varying percentages, because significant loss of statistical information can occur with these simplified schemes. It would, however, be superfluous to list percentages based on very few strains. As systematic bacteriology advances, it will be increasingly important to publish the actual data on individual strains or deposit it in archives; such data will show its full value when test methods become very highly standardized.

It is evident that numerical taxonomy (and also numerical identification; see the chapter on "Identification of Bacteria" in this *Manual*) place considerable demands on laboratory expertise. New test methods are continually being devised. New information is continually being accumulated. It is important that progress should be made toward agreed data bases (Krichevsky and Norton, 1974), as well as toward improvements in standardization of test methods in determinative bacteriology, if the full potential of numerical methods is to be achieved.

BACTERIAL CLASSIFICATION III

Nucleic Acids in Bacterial Classification

John L. Johnson

Historically, classification of bacteria has been based on similarities in phenotypic characteristics. Although this method has been quite successful, it has not been precise enough for distinguishing superficially similar organisms or for determining phylogenetic relationships among the bacterial groups. Nucleic acid studies were first applied to such problems in bacterial classification more than 20 years ago and have since become of major importance. There are several advantages to be gained by basing classification on genetic relatedness:

1. A more unifying concept of a bacterial species is possible,
2. Classifications based on genetic relatedness tend not to be subject to frequent or radical changes,

3. Reliable identification schemes can be prepared after organisms have been classified on the basis of genetic relatedness,
4. Information can be obtained that is useful for understanding how the various bacterial groups have evolved and how they can be arranged according to their ancestral relationships.

The purpose of this chapter is to provide an overview of the principles involved in nucleic acid methodology, to give a brief description of the procedures being used, to compare the results obtained by one procedure with those obtained by another, and to indicate how the results are being used in bacterial classification.

PROPERTIES OF NUCLEIC ACIDS

DNA Base Composition

The first unique feature of DNA that was recognized as having taxonomic importance was its mole percent guanine plus cytosine content (mol% G + C). Among the bacteria, the mol% G + C values range from ca. 25–75 and the value is constant for a given organism. Closely related bacteria have similar mol% G + C values. However, it is important to recognize that two organisms that have similar mol% G + C values are not necessarily closely related; this is because the mol% G + C values *do not take into account the linear arrangement of the nucleotides in the DNA.*

Mol% G + C values were initially determined by acid-hydrolyzing the DNA, separating the nucleotide bases by paper chromatography, and then eluting and quantifying the individual bases. Other methods have since become more popular.

Thermal Denaturation Method. During the controlled heating of a preparation of double-stranded DNA in an ultraviolet spectrophotometer, the absorbance increases by ∼ 40%. This is due to the disruption of the hydrogen bonds between the base pairs that link the two DNA strands. *The temperature at the midpoint of the curve obtained by plotting temperature versus absorbance is called the "melting temperature," or T_m.* The T_m is correlated in a linear manner with the mol% G + C content of the DNA (Marmur and Doty, 1962). The higher the T_m, the higher the mol% G + C of the DNA (see Johnson, J.L. (1981) for further details).

Buoyant Density Method. When DNA is subjected to centrifugation in a cesium chloride density gradient (isopycnic centrifugation), it will become located in the form of a band at a position where its density

exactly matches that of the cesium chloride solution. The higher the density of cesium chloride where the DNA forms a band, the higher is the mol% G + C value of the DNA (Schildkraut et al. (1962), also see Mandel et al. (1968) for further details).

Although these methods are widely used for estimating DNA base composition, technical problems occasionally do arise because of contamination of the DNA preparation by polysaccharides or pigments, or because of excessive fragmentation of the DNA during its purification. Recent developments in high pressure liquid chromatography have resulted in methods that will accurately and rapidly quantify the free bases, nucleosides, or nucleotides of DNA (e.g. see Ko et al. 1977).

DNA Denaturation and Renaturation

A unique physical property of double-stranded (native) DNA is that under certain conditions (high temperature or high pH) the complementary strands will dissociate (denature). When the resulting single-stranded DNA is then subjected to a somewhat lower temperature and a rather high salt concentration, the complementary strands will reassociate (renature) to form double-stranded DNA/DNA structures (duplexes) that are very similar if not identical to the native DNA (Marmur and Doty, 1961). The renaturation rate is inversely proportional to the genome size (see the following references for further details: Wetmer and Davidson, 1968; Wetmer, 1976.

RNA/DNA Hybrids

Since only one strand of DNA is used by a cell as a template for RNA synthesis, RNA is complementary only to that strand. Since RNA is single-stranded, RNA molecules do not associate with other RNA molecules; however, when mixed with denatured DNA they can pair with a complementary DNA strand (hybridization) (see Galau et al., 1977 for further details).

Heterologous DNA Duplexes or RNA Hybrids

If denatured DNA from one organism is mixed with denatured DNA from a second organism, heterologous duplexes may form (i.e. duplexes consisting of one strand from the first DNA hybridized with one strand from the second DNA). Similarly, heterologous RNA duplexes may be formed when RNA from one organism is mixed with denatured DNA from a second organism. However, in order for heterologous DNA duplexes or RNA hybrids to occur, the two strands must be complementary in their nucleotide base sequence. A perfect match is not required, and estimates of the amount of base pair mismatch that is tolerated range from ~ 8–10% (Ullman and McCarthy, 1973). The thermal stability is usually determined by measuring strand separation during stepwise increases in temperature, and the results mimic the optical melting profile previously discussed under DNA Base Composition. The thermal stability is usually represented by the term "$T_{m(e)}$," which is the midpoint of the thermal stability profile (i.e. analogous to T_m of native DNA). The difference between the $T_{m(e)}$ of a heterologous duplex and that of a homologous duplex is referred to as the $\Delta T_{m(e)}$ and is used as a measure of the degree of base-pair mismatching in the heterologous duplex. The $\Delta T_{m(e)}$ values for heterologous duplexes range from 0 (no mismatching) to 18°C (considerable mismatching). In general, as the fractions of the genomes which can form heterologous duplexes decrease, the thermal stabilities of the duplexes that do form also decrease.

DNA AND RNA HOMOLOGY EXPERIMENTS

Such experiments attempt to answer one question: does DNA or RNA from organism A have a base sequence that is sufficiently similar to that from organism B to allow the formation of DNA heteroduplexes or heterologous RNA hybrids?

DNA Homology Values

These are average measurements of similarity in which the *entire genome of one organism is compared with that of another*.

RNA Homology Values

These values are specific for each type of RNA:

Messenger RNA (mRNA) Homology Values. These are similar to those obtained by DNA homology (at least for bacteria) because a large portion of the genome is used for transcribing the mRNA molecules. For this reason, and because mRNA is difficult to label, mRNA homology has not been widely used in bacterial taxonomy.

Ribosomal RNA (rRNA) and Transfer RNA (tRNA) Homology Values. In contrast to mRNA, rRNA and tRNA are coded for by *only a small fraction of the bacterial genome*; therefore, in homology experiments using either of these two types of RNA, only those fractions of the genome are being compared, not the entire genomes. In all groups of bacteria so far studied, the arrangement of nucleotides in the rRNA and tRNA cistrons of the DNA appears to have evolved less rapidly than the bulk of the cistrons in the DNA. This is probably due to their role in determining the structural and functional aspects of the ribosome (Woese et al., 1975).

Therefore, DNA homology experiments are used to detect similarities between *closely related* organisms, whereas RNA homology experiments are used to detect similarities between *more distantly related* organisms.

METHODS FOR HOMOLOGY EXPERIMENTS

Many procedures have been developed for detecting heterologous DNA duplexes or RNA hybrids. A brief description of some of these follows.

Heavy Isotopes

The earliest efforts to quantify the formation of heteroduplexes were made by incorporating a heavy base (5-bromouracil) or a heavy isotope (^{15}N) into one of the DNA preparations. After the labeled and unlabeled DNA preparations were mixed and allowed to reassociate, the mixture was subjected to ultracentrifugation with cesium chloride. This allowed the separation of heteroduplexes (which had an intermediate buoyant density) from the homologous duplexes (which had either a light or heavy density). These experiments were time-consuming and worked best only for small genomes such as those of viruses.

Agarose Gels

In 1963, McCarthy and Bolton immobilized high molecular weight denatured DNA in an agarose gel. The gel was then cut into small particles by forcing the agar through a small mesh screen. The agar particles were then incubated with radioactive-labeled RNA or fragmented DNA. The smaller RNA molecules or DNA fragments could diffuse through the agar and form hybrids or duplexes with complementary immobilized DNA. The immobilization of the high molecular weight DNA prevented it from reassociating with other high molecular weight DNA and also provided a means for washing unreacted labeled nucleic acid fragments away from those that had formed hybrids or duplexes with the immobilized DNA. The results from such experiments were quantitative and could be readily applied to broad taxonomic studies (Hoyer et al., 1964).

Binding to Nitrocellulose

In 1963, Nygaard and Hall found that native DNA, denatured DNA, and RNA/DNA hybrids would bind to nitrocellulose whereas RNA would not. This provided another means for immobilizing denatured DNA for use in RNA/DNA hybridization experiments and also for separating RNA/DNA hybrids from free RNA. The parameters for these experiments were worked out in detail by Gillespie and Spiegelman (1965).

In 1966, Denhardt described a procedure for covering the DNA binding sites on nitrocellulose membranes. This made it possible to first immobilize a given amount of denatured DNA on the membrane and then treat the membrane with a mixture that prevented additional DNA from binding to the membrane (unless it was complementary to the immobilized DNA on the membrane). Thus the membrane procedure became readily applicable to DNA homology experiments and has completely replaced the agarose gel method.

By the use of nitrocellulose membranes, DNA or RNA homology values can be determined by either *direct binding* or by *competition* experiments.

Direct Binding Method. In the direct binding method, a given amount of denatured labeled DNA or RNA is incubated under standardized conditions with various single-stranded DNA preparations that have been immobilized on nitrocellulose membranes. After incubation the unbound labeled nucleic acid is washed away and the radioactivity remaining on the membrane (due to duplex or hybrid formation) is measured. The *percent homology* is expressed as the *amount of heterologous binding divided by the amount of homologous binding × 100*. The results are somewhat variable because it is difficult to consistently get the same amounts of DNA on the membranes. This problem is circumvented with the competition method.

Competition Method. In the competition method, unlabeled denatured reference DNA is fixed onto nitrocellulose membranes. A direct binding reaction, used for a reference point, is performed between the homologous denatured labeled DNA in solution and membrane-bound reference DNA. The competitive reactions have the same components as the direct binding reaction but additionally contain high concentrations of unlabeled denatured DNA fragments in solution. If the competitor DNA is homologous to the labeled DNA in solution and to the unlabeled DNA bound to the membrane, the competitor DNA will form duplexes with both the labeled DNA and the immobilized DNA: consequently, the amount of labeled DNA that forms duplexes with the immobilized DNA will be much lower than that occurring in the direct binding reaction. The homologous competition will be ~90% effective. On the other hand, if the competitor DNA is not related, it will not form duplexes with the labeled DNA and immobilized DNA and there will be no competition. The percent homology is the ratio of the heterologous competition to the homologous competition × 100. Such competition experiments give very reproducible results but do require relatively large quantities of DNA (Johnson, J.L., 1981).

Free Solution Reassociation

In this method all of the component nucleic acids are in solution rather than being immobilized in some manner. Reassociation of DNA may be monitored optically by ultraviolet spectrophotometry or by means of a labeled probe.

Optical Procedure. In the optical procedure, the rates of reassociation are determined. Since DNA reassociation is a 2nd-order reaction, the rate will be proportional to the square of the concentration. The general procedure for comparing the DNAs from two organisms is to measure the reassociation rates of equivalent concentrations from each of the organisms separately and compare those rates with that of an equal mixture of the two DNA preparations. If the two organisms are identical, the reassociation rates in the three cuvettes will be the same. If the two organisms are unrelated, then each kind of DNA in the mixture will reassociate independently of the other and, since they are each at half the concentration as that used in the cuvettes with a single DNA component, the overall rate will be one-half. De Ley et al. (1970) have studied the parameters of the method in detail and have derived equations for calculating the homology values.

Labeled DNA Probe. The most popular procedure for free solution reassociation involves the use of a labeled DNA probe. As discussed above, the rate of DNA reassociation is a function of DNA concentration and, because the labeled probe DNA is used at a very low concentration, very little of it will reassociate. The unlabeled test DNA with

which the probe DNA is incubated is at a much higher concentration and most of it will reassociate. Therefore, if the probe DNA is identical to the unlabeled test DNA, it will reassociate with the unlabeled DNA at the rate at which the unlabeled DNA is reassociating. On the other hand, if the two DNAs are unrelated the unlabeled DNA will reassociate but most of the probe DNA will remain single-stranded. To determine the amount of probe DNA that has duplexed with the unlabeled DNA, either *hydroxylapatite* or *S1 nuclease* is usually used.

Hydroxylapatite is used to separate single-stranded (denatured) DNA from double-stranded DNA. At a phosphate concentration of 0.14 M, only double-stranded DNA will adsorb to hydroxylapatite and single-stranded DNA can be washed away. The double-stranded DNA can then be desorbed by increasing the phosphate concentration. Although originally used as a column chromatography procedure (Bernardi, 1969a, b; Miyazawa and Thomas, 1965), the batch procedure described by Brenner et al. (1969) has been widely used.

Under suitable conditions, S1 nuclease will have little effect on double-stranded DNA but will hydrolyze single-stranded DNA. Consequently, the extent of duplex formation by the probe DNA can be determined by the amount of S1 nuclease-resistant (i.e. acid-precipitable) radioactivity (Crosa et al., 1973).

Comparison of the Various Homology Methods

In spite of the diversity of the DNA homology methods, they are all used to measure the same phenomenon and so it is comforting to find that, for the most part, they all give similar results. The major experimental parameters that affect homology results are the sodium ion concentration and the reassociation temperatures. The most commonly used sodium ion concentration is about 0.4 M although concentrations up to 1 M do not alter the results significantly. The reassociation temperature can have a profound effect on the homology values and therefore a standardized temperature of about 25°C below the T_m (T_m − 25 °C) is most commonly used (Marmur and Doty, 1961). The reassociation temperature effect is approximately linear for the membrane competition and the hydroxylapatite procedures: for organisms having less than 50% homology, the homology values will increase by about 20% at 10 C below the T_m − 25°C temperature and decrease by about 20% at 10°C above the T_m − 25°C temperature. Reassociation temperature differences do not have as great an effect on the optical (De Ley et al., 1970) or the S1 nuclease methods (Grimont et al., 1980).

Under similar conditions of reassociation, the hydroxylapatite, membrane competition and spectrophotometric methods give very similar results (Kurtzman et al., 1980). The S1 nuclease procedure results in somewhat lower (15–20%) homology values, particularly between organisms having less than 50% homology.

The rRNA cistrons have been found to be very conserved in all groups of organisms that have been investigated. The nitrocellulose membrane procedures, such as competition, direct binding and thermal stability of hybrids have been used for most of the rRNA homology studies. Results from these experiments appear to reflect nucleotide sequence differences that are similar to those found in the DNA homology experiments discussed above.

rRNA OLIGONUCLEOTIDE CATALOGUES

Besides the use of RNA/DNA homology experiments for comparison of the rRNA cistrons from various bacteria, rRNA molecules have been compared directly by determining the nucleotide sequences in oligonucleotides. The rRNA preparation is first digested with T1 ribonuclease which cleaves between the 3′-guanylic acid and the 5′hydroxyl group of the adjacent nucleotide. This results in a guanine residue at the 3′ end of each oligonucleotide. The oligonucleotides are then separated by two-dimensional electrophoresis (Sanger et al., 1965; Uchida et al., 1974). The first dimension is on cellulose acetate at pH 3.5. The oligonucleotides are then transferred from the cellulose acetate

strip onto DEAE cellulose and electrophoresed in the second dimension in 6.5% formic acid. The oligonucleotide spots form three-to-four series of wedge-shaped patterns (Sanger et al., 1965). Within each pattern the oligonucleotides contain a constant number of uracil residues and the locations of the spots within a pattern indicate the number of adenine and cytosine residues. Therefore, by inspecting the pattern one can predict the nucleotide sequence of the shorter oligonucleotides and the base compositions for the longer ones. The spots containing the longer nucleotides are then cut out for secondary analysis. After digestion with other ribonucleases they are again electrophoresed on

DEAE cellulose. If the nucleotide sequence still is not clear, a tertiary analysis is required. The unique oligonucleotides (usually only one per rRNA molecule) of each organism are entered (catalogued) into computer storage. The oligonucleotide catalog from one organism can then be compared with that of another. The similarity values between two organisms is the number of unique oligonucleotides (in each of their rRNA molecules) that they both share divided by the average total number of unique oligonucleotides. This procedure compares the sequence for a rather large portion of the rRNA molecules.

Most recently, procedures have been developed for rapidly sequencing long segments of DNA and RNA (Maxam and Gilbert, 1977; Peattie, 1979; Sanger et al., 1977). DNA from several viruses have been sequenced. Sequencing all of the DNA of a bacterium would generate a rather formidable amount of data; however, specific cistrons have been compared by sequence analysis, such as the genes of the tryptophan operon of *Escherichia coli* and *Salmonella typhimurium* (Crawford et al., 1980).

CONTRIBUTIONS OF NUCLEIC ACID STUDIES TO BACTERIAL TAXONOMY

Concept of a Bacterial Species

A major contribution of DNA homology studies has been to provide a more unifying concept of a bacterial species. Although the exact level of DNA homology above which one considers organisms as belonging to the same species is arbitrary, similar homology clusters have been found in all bacterial groups that have been investigated. I have previously suggested what seemed to be reasonable cut-off points for delineating subspecies, species, and closely related species (Johnson, 1973). These are illustrated in Figure III.1. DNA heterogeneity in the species range (*A*) has been found for many bacterial groups that are phenotypically very similar. In some instances the homology values will tend to cluster in the 80–90% homology range (*B*). Examples of this are the clustering of *Propionibacterium acnes* (Johnson and Cummins, 1972) and *Bacteroides uniformis* (Johnson, 1978). In other instances there may also be clustering at the lower end of the species range (*C*). *Bacteroides fragilis*, for example, clusters into two groups where the intergroup homology values are in the range of 60–70% and the intragroup homology values in the 80–90% range (Johnson, 1978). It is important to note that the thermal stabilities of heteroduplexes between organisms in the 80–80% DNA homology range will be very similar to those of homoduplexes ($\Delta T_{m(e)}$ values of 0–3°C), whereas with heteroduplexes between 60–70% homology they will be substantially lower ($\Delta T_{m(e)}$ values of 6–9°C). Therefore, it appears that 60–70% homology is a transitional point between genetic events that may be largely cistron-rearranging in nature and genetic events where there are also many changes in the base sequences (Johnson, 1973). In other instances, e.g. *Bacteroides ovatus* (Johnson, 1978), multiple groups within the 60–70% homology range make subgrouping at this level rather complicated so that, unless there are other important considerations, such as pathogenicity (Krych et al., 1980), it may not be justified.

The DNA homology groups in the lower homology range (*D* in Fig. III.1) often are quite distinct phenotypically from the species with which they are being compared, although in some instances they may differ only in a few characters (Johnson and Ault, 1978; Johnson, 1981; Mays et al., 1982).

It is important to remember that few bacteria have read Fig. III.1; therefore, the exact limits chosen for a given group of organisms will have to remain at the discretion of the individual investigators.

Identification Schemes

A major practical use of DNA homology data is for correlation with individual phenotypic tests. It is common to find variability for a trait

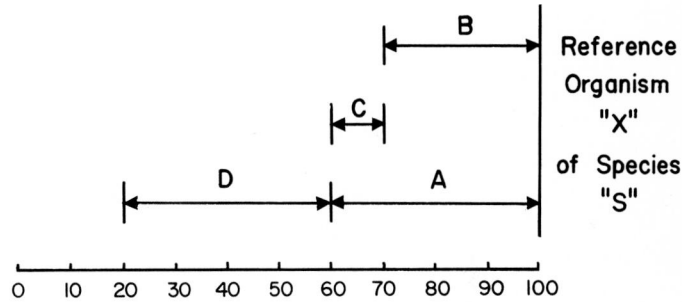

PERCENT DNA HOMOLOGY

Figure III.1. Proposed taxonomic groupings based upon DNA homology data. *A*, organisms belonging to Species "*S*"; *B*, varieties within subspecies to which "*X*" belongs; *C*, other subspecies that belong to "*S*"; *D*, species that are closely related to Species "*S*".

among strains within a DNA homology group as well as distinct DNA homology groups that differ from each other by only a few traits (Johnson and Ault, 1978; Johnson, 1980; Mays et al., 1982; Holdeman et al., 1982). Correlating phenotypic test results with DNA homology groups enables investigators to select phenotypic tests that are required for the accurate identification of organisms belonging to these groups.

Concept of a Bacterial Genus and Higher Taxa

Comparisons of rRNA cistrons by rRNA homology experiments and by 16S oligonucleotide catalogue similarities are providing data from which a more unifying phylogenetic concept for higher bacterial taxa is possible. De Ley and his associates (De Ley et al., 1978; De Smedt et al., 1980) have proposed the establishment of several genera on the basis of rRNA homology results. On the basis of 16S rRNA oligonucleotide similarity, Woese (in Fox et al., 1980) has proposed the reestablishing of the higher bacterial taxa which were dropped from the eighth edition of *Bergey's Manual* because it was thought that the higher taxa listed in the seventh edition did not represent phylogenetic relationships. As examples, the 16S rRNA oligonucleotide similarity values have contributed greatly to the present taxonomic scheme of the methanogenic bacteria (Balch et al., 1979) and to the establishment of Division IV *Mendosicutes* in the Kingdom *Procaryotae* (see the chapter on The Higher Taxa by Murray in this *Manual*).

BACTERIAL CLASSIFICATION IV

Genetic Methods

Dorothy Jones

The use of genetic characteristics in bacterial classification is comparatively recent. It dates from the mid 1950s when bacterial gene transfer was discovered and Watson and Crick demonstrated the molecular basis of genetic information in the sequence of bases on the deoxyribonucleic acid (DNA) molecule. Since that time the development of physicochemical techniques for the analysis of the genetic material, together with the exploitation of bacteria as genetic tools, has resulted in the accumulation of material which has proved significant for bacterial systematics.

In the last 2 decades it has become clear that the genetic complement of a bacterial cell lies not only in the main chromosome but, in many cases, also in extrachromosomal elements such as plasmids, transposons and lysogenic or temperate phages. All these elements carry genetic material capable of phenotypic expression. What contribution such extrachromosomal entities make to a particular bacterial phenotype, either by direct expression or interaction with the chromosomal DNA of the cell, is only just beginning to be understood (see Broda, 1979; Harwood, 1980; Hardy, 1981).

For the bacterial taxonomist the genetic approach to systematics has great appeal both for its potential to reveal biologically significant, stable groupings (taxa) and for the elucidation of bacterial evolutionary relationships (phylogeny). Consequently several of the newer taxonomic methods have been and are being directed towards the characterization of the genetic complement of bacteria.

Physicochemical methods for the analysis of bacterial genomes have been discussed in the previous chapter. The present chapter is concerned with genetic methods used in bacterial classification, i.e. methods based on the transfer of genes between bacteria.

CHROMOSOMAL GENE EXCHANGE

The three main classes of chromosomal gene exchange are: (a) those in which genes are transferred as soluble DNA molecules, i.e. **transformation**; (b) those involving transfer by bacteriophage i.e. **transduction**; and (c) those involving cell contact followed by transfer of whole or part of the bacterial chromosome, i.e. **conjugation**. Of these classes, transformation studies have so far proved the most useful for determining relationships between bacteria.

Transformation

Transformation has been demonstrated usually between different taxospecies and only rarely between taxa presently recognized as different genera. Interspecific transformation has revealed three distinct homology groups amongst neisseriae and moraxellae. Transformation studies have indicated a close relationship between *Rhizobium leguminosarum* and *Agrobacterium tumefaciens*. Studies with the micrococci have shown a close relationship between *Micrococcus luteus* and *Micrococcus lylae* and, in this case, was confirmed by DNA reassociation studies in vitro. Similar studies have shown a low rate of transformation between *Pasteurella multocida*, *P. haemolytica*, *P. ureae* and *P. pneumotropica*, taxa which are also closely related on phenetic and DNA reassociation criteria. A great deal of transformation work has been done on the genus *Haemophilis*. *Haemophilus influenzae*, *H. aegyptius* and *H. parainfluenzae* all appear to be closely related.

Transformation of chromosomal DNA has been demonstrated also among other taxa and there is no doubt that it is a good indication of the degree of relatedness between different taxospecies and can highlight areas of taxonomic homogeneity and heterogeneity (Jones and Sneath, 1970; Bøvre, 1980).

Transduction

In transduction, host chromosomal material is incorporated into a bacteriophage by several mechanisms and transmitted from one bacteriophage host to another by phage-mediated transduction. Only a small range of bacterial groups are presently known to be susceptible to transduction, e.g. the *Enterobacteriaceae*, the genus *Bacillus*, pseudomonads and some streptococci. Not much is known about how readily strains of the same species can be transduced, but the host range pattern of the transducing bacteriophages is probably a major limiting factor. It has been suggested also that the greater difficulties associated with transduction are due to the larger sizes of the DNA fragments involved in transduction as compared with transformation, the larger fragments being less easily integrated into the recipient chromosome. Again this mechanism of genetic transfer appears to have significance for bacterial classification only at the taxospecies level, and its usefulness is further restricted by the host range of bacteriophages (see Jones and Sneath, 1970).

It is appropriate here to mention the other roles of bacteriophages in bacterial classification. As noted earlier, a temperate bacteriophage can lysogenize in a host bacterium and express its genetic information as phenotypic characters different from those typical of the bacterium devoid of phage. The consequences of this for bacterial classification will be dealt with later (see Extrachromosomal Elements). Additionally, and this is perhaps their best known feature, virulent bacteriophages infect and lyse bacteria. The process is referred to as **phage lysis**.

The inclusion of phage lysis in a section dealing with genetic methods may cause the reader some surprise. Phage lysis of the bacterial cell (as distinct from the much less specific phage adsorption, or killing of the cell followed by lysis from without) involves phage infection with phage multiplication but without lysogenization. In bacteriophage infection, the genes of the virulent phage are transferred and expressed even though they are not integrated into the host chromosome nor of course in the lineage of the recipient. Specific phage receptors are necessary for the adsorption of virulent phage on to the recipient bacterial cell; once in the cell the phage may be repressed if the bacterium is carrying a homologous prophage or it may be restricted enzymically. The ability of two bacterial strains to support the growth of a given virulent phage may reflect similarity in only one or two host genes. Therefore, the technique has little value for *bacterial classification*. However, the value of phage lysis cross-reactions for *bacterial identification* is high. The reported host range of bacteriophages extends from those specific for very few strains of one taxospecies, to those that can lyse bacteria which are currently placed in different bacterial genera, families and even orders. However, most reports in the literature show that most phages lyse a significant proportion of strains belonging to the same taxospecies as the propagating strain. Phage-typing schemes are playing increasingly important epidemiological and identification roles among a number of bacterial groups, e.g. some pyogenic streptococci, staphylococci and enterobacteria.

Conjugation

This method of gene exchange refers to the transfer of the whole or a portion of the bacterial chromosome following cell-to-cell contact. The conjugation system is best understood among the coliforms (Curtiss, 1969). Similar systems have been noted amongst other genera such as *Pseudomonas*, *Vibrio*, *Pasteurella* and *Rhizobium* and are known to occur among other groups, but the mechanism is less well understood. In the streptococci there is evidence that in some cases the bacteria make use of sex pheromones to generate cell-to-cell contact (Clewell, 1981). Transfer of bacterial chromosomal material by conjugation has not been reported so frequently as by transformation or transduction. However, evidence suggests that it takes place only between closely related taxa.

Bacterial taxonomy has not, to date, benefited greatly from studies involving genetic exchange of chromosomal material and the concept of a bacterial genospecies is far from being realised. However, bacteriologists no longer believe that gene transfer is so rare among bacteria that it is of no consequence for natural bacterial populations. In the past 2 decades it has been recognized that gene transfer particularly involving phages, plasmids and transposons, their interaction with each other, and the bacterial chromosome together with the gene transfer mediated by insertion sequences, can be a significant factor in bacterial variation. This variation has obvious consequences for bacterial systematics.

EXTRACHROMOSOMAL ELEMENTS

Plasmids, transposons and phages are collectively referred to as extrachromosomal elements (Novick, 1969; Broda, 1979; Hardy, 1981). Their transfer between bacteria is essentially by the same mechanisms as those described under Genetic Methods. Phages play a role in the transduction of all genetic material between bacteria and it is now recognized that the F′ factor is a plasmid. It is therefore probably artificial to make too clear a distinction between the transfer of chromosomal DNA and that of extrachromosomal elements between bacteria. Transformation by chromosomal DNA may or may not be accompanied by plasmid DNA. In transduction phages can carry a portion of the chromosomal or plasmid DNA, and in conjugation plasmid and chromosomal DNA can be transferred at the same time. The situation is far more complex than was previously realized (Novick, 1969; Hardy, 1981; Clewell, 1981).

A range of methods now exists for the isolation of extrachromosomal genetic elements from bacteria and for their analyses by physicochemical methods. A good account is given by Hardy (1981).

The two aspects of extrachromosomal elements which are of prime interest to the bacterial taxonomist are their ability to code for phenotypic traits in a range of bacteria, and their significance in evolution.

Phenotypic Traits

Plasmids have been observed in virtually every bacterial genus examined. Many plasmids detected by physical screening methods are not known to code for any phenotypic trait in the host bacterium. They are called **cryptic plasmids**. The fact that their presence has not been correlated with a phenotypic characteristic does not mean that they do not code for such a trait. It may be that their particular phenotypic traits have not been identified.

Phenotypic traits known to be coded for by plasmids include resistance to a variety of antibiotics, heavy metal ions and ultraviolet light; production of enterotoxin, exfoliate toxin, the surface antigens K88 and K89, hemolysins, proteases, bacteriocins, urease and H_2S; metabolism of lactose, sucrose, raffinose and citrate; degradation of a variety of organic compounds such as camphor, octanol and toluene (at least part of the remarkable diversity shown by pseudomonads in the degradation of organic compounds is due to the presence of degradative

plasmids); and nitrogen fixation. Preliminary evidence suggests that the production of gas vacuoles in *Halobacterium* is controlled by a plasmid. There is also evidence that pigment, coagulase and fibrolysin production in staphylococci are plasmid determined. It also seems highly probable that among the streptococci the production of serum opacity factor, M protein production, nisin production and the ability to ferment galactose and xylose are plasmid-coded.

Transposons found on the plasmids of Gram-negative bacteria have been shown to code for resistance to a number of antibiotics, lactose fermentation (in *Yersinia enterocolitica*), heat stable toxin (in *E. coli*) and doubtless others coding for other phenotypic traits will be found. Full accounts of the phenotypic traits conferred on bacteria by plasmids and transposons are given by Harwood (1980), Clewell (1981) and Hardy (1981).

The classic example of a phage-encoded phenotypic trait is the diphtheria toxin which was shown by Freeman in 1950 to be produced only when *Corynebacterium diphtheriae* is lysogenized by a particular phage. The structural gene for the protein toxin is on the phage chromosome. This phage can lysogenize and synthesize toxin in a number of closely related corynebacteria viz. *C. diphtheriae*, *C. ulcerans* and *C. ovis* (Barksdale, 1970).

Effect of Extrachromosomal Elements on Classification and Identification

Since these elements confer extra phenotypic traits on their hosts they could have a marked effect on bacterial classification if those characters were ones on which the classification was based. Two examples of the presence of plasmids which relate to species nomenclature are the plasmid-coded hemolysin of *Streptococcus faecalis* which resulted in the naming of such plasmid-bearing strains as *Streptococcus faecalis* var. *zymogenes*, and the plasmid-determining citrate utilization in *Streptococcus lactis* which appears to be responsible for the name *Streptococcus lactis* subsp. *diacetylactis*. However, the effect of an extra chromosomal-coded trait on a classification based on a large number of characters would normally be expected to be small and this has proved to be the case in the few preliminary studies so far conducted.

Such characters can, however, affect the identification of bacteria when the identification is based on a small number of characters and considerable weight is placed on individual features, e.g. lactose fermentation in the identification of enterobacteria. It is best therefore if identification schemes are based on stable features chosen as a result of a taxonomic study where a large number of characters have been employed, e.g. computer-assisted classifications (numerical taxonomy). Ideally, computer-based identification matrices derived from such studies should be employed. The risk of a misidentification due to the loss or gain of one or two phenotypic characters is thereby reduced to a minimum.

It has been suggested that strains known to carry extrachromosomal elements should be excluded from taxonomic studies. Such a policy is not practical because present methods do not always detect such strains; further, it is believed that many bacterial populations depend on the presence of these elements for their survival. It has also been suggested that known or suspected extrachromosomal coded characters should be excluded when classifications are constructed. Again, present methods do not allow all such characters to be determined, besides, such characters may have taxonomic relevance.

Bacteriologists should accept that extrachromosomal elements do contribute to bacterial variation. This variation should therefore be recognized, and due allowance made, when bacterial taxa are described and when identification schemes are constructed.

Extrachromosomal Elements and Evolution

At the present time the relative contributions of mutation and recombination to bacterial evolution is difficult to assess. Mutation results in changes in the protein structure of the organism. Recombination leads to the rearrangement of existing genes. Until recently little attention was paid to the possible involvement of gene rearrangement in evolution. The recognition that gene transfer involving extrachromosomal elements can be a significant factor in bacterial variation has led to a view that these elements have played an important role in bacterial evolution. Whether or not the role which these elements play in contemporary bacterial variation and adaptation is one of the major ways in which bacteria have evolved from the earliest times is still not resolved (Cullum and Saedler, 1981; Hardy, 1981; Koch, 1981; Reanney, 1976).

BACTERIAL CLASSIFICATION V

Serology and Chemotaxonomy

Dorothy Jones and Noel R. Krieg

Serology and chemotaxonomy are both methods for investigating the molecular architecture of the bacterial cell although the methodologies used in the two techniques are quite different.

SEROLOGY

Serological techniques depend on the ability of the chemical constituents of bacterial cells to behave as antigens, i.e. to elicit the production of antibodies in vertebrate animals. The antibodies used in serological studies are the humoral antibodies found in the blood serum and referred to as antiserum.

Serological techniques used include agglutination, precipitation (including many refinements, e.g. use of gels and electrophoretic techniques), complement fixation and immunofluorescence. Details of the techniques may be found in a number of immunological or microbiological text books.

Serological studies of value in bacterial taxonomy can be divided into two broad classes: (a) those concerned with detecting differences or similarities between bacteria **on the basis of their cell surface and associated antigenic complement** (e.g. flagella, pili, cell walls, cytoplasmic membranes, capsules and slime layers) and (b) the use of antisera raised against purified enzymes to assess **structural similarities between homologous proteins from different bacteria**.

Cell Surface and Associated Antigens

On the basis of the antigenic complexity of their surface antigens (cell wall lipopolysaccharide, flagella and capsule constituents), the genera of the family *Enterobacteriaceae* can be divided into many serovars; e.g. more than 1000 serovars have been detected within the genus *Salmonella* (Kauffmann, 1966). Contrary to the view of Kauffmann (1966) these serovars do not represent separate taxospecies. The information derived from serological studies of a group such as the enterobacteria is now so large, and so many cross-reactions occur that, in the absence of any methods (e.g. computer programs) for analyzing the plethora of data in an objective fashion, it is generally accepted that these techniques are of little value in classification but are valuable in epidemiological studies.

Serological studies of the streptococci based on the use of acid-extracted polysaccharide antigens (Lancefield, 1933, 1934) have resulted in the division of the genus *Streptococcus* into a number (now approaching 30) of serological groups labeled A, B, C, etc. Until fairly recently, very great emphasis was placed on the serological grouping of streptococci for purposes of both classification and identification.

Although some serological groups correspond to distinct taxospecies (e.g. serological group A (*S. pyogenes*) and serological group B (*S. agalactiae*)) other serological groups comprise more than one taxo-species (e.g. serological groups C, D and N), while serological groups G, H and K do not serve to define any good taxa (see Jones, 1978).

Other serological studies of this kind have been based on the use of different classes of antigenic material, e.g. cell walls, spore suspensions, etc. A review of the serochemical specificity and location of antigens in the bacterial cell together with observations on the significance of such serological studies for bacterial classification has been provided by Cummins (1962).

Use of Antisera Raised Against Purified Proteins

The basis of this approach is that one antiserum raised against a purified enzyme can be used to detect the serological cross-reactions of homologous proteins in crude extracts of other bacteria if the bacteria possess the same enzyme. The use of microcomplement fixation techniques makes this approach a very sensitive one. Comparative studies on purified proteins of known primary structure have indicated that there is a very high correlation between the amino acid sequence of the proteins and the degree of serological similarity (see later section on Amino Acid Sequences). Examples of this approach include studies of the muconate-lactonizing enzymes of the *Pseudomonadaceae* (Stanier et al., 1970), the fructose diphosphate aldolases of the lactic acid bacteria (London and Kline, 1973) and the catalases of staphylococci and micrococci. In the instance of the staphylococci, a very high correlation has been shown to exist between the serological relationships of their catalases and genetic relatedness based on DNA/DNA homology data (see Kandler and Schleifer, 1980).

Similar studies on the transaldolases of several species of bifidobacteria (Sgorbati and Scardovi, 1979; Sgorbati, 1979) indicate that the genus *Bifidobacterium* contains several distinct clusters based on the index of dissimilarity of their respective aldolases. In some instances there was good correlation between the clusters so obtained and clusters formed on the basis of other criteria, but in others the correlation was not so high.

Baumann et al. (1980) and Bang et al. (1981) found the immunological relationships among glutamine synthetases and superoxide dismutases of *Vibrio* and *Photobacterium* species were in good agreement with relationships based on rRNA/DNA homology experiments. The amino acid sequence of the glutamine synthetases was conserved to a greater extent than that of the superoxide dismutases, and this supports the idea that the study of proteins having different evolutionary rates

15

can permit the resolution of close, intermediate, and distant relationships among organisms.

It should be noted that serological homology studies of proteins, like many other techniques, have their limitations. There is evidence that the approach is useful only for the study of proteins with relatively high (70% or greater) sequence homologies. Further, serological techniques measure similarities only at the surface of proteins and it is at the protein surface that the greatest number of amino acid changes occur. The results can be influenced also by the number of antigenic sites per protein molecule. Nevertheless, serological techniques of this kind provide a rapid and convenient method for assessing structural similarities between homologous proteins and are useful in the classification of bacteria and can also cast some light on possible phylogenetic relationships.

CHEMOTAXONOMY

During the past 20 years or so, the application of chemical and physical techniques to elucidate the chemical composition of whole bacterial cells or parts of cells has produced information of great value in the classification and identification of bacteria. Indeed, so useful have some of the data generated proved to be, that the word "chemotaxonomy" used to describe the classification of bacteria on the basis of their chemical composition, is now firmly entrenched in the literature.

In addition, techniques such as gas chromatography have allowed the more precise analysis of the products of fermentation and there is a growing awareness of the taxonomic significance of enzyme systems and their regulation as opposed to the detection of individual enzymes.

Cell Wall Composition

The characteristic cell wall polymer of many procaryotes, present in Gram-negative and Gram-positive bacteria and in the cyanobacteria, is peptidoglycan. Peptidoglycan is not found in the mycoplasmas nor in the archaebacteria. The chemical structure of the peptidoglycan of Gram-negative bacteria is, with few exceptions, reasonably uniform. However, the variation in qualitative amino acid and/or sugar composition, especially the variation in the primary structure of the peptidoglycans of various Gram-positive bacteria has provided information of enormous taxonomic value.

The cell wall composition of Gram-positive bacteria was one of the earliest useful chemotaxonomic characters. On the basis of the analysis of the purified cell walls of Gram-positive bacteria, Cummins and Harris (1956) suggested that the cell wall amino acid composition might prove to be an important taxonomic criterion at the generic level and that the sugar composition might help to distinguish between species. Subsequent studies have indicated this to be the case. The amino acids present in the cell wall are now an accepted important part of the generic description. Information from cell wall analysis of the type done by Cummins and Harris (1956) has proved of especial value among the coryneform group of bacteria (see Keddie and Bousfield, 1980, for a comprehensive review).

Information of even greater taxonomic value has resulted from the methods devised by Schleifer and Kandler (1967) and used by them and their associates to determine the peptidoglycan types of a wide range of Gram-positive bacteria (see Schleifer and Kandler, 1972; Kandler and Schleifer, 1980). This approach has revealed differences between bacteria which could not possibly have been detected by qualitative cell wall analysis. The methods for determining differences in peptidoglycan types are, however, quite specialized and cannot be used routinely to screen large numbers of bacteria.

More recently a number of "rapid" methods have been developed for the routine screening of bacteria to determine those cell wall components which have been shown to be of the greatest discriminatory value in bacterial classification and identification. The review of Keddie and Bousfield (1980) contains references to the pertinent literature as does that of Kandler and Schleifer (1980).

A novel peptidoglycan, the so-called pseudomurein, characterized by the replacement of muramic acid by talosaminouronic acid, has been found to be the typical cell wall constituent of the genus *Methanobacterium*, which taxon is now recognized as a member of the archaebacteria (see Kandler and Schleifer, 1980).

Lipid Composition

Among the procaryotes there are two quite distinct lipid categories. The eubacteria possess acyl lipids (ester-linked) while the archaebacteria possess ether-linked lipids. Thus, the presence of ether-linked lipids serves to distinguish the archaebacteria. Further details of the archaebacterial lipids are given by Kates (1978).

Lipids occur in the cytoplasmic membranes of all eubacteria and in the cell wall complex of Gram-negative bacteria and certain Gram-positive bacteria such as the genera *Corynebacterium* and *Mycobacterium*. The eubacterial lipids comprise a number of different classes and, in the last decade, it has become increasingly clear that at least some of these lipids have chemotaxonomic potential (see Lechevalier, 1977).

The fatty acid composition of the bacterial cell has proved useful in the classification of certain bacteria and in some cases the fatty acid pattern may be characteristic for a particular taxon (see Lechevalier, 1977). However, it should be noted that the fatty acid patterns obtained may be influenced by a number of factors: composition of growth medium, temperature of incubation, age of culture and by the techniques employed to analyze the sample.

A special category of fatty acids free from the aforementioned limitations are the mycolic acids. These long-chain 3-hydroxy 2-branched acids have been found, so far, only in the taxa *Bacterionema, Corynebacterium, Micropolyspora, Mycobacterium, Nocardia* and *Rhodococcus*. Differences in the structure of their component mycolic acids have proved to be a valuable criterion in the classification and identification of members of these taxa (see Minnikin and Goodfellow, 1980; Collins et al., 1982).

Another class of lipids of recognized chemotaxonomic potential are the polar lipids which occur in all bacteria. The most common polar lipid types are the phospholipids and the glycolipids. Phospholipids occur in many bacteria but certain actinomycetes and coryneform bacteria contain very characteristic phospholipids, the phosphatidylinositol mannosides. Other highly characteristic phospholipids include phosphosphingolipids found in certain Gram-negative taxa, e.g. *Bacteroides* (see Lechevalier, 1977). Glycolipids (glycosyl diacylglycerols) are widely distributed amongst Gram-positive bacteria and can also be used as chemotaxonomic markers (see Shaw, 1975).

Other lipids with chemotaxonomic potential include hopanoids, hydrocarbons, and carotenoids.

Isoprenoid Quinones

Isoprenoid quinones are a class of terpenoid lipids located in the cytoplasmic membranes of many bacteria. They play important roles in electron transport, oxidative phosphorylation and possibly active transport. Their potential as an aid to the classification of bacteria was recognized by Jeffries et al. (1969), Yamada et al. (1976) and others (see Collins and Jones, 1981). Representatives of one, or more than one, of the three main types, ubiquinones, menaquinones and demethylmenaquinones are present in the majority of procaryotes so far examined. The cyanobacteria contain neither ubiquinones nor menaquinones. However, they do contain phylloquinones and plastoquinones which are indigenous to the plant kingdom but not normally found in bacteria. All the mycoplasmas so far examined contain menaquinones only. Among the archaebacteria no isoprenoid quinones have been detected in the fastidious anaerobic species *Methanobacterium thermoautotrophicum*, a situation in keeping with that most commonly, but not invariably, found among the strictly anaerobic eubacteria. An unusual terpenoid, caldariellaquinone, has been detected in the extreme acidophile "*Caldariella acidophila*." The other archaebacteria examined possess menaquinones.

The majority of the strictly aerobic, Gram-negative bacteria produce only ubiquinones, with the exception of cytophagas and myxobacters which produce only menaquinones. Facultatively anaerobic, Gram-negative bacteria contain ubiquinones, menaquinones or demethylmenaquinones or a combination of the three. Strictly anaerobic, Gram-negative bacteria (e.g. the genus *Bacteroides*) produce only menaquinones.

The majority of the aerobic and facultatively anaerobic Gram-positive bacteria produce only menaquinones. Most streptococci do not contain any isoprenoid quinones but demethylmenaquinones are present in *Streptococcus faecalis* and menaquinones have been detected in "*S. faecium* subsp. *casseliflavus*" and *S. lactis*. Similarly members of the genus *Lactobacillus* generally lack isoprenoid quinones but recently low levels of an uncharacterized menaquinone have been detected in one strain of *L. brevis*. Uncharacterized menaquinones have also been reported in some strains of the strictly anaerobic genus *Clostridium* although, in general, this genus lacks quinones.

The current data on isoprenoid quinone structural types in bacteria and their implications for taxonomy are reviewed by Collins and Jones (1981). From the available data on procaryotes in general, it appears that menaquinones have far greater discriminatory value that ubiquinones. Menaquinones possess not only a greater range of isoprenologues but additional modifications such as ring demethylation and partial hydrogenation of the polyprenyl side chain occur. The available data strongly suggest that these compounds will be of considerable value in the classification of micrococci, staphylococci, coryneform bacteria and certain actinomycetes (see Collins and Jones, 1981).

Cytochrome Composition

Cytochromes are specialized forms of hemoproteins which are involved in a variety of redox processes in the procaryote cell. They can be assigned to four main classes, *a, b, c* and *d* according to the structures of their heme prosthetic groups. Cytochrome *o* is an autoxidizable *b*-type cytochrome.

Two basic methods are available which use cytochromes as an aid to classification and identification of bacteria; the "pattern" and the "structure" approach. The former compares the cytochrome patterns of different bacterial species as compared by conventional difference spectrophotometry (see Meyer and Jones, 1973); the latter compares primary structures and, where possible, the tertiary structures of easily purified cytochrome *c* as determined by amino acid sequence and x-ray diffraction (see Ambler, 1976, and see later section on Amino Acid Sequences).

Cytochrome patterns show greater variation among the procaryotes than among eucaryotes and can therefore be a useful aid in bacterial classification. Qualitative analyses of the cytochrome composition of over 200 species of bacteria have now been done. The results indicate that the heterotrophic Gram-positive bacteria comprise a rather homogeneous grouping with cytochromes $bcaa_3o$ forming the predominant pattern. There are some variations however and cytochrome *c* is often absent from facultatively anaerobic Gram-positive bacteria. Some lactic acid bacteria contain only cytochrome *b* when grown on a heme-containing medium. Propionibacteria exhibit a cytochrome bda_1 pattern. The genus *Clostridium* lacks cytochromes. It is of interest that the cytochrome $bcaa_3o$ pattern of logarithmic growth-phase cells of the aerobe *Arthrobacter globiformis* changes to $bcaa_3od$ when the cells become oxygen-limited and lose their ability to retain the crystal violet-iodine complex in the Gram stain. Cytochrome *d* is characteristic of many Gram-negative bacteria.

In contrast, the Gram-negative heterotrophic bacteria form a much less homogeneous group on the basis of cytochrome composition. The majority have the basic pattern $bcdoa_1$ from which *c* may often be absent. Cytochrome c_{co} appears to be characteristic of methylotrophs; however, it is also present in the nonmethylotrohpic genus *Chromobacterium*. The phototrophic bacteria contain cytochromes *b* and *c* when grown photosynthetically but there are differences between the taxa when grown aerobically in the absence of light. The obligately aerobic chemolithotrophs exhibit the cytochrome pattern $bcaa_3oa$, with some occasional omissions. Neither the phototrophs nor the chemolithotrophs have been shown to produce cytochrome *d*.

There is now sufficient evidence available to indicate that cytochrome patterns, in conjunction with other evidence, are useful guides in bacterial classification. There is, however, little evidence that cytochrome patterns will be useful for purposes of identification mainly because bacteria contain relatively few types of spectrally distinct cytochromes. A comprehensive review of the use of cytochrome patterns in bacterial classification is that of Jones (1980).

It should be stressed that when cytochrome patterns are used for taxonomic purposes, the influence of the growth environment should be taken into account. Growth conditions can influence the quantitative and, to a lesser extent, the qualitative cytochrome content of bacteria.

Amino Acid Sequences of Various Proteins

Comparison of the amino acid sequence of specific kinds of proteins, or of properties such as antigenic reactivity which reflect the amino acid sequence of these proteins, have been used as measures of phylogenetic relationships among organisms. The fundamental concept involved is that most extant proteins are likely to have evolved from a very small number of archetypal proteins by the processes of genetic duplication and modification. In comparing the proteins of any particular group (such as cytochrome *c*, superoxide dismutase, ferredoxin or other enzymes), the greater the difference in amino acid sequence in the protein of one organism from the corresponding protein of another organism, the greater is believed to be the evolutionary divergence between the two organisms. Conversely, if the amino acid sequence of corresponding proteins from two organisms is very similar, the two organisms are believed to be closely related phylogenetically. Even distant relationships between organisms can be deduced by this approach, and various phylogenetic schemes have been constructed to reflect the perceived evolutionary development of a great variety of organisms, both procaryotic and eucaryotic (see the review by Schwartz and Dayhoff, 1978). Among the proteins that have been used for such studies are ferredoxins, flavodoxins, azurins, plastocyanins, and cytochrome *c*. For example, a remarkable similarity in the structure of cytochrome *c* exists between certain nonsulfur purple photosynthetic bacteria (i.e. *Rhodopseudomonas capsulatus* and *R. sphaeroides*), the nonphotosynthetic respiring bacterium *Paracoccus denitrificans*, and the mitochondria of eucaryotic organisms; this and other kinds of congruent data have led to the view that *P. denitrificans* descended from nonsulfur purple bacteria by loss of photosynthetic properties and that this species is the procaryote that most closely resembles the putative procaryotic ancestor of mitochondria (see the exposition by Dickerson (1980) as well as a critical review of the various theories for the endosymbiont origin of mitochondria and chloroplasts by Gray and Doolittle (1982). Reservations about many of the conclusions based on cytochrome *c* sequences have been expressed by Ambler et al. (1979 a, b), and an analysis of some of the limitations involved in comparing amino acid sequences of proteins has been given by Doolittle (1981).

Protein Profiles

The basic premise here is that closely related organisms should have similar or identical kinds of cellular proteins. Two-dimensional electrophoretic and isoelectric focusing procedures (O'Farrell, 1975) have made it possible to resolve several hundred proteins from a cell extract. The protein "fingerprint" so obtained for one bacterial strain is a reflection of the genetic background of that strain and can be compared with the "fingerprints" from other strains as a measure of relatedness.

For examples of the application of this method, see the comparison of *Rhizobium* strains made by Roberts et al. (1980), and the comparison of *Spiroplasma* strains made by Mouches et al. (1979). The method requires a considerable degree of standardization in order to yield optimum results.

One-dimensional polyacrylamide gel electrophoresis (PAGE) of cellular proteins can yield patterns of up to ca. 30 bands and, although it is not comparable in resolving power to the two-dimensional separation method, it can distinguish related organisms from unrelated organisms. In general, whole cells or cellular membrane fractions are used, and the proteins are solubilized by means of a detergent such as sodium dodecyl sulfate (SDS); however, many studies have employed merely the water-soluble proteins ("soluble" fraction) from disintegrated cells. A few examples of the application of PAGE are: identification of mycoplasmas (Razin and Rottem, 1967), taxonomy of *Haemophilus* strains (Nicolet et al., 1980), comparison of isolates from gingival crevice floras (Moore et al., 1980), and differentiation of isolates of indigenous *Rhizobium* populations (Noel and Brill, 1980). By use of rigorously standardized conditions, extremely reproducible protein patterns can be obtained which are amenable to rapid, computerized, numerical analysis (Kersters and De Ley, 1975).

In an analysis of the patterns of soluble cellular proteins from strains of 70 *Clostridium* species, Cato et al. (1982) found that strains having greater than 80% DNA/DNA homology usually produced identical patterns, strains related by ca. 70% homology showed overall similarity of the total patterns but also showed minor differences, and strains unrelated by DNA homology showed major differences. In many instances, the patterns obtained within 24 h of isolating an organism were sufficiently distinctive so that the identity of the organism could be strongly suspected.

Enzyme Characterization

It is now recognized that the functional and structural patterns displayed by certain bacterial enzymes provide data of use in classification. Good examples are the diverse regulatory and molecular size patterns exhibited by bacterial citrate synthases and succinate thiokinases. Both these are enzymes of the citric acid (Krebs) cycle and the near universal occurrence of this cycle in living cells makes it a very suitable pathway for comparative studies between different organisms (see Weitzman, 1980).

In general, the citrate synthases of Gram-negative bacteria are inhibited by reduced nicotinamide adenine dinucleotide (NADH) while those of Gram-positive bacteria are not. The citrate synthases of Gram-negative bacteria can be further divided into two classes on the basis of whether or not their NADH sensitivity is overcome by adenosine monophosphate (AMP). Citrate synthases from the majority of strictly aerobic Gram-negative bacteria are reactivated by AMP while those of the facultatively anaerobic Gram-negative bacteria are not. Citrate synthases of the Gram-negative facultative anaerobes are also inhibited by α-oxoglutarate but the enzymes from the aerobic Gram-negative bacteria and from Gram-positive bacteria are not. Citrate synthases of the cyanobacteria are not inhibited by NADH but they are inhibited by α-oxoglutarate and by succinyl-CoA.

Bacterial citrate synthases fall into two groups, "large" and "small," on the basis of molecular size. The majority of Gram-negative bacteria possess large citrate synthases (mol. wt. = $\sim$250,000) while the majority of Gram-positive bacteria produce citrate synthases of the small type (mol. wt. = $\sim$100,000).

Exceptions to the broad general pattern occur. The citrate synthases of the Gram-negative genus *Acetobacter* do not appear to be inhibited by NADH although the enzyme is of the large type. On the other hand, the citrate synthase of *Thermus aquaticus* is both insensitive to NADH and is of the small type. In both respects it resembles the citrate synthases of the archaebacterial genus *Halobacterium* and the majority of the citrate synthases of Gram-positive bacteria.

Similar molecular size patterns occur among bacterial succinate thiokinases. All succinate thiokinases from the Gram-positive bacteria so far studied are of the small type (mol. wt. = 70,000–75,000) whereas those of Gram-negative bacteria, cyanobacteria and *Halobacterium* species are of the large type (mol. wt. = 140,000–150,000). Bacterial succinate thiokinases can be further subdivided on the basis of their specificity for nucleotide substrates (guanosine diphosphate or inosine diphosphate) and preliminary results point to interesting patterns of enzyme diversity of possible potential in bacterial classification.

Rapid methods are now available for the routine laboratory screening of bacterial citrate synthases and as further such methods are developed it is likely that the regulatory and molecular properties of these and other enzymes will prove useful in the classification of bacteria (see Weitzman, 1980).

Fermentation Product Profiles

The use of gas liquid chromatographic methods to analyze the fatty acids formed as end products of protein or carbohydrate metabolism is particularly useful in the classification and identification of the anaerobic genera *Clostridium*, *Bacteroides*, *Eubacterium*, etc. (see Holdeman et al., 1977).

Bacterial Nomenclature

Peter H. A. Sneath

THE SCOPE OF NOMENCLATURE

Nomenclature has been called the handmaid of taxonomy. The need for a stable set of names for living organisms, and rules to regulate them, has been recognized for over a century. The rules are embodied in *international codes of nomenclature*. There are separate codes for animals, noncultivated plants, cultivated plants, bacteria and viruses. But partly because the rules are framed in legalistic language (so as to avoid imprecision) they are often difficult to understand. Useful commentaries are found in Ainsworth and Sneath (1962), Cowan (1978) and Jeffrey (1977).

The nomenclature of the different kinds of living creatures falls into two parts: (a) informal or vernacular names, or very specialized and restricted names, and (b) scientific names of taxonomic groups (taxon, plural taxa).

Examples of the first are vernacular names from a disease, strain numbers, the symbols for antigenic variants, and the symbols for genetic variants. Thus one can have a vernacular name like the tubercle bacillus, a strain with the designation K12, a serological form with the antigenic formula Ia, and a genetic mutant requiring valine for growth labeled *val*. These names are usually not controlled by the codes of nomenclature, although the codes may recommend good practice for them.

Examples of scientific names are the names of species, genera, and higher ranks. Thus *Mycobacterium tuberculosis* is the scientific name of the tubercle bacillus, a species of bacterium.

These scientific names are regulated by the codes (with few exceptions) and have two things in common: (a) they are all Latinized in form so as to be easily recognized as scientific names, and (b) they possess definite positions in the taxonomic hierarchy. These names are international: thus microbiologists of all nations know what is meant by *Bacillus anthracis*, but few would know it under vernacular names like Milzbrandbacillus or Bactéridie de charbon.

The scientific names of bacteria are regulated by the *International Code of Nomenclature of Bacteria* (Lapage et al., 1975). This is the most recent edition, and is also referred to as the *Revised Code*. This edition authorized a new starting date for names of bacteria on January 1, 1980, and the starting document is the *Approved Lists of Bacterial Names* (Skerman et al., 1980), which contains all the scientific names of bacteria that retain their nomenclatural validity from the past. The operation of these Lists will be referred to later. The Code and the Lists are under the aegis of the International Committee on Systematic Bacteriology, which is a constituent part of the International Union of Microbiological Associations. The Committee is assisted by a number of Taxonomic Subcommittees on different groups of bacteria, and by the Judicial Commission which considers amendments to the Code and any exceptions that may be needed to specific Rules.

LATINIZATION

Since scientific names are in latinized form, they obey the grammar of classic or medieval Latin. Fortunately the necessary grammar is not very difficult, and the commonest point to watch is that adjectives agree in gender with the substantives they qualify. Some examples are given later. The names of genera and species are normally printed in italics (or underlined in manuscripts to indicate italic font). For higher categories conventions vary: in Britain they are often in ordinary roman type, but in America they are usually in italics, which is preferable because this reminds the reader they are latinized scientific names.

THE TAXONOMIC HIERARCHY

The taxonomic hierarchy is a conventional arrangement. Each level above the basic level of species is increasingly inclusive. The names belong to successive **categories**, each of which possesses a position in the hierarchy called its **rank**. The lowest category ordinarily employed is that of species, though sometimes these are subdivided into subspe-cies. The main categories in decreasing rank, with their vernacular and latin forms, and examples, are shown in Table VI.1.

Additional categories may sometimes be intercalated (e.g. subclass below class, and tribe below family).

Table VI.1.
The ranking of taxonomic categories

Category	Example[a]
Kingdom (*Regnum*)	*Procaryotae*
Phylum (*Phylum*) in zoology or Division (*Divisio*) in botany and bacteriology	*Gracilicutes*
Class (*Classis*)	*Scotobacteria*
Order (*Ordo*)	*Rickettsiales*
Family (*Familia*)	*Rickettsiaceae*
Genus (*Genus*)	*Coxiella*
Species (*Species*)	*Coxiella burnetii*

[a] Based on the classification given by Murray in the chapter on The Higher Taxa, in this *Manual.*

FORM OF NAMES

The form of latinized names differs with the category. The species name consists of two parts. The first is the **genus name**. This is spelled with an initial capital letter, and is a latinized substantive. The second is the **specific epithet**, and is spelled with a lower case initial letter. The epithet is a latinized adjective in agreement with the gender of the genus name, or a Latin word in the genitive case, or occasionally a noun in apposition. Thus in *Mycobacterium tuberculosis*, the epithet *tuberculosis* means "of tubercle," so the species name means the mycobacterium of tuberculosis. The species name is called a **binominal name**, or **binomen**, because it has two parts. When subspecies names are used, a trinominal name results, with the addition of an extra **subspecific epithet**. An example is the subspecies of *Lactobacillus casei* that is called *Lactobaccillus casei* subsp. *rhamnosus*. In this name, *casei* is the specific epithet and *rhamnosus* is the subspecific epithet. The existence of a subspecies such as *rhamnosus* implies the existence of another subspecies, in which the subspecific and specific epithets are identicial, i.e. *Lactobacillus casei* subsp. *casei*.

One problem that frequently arises is the scientific status of a species. It may be difficult to know whether an entity differs from its neighbors in certain specified ways. A useful terminology was introduced by Ravin

(1963). It may be believed, for example, that the entity can undergo genetic exchange with a nearby species, in which event they could be considered to belong to the same **genospecies**. It may be believed that the entity is not phenotypically distinct from its neighbors, in which event they could be considered to belong to the same **taxospecies**. Yet the conditions for genetic exchange may vary greatly with experimental conditions and the criteria of distinctness depend on what properties are considered, so that it may not be possible to make clear-cut decisions on these matters. Nevertheless, it may be convenient to give the entity a species name and to treat it in nomenclature as a separate species, a **nomenspecies**. It follows that all species in nomenclature should strictly be regarded as nomenspecies.

Genus names, as mentioned above, are latinized nouns, and so are subgenus names (now rarely used) which are conventionally written in parentheses after the genus name, e.g. *Bacillus* (*Aerobacillus*) indicates the subgenus *Aerobacillus* of the genus *Bacillus*. As in the case of subspecies, this implies the existence of a subgenus *Bacillus* (*Bacillus*).

Above the genus level most names are plural adjectives in the feminine gender, agreeing with the word *procaryotae*, so that *Brucellaceae* means *procaryotae brucellaceae*, for example.

PURPOSES OF THE CODES OF NOMENCLATURE

The codes have three main aims:

1. Names should be stable,
2. Names should be unambiguous,
3. Names should be necessary.

These three aims are sometimes contradictory, and the rules of nomenclature have to make provision for exceptions where they clash. The principles are implemented by three main devices: (a) priority of publication to assist stability, (b) establishment of nomenclatural types to ensure the names are not ambiguous, and (c) publication of descriptions to indicate that different names do refer to different entities. These are supported by subsidiary devices such as the latinized forms of names, and the avoidance of synonyms for the same taxon.

PRIORITY OF PUBLICATION

In order to achieve stability the first name given to a taxon (provided the other rules are obeyed), is taken as the correct name. This is the **principle of priority**. But to be safeguarded in this way a name obviously has to be made known to the scientific community: one cannot use a name that has been kept secret. Therefore names have to be published in the scientific literature, together with sufficient indication of what they refer to. This is called **valid publication**. If a name is merely published in the scientific literature it is called **effective publication**: to be valid it also has to satisfy additional requirements, which are summarized later.

The earliest names that must be considered are those published after an official starting date. For many groups of organisms this is Linnaeus' *Species Plantarum* of 1753, but the difficulties of knowing to what the early descriptions refer, and of searching of voluminous and growing literature, have made the principle of priority increasingly hard to obey.

The Code of nomenclature for bacteria, therefore, has established a new starting date of 1980, with a new starting document, the *Approved Lists of Bacterial Names* (Skerman et al., 1980). This list contains names of bacterial taxa that are recognizable and in current use. Names not on the lists lost standing in nomenclature on January 1, 1980, although there are provisions for reviving them if the taxa are subsequently rediscovered or need to be reestablished. In order to prevent the need to search the voluminous scientific literature, the new provisions for bacterial nomenclature require that for valid publication new names (including new names in patents) must be published in certain official publications. Alternatively, if the new names were effectively published in other scientific publications they must be announced in the official publications to become validly published. Priority dates from the official publication concerned. At present the only official publication is the *International Journal of Systematic Bacteriology*.

NOMENCLATURAL TYPES

In order to make clear what names refer to, the taxa must be recognizable by other workers. In the past it was thought sufficient to publish a description of a taxon. This has been found over the years to be inadequate. Advances in techniques and in knowledge of the many undescribed species in nature have shown that old descriptions are usually insufficient. Therefore an additional principle is employed, that of **nomenclatural types**. These are actual specimens (or names of subordinate taxa that ultimately relate to actual specimens). These type specimens are deposited in museums and other institutions. For bacteria (like some other microorganisms that are classified according to their properties in artificial culture) instead of type specimens, **type strains** are employed. The type specimens or strains are intended to be typical specimens or strains which can be compared with other material when classification or identification is undertaken, hence the word "type." However, a moment's thought will show that if a type specimen has to be designated when a taxon is *first* described and named, this will be done at a time when little has yet been found out about the new group. Therefore it is impossible to be sure that it is indeed a typical specimen. By the time a completely typical specimen can be chosen the taxon may be so well known that a type specimen is unnecessary: no one would now bother to designate a type specimen of a bird so well known as the common house sparrow.

The word "type" thus does *not* mean it is typical, but simply that it is a **reference specimen for the name**. This use of the word, type, is a very understandable cause for confusion that may well repay attention by the taxonomists of the future.

In recent years other type concepts have been suggested. Numerical taxonomists have proposed the hypothetical median organism (Liston et al., 1963), or the centroid: these are mathematical abstractions, not actual organisms. The most typical strain in a collection is commonly taken to be the **centrotype** (Silvestri et al., 1962), which is broadly equivalent to the strain closest to the center (centroid) of a species cluster. Some workers have suggested that several type strains should be designated. Gordon (1967) refers to this as the "population concept." One strain, however, must be the official nomenclatural type in case the species must later be divided. Gibbons (1974) proposed that the official type strain should be supplemented by reference strains that indicated the range of variation in the species, and that these strains could be termed the "type constellation." It may be noted that some of these concepts are intended to define not merely the center but, in some fashion, the limits of a species. Since these limits may well vary in different ways for different characters, or classes of characters, it will be appreciated that there may be difficulties in extending the type concept in this way. The centrotype, being a very typical strain, has often been chosen as the type strain, but otherwise these new ideas have not had much application to bacterial nomenclature.

Type strains are of the greatest importance for work on both classification and identification. These strains are preserved (by methods to minimize change to their properties) in culture collections from which they are available for study. They are obviously required for new classificatory work, so that the worker can determine if he has new species among his material. They are also needed in diagnostic microbiology, because one of the most important principles in attempting to identify a microorganism that presents difficulties is to compare it with authentic strains of known species. The drawback that the type strain may not be entirely typical is outweighed by the fact that the type strain is by definition authentic.

Not all microorganisms can be cultured, and for some the function of a type can be served by a preserved specimen, a photograph, or some other device. In such instances, these are the nomenclatural types, though it is commonly considered wise to replace them by type strains when this becomes possible.

Sometimes types become lost, and new ones (**neotypes**) have to be set up to replace them: the procedure for this is described in the Code. In the past it was necessary to define certain special classes of types, but most of these are now not needed.

Types of species and subspecies are type specimens or type strains. For categories above the species the function of the type—to serve as a point of reference—is assumed by a *name*, e.g. that of a species or subspecies. The species or subspecies is, of course, tied to its type specimen or type strain.

Types of genera are **type species** (one of the included species) and types of higher names are usually **type genera** (one of the included genera). This principle applies up to and including the category of Order. This can be illustrated by the types of an example of a taxonomic hierarchy shown in Table VI.2.

Just as the type specimen, or type strain, must be considered a member of the species whatever other specimens or strains are excluded, so the **type species of a genus must be retained in the genus even if all other species are removed from it**. A type, therefore, is sometimes called a **nominifer** or **name bearer**: it is the reference point for the name in question.

Table VI.2.

An example of taxonomic types.

Category	Taxon	Type
Family	*Pseudomonadaceae*	*Pseudomonas*
Genus	*Pseudomonas*	*Pseudomonas aeruginosa*
Species	*Pseudomonas aeruginosa*	American Type Culture Collection strain number 10145

DESCRIPTIONS

The publication of a name, with a designated type, does in a technical sense create a new taxon—insofar as it indicates that the author believes he has observations to support the recognition of a new taxonomic group. But this does not afford evidence that can be readily assessed from the bald facts of a name and designation of a type. From the earliest days of systematic biology it was thought important to describe the new taxon for two reasons: (a) to show the evidence in support of a new taxon, and (b) to permit others to identify their own material with it—indeed this antedated the type concept (which was introduced later to resolve difficulties with descriptions alone).

It is, therefore, a requirement for valid publication that a description of a new taxon is needed. However, just how full the description should be, and what properties must be listed, is difficult to prescribe.

The codes of nomenclature recognize that the most important aspect of a description is to provide a list of properties that distinguish the new taxon from others that are very similar to it, and that consequently fulfill the two purposes of adducing evidence for a new group and allowing another worker to recognize it. Such a brief differential description is called a **diagnosis**, by analogy with the characteristics of diseases that are associated with the same word. Although it is difficult to legislate for adequate diagnoses, it is usually easy to provide an acceptable one: inability to do so is often because insufficient evidence has been obtained to support the establishment of the new taxon.

The Code provides guidance on descriptions, in the form of recommendations. Failure to follow the recommendations does not of itself invalidate a name, though it may well lead later workers to dismiss the taxon as unrecognizable or trivial. The code for bacteria recommends that as soon as minimum standards of description are prepared for various groups, workers should thereafter provide that minimum information; this is intended as a guide to good practice, and should do much to raise the quality of systematic bacteriology. For an example of minimum standards, see the report of the International Committee on Systematic Bacteriology Subcommittee on the Taxonomy of *Mollicutes* (1979).

CLASSIFICATION DETERMINES NOMENCLATURE

The student often asks how an organism can have two different names. The reason lies in the fact that a name implies acceptance of some taxonomy, and on occasion no taxonomy is generally agreed. Scientists are entitled to their own opinions on taxonomies: there are no rules to force the acceptance of a single classification.

Thus opinions may be divided on whether the bacterial genus *Pectobacterium* is sufficiently separate from the genus *Erwinia*. The soft-rot bacterium was originally called *Bacterium carotovorum* in the days when most bacteria were placed in a few large genera such as *Bacillus* and *Bacterium*. As it became clear that these unwieldy genera had to be divided into a number of smaller genera, which were more homogeneous and convenient, this bacterium was placed in the genus *Erwinia* (established for the bacterium of fireblight, *Erwinia amylovora*) as *Erwinia carotovora*. When further knowledge accumulated, it was considered by some workers that the soft-rot bacterium was sufficiently

distinct to merit a new genus, *Pectobacterium*. The same organism, therefore, is also known as *Pectobacterium carotovorum*. Both names are correct in their respective positions. If one believes that two separate genera are justified, then the correct name for the soft-rot bacterium is *Pectobacterium carotovorum*. If one considers that *Pectobacterium* is not justified as a separate genus, the correct name is *Erwinia carotovora*.

Classification, therefore, determines nomenclature, not nomenclature classification. Although unprofitable or frivolous changes of name should be avoided, the freezing of classification in the form it had centuries ago is too high a price to pay for stability of names. Progress in classification must reflect progress in knowledge (for example, no one now wants to classify all rod-shaped bacteria in *Bacillus* as was popular a century ago). Changes in name must reflect progress in classification: some changes in name are thus inevitable.

CHANGES OF NAME

Most changes in name are due to moving species from one genus to another or dividing up older genera. Another cause, however, is the rejection of a commonly-used name because it is incorrect under one or more of the Rules. A much-used name, for example, may not be the earliest, because the earliest name was published in some obscure journal and had been overlooked. Or there may already be another identical name for a different microorganism in the literature. Changes can be very inconvenient if a well-established name is found to be **illegitimate** (contrary to a Rule) because of a technicality. The codes of nomenclature therefore make provision to allow the organizations that are responsible for the codes to make exceptions if this seems necessary. A name thus retained by international agreement is called

a **conserved name**, and when a name is conserved the type may be changed to a more suitable one.

When a species is moved from one genus into another, the specific epithet is retained (unless there is by chance an earlier name which forms the same combination, when some other epithet must be chosen), and this is done in the interests of stability. The new name is called a **new combination**. An example has been given above. When the original *Bacterium carotovorum* was moved to *Erwinia*, the species name became *Erwinia carotovora*. The gender of the species epithet becomes the same as that of the genus *Erwinia*, which is feminine, so the feminine ending, -*a*, is substituted for the neuter ending, -*um*.

NAMES SHOULD BE NECESSARY

The codes require that names should be necessary, that is **there is only one correct name for a taxon** in a given or implied taxonomy. This is sometimes expressed by the statement that an organism with a

given position, rank and circumscription can have only one correct name.

NAMES ARE LABELS, NOT DESCRIPTIONS

In the early days of biology there was no regular system of names, and organisms were referred to by long Latin phrases which described them briefly, such as *Tulipa minor lutea italica folio latiore*, "the little yellow Italian tulip with broader leaves." The Swedish naturalist Linnaeus tried to reduce these to just two words for species, and in doing

so he founded the present *binominal system* for species. This tulip might then become *Tulipa lutea*, just "the yellow tulip." Very soon it would be noted that a white variant sometimes occurred. Should it then still be named "the yellow tulip"? Why not change it to "the Italian tulip"? Then someone would find it in Greece and point out that the record from Italy was a mistake anyway. Twenty years later an orange form would be found in Italy after all. Soon the nomenclature would be confused again.

After a time it was realized that the original name had to be kept, even if it was not descriptive, just as a man keeps his name of Fairchild Goldsmith as he grows older, and even if he becomes a farmer. The scientific names or oganisms are today only **labels**, to provide a means of referring to taxa, just like personal names.

A change of name is therefore only **rarely** justified, even if it sometimes seems inappropriate. Provisions exist for replacement when the name causes great confusion.

CITATION OF NAMES

A scientific name is sometimes amplified by a *citation*, i.e., by adding after it the author who proposed it. Thus the bacterium that causes crown-galls is *Agrobacterium tumefaciens* (Smith and Townsend) Conn. This indicates that the name refers to the organism first named by Smith and Townsend (as *Bacterium tumefaciens*, in fact, though this is not evident in the citation) and later moved to the genus *Agrobacterium* by Conn, who therefore created a **new combination**. Sometimes the citation is expanded to include the date (e.g. *Rhizobium* Frank 1889), and more rarely to include also the publication, e.g. *Proteus morganii* Rauss 1936 *Journal of Pathology and Bacteriology* Vol. 42, p. 183.

It will be noted that citation is only necessary to provide a suitable reference to the literature or to distinguish between inadvertent duplication of names by different authors. A citation is *not* a means of giving credit to the author who described a taxon: the main functions of citation would be served by the bibliographic reference without mentioning the author's name. Citation of a name is to provide a **means of referring** to a name, just as a name is a means of referring to a taxon.

SYNONYMS AND HOMONYMS

A homonym is a name identical in spelling to another name but based on a different type, so they refer to different taxa under the same name. They are obviously a source of confusion, and the one that was published later is suppressed. The first published name is known as the **senior homonym**, and later published names are **junior homonyms**. Names of higher animals and plants that are the same as bacterial names are not treated as homonyms of names of bacteria, but to reduce confusion among microorganisms, bacterial names are suppressed if they are junior homonyms of names of fungi, algae, protozoa or viruses.

A synonym is a name that refers to the same taxon under another scientific name. Synonyms thus come in pairs or even swarms. They are of two kinds:

1. **Objective synonyms** are names with the same nomenclatural type, so that there is no doubt that they refer to the same taxon. These are often called nomenclatural synonyms. An example is *Erwinia carotovora* and *Pectobacterium carotovorum*: they have the same type strain, American Type Culture Collection strain 15713.

2. **Subjective synonyms** are names that are believed to refer to the same taxon but which do not have the same type. They are matters of taxonomic opinion. Thus *Pseudomonas geniculata* is a subjective synonym of *Pseudomonas fluorescens* for a worker who believes that these taxa are sufficiently similar to be included in one species, *P. fluorescens*. They have different types, however, (American Type Culture Collection strains #19374 and 13525, respectively) and another worker is entitled to treat them as separate species if he so wishes.

There are senior and junior synonyms, as for homonyms. The synonym that was first published is known as the **senior synonym**, and those published later are **junior synonyms**. Junior synonyms are normally suppressed.

PROPOSAL OF NEW NAMES

The valid publication of a new taxon requires that it be named. The Code insists that an author should make up his mind about the new taxon: if he feels certain enough to propose a new taxon with a new name then he should say he does so propose; if he is not sure enough to make a definite proposal then his name will not be afforded the protection of the Code. He cannot expect to suggest provisional names—or possible names, or names that one day might be justified—and then expect others to treat them as definite proposals at some unspecified future date: how can a reader possibly know when such vague conditions have been fulfilled?

If a taxon is too uncertain to receive a new name it should remain with a vernacular designation (e.g. the marine form, group 12A). If it is already named, but its affinities are too uncertain to move it to another genus or family, it should be left where it is. There is one exception, and that is that a new species should be put into some genus even if it is not very certain which is the most appropriate, or if necessary a new genus should be created for it. Otherwise, it will not be validly published, it will be in limbo, and it will be generally overlooked, because no one else will know how to index it or whether they should consider it seriously. If it is misplaced it can later be moved to a better genus.

The basic needs for publication of a new taxon are four: (a) the publication should contain a new name in proper form that is not a homonym of an earlier name of bacteria, fungi, algae, protozoa or viruses; (b) the taxon should not be a synonym of an earlier taxon; (c) a description or at least a diagnosis should be given; (d) the type should be designated. A new species is indicated by adding the Latin abbreviation *sp. nov.*, a new genus by *gen. nov.*, and a new combination by *comb. nov.* The most troublesome part is the search of the literature to cover the first two points. This is now greatly simplified for bacteria, because the new Starting Date means that one need only search the *Approved Lists of Bacterial Names* and the issues of the *International Journal of Systematic Bacteriology* from January 1980 onwards for all validly published names that have to be considered. However, the new name has to be published in that journal, with its description and designation of type, or—if published elsewhere—the name must be announced in that journal to render it validly published.

Identification of Bacteria

Noel R. Krieg

THE NATURE OF IDENTIFICATION SCHEMES

Identification schemes are not classification schemes, although there may be a superficial similarity. An identification scheme for a group of organisms can be devised only **after** that group has first been classified (i.e. recognized as being different from other organisms); it is based on one or more characters, or on a pattern of characters, which all the members of the group have and which other groups do not have. The characters used are often not those that were involved in classification of the group; for example, classification might be based on a DNA/DNA hybridization study, whereas identification might be based on a phenotypic character that is found to correlate well with the genetic information. In general, the characters chosen for an identification scheme should be **easily determinable**, whereas those used for classification may be quite difficult to determine (such as DNA homology values). The characters should also be **few in number**, whereas classification may involve large numbers of characters, such as in a numerical taxonomy study. These ideal features of an identification scheme may not always be possible, particularly with genera or species that are not susceptible to being characterized by traditional biochemical or physiological tests. In such cases, one may need to resort to relatively difficult procedures in order to achieve an accurate identification—procedures such as polyacrylamide gel electrophoresis (PAGE) of cellular proteins, cellular lipid patterns, genetic transformation, or even nucleic acid hybridization.

Serological reactions, which generally have only limited value for classification, often have enormous value for identification. Slide agglutination tests, fluorescent antibody techniques, and other serological methods can be performed simply and rapidly and are usually highly specific; therefore, they offer a means for achieving quick, presumptive identification of bacteria. Their specificity is frequently not absolute, however, and confirmation of the identification by additional physiological or biochemical tests is usually required.

With many genera and species, identification may not be based on only a few tests, but rather on the pattern given by applying a whole battery of tests. The members of the family *Enterobacteriaceae* represent one example of this. To alleviate the need for inoculating large numbers of tubed media, a variety of convenient and rapid multitest systems have been devised and are commercially available for use in identifying various taxa, particularly those of medical importance. A summary of some of these systems has been given by Smibert and Krieg (1981), but new systems are being developed continually. Each manufacturer provides charts, tables, coding systems, and characterization profiles for use with the particular multitest system being offered.

NEED FOR STANDARDIZED TEST METHODS

One difficulty in devising identification schemes is that the results of characterization tests may vary depending on the size of the inoculum, incubation temperature, length of the incubation period, composition of the medium, the surface-to-volume ratio of the medium, and the criteria used to define a "positive" or "negative" reaction. Therefore, the results of characterization tests obtained by one laboratory often do not match exactly those obtained by another laboratory, although the results within each laboratory may be quite consistent. The blind acceptance of an identification scheme without reference to the particular conditions employed by those who devised the scheme can lead to error (and, unfortunately, such conditions are not always specified). Ideally, it would be desirable to standardize the conditions used for

testing various characteristics, but this is easier said than done, especially on an international basis. The use of commercial multitest systems offers some hope of increasing the standardization among various laboratories because of the high degree of quality control exercised over the media and reagents, but no one system has yet been agreed on for universal use for any given taxon. **It is therefore always advisable to include strains whose identity has been firmly established** (type or reference strains, available from national culture collections) **for comparative purposes when making use of an identification scheme**, to make sure that the scheme is valid for the conditions employed in one's own laboratory.

NEED FOR DEFINITIONS OF "POSITIVE" AND "NEGATIVE" REACTIONS

Some tests may be found to be based on plasmid- or phage-mediated characteristics; such characteristics may be highly mutable and there-

fore unreliable for identification purposes. Even with immutable characteristics, certain tests may not be well suited for use in identification schemes because they may not give highly reproducible results (e.g. the catalase test, oxidase test, Voges-Proskauer test, and gelatin liquefaction are notorious in this regard). Ideally, a test should give reproducible results that are clearly either positive or negative, without equivocal reactions. In fact, no such test may exist. The Gram reaction of an organism may be "Gram-variable," the presence of endospores in a strain that makes only a few may be very difficult to determine by staining or by heat-resistance tests, acid production from sugars may be difficult to distinguish from no acid production if only small amounts of acid are produced, and a weak growth response may not be clearly distinguishable from "no growth." A precise (although arbitrary) definition of what constitutes a "positive" and "negative" reaction is often important in order for a test to be useful for an identification scheme.

PURE CULTURES

Although a few bacteria are so morphologically remarkable as to make them identifiable without isolation, pure cultures are nearly always a necessity before one can attempt identification of an organism. **It is important to realize that the single selection of a colony from a plate does not assure purity.** This is especially true if selective media are used; live but non-growing contaminants may often be present in or near a colony and can be subcultured along with the chosen organism. It is for this reason that **non-selective media are preferred for final isolation**, because they allow such contaminants to develop into visible colonies. Even with non-selective media, apparently well-isolated colonies should not be isolated too soon; some contaminants may be slow growing and may appear on the plate only after a longer incubation. Another difficulty occurs with bacteria that form extracellular slime or that grow as a network of chains or filaments; contaminants often become firmly embedded or entrapped and are difficult to penetrate. In the instance of cyanobacteria, contaminants frequently penetrate and live in the gelatinous sheaths that surround the cells, making pure cultures difficult to obtain.

In general, colonies from a pure culture that has been streaked on a solid medium are similar to one another, providing evidence of purity. Although this is generally true, there are exceptions, as in the case of S → R variation, capsular variants, pigmented or nonpigmented variants, etc., which may be selected by certain media, temperatures, or other growth conditions. Another criterion of purity is morphology: organisms from a pure culture generally exhibit a high degree of morphological similarity in stains or wet mounts. Again, there are exceptions, depending on the age of the culture, the medium used, and other growth conditions: coccoid body formation, cyst formation, spore formation, pleomorphism, etc. For example, examination of a broth culture of a marine spirillum after 2 or 3 days may lead one to believe the culture is highly contaminated with cocci unless one is previously aware that such spirilla generally develop into thin-walled coccoid forms following active growth.

APPROACHES TO IDENTIFICATION OF AN ISOLATE

The vernacular headings of the various sections of *Bergey's Manual* indicate major categories of the procaryotes and are a good starting point for identification. The categories are concerned with such phenotypic characteristics as the Gram staining reactions, morphology, and general type of metabolism. It is therefore important to establish whether the new isolate is a chemolithotrophic autotroph, a photosynthetic organism or a chemoheterotrophic organism. Living cells should be examined by phase-contrast microscopy and Gram-stained cells by light microscopy; other stains can be applied if this seems appropriate. If some outstanding morphological property, such as endospore production, sheaths, holdfasts, acidfastness, cysts, stalks, fruiting bodies, budding division, or true branching, is obvious, then further efforts in identification can be confined to those groups having such a property. Whether or not the organisms are motile, and the type of motility (swimming, gliding) may be very helpful in restricting the range of possibilities. Gross growth characteristics, such as pigmentation, mucoid colonies, swarming, or a minute size, may also provide valuable clues to identification. For example, a motile, Gram-negative rod that produces a water-soluble fluorescent pigment is likely to be a *Pseudomonas* species, whereas one that forms bioluminescent colonies is likely to belong to the *Vibrionaceae*.

The source of the isolate can also help to narrow the field of possibilities. For example, a spirillum isolated from coastal sea water is likely to be an *Oceanospirillum*, whereas Gram-positive cocci occurring in grape-like clusters and isolated from the human nasopharynx are likely to belong to the genus *Staphylococcus*.

The relation of the isolate to oxygen (i.e. whether it is aerobic, anaerobic, facultatively anaerobic, or microaerophilic) is often of fundamental importance in identification. For example, a small, microaerophilic vibrio isolated from a case of diarrhea is likely to be a *Campylobacter*, whereas an anaerobic, Gram-negative rod isolated from a wound infection is probably a member of the *Bacteroidaceae*. Similarly, it is important to test the isolate for its ability to dissimilate glucose (or other simple sugar) to determine if the type of metabolism is oxidative or fermentative, or whether sugars are catabolized at all.

Above all, common sense should be used at each stage where the possibilities are narrowed in deciding what additional tests should be performed. There should be a reason for the selection of each test, in contrast to a "shotgun" type of approach where many tests are used but most provide little pertinent information for the particular isolate under investigation. As the category to which the isolate belongs becomes increasingly delineated, one should follow the specific tests indicated in the particular diagnostic tables or keys that apply to that category.

The following summary is taken from "The Mechanism of Identification" by S. T. Cowan and J. Liston in the eighth edition of the *Manual*, with some modifications:

1. Make sure that you have a pure culture.
2. Work from broad categories down to a smaller, specific category of organism.
3. Use all the information available to you in order to narrow the range of possibilities.
4. Apply common sense at each step.
5. Use the minimum number of tests to make the identification.
6. Compare your isolate to type or reference strains of the pertinent taxon to make sure the identification scheme being used actually is valid for the conditions in your particular laboratory.

If, as may well happen, you cannot identify your isolate from the information contained in the *Manual*, neither despair nor immediately assume that you have isolated a new genus or species; many of the problems of microbial classification are the result of people jumping to this conclusion prematurely. When you fail to identify your isolate, check (a) its **purity**, (b) that you have carried out the **appropriate tests**, (c) that your **methods are reliable**, and (d) that you have used correctly the various keys and tables of the *Manual*. It has been said that the most frequent cause of mistaken identity of bacteria is error in the determination of shape, Gram-staining reaction and motility. In most cases, you should have little difficulty in placing your isolate into a genus; allocation to a species or subspecies may need the help of a specialized reference laboratory.

On the other hand, it is always possible that you have actually isolated a new genus or species. A comparison of the present edition of the *Manual* with the previous edition indicates that a number of new genera and species have been added. Some prime examples can be found in the family *Legionellaceae, Other Genera of the Family Enterobacteriaceae*, the genus *Azospirillum, Dissimilatory Sulfate- or Sulfur-Reducing Bacteria*, the genus *Meniscus*, etc. Undoubtedly, there exist in nature a great number of bacteria that have not yet been classified, and therefore cannot yet be identified by existing schemes. Yet, before describing and naming a new taxon, one must be **very sure that it is really a new taxon** and not merely the result of an inadequate identification.

Further Readings for Bacterial Identification

Goodfellow, M. and R.G. Board (Editors). 1980. Microbiological classification and identification. *Society for Applied Bacteriology Symposium Series No. 8.* Academic Press, London.
Hedén, C. and T. Illéni (Editors). 1975. *New Approaches to the Identification of Microorganisms.* John Wiley & Sons, New York.
Holding, J.A. and J.G. Colee. 1971. Routine biochemical tests. *In* Norris and Ribbons (Editors), *Methods in Microbiology*, Vol. 6A, Academic Press, New York, pp. 1–32.
Mitruka, B.J. 1976. *Methods of Detection and Identification of Bacteria.* CRC Press, Cleveland, Ohio.
Skerman, V.B.D. 1967. *A Guide to the Identification of the Genera of Bacteria*, 2nd Ed., Williams & Wilkins, Baltimore.
Skerman, V.B.D. 1969. *Abstracts of Microbiological Methods.* Wiley-Interscience, New York.
Skerman, V.B.D. 1974. A key for the determination of the generic position of organisms listed in the *Manual. In* Buchanan and Gibbons (Editors), *Bergey's Manual of Determinative Bacteriology*, 8th Ed., Williams & Wilkins, Baltimore, pp. 1098–1146.
Skinner, F.A. and D.W. Lovelock. 1979. Identification methods for microbiologists. *Society for Applied Bacteriology Technical Series No. 14.* Academic Press, New York.
Smibert, R.M. and N.R. Krieg. 1981. General characterization. *In* Gerhardt, Murray, Costilow, Nester, Wood, Krieg and Phillips (Editors), *Manual of Methods for General Bacteriology*, American Society for Microbiology, Washington, D.C, pp. 409–443.

NUMERICAL IDENTIFICATION
Peter H. A. Sneath

The success of numerical taxonomy has in recent years led to the development of a new diagnostic method based upon it, called **numerical identification**. The rapidly growing field is well reviewed by Lapage et al. (1973), and Willcox et al. (1980). The essential principles can be illustrated geometrically (Sneath, 1978) by considering the columns of percent positive test reactions in a new table, a table of q taxa for m diagnostic characters. If an object is scored for two variables, its position can be represented by a point on a scatter diagram. Three variables determines a position in a three-dimensional model. Objects that are very similar on the variables will be represented by clusters of points in the diagram or the model, and a circle or sphere can be drawn round each cluster so as to define its position and radius. The same principles can be extended to many variables or tests, which then represent a multidimensional space or "hyperspace." A column representing a species defines, in effect, a region in hyperspace, and it is useful to think of a species as being represented by a hypersphere in that space, whose position and radius are specified by the numerical values of these percentages.

The operation of numerical identification is to compare an unknown strain with each column of the table in turn, and to calculate a distance (or its analogue) to the center of each taxon hypersphere. If the unknown lies well within a hypersphere, this will identify it with that taxon. Further, such systems have important advantages over most other diagnostic systems. The numerical process allows a likelihood to be attached to an identification, so that one can know to some order of magnitude the certainty that the identity is correct. The results are not greatly affected by an occasional aberrant property of the unknown, or an occasional experimental mistake in performing the tests. Furthermore, the system is robust toward missing information, and quite good identifications can be obtained if only a moderate proportion of the tests have been performed.

Further Readings for Numerical Identification

Lapage, S.P., S. Bascomb, W.R. Willcox and M.A. Curtis. 1973. Identification of bacteria by computer. I. General aspects and perspectives. J. Gen. Microbiol. 77: 273–290.
Sneath, P.H.A. 1978. Identification of microorganisms. *In* Norris and Richmond (Editors), *Essays in Microbiology.* John Wiley, Chichester, England, pp. 10/1–10/32.
Willcox, W.R., S.P. Lapage and B. Holmes. 1980. A review of numerical methods in bacterial identification. Antonie van Leeuwenhoek J. Microbiol. Serol. *46:* 233–299.

Reference Collections of Bacteria—
The Need and Requirements
for Type Strains

the late N. E. Gibbons
Revised by Peter H. A. Sneath and Stephen P. Lapage

As it became possible to grow bacteria in liquid and solid media, microbiologists began to exchange cultures with their colleagues for information and comparison. Each investigator kept his own isolates, added those received from others and in this way built up his own reference and working collection.

About the turn of the century, Professor František Král of Prague realized the value of a central collection and began to collect cultures which he made available for a fee to other workers. After Král's death in 1911, the collection was acquired by Professor Ernst Pribram and transferred to the University of Vienna in 1915. Pribram brought part of the collection to Loyola University in Chicago some years before the Second World War. He was killed in a car accident in 1940, but the fate of his collection is not known. The cultures left in Vienna were destroyed during World War II.

The next oldest collection—Centraalbureau voor Schimmelcultures—was founded in 1906 by the Association internationale des Botanistes. Although the founding association did not survive the First World War, the collection is still in existence at Baarn under the auspices of The Royal Netherlands Academy of Sciences. This collection provides a holding and distribution center for fungi and an identification service.

Since then many other collections have developed, some general, some specialized, some oriented to service. A full account of the history of culture collections is given by Porter (1976). Some salient developments may be mentioned briefly. About 1946, Professor P. Hauduroy established a centralized information facility at Lausanne, the "Centre de Collections de Types Microbiens" which provided information on which collections held cultures of various bacterial species. In 1947, the Lausanne Centre became associated with the International Association of Microbiological Societies (IAMS, now the International Union of Microbiological Societies, IUMS) and, in cooperation with it, an International Federation of Type Culture Collections was formed. This Federation had ambitious plans which were never realized, and the Federation went out of existence within a few years.

In 1962, therefore, a Conference on Culture Collections (Martin, 1963), held after the VIIth International Congress for Microbiology, asked IAMS to form a Section on Culture Collections. The Section was set up in 1963 and, on the reorganization of IAMS in 1970, became the World Federation of Culture Collections (WFCC). The WFCC is also a multidisciplinary Commission of the International Union of Biological Sciences in the Divisions of Botany and Zoology, linking it with other organizations concerned with problems of biological preservation, such as herbaria, zoological gardens and museums. It has collected information on several hundred collections throughout the world and the *World Directory of Collections of Cultures of Microorganisms* (Martin and Skerman, 1972) has been published. This has recently been updated (V. F. McGowan and V. B. D. Skerman (Eds.), 1982, *World Directory of Collections of Cultures of Microorganisms*, 2nd Ed., World Data Centre on Microorganisms, University of Queensland, Brisbane, Australia). Pridham (1974) has also compiled a useful list of the acronyms and abbreviations for numerous culture collections. A number of national Federations of Culture Collections have also been formed which are affiliated to the WFCC. The aims of the WFCC include the collection of information on strains held by the collections and more detailed information on the strains themselves.

In the preservation of cultures, satisfactory methods of maintenance, with minimal change in the cultures, are essential, and an accepted system of taxonomy should be used. Particularly stringent standardization is required in the case of cooperative and comparative studies. In pursuit of these and similar aims, the WFCC has held training courses for curators and workers in culture collections at which other important functions of culture collections are also discussed. The WFCC also works in close cooperation with the International Committee on Systematic Bacteriology (ICSB) and its Judicial Commission, and other related national or international bodies dealing with all aspects of the preservation of various groups of microorganisms. It has also sponsored a number of international conferences on culture collections. A list of these, together with a summary of other WFCC activities is given by Lapage (1975). The WFCC supports the development of the World Data Centre at the University of Queensland, Brisbane, Australia, which is collecting cultural, physiological and other data on strains of microorganisms, and is exploring methods of recording such information in a standard format.

THE NEED FOR CULTURE COLLECTIONS

It is essential for the orderly development of bacteriology that cultures of organisms described or mentioned in publications be available for independent study. Because microbiologists are mortal and their interests vary during their working life, collections are necessary to provide an element of stability and continuity.

While some microbiologists spend a lifetime on one or two groups of organisms and build up large specialized collections, others move from one organism to another, abandoning old favorites. Both approaches generate problems in the preservation of organisms. The specialized collection may become so large and so specialized that it is hard to find a willing successor to the original enthusiastic curator. The worker whose interests are more fickle seldom worries about the systematic aspects, which make the preservation of cultures so desirable to the taxonomist.

Until the 1920s, the main reason for the existence of collections was their value for taxonomic and epidemiological studies. In the 1930s, the burgeoning interest in microbial physiology and biochemistry gave rise to a need for preserving organisms that produced or gave better yields of specific compounds. This greatly increased the value of culture collections.

More recently, studies on bacterial genetics have resulted in the isolation of numerous mutants which have, in turn, necessitated specialized collections. Some of these mutants are concerned with genetic loci useful in studies of nutrient and of biochemical pathways. The 1972 Stockholm Conference on the Environment recognized the importance of genetic pools and of collections of microorganisms.

Current developments in culture collections are diverse. Reviews of these can be found in Lapage (1971) and the volume edited by Colwell (1976). Methods of preservation are undergoing change, with increasing use of storage at very low temperatures to reduce the risk of genetic change. Recent legislation on patents has led to the need for deposition of strains used in industry. Cultures are also needed for teaching of microbiology and for quality control in many fields. The growth of numerous new diagnostic aids requires that large sets of strains from numerous species shall be available for establishing the data bases that are needed (Sneath, 1977). Culture collections may also in the future expand associated activities, such as storage and supply of dried material of microbial origin, standard antisera, nucleic acid preparations, and the like.

TYPE STRAINS

A particularly important function of culture collections is to preserve type strains and make them available to microbiologists who are undertaking taxonomic revisions. The nomenclatural aspects of type strains are discussed in the article on Bacterial Nomenclature, but some related points are briefly summarized here.

Type strains of bacterial species and subspecies are essential for the advance of taxonomy. They are required for comparison with strains that an author may believe belong to a new species or subspecies. Descriptions have never proved to be sufficient, because new techniques in systematics are continually being devised, and there is no substitute for an authentic strain when one wishes to make a critical comparison.

Type strains are of such taxonomic and nomenclatural importance that in this edition as many of them as possible are listed by their designation and catalog number in the main collections; a list of collections mentioned is given in the next article.

The new *International Code of Bacteriological Nomenclature* (Lapage et al., 1975; *Revised Code*) has made several special provisions for types. It is now a requirement for valid publication of a cultivable new species or subspecies of bacteria that a type strain shall be designated (alternative provisions exist for noncultivable bacteria). The Code urges that a type strain should be deposited in one or more of the permanently established culture collections. The numerous problems caused in the past by taxa for which there were no type strains should thus be largely overcome. Type cultures should, in the future, be available for all cultivable species of bacteria.

In the past it was frequently necessary to distinguish between different classes of type culture, in particular between types and neotypes, but the *Revised Code* has made most of these distinctions unnecessary. The new starting document for bacterial nomenclature, which came into force on January 1, 1980 (*Approved Lists of Bacterial Names*; Skerman, McGowan and Sneath, 1980) lists the type strains for the names of bacterial species that are currently recognized and given in the *Lists*. In the past, when many species had no type strains, it was necessary to establish **neotypes** for taxa where no type existed or had been lost. A neotype was thus a replacement for a type, and there should rarely be a need in the future for neotypes, except in the case of loss of the types. The procedure for establishing a neotype is given in the *Revised Code* and, needless to say, it should be deposited in one, or preferably several of the main culture collections.

While culture collections maintain type strains and neotypes as described above, they also keep typical and atypical strains, reference strains, and strains with particular properties of interest to biochemistry, genetics, serology, bacteriophage studies and the like; they also carry out many other functions, of which a general account can be found in Lapage (1971). Culture collections are therefore of great value not only to systematists but to all bacteriologists, and are essential to the development of the subject.

List of Culture Collections

There are several hundred culture collections in the world, the majority being small specialized collections, often collected by one individual. Details of most of these may be found in the *World Directory of Collections of Cultures of Microorganisms* (edited S. M. Martin and V. B. D. Skerman, 1972). This has recently been updated (V. F. McGowan and V. B. D. Skerman (Eds.), 1982, *World Directory of Collections of Cultures of Microorganisms*, 2nd Ed., World Data Centre on Microorganisms, University of Queensland, Brisbane, Australia). A smaller number of collections are frequently referred to in bacteriological work, and a selection of these is given below with commonly used abbreviations.

AMRC	FAO-WHO International Reference Centre for Animal Mycoplasmas, Institute for Medical Microbiology, University of Aarhus, Aarhus, Denmark.
ATCC	American Type Culture Collection, 12301 Parklawn Drive, Rockville, Maryland 20852, U.S.A.
BKM	All-Union Collection of Microorganisms, Institute of Microbiology, U.S.S.R. Academy of Sciences, Profsojunznaja 7, B-133 Moscow, U.S.S.R.
BKMW	Culture Collection, Institute of Microbiology, U.S.S.R. Academy of Sciences, Moscow, U.S.S.R.
CBS	Centraalbureau voor Schimmelcultures, Oosterstraat 1, Baarn, The Netherlands.
CCEB	Culture Collection of Entomophagous Bacteria, Institute of Entomology, Czechoslovak Academy of Sciences, Flemingovo N2, Prague 6, Czechoslovakia.
CCM	Czechoslovak Collection of Microorganisms, J. E. Purkyne University, Tr. Obr. Miru 10, Brno, Czechoslovakia.
CDC	Centers for Disease Control, Atlanta, Georgia, 30333 U.S.A.
CIP	Collection of the Institut Pasteur, Rue du Dr. Roux, Paris 15, France.
CNC	Czechoslovak National Collection of Type Cultures, Institute of Epidemiology and Microbiology, Srobarova 48, Prague 10, Czechoslovakia.
DSM	Deutsche Sammlung von Mikroorganismen, Schnitsphanstrasse, Darmstadt, Federal Republic of Germany.
IAM	Institute of Applied Microbiology, University of Tokyo, Bunkyo-ku, Tokyo, Japan.
ICPB	International Collection of Phytopathogenic Bacteria, University of California, Davis, California 95616, U.S.A.
IFO	Institute for Fermentation, 4-54 Jusonishinocho, Osaka, Japan.
IMRU	Institute of Microbiology, Rutgers—The State University, New Brunswick, New Jersey 08903, U.S.A.

IMV	Institute of Microbiology and Virology, Academy of Sciences of the Ukrainian S.S.R., Kiev, U.S.S.R.
INA	Institute for New Antibiotics, Nagatinskaya 3A, Moscow, U.S.S.R.
INMI	Institute for Microbiology, U.S.S.R. Academy of Sciences, Moscow, U.S.S.R.
IPV	Istituto di Patologia Vegetale, Milan, Italy.
KCC	Kaken Chemical Company Ltd., 6-42 Jujodai-1-Chome, Tokyo 114, Japan.
LIA	Museum of Cultures, Leningrad Research Institute of Antibiotics, 23 Ogorodnikov Prospect, Leningrad L-20, U.S.S.R.
LSU	Louisiana State University, Baton Rouge, Louisiana 70803, U.S.A.
LMD	Laboratorium voor Microbiologie, Technische Hogeschool, Julianalaan 67a, 2623 BC Delft, The Netherlands.
NCDO	National Collection of Dairy Organisms, National Institute for Research in Dairying, University of Reading, Shinfield, Reading, England, U.K.
NCIB	National Collection of Industrial Bacteria, Torry Research Station, Aberdeen AB9 8DG, Scotland, U.K.
NCPPB	National Collection of Plant Pathogenic Bacteria, Plant Pathology Laboratory, Hatching Green, Harpenden, England, U.K.
NCTC	National Collection of Type Cultures, Central Public Health Laboratory, Colindale, London NW9 5HT, England, U.K.
NIAID	National Institute of Allergy and Infectious Diseases, Hamilton, Montana 59840, U.S.A.
NIHJ	National Institute of Health, Tokyo, Japan.
NRC	National Research Council, Sussex Drive, Ottawa 2, Canada.
NRL	Neisseria Reference Laboratory, U.S. Public Health Service Hospital, Seattle, Washington 98114, U.S.A.
NRRL	Northern Utilization Research and Development Division, U.S. Department of Agriculture, Peoria, Illinois 61604, U.S.A.
PDDCC	Culture Collection of Plant Diseases Division, New Zealand Department of Scientific and Industrial Research, Auckland, New Zealand
TC	Thaxter Collection, Farlow Herbarium, Harvard University, Cambridge, Massachusetts 02138, U.S.A.
TPH	Microbiological Culture Collection, Public Health Laboratory, Ontario Department of Health, Toronto 116, Canada.
UMH	University of Missouri Herbarium, Columbia, Missouri 65201, U.S.A.

UQM	Culture Collection, Department of Microbiology, University of Queensland, Herston, Brisbane 4006, Australia.	VPI	Anaerobe Laboratory, Virginia Polytechnic Institute and State University, Blackburg, Virginia 24061, U.S.A.
VKM	Institute of Microbiology, Academy of Sciences of the U.S.S.R., Moscow, U.S.S.R.	WINDSOR	Culture Collection, University of Windsor, Windsor, Ontario, Canada

The Higher Taxa, or,
A Place for Everything . . . ?

R. G. E. Murray

"Quot homines tot sententiae; suo quoque mos." (so many men, so many opinions; each to his own taste).

Terence, Phormio.

When the eighth edition of *Bergey's Manual* was in preparation, a major taxonomic concern was the provision of a clear statement of where the bacteria fitted among living things. This was set out in *A Place for Bacteria in the Living World* (Murray, 1974), which summarized the reasons for recognizing the Kingdom *Procaryotae*, inclusive of the bacteria and the "blue-green algae" (Cyanobacteria). This concept, based on cellular organization, is now a part of the fundamental training of all biologists and a formal repetition is no longer a necessity. The student who wishes to relive that era should consult the major essays for details and references (Stanier, 1961; Stanier and van Niel, 1962; Murray, 1962; Allsopp, 1969; Stanier, 1970). Taxonomy is not static, however, and new horizons are being explored that give perspective and greater definition to higher taxa as well as the lower categories of genus and species.

The prefatory chapter mentioned above included a tentative proposal of appropriate higher taxa. The arguments and proposals that have arisen since then concern the levels of dissection of the kingdoms of the living world (Whittaker and Margulis, 1978; Woese and Fox, 1977a), the definition and levels of dissection of the major procaryotic groups (Gibbons and Murray, 1978; Whittaker and Margulis, 1978) and the integration of evolutionary information (Stackebrandt and Woese, 1981). There is a renewal of interest in bacterial taxonomy stimulated by the recognition of novel groups of bacteria that do not fit comfortably into current systematic schemes and by new understanding of the taxonomic utility and phylogenetic significance of molecular and genetic data. The system of "superphyla" and phyla proposed by Whittaker and Margulis (1978) is not sensitive to current interpretations of biochemical relatedness based on wall chemistry or other unique features of well-established groups of procaryotes. There is no advantage, at this stage of our understanding, in debating the relative value of recognizing "super kingdoms" (Whittaker and Margulis, 1978) or "primary kingdoms" (the Urkingdoms of Woese and Fox, 1977a) to accommodate views of cellular organization in protists, plants and animals as well as speculations about the nature of putative progenitors. We should be content for now to deal with the *Procaryotae* and the systematic problems that arise within that circumscription; there must be sufficient time for assimilation and consolidation of the burgeoning data. There are attractive features in the dendrograms generated by C. R. Woese and his colleagues; their time will come when the patterns of associations are less fragmentary. For now it would appear sensible to look to the Gibbons and Murray (1978) proposal as capable of modifi-

cation as an interim broad classification with a few areas of taxonomic validity.

The conceptual changes deriving from genetic and molecular studies are making inroads into the cherished beliefs of taxonomists and bacteriologists. We have to agree with the moderate statement taken from Stackebrandt and Woese (1981): "... what bacterial classification we have (say up through the eighth edition of *Bergey's Manual*, 1923–1974) is probably not in very good accord with the natural relationships that exist among organisms." This is true enough because it is only in the past decade that sequencing of biopolymers and molecular genetics has provided convincing data on relatedness and because the intent and role of *Bergey's Manual* has always been to provide a basis for the determination of the identity of a pure culture. It is unfortunate that the expression of nomenclatural decisions, hierarchical arrangements (even if all but abandoned in the eighth edition), and the mask of authority has tended to induce undue confidence in the relationships implied in earlier editions.

Even if our perception of "natural relationships" is flawed by ignorance as well as inadequate information, the practical bacteriologist needs a simple scheme of classification as a framework for recognition. At this stage, for example, the possibility that some micrococci are more closely related to *Arthrobacter* than to other spherical Gram-positive cocci and the implication for a further splitting of the genus *Micrococcus* (Stackebrandt and Woese 1979, 1981) would be confusing to the practical bench worker or the physician and is unhelpful. At the higher taxonomic levels some sort of serviceable scheme that can recognize and accomodate to a degree the possibilities will be of service, will cushion the shocks to come, and will stimulate appropriate research. This sort of practicality is almost realized in the proposal by Gibbons and Murray (1978). But any scheme will require future modification because it will take time to attain a complete reassessment of the taxonomic significance and validity of those characters that are reasonably easy to determine and apply effectively to each level of identification. The alternative possibility is that we should maintain two entirely independent schemes: a practical taxonomy and an academic (phylogenetic) taxonomy. The dichotomy of these phenotypic and genotypic approaches is with us because people of such persuasions work in semi-isolation and because, as argued by Stackebrandt and Woese (1981), "... the classically defined taxonomic categories, the genera and families, do not correspond to fixed (minimal) S_{AB} values." Furthermore, the genetics of today is beginning to clarify the mechanisms operating in the grand evolutionary experiment, blessed with minimal constraints of time and circumstance, conducted in nature's laboratory. It is clear that point mutations are less important than effective reassociations of determinants with their modifying segments and the mechanisms allowing the exchange, chromosomal integration and amplification of operative sets of determinants (Campbell, 1981; Cullum and Saedler, 1981). It is conceivable, nowadays, that major complex characters of physiological and taxonomic significance (involving a considerable number of genes) could be transferred between organisms both closely and distantly related in the clonal arborizations of the phylogenetic tree. All that is required is an occasional evolutionarily successful experiment in a time frame measured in thousands, millions or billions of years (or cell divisions, for that matter).

An overall taxonomic scheme that is capable of incorporating phylogenetic data (as well as providing a primary key) would be helpful in minimizing the dichotomy of interests and understanding among bacteriologists. It is desirable to bridge the growing gap between the practical applied fields and the academic substratum with something more than the perfidy of plasmids and technological legerdemain. The perpetual quandary is how and when to incorporate into systematic bacteriology the generalizations derived by intensive and expensive study of "model" organisms or of a limited set.

The most exciting and evocative of recent explorations of the possibilities of extracting phylogenetic information from highly conserved biopolymers (the "semantides" of Zuckerkandl and Pauling, 1965) are the comparisons of 16S ribosomal RNA catalogs undertaken by Carl Woese and his colleagues (Woese and Fox 1977b). This approach together with that involving the functional structural homologies and interchangeability of whole ribosome parts and some protein components shows promise of providing comparative data with evolutionary significance for the whole living world (Brimacombe et al., 1978; Kandler, 1981). Of most interest to us is the capability of the technique of RNA nucleotide analysis, expensive and slow though it may be, in spanning the widest range of procaryotic clones. The stability of most of this fairly large (1540 residues) molecule is the basis of the application to assessment of taxa at the level of family and higher. The fact that a few variable domains exist in the molecule allows for the partially realized possibility of contributing a degree of resolution at the level of genus. This could add to the data on relations within genus and species generated by utilizing DNA/DNA (Johnson, 1973) and RNA/DNA hybridization (De Smedt and DeLey, 1977). The figures that are generated (either the number of shared oligonucleotides or an association coefficient, the S_{AB} value) are based on a computer comparison of the catalogs of oligonucleotides (liberated from the 16S RNA by ribonuclease T_1) large enough to show individuality (larger than pentamers). This means that a considerable portion of the molecule yielding oligonucleotides smaller than hexamers is not taken into account and, therefore, the sequence data cannot be directly related to all other hybridization data. Nevertheless the results of comparing some 200 representative bacteria, surprises and all, support the directing thesis that most of the 16S rRNA sequence has drifted only slowly with time. We surmise (Stackebrandt and Woese, 1981) that the comparison provides a measure of the "depth" of the separation between the phylogenetic units or branches of the phylogenetic tree. It is an article of faith that the degree of cleavage (a low S_{AB} value) is proportional to time and is reasonable for initial purposes, although the time scale may be different for different major taxa.

Clones giving rise to unique and now recognizable groups of bacteria must have separated at various stages of procaryotic evolution (Fox et al., 1980). Earliest among these departures from the main stem so far detected by this oligonucleotide cataloging are the *Archaeobacteria**, which comprise the methanogens, halobacteria and thermoacidophiles; these, it is now known (Kandler 1981), possess peculiar lipids and either no murein or a pseudomurein in their cell walls. The eight or so major groups of photosynthetic and chemosynthetic bacteria (all designated "Eubacteria" in the papers cited) arose somewhat later in this imprecise evolutionary time scale. The data suggest that some genera are truly ancient (e.g. *Clostridium*, *Spirochaeta*) and older, in fact, than some very complex associations of genera (e.g. the actinomycetes).

Those involved in the comparative studies of 16 S rRNA, ribosomes and ribosomal proteins have come to the enthusiastic conclusion that the procaryotes are made up of two kingdoms, the *Eubacteria* and the *Archaeobacteria*. The molecular and biochemical basis for the separation have been summarized by Woese (1981) and Kandler (1981). There is no doubt whatever that the *Archaeobacteria* are distinguished by a number of specialized characters from the rest of the procaryotes ("Eubacteria" has had too many meanings in the past to be a useful term). They include the number of ribosomal proteins, the size and shape of the ribosomal S unit, the proportion of acidic ribosomal proteins, the constitution of tRNA initiator, the presence of ether-linked rather than ester-linked lipids, and the absence of muramic acid or the normal form of peptidoglycan from cell walls. These and some other intimate features make for interesting thoughts about eucaryotes, mitochondria and chloroplasts, as well as the procaryotes. But these distinctions are not suitable to kingdom status. Stackebrandt and Woese (1981) sum up the situation as follows: ". . . the general conclusion that seems to be emerging with regard to the differences among archaebacteria, true bacteria and eucaryotes is that all are identical in the basic aspects of their basic processes, yet all differ from one another in the details of these processes." An examination of any of the methanogens, strict halophiles or thermoacidophiles would place them in the kingdom *Procaryotae* as presently defined. It is not appropriate to separate kingdoms on any basis but a major, reasonably easily determined, difference in organization. Therefore, it seems sensible to treat the *Archaeobacteria* as a major taxon within the *Procaryotae* and, if necessary amend its status at some later date when more of the evidence is collected and digested. Perhaps we will soon recognize other equally distinctive clones that diverged very early from the stem clones of primitive microbes.

There is a comforting sequel to pondering compilations of articles regarding biochemical evolution (Wilson et al., 1977; Carlile et al., 1981). Although these studies suggest that "strange bedfellows" may be assigned to some of the more complex groups or point to unexpected separations (e.g. among the photosynthetic bacteria, Gibson et al., 1979; among the micrococci, Stackebrandt and Woese, 1979) there are concordant features. Gram-positiveness and Gram-negativeness are still unassailable characters except in what are now known to be phylogenetically and biochemically separate groups, the *Archaeobacteria* (Balch et al., 1979), the radiation-resistant cocci (Brooks et al., 1980) and, of course, the wall-less *Mollicutes*. Among the Gram-positives, it is encouraging to see that the many peptidoglycan types form consistent patterns in the branches of the dendrograms generated by comparison of the oligonucleotide catalogs (Schleifer and Kandler, 1972; Kandler and Schleifer, 1980; Kandler, 1981).

The infinitely diverse groupings of Gram-negative bacteria have yet to be surveyed to an extent that allows of any decisive taxonomic proposals and many of those included up to now exhibit phylogenetic and phenotypic incoherence. A major surprise arising from the analysis of rRNA oligonucleotides has been that each of the three coherent phylogenetic groupings of anoxygenic photosynthesizers contain a variety of seemingly related, diverse and well-known nonphotosynthetic genera showing subordinate S_{AB} values (Gibson et al., 1979; Stackebrandt and Woese, 1981). The implication is that photosynthetic clones may spawn apochlorotic derivatives; a thesis often directed to the cyanobacteria in the past (Pringsheim, 1967) but not yet subjected to this sort of phylogenetic analysis. It would seem wise not to make phototrophism an overriding taxonomic unit until the situation clarifies. This makes for difficulties in more conventional schemes such as that of Gibbons and Murray (1978), which separated in simplistic fashion the photosynthetic (*Photobacteria*) and the nonphotosynthetic (*Scotobacteria*) included in the Gram-negative bacteria (*Gracilicutes*). But it may be too early to be either discouraged or encouraged and there is still room for the exercise of one's prejudices.

Classifications cannot be final and there are many ways in which

---

* Equivalent to the term *Archaebacteria*. Because the word is formed by a combination of two Greek words (*archaios*, ancient, and *bakterion*, a small rod) the letter *o* should be used as the combining vowel, hence, *Archaeobacteria*.

bacteria can be classified (Cowan, 1968); this one has no more permanence than those that went before it. But the doubts and criticisms that come to the mind of the reader are the stimuli to further work and a deeper consideration of the taxonomic implications of the new mix of biochemical and phylogenetic data. Changes from the earlier versions (Murray, 1974; Gibbons and Murray, 1978) were inevitable. The hierarchical levels needed to be raised to give greater scope for classifying the range of organisms included at each major level. For example, students of the cyanobacteria, even those most sympathetic to their incorporation into bacterial taxonomy, despair of being able to accommodate their charges within the single order assigned by Gibbons and Murray (1978). As already indicated, it would take more than ordinary taxonomic agility to make a phylogenetically sensitive classification of the phototrophic bacteria and their derivatives in our present state of understanding.

The following is proposed as an arrangement of higher taxa which can serve during this time of taxonomic transition. It involves some amendments of rank and new names.

Kingdom *Procaryotae* Murray 1968, 252.
   Division I.   *Gracilicutes* Gibbons and Murray 1978, 3.
      Class I.   *Scotobacteria* Gibbons and Murray 1978, 4.
      Class II.   *Anoxyphotobacteria* (Gibbons and Murray) classis nov. (Subclassis *Anoxyphotobacteria* Gibbons and Murray 1978, 4.)
      Class III.   *Oxyphotobacteria* (Gibbons and Murray) classis nov. (Subclassis *Oxyphotobacteria*, Gibbons and Murray 1978, 3.)
   Division II.   *Firmicutes* Gibbons and Murray 1978, 5. (*Firmacutes* (sic) Gibbons and Murray 1978, 5.)
      Class I.   *Firmibacteria* classis nov.; L. adj. *firmus* strong; Gr. dim.n. *bakterion* a small rod; M.L. fem.pl.n. *Firmibacteria* strong bacteria, indicative of simple Gram-positive bacteria.
      Class II.   *Thallobacteria* classis nov.; Gr. n. *thallos* branch; Gr. dim.n. *bakterion* a small rod; M.L. fem.pl.n. *Thallobacteria* branching bacteria.

         (These new names are proposed to express the general basis of splitting the division into the simple Gram-positive bacilli and those Gram-positive bacteria showing a branching habit, the actinomycetes and related organisms.)
   Division III.   *Tenericutes* div. nov.; L. adj. *tener* soft, tender; L. fem.n. *cutis* skin; M.L. fem.n. *Tenericutes* procaryotes of pliable, soft nature, indicative of lack of a rigid cell wall.
      Class I.   *Mollicutes* Edward and Freundt 1967, 267.

         (The *Mollicutes* are a distinctive group of wall-less procaryotes of sufficiently diverse phylogenies that separate classes may well be required in the future.)
   Division IV.   *Mendosicutes* Gibbons and Murray 1978, 2. (*Mendocutes* (sic) Gibbons and Murray 1978, 2.)
      Class I.   *Archaeobacteria* (Woese and Fox) classis nov. (Kingdom *Archaebacteria* (sic) Woese and Fox 1977a, 5089.)

         (The *Archaeobacteria* are defined in terms of being procaryotes with unusual walls, membrane lipids, ribosomes and RNA sequences (Kandler, 1981). The future may bring further classes into the *Mendosicutes* when (and if) truly primitive organisms as envisioned by Woese and Fox (1977b) are isolated and recognized as related to a "universal ancestor" or progenote.)

This arrangement of the procaryotes continues to recognize the absence or presence and nature of cell walls as determinative at the highest level. The omission of Photobacteria (Gibbons and Murray, 1978) as a class and the elevation of *Oxyphotobacteria* and *Anoxyphotobacteria* to class status is intended to provide more scope than offered by Gibbons and Murray (1978) for the inevitable arrangement and rearrangement of the lower taxa within these categories as new under-

standing of lineage and relationships is brought to bear. For instance, the groups of phototrophic bacteria are formed of several major subgroups; separation of these from the level of class would be appropriate to the deep phylogenetic clefts that may be established in both the *Oxyphotobacteria* (the Cyanobacteria and the *Prochlorales*) and the *Anoxyphotobacteria* as pointed out by Stackebrandt and Woese (1981). Furthermore, it may not be appropriate to place all the nonphotosynthetic, Gram-negative bacteria in the *Scotobacteria* if the molecular evidence points clearly to derivation from photosynthetic ancestors; undoubtedly this would require separation at a high level within the class. The same need for broad scope is apparent in the *Firmicutes* with its two major divisions, the simple Gram-positive bacteria (*Clostridium* and relatives) and the actinomycetes.

The *Tenericutes* would have less support from the molecular phylogenetic evidence as a taxon at the highest level because of their probable origin from Gram-positive bacteria and the possibility that they may not have a single common ancestry (Woese et al., 1980). However, they form a stable and distinctive group; they are not obviously a subset of the *Firmicutes*, and their wallless state puts them clearly in a division by themselves as long as we base our classification on the presence or absence and character of the cell wall. On the other hand, we must recognize that an organism may lose a component of a very complex wall and still merit consideration as a member of that class; e.g. the members of the genus *Chlamydia* have no muramic acid but other characters, including some concerned with the relict wall, suggest a relationship to organisms that are definitive members of the *Gracilicutes*.

The *Archaeobacteria*, for their part, are a very diverse group in terms of cell-wall attributes (all the way from a complex wall including pseudomurein to wall-less) and there are at least five groupings with S_{AB} values of less than 0.3. There is sufficient scope within a division for the apparent complexity laid out by Balch et al., (1979) and any extraordinary "primitive" organisms that may be isolated. The sensible approach would seem to be maintenance of the consistency of the higher taxa by including in the class *Mendosicutes* all those procaryotes with a cell-wall composition inconsistent with that defined for *Gracilicutes* and *Firmicutes* (e.g. in simplest terms, not possessing muramic acid). This view is supported by Starr and Schmidt (1981). The inclusion of wall-less thermoacidophiles among the *Archaeobacteria* will be necessary even if seemingly inconsistent. There will come a time, without doubt, when we can set up taxa that are precisely defined in terms of molecular genetics; but that time has not yet arrived.

We cannot assume that all possible procaryotic organisms have been observed and isolated for study. The possibility exists that organisms will be found that are even more "primitive" (i.e. separated from the main stem even earlier) than the *Archaeobacteria* and have a constitution revealing more of the nature of the "universal common ancestor" (Woese and Fox, 1977; Stackebrandt and Woese, 1981). A class can be formed in the future as a suitable home for any organism whose proteins and genetic translation apparatus do not fit into the line of evolution represented by the procaryotes and eucaryotes studied up to now. They may have characteristics that foreshadow the fundamental eucaryotic cell.

No place is provided in our scheme for the fossil microbes being described from specimens of precambrian and possibly archean cherts (see Walter, 1977). This is because they can only be described in terms of size, shape and associations. The oldest are in stratified structures closely resembling the stromatolites and "algal mats" that can be found today, which are complex consortia of cyanobacteria, algae and bacteria with equally complex layering of metabolic and physiological characteristics. There are several attractive morphological resemblances. Size is about all that distinguishes the interpretation of forms as bacterial or algal. A further interpretation that a particular form is photosynthetic is entirely circumstantial and assumptive. Names have been assigned in binominal form (usually based on the Botanical Code) as a means of classifying the varied types being observed. This is a legitimate and stimulating activity but it is not yet helpful in terms of procaryotic classification or evolutionary taxonomy. Because of the uncertainties of alignment, they should be classified for determinative purposes in a separate group of microbiota.

Our view of classification and the taxonomic edifice is based on a century of experience supporting the contention that there is a reasonable degree of fixity in the characters describing a species. Such variation as there is within the clusters (as we now see in computer-assisted studies) can be included in the circumscription offered in describing the species or other categories. Certainly the species is a concept and not an entity (Cowan, 1968); but phenetic studies utilizing large numbers of strains and the widest possible range of characters has, if anything, clarified our concept of taxonomic groups (Sneath, 1978). Despite a growing appreciation of the Adansonian approach to species, the attitude to the definition of higher taxa involves the selection of seemingly single, but very complex characters as exemplified by this essay.

The alternative and extreme view that the bacteria are so pleomorphic that the species concept has no reality, is misleading, and draws the attention away from the important facts of life in clonal populations. Views of this character have been put forward by Sonea (1971) and Sonea and Panisset (1976, 1980). They believe, along with the rest of us, that the procaryotic clones are united by their lineage but differ in their interpretations of the stability of taxonomic units. They argue that genetic exchange (and "communications" of all sorts) between diverse clones make nonsense of the species concept. The extension of their analysis into a concept of unity for the entire world population of bacteria and their interactions with the environment (a sort of global organism as real as a horse or an elephant) is an interesting but philosophical curiosity. The approach stimulates thought but it seems evident that modern pleomorphists will have to modify and adapt their views to practical necessities dictated by new knowledge as much as ours views, expressed in this chapter, will have to bend with the winds of change. A century of bacteriological analysis and studies of cultures can convince one that there is sufficient stability to allow of the recognition of most taxonomic clusters.

No doubt, there are many and variant attitudes to the details of bacterial taxonomy. But the substratum of fact is now beginning to be revealed beneath the veneer of fancy or prejudice and to allow judgement to operate. Perhaps bacteriology is maturing after about 150 years of seeking a stable basis for classification.

"For now we see through a glass, darkly; but then face to face: now I know in part . . . "

The first Epistle of Paul
to the Corinthians (XIII:12)

That we can now perceive, albeit dimly, aspects of classification and phylogeny in the macromolecules of the *Procaryotae* is the legacy of many more scientists than have been cited so far in the prefatory chapters. The views that we espouse today still reflect the prejudices and enthusiasms of our teachers, our teachers' teachers and the influential observers and exponents of each stage in the maturation of bacteriology. The praise and the blame cannot easily be apportioned but undoubtedly we can recognize the pervasive influence of strong minded people of the modern era who enjoyed trying to make order out of chaos. These included D. H. Bergey, R. S. Breed, R. E. Buchanan and N. E. Gibbons whose efforts have brought *Bergey's Manual* into print in its various editions. But they, like others before and after them and despite their special interests, not to say prejudices, were collectors of the intellections and arrangements of "authorities." Despite the arguments concerning validity that undoubtedly surfaced at the time, their efforts could not prevent the perpetuation of numerous unstable taxa that we still struggle with today: form genera, color genera, physiological genera, etc., encompassing diverse and probably unrelated species, as genetical and biochemical criteria now force us to realize. But it is more important, perhaps, to realize that the most pervasive influences on all manner of approaches have been the writings of the "Delft school" (M. W. Beijerinck and A. J. Kluyver) and the product of that school, C. B. van Niel. The discussions accompanying "the van Niel course" (attended by an equally remarkable collection of microbiologists) started many thinking about microbes in new ways and set them and their students on productive lines of work. The comparative studies that resulted from these stimuli were of great significance in microbial biochemistry, physiology and ecology; a high proportion of the studies made significant contributions to systematic bacteriology. Not least among those influenced by van Niel was R. Y. Stanier, whose death came just as this volume was being readied for the press. It is obvious that we owe a particular debt to this lineage of bacteriologists. Stanier's arguments in our meetings, when he was a member of the Board of Trustees, were largely responsible for major changes in attitude and format expressed in the eighth and in this edition of the *Manual*; we sharpened our judgements with the help of other well established heretics such as S. T. Cowan (1970). And now, as strongly expressed and supported in this essay, a new breed of heretic is influencing bacterial systematics, as we had been warned would be the case by another former trustee, A. W. Ravin.

In the end, a reassessment of the diverse characters used in classifications will have as important consequences as the major generalization based on cellular organization used in defining the kingdom. We must identify characters of proven reliability and validity encompassing more of the genome than seems to be the case today. The techniques must allow the comparison of groups whatever their ecological niche or the professional proclivities of those that study them. Happily, we can echo the Rabbi ben Ezra and proclaim: "The best is yet to be."

Editorial Note

A publication presenting much of the data and examining the perspectives revealed in research on the *Archaeobacteria* appeared after the completion of this essay. It includes contributions from most if not all of the laboratories engaged in this thrilling task and starts with "an overview" by C. R. Woese. This comprehensive collection of papers is in Zentralblatt für Bakteriologie, Mikrobiologie und Hygiene, I Abt. Orig. C.*3*(1/2): 1–345, March/May 1982.

Kingdom Procaryotae
Murray, 1968, 252^{AL}*

R. G. E. Murray

Pro.car.y.o'tae. Gr. pref. *pro* before (primordial); Gr. n. *karyon* nut, kernel (nucleus); M.L. fem.pl.n. *Procaryotae* organism with primordial nucleus.

Single cells or simple associations of similar cells (0.2–10.0 μm in smallest dimension) forming a kingdom defined by cellular, not organismal, properties. The nucleoplasm (genophore) is never separated from the cytoplasm by a unit-membrane system (nuclear membrane) and is not associated with a basic protein. Cell division is not accompanied by cyclical changes in the texture or staining properties of either nucleoplasm or cytoplasm; a microtubular (spindle) system is not formed. The plasma membrane is frequently complex in topology and forms vesicular, lamellar or tubular intrusions into the cytoplasm; vacuoles and replicating cytoplasmic organelles enclosed by unit membranes are absent. Cytoplasmic organelles independent of the plasma membrane system (chlorobium vesicles, gas vacuoles) are relatively rare, and are enclosed by nonunit membranes. Respiratory and photosynthetic functions are associated with the plasma-membrane system in those members possessing these physiological attributes, although in the cyanobacteria there may be an independence of plasma and thylakoid membranes. Ribosomes of the 70S type (except for one division with slightly higher S values) are dispersed in the cytoplasm; an endoplasmic reticulum with attached ribosomes is not present. The cytoplasm is immobile; cytoplasmic streaming, pseudopodial movement, endocytosis and exocytosis are not observed. Nutrients are acquired in molecular form. Enclosure of the cell by a rigid wall is common but not universal. The cell may be nonmotile or exhibit swimming motility (mediated by flagella of bacterial type) or gliding motility on surfaces.

In organismal terms, these ubiquitous inhabitants of moist environments are predominantly unicellular microorganisms, but filamentous, mycelial or colonial forms also occur. Differentiation is limited in scope (holdfast structures, resting forms and modifications in cell shape). Mechanisms of gene transfer and recombination occur but these processes never involve gametogenesis and zygote formation.

DIVISION I. **GRACILICUTES** GIBBONS AND MURRAY 1978, 3^{AL}

Gra.cil.i.cu'tes. or Gra.cil.i' cu.tes. L. adj. *gracilis* thin; L. fem.n. *cutis* skin; M.L. fem.pl.n. *Gracilicutes* procaryotes with thinner cell walls, implying a Gram-negative type of cell wall.

Procaryotes that have a complex (Gram-negative type) cell-wall profile consisting of an outer membrane and an inner, thin peptidoglycan layer (which contains muramic acid and is present in all but a few organisms that have lost this portion of wall) and a variable complement of other components outside or between these layers. Usually stain Gram-negative. Cell shapes may be spheres, ovals, straight or curved rods, helices or filaments; some of these forms may be sheathed or capsulated. Reproduction is by binary fission but some groups show budding and a rare group (*Pleurocapsales*) shows multiple fission. Endospores are not formed. Fruiting bodies and myxospores may be formed by *Myxobacterales*. Swimming motility, gliding motility and non-motility are commonly observed. Members of the division may be phototrophic or non-phototrophic (both lithotrophic and heterotrophic) bacteria and include aerobic, anaerobic and facultatively anaerobic species; some members are obligate intracellular parasites.

DIVISION II. **FIRMICUTES** GIBBONS AND MURRAY 1978, 5^{AL}
(*Firmacutes* (sic) Gibbons and Murray 1978, 5)

Fir.mi.cu′ tes. or Fir.mi′cu.tes. L. adj. *firmus* strong, durable; L. fem. noun *cutis* skin; M.L. fem.pl.n. *Firmicutes* procaryotes with thick and strong skin, indicative of a Gram-positive type of cell wall.

Procaryotes with a cell-wall profile of Gram-positive type; reaction with Gram's stain generally, but not always, positive. Cells may be spheres, rods or filaments; the rods and filaments may be nonbranching but many show true branching. Cellular reproduction generally by binary fission; some produce spores as resting forms (endospores or spores on hyphae). Not photosynthetic; generally they are chemosynthetic heterotrophs and include aerobic, anaerobic and facultatively anaerobic species. The members of this division include simple asporogenous and sporogenous bacteria as well as the actinomycetes and their relatives.

DIVISION III. **TENERICUTES** DIV. NOV.
(Division *Mollicutes* Gibbons and Murray 1978, 5)

Te.ner.i.cu′tes. or Te.ner.i′cu.tes. L. adj. *tener* soft, tender; L. fem.n. *cutis* skin; M.L. fem.pl.n. *Tenericutes* procaryotes of pliable soft nature, indicative of lack of a rigid cell wall.

Procaryotes that lack a cell wall (commonly called the mycoplasmas and including the Class *Mollicutes*) and do not synthesize the precursors of peptidoglycan. They are enclosed by a unit membrane, the plasma membrane. The cells are highly pleomorphic and range in size from large deformable vesicles to very small (0.2 μm), filterable elements. Filamentous forms with branching projections are common. Reproduction may be by budding, fragmentation and/or by binary fission. Some groups show a degree of regularity of form due to placing of internal structures. Usually nonmotile but some species show a form of gliding motility. No resting forms are known. Stain Gram-negative. Most require complex media for growth (high osmotic pressure surroundings) and tend to penetrate the surface of solid media forming characteristic "fried egg" colonies. The organisms resemble the naked L-forms that can be generated from many species of bacteria (notably *Firmicutes*) but differ in that the mycoplasmas are unable to revert and make cell wall. Most species are further distinguished by requiring both cholesterol and long chain fatty acids for growth; unesterified cholesterol is a unique component of the membranes of both sterol-requiring and nonrequiring species if present in the medium. The guanine and cytosine content of ribosomal RNA is 43–48 mol% (lower than the 50–54 mol% of walled *Gracilicutes* or *Firmicutes*); the guanine and cytosine content of the DNA is also comparatively low, 23–46 mol%, and the genome size of the mycoplasmas is less than that of other procaryotes at 0.5–1.0 × 10^9 daltons. All are completely resistant to β-lactam antibiotics. The mycoplasmas may be saprophytic, parasitic or pathogenic, and the pathogens cause diseases of animals, plants and tissue cultures.

DIVISION IV. **MENDOSICUTES** GIBBONS AND MURRAY 1978, 2$^{VP^*}$
(*Mendocutes* (sic) Gibbons and Murray 1978, 2)

Men.dos.i.cu′tes. or Men.dos.i′cu.tes. L. Adj. *mendosus* having faults; L.fem.n. *cutis* skin; M.L. fem.pl.n. *Mendosicutes* procaryotes having faulty cell walls, suggesting the lack of conventional peptidoglycan.

Procaryotes that give evidence of an earlier phylogenetic origin from ancestral forms than the groups included in *Gracilicutes* and *Firmicutes*. The phylogenetic distinction (based on ribosomal RNA oligonucleotide catalogs) is correlated with unique sequences in transfer RNA, and with ether-linked polyisoprenoid branched-chain lipids. There may be systematic alterations in the form of the ribosomes (with higher S values because of a 10–12% increase in the number of component proteins) and in the proportion of acidic proteins. Most, but not all members possess some form of cell wall but this does not contain muramic acid and, therefore, they have no conventional peptidoglycan (murein); some have walls made purely of protein macromolecules and others of heteropolysaccharides; and some are lacking in wall material. Morphological forms include cocci, rods, filaments and irregular, mycoplasma-like elements; many are pleomorphic. Some stain Gram-positive, others stain Gram-negative. Endospores and resting forms have not been observed. Most are strict anaerobes but some groups are aerobic. Many are motile (bacterial flagella). The members are ecologically and metabolically diverse and the known members (including methanogens, strict halophiles and thermoacidophiles) live in somewhat extreme environments.

---

*$^*VP$, denotes that this name has been validly published in the official publication, International Journal of Systematic Bacteriology.

Important Notes for Users of this Edition

1. Always read both generic and species descriptions because characters listed in the generic description are not usually listed in the species descriptions.

2. Unless otherwise indicated in footnotes to tables, the meanings of symbols are as follows:

+ 90% or more of strains are positive

− 90% or more of strains are negative

d 11–89% of strains are positive

v strain instability (*not* equivalent to "d")

D different reactions in different taxa (species of a genus or genera of a family)

3. All other symbols are defined in footnotes to tables.

SECTION 1

The Spirochetes

ORDER I. **SPIROCHAETALES** BUCHANAN 1917, 163[AL]*

E. CANALE-PAROLA

Spi.ro.chae.ta′les. M.L. fem. n. *Spirochaetaceae* type family of order; *-ales* ending to denote an order; M.L. fem. pl. n. *Spirochaetales* the *Spirochaetaceae* order.

Helically shaped, motile bacteria, 0.1–3.0 μm and 5–250 μm in length. Unicellular, with one possible exception (*Spirochaeta plicatilis*). The outermost structure of the helical cell is a multilayered membrane referred to as "outer sheath" or "outer cell envelope" (Fig. 1.1). The **outer sheath** completely surrounds the **protoplasmic cylinder,** which consists of the cytoplasmic and nuclear regions enclosed by the cytoplasmic membrane-cell wall complex (Fig. 1.1). Around the helical protoplasmic cylinder are wound **periplasmic flagella** (which have also been called axial fibrils, axial filaments, flagella, and endo-flagella, and periplasmic fibrils) (Canale-Parola, 1978; Holt, 1978). The periplasmic flagella are enclosed by the outer sheath and, thus, are located between this membrane and the protoplasmic cylinder (Fig. 1.1). The number of periplasmic flagella ranges from 2 to more than

Figure 1.1. Schematic representation of a spirochete. The *broken line* indicates the outer sheath (outer cell envelope). The area delimited by the *thick, solid line* adjacent to the broken line represents the protoplasmic cylinder. The *circles* near the ends of the protoplasmic cylinder indicate the insertion points of the periplasmic flagella. The *solid thin lines* wound around the protoplasmic cylinder are the periplasmic flagella. (Reproduced with permission from E. Canale-Parola, *Bacteriological Reviews 41:* 181–204, 1977, © American Society for Microbiology.)

100/cell, depending on the species. One end of each flagellum is inserted near a pole of the protoplasmic cylinder, and the other end is not inserted (Fig. 1.1). The periplasmic flagella are components of the motility apparatus of the cell and perform a function(s) essential for locomotion and for the other movements typical of spirochetes (Paster and Canale-Parola, 1980). The periplasmic flagella of spirochetes are similar to other bacterial flagella in ultrastructure and in certain chemical characteristics. However, unlike other bacterial flagella, the periplasmic flagella are (a) permanently wound around the cell body, and (b) entirely endocellular, being enclosed by the outer sheath. Thus, **the motility of spirochetes differs from that of other bacteria because spirochetes suspended in liquids are able to locomote even though their cells do not have flagella that propel them by rotating in direct contact with the external environment** (see Further Comments on *Spirochaetales* for comparison with motility of spiroplasmas).

Spirochetes have three main types of movements in liquid environments: locomotion, rotation about their longitudinal axis, and flexing motions (Canale-Parola, 1978). Cells of spirochetes **remain locomotory in environments of relatively high viscosity** (e.g. 250 to almost 1000 centipoise for strains that have been tested), whereas other flagellated bacteria usually become immotile at approximately 60 centipoise. Spirochetes perform translational motility through agar media (e.g. in media containing 1% agar, w/v). **Creeping or crawling movements** of spirochetes in contact with solid substrata (e.g. glass surfaces) have been reported.

Chemoheterotrophic. Carbohydrates, amino acids, long chain fatty acids, or long chain fatty alcohols serve as carbon and energy sources (Canale-Parola, 1977; Johnson, 1977). **Anaerobic, facultatively anaerobic, or aerobic.** Gram-negative.

Free living or in association with animal and human hosts. Some species are pathogenic.

The mol% G + C of the DNA ranges from 25–65 (Bd).

Further Comments on **Spirochaetales**

In spite of some similarities, the motility of spirochetes differs from that of spiroplasmas. Spiroplasmas, which lack both flagella and periplasmic flagella, exhibit translational motility in semisolid agar media and in liquids of elevated viscosity, but do not show significant translational movement in liquids of viscosity comparable to that of water (e.g. in ordinary liquid culture media) (Davis, 1979). In contrast, spirochetes not only locomote in viscous environments, but also perform vigorous translational motility when the viscosity of the liquid in which they swim is not elevated.

Cells of spirochetes as well as cells of spiroplasmas exhibit translational movement in contact with solid surfaces.

---

*AL denotes the inclusion of this name on Approved Lists of Bacterial Names.

Key to the familes of the order **Spirochaetales**

I. Cell diameter 0.1–3.0 μm. Ends of cells are usually not hooked. The diamino acid present in the peptidoglycan is L-ornithine. Anaerobic, facultatively anaerobic, or microaerophilic. Use carbohydrates and/or amino acids as carbon and energy sources.

Family I. *Spirochaetaceae*

II. Cell diameter 0.1 μm. Ends of cells are usually hooked. The diamino acid present in the peptidoglycan is diaminopimelic acid. Aerobic. Use long chain fatty acids or long chain fatty alcohols as carbon and energy sources.

Family II. *Leptospiraceae*

FAMILY I. **SPIROCHAETACEAE** SWELLENGREBEL 1907, 581[AL]

E. CANALE-PAROLA

Spi.ro.chae.ta′ce.ae M.L. fem. n. *Spirochaeta* type genus of the family; -*aceae* ending to denote a family; M.L. fem. pl. n. *Spirochaetaceae* the *Spirochaeta* family.

Helical cells, 0.1–3.0 μm in diameter and 5–250 μm in length. **Ends of cells usually not hooked.** Individual periplasmic flagella extend along most of the length of the cell. Thus, flagella inserted near opposite poles of the cell overlap and are in close apposition to one another in the central region of the cell (Fig. 1.1). **The diaminoamino acid present in the peptidoglycan is L-ornithine.** Motile.

Anaerobic, facultatively anaerobic, or microaerophilic. Chemoorganotrophic. **Utilize carbohydrates and/or amino acids as carbon and energy sources. Do not utilize long chain fatty acids or long chain fatty alcohols as energy sources.**

Free living or in association with animal and human hosts. Some species are pathogenic.

The mol% G + C of the DNA is 25–65 (Bd). Species that have been examined by means of 16S rRNA cataloging are phylogenetically distant from *Leptospiraceae*.

Type genus: *Spirochaeta* Ehrenberg 1835, 313.

Key to the genera of the family **Spirochaetaceae**

I. Cells are 0.2–0.75 μm in diameter and 5–250 μm in length. Obligately anaerobic and facultatively anaerobic. Carbohydrates serve as energy and carbon sources. Amino acids are not utilized as growth substrates. Free living in aquatic environments such as the sediments, mud and water of ponds, marshes, lakes and rivers. Present in freshwater and in marine environments. The mol% G + C of the DNA is 51–65 (Bd).

Genus I. *Spirochaeta*, p. 39

II. Cells are 0.5–3.0 μm in diameter and 30–180 μm in length. The periplasmic flagella are present as a bundle that distends the outer sheath to form a ridge (crista). Inhabit the crystalline style or the fluid of the digestive tract of marine and fresh water molluscs. Not grown in pure culture.

Genus II. *Cristispira*, p. 46

III. Cells usually are 0.1–0.4 μm in diameter and 5–20 μm in length. Species that have been grown in pure culture are obligate anaerobes. Carbohydrates and amino acids are utilized as fermentable substrates. Indigenous to the mouth, intestinal tract, and genital areas of humans and animals. Some species are pathogenic. The mol% G + C of the DNA is 25, 36–43, and 53.

Genus III. *Treponema*, p. 49

IV. Cells usually measure 0.2–0.5 μm in diameter and 3–20 μm in length. Probably microaerophilic. Pathogens, causative agents of relapsing fever. The mol% G + C of the DNA has not been reported.

Genus IV. *Borrelia*, p. 57

Genus I. **Spirochaeta** Ehrenberg 1835, 313[AL]

E. CANALE-PAROLA

(*Spirochoeta* Dujardin 1841, 225, and *Spirochaete* Cohn 1872, 180 (orthographic variants of *Spirochaeta*); *Ehrenbergia* Gieszczkiewicz 1939, 24.)

Spi.ro.chae′ta. Gr. n. *spira* a coil; Gr. n. *chaete* hair; M.L. fem. n. *Spirochaeta* coiled hair.

Helical cells 0.2–0.75 μm in diameter and 5–250 μm in length. All species have 2 periplasmic flagella/cell except *Spirochaeta plicatilis*, which has many periplasmic flagella. Under unfavorable conditions spherical cells or structures 0.5–2.0 μm (occasionally up to 10 μm) in diameter are formed. Cells locomote when suspended in liquids and crawl or creep when in contact with solid surfaces. **Obligately anaerobic or facultatively anaerobic.** Under aerobic growth conditions the **facultatively anaerobic species usually produce carotenoid pigments** that give a yellow, yellow-orange or red coloration to colonies. Optimum temperature, 25–40°C. **Chemoorganotrophic, using a variety of carbohydrates as carbon and energy**

sources. The main products of anaerobic carbohydrate metabolism are ethanol, acetate, CO_2 and H_2, except for one species (*Spirochaeta zuelzerae*) that produces succinate and lactate instead of ethanol. Facultatively anaerobic species oxidize carbohydrates aerobically yielding primarily CO_2 and acetate. Indigenous to aquatic environments such as the sediments, mud and water of ponds, marshes, swamps, lakes and rivers. Occur commonly in H_2S-containing environments. Present in freshwater and marine environments. **Free living.** None reported to be pathogenic. The mol% G + C of the DNA is 51–65 (Bd).

Type species: *Spirochaeta plicatilis* Ehrenberg 1835, 313.

Further Descriptive Information

Cells of all species are helical in shape. The helical shape may be lost through mutation (Hel⁻ mutants). These mutants are rod shaped, frequently with one or both cell ends bent, coiled or wavy (Greenberg and Canale-Parola, 1977b).

The nature of the spherical structures, called "spherical bodies" that are formed under unfavorable growth conditions has not been determined. Spherical bodies occur either in physical association with helical cells or free.

Cells of strains that have been studied locomote suspended in liquids "in straight lines or nearly straight lines, and they appear to spin rapidly about their longitudinal axis ... Occasionally a cell stops momentarily and flexes, and then resumes spinning and translational motility. However, when translation resumes, the direction of movement is usually altered ..." and frequently the previously leading cell end becomes the trailing end (Greenberg and Canale-Parola, 1977a). Cells in motion usually retain their basic helical configuration, but they assume a variety of shapes as a result of flexing, undulating, and contracting movements, as well as wave propagation. Broad secondary coils or waves superimposed on the smaller primary coils are formed frequently (Canale-Parola, 1977, 1978). During creeping movements of *S. plicatilis* on solid surfaces (Blakemore and Canale-Parola, 1973) "... the rear coils follow the tortuous path of the anterior cell end almost exactly" (Canale-Parola, 1978).

Cells retain translational motility in environments of relatively high viscosity, usually becoming immotile at viscosities ranging from 300–1000 centipoise, depending on the strain (Greenberg and Canale-Parola, 1977b, 1977c).

Strains of *S. aurantia* that have been tested exhibit chemotaxis toward carbohydrates, but not toward amino acids (Breznak and Canale-Parola, 1975; Greenberg and Canale-Parola, 1977a). Effective attractants for *S. aurantia* strain M1 are: D-glucose, 2-deoxy-D-glucose, α-methyl-D-glucoside, D-galactose, D-fucose, D-mannose, D-fructose, D-xylose, maltose, cellobiose, and D-glucosamine (Greenberg and Canale-Parola, 1977a). Taxis toward D-galactose and D-fucose is induced by the presence of D-galactose in the growth medium.

The helical shape of the cells is maintained by the peptidoglycan layer (Joseph and Canale-Parola, 1972). L-ornithine is the only diaminoamino acid in the peptidoglycan of *S. stenostrepta*, *S. litoralis*, *S. aurantia*, and *S. halophila* (Joseph et al., 1973; B.J. Paster and E. Canale-Parola, unpublished data). The peptidoglycans of *S. zuelzerae* and *S. plicatilis* have not been tested for the presence of L-ornithine.

The peptidoglycan of *S. stenostrepta* is composed of acylglucosamine, acylmuramic acid, L-alanine, D-glutamic acid, L-ornithine, and D-alanine (Joseph et al., 1973; Schleifer and Joseph, 1973). Peptidoglycan of similar composition is present in *S. litoralis*. At least 50% of the peptide subunits of *S. stenostrepta*'s peptidoglycan consist of the tripeptide *N*-acyl-muramyl-L-alanyl-α-D-glutamyl-L-ornithine. Cross-linkage (30%) is between the δ-amino group of L-ornithine and the carboxyl group of D-alanine present in the remaining peptide subunits of sequence *N*-acyl-muramyl-L-alanyl-α-D-glutamyl-L-ornithyl-D-alanine (Schleifer and Joseph, 1973).

A lipoprotein layer, adjacent and external to the peptidoglycan, has been detected in *S. stenostrepta* (Joseph et al., 1970). This layer consists of a fine array of tightly packed, longitudinally oriented helices measuring 2.5 nm in diameter (Holt and Canale-Parola, 1968).

Colonies of *Spirochaeta* diffuse or spread through the agar medium in which they are growing. This phenomenon is especially apparent in agar media containing low substrate concentrations and 1% or less agar. Diffusion of colonies is due to migration of the growing cells through the agar medium. Migration of the cells is the result of chemotaxis toward the growth substrate and of the ability of spirochetes to locomote through agar gels (Canale-Parola, 1977, 1978).

S. stenostrepta may be grown in medium GYPT, *S. zuelzerae* in medium SZ, *S. litoralis* in medium MGTY, and *S. aurantia* in medium GTY (see Table 1.3). Chemically defined growth media have been described for *S. litoralis* (Hespell and Canale-Parola, 1970b) and *S. aurantia* (Breznak and Canale-Parola, 1975).

S. halophila may be grown in medium ISM which contains 0.2 g of peptone (Difco) and 0.4 g of yeast extract (Difco)/98 ml of an inorganic salt solution. This solution has the following composition: $CaCl_2$, 0.01 M; NaCl, 0.75 M; $MgSO_4$, 0.2 M. In preparing the inorganic salts solution the dihydrate form of $CaCl_2$ and the heptahydrate form of $MgSO_4$ are used, and the salts are added in the order in which they are listed, to prevent formation of a precipitate. The pH of the medium is adjusted to 7.5 with KOH. After autoclaving, the volume of the medium is brought to 100 ml by adding a separately sterilized solution of maltose to obtain a final concentration of 0.5% (Greenberg and Canale-Parola, 1976).

The obligately anaerobic species of *Spirochaeta* grow readily in media gelled by the addition of 1.0 or 1.5 g of agar/100 ml, whereas the growth of some strains of facultatively anaerobic species of *Spirochaeta* is inhibited in media containing more than 1.0% agar. These strains grow abundantly, however, when the agar concentration in the medium is 1% or lower.

Under anaerobic conditions *S. stenostrepta*, *S. litoralis*, *S. aurantia*, and *S. halophila* ferment carbohydrates to pyruvate via the Embden-Meyerhof pathway (Canale-Parola, 1977; Greenberg and Canale-Parola, 1976). Pyruvate is metabolized to acetyl-CoA, CO_2, and H_2 by means of a clostridial-type clastic reaction (Fig. 1.2). Acetyl-CoA is converted to acetate in reactions catalyzed by phosphotransacetylase and acetate kinase, and to ethanol through a double reduction involving aldehyde and alcohol dehydrogenase activities (Fig. 1.2; see Table 1.5) (Canale-Parola, 1977). The pathways in Figure 1.2 constitute the major anaerobic energy-yielding mechanisms utilized by the four *Spirochaeta* species mentioned above (Canale-Parola, 1977). The pathways of carbohydrate catabolism utilized by *S. zuelzerae* have not been elucidated. This spirochete does not form ethanol but produces succinate and larger amounts of lactate than other species (see Table 1.5).

When growing aerobically *S. aurantia* and *S. halophila* derive energy by performing an incomplete oxidation of carbohydrates, with CO_2 and acetate being the main dissimilatory products. The tricarboxylic acid cycle either is not present or serves in a minor catabolic capacity in these two species. Determinations of molar growth yields and other

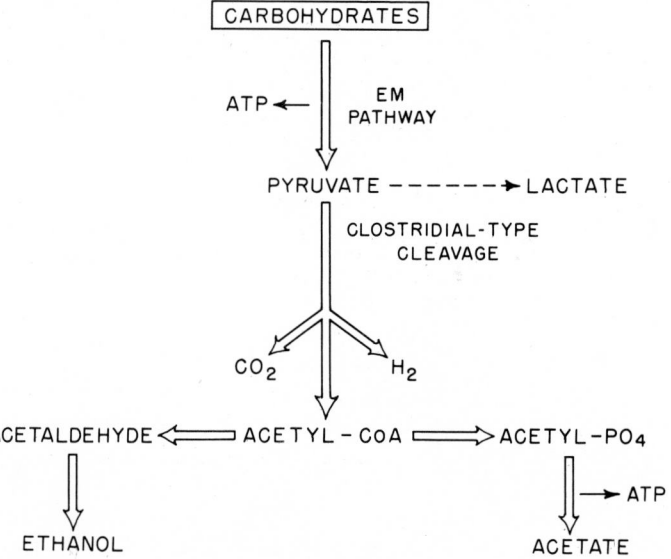

Figure 1.2. Pathways for anaerobic dissimilation of carbohydrates by *Spirochaeta stenostrepta*, *S. litoralis*, *S. aurantia* and *S. halophila*. The *broken line* indicates a minor pathway. (Reproduced with permission from E. Canale-Parola, Bacteriological Reviews *41*: 181–204, 1977, © American Society for Microbiology.)

studies indicate that, when growing aerobically, *S. aurantia* and *S. halophila* generate ATP via oxidative phosphorylation as well as by substrate level phosphorylation (Breznak and Canale-Parola, 1972b; Greenberg and Canale-Parola, 1976). Cytochromes b_{558} and cytochrome o are present in *S. aurantia* (Breznak and Canale-Parola, 1972b).

Spirochaeta species are able to synthesize all of their cell lipids de novo. The chain length of cellular fatty acids varies from 12–18 carbons (Livermore and Johnson, 1974). *S. aurantia* and *S. zuelzerae* synthesize unsaturated fatty acids, whereas *S. litoralis* and *S. stenostrepta* do not. Anteiso-branched chain fatty acids are synthesized by *S. stenostrepta* and *S. zuelzerae*, but not by *S. litoralis* and *S. aurantia* (Livermore and Johnson, 1974).

Spirochaeta species are resistant to the antibiotic rifampin (rifampicin) at concentrations ranging from 1–50 μg rifampin/ml. Resistance to rifampin may be due to low affinity of the spirochetes' RNA polymerase for the antibiotic.

Isolation and Enrichment Procedures

Free living spirochetes capable of anaerobic growth (genus *Spirochaeta*) are readily isolated by means of a method that involves the antibiotic rifampin as the selective agent (Stanton and Canale-Parola, 1979; Leschine and Canale-Parola, 1980; Weber and Greenberg, 1981). In this method, the isolation medium for marine strains is identical to medium MGTY (see Table 1.3) except that it includes cellobiose (0.2 g/100 ml of medium) instead of glucose, and that rifampin is added to it (2 μg/ml of medium). Rifampin is added as a filter-sterilized solution to the autoclaved medium. A similar medium may be used for the isolation of freshwater strains, except that distilled water is included in the medium instead of sea water. Medium GYPT (see Table 1.3), to which rifampin is added as indicated above, may serve as isolation medium for freshwater strains. All isolation media contain 1.0 g of agar/100 ml. When the isolation media are prereduced, 0.1 mg of resazurin is included/100 ml of medium.

The spirochetes are isolated as follows by means of the rifampin selection method. A small amount of freshwater or marine mud is serially diluted into melted (45°C) deeps of isolation medium. When the isolation medium is prereduced either a N_2 or Ar atmosphere is used. After incubation (22–30°C), samples of spherical colonies typical of spirochetes (see below) are removed from the agar deeps by stabbing a Pasteur pipette through each selected colony. Spirochete cells from the sampled colonies are cloned by transferring them 3 times (serial dilutions) in isolation medium. Spirochete colonies in deeps of isolation media are readily recognizable because they tend to diffuse through the agar gel and resemble in appearance a "cotton ball," a "transparent bubble" or "veil-like growth with a denser center" (Stanton and Canale-Parola, 1979).

Other selective procedures for the isolation of *Spirochaeta* species are (a) the filtration method; (b) the migration through agar media method; and (c) the filter disk-on-plate method (Canale-Parola, 1973).

The filtration method has been used for the isolation of thin spirochetes present in mud. These organisms are separated from most of the accompanying microbiota by filtering mud suspensions through cellulose-ester filter disks (Millipore, pore dia., 0.3 or 0.45 μm). Thin spirochetes pass through the pores of the filter disks and are found in the filtrate, whereas most other bacteria are retained on the filter disks. The filtrate is inoculated into appropriate liquid media to allow growth of the spirochetes (Canale-Parola, 1973). After incubation of these cultures, cloning of the spirochetes is accomplished by standard techniques involving serial dilutions through agar media. The type strains of *S. stenostrepta*, *S. litoralis*, and *S. halophila* were isolated by means of the filtration method (Canale-Parola et al., 1967, 1968; Hespell and Canale-Parola, 1970b; Greenberg and Canale-Parola, 1976).

As mentioned previously, spirochetes have the ability to migrate through agar media. This property of spirochetes may be used for their selective isolation from natural environments (Canale-Parola, 1973). A minute amount of spirochete-containing mud is placed in a small well (approximately 3 mm deep) melted through the surface of solidified agar medium by touching the agar gel with the heated tip of a pipette. The medium has been previously solidified in a slanted position in a 60-ml bottle sealed with a rubber stopper. Alternatively, the mud may be placed in a well in the center of an agar medium plate. During anaerobic incubation (N_2 atmosphere) of the well enrichments, the spirochetes multiply and migrate through the agar medium forming a characteristic subsurface growth veil that extends away from the well. The growth of most other bacteria is restricted to the well. Pure cultures of spirochetes from the growth veil are obtained by standard techniques. This procedure has been used successfully for the isolation of strains of anaerobic marine spirochetes (Canale-Parola, 1973).

The filter disk-on-plate method involves both filtration and migration of spirochetes. This method is suitable for the isolation of facultatively anaerobic strains of *Spirochaeta*. A sterile cellulose-ester filter disk (Millipore, 47-mm disk dia., 0.3 or 0.45 μm pore dia.) is placed on the surface of an agar medium plate. A drop of pond water or water-mud slurry is added near the center of the filter disk. During incubation of this plate enrichment (22–30°C, 12–24 h, in air) thin spirochetes in the inoculum move through the pores of the filter disk onto the surface of the medium. Then the filter disk is removed from the plate enrichment and incubation of the plate is continued in order to allow growth and migration of the spirochetes. The spirochetes form a characteristic subsurface growth veil which extends toward the periphery of the plate away from colonies of contaminating bacteria that have passed through the filter disk. Cloning of spirochetes from the growth veil is accomplished by means of standard techniques. The type strain and other strains of *S. aurantia* have been isolated by means of the filter disk-on-plate method (Breznak and Canale-Parola, 1969, 1975).

Maintenance Procedures

Species of *Spirochaeta* remain viable for many years when stored in the frozen state at the temperature of liquid nitrogen. Methods for liquid nitrogen storage of *Spirochaeta* species and for the preparation of other types of stock cultures of these bacteria have been described (Canale-Parola, 1973).

Differentiation of the genus **Spirochaeta** from other genera

Characteristics that distinguish the genus *Spirochaeta* from other genera of spirochetes are presented in Table 1.1. Other characteristics are listed elsewhere in Tables 1.2, 1.4, and 1.5.

Taxonomic Comments

The genus *Spirochaeta* represents one of the major eubacterial groups so far identified by means of 16 S rRNA oligonucleotide cataloging (Fox et al., 1980).

The genus includes two groups of organisms distinguishable on the basis of their relation to molecular oxygen. One of these groups comprises the obligate anaerobes (*S. stenostrepta*, *S. zuelzerae*, *S. litoralis*), the other the facultative anaerobes (*S. aurantia*, *S. halophila*) (see Table 1.2). It has been suggested that each of these groups should constitute a separate genus (Canale-Parola et al., 1968).

Revision of the existing classification of organisms presently assigned to the genus *Spirochaeta* is hampered by the lack of information on *S. plicatilis* (the type species), which has not been obtained in pure culture. For example, the relation of this organism to molecular oxygen is not known. Recently the Judicial Commission of the ICSB* has been asked

---

* International Committee on Systematic Bacteriology.

Table 1.1.

Differentiation of the genus **Spirochaeta** *from other genera of spirochetes*[a]

Characteristic	Spirochaeta	Cristispira	Treponema	Borrelia	Leptospira
Free living	+	−	−	−	+
Host associated	−	+	+	+	+
Obligate aerobes	−		−[b]	+[c]	+
Obligate anaerobes	+		+[b]	−	−
Facultative anaerobes	+		−[b]	−	−
Energy and carbon sources:					
Carbohydrates	+		+	+	−
Amino acids	−[d]		+		−
Long chain fatty acids	−		−		+
De novo cellular fatty acid synthesis	+		−		Rare
Ability to synthesize[e]					
Monoglycosyl diglyceride	+		+		−
Phosphatidyl choline	−		+		−
Mol% G + C of DNA (Bd, T_m)	51–65		25–54		35–53

[a] Symbols: see standard definitions.

[b] *Treponema pallidum* may be a microaerophile or a facultative anaerobe.

[c] *Borrelia* may be microaerophilic.

[d] Some obligately anaerobic marine strains of *Spirochaeta* ferment L-valine, L-isoleucine, and L-leucine with ATP generation, but do not utilize these amino acids as growth substrates.

[e] Data from Livermore and Johnson, 1974.

Table 1.2.

Differential characteristics of the species of the genus **Spirochaeta**[a]

Characteristics	1. S. plicatilis	2. S. stenostrepta	3. S. zuelzerae	4. S. litoralis	5. S. aurantia	6. S. halophila
Cultivation in pure culture	−	+	+	+	+	+
Cell diameter, μm	0.75	0.20–0.30	0.20–0.35	0.40–0.50	0.30	0.40
Number of periplasmic flagella:						
Two flagella/cell	−	+	+	+	+	+
Many flagella/cell	+	−	−	−	−	−
Relationship to oxygen:						
Obligate anaerobe		+	+	+	−	−
Facultative anaerobe		−	−	−	+	+
High NaCl concentration required for growth		−	−	+	−	+
Ethanol formed		+	−	+	+	+
Succinate formed		−	+	−	−	−
Optimum growth temperature, °C		30–37	37–39	30	25–30	35–40
Pigmentation:						
Yellow-orange		−	−	−	+	−
Red		−	−	−	−	+
Mol% G + C of DNA (Bd)		60	56	51	61–65[b]	62

[a] Symbols: see standard definitions.

[b] *S. aurantia* subsp. *aurantia* = 62–65; *S. aurantia* subsp. *stricta* = 61.

to issue an opinion recognizing *S. stenostrepta* Zuelzer 1912, 17 as the type species of the genus *Spirochaeta* (Canale-Parola, 1981). Recognition of *S. stenostrepta* as the type species would allow a taxonomic reorganization of the species presently assigned to the genus *Spirochaeta*. *Editorial Note*: The Judicial commission recently rejected the proposal to recognize *S. stenostrepta* as the type species of the genus (L. G. Wayne, Int. J. Syst. Bacteriol. *32:* 464–465, 1982).

Further Reading

Canale-Parola, E. 1973. Isolation, growth, and maintenance of anaerobic free living spirochetes. *In* J.R. Norris and D.W. Ribbons (Editors), Methods in Microbiology, Vol. 8, Academic Press, New York, pp. 61–73.

Canale-Parola, E. 1977. Physiology and evolution of spirochetes. Bacteriol. Rev. *41:* 181–204.

Canale-Parola, E. 1978. Motility and chemotaxis of spirochetes. Annu. Rev. Microbiol. *32:* 69–99.

Holt, S.C. 1978. Anatomy and chemistry of spirochetes. Microbiol. Rev. *42:* 114–160.

Differentiation of the species of the genus **Spirochaeta**

Characteristics useful for distinguishing the various species of the genus *Spirochaeta* are indicated in Table 1.2.

Table 1.3.

Growth media for species of **Spirochaeta**[a]

Medium component[b]	GYPT Medium	SZ Medium	MGTY Medium	GTY Medium
Distilled water (ml)	100	100	20	99
Sea water (ml)			75	
Glucose	0.5	0.2	0.2	0.2
Peptone	0.2			
Trypticase (BBL)			0.1	0.5
Yeast extract	0.2	0.4	0.1	0.2
L-cysteine (or Na thioglycolate)	0.05	0.05	0.05	
$CaCl_2 \cdot 2H_2O$		0.004		
$MgSO_4 \cdot 7H_2O$		0.05		
$NaHCO_3$		0.1		
KH_2PO_4		0.1		
Potassium phosphate buffer, 1 M, pH 7 (ml)				1
Tris-HCl buffer, 1 M (ml)			5	

[a] Numbers in table indicate grams unless otherwise specified.

[b] When preparing prereduced media (GYPT, SZ, or MGTY) 0.1 ml of a 0.1% resazurin solution may be added. A N_2 atmosphere is used for prereduced media. $NaHCO_3$, KH_2PO_4, and the K-phosphate buffer are added to the sterile media as separately sterilized solutions. The Tris-HCl buffer is prepared as follows: to 1865 ml of distilled water are added 242.2 g of Trizma (2-amino-2-hydroxymethyl-1,3-propanediol) base, reagent grade (Sigma T-1503) and 133.3 ml of concentrated HCl. The final concentration of this buffer in MGTY medium is 50 mM. A 50-mM solution of the Tris-HCl buffer has a pH of 7.5 at 30°C. Prior to sterilization, the pH of the media is adjusted as follows: medium GYPT, pH 7.4; medium SZ, pH 7.2; medium GTY, pH 7.5. GYPT medium is for *S. stenostrepta*, SZ medium for *S. zuelzerae*, MGTY medium for *S. litoralis*, and GTY medium for *S. aurantia*.

List of the species of the genus **Spirochaeta**

1. **Spirochaeta plicatilis** Ehrenberg 1835, 313.[AL]

pli.ca′ti.lis. L. adj. *plicatilis* flexible.

Helical cells, 0.75 μm in diameter and usually 80–250 μm in length. Cells have regular primary coils which are stable (they persist both in the presence and absence of movement). Cells in motion may exhibit broad secondary coils superimposed on the smaller primary coils. Moving cells suspended in liquids display rotation about the longitudinal axis and wide waves traveling along the length of the organism. Cells creep in contact with solid surfaces (Blakemore and Canale-Parola, 1973).

Regularly spaced cross-walls or transverse septa are present (Blakemore and Canale-Parola, 1973). Long specimens may consist of chains of multicellular spirochetes. Many periplasmic flagella are present, occurring as a bundle wound around the protoplasmic cylinder. Phase-contrast photomicrographs and electron micrographs of the cells have been published (Blakemore and Canale-Parola, 1973).

Not cultivated in pure culture. Presumed to be either a microaerophile or an anaerobe that can tolerate low O_2 tensions. Present in H_2S-containing freshwater, brackish and marine mud, frequently in association with *Beggiatoa* trichomes.

2. **Spirochaeta stenostrepta** Zuelzer 1912, 17.[AL]

ste.no.strep′ta. Gr. adj. *stenus* narrow; Gr. adj. *streptus* pliant, easily bent; M.L. adj. *stenostrepta* tightly coiled.

Helical cells, 0.2–0.3 μm in diameter and 15–45 μm in length. Some of the cells in cultures are shorter than 15 μm. In the late exponential and stationary phases the organisms increase in length (up to 300 μm). Long organisms occasionally pair and become entwined, or a single organism becomes partially wrapped around itself. The cells have regular, stable primary coils. Cells in motion occasionally exhibit broader secondary coils or waves superimposed on the smaller primary coils. Spherical bodies generally 1–3 μm in diameter are occasionally observed in cultures. The spherical bodies occur either free or in association with helical cells. Each cell has two subterminally inserted periplasmic flagella that overlap in the central region of the cell (1–2–1 arrangement). Phase-contrast photomicrographs and electron micrographs of the cells have been published (Canale-Parola et al., 1967, 1968; Holt and Canale-Parola, 1968).

Subsurface colonies (in GYPT medium containing 1.5 g agar/100 ml; Table 1.3) are white, spherical, fluffy, approximately 2–3 mm in diameter when fully developed. Smaller spherical colonies lacking the characteristic fluffiness are present occasionally.

Obligately anaerobic, having a fermentative type of metabolism. Various carbohydrates are fermented (Table 1.4) (Hespell and Canale-Parola, 1970a). The main products of glucose fermentation are ethanol, acetic acid, CO_2, H_2, and smaller amounts of lactic acid (Table 1.5) (Canale-Parola et al., 1967, 1968; Hespell and Canale-Parola, 1970a). Catalase-negative.

Growth reported only on complex media. Minimal growth requirements are unknown. Growth occurs between 15 and 40°C; optimum temperature is 35–37°C. Optimum growth yields result when the initial pH of the medium is between 7.0 and 7.5.

Originally isolated from H_2S-containing mud of a freshwater pond (Canale-Parola et al., 1967, 1968).

The mol% G + C of the DNA is 60 (Bd) (strain Z1).

Type strain: ATCC 25083 (DSM 2028; Z1).

3. **Spirochaeta zuelzerae** Canale-Parola 1980, 594.[VP*] (*Treponema zuelzerae* Veldkamp 1960, 122.)

zuel′ze.rae. M.L. gen. n. *zuelzerae* of Zuelzer; named after Margarete Zuelzer, who described the occurrence of morphologically diverse spirochetes in sulfide-containing environments.

Helical cells, 0.20–0.35 μm in diameter and 8–16 μm in length. Shorter cells (as short as 2–3 μm) are occasionally observed in cultures.

---

* *VP*, denotes that this name has been validly published in the official publication, International Journal of Systematic Bacteriology.

Table 1.4.
Utilization of carbohydrates as energy sources for growth by species of the genus **Spirochaeta**[a, b]

Carbohydrates	2. S. stenostrepta	3. S. zuelzerae	4. S. litoralis	5. S. aurantia	6. S. halophila
L-arabinose	+	+	+	+	+
Cellobiose	+	+	+	+	+
Dextrin				+	+
Dulcitol				−	−
D-Fructose	+	−	+	+	+
D-Galactose	+	+	+	+	+
D-Glucose	+	+	+	+	+
Inulin		−	+	+	+
Lactose	+	−	+	+	+
Maltose	+	+	+	+	+
Mannitol		−		+	−
D-Mannose	+	+	+	+	+
Raffinose		−	+	−	−
L-Rhamnose		−	+	+	+
D-Ribose	+		−	−	
Sorbitol		−		−	−
Sucrose	+	−	+	+	+
Trehalose		+	+	+	+
D-Xylose	+	+		+	+

[a] Symbols: see standard definitions.
[b] Data are for the type strains.

Table 1.5.
Fermentation products of **Spirochaeta** *species*[a]

Products[b]	2. S. stenostrepta		3. S. zuelzerae	4. S. litoralis		5. S. aurantia		6. S. halophila
	A[c]	B	A	A	B	A	B	A
Acetate	93	20.4	82	37.5	57.0	69.2	50.3	52.4
Ethyl alcohol	84	146.2	ND	109.5	140.5	151.0	78.4	132.0
CO_2	140	187.5	68	127.5	201.8	165.3	128.2	176.1
H_2	180	27.2	164	74.0	74.4	107.2	79.5	130.3
Lactate	10	8.2	87	6.5	Trace	1.0	17.2	1.8
Formate	Trace	NR	ND	2.8	Trace	5.2	NR	ND
Pyruvate	NR	NR	NR	0.3	Trace	NR	3.1	ND
Succinate	ND	ND	13	ND	ND	ND	NR	ND
Glycerol	NR	NR	ND	NR	NR	NR	4.4	NR
Acetoin, diacetyl	ND	NR	ND	ND	ND	Trace	NR	ND

[a] Data from Canale-Parola (1977) and references therein.
[b] Products of the following strains: *S. stenostrepta* Z1, *S. zuelzerae* ATCC 19044, *S. litoralis* R1, *S. aurantia* J1, *S. halophila* RS1.
[c] The abbreviations used are: A, Products of growing cells; B, products of cell suspensions; ND, not detected; NR, not reported.

Long organisms (up to 80 μm) are present in old cultures. Exponentially-growing cells have fairly regular, stable primary coils. Secondary coils or waves are present infrequently. Spherical bodies, generally not exceeding 3–4 μm in diameter, are formed usually at the ends of the cells in the stationary phase of growth. Two subterminally inserted periplasmic flagella are present in a 1–2–1 arrangement. Phase-contrast photomicrographs and electron micrographs of the cells have been published (Canale-Parola et al., 1968; Joseph and Canale-Parola, 1972).

Subsurface colonies in agar media (for composition of medium see Veldkamp, 1960, and also SZ medium in Table 1.3) are white, fluffy, spherical, with a tendency to diffuse in the agar medium. Disc-shaped colonies are present occasionally.

Obligately anaerobic, having a fermentative type of metabolism. Various carbohydrates are fermented (Table 1.4) (Veldkamp, 1960).

Cells growing in media containing 0.05% $NaHCO_3$ ferment glucose mainly to acetic, lactic and succinic acids, CO_2 and H_2 (Table 1.5) (Veldkamp, 1960). Catalase-negative.

Growth occurs at 20°C but not at 45°C. The optimum temperature ranges from 37–40°C. Optimum growth yields result when the initial pH of the medium is between 7 and 8. Inorganic ammonium salts or nitrates are not utilized as sole nitrogen sources. Added CO_2 is an absolute requirement for growth (Veldkamp, 1960). Growth has been reported only on complex media. The minimal growth requirements are unknown.

Cells have a protein antigen that gives a positive complement-fixation reaction with syphilitic serum.

The mol% G + C of the DNA is 56 (Bd) (strain ATCC 19044).

Originally isolated from an enrichment culture for green photosynthetic bacteria that had been inoculated with sulfide-containing mud from a freshwater pond (Veldkamp, 1960).

Type strain: ATCC 19044 (DSM 1903).

4. **Spirochaeta litoralis** Canale-Parola 1980, 594.[VP]

li.to.ra′lis. L. adj. *litoralis* of the shore.

Helical cells, 0.4–0.5 μm in diameter and 5.5–7.0 μm in length. The cells are regularly and tightly coiled during the exponential phase of growth. Spherical bodies (2.0–3.5 μm in diameter) are present in the stationary growth phase under unfavorable growth conditions (e.g. in the presence of O_2). Two subterminally inserted periplasmic flagella are present in a 1–2–1 arrangement. Phase-contrast photomicrographs and electron micrographs of the cells have been published (Hespell and Canale-Parola, 1970b; Joseph and Canale-Parola, 1972).

Subsurface colonies in agar media are spherical, fluffy, cream colored, 1–5 mm in diameter. Surface colonies (anaerobic) are round, growing partially within the agar medium, cream colored, 2–5 mm in diameter.

Obligately anaerobic, having a fermentative type of metabolism. Various carbohydrates are fermented (Table 1.4) (Hespell and Canale-Parola, 1970b). Main products of glucose fermentation are ethanol, acetic acid, CO_2, H_2, and trace amounts of lactic, formic and pyruvic acids (Table 1.5) (Hespell and Canale-Parola, 1970b, 1973). Nitrite is not accumulated in the medium by cells growing in the presence of nitrate. Catalase-negative.

Cells grow in media prepared with seawater, but do not grow in media prepared with freshwater unless NaCl is added (minimum concentration, 0.05 M; optimum, 0.35 M). Cells have specific requirements for Na^+ and Cl^-. Exogenous supplements of biotin, niacin, and coenzyme A are required for growth. Coenzyme A may be replaced by pantothenate, but the resulting cell yields are low. Added thiamine is stimulatory for growth. A reducing agent (e.g. sulfide or cysteine) is required for growth in laboratory media. Cells grow in chemically defined media containing glucose, $(NH_4)_2SO_4$ or amino acids, sulfide, NaCl, vitamins, coenzyme A, and inorganic salts (Hespell and Canale-Parola, 1970b).

Optimum temperature, near 30°C. Growth occurs slowly at 15°C and not at all at 5 or 40°C. Optimum growth yields result when the initial pH of the medium is between 7.0 and 7.5.

Isolated from sulfide-containing marine mud.

The mol% G + C of the DNA is 51 (Bd) (strain R1).

Type strain: ATCC 27000 (DSM 2029; R1).

5. **Spirochaeta aurantia** Canale-Parola 1980, 594.[VP]

au.ran′tia. M.L. n. *aurantium* the orange; M.L. adj. *aurantia* orange colored.

Helical cells, 0.3 μm in diameter and 5–50 μm in length. Most cells in cultures measure 10–20 μm in length during exponential growth. Spherical bodies 0.5–2.0 μm in diameter are present, especially in the stationary phase of growth or when the cells are incubated at temperatures unfavorable for growth (e.g. 37°C). The spherical bodies are either in association with cells or free. Each cell has two subterminally inserted periplasmic flagella in a 1–2–1 arrangement. Phase-contrast photomicrographs and electron micrographs of the cells have been published (Canale-Parola et al, 1968; Breznak and Canale-Parola, 1969, 1975).

Colonies on aerobic plates (in media containing 100 g of agar/1·0 ml; see Breznak and Canale-Parola, 1975) are 1–4 mm in diameter, yellow-orange to orange, round with slightly irregular edges, growing primarily within the agar medium just under the surface, sometimes with a slightly raised center. At low carbohydrate concentrations (see Breznak and Canale-Parola, 1975) the colonies are larger, and they diffuse through the agar medium in the shape of almost perfect circles. Under these growth conditions the colonies have a lower cell density and their pigmentation is not readily apparent. Anaerobically grown colonies are white. Subsurface anaerobic colonies are spherical, fluffy, 1–3 mm in diameter.

Facultatively anaerobic, having both fermentative and respiratory types of metabolism. Carbohydrates, but not amino acids, are utilized

as energy sources for growth (Table 1.4) (Breznak and Canale-Parola, 1969, 1975). Amino acids serve as sole nitrogen sources; inorganic ammonium salts or nitrates usually do not. Exogenous thiamine is required by all strains tested, and riboflavin is required by most strains. Exogenous biotin is required for growth of the type strain and is stimulatory to the growth of other strains (Breznak and Canale-Parola, 1975). Nitrate is reduced to nitrite anaerobically. Oxidase-negative. Weakly catalase-positive. Superoxide dismutase (SOD) is present, and aerobically grown cells have higher levels of SOD than do anaerobically grown cells (F. Austin, personal communication).

Optimum growth occurs between 25 and 30°C. Slow growth occurs at 15°C and usually no growth at 5°C. There is poor or no growth at 37°C. Optimum growth yields result when the initial pH of the medium is 7.0–7.3.

Cells grown anaerobically ferment glucose primarily to ethanol, acetic acid, CO_2 and H_2 (Table 1.5) (Breznak and Canale-Parola, 1969, 1972a). Under aerobic conditions, growing cells oxidize glucose mainly to CO_2 and acetic acid (Breznak and Canale-Parola, 1972b).

Cells growing aerobically produce carotenoid pigments responsible for the yellow-orange to orange color of colonies. The major carotenoid pigment is 1′,2′-dihydro-1′-hydroxytorulene (Greenberg and Canale-Parola, 1975). Nonpigmented mutants have been isolated.

Chemotactic toward carbohydrates but not toward amino acids.

Isolated from water and mud of freshwater ponds and swamps.

The mol% G + C of the DNA is 61–65 (Bd).

Type strain: ATCC 25082 (DSM 1902; J1).

5a. **Spirochaeta aurantia** subspecies **aurantia** Canale-Parola 1980, 594.[VP]

The characteristics are as described for the species. Distinguished from the subspecies *stricta* by having a cell wavelength of 2.0–2.8 μm, a wave amplitude of 0.5 μm (the cells have loose coils), and a mol% G + C of 62–65 (Bd).

Type strain: ATCC 25082.

5b. **Spirochaeta aurantia** subspecies **stricta** subsp. nov.

stric′ta. L. v. *stringere* to draw tight, compress; L. past part. *strictus* drawn tight.

The characteristics are as described for the species. Distinguished from the subspecies *aurantia* by having a cell wavelength of 1.1–1.5 μm, a wave amplitude of 0.35 μm (cells have tight coils), and a mol% G + C of 61 (Bd).

Type strain: J4T (Breznak and Canale-Parola, 1975).

6. **Spirochaeta halophila** Greenberg and Canale-Parola 1976, 185.[AL]

ha.lo.phi′la. Gr. n. *hals*, *halos* salt; Gr. adj. *philus* loving; M.L. fem. adj. *halophila* salt-loving.

Helical cells, 0.4 μm in diameter and 15–30 μm in length. Some of the cells in cultures are as short as 5 μm and as long as 60 μm. Cells have regular, stable primary coils. Spherical bodies 1–2 μm in diameter occur in cultures, especially in the stationary phase of growth or during growth at unfavorable temperatures (e.g. 45°C). Each cell has two subterminally inserted periplasmic flagella that overlap in the central region of the cell (1–2–1 arrangement). Phase-contrast micrographs and electron micrographs of the cells have been published (Greenberg and Canale-Parola, 1976).

Colonies growing aerobically on ISM plates (0.75 g of agar/100 ml of medium; see earlier section on Further Descriptive Information) are red, round, with areas of diffuse growth at their periphery, and usually 2–6 mm in diameter (after 5 days at 35°C). Each colony grows partially above and partially below the surface of the agar medium. Anaerobically grown colonies are white. When cells are streaked onto agar medium plates and incubated anaerobically, the colonies grow below the surface of the medium and are spherical, diffuse and white.

Facultatively anaerobic, having both respiratory and fermentative types of metabolism. Carbohydrates, but not amino acids, are utilized as energy sources for growth (Table 1.4) (Greenberg and Canale-Parola,

1976). Cells have specific growth requirements for relatively high concentrations of Na^+, Cl^-, Ca^{2+} and Mg^{2+} (Greenberg and Canale-Parola, 1976). Optimum cell yields result when 0.75 M NaCl, 0.2 M $MgSO_4$ and 0.01 M $CaCl_2$ are included in growth media containing (g/100 ml) a carbohydrate (0.5), peptone (0.2) and yeast extract (0.4). No growth occurs when any one of the three inorganic salts is omitted from the medium (e.g. ISM medium). Nitrate is reduced to nitrite anaerobically. Catalase-negative.

Cells growing anaerobically ferment glucose primarily to ethanol, acetic acid, CO_2 and H_2 (Table 1.5) (Greenberg and Canale-Parola, 1976). Under aerobic conditions, growing cells oxidize glucose mainly to CO_2 and acetic acid (Greenberg and Canale-Parola, 1976).

Optimum temperature, 35–40°C. Poor growth occurs at 45°C and no growth occurs at 22°C.

Cells growing aerobically produce carotenoid pigments responsible for the red color of the colonies. The major carotenoid pigment is 4-keto-1′,2′-dihydro-1′-hydroxytorulene (Greenberg and Canale-Parola, 1975). Nonpigmented mutants, occurring spontaneously in cultures, have been isolated.

Isolated from H_2S-containing mud of a high salinity pond (Solar Lake) located on the Sinai shore of the Gulf of Elat.

The mol% G + C of the DNA is 62 (T_m, Bd) (strain RS1).

Type strain: ATCC 29478 (RS1).

Other organisms

Among the free living spirochetes capable of anaerobic growth there exists greater diversity than is reflected by the species presently recognized in the genus *Spirochaeta*. A free living, strictly anaerobic spirochete (strain Z4), resembling *S. zuelzerae* morphologically, but differing in certain physiological properties, has been isolated from freshwater mud (Canale-Parola et al., 1968). The spirochete ferments glucose to acetic, lactic, succinic, and formic acids, ethanol, CO_2, and H_2. Unlike *S. zuelzerae*, it does not require an exogenous source of CO_2, and the mol% G + C of its DNA is 59.2 (Bd).

In addition, various strains of facultatively and obligately anaerobic spirochetes have been isolated from intertidal marine muds. An obligately anaerobic strain was isolated from water collected at a depth of 2550 meters near the Galápagos hydrothermal vents in the Pacific Ocean. All these isolates are indigenous to marine environments inasmuch as they have Na^+ requirements typical of marine bacteria. The

facultatively anaerobic marine isolates form either white or yellow colonies. Thus, they differ both in salt requirements and in pigmentation from *S. aurantia* which is a freshwater species and forms orange colonies, and from *S. halophila* which requires high concentrations of Ca^{++} and Mg^{++} and forms red colonies.

One of the obligately anaerobic marine isolates (strain MA-2) has been studied in some detail. This spirochete ferments glucose to products similar to those formed by *S. litoralis* (Table 1.5), but differs from the latter species because it has the ability to generate ATP by catabolizing L-leucine, L-isoleucine and L-valine with formation of branched-chain fatty acids as end products. ATP thus formed is not utilized for growth processes, but serves as a source of maintenance energy for strain MA-2 during periods of starvation (Harwood and Canale-Parola, 1981). The mol% G + C of the DNA is 64.5 (T_m).

Genus II. **Cristispira** *Gross 1910, 44* [AL]

JOHN A. BREZNAK

Cris·ti·spi′ra. L. fem. n. *crista* a crest; G. fem. n. *spira* a coil; M.L. fem. n. *Cristispira* a crested coil.

Helical or undulate cells 0.5–3.0 μm in diameter and 30–180 μm in length, generally displaying 2–10 complete helical turns. Ends of cells are blunt, rounded or tapered; in fixed and stained preparations a filament or spicule may emanate from one or both ends. Stained preparations reveal a series of **ovoid inclusions** of unknown composition which impart a chambered appearance to the protoplasmic cylinder. Electron microscopy of thin sections reveals multiple cytoplasmic vesicles bounded by a double membrane. **Cell division is by transverse fission. A bundle of 100 or more periplasmic flagella** (which also have been termed axial fibrils, endoflagella or periplasmic fibrils) **is intertwined with the protoplasmic cylinder and may distend the outer sheath to form a ridge or crest (the so-called "crista") on the protoplasmic cylinder.** The crista is not always obvious on live cells but may be conspicuous when the cells stop moving. It is frequently seen by light microscopy of stained cells. Motility is parallel to the cell's long axis and individual cells move forward or backward with no anterior-posterior polarity. Translocation may include rotation about the cell's long axis or may take the form of an irrotational traveling helical wave. **Flexing movements are common.** When removed from its habitat, degenerative changes readily occur and accompany the loss of motility. Cristae may become markedly distended, multiple swellings may appear on the cell body, and cells may lyse or form spherical bodies. Strains are **widely distributed among marine and freshwater molluscs** (clams, mussels, and oysters) and inhabit the crystalline style (a mucoproteinaceous rod-shaped organ) or fluid of the digestive tract. **They are probably commensals.** Also found in gastropods and may occur in nonmollusc species as well. No strains of *Cristispira* have been grown in pure culture.

Type species: *Cristispira pectinis* Gross 1910, 44.

Further Descriptive Information

Observations of cristispires, such as depicted in Fig. 1.3, are customarily made by examining the crystalline style of the host, probably because this structure is easy to remove (Breznak, 1973). The spirochetes may occur in the firm cortical region of the style or the more fluent inner zone, being more numerous and motile in the latter region (Perrin, 1906). However, the overall texture of a crystalline style may govern its susceptibility to colonization by *Cristispira*: molluscan species with soft textured styles appear to harbor the spirochetes more frequently (Noguchi, 1921). Cristispires have also been observed to protrude through, or adhere to, the outer layer of the crystalline style, and such cells appear to have a longer wavelength than those within the style matrix (Bernard, 1970; Tall and Nauman, 1981). When present in a crystalline style, cristispires move freely within it, but they avoid the anterior end which usually contains a tassel of ground food (plankton) formed by the pulverizing action of the style as it impinges on the gastric shield region of the stomach (Breznak, 1973). The anterior end of the style may contain substances toxic to cristispires (Berkeley, 1933, 1959, 1962).

Cristispires have also been observed in organs other than the crystalline style, including the style pouch, gastric shield, and fluid of the pallial cavity, stomach, anterior intestine, cecum, and rectum (Bernard, 1970; Breznak, 1973). They may be observed in aquaria water used to incubate their hosts (Fantham, 1908; Bosanquet, 1911). Environmental factors, as well as host physiological status, may bear on the ability of *Cristispira* to colonize a mollusc. Not all species of molluscs examined harbor *Cristispira*; neither do all specimens of a recognized host species (Bernard, 1970; Breznak, 1973).

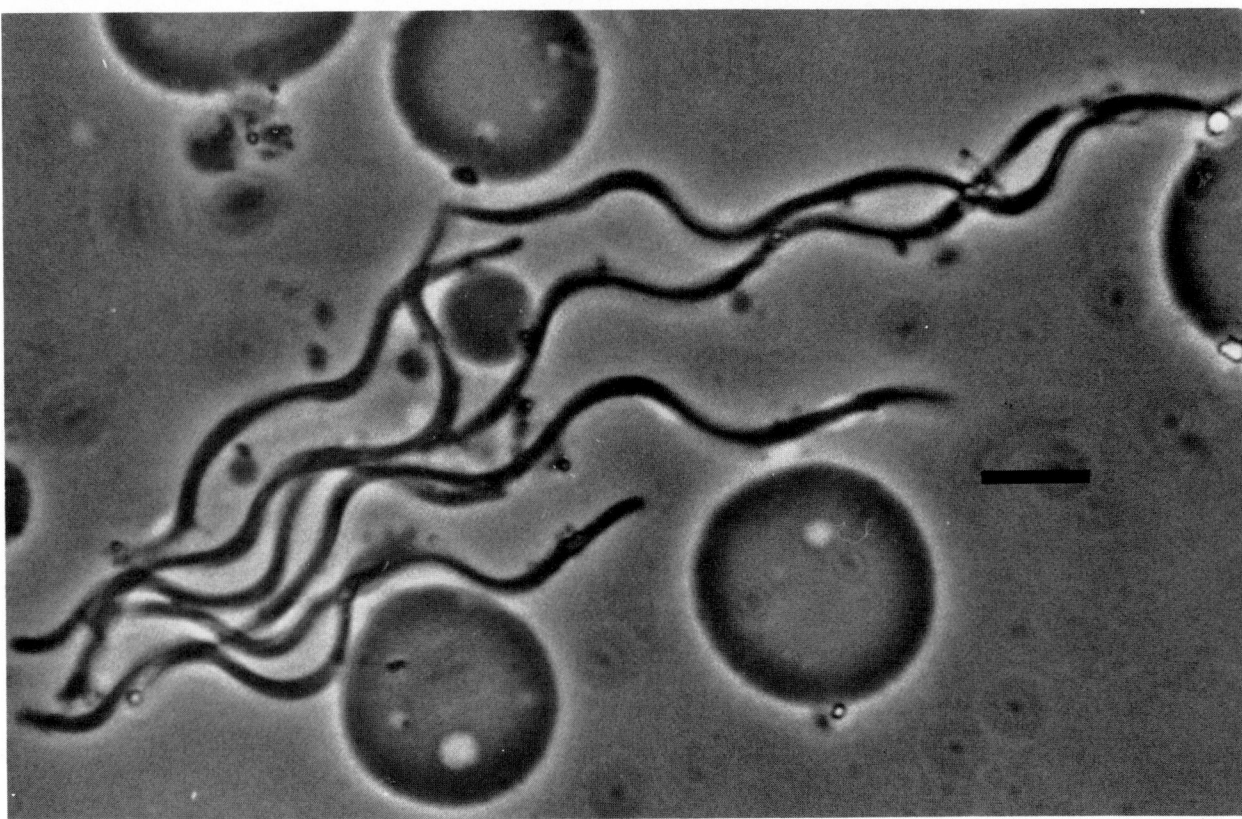

Figure 1.3. Phase-contrast micrograph of a cluster of *Cristispira* cells from the crystalline style of *Ostrea (Crassostrea) virginica*. The *bar* = 10 μm. (Reproduced with permission from R. K. Nauman.)

Information on the physiology of *Cristispira* is fragmentary and based mainly on observations of the cells' motility after removal from the crystalline style. Cells remain motile longer at 5–10°C than at temperatures above 20°C (Perrin, 1906; Fantham, 1908; Noguchi, 1921); glucosone inhibits motility (Berkeley, 1962); motility occurs under aerobic and putative anaerobic conditions (Berkeley, 1959); and suspension of cells in distilled water may induce lysis (Perrin, 1906; Fantham, 1908). Proteinaceous substrates (Noguchi, 1921) and fructose (Kubomura, 1969) appear to support limited growth of *Cristispira* on primary culture.

At present there is no reason to believe that *Cristispira* is pathogenic to its host. It is usually found in healthy, palatable univalve and bivalve molluscs obtained from well aerated beds (Berkeley, 1959). If the normal physiological activities of the host (e.g. siphoning and feeding) are disrupted, the crystalline style generally dissolves and the normal populations of *Cristispira* disappear or decrease drastically (Kuhn, 1974). If oysters are subjected to drying at cold temperatures, *Cristispira* disappears rapidly (Dimitroff, 1926).

The range of molluscan hosts reported to harbor *Cristispira* is listed in Table 1.6. Large spirochetes resembling *Cristispira* have been observed in starfish (Collier, 1921), tunicates (Hellmann, 1913), and termites (Hollande, 1922): however, their natural relationships to the molluscan cristispires are unknown.

Differentiation of the genus **Cristispira** from other genera

Three main criteria, in aggregate, serve to distinguish *Cristispira* from other genera of spirochetes: (a) their habitat, (b) the crista, and (c) their relatively large size.

Taxonomic Comments

The genus *Cristispira* is presently monospecific. However, the early literature advanced many different specific epithets intended to acknowledge the molluscan host as well as differences (often subtle) in the morphology between cristispires (Breznak, 1973; Kuhn, 1974). In the eighth edition of the *Manual*, Kuhn (1974) grouped such species as *species incertae sedis*. It would seem best to avoid reviving such names until pure cultures are obtained and the extent of morphological and physiological variation among critispires can be ascertained. A similar strategy seems desirable for molluscan spirochetes of smaller size than *Cristispira*, many of which do not possess the multitude of periplasmic flagella typical of cristispires. Such spirochetes have been observed in *Anodonta mutabilis* (Schellack, 1909), *Pachelebra* (probably *Pachylabra*) *moesta* (a snail; deMello, 1921), *Pinna squamosa* (Gonder, 1908), *Scrobicularia piperata* (Pillot and Ryter, 1965), planorbid molluscs (Richards, 1978) and *Polydora flava* (a marine polychete annelid; Mesnil and Caullery, 1916). In the eighth edition of the *Manual*, Kuhn (1974) treated many of these as *species inquirendae*.

Table 1.6.

Molluscan hosts of **Cristispira**

Host	Geographical Source	Reference
Amphidesma australe (Gm.)	Cheltenham Beach, Auckland, New Zealand (marine)	Judd, 1979
Anodonta cygnea Linn.	River Cam, England (freshwater)	Dobell, 1912; Fantham, 1908
A. grandis Say	Wisconsin, U.S.A. (freshwater)	Nelson, 1918
A. mutabilis Cless.	Germany (freshwater)	Keysselitz, 1906; Schellack, 1909
Anodonta (species unknown)	England (freshwater)	Bosanquet, 1911
Cardium papillosum Poli	Rovigno, Adriatic Sea (marine)	Schellack, 1909
Chama gryphoides Linn.	Rovigno, Adriatic Sea (marine)	Schellack, 1909
C. sinistrorsa Brocchi	Rovigno, Adriatic Sea (marine)	Schellack, 1909
Clinocardium nuttallii (Con.)	British Columbia, Canada (marine)	Bernard, 1970
Crassostrea gigas Thun.	British Columbia, Canada (marine)	Berkeley, 1959; Bernard, 1970
Cyclas (species unknown)	Rovigno, Adriatic Sea (marine)	Schellack, 1909
Diplodonta orbella (Gould)	British Columbia, Canada (marine)	Bernard, 1970
Entodesma saxicola (Baird)	British Columbia, Canada (marine)	Bernard, 1970
Gastrochaena dubia Penn.	Rovigno, Adriatic Sea (marine)	Schellack, 1909
Lampsilis anodontoides Lea	Wisconsin, USA (freshwater)	Nelson, 1918
Lima hyans Gm.	Rovigno, Adriatic Sea (marine)	Schellack, 1909
L. inflata Lam.	Rovigno, Adriatic Sea (marine)	Schellack, 1909
Lyonsia pugetensis Dall	British Columbia, Canada (marine)	Bernard, 1970
Macoma (species unknown)	California, U.S.A. (marine)	Berkeley, 1959
Mactra sulcataria Desh.	Fukuoka, Japan (marine)	von Prowazek, 1910
Modiola barbata Linn.	Rovigno, Adriatic Sea (marine)	Schellack, 1909
M. modiolus Linn.	Woods Hole, Mass., U.S.A. (marine)	Noguchi, 1921
Ostrea angulata Lam.	Arcachon and LaRochelle, France (marine)	Certes, 1882
O. edulis Linn.	France and Adriatic Sea (marine)	Certes, 1882; Fantham, 1911; Ryter and Pillot, 1965; Schellack, 1909; Swellengrebel, 1907; von Prowazek, 1910
O. lurida Carp.	British Columbia, Canada (marine)	Bernard, 1970
O. talienwhaneensis Cross	Fukuoka, Japan (marine)	von Prowazek, 1910
O. (Crassostrea) virginica Gm.	Tuckerton, N.J., U.S.A.; Baltimore, Md., U.S.A.; Woods Hole, Mass., U.S.A. (marine)	Nelson, 1918; Dimitroff, 1926; Noguchi, 1921; Tall and Nauman, 1981
Panope generosa Gould	British Columbia, Canada (marine)	Bernard, 1970
Paphia staminea[a]	British Columbia, Canada (marine)	Berkeley, 1959
Pecten jacobaeus Linn.	Gulf of Naples, Italy (marine)	Gross, 1910
Pinna nobilis Linn.	Rovigno, Adriatic Sea; Gulf of Naples, Italy (marine)	Gonder, 1908; Schellack, 1909
Protothaca staminea (Con.)	British Columbia, Canada (marine)	Bernard, 1970
Saxicava arctica (Linn.)	Rovigno, Adriatic Sea (marine)	Schellack, 1909
Saxidomus giganteus Desh.	British Columbia, Canada (marine)	Berkeley, 1959; Bernard, 1970
Semisulcospira libertina Gould (a snail)	Hiroshima, Japan (freshwater)	Terasaki, 1958
Siliqua patula Dixon	California, U.S.A. (marine)	Berkeley, 1959
Solen ensis Linn.	(?)	Fantham, 1911
Soletellina acuminata Desh.	Tamblegam Lake, Ceylon (saltwater)	Dobell, 1911, 1912
Strophitus (species unknown)	Wisconsin River, Wis., U.S.A. (freshwater)	Breznak, 1973
Tapes aureus Gm.	France (marine)	Fantham, 1911; Ryter and Pillot, 1965
T. decussatus (Linn.)	Rovigno, Adriatic Sea and France (marine)	Ryter and Pillot, 1965; Schellack, 1909
T. laeta Wkff.	Rovigno, Adriatic Sea (marine)	Schellack, 1909
T. (Venerupis) philippinarum (Adams and Reeve)	Fukuoka, Urayasu, and Inage, Japan (marine)	von Prowazek, 1910; Kubomura, 1969
T. pullastra (Mont.)	France (marine)	Ryter and Pillot, 1965
Tivela stultorum Mawe	California, U.S.A. (marine)	Berkeley, 1959; Jahn and Landman, 1965
Tresus capax (Gould)	British Columbia, Canada (marine)	Bernard, 1970
T. nuttallii (Con.)	British Columbia, Canada (marine)	Bernard, 1970
Unio pictorum Linn.	Rovigno, Adriatic Sea (marine)	Schellack, 1909

Table 1.6.—*continued*

Host	Geographical Source	Reference
Venerupis japonica (Desh.)	British Columbia, Canada (marine)	Bernard, 1970
Venus casta Chem.	Tamblegam Lake, Ceylon (saltwater)	Dobell, 1911, 1912
V. mercenaria Linn.	Woods Hole, Mass., U.S.A. (marine)	Noguchi, 1921
V. verrucosa Linn.	France (marine)	Ryter and Pillot, 1965
Others		
"Schleswig Holstein oyster"	North Germany (marine)	Möbius, 1883
"Adriatic oyster"	Adriatic Sea (marine)	Perrin, 1906
"French," "English," and "Abervrach" oysters	France and England (marine)	Fantham, 1908

[a] Authority unknown

List of the species of the genus **Cristispira**

1. **Cristispira pectinis** Gross 1910, 44.[AL]

pec′ti.nis. M.L. masc. n. *Pecten* a genus of molluscs; M.L. gen. n. *pectinis* of *Pecten*.

Helically coiled, flexible cells, 1.5 µm in diameter. Length of helix, 36–72µm. Possess no more than four complete turns of the helix. Ends of the cells are round or tapered with no terminal appendages. Stained preparations reveal cross-striations, polar granulation, and multiple inclusions. In fixed and stained preparations a crista extends along the side of the cell. Division is by transverse fission.

Found in the crystalline style and intestinal fluid of *Pecten jacobaeus* from the Gulf of Naples.

Type strain: no culture isolated.

Genus III **Treponema** Schaudinn 1905, 1728 [AL]

ROBERT M. SMIBERT

(*Spironema* Vuillemin 1905, 1568; *Microspironema* Stiles and Pfender 1905, 936). Tre.po.ne ′ ma. Gr. v. *trepo* turn; Gr. n. *nema* a thread; M.L. neut. n. *treponema* a turning thread.

Helical rods 0.1–0.4 µm diameter and 5–20 µm in length. Cells have tight regular or irregular spirals. They have one or more periplasmic flagella (axial fibrils or axial filaments) inserted at each end of the protoplasmic cylinder. Cytoplasmic fibrils (microtubules and intracytoplasmic tubules) are seen in the protoplasmic cylinder just under the cytoplasmic membrane and positioned under the periplasmic flagella. Under unfavorable cultural or environmental conditions spherical cells or spirochetal spheres are formed. These can also be seen in old cultures. Gram-negative. Cells stain well with silver impregnation methods. Most species stain poorly if at all with Gram's or Giemsa's stain. Best observed with darkfield or phase contrast microscopy. **Motile. Cells have both rotational and translational movement in liquid media.** In a semisolid or solid medium, cells exhibit a serpentine type movement. **Strictly anaerobic or microaerophilic.** Human pathogenic species are now considered to be microaerophiles and have not been cultivated in artificial media or in tissue culture. **Chemoorganotrophs**, using a variety of **carbohydrates or amino acids for carbon and energy sources.** Cultivated anaerobic species are catalase- and oxidase-negative. Some require long chain fatty acids found in serum for growth, while other cultivated species require short chain volatile fatty acids for growth. **Found in the oral cavity, intestinal tract, and genital areas of man and animals.** Host-associated. Some species are pathogenic. The mol% G + C of the DNA ranges from 25–54.

Type species: *Treponema pallidum* (Schaudinn and Hoffman 1905) Schaudinn 1905, 1728.

Further Descriptive Information

Cells of all species are helical (Stepan and Johnson, 1981) and vary in cell diameter from 0.1–0.4 µm. The outer envelope of treponemes contains lipid, protein and carbohydrates and is similar to the outer membrane of Gram-negative bacteria. The lipid is mainly phospholipid and glycolipid. The cell walls contain muramic acid, glucosamine and ornithine. Peptidoglycan represents 1% of the dry weight of cells. Endotoxin in significant amounts has not been detected from *Treponema phagedenis* (Johnson, 1976).

The insertion apparatus of the periplasmic flagella has a proximal hook and insertion disks similar to those found in other bacteria. The molecular weight of the monomers of dissociated periplasmic flagella of *T. phagedenis* is 32,000–36,000 (Holt, 1978).

Treponemes have cytoplasmic fibrils (microtubules or intracytoplasmic tubules) that extend along the inner layer of the cytoplasmic membrane; they are about 7 nm in diameter and occur in clusters of 6–8. The clusters occur at the ends of the cell and are in close association with the insertion apparatus of the periplasmic flagella. Most *Treponema* species examined have been found to contain cytoplasmic fibrils.

Treponemes are host-associated and are found in the flora of man and animals; some are pathogenic for man. They occur in the oral cavity, rumen, intestinal tract and genitals of man and animals. The pathogenic treponemes causing syphilis, yaws, pinta and nonvenereal endemic syphilis in man have not been cultivated. These organisms are usually propagated in laboratory animals, e.g. in the testes of rabbits. Most investigators now consider *T. pallidum* to be a microaerophile requiring low concentrations of oxygen (1.5–5.0%).

The cultivated treponemes are strict anaerobes and require good anaerobic conditions for growth. Some of these treponeme species ferment glucose and require a carbohydrate as an energy source, whereas other species gain energy from the fermentation of amino acids. Treponemes also differ in their requirement for fatty acids. Some species require long chain fatty acids found in serum, while others require short chain volatile fatty acids found in rumen fluid. All cultivated treponeme species require either long or short chain fatty acids. The oral organisms *Treponema denticola*, *T. vincentii* and *T. scoliodontum* require thiamine pyrophosphate in addition to serum.

Cultivated treponemes (*T. phagedenis*, *T. refringens*, *T. denticola* and *T. vincentii*) are inhibited by penicillin (0.1–1.0 U/ml), ampicillin (0.1–1.0 µg/ml), oxacillin (0.1–10 µg/ml), cloxacillin (0.1–1.0 µg/ml), cephalothin (0.1–10 µg/ml), vancomycin (0.1–10 µg/ml), bacitracin (0.1–1.0 µg/ml), erythromycin (0.1–1.0 µg/ml), novobiocin (10–500 µg/ml), tetracycline (1 µg/ml), doxycycline (0.1–1.0 µg/ml), chloramphenicol (100–500 µg/ml), kanamycin (100–1000 µg/ml), and viomycin (10–1000 µg/ml). All are resistant to cycloserine (500–1000 µg/ml), polymyxin B

(500–1000 μg/ml) and nalidixic acid (500–1000 μg/ml) (Abramson and Smibert, 1971). Bactericidal concentrations of antibiotics are usually much higher than inhibitory concentrations (Abramson and Smibert, 1971). Oral and rumen treponemes are resistant to rifampin at concentrations of 1–50 μg/ml (Stanton and Canale-Parola, 1979; Leschine and Canale-Parola, 1980).

Isolation and Enrichment Procedures

Treponemes can be isolated by two general methods. The first uses membrane filters placed on the surface of agar media. The second method uses rifampin as a selective agent (Stanton and Canale-Parola, 1979; Leschine and Canale-Parola, 1980).

A membrane filter with a pore size of 0.2 μm is placed on an agar medium containing either 10% inactivated animal serum or 30% rumen fluid. The agar concentration should be 1.3–1.4%. The sample is placed on the filter and the Petri dish quickly placed in an anaerobic jar or chamber and incubated. Incubation usually requires 1–2 weeks at 37°C. After incubation the filter is removed and a white haze can be seen in the agar. Treponemes migrate through the filter and grow into the agar. A plug of agar is removed and inoculated into a prereduced culture medium.

In the second method, filter-sterilized rifampin is added to either prereduced broth or molten agar medium in tubes to a final concentration of 1.0–2.0 μg/ml. The tubes are inoculated and incubated at 37°C.

Serial dilutions of samples can be made in the selective medium. The medium contains 1% agar and is allowed to solidify in the form of agar deeps. After incubation white cottony colonies of treponemes are seen within the medium. These colonies can be removed with a Pasteur pipette and inoculated into prereduced medium.

The selective broth medium will contain treponemes and a few other kinds of bacteria because the rifampin does not inhibit all other bacteria. Treponemes can be isolated from the enriched selective culture by inoculating prereduced agar medium (1.3–1.4% agar) in 6- or 8-oz prescription bottles. The "bottle plates" can be streaked or the culture inoculated into melted agar (45°C) in the bottles, mixed, and the agar allowed to solidify on the flat side of the bottle. The gas phase in the bottle plates should be 90% N_2 and 10% CO_2.

Other methods used for isolation of species of the genus *Spirochaeta* may be used for treponemes. For additional methods and details see Smibert, 1981.

Maintenance Procedures

Cultures of all species of treponemes can be stored in the frozen state using liquid nitrogen or mechanical freezers at −80°C. For long term preservation cryoprotective agents such as 10% glycerol or dimethyl sulfoxide are added to cultures to be frozen. Lyophilization has generally not been successful for preservation of treponemes.

Differentiation of the genus **Treponema** from other closely related genera

Characteristics useful for distinguishing *Treponema* from the other members of the family *Spirochaetaceae* are indicated in the key to the family.

Taxonomic Comments

The genus *Treponema* contains organisms with a mol% G + C of the DNA ranging from 25–54. It also includes organisms that produce acetate and succinate, acetate alone, and acetate and butyrate. The wide range of mol% G + C values, the differences in metabolic products and cell diameter, and the differences of some species in relationship to oxygen might indicate that various groups could be considered separate genera rather than a single genus.

Revision of the existing classification of treponemes is hampered by the inability to grow the type species, *T. pallidum*, in pure culture. Even though it is now believed that *T. pallidum* is a microaerophile requiring 1–5% oxygen, the question will not be answered until *T. pallidum* has been cultured for more than one transfer.

Studies involving DNA/DNA hybridization with *T. pallidum* and *T. pertenue* show that there is 100% homology between these two organisms (Miao and Fieldsteel, 1980). In view of these results, these two species have been combined in this edition of *Bergey's Manual* and are considered as subspecies of *T. pallidum*. Differences that exist between them concern the mode of infection, severity of infection, and infectivity for different laboratory animals. The organism causing nonvenereal endemic syphilis has never been named. It is considered by most

investigators as a variant of *T. pallidum*. Therefore, a third subspecies of *T. pallidum* has been proposed in this edition of the *Manual* to correct this omission.

Acknowledgments

Our research has been supported by grants GM-14604, AI-12726, DE-05054, and DE-05139 from the National Institutes of Health and by project 2022820 from the Commonwealth of Virgina.

Further Reading

Canale-Parola, E. 1977. Physiology and evolution of spirochetes. Bacteriol. Rev. *41:* 181–204.
Canale-Parola, E. 1978. Motility and chemotaxis of spirochetes. Annu. Rev. Microbiol. *32:* 69–99.
Holt, S.C. 1978. Anatomy and chemistry of spirochetes. Microbiol. Rev. *42:* 114–160.
Hovind-Hougen, K. 1976. Determination by means of electron microscopy of morphological criteria of value for classification of some spirochetes, in particular treponemes. Acta Path. Microbiol. Scan. Sect. B, Supp, No. 255, 1–41.
Johnson, R.C. 1977. The spirochetes. Annu. Rev. Microbiol. *31:* 89–106.
Johnson, R.C. (editor) 1976. *The Biology of Parasitic Spirochetes*. Academic Press, New York.
Smibert, R.M. 1973. The *Spirochaetales*, a review. Crit. Rev. Microbiol. *2:* 491–552.
Smibert, R.M. 1981. The Genus *Treponema. In* Starr, Stolp, Truper, Balows, and Schlegel (Editors). *The Prokaryotes, a Handbook on Habitats, Isolation and Identification of Bacteria*. Springer-Verlag, New York, pp. 564–577.

Differentiation of the species of the genus **Treponema**

Differentiation of the noncultivable species of *Treponema* is presented in Table 1.7. Characteristics useful for differentiating the cultivable species are given in Tables 1.8–1.11.

List of the species of the genus **Treponema**

Species 1 to 3 have not been cultivated.

1. **Treponema pallidum** (Schaudinn and Hoffman 1905) Schaudinn 1905, 1728.[AL] (*Spirochaeta pallida* Schaudinn and Hoffman 1905, 528). pal′li.dum. L. adj. *pallidum* pale, pallid.

Tightly coiled, 0.10–0.18 μm in diameter by 6–20 μm long. The average diameter is 0.13–0.15 μm and the average length is 10–13 μm. The wavelength of coils is 1.1 μm and the amplitude is 0.2–0.3 μm. The

ends of the cells are pointed and covered with a sheath. Three periplasmic flagella are inserted into each end of the cell. Motile with graceful flexuous movements. Probably microaerophilic.

Pathogenic for man. Causes infection and skin lesions in various animals.

Isolated from man.

The mol% G + C of the DNA for those strains that have been tested is 52.4–53.7 (T_m). A recent report indicates that 100% DNA/DNA

Table 1.7.
Differential characteristics of the noncultivable species of the genus **Treponema**[a]

Characteristic	1. T. pallidum			2. T. cara-teum	3. T. paraluis-cuniculi
	subsp. pallidum	subsp. pertenue	subsp. endemicum		
Natural host					
Humans	+	+	+	+	−
Rabbits	−	−	−	−	+
Nature of infection in natural host					
Systemic; may affect most internal organs	+	−	−	−	−
Usually restricted to cutaneous lesions	−	+	+	+	+
Sexually transmitted	+	−	−	−	+
Geographical distribution:					
Worldwide	+	−	−	−	
Found only in tropical countries in both hemispheres	−	+	−	−	
Found only in tropical countries in the Western hemisphere	−	−	−	+	
Restricted to the Middle East, Africa, southeast Asia and Yugoslavia	−	−	+	−	
Cutaneous lesion produced in:					
Rabbits	+	+	+	−	
Hamsters	−	+	+	−	
Mice	−	−	−	−	
Guinea pigs	−[b]	−	+	−	

[a] Symbols: see standard definitions.
[b] A slight lesion is occasionally seen at the point of injection of guinea pigs.

homology exists between *T. pallidum* and *T. pertenue* (Miao and Fieldsteel, 1980). It seems unlikely from these data that the organisms represent two distinct species, and it is more logical to consider them as the same species but with different degrees of virulence and clinical symptoms in man and with a different ability to infect various laboratory animals. Therefore *T. pertenue* will be considered as a subspecies of *T. pallidum*.

Type strain: none designated.

la. Treponema pallidum subspecies **pallidum** (Schaudinn and Hoffman 1909) Schaudinn 1905, 1728.

The morphology and characteristics are as described for the species and as listed in Table 1.7.

The cause of venereal and congenital syphilis is man. Has not been successfully cultivated in artificial media or in tissue culture. Propagated by intratesticular inoculation of rabbits.

Successful replication of *T. pallidum* (virulent Nichols strain) has been reported to occur on the surface of tissue culture cells of cottontail rabbit epithelium (SflEP) growing in a monolayer in an atmosphere of 1.5% O_2. A 49-fold increase (mean value) in cell numbers was reported in a single passage of *T. pallidum* (Fieldsteel et al., 1981).

T. pallidum cells show rapid polar attachment to cultured mammalian cells (Fitzgerald et al., 1977; Hayes et al., 1977). Mucopolysaccharidase activity was found on *T. pallidum* that was active against acid mucopolysaccharides (Fitzgerald and Johnson, 1979). Mucopolysaccharides have been found on the surface of *T. pallidum* and in syphilitic lesions (Fitzgerald and Johnson, 1979a, b).

Pathogenic for man and monkeys. Virulent strains, e.g. the Nichols pathogenic strain, are propagated by intratesticular inoculation of rabbits. Cutaneous inoculation of rabbits produces skin lesions. Cutaneous inoculation of hamsters, mice and guinea pigs produces no apparent infection or visible lesions. A slight lesion is occasionally seen at the point of injection of guinea pigs.

Probably microaerophilic. Survives in artificial media or tissue culture longest when incubated in an atmosphere of 3–5% O_2 (Fieldsteel et al., 1977; Fitzgerald et al., 1977; Norris et al., 1978; Sandok et al., 1978). Respiration and oxidative phosphorylation has been demonstrated (Lysko and Cox, 1978). Cytochromes of the *b* and *c* types have

been reported (Lysko and Cox, 1977). Glucose is metabolized by way of the Embden-Meyerhof-Parnas and hexose monophosphate pathways (Schiller and Cox, 1977). Oxygen uptake by *T. pallidum* has been reported and is glucose-dependent (Cox and Barber, 1974; Barbieri and Cox, 1981). Oxidation of pyruvate occurs only when oxygen is present (Barbieri and Cox, 1979). Major fermentation products of glucose are acetate and CO_2 (Nichols and Baseman, 1975).

Isolated from syphilitic lesions in man.

The mol% G + C of the DNA is 52–53.7. Shows 100% DNA/DNA homology with *T. pertenue* (Gauthier strain) but no homology to *T. phagedenis* or *T. refringens* (Miao and Fieldsteel, 1978 and 1980).

Type strain: None designated. Reference strain: Nichols pathogenic.

1b. Treponema pallidum subspecies **pertenue** (Castellani 1905) subsp. nov. (*Spirochaeta pertenuis* Castellani 1905, 54.)

per.ten'u.e. L. neut. adj. *pertenue* very thin, slender.

The morphology and characteristics as described for the species and as listed in Table 1.7.

Pathogenic. The cause of yaws in man, a contagious disease that is spread by contact.

Virulent strains have not been cultivated in artificial media or in tissue culture. Cutaneous lesions are produced at the point of inoculation in rabbits and Syrian hamsters but not in guinea pigs.

Sera from patients with yaws gives positive results with serologic tests for syphilis. Attachment of *T. pallidum* subsp. *pertenue* to 5 different mammalian cell lines was compared to that of *T. pallidum* subsp. *pallidum* (Fieldsteel et al., 1979). *T. pallidum* subsp. *pertenue* attached to all 5 cell lines as did *T. pallidum* subsp. *pallidum*. No preferential attachment was found with *T. pallidum* subsp. *pertenue* for nude mouse ear and cottontail rabbit epithelial (SflEp) cells, but preferential attachment did occur with *T. pallidum* subsp. *pallidum*.

Inbred hamsters (LSH/Ss LAK LSH) infected with *Treponema pallidum* subsp. *endemicum* (Bosnia A strain) were resistant to reinfection with both *T. pallidum* subsp. *pertenue* and *T. pallidum* subsp. *pallidum* (Schell et al., 1980).

Found in lesions from cases of yaws. Common in tropical countries such as Africa, Southeast Asia, the western Pacific Islands and tropical countries of South and Central America.

The mol% G + C of the DNA is 52–53.7. Strain Gauthier shows 100% DNA homology to *T. pallidum* (Nichols and KKJ strains) and no homology to *T. phagedenis* and *T. refringens* (Miao and Fieldsteel, 1980).

Type strain: None designated. Reference strains: Gauthier or Haiti B.

1c. Treponema pallidum subspecies endemicum subsp. nov.

en.de′mi.cum. Gr. m. adj. *endemos*, native, dwelling in place. M.L. neut. adj. *endemicum* endemic.

The morphology and characteristics as are described for the species and as listed in Table 1.7.

Pathogenic. The cause of nonvenereal endemic syphilis in man, a contagious disease spread by contact.

Has not been successfully cultivated in artificial media or tissue culture. Propagated by intratesticular inoculation of rabbits or by intradermal inoculation of hamsters. The organisms can be isolated from inguinal lymph nodes 3–4 weeks after intradermal infection. Inbred hamsters, LSH/Ss LAK LSH, are particularly useful for study of this organism (Schell et al., 1980).

Produces cutaneous lesions in rabbits, hamsters, and guinea pigs but not in mice.

Sera from patients with nonvenereal epidemic syphilis give positive results with serologic tests for syphilis.

Found in Africa, the middle east, some areas of southeast Asia and certain areas of Yugoslavia. Isolated from lesions from patients with nonvenereal endemic syphilis, a contagious disease spread by contact.

This subspecies is created because the organism is considered a variant of *T. pallidum* and has its own clinical symptoms in human infection as well as the ability to infect and produce skin lesions in different laboratory animals than does the organism of venereal syphilis.

Type strain: None designated. Reference strain: Bosnia A.

2. Treponema carateum (ex Brumpt 1939) nom. rev.

ca.ra′te.um. M.L. n. *carate* name of a South American disease, pinta. M.L. neut. adj. *carateum* of carate.

The cause of pinta or carate, a contagious disease of man.

Morphologically similar to *Treponema pallidum*. Virulent strains have not been grown in vitro. Experimental transmission of the disease has been accomplished in man as well as in chimpanzees by intradermal inoculation and by direct exposure of scarified areas of skin to abraded human lesions. Not proven virulent for rabbits, hamsters or guinea pigs.

Found in the lymph fluid of cutaneous lesions of pinta. Occurs only in Mexico, Central America and parts of subtropical South America, the West Indies and Cuba.

Type strain: None designated.

3. Treponema paraluiscuniculi (Jacobsthal 1920) Smibert 1974, 177.[AL] (*Spirochaeta paraluis-cuniculi* Jacobsthal 1920, 571.)

par′a.lu.is-cu.ni′cu.li. Gr. pref. *para* resembling; L. n. *luis* pestilences, syphilis; L. gen. n. *cuniculi* of a rabbit; M.L. n. *paraluis-cuniculi* of a syphilis-like (disease) of rabbits.

Produces benign venereal spirochaetosis (rabbit spirochaetosis or rabbit syphilis) in rabbits.

Morphologically similar to *Treponema pallidum*. Has not been cultivated in vitro. The organism can be propagated by intratesticular inoculation of rabbits. Causes a latent infection of mice, guinea pigs and hamsters. Treponemes are found in the lymph nodes of these animals. Cutaneous lesions are found only in guinea pigs and rabbits.

The organisms can be found in lesions in the genitoperineal area of rabbits. Primarily involves the genitalia although cutaneous lesions often occur around the face, eyes, ears and nose. For additional information see Smith and Persetsky (1967).

Type strain: None has been designated.

Species 4 to 13 are anaerobes and have been cultivated. Characteristics of these species are in Tables 1.8–1.11.

4. Treponema denticola (ex Brumpt 1922) nom. rev.

den.ti′co.la. L. masc. n. *dens, dentis* tooth; L.v. suff. *cola* from L.v. *colo* to dwell; M.L. n. *denticola* tooth dweller.

Many characteristics are listed in Tables 1.8 and 1.9.

Two to three periplasmic flagella are inserted into each end of the cell.

Motile with a jerky but fairly rapid motion.

Surface and subsurface colonies are 0.3–1.0 mm in diameter, white, diffuse, and appear after 2 weeks incubation.

Grows well in peptone-yeast extract-serum medium under anaerobic conditions.

End products of fermentation of amino acids in peptone-yeast extract-serum medium are major amounts of acetic acid, moderate amounts of lactic acid and minor amounts of succinic and formic acids. Trace amounts of propionic acid, *n*-butyric acid, ethanol, *n*-propanol and *n*-butanol may occasionally be found. No additional end products are produced in the presence of glucose.

Glucose can be degraded by the Embden-Meyerhof pathway; however, *T. denticola* grown in a peptone-rich medium containing glucose is primarily an amino acid fermenter and does not use the glycolytic pathway as a major source of energy. Only 10% of the end products are from glucose. Alanine, cysteine, glycine and serine are fermented. Pyruvate is oxidized with a clostridium-type clastic pathway (Hespell and Canale-Parola, 1971). Arginine is metabolized to citrulline, NH_3, CO_2, proline and small amounts of ornithine. Arginine iminohydrolase and ornithine carbamoyltransferase activity have been reported (Blakemore and Canale-Parola, 1976). Arginine can act as an energy source. Ornithine can be converted to putrescine and to proline (Leschine and Canale-Parola, 1980).

Methyl red-negative. Produces ammonia from amino acids. Chopped meat-serum medium is neither blackened nor digested. No action on milk.

Does not grow at a pH of 6.0 or 9.6. Grows at a pH range of 6.5–8.0. Grows in a temperature range of 30 to 42°C. Very slight or no growth of a few strains at 25 and 45°C.

Found in the oral cavity of man and chimpanzees, usually in the deposit at the juncture between the teeth and gums.

The mol% G + C content of the DNA is 37–38 (T_m). DNA from strains of *T. denticola* show a high homology with DNA from a culture labeled *T. comandonii*.

Type strain: None designated. Reference strain: FM.

5. Treponema vincentii (ex Brumpt 1922) nom. rev.

vin.cen′ti.i. M.L. gen. n. *vincentii* of Vincent; named after Dr. H. Vincent, a French bacteriologist.

Many characteristics are listed in Tables 1.8 and 1.9.

Cells may have shallow and irregular spirals. Four to six periplasmic flagella are inserted at each end of cell.

Motile with a rapid, jerky, vibratory motion.

Colonies of strain N-9 are visible after incubation for 2 weeks. The colonies are small, white, 12–15 mm in diameter, appearing as a slight haze in the agar.

Grows in a peptone-yeast extract medium under anaerobic conditions. Requires animal serum or ascitic fluid for growth.

End products of fermentation of amino acids are major amounts of acetic and *n*-butyric acids, moderate amounts of lactic acid and smaller amounts of succinic and formic acids. Trace amounts of propionic acid may also be found. Trace amounts of ethanol, *n*-propanol and *n*-butanol may also be found. No additional end products are produced in the presence of glucose.

Methyl red-negative. Skim milk is not changed. Ammonia is produced in cultures. Chopped meat is neither blackened nor digested. A slight fetid odor is produced in cultures.

Grows in a pH range of 6.5–7.5. Grows in a temperature range of 25–45°C.

Meyer and Hunter (1967) reported that *T. vincentii* strain N-9 was antigenically distinct from *T. denticola* (FM) and the Nichols and

Table 1.8.
Differential characteristics of the cultivable species of the genus **Treponema**[a]

Characteristic	Species not Fermenting Carbohydrates					Species Fermenting Carbohydrates				
	4. *T. denticola*	5. *T. vincentii*	6. *T. scoliodontum*	7. *T. refringens*	8. *T. minutum*	9. *T. phagedenis*	10. *T. succinifaciens*	11. *T. bryantii*	12. *T. hyodysenteriae*	13. *T. innocens*
Cell diameter, μm										
0.15–0.20	+	–	+	–	+	–	–	–	–	–
0.20–0.25	–	+	–	+	–	+	–	–	–	–
0.30	–	–	–	–	–	–	+	+	–	–
0.36–0.38	–	–	–	–	–	–	–	–	+	+
Growth requirements										
Serum	+	+	+	+	+	+	–	–	+	+
Thiamine pyrophosphate	+	+	+	–	–	–	–	–	–	–
Volatile fatty acids[b]	–	–	–	–	–	–	+	+	–	–
Glucose	–	–	–	–	–	–	+	+	–	–
Fermentation of[c]										
Glucose	–	–	–	–	–	+	+	+	d	d
Fructose	–	–	–	–	–	+	–	–	–	+
Lactose	–	–	–	–	–	+	+	+	–	–
Maltose	–	–	–	–	–	–	+	–	+	–
Mannitol	–	–	–	–	–	+	–	–	+	–
Starch	–	–	–	–	–	–	+	–	–	–
Sucrose	–	–	–	–	–	+	–	+	–	–
Ribose	–	–	–	–	–	+	–	–	–	–
Esculin hydrolysis	+	d	–	+	+	d[d]			+	+
Indole production	d[e]	weak	–	+	+	+			+	–
1% Glycine (growth)	d	–	–	d[f]	+	+				
Phosphatase	+	–	–	–	–	+				
Convert fumarate to succinate	–	–	–	+	–	+				
β-Hemolysis	–	–	–	–	–	–			+	weak

[a] For symbols see standard definitions.
[b] As occur in rumen fluid.
[c] Acid production in weakly buffered medium containing carbohydrate.
[d] Distinguishes between biovars: biovar *reiter* does not hydrolyze esculin, whereas biovar *kazan* does.
[e] Distinguishes between biovars: biovar *denticola* is indole-positive, biovar *comondonii* is indole-negative.
[f] Distinguishes between biovars: biovar *refringens* does not grow with 1% glycine, biovar *calligyrum* does grow in 5–6 days.

Noguchi strains of *T. refringens*. Antigens were shared with *S. zuelzerae* and *T. phagedenis* (Reiter and Kazan strains).

Found in the oral cavity of man.

The mol% G + C of the DNA is not known.

Type strain: None designated. Reference strain: *T. vincentii* strain N-9.

6. **Treponema scoliodontum** (*ex* Noguchi 1928) nom. rev.

sco.li o.dont′.um. Gr. adj. *scotios* crooked, bent; Gr. n. *odous, odontos* tooth; M.L. neut. n. *scoliodontum* crooked tooth.

Many characteristics are listed in Tables 1.8 and 1.9.

Very tightly coiled cells.

Motile with a jerky but fairly rapid motion.

Grows in peptone-yeast extract-serum medium under anaerobic conditions. Requires animal serum or ascitic fluid.

End products of fermentation of amino acids in a peptone-yeast extract-serum medium are moderate amounts of acetic acid, and small amounts of formic, succinic, lactic, propionic and *n*-butyric acids. No additional end products are produced in the presence of glucose.

Methyl red-negative. Ammonia is not produced from amino acids. No action on milk. Chopped meat serum medium is neither blackened nor digested. Produces a slight fetid odor.

Does not grow at a pH of 6.0 or 9.6. Grows at a pH range of 6.5–8.0. Grows in a temperature range of 30–42°C.

From the oral cavity of man.

The mol% G + C of the DNA is not known.

Type strain: None designated. Reference strain: *T. scoliodontum* Institut Pasteur, Paris.

7. **Treponema refringens** (*ex* Castellani and Chalmers 1919) nom. rev.

re.frin′gens. L. part. adj. *refringens* refringent, refractive.

Many characteristics are listed in Tables 1.8 and 1.9.

The average cells are 5–8 μm long and 0.24 μm wide. Some cells may appear loosely coiled. Two to four periplasmic fibrils are inserted at each end of the cell.

Motile, with a slow, sluggish movement. Rotation of cells is rare. The cells usually rotate slowly.

Colonies on prereduced anaerobic peptone-yeast extract-serum agar (1.4%) are visible in 9–15 days. They are small, white, round, pinpoint surface colonies 0.5–1 mm in diameter. Some colonies after longer incubation are white, fluffy and up to 1.5 mm in diameter. Colonies grow on the surface of the medium as well as into the medium. Size and texture of colonies varies with the concentration of agar in the medium.

Grows well in peptone-yeast extract-serum medium under anaerobic conditions. Requires animal serum (inactivated at 56–60°C for 1 h) for growth.

End products of fermentation of amino acids in serum medium are major amounts of acetic acid, moderate amounts of succinic, and smaller amounts of lactic acid. Trace amounts of propionic, *n*-butyric and formic acids produced by some strains. Trace amounts of ethanol,

Table 1.9.

Other characteristics of the species of the genus **Treponema** *that do not ferment carbohydrates[a]*

Characteristic	4. T. denticola	5. T. vincentii	6. T. scoliodontum	7. T. refringens	8. T. minutum
Cell length, μm	6–16	5–16	6–16	5–16	9–12
Wavelength, μm	0.9	1.3		1.8	1.3
Amplitude, μm	0.15	0.2–0.3		0.2–0.3	0.2
Ends of cell					
Blunt	+	+	–	–	–
Pointed	–	–	+	–	–
Tapered	–	–	–	+	+
Amino acids fermented	+	+	+	+	+
Final pH of glucose broth	6.7	6.9–7.2	6.9–7.1	6.5–6.7	6.5–6.7
Hydrolysis of					
Cellulose	–			–	–
Hippurate	–		–	+	+
Dextrin, glycogen, starch, gelatin	+	+	+	+	+
Medium containing 1% mucin is blackened	+	+	–	+	+
Utilization of					
Lactate	–	–	–	–	–
Pyruvate	+	–	–	+	+
Growth in presence of					
3% NaCl	–	–	–	–	–
1% Bile	–	–	–	–	–
Optimum pH for growth	7.0	7.0	7.0	7.0	7.0
Optimum temperature for growth, °C	37	37	35	37	37
H_2S production	+	+	–	+	+
Phosphatase activity	+	–		–	–
Fumarate converted to succinate	–	–		+	–
Produce acetate, trace of propionate	+	–		–	–
Produce acetate and butyrate	–	+		–	–
Produce acetate only	–	–		+	+
Mol% G + C of DNA	37–38			39–43	37

[a] For symbols see standard definitions.

n-propanol and n-butanol are also produced by some strains. No additional end products are produced in the presence of glucose.

Methyl red-negative. Skim milk is only slightly curdled. Ammonia produced by most strains.

Grows in a pH range of 6.5–8.0. Does not grow at a pH of 6.0 or 9.6. Grows in a temperature range of 30–42°C. Only very slight or no growth at 25 or 45°C. Chopped meat serum medium is neither blackened nor digested. Only a very slight putrid odor is detectable.

Dupouey (1963) reported that strains labeled *T. refringens* and *T. calligyrum* were closely related antigenically, sharing 4–5 common antigens, but *T. minutum* was only slightly related to them antigenically. The three strains had only one antigen in common with *T. pallidum*.

Not pathogenic. Isolated from condyloma acuminata lesions, occasionally from syphilitic lesions. Part of normal flora of male and female genitalia of man and animals.

The mol% G + C of the DNA is 39–43 (T_m). DNA from *T. refringens* shows a high homology with DNA from Nichols, Noguchi strains and *T. calligyrum*, and a very low homology with DNA from strains of *T. denticola* and *T. phagedenis* (data obtained with the cooperation of J. Johnson). No DNA homology to *T. phagedenis* or *T. pallidum* (Nichols) and *T. pertenue* (Miao and Fieldsteel, 1978 and 1980).

Type strain: None designated. Reference strain: *T. refringens*, Institut Pasteur, Paris.

8. **Treponema minutum** Dobell 1912, 117[AL]

mi.nu′tum. L. neut. adj. *minutum*, small, tiny.

Many characteristics are listed in Tables 1.8 and 1.9.

Two to three periplasmic flagella are inserted into each end of the cell.

Motile with sluggish movement.

Colonies on prereduced peptone-yeast extract-serum agar (1.4%) are visible in 9–15 days. They are small, white, round, pinpoint surface colonies 0.5–1 mm in diameter. Some colonies after longer incubation are white, fluffy and up to 1.5 mm in diameter. Colonies grow on the surface of the medium as well as into the medium. Size and texture of colonies will vary with the concentration of agar in the medium.

Grows well in peptone-yeast extract-serum medium under anaerobic conditions. Requires animal serum (inactivated at 56–60°C for 1 h) for growth.

End products of fermentation of amino acids in peptone-yeast extract-serum medium are major amounts of acetic acid, moderate amounts of succinic, and smaller amounts of lactic acid. Trace amounts of propionic, n-butyric and formic acids produced by most strains. Trace amounts of ethanol, n-propanol and n-butanol are also produced by most strains. No additional end products are produced in the presence of glucose.

Methyl red-negative. Skim milk is only slightly curdled. Ammonia produced by most strains.

Grows in a pH range of 6.5–8.0. Does not grow at a pH of 6.0 or 9.6. Grows in a temperature range of 34–40°C. Chopped meat serum medium neither blackened nor digested. Slight putrid odor.

Dupouey (1963) reported that a strain labled *T. minutum* was only slightly related antigenically to *T. refringens*.

Not pathogenic.

From male and female genitalia.

The mol% G + C of the DNA is 37 (T_m).

Type strain: CIP 5162.

9. **Treponema phagedenis** (ex Brumpt 1922) nom. rev.

phag.e.den′is. Gr. *phagedena*, *phagedenis* of a cancerous sore.

Many characteristics are listed in Tables 1.8, 1.10 and 1.11.

Table 1.10.

Fermentation characteristics of carbohydrate-fermenting species of the genus **Treponema**[a,b]

Carbohydrate	9. T. phagedenis		10. T. succinifaciens	11. T. bryantii	13. T. hyodysenteriae	14. T. innocens
	biovar *reiter*	biovar *kazan*				
Amygdalin	−	−			−	−
Arabinose	−	−	+	+	−	−
Cellobiose	−	−	+	+	−	−
Erythritol	−	−			−	−
Esculin	−	−			−	−
Fructose	+	+	−	−	−	+
Galactose	d	+	+	+		
Glucose	+	+	+	+	+	+
Glycogen	−	−			−	−
Lactose	+	+	+	+	−	d
Maltose	−	−	+	−	+	d
Mannitol	+	+	−	−	−	−
Mannose	+	+	+	+	−	−
Melizitose	−	−			−	−
Melibiose	−	−				
Ribose	+	d	−	−	−	−
Salicin	−	−			−	−
Starch	−	−	+	−	−	−
Sucrose	−	−	−	+	−	−
Trehalose	d	d	−		−	−
Xylose	−	−	+	+	−	−

[a] For symbols see standard definitions.

[b] Carbohydrates not fermented by any species include: inulin, glycerol, cellulose, sorbose, dulcitol, inositol, pectin, raffinose, rhamnose and sorbitol.

Table 1.11.

Other characteristics of carbohydrate-fermenting species of the genus **Treponema**[a]

Characteristic	9. T. phagedenis	10. T. succinifaciens	11. T. bryantii	12. T. hyodysenteriae	13. T. innocens
Average cell length, μm	6–12	4–8	3–8	7–9	7–9
Wavelength, μm	1.4–1.6				
Amplitude, μm	0.2–0.3				
Carbohydrates required for growth	−	+	+	−	−
Amino acids fermented	+	−	−		
Final pH of glucose medium	5.8–6.3			5.9	5.9
Cellulose digestion	−				
Utilization of					
Lactate	−	−		−	−
Pyruvate	+	+		+	+
Optimum temperature, °C	37	35–39	37	42	37
Growth in presence of					
3% NaCl	−				
6.5% NaCl	−	−		+	+
1% Bile	−				
Fumarate converted to succinate	+				
H_2S production	d			−	−
Urease activity	−			−	−
Hydrolysis of					
Gelatin	+			weak	−
Hippurate	−			weak	−
H_2 produced	−	−	−	+	+
Makes acetate, *n*-butyrate	−	−	−	+	+
Makes acetate, propionate, *n*-butyrate	+	−	−	−	−
Makes acetate, lactate, formate, succinate	−	+	−	−	−
Makes acetate, formate, succinate	−	−	+	−	−
Nitrate reduction	−			−	−
Acetyl methyl carbinol produced	−			−	−
Mol% G + C on DNA	38–39	36	36	25–26	25–26

[a] For symbols see standard definitions.

Widest cells show double contours with dark-field microscopy. Ends of the cells are blunt with no covering sheath. Three-to-eight periplasmic flagella are inserted into each end of the cell. In old cultures, the flagella may be seen trailing from the ends of the cells.

Motility in culture media is jerky with slow rotational movement.

Colonies in prereduced anaerobic peptone-yeast extract-serum medium containing 1.3–1.4% agar are small, white, annular, 0.5–1 mm in diameter with a dense center. They can be seen after incubation for 2–5 days at 37°C. Colonies grow on the surface but mainly in the agar. The description and time of appearance of colonies varies with the concentration of agar used in the medium.

Requires animal serum (heat inactivated at 56–60°C for 30 min to 1 h) for growth. Bovine serum albumin supplemented with a pair of fatty acids can substitute for serum (Johnson and Eggebraten, 1971). Fatty acids required are (a) an unsaturated one such as oleic acid and (b) a saturated one such as palmitic acid. No growth with short chain fatty acids or α- β- γ-globulins.

Fermentation of glucose is by the Embden-Meyerhof-Parnas pathway. Contains ferredoxin.

End products of fermentation in a serum medium without glucose are major amounts of acetic and n-butyric acids and moderate-to-small amounts of propionic and formic acids and usually small-to-trace amounts of lactic and succinic acids. Trace amounts of alcohols are also produced. In a medium containing glucose, large amounts of ethanol and n-butanol and smaller amounts of n-propanol are produced.

A very slight curd is formed in skim milk. Weakly methyl red-positive. Does not grow at a pH of 6.0 or 9.6. Temperature growth range is from 30–42°C. No growth to slight growth at 25 and 45°C. Reduces neutral red.

Chopped meat serum medium is neither blackened nor digested. A slight fetid odor is produced in cultures.

Meyer and Hunter (1967) showed that the Reiter, Kazan and English Reiter strains are antigenically closely related. Reiter and English Reiter contained the same antigens while the Kazan strain contained an antigen not shared by the Reiter treponemes. Cultivated Nichols strain of *Treponema refringens* was antigenically unrelated. Christiansen (1964) also reported that the Reiter and Kazan II strains were closely related but not identical. Dupouey (1963) reported that *T. phagedenis* and Reiter strain were closely related antigenically, sharing at least 6 common antigens. The Reiter strain has a large amount of an antigen that may be shared with a number of other species including *T. pallidum*.

More than 40 water-soluble antigens have been demonstrated in the Reiter treponeme by crossed immunoelectrophoresis. Five antigens cross-reacted with antibodies in syphilitic sera (Strandberg Pederson et al., 1980, 1981).

Nonpathogenic. Isolated from phagedenic ulcer on human external genitalia. Reiter treponeme was isolated from a case of primary syphilis in man. Also found as normal flora on the anal and genital areas of normal male and female chimpanzees.

The mol% G + C of the DNA is 38–39 (T_m) (Smibert, 1974; Miao and Fieldsteel, 1978). The data from Smibert were determined with the cooperation of Dr. J. Johnson. Reiter and Kazan strains have high DNA/DNA homology to each other and no homology to *T. refringens* (Smibert, 1974; Miao and Fieldsteel, 1978). There is no DNA/DNA homology to *T. denticola* (Smibert, 1974). There is no detectable DNA homology between *T. phagedenis* (Reiter and Kazan 5) and pathogenic Nichols strain of *T. pallidum* (Miao and Fieldsteel, 1978).

Type strain: None designated. Reference strain: Reiter treponeme.

Additional information on the Reiter treponeme can be found on an excellent review by Wallace and Harris (1967).

10. **Treponema succinifaciens** Cwyk and Canale-Parola 1981, 383.[VP] (Effective publication: Cwyk and Canale-Parola 1979, 231.)

suc.ci.ni.fa′ci.ens. M.L. noun *acidum succinicum*, succinic acid; L.v. *facio* make, produce; M.L. part. adj. *succinifaciens* succinic acid-producing.

Some cells may be up to 16 μm long. May form chains. Two periplasmic flagella are inserted at each end of cell. No transitional movement at 25°C. Motile at 37°C. Requires carbon dioxide.

Colonies in rumen fluid agar deeps (1% agar) are spheroid with an opaque center and diffuse peripheral growth. Colonies are 4–8 mm in diameter after 2 days growth at 37°C. In broth, cell yields are 1.5×10^9 cells/ml with an average generation time of 3.5 h.

The pH of a glucose culture after 48 h is about 6.0. Ferments glucose by the Embden-Meyerhof pathway. Pyruvate is metabolized by a coliform-type clastic reaction. Does not ferment acetate, formate, succinate, xylitol, pyruvate, lactate, gluconate, and Tween 80. End products are (μmol/100 μmol glucose and 51 μmol of CO_2 utilized) acetate, 82; formate, 81; succinate, 58; lactate, 30; 2,3-butanediol, 5; pyruvate, 4; and acetoin, 3.

Catalase-negative. Poor growth occurs at 22 and 43°C.

Inhibited by penicillin G (4 U/ml), cephalothin (4 mg/ml), and chloramphenicol (4 μg/ml). Not inhibited by erythromycin (4 μg/ml), oxytetracycline (4 μg/ml), polymyxin B (40 U/ml), rifampin (4 μg/ml), streptomycin (4 μg/ml), tetracycline (4 μg/ml) and vancomycin (4 μg/ml).

Shares common antigens(s) with *T. pallidum*.

From colon of swine.

The mol% G + C of the DNA is 36 (T_m).

Type strain: ATCC 33096.

11. **Treponema bryantii** Stanton and Canale-Parola 1981, 676.[VP] (Effective publication; Stanton and Canale-Parola 1980, 145.)

bry.an′ti.i. M.L. gen. n. *bryantii* of Bryant; named after M. P. Bryant. One periplasmic flagellum is inserted at each end of the cell. No translational motility occurs at 22°C. Motile at 37°C. Requires carbon dioxide. Grows in chemically defined medium containing isobutyrate, DL-2-methyl butyrate, pyridoxal, folic acid, niacinamide, biotin, thiamine, glucose, CO_2, salts and ammonium sulfate. Riboflavin is stimulatory.

Subsurface, colonies in agar deeps with 0.7% Noble agar (Difco) are subsurface, spherical, 0.5–1.0 mm in diameter, and white resembling cotton balls after 24–36 h of incubation. Colonies usually visible after 24 h. Colonies increase in size to 2–3 mm in diameter after further incubation.

In broth medium containing rumen fluid, glucose and sodium bicarbonate, the final yield of cells is 1.9×10^9 cells/ml.

Does not attack gluconate, succinate, acetate, formate, fumarate, sugar alcohols, or Tween 80. Grows in a medium containing cellulose when cocultured with a cellulolytic bacterium such as *Bacteroides succinogenes* or *Ruminococcus albus*.

No growth at 22 or 45°C.

End products of fermentation (μmol/100 μmol of glucose and 84 μmol of CO_2 utilized) are acetate, 100; formate, 119; and succinate 53. About 15% of the glucose carbon used is assimilated into cell material.

Growth in an antibiotic disk assay method was inhibited by penicillin (10-U disk), cephalothin (30-μg disk), tetracycline (30-μg disk), chloramphenicol (30-μg disk), erythromycin (15-μg disk), and vancomycin (30-μg disk). There is slight inhibition by polymyxin B (100-U/disk). Resistant to rifampin (5-μg disk) and up to 10 μg/ml.

Isolated from bovine rumen.

The mol% G + C of the DNA is 36 ± 1% (T_m).

Type strain: ATCC 33254, DSM 1788.

12. **Treponema hyodysenteriae** Harris, Glock, Christensen and Kinyon 1972, 61.[AL]

hy.o.dys.en.te′.ri.ae. Gr. n. *hys, hyos*, a hog; Gr. n. *dysenteria*, dysentery; M.L. gen. n. *hyodysenteriae* of hog dysentery.

Eight to nine periplasmic flagella are inserted at each end of cell. Either fetal calf serum or rabbit serum can be used to satisfy the serum requirement. Colonies on blood agar plates incubated anaerobically are small and translucent.

Strongly β-hemolytic. The hemolysin is soluble. Grows in a liquid medium composed of prereduced trypticase soy broth supplemented with 10% fetal calf serum and incubated in an atmosphere of CO_2 (or N_2 and 10% CO_2). Grows best in broth media with a redox potential not higher than −125 mV and a pH of 6.9. Grows at 36 and 42°C. The generation time at 37°C is 5.2 h and at 42°C is 3.3 h (Lemcke et al., 1979). Growth is inhibited by digitonin. Suggested to require cholesterol for growth in a medium containing bovine serum albumin and cholesterol (Lemcke and Burrows, 1980). Sitosterol and cholestanol produced a growth response similar to that of cholesterol.

Weakly fermentative. An acid reaction is considered to occur when the pH is 0.25 units lower than the pH of the control. Produces H_2, CO_2, and small amounts of acetic and butyric acids.

Negative for lipase and lecithinase. Does not grow in media with 1% glycine.

Four serovars have been proposed based on soluble antigens found in the water phase of cells extracted with hot phenol water (Baum and Joens, 1979).

Resistant to spectinomycin. Trypticase soy blood agar plates containing 400 μg/ml of spectinomycin (TSA-S400) incubated anaerobically can be used as a selective isolation medium (Songer et al., 1976).

Enteropathogenic for swine. Pathogenic for CF1 mice, producing caecitis (Joens et al. 1980). Found in the intestinal tract of swine on the luminal surface and in mucosal crypts of the large intestine of swine with swine dysentery (Glock et al., 1974).

The mol% G + C of the DNA is 25.7–25.9. Does not show DNA homology with *T. innocens, T. pallidum, T. phagedenis* or *T. refringens* (Miao et al., 1978).

Type strain: ATCC 27168 (strain B78).

13. **Treponema innocens** Kinyon and Harris 1979, 102.[VP]

in′.no.cens. L. adj. *innocens* doing no harm.

Eight or nine periplasmic flagella are inserted at each end of the cell.

Colonies on blood agar plates incubated in an anaerobic jar are small and translucent. Weakly β hemolytic. Grows in a liquid medium composed of prereduced trypticase soy broth supplemented with 10% fetal calf serum and incubated in an atmosphere of CO_2 (or N_2 and CO_2). Grows at 37 and 42°C. Does not grow at 25 and 30°C.

Negative for lipase and lecithinase. Does not grow in media with 1% glycine.

Weakly fermentative. An acid reaction is considered to occur when the pH is 0.25 units lower than the pH of an uninoculated control medium. Produces H_2, CO_2 and small amounts of acetic and butyric acids.

Resistant to spectinomycin. Trypticase soy blood agar plates containing 400 μg/ml of spectinomycin (TSA-S400) incubated anaerobically serves as a selective isolation medium (Songer et al., 1976).

Not enteropathogenic for swine. Isolated from intestinal contents and feces of swine and dogs.

The mol% G + C of the DNA is 25.7–25.9. Shows only a low DNA homology (28%) with *T. hyodysenteriae, T. pallidum, T. phagedenis* and *T. refringens* (Miao et al., 1978).

Type strain: ATCC 29796 (strain B256).

Other organisms

The organisms listed below can be found in the literature. However there are no known cultures of these species. They are listed so that if they are isolated again, the description can be used to aid in their identification. The names presently have no standing in nomenclature.

a. **"Treponema macrodentium"** Noguchi 1912, 82. (*Spirochaeta macrodentium* (Noguchi 1912) Pettit 1928, 182.)

mac.ro.den′ti.um. Gr. adj. *macrus* long; L. n. *dens, dentis* tooth; M.L. gen. pl. n. *macrodentium* (*sic*) of large teeth.

Slender helical rods, 5–16 μm long and 0.1–0.25 μm wide. The ends of the cell are pointed. One periplasmic flagellum is inserted into each end of the cell.

Motile with a fairly rapid motion. Young cells rotate rapidly on their long axis.

Grows in peptone-yeast extract-medium or PPLO medium (BBL) containing 10% serum or ascitic fluid with cocarboxylase (5 μg/ml), glucose (1 mg/ml) and cysteine (1 mg/ml). Requires animal serum for growth. This requirement can be replaced by isobutyric acid (20 μg/ml), spermine (150 μg/ml) and nicotinamide (400 μg/ml). Will also grow in a medium supplemented with rumen fluid and cocarboxylase. Requires a fermentable carbohydrate as an energy source.

Carbohydrates are fermented. Acid but no gas is produced. The final pH in glucose broth is 5.0–5.4. Ferments fructose, glucose, maltose, ribose and sucrose. May ferment cellobiose, galactose, and xylose. Does not ferment mannose, rhamnose, sorbose, lactose, arabinose, trehalose, mannitol, inulin, sorbitol or salicin. Starch is not hydrolyzed.

End products of fermentation of glucose are major amounts of lactic acid, moderate amounts of acetic and formic acids and traces of succinic acid.

Gelatin is hydrolyzed. Indole-negative. Hydrogen sulfide is produced. Lactate is not used. Ammonia is not produced.

Optimum temperature, 37°C. Grows at a pH of 7.0.

From the gingival crevice of man.

The mol% G + C of the DNA is 39 (T_m).

Type strain: no culture available.

b. **"Treponema orale"** Socransky, Listgarten, Hubersak, Cotmore and Clark 1969, 881. (*Treponema oralis* (*sic*) Socransky, Listgarten, Hubersak, Cotmore and Clark 1969, 881.)

o.ra′le. L. noun *os, oris* the mouth; M.L. neut. adj. *orale* of the mouth.

Slender helical cells, 6–16 μm long and 0.10–0.25 μm wide. Occasional chains are formed. One periplasmic flagellum is inserted into each end of the cell. Frequently end granules are seen in broth cultures.

Motile with a jerky but fairly rapid motion.

Grows in either PPLO medium without crystal violet (BBL) or peptone-yeast extract medium. Each medium contains glucose (1 mg/ml), cysteine (1 mg/ml), nicotinamide (500 μg/ml), cocarboxylase (5 μg/ml), spermine tetrahydrochloride (150 μg/ml) and sodium isobutyrate (20 μg/ml), and each is further supplemented with 10% inactivated rabbit serum or ascitic fluid, or 0.05% α-globulin. Uniform turbidity occurs in liquid media. Does not grow well on surface cultivation. Does not require carbohydrates as an energy source. Carbohydrates not fermented. Amino acids are fermented. The final pH in glucose broth is 6.8–7.2. End products of fermentation of amino acids are acetic and propionic acids.

Hydrolyzes gelatin but not starch. Indole-positive. H_2S produced. Utilizes lactate. Does not produce ammonia in cultures.

Grows at a pH of 7.0 and at a temperature of 37°C.

Found in the gingival crevice of man.

The mol% G + C of the DNA is 37.

Type strain: No culture available.

Genus IV. **Borrelia** *Swellengrebel 1907, 582*[AL]

RICHARD T. KELLY

Bor. re′ li.a. M.L. fem. n. *Borrelia* named after A. Borrel.

Helical cells 0.2–0.5 by 3–20 μm, composed of 3–10 loose coils. The cells are surrounded by a surface layer, an outer membrane, and a cytoplasmic membrane. Fifteen to 20 periplasmic flagella (which also have been termed axial fibrils, periplasmic fibrils or endoflagella) originate at each end of the cell and wind about the protoplasmic cylinder to overlap in the middle of the cell. The cells are actively motile with frequent reversal of the direction of translational movement. Gram-negative. Stain well with Giemsa's stain. Species which have been grown in vitro are **microaerophilic.** Nutritional requirements for in vitro growth are complex. Pathogens of man, other mammals, and birds. **The causative agents of tick-borne and louse-borne relapsing fever in man.** The mol% G + C of the DNA is not known.

Type species: *Borrelia anserina* (Sakharoff 1891) Bergey, Harrison, Breed, Hammer and Huntoon 1925, 435.

Further Descriptive Information

The appearance of borreliae in stained blood films is illustrated in Figure 1.4.

Chemical analysis of *Borrelia hermsii* cells that were cultivated in vitro has demonstrated the presence of muramic acid and ornithine in the whole cells and in the protoplasmic cylinders, but not in the outer envelope preparations (Klaviter and Johnson, 1979). These data suggest that ornithine is a component of the cell wall. *B. hermsii* contains cholesterol glucoside and its acylated derivatives (Livermore et al., 1978), and the lipid composition and metabolism of this species of *Borrelia* is remarkably similar to that of several species of mycoplasmas.

Only a few *Borrelia* species have been cultivated in vitro, but some essential components of the media for those species that have been cultivated include rabbit serum, bovine serum albumin, peptones, pyruvate, citrate and *N*-acetylglucosamine. The limited data regarding the in vitro nutritional requirements of the various *Borrelia* species indicate some similarities as well as specific differences. The culture medium developed for in vitro propagation of *B. hermsii* (medium "A", see Table 1.13) was subsequently found to support the growth of *B. parkeri*, *B. turicatae* and *B. duttonii*, but not *B. hispanica* or *B. recurrentis* (Kelly, 1971, 1976). A subsequent culture medium which supported the growth of *B. hispanica* did not support the growth of *B. recurrentis*.

Smibert (1976) has studied three species of *Borrelia* grown in vitro to determine their ability to utilize a variety of carbohydrates. *B. hermsii* and *B. parkeri* fermented glucose, maltose, trehalose, starch, dextrin and glycogen, but not raffinose. In contrast, *B. turicatae* was able to ferment only glucose, raffinose and dextrin. From these data, Smibert concluded that *B. hermsii* and *B. parkeri* may be identical, but that *B. turicatae* is probably another species.

Glucose is fermented via the Embden-Meyerhof pathway to DL-lactic acid. Lysolecithinase, glycerophosphorylcholine diesterase and acid phosphatase have been demonstrated in sonic extracts of three species (Pickett and Kelly, 1974). One species (*B. hermsii*) has been reported to contain superoxide dismutase, but lacks catalase and peroxidase (Austin et al., 1981), which may be relevant to the microaerophilic behavior of the organisms.

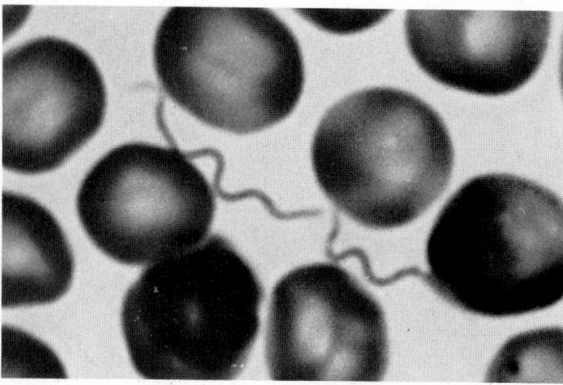

Figure 1.4. Appearance of *Borrelia hermsii* in blood. Giemsa stain. × 2000.

Borreliae are antigenically quite labile, with the result that a given species may exist in eight or more entirely different antigenic phases. This property is responsible for the numerous relapses which occur in the infected host. As immunity develops with termination of infection by one antigenic phase, there is mutation and an antigenetically different phase proliferates.

Identification of the various species of *Borrelia* by serological or biochemical reactions has not yet been accomplished.

Borreliae are pathogens of man, other mammals, and birds. Infections are acquired from ticks or lice which are parasitized with borreliae. Infections in man produce a severe septicemic illness of which two varieties are recognized: tick-borne and louse-borne relapsing fever.

The causative agent of louse-borne relapsing fever, *Borrelia recurrentis*, is transmitted from human-to-human by the body louse *Pediculus humanus* subsp. *humanus*. Man acquires infection by crushing infected lice during scratching, thereby liberating borreliae which enter the body at the site of the louse bite or through skin abraded by scratching. Lice acquire *Borrelia recurrentis* by feeding upon infected humans and remain infectious for the remainder of their relatively short life span. Congenital transmission of borreliae does not take place.

Tick-borne relapsing fever is transmitted by a number of different species of *Ornithodoros* ticks. Nine different species of tick-borne borreliae have been recognized, many of which are named after the species of *Ornithodoros* tick transmitting the infection. Numerous small animals serve as natural reservoirs for tick-borne borreliae. Borreliae are frequently vertically transmitted in infected ticks, resulting in a variable percentage of infectious offspring.

The geographical distribution of the *Borrelia* species, and also the specific arthropod vectors and their hosts, are listed later in Table 1.12.

Enrichment and Isolation

In vitro Cultivation

Six species have been grown in vitro. *B. hermsii*, *B. parkeri*, *B. turicatae* and *B. duttonii* are grown in medium A (Table 1.13). Other growth media have been developed for *B. hispanica* and *B. recurrentis*; however, low levels of growth are obtained (Kelly, 1976). In medium A the maximal yield of $3–5 \times 10^7$ organisms/ml is obtained after 6–7 days of incubation. The calculated generation time for the various borreliae species is from 12–18 h.

Cultured borreliae have retained infectivity for mice for over 100 subcultures. With prolonged in vitro cultivation (over 150 subcultures), there is attenuation and infectivity is lost.

Isolation by Animal Inoculation

Because experience with isolation of borreliae by in vitro cultivation has been limited to only a few species, animal inoculation remains the method of choice for isolation of borreliae. A number of animals have been used for isolation of nonavian borreliae including mice, rats, rabbits, guinea pigs and monkeys. Results have been variable depending on the species of borrelia and the age and species of animals employed. The most consistent results have been obtained with suckling mice. Citrated blood or triturates of ticks or lice are inoculated in a volume of 0.1–0.2 ml via the subcutaneous or intraperitoneal route. At daily intervals, the mice are bled by clipping the tip of the tail with scissors to obtain a small drop of blood which is examined by darkfield microscopy or in smears stained by the Wright or Giemsa method. For darkfield microscopy, a slide is touched to the cut surface of the tail. A coverslip is placed over the blood thereby spreading the blood into a thin layer, and the edges of the coverslip are sealed with either lanolin or liquid paraffin. Specimens are examined at a magnification of about × 400 (high dry). Borreliae are frequently first detected by observing movement of erythrocytes which are disturbed as the organisms move about. When these areas are focused upon, the characteristic large spirochetes are readily observed.

For isolation of *B. anserina*, newly hatched chicks are inoculated via the intraperitoneal route.

Maintenance Procedures

Suspensions of borreliae containing 10% glycerol retain viability for years when stored at or below −70°C. Limited attempts to preserve borreliae by lyophilization have not been successful.

Tick-borne borreliae may also be maintained in the laboratory for several years in infected ticks with only occasional feedings on experimental animals.

Differentiation of the genus **Borrelia** from other genera

Two other genera of spirochetes which may be present in biological fluids or tissues are species of *Treponema* and *Leptospira*. Borreliae stain readily and can be observed with conventional microscopy whereas leptospires and treponemes are not visualized. The coils of borreliae are loose, coarse and irregular. Leptospires are coiled so tightly that they are difficult to observe with darkfield microscopy. Treponemal species are regularly and rigidly coiled and have a more uniformly helical structure than do borreliae species.

Taxonomic Comments

Borreliae have been classified into species according to their arthropod vectors. Those borreliae transmitted by human body lice presently belong to the species *B. recurrentis*; those transmitted by ticks are differentiated on the basis of the tick-spirochete specificity theory which states that borreliae carried by a given species of tick are specific for that vector and, therefore, constitute individual species of borreliae. Thus, *Borrelia parkeri* is classified on the basis of its transmission by

the tick *Ornithodoros parkeri*; *Borrelia duttonii* is transmitted by the tick *Ornithodoros moubata*, etc. (Table 1.12). This theory, however, is flawed by the fact that some borreliae isolated from one species of tick are capable of infecting other species of ticks. Moreover, many species of tick-borne borreliae have been shown to develop in experimentally infected body lice. Consequently, classification according to the traditional scheme may or may not be correct, and for this reason the currently accepted species of *Borrelia* are merely listed alphabetically in the present edition of the *Manual*. Since biochemical characterization of borreliae has been limited to only a few species, and since nucleic acid hybridization studies have not yet been performed, additional studies will be necessary to determine if the present classification is in fact valid.

Further Reading

Felsenfeld, O., 1971. *Borrelia*. W.H. Green, St. Louis, Missouri.
Johnson, R.C. (Editor). 1976. The biology of the parasitic spirochetes. Academic Press, New York.

Differentiation of the species of the genus **Borrelia**

The characteristic features of the various species of *Borrelia* are indicated in Table 1.12.

List of species of the genus **Borrelia**

1. **Borrelia anserina** (Sakharoff 1891) Bergey, Harrison, Breed, Hammer and Huntoon 1925, 435.[AL] (*Spirochaeta anserina* Sakharoff 1891, 565; *Spiroschaudinnia anserina* (Sakharoff 1891) Sambon 1907, 834.)

an.se.ri′na. L. adj. *anserina* pertaining to geese.

The organisms are 0.2–0.3 μm wide and 8–20 μm in length, and consist of 5–8 coils.

No reliable data are available concerning in vitro cultivation. The organisms can be grown in embryonated duck or chicken eggs. They can also be maintained in young chickens or ducks. Not infective for mice, rats or rabbits.

There are a number of antigenically different strains. The most reliable distinguishing characteristic of this species is that it is infectious for birds but not for rodents. Biochemical characteristics have not been determined.

The mol% G + C of the DNA is not known.

Type strain: no culture available.

2. **Borrelia brasiliensis** Davis 1952, 476.[AL]

bra.si.li.en′sis. M.L. adj. *brasiliensis* the specific epithet of the tick vector, *Ornithodoros brasiliensis*.

Resembles *B. recurrentis* in morphology.

In vitro cultivation has not been reported. Isolated and maintained in mice, rats, and guinea pigs.

Distinguishing characteristics: not well characterized. No biochemical data available.

The mol% G + C of the DNA is not known.

Type strain: no culture available.

3. **Borrelia caucasica** (Kandelaki 1945) Davis 1957, 901.[AL] (*Spirochaeta caucasica* Kandelaki according to Maruashvili 1945, 24.)

cau.ca′si.ca. M.L. adj. *caucasica* pertaining to the Caucasus.

Resembles *B. recurrentis* in morphology.

In vitro cultivation has not been reported. Isolated and maintained in mice, rats and guinea pigs.

Distinguishing characteristics: not well characterized. No biochemical data available.

The mol% G + C of the DNA is not known.

Type strain: no culture available.

4. **Borrelia crocidurae** (Leger 1971) Davis 1957, 903.[AL] (*Spirochaeta crocidurae* Leger 1917, 281.)

cro.ci.du′rae. M.L. gen. n. *crocidurae* of *Crocidura*, a genus of Insectivora.

Resembles *B. recurrentis* in morphology.

In vitro cultivation has not been reported. Isolated and maintained in young mice and rats.

Four serovars have been recognized; however, the species has not been well characterized. No biochemical data are available.

The mol% G + C of the DNA is not known.

Type strain: no culture available.

5. **Borrelia dugesii** (Mazzotti 1949) Davis 1957, 902.[AL] (*Spirochaeta dugesii* Mazzotti 1949, 278.).

du.ge′si.i. M.L. gen. n. *dugesii* of dugesi, the specific epithet of the tick vector, *Ornithodoros dugesi*.

Resembles *B. recurrentis* in morphology.

In vitro cultivation has not been reported. Isolated and maintained in young mice or rats.

Distinguishing characteristics: not well characterized. No biochemical data available.

The mol% G + C of the DNA is not known.

Type strain: no culture available.

6. **Borrelia duttonii** (Novy and Knapp 1906) Bergey, Harrison, Breed, Hammer and Huntoon 1925, 434.[AL] (*Spirillum duttoni (sic)* Novy and Knapp 1906, 296; *Spirochaeta duttoni* (Novy and Knapp 1906) Breinl 1906, 1691.))

dut.to′ni.i. M.L. gen. n. *duttonii* of Dutton; named after J.E. Dutton.

Resembles *B. recurrentis* in morphology.

Table 1.12.

Differential characteristics of the species of the genus **Borrelia**

Species	Vector	Host	Distribution	Disease
1. *B. anserina*	*Argas miniatus* *Argas persica* *Argas reflexus*	Numerous birds	Worldwide	Avian borreliosis
2. *B. brasiliensis*	*Ornithodoros brasiliensis*	Rodents	Brazil	
3. *B. caucasica*	*Ornithodoros verrucosus*	Rodents, man	Caucasus	Tick-borne relapsing fever
4. *B. crocidurae*	*Ornithodoros erraticus* (small variety)	Rodents, man	Africa, Near East, Central Asia	Tick-borne relapsing fever
5. *B. dugesii*	*Ornithodoros dugesi*	Probably rodents	Mexico	
6. *B. duttonii*[a]	*Ornithodoros moubata*	Man	Africa	Tick-borne relapsing fever
7. *B. graingeri*	*Ornithodoros graingeri*	Rodents, man	East Africa	Tick-borne relapsing fever
8. *B. harveyi*	Unknown	Monkeys	Africa	
9. *B. hermsii*[a]	*Ornithodoros hermsi*	Rodents, man	Western U.S. and Canada	Tick-borne relapsing fever
10. *B. hispanica*[b]	*Ornithodoros erraticus* (large variety)	Rodents, man	Spain, Portugal, Morocco, Algeria, Tunisia	Tick-borne relapsing fever
11. *B. latyschewii*	*Ornithodoros tartakovskyi*	Rodents, reptiles, man	Iran, Central Asia	Tick-borne relapsing fever
12. *B. mazzottii*	*Ornithodoros talaje*	Rodents, armadillos, monkeys, man	Mexico and Guatemala	Tick-borne relapsing fever
13. *B. parkeri*[a]	*Ornithodoros parkeri*	Rodents, man	Western U.S.A.	Tick-borne relapsing fever
14. *B. persica*	*Ornithodoros tholozani*	Rodents, bats, man	Middle East, Central Asia	Tick-borne relapsing fever
15. *B. recurrentis*[b]	*Pediculus humanus,* subsp. *humanus*	Man	South America, Europe, Africa, Asia	Louse-borne relapsing fever
16. *B. theileri*	*Rhipicephalus decoloratus* *Rhipicephalus evertsi* *Boophilus micropus*	Ruminants, horses	South Africa, Australia	Cattle and horse borreliosis
17. *B. tillae*	*Ornithodoros zumpti*	Rodents	South Africa	
18. *B. turicatae*[a]	*Ornithodoros turicatae*	Rodents, man	U.S.A. and Mexico	Tick-borne relapsing fever
19. *B. venezuelensis*	*Ornithodoros rudis*	Rodents, man	Central and South America	Tick-borne relapsing fever

[a] Grow well in vitro.

[b] Grow in vitro but only low cell numbers are obtained.

Grows well in medium A (Table 1.13).

Isolated and maintained in young mice and rats.

Distinguishing characteristics: not well characterized. No biochemical data available.

The mol% G + C of the DNA is not known.

Type strain: no strain designated.

7. **Borrelia graingeri** (Heisch 1953) Davis 1957, 903.[AL] (*Spirochaeta graingeri* Heisch 1953, 133.)

grain'ger.i. M.L. gen. n. *graingeri* the specific epithet of the tick vector, *Ornithodoros graingeri.*

Resembles *B. recurrentis* in morphology.

In vitro cultivation has not been reported. Animal pathogenicity: present in the blood of mice and rats only transiently after inoculation.

Distinguishing characteristics: not well characterized. No biochemical data available.

The mol% G + C of the DNA is not known.

Type strain: no culture available.

8. **Borrelia harveyi** (Garnham 1947) Davis 1948, 316.[AL] (*Spirochaeta harveyi* Garnham 1947, 49.)

har'vey.i. M.L. gen. n. *harveyi* of Harvey; named after A. E. C. Harvey.

Resembles *B. recurrentis* in morphology.

In vitro cultivation has not been reported. May be maintained in mice, rats and monkeys.

Distinguishing characteristics: not well characterized. No biochemical data available.

The mol% G + C of the DNA is not known.

Type strain: no culture available.

9. **Borrelia hermsii** (Davis 1942) Steinhaus 1946, 453.[AL] (*Spirochaeta hermsi (sic)* Davis 942, 46.)

herm'si.i. M.L. gen. n. *hermsii* of hermsi, the specific epithet of the tick vector, *Ornithodoros hermsi.*

Table 1.13.

Composition of **Borrelia** *growth medium "A"*

Component	g/liter of Double-Distilled Water
Stock solution 1 (store at −20°C):	
$Na_2HPO_4 \cdot 7H_2O$	26.52
$NaH_2PO_4 \cdot H_2O$	1.03
NaCl	1.20
KCl	0.85
$MgCl_2 \cdot 6H_2O$	0.68
Glucose	12.75
Proteose peptone No. 2 (Difco)	5.95
Tryptone (Difco)	2.55
Sodium pyruvate	1.06
Sodium citrate · $2H_2O$	0.47
N-Acetylglucosamine	0.53

	g/100 ml of Double-Distilled Water
Stock solution 2 (adjust pH to 7.8 with NaOH; store at −20°C):	
Bovine albumin, fraction V	10.0
Stock solution 3 (freshly prepared):	
$NaHCO_3$	4.5
Stock solution 4 (autoclave at 115°C for 115 min; store at 4°C):	
Gelatin	7.0
Stock solution 5 (store at 4°C):	
Phenol red	0.5

1. to 80 ml of solution 1, add 34 ml of solution 2, 4.0 ml of solution 3, 0.7 ml of solution 5, and 1.3 ml of distilled water.
2. Sterilize by filtration through a membrane filter (0.22 μm pore size).
3. Dispense in 6-ml portions into sterile borosilicate screw-cap tubes (13 × 100 mm) with Teflon liners.
4. Liquefy solution 4 by immersion in warm water and add 2.0 ml to each tube.
5. Add 0.5 ml of pooled, sterile rabbit serum to each tube and mix by inversion.
6. Store the complete medium at room temperature. Use within 30 days.
7. To initiate cultures, add 0.05 ml of citrated, infected blood to each tube and incubate at 35°C.

Resembles *B. recurrentis* in morphology.

Grows well in borrelia medium A (Table 1.13). Readily isolated and maintained in young mice and rats.

Distinguishing characteristics: ferments glucose, maltose, trehalose, starch, dextrin and glycogen, but not raffinose.

The mol% G + C of the DNA is not known.

Type strain: none designated.

10. **Borrelia hispanica** (de Buen 1926) Steinhaus 1946, 453.[AL] (*Spirochaeta hispanica* de Buen 1926, 185.)

hi.spa′ni.ca. L. adj. *hispanica* Spanish.

Resembles *B. recurrentis* in morphology.

May be cultivated in vitro, but the yield of organisms (7×10^6/ml) is low (Kelly, 1976). Animal pathogenicity: causes a mild infection in young mice and rats. A strain in the author's laboratory is virulent for young guinea pigs but is not infectious for mice.

Distinguishing characteristics: not well characterized. No biochemical data available.

The mol% G + C of the DNA is not known.

Type strain: none designated.

11. **Borrelia latyschewii** (Sofiev 1941) Davis 1948, 315.[AL] (*Spirochaeta latyschewi (sic)* Sofiev 194, 271.)

la.ty.sche′wi.i. M.L. gen. n. *latyschewii* of Latyschew: named after Latyschew (Latyshev).

Resembles *B. recurrentis* in morphology.

In vitro cultivation has not been reported. Maintained in young mice.

Distinguishing characteristics: not well characterized. No biochemical data available.

The mol% G + C of the DNA is not known.

Type strain: no culture available.

12. **Borrelia mazzottii** Davis 1956, 17.[AL]

maz.zot′ti.i. M.L. gen. n. *mazzottii* of Mazzotti; named after L. Mazzottii.

Resembles *B. recurrentis* in morphology.

In vitro cultivation has not been reported. Isolated and maintained in young mice and rats.

Distinguishing characteristics: not well characterized. No biochemical data availale.

The mol% G + C of the DNA is not known.

Type strain: no culture available.

13. **Borrelia parkeri** (Davis 1942) Steinhaus 1946, 453.[AL] (*Spirochaeta parkeri* Davis 1942, 46).

par′ker.i. M.L. gen. n. *parkeri* the specific epithet of the tick vector, *Ornithodoros parkeri*.

Resembles *B. recurrentis* in morphology.

Grows well in borrelia medium A (Table 1.13). Animal pathogenicity: isolated and maintained in young mice and rats.

Distinguishing characteristics: fermentation of carbohydrates is identical to that of *Borrelia hermsii*. This suggests that the two species are in fact variants of a single species.

The mol% G + C of the DNA is not known.

Type strain: none designated.

14. **Borrelia persica** (Dschunkowsky 1913) Steinhaus 1946, 453.[AL] (*Spirochaeta persica* Dschunkowsky 1913, 419.)

per′si.ca. L. adj. *persica* Persian.

Resembles *B. recurrentis* in morphology.

In vitro cultivation has not been reported. Animal pathogenicity varies according to the strain. Most isolates can be maintained in mice or guinea pigs.

Distinguishing characteristics: not well characterized. No biochemical data available.

The mol% G + C of the DNA is not known.

Type strain: no culture available.

15. **Borrelia recurrentis** (Lebert 1874) Bergey, Harrison, Breed, Hammer and Huntoon 1925, 433.[AL] (*Spirochaeta recurrentis* Lebert 1874, 273; *Spiroschaudinnia recurrentis* (Lebert 1874) Sambon 1907, 833.)

re.cur.ren′tis. L. part. adj. *recurrens, recurrentis* recurring.

The organisms are 0.3–0.6 μm wide and 8–18 μm in length, and consist of 3–8 coils.

May be cultivated in vitro, but the yield is poor (Kelly, 1976). Most strains may be maintained in young rats. Guinea pigs are not infected. Recent isolates from Ethiopia do not infect any rodents but are infectious for baboons.

Distinguishing characteristics: not well characterized. No biochemical data available.

The mol% G + C of the DNA is not known.

Type strain: no culture available.

16. **Borrelia theileri** (Laveran 1903) Bergey, Harrison, Breed, Hammer and Huntoon 1925, 435.[AL] (*Spirochaeta theileri* Laveran 1903, 941.)

thei′le.ri. M.L. gen. n. *theileri* of Theiler; named after A. Theiler.

The organisms are 0.25–0.30 μm wide and 20–30 μm in length in cattle; they are reported to be shorter in horses.

In vitro cultivation has not been reported. May be maintained in cattle and horses.

Distinguishing characteristics: not well characterized. No biochemical data available.

The mol% G + C of the DNA is not known.

Type strain: no culture available.

17. **Borrelia tillae** Zumpt and Organ 1961, 33.[AL]

till'ae. M.L. gen. n. *tillae* of Till; named after Dr. W. Till.

Resembles *B. recurrentis* in morphology.

In vitro cultivation has not been reported. Isolated and maintained in young mice and rats.

Distinguishing characteristics: not well characterized. No biochemical data available.

The mol% G + C of the DNA is not known.

Type strain: no culture available.

18. **Borrelia turicatae** (Brumpt 1933) Steinhaus 1946, 453.[AL] (*Spirochaeta turicatae* Brumpt 1933, 1369.)

tu.ri.ca'tae. M.L. gen. n. *turicatae* of turicata, the specific epithet of the tick vector, *Ornithodoros turicata.*

Resembles *B. recurrentis* in morphology.

Grows well in borrelia medium "A" (Table 1.13). Isolated and maintained in young mice and rat.

Distinguishing characteristics: Ferments only glucose, raffinose and dextrin, thus differing from *B. hermsii* and *B. parkeri.*

The mol% G + C of the DNA is not known.

Type strain: none designated.

19. **Borrelia venezuelensis** (Brumpt 1921) Brumpt 1922, 495.[AL] (*Treponema venezuelense* Brumpt 1921, 207.)

ve.ne.zue.len'sis. M.L. adj. *venezuelensis* the specific epithet of the tick vector, *Ornithodoros rudis (O. venezuelensis).*

Resembles *B. recurrentis* in morphology.

In vitro cultivation has not been reported. Isolated and maintained in young mice or rats.

Distinguishing characteristics: not well characterized. No biochemical data available.

The mol% G + C of the DNA is not known.

Type strain: No culture available.

Species Incertae Sedis

Borrelia baltazardii Karami, Hovind-Hougen, Birch-Andersen and Asmar 1983, 438.[VP] (*Borrelia baltazardi* (sic) Karami, Hovind-Hougen, Birch-Andersen and Asmar 1979, 157).

Isolated from the blood of a patient with thrombocytopenic purpura in Iran.

FAMILY II. **LEPTOSPIRACEAE** HOVIND-HOUGEN 1979, 245.[AL]

RUSSELL C. JOHNSON AND S. FAINE

Lep.to.spi.ra'ce.ae. M.L. fem. n. *Leptospira* type genus of the family; *-aceae* ending to denote a family; M.L. fem. pl. n. *Leptospiraceae* the *Leptospira* family.

Flexible helical cells, 0.1 μm in diameter and 6 to over 12 μm in length. The helical conformation is right-handed (clockwise coiling). One or both ends of the cells are typically hooked. Two periplasmic flagella (which have also been termed flagella, endoflagella or axial fibrils) occur per cell and rarely overlap in the central region of the cell. **The diamino acid present in the peptidoglycan is α,ε-diaminopimelic acid.** Motile. **Aerobic.** Chemoorganotrophic. **Utilize long chain fatty acids or long chain fatty alcohols as carbon and energy sources.** Do not utilize carbohydrates or amino acids as energy sources. Free living or in association with animal and human hosts. Some species are pathogenic for man and animals. The mol% G + C of the DNA is 35–53 (T_m).

Type genus: *Leptospira* Noguchi 1917, 755.

Taxonomic Comments

The family *Leptospiraceae* has been included as a second family in this edition of the *Manual.* In addition to the genus *Leptospira,* a second genus, *Leptonema,* was proposed to be included in this family (Hovind-Hougen, 1979). The type species of *Leptonema* was designated as *L. illini.* However, the Subcommittee on the Taxonomy of *Leptospira* (ICSB) considers that the information currently available is insufficient to define satisfactorily the genus *Leptonema,* and the species *illini* has been retained in the genus *Leptospira* as a *species incertae sedis.*

Genus I. **Leptospira** Noguchi 1917, 755.[AL]

RUSSELL C. JOHNSON AND S. FAINE

Lep.to.spi'ra. Gr. adj. *lepto* thin, narrow, fine; Gr. n. *spira* a coil; M.L. fem. n. *Leptospira* a fine coil.

Flexible helicoidal rods, 0.1 μm in diameter and 6 to over 12 μm in length. Resting stages not known. Gram-negative; appear faintly stained with aniline dyes. Unstained cells are **not visible by bright-field microscopy** but are visible by dark-field illumination and phase-contrast microscopy. **Motile. Obligately aerobic,** having a respiratory type of metabolism with oxygen as the terminal electron acceptor. Optimum temperature, 28–30°C. Generation time, 6–16 h. Diffuse-to-discrete subsurface colonies are formed in 1% agar and turbid-to-clear surface colonies on 2% agar; some strains form both subsurface and surface colonies on 1% or 2% agar. **Oxidase-positive.** Catalase- and/or peroxidase-positive. **Chemoorganotrophic, using fatty acids or fatty alcohols (15 carbons or more) as energy and carbon sources. Some strains are parasitic and may be pathogenic for man and animals, while other strains are free living in soil, freshwater or marine habitats.** The mol% G + C of the DNA is 35–41 (T_m) with the exception of one strain which is 53.

Type species: *Leptospira interrogans* (Stimson 1907) Wenyon 1926, 1281.

Further Descriptive Information

The cells of *Leptospira* have 18 or more coils/cell. The helical conformation is right-handed (clockwise coiling) (Carleton et al., 1979) and the coils have an amplitude of 0.10–0.15 μm and a wavelength (pitch) of approximately 0.5 μm (Fig. 1.5). Nonviable spherical forms occur in old cultures or when cells are exposed to adverse conditions such as hypertonicity. One or both ends of the cell are typically hooked in liquid environments and the characteristic movements appear as an alternating rotation around the long axis and translation in the direction of the unhooked cell end. In environments of higher viscosity flexuous, boring, and serpentine movements also take place. A few strains lack the ability to form hooked ends and have little or no translational motility (but do exhibit rotational motility).

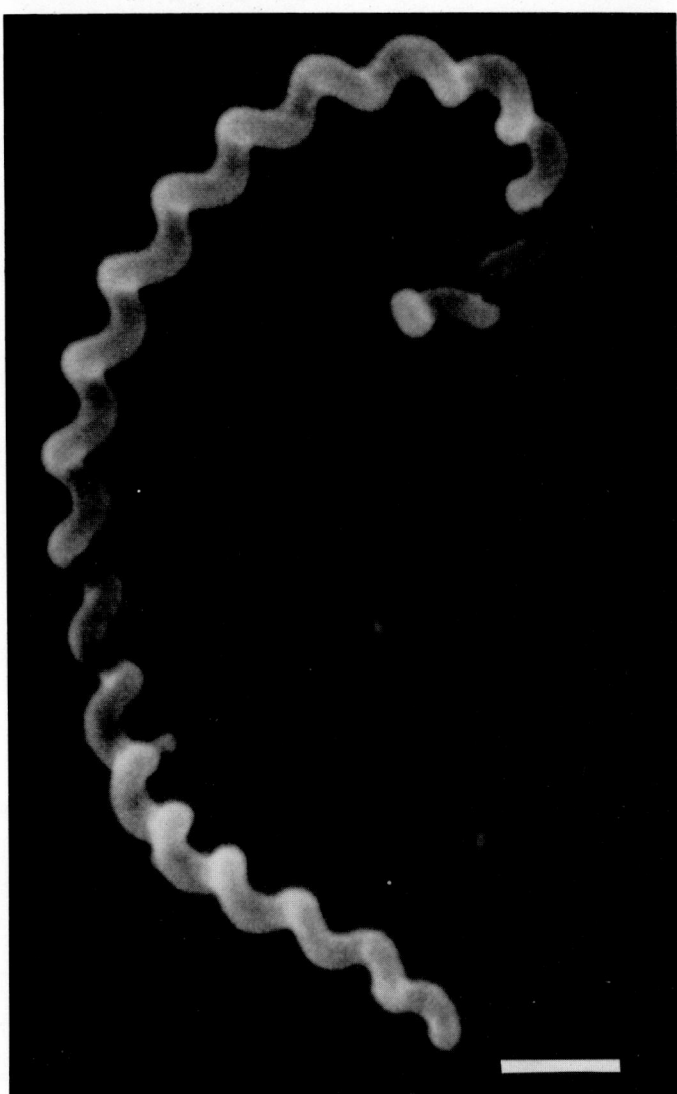

Figure 1.5. Scanning electron micrograph of *Leptospira interrogans* serovar *canicola* strain Moulton. *Bar* = 0.5 μm.

The peptidoglycan contains the diamino acid α,ε-diaminopimelic acid (Azuma et al., 1975).

Leptospira cells have the same basic morphological features as described for the order *Spirochaetales*. All species have 2 periplasmic flagella (which have also been termed endoflagella or axial fibrils) which rarely overlap in the central region of the cell. The basal bodies of the periplasmic flagella are similar to those of the flagella of Gram-negative bacteria with the exception of those of *"Leptospira illini"* (see Species Incertae Sedis) which are similar to the basal bodies of Gram-positive bacteria. In *"L. illini"* a bundle of cytoplasmic tubules (diameter ~7 nm) extends from each region of the insertion of the periplasmic flagellum and are approximately equivalent to the periplasmic flagellum in length. These cytoplasmic tubules have not been observed on other leptospires (Hovind-Hougen, 1976).

Nonpigmented diffuse to discrete subsurface colonies are commonly formed on 1% agar (Cox and Larson, 1957) and clear to turbid surface colonies occur on 2% agar. Some strains may grow as both surface and subsurface colonies on 1% or 2% agar (Wood et al., 1981; Wood and Johnson, personal communication). Colonial morphology has not been a useful differentiating characteristic to date.

The nutritional requirements of *Leptospira* are relatively simple (Johnson, 1981). Organic compounds required are long chain fatty acids (15 carbon or greater) and the vitamins thiamine (B_1) and cyanocobalamin (B_{12}). Fatty acids are the major carbon and energy source. In addition, they serve as the source of cellular fatty acids because *Leptospira* cannot synthesize fatty acids *de novo*. Some pathogenic strains require both saturated and unsaturated fatty acids for growth while others require only saturated fatty acids. Due to the inherent toxicity of free fatty acids, they are presented to the leptospires bound to albumin or in a detoxified esterified form. Neither carbohydrates nor amino acids are suitable sources of carbon and energy. Ammonium salts, but not amino acids provide an effective source of nitrogen (Johnson, 1981). The initiation of the growth of parasitic leptospires is enhanced by pyruvate, a nonessential nutrient (Johnson et al., 1973). Purines, but not pyrimidines are incorporated by leptospires. Because of this, leptospires are resistant to the pyrimidine analogue, 5'-fluorouracil, which inhibits the growth of most other bacteria, and this compound is used in the formulation of a selective isolation medium (Johnson and Rogers, 1964a).

Media used for the cultivation of leptospires are those enriched with rabbit serum (Fletcher, 1928; Stuart, 1946) or bovine serum albumin (Ellinghausen and McCullough, 1965; Johnson and Harris, 1967) and those that are protein free (Shenberg, 1967; Bey and Johnson, 1978).

Leptospires metabolize fatty acids by beta oxidation and the metabolic end products are acetate and carbon dioxide. Enzymes of the citric acid cycle, glycolytic and pentose have been demonstrated (Baseman and Cox, 1969a). Terminal electron acceptors are cytochromes *a*, *c*, and c_1, (Baseman and Cox, 1969b). Catalase (Faine, 1959) and peroxidase (Corin et al., 1978) have been described.

Seven distinct genetic groups have been identified in the genus *Leptospira* (Brendle et al., 1974). Neither bacteriophages nor plasmids have been isolated from leptospires. Leptospires are not known to produce bacteriocins.

The antigenic composition of leptospires is used for taxonomic purposes. The basic taxon is the serovar. The standard method for determining the serovars is the microscopic agglutination test, with cross-agglutinin absorption (Alexander, 1980b). For convenience, antigenically related serovars are organized into serogroups.

In vitro, leptospires are sensitive to many antibiotics. Penicillin, streptomycin, and tetracycline are commonly used for the treatment of leptospirosis.

Serovars of *L. interrogans* are the cause of leptospirosis, a zoonosis primarily infecting wild and domestic animals, which may act as reservoirs. Humans are accidental hosts and the clinical manifestations vary from an influenza-like illness to a severe icteric form. Leptospires and leptospirosis occur throughout the world.

Leptospira interrogans strains are parasitic or pathogenic for animals and man. The mechanism of pathogenicity is not known. The proximal convoluted tubule of the kidney is the natural habitat for these spirochetes. Multiplication occurs at this site and leptospires are shed in the urine. The primary mode of transmission of leptospirosis is by contact with the urine of leptospiruric animals directly or through contact with urine-contaminated soil and water. The reservoirs of *L. interrogans* are rodents and other feral or domestic animals. *Leptospira biflexa* strains are free living and can be found in moist soil and fresh surface waters. They are frequently referred to as "water leptospires" even though they survive better in the moist soil environment (Henry and Johnson, 1978). Several halophilic strains have been isolated from seawater (Cinco et al., 1975). None of the *L. biflexa* strains have been shown to be infective for experimental animals.

Isolation of Leptospires

*Isolation of **L. interrogans** strains (pathogenic leptospires).* The specimen to be collected for the isolation of *L. interrogans* from the infected host will depend on the phase of the disease. Leptospires are present in the blood and spinal fluid during the leptospiremic phase of the disease which occurs for approximately 1 week after the onset of symptoms. Blood specimens treated with heparin or sodium oxalate as anticoagulant are best collected prior to antimicrobial therapy.

After the first week of the disease, leptospires may be isolated from the urine. Leptospiruria usually lasts from a few weeks to months in man and domestic animals, respectively. Tissues (kidney, liver) may be cultured for leptospires at necropsy. Because clinical specimens submitted for isolation of leptospires may be contaminated with other bacteria, 5'-fluorouracil (100–200 μg/ml) may be incorporated into the isolation medium to control the growth of these contaminants.

Leptospires can also be isolated from contaminated materials by injecting the material intraperitoneally into weanling hamsters or guinea pigs and culturing the heart blood or kidneys. This method is used to isolate pathogenic leptospires from soil and water (Alexander, 1980a, Sulzer and Jones, 1976).

Isolation of **L. biflexa** *strains (free living leptospires). Leptospira biflexa* strains are present in surface waters and moist soil. Water samples are passed through 0.45- or 0.22-μm pore filters and the filtrate inoculated into semisolid culture medium containing 5'-fluorouracil. When needed, leptospires can be separated from other bacteria by placing a drop of the filtrate in the center of a Petri dish containing medium with 1% agar and 5'-fluorouracil. Leptospires readily migrate through the agar while most other bacteria remain at the site of inoculation. Leptospiral growth is visualized as a veil of growth extending toward the periphery of the plate.

Soil samples (25 g) are suspended in 100 ml of sterile distilled water, the suspension allowed to settle and the upper portion centrifuged at 800 × g for 5 min. The supernatant is processed in the same manner as the water sample (Henry and Johnson, 1978).

Maintenance Procedures

Screw-capped tubes containing semisolid media (0.2% agar) are commonly used for the maintenance of stock cultures. They are stored at 30°C or room temperature and transferred every 2–6 months.

Serovars such as *icterohaemorrhagiae* tend to decrease in virulence and immunogenicity after continuous transfer in media. Maximal virulence and immunogenicity of the pathogenic leptospires can be maintained by animal passage.

For long term preservation, cryoprotective agents such as 10% glycerol or dimethyl sulfoxide are added to the cultures which are then maintained at liquid nitrogen temperatures (Alexander et al., 1972). Lyophilization generally has not been a successful method for the preservation of leptospires.

Differentiation of the genus **Leptospira** *from other genera*

Leptospira strains are readily differentiated from other morphologically similar spirochetes by being obligately aerobic, tightly coiled, hook-ended organisms that require long fatty acids or alcohols as their source of carbon and energy. In addition, they are resistant to the inhibitory action of 5'-fluorouracil to which other spirochetes are sensitive. Also their cellular lipids do not contain glycolipids (Livermore and Johnson, 1974) and their cell walls contain the diamino acid, diaminopimelic acid rather than ornithine.

Taxonomic Comments

In the eighth edition of the *Manual*, only one species of *Leptospira* was listed. This has now been increased to two species, *L. interrogans* and *L. biflexa*. A third species, "*L. illini*", has an uncertain taxonomic status (see below) and is listed as a *species incertae sedis*.

Genetic studies involving DNA/DNA hybridization and mol% G + C content indicate that there should be at least seven species of *Leptospira* (Brendle et al., 1974). Genetic relatedness was generally correlated with antigenic composition; however, some antigenically related serovars were genetically dissimilar. This type of heterogeneity has also been observed with cross-immunity studies (Plesko and Hlavata, 1971) and the biological grouping of leptospires (Johnson and Harris, 1968) (see Table 1.15). Furthermore, DNA oligonucleotide analyses point to a greater genetic diversity within the genus (Marshall et al., 1981). Not more than two species can be defined on the basis of the known phenotypic characteristics.

"*Leptospira illini*" was given the taxonomic status of *species incertae sedis* (Subcommittee on the Taxonomy of *Leptospira* ICSB, Munich, 1978). This leptospire is phenotypically similar to *L. biflexa* (see Table 1.15); however, it differs morphologically from *L. interrogans* and *L. biflexa* in possessing cytoplasmic tubules which are also present in treponemes but not in the borrelias or members of the genus *Spirochaeta*. Also the insertion organelles of the periplasmic flagella of "*L. illini*" are similar to those of Gram-positive bacteria, borrelias, treponemes, and species of *Spirochaeta* whereas those of other *Leptospires* resemble those of other Gram-negative bacteria (Hovind-Hougen, 1976). Culturally "*L. illini*" strains are the only leptospires that are able to grow in trypticase soy broth without the addition of serum or serum albumin (Hanson et al., 1974); however, they do have the same nutritional requirements as other leptospires, but are more resistant to free fatty acids (Steiner and Johnson, personal communciation). Additionally, "*L. illini*" strains have DNA with a mol% G + C content of 51.2–53, which is considerably higher than that of other leptospires (35.2–41.2) (Brendle et al., 1974).

The two species of leptospires are currently arranged into serovars based on their antigenic composition. This antigen analysis is accomplished by microscopic agglutination and cross-agglutinin absorption tests which require a considerable amount of expertise and is restricted largely to leptospirosis reference laboratories. An added complication to this serological grouping is the variable presence of heat-labile antigens in some strains of *Leptospira* (Borg-Peterson, 1971, 1974, Kmety, 1972). Kmety (1967) has identified serovars of *L. interrogans* based on the presence and absence of major and minor antigens by a method that requires selectively absorbed antisera.

Acknowledgments

We gratefully acknowledge the help of the members of the Subcommittee on the Taxonomy of *Leptospira* (ICSB) in defining the genus and species of *Leptospira*.

Further Readings

Canale-Parola, E. 1977. Physiology and evolution of spirochetes Bacteriol. Rev. *41:* 181–204.
Holt, S.C. 1978. Anatomy and chemistry of spirochetes. Microbiol. Rev. *42:* 114–160.
Johnson, R.C. (Editor). 1976. *The Biology of the Parasitic Spirochetes.* Academic Press, New York.
Johnson, R.C. 1977. The spirochetes. Annu. Rev. Microbiol. *31:* 89–106.
Johnson, R.C. 1981. Introduction to the spirochetes. In Starr, Stolp, Trüper, Balows and Schlegel (Editors), *The Prokaryotes, a Handbook on Habitats, Isolation and Identification of Bacteria.* Springer-Verlag, New York, pp. 533–537.
Johnson, R.C. 1981. Aerobic spirochetes: the genus *Leptospira.* In Starr, Stolp, Trüper, Balows and Schlegel (Editors), *The Prokaryotes, a Handbook on Habitats, Isolation and Identification of Bacteria.* Springer-Verlag, New York, pp. 582–591.
Smibert, R.M. 1973. Spirochaetales, a review. Crit. Rev. Microbiol. *2:* 491–552.

Differentiation and characteristics of **Leptospira** *species*

The differential characteristics of the two species of *Leptospira* are indicated in Table 1.14. Table 1.15 lists characteristics that separate members of the genus into five biological groups.

List of the species of the genus **Leptospira**

1. **Leptospira interrogans** (Stimson 1907) Wenyon 1926, 1281.[AL] (*Spirochaeta interrogans* Stimson 1907, 541.)

in.ter'ro.gans. M.L. part. adj. *interrogans* interrogation, here meaning shaped like a question mark.

Morphological, cultural and physiological characteristics are generally those of the genus *Leptospira*.

Cytochrome enzymes, catalase, peroxidase and enzymes of the citric acid cycle, glycolytic and pentose pathways and an acylcoenzyme A dehydrogenase for beta-oxidation of fatty acids have been demonstrated.

Growth usually inhibited by 8-azaguanine, 225 µg/ml and at 13°C. Some strains of specific serological types produce a hot-cold hemolysin specific for ruminant cells. Lipase is found in some but not all strains.

Table 1.14.

Characteristics which differentiate species of **Leptospira**[a]

Characteristic	1. *L. interrogans*	2. *L. biflexa*
Pathogenicity	+	−
Growth at 13°C	−	+
Inhibition of growth by 8-azaguanine (225 µg/ml)	+	−
Conversion of cells to spherical forms by 1 M NaCl[b]	+	−
Lipase[c]	d	+
Mol% G + C of DNA	35.3–39.9	38–41

[a] Symbols: +, all strains positive for characteristic; −, all strains negative; d, differs among strains.
[b] Conversion of 75% of washed cells to spherical forms within 2 h in 1 M NaCl at between 20–30°C.
[c] See Table 1.15.

Serologically heterogeneous as determined by cross-agglutination and agglutinin-adsorption tests. The basic taxon is a serovar. Approximately 180 different serovars have been reported. It has been convenient to assemble antigenically related serovars into serogroups (Table 1.16).

Parasitic in man and animals, manifesting frank to subclinical infections. Produce lethal to subclinical infections in a wide variety of animals. In naturally occurring infections, leptospires localize in the kidneys of their hosts, whence they may be shed in the urine.

Genetically, three different groups of parasitic strains have been described on the basis of DNA/DNA annealing tests. Between groups, strains share partial sequences of nucleic acids. The serological relationships do not necessarily correlate with genetic relationships, although serologically closely related strains appear to be genetically homologous. Genetic groups may contain serologically diverse serovars.

Type strain: ATCC 23581. This strain is also the reference strain for the serovar *icterohaemorrhagiae*.

2. **Leptospira biflexa** (Wolbach and Binger 1914) Noguchi 1918, 585.[AL] (*Spirochaeta biflexa* Wolbach and Binger 1914, 23.)

bi·fle'xa. L. adj. *bis* twice; L. part adj. *flexus* bent; M.L. adj. *biflexus* twice bent.

Morphological, cultural and physiological characteristics are generally those of the genus *Leptospira* except that strains can grow at 13°C.

Usually resistant to 8-azaguanine, 225 µg/ml. Some strains isolated from seawater are sodium dependent and will not grow at Na⁺ concentrations below 0.24 M. A hemolysin may be produced which acts on erythrocytes containing lecithin (e.g. rats and mice). All strains produce a lipase.

Serologically heterogeneous as determined by agglutination and cross-agglutinin adsorption tests. The basic taxon is the serovar. More than 60 different serovars have been reported. Antigenically related serovars have been assembled into serogroups for convenience (Table 1.17).

Table 1.15.

Differential characteristics of biological groups of leptospires[a]

Genetic Group	Representative Serovar and Strain	Mol% G + C	Pathogenicity	Conversion to Spherical Forms[b]	Periplasmic Flagellar Basal Bodies 1 ring	Periplasmic Flagellar Basal Bodies 2 rings	Cytoplasmic tubules	Lipase activity	13°C	8-Azaguanine	2,6-Diaminopurine	Trypticase Soy Broth	1–2% NaCl Required for Growth
I[d]	*bataviae* Van Tienen	35.3	+	+	−	+	−	+	−	−	−	−	−
II[d]	*javanica* Veldrat Bataviae 46	39.9	+	+	−	+	−	−	−	−	+	−	−
III[d]	*patoc* Patoc I	38.3	−	−	−	+	−	+	+	+	+	−	−
IV[d]	*codice* CDC	38.0	−	−	−	+	−	+	+	+	+	−	−
V[d]	*ranarum* Iowa City Frog	41.2	−	−	−	+	−	+	+	+	+	−	−
VI[c]	*illini* 3055	53.0	−	−	+	−	+	+	+	+	+	+	−
	muggia[e]		−					+			−		+

[a] For symbols see Table 1.14.
[b] Conversion of 75% of washed cells to spherical forms within 2 h in 1 M NaCl.
[c] Johnson and Harris (1968).
[d] Brendle et al. (1974).
[e] Cinco et al. (1975).

Table 1.16.

Serovars of Leptospira interrogans[a]

Serogroup	Serovar		Serogroup	Serovar	
Australis	*australis*	*lora*		*dakota*	*sarmin*
	bangkok	*muenchen*		*gem*	*smithi*
	bratislava	*nicaragua*		*icterohaemorrhagiae*	*tonkini*
	fugis	*peruviana*		*mankarso*	*weaveri*
	hawain	*ramusi*		*monymusk*	
	jalna		Javanica	*an hoa*	*poi*
Autumnalis	*alice*	*louisiana*		*ceylonica*	*rio*
	autumnalis	*mooris*		*coxi*	*sofia*
	bangkinang	*orleans*		*fluminense*	*sorex-jalna*
	bulgarica	*rachmati*		*javanica*	*vargonica*
	erinacei-auriti	*srebarna*		*menoni*	*waskurin*
	fort-bragg	*sumatrana*	Panama	*cristobali*	*panama*
	lanka	*tingo maria*		*mangus*	
Ballum	*arboreae*	*castellonis*	Pomona	*monjakov*	*proechimys*
	ballum			*mozdok*	*tropica*
Bataviae	*argentiniensis*	*claytoni*		*pomona*	
	balboa	*djatzi*	Pyrogenes	*abramis*	*manilae*
	bataviae	*kobbe*		*alexi*	*myocastoris*
	brasiliensis	*paidjan*		*biggis*	*princestown*
Butembo	*butembo*			*camlo*	*pyrogenes*
Canicola	*bafani*	*jonsis*		*guaratuba*	*robinsoni*
	benjamin	*kamituga*		*hamptoni*	*varela*
	bindjei	*malaya*		*kawali*	*zanoni*
	broomi	*portland-vere*	Sejroe	*balcanica*	*nyanza*
	canicola	*schueffneri*		*caribe*	*polonica*
	galtoni	*sumneri*		*dikkeni*	*recreo*
Celledoni	*celledoni*	*whitcombi*		*geyaweera*	*ricardi*
Cynopteri	*cynopteri*	*tingo mariensis*		*gorgas*	*roumanica*
Djasiman	*djasiman*	*sentot*		*guaricurus*	*rupa rupae*
	gurungi			*haemolytica*	*saxkoebing*
Grippotyphosa	*canalzonae*	*ratnapura*		*hardjo*	*sejroe*
	grippotyphosa	*valbuzzi*		*istrica*	*trinidad*
	muelleri	*vanderhoedeni*		*medanensis*	*wolffi*
Hebdomadis	*beye*	*kremastos*	Shermani	*babudieri*	*shermani*
	borincana	*maru*	Tarassovi	*atchafalaya*	*kisuba*
	georgia	*mini*		*atlantae*	*langati*
	goiano	*nona*		*bakeri*	*luis*
	hebdomadis	*perameles*		*bravo*	*moldaviae*
	jules	*szwajizak*		*carimagua*	*navet*
	kabura	*tabaquite*		*chagres*	*osetica*
	kambale	*worsfoldi*		*darien*	*rama*
Icterohaemorrhagiae	*birkini*	*mwogolo*		*gatuni*	*tarassovi*
	bog-vere	*naam*		*guidae*	*tunis*
	budapest	*ndahambukuje*		*kanana*	*vughia*
	copenhageni	*ndambari*		*kaup*	

[a] Provisional List.

Genetically, three different groups of strains have been demonstrated on the basis of DNA/DNA annealing tests. Between groups, strains share partial sequences of nucleic acids. No genetic homology with strains of *Leptospira interrogans*.

Strains are commonly found in natural freshwater, tap water and occasionally in seawater and animals. Not pathogenic in experimental animals commonly susceptible to *Leptospira interrogans*.

Type strain: ATCC 23582. This strain is also the reference strain for the serovar *patoc*.

Species Incertae Sedis

a. "*Leptospira illini*" Hanson, Tripathy, Evans and Alexander 1974, 355. (*Leptonema illini* Hovind-Hougen 1983, 439.[VP])

il.li'ni. M.L. gen. n. *illini* of Illinois; named after the state of Illinois, U.S.A. where it was isolated.

Morphological characteristics are generally those of genus *Leptospira* except that the basal bodies of the periplasmic flagella are analogous to those of the flagella of Gram-positive organisms. A bundle of cytoplasmic tubules, each with a diameter of about 7 nm, extends from each end of the cell from the region of the insertion points of the periplasmic flagella. These bundles are approximately the same length as the periplasmic flagella.

Metabolism is respiratory using fatty acids as a major source of carbon. Growth in media the same as for *Leptospira biflexa*. In addition, can grow on trypticase broth, unlike other leptospires.

Only one serovar described. Genetically unrelated to *Leptospira interrogans* and *Leptospira biflexa* by DNA/DNA annealing tests. The mol% G + C of the DNA is 53.

Nonpathogenic for hamsters, mice, gerbils, guinea pigs or cattle.

Type strain: NCTC 11301 (strain 3055 isolated from a bull in Illinois, U.S.A. in 1965 (Hanson, Tripathy, Evans and Alexander, 1974).

Table 1.17.
Serovars of **Leptospira biflexa**

Serogroup	Serovar		Serogroup	Serovar	
Abaete	*abaete*		Lazio	*lazio*	
Ancona	*ancona*		Malomirovo	*malomirovo*	
Andamana	*andamana*	*bovedo*	Maritza	*maritza*	*valderio*
Aurisina	*aurisina*	*farneti*	Muggia	*muggia*	
Babrich	*babrich*	*lugo*	Nomentano	*nomentano*	
Basovizza	*basovizza*	*sangiusto*	Ondina	*ondina*	
Bessemans	*bessemans*		Orvenco	*orvenco*	
Botanica	*botanica*	*montefiascone*	Parapatan	*parapatan*	
Cadore	*cadore*		Percedol	*percedol*	
Camtchia	*camtchia*		Poona	*poona*	
Cau	*cau*	*khera*	Pulpudeva	*canela*	*jequitaia*
Codice	*cdc*	*piraja*		*fons*	*pulpudeva*
Dindio	*dindio*		Ranarum	*evansi*	*ranarum*
Doberdo	*doberdo*	*isolasacra*	Semaranga	*montevalerio*	*sao paulo*
	drahovce	*rupino*		*patoc*	*semaranga*
	eleven		Sidonia	*sidonia*	
Garcia	*garcia*		Sobradinho	*sobradinho*	
Holland	*acquamarcia*	*roma*	Tevere	*tevere*	
	holland	*tredici*	Thracia	*thracia*	
	lucaia	*wild*	Tororo	*tororo*	
	piatan	*zoo*	Udine	*udine*	
Iran	*iran*		Vinzent	*vinzent*	
Khoshamian	*compton*	*khoshamian*			

OTHER ORGANISMS

Hindgut Spirochetes of Termites and *Cryptocercus punctulatus*

JOHN A. BREZNAK

The hindgut microbiota of termites (Table 1.18) and of the wood-eating cockroach *Cryptocercus punctulatus* Scudder (Grimstone, 1963) includes an abundance of motile, spiral-shaped organisms (Breznak, 1973). Based on light microscopy observations of living or stained specimens, early workers referred to the microbes as spirilla (Leidy, 1877) or vibrios (Leidy, 1881), assigned them to currently recognized spirochete genera (see Taxonomic Comments), or simply termed them "spirochetes" (Damon, 1926). However, not until recently have ultrastructural studies shown some of the organisms to be true spirochetes. Hindgut spirochetes occur free in the gut fluid as well as attached to the surface of polymastigote and hypermastigote protozoa which may also inhabit the hindgut. Unfortunately, none have been isolated and grown in pure culture.

The size of free hindgut spirochetes ranges from about 0.2 μm in diameter × 3 μm long (Breznak and Pankratz, 1977) to as large as 1.0 μm in diameter × 100 μm long (Hollande and Gharagozlou, 1967). Likewise, the number of periplasmic flagella ranges from a few per cell to as many as 100 or more in the larger forms (To et al., 1978). However, the multitudinous periplasmic flagella in the large forms do not generally occur in a tight bundle as in *Cristispira*. It is common for termites to possess numerous morphotypes of spirochetes in their hindgut, distinguishable on the basis of size as well as wavelength and wave amplitude of the primary coils. To et al. (1980) were able to discriminate more than 15 morphological types of spirochetes present in the hindgut of *Pterotermes occidentis*. In fresh samples of termite hindgut fluid, the spirochetes exhibit vigorous spinning, lashing, and locomotory movements typical of this group of organisms (Canale-Parola, 1978).

Some hindgut spirochetes possess distinctive ultrastructural features which include (a) a crenulated outer sheath (Hollande and Gharagozlou,

1967; Fig. 1.6), (b) a helicoidal groove or "sillon" in the outer sheath which brings it in close apposition to the protoplasmic cylinder (Hollande and Gharagozlou, 1967; Gharagozlou, 1968; Fig. 1.6 and 1.7), (c) an extension of an electron-dense coat on the inner surface of the outer sheath so as to partially envelop the surface of the protoplasmic cylinder (Gharagozlou, 1968; Fig. 1.7), and (d) microtubule-like structures 20 nm in diameter which are aligned longitudinally within the protoplasmic cylinder (Margulis et al., 1979). The latter structures may be responsible for reactivity of certain of the spirochetes to antitubulin antibody (Margulis et al., 1978).

Virtually nothing is known about the physiology of free hindgut spirochetes, although casual observations suggest they are anaerobic since they usually lose their motility and morphological integrity when exposed to air. This is commonly seen microscopically when spirochetes in wet mount preparations swim near air bubbles or the edge of the coverslip. Anaerobicity of the organisms would be a property consistent with that of their natural habitat which appears to contain little or no O_2 and be of low E_0' (Bignell and Anderson, 1980; Veivers et al., 1980). Limited success at initial cultivation of the spirochetes was achieved by Ghidini and Archetti (1939) who used a serum-containing agar medium. Although the spirochetes could not be completely separated from contaminating organisms, it appeared that their optimum temperature for growth was 25–28°C and the optimum pH was 7.2–7.7.

Some hindgut spirochetes occur as ectosymbionts attached end on to the surface of certain flagellate protozoa found only in the lower termites (i.e. families Mastotermitidae, Kalotermitidae, Hodotermitidae, Serritermitidae, and Rhinotermitidae) and *Cryptocercus punctulatus*. Depending on the particular species of protozoan, ectosymbiotic spirochetes may be uniformly distributed over the surface or localized in specific regions (Kirby, 1941; Ball, 1969). Ultrastructural studies

Table 1.18.

Occurrence of spirochetes or spirochete-like organisms in the hindgut of termites

Termite	Geographical Source	EM[a]	Reference
Bifiditermes sp.	New South Wales, Australia	−	To et al., 1978.
Calcaritermes (Kalotermes) nigriceps (Emerson)	British Guinea	−	Damon, 1926.
Ceratokalotermes spoliator (Hill)	Canberra, Australia	−	To et al., 1978.
Coptotermes acinaciformis (Froggatt)	Australia	−	To et al., 1978.
C. formosanus Shiraki	Lake Charles, LA. and Honolulu, HA., U.S.A.	+	Breznak and Pankratz, 1977; To et al., 1978.
C. lacteus (Froggatt)	Sydney, Australia	−	Eutick et al., 1978.
Cryptotermes brevis (Walker)	Puerto Rico, and Miami, FL U.S.A.	−	Damon, 1926; To et al., 1978.
C. cavifrons Banks	South Florida, U.S.A.	−	To et al., 1978.
C. gearyi (Hill)	S. Queensland, Australia	−	To et al., 1978.
Glyptotermes iridipennis Froggatt	Melbourne, Australia	−	Duboscq and Grassé, 1927.
G. neotuberculatus (Hill)	Canberra, Australia	−	To et al., 1978.
Heterotermes aureus (Snyder)	Tucson, AZ., U.S.A.	−	To et al., 1978.
Incisitermes milleri (Emerson)	Florida Keys, U.S.A.	−	To et al., 1978.
I. minor (Hagen)	Tucson, AZ., U.S.A.	−	To et al., 1978.
I. schwarzi (Banks)	South Florida, U.S.A.	+	Damon, 1926; To et al., 1978.
Kalotermes approximatus Synder	North Florida, U.S.A.	−	To et al., 1978.
K. banksiae Hill	Cape Bonran, Victoria, Australia.	−	To et al., 1978.
K. flavicollis (Fabricius)	Barcelona and Elche, Alicante, Spain; Banyuls-sur-Mer, France.	+	Grassi and Sandias, 1896–1897; Gharagozlou, 1968; To et al., 1978.
Leucotermes lucifugus Sjöstedt	Bastia, Corsica	−	Hollande, 1922
L. tenuis Snyder	British Guinea	−	Damon, 1926.
Marginitermes hubbardi (Banks)	Tucson, AZ., U.S.A.	−	To et al., 1978.
Mastotermes darwiniensis Froggatt	Australia (various sites)	+1	Cleveland and Grimstone, 1964: To et al., 1978.
Nasutitermes costalis (Holmgren)	Paquera, Puerto Rico	−	To et al., 1978.
N. exitiosus (Hill)	Sydney, Australia	−	Eutick et al., 1978.
N. morio Banks	Puerto Rico	−	Damon, 1926.
Neotermes castaneus (Burmeister)	South Florida, U.S.A.	−	To et al., 1978.
N. insularis (White)	Australia	−	To et al., 1978.
N. jouteli (Banks)	South Florida, U.S.A.	−	To et al., 1978.
Paraneotermes simplicicornis (Banks)	California and Tucson, AZ., U.S.A.	−	To et al., 1978.
Porotermes adamsoni (Froggatt)	Australia	−	To et al., 1978.
Postelectrotermes (Kalotermes) praecox (Hagen)	Island of Madeira, Portugal	+	Hollande and Gharagozlou, 1967.
P. militaris (Desneux)	Paradeniya, Sri Lanka (Ceylon)	−	Dobell, 1910, 1912.
Pterotermes occidentis (Walker)	Sonoran Desert, AZ., U.S.A.	+	To et al., 1978, 1980.
Reticulitermes flavipes (Kollar)	New Jersey; Maryland; Janesville, WI.; Spring Arbor and Dansville, MI.; Gulfport, MS.; Naples, FL.; and Woods Hole, MA. (all in U.S.A.).	+	Leidy, 1877, 1881; Damon, 1926; Breznak and Pankratz, 1977; To et al., 1978; Smith and Arnott, 1974; Smith, Buhse and Stamler, 1975; Smith, Stamler, and Buhse, 1975.
R. hageni Banks	Maryland, U.S.A.	−	Damon, 1926.
R. hesperus Banks	San Diego, CA., U.S.A.	+	To et al., 1978.
R. lucifugus (Rossi)	Japan and Italy	−	von Prowazek, 1910; Ghidini and Archetti, 1939.
R. tibialis Banks	Colorado, U.S.A.	+	Bloodgood and Fitzharris, 1976
R. virginicus (Banks)	Maryland, U.S.A.	−	Damon, 1926.
Zootermopsis angusticollis (Hagen)	California, U.S.A.	−	Damon, 1926; To et al., 1978.
Z. nevadensis (Hagen)	Oregon and other sites in western U.S.A.	−	Damon, 1926; To et al., 1978.

[a] Verification of spirochetal ultrastructure by electron microscopy (+); or no verification made (−).

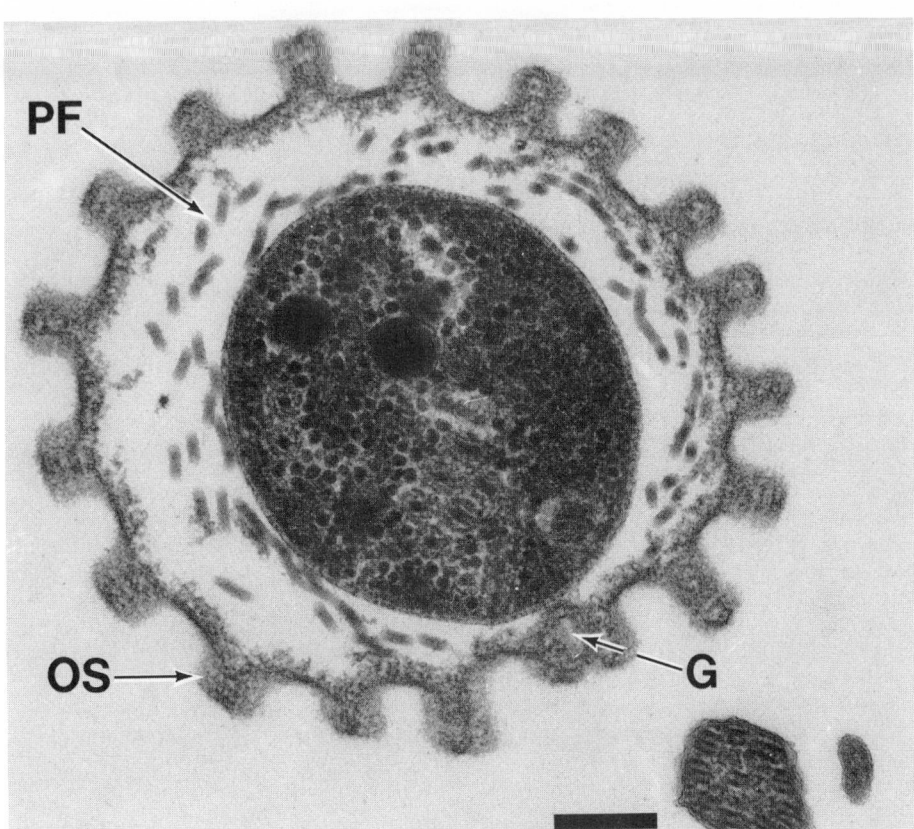

Figures 1.6 and 1.7. Transmission electron micrographs of transverse sections of spirochetes present in the hindgut of *Reticulitermes flavipes*. New genera have been proposed for spirochetes of the morphology shown in Figure 1.6 (Hollande and Gharagozlou, 1967) and in Figure 1.7 (Gharagozlou, 1968) (see Taxonomic Comments). *OS*, outer sheath; *G*, groove; *PF*, periplasmic flagella; *marker bars* = 0.2 μm (Prepared by H. S. Pankratz and J. A. Breznak).

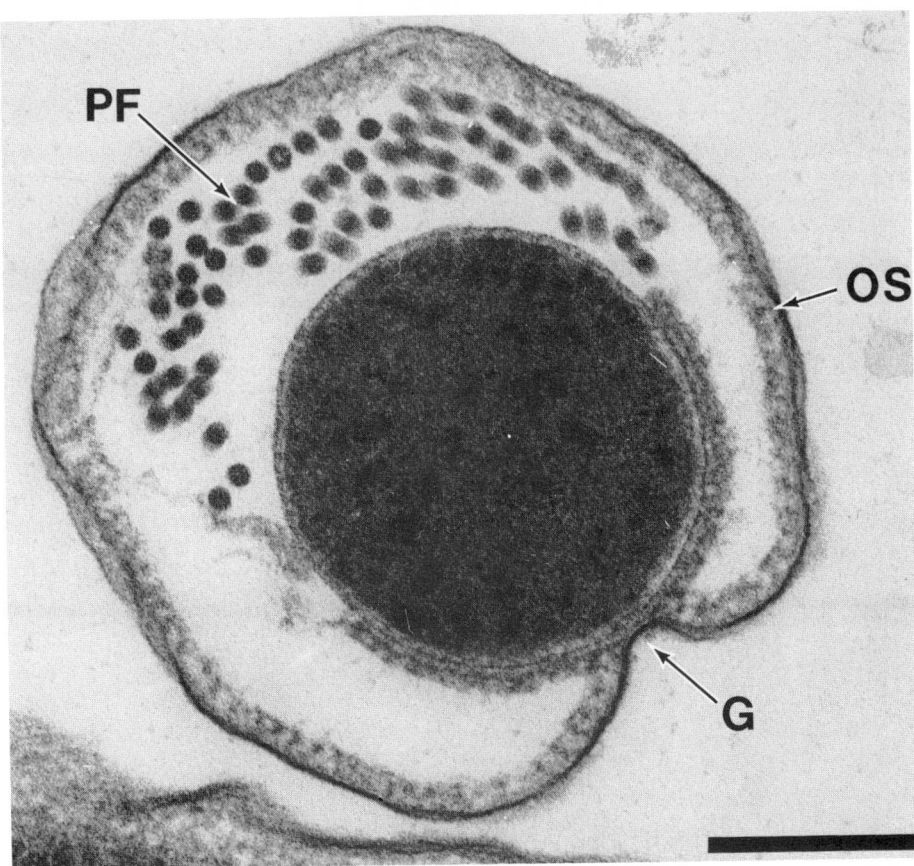

Figure 1.7.

suggest that attachment of some spirochetes is accomplished through a structural modification of an end of the spirochetal cell, with little or no complementary modification of the protozoan plasma membrane. This appears to be the case for spirochetes attached to *Pyrsonympha* (from *Reticulitermes flavipes* and *R. tibialis*) and *Barbulanympha* (from *C. punctulatus*) wherein the spirochetal attachment organelle appears as a flattened or rounded structure or a noselike extension of the cell pole (Bloodgood et al., 1974; Bloodgood and Fitzharris, 1976). Other studies indicate that ectosymbiotic spirochetes are accommodated through structural modification of the protozoa alone. For example, in *Mixotricha paradoxa* (from *Mastotermes darwiniensis*) the plasma membrane is modified to form bracket-like elements (Cleveland and Grimstone, 1964), whereas in polymastigotes from *R. flavipes* the protozoan membrane is differentiated into screwlike structures into which spirochetes engage (Smith, Buhse and Stamler, 1975; Smith, Stamler and Buhse, 1975). Still other studies indicate that both the protozoan plasma membrane and the spirochetal poles are modified to form an attachment complex (Smith and Arnott, 1974). Besides distinct attachment to protozoa, free swimming spirochetes are often seen to course in and out of the flagellar vestment of the eucaryotes: the cause of such behavior is unknown.

Occasionally, motile spirochetes are observed within the cytoplasm of hindgut protozoa (Kirby, 1941; Margulis et al., 1979; To et al., 1980). However, it is not clear whether such intracellular spirochetes coexist stably with the protozoa (i.e. are true endosymbionts) or were merely endocytosed into food vacuoles.

Aside from the classic study of Cleveland and Grimstone (1964), in which it was shown that adherent spirochetes serve a locomotory function for *M. paradoxa*, nothing is known about the role of spirochetes attached to other protozoa and for which a propulsive function is doubtful. Nevertheless, this has remained a topic of much speculation (Breznak, 1973; Bloodgood and Fitzharris, 1976), which includes the notion that eucaryotic flagella and cilia evolved from ectosymbiotic spirochetes (Margulis et al., 1979). Similarly, little is known about the function of free hindgut spirochetes (Breznak, 1973), although there is

no reason to believe they are pathogenic to their hosts. They do not invade the hindgut epithelium, and the insects harboring them appear vigorous and healthy. In fact, there is some evidence to suggest they may benefit the host. Eutick et al. (1978) observed a reduced life span of *Nasutitermes exitiosus* termites when spirochete-like organisms were eliminated from the hindgut.

Taxonomic Comments

Spirochete-like organisms in termite hindguts have been assigned by some workers to recognized spirochete genera, for example: "*Spirochaeta termitis*" (Dobell, 1910), "*S. minei*" (von Prowazek, 1910), "*S. leucotermitis*" (Hollande, 1922), and "*S. staphylina*" (Ghidini and Archetti, 1939); "*Treponema termitis*" and "*T. minei*" (Dobell, 1912); and "*Cristispira termitis*" (Hollande, 1922). However, these names have not been validly published or revived in accord with rules 27 and 28a of the *International Code of Nomenclature of Bacteria* (Lapage et al., 1975). Moreover, although the general morphology of the organisms was similar to that of spirochetes, the original assignments were made without evidence that the organisms possessed ultrastructural features of true spirochetes (i.e. protoplasmic cylinder, periplasmic flagella, and outer sheath). By contrast, based on electron microscopic studies, other workers have proposed new generic and specific epithets for true spirochetes of termite hindguts. These epithets include "*Pillotina calotermitidis*" (Hollande and Gharagozlou, 1967; Fig. 1.6), "*Diplocalyx calotermitidis*" (Gharagozlou, 1968; Fig.1.7), "*Hollandina pterotermitidis*" (To et al., 1978), and "*Clevelandina*" (Margulis et al., 1979). However, none of these names has been validly published either. Because hindgut spirochetes of termites and *C. punctulatus* have not yet been isolated in pure culture, it is debatable whether any of the aforementioned names should be revived. It seems desirable to postpone further efforts at nomenclature until pure cultures are obtained and the extent of physiological and morphological variation of the organisms can be assessed, as well as their host range and natural relationships to representatives of existant spirochete genera.

Important Notes for Users of this Edition

1. Always read both generic and species descriptions because characters listed in the generic description are not usually listed in the species descriptions.

2. Unless otherwise indicated in footnotes to tables, the meanings of symbols are as follows:

+ 90% or more of strains are positive

− 90% or more of strains are negative

d 11–89% of strains are positive

v strain instability (*not* equivalent to "d")

D different reactions in different taxa (species of a genus or genera of a family)

3. All other symbols are defined in footnotes to tables.

SECTION 2

Aerobic/Microaerophilic, Motile, Helical/Vibrioid Gram-Negative Bacteria*

Noel R. Krieg

Helical or vibrioid cells, having from less than one complete helical turn to many turns. Many species form intracellular poly-β-hydroxybutyrate. Many possess a "polar membrane" attached to the inside of the cytoplasmic membrane by barlike links; this generally occurs underlying the region of flagellar insertion. Many of the organisms become converted into thin-walled "coccoid bodies" in old cultures. One genus (*Azospirillum*) produces enlarged, encapsulated forms that may superficially resemble cysts. Gram-negative; one genus (*Azospirillum*) may be Gram-variable. **Motile in broth cultures by polar flagella. Swim in straight lines with a characteristic corkscrew-like motion. Aerobic or microaerophilic, having a respiratory type of metabolism with oxygen as the normal electron acceptor.** Growth under anaerobic conditions is absent or very poor, or may occur only with special electron acceptors such as nitrate or fumarate. Most members of the group are incapable of catabolizing carbohydrates, but some can oxidize a limited variety. A few species can acidify certain sugar media under anaerobic conditions, but they grow very poorly and much prefer aerobic conditions. Typically **oxidase-positive. Indole is not produced.** Chemoorganotrophic; however, some species may grow autotrophically in a mixture of $H_2 + CO_2 + O_2$. Some can fix N_2 under microaerobic conditions. Occur in soil, fresh water or marine environments, within plant roots, or in the reproductive organs, intestinal tract and oral cavity of man and animals. Some are predatory on other microorganisms. The mol% G + C of the DNA ranges from 30 to 70.

Further Comments

In the eighth of *Bergey's Manual*, two genera—*Spirillum* (now divided into the three genera *Spirillum*, *Aquaspirillum* and *Oceanospirillum*) and *Campylobacter*—were included in the family *Spirillaceae* Ehrenberg 1832. This was merely a family of convenience, since the genetic relationships between these genera were unknown. Recent results from 16S rRNA oligonucleotide cataloging Woese et al., 1982 and DNA/rRNA hybridization studies (DeSmedt et al., 1980) have provided some useful preliminary indications based on a limited number of species. *Aquaspirillum itersonii* is phylogenetically related to *Aquaspirillum polymorphum*, to the genus *Azospirillum*, and to the phototrophic species *Rhodospirillum rubrum*, but has little relatedness to *Aquaspirillum gracile*, *Aquaspirillum serpens*, or *Spirillum volutans*. *A. serpens* is related to *Rhodospirillum tenue* and, to a lesser extent, to *S. volutans* and *A. gracile*. *A. gracile* is related to *Rhodopseudomonas*

gelatinosa, and to a lesser extent to *A. serpens*, *R. tenue* and *S. volutans*. The *Oceanospirillum* species so far studied do not appear to be closely related to *Aquaspirillum*, *Azospirillum*, or *Spirillum*. It is becoming evident that there is considerable phylogenetic heterogeneity within the genus *Aquaspirillum* and the genus *Oceanospirillum* and, therefore, these genera may not be genetically valid taxa. To date, nothing is known of the phylogenetic relationships of *Campylobacter* and *Bdellovibrio* to other helical organisms. Although some information is available for the other genera, phylogenetic studies are presently not comprehensive enough to permit the grouping of these genera (which themselves may not be genetically valid taxa) at a family level.

From a purely practical point of view, it is difficult to decide which genera should presently be included in the family *Spirillaceae*. Although the genera *Aquaspirillum*, *Spirillum*, *Azospirillum*, *Oceanospirillum*, *Campylobacter* and *Bdellovibrio* exhibit certain similarities to one another, they also exhibit a sufficient number of differences so that one would be hard pressed to arrive at a satisfactory useful definition of a family that could include all six genera. Yet to exclude one or more of the genera would seem to be unreasonable and arbitrary. For now, the best policy may be not to assign the genera to any family, but rather to consider them merely as forming a loose assemblage of taxa that exhibit some morphological and/or physiological similarities. It is anticipated that future studies of genetic relatedness will help to provide a more satisfactory and stable arrangement.

One difficulty in dealing with helical/vibrioid organisms is the great taxonomic weight placed on the cell morphology. The importance of a helical/vibrioid morphology loses some of its force when one considers (a) the finding of straight rod variants in certain spirillum species, and (b) certain naturally occurring straight rods (*Aquaspirillum fasciculus* and *Serpens flexibilis*†) which have many of the typical characteristics of aquaspirilla, including bipolar tufts of flagella and an inability to catabolize carbohydrates. It is difficult to know just where to place organisms such as *A. fasciculus* or *S. flexibilis* on a practical basis, although *S. flexibilis* is probably related to the family *Pseudomonadaceae* (see Taxonomic Comments, the genus *Serpens*). It is also difficult at present to know whether the difference between rods that are curved in one plane (such as certain pseudomonads) vs. those that are curved with a twist (vibrioid organisms, such as *Aquaspirillum delicatum*) is an important difference. Studies of genetic relatedness and of the determinants of shape and form can undoubtedly resolve many of these problems.

---

* One species of **straight rods** is also included in this section; see Partial Key to Section 2.
† See Section 4.

Partial key to section 2

I. Helical or vibrioid cells.
   A. Not normally associated with man or animals. Not predacious on other bacteria or on algae.
      1. Fresh water habitat.
         a. Helical or vibrioid cells 0.2–1.4 μm in diameter. Typically possess bipolar tufts of flagella; some have a single flagellum at one or both poles. Aerobic to microaerophilic. Inhibited by 3% NaCl. Typically do not catabolize carbohydrates; some catabolize a very limited variety. The mol% G + C of the DNA is 49–66.
                    Genus *Aquaspirillum*, p. 72
         b. Helical cells 1.4–1.7 μm in diameter and up to 60 μm in length. Possess large bipolar flagellar fascicles easily visible by phase-contrast microscopy. Microaerophilic. Carbohydrates not catabolized. The mol% G + C of the DNA is ~38.
                    Genus *Spirillum*, p. 90
      2. Occur in or on the roots of plants and free living in soil. Vibrioid cells ~1.0 μm in diameter. In liquid media they possess a single polar flagellum; on solid media they additionally possess numerous lateral flagella of shorter wavelength. Fix nitrogen under microaerobic conditions; with a source of fixed nitrogen they grow as aerobes. Grow well on malate, succinate, lactate or pyruvate. Fructose and certain other monosaccharides are catabolized; disaccharides are not attacked. The mol% G + C of the DNA is ~70.
                    Genus *Azospirillum*, p. 94
      3. Marine habitat (coastal waters). Seawater or NaCl required for growth. Helical cells 0.3–1.4 μm in diameter. Typically possess bipolar tufts of flagella; some have a single flagellum at each pole. Aerobic. Carbohydrates are not catabolized. The mol% G + C of the DNA is 42–51.
                    Genus *Oceanospirillum*, p. 104
   B. Associated with man or animals. Not predacious on other bacteria or on algae.
      1. Found in the reproductive organs, intestinal tract and oral cavity of man and animals. Vibrioid or helical cells 0.2–0.5 μm in diameter. Possess a single flagellum at one or both poles. Microaerophilic. Carbohydrates are not catabolized. The mol% G + C of the DNA is 30–38.
                    Genus *Campylobacter*, p. 111
      2. Causative agent of one form of rat-bite fever in man. Occur in patient's blood. Natural parasites of rats. Helical cells ~0.2 μm in diameter with one or more flagella at each pole. Probably have never been successfully cultured in artificial media.
                    "*Spirillum minus*" (see Species Incertae Sedis, the genus *Aquaspirillum*), p. 89
      3. Causative agent of a diphtheroid stomatitis in chickens. Helical cells ~1 μm in diameter. Motile by a single polar flagellum at each pole. Have not been cultured in artificial media.
                    "*Spirillum pulli*" (see Species Incertae Sedis, the genus *Aquaspirillum*), p. 89
      4. Found in the intestinal contents of tadpoles. Enormous helical cells 1.8–4.8 μm in diameter and 40–100 μm in length. Contain endospore-like structures. Motile by an unknown mechanism. Possibly anaerobic. Have never been isolated in artificial media.
         Genus "*Sporospirillum*" (see Genus Incertae Sedis, in the chapter on the genus *Aquaspirillum*), p. 89
   C. Not associated with man or animals. Predacious on other bacteria or on algae.
      1. Predacious on other Gram-negative bacteria. Occur in soil, sewage, freshwater and marine habitats. Small vibrioid cells 0.2–0.5 μm in diameter. Motile by a single sheathed polar flagellum. Have a morphologically and physiologically biphasic growth cycle, alternating between a nongrowing predatory phase and an intracellular reproductive growth phase. Aerobic. The mol% G + C of the DNA is 33–51.5.
                    Genus *Bdellovibrio*, p. 118
      2. Predacious on the eukaryotic alga *Chlorella*. Vibrioid cells 0.3 μm in diameter. Motile by a single nonsheathed flagellum. Do not penetrate the host cell. The mol% G + C of the DNA is 50.
                    Genus *Vampirovibrio*, p. 124
II. Straight rods, possessing bipolar tufts of flagella. Carbohydrates are not catabolized.
   A. Cell diameter 0.7–0.9 μm. Bipolar flagellar fascicles easily visible by phase-contrast microscopy and can coil up into loops. Exhibit an ineffectual "floundering" type of motility in ordinary media, but can swim in straight lines in media of high viscosity. Nitrogen is fixed under microaerobic conditions. Grow aerobically in complex media. Occur in stagnant fresh water. The mol% G + C of the DNA is 62–65.
      Aquaspirillum fasciculus (see the genus *Aquaspirillum*), p. 86

Genus **Aquaspirillum** *Hylemon, Wells, Krieg and Jannasch 1973, 361*[AL]·

NOEL R. KRIEG

Aq.ua.spi.ril'lum. L. *aqua* water; Gr. n. *spira* a spiral; M.L. dim. neut. n. *spirillum* a small spiral; *Aquaspirillum* a small water spiral.

Rigid, generally helical cells, 0.2–1.4 μm in diameter; however, one species is vibrioid, another consists of straight rods. A **polar membrane** underlies the cytoplasmic membrane at the cell poles in all species so far examined for this characteristic by electron microscopy. Intracellular **poly-β-hydroxybutyrate** is usually formed. Some species form thin-walled **coccoid bodies** which predominate in old

---

cultures. Gram-negative. **Motile by polar flagella,** generally **bipolar tufts;** one species is monotrichous, others have a single flagellum at each pole. **Aerobic** to microaerophilic, having a respiratory type of metabolism with oxygen as the terminal electron acceptor; a few species can grow anaerobically with nitrate. The optimum temperature for most species is 30–32°C. Chemoorganotrophic; however, one species is a facultative hydrogen autotroph. **Oxidase-positive.** Usually catalase- and phosphatase-positive. Indole- and sulfatase-negative. Casein, starch and hippurate are not hydrolyzed. **No growth occurs in the presence of 3% NaCl.** A few species can denitrify. Nitrogenase activity occurs in some species, but only under microaerobic conditions. **Carbohydrates are not usually catabolized,** but a few species can attack a limited variety. Amino acids or the salts of organic acids serve as carbon sources. Vitamins are not usually required. Usually occur in **stagnant, freshwater environments.** The mol% G + C of the DNA ranges from 49–66 (T_m).

Type species: *Aquaspirillum serpens* (Müller 1786) Hylemon, Wells, Krieg and Jannasch 1973, 366.

Further Descriptive Information

Most species of *Aquaspirillum* have helical cells; however, *A. delicatum* is vibrioid (has less than one complete turn or twist) and *A. fasciculus* consists of straight rods. Variants that are nearly straight rods have been obtained from helical species after prolonged transfer (Terasaki, 1972). For helical aquaspirilla, the cells within a given species have a constant type of helix—clockwise (right-handed) or counterclockwise (left-handed) (Terasaki, 1972). Photographs showing the comparative size and shape of various aquaspirilla are presented in Figure 2.1.

Although aquaspirilla are more rigid than spirochetes, they do have a certain degree of flexibility. For example, during rapid swimming the helical cells tend to become straighter. Also, cells embedded in glycerol gelatin can be stretched to 3 times their original length (Isaac and Ware, 1974).

As unusual elaboration of the plasma membrane, the "polar membrane," occurs in all of 15 species so far examined (Beveridge and Murray, unpublished results). It is attached to the inside of the plasma membrane by barlike links and is located, most commonly, in the region surrounding the polar flagella (Murray and Birch-Andersen, 1963; and Fig. 2.2). Such a membrane has been found mainly in genera of helical bacteria, such as *Spirillum, Oceanospirillum, Campylobacter, Ectothiorhodospira* and *Rhodospirillum.*

Intracellular poly-β-hydroxybutyrate occurs in all species except *A. gracile* and *A. psychrophilum.* The granules of this polymer stain with metachromatic dyes such as toluidine blue (Martinez, 1963) as well as with lipid-soluble dyes such as Sudan black.

In certain species of *Aquaspirillum* the cells develop into thin-walled coccoid bodies (sometimes termed "microcysts") within several days to several weeks. All species may show a few such forms in old cultures, but in *A. itersonii, A. peregrinum* subsp. *peregrinum, A. polymorphum* and *A. fasciculus* (Fig. 2.3) they become greatly predominant. Such coccoid bodies also are formed by members of the genus *Oceanospirillum* and the genus *Campylobacter.* In *A. itersonii,* the development of the helical cells into coccoid bodies can be greatly accelerated by treatment with mitomycin or ultraviolet light; this effect has been correlated with the induction of a defective bacteriophage (Clark-Walker, 1969). Whether the coccoid bodies of aquaspirilla have resistance to desiccation, or whether they are viable, is not known.

Most species of *Aquaspirillum* are motile by means of bipolar tufts or fascicles of flagella. However, *A. delicatum* has mainly 1–2 flagella at a single pole, and *A. polymorphum* has bipolar single flagella.

Aquaspirilla generally have flagella that are crescent shaped or that have a long wavelength (over 3 μm) with less than one complete wave. Such flagella are especially likely to occur with the larger aquaspirilla such as *A. serpens, A. metamorphum* and *A. putridiconchylium,* and the motility of such spirilla is similar to that described for *Spirillum volutans* (i.e. the flagellar fascicles form cones of revolution). However, some aquaspirilla, especially those that are small or medium in cell diameter such as *A. dispar* or *A. delicatum,* have more conventional, helical flagella (see Hylemon et al., 1973). The rod-shaped *A. fasciculus* (Fig. 2.4) has bipolar flagellar fascicles that exhibit unusual behavior (Strength et al., 1976), and the cells can swim only when suspended in a medium of high viscosity.

A. serpens has been studied extensively with regard to its flagella cell-wall association (Coulton and Murray, 1978), cell-wall ultrastructure and cell-wall chemical composition (Murray et al. 1965; Chester and Murray, 1975, 1978). A protein layer consisting of a regular array or mosaic of subunits surrounds the cell walls of certain species of aquaspirilla (Beveridge and Murray, 1976; Buckmire and Murray, 1970; Stewart et al, 1980); such protein layers can be dissociated by agents such as sodium dodecyl sulfate or guanidine, and can subsequently be reassembled onto templates in vitro in the presence of Ca^{2+}. One function of the protein layer of *A. serpens* is a protective one against attack by bdellovibrios (F. L. A. Buckmire, 1971, Bacteriological Proceedings, p. 43). The lipopolysaccharide of the cell wall of *A. serpens* differs from the majority of other Gram-negative bacteria in that it lacks 2-keto-3-deoxyoctonic acid (Chester and Murray, 1975); this compound is present in the lipopolysaccharide of *A. itersonii* and *A. peregrinum.* In *A. serpens,* lipid A of the lipopolysaccharide differs from that found in members of *Enterobacteriaceae* in that 3-hydroxydodecanoic acid is the N-acylating acid rather than 3-hydroxytetradecanoic acid.

All species of *Aquaspirillum* are aerobic. Although *A. itersonii* and *A. peregrinum* can acidify fructose media sealed with a layer of oil or petrolatum, no significant degree of turbid growth occurs and these species should be considered to have an essentially oxidative type of metabolism. *A. itersonii, A. dispar, A. psychrophilum* and *A. fasciculus* can grow anaerobically with nitrate and possess a dissimilatory nitrate reductase. *A. itersonii, A. dispar* and *A. psychrophilum* can reduce nitrate beyond nitrite, but only *A. psychrophilum* appears to form visible amounts of gas from nitrate (Terasaki, 1972, 1979).

The respiratory chain of *A. itersonii* has been studied in detail and appears to be an unbranched, membrane-bound electron transport chain from NADH and succinate to oxygen (Dailey, 1976). Cytochromes of both the *b* and *c* type, but not of the *a* type, are present, and their biosynthesis and properties have been investigated (Clark-Walker and Lascelles, 1970; Clark-Walker et al., 1967; Dailey and Lascelles, 1974; Ho and Lascelles, 1971). *A. itersonii* synthesizes higher levels of cytochromes *b* and *c* under semianaerobic conditions with nitrate in the medium than it does anaerobically without nitrate (Clark-Walker et al., 1967), and much of the cytochrome *c* is present in a soluble form in the periplasmic space (Gauthier et al., 1970). This soluble cytochrome *c*, and other periplasmic proteins, can be selectively liberated from the cells by the use of a mixture of Tris buffer and EDTA (Garrard, 1971). Biosynthesis of the soluble cytochrome *c* has been investigated by Garrard (1972).

Aquaspirillum species cannot grow in the presence of 3% NaCl, and many species cannot tolerate even 1% NaCl. This lack of salt tolerance distinguishes *Aquaspirillum* from *Oceanospirillum,* because the latter genus requires seawater or Na^+ for growth.

Commonly used culture media for aquaspirilla are PSS broth,* MPSS broth* and nutrient broth.* Aquaspirilla generally produce moderate to abundant turbid growth in 2 or 3 days in PSS broth (Hylemon et

---

* *PSS broth* (g/liter): Bacto peptone (Difco), 10.0; succinic acid (free acid), 1.0; $(NH_4)_2SO_4$, 1.0; $MgSO_4 \cdot 7H_2O$, 1.0; $FeCl_3 \cdot 6H_2O$, 0.002; and $MnSO_4 \cdot H_2O$, 0.002. The pH is adjusted to 6.8 with KOH before autoclaving the medium. For *PSS agar,* 15.0 g of agar is added/liter; for *PSS semisolid medium,* 1.5 g of agar is added/liter. For *MPSS media* use 5.0 g of peptone rather than 10.0 g. *Nutrient broth* (g/liter): peptone, 5.0; meat extract, 3.0 g; the pH is adjusted to 7.0–7.2 with KOH before autoclaving. For *nutrient agar,* 15.0 g of agar is added/liter.

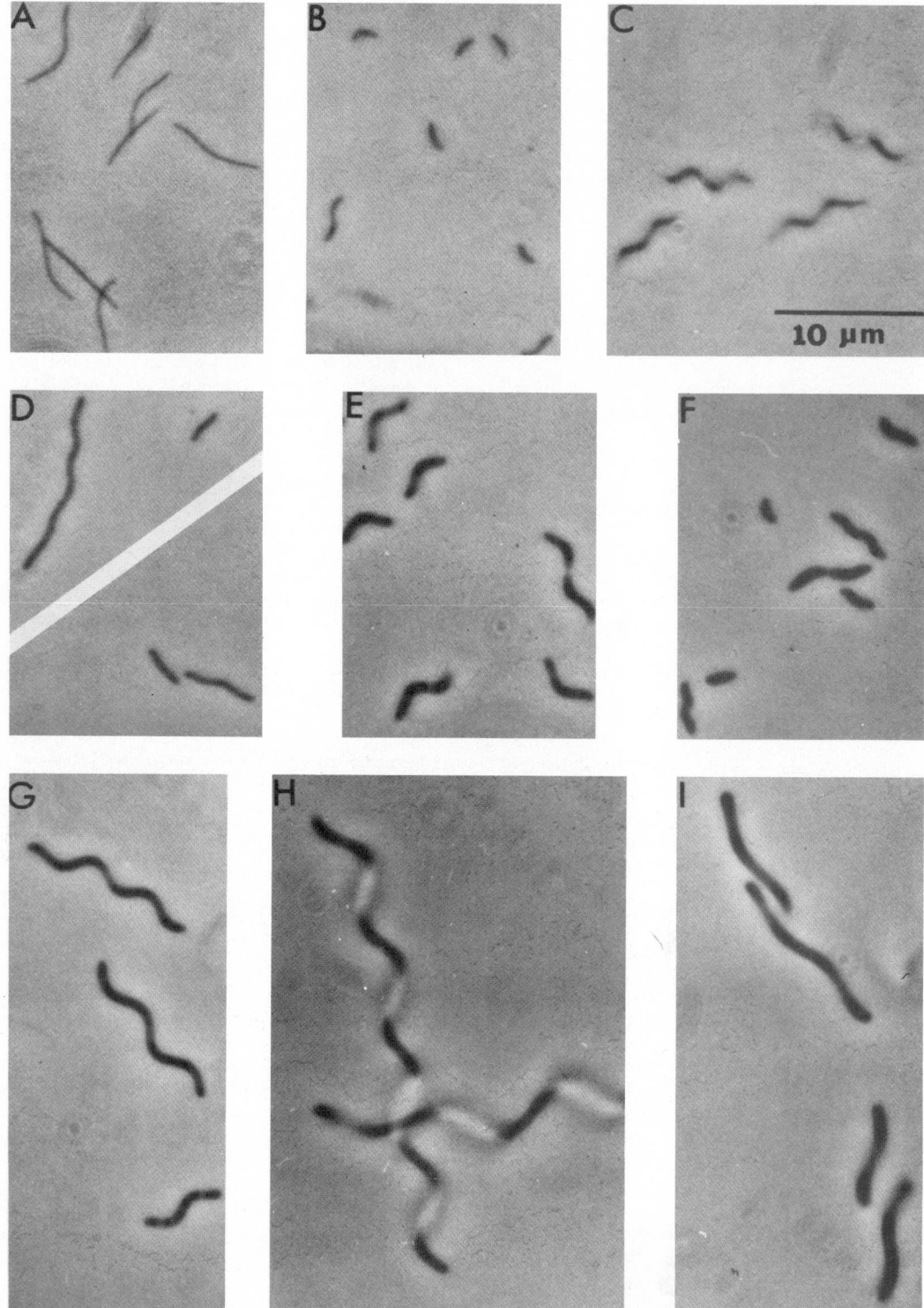

Figure 2.1. Phase contrast photomicrographs of several species of the genus *Aquaspirillum* and of *Spirillum volutans*. The spirilla were cultured in MPSS broth for 24–48 h at 30°C; however, *A. bengal* was incubated at 37°C, and *A. delicatum* was cultured in nutrient broth since its morphology and motility are more characteristic in this medium. All photomicrographs were taken at the same magnification. *A, A. gracile* ATCC 19624. *B, A. delicatum* ATCC 14667. *C, A. polymorphum* NCIB 9072. *D, A. aquaticum* ATCC 11330. *E, A. itersonii* ATCC 12639. *F, A. dispar* ATCC 27510. *G, A. peregrinum* ATCC 15387. *H, A. sinuosum* ATCC 9786. *I, A. putridiconchylium* ATCC 15279. *J, A. serpens* strain VH. *K, A. serpens* ATCC 12638. *L, A. bengal* ATCC 27641. *M, A. metamorphum* ATCC 15280. *N, A. anulus* NCIB 9012. *O, A. giesbergeri* NCIB 8320. *P, S. volutans.* (Reproduced with permission from N. R. Krieg, Bacteriological Reviews *40:* 55–115, 1976, © American Society for Microbiology.)

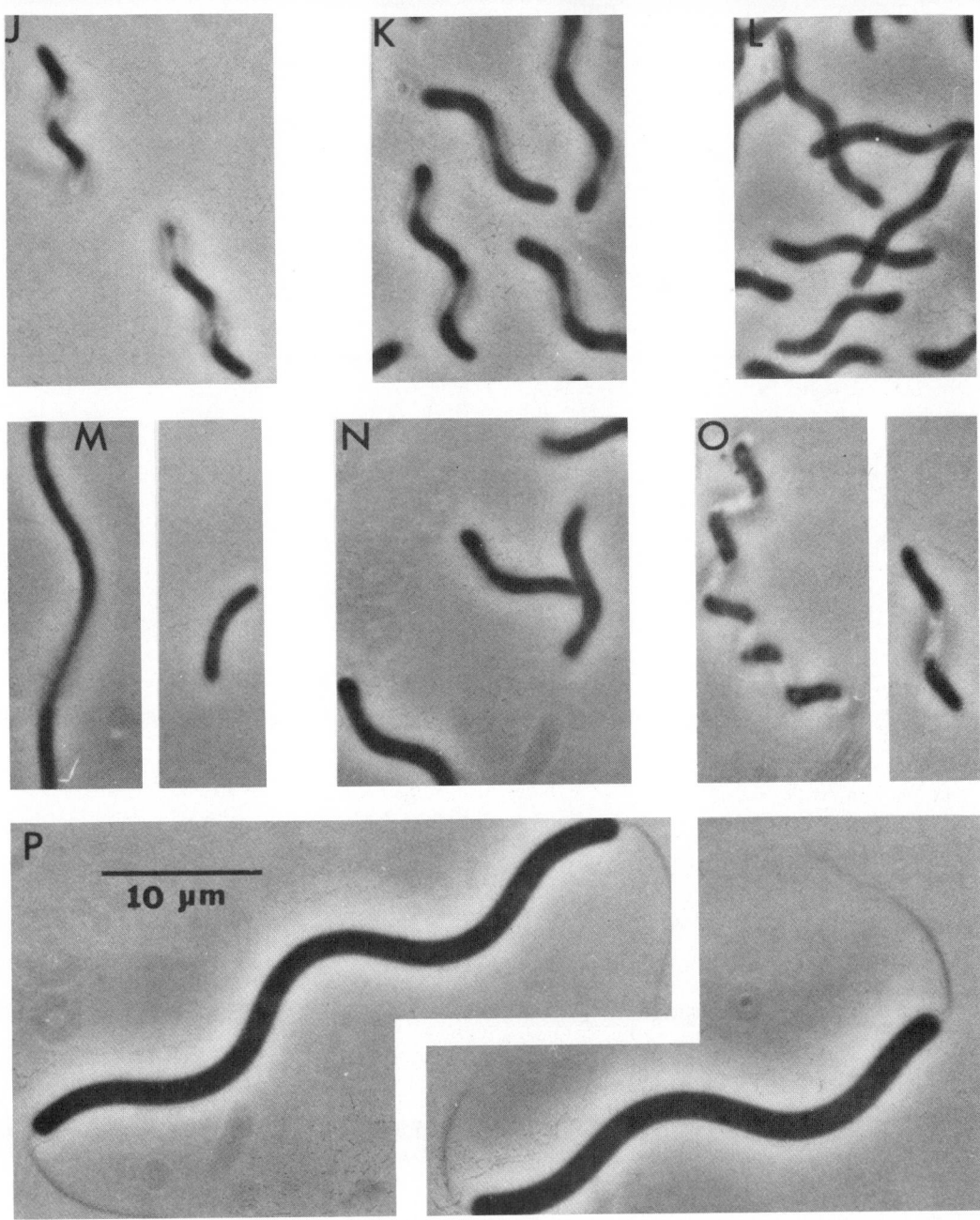

Figure 2.1 *J–P.*

al., 1973). In nutrient broth, membranous masses are often formed at the surface and can be dispersed with shaking to yield turbid cultures (Terasaki, 1972).

Colonies of aquaspirilla generally develop within 2 or 3 days on PSS agar and are usually white, circular and convex, ranging from pinpoint to 1.5 mm in diameter (Hylemon et al., 1973). Colonies on nutrient agar are generally pinpoint in size at 48 h but become larger (up to 2.0 mm in diameter) at 7 days; they are usually convex or umbonate, glistening, opaque, pale yellow and butyrous (Terasaki, 1972). S-R variation has been found in several species (Terasaki, 1972). Some species produce a water soluble, yellow-green fluorescent pigment on PSS agar.

Most species grow best at a temperature of 30–32°C. However, *A. psychrophilum* grows best at 20°C and cannot grow above 26°C, and *A. bengal* grows best at 41°C. The optimum pH for most species is 6.5–7.5, but many species can grow at pH values as high as 8.5 or 9.0 (Terasaki, 1972).

The nutrition of aquaspirilla is generally simple. Most species grow in simple defined media with amino acids or the salts of organic acids as carbon sources and ammonium salts as the nitrogen source. Only two species are known to require vitamins: *A. gracile* requires biotin and *A. aquaticum* requires niacin. Few species can catabolize sugars, but a limited variety can be catabolized by *A. gracile*, *A. itersonii* and *A. peregrinum*. Acidification of sugar media by these species occurs only when the peptone level is kept low (0.2% or less). A listing of the carbon sources for aquaspirilla is given later in Table 2.4. From this table, some contradictions can be seen to occur between the results obtained from different laboratories, although the results within each laboratory are reproducible. These contradictions are likely due to the differences in methodology and in definition of what constitutes a positive growth response and, in some cases, to the use of different strains.

A. autotrophicum is the only member of the genus known to be a facultative hydrogen autotroph, i.e. capable of growing with CO_2 as a

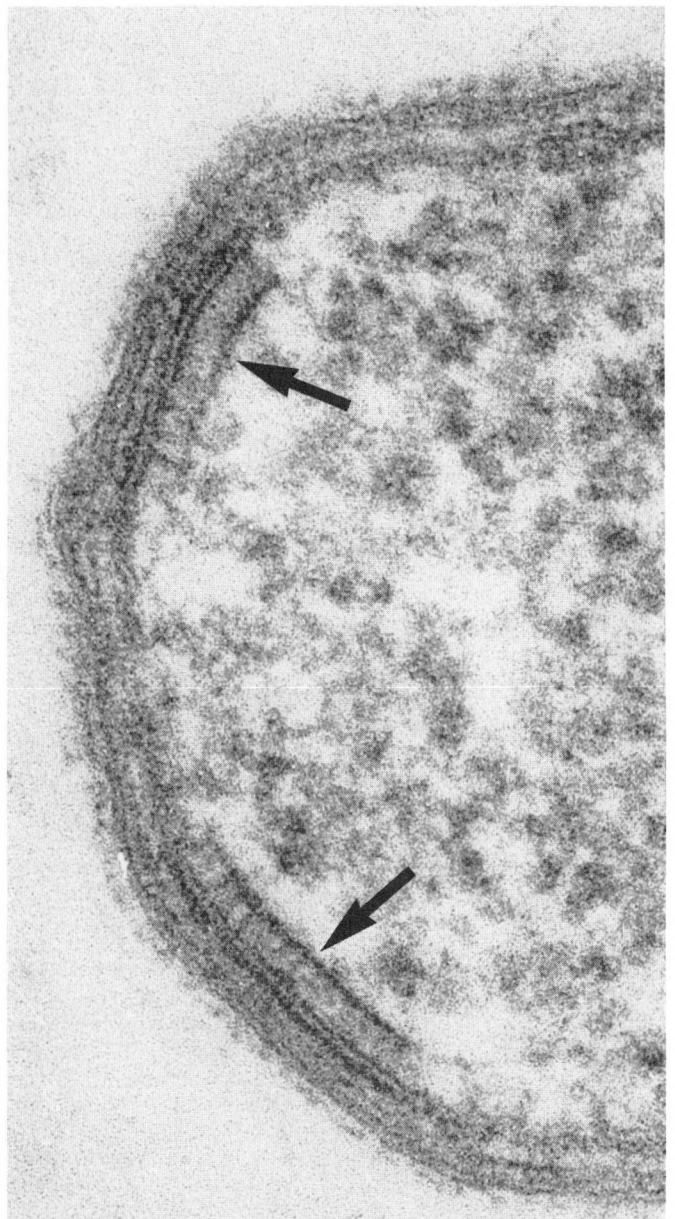

Figure 2.2.   Thin-section through the polar region of a cell of *Aquaspirillum serpens* strain VHA, showing the polar membrane (*arrows*). A protein layer can also be seen external to the outer wall membrane. × 262,000. (Reproduced with permission from Dr. R. G. E. Murray, University of Western Ontario, London, Canada.)

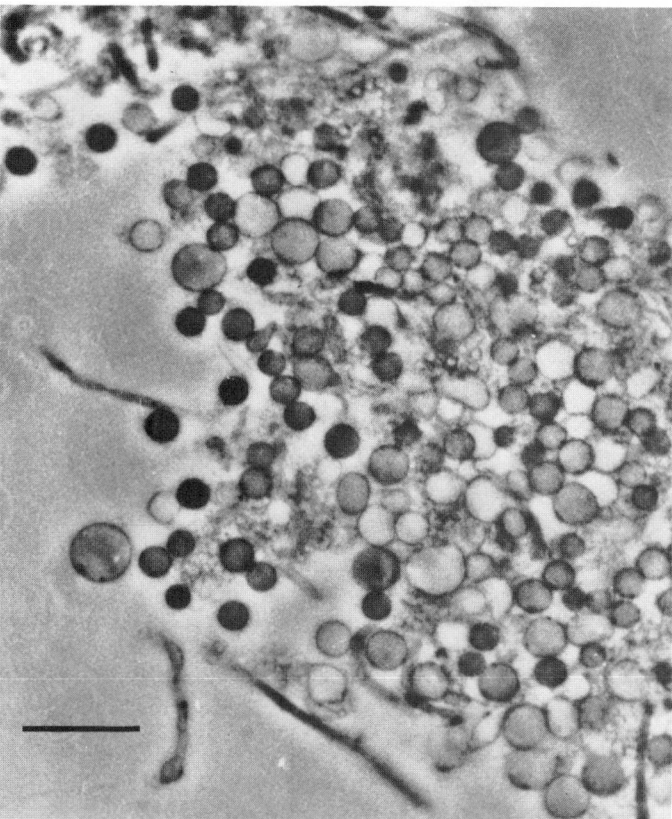

Figure 2.3.   Coccoid bodies of *Aquaspirillum fasciculus* in peptone-fumarate-salts broth cultures incubated at 30°C for 48 h on a shaking machine. Phase contrast microscopy. The *bar* indicates 10 μm.

sole carbon source under an atmosphere containing H_2, O_2 and CO_2 (Aragno and Schlegel, 1978). However, hydrogen autotrophy may not have been tested in other species.

A. *peregrinum* and *A. fasciculus* exhibit nitrogenase activity, but only under microaerobic conditions (Strength et al., 1976). In this respect they are similar to members of the genus *Azospirillum*. Atmospheric nitrogen fixation has also been reported in certain strains of *Aquaspirillum itersonii* (S. D. Ketkar, M. S. Thesis, University of Bombay, 1967; S. D. Ketkar and S. A. Dhala, XIX Annual Conference, Association of Microbiologists of India, Baroda, 1978).

The intermediary metabolism of sugars has been studied in *A. itersonii* and *A. gracile. A. itersonii* can acidify glucose media under semianaerobic conditions but not under aerobic conditions (Terasaki, 1972, 1979); this has not yet been explained. Under aerobic conditions

A. *itersonii* is impermeable to glucose despite the occurrence of high levels of glucokinase activity (Hylemon et al., 1974). Fructose is transported and phosphorylated by means of a fructose-specific phosphoenolpyruvate phosphotransferase system (P. V. Phibbs, unpublished results). The Embden-Meyerhof-Parnas and Entner-Doudoroff pathways occur in *A. itersonii* and *A. gracile*, but the hexose monophosphate pathway is absent (Hylemon et al., 1974; Laughon and Krieg, 1974). *A. gracile* acidifies sugar media by formation of sugar acids, such as gluconic acid; other organic acids are not formed (Laughon and Krieg, 1974).

The tricarboxylic acid cycle has been demonstrated in *A. serpens* and *A. itersonii* (Cole and Rittenberg, 1971). Whether the glyoxylate shunt occurs is not known.

Serological differentiation between most species of aquaspirilla has been reported by McElroy and Krieg (1972). Antisera are prepared against whole cells and adsorbed with heated cells, leaving only antibodies against thermolabile cell components. The use of such antisera in agglutination tests with a limited number of strains suggests that most species can be distinguished from one another and from organisms of other genera.

There has been only one report of bacteriophages for aquaspirilla. An icosahedral, double-stranded DNA phage specific for a particular strain of *A. itersonii* was isolated from raw sewage in Australia by Clark-Walker and Primrose (1971). It could produce plaques on plate cultures but was unable to lyse broth cultures. Oddly, the host strain of *A. itersonii* was isolated originally from Lake Erie, U.S.A., rather than Australia.

Aquaspirilla are considered to be nonpathogenic for humans or animals. An organism known as "*Spirillum minus*" (see Species Incertae Sedis) is the cause of one of the two forms of rat-bite fever in man, and another organism, "*Spirillum pulli*" (see Species Incertae Sedis), is

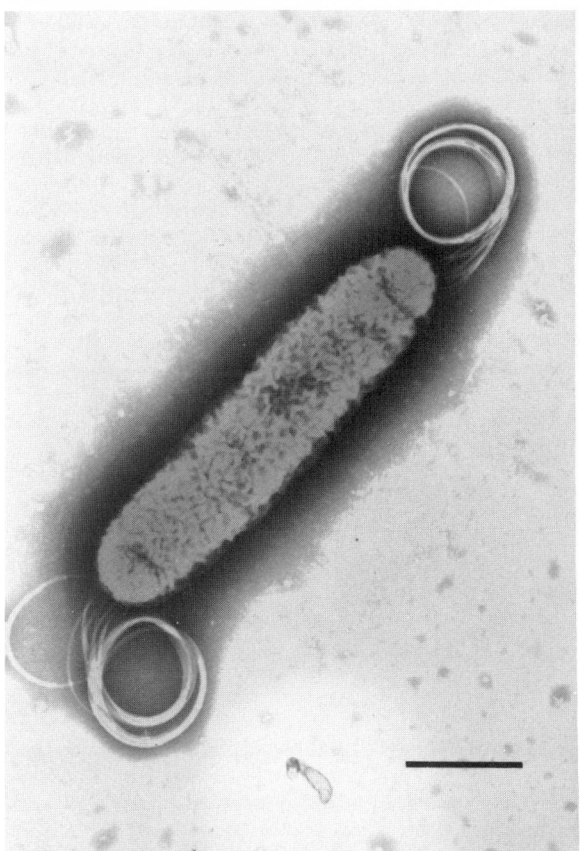

Figure 2.4. Formalin-fixed cells of *Aquaspirillum fasciculus* showing the bipolar fascicles of flagella coiled into loops. When the flagellar fascicles are extended, they have a helical configuration with several waves. The *bar* indicates 1.0 μm. (Reproduced with permission from Strength and Krieg, Canadian Journal of Microbiology *17:* 1133–1137, 1971, © National Research Council of Canada.)

apparently the cause of a diphtheritic stomatitis in chickens. Neither of these species belong to the genus *Spirillum* or to the genus *Aquaspirillum*, and their affiliation with other established genera is uncertain. A few cases of human infection have been reported to be caused by, or associated with, organisms resembling aquaspirilla (e.g. see Kowal, 1961; and Edwards and Kraus, 1960); the identification is uncertain, however. Spirillum-like organisms have also been detected in diseased mosquito larvae (Fulton et al., 1974), but whether they are aquaspirilla is uncertain. Giant spore-forming spirilla have been described in the intestinal contents of tadpoles by Delaporte (1964a) (see Genus Incertae Sedis); little is known of these organisms, but they do not appear to be aquaspirilla.

Aquaspirilla have been isolated from a wide variety of fresh water sources, especially those which are stagnant or contain organic matter: ditch water, canal water, stagnant ponds, primary oxidation ponds and eutrophic lakes. They have also been isolated from storage tanks of distilled water in laboratories, where the organisms and the nutrients to support their growth or survival apparently come from the surrounding air. Aquaspirilla have also been isolated from hay infusions made with pond water (the water is probably the source of the organisms) and from putrid infusions of freshwater mussels (where the mud adherent to the shellfish is probably the source).

Although widely distributed in nature, aquaspirilla comprise only a very small proportion of the total flora of natural habitats. Helical bacteria in general represent only 0.1–0.6% of the flora of pond mud, surface water, slime on stones, or trickling filter effluents, and less than 0.01% of most other habitats (Scully and Dondero, 1973). Con-

sequently, an enrichment procedure is usually necessary before spirilla can be isolated from these habitats.

Enrichment and Isolation Procedures

A number of enrichment methods have been used and advantage is usually taken of the ability of aquaspirilla to grow with levels of nutrients low enough to discourage active growth of many other organisms. Two methods employed by Williams and Rittenberg (1957) have yielded excellent results.

(a) To a sample of the source water is added 1% peptone or yeast autolysate. The samples are incubated at room temperature for ~1 week or until the spirilla become numerous. A portion of this culture is then added to an equal quantity of the source water and the mixture is sterilized by autoclaving. It is then inoculated from the unsterilized portion of the initial culture. After 1–3 transfers through successively nutrient-exhausted medium the spirilla predominate.

(b) A second method is to enrich the initial sample of source water with 1% calcium malate or lactate and incubate for ~1 week. A serial transfer is then made into more source water similarly supplemented with malate or lactate. Spirilla predominate after 3 or 4 such transfers.

For isolation, the enrichments are diluted 1:100 to 1:100,000 with sterile tap water. The dilution bottles are shaken vigorously and allowed to stand at room temperature for 20 min, to allow migration of spirilla to the surface of the diluent. Isolation is then accomplished by streaking the surface water onto a suitable medium such as PSS agar or nutrient agar.

For enrichment by use of putrid infusions of mussels or mud and sand samples, see Terasaki (1963, 1970, 1980). Other general methods have been summarized by Jannasch (1965). Special methods have been used for the following organisms. *A. gracile*, see Canale-Parola et al. (1966); *A. faciculus*, see Strength et al. (1976); *A. autotrophicum*, see Aragno and Schlegel (1978); *A. bengal*, see Kumar et al. (1974).

Maintenance Procedures

Aquaspirilla may be maintained in semisolid PSS medium at 30°C (except for *A. psychrophilum* which is maintained at 15°C) with weekly transfer (Hylemon et al., 1973). Cultures may also be maintained as nutrient agar stabs at room temperature (except for *A. psychrophilum* which is maintained in a refrigerator) with monthly transfer (Terasaki, 1972).

Preservation is most easily accomplished by suspending a dense suspension of cells in nutrient broth containing 10% (v/v) dimethyl sulfoxide, with subsequent freezing in liquid nitrogen. A method for freeze-drying spirilla has been reported by Terasaki (1975).

Procedures for Testing Special Characters

Characterization methods for aquaspirilla have been described in detail by Terasaki (1972, 1979) and by Hylemon et al. (1973). The following comments refer to certain aspects of these procedures. Cell dimensions are best measured in wet mounts of broth cultures by phase-contrast microscopy rather than by dark-field microscopy or by light microscopy of stained smears. To determine whether the cells have a clockwise or counterclockwise type of helix, refer to footnote *d* of Table 2.2, shown later. The presence of intracellular poly-β-hydroxybutyrate is best determined by chemical analysis; for example, *A. delicatum* has no visible granules but does make the polymer. The type and number of flagella is best determined by electron microscopy rather than by flagella staining (Williams, 1960). With regard to coccoid bodies, all strains have a few such forms in old cultures; however, it is only in certain species that coccoid bodies become predominant in old cultures and have taxonomic significance. For determination of acidification of sugar media it is important to use a low concentration of peptone (0.2% or less). For the urease test, cells should be cultured in PSS broth for 24 h, centrifuged and suspended in sterile water to a dense concentration; 0.5 ml of this suspension is then added to 2.0 ml of a medium consisting of 0.1% BES buffer (*N,N*-bis(2-hydroxyethyl)-2-aminoethanesulfonic acid), 2% urea and 0.001% phenol red; pH = 7.0. (This medium must be sterilized by filtration because of the thermolability of the urea). A red or magenta color after incubation at 30°C for 24 h indicates a positive reaction, provided that controls in similar media lacking urea do not develop color. For determination of a water-soluble fluorescent pigment, cultures are streaked in a line across plates of PSS agar and incubated for 48–72 h; the covers of the plates are removed and the plates examined with an ultraviolet lamp of the type used for mineralogical specimens (254-nm wavelength). The occurrence of a distinct yellow-green fluorescent zone in the agar medium surrounding the growth constitutes a positive test. Cultures to be tested for nitrogenase activity should be cultured in nitrogen-deficient semisolid malate medium (see the genus *Azospirillum*) supplemented with 0.005% yeast extract. Cultures are incubated for 3 days at 30°C and then sealed with rubber vaccine bottle stoppers. Acetylene

is injected to a final concentration of 10% (v/v) and the cultures are tested for ethylene production by gas chromatography afte 1 h of further incubation. Controls using liquid rather than semisolid medium and semisolid medium containing 0.1% $(NH_4)_2SO_4$ should be negative for ethylene production. For testing hydrogen autotrophy, the mineral medium of Aragno and Schlegel (1978) is used*; cultures are incubated under an atmosphere of 5% O_2, 10% CO_2 and 85% H_2). A requirement for both H_2 and CO_2 should be demonstrated. For testing sole carbon sources, the procedures of Terasaki (1972, 1979) or Hylemon et al. (1973b) should be followed for most species; see Table 2.5 for additional methods. It is recommended that the type strain of the suspected species be subjected to the same battery of characterization tests as used for the new isolate in order to confirm an identification.

Differentiation of the genus **Aquaspirillum** from other genera

Table 2.1 indicates the characteristics of *Aquaspirillum* that distinguish it from other genera of morphologically or physiologically similar genera.

Taxonomic Comments

In the eighth edition of the *Manual* a single genus, *Spirillum*, contained all of the various aerobic and microaerophilic spirilla. However, the DNA base composition for this genus ranged from 38–65 mol% G + C, a range that was much greater than is usually the case for a bacterial genus. Three groups within the genus were evident: (a) the aerobic, freshwater spirilla that could not tolerate 3% NaCl (mol% G + C is 49–66); the aerobic marine spirilla that required seawater for growth (mol% G + C is 42–51); and the large microaerophilic spirilla that belonged to the species *Spirillum volutans* (mol% G + C is 38). Accordingly, Hylemon et al. (1973) divided the genus into the three genera *Aquaspirillum*, *Oceanospirillum* and *Spirillum*, respectively.

Although this scheme is useful for practical purposes, there is little genetic evidence to support the concept that the various species of *Aquaspirillum* are phylogenetically related to one another. Indeed, analysis of the oligonucleotide catalogs of the 16S rRNA of three species indicates that there is considerable phylogenetic heterogeneity within the genus (Woese et al., 1982). *A. serpens* belongs to group II of the phototrophic bacteria as defined by Gibson et al. (1979) and is related to *Rhodospirillum tenue* and also to a lesser extent to *Spirillum volutans*. *A. gracile* is also a member of group II, but is more closely related to *Rhodopseudomonas gelatinosa* than to *A. serpens*, *S. volutans*, or *R. tenue*. *A. itersonii* is a member of group I and is related to *Rhodospirillum rubrum* and *Azospirillum brasilense*. The relation of *A. itersonii*, and also *A. polymorphum*, to the genus *Azospirillum* has also been indicated by rRNA/DNA hybridization studies (DeSmedt et al., 1980).

Except for the restriction of a mol% G + C range of 49–66, the genus *Aquaspirillum* is at present based largely on a pattern or core of phenotypic characteristics, considered to be the typical characteristics of the genus. These include: a helical shape; bipolar tufts of flagella; poly-β-hydroxybutyrate formation; a strictly respiratory type of metabolism; positive oxidase, catalase and phosphatase reactions; an inability to attack sugars, starch or casein; a negative indole test; lack of tolerance to 3% NaCl; an optimum temperature of 30°C: and a simple chemoheterotrophic nutrition with amino acids or the salts of organic acids serving as carbon sources. Species have been assigned to the genus on the basis of a similarity of their characteristics to this pattern of core characteristics, with the recognition that exceptional characteristics may occur. Such exceptional characteristics have included: a

vibrioid shape or a straight rod shape; nitrogen-fixing ability; hydrogen autotrophy; lack of poly-β-hydroxybutyrate; high or low temperature optima; a single flagellum at one or both poles; catabolism of a limited variety of sugars; and vitamin requirements.

Some examples can illustrate the taxonomic problems one may encounter with aquaspirilla. *A. delicatum* is vibrioid rather than helical, and has one or two flagella at only one pole rather than bipolar flagellar tufts. This morphology differs from that of all other aquaspirilla. Although Leifson (1962) placed the organism in the genus *Spirillum* (*Aquaspirillum*), he recognized that it differed morphologically from typical spirilla and suggested the possibility of creating a new genus for such noncarbohydrate-utilizing vibrios. Such a genus might have been "*Comamonas*," established by Davis and Park (1962) for curved rods having single or multiple flagella at one pole, a high mol% G + C value, and an oxidative metabolism. However, this genus was abandoned when it was discovered that the type species "*Comamonas terrigena (Vibrio percolans)*" was apparently synonymous with *Pseudomonas testosteroni* (Hugh, 1965). Because *A. delicatum* has a strictly respiratory type of metabolism, is incapable of catabolizing any carbohydrates, does not grow with 3% NaCl, has poly-β-hydroxybutyrate, does not require vitamins, and has a mol% G + C of 63, Hylemon et al. (1973) included it in the genus *Aquaspirillum*. Because certain pseudomonads such as *P. lemoignei* are slightly curved rods that cannot use carbohydrates, it is possible that *A. delicatum* could be placed in the genus *Pseudomonas*. However, *A. delicatum* is vibrioid (curved with a twist), whereas pseudomonads, if curved, are curved in one plane. It is difficult at present to evaluate the importance of this distinction. Meanwhile, it seems advisable to retain *A. delicatum* in the genus *Aquaspirillum* until genetic evidence contraindicates this arrangement.

Another example of taxonomic difficulty concerns *A. fasciculus*, which has a straight-rod shape, forms viscous flocs, fails to swim except in a viscous medium, and exhibits nitrogenase activity. These characteristics are not typical of aquaspirilla, and it seems especially heretical to include a straight rod in a genus of spirilla. However, *A. fasciculus* does possess bipolar tufts of flagella, forms poly-β-hydroxybutyrate, has a strictly respiratory type of metabolism, attacks no sugars, cannot grow with 3% NaCl (or even 1% NaCl), has a simple nutrition, grows best at 30°C, is indole-negative, does not hydrolyze starch or casein, possesses oxidase, catalase and phosphatase activity, and has a mol% G + C from 62–65. Moreover, *A. fasciculus* forms coccoid bodies in abundance, which is characteristic of certain aquaspirilla (*A. itersonii*, *A. polymorphum* and *A. peregrinum* subspecies *peregrinum* and other helical or vibrioid bacteria (*Oceanospirillum*, *Campylobacter*, *Desulfovibrio* and *Vibrio*; see Baker and Park, 1975; Felter et al., 1969; Levin and Vaughn, 1968; Ogg, 1962; Williams and Rittenberg, 1957). Fur-

---

* Mineral medium (g/liter); $Na_2HPO_4 \cdot 12H_2O$, 9.0; KH_2PO_4, 1.5; $MgSO_4 \cdot 7H_2O$, 0.2; NH_4Cl, 1.0; ferric ammonium citrate, 0.005; $CaCl_2 \cdot 2H_2O$, 0.01; trace element solution (see below), 3.0 ml; pH = 7.1. For a solid medium, 17.0 g of agar are added. For both liquid and solid media, 0.05% $NaHCO_3$ should be incorporated aseptically into the sterilized medium to buffer against changes in pH caused by the CO_2 of the gas atmosphere. *Trace element solution* (mg/liter): $ZnSO_4 \cdot 7H_2O$, 10.0; $MnCl_2 \cdot 4H_2O$, 3.0; H_3BO_3, 30.0; $CoCl_2 \cdot 6H_2O$, 20.0; $CuCl_2 \cdot 6H_2O$, 0.79; $NiCl_2 \cdot 6H_2O$, 2.0; $Na_2MoO_4 \cdot 2H_2O$, 3.0.

Table 2.1.

Differential characteristics of the genus **Aquaspirillum** *and other genera of oxidase-positive, motile, curved, vibrioid or helical Gram-negative rods[a]*

Characteristics	Aquaspirillum	Spirillum	Oceano-spirillum	Campylo-bacter	Azospiril-lum	Bdellovibrio	Pseudomonas	Vibrio	Alteromonas
Predominant shape:									
Helical	$+^b$	+	+	$-^c$	−	−	−	−	−
Vibrioid or curved	$-^b$	−	−	$+^c$	$+^d$	+	$-^e$	D	D
Straight	$-^b$	−	−	−	$-^d$	−	$+^e$	D	D
Cell diameter, μm	0.2–1.4	1.4–1.7	0.3–1.4	0.2–0.5	1.0	0.2–0.5	0.5–1.0	0.5–0.8	0.7–1.5
Polar membrane present	+	+	+	+	+	−	−	−	
Usual arrangement of polar flagella:									
Bipolar tufts	$+^f$	+	$+^g$	−	−	−	−	−	−
Tuft at one pole	$-^f$	−	−	−	−	−	D	$-^h$	−
Single flagellum at one or both poles	$-^f$	−	$-^g$	+	+	+	D	$+^h$	+
Lateral flagella also formed under certain conditions	−	−	−	−	+	−	D	D	−
Intracellular poly-β-hydroxybutyrate formed	$+^i$	+	+	−	+	−	D	D	−
Relation to oxygen:									
Aerobic	$+^j$	−	+	−	$+^k$	+	+	−	+
Facultative	$-^j$	−	−	−	$-^k$	−	−	+	−
Microaerophilic	$-^j$	+	−	+	$-^k$	−	−	−	−
CO$_2$ required for growth	−	−	−	+	−	−	−	−	−
Sea water or Na$^+$ required for growth	−	−	+	−	−	D	D	$+^l$	+
Tolerant to 3% NaCl	−		+	D	D	D		+	+
Indole test	−	−	−	−	−			D	
Carbohydrates are catabolized	$-^m$	−	−	−	+	−	D	+	+
Habitat:									
Freshwater	+	+	−	−	−	D	D	D	−
Marine	−	−	+	−	−	D	D	D	+
Soil	−	−	−	−	+	D	D	−	−
Humans or mammals	−	−	−	+	−	−	D	D	−
Capable of multiplying in the periplasmic space of other bacteria	−	−	−	−	−	+	−	−	−
Nitrogenase activity	$-^j$	−			$+^k$		−	−	−
Mol% G + C of DNA	49–66	38	42–51	30–38	69–71	33–37 42–51	57–70	38–51	38–50

[a] Symbols: +, all species positive except where noted; −, all species negative except where noted; D, differs among species.

[b] *A. delicatum* is mainly vibrioid; *A. fasciculus* is a straight rod.

[c] Cells in chains resemble spirilla.

[d] In pure cultures, a proportion of the cells may be straight rods.

[e] The genus *Pseudomonas* contains straight rods and rods that are curved in one plane, but not helically curved rods.

[f] *A. delicatum* has mainly a single flagellum at one pole; *A. polymorphum* has mainly a single flagellum at each pole.

[g] *O. pusillum* has mainly a single flagellum at each pole.

[h] Most species of *Vibrio* have a single polar flagellum, but two species (*V. fischeri* and *V. logei*) have a tuft of polar flagella.

[i] *A. gracile* and *A. psychrophilum* lack this polymer.

[j] Some species (*A. peregrinum* and *A. fasciculus*) can fix nitrogen under microaerophilic conditions. With a source of fixed nitrogen they grow as aerobes. Some species (*A. itersonii* and *A. peregrinum*) can acidify certain sugar media under anaerobic conditions, suggesting that some fermentative ability may be present; however, these species do not exhibit visibly turbid growth under these conditions and should be considered to have mainly a respiratory type of metabolism.

[k] All species fix nitrogen under microaerophilic conditions. With a source of fixed nitrogen they grow as aerobes. *A. lipoferum* has weak fermentative ability and can exhibit slight growth under anaerobic conditions with certain sugars; however, the organisms have a mainly respiratory type of metabolism and should not be considered as facultative bacteria.

[l] Growth of all species is stimulated by Na$^+$, and most species have an absolute requirement for Na$^+$.

[m] *A. gracile*, *A. itersonii* and *A. peregrinum* can catabolize a very restricted number of sugars. All other species are incapable of catabolizing any carbohydrates.

thermore, the presence of a polar membrane in *A. fasciculus* is also a characteristic associated mainly with helical bacteria (Hickman and Frenkel, 1965a, b; Keeler et al., 1966; Murray and Birch-Andersen, 1963; Remson et al., 1963; Ritchie, Keeler and Bryner, 1966) although it has also been found in the rod-shaped organism *Chromatium* (Murray and Birch-Andersen, 1963). Nitrogenase activity, although unusual for aquaspirilla, does occur in two helical species. With regard to the rod shape of *A. fasciculus*, Strength et al. (1976) reported the isolation of a strain that was at first morphologically identical to *A. fasciculus* but later exhibited many cells that were curved or even S-shaped after prolonged transfer. Moreover, some helical species of *Aquaspirillum* develop variants that are nearly straight rods upon prolonged transfer (Terasaki, 1973; Williams, 1959). All of this evidence supports the idea that *A. fasciculus* is a nonhelical aquaspirillum. To exclude it from the genus merely because of its rod shape does not seem reasonable.

Still another example of taxonomic difficulty can be found with *A. autotrophicum*, which can grow autotrophically under an atmosphere of oxygen, carbon dioxide and hydrogen. Other members of the genus appear not to be facultative hydrogen autotrophs, although it is likely that many have not been tested for this character. Although hydrogen autotrophs were once all placed into a single genus, "*Hydrogenomonas*," it has since been recognized that this genus is taxonomically heterogeneous and the genus has been abandoned (Davis et al., 1969). The Gram-negative, polar-flagellated members of "*Hydrogenomonas*" have generally been reclassified into the genus *Pseudomonas*, but in the case of *A. autotrophicum* this would not be suitable. Aside from the facultative hydrogen autotrophy, the characteristics of *A. autotrophicum* are completely consistent with the typical characteristics of the genus *Aquaspirillum*.

The ability of nearly half the species in the genus to form a water-soluble fluorescent pigment suggests a possible relationship of aquaspirilla to the fluorescent pseudomonads. However, whether the pigment is the same as that formed by pseudomonads is not known and, in the case of *A. serpens* (where most of the strains form a fluorescent pigment) the mol% G + C of the strains is ~50, well below the range of 57–70 for the genus *Pseudomonas*. Moreover, other genera have also been found to form fluorescent pigments, e.g. *Azotobacter*, *Azomonas* and *Beijerinckia*. Consequently, the taxonomic significance of fluorescent pigment production remains uncertain at the genus level.

Although 17 species are presently recognized within the genus *Aquaspirillum*, no DNA/DNA hybridization studies have been done to support or deny this number of species (except in the case of *A. autotrophicum* vs. *A. dispar* (Aragno and Schlegel, 1978). Speciation has been based mainly on differences in morphology, nutrition and DNA base composition. In some case species are obviously different; for example, a small spirillum such as *A. polymorphum* would not belong in the same species as a large spirillum (such as *A. serpens*), and a spirillum having a mol% G + C of 50 would not be placed in the same species as one with a mol% G + C of 60. However, species distinctions are less firmly based within a particular morphological group of strains having a similar DNA base composition. One other difficulty is that many species are represented by only one or two strains, and the limits of variation within such a species may be broader than is presently assumed. Therefore, it is possible that some species may not deserve separate species status (for example, the differences between *A. serpens*, *A. bengal* and *A. putridiconchylium* may not be a strong basis for separation of the group into three species). On the other hand, it is also possible that some of the species may represent more than a single species; for example, the variation in a number of physiological or antigenic characters occuring within the species *A. serpens* perhaps may be indicative of the occurrence of more than one species. It is likely that DNA/DNA hybridization studies could resolve many of these questions.

Further Reading

Krieg, N.R. 1976. Biology of the chemoheterotrophic spirilla. Bacteriol. Rev. *40:* 55–115.

Krieg, N.R. and P.B. Hylemon. 1976. The taxonomy of the chemoheterotrophic spirilla. Annu. Rev. Microbiol. *30:* 303–325.

List of the species of the genus **Aquaspirillum**

1. **Aquaspirillum gracile** (Canale-Parola, Rosenthal and Kupfer 1966) Hylemon, Wells, Krieg and Jannasch 1973, 369.[AL] (*Spirillum gracile* Canale-Parola, Rosenthal and Kupfer 1966, 124.)

gra.ci′le. L. neut. adj. *gracile* slender or thin.

The smallest of the aquaspirilla. The morphological characters are as depicted in Figure 2.1 and described in Tables 2.2 and 2.3.

When originally isolated, all strains formed subsurface, spreading, semitransparent colonies on a medium containing 1.0% agar, the spreading occurring within the medium (Canale-Parola et al., 1966). After prolonged subculturing, some of the spirilla in each strain lost the ability to diffuse through 1.0% agar and formed small, nonspreading colonies. Although *A. gracile* was originally described as being microaerophilic, based on the growth of the organisms a few millimeter below the surface of semisolid media (Canale-Parola et al., 1966), more recent analysis of the type strain indicates that in liquid media a maximum growth response occurs under an air atmosphere (B. E. Laughon, 1973, M.S. thesis, Virginia Polytechnic Institute and State University).

Optimum temperature 30°C. Scanty growth in PSS broth at 25°C; no growth at 10 or 42°C.

Physiological characters are listed in Tables 2.2 and 2.3. Carbon sources are listed in Table 2.4. Ammonium chloride can be used as a nitrogen source but potassium nitrate cannot (Canale-Parola et al., 1966). A chemically defined medium has been devised by B. E. Laughon (1973, M.S. thesis; see also Krieg, 1976). Biotin is required for growth.

Isolation is accomplished by allowing the spirilla to pass through a membrane filter disk (0.45 μm pore size) to underlying agar medium (Canale-Parola et al., 1966).

Isolated from pond or stream water.

The mol% G + C of the DNA is 64–65 (T_m).

Type strain: ATCC 19624. *Reference strains:* ATCC 19625 and 19626.

2. **Aquaspirillum delicatum** (Leifson 1962) Hylemon, Wells, Krieg and Jannasch 1973, 371.[AL] (*Spirillum delicatum* Leifson 1962, 164.)

de.li.ca′tum. L. neut. adj. *delicatum* delicate.

The only species of the genus to have a vibrioid shape. Chains of cells may resemble spirilla. The morphological characters are as depicted in Figure 2.1 and described in Tables 2.2 and 2.3, and are most characteristic for cultures grown in nutrient broth rather than PSS broth. Intracellular granules are not evident, but chemical tests indicate the presence of this polymer.

Optimum temperature 30–32°C.

Physiological characters are described in Tables 2.2 and 2.3. Sole carbon sources are listed in Table 2.4. Growth is poor in defined media; the best growth occurs with malate as the carbon source and glutamine as the nitrogen source. Ammonium salts or potassium nitrate are used poorly as nitrogen sources.

Isolated from stored distilled water.

The mol% G + C of the DNA is 63 (T_m).

Type strain: ATCC 14667.

3. **Aquaspirillum polymorphum** (Williams and Rittenberg 1957) Hylemon, Wells, Krieg and Jannasch 1973, 371.[AL] (*Spirillum polymorphum* Williams and Rittenberg 1957, 85.)

Table 2.2

Differential characteristics of species of the genus Aquaspirillum[a]

Characteristics	1. A. gracile	2. A. delicatum	3. A. polymorphum	4. A. aquaticum	5. A. itersonii	6. A. peregrinum	7. A. dispar	8. A. autotrophicum	9. A. sinuosum	10. A. psychrophilum	11. A. fasciculus	12. A. serpens	13. A. putridiconchylium	14. A. bengal	15. A. metamorphum	16. A. giesbergeri	17. A. anulus
Cell diameter, μm[b]	0.2–0.3	0.3–0.4	0.3–0.5	0.5–0.6	0.4–0.8[c]	0.5–0.7	0.5–0.7	0.6–0.8	0.6–0.9	0.7–0.9	0.7–0.9	0.6–1.1	0.7–1.2	0.9–1.2	0.7–1.3	0.7–1.4	0.8–1.4[c]
Shape	H	V	H	H	H	H	C	H	H	H	SR	H	H	H	H	H	H
Type of helix[d]	C		CC	C	CC	CC	C	C	C	C		C	C	C	C	C	C
Polar membrane present	+	+	+	+	+	+	+	+	+	–	+	+	+	+	+	+	+
Poly-β-hydroxybutyrate formed	–	+	+	+	+	+	+	+	+	–	+	+	+	+	+	+	+
Flagellar arrangement	BT	U (1–2)	BS	BT	BT	BT	BT	BT	BT	BT	BT	BT	BT	BT	BT	BT	BT
Coccoid bodies predominant at 3–4 weeks	–	–	+	–	+	d[e]	–	–	–	–	+	–	–	–	–	–	–
Acid produced from sugars[f]	+	–	–	–	+	+	+	–	–	+	–	–	–	–	–	–	–
Anaerobic growth with KNO₃	–	–	–	–	+	–	+	–	–	+	+	d	–	–	–	–	–
KNO₃ reduced only to KNO₂	+	+	+	+	+	–	+	–	–	+	+	–	–	–	–	–	–
Denitrification	–	–	–	+	+	+	+	+	+	–	–	+	+	+	+	+	+
Phosphatase test	+	+	+	–	+	+	–	–	+	+	–	–	–	–	–	–	–
Esculin hydrolysis	+	+	+	–	–	–	–	–	+	–	+	d	–	–	–	+	–
Urease[g]	–	–	–	+	+	+	+	+	–	+	+	–	–	–	–	+	–
Optimum temperature = 20°C; no growth at >26°C	+	–	–	–	–	–	–	–	–	+	–	–	–	–	–	–	–
Optimum temperature = 41°C	–	–	–	–	–	–	–	–	–	–	–	–	–	+	–	–	–
Growth factors required[h]	+	–	+	+	–	+	d	–	–	–	+	–	+	+	–	–	+
Glutamate as sole C source[i]	–	+	–	+	+	+	–	d	–	–	–	+	+	–	–	–	–
Histidine as sole C source[i]	–	–	–	–	+	–	–	–	–	–	–	–	+	–	–	–	–
Tryptophan and glycine as sole C sources[i]	–	–	–	–	–	–	–	+	–	–	+	–	–	–	–	–	–
Nitrogenase activity[g]	–	–	–	–	d	+	–	+	–	–	–	–	–	–	–	–	–
Hydrogen autotrophy[g]	–	–	–	–	–	–	–	+	–	–	–	–	–	–	–	–	–
Mol% G + C of DNA	64–65	63	61–62	64–65	60–66[j]	60–64[k]	63–65	60–62	57–59	65	62–65	49–51	52	52	63	57–58	58–59

[a] Symbols: +, positive for all strains; –, negative for all strains; blank space, not determined; H, helical (one or more complete turns or twists); V, vibrioid (less than one complete turn or twist); SR, straight rod; C, clockwise helix; CC, counterclockwise helix; BT, bipolar tufts; U (1–2), 1 or 2 flagella at only one pole; BS, single flagellum at each pole; d, differs among strains.

[b] By phase-contrast microscopy of 24- to 48-h-old broth cultures.

[c] The range for the subspecies nipponicum is 0.5–0.8 μm; for the subspecies itersonii the range is 0.4–0.6 μm.

[d] Determined by focusing on the bottom of the cells. The pattern //// indicates a clockwise (right-handed) helix, whereas the pattern \\\\ indicates a counterclockwise (left-handed) helix.

[e] The subspecies integrum fails to form coccoid bodies, whereas the subspecies peregrinum forms them readily in old cultures.

[f] For A. gracile, acid from glucose, galactose and arabinose (aerobically). For A. itersonii, acid from glycerol (aerobically), fructose (aerobically and anaerobically) and glucose (aerobically only). Although A. itersonii and A. peregrinum, acid from fructose (aerobically and anaerobically). Peptone concentrations must be kept low (0.2% or less) in order to detect change in pH indicator. A. peregrinum acidify sugar media anaerobically, turbid growth does not occur and the organisms should be considered to have mainly a respiratory rather than fermentative type of metabolism.

[g] See Procedures for Testing Special Characters.

[h] See Table 2.5.

[i] A. gracile requires biotin and A. aquaticum requires niacin.

[j] The mol% G + C for the subspecies nipponicum is 66, whereas it is 60–64 for the subspecies itersonii.

[k] The mol% G + C for the subspecies integrum is 64, whereas it is 60–62 for the subspecies peregrinum.

Table 2.3.

Other characteristics of **Aquaspirillum** *species 1–8[a]*

Characteristic	1. A. gracile	2. A. delicatum	3. A. polymorphum	4. A. aquaticum	5. A. itersonii	6. A. peregrinum	7. A. dispar	8. A. autotrophicum
Wavelength of helix, μm	2.8–3.5	−[b]	4.0–5.0	2.0–5.0	2.5–6.0	3.0–4.5	2.0–3.5	3.0–4.0
Helix diameter, μm	0.5–2.1	0.4–0.7[b]	1.0–1.5	0.8–1.0	1.0–2.2	1.4–2.0	1.0–2.1	
Length of helix, μm	3.5–14.0	3.0–5.0[b]	3.5–8.4	2.5–13.0	2.0–10.0	1.5–22.0	2.1–6.5	2.0–5.0
Catalase	+	+	+	+	+	+	+	+
Oxidase	+	+	+	+	+	+	+	+
Growth in presence of:								
1% Oxgall	+	−	+	+	+[c]	+[d]	+	+
1% Glycine	−	−	−	+	−[c]	−[d]	+	−
3% NaCl	−	−	−	−	−	−	−	−
Water-soluble brown pigment formed in presence of:								
0.1% Tyrosine	−	−	−	−	+[c]	−[d]	−	
0.1% Tryptophan	−	−	−	−	+[c]	−[d]	−	
Alkaline reaction in litmus milk	−	−	−	−	−[c]	+[d]	−	
Water-soluble yellow-green fluorescent pigment	d	−	+	+	+[c]	+[d]	+	−
Deoxyribonuclease (DNase)	+	−	−	+	+[c]	−[d]	−	
Ribonuclease (RNase)	+	+	−	+	+[c]	−[d]	−	
Growth on:								
Eosin methylene blue agar	+	−	+	+	+[c]	+[d]	+	
MacConkey agar	−	−	+	+	+[c]	−[d]	+	
Triple-sugar iron agar	−	+	−	+	+[c]	+[d]	+	
Seller agar	−	−	−	+	+[c]	+[d]	+	
Methyl red-Voges Proskauer broth	−	+	−	+	+[c]	−[d]	+	
Reduction of 0.3% H_2SeO_3	−	−	+	+	d[c]	+[d]		
H_2S from 0.2% cysteine in PSS broth, 7 days	+	+	+	+	+	+	+	−
H_2S from 0.01% cystine in nutrient broth, 7 days		+	+	+	+	+		
Indole test	−	−	−	−	−	−	−	−
Hydrolysis of casein and starch	−	−	−	−	−	−	−	−
Hydrolysis of hippurate	−	−	−	−	−	−	−	
Gelatin hydrolysis, 30°C, 4 days[e]	−	−	−	−	−	−	−	−
Gelatin liquefaction, 20°C:[f]								
7 days		−	−		−	−		
28 days		−	−		−	+		
Sulfatase (0.01% phenolphthalein disulfate)	−	−	−	−	−	−	−	−
Protocatechuate cleavage of the *ortho* type in *p*-hydroxybenzoate metabolism								+
Temperature range for growth, °C	9–4.0	14–3.6	12–42	12–42	11–40			10–35
pH range for growth	5.5–8.5	6.0–8.5	5.5–9.0	5.5–9.0	5.5–9.0			50–8.0

[a] Symbols: +, all strains positive; −, all strains negative; d, differs among strains; blank space, not determined.

[b] *A. delicatum* is vibrioid rather than helical; thus the wavelength cannot be determined. The helix diameter refers to the width of the vibrio, and the length of helix refers to the length of the vibrio.

[c] Has not been determined for the subspecies *nipponicum*.

[d] Has not been determined for the subspecies *integrum*.

[e] Method of Hylemon et al. (1973).

[f] Method of Terasaki (1972).

po.ly.mor′phum. Gr. adj. *poly* many; Gr. n. *morphus* form, shape; M.L. neut. adj. *polymorphum* of many shapes.

Although originally reported to have bipolar tufts of flagella (Williams and Rittenberg, 1957), this species appears to have only a single flagellum at each pole (Terasaki, 1972; Hylemon et al., 1973b). The morphological characters are as depicted in Figure 2.1 and described in Tables 2.2 and 2.3.

Optimum temperature 30°C.

Physiological characteristics are described in Tables 2.2 and 2.3. Although originally reported to grow anaerobically with nitrate (Williams and Rittenberg, 1957), more recent studies have not confirmed this (Terasaki, 1972; Hylemon et al., 1973). Sole carbon sources are listed in Table 2.4. The best sole carbon sources are glutamate and aspartate; they are also the best sole nitrogen sources. Ammonium salts can serve as nitrogen sources; there are conflicting results concerning the utilization of nitrate as a sole nitrogen source (Terasaki, 1972; Hylemon et al., 1973).

Growth in PSS broth abundant, cloudy. Colonies on PSS agar are circular, convex, translucent, pinpoint. Optimum temperature 30°C.

Isolated from pond water.

The mol% G + C of the DNA is 61–62 (T_m).

Type strain: ATCC 11332.

4. **Aquaspirillum aquaticum** Hylemon, Wells, Krieg and Jannasch 1973, 372.[AL]

a.qua′ti.cum. L. neut. adj. *aquaticum* living in water.

The morphological characters are as depicted in Figure 2.1 and described in Tables 2.2 and 2.3.

Optimum temperature 35°C.

Physiological characters are listed in Tables 2.2 and 2.3. Carbon sources are listed in Table 2.4. No growth occurs in the absence of niacin (Kropinksi, 1975).

Isolated from freshwater sources.

The mol% G + C of the DNA is 64–65 (T_m).

Type strain: ATCC 11330.

5. **Aquaspirillum itersonii** (Giesberger 1936) Hylemon, Wells, Krieg and Jannasch 1973, 370.[AL] (*Spirillum itersonii* Giesberger 1936, 68.)

i.ter.so′ni.i. M.L. gen. n. *itersonii* of Iterson; named for G. Van Iterson, a Dutch bacteriologist.

The morphological characters are as depicted in Figure 2.1 and described in Tables 2.2 and 2.3.

Optimum temperature 32–35°C.

Physiological characters are as described in Tables 2.2 and 2.3. Sole carbon sources are listed in Table 2.4. Ammonium salts can serve as sole nitrogen sources. Nitrate supports either no growth or very scanty growth. In 1957 Williams and Rittenberg established a subspecies, *vulgatum*, to include those strains that use nitrate; however, this could not be confirmed by Terasaki (1979) for the type strain of the subspecies (ATCC 11331) using the methods of Williams and Rittenberg.

Isolated from pond water and from putrid infusions of freshwater mussels.

The mol% G + C of the DNA is 60–66 (T_m).

Type strain: ATCC 12639.

5a. **Aquaspirillum itersonii** subspecies **itersonii** (Giesberger 1936) Hylemon, Wells, Krieg and Jannasch 1973, 370.[AL] (*Spirillum itersonii* Giesberger 1936, 68.)

Morphology and characteristics as for the species. Differs from the subspecies *nipponicum* by having a cell diameter of 0.4–0.6 μm, a mol% G + C of 60–64, and rapid formation of coccoid bodies (within 7 days).

Type strain: ATCC 12639.

5b. **Aquaspirillum itersonii** subspecies **nipponicum** (Terasaki 1973) Terasaki 1979, 140.[AL] (*Spirillum itersonii* subspecies *nipponicum* Terasaki 1973, 58.)

nip.po′ni.cum. M.L. neut. adj. *nipponicum* pertaining to the country of Japan.

Morphology and characteristics as for the species. Differs from the subspecies *itersonii* by having a cell diameter of 0.5–0.8 μm, a mol% G + C of 66, and a delayed formation of coccoid bodies (~2 weeks before they become predominant).

Type strain: IFO 13615.

6. **Aquaspirillum peregrinum** (Pretorius 1963) Hylemon, Wells, Krieg and Jannasch 1973, 370.[AL] (*Spirillum peregrinum* Pretorius 1963, 407.)

pe.re.gri′num. L. neut. adj. *peregrinum* strange, foreign.

The morphological characters are as depicted in Figure 2.1 and described in Tables 2.2 and 2.3.

Optimum temperature 32°C.

Physiological characters are as described in Tables 2.2 and 2.3. Sole carbon sources are listed in Table 2.4. Ammonium salts can be used as sole nitrogen sources. There are conflicting reports concerning the ability to use nitrate as a sole nitrogen source (Terasaki, 1972, 1979; Hylemon et al., 1973).

Isolated from a primary oxidation pond and from the putrid infusion of a freshwater mussel.

The mol% G + C of the DNA is 60–64 (T_m).

Type strain: ATCC 15387.

6a. **Aquaspirillum peregrinum** subspecies **peregrinum** (Pretorius 1963) Hylemon, Wells, Krieg and Jannasch 1973, 370.[AL] (*Spirillum peregrinum* subspecies *peregrinum* Pretorius 1963, 407.)

Morphology and characteristics as for the species. Differs from the subspecies *integrum* by forming coccoid bodies and by having a mol% G + C of 60–62 (T_m).

Type strain: ATCC 12639.

6b. **Aquaspirillum peregrinum** subspecies **integrum** (Terasaki 1973) Terasaki 1979, 141.[AL] (*Spirillum peregrinum* subspecies *integrum* Terasaki 1973, 60.)

in′te.grum. L. neut. adj. *integrum* unchanged (referring here to failure to form coccoid bodies).

Morphology and characteristics as for the species. Differs from the subspecies *peregrinum* by failing to form coccoid bodies as a predominant form even after 28 days of incubation, and by having a mol% G + C of 64 (T_m).

Type strain: IFO 13617.

7. **Aquaspirillum dispar** Hylemon, Wells, Krieg and Jannasch 1973, 372.[AL]

dis′par. L. neut. adj. *dispar* unlike.

The morphological characters are as depicted in Figure 2.1 and described in Tables 2.2 and 2.3.

Optimum temperature 30°C. Moderate growth at 25 and 37°C; no growth at 10 or 45°C.

Physiological characters are as described in Tables 2.2 and 2.3. Although originally described as being unable to grow anaerobically with nitrate (Hylemon et al., 1973), the species *does* grow well under these conditions (Aragno and Schlegel, 1978). Sole carbon sources are listed in Table 2.4. Ammonium salts are used well as sole nitrogen sources; nitrate is not used.

Isolated from fresh water.

The mol% G + C of the DNA is 63–65 (T_m).

Type strain: ATCC 27510.

8. **Aquaspirillum autotrophicum** Aragno and Schlegel 1978, 116.[AL]

au.to.tro′phi.cum. Gr. n. *autos* self; Gr. adj. *trophikos* nursing, tending or feeding; M.L. neut. adj. *autotrophicum* self-nursing or self-feeding.

The morphological characters are as described in Tables 2.2 and 2.3.

Optimum temperature 28°C.

Table 2.4.

Carbon sources used by the species of **Aquaspirillum**

Species	Citrate	Aconitate	Isocitrate	α-Ketoglutarate	Succinate	Fumarate	Malate	Oxaloacetate	Pyruvate	Lactate	Malonate	Tartrate	Acetate	Propionate	Butyrate	Caproate	β-Hydroxybutyrate	p-Hydroxybenzoate	Ethanol	n-Propanol	n-Butanol	Glycerol	D-Fructose	D-Glucose	D-Xylose	L-Arabinose	L-Histidine	L-Tyrosine	L-Phenylalanine	L-Alanine	L-Glutamate	L-Aspartate	L-Glutamine	L-Asparagine	L-Proline	L-Hydroxyproline	L-Ornithine	L-Citrulline	L-Arginine	L-Lysine	Putrescine	L-Methionine	L-Serine	L-Cysteine	Glycine	L-Leucine	L-Isoleucine	L-Valine	L-Tryptophan
1. *A. gracile*[a]	-	+	+	+	+	-	-		+	+			+	-		-	-	-	-			+	-	+	+	+	-	-	-	-	+	+	-	-	-	-	-	-	-	-	-	-	-	-	-	-	-	-	-
2. *A. delicatum*																																																	
Method A[b]	-	-	+	+	w	+	+	+	w	w	+	+	+	-		-	-	-	w	-	-	-	-	-	-	-	-	-	-	-	-	+	+	-	+	-	-	-	-	-	+	-	-	-	-	-	-	-	-
Method B[c]	-	-	-	+	w	w	w	+	w	w	w	-	w	-	w	-	-	-	w	-	w	w	-	-	-	-	-	-	-	-	-	+	+	-	+	-	-	-	-	-	-	-	-	-	-	-	-	-	-
3. *A. polymorphum*																																																	
Method A[b]	-	-	-	-	-	+	+	-	-	-	+	-	+	-	-	-	-	-	-	-	-	-	-	-	-	-	-	-	-	-	+	+	+	+	+	-	-	-	-	-	-	-	-	-	-	-	-	-	-
Method B[c]	-	-	-	+	+	+	+	+	+	+	w	+	w	w	-	-	+	-	-	-	-	-	-	-	-	-	-	-	-	-	+	+	+	+	+	+	-	+	-	-	-	-	-	-	-	-	-	-	-
4. *A. aquaticum*[d]	-	+	+	+	+	+	+	+	+	+	+	+	+	+	-	-	+	-	-	-	-	-	+	-	-	-	-	-	-	-	+	+	+	+	+	+	-	-	-	-	-	-	-	-	-	-	-	-	-
5. *A. itersonii*																																																	
Method A[b,e]	-	+	-	+	+	+	+	d	d	d	d	-	-	d	-	-	+	-	+	+	d	d	+	-	-	-	+	-	+	d	+	+	+	+	+	d	d	d	d	d	d	-	-	d	-	-	d	d	-
Method B[c]	d	+	-	d	d	d	d	d	d	d	d	-	+	d	d	-	+	-	d	d	d	d	+	-	-	-	-	-	+	d	d	+	+	+	+	d	d	d	d	d	d	-	-	d	-	-	d	d	d
6. *A. peregrinum*																																																	
Method A[b,f]	-	+	-	+	+	+	+	+	+	+	+	-	+	+	+	+	+	-	+	+	+	-	+	-	-	-	-	-	-	+	+	+	+	+	+	+	-	+	-	-	-	-	-	-	-	-	-	-	-
Method B[c]	+	+	+	+	+	+	+	+	+	+	+	+	+	+	+	+	+	+	+	+	+	d	+	-	-	-	-	+	-	+	+	+	+	+	+	-	+	+	-	-	-	-	-	-	+	+	+	-	+
7. *A. dispar*[b]	+	+	+	+	+	+	+	+	+	+	+	-	+	+	-	-	-	-	d	-	-	-	-	-	-	-	-	+	+	+	d	-	-	-	-	-	-	-	-	-	-	-	-	-	-	-	-	-	-
8. *A. autotrophicum*[g]																												+	+	+	d	-	+	+	+	-	-	-	-	-	-	-	-	-	+	+	+	-	+
9. *A. sinuosum*																																																	
Method A[b]	-	-	-	-	w	w	w	+	w	-	-	-	-	-	-	-	+	-	-	-	-	-	-	-	-	-	-	-	-	-	-	+	+	+	+	-	-	-	-	-	-	-	-	-	-	-	-	-	-
Method B[c]	-	-	-	-	w	-	w	+	w	w	-	-	w	-	d	-	-	-	-	-	-	-	-	-	-	-	-	-	-	-	-	-	-	-	-	-	-	-	-	-	-	-	-	-	-	-	-	-	-
10. *A. psychrophilum*[c,h]	-	-	d	-	-	-	-	+	-	-	-	-	-	-	-	-	-	-	-	-	-	-	-	-	-	-	-	-	-	-	-	-	-	-	+	-	-	-	+	-	-	-	-	-	-	-	-	-	-
11. *A. fasciculus*[i]	-	-	-	-	+	-	-	+	+	+	-	-	-	-	-	-	+	-	-	-	-	-	-	-	-	-	-	-	-	-	-	+	+	+	+	-	-	-	+	-	-	-	-	-	-	-	-	-	-
12. *A. serpens*																																																	
Method A[b]	-	-	d	d	d	-	-	d	d	d	-	-	-	-	-	-	-	-	-	-	-	-	-	-	-	-	-	-	-	d	d	d	-	+	+	d	-	-	-	-	-	-	-	-	d	-	-	-	-
Method B[c]	-	-	-	-	+	+	+	+	+	+	-	-	+	d	+	-	-	-	-	-	-	-	-	-	-	-	-	-	-	+	+	+	d	d	d	d	-	-	-	-	-	-	d	-	d	-	-	-	-

13. A. putridiconchylium		
Method A[b]		
Method B[c]		
14. A. bengal[i]		
15. A. metamorphum		
Method A[b]		
Method B[c]		
16. A. giesbergeri		
Method A[b]		
Method B[c]		
17. A. anulus		
Method A[b]		
Method B[c]		

[a] As determined by the method of Canale-Parola et al. (1966). A complex growth-limiting medium was used containing the carbon sources at 0.05%. Growth in the presence of the test compounds was compared turbidimetrically or by microscopic count to that occurring in the absence of the compounds. Symbols: +, >10% increase in the growth of all strains in the presence of the test compound; −, 10% or less increase in growth; blank space, not determined.

[b] As determined by the method of Hylemon et al. (1973). A turbidimetrically standardized cell suspension in physiological saline was inoculated into a defined, vitamin-free medium containing the carbon sources (0.1%) and ammonium sulfate as the nitrogen source. Growth responses were measured turbidimetrically after one 72-h serial transfer from the initial cultures, using a Klett colorimeter with the blue (420 nm) filter and 16-mm cuvettes. Symbols: +, 10 or more Klett units of turbidity for all strains tested; −, less than 10 Klett units of turbidity; d, differs among strains; blank space, not determined.

[c] As determined by the method of Terasaki (1972, 1979). A cell suspension washed in basal, defined, vitamin-free medium (Williams and Rittenberg, 1957) lacking carbon sources was inoculated into similar media containing the test compounds (0.05%) and ammonium chloride as the nitrogen source. After 7 days growth was estimated turbidimetrically. Symbols: +, a turbidity of 0.025 absorbance units or greater for all strains tested; W, a turbidity of less than 0.025; −, no growth (turbidity equals the same as the appearance of controls without a carbon source); d, differs among strains; blank space, not determined.

[d] As determined by the method of Kropinski (1975). Samples (0.1 ml) of a washed cell suspension were spread on plates of a minimal agar medium containing ammonium sulfate and niacin. Approximately 8 mg of the test compounds were placed in small areas on the plates. After incubation for 48 h the growth response was estimated. Symbols: +, growth in the area around the test compound; −, no growth; blank space, not determined.

[e] Strains of the subspecies nipponicum were not tested.

[f] Strains of the subspecies integrum were not tested.

[g] The utilization of compounds as sole carbon sources was tested on agar plates as described by Stanier et al. (1966) using a velvet-disk replicator. The medium was the basal mineral agar described under Procedures in Testing Special Characters, supplemented with 0.2% carbohydrates or 0.1% of other compounds. Symbols: +, growth greater than on control plate with no carbon source; −, growth no greater than on control plate; blank space, not determined.

[h] The nutritional requirements of A. psychrophilum have not yet been determined.

[i] As determined by the method of Strength et al. (1976). A washed, turbidimetrically standardized suspension was inoculated into a defined, vitamin-free, semisolid medium containing the test compounds (on an equal carbon basis relative to 0.2% fumaric acid) and ammonium sulfate as the nitrogen source. Growth responses were measured turbidimetrically at 36 h after gently inverting the semisolid cultures several times to obtain an even distribution of cells. A Klett colorimeter was used (blue filter, 420 nm) with 16-mm cuvettes. Symbols: +, production of at least 10 Klett units of turbidity; −, less than 10 Klett units; blank space, not determined.

[j] As determined by Kumar et al. (1974), using a modification of the method of Hylemon et al. (1973b). Symbols: +, a turbidity of 0.03 absorbance units or more, using a green filter and 16-mm cuvettes; −, turbidity less than 0.03; blank space, not determined.

Physiological characters are as described in Tables 2.2 and 2.3. Sole carbon sources are listed in Table 2.4. Ammonium salts and nitrate can be used as sole nitrogen sources.

Isolated from a eutrophic lake in Switzerland.

The mol% G + C of the DNA is 60–62 (T_m).

Type strain: DSM 732.

9. **Aquaspirillum sinuosum** (Williams and Rittenberg 1957) Hylemon, Wells, Krieg and Jannasch 1973, 368.[AL] (*Spirillum sinuosum* Williams and Rittenberg 1957, 94.)

si.nu.o′sum. L. neut adj. *sinuosum* full of curves.

The morphological characters are as depicted in Figure 2.1 and described in Tables 2.2 and 2.5.

Optimum temperature 30°C.

Physiological characters are as described in Tables 2.2 and 2.5. Sole carbon sources are listed in Table 2.4. Ammonium salts can be used as a sole nitrogen source; nitrate is not used.

Isolated from freshwater.

The mol% G + C of the DNA is 57–59 (T_m).

Type strain: ATCC 9786.

10. **Aquaspirillum psychrophilum** (Terasaki 1973) Terasaki 1979, 138.[AL] (*Spirillum psychrophilum* Terasaki 1973, 57.)

psy.chro′phi.lum. Gr. adj. *psychros* cold; Gr. adj. *philus* liking, preferring; M. neut. adj. *psychrophilum* preferring cold.

The morphological characters are as described in Tables 2.2 and 2.5.

Grows in nutrient broth or PSS broth. Optimum temperature 20°C. No growth above 26°C. The organism may best be considered as psychrotolerant or psychrotrophic rather than psychrophilic (Morita, 1975).

Physiological characters are as described in Tables 2.2 and 2.5. This species has not been cultured in vitamin-free defined media and most likely has a growth factor requirement.

Isolated from antarctic mosses.

The mol% G + C of the DNA is 65 (T_m).

Type strain: IFO 13611.

11. **Aquaspirillum fasciculus** Strength, Isani, Linn, Williams, Vandermolen, Laughon and Krieg 1976, 266.[AL]

fas.ci′cu.lus. L. mas. dim. n. *fasciculus* a small bundle.

The only species in the genus that consists of straight rods. Curved or S-shaped variants have been reported to occur in one strain after prolonged serial transfer (Strength et al., 1976). Cells from broth cultures have bipolar flagellar fascicles composed of up to 11 flagella (Fig. 2.4). The fascicles can be clearly seen by dark-field microscopy and show an unusual and distinctive behavior when the cells are suspended in ordinary, nonviscous media; helical wave propagation with waves progressing from base to tip (Fig. 2.5*A*), an ability to coil up like springs (Fig. 2.5*B*) and basal bending accompanied by a change in wavelength (Fig. 2.5*C*). The behavior of one fascicle is not coordinated with that of the fascicle at the opposite pole. In ordinary media the cells do not swim and, instead, exhibit an ineffectual "floundering about" movement. When cells from broth cultures are suspended in a medium of high viscosity (10–200 centipoise, obtained by the use of agents such as methylcellulose "400 centipoise," they swim steadily in straight lines (Strength et al., 1976). When viscous cell flocs formed by freshly isolated strains are crushed with a glass rod and homogenized in a small quantity of distilled water, free-swimming cells can also be seen moving in straight lines; here, the tailing flagellar fascicle is extended behind each cell, while the leading fascicle is either coiled into a polar loop or is coiled around the cell (Fig. 2.5*D*). The ability to swim only in viscous media may represent an adaptation to the viscous conditions that occur within cell flocs.

Initial isolation cannot be achieved by ordinary streaking onto agar media, and is instead achieved after preliminary enrichment by the pour-plate method, using L-proline as the sole carbon and nitrogen

source (see Strength et al. (1976) for details). On initial isolation, *A. fasciculus* forms highly viscous flocs. This floc-forming ability is gradually lost during subsequent transfers, and eventually the strains exhibit homogeneous, turbid growth. A suitable liquid medium containing peptone, fumarate and minerals has been described by Strength et al. (1976). Optimum temperature, 30°C; no growth at 20 or 40°C.

Physiological characters are as described in Tables 2.2 and 2.5. Sole carbon sources are listed in Table 2.4; pyruvate and proline are the most effective. Sole nitrogen sources include nitrate, ammonium salts, and 10 amino acids, with L-proline and L-alanine being the most effective. In defined media, growth occurs best when 0.15% agar is added to give a semisolid consistency (Strength et al., 1976).

Habitat: pond water.

The mol% G + C of the DNA is 62–65 (T_m).

Type strain: ATCC 27740.

12. **Aquaspirillum serpens** (Müller 1786) Hylemon, Wells, Krieg and Jannasch 1973, 366.[AL] (*Vibrio serpens* Müller 1786, 48.)

ser′pens. L. v. *serpo* to crawl or creep; L. part. adj. *serpens* creeping.

The morphological characters are as depicted in Figure 2.1 and described in Tables 2.2 and 2.5.

Optimum temperature 35°C.

Physiological characters are as described in Tables 2.2 and 2.5. Sole carbon sources are listed in Table 2.4; the best growth occurs with glutamate. Nitrate is not used as a sole nitrogen source; results with ammonium salts have been conflicting (Hylemon et al., 1973; Terasaki, 1972). A defined medium suitable for batch and continuous cultures has been described by Whitby and Murray (1980).

Habitat: pond water.

The mol% G + C of the DNA is 62–65 (T_m).

Type strain: ATCC 27740.

13. **Aquaspirillum putridiconchylium** (Terasaki 1961) Hylemon, Wells, Krieg and Jannasch 1973, 367.[AL] (*Spirillum putridiconchylium* Terasaki 1961, 80.)

pu′tri.di.con.chy.li.um. L. adj. *putridus* putrid, decayed; L. n. *conchylium* a shellfish; L. n. *putridiconchylium* decayed shellfish.

The morphological characters are as depicted in Figure 2.1 and described in Tables 2.2 and 2.5.

Optimum temperature 32°C.

Physiological characters are as described in Tables 2.2 and 2.5. Sole carbon sources are listed in Table 2.4. Ammonium salts can be used as a sole nitrogen source; nitrate is not used.

Isolated from the putrid infusion of a fresh water mussel.

The mol% G + C of the DNA is 52 (T_m).

Type strain: ATCC 15279.

14. **Aquaspirillum bengal** Kumar, Banerjee, Bowdre, McElroy and Krieg 1974, 457.[AL]

ben′gal. M.L. n. *bengal* Bengal.

The morphological characters are as depicted in Figure 2.1 and described in Tables 2.2 and 2.5.

Growth in PSS broth abundant, cloudy. Mechanical shaking of cultures fails to increase the rate of growth, and a microaerophilic band of cells forms after ~5 min in wet mounts; these results suggest that the species may possibly prefer microaerobic conditions for growth, although good growth does occur under an air atmosphere. Optimum temperature 41°C.

Physiological characters are as described in Tables 2.2 and 2.5. Sole carbon sources are listed in Table 2.4. Ammonium salts can be used as a sole nitrogen source; nitrate is not used.

Isolated from a freshwater pond in West Bengal. Occurs in greatest numbers during months when the temperature of the pond water is 30°C or higher.

The mol% G + C of the DNA is 52 (T_m).

Type strain: ATCC 27641.

Table 2.5.

Other characteristics of Aquaspirillum species 9–17[a]

Characteristics	9. A. sinuosum	10. A. psychrophilum	11. A. fasciculus	12. A. serpens	13. A. putridiconchylium	14. A. bengal	15. A. metamorphum	16. A. giesbergeri	17. A. anulus
Wavelength of helix, μm	8.6–10.5	5.5–6.5	–[b]	3.5–12.0	4.5–7.0	4.6–8.1	7.5–12.0	4.5–8.4	5.0–13.0
Helix diameter, μm	1.4–3.5	1.0–1.4	–[b]	1.2–4.2	1.2–2.0	1.7–2.3	2.2–3.5	1.2–5.0	1.7–4.5
Length of helix, μm	5.0–42.0	1.5–14.0	3.6–43.0[b]	3.5–42.0	4.0–23.0	5.2–22.0	3.5–11.0	4.0–40.0	4.0–52.0
Catalase	+	+	+	+	W or –	+	+	+	+
Oxidase	+	+	+	+	+	+	+	+	+
Growth in presence of:									
1% Oxgall	+		+	+	+	+	+	+	–
1% Glycine	–		–	–	–	–	–	–	–
3% NaCl	–	–	–	–	–	–	–	–	–
Water-soluble brown pigment formed in the presence of:									
0.1% Tyrosine	–		–	–	–	+	–	–	–
0.1% Tryptophan	–		–	–	–	+	–	–	–
Water-soluble yellow-green fluorescent pigment	–		W	d	–	–	+	–	–
Alkaline reaction in litmus milk	–		–	d	–	–	–	–	–
Deoxyribonuclease (DNase)	+			d	–	+[c]	+	+	d
Ribonuclease (RNase)	–			d	+	–	+	+	–
Growth on:									
Eosin methylene blue agar	–			+	+	–	+	–	–
MacConkey agar	–			d	–	–	–	–	–
Triple sugar iron agar	–			d	+	+	+	–	–
Sellers agar	–			d	–	+	+	+	–
Methyl red-Voges Proskauer broth	–			d	–	–	–	–	–
Reduction of 0.3% H$_2$SeO$_3$	+			+	+	+	+	d	+
H$_2$S from 0.2% cysteine in PSS broth, 7 days	+		– or W	+	+	+	+	d	+
H$_2$S from 0.01% cystine in nutrient broth, 7 days	–	–		+	+	+	+	W	W
Indole test	–	–	–	–	–	–	–	–	–
Hydrolysis of casein and starch	–	–	–	–	–	–	–	–	–
Hydrolysis of hippurate	–	–	–	–	–	–	–	–	–
Gelatin hydrolysis, 30°C, 4 days[d]	–	–	–	d	–	–	+	–	–
Gelatin liquefaction, 20°C[e]									
7 Days	+	+		d	–		+	+	–
28 Days	+	+		d	+		+	+	–
Sulfatase (0.01% phenolphthalein disulfate)	–						–	–	–
Temperature range for growth, °C	9–37	2–26		12–44	8–40	15–42	3–38	9–36	3–36
pH range for growth	6.0–9.0	5.5–9.0	5.5–8.5	6.0–9.0	5.5–8.5	6.0–8.4	6.0–9.0	6.0–9.0	6.0–8.5

[a] Symbols: +, all strains positive; –, all strains negative; W, weak; d, differs among strains; blank space, not determined.
[b] A. fasciculus is a straight rod. The helix length refers to the length of the rod.
[c] Positive at 37°C but not at 41°C.
[d] Method of Hylemon et al. (1973b).
[e] Method of Terasaki (1972).

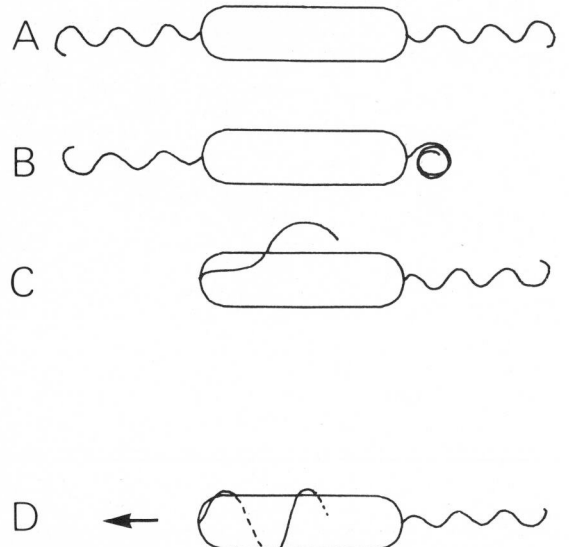

Figure 2.5 Flagellar behavior in *A. fasciculus*. *A–C*, flagellar orientations observed in nonviscous media: *A*, flagellar fascicles extended; *B*, coiling into a polar loop; and *C*, basal bending accompanied by a change in wavelength. *D*, orientation of fascicles in motile cells suspended in a viscous medium; the *arrow* indicates the direction of swimming.

15. **Aquaspirillum metamorphum** (Terasaki 1961) Hylemon, Wells, Krieg and Jannasch 1973, 366.[AL] (*Spirillum metamorphum* Terasaki 1961, 220.)

me.ta.mor′phum. Gr. neut. adj. *metamorphum* changing.

The morphological characters are as depicted in Figure 2.1 and described in Tables 2.2 and 2.5.

Optimum temperature 30°C.

Physiological characters are as described in Tables 2.2 and 2.5. Sole carbon sources are listed in Table 2.4. Ammonium salts can be used as a sole nitrogen source; nitrate is not used.

Isolated from the putrid infusion of a freshwater mussel.

The mol % G + C of the DNA is 63 (T_m).

Type strain: ATCC 15280.

16. **Aquaspirillum giesbergeri** (Williams and Rittenberg 1957) Hylemon, Wells, Krieg and Jannasch 1973, 368.[AL] (*Spirillum giesbergeri* Williams and Rittenberg 1957, 88.)

gies′ber.ger.i. M.L. n. *giesbergeri* of Giesberger, the first investigator to study certain physiological characteristics of spirilla.

The morphological characters are as depicted in Figure 2.1 and described in Tables 2.2 and 2.5.

Optimum temperature 30°C.

Physiological characters are as described in Tables 2.2 and 2.5. Sole carbon sources are listed in Table 2.4. Ammonium salts can be used as sole nitrogen sources; nitrate is not used.

This species includes organisms previously assigned to the two species *Spirillum giesbergeri* and *Spirillum graniferum* by Williams and Rittenberg (1957). The two species were combined into a single species by Hylemon et al. (1973) on the basis of a high degree of similarity in phenotypic characters and in DNA base composition.

Isolated from pond water.

The mol% G + C of the DNA is 57–58 (T_m).

Type strain: ATCC 11334.

17. **Aquaspirillum anulus** (Williams and Rittenberg 1957) Hylemon, Wells, Krieg and Jannasch 1973, 368.[AL] (*Spirillum anulus* Williams and Rittenberg 1957, 86.)

a′nu.lus. L. masc. n. *anulus* a ring.

The morphological characters are as depicted in Figure 2.1 and described in Tables 2.2 and 2.5.

Optimum temperature 27–30°C.

Physiological characters are as described in Tables 2.2 and 2.5. Sole carbon sources are listed in Table 2.4. Ammonium salts can be used as sole nitrogen sources; nitrate is not used.

Isolated from pond water and from putrid infusions of fresh water mussels.

The mol% G + C of the DNA is 58–59 (T_m).

Type strain: NCIB 9012.

Other spirilla possibly belonging to the genus **Aquaspirillum**

According to Morita (1975) psychrophiles are defined as organisms having an optimum temperature for growth of ~15°C or lower, a maximum temperature for growth of ~20°C, and a minimum growth temperature of 0°C or below. A spirillum conforming to this definition has been isolated from Antarctica and is described below. The species may belong to the genus *Aquaspirillum*; however, it is atypical in that it possesses only a single polar flagellum rather than bipolar tufts and produces acid from a carbohydrate (xylose); moreover, the tolerance to 3% NaCl has not yet been reported.

a. "*Spirillum pleomorphum*" Inoue and Komagata 1976, 170.

ple.o.mor′phum. L. adj. *pleomorphum* pleomorphic.

Helical cells, curved rods, crescent-shaped cells, U-form cells and nearly ringlike forms occur on peptone-yeast extract-glucose (PYG) agar. Cell size: 0.7–1.0 μm by 2.0–4.5 μm. Motile by a single polar flagellum.

Growth in PYG broth is turbid with sediment. Colonies on PYG agar are circular, smooth, convex, entire, opaque and pale brown. Optimum temperature, 9°C; maximum, 20°C; minimum below 0°C.

Aerobic. Oxidase- and catalase-positive. Indole, Voges-Proskauer and methyl red tests are negative. No growth with 5% NaCl. Acid but no gas from xylose (aerobically). No acid or gas from glucose, lactose, sucrose, maltose, arabinose or glycerol aerobically or anaerobically. No change in litmus milk. H_2S is not produced. Starch and cellulose are

not degraded. Nitrate is reduced to nitrite. No growth occurs anaerobically with nitrate.

Succinate, formate, acetate, fumarate and propionate are assimilated. Citrate, lactate, protocatechuate, *p*-hydroxybenzoate and hippurate are not assimilated.

Isolated from antarctic soil.

The mol% G + C of the DNA is 63 (T_m).

Type strain: IAM 12028 (strain 22-o-d).

b. Magnetotactic spirilla. A microaerophilic, magnetotactic spirillum has been isolated by Blakemore et al. (1979) from sediments from a freshwater swamp in Massachusetts. This organism may belong to the genus *Aquaspirillum*; however, it differs from the typical characteristics of the genus with respect to having a single flagellum at each pole rather than bipolar flagellar tufts, being an obligate microaerophile, and being oxidase negative, in addition to its magnetotactic properties. The name *Aquaspirillum magnetotacticum* will be proposed in the near future for the organism by D. Maratea and R. Blakemore (personal communication). The characteristics of the organism are as follows.

Helical (clockwise) cells, 0.28–0.36 μm in diameter by 2.3–11.1 μm in length. Motile by a single flagellum at each pole. Intracellular poly-β-hydroxybutyrate is present. A chain of 5–41 electron-dense inclusion bodies ("magnetosomes") containing magnetite (Fe_3O_4) is present within each cell and is responsible for the tactic response of the spirilla to magnetic fields. The magnetosomes are cubic with rounded edges, ~42 nm wide (Balkwill et al., 1980). In an artificial magnetic field the motile spirilla orient within several seconds with their long axis in the north-south axis of the field. In strong magnetic fields, such as produced by attaching a small permanent magnet to the side of the culture tube, the cells accumulate near the poles of the magnet. In nature, the role of such magnetotaxis in response to the earth's magnetic field may be to direct the motile cells downward in aquatic environments to regions of low oxygen tension (Blakemore et al., 1979).

Gram-negative. Coccoid bodies are formed in old cultures.

Microaerophilic, growing best under an atmosphere of 1–3% oxygen; unable to grow anaerobically or under an air atmosphere, although aerobic growth can occur when catalase is added to the medium (Blakemore et al., 1979). When cultured at initial oxygen levels of 6% or higher, cells are not magnetic. In semisolid media incubated under an air atmosphere, growth occurs initially as a thin band some distance from the surface; as growth becomes more dense this band migrates progressively upwards toward the agar-air interface.

No growth under strict anaerobic conditions; however, at low oxygen levels nitrate is dissimilated to nitrous oxide and nitrogen gas (Escalante-Semerena et al., 1980).

Chemoheterotrophic, growing best in a simple, chemically defined medium (Blakemore et al., 1979). Oxidase-negative. Catalase-negative. Fumarate, tartrate, malate, succinate, lactate, pyruvate, oxaloacetate, β-hydroxybutyrate and maleate are used as sole carbon sources. Sugars, sugar alcohols, glycerol, gluconate, citrate, isocitrate, aconitate, α-ketoglutarate, benzoate, quinate, pimelate, phthallate, mandelate, sulfanilate, methanol, butanol, sec-butanol, t-butanol, ethanol, propanol, isopropanol, benzene, xylene, toluene, phenylalanine, glutamate, alanine, serine and lysine are not used. Butyrate, isobutyrate, propionate, protocatechuate, formate, oxalate, caproate, L-epinephrine, L-β-3, 4-dihydroxyphenylalanine, catechol, L-arterenol, phenol, methionine and cysteine are inhibitory. Ammonium ions or nitrate can serve as sole nitrogen sources.

A nonmagnetotactic variant, which lacks magnetosomes, has been obtained (Blakemore et al., 1979; Balkwill et al., 1980).

The mol% G + C of the DNA is 65 (T_m).

Suggested type strain: Strain MS-1; ATCC 31632 (D. Maratea and R. Blakemore, personal communication).

Editorial Note.

After completion of the manuscript for the present article, the name *Aquaspirillum magnetotacticum* Maratea and Blakemore 1981, 454 was validly published in the International Journal of Systematic Bacteriology, together with a complete description. ATCC strain 31632 was designated the type strain. Although oxidase-negative by ordinary tests, the species has been found to give a faintly positive delayed test when toluene-treated cells are employed (Maratea and Blakemore, 1981).

Species Incertae Sedis

The following two species do not belong to either the genus *Aquaspirillum*, *Oceanospirillum* or *Spirillum*, and their placement is uncertain. Studies of these species have been hampered by lack of reproducible in vitro cultivation methods. They do not appear on the Approved Lists of Bacterial Names because no type or reference strains are available and the organisms are not well characterized. The disease syndromes caused by these species are distinct and recognizable, however. If possible, neotype strains should be designated and either maintained by animal passage or preserved in a recognized culture collection.

a. "*Spirillum minus*" Carter 1888, 47. (*Spirillum minor* (*sic*) Carter 1888, 47; the specific epithet *minor* is grammatically incorrect as noted by Robertson (1924) and the correct form is *minus*.)

mi'nus. L. neut. adj. *minus* less, smaller.

Rigid cells; usually described as spiral with two or three turns, although the waves have been reported to be planar (McDermott, 1928).

The ends of the cell may be blunt or pointed. Cell diameter, ~0.2 μm; cell length, 3–5 μm; wavelength, 0.8–1.0 μm. Actively motile by one or more flagella at each pole.

Causes one of the two forms of rat-bite fever in man. The disease caused by "*S. minus*" is often termed "Sodoku"; it occurs worldwide but has its greatest frequency in the Far East. The organisms are usually transmitted to humans through the bite of an infected rat, although mice, squirrels and rodent-ingesting animals such as cats, dogs, ferrets and weasels have also been implicated. "*S. minus*" appears to be a natural parasite of rats, which act as carriers; the infection is usually not lethal in rats. The natural infection frequency for rats varies from country to country but may be as high as 25% (see Babudieri, 1973, for pertinent literature).

The clinical aspects of rat-bite fever and the distinctions between the form caused by "*S. minus*" and that caused by *Streptobacillus moniliformis* have been summarized by Joklik et al. (1980) and by Rogosa (1980). Experimental infections of humans and animals by "*S. minus*" have been described by Babudieri (1973).

"*S. minus*" is best observed in blood or exudates from patients by dark-field or phase-contrast microscopy of wet mounts; staining with Giemsa or Wright's stain or by silver impregnation is also useful.

"*S. minus*" is cultured in vivo by intraperitoneal inoculation of patients' blood or exudates from lesions, or blood from naturally infected rats, into spirillum-free mice or guinea pigs (Rogosa, 1980); mice are the animals most susceptible to "*S. minus*" infection (Babudieri, 1973). It is questionable whether the organism has ever been cultured successfully in artificial media. Numerous attempts have failed, and various claims of successful cultivation have been unable to be confirmed. One report that may indicate successful cultivation is that by Hitzig and Liebesman (1944), who inoculated blood from a patient into 2% dextrose-veal infusion broth and into 10% tomato extract-veal infusion broth. The addition of citrated human or rabbit blood was required for successful subculturing; also, the organisms initially required incubation in a candle jar but eventually were able to grow aerobically after 5 months of serial transfer. Confirmation of this report is needed. Considering the morphology, pathogenicity and sources of "*S. minus*," serious attention should be given to the possibility that the organism might belong to, or be related to, the genus *Campylobacter*, and the microaerophilic techniques employed for campylobacters might also prove useful for "*S. minus*."

Type strain: none. *Reference strains:* none.

b. "*Spirillum pulli*" Mathey 1956, 745.

pul'li. L. gen. n. *pulli* of a young chicken.

Rigid spiral cells. By dark-field microscopy the cell diameter is ~1 μm and the cell length is from 5–12 μm. Actively motile by means of a single flagellum at each end of the cell.

Believed to be the cause of a diphtheroid stomatitis in the mouths of adult chickens. The lesions are yellowish white, rather firm, and adherent to the underlying tissue; often they are symmetrical ovoids, one at each side of the lower jaw. Lesions also occur on the palate, the lower surface of the tongue, on the floor of the mouth, between the larynx and the transverse row of papillae on the tongue, around the larynx, and on the walls of the pharynx. The lesions vary in size from ~2–20 mm.

Attempts to culture "*S. pulli*" in artificial media have been unsuccessful. Experimental passage of the disease in chickens has been accomplished by contact and by experimental inoculation.

Type strain: none. *Reference strains:* none.

Genus Incertae Sedis

a. "*Sporospirillum*" Delaporte 1964, 257.

spor.o.spi.ril'lum. Gr. n. *sporos* a seed (spore); Gr. n. *spira* a spiral; M.L. dim. n. *spirillum* a small spiral; M.L. neut. n. *Sporospirillum* a small spore (-forming) spiral.

Rigid, helical bacteria of enormous size, 1.8–4.8 μm in diameter and 40–100 μm in length. Structures that morphologically resemble endospores occur within the cells, but their thermal resistance has not been determined. The sporelike structures have the ability to rotate and to migrate within the cytoplasm of the bacteria. They initially develop near the cell poles and later migrate to the center where they are released after the cell ruptures and disintegrates. The Gram-reaction has not been reported. The cells are motile, but no organs of locomotion are evident. The relationship of the cells to oxygen is unknown. Occur in the intestinal contents of tadpoles. Have not been isolated.

Type species: none designated.

List of the Species of the Genus "Sporospirillum"

aa. "*Sporospirillum praeclarum*" (Collin 1913) Delaporte 1964, 259. (*Spirillum praeclarum* Collin 1913, 62.)

prae.cla′rum. L. adj. *praeclarum* distinguished, famous.

Cell diameter, 3.0–4.0 μm. Length, 50–100 μm. Diameter of helix, 5–10 μm. Wavelength, 17–23 μm. A single endospore is present, 3–4 μm by 9–12 μm.

bb. "*Sporospirillum gyrini*" Delaporte 1964, 259.

gy.ri′ni. L. n. *gyrinus* a tadpole; L. gen. n. *gyrini* of a tadpole.

Cell diameter, 1.8–2.6μm. Length, 40–100 μm. Diameter of helix, 3–6 μm. Wavelength, 13–20 μm. A single endospore is present, 2 μm by 5–7 μm.

cc. "*Sporospirillum bisporum*" Delaporte 1964, 260.

bi.spo′rum. L. adv. *bis* twice; G. n. *sporos* a seed; M.L. gen. pl. n. *bisporum* of two seeds (spores).

Cell diameter, 3.5–4.8 μm. Length, 50–90 μm. Diameter of helix, 11–15 μm. Wavelength, 27–35 μm. At each pole an endospore occurs, 2–4 μm by 10–14 μm.

Genus **Spirillum** Ehrenberg 1832, 38[AL]

NOEL R. KRIEG

Spi.ril′lum. Gr. n. *spira* a spiral; M.L. dim. neut. n. *Spirillum* a small spiral.

Rigid, helical cells, 1.4–1.7 μm in diameter by 14–60 μm in length. A **polar membrane** underlies the cytoplasmic membrane at the cell poles and is visible in ultrathin sections. Intracellular **poly-β-hydroxybutyrate** granules are formed. Coccoid bodies are not formed. Gram-negative. Motile by large **bipolar tufts of flagella** having a long wavelength and about one helical turn; these are easily visible by dark-field or phase-contrast microscopy. **Microaerophilic** in ordinary liquid media, but can grow aerobically in special media or with certain supplements. Colonies on solid media can be obtained only under special conditions. Have a **strictly respiratory type of metabolism** with oxygen as the terminal electron acceptor. Growth does not occur anaerobically with nitrate. Optimum temperature, 30°C. **Oxidase- and phosphatase-positive. Catalase-negative.** Indol- and sulfatase-negative. Casein, starch, esculin, gelatin, DNA and RNA are not hydrolyzed. Inhibited by extremely low levels of hydrogen peroxide in the culture medium. NaCl levels above 0.02% are inhibitory. Phosphate levels greater than 0.01 M are inhibitory. **Carbohydrates are not catabolized.** The salts of certain organic acids, particularly succinate, are used as carbon sources. Vitamins are not required. Occur in **stagnant, freshwater environments.** The mol% G + C of the DNA is 38 (T_m) or 36 (Bd).

Type species: *Spirillum volutans* Ehrenberg 1832, 38.

Further Descriptive Information

The genus presently contains only a single species, *S. volutans*. This species consists of very large, helical cells (for a photomicrograph see the genus *Aquaspirillum*, Fig. 2.1P), and the only other spirilla that are of comparable size are members of the phototrophic genus *Thiospirillum*.

An unusual elaboration of the plasma membrane, the "polar membrane" occurs in *S. volutans* (Coulton and Murray, 1978). It is attached to the inside of the plasma membrane by barlike links and is located in the region surrounding the polar flagella. Such a membrane has been found mainly in genera of helical bacteria, such as *Aquaspirillum*, *Campylobacter*, *Ectothiorhodospira* and *Rhodospirillum*.

Intracellular poly-β-hydroxybutyrate occurs in the form of prominent granules which are refractile by phase-contrast microscopy and which stain with metachromatic stains such as Ponder's stain (Wells and Krieg, 1965) or with lipid stains such as Sudan black.

The cells are actively motile and swim in straight lines with frequent reversal of direction. The bipolar flagellar fascicles are exceptionally large and consist of many individual flagella (Fig. 2.6). As noted by Metzner (1920), during its rotation the fore fascicle appears to describe a wide bell which is opened toward the rear of the cell; the aft fascicle extends behind the cell and appears to describe a wide goblet (Fig. 2.7). There is no true anterior cell pole: when the fascicles change their direction of rotation and their orientation the cell reverses its direction of swimming (Fig. 2.7). Because of the normally wide zones of rotation of the fascicles, Metzner believed that the mechanical effect of the flagella was mainly indirect, i.e. to cause an opposite rotation of the cell body which, because of its helical shape, would then screw through the medium. This type of propulsion would differ markedly from that of most other kinds of bacteria and might be taxonomically important; however, Metzner also noted that the aft fascicle sometimes described the form of a narrow bell that provided a direct screwlike thrust. Padgett et al. (1983) reported that straight mutant cells could swim at nearly the same speed as helical cells. This indicated that direct thrust was operative and seemed to contradict the concept that both flagellar fascicles normally act mainly by causing indirect thrust. However, further observations of this mutant (N. R. Krieg and M. W. Friedman, personal communication) have indicated that the zone of rotation of the aft fascicles is narrower than that of the wild type cells; therefore, the mutant may not have provided a sufficient test. Winet and Keller (1976) provided evidence that the aft fascicle of helical cells beats in a helical fashion just as other bacterial flagella do. Swan (1982) reported that helical cells flagellated at only one pole were capable of reversing swimming direction, indicating that a fore fascicle alone could propel the cells; just how it does this is not yet clear.

S. volutans is a microaerophile but can be cultivated easily in a semisolid medium, such as MPSS broth or CHSS broth* prepared with 0.15% agar, incubated under an air atmosphere. Growth is initiated as a thin band or disc several millimeters or centimeters below the surface of the medium, where the respiratory rate of the cells matches the rate of diffusion of oxygen to the cells. As the cell numbers increase, the band becomes more dense and migrates closer to the surface. Dense growth just beneath the surface occurs in 48 h.

In MPSS broth, *S. volutans* grows only when incubated under atmospheres of from 1–12% oxygen, despite the occurrence of a superoxide dismutase (SOD) of the iron type in the cells (W. H. Cover, M.S. thesis, Virginia Polytechnic Institute, 1978; P. J. Padgett, Ph.D. dissertation, Virginia Polytechnic Institute, 1981). Growth does not occur under anaerobic conditions or under an air atmosphere. Addition of

---

* MPSS broth (g/liter): Bacto peptone (Difco), 5.0; succinic acid (free acid), 1.0; (NH$_4$)$_2$SO$_4$, 1.0; MgSO$_4$·7H$_2$O, 1.0; FeCl$_3$·6H$_2$O, 0.002; and MnSO$_4$·H$_2$O, 0.002. The pH is adjusted to 6.8 with 2 N KOH before autoclaving. For CHSS broth, 2.5 g of "vitamin-free, salt-free acid-hydrolyzed casein" (ICN Nutritional Biochemicals, Cleveland, Ohio) is substituted for the peptone component, and 0.1 g NaCl is added. For semisolid media, 1.5 g agar is added.

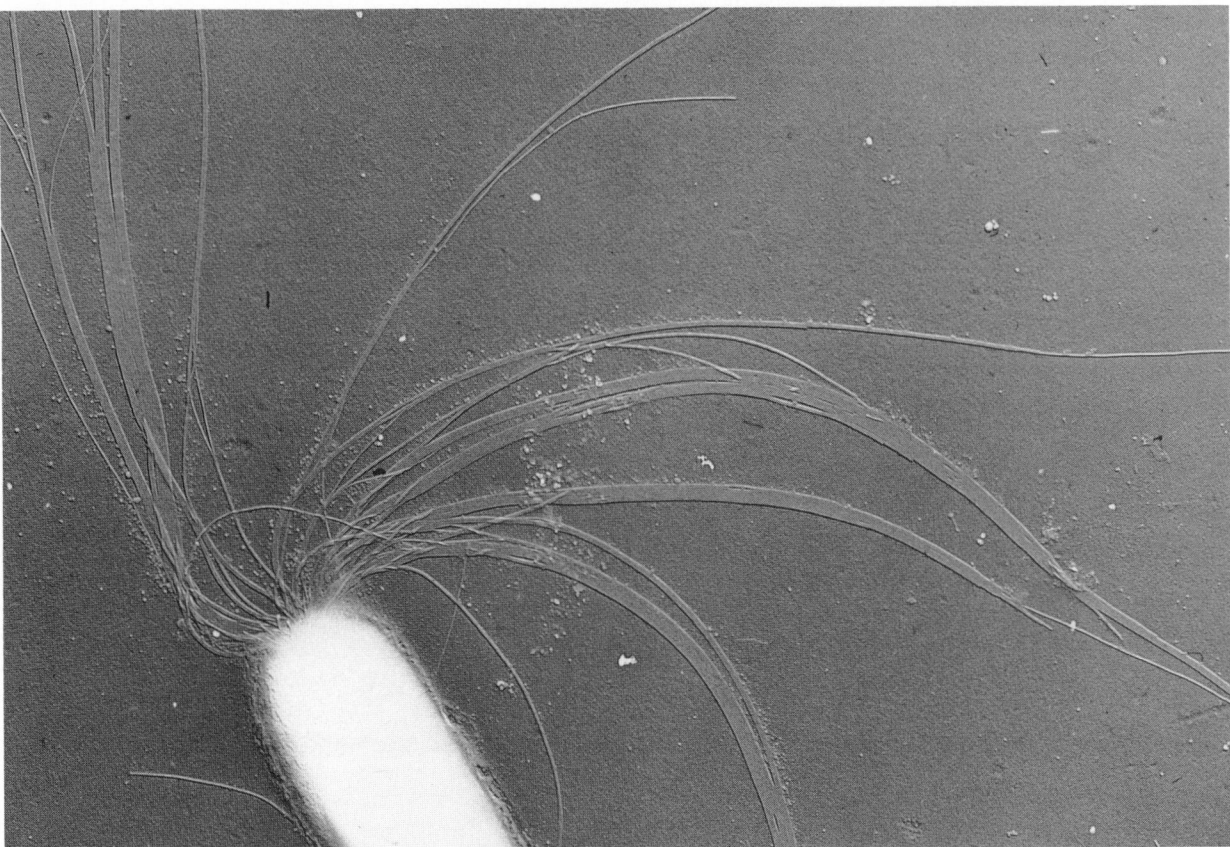

Figure 2.6. Flagellar fascicle of *Spirillum volutans* (× 16,000).

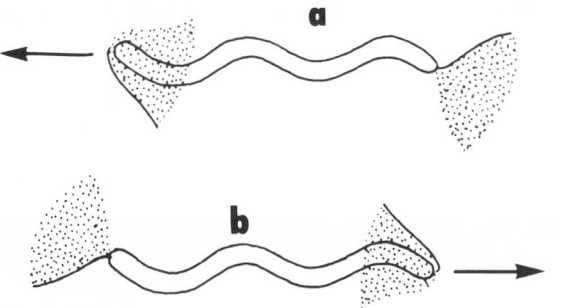

Figure 2.7. Diagram of *Spirillum volutans* showing the orientation of the bipolar flagellar fascicles. *a*, during swimming the fascicles form oriented cones of revolution. *b*, during reversal of swimming direction both fascicles reorient simultaneously. *Arrows* in *a* and *b* indicate the direction of swimming. (Reproduced with permission from J. H. Bowdre and N. R. Krieg, Virginia Polytechnic Institute and State University Water Resources Research Center Bulletin No. 69, 1974, © Virginia Water Resources Research Center.)

catalase (0.8 U/ml) or bovine erythrocyte SOD (4 U/ml) to MPSS broth allows growth to occur under an air atmosphere with static incubation; the two enzymes are most effective when used in combination (0.08 U/ml each) and exert a synergistic effect (Padgett et al., 1982). Potassium metabisulfite (0.005%), acting in conjunction with the $FeCl_3$ component of the MPSS broth, also permits aerobic growth to occur (Padgett et al., 1982). *S. volutans* lacks catalase and is extraordinarily sensitive to hydrogen peroxide; inhibition of growth occurs by addition of as little as 0.29 μM H_2O_2 to culture media (Padgett et al., 1982). Exposure to MPSS broth to moderate or strong illumination causes the generation of H_2O_2 in MPSS broth at levels sufficient to inhibit growth, and culture media should be protected from illumination. Potassium bisulfite, and especially a combination of SOD and catalase, can help to prevent the inhibitory effects caused by illumination.

Aerobic growth can also be obtained by the use of CHSS broth; turbid cultures are obtained in this medium under an air atmosphere (with static incubation) in 24–48 h. Also, a chemically defined medium for *S. volutans* devised by Bowdre et al. (1976) supports growth under a 6% oxygen atmosphere. When supplemented with norepinephrine (10^{-5} to 10^{-6} M), this medium also supports growth under an air atmosphere.

Growth of *S. volutans* on solid media (MPSS or CHSS broth solidified with 15.0 g agar/liter) is difficult to obtain and depends on several factors: (a) protection of the plates from exposure to illumination during preparation and incubation; (b) addition of potassium bisulfite (0.002%), catalase (130 U/ml) or SOD (30 U/ml) to the medium (the enzymes must be added aseptically to the molten medium just before dispensing into Petri dishes); and (c) incubation of the plates in an atmosphere of high humidity for 24 h prior to inoculation and for 5 days after inoculation (Padgett et al., 1982). Even with these precautions, colonies develop only under oxygen atmospheres of 12% or less, and only 22–72% of the cells spread onto the surface of the plates develop into colonies.

Catalase activity in *S. volutans* is not detectable by spectrophotometric methods (Padgett et al., 1982). The formation of a few bubbles from 3% hydrogen peroxide after 30 min by cultures in semisolid MPSS medium is probably attributable to the high pH developed during growth from oxidation of succinate, although Cole and Rittenberg (1971) have reported very low catalase levels in cell suspensions by use of an oxygen electrode.

Respiration rates of *S. volutans* suspended in 0.05 M phosphate buffer are very low (Cole and Rittenberg, 1971), but are higher—and, in fact, comparable to those for aerobes—when a less inhibitory buffer is used, such as BES buffer (N, N-bis(2-hydroxyethyl)-2-aminoethanesulfonic acid) (Caraway and Krieg, 1974). Succinate supports the highest rate of oxygen uptake; fumarate, malate, oxaloacetate and pyruvate are also readily oxidized. Lactate, butyrate and β-hydroxybutyrate are oxidized to a lesser extent, and citrate, aconitate, isocitrate, α-ketoglutarate, aspartate, glutamate, casein hydrolysate and carbohydrates are oxidized only very slightly or not at all. When placed in a non-nutritive medium, S. volutans continues to retain motility for up to 24 h, with intracellular poly-β-hydroxybutyrate serving as an endogenous energy source (Caraway and Krieg, 1974).

Cytochromes of the b, c and o types have been detected in S. volutans, as well as cytochrome oxidase, NADH oxidase, and various tricarboxylic acid cycle enzymes (Cole and Rittenberg, 1971).

Aerotactic responses of S. volutans to self-created oxygen gradients have been described by Wells and Krieg (1965) and Caraway and Krieg (1974).

The nutritional requirements of S. volutans are not well understood, since the defined medium of Bowdre et al. (1976) may not necessarily be a minimal medium. If succinate is omitted from this medium, however, no growth occurs, and omission of any of the amino acids present (threonine, methionine, histidine, isoleucine and cystine) give a decreased growth response. A very low level of NaCl is required. The phosphate concentration must be no higher than 10^{-4} M for aerobic growth to occur, although growth under microaerobic conditions will occur if the level is increased to 10^{-2} M. Because of the toxic effects of low levels of heavy metals, glassware used for cultivation of S. volutans must be cleaned with acid and washed extensively in tap and distilled water; even growth in CHSS broth is dependent upon the use of exceptionally clean glassware.

S. volutans has been isolated from stagnant pond water in Virginia and from the cooling water of a sugar beet refinery in England (Wells and Krieg, 1965; Rittenberg and Rittenberg, 1962); however, the organism is widely distributed in stagnant freshwater sources and can be demonstrated in nearly any hay infusion prepared from such sources. In hay infusions the organism occurs in greatest numbers just beneath the surface scum (composed of aerobic organisms), presumably at a location where the dissolved oxygen level is most suitable or where hydrogen peroxide is being destroyed by other bacteria.

Enrichment and Isolation Procedures

Initial enrichment from hay infusion is accomplished by inoculation of Pringsheim's soil medium (Rittenberg and Rittenberg, 1962). This medium is prepared by placing one wheat or barley grain in a large test tube, covering the grain with 3–4 cm of garden soil, filling the tube nearly to the top with tap water, and sterilizing in an autoclave. Even with this enrichment, S. volutans is vastly outnumbered by other bacteria. At present, the only successful method for isolation is a mechanical method based on the ability of S. volutans to out-swim other bacteria. This method, first devised by Giesberger (1936). for isolation of other spirilla, has been successfully used for S. volutans by Rittenberg and Rittenberg (1962) and Wells and Krieg (1965).

A capillary is prepared by heating sterile 5-mm glass tubing in a flame, pinching the softened portion with square-ended forceps until almost closed, then reheating the flattened portion and drawing it out. The resulting oval capillary should be 15–30 cm long and 0.1–0.3 mm

in diameter. After cooling, the capillary is broken with sterile forceps at the tip and 10–20 cm of sterile medium (the supernatant fluid from Pringsheim's soil medium) are drawn into it, followed by 2–4 cm of enrichment culture. There should be no air space between the sterile medium and the culture. The tip is now sealed in a flame, leaving a small air space at the tip of the capillary. The capillary is mounted horizontally on the stage of a × 100 microscope. Because of its rapid motility, S. volutans will often be able outdistance the contaminants and be the first to arrive in the distal portion of the sterile medium. (S. volutans will frequently form a band of cells that migrates along the capillary in response to a self-created oxygen gradient; however, the migration rate of such a band is relatively slow, and it is more fruitful to watch for faster-swimming cells well in advance of such a band.) As soon as some spirilla have entered the distal regions of the sterile medium, the capillary is broken behind them and sealed in a flame. After the outside of the capillary has been sterilized with strong hypochlorite, and the latter removed with sterile thiosulfate solution, the tip is broken and the spirilla are expelled into a suitable medium. The medium used by Rittenberg and Rittenberg (1962) and Wells and Krieg (1965) was sterile Pringsheim's soil medium contained in a dialysis sac which was suspended in a mixed culture of other bacteria; however, it seems likely that simply expelling the spirilla into a tube of semisolid MPSS or CHSS medium would be a satisfactory alternative. Purity of the cultures is verified by microscopic examination.

Maintenance Procedures

S. volutans may be maintained in semisolid MPSS or CHSS medium at 30°C with transfer every 4–5 days. Cultures may also be maintained in CHSS broth (incubated statically under an air atmosphere) at 30°C with daily transfer.

Preservation by lyophilization has not yet been possible, but cultures may be preserved indefinitely in liquid nitrogen (Pauley and Krieg, 1974). Cells from a broth culture are harvested at 3500 × g, washed once with sterile nutrient broth, and suspended to a dense concentration in nutrient broth containing 10% (v/v) dimethyl sulfoxide. After incubation for 30 min to allow penetration of the cells by the cryoprotective agent, the suspension is dispensed into vials. After sealing, the vials are frozen in a mixture of dry ice and alcohol and stored by submersion in liquid nitrogen.

Procedures for Testing Special Characters

The characteristics listed later in Table 2.6 were determined by methods described by Hylemon et al. (1973a). These authors employed PSS broth (containing 1.0% peptone) rather than MPSS broth, but the latter medium is more satisfactory. Cultures in liquid media are incubated under an atmosphere containing 6% oxygen; this oxygen level is most easily obtained by exhausting the air from the culture vessel and refilling the vessel several times with a mixture of 6% oxygen: 94% nitrogen. An alternative procedure is to exhaust the air in the culture vessel until the pressure becomes 0.29 atm, then fill the vessel with nitrogen to 1 atm. For certain biochemical tests such as phosphatase activity, the surface of plates of PSS (or MPSS) medium containing 0.7% agar is inoculated heavily (to give confluent growth) and the plates are incubated in a humid atmosphere under 6% oxygen.

The comments given in the Manual for the genus Aquaspirillum for performance of biochemical tests also apply to the genus Spirillum as well.

Differentiation of the genus **Spirillum** from other genera

See the genus Aquaspirillum, Table 2.1, for characteristics of Spirillum that distinguish the genus from other morphologically or physiologically similar genera.

Taxonomic Comments

In the eighth edition of the Manual, a single genus, Spirillum, contained all of the various aerobic and microaerophilic spirilla, in-

cluding freshwater and marine species. However, the DNA base composition for this genus ranged from 38–65 mol% G + C and appeared to be unusually broad for a bacterial genus. Moreover, three groups were evident within the genus: the aerobic freshwater spirilla that could not tolerate 3% NaCl (mol% G + C is 49–66); the aerobic marine spirilla that required seawater for growth (mol% G + C is 42–51); and the large, microaerophilic freshwater spirilla that belonged to the

Table 2.6.
Characteristics of **Spirillum volutans**

Characteristics	Reaction or Result	Characteristics	Reaction or Result
Cell diameter, μm^a	1.4–1.7	Hydrolysis of casein (single-strength milk)[d]	−
Wavelength, μm^a	16–28	Hydrolysis of esculin[a]	−
Diameter of helix, μm^a	5–8	Hydrolysis of 0.1% DNA or RNA (clear zone after acidification)[d]	−
Length of helix, μm^a	14–60	Indol production from 0.1% tryptophan[a]	−
Number of turns[a]	1–5	Hydrolysis of 10% soluble starch[d]	−
Intracellular poly-β-hydroxybutyrate granules present[a]	+	Aerobic reduction of 0.1% KNO_3[a]	−
Coccoid bodies present in older cultures[a]	−	Aerobic reduction of 0.1% H_2SeO_2 (by pink color)[a]	−
Bipolar tufts of flagella present[a]	+	Visible growth with 1% bile or 1% glycine[a]	−
Oxidase (moistened test disc inoculated with centrifuged cells)[a]	+	Anaerobic growth with 0.1% KNO_3 (sealed with petrolatum)[b]	−
Catalase[b]	−[c]	Acid reaction from carbohydrates (38 compounds tested)[e]	−
Phosphatase (0.01% phenolphthalein diphosphate)[d]	+	Urease[f]	−
Sulfatase (0.01% phenolphthalein disulfate)[d]	−	Mol% G + C of DNA (T_m)	38
H_2S from 0.2% cysteine (detector strip)[a]	+		
Liquefaction of 12% gelatin[a]	−		

[a] Basal medium is PSS broth.

[b] Basal medium is PSS broth + 0.15% agar.

[c] When a few drops of 3% H_2O_2 are added, a few bubbles of oxygen form after 30-min incubation. However, this reaction is probably attributable to the alkalinity of the cultures rather than to catalase activity.

[d] Basal medium is PSS broth + 0.7% agar.

[e] Basal medium is PSS broth lacking succinate and with peptone decreased to 0.2%; 0.0018% phenol red indicator added.

[f] Cells suspended in distilled water to a dense, milky concentration; 0.5 ml added to 2.0 ml of the following medium: BES buffer, 0.1065%; urea, 2.0%; phenol red, 0.001%; pH 7.0. Controls without urea are used. The test is incubated for 24 h.

species *S. volutans* (mol% G + C is 38). Accordingly, Hylemon et al. (1973b) divided the genus into the three genera *Aquaspirillum, Oceanospirillum* and *Spirillum*, respectively, with the latter genus containing the type species of the original genus.

Analysis of the oligonucleotide catalog of the 16S rRNA of *S. volutans* (Woese et al., 1982) indicates that the species belongs to group II of the phototrophic bacteria as defined by Gibson et al. (1979), and is distantly related to *Aquaspirillum serpens* and *Rhodospirillum tenue*.

The species *Campylobacter fetus* exhibits certain similarities to *S. volutans*, in that both contain cells with polar flagella and are microaerophilic, with a strictly respiratory type of metabolism. Neither species can catabolize carbohydrates. The DNA base composition is similar (32–35 mol% G + C for *C. fetus*: 38 mol% G + C for *S. volutans*). Moreover, *C. fetus*, although nominally vibrioid in shape, can often exhibit a spirillum-like appearance, and, like *S. volutans*, *Campylobacter* species have a polar membrane. Despite these similarities, there are marked differences between the two species: *S. volutans* has a much larger cell diameter, has large bipolar tufts of flagella rather than a single flagellum at one or both poles, does not form coccoid bodies, and is not associated with animals or humans. Studies involving rRNA/ DNA hybridization or rRNA oligonucleotide cataloging would be helpful in determining if a phylogenetic relationship exists.

Distinctive morphological similarities exist between *S. volutans* and members of the phototrophic genus *Thiospirillum*. *Thiospirillum jenense* has flagellar fascicles which are remarkably similar to those of *S. volutans*, and both species have cells of large diameter. However, *T. jenense* is obligately phototrophic and anaerobic, and also has a higher mol% G + C for its DNA (45%). Phylogenetic studies based on rRNA would be helpful in determining if any relatedness exists.

Further Reading

Krieg, N.R. 1976. Biology of the chemoheterotrophic spirilla. Bacteriol. Rev. *40:* 55–115.

Krieg, N.R. and P.B. Hylemon. 1976. The taxonomy of the chemoheterotrophic spirilla. Annu. Rev. Microbiol. *30:* 303–325.

List of the species of the genus **Spirillum**

1. **Spirillum volutans** Ehrenberg 1832, 38.[AL]

vo'lu·tans. L. v. *voluto* to tumble about; 1. part. adj. *volutans* tumbling about.

The morphological characteristics are as listed in Table 2.6 and depicted in Figure 2.6 and 2.7. See also the phase-contrast photomicrograph of *S. volutans* presented in Figure 2.1*P* under the genus *Aquaspirillum*.

Other characteristics are as described for the genus or as listed in Table 2.6.

Isolated from stagnant freshwater sources.

The mol% G + C of the DNA is 38 (T_m) (Hylemon et al., 1973a) or 36 (Bd) (Cole, 1972).

Type strain: ATCC 19554. *Reference strain*: ATCC 19553.

Species Incertae Sedis

The two species "*Spirillum minus*" and "*Spirillum pulli*" do not belong to the genus *Spirillum* and are described in *Species Incertae Sedis* following the treatment of the genus *Aquaspirillum* (see p. 89).

Genus **Azospirillum** *Tarrand, Krieg and Döbereiner 1979, 79[AL] (Effective publication:* Tarrand, Krieg and Döbereiner 1978, 978)

NOEL R. KRIEG AND JOHANNA DÖBEREINER

A.zo.spi.ril'lum. Fr. n. *azote* nitrogen; Gr. n. *spira* a spiral; M.L. dim. neut. n. *spirillum* a small spiral; *Azospirillum* a small nitrogen spiral.

Plump, slightly-curved and straight rods, about 1.0 μm in diameter and 2.1–3.8 μm in length, often with pointed ends. Intracellular granules of **poly-β-hydroxybutyrate** present. Enlarged, pleomorphic forms may occur in old, alkaline cultures or under conditions of excess oxygen. **Gram-negative to Gram-variable. Motile** in liquid media by a **single polar flagellum;** on solid media at 30°C numerous **lateral flagella** of shorter wavelength are also formed. **Nitrogen fixers,** exhibiting N_2-dependent growth under **microaerobic conditions.** Grow well under an air atmosphere in the presence of a source of fixed nitrogen such as an ammonium salt. **Possess mainly a respiratory type of metabolism** with oxygen or nitrate as the terminal electron acceptor, but weak fermentative ability may also occur. Under severe oxygen limitation **nitrate is dissimilated** to nitrite or to nitrous oxide and nitrogen gas. Optimum temperature, 35–37°C. Colonies on potato agar are typically **light or dark pink,** often wrinkled and nonslimy. **Oxidase-positive.** Chemoorganotrophic; some strains are facultative hydrogen autotrophs. **Grow well on the salts of organic acids such as malate, succinate, lactate or pyruvate.** Fructose and certain other sugars can also be used as carbon sources. Disaccharides are not used. Some strains require biotin. **Occur free living in the soil or associated with the roots of cereal crops, grasses and tuber plants.** Root nodules are not induced. The mol% G + C of the DNA is 69–71 (T_m).

Type species: *Azospirillum lipoferum* (Beijerinck 1925) Tarrand, Krieg and Döbereiner 1979, 79. (Effective publication: Tarrand, Krieg and Döbereiner 1978, 978.)

Further Descriptive Information

In complex media such as MPSS broth*, azospirilla grow as plump, slightly curved rods and straight cells having a diameter of ~1.0 μm (Figs. 2.8 and 2.9). Many of the cells have pointed ends. In semisolid nitrogen-free malate (Nfb) medium,† *A. lipoferum* develops predominantly into pleomorphic cells within 48 h (Fig. 2.10), in contrast to *A. brasilense* which retains mainly the vibrioid form (Fig. 2.11). *A. lipoferum* grows as elongated cells (1.4–1.7 μm × 5 to over 30 μm long) which are nonmotile and have an S shape or helical shape (Fig. 2.10). These forms eventually seem to fragment into shorter ovoid forms, many of which beome very large and rounded and may contain several cells filled with phase-refractile granules (probably poly-β-hydroxybutyrate). Alkalinization of the malate medium, due to oxidation of the malate, may be related to development of pleomorphism in *A. lipoferum.* Pleomorphism fails to occur when the organisms are cultured in semisolid nitrogen-free glucose medium, which does not become alkaline (Fig. 2.12).

In semisolid nitrogen-free malate medium, *A. brasilense* grows mainly as motile, vibrioid cells (Fig. 2.11). Nonmotile, enlarged, pleomorphic forms (C forms) may also occur (Fig. 2.13), especially in older cultures (Eskew et al., 1977; Tarrand et al., 1978), on the surface of nitrogen-free agar media (Berg et al., 1980), in association with plant callus

cultures (Berg et al., 1979), or in association with the roots of grass seedlings (Umali-Garcia et al., 1980). A capsule is formed external to the outer wall membrane of C forms (Fig. 2.14) and may be a protective mechanism against unfavorable levels of oxygen under nitrogen-fixing conditions (Berg et al., 1980). Very large, rounded forms containing several cells may be formed (Fig. 2.14). The ultrastructure of the C forms indicates little similarity to the cysts of *Azotobacter.* R. B. Lamm and C. A. Neyra (Abstract Annual Meeting American Society Microbiology I-65, 1981) have reported that strains of *A. brasilense* and *A. lipoferum* enriched in C forms exhibit a resistance to desiccation and temperature not found in cultures lacking these forms.

Encapsulation may be related to resistance to Gram-decolorization exhibited by a small proportion of cells when cultured on MPSS agar at 37°C for 48–72 h. Gram-variability appears to be more pronounced with *A. brasilense* than with *A. lipoferum.* Cultures grown in MPSS broth appear not to contain encapsulated forms, at least in young cultures, and the cell stain uniformly Gram-negative.

An unusual elaboration of the plasma membrane, the "polar membrane," has been found in thin sections of the type strain of *A. brasilense* (R. G. E. Murray, unpublished results). This structure is attached to the inside of the plasma membrane by barlike links and is located, most commonly, in the region surrounding the polar flagellum. Such a membrane has been found mainly in genera of helical bacteria, such as *Spirillum, Aquaspirillum, Oceanospirillum, Campylobacter, Ectothiorhodospira* and *Rhodospirillum.* (See the genus *Aquaspirillum* for further information about polar membranes.)

Poly-β-hydroxybutyrate may constitute from 25–50% of the dry weight of cells cultured in nitrogen-free media. In cells cultured with an ammonium salt as the nitrogen source, the polymer constitutes only 0.5–1.0% of the cell weight (Okon et al., 1976).

Azospirilla possess a single polar flagellum when cultured in MPSS broth; however, when cultured on MPSS agar at 30°C numerous lateral flagella are formed in addition to the polar flagellum (see Fig. 2.15). The polar flagellum appears to be thicker than the lateral flagella and also has a longer wavelength. The function of the lateral flagella is unknown.

On BMS agar§ after 1–2 weeks of incubation at 33–35°C, colonies of azospirilla are pink, opaque, irregular or round, often wrinkled, and typically have umbonate elevations (Döbereiner et al., 1976; Döbereiner and Baldani, 1979). Pigmentation is best on BMS agar incubated in the light. Certain strains and variants of *A. brasilense* form colonies that have a very deep pink color (Eskew et al., 1977; Tarrand et al., 1978). In one such strain (ATCC 29729), this intense color is attributable to the formation of several carotenoid pigments which occurs only under aerobic conditions and may be related to protection of the nitrogenase from oxidative damage (Nur et al., 1981). The more typical pink color of other strains of *A. brasilense* may possibly be due to their content of cytochrome *c* (Nur et al., 1981).

In nitrogen-free malate media, nitrogen fixation occurs only under

---

* MPSS broth (g/liter): peptone (Difco), 5.0; succinic acid (free acid), 1.0; $(NH_4)_2SO_4$, 1.0; $MgSO_4 \cdot 7H_2O$, 1.0; $FeCl_3 \cdot 6H_2O$, 0.002; $MnSO_4 \cdot H_2O$, 0.002; pH adjusted to 7.0 with KOH. For solid media add 15.0 g agar/liter.

† Nfb medium (g/liter): L-malic acid, 5.0; K_2HPO_4, 0.5; $MgSO_4 \cdot 7H_2O$, 0.2; NaCl, 0.1; $CaCl_2$, 0.02; trace element solution ($Na_2MoO_4 \cdot 2H_2O$, 0.2 g; $MnSO_4 \cdot H_2O$, 0.235 g; H_3BO_3, 0.28 g; $CuSO_4 \cdot 5H_2O$, 0.008 g; $ZnSO_4 \cdot 7H_2O$, 0.024 g; distilled water, 1000 ml), 2.0 ml; bromthymol blue (0.5% aqueous solution (dissolve in 0.2 N KOH)), 2.0 ml; Fe EDTA (1.64% solution), 4.0 ml, vitamin solution (biotin, 0.01 g; pyridoxin, 0.02 g; distilled water, 1000 ml), 1.0 ml; KOH, 4.0; pH adjusted to 6.8 with KOH. For a semisolid medium add 1.75 g agar/liter; for a solid medium add 15.0 g agar/liter.

§ BMS agar: washed, peeled, sliced potatoes, 200 g; L-malic acid, 2.5 g; KOH, 2.0 g; raw cane sugar, 2.5 g; vitamin solution (biotin, 0.01 g; pyridoxin, 0.02 g; distilled water, 1000 ml), 1.0 ml; bromthymol blue (0.5% alcoholic solution), 2 drops; agar, 15.0 g. The potatoes are placed in a gauze bag, boiled in 1 liter of water for 30 min, then filtered through cotton, saving the filtrate. The malic acid is dissolved in 50 ml of water and the bromthymol blue added. KOH is added until the malic solution is green (pH 7.0). This solution, together with the cane sugar, vitamins and agar, is added to the potato filtrate. The final volume is made up to 1 liter with distilled water. The medium is boiled to dissolve the agar, then sterilized by autoclaving.

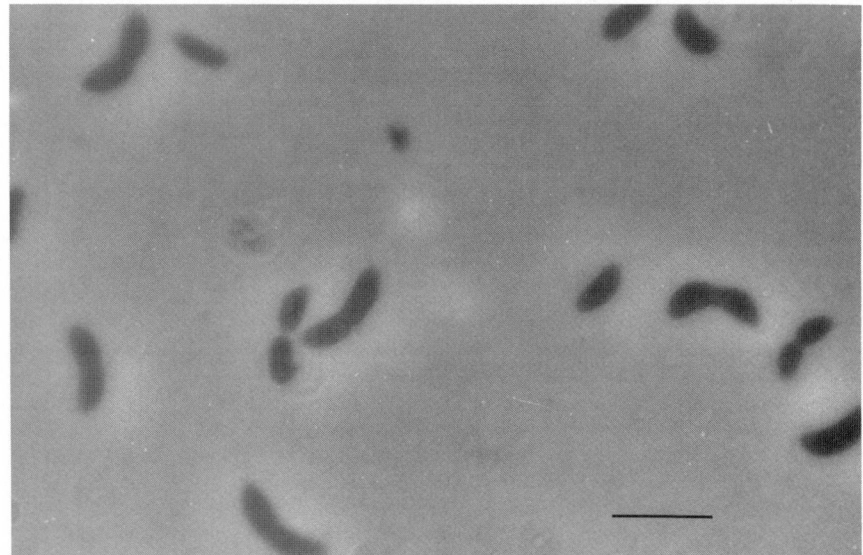

Figure 2.8. *Azospirillum brasilense* ATCC 29145 cultured in MPSS broth at 37°C for 24 h. Phase-contrast microscopy. *Bar*, 5 μm.

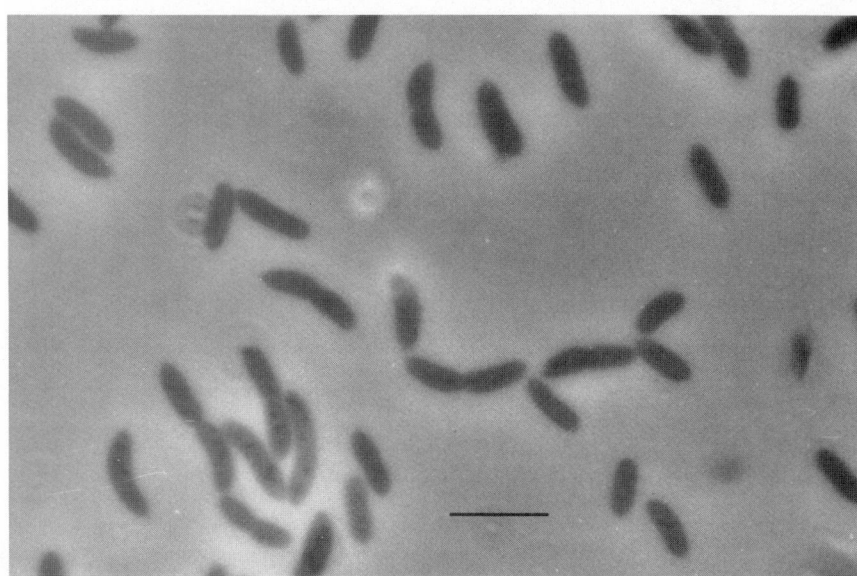

Figure 2.9. *Azospirillum lipoferum* ATCC 29707 cultured in MPSS broth at 37°C for 24 h. Phase-contrast microscopy. *Bar*, 5 μm.

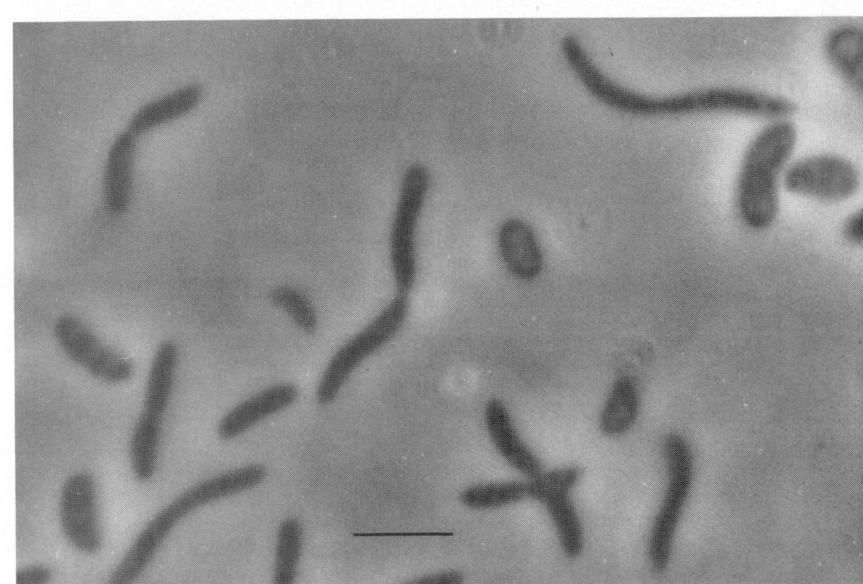

Figure 2.10. *Azospirillum lipoferum* ATCC 29707 cultured in semisolid nitrogen-free malate medium at 37°C for 48 h, showing characteristic elongated S-shaped forms and enlarged ovoid forms. Phase-contrast microscopy. *Bar*, 5 μm.

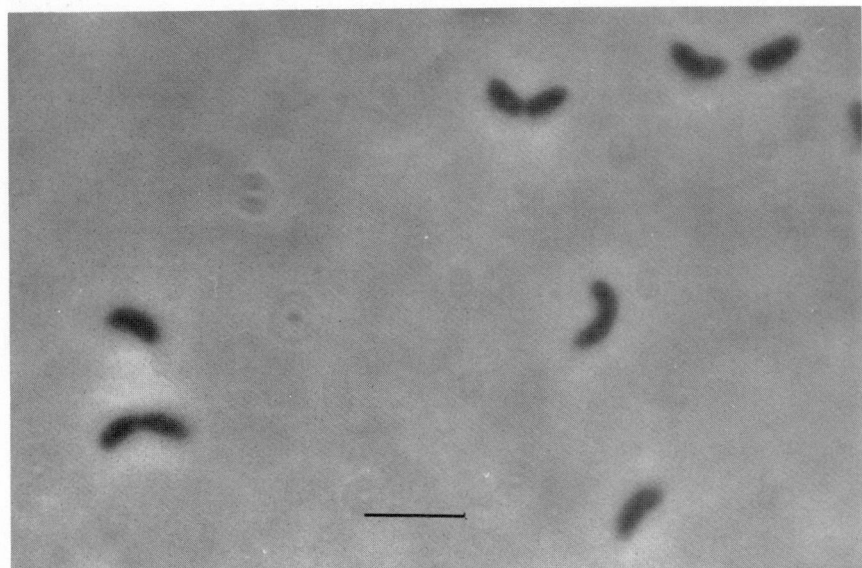

Figure 2.11. *Azospirillum brasilense* ATCC 29145 cultured in semisolid nitrogen-free malate medium at 37°C for 48 h, showing predominant vibrioid forms. Phase-contrast microscopy. *Bar*, 5 μm.

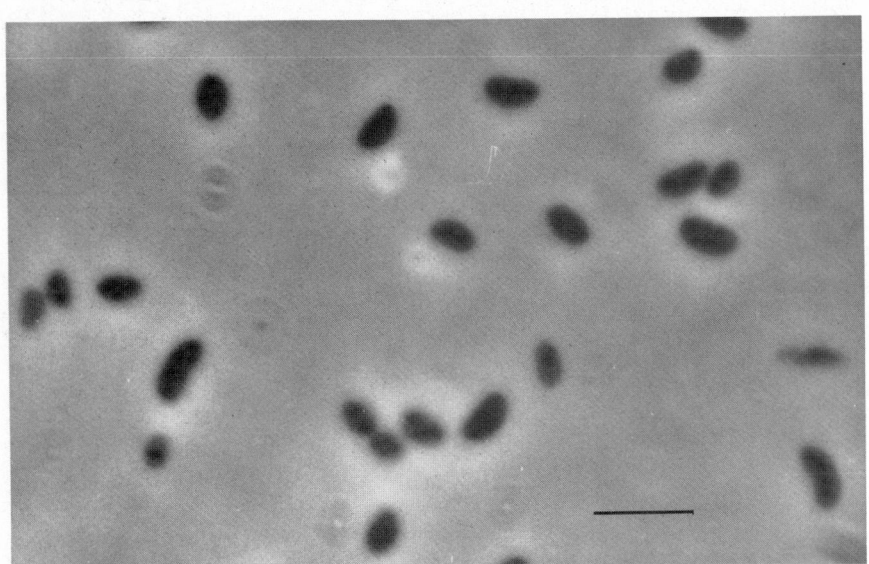

Figure 2.12. *Azospirillum lipoferum* ATCC 29707 cultured in semisolid nitrogen-free glucose medium at 37°C for 48 h, showing the characteristic vibrioid forms. (Compare with Fig. 2.10). Phase-contrast microscopy. *Bar*, 5μm.

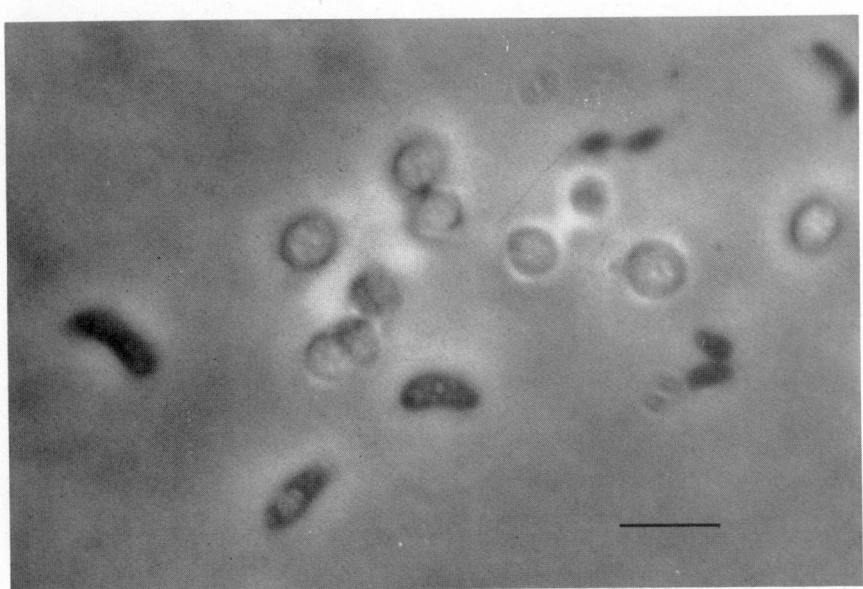

Figure 2.13. *Azospirillum brasilense* ATCC 29145 cultured in semisolid nitrogen-free malate medium at 37°C for 48 h, showing enlarged vibrioid forms and several refractile, ovoid C forms. Phase-contrast microscopy. *Bar*, 5 μm.

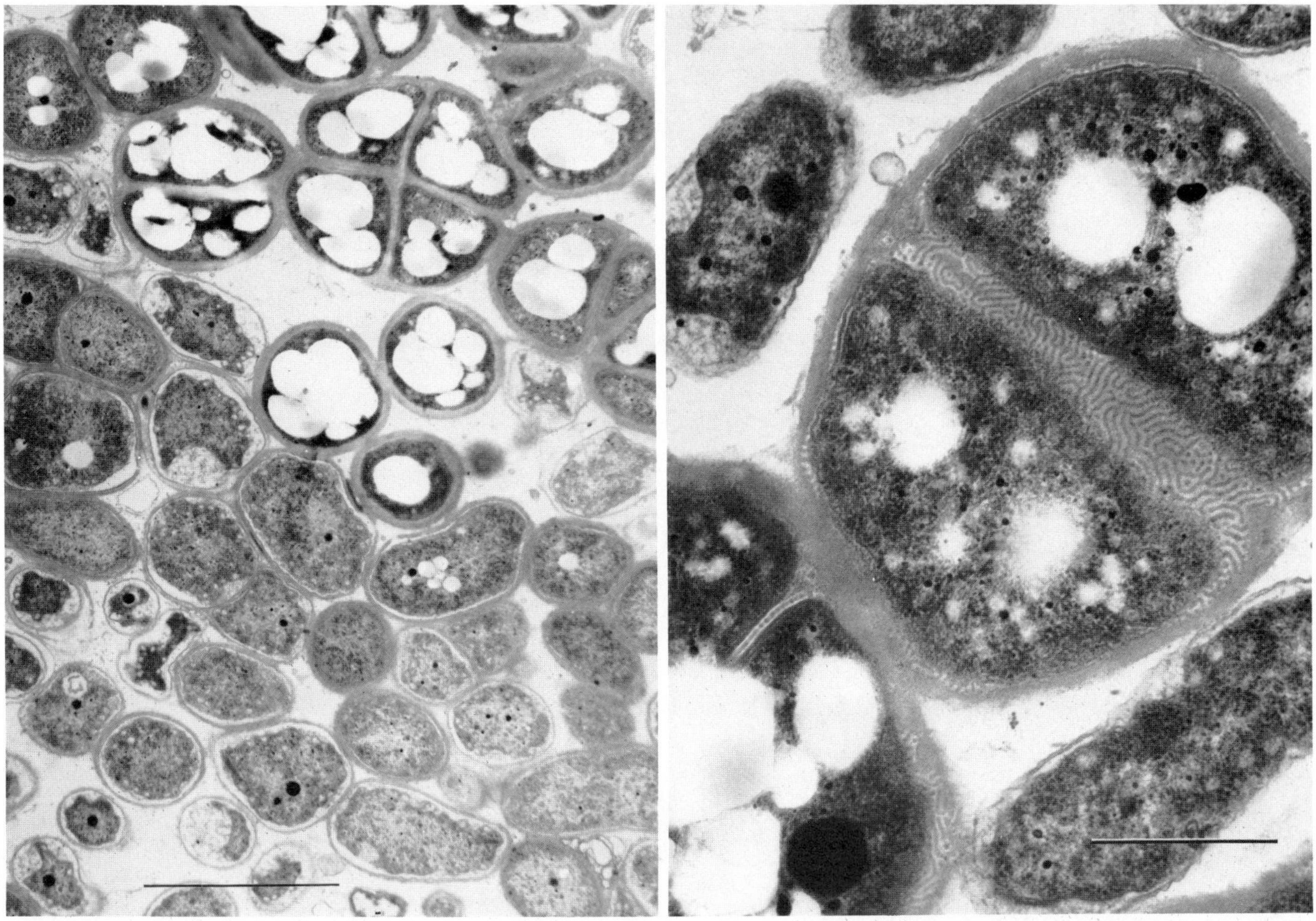

Figure 2.14. Ultrastructure of vibrioid and encapsulated (C) forms of *Azospirillum brasilense* ATCC 29145 grown in association with a plant callus (sugarcane). *Left:* gradient of vibrioid forms (*lower*) to C forms (*upper*). Multicellular C forms filled with poly-β-hydroxybutyrate granules (*white areas*) are seen at the top. *Bar,* 1.5 μm. *Right:* comparative fine structure of vibrioid forms and C forms. *Bar,* 0.5 μm. (Reproduced with permission from R.H. Berg, V. Vasil, and I. Vasil, Protoplasma *101:* 143–163, 1980, Springer-Verlag.)

Figure 2.15. Electron micrograph of *Azospirillum brasilense* ATCC 29145 cultured on MPSS agar at 30°C for 24 h. Both the single polar flagellum and the numerous lateral flagella can be seen. Shadowed with tungsten oxide (~X 15,000).

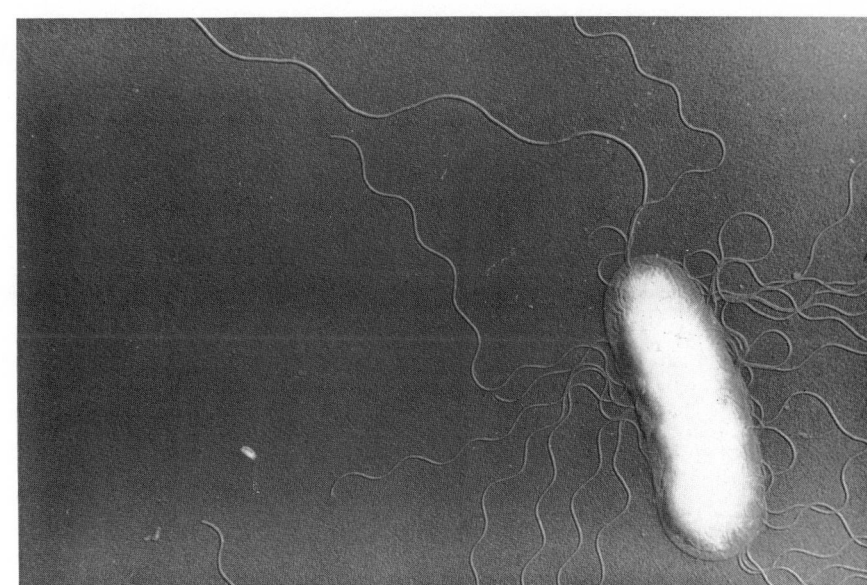

microaerobic conditions ($P_{O2} \simeq 0.003$ atm), due to lack of oxygen protection mechanisms for the nitrogenase. In liquid nitrogen-free media there is no growth or nitrogen fixation under an air atmosphere. The fastest growth rates are obtained in continuous or batch cultures when the oxygen input is in equilibrium with bacterial respiration. Depending on the cell density, a gas mixture containing from 1–25% oxygen can be bubbled through the culture to achieve this. A simpler way to obtain N_2-dependent growth is by culturing the organisms in semisolid nitrogen-free media incubated under an air atmosphere. Here, growth is initiated as a thin veil or disc several millimeters or centimeters below the surface of the medium at a point where the rate of diffusion of oxygen into the medium corresponds to the respiration rate of the organisms so that no excess oxygen remains in solution. As the bacteria multiply the disc of growth migrates closer to the surface until finally it is just below the surface. Okon et al. (1980) have found a similar veil or disc formation even in media containing a source of fixed nitrogen, suggesting that the organisms may prefer microaerobic conditions even when not fixing nitrogen.

When supplied with a source of fixed nitrogen, such as an ammonium salt or nitrate, azospirilla can grow under aerobic conditions. Nitrate is assimilated by an assimilatory nitrate reductase (Neyra and van Berkum, 1977.) All strains reduce nitrate to nitrite either by an aerobic assimilatory pathway or an anaerobic dissimilatory or respiratory pathway. Approximately half of the strains of both species so far isolated can dissimilate nitrite further to nitrous oxide and nitrogen gas (Neyra et al., 1977). The nitrite reductase (*nir*) is the key enzyme for denitrification (Magalhães et al., 1978). Plant roots are infected almost exclusively by nondenitrifying strains (*nir⁻*). Chlorate-resistant spontaneous mutants lacking the dissimilatory nitrate reductase (*nr⁻* mutants) and/or the nitrite reductase (*nir⁻* mutants) have been isolated from both species of *Azospirillum*. Nitrogenase activity is retained by such mutants and proceeds even in the presence of 10 mM nitrate.

Steady state cultures of denitrifying strains of *A. brasilense* have been obtained under anaerobic conditions in malate containing 100 mM nitrate. When the level of nitrate is lowered to 20 mM, nitrogenase activity occurs but growth is severely decreased (Nelson and Knowles, 1978). Anaerobic suspensions of nitrogenase-containing cells in Nfb medium containing 10 mM nitrate exhibit nitrate-dependent nitrogenase activity after 30–60 min, but no growth occurs (Neyra and van Berkum, 1977; Scott et al. 1979); during this period nitrite is accumulated in the medium. After 4 h the cells start assimilating the nitrite, start growing again, and stop fixing nitrogen (H. Bothe, M. P. Stephan and J. Döbereiner, unpublished results).

In peptone-based media, azospirilla grow abundantly under anaerobic conditions due to the dissimilatory nitrate reductase, with the nitrate being reduced to nitrite or to nitrous oxide and nitrogen gas.

The nitrogenase system of *Azospirillum* consists of 3 components: the Mo-Fe protein, the Fe protein, and an activating factor for the Fe protein (Ludden et al. 1978). The activating factor is interchangeable with that from *Rhodospirillum rubrum*. The *Azospirillum* nitrogenase system in vivo does not liberate H_2 because the latter is recycled by hydrogenase (Berlinger and Lespinat, 1980; Chan et al., 1980; Volpon et al., 1981).

A. lipoferum has the potential for H_2-dependent autotrophic growth under nitrogen-fixing or nonnitrogen-fixing conditions. Strain 208 was able to grow autotrophically under an atmosphere of H_2:CO_2:O_2:N_2 (20:5:2:73) in liquid Nfb medium (Sampaio et al., 1981). Both ribulose-1,5-biphosphate carboxylase (RubPcase) and hydrogenase activities were present and were much higher in autotrophically grown cells than in lactate-grown cells. In contrast to *A. lipoferum*, *A. brasilense* ATCC 29145 showed only marginal growth with H_2 under autotrophic conditions. Malik and Schlegel (1981) found that all of the strains of *A. lipoferum* they tested (including the type strain) were capable of hydrogen autotrophy provided the oxygen of the gas atmosphere was less than 2% (v/v). RubPcase and hydrogenase activities were present. In the absence of a source of combined nitrogen, the growth rate was very slow under autotrophic conditions and ceased at a cell concentration

of 0.2 g/liter; with an ammonium salt present the growth rate was much faster. None of the strains of *A. brasilense* tested were capable of autotrophic growth.

Both *A. brasilense* ATCC 29145 and *A. lipoferum* Sp 208 seem to grow well on methane, methanol or formate as sole energy sources (Sampaio et al., 1981). *A. lipoferum* cells derepressed under CH_4 were devoid of RubPcase and hydrogenase activities; however, RubPcase, but not hydrogenase, could be induced by CO_2 in the presence of methane. These results strongly suggest that *A. brasilense* and *A. lipoferum* have the potential for methylotrophic growth; however, ^{14}C-incorporation studies have not yet been performed to confirm this.

Compounds that can serve as sole carbon and energy sources for N_2-dependent growth include malate, succinate, pyruvate, lactate and fructose, and, in the case of *A. lipoferum*, glucose and α-ketoglutarate. A combination of a tricarboxylic acid cycle and a pentose has been reported to stimulate nitrogenase activity (Child and Kurz, 1978).

The type strains of *A. brasilense* and *A. lipoferum* have been found to phosphorylate fructose to fructose-1-phosphate by means of a phosphoenolpyruvate phosphotransferase system (E. M. Goebel, Ph.D. dissertation, Virginia Polytechnic Institute and State University, 1980). Although both type strains possessed glucokinase, only *A. lipoferum* was permeable to glucose. Hexokinase activity occurred only in *A. lipoferum*. Enzymes of the Embden-Meyerhof-Parnas pathway and the Entner-Doudoroff pathway occurred in fructose-grown cells of both strains; however, *A. brasilense* lacked glucose-6-phosphate dehydrogenase activity and therefore probably does not catabolize fructose by the Entner-Doudoroff pathway. *A. lipoferum* did possess glucose-6-phosphate dehydrogenase activity. Neither strain possessed the hexose monophosphate pathway. Both strains did possess pyruvate dehydrogenase, a complete tricarboxylic acid cycle, a glyoxylate shunt, and malic enzyme.

With glucose or fructose as a carbon source, *A. lipoferum* exhibits weak fermentative ability in media containing a source of fixed nitrogen; however, the organisms grow far better aerobically than anaerobically. *A. lipoferum* is capable of acidifying glucose or fructose media anaerobically, of forming very small amounts of gas in Durham vials, of exhibiting slight growth in glucose or fructose broth under anaerobic conditions, and of forming minute colonies on glucose or fructose agar anaerobically. Variable results for these tests occur within some strains, and variants with decreased fermentative ability can be selected. Such variants continue to require biotin, use glucose as a sole carbon source for N_2-dependent growth, and exhibit the characteristic pleomorphic changes associated with this species.

By an auxanographic method using media containing ammonium sulfate as the nitrogen source (Tarrand et al., 1978), the following compounds serve universally as sole carbon sources for all *Azospirillum* strains: malate, succinate, lactate, pyruvate, fumarate, β-hydroxybutyrate, gluconate, glycerol and fructose. Whether all of these compounds can support N_2-dependent growth has not yet been ascertained. Glucose is used by all strains of *A. lipoferum* and by a very few strains of *A. brasilense*; however, *A. brasilense* cannot use glucose as a sole carbon source for N_2-dependent growth, and it also produces a lower degree of acidification of glucose than does *A. lipoferum* (Tarrand et al. 1980).

The following tests are universally positive for the genus: oxidase, phosphatase, urease, esculin hydrolysis, ability to grow anaerobically with nitrate in peptone-based media, and ability to dissimilate nitrate to either nitrite or to nitrous oxide and nitrogen gas. The following tests are universally negative: starch and gelatin hydrolysis, production of water-soluble pigments, indol production and acidification of media containing lactose, sucrose, rhamnose, cellobiose, erythritol, dulcitol or melibiose. Catalase activity ranges from strong to undetectable.

Pectolytic activity (pectic lyase and endogalacturonase (Umali-Garcia et al., 1980); polygalacturonic acid transeliminase (Tien et al. 1981) has been reported in some strains of *Azospirillum* and may have significance for the colonization of the middle lamella of root tissue (Umali-Garcia et al., 1980); however, the enzymatic activity appears to be weaker in comparison with that of *Erwinia* (Tien et al., 1981). The production of the plant growth substance indoleacetic acid, and also indolelactic acid, gibberellin and cytokinin-like substances, has been reported in strains of *A. brasilense* (Reynders and Vlassak, 1979; Tien et al., 1979).

Genetic transformation has been reported with *A. brasilense* ATCC 29145 (Mishra et al., 1979). Maximum competence was achieved by culturing the recipient cells for 4 h at 30°C in nutrient broth.

Genetic recombination has been demonstrated in *A. brasilense* ATCC 29145, using plasmid R68-45 (derived from *Pseudomonas aeruginosa*) to mobilize the chromosome (Franche et al., 1981). The mode of gene transfer promoted by the plasmid appeared to be unpolarized, suggesting the existence of multiple origins of transfer.

A. G. Wood and D. E. Duggan (Abstracts Annual Meeting American Society Microbiology, 1981, H-106) reported the occurrence of multiple large plasmids in *A. brasilense* and *A. lipoferum*. Six strains of *A. brasilense* and two of *A. lipoferum* appeared to possess as many as eight differently sized plasmids ranging from 60 to well over 400 megadaltons. Similarly, Franche and Elmerich (1981) found that four strains of *A. brasilense* and five strains of *A. lipoferum* harbored 1–5 plasmids of sizes ranging from 3.5 to over 300 megadaltons. E. M. Menezes and D. E. Duggan (Abstracts Annual Meeting American Society Microbiology, 1981, H-107) reported that auxotrophs were transformed to prototrophy at high frequency by plasmid DNA. Various antibiotic resistances and also biosynthetic markers could be transferred by the plasmid DNA. These findings suggested that *Azospirillum* might be a multilinkage group procaryote.

Six of nine strains studied by Franche and Elmerich (1981) were found to be lysogenic. Phage production was inducible by mitomycin C.

Antisera prepared against whole cells have been used by the indirect fluorescent antibody technique to distinguish between the species and also between groups of strains of azospirilla (Schank et al., 1979; De Polli et al., 1980). The technique is useful for specific visualization of azospirilla in root sections and supports differences between *nir* groups and host plant specificity groups.

Azospirilla appear to have a worldwide distribution and occur in large numbers (up to 10^7/g) in rhizosphere soils and in association with the roots of a variety of C3 and C4 plants. Most isolates have been obtained from tropical forage grasses and from cereal plants. There appears to be a greater occurrence of azospirilla in regions with a tropical rather than a temperate climate (Döbereiner et al., 1976), although some researchers have found little difference (Tyler et al., 1979).

In field-grown maize, azospirilla occur on the surface of roots, in the outer cortex, inner cortex and in the stele (Patriquin and Döbereiner, 1978). Infection of the inner cortex and stele occurs in the absence of significant bacterial colonization or collapse of outer lying tissues. Paraxylem vessels can be completely plugged with bacteria. Infection occurs initially in root branches and spreads longitudinally into main roots. In monoaxenic cultures of pearl millet and guinea grass, azospirilla are found within the mucigel layer of roots and become firmly attached to root hairs. The bacteria enter the roots through lysed root hairs and void spaces created by epithelial desquamation and lateral root emergence (Umali-Garcia et al., 1980). Azospirilla have been observed in intercellular locations within the middle lamella of root tissues; they have also been observed intracellularly, sometimes in very large numbers, but this may represent invasion of dead plant cells where the wall has undergone autolysis.

Host plant specificity in the infection of many grasses and cereals by azospirilla has been demonstrated. C4 plants are selectively infected by *A. lipoferum*, whereas C3 cereals (wheat, oats, barley and rye) are selectively infected by *nir⁻* strains of *A. brasilense* (Baldani and Döbereiner, 1980; Döbereiner and Baldani, 1979; Döbereiner and De Polli, 1980). Depending on the soil, low level (10–20 μg/ml) streptomycin resistance is often required for infection (Döbereiner and Baldani, 1979).

Enrichment and Isolation Procedures

Soil samples or washed root pieces (5–8 mm long) macerated with a forceps are placed into semisolid Nfb medium (4 ml in 6-ml cotton-stoppered serum vials). After ~40 h of incubation at 32–35°C the vials are sealed with rubber stoppers, 12% of the air is replaced by acetylene, and nitrogenase activity is tested. For vials exhibiting acetylene reduction, a second transfer is made to semisolid Nfb medium. After 24 h, the growth is streaked onto plates of solidified Nfb medium containing 0.02 g yeast extract/liter. After 1 week, typical small, white, dense, single colonies are subcultured to semisolid Nfb medium, where subsurface growth in the form of a veil or disc is a presumptive indication of successful enrichment. For final purification, cultures are streaked onto BMS agar and the typical pink, often wrinkled, colonies are transferred for storage and identification.

The following method has been used for specific enrichment of *A. lipoferum* (K. A. Malik, personal communication). A nitrogen-free mineral medium is inoculated with soil samples or washed root pieces and incubated under a gas atmosphere of 5% (v/v) O_2, 10% CO_2, 30% H_2, and 55% N_2. After 1 week, 0.5 ml of the suspension is transferred to fresh medium and incubated as before. The procedure is repeated 2 or 3 times before serial dilutions are plated onto nitrogen-free mineral media containing 1.5% (w/v) agar. Colonies are selected and propagated under autotrophic conditions.

Maintenance Procedures

Stock cultures may be maintained in semisolid Nfb medium at 8–30°C with monthly transfers for *A. brasilense* and biweekly transfers for *A. lipoferum*. Stocks may also be maintained on trypticase soy agar with monthly transfer (Tyler et al., 1979). In sterile vermiculite moistened with potato broth, *A. brasilense* remains viable for several years and *A. lipoferum* for 3–6 months; drying should be avoided by tightly sealing the vials with screw caps. A combination of soil and farmyard manure (1:1) can maintain a population of *A. brasilense* at high levels for up to 6 months (Tilak et al., 1979).

For preservation, heavy suspensions of cells harvested from MPSS broth or MPSS agar plates are prepared in nutrient broth containing 10% (v/v) dimethyl sulfoxide (Tarrand et al., 1978) or in TSS broth (1% trypticase, 0.5% succinate with salts as in Nfb medium) in which 15% of the water is replaced by glycerol (Tyler et al., 1979). The suspensions are placed in vials and preserved in liquid nitrogen.

Procedures for Testing Special Characters

Glucose used as sole carbon source for growth in semisolid nitrogen-free medium. A loopful of culture grown in semisolid Nfb malate medium is inoculated into a tube of semisolid nitrogen-free glucose medium (malate replaced by 1.0% glucose which has been sterilized by filtration). *A. lipoferum* forms a veil or disc of growth in the depths of the medium; within 3 days at 37°C this disc migrates close to the surface of the medium and becomes very dense. *A. brasilense* either gives no response or forms a slight pellicle in the depths of the medium that later disperses. The difference in response between the two species is very pronounced.

Biotin requirement. Glassware should be rinsed copiously with distilled water and subsequently baked in an oven to destroy traces of biotin. A medium of the following composition is used (g/liter): K_2HPO_4, 0.5; succinic acid (free acid), 5.0; $FeSO_4 \cdot 7H_2O$, 0.01; $Na_2MoO_4 \cdot 2H_2O$, 0.002; $MgSO \cdot 7H_2O$, 0.2; NaCl, 0.1; $CaCl_2 \cdot 2H_2O$, 0.026; $(NH_4)_2SO_4$, 1.0. The pH is adjusted to 7.0 with KOH pellets. Biotin (0.0001 g/liter) is added to one portion. The biotin-free and biotin-containing media are sterilized in 5.0-ml amounts in screw-capped tubes by autoclaving. Cultures grown in MPSS broth are inoculated by a loop into 25 ml of one quarter strength nutrient broth (Difco) and incubated at 37°C for 24 h. The cells are harvested by centrifugation, washed twice with 10-ml portions of sterile distilled water, and suspended in water to a turbidity of 20 Klett units (blue filter, 16-mm tubes). One-tenth ml of this suspension is used to inoculate each 5.0 ml of medium (with and without biotin). The cultures are incubated for 48 h at 37°C. In cases where growth occurs in the absence of biotin, a second serial transfer is made to media with and without biotin, using 0.1 ml from the first culture.

Difference in pleomorphism between A. lipoferum and A. brasilense. A loopful of growth from a 24-h-old MPSS broth culture is inoculated into a tube of semisolid nitrogen-free malate medium (containing 0.05 g yeast extract or 0.0001 g biotin/liter). The cultures are examined by phase-contrast microscopy after incubation at 37°C for 24–48 h (Fig. 2.8). *A. brasilense* remains mainly vibrioid and motile, wheras *A. lipoferum* becomes wider, longer, nonmotile, and S shaped or helical. These forms of *A. lipoferum* eventually undergo fragmentation into shorter, ovoid cells, many of which become very large and pleomorphic and are filled with refractile granules.

Auxanographic method for sole carbon sources. The following ingredients are dissolved in 50 ml of distilled water passed through a Bantam multibed resin cartridge (Barnstead Co., Boston, MA): $(NH_4)_2SO_4$, 1.0 g; $MgSO_4 \cdot 7H_2O$, 1.0 g; K_2HPO_4, 2.0 g; $FeCl_3 \cdot 6H_2O$, 0.0047 g; $MnSO_4 \cdot H_2O$, 0.0025 g; $ZnSO_4 \cdot 7H_2O$, 0.00072 g; $CuSO_4 \cdot 5H_2O$, 0.000125 g; $CoSO_4 \cdot 7H_2O$, 0.00014 g; H_3BO_3, 0.000031 g; and $Na_2MoO_4 \cdot 2H_2O$, 0.000245 g. The pH is adjusted to 2.5 with HCl to dissolve precipitates and then adjusted to 7.0 with KOH. Biotin (0.0001 g) and $CaCO_3$ (0.001 g) are added and the medium is sterilized by autoclaving. The sterile medium is added aseptically to an equal volume of sterile agar solution (15.0 g purified agar (Difco) in 500 ml of distilled water, sterilized by autoclaving) at 45–50°C. Cells are prepared as for the determination of biotin requirement (previous paragraph) except that the final suspension is adjusted to 30 Klett units instead of 20. Two ml of cell suspension are used to seed 20 ml of molten medium at 45–50°C in a Petri dish. After solidification of the medium, sterile 7-mm diameter paper discs (punched from Beckman electrophoresis filter paper, Cat. No. 319328) are dipped into 5% (w/v) aqueous solutions of carbon sources sterilized by filtration. (Solutions of organic acids are adjusted to pH 7.0 with KOH before sterilization.) The saturated discs are then placed near the periphery of the seeded agar plates (3 discs/plate). The plates are incubated at 37°C for 72 h. Any visible zone of turbidity around the discs, as judged by holding the plates against a black background and illuminating them with oblique lighting from the rear, constitutes a positive growth response.

Acidification of peptone-based glucose medium. The following medium is used (g/liter): peptone (Difco), 2.0; $MgSO_4 \cdot 7H_2O$, 1.0; $(NH_4)_2SO_4$, 1.0; $FeCl_3 \cdot 6H_2O$, 0.002; $MnSO_4 \cdot H_2O$, 0.002; bromthymol blue (dissolved in dilute KOH), 0.025. The medium is made up to a volume of 950 ml, adjusted to pH 7.0 and sterilized by autoclaving. After it has cooled, 50 ml of a 20% (w/v) solution of glucose (sterilized by filtration) is added aseptically. The development of a yellow color during incubation for 96 h at 37°C indicates acidification.

Fermentative ability. Tubes of peptone-glucose or fructose media, or defined glucose or fructose broth (Tarrand et al., 1978), are inoculated and placed in GasPak jars (BBL) containing fresh catalyst. The jars are evacuated, filled once with N_2 and three times with H_2, sealed and incubated at 37°C for up to 2 weeks. Development of a yellow color indicates acidification. Slight growth in the form of a sediment can be detected by agitating the tubes.

Test for autotrophic growth (Malik and Schlegel, 1981). A mineral medium is used with the following composition (per liter): KH_2PO_4, 2.3 g; $Na_2HPO_4 \cdot 2H_2O$, 2.9 g; NH_4Cl, 1.0 g; $MgSO_4 \cdot 7H_2O$, 0.5 g; $CaCl_2 \cdot 2H_2O$, 0.01 g; ferric ammonium citrate, 0.05 g; and trace element solution*, 6.0 ml. When necessary, the medium is supplemented with the appropriate growth factors. For testing in liquid culture, 20-ml portions of medium are distributed into 100- or 250-ml Erlenmeyer flasks. After inoculation, the flasks are placed in desiccator jars or anaerobic jars. The jars are then evacuated and filled with the following gas mixture: 2% (v/v) O_2, 10% CO_2, 60% H_2 and 28% N_2. The jars are incubated at 30–37°C with shaking and are periodically evacuated and refilled with fresh gas mixture. For testing the growth on agar plates, the mineral medium is solidified with 1.5% agar.

Differentiation of the genus **Azospirillum** *from other genera*

Table 2.7 indicates the characteristics of *Azospirillum* that distinguish it from other genera of morphologically or physiologically similar genera.

Taxonomic Comments

An analysis of the oligonucleotide catalog obtained from the 16S rRNA of *A. brasilense* indicates that this species belongs to group Ic of the purple phototrophic bacteria as defined by Gibson et al. (1979). In this group the closest relatives now known of *A. brasilense* appear to be *Aquaspirillum itersonii* and *Rhodospirillum rubrum* (C. Woese and collaborators, unpublished data). Since *A. lipoferum* is closely related to *A. brasilense* on the basis of DNA/DNA and rRNA/DNA hybridization studies (see below), the rRNA oligonucleotide data for *A. brasilense* almost certainly apply to *A. lipoferum* as well.

Ribosomal RNA/DNA hybridization studies by De Smedt et al. (1980) also support a phylogenetic relationship of *Azospirillum* to *Aquaspirillum* (*Rhodospirillum* has not yet been tested). The 32 S rRNA cistrons of *A. brasilense* closely resemble those of *A. lipoferum* ($T_{m(e)}$ is 80–82.5°C) and also exhibit a lower degree of similarity to those of *Aquaspirillum itersonii* ($T_{m(e)}$ is 72.5°C) and *Aquaspirillum polymorphum* ($T_{m(e)}$ is 71.5°C). The studies by De Smedt et al. (1980) indicate that azospirilla belong to the same rRNA superfamily as *Beijerinckia, Rhizobium, Agrobacterium, Acetobacter, Gluconobacter, Xanthobacter* and *Aquaspirillum*, with members of the latter genus presently being the closest known relatives of *Azospirillum*. Although azospirilla and aquaspirilla belong in one RNA group, the inclusion of azospirilla in a separate genus seems justified both on a phenotypic and phylogenetic basis. With regard to rRNA cistrons, azospirilla differ as much from aquaspirilla as azotobacters differ from *Alteromonas communis* and *Alteromonas vaga* (De Smedt et al., 1980).

A relationship between *Azospirillum* and *Rhodospirillum*, as indicated by oligonucleotide catalogs from 16 S rRNA, is supported by certain phenotypic evidence. The most striking similarity concerns the activation factor for the Fe protein of nitrogenase. This factor has so far been found only in *Azospirillum* and *Rhodospirillum* (Ludden et al., 1978). *A. lipoferum* and *R. rubrum* share a growth requirement for biotin. Both *Azospirillum* and *R. rubrum* form a pink or red pigment when grown in the dark under aerobic conditions; whether the pigments are similar has not been investigated. Both genera form intracellular poly-β-hydroxybutyrate. *Azospirillum* appears not to be phototrophic, however, and also *R. rubrum* possesses bipolar tufts of flagella and appears to have cells that are thinner and more helical than those of *Azospirillum*. The mol% G + C of the DNA of *R. rubrum* is 64–66, in contrast to the value of 69–71 for *Azospirillum*.

Certain species of the genus *Aquaspirillum* exhibit phenotypic characters in common with *Azospirillum*: (a) *Aquaspirillum peregrinum* and *Aquaspirillum fasciculus*, and, putatively, some strains of *Aquaspirillum itersonii*, exhibit nitrogenase activity under microaerophilic conditions; (b) *A. itersonii, Aquaspirillum dispar* and *Aquaspirillum psychrophilum* are denitrifiers; (c) *Aquaspirillum delicatum* has mainly a vibrioid shape and possesses a single polar flagellum; (d) *A. itersonii, A. peregrinum* and *Aquaspirillum gracile* can oxidize a limited variety of sugars, and the former two species may have weak fermentative ability; (e) *A. gracile* requires biotin for growth; and (f) aquaspirilla

---

* Trace element solution SL-6 (Pfennig, 1974) contains (g/liter of distilled water): $ZnSO_4 \cdot 7H_2O$, 0.1; $MnCl_2 \cdot 4H_2O$, 0.03; H_3BO_3, 0.3; $CoCl_2 \cdot 6H_2O$, 0.2; $CuCl_2 \cdot 2H_2O$, 0.01; $NiCl_2 \cdot 6H_2O$, 0.02; and $Na_2MoO_4 \cdot 2H_2O$, 0.03.

Table 2.7.

Differential characteristics of the genus **Azospirillum** *and other morphologically or physiologically similar genera[a]*

Characteristics	Azospirillum	Derxia	Xanthobacter	Azotobacter	Azomonas	Beijerinckia	Rhizobium/Bradyrhizobium	Vibrio	Aquaspirillum	Pseudomonas	Rhodospirillum
Have mainly a vibrioid shape	+	−	−	−	−	−[b]	−	+	−[c]	−[d]	−
Gram variability may occur	+[e]	−	+	+	+	−	−	−	−	−	−
Cell diameter 2.0 μm or more	−	−	−	+	+	−	−	−	−	−	+
Motility	+	+	−	D	+	D	+	+	+	+	+
Flagellar arrangement											
Monotrichous	+[f]	+		−	D	−	D	+[g]	−[h]	D	−
Lophotrichous	−	−		−	D	−	−	−	+	D	+
Peritrichous	−[f]	−		+	D	+	D	−[g]	−	−[i]	−
Cysts formed	−[j]	−	−	+	−	−[k]	−	−	−[l]	−	−
Nitrogenase activity occurs	+	+	+	+	+	+	+	−	D	−	+
Nitrogen fixed only under microaerophilic conditions	+	−	+	−	−	−	+	−	+	−	+[m]
Known to be root-associated nitrogen fixers	+	−	−	D	−	−	+	−	−	−	−
Root hypertrophies produced	−	−	−	−	−	−	+	−	−	−	−
Phototropic	−	−	−	−	−	−	−	−	−	−	+
Colonies are pink or red	+	−	−	−	−	D	−	−	−	−	+
Some species or strains are denitrifiers	+	−	−	−	−	−	+	−	+	+	
Fermentative ability present	D[n]	−	−	−	−	−	−	+	D[o]	−	
Sucrose can be used as a carbon source	−	−	+	+		+	D	+	−	D	
Mol% G + C of DNA	69–71	69–73	65–70	63–67.5	52–59	55–61	57–65	38–51	49–66	57–70	62–66

[a] Symbols: +, typically positive; −, typically negative; D, differs among species.

[b] *Beijerinckia* may be straight, slightly curved or pear shaped.

[c] Only *A. delicatum* is vibrioid.

[d] Curved rods are not excluded from the genus if they are curved in only one plane.

[e] Best seen in *A. brasilense* strains cultured on MPSS Agar for 48–72 h.

[f] In liquid medium the cells possess a single polar flagellum; on agar media at 30°C numerous lateral flagella of shorter wavelength occur in addition to the polar flagellum.

[g] Some species exhibit the same flagellar arrangement as described in footnote *f*.

[h] *A. delicatum* is predominantly monotrichous; *A. polymorphum* has predominantly bipolar single flagella.

[i] *P. stutzeri* and *P. mendocina* exhibit the same flagellar arrangement as described in footnote *f*.

[j] Large, encapsulated, ovoid or pleomorphic forms can occur under certain conditions.

[k] *B. fluminensis* forms zoogloea-like forms filled with several cells, somewhat resembling the encapsulated or C forms of *A. brasilense*.

[l] A few species of *Aquaspirillum*, such as *A. itersonii*, form thin-walled coccoid bodies which have sometimes been called "microcysts."

[m] In the dark. *Rhodospirillum* can also fix nitrogen anaerobically in the light.

[n] Weak fermentative ability occurs in *A. lipoferum*. The organisms have mainly a respiratory type of metabolism and grow far better aerobically than anaerobically.

[o] *A. itersonii* and *A. peregrinum* acidify certain sugar media anaerobically but do not grow significantly under fermentative conditions. They may have weak fermentative ability.

grow best on the salts of organic acids. However, on the whole, the genus *Aquaspirillum* differs from *Azospirillum* by (a) having a mol% G + C of 49–66; (b) usually having helical cells with bipolar tufts of flagella; (c) typically failing to oxidize or ferment carbohydrates; and (d) not being associated with plants and having mainly an aquatic habitat.

There is little evidence, genetic or phenotypic, that *Azospirillum* is closely related to the genus *Azotobacter*, although *Azotobacter paspali* is, like *Azospirillum*, associated with the roots of plants. Azospirilla, however, lack the respiratory protection against oxygen inactivation of nitrogenase that is found in *Azotobacter*, and the ultrastructure of the C forms of azospirilla indicates little similarity to the cysts of *Azoto-*

bacter. Moreover, azotobacters are not vibrioid in shape and are not motile by a single polar flagellum in liquid media.

Although the mol% G + C of the DNA of the nitrogen-fixing genus *Derxia* is similar to that of *Azospirillum*, *Derxia* differs from *Azospirillum* in (a) being a straight rod, (b) failing to grow well on malate or succinate, and (c) forming very large, distinctive, very dense colonies which enable them to fix nitrogen under an air atmosphere. The genus *Xanthobacter* resembles *Azospirillum* in being a microaerophilic nitrogen fixer and in having a mol% G + C of 69–70; however, the cells are smaller and under certain conditions exhibit branching, the cells are nonmotile, a yellow pigment is formed, and the disaccharide sucrose can be used as a carbon source.

Table 2.8.

Characteristics differentiating **Azospirillum lipoferum** *and* **Azospirillum brasilense**[a]

Characteristics	1. *A. lipoferum*	2. *A. brasilense*
Biotin required for growth	+	−
Glucose, α-ketoglutarate and mannitol used as sole carbon sources for growth in nitrogen-free semisolid media	+	−
Acidification of peptone-based glucose medium, 96 h	+	−
In semisolid nitrogen-free malate medium nearly all cells become wider (1.4–1.7 μm), longer (5 μm to over 30 μm) and nonmotile in 24–48 h. Eventually many cells become ovoid or pleomorphic and filled with highly refractile granules. Multicellular forms may also occur.	+	−
Acid from glucose or fructose anaerobically; slight growth also occurs; a small amount of gas is sometimes formed.	+ or v	−

[a] All tests done with incubation at 37°C.

The nitrogen-fixing genus *Rhizobium* resembles *Azospirillum* in being plant associated, but the host-bacterium interaction is far more highly specialized for symbiotic nitrogen fixation than is the case with *Azospirillum*.

The vibrioid shape, occurrence of lateral flagella when grown on solid media, and occurrence of some fermentative ability suggest a relationship between *Azospirillum* and the genus *Vibrio*; however, the mol% G + C of the DNA of *Vibrio* is 38–51, well below that of *Azospirillum*.

Pseudomonas stutzeri possesses a flagellar arrangement similar to that of *Azospirillum* and, like *Azospirillum*, it is a denitrifier. The mol% G + C for the genus *Pseudomonas* is 57–70, and vibrioid forms are not excluded by the definition of the genus. Yet the occurrence of some fermentative ability in *A. lipoferum* would appear to exclude azospirilla from the genus *Pseudomonas*. Moreover, nitrogen-fixing ability has not been reported for members of the genus *Pseudomonas*. Also, unlike plant-associated pseudomonads, azospirilla are not pathogenic for plants.

From hybridization studies using a membrane-filter competition method, *A. lipoferum* and *A. brasilense* exhibit an interspecies DNA/DNA homology value of 30–50% (Tarrand et al., 1978). Within the species *A. brasilense*, strains exhibit a continuum of DNA/DNA homology values ranging from ~70–100% with the type strain. Within the species *A. lipoferum*, strains exhibit homology values of 70–76% with the type strain. The occurrence of two homology groups is highly

correlated with the differential characteristics indicated in Table 2.8 and appears to provide a firm basis for speciation within the genus.

In 1921–1922, Beijerinck observed extensive development of a spirillum-like bacterium in nitrogen-deficient glucose and mannitol solutions that had been inoculated heavily with garden soil or soil from a sand bed. Although the organism grew well at first, it was later displaced by competitive growth of *Azotobacter* and *Clostridium*. However, when calcium malate or lactate was employed as the carbon source instead of the carbohydrates, the organism grew well and was not overgrown by other nitrogen fixers. Beijerinck found that partially purified cultures of the spirillum exhibited increases in nitrogen at the expense of malate, whereas cultures lacking the spirillum failed to show such increases. Pure cultures of the spirillum failed to grow in the absence of a source of fixed nitrogen, and Beijerinck suggested that the good growth in partially pure cultures might be attributable to the microaerophilic nature of the spirillum, as suggested by microaerotactic band formation in wet mounts. In general, cells cultured in sugar media were plump, curved rods containing many lipoidal droplets which sometimes distorted the shape of the cells. On malate or lactate agar, the cells tended to be thinner and straighter, while in dilute bouillon they exhibited a distinct spirillum shape with one or more helical turns. Because of the ease of cultivation on salts of organic acids, as well as the spirillum shape exhibited under certain conditions, Beijerinck considered the organism to be a member of the genus *Spirillum* and to be a bridging organism linking the genus *Spirillum* with the genus *Azotobacter*. He initially named the organism *Azotobacter spirillum* (Beijerinck, 1922), but later renamed it *Spirillum lipoferum* (Beijerinck, 1925).

Later studies of *S. lipoferum* by Schröder (1932) also failed to demonstrate N₂ fixation by pure cultures, and the organism was forgotten for many years except for a few scattered reports. However, in 1963 Becking isolated an organism resembling *S. lipoferum* that showed uncontestable nitrogenase activity. Finally, the discovery of the association of such organisms with plant roots (Döbereiner and Day, 1976) caused much interest in *S. lipoferum* and led to detailed taxonomic studies.

Tarrand et al. (1978) reserved Beijerinck's specific epithet *lipoferum* for their DNA homology group II (i.e. *A. lipoferum*), because this group seemed to correspond in more ways to Beijerinck's description of *S. lipoferum*, particularly with regard to growth with glucose or mannitol and to formation of spirillum-shaped cells under certain conditions. Since Beijerinck's strains no longer exist, however, their correspondence with homology group II cannot be estimated with certainty.

Further Reading

Döbereiner, J. and H. De-Polli. 1980. Diazotrophic rhizocoenoses. *In* Stewart and Gallon (Editors), *Nitrogen Fixation*, Academic Press, London, pp. 301–333.

Tarrand, J.J., N.R. Krieg and J. Döbereiner. 1978. A taxonomic study of the *Spirillum lipoferum* group, with descriptions of a new genus, *Azospirillum* gen. nov. and two species, *Azospirillum lipoferum* (Beijerinck) comb. nov. and *Azospirillum brasilense* sp. nov. Can. J. Microbiol. *24:* 967–980.

Differentiation and characteristics of species of **Azospirillum**

The differential characteristics of the species of *Azospirillum* are indicated in Table 2.8. Other characteristics of the species are presented in Table 2.9.

List of the species of the genus **Azospirillum**

1. **Azospirillum lipoferum** (Beijerinck 1925) Tarrand, Krieg and Döbereiner 1979, 79.[AL] (*Effective publication:* Tarrand, Krieg and Döbereiner 1978, 978.) (*Spirillum lipoferum* Beijerinck 1925, 353.)

li.po′fe.rum. Gr. n. *lipus* fat; L. v. *fero* to carry; M.L. adj. *lipoferus* fat bearing.

Morphology is as depicted in Figs. 2.9, 2.10 and 2.12 and described in Table 2.8. The characteristic pleomorphism seen in malate medium does not occur in glucose medium.

All strains stain uniformly Gram-negative when cultured in MPSS

broth. Most strains stain uniformly Gram-negative when cultured for 48–72 h on MPSS agar but a few strains exhibit a small proportion of cells that are resistant to Gram-decolorization.

Colonies are not slimy on BMS agar; when 0.5% glucose is added to the medium, some strains form large, slimy, white colonies.

Aerobic growth in liquid media containing a source of fixed nitrogen often shows extensive clumping, particularly in defined media. No growth occurs anaerobically in the absence of a source of fixed nitrogen.

Strains of *A. lipoferum* which are able to grow lithoautotrophically

Table 2.9.

Other characteristics of **Azospirillum lipoferum** *and* **Azospirillum brasilense**[a]

Characteristics	1. *A. lipoferum*	2. *A. brasilense*
Oxidase	+	+
Catalase	d	+
Esculin hydrolysis	+	+
Phosphatase	+	+
Pink pigment on BMS agar	+	+
Anaerobic growth with nitrate in peptone-based medium	+	+
NO_3^- to NO_2^-	+	+
NO_2^- to N_2O	d	d
N_2O to N_2	+	+
Starch hydrolysis	−	−
Urease	+	+
Gelatin hydrolysis	−	−
Indole	−	−
Growth in presence of 1% oxgall	+	+
Growth in presence of 3% NaCl	d	−
Voges-Proskauer, 2% glucose, 5 days	d	−
Acid produced aerobically from carbohydrates:		
Fructose	+	+
Galactose, arabinose	+	d
Mannitol, sorbitol, ribose, *i*-inositol, xylose	d	−
Rhamnose, dulcitol, erythritol, maltose, sucrose, cellobiose, melibiose, lactose	−	−
Sole carbon sources (auxanographic method)[b] with $(NH_4)_2SO_4$ as nitrogen source:		
Succinate, malate, lactate, pyruvate, oxaloacetate, fumarate, gluconate, β-hydroxybutyrate, glycerol, fructose, propionate, citrate	+	+
α-Ketoglutarate	+	−
Glucose	+	d[c]
Galactose, arabinose	d	d
Mannitol, sorbitol, ribose	d	−
Malonate	−	−

[a] By methods of Tarrand et al., 1978.

[b] See Procedures for Testing Special Characters.

[c] Only a few strains show this character under the conditions of the test (see Procedures for Testing Special Characters) and the growth response is weak. The strains are unable to use glucose as a sole carbon source for N_2-dependent growth in nitrogen-free semisolid medium.

with H_2 contain an uptake hydrogenase as well as RubPcase. They can be categorized as aerobic hydrogen-oxidizing bacteria.

Physiological and nutritional characteristics of the species are presented in Tables 2.8 and 2.9. Variants with decreased fermentative ability may arise but continue to require biotin, use glucose as a sole carbon source for N_2-dependent growth, and exhibit the characteristic pleomorphism associated with this species in malate medium.

Maize and other C4 plants are selectively infected by this species rather than by *A. brasilense*. Depending on the soil, low level streptomycin resistance can be required for infection of plants.

The mol% G + C of the DNA is 69–70 (T_m).

Neotype strain: ATCC 29707 (strain Sp 59b of Tarrand, Krieg and Döbereiner, 1978). *Reference strains:* ATCC 29708 (strain Sp Rg 20a), ATCC 29709 (strain Sp Br 17), ATCC 29731 (strain Sp Rg 6xx).

2. **Azospirillum brasilense** Tarrand, Krieg and Döbereiner 1979, 79.[AL] (*Effective publication:* Tarrand, Krieg and Döbereiner 1978, 979.) bra.si.len'se. M.L. adj. *brasilensis* pertaining to the country of Brazil, South America.

Morphology is as depicted in Figures 2.8, 2.11 and 2.13–2.15. In semisolid nitrogen-free malate medium cells remain mainly vibrioid even when the cultures become alkaline, in contrast to *A. lipoferum*. Some encapsulated (C) forms (Figs. 2.13 and 2.14) may occur, especially in older cultures. C forms occur in abundance in colonies grown on the surface of nitrogen-free media, in association with plant callus cultures, or in grass seedlings inoculated with this species.

All strains stain uniformly Gram-negative when cultured in MPSS broth. Most strains exhibit a small proportion of cells showing resistance to Gram-decolorization when cultured on MPSS agar for 48–72 h; this Gram-variability may be related to the occurrence of encapsulated forms.

Certain strains and variants of *A. brasilense* form colonies that are a much deeper shade of pink than is usually the case.

Aerobic growth in liquid media containing a source of fixed nitrogen is usually homogeneous and turbid, without clumping. No growth occurs aerobically in the absence of a source of fixed nitrogen.

Physiological and nutritional characteristics of the species are presented in Tables 2.8 and 2.9.

Wheat, barley, rice, oats and rye are selectively infected by *nir*⁻ strains of *A. brasilense* rather than by *A. lipoferum*.

The mol% G + C of the DNA is 70–71 (T_m).

Type strain: ATCC 29145 (strain Sp 7 of Tarrand, Krieg and Döbereiner 1978). *Reference strains:* ATCC 29710 (strain Cd; deep pink), ATCC 29711 (Sp 35).

Other organisms

A noncellulolytic bacterium (strain Mc-2s) having the characteristics of the genus *Azospirillum* was isolated from cellulolytic nitrogen-fixing mixed cultures by Wong et al. (1980). The strain exhibited a biotin requirement, was pleomorphic in nitrogen-deficient malate medium and could utilize mannitol and α-ketoglutarate. Although these are characteristics of *A. lipoferum*, the strain failed to use glucose as a sole carbon source for N_2-dependent growth and did not acidify glucose or ribose media. The strain may be a variant of *A. lipoferum*. In another study, Franche and Elmerich (1981) found *Azospirillum* strains K67 and KR77 to be capable of glucose utilization, yet they had no biotin requirement; they were provisionally classified as *A. brasilense*. A third strain, SpN, appeared to be an *A. lipoferum* strain but seemed to require other growth factors in addition to biotin.

Nur et al. (1980) isolated microaerophilic, nitrogen-fixing vibrioid bacteria having certain characteristics in common with *Azospirillum*.

The organisms were associated with grass roots and were vibrioid cells with a single polar flagellum, and could use malate, lactate, arabinose and galactose, but not mannitol, as sole carbon sources; one isolate could use glucose. Biotin was not required. These strains differed from *Azospirillum* by having a smaller cell diameter (0.5–0.6 μm), by having no lateral flagella, and by forming yellow-pigmented colonies. Their taxonomic placement is uncertain.

Two strains of *Azospirillum* (Am-53 and A1-3) were isolated by Malik and Schlegel (1980) from soil and plant litter in Göttingen, Germany, as nitrogen-fixing, hydrogen-oxidizing bacteria. These were Gram-negative but showed Gram-positive granules and contained lipid droplets. Both had polar flagella and exhibited an active spirillum-like motility. Strain Am-53 (DSM 1727) required biotin, was pleomorphic in nitrogen-deficient malate medium, and in several other characters resembled *A. lipoferum*. Like *A. lipoferum*, it could grow chemolithotrophically on H_2 (Malik and Schlegel, 1981). However, unlike *A. lipoferum*, it could not utilize glucose, malonate, mannitol, or α-ketoglutarate as a sole source of carbon. Strain Al-3 (DSM 1726) was shorter than *A. lipoferum*. The cells were pleomorphic, but in the presence of 2–3% (w/v) NaCl the cells developed an active spinning motility (around the axis). Morphologically, and in several other respects, the strain resembled *A. brasilense*; however, it could grow chemolithotrophically on H_2, could utilize glucose and α-ketoglutarate as a sole carbon source, and could produce acid from glucose aerobically, like *A. lipoferum*. Both these strains may be variants of *A. lipoferum*.

Species Incertae Sedis

The following organism resembles *Azospirillum* in certain respects but appears to fix nitrogen under aerobic conditions and has an aquatic habitat. It has not yet been well characterized and its taxonomic placement is uncertain. It should not be included in the genus *Spirillum*, but may possibly be related to *Aquaspirillum* or *Azospirillum*.

a. "*Spirillum azotocolligens*" Rodina 1956, 149.

a.zo.to.coll´i.gens. Fr. n. *azote* nitrogen; L. part. adj. *colligens* collecting; *azotocolligens* nitrogen-collecting.

When cultured in liquid nitrogen-free medium (Fyodorov's), the cells are mainly vibrioid, 1.3–2.0 μm in diameter and 3.9–6.6 μm in length. Longer cells occur on Fyodorov's agar, some of which are S shaped. On media containing peptone, involution forms occur even in young cultures; these cells are larger, more severely twisted and are often united in pairs. Intracellular lipid inclusions, probably poly-β-hydroxybutyrate, are present. Motile, even in old cultures.

Grow on liquid or solid nitrogen-free medium (Fyodorov's) under aerobic conditions, fixing 4–12 mg nitrogen/g of mannitol. In Fyodorov's liquid medium they produce a copious amount of sediment (the medium becomes whitish). On Fyodorov's agar the colonies are small, whitish, slightly convex and somewhat dry.

Isolated from rivers, water receptacles and fishery ponds in the southern region of the U.S.S.R.

Type strain: none designated.

Genus Oceanospirillum Hylemon, Wells, Krieg and Jannasch 1973, 361[AL]

NOEL R. KRIEG

O.ce.an.o.spi.ril´ lum. M.L. n. *oceanus* ocean; Gr. n. *spira* a spiral; M.L. dim. neut. n. *spirillum* a small spiral; *Oceanospirillum* a small spiral (organism) from the ocean (sea water).

Rigid, helical cells 0.3–1.4 μm in diameter. **A polar membrane** underlies the cytoplasmic membrane at the cell poles in all species so far examined by electron microscopy. Intracellular **poly-β-hydroxybutyrate** is formed. Most species form thin walled **coccoid bodies which predominate in old cultures**. Gram-negative. Motile by **bipolar tufts of flagella or by a single flagellum at each pole**. **Aerobic**, having a strictly respiratory type of metabolism with oxygen as the terminal electron acceptor. Nitrate respiration does not occur. Optimum temperature, 25–32°C. **Oxidase-positive**. Indole- and arylsulfatase-negative. Casein, starch, hippurate and esculin are not hydrolyzed. **Seawater is required for growth. Carbohydrates are neither oxidized nor fermented**. Amino acids or the salts of organic acids serve as carbon sources. Growth factors are not usually required. Isolated from coastal sea water, decaying seaweed and from putrid infusions of marine mussels. The mol% G + C of the DNA ranges from 42–51 (T_m).

Type species: Oceanospirillum linum (Williams and Rittenberg, 1957) Hylemon, Wells, Krieg and Jannasch 1973, 374.

Further Descriptive Information

All species of *Oceanospirillum* consist of helical cells; however, variants having less curvature may arise after prolonged transfer. For example, the type strain of *O. japonicum* consisted initially of long, helical cells with several turns (Watanabe, 1959), but now consists of slightly curved or S-shaped cells. The cells within a given species have a constant and characteristic type of helix—clockwise (right-handed) or counterclockwise (left-handed) (Terasaki, 1972). Photographs showing the size and shape of various species of oceanospirilla are presented in Fig. 2.16.

An unusual elaboration of the plasma membrane, the "polar membrane," occurs in all of four species so far examined (Beveridge and Murray, unpublished results). It is attached to the inside of the plasma membrane by barlike links and is located, most commonly, in the region surrounding the polar flagella (Murray and Birch-Andersen, 1963). Such a membrane has been found mainly in genera of helical bacteria, such as *Spirillum*, *Campylobacter*, *Aquaspirillum*, *Ectothiorhodospira* and *Rhodospirillum*.

All species have intracellular poly-β-hydroxybutyrate, but granules may not be evident in cells having a small diameter (such as *O.*

minutulum) and chemical analysis may be required to demonstrate the polymer.

All species have bipolar tufts of flagella except *O. pusillum*, which has a single flagellum at each pole. The flagella of *O. japonicum* appear to be crescent shaped with less than one helical turn, whereas those of other species have one or more helical turns.

All species except *O. japonicum* show extensive formation of coccoid bodies (sometimes termed "microcysts") in old cultures. These bodies have thin walls and resemble spheroplasts; however, they are resistant to lysis in distilled water (B. L. Kelly, 1959, Ph.D. dissertation, University of Southern California). Whether coccoid bodies are resistant to desiccation is not known. Three main modes of formation of coccoid bodies have been described by Williams and Rittenberg (1957). (a) Two cells may entwine and apparently fuse. The cells become shorter and thicker and a protuberance develops at the point of fusion. This gradually enlarges and absorbs the organisms to form the coccoid body. More than one coccoid body may develop from a pair of entwined spirilla. (b) A spirillum may become shorter and thicker and a protuberance arises from the center of the cell or from each end of the cell. The protuberances enlarge and eventually merge into a single coccoid body as the helical cell is absorbed. (c) A spirillum may undergo a gradual shortening and rounding to form a coccoid body. The majority of coccoid bodies present in old cultures appear to be viable and can "germinate" when placed into a fresh medium (Williams and Rittenberg, 1956). Germination is by unipolar or bipolar growth of a helical cell from the coccoid body, with the latter being absorbed into the developing helical cell.

Seawater is required for the growth of all species. Media prepared with natural seawater or with 2.75% NaCl have been used for enrichment and isolation (Williams and Rittenberg, 1957; Terasaki, 1963, 1970, 1980). Commonly used culture media for oceanospirilla are nutrient broth* prepared with natural seawater and PSS or MPSS broth* prepared with artificial seawater†.

Oceanospirilla generally produce moderate to abundant turbid growth in 2 or 3 days in PSS seawater broth (Hylemon et al., 1973). In seawater-nutrient broth membranous masses are often formed at the surface and can be dispersed with shaking to yield turbid cultures (Terasaki, 1972).

Colonies of oceanospirilla generally develop within 2 or 3 days on

---

*See the genus *Aquaspirillum* for recipes for these media.
†Artificial sea water for use in PSS broth, g/liter of distilled water: NaCl, 27.5 g; $MgCl_2$, 5.0; $MgSO_4$, 2.0; $CaCl_2$, 0.5; KCl, 1.0; and $FeSO_4$, 0.01.

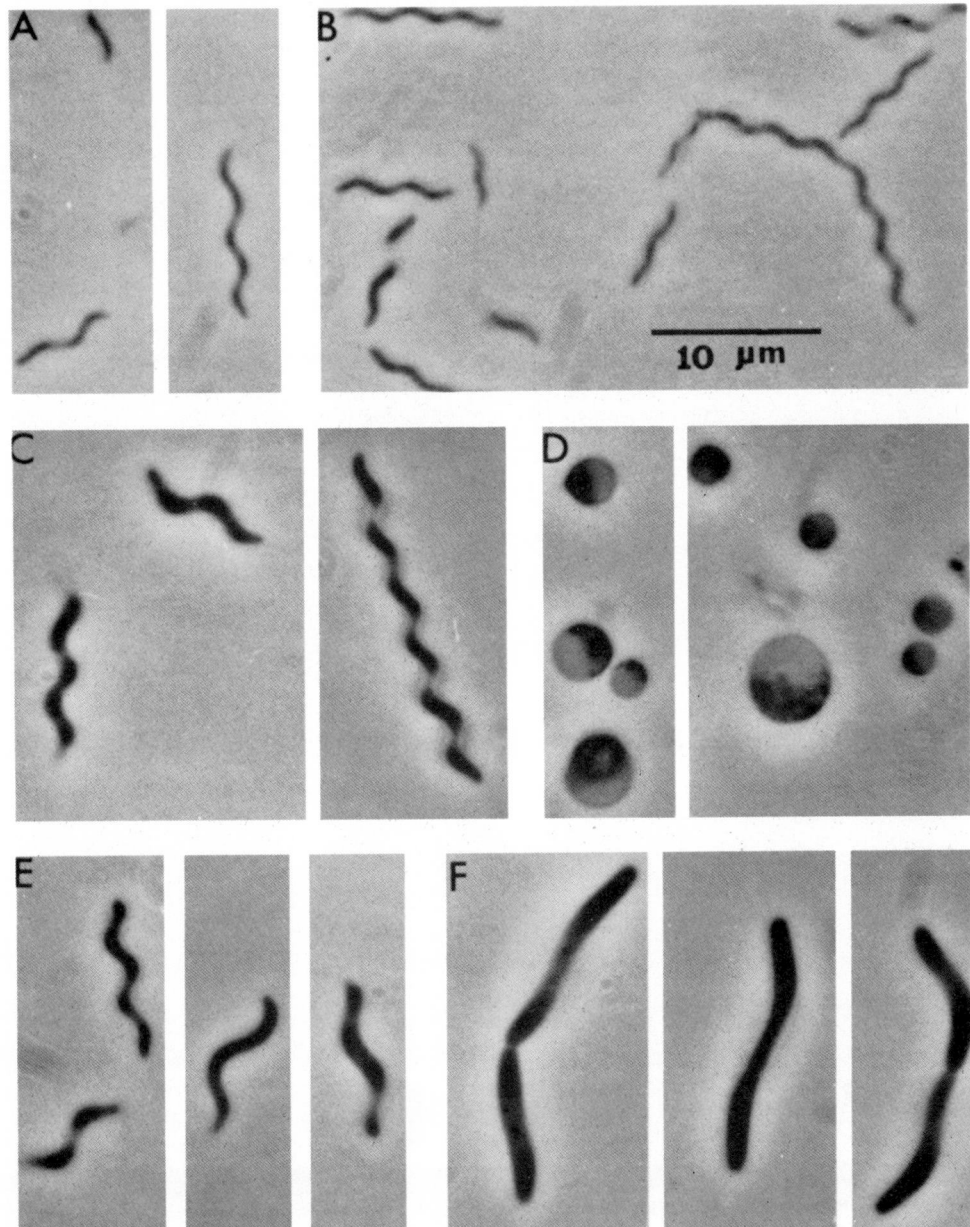

Figure 2.16. Phase-contrast photomicrographs of several species of the genus *Oceanospirillum*. All photomicrographs were taken at the same magnification. *A, O. minutulum* ATCC 19193; *B, O. linum* ATCC 11336; *C, O. maris* ATCC 27509; *D,* coccoid bodies of *O. maris* formed after 7 days of incubation; *E, O. beijerinckii* ATCC 12754; and *F, O. japonicum* ATCC 19191. (Reproduced with permission from N. R. Krieg, Bacteriological Reviews *40:* 55–115, 1976, ©American Society for Microbiology.)

PSS seawater agar and are usually white, circular and convex, ranging from pinpoint to 1.5 mm in diameter (Hylemon et al., 1973). Colonies on seawater-nutrient agar are generally pinpoint in size at 48 h but become larger (up to 2.0 mm in diameter) at 7 days; they are usually convex or umbonate, glistening, opaque, pale yellow and butyrous (Terasaki, 1972). Rough or "R" colonies may arise on prolonged transfer; for example, the colonies of the type strain of *O. japonicum* are presently of the R type. Some species produce a water-soluble, yellow-green fluorescent pigment on PSS seawater agar.

Most species grow best at a temperature of 30–32°C; however, *O. hiroshimense* grows best at 25°C and *O. japonicum* grows best at 35–37°C (Terasaki, 1972).

The nutrition of oceanospirilla is generally simple. Most species grow in simple defined media with amino acids or the salts of organic acids as carbon sources and ammonium ions as the nitrogen source. However, *O. linum* is specifically stimulated by methionine in a medium containing succinate and malate as carbon sources, and *O. maris* subsp. *williamsae* has a growth factor requirement that has not yet been identified. A listing of the carbon sources for oceanospirilla is given later in Table 2.12. Some apparent contradictions occur between the results obtained from different laboratories, although the results within each laboratory are reproducible. These differences are likely attributable to differences in methodology and in definitions of what constitutes a positive growth response, and in some cases to the use of different strains.

The use of antisera in agglutination tests with a limited number of strains has indicated that the species of *Oceanospirillum* can be distinguished serologically (McElroy and Krieg, 1972). The antisera were prepared against whole cells and adsorbed with heated cells, leaving only antibodies against thermolabile antigens.

Oceanospirilla have been isolated from coastal sea water (Williams and Rittenberg, 1957), decaying sea weed (Jannasch, 1963) and putrid infusions of marine mussels (Terasaki, 1963, 1970, 1980). By direct microscopic counts of the bacteria present in clear and turbid sea waters near Port Aransas, Texas, Oppenheimer and Jannasch (1962) found that spirilla comprised only from 0.1–2.5% of the total bacteria present. Whether oceanospirilla occur in the open sea is not known. On the basis of chemostat experiments, Jannasch (1963) suggested that the growth of oceanospirilla may be restricted to environments of higher nutrient concentration than found in ordinary sea water, such as in zones surrounding decaying particulate matter. With regard to occurrence of oceanospirilla in putrid infusions of marine mussels, the source is most likely marine mud adherent to the mussels (Terasaki, 1970).

Enrichment and Isolation Procedures

The enrichment and isolation method used by Williams and Rittenberg (1957) is as follows. A seawater sample is mixed with an equal volume of Giesberger's base medium (NH$_4$Cl, 0.1%; K$_2$HPO$_4$, 0.05%; MgSO$_4$, 0.05%) plus 1.0% calcium lactate. After incubation and appearance of spirilla, a portion of the initial culture is sterilized and mixed with an equal volume of sterile Giesberger's medium lacking NH$_4$Cl. This mixture is then inoculated from the unsterilized portion of the initial culture. One to three subcultures done in this manner are sufficient to establish the spirilla as the predominant type. For isolation, the enrichment is diluted 1:100 to 1:100,000 with sterile seawater. The dilution bottles are shaken vigorously and allowed to stand at room temperature for 20 min to allow migration of spirilla to the surface of the dilution. Isolation is then accomplished by streaking the surface water onto a suitable agar medium such as nutrient agar prepared with seawater.

The method of Terasaki (1970) has yielded excellent results. Marine mussels are smashed with a hammer and placed in a Petri dish with a teaspoon of marine mud. Sterilized seawater is poured into the dish until the mussels sink completely in the solution. The infusion is incubated at 27–28°C and examined for the development of spirilla after 1, 2, 4 and 7 days. Isolation is accomplished by streaking dilutions onto suitable agar media.

For enrichment by use of continuous cultures, see Jannasch (1967).

Maintenance Procedures

Oceanospirilla may be maintained in semisolid PSS seawater medium (containing 0.15% agar to give a jelly-like consistency) at 30°C with weekly transfer (Hylemon et al., 1973). Cultures may also be maintained as stabs in seawater-nutrient agar at room temperature with monthly transfer (Terasaki, 1972).

Preservation is most easily accomplished by suspending a dense concentration of cells in seawater-nutrient broth containing 10% (v/v) dimethyl sulfoxide, with subsequent freezing in liquid nitrogen. A method for freeze-drying oceanospirilla has been reported by Terasaki (1975).

Procedures for Testing Special Characters

Characterization methods for oceanospirilla have been described in detail by Terasaki (1972, 1979) and Hylemon et al. (1973). The comments given in this *Manual* for the genus *Aquaspirillum* also apply to the genus *Oceanospirillum* as well, except that media containing natural or artificial seawater must be used for all characterization tests.

Differentiation of the genus **Oceanospirillum** from other genera

See the genus *Aquaspirillum*, Table 2.1 for characteristics of *Oceanospirillum* that distinguish the genus from other morphologically or physiologically similar genera.

Taxonomic Comments

In the eighth edition of the *Manual* a single genus, *Spirillum*, contained all of the various aerobic and microaerophilic spirilla, including freshwater and marine species. However, the DNA base composition for the genus ranged from 38–65 mol% G + C and appeared to be unusually broad for a bacterial genus. Moreover, three groups were evident within the genus: (a) the aerobic, freshwater spirilla that could not tolerate 3% NaCl (mol% G + C is 50–65); the aerobic marine spirilla that required seawater for growth (mol% G + C is 42–48); and the large, microaerophilic spirilla that belong to the species *S. volutans* (mol% G + C is 38). Accordingly, Hylemon et al. (1973) divided the genus into the three genera *Spirillum*, *Aquaspirillum* and *Oceanospirillum*, with the marine organisms comprising the latter genus.

Although this scheme is useful for practical purposes, it is only recently that the phylogenetic relationships between the species of *Oceanospirillum* have begun to be investigated. In an analysis of the oligonucleotide catalogs of the 16S rRNA of the two species *O. maris* and *O. minutulum*, Woese et al. (1982) found that both organisms belonged to group III of the phototrophic bacteria as defined by Gibson et al. (1979) but were not closely related to each other. Additional species of *Oceanospirillum* have not yet been tested.

Except for the restriction of a mol% G + C range of 42–48 (the recent inclusion of *O. pusillum* now extends this range to 51), the genus *Oceanospirillum* is at present based on a pattern or core of phenotypic characteristics that are considered to be the typical characteristics of the genus. These include a helical shape; bipolar tufts of flagella; poly-β-hydroxybutyrate formation; a predominance of coccoid bodies in old cultures; a strictly respiratory type of metabolism with oxygen as the terminal electron acceptor; a positive oxidase reaction; an inability to catabolize sugars, starch or casein; a negative indole test; a requirement for seawater; an optimum temperature of 30–32°C; and a simple heterotrophic nutrition. Species have been assigned to the genus on the basis of the correspondence of their characteristics to the core characteristics, with the recognition that certain exceptional characteristics may occur. Such exceptional characteristics have so far included the following: a single flagellum at each pole instead of bipolar flagellar tufts, inability to form coccoid bodies, and an optimum temperature of 25°C rather than 30–32°C.

Although 9 species are presently recognized in the genus, no DNA/DNA hybridization studies have been done to confirm this number of species. Speciation has been based mainly on differences in morphology, nutrition and DNA base composition. In some cases, species are obviously different: for example, a small spirillum such as *O. minutulum* would not belong to the same species as a large spirillum such as *O. maris* or *O. japonicum*, and a spirillum having a mol% G + C of 42–44 would not belong to the same species as one with a mol% G + C of 48–50 such as *O. linum*. However, species distinctions are less firmly based within a particular morphological group of strains, especially if they have a similar DNA base composition. Therefore, it is possible that some presently recognized species may not deserve separate species status (for example, *O. maris* vs. *O. beijerinckii*). It is likely that DNA/DNA hybridization experiments could resolve many such questions.

Another difficulty with speciation in the genus is that most species are represented by only one or two strains. Consequently, the limits of variation of characters within these species may be broader than is presently assumed.

An organism known as "*Spirillum lunatum*" (Williams and Rittenberg, 1957) was included in the genus *Oceanospirillum* by Hylemon et al. (1973) but has posed taxonomic difficulties. The characteristics of the type strain (ATCC 11337 or NCMB 54) did not fit the original description of the species, and Linn and Krieg (1978) found that NCMB strain 54 consisted of a mixture of two dissimilar organisms. The first type was a short, vibrioid rod that possessed a single polar flagellum, grew in either the presence or absence of seawater, catabolized sugars, did not form coccoid bodies, and had a mol% G + C of 63–64. The second type was a larger, helical organism that possessed bipolar flagellar tufts, required seawater, failed to attack sugars, formed coccoid bodies, and had a mol% G + C of 45. The smaller organism did not appear to belong to either *Oceanospirillum* or *Aquaspirillum* and it

remains unclassified. The larger organism had characteristics more in accord with the original description of "*S. lunatum*" but differed in certain respects; it has been classified as a new subspecies of *O. maris*, viz., *O. maris* subsp. *williamsae*.

Further Reading

Krieg, N.R. 1976. Biology of the chemoheterotrophic spirilla. Bacteriol. Rev. *40:* 55–115.
Krieg, N.R. and P.B. Hylemon. 1976. The taxonomy of the chemoheterotrophic spirilla. Annu. Rev. Microbiol. *30:* 303–325.

Differentiation and characteristics of the species of **Oceanospirillum**

The differential characteristics of the species of *Oceanospirillum* are indicated in Table 2.10. Other characteristics of the species are indicated in Table 2.11.

List of the species of the genus **Oceanospirillum**

1. **Oceanospirillum minutulum** (Watanabe 1959) Hylemon, Wells, Krieg and Jannasch 1973, 373.[AL] (*Spirillum minutulum* Watanabe 1959, 83.)

mi.nu′tu.lum. L. dim. neut. adj. *minutulum* very little.

The smallest of the oceanospirilla. The morphological characters are depicted in Fig. 2.16 and listed in Tables 2.10 and 2.11.

The colonial characteristics on sea water-nutrient agar differ from

other oceanospirilla in that the colonies are flat and grow into the agar medium, with the growth reaching to the bottom of the plate (Terasaki, 1972).

The physiological characters are listed in Tables 2.10 and 2.11. Sole carbon sources are listed in Table 2.12. Ammonium ions can be used as a sole nitrogen source; nitrate is not used.

This species includes organisms previously assigned to the two spe-

Table 2.10
Differential characteristics of species of the genus **Oceanospirillum**[a]

Characteristics	1. *O. minutulum*	2. *O. pusillum*	3. *O. linum*	4. *O. multiglobuliferum*	5. *O. hiroshimense*	6. *O. pelagicum*	7. *O. beijerinckii*	8. *O. maris*	9. *O. japonicum*
Cell diameter, μm	0.3–0.4	0.3–0.5	0.4–0.6	0.5–0.9	0.6–1.1	0.6–1.2	0.7–1.0	0.7–1.0	0.8–1.4
Type of helix[b]	C	CC	C	C	C	C	C	C	C
Flagellar arrangement	BT	BS	BT	BT	BT	BT	BT	BT	BT
Coccoid bodies predominant at 3–4 weeks	+	+	+	+	+	+	+	+	−
Coccoid bodies predominant at 24–48 h	−	−	−	+	−	−	−	−	−
Maximum salt tolerance is low (4% NaCl)	−	−	−	+	−	−	−	−	−
Optimum temperature: 25°C rather than 30–32°C	−	−	−	−	+	−	−	−	−
Catalase	+	W or−	+ or W	+	W or −	+	+ or W	+ or W	W or −
Phosphatase	−	W	+	+	+	+	+	−	W
Nitrate reduced to nitrite	+	+	−	−	−	−	−	−	−
Auxotrophic growth requirement	−	−	+[c]	−	−	−	−	d[c]	−
Mol% G + C of DNA	42–44	51	48–50	46	47	49	47	45–46	45

[a] Symbols: +, positive for all strains; −, negative for all strains; d, differs among strains; C, clockwise helix; CC, counterclockwise helix; BT, bipolar tufts; BS, bipolar single; and W, weak reactions.
[b] Determined by focusing on the bottom of the cells. The pattern ⁄ ⁄ ⁄ ⁄ indicates a clockwise helix, whereas the pattern \ \ \ \ indicates a counterclockwise helix.
[c] *O. linum* grows poorly or not at all in defined media with single carbon sources and ammonium ions as the nitrogen source; however, abundant growth occurs in a defined medium containing succinate plus malate as carbon sources and methionine as the nitrogen source. (See also footnote d, Table 2.12) *O. maris* subspecies *williamsae* failes to grow in vitamin-free defined media and has a growth factor requirement that has not yet been identified.

Table 2.11
Other characteristics of the species of the genus **Oceanospirillum**[a]

Characteristics	1. O. minutulum	2. O. pusillum	3. O. linum	4. O. multiglobuliferum	5. O. hiroshimense	6. O. pelagicum	7. O. beijerinckii	8. O. maris	9. O. japonicum
Wavelength of helix, μm	2.0–2.8	1.7–2.0	1.8–4.0	3.5–5.0	3.0–4.0	3.0–6.0	6.3–7.2	3.5–7.0	7.0–20.0
Helix diameter, μm	0.6–1.5	1.0–1.2	0.8–1.4	1.0–2.0	1.2–2.5	1.0–2.0	1.5–3.0	1.4–2.8	2.0–5.0
Length of helix, μm	3.0–8.0	1.2–4.0	4.0–30.0	2.0–10.0	2.5–40.0	2.0–14.0	7.0–15.5	7.0–21.0	5.0–75.0
Polar membrane present	+						+	+	+
Intracellular poly-β-hydroxybutyrate formed	+[b]	+	+	+	+	+	+	+	+
Oxidase	+	+	+	+	+	+	+	+	+
Anaerobic growth with nitrate	−	−	−	−	−	−	−	−	−
Denitrification	−	−	−	−	−	−	−	−	−
Acid produced from sugars	−	−	−	−	−	−	−	−	−
Hydrolysis of esculin, hippurate, starch or casein	−	−	−	−	−	−	−	−	
Indole test	−	−	−	−	−	−	−	−	−
Sulfatase (0.01% phenolphthalein disulfate)	−	−					−	−	−
Temperature range for growth, °C	11–37	6–40	11–39	6–37	2–35	8–41	14–37		10–43
Gelatin liquefaction, 20°C[c]									
7 Days	−	−	−	−	−	d	+		d
28 Days	−	−	+	−	−	d	+		d
42 Days	−	−	+	−	−	d	+		d
Gelatin hydrolysis, 30°C., 4 Days[d]	−		−				−		−
Range of NaCl (%) for growth in peptone water, 7 days[c]	0.5–8.0	0.5–8.0	0.5–8.0	0.5–4.0	0.5–8.0	0.5–8.0	0.5–6.0		0.5–8.0
Growth in PSS-seawater broth containing:[d]									
9.75% total NaCl	+		+				+	+	+
12.75% total NaCl	d		−				−	−	−
Reduction of 0.3% H_2SeO_3[d]	+		+				−	−	+
Water-soluble brown pigment from:[d]									
0.2% Tyrosine	−		+				+	d	−
0.2% Phenylalanine	−		−				−	−	−
0.2% Tryptophan	d		d				−	d	−
Growth in presence of:[d]									
1% Oxgall	+		+				+	+	+
1% Glycine	+		+				−	+	−
Growth on:[d]									
Eosin-methylene blue agar	−		−				−	d	−
MacConkey agar	−		−				−	−	−
Triple-sugar iron agar	+		+				−	+	−
Sellers agar	−		−				−	−	−
Methyl red-Voges Proskauer broth	−		+				−	+	−
Water-soluble fluorescent pigment[d]	−		+				−	d	−
Deoxyribonuclease[d]	−		−				+	−	−
Ribonuclease[d]	−		d				+	−	−
Urease[d]	−		−				−	−	−

[a] Symbols: +, positive for all strains; −, negative for all strains; d, differs among strains; blank space, not determined.
[b] No granules are visible microscopically in *O. minutulum* but chemical analysis indicates presence of the polymer.
[c] Data from Terasaki (1972, 1979).
[d] Data from Hylemon et al. (1973).

cies *Spirillum minutulum* and *Spirillum halophilum* by Watanabe (1959). The two species were combined into the single species *O. minutulum* by Hylemon et al. (1973) on the basis of a high degree of similarity in phenotypic characters and DNA base composition.

Isolated from putrid infusions of marine mussels.

The mol% G + C of the DNA is 42–44 (T_m).
Type strain: ATCC 19193.

2. **Oceanospirillum pusillum** (Terasaki 1973) Terasaki 1979, 142.[AL] (*Spirillum pusillum* Terasaki 1973, 67.)

Table 2.12.
Carbon sources used by the species of Oceanospirillum

Specie	Citrate	Aconitate	Isocitrate	α-Ketoglutarate	Succinate	Fumarate	Malate	Oxaloacetate	Pyruvate	Lactate	Malonate	Tartrate	Acetate	Propionate	Butyrate	Caproate	β-Hydroxybutyrate	p-Hydroxybenzoate	Ethanol	n-Propanol	n-Butanol	Glycerol	L-Histidine	L-Tyrosine	L-Phenylalanine	L-Alanine	L-Glutamate	L-Aspartate	L-Glutamine	L-Asparagine	L-Proline	L-Hydroxyproline	L-Ornithine	L-Citrulline	L-Arginine	L-Lysine	Putrescine	L-Methionine	L-Serine	L-Cysteine	Glycine	L-Leucine	L-Isoleucine	L-Valine	L-Tryptophan
1. O. minutulum																																													
Method A[a]	−	−	−	d	d	+	+	+	+	+	+	d	+	+	+	−	−	−	−	−	−	−	−	−	−	−	+	−	d	−	+	+	−	−	−	−	−	−	−	−	−	−	−	−	−
Method B[b]	−	−	−	−	+	+	+	−	+	+	+	−	+	+	+	−	−	−	−	−	−	+	−	−	−	−	+	−	−	−	−	−	−	−	−	−	−	−	−	−	−	−	−	−	−
2. O. pusillum	+	−	−	−	+	+	+	−	+	+	+	−	+	+	+	−	−	−	+	+	−	+	−	−	−	−	−	−	−	−	−	−	−	−	−	−	−	−	−	−	−	−	−	−	−
3. O. linum																																													
Method A[a,c]	−	−	−	−	−	−	−	−	−	−	−	−	d	−	−	−	−	−	−	−	−	+	−	−	−	−	−	−	−	−	−	−	−	−	−	−	−	−	−	−	−	−	−	−	−
Method B[b,d]	−	−	−	−	−	+	+	−	+	+	−	−	+	+	+	−	−	−	−	−	−	+	−	−	−	−	−	−	−	−	−	−	−	−	−	−	−	−	−	−	−	−	−	−	−
4. O. multiglobuliferum[b]	+	−	−	−	+	+	+	−	+	+	−	−	+	+	−	−	−	−	−	−	−	−	−	−	−	−	−	−	−	−	−	−	−	−	−	−	−	−	−	−	−	−	−	−	−
5. O. hiroshimense[b]	−	−	−	−	+	+	+	+	+	+	−	+	+	+	−	−	d	d	d	d	−	−	−	−	−	−	−	−	−	−	−	−	−	−	−	−	−	−	−	−	−	−	−	−	−
6. O. pelagicum[b]	d	−	−	−	+	+	+	+	+	+	−	+	+	+	−	−	d	d	d	d	−	−	−	−	−	−	−	−	−	−	−	−	−	−	−	−	−	−	−	−	−	−	−	−	−
7. O. beijerinckii																																													
Method A[a]	−	−	−	−	−	−	+	−	+	−	−	−	−	−	−	−	−	−	−	−	−	−	−	−	−	−	−	−	−	−	−	−	−	−	−	−	−	−	−	−	−	−	−	−	−
Method B[b]	−	−	−	−	+	+	+	−	−	−	−	−	−	−	−	−	−	−	−	−	−	−	−	−	−	−	−	−	−	−	−	−	−	−	−	−	−	−	−	−	−	−	−	−	−
8. O. maris[a,e]	−	−	−	−	−	−	d	+	−	−	−	−	−	−	−	−	−	−	−	−	−	−	−	−	−	−	+	−	−	−	d	−	−	−	−	−	−	−	−	−	−	−	−	−	−
9. O. japonicum																																													
Method A[a]	−	−	−	−	−	−	+	+	+	+	−	−	+	−	−	−	−	−	−	−	−	−	−	−	+	+	+	−	+	−	−	−	−	−	−	−	−	−	−	−	−	−	−	−	−
Method B[b]	−	−	−	−	+	+	+	+	+	−	−	−	+	+	+	−	−	−	−	−	−	−	−	−	+	+	+	−	−	−	−	−	−	−	−	−	−	−	−	−	−	−	−	−	−

[a] As determined by the method of Hylemon et al. (1973). A turbidimetrically standardized cell suspension in synthetic seawater was inoculated into a defined, vitamin-free medium containing the carbon source (0.1%) and ammonium sulfate as the nitrogen source. Growth responses were measured turbidimetrically after one 72-h serial transfer from the initial cultures, using a Klett colorimeter with the blue (420 nm) filter and 16-mm cuvettes. Symbols: +, 10 or more Klett units of turbidity for all strains tested; −, less than 10 Klett units of turbidity for all strains tested; d, differs among strains; blank space, not determined.

[b] As determined by the method of Terasaki (1972, 1979). A cell suspension was washed in basal, defined, vitamin-free medium (Williams and Rittenberg, 1957) containing natural seawater and lacking carbon sources. The cells were inoculated into similar media containing the test compound (0.05%) and ammonium chloride as the nitrogen source. After 7 days, the growth was estimated turbidimetrically. Symbols: +, a turbidity of 0.025 absorbance units or greater for all strains tested; −, a turbidity of less than 0.025; d, differs among strains; blank space, not determined.

[c] ATCC strain 12753 failed to grow with any sole carbon source, while strain 11336 grew only with acetate. Both strains grew abundantly when succinate plus malate were supplied as carbon sources and L-methionine as the nitrogen source.

[d] Strain OF3 (Terasaki, 1972, 1973) differs from the results given in the table in that it grows with a large variety of sole carbon sources: citrate, succinate, fumarate, malate, pyruvate, lactate, acetate, propionate and butyrate. Whether this strain should be included in the species O. linum is uncertain.

[e] The results are given only for O. maris subsp. maris. O. maris subsp. williamsae fails to grow with any sole carbon (or sole nitrogen) source and, therefore, appears to have an auxotrophic requirement. This requirement has not yet been identified.

[f] Corrected from the original study.

109

pu.sil'lum. L. dim. neut. adj. *pusillum* very small.

The morphological and physiological characters are listed in Tables 2.10 and 2.11. Sole carbon sources are listed in Table 2.12. Ammonium ions can serve as a sole nitrogen source; nitrate is not used.

Isolated from putrid infusions of marine mussels.

The mol% G + C of the DNA of the type strain is 51.

Type strain: IFO 13613.

3. **Oceanospirillum linum** (Williams and Rittenberg 1957) Hylemon, Wells, Krieg and Jannasch 1973, 374.[AL] (*Spirillum linum* Williams and Rittenberg 1957, 82.)

li'num. L. n. *linum* flax, thread.

The morphological characters are depicted in Figure 2.16 and listed in Tables 2.10 and 2.11. The physiological characters are indicated in Tables 2.10 and 2.11. Sole carbon sources are listed in Table 2.12. Growth in defined media is usually poor; however, abundant growth occurs in defined media containing malate plus succinate as carbon sources and *methionine* as the nitrogen source. Nitrate is not used.

Strain OF3, isolated by Terasaki (1972, 1973) differs from other strains of *O. linum* in that it can grow well in defined media with a variety of sole carbon sources and ammonium ions as the nitrogen source (see footnote d, Table 2.12). Other characteristics of this strain are similar to those of the type strain (Terasaki, 1973), but whether it should be included in this species is uncertain.

This species includes organisms previously assigned to the two species *Spirillum linum* and *Spirillum atlanticum* by Williams and Rittenberg (1957). The two species were combined into the single species *O. linum* by Hylemon et al. (1973) on the basis of a high degree of similarity in phenotypic characters and in DNA base composition.

Isolated from coastal sea water.

The mol% G + C is 48–50 (T_m).

Type strain: ATCC 11336.

4. **Oceanospirillum multiglobuliferum** (Terasaki 1973) Terasaki 1979, 143.[AL] (*Spirillum multiglobuliferum* Terasaki 1973, 69.)

mul.ti.glo.bu.li'fe.rum. L. adj. *multus* much, many; L. dim. n. *globulus* a small sphere, globule; L. v. *fero* to bear, carry; M.L. neut. adj. *multiglobuliferum* bearing many globules.

The morphological and physiological characters are listed in Tables 2.10 and 2.11. Differs from other species by forming unusually large numbers of coccoid bodies even in 24- to 48-h-old broth cultures. Sole carbon sources are listed in Table 2.12. Ammonium ions can serve as a sole nitrogen source; nitrate is not used.

Isolated from putrid infusions of marine mussels.

The mol% G + C of the DNA of the type strain is 46.

Type strain: IFO 13614.

5. **Oceanospirillum hiroshimense** (Terasaki 1973) Terasaki 1979, 141.[AL] (*Spirillum hiroshimense* Terasaki 1973, 62.)

hi.ro.shi.men'se. M.L. neut. adj. *hiroshimense* pertaining to Hiroshima, Japan.

The morphological and physiological characters are listed in Tables 2.10 and 2.11. Sole carbon sources are listed in Table 2.12. Ammonium ions can serve as a sole nitrogen source; nitrate is not used.

Isolated from putrid infusions of marine mussels.

The mol% G + C of the type strain is 47 (T_m).

Type strain: IFO 13616.

6. **Oceanospirillum pelagicum** (Terasaki 1973) Terasaki 1979, 142.[AL] (*Spirillum pelagicum* Terasaki 1973, 65.)

pe.la'gi.cum. L. neut. adj. *pelagicum* belonging to the sea.

The morphological and physiological characters are listed in Tables 2.10 and 2.11. Sole carbon sources are listed in Table 2.12. Ammonium ions can serve as a sole nitrogen source; nitrate is not used.

Isolated from putrid infusions of marine mussels.

The mol% G + C of the DNA of the type strain is 49.

Type strain: IFO 13612.

7. **Oceanospirillum beijerinckii** (Williams and Rittenberg 1957) Hylemon, Wells, Krieg and Jannasch 1973, 375.[AL] (*Spirillum beijerinckii* Williams and Rittenberg 1957, 90.)

bei.jer.inck'i.i. M.L. gen. n. *beijerinckii* of Beijerinck; named for Prof. M. W. Beijerinck of Delft, Holland.

The morphological characters are depicted in Fig. 2.16 and listed in Tables 2.10 and 2.11. Physiological characters are indicated in Tables 2.10 and 2.11. Sole carbon sources are listed in Table 2.12. Ammonium ions can serve as a sole nitrogen source; nitrate is not used.

Isolated from coastal sea water.

The mol% G + C of the DNA of the type strain is 47.

Type strain: ATCC 12754.

8. **Oceanospirillum maris** Hylemon, Wells, Krieg and Jannasch 1973, 376.[AL]

ma'ris. L. n. *mare* the sea; L. gen n. *maris* of the sea.

The morphological characters are depicted in Fig. 2.16 and listed in Tables 2.10 and 2.11. Physiological characters are indicated in Tables 2.10 and 2.11. Sole carbon sources are listed in Table 2.12.

Isolated from coastal sea water (Jannasch, 1967).

The mol% G + C of the DNA is 45–46 (T_m).

Type strain: ATCC 27509.

8a. **Oceanospirillum maris** subspecies **maris** Hylemon, Wells, Krieg and Jannasch, 1973, 376.

Characters are as described for the species. Differs from the subspecies *williamsae* in having strong catalase activity, growing in the presence of 1% glycine, lacking deoxyribonuclease and ribonuclease activity, and in having no auxotrophic growth requirements.

Type strain: ATCC 27509.

8b. **Oceanospirillum maris** subspecies **williamsae** Linn and Krieg 1978, 137.[VP*]

will'iam.sae. M.L. gen. n. *williamsae* of Williams; named for Marion A. Williams, who was the first to describe species of marine spirilla.

Characters are as described for the species. Differs from the subspecies *maris* in having very weak catalase activity, failing to grow in the presence of 1% glycine, possessing deoxyribonuclease and ribonuclease activity, and failing to grow in defined vitamin-free media with any sole carbon or nitrogen source. The auxotrophic growth requirement has not yet been identified.

Isolated from a mixture of organisms comprising NCMB strain 54 by Linn and Krieg (1978).

Type strain: ATCC 29547.

9. **Oceanospirillum japonicum** (Watanabe 1959) Hylemon, Wells, Krieg and Jannasch 1973, 375.[AL] (*Spirillum japonicum* Watanabe 1959, 78.)

ja.pon'i.cum. M.L. neut. adj. *japonicum* pertaining to Japan.

The morphological characters are listed in Tables 2.10 and 2.11. A photomicrograph of the type strain is presented in Fig. 2.16; however, this strain presently has morphological features that differ from the original description in that the cells are no longer helical with several waves but instead are curved, straight or S shaped; moreover, colonies of this strain are presently of the R (rough) type. Therefore, it is likely that the type strain has undergone alteration since its isolation in 1959. Three reference strains isolated by Terasaki (1972, 1973) have morphological features that more nearly correspond to those given in the original description (strains IF4, IF8, and UF3), and also they form colonies of the S (smooth) type.

The physiological characters are listed in Tables 2.10 and 2.11. Sole carbon sources are listed in Table 2.12.

Isolated from putrid infusions of marine mussels.

The mol% G + C of the DNA of the type strain is 45.

Type strain: ATCC 19191.

---

*VP denotes that this name has been validly published in the official publication, International Journal of Systematic Bacteriology.

Genus **Campylobacter** Sebald and Véron 1963, 907[AL]

ROBERT M. SMIBERT

Cam.py′.lo.bac.ter. Gr. adj. *campylo* curved; Gr. n. *bacter* rod; M.L. masc. n. *Campylobacter* a curved rod.

Slender, spirally curved rods, 0.2–0.5 μm wide and 0.5 to 5 μm long. The rods may have one or more spirals and can be as long as 8 μm. They also appear S shaped and gull-winged when two cells form short chains. Nonsporeforming. Cells in old cultures may form spherical or coccoid bodies. Cells have a multilaminar polar membrane at both ends of the cell that is located under the cytoplasmic membrane. Gram-negative. **Motile with a characteristic corkscrew-like motion by means of a single polar flagellum at one or both ends of the cell.** The flagella may be 2 to 3 times the length of the cells. **Microaerophilic, with a respiratory type of metabolism.** Require an oxygen concentration of between 3 and 15% and a carbon dioxide concentration of 3–5%. Occasionally a few strains may grow slightly under aerobic conditions (20% O_2). Some species can grow under anaerobic conditions with either fumarate, formate and fumarate, or oxygen and fumarate in the medium. **Chemoorganotrophs. Carbohydrates are neither fermented nor oxidized.** No acid or neutral end products produced. Does not require serum or blood for growth. Energy is obtained from amino acids or tricarboxylic acid cycle intermediates, not carbohydrates. Gelatin and urea are not hydrolyzed. Methyl red- and Voges Proskauer-negative. No lipase activity. **Oxidase-positive.** Nitrate reduced. Pigments are not produced. Some species are pathogenic for man and animals. **Found in the reproductive organs, intestinal tract and oral cavity of man and animals.** The mol% G + C of the DNA ranges from 30–38.

Type species: *Campylobacter fetus* (Smith and Taylor 1919) Sebald and Véron 1963, 907.

Further Descriptive Information

Cells are helically curved and have a very characteristic corkscrew-darting type of motility which is observed with phase contrast or dark-field microscopy. The growth medium becomes alkaline (pH 8.5–9.0) and coccoid forms occur under these unfavorable conditions. They are considered by many to be degenerative forms and not viable.

The outer cells membrane is double layered, loosely fitted over the cell wall, and has a wavy morphology. The cytoplasmic membrane is thickened at the polar region. This polar membrane can be seen at both ends of the cell and a similar structure has been reported in other helical bacteria such as *Aquaspirillum, Rhodospirillum, Ectothiorhodospira mobilis* and the straight rod *Chromatium* and *Selenomonas* (Smibert, 1978).

Campylobacter species have a respiratory metabolism. *C. fetus* oxidizes citrate, *cis*-aconitate, isocitrate, α-ketoglutarate, succinate, fumarate, malate and oxaloacetate. A complete tricarboxylic acid (TCA) cycle has been demonstrated. There is no oxidation or fermentation of carbohydrates. Energy for *C. fetus* is obtained from TCA intermediates and from amino acids such as glutamate and aspartate that can be deaminated to TCA intermediates.

Campylobacters are microaerophilic, requiring a low oxygen tension (3–6% O_2) for growth. This is especially true with small inocula containing few cells. The oxygen toxicity can be overcome in solid or liquid media with very large inocula containing large numbers of cells. *C. fetus, C. jejuni* and *C. coli* have catalase, oxidase and superoxide dismutase activity (Hoffman et al. 1979a, b).

Addition of 0.025% each of ferrous sulfate, sodium metabisulfite and sodium pyruvate (FBP) to culture media will increase the aerotolerance of cultures of these species and allow growth at oxygen concentrations of 15–20% (George et al., 1978; Hoffman et al., 1979a, b). The FBP mixture destroys hydrogen peroxide and superoxide anions that appear in the medium when exposed to air and light. *C. fetus, C. jejuni* and *C. coli* appear to be extremely sensitive to these toxic forms of oxygen. Bovine superoxide dismutase and catalase added to culture media greatly enhance oxygen tolerance (Hoffman et al., 1979a). Campylo-

bacters have been shown to be extremely sensitive to hydrogen peroxide, with inhibition of growth occurring with as little as 0.00124% hydrogen peroxide. The use of agar media containing blood also enhances the aerotolerance of campylobacters because blood contains catalase and superoxide dismutase.

Cytochromes *c*, *b*, and *d* were found in *C. jejuni* (Hoffman et al. 1979b). The cytochrome *c* is membrane-bound and soluble. Cytochromes *c* and *b* were found in *C. fetus* subsp. *fetus* by Harvey and Lascelles (1980). They suggested that low and high potential forms of cytochromes *c* and *b* exist in the respiratory chain and that the carbon monoxide binding form of cytochrome *c* may be the terminal oxidase.

The aerotolerance of the catalase-negative organism *C. sputorum* subsp. *bubulus* is not affected by FBP supplements to culture media (George et al., 1978). One site of oxygen damage to *C. sputorum* subsp. *bubulus* is thought to be lactic dehydrogenase (Niekus et al., 1977). It is thought that low levels of oxygen are needed to prevent the accumulation of hydrogen peroxide when cells are grown in the presence of oxygen in a medium containing formate. The enzyme, formate oxidase, which is associated with the cytoplasmic membrane, produces hydrogen peroxide. Cultivation in high oxygen concentrations results in a slowing of growth and loss of lactic and formate dehydrogenase activity because of local accumulation of hydrogen peroxide that the cellular peroxidase cannot handle. Peroxidase activity does not increase with growth in higher oxygen environments (Niekus et al., 1980a, b).

Bacteriophages have been demonstrated in *C. coli, C. fetus* and *C. jejuni* (Fletcher, 1965; Bryner et al., 1970). In *C. coli* the phage was specific for *C. coli* and did not infect *C. fetus* or *C. jejuni* (Fletcher, 1965). These are temperate bacteriophages. Phage C was lytic for 30 of 41 strains of *C. jejuni*. Phages I, II, III and IV are considered specific for *C. fetus* (Bokkenheuser et al., 1979). Transduction of streptomycin resistance in *C. fetus* and glycine tolerance has been shown (Chang and Ogg, 1970, 1971). Development of phage typing of these organisms is being studied (Bryner et al., 1973).

Plasmids have also been found in *C. jejuni* (Austen and Trust, 1980; Taylor et al., 1981). Tetracycline resistance was shown to be mediated through a plasmid with a molecular weight of 38×10^6 (Taylor et al., 1981). The plasmid could be transferred to *C. fetus* subsp. *fetus* but not to *Escherichia coli*. Plasmids were found in 19% of the strains (Austen and Trust, 1980).

C. fetus causes infection in animals and man. *C. fetus* subsp. *fetus* is associated with sporadic abortion in cattle and abortion in sheep. In man it causes fever and other symptoms. It is usually isolated from blood cultures from patients with some other debilitating condition. *C. fetus* subsp. *venerealis* causes abortion and reproductive problems only in cattle. Human infection has not been reported. *C. jejuni* causes abortion in sheep and fever and enteritis in man. Campylobacter enteritis has in the last few years been found to cause as much enteritic disease in man as *Salmonella* and *Shigella* and thus has emerged as an important human intestinal pathogen. *C. jejuni* infects people of all ages. It is more frequently diagnosed in children than adults and the seasonal incidence shows it to be higher in the summer and fall than in winter or spring (Butzler and Skirrow, 1979). The organism is found in the intestinal tract of poultry, dogs, cats, sheep and cattle (Smibert 1965, 1969, 1978 and Suedham and Kaijer, 1981). Enteric infection in animals has been reported. Transmission is most likely oral. Food, milk and water contamination presents a problem needing investigation. Outbreaks of *C. jejuni* infection associated with the drinking of unpasteurized milk have been reported (Porter and Reid, 1980). *C. coli* is found in the intestinal tract of pigs and poultry, and the incidence of infection in man is probably low. There are no laboratory animals that can be used at the present time as models for studies of *Campylobacter* infection. Thus the question of what percentage of *Campylobacter* isolates from intestinal material of various animals or food products are capable of causing enteritis in man cannot be answered at the present time.

Isolation and Enrichment Procedures

Isolation of *Campylobacter* species can be accomplished by two methods. The first involves filtration of the cells through membrane filters with a pore size of 0.45, 0.65 or 0.8 μm. The filtrate is streaked onto agar medium or inoculated into broth medium. All agar plates or broth cultures must be incubated in a microaerophilic atmosphere containing 5% O_2, 10% CO_2 and 85% N_2. Commercial products are available for establishing microaerophilic environments.

The second method for isolating campylobacters is the use of selective agar media. *C. fetus* subsp. *fetus* and *C. fetus* subsp. *venerealis* can be isolated from sheep and cattle using blood agar or brucella agar containing antibiotics such as bacitracin (2 U/ml) and novobiocin (2 μg/ml). Isolation of *C. fetus* subsp. *fetus* from man is best accomplished by blood culture using commercial blood culture products.

Isolation of *C. jejuni* and *C. coli* is best accomplished from culture of fecal material or fecal swabs. Selective media are available commercially for isolation of these organisms. The medium of Skirrow contains blood agar base, 5 to 7% lysed horse blood, vancomycin (10 mg/liter), polymyxin (2500 U/liter) and trimethoprim (5 mg/liter). The medium of Butzler contains blood agar base, 5–7% blood, bacitracin (2500 U/liter), cycloheximide (50 mg/liter), colistin sulfate (10,000 U/liter), cefazolin (15 mg/liter) and novobiocin (5 mg/liter). Campy-BAP contains brucella agar, 5% sheep erythrocytes, vancomycin (10 mg/liter), trimethoprim (5 mg/liter), polymyxin B (2500 U/liter), cephalothin (15 mg/liter) and amphotericin B (2 mg/liter).

These selective media are incubated at 42–43°C in a microaerophilic atmosphere. Isolation rates with incubation at 42–43°C are higher than with incubation at 37°C. Selective media do not eliminate all the other intestinal organisms but only limit growth of these organisms to favor isolation of campylobacters. A word of caution is needed. The commercial media designed for *C. jejuni* and *C. coli* cannot be used for isolation of *C. fetus*. Cefazolin and cephalothin inhibit growth of *C. fetus* subsp. *fetus* and *C. fetus* subsp. *venerealis*. Incubation at 42–43°C, which is best for isolation of *C. jejuni* and *C. coli*, will not support growth of *C. fetus*. Therefore special media and incubation at 37°C are required for isolation of *C. fetus*. A combination of filtration and selective media can be used for isolation of campylobacters.

C. sputorum subsp. *mucosalis* can be isolated from mucosal tissue of infected pigs using horse blood agar containing brilliant green (1:60,000) and novobiocin (5 μg/ml) (Lawson 1974). Growth of this organism requires an atmosphere containing hydrogen (3% oxygen and 97% hydrogen, or anaerobic conditions with hydrogen and fumarate in the medium).

C. concisus also requires hydrogen for growth in an atmosphere with 5% O_2, 10% CO_2 and 10–85% H_2.

For more details on isolation methods and media see Smibert, 1981; Ullmann, 1979; Kaplan, 1980 and Tanner et al., 1981.

Maintenance Procedures

Stock cultures of *Campylobacter* species can be maintained under aerobic conditions by weekly transfer in a semisolid medium such as brucella broth with 0.16% agar. Addition of blood to media may increase survival. Cultures may be stored for many years by lyophilization, freezing at −80°C, or in liquid nitrogen. Cryoprotective agents such as 10% glycerol or dimethyl sulfoxide should be added to cultures before freezing, and heavy cell concentrations should be used.

Taxonomic Comments

The genus *Campylobacter* represents a well defined genus. Taxonomic problems are at the species level.

The nomenclature used in this edition of *Bergey's Manual* is that of Véron and Chatelain, 1973. The manuscript for the eighth edition of *Bergey's Manual* was started in 1970 and finished by 1971. The nomenclature and classification developed was an improvement over previous ones. A conservative approach was used in the eighth edition classification with full appreciation that it was temporary and that with more interest and work in the field the taxonomy would be changed. The basic view was that taxonomy is a dynamic and not a static field. Because of some problems, the publication of *Bergey's Manual* was delayed until 1974. In 1973, a study on *Campylobacter* was published by Véron and Chatelain. Thus the earlier date of publication made the nomenclature of Véron and Chatelain the valid one. Therefore, the nomenclature used in this current edition of Bergey's Manual is that of Véron and Chatelain (1973).

The main taxonomic problem is with the separation of *C. jejuni* and *C. coli*. Most characteristics such as brilliant green sensitivity, tetrazolium sensitivity and growth in 8% glucose are not very reliable tests (Skirrow and Benjamin, 1980). The most reliable test is the hippurate hydrolysis test done according to the method of Hwang and Ederer (1975) as described by Harvey (1980). With DNA homology studies, *C. jejuni* and *C. coli* isolates from different animal sources may eventually represent more than two species. The problem will be to find simple and reliable tests that can be used to separate these homology groups.

One additional taxonomic problem is the status of *C. sputorum* subsp. *mucosalis*. The question would be whether this subspecies, which requires hydrogen and oxygen for growth, is a microaerophilic subspecies of *Campylobacter sputorum* or a separate species of *Campylobacter*.

C. concisus is not related to *C. sputorum* subsp. *sputorum* or subsp. *bubulus* by DNA homology. The relationship of *C. concisus* to *C. sputorum* subsp. *mucosalis* should also be investigated, because both of these organisms require H_2 for microaerophilic growth.

Further Reading

Butzler, J.P. and M.B. Skirrow. 1979. *Campylobacter* enteritis. Clin. Gastroenterol. *8:* 737–765.
Kaplan, R. 1980. *Campylobacter. In* Lennette, Balows, Hausler and Truant (Editors), *Manual of Clinical Microbiology*, 3rd Ed., American Society for Microbiology, Washington, D.C., pp. 235–241.
Skirrow, M.B. and J. Benjamin. 1980. "1001" Campylobacters: cultural characteristics of intestinal campylobacters from man and animals. J. Hyg. (Camb.) *85:* 427–442.
Smibert, R.M. 1978. The genus *Campylobacter*. Annu. Rev. Microbiol. *32:* 673–709.
Smibert, R.M. 1981. The Genus *Campylobacter. In* Starr, Stolp, Truper, Balows and Schleger (Editors), *The Prokaryotes, a Handbook on Habitats, Isolation and Identification of Bacteria*, Springer-Verlag, New York, pp. 609–617.
Ullmann, U. 1979. Methods in *Campylobacter. In* Bergan and Norris (Editors), *Methods in Microbiology*, Vol. B, Academic Press, New York, pp. 439–452.

Differentiation of the species of the genus **Campylobacter**

Characteristics useful in distinguishing the various species of the genus *Campylobacter* are listed in Table 2.13. Additional characteristics are given in Table 2.14.

List of the species of the genus **Campylobacter**

1. **Campylobacter fetus** (Smith and Taylor 1919) Sebald and Véron 1963, 907.[AL] (*Vibrio fetus* Smith and Taylor 1919, 301.)
fe′.tus. L. n. *foetus* fetus.
Slender curved rods that are 0.2–0.3 μm in diameter and 1.5–5 μm long. They appear comma-, S-, and gull-shaped. The ends of the cells are pointed. Loosely wound spiral filaments up to 8 μm long appear in old cultures. Spherical or coccoid forms are also found in old cultures especially when grown on agar plates.

• Very actively motile with a characteristic darting and corkscrew-like motion. Motility and rotation of the cells are so rapid the curvation of

Table 2.13.

Differential characteristics of the species and subspecies of the genus **Campylobacter**[a]

Characteristics	1. C. fetus		2. C. jejuni	3. C. coli	4. C. sputorum			5. C. concisus
	1a. subsp. fetus	1b. subsp. venerealis			4a. subsp. sputorum	4b. subsp. bubulus	4c. subsp. mucosalis	
H_2 required for microaerophilic growth	−	−	−	−	−	−	+	+
Catalase	+	+	+	+	−	−	−	−
Nitrite reduction	−	−	−	−	+	+	+	+
H_2S production (TSI or SIM)[a]	−	−	−	−	+	+	+	+
Growth in presence of:								
1% Glycine	+	−	+	+	+	+	−	−
3.5% NaCl	−	−	−	−	−	+	−	−
Growth at:								
25°C	+	+	−	−	−	d	−	−
42°C	−	−	+	+	+	−	+	−
Inhibited by:								
Nalidixic acid[b]	−	−	+	+	−	d	d	−
Cephalothin[b]	+	+	−	−	+	+	+	
Hippurate hydrolysis[c]	−	−	+	−	−	−	−	
Anaerobic growth with fumarate in the absence of H_2 or formate	−	−	−	−	+	+	−	−
Anaerobic growth requires both fumarate and H_2 or formate[d]	−	−	−	−	−	−	+	+
Colonies are a dirty yellow color	−	−	−	−	−	−	+	−
Mol% G + C of DNA	33–36	33–36	31	32–34	29–31	29–31	34	34–38

[a] Symbols: see standard definitions; and TSI, triple sugar iron agar; SIM, sulfide-indole-motility medium.
[b] 30 µg/disk.
[c] Using the test described by Harvey (1980).
[d] Using brucella broth as the basal medium.

Table 2.14.

Additional characteristics of the species and subspecies of the genus **Campylobacter**[a]

Characteristics	1. C. fetus		2. C. jejuni	3. C. coli	4. C. sputorum			5. C. concisus
	1a. subsp. fetus	1b. subsp. venerealis			4a. subsp. sputorum	4b. subsp. bubulus	4c. subsp. mucosalis	
Ferment or oxidize glucose	−	−	−	−	−	−	−	−
Nitrate reduction	+	+	+	+	+	+	+	+
Oxidase test	+	+	+	+	+	+	+	+
Indole production	−	−	−	−	−	−	−	−
H_2S, lead acetate strip method[b]	d	d	+	+	+	+	+	+
Growth in presence of:								
1.0% Bile	+	+	+	+	+	−	−	+
1.5% NaCl	d	d	−	−	+	+	+	+
Brilliant green, 1:100,000	+	+	−	+			+	
Brilliant green, 1:33,000	+	+	−					
Growth at:								
25°C	+	+	−	−	−	d	−	−
30.5°C	+	+	−	+	+			
43°C	−	−	+	+	+		−	−
45.5°C	−	−	d	d	−	−	−	−
Selenite reduced	d	−	+	+	+	d		
Inhibited by metronidazole[c]	−	−	d	d	+	+		+
Fumarate converted to succinate	+	+	+	+	+	+	+	+
Cell lipid contains C-19 fatty acid[d]	−	−	+					

[a] Symbols: see standards definitions.
[b] Lead acetate-impregnated strips of paper hung over lip of tube of Brucella broth containing 0.16% agar and cysteine-HCl.
[c] 5 µg/disk.
[d] C-19, cyclopropane-C-19:0 fatty acid.

the cells may be overlooked. Best observed with a phase-contrast microscope.

Whole cell hydrolysates contain small amounts of *meso*-diaminopimelic acid (DAP). DAP was not found in isolated cell walls (Smibert, 1970).

Cell lipid contains C-14:0, C-16:0, C-16:1 and C-18:1 fatty acids. Cyclopropane C-19:0 fatty acid not found (Blaser et al., 1980).

Several types of colonies are found on agar on primary isolation (Bryner et al., 1962). Smooth colonies, the most frequently found, are small, 0.5 mm in diameter, round, slightly raised, smooth, colorless and slightly translucent. "Cut glass" colonies are 1 mm in diameter, round, raised, translucent and granular with reflecting facets. Rough colonies are rare and similar to smooth colonies with the exception of being granular and more opaque. Mucoid colonies are similar to smooth and cut glass colonies but are viscid. Sometimes on primary isolation, colonies occur as a thin veil of confluent growth that is translucent and a very light gray or tan color. Colonies on blood agar are nonhemolytic, round, 1 mm in diameter, smooth, raised, convex and grayish white in appearance.

Slight uniform turbidity in broth.

Many physiological characteristics are listed in Tables 2.13 and 2.14.

Potassium gluconate is not oxidized. Final pH of glucose semisolid medium (0.16% agar) is 8.2–8.7 after 3 weeks.

Fatty acids are not produced from glucose as end products of fermentation. However trace amounts of ethanol and *n*-propanol may be produced in cultures. Both pyruvate and lactate are utilized, but fatty acids do not accumulate in the medium as end products.

Esculin, casein, ribonucleic acid and deoxyribonucleic acid are not hydrolyzed.

Glutamic acid, glutamine, aspartic acid and asparagine are deaminated and ammonia is produced in cultures. Phenylalanine, tyrosine and tryptophan are not deaminated. Lysine, ornithine and arginine are not decarboxylated. Malonate is not utilized.

Hydrogen sulfide is not produced on triple sugar iron agar slants; there is only an alkaline reaction on this medium.

Does not grow in litmus milk. Peptone-supplemented litmus milk is reduced and turned slightly alkaline. Only slight growth occurs on Sellers' medium (Sellers, 1964); the slant is slightly alkaline and the butt is not changed.

Grows in a semisolid medium containing 0.12–0.20% agar when incubated under aerobic conditions. Growth in a semisolid medium occurs only within the first few millimeters below the surface of the medium. There is little or no growth under strict anaerobic conditions. No growth anaerobically with fumarate or formate and fumarate. There is no growth in deep stab cultures. Slight growth on MacConkey agar.

Optimum temperature, 37°C; optimum pH, 7.0.

Growth is not inhibited by 0.001 M iodoacetate or the vibriostatic agent 0/129.

The mol% G + C of the DNA ranges from 32–36.

Type strain: CIP 5396 (ATCC 27374; NCTC 10842).

1a. Campylobacter fetus subspecies **fetus** Véron and Chatelain 1973, 126.[AL] (*Vibrio fetus* var. *intestinalis* Florent 1959, 955; *Vibrio foetus-ovis* Buxton 1929, 47; *Campylobacter fetus* subsp. *intestinalis* Smibert, 1974, 209).

Morphology and characteristics as for species except as noted.

Intermediate sized spirals with an average wavelength of 1.8 μm and an average amplitude of coils 0.55 μm (Karmali et al., 1981).

Cell walls contain galactose only, galactose and mannose, galactose and rhamnose, or galactose, glucose and mannose (Smibert, 1970).

Several types of colonies are found on agar on primary isolation (Bryner et al., 1962). Smooth colonies are 1 mm in diameter, colorless to slightly cream colored. Rough colonies are small, round, finely granular, opaque and white to cream or tan colored. They are 1–2 mm in diameter. "Cut glass" colonies do not develop in primary cultures. Smooth colonies incubated for 6–8 days become mucoid. Upon subculture, smooth cut glass and rough cut glass colonies appear, as well as smooth colonies. Frequently, on primary isolation colonies are low, flat, grayish to tan colored and translucent with an irregular edge. They spread along the direction of the streak and coalesce. They may also form a thin veil of confluent growth on agar plates. Colonies on blood agar are nonhemolytic, round, 1–2 mm in diameter, smooth, convex and greyish white or light tan colored.

Uniform turbidity is produced in Brucella broth. A butyrous sediment may be seen in some broth cultures. Twenty-five percent of the strains may be arylsulfatase positive. No phosphatase activity.

Many physiological characteristics are given in Tables 2.13 and 2.14. Grows on slants with 0.1% sodium selenite; selenite is usually reduced.

Grows at 25°C, but usually not at 42°C. A few strains may grow at both 25 and 42°C.

Sensitive to chloramphenicol (4 μg/ml), dihydrostreptomycin (4 μg/ml), erythromycin (2 μg/ml), neomycin (8 μg/ml), oxytetracycline (8 μg/ml), streptomycin (4 μg/ml), and tetracycline (1 μg/ml). Moderately sensitive to penicillin (32 U/ml). Resistant to bacitracin (128 μg/ml), novobiocin (128 μg/ml), and polymyxin B, (1024 U/ml).

Also sensitive to *N*-formimodoylthienamycin (< 0.19 μg/ml), gentamicin (< 0.19 μg/ml), ampicillin (0.19–0.78 μg/ml), moxalactam (1.56–6.25 μg/ml), cefotaxime (3.12–12.5 μg/ml), cephalothin (3.12–12.5 μg/ml), and cefoperazone (50–100 μg/ml) (Spelhaug et al., 1981).

Serologic studies with *C. fetus* subsp. *fetus* using heated cells (100°C for 2 h) divide this subspecies into two serovars: serovar A-2 and B. Strains with both antigens (A-B-2) occur. These serovars contain surface heat-stable antigens A and B (Berg et al., 1971). Serovar A-2 is the same as serovar A of Morgan, serovar 1 of Mitscherlich and Liess, and serovar III and V of Marsh and Firehammer. Serovar B is the same as serovar B of Morgan, serovar 2 of Mitscherlich and Liess and serovar II of Marsh and Firehammer (Smibert, 1978; Berg et al., 1971).

An antiphagocytic antigen (a) which makes up the microcapsule is found on *C. fetus* subsp. *fetus*. It is a glycoprotein with about 4% carbohydrate consisting of hexose, pentose and methylpentose and has a molecular weight of ~98,000 (McCoy et al., 1975; Winter et al., 1978).

Pathogenic. Cause of abortion in sheep and sporadic abortion in cattle, as well as a cause of human infections. Transmitted orally.

Isolated from the placentas and stomach content of fetuses from aborted sheep and cattle; and from the blood, intestinal content and bile of infected ewes, and cattle. Isolated from blood, spinal fluid and abscesses from most parts of the body of man. This organism will grow in the intestinal tract and gallbladder of man and animals (Bryner et al., 1964).

The mol% G + C of the DNA is 33–36.

Type strain: CIP 5396 (ATCC 27374; NCTC 10842).

1b. Campylobacter fetus subspecies **venerealis** (Florent 1959) Véron and Chatelain 1973, 126.[AL] (*Vibrio fetus* var. *venerealis* Florent 1959, 955; *Campylobacter fetus* subsp. *fetus* Smibert 1974, 209.)

ve.ne′.re.al.is. L. *venereus* from Venus goddess of Love, L. adj. *venereal*, L. gen. n. *venerealis* of Venus goddess of love.

Morphology and characteristics are for species except as noted.

Large spirals with an average wavelength of 2.43 μm and an average amplitude of 0.73 μm (Karmali et al., 1981).

Cell walls contain galactose and either mannose or glucose (Smibert, 1970).

Many physiological characteristics are listed in Tables 2.13 and 2.14.

Hydrogen sulfide usually not produced in a medium containing cysteine with lead acetate-impregnated paper strips as the detection system. A few isolates (3–4%) will be slightly H₂S-positive after 5 days of incubation. These have been reported as subtype I (Bryner et al., 1962). These workers called the H₂S-negative strains type I. H₂S-positive strains are also called *C. fetus* subsp. *venerealis* biovar *intermedius* (Véron and Chatelain, 1973). No growth on 0.1% sodium selenite medium and no reduction of selenite. No phosphatase or arylsulfatase activity occur.

Grows at 25°C but not at 42 or 45°C. A few strains have been reported that grow at 42°C.

Sensitive to chloramphenicol (2 μg/ml), dihydrostreptomycin (4 μg/ml), erythromycin (2 μg/ml), neomycin (8 μg/ml), oxytetracycline (4 μg/ml), streptomycin (4 μg/ml), and tetracycline (0.5 μg/ml). Moderately sensitive to novobiocin (64 μg/ml) and penicillin (1 U/ml). Resistant to bacitracin (256 μg/ml) and polymyxin B (512 U/ml).

Serologic studies with *C. fetus* subsp. *venerealis* using heated cells (100°C for 2 h) divide this subspecies into two groups; serovars A-1 (serovar A-biovar 1) and A-sub 1 (serovar A-biovar-sub 1). These serovars contain surface heat stable antigen A. Serovar A-sub 1 differs from serovar A-1 in producing H₂S using the lead acetate strip method of detection. Serovar A-1 is the same as serovar A of Morgan, serovar 1 of Mitscherlich and Liess and serovars III-II, III and III-V of Marsh and Firehammer (Berg et al., 1971; Smibert, 1978).

Pathogenic. A cause of abortion and infertility in cattle. Transmitted venereally.

Found in the vaginal mucus of infected cows, the semen and prepuce of bulls and in the placenta and tissues of aborted bovine fetuses. Pathogenic for cattle, guinea pigs, hamsters and embryonated chicken eggs. Not pathogenic for rabbits, mice or rats when injected intraperitoneally. Will not multiply in the intestinal tract of man and animals (Bryner et al., 1964).

The mol% G + C of the DNA is 33–36.

Type strain: ATCC 19438.

2. **Campylobacter jejuni** (Jones, Orcutt and Little 1931) Véron and Chatelain 1973, 128.[AL] (*Vibrio jejuni* Jones, Orcutt and Little 1931, 861; *Vibrio hepaticus* Mathey and Rissberger 1964; 1339; *Campylobacter fetus* subsp. *jejuni* Smibert 1974, 209.)

je.ju'ni. L. adj. *jejunus* insignificant, meagre; M.L. gen. n. *jejuni* of the jejunum.

Small tightly coiled spirals. Average wavelength is 1.12 μm and average amplitude of coils is 0.48 μm. Rapid formation of coccoid bodies occurs when exposed to air. Swarming growth occurs on moist agar plates (Karmali et al., 1981).

Cell walls contain galactose only, galactose and glucose, or galactose, glucose and mannose (Smibert, 1970).

Cell lipid contains a cyclopropane C-19:0 fatty acid (Blaser et al., 1980). Also has C-14:0, C-16:0, C-16:1 and C-18-1 fatty acids.

Two types of colonies are found on primary isolation (Smibert, 1965, 1969). One is low, flat, grayish, finely granular and translucent with an irregular edge. It spreads along the direction of the streak and tends to swarm and coalesce. The other is round, 1–2 mm in diameter, raised, convex, smooth and glistening, with an entire edge. It has a translucent edge and a darker, dirty brownish, slightly opaque center. Colonies on blood agar are nonhemolytic. Growth in broth usually has a butyrous sediment.

Many physiological characteristics are listed in Tables 2.13 and 2.14.

Casein, ribonucleic acid and deoxyribonucleic acid are hydrolyzed by about 40% of strains. Ninety to ninety-five percent have phosphatase activity and 6% are arylsulfatase positive. Grows on slants with 0.1% sodium selenite; selenite is reduced. Slight growth on MacConkey agar. Does not grow at 25°C. Grows at 42°C and usually at 45°C.

Sensitive to chloramphenicol (4 μg/ml), dihydrostreptomycin (2 μg/ml), erythromycin (8 μg/ml), neomycin (8 μg/ml), oxytetracycline (4 μg/ml), streptomycin (2 μg/ml), and tetracycline (1 μg/ml). Moderately sensitive to novobiocin (64 μg/ml) and penicillin (64 U/ml), bacitracin (128 μg/ml), and polymyxin B (1024 U/ml).

A study by Karmali et al. (1981) showed that 90% of strains were sensitive to erythromycin (1 μg/ml), clindamycin (1 μg/ml), tetracycline (64 μg/ml), metronidazole (64 μg/ml), nalidixic acid (16 μg/ml), nitrofurantoin (1 μg/ml), rifampin (64 μg/ml), gentamicin (0.5 μg/ml), kanamycin (8 μg/ml), novobiocin (512 μg/ml), bacitracin (512 μg/ml), vancomycin (512 μg/ml), trimethoprim (512 μg/ml), polymyxin B (16 μg/ml), penicillin (64 μg/ml), ampicillin (64 μg/ml), carbenicillin (64 μg/ml), cloxacillin (512 μg/ml), cephalosporin C (16 μg/ml), cephaloridine (64 μg/ml), cephalexin (256 μg/ml), cephalothin (512 μg/ml), cefamandole (512 μg/ml), cefazolin (512 μg/ml), cefotaxime (8 μg/ml) and cefoxitin (256 μg/ml).

Serologic studies using heated coils (100°C for 2 h) showed one serovar, C. Strains in this serovar contained surface heat-stable antigen C (Berg et al., 1971). Serovar C is the same as serovar 13 of Mitscherlich and Liess and serovar I of Marsh and Firehammer (Berg et al., 1971 and Smibert, 1978). Additional serovars of this species probably exist.

An antigen that sensitizes sheep red blood cells can be extracted from cells heated at 100°C. Using passive hemagglutination, 23 serovars from 114 strains were found indicating great antigenic diversity. Serovars 1, 2, and 3 were most frequently found in human infections (Penner and Hennessy, 1980).

Using a slide agglutination test to detect heat-labile antigens, 22 serogroups were found (Lior et al., 1981). Eighty-five percent of human isolates and 80% animal isolates were typable with the 21 antisera. Most human isolates were in serogroups 1, 2, 4, 7 and 11.

Pathogenic. Causes abortion in sheep and fever and enteritis in man.

From placentas and stomach content of fetuses from aborted sheep. From blood and fecal material from man with enteritis. May also be found in food, water and unpasteurized milk. May be associated with enteric disease of calves, lambs and other animals. Also found as normal intestinal flora of young cattle, sheep, goats, dogs, rabbits, monkeys, cats, chickens, turkeys, ducks, seagulls, pigeons, blackbirds, starlings and sparrows. Isolates from seagulls have been called NARTC (nalidixic acid resistant thermophilic campylobacter) strains. The NARTC strains are probably not *C. jejuni*. Transmitted orally. Can grow in the intestinal tract of man and animals.

The mol% G + C of the DNA is 31.

Type strain: CIP 702 (NCTC 11351; ATCC 33560).

3. **Campylobacter coli** (Doyle 1948) Véron and Chatalain 1973, 127.[AL] (*Vibrio coli* Doyle 1948, 50.)

co'.li. Gr. n. *colon* large intestine, colon. M.L. gen. n. *coli* of the colon.

Cells tightly coiled 0.2–0.3 μm in diameter and 1.5–5.0 μm long. Coccoid bodies formed. Fairly aerotolerant.

Colonies are round, 1–2 mm in diameter, raised convex, smooth and glistening. White to tan colored. On moist medium, colonies are low flat, grayish colored and spread in the direction of the streak. Colonies on blood agar are nonhemolytic. Growth in broth usually has a butyrous sediment. Usually no growth occurs aerobically; no growth occurs under strict anaerobic conditions. No growth in deep stab cultures.

Gelatin and casein not hydrolyzed. Arylsulfatase-negative. Most strains (75%) phosphatase-positive. DNase activity usually present (65%). RNase activity usually present. Lysine and ornithine are not decarboxylated. Phenylalanine and tyrosine deaminase-negative. Lipase-negative. Methyl red-negative. Acetylmethylcarbinol not produced. Esculin not hydrolyzed. Slight growth occurs on MacConkey agar. Grows on selenite (0.1%) agar; selenite is reduced. Litmus milk is reduced; no other change occurs. Grows at 42°C but not at 25°C. Grows in medium with 8% glucose. Sensitive to nalidixic acid (30 μg disk). Resistant to cephalothin (30 μg/disk). Grows at 30.5°C and 43°C but not at 25°C. Hippurate not hydrolyzed (Harvey, 1980, and Skirrow and Benjamin, 1981).

It is difficult to distinguish *C. coli* from *C. jejuni*. Growth in 8% glucose and resistance to tetrazolium chloride and brilliant green are not reliable characteristics (Skirrow and Benjamin, 1980). Hippurate hydrolysis seems to be a more reliable test.

May be pathogenic.

Isolated from the intestinal tract of swine, poultry and man.

The mol% G + C of the DNA ranges from 32–34.

Type strain: NCTC 11366 (CIP 7080; ATCC 33559).

4. **Campylobacter sputorum** (Prévot) Véron and Chatelain 1973, 128.[AL] (*Vibrio sputorum* Prévot 1940, 85.)

spu.to'.rum. L. n. *sputum* spit, sputum; L. gen. pl. n. *sputorum* of sputa.

Slender, curved rods 0.3–0.5 μm wide and 2–4 μm long. They appear comma-shaped and gull-winged and occasionally occur as filaments up to 8 μm long. The ends of the cells are usually rounded. Cultures 10- to 14-h-old have some motile cells that have a characteristic darting and corkscrew-like motion; other cells appear nonmotile.

Colonies on blood agar are gray, 1–2 mm in diameter, smooth, shiny, low convex and round with thin irregular spreading edges. Some may be weakly α-hemolytic. Growth in broth is light and evenly dispersed. Growth in a medium with 0.12–0.20% agar occurs in the upper third of the medium as a moderate evenly dispersed turbidity.

The final pH of glucose semisolid medium (0.16% agar) is 6.7–7.0 after 3 weeks. Fatty acids are not produced from glucose as end products or from peptone medium without glucose. No fatty acids or alcohols detected when grown under microaerophilic conditions. When grown under anaerobic conditions with fumarate in the medium, small amounts of acetate and large amounts of succinate are produced.

Many physiological characteristics are listed in Tables 2.13 and 2.14.

Nitrate in a medium enhances growth. A small amount of ammonia and CO_2 produced by cultures. No change in skim milk. Litmus milk is reduced.

Microaerophilic to anaerobic. Microaerophilic growth on the surface of agar plates or in broth requires the oxygen concentration of the atmosphere to be decreased to 5%. No growth occurs in broth or on agar under aerobic conditions. Grows in medium containing 0.12–0.20% agar when incubated under aerobic conditions. Grows in deep stab cultures. Anaerobic growth occurs in media containing fumarate alone, formate and fumarate, or hydrogen gas and fumarate. Reduces fumarate to succinate.

The mol% G + C of the DNA is 29–31.

Type strain: Forsyth ER33. (The location of this strain has not yet been determined.) Reference strain: VPI S-17.

4a. Campylobacter sputorum subspecies **sputorum** Véron and Chatelain 1973, 128.[AL]

Characteristics differentiating this subspecies from other subspecies of *C. sputorum* are indicated in Table 2.13. Additional characteristics are listed in Table 2.14.

Grows in semisolid medium containing 1 or 10% bile and in media with 2%, but not 3.5% NaCl.

Anaerobic growth occurs in a medium containing fumarate. End products are some acetate and large amounts of succinate. Microaerophilic growth occurs in an atmosphere of 5% O_2, 10% CO_2 and 85% N_2.

Found in the gingival crevice flora of man.

Type strain: Forsyth ER33 (location not yet determined). Reference strain: VPI S-17.

4b. Campylobacter sputorum subspecies **bubulus** (Loesche, Gibbons and Socransky 1965*) Véron and Chatelain 1973, 128.[AL] (*Vibrio bubulus* Thouvenot and Florent 1954, 237; *Campylobacter bubulus* (Thouvenot and Florent 1954) Sebald and Véron 1963, 907; *Vibrio sputorum* var. *bubulum* (Prévot 1940) Loesche, Gibbons and Socransky 1965, 1109.)

bub′.u.lus. L. adj. *bubulus* pertaining to cattle.

Characteristics differentiating this subspecies from the other subspecies of *C. sputorum* are indicated in Table 2.13. Additional characteristics are given in Table 2.14.

In cultures grown microaerophilically, fatty acids and alcohols are not produced from glucose as end products of fermentation, although a few strains have been reported to produce a trace of acetic acid or acetic and lactic acids.

Anaerobic growth with fumarate in the medium yields some acetate and large amounts of succinate.

Grows in medium with 0.1% selenite; selenite reduction is variable. Optimum temperature 37°C.

Under microaerophilic conditions lactate is used and acetate produced. Formate is used and stimulates growth under microaerophilic conditions. A membrane-bound formate dehydrogenase is present. Activity of this enzyme is lost in air. A hydrogenase has been found. Cytochromes *b* and *c* and a CO_2-binding pigment of the *c*-type has been reported. The *b*-cytochrome is membrane-bound, while the *c*-type cytochrome is mainly soluble. Lactate and formate can be electron donors to oxygen under microaerophilic conditions. Grows anaerobically with nitrate, fumarate or malate as electron acceptors. Hydrogen or formate may serve as electron donors. However, the organism is not dependent on hydrogen or formate for growth and can grow anaerobically with only fumarate in the medium (Niekus et al., 1977, 1980a, b). Microaerophilic growth occurs in an atmosphere of 5% O_2, 10% CO_2 and 85% N_2.

Sensitive to chloramphenicol (1 μg/ml), dihydrostreptomycin (8 μg/ml), erythromycin (2 μg/ml), oxytetracycline (4 μg/ml), penicillin (0.25 U/ml), polymyxin (4 U/ml), streptomycin (8 μg/ml), and tetracycline (1 μg/ml). Moderately sensitive to neomycin (16 μg/ml). Resistant to bacitracin (64 μg/ml) and novobiocin (256 μg/ml).

Found in the genital tract of male and female cattle and sheep. Can be isolated from semen, preputial and vaginal mucus of normal animals.

The mol% G + C of the DNA ranges from 29–31. *C. sputorum* subsp. *bubulus* shows high DNA homology (93%) to *C. sputorum* subspecies *sputorum* and low DNA homology (15–26%) to *C. concisus* (Tanner et al., 1981).

Type strain: CIP 53103 (ATCC 33562).

4c. Campylobacter sputorum subspecies **mucosalis** Lawson, Leaver, Pettigrew and Rowland 1981, 385.[VP] (*Campylobacter sputorum* subspecies *mucosalis* Lawson, Rowland and Wooding 1975, 121.)

mu.co′.sal.is. L. n. *mucosus*, mucus, L. f. adj. *mucosal*, L. gen. n. *mucosalis* of mucus, pertaining to mucus.

Characteristics differentiating this subspecies from the other subspecies of *C. sputorum* are listed in Table 2.13. Additional characteristics are given in Table 2.14.

Cells short, irregularly curved, 0.25–0.30 μm in diameter and 1.0–2.8 μm long. In old cultures, coccoid cells and filamentous forms 7–8 μm long are seen.

Colonies are 1.5 mm in diameter, circular, raised with a flat surface, and have a dirty yellowish color. On moist agar colonies, tend to swarm along the line of inoculation.

Does not grow microaerophilically in an atmosphere containing 5% O_2, 10% CO_2 and 85% N_2. Grows microaerophilically only with 3–5% O_2, 10% CO_2 and H_2 (Lawson et al., 1974). Grows anaerobically in an atmosphere containing hydrogen and fumarate. Requires hydrogen as an electron donor. Formate may replace hydrogen for growth of most strains (Lawson et al., 1981). Anaerobic growth requires fumarate as an electron acceptor. Converts fumarate to succinate. Oxygen utilization by cell suspensions is greatly increased in the presence of hydrogen and formate. Utilization of oxygen with these substrates is unaffected by cyanide (0.4 mM). Cells contain large amounts of cytochrome c_{553} which is reduced when cell suspensions are incubated with hydrogen or formate. When exposed to air the reduced cytochrome *c* is reoxidized. Lactate, succinate, and NADH give slight reduction of cytochrome *c* while methanol, malate, glutamate and serine are inactive. It is not affected by cyanide (Lawson et al., 1981).

Urease, lipase and lecithinase activity not found. DNase activity is variable. Gelatin not hydrolyzed. Indole and acetylmethylcarbinol are not produced. Does not grow on MacConkey agar.

Grows in media with 0.5% sodium deoxycholate. Grows on agar media with brilliant green (1:100,000). Grows with 6% ox bile and in a medium with triphenyltetrazolium chloride (0.75 mg/ml). Not inhibited by novobiocin (5 μg/ml).

Antigenic analysis shows that strains of this organism are closely related antigenically (Lawson et al., 1975). Antisera prepared against 5 strains agglutinated homologous and heterologous cells. When antisera was absorbed with one strain, the antisera reacted with only homologous cells. An antigenic analysis has been reported (Lawson et al., 1977) and three serovars, A, B and C, have been reported (Lawson et al., 1981).

Pathogenic for pigs.

Isolated from the intestinal mucosa of pigs with porcine intestinal adenamatosis, necrotic enteritis, regional ileitis and proliferative hemorrhagic enteropathy; also isolated from the porcine oral cavity.

The mol% G + C of the DNA is 34.

Type strain: NCTC 11000 (FS253/72).

5. Campylobacter concisus Tanner, Badger, Lai, Listgarten, Visconti and Socransky 1981, 442.[VP]

con.cis′us. L. part. adj. *concisus* brief, concise.

---

*Incorrectly cited as Florent 1953 in the Approved Lists of Bacterial Names.

Cells are small and curved, 0.5 μm in diameter and 4 μm long, with rounded ends. A membrane-like polar cap occurs at the ends of the cells.

Colonies are convex, translucent, 1 mm in diameter, with entire edges. The agar is not pitted by the colonies.

Grows microaerophilically with 5% O_2, 10% CO_2 and 10–85% H_2. Requires H_2 for microaerophilic growth. Will not grow in semisolid medium (0.16% agar), in air or in an atmosphere containing 5% O_2, 10% CO_2 and 85% N_2. Anaerobic growth occurs with formate and fumarate in the medium. Formate is oxidized to hydrogen and CO_2; and fumarate is reduced to succinate, which accumulates in medium. Growth is stimulated by nitrate, formate and fumarate. End products of metabolism when grown in a medium with formate and fumarate are acetate, succinate and H_2.

Many physiological characteristics are listed in Tables 2.13 and 2.14.

Benzidine test positive. Lysine, ornithine and arginine are not decarboxylated. Starch, dextrin, esculin, casein, gelatin and DNA are not hydrolyzed. Urease, lecithinase and lipase activities are not found. Acetylmethylcarbinol, H_2O_2 and ammonia are not produced. Neutral red and benzyl viologen are reduced. Gas is not formed.

Growth not inhibited by sodium fluoride, 0.05%; oxgall, 1.0%; sodium deoxycholate, 0.1%; janus green, 0.01%; basic fuchsin, 0.005%; crystal violet, 0.0005%; safranin, 0.01%; alizarin red S, 0.032%; azure II, 0.005%; potassium cyanide, 0.1%; methyl orange, 0.025% and brilliant green, 0.00125%.

Minimum inhibitory concentrations of antibiotics are (μg/ml): bacitracin, 128; chloramphenicol, 4.0; clindamycin, 2–4; colistin, 0.5–1.0; erythromycin, 4.0; gentamicin, 2–4; kanamycin, 1–2; metronidazole, 0.5–2.0; minocycline, 2; nalidixic acid, 64–128; neomycin, 16–32; penicillin, 0.5–4.0; polymyxin, 0.25–1.0; rifampin, 16–64; streptomycin, 1–2; tetracycline, 1–2; and vancomycin, 128.

Not known to be pathogenic.

Found in the gingival crevices of man with gingivitis, periodontitis and periotontosis.

The mol% G + C of the DNA is 34–38. Shows low homology (23–26%) with *C. sputorum* subsp. *sputorum* and *C. sputorum* subsp. *bubulus* and also with *C. fetus* and *Wolinella recta* (Tanner et al., 1981). While *C. concisus* shows no DNA homology to *C. sputorum* subsp. *sputorum* and *C. sputorum* subsp. *bubulus* and is a separate species, there are no known phenotypic characteristics that can be used to separate the species.

Type strain: ATCC 33237 (FDC 484).

Other organisms

Several organisms have been reported in the literature that may be *Campylobacter*-like organisms. These organisms are briefly described in this section.

a. "*Campylobacter fecalis*" (Firehammer 1965) Smibert, 1974, 211. ("*Vibrio fecalis*" Firehammer 1965, 493.)

fe.cal'is. L. n. *faex, faccis*, dregs; M.L. adj. *faecalis* pertaining to feces.

Slender, curved, rods 0.3–0.6 μm wide and 2–4 μm long. Ends of the cells are rounded. Morphology and motility otherwise similar to *C. fetus*.

Colonies on blood agar are pinpoint to 3.5 mm in diameter. They are shiny, smooth, convex and round, with entire edges.

Growth in a semisolid medium (0.12–0.20% agar) occurs in the upper third of the medium.

Indole-negative. Catalase- and oxidase-positive. Hydrogen sulfide produced in peptone iron agar. Also H_2S-positive with lead acetate strips as the method of detection. Nitrate reduced to nitrite.

Microaerophilic to anaerobic. Growth on the surface of agar plates or in broth requires the oxygen concentration to be reduced below 10%. Only traces of growth found on agar plates incubated under aerobic conditions. Poor growth under strict anaerobic conditions. Light growth in deep stab cultures.

Grows in semisolid medium containing 1% glycine. Growth in 1% bile is variable. Grows in medium with 0.1% selenite and reduces selenite. Grows in 2% NaCl. Some strains fail to grow in a medium with 4% NaCl. Some strains require at least 0.5% NaCl in the medium. Grows at 42°C but not at 25°C. Optimum temperature is 37°C.

Sensitive to chloramphenicol (1 μg/ml), dihydrostreptomycin (4 μg/ml), erythromycin, (4 μg/ml), oxytetracycline (1 μg/ml), penicillin (0.25 U/ml), streptomycin (4 μg/ml), and tetracycline (0.25 μg/ml). Resistant to bacitracin (74 μg/ml), neomycin (32 μg/ml), and novobiocin (512 μg/ml). Moderately sensitive to polymyxin (16 U/ml).

Isolated from sheep feces. May also be found in bovine semen and vagina.

Reference strain: ATCC 33709.

b. A nitrogen-fixing *Campylobacter*-like organism was isolated from the roots of *Spartina alterniflora* growing in a Nova Scotia salt marsh (McClung and Patriquin, 1980). It is a small curved rod (0.3–0.6 μm wide × 1–3 μm long), motile with a single polar flagellum. Coccoid forms develop in old cultures.

The metabolism is respiratory. Organic acids and amino acids serve as energy sources. Carbohydrates are not used. Catalase- and oxidase-positive. Phosphatase- and sulfatase-negative. Gelatin, esculin, casein and starch are not hydrolyzed. Indole-negative. H_2S-positive; nitrogenase- and urease-positive. Nitrate is reduced; nitrite is not reduced. Does not grow with 1% bile, 1% glycine or 0.1% 2,3,5-triphenyltetrazolium chloride. Microaerophilic. Grows anaerobically with fumarate in the medium. Grows at 6–25°C but not at 37°C or 42°C. Optimum temperature, 30°C. Requires salt. Grows in media with a range of 2–70 g/liter NaCl. Optimum growth with 10–40 g/liter NaCl. Anaerobic growth and acetylene reduction are detected in an anaerobic medium supplemented with fumarate, but not in a medium with fumarate and nitrate.

The mol% G + C of the DNA of strain C1 is 32.

Reference strain: ATCC 33309 (strain C1).

c. *Campylobacter*-like organism from porcine and ovine abortions (Neill et al., 1979). Curved cells 0.4 μm wide × 1.8 μm long. Catalase- and oxidase-positive. Not fermentative. A microaerophilic atmosphere is required for primary isolation. Grows aerobically after several subcultures. Grows at 25 but not at 42°C. H_2S-negative. Selenite (0.1%) not reduced. Growth in 1% glycine is variable. Sensitive to nalidixic acid (40 μg/ml) and tetrazolium chloride (0.01%). Resistant to vibriostatic agent 0/129.

Colonies are 1 mm in diameter, convex with an entire edge. Not pigmented.

Isolated from fetal tissue and fluids from porcine and bovine abortions in EMJH medium (leptospira isolation medium) incubated at 30°C.

The mol% G + C of the DNA ranges from 29–34.

d. A small curved rod (0.5 μm wide × 2.5–20 μm long) called "spirillum 5175" has been isolated from an anaerobic enrichment culture for *Desulfuromonas* in which the major constituents were acetate and elemental sulfur (Wolfe and Pfennig, 1977). Colonies in agar-shake cultures are lens shaped, 1–3 mm in diameter with a yellowish color. Anaerobic growth can occur with malate or fumarate as well as elemental sulfur as electron acceptors and with hydrogen as the electron donor. Formate can also serve as an electron donor.

The mol% G + C of the DNA is 38.

e. A *Campylobacter*-like organism with many characteristics similar to spirillum 5175 has been isolated from an anaerobic digester fed waste water from a potato processing factory (Laanbroek, 1977, 1978). Cells are 0.3–0.5 μm wide and 1–3 μm long and are motile by means of a single polar flagellum. Colonies on blood agar plates in 5% O_2 are 1 mm in diameter, translucent, and round with an entire edge.

Catalase-positive; indole-, urease-, lipase-, and gelatinase-negative.

Does not hydrolyze casein. Carbohydrates are not fermented. Grows at 25 but not at 42°C. H₂S-positive. No growth in 1% glycine and 3.5% NaCl.

Aspartate, malate and fumarate used and converted to acetate and succinate. Grows in brucella broth with 0.5% L-aspartate and 0.2% formate. Grows anaerobically with hydrogen or formate as electron donors and nitrate, thiosulfate, elemental sulfur, or oxygen as electron acceptors, with acetate as a carbon source.

Resting cells reduce aspartate, fumarate, malate, nitrate, nitrite, hydroxylamine, sulfite and sulfur with hydrogen.

The mol% G + C of the DNA is 41.6.

Genus *Bdellovibrio* Stolp and Starr 1963, 243[AL]

JEFFREY C. BURNHAM AND SAMUEL F. CONTI

Bdel.lo.vib'ri.o. Gr. n. *bdella* leech, sucker; M.L. masc. n. *Vibrio* a generic name; M.L. masc. n. *Bdellovibrio* a leechlike vibrio.

This genus consists of a group of remarkable bacteria which are **predacious upon other Gram-negative bacteria**. The cells are **comma-shaped rods, 0.2–0.5 μm in diameter** and 0.5–1.4 μm in length. They are **motile by means of a single polar flagellum** that is surrounded by a sheath which is continuous with the outer membrane of the cell wall. **Bdellovibrios exhibit a morphologically and physiologically biphasic life cycle, alternating between a nongrowing predatory phase and an intracellular reproductive phase** (Fig. 2.17). The highly motile bdellovibrio appears to locate its prey by means of chance collision; it forceably strikes and attaches to the generally much larger prey cell, then rapidly penetrates into the prey's periplasmic space. The prey cell containing the invading bdellovibrio usually rounds up and swells into a spherical form (bdelloplast). The bdellovibrio kills the prey cell very early in the attack; indeed, the prey cell loses essentially all metabolic potential, including energy generation, biosynthesis, and the activity of degradative enzymes. The prey cell, soon after attack, is functionally a substrate for bdellovibrio development. The developing bdellovibrio elongates into a snake form at the expense of the prey's protoplast. The spiral-shaped nonmotile cell then fragments into motile, unit-sized predacious vibrios which leave the prey (now a "ghost" cell) to begin the cycle anew.

Obligately **aerobic**, having a strictly respiratory type of metabolism with oxygen as the terminal electron acceptor. Optimum temperature, generally 28–30°C; growth is poor above 37°C and below 10°C. **All wild-type strains upon initial isolation are dependent on intraperiplasmic growth in susceptible prey**; the reason for the dependence is unknown. Growth of isolates in the absence of prey has been achieved in some cases with media supplemented with high concentrations of bacterial cell extracts. Only predacious bdellovibrios have been obtained on initial isolation from nature, but mutants capable of axenic growth (growth in the absence of prey cells, "**prey-independent**" strains) have been derived from the predacious strains. Some strains are **facultative**, i.e. capable of growth in the presence or absence of prey cells. Habitats of bdellovibrios include soil, sewage, freshwater and marine environments. The mol% G + C of the DNA ranges from 33.4–51.5 (Bd, T_m).

Type species: *Bdellovibrio bacteriovorus* Stolp and Starr 1963, 243.

Further Descriptive Information

The morphology of *Bdellovibrio* varies in accordance with the phase of the life cycle. The free living predacious cell is a small curved or comma-shaped rod (Fig. 2.18) which can move rapidly (up to 100 cell lengths/sec) with a characteristic darting pattern. The single polar flagellum is ensheathed (Fig. 2.19) by an apparent extension of the outer cell membrane (Burnham et al., 1968; Seidler and Starr, 1968). The cell envelope at the attachment pole possesses distinct ring-like structures, with fibers radiating from these structures (Abram and Davis, 1970).

The ultrastructure of the cell envelope is characteristic of Gram-negative bacteria. Thomashow and Rittenberg (1978a) have shown that the chemical composition of the peptidoglycan is similar to that of other Gram-negative bacteria. These authors also demonstrated a high rate of peptidoglycan turnover, which accounts for the sensitivity of nongrowing bdellovibrios to antibiotics with activities directed at peptidoglycan biosynthesis; they also reported that antibiotic-induced spheroplasts are osmotically stable.

A schematic representation of the life cycles of predacious, prey-independent, and facultative strains is presented in Figure 2.17. The axenic cycle illustrates cell growth and division in a complete medium devoid of prey bacteria. The predacious cycle represents the morphological events associated with predation of susceptible bacteria. The facultative cycle reflects the observations that some bdellovibrios (e.g. the type strain of *B. stolpii*, initially designated *B. bacteriovorus* UKi2) are capable of growing and multiplying in either the predacious or axenic modes (Diedrich et al., 1970).

The predacious bdellovibrios attach to the prey cell (Fig. 2.20, *p*) and penetrate through both the outer cell membrane and peptidoglycan layer of the prey cell to gain entrance into the periplasmic space. During the initial attack, the prey cells are converted to osmotically stable spheres (bdelloplasts) (Fig. 2.20, *i*) (Starr and Baigent, 1966; Burnham et al., 1968; Abram et al., 1974).

The elongation phase is similar for both axenic and predacious bdellovibrios. The lengthening is unidirectional and approaches a 20-fold increase over the original cell length (Eksztejn and Varon, 1977). The ultimate length of an elongating predacious bdellovibrio, and the number of progeny, is a function of the prey cell volume (Kessel and Shilo, 1976). Cell division of the spiral-shaped cells takes place as concurrent multiple septations along their length (Fig. 2.17). This type of cell division occurs both in axenic cultures and in cultures where the prey cell volume places constraints upon the length of the spiral (Burnham et al., 1970; Eksztejn and Varon, 1977). The spiral-shaped cells fragment into unit-sized cells followed by lysis of the bdelloplast wall and release of the progeny bdellovibrios.

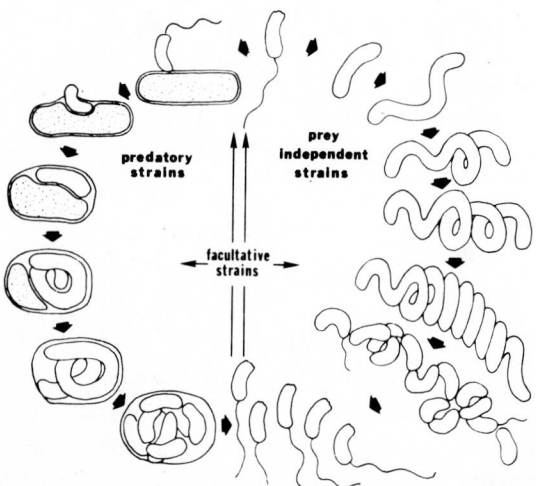

Figure 2.17. A schematic representation of the life cycles of prey-dependent (predatory), prey-independent and facultative strains of the genus *Bdellovibrio*. (Reproduced with permission from J. C. Burnham, T. Hashimoto and S. F. Conti, Journal of Bacteriology *101*: 997–04, 1970, ©American Society for Microbiology.)

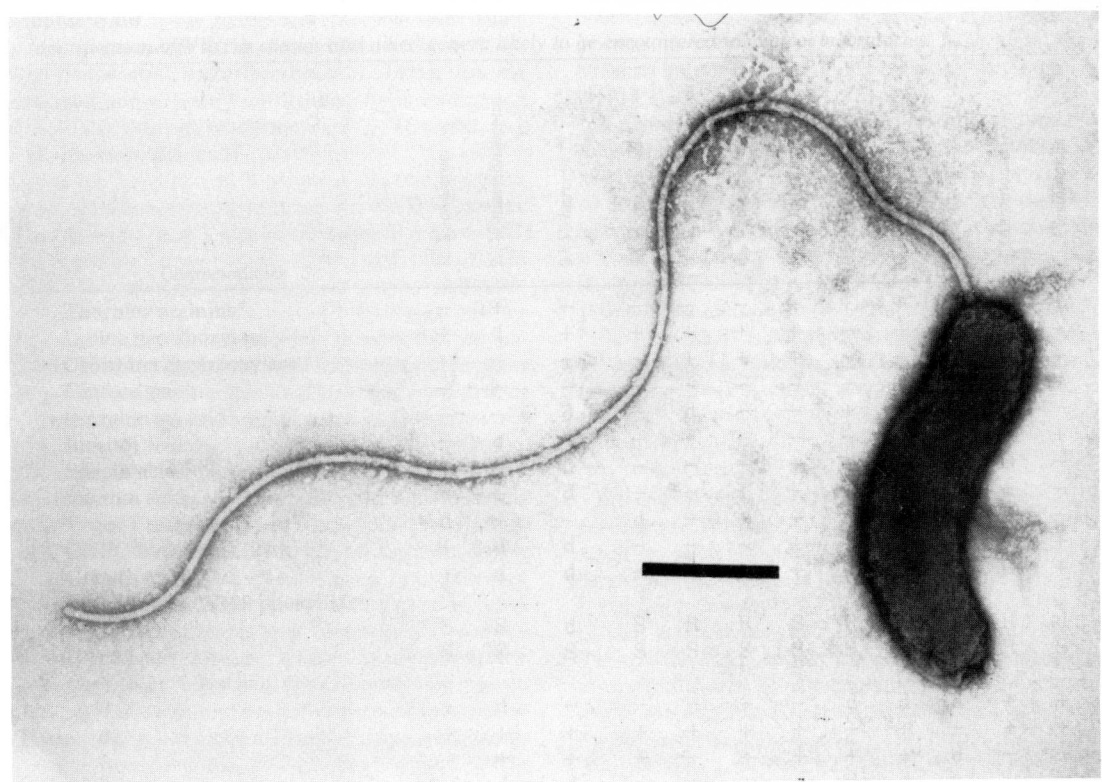

Figure 2.18. Negatively stained preparation of *Bdellovibrio bacteriovorus* ATCC 15143 showing the comma-shaped cell with its ensheathed polar flagellum. *Bar*, 0.2 μm. (Reproduced with permission from J. C. Burnham, T. Hashimoto and S. F. Conti, Journal of Bacteriology *96:* 1366–1381, 1968, ©American Society for Microbiology.)

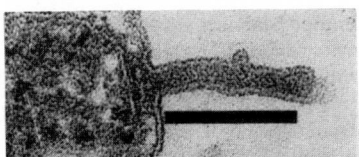

Figure 2.19. A thin section of *Bdellovibrio bacteriovorus* showing the polar flagellum to be ensheathed by a continuation of the cell wall. *Bar*, 0.1 μm. (Reproduced with permission from J. C. Burnham, T. Hashimoto and S. F. Conti, Journal of Bacteriology *92:* 1366–1381, 1968, ©American Society for Microbiology.)

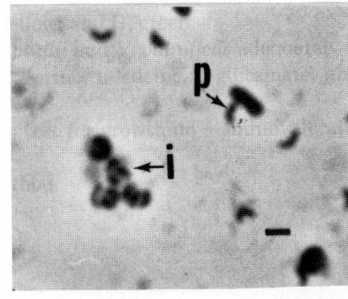

Figure 2.20. Phase-contrast photomicrograph of *Bdellovibrio bacteriovorus* ATCC 15143 and cells of *Escherichia coli* ATCC 15144. Shown are a bdellovibrio (*p*) penetrating an *E. coli* prey cell and several *E. coli* cells already infected with bdellovibrios (*i*). *Bar*, 1 μm.

Several investigators have noted that the nonflagellated pole of the bdellovibrio cell is differentiated into a swollen "holdfast" and that often a mesosome-like structure is present in this region of the cell (Burnham et al., 1968; Burnham et al., 1970; Starr and Baigent, 1966). Possibly this holdfast is the result of disruption by the fixatives and stains used for electron microscopy; nevertheless, its presence indicates a chemical differentiation of this region. Bdellovibrios are known to contain phosphonosphingolipids (Steiner et al., 1973) which are rarely found in bacteria. Although their localization in the cell and their function have not been determined, they may be concentrated in the holdfast region and serve to protect the bdellovibrio from the enzymatic activities which are localized at this site during the initial predator-prey interaction.

The range of susceptible prey varies with the strains of *Bdellovibrio* (Stolp, 1973) but is confined to Gram-negative bacteria. Bdellovibrios may attack prey from several, but not all, genera of Gram-negative bacteria. A study of the prey range of 12 strains revealed that only *Aquaspirillum serpens* strain MW5 was lysed by all the strains (Torella et al., 1978), and the authors suggested that prey cell specificity is an unreliable taxonomic characteristic in the absence of standardized experimental conditions. In general, enteric bacteria are susceptible prey for most strains of *Bdellovibrio*; these prey all have similar lipopolysaccharide (LPS) structures (Luderlitz et al., 1966), and there is some evidence suggesting that receptor sites on the prey are a vital feature of the prey and may reside, at least in part, in the core of the LPS component of the cell wall. Recent studies suggest that the outer membrane proteins of susceptible prey are involved in the attachment of predator to prey (S. F. Conti and D. Diedrich, unpublished observations). Presumably, the failure of bdellovibrios to grow on Gram-positive cells is due to the great difference in structure of their cell wall as compared to that of Gram-negative bacteria.

A variety of enzymatic activities is expressed at the time of attachment of a bdellovibrio to its substrate cell, thereby enabling the predator to produce an entry pore and to stabilize the bdelloplast wall. Enzymatic activities detected to date include a "glycanase," a "peptidase," an activity directed against LPS, an *N*-deacetylase, an "acylase," and an activity that removed the Braun lipoprotein from the peptidoglycan (Thomashow and Rittenberg, 1978b–d). These enzymatic activities act on the substrate cell in a coordinated and highly regulated manner which enables the bdellovibrio to penetrate the prey cell wall and to modify the substrate cell wall in a significant way. The production of a spherical prey cell provides the maximum space for the bdellovibrio's intraperiplasmic growth. The conversion of the peptidoglycan sacculus to a lysozyme-insensitive form leads to the exclusion of other bdellovibrios and thus provides the predator with a protected environment in which there is no competition with other predators for nutrients (Thomashow and Rittenberg, 1978d). These advantages of the predator are abolished when there is a high multiplicity of infection (e.g. 20 bdellovibrios/prey cell). Under these conditions a large number of bdellovibrios may attach simultaneously to the prey (Fig. 2.21); this can rapidly cause prey cell disruption prior to predator penetration.

Intraperiplasmic growth is highly efficient (Rittenberg and Hespell, 1975) and is primarily due to the ability of the bdellovibrios to construct their DNA (Matin and Rittenberg, 1972), RNA (Hespell et al., 1975),

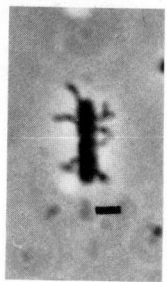

Figure 2.21. Phase-contrast photomicrograph of two *Escherichia coli* B/r prey cells being attacked by multiple cells of *Bdellovibrio bacteriovorus* ATCC 15143. *Bar*, 1 μm.

lipids (Kuenen and Rittenberg, 1975) and proteins (Rittenberg and Langley, personal communication) from monomeric units derived from the substrate cell. The conservation of phosphate-ester bonds from the substrate cell, including those from the phospholipids (Rittenberg and Langley, 1975), also appears to be involved in the unusual energy efficiency. The bdellovibrios also incorporate long chain fatty acids from the substrate cell into their lipids; thus their lipid composition reflects the lipid composition of the prey cells (Kuenen and Rittenberg, 1975). A comparison of some of the general features of obligate intracellular parasites and the unique intraperiplasmic mode of existence of the bdellovibrios yields some new and important insights into the nature of bdellovibrios (Rittenberg and Thomashow, 1979).

Prey independent mutants of predacious strains often are nonmotile and possess an increased cell diameter; colonies of these variants generally have a yellow pigmentation (Seidler and Starr, 1969). Mutants apparently arise from obligately predacious strains by a mutational event which occurs at a conversion frequency of 5×10^{-6} (Seidler and Starr, 1969). Prey-independent strains are uniformly nonfermentative, oxidase-positive, and possess cytochromes *a* and *c* (Seidler and Starr, 1969) and cytochrome *b* (Torella et al., 1978). Catalase activity is variable (Seidler and Starr, 1969).

Bdellovibrio strain W expresses an additional life-cycle variation that is not illustrated in Figure 2.17. This strain can produce encysted, resting cells termed "bdellocysts." These bdellocysts are produced within the bdelloplast and possess enhanced resistance to high temperature, desiccation and disruption. Bdellocysts contain increased amounts of DNA, RNA, protein and carbohydrate per cell when compared to vegetative cells (Tudor and Conti, 1977a). Encystment results in the addition of a 25- to 30-nm electron-dense wall layer to the bdellovibrio envelope. This layer appears to be associated intimately with the cell envelope of the host (Fig. 2.22) (Tudor and Conti, 1977b) and is composed primarily of peptidoglycan (Tudor, 1980).

Marine bdellovibrios (24 strains studied) exhibit a requirement for NaCl (Reichelt and Baumann, 1974; Taylor et al., 1974; Miyamoto and Kuroda, 1975; Marbach et al., 1976). Some of the marine bdellovibrios

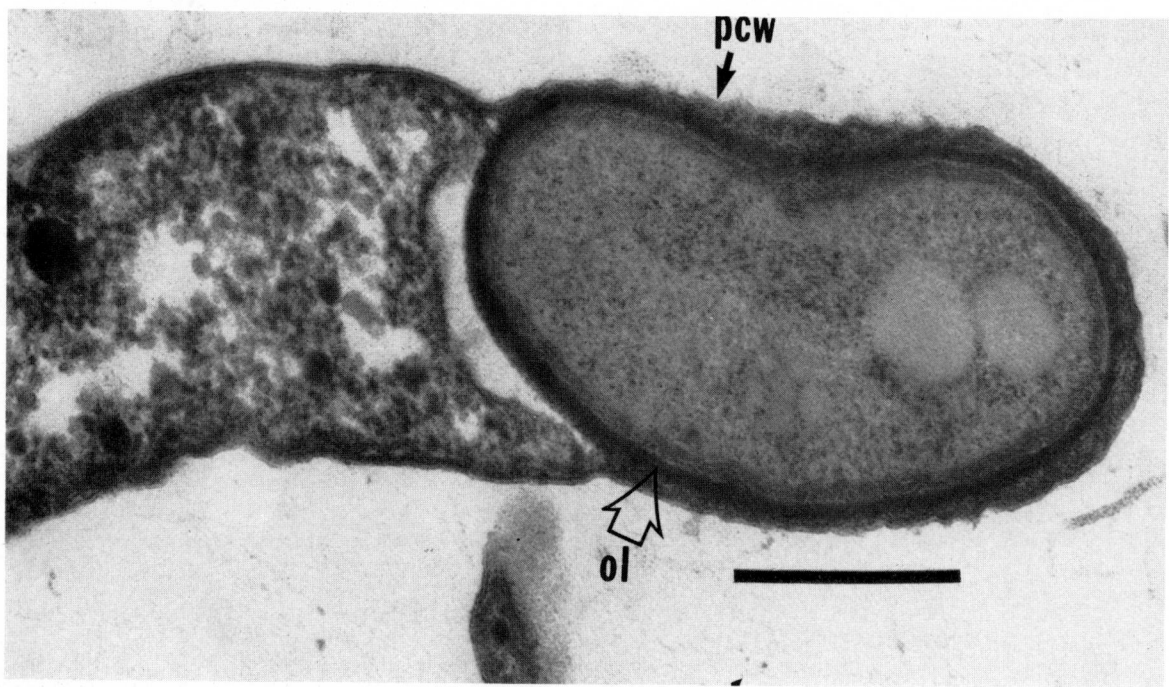

Figure 2.22. Electron micrograph of *Bdellovibrio* sp. strain W during encystment in *Rhodospirillum rubum*. The outer layer (*ol*) of the bdellovibrio has increased in thickness and is intimately associated with the prey cell wall (*pcw*). *Bar*, 0.2 μm. (Reproduced with permission from J. J. Tudor and S. F. Conti, Journal of Bacteriology *131*: 323–330, 1977, ©American Society for Microbiology.)

also require KCl, MgCl$_2$ and CaCl$_2$ (Marbach et al., 1976). The marine strains generally require at least 75 mM NaCl for growth, and this requirement cannot be replaced by KCl. A comparison of cell yields has shown that natural seawater is 10-fold superior to artificial seawater in the growth medium for marine strains (Miyamoto and Kuroda, 1975). This suggests that unknown factors in natural seawater may be important to the development of marine bdellovibrios (Varon and Shilo, 1980).

The genome sizes of five tested *Bdellovibrio* strains are all smaller than that of *E. coli* and range from 1.32×10^9 to 1.70×10^9 daltons (Torella et al., 1978).

Bacteriophages have been isolated for bdellovibrios (Hashimoto et al., 1970; Varon and Levisohn, 1972; Althauser et al., 1972) and have been used to divide the genus into six phage-susceptibility groups. Groups I to IV include strains presently classified as *B. bacteriovorus*; this suggests that further subdivision of this species is possible. *B. stolpii* (Group V) and *B. starrii* (Group VI) were unrelated. A phage capable of attacking *B. starrii* has not been isolated. No difference in phage susceptibility has been found between prey-dependent strains and their prey-independent derivatives (Varon and Levisohn, 1972).

Serological studies on bdellovibrios have revealed at least one common antigen (Kramer and Westergaard, 1977; Schelling, J. E., C. Anderson and S. F. Conti, Abst. Annu. Mtg. Amer. Soc. Microbiol., 1977, I-146). *B. starrii* and *B. stolpii* are antigenically distinct from each other and both are distinct from the group of strains placed in the species *B. bacteriovorus* (Schelling et al., see above). Subsequent work on 21 strains has led to the division of *B. bacteriovorus* into nine serovars (Schelling and Conti, unpublished data).

The bdellovibrios are found worldwide in diverse habitats. Strains have been found in the oceans, brackish and freshwater ponds, rivers and estuaries, sewage systems, and soil (Varon and Shilo, 1980). The concentration of bdellovibrios has been correlated with the extent of pollution present (Fry and Staples, 1976). It has been suggested that the bdellovibrios are not distributed uniformly in these environments but are concentrated in the ecosystems that provide the necessary prey densities (Varon and Shilo, 1980). Although bdellovibrios are chemotactic (Straley and Conti, 1974), the positive chemotactic response is toward the ionic environment favored by prey bacteria, and the response is not affected by prey density (Straley, 1977). This maximizes the probability of *Bdellovibrio* contact with susceptible prey. A high prey concentration in the bdellovibrios' habitat appears to be critical for the survival of the bdellovibrios. Hespell et al. (1974) and Varon and Ziegler (1978) have calculated that 1.5×10^5 to 10^6 prey/ml are required to sustain a *Bdellovibrio* population.

Enrichment and Isolation Procedures

Freshwater specimens. Water samples may be used directly or after filtration through a membrane filter (1.2 μm pore size) to remove the majority of larger bacteria present (Varon and Shilo, 1970). In the "double layer" procedure of Staples and Fry (1973), either 0.1 or 1.0 ml of the sample is added to a tube containing 5 ml of soft agar (NB/500* containing 0.6% agar) that has been melted and cooled to 45°C. One ml of a suspension of suitable prey cells is added to the soft agar. (Suitable bacterial prey are grown in nutrient broth or other appropriate medium for 20–24 h, harvested under aseptic conditions, washed

once with sterile NB/500, and suspended in NB/500 to yield a dense concentration, ~ 1/10 the original volume.) The soft agar is then poured over agar plates (NB/500 containing 1.2% agar, or other appropriate medium) which have been prewarmed to 37°C. After the overlay has gelled, the plates are inverted and incubated at 30°C for a maximum of 6 days. "*Achromobacter*" sp. (NCIB 8250) is recommended as an effective prey for freshwater enumerations since it consistently permitted the highest number of plaques to form (Staples and Fry, 1973). Plaques of bdellovibrios usually appear in 2–6 days and will gradually increase in size. Phage plaques will develop more rapidly and may be scored after 24 h. The phage plaques, in contrast to the bdellovibrio plaques, do not continue to increase in size with continued incubation. Microscopic examination will identify plaques formed by protozoa or myxobacters. Serial transfers from the edge of the bdellovibrio plaques to fresh lawns of prey bacteria usually will provide pure cultures.

Soil specimens. A slight modification of the procedure of Klein and Casida (1967) is recommended. A 10-fold dilution of a soil sample is prepared by shaking the soil in tap water for 20 min. A 1-ml sample is further diluted 10-fold and added to molten soft agar tubes containing prey bacteria in NB/500 as described above for freshwater specimens. (*Note*: Klein and Casida (1967) recommended *E. coli* as the initial prey and reported that over 10 bdellovibrios were present in 21 of the 23 samples tested.) The contents of each soft agar tube is rapidly added to a prewarmed base-agar (NB/500) plate. The plates should be inverted and incubated at 30°C for up to 6 days. *Bdellovibrio* plaques may be distinguished as described above.

Marine specimens. The procedures of Marbach et al. (1976) are recommended. Samples (250 ml) are filtered through membrane filters (0.8 μm pore size) and centrifuged at $27,000 \times g$ for 10 min. The pellet is suspended in 2.5 ml of MPY/10 medium.† Portions of this concentrate are plated by the double layer technique described for freshwater samples, with MPY/10 medium being substituted for NB/500. Plates are incubated at 25°C and examined daily for 10 days. Multiple transfers of single plaques lead to the isolation of pure cultures.

Maintenance Procedures

Freshwater and soil strains. Bdellovibrios may be maintained by serial transfer on lawns of prey bacteria as described under *Enrichment and Isolation*, or they may be transferred to PYE broth§ containing 24- to 48-h-old cultures of a suitable prey (Stolp and Starr, 1963).

Cultures of bdellovibrios can be preserved indefinitely by centrifuging growing cells at $10,000 \times g$ for 10 min, suspending the pellet to the original volume either in skim milk or 12% sucrose, and freeze-drying this suspension in ampoules. To revive cultures, the freeze-dried cells are suspended in NB/500 to their original volume, and both lawns and small broth cultures of the appropriate bacterial prey are inoculated. Active bdellovibrios should be visible by phase-contrast microscopy in 72–96 h.

Marine strains. The organisms may be maintained by serial transfer on double-layer plates as described under *Enrichment and Isolation*, or cultures may be grown at 25°C in broth as described for freshwater strains, using MPY medium containing a culture of the suitable prey (Taylor et al., 1974; Marbach et al., 1976).

Differentiation of the genus **Bdellovibrio** from other taxa

It is our opinion that plaque-forming bacterial strains tentatively identified as members of the genus *Bdellovibrio* must be confirmed to

grow in the periplasmic space of bacteria. This is the primary characteristic that binds this diverse group of comma-shaped bacteria. This

---

* NB/500 is nutrient broth (Difco) at 1/500 normal strength (equals 0.016 g of dehydrated broth/liter).

† MPY medium consists of (g/liter): peptone (Difco), 5.0; yeast extract (Difco), 3.0; NaCl, 28.15; MgSO$_4 \cdot$7H$_2$O, 6.92; MgCl$_2 \cdot$6H$_2$O, 5.51; CaCl$_2 \cdot$H$_2$O, 1.45; and KCl, 0.67. The pH is adjusted to 7.4 with 1 N NaOH. *Medium MPY/10* is similar but contains 1/10 the level of peptone and yeast extract.

§ The recommended prey-independent maintenance medium is PYE medium (Seidler and Starr, 1969) which consists of 10.0 g of Bacto-peptone (Difco), 3.0 g of yeast extract (Difco), and 1000 ml of distilled water.

can be accomplished by phase-contrast light microscopy or transmission electron microscopy. These techniques should confirm the location of the bdellovibrios *within* the host cell. An acceptable procedure is described in Burnham et al., (1968). Negatively stained specimens examined with the electron microscope will also aid in this identification by confirming the presence of a polar ensheathed flagellum.

Taxonomic Comments

Since the publication of the eighth edition of *Bergey's Manual*, the lower extreme of the mol% G + C range of the strains assigned to the genus *Bdellovibrio* has been adjusted from 42 to 33.4, resulting in a broad range of 17 mol% G + C. The marine strains possess a mol% G + C of 33.4–43.5 (Taylor et al., 1974; Marbach et al., 1976). The mol% G + C values for *B. stolpii* and *B. starrii* are 41.8 and 43.5, respectively, and the range for *B. bacteriovorus* is 49.5–51 (Torrella et al., 1978). The amount of DNA hybridization between *Bdellovibrio* strains (see below) also reflects the genetic diversity within the genus. This diversity places considerable stress upon the concept of a single genus *Bdellovibrio* for the intraperiplasmic, predatory bacteria isolated to date.

DNA homology studies have shown that considerable differences exist between *B. stolpii* and *B. starrii* (16% homology) despite only a 2% difference in the mol% G + C of their DNA. *B. bacteriovorus* strain 100 exhibited a 1% or less homology with either *B. stolpii* or *B. starrii*, but it possessed high homologies with *B. bacteriovorus* strains 109, 114 and 118 (Seidler et al., 1972). *Bdellovibrio* strain W was 23 and 28.5% homologous to *B. starrii* and *B. stolpii*, respectively. Marine *Bdellovibrio* strain N6804 was compared to other strains (Torrella et al., 1978) and was found to possess homologies of 0, 37 and 32% to *B. starrii*, *B. stolpii* and *Bdellovibrio* strain W, respectively.

The evolutionary background of the *Bdellovibrio* strains has been questioned, and it is suggested that the organisms that we describe as "bdellovibrios" originated from diverse organisms which, through converging evolutionary patterns, have arrived at the predacious way of life (Varon and Shilo, 1980; Shilo, personal communication). This predatory ability dictates that these bacteria must possess those characters we have indicated for the bdellovibrios under Further Descriptive Information. Although we recognize the validity of the concept of varied ancestries for the present strains of "bdellovibrios," we choose to continue the recognition of the group of "bdellovibrios" as a single genus, since a single genus affords a practical approach to characterizing and categorizing these unique bacteria.

The *Bdellovibrio* literature has addressed numerous significant physiological and biochemical processes which could be important in further subdividing the species. Unfortunately, comparative studies of these processes employing representatives of *Bdellovibrio* strains have not been performed. Thus, we are continuing to present the three species described in the eighth edition of the *Manual* (Burnham and Robinson, 1974) and we are adding one additional listing for the marine bdellovibrio strains under the designation "*Bdellovibrio sp.* (marine strains)," leaving the choice of species name or names to those who wish to formally propose a marine species taxonomy.

Bdellovibrio bacteriovorus remains the type species, with strain 100 (NCIB 9529) as its type strain. Unless the mol% G + C of soil or freshwater strains has been determined to lie between 49 and 51, we recommend that investigators refrain from using the *B. bacteriovorus* designation and more safely refer to a strain as a *Bdellovibrio sp.* (specify), with the strain designation preferably expressing letters and numbers signifying the site of isolation and some property of the isolate, e.g. UKi1 (University of Kentucky, prey-*i*ndependent strain, *1*st isolate).

Further Reading

Burnham, J.C., T. Hashimoto and S.F. Conti. 1970. Ultrastructure and cell division of a facultatively parasitic strain of *Bdellovibrio bacteriovorus*. J. Bacteriol. *101:* 997–1004.

Rittenberg, S.C. and M.F. Thomashow. 1979. Intraperiplasmic growth—life in a cozy environment. *In* Schlessinger (Editor), *Microbiology—1979*, American Society for Microbiology, Washington, D.C., pp. 80–86.

Starr, M.P. and R.J. Seidler. 1971. The bdellovibrios. Annu. Rev. Microbiol. *25:* 649–678.

Stolp, H. 1973. The bdellovibrios: bacterial parasites of bacteria. Annu. Rev. Phytoplathol. *11:* 53–76.

Varon, M. and M. Shilo. 1980. Ecology of aquatic bdellovibrios. *In* Droop and Jannasch (Editors), *Advances in Aquatic Microbiology*, Vol. 2, Academic Press, New York, pp. 1–48.

Differentiation of the species of the genus **Bdellovibrio**

Table 2.15 presents the characteristics that can be used to differentiate the species of *Bdellovibrio*. Table 2.16 lists other characteristics of the species.

Table 2.15.
Differential characteristics of the species of the genus **Bdellovibrio**[a]

Characteristics	1. *B. bacteriovorus*	2. *B. stolpii*	3. *B. starrii*	4. Unnamed Species (Marine Strains)
Facultatively predatory[b]	−	+	−	
Na+ required for growth (75 mM or higher)	−	−	−	+
Source:				
Seawater	−	−	−	+
Freshwater or terrestrial	+	+	+	−
Sensitive to vibriostatic agent 0/129[c]	+	+	−	
Bacteriophage susceptibility[d]	+	+	−	
Protease production	low	high	moderate	
Mol% G + C of DNA	50.4	42	43.5	33.4–38.6; 43.5[e]

[a] Symbols: see standard definitions.

[b] Defined to mean that any individual comma-shaped cell may complete its life cycle either prey-dependently or prey-independently.

[c] 0/129 = 2.4-diamino-6,7-diisopropylpteridine phosphate.

[d] Althauser et al., 1972; Varon and Levisohn, 1972.

[e] Only one marine strain has been found to have a mol% G + C of 43.5.

Table 2.16.
Other characteristics of the species of the genus **Bdellovibrio**[a]

Characteristics	1. B. bacteriovorus	2. B. stolpii	3. B. starrii	4. Unnamed Species (Marine Strains)
Comma-shaped predacious cells, 0.35 × 1.2 μm	+	+	+	+
Motility	+	+	+	+
Single, polar, sheathed flagellum	+	+	+	+
Elongated to spiral-shaped forms develop during growth	+	+	+	+
Grow and divide within prey cell	+	+	+	+
Prey-dependent strains occur	+	+	+	+
Prey-independent variants occur	+	+	+	+
Gram-negative, but not Gram-positive, bacteria can serve as hosts	+	+	+	+
Circular, plaque-like zones of clearing occur on agar-grown lawns of host bacteria	+	+	+	+
Aerobic	+	+	+	+
Oxidase test	+			+
Catalase test[b]	+	+	−	+
Fermentative ability present	−	−	−	−
Nitrate reduced to nitrite	−	+	−	+
Gelatinase activity	+			+

[a] Symbols: see standard definitions.
[b] Cultures may be initially positive but become variable with subsequent transfer (Seidler and Starr, 1969).

List of the species of the genus Bdellovibrio

1. Bdellovibrio bacteriovorus Stolp and Starr 1963, 243.[AL]
bac.te.ri.o′vo.rus. Gr. dim. n. *bacterium* a small rod; L. v. *voro* to devour; M.L. adj. *bacteriovorus* bacteria-devouring.

The characteristics are as described for the genus and as listed in Tables 2.15 and 2.16.

The number of progeny produced depends upon the strain and the volume of the prey bacterium.

Prey-independent strains have been isolated and appear in culture to parallel the developmental cycle of the predacious form (Fig. 2.17). Colonies on media containing yeast extract and peptone are small to pinpoint and pale yellow to orange after incubation for 5 or 6 days.

Isolated from soil, sewage and freshwater sources.

The mol% G + C of the DNA is 50.4 ± 0.9 (Bd).

Type strain: NCIB 9529 (strain 100).

2. Bdellovibrio stolpii Seidler, Mandel and Baptist 1972, 216.[AL]
stolp′i.i. M.L. gen. n. *stolpii* of Stolp; named after Heinz Stolp, discoverer of the bdellovibrios.

The characteristics are as described for the genus and as listed in Tables 2.15 and 2.16.

The prey range resembles that of *B. bacteriovorus*.

The original isolate was obtained from the sewage plant in Lexington, Kentucky.

The mol% G + C of the DNA is 41.8 (Bd).

Type strain: ATCC 27111 (DSM 50722; strain UKi2). The ATCC distributes this strain with prey *E. coli* B (ATCC 11303).

3. Bdellovibrio starrii Seidler, Mandel and Baptist 1972, 216.[AL]
starr′i.i. M.L. gen. n. *starrii* of Starr; named after M.P. Starr, a contemporary investigator of the bdellovibrios.

The characteristics are as described for the genus and as listed in Tables 2.15 and 2.16.

The host range appears unique in that no *Enterobacteriaceae* are lysed by this species (Starr and Seidler, 1971; Torrella et al., 1978).

The mol% G + C of the DNA is 43.5 (Bd).

Type strain: ATCC 15145 (strain A3.12). The ATCC distributes this strain with prey *Pseudomonas putida* ATCC 12633.

Unnamed species of Bdellovibrio

4. Bdellovibrio sp. (Marine strains)
The characteristics are as described for the genus and as listed in Tables 2.15 and 2.16.

The temperature range is similar to that of other bdellovibrios with the exception that plaques may form at a lower minimum temperature (10°C).

The prey range varies for Gram-negative marine bacteria. Nonmarine bacterial strains appear to be poor prey (Taylor et al., 1974).

Na⁺ (75 mM or greater) is required for growth.

The mol% G + C of the DNA is 33.4–38.6 (Bd). Taylor et al. (1974) reported one strain with a G + C content of 43.5, but all strains isolated by Marbach et al. (1976) were within the 33–39% range.

Comments on other organisms possibly belonging to the genus Bdellovibrio

a. Marine microvibrios (called "*Microvibrio marinus roscoffensis*" by Guelin et al., 1977). These small Gram-negative bacteria resemble the bdellovibrios by possessing a biphasic morphology: a small comma-shaped form (diameter less than 0.65 μm) which is motile, and elongated spirillar forms which always are seen preceding the presence of the comma-shaped bacteria. This implies a division form analogous to that of the bdellovibrios (Guelin et al., 1978). Although growth and multiplication of these microvibrios depends upon the presence of suitable

Gram-negative bacteria, they do not penetrate the prey cell wall, and therefore cannot be considered to be bdellovibrios. The attacked bacteria (*E. coli*) lose their viability as a result of interaction with the microvibrios but they are not lysed. Guelin et al. (1978) reported that the fatty acid composition of the microvibrios was qualitatively similar to that of the bdellovibrios and distinct from the genera *Vibrio, Pseudomonas*, or members of the *Enterobacteriaceae*. The overall similarities of microvibrios to the bdellovibrios led those authors to postulate an evolutionary relationship between these bacteria.

b. Bacterial endoparasite of *Scenedesmus*. A strain of a spirillum-like bacterium which attacks, penetrates and grows within the alga *Scenedesmus acutus* was described by Schnepf et al. (1974). The parasite appears to enter the cytoplasm of the cell causing the lysis of the cell membrane. This bacterium develops endoparasitically into a long thin spirillar form which digests the cytoplasm of the host.

Genus **Vampirovibrio** Gromov and Mamkaeva 1980, 676[VP]
(*Effective publication*: Gromov and Mamkaeva 1980, 165)

JEFFREY C. BURNHAM AND SAMUEL F. CONTI

Vam.pi.ro.vib′ri.o. Fr. n. *vampire* vampire; M.L. masc. n. *Vibrio* a genus of curved bacteria; M.L. masc. n. *Vampirovibrio* a vampire-like vibrio.

Comma-shaped rods, 0.3 μm in diameter. Larger coccoid forms (0.6 μm) have also been described (Coder and Starr, 1978). Gram-negative. Motile by means of a single, polar, nonsheathed flagellum. The bacterium does not develop into a spirillar form as part of its life cycle. Reproduction is by binary fission. Cells attach to strains of the alga *Chlorella* and require viable cells of the alga for development. Penetration of the host cells has not been reported. The mol% G + C of the DNA is 50 (cited by Coder and Starr, 1978).

Type species: *Vampirovibrio chlorellavorus* Gromov and Mamkaeva 1980, 676.

Taxonomic Comments

Although *Vampirovibrio* resembles *Bdellovibrio* in certain respects (morphology, DNA base composition, requirement for susceptible cells), there are important differences: (a) the host is eucaryotic, not procaryotic; (b) no elongated spirillar forms occur during growth; (c) growth occurs outside the host cell and the latter is not penetrated; and (d) the flagellum has no sheath. For these reasons a new genus was created to contain the "chlorellavorus bacterium."

List of the species of the genus **Vampirovibrio**

1. **Vampirovibrio chlorellavorus** Gromov and Mamkaeva 1980, 676.[VP] (Effective publication: Gromov and Mamkaeva 1980, 165.) (*Bdellovibrio chlorellavorus* Gromov and Mamkaeva 1972, 259.)

chlo.rel.la′vo.rus. M.L. fem. n. *Chlorella* a genus of algae; L. v. *voro* to devour; M.L. adj. *chlorellavorus* *Chlorella*-devouring.

The characteristics are as described for the genus.
Type strain: ATCC 29753.

Important Notes for Users of this Edition

1. Always read both generic and species descriptions because characters listed in the generic description are not usually listed in the species descriptions.

2. Unless otherwise indicated in footnotes to tables, the meanings of symbols are as follows:
   - + 90% or more of strains are positive
   - − 90% or more of strains are negative
   - d 11–89% of strains are positive
   - v strain instability (*not* equivalent to "d")
   - D different reactions in different taxa (species of a genus or genera of a family)

3. All other symbols are defined in footnotes to tables.

SECTION 3

Nonmotile (or Rarely Motile), Gram-Negative Curved Bacteria

Partial key to section 3

I. Strictly respiratory type of metabolism. Curved or C-shaped rods. Rings may be formed by overlapping of the ends of a cell. Coils and helical spirals may be present. Oxidase-positive.

    A. Rings commonly formed. Coils, helical forms, and long, sinuous filaments may be present. Gas vacuoles are not formed. Nonmotile. Colonies are yellow or pink. The mol% G + C of the DNA is 34–53. (Family *Spirosomaceae*, p. 125

        Genus I. *Spirosoma*, p. 126
        Genus II. *Runella*, p. 128
        Genus III. *Flectobacillus*, p. 129

    B. Rings formed occasionally. Coils, helical forms and filaments are not formed. Gas vacuoles may be present. Rarely motile. Colonies are white to cream colored. The mol% G + C of the DNA is 66–69.

        Genus IV. *Microcyclus*, p. 133

II. Strictly fermentative type of metabolism. Straight or vibrioid cells. Gas vacuoles formed. Nonmotile. Oxidase-negative. The mol% G + C of the DNA is 45.

        Genus V. *Meniscus*, p. 135

III. Bow-shaped cells with gas vacuoles. Cells arranged in coenobia of two rings or four rings (cloverleaf appearance). Pretzel-shaped cells may occur. Can attach to surfaces by means of a mucoid substance. Nonmotile. Have not been isolated in pure culture. Found in ponds and lakes under conditions where sulfide is present and oxygen absent. Gram-reaction not reported.

        Genus VI. "*Brachyarcus*," p. 137

IV. Slender S-shaped cells arranged side-by-side in flat, sigmoid aggregates of four or multiples of four. Aggregates are occasionally motile, possibly by a tuft of flagella. Have not been isolated in pure culture. Occur in and on mud in fresh and brackish waters where sulfide is present and oxygen absent. Gram-reaction not reported.

        Genus VII. "*Pelosigma*," p. 138

FAMILY I. SPIROSOMACEAE LARKIN AND BORRALL 1978, 595[AL]*

JOHN M. LARKIN AND RENEÉ BORRALL

Spi.ro.so.ma'ce.ae. M.L. noun *Spirosoma* type genus of the family; -*aceae* ending to denote family; M.L. fem. pl. n. *Spirosomaceae* the Spirosoma family.

Rigid straight to curved rods, the degree of curvature varying among individuals within a culture. The cells measure 0.3–1.0 μm wide × 1.5–6.0 μm long. Long sinuous filaments up to 50 μm long may be present. **Rings 0.8–10.0 μm in outer diameter are formed by overlapping of the ends of a cell.** Coils and helical spirals may be present. Gram-negative. **Nonmotile.** Resting stages are not known. **Obligate aerobes,** possessing a strictly respiratory type of metabolism with oxygen as the terminal electron acceptor. Acid is produced aerobically from some carbohydrates but not from sugar alcohols. Optimum temperature, 20–30°C. Colonies contain a **pink or yellow water-insoluble pigment. Catalase- and oxidase-positive.** Chemoorganotrophic. Inhibitants of soil, freshwater and marine water. The mol% G + C of the DNA ranges from 34–53 (T_m, Bd or absorbance ratio).

Type genus: *Spirosoma* Migula 1894, 235.

---

* *AL*, denotes the inclusion of this name on the Approved Lists of Bacterial Names (1980).

Key to the genera of the family **Spirosomaceae**

I. Pigmentation on MS agar* is yellow.

Genus I. *Spirosoma*, p. 126

II. Pigmentation on MS agar is pink.

A. Acid is not produced aerobically from most carbohydrates, including fructose, lactose and xylose.

Genus II. *Runella*, p. 128

B. Acid is produced aerobically from most carbohydrates, including fructose, lactose and xylose.

Genus III. *Flectobacillus*, p. 129

Genus I. **Spirosoma** Migula 1894, 237[AL]

JOHN M. LARKIN AND RENEÉ BORRALL

Spi.ro.so′ma. Gr. n. *spira* coil; Gr. n. *soma* body; M.L. neut. n. *Spirosoma* coiled body.

Rigid **straight to curved rods**, the degree of curvature varying among individual cells within a culture. The cells measure 0.5–1.0 μm by 1.5–6.0 μm. Long sinuous filaments up to 50 μm long may be present. **Rings 1.5–3.0 μm in outer diameter are formed by overlapping of the ends of a cell. Coils and helices may be present.** Gram-negative. **Nonmotile.** Resting stages are not known. **Obligate aerobes**, possessing a strictly respiratory type of metabolism with oxygen as the sole terminal electron acceptor. Acids are produced aerobically from a variety of carbohydrates. Optimum temperature, 20–30°C. **Colonies contain a pale to light yellow, water-insoluble pigment. Catalase- and oxidase-positive.** Chemoorganotrophic. Isolated from soil and fresh water. The mol% G + C of the DNA is 51–53 (T_m).

Type species: *Spirosoma linguale* (Eisenberg 1891) Migula 1894, 235.

Further Descriptive Information

Cells of *Spirosoma* typically appear as helices which may be loosely or tightly coiled and which are bent in more than one plane (Fig. 3.1); they also may appear as long serpentine and undulating filaments which may or may not have coiled portions.

Colonies on MS agar* produce a yellow water-insoluble, nonfluorescent pigment. The colonies are circular and convex with an entire margin. The organisms grow poorly or not at all on rich media such as chocolate agar or blood agar, or on enteric-selective media such as eosin methylene blue agar, MacConkey agar or salmonella-shigella agar. Only one of the four available strains grows on Trypticase soy agar.

Spirosoma is chemoorganotrophic and is active in the acidification of carbohydrate media. Acidification occurs with all but one (sorbose) of the 20 carbohydrates tested using the medium of Hugh and Leifson (1953) and aerobic incubation, although up to 3 weeks is sometimes required. Acidification does not occur with any sugar alcohols.

Cellulose and chitin are not hydrolyzed, but some strains hydrolyze esculin and (weakly) casein. All strains hydrolyze gelatin, tributyrin and (weakly) starch. Three of the available strains of *Spirosoma* produce a soft curd in litmus milk, accompanied by the reduction and then reoxidation of the litmus. A fourth strain produces only an increased alkalinity in litmus milk.

Spirosoma strains may be grown in the defined medium of Gordon and Mihm† (1957) with various sole carbon sources. Of 11 substrates that have been tested, only glycerol phosphate, succinate, tartrate and malonate are utilized as sole carbon sources.

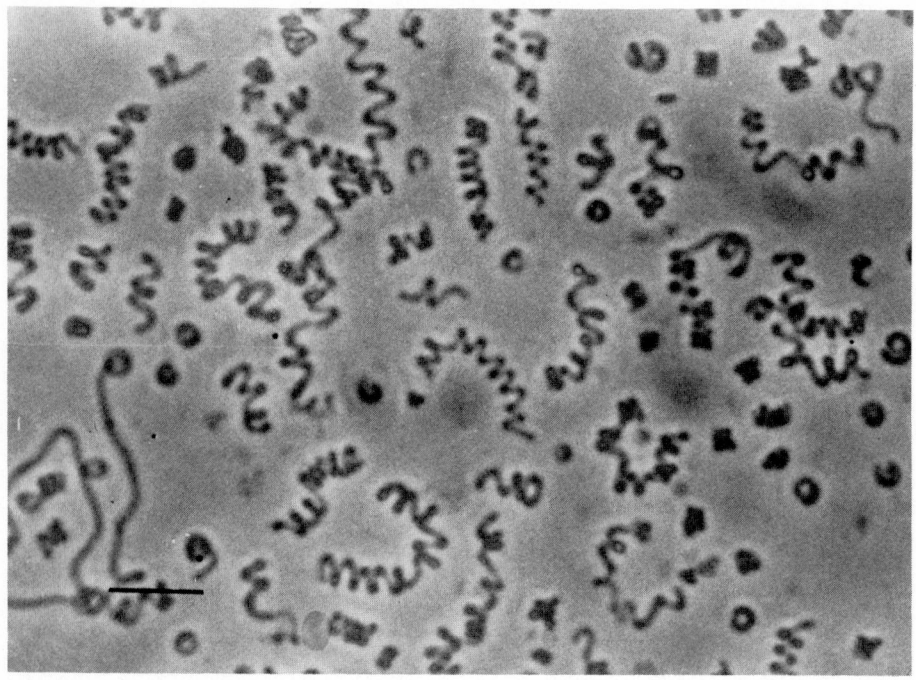

Figure 3.1. Phase-contrast micrograph of *Spirosoma linguale* cultured in MS broth. *Bar,* 10 μm. (Reproduced with permission from J. M. Larkin, P. M. Williams and R. Raylor, International Journal of Systematic Bacteriology *27:* 147–156, 1977, © International Association Microbiology Societies.)

---

* MS agar (g/liter): peptone, 1.0; yeast extract, 1.0; glucose, 1.0; agar, 15.0. For marine organisms add 30.0 g NaCl.

† Gordon and Mihm's medium (g/liter): $MgSO_4$, 0.2; $(NH_4)_2HPO_4$, 1.0; KH_2PO_4, 0.5; NaCl, 1.0; carbon source, 2.0; agar, 15.0; and bromthymol blue, 0.08.

Radiorespirometric studies by Kottel and Raj (1973) reveal that *Spirosoma linguale* ATCC 23276 (formerly "*Microcyclus flavus*" Raj 1970) catabolizes glucose almost entirely (96%) by the Embden-Meyerhof pathway with a small amount (4%) being catabolized by the pentose phosphate pathway. The Entner-Doudoroff pathway is inducible, with 75% of radiolabeled gluconate being metabolized by that pathway. Gluconate recovery patterns of this strain show evidence for the possibility of a 2-5-diketogluconate pathway or some other unorthodox pathway (Raj, 1977). Radiorespirometric and enzymatic data indicate that a functional tricarboxylic acid cycle occurs in this strain (Kottel and Raj, 1973).

Spirosoma is not known to be pathogenic.

Although a culture of *Spirosoma* typically shows a ringlike morphology, it exhibits variations of shape and size under certain cultural conditions (Raj 1979). Occasionally a culture of *Spirosoma* may be composed mainly of relatively straight cells, especially after prolonged subculturing on MS agar. The curly form may be reobtained by examining colonies from agar streaked for isolation (Larkin, unpublished observation). S. R. Maloy, L. A. Anderson and H. D. Raj (1978, Abst. Annu. Mtg. Amer. Soc. Microbiol., I-136) showed that there is a relationship between morphology and the phosphate content of the medium. The rings, coils and helices are produced when the phosphate level is below 20 mM at pH 7.2. At 20–60 mM phosphate, long nonseptate filaments develop and the ratio of diphosphatidyl glycerol to phosphatidyl glycerol is 2-fold higher in the filaments. Normal cells contain twice the amount of muramic acid in the peptidoglycan layer as do the filaments; furthermore, this layer in normal cells contains equal amounts of N-acetylglucosamine and O-acetylglucosamine, while that of filamentous cells has nearly twice as much N-acetylglucosamine as O-acetylglucosamine (J. C. Miller and H. D. Raj, 1978, Abst. Annu. Mtg. Amer. Soc. Microbiol., K-180). The filamentous forms appear to be multinucleate (M. A. Redell, S. R. Maloy and H. D. Raj, 1981, Abst. Annu. Mtg. Amer. Soc. Microbiol., I-63).

Isolation Procedures

Spirosoma may be isolated by repeated streaking of water or diluted soil samples onto MS agar (Larkin et al., 1977) or tryptone glucose extract agar plus 0.1% yeast extract (TGEY) (Raj 1970) with incubation at room temperature for up to 2 weeks. The yellow pigmentation of the colonies aids detection of *Spirosoma*.

Maintenance Procedures

S. linguale is grown on MS agar, TGEY or nutrient agar at room temperature until abundant growth occurs (usually 1–4 days). Cultures may then be stored at 4°C for at least 4 weeks. Preservation by lyophilization is effective for several years.

Differentiation of the genus **Spirosoma** from other genera

Table 3.1, provides the primary characteristics that can be used to differentiate this genus from the morphologically similar genera *Flectobacillus*, *Runella* and *Microcyclus*.

Taxonomic Comments

In the eighth edition of the *Manual* the genus *Microcyclus* consisted of three species which were placed together primarily because of their ability to form rings during growth. The mol% G + C values were quite different, being ~39.5% for "*M. major*," 51% for "*M. flavus*" and 67% for *M. aquaticus* (the type species). Claus (1967) and Claus et al. (1968) suggested that "*M. flavus*" and two additional isolates corresponded to the description of the forgotten genus *Spirosoma*. Moreover, Staley (1974) suggested that the grouping of the three species was unsatisfactory, as did Konopka et al. (1976) who found only a 0–14% binding of the DNA from three *Spirosoma* strains to that of *M. aquaticus*. Larkin et al. (1977) proposed the reintroduction of the genus *Spirosoma* and emended its description to include "*Microcyclus flavus*" and three other isolates.

The formation of rings is not a sufficiently restrictive character to delineate a single taxonomic group. Ring formation is a characteristic that occurs in several bacteria including *Rhodocyclus* (Pfennig, 1978), *Clostridium cocleatum* and *C. spiroforma* (Kaneuchi et al., 1979), as well as certain strains of *Bacillus subtilis* (Mendelson, 1976), *B. stearothermophilus* (Agnostopolous and Sidhu; 1978) and other organisms (Raj, 1977).

The literature concerning the utilization of sole carbon sources by *S. linguale* is conflicting. Raj (1970) reported an inability of the species to use malonate, succinate or tartrate, in contrast to results obtained by Larkin et al. (1977). The discrepancy may be attributable to a difference in the media employed or to the method of detecting growth. Raj (1970) used Simmons citrate agar with various organic substrates substituted for citrate and with 0.2% yeast extract added; growth responses were indicated by a color change. Larkin et al. (1977) used the agar medium of Gordon and Mihm (1957) and estimated visible growth after four successive transfers even in the absence of a color change.

Table 3.1.

Differential characteristics of the genus **Spirosoma** *and other morphologically similar genera[a]*

Characteristics	Spiro-soma	Runella	Flecto-bacillus	Micro-cyclus
Cell shape:				
Rings formed	+	+	+	−
Coils and helices formed	+	+	+	−
Pigmentation				
White to cream	−	−	−	+
Yellow	+	−	−	−
Pink	−	+	+	−
Methanol utilized	−	−	−	+
Mol% G + C of DNA	51–53	49–50	34–40	66–69

[a] For symbols see standard definitions.

Four strains of *Spirosoma linguale* have been isolated and characterized: one from garden soil and three from fresh water. *Spirosoma*-like organisms have been seen in gelatinous deposits on wet planks from a mine (Krapelin and Passern, 1980).

List of the species of the genus **Spirosoma**

1. **Spirosoma linguale** (Eisenberg 1891) Migula 1894, 235.[AL] (*Vibrio lingualis* Eisenberg 1891, 212, *Microcyclus flavus* Raj 1970, 62.)

lin.gua′le. L. adj. *linguale* of the tongue;

The description is as for the genus. Morphological features are depicted in Figure 3.1. Physiological and nutritional characters are presented in Tables 3.1 and 3.2.

Found in soil and fresh water.

The mol% G + C of the DNA ranges from 51 to 53 (T_m).

Type strain: DSM 74.

Table 3.2.
Other characteristics of **Spirosoma linguale** *and* **Runella slithyformis.**[a]

Characteristics	Spirosoma linguale	Runella slithyformis	Characteristics	Spirosoma linguale	Runella slithyformis
Enzyme activity:			Hydrolytic activity on:		
Oxidase, β-galactosidase	+	+	Tributyrin	+	+
Catalase	+	W	Starch	W	W
Phosphatase	d	+	Gelatin	+	−
Urease, lecithinase, lysine decarboxylase, ornithine decarboxylase, phenylalanine deaminase, hemolysin, indole, methyl red test, Voges-Proskauer test, NO₃ reduction, H₂S production from peptone	−	−	Esculin	d	−
			Casein	d	NG
			Agar, cellulose, chitin	−	−
Acid produced aerobically from sugars and sugar alcohols:			Sensitivity to antimicrobial agents:		
			Agent and concentration per disk		
Glucose, inulin, maltose, sucrose	+	+	Ampicillin, 10 µg	S	S
Galactose, mannose, raffinose, rhamnose	+	d	Carbenicillin, 50 µg	S	S
Arabinose, α-methyl-D-glucoside, cellobiose, dextrin, fructose, lactose, melibiose, ribose, salicin, trehalose, xylose	+	−	Cephalothin, 30 µg	S	S
			Chlortetracycline, 15 µg	S	S
			Colistin, 10 µg	R	R
Sorbose	−	−	Erythromycin, 15 µg	S	S
Dulcitol, erythritol, glycerol, mannitol, sorbitol	−	−	Gentamicin, 10 µg	S	S
Utilization of single carbon sources:			Kanamycin, 30 µg	S	d
Glycerol phosphate, malonate, succinate, tartrate	+	−	Neomycin, 30 µg	d	d
			Nitrofurantoin, 300 µg	S	S
Acetate, benzoate, citrate, formate, methanol, methylamine, propionate	−	−	Penicillin G, 10 U	S	S
			Polymyxin B, 300 U	R	R
Litmus milk:			Streptomycin, 10 µg	d	S
Acid production	−	−	Sulfamethoxazole/trimethoprim, 25 µg	S	S
Hard curd	−	−	Sulfathiazole, 300 µg	S	
Alkali production	−	−	Tetracycline, 30 µg	S	S
Soft curd	d	−	Triple sulfa, 1 mg	R	R
Peptonization	−	−	*Agent and concentration per milliliter*		
Litmus reduced	d	−	Actinomycin D, 100 µg	S	S
Litmus reoxidized	d	−	Mitomycin C, 1 µg	S	S

[a] Symbols: see standard definitions; and W, weak reaction; NG, no growth on medium employed; S, susceptible; and R, resistant.

Genus II. **Runella** *Larkin and Williams 1978, 35*[AL]

John M. Larkin and Renée Borrall

Ru.nel′la. M.E. n. *rune* an ancient alphabet; M.L. dim. ending *-ella*; M.L. fem. n. *Runella* that which resembles figures of the runic alphabet.

Rigid **straight to curved rods**, the degree of curvature varying among cells within a culture. The cells measure 0.5–0.9 by 2.0–4.5 µm. The ends of a cell may overlap, **producing a ringshaped structure with an outside diameter of 2.0–3.0 µm.** Filaments up to 14 µm long may be produced. On rare occasions, a coil of two to three turns may be produced. Gram-negative. **Nonmotile.** Resting stages are not known. **Obligate aerobe**, possessing a strictly respiratory metabolism with oxygen as the terminal electron acceptor. **Acid is produced aerobically from only a few carbohydrates.** Optimum temperature, 20–30°C. Colonies contain a **pale pink, water-insoluble pigment.** Catalase weakly positive. **Oxidase-positive.** Chemoorganotrophic. Isolated from fresh water. The mol% G + C of the DNA is 49–50 (T_m, absorbance ratio).

Type species: *Runella slithyformis* Larkin and Williams 1978, 35.

Further Descriptive Information

Cells of *Runella* typically appear as rods whose degree of curvature varies from nearly straight to crescent shape (Fig. 3.2). An individual cell may be bent in more than one plane.

Colonies on MS agar* produce a pale pink, water insoluble, nonfluorescent pigment. The colonies are circular and convex with an entire margin. Abundant growth occurs on MS agar and on nutrient agar. Scant growth occurs on chocolate agar, peptonized milk agar and yeast extract-acetate-tryptone agar. No growth occurs on blood agar, eosin methylene blue agar, nutrient agar containing 5% sucrose, phenol red mannitol salt agar, phenylethyl alcohol agar, trypticase soy agar (with or without 5% sucrose or 3% glucose), MacConkey agar, bismuth sulfite agar or salmonella-shigella agar (Larkin and Williams, 1978).

Runella is chemoorganotrophic. By the technique of Hugh and Leifson (1953) acid is produced aerobically only from glucose, maltose, sucrose and inulin. Strains differ in their ability to produce an acid reaction from rhamnose, galactose, mannose and raffinose. Sugar alcohols are not acidified.

Starch and tributyrin are hydrolyzed; esculin, cellulose, agar, chitin and casein are not hydrolyzed.

None of 11 compounds tested in the medium of Gordon and Mihm (1957)* are utilized as sole carbon sources.

Runella is not known to be pathogenic.

* See p. 126 for recipes for these media.

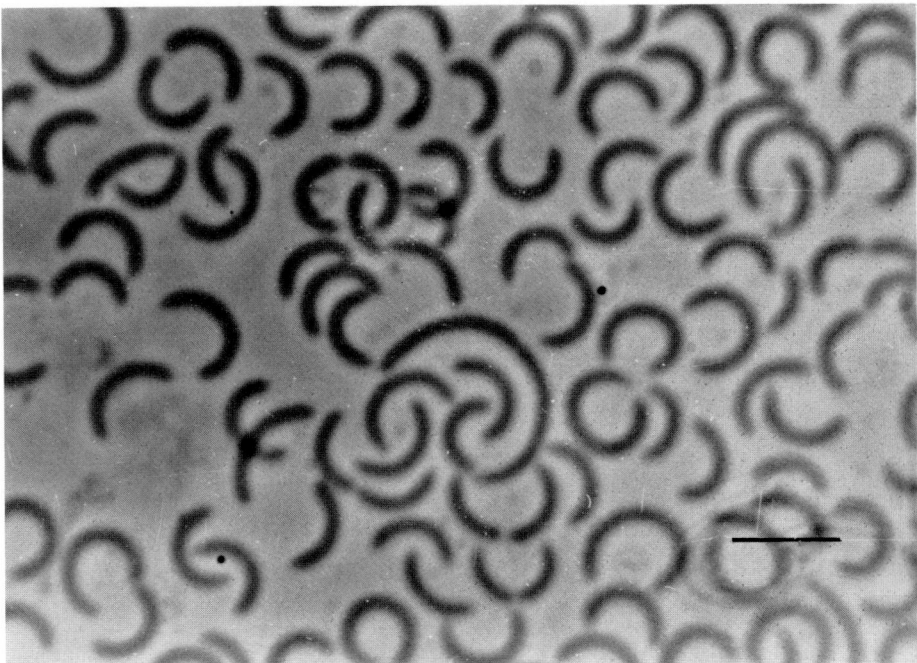

Figure 3.2. A phase contrast micrograph of *Runella slithyformis* cultured in MS medium. *Bar, 5 μm.*

Isolation Procedures

Both known isolates were obtained by repeated streaking of water samples onto MS agar with incubation at room temperature for up to two weeks. The pale pink pigmentation of the colonies aids detection.

Maintenance Procedures

Strains of *Runella* are grown on MS agar or nutrient agar at room temperature for several days to allow abundant growth. They will then survive refrigeration (4°C) for at least 3 weeks. They may also be preserved indefinitely by lyophilization.

Procedures for Testing Special Characters

Utilization of carbohydrates. The production of acid from carbohydrates by aerobic or anaerobic means is determined by the method of Hugh and Liefson (1953), in which MS agar is used but with the glucose replaced by 1% of the substrate to be tested and the agar concentration lowered to 0.3%. Incubation is continued for 8 weeks for cultures giving negative results.

Utilization of single carbon sources. The basal medium used is that of Gordon and Mihm (1957), to which is added 0.2% of the substrate or the sodium salt of the substrate. If growth occurs through four successive subcultures, the results are considered positive even in the absence of a color change in the bromthymol blue indicator.

Differentiation of the genus **Runella** from other genera

Table 3.1 of the previous genus, *Spirosoma*, provides the primary characteristics that can be used to differentiate this genus from the morphologically similar genera *Spirosoma*, *Flectobacillus* and *Microcyclus*.

Taxonomic Comments

Only two strains of *R. slithyformis* have been isolated and characterized. Both were isolated from eutrophic fresh waters. Similar organisms

were seen in gelatinous deposits on wet planks from a mine (Kraepelin and Passern, 1980), in marine waters (Overbeck, 1974; Sieburth, 1978) and in fresh waters (Larkin, unpublished observations).

List of the species of the genus **Runella**

1. **Runella slithyformis** Larkin and Williams 1978, 35[AL]
slith.y.form′is. *slithy* a nonsense word from Lewis Carroll's *Jabberwocky* for a fictional organism that is "slithy" (presumably a combination of slinky and lithe); L. n. *forma* shape, form; M.L. adj. *slighyformis*, slithy in form.

The description is as for the genus. Morphological features are depicted in Figure 3.2. Physiological and nutritional characters are presented in Tables 3.1 and 3.2 of the genus *Spirosoma*.

Found in fresh water.

The mol% G + C of the DNA is 49 to 50 (T_m, absorbance ratio).

Type strain: ATCC 29530.

Genus III. **Flectobacillus** Larkin, Williams and Taylor 1977, 152[AL]

JOHN M. LARKIN AND RENEÉ BORRALL

Flec.to.ba.cil′lus. L. v. *flecto* to curve; L. n. *bacillus* a little staff, rod; M.L. masc. n. *Flectobacillus* curved rod.

Rigid **straight to curved rods**, the degree of curvature varying among individual cells with a culture. **The most abundantly occur-** **ring are cells in the shape of the letter C or a closed ring.** The cells measure 0.3–1.0 μm wide by 1.5–5.0 μm long. Long sinuous

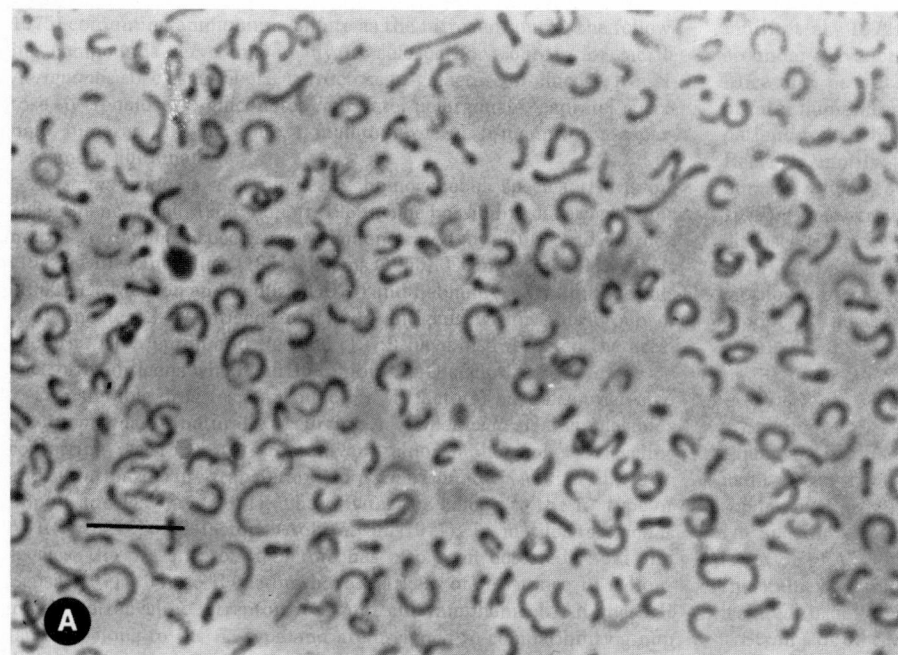

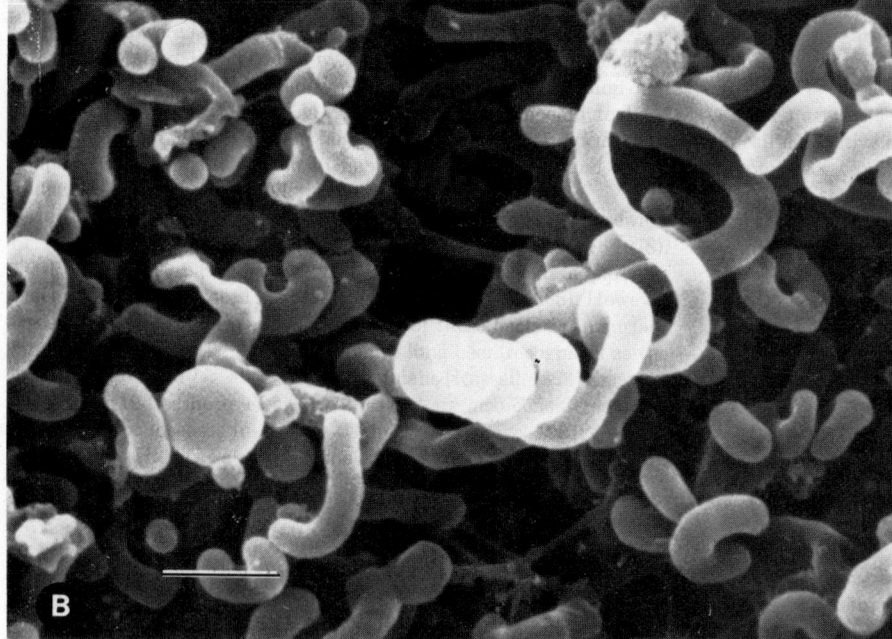

Figure 3.3. Appearance of two species of *Flectobacillus*. *A*, *F. major* cultured in MS broth. *Bar*, 10 μm. (Reproduced with permission from J. M. Larkin, P. M. Williams and R. Taylor, International Journal of Systematic Bacteriology 27: 147–156, 1967, ©American Society for Microbiology.) *B*, *F. marinus* cultured in mZ broth. *Bar*, 1 μm. (Reproduced with permission from H. Raj.)

filaments up to 50 μm long may be present. **Rings 0.8–10.0 μm in outer diameter are formed by overlapping of the ends of a cell.** Coils or helices may be present. Gram-negative. **Nonmotile.** Resting stages are not known. **Obligate aerobes,** possessing a strictly respiratory type of metabolism with oxygen as the terminal electron acceptor. Acids are produced aerobically from a variety of carbohydrates. Optimum temperature, 20–30°C. Colonies contain a **pale pink or a rose-colored, water-insoluble pigment. Catalase- and oxidase-positive.** Chemoorganotrophic. Isolated from fresh and marine waters. The mol% G + C of the DNA is 34–40 (T_m, Bd).

Type species: *Flectobacillus major* (Gromov 1963) Larkin, Williams and Taylor 1977, 155.

Further Descriptive Information

Cells of *Flectobacillus* typically appear as curved rods whose degree of curvature varies from nearly straight to crescent- or horseshoe-shaped, to rings. An individual cell may be bent in more than one plane (Fig. 3.3). Regular and irregular coils and helices as well as long sinuous filaments may be produced commonly by *F. marinus* and rarely by *F. major*.

Colonies on MS agar* (supplemented with 3.0% NaCl for *F. marinus*) produce a pink, water-insoluble, nonfluorescent pigment. The colonies are circular and convex with an entire margin. *F. major* does not grow on eosin methylene blue agar, phenol red mannitol salt agar, phenyle-

---

* See Key to the Genera of the Family *Spirosomaceae*.

thyl alcohol agar, MacConkey agar, bismuth sulfite agar or salmonella-shigella agar. Abundant growth occurs on MS agar, peptonized milk agar, yeast extract-acetate-tryptone agar and nutrient agar (Larkin et al., 1977). *F. marinus* grows on brain heart infusion agar, tryptone glucose yeast extract agar, Mueller-Hinton agar, cystine trypticase agar with glucose, and eosin methylene blue agar, when 3.0% NaCl is added. It also grows abundantly on modified Zobell 2216 (mZ) agar†. Poor growth occurs on nutrient agar and there is no growth on Lowenstein-Jensen agar (Raj, 1976).

Flectobacillus is chemoorganotrophic and produces an acid reaction from a wide variety of carbohydrates including pentoses, hexoses and disaccharides when incubated aerobically in the medium of Hugh and Leifson (1953); NaCl (3.0%) must be added to the medium for growth of *F. marinus*. Acidification does not occur with any sugar alcohols.

Casein, cellulose and chitin are not hydrolyzed. *F. major*, but not *F. marinus*, hydrolyzes gelatin, tributyrin and starch. Litmus milk is unchanged by *F. major*, but *F. marinus* produces an acid curd with litmus reduction. *F. marinus* is more versatile than *F. major* in its ability to utilize single carbon sources.

The pathway for carbohydrate catabolism by *F. marinus* has been examined by radiorespirometric techniques (Raj, 1976). Approximately 80% of respired glucose was metabolized via the Embden-Meyerhof pathway, 13% via the Entner-Doudoroff pathway, and only 7% via the pentose phosphate pathway. When given radiolabeled gluconate instead of glucose, *F. marinus* respired 75% of the substrate via the Entner-Doudoroff pathway; moreover, gluconate recovery patterns of *F. marinus* suggest the possibility of a 2,5-diketogluconate pathway or some other unorthodox pathway (Raj, 1977).

Radiorespirometric and enzymatic data indicate the occurrence of a functional tricarboxylic acid cycle in *F. marinus* (Raj, 1976).

Flectobacillus is not known to be pathogenic.

Isolation Procedures

Both species of *Flectobacillus* can be isolated by repeated streaking of samples onto MS agar (containing 3.0% NaCl for *F. marinus*) with incubation at room temperature for up to 2 weeks. The pink pigmentation of the colonies aids detection. *F. marinus* may also be isolated on modified Zobell 2216 (mZ) medium (Raj, 1976).

Maintenance Procedures

Strains of *F. major* are grown on MS or nutrient agar. *F. marinus* may be grown on MS agar with 3.0% NaCl added, or on mZ medium. Incubation is at room temperature (25°C) for several days to allow abundant growth. The cultures will then survive refrigeration at 4°C for at least 3 weeks. They may also be preserved indefinitely by lyophilization.

Procedures for Testing Special Characters

Utilization of single carbon sources. The following medium (g/liter) is used for *F. major*: $MgSO_4$, 0.2; $(NH_4)_2HPO_4$, 1.0; KH_2PO_4, 0.5; NaCl, 1.0; carbon source, 2.0; bromthymol blue, 0.08; agar, 20.0. The appearance of growth through four successive subcultures is considered positive even in the absence of a color change in the pH indicator (Larkin et al., 1977). The same medium supplemented with 3.0% NaCl may be used for *F. marinus*.

Differentiation of the genus **Flectobacillus** *from other closely related taxa*

Table 3.1 in the description of the genus *Spirosoma* presents characteristics that can be used to differentiate *Flectobacillus* from the morphologically similar genera *Spirosoma*, *Runella* and *Microcyclus*.

Taxonomic Comments

In the eighth edition of the *Manual*, *F. major* was a member of the genus *Microcyclus* together with *M. aquaticus* and "*M. flavus*." The only similarity among these species was their ability to form rings, a feature which is not sufficiently restrictive to delineate a genus (see taxonomic comments for the genus *Spirosoma*). Moreover, the mol% G + C values for the DNA of the three species of *Microcyclus* were quite different. It was suggested that the grouping of the three species into a single genus was unsatisfactory (Claus, 1967; Claus et al., 1968; Staley, 1974; Konopka et al., 1976). Larkin et al. (1977) examined all known isolates of the three species and determined that separate generic status for each species was appropriate. Thus, "*M. flavus*" became *Spirosoma linguale*, "*M. major*" became *Flectobacillus major* and *M. aquaticus* remained within the genus as the only species. While the latter study was in press, Raj (1976) described an additional species of *Microcyclus*, "*M. marinus*." In many features it resembled *F. major*, and after studying both species Borrall and Larkin (1978) assigned it to the genus *Flectobacillus* as *F. marinus*.

H. D. Raj (Abst. Annu. Meet. Amer. Soc. Microbiol., I-31, 1979)

reported that *F. major* and *F. marinus* were sufficiently distinct and that *F. marinus* should be accorded status as a separate genus. The name "*Cyclobacterium marinus*" was suggested (Raj, personal communication). However, at present it seems best to retain it within the genus *Flectobacillus* because hybridization between the DNA of *F. major* and *F. marinus* shows a 70% homology (Borrall and Larkin, unpublished results).

Until recently, only two strains of *F. major*, both from fresh water, have been isolated. However, Metcalf and Krueger (personal communication) have located two shallow lakes near Sacramento, California which have relatively large populations of *F. major*. Counts on MS agar ranged from zero to several thousand per milliliter and over 60 isolates have now been obtained from separate samplings. The isolates are strikingly similar to the type strain of *F. major* in their morphological and physiological characteristics, but differ from it in being unable to hydrolyze urea or gelatin.

Only a single strain of *F. marinus* has been isolated. It is from a *Dendraster* found off the Pacific Coast in Newport Beach, California. Organisms similar in appearance to *Flectobacillus* have been seen in gelatinous deposits on wet planks from a mine (Kraepelin and Passern, 1980), in marine waters (Overbeck, 1974; Sieburth, 1978) and fresh waters (Larkin et al., 1977).

Differentiation and characteristics of species of **Flectobacillus**

Table 3.3 presents the characteristics which differentiate *F. major* from *F. marinus*. Table 3.4 lists other characteristics of the two species.

List of the species of the genus **Flectobacillus**

1. **Flectobacillus major** (Gromov 1963) Larkin, Williams and Taylor 1977, 155.[AL] (*Microcyclus major* Gromov 1963, 733.)

ma′jor. L. adj. *major* larger.

The morphology is as described for the genus (see also Table 3.3 and Fig. 3.3*A*).

Physiological and nutritional characteristics are as described for the genus and as listed in Table 3.4.

Found in fresh water.

The mol% G + C of the DNA ranges from 39.5–40.3 (T_m).

Type strain: BKM 859 (DSM 103).

† Modified Zobell 2216 (mZ) medium (g/liter of sea water): peptone, 5.0; yeast extract, 1.0; ferrous sulfate, 0.2; agar, 20.0 (Raj, 1976).

Table 3.3.

Characteristics differentiating **Flectobacillus major** *and* **Flectobacillus marinus.**[a]

Characteristics	1. *F. major*	2. *F. marinus*
Cell diameter (μm)	0.6–1.0	0.3–0.7
Outer diameter of rings (μm)	5.0–10.0	0.8–2.0
Growth in 5.0% NaCl	−	+

[a] For symbols, see standard definitions.

2. **Flectobacillus marinus** (Raj 1976) Borrall and Larkin 1978, 342.[AL] (*Microcyclus marinus* Raj 1976, 540.)

ma.ri′nus. L. adj. *marinus* of the sea, marine.

The morphology is as described for the genus (see also Table 3.3 and Fig. 3.3*B*).

Sea water or Na^+ is required for growth.

Physiological and nutritional characteristics are as described for the genus and as listed in Table 3.4.

Found in marine environments.

The mol% G + C of the DNA ranges from 34 (T_m) to 38 (Bd).

Type strain: ATCC 25205.

Table 3.4.

Other characteristics of **Flectobacillus major** *and* **Flectobacillus marinus**[a]

Characteristics	1. *F. major*	2. *F. marinus*	Characteristics	1. *F. major*	2. *F. marinus*
Enzyme activity:			Sensitivity to antimicrobial agents:		
Oxidase	+	+	*Agent and concentration per disk*		
Catalase	W	+	Ampicillin, 10 μg	S	S
Urease	+	−	Aureomycin, 15 μg	S	R
β-galactosidase (ONPG), phosphatase	+		Bacitracin, 10 μg		R
			Carbenicillin, 10 μg	S	
Lecithinase, lysine decarboxylase, ornithine decarboxylase, phenylalanine deaminase	−		Cephalothin, 30 μg	S	S
			Chloramphenicol, 30 μg		S
			Colistin, 10 μg	R	R
Hemolysin, indole, methyl red test, Voges-Proskauer test, NO_3 reduction, H_2S production from peptone	−	−	Declomycin, 30 μg		R
			Dihydrostreptomycin, 10 μg		R
			Doxycycline, 30 μg		R
			Erythromycin, 15 μg	S	S
Acid production from carbohydrates:			Gentamicin, 10 μg	S	
Arabinose, fructose, galactose, glucose, lactose, maltose, mannose, melibiose, raffinose, rhamnose, salicin, sucrose, trehalose, xylose	+	+	Kanamycin, 30 μg	d	R
			Lincomycin, 2 μg		S
			Methicillin, 5 μg		R
			Nalidixic acid, 30 μg		S
			Neomycin, 30 μg	d	R
α-methyl-D-glucoside, cellobiose, dextrin	+		Nitrofurantoin, 300 μg	S	S
			Novobiocin, 30 μg		S
Inulin	d	+	Oleandomycin, 15 μg		R
Melizitose		+	Oxacillin, 1 μg		R
Ribose, sorbose	−		Oxytetracycline, 30 μg		R
Acid produced aerobically from alcohols:			Penicillin G, 10 U	S	R
			Polymyxin B, 50 U	R	R
Dulcitol, glycerol, mannitol, sorbitol	−	−	Ristocetin 30 μg		S
			Streptomycin, 10 μg	S	R
Erythritol	−		Sulfadiazine, 1 mg		R
Adonitol, inositol		−	Sulfamerazine, 50 μg		R
Utilization of single carbon sources:			Sulfamethizole, 50 μg		R
Succinate	+	+	Sulfamethoxazole/trimethoprim, 25 μg	S	R
Acetate, citrate, malonate, tartrate	−	+	Sulfamethoxypyridazene, 1 μg		R
Fumarate, malate, pyruvate		+	Sulfathiazole, 300 μg	S	
Gluconate, lactate, oxalacetate		−	Sulfathiazole, 1 mg		S
Glycerol phosphate, methanol, methylamine		−	Sulfioxazole, 250 μg		R
			Sulfisomidine, 50 μg		R
Propionate	−		Tetracycline, 30 μg	S	S
Benzoate, formate	−	−	Triple sulfa, 250 μg	R	R
Hydrolytic activity on:			*Agent and concentration per milliliter*		
Gelatin, starch, tributyrin	+	−	Actinomycin D, 100 μg	S	
Esculin	+		Mitomycin C, 1 μg	R	
Agar, casein, cellulose, chitin	−	−			

[a] Symbols: see standard definitions; and W, weak reaction; S, susceptible; and R, resistant.

OTHER GENERA

Genus **Microcyclus*** Ørskov 1928, 183[AL]

JAMES T. STALEY AND A. E. KONOPKA

Mi.cro.cy′clus. Gr. adj. *micros* small; Gr. n. *cyclos*, circle; M.L. masc. n. *Microcyclus* a small circle.

Curved rods, 0.3–1.0 μm in diameter and 1.0–3.0 μm in length. **Rings** (0.9–3.0 μm outer diameter) formed occasionally prior to cell separation. Coils, helical, or filamentous forms are not produced. Cells are encapsulated. Resting stages not known. **Some strains produce gas vacuoles.** Gram-negative. **Generally non-motile,** but motility occurs in one gas vacuolate strain by means of a single polar flagellum. **Obligately aerobic,** possessing a strictly respiratory type of metabolism with oxygen as the terminal electron acceptor. Optimum temperature, 22–37°C. Colonies are translucent opaque and white to cream colored. Pellicles are produced in liquid media. **Oxidase-positive.** Catalase-positive. Chemoorganotrophic, using a variety of sugars or salts of organic acids as carbon sources. Chemolithotrophic growth on molecular hydrogen has been reported. Strains that have been tested can use methanol and formate (**facultatively methylotrophic**). Occur in soil and freshwater sources. The mol% G + C of the DNA is 66–69 (Bd).

Further Descriptive Information

Cells of *Microcyclus aquaticus* typically appear as curved rods (Figs. 3.4 and 3.5). Ringlike forms occur when the ends of a curved cell overlap prior to cell separation; this occurs infrequently under normal growth conditions. Pleomorphic forms have been reported under certain cultural conditions. For example, older cultures of the type strain contain swollen cells and other involution forms (Raj, 1970, 1977).

A number of gas vacuolate strains of *M. aquaticus* have been isolated (Van Ert and Staley, 1971; Nikitin, 1971; Konopka et al., 1976). These appear to be very similar to the type strain so no new species have been proposed for them (see Taxonomic Comments). We consider the type strain of *M. aquaticus* to be avacuolate, despite a claim to the contrary (Raj, 1977). In addition to a lack of convincing microscopic evidence for vacuoles in this strain in our laboratory, we have found that antiserum prepared against gas vacuoles is incapable of causing precipitation in lysates of the type strain, whereas precipitation does occur with the lysates of all gas-vacuolated strains.

Cells are encapsulated.

Motility has not been detected in *Microcyclus* except in the case of one of the gas vacuolate strains of *M. aquaticus* (Konopka et al., 1976). In this strain (strain M6) the cells possess a single polar flagellum. The occurrence of flagella is best seen in cells cultivated at higher growth temperatures, i.e. above 20°C, a condition which precludes extensive gas vacuole formation. The strain forms numerous gas vacuoles when cultivated at 20°C but flagella are rarely formed at this temperature.

The maximum temperature at which growth occurs is usually 37°C, although one strain can grow at 43°C. The minimum temperature for growth is 5°C.

Colonies of *M. aquaticus* are nonpigmented to cream colored and are circular and convex with an entire margin. In gas vacuolate strains, the colonies appear in gradations from translucent (if few vacuoles are formed) to opaque, chalky-white (if vacuoles are abundant).

Acid is produced from a variety of carbon sources when the medium of Hugh and Leifson (1953) is incubated aerobically (Raj, 1970; Van Ert and Staley, 1971; Larkin et al. 1977). All strains of *Microcyclus aquaticus* form acid from arabinose, xylose, glucose, fructose, galactose, mannose, melibiose, sucrose, mannitol, sorbitol, and inulin. Gelatin is not liquefied. Starch is not hydrolyzed. Additionally, some of the strains, including the type strain, have been shown to produce acid

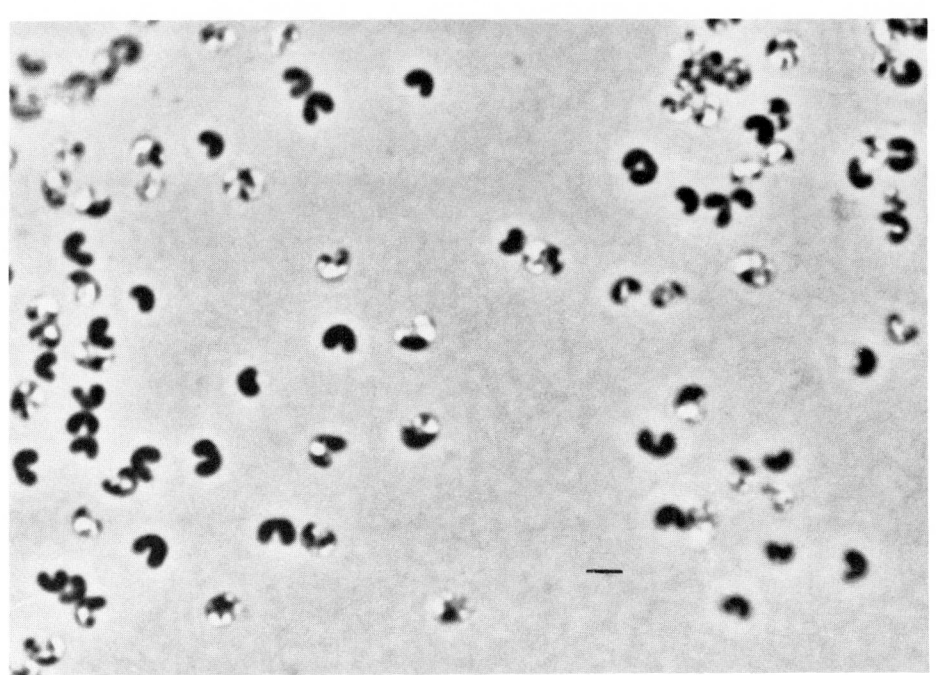

Figure 3.4. A phase-contrast photomicrograph showing numerous cells of *Microcyclus aquaticus.* The refractile intracellular areas of some cells are gas vacuoles. *Bar,* 2.0 μm.

---

* *Editorial Note:* The generic name *Microcyclus* Ørskov 1928 has recently been found to be illegitimate because of the precedence of the fungal genus *Microcyclus* Saccardo 1904. A new generic name, *Ancylobacter*, has been proposed as a substitute, with *A. aquaticus* as the type species (H. D. Raj, 1983, Int. J. Syst. Bacteriol. *33*: 397–398).

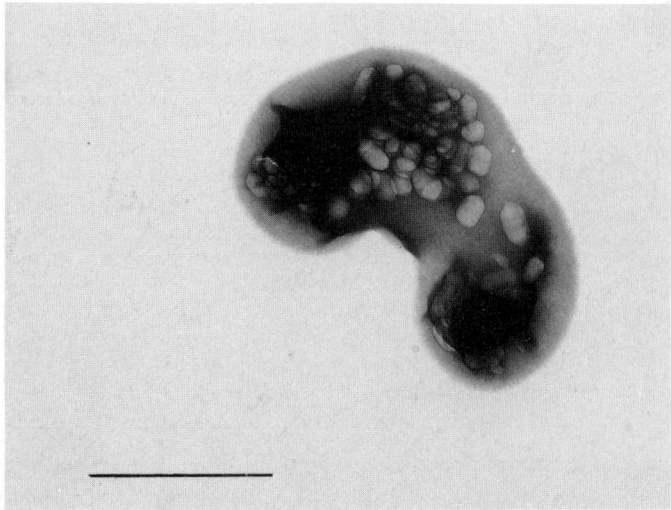

Figure 3.5. An electron micrograph of a cell of gas-vacuolate strain of *Microcyclus aquaticus*. Individual gas vesicles can be seen. *Bar*, 1.0 μm. (Courtesy of Dr. J. C. Lara)

oxidatively from glycerol and ribose, reduce nitrate to nitrite and ammonia, curdle and peptonize litmus milk by rendering it alkaline, and possess catalase, β-galactosidase, lipase, ornithine decarboxylase, oxidase and urease. However, they have been shown not to hydrolyze agar, chitin, casein, cellulose or esculin, and not to produce indole, acetyl methyl carbinol and H_2S (Larkin et al., 1977; Raj, 1970, 1977).

Strains of *M. aquaticus* may be grown in a defined medium* containing an ammonium salt as a sole nitrogen source; individual carbon sources may be tested in this medium as sole carbon sources for growth. Lactate and pyruvate are used by all strains of *M. aquaticus* under these conditions. In addition, the type strain and some others utilize acetate, citrate, formate, gluconate, malonate, oxaloacetate and succinate as sole carbon sources (Raj, 1977). All strains of *M. aquaticus* so far tested have been found to be facultative methylotrophs being capable of utilizing 1-carbon compounds such as methanol and formate (Van Ert and Staley, 1971; Namsaraev, 1973; Larkin et al., 1977). The one strain (Z-238) that has been analyzed for its enzyme content has been found to have ribulose biphosphate carboxylase activity when grown on methanol (Loginova et al., 1978). Due to induced dehydrogenases methanol is oxidized sequentially to formaldehyde, formate and finally carbon dioxide. Carbon dioxide is assimilated by ribulose bisphosphate carboxylase and used in the Calvin cycle. Autotrophic

growth of several strains has been achieved with hydrogen (Namsaraev and Nozhevnikova, 1978; Malik and Schlegel, 1981).

The pathways of glucose dissimilation have been determined by the use of position radiolabeled glucose and gluconate (Kottel and Raj, 1973; Raj, 1976). In *M. aquaticus* Ørskov the Entner-Doudoroff pathway is the primary pathway and the pentose phosphate and Embden-Meyerhof-Parnas pathways are of lesser importance. Also a tricarboxylic acid cycle is present as a secondary pathway as indicated by studies using specifically labeled acetate, glutamate, and pyruvate. The glyoxylic acid cycle does not appear to be operative in this bacterium. Activities of key enzymes from each of these pathways have been demonstrated by assays of cell extracts (Kottel and Raj, 1973).

Strains that have been tested grow well on nitrogen-free media (Nikitin, 1971; Malik and Schlegel, 1981). Attempts to demonstrate nitrogenase activity by the acetylene reduction method or by $^{15}N_2$ incorporation, however, have been unsuccessful (Malik and Schlegel, 1981).

The type strain of *M. aquaticus* is sensitive to the following antibiotics (concentrations, μg/disk): ampicillin, 10; cephalothin, 30; chlortetracycline, 30; demeclocycline, 30; dihydrostreptomycin, 10; doxycycline, 30; erythromycin, 15; kanamycin, 30; neomycin, 30; nalidixic acid, 30; nitrofurantoin, 300; novobiocin, 30; oxytetracycline, 30; streptomycin, 5; tetracycline 5 (Raj, 1970, 1976; Larkin et al., 1977).

Microcyclus is not known to be pathogenic to humans.

M. aquaticus occurs in freshwater habitats including ponds, creeks and lakes and also in soil environments. A number of strains have also been isolated from pulp mill oxidation lagoons.

Enrichment and Isolation Procedures

The original strain of *M. aquaticus* was isolated on a medium consisting of agar (2%) and water (Ørskov, 1928). Gas vacuolate strains can be isolated from enrichment cultures prepared with 100 ml of a freshwater source such as creek water added to a sterile aluminum foil-covered beaker containing 10 mg Bacto-peptone (Difco). The enrichments are incubated at room temperature for 2 weeks and dilutions are plated onto a casein hydrolysate medium† containing glucose (Van Ert and Staley, 1971). These plates are incubated at 30°C for 1 week and examined for the development of chalky white colonies which are indicative of gas vacuolate strains. Cells from such colonies should be observed by phase-contrast and electron microscopy to confirm that curved rods with refractile areas typical of gas vacuoles are present.

Maintenance Procedures

Strains of *M. aquaticus* are normally cultured on glucose-casamino acids medium or TGEY medium.§ After incubation at 20–30°C to allow abundant growth, the cultures may be maintained in a refrigerator (5°C) for at least 3 weeks. They may also be preserved indefinitely by lyophilization.

Differentiation of the genus **Microcyclus** from other genera

Table 3.5 provides the primary characteristics that can be used to differentiate this genus from morphologically similar, aerobic, nonmotile, nonphotosynthetic bacteria of the genera *Flectobacillus*, *Spirosoma* and *Runella*. Two other genera of gas vacuolate curved rods that may be confused with *Microcyclus* are *Brachyarcus* and *Meniscus*. *Brachyarcus*, which has never been isolated, differs from *Microcyclus* in having cells arranged in groups (coenobia) consisting of two, four or

more rings (Skuja, 1964). *Meniscus* is differentiated from *Microcyclus* by being an aerotolerant anaerobe rather than an aerobe and by having a mol% G + C of 45 (Irgens, 1977).

Taxonomic Comments

A number of changes have been made since the eighth edition of the *Manual* in which it was indicated that the genus *Microcyclus* was

---

* Defined medium has the following composition (per liter of distilled water): $(NH_4)_2SO_4$, 0.25 g; glucose or other carbon source, 0.25 g; Na_2HPO_4, 0.071 g; modified Hutner's salt solution, 20 ml; and vitamin solution, 10 ml. The salt solution is as described by Van Ert and Staley (1971) except that the amount of sodium molybdate is 12.67 mg. The vitamin solution is as described by Staley (1968).
† Glucose-casamino acids medium (per liter of distilled water): glucose, 1.0 g; Bacto-casamino acids (Difco), 1.0 g; modified Hutner's salt solution (see defined medium), 20 ml; vitamin solution (see defined medium), 10 ml; agar, 15.0 g.
§ TGEY medium is Bacto-tryptone glucose extract agar (Difco) supplemented with 0.1% yeast extract (Raj, 1970).

Table 3.5.
*Differential characteristics of the genus **Microcyclus** and other genera of curved, nonmotile, aerobic, nonphotosynthetic bacteria[a]*

Characteristics	Micro-cyclus	Spirosoma	Flecto-bacillus	Runella
Shape:				
Curved or C-shaped rods	+	+	+	+
Rings	Rare	+	+	+
Coils or helical forms	−	+	Rare	Rare
Sinuous filaments	−	+	+	+
Gas vacuoles present	d[b]	−	−	−
Motility	−[c]	−	−	−
Pigmentation of colonies:				
White to cream	+	−	−	−
Yellow	−	+	−	−
Pink to rose	−	−	+	+
Alkaline reaction and peptonization in litmus milk	+[d]	−	−	−
NO₃ reduction	+[d]	−	−	−
Ornithine decarboxylase activity	+[d]	−	−	−
Utilization of single carbon sources:				
Formate	+[d]	−	−	−
Methanol	+[d]	−	−	−
Glycerol phosphate	−[d]	+	−	−
Acid production (oxidatively) from:				
Glycerol	+[d]	−	−	−
Mannitol	+[d]	−	−	−
Ribose	+[d]	+	−	−
Sorbitol	+[d]	−	−	−
Mol% G + C of DNA	66–69	51–53	34–41	49–50

[a] Symbols: see standard definitions.
[b] Most strains of *M. aquaticus*.
[c] Motility has been found only in 1 gas vacuolated strain of *M. aquaticus*.
[d] Found in those strains, including the type strain, that have been tested.

composed of at least three distinct groups (i.e. genera) with widely disparate DNA base compositions (see also Claus et al., 1968). To accommodate two of these groups, the genus *Spirosoma* Migula 1894, which had been in disuse, was reinstituted and emended to include "*Microcyclus flavus*" and similar strains (Larkin et al., 1977), and a new genus, *Flectobacillus* Larkin et al. 1977 was created to include "*M. major*" (Larkin et al., 1977). Removal of these species left *Microcyclus*

aquaticus as the only species remaining in the genus *Microcyclus*, and the genus now seemed to be distinct from *Flectobacillus* and *Spirosoma* because of its higher mol% G + C, lack of pigmentation, its mostly vibrioid morphology, and its inability to form helical coils or long, sinuous filaments. A new species, *Microcyclus marinus*, was subsequently described by Raj (1976). This species had a mol% G + C nearly 30% lower than *M. aquaticus*, produced pink colonies, developed helical forms and was isolated from a marine rather than a freshwater habitat. These characteristics were clearly not in accordance with the characteristics of *M. aquaticus*; moreover, DNA homology studies indicated that there was no significant DNA/DNA hybridization between the two species (Konopka et al., 1976). Borral and Larkin (1978) proposed that this species be included in the genus *Flectobacillus* and renamed *F. marinus* because of its similarity to *F. major*. An Adansonian analysis of nonmotile ring-forming bacteria was recently conducted (H.D. Raj, Abstr. Annu. Mtg. Amer. Soc. Microbiol., 1979, I-31). H. D. Raj (personal communication) believes that *F. marinus* should be placed in a separate genus because its similarity value with *F. major* was "only 58%." The name "*Cyclobacterium marinus*" Raj. comb. nov. was suggested (H. D. Raj, Abstr. Annu. Mtg. Amer. Soc. Microbiol. 1980, K-49). Unpublished DNA homology data indicate, however, there is significant homology between *F. marinus* and *F. major* (J. Larkin, personal communication) and these results are therefore in support of the present taxonomy.

The DNA/DNA homology studies of *M. aquaticus* strains B, H, M, M₆ and W with the type strain showed a much lower genetic relatedness (28–45%) than expected of strains of the same species, even though their mol% G + C values were very close (Konopka et al., 1976).

The finding that many strains of *M. aquaticus* are facultative methylotrophs with a Calvin cycle, and the report that at least one strain can grow as a hydrogen autotroph, indicate these bacteria may be related to methylotrophs and chemoautotrophic bacteria. Furthermore, it may be significant that certain phototrophic bacteria, i.e. those of the nonsulfur purple genus *Rhodocyclus*, are the morphological counterparts of *M. aquaticus*. Studies involving rRNA/DNA hybridization, rRNA oligonucleotide cataloging and genome size may prove useful for investigating such potential relationships.

"*Renobacter vacuolatum*" Nikitin 1971, a new genus and species proposed for nonmotile, gas vacuolate bacteria, has been regarded as being identical to *M. aquaticus* (Namsaraev, 1973; J. Larkin, personal communication). Also, a new species, "*Vibrio alternans*" Hallock 1960, was named for 20 isolates that morphologically as well as biochemically seem to be identical or strongly similar to *M. aquaticus* Ørskov (Claus, 1967). Indeed, it is possible that "*V. alternans*" and *M. aquaticus* may justify placement in the same genus and possibly within the same species (Raj, 1977).

The original strain described by Ørskov as the type strain of *M. aquaticus* was lost (Ørskov, 1953). Subsequently, new strains were isolated by Ørskov (1953), and Larkin and Borrall (1979) proposed that one of these strains (ATCC 25396) be the neotype strain.

List of the species of the genus **Microcyclus**

1. **Microcyclus aquaticus** Ørskov 1928, 183.[AL]
a.qua′ti.cus. L. adj. *aquaticus* living in water.
The characteristics are as described for the genus.

Occur in soil and freshwater.
Type strain: ATCC 25396 (Larkin and Borral, 1979).

Genus **Meniscus** Irgens 1977, 42[AL]

Roar L. Irgens

Me.nis′cus. Gr. n. *meniskos* crescent moon.

Curved or straight rods, 0.7–1.0 μm in diameter and 2.0–3.0 μm in length. Cultures may show single cells, pairs, tightly coiled spirals, S shapes (two cells, one inverted), and doughnut-shaped cells, where the ends are overlapping before division by binary fission has occurred. "Rings" have outer diameters of about 3.0 μm. **Gram-negative. Nonmotile. Resting stages not known. Encapsulated. Gas vacuoles**

arranged at random within cells. **Colonies chalky white.** Chemoorganotrophic; strictly **fermentative** metabolism with no gas production. **Catalase- and oxidase-negative. Aerotolerantly anaerobic;** capable of growth under an air atmosphere provided at least 1% CO_2 is present. Vitamin B_{12}, thiamin and CO_2 required for growth. Optimum temperature, 30°C. No growth at 10 or 40°C. Isolated from anaerobic digester sludge. The mol% G + C of the DNA is 44.9 (Bd).

Types species: *Meniscus glaucopis* Irgens 1977, 42.

Further Descriptive Information

Cells of *Meniscus* appear as curved or straight rods. The curved rods often form ringlike shapes when the ends of a cell overlap prior to cell separation. Helical filamentous forms may also be seen.

Gas vacuoles (Figs. 3.6 and 3.7) may be observed by phase contrast microscopy and the individual gas vesicles are resolved by observing whole cells or thin sections in the transmission electron microscope.

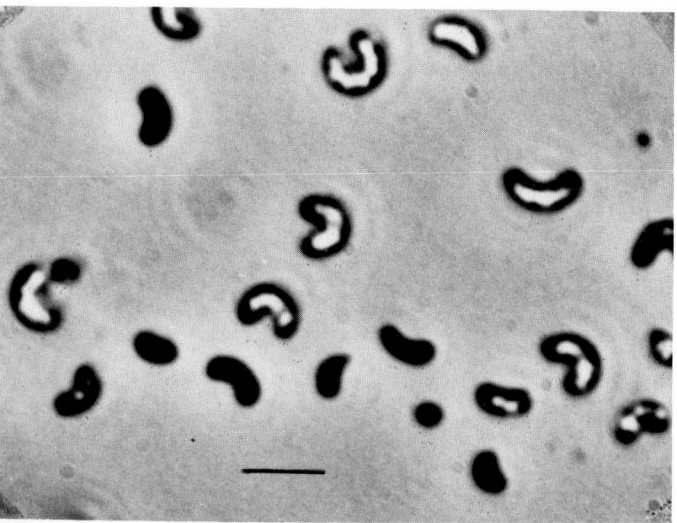

Figure 3.6. *Meniscus glaucopis* ATCC 29398. Phase-contrast. Note gas vacuoles. *Bar,* 2.0 μm.

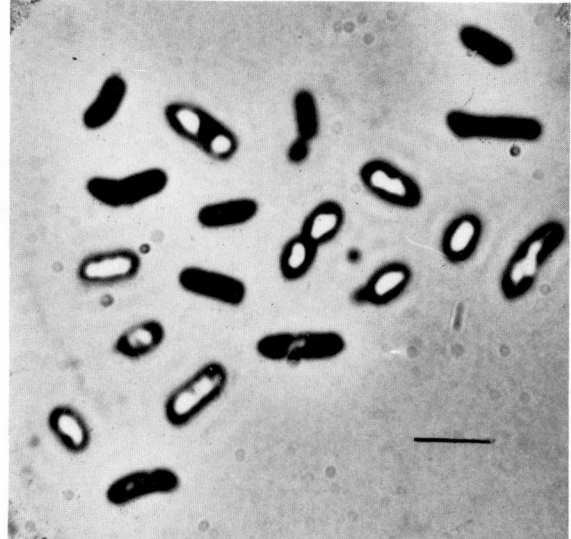

Figure 3.7. *Meniscus glaucopis,* strain R. Phase-contrast. Note gas vacuoles. *Bar,* 2.0 μm.

Colonies are circular, convex in elevation, with an entire margin and smooth, glistening surface. The colonies may appear translucent or opaque, chalky white. The larger the number of gas vacuoles within the cells of the colony, the whiter the colony. The consistency of the colonies is buttery when grown anaerobically and rubbery when grown aerobically.

When grown stationary in test tubes, the cells often buoy to the surface where they form a white band.

The cells apparently do not carry out respiration as aerobic and anaerobic cultures yield the same number of cells. Fermentation end products when grown on maltose are acetic, propionic, and succinic acids (Irgens, 1977).

Optimum growth occurs around 30°C at pH 7.0. Growth occurs at 15 and 35°C, but not at 10 or 40°C. Cobalamin (vitamin B_{12}), thiamin, and CO_2 are required for growth. Good growth in defined medium occurs with ammonium as the nitrogen source.

The following characteristics are negative: deaminase (casamino acids), urease, acetyl methyl carbinol, indole, H_2S production and nitrate reduction.

Do ferment agar (weakly), dextrin, melezitose, raffinose, cellobiose, sucrose, lactose, maltose, melibiose, trehalose, fructose, galactose, glucose, rhamnose (weakly), CH_3-α-D-glucoside, esculin, salicin, D-ribose, D-xylose, and arabinose.

Do not ferment mannose, sorbose, glycerol, lactate, mannitol, sorbitol, adonitol, dulcitol, inositol, or amino acids.

Do not hydrolyze starch, cellulose, DNA, gelatin, casein, pectin, inulin, gum arabic, tributyrin, chitin, xylan, or glycogen.

Isolated from anaerobic digester sludge, but probably also present in anaerobic hypolimnion of lakes.

Enrichment and Isolation Procedures

Members of this genus may be isolated on a complex medium (BGM) having the following composition (per liter): yeast extract, 1.0 g; KH_2PO_4, 0.5 g; NaCl, 0.4 g; NH_4Cl, 0.4 g; $CaCl_2 \cdot 2H_2O$, 0.01 g; sodium thioglycolate, 0.3 g; $MgSO_4 \cdot 7H_2O$, 0.2 g; $FeSO_4 \cdot 7H_2O$, 0.001 g; at least 1% CO_2; trace elements solution (TES), 1.0 ml. The TES, modified from Pfennig's formula (personal communication), contains (per liter): $ZnSO_4 \cdot 7H_2O$, 0.10 g; $MnCl \cdot 4H_2O$, 0.03 g; H_3BO_3, 0.3 g; $CoCl_2 \cdot 6H_2O$, 0.2 g; $CuCl_2 \cdot 2H_2O$, 0.01 g; $NiCl_2 \cdot 6H_2O$, 0.02 g; $Na_2MoO_4 \cdot 2H_2O$, 0.03 g; pH 3.0–4.0.

The pH of the medium is adjusted to 7.3 with 10% Na_2CO_3 before autoclaving, or the medium may be autoclaved with the pH unadjusted and then adjusted to about pH 7.0 after autoclaving by the addition of 2.0 ml of 5.0% $NaHCO_3$, sterilized by filtration, per 100 ml of medium. All solid media contain 1.0% $CaCO_3$ and 1.5% agar. Stock media contain 0.3% agar. Maltose, at a concentration of 0.2–0.5%, is used as the carbon source for growth studies and stock cultures. Pour plates are prepared using anaerobic digester sludge as the inoculum. The plates are incubated in a GasPak anaerobic jar (BBL) at 25–30°C. After 10 days chalky white colonies indicative of the presence of gas vacuoles may be observed.

Maintenance Procedures

Stock cultures are maintained in test tubes rendered anaerobic by the pyrogallic acid technique or they may be maintained on slants in cotton-stoppered test tubes in anaerobic jars. Lyophilized cultures may be stored indefinitely.

Procedures for Testing Special Characters

The presence of gas vacuoles may be demonstrated by the disappearance of the vacuoles upon the application of a sharp blow with a hammer to a wet mount of the cells. The cover slip is protected with a 3- to 4-mm-thick rubber pad. Phase-contrast microscopic examination of wet mounts made with dilute India ink is used to demonstrate capsules. The hydrolysis of glycogen, inulin, pectin, gum arabic, and dextrin is tested in BGM without maltose in liquid culture. Hydrolysis is considered positive if growth occurs and the pH drops.

Differentiation of the genus **Meniscus** from other genera

Two genera of gas-vacuolated curved rods that may be confused with *Meniscus* are "*Brachyarcus*" and *Microcyclus*. "*Brachyarcus*" has never been isolated and differs from *Meniscus* in having cells arranged in groups (coenobia) consisting of two, four or more rings (Skuja, 1964).

Microcyclus is differentiated from *Meniscus* as being a catalase-positive obligate aerobe rather than an aerotolerant anaerobe and by having a mol % G + C of 66–69 (van Ert and Staley, 1971).

Taxonomic Comments

Gas vacuoles, being present in both phototrophs and morphologically and physiologically unrelated heterotrophs, cannot be used as a unifying taxonomic characteristic. One must therefore consider the possible relationship of *Meniscus* to nonvacuolated vibrioid organisms such as *Aquaspirillum* and *Vibrio*, and possibly to other Gram-negative aerotolerant anaerobes such as *Zymomonas* and *Eikenella*. As the members of the genus *Aquaspirillum* are oxidase- and catalase-positive and incapable of fermenting carbohydrates, *Meniscus* must be excluded from this genus. The members of the genus *Vibrio*, although facultative anaerobes, are all catalase-positive and most are oxidase-positive, whereas *Meniscus* is catalase- and oxidase-negative. *Zymomonas*, an oxidase-negative and catalase-positive aerotolerant bacterium with fermentative metabolism produces mainly ethanol and CO_2 as end products, whereas *Meniscus* produces all acid end products. *Eikenella*, on the other hand, is oxidase-positive and catalase-negative. Based on the above facts, *Meniscus* was accorded separate generic status.

Meniscus glaucopis includes both a vibrioid strain and a straight rod strain. This is justified by the fact that all metabolic and physiological characteristics tested are identical for the two strains. These strains are merely morphovars of the same species.

List of species of the genus **Meniscus**

1. **Meniscus glaucopis** Irgens 1977, 42.[AL]
glau.co'pis. Gr. adj. *glaucopis* gleaming-eyed.
The description of the species is as given for the genus.

Isolated from anaerobic digester sludge.
The mol% G + C of the DNA is 44.9 (Bd).
Type strain: ATCC 29398.

Genus "**Brachyarcus**" Skuja 1964, 19

PETER HIRSCH

Bra.chy.ar'cus. Gr. adj. *brachys* short; L. n. *arcus* a bow; M.L. masc. n. *Brachyarcus* a short bow.

Rod-shaped cells bent like a bow, 1.0 × 1.5 wide and to 2.5 μm long, colorless, usually with several cylindrical **gas vesicles** which may appear reddish if viewed with a phase-contrast microscope. The cells may occasionally contain some minute **sulfur granules.** Cells are arranged in groups (coenobia) as a result of polar growth and median cross-division combined with tight attachment to a surface by means of a mucoid substance. Occasionally cells are merely embedded in the polymer and are thus free floating as a flat coenobium. After division, **cells form coenobia of two (rings) or four (clover-leaf appearance)** or more cells. Delayed cell division results in the formation of **pretzel-shaped cells.** Division in the coenobia may or may not be synchronous. Coenobia of 2–10 cells may measure 3–6 μm in diameter; secondary families of irregularly humped agglomerates of up to 100 μm or more in size can be found. Often the mucoid capsule is thin and not distinctly delineated. The Gram reaction has not been reported. Nonmotile. Microaerophilic or **probably anaerobic.** Has not been obtained in pure culture. Originally found in Lake Vuolep Njakajaure, Swedish Lappland, in April, at a depth of 12–13.5 m. The mol% G + C of the DNA is not known.

Type species: "*Brachyarcus thiophilus*" Skuja 1964, 20.

Further Descriptive Information

"*Brachyarcus*" strains have not been grown in laboratory culture. Hirsch (1981) collected observations of "*Brachyarcus*" occurrence in various lakes and a pond. These arc-shaped bacteria were found between April and October at depths of 0.25 m (pond) to 27 m (Lake Plußsee, Germany). The temperature ranged from 2–20°C but was usually below 16.5°C. The pH ranged from 7.12–7.8. Sulfide was always present; oxygen was always absent.

Table 3.6.

Differential characteristics of the genus "**Brachyarcus**" *and other morphologically similar genera*[a]

Characteristics	"Brachyarcus"	Microcyclus	Chlorobium	Rhodocyclus	"Renobacter"
Cells colorless	+	+	−	−	+
Cells with photosynthetic pigments	−	−	+	+	−
Gas vesicles present	+	D	−	−	+
Cells arranged symmetrically in a plane in multiples of two	+	−	−	−	−
Intracellular sulfur globules present	+[b]	−	−[c]	−	−
Cells before division often 3-shaped or pretzel-shaped	+	+	−	−	−
Cells before division S-shaped	−	−	−	−	+
Anaerobic	+[d]	−	+	+	−
Psychrophilic	+	−	−	−	−

[a] Symbols: see standard definitions.
[b] Sulfur globules were only seen by Skuja (1964).
[c] External sulfur globules may be present.
[d] Probably anaerobic, based on the habitat.

Differentiation of the genus "**Brachyarcus**" from other genera

Table 3.6 indicates the characteristics that differentiate "*Brachyarcus*" from other morphologically similar genera.

Taxonomic Comments

Until "*Brachyarcus*" can be cultivated in the laboratory and pure cultures obtained, it will continue to be difficult to determine the proper taxonomic placement of the genus.

The generic name and species name were not included in the Approved Lists of Bacterial Names in 1980 and presently have no standing in nomenclature.

List of the species of the genus **"Brachyarcus"**

1. **"Brachyarcus thiophilus"** Skuja 1964, 20.
thi.o'phi.lus. Gr. n. *thium* sulfur; Gr. adj. *philus* loving; M.L. masc. adj. *thiophilus* sulfur-loving.

The characteristics are as described for the genus. The mol% G + C of the DNA is not known.
Type strain: none isolated.

Genus **"Pelosigma"** *Lauterborn 1913, 100*

PETER HIRSCH

Pe.lo.sig'ma. Gr. adj. *pelos* dark-colored, hence anaerobic mud; Gr. n. *sigma* the letter S; M.L. neut. n. *Pelosigma* S-shaped mud bacterium.

S-shaped, slender filaments, 0.23–0.35 μm wide × 9–30 μm long, colorless or pale grey, usually with a slight spiral twist. **Generally arranged side by side in sigmoid aggregates of four or multiples of four;** multiple aggregates of considerable thickness have been observed. The cells may be held together throughout their length by a mucoid substance, or only at one end forming a point, the other end being wider and spread out to reveal the individual filaments. Multiplication is presumably by synchronous cross-division of the aggregate and sudden separation of the daughter aggregates. The Gram reaction has not been recorded. **The aggregate may be motile** by an organelle visible by light microscopy and located at the pointed end. This organelle is interpreted by the author as a tuft of flagella. "*Pelosigma*" species have not yet been obtained in pure culture. **They occur in and on mud in fresh and brackish waters.** The mol% G + C of the DNA is not known.

Type species: "*Pelosigma cohnii*" (Warming 1875) Lauterborn 1913, 100.

Further Descriptive Information

To date, "*Pelosigma*" species have never been cultivated in the laboratory. Recently, Hirsch (1981) has listed several locations where "*Pelosigma*" has been sighted and has given some environmental parameters for its occurrence in lakes and ponds. "*Pelosigma*" species were found from May to August at depths ranging from 0.7 m (a forest pond in Michigan) to 31 m (Lake Plußsee, Germany). The temperature of these habitats varied from 4–19°C, but generally was below 11°C. The pH varied from 7.25–7.9. Oxygen was always absent; sulfide was always present.

The lengths of the aggregates ranged in all but one instance from 18.7–20.3 μm; the widths were 5.1 or 10.2 μm. There were usually four or multiples of four cells per aggregate, and the width of the individual cells was 0.23–0.31 μm.

Differentiation of the genus **"Pelosigma"** *from other genera*

Table 3.7 presents the characteristics of "*Pelosigma*" that distinguish it from other genera of morphological similarity. The distinct form of the aggregates and their exclusive occurrence in anaerobic waters allow an easy recognition.

Taxonomic Comments

Taxonomic placement of this morphological genus is presently difficult and will continue to be so until the organisms can be cultivated

in the laboratory and studied in pure culture. The question of species separation may also have to be reinvestigated, since the present basis of differentiation is entirely morphological and ecological.

The generic name and species names were not included in the Approved Lists of Bacterial Names in 1980 and presently have no standing in nomenclature.

Differentiation of the species of the genus **"Pelosigma"**

Table 3.8 presents the characteristics differentiating the two species of "*Pelosigma*."

List of the species of the genus **"Pelosigma"**

1. **"Pelosigma cohnii"** (Warming 1875) Lauterborn 1913, 100. (*Spiromonas cohnii* Warming 1875, 370.)
cohn'i.i. M.L. gen. n. *cohnii* of Cohn; named after F. Cohn, a German microbiologist.

The cell morphology is as described for the genus. The cells are wound around an imaginary axis in an S-curve with 1¼ turns. They are arranged in aggregates ranging in width from 1.2–4.0 μm and in length from 9–20 μm.

Originally found in actively decaying mud ("Faulschlamm") of brackish water habitats near Kalundborg, Vejle, Hofmansgave, and from the Limfjord, Denmark.

The mol% G + C of the DNA is unknown.
Type strain: none isolated.

2. **"Pelosigma palustre"** Lauterborn 1915, 418.
pa.lus'tre. L. neut. adj. *palustre* marshy, swampy.

The cell morphology is as described for the genus. Aggregates are flat and bandlike, 8–10 μm wide and 20–25 μm in length; one end is pointed, the other is spread out like a fan. The pointed end has a flagellum or flagella (a composite structure). Movements are slow, trembly, and connected with a rotation around the longitudinal axis; the aggregate may remain motionless for long periods.

Originally found in "Faulschlamm" (decaying mud) of ponds with *Chara* spp. near Ludwigshafen, Germany.

The mol% G + C of the DNA is unknown.
Type strain: none isolated.

Table 3.7.

Differential characteristics of the genus "Pelosigma" and other morphologically or physiologically similar genera[a]

Characteristics	*"Pelosigma"*	*Peloploca*	*Saprospira*
Cells aggregated in sigmoid bundles of four or multiples of four	+	+[b]	−
Cells S-shaped and of constant form	+	+	−
Cells usually aggregating with one end, the other end spreading out like a fan	+	−	−
Cell aggregates occasionally motile	+	−	+[c]
Cells with gas vesicles	−	+	−
Found in anaerobic habitats	+	+	−

[a] The genera *Peloploca* and *Saprospira* are described more fully in Volume 3 of the *Manual*. For symbols, see standard definitions.
[b] *Peloploca* spp. usually form flat ribbons. There may be more than four filaments in a *Peloploca* aggregate.
[c] Motility is by gliding.

Table 3.8.

Characteristics differentiating "Pelosigma cohnii" and "Pelosigma palustre"[a]

	1. "P. cohnii"	2. "P. palustre'
Aggregates helically curved	+	−
Aggregates flat and band-like	−	+
Aggregate width, μm	1.2–4.0	8–10
Aggregate length, μm	9–20	20–25
Occurrence in brackish water	+	−
Found in freshwater	−	+

[a] For symbols, see standard definitions.

Important Notes for Users of this Edition

1. Always read both generic and species descriptions because characters listed in the generic description are not usually listed in the species descriptions.

2. Unless otherwise indicated in footnotes to tables, the meanings of symbols are as follows:
- \+ 90% or more of strains are positive
- − 90% or more of strains are negative
- d 11–89% of strains are positive
- v strain instability (*not* equivalent to "d")
- D different reactions in different taxa (species of a genus or genera of a family)

3. All other symbols are defined in footnotes to tables.

SECTION 4

Gram-Negative Aerobic Rods and Cocci

Table 4.1.

Some differential characteristics of the various families of Section 4[a]

Characteristics	Family I. PSEUDOMONADACEAE, p. 141	Family II. AZOTOBACTERACEAE, p. 219	Family III. RHIZOBIACEAE, p. 234	Family IV. METHYLOCOCCACEAE, p. 256	Family V. HALOBACTERIACEAE, p. 261	Family VI. ACETOBACTERACEAE, p. 267	Family VII. LEGIONELLACEAE, p. 279	Family VIII. NEISSERIACEAE, p. 288
Morphology:								
Rods	+	+	+	D	D	+	+	D
Cocci	−	−	−	D	D	−	−	D
Cell diameter, 2.0 μm or greater	−	+	−	−	D	−	−	−
Muramic acid-containing peptidoglycan present	+	+	+	+	−	+	+	+
Nitrogenase activity present	−	+	D	D	−	−	−	−
Facultative hydrogen autotrophy	D	−	D	−	−	−	−	−
Methane used as a sole carbon source	−	−	−	+	−	−	−	−
Need at least 12–15% NaCl for growth	−	−	−	−	+	−	−	−
Cysts (*sensu stricto*) are formed	−	D	−	D	−	−	−	−
Parasites of warm-blooded hosts	−[b]	−	−	−	−	−	−	D
Pathogenic for man	D	−	−	−	−	−	+	D
Ethanol oxidized to acetic acid in neutral or acid (pH 4.5) media	−	−	−	−	−	+	−	−
Motility (swimming)	+	D	D	D	D	D	+	−
Flagellar arrangement:								
Polar	+	D	D	+	+	D	+	
Lateral	−[c]	D	D	−	−	D	+	
Plant pathogenicity	D[d]	−	D[e]	−	−	D[f]	−	−
N₂ fixed under atmospheric pO₂	−	+	−	−	−	−	−	−
N₂ fixed in root nodules of plants	−	−	D	−	−	−	−	−
Oxidase test	D	D			+	−	D	D
Catalase test	+	+			+	+	+	D
Some sugars can be utilized	+[g]	+	+		D	+	−	D
Cysteine + iron salts required for growth	−	−	−	−	−	−	+	−
Zoogloeae (flocs with dendritic outgrowths in broth) are formed	D	−	−	−	−	−	−	−
Marine organisms; required seawater-based media	D	−	−	−	−	−	−	−
Yellow colonies are formed	D	D	−	−	−	−	−	D
Water-soluble fluorescent pigment formed	D	D	−	−	−	−	−	−
Mol% G + C of DNA	58–71	52–68	57–66			51–65	39–43	38–55

[a] Symbols: see standard definitions.

[b] *Pseudomonas mallei* is an exception.

[c] A few species may also form lateral flagella in addition to polar flagella when cultured on a solid medium.

[d] Leaf stripe, leaf spot, wilt, necrosis, and similar diseases.

[e] Gall hypertrophies on stems and roots of plants; nodules incited on leaves of plants.

[f] Pink disease in pineapple fruit and rot in apples and pears.

[g] There are some rare exceptions.

FAMILY I. **PSEUDOMONADACEAE** WINSLOW, BROADHURST, BUCHANAN KRUMWIEDE, ROGERS AND SMITH 1917, 555[AL*]

NORBERTO J. PALLERONI

Pseu.do.mo.na.da′ce.ae. M.L. fem. n. *Pseudomonas* type genus of family; *-aceae* ending to denote family; M.L. fem. pl. n. *Pseudomonadaceae* the *Pseudomonas* family.

Straight or curved Gram-negative rods, motile by polar flagella. Chemoorganotrophs. Strictly aerobic. The metabolism is respiratory, never fermentative. Growth occurs from 4°C or lower to 43°C. Chemoorganotrophic, able to use other than one-carbon organic compounds as sole carbon and energy sources. Catalase-positive; usually oxidase-positive. The mol% G + C of the DNA ranges from 58–71.

Type genus: *Pseudomonas* Migula 1894, 237.

Four genera, *Pseudomonas, Xanthomonas, Frateuria* and *Zoogloea*, are included in the family. A few selected characteristics permitting their differentiation are given in Table 4.2.

Table 4.2.

Differential characteristics of the four genera of **Pseudomonadaceae**[a]

Characteristics	I. *Pseudo-monas*, p. 141	II. *Xantho-monas*, p. 199	III. *Fra-teuria*, p. 210	IV. *Zoo-gloea*, p. 214
Requirement for growth factors	−	+	−	+
Growth at pH 3.6	−	−	+	−
Formation of flocs with dendritic out-growths	−	−	−	+
Production of xantho-monadins	−	+[b]	−	−
Plant pathogenicity	D	+	−[c]	−

[a] For symbols see standard definitions.

[b] One species is negative.

[c] Browning and rot are caused when injected in apples or pears.

Comments on the Circumscription of the Family

In the past many bacterial genera were included in the family *Pseudomonadaceae*. For example, Kluyver and van Niel (1936) proposed the inclusion of no less than 25 genera grouped into three tribes. The hypertrophy was relieved from time to time by the application of criteria whose phylogenetic meaning was difficult to evaluate, but in recent years the reduction in the number of genera has become more substantial as our knowledge of the natural relationships among the various taxa has become more precise. In the eighth edition of *Bergey's Manual* the number of genera was reduced to four, as in the present treatment, but one of the four genera, *Gluconobacter*, has now been assigned to the family *Acetobacteraceae*, and its place has been taken by the recently created genus *Frateuria*.

The type genus of the family is *Pseudomonas*, one of the most complex groups of Gram-negative bacteria, with phenotypic similarities to many other genera. Members of this genus are characterized by their ability to grow in simple media of neutral reaction at the expense of a great variety of simple organic compounds. The oxidase reaction is usually positive, and the mol% G + C range is 58–70. The strains of *Xanthomonas* are plant pathogens. They are nutritionally more fastidious than *Pseudomonas*, they produce characteristic yellow cellular pigments (xanthomonadins), and they give a weak or negative oxidase reaction. Their G + C range is 63–71 mol%. Whereas *Pseudomonas* and *Xanthomonas* strains are not acid tolerant, *Frateuria* strains can grow at pH 3.6. The G + C range for this genus is 62–64 mol%. Finally, the genus *Zoogloea* produces dendritic slimy masses attached to solid surfaces under natural conditions and is capable of floc formation under certain laboratory conditions. The mol% G + C of the DNA of *Zoogloea* is 65.

Natural relationships among the genera of *Pseudomonadaceae* were reported for *Xanthomonas* and *Pseudomonas* by Palleroni et al. (1973) on the basis of rRNA/DNA hybridization data. More recently, Swings et al. (1980) have demonstrated by similar techniques a closer relationship between *Frateuria* and *Xanthomonas* than between *Frateuria* and *Gluconobacter* and *Acetobacter*, even though these last three genera have many phenotypic characters in common. The genus *Frateuria* can be differentiated from *Acetobacter* and *Gluconobacter* by H_2S formation, growth in 30% glucose and on D-mannose, L-arabinose, D-lyxose and L-lyxose as carbon sources (see Table 4.41 under *Frateuria*). Other characters of *Frateuria* usually not possessed by the other two genera include the formation of a yellow-brown, soluble pigment on glucose-yeast extract-$CaCO_3$ medium, the formation of 2,5-diketogluconate from glucose, and the ability to grow in Frateur's Hoyer mannitol medium. Some of the strains of *Gluconobacter* and/or *Acetobacter* also exhibit these properties. With regard to *Zoogloea*, no attempt at determining its phylogenetic relationships to other genera of the family has been reported so far, and inclusion of the genus in the family is only tentative (Palleroni, 1981).

In summary, the family *Pseudomonadaceae* presently includes three genera possessing the basic properties of aerobic pseudomonads. This, however, may not represent the most satisfactory arrangement from a phylogenetic viewpoint. Some of the species included in these genera may be more closely related to members of other families than to other species of *Pseudomonadaceae* (De Smedt et al., 1980). A fourth genus, *Zoogloea*, is placed in the family on a tentative basis. It must be admitted that, from a practical determinative standpoint, a clear-cut circumscription of the family by classical phenotypic criteria has become more difficult than ever before, and differentiation of the genera may require specialized techniques which are still beyond the reach of many laboratories involved in problems of determinative bacteriology.

* *AL*, denotes the inclusion of this name on the Approved Lists of Bacterial Names (1980).

Genus I. **Pseudomonas** *Migula 1894, 237*[AL] (Nom. cons. Opin. 5, Jud. Comm. 1952, 237)

NORBERTO J. PALLERONI

Pseu.do′mo.nas or Pseu.do.mo′nas. Gr. adj. *pseudes* false; Gr. n. *monas* a unit, monad; M.L. fem. n. *Pseudomonas* false monad.

Straight or slightly curved rods but not helical, 0.5–1.0 µm in diameter by 1.5–5.0 µm in length. Many species accumulate poly-β-hydroxybutyrate as carbon reserve material, which appears as sudanophilic inclusions. Do not produce prosthecae and are not surrounded by sheaths. No resting stages are known. Gram-negative. **Motile by one or several polar flagella;** rarely nonmotile. In some species lateral flagella of shorter wavelength may also be formed. **Aerobic, having a strictly respiratory type of metabolism with oxygen as the terminal electron acceptor; in some cases nitrate can be used as an alternate electron acceptor,** allowing growth to occur anaerobically. Xanthomonadins are not produced. Most, if not all, species fail to grow under acid conditions (pH 4.5). Most species do not require organic growth factors. Oxidase-positive or -negative. Catalase-positive. Chemoorganotrophic; **some species are facultative chemolithotrophs, able to use H₂ or CO as energy sources**. Widely distributed in nature. Some species are pathogenic for humans, animals or plants. The mol% G + C of the DNA is 58–70 (Bd).

Type species: *Pseudomonas aeruginosa* (Schroeter 1872) Migula 1900, 884.

Cell morphology and fine structure. The cells of *Pseudomonas* strains occasionally differ substantially in size and shape from the general definition. In some species ("*P. ovalis*," a synonym of *P. putida*) the cells are oval, and some plant pathogenic species have cells exceeding 4 μm in length.

Many species accumulate large amounts of poly-β-hydroxybutyrate (PHB) when the cells are grown under conditions of nitrogen limitation. In cases where PHB accumulation is very low (e.g. in *P. pseudoalcaligenes*), extraction and hydrolysis by specific enzymes (depolymerases) may be necessary for its detection.

Some species (*P. vesicularis, P. pseudoflava*) accumulate a glucose polysaccharide which resembles glycogen (Ballard et al., 1968; Auling et al., 1978). Because glycogen is less conspicuous than PHB, its demonstration requires isolation and chemical characterization, and no extensive survey of its occurrence in the genus has yet been made.

By electron microscopy, *Pseudomonas* species have cell walls and membranes typical of Gram-negative bacteria. In freeze-etched preparations, as many as nine layers can be defined, accounting for all of the electron-dense and electron-transparent layers that can be seen in thin sections (Lickfield et al., 1972; Gilleland et al., 1973). Gilleland et al. (1973) showed that the outer membrane of the cell wall could be split down the center when the cells were freeze-etched in the presence of a cryoprotective agent, thus confirming an earlier interpretation by Lickfield et al. (1972) of their own freeze-etch studies. The outer layer thus separated from the outer membrane had in its inner face numerous spherical units and small rods composed of the spherical units. These elements could be removed by previous exposure of the cells to ethylenediaminetetraacetate (EDTA), resulting in osmotically fragile cells. The process could be reversed by addition of magnesium ions; the spherical units were able to reaggregate and osmotic stability was restored. The spherical units were shown to be protein.

Four components of the cell envelopes of *P. aeruginosa* could be defined by Hancock and Nikaido (1978) after sucrose-gradient centrifugation of preparations in the absence of EDTA. The two densest bands were outer membrane components, differing in their phospholipid content. The lightest of the four components was mainly inner (or cytoplasmic) membrane material. The densest band very likely corresponded to the spherical units in the concave surface of the outer membrane observed by Gilleland et al. (1973). Vesicles prepared with the addition of either one of the two heaviest components of the outer membrane were able to retain saccharides of molecular weight >9000, suggesting an exclusion limit much larger than the figure obtained for *Escherichia coli* and *Salmonella typhimurium* (550–650). The much larger pores of the *P. aeruginosa* outer membrane may favor the utilization of large peptides and hydrophobic compounds.

Mesosome-like structures have been detected in cells of *P. aeruginosa* and *P. saccharophila* (Carrick and Berk, 1971; Hoffman et al., 1973; Young et al., 1972).

The microtubular structures known as rhapidosomes occur in a number of *Pseudomonas* species (Yamamoto, 1967). The biological meaning of the rhapidosomes is at present obscure. Evidence favoring the identity of rhapidosomes with the polymerized sheaths of bacteriocins in *P. fluorescens* has been presented by Amako et al. (1970). Such a relationship is not clear in the case of *P. aeruginosa* rhapidosomes, and Baechler and Berk (1972) suggested that the structures may involve their origin in rearrangement of membranes and mesosomes.

Flagella and pili. Typically, *Pseudomonas* cells have polar flagella (Figs. 4.1 and 4.2), although the flagellar insertion in some cases is subpolar. In practice it is often difficult to differentiate the subpolar insertion from the so-called degenerately peritrichous type observed in

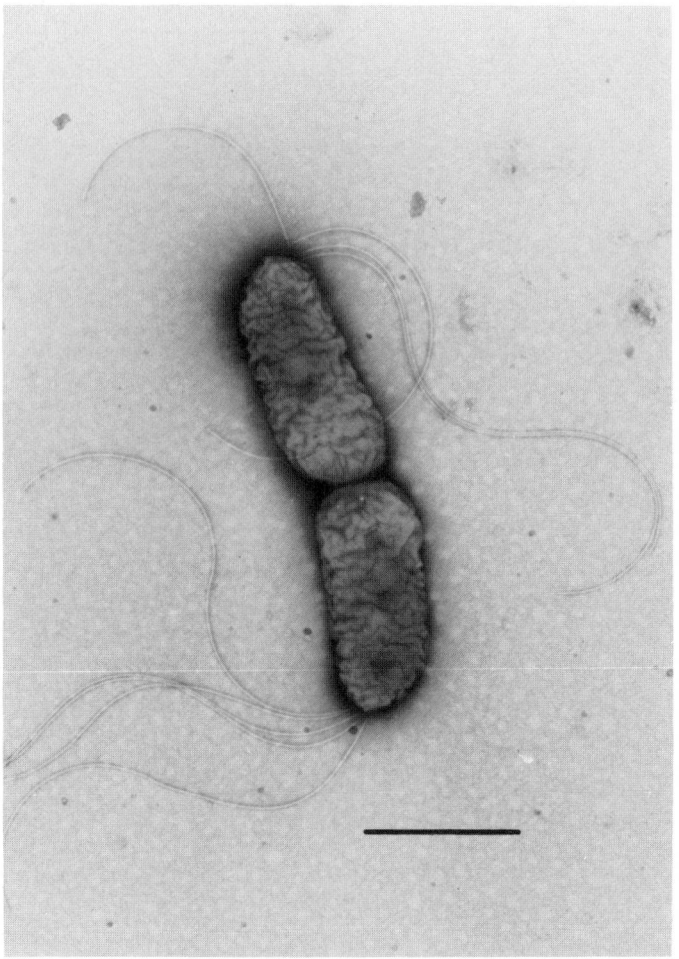

Figure 4.1. *P. rubrisubalbicans* with several polar flagella. Negative staining, 1% phosphotungstate in bovine serum albumin. *Bar*, 1 μm. (Courtesy of Dr. A. C. Hayward.)

members of other genera when using current staining methods. In addition, lateral flagella of short wavelength may be produced by some species (*P. stutzeri, P. mendocina*), and they are shed much more easily than the polar ones. Growth on solid media favors the formation of these lateral flagella, which suggests that perhaps they may be involved in swarming of the population on solid surfaces (Shinoda and Okamoto, 1977). In *P. testosteroni* both polar and lateral flagella are formed when the cells grow at low temperatures (H. Lautrop, personal communication; Davis, 1967). The capacity for production of lateral flagella throws some doubts on the validity of the mode of flagellar insertion as an indicator of evolutionary relationships among bacterial groups and, in fact, some of the species presently assigned to *Pseudomonas* appear to be less closely related to one another than they are to some species of other genera having peritrichous flagella.

At least one species, *P. andropogonis* (syn. "*P. stizolobii*") (Goto and Starr, 1971; Hayward, 1972) is characterized by the production of sheathed flagella, i.e. flagella surrounded by a membrane that is a continuation of the cell wall (Fig. 4.2) (Fuerst and Hayward, 1969b).

Nonmotile strains of various species are occasionally isolated from nature, but the genus includes one species (*P. mallei*) whose members permanently lack flagella. Inclusion of this species in the genus has been decided only after the demonstration of a close relationship to normally flagellated species (Redfearn et al., 1966), but for other nonmotile species such a demonstration is lacking and they are excluded from this treatment.

The number of polar flagella is an important taxonomic character,

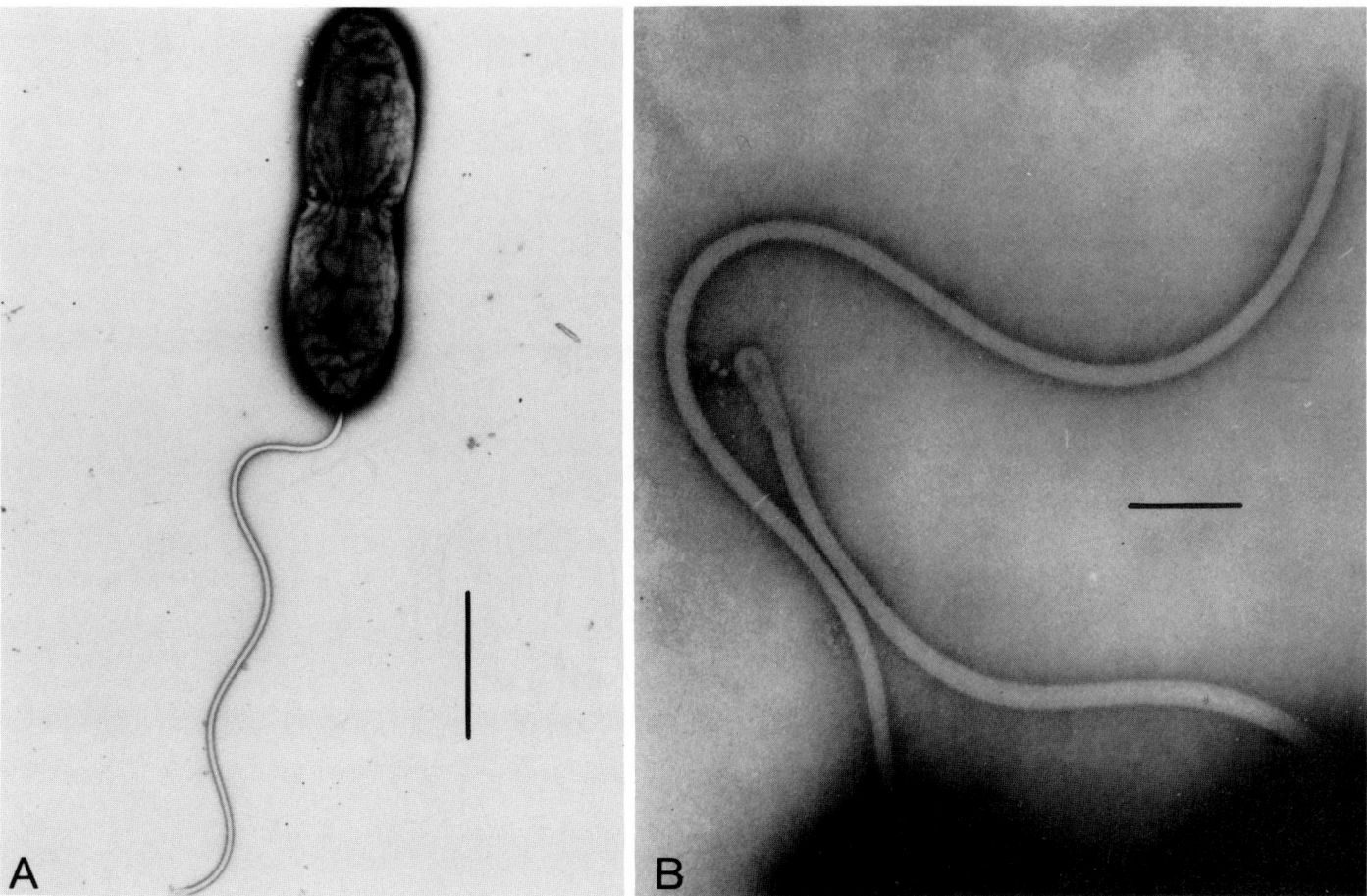

Figure 4.2. *A, P. andropogonis* (syn. *P. stizolobii*), with single sheathed polar flagellum. Negative staining, 1% uranyl acetate, 0.4% sucrose. *Bar*, 1 μm. *B*, Sheathed flagella of *P. andropogonis*. Same treatment as in *A*. *Bar*, 0.2 μm. (Courtesy of Dr. J. A. Fuerst.)

provided that its determination is performed under well controlled conditions and the results are expressed on a statistical basis (Lautrop and Jessen, 1964). Unfortunately, these recommendations have not been followed in the description of most species.

Pili or fimbriae have been reported for a number of *Pseudomonas* species by Fuerst and Hayward (1969a). Their insertion can be polar (*P. aeruginosa*, *P. acidovorans*, *P. testosteroni*, *P. maltophilia*, *P. alcaligenes*, *P. solanacearum*) or peritrichous (*P. cepacia*, *P. fragi*). Auling et al. (1978) described pili 6–9 nm in diameter and 4–5 times the length of the cells in the species *P. pseudoflava*. No fimbriae have been observed in strains of *P. fluorescens*, *P. aureofaciens*, *P. chlororaphis*, *P. putida*, *P. oleovorans*, and *P. andropogonis* (syn. "*P. stizolobii*").

The polar pili of *P. aeruginosa* are ~6 nm wide, thinner than those of *Enterobacteriaceae* (Fig. 4.3). They act as receptors for various phages and are retractile (Bradley, 1972a, b), but not all the pili of this species have the property of withdrawing into the cell (Bradley, 1974). The retractile polar pili have chromosomal determinants which can be mobilized by FP plasmids (Bradley, 1980a). Interestingly, the amino acid sequences of pili seem to have been conserved through evolution, which is shown by the fact that the amino terminal sequence of *P. aeruginosa* pili has marked homology with that of *Neisseria meningitidis* and *Moraxella nonliquefaciens* (Buchanan and Pearce, 1979).

Pili are found in many Gram-negative pathogenic organisms including *P. aeruginosa*, helping in the attachment of the pathogen to cell surfaces and affecting phagocytosis (Buchanan and Pearce, 1979). The pili of *P. rhodos* and *P. echinoides* are reported to be involved in the formation of rosettes or star-shaped aggregates characteristic of these species (Heumann, 1962).

Active swimming motility due to the presence of polar flagella is one of the characteristics of most *Pseudomonas* species. However, a peculiar type of translocation on solid surfaces known as "twitching motility" was reported initially by Lautrop (1965) for strains of *P. aeruginosa* lacking flagella, and was confirmed by Henrichsen (1972) who also observed it in various fluorescent pseudomonads which also lacked flagella. Twitching motility has not been observed in *P. mallei*, a species whose cells are devoid of flagella. A correlation between twitching translocation and the presence of retractile polar pili was reported for *P. aeruginosa* strain PAO by Bradley (1980b).

Composition of cell envelope. Cell walls of *Pseudomonas* species are very similar in structure to those of other Gram-negative bacteria, but important differences can be found in the chemical composition. Chemical differences are also apparent among the *Pseudomonas* species and even among the strains within a species.

The species most studied has been *P. aeruginosa*. Cells of this species are lysed rapidly by EDTA unless osmotic protection is provided in the medium. EDTA treatment results in the release of proteins from complexes present in the outer membrane, and also in the release of lipopolysaccharide (LPS) and loosely bound lipid. After loss of these components, the mucopeptide seems unable to maintain cell integrity (Wilkinson, 1970). The murein layer of *P. aeruginosa* is only loosely connected to the outer membrane and, in view of recent studies of a mutant of *E. coli* (Hirota et al., 1977), this might be related to the high EDTA sensitivity. Also, Wilkinson (1970) has found a correlation between EDTA sensitivity and a high phosphorous content in the outer membrane (as occurs in *P. aeruginosa* and *P. alcaligenes*). Much of this phosphorous is acid labile and its metal-binding capacity may have the

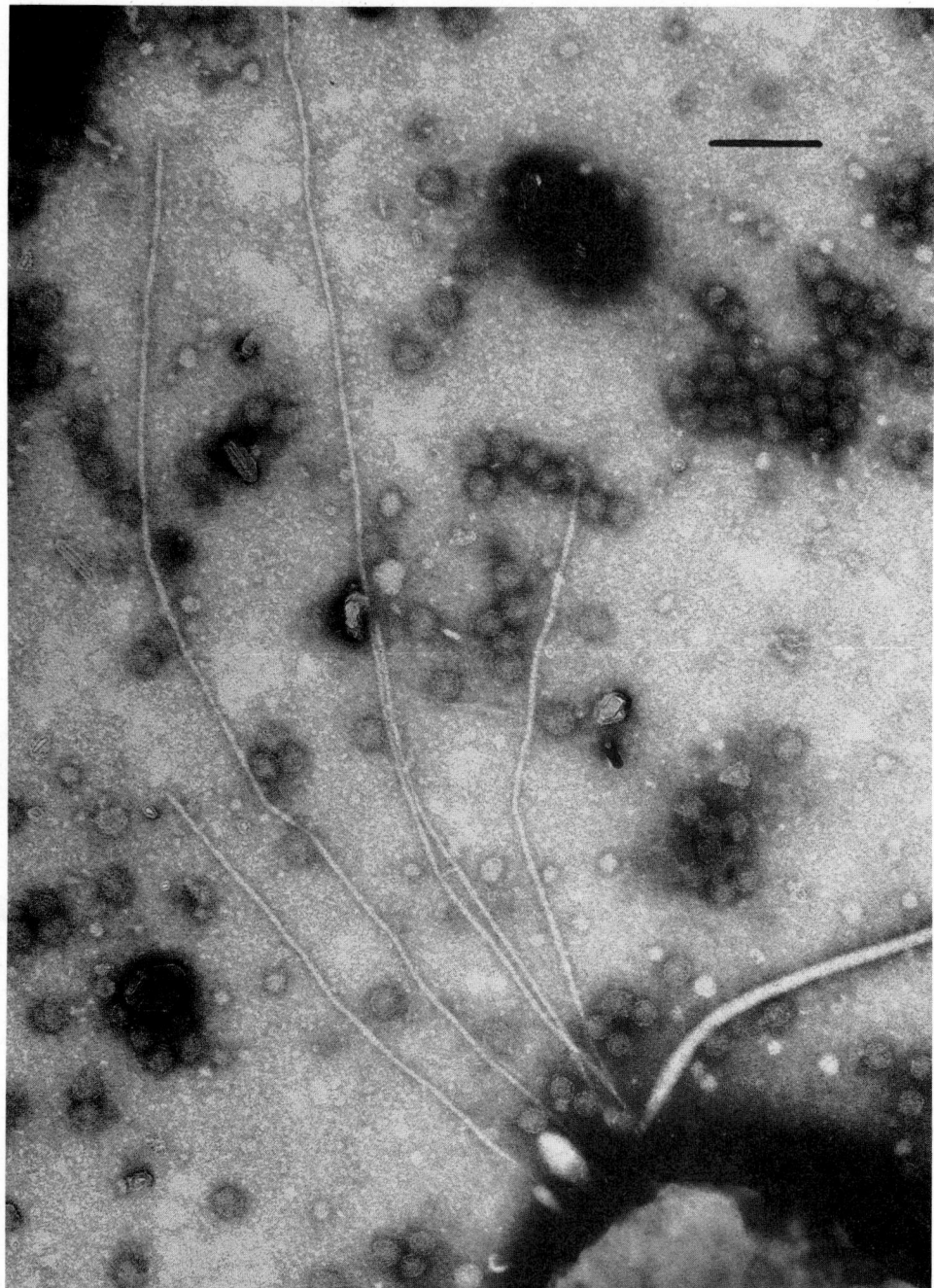

Figure 4.3. *Pseudomonas aeruginosa* strain PAC5 polar pili, which have been prevented from retracting by the adsorption of a pilus-specific bacteriophage; this can be seen scattered over the field. *Bar*, 0.1 μm. (Reproduced with permission from D. E. Bradley, Canadian Journal of Microbiology *26:* 155–160, 1980, © National Research Council of Canada.)

function of linking the LPS to other wall components (Wilkinson et al., 1973). Less EDTA-sensitive species (containing less phosphorus) include *P. aureofaciens, P. chlororaphis,* "*P. denitrificans.*" *P. fluorescens, P. fragi, P. mucidolens,* "*P. ovalis,*" *P. putida, P. stutzeri,* "*P. syncyanea,*" *P. synxantha,* and *P. taetrolens* (Wilkinson, 1970). Of these species, those that have been thoroughly characterized belong to the RNA homology group I. Since this homology group also includes the sensitive species *P. aeruginosa* and *P. alcaligenes,* two subgroups can thus be differentiated on the basis of EDTA sensitivity. *P. diminuta* and *P. maltophilia,* which are EDTA resistant, are only distantly related to group I.

Little is known about rough colony mutants defective in the production of polysaccharide in Gram-negative bacteria outside of the *Enterobacteriaceae*; however, in a single strain of *P. aeruginosa* Koval and Meadow (1977) have found four different types of LPS, one of which consists of the core polysaccharide joined to short antigenic chains, as in the semirough mutants of *Enterobacteriaceae*. Some rough *P. aeruginosa* strains have high molecular weight LPS, while other mutants defective in LPS composition have a typical smooth colony appearance. No clear-cut correlation is evident between LPS composition and resistance to carbenicillin and bacteriocins.

Evidence for the sequential incorporation of sugars from the nucleotide derivatives has been derived from the study of some defective mutants (Koval and Meadow, 1977) and from the patterns of R-type pyocins and phage sensitivity (Meadow and Wells, 1978). The core of the LPS of *P. aeruginosa* contains glucose, rhamnose, galactosamine, heptose, 2-keto-3-deoxyoctonate (KDO) and alanine (Meadow, 1975). The synthesis includes the sequential addition of glucose followed by incorporation of rhamnose (Asonganyi and Meadow, 1980).

The analytical data of Moss et al. (1973) on the lipids of whole cells of *P. maltophilia*, which reflect the fatty acid composition of lipid A, exhibit a marked similarity to the data reported by Rietschel et al. (1975) for several *Xanthomonas* species. This is in accord with the taxonomic relationship between *P. maltophilia* and *Xanthomonas* species as shown by rRNA/DNA hybridization (Palleroni et al., 1973). The major fatty acids in the LPS of *P. maltophilia* include 9-methyldecanoic, 2-hydroxy-9-methyldecanoic, 3-hydroxy-9-methyldecanoic, 3-hydroxydodecanoic, and 3-hydroxy-11-methyldecanoic (S. Wilkinson, personal communication). Both *P. maltophilia* and *Xanthomonas* have a lipid A based on glucosamine (Hase and Rietschel, 1976). Similarities in the LPS composition also include the presence of a high content of rhamnose, the presence of D-galacturonic acid, the absence of a heptose, and a low content of KDO (Neal and Wilkinson, 1979). An additional interesting feature of the LPS of *P. maltophilia* is the presence of a pentose derivative that has been identified as 3-*O*-methyl-L-xylose, thus far not found in other bacteria with the exception of *Rhodopseudomonas viridis* (Weckesser et al., 1974).

Lipid A and the inner core of LPS are rather conservative structures in Gram-negative bacteria (Wilkinson and Taylor, 1978). The structure of lipid A in pseudomonads, as in the case of enteric bacteria, usually contains a phosphorylated disaccharide of glucosamine and fatty acids (Rietschel et al., 1977), but in the case of *P. diminuta* and perhaps also in *P. vesicularis*, glucosamine is replaced by 2,3-diamino-2,3,-dideoxy-D-glucose (Wilkinson et al., 1973; Wilkinson and Taylor, 1978). This diamine is also present in species of *Rhodopseudomonas* (Roppel et al., 1975; Keilich et al., 1976). The core oligosaccharide of *P. diminuta* has KDO, glucose and heptose (almost certainly the usual L-glycero-D-manno-heptose), but no phosphorus or amino components (S. G. Wilkinson, personal communication).

P. diminuta and *P. vesicularis* have marked similarities in lipid composition. Both species have phosphatidylglycerol, a phosphatidylglucosyldiacyl-glycerol, and three major glucosyldiacylglycerols (Wilkinson and Galbraith, 1979). The sugar-containing lipids are prominent in the composition of both species. Glucuronosyldiglycerides are produced by typical pseudomonads in small amounts (Wilkinson, 1970), but the lipids of the above two species differ from those of all other Gram-negative species analyzed thus far, including *Gluconobacter* (Wilkinson and Galbraith, 1979). These findings are of interest in view of the rather isolated taxonomic position of the two species. The resemblance to *Gluconobacter* in some of the physiological properties (Ballard et al., 1968) does not extend to polar lipid composition (Wilkinson and Galbraith, 1979).

Wilkinson found that the fatty acid composition of the cell walls of *P. cepacia* differed from that of other species. The major components are myristic, 3-hydroxymyristic, and 3-hydroxypalmitic acids (Wilkinson, personal communication). These results agree with the fatty acid composition of whole cells (Samuels et al., 1973). The core polysaccharide contains glucose, rhamnose and heptoses but no phosphorus. The LPS has a low content of KDO, and the side chain is basically a mannan. One added strange feature is the presence of an acid-labile amino sugar phosphate presumably associated with the lipid A. All these properties, according to Wilkinson, set apart *P. cepacia* from all other species that have been examined thus far. Manniello et al. (1979) have confirmed some of these results. Their analyses of *P. cepacia* LPS composition revealed the presence of rhamnose, glucose, heptose, and hexosamine, but no KDO, and the phosphorus content was about one-third that of *P. aeruginosa* LPS.

Many procaryotes have hopanes in their membranes. These compounds are triterpene derivatives that are similar to sterols in size, rigidity and amphiphilic character, and may play in the membranes a role similar to that of the sterols in eucaryotes. In fact, hopanes are probably chemical ancestors of the sterols. Among the *Pseudomonas* species that have been examined, only *P. cepacia* (of RNA group II) appears to contain hopanes, while *P. aeruginosa*, *P. fluorescens*, *P. chlororaphis*, *P. stutzeri* (all of RNA group I), *P. acidovorans* (RNA group III), *P. diminuta* (RNA group IV), and *P. maltophilia* (RNA group V) do not contain these compounds (Rohmer et al., 1979; M. Rohmer, personal communication).

Pigmentation. Early treatments of the genus *Pseudomonas* included pigmentation as a generic character, but at present the genus includes many so-called nonpigmented species. Actually, the colonies and other cell masses always display some colors due to normal cellular components which, in some cases, become quite apparent. Thus, *P. stutzeri* is grouped with the nonpigmented species, but the colonies of many strains become reddish brown with age due to high concentration of cytochrome *c* in the cells.

Among the most notorious of *Pseudomonas* soluble pigments are the fluorescent (pyoverdin) and pyocyanin, which have become familiar to bacteriologists for many years. While the structure of pyocyanin is well known, the pyoverdins have been only partially characterized. Fluorescent pigments are produced abundantly in media of low iron content, and the fluorescence varies from white to blue-green upon excitation with ultraviolet radiation. The wavelength for maximum excitation is around 400 nm, but the pigments are still able to fluoresce intensely at lower wavelengths, a property that is useful for a differentiation of pyoverdins from other fluorescent compounds of different chemical nature produced by pseudomonads of other groups (Stanier et al., 1966; Hildebrand et al., 1973).

Pyoverdins are unstable, and various compounds have been described in the past which were probably decomposition products of lower molecular weight. Recently, the work of Meyer and his collaborators (Meyer, 1977; Meyer and Abdallah, 1978; Meyer and Hornsperger, 1978) has substantially advanced our knowledge of the pyoverdin of *P. fluorescens*, and it is to be hoped that the structure of the molecule will be known in the near future. The main components are a quinoline chromophore linked to a cyclic peptide whose composition may be different according to the species. The chemical studies have been complicated by the chemical instability of pyoverdin, although this can be lowered by formation of the ferric complex. Meyer and Hornsperger (1978) have demonstrated the function of pyoverdin in iron transport, but other fluorescent compounds of unknown structure have also been postulated for this role (Garibaldi, 1971; Kloepper et al., 1980). Kloepper et al. (1980) described a fluorescent chelator under the name of pseudobactin, to which they attribute the role of promoting plant growth by inhibition of plant-pathogenic bacteria in the rhizosphere of some cultivated plants. In common with other siderophores, the pyoverdin of *P. fluorescens* contains the unusual amino acid N^5-hydroxyornithine (Meyer and Hornsperger, 1978).

By growing *P. stutzeri*, a species related to the fluorescent pseudomonads, under iron-deficient conditions, Meyer and Abdallah (1980) were able to show the excretion of a colorless siderophore that was identified as nocardamine, a compound produced by actinomycetes under similar growth conditions. But even though a number of siderophores of microbial origin are common in soil, it is unknown whether these compounds represent a source of iron available for higher plants (Powell et al., 1980).

Other fluorescent compounds have been isolated from fluorescent pseudomonads. These include four fluorescent pteridine derivatives isolated by Suzuki and Goto (1971) from "*P. ovalis*" (presumably *P. putida*), the two fluorescent antibiotics isolated from *P. fluorescens* by Shirarata et al. (1970), who named them fluopsin C and fluopsin F; 6-hydroxymethylpterine, an intermediate in folic acid synthesis (Viscontini and Frater-Schröder, 1968), and erithroneopterine of *P. putida* (Suzuki and Goto, 1972).

Pyocyanin of *P. aeruginosa* is the most familiar of a group of pigments found among various fluorescent and nonfluorescent pseudomonads, the phenazine pigments. Pyocyanin has a blue color and diffuses freely into the medium, like the fluorescent pigments. Other phenazine pigments are less soluble and in some instances they crystallize around the colonies. Such is the case of the green pigment chloraphin, produced by the species *P. chlororaphis*. Phenazine pigments are also produced by nonfluorescent species (Morris and Roberts, 1959).

Pigments that are not soluble in water and remain associated with cellular structures are found in many species: *P. alcaligenes, P. mendocina, P. vesicularis, P. flava, P. pseudoflava, P. palleronii, P. rhodos, P. echinoides, P. radiora, P. mesophilica*, etc. These pigments are usually called "carotenoids" but rigorous chemical characterization is lacking in some cases. The carotenoid pigments of *P. rhodos* are unusual compounds that have been identified as 4,4′-diapocarotene-4-oic acid, di(β-D-glucosyl-4,4′-diapocarotene-4,4′-dioate), and β-D-glucosyl-4,4′-diapocarotene-4-oate-4′-oic acid. The glucosyl residue may be esterified to one of several fatty acids (Kleinig et al., 1979). *P. echinoides* contains, instead, derivatives of β,β-carotene (Czygan and Heumann, 1967). Another recently described species, *P. paucimobilis*, was found to contain the carotenoid nostoxanthin (Jenkins et al., 1979).

Strains of the *P. fluorescens* biovar IV, which were originally assigned to the species "*P. lemonnieri*", produce an intracellular insoluble pigment of structure related to that of indigoidine, the purple pigment of *P. indigofera*. Both pigments are derivatives of 3,3′-bipyridyl (Kuhn et al., 1965). The pigment of "*P. lemonnieri*" has been recently reexamined by Ferguson et al. (1980) and by Jain and Whalley (1980), who have named it lemonnierin and have determined the correct chemical structure of the compound.

P. indigofera has been referred to as one of the "indigo bacteria" (Sneath, 1960), but it is important to point out here that its pigment, indigoidine, is not the same as indigo, or indigotine (Kuhn et al., 1965), a blue, water-soluble compound which accumulates in and around the colonies of "*P. indoloxidans*" (renamed *P. acidovorans* by Stanier et al., 1966) growing in a medium containing indole (Gray, 1928).

Pigments are in general very conspicuous and in many instances have a definite taxonomic significance, since their production by pseudomonads correlates with other group properties. However, their production may cease or become erratic even in the recommended media, especially after prolonged cultivation in laboratory conditions. The taxonomic importance of pigments is also affected in some cases by the fact that different species of the same genus or even species of unrelated genera may be capable of producing the same pigment.

Colonial characters. Colonial and various cultural characters have been included in early descriptions of *Pseudomonas* species, but in our opinion most of these properties lack diagnostic value. Therefore, they will be used in the present treatment only where they appear to be highly characteristic of the taxon under discussion.

Nutrition and growth conditions. Most *Pseudomonas* species can grow in mineral media with ammonium ions or nitrate and a single organic compound as the sole carbon and energy source. Only a few species require the addition of organic growth factors. Some of these species belong to groups bearing only a remote relationship to other *Pseudomonas* groups. *P. maltophilia*, which is related to *Xanthomas* species more closely than to species of *Pseudomonas*, requires methionine, and the species of RNA group IV (*P. diminuta* and *P. vesicularis*) require pantothenate, biotin and cyanocobalamin. In addition, *P. diminuta* also needs cystine or methionine. Species of carboxydobacteria (i.e. bacteria capable of using CO as an energy source) also may require growth factors. Thus, "*P. compransoris*" requires thiamine and "*P. gasotropha*" does not grow in the absence of vitamin B_{12}. The natural relationships of these species to other species of the genus are at present obscure.

Growth factor-requiring strains of species of other *Pseudomonas* groups are occasionally encountered in nature (for instance, strains of *P. caryophylli* and of *P. syringae*). Such strains often can be assigned to the proper species on the basis of various phenotypic characters.

In view of the above considerations, it is likely that the genus *Pseudomonas* will be reserved in the future for species not requiring growth factors. A step in this direction has been taken in the present treatment, as we shall discuss in a later section, by suggesting the assignment of *P. maltophilia* to the genus *Xanthomonas* and the crea-

tion of a new genus for *P. diminuta* and *P. vesicularis*. *P. maltophilia* and *P. vesicularis* also differ from most other *Pseudomonas* species by their inability to assimilate nitrate.

From time to time the capacity for nitrogen fixation has been claimed for some species which later on proved to be unable to perform this function under strictly controlled conditions. Of the species described in the present treatment, *P. glathei* comes close to being a nitrogen fixer, but several facts point to the contrary. *P. glathei* grows in nitrogen-deficient media even after precautions have been taken to prevent contact with ammonia from the laboratory atmosphere. However, the cells grown under these conditions fail to reduce acetylene and, in addition, growth is not stimulated by low oxygen tensions (Zolg and Ottow, 75). For the moment we must assume that none of the *Pseudomonas* species can be considered a legitimate nitrogen fixer.

The capacity for growth in very simple mineral media at the expense of a single carbon compound has served as the basis for extensive nutritional characterization of *Pseudomonas* species. These may differ from one another in the number and types of organic compounds that can support growth when used individually in chemically defined media, and the nutritional pattern of a given strain is of considerable help for its assignment to a given species. Nutrition in *Pseudomonas* is of the chemoorganotrophic type, but some species can also live under autotrophic conditions using CO and/or H_2 as energy sources.

The best growth temperature for most strains is around 28°C. None of the well characterized species (with the exception of members of RNA group IV) tolerate acid conditions and growth is invariably negative at pH 4.5.

Metabolism and metabolic pathways. The metabolism of *Pseudomonas* is typically respiratory with oxygen as the terminal electron acceptor, but many species can also use nitrate as an alternate electron acceptor and can carry out oxygen-repressible denitrification (dissimilatory reduction of nitrate to N_2O or N_2). Some cytochromes are involved in denitrification through the participation of a special cytochrome oxidase which, according to Yamanaka (1964) may represent a very primitive mechanism dating from the preoxygen era of the planet.

Assimilation of nitrate occurs aerobically through reduction to ammonia. Although there may be common elements in the assimilatory and dissimilatory pathways for nitrate reduction (Hartinsgvelt et al., 1971), Sias et al. (1980) have shown that the assimilatory and dissimilatory nitrate reductases in *P. aeruginosa* are encoded by different sets of genes, and that the apparent pleiotropic effects of some mutations can be attributed to lesions in a gene affecting molybdenum incorporation into both enzymes.

The cytochrome composition was determined by Stanier et al. (1966) for various species, by Sands et al. (1967) for plant pathogenic nomenspecies, by Davis (Ph.D. Thesis, University of California, Berkeley, 1967) and Auling et al. (1978) for hydrogen pseudomonads, and by Ballard et al. (1968) for the diminuta group (RNA group IV). The plant pathogens studied by Sands et al. (1967) all showed the absence of a cytochrome c peak, as is also the case for *P. maltophilia*. These species also gave a negative oxidase reaction, thus confirming the observation that the "oxidase" test may be diagnostic for the presence of cytochrome c in the electron transport system of Gram-negative bacteria (Stanier et al., 1966; Baumann et al., 1968).

The oxidative degradation of some organic substrates or their intermediates by *Pseudomonas* occasionally involves the participation of oxygenases. Both mono- and dioxygenases coupled to a variety of electron donors are well represented in various *Pseudomonas* species. Some of the oxygenases have a rather broad specificity, contributing to the nutritional versatility of the strains. Oxygenases acting on aliphatic compounds such as alkanes may be part of complex oxidative systems. This is also true of the system described by Gunsalus et al. (1971) for the oxidation of camphor by *P. putida*, with oxygenase steps of considerable complexity, one of which includes the Fe-S protein putidaredoxin and cytochrome $P450_{CAM}$. Other systems include the iron protein rubredoxin (Lode and Coon, 1971) and cytochrome o (Peterson, 1970).

The classical reactions of the tricarboxylic acid cycle are found in all species of *Pseudomonas* that have been examined. A key reaction is the synthesis of citrate from oxaloacetate and acetyl-CoA, which is under a control system typical of absolute aerobic organisms (Weitzman and Jones, 1968). The central position of the tricarboxylic acid cycle is clearly manifested in the repression control of peripheral catabolic enzymes (amidase, histidase, enzymes of aromatic compounds and camphor metabolism) by the tricarboxylic acid cycle intermediates. For biosynthetic purposes, tricarboxylic acid cycle intermediates can be replenished by carboxylation of pyruvate and also by the action of the enzymes isocitrate lyase and malate synthase, which are part of the anaplerotic system known as the glyoxylate cycle (Kornberg and Madsen, 1958).

In addition to low molecular weight compounds, a variety of macromolecules can be degraded by some strains by means of extracellular enzymes. Hydrolytic enzymes that have been studied in detail include the proteases of *P. aeruginosa* (Morihara, 1964; Morihara et al., 1965), α-amylase from *P. saccharophila* (Markowitz et al., 1956) and from *P. stutzeri* (Robyt and Ackerman, 1971), and poly-β-hydroxybutyrate depolymerase from *P. lemoignei* (Delafield et al., 1965; Lusty and Doudoroff, 1966). Proteases from *P. aeruginosa* may play a role in pathogenesis and the development of symptoms, and the same is true of the enzymes produced by plant pathogenic bacteria that are capable of degrading high molecular weight plant components. Hydrolysis of pectin has been reported by Hildebrand (1971), Wilkie et al. (1973), Ohuchi and Tominaga (1973, 1975), and attack of xylan has been described by Maino et al. (1974) and glycosides by Hayward (1977). Cellulase activity has been reported for phytopathogens (Gehring, 1962; Lange and Knösel, 1970) and for saprophytic species. Ueda et al. (1952) have described four organisms reputed to be cellulose decomposers under the names "*P. alboflava*," *P. viscosissima*," "*P. rubeorifaciens*," and "*P. fluorescens* var. *cellulosa*." Of these, the last two are capable of gas production in the presence of carbohydrates, a character that excludes them from the genus *Pseudomonas*. We have not examined the other two species, and to our knowledge representative strains have not been deposited in any of the main culture collections. Therefore, they will not be further considered in this treatment.

One-carbon compounds can be utilized by a variety of microorganisms, some of which have been assigned to the genus *Pseudomonas* mainly on the basis of their morphological properties and aerobic behavior. These pseudomonads belong to a special physiological group differing from members of the genus *Pseudomonas* as presently described in their nutritional properties, their fine structure and the DNA base composition, and will not be further treated here. Good reviews on one-carbon-utilizing pseudomonads include those by Quayle (1972) and by Whittenbury et al. (1975).

Sugar catabolism. Some species of *Pseudomonas* have received names suggesting a preference for carbohydrates (e.g. *P. saccharophila*, *P. maltophilia*), but these compounds are in general not the best carbon and energy sources for many species. Some species grow slowly, while others fail altogether to grow at the expense of sugars, as is the case with *P. acidovorans*, *P. testosteroni*, *P. alcaligenes*, *P. pseudoalcaligenes*, *P. lemoignei*, *P. palleronii*, *P. stutzeri*, *P. mendocina*, etc. *P. saccharophila* grows well on complex saccharides, and uses glucose, fructose, mannose or D-arabinose only after mutation.

Most hexoses and related compounds are degraded by the Entner-Doudoroff pathway, first discovered in *P. saccharophila* (Entner and Doudoroff, 1952). The metabolism of glucose and other sugars by this species is depicted in Figure 4.4.

The fluorescent pseudomonads (*P. aeruginosa*, *P. fluorescens*, *P. putida*) have multiple peripheral pathways for glucose oxidation that converge to the synthesis of 6-phosphogluconate, which is further degraded by the Entner-Doudoroff pathway (Eisenberg et al., 1974). Of these routes, one involves direct oxidation of the sugar (oxidative pathway) and either gluconate or 2-ketogluconate can serve as precursor of 6-phosphogluconate. However, in *P. putida*, 6-phosphogluconate is synthesized preferentially from 2-ketogluconate (Vicente and Cánovas, 1973). Induction of the oxidative pathway in *P. aeruginosa* can only occur in the presence of oxygen. Under denitrifying conditions, only the so-called phosphorylative pathway (which starts with the phosphorylation of glucose) is operative in this organism (Hunt and Phibbs, 1981).

The metabolism of fructose by various species (*P. aeruginosa*, *P. putida*, *P. fluorescens*, *P. stutzeri*, *P. mendocina*, *P. acidovorans*, *P.

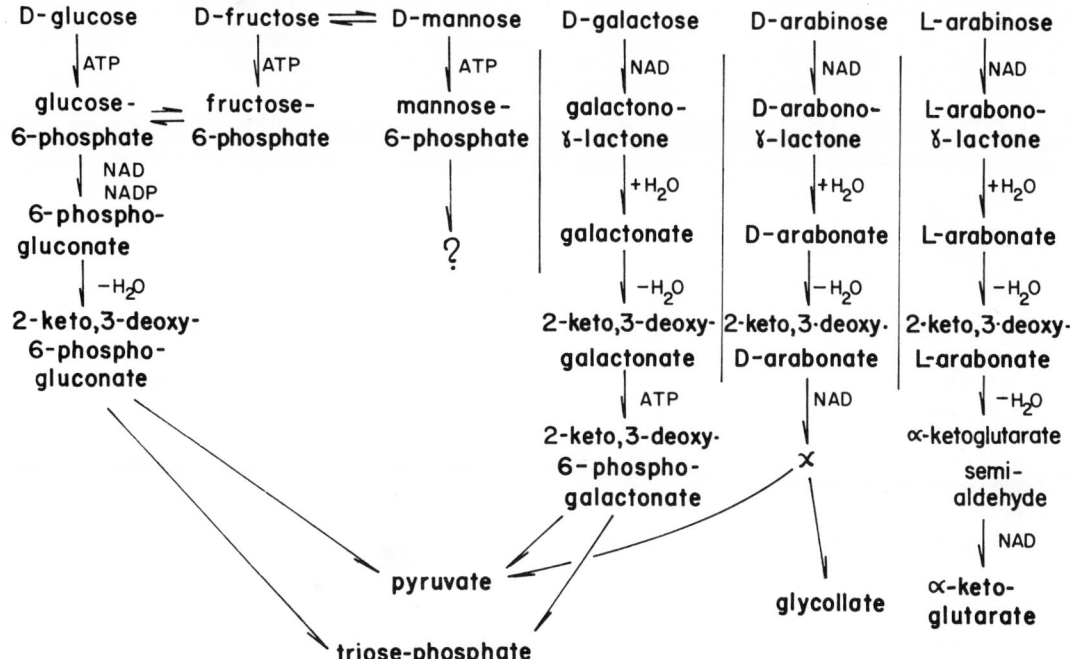

Figure 4.4. Metabolism of monosaccharides by *P. saccharophila*. (Reproduced with permission from N. J. Palleroni, *The Pseudomonas Group*, 1978, © Meadowfield Press, Shildon, Co. Durham, England.)

maltophilia) occurs by means of a phosphoenolpyruvate (PEP) phosphotransferase system. This gives fructose-1-phosphate, which may be further phosphorylated to fructose-1,6-diphosphate and cleaved by an aldolase. Further metabolism involves mainly the Entner-Doudoroff pathway, but the Embden-Meyerhof pathway can also operate as demonstrated by the metabolism in mutants blocked in the first route (Baumann and Baumann, 1975a, b; Sawyer et al., 1977a, b). No mention is made, however, of an initial reaction which is found in several of the above mentioned species. Fructose and mannose can be interconverted as free sugars through the action of mannose isomerase (Palleroni and Doudoroff, 1956a; Palleroni and Palleroni, unpublished observations). In *P. saccharophila* mannose is phosphorylated, but the cell does not seem to make further use of the resulting mannose-6-phosphate. This point perhaps deserves investigation in other species of the genus.

Catabolism of organic acids. Many fatty acids are used by pseudomonads for growth. Acetate can enter the tricarboxylic acid cycle and the higher members of the series can be degraded to acetyl-CoA by a beta-oxidation system.

In *P. aeruginosa* mutants with a defective isocitric lyase behave differently in their ability to metabolize mono- and dicarboxylic acids of even or odd numbers of carbon atoms, in confirmation of the essential role of this enzyme in the glyoxylate cycle (Chapman and Duggleby, 1967).

Interesting details of the metabolism of other organic acids have been reported by Shilo and Stanier in 1957, Hurlbert and Jacoby (1965), Dagley and Trudgill (1965) for the tartrates, and by Cooper and Kornberg (1964) for itaconate.

Acetamide and biochemical evolution. Acetamide is used for growth by several *Pseudomonas* species, including *P. aeruginosa* (Stanier et al., 1966). It is hydrolyzed with liberation of ammonia and this is probably the only reaction of biological significance, although the amidase can also catalyze acyl transfer reactions. Amides hydrolyzed by the wild type enzyme include propionamide, acetamide, formamide and butyramide, in order of decreasing rates of hydrolysis. Clarke and her collaborators found this system to be admirably suited for studies of experimental evolution during which it became possible to evolve amidases with altered substrate specificities (Brown et al., 1969; Brown and Clarke, 1972; Betz and Clarke, 1972; Betz et al., 1974). The work of Clarke and collaborators stands out as one of the most beautifully documented examples of biochemical evolution achieved in the laboratory.

Metabolism of aromatic compounds. Many aromatic compounds can be used for growth by *Pseudomonas* species. A number of these compounds (benzoate, *p*-hydroxybenzoate, mandelate, tryptophan, phthalate, salicylate) may be metabolized by pseudomonads of RNA group I following pathways that converge to a common intermediate, β-ketoadipate. This intermediate is formed soon after the last aromatic intermediate is cleaved by a 1,2-dioxygenase (Fig. 4.5). The type of ring fission catalyzed by this enzyme is frequently referred to as an *ortho* cleavage. Members of RNA group II also use this mechanism of ring fission. In contrast, species of RNA group III are able to split the aromatic substrates using divergent pathways, without participation of β-keto-adipate as an intermediate (Figs. 4.6 and 4.7).

The two types of ring opening appear to be of taxonomic importance as long as the experimental conditions are precisely specified. For instance, cleavage of a given diphenolic intermediate such as catechol by either a 1,2- or a 2,3-dioxygenase may be carried out by the same strain depending on whether the growth substrate was benzoate or phenol, respectively (Feist and Hegeman, 1969). Genes encoding for *meta* cleavage enzymes in these cases may be located in plasmids (Austen and Dunn, 1980), but this is not the rule. Hewetson et al. (1978) have presented evidence for a plasmid in *P. putida* carrying genes for the degradation of *p*-cresol via the protocatechuate *ortho* pathway.

The pathways of degradation of aromatic compounds have been extensively analyzed in many aspects, including the regulatory mechanisms, the immunological properties and amino acid sequences of

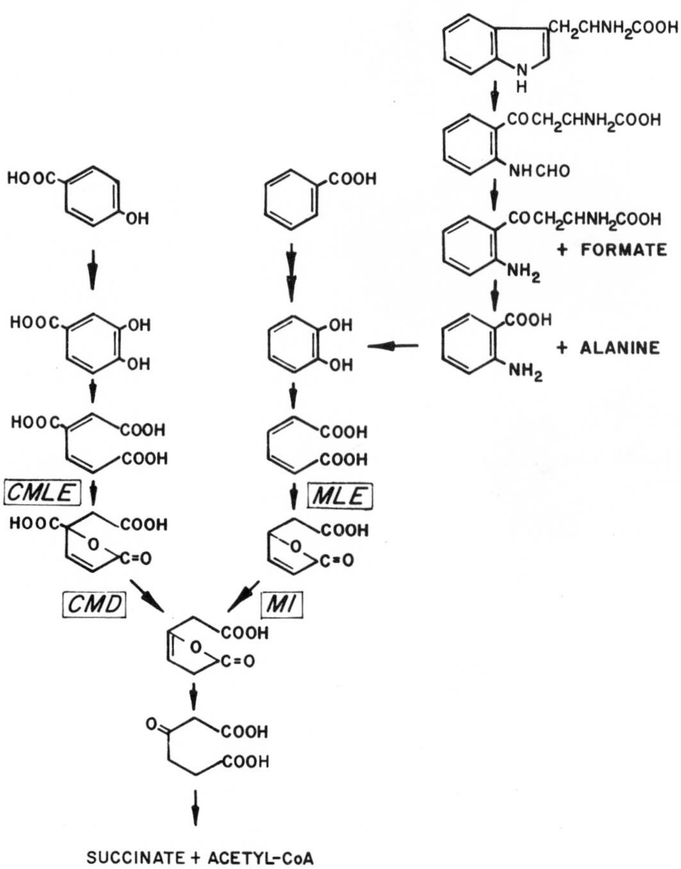

Figure 4.5. Convergent pathways for the degradation of *p*-hydroxybenzoate, benzoate and tryptophan by *P. putida*. Four enzymes of the pathways are indicated with abbreviated names. *CMLE*, carboxymuconate-lactonizing enzyme; *CMD*, carboxymuconate decarboxylase; *MLE*, muconate-lactonizing enzyme; *MI*, muconolactone isomerase. The three pathways generate β-ketoadipate, which is converted to succinate and acetyl-CoA. (Reproduced with permission from N. J. Palleroni, *The Pseudomonas Group*, 1978, © Meadowfield Press, Shildon, Co. Durham, England.)

selected enzymes, and the genetics; see the review by Clarke and Ornston (1975). The phylogenetic conclusions of such investigations are of interest and help one to understand the relationships of *Pseudomonas* to other bacterial genera, and often confirm the taxonomic criteria on which is based the present internal subdivision of the genus.

Amino acid catabolism. Most amino acids can be used as carbon, nitrogen and energy sources by many *Pseudomonas* species. The exception is methionine, which can be used by *P. aeruginosa* as a nitrogen source only (Kay and Gronlund, 1969).

Permeases for various amino acids have been identified mainly in *P. aeruginosa*, *P. fluorescens*, *P. putida* and *P. acidovorans*. The specificity of some permeases is high, as in the case of L-tryptophan permease of *P. acidovorans* (Rosenfeld and Feigelson, 1969), but the opposite may also be true. In *P. fluorescens*, L-alanine and L-proline can be transported by their respective specific permeases or by a less specific system that also mediates β-alanine transport (Hechtman and Scriver, 1970).

A few examples of amino acid catabolism will be cited here. Threonine is not used by fluorescent pseudomonads. In *P. cepacia* Lessie and Whiteley (1969) demonstrated the induction of threonine dehydrogenase, which appears to be indispensable for growth at the expense of this amino acid.

Lysine catabolism can occur in *Pseudomonas* by at least three different pathways eventually converging to glutarate, which generates ace-

which passes through quinoline derivatives (Behrman, 1962; Stanier et al., 1966). Interesting regulatory characteristics have been discovered in the first pathway. In a strain labelled *P. fluorescens*, L-kynurenine (the second intermediate of the inducible pathway) and not tryptophan is the inducer (Palleroni and Stanier, 1964). This was the first documented case of "metabolite induction," and also the first example of sequential groups of enzymes that are coordinately regulated. Later this mechanism was found to be operative in the tryptophan catabolic pathway of *P. acidovorans* (Rosenfeld and Feigelson, 1969), but not in that of *P. aureofaciens* (Salcher and Lingens, 1980).

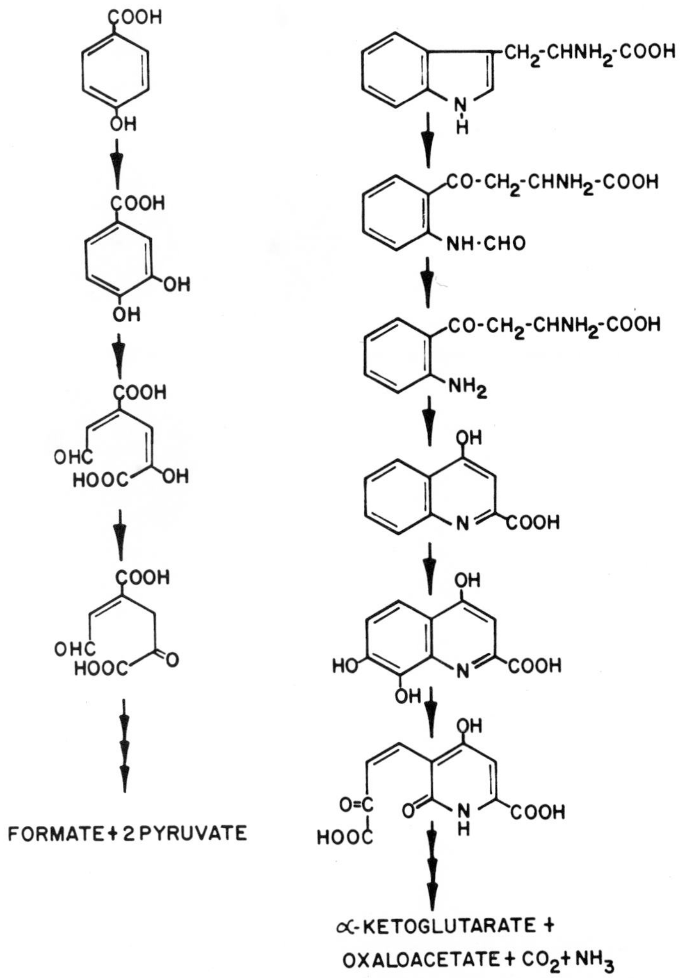

Amino acid biosynthesis. The biosynthesis of amino acids has been

Figure 4.6. Divergent pathways for the metabolism of *p*-hydroxybenzoate and tryptophan by *P. acidovorans*. (Reproduced with permission from N. J. Palleroni, *The Pseudomonas Group*, 1978, © Meadowfield Press, Shildon, Co. Durham, England.)

tyl-CoA. For short, these pathways may be called the "oxygenase," the "pipecolate" and the "cadaverine" routes (Miller and Rodwell, 1971; Chang and Adams, 1971; Fothergill and Guest, 1977), and their distribution in various *Pseudomonas* species is presented in Table 4.3.

For the branched-chain amino acids (leucine, isoleucine, valine) Marshall and Sokatch (1972) have described a common catabolic pathway in *Pseudomonas* with a common point of convergence in α-ketovalerate.

Arginine can be used by many pseudomonads for growth. The first two intermediates of its degradation are citrulline and ornithine. The enzymes involved are arginine deiminase and ornithine transcarbamylase and, in the last reaction, carbamyl phosphate is also formed, which can serve as a source of high energy phosphate for ATP synthesis (Stalon et al., 1972). The reactions constitute the commonly called "arginine dihydrolase" system, by which aerobic pseudomonads can obtain ATP under anaerobic conditions (Sherris et al., 1959).

A second pathway of arginine catabolism was described by Miller and Rodwell (1971) in *P. putida*, where the successive intermediates are α-ketoarginine, γ-guanidinobutyrate, and γ-aminobutyrate *plus* urea.

Tryptophan can be catabolized via anthranilate, catechol and the β-ketoadipate pathway in the *Pseudomonas* species of RNA group I. *P. acidovorans*, a member of RNA group III, follows a different pathway

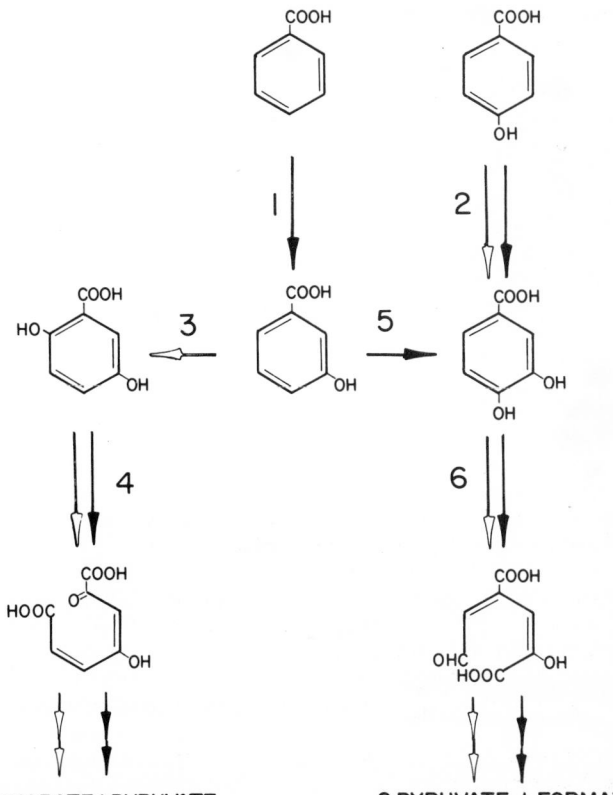

Figure 4.7. Metabolism of benzoate and *p*-hydroxybenzoate by members of the acidovorans group (RNA group III). Numbers represent various enzymes: *1*, benzoate-3-hydroxylase; *2*, *p*-hydroxybenzoate-3-hydroxylase; *3*, *m*-hydroxybenzoate-6-hydroxylase; *4*, gentisate oxygenase; *5*, *m*-hydroxybenzoate-4-hydroxylase; and *6*, protocatechuate-4,5-oxygenase. *White arrows*, *P. acidovorans*; and *black arrows*, *P. testosteroni*. (Reproduced with permission from N. J. Palleroni, *The Pseudomonas Group*, 1978, © Meadowfield Press, Shildon, Co. Durham, England.)

Table 4.3.
Distribution of lysine catabolic pathways in **Pseudomonas** *species[a]*

Species	Lysine Catabolic Pathway		
	Oxygenase	Pipecolate	Cadaverine
P. aeruginosa	−	+	+
P. fluorescens	+	+	+
P. putida biovar A	+	+	−
P. putida biovar B	+	−	−
P. cepacia	−	+	−

[a] Symbols: see standard definitions.

investigated mainly in *P. aeruginosa* and *P. putida*, and the pathways appear to be essentially similar to those of other groups of bacteria. However, the regulatory mechanisms in *Pseudomonas* may vary considerably, and the differences often have interesting taxonomic implications. From the genetic standpoint, the biosynthetic pathways in *Pseudomonas* may differ from those of other Gram-negative bacteria in the degree of clustering of the genes involved. Thus, while the histidine genes constitute a single operon of nine genes in *Salmonella typhimurium* (Hartman et al., 1971), such clustering does not occur in *P. aeruginosa* (Mee and Lee, 1967).

As we have seen, pseudomonads are capable of using most amino acids for growth, and the catabolic schemes, with their own regulatory peculiarities, often introduce formidable complications in the study of biosynthetic routes.

Here, again, only a few selected examples will be cited. In the biosynthetic pathways of amino acids of the aspartate family, *Pseudomonas* species have a single aspartokinase, which is sensitive to concerted feedback inhibition by lysine and threonine. High concentrations of either amino acid inhibit the enzyme in the organisms of the fluorescent group and in *P. cepacia*, whereas in the acidovorans group (RNA group III), the sensitivity to single amino acids is markedly lower. Homoserine dehydrogenase in the members of this last group is absolutely specific for NAD, but in all other pseudomonads that have been examined, NADP is more effective than NAD (Cohen et al., 1969).

Arginine biosynthesis proceeds in *Pseudomonas* by a sequence of reactions that start with glutamate and lead to ornithine through acetylated intermediates (Udaka, 1966; Leisinger et al., 1972). Exogenous acetylation of glutamate is only needed at the beginning, and subsequently acetyl groups are transferred from acetylornithine back to glutamate without further energy expenditure, whereas, in other bacterial groups, acetylornithine is merely deacylated and de novo acetylation of glutamate is required every time.

The last stages of the arginine biosynthetic pathway have a marked similarity to those of the catabolic system, but the work of Wiame and his collaborators in Belgium, and later more detailed investigations of Stalon et al. (1972), have clearly shown that the catabolic enzymes are different from the biosynthetic ones, and that they can be easily differentiated by their physicochemical and regulatory properties. For example, the catabolic ornithine transcarbamylase (OTC) is induced by arginine, while the anabolic one is repressed. As shown by Stalon et al. (1972) the two enzymes have regulatory properties so that, under physiological conditions, they work in opposite directions. From strains lacking the biosynthetic OTC, these workers could obtain mutants with a catabolic OTC modified to the extent of being capable of performing a biosynthetic function.

Jensen and his collaborators (1967) have reported interesting comparative studies on the control of a branch-point enzyme, 3-deoxy-D-arabinoheptulosonate-7-phosphate (DAHP) synthetase in different organisms, including a number of the strains of *Pseudomonas* from the Berkeley collection. In tests with crude cell extracts of *P. aeruginosa*, *P. putida*, *P. fluorescens*, *P. alcaligenes*, *P. pseudoalcaligenes*, *P. stutzeri*, and *P. cepacia*, it was found that there was feedback inhibition of a single enzyme species by only one of the three final products of the diverging pathways, namely tyrosine. The DAHP synthetase of *P. acidovorans* and of *P. testosteroni*, on the other hand, appeared to be also sensitive to phenylalanine inhibition.

More recently Whitaker et al. (1981a) have expanded these studies considerably by including 90 species of the genera *Pseudomonas*, *Xanthomonas* and *Alcaligenes*. The results obtained with the pseudomonads are of particular interest to us. The comparative allostery of DAHP synthetase has permitted the unmistakable identification of organisms of two of the RNA groups, groups IV and V of Palleroni et al. (1973). The species of group IV appear to be distinguished by the retrotryptophan pattern of control of DAHP synthetase, while those of group V have a synthetase that is not stimulated by divalent cations and is sensitive to feedback inhibition by chorismate. RNA groups I, II and III present overlapping patterns where DAHP synthetase appears to

be sensitive to both phenylalanine and tyrosine. No clear-cut differentiation of organisms belonging to these three groups is possible, however, the study of the control mechanisms of DAHP synthetase can be very useful as a confirmation of species assignments when used in conjunction with the regulatory patterns of tyrosine biosynthesis.

The specific reactions of tyrosine biosynthesis in *Pseudomonas* participate in one of two alternative routes. Prephenic acid (PPA) can be converted by a specific dehydrogenase to 4-hydroxyphenylpyruvate, which can be aminated to tyrosine. Alternatively, PPA can first be aminated to arogenate (AGN) and subsequently converted to tyrosine by a dehydrogenase. Species of RNA groups II and III have NADP-linked AGN dehydrogenase activity, and this is lacking in members of RNA groups I, IV and V. The AGN dehydrogenase of species of group II responds to feedback inhibition by tyrosine, while the dehydrogenase of species of group III is insensitive. Members of group IV have PPA dehydrogenase linked to NADP, and this is absent in groups I and V. The AGN-NAD dehydrogenase of group I is sensitive to inhibition by tyrosine, a property absent in the NAD-linked PPA dehydrogenase of the same group. This particular pattern is reversed in organisms of group V. These observations are summarized in Table 4.4 (Byng et al., 1980).

Earlier experiments by Baumann and Baumann (1978) on a smaller scale of *Pseudomonas* strains supported similar conclusions on the differences among the various *Pseudomonas* groups. Using antibody prepared against glutamine synthetase of *E. coli* and extracts of bacteria of various genera in double diffusion and quantitative microcomplement fixation experiments, Baumann and Baumann (1978) concluded that the primary structure of this enzyme was highly conserved and, as might be expected, congruence of the results with rRNA homologies was observed. The data also supported the conclusion (later confirmed by the work of Jensen and his collaborators) that some species of *Alcaligenes* are more closely related to some *Pseudomonas* species than to each other.

The enzymatic and regulatory patterns of L-phenylalanine biosynthesis have, in addition, provided a "fine-tuned probe" whose use has further confirmed the currently accepted internal subdivision of the genus *Pseudomonas* (Whitaker et al. 1981b). The five groups defined by rRNA homology can be further subdivided into eight DNA homology groups, as demonstrated by the Berkeley workers. Thus, DNA/DNA hybridization studies have permitted a differentiation of *P. pickettii* and *P. solanacearum* from all other species of RNA group II, and likewise, RNA group III could be subdivided into DNA subgroup IIIa (*P. acidovorans* and *P. testosteroni*), IIIb (*P. delafieldii* and *P. facilis*), and IIIc (*P. saccharophila*). This internal subdivision has been confirmed by the phenylalanine enzymatic studies of Whitaker et al. (1981b) and, in addition, these experiments could discriminate the phenotypic cluster of fluorescent organisms included in RNA group I (subgroup Ib) (Table 4.5).

Further evidence for the differences separating two important *Pseu-*

Table 4.4.

Distribution of tyrosine biosynthetic pathways and feedback inhibition patterns in various **Pseudomonas** *natural groups[a]*

Pathway	Enzyme	Coenzyme	Sensitivity to Tyrosine	RNA Homology Group
4-Hydroxyphenylpyruvate	Prephenate dehydrogenase	NADP	?	IV
		NAD	−	I
			+	V
Arogenate	Arogenate dehydrogenase	NADP	+	II
			−	III
		NAD	+	I
			−	III

[a] Symbols: see standard definitions.

Table 4.5.

*Enzymatic and regulatory patterns in L-phenylalanine biosynthesis by various **Pseudomonas** subgroups[a]*

DNA Homology Subgroup	Prephenate Dehydratase Activity	Arogenate Dehydratase Activity	Tyrosine Activation Factor[b]
Ia	+	−	1
Ib	+	+	1
IIa	+	+	2–3
IIb	+	+	1
IIIa	+	−	6–12
IIIb	+	+	3–4
IIIc	+	+	>15
IV	−	+	
V	+	+	1

[a] Data from Whitaker et al. (1981b). See standard definitions.

[b] Ratio of activity of prephenate dehydratase activity in the presence and in the absence of L-tyrosine.

domonas RNA groups has been presented by Queener and Gunsalus (1970) from their studies on anthranilate synthetase, one of the enzymes of tryptophan biosynthesis. This enzyme can be resolved into two subunits and hybrid molecules can be formed with subunits of different bacterial species. Hybrid molecules made with subunits from *P. aeruginosa* and *P. putida*, two species of RNA group I, are fully active, and the same is true of the hybrids between subunits from *P. acidovorans* and *P. testosteroni*, both of RNA group III. When subunits from different groups are hybridized, however, the resulting molecules have lower activity than either native enzyme. *P. stutzeri*, of group I, has a synthetase resembling the putida-aeruginosa enzyme, while *P. cepacia* (of group II) enzymes is similar to the acidovorans-testosteroni class.

Genetics. Pseudomonas species have attracted increasing attention from bacterial geneticists. The main reasons are their widespread occurrence, their biological and medical importance, their nutritional and biochemical versatility, and the simplicity of the conditions required for their cultivation in the laboratory. For general information in this field, see the reviews by Holloway (1969) and Holloway et al. (1971, 1979).

The general principles and methodology for isolation of regulatory and structural gene mutations for both catabolic and biosynthetic pathways have been discussed by Clarke (1976) and, for the particular case of *P. aeruginosa*, by Holloway (1969, 1974, 1975). The choice of mutagen may be a crucial point for success in mutant isolation (Holloway, 1979a).

Of all *Pseudomonas* species, by far the best known from the genetical point of view, is *P. aeruginosa*, of which strain PAO has been the most extensively studied. A detailed chromosome map for this strain including over 90 genes is available (Holloway et al., 1979; Royle et al., 1981). A second strain, PAT, is also familiar to geneticists, and its map closely resembles that of PAO (Watson and Holloway, 1978a). Since these two strains have different geographic origins, the genetic findings once more confirm the remarkable homogeneity of the species already noted by taxonomists.

The three common systems of gene recombination found in other bacterial groups, namely, conjugation, transduction and transformation, have been observed in members of the genus. For *P. aeruginosa*, the first two processes are practically much more important than the third. Studies on chromosome composition in *P. aeruginosa* have depended heavily on conjugation, which in turn depends in most instances on the chromosome donor ability (*cda*) or chromosome mobilizing activity (*cma*) of some plasmids. In *P. aeruginosa*, plasmid FP2 has been used extensively for mapping purposes, but other plasmids, such as FP5, FP39 and FP110 have also been used. Both FP2 and FP5 have a mercury resistance marker (Loutit, 1971; Matsumoto and Tazaki,

1973). Each of these plasmids has one predominant attachment site, which has been an obstacle for the demonstration of chromosome circularity in strain PAO, but this demonstration has been possible by a series of 2- and 3-factor crosses with *cma* plasmids FP2, FP5, FP110 and R68.45 (Royle et al., 1981). Chromosome circularity for strain PAT was reported earlier by Watson and Holloway (1978a, b).

Plasmid R68.45 derives from R68, a plasmid of incompatibility group 1 (IncP-1), one of the groups into which *Pseudomonas* plasmids are currently classified (Jacoby and Shapiro, 1977). R68 has a much higher *cma* in strain PAT than in PAO, but in R68.45 this capacity for PAO is increased considerably (Haas and Holloway, 1976, 1978; Jacob et al., 1977). In cases like this, the plasmid is said to possess enhanced chromosome mobilization (ECM) ability (Holloway et al., 1979). R68.45 was shown to carry a 2.1 kb insertion sequence which has been named IS21, formed by duplication of a segment already present in R68 (Willetts et al., 1981). Presumably, this structure is capable of interacting with the bacterial chromosome triggering its transfer from donor to recipient. This has been shown to occur not only in *P. aeruginosa*, but also in *P. putida* (Martinez and Clarke, 1975), *P. glycinea* (Fulbright and Leary, 1978), and even in species of other genera (Holloway, 1979b). Transfer in nature may be more common than it is generally assumed, but in spite of active search, plasmids with *cma* have been reported in only a few instances.

P. aeruginosa is notorious for the common occurrence of lysogeny, and the general opinion is that there may be very few strains of the species where this phenomenon does not occur, although some workers believe that the term lysogeny may have been applied rather loosely at times (Barksdale and Arden, 1974). Among the numerous *P. aeruginosa* phages that have been described, F116, F116L (capable of transducing larger segments of chromosome than F116 (Holloway, 1974) and G101 have been used extensively for fine mapping purposes (Holloway and Krishnapillai, 1975). Phage F116 is normally present in the cells as an extrachromosomal element (Miller et al., 1977).

In comparison with conjugation and transduction, transformation has not been an efficient method of chromosome study in *P. aeruginosa*, but plasmid transformation after CaCl₂ treatment has been achieved (Sano and Kageyama, 1977; Sinclair and Morgan, 1978). Recently, Hara et al. (1981) reported CaCl₂-induced transformation of *P. aeruginosa* with chromosomal DNA.

Pseudomonas species other than *P. aeruginosa* have been less rewarding subjects for genetic research. This is unfortunate in the case of a species like *P. putida*, which has been extensively investigated from the biochemical viewpoint. The species presents serious problems for the development of a satisfactory chromosome transfer system and, consequently, a less detailed chromosome map is available (Mylroie et al., 1977). Recently, circularity of the map of *P. putida* strain PPN (related to strain Stanier 90, with which many metabolic studies have been carried out) has been demonstrated by Dean and Morgan (1983).

Even though *P. aeruginosa* and *P. putida* are members of the same natural *Pseudomonas* group (RNA group I), the two species appear taxonomically as well as genetically quite different; however, many amino acid auxotrophic markers of *P. aeruginosa* PAO and *P. putida* PPN have been found to be functionally equivalent, as evidenced by complementation analysis using R primes generated from R68.45 (B. W. Holloway, personal communication).

Lysogeny has been seldom reported for species other than *P. aeruginosa*, where the phenomenon is extremely common. Temperate phages have been found in *P. acidovorans*, where generalized transduction is possible (H. E. Heath, personal communication), and also in *P. maltophilia* (Moillo, 1973). Phage pf16 has been used for transductional analysis in *P. putida* (Chakrabarty et al., 1968) but, in general, transduction is not a very promising mechanism for genetic analysis in most *Pseudomonas* species, and the same can be said of transformation in spite of the fact that chromosomal DNA could be used for transformation in *P. putida* (Mylroie et al., 1978) and in *P. solanacearum* (Boucher and Sequeira, 1978).

R′ plasmids are attractive vectors for interspecific and intergeneric transfer. Derivatives of R68.45 have been used in various laboratories in crosses involving *recA* recipients, or strains of different species or genera (Hedges and Jacob, 1977; Holloway, 1978; Morgan, 1982). Complement of auxotrophic markers has been achieved, although not always has the enzymological evidence been sufficient to decide the nature of the enzyme responsible for the restored activity. TOL plasmid of *P. putida* carrying a carbenicillin resistance marker has been transferred to *P. aeruginosa* and *E. coli* but expression has only been observed in *Pseudomonas* (Benson and Shapiro, 1978). In contrast, genes from *E. coli* have less difficulty for expression in *Pseudomonas* (Mergeay and Gerits, 1978), which suggests the convenience of using some members of this last genus for cloning purposes in preference to *E. coli* (Holloway et al., 1979).

Studies on gene arrangement in *Pseudomonas* species have revealed interesting interspecific differences. In general, genes of individual biosynthetic pathways are grouped into small cotransducible clusters or not clustered at all. The absence of clustering of functionally related genes was reported years ago by Fargie and Holloway (1965), but this is not invariably so. Various genes coding for enzymes in catabolic pathways have been found to be clustered in *P. putida* (Chakrabarty et al., 1968; Wheelis, 1975). The degree of clustering may vary considerably for the same pathway in different species and this variation is at times quite striking for some of the peripheral catabolic pathways present in different species of the same natural group (Rosenberg and Hegeman, 1969; Wheelis and Stanier, 1970). Heath and collaborators (cited by Holloway et al., 1979) have shown that genes for histidine utilization in *P. acidovorans* are not part of an operon, and that the gene cluster is similar to that found in *P. putida* (Leidigh and Wheelis, 1973).

Plasmids, phages, bacteriocins. As mentioned before, plasmids are important components of the genetic make up of *Pseudomonas*. Some of them act as fertility factors, some may confer resistance to various agents (R plasmids), or some may give to the cell the capacity for degradation of unusual carbon sources, thus contributing to the nutritional versatility of many members of the genus. A large number of *Pseudomonas* plasmids have been described in recent years. Table 4.6, prepared with data taken from Jacoby (1979; and personal communication) presents some of the salient properties governed by genes located in plasmids.

Classification of *Pseudomonas* plasmids can be most effectively achieved by compatibility, that is, by the capacity of a given plasmid to coexist with other plasmids in the same cell. In *Pseudomonas*, at least 10 incompatibility groups have been defined (Jacoby, 1977; Jacoby and Shapiro, 1977; Korfhagen et al., 1978). Various aspects of *Pseudomonas* plasmid research are discussed by Jacoby (1979). Plasmids of the various incompatibility groups have different host ranges. The widest range is that of IncP-1 plasmids, while those of groups 2, and 5–9 are more specific. Plasmids of IncP-2 include R plasmids and plasmids carrying genes for the degradation of unusual carbon compounds. Some of them are among the largest of *Pseudomonas* plasmids, exceeding 300 megadaltons, while the majority of plasmids in general range from 10–60 megadaltons. Outside of IncP-2, degradative plasmids are also found in IncP-9, but several remain unclassified as to compatibility. Degradative plasmids named CAM, OCT, SAL, NAH, TOL, XYL, are involved in the degradation of camphor, *n*-octanol, salicylate, naphthalene, toluates, and xylene, respectively.

Unique *Pseudomonas* strains capable of degrading many unusual compounds by means of enzymatic systems encoded in plasmids, can be developed in the laboratory. A notorious example, which has involved a U.S. Supreme Court decision authorizing patenting of such laboratory developed organisms, has consisted in the transfer to *P. aeruginosa* and *P. putida* strains of various plasmids carrying genes for the degradation of complex hydrocarbons and other compounds present in crude oil (Chakrabarty, 1981). The strains are also active on compounds present in the residual oil after the useful components have been removed from crude oil. The plasmids carrying the genes for these energy-generating peripheral catabolic pathways may be incompatible, but in the present case such problem has been solved by irradiating the cells carrying the incompatible plasmids with ultraviolet, a treatment that may result in the fusion of the plasmids and a greater stability of these elements in the cells.

Evidence for plasmid involvement in the oxidative degradation of tryptophan to the plant hormone indoleacetic acid (IAA) by "*P. savastanoi*" (now named *P. syringae* pv. *savastanoi*) has been reported by Comai and Kosuge (1980). Concentrations of IAA higher than normally found in plant tissues cause gall formation.

The range of antibiotic resistance determined by plasmids and the mechanisms of the resistance in *Pseudomonas* are in general similar to those found in enteric bacteria. R factors in some cases may be nontransmissible, but they may be mobilized by other plasmids which are transmissible but do not carry resistance genes (Holloway, 1979a).

General discussions on *Pseudomonas* phages and bacteriocins have been presented by Bradley (1967) and by Holloway and Krishnapillai (1975). Many different lytic and temperate phages have been identified in *Pseudomonas*, and the morphological diversity of the phages is at least as great as for other bacterial genera. As we have mentioned earlier, lysogeny is a very common phenomenon in *P. aeruginosa*, and transducing phages have been very useful in linkage studies. Most *Pseudomonas* phages contain double-stranded DNA, but some are RNA phages, including one with double-stranded RNA (Semancik et al., 1973; Vidaver et al., 1973).

Even though host specificity is the rule, some phages can plate on different but related species, with interesting taxonomic implications (M. P. Starr, personal communication). Billing (1963, 1970a) has attempted to improve the methodology used for the differentiation of phytopathogenic pseudomonads by inclusion of phage sensitivity tests. Satisfactory differentiation could be achieved among the nomenspecies "*P. phaseolicola*" (*P. syringae* pv. *phaseolicola*), *P. syringae* and "*P. glycinea*" (*P. syringae* pv. *glycinea*), while in the cases of "*P. tabaci*" (*P. syringae* pv. *tabaci*), "*P. garcae*" (*P. syringae* pv. *garcae*) and "*P. morsprunorum*" (*P. syringae* pv. *morsprunorum*), the sensitivity tests were useful only when used in combination with biochemical tests. In many cases the lytic reaction correlated with the host of origin, regardless of the country where the pathogen was isolated. *P. syringae* isolated from pear and "*P. morsprunorum*" (*P. syringae* pv. *morsprunorum*) from cherry could be distinguished by phage tests by Crosse and Garrett (1963) and also by Billing (1970a), although serological tests could not differentiate the two nomenspecies (Lovrekovich et al., 1963). On the other hand, Crosse and Garrett (1963) could not readily distinguish pathogenic pseudomonads from saprophytic ones by their lysotypes. Other taxonomically interesting findings have been reported by several authors, among them Taylor (1972) and Sato and Takahashi (1972).

Table 4.6.
Properties of **Pseudomonas** *coded by plasmid genes*[a]

Resistance characters to:
Antibiotics: carbenicillin, chloramphenicol, gentamicin, kanamycin, streptomycin, tetracycline, tobramycin, sulfonamide.
Chemical and physical agents: borate, chromate, mercury ion, organomercurials, tellurite, ultraviolet radiation.
Bacteriophage: interference with bacteriophage propagation; interference to lysogenization by some temperate phages; DNA restriction and modification.
Bacteriocins.
Other characters
Chromosome donor ability
Donor-specific phage susceptibility
Inhibition of bacteriocin production
Fertility inhibition
Incompatibility with other plasmids

[a] Information taken from Jacoby (1979), with permission from the author and Academic Press, Inc.

Phage sensitivity can also be used successfully at the subspecies level. Typing of *P. aeruginosa* isolates by their sensitivity to lytic phages is currently in use, although it is not considered a method to be used by itself as an epidemiological tool without comparison to other typing procedures (Brokopp and Farmer, 1979). A comprehensive discussion of various phage typing procedures for *P. aeruginosa* is given by Bergan (1978). See also the chapter cited in the reference by Brokopp and Farmer (1979) for useful information on phage typing in general.

Some factors required for phage nucleic acid replication appear to have been conserved through evolution in some groups of Gram-negative bacteria. One of the elements required for in vitro replication of sex-specific single stranded RNA coliphage Qβ is the so-called host factor (HF). This is a heat-resistant RNA-binding protein of molecular weight 12,000, usually present in the *E. coli* cells as a hexamer (Franze de Fernández et al., 1972). In *P. putida*, DuBow and Blumenthal (1975) have identified a polypeptide of molecular weight 11,000 which gives an immunological cross-reaction with *E. coli* HF and allows Qβ replication in vitro. In a survey on various species of pseudomonads, DuBow and Ryan (1977) were able to detect material cross-reacting with *E. coli* HF antiserum in the extracts of *P. aeruginosa*, *P. fluorescens*, *P. putida*, *P. stutzeri*, *P. maltophilia*, "*Xanthomonas vesicatoria*" (now named *Xanthomonas campestris* pv. *vesicatoria*), *P. delafieldii*, *P. facilis*, *P. acidovorans*, *P. cepacia*, and *P. solanacearum*, but not in the extracts of *P. diminuta* and *P. vesicularis*. Strains of the first six species (RNA groups I and V) gave extracts which after heat treatment had HF activity. Heat-treated extracts of *P. delafieldii*, *P. facilis*, *P. acidovorans*, *P. cepacia*, and *P. solanacearum* (RNA groups II and III) did not stimulate replication of Qβ RNA and, in addition, they inhibited the activity of systems containing saturating amounts of *E. coli* HF. Species of RNA groups II and III therefore produce a heat-stable inhibitor of Qβ RNA replication. *P. diminuta* and *P. vesicularis* constitute RNA group IV, which is only remotely related to the other RNA groups (Palleroni et al., 1973), and the findings of DuBow and Ryan (1977) are a confirmation of the unique properties of this natural group.

Bacteriocins are proteins produced by some bacterial strains, which have a lethal action on other strains of the same species. They have been frequently found in strains of *Pseudomonas*. In *P. aeruginosa* the names pyocin (from the old name *P. pyocyanea*) and aeruginosin are in use, but both names are objectionable. Pyocin can be mistaken for pyocyanin, the blue diffusible pigment characteristic of the species, while the name aeruginosin has been used for two red pigments produced by some strains (aeruginosin A and B) (Holliman, 1957; Herbert and Holliman, 1964).

Bacteriocins of *P. aeruginosa* have been classified by Bradley (1967) into an S type, of amorphous appearance and sensitive to proteolytic enzymes, and an R type, which has the structure of phage components and is insensitive to proteolysis.

The bacteriocinogenic condition can be detected and assayed very easily, and in practice it can be used with advantage in methods of typing of *P. aeruginosa* strains for epidemiological purposes, as first suggested by Holloway (1960). Pyocin typing can be performed in two ways. One is by determining the type of pyocin produced by the strain under study against a collection of sensitive indicator strains, and the other, by assessing the sensitivity of the strain to known pyocins. The first method is commonly used, although the second is the easier. The methodology for pyocin typing is described by Govan (1978) and Brokopp and Farmer (1979).

Bacteriocin production in species other than *P. aeruginosa* has been reported in a few instances. Bacteriocin typing of unknown *P. syringae* isolates was found to be of limited value in assessing the host plant of origin. Bacteriocins of "*P. glycinea*" (*P. syringae* pv. *glycinea*) ("glycins") are more restricted in host range, and the highest specificity was shown by the phaseolicins (from "*P. phaseolicola*" (*P. syringae* pv. *phaseolicola*)) (Vidaver et al., 1972). Vidaver and Buckner (1978) extended these studies to a larger number of isolates, with results that did not appear as clear-cut as before. Correlation between bacteriocin type and host plant of origin or phytotoxin production was rather poor.

Typing of *P. syringae*, however, still can be performed, since 86% of the strains examined were able to produce bacteriocins.

The possibility of using phages and bacteriocins as control agents for plant diseases caused by various bacterial pathogens including *Pseudomonas*, has been discussed by Vidaver (1976).

In summarizing the findings on the location of prophages and determinants of bacteriocins in the *P. aeruginosa* chromosome, Holloway et al. (1979) point out that there are three main regions where these elements are found most frequently, and that these regions should be taken into consideration for the development of specialized transduction systems.

Antigenic structure. The antigenic properties of *P. aeruginosa* have been studied more thoroughly than those of other *Pseudomonas* species, and here again the interest of workers reflects the medical importance of this species.

Agglutination of intact *P. aeruginosa* cells can be caused by specific antibodies elicited in animals by cell components that are very similar to the O-antigens of other Gram-negative bacteria. The heat-stable O-antigen, considered to be the most stable marker for *P. aeruginosa*, is represented by the lipopolysaccharide (LPS). Mild heating of a cell suspension reduces its agglutinability by specific antisera, but the property can be fully recovered by a more intense treatment (100–120°C for 2–2.5 h). Mucous strains of *P. aeruginosa* can be selected under certain artificial conditions (Govan and Fyfe, 1978), or they can be directly isolated from patients with cystic fibrosis (Doggett, 1969). The mucus, which is similar to alginate, does not interfere with the O-antigenicity.

In addition to the immune response against *P. aeruginosa* LPS, which in its early manifestation consists of IgM antibodies (Høiby, 1979), there is a humoral response to cross-reactive antigens which are present in many other, mostly Gram-negative, bacteria (Høiby, 1975). A heat labile antigen common to a wide range of bacteria has been isolated from *P. aeruginosa* and shown to be an acidic protein composed of subunits of molecular weight 62,000, present in the cytoplasmic fraction of the cells (Sompolinsky et al., 1980a, b).

Various other antigens can be identified in *P. aeruginosa*. Heat-labile surface antigens are represented by flagella and fimbriae (Pitt and Bradley, 1975; Bradley and Pitt, 1975). An extracellular slime can elicit the production of an agglutinin, and exoenzymes such as phosphatases, proteases and phospholipases can also act as antigens.

We have already mentioned typing procedures for *P. aeruginosa* strains based on bacteriocin production and sensitivity to lytic phages. Other methods such as biotyping (or characterization of strains by biochemical and physiological reactions) and antibiograms (or spectra of sensitivity to antibiotics) can also be used. However, the most important method for epidemiological purposes is serotyping which, as usually practiced, can be described as the reaction of a cell suspension toward a standard set of antibody preparations.

Many schemes of serotyping have been proposed, but the system of Habs (1957) has gained wide acceptance and, with some modifications, is at present in general use. Habs defined 12 somatic groups which could be identified by agglutination tests. Different additional proposals and modifications have made it hard to develop a standard procedure, but the situation may improve in the near future, since an international antigenic typing scheme based on Habs' 12 original O-groups plus 5 additional ones, is under consideration by the Subcommittee on *Pseudomonas* of the International Committee on Systematic Bacteriology. A table comparing the O-antigens of the major *P. aeruginosa* serotyping schemes can be found in Brokopp and Farmer (1979). Standard strains and typing sera are at present commercially available.

In the opinion of Brokopp and Farmer (1979), serological typing of *P. aeruginosa* based on the O-antigens produces more reliable evidence for relatedness than can be obtained by other, less specific typing methods. Thermolabile surface antigens can also be used for serotyping, and various methods based on flagellar (H) antigens have been proposed (Verder and Evans, 1961; Lányi, 1970). These approaches to serotyping have not been widely accepted, the main reason being the difficulties encountered in the preparation of specific flagellar antisera (Brokopp and Farmer, 1979).

Comprehensive discussions on the antigenic composition of *P. aeruginosa* and its relationship to virulence as well as to the practical aspects of serotyping have been presented by Lányi and Bergan (1978) and by Véron and Berche (1976).

Serological studies on the agents of melioidosis and of glanders (*P. pseudomallei* and *P. mallei*, respectively) have been described by Murase et al. (1952), Levine and Maurer (1958), Nigg (1963), Redfearn (1964), Gangulee et al. (1966), Sen et al. (1968), and Strauss et al. (1969). Serological relationships between the two bacterial species were reported for the first time by Stanton and Fletcher (1921). Occult or latent forms of glanders disease can be revealed by the intrapalpebral test based on the antigenicity of the endotoxic substance mallein, but other serological reactions such as agglutination, complement fixation and precipitin tests can also be used (Sen et al., 1968).

From *P. facilis* cell envelopes, Rittenhouse et al. (1973) have isolated a phospholipoprotein (PLP) with which they could produce an antiserum that agglutinates a wide range of Gram-negative bacteria including not only *Pseudomonas* species but also species of many other genera. The protein component of the PLP is similar to the protein isolated from *P. aeruginosa* cell envelopes by Homma and Suzuki (1966), and considered to be part of the lipopolysaccharide-containing endotoxin.

Immunological cross-reactions have also been demonstrated by Detrick-Hooks and Kennedy (1974) among strains of "*Hydrogenomonas*," *Pseudomonas* and *Alcaligenes*. The authors have demonstrated the presence of three antigens common to "*H. facilis*" (*P. facilis*) and "*H. eutropha*" (*Alcaligenes eutrophus*). One of these antigens was related to antigens present in six other *Pseudomonas* species. The authors used the immunodiffusion technique and could not confirm the same antigenic relationships by agglutination. One conclusion of the work is that members of the three genera examined represent a continuous series of serologically interrelated organisms. In contrast, the work of Pootjes (1964) and Pootjes et al. (1966) has demonstrated the sharp specificity of *P. facilis* bacteriophages, by means of which this species appears unrelated to many other species of the genus.

In the field of plant pathogenic pseudomonads, the serological approach has had variable success, and has not helped substantially in the circumscription of nomenspecies. The lack of a solid taxonomic frame of reference has often made it very difficult to interpret the resolving power of the serological techniques. Otta and English (1971) were unable to define a precise serological differentiation of virulent strains from nonvirulent ones. On the other hand, species-specific antigens could be identified in "*P. phaseolicola*" (*P. syringae* pv. *phaseolicola*) (Guthrie, 1968) and in "*P. lachrymans*" (*P. syringae* pv. *lachrymans*) (Lucas and Grogan, 1969a, b). Otta (1977) succeeded in distinguishing *P. syringae* strains isolated from different groups of host plants. In general, however, reproducibility of serological data by different authors using different strains of the same nomenspecies, appears to be rather poor.

Antibiotic sensitivity. Many *Pseudomonas* species are resistant to a number of antibacterial agents. Antibiotic and drug susceptibility may have important practical implications, and is occasionally included in the description of new species. However, research on *P. aeruginosa* has been more systematic and thorough than on other species of the genus, and in this section we shall refer almost exclusively to this species. For a more comprehensive discussion see Bryan (1979).

Few penicillins are active against *Pseudomonas*, and even though a MIC of 5 µg/ml for penicillin G has been reported for *P. saccharophila* (Palleroni, 1980), the resistance of most other species is much higher. A β-lactam commonly used against *P. aeruginosa* infections is carbenicillin, either by itself or in mixture with aminoglycosides. Carbenicillin produces cell enlargement and filament formation, but the sensitivity is in general not very high and the MIC ranges from 25–200 µg/ml. The emergence of carbenicillin-resistant mutants in response to carbenicillin treatment has been reported by Gaman et al. (1976) and Lowbury et al. (1969). Mucoid strains of *P. aeruginosa* are somewhat more resistant than the other colony variants.

Resistance to carbenicillin may be due to β-lactamases whose genes are carried by plasmids. The plasmids may belong to IncP-1 (Sykes and Richmond, 1970) and the resistance can be transferred to enteric bacteria (Lowbury et al., 1969), but resistance genes can also be found in IncP-2 plasmids, which are host restricted (Bryan, 1979). Seven different kinds of β-lactamases have been found by Jacoby and Matthew (1979) in 24 *Pseudomonas* plasmids belonging to at least eight of the incompatibility groups. In addition to β-lactamases, resistance may also be due to difficulty of penetration of the antibiotic to the target site (Zimmermann, 1979, 1980). The above mechanisms of resistance, however, fail to account for the fact that success in the treatment of *P. aeruginosa* with carbenicillin has been quite variable, since the failure is often independent of the development of carbenicillin-resistant populations.

β-Lactamase genes are not invariably carried by plasmids and, in fact, there is in *P. aeruginosa* a constitutive penicillinase ("Dalgleish enzyme") (Furth, 1975) and an inducible cephalosporinase also able to act on penicillins (Sabbath et al., 1965), whose genetic determinants are chromosomic (Sykes and Matthew, 1976).

Mutants of *P. aeruginosa* sensitive to β-lactam antibiotics without substantial change in their β-lactamase activity have been reported by Sykes and Matthew (1976), but in other instances increased sensitivity (up to several thousand-fold) correlates with much lower β-lactamase activity (unpublished observations).

Response to β-lactam antibiotics in *Pseudomonas* can also result in their utilization as substrates for growth. Johnsen (1977) isolated from soil a strain of *P. fluorescens* which was able to utilize benzylpenicillin as carbon, nitrogen and energy source. All 16 strains of *P. cepacia* and some strains of "*P. marginata*" (*P. gladioli*) and *P. caryophylli* tested by Beckman and Lessie (1980) appeared to be able to utilize penicillin G for growth. The benzylpenicillin derivatives ampicillin and carbenicillin were not utilized, although the strains were resistant to these antibiotics. The penicillin-utilizing strains had high levels of β-lactamase activity, but the enzyme did not appear to be necessary for growth on penicillin. Mutants lacking β-lactamase and the capacity for growth on β-lactam were used as recipients of plasmid RP1 from *P. aeruginosa*. β-Lactamase activity was restored, but the transconjugants did not regain the ability to use penicillin for growth.

Gentamicin, tobramycin and amikacin are among the most effective aminoglycosides against *P. aeruginosa*. The sensitivity to these antibiotics cannot be explained in simple terms, even though the predominant effect is the inhibition of protein synthesis, and the estimation of the antibiotic activity in vitro is often difficult to extrapolate to situations in vivo. The effect of pH and ionic composition of the test medium (particularly the concentration of divalent cations) are factors to be controlled closely (Garrod and Waterworth, 1969).

The most common mechanisms of gentamicin resistance in *P. aeruginosa* are enzymatic modifications of the antibiotic consisting of *N*-acetylation and O-adenylylation and, to a much lesser extent, phosphorylation. These properties are controlled by genes carried mainly by IncP-2 plasmids, although IncP-3 plasmids are also involved (Bryan et al., 1972, 1973, 1974; Jacoby, 1974a, b). Permeability factors may also be involved in gentamicin resistance (see Bryan et al., 1975; Bryan, 1979). Resistance to aminoglycosides other than gentamicin may also be based on modification of the antibiotic and occasionally on diminished permeability (Mathias et al., 1976).

Polymyxins are very active against *P. aeruginosa* and other pseudomonads when tested in vitro, but their efficacy in vivo is limited for a number of practical reasons. Resistant strains can be obtained, but the condition is unstable in *P. aeruginosa* and it seems to correlate with an increased EDTA tolerance. Gilleland and Murray (1976) noted in these variant strains the disappearance of the particles characteristic of the concave cell wall layer separated by freeze-etching (see Cell Morphology and Fine Structure, above), but the particles—and the sensitivity to EDTA—reappeared in cells of the same strains grown in the absence of polymyxin. Acquisition of a temporary resistance to polymyxin by the population is accompanied by a decrease in the phosphorus content of the outer membrane, in line with Wilkinson's (1970) observation on the relationship between EDTA sensitivity and the phosphorus content of the LPS of pseudomonads (see Composition of Cell Envelope, above). Growth under conditions of limiting Mg concentrations has a similar effect (Brown and Melling, 1969).

About one-half of *P. aeruginosa* plasmids confer resistance to mercuric ions (Jacoby and Shapiro, 1977). Resistance to chromium, boron and tellurium is also determined by some *Pseudomonas* plasmids (Summers et al., 1978).

Pathogenicity for plants. Phytopathogenic pseudomonads are allocated in three of the five RNA homology groups, groups I, II and V. In this last group, the pathogenic species are currently assigned to the genus *Xanthomonas.*

Various symptoms produced in plants by pathogenic pseudomonads, such as tumorous outgrowth, rot, blight or chlorosis, and necrosis, are caused by alteration of the normal metabolism of plant cells by substances excreted by the pathogen. Among these excretions there are enzymes capable of attacking various components of plant tissues, toxins, and plant hormones.

Some hydrolytic enzymes that are found in plant pathogenic pseudomonads and not in saprophytes include β-glucosidases (Hildebrand and Schroth, 1964; Hayward, 1977), xylosidases or xylanases (Hayward, 1977; Maino et al., 1974), pectinases (Hildebrand, 1971; Wilkie et al., 1973), of which the most common sort is polygalacturonic transeliminase (Ohuchi and Tominaga, 1973, 1975), and cellulases (Gehring, 1962; Lange and Knösel, 1970). Rots are mainly produced by the nonfluorescent plant pathogens (*P. cepacia, P. caryophylli, P. gladioli*) and these alterations are due to active pectinolytic enzymes and cellulases (Hildebrand, 1971; Gehring, 1962). Some of the hydrolytic enzymes may be involved in the nutrition of the pathogen (Hildebrand, 1971), but the products of hydrolysis by β-glucosidases may not be utilized for growth (Joubert et al., 1970).

Various nomenspecies of plant-pathogenic bacteria described in this treatment as pathovars of *P. syringae (syringae, tabaci, phaseolicola, coronafaciens, glycinea, tomato)* have been found to produce phytotoxins capable of provoking disease symptoms, principally chlorosis, in susceptible plants. *P. syringae* pv. *phaseolicola*, the agent of halo blight in beans, produces several toxins of which the best characterized is known by the trivial name of phaseolotoxin (Mitchell, 1976). *P. syringae* pv. *glycinea* secretes glyotoxin (Strobel, 1977), *P. syringae* produces syringomycin (Gross and De Vay, 1977), and tabtoxin is characteristic of *P. syringae* pv. *tabaci* and *P. syringae* pv. *coronafaciens* (Sinden and Durbin, 1970). Several phytotoxins have an amino acid or peptide nature (Leisinger and Margraff, 1979). Interestingly, tabtoxin has a β-lactam structure, a rare example among secondary metabolites of *Pseudomonas* (Stewart, 1971; Durbin et al., 1978). However, recently two new β-lactam antibiotics were found to be produced by acidophilic pseudomonads (Imada et al., 1981).

Toxins do not necessarily parallel the host specificity of the respective phytopathogen (Patil, 1974). Their importance as taxonomic tools is very limited, but their practical significance is considerable. In certain instances their production correlates with the presence of plasmids in the cells, and this has provoked renewed interest in these compounds in recent years (Leisinger and Margraff, 1979). Loss of a plasmid by treatment with acridine orange was accompanied by loss of syringomycin production (González and Vidaver, 1977), but no correlation was observed between phaseolotoxin production and the presence of plasmids in *P. syringae* pv. *phaseolicola* (Jamieson et al., 1981).

High concentrations of the plant hormone indole-3-acetic acid produced by "*P. savastanoi*" (now *P. syringae* pv. *savastanoi*) are thought to be the cause of tumors in plants of the family *Oleaceae.*

Pathogenicity for humans and animals. Pseudomonas strains can frequently be isolated from assorted clinical materials and they may be the cause of nosocomial infections, particularly in patients in which the normal host defenses are depressed (neoplasias, burns, cystic fibrosis, etc.). Useful descriptions of sources of isolation and diagnostic procedures for *Pseudomonas* species found in clinical laboratories are given by Gilardi (1971) and by Hugh and Gilardi (1980).

Among the *Pseudomonas* species, the most serious animal pathogens are *P. mallei* and *P. pseudomallei*, of RNA group II. Strains of these two species are the agents of glanders and melioidosis, respectively, two related diseases that can be transmitted from animals to man. *P. mallei* is an nonmotile pseudomonad that is found only in animal hosts, and therefore its distribution is coterminous with that of the natural hosts. Nearly all warm-blooded animals are susceptible to glanders, with the exception of cows, pigs and pigeons. Horses, mules, asses, guinea pigs, hamsters and field mice are very susceptible, and less so are monkeys, sheep, camels, rabbits, dogs, rats, white mice, goats, cats, ferrets and moles. Melioidosis is also remarkably ubiquitous, and the list of animals includes the ones mentioned above and several more (pigs, tree-climbing kangaroos and parrots) (Redfearn and Palleroni, 1975), but its etiologic agent, *P. pseudomallei*, is a free living soil organism restricted, with a few exceptions, to tropical and subtropical zones (Redfearn et al., 1966). The two species are closely related to each other (Redfearn et al., 1966; Rogul et al., 1970), and to other *Pseudomonas* species that are now included in RNA group II (Rogul et al., 1970; Ballard et al., 1970). This natural relationship between the two agents could not be suspected at the time of discovery of melioidosis, but the first description of the disease by Whitmore (1913) indicated its similarity to glanders.

Clinically active cases of human melioidosis are relatively rare but, of those that are brought to the physician's attention, the mortality rate is very high. Treatment with tetracycline, occasionally combined with chloramphenicol, is at present the most effective, but it must be continued for long periods of time. Most episodes, however, can be considered to be of a clinically inactive or subclinical type, which in its mildest manifestations can only be recognized by antibody response. Two toxins have been found in filtrates from broth cultures of *P. pseudomallei*. One of them is a lethal factor with anticoagulant activity, and the other a skin-necrotizing proteolytic agent (Piggot and Hochholzer, 1970).

Human susceptibility to glanders is high, but here again clinical cases are relatively uncommon. However, misdiagnoses may be suspected due to the variety of clinical manifestations of the disease, and numerous subclinical forms may also be suspected, as in melioidosis. *P. mallei* produces an endotoxic substance called *mallein*, which is used as the basis of a confirmatory test for the diagnosis.

Further details on melioidosis and glanders can be found in the chapter cited in the reference by Redfearn and Palleroni (1975).

Other *Pseudomonas* species can be classified as opportunistic human pathogens because of their low pathogenic potential towards normal individuals. In recent years, due to the widespread use of antibiotics and other antibacterial agents to which *Pseudomonas* species are notoriously resistant, these have become important agents of infections in humans and animals. *P. aeruginosa* is at present the most significant of these opportunistic pathogens for the immunocompromised host. Various aspects of the clinical manifestations of *P. aeruginosa* infections and the current trends in therapy are covered in numerous papers and in the book edited by Doggett (1979).

The specific factors responsible for the virulence of *P. aeruginosa* have not yet been defined and various workers differ in their opinion on the relative significance of the various factors identified so far. The role of LPS is still being debated. Liu (1979) attributes pathogenicity of *P. aeruginosa* mainly to its extracellular products, such as the pigments, which may suppress the action of other microorganisms; proteases of localized activity; hemolytic toxins; enterotoxin; and the lethal toxin (exotoxin A), an acidic protein capable of inhibiting protein synthesis at the ribosomal level. Antisera prepared against the latter toxin have been able to protect mice against infection, without need of other antibodies against cellular components (Liu and Hsieh, 1973).

Other *Pseudomonas* species are less frequently found as opportunistic human or animal pathogens. *P. maltophilia*, a species which in the present treatment we propose to exclude from the genus *Pseudomonas*, is the second most prevalent pseudomonad in hospitals and clinical laboratories (Hugh and Gilardi, 1980) and at times is found associated with serious infective processes. Chloramphenicol and colistin are among the compounds most effective against this organism.

P. cepacia is also an important potential pathogen. Strains of this species have been isolated from many clinical materials and from infected tissues in humans, but until recently the isolates were incorrectly identified and even assigned to other genera (Sorrell and White, 1953; Schiff et al., 1961; Hardy et al., 1970). *P. cepacia* strains are resistant to polymyxin B, colistimethate and gentamicin, and sensitive to chloramphenicol, sulfonamides, and carbenicillin (although some resist this β-lactam antibiotic (Hardy et al., 1970)). Trimethoprim-sulfamethoxazole combinations are usually recommended for treatment.

Group Va-1, one of the two biovars recognized by Pickett and Greenwood (1980) within the species *P. pickettii*, has been tentatively considered by these authors to be etiologically significant in human disease. Other well characterized species of *Pseudomonas* (*P. alcaligenes*, *P. stutzeri*, *P. mendocina*, *P. fluorescens*, *P. putida*) are far less prevalent in clinical laboratories and hospitals. Some species lack the capacity for growth at 37°C, but they can grow at refrigerator temperatures and may be troublesome contaminants of clinical materials where they can be the source of endotoxic reactions.

Ecology—habitats and niches. Strains of many *Pseudomonas* species are ubiquitous, and isolation data often throw little light on their ecology. Numerous natural materials are good sources for isolation. In some of the habitats, pseudomonads may represent a minority of the total microbial flora, but under certain conditions (pH close to neutrality, organic matter in solution, temperature in the mesophilic range, good supply of dissolved oxygen), their capacity for rapid growth in the absence of complex growth factors decides their predominance. Even in media of extremely low nutrient content, pseudomonads occasionally multiply to a considerable extent. Thus, *P. aeruginosa* has been found capable of growth at the expense of minor impurities present in hospital distilled water (Favero et al., 1971).

When dealing with organisms of such versatility, ecological conclusions are particularly difficult to draw. Of the 57 strains of fluorescent pseudomonads isolated by den Dooren de Jong (1926), 23 had their origin in soil and, with one exception, all could be classified as *P. putida* on the basis of their incapacity to attack gelatin. All other strains were isolated from water, appeared to be gelatin-liquefiers, and were classified as *P. fluorescens*. The conclusion that *P. putida* is a soil organism while *P. fluorescens* predominates in water could not be supported by an analysis of isolation data of many other strains from Berkeley and from Jessen's collection in Denmark (Jessen, 1965). Careful sample collection is here mandatory, due to the ample opportunities for cross-contamination between the two habitats and the obvious similarities in nutritional versatility of the species involved.

P. spinosa was isolated by Leifson (1962a) from river water, and *P. huttiensis* and *P. lanceolata* from distilled water (Leifson, 1962b). The normal habitat of the Na^+-requiring species *P. marina*, *P. nautica* and *P. doudoroffii* is the ocean (Baumann et al., 1972). Wet soil and mud are recommended as the best sources for the isolation of carboxydobacteria, while dry soils are totally inadequate (Zavarzin and Nozhevnikova, 1977).

Saprophytic fluorescent pseudomonads are very common in soils, and they also abound in plant rhizospheres, where they seem to have a stimulating effect on plant growth. This may be due in part to inhibition of plant pathogenic organisms by iron starvation (Kloepper et al., 1980), although utilization of the iron-siderophore complex by the plant may also be involved (Powell et al., 1980). The predominant fluorescent pseudomonad in wheat rhizosphere is *P. fluorescens* biovar G (Sands and Rovira, 1971).

The rhizosphere of alder trees has been the source of isolation of *P. rhodos* (Heumann, 1962). Debette and Blondeau (1980) have found a predominance of *P. maltophilia* over other pseudomonads in the rhizospheres of several cultivated plants, such as cabbage, rape, mustard, corn and beet. Some of these plants have proteins of high sulfur content, and the abundance of *P. maltophilia*, which requires methionine for

growth, may be due to excretion of S-containing amino acids by the roots. However, methionine could not be detected in significant amounts in the rhizosphere.

Many species of *Pseudomonas* have been isolated from *Lolium* leaves (Stout, 1960), and Austin et al. (1978) have often found *P. fluorescens* and less frequently *P. maltophilia* and *P. cepacia*, as well as groups of "pink chromogens," one of which has been described as a new species under the name *P. mesophilica* (Austin and Goodfellow, 1979). In their review of the literature, these authors mention isolation of *P. fluorescens* from leaves of *Phaseolus vulgaris*, *Fagus*, and *Pinus* by various workers, and consider this species of *Pseudomonas* indigenous to leaf surfaces. Whether the strains can become opportunistic plant pathogens is an open question, but some of the epiphytic pseudomonads have definite pathogenic potentialities, as was suggested by Billing (1970b) for *P. viridiflava*.

Plant pathogenic pseudomonads are normally isolated from lesions in plant hosts, which are natural enrichment cultures. Because of their ecological niches, these bacteria may offer some of the most interesting materials for the study of bacterial speciation (Palleroni and Doudoroff, 1972). In general, animals are not as good sources of pseudomonads, since these are frequently opportunistic pathogens, with the exception of *P. mallei*, which is isolated from equids affected with glanders (Redfearn and Palleroni, 1975).

In different areas, speciation of plant pathogens may proceed at different rates according to the selective forces imposed by various environmental factors. While the nomenspecies *P. syringae* and "*P. morsprunorum*" appear as a single species by many criteria, Garrett et al. (1966) consider "*P. morsprunorum*" as a distinctive ecotype derived from *P. syringae* under the selective pressures of the horticultural regions of southeast England.

Continued association with living hosts is important to the survival of members of the *P. syringae* group (Schroth et al., 1981), and the association does not necessarily have to involve lesions, since the pathogen may be able to survive as an epiphyte on hosts and nonhosts (Ercolani et al., 1974; Leben, 1974). Survival of the *P. syringae* pathovars in soil may not be long, in contrast with other fluorescent species and the plant pathogens of RNA group II (Schroth et al., 1981). Indeed, *P. cepacia* has been frequently isolated from soil as well as from many clinical specimens. In soils it is not restricted to tropical and subtropical areas as *P. pseudomallei* (Redfearn et al., 1966), another member of the same group with which *P. cepacia* has marked similarities, including a high maximal temperature. In view of its extraordinary nutritional versatility, and its capacity for surviving in soil, it is also hard to understand why *P. cepacia* does not predominate in soils. For reasons that are not yet clear, it is outgrown by fluorescent pseudomonads at temperatures close to the optimum for both groups.

Enrichment and Isolation Procedures

Direct isolation of pseudomonads can often be achieved from many natural materials, especially soil and water. Extensive lists of sources for several species are given by Hugh and Gilardi (1980), with particular emphasis on clinical sources, and by Palleroni (1981).

A medium that is frequently used for direct isolation, especially by plant pathologists, is medium B of King et al. (1954),* which enhances pyoverdin production and is also satisfactory for general purposes. Some selective solid media have been developed on the basis of medium B. An example is the medium proposed by Sands and Rovira (1970) which contains penicillin G, novobiocin and cycloheximide. These compounds do not inhibit the fluorescent pseudomonads, and their colonies can be identified on the plates by the characteristic diffusible pigment. Another modification of medium B was later published by Sands et al. (1972) for the isolation of pectolytic pseudomonads. The medium contains 0.5% sodium polypectate, and the pectolytic activity can be detected by flooding the plates with a 1% solution of hexadecyltrimethylammonium bromide, which precipitates the intact pectin;

---

*Medium B (g/liter): Proteose-peptone (Difco), 20.0; Bacto-agar (Difco), 15.0; glycerol, 10.0; KH_2PO_4, 1.5; and $MgSO_4 \cdot 7H_2O$, 1.5. The pH is adjusted to 7.2.

colonies are isolated soon after identification, since the compound is toxic to the cells.

Selective media for oxidase-negative fluorescent plant pathogens have been described by Sands et al. (1980). Oxidase-positive, nonfluorescent phytopathogens can be isolated from various sources, including soils, using the medium described by Sumner and Schaad (1977). The isolation and characterization of *P. solanacearum* is facilitated by the use of Kelman's (1954) medium, containing 2,3,5-triphenyltetrazolium chloride, which allows a differentiation of virulent and avirulent mutants of the species. Various other media and procedures for the isolation of phytopathogenic bacteria can be found in the laboratory guide edited by Schaad (1980) and in the chapter by Schroth et al. (1981).

A solid medium composed of a mineral agar base and an overlayer with a suspension of poly-β-hydroxybutyrate granules (see Procedures for Testing for Special Characters, below) has been successfully used by Delafield et al. (1965) for the direct isolation of *P. lemoignei*.

Practical methods for the isolation of certain species of the genus include enrichment. The lesions produced by plant pathogenic pseudomonads on various plant organs represent a natural enrichment, and isolation of pure cultures from these lesions is common practice. The lesions selected for isolation should not be too old, when other organisms may have gained access to the diseased tissues.

Strains of the acidovorans group (*P. acidovorans* and *P. testosteroni*) are nutritionally versatile and can be found in many soil samples, often predominating in enrichments with higher dicarboxylic acids or with kynurenate at 30°C (Stanier et al., 1966). Maleate and D-tryptophan should be equally effective, since they are good carbon sources for strains of this group and are not used by many other pseudomonads. Use of *cis, cis-* or *cis, trans*-muconic acid as the sole carbon source has been found very effective for the enrichment of *P. acidovorans* (Robert-Gero et al., 1969), while imidazole-propionate or imidazole-lactate gives a population enriched in *P. testosteroni* (Coote and Hassal, 1973; Hassal, 1966; Hassal and Rabie, 1966).

Denitrification enrichments with various carbon sources can be recommended for the isolation of *P. aeruginosa*, *P. stutzeri*, *P. mendocina* and *P. pseudomallei*, particularly when the incubation is carried out at 37°C. Denitrification using citrate as the carbon source at the same temperature has allowed the isolation of *P. indigofera* (Elazari-Volcani, 1939).

The nitrogen-deficient medium at pH 5 developed by Becking (1961) inoculated with a sample of soil of the fossil, lateritic type, gave enrichment cultures from which *P. glathei* was isolated (Zolg and Ottow, 1975).

Enrichments in a mineral medium devoid of organic carbon sources, incubated under an atmosphere of hydrogen, oxygen and carbon dioxide, easily yield hydrogen pseudomonads. Some of these organisms are sensitive to oxygen when grown under autotrophic conditions, and therefore a general purpose enrichment should be carried out under an atmosphere with no more than 10% oxygen. Hydrogen and carbon dioxide can be replaced by carbon monoxide as both carbon and energy source for the enrichment of carboxydobacteria.

Enrichment cultures can sometimes be designed in more than one step, including a succession of cultural conditions highly selective for a particular group of organisms. However, starting from sources of formidable microbiological complexity such as soil, these enrichments often fail to give the desired results, a reflection of our ignorance of the proper conditions under which such experiments should be conducted.

Maintenance Procedures

Most *Pseudomonas* strains can be maintained on slants of common bacteriological media (nutrient agar or other standard complex media, or various chemically defined media with the addition of 0.5% yeast extract and 0.1% lactate or glycerol), with transfers every 1 or 2 months.

Slants can be kept at 4–8°C. The collection examined by Stanier et al. (1966) included a number of strains of fluorescent and nonfluorescent pseudomonads that had been kept on slants of ordinary media with periodic transfers since the 1920s, and their phenotypic properties appeared to have remained essentially unchanged.

Many strains are not so easily maintained. For example, many of the species of RNA group II die in a relatively short time when kept on agar slants in the refrigerator. Examples of the conflicting statements of different authors on viability of one of the species of this group, *P. solanacearum*, have been discussed by Kelman (1953). The discrepancies are attributed to differences in storage temperatures, rate of desiccation and composition of culture medium. The author recommends covering the slants with sterile mineral oil, which appears to be satisfactory for the maintenance of both viability and virulence.

P. saccharophila has been kept for long periods of time at room temperature in liquid mineral medium with sucrose, but did not survive for very long on the same medium solidified with agar either at room temperature or in the refrigerator. This maintenance procedure, however, cannot be recommended for strains of most other species.

Lyophilization or liquid nitrogen techniques appear to be the safest for the preservation of all strains of the genus. Cell suspensions of many strains can be dried under vacuum without prior freezing. A practical method consists in placing a labeled, sterile strip of thick filter paper in a tube, soaking the paper with the cell suspension, and pushing a cotton plug into the tube to about 1 cm from the paper strip. The tube is immediately attached to the high vacuum line of any freeze-drying machine. After a few hours, the tube is sealed under vacuum. For reactivation of the culture, the tube is opened after marking with a file just above the cotton plug, which will filter the air rushing into the tube, and the paper strip is removed with tweezers and dropped into a liquid culture medium.

Procedures for Testing Special Characters

Screening of nutritional and other physiological properties. The medium described by Stanier et al. (1966) has been used extensively for nutritional screenings and for testing various physiological properties of *Pseudomonas* strains. More satisfactory results are obtained using the simpler medium recommended by Palleroni and Doudoroff (1972),* which is most satisfactory for autotrophic and heterotrophic enrichment and cultivation of all *Pseudomonas* species. Mineral media of similar composition have been described by Schlegel and Lafferty (1971) and by Zavarzin and Nozhevnikova (1977). These last authors recommend supplementation of the medium with minor elements for the growth of the carboxydobacteria, but no such supplementation seems to be necessary for the strains of other *Pseudomonas* groups. *Pseudomonas* strains also grow very well in the mineral medium described for *Arthrobacter* by Owens and Keddie (1969).

Good heterotrophic growth in all these mineral media can be obtained by addition of a single organic compound as carbon and energy source of 0.1%. Experimental details of the nutritional analysis have been extensively discussed by Stanier et al. (1966) and by Palleroni and Doudoroff (1972). Ideally, the analysis should be performed under conditions which prevent cross-feeding or competition among the various strains when the medium is solidified with agar and several strains are patched on a single plate. Organic impurities in the agar may also be a serious problem. For critical tests liquid media may be preferable, although the amount of work and the facilities required may be considerably greater. Carbon compounds to be tested in different concentrations or requiring special treatments for their use in nutritional screenings have been discussed by Palleroni and Doudoroff (1972).

Pigment production. Perhaps the most widely used medium for pyoverdin production is the medium B of King et al. (1954). Paton (1959) has described a medium whose preparation involves extensive

---

*Medium of Palleroni and Doudoroff (1972) (g/liter of 0.33 M Na-K phosphate buffer (pH 6.8)): NH_4Cl, 1.0; $MgSO_4 \cdot 7H_2O$, 0.5; ferric ammonium citrate, 0.05; and $CaCl_2$, 0.005. The first two ingredients are added to the buffer and sterilized by autoclaving. The ferric ammonium citrate and $CaCl_2$ are added aseptically from a single stock solution that has been sterilized by filtration.

treatment of the solution and the agar with iron-chelating agents such as oxine (8-hydroxyquinoline) and EDTA. This procedure has been criticized by Garibaldi (1967) because the removal of iron is so complete that growth of bacteria is often inhibited. Garibaldi recommends a medium which is based on the well known iron-chelating capacity of conalbumin, a component of egg white.* Luisetti et al. (1972) recommended a medium† for the enhancement of fluorescent pigment production by plant pathogenic and other pseudomonads which fail to fluoresce in the medium B of King et al. (1954).

A point that is frequently incorrectly stated or disregarded altogether in descriptions of procedures for the production of pyoverdins by fluorescent pseudomonads is the type of ultraviolet lamp to be used for the observation of the fluorescence. Some *Pseudomonas* species (*P. cepacia, P. gladioli, P. caryophylli*) produce diffusible yellow-green pigments that are sometimes mistaken for the fluorescent pigments; these can be readily distinguished from the pyoverdins by examination of the cultures on solid media under a source of ultraviolet light of short wavelength (~254 nm), under which only the pyoverdins will fluoresce.

Medium A of King et al. (1954)§ is generally recommended for the production of phenazine pigments. In our experience this medium is not always effective, but unfortunately no other medium can be recommended strongly as an alternative. Phenazine pigment production often appears erratic, particularly with cultures that have been kept for a long time under ordinary laboratory conditions.

The blue pigment of *P. fluorescens* biovar IV ("*P. lemonnieri*") can be produced abundantly by fresh cultures in the potato medium¶ familiar to mycologists. As mentioned earlier, the pigment of this organism is related to indigoidine (Starr et al., 1967), and it is interesting to mention that Starr (1958) had already noticed the convenience of using a medium based on potato for the production of indigoidine by a totally unrelated organism, *Corynebacterium insidiosum*.

Indigoidine is also characteristic of *P. indigofera*. A good medium for the enhancement of indigoidine production by this species has been formulated by Sommer et al. (1961). ‖ A mineral medium supplemented with methionine, glutamine, aspartic acid and arginine has also been recommended by Sommer et al. (1961).

The oxidation of indole to indigotine by some *Pseudomonas* strains can be observed in the medium described by Gray (1928).**

Accumulation of poly-β-hydroxybutyrate (PHB). Accumulation of PHB can be enhanced in the liquid medium of Palleroni and Doudoroff (1972) already described, but modified to contain one-fifth of its ammonium chloride content. The medium is supplemented with a single carbon source, DL-β-hydroxybutyrate, at a concentration of 0.5%. The low nitrogen content of the medium reduces protein synthesis and the carbon compound is preferentially converted into polymer.

Hydrolysis of PHB. Preparation of PHB granules has been described by Delafield et al. (1965). Good yields of granules can be obtained from *Bacillus megaterium* strain KM (ATCC 13632). The cells are grown in the medium described by Macrae and Wilkinson (1958)†† for 18 h at 30°C with aeration, when the culture will be at the end of the exponential phase of growth. The cells are collected by centrifugation and the pellets are suspended in cold phosphate buffer, pH 7.2. The suspension is sonicated for 10 min in a 10-kc Raytheon oscillator and then centrifuged. A 5% (wet weight/volume) of the pellet in alkaline hypochlorite (see below) is incubated at 37°C for 2 h with continuous stirring. The polymer granules are suspended in water and dialyzed against running water until no hypochlorite reaction is observed with the starch-iodide reagent. Last traces of chlorine can be eliminated by washing the granules with a $Na_2S_2O_3$ solution. After further dialysis vs. distilled water, the polymer is centrifuged, washed with a little acetone and finally placed in a Soxhlet (or similar solid/liquid) extracting apparatus, in which it is continuously extracted for 3 days with acetone-ether (2:1, v/v). The semidry powder is finally pulverized, dried in a desiccator, and stored at room temperature. Based on the dry weight of bacteria, the final yield of PHB is about 25%.

Strain KM of *B. megaterium* is an asporogenous mutant, and thus disappearance of PHB as the cells become committed to sporulation in this species is eliminated, and the moment of harvesting the cells is not too critical. We have also used various strains of *Pseudomonas* of the species *P. acidovorans, P. testosteroni* and *P. cepacia*, as sources of polymer granules, and the medium recommended above for the test of PHB accumulation is very convenient.

The alkaline hypochlorite reagent solution can be prepared from fresh bleaching powder according to the method described by Williamson and Wilkinson (1958), or sodium hypochlorite solution (reagent grade, ~7% (w/v)) can be used.

For the test of PHB hydrolysis, the method described by Delafield et al. (1965) is most convenient. Homogeneous suspensions of PHB granules in water are obtained with the aid of a Potter-Elvehjem homogenizer and a sonic oscillator. PHB overlay plates are prepared by pouring 4 ml of melted 2% agar containing 5 mg of dispersed polymer onto a solidified mineral agar plate that has been prewarmed to 50°C. The polymer agar for the overlay can be sterilized in the autoclave.

Testosterone degradation. A test system similar to that for PHB degradation is used here, but testosterone cannot be autoclaved without decomposition. A suspension of testosterone in distilled water is sonicated extensively to achieve good dispersion and also to reduce the number of contaminants. An aliquot of the suspension is added to the agar for the overlay to a concentration of 0.2%. Contaminants are not a problem in a mineral medium devoid of carbon sources other than testosterone and the minor impurities present in purified agar.

Ring fission mechanisms. The *ortho* or *meta* cleavage of two central intermediates in the metabolism of aromatic compounds (catechol and protocatechuate) can be tested by the method originally suggested by Dr. K. Hosokawa (Stanier et al., 1966). After growth in a chemically defined medium with an aromatic substrate, the cells are suspended in 0.02 M tris buffer (pH 8). To 2-ml portions are added a drop of toluene

---

*Garibaldi's medium. Sterile egg white is drawn aseptically from eggs that have been sterilized externally by immersion in 70% ethanol for 5 min, drained and flamed to remove the residual alcohol. The egg white is warmed to 45°C and added aseptically to give a final concentration of 10% (v/v) to any commercial complex solid medium which has been autoclaved and cooled to 45°C. After mixing, the medium is dispensed into Petri dishes. A solution of the protein conalbumin (available commercially) which has been sterilized by filtration can be used instead of egg white at a final concentration of 1.7 mg/ml of medium.

†Medium of Luisetti et al. (1972) (g/liter): vitamin-free casamino acids, 10.0; K_2HPO_4, 1.0; $MgSO_4 \cdot 7H_2O$, 1.0; sucrose, 10.0; gelatin, 30.0; and agar, 20.0; pH 7.0.

§Medium A of King et al. (1954) (g/liter): Bacto-peptone (Difco), 20.0; Bacto-agar (Difco), 15.0; glycerol, 10.0; K_2SO_4, 10.0; and $MgCl_2$, 1.4; pH 7.2.

¶ A formula recommended by Dr. D. C. Hildebrand, who has isolated a number of strains of this taxon from soil, is as follows. Sliced potatoes (250 g) are steeped in 1 liter of water at 65°C for 1 h. The preparation is filtered through three layers of cheesecloth, and 2.0 g of glucose and 20.0 g of agar are added to the solution. The medium is sterilized for 30 min at 20 psi.

‖ Medium of Sommer et al. (1961) (g/liter): Trypticase (BBL), 15.0; Phytone, soy (BBL), 5.0; NaCl, 5.0; Bacto-agar (Difco), 20.0; Bacto-yeast extract (Difco), 1.0; and $CaCO_3$, 10.0.

** Gray's medium (g/liter): $(NH_4)_2SO_4$, 0.5; K_2HPO_4, 1.0; $MgSO_4 \cdot 7H_2O$, 0.2; NaCl, 0.1; $CaCl_2$, 0.1; $FeCl_3$, 0.02; glycerol, 1.0; indole, 0.1; and agar, 18.0. For organisms unable to grow on glycerol, other carbon sources can be used. The medium is sterilized by autoclaving.

†† Medium of Macrae and Wilkinson (1958) (g/liter): Na_2HPO_4, 6.0; KH_2PO_4, 3.0; NaCl, 3.0; NH_4Cl, 1.0; Na_2SO_4, 0.1; $MgCl_2 \cdot 6H_2O$, 0.1; $MnCl_2 \cdot 4H_2O$, 0.1; Casamino acids (Difco), 0.1; glucose, 2.0; and sodium acetate, 0.1 M.

and 0.2 ml of either 0.1 M catechol or 0.1 M protocatechuate. A bright yellow color appearing in a short time (usually within a few seconds) indicates a *meta* cleavage. The tubes are shaken for 1 h at 30°C, and then tested for the appearance of β-ketoadipate (indicative of an *ortho* cleavage) by the Rothera reaction, as follows. Solid ammonium sulfate is added to saturation and the pH is brought up to ~10 by the addition of 2 drops of 5 N ammonium hydroxide. One drop of freshly prepared 25% sodium nitroprusside is added. A deep purple color indicates a positive reaction.

This method has been criticized by Ottow and Zolg (1974) because of alleged weak reactions due to a very low concentration of the aromatic intermediates. The authors propose instead to increase the final concentration of catechol and protocatechuate to 2 mM, apparently failing to realize that in the above method the final concentration is ~9 mM, which is quite appropriate to give quick, clear and reproducible results. It is unnecessary to prolong the incubation to 18 h, as advocated by Ottow and Zolg, in order to obtain a more intense color in the Rothera reaction.

Arginine dihydrolase reaction. The method giving the most unequivocal results is the direct one, consisting of incubating a bacterial suspension (200 Klett units, as measured with the green filter No. 54) in the presence of arginine (2.5×10^{-4} M) under anaerobic conditions. A similar suspension without arginine is used as control. After incubation for 2 h at 30°C, the tubes are immersed in a boiling water bath for 15 min, and arginine is determined quantitatively by the method of Rosenberg et al. (1956) in a sample of the supernatant after removal of the cells.

This procedure is cumbersome and time-consuming, and other methods appear much simpler and quite reliable. Among them, the method of Thornley (1960) is perhaps the most convenient. As recommended by Lelliott et al. (1966), Thornley's medium* can be conveniently dispensed in 3-ml portions into 5-ml screw cap vials. After autoclaving, the vials are inoculated by stabbing and sealed with melted paraffin. Anaerobic formation of ammonia from arginine can be detected by a change of color of the indicator within 3 days.

Acid production from carbohydrates. As pointed out by Palleroni and Doudoroff (1972), the reaction of acid production from carbohydrates is not necessarily correlated with nutritional data, since the acid produced by oxidation of a sugar may not always be a suitable carbon source for growth. This, however, should not diminish the taxonomic value of the test. More serious is the objection of redundancy, in that a single enzyme may effect the oxidation of several different sugars (Weimberg, 1962; Baumann et al., 1968).

Acid production from carbohydrates has been used very extensively for descriptive purposes since the last century and is, therefore, mentioned here among the recommended procedures, provided that the results are interpreted with proper reservations. The method of choice is that described by Hugh and Leifson (1953) and gives clear and reproducible results. The main advantage of the Hugh-Leifson medium† is its low concentration of peptone. When the peptone concentration is high, its deamination can neutralize a positive acid reaction from the sugar.

Hydrolysis of Tween 80. This reaction is an indication of lipolytic activity, and is carried out in a medium that has been proposed by Sierra (1957).§ The medium is usually dispensed into Petri dishes and is spot-inoculated with the cultures. During the incubation the plates are observed daily for opacity around the patches, due to formation of insoluble calcium soaps. If the reaction has been negative for 10 days, the bacterial mass is scraped off and the medium under the patch is examined for the presence of precipitate. However, the interpretation of such a positive reaction is not easy. In the lower central area of a bacterial patch anaerobic conditions may occur and endocellular lipases may be released after lysis of part of the population.

Reduction of nitrate and denitrification. Reduction of nitrate to nitrite can be tested according to Lelliott et al. (1966), and denitrification as recommended by Stanier et al. (1966). Some of the problems encountered in the denitrification test have been discussed by Palleroni and Doudoroff (1972).

Tests for chemolithotrophic growth. Petri plates with a solid mineral medium can be patch-inoculated with the organism to be tested and incubated in a jar or chamber which can be evacuated and refilled with the appropriate gas mixture. Since evacuation is never complete, the operation is repeated three or more times to insure an atmosphere of the proper composition. Lithotrophic growth with H_2 should be tested at initial oxygen concentrations of 5–10%; e.g. a gas mixture containing 50% H_2, 9% O_2, 5% CO_2 and 36% N_2 would be satisfactory. Hydrogen lithotrophy should be confirmed by demonstrating that *both* H_2 and CO_2 are required for growth.

For carboxydobacteria, CO can be prepared by adding formic acid to hot sulfuric acid and subsequently washing the gas with an alkaline solution. As indicated by Zavarzin and Nozhevnikova (1977), a convenient gas mixture is 80% CO:20% O_2.

Miscellaneous tests. Bibliographic sources for other methods are given below. *Oxidase test*: Stanier et al. (1966); Kovacs (1956). Baumann et al. (1972) have noted that the speed of the color change is much accelerated by the addition of one drop of toluene to the cell suspension prior to addition of the N,N'-dimethyl-p-phenylenediamine. *Levan formation from sucrose*: Stanier et al. (1966); Lelliott et al. (1966). *Gelatinase*: Skerman (1967). *Egg-yolk reaction*: Stanier et al. (1966); Lelliott et al. (1966). *Catalase*: ordinary methods are usually satisfactory, but occasionally the reaction may be so weak as to require sensitive detection methods involving the use of an oxygen electrode (Auling et al., 1978).

DNA/DNA hybridization has been performed extensively with pseudomonads by the competition technique originally described by Johnson and Ordal (1968). The technique is also described in detail by Ballard et al. (1970) and by Johnson (1981). Direct binding has been used to a limited extent (Palleroni and Doudoroff, 1971). The methodology for rRNA/DNA hybridization as applied to pseudomonads is described by Palleroni et al. (1973).

Differentiation of the genus **Pseudomonas** *from other closely related Taxa*

Table 4.2 of the family *Pseudomonadaceae* indicates characteristics useful for distinguishing the genus *Pseudomonas* from *Xanthomonas*, *Frateuria* and *Zoogloea*.

The particular combination of properties that presently circumscribes the genus *Pseudomonas* permits a differentiation from other genera of Gram-negative, nonsporeforming, aerobic bacteria. For example, the prosthecate genera *Caulobacter* and *Asticcacaulis*, and the sheathed genera *Spherotilus* and *Leptothrix*, are separated on the basis of morphological characters, while the aerobic organisms belonging to the genera *Nitrobacter*, *Nitrosomonas* and *Thiobacillus* are distinguished by their unique type of energy-yielding metabolism.

Differentiation of *Pseudomonas* from *Gluconobacter* may not be clearcut as long as the former genus includes species of marked acid tolerance, such as *P. diminuta* and *P. vesicularis*. These species, in our opinion, should be excluded from *Pseudomonas* (see General Taxonomic Comments, below), but their phylogenetic relationships to other

---

*Thornley's medium (g/liter): peptone, 1.0; NaCl, 5.0; K_2HPO_4, 0.3; agar, 3.0; phenol red, 0.01; and L-arginine-HCl, 10.0; pH 7.2.
†Hugh-Leifson medium (g/liter): peptone, 2.0; NaCl, 5.0; K_2HPO_4, 0.3; agar, 3.0; brom thymol blue, 0.03; and carbohydrate, 10.0; pH 7.1.
§Sierra's medium (g/liter): Bacto-peptone (Difco), 10.0; NaCl, 5.0; $CaCl_2 \cdot H_2O$, 0.1; and agar, 17.0. Tween 80 is sterilized separately and added to the medium after autoclaving to give a final concentration of 10.0 g/liter.

genera are presently obscure and the solution to the problem may be the creation of a separate genus. In section V of the taxonomic treatment we include in the genus *Pseudomonas* the species *P. glathei*, which has marked acid tolerance. This species has uncertain taxonomic affiliations, but it is not possible to foresee at this moment whether further study will show the convenience of assigning it to a different genus.

Members of the genus *Agrobacterium* occasionally have polar flagella, and *Alcaligenes* strains may have "degenerately peritrichous" flagella. These organisms may be differentiated from *Pseudomonas* by characters of questionable phylogenetic value. Indeed, the results of immunological studies by Baumann and Baumann (1978) and of enzymological experiments on pathways of aromatic acid biosynthesis (Byng et al., 1980; Whitaker et al., 1981a, b), suggest that some species of *Alcaligenes* are more closely related to *Pseudomonas* species than to each other.

In their study on the aerobic marine bacteria, Baumann et al. (1972) examined a number of polarly flagellated organisms sharing the basic morphological and physiological properties characteristic of *Pseudomonas*. The mol% G + C of the DNA of the polarly flagellated strains ranged from 30.5–64.7, with discontinuities defining three clusters of G + C values: 30.5, 43.2–48, and 52–64.7. The strains in the third cluster with G + C contents within the range of *Pseudomonas* (57–70 mol%) were placed in this genus, while those having G + C contents between 43.2 and 48 mol% were collected under the new genus *Alteromonas*. Finally, the strains having from 52–56.4 mol% G + C were left unnamed.

Strictly speaking, the reasons for restricting *Pseudomonas* to organisms with G + C contents within the range of values observed in a limited collection, are purely arbitrary, and in the absence of solid criteria for the segregation of marine and terrestrial genera of microorganisms, we have no good phenotypic arguments to exclude the marine strains with less than 57 mol% G + C from *Pseudomonas*.

The relationship between *Pseudomonas* and *Xanthomonas* also merits special discussion, since some members of these two genera share a substantial degree of rRNA/DNA homology. Plant pathogenicity and the production of characteristic yellow cellular pigments (xanthomonadins) are two important disctinctive properties of the xanthomonads, but otherwise these organisms are typical pseudomonads. Whether these properties are sufficient for a generic differentiation has been the subject of much debate, but the preservation of *Xanthomonas* at least has helped to reduce the considerable hypertrophy of *Pseudomonas*. In the present treatment we propose to place *P. maltophilia* and *Xanthomonas* species in the same natural group. The nomenclatural problem could be solved in one of two ways. *P. maltophilia* could be assigned to the genus *Xanthomonas* after an adequate generic redefinition or, perhaps more rationally, a new genus could be created for *P. maltophilia* and placed together with *Xanthomonas* in a separate group at the family level. *Editorial Note*: J. Swings, P. De Vos, M. Van den Mooter and J. De Lay have recently proposed transfer of *P. maltophilia* to the genus *Xanthomonas* as *X. maltophilia* (Int. J. Syst. Bacteriol. *33*: 409–413, 1983).

The genus *Zoogloea*, which is provisionally included in the family *Pseudomonadaceae*, is characterized by the production of extracellular slime and the aggregation of the cells into free floating flocs (Crabtree and McCoy, 1967), but the capsule formation depends on the growth conditions, and it may not represent a solid generic character. The fact that *Zoogloea* cells accumulate poly-β-hydroxybutyrate suggests a relationship to some *Pseudomonas* group other than section I, but our present knowledge of these pseudomonads is rather incomplete to attempt to determine their present phylogenetic relationships.

The comparison of catalogs of 16S rRNA oligonucleotide sequences has been recently introduced for the evaluation of phylogenetic relationships among procaryotic organisms (Fox et al., 1980). This approach has resulted in genealogic schemes at the generic or suprageneric level, where most Gram-negative bacteria, including *Pseudomonas*, are arranged in various groups whose ancestors are the purple photosynthetic bacteria. At a different level, rRNA/DNA hybridization studies

have also been of great value for intrageneric studies and form the basis of the current classification of *Pseudomonas* species groups. Refinements of the present techniques for evaluation of sequence homologies of conservative areas of the genome will no doubt be very important for a better understanding of the relationships among pseudomonads.

General Taxonomic Comments

The main body of the present classification of *Pseudomonas* species will follow the groupings proposed by Palleroni et al. (1973) on the basis of rRNA/DNA homology studies (Fig. 4.8). This internal subdivision of the genus is, with some exceptions, in agreement with the phenotypic groupings presented in the eighth edition of the *Manual*.

Pseudomonas species were originally grouped on the basis of the visual evaluation of the results of the extensive phenotypic characterization of a large collection of strains. General comments on the phenotypic classification have appeared in various reviews (Palleroni and Doudoroff, 1972; Palleroni, 1975; Palleroni, 1978), and practical determinative keys include those proposed by Palleroni (1977), Bergan (1981) and Stolp and Gadkari (1981). Even though the formidable body of information collected by the Berkeley workers constituted a very appropriate subject for numerical taxonomy, this type of data treatment was reported for only a fraction of the strains. Numerical analyses of strains of section I have been carried out by Sands et al. (1970), Palleroni et al. (1972), Ralston (Ph.D. Thesis, University of California, Berkeley, 1972), and Champion et al. (1980), and a limited analysis of the nonfluorescent phytopathogenic species was reported by Ballard (Ph.D. Thesis, University of California, Berkeley, 1970). Recently, a more extensive numerical treatment of the data of the Berkeley collection by Sneath et al. (1982) has given results in very good agreement with the groupings of Stanier, Doudoroff, Palleroni and their collaborators.

The definitive groupings of the species to be presented here and summarized in Figure 4.8 was mainly achieved on the basis of rRNA/DNA hybridization studies (Palleroni et al., 1973). Confirmation of this system of classification was reported in numerous papers from several laboratories, following various other approaches to bacterial phylogeny which included investigations on metabolic pathways and their regulatory mechanisms, determination of amino acid sequences of selected proteins, immunological studies, and cell wall composition. The results reported by Byng et al. (1980) and by Whitaker et al. (1981a) deserve special mention. Work performed on a large number of species on the regulation of tyrosine biosynthesis, on the comparative allostery of 3-deoxy-D-arabino-heptulosonate-7-phosphate synthetase, as well as the studies on phenylalanine biosynthesis by Whitaker et al. (1981b), amply support the division of *Pseudomonas* into various RNA groups, and at the same time suggest that the analysis of regulatory mechanisms may represent a useful tool for the tentative assignment of new species to RNA homology groups.

The proposed internal subdivision of *Pseudomonas* has also received confirmation from the studies of Woese and his collaborators on the sequence analysis of 16S ribosomal RNA component. The different RNA homology groups can be clearly differentiated by the results of such analysis and, in addition, they have been found to be related to different groups of purple sulfur and purple nonsulfur photosynthetic bacteria, thus confirming the hypothesis of the multigeneric nature of *Pseudomonas* as presently defined (Stackebrandt and Woese, 1981).

Unfortunately, there is still a large number of species which deserve further study but are not included in these groups, and their natural relationships are therefore unknown at present. The phenotype of some of these species are known in considerable detail, while other species have been partially characterized. Representative strains of the incompletely described species can be found in culture collections, and one important task of *Pseudomonas* taxonomists in the future will be to attempt to analyze their natural relationships in order to decide their most appropriate allocation in a phylogenetic system of classification. The description of many of these partially characterized species have often departed from standard procedures, which makes the construc-

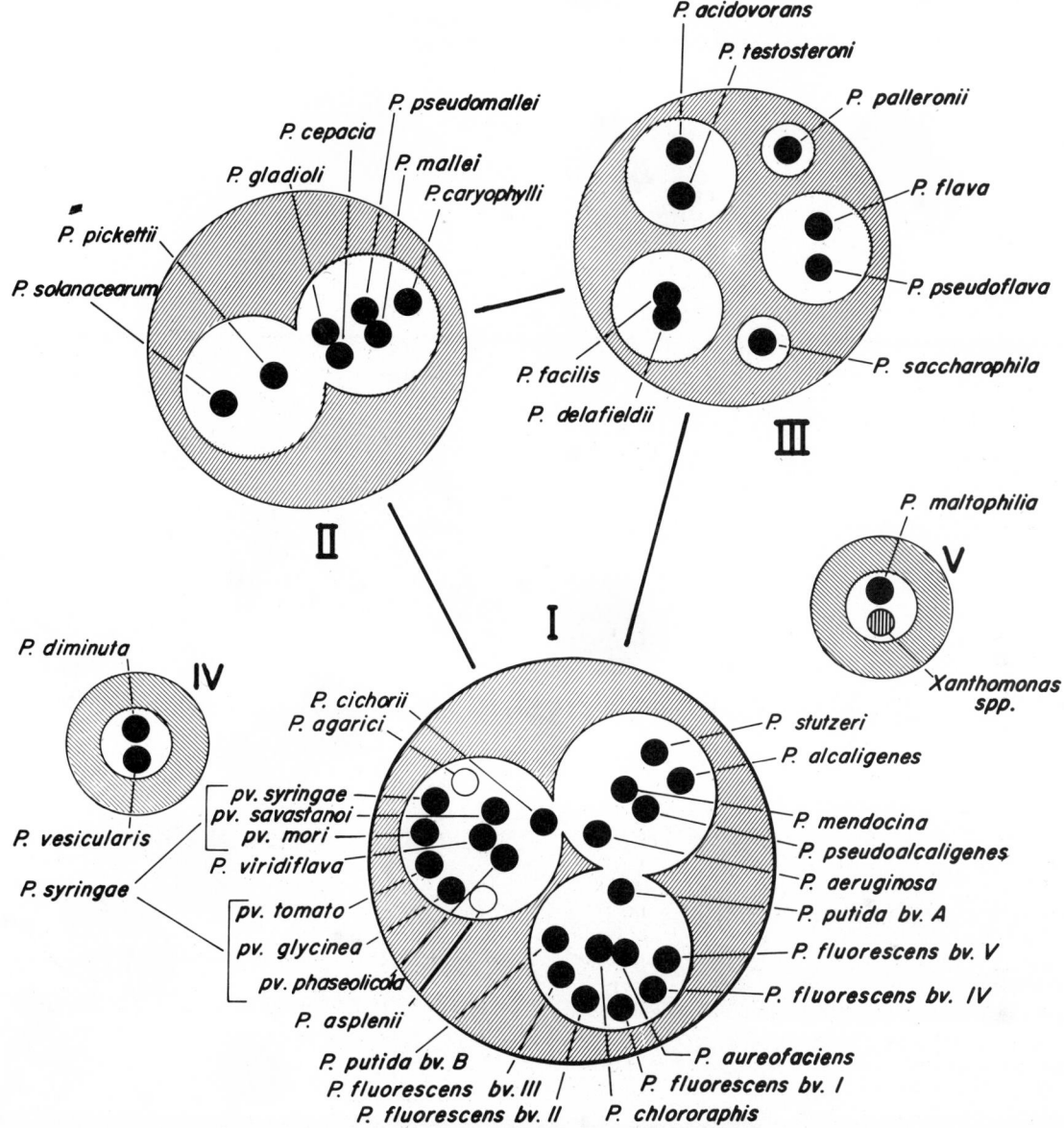

Figure 4.8. *Pseudomonas* species, biovars and pathovars arranged according to rRNA and DNA homologies. The *large circles* represent rRNA homology groups, within which DNA homology groups are indicated by *white circles*. The diminuta group (group IV) and the *P. maltophilia-Xanthomonas* group (group V) are included for reference, but appear unconnected to the *Pseudomonas* groups. Assignment of *P. agarici* and *P. asplenii* to the "fluorescent" DNA homology group within RNA group I is only tentative. (Modified from Palleroni et al. (1973).)

tion of determinative keys particularly difficult. Therefore, their classification in the present edition of the *Manual* will be based on purely empirical principles.

Acknowledgments

The generous and valuable help of several colleagues in the preparation of this manuscript is acknowledged with gratitude. B. Austin, D. E. Bradley, H. Dean, J. A. Fuerst, R. L. Gherna, M. Goodfellow, A. C. Hayward, D. C. Hildebrand, B. W. Holloway, G. A. Jacoby, R. A. Jensen, T. Kodama, M. Mandel, O. Meyer, A. Morgan, M. N. Schroth,

B. G. Shaw, M. Véron, and S. G. Wilkinson have contributed with some of the figures, unpublished information and helpful discussion.

Further Reading

Clarke, P.H. and M.H. Richmond (Editors). 1975. *Genetics and Biochemistry of Pseudomonas.* John Wiley, London, pp. 1–366.
Palleroni, N.J. 1978. *The Pseudomonas Group.* Meadowfield Press, Shildon, Durham, England, pp. 1–80.
Starr, M.P., H. Stolp, H.G. Trüper, A. Balows and H.G. Schlegel (Editors). 1981. *The Prokaryotes: a Handbook on Habitats, Isolation and Identification of Bacteria.* Springer-Verlag, Berlin-Heidelberg-New York, pp. 656–763.

Lists of the major sections and the species of the genus **Pseudomonas**

In the following part of this article, a general classification of *Pseudomonas* groups and species will be presented in five sections. The first four sections will not differ significantly from the four sections of the

eighth edition of the *Manual*. The fifth section will include species whose relationships to the above groups are, for the most part, unknown at present.

General key for the species of rRNA groups I, II and III
(which will be treated in sections I, II and III, respectively)

A. Poly-β-hydroxybutyrate (PHB) is not accumulated as a carbon reserve material, except for some strains of *P. pseudoalcaligenes* . . . . RNA group I (section I)
  1. Fluorescent pigment produced by most strains
    a. Arginine dihydrolase present; saprophytic or opportunistic animal pathogens
          1. *P. aeruginosa*
          2. *P. fluorescens* biovars
          3. *P. chlororaphis*
          4. *P. aureofaciens*
          5. *P. putida* biovars
    b. Arginine dihydrolase absent; phytopathogenic
      i. Oxidase-negative
          6. *P. syringae* pathovars
          7. *P. viridiflava*
      ii. Oxidase-positive
          8. *P. cichorii*
  2. Fluorescent pigment not produced
    a. Use glucose for growth
          9. *P. stutzeri*
         10. *P. mendocina*
    b. Do not use glucose for growth
         11. *P. alcaligenes*
         12. *P. pseudoalcaligenes*
B. PHB is accumulated as carbon reserve material.
  1. Pathogenic, except for *P. pickettii*. Most species grow at 40°C (the exception is *P. solanacearum*). Capable of *ortho* cleavage of protocatechuate . . . . RNA group II (section II)
    a. Arginine dihydrolase present
         13. *P. mallei*
         14. *P. pseudomallei*
         15. *P. caryophylli*
    b. Arginine dihydrolase absent
      i. Do not denitrify
         16. *P. cepacia*
         17. *P. gladioli*
      ii. Denitrify
         18. *P. pickettii*
         19. *P. solanacearum*
  2. Nonpathogenic. Only *P. pseudoflava* can grow at 40°C. Capable of *meta* cleavage of protocatechuate . . . . RNA group III (section III)
    a. Do not grow autotrophically with hydrogen
         20. *P. acidovorans*
         21. *P. testosteroni*
         22. *P. delafieldii*
    b. Grow autotrophically with hydrogen
         23. *P. facilis*
         24. *P. saccharophila*
         25. *P. flava*
         26. *P. pseudoflava*
         27. *P. palleronii*

SECTION I

Characteristics useful for the differentiation of the species of section I (RNA group I) are given in Table 4.7. Other characteristics of the species are presented later in Table 4.8–4.16.

rRNA Group I

Taxonomic Comments

The species of this group include the type species *P. aeruginosa*, other saprophytic and phytopathogenic fluorescent pseudomonads, the nonpigmented denitrifying species of the "stutzeri group" (Palleroni et al., 1970), and the nonpigmented strains that constitute the "alcaligenes group" (Ralston-Barrett et al., 1976).

The assignment of species and species clusters within this large group was originally decided on the basis of a more or less subjective analysis of phenotypic characters. Later, this criterion was amply confirmed by nucleic acid hybridization experiments and by numerical analysis of the phenotypic data. DNA/DNA hybridization experiments allowed the circumscription of a so-called "*P. fluorescens* homology group" (Palleroni et al., 1972). The competition technique of DNA hybridization demonstrated some level of DNA homology among the various members of the group either directly or indirectly, and at the

Table 4.7.
Characteristics differentiating species 1–12 (section I) of
Pseudomonas[a]

Characteristics	1. P. aeruginosa	Species 2, 3 and 4; see Table 4.8	5. P. putida	6. P. syringae pvs.	7. P. viridiflava	8. P. cichorii	9. P. stutzeri	10. P. mendocina	11. P. alcaligenes	12. P. pseudoalcaligenes
No. of flagella	1	>1	>1	>1	1–2	>1	1[b]	1[b]	1	1
Fluorescent pigments	d	+	+	+	+	+	−	−	−	−
Pyocyanin	d	−	−	−	−	−	−	−	−	−
Carotenoids	−	−	−	−	−	−	−	+	d	−
Growth at 41°C	+	−	−	−	−	−	d	+	+	+
Levan formation from sucrose	−	d	−	d	−	−	−	−	−	−
Arginine dihydrolase	+	+	+	−	−	−	−	+	+	d
Oxidase reaction	+	+	+	−	−	+	+	+	+	+
Denitrification	+	d	−	−	−	−	+	+	+	d
Gelatin hydrolysis	+	+	−	d	+	−	−	−	d	d
Starch hydrolysis	−	−	−	−	−	−	+	−	−	−
Utilization of:										
Glucose	+	+	+	+	+	+	+	+	−	−
Trehalose	−	+	−							
2-Ketogluconate	+	+	+	−	−	−	−	−		
meso-Inositol	−	+	−	d	+	d	−	−	−	−
Geraniol	+	−	−	−	−	−	−	+		
L-Valine	d	+	+	−	−	−	+	+	−	−
β-Alanine	+	+	+	−	−	−	−	+	d	d
DL-Arginine	+	+	+	d	+	+	−	+	+	+

[a] Symbols: see standard definitions.
[b] Lateral flagella of short wavelength may also be produced under certain conditions.

same time, a negligible level of homology with species outside the group. Moreover, when the experiments were repeated under more stringent conditions, the large homology group fell apart into its constituent subgroups or species, while appreciable DNA reannealing could still be observed among the members of a given subgroup. Application of rRNA/DNA hybridization techniques (Palleroni et al., 1973) further supported the clear-cut separation of this group from all other *Pseudomonas* species.

The fluorescent species of section I are characterized by the production of water-soluble pigments (pyoverdins), and, in some species, phenazine pigments. Many freshly isolated strains do not produce the characteristic pigments, and others lose the capacity after repeated subcultivation under laboratory conditions, but nonpigmented strains often can be identified with the appropriate species by their nutritional and physiological properties.

The saprophytic species of fluorescent pseudomonads can be distinguished from the phytopathogenic species, *P. syringae*, *P. viridiflava* and *P. cichorii*, by their positive arginine dihydrolase reaction, a more rapid growth in most media, and the ability to utilize certain substrates. The general outline of the taxonomic proposals for this group made by Stanier et al. (1966) is followed, and this reference should be consulted for a discussion of the problems of the internal subdivision of the group and for a more complete characterization of some of the species and biovars (biotypes) than is presented here. A majority of strains encountered in nature and studied in the past fall into fairly well defined categories, but some strains cannot be reasonably assigned to any well characterized species. In fact, it may not be a gross exaggeration to state that perhaps this is the most complex of all groups of aerobic pseudomonads, particularly when considered together with the fluorescent plant pathogens, and that we are still far from having achieved a satisfactory taxonomic solution for its internal subdivision into species.

The complexity of the fluorescent saprophytes other than *P. aeruginosa* has been well illustrated by the extensive studies of Jessen (1965), Stanier et al. (1966), Palleroni et al. (1972), and Champion et al. (1980). In total, 81 biovars were recognized by Jessen (1965). This number was reduced to seven *P. fluorescens* biovars and two biovars of *P. putida* by Stanier et al. (1966). A recent report by Champion et al. (1980) defined the relationships among strains of fluorescent pseudomonads in phenotypic properties, DNA hybridization and quantitative microcomplement fixation using antibodies directed against the protein azurin. The immunological studies could define clusters showing 98% agreement with assignment based on DNA homology, suggesting that the evolution pattern probably holds true for many genes. Clustering on the basis of phenotypic properties agreed with the DNA/azurin clustering for the majority of the strains and, in a more limited scale, with the data from studies by Ambler (1974) on cytochrome amino acid sequences.

An important conclusion of these studies has been the suggestion of very limited gene exchange between the clusters, although this may be significant at the intracluster level.

The above studies, among others, point to the considerable heterogeneity of the two fluorescent species, *P. fluorescens* and *P. putida*. Within *P. fluorescens*, however, the cluster constituted by biovars D and E of Stanier et al. (1966) appears by all criteria quite clearly outlined and distinct, supporting the temperament adopted in the eighth and in the present edition of the *Manual* of separating these two biovars under their original species names, *P. chlororaphis* and *P. aureofaciens*, respectively.

Of the fluorescent species described in the present treatment, only *P. aeruginosa* is characterized by the capacity for growth at 41°C. However, Ajello and Hoadley (1976) have examined a collection of fluorescent organisms capable of growing at this temperature, but distinct from *P. aeruginosa*. The cellular fatty acids composition of some of these unidentified pseudomonads suggests a relationship with the "stutzeri" and the "alcaligenes" groups, and some of the strains have phenotypic similarity and a substantial level of DNA homology with *P. mendocina*.

Phytopathogenic fluorescent pseudomonads also present particular taxonomic problems. Many strains isolated from various diseased plants have been named on the assumption that each organism was highly specific for its host and in the type of lesion that it produced. Many of these species are unrecognizable from their original descriptions, their host ranges have not been determined, and the type strains have been lost. It is therefore understandable that the taxonomy of the phytopathogenic pseudomonads should be qualified as "abstruse and unsettled" (Schroth et al., 1981). The problem is particularly acute for the group of fluorescent phytopathogens. Solid bacteriological criteria for the differentiation of species and pathovars from one another and from saprophytic forms are still needed and, consequently, strains isolated from sources other than plant lesions seldom can be assigned with certainty to described species or pathovars.

Even though some plant pathologists enthusiastically emphasize the diagnostic potentialities of the plant itself as a complex test system (Schroth et al., 1981), it cannot be denied that practical considerations often preclude the use of the plants for diagnostic purposes, and it seems reasonable to insist in the search for appropriate bacteriological characterization of the taxa. In Skerman's (1967) opinion "... one can either test pathogenicity over the entire plant world or, more rationally, attempt classification by other means."

In 1966 a set of methods proposed by Lelliott, Billing and Hayward under the name of LOPAT (levan formation, oxidase, potato rotting capacity, arginine dihydrolase, and tobacco hypersensitivity) provided a firm basis for clustering most fluorescent species under two names, *P. syringae* and *P. cichorii*, while *P. marginalis* was found to resemble the saprophytic forms of the *P. fluorescens*-*P. putida* group. Work performed at Berkeley by Stanier and collaborators (Stanier et al., 1966) stimulated projects centered on extensive phenotypic characterization of plant pathogenic species, with particular emphasis on the nutritional analysis (Misaghi and Grogan, 1969; Sands et al., 1970) and on their relationships by nucleic acid hybridization studies (Pecknold and Grogan, 1973).

The results obtained by Sands et al. (1970) allowed the grouping of most fluorescent plant pathogens in one cluster characterized by a negative arginine dihydrolase reaction, a positive hypersensitive reaction on tobacco leaves, and a slow growth rate as compared to that of the saprophytic pseudomonads. All organisms of this group were oxidase-negative, with the exception of *P. cichorii*; they were unable to denitrify and to grow at 37°C, and their nutritional versatility was markedly lower than that of the saprophytes. The slow growth and lower versatility of the fluorescent pathogens had already been noticed by Jessen (1965), who placed them in a separate group, and by Stanier et al., (1966). As pointed out by Misaghi and Grogen (1969), loss of pathogenicity does not transform a fluorescent plant pathogen into a known saprophytic form, which in turn implies that it is highly unlikely that pathogenic mutants could be isolated from a "statistically avirulent" population.

It was mainly on the basis of the work performed by various groups on the phenotypic characterization of the fluorescent plant pathogens that the treatment presented in the eighth edition of the *Manual* (Doudoroff and Palleroni, 1974) recognized only four species in this group. The two main species were *P. syringae* (including many synonyms) and *P. cichorii*, differing from one another mainly in the oxidase reaction. A third plant pathogenic species (*P. marginalis*) was included in one of the *P. fluorescens* biovars (biotypes) and the fourth was *P. aeruginosa*, whose role as a plant pathogen is not yet well established.

This treatment was found ". . . inadequate from a plant pathologist's viewpoint because it does not provide a nomenclature that expresses the phytopathogenic individuality of these bacteria" (Dye et al., 1975). The recommendation was to retain "the names that are nomenclaturally legitimate until a satisfactory rationalization can be made."

In attempting to satisfy the need for a special purpose classification of plant pathogenic pseudomonads, particularly those belonging to the fluorescent group, most of the names listed by Young et al. (1978) and by Dye et al. (1980) are included here in the infrasubspecific category of pathovars. Preservation of these names is thought to "meet the demands of plant pathologists who wish to refer to pathogenic ability" (Young et al., 1978). In the absence of solid criteria based on overall similarities for the differentiation of these various pathovars, and a fragmentary knowledge of host ranges and specificities, it is debatable whether the present classification will serve as a guide for determinative purposes, particularly for strains isolated from sources other than the diseased plants or from hosts that are marginally located in the respective host ranges.

Internal subdivision of the fluorescent plant pathogens by Sands et al. (1970) could define two clusters corresponding to the nomenspecies "*P. savastanoi*" and "*P. phaseolicola*," but the characters of diagnostic value for these clusters were possessed in various combinations by strains of other nomenspecies now included as pathovars of *P. syringae*. The experiments on DNA/DNA hybridization carried out on a limited number of strains by Palleroni et al. (1972) and more extensively by Pecknold and Grogan (1973) permitted the definition of the following groups:

1. The syringae group, including the nomenspecies "*P. aptata*," *P. syringae*, "*P. panacis*," and "*P. pisi*";
2. The morsprunorum group, with the nomenspecies "*P. morsprunorum*," "*P. phaseolicola*," "*P. tabaci*," "*P. mori*," "*P. lachrymans*," "*P. glycinea*" and "*P. savastanoi*";
3. The tomato group, with the species "*P. tomato*," "*P. helianthi*," "*P. delphinii*," and "*P. coronafaciens*";
4. The viridiflava group, with *P. viridiflava* strains;
5. The cichorii group, with *P. cichorii* strains, and
6. The marginalis group, with *P. marginalis* strains.

The first two groups appear to be very "tight," with intragroup homology values in the range of the best defined species of the genus. Strains of a given species within these two groups have higher homology among them than with strains of other nomenspecies of the same group, and the range of values do not overlap. The separation of the syringae and morsprunorum groups is interesting in view of the different opinions on the convenience of separation of these two nomenspecies.

The results with the tomato group were less clear, and this group has been left undefined. An analogous situation holds true for the so-called viridiflava group, whose strains are linked by a relatively low level of homology. As for the cichorii group, this cluster of oxidase-positive strains is very well defined and deserves independent status, but the results on *P. marginalis* once more support the opinion that this species is more closely related to *P. fluorescens* than to the other plant pathogenic species. Both *P. marginalis* and *P. viridiflava* are species that resemble the saprophytic species and are not yet fully specialized as phytopathogens. Therefore, it is our feeling that *P. marginalis* is better kept for the moment in the *P. fluorescens* group until further work demonstrates the necessity for its separation as an independent taxon. As for *P. viridiflava*, its close relationship to other oxidase-negative phytopathogens and its occurrence as, an epiphyte with pathogenic potentialities (Billing, 1970b) are reasons for recommending its preservation as a separate species, even though its natural relationships are not yet well known.

Four species of nonfluorescent denitrifying *Pseudomonas* are included in section I. Two species, *P. stutzeri* and *P. mendocina*, constitute the so-called "stutzeri group." *P. mendocina* is a very homogeneous species, while *P. stutzeri* is heterogeneous with respect to many phenotypic characters and DNA composition. The nomenclatural problems were discussed by Palleroni et al. (1970). The last two species of section I are *P. alcaligenes* and *P. pseudoalcaligenes*. A comprehensive characterization of these species can be found in the papers by Stanier et al. (1966) and by Ralston-Barrett et al. (1976).

List of the species of section I

1. **Pseudomonas aeruginosa** (Schroeter 1872) Migula 1900, 884.[AL] (*Bacterium aeruginosum* Schroeter 1872, 126.)

ae.ru.gi.no'sa. L. fem. adj. *aeruginosa* full of copper rust or verdigris, hence green.

Characteristics of the species are given in Tables 4.8 and 4.9. Characters useful in the differentiation from other species of the genus are given in Table 4.7 and later in Table 4.30.

Two main colony types can be observed on common solid media. One is large, smooth, with flat edges and elevated center ("fried egg" appearance), and the other is small, rough, convex. Clinical materials are, in general, good sources of the large colony type, while the small is commonly obtained from natural sources (Véron and Berche, 1976). Variation of the large type to the small is easy to observe, but the reverse variation is extremely rare. A third colony type (mucous) often can be obtained from respiratory and urinary tract secretions, and was first observed by Sonnenshein (1927). Mucoid mutants of *P. aeruginosa* can be divided into two groups according to whether the mucus (alginate) is produced in chemically defined media (Fyfe and Govan, 1980).

Aside from pyoverdin and pyocyanin, other pigments may be produced by some strains, including a dark red pigment.

Optimum temperature, 37°C.

Can be isolated from soil and water, particularly from enrichment cultures for denitrifying bacteria. Commonly isolated from clinical specimens (wound, burn and urinary tract infections). Causative agent of "blue pus," origin of the synonym *pyocyaneus*. Occasionally pathogenic for plants. Strains isolated from leaf spot of tobacco, identical with or similar to *P. aeruginosa* have been named "*P. polycolor*" (Clara, 1930).

Table 4.8.
General characteristics of species 1–5 of **Pseudomonas** *(section I)[a]*

Characteristics	1. P. aeruginosa	2. P. fluorescens biovar I	2. P. fluorescens biovar II	2. P. fluorescens biovar III	2. P. fluorescens biovar IV	2. P. fluorescens biovar V	3. P. chlororaphis	4. P. aureofaciens	5. P. putida biovar A	5. P. putida biovar B
Cell diameter, μm	0.5–0.7	0.7–0.8	0.7–0.8	0.8	0.7	0.8	0.7–0.8	0.7–0.8	0.7–1.1	0.7–1.1
Cell length, μm	1.5–3.0	2.3–2.8	2.0–2.8	2.0–2.8	2.0–2.5	2.0–3.0	1.5–3.6	1.9–2.8	2.0–4.0	2.0–4.0
Flagellar number	1	>1	>1	>1	>1	>1	>1	>1	>1	>1
Pyocyanin production	+	–	–	–	–	–	–	–	–	–
Pyoverdin production	+	+	d	+	+	d	d	+	+	d
Chlororaphin production	–	–	–	–	–	–	+	–	–	–
Phenazine monocarboxylate production	–	–	–	–	–	–	–	+	–	–
Other pigments (not carotenoids)	+	–	–	–	d	–	–	–	–	–
Yellow-orange cellular pigments	–	–	–	–	–	–	–	–	–	–
Oxidase	+	+	+	+	+	+	+	+	+	+
PHB accumulation	–	–	–	–	–	–	–	–	–	–
Levan formation from sucrose	–	+	+	–	+	–	+	+		
Gelatin liquefaction	+	+	+	+	+	+	+	+		
Starch hydrolysis	–	–	–	–	–	–	–	–		
Autotrophic growth with H₂	–	–	–	–	–	–	–	–		
Lecithinase (egg yolk)	–	+	±	+	+	d	+	d	–	–
Lipase (Tween 80 hydrolysis)	±	d	–	d	d	d	+	d	d	d
Extracellular PHB hydrolysis	–	–	–	–	–	–	–	–		
Growth at 4°C	–	+	+	+	+	d	+	+	d	+
Growth at 41°C	+	–	–	–	–	–	–	–	–	–
Denitrification	+	–	+	+	+	–	+			
Arginine dihydrolase	+	+	+	+	+	+	+	+	+	+
Catechol, *ortho* cleavage	+	+	+	+	+	+	+	+	+	+
Protocatechuate, *ortho* cleavage	+	+	+	+	+	+	+	+	+	+
Mol% G + C of DNA (Bd)	67.2	60.5	61.3	60.6	59.4	60.5	63.5	63.6	62.5	60.7

[a] For symbols, see standard definitions.

The species can be internally divided into a number of subgroups useful for epidemiological purposes.

The mol% G + C of the DNA is 67.2 (Bd).

Type strain: ATCC 10145 (NCIB 8295; NCTC 10332).

2. **Pseudomonas fluorescens** (Trevisan 1889) Migula 1895, 29.[AL] (*Bacillus fluorescens* Trevisan 1889, 18.)

flu.o.res′cens. L. n. *fluor* a flux; M.L. v. *fluoresco* to fluoresce; M.L. part. adj. *fluorescens* fluorescing.

Characteristics of the species and its five biovars are given in Tables 4.8 and 4.9. Most of the information has been taken from Stanier et al. (1966), but data on a larger number of strains of biovars III (68 strains) and V (89 strains) are presented in the tables. Characteristics of diagnostic value are given in Tables 4.7 and 4.10.

Optimum temperature, 25–30°C.

Found in soil and water, from which it can be isolated after enrichment in media containing various carbon sources, incubated aerobically; strains of the denitrifying biovars can be enriched in similar media containing nitrate, incubated under anaerobic conditions. Commonly associated with spoilage of foods (eggs, cured meats, fish and milk). Often isolated from clinical specimens. Some strains assigned to this species (biovar II) have been isolated from diseased plants (e.g. lettuce), and identified as *Pseudomonas marginalis* (Brown 1918) Stevens 1925.[AL] This taxon has been subdivided by Young et al. (1978) into three pathovars: (a) *P. marginalis* pv. *marginalis* (Brown 1918) Stevens 1925. Designated type strain is ATCC 10844 (PDDCC 3553; NCPPB 667); (b) *P. marginalis* pv. *alfalfa* Shinde and Lukezic 1974. This pathovar is associated with discolored alfalfa (*Medicago sativa*, fam. *Leguminosae*) roots. Reference strain is PDDCC 5708 (NCPPB 2644). (c) *P. marginalis* pv. *pastinacae* (Burkholder 1960) comb. nov., a pathogen of cultivated parsnip (*Pastinaca sativa*, fam. *Umbelliferae*). Reference strain is ATCC 13889 (PDDCC 5709; NCPPB 806).

Biovar I (Biotype A of Stanier et al., 1966) is considered to be typical of *P. fluorescens*, and the type strain of the species belongs to this group. Aside from *P. marginalis* strains, biovar II also includes saprophytic organisms. Within biovar III (biotype C of Stanier et al., 1966) at least two subgroups, which differ from each other in their capacity for utilization of dicarboxylic acids, can be defined. Biovar IV (biotype F of Stanier et al., 1966) contains the type strain of "*P. lemonnieri*" (Lasseur 1913) Breed 1948, 178. Several strains are known at present, and these can be grouped into at least two clusters (Mark Lipstein, personal communication). The group of miscellaneous strains assigned to *P. fluorescens* biotype G by Stanier et al (1966) constitute the last (V) biovar of the species. The biovar is very heterogeneous in its nutritional properties, and may consist of strains that have lost one or more of the properties considered to be of diagnostic importance in differentiating among the better characterized biovars. Among the nonauthentic strains assigned to this biovar are strains labeled "*P. schuylkilliensis*" and *P. geniculata* (Wright 1895) Chester 1901, 313[AL] (see Stanier et al., 1966). Strains of biovar V are very common in soils (Sands and Rovira, 1971).

The mol% G + C of the DNA is 59.4–61.3 (Bd).

Type strain: ATCC 13525 (NCIB 9046; NCTC 10038).

3. **Pseudomonas chlororaphis** (Guignard and Sauvageau 1894) Bergey, Harrison, Breed, Hammer and Huntoon 1930, 166.[AL] (*Bacillus chlororaphis* Guignard and Sauvageau 1894, 841.)

Table 4.9.
Nutritional characteristics of species 1–5 of **Pseudomonas** *(section I)*[a]

Characteristics	1. P. aeruginosa	2. P. fluorescens biovar I	2. P. fluorescens biovar II	2. P. fluorescens biovar III	2. P. fluorescens biovar IV	2. P. fluorescens biovar V	3. P. chlororaphis	4. P. aureofaciens	5. P. putida biovar A	5. P. putida biovar B
Utilization of:										
Acetate, heptanoate, caprylate, pelargonate, caprate, succinate, fumarate, glutarate, L-malate, β-hydroxybutyrate, lactate, citrate, α-ketoglutarate, pyruvate, glycerol, L-alanine, β-alanine, L-aspartate, L-glutamate, L-arginine, γ-aminobutyrate, L-proline, L-tyrosine, putrescine, spermine, betaine	+	+	+	+	+	+	+	+	+	+
D-Arabinose, D-fucose, maltose, cellobiose, lactose, starch, inulin, oxalate, maleate, ethylene glycol, phthalate, isopropanol, L-threonine, L-norleucine, poly-β-hydroxybutyrate	−	−	−	−	−	−	−	−	−	−
D-Ribose, mannitol	+	+	+	d	+	d	+	+	d	d
D-Xylose	−	+	d	d	d	d	−	−	d	d
L-Arabinose	−	+	+	d	+	d	−	+	d	+
L-Rhamnose	−	−	d	d	−	d	−	−	−	−
Glucose	+	+	+	+	+	+	+	d	+	+
D-Mannose	−	+	+	+	+	d	+	d	d	d
D-Galactose	−	+	+	d	+	d	d	+	−	d
D-Fructose	+	+	d	+	+	+	d	d	+	+
Sucrose	−	+	+	−	+	d	+	d	−	d
Trehalose	−	+	+	d	+	d	+	d	−	−
Gluconate	+	+	+	+	+	+	+	d	+	+
2-Ketogluconate	+	+	+	+	d	+	+	d	d	+
Saccharate	−	+	+	d	+	d	+	+	+	+
Mucate	−	+	+	d	+	+	+	+	d	+
Propionate	+	+	+	d	+	+	+	+	+	+
Butyrate	+	−	d	d	+	d	+	d	+	+
Isobutyrate	+	−	d	d	−	d	−	d	d	d
Valerate, isovalerate	+	d	d	d	−	d	+	+	+	+
Caproate	+	+	d	+	+	+	d	+	+	+
Malonate	+	+	+	d	+	d	+	+	d	+
Adipate, sebacate	+	−	−	d	−	−	−	−	−	d
Pimelate, suberate	d	−	−	d	−	−	−	−	−	−
Azelate	+	−	−	d	−	−	−	−	−	−
D-Malate	d	−	d	d	+	d	d	−	d	d
D(−)-Tartrate	−	−	d	−	−	d	−	−	d	d
L(+)-Tartrate	−	−	−	−	+	−	d	−	d	d
m-Tartrate	−	−	−	d	−	d	−	−	d	−
Glycolate	−	−	−	−	−	−	−	−	d	−
Glycerate	+	+	+	d	+	d	+	+	d	+
Hydroxymethylglutarate	−	d	d	−	−	d	−	−	−	−
Aconitate	+	+	+	d	+	d	d	+	+	+
Levulinate	+	d	−	d	−	d	d	d	d	d
Citraconate	−	d	d	d	+	d	−	−	−	d
Itaconate, mesaconate	+	+	d	d	−	d	+	+	d	−
Erythritol	−	d	d	+	−	d	−	−	−	−
Sorbitol	−	+	+	d	+	d	−	−	−	d
m-Inositol	−	d	+	d	+	d	+	+	−	−
Adonitol	−	+	−	d	−	d	−	−	−	−
Propylene glycol	+	−	+	d	−	d	−	−	d	+

chlo.ro.ra′phis. Gr. adj. *chlorus* green; Gr. n. *raphis* a needle; M.L. fem. n. *chloraphis* a green needle.

The characteristics of the species are presented in Tables 4.8 and 4.9. Characteristics useful for differentiation from other species of section I are given in Tables 4.7 and 4.10.

Optimum temperature, ~30°C.

Isolated from dead larvae of the cockchafer (a large European beetle) and from water.

The mol% G + C of the DNA is 63.5 (Bd).

Type strain: ATCC 9446 (NRRL B-560; NCIB 9392; IFO 3904).

Table 4.9.—*continued*

	1. P. aeruginosa	2. P. fluorescens biovar I	2. P. fluorescens biovar II	2. P. fluorescens biovar III	2. P. fluorescens biovar IV	2. P. fluorescens biovar V	3. P. chlororaphis	4. P. aureofaciens	5. P. putida biovar A	5. P. putida biovar B
2,3-Butylene glycol	+	d	+	d	+	d	d	d	d	d
Ethanol, n-propanol	+	−	+	d	−	d	d	−	d	d
n-Butanol	+	d	d	d	+	d	d	−	+	+
Isobutanol	+	−	d	d	−	d	−	−	d	d
Geraniol	+	−	−	−	−	−	−	−	−	−
D-Mandelate	−	−	−	−	−	−	−	−	d	d
L-Mandelate	+	−	−	−	−	d	−	−	−	d
Benzoylformate	+	−	−	−	−	−	+	+	d	d
Benzoate	+	d	d	d	+	d	+	d	d	+
o-Hydroxybenzoate	−	−	−	−	−	−	d	−	d	d
m-Hydroxybenzoate	−	−	−	−	−	−	d	d	d	d
p-Hydroxybenzoate	+	+	+	d	+	d	+	+	+	+
Phenylacetate	−	−	−	d	−	d	d	+	d	+
Phenylethanediol, naphthalene, α-aminobutyrate, D-tryptophan	−	−	−	−	−	−	−	−	−	d
Phenol, creatine	−	−	−	d	−	−	−	−	d	d
Quinate	d	+	+	+	+	d	+	+	+	+
Testosterone	−	−	−	−	d	−	−	−	−	+
Glycine	d	−	−	−	d	−	−	−	d	+
D-Alanine	+	+	+	+	+	−	+	−	+	+
L-Serine	d	+	d	+	+	d	d	+	d	d
L-Leucine	+	+	d	+	+	+	+	+	+	+
L-Isoleucine, L-valine	d	+	+	+	+	+	+	+	+	+
L-Lysine	+	+	d	d	+	d	d	d	+	d
L-Ornithine	+	+	d	d	d	d	d	+	+	+
L-Citrulline	d	d	d	d	−	d	d	d	d	d
α-Aminovalerate	−	−	−	d	−	−	−	−	d	−
δ-Aminovalerate	+	d	+	+	+	+	d	d	+	+
L-Histidine	+	+	d	+	+	d	+	+	+	+
L-Phenylalanine	d	d	d	d	+	d	d	+	+	+
L-Tryptophan	d	+	d	d	−	d	+	+	−	+
L-Kynurenine, anthranilate	+	d	d	d	−	d	+	+	−	+
Kynurenate	d	d	−	d	−	−	d	d	−	d
Ethanolamine	d	+	d	d	+	d	d	+	d	d
Benzylamine	−	−	−	d	−	d	−	−	d	+
Histamine	+	d	−	d	−	d	d	d	d	+
Tryptamine	−	−	d	d	−	−	−	−	d	+
Butylamine	−	−	−	−	−	d	−	d	+	+
α-Amylamine	−	−	d	d	−	d	−	+	d	+
Sarcosine	d	+	d	d	+	d	d	+	+	+
Hippurate	−	−	−	d	−	−	−	−	d	d
Pantothenate	−	−	−	d	−	−	−	−	−	−
Acetamide	+	−	−	−	−	−	−	−	d	d
Nicotinate	−	−	−	−	−	d	−	−	d	+
Trigonelline	−	d	d	d	−	d	−	−	d	+
Dodecane, hexadecane	d	−	−	−	−	−	−	−	−	−

a For symbols see standard definitions.

4. **Pseudomonas aureofaciens** Kluyver 1956, 406.[AL]

au.re.o.fa′ci.ens. L. adj. *aureus* golden; L. v. *facio* to make; M.L. part. adj. *aureofaciens* making golden.

The characteristics of the species are given in Tables 4.8 and 4.9. Characteristics useful for differentiation from other species of section I are given in Tables 4.7 and 4.10.

Optimum temperature, ∼30°C.

Occurs in soil and water. The type strain was isolated from clay suspended in kerosene for 3 weeks.

The mol% G + C of the DNA is 63.6 (Bd).

Type strain: ATCC 13985 (NCIB 9030).

5. **Pseudomonas putida** (Trevisan 1889) Migula 1895, 29.[AL] (*Bacillus putidus* Trevisan 1889, 18.)

Table 4.10.

Characteristics of species 2 (biovars I–V), 3 and 4 of **Pseudomonas** *(section I)[a]*

Characteristics	Biovar I	Biovar II	Biovar III	Biovar IV	Biovar V	3. *P. chlororaphis*	4. *P. aureofaciens*
			2. *P. fluorescens*				
P. fluorescens biovars as designated by Stanier et al. (1966)	A	B	C	F	G	D	E
Nonfluorescent pigments:							
Green (chlororaphin)	–	–	–	–	–	+	–
Orange (phenazine-1-carboxylate)	–	–	–	–	–	–	+
Blue, nondiffusible	–	–	–	+	–	–	–
Levan formation from sucrose	+	+	–	+	–	+	+
Denitrification	–	+	+	+	–	+	d
Carbon sources used for growth:							
L-Arabinose	+	+	d	+	d	–	+
Sucrose	+	+	–	+	d	+	d
Saccharate	+	+	d	+	d	+	+
Propionate	+	+	d	+	+	+	+
Butyrate	–	d	d	+	d	+	d
Sorbitol	+	+	d	+	d	–	–
Adonitol	+	–	d	–	d	–	–
Propylene glycol	–	+	d	–	d	–	–
Ethanol	–	+	d	–	d	d	–

[a] For symbols see standard definitions.

pu′.ti.da. L. fem. adj. *putida* stinking, fetid.

Characteristics of the species (103 strains of biovar A and 9 strains of biovar B) are given in Tables 4.8 and 4.9. Characteristics of diagnostic value are presented in Table 4.7.

Optimum temperature, 25–30°C.

Isolated from soil and water after enrichment in mineral media with various carbon sources.

The majority of the strains can be assigned to biovar A (biotype A of Stanier et al., 1966), which is considered to be typical. This biovar has a G + C content of 62.5 mol%. Biovar B, which has a G + C content of 60.7, differs from biovar A only in a few phenotypic properties: all known strains of this biovar utilize L-tryptophan, kynurenine and anthranilate, and most use D-galactose as carbon sources. None of the strains of biovar B uses nicotinate.

Type strain: ATCC 12633 (NCIB 9494).

6. **Pseudomonas syringae** van Hall 1902, 141.[AL] (*Phytomonas syringae* (van Hall 1902) Bergey, Harrison, Breed, Hammer and Huntoon 1930, 257; *Pseudomonas barkeri* (Berridge 1924) Clara 1934, 11; *Pseudomonas citrarefaciens* (Lee 1917) Stapp 1928, 190; *Pseudomonas citriputealis* (Smith 1913) Stapp 1928, 190; *Pseudomonas hibisci* (Nakada and Takimoto 1923) Stapp 1928, 203; *Pseudomonas prunicola* Wormald 1930, 742; *Pseudomonas punctulans* (Bryan 1933) Savulescu 1947, 12; *Pseudomonas rimaefaciens* Koning 1938, 11; *Pseudomonas spongiosa* (Aderhold and Ruhland 1905) Braun 1927, 2; *Pseudomonas tonelliana* (Ferraris 1926) Burkholder 1948, 132; *Pseudomonas trifoliorum* (Jones, Williamson, Wolf, and McCulloch 1923) Stapp 1928, 177; *Pseudomonas utiformica* Clara 1932, 111; *Pseudomonas vignae* Gardner and Kendrick 1923, 275; and *Pseudomonas viridifaciens* Tisdale and Williamson 1923, 150.) Some of these nomenspecies may be synonyms, biovars or pathovars of *P. syringae*, and others may deserve independent species rank. Other names considered to be synonyms have been omitted following the recommendation of Dye et al. (1975) since they are not represented by authentic cultures or have been very inadequately described.

sy.rin′gae. M.L. fem. n. *Syringa* generic name of lilac; M.L. fem. gen. n. *syringae* of the lilac.

Characteristics of the species and some of its pathovars are described

Table 4.11.

General characteristics of species 6–8 of **Pseudomonas** *section I[a]*

Characteristics	6. *P. syringae* pathovars	7. *P. viridiflava* (Sands et al., 1970)	7. *P. viridiflava* (Billing, 1970b)	8. *P. cichorii*
Cell diameter, µm	0.7–1.2			0.8
Cell length, µm	1.5			1.2–3.5
Flagellar number	>1	1–2[b]		>1
Pyoverdin production	+	+	+	+
Phenazine pigment production	–	–	–	–
Yellow or orange cellular pigments	–	–	–	–
Oxidase	–	–	–	+
PHB accumulation	–	–	–	–
Levan formation from sucrose	d	–	–	–
Gelatin liquefaction	d	+[b]	+	–
Lecithinase (egg yolk)	d	d	–	+
Lipase (Tween 80 hydrolysis)	d	d	–	–
Denitrification	–	–	–	–
Arginine dihydrolase	–	–	–	–
Starch hydrolysis	–[b]	–		
Growth at 4°C	d			–
Protocatechuate, *ortho* cleavage	+	+		+
Mol% G + C of DNA (Bd)	59–61			59

[a] For symbols see standard definitions.
[b] Modified from Clara (1934).

in Tables 4.7 and 4.11–4.14. The pattern of acid production from sugars will be found in Haynes and Burkholder (1957) and in the original papers describing the nomenspecies here included as pathovars.

Cytochrome *c* is not detectable.

Table 4.12.
Nutritional characteristics of species 6–8 of **Pseudomonas** *(section I)[a]*

Characteristics	6. *P. syringae* pathovars	7. *P. viridiflava* (Sands et al., 1970)	7. *P. viridiflava* (Billing, 1970b)	8. *P. cichorii*	Characteristics	6. *P. syringae* pathovars	7. *P. viridiflava* (Sands et al., 1970)	7. *P. viridiflava* (Billing, 1970b)	8. *P. cichorii*
Utilization of:[b]					Erythritol, sorbitol	d	+	+	−
Glucose, mucate, succinate,[c] glycerol, L-aspartate, L-glutamate, L-glutamine,[c] γ-aminobutyrate	+	+	+	+	L-Alanine	+	d	+	+
					D-Alanine	d	+		d
D-Ribose, D-xylose, acetate, propionate, β-hydroxybutyrate	d	d	d	+	L-Leucine	−	−	+	−
					L-Histidine	d	+	d	d
L-Arabinose, gluconate, L-malate, citrate, aconitate	d	d	+	+	L-Tyrosine	−	d	−	+
					L-Tryptophan	−	−	d	−
D-Mannose, D-galactose, caproate, L-arginine, betaine	d	+	d	+	Putrescine	d	d	−	−
					Sarcosine	d	−	d	d
D-Fructose, caprylate, pelargonate, lactate, mannitol, m-inositol, p-hydroxybenzoate, quinate, L-serine, L-proline	d	+	+	+	Laurate			d	
					L-Sorbose, melezitose, amygdalin, dextrin, formate, dulcitol, isophthalate, L-methionine, m-aminobenzoate, p-aminobenzoate, methylamine			−	
Raffinose	d	−	−	d					
Fumarate	+	+	−	+	Linolenate	d	+		d
Sucrose, glutarate	d	+	−	+	Triacetin	d	+		+
Saccharate	d	d	+	+	Tripropionin	d	+		+
Valerate	d	d	−	d	Tricaproin	d	+		d
Caprate	+	+	d	d	Ascorbate	d	−		−
Malonate, m-tartrate	d	d		+	Isoascorbate	d	+		−
D-Malate, glycerate, trigonelline	d	+		+	Lecithin	d	d		+
D(−)-Tartrate	−	+	+	−	Asparagine	d	+		+
L(+)-Tartrate, α-ketoglutarate	d	−	−	+					
Hydroxymethylglutarate	d	d		−					
Pyruvate	+	+	d	+					

[a] For symbols see standard definitions.

[b] The following compounds are not utilized by any species: D-arabinose, D-fucose,[c] L-rhamnose, trehalose, maltose, cellobiose, lactose, meliobiose, methylglucoside,[b] starch, inulin, 2-ketogluconate,[c] salicin,[c] N-acetylglucosamine,[c] isobutyrate, isovalerate, linoleate,[c] laurylsulfate,[c] tannate,[c] oxalate, maleate, adipate, pimelate, suberate, azelate, sebacate, glycolate, thioglycolate,[c] levulinate, citraconate, itaconate, mesaconate, 3-phosphoglycerate,[c] hydroxymethylbutyrate,[c] adonitol, ethylene glycol, propylene glycol, 2,3-butylene glycol, methanol,[c] ethanol, n-propanol, isopropanol, n-butanol, isobutanol, geraniol,[c] D-mandelate, L-mandelate,[c] benzoylformate,[c] benzoate, o-hydroxybenzoate, m-hydroxybenzoate, phthalate, phenylacetate, phenylethanediol,[c] eicosenedioate,[c] naphthalene, phenol, testosterone, glycine, β-alanine, L-threonine, L-isoleucine, L-norleucine, L-valine, L-lysine, L-ornithine, L-citrulline, α-aminobutyrate, δ-aminovalerate,[c] L-phenylalanine, L-hydroxyproline,[c] D-tryptophan,[c] indoleacetic acid,[c] L-kynurenine,[c] kynurenate,[c] anthranilate, methylamine, ethanolamine, benzylamine, spermine, histamine,[c] tryptamine,[c] butylamine, α-amylamine, creatine, choline,[c] hippurate, urate,[c] pantothenate, acetamide, nicotinate, dodecane, hexadecane, poly-β-hydroxybutyrate,[c] pectate,[c] chlorogenate,[c] and uridine.[c]

[c] Not reported for *P. viridiflava* by Billing (1970b).

Rare strains require organic growth factors. Growth of most strains slow in mineral media with a single carbon source and are relatively slow in complex media. Nutritional spectrum is less extensive and more heterogeneous than that of the saprophytic fluorescent pseudomonads.

Optimum temperature, ~25–30°C.

The original strain was isolated from lilac (*Syringa vulgaris*, fam. *Oleaceae*) but strains conforming to the original description are pathogenic for many unrelated plants.

Type strain: ATCC 19310 (PDDCC 3023; NCPPB 281).

The following is a list of pathovars of *P. syringae* and of the respective neopathotypes proposed by Dye et al. (1980). The names of the most important host plants and of their botanical families are also included. The list includes pathovar *mori*, although this name has priority over *syringae*. As recommended by Young et al. (1978), *P. syringae* is widely known and it should be conserved over *P. mori*. Many of the pathovars in the list may not be distinguishable from *P. syringae* except for their host range. Few phenotypic characters are taxonomically useful.

P. syringae pv. *syringae* van Hall 1902.
Hosts: lilac (*Syringa vulgaris*, fam. *Oleaceae*) and several unrelated plants.
Neopathotype strain: ATCC 19310 (PDDCC 3023; NCPPB 281).

P. syringae pv. *aceris* (Ark 1939) Young et al. 1978.
Hosts: *Acer macrophyllum, Acer* spp. (fam. *Aceraceae*)
Neopathotype strain: ATCC 10853 (PDDCC 2802; NCPPB 958) (Sneath and Skerman, 1966).

P. syringae pv. *antirrhini* (Takimoto 1920) Young et al. 1978.
Host: *Antirrhinum majus* (fam. *Scrophulariaceae*).
Neopathotype strain: PDDCC 4303 (NCPPB 1817).

P. syringae pv. *apii* (Jagger 1921) Young et al. 1978.
Host: *Apium graveolens* (fam. *Umbelliferae*).
Neopathotype: ATCC 9654 (PDDCC 2814; NCPPB 1626).

P. syringae pv. *aptata* (Brown and Jamieson 1913) Young et al. 1978.
Hosts: Sugar beet (*Beta vulgaris*, fam. *Chenopodiaceae*), *Nasturtium* sp. (fam. *Cruciferae*); lettuce (*Lactuca sativa*, fam. *Compositae*).
Neopathotype strain: PDDCC 459 (NCPPB 871).

P. syringae pv. *atrofaciens* (McCulloch 1920) Young et al. 1978.

Table 4.13.

Characters differentiating various **P. syringae** *pathovars,* **P. viridiflava** *and* **P. cichorii** *(section I)[a]*

| | 6. *P. syringae* pathovars | | | | | | 7. *P. viridiflava* | 8. *P. cichorii* |
Characteristics	*savastanoi*	*phaseolicola*	*mori*	*coronafaciens*	*tomato*	*syringae*		
Utilization of:								
D-Xylose	3	0	100	100	100	100	33	100
D-Gluconate	94	100	100	0	100	100	33	100
Acetate	35	0	86	25	33	58	66	100
Propionate	0	50	29		33	0	66	100
Linolenate	0	0	0	0	66	67	100	50
Triacetin	0	100	100	100	100	92	100	100
Tripropionin	0	100	100	75	66	67	100	100
Tricaproin	0	37	71	62	0	75	100	50
Glutarate	68	0	43	62	33	100	100	100
L-Malate	85	63	71	0	66	100	66	100
L(+)-Tartrate	71	0	14	0	0	8	0	100
D(−)-Tartrate	0	0	14	12	66	0	66	0
meso-Tartrate	6	0	57	62	100	100	66	100
DL-β-Hydroxybutyrate	0	37	14	38	66	75	66	100
DL-Lactate	0	0	0	25	0	67	100	100
DL-Glycerate	59	0	14	25	33	100	100	100
L-Ascorbate	70	0	71	100	0	25	0	0
Isoascorbate	47	0	(50)	71	100	100	100	0
α-Ketoglutarate	94	75	100	12	100	100	0	100
L-Histidine	32	0	43	0	0	67	100	50
L-Tyrosine	0	0	(17)	12	100	8	66	100
Betaine	76	0	57	62	100	100	100	100
Erythritol	0	0	0	87	0	83	100	0
Sorbitol	88	0	57	75	100	100	100	0
m-Inositol	53	0	100	50	100	100	100	100
Quinate	0	87	29		100	92	100	100
α-Lecithin	6	63	0	0	100	42	66	100
DL-Asparagine	76	0	(50)	62	100	83	100	100
Miscellaneous tests:								
Oxidase	0	0	0	0	0	0	0	100
Egg yolk	6	12	14	50	0	67	66	100
Ribonuclease	74	100	100	62	66	83	100	0

[a] Data with permission from Sands et al. (1970). The numbers in the table represent percentages of positive strains. Parentheses indicate that not all strains have been tested.

Hosts: Wheat (*Triticum* sp.) and other plants of the family *Gramineae*.

Neopathotype strain: PDDCC 4394 (NCPPB 2612).

P. syringae pv. *atropurpurea* (Reddy and Godkin 1923) Young et al. 1978.

Hosts: *Bromus inermis* and many other *Bromus* spp., *Agropyron repens* and many other plants of the family *Gramineae*.

Neopathotype strain: PDDCC 4457 (NCPPB 2397).

P. syringae pv. *berberidis* (Thornberry and Anderson 1931a) Young et al. 1978.

Hosts: Barberry (*Berberis thunbergii, B. inermis,* fam. *Berberidaceae*).

Neopathotype strain PDDCC 4116 (NCPPB 2724).

P. syringae pv. *cannabina* (Šutić and Dowson 1959) Young et al. 1978.

Hosts: *Cannabis sativa* (fam. *Moraceae*); *Phaseolus vulgaris, Vicia sativa* (fam. *Leguminosae*).

Neopathotype strain: PDDCC 2823 (NCPPB 1437).

P. syringae pv. *ciccaronei* (Ercolani and Caldarola 1972) Young et al. 1978.

Host: Carob (*Ceratonia siliqua,* fam. *Leguminosae*).

Pathotype strain: PDDCC 5710 (NCPPB 2355).

P. syringae pv. *coronafaciens* (Elliott 1920) Young et al. 1978.

Hosts: *Avena sativa, Bromus inermis, Agropyron repens* and various other wild and cultivated plants of the family *Gramineae*.

Neopathotype strain: PDDCC 3113 (NCPPB 600).

P. syringae pv. *delphinii* (Smith 1904) Young et al. 1978.

Host: *Delphinium* sp. (fam. *Ranunculaceae*).

Neopathotype strain: PDDCC 529 (NCPPB 1879).

P. syringae pv. *dysoxyli* (Hutchinson 1949) Young et al. 1978.

Host: *Dysoxylum spectabile* (fam. *Meliaceae*).

Neopathotype strain: ATCC 19863 (PDDCC 545; NCPPB 225).

P. syringae pv. *eriobotryae* (Takimoto 1931) Young et al. 1978.

Host: Loquat (*Eriobotrya japonica,* fam. *Rosaceae*).

Neopathotype strain: PDDCC 4455 (NCPPB 2331).

P. syringae pv. *garcae* (Amaral, Teixeira and Pinheiro 1956) Young et al. 1978.

Hosts: Coffee (*Coffea arabica,* fam. *Rubiaceae*) and many other unrelated plants.

Pathotype strain: PDDCC 4323 (NCPPB 588; ATCC 19864).

P. syringae pv. *glycinea* (Coerper 1919) Young et al. 1978.

Host: Soybean (*Glycine max,* fam. *Leguminosae*).

Neopathotype strain: PDDCC 2189 (NCPPB 2411).

Table 4.14.

Characterics of **Pseudomonas syringae** *pathovars (section I)[a]*

Characteristics	tabaci	lachrymans	syringae	aptata	pisi	antirrhini	morsprunorum	delphinii	tomato	eriobotryae	sesami	savastanoi	coronafaciens	striafaciens	mori	passiflorae	glycinea	phaseolicola	persicae	cannabina
Levan formation	+	+	+	+	+	+	+	d	+	+	+	−	+	+	+	−	+	+	+	+
Pectate gel pitting[b]	4, 8	4, 8	−	−	−	4	4	4	4	4	4, 8	4	−	4		4	4	4	4	
β-Glucosidase	+	+	+	+	d	+	−	+	+	+	−	−	+	+	−	+	−	−	−	−
Growth on:																				
Mannitol	+	+	+	+	+	+	+	+	+	+	−	+	+	+	+	+	d	−	+	−
Betaine	+	−	+	+	+	+	+	+	+	+	+	+	+	−	+	−	d	−		+
Inositol	+	+	+	+	+	+	+	+	+	−	+	d	+	+	+	+	+	−	−	−
Sorbitol	+	+	+	+	+	+	+	+	+	−	d	+	+	d	+	−	−	+	−	
Trigonelline	+	+	+	+	+	+	+	+	+	+	d	−	−	+	−	+	+	−	+	
Quinate	+	+	+	+	+	+	+	+	+	+	−	+	−	−	+	+	+	−	−	
Erythritol	d	+	+	+	d	−	d	+	−	+	+	−	+	+	−	−	−	−	−	−
L-Tartrate	+	+	−	−	−	−	+	−	−	+	+	d	−							
D-Tartrate	−	−	d	+	−	+	−	−	+	−	−	−								
L-Lactate	−	−	+	+	d	−	−	−	−	−	−	−								
Anthranilate	−	−	−	−	−	−	+	−	−	−	−	d	−							
Homoserine	−	−	−	−	+	+	−	−	−	−	−	−								

[a] Data with permission from Sands et al. (1980). For symbols see standard definitions.

[b] Method of Hildebrand (1971); numbers represent pH values at which pitting occurs.

P. syringae pv. *helianthi* (Kawamura 1934) Young et al. 1978.
  Host: sunflower (*Helianthus debilis*, fam. *Compositae*).
  Neopathotype strain: PDDCC 4531 (NCPPB 2640).

P. syringae pv. *japonica* (Mukoo 1955) Dye et al. (1980) (syn. *P. striafaciens* var. *japonica* Mukoo 1955).
  Hosts: rye (*Secale cereale*), barley (*Hordeum sativum*), wheat (*Triticum vulgare*), rice (*Oryza sativa*) and many other *Gramineae* of the genera *Setaria, Panicum, Bromus, Lolium, Andropogon, Alopecurus*, etc., as well as plants of other botanical families *Solanaceae, Chenopodiaceae, Oxalidaceae*, etc.
  Neopathotype strain: PDDCC 6305 (NCPPB 3093).

P. syringae pv. *lachrymans* (Smith and Bryan 1915) Young et al. 1978.
  Host: Cucumber (*Cucumis sativus*, fam. *Cucurbitaceae*).
  Neopathotype strain: ATCC 7386 (PDDCC 3988; NCPPB 537).

P. syringae pv. *lapsa* (Ark 1940) Young et al. 1978.
  Hosts: corn (*Zea mays*), sugarcane (*Saccharum officinarum*) (fam. *Gramineae*).
  Neopathotype strain: PDDCC 3947 (NCPPB 2096).

P. syringae pv. *maculicola* (McCulloch 1911) Young et al. 1978.
  Hosts: cabbage, cauliflower (*Brassica oleraceae*, fam. *Cruciferae*).
  Neopathotype strain: PDDCC 3935 (NCPPB 2039).

P. syringae pv. *mellea* (Johnson 1923) Young et al. 1978.
  Host: tobacco (*Nicotiana tabacum*, fam. Solanaceae).
  Neopathotype strain: PDDCC 5711 (NCPPB 2356).

P. syringae pv. *mori* (Boyer and Lambert 1893) Young et al. 1978.
  Host: Mulberry (*Morus* spp., fam. *Moraceae*).
  Neopathotype strain: ATCC 19873 (PDDCC 4331; NCPPB 1034).

P. syringae pv. *morsprunorum* (Wormald 1931) Young et al. 1978.
  Hosts: *Prunus* spp. (fam. Rosaceae).
  Neopathotype strain: ATCC 19322 (PDDCC 5795; NCPPB 2995) (Sneath and Skerman, 1966).

P. syringae pv. *panici* (Elliott 1923) Young et al. 1978.
  Host: Proso or broom-corn millet (*Panicum miliaceum*, fam. *Gramineae*).
  Neopathotype strain: ATCC 19875 (PDDCC 3955; NCPPB 1498).

P. syringae pv. *papulans* (Rose 1917) Dhanvantari 1977.
  Host: Apple (*Pyrus malus*, fam. *Rosaceae*).
  Neopathotype strain: PDDCC 4048 (NCPPB 2848).

P. syringae pv. *passiflorae* (Reid 1938) Young et al. 1978.

Host: *Passiflora edulis* (fam. *Passifloraceae*).
  Neopathotype strain: PDDCC 129 (NCPPB 1387).

P. syringae pv. *persicae* (Prunier et al. 1970) Young et al. 1978.
  Host: Peach (*Prunus persicae*, fam. *Rosaceae*).
  Neopathotype strain: PDDCC 5846 (NCPPB 2761).

P. syringae pv. *phaseolicola* (Burkholder 1926) Young et al. 1978.
  Hosts: Beans (*Phaseolus vulgaris*), kudzu vine (*Pueraria hirsuta*), (fam. *Leguminosae*).
  Neopathotype strain: ATCC 19304 (PDDCC 2740; NCPPB 52).

P. syringae pv. *pisi* (Sackett 1916) Young et al. 1978.
  Host: Pea (*Pisum sativum*, fam. *Leguminosae*).
  Neopathotype strain: PDDCC 2452 (NCPPB 2585).

P. syringae pv. *primulae* (Ark and Gardner 1936) Young et al. 1978.
  Host: *Primula polyantha* (fam. *Primulaceae*).
  Neopathotype strain: ATCC 19306 (PDDCC 3956; NCPPB 133).

P. syringae pv. *ribicola* (Bohn and Maloit 1946) Young et al. 1978.
  Host: Golden currant (*Ribes aureum*, fam. *Saxifragaceae*).
  Neopathotype strain: ATCC 13456 (PDDCC 3882; NCPPB 963) (Sneath and Skerman, 1966).

P. syringae pv. *savastanoi* (Smith 1908) Young et al. 1978.
  Hosts: olive tree (*Olea europaea*), *Fraxinus* spp. (fam. *Oleaceae*).
  Neopathotype strain: ATCC 13522 (PDDCC 4352; NCPPB 639).

P. syringae pv. *sesami* (Malkoff 1906) Young et al. 1978.
  Host: Sesame (*Sesamum indicum*, fam. *Pedaliaceae*).
  Neopathotype strain: ATCC 19879 (PDDCC 763; NCPPB 1016).

P. syringae pv. *striafaciens* (Elliott 1927) Young et al. 1978.
  Hosts: Oats (*Avena sativa*), barley (*Hordeum sativum*) (fam. *Gramineae*).
  Pathotype strain: ATCC 10730 (PDDCC 3961; NCPPB 1898). Pathogenic reference strain: PDDCC 4483 (NCPPB 2394).

P. syringae pv. *tabaci* (Wolf and Foster 1917) Young et al. 1978.
  Host: Tobacco (*Nicotiana tabacum*, fam. *Solanaceae*).
  Neopathotype strain: PDDCC 2835 (NCPPB 1427).

P. syringae pv. *tagetis* (Hellmers 1955) Young et al. 1978.
  Host: *Tagetes erecta* (fam. *Compositae*).
  Neopathotype strain: ATCC 4091 (NCPPB 2488).

P. syringae pv. *theae* (Hori 1915) Young et al. 1978.
  Host: Tea plant (*Thea sinensis*, fam. *Theaceae*).
  Neopathotype strain: PDDCC 3923 (NCPPB 2598).

P. syringae pv. *tomato* (Okabe 1933) Young et al. 1978.
  Host: Tomato (*Lycopersicon esculentum*, fam. *Solanaceae*).
  Neopathotype strain: PDDCC 2844 (NCPPB 1106).

P. syringae pv. *ulmi* (Šutić and Tešić 1958) Young et al. 1978.
  Host: Elm (*Ulmus* sp., fam. *Ulmaceae*).
  Neopathotype strain: ATCC 19883 (PDDCC 3962; NCPPB 632).

P. syringae pv. *viburni* (Thornberry and Anderson 1931b) Young et al. 1978.
  Host: *Viburnum opulus* (fam. *Caprifoliaceae*).
  Neopathotype strain: ATCC 13458 (PDDCC 3963; NCPPB 1921).

7. Pseudomonas viridiflava (Burkholder 1930) Dowson 1939, 177.[AL] (*Phytomonas viridiflava* Burkholder 1930, 63.)

vi.ri.di.fla'va. L. *viridis* green; L. *flavus* yellow; M.L. adj. *viridiflavus* greenish yellow.

Characteristics of the species are presented in Tables 4.11 and 4.12. Characters useful for the differentiation from other species of section I and from various pathovars of *P. syringae* are given in Tables 4.7 and 4.13.

Further information is given by Haynes and Burkholder (1957), Clara (1934), and Billing (1970b). The following characteristics are mentioned in this last paper. The bacterial mass usually has a yellow tinge in media with 5% sucrose and olive to golden brown in media with yeast extract and glycerol. A blue-green insoluble pigment is produced by some strains. Aside from the characters described in the tables, the potato rot and esculin reactions are positive. From her extensive nutritional screening of strains of the species, Billing (1970b) has concluded that few substrates have diagnostic value. The only ones distinguishing *P. viridiflava* from most other oxidase-negative plant pathogens are the inability to use sucrose and the capacity for use of D(−)-tartrate. This substrate is used by *P. viridiflava*, *P. syringae* pv. *tomato*, and rarely by other species or pathovars. In addition to these characters Billing (1970b) mentions that the reaction in beans was similar to that produced by *P. syringae* and both were different from the water-soaked lesions given by "*P. phaseolicola*."

Pathogenic on bean (*Phaseolus vulgaris*).

The mol% G + C of the DNA is not known.

Type strain: ATCC 13223 (PDDCC 2848; NCPPB 635).

8. Pseudomonas cichorii (Swingle 1925) Stapp 1928, 291.[AL] (*Phytomonas cichorii* Swingle 1925, 730.)

ci.cho'ri.i. Gr. *cichora* (pl.) succory, chicory; L. n. *cichorium* chicory; M.L. gen. n. *cichorii* of chicory.

Characteristics of the species are given in Tables 4.11 and 4.12. Characteristics useful for the differentiation from other species of section I and from various pathovars of *P. syringae* are given in Tables 4.7 and 4.13.

Optimum temperature, ~30°C.

The oxidase reaction is weak and slow.

Isolated from *Cichorium intybus* and *C. endivia*, for which it is pathogenic.

The mol% G + C of the DNA is 59 (Bd).

Type strain: ATCC 10857 (PDDCC 5707; NCPPB 943).

9. Pseudomonas stutzeri (Lehmann and Neumann 1896) Sijderius 1946, 115.[AL] (*Bacterium stutzeri* Lehmann and Neumann 1896, 237.)

stut'ze.ri. Stutzer patronymic; *stutzeri* of Stutzer.

Characteristics of the species are given in Tables 4.15 and 4.16. Characters differentiating the species from other species of section I and from other species of denitrifying pseudomonads are given in Tables 4.7 and later in Table 4.30.

Table 4.15.

General characteristics of species 9–12 of **Pseudomonas** (*section I*)[a]

Characteristics	9. *P. stutzeri*	10. *P. mendocina*	11. *P. alcaligenes*	12. *P. pseudoalcaligenes*
Cell diameter, μm	0.7–0.8	0.7–0.8	0.5	0.7–0.8
Cell length, μm	1.4–2.8	1.4–2.8	2.0–3.0	1.2–2.5
Number of flagella	1[b]	1[b]	1	1
Production of pyoverdins or phenazine pigments	−	−	−	−
Yellow or orange cellular pigments	−	+	d	−
Autotrophic growth with H_2	−	−	−	−
Oxidase	+	+	+	+
PHB accumulation	−	−	−	d
Gelatin liquefaction	−	−	d	d
Starch hydrolysis	+	−	−	−
Lecithinase (egg yolk)	−	−	−	−
Lipase (Tween 80 hydrolysis)	+	+	d	−
Extracellular PHB hydrolysis	−	−	−	−
Growth at 4°C	−	−	−	−
Growth at 41°C	+	+	+	+
Denitrification	+	+	+	+
Arginine dihydrolase	−	+	+	d
Levan formation from sucrose	−	−	−	−
Catechol, *ortho* cleavage	+	+		
Protocatechuate, *ortho* cleavage	+	+		
Mol% G + C of DNA (Bd)	60.6–66.3	62.8–64.3	64–68	62–64

[a] For symbols see standard definitions.

[b] Lateral flagella of short wavelength may be produced under certain conditions.

Table 4.16.
Nutritional characteristics of species 9–12 of **Pseudomonas** (*section I*)[a]

Characteristic	9. P. stutzeri	10. P. mendocina	11. P. alcaligenes	12. P. pseudoalcaligenes
Utilization of:[b]				
Acetate, caprylate, caprate, succinate, fumarate, lactate, α-ketoglutarate, L-alanine, L-glutamate, L-proline	+	+	+	+
Glucose, glycolate	+	+	−	−
Fructose	d	d	−	+
Maltose, starch, azelate, sebacate	+	−	−	−
Gluconate, glutarate, L-serine	d	+	−	d
Saccharate, mucate, isovalerate, malonate, L-isoleucine, L-valine, L-aspartate, α-aminobutyrate	d	+	−	−
Propionate, pelargonate, spermine	d	+	+	+
Butyrate, propylene glycol, ethanol	+	+	d	+
Isobutyrate, L(+)-tartrate, citraconate, tryptamine	−	d	−	−
Valerate, caproate, heptanoate, aconitate, L-tyrosine	d	+	d	d
Adipate, mannitol, p-hydroxybenzoate, 2,3-butylene glycol	d	−	−	−
D-Malate, glycine, L-phenylalanine	d	d	−	d
β-Hydroxybutyrate, mesaconate	+	+	−	+
L-Malate	+	−	+	+
Glycerate	d	+	−	+
Hydroxymethylglutarate, benzoate	d	d	−	−
Citrate, D-alanine	+	+	d	d
Pyruvate	+	d	+	+
Levulinate, geraniol, quinate, α-aminovalerate	−	+	−	−
Itaconate, ethylene glycol	+	+	−	d
Sorbitol, creatine	−	−	−	d
Glycerol	+	d	−	d
n-Propanol, butanol, putrescine	d	+	d	+
Isobutanol	d	+	d	−
β-Alanine, L-histidine	−	+	d	d
L-Leucine	+	+	+	d
L-Arginine	−	+	+	+
γ-Aminobutyrate	−	−	+	+
δ-Aminovalerate	d	−	d	d
Ethanolamine	d	−	−	+
Histamine	−	−	d	d
Butylamine, α-amylamine, dodecane	−	−	d	−
Betaine	−	+	−	+
Sarcosine	−	+	−	d

[a] For symbols see standard definitions.

[b] The following compounds are not used by any species: D-ribose, D-xylose, D-arabinose, L-arabinose, L-rhamnose, D-mannose, D-galactose, sucrose, trehalose, cellobiose, lactose, 2-ketogluconate, salicin, oxalate, maleate, pimelate, suberate, D(−)-tartrate, m-tartrate, erythritol, m-inositol, adonitol, isopropanol, D-mandelate, L-mandelate, benzoylformate, o-hydroxybenzoate, m-hydroxybenzoate, phthalate, phenylacetate, phenylethanediol, naphthalene, phenol, testosterone, L-threonine, L-norleucine, L-lysine, L-ornithine, L-citrulline, L-tryptophan, D-tryptophan, kynurenine, kynurenate, anthranilate, benzylamine, hippurate, pantothenate, acetamide, nicotinate, trigonelline, hexadecane, poly-β-hydroxybutyrate.

Freshly isolated colonies are adherent and have a characteristic wrinkled appearance. After repeated transfers in laboratory media, the colonies may become smooth, butyrous and pale in color.

Some strains grow at 43°C. Optimum temperature, ~ 35°C.

The species is markedly heterogeneous in nutritional properties and in DNA base composition (mol% G + C of the DNA is 60.6–66.3 (Bd)).

Further descriptive information: (van Niel and Allen, 1952), Sijderius (1946) and Palleroni et al. (1970).

Found in soil and water, from which it can be isolated after enrichment in media with nitrate under anaerobic conditions at 30°C using various carbon sources. L(+)-tartrate gives excellent results in the enrichments (Van Niel and Allen, 1952) although, paradoxically, strains obtained in this manner may not grow with tartrate in pure culture. Many strains are isolated from clinical specimens.

The name "*P. stanieri*" Mandel (1966) has been proposed for strains with a mol% G + C of ~62%. However, this species is not clearly differentiated from *P. stutzeri* on the basis of phenotypic characters (Palleroni et al., 1970).

Type strain: ATCC 17588 (Lautrop AB 201; Stanier 221 (Stanier et al., 1966)).

10. **Pseudomonas mendocina** Palleroni, *in* Palleroni, Doudoroff, Stanier, Solanes and Mandel 1970, 220.[AL]

men.do.ci'na. Spanish fem. n. *mendocina* native of Mendoza.

Characteristics of the species are presented in Tables 4.15 and 4.16. Characters for the differentiation from other species of section I and from other denitrifying pseudomonads are given in Tables 4.7 and later in Table 4.30.

Colonies are yellowish as a result of production of carotenoid pigment; not adherent or wrinkled in appearance.

Optimum temperature, ~35°C.

For further descriptive information see Palleroni et al. (1970).

Found in soil and water; isolated by enrichment in media with nitrate under anaerobic conditions, especially at 40°C. Ethanol and L(+)-tartrate can be used as carbon sources for the enrichments. It has also been isolated from urine (Hugh and Gilardi, 1980)

The mol% G + C is 62.8–64.3 (Bd).

Type strain: ATCC 25411.

11. **Pseudomonas alcaligenes** Monias 1928, 332.[AL]

al.ca.li'ge.nes. M.L. adj. *alcaligenes* alkali-producing.

Characteristics of the species are given in Tables 4.15 and 4.16. Characters for differentiation from other species of section I and from other denitrifying pseudomonads are presented in Tables 4.7 and later in Table 4.30.

Optimum temperature, ~35°C.

One of the three known strains produces yellow cellular pigments.

For further descriptive information see Ralston-Barrett et al. (1976) and Stanier et al. (1966).

The type strain was isolated from swimming pool water (Hugh and Ikari, 1964).

The mol% G + C is 64–68 (Bd).

Type strain: ATCC 909 (Stanier 142; NCTC 10367; NCIB 9945).

12. **Pseudomonas pseudoalcaligenes** Stanier, *in* Stanier, Palleroni and Doudoroff 1966, 247.[AL]

pseu.do.al.ca.li'ge.nes. Gr. adj. *pseudes* false; M.L. adj. *alcaligenes* alkali-producing; M.L. adj. *pseudoalcaligenes* false alkali-producing.

Characteristics of the species are given in Tables 4.15 and 4.16. Differentiation from other species of section I and other denitrifiers is given in Tables 4.7 and later in Table 4.30.

Optimum temperature, 35°C.

Further descriptive information can be found in Ralston-Barrett et al. (1976) and Stanier et al. (1966).

The species is rather heterogeneous. The collection examined by Stanier *et al.* (1966) included at least two groups. Strains of one of these groups were capable of PHB accumulation and gave a positive arginine dihydrolase reaction, while the strains of the second group were negative for the two properties and also differed in some nutritional characteristics (Doudoroff and Palleroni, 1974). Recently the creation of a new subspecies, *P. pseudoalcaligenes* subspecies *citrulli* Schaad, Sowell, Goth, Colwell and Webb 1978, 123,[AL] has been proposed to include a group of strains that are pathogenic for watermelon (*Citrullus lanatus*). The phenotypic resemblance of these organisms to *P. pseudoalcaligenes* is evident, although the mol% G + C of the DNA is higher (65–67) than that of *P. pseudoalcaligenes* (62–64). The property of poly-β-hydroxybutyrate accumulation in the watermelon strains is not universal, and none of the strains give an arginine dihydrolase

reaction, thus differing from *P. pseudoalcaligenes* in the correlation between these two properties. Other differences have been described by Schaad et al. (1978). There is no question, however, that the watermelon isolates belong to the same phenotypic group as *P. pseudoalcaligenes*, but we believe that their acceptance as a new subspecies may have to be postponed until more is known about their natural relationships to *P. pseudoalcaligenes*.

Isolated from various natural materials, including clinical specimens. *Type strain*: ATCC 17440 (NCIB 9946).

SECTION II

The differential characteristics of *Pseudomonas* species included in this section are given in Table 4.17. Descriptions of the various species are presented later in Tables 4.18–4.23.

rRNA Group II

Taxonomic Comments

No clear-cut phenotypic differentiation can be drawn between this group and RNA group III (section III). With the exception of *P. solanacearum*, all strains of group II grow at 40°C. Arginine and betaine are used by all strains except by those of *P. solanacearum* and *P. pickettii*, which thus resemble members of group III, but strains of this group are incapable of denitrification, while those of *P. pickettii* and *P.*

solanacearum do denitrify. Protocatechuate is cleaved by an *ortho* mechanism (1,2-dioxygenase) by all strains of group II; members of group III degrade this aromatic intermediate starting with a *meta* cleavage (2,3-dioxygenase).

The most outstanding property of RNA group II is the fact that, with the exception of *P. pickettii*, the species included are animal or plant pathogens. However, Pickett and Greenwood (1980) have recognized two subgroups within *P. pickettii*, one of which can be considered

Table 4.17.
Characteristics differentiating species 13–19 (Section II) of **Pseudomonas**[b]

Characteristics	13. *P. mallei*	14. *P. pseudomallei*	15. *P. caryophylli*	16. *P. cepacia*	17. *P. gladioli*	18. *P. pickettii*	19. *P. solanacearum*
Number of flagella	0	>1	>1	>1	>1	1	>1
Diffusible pigments	–	–	+[b]	+[b]	+[b]	–	d[c]
Arginine dihydrolase	+	+	+	–	–	–	–
Denitrification	+	+	+	–	–	+	+
Growth at 40°C	+	+	+	+	+	+	–
Gelatin hydrolysis	+	+	–	d	+	–	–
Starch hydrolysis	d	+	–	–	–	–	–
Extracellular PHB hydrolysis	d	+	–	–	–	–	–
Carbon sources used for growth:							
D-Xylose	+	–	+	d	+	+	–
D-Ribose	–	+	+	+	+	–	d
L-Rhamnose	–	–	+	d	–	–	–
Saccharate	–	–	+	+	+	+	+
Levulinate	–	+	–	+	–	+	d
Citraconate	–	–	–	+	+	+	–
Mesaconate	–	–	–	–	+	–	–
D(–)-Tartrate	–	–	–	–	+	–	–
meso-Tartrate	–	–	+	+	+	+	d
Erythritol	–	+	–	–	–	–	–
Adonitol	–	d	–	+	+	–	–
2,3-Butylene glycol	–	–	+	+	–	–	–
m-Hydroxybenzoate	–	–	–	+	–	–	–
Tryptamine	–	–	–	+	–	–	–
α-Amylamine	+	+	–	+	–	–	–

[a] For symbols, see standard definitions.

[b] Strains of *P. cepacia* may produce nonfluorescent pigments of various colors; strains of *P. gladioli* and *P. caryophylli* may produce yellow-green nonfluorescent pigments.

[c] Brown diffusible pigment produced by some strains.

as potentially pathogenic for humans (see below). Moreover, *P. pickettii* strains have been isolated most frequently from hospital specimens including urine, blood, cerebrospinal fluid, nasopharynx, abscesses, wounds (Ralston et al., 1973; Hugh and Gilardi, 1980).

In the eighth edition of the *Manual P. solanacearum* was placed in rRNA group III, but subsequently *P. pickettii* strains were found to have some level of DNA homology with the former species and also with strains of *P. cepacia* and *P. mallei* (Ralston et al., 1973), thus representing natural links between *P. solanacearum* and the other pathogenic species of group II. This finding was supported by the results of rRNA/DNA hybridization experiments (Palleroni et al., 1973), which clearly showed that all these organisms belong to the same RNA homology group.

The enzymological studies of Jessen and collaborators (Byng et al.,

1980; Whitaker et al., 1981a, b) suggest the assignment of the species *P. pyrrocinia* and "*P. methanolica*" to rRNA group II. Since these species have not been extensively characterized, it is impossible at this moment to include them in comparative tables with the rest of the species of group II, and their assignment to the group should also require confirmation by nucleic acid hybridization studies. The same authors have also confirmed the conclusions of Baumann and Baumann (1978), based on the immunological properties of glutamine synthetase, of a relationship between *Alcaligenes eutrophus* and members of *Pseudomonas* group II. This is rather surprising, in view of the general phenotypic properties of *A. eutrophus*, which are closer to those of members of group III, particularly the so-called "acidovorans group" (Davis, 1967).

List of the species of section II

13. **Pseudomonas mallei** (Zopf 1885) Redfearn, Palleroni and Stanier 1966, 305.[AL] (*Bacillus mallei* Zopf 1885, 89.)

mal'le.i. L. n. *malleus* the disease glanders; L. gen. n. *mallei* of glanders.

Characteristics of the species are listed in Tables 4.18 and 4.19. Characteristics useful for differentiation from other species of section II and from other denitrifying pseudomonads are presented in Tables 4.17 and later in Table 4.30.

Optimum temperature, ~37°C.

For further descriptive information, see Redfearn et al. (1966) and Redfearn and Palleroni (1975).

Parasitic on horses and donkeys, in which it causes glanders and farcy. The infection is transmissible to man and to other animal species.

The mol% G + C of the DNA is 69 (Bd).

Type strain: ATCC 23344 (NBL 7) (Redfearn et al., 1966).

14. **Pseudomonas pseudomallei** (Whitmore 1913) Haynes 1957, 100.[AL] (*Bacillus pseudomallei* Whitmore 1913, 9.)

pseu.do.mal'le.i. Gr. adj. *pseudes* false; L. n. *malleus* the disease glanders; M.L. gen. n. *pseudomallei* of false glanders.

Characteristics of the species are given in Tables 4.18 and 4.19. Properties useful for the differentiation from other species of section II and from other species of denitrifying pseudomonads are presented in Tables 4.17 and later in Table 4.30.

Colonies can range in structure from extreme rough to mucoid, and in color from cream to bright orange.

Optimum temperature, ~37°C.

For further descriptive information see Redfearn et al. (1966) and Redfearn and Palleroni (1975).

Isolated from human and animal cases of melioidosis and from soil and water in tropical regions, particularly Southeast Asia. Probably a soil organism and accidental pathogen, causing melioidosis.

The mol% G + C of the DNA is 69.5 (Bd).

Type strain: ATCC 23343 (WRAIR 286; NBL 121) (Redfearn et al., 1966).

15. **Pseudomonas caryophylli** (Burkholder 1942) Starr and Burkholder 1942, 601.[AL] (*Phytomonas caryophylli* Burkholder 1942, 143.)

ca.ry.o'phyl.li. M.L. masc. n. *caryophyllus* specific epithet of *Dianthus caryophyllus*, carnation; M.L. gen. n. *caryophylli* of the carnation.

Characteristics of the species are presented in Tables 4.20 and 4.21. Differentiation from other species of section II and from other denitrifying species of *Pseudomonas* is indicated in Tables 4.17 and later in Table 4.30.

Optimum temperature, ~30–33°C.

For further descriptive information see Ballard et al. (1970).

Isolated from diseased carnations, for which the species is pathogenic.

Table 4.18.

General characteristics of **Pseudomonas mallei** *and* **Pseudomonas pseudomallei** *(*Pseudomonas *section II)*[a]

Characteristics	13. *P. mallei*	14. *P. pseudomallei*
Cell diameter, μm	0.5	0.8
Cell length, μm	1.4–4.0	1.5
Number of flagella	0	>1
Diffusible pigment production	−	−
Yellow or orange cellular pigments	−	+
Autotrophic growth on H_2	−	−
Oxidase reaction	+	+
Poly-β-hydroxybutyrate accumulation	+	+
Gelatin liquefaction	+	+
Starch hydrolysis	d	+
Lipase (Tween 80 hydrolysis)	d	+
Extracellular poly-β-hydroxybutyrate hydrolysis	d	+
Denitrification	+	+
Arginine dihydrolase	+	+
Catechol, *ortho* cleavage	+	+
Protocatechuate, *ortho* cleavage	+	+
Growth at 4°C	−	−
Growth at 41°C	+	+
Mol% G + C of DNA (Bd)	69	69.5

[a] For symbols see standard definitions.

The mol% G + C of the DNA is 65.3 (Bd).

Type strain: ATCC 25418 (PDDCC 512; NCPPB 2151; ICPB PC113; Ballard 720 (Ballard et al., 1970).)

16. **Pseudomonas cepacia** (*ex* Burkholder 1950) Palleroni and Holmes 1981, 479.[VP]* Note: This species was unaccountably omitted from the Approved Lists of Bacterial Names, but the name has been revived by Palleroni and Holmes (1981).

ce.pa'ci.a. L. fem. n. *caepa* or *cepa* onion; M.L. fem. adj. *cepacia* of or like an onion.

Characteristics of the species are presented in Tables 4.20 and 4.21. Characteristics useful for differentiation from other species of section II are given in Table 4.17.

For further descriptive information see Ballard et al. (1970) and Palleroni and Holmes (1981).

Optimum temperature, ~30–35°C.

Many strains isolated from rotten onions; others from soil and from clinical specimens. The species is considered to be an opportunistic

---

* *VP*, denotes that this name has been validly published in the official publication, International Journal of Systematic Bacteriology.

Table 4.19.
Nutritional characteristics of **P. mallei** *and* **P. pseudomallei**
(**Pseudomonas** *section II*)[a]

Characteristics	13. P. mallei	14. P. pseudomallei
Utilization of[b]:		
D-Arabinose, D-fucose, D-glucose, D-mannose, D-galactose, sucrose, trehalose, cellobiose, acetate, propionate, succinate, fumarate, adipate, L-malate, β-hydroxybutyrate, lactate, glycerate, α-ketoglutarate, pyruvate, mannitol, sorbitol, m-inositol, glycerol, benzoate, p-hydroxybenzoate, phenylacetate, quinate, L-alanine, D-alanine, β-alanine, L-threonine, L-aspartate, L-glutamate, L-arginine, γ-aminobutyrate, L-histidine, L-tyrosine, L-tryptophan, betaine, hippurate, poly-β-hydroxybutyrate	+	+
D-Ribose, isobutyrate, valerate, isovalerate, caproate, heptanoate, pelargonate, caprate, sebacate, aconitate, levulinate, erythritol, benzoylformate, L-isoleucine, L-lysine, L-kynurenine, kynurenate, ethanol, ethanolamine, butylamine, α-amylamine	−	+
D-Xylose, glycine	+	−
L-Arabinose, malonate, α-aminobutyrate	d	−
D-Fructose, maltose, starch, gluconate, 2-ketogluconate, salicin, butyrate, caprylate, suberate, citrate, L-serine, L-valine, δ-aminovalerate, L-proline, L-phenylalanine, anthranilate, putrescine	d	+
Glutarate, pimelate, azelate, D-malate, sarcosine, trigonelline	d	d
Adonitol, ethanol, L-mandelate, spermine, hexadecane	−	d

[a] For symbols see standard definitions.
[b] The following compounds are not used by either species: L-rhamnose, lactose, inulin, saccharate, mucate, oxalate, maleate, D(−)-tartrate, L(+)-tartrate, m-tartrate, glycolate, hydroxymethylglutarate, citraconate, itaconate, mesaconate, ethylene glycol, propylene glycol, 2,3-butylene glycol, methanol, n-propanol, isopropanol, n-butanol, isobutanol, geraniol, D-mandelate, o-hydroxybenzoate, m-hydroxybenzoate, phthalate, isophthalate, terephthalate, phenylethanediol, phenol, testosterone, L-leucine, L-norleucine, L-ornithine, L-citrulline, α-aminovalerate, D-tryptophan, methylamine, benzylamine, histamine, tryptamine, creatine, pantothenate, acetamide, nicotinate, dodecane, p-aminobenzoate, m-aminobenzoate.

human pathogen, and it has been found associated with various types of infections of nosocomial origin.

The mol% G + C of the DNA is 67.4 (Bd).

Type strain: ATCC 25416 (Burkholder 717; ICPB 25) (Ballard et al., 1970).

17. Pseudomonas gladioli Severini 1913, 420.[AL] (*Pseudomonas marginata* (McCulloch 1921) Stapp 1928, 56.)

gla.di'o.li. L. n. *gladiolus* a small sword lily; M.L. masc. n. *Gladiolus* generic name of gladiolus; M.L. gen. n. *gladioli* of gladiolus.

Characteristics of the species are presented in Tables 4.20 and 4.21. Characteristics useful for differentiation from other species of section II are given in Table 4.17.

Optimum temperature, ~30–35°C.

For further descriptive information see Ballard et al. (1970), where

the species appears under the name "*P. marginata*" (see also Hildebrand et al., 1973).

Isolated from decayed onions, *Gladiolus* spp. and *Iris* spp., for which the species is believed to be pathogenic.

Young et al. (1978) have proposed to include in this species the pathovars *gladioli* and *alliicola*, since it seems probable that the species comprises organisms of different pathogenic ability.

The mol% G + C of the DNA is 68.5 (Bd).

Type strain: NCPPB 1891 (ATCC 10248; PDDCC 3950). This is also the reference strain for *P. gladioli* pv. *gladioli*. The reference strain for *P. gladioli* pv. *alliicola* is ATCC 19302 (PDDCC 2804; NCPPB 947).

18. Pseudomonas pickettii Ralston, Palleroni and Doudoroff 1973, 18.[AL]

pick.et'ti.i. Pickett patronymic; M.L. gen. n. *pickettii* of Pickett; named after M. J. Pickett.

Characteristics of the species are given in Tables 4.22 and 4.23. Differentiation from other species of section II and from other denitrifying species is presented in Tables 4.17 and later in Table 4.30.

Optimum temperature, ~35°C.

Further descriptive information is given by Ralston et al. (1973) and by King et al. (1979).

According to Pickett and Greenwood (1980), *P. pickettii* is unique among nonfermenting bacilli because of the following properties: resistance to polymyxin B, production of acid from glucose but not from ethanol, mannitol or sucrose; formation of alkali from urea and malo-

Table 4.20.
General characteristics of **Pseudomonas caryophylli,**
Pseudomonas cepacia *and* **Pseudomonas gladioli** (*section II*)[a]

Characteristics	15. P. caryophylli	16. P. cepacia	17. P. gladioli
Cell diameter, μm		0.8–1.0	0.8
Cell length, μm		1.6–3.2	2.0
Number of flagella	>1	>1	>1
Diffusible pigment production	+	+	+
Organic growth factor requirement	d	−	−
Autotrophic growth with H₂	−	−	−
Oxidase reaction	+	+	+
Poly-β-hydroxybutyrate accumulation	+	+	+
Slime production from sucrose	−	d	+
Gelatin liquefaction	−	d	+
Starch hydrolysis	−	−	−
Lecithinase (egg yolk)	d	d	+
Lipase (Tween 80 hydrolysis)	d	+	+
Extracellular poly-β-hydroxybutyrate hydrolysis	−	−	−
Denitrification	+	−	−
Reduction of nitrate to nitrite		+	
Arginine dihydrolase	+	−	−
Growth at 4°C	−	−	−
Growth at 41°C	+	d	+
Protocatechuate, *ortho* cleavage	+	+	+
Mol% G + C of DNA (Bd)	65.3	67.4	68.5

[a] For symbols see standard definitions.

Table 4.21.
Nutritional characteristics of **Pseudomonas caryophylli, Pseudomonas cepacia,** *and* **Pseudomonas gladioli,** *(section II)[a]*

Characteristics	15. *P. caryophylli*	16. *P. cepacia*	17. *P. gladioli*	Characteristics	15. *P. caryophylli*	16. *P. cepacia*	17. *P. gladioli*
Utilization of[b]:				Heptanoate, caproate, caprylate, pelargonate, caprate, adipate, azelate, sebacate, citraconate, adonitol, benzoate, D-alanine, ornithine, kynurenate, ethanolamine	−	+	+
D-Ribose, D-arabinose, L-arabinose, D-fucose, D-glucose, D-mannose, D-galactose, D-fructose, sucrose, cellobiose, gluconate, 2-ketogluconate, saccharate, mucate, salicin, acetate, propionate, butyrate, isobutyrate, valerate, malonate, succinate, fumarate, D-malate, L-malate, *m*-tartrate, β-hydroxybutyrate, lactate, glycerate, hydroxymethylglutarate, citrate, α-ketoglutarate, pyruvate, aconitate, mannitol, sorbitol, *m*-inositol, glycerol, *p*-hydroxybenzoate, phenylacetate, quinate, L-alanine, β-alanine, L-serine, L-aspartate, L-glutamate, L-arginine, γ-aminobutyrate, L-histidine, L-proline, L-tyrosine, L-phenylalanine, L-tryptophan, betaine, hippurate	+	+	+	Pimelate, suberate, levulinate, *m*-hydroxybenzoate, δ-aminovalerate, putrescine, spermine, butylamine, tryptamine, α-amylamine	−	+	−
				D(−)-Tartrate, mesaconate	−	−	+
				L(+)-Tartrate, ethanol, L-isoleucine, nicotinate, trigonelline	−	d	+
				Itaconate, propylene glycol, glycine, norleucine, α-aminobutyrate, α-aminovalerate	−	−	d
				2,3-Butylene glycol	+	+	−
				n-Butanol	+	d	d
				Isobutanol	−	d	d
				L-Mandelate, benzoylformate	d	d	−
				o-Hydroxybenzoate, testosterone, benzylamine, histamine, acetamide	−	d	−
D-Xylose, *n*-propanol	+	d	+	L-Leucine, L-valine	d	d	+
L-Rhamnose, glycolate	+	d	−	Dodecane, hexadecane		d	
Trehalose, L-threonine	d	+	+				
Isovalerate, glutarate, citrulline, anthranilate, sarcosine	−	+	d				

[a] For symbols see standard definitions.
[b] The following compounds are not utilized by any species: maltose, lactose, starch, inulin, oxalate, maleate, erythritol, ethylene glycol, isopropanol, geraniol, D-mandelate, phthalate, phenylethanediol, naphthalene, phenol, D-tryptophan, creatine, pantothenate, poly-β-hydroxybutyrate.

nate, and formation of gas from nitrate at temperatures between 20°C and 30°C. Two biovars may be identified within the species. These clusters were first outlined by Tatum et al. (1974) within group Va described by King (1964), and could be defined more precisely on the basis of the data of Ralston et al. (1973) on DNA homology and nutritional characteristics (utilization of caproate, histamine, L-histidine, L-phenylalanine, and valerate). Strains of one of the biovars (Va-1) may be significant as potential human pathogens.

The cluster analysis reported by King et al. (1979) has supported the conclusion that *P. pickettii* comprises several biovars, among which could be included a group of strains named "*P. thomasii*" by Phillips and Eykyn (1972). This name has been proposed at an earlier date than *P. pickettii*, but has no priority due to the fact that it has not been validly published.

Isolated from natural materials and from diverse clinical specimens. The mol% G + C of the DNA is 64 (Bd).

Type strain: ATCC 27511 (K-288 (Ralston et al., 1973).)

19. **Pseudomonas solanacearum** (Smith 1896) Smith 1914, 178.[AL] (*Bacillus solanacearum* Smith 1896, 10.)

so.la.na.ce.a′rum. M.L. fem. pl. n. *Solanaceae* the nightshade family; M.L. fem. pl. gen. n. *solanacearum* of the *Solanaceae*.

Characteristics of the species are given in Tables 4.22 and 4.23. Differentiation for other species of *Pseudomonas* is indicated in Tables 4.17 and later in Table 4.30.

At least two different kinds of colonies are produced on complex media; one type is smooth, fluid, elevated; the other is somewhat rough, dry and flat. Some strains produce a diffusible brown pigment in complex culture media; the pigment has not been chemically identified.

The species has been divided into four biovars by Hayward (1964) mainly by acid formation from sugars and denitrification. The nutritional spectra of these biovars are fairly similar, but biovars I and II can be separated from III and IV by being unable to use galactose, lactate, mannitol, sorbitol, and *p*-hydroxybenzoate as carbon sources for growth (Palleroni and Doudoroff, 1971). On the basis of biochemical

Table 4.22.

General characteristics of **Pseudomonas pickettii** *and* **Pseudomonas solanacearum** *(section II)*[a]

Characteristics	18. *P. pickettii*	19. *P. solanacearum*
Cell diameter, μm	0.5–0.6	0.5–0.7
Cell length, μm	1.5–3.0	1.5–2.5
Number of flagella	1	>1
Pigment production	–	d
Autotrophic growth with H_2	–	–
Oxidase reaction	+	+
Poly-β-hydroxybutyrate accumulation	+	+
Levan formation from sucrose	–	–
Gelatin liquefaction	–	–
Starch hydrolysis	–	–
Lecithinase (egg yolk)	–	–
Lipase (Tween 80 hydrolysis)	+	d
Hydrolysis of extracellular poly-β-hydroxybutyrate	–	–
Growth at 4°C	–	
Growth at 41°C	+	–
Denitrification	+	+
Arginine dihydrolase reaction	–	–
Protocatechuate, *ortho* cleavage	+	+
Mol% G + C of DNA (Bd)	64	66.5–68

[a] For symbols see standard definitions.

properties and serological, phage and bacteriocin typing, Okabe and Goto (1952) could subdivide the species in about 40 groups, but there is no clear-cut relationship of these groups to the 13 pathovars later defined by the same authors (Okabe and Goto, 1961). In general, there is no agreement among different authors as to the internal division of this complex species.

For further descriptive information see Kelman (1953), Buddenhagen and Kelman (1964), Hayward (1964), and Palleroni and Doudoroff (1971).

Important as a plant pathogen, particularly in warm and humid climates, causing wilt of many cultivated plants. Isolated from a wide variety of diseased plants, including several *Solanaceae* (particularly potato, tomato, tobacco), *Casuarina*, *Strelitzia*, ginger, banana, *Heliconia*, peanut, *Pelargonium*. Believed to be transmitted through soil (for a discussion see Buddenhagen, 1965) and occasionally by insects (Buddenhagen and Elsasser, 1962). The isolated cultures easily lose their pathogenicity upon cultivation on laboratory media, and this phenomenon has been correlated with colony variation (Buddenhagen and Kelman, 1964).

The mol% G + C of the DNA is 66.5–68 (Bd).

Type strain: ATCC 11696 (PDDCC 5712; NCPPB 325) (Sneath and Skerman, 1966).

Table 4.23.

Nutritional characteristics of **Pseudomonas pickettii** *and* **Pseudomonas solanacearum** *(section II)*[a]

Characteristics	18. *P. picketti*	19. *P. solanacearum*
Utilization of [b]:		
D-Glucose, gluconate, saccharate, mucate, acetate, propionate, butyrate, isobutyrate, succinate, fumarate, L-malate, β-hydroxybutyrate, citrate, α-ketoglutarate, pyruvate, aconitate, glycerol, L-alanine, D-alanine, L-aspartate, L-glutamate, γ-aminobutyrate, L-proline	+	+
D-Ribose, mannitol, sorbitol, *m*-inositol, L-ornithine, L-citrulline, L-phenylalanine, sarcosine	–	d
D-Xylose, L-arabinose, 2-ketogluconate, isovalerate, heptanoate, caprylate, caprate, glycerate, hydroxymethylglutarate, malonate, maleate, adipate, suberate, citraconate, L-mandelate, benzoylformate, glycine, L-valine, L-tryptophan, kynurenate	+	–
D-Galactose, pimelate, azelate, sebacate, L(+)-tartrate, *m*-tartrate, lactate, glycolate, levulinate, *p*-hydroxybenzoate, β-alanine, L-serine, L-threonine, L-tyrosine, hippurate, quinate	+	d
Sucrose, trehalose, benzoate	–	+
Raffinose	–	
Inulin		–
Valerate, caproate, D-malate, L-kynurenine	d	d
Glutarate, 2,3-butylene glycol, *n*-propanol, n-butanol, isobutanol, D-tryptophan, histamine	d	–
cis, cis-Muconate		–
Asparagine		+
L-Histidine	d	+

[a] For symbols see standard definitions.
[b] The following compounds are not utilized by either species: D-arabinose, D-fucose, L-rhamnose, D-mannose, maltose, cellobiose, lactose, starch, salicin, pelargonate, D(–)-tartrate, itaconate, mesaconate, erythritol, adonitol, ethylene glycol, propylene glycol, ethanol, isopropanol, geraniol, D-mandelate, o-hydroxybenzoate, m-hydroxybenzoate, phthalate, phenylacetate, naphthalene, phenol, testosterone, L-leucine, L-isoleucine, L-norleucine, L-lysine, L-arginine, α-aminobutyrate, α-aminovalerate, δ-aminovalerate, anthranilate, ethanolamine, benzylamine, putrescine, spermine, tryptamine, butylamine, α-amylamine, betaine, creatine, acetamide, nicotinate, trigonelline, dodecane, hexadecane, poly-β-hydroxybutyrate.

SECTION III

The properties of the species of section III are given in Table 4.24 and later in Tables 4.25–4.28.

rRNA Group III

Taxonomic Comments

All strains of the species listed in rRNA group III accumulate poly-β-hydroxybutyrate as carbon reserve material, are unable to use arginine or betaine as carbon sources for growth (growth of one of the species, *P. pseudoflava*, has been recorded as scanty or variable) and, with the exception of *P. pseudoflava*, are unable to grow at 41°C or to denitrify. These last two properties are almost universally positive in group II.

All strains of rRNA group III are able to perform a *meta* cleavage of protocatechuate, but this property has not been recorded for *P. pseudoflava*. As mentioned before, strains of the species of group II cleave

Table 4.24.
Characteristics differentiating species 20–27 (section III) of **Pseudomonas**[a]

Characteristics	20. P. acidovorans	21. P. testosteroni	22. P. delafieldii	23. P. facilis	24. P. saccharophila	25. P. flava	26. P. pseudoflava	27. P. palleronii
Number of flagella	>1	>1	1	1	1	1[b]	1[c]	1[b]
Carotenoid pigments	−	−	−	−	−	+	+	+
Autotrophic growth with H_2	−	−	−	+	+	+	+	+
Gelatin hydrolysis	−	−	−	+	+	−	−	−
Starch hydrolysis	−	−	−	−	+	−	−	−
Extracellular PHB hydrolysis	−	d	+	+	−	−	−	−
Denitrification	−	−	−	−	−	−	+	−
Carbon sources used for growth:								
Glucose	−	−	+	+	+[d]	+	+[d]	+
Fructose	+	−	+	+	+[d]	+	+	−
L-Arabinose	−	−	+	+	+	+	+	−
Sucrose	−	−	−	−	+	+	+	−
Malonate	+	−	+	+	−	−	−	−
Ethanol	+	−	−	−	−	+	−	+
Glycolate	+	+	−	−	−	−	−	−
p-Hydroxybenzoate	+	+	−	−	−	−	+	+
Testosterone	−	+	−	−	−	−	−	−

[a] For symbols see standard definitions.
[b] With a tendency toward multitrichous flagella and subpolar flagellar insertion.
[c] Polar or subpolar insertion.
[d] Positive for all strains studied, but may require mutation in strains isolated from nature.

protocatechuate by the *ortho* mechanism. The cleavage reaction is, therefore, a very useful tool to distinguish members of the two groups, but it must be remembered that the conditions of induction must be strictly specified. In special instances a given organism may be able to split the same aromatic intermediate in different ways depending on the substrate used for the growth of the cells.

The species of rRNA group III can be subdivided into two major subgroups. Three of the species (*P. acidovorans, P. testosteroni* and *P. delafieldii*) are unable to grow chemolithotrophically at the expense of H_2 oxidation, while the other species (*P. facilis, P. saccharophila, P. flava, P. pseudoflava* and *P. palleronii*) are typical hydrogen pseudomonads. In section V of this treatment, other species of hydrogen pseudomonads ("*P. hydrogenovora*" and "*P. hydrogenothermophila*") will also be described, as well as five species of CO-utilizing pseudomonads that can also live autotrophically using hydrogen as the source of energy. The natural relationships of these species to other species of the genus are at present unknown.

Three of the species here included in rRNA group III (*P. flava, P. pseudoflava* and *P. palleronii*) have not been examined by rRNA/DNA annealing techniques by Palleroni et al. (1973). Their inclusion in this natural group is based on the results reported by De Vos (1980) on *P. flava* and *P. palleronii* and their homology with other members of the group, and on the results of Auling et al. (1978) on the comparison of *P. flava* and *P. pseudoflava*. The results reported by the two laboratories prove that the three species mentioned above and the other species of group III are part of the same natural cluster.

The inclusion of *P. palleronii* in group III is also supported by the work of Jensen and his collaborators (Byng et al., 1980; Whitaker et al., 1981a, b). These workers have also suggested that *P. andropogonis* (syn. "*P. stizolobii*") and "*P. alboprecipitans*" (a synonym of *P. avenae*) also belong in group III. Since our knowledge of these species is rather incomplete, they are here included in section V of this treatment among other species of uncertain relationships.

List of the species of section III

20. **Pseudomonas acidovorans** den Dooren de Jong 1926, 106.[AL]
a.ci.do′vo.rans. L. neut. n. *acidum* acid; L. v. *voro* to devour; M.L. part. adj. *acidovorans* acid-devouring.

Characteristics of the species are presented in Tables 4.25 and 4.26. Characteristics useful for differentiation from other species of section III are presented in Table 4.24.

For further descriptive information see den Dooren de Jong (1926) and Stanier et al. (1966).

Optimum temperature, ~30°C.

Occurs in soil; isolated after aerobic enrichment with a variety of organic compounds as sole carbon sources. Occasionally found in clinical specimens.

The mol% G + C of the DNA is 67 (Bd).

Type strain: ATCC 15668 (den Dooren de Jong 7; Stanier 14; NCIB 9681) (Stanier et al., 1966).

21. **Pseudomonas testosteroni** Marcus and Talalay 1956, 661.[AL]
tes.tos.te.ro′ni. M.L. gen. n. *testosteroni* of testosterone, a chemical compound.

Characteristics of the species are presented in Tables 4.25 and 4.26. Characteristics for differentiation from other species of section III are given in Table 4.24.

For further descriptive information see Stanier et al. (1966).

Optimum temperature, ~30°C.

Occurs in soil. Isolated after aerobic enrichment with a variety of organic compounds as sole carbon sources, including testosterone.

The mol% G + C of the DNA is 62 (Bd).

Type strain: ATCC 11996 (NCIB 8955).

22. **Pseudomonas delafieldii** Davis, in Davis, Stanier, Doudoroff and Mandel 1970, 12.[AL]

Table 4.25.

General characteristics of **Pseudomonas acidovorans** *and* **Pseudomonas testosteroni** *(section III)[a]*

Characteristics	20. *P. acido-vorans*	21. *P. testo-steroni*
Cell diameter, μm	0.8–1.1	0.7–0.8
Cell length, μm	2.5–4.1	2.1–2.9
Number of flagella	>1	>1
Pigment production	−	−
Oxidase reaction	+	+
Poly-β-hydroxybutyrate accumulation	+	+
Levan formation from sucrose	−	−
Gelatin liquefaction	−	−
Starch hydrolysis	−	−
Lipase (Tween 80 hydrolysis)	+	+
Extracellular poly-β-hydroxybutyrate hydrolysis	−	d
Autotrophic growth with H_2	−	−
Denitrification	−	−
Arginine dihydrolase	−	−
Growth at 4°C	−	−
Growth at 41°C	−	−
Protocatechuate, *meta* cleavage	+	+
Mol% G + C of DNA (Bd)	67	62

[a] For symbols see standard definitions.

de.la.fiel′di.i. M.L. gen. n. *delafieldii* of Delafield; named after F. P. Delafield, who first isolated the organism.

Characteristics of the species are presented in Tables 4.27 and 4.28. Differentiation from other species of section III is indicated in Table 4.24.

For further descriptive information see Davis et al. (1970).

Optimum temperature, 30°C.

Isolated from soil by enrichment with poly-β-hydroxybutyrate as the sole carbon source.

The mol% G + C of the DNA is 65–66 (Bd).

Type strain: ATCC 17505 (Delafield FD-6).

23. **Pseudomonas facilis** (Schatz and Bovell 1952) Davis, in Davis, Stanier, Doudoroff and Mandel 1969, 385.[AL] (*Hydrogenomonas facilis* Schatz and Bovell 1952, 88.)

fa′ci.lis. L. adj. *facilis* ready, quick.

Characteristics of the species are given in Tables 4.27 and 4.28. A comparison with other species of section III and with other chemolithotrophic pseudomonads is given in Tables 4.24 and 4.29.

For further descriptive information see Davis et al. (1969, 1970).

Optimum temperature, ~28°C.

Isolated from soil by enrichments in mineral media with incubation under an atmosphere containing H_2, O_2 and CO_2.

The mol% G + C of the DNA is 61.7–63.8 (Bd).

Type strain: ATCC 11228.

24. **Pseudomonas saccharophila** Doudoroff 1940, 59.[AL]

sac.cha.ro.phi′la. Gr. n. *sacchar* sugar; Gr. adj. *philus* loving; M.L. fem. adj. *saccharophila* sugar-loving.

Characteristics of the species are given in Tables 4.27 and 4.28. A comparison with other species of section III and with other chemolithotrophic pseudomonads is given in Tables 4.24 and 4.29.

For further descriptive information see Doudoroff (1940), Davis et al. (1970), Palleroni and Doudoroff (1965) and Palleroni (1978).

Optimum temperature, ~30°C.

Only one strain of the species is known. It was isolated by enrichment in mineral medium inoculated with mud from a stagnant pool, and incubated in an atmosphere containing 83% hydrogen, 2% oxygen and

15% CO_2. A similar strain was recently described (Palleroni, 1980), which differs from *P. saccharophila* in some phenotypic properties and in DNA sequence.

The mol% G + C of the DNA is 68.9 (Bd).

Type strain: ATCC 15946.

25. **Pseudomonas flava** (Niklewski 1910) Davis, *in* Davis, Stanier, Doudoroff and Mandel 1969, 385.[AL] (*Hydrogenomonas flava* Niklewski 1910, 123.)

fla′va. L. fem. adj. *flava* yellow.

Characteristics of the species are given in Tables 4.27 and 4.28.

Table 4.26.

Nutritional characteristics of **Pseudomonas acidovorans** *and* **Pseudomonas testosteroni** *(section III)[a]*

Characteristics	20. *P. acido-vorans*	21. *P. testo-steroni*
Utilization of [b]:		
Gluconate, saccharate, mucate, acetate, propionate, butyrate, valerate, isovalerate, caproate, succinate, fumarate, glutarate, adipate, pimelate, suberate, azelate, sebacate, D-malate, L-malate, β-hydroxybutyrate, lactate, glycolate, glycerate, hydroxymethylglutarate, citrate, α-ketoglutarate, pyruvate, aconitate, levulinate, citraconate, itaconate, butanol, m-hydroxybenzoate, p-hydroxybenzoate, glycine, D-alanine, L-leucine, L-isoleucine, L-norleucine, L-aspartate, L-glutamate, L-histidine, L-proline, L-tyrosine, L-phenylalanine, L-kynurenine, kynurenate	+	+
Isobutyrate, mesaconate, m-tartrate, n-propanol, L-alanine, nicotinate, trigonelline, hippurate	+	d
Caprate, m-inositol, glycerol, 2,3-butylene glycol, isobutanol, β-alanine	d	−
D(−)-Tartrate, geraniol, benzoate, phthalate, L-valine, poly-β-hydroxybutyrate, benzoylformate	−	d
D-Fructose, malonate, maleate, L(+)-tartrate, mannitol, ethanol, quinate, phenylacetate, α-aminobutyrate, δ-aminovalerate, L-tryptophan, D-tryptophan, acetamide	+	−
Testosterone	−	+
α-Aminovalerate, γ-aminobutyrate	d	d

[a] For symbols see standard definitions.

[b] The following compounds are not utilized by either species: D-ribose, D-xylose, D-arabinose, L-arabinose, D-fucose, L-rhamnose, D-glucose, D-mannose, D-galactose, sucrose, trehalose, maltose, cellobiose, lactose, starch, 2-ketogluconate, salicin, heptanoate, caprylate, pelargonate, oxalate, erythritol, sorbitol, adonitol, ethylene glycol, propylene glycol, D-mandelate, L-mandelate, o-hydroxybenzoate, phenylethanediol, naphthalene, phenol, L-serine, L-threonine, L-lysine, L-arginine, L-ornithine, L-citrulline, anthranilate, ethanolamine, benzylamine, putrescine, spermine, histamine, tryptamine, butylamine, α-amylamine, betaine, sarcosine, creatine, pantothenate, dodecane, hexadecane.

Table 4.27.
General characteristics of species 22–27 of **Pseudomonas** *(section III)[a]*

Characteristics	22. P. delafieldii	23. P. facilis	24. P. saccharophila	25. P. flava	26. P. pseudoflava	27. P. palleronii
Cell diameter, μm	0.5	0.3–0.5	0.5	1.5	0.5	0.4
Cell length, μm	1.8–2.6	1.8–2.8	3.0–4.0		1.0–2.0	1.5–2.6
Number of flagella	1	1	1	1[b]	1[c]	1[b]
Production of pyoverdins or phenazine pigments	–	–	–	–	–	–
Yellow or orange cellular pigments	–	–	–	+	+	+
Autotrophic growth with H_2	–	+	+	+	+	+
Oxidase reaction	+	+	+	+	+	+
Poly-β-hydroxybutyrate accumulation	+	+	+	+	+	+
Glycogen accumulation				+		
Gelatin liquefaction	–	+	+	–	–	–
Starch hydrolysis	–	–	+	–	–	–
Lipase (Tween 80 hydrolysis)	–	–	–	–	+	+
Extracellular poly-β-hydroxybutyrate hydrolysis	+	+	–	–	–	–
Growth at 41°C	–	–	–	–	+	–
Denitrification	–	–	–	–	+	–
Arginine dihydrolase	–	–	–	–		–
Protocatechuate, *meta* cleavage	+	+	+	+		+
Mol% G + C of DNA (Bd)	65–66	61.7–63.8	68.9	67.3	66.5–68	66.8

[a] For symbols see standard definitions.
[b] With a tendency toward multitrichous flagella and subpolar flagellar insertion.
[c] Polar or subpolar insertion.

Differentiation from other chemolithotrophic pseudomonads is presented in Tables 4.24 and 4.29.

For further descriptive information see Davis et al. (1970) and Kluyver and Manten (1942).

Optimum temperature, ~30°C.

The strain described by Kluyver and Manten (1942) is sensitive to oxygen, but insensitive strains can also be isolated from nature (Davis et al., 1970). The species may be confused with *Alcaligenes paradoxus* (Davis et al., 1969), some strains of which are initially microaerophilic after isolation. It can also be confused with *P. palleronii*. The nutritional characters presented in the description, however, permit a clear differentiation from this last species.

The mol% G + C of the DNA is 67.3 (Bd).

Type strain: DSM 619 (Davis et al., 1969).

26. **Pseudomonas pseudoflava** Auling, Reh, Lee and Schlegel 1978, 93.[AL]

pseu.do.fla′va. Gr. adj. *pseudes* false; L. fem. adj. *flava* yellow; M.L. fem. adj. *pseudoflava* not the true (*Pseudomonas*) *flava*.

Characteristics of the species are presented in Tables 4.27 and 4.28. Differentiation from other species of section III and from other chemolithotrophic pseudomonads, as well as from denitrifying species, is given in Tables 4.24, 4.29 and 4.30.

For further descriptive information see Auling et al. (1978).

The colonies are yellow, and this may be due to the presence of a carotenoid pigment in the cells. However, the structure of the pigment is unknown. The cells are capable of accumulating poly-β-hydroxybutyrate, glycogen and polyphosphate.

Optimum temperature, ~35–38°C.

DNA/DNA hybridization experiments demonstrated high homology within the strains of the species, and a moderate homology with *P. flava*. Immunological cross-reactions have been detected with *P. facilis*, *P. flava* and two strains of *Alcaligenes paradoxus* (Auling et al., 1978).

Isolated from soil, mud, or water by liquid enrichment for hydrogen bacteria and by the use of a gas atmosphere consisting of 10% oxygen, 10% CO_2 and 80% hydrogen.

The mol% G + C of the DNA is 66.5–68 (Bd).

Type strain: DSM 1034 (Auling et al., 1978).

27. **Pseudomonas palleronii** Davis, *in* Davis, Stanier, Doudoroff and Mandel 1970, 11.[AL]

pal.le.ro′ni.i. M.L. gen. n. *palleronii* of Palleroni; named after N. J. Palleroni, who first isolated the organism.

Characteristics of the species are given in Tables 4.27 and 4.28. Differentiation from other species of section III and from other chemolithotrophic pseudomonads is indicated in Tables 4.24 and 4.29.

For further descriptive information see Davis et al. (1970).

Optimum temperature, ~30°C.

All strains can tolerate a partial pressure of 0.2 atm of O_2 for autotrophic growth with H_2.

The species is easily confused with *Alcaligenes paradoxus* Davis et al. (1969) and with *P. flava* (Niklewski) Davis et al. (1970) and related strains. The distinguishing characters are listed by Davis et al. (1970).

Isolated from soil and water by enrichment in minimal media and atmospheres containing H_2, O_2 and CO_2.

The mol% G + C of the DNA is 66.8 (Bd).

Type strain: ATCC 17724 (Davis 1-6-1; Stanier 366) (Davis et al., 1970).

Table 4.28.

Nutritional characteristics of species 22–27 of **Pseudomonas** *(section III)[a]*

Characteristics	22. P. delafieldii	23. P. facilis	24. P. saccharophila	25. P. flava	26. P. pseudoflava	27. P. palleronii
Utilization of[b]:						
Glucose	+	+	+[c]	+	+	+[c]
Gluconate, acetate, succinate,[d] fumarate, L-malate,[d] β-hydroxybutyrate, lactate, pyruvate, L-alanine, L-aspartate, L-glutamate, L-proline, L-phenylalanine	+	+	+	+	+	+
D-Ribose, propionate	+	+	+	−	−	−
D-Arabinose, ethylene glycol	−	−	+[c]	−	−	−
D-Xylose	+	d	+	−	+	−
L-Arabinose, D-galactose	+	+	+	+	+	−
L-Rhamnose, butylamine	−	−	−	+		−
D-Fructose, D-mannose	+	+	+[c]	+	+	−
Sucrose, trehalose, maltose, cellobiose	−	−	+	+	+	−
Starch, isobutyrate	−	−	+	−	−	−
2-Ketogluconate, citraconate	+	−	−	−	−	−
Mucate, L-mandelate, kynurenate	−	−	−	−	−	+
Saccharate	+	−	−	−	s	+
Butyrate	+	+	+	−	s	+
Valerate, isovalerate	−	−	+	−	s	−
Malonate, poly-β-hydroxybutyrate	+	+	−	−	−	−
Maleate, hydroxymethylglutarate, hippurate	+	−	−	−	−	−
Glutarate	+	+	+	−		−
Adipate	+	−	+[c]	−		d
Pimelate	d	−	+[c]	−	s	−

Characteristics	22. P. delafieldii	23. P. facilis	24. P. saccharophila	25. P. flava	26. P. pseudoflava	27. P. palleronii
Suberate	+	+	+[c]	−	s	+[c]
Azelate	+	+	+[c]	−		+
Sebacate	d	−	+	−		−
D-Malate	−	+	−	+	+	+
Glycolate, D(−)-tartrate, n-propanol	−	−	−	+	s	+
L-(+)-Tartrate	+	−	+[c]	+[c]	s	+
α-Ketoglutarate	+	+	+	+	s	+
Citrate, aconitate	−	−	+	−	+	+
Levulinate	−	−	+	−		
Citraconate	+	−	−	−	−	−
Mannitol	+	+	−	+	+	−
Sorbitol	+	+	−	+	−	−
m-Inositol	−	−	−	+	+	+
Glycerol, L-tyrosine	+	+	−	+	+	+
Propylene glycol, m-hydroxybenzoate	−	−	−	+		+
Ethanol, DL-glycerate	−	−	−	+	s	+
n-Butanol	−	−	+[c]	−	s	+
Isobutanol	−	−	+[c]	−		+
Benzoylformate	−	−	−	−		d
p-Hydroxybenzoate, phenol, L-isoleucine	−	−	−	−	+	d
D-Alanine, L-tryptophan	−	−	−	−	+	+
Quinate	+[c]	+[c]	+	−	+	−
Glycine, L-valine, L-arginine	−	−	−	−	s	−
β-Alanine, L-serine	+	+	−	−		−
L-Leucine	+	+	−	−	+	+
L-Lysine, L-ornithine	−	−	−	−	+	−
L-Histidine	+	−	−	−	+	−
Ethanolamine	−	−	−	+	+	−

[a] Symbols: see standard definitions; and s, growth scanty or variable.

[b] The following compounds cannot be utilized by any species:

D-fucose, lactose,[d] salicin,[d] inulin,[d] formate,[d] caproate, heptanoate,[d] caprylate,[d] pelargonate,[d] caprate,[d] oxalate, itaconate[d], mesaconate, erythritol,[d] adonitol,[d] 2,3-butylene glycol, methanol,[d] isopropanol,[d] D-mandelate,[d] benzoate, o-hydroxybenzoate, phthalate,[d] isophthalate,[d] terephthalate,[d] phenylacetate,[d] phenylethanediol,[d] eicosenedioate,[d] testosterone,[d] threonine, norleucine,[d] citrulline, α-aminobutyrate,[d] γ-aminobutyrate,[d] α-aminovalerate,[d] δ-aminovalerate,[d] D-tryptophan,[d] L-kynurenine, anthranilate, methylamine,[d] benzylamine,[d] putrescine,[d] spermine,[d] histamine,[d] tryptamine,[d] α-amylamine,[d] betaine,[d] sarcosine,[d] creatine,[d] pantothenate, acetamide,[d] nicotinate, trigonelline,[d] m-aminobenzoate,[d] p-aminobenzoate.[d]

[c] Positive for all strains studied, but may require mutation in strains isolated from nature.

[d] Results not reported for P. pseudoflava.

SECTION IV

The three *Pseudomonas* species included in this section (*P. diminuta*, *P. vesicularis* and *P. maltophilia*) constitute RNA groups IV and V of the classification based on rRNA homology data. *P. diminuta* and *P. vesicularis* are the members of group IV and of the phenotypic cluster known as the diminuta group. *P. maltophilia* belongs to group V, together with *Xanthomonas* species.

Many properties of these species set them apart from all other species of the genus *Pseudomonas*. In addition to properties already discussed, the species share two characters that are absent in the vast majority of the fully characterized species of *Pseudomonas*. One is a growth factor requirement, which, outside of groups IV and V, is only present in exceptional strains of well characterized species, or in species whose natural relationships are still largely unknown. The other character is the inability of *P. maltophilia* and *P. vesicularis* to use nitrate as a nitrogen source. The exclusion of the growth factor requirement and the inability to use nitrate nitrogen would indeed contribute to a more precise circumscription of the genus *Pseudomonas*.

Table 4.29.
Characteristics differentiating various hydrogen and CO pseudomonads[a]

Characteristics	Yellow cellular pigment	Gelatin	Starch	PHB	Tween 80	Denitrification	Glucose	Fructose	Oxalate	Succinate	Malonate	β-Hydroxybutyrate	p-Hydroxybenzoate	L-Tyrosine	L-Proline	Mol% G + C of DNA
Hydrogen bacteria:																
P. saccharophila	−	−	+	−	−	−	+[b]	+[b]	−	+	−	+	−	−	+	68.9[c]
P. facilis	−	+	−	+	−	−	+	+	−	−	+	+	−	+	+	61.7–63.8[c]
P. flava	+	−	−	−	−	−	+	+	−	+	−	+	−	−	+	67.3[c]
P. pseudoflava	+	−	−	−	−	−	+	+	−	+	−	+	−	+	+	66.5–68[d]
P. palleronii	+	−	−	−	+	−	+[b]	−	−	−	−	+	+	+	+	66.8[c]
"P. hydrogenovora"	−	−	+	−	+	−	+	+	−	+	−	+	−	+		62.5[e]
"P. hydrogenothermophila"	+	−	−				−	−	−	+	−	−				63.5[e]
Carboxydobacteria:																
"P. carboxydoflava"	+	−	−	−	−	+	+	−	−	+	−	−	−	−	+	68.0[e]
P. carboxydohydrogena	−	−	−	−	−	−	−	+	+	+	−	−	−	−	−	58.2[e]
"P. compransoris"	−	−	−	−	−	−	−	−	−	−	−	+	−	−	−	63.3[e]
"P. carboxydovorans"	−	−	−	−	−	−	−	−	+	−	−	−	−	−	−	62.2[e]

[a] For symbols see standard definitions.
[b] Positive for all strains studied, but may require mutation in strains isolated from nature.
[c] Bd.
[d] Bd and T_m.
[e] T_m.

Table 4.30.
Characteristics useful for differentiation of the denitrifying pseudomonads[a]

Characteristics	P. aeruginosa	P. fluorescens II, III, P. aureofaciens	P. stutzeri	P. mendocina	P. alcaligenes	P. pseudoalcaligenes[b]	P. mallei	P. pseudomallei	P. caryophylli	P. solanacearum	P. pickettii	P. pseudoflava
RNA group	I	I	I	I	I	I	II	II	II	II	II	III
Mol% G + C of DNA	67.2	59.4	60.7–66.3	62.8–64.3	64–68	62–64	69	69.5	65.3	66.5–68	64	66.5–68
Number of flagella	1	>1	1	1	1	1	0	>1	>1	>1	1	1
PHB accumulation	−	−	−	−	−	d	+	+	+	+	+	+
H₂ autotrophy	−	−	−	−	−	−	−	−	−	−	−	+
Growth at 40°C	+	−	+	+	+	+	+	+	+	−	+	+
Pyoverdin production	+	+	−	−	−	−	−	−	−	−	−	−
Pyocyanin production	+	−	−	−	−	−	−	−	−	−	−	
Yellow cellular pigment	−	−	+	−	d	−	−	d	−	−	−	+
Arginine dihydrolase	+	+	−	+	+	d	+	+	+	−	−	
Starch hydrolysis	−	−	+	−	−	−	d	+	−	−	−	−
PHB hydrolysis	−	−	−	−	−	−	d	+	−	−	−	−
Growth on:												
D-Xylose	−	d	−	−	b	−	+	−	+	−	+	+
Maltose	−	−	+	−	−	−	d	+	−	−	−	+
Saccharate	−	d	d	+	−	−	−	−	+	+	+	d
Mannitol	+	+	d	−	−	−	+	+	+	d	−	+
Ethylene glycol	−	−	+	+	−	d	−	−	+	−	−	−
2,3-Butylene glycol	+	d	d	−	−	−	−	−	+	−	d	−
Geraniol	+	−	−	+	−	−	−	−	−	−	−	
Azelate	+	d	−	+	−	−	−	+	−	d	+	
Levulinate	+	d	−	+	−	−	−	+	−	d	+	
Glycolate	−	−	+	+	−	−	−	−	+	d	+	d
L-Serine	d	d	d	+	−	d	d	+	+	d	+	
L-Arginine	+	+	−	+	+	+	+	+	+	−	−	
L-Histidine	+	+	−	+	d	d	+	+	+	+	d	+
Betaine	+	+	−	+	−	+	+	+	+	−	−	
Sarcosine	+	+	−	+	−	d	d	d	−	d		

[a] For symbols see standard definitions.
[b] Not all strains of P. pseudoalcaligenes are denitrifiers.

Table 4.31.

Characteristics for the differentiation of **Pseudomonas diminuta, Pseudomonas vesicularis** *and* **Pseudomonas maltophilia** *(section IV)[a]*

Characteristics	RNA Group IV		RNA Group V
	a. *P. diminuta*	b. *P. vesicularis*	c. *P. maltophilia*
Number of flagella	1	1	>1
Colonies are yellow	−	+	+
Growth factors required:			
Pantothenate	+	+	−
Biotin	+	+	−
Cyanocobalamin	+	+	−
Methionine or cystine	+	−	+
Oxidase reaction	+	W	−
Hydrolysis of:			
Gelatin	−	−	+
Tween 80	−	−	+
Carbon sources used for growth:			
Glucose	−	+	+
Cellobiose	−	+	+
β-Hydroxybutyrate	+	+	−
L-Histidine	+	−	+
Pantothenate	+	−	−

[a] For symbols see standard definitions; and W, weak reaction.

Key for the differentiation of rRNA Groups IV and V

A. Accumulate poly-β-hydroxybutyrate as carbon reserve material
        a. *P. diminuta* (group IV)
        b. *P. vesicularis* (group IV)
B. Do not accumulate poly-β-hydroxybutyrate
        c. *P. maltophilia* (group V)

Characteristics useful for the differentiation of the species of the two groups are presented in Table 4.31.

rRNA Group IV

Taxonomic Comments

As we have already discussed, the superficial resemblance of *P. diminuta* and *P. vesicularis* to some of the acetic acid bacteria is not supported by other criteria, and the unique properties of this group set it apart not only from other *Pseudomonas* groups but also from other genera of Gram-negative bacteria. Therefore, although the two species are temporarily preserved in the genus *Pseudomonas* in the present treatment, the most satisfactory solution may be the creation of a new genus for this natural group.

List of the species of RNA group IV

a. **Pseudomonas diminuta** Leifson and Hugh 1954, 68.[AL]
di.mi.nu′ta. L. adj. *minutus* small; M.L. fem. adj. *diminuta* defective, minute.
Characteristics of the species are given in Tables 4.31–4.33.
Optimum temperature, ~30°C.
For further descriptive information see Ballard et al. (1968).
Isolated from water and clinical specimens.
The mol% G + C of the DNA is 66.3–67.3 (Bd).
Type strain: ATCC 11568 (RH342; NCTC 8545; NCIB 9393).

b. **Pseudomonas vesicularis** (Büsing, Doll and Freytag 1953) Galarneault and Leifson 1964, 167.[AL] (*Corynebacterium vesiculare* Büsing,
Doll and Freytag 1953, 76.)
ve.si.cu.la′ ris. M.L. fem. adj. *vesicularis* pertaining to a vesicle.
Characteristics of the species are given in Tables 4.31–4.33.
Optimum temperature, 30°C. Growth occurs at 37°C.
Growth occurs on ethanol with acid production. Isopropanol is oxidized to acetone but is not used as a carbon source. Ammonium salts, but not nitrates, are used as a nitrogen source (Ballard et al., 1968).
For further descriptive information see Ballard et al. (1968).
The type strain was isolated from a medicinal leech (*Hirudo medicinalis*). Other strains have been isolated from streams.
The mol% G + C of the DNA is 65.8 (Bd).
Type strain: ATCC 11426 (NCMB 1945).

rRNA Group V

Taxonomic Comments

Due to the fact that *P. maltophilia* and species of *Xanthomonas* share a substantial level of rRNA homology (Palleroni et al., 1973), it appears appropriate to assign the former to the genus *Xanthomonas*. *P. maltophilia* strains have a weak oxidase reaction, and this is also weak or negative in *Xanthomonas*. Methionine is required as a growth factor by *P. maltophilia*, and this amino acid is also required, in addition to other factors, by many *Xanthomonas* strains. Other physiological similarities have been mentioned by Palleroni (1981), and striking resemblances between the two groups are also found in the composition of the cell envelopes. Admittedly, inclusion of *P. maltophilia* in the genus

Table 4.32.

General characteristics of **Pseudomonas diminuta** *and* **Pseudomonas vesicularis** *(section IV)*[a]

Characteristics	a. *P. diminuta*	b. *P. vesicularis*
Cell diameter, μm	0.5	0.5
Cell length, μm	1.0–4.0	1.0–4.0
Number of flagella	1	1
Flagellar wavelength, μm	0.6–1.0	0.6–1.0
Soluble pigment production	−	−
Yellow or orange cellular pigments	−	+
Organic growth factor requirements	+[b]	+[c]
Autotrophic growth with H$_2$	−	−
Oxidase reaction	+	W
Nitrate used as a nitrogen source	−	−
Poly-β-hydroxybutyrate accumulation	+	+
Accumulation of glucose polysaccharide	−	+
Gelatin liquefaction	−	−
Lecithinase (egg yolk)	−	−
Lipase (Tween 80 hydrolysis)	−	−
Extracellular poly-β-hydroxybutyrate hydrolysis	−	−
Starch hydrolysis	−	−
Denitrification	−	−
Reduction of NO$_3^-$ to NO$_2^-$	d	−
Growth at 4°C	−	−
Growth at 41°C	d	−
Mol% G + C of DNA (Bd)	66.3–67.3	65.8

[a] For symbols see standard definitions; and W, weak reaction.
[b] Pantothenate, biotin, cyanocobalamin, and cystine or methionine required.
[c] Pantothenate, biotin and cyanocobalamin required.

Xanthomonas requires a redefinition of this latter genus to accommodate organisms which lack two important properties such as plant pathogenicity and the presence of characteristic pigments. Alternatively, *P. maltophilia* could be assigned to a new genus which, together with *Xanthomonas*, could constitute a separate new taxon.

Table 4.33.

Nutritional characteristics of **Pseudomonas diminuta** *and* **Pseudomonas vesicularis** *(section IV)*[a]

Characteristics	a. *P. diminuta*	b. *P. vesicularis*
Utilization of[b]:		
Acetate, butyrate, β-hydroxybutyrate, pyruvate, L-alanine, D-alanine, L-aspartate, L-glutamate, L-proline	+	+
D-glucose, D-galactose, maltose, cellobiose, α-ketoglutarate	−	+
Propionate, hydroxymethylglutarate, L-serine, L-leucine, L-isoleucine	d	−
Succinate, L-malate, ethanol, *n*-propanol	d	+
Fumarate	d	d
Aconitate	−	d
L-Histidine, pantothenate	+	−

[a] For symbols see standard definitions.
[b] The following compounds are not utilized by either species: D-ribose, D-xylose, D-arabinose, L-arabinose, D-fucose, L-rhamnose, D-mannose, D-fructose, sucrose, trehalose, lactose, starch, inulin, gluconate, 2-ketogluconate, saccharate, mucate, salicin, isobutyrate, valerate, isovalerate, caproate, heptanoate, caprylate, pelargonate, caprate, oxalate, malonate, maleate, glutarate, adipate, pimelate, suberate, azelate, sebacate, D-malate, D(−)-tartrate, L(+)-tartrate, *m*-tartrate, lactate, glycolate, glycerate, citrate, levulinate, citraconate, itaconate, mesaconate, erythritol, mannitol, sorbitol, *m*-inositol, adonitol, glycerol, ethylene glycol, propylene glycol, 2,3-butylene glycol, isopropanol, *n*-butanol, geraniol, D-mandelate, L-mandelate, benzoylformate, benzoate, *o*-hydroxybenzoate, *m*-hydroxybenzoate, *p*-hydroxybenzoate, phthalate, phenylacetate, phenylethanediol, naphthalene, phenol, quinate, testosterone, glycine, β-alanine, L-threonine, L-norleucine, L-valine, L-lysine, L-arginine, L-ornithine, L-citrulline, α-aminobutyrate, γ-aminobutyrate, α-aminovalerate, δ-aminovalerate, L-tyrosine, L-phenylalanine, L-tryptophan, D-tryptophan, L-kynurenine, kynurenate, anthranilate, ethanolamine, benzylamine, putrescine, spermine, histamine, tryptamine, butylamine, α-amylamine, betaine, sarcosine, creatine, hippurate, acetamide, nicotinate, trigonelline, dodecane, hexadecane, poly-β-hydroxybutyrate.

List of the Species of RNA group V

c. **Pseudomonas maltophilia** (*ex* Hugh and Ryschenkow 1960) Hugh 1980, 195.[VP] This species was omitted from the Approved Lists of Names, but the name has been revived recently (Hugh, 1981).

mal.to.phi′li.a. Anglo-Saxon noun *malt*; Gr. n. *philia* friend; M.L. fem. n. *maltophilia* friend of malt.

Straight or slightly curved rods, 0.5 × 1.5 μm, singly or in pairs. Do not accumulate PHB as an intracellular carbon reserve. Motile by means of polar multitrichous flagella.

Colonies may be yellowish; the yellow color is not due to carotenoid pigments or to xanthomonadins.

Denitrification does not occur.

Strongly lipolytic.

Methionine or cystine is required for growth, although strains which do not have this requirement may be isolated (Ikemoto et al., 1980).

Limited in its nutritional spectrum. Individual strains used between 24 and 28 of 146 organic compounds tested as principal carbon sources (Stainer et al., 1966). The utilizable compounds were: glucose, mannose, sucrose, trehalose, maltose, cellobiose, lactose, salicin, acetate, propionate, valerate, malonate, succinate, fumarate, L-malate, lactate, citrate, α-ketoglutarate, pyruvate, L-alanine, D-alanine, L-glutamate, L-histidine, and L-proline. A variable number of strains also used fructose, isobutyrate, aconitate and *n*-propanol. Under aerobic conditions acid is readily produced in complex media with maltose but not with glucose (Hugh and Ryschenkow, 1960). Nitrate is not used as nitrogen source. A comparison with *P. diminuta* and *P. vesicularis* is presented in Table 4.31.

Obligately aerobic. No growth at 4 or 41°C. Optimum temperature, 35°C.

Most strains are isolated from clinical specimens. According to Hugh and Gilardi (1980), *P. maltophilia* is the second most frequently isolated *Pseudomonas* species, after *P. aeruginosa*, in the clinical laboratory. Strains of the species appear to be opportunistic human pathogens. Also found in water, milk and frozen food. Debette and Blondeau (1980) were able to isolate more strains of *P. maltophilia* than of any other species of the genus from the rhizosphere of some cultivated plants.

Recently, the possibility of association of this ubiquitous species with plant diseases has been suggested by the results of a re-examination of the properties of the type strain of *P. hibiscicola* (ATCC 19867), which seems to be identical with *P. maltophilia* (R. L. Gherna, personal communication). *P. hibiscicola* is a plant pathogenic species that has been excluded from the present treatment because of its atypical characteristics (Young et al., 1978), but some interesting consequences of the above findings deserve consideration. First of all, a confirmation of the phytopathogenic capabilities of *P. maltophilia* would add further support to the hypothesis of a close relationship of this species to members of *Xanthomonas*. In the second place, some nomenclatural complications seem to be unavoidable. As noted by V. B. D. Skerman (personal communication), even though the description of *P. malto-*

philia precedes that of *P. hibiscicola*, the former has not been included in the approved lists of bacterial names (Skerman et al., 1980) and therefore the name *P. hibiscicola* should take priority.

Type strain: ATCC 136237 (Hugh 810-2; NCIB 9203; NCTC 10257; NRC 729).

SECTION V

This section includes a number of species of *Pseudomonas* whose natural relationships with well characterized species of the genus are largely unknown. The main reason for considering them in this treatment is that their names have been temporarily preserved in the Approved Lists of Bacterial Names (Skerman et al., 1980).

In addition, this section also contains some species that have not been considered in the preparation of the Approved Lists even though they have been subjects of considerable taxonomic and biochemical research.

The descriptions in this section are of very uneven quality, ranging from almost nonexistent to very extensive. However, a common denominator is that representative cultures are available in all cases from

culture collections and, therefore, a definitive judgment on their natural relationships can be achieved by future research. In fact, for some of the species a tentative allocation has already been suggested by Byng et al. (1980) and Whitaker et al. (1981a, b). Thus *P. agarici*, *P. asplenii*, and *P. fragi* are related to species of RNA group I; *P. pyrrocinia* and *P. huttiensis* to species of group II, and *P. andropogonis* to members of group III.

Since natural relationships are either unknown or tentative, we thought that the classification of the species of this section could be conveniently presented following some utilitarian criterion such as the source of isolation of representative strains.

Key for the sources of strains of the **Pseudomonas** *species of section V*

A. Isolated from plants
   I. Diseased plants and cultivated mushrooms
     a. Fluorescent organisms
                               i. *P. agarici*
                             ii. *P. asplenii*
                            iii. *P. caricapapayae*
                            iv. *P. tolaasii*
     b. Nonfluorescent organisms
                             v. *P. amygdali*
                           vi. *P. andropogonis*
                         vii. *P. avenae*
                       viii. *P. cattleyae*
                         ix. *P. cissicola*
                          x. *P. corrugata*
                         xi. *P. glumae*
                       xii. *P. rubrilineans*
                      xiii. *P. rubrisubalbicans*
                      xiv. *P. woodsii*
  II. Normal plants
     a. Leaf surfaces*
                          xv. *P. mesophilica*
     b. Rhizosphere†
                        xvi. *P. rhodos*
B. Isolated from animals
                       xvii. *P. anguilliseptica*
C. Isolated from soils
    I. By enrichment in an atmosphere of H_2, O_2 and CO_2
                    xviii. "*P. hydrogenovora*"
                    xix. "*P. hydrogenothermophila*"
    II. By enrichment in an atmosphere of CO and O_2§

---

* *Pseudomonads isolated from normal plant leaf surfaces.* These include many strains of *P. fluorescens*, a species which seems to be indigenous to this habitat. For a discussion of the literature and a list of plants from which strains of this species have been isolated, see Austin et al. (1978). These authors find *P. fluorescens* as an important component of the normal microflora of *Lolium perenne* leaf blades. A group of "pink chromogens" (phenon 1) isolated from the same material was later described under the new name *P. mesophilica* (see key), while another group of "pink chromogens" has been left unnamed. As mentioned in an earlier section, *P. viridiflava* has also been observed on normal plant tissues as an epiphyte (Billing, 1970b).
† Strains of various *Pseudomonas* species have been isolated from plant rhizospheres. Among the most prominent instances are the isolation of *P. fluorescens* biotype G (our biovar V), the dominant fluorescent pseudomonad of wheat rhizospheres and South Australian soils (Sands and Rovira, 1971), and the isolation of *P. maltophilia* from the rhizosphere of cultivated plants of the family *Cruciferae* (Debette and Blondeau, 1980). The species listed in the key, *P. rhodos*, was isolated from the rhizosphere of alder.
§ *Pseudomonas species isolated from CO-O₂ enrichments.* The preferred source for the isolation of these bacteria, collectively named carboxydobacteria, appears to be mud or river sludge. The information on the species listed has been taken mainly from the thesis work of Söder (1980), which was kindly made available by Dr. O. Meyer. Additional information was also taken from Zavarzin and Nozhevnikova (1977), Nozhevnikova and Zavarzin (1974), Sanjieva and Zavarzin (1971), Meyer and Schlegel (1978), and Meyer et al. (1980).

xx. *"P. carboxydoflava"*
xxi. *P. carboxydohydrogena*
xxii. *"P. compransoris"*
xxiii. *"P. carboxydovorans"*
xxiv. *"P. gasotropha"*

III. By direct isolation on PHB agar

xxv. *P. lemoignei*

IV. By denitrification enrichment with calcium citrate

xxvi. *P. indigofera*

V. By enrichment with amines

xxvii. *P. aminovorans*

VI. By enrichment with isoprenoid compounds

xxviii. *P. citronellolis*

VII. By enrichment with natural resins

xxix. *P. resinovorans*

VIII. By enrichment with aromatic compounds

xxx. *P. boreopolis*
xxxi. *P. pictorum*

IX. In antibiotic screenings

xxxii. *"P. acidophila"*
xxxiii. *P. mesoacidophila"*
xxxiv. *P. pyrrocinia*

X. By the use of acid, low nitrogen media

xxxv. *P. glathei*

XI. By direct isolation

xxxvi. *P. aurantiaca*

D. Isolated from water
  I. Fresh and distilled water

xxxvii. *P. huttiensis*
xxxviii. *P. lanceolata*
xxxix. *P. spinosa*

  II. Sea water

xl. *P. doudoroffii*
xli. *P. marina*
xlii. *P. nautica*
xliii. *P. elongata*
xliv. *P. gelidicola*

E. Isolated from foods
  I. Milk and dairy products

xlv. *P. fragi*
xlvi. *P. mephitica*
xlvii. *P. synxantha*

  II. Eggs

xlviii. *P. mucidolens*
xlix. *P. taetrolens*

  III. Meats[¶]

xlv. *P. fragi*[‖]

[¶] *Pseudomonads isolated from meats.* In a recent study of 110 strains of pseudomonads responsible for low temperature aerobic spoilage of meats, and 10 strains of well characterized species of *Pseudomonas*, using 161 unit characters, B. G. Shaw and J. B. Latty (B. G. Shaw, personal communication) could define four clusters which were clearly distinct from the known species, with the exception of *P. fragi*, which fell in one of the clusters.

The new strains could be generally characterized as follows. Two clusters included monotrichous nonfluorescent strains; the third cluster had multitrichous fluorescent strains, and the fourth, monotrichous fluorescent strains. Characters useful in the differentiation of the four clusters are presented later in Table 4.38, p. 196, which was prepared with data from the original manuscript by Shaw and Latty. These data have been presented at the XI International Symposium of the International Association of Microbiological Societies (IAMS) Committee on Food Microbiology and Hygiene (Aalborg, Denmark, July 1980).

As mentioned by the authors, many workers have identified pseudomonads among the components of the aerobic spoilage microflora of meats, but only occasionally could the identification reach the species level through the use of the various keys and descriptions presented in current editions of *Bergey's Manual*. Among the identifiable taxons, *P. fragi* appears to be one of the species most frequently found in the studies from various laboratories, and in the investigations of Shaw and Latty this species is included in the largest of four clusters, containing monotrichous, nonfluorescent pseudomonads. *P. geniculata* and "*P. rugosa*" are also frequently reported, although their present taxonomic position is quite uncertain. In the numerical analysis, *P. taetrolens* joined the first two clusters at a relatively high matching value.

Similar conclusions about the predominance of *P. fragi* among strains isolated from meats have been recently reported by workers from the Swedish Meat Research Institute (Molin and Ternström, 1982). Of 200 psychrotropic strains, 112 were classified as *P. fragi*, constituting one of the major clusters out of 15 groups defined by numerical analysis. Other pseudomonads identified in this work were various biotypes of *P. fluorescens* and *P. putida*.

[‖] See previous listing under *Milk and Dairy Products*.

List of the Species of Section V

i. **Pseudomonas agarici** Young 1970, 985.

a.gar'i.ci. M.L. n. *Agaricus* a genus of fungi; M.L. gen. n. *agarici* of Agaricus.

The description is a summary of the one presented by Young (1970).

Short rods (no dimensions given), motile by one or, rarely, two polar flagella. Green diffusible pigment with weak fluorescence under ultraviolet light of unspecified wavelength. Acid is produced from arabinose, glucose and mannitol; little acid from fructose, galactose and ribose, and no acid from rhamnose, xylose, mannose, lactose, sucrose, maltose, trehalose, melibiose, cellobiose, raffinose, starch, inulin, dextrin, glycogen, adonitol, sorbitol, inositol and salicin.

Acetate, benzoate, citrate, formate, fumarate, gluconate, lactate, propionate, and succinate are utilized. Galacturonate, oxalate and tartrate are not.

Oxidase and catalase reactions positive. Nitrate reduction, pectate liquefaction, starch hydrolysis, esculin hydrolysis, levan production, and growth factor requirements are all negative.

The organism causes drippy gill of mushrooms, and one of the main differences with another mushroom pathogen, *P. tolaasii*, is in the utilization of benzoate.

Even though the organism was described in 1970, no comparative studies have been made with any of the numerous fluorescent organisms of rRNA group I which were already known to use benzoate as a carbon source. Many strains producing an excess of catechol from benzoate are also known; catechol is oxidized to a black pigment which diffuses into the medium, and this character is also present in *P. agarici*.

The species has been tentatively assigned to rRNA group I by Byng et al. (1980).

Type strain: ATCC 25941.

ii. **Pseudomonas asplenii** (Ark and Tompkins 1946) Savulescu 1947, 11.[AL] (*Phytomonas asplenii* Ark and Tompkins 1946.)

a.sple'ni.i. M.L. neut. n. *Asplenium* genus of ferns, spleenworts; M.L. gen. n. *asplenii* of *Asplenium*.

A description is given by Haynes and Burkholder (1957) in the seventh edition of the *Manual* (p. 124). Tentatively placed in rRNA group I by Byng et al. (1980). Pathogenic for the bird's-nest fern (*Asplenium nidus*.)

Type strain: ATCC 23835.

iii. **Pseudomonas caricapapayae** Robbs 1956, 74.[AL]

ca.ri.ca.pa'pay.ae. M.L. gen. n. *caricapapayae* of *Carica papaya*, paw-paw.

The description is taken from Robbs (1956).

Straight or slightly curved rods, 0.86–1.08 × 1.29–3.01 μm. Motile by 3–6 polar flagella. Green fluorescent pigment produced. Gelatin is liquefied. Starch is not hydrolyzed. Nitrates are not reduced. Acid is produced from glucose, mannose, sucrose, glycerol and mannitol, but not from lactose, maltose, salicin, or starch. Citrate and tartrate are utilized; lactate is not.

Optimum growth between 23°C and 29°C; no growth at temperatures below 7°C and above 45°C.

Isolated from water-soaked, angular spots on leaves of pawpaw.

Type strain: NCPPB 1873.

iv. **Pseudomonas tolaasii** Paine 1919, 210.[AL]

to.laa'si.i. Tolaas patronymic; M.L. gen. n. *tolaasii* of Tolaas.

Description is given by Haynes and Burkholder (1957) in the seventh edition of the *Manual* (p. 136). Rapid identification tests have been described (Wong and Preece, 1979).

Pathogenic for cultivated mushrooms.

Type strain: NCPPB 2192.

v. **Pseudomonas amygdali** Psallidas and Panagopoulos 1975, 105.[AL]

a.myg'da.li. L. n. *amygdalum* almond; L. gen. n. *amygdali* of the almond.

Description taken from Psallidas and Panagopoulos (1975).

Rods, 0.7 × 1.7 μm or much longer (filaments 10–15 times the length of the normal cells). Motile by means of 1–6 polar flagella. No PHB accumulated.

It grows better in potato-dextrose than in nutrient agar. Growth range 3°C to 32°C. No growth below pH 5. No fluorescent pigment produced.

Acid is formed from D-ribose, L-arabinose, glucose, mannose, galactose, fructose, sucrose, mannitol, and sorbitol. No utilization of xylose, L-rhamnose, L-sorbose, cellobiose, lactose, maltose, melibiose, trehalose, raffinose, inulin, esculin, amygdalin, arbutin, salicin, dulcitol, erythritol, glycerol, inositol, dextrin, and α-methyl-D-glucoside. Malate, citrate, succinate and fumarate are utilized. Gluconate is slowly assimilated. Acetate, propionate, oxalate, maleate, malonate, tartrate, lactate, sulfanilic acid, picrate, hippurate and benzoate are not utilized. Among the amino acids, serine, aspartate, glutamate, arginine, asparagine, proline and histidine are utilized. Not used as carbon and/or nitrogen sources are glycine, β-alanine, leucine, isoleucine, valine, lysine, orni-

** See previous listing under *Grains.*

thine, tyrosine, phenylalanine, tryptophan, cystine, cysteine, methionine and creatine.

Some isolates are urease-positive. Tween 80 and tributyrin are rapidly hydrolyzed. Lecithinase- and arginine dihydrolase-negative. Gelatin, casein, esculin, arbutin and starch are not hydrolyzed. Nitrates are not reduced.

Rotting of potato slices does not occur, but the organism is positive in the hypersensitivity test on tobacco leaves. Further details are given in the original paper.

Pathogen for the almond tree (*Prunus dulcis*, fam. *Rosaceae*) in which it produces a hyperplastic bacterial canker. Not pathogenic for other fruit trees.

Type strain: NCPPB 2607.

vi. **Pseudomonas andropogonis** (Smith 1911) Stapp 1928, 27.[AL] (*Pseudomonas stizolobii* (Wolf 1920) Stapp 1935, 407; *Aplanobacter stizolobii* Wolf 1920, 75.)

an.dro.po'go.nis. M.L. n. *Andropogon* genus of widely distributed grasses; M.L. gen. n. *andropogonis* of the genus *Andropogon*.

The following description is a summary of that of Goto and Starr (1971).

Slender rods with round ends, 0.5–0.7 × 1–2 µm, with one or rarely two polar flagella. PHB is accumulated. Colonies change from butyrous to viscid with age. No fluorescent pigment is produced.

Most strains are oxidase-negative. Gelatin liquefaction, nitrate reduction, lipolysis and arginine dihydrolase reaction negative. Glucose, fructose, mannose, galactose, adonitol, glycerol, sorbitol, mannitol, succinate, citrate and malonate are utilized as carbon sources. Sucrose, maltose, raffinose, glycogen, salicin, dulcitol, and tartrate are not utilized.

The species has been tentatively assigned to rRNA group III (Byng et al., 1980).

Fuerst and Hayward (1969b) have reported the production of sheathed flagella by strains of this species (Fig. 4.2).

Pathogenic for sorghum, corn, clover and velvet bean (*Stizolobium deeringianum*). The species may be subdivided into two specialized pathovars, namely, pv. *andropogonis*, the agent of a stripe disease of sorghum, and pv. *stizolobii*, which has been described as the cause of leaf spot of velvet bean (Fuerst and Hayward, 1980).

Type strain: ATCC 23061.

vii. **Pseudomonas avenae** Manns 1909, 1933.[AL] (*Pseudomonas alboprecipitans* Rosen 1922, 383.)

a've.nae. M.L. n. *Avena* genus of plants; M.L. gen. n. *avenae* of *Avena*.

The following brief description is taken from the paper by Schaad et al. (1975), who established the synonymy of *P. avenae* and "*P. alboprecipitans*."

Rods, 0.6 × 1.6 µm, having polar flagella. Old colonies are sticky and adherent to the agar. Optimal growth temperature, 36°C. No fluorescent pigment is produced. Nitrate reduction-positive; denitrification-negative. Leucine, xylose, and sorbitol are utilized as carbon sources; threonine, inositol, D-arabinose and trehalose are not utilized. Acid is produced from dextrose, galactose and mannitol and more slowly from glycerol; no acid is produced from sucrose, lactose or maltose. Gelatin liquefaction is variable. Oxidase-negative. Slight starch hydrolysis.

Schaad et al. (1975) have included in their discussion the comments of Rosen (1922) and of Manns (1909) on the stainable granules seen in the cytoplasm of *P. avenae* and of "*P. alboprecipitans*." In both instances, the idea that these bodies represent spores was rejected. On the basis of the modern knowledge of *Pseudomonas* cytology, it appears now rather obvious that both authors have observed poly-β-hydroxybutyrate (PHB) granules. As a matter of fact, the name *alboprecipitans* may have its origin in the striking appearance of the cell mass as a consequence of copious accumulation of the polymer granules. Our own unpublished observations support this interpretation. Indeed, the type strain of *P. avenae* is capable of abundant PHB accumulation under conditions of high carbon/low nitrogen in the culture medium. In confirmation of this finding, Byng et al. (1980) have tentatively placed

strain ATCC 19860 of "*P. alboprecipitans*" in rRNA group III of Palleroni et al. (1973).

Pathogenic for oats (*Avena sativa*) and foxtail (*Chaetochloa lutescens*) (fam. *Gramineae*).

The mol% G + C of the DNA is 70.2 (T_m).

Type strain: NCPPB 1011 (ATCC 19860).

viii. **Pseudomonas cattleyae** (Pavarino 1911) Savulescu 1947, 11.[AL] (*Bacterium cattleyae* Pavarino 1911, 234.)

catt'ley.ae. M.L. fem. n. *Cattleya* a generic name; M.L. gen. n. *cattleyae* of *Cattleya*.

The description of this species, as given by Ark and Thomas (1946), has been published in the seventh edition of the Manual (Haynes and Burkholder, 1957, p. 148).

Pathogenic for *Cattleya* sp. and *Phalaenopsis* sp. (fam. *Orchidaceae*).

Type strain: NCPPB 961.

ix. **Pseudomonas cissicola** (Takimoto 1939) Burkholder 1948, 134.[AL] (*Aplanobacter cissicola* Takimoto 1939, 43.)

cis.si'co.la. Gr. n. *cissus* ivy; M.L. fem. n. *Cissus* generic name of flowering plant; L. suffix -*cola* dweller; M.L. fem. n. *cissicola* Cissus-dweller.

A description of this species appears in the seventh edition of the Manual (Haynes and Burkholder, 1957, p. 140). The following additional information has been taken from Goto and Makino (1977).

Immediately after isolation the cells were nonmotile, but always motile clones appeared after subculturing. A polar flagellum can be observed. PHB granules are accumulated as reserve material. No fluorescent pigment.

Colony variants (smooth and transparent) appear with high frequency. The original type (raised, translucent) accumulates levan from sucrose, but not the variants. Gelatin is liquefied and starch is strongly hydrolyzed. Arginine dihydrolase reaction negative. No reduction of nitrate to nitrite.

Optimum growth temperature, 28–30°C. No growth below 5°C or above 37°C.

Most isolates produce acid from xylose, ribose, arabinose, rhamnose, glucose, galactose, mannose, fructose, sucrose, maltose, lactose, trehalose, cellobiose, melibiose, inositol, adonitol, mannitol, sorbitol, glycerol, dextrin, starch, and glycogen. No acid is produced from dulcitol or inulin. Malate, succinate, citrate and tartrate are utilized.

The organism is pathogenic for *Cissus japonica* (fam. *Vitaceae*).

The mol% G + C of the DNA is 60.9 (Bd).

Type strain: NCPPB 2982.

x. **Pseudomonas corrugata** Roberts and Scarlett 1981, 216.[VP] (*Pseudomonas corrugata* Scarlett et al. 1978, 109.)

cor.ru'ga.ta. L. v. *corrugare* to wrinkle up; part. adj. *corrugatus* wrinkled up.

The description is summarized from the original paper by Scarlett et al. 1978.

Rods motile by multitrichous polar flagella. Accumulate PHB as carbon reserve. Colonies are wrinkled (which is the reason for the species name), yellowish, sometimes with green center. With age, the color may change to khaki or fawn, depending on the medium. Yellow to yellow-green diffusible, nonfluorescent pigment is produced.

Gelatin is hydrolyzed, but starch is not. Egg yolk (lecithinase) reaction positive. Levan production negative. There is growth at 37°C but not at 41°C.

Among the characters that differentiate this species from *P. cepacia* and *P. gladioli* ("*P. alliicola*") are the absence of pectate hydrolysis and rot of onion slices, and the lack of utilization of D-arabinose, cellobiose, adipate, *meso*-tartrate and citraconate. All these characters were positive for the strains of the last two species included in the study.

Isolated from tomato pith necrosis.

Type strain: NCPPB 2445.

xi. **Pseudomonas glumae** Kurita and Tabei 1967, 111.[AL]

glu'mae. L. n. *gluma* husk, hull; L. gen. n. *glumae* of a husk.

Description taken from the original paper.

Rods, 0.5–0.7 × 1.5–2.5 μm, motile by means of 2–4 polar flagella. Fluorescent pigment produced in potato agar. Nitrate reduction, starch hydrolysis and H₂S production are negative. Gelatin liquefaction is not reported.

Temperature limits for growth: 11–40°C. Optimum: 30–35°C.

Acid is produced from arabinose, glucose, fructose, galactose, mannose, xylose, glycerol, mannitol, inositol. No acid is produced from rhamnose, sucrose, maltose, lactose, raffinose, dextrin, starch, inulin, or salicin.

Milk is coagulated and peptonized.

Pathogenic for the rice plant (*Oryza sativa*, fam. *Gramineae*).

Type strain: NCPPB 2981.

xii. **Pseudomonas rubrilineans** (Lee, Purdy, Barnum and Martin 1925) Stapp 1928, 35.^{AL} (*Phytomonas rubrilineans* Lee, Purdy, Barnum and Martin 1925, 73.)

ru.bri.li′ne.ans. L. adj. *ruber* red; L. v. *lineo* to make a straight line; M.L. part. adj. *rubrilineans* making red stripes.

The following is a summary of the description given by Hayward (1962).

Straight rods motile by means of one polar flagellum. PHB is accumulated. Colonies are nonmucoid in 2% glucose-peptone agar. No pigments are produced.

Oxidase-positive; H₂S produced; Tween 80 is hydrolyzed. Nitrates are reduced to nitrites, but denitrification is negative. Starch hydrolysis is negative and gelatin liquefaction is weak. Capable of growth at 40°C.

Acid is produced from glucose, fructose, galactose, arabinose, mannitol, glycerol, and sorbitol. No acid from sucrose, raffinose, salicin, lactose, maltose, cellobiose, or *m*-inositol.

The organism is the agent of the red stripe of sugarcane.

Type strain: ATCC 19307.

xiii. **Pseudomonas rubrisubalbicans** (Christopher and Edgerton 1930) Krasil′nikov 1949, 379.^{AL} (*Phytomonas rubrisubalbicans* Christopher and Edgerton 1930, 266.)

ru.bri.sub.al′bi.cans. L. adj. *ruber* red; L. adj. *subalbicans* whitish; M.L. adj. *rubrisubalbicans* red-whitish.

The following is taken from the description by Hayward (1962).

Slightly curved rods, motile by means of several polar flagella. PHB is accumulated. Colonies in 2% glucose-peptone agar are mucoid (not so in 2% sucrose-peptone agar). No pigments are produced.

Oxidase-positive. Do not produce H₂S. No hydrolysis of gelatin, starch, and Tween 80. Most strains reduce nitrate to nitrite. Denitrification is negative. Growth occurs at 40°C.

Acid is produced from glucose, fructose, galactose, arabinose, mannitol, lactose, glycerol, and sorbitol. No acid from sucrose, raffinose, salicin, maltose, cellobiose or *m*-inositol.

Agent of the mottled stripe of sugarcane.

Type strain: ATCC 19308.

xiv. **Pseudomonas woodsii** (Smith 1911) Stevens 1925, 39.^{AL} (*Bacterium woodsii* Smith 1911, 62.)

wood′si.i. M.L. gen. n. *woodsii* of Wood; named after A. F. Woods, an American plant pathologist.

A description taken from Burkholder and Guterman (1935) has been included by Haynes and Burkholder (1957) in the seventh edition of the *Manual* (p. 150).

Pathogenic for carnation (*Dianthus caryophyllus*, fam. *Caryophyllaceae*)

Type strain: ATCC 19311.

xv. **Pseudomonas mesophilica** Austin and Goodfellow 1979, 377.^{AL}

me.so.phi′li.ca. Gr. n. *mesos* middle; Gr. adj. *philus* loving; M.L. fem. adj. *mesophilica* middle (temperature)-loving, i.e. mesophilic.

The following description is a summary of that presented in the original paper.

Rods, 1 × 3–4 μm. The cells are vacuolated. Motile by means of 1–3 polar flagella.

The colonies are pink due to the presence of an insoluble pigment associated with the cells. No growth at 5°C or 37°C. Nitrate reduction negative. No degradation of esculin, blood, casein, cellulose, chitin, gelatin, lecithin, Tweens 20 and 80, tyrosine, xanthine or urea. No accumulation of PHB. Oxidase reaction is negative.

The following compounds are utilized for growth: arabinose, galactose, glucose, glycerol, citrate, glutamate, malate, malonate, pyruvate, succinate. None of the following compounds is utilized: *cis*-aconitate, adonitol, alanine, arginine, lactate, cellobiose, erythritol, fructose, *p*-hydroxybenzoate, *m*-inositol, inulin, lactose, leucine, maltose, mannitol, mannose, melezitose, norleucine, proline, raffinose, ribose, salicin, serine, benzoate, oxalate, tartrate, sorbitol, sucrose, trehalose, tryptophan, valine, and xylose.

The mol% G + C of the DNA is 65.8 (T_m).

Type strain: ATCC 29983.

xvi. **Pseudomonas rhodos** Heumann 1962, 342.^{AL}

rho′dos. Gr. n. *rhodon* rose; M.L. adj. *rhodos* rose colored.

The following is a summary of the description given by Heumann (1962).

Plump rods, 1.4 × 2.7 μg. Motile by means of a single polar flagellum of wavelength ~2.4 μm. Star-shaped cell aggregates (rosettes) are formed in nutrient broth and in solid media.

Colonies are rough and red. The color is due to the production of carotenoid pigments. Mutants of smooth colonies or lower pigmentation can be obtained.

Nitrates are reduced to nitrites and small amounts of ammonia. Gelatin is not liquefied and starch is slowly hydrolyzed. Optimum growth temperature, ~28°C.

Austin and Goodfellow (1979) have supplemented the above description, and the following data have been taken from their paper and from personal communication.

Vacuoles may be observed in the cells. Cells with 2–4 polar flagella are also present. The colonies are small, reaching 1 mm in diameter after 3 days at 25°C on yeast extract-glucose agar. No growth at 4°C or 37°C.

Lipase (Tween 20 hydrolysis)-positive. Fructose, citrate and inulin are utilized, but L-arabinose, galactose, lactate, glutamate and malonate are not. The following properties are negative: oxidase, β-galactosidase, nitrate reduction, gluconate oxidation, production of H₂S and degradation of esculin, gelatin, casein, chitin, lecithin, starch, tyrosine, urea and xanthine.

Growth with glucose, fructose, glycerol, inulin, malate, pyruvate, and citrate. No growth with L-arabinose, xylose, ribose, sucrose, cellobiose, trehalose, melezitose, acetate, malonate, lactate, malate, tartrate, adonitol, inositol, mannitol, proline or glutamate.

Isolated from the rhizosphere of alder (*Alnus* sp., fam. *Betulaceae*).

Type strain: ATCC 14821.

xvii. **Pseudomonas anguilliseptica** Wakabayashi and Egusa 1972, 589.^{AL}

an.guil.li.sep′ti.ca. L. n. *anguilla* eel; Gr. adj. *septica* putrefactive; M.L. adj. *anguilliseptica* pertaining to diseased eels.

The description below is taken from the original paper.

Rods, 0.4 × 2 μm, with a tendency to become filamentous. Motile by a single polar flagellum; motility is better at 15°C than at 25°C.

Catalase and oxidase reactions positive. Gelatin is liquefied. Nitrate reduction, urease, fluorescent pigment production, and starch hydrolysis are all negative. There is no acid production from arabinose, xylose, rhamnose, fructose, galactose, glucose, mannose, sorbose, lactose, maltose, raffinose, sucrose, starch, dextrin, salicin, glycerol, mannitol or inositol.

Isolated from diseased pond-cultured eels (*Anguilla japonica*)

Type strain: NCMB 1949.

xviii. **"Pseudomonas hydrogenovora"** Igarashi, Kodama and Minoda 1980, 1278. (As of this writing, this species has not been validly published.)

hy.do.gen.o′vo.ra. Gr. n. *hydro* water; Gr. n. *genus* race, offspring;

whence M.L. *hydrogenum* hydrogen, that which produces water; L. v. *voro* to devour; M.L. adj. *hydrogenovora* hydrogen devouring.

The description is taken from the paper by Igarashi et al. (1980).

Rods, 0.4–0.6 by 1.0–1.5 μm, motile with several polar flagella.

The colonies are cream-yellow.

The following properties are positive: nitrate reduction, oxidase, catalase, starch and gelatin hydrolysis, and accumulation of PHB. Among the negative characters are the following: denitrification, arginine dihydrolase, nitrogen fixation and growth factor requirement.

The cells can grow at the expense of glucose, fructose, galactose, L-arabinose, D-arabinose, lactose, L-rhamnose, cellobiose, lactate, arginine, betaine, Tween 80, acetate, and succinate. Citrate, gluconate, xylose, sucrose, β-hydroxybutyrate and starch are more poorly utilized. Malonate, trehalose, p-hydroxybenzoate, ethanol, mesaconate, testosterone, mannitol and raffinose are not utilized. A comparison with other chemolithotrophic pseudomonads is given in Table 4.29.

Aside from *P. saccharophila*, this is the only species of hydrogen pseudomonad that can use starch for heterotrophic growth.

Additional information can be found in the papers by Igarashi et al. (1980) and Kodama et al. (1975).

The mol% G + C of the DNA is ~62.5 (T_m).

Type strain: strain 9-5 (Igarashi et al., 1980). According to these authors, the strain has been deposited in the Deutsche Sammlung von Mikroorganismen (DSM), but no number is given.

xix. "**Pseudomonas hydrogenothermophila**" Goto, Kodama and Minoda 1978, 1307. (As of this writing, this species has not been validly published.)

hy.dro.gen.o.ther.mo′phi.la. Gr. n. *hydro* water; Gr. n. *genus* race, offspring; whence M.L. *hydrogenum* hydrogen, that which produces water; Gr. n. *thermus* heat; Gr. adj. *philus* loving; M.L. fem. adj. *hydrogenothermophila* (presumably intended to mean) hydrogen and heat loving.

A description taken from the original paper follows.

Rods 0.5–0.6 × 2–3 μm, motile by polar flagella. Colonies dull yellow. Optimum growth temperature, 52°C.

Acetate, propionate, butyrate, succinate, fumarate, DL-malate, DL-lactate, pyruvate and α-ketoglutarate are utilized for growth. Compounds not utilized are: xylose, D-arabinose, L-arabinose, D-galactose, sucrose, glucose, mannose, fructose, L-sorbose, maltose, L-rhamnose, cellobiose, lactose, trehalose, xylan, raffinose, soluble starch, glycogen, dextrin, inulin, dextran, salicin, sorbitol, mannitol, inositol, benzoate, oxalate, malonate, maleate, glutarate, DL-tartrate, citrate, formate, methanol, ethanol, arginine, betaine, glycolate, p-hydroxybenzoate, testosterone.

The organism was isolated from soil taken in the vicinity of hot springs.

For further information consult the papers by Goto et al. (1977, 1978).

A comparison of "*P. hydrogenothermophila*" with other chemolithotrophic pseudomonads is presented in Table 4.29.

The mol% G + C of the DNA is ~ 63.5 (T_m)

Type strain: TH1 (Goto et al., 1978).

xx. "**Pseudomonas carboxydoflava**" Nozhevnikova and Zavarzin 1974, 438. (As of this writing, this species has not been validly published.)

car.box.y.do.fla′va. L. n. *carbo* charcoal, carbon; Gr. adj. *oxys* sour, acid; L. adj. *flavus* yellow; M.L. fem. adj. *carboxydoflava* yellow carbonic acid-(forming).

Characteristics are presented in Tables 4.34 and 4.35, and a comparison with other chemolithotrophic pseudomonads is given in Table 4.29.

Type strain: DSM 1084.

xxi. **Pseudomonas carboxydohydrogena** (Sanjieva and Zavarzin 1971) Meyer, Lalucat and Schlegel 1980, 194.[VP] (*Seliberia carboxydohydrogena* Sanjieva and Zavarzin 1971, 956.)

car.box.y.do.hy.dro′ge.na. L. n. *carbo* charcoal, carbon; Gr. adj. *oxys*

sour, acid; Gr. n. *hydro* water; Gr. n. *genus* race, offspring; whence M.L. *hydrogenum* hydrogen; M.L. adj. *carboxydohydrogena* hydrogen carbonic acid-(forming).

Characteristics are presented in Tables 4.34 and 4.35, and a comparison with other chemolithotrophic pseudomonads is given in Table 4.29.

Type strain: DSM 1083.

xxii. "**Pseudomonas compransoris**" (Nozhevnikova and Zavarzin 1974) Söder 1980, 73. (*Comamonas compransoris* Nozhevnikova and Zavarzin 1974, 438.) (As of this writing, this species has not been validly published.)

com.pran′so.ris. L. gen. n. *compransoris* of a dinner companion.

Characteristics are presented in Tables 4.34 and 4.35, and a comparison with other chemolithotrophic pseudomonads is given in Table 4.29.

Type strain: DSM 1231.

xxiii. "**Pseudomonas carboxydovorans**" (Kistner 1954) Meyer and Schlegel 1978, 42. (*Hydrogenomonas carboxydovorans* Kistner 1954, 186.) (As of this writing, this species has not been validly published.)

car.box.y.do′vor.ans. L. n. *carbo* charcoal, carbon; Gr. adj. *oxys* sour, acid; L. v. *voro* devour; M.L. part. adj. *carboxydovorans* carbon-acid devouring.

Characteristics are presented in Tables 4.34 and 4.35, and a comparison with other chemolithotrophic pseudomonads is given in Table 4.29.

Table 4.34.

General characteristics of CO-oxidizing **Pseudomonas** *species (section V)[a]*

Characteristics	"*P. carboxydoflava*"	*P. carboxydohydrogena*	"*P. compransoris*"	"*P. carboxydovorans*"
Cell diameter, μm	0.6–0.9	0.3	0.7–0.9	0.5–0.7
Cell length, μm	1.1–2.4	2.0	1.5–2.1	1.0–2.0
Number of flagella	0[b]	1[c]	1	1[c]
Yellow cellular pigment	+	−	−	−
Oxidase reaction	+	+	+	+
Growth factor requirement	−	−	+[d]	−
Denitrification	+	−	−	−
Nitrate reduction (in the absence of NH_4^+)	+	+	−	+
Nitrate reduction (in the presence of NH_4^+)	+	+	−	−
Autotrophic nitrate respiration	−	−	−	−
Growth in presence of 2.5% NaCl or more	−	−	−	−
Hydrolysis of starch, cellulose, poly-β-hydroxybutyrate, gelatin, casein or Tween 80	−	−	−	−
Lecithinase (egg yolk)	+	−	−	−
Urease	−	−	−	−
Type of hydrogenase	m	m	m	m
Type of CO-oxidase	s	s	s	s
Mol% G + C of DNA (T_m)	68.0	58.2	63.3	62.2

[a] Symbols: standard definitions; and m, membrane-bound; and s, soluble.

[b] No flagella observed in stained or electron microscope preparations; motility observed in a minority of cells.

[c] Subpolar insertion.

[d] Thiamine required for growth.

Table 4.35.
Nutritional characteristics of CO-oxidizing **Pseudomonas** *species (section V)[a]*

Characteristics	"P. carboxydoflava"	P. carboxydohydrogena	"P. compransoris"	"P. carboxydovorans"
Utilization of[b]:				
Fumarate, α-ketoglutarate, pyruvate	+	+	+	+
D-Xylose, D-glucose, D-mannose, D-galactose, sucrose, trehalose, maltose, cellobiose, gluconate, mannitol, sorbitol, glutamine, L-lysine, L-ornithine, L-proline	+	−	−	−
L-Arabinose, lactose, DL-tryptophan	−(+)[c]	−	−	−
D-Fructose, citrate	−(+)[c]	+	−	−
Acetate, succinate, DL-malate	−(+)[c]	+	+	+
Propionate	−	−	−(+)[c]	−
Isobutyrate, β-hydroxybutyrate, γ-aminobutyrate	−	−	+	−
Valerate	(−)[c]		(+)[c]	
Formate, oxalate, glyoxylate	−	+	−	+
L-Tartrate	−	+	−	−
Lactate	+	+	+(−)[c]	+
Ascorbate	−	−	−	+
Glycolate, butanol, L-histidine	(+)[c]		(−)[c]	
Ethanol	−(+)[c]	−	−(+)[c]	−
Propanol	+	−	−(+)[c]	−
DL-Alanine, L-serine, L-glutamate, L-arginine	+	−	+	−
L-Aspartate	+	+	−	−
Asparagine	+	−	+(−)[c]	−

[a] For symbols see standard definitions.

[b] The following compounds are not utilized by any species: D-ribose, D-fucose, L-rhamnose, L-sorbose, 6-deoxyglucose, tagatose, raffinose, starch, cellulose, glycogen, glucosamine, 2-ketogluconate, malonate, thioglycolate, citraconate, mesaconate, methanol, DL-mandelate, benzoate, m-hydroxybenzoate, p-hydroxybenzoate, glycine, glycyl-glycine, L-threonine, L-leucine, L-isoleucine, L-methionine, L-cysteine, L-valine, L-tyrosine, L-phenylalanine, methylamine, diethylamine, betaine, creatine, pantothenate, urea, uric acid, p-aminobenzoate, sulfanilic acid, tris-aminomethane, trichloroacetate.

[c] Data in parentheses are from Zavarzin and Nozhevnikova (1977).

Type strain: DSM 1227 (OM5; Meyer and Schlegel, 1978).

xxiv. **"Pseudomonas gasotropha"** Nozhevnikova and Zavarzin 1974, 438. (As of this writing, this species has not been validly published.)

gas.o.tro′pha. Gr. n. *chaos* gas, whence *gas* (coined word); Gr. n. *trophe* food; M.L. n. *gasotropha* gas-food.

No information is given in the tables on this species; it was not examined by Söder (1980), and it has not been possible to make a full comparison with other CO-oxidizing species. The following brief description is taken from Nozhevnikova and Zavarzin (1974) and from Zavarzin and Nozhevnikova (1977), who have made a nutritional comparison of "*P. gasotropha*" with other carboxydobacteria.

Rods, 0.7 × 1.1 μm, motile by means of a single subpolar flagellum. Slime is produced. Colonies are very small. Little or no growth is observed on meat peptone agar. Cyanocobalamin (vitamin B_{12}) is required. Only organic acids and alcohols (including methanol) are utilized heterotrophically. Nitrate reduced to nitrite. Starch, cellulose

and gelatin are not hydrolyzed. Cytochromes of the *a*, *b*, and *c* types are present.

There is a slight resemblance of this species to *P. facilis,* but the nutritional spectra of the two species are quite different. "*P. gasotropha*" is also different from other methanol-utilizing pseudomonads.

The mol% G + C of the DNA is 66.9 (T_m).

Type strain: Z-1156 (Nozhevnikova and Zavarzin, 1974).

xxv. **Pseudomonas lemoignei** Delafield, Doudoroff, Palleroni, Lusty and Contopoulou 1965, 1460.[AL]

le.moig′ne.i. M.L. gen. n. *lemoignei* of Lemoigne; named after M. H. Lemoigne, a French bacteriologist.

A description taken from Delafield et al. (1965) has been included by Doudoroff and Palleroni (1974) in the eighth edition of the *Manual* (p. 231).

Only one strain of this species (ATCC 17989) is known. The natural relationships to other species of the genus is at present not clear, although De Vos and De Ley (1982) have found that the species is marginally related to members of rRNA group III. The unique phenotypic properties of *P. lemoignei,* which is by far the nutritionally least versatile of all the well characterized *Pseudomonas* species, suggest no resemblance whatsoever to species of group III in morphology or in physiological characteristics, as well as in DNA base composition.

Type strain: ATCC 17989; NCIB 9947.

xxvi. **Pseudomonas indigofera** (Voges 1893) Migula 1900, 950.[AL] (*Bacillus indigoferus* Voges 1893, 307.)

in.di.go′fe.ra. Fr. *indigo* the dye indigo (from India); L. suffix *-fer* (*-ferus*) from L. v. *fero* to bear; M.L. adj. *indigofera* bearing (producing) indigo.

The following is a summary of the description given by Elazari-Volcani (1939).

Rods, 0.5 × 1–4 μm, occurring singly, in pairs or more rarely in chains. Motile by means of a single polar flagellum. A strain lacking flagella has been described by Elazari-Volcani (1939) as "*P. indigofera* var. *immobilis.*"

Colonies blue with a metallic sheen; margin thin, of yellow color. A dark blue print remains in the agar after a colony is scraped off. The blue color is abundantly produced in media with sucrose.

Gelatin is not liquefied. No acid is produced from several carbohydrates. A strong indole reaction is reported. However, in our hands this reaction has been negative. However, upon acidification, the pigment changes to a color that resembles that of indole under similar conditions.

Optimum temperature for pigment production is 18–20°C. The organism grows without pigmentation at 37°C.

The property of denitrification deserves some discussion. Elazari-Volcani (1939) isolated his strains by enrichment in a mineral medium with nitrate and calcium citrate under anaerobic conditions at 37°C. The individual cultures could grow under these conditions, presumably producing gas from the nitrate, although this property is not mentioned again in the paper for the pure cultures. McFadden and Howes (1961) reported gas formation under anaerobic conditions with nitrate, but Sommer et al. (1961) describe a new strain of the species as a nondenitrifier, even though they claim that this strain was both morphologically and physiologically identical with that of Elazari-Volcani.

Further descriptive information is found in the papers mentioned above.

Our unpublished nutritional experiments have shown that the type strain of *P. indigofera* uses at least 40 organic compounds for growth. These compounds are: glucose, fructose, sucrose, trehalose, maltose, acetate, propionate, butyrate, succinate, fumarate, D-malate (after mutation), L-malate, β-hydroxybutyrate, lactate, pyruvate, *m*-inositol, glycerol, ethanol, propanol, p-hydroxybenzoate, quinate, L-alanine, D-alanine, β-alanine, L-threonine, L-leucine, L-isoleucine, asparagine, L-aspartate, L-glutamate, L-lysine, L-arginine, L-ornithine, L-citrulline, γ-aminobutyrate, δ-aminovalerate, L-histidine, L-proline, L-tyrosine, putrescine.

We have also observed accumulation of PHB as a carbon reserve material.

The mol% G + C of the DNA is 64.3 (Bd) (M. Mandel, personal communication).

Type strain: ATCC 19706.

xxvii. **Pseudomonas aminovorans** den Dooren de Jong 1926, 161.[AL]
am.i.no′vo.rans. M.L. n. *aminum* amine; L. part. adj. *vorans* devouring, digesting; M.L. part. adj. *aminovorans* amine digesting.

In his classical work, den Dooren de Jong (1926) isolated from soil enrichments with various amines (methylamine, trimethylamine, tetramethylammonium, ethylamine, and ethylurea) a group of organisms among which seven strains were placed in the genus *Pseudomonas*. Den Dooren de Jong created the name *Pseudomonas aminovorans* for the group because of the capacity of the strains to grow at the expense of various amines as sole carbon and energy sources. The group was divided internally into four subgroups named by den Dooren de Jong with the first letters of the Greek alphabet. In only two of the strains, belonging to the subgroups α and γ could active motility and the presence of polar flagella be demonstrated, but the remaining five strains were assigned to the same species on the basis of their overall similarity.

P. aminovorans is a species whose relationship to other species of the genus is at present obscure. It is likely that future studies will suggest its elimination from the genus. Aside from their capacity for growth at the expense of amines, the strains are able to utilize one-carbon organic compounds, a property absent from other species of *Pseudomonas*. This characteristic places *P. aminovorans* closer to the special group of one-carbon utilizers but, as we shall see below, the nutritional versatility of the species is quite high.

The following information summarizes the nutritional studies of den Dooren de Jong on his seven strains of the species.

Substrates utilized by all strains: glucose, acetate, lactate, succinate, malate, citrate, glucosamine, sarcosine, tyrosine, methylamine, trimethylamine.

Substrates not used by any strain: adipate, phenylaminoacetate, triethylamine, tetraethylamine, tetrapropylamine, triisobutylamine, diamylamine, triamylamine, hexylamine, heptylamine, diethanolamine, allylamine, ethylenediamine, hexamethylenetetramine, α-phenylethylamine, piperidine, piperazine, pyrrol, pyridine, urea, *asym*-diethylamine, tetraethylamine.

Substrates used by some of the strains (in parentheses: numbers of positive strains): valerate (3); α-crotonate (5); undecylenate (2); β-hydroxybutyrate (5); glycerate (6); pyruvate (6); malonate (3); methylsuccinate (2); glutarate (3); maleate (2); fumarate (6); tartrate (2); β-phenylpropionate (2); quinate (3); ethanol (3); betaine (5); hippurate (4); α-alanine (4); phenylalanine (3); δ-aminovalerate (3); α-aminocaproate (3); leucine (4); aspartate (5); propionamide (5); capronamide (3); lactamide (5); succinamide (4); asparagine (5); creatine (6); allantoin (3); urate (4); dimethylamine (4); tetramethylamine (1); ethylamine (4); diethylamine (3); propylamine (3); isopropylamine (1); dipropylamine (1); tripropylamine (1); butylamine (3); isobutylamine (5); diisobutylamine (1); amylamine (3); ethanolamine (3); choline (3); neurine (3); pentamethylendiamine (2); benzylamine (2); histamine (1); methylurea (3); ethylurea (2); and *sym*-diethylurea (1).

To our knowledge, only one strain (strain 26 of den Dooren de Jong) has been analyzed for DNA base composition. The mol% G + C is 63.2 (Bd), well within the range of *Pseudomonas* (Mandel, 1966).

Type strain: ATCC 23314.

xxviii. **Pseudomonas citronellolis** Seubert 1960, 428.[AL]
cit.ro.nel′lo.lis. M.L. gen. n. *citronellolis* of citronellol.

The description is summarized from the original paper by Seubert (1960).

Rods, 0.5 × 1.0–1.5 μm, motile by means of a single polar flagellum. Optimum growth temperature, ~ 31°C.

Gelatin is not liquefied. Nitrates are reduced to nitrites. Growth can occur anaerobically in the presence of nitrate. H_2S is not produced.

Besides citronellol, glucose, acetate, farnesol and ionone can support growth. No growth was observed with squalene or camphoric acid. Ammonium salts, nitrate, peptone or yeast extract can be used as nitrogen sources.

Acid is produced from glycerol, but not from glucose, galactose, arabinose, fructose, sucrose, maltose, lactose, dulcitol, inositol, mannitol, inulin or dextrin.

Type strain: ATCC 13674.

xxix. **Pseudomonas resinovorans** Delaporte, Raynaud and Daste 1961, 1075.[AL]
re.si.no′vor.ans. L. n. *resina* resin; L. v. *voro* to devour, digest; M.L. part. adj. *resinovorans* resin digesting.

Rods, 0.6–0.7 × 2.0–2.5 μm. Motile by means of a polar flagellum.

Fluorescent pigment is produced. Gelatin is not liquefied. Nitrate reduction is weak, and denitrification is negative. Oxidase reaction positive. Optimum growth temperature, 28–30°C; no growth at 5°C or 42°C.

No acid is produced from arabinose, xylose, rhamnose, glucose, fructose, galactose, mannose, lactose, maltose, sucrose, raffinose, inulin, salicin, dextrin, glycerol, mannitol, inositol or dulcitol. Starch hydrolysis very weak.

Growth occurs at the expense of colophony, Canada balsam or abietic acid. Phenol, phenanthrene, salicylic acid, *m*-cresol and naphthalene can also be used as carbon and energy sources for growth.

Further information may be found in the original paper by Delaporte et al. (1961).

Type strain: ATCC 14235.

xxx. **Pseudomonas boreopolis** Gray and Thornton 1928, 92.[AL]
bo.re.o′po.lis. Gr. *boreas* north; Gr. *polis* a city; M.L. gen. fem. n. *boreopolis* of North City.

Description taken from the original paper.

Rods, 0.5–1.0 × 2–3 μm. Paired cells up to 6 μm long are common. Motile.

Fluorescent pigment is produced. Gelatin is liquefied. Some strains produce a red pigment in the gelatin medium. Acid is produced from glucose by most strains. Nitrate reduced to nitrite by some strains. Starch is not hydrolyzed.

Naphthalene is used by all strains.

Some additional details on this species are given by Haynes and Burkholder (1957). The cells may have from 1–5 flagella. Growth occurs at 35–37°C.

Type strain: NCIB 9401.

xxxi. **Pseudomonas pictorum** Gray and Thornton 1928, 89.[AL]
pic.to′rum. M.L. gen. n. *pictorum* of the Picts; names after the Picts, a Scottish tribe.

The description of the single strain of this species is taken from the original paper.

Rods, 0.5–0.8 × 1.5–5 μm (the majority about 3 μm), motile by means of one short polar flagellum.

Colonies are yellow. Gelatin not hydrolyzed. Nitrate reduced to nitrite. Starch is not hydrolyzed. Acid is produced from glucose and maltose.

Phenol is utilized.

Type strain: ATCC 23328.

xxxii. **"Pseudomonas acidophila"** Imada, Kitano and Asai 1980b, 1. (As of this writing, this species has not been validly published.)
a.ci.do′phi.la. L. adj. *acidus* sour; M.L. neut. n. *acidum* acid; Gr. adj. *philus* loving; M.L. adj. *acidophila* acid loving.

Description taken from the U.S. patent issued to Imada et al. (1980b).

Rods, 0.7–1.0 × 0.7–3.5 μm, motile with one to several polar flagella.

PHB is not accumulated as carbon reserve material. No diffusible pigment is produced. No liquefaction of gelatin. Nitrate reduction negative; denitrification negative. Hydrolysis of starch negative. Poor utilization of nitrate as nitrogen source. Tween 80 is hydrolyzed. The oxidase reaction is negative.

Growth occurs between pH 4 and 8, and the optimum is 4.5–6.0. Optimum growth temperature 25–30°C; limits of growth: 2–37°C. No growth factor requirement.

Weak acid production from glucose, mannose, galactose, maltose, sucrose, trehalose, sorbitol, mannitol, inositol and glycerol. The following compounds are used as carbon sources for growth: L-arabinose, xylose, glucose, mannose, fructose, galactose, maltose, sucrose, trehalose, sorbitol, mannitol, inositol, glycerol, α-methylglucoside, adonitol, raffinose, cis-aconitate, citrate, isocitrate, gluconate, acetate, fumarate, malate, tartrate, p-hydroxybenzoate and L-alanine. Lactose, starch and dulcitol are not utilized.

The mol% G + C of the DNA is 59.5.

There are some discrepancies between the above description and that given by Imada et al. (1981), where the flagellation is described as monotrichous and the G + C content is reported to be 64 mol%. Neither source indicates the method of DNA base analysis.

The organism produces the new β-lactam antibiotic sulfazecin (Imada et al., 1981).

Type strain: ATCC 31363 (FERM-P 4344; IFO 13774; G-6302) (Imada et al., 1981).

xxxiii. **"Pseudomonas mesoacidophila"** Imada, Kintaka and Haibara 1980a, 1. (As of this writing, this species has not been validly published.)

me.so.a.ci.do'phi.la. Gr. n. *mesos* middle; L. adj. *acidus* sour; M.L. neut. n. *acidum* acid; Gr. adj. *philus* loving; M.L. adj. *mesoacidophila* loving middle (range) acidity.

Description taken from the U.S. patent issued to Imada et al. (1980a).

Rods, 0.8–1.1 × 1.6–4.1 μm. Motile by means of one to several polar flagella. PHB is accumulated in the cells. No pigments are produced.

Gelatin is liquefied. Nitrate is not reduced; no denitrification. Starch is not hydrolyzed. Hydrolysis of Tween 80 is positive. The arginine dihydrolase reaction is positive.

Growth occurs between pH 4 and 8.85, optimal pH 4.5–7.0. Optimum growth temperature, 24–36°C; temperature limits of growth, 8–42°C.

Weak acid production from L-arabinose, xylose, glucose, mannose, fructose, galactose, maltose, sucrose, trehalose, sorbitol, mannitol, inositol and glycerol.

The following compounds can be used as the sole carbon and energy sources: L-arabinose, xylose, glucose, mannose, fructose, galactose, maltose, sucrose, trehalose, sorbitol, mannitol, inositol, glycerol, raffinose, citrate, acetate, L-alanine, β-alanine, succinate, 2-ketogluconate, L-arginine and betaine. Lactose and starch are not utilized.

Antibiotic SB-72310, isosulfazecin, a novel monocyclic β-lactam compound, is produced by this organism (Imada et al., 1981).

The mol% G + C of the DNA is given as 64.3 by Imada et al. (1980a) and as 70 by Imada et al. (1981). Neither source indicates the method used for the determination of DNA base composition.

Type strain: ATCC 31433 (FERM 4653; IFO 13884; SB-72310) Imada et al., 1980a).

xxxiv. **Pseudomonas pyrrocinia** Imanaka, Kousaka, Tamura and Arima 1965, 205[AL]

pyr.ro.ci'ni.a. Etymology uncertain. Possibly a M.L. adj. based on "pyrrocin," referring to the antibiotic properties of pyrrolnitrin.

The following description is taken from the original paper.

Rods, 0.5–0.8 × 1.2–2.0 μm, occurring singly; motile with polar flagella. No pigment is produced. Oxidase reaction negative. Nitrate reduction and denitrification negative. Starch not hydrolyzed. H₂S is produced. Optimum temperature for growth, 26–30°C. Growth scanty at 37°C and negative at 42°C.

Acid produced from glucose, galactose, lactose, sucrose, glycerol, but not from maltose, trehalose, mannose, raffinose, starch or inulin. 2-ketogluconate is produced from gluconate.

Growth on glucose, gluconate, 2-ketogluconate, and p-hydroxybenzoate as sole carbon sources. 5-ketogluconate, citrate, ethanol, phenol, succinate, benzoate, salicylate, m-hydroxybenzoate, protocatechuate, gentisate, anthranilate, and p-aminobenzoate are not utilized for growth.

The name of the species refers to its capacity for the synthesis of the antibiotic pyrrolnitrin. This compound was also found to be produced by some strains of *P. chlororaphis*, *P. aureofaciens*, *P. cepacia* (*P. multivorans*) (Elander et al., 1968) suggesting that its production may be restricted to a few species of the genus. It is interesting to note here that Byng et al. (1980) have tentatively allocated *P. pyrrocinia* to rRNA group II, which includes *P. cepacia*.

Type strain: ATCC 15958.

xxxv. **Pseudomonas glathei** Zolg and Ottow 1975, 296.[AL]

gla'the.i. M.L. gen. n. *glathei* of Glathe; named after H. Glathe of Giessen, Germany.

The description is taken from Zolg and Ottow (1975).

Rods to oval cocci, 0.5–0.7 × 1.5 μm, motile by a polar flagellum

Optimum growth temperature 30–37°C. Oxidase reaction positive. Arginine dihydrolase negative. Starch, gelatin, lecithin (egg yolk reaction), esculin and polypectate are not hydrolyzed, but tributyrin, urea and hippurate are. Nitrate reduced to nitrite. No denitrification. H₂S is not produced.

No growth factor requirement has been found. The organism is capable of growth in nitrogen-deficient media, but acetylene reduction by cells grown under those conditions is negative. Acid tolerant (pH 4.5).

At least 68 organic compounds have been found to be utilized as sole carbon and energy sources for growth. These include aldoses, ketoses, deoxysugars, sugar-alcohols, and sugar-acids. Except for lactose, melibiose and melezitose, none of the di-, tri- and polysaccharides are utilized. The only amino acids that are not utilized are glycine, L-serine, L-isoleucine, L-methionine, and β-alanine. Of 28 aliphatic organic acids, 23 are utilized. The list of utilizable acids includes oxalate.

A 1,2-oxygenase responsible for the *ortho* cleavage of aromatic compounds is produced constitutively.

This species differs in several important characteristics from other species of the genus. The main differences include the capacity for growth in nitrogen-deficient media, the acid tolerance, and the utilization of oxalate as substrate for growth.

The mol% G + C of the DNA of the type strain is 64.8 (T_m).

Type strain: ATCC 29195 (N15; Zolg and Ottow, 1975).

xxxvi. **Pseudomonas aurantiaca** Nakhimovskaya 1948, 64.[AL]

au.ran.ti'a.ca. M.L. adj. *aurantiaca* orange colored.

The description is taken from the original paper.

Rods, 0.3–0.5 × 0.8–2.0 μm; motile, lophotrichous, with 4–6 flagella.

Two main diffusible pigments, green and orange, are produced. The colony may remain orange, while the medium is stained green and fluoresces. Colorless colonies are also produced, presumably by mutation.

Gelatin liquefaction is rapid. Starch is not hydrolyzed. Growth is good in complex and synthetic media. Nitrogen sources assimilated include ammonia, nitrate, amino acids and peptone. Oligonitrophile. Optimum temperature 25°C.

Capable of growth and acid formation from arabinose, xylose, glucose, galactose, sucrose, raffinose, glycerol, and mannitol. Lactose, maltose and starch are not utilized.

Comparison of this species with *P. aureofaciens* deserves consideration.

Type strain: NCIB 10068.

xxxvii. **Pseudomonas huttiensis** Leifson 1962b, 167.[AL]

hut.ti.en'sis. M.L. adj. *huttiensis* pertaining to Lower Hutt, New Zealand.

Description taken from the original paper.

Curved rods with blunt ends, 0.4 × 1.8 μm, with spherical inclusions. Polar multitrichous flagellation (1–3 flagella), wavelength 2.1 μm.

Growth is good in common peptone or yeast extract media. Mesophilic.

Acid is produced from glucose and xylose; slight acidification from mannose and sorbitol, and no acid from sucrose, maltose, lactose and raffinose.

Nitrate is reduced, and hydrolysis of gelatin, starch and cellulose are all negative. No growth factors are required.

Isolated from distilled water.

Type strain: ATCC 14670.

xxxviii. **Pseudomonas lanceolata** Leifson 1962b, 166.[AL]

lan.ce.o.la'ta. L. adj. *lanceolata* lancet shaped.

Description taken from the original paper.

Cells oval, of small (0.6 × 1.2 μm) or large (0.9 × 1.8 μm) size; single or in pairs. Motile by a polar flagellum. Flagellum wavelength is 1.9 μm.

Growth is good in peptone or yeast extract media. Organic growth factors are required by some of the strains.

Good growth occurs at 20–30°C; poor growth at 37°C. No growth with 1% NaCl.

Nitrate reduction is positive. No hydrolysis of gelatin, starch, or cellulose.

Slight amount of acid is produced from glucose. No acid from sucrose, maltose, xylose, lactose, sorbitol, mannose, raffinose.

Isolated from distilled water.

Type strain: ATCC 14669. A type culture of an infraspecific form has also been deposited (ATCC 14668).

xxxix. **Pseudomonas spinosa** Leifson 1962a, 90.[AL]

spi.no'sa. L. adj. *spinosa* thorny, spiny.

Description taken from the original paper.

Rods, 0.6–0.8 × 4–6 μm, with one, two, or occasionally, three long polar flagella. Under unfavorable growth conditions the rods may be curved. In their natural habitat (river water) the curvature is more marked, and no flagella are observed. Instead, numerous "spines" (fimbriae?) may be seen in the stained preparations. The property of forming these spines under unfavorable conditions has been lost in the strain preserved in collections. Spherical inclusions may be seen in the cells.

Growth is negative at 37°C or in media with 1% NaCl.

Weak acidity is produced from glucose, sucrose and maltose. No acid is produced from mannose, sorbitol, xylose, raffinose or lactose.

Nitrate is reduced to nitrite. Gelatin, cellulose or starch are not hydrolyzed. Growth factors are required.

Type strain: ATCC 14606.

xl. **Pseudomonas doudoroffii** Baumann, Baumann, Mandel and Allen 1972, 422.[AL]

dou.do.rof'fi.i. M.L. gen. n. *doudoroffii* of Doudoroff; named after M. Doudoroff.

Characteristics are presented in Tables 4.36 and 4.37.

Sodium is required for growth. Growth factors are not required. Nitrogen is not fixed. Nonluminescent.

Consult Baumann et al. (1972) for further information.

The mol% G + C of the DNA is 59.1 (Bd).

Type strain: ATCC 27123.

xli. **Pseudomonas marina** (Cobet, Wirsen and Jones 1970) Baumann, Baumann, Mandel and Allen 1972, 423.[AL] (*Arthrobacter marinus* Cobet, Wirsen and Jones 1970, 159.)

ma.ri'na. L. fem. adj. *marina* of the sea, marine.

Characteristics are presented in Tables 4.36 and 4.37.

Sodium is required for growth. Growth factors are not required. Nitrogen is not fixed. Nonluminescent.

Consult Baumann et al. (1972) for further descriptive information.

The mol% G + C of the DNA is 63 (Bd).

Type strain: ATCC 25374.

xlii. **Pseudomonas nautica** Baumann, Baumann, Mandel and Allen 1972, 423.[AL]

nau'ti.ca. L. fem. adj. *nautica* nautical.

Table 4.36.

General characteristics of **Pseudomonas doudoroffii, Pseudomonas marina** *and* **Pseudomonas nautica** (*section V*)[a]

Characteristics	P. doudoroffii	P. marina	P. nautica
Number of flagella	>1	0–1	1[b]
Pyroverdin production	−	−	−
Phenazine pigment production	−	−	−
Yellow cellular pigments	−	−	−
Oxidase reaction	+	−	+
Poly-β-hydroxybutyrate accumulation	+	+	−
Levan production from sucrose	−	−	−
Gelatin liquefaction	−	−	d
Starch hydrolysis	−	−	d
Lipase (Tween 80 hydrolysis)	−	−	+
Denitrification	−	−	d
Nitrate reduced to nitrite	d	−	d
Arginine dihydrolase	−	−	−
Autotrophic growth with H$_2$	−	−	−
Growth at 4°C	−	d	−
Growth at 41°C	+	d	d
Catechol, *ortho* cleavage	+		+
Protocatechuate, *ortho* cleavage	+		
Mol% G + C of DNA (Bd)	59.1	63	57.8–61.1

[a] For symbols see standard definitions.

[b] Lateral flagella are produced by some strains.

Characteristics are presented in Tables 4.36 and 4.37.

Sodium is required for growth. Growth factors are not required. Nitrogen is not fixed. Nonluminescent.

Only one strain of *P. nautica* liquefies gelatin.

For further descriptive information consult Baumann et al. (1972).

The mol% G + C of the DNA is 57.8–61.1 (Bd).

xliii. **Pseudomonas elongata** Humm 1946, 60.[AL]

e.lon'ga.ta. L. fem. part. adj. *elongata* elongated, stretched out.

Description taken from the original paper by Humm (1946).

Cells single, in chains or filaments, 0.3–0.4 × 3–6 μm. Both coccoid and longer cells are also present. Colonies in agar are sunken (due to agar liquefaction). Yellowish brown pigment diffusing into the medium is produced from peptone. Sodium ions are required (optimum concentration of NaCl, 2–3%).

Gelatin is slowly liquefied. Nitrate reduction negative. Starch, cellulose, chitin, and alginic acid are hydrolyzed. Acid is produced from arabinose, xylose, cellobiose, maltose, sucrose and salicin.

Ammonia, nitrate, nitrite or amino acids can be used as nitrogen sources. Fructose, galactose, glucose, mannose, lactose, acetate, lactate, malate, propionate, succinate, arginine, glutamate, and leucine are used as carbon sources.

Rhamnose, raffinose, inulin, ethanol, ethylene glycol, glycerol, mannitol, dulcitol, inositol, sorbitol, creatine, methionine, phenylalanine, serine, valine, butyrate, citrate, gluconate, malate, maleate, malonate, mucate, oxalate, tartrate, and isovalerate are not utilized. Further descriptive information in Humm (1946).

Isolated from intertidal sand, sea water and bottom sediments.

Type strain: ATCC 10144.

Further Comments. Various other species of marine pseudomonads are described in the paper by Humm (1946): "*P. atlantica,*" "*P. beaufortensis,*" "*P. corallina,*" "*P. droebachense,*" "*P. floridana,*" "*P. inertia,*"

Table 4.37.
Nutritional characteristics of **Pseudomonas doudoroffii, Pseudomonas marina** *and* **Pseudomonas nautica** (*section V*)[a]

Characteristics	P. doudoroffii	P. marina	P. nautica
Utilization of [b]:			
Acetate, succinate, fumarate, lactate, pyruvate	+	+	+
D-Ribose	d	+	−
D-Glucose, D-galactose, gluconate, mannitol, L-leucine, L-tyrosine	−	+	−
D-Fructose, glycerate, α-ketoglutarate, aconitate, β-alanine, L-serine, L-aspartate, γ-aminobutyrate	+	+	−
Propionate, caprate	d	+	+
Butyrate, valerate, caprylate, isobutyrate, caproate	−	d	+
Heptanoate, pelargonate	−	+	+
Isovalerate	−	+	d
Malonate, L(+)-tartrate	d	d	−
Glutarate, glycolate, glycine, γ-aminovalerate, L-histidine, putrescine, creatine, allantoin	+	−	−
Adipate, pimelate, suberate, azelate, sebacate, levulinate, propylene glycol, isobutanol, trigonelline, hexadecane	−	−	d
DL-Malate, β-hydroxybutyrate, citrate, L-alanine, D-alanine, L-glutamate, L-proline	+	+	d
m-Inositol, L-isoleucine, L-lysine, L-phenylalanine	−	d	−
2,3-Butylene glycol, *n*-propanol	d	d	d
Ethanol	+	d	d
n-Butanol	−	d	d
Benzoate	+	−	d
Glucuronate, *p*-hydroxybenzoate, L-threonine, kynurenate, spermine, histamine, hippurate, adenine	d	−	−
L-Arginine, L-ornithine, betaine, sarcosine	+	d	−

[a] For symbols see standard definitions.

[b] The following compounds are not utilized by any species: D-xylose, D-arabinose, L-arabinose, D-fucose, L-rhamnose, D-mannose, sucrose, trehalose, maltose, cellobiose, lactose, melibiose, saccharate, mucate, galacturonate, *N*-acetylglucosamine, salicin, inulin, cellulose, agar, formate, oxalate, maleate, D(−)-tartrate, *m*-tartrate, citraconate, itaconate, mesaconate, erythritol, sorbitol, adonitol, ethylene glycol, methanol, isopropanol, geraniol, D-mandelate, L-mandelate, benzoylformate, *o*-hydroxybenzoate, *m*-hydroxybenzoate, phenylacetate, phenylethanediol, naphthalene, quinate, L-valine, L-norleucine, L-citrulline, α-aminobutyrate, L-tryptophan, D-tryptophan, kynurenine, anthranilate, methylamine, ethanolamine, benzylamine, tryptamine, butylamine, α-amylamine, 2-amylamine, pantothenate, acetamide, nicotinate, nicotinamide, cytosine, thymine, uracil, guanine, *m*-aminobenzoate, *p*-aminobenzoate, poly-β-hydroxybutyrate.

Table 4.38.
Differentiating characteristics of clusters of pseudomonads isolated from meats (*section V*)[a,b]

Characteristics	Cluster 1	Cluster 2	Cluster 3	Cluster 4
Number of strains	32	53	18	5
Fluorescent pigment	−(3)	−(4)	d(72)	+
Multitrichous flagella	−	−(2)	d(78)	−
Gelatin hydrolysis	d(31)	d(24)	+	−
Tween 80 hydrolysis	−(3)	−(4)	d(61)	−
Egg yolk reaction	−	−	d(72)	−
Nitrate as *N*-source	d(34)	d(89)	d(22)	+
Arginine dihydrolase	d(53)	+(94)	+	+
Utilization of:				
Mesaconate	−	+	−	−
Itaconate	−(9)	+(98)	−(6)	−
Trehalose	d(18)	+	d	+
Meso-tartrate	d(13)	+	−	d
Mucate	+	+	−	d
Saccharate	+(97)	+(98)	−	d
Quinate	+(97)	+(98)	d(17)	+
p-Hydroxybenzoate	+	+(98)	−(6)	+
Hippurate	d(78)	+(98)	−	+
Erythritol	−	−	−	+
Pimelate	−	−	−	+
Suberate	−	−	−	+
L-Tryptophan	−	−	−	+

[a] For symbols see standard definitions. Values given in parentheses represent the percent of the strains giving a positive reaction.

[b] Data from Dr B. G. Shaw (personal communication). See also p. 187.

"*P. iridescens*," "*P. roseola*," as well as species reported in other bibliographic sources.

xliv. **Pseudomonas gelidicola** Kadota 1951, 58[AL]

ge.li.di′co.la. L. noun *gelida* cold water; L. substantive ending *-cola* inhabitant; M.L. n. *gelidicola* cold water inhabitant.

The description of this species is taken from the original paper.

Rods, 0.4–0.5 × 1.5–2.5 µm, occurring singly or in pairs, with monotrichous flagella.

Colonies are yellow. Optimum NaCl concentration for growth, 3%. Hydrolyzes agar but not gelatin. Nitrates are not reduced; however, the preferred nitrogen source appears to be nitrate.

Utilizes glucose, galactose, mannose, sucrose, lactose, raffinose, starch, and agar for growth.

Isolated from sea water, algae and rotted straw submerged in sea water.

Type strain: IAM 1127.

xlv. **Pseudomonas fragi** (Eichholz 1902) Gruber 1905, 122.[AL] (*Bacterium fragi* Eichholz 1902, 425.)

fra′gi. L. neut. n. *fragum* strawberry; L. gen. n. *fragi* of the strawberry.

The description of this species is given in the seventh edition of the *Manual* (Haynes and Burkholder, 1957, p. 110). This species was tentatively placed in rRNA group I by Byng et al. (1980).

Type strain: ATCC 4973.

xlvi. **Pseudomonas mephitica** Claydon and Hammer 1939, 254.[AL]

me.phi′ti.ca. L. adj. *mephitica* pestilential (skunklike) odor.

The description of this species is given by Haynes and Burkholder (1957) in the seventh edition of the *Manual* (p. 111).

Type strain: NCIB 9672.

xlvii. **Pseudomonas synxantha** (Ehrenberg 1840) Holland 1920, 220.[AL] (*Vibrio synxanthus* Ehrenberg 1840, 202.)

syn.xan′tha. Gr. pref. *syn-* along with, together; Gr. adj. *xanthus* yellow; M.L. adj. *synxanthus* with yellow.

The description of this species is given by Haynes and Burkholder (1957) in the seventh edition of the *Manual* (p. 104).

Type strain: ATCC 9890.

xlviii. **Pseudomonas mucidolens** Levine and Anderson 1932, 344.[AL]

mu.ci'do.lens. L. adj. *mucidus* musty; L. v. *oleo* to smell of; L. part. adj. *mucidolens* musty smelling.

The description is taken from the original paper as follows.

Rods, rounded ends, occurring singly and in pairs, actively motile. Filamentous cells are frequently observed. Fluorescent pigment is produced.

Gelatin is liquefied. Nitrates reduced to nitrites and gas.

Optimum growth temperature 23–25°C; slight growth at 37°C and at 10°C.

Acid produced from glucose, rhamnose, arabinose, erythritol, sorbitol, trehalose, galactose, fructose, mannose. Moderate amount of acid from glycerol, mannitol, dulcitol, glycogen, inulin, maltose, melezitose, pectin, raffinose, salicin, starch, sucrose, or xylan. Lactate, citrate and urate can be used for growth.

Acetate, oxalate, sulfanilate, tartrate, salicylate and formate do not support growth.

A physiologically less active form of this species is described in the same paper under the name "*P. mucidolens* var. *tarda.*"

P. mucidolens is one of the species causing mustiness in eggs, which were the source for its isolation. The general characters as described above roughly conform to those of one of the denitrifying biovars of *P. fluorescens*, but a decision on this point should require comparative studies.

Type strain: ATCC 4685.

xlix. **Pseudomonas taetrolens** Haynes 1957, 108.[AL]

taet'ro.lens. L. adj. *taeter* offensive; L. part adj. *olens* having an odor; M.L. part. adj. *taetrolens* foul smelling.

The description of this species is given by Haynes and Burkholder (1957) in the seventh edition of the *Manual*.

Type strain: ATCC 4683.

l. **Pseudomonas azotoformans** Iizuka and Komagata 1963, 137.[AL]

a.zo.to.for'mans. Fr. n. *azote* nitrogen; L. v. *formo* to fashion, form; M.L. part. adj. *azotoformans* nitrogen forming (by denitrification).

The description is taken from the original paper.

Rods, 0.6–0.8 × 1.4–2.0 μm, motile with polar flagella. Fluorescent pigment is produced. Oxidase reaction positive.

Gelatin is liquefied. Nitrate reduction positive. Denitrification positive. H_2S not produced. Starch not hydrolyzed. Acid is produced from glucose, glycerol, xylose, sucrose; no acid from lactose or starch.

Optimum growth temperature, 25–30°C. No growth at 37°C.

Glucose, gluconate, 2-ketogluconate, citrate, succinate, ethanol, *p*-hydroxybenzoate, protocatechuate, and anthranilate are assimilated. Phenol, benzoate, salicylate, *m*-hydroxybenzoate, gentisate, *p*-aminobenzoate, and 5-ketogluconate are not assimilated.

More details are given in the original paper. The organism appears from the description to be similar to some of the dentrifying biovars of *P. fluorescens*, but no comparative studies have been performed.

Isolated from Japanese rice paddies.

Type strain: IAM 1603.

li. **Pseudomonas fulva** Iizuka and Komagata 1963, 138.[AL]

ful'va. L. adj. *fulva* tawny, yellowish brown.

Description taken from the original paper.

Rods, 0.6–0.8 × 1.2–1.8 μm. Motile with one to three polar flagella. Fluorescent pigment is produced. Oxidase reaction feebly positive.

Gelatin is not liquefied. Nitrate not reduced to nitrite or denitrified. H_2S not produced. Starch is not hydrolyzed.

Acid is produced from glucose. No acid from glycerol, xylose, sucrose, lactose or starch.

Glucose, gluconate, 2-ketogluconate, citrate, succinate, ethanol, *p*-hydroxybenzoate and protocatechuate are assimilated. Phenol, benzoate, *m*-hydroxybenzoate, gentisate, *p*-aminobenzoate, and 5-ketogluconate are not assimilated.

Optimum temperature for growth, 25–30°C. Poor growth at 37°C and no growth at 42°C.

Isolated from Japanese rice paddies.

Type strain: IAM 1529.

lii. **Pseudomonas radiora** Ito and Iizuka 1971, 1568.[AL]

ra.di.o'ra. The etymology is uncertain.

Description taken from the original paper.

Rods, 0.6–0.8 × 1.4–2.5 μm, motile with a single polar flagellum. Fat globules are present in the cells.

The colonies are pink to red due to the production of a carotenoid pigment (α-bacterioruberin).

Gelatin is not liquefied. Nitrate is reduced to nitrite. Denitrification negative. Starch is slightly hydrolyzed. H_2S is produced. Urease strongly positive.

Acid is produced from xylose, glucose and glycerol, but not from sucrose, lactose and starch. Glucose, gluconate, 2-ketogluconate, citrate, succinate, ethanol, methanol, benzoate, salicylate, and *p*-hydroxybenzoate are assimilated. Phenol is not assimilated. Oxidase reaction strongly positive. Salt tolerance, up to 0.5%.

The organism has been isolated from Japanese unhulled old rice.

The radio-resistance of this organism was 10–40 times higher than that of ordinary species of the genus *Pseudomonas*.

Type species: ATCC 27329.

liii. **Pseudomonas straminea** Iizuka and Komagata 1963, 139.[AL]

stra.mi'ne.a. L. adj. *straminea* made of straw.

Description taken from the original paper.

Slender rods, 0.3 × 3.0 μm. Motile with a polar flagellum.

Colonies are yellow. Fluorescent pigment is produced. Oxidase reaction positive.

Gelatin is liquefied. Nitrates are not reduced to nitrites. Denitrification is negative. H_2S is produced. Starch is not hydrolyzed. Acid produced from glucose. Glucose, gluconate, citrate, succinate, and ethanol are assimilated. Phenol, benzoate, salicylate, *m*-hydroxybenzoate, *p*-hydroxybenzoate, protocatechuate, gentisate, anthranilate, *p*-aminobenzoate, 2-ketogluconate and 5-ketogluconate are not assimilated.

Optimum growth temperature, 25–30°C. No growth at 37°C.

The organism has been isolated from Japanese rice paddies.

Type strain: IAM 1598.

liv. **Pseudomonas beijerinckii** Hof 1935, 152.[AL]

bei.jer.inck'i.i. M.L. gen. n. *beijerinckii* of Beijerinck; named after M. W. Beijerinck of Delft, Holland.

The description of this species is given by Haynes and Burkholder (1957) in the seventh edition of the *Manual*.

Type strain: ATCC 19372.

lv. **Pseudomonas cocovenenans** van Damme, Johannes, Cox and Berends 1960, 255.[AL]

co.co.ve'ne.nans. M.L. n. *Cocos* genus of coconut; L. v. *veneno* to poison; M.L. part. adj. *cocovenenans* coconut poisoning.

To our knowledge, no description of this species is available and the first time this name was proposed (van Damme et al., 1960) was in conjunction with the description of toxoflavin, the yellow poisonous compound produced by this organism. Even though the most appropriate position in such a situation is the one adopted in the eighth edition of the *Manual*, where the species is placed among the *species incertae sedis*, the name is mentioned here for the potential interest of the organism whose type strain is available from collections (NCIB 9450).

lvi. "**Pseudomonas butanovora**" Takahashi, Ichikawa, Sagae, Komura, Kanou and Yamada 1980, 1837. (As of this writing, this species has not been validly published.)

but.an.o'vo.ra. L. *butyricum* butter, whence *butane* (coined word); L. v. *voro* to devour; M.L. fem. adj. *butanovoro* butane devouring.

The following is a summary of the description presented in the paper by Takahashi et al. (1980).

Rods, 0.6–0.8 × 1.1–2.4 μm, occurring singly. Motile with a single polar flagellum. The cells accumulate PHB as carbon reserve material.

Colonies are pale yellow to brownish yellow. Growth factors are not required. Maximum growth temperature, 42.5°C.

Catalase and oxidase reactions positive. Arginine dihydrolase reaction negative. Nitrate reduction and denitrification are positive. Hydrolysis of gelatin and of starch, negative. Arginine and betaine are not assimilated. C_2–C_9 normal hydrocarbons are assimilated, but the C_{10}–C_{16} n-alkanes and methane are not. Ethanol, n-propanol, n-butanol, acetate, propionate, butyrate, lactate, 1,2-propanediol, 2,3-butanediol, glycerol and valine are utilized. Glucose, fructose, xylose, L-arabinose, sucrose, geraniol, 1,2-ethanediol, glycolate, arginine, ethylene, propylene, 1-butene, 2-butene, 1,3-butadiene, sorbitol and erythritol are not assimilated. Since methane and methanol are not utilized. the organism is unrelated to the special group of C_1 utilizers. In their paper, Takahashi et al. (1980) have included a comparison with other dentrifying species of the genus *Pseudomonas*, although there is no mention of the strains used for the comparison. *P. pseudoalcaligenes*, a species characterized for the denitrifying capacity and the inability to use glucose, was not included in the comparison. This species resembles "*P. butanovora*" in the accumulation of PHB, absence of pigment, growth at 41°C, denitrification, and incapacity for growth on glucose, xylose, glycolate, geraniol, and for the hydrolysis of gelatin and of starch, although some of these properties are not universal in *P. pseudoalcaligenes*. Differences include the utilization of arginine, 2,3-butyleneglycol, valine, and the DNA base composition.

The mol% G + C of the DNA is 67.3 (T_m).

Type strain: IAM 12574.

lvii. **Pseudomonas oleovorans** Lee and Chandler 1941, 377.[AL]

o.le.o′vor.ans. L. n. *oleum* oil; L. v. *voro* to destroy, consume; M.L. part. adj. *oleovorans* oil consuming.

Description taken from the original paper.

When grown on agar, the cells are almost coccoid (0.5 × 0.8 μm), but the length increases to about 1.5 μm during the exponential phase in broth.

The colonies have a typical fluorescence that is not imparted to the medium.

Nitrate is reduced to nitrite. Gelatin not liquefied. Starch is hydrolyzed.

Isolated from oil-water emulsions used as lubricants and cooling agents in the cutting and grinding of metals. Apparently, the organism lives on some normal constituent of the cutting compound, probably the naphthenic acids which act as emulsifying agents. Consequently, the name *oleovorans* may not be justified.

Type strain: ATCC 8062.

lviii. **Pseudomonas iners** Iizuka and Komagata 1964b, 225.[AL]

i′ners. L. adj. *iners* inactive.

The following description is taken from the original paper.

Rods, 0.6–0.8 × 1.2–2.0 μm, occurring singly; motile with a single polar flagellum. No pigments are produced.

Oxidase reaction negative; catalase reaction weak. Gelatin is not liquefied. Nitrates are not reduced. Denitrification negative. Starch is not hydrolyzed. The strains fail to grow in synthetic media, suggesting growth factor(s) requirement.

No acid is formed in peptone media from glycerol, xylose, glucose, sucrose, lactose, and starch. A number of compounds are listed as unable to support growth although the authors do not report experiments designed to define the conditions necessary to allow growth in chemically defined media.

Optimum growth temperature, 25–30°C. Growth scanty at 37°C and no growth at 42°C.

Source: oil brines in Japan.

Type strain: IAM 1419.

lix. **Pseudomonas nitroreducens** Iizuka and Komagata 1964a, 214.[AL]

ni.tro.re.du′cens. L. n. *nitrum* nitre, nitrate; L. v. *reduco* to draw backwards, bring back to a state or condition; M.L. part. adj. *nitroreducens* nitrate reducing.

Description taken from the original paper.

Rods, 0.4–0.6 × 1.4–1.8 μm, occurring singly, rarely in pairs; motile with polar flagella.

Fluorescent pigment produced. Gelatin is not liquefied. Nitrates are reduced to nitrites and to gas. Oxidase reaction positive.

H_2S not produced. Starch is not hydrolyzed. Acid is produced from glucose.

Glucose, gluconate, 2-ketogluconate, citrate, succinate, ethanol, p-hydroxybenzoate and protocatechuate can be utilized as sole sources of carbon. 5-ketogluconate, benzoate, salicylate, gentisate, anthranilate, and p-aminobenzoate are not utilized. The fresh isolate could utilize kerosene, but the activity was lost after subcultivation in the laboratory.

Optimum growth temperature, 25–30°C. No growth at 37°C.

The source of isolation has been oil brine in Japan.

This species has two characteristics never found in association in any of the fluorescent species described in RNA group I, namely, the capacity for denitrification and the inability to liquefy gelatin.

Type strain: IAM 1439.

lx. **Pseudomonas paucimobilis** Holmes, Owen, Evans, Malnick and Willcox 1977, 144.[AL]

pau.ci.mo′bi.lis. L. adj. *paucus* few; L. adj. *mobilis* mobile; M.L. fem. adj. *paucimobilis* intended to mean a few cells motile.

Description is taken from the original paper.

Rods, 0.7 × 1.4 μm, occurring singly or in pairs, motile by means of a single polar flagellum. Motility is observed at 18–22°C, but not at 37°C. A yellow pigment is produced, which is insoluble in water, is not fluorescent, and has absorption maxima in methanol at 452 and 479 nm, with a slight inflection at 425 nm. As mentioned in an earlier section, this pigment has been identified as a carotenoid (nostoxanthin) (Jenkins et al., 1979).

Growth occurs at 37°C but not at 5°C or 42°C. Optimum temperature for growth is ∼ 30°C.

Oxidase reaction positive. Catalase and deoxyribonuclease are produced. Tween 20 and 80 are hydrolyzed. Lecithinase reaction negative.

Nitrate is not reduced to nitrite; nitrate is not reduced.

Lipid inclusions presumed to be poly-β-hydroxybutyrate are produced.

Gelatin is not liquefied; casein is not digested. Starch is hydrolyzed. Arginine dihydrolase, arginine desimidase, lysine decarboxylase, and ornithine decarboxylase are not produced. 3-ketolactose is not produced.

Acid is produced in mineral medium from arabinose, cellobiose, ethanol, fructose, glucose, lactose, maltose, raffinose, salicin, sucrose, trehalose, and xylose. Acid is not produced in mineral medium from adonitol, dulcitol, inositol, mannitol, sorbitol, glycerol, or rhamnose.

Isolated from a respirator.

The mol% G + C of the DNA is 65 (T_m).

Type strain: ATCC 29837 (NCTC 11030).

lxi. **Pseudomonas echinoides** Heumann 1962, 343.[AL]

e.chi.noi′des. Gr. adj. *echinos* spiny appearance; Gr. n. *eidus* form, shape; M.L. adj. *echinoides* spiny shape.

The description is taken from the original paper.

Rods, 0.8 × 1.9 μm, slightly curved, with sharp ends. Monotrichous flagella. Flagella wavelength, 1.9 μm. Yellow intracellular pigments (carotenoids) are produced. Nonpigmented and egg yolk-yellow mutants can be obtained.

The surface of the colonies is smooth, but rough mutants are frequent. Cell aggregates (rosettes) form in broth and also in agar media. In liquid media they can be observed as small flocs.

Nitrate is reduced to nitrite, not to ammonia. No H_2S is produced. Starch is slowly hydrolyzed. Gelatin is not liquefied.

Optimum temperature, ∼ 28°C.

Isolated as a laboratory contaminant.

Type strain: ATCC 14820.

lxii. **Pseudomonas pertucinogena** Kawai and Yabuuchi 1975, 318.[AL]

per.tu.ci.no′ge.na. *pertucin* (coined word), a bacteriocin active against *Bordetella pertussis*; L. v. *gigno* to produce; M.L. fem. adj. *pertucinogena* intended to mean pertucin producing.

The description is from the original paper, referring to two strains of the species.

Rods, 0.4 × 1.1 μm, motile by means of single polar flagella. Do not accumulate PHB. Oxidase reaction positive. No pigments are produced.

Hydrolysis of gelatin, starch or Tween 80, negative. Arginine dihydrolase negative.

Acid production from sugars is in general weak or negative (glucose, D-arabinose and galactose appear to be positive). Pyruvate, succinate, oxaloacetate, β-hydroxybutyrate and L-alanine are utilized for growth. A number of amino acids, carbohydrates, and alcohols are not utilized. Further details can be obtained from the original paper.

The two strains produce pertucin, a bacteriocin active against *Bordetella pertussis*, and were kept for many years in the American Type Culture Collection (ATCC) as members of this species.

The mol% G + C of the DNA is ∼ 60 for the two strains, and their DNA homology is very high.

Type strain: ATCC 190.

Genus II. **Xanthomonas** Dowson 1939, 187.[AL]

J. F. BRADBURY

Xan.tho′mo.nas according to strict Latin usage; Xan.tho.mo′nas is commonly used; Gr. adj. *xanthus* yellow; Gr. fem. n. *monas* unit, monad; M.L. fem. n. *Xanthomonas* yellow monad.

Cells are **straight rods, usually within the range 0.4–0.7 wide × 0.7–1.8 μm long,** predominantly single. Do not produce poly-β-hydroxybutyrate inclusions. Do not have sheaths or prosthecae. No resting stages known. Gram-negative. **Motile by a single polar flagellum. Obligately aerobic,** having a strictly respiratory type of metabolism with oxygen as the terminal electron acceptor. **No denitrification or nitrate reduction occurs.** Optimum temperature, 25–30°C. Colonies are usually **yellow,** smooth and butyrous or viscid. **The pigments are highly characteristic brominated aryl polyenes, or "xanthomonadins."** The oxidase test is negative or weak. Catalase-positive. Chemoorganotrophic; able to use a variety of carbohydrates and salts of organic acids as sole carbon sources. Small amounts of acid are produced from many carbohydrates, but not from rhamnose, inulin, adonitol, dulcitol, sorbitol, *meso*-inositol or salicin. Acid is not produced in purple milk or litmus milk. **Asparagine is not used as a sole source of carbon and nitrogen. Growth is inhibited by 0.1% (and usually by 0.02%) triphenyltetrazolium chloride. Growth factors required usually include methionine, glutamic acid, nicotinic acid, or a combination of these. Plant pathogens;** occur in association with plants. The mol% G + C of the DNA is 63–71 (T_m, Bd).

Type species: *Xanthomonas campestris* (Pammel 1895) Dowson 1939, 190.

Further Descriptive Information

Morphology. Cells examined directly from plant tissues or from old cultures may show irregularities in size and shape, but those from actively growing cultures are reasonably uniform rods with rounded ends. The vast majority of the cells are within the ranges of width given above. The greater widths reported in a few early descriptions seem to be erroneous. For example "*Xanthomonas phaseoli* var. *fuscans*," readily distinguished by its production of brown diffusible pigment, was originally reported to produce cells up to 1.35 μm wide and 4 μm long (Burkholder, 1930). However, no cells larger than the limits given above for the genus have been found in the many cultures examined from various countries in our laboratory. This organism is now considered to be synonymous with *X. campestris* pathovar *phaseoli*. Cell length is a more variable parameter. It is also frequently difficult to decide whether one is viewing a single long cell, or two or more together. Filamentous cells are occasionally seen in cultures of *X. ampelina* (Panagopoulos, 1969). These are of normal width, but up to 10 times the normal length.

Capsules. No special structures or inclusions are visible by light microscopy with simple staining techniques; however, capsular staining, e.g. with India ink, has led to reports of capsules in many isolates of *X. campestris* and in *X. axonopodis*. Results are variable with the technique and depend on how the cells are handled prior to staining,

because the capsular polysaccharides are often quite loosely associated with the cells. Production of large amounts of these extracellular polysaccharides or xanthan gums result in the slimy or mucoid colonies of *X. fragariae* and *X. campestris* when grown on media containing usable carbohydrates. Xanthans have high molecular weights and acidic properties (Leach et al., 1957). Those synthesized by *X. campestris* pv. *campestris* are anionic heteropolysaccharides composed of D-glucose, D-mannose, and D-glucuronic acid in a ratio of 2:2:1, with small amounts of pyruvic and acetic acids that vary with strains and conditions (Cadmus et al., 1978; Whitfield et al., 1981).

Changes in viscosity and spectroscopic properties occur in aqueous solutions of xanthans. These are the result of changes of molecular configuration from disordered to ordered, which are thought to be related to the covalent structure and fiber conformations (Morris et al., 1977). Holzworth and Prestridge (1977) examined the structure of the polysaccharide by electron microscopy and found that in nature it has unbranched, probably double-stranded fibers 4 nm wide by 2–10 μm long. Denatured xanthan consists of single strands 2 nm wide by 0.3–1.8 μm long. The renatured material shows short unravelled regions with two or three strands arranged in a right-handed twist. The ordered form can bind cooperatively to certain polysaccharides present in plant cell walls. This suggests a role in the early stages of colonization of a plant by a bacterial pathogen (Morris et al., 1977).

El Banoby and Rudolph (1979) obtained specific fractions of extracellular polysaccharide by column chromatography from several plant pathogenic pseudomonads and xanthomonads. These fractions, when infiltrated into leaves, produced persistent watersoaked spots only in the host plants of the bacteria from which they were derived. The role of extracellular polysaccharides in host-pathogen interactions is still far from understood.

The material is also thought to play a protective role that helps to maintain viability of bacterial cells in exudate and plant material under dry conditions.

Because of its physical properties and lack of toxicity to man and animals xanthans have a wide range of uses in industry. They are gelling agents, emulsifiers, stabilizing agents and plasticizers, particularly in the food industry (Jeannes, 1974; Sutherland and Ellwood, 1979).

Cell walls. The structure of the cell envelope in *Xanthomonas* appears to be similar to that of other Gram-negative cells as reviewed by Costerton et al. (1974). Schnaitman (1970) found that the morphology of the cell wall of a *X. campestris* isolate, as viewed with the electron microscope, was similar to that described for *Escherichia coli* by De Petris (1967). The electrophoretic protein pattern he obtained was broadly similar to *E. coli*, but the major protein constituent had a slightly lower mobility.

The lipopolysaccharides of the cell wall consist of a heteropolysaccharide portion to which "lipid A" is covalently bonded. The backbone of lipid A in a number of isolates from various genera, including one from *Xanthomonas*, was shown to contain β-1',6-linked glucosamine disaccharides carrying two phosphate groups, one glycosidically linked and one with ester linkage (Hase and Rietschel, 1976). Long chain fatty acids are joined to the backbone through ester and amide linkages. The fatty acids found in several isolates of *X. campestris* included the branched-chain acids D-3-hydroxy-9-methyldecanoic, D-3-hydroxy-11-methyldodecanoic, and smaller amounts of 2-hydroxy-9-methyldecanoic (Rietschel et al., 1975). Branched hydroxy fatty acids are most unusual in bacteria, but the three that occur in *Xanthomonas* also occur in *Pseudomonas maltophilia* (Moss et al., 1973). This latter species shows many similarities to the xanthomonads, as will be seen in the following pages. It should probably be included in the genus *Xanthomonas*, even though it does not agree completely with the present generic description (see Taxonomic Comments).

The polysaccharides of the cell wall of *Xanthomonas* have been examined by Volk (1966, 1968a, b), who obtained D-galacturonic acid-1-phosphate, glucose, mannose and 2-keto-3-deoxyoctonate. The latter occurs in larger amounts in enteric organisms. Studying 17 isolates of *X. campestris*, Volk (1968b) found two different ratios of uronic acid: mannose. Five isolates had 1.2:1 while the remainder had 0.66:1 i.e., about half. Most also contained xylose or fucose, but not both. High levels of rhamnose were present in 15 isolates, low levels in two. In addition 4,7-anhydro- and 4,8-anhydro-3-deoxyoctulosonic acids (Volk et al., 1972) and 3-acetamido-3,6-dideoxy-D-galactose (Hickman & Ashwell, 1966) have been isolated. All the isolates examined so far belong to *X. campestris*. Results for other species would be interesting.

The high content of rhamnose, low content of 2-keto-3-deoxyoctonate, presence of D-galacturonic acid, and the absence of heptoses are also found in *Pseudomonas maltophilia* (Wilkinson, 1968; Neal and Wilkinson, 1979).

Flagella and pili. The single polar flagellum may be demonstrated by flagella staining or by electron microscopy of cells from the syneresis water of young cultures. The wavelength of normal flagella averages 1.79 μm, but occasionally variants occur with wavelengths of 2–3 μm. Very rarely cells occur with two polar flagella (Leifson, 1960).

No reports of fimbriae or pili in *Xanthomonas* have been found, but it is not certain whether this indicates absence of the structures, or lack of a search.

Pigments. Yellow pigments are present in all species of *Xanthomonas*, but nonpigmented strains sometimes occur. *X. campestris* pv. *manihotis* and some strains of pv. *ricini* occur naturally as nonpigmented organisms. Occasionally nonpigmented mutants arise in culture (Bryan, 1932; Durgapal, 1977). Although presence of yellow pigment is an important characteristic for identification, its absence does not exclude an organism from the genus if other characteristics are in agreement.

Starr and Stephens (1964), examining a number of isolates belonging to *X. campestris*, found that all contained one or more pigments, thought to be carotenoids, that showed absorption maxima at 418 (a shoulder), 437 and 463 nm when dissolved in petroleum ether. The pigments were shown to be released from disrupted cells at the same rate as a component of the cytoplasmic membrane, strong evidence for their attachment to the cell envelope (Stephens and Starr, 1963). Later it was found that they are not carotenoids but brominated aryl polyenes (Andrewes et al., 1973; Starr et al., 1977). They were given the trivial name "xanthomonadins." Xanthomonadin I, the main component of the pigment of *X. campestris* pv. *juglandis*, was found to be 17-(4-bromo-3-methoxyphenyl-17-bromo-heptadeca-2,4,6,8,10,12,14,16-octaenoic acid (Andrewes et al., 1976).

From two to five pigments with this type of structure were found in isolates of *X. albilineans*, *X. axonopodis*, *X. fragariae* and *X. campestris* by Starr et al. (1977). The pigments of *X. ampelina*, however, were more difficult to dissolve and purify, and showed different absorption spectra.

The only other pigments approaching these in structure and known to occur in microorganisms are not brominated and occur in *Flexibacter*

elegans and the fungus *Corticium salicinum* (Starr et al., 1977). Selective production of bromine compounds in the presence of excess chlorine is found in some marine organisms, but is rare in land organisms. There are, however, no indications of a recent marine origin for the genus *Xanthomonas*.

Colonial and cultural characteristics. Colonies of all species of *Xanthomonas* are normally smooth, round, entire, and butyrous, at least when young, but may show surface markings such as striations and become lobed when older. Mutant colonies of different appearance occur occasionally. These may be less mucoid and tend toward rough (Corey and Starr, 1957a) or may be crenated (Whitfield et al., 1981).

When they are first visible, colonies are transparent and buff to very pale yellow, but the color soon deepens and then varies according to the species and the medium used. Pathovars of *X. campestris* and *X. axonopodis* darken towards a buttercup yellow or light cadmium, especially on nutrient agar supplemented with 5% glucose or sucrose. *X. fragariae* remains small and pale on nutrient agar, but gives abundant, spreading, orange-yellow growth on addition of 1% glucose. *X. ampelina* grows slowly on nutrient agar and is pale yellow; better growth is obtained when glucose or sucrose is added. The best medium consists of 2% galactose, 1% yeast extract, 2% $CaCO_3$ and 2% agar; on this medium the organism produces deep yellow colonies and a brown diffusible pigment. *X. albilineans* does not grow on nutrient agar but does grow on an agar medium containing 2% sucrose and 1% peptone, and is buff yellow or honey colored.

Growth rates vary widely in the genus. The faster-growing pathovars of *X. campestris* will produce visible colonies from single cells in 24–36 h at 25°C. Slower-growing isolates of this species may take 2 or 3 days for single colonies to become visible. The time taken to produce colonies about 1 mm in diameter at this temperature varies from 2 or 3 days to a week or 10 days. The other four species are all much slower-growing than this, all taking about a week to appear, and often longer on isolation plates.

Nutrition and growth conditions. *Xanthomonas* species are chemoorganotrophic. In purely synthetic media containing minerals, ammonium nitrogen, and a suitable carbon source such as glucose, amino acid supplements are required by most strains of all species. *X. ampelina* requires 0.1% glutamate, and *X. albilineans* requires glutamate and methionine for growth (Panagopoulos, 1969). *X. axonopodis* and *X. fragariae* will grow when vitamin-free casein hydrolysate is added, but their exact amino acid requirements are unknown. For the former, supplementation with glutamate, methionine or both is insufficient (Starr and Garces, 1950), while for the latter methionine is without effect and glutamate is only partially effective (Kennedy and King, 1962). Most pathovars of *X. campestris* will grow with added glutamate or methionine, but occasionally both are needed, and *X. campestris* pv. *pruni* also requires nicotinic acid (Starr, 1946). Experiments have shown that in this pathovar nicotinic acid can be replaced by tryptophan, which this organism can convert to nicotinic acid (Wilson and Henderson, 1963).

Ammonium salts, e.g. ammonium phosphate, can serve as nitrogen sources, but nitrates are less often acceptable. Patel and Kulkarni (1949) reported that *X. campestris* pv. *malvacearum* will use potassium nitrate, but not sodium nitrate. Kotasthane et al. (1965) examined 16 pathovars of *X. campestris* and found that all would accept DL-alanine, L-glutamate, and L-proline as sole carbon sources. Glycine, L-leucine, DL-methionine, DL-serine and DL-norleucine were not used. As sole nitrogen sources, DL-alanine, L-glutamate, L-proline, DL-methionine, DL-threonine, DL-aspartate, L-asparagine, L-hydroxyproline and L-histidine were accepted by all; DL-serine, DL-norleucine and L-tyrosine by none; while results with glycine, DL-valine, L-tryptophan, L-leucine, DL-isoleucine, L-arginine, DL-lysine and L-cystine varied with the pathovar. Asparagine can serve as a nitrogen source if glucose is supplied, but it cannot serve as a simultaneous source of both carbon and nitrogen (Starr and Weiss, 1943; Starr, 1946). This is used as a diagnostic test for *Xanthomonas* (Dye, 1962), since yellow *Enterobacteriaceae* and many *Pseudomonas* species will grow with asparagine as sole source of both elements. Glutamate or alanine can serve in this way for *Xanthomonas* (Lewis, 1930).

The optimum growth temperature is usually in the range 25–30°C, with the minimum above 5°C and the maximum varying from 30–39°C.

Respiratory chain. Xanthomonads are obligately aerobic and use oxygen as the terminal electron acceptor. They are oxidase-negative by the Kovacs' test.

Hochster and Nozzolillo (1960) postulated that the respiratory chain in *X. campestris* pv. *phaseoli* is:

Flavoprotein → cytochrome b_1
$$\rightarrow \text{cytochrome } a_1 \rightarrow \text{cytochrome } a_2 \rightarrow O_2$$

The difference spectra that they found at liquid air temperature showed a slight shoulder at 549 nm, suggesting a very small content of cytochrome *c*, and with cell-free extracts they showed some reduced nicotinamide adenine dinucleotide (NADH)-cytochrome *c* reductase activity. The difference spectra of Sands et al. (1967) for *Pseudomonas syringae*, and of Stanier et al. (1966) for *P. maltophilia* are very similar, but do not show any noticeable peak or shoulder at or near 549 nm. Both of these latter species are oxidase-negative and thought to contain no cytochrome *c*.

Carbohydrate metabolism. Only nomenspecies now included as pathovars of *X. campestris* have been examined with respect to pathways of glucose catabolism. Glucose is oxidized readily, but not gluconate. No gluconate or 2-keto-gluconate is produced from glucose (Lockwood et al., 1941). The Entner-Doudoroff pathway is present (Katznelson, 1955, 1958) and is the predominant pathway of glucose catabolism, with only 8–16% of the glucose following the pentose phosphate pathway. The hexose cycle is not significantly used by the pathovars examined (Zagallo and Wang, 1967). The tricarboxylic acid and glyoxylate cycles are present (Madsen and Hochster, 1959). Two enzymes of the tricarboxylic acid cycle that show interesting variations of taxonomic significance are citrate synthase and succinate thiokinase (Weitzman, 1980). *X. campestris* pv. *hyacinthi* has a citrate synthase that is typical of Gram-negative aerobes: it is a large molecule (molecular weight is ~250,000), is inhibited by NADH, and is reactivated by adenosine monophosphate. The succinate thiokinase is also typical of Gram-negative bacteria, being large in size (molecular weight is 140,000–150,000).

With respect to the breakdown of oligosaccharides, Hayward (1977) assessed the activity of α-glucosidase, β-glucosidase, β-galactosidase and β-xylosidase in various bacteria, mostly plant pathogens; of 39 strains examined, most showed the activity of all 4 glycosidases, as did *Pseudomonas maltophilia* alone among the *Pseudomonas* species. α-Glucosidase activity was weak in *X. albilineans* and was absent for all isolates of *X. campestris* pvs. *holcicola, pruni* and *vitians*. Dekker and Candy (1979) showed that β-mannanases and carboxymethylcellulases produced by *X. campestris* were present as membrane-bound constituents and also extracellularly. Pectolytic activity has been shown for xanthomonads, but varies among strains. Small variations in methodology can also cause variation of results. Dye (1960) examined 142 isolates of various pathovars of *X. campestris* and 3 isolates of *X. albilineans*. Some of the *X. campestris* strains had as much activity as *Erwinia carotovora*, but many showed little or no activity. Two of the *X. albilineans* strains showed only weak polygalacuronase activity, while one also showed some pectin methylesterase activity. Using pectate gels at various pH levels, Hildebrand (1971) found that the pitting produced by two pathovars of *X. campestris* was unaffected by pH, whereas with six others there was greatest activity at pH 5. Three isolates of *X. fragariae* showed slight activity. Lange and Knösel (1970) examined eight isolates belonging to three pathovars of *X. campestris* and detected strong pectin methylesterase, pectic transeliminase, and some polygalacturonase activity in pathovar *vesicatoria* (one isolate); the remainder showed slight or no activity. These authors also demonstrated significant $C_x$cellulase activity and proteolytic activity in *Xanthomonas*.

Metabolism of aromatic compounds. Degradation of protocatechuate,

phenylalanine and synephrine by 10 pathovars of *X. campestris* was found to occur by *meta* cleavage (William and Mahadevan, 1980). These compounds could serve as carbon sources for growth, but benzoate and *o*-hydroxybenzoate could not be used.

In the synthesis of aromatic amino acids by *Xanthomonas*, the first step is the combination of the multifunctional metabolites D-erythrose-4-phosphate and phosphoenol pyruvate to form 3-deoxy-D-arabino-heptulosonate-7-phosphate (DAHP), the first metabolite that is specific to the pathway to aromatic amino acids. The step is catalyzed by the enzyme DAHP synthetase. The subsequent pathway is multibranched and the activity of DAHP synthetase may be governed by various regulatory systems. Thirty-two genera were examined by Jensen et al. (1967) and found to have one of six different control patterns. The control in *X. campestris* (pvs. *campestris* and *hyacinthi*) was unusual. It was identified as sequential feedback inhibition by chorismate. Arogenate also produced some inhibition, but prephenate, phenylalanine and tyrosine had little or no effect. There was also some inhibition by tryptophan in pv. *hyacinthi* but little or none in pv. *campestris*.

Recently Whitaker et al. (1981a) extended this work. They examined, among other species, five isolates of *X. campestris* (pathovars not specified), one of *X. albilineans*, one of *X. axonopodis*, two of *Pseudomonas maltophilia*, one of "*P. gardneri*," and the type strain of *P. geniculata*, and found that all had DAHP synthetases that showed sequential feedback inhibition by chorismate and lack of stimulation by divalent cations. These properties have not been found in any other groups of bacteria. It is interesting to note that Dye (1966b) found an isolate of "*P. gardneri*" to be a synonym of *X. campestris* pv. *vesicatoria*, and De Ley (1978) considered it to be a *Xanthomonas* on the basis of DNA/rRNA hybridization. Whitaker et al. (1981a) point out that the isolate of *P. geniculata* was originally deposited in the ATCC as a strain of *P. maltophilia*.

X. ampelina was found to be different, showing less inhibition by chorismate but strong inhibition by arogenate, phenylalanine and tyrosine, and also considerable inhibition by tryptophan and prephenate. Its enzyme also requires addition of cobalt for maximum activity, unlike all of the other organisms studied.

The enzymes involved in the biosynthesis of tyrosine and phenylalanine from prephenate have also recently been studied and found to have taxonomic importance. The dehydrogenases of the two pathways to tyrosine were studied by Byng et al. (1980). They found that, although both dehydrogenase activities (i.e. prephenate dehydrogenase and arogenate dehydrogenase) were shown by nearly all species of *Pseudomonas* and *Xanthomonas* studied, the enzymes varied in the cofactor required and in their inhibition by tyrosine (feedback control). Four isolates of *X. campestris* (pvs. *malvacearum, hyacinthi, pelargonii* and *phaseoli*), one of *X. albilineans*, one of *X. axonopodis* and one of *Pseudomonas maltophilia* all showed an NAD-dependent prephenate dehydrogenase that was inhibited by tyrosine, and all showed an NAD-dependent arogenate dehydrogenase except *X. axonopodis*, which showed no activity. The arogenate dehydrogenase was not inhibited by tyrosine. *X. ampelina* was found to be very different, having dehydrogenases that were completely insensitive to tyrosine.

The dehydratases of the two pathways to phenylalanine (i.e. prephenate dehydratase and arogenate dehydratase) were studied by Whitaker et al. (1981b). They found that not only were one or the other of these enzymes absent in some groups, but that the degree of activation of prephenate dehydratase by tyrosine varied considerably between groups. The combination of these characteristics gave a fine taxonomic tool for distinguishing not only the five rRNA/DNA hybridization groups, but also two subgroups of group II and three subgroups of group III, along the same lines as found with DNA homology. Group V organisms contained arogenate dehydratase and prephenate dehydratase which was not activated by tyrosine. They could not be distinguished from group Ib using these enzymes. Assigned to group V were *P. maltophilia*, *X. campestris* (three isolates), *X. albilineans*, *X. axonopodis*, "*P. gardneri*" and *P. geniculata*.

A further metabolic pathway involving aromatic amino acids is relatively well known, at least in one pathovar of *X. campestris*. *X. campestris* pv. *pruni* requires nicotinic acid for growth, but can accept tryptophan or 3-hydroxyanthranilic acid as a substitute. This it does by synthesis of nicotinic acid from these substances by a pathway that leads to synthesis of nicotinamide adenine dinucleotide (NAD) from tryptophan (Wilson and Henderson, 1963). This pathway has been found in animals, the fungus *Neurospora crassa*, yeasts, and some *Streptomyces* species, but is absent from various bacteria that have been examined (Yanofsky, 1954; Lingens et al., 1966). The first three enzymes involved, tryptophan pyrrolase, kynurenine formamidase and kynureninase, are coordinately induced by L-tryptophan (Brown and Wagner, 1970). Wagner and Brown (1970) presented evidence that the pathway is regulated by the first step from tryptophan. The enzyme involved, tryptophan pyrrolase, is an allosteric enzyme whose activity is under feedback control (inhibition) by the end products of the pathway, NADH and NADH phosphate.

Van Eys (1960) has shown that nicotinic acid may also be replaced by NAD for growth of *X. campestris* pv. *pruni*. The final step in degradation of NAD to nicotinic acid is catalyzed by a pyridine riboside that shows unusual preference for the pyridine-riboside linkage over the purine-riboside linkage.

Catabolism of nucleoside triphosphates. Huang et al. (1975) found that ATP is bound to the cell membrane of *X. campestris* pv. *oryzae* by a mechanism that involves magnesium ions, which also stabilize this enzyme. Nucleoside tri- and diphosphates are hydrolyzed but not monophosphates. The pathway for degradation of ATP has been elucidated by Hochster and Madsen (1959). Working with *X. campestris* pv. *oryzae*, Yang et al. (1975) obtained evidence for the extracellular breakdown of deoxycytidine triphosphate to the diphosphate, monophosphate, and finally to deoxycytidine. The material was taken up by the cell and incorporated into DNA, probably by stepping up again through mono-, di-, and triphosphates.

Genetics, plasmids and bacteriophages. Genetic transformation in *X. campestris* pv. *phaseoli* has been shown to occur with DNA from strains differing in colony type or in streptomycin resistance (Corey and Starr, 1957b, c).

Lai et al. (1977a) obtained transmission of plasmids RP4 and RK2 from *Escherichia coli* to *X. campestris* pv. *vesicatoria*, conferring resistance to various antibiotics and production of penicillinase. The transconjugants were able to transmit the plasmids to other strains of the same pathovar, as well as seven or possibly eight other pathovars, and also to bacterial phytopathogens belonging to other Gram-negative genera. Stabilities of the plasmids in the transconjugants varied. Plasmids were retained by a high proportion of bacteria in leaf lesions on plants, in detached leaves kept moist at 5°C, and in dried leaves, but were rapidly lost from cells growing on agar medium (Lai et al., 1977b). Indigenous plasmids have been found in *X. campestris* pv. *manihotis* (Lin et al., 1979). Three different sizes of supercoiled DNA were obtained. Melting temperatures were identical for both plasmid and chromosomal DNA. No associated phenotypic traits were found. The same method of isolation was used on other pathovars without success.

Recently, Kado and Liu (1981), using a rapid technique, have detected "cryptic" plasmids in *X. campestris* pvs. *pruni* and *vitians*, and confirmed the findings of Lin et al. (1979) with *X. campestris* pv. *manihotis*.

Bacteriophages that attack xanthomonads have been known for many years, but work has concentrated on a few pathovars of *X. campestris*. Thus only nine nomenspecies of *X. campestris* are listed as primary phage hosts by Okabe and Goto (1963) and Vidaver (1976) in their reviews. To be useful for diagnosis and identification a phage must have the right degree of specificity and its bacterial host range must correlate with the feature of interest shown by the bacteria, e.g. pathovar, species, host range, geographical area of the bacterial host. Unfortunately this is very rarely the case. Phage typing may, however, be of interest in epidemiological studies. Dye et al. (1964) examined over 80 isolates of *X. campestris* pv. *vesicatoria* from various plant

hosts and countries. They found 17 phage groups by the use of 7 phages. Group I, containing 10 bacterial isolates, did not react with any of the phages. Hence for diagnostic purposes, even if all the phages are used, more than 10% of the pathogens will be undetected. Their results do, however, show some geographical distributions of phage types which might be of interest to an epidemiologist. Hayward (1964) obtained good correlation between physiological tests and phage reactions in *X. campestris* pv. *malvacearum*. Two clearcut groups emerged. Phages with high specificity for *X. campestris* pv. *oryzae* have been used to predict the outbreak of bacterial blight of rice (Wakimoto and Mew, 1979).

Serology. The use of serology as a taxonomic tool has not been very successful in *Xanthomonas*. Work with large numbers of different pathovars of *X. campestris* has mostly resulted in multiple groupings and often many cross-reactions. Thus Elrod and Braun (1947a), using whole cells as antigens for agglutination, found that 36 nomenspecies fell into five groups. Cross-reactions occurred, although these were fewer if the mucoid exopolysaccharide was removed by thorough washing or by growing the cells on medium low in carbohydrate. Further work on one group (Elrod and Braun, 1947b), using cross-absorbed antisera, separated two "serologically good species," "*X. juglandis*" (four isolates) and "*X. carotae*" (three isolates). The remainder fell into two groups. More recently Yano et al. (1979) examined 31 isolates belonging to 14 pathovars of *X. campestris*. They used whole cells washed in formalized saline for injection, preparations of extracellular polysaccharide as antigens, and indirect hemagglutination. They found nine serovars, but cross-reactions were evident.

When single taxa are considered, *X. campestris* pv. *vesicatoria* has received most attention, and the findings are of some interest. Lovrekovich and Klement (1965) found 2 serovars among 34 isolates. The serovars correlated with phagovars and with host of origin (tomato or pepper). Charudattan et al. (1973) also found two serovars in 72 isolates examined, but there was no correlation with pathogenicity. Schaad (1976) serologically examined ribosome preparations from 25 isolates. He was able to separate three serovars, although the third was a single isolate. There was an interesting correlation with ability to hydrolyze starch, but again no correlation with pathogenicity. Out of 16 other *Xanthomonas* isolates he included in the tests, only two isolates of *X. campestris* pv. *campestris* cross-reacted with *X. campestris* pv. *vesicatoria*.

Schaad (1978) again used ribosomes as antigens in identification of *X. campestris* pv. *campestris* by immunofluorescence. Ribosomes were chosen because they are very specific. This reduces nonspecific fluorescence, which can interfere, especially with the indirect method. The use of such preparations of cell constituents as antigens for the production of antisera may lead to taxonomically useful results in the future.

Pathogenicity. All organisms currently included in the genus *Xanthomonas* are plant pathogens. The symptoms produced vary widely among the various species and pathovars and are very helpful, sometimes essential, for identification.

X. albilineans, *X. ampelina* and *X. axonopodis* all produce systemic diseases. With *X. albilineans* there are two distinct phases: the acute, in which stems rapidly wilt and die, and the chronic, in which long narrow white "pencil lines" are seen on the leaves. Occasionally chronically diseased plants showing very little signs of disease will suddenly show acute symptoms with dramatic effect. The reasons for the change are not known, but are likely to be connected with stress to the plants as may occur when a dry spell follows a time of plentiful rain. Infection of grape with *X. ampelina* produces cracks and cankers on stems, and slight swellings may be seen due to hyperplasia of the cambial tissues (Panagopoulos, 1969). Leaf necrosis is seen occasionally, but ooze is seldom found. *X. axonopodis* produces a typical gummosis. Diseased stems look pale and, when cut, bacterial ooze is usually seen. The pathovars of *X. campestris* produce widely different diseases, sometimes on the same host species as, e.g. *X. campestris* pv. *cassavae* and pv. *manihotis* on cassava. The former produces a leaf spot that does not normally become systemic, whereas the latter readily spreads into the vascular system and causes shoots to die back.

The extent of development of symptoms and the type of symptom shown usually depends on the environmental conditions under which the plant grows and may vary considerably with host plant cultivar. The mechanisms involved in symptom production remain largely unknown. Toxins, enzymes and other metabolic products have been suggested, and a small amount of circumstantial evidence has been accumulated. Of special interest was the discovery of El Banoby and Rudolph (1979) that specific fractions of extracellular polysaccharides from several xanthomonads produced persistent water-soaked spots when introduced into the leaves of the host plants from which the bacteria came. Nonhomologous plants failed to react in this way. Hammerschlag (1979) found a toxin in a low molecular weight culture filtrate of *X. campestris* pv. *pelargonii* that is active on geranium cells in tissue culture. It may also be present in diseased plants.

Host specificity is high in pathovars of *X. campestris*. In the list of pathovars below only about half a dozen have been found to have hosts in more than one plant family. On the other hand, Schnathorst (1966) found that *X. campestris* pv. *malvacearum*, when inoculated onto *Phaseolus* sp., produced very similar early stages of infection to pv. *phaseoli*. The two pathovars were also serologically related according to earlier experiments. His experiments helped to disprove earlier ideas that host specificity could be changed by artificial passage through nonhost plants.

A critical point in the life cycle of xanthomonads, as with other plant pathogens, is transmission to a new host, particularly if a period of survival in the absence of the host is necessary. Such survival may be achieved in many ways, such as with seed, plant residues, perennial hosts, epiphytically, saprophytically in soil, and in insects (Schuster and Coyne, 1974). Many xanthomonads solve this problem by transmission with the seed of their host. These organisms are listed by Richardson (1979). The bacteria may be carried in detritus with the seed, on or in the seed coat, or deeper inside the tissues of the seed itself. Epiphytic survival is becoming well known, e.g. *X. campestris* pv. *manihotis* and *X. campestris* pv. *pruni* are both known to spend the interseasonal time epiphytically (Persley, 1978; Young, 1978). In agriculture many pathogens are most effectively carried by the activities of man, either in infected planting material, which may be symptomless (Hayward, 1974), or on tools, wheeled vehicles, and even grazing animals, as for *X. axonopodis* (Castano et al., 1964). Survival in soil saprophytically is unusual, but has been suggested by the work of Goto et al. (1978) for *X. campestris* pv. *citri*.

Xanthomonads have occasionally been detected in run-off water and ditches around fields of infected plants. Results suggest that survival in this situation would be short (Steadman et al., 1975).

Enrichment and Isolation Procedures

With plant pathogenic bacteria, the plant itself is probably the ideal "enrichment medium" and isolation is best made from plant material, using a dilution procedure unless contraindicated, on to a nonselective medium. Nutrient agar is usually satisfactory for isolation because it is not sufficiently rich to encourage growth of saprophytes. Dye (1980) recommends supplementing Difco nutrient agar with yeast extract (5 g/liter).

For better production of pigment Dye's GYCA medium* is useful. It is also a good medium for *X. campestris* pv. *cassavae*, which grows poorly and is short-lived on media containing peptone. GYCA is also good for keeping stock cultures.

For isolation of *X. albilineans* a sucrose peptone agar is best. Dye (1980) recommends YSP,† which is simpler to prepare than the various modifications of Wilbrink's agar that have been used.

For isolation of *X. ampelina* Panagopoulos (1969) recommends Difco nutrient agar. Growth on this medium is very slow and is better on a medium containing yeast extract (1.0%), galactose (2.0%), CaCO₃ (2.0%) and agar (2.0%); this medium might be used in addition to nutrient agar.

Maintenance Procedures

Stock cultures may be maintained on GYCA for periods up to about 6 months with regular subculturing at approximately monthly intervals. Less frequent subculturing is possible with most isolates at lower-than-room temperature, e.g. in a refrigerator at ~7–10°C, but care should be taken to ensure that the isolate will tolerate the temperature before entrusting all one's isolates to this treatment.

For longer periods of storage, most isolates can be freeze-dried and will keep for years (Lelliott, 1965). *Xanthomonas albilineans*, which does not keep well in the lyophilized condition can be kept on an agar slope under sterile mineral oil for periods up to ~2 years.

Differentiation of the genus **Xanthomonas** from closely related genera

Characteristics useful for differentiating *Xanthomonas* from the other genera of the family *Pseudomonadaceae* are listed earlier in Table 4.2 of the family description.

Taxonomic Comments

The rRNA/DNA hybridization carried out by Palleroni et al. (1973) showed close relationships between *Xanthomonas campestris* and *Pseudomonas maltophilia*. Later De Vos (1981) included all five species of *Xanthomonas* in such studies. His similarity map obtained with ¹⁴C-labeled 23S rRNA from the type strain of *X. campestris* showed close grouping of all eight isolates of "*Xanthomonas populi*" with all the isolates of four of the accepted species of *Xanthomonas*. The six isolates of *X. ampelina*, however, were grouped together in a distant part of the map. The four isolates of *P. maltophilia* were grouped very close to the *Xanthomonas* group, closer than to any other isolates.

Based on phenotypic characteristics, the recent proposal by Ridé and Ridé (1978, 1979) that the species previously known as "*Aplanobacterium populi*" (Ridé, 1958) be transferred to the genus *Xanthomonas* as

"*X. populi*" seems strongly supported, and there is little doubt that this organism belongs in the genus. It is to be hoped that the name will soon be validly published.

Various results and observations support the exclusion of *X. ampelina* from the genus. Its occasional production of filamentous cells, the presence of urease, utilization of tartrate and the failure to produce acid from glucose are atypical of the genus. The pigments of *X. ampelina* are also different (Starr et al., 1977), and the enzymes of aromatic amino acid synthesis also show important differences (see previous discussion on metabolism). The organism does not fit well into *Xanthomonas*, but as a monotrichous, Gram-negative, aerobic plant pathogen it can be retained at present as an anomalous species until a more suitable classification can be arranged.

There are strong reasons for inclusion of *Pseudomonas maltophilia* in the genus *Xanthomonas*, but it would necessitate alteration of the genus description. This organism is multitrichous and not a plant pathogen, normally sufficient reasons for excluding it. However, like many xanthomonads and unlike other pseudomonads it requires me-

---

* GYCA medium consists of (g/liter): yeast extract, 5.0; glucose, 5.0, CaCO₃, 40.0, and agar, 15.0. The carbonate is mixed well into the medium just before solidification.
† YSP medium consists of (g/liter): yeast extract, 5.0; sucrose, 20.0; peptone, 10.0; and agar 15.0.

thionine and is oxidase-negative. The lipid A of the cell walls contains the same three very unusual branched hydroxy fatty acids that occur in *X. campestris* (Moss et al., 1973; Rietschel et al., 1975). These do not occur in other bacteria examined so far. The polysaccharides of the cell walls of the two organisms show similar high levels of rhamnose, presence of D-galacturonic acid, low levels of 2-keto-3-deoxyoctonate and absence of heptoses (Neal and Wilkinson, 1979).

The two organisms also have strong similarities in their enzymes. Both have sequential feedback inhibition of DAHP synthetase by chorismate which lacks stimulation by divalent cations, a property not so far found in any other bacteria (Whitaker et al., 1981a). They also show the same combination of an NAD-dependent prephenate dehydrogenase that is inhibited by tyrosine and an NAD-dependent arogenate dehydrogenase that is not (Byng et al., 1980). Like xanthomonads, but unlike the other pseudomonads, *P. maltophilia* shows activity of α- and β-glucosidase, β-galactosidase and β-xylosidase (Hayward, 1977).

These similarities, some highly unusual, taken together with the nucleic acid hybridization results make a strong argument in favor of removing *P. maltophilia* from the genus *Pseudomonas* and placing it either in a modified genus *Xanthomonas* or in a new, closely related genus. *Editorial Note:* J. Swings, P. De Vos, M. Van den Motter and J. De Ley have recently proposed transfer of *P. maltophilia* to the genus *Xanthomonas* as *X. maltophilia* (Int. J. Syst. Bacteriol. *33*: 409–413, 1983).

The condensation of over 100 different pathogens into a single species, *Xanthomonas campestris*, in the 1974 edition of the *Manual* was taxonomically sound. It was, however, a shock for many plant pathologists, who needed the names for practical reasons. However, since the nomenspecies were separated almost exclusively on characteristics of pathogenicity it was logical to retain their names in the rank of pathovar. This has been done, and a list of acceptable pathovars, together with their pathotype culture numbers was published (Dye et al., 1980). These names, which are in current use by plant pathologists, should not be reused, or much confusion will result. The pathovars of *X. campestris* are listed here under the species, together with their major hosts and pathotype strains.

Further Reading

Bradbury, J.F. 1983 (in press). *Guide to plant pathogenic bacteria.* Commonwealth Mycological Institute, Kew, England. (*Note:* this reference provides host ranges and geographical distributions.)

De Vos, P. 1981. Intra- and intergeneric similarities of ribosomal ribonucleic acid cistrons in and with the genus *Pseudomonas.* Meded. Acad. Wat. Lett. Schonc Kunsten Belg. Kl. Wet. *43:* 23–60.

Murata, N. and M.P. Starr. 1973. A concept of the genus *Xanthomonas* and its species in the light of segmental homology of deoxyribonucleic acids. Phytopathol. Z. 77: 285–323.

Palleroni, N.J. 1978. *The Pseudomonas group.* Meadowfield Press, Shildon, England.

Stolp, H., M.P. Starr and N.L. Baigent. 1965. Problems in speciation of phytopathogenic pseudomonads and xanthomonads. Annu. Rev. Phytopathol. *3:* 231–264.

Young, J.M., D.W. Dye, J.F. Bradbury, C.G. Panagopoulos and C.F. Robbs. 1978. A proposed nomenclature and classification for plant pathogenic bacteria. N.Z. J. Agric. Res. *21:* 153–157. (*Note:* this reference gives the pathovar system.)

Differentiation of the species of the genus **Xanthomonas**

Table 4.39 gives the main characteristics that are of use in differentiating the species of this genus, including a tentative new member, "*Xanthomonas populi*," whose name has not yet been validated by publication in the International Journal of Systematic Bacteriology.

List of the species of the genus **Xanthomonas**

1. **Xanthomonas campestris** (Pammel 1895) Dowson 1939, 190.[AL] (*Bacillus campestris* Pammel 1895, 284.)

cam.pes′tris. L. gen. n. *campestris* of a level field; also the specific epithet of *Brassica campestris*, a host plant.

The characteristics are as described for the genus and as listed in Tables 4.39 and 4.40.

Many strains require growth factors, usually amino acids, often methionine and/or glutamic acid. Nicotinic acid is occasionally required.

The original isolates caused a vascular disease of *Brassica* spp., but the species now includes, and is subdivided into, a large number of pathovars that cause diseases on many plants. They are not distinguishable with certainty by phenotypic characterization without knowledge of their hosts. Some characteristics, such as starch and gelatin hydrolysis, are of limited use in distinguishing between groups of pathovars, but variations are usually found when larger numbers of strains, collected worldwide, are examined.

The mol% G + C of the DNA ranges from 63.5–69.2 (T_m, Bd).

Type strain: NCPPB 528.

Further comments

In the following list of currently acceptable pathovars of *X. campestris* and their pathotype strains (Dye et al., 1980) the names of host plants and their respective families are given.

Xanthomonas campestris pv. *campestris* (Pammel 1895) Dowson 1939.
    Hosts: *Brassica* spp., *Capsella bursa-pastoris, Lepidium sativum, Mat-*
thiola spp., *Raphanus sativus, Rorippa armoracia*, (fam. *Cruciferae*); *Boerhaavia erecta* (fam. *Nyctaginaceae*).
    Neopathotype strain: PDDCC 13 (NCPPB 528).

X. campestris pv. *aberrans* (Knösel 1961) Dye 1978.
    Host: *Brassica oleracea* var. *botrytis*, (fam. *Cruciferae*).
    Neopathotype strain: PDDCC 4805 (NCPPB 2986).

X. campestris pv. *alangii* (Padhya and Patel 1962) Dye 1978.
    Host: *Alangium lamarckii*, (fam. *Cornaceae*).
    Neopathotype strain: PDDCC 5717 (NCPPB 1336).

X. campestris pv. *alfalfae* (Riker, Jones and Davis 1935) Dye 1978.
    Hosts: *Medicago sativa, Melilotus indica, Pisum sativum, Phaseolus vulgaris, Trigonella foenum-graecum* (fam. *Leguminosae*).
    Neopathotype strain: PDDCC 5718 (NCPPB 2062).

X. campestris pv. *amaranthicola* (Patel, Wankar and Kulkarni 1952) Dye 1978.
    Hosts: *Amaranthus* spp. (fam. *Amaranthaceae*).
    Neopathotype strain: PDDCC 441 (NCPPB 570; ATCC 11645).

X. campestris pv. *amorphophalli* (Jindal, Patel and Singh 1972) Dye 1978.
    Host: *Amorphophallus campanulatus* (fam. *Araceae*).
    Neopathotype strain: PDDCC 3033 (NCPPB 2371).

X. campestris pv. *aracearum* (Berniac 1974) Dye 1978.
    Host: *Xanthosoma sagittifolium* (fam. *Araceae*).
    Neopathotype strain: PDDCC 5381 (NCPPB 2832).

X. campestris pv. *arecae* (Rao and Mohan 1970) Dye 1978.
    Host: *Areca catechu* (fam. *Palmaceae*).
    Neopathotype strain: PDDCC 5719 (NCPPB 2649).

Table 4.39.

Characteristics differentiating the species of the genus **Xanthomonas**[a]

Characteristics	1. X. campestris	2. X. fragariae	3. X. albilineans	4. X. axonopodis	5. X. ampelina	"X. populi"
Mucoid growth on nutrient agar + 5% glucose	+	+	−	−	−	+
Xanthomonadins produced	+	+	+	+	−[b]	
Hydrolysis of:						
Gelatin	d	+	d	−	−	−
Esculin	+	−	+	+	−	
Starch	d	+	−	+	−	
Milk proteolysis	+	−	−	−	−	Slow
H_2S from peptone	+	−	−	+	d	−
Urease activity	−	−	−	−	+	−
Maximum growth temperature, °C	35–39	33	37	35–37	30	27.5
Maximum NaCl tolerance, %	2.0–5.0	0.5–1.0	0.5	1.0	1.0	0.4–0.6
Acid production within 21 days on Dye's medium C from:						
Arabinose	+	−	−	−	+	−
Glucose, sucrose	+	+	+	+	−	+
Mannose	+	+	+	−	−	+
Galactose	+	−	d	−	+	+
Trehalose	+	−	−	+	−	+
Cellobiose	+	−	−	−	−	−
Fructose	+	+	−	−	−	+

[a] Information from Dye (1962, 1966a), Panagopoulos (1969), Ridé and Ridé (1978), and Starr et al. (1977). Symbols: see standard definitions.

[b] The pigments of *X. ampelina* have different absorption spectra from other xanthomonads and probably are not xanthomonadins.

Important Notes for Users of this Edition

1. Always read both generic and species descriptions because characters listed in the generic description are not usually listed in the species descriptions.

2. Unless otherwise indicated in footnotes to tables, the meanings of symbols are as follows:

  +   90% or more of strains are positive

  −   90% or more of strains are negative

  d   11–89% of strains are positive

  v   strain instability (*not* equivalent to "d")

  D   different reactions in different taxa (species of a genus or genera of a family)

3. All other symbols are defined in footnotes to tables.

Table 4.40.

Other characteristics of the species of the genus **Xanthomonas**[a]

Characteristics	1. *X. campestris*	2. *X. fragariae*	3. *X. albilineans*	4. *X. axonopodis*	5. *X. ampelina*
Acid production within 21 days from:[b]					
Lactose, maltose	d	−	−[c]	−[d]	−
Xylose	d	−	+		−
Ribose	d	−			−
Melibiose	d	−	−		−
Raffinose	d	−	−[c]	−[d]	−
Melezitose	d	−			
Dextrin	d	−		−[d]	
Glycogen	d	−			−
Glycerol	d	−		−[d]	d
Adonitol, mannitol, sorbitol, dulcitol, rhamnose, salicin, *meso*-inositol, inulin, α-methylglucoside	−	−	−	−	−
Utilization of:					
Acetate, citrate, malate	+				+
Propionate	+	−[e]			−[e]
Succinate	+	+			+
Lactate	+				d
dl-Tartrate	−	−	−		+
Benzoate	−[e]	−	−[e]		−[e]
Growth on nutrient agar:					
Good	+	−	−	−	−
Poor to very poor	−	+	−	+	+
No growth	−	−	+	−	−
Growth rate in culture:					
Moderate	+	−	−	−	−
Slow to very slow	−	+	+	+	+
Single polar flagellum occurs	+	+	+	+	+
Catalase	+	+	+	+	+
Nitrate reductase	−	−	−	−	−
Indole production	−	−	−	−	−
Voges-Proskauer test	−	−	−	−	−
Growth in Dye's asparagine medium	−	−	−	−	−
Arginine dihydrolase (Thornley's)	−	−	−	−	−
Lysine and ornithine decarboxylase	−	−	−	−	−

[a] For symbols, see standard definitions.

[b] In Dye's medium C (Dye, 1962) unless otherwise stated. Medium C consists of (per liter): $NH_4H_2PO_4$, 0.5 g; K_2HPO_4, 0.5 g; $MgSO_4 \cdot 7H_2O$, 0.2 g; NaCl, 5.0 g; yeast extract, 1.0 g; carbon source, 5.0 g; bromocresol purple (1.5% alcoholic solution), 0.7 ml; and agar, 12.0 g.

[c] In the medium of Hayward (1964). Hayward's medium consists of (per liter): $NH_4H_2PO_4$, 1.0 g; KCl, 0.2 g; $MgSO_4 \cdot 7H_2O$, 0.2 g; peptone, 1.0 g; carbon source, 10.0 g; bromothymol blue (1.0% aqueous solution), 0.3 ml; and agar, 15.0 g.

[d] In the medium of Starr and Garces (1950), which contains the basal mineral medium of Starr (1946), 0.5% casein hydrolysate, 0.1% yeast extract, 0.1% carbon source, and a mixture of cresol red and bromocresol purple as indicator.

[e] Inhibition occurs.

X. campestris pv. *argemones* (Srinivasan, Patel and Thirumalachar 1961a) Dye 1978.
  Host: *Argemone mexicana* (fam. *Papaveraceae*).
  Neopathotype strain: PDDCC 1617 (NCPPB 1593).

X. campestris pv. *armoraciae* (McCulloch 1929) Dye 1978.
  Hosts: *Armoracia rusticana*, *Brassica oleracea* var. *botrytis*, *B. oleracea* var. *capitata* (fam. *Cruciferae*), *Phaseolus vulgaris* (fam. *Leguminosae*).
  Neopathotype strain: PDDCC 7 (NCPPB 347).

X. campestris pv. *arracaciae* (Pereira, Paradella and Zagatto 1971) Dye 1978.
  Host: *Arracacia xanthorrhiza* (fam. *Umbelliferae*).
  Pathotype strain: PDDCC 3158 (NCPPB 2436).

X. campestris pv. *azadirachtae* (Desai, Gandhi, Patel and Kotasthane 1966) Dye 1978.
  Host: *Azadirachta indica* (fam. *Meliaceae*).
  Neopathotype strain (nonpigmented): PDDCC 3102 (NCPPB 2388).

X. campestris pv. *badrii* (Patel, Kulkarni and Dhande 1950) Dye 1978.
  Host: *Xanthium strumarium* (fam. *Compositae*), *Pisum sativum* (fam. *Leguminosae*).
  Neopathotype strain: PDDCC 571 (NCPPB 571; ATCC 11672).

X. campestris pv. *barbareae* (Burkholder 1941) Dye 1978.
  Host: *Barbarea vulgaris* (fam. *Cruciferae*).
  Neopathotype strain: PDDCC 438 (NCPPB 983; ATCC 13460).

X. campestris pv. *bauhiniae* (Padhya, Patel and Kotasthane 1965a) Dye 1978.

Host: *Bauhinia racemosa* (fam. *Leguminosae*).
Neopathotype strain: PDDCC 5720 (NCPPB 1335).

X. campestris pv. *begoniae* (Takimoto 1934) Dye 1978.
  Hosts: *Begonia* spp. (fam. *Begoniaceae*).
  Neopathotype strain: PDDCC 194 (NCPPB 1926).

X. campestris pv. *betlicola* (Patel, Kulkarni and Dhande 1951) Dye 1978.
  Hosts: *Piper betle, P. hookeri, P. longum* (fam. *Piperaceae*).
  Neopathotype strain: PDDCC 312 (NCPPB 2972; ATCC 11677).

X. campestris pv. *biophyti* (Patel, Chauhan, Kotasthane and Desai 1969) Dye 1978.
  Host: *Biophytum sensitivum* (fam. *Oxalidaceae*).
  Neopathotype strain: PDDCC 2780 (NCPPB 2228).

X. campestris pv. *blepharidis* (Srinivasan and Patel 1956) Dye 1978.
  Hosts: *Blepharis boerhaavifolia, B. molluginifolia* (fam. *Acanthaceae*).
  Neopathotype strain: PDDCC 5722 (NCPPB 1757; ATCC 17995).

X. campestris pv. *cajani* (Kulkarni, Patel and Abhyankar 1950) Dye 1978.
  Host: *Cajanus cajan* (fam. *Leguminosae*).
  Neopathotype strain: PDDCC 444 (NCPPB 573; ATCC 11639).

X. campestris pv. *cannabis* Severin 1978.
  Host: *Cannabis sativa* (fam. *Moraceae*).
  Pathotype strain: PDDCC 6570 (NCPPB 2877).

X. campestris pv. *carissae* (Moniz, Sabley and More 1964) Dye 1978.
  Hosts: *Carissa congesta, C. carandas, Thevetia nerifolia* (fam. *Apocynaceae*), *Cestrum nocturnum* (fam. *Solanaceae*).
  Neopathotype strain: PDDCC 3034 (NCPPB 2373).

X. campestris pv. *carotae* (Kendrick 1934) Dye 1978.
  Host: *Daucus carota* (fam. *Umbelliferae*).
  Neopathotype strain: PDDCC 5723 (NCPPB 1422).

X. campestris pv. *cassavae* (Wiehe and Dowson 1953) Maraite and Weyns 1979.
  Hosts: *Manihot* spp. (fam. *Euphorbiaceae*).
  Neopathotype strain: PDDCC 204 (NCPPB 101).

X. campestris pv. *cassiae* (Kulkarni, Patel and Dhande 1951) Dye 1978.
  Hosts: *Cassia tora, C. occidentalis, Cicer arietinum, Pisum sativum* (fam. *Leguminosae*).
  Neopathotype strain: PDDCC 358 (NCPPB 2973; ATCC 11638).

X. campestris pv. *celebensis* (Gaumann 1923) Dye 1978.
  Hosts: *Musa* spp. (fam. *Musaceae*).
  Neopathotype strain: PDDCC 1488 (NCPPB 1832; ATCC 19045).

X. campestris pv. *centellae* Basnyat and Kulkarni 1979.
  Host: *Centella asiatica* (fam. *Umbelliferae*).
  Pathotype strain: PDDCC 6746.

X. campestris pv. *cerealis* (Hagborg 1942) Dye 1978.
  Hosts: *Agropyron* spp., *Avena* spp., *Bromus* spp., *Hordeum* spp., *Secale cereale, Triticum* spp. (fam. *Gramineae*).
  Neopathotype strain: PDDCC 1409 (NCPPB 1944).

X. campestris pv. *citri* (Hasse 1915) Dye 1978.
  Hosts: *Aegle marmelos, Atalantia* spp. *Balsamocitrus paniculata, Casimiroa edulis, Chaetospermum glutinosa, Citropsis schweinfurthii, Citrus* spp. and hybrids, *Clausena lansium, Eremocitrus glauca, Evodia* spp. *Feronia* spp., *Feroniella* spp., *Fortunella* spp., *Hesperethusa crenulata, Limonia* spp., *Melicope triphylla, Microcitrus* spp., *Murraya erotica, Paramigyna longipedunculata, Poncirus trifoliata* and hybrids, *Severina buxifolia, Toddalia asiatica, Zanthoxylum* spp. (fam. *Rutaceae*).
  Neopathotype strain: PDDCC 24 (NCPPB 409).

X. campestris pv. *clerodendri* (Patel, Kulkarni and Dhande 1952a) Dye 1978.
  Host: *Clerodendron phlomoides* (fam. *Verbenaceae*).
  Neopathotype strain: PDDCC 445 (NCPBB 575; ATCC 11676).

X. campestris pv. *clitoriae* Pandit and Kulkarni 1979
  Host: *Clitoria biflora.* (fam. *Leguminosae*).
  Neopathotype strain: PDDCC 6574 (NCPPB 3092).

X. campestris pv. *convolvuli* (Nagarkoti, Banerjee and Swarup 1973) Dye 1978.
  Host: *Convolvulus arvensis* (fam. *Convolvulaceae*).
  Neopathotype strain: PDDCC 5380 (NCPPB 2498).

X. campestris pv. *coracanae* (Desai, Thirumalachar and Patel 1965) Dye 1978.
  Host: *Eleusine coracana* (fam. *Gramineae*).
  Neopathotype strain: PDDCC 5724 (NCPPB 1786).

X. campestris pv. *coriandri* (Srinivasan, Patel and Thirumalachar 1961b) Dye 1978.
  Hosts: *Coriandrum sativum, Foeniculum vulgare.* (fam. *Umbelliferae*).
  Pathotype strain: PDDCC 5725 (NCPPB 1758; ATCC 17996).

X. campestris pv. *corylina* (Miller, Bollen, Simmons, Gross and Barss 1940) Dye 1978.
  Hosts: *Corylus* spp. (fam. *Betulaceae*).
  Neopathotype strain: PDDCC 5726 (NCPPB 935; ATCC 19313).

X. campestris pv. *cucurbitae* (Bryan 1926) Dye 1978.
  Hosts: *Citrullus vulgaris, Cucumis sativus, Cucurbita* spp. (fam. *Cucurbitaceae*).
  Neopathotype strain: PDDCC 2299 (NCPPB 2597).

X. campestris pv. *cyamopsidis* (Patel, Dhande and Kulkarni 1953) Dye 1978.
  Host: *Cyamopsis tetragonoloba* (fam. *Leguminosae*).
  Neopathotype strain: PDDCC 616 (NCPPB 637).

X. campestris pv. *desmodii* (Patel 1949) Dye 1978.
  Host: *Desmodium diffusum* (fam. *Leguminosae*).
  Neopathotype strain: PDDCC 315 (NCPPB 481; ATCC 11640).

X. campestris pv. *desmodiigangetici* (Patel and Moniz 1948) Dye 1978.
  Host: *Desmodium gangeticum* (fam. *Leguminosae*).
  Neopathotype strain: PDDCC 577 (NCPPB 577; ATCC 11671).

X. campestris pv. *desmodiilaxiflori* Pant and Kulkarni, 1976a).
  Hosts: *Desmodium laxiflorum, Tamarindus indica* (fam. *Leguminosae*).
  Pathotype strain: PDDCC 6502 (NCPPB 3086; ITCC 29).

X. campestris pv. *desmodiirotundifolii* (Desai and Shah 1960) Dye 1978.
  Host: *Desmodium rotundifolium* (fam. *Leguminosae*).
  Neopathotype strain: PDDCC 168 (NCPPB 885).

X. campestris pv. *dieffenbachiae* (McCulloch and Pirone 1939) Dye 1978.
  Hosts: *Aglaonema robellinii, Anthurium andraeanum, Dieffenbachia* spp. (fam. *Araceae*), *Dracaena fragrans* (fam. *Liliaceae*).
  Neotype strain: PDDCC 5727 (NCPPB 1833).

X. campestris pv. *durantae* (Srinivasan and Patel 1957) Dye 1978.
  Host: *Duranta repens* (fam. *Verbenaceae*).
  Neopathotype strain: PDDCC 5728 (NCPPB 1456).

X. campestris pv. *erythrinae* (Patel, Kulkarni and Dhande 1952b) Dye 1978.
  Host: *Erythrina indica* (fam. *Leguminosae*).
  Pathotype strain: PDDCC 446 (NCPPB 578; ATCC 11679).

X. campestris pv. *esculenti* (Rangaswami and Easwaran 1962) Dye 1978.
  Host: *Hibiscus esculentus* (fam. *Malvaceae*).
  Neopathotype strain: PDDCC 5729 (NCPPB 2190).

X. campestris pv. *eucalypti* (Truman 1974) Dye 1978.
  Hosts: *Eucalyptus citriodora, E. maculata* (fam. *Myrtaceae*).
  Pathotype strain: PDDCC 5382 (NCPPB 2337).

X. campestris pv. *euphorbiae* (Sabet, Ishag and Khalil 1969) Dye 1978.
  Host: *Eurphorbia acalyphoides* (fam. *Euphorbiaceae*).
  Neopathotype strain: PDDCC 5730 (NCPPB 1828).

X. campestris pv. *fascicularis* (Patel and Kotasthane 1969a) Dye 1978.
  Host: *Corchorus fascicularis* (fam. *Tiliaceae*).
  Neopathotype strain: PDDCC 5731 (NCPPB 2230).

X. campestris pv. *fici* (Cavara 1905) Dye 1978.
  Host: *Ficus carica* (fam. *Moraceae*).
  Neopathotype strain: PDDCC 3036 (NCPPB 2372).

X. campestris pv. *glycines* (Nakano 1919) Dye 1978.
  (Syn. *X. phaseoli* var. *sojensis* (Hedges 1922) Starr and Burkholder 1942).
  Hosts: *Brunnichia cirrhosa* (fam. *Polygonaceae*), *Dolichos uniflorus, Glycine* spp. *Phaseolus lunatus, P. vulgaris* (fam. *Leguminosae*).
  Neopathotype strain: PDDCC 5732 (NCPPB 554).

X. campestris pv. *graminis* (Egli, Goto and Schmidt 1975) Dye 1978.*

---

* At the recent Fifth International Conference on Plant Pathogenic Bacteria, Egli and Schmidt presented evidence that this organism should be divided into at least four pathovars. Their proposals are currently in press.

Hosts: *Arrhenatherum elatius, Alopecurus pratensis, Dactylis glomerata, Festuca* spp., *Lolium multiflorum, L. perenne, Phleum pratense* (fam. *Gramineae*).
Pathotype strain: PDDCC 5733 (NCPPB 2700).

X. campestris pv. *guizotiae* (Yirgou 1964) Dye 1978.
Host: *Guizotia abyssinica* (fam. *Compositae*).
Neopathotype strain: PDDCC 5734 (NCPPB 1932).

X. campestris pv. *gummisudans* (McCulloch 1924) Dye 1978.
Host: *Gladiolus* sp. (fam. *Iridaceae*).
Neopathotype strain: PDDCC 5780 (NCPPB 2182).

X. campestris pv. *hederae* (Arnaud 1920) Dye 1978.
Host: *Hedera helix* (fam. *Araliaceae*).
Neopathotype strain: PDDCC 453 (NCPPB 939).

X. campestris pv. *heliotropii* (Sabet, Ishag and Khalil 1969) Dye 1978.
Host: *Heliotropium aegypticum, H. sudanicum* (fam. *Boraginaceae*).
Neopathotype strain: PDDCC 5778 (NCPPB 2057).

X. campestris pv. *holcicola* (Elliott 1930) Dye 1978.
Hosts: *Sorghum halepense, S. vulgare* (fam. *Gramineae*).
Neopathotype strain: PDDCC 3103 (NCPPB 2417).

X. campestris pv. *hordei* (Hagborg 1942) Dye 1978.
Hosts: *Bromus inermis, Hordeum* spp. (fam. *Gramineae*).
Neopathotype strain: PDDCC 5735 (NCPPB 2389).

X. campestris pv. *hyacinthi* (Wakker 1883) Dye 1978.
Hosts: *Hyacinthus orientalis* (fam. *Liliaceae*).
Neopathotype strain: PDDCC 189 (NCPPB 599; ATCC 19314).

X. campestris pv. *incanae* (Kendrick and Baker 1942) Dye 1978.
Host: *Matthiola incanae* (fam. *Cruciferae*).
Neopathotype strain: PDDCC 574 (NCPPB 937; ATCC 13462).

X. campestris pv. *ionidii* (Padhya and Patel 1963a) Dye 1978.
Host: *Ionidium heterophyllum* (fam. *Violaceae*).
Neopathotype strain: PDDCC 5736 (NCPPB 1334).

X. campestris pv. *juglandis* (Pierce 1901) Dye 1978.
Hosts: *Juglans* spp. (fam. *Juglandaceae*).
Neopathotype strain: PDDCC 35 (NCPPB 411).

X. campestris pv. *khayae* (Sabet 1959) Dye 1978.
Hosts: *Khaya senegalensis, K. grandifoliola* (fam. *Meliaceae*).
Neopathotype strain: PDDCC 671 (NCPPB 536).

X. campestris pv. *lantanae* (Srinivasan and Patel 1957) Dye 1978.
Host: *Lantana camara* var. *aculeata* (fam. *Verbenaceae*).
Neopathotype strain: PDDCC 5737 (NCPPB 1455).

X. campestris pv. *laureliae* (Dye 1963) Dye 1978.
Host: *Laurelia novae-zelandiae* (fam. *Monimiaceae*).
Neopathotype strain: PDDCC 84 (NCPPB 1155).

X. campestris pv. *lawsoniae* (Patel, Bhatt and Kulkarni 1951) Dye 1978.
Host: *Lawsonia alba* (fam. *Lythraceae*).
Pathotype strain: PDDCC 319 (NCPPB 579; ATCC 11674).

X. campestris pv. *leeana* (Patel and Kotasthane 1969b) Dye 1978.
Host: *Leea edgeworthii* (fam. *Vitaceae*).
Neopathotype strain: PDDCC 5738 (NCPPB 2229).

X. campestris pv. *lespedezae* (Ayres, Lefebvre and Johnson 1939) Dye 1978.
Host: *Lespedeza* spp. (fam. *Leguminosae*).
Neopathotype strain: PDDCC 439 (NCPPB 993; ATCC 13463).

X. campestris pv. *maculifoliigardeniae* (Ark and Barrett 1946) Dye 1978.
Hosts: *Gardenia* spp., *Ixora coccinea* (fam. *Rubiaceae*).
Neopathotype strain: PDDCC 318 (NCPPB 971).

X. campestris pv. *malvacearum* (Smith 1901) Dye 1978.
Hosts: *Gossypium* spp., (fam. *Malvaceae*), *Ceiba pentandra* (fam. *Bombacaceae*), *Thespesia lambas* (fam. *Malvaceae*).
Neopathotype strain: PDDCC 5739 (NCPPB 633).

X. campestris pv. *mangiferaeindicae* (Patel, Moniz, and Kulkarni 1948) Robbs, Ribeiro and Kimura 1974.
Hosts: *Anacardium occidentale, Mangifera indica, Spondias mangifera* (fam. *Anacardiaceae*).
Neopathotype strain: PDDCC 5740 (NCPPB 490; ATCC 11637).

X. campestris pv. *manihotis* (Berthet and Bondar 1915) Dye 1978.
Hosts: *Manihot* spp. (fam. *Euphorbiaceae*).
Neopathotype strain (nonpigmented): PDDCC 5741 (NCPPB 1834).

X. campestris pv. *martyniicola* (Moniz and Patel 1958) Dye 1978.
Host: *Martynia diandra* (fam. *Martyniaceae*).
Neopathotype strain: PDDCC 82 (NCPPB 1148).

X. campestris pv. *melhusii* (Patel, Kulkarni and Dhande 1952b) Dye 1978.
Host: *Tectona grandis* (fam. *Verbenaceae*).
Neopathotype strain: PDDCC 619 (NCPPB 994; ATCC 11644).

X. campestris pv. *merremiae* Pant and Kulkarni 1976b.
Host: *Merremia gangetica* (fam. *Convolvulaceae*).
Neopathotype strain: PDDCC 6747 (NCPPB 3114; ITCC 30).

X. campestris pv. *musacearum* (Yirgou and Bradbury 1968) Dye 1978.
Hosts: *Ensete ventricosum, Musa* spp. (fam. *Musaceae*).
Pathotype strain: PDDCC 2870 (NCPPB 2005).

X. campestris pv. *nakataecorchori* (Padhya and Patel 1963b) Dye 1978.
Host: *Corchorus acutangulus* (fam. *Tiliaceae*).
Neopathotype strain: PDDCC 5742 (NCPPB 1337).

X. campestris pv. *nigromaculans* (Takimoto 1927) Dye 1978.
Host: *Arctium lappa* (fam. *Compositae*).
Neopathotype strain: PDDCC 80 (NCPPB 1935; ATCC 23390).

X. campestris pv. *olitorii* (Sabet 1957) Dye 1978.
Host: *Corchorus olitorius* (fam. *Tiliaceae*).
Neopathotype strain: PDDCC 359 (NCPPB 464).

X. campestris pv. *oryzae* (Ishiyama 1922) Dye 1978.
Hosts: *Isachne globosa, Leersia* spp., *Leptochloa* spp., *Oryza sativa, Oryza* spp., *Phalaris arundinacea, Phragmites communis, Zizania aquatica* (fam. *Gramineae*).
Neopathotype strain: PDDCC 3125 (NCPPB 3002).

X. campestris pv. *oryzicola* (Fang, Ren, Chen, Chu, Faan and Wu 1957) Dye 1978.
Hosts: *Oryza sativa, Oryza* spp., *Leersia hexandra* (weak) (fam. *Gramineae*).
Neopathotype strain: PDDCC 5743 (NCPPB 1585).

X. campestris pv. *papavericola* (Bryan and McWhorter 1930) Dye 1978.
Hosts: *Meconopsis baileyi, Papaver* spp. (fam. *Papaveraceae*).
Neopathotype strain: PDDCC 220 NCPPB 2970; ATCC 14179

X. campestris pv. *passiflorae* (Pereira 1969) Dye 1978.
Host: *Passiflora edulis* (fam. *Passifloraceae*).
Neopathotype strain: PDDCC 3151.

X. campestris pv. *patelii* (Desai and Shah 1959) Dye 1978.
Host: *Crotalaria juncea* (fam. *Leguminosae*).
Neopathotype strain: PDDCC 167 (NCPPB 840).

X. campestris pv. *pedalii* (Patel and Jindal 1972) Dye 1978.
Host: *Pedalium murex* (fam. *Pedaliaceae*).
Neopathotype strain (nonpigmented): PDDCC 3030 (NCPPB 2368).

X. campestris pv. *pelargonii* (Brown 1923) Dye 1978.
Host: *Geranium* spp., *Pelargonium* spp. (fam. *Geraniaceae*).
Neopathotype strain: PDDCC 4321 (NCPPB 2985).

X. campestris pv. *phaseoli* (Smith 1897) Dye 1978.
Hosts: *Lablab purpureus* (syn. *Dolichos lablab*), *Lupinus polyphyllus, Phaseolus lunatus, Phaseolus vulgaris* (fam. *Leguminosae*).
Neopathotype strain: PDDCC 5834 (NCPPB 3035; ATCC 9563).

X. campestris pv. *phleipratensis* (Wallin and Reddy 1945) Dye 1978.
Host: *Phleum pratense* (fam. *Gramineae*).
Neopathotype strain: PDDCC 5744 (NCPPB 1837).

X. campestris pv. *phormiicola* (Takimoto 1933) Dye 1978.
Host: *Phormium tenax* (fam. *Liliaceae*).
Neopathotype strain: PDDCC 4294 (NCPPB 2983).

X. campestris pv. *phyllanthi* (Sabet, Ishag and Khalil 1969) Dye 1978.
Host: *Phyllanthus niruri* (fam. *Euphorbiaceae*).
Neopathotype strain: PDDCC 5745 (NCPPB 2066).

X. campestris pv. *physalidicola* (Goto and Okabe 1958) Dye 1978.
Host: *Physalis alkekengi* var. *francheti* (fam. *Solanaceae*).
Neopathotype strain: PDDCC 586 (NCPPB 761).

X. campestris pv. *physalidis* (Srinivasan, Patel and Thirumalachar 1962) Dye 1978.
Hosts: *Physalis minima, P. peruviana* (fam. *Solanaceae*).
Pathotype strain: PDDCC 5746 (NCPPB 1756; ATCC 17994).

X. campestris pv. *pisi* (Goto and Okabe 1958) Dye 1978.
Host: *Pisum sativum* (fam. *Leguminosae*).
Neopathotype strain: PDDCC 570 (NCPPB 762).

X. campestris pv. *plantaginis* (Thornberry and Anderson 1937) Dye 1978.
Hosts: *Plantago* spp. (fam. *Plantaginaceae*).
Reference strain (nonpathogenic): PDDCC 1028 (NCPPB 1061; ATCC 23382).

X. campestris pv. *poinsettiicola* (Patel, Bhatt and Kulkarni 1951) Dye 1978.
Hosts: *Euphorbia pulcherrima, E. milii, Manihot esculenta* (fam. *Euphorbiaceae*).
Neopathotype strain: PDDCC 5779 (NCPPB 581; ATCC 11643).

X. campestris pv. *pruni* (Smith 1903) Dye 1978.
Hosts: *Prunus* spp., *Sorbus japonica* (fam. *Rosaceae*).
Neopathotype strain: PDDCC 51 (NCPPB 416; ATCC 19316).

X. campestris pv. *punicae* (Hingorani and Singh 1959) Dye 1978.
Host: *Punica granatum* (fam. *Punicaceae*).
Neopathotype strain: PDDCC 360 (NCPPB 466).

X. campestris pv. *raphani* (White 1930) Dye 1978.
Hosts: *Brassica* spp., *Raphanus sativus*, (fam. *Cruciferae*), *Capsicum annuum, Lycopersicon esculentum, Nicotiana tabacum* (fam. *Solanaceae*).
Neopathotype strain: PDDCC 1404 (NCPPB 1946).

X. campestris pv. *rhynchosiae* (Sabet, Ishag and Khalil 1969) Dye 1978.
Hosts: *Lupinus termis, Mucuna pruriens* (syn. *Stizolobium alterrimum*), *Rhynchosia memnonia* (fam. *Leguminosae*).
Neopathotype strain: PDDCC 5748 (NCPPB 1827).

X. campestris pv. *ricini* (Yoshi and Takimoto 1928) Dye 1978.
Hosts: *Ricinus communis* (fam. *Euphorbiaceae*).
Neopathotype strain (nonpathogenic): PDDCC 5747 (NCPPB 1063; ATCC 19317).

X. campestris pv. *secalis* (Reddy, Godkin and Johnson 1924) Dye 1978.
Hosts: *Secale cereale; Hordeum* and *Triticum* spp. may become artificially infected (fam. *Gramineae*).
Neopathotype strain: PDDCC 5749 (NCPPB 2822).

X. campestris pv. *sesami* (Sabet and Dowson 1960) Dye 1978.
Host: *Sesamum orientale* (fam. *Pedaliaceae*).
Neopathotype strain: PDDCC 621 (NCPPB 631).

X. campestris pv. *sesbaniae* (Patel, Kulkarni and Dhande 1952a) Dye 1978.
Host: *Sesbania aegyptiaca* (fam. *Leguminosae*).
Neopathotype strain: PDDCC 367 (NCPPB 582; ATCC 11675).

X. campestris pv. *spermacoces* (Srinivasan and Patel 1956) Dye 1978.
Host: *Spermacoce hispida* (fam. *Rubiaceae*).
Pathotype strain: PDDCC 5751 (NCPPB 1760; ATCC 17998).

X. campestris pv. *tamarindi* (Patel, Bhatt and Kulkarni 1951) Dye 1978.
Hosts: *Caesalpina sepiaria, Tamarindus indica* (fam. *Leguminosae*).
Pathotype strain: PDDCC 572 (NCPPB 584; ATCC 11673).

X. campestris pv. *taraxaci* (Niederhauser 1943) Dye 1978.
Host: *Taraxacum bicorne* (fam. *Compositae*).
Neopathotype strain: PDDCC 579 (NCPPB 940; ATCC 19318).

X. campestris pv. *tardicrescens* (McCulloch 1937) Dye 1978.
Hosts: *Belamcanda* sp., *Iris* spp. (fam. *Iridaceae*).
Neopathotype strain: PDDCC 4295 (NCPPB 2984 (Goto 1977)).

X. campestris pv. *theicola* Uehara and Arai 1980.
Host: *Camellia sinensis* (fam. *Theaceae*).
Pathotype strain: PDDCC 6774.

X. campestris pv. *thirumalacharii* (Padhya and Patel 1964) Dye 1978.
Host: *Triumfetta pilosa* (fam. *Tiliaceae*).
Neopathotype strain: PDDCC 5852 (NCPPB 1452).

X. campestris pv. *translucens* (Jones, Johnson and Reddy 1917) Dye 1978.
Hosts: *Hordeum* spp. and other cereals. (fam. *Gramineae*).
Neopathotype strain: PDDCC 5752 (NCPPB 973; ATCC 19319).

X. campestris pv. *tribuli* (Srinivasan and Patel 1956) Dye 1978.
Host: *Tribulus terrestris* (fam. *Zygophyllaceae*).
Neopathotype strain: PDDCC 5753 (NCPPB 1454).

X. campestris pv. *trichodesmae* (Patel, Kulkarni and Dhande 1952b) Dye 1978.
Host: *Trichodesma zeylanicum* (fam. *Boraginaceae*).
Pathotype strain: PDDCC 5754 (NCPPB 585; ATCC 11678).

X. campestris pv. *undulosa* (Smith, Jones and Reddy 1919) Dye 1978.
Hosts: *Triticum* spp. (fam. *Gramineae*), *Secale cereale, Hordeum* spp. by inoculation.
Neopathotype strain: PDDCC 5755 (NCPPB 2821).

X. campestris pv. *uppalii* (Patel 1948) Dye 1978.
Hosts: *Ipomoea muricata* (fam. *Convolvulaceae*), *Tropaeolum majus* (fam. *Tropaeolaceae*).

Neopathotype strain: PDDCC 5756 (NCPPB 586; ATCC 11641).

X. campestris pv. *vasculorum* (Cobb 1893) Dye 1978.
Hosts: *Bambusa vulgaris, Brachiaria mutica, Coix lacryma-jobi, Panicum maximum, Pennisetum purpureum, Saccharum officinarum, Sorghum* spp., *Thysanolaena maxima, Zea mays* (fam. *Gramineae*), *Cocos nucifera, Dictyosperma alba* (fam. *Palmae*).
Neopathotype strain: PDDCC 5757 (NCPPB 796).

X. campestris pv. *vernoniae* (Patel, Desai and Patel 1968) Dye 1978.
Host: *Vernonia cinerea* (fam. *Compositae*).
Neopathotype strain: PDDCC 5758 (NCPPB 1787).

X. campestris pv. *vesicatoria* (Doidge 1920) Dye 1978.
Hosts: *Capsicum* spp., *Datura stramonium, Hyoscyamus* spp., *Lycium* spp., *Lycopersicon* spp., *Nicotiana rustica, Nicandra physalodes, Physalis minima, Solanum* spp. (fam. *Solanaceae*).
Neopathotype strain: PDDCC 63 (NCPPB 422).

X. campestris pv. *vignaeradiatae* (Sabet, Ishag and Khalil 1969) Dye 1978.
Hosts: *Lablab purpureus* (syn. *Dolichos lablab*), *Vigna radiata* (fam. *Leguminosae*).
Neopathotype strain: PDDCC 5759 (NCPPB 2058).

X. campestris pv. *vignicola* (Burkholder 1944) Dye 1978.
Hosts: *Phaseolus vulgaris, Vigna pubigera, V. unguiculata* (fam. *Leguminosae*).
Neopathotype strain: PDDCC 333 (NCPPB 1838; ATCC 11648).

X. campestris pv. *vitians* (Brown 1918) Dye 1978.
Hosts: *Lactuca* spp. (fam. *Compositae*).
Neopathotype strain: PDDCC 336 (NCPPB 976; ATCC 19320).

X. campestris pv. *viticola* (Nayudu 1972) Dye 1978.
Hosts: *Azadirachta indica* (fam. *Meliaceae*), *Phyllanthus maderaspatensis* (fam. *Euphorbiaceae*), *Vitis vinifera* (fam. *Vitaceae*).
Neopathotype strain (nonpigmented): PDDCC 3867 (NCPPB 2475).

X. campestris pv. *vitiscarnosae* (Moniz and Patel 1958) Dye 1978.
Hosts: *Vitis carnosa* (fam. *Vitaceae*).
Neopathotype strain: PDDCC 90 (NCPPB 1149).

X. campestris pv. *vitistrifoliae* (Padhya, Patel and Kotasthane 1965b) Dye 1978.
Host: *Vitis trifolia* (fam. *Vitaceae*).
Neopathotype strain: PDDCC 5761 (NCPPB 1451).

X. campestris pv. *vitiswoodrowii* (Patel and Kulkarni 1951) Dye 1978.
Host: *Vitis woodrowii* (fam. *Vitaceae*).
Neopathotype strain (nonpigmented): PDDCC 3965 (NCPPB 1014; ATCC 11636).

X. campestris pv. *zantedeschiae* (Joubert and Truter 1972) Dye 1978.
Host: *Zantedeschia aethiopica* (fam. *Araceae*).
Neopathotype strain: PDDCC 2372 (NCPPB 2978).

X. campestris pv. *zinniae* (Hopkins and Dowson 1949) Dye 1978.
Host: *Zinnia elegans* (fam. *Compositae*).
Neopathotype strain: PDDCC 5762 (NCPPB 2439).

2. **Xanthomonas fragariae** Kennedy and King 1962, 875.[AL]
fra.gar′i.ae. M.L. noun *Fragaria* generic name of strawberry; M.L. gen. n. *fragariae* of strawberry.

The characteristics are as described for the genus and as listed in Tables 4.39 and 4.40.

Amino acids are required for growth, but the exact requirement is unknown.

Causes a leaf spot disease of strawberry.

The mol% G + C of one strain is 62.6 (Bd); for a second strain it is 63.3 (T_m).

Type strain: NCPPB 1469 (PDDCC 5715).

3. **Xanthomonas albilineans** (Ashby 1929) Dowson 1943, 11.[AL]
(*Bacterium albilineans* Ashby 1929, 135.)
al.bi.lin′e.ans. L. adj. *albus* white; L. part. adj. *lineans* striping; M.L. adj. *albilineans* white striping.

The characteristics are as described for the genus and as listed in Tables 4.39 and 4.40.

Colonies are yellowish buff on Dye's YSP medium or similar sucrose-peptone media.

Glutamic acid and methionine are required for growth.

Causes leaf scald disease of sugar cane (*Saccharum officinarum*).

Other natural hosts are: *Brachiaria piligera, Imperata cylindrica* var. *major, Paspalum conjugatum* (fam. *Gramineae*). The following can be inoculated successfully, but are probably not natural hosts: *Bambusa vulgaris, Coix lacryma-jobi, Cymbopogon citratus, Panicum maximum, Paspalum dilatatum, P. paniculatum, P. scrobiculatum* var. *commersonii, Pennisetum purpureum, Sorghum halepense, S. verticilliflorum, Thysanolaena agrostis, T. maxima, Zea mays* (fam. *Gramineae*).

The mol% G + C of the DNA is 63.1 and 63.5 (Bd) for two strains, and 64.5 (T_m) for a third strain.

Type strain: PDDCC 196 (NCPPB 2969).

4. **Xanthomonas axonopodis** Starr and Garces 1950, 81.[AL] (*Xanthomonas axonoperis* (*sic*) Starr and Garces 1950, 81 (an orthographic variant).)

ax.on.o′pod.is. M.L. n. *Axonopus* generic name of a grass; M.L. gen. n. *axonopodis* of *Axonopus*.

The characteristics are as described for the genus and as listed in Tables 4.39 and 4.40.

Amino acids are required for growth, but the exact requirement is unknown.

Causes gummosis of *Axonopus scoparius, A. micay, A. compressus*

and *A. affinis*. Hosts by inoculation are *Digitaria decumbens, Hypharrhenia rufa, Panicum* sp. and *Saccharum officinarum* (fam. *Gramineae*).

The mol% G + C of two isolates is 62.6 and 64.4 (Bd); a third isolate has a value of 65 (T_m).

Type strain: ATCC 19312 (PDDCC 50; NCPPB 457).

5. **Xanthomonas ampelina** Panagopoulos 1969, 75.[AL]

am.pe.li′na. Gr. n. *ampelos* the grape vine; M.L. fem. adj. *ampelina* of the vine.

The characteristics are as described for the genus and as listed in Tables 4.39 and 4.40.

A brown diffusible pigment is produced on yeast extract-galactose-chalk agar (1.0%, 2.0%, 2.0% and 2.0%, respectively). Growth is better on this medium than on most other media.

Glutamic acid is required for growth.

Causes bacterial blight and canker of grape vine (*Vitis vinifera*). *Erwinia vitivora*, a synonym of *E. herbicola*, was for many years wrongly thought to be the causal agent.

The mol% G + C of the DNA of 4 strains is 68.1–68.5 (T_m); the type strain has a value of 70.8 (T_m).

Type strain: NCPPB 2217 (PDDCC 4298).

Other organisms probably belonging to the genus **Xanthomonas**

"**Xanthomonas populi**" (Ridé 1958) Ridé and Ridé 1978, 310. (*Aplanobacter populi* Ridé 1958, 2797; *Aplanobacterium populi* Ridé 1958, 2797, an orthographic variant.)

pop′u.li. M.L. n. *Populus* generic name of poplar tree; M.L. gen. noun *populi* of poplar.

This species seems to conform to the definition of the genus, although it is not known whether xanthomonadins are produced. See Table 4.39.

Colonies on Ridé's LPGA medium* are cream colored, mucoid, and slow growing. On medium S† growth is very slow but colonies are a definite yellow. No growth occurs on nutrient agar.

Flagellated cells are few, and the single polar flagellum was detected by immunofluorescence.

Oxidase-negative. Catalase-positive. Nitrate reductase negative.

A hypersensitive reaction has been reported on tobacco.

Causes a canker of poplar (*Populus* spp.) Pathogenic to all species in section *Aigeros, Deltoides, Tacamahaca, Leuce* (subsections *Albidae* and *Trepidae*) and their hybrids.

The mol% G + C of the DNA is 62–65 (T_m).

Reference strain: NCPPB 2959.

Genus III. **Frateuria** *Swings, Gillis, Kersters, De Vos, Gosselé and De Ley, 1980, 547*[VP]

JEAN SWINGS, JOZEF DE LEY AND MONIQUE GILLIS

Fra.teur′i.a. M.L. fem. n. *Frateuria* named after the late Joseph Frateur (1903–1974) eminent Belgian microbiologist.

Regular straight rods, 0.5–0.7 μm in diameter and 0.7–3.5 μm in length, occurring singly or in pairs. Gram-negative. Generally **motile by polar flagella** or nonmotile. **Obligately aerobic.** Optimum temperature for growth 25–30°C. Colonies on mannitol-yeast extract-peptone (MYP) agar§ are yellow to orange. On glucose-yeast extract-CaCO₃ (GYC) agar¶ most strains produce a typical brown water-soluble pigment. **Oxidase-negative. Grows at pH 3.6.** No nitrate reduction. Starch and gelatin are not hydrolyzed. **H₂S is produced.** Chemoorganotrophic. Acid is produced from ethanol and a number of C sources. On D-glucose and D-xylose, the pH drops below 4.0. From D-glucose, 2-keto- and 2,5-diketogluconic acids are formed, but not 5-ketogluconic acid. **No requirements for growth factors.** Isolated from *Lilium auratum* and from the fruit of *Rubus parvifolius* (raspberry) in Japan. The mol% G + C of the DNA is 62–64 (T_m).

Type species: *Frateuria aurantia* Swings, Gillis, Kersters, De Vos, Gosselé and De Ley 1980, 555.

Further Descriptive Information

On GYC agar, cells of *Frateuria* appear mostly as rods, rarely as ovoids, never in chains. Some strains form filaments. *Frateuria* grows luxuriantly on this medium and forms typical dark coffee-brown colonies which measure 2–5 mm in diameter with a dark center, and are glistening or rough, flat or raised and circular with a regular edge. Colonies on MYP agar are yellow to orange, glistening or rough, regular or irregular, highly convex, measuring 1–3 mm in diameter. All strains can be grown in a defined Hoyer medium‖ with ammonium sulfate as the sole nitrogen source and mannitol as the carbon source. Luxuriant growth is obtained when mixtures of amino acids are supplied instead

---

* LPGA medium (Ridé and Ridé, 1978) consists of (g/liter): yeast extract (Difco), 5.0; Bacto peptone (Difco), 5.0; glucose, 10.0; and agar, 20.0. The pH is adjusted to 7.2.

† S medium of Ridé and Ridé (1978) consists of (g/liter): yeast extract (Difco), 2.0; Bacto peptone (Difco), 5.0; NaCl, 5.0; sucrose, 50.0; and agar, 20.0. The pH is adjusted to 7.2.

§ Mannitol-yeast extract-peptone (MYP) agar has the following composition (g/liter): mannitol, 25.0; yeast extract, 5.0; peptone, 3.0 and agar, 25.0.

¶ Glucose-yeast extract-CaCO₃ (GYC) agar consists of (g/liter): D-glucose, 50.0; yeast extract, 10.0; CaCO₃, 30.0 and agar, 25.0.

‖ The Hoyer medium consists of (g/liter): mannitol, 30.0; (NH₄)₂SO₄, 1.0; K₂HPO₄, 0.1; KH₂PO₄, 0.9; MgSO₄·7H₂O, 0.25 and FeCl₃, 0.005.

of NH_4^+. Some strains grow in yeast extract broth or in peptone broth, but for routine cultivation of *Frateuria* D-mannitol and D-glucose are incorporated in the media as carbon sources. In a mannitol-salts-vitamins medium,† several single amino acids are used by *Frateuria* as sole nitrogen sources. L-Alanine, L-aspartic acid, L-glutamic acid or L-proline are not utilized as carbon and nitrogen sources by *Frateuria*.

The following physiological tests are positive for all the strains: catalase, production of H_2S, ketogenesis on glycerol and on D-mannitol, and oxidation of DL-lactate to CO_2 and water. The following tests are universally negative: oxidase, reduction of nitrates, indole formation, gelatin hydrolysis and oxidation of acetate.

Injected into apples or pears, *Frateuria* causes browning and rotting symptoms (Vanden Abeele et al., 1980).

Nine authentic *Frateuria* strains have been mentioned so far in the literature; they were all isolated in Japan.

Enrichment and Isolation Procedures

Yamada et al. (1976) used the following medium for the enrichment of *Frateuria* from raspberries (per liter of 10% potato extract): D-glucose, 10.0 g; ethanol, 5.0 ml; yeast extract, 5.0 g; peptone, 3.0 g and acetic acid, 0.3 ml. The pH was 4.5 and the incubation temperature 30°C. Isolation of *Frateuria* is possible on a $CaCO_3$—containing medium, where the colonies dissolve the $CaCO_3$.

Maintenance Procedures

Originally *Frateuria* was grown and maintained on a medium containing (per liter of 20% potato extract): D-glucose, 5.0 g; glycerol, 15.0 g; yeast extract, 30.0 g; peptone, 20.0 g and $CaCO_3$, 10.0 g. *Frateuria* can be maintained for 5 months at 4°C on slants of this medium contained in screw-capped vials. *Frateuria* survives the ordinary lyophilization procedure for many years.

Differentiation of the genus **Frateuria** from other genera

Frateuria resembles the acetic acid bacteria in its ability to oxidize ethanol to acetic acid and to grow at pH 3.6. Table 4.41 shows the characteristics that can be used to differentiate the genus *Frateuria* from the genera *Acetobacter* and *Gluconobacter*. *Frateuria* resembles mostly the polarly-flagellated *Gluconobacter* but it oxidizes lactate, requires no vitamins and does not produce 5-ketogluconic acid. *Acetobacter* can be differentiated from *Frateuria* by its peritrichous flagellation and its oxidation of acetate. *Frateuria* makes the ubiquinone Q_8, whereas *Gluconobacter* possesses ubiquinone Q_{10} and *Acetobacter* Q_9 or Q_{10}.

Taxonomic Comments

Until recently 11 "*Acetobacter aurantius*" strains were considered as polarly flagellated acetic acid bacteria intermediate between *Acetobacter* and *Gluconobacter* (Asai, 1968). Yamada et al. (1976) already suspected these strains to form a separate new genus mainly because of its typical ubiquinone Q_8.

DNA/rRNA hybridization studies, examination of protein electrophoregrams, and phenotypic analysis show that these organisms belong in three taxa (Gillis and De Ley, 1970; Swings et al., 1980). Strain IFO 3248 belongs in *Acetobacter*; strain IFO 3246 is located in *Gluconobacter*.

The remaining nine strains form the tight cluster of *Frateuria*. Their phenotypic features and protein gel electrophoregrams are very similar, suggesting a high genetic homogeneity. DNA/rRNA hybridizations show that *Frateuria* is not related to either *Acetobacter* or *Gluconobacter*; moreover, it does not belong in the rRNA superfamily IV (*sensu* De Ley, 1978) of which the family *Acetobacteraceae* constitutes a separate branch (Gillis and De Ley, 1980; Swings et al., 1980). Rather, *Frateuria* belongs in the rRNA superfamily II (*sensu* De Ley 1978) together with the genera *Xanthomonas, Azotobacter, Azomonas* and the pseudomonads of section I (De Vos, 1980; Swings et al., 1980). Within this rRNA superfamily the *Frateuria* strains constitute a separate genus with $T_{m(e)}$ values of about 72°C when hybridized with [^{14}C]rRNA from *Xanthomonas campestris*. With this [^{14}C]rRNA all members of the genus *Xanthomonas* form DNA/rRNA hybrids with $T_{m(e)}$ values above 79°C. Thus, although *Frateuria* seems to be closely related to *Xanthomonas*, it is nevertheless clearly distinct from that genus.

Further Reading

Swings, J., M. Gillis, K. Kersters, P. De Vos, F. Gosselé and J. De Ley. 1980. *Frateuria*, a new genus for "*Acetobacter aurantius*." Int. J. Syst. Bacteriol. *30:* 547–556.

List of the species of the genus **Frateuria**

1. **Frateuria aurantia** Swings, Gillis, Kersters, De Vos, Gosselé and De Ley 1980, 547.^{VP}

au.ran'tia. L. v. *aurare* to overlay with gold; M.L. adj. *aurantius* gold colored; refers to the gold-yellow color of the strains on MYP agar.

The description of the species is as for the genus. See Tables 4.41

and 4.42 for additional characteristics.

Isolated from *Lilium auratum* and from the fruit of *Rubus parvifolius* in Japan.

The mol% G + C of the DNA is 62–64 (T_m).

Type strain: IFO 3245.

---

† The medium for testing the utilization of L-amino acids as sole nitrogen sources consists of D-mannitol, 30.0 g; *d*-biotin, 0.001 g; Ca-D-panthothenate, 0.001 g; thiamine, 0.001 g; pyridoxal hydrochloride, 0.0015 g; niacin, 0.0015 g; riboflavin, 0.0015 g; *p*-aminobenzoic acid, 0.001 g; vitamin B_{12}, 0.001 g; folic acid, 0.001 g; salt solution A, 5 ml; salt solution B, 5 ml; L-amino acid, 1 g in 1 liter of 0.2 M Tris-maleate buffer, pH 5.4. Salt solution A contains KH_2PO_4, 10.0 g and K_2HPO_4, 10.0 g/liter of distilled water. Salt solution B contains $MgSO_4 \cdot 7H_2O$, 4.0 g; NaCl, 0.2 g; $FeSO_4 \cdot 7H_2O$, 2.0 g; $MnSO_4 \cdot H_2O$, 0.15 g/liter water.

Table 4.41.

Characteristics differentiating the genera **Frateuria, Gluconobacter** *and* **Acetobacter**[a]

Characteristics	Frateuria	Glucono-bacter	Aceto-bacter
Flagellar arrangement in motile strains:			
Polar	+	+	−
Peritrichous	−	−	+
Overoxidation of ethanol	−	−	+
Oxidation of DL-lactate to CO_2 and H_2O	+	−	+
Oxidation of acetate to CO_2 and H_2O	−	−	+
Ketogenesis	+	+	D
Formation of brown water-soluble pigments on GYC agar	+[b]	−	−
Growth factors required	−	+	D
Formation of H_2S	+	−	−
Products formed from D-glucose:			
2-Ketogluconic acid	+	+	D
5-Ketogluconic acid	−	+	D
2,5-Diketogluconic acid	+	D	D
Acetylmethylcarbinol (Voges-Proskauer)	−	D	D
Type of ubiquinone formed:			
Q_8	+	−	−
Q_9	−	+	D
Q_{10}	−	−	D
	−		
Growth in presence of 30% D-glucose	+[b]	−	−
Growth on Frateur's Hoyer mannitol medium	+	−	−
Acid produced from:			
D-Arabinose	+[b]	+	−
i-Inositol	+	D	−
Maltose	−	D	−
m-Erythritol, D-fructose	D	+	−
Carbon sources for growth:			
D-Mannose, L-arabinose, D-lyxose, L-lyxose	+	−	−
n-Propanol	−	−	D
Acetate, glycerate, lactate	+	−	D
Mol% G + C of DNA	62–64	57–64	51–65

[a] Symbols: see standard definitions.

[b] Some strains are negative.

Table 4.42.

Other characteristics of **Frateuria aurantia**[a]

Characteristic	Result or Reaction	Characteristic	Result or Reaction
Growth in 0.5% yeast extract broth or 0.5% peptone broth	d	L-Amino acids as sole nitrogen source:	
Growth in Frateur's Hoyer medium with		Arginine, asparagine, glutamine, glutamate, isoleucine, leucine, phenylalanine	+
Ethanol	−	Alanine, aspartate, histidine, methionine, proline, serine, valine	d
Glucose	d	Cysteine, glycine, lysine, threonine, tryptophan	−
Mannitol	+	L-Amino acids as sole carbon and nitrogen sources:[f]	
Growth in SM[b] containing 0.5–2% NaCl	d	Alanine, aspartate, glutamate, proline	−
Growth in SM at pH 3.6–8.1	+	Antimicrobial agents (amount per disk):	
Temperature range, growth in SM at:		Ampicillin, 10 μg	d
4°C	−	Bacitracin, 10 U	R
30–37°C	+	Cephaloridine, 25 μg	d
Ethanol tolerance, growth in SM containing:		Chloramphenicol, 30 μg	R
1% Ethanol	+	Colistin sulfate, 10 μg	R
2–5% Ethanol	d	Erythromycin, 10 μg	R
10% Ethanol	−	Fusidic acid, 10 μg	R
Glucose tolerance, growth in 0.5% yeast extract containing:		Gentamicin, 10 μg	R
20% Glucose	+	Kanamycin, 30 μg	S
25–30% Glucose	d	Lincomycin, 2 μg	R
35% Glucose	−	Methicillin, 10 μg	R
Tolerance to metals and dyes, growth in presence of:		Nalidixic acid, 30 μg	d
Malachite green, 0.001%; crystal violet, 0.0001%; brilliant green, 0.001%	+	Neomycin, 30 μg	R
$Cd(CH_3COO)_2 \cdot 2H_2O$, 0.01%; $CoSO_4 \cdot 7H_2O$, 0.01%	−	Nitrofurantoin, 200 μg	R
$HgCl_2$, 0.001%	+	Novobiocin, 30 μg	S
CH_3COOTl	d	Penicillin G, 10 U	R
Vitamin requirements:		Polymyxin B, 300 U	R
p-Aminobenzoic acid, thiamine, niacin, pantothenate	−	Streptomycin, 10 μg	S
		Sulfafurazole, 100 μg	R
Final pH in D-glucose, D-xylose or D-galactose media	≦4.5	Tetracycline, 30 μg	S
Acid produced from:[c]		Catalase	+
Ethanol, D-ribose, D-mannose, n-propanol, i-inositol	+	Oxidase	−
m-Erythritol, D-arabinose,[d] D-fructose, sucrose, n-butanol	d	Nitrate reduction	−
		Indole production	−
Glycerol,[d] L-rhamnose,[d] L-sorbose, D-mannitol, sorbitol, D-cellobiose,[d] D-lactose,[d] maltose, raffinose, dextrin, starch, n-amyl alcohol	−	Gelatin hydrolysis	−
		Change in litmus milk	−
Carbon sources for growth:[e]		Ferric chloride reaction on media containing:	
Glycerol, m-erythritol, D-ribose, D-fructose, D-glucose, D-mannose, D-mannitol, D-sorbitol, L-arabinose, D-lyxose, L-lyxose, adonitol, i-inositol, acetate, glycerate, lactate	+	D-Glucose	d
		D-Fructose	+
Ethanol, D-arabinose, D-xylose, L-rhamnose, D-galactose, L-sorbose, D-cellobiose, D-lactose, maltose, sucrose, raffinose, starch, methanol, ethanediol, L-arabitol, m-xylitol, oxalate, malonate, tartrate, malate, gluconate, dextrin	d	D-Galactose	−
n-Propanol, dulcitol, formate, citrate	−		

[a] For symbols see standard definitions; also R, resistant; and S, susceptible.

[b] SM medium contains (g/liter): D-glucose, 50.0; yeast extract, 5.0.

[c] Acid formation was tested in the following medium (g/liter): yeast extract, 5.0; bromocresol purple, 0.02; carbon source, 10.0. A final pH below 5.9 was considered as positive for acid formation.

[d] After retesting, some corrections of our previous results (Swings et al., 1980) were necessary.

[e] Growth was tested on the following medium (g/liter): carbon source, 3.0; yeast extract, 0.5; vitamin-free casamino acids, 3.0; agar (Oxoid No. 1), 25.0.

[f] Frateur's Hoyer medium was employed, with D-mannitol omitted.

Genus IV. **Zoogloea** *Itzigsohn 1868, 30.*[AL]

RICHARD F. UNZ

Zo.o.gloe′a. Gr. adj. *zoos* living; Gr. noun *gloia* glue; M.L. fem. noun *Zoogloea* living glue.

Straight to slightly curved, plump rods, 1.0–1.3 μm in diameter and 2.1–3.6 μm in length, with rounded ends; sometimes tapered to a blunt point at one or both poles. **Nonsporeforming and noncystforming.** Cells in older cultures are demonstrably encapsulated. Gram-negative. **Actively motile**, especially in young cultures, by means of a **single polar flagellum.** Intracellular granules of poly-β-hydroxybutyrate are formed on media containing the salts of organic acids. Cultures enter into formation of **flocs and films** in liquid media at late growth stages; **the cells become embedded in gelatinous matrices to form zoogloeae, which are distinguished by a "tree-like" or "finger-like" morphology.** Young colonies on solid media under a normal air atmosphere are translucent and punctiform but may increase to 1 or 2 mm in diameter and exhibit opaque centers. Nonpigmented. **Aerobic**, having a strictly respiratory type of metabolism with oxygen as the terminal electron acceptor; **growth can also occur anaerobically in the presence of nitrate** (nitrate respiration). Denitrification occurs with formation of N_2. Optimum temperature for growth, 28–37°C. Optimum pH, 7.0–7.5. **Oxidase-positive.** Weakly catalase-positive. Chemoorganotrophic. **Acid is not formed from carbohydrates** except xylose, glycerol and ethanol, which are attacked oxidatively by a few strains. **Proteolytic on gelatin.** Most strains are **urease-positive.** Litmus milk is unchanged. Hydrogen sulfide is not usually produced from cysteine. Major carbon sources include salts of several organic acids (e.g. lactate, pyruvate and fumarate), dicarboxylic amino acids (e.g. aspartate, glutamate, and asparagine), alcohols, and salts of certain aromatic acids (e.g. benzoate and *m*-toluate). **Benzene derivatives are attacked by *meta* cleavage of the ring structure.** Organic nitrogen compounds (e.g. dicarboxylic amino acids) and ammonia serve as nitrogen sources; nitrate is unsuitable. Specific growth factor requirements, if any, are unknown. **Occur free living in organically polluted fresh waters and in wastewaters at all stages of treatment.** The mol% G + C of the DNA is 65.3 (Bd).

Type species: *Zoogloea ramigera* Itzigsohn 1868, 30.

Further Descriptive Information

Zoogloea strains form flocculent masses of zoogloeae in both complex and defined media containing suitable carbon sources (Fig. 4.9). Arrangement of the bacteria into sharply demarcated columns or "fingers" (Figs. 4.10 and 4.11) which protrude from a cluster or aggregate of cells constitutes the historically recognized growth form of *Z. ramigera* (Koch, 1877; Butterfield, 1935; Dugan and Lundgren, 1960; Unz and Farrah, 1976a). However, taken by itself, zoogloea morphology is an unreliable character upon which to base the *identification* of *Z. ramigera* since (a) fragmented portions of flocs and pellicles, and artifacts created by random coalescence of bacteria, may be mistaken under microscopic observation for finger-like zoogloeae, and (b) *Z. ramigera* may form amorphous rather than finger-like zoogloeae (Fig. 4.12). The extent of zoogloea production varies among strains and with the culture conditions and may diminish greatly or be lost, especially during frequent transfer of strains in rich culture media.

Cells of *Zoogloea* are plump rods which are motile by means of a single, monopolar flagellum (Fig. 4.13). Chains of cells are rare. Cells in very old cultures may appear elongated.

Colonies produced on CY agar* are initially punctiform. After 3 or 4 days, the colonies reach 1 mm in diameter and appear circular, slightly raised and translucent with opaque centers (Fig. 4.14) or completely grey-white. Colony edges are entire or lobate. Mature colonies are distinctively tenacious and cohesive and may be lifted intact from the agar surface with a needle. Colonies develop poorly on ordinary nutrient agar.

Growth of *Zoogloea* strains is slow at 9°C and nonexistent at 45°C. The pH range for growth lies between pH 5.0 and 10.0, with poor development occurring in the region of extreme pH values.

Strains survive but do not grow under strictly anaerobic conditions in the absence of nitrate. They exhibit a microaerophilic tendency in semisolid agar deeps as evidenced by the appearance of culture bands 3–5 mm below the agar surface (Unz and Dondero, 1967b). Approximately 50% of strains tested reduce trimethylamine oxide to trimethylamine (Unz and Dondero, 1967b).

Acid was produced oxidatively from xylose by 3 of 65 strains in one study (Unz and Dondero, 1967a), and by 5 of 37 strains in another study (Unz and Farrah, 1972). Acid was also produced oxidatively from glycerol and/or ethanol by 10 of 65 strains (Unz and Dondero, 1967a).

No growth on citrate occurred in one study (Unz and Dondero, 1967a), but growth by 20 of 37 strains occurred on Koser's citrate in another study (Unz and Farrah, 1972).

Zoogloea strains are not fastidious nutritionally and may be cultured on a variety of organic carbon sources in a simple defined medium.† Carbon sources which support growth of at least 90% of *Zoogloea* strains are lactate, pyruvate, α-hydroxybutyrate, fumarate, ethanol, butanol, benzoate, glutamate, aspartate and asparagine (Unz and Dondero, 1967b).

Resistance to 2.5 U of penicillin G occurs in 67% of the strains tested (Unz and Dondero, 1967a).

Zoogloea strains are found principally in organically polluted fresh waters, wastewaters and aerobic biological wastewater treatment systems, e.g. activated sludge and trickling filter units.

Enrichment and Isolation Procedures

Finger-producing strains of *Zoogloea* may be cultivated best in enrichment media inoculated with activated sludge or the films from aerobic wastewater treatment devices as follows. Approximately 2 ml of an inoculum are added to 15 ml of mineral salts solution§ which overlays 5 ml of a nutrient-enriched agar¶ contained at the bottom of a metal-capped test tube (20 mm O.D. × 150 mm). The enrichment culture is incubated at 28°C until a pellicle develops, usually within 2 or 3 days with activated sludge as the inoculum. A simpler method for enrichment of fingered zoogloeae involves storage of activated sludge directly in covered glass containers at room temperature until a surface film appears (Amin and Ganapati, 1967). The latter method is not always reliable and success may depend on obtaining a proper ratio of

---

* CY agar (g/liter of distilled water): Casitone (Difco), 5.0; yeast autolysate, 1.0; agar, 15.0.

† Defined medium (g/liter of distilled water): carbon source, 0.25–0.5; $(NH_4)_2SO_4$, 0.375; $MgSO_4\cdot7H_2O$, 0.2; $CaCl_2$, 0.2; K_2HPO_4, 0.1; and $FeSO\cdot7H_2O$, 0.005. The pH is adjusted to 7.2 with 0.1 N NaOH. Yeast autolysate (0.01 g) and vitamin B_{12} (1×10^{-6} g) may be included as sources of growth factors to decrease growth lag.

§ Mineral salts solution (g/liter): $(NH_4)_2SO_4$, 0.3; NaCl, 5.85; $CaCl_2\cdot2H_2O$, 0.2; K_2HPO_4, 0.1; $MgSO_4\cdot7H_2O$, 0.14; $FeSO_4\cdot7H_2O$, 0.0003; $MnCl_2\cdot4H_2O$, 0.0063; $CoSO_4\cdot7H_2O$, 0.00011; H_3BO_4, 0.0006; $ZnCl_2$, 0.00022; and $CuSO_4\cdot5H_2O$, 0.00008. The medium components are prepared from stock solutions and the pH of the medium is adjusted to 8.5 ± 0.1 pH unit with 0.5 N NaOH.

¶ Nutrient-enriched agar: to mineral salts solution containing agar (20 g/liter) is added any of the following sole carbon sources in the amount shown per liter: starch, 2.4 g (w/v); *m*-toluic acid (neutralized), 1.35 g (w/v); *n*-butanol, 1.5 ml (v/v); lactic acid (85%), 1.35 g (w/v); ethanol (95%), 1.5 ml (v/v); or glucose, 2.4 g (w/v). After adding the carbon source to the liquid mineral salts-agar mixture, the pH is adjusted to 8.5 ± 0.1 pH unit with 0.5 N NaOH.

Figure 4.9. Flocculent growth habit of *Zoogloea ramigera*. *Left*, ATCC strain 19544. *Right*, strain G4, freshly isolated and exhibiting the development of a thick, straggly pellicle. Casitone-yeast autolysate medium, 28°C, 72 h.

the height of activated sludge solids to that of the total volume of liquid stored in the container.

A wet mount of enrichment culture surface film is examined at × 100 magnification by phase contrast microscopy to confirm the presence of fingered zoogloeae. Several loopfuls of film are transferred to a 2-ml droplet of CY broth contained in a Petri plate. The fingered zoogloeae may be located easily in the droplet under × 45–60 magnification with the aid of a dissecting microscope. Approximately 10–12 of the fingered zoogloeae are transferred individually and successively by a micropipette through each of four 0.7-ml droplets of CY medium in

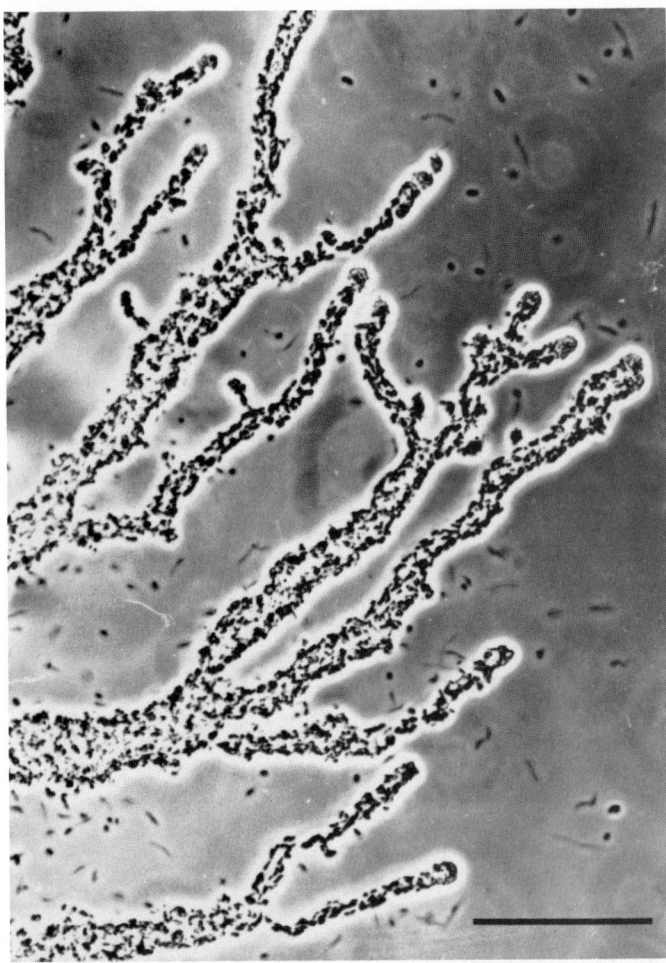

Figure 4.10. Finger-like zoogloea. *Zoogloea ramigera* ATCC strain 19544 grown in lactate-mineral salts medium, 28°C, 60 h. Phase-contrast. *Bar*, 50 μm.

order to free loosely attached debris and microorganisms from the zoogloeae. Finally, the washed zoogloeae are collectively transferred to 3 ml of CY medium and subjected briefly to sonic oscillation (e.g. 30 s at 50 watts) to release the cells. A loopful of the sonicate is streaked onto solid CY medium and incubated at 28°C. Typically cohesive colonies of *Zoogloea* are large enough in 3 or 4 days to be transferred intact to CY broth. Inoculation of CY broth with a single colony usually results in appearance of slight turbidity and a slippery, glistening pellicle after 3 days at 28°C. Pellicles may be composed entirely of fingered zoogloeae or amorphous zoogloeae which, following detachment, settle and give rise to a flocculent sediment.

Maintenance Procedures

Zoogloea strains may be maintained in half-strength CY medium at 20°C for at least 2 months between subculturing. The formation of zoogloeal flocs is visibly reduced upon continuous transfer in laboratory culture media. The zoogloeal growth habit may be restored by plating a broth culture and transferring a colony back to liquid medium.

Strains do not survive prolonged refrigeration; however, they may be preserved indefinitely by lyophilization.

Differentiation of the genus **Zoogloea** from other genera

The taxonomic status of the genus *Zoogloea* is uncertain. Its current position among the genera of the family *Pseudomonadaceae* is based on the chemoorganotrophic, nonfermentative and monotrichous nature of the type species. Some differentiation of the genus *Zoogloea* from

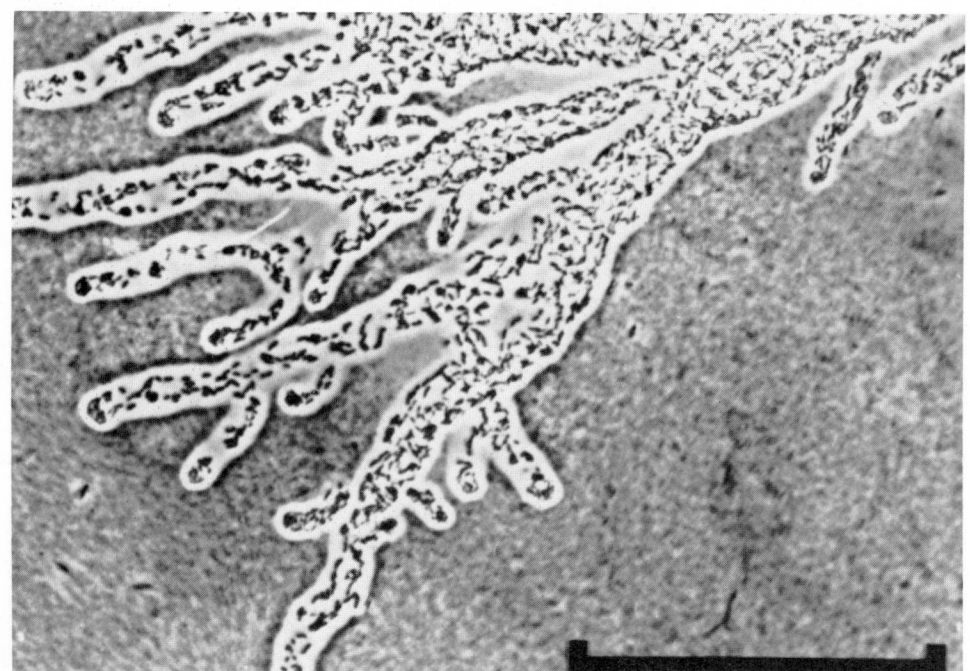

Figure 4.11. Finger-like zoogloea treated in wet mount with skim milk to accentuate the exopolymer in which the cells are embedded. *Zoogloea ramigera* ATCC strain 19544. Lactate-mineral salts medium, 28°C, 60 h. Phase-contrast. *Bar,* 100 μm. (Reproduced with permission from R. F. Unz, International Journal of Systematic Bacteriology *21:* 91–99, 1971, © International Association of Microbiological Societies.)

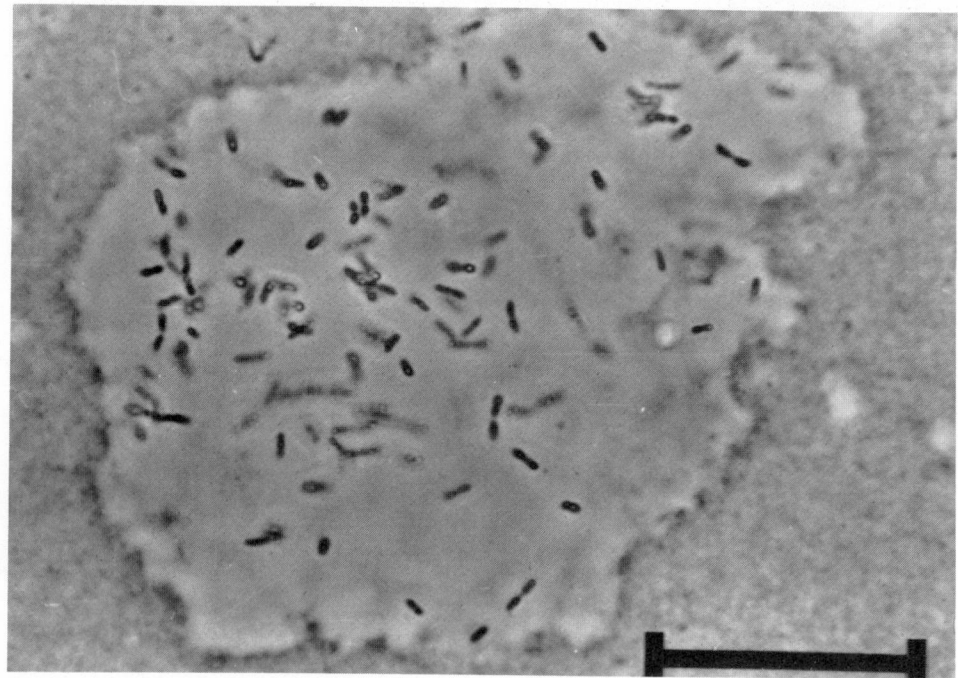

Figure 4.12. Amorphous zoogloea treated in wet mount with skim milk to accentuate the exopolymer in which the cells are embedded. *Zoogloea ramigera* ATCC strain 19544. Lactate-mineral salts medium, 28°C, 60 h. Phase-contrast. *Bar,* 30 μm. (Reproduced with permission from R. F. Unz, International Journal of Systematic Bacteriology, *21:* 91–99, 1971, © International Association of Microbiological Societies.)

other genera of *Pseudomonadaceae* family is possible (Table 4.43); however, definitive relationships cannot be established in the absence of critical information on the genetic and antigenic properties of strains. In most cases, the mol% G + C of the DNA provides little basis for comparison. Floc formation is too nonspecific to be very useful in the delineation of *Zoogloea* from other pseudomonads and several eubacteria, particularly in the absence of information on exopolymer production or its chemical and structural composition. The qualitative chemistry of exopolymers—e.g. hexosamine, glucose, xylose and arabinose (Crabtree et al., 1966); glucose, mannose and galactose (Wallen and Davis, 1972); glucose and galactose (Parsons and Dugan, 1971); and glucosamine and fucosamine (Tezuka, 1973)—is diverse among

bacteria designated *Zoogloea* and is in keeping with general phenotypic dissimilarities. The mucopolysaccharide (Farrah and Unz, 1976) produced by the neotype strain of *Z. ramigera* (ATCC 19544) appears to be chemically related to the exopolymer described by Tezuka; moreover, it exhibits a fine, strandlike mesh (Fig. 4.15) rather than the coarse, fibrillar network of cellulose-like glycans observed for other zoogloeal and nonzoogloeal bacteria (Friedman et al., 1968; Friedman et al., 1969) or the cellulose of *Acetobacter* species (Ohad et al., 1962). The exocellular homo- and heteropolysaccharide exopolymers which have been characterized for species of *Xanthomonas, Pseudomonas, Arthrobacter* and *Alcaligenes* are variably water-soluble and may increase the consistency of the culture medium (Sutherland, 1979; Powell, 1979). In

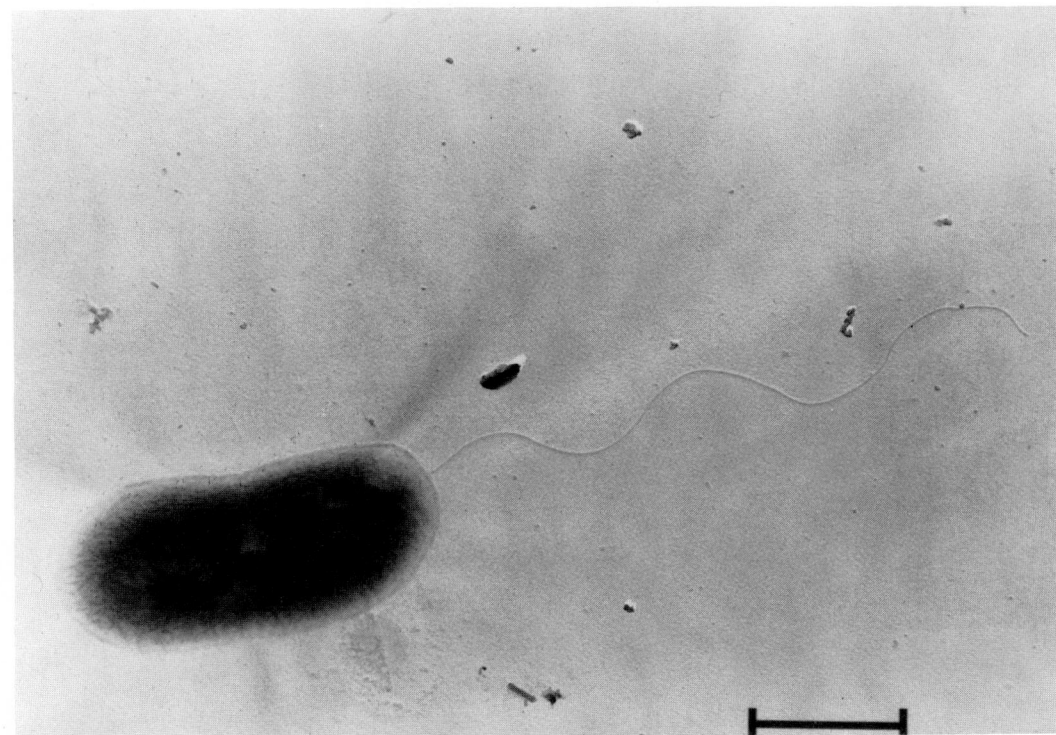

Figure 4.13. *Zoogloea ramigera* ATCC strain 19544 showing the single, polar flagellum. Casitone-yeast autolysate medium, 28°C, 24 h. Platinum-carbon shadowed. *Bar*, 1 μm. (Reproduced with permission from R. F. Unz, International Journal of Systematic Bacteriology, *21:* 91–99, 1971, © International Association of Microbiological Societies.)

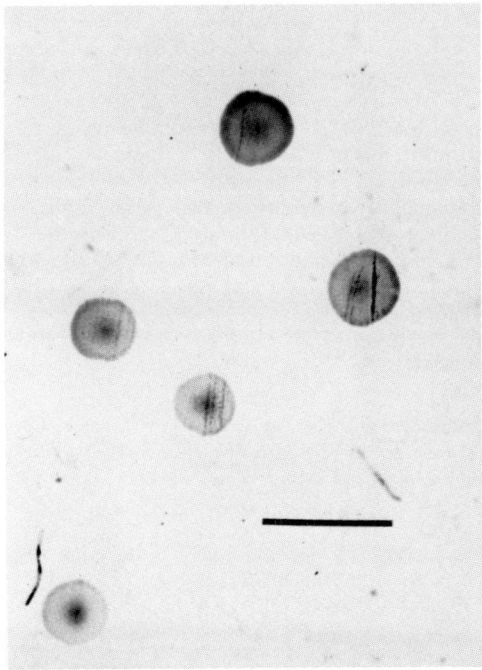

Figure 4.14. Colonies of *Zoogloea ramigera* ATCC strain 19544. Casitone-yeast autolysate agar, 28°C, 72 h. Photographed by reflected light. *Bar*, 2 mm. (Reproduced with permission from R. F. Unz, International Journal of Systematic Bacteriology, *21:* 91–99, 1971, © International Association of Microbiological Societies.)

Table 4.43.
Differential characteristics of the genus **Zoogloea** *and other genera of the family* **Pseudomonadaceae**[a]

Characteristics	Zoogloea	Xanthomonas	Pseudomonas	Frateuria
Cell diameter typically 1.0 μm or slightly larger	+	−	−	−
Pigment formed by colonies	−	+	D	+
Denitrification (to N_2)	+	−	D	−
Urease	+	−		
Oxidase	+	−	+	−
Acid from glucose	−	+	+	+
Gelatinase	+	D	D	−
H_2S from cysteine	−	+	D	+
Acetic acid oxidized	+	+	D	−
Meta-cleavage of aromatic rings	+		D	
Mol% G + C of DNA	65.3	63–71	58–70	62–64

[a] Symbols: see standard definitions.

contrast, the mucopolysaccharide-producing neotype strain of *Z. ramigera* produces a water-insoluble exopolymer at room temperature and the culture medium is not visibly thickened.

The small cell size of members of the family *Acetobacteraceae*, as well as the occurrence of polytrichous flagella, ability to grow at pH 4.5, and the ability of at least some strains to oxidize gluconate and several sugars, provides a measure of difference between this family and the genus *Zoogloea*.

The genus *Comamonas* (Davis and Park, 1962)—a *genus incertae sedis*—has been used for classifying nonfloc-forming bacteria to distinguish them from otherwise similar floc-forming bacteria considered to be *Zoogloea* strains (Dias and Bhat, 1964).

Taxonomic Comments

Two important developments which have occurred in the genus *Zoogloea* since the last edition of the *Manual* are (a) replacement of

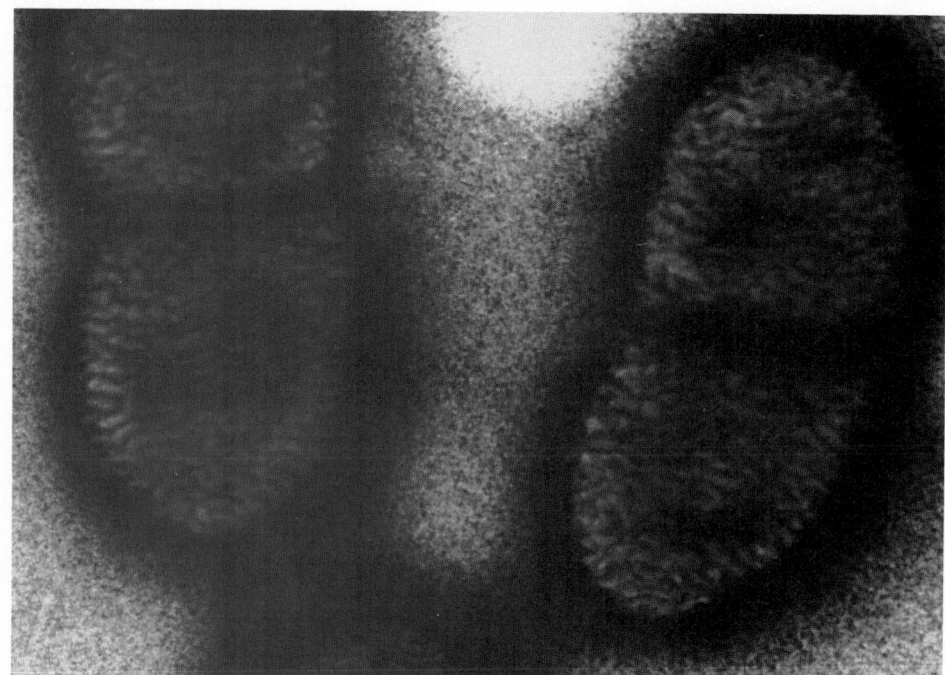

Figure 4.15. Cell of *Zoogloea ramigera* ATCC strain 19544 embedded in a fine mesh-work of exopolymer. Casitone-yeast autolysate medium, 28°C, 72 h. Preparation negatively stained with 2% phosphotungstic acid for 30 s. ×44,400.

ATCC strain 19623 (strain I-16-M; Crabtree and McCoy, 1967) with ATCC 19544 (strain 106; Unz, 1971) as the neotype strain of the species *Z. ramigera* (Judicial Commission, 1979), and (b) removal of *Z. filipendula* as a valid species of the genus (Skerman et al., 1980). The genus *Zoogloea* perpetually suffers an ambiguous taxonomic status owing largely to nomenclatural difficulties arising from (a) uncertainties about the type species which was originally described on the basis of a growth form of an organism in mixed culture, (b) acceptance of the flocculent growth habit as the principal characteristic for the identification of sundry aggregative bacteria as strains of *Zoogloea* or of *Z. ramigera*, and (c) conflicting descriptions of bacteria stated to be *Z. ramigera*. Nucleic acid homology studies are not known for *Zoogloea* but would be most helpful in resolving these taxonomic difficulties.

Notwithstanding the above comments, *Z. ramigera* ATCC 19544 is one of 147 *Zoogloea* strains of certain origin, in that the bacteria were isolated as single cells during micromanipulation of natural, finger-like zoogloeae while under microscopic observation (Unz and Dondero, 1967a). *Z. ramigera* ATCC 19544 forms fingered zoogloeae in liquid culture which are similar in appearance to the natural type from which it came and to the zoogloeae referred to as "tree-like" ramifications of spherical gelatinous masses in the original description of the species (Itzigsohn, 1868).

At least 90–100% of *Zoogloea* strains denitrify, demonstrate positive oxidase and catalase reactions, hydrolyze gelatin, carry out *meta*-cleavage of aromatic ring structures, and fail to form acid from carbohydrates (Unz and Dondero, 1967a; Unz and Farrah, 1972). Approximately 82% of strains demonstrate urease activity. To define the salient characteristics of the genus is difficult because minor attributes, such

as acid formation from ethanol and glycerol, are often located in strains of a common natural provenance—i.e. progeny of a single zoogloea or of several zoogloeae found in a specific wastewater treatment plant or polluted environment.

Floc formation (Finstein, 1967) and major exopolymer production (Unz and Farrah, 1976b) by *Zoogloea* strains take place during the decreasing and stationary growth phases. Consequently, bioflocculation and exopolymer formation are to be regarded as conditional characters of *Zoogloea* strains, albeit ones that are exhibited reliably by freshly-isolated organisms.

Degradation of the benzenoid ring structure by a *meta*-cleavage mechanism is a distinctive character of *Zoogloea* strains and is also considered taxonomically significant for the acidovorans group of *Pseudomonas* (Stanier et al., 1966). *Z. ramigera* differs from *P. acidovorans* and *P. testosteroni* by its ability to hydrolyze gelatin and to denitrify, and by its inability to hydrolyze starch or to utilize glycolate, caproate and several amino acids.

Acknowledgments

The mol% G + C of the DNA of *Z. ramigera* strains was determined by Dr. Manley Mandel, University of Texas Medical Center. Figures 4.9 and 4.15 were prepared by Paul Wichlacz and Terry Williams, The Pennsylvania State University.

Further Reading

Zvirbulis, E. and H.D. Hatt. 1967. Status of the generic name *Zoogloea* and its species. Int. J. Syst. Bacteriol. **17:** 11–21.

Differentiation of strains of **Zoogloea ramigera**

Because of the taxonomic difficulties described previously, the characteristics of the neotype strain of *Z. ramigera* are presented for comparison with those of two other strains whose inclusion in the

species is considered by this author to be uncertain (Table 4.44). See also the section on *Other Organisms*.

List of the species of the genus **Zoogloea**

1. **Zoogloea ramigera** Itzigsohn 1868, 30.[AL]
ra.mi′ge.ra. L. n. *ramus* branch; *L. v. gero* to bear; M.L. adj. *ramigera* branch-bearing.

The morphological, cultural, physiological and nutritional characters are given in the description of the genus and in the description of the neotype strain (Table 4.44). No growth occurs on Koser's citrate. H₂S

Table 4.44.
Characteristics of **Zoogloea ramigera** *ATCC 19544 and two strains of uncertain affiliation*[a]

Characteristics	ATCC 19544 (Neotype strain; strain 106)	ATCC 19623 (Strain I-16-M)	ATCC 25935 (Strain 115)	Characteristics	ATCC 19544 (Neotype strain; strain 106)	ATCC 19623 (Strain I-16-M)	ATCC 25935 (Strain 115)
Cell diameter is 1.0 μm or slightly larger	+	−		Galactose, maltose	−	+[d]	+
Flagellar arrangement:				Cellobiose, glycogen, lactose, mannose	−	−[d]	+
Monotrichous only	+	−	+	Glucose, xylose, rhamnose, mannitol	−	+[b]	−
Monotrichous and polytrichous (lateral)	−	+	−	Ribose	−	+[b]	
Zoogloeae are produced	+	−[b]	+	Utilized as sole carbon source:			
Straw-colored colonies	−	+	+	Acetate	+	+	+
Growth on potato	−	+		Citrate	+	−	+[e]
Hydrolysis of gelatin and casein	+	−	+	Malate, pyruvate, fumarate	+	+	
Tyrosine agar cleared	−	+		Butyrate, α-ketoglutarate, propionate	+		+[e]
Arginine dihydrolase	−	+		n-Propanol, ethanol, n-butanol	+	+[d]	+
Sensitivity to 0/129[c]	+	−[b]		Acetaldehyde, succinate, oxalacetate, lactate, α-hydroxybutyrate, palmitate, myristate	+		
Growth in presence of 3% NaCl	−	+		Benzoate, m-toluate, p-toluate, phenol, o-cresol, m-cresol, p-cresol	+[f]	−[f]	
Litmus milk:							
Alkalinity produced	−	+	+	Methanol	−	+[d]	+
Reduction occurs	−	+	−	Mol% G + C of DNA[g]	65.3	65.3	
Denitrification (to N_2)	+	−	−				
Urease activity	+	−	+				
Hydrolysis of starch	−	−	+				
Acid formed oxidatively from:							
Arabinose, fructose, sucrose	−	+[b]	+				

[a] Unless otherwise indicated, the sources of data for strains are as follows: ATCC 19544 (Unz, 1971); ATCC 19623 (Crabtree and McCoy, 1967); and ATCC 25935 (Friedman and Dugan, 1968). For symbols see standard definitions.
[b] Unz, 1971.
[c] Vibriostatic agent 0/129 is 2,4-amino-6,7-di-isopropylpteridine.
[d] Friedman and Dugan, 1968.
[e] Joyce and Dugan, 1970.
[f] Unz and Farrah, 1972.
[g] M. Mandel, personal communication.

is not produced from cysteine or on Kligler iron agar. Indole is not produced. Ammonia is produced from asparagine. Lipolytic activity occurs.

Growth on benzoate, *m*-toluate, and *o*-cresol occurs with *meta* cleavage of the aromatic ring.

Methanol, formate, and formaldehyde are not utilized as sole carbon sources.

Neotype strain: ATCC 19544 (strain 106 of Unz, 1971). This strain was cultivated from a single cell isolated by micromanipulation of a finger-like zoogloea obtained in a trickling filter sample collected in Freehold, New Jersey.

Other organisms

In addition to the neotype strain of *Z. ramigera*, there exist two other strains which have attained some prominence through experimental use: strain I-16-M (Crabtree and McCoy, 1967) and strain 115 (Friedman and Dugan, 1968). The three strains all share the property of floc formation and each was obtained from polluted environments; however, dissimilarities between the strains are obvious in several respects (see Table 4.44), and it may be that strains I-16-M and 115 do not belong to the genus *Zoogloea*. Additional evidence for this view is that cross-reactions are not observed between strains I-16-M or 115 and fluores-cein-labeled, whole-cell antiserum against the neotype strain (Farrah and Unz, 1975). Weak, whole-cell antigen-antibody reactions do occur between strain 115 and strain I-16-M, although strain 115 bears a greater antigenic relationship to *Gluconobacter oxydans* subsp. *suboxydans* (ATCC strain 621) (Chorpenning et al., 1978).

Nonextant strains of *Z. ramigera*, which bear some resemblance to the neotype strain according to published descriptions, are strain Z-1 (Butterfield, 1935) and an early isolate of questionable purity (Bloch, 1918).

FAMILY II. **AZOTOBACTERACEAE** PRIBRAM 1933, 5[AL]

YAO-TSENG TCHAN

A.zo.to.bac.ter.a′ce.ae. Fr. n. *azote* nitrogen; M.L. masc. n. *bacter* equivalent of Gr. neut. n. *bactrum* a rod or staff; M.L. masc. n. *Azotobacter* a nitrogen rod; *-aceae* ending to denote family; M.L. fem. n. *Azotobacteraceae* the *Azotobacter* family.

Blunt rods to oval cells, 2 μm or more in diameter, of various length. The morphology changes with various growth conditions. Cells are often in pairs, also in chains. **Motile by peritrichous or polar flagella or nonmotile. Intracellular poly-β-hydroxybutyrate present. Gram-negative to Gram-variable.** Endospores not produced. **Cysts formed in one genus (*Azotobacter*). Chemoheterotrophic. Obligately aerobic.** Catalase-positive. **Capable of fixing molecular nitrogen.** Normally fix 10 mg of nitrogen/g of a suitable carbohydrate consumed under laboratory conditions on synthetic media, with or without noticeable production of homopolysaccharides. Organic growth factors not required. Trace elements involved in nitrogenase activity (e.g. molybdenum) are required. Can utilize various sources of combined nitrogen; some species do not, or poorly utilize nitrate. Water-insoluble and water-soluble pigments and fluorescent pigments produced by some species. **Habitat: soil, water and the plant rhizosphere.** The mol% G + C of the DNA ranges from 52–67.5.

Type genus: *Azotobacter* Beijerinck 1901, 567.

Taxonomic Comments

The concept of a family *Azotobacteraceae* as defined in the eighth edition of *Bergey's Manual* (Becking, 1974) is no longer adequate. Traditionally this family has contained various Gram-negative, aerobic, heterotrophic bacteria capable of fixing nitrogen nonsymbiotically under a normal atmospheric partial pressure of oxygen.

The heterogeneity of this family was noted by Tchan (1953). More recently, based on numerical analysis, Thompson and Skerman (1979) concluded that the family *Azotobacteraceae* is composed of "a collection of bacterial genera that fix nitrogen nonsymbiotically, that are Gram-negative, nonsporeforming. . .and that cannot be readily identified with genera in other families."

De Smedt et al. (1980) suggested the removal of the genus *Beijerinckia* from the family *Azotobacteraceae* based on an analysis of rRNA cistrons. *Beijerinckia* is not closely related to *Azotobacter*, *Azomonas* or *Derxia*. It belongs instead to the fourth rRNA super family defined by De Smedt et al., which contains *Agrobacterium* and *Rhizobium*. The rRNA similarity map prepared by De Smedt et al. shows no genera in the immediate vicinity of *Beijerinckia*, which suggests that this genus has a separate phylogenetic position within the super family; moreover, it is also phenotypically distinguishable from all other genera of bacteria. These recent findings support earlier arguments for separating *Beijerinckia* from the family *Azotobacteraceae*. Tchan (1953) previously noted that *Beijerinckia* was morphologically and physiologically different from *Azotobacter* and that its place in the family *Azotobacteraceae* was only temporary.

De Smedt et al. (1980) advocated the removal of the genus *Derxia* from the family *Azotobacteraceae* based on rRNA cistron analysis. Their results showed that the rRNA cistrons of *Derxia* closely resemble those of their third super family, which contains *Pseudomonas*, *Chromobacterium*, *Janthinobacterium* and *Alcaligenes*, and differ from *Azotobacter*, *Azomonas* and *Beijerinckia*.

Derxia has the capability of autotrophically fixing N_2 using H_2 (Pedrosa et al., 1980). This property is not found in *Azotobacter*, *Azomonas* or *Beijerinckia*. The nitrogen-fixing ability of *Derxia* can be inhibited very easily by excessive aeration (in liquid medium). Because of its oxygen sensitivity, *Derxia* exhibits a characteristic variation in colony morphology not shared by *Azotobacter*, *Azomonas* and *Beijerinckia* (Tchan and Jensen, 1963).

Using numerical taxonomic analysis, Thompson and Skerman (1979) indicated that both *Beijerinckia* and *Derxia* should be hierarchically classified away from *Azotobacter* and *Azomonas*. Recent immunological analysis lends further support to this idea. Unlike *Azotobacter* and *Azomonas*, which exhibit a strong immunological cross-reaction, *Derxia* and *Beijerinckia* do not generally form precipitation complexes with antisera produced against *Azotobacter* and *Azomonas* (Tchan et al., 1980; Tchan, unpublished results; Ann M. Simpson, Bachelor of Science (Hons) Dissertation, University of Sydney, 1980).

After removal of *Beijerinckia* and *Derxia* from the family *Azotobacteraceae*, the rRNA similarity of the remaining genera would lead to their inclusion in the same rRNA super family as *Pseudomonas* (De Smedt et al., 1980). This is in agreement with the finding by Ambler (1973) that *Azotobacter* is very similar to *Pseudomonas* with respect to the amino acid sequence of its cytochrome c_{551}. However *Azotobacter* and *Azomonas* are related closely to one another and only distantly to *Pseudomonas*, as shown in rRNA similarity maps (De Smedt et al.) as well as their nitrogen fixing ability.

Thus the family *Azotobacteraceae* presently consists of two well accepted genera, *Azotobacter* and *Azomonas*, which can be differentiated from each other by their cyst formation and DNA base composition. Such a grouping is further supported by bacteriophage typing (Hegazi and Jensen, 1973).

Key to the genera of the family **Azotobacteraceae**

I. Cysts formed.

Genus I. *Azotobacter*, p. 220

II. Cysts not formed.

Genus II. *Azomonas*, p. 230

Genus I. **Azotobacter** Beijerinck 1901, 567.[AL]

YAO-TSENG TCHAN AND PETER B. NEW

A.zo.to.bac'ter. Fr. n. *azote* nitrogen; M.L. masc. n. *bacter* the equivalent of Gr. neut. n. *bactrum* a rod or staff. M.L. masc. n. *Azotobacter* a nitrogen rod.

Large ovoid cells 1.5–2.0 μm or more in diameter. Pleomorphic, ranging from rods to coccoid cells. Occur singly, in pairs or irregular clumps, and sometimes in chains of varying lengths. Do not produce endospores, **but form cysts.** Gram-negative. **Motile** by peritrichous flagella, or nonmotile. **Aerobic**, but can also grow under decreased oxygen tensions. Water-soluble and water-insoluble pigments are produced by some strains of all species. Chemoorganotrophic, using sugars, alcohols and salts of organic acids for growth. **Nitrogen fixers; generally fix nonsymbiotically at least 10 mg of atmospheric nitrogen/g of carbohydrate (usually glucose) consumed. Molybdenum is required for nitrogen fixation** but may be partially replaced by vanadium. Nonproteolytic. Can utilize nitrate and

ammonium salts (all but one species) and certain amino acids as sources of nitrogen. **Catalase-positive.** The pH range for growth in the presence of combined nitrogen is 4.8–8.5; the optimum pH for growth and nitrogen fixation is 7.0–7.5. Occur in soil and water; one species occurs in association with plant roots. The mol% G + C of the DNA is 63.2–67.5 (T_m).

Type species: *Azotobacter chroococcum* Beijerinck 1901, 567.

Further Descriptive Information

In nitrogen-free medium* with glucose as the carbon source, the young cells of different species are remarkably similar in appearance

---

* Winogradsky's nitrogen-free mineral medium is prepared from a stock solution which has the following composition (g/liter): KH_2PO_4, 50.0; $MgSO_4 \cdot 7H_2O$, 25.0; NaCl, 25.0; $FeSO_4 \cdot 7H_2O$, 1.0; $Na_2MoO_4 \cdot 2H_2O$, 1.0; and $MnSO_4 \cdot 4H_2O$, 1.0; the pH is adjusted to 7.2 with NaOH. (This stock solution can be stored indefinitely at room temperature.) The medium is prepared by using 5.0 ml of the stock solution and 0.1 g of $CaCO_3$/liter

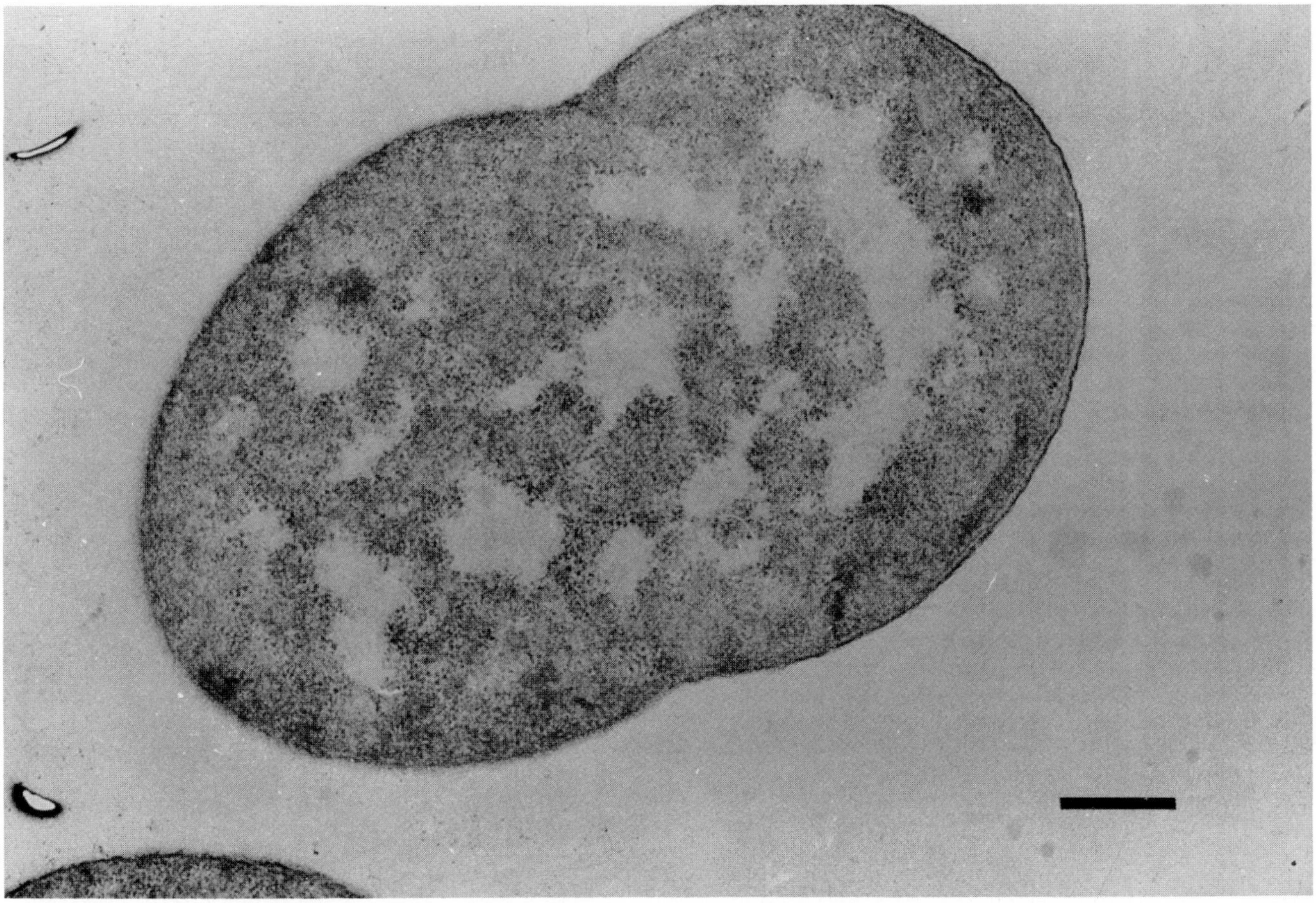

Figure 4.16. Electron micrograph of a vegetative cell of *A. chroococcum*. *Bar*, 0.2 μm.

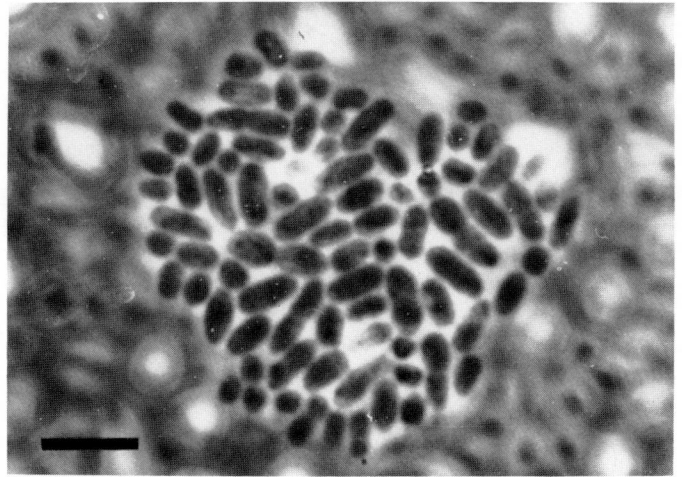

Figure 4.17. Vegetative cells of *A. chroococcum* (phase-contrast microscopy). *Bar*, 10 μm.

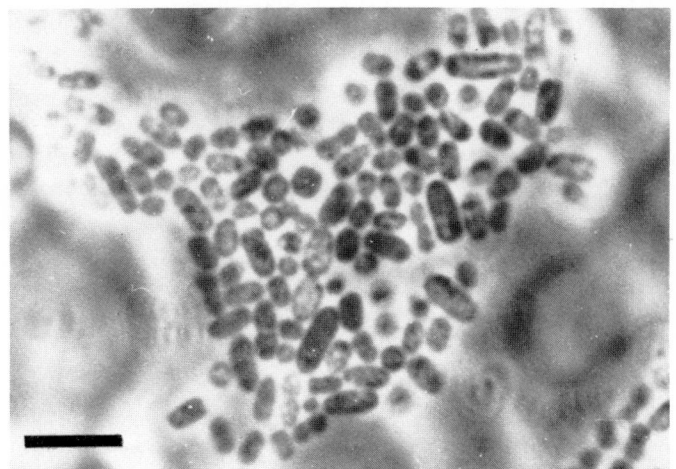

Figure 4.18. Vegetative cells of *A. nigricans* (phase-contrast microscopy). *Bar*, 10 μm.

(Figs. 4.16–4.18), namely rods with rounded ends, 1.3–2.7 μm in diameter and 3.0–7.0 μm in length. In older cultures the cells tend to be ellipsoidal; chains and filamentous forms become more common and metachromatic and sudanophilic granules are observed. *A. paspali* produces long filamentous forms even in young cultures (Fig. 4.19), which differentiates this species from all of the others. In peptone-yeast extract agar† all members of the genus produce distorted cells.

Cysts are formed in old cultures grown with sugar as the carbon

of distilled water. The medium is sterilized at 120°C for 20 min. To make nitrogen-free organic medium, a suitable quantity of organic substrate is added to the mineral medium. (Certain sugars, including glucose, must be sterilized separately before addition to the sterilized mineral medium.) For a solid medium, 15.0 g of agar is added/liter.

† Peptone-yeast extract agar (g/liter): Bacto peptone (Difco), 1.0; yeast extract (Difco), 0.5; NaCl, 0.5; Bacto agar (Difco), 1.5. The pH is adjusted to 7.2 with NaOH prior to autoclaving, giving a final pH of 7.0.

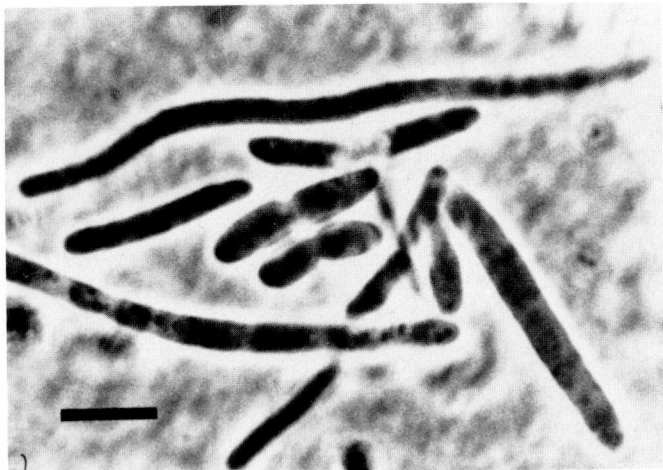

Figure 4.19. Vegetative cells of *A. paspali* (phase-contrast microscopy). *Bar, 10 μm.*

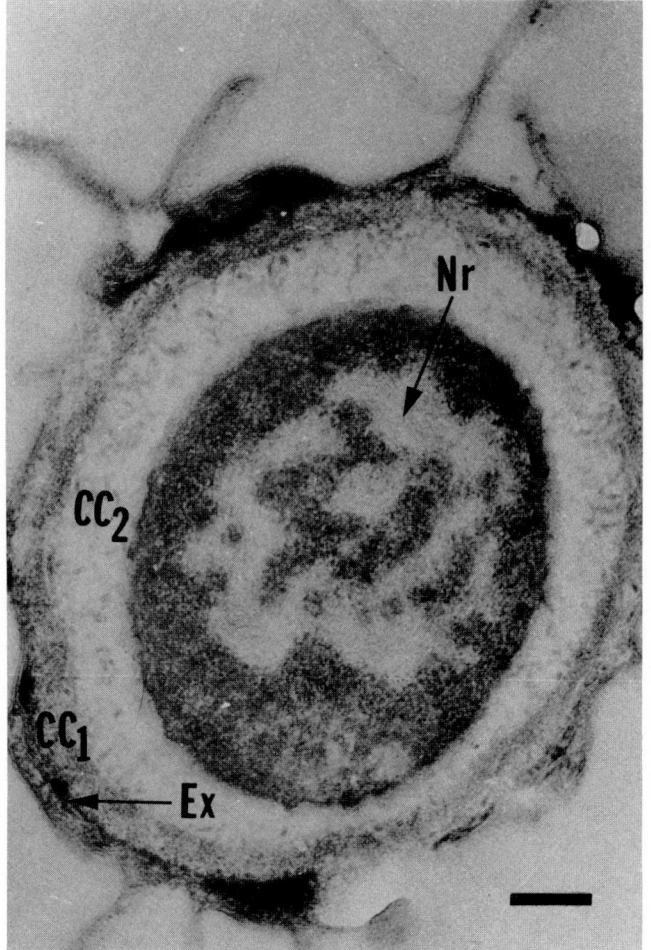

Figure 4.20. Ultrastructure of a cyst. The exocystorium (*Ex*) and the two layers of exine (*CC$_1$, CC$_2$*) are visible. A nuclear region (*Nr*) showing typical vegetative structure and ribosomes are observable within the central body. *Bar, 0.2 μm.*

source; for some species a medium* containing butan-1-ol as the organic substrate enhances cyst formation. The cyst may be distinguished from an endospore by its characteristic structure: a central body surrounded by a cyst coat, consisting of an exocystorium and an exine (Figs. 4.20 and 4.21). Unlike a spore, the cell inside the cyst coat is similar to the vegetative form and there are no cytological changes in the cell prior to its germination. During germination (Fig. 4.22) the cyst exocystorium is ruptured at one point and the cell which emerges may already be in a dividing state (Tchan et al., 1962; Socolofsky and Wyss, 1961).

All members of the genus are Gram-negative, although they have been described as Gram-variable by some authors (e.g. Jensen and Petersen, 1954; Kirakosyan and Melkonyan, 1964; Norris and Kingham, 1968; Johnstone, 1974). The Gram-variable results are probably due to incomplete decolorization of thick smears (Thompson and Skerman, 1979).

A. beijerinckii and *A. nigricans* are nonmotile. Other species are motile by peritrichous flagella (Figs. 4.23 and 4.24).

On nitrogen-free agar medium with sugar as the carbon source, colonies appear within 48 h at 30°C and reach a diameter of 2–6 mm in a week. The colonies are generally smooth, glistening, opaque, low convex, and viscid; however, colonial variations may occur. For *A. vinelandii*, smaller variant colonies may appear due to decreased production of extracellular polysaccharide. *A. armeniacus* sometimes produces translucent colonies and *A. paspali* forms undulate edged and unevenly convex colonies with a dull or rough surface. On sucrose or raffinose agar the production of diffusible homopolysaccharides, resulting in formation of a diffuse halo around the colony, is species dependent. *A. vinelandii* and *A. paspali* do not form diffusible homopolysaccharides. *A. chroococcum* and some strains of *A. beijerinckii* and *A. nigricans* produce diffusible homopolysaccharides from both sugars, whereas *A. armeniacus* and some other strains of *A. beijerinckii* and *A. nigricans* produce them only from sucrose (Thompson and Skerman,

1979). No cultures of *Azotobacter* produce colony-retained homopolysaccharides (defined as an increase in size, transparency and viscosity of the colony on sucrose or raffinose, compared with glucose).

On nitrogen-free medium, *A. chroococcum* produces a nondiffusible brown-black pigment, which is medium dependent. Many strains of this species produce gray-to-brown pigments on the medium of Stanier et al. (1966)† with 0.2% gluconate. Production of diffusible brown-black pigment in the presence of benzoate is variable for *A. chroococcum*. *A. armeniacus* produces a brown-black nondiffusible pigment which also is produced by one strain of *A. nigricans*. Other strains of *A. nigricans* produce a yellow nondiffusible pigment or no nondiffusible pigment. No brown-black diffusible pigment is formed by these two

---

* To 100 ml of molten Winogradsky's nitrogen-free mineral agar medium (see preceding footnote *), 0.1–0.2 ml of butan-1-ol is added prior to pouring the agar plates.

† Stanier's basal medium (modified slightly from that described by Stanier et al., 1966). Prepare stock solutions A, B and C as follows. *Solution A* (g/100 ml): disodium-EDTA (ethylenediamine tetra-acetate), 0.315; ZnSO$_4$·7H$_2$O, 1.095; FeSO$_4$·7H$_2$O, 0.5; MnSO$_4$·H$_2$O, 0.154; CuSO$_4$·5H$_2$O, 0.0392; Co(NO$_3$)$_2$·6H$_2$O, 0.0248; Na$_2$B$_4$O$_7$·10H$_2$O, 0.0177. A few drops of 25% H$_2$SO$_4$ are added to retard precipitation. *Solution B* is prepared by adding the following ingredients to 50 ml of solution A: nitrilotriacetic acid, 10.0 g; MgSO$_4$·7H$_2$O, 29.0 g; CaCl$_2$·2H$_2$O, 3.335 g; (NH$_4$)$_6$Mo$_7$O$_{24}$·4H$_2$O, 0.0093 g; FeSO$_4$·7H$_2$O, 0.099 g. Dissolve the nitrilotriacetic acid, neutralize with KOH ($\approx$ 7.3 g), add the remaining ingredients, adjust the pH to 6.6–6.8 and increase the volume to 1000 ml with distilled water. *Solution C:* mix equal volumes of 1.0 M Na$_2$HPO$_4$ (142 g/liter) and 1.0 M KH$_2$PO$_4$ (136 g/liter); the pH of the mixture should be 6.8. To prepare the complete medium, combine 40 ml of solution B, 66 ml of solution C, 2.0 g of (NH$_4$)$_2$SO$_4$ and 1000 ml of distilled water. Mix with an equal volume of 2% (w/v) Ionagar (Oxoid). After sterilizing the medium at 120°C for 15 min, add sufficient sterile organic substrate (e.g. gluconate or, if this is not utilized, glucose) to give a final concentration of 0.2% (1.0% in the case of glucose). The required pH can be obtained by addition of appropriate amounts of sterile 1.0 M KOH or 0.5 M H$_2$SO$_4$.

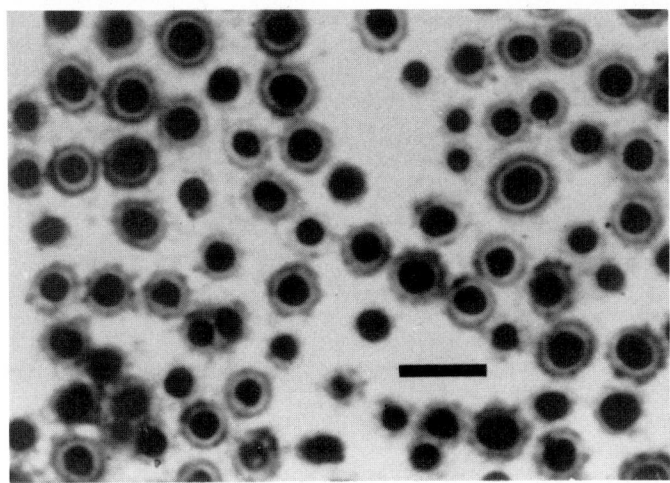

Figure 4.21. Violamine-stained cysts. The central body is surrounded by the cyst coat. *Bar,* 5 μm.

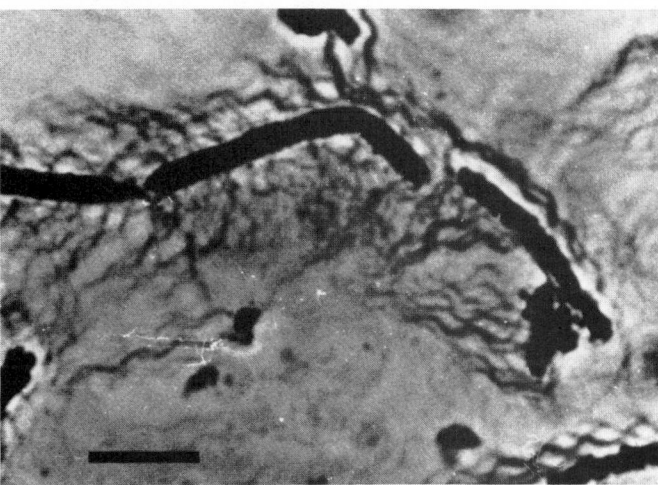

Figure 4.23. Flagella stain of *A. paspali. Bar,* 10 μm.

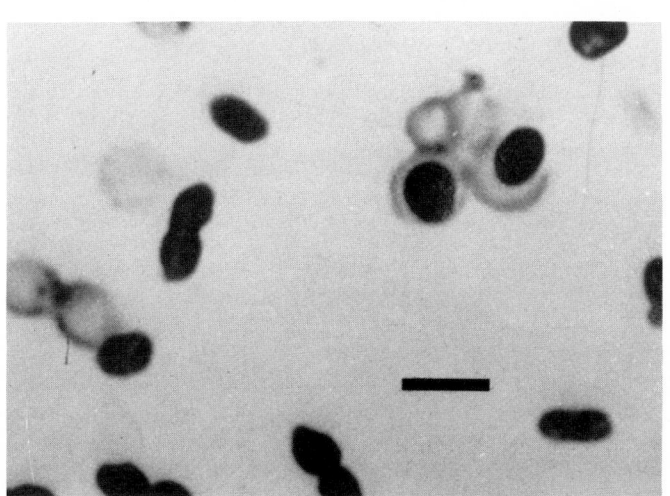

Figure 4.22. Germinating cysts and vegetative cells (violamine stained). *Bar,* 5 μm.

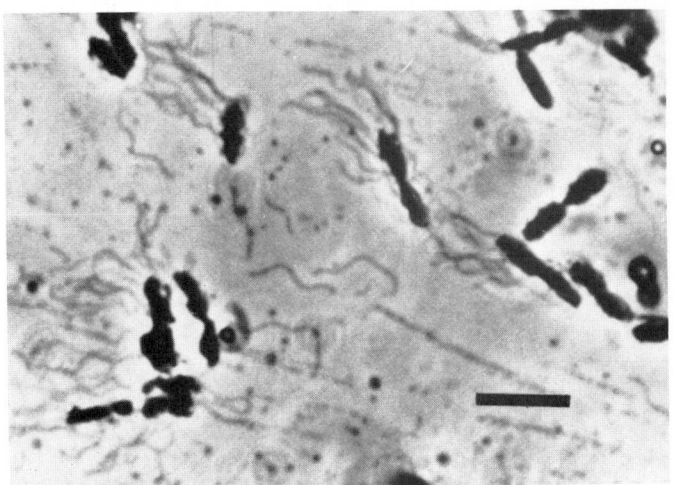

Figure 4.24. Flagella stain of *A. vinelandii. Bar,* 10 μm.

species in the presence of benzoate. *A. beijerinckii* produces a yellow to light brown nondiffusible pigment. In the presence of benzoate it produces a brown diffusible pigment. *A. vinelandii* and *A. paspali* do not produce nondiffusible pigment. In the presence of benzoate *A. vinelandii* usually produces brown-black pigment but pigment is rarely produced by *A. paspali.* The production of diffusible pigments on iron-deficient medium* and other media is summarized later in Table 4.46.

All members of the genus initially produce turbidity in liquid culture, especially near the surface of an undisturbed medium. As the cultures age, a pellicle is formed. Dispersible sediment is the rule with older cultures.

Most strains of the different species use the following substances as sole carbon and energy source: fructose, glucose, sucrose, ethanol, acetyl methyl carbinol, acetate, fumarate, pyruvate, α-oxoglutarate, gluconate, succinate. The ability of other organic substances to support the growth of different species has been exhaustively investigated by Thompson and Skerman (1979). Table 4.47 provides examples to illustrate the nutritional versatility of species of this genus.

Gas is not produced from glucose. In buffered glucose media† a titratable acidity of >2 microequivalents/ml is produced by *A. paspali* and by one strain of *A. nigricans.*

Most members of *Azotobacter* require a minimum temperature of 14°C for growth, but some strains of *A. nigricans* and *A. beijerinckii* are capable of multiplication at a temperature as low as 9°C. One strain of *A. armeniacus* has been reported to have a minimum growth temperature of 28°C. The optimum temperature for growth is near 32°C, except for *A. vinelandii, A. paspali,* and some strains of *A. chroococcum* which show optimum growth at 37°C. Differences in temperature range for growth and ability to utilize glycerol have also been found between strains isolated from subtropical and temperate environments (Thompson and Skerman, 1979). All species resist 50°C for 10 min; none can survive 60°C for 10 min.

---

* Iron-deficient medium: see Procedures for Testing Special Characters for formulation.
† Titratable acidity media: any one of the following media can be used (Thompson and Skerman, 1979). *Medium A* (g/liter): K₂HPO₄, 5.0; CaCl₂·2H₂O, 0.1; MgSO₄·7H₂O, 0.2; FeSO₄·7H₂O, 0.05; Na₂MoO₄·2H₂O, 0.005; glucose, 10.0; pH 7.8. *Medium B* (g/liter): K₂HPO₄, 1.5; KH₂PO₄, 3.5; MgSO₄·7H₂O, 0.2; FeSO₄·7H₂O, 0.05; Na₂MoO₄·2H₂O, 0.005; glucose, 10.0; pH 6.8. *Medium C* consists of medium A supplemented with either 2.0 g of tryptone (Difco) or yeast extract (Difco)/liter.

Positive oxidase and peroxidase reactions are given by some, but not all, strains. Only *A. paspali* is always peroxidase-negative. All species except *A. armeniacus* and *A. paspali* reduce nitrate to nitrite, but none denitrifies. H_2S is produced from thiosulfate by representatives of all species except *A. armeniacus*. H_2S is produced from cysteine by *A. paspali*, but only rarely by *A. vinelandii*.

Nitrogenase activity is positive in vegetative cells of all normal cultures on nitrogen-free medium. The activity is decreased in older cultures and practically nonexistent when near encystment. During germination, the cell inside the cyst expresses nitrogenase activity before the rupture of the exocystorium coat (Funnell and Tchan, 1977). The optimum pH requirement for nitrogen fixation is close to neutrality. *A. beijerinckii* is more acid tolerant, with a pH range of 5–10 (Tchan, 1953; Jensen and Petersen, 1955).

Spontaneous mutants of *A. vinelandii* unable to fix molecular nitrogen (Nif⁻) have been isolated from cultures (Wyss and Wyss, 1950). However, *Azotobacter* mutants are often unstable and some classes of mutation have been difficult to obtain. This problem is not due to an inherent resistance to mutagenic agents, but appears to reside in segregation or stabilization of mutated DNA (Terzaghi, 1980a). Nif⁻ strains were used by Gordon and Brill (1972) to obtain Nif-derepressed strains (Nif-Drd), and Bishop et al. (1977) used Nif-Drd strains to prepare mutants capable of reducing nitrogen and excreting ammonia in the presence of NH_4^+. Ammonia-excreting mutants have also been obtained directly by ultraviolet mutagenesis of wild type *A. vinelandii* (Terzaghi, 1980a, b).

A. beijerinckii has been manipulated in similar ways, but with less success.

Bacteriophages isolated from soil using *A. vinelandii* and *A. chroococcum* as indicator strains are capable of lysing cultures of *A. chroococcum*, *A. vinelandii* and, to a lesser degree, *A. beijerinckii*. None of the phages are capable of affecting *Azomonas* (Hegazi and Jensen, 1973). No information on phage is available for *A. nigricans*, *A. armeniacus* and *A. paspali*.

Despite serologic studies involving agglutination and immunodiffusion techniques, there is no consensus regarding the serological relationship between *A. vinelandii*, *A. chroococcum* and *A. beijerinckii*; Norris and Kingham (1968) found some cross-reaction between *A. chroococcum* and *A. beijerinckii*, but not between *A. chroococcum* and *A. vinelandii*, whereas E. Chia (M.Sc. dissertation, University of Sydney, 1970) found that *A. vinelandii* cross-reacted with *A. chroococcum*, but that *A. beijerinckii* gave little or no cross-reaction with these two species. Chia did not record intergeneric cross-reactions but the data obtained by Norris and Kingham indicate that antiserum against *Azomonas agilis* reacted with *A. vinelandii*. Furthermore, serum produced against *A. paspali* cross-agglutinated *A. vinelandii* and *Azomonas macrocytogenes* strain 10EM, but the identification of the latter strain is in some doubt; it is serologically indifferentiable from *A. vinelandii*, but distinct from another strain (LH2) of *A. macrocytogenes* (Norris and Kingham, 1968; Tchan et al., 1980).

The immunofluorescence method has not detected any interspecific or intergeneric cross-reactions (Tchan and de Ville, 1970). More recent studies using rocket line immunoelectrophoresis have shown that some precipitation antigens are strain-specific, while others are species-specific. A thermoresistant antigen (γtr) is family specific and shared by *Azotobacter* and *Azomonas* (Tchan et al., 1980).

All species of *Azotobacter* are sensitive to streptomycin (0.2 μg/ml). Their sensitivities to chlortetracycline, oxytetracycline, polymyxin, sulfanilamide, penicillin and neomycin are variable depending upon strain.

A. chroococcum and *A. vinelandii* are resistant to phenol (0.05%). Benzoate up to 0.5% is tolerated by many strains of different species. No culture is resistant to 1 mM iodoacetate (Thompson and Skerman, 1979). Resistance to other chemicals is summarized in Table 4.47.

Azotobacter species are generally found in soil of slightly acid to alkaline pH, although in acid soil they can survive in niches of neutral reaction. Because of their relatively high requirement for phosphorus,

they are more commonly found in fertile soils. Only *A. paspali* is known to have a constant association with the root system of a plant, *Paspalum notatum* (Döbereiner, 1970), where availability of organic substances and a more suitable pH is conditioned by the plant in its rhizosphere. The restriction of *A. paspali* to the rhizosphere of *Paspalum* (it has never been isolated from other habitats) may be related to its reduced ability to utilize many organic substances commonly occurring as end products of fermentation in the soil environment. In addition, *A. paspali* is the only species antagonistic to Gram-positive bacteria, a property which may be advantageous for life in the rhizosphere.

In the experience of the authors, *A. chroococcum* is the most common *Azotobacter* species in soil. *A. vinelandii* is best known to biochemists and geneticists but its occurrence in nature seems restricted. Acid soils favor *A. beijerinckii* because of its tolerance to lower pH; however there is uncertainty regarding the identification of this species in past general surveys on the occurrence of *Azotobacter* in nature which should be remedied in future studies. Little information is available for *A. nigricans* and *A. armeniacus*, and the distribution of these two species in their ecological niches can not be stated with accuracy.

The presence of cysts in this genus confers resistance to desiccation, thus insuring survival in soil during dry periods; longevity under dry conditions may exceed 2300 years (Abd-al-Malek and Ishac, 1966).

Enrichment and Isolation Procedures

The enrichment and isolation techniques depend upon the ability of *Azotobacter* to grow by fixing nitrogen, using organic matter as the energy source.

Nonselective enrichment methods (Tchan, 1952). Five grams of soil from the soil surface or the rhizosphere is added to 45 ml of liquid Winogradsky's medium containing 1.0% glucose, and 5 ml of this suspension is diluted in 45 ml of the same medium. Both dilutions are incubated at 30°C.

Selective enrichment methods (see also Thompson and Skerman, 1979). Selective enrichment for various species of *Azotobacter* is based upon addition of selectively stimulatory or toxic substances to liquid Winogradsky's medium.

Azotobacter vinelandii can be selectively enriched by using L-rhamnose, ethylene glycol, erythritol or D-arabitol as the carbon source. Alternatively, 1.0% sodium benzoate or 0.1% phenol can be used to inhibit the growth of other *Azotobacter* species. Incubation at 37°C will also favor the development of *A. vinelandii*.

Azotobacter beijerinckii can be selectively enriched using L-tartrate, o-hydroxybenzoate, D-glucuronate or D-galacturonate. A pH of 6 or slightly less favors the growth of this species.

Azotobacter chroococcum is the most common species and needs no special enrichment for its isolation.

Azotobacter armeniacus can be enriched with caprylate but *Azotobacter vinelandii* will also grow in this medium. However, further isolation will differentiate these two species.

Azotobacter paspali can be isolated from the rhizosphere of *Paspalum notatum*. There is no selective medium for this species but an incubation temperature of 35–37°C will favor its growth.

There is no specific method to isolate *Azotobacter nigricans*.

Isolation Methods. The enrichment cultures are inspected daily. When macroscopic growth becomes apparent, the culture is examined using phase-contrast microscopy to detect the presence of *Azotobacter* cells, which are recognizable by their general morphology. Usually positive growth will be obtained within 2–5 days. Positive cultures are streaked onto nitrogen-free agar medium with glucose as the carbon source. For selective isolation, nitrogen-free agar media with appropriate selective substances should also be used. Sometimes nitrogen-free iron-deficient agar media can be used to advantage to encourage pigment production by some species. For soils containing high enough numbers of *Azotobacter*, the different species can be isolated directly on agar plates: 0.1–0.5 g dry soil or an appropriate dilution is spread on the surface of nitrogen-free agar medium.

Maintenance Procedures

For routine maintenance, *Azotobacter* should be subcultured at monthly or bimonthly intervals on Winogradsky's agar medium with sucrose, except that glucose should be substituted for sucrose in the case of *A. armeniacus* and certain other strains which prefer glucose.

Paraffin oil preservation method (Tchan, unpublished data). Ten milliliters of a suitable nitrogen-free agar medium is poured into a 28-ml capacity McCartney bottle and allowed to solidify to form a slope approximately 3 cm in length with a base 2 cm in height. The culture is inoculated in the middle of the surface of the slope with a single stroke. When the growth reaches about 1 mm in thickness, the culture is checked for purity. Then 5–7 ml of sterile paraffin (mineral) oil is poured into the bottle to completely cover the slope. (It is important that no agar medium should be in direct contact with air, otherwise the culture may dry out and some cultures may die.) Usually *Azotobacter* survives for many months (even years). Routine subculture at 6-month intervals is adequate. This method is particularly advantageous for frequently used stock cultures.

Lyophilization. Dense culture suspensions in 10% sucrose solution (Bascombe and Jackson, 1965) can be lyophilized and stored at 5°C. In the authors' experience, this method is not suitable for all cultures.

Procedures for Testing Special Characters

Cell morphology. Cultures grown for 24–48 h on both agar and liquid media should be used for the study of general morphology using phase-contrast microscopy. Motility may be observed in wet mounts, but it is advisable to incorporate some air bubbles in the mount to insure an adequate supply of oxygen. Motile cells are usually found near the edge of an air bubble.

Gram stain. The method of Tchan (Bunt and Tchan, 1955) is recommended.

Flagellar arrangement. To study the arrangement of the flagella on cells cultured on solid media, a sterile Pasteur pipette with the tip sealed is used to pick up the culture, so that the terminal 1 cm is covered with bacteria. The Pasteur pipette is placed in a test tube containing a 7- to 8-mm depth of distilled water and left standing, so that motile cells will swim out and be washed free of organic matter carried over from the culture medium. Small droplets of the cell suspension are then carefully deposited on a slide with minimal disturbance and examined for motility using phase contrast microscopy (Tchan, unpublished data). For examination of flagella, the bacteria are fixed and stained by the method of Rhodes (1958). Alternatively,

the flagella can be studied by electron microscopy using negative staining. In the case of liquid media, no special preparation is required before carefully transferring the culture to the slides, although sometimes difficulties with the flagella stain may occur due to the presence of organic matter in the medium.

Cysts. These resting forms are best studied with cultures at least 2 weeks old, grown on nitrogen-free medium containing 0.2% butanol. If the culture will not grow on butanol (e.g. *A. paspali*), nitrogen-free medium with 0.2% glucose can be used. The cysts may be stained with violamine (Winogradsky, 1938),* acridine orange† (Tchan, unpublished data) or with a mixture of neutral red and light green SF yellowish (Vela and Wyss, 1964). Examination by phase-contrast microscopy is sometimes helpful when violamine stain is used.

Production of homopolysaccharides. Any one of three basal agar media§ (Thompson and Skerman, 1979) is sterilized and enriched with an appropriate volume of 50% glucose, sucrose or raffinose solution (sterilized by filtration) to give a final sugar concentration of 5%. Production of diffusible homopolysaccharide is indicated by the development of a white opaque halo on sucrose and raffinose, but not on glucose. Production of colony-retained homopolysaccharide is recognized by the production of larger, more transparent and viscid colonies on sucrose or raffinose, than on glucose.

Pigment production. The production of diffusible pigments in the presence or absence of benzoate is not constant on iron-containing media, and is unreliable for strain differentiation. Diffusible pigments are readily produced on the following iron-deficient media (Thompson and Skerman, 1979). Any one of the three basal agar media§ is modified by omission of $FeSO_4 \cdot 7H_2O$ and reduction of $Na_2MoO_4 \cdot 2H_2O$ to 1 µg/ml. Glucose is added to give a concentration of 10 g/liter. The medium is inoculated by depositing a small drop of a suspension of culture on the surface of the agar. The plates are examined in daylight for diffusible pigment and under ultraviolet light (wavelength 364 nm) for fluorescent pigments. The medium of Stainer et al. (1966) is used for the production of nondiffusible pigments or, for those strains which will not grow on this medium, the basal medium of Thompson and Skerman is enriched with sodium gluconate (2.0 g/liter).

Acid production from sugars and alcohols. It is essential that the buffering capacity of the medium be low, in order not to mask the production of the small quantity of acid. The bacterial culture is inoculated onto the surface of the test medium.¶ The production of acid is indicated by a change in color of the medium to a definite yellow after 2, 5, 7 or 14 days.

Differentiation of the genus **Azotobacter** from other genera

Table 4.45 provides the characteristics that can be used to differentiate the genus *Azotobacter* from morphologically or physiologically similar genera.

Taxonomic Comments

Thompson and Skerman (1979), using numerical taxonomic analysis, indicated that *Azotobacter paspali* showed significant differences from the other species. These authors recognized that this species is similar to *A. vinelandii* in the production of diffusible and ultraviolet fluorescent pigments and its temperature range for growth. However the ability of *A. paspali* to use organic compounds (17 out of 159 tested) is

---

* Violamine stain: violamine "2 R," 1.0 g; phenol, 5.0 g; water, 100 ml. An air-dried and fixed smear is treated with violamine stain for 45 sec, washed and counterstained with dilute crystal violet (0.05% or less) for 1 min. After rinsing with water, the preparation is mounted in water for microscopic examination.
† Acridine orange stain: Cells are suspended in a 0.01% aqueous solution of acridine orange. One drop of the suspension is deposited on a slide, then covered with 2 drops of paraffin (mineral) oil. The drops are then spread in the same manner as for a blood smear. The preparation is mounted with a cover glass. A very thin monolayer of cell suspension can be found in the preparation where the depth of liquid is such that motile cells are immobilized to facilitate microscopic examination. When examined by fluorescence microscopy, the vegetative cells are green and the cysts have a green central body and a red coat.
§ Basal agar media (Thompson and Skerman, 1979). *Medium of Norris and Jensen (1958)* (g/liter): K_2HPO_4, 1.0; $CaCl_2 \cdot 2H_2O$, 0.1; $MgSO_4 \cdot 7H_2O$, 0.2; $FeSO_4 \cdot 7H_2O$, 0.05; $Na_2MoO_4 \cdot 2H_2O$, 0.005; Ionagar (Oxoid), 10.0; pH 7.3. *Acidic modification* (g/liter): K_2HPO_4, 0.3; KH_2PO_4, 0.7; $MgSO_4 \cdot 7H_2O$, 0.2; $FeSO_4 \cdot 7H_2O$, 0.05; $Na_2MoO_4 \cdot 2H_2O$, 0.005; Ionagar (Oxoid), 10.0; pH 6.2. *Enriched modification:* Norris and Jensen's medium plus 0.5% yeast extract (Difco).
¶ Test medium for acid production (g/liter): K_2HPO_4, 0.3; $CaCl_2 \cdot 2H_2O$, 0.1; $MgSO_4 \cdot 7H_2O$, 0.2; $FeSO_4 \cdot 7H_2O$, 0.05; $Na_2MoO_4 \cdot 2H_2O$, 0.005; bromthymol blue, 0.08; Ionagar No. 2 (Oxoid), 10.0; organic substrate being tested (added aseptically from a separately sterilized 10% solution), 10.0. The pH is adjusted to 7.1.

Table 4.45.

Differential characteristics of **Azotobacter** *and other morphologically or physiologically similar genera.*[a]

Characteristics	Azotobacter	Azomonas	Beijerinckia	Derxia	Azospirillum	Rhizobium/ Bradyrhizobium	Klebsiella	Pseudomonas	"Azotomonas"
Cell morphology:									
Pleomorphic: ovoid to rod shaped	+	+	−	−	−	−	−	−	−
Dumbell shaped with granule positioned at poles	−	−	+	−	−	−	−	−	−
Rod containing granules which give beaded appearance to the cell	−	−	−	+	−	−	−	−	−
Cell $\geq 2\ \mu$m diameter $\times \geq 3\ \mu$m length	+	+	−	−	−	−	−	−	−
Motility	D	+	+	+	+	+	−	+	+
Flagellar arrangement:									
Monotrichous	−	D	−	+	+[b]	D	−	D	−
Lophotrichous	−	D	−	−	−	−	−	D	−
Peritrichous	D	D	+	−	+[b]	D	−	−	+
Cysts (*sensu stricto*) produced	+	−	−	−	−	−	−	−	−
Nitrogen fixation under atmospheric partial pressure of O_2	+	+	+	+	−	−	−	−	−
Nitrogen fixation only under anaerobic or microaerophilic conditions	−	−	−	−	+	+	+[c]	−	−
Autotrophic use of H_2 to fix N_2	−	−	−	+	D[d]	D	−	−	−
Root-associated nitrogen fixation:									
Not producing root hypertrophy	D	−	−	−	+	−	−	−	−
Producing root hypertrophy	−	−	−	−	−	+	−	−	−
Mol% G + C of DNA	63.2–67.5	52–58.6	55–61	69–73	69–71	59–65	53–58	58–70	58–60.5

[a] Symbols: see standard definitions.

[b] When cultured in liquid media, *Azospirillum* strains have a single polar flagellum, but when cultured on solid media at 30°C they also form numerous lateral flagella having a shorter wavelength.

[c] Not all strains fix nitrogen.

[d] Some strains of *A. lipoferum* appear to be capable of hydrogen autotrophy (see the earlier article on the genus *Azospirillum* for additional information).

considerably less than that of other species (48 for *A. chroococcum* and 60 for *A. vinelandii*). The typical production of exceptionally long rods is a significant morphological difference from other *Azotobacter* species. In the numerical analysis, *A. paspali* fused at a low hierarchical level with *Azomonas macrocytogenes*, and the *A. paspali—A. macrocytogenes* group fused with the two other *Azomonas* species before fusing with *Azotobacter*, suggesting that *A. paspali* is not closely related to other *Azotobacter* species (Fig. 4.25). Further grounds of differentiation such as antagonism to Gram-positive bacteria and restricted habitat of the rhizosphere were used by these authors as support for the separation of *A. paspali* from other *Azotobacter* species.

Thompson and Skerman (1979) proposed a new genus, *Azorhizophilus* Thompson and Skerman 1979, to accommodate *A. paspali*. This view was opposed by De Smedt et al. (1980) who found that the rRNA cistron of *A. paspali* is almost identical with those of *A. chroococcum*, *A. beijerinckii*, *A. vinelandii* and *A. nigricans*; therefore *A. paspali* should be considered a normal member of the genus *Azotobacter*. Recent evidence obtained by rocket line immunoelectrophoresis supports this view, showing that *A. paspali* is not immunologically separate from other members of the genus (Tchan et al., 1983). In view of this, it is not considered desirable to place *A. paspali* in a new genus.

The genus *Azotobacter* is differentiated from *Azomonas* by the formation of drought-resistant cysts. This character also differentiates it from *Derxia* and *Beijerinckia*. Furthermore *Derxia* and *Beijerinckia* are morphologically and physiologically different from *Azotobacter*. *Azotobacter* is different from *Azospirillum* although *A. paspali*, like *Azospirillum*, is associated with the roots of plants. *A. paspali*, however, differs

from *Azospirillum* by its morphology, flagellar arrangement, its tolerance to atmospheric pO_2, and its ability to use sucrose as a carbon source for nitrogen fixation.

Some rhizobia can fix nitrogen nonsymbiotically but, unlike *Azotobacter*, can only do so under conditions of very low oxygen tension. Cells of *Rhizobium* are also much smaller than *Azotobacter* cells. *A. paspali* is superficially similar to the rhizobia in its association with the root system of a plant, but the symbiosis involving *A. paspali* is a loose system compared with the *Rhizobium* or even the *Azospirillum* symbiosis. Immunoprecipitation tests have shown no cross-reaction between *Azotobacter* and rhizobia using antiazotobacterian sera (Tchan et al., 1980; Ann M. Simpson, Bachelor of Science (Honors) dissertation, University of Sydney, 1980). Based on nucleic acid comparisons, *Rhizobium* belongs to the fourth super-family of De Smedt et al. (1980), whereas *Azotobacter* is part of the second super-family.

Although some members of the *Enterobacteriaceae*, including some *Klebsiella* strains, are capable of fixing nitrogen, *Azotobacter* cells are much larger. Unlike the enterobacteria, *Azotobacter* is unable to utilize lactose, and can fix nitrogen under a normal atmospheric partial pressure of oxygen.

"*Azotomonas*" can be separated from *Azotobacter* by its inability to fix molecular nitrogen. The cell is usually smaller than that of the members of *Azotobacter*. The mol % G + C of the DNA (T_m) of these organisms (58–60.5) also differs from *Azotobacter* (63.2–67.5) (De Smedt et al., 1980).

The inability of *Pseudomonas* to form cysts and its lack of nitrogen-fixing ability differentiate it from the members of *Azotobacter*. Unlike

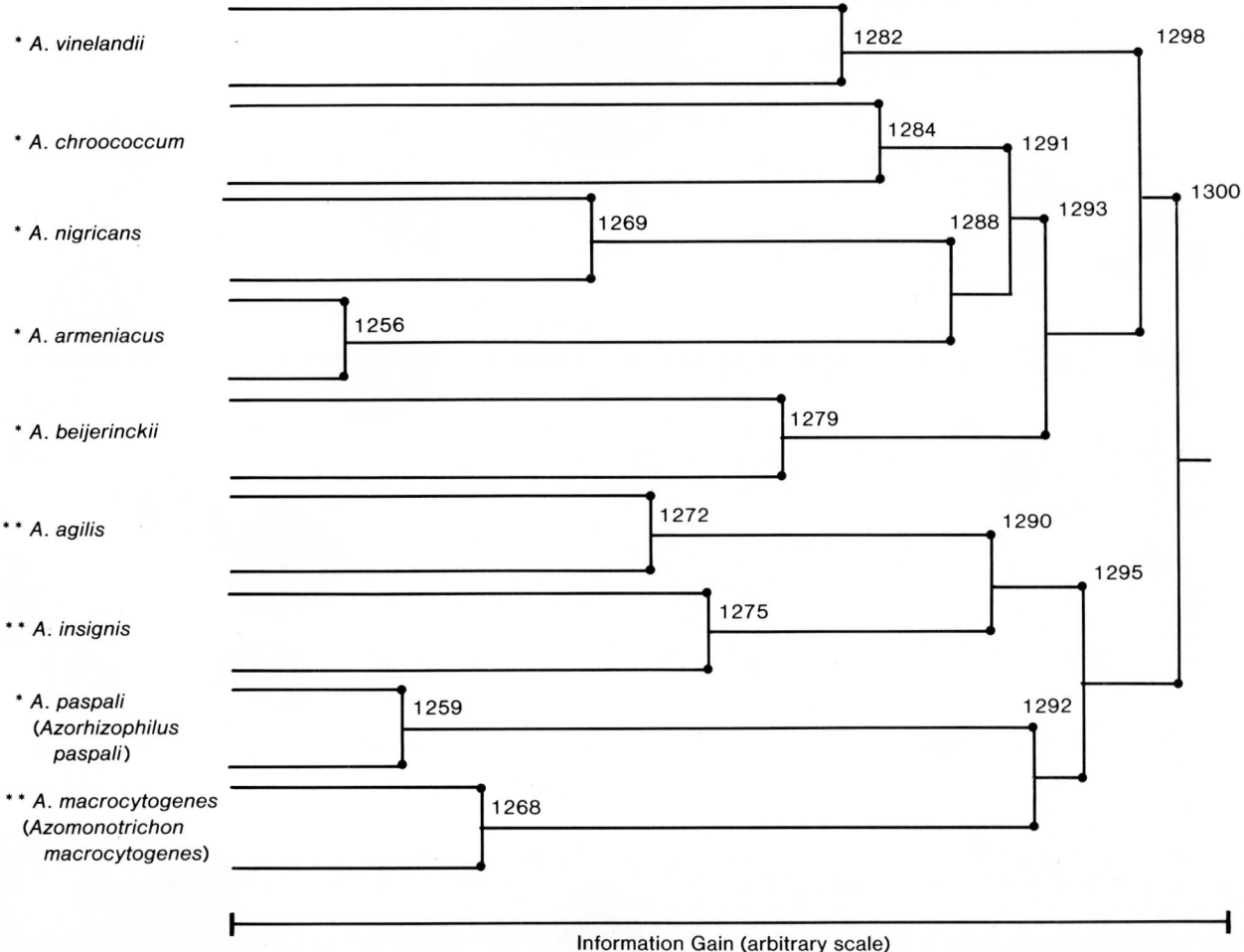

Figure 4.25. Hierarchical classification of *Azotobacteraceae*. The simplified dendrogram according to Thompson and Skerman is prepared to indicate the degree of similarity between different members of the family. The number at the first vertical bar of a species is related to the degree of similarity of different strains within the species (expressed as *Information Gain*, which increases as similarity decreases), e.g. 1281 for *A. vinelandii*. Other numbers at subsequent vertical bars denote the degree of similarity between different species (e.g. 1288 for *A. nigricans* and *A. armeniacus*) or between genera (e.g. 1300 for *Azotobacter* and *Azomonas*). Note the unusual position of *A. paspali*. It fuses with *A. macrocytogenes* at 1292 before this group fuses with the remaining members of *Azomonas* at 1295. (* denotes *Azotobacter*; ** denotes *Azomonas*). (Reproduced with permission from J. P. Thompson and V. B. D. Skerman, *Azotobacteraceae: The Taxonomy and Ecology of the Aerobic Nitrogen-Fixing Bacteria*, 1979, Academic Press, London.)

some *Pseudomonas* spp., the root association of *A. paspali* is not pathogenic.

Acknowledgments

The authors wish to acknowledge the technical assistance of Mrs. Z. Wyszomirska-Dreher for immunological analysis and photomicrography. Our thanks are extended to Miss M. Hailwood for her indefatigable efforts in helping to prepare the manuscript.

Further Reading

Döbereiner, J. 1970. Further research on *Azotobacter paspali* and its variety specific occurrence in the rhizosphere of *Paspalum notatum* Flügge. Zentralbl. Bakteriol. Parasitenkd. Infektionskr. Hyg. Abt. II. *124:* 224–230.
Jensen, H.L. 1954. The *Azotobacteriaceae*. Bacteriol. Rev. *18:* 195–214.
Tchan, Y.T. 1968. Importance of systematics of *Azotobacteriaceae* in the study of its ecology. Trans. 9th Int. Congr. Soil Sci., Adelaide *2:* 115–124.
Winogradsky, S. 1938. Etudes sur la microbiologie du sol et des eaux. Sur la morphologie et l'écologie des *Azotobacter*. Ann. Inst. Pasteur (Paris) *60:* 351–400.

Differentiation and characteristics of the species of the genus **Azotobacter**

The differential characteristics of the species of *Azotobacter* are presented in Table 4.46. Other characteristics of the species are indicated in Table 4.47.

List of the species of the genus **Azotobacter**

1. **Azotobacter chroococcum** Beijerinck 1901, 567.[AL]
chro.o.coc′cum. Gr. n. *chroa* color; Gr. n. *coccus* a grain; M.L. neut. n. *chroococcum* colored coccus.

The morphology is as depicted in Figs. 4.16 and 4.17. Motile cells from 1- to 2-day-old cultures have peritrichous flagella with a wavelength of 2.5–3.3 μm and an amplitude of 0.37–0.76 μm.

Table 4.46.
Differential characteristics of the species of the genus **Azotobacter**[a]

Characteristics	1. A. chroococcum	2. A. vinelandii	3. A. beijerinckii	4. A. nigricans	5. A. armeniacus	6. A. paspali
Motility	+	+	−	−	+	+
Long filaments in young culture	−[b]	−[b]	−[b]	−[b]	−[b]	+
Water soluble pigments:						
Yellow-green fluorescent[c]	−	+	−	−	−	+
Green	−	d	−	−	−	−
Brown-black	−	−	−	d	−	−
Brown-black to red-violet	−	−	−	+	+	−
Red-violet	−	d	−	d	+	+
Utilization as carbon source:						
Rhamnose	−	+	−	−	−	−
Caproate	+	+	−	−	−	−
Caprylate	−	+	−	−	+	−
meso-inositol	−	+	d	−	d	−
Mannitol	+	+	d	d	+	−
Malonate	d	+	+	d	−	−

[a] For symbols see standard definitions.
[b] These species may sporadically produce filamentous forms of different lengths.
[c] On iron-deficient medium: see Procedures for Testing Special Characters.

Colonies are not slimy but variant colony forms may arise due to the quantity of extracellular polysaccharides produced.

Physiological and nutritional characteristics of the species are presented in Tables 4.46 and 4.47. Ammonium and nitrate are used as nitrogen sources with inhibition of nitrogen fixation. Glutamate is not utilized.

Starch is hydrolyzed either by α-amylase or β-amylase.

The mol% G + C of the DNA is 65.8–67.5 (T_m).

Type strain: ATCC 9043 (R. L. Starkey's strain 43; WR-88).

2. **Azotobacter vinelandii** Lipman 1903, 238.[AL] (*Azotobacter miscellum* Pshenin 1964, 684.)

vine.lan'di.i. M.L. gen. n. *vinelandii* of Vineland; named after Vineland, New Jersey, where the species was first isolated.

The morphology is similar to that of *A. chroococcum*. Motile cells from 1- to 2-day-old cultures have peritrichous flagella with a wavelength of 2.4–2.9 μm and an amplitude of 0.39–0.55 μm (Fig. 4.24).

Colonies are not slimy but variant colony forms may arise due to the quantity of extracellular polysaccharides produced. Diffusible pigments are produced.

Physiological and nutritional characteristics of the species are presented in Tables 4.46 and 4.47.

Ammonium and nitrate are used as nitrogen source with inhibition of nitrogen fixation.

Starch is not hydrolyzed.

The mol% G + C of the DNA is 64.9–66.5 (T_m).

Type strain: ATCC 478 (WR-100).

Further comments. Nonmotile strains have been reported. They are similar to "*Azotobacter non-vinelandii*" (Derx, 1951), which is not considered to have been validly published.

3. **Azotobacter beijerinckii** Lipman 1904, 248.[AL] (*Azotobacter beijerinckii* subsp. *acidotolerans* Tchan 1953, 83.)

bei.jer.inck'i.i. M.L. gen. n. *beijerinckii* of Beijerinck; named after M. W. Beijerinck of Delft, the Dutch microbiologist.

The morphology is similar to that of *A. chroococcum*. The cells are nonmotile.

Colonies are smooth but variants may arise due to the quantity of extracellular polysaccharide produced.

Physiological and nutritional characteristics of the species are presented in Tables 4.46 and 4.47.

Ammonium and nitrate are used as nitrogen sources with inhibition of nitrogen fixation.

Some strains hydrolyze starch by β-amylase. All strains hydrolyze urea.

The mol% G + C of the DNA is 66.0–66.2 (T_m).

Type strain: ATCC 19360 (WR-135). The type strain originated from a subculture of NCIB 8948 (WR-147).

Further comments. Two sub-groups of *A. beijerinckii* have been delineated. The acid-tolerant subgroup corresponds with "*A. beijerinckii* subsp. *acidotolerans*" Tchan. The other subgroup is distinguished by its sensitivity to 0.05% phenol or 40 μg/ml of diamond fuchsin, its inability to use sorbitol or aconitate, and its failure to produce diffusible homopolysaccharides.

4. **Azotobacter nigricans** Krasil'nikov 1949, 506.[AL] (*Azotobacter beijerinckii* subsp. *achromogenes* Jensen and Petersen 1954, 106.)

ni'gri.cans. L. adj. *nigricans* black.

The morphology is as depicted in Figure 4.18. The cells are nonmotile.

Colonies are smooth but variants occur due to the quantity of extracellular polysaccharide produced.

Physiological and nutritional characteristics of the species are presented in Tables 4.46 and 4.47.

The production of diffusible homopolysaccharide from sucrose or raffinose is strain dependent.

Ammonium and nitrate are used as nitrogen source in preference to nitrogen.

The mol% G + C of the DNA is 64.5 (T_m).

Type strain: UQM 1967.

5. **Azotobacter armeniacus** Thompson and Skerman 1981, 215.[VP] (Effective publication: Thompson and Skerman 1979, 272.) (*Azotobacter agilis* subsp. *armeniae* Kirakosyan and Melkonyan 1964, 41; *Azotobacter vitreus* subsp. *armeniae* Kirakosyan and Melkonyan 1964, 41.)

ar.me.ni.a'cus. M.L. neut. n. *armeniacus* named after Armenia, U.S.S.R., where the species was isolated.

The morphology is as described for the genus. Motile with peritrichous flagella having a wavelength of 2.4–2.9 μm and an amplitude of 0.32–0.51 μm.

Colonies are smooth, high convex, glistening and viscid.

Physiological and nutritional characteristics of the species are presented in Tables 4.46 and 4.47.

Diffusible homopolysaccharides are produced by some strains when grown on sucrose but not when grown on raffinose. Starch is hydrolyzed by either α-amylase or β-amylase, depending on the strain.

Ammonium, nitrate and glutamate are not used as a sole source of nitrogen for growth.

Table 4.47.
Other characteristics of the species of the genus **Azotobacter**[a]

Characteristics	1. A. chroococcum	2. A. vinelandii	3. A. beijerinckii	4. A. nigricans	5. A. armeniacus	6. A. paspali	Characteristics	1. A. chroococcum	2. A. vinelandii	3. A. beijerinckii	4. A. nigricans	5. A. armeniacus	6. A. paspali
Peritrichous flagella	+	+	−	−	+	+	β-phenylpropionate	d	+	d	−	−	−
Nitrogen fixation occurs at pH:[b]							n-Butyrate	+	+	+	d	+	−
5.0–5.5	−	−	d	d	−	−	Glutarate	−	+	−	−	d	−
6.0	−	+	+	d	−	d	Itaconate	d	d	d	d	−	−
6.5–9.5	+	+	+	+	+	+	Oxaloacetate	d	+	d	+	+	+
10.0	+	+	+	+	d	+	DL-β-Hydroxybutyrate	d	d	+	d	+	−
Growth at a temperature of:							Benzoate	+	d	+	−	−	−
9°C	−	−	d	d	−	−	D-Glucuronate	−	d	+	−	−	−
14°C	d	+	+	d	−	+	D-Galacturonate	−	d	+	−	+	−
18°C	+	+	+	+	d	+	Glycolate	+	−	−	−	−	−
32°C	+	+	+	+	+	+	D-Galactose	d	+	d	d	−	−
37°C	d	+	−	−	d	+	Trehalose	d	−	d	+	+	−
Acid production from carbohydrates in weakly buffered media[c]	d	d	d	d	+	+	Maltose	+	+	d	d	+	−
Diffusible homopolysaccharides produced[c]	+	−	d	+	d	−	Raffinose	+	d	d	d	d	−
Colony-retained homopolysaccharides produced[c]	−	−	−	−	−	−	L-Fucose	−	−	−	−	−	−
Peroxidase	d	+	d	d	d	−	Dulcitol	d	d	−	−	−	−
Oxidase	+	+	+	+	d	+	Propan-1-ol	d	+	d	d	−	d
Nitrate reduced to nitrite	+	+	+	+	−	−	Butan-1-ol	+	+	+	d	d	−
Production of H₂S from:							Glycerol	d	+	d	−	−	−
Thiosulfate	d	+	d	d	−	+	Sorbitol	+	+	d	d	+	−
Cysteine	−	d	−	−	−	d	Butan-2-ol	−	−	−	−	−	−
Growth in presence of 1% NaCl	+	+	d	d	+	d	Susceptibility to antimicrobial agents:						
Habitat:							Streptomycin, 0.2 µg/ml	S	S	S	S	S	S
Soil	+	+	+	+	+	+	Chlortetracycline, 5 µg/ml	d	R	d	d	R	R
Rhizosphere	−	−	−	−	−	+	Oxytetracycline, 0.2 µg/ml	R	R	R	R	R	R
Utilization as sole carbon source:[d]							Chloramphenicol, 25 µg/ml	d	R	d	R	d	d
Fructose, glucose, sucrose, ethanol, acetyl methyl carbinol, acetate, fumarate, lactate, DL-malate, pyruvate, α-oxoglutarate, gluconate, or succinate	+	+	+	+	+	+	Sulfanilamide, 25 µg/ml	d	d	d	d	R	S
							Penicillin G, 5 U/ml	d	d	d	R	S	R
							Phenol, 0.05%	R	R	d	d	S	d
							Sodium benzoate, 0.5%	d	R	d	R	R	d
							Sodium fluoride, 0.01 M	d	R	R	d	R	R
							Mercuric chloride, 10 µg/ml	d	d	d	d	R	R
Propionate	+	+	+	−	d	−	Iodoacetate, 1 mM	S	S	S	S	S	S

[a] For symbols see standard definitions; also S, susceptible, and R, resistant.

[b] Results of Thompson and Skerman (1979) on solid media. Machado and Döbereiner (1969) found essentially no growth of *A. paspali* above pH 8.0, in their liquid medium.

[c] See Procedures for Testing Special Characters.

[d] See also Table 4.46 for additional compounds. The following substances are used by at least 11% of strains of *Azotobacter vinelandii*: ethylene glycol, erythritol, adipate, pimelate, suberate and sebacate; they are used by no other species. L-sorbose, D-arabitol and p-cresol are only used by one strain of *A. chroococcum* or *A. beijerinckii*.

The mol% G + C of the DNA is 63.5–65.0 (T_m).

Type strain: N28 of Kirakosyan (equals WR-136).

6. **Azotobacter paspali** Döbereiner 1966, 364.[AL] (*Azorhizophilus paspali* (Döbereiner 1966) Thompson and Skerman 1981, 215.[VP])

pas.pal′i. M.L. gen. n. *paspali* of *Paspalum*; named after *Paspalum*, the generic name of a grass.

The morphology is depicted in Figure 4.19. Motile by peritrichous flagella having a wavelength of 2.3–2.8 µm and an amplitude of 0.38–0.52 µm (Fig. 4.23). Some strains have curly flagella with a wavelength of 1.1–1.4 µm and an amplitude of 0.25–0.34 µm.

Colonies are opaque with an entire or undulate edge, raised, usually not evenly convex, butyrous with a dull or rough surface.

Physiological and nutritional characteristics are presented in Tables 4.46 and 4.47.

Diffusible homopolysaccharides are not produced from sucrose and raffinose.

Starch is not hydrolyzed.

Ammonium and nitrate, but not glutamate, can be used as the sole nitrogen source for growth (Thompson and Skerman, 1979), but nitrogenase activity is not affected by the presence of nitrate (Döbereiner and Day, 1975).

The organism has never been isolated from environments other than the rhizosphere of the grass *Paspalum notatum*.

The mol% G + C of the DNA is 63.2–64.6 (T_m).

Type strain: ATCC 23833 (Ax-8; WR-129).

Genus II. Azomonas Winogradsky 1938, 391[AL]

YAO-TSENG TCHAN AND PETER B. NEW

A.zo.mon'as. Gr. adj. *a* not; Gr. n. *zoê* life; Gr. n. *azoê* not sustaining life, nitrogen; Gr. n. *monas* a unit, monad; M.L. fem. n. *Azomonas* nitrogen monad.

Cells 2.0 μm or more in diameter and of various lengths, ranging from rods to ovoid to coccoid in shape. Occur singly, in pairs or in clumps. Pleomorphism is generally present. **Generally Gram-negative, sometimes Gram-variable. Do not produce endospores or cysts. Motile** by peritrichous or polar flagella. **Aerobic,** but can also grow under decreased oxygen tensions. **Water-soluble pigments and fluorescent pigments** are produced by nearly all strains. Chemoorganotrophic, using sugars, alcohols and the salts of organic acids for growth. **Nitrogen fixers; generally fix nonsymbiotically at least 10 mg of molecular nitrogen/g of carbohydrate (usually glucose) consumed. Molybdenum is required for nitrogen fixation.** Nonproteolytic. Can utilize ammonium salts and certain amino acids as sources of nitrogen. **Catalase-positive.** The optimum pH for nitrogen fixation is close to neutrality, but certain strains can also fix nitrogen at a pH of 4.6–4.8. Occur in soil and water. The mol% G + C of the DNA is 52–58.6 (T_m).

Type species: *Azomonas agilis* Winogradsky 1938, 391.

Further Descriptive Information

On nitrogen-free glucose medium* young cells of all species are ellipsoidal, 2.4–3.9 μm × 1.6–2.8 μm, occurring singly or in pairs. Rod-shaped cells sometimes occur. In older cultures *A. agilis* exhibits some variation in shape. *A. insignis* retains its shape but shows changes in size. *A. macrocytogenes* produces very large cells and sometimes spindle-shaped or filamentous forms, especially when ethanol is used as the carbon source. On peptone-yeast extract medium† distortion of cells is apparent in *A. macrocytogenes* with irregular, filamentous forms predominating. *A. insignis* does not show any distortion, but the cells appear thinner than when grown in glucose medium. *A. agilis* is less affected and does not show distortion (Thompson and Skerman, 1979).

Metachromatic and sudanophilic granules can be observed in *A. agilis* and *A. macrocytogenes*. No cysts are formed by any species; however, in *A. macrocytogenes* cells emerging from capsular material may be found and should not be mistaken for germinating cysts (Jensen, H., 1955; Tchan, 1968).

A. agilis and *A. insignis* are uniformly Gram-negative. *A. macrocytogenes* tends to be Gram-variable; usually both Gram-positive and Gram-negative cells may be seen in the same film, and on lactate agar Gram-negative cells with distinct Gram-positive granules are seen. The enlarged cells are often markedly Gram-positive (H. Jensen, 1955).

All members of the genus are motile, with the flagellar arrangement being dependent on the species (see Table 4.48).

All species are obligate aerobes, but nitrogen fixation can be enhanced by a decreased oxygen tension (0.04 atm).

On nitrogen-free agar medium with a sugar as the carbon source, *A. agilis* grows at a relatively rapid rate, forming colonies 2–6 mm in diameter after incubation for 7 days at 30°C. The colonies are opaque, glistening, smooth and convex, and variants are rarely observed. *A. insignis* produces similarly sized, translucent, glistening, smooth colonies. *A. macrocytogenes* forms colonies ranging from 1–6 mm in diameter after 1 week of incubation; initially, they may be coarsely wrinkled but later they become smooth. On sucrose agar‡ *A. macrocytogenes* produces voluminous and moist colonies due to the formation of colony-retained homopolysaccharides. No species of *Azomonas* produces diffusible homopolysaccharides; i.e. there is no formation of a diffuse halo around the colonies.

Azomonas species do not produce insoluble pigments. On iron-deficient agar medium* a yellow-green diffusible pigment is produced. On other media red-violet diffusible pigments are produced by all species. *A. insignis* differs from other azomonads by its production of a brown-black pigment in the presence of benzoate; it is also the only species that fails to form a fluorescent pigment.

In liquid media *A. agilis* produces a pellicle and a dispersible sediment, but no turbidity; in contrast, *A. macrocytogenes* and *A. insignis* produce turbidity in addition to a pellicle and bottom sediment.

Molybdenum is required for nitrogen fixation, but lower levels are needed than by *Azotobacter* (Jensen, H., 1955). The optimum pH requirement for nitrogen fixation is near neutrality, but *A. macrocytogenes* is acid tolerant (Jensen, H., 1955).

The optimum temperature for growth is approximately 30°C, although some strains of *A. agilis* prefer 37°C. The minimum temperature for growth is generally close to 14°C; however, some strains, particularly of *A. insignis*, are able to grow at 9–10°C.

No acid or gas is produced on buffered glucose medium. Only *A. macrocytogenes* produces titratable acidity (>10 microequivalents/ml within 5 days).

The following organic substances are used as sole carbon and energy sources by all species of the genus: D-fructose, D-glucose, ethanol, acetate, succinate, fumarate, lactate, malate, pyruvate, gluconate, α-oxoglutarate. Only *A. macrocytogenes* can use D-galactose, melizitose, mannitol and sorbitol. Maltose and sucrose are used by *A. macrocytogenes* and most strains of *A. agilis*. Caproate, malonate and mucate are used by *A. agilis* and *A. insignis* but not by *A. macrocytogenes*.

Catalase is present in all species. A positive oxidase reaction is given by the majority of *A. agilis* and *A. insignis* but rarely by *A. macrocytogenes*. Peroxidase is produced by the other species, but not by *A. macrocytogenes*.

Nitrate is reduced to nitrite by many strains of *A. insignis*, rarely by *A. agilis* and never by *A. macrocytogenes* (Becking, 1962; Thompson and Skerman, 1979). None of the species denitrifies. H_2S can be produced from thiosulfate by nearly all the cultures of *A. macrocytogenes* and *A. agilis* but by only a few strains of *A. insignis*. The production of H_2S from cysteine by *A. macrocytogenes* and *A. agilis* is variable. *A. insignis* fails to produce H_2S from cysteine.

When one strain each of *A. agilis* and *A. macrocytogenes* was tested by Terzaghi (1980) for derepression of nitrogenase (ability to fix nitrogen in the presence of combined nitrogen), the *A. agilis* strain only was partially derepressed (Nif-Drd).

Hegazi and Jensen (1973) reported that bacteriophages isolated using *A. chroococcum* and *A. vinelandii* as hosts were unable to affect *A. agilis*, *A. insignis* and *A. macrocytogenes*.

Antisera produced against *A. agilis* and *A. insignis* do not cross-agglutinate with members of *Azotobacter* (Tchan, unpublished data) but Norris and Kingham (1968) indicated that antisera against *A. agilis* cross-react with *A. vinelandii*. This discrepancy may be due to strain differences or to variability in immune response between animals.

Within the genus *Azomonas*, serum against *A. insignis* cross-agglutinates with *A. agilis* (Tchan, unpublished data).

Immunodiffusion produces numerous cross-precipitation lines. Rocket-line immunoelectrophoresis indicated that *Azomonas* and *Azotobacter* can be differentiated by the unshared precipitation peaks using an appropriate antiserum (Tchan et al., 1980; Ann M. Simpson, Bach-

---

* See the genus *Azotobacter*, footnote * on p. 220.

† See the genus *Azotobacter*, footnote † on p. 221.

‡ See the genus *Azotobacter*, *Procedures for Testing Special Characters*.

elor of Science (Honors) dissertation, University of Sydney, 1980). The thermoresistant antigen γ_{tr} is shared by *Azomonas* and *Azotobacter* (Tchan et al., 1980).

All species of the genus are sensitive to streptomycin (0.2 μg/ml). Their sensitivities to chlortetracycline, oxytetracycline, polymyxin, sulfanilamide, penicillin and neomycin vary depending upon strain. Only *A. insignis* is entirely sensitive to penicillin (Thompson and Skerman, 1979).

The members of this genus are resistant to phenol, benzoate and mercuric chloride to varying degrees. Most strains tolerate sodium fluoride up to 0.01 M. The resistance of *A. agilis* to iodoacetate can be used for selective isolation of this species (Thompson and Skerman, 1979).

A. agilis and *A. insignis* are aquatic bacteria and *A. macrocytogenes* is found in soil, although no information is available as to whether *A. macrocytogenes* also has an aquatic habitat. Thus it appears that the genus is ecologically heterogeneous. Only *A. macrocytogenes* is resistant to desiccation (five out of seven strains survived desiccation for at least 1 month (Thompson and Skerman, 1979)), as the other members have no such need. All species have a salt tolerance of up to 1%, suggesting that the aquatic members are capable of living in contaminated water where concentrations of organic matter and mineral salts can be relatively high.

Enrichment and Isolation Procedures

The enrichment and isolation techniques depend upon the ability of *Azomonas* to grow by fixing nitrogen, using organic matter as the energy source.

Nonselective enrichment methods (Tchan, 1952). These methods are the same as those described for the genus *Azotobacter.*

Selective enrichment methods (Thompson and Skerman, 1979). Media containing either 1% sodium benzoate (Derx, 1951b; V. Jensen, 1955) or 1 mM iodoacetate are selective for *A. agilis.* Incorporation of 1% benzoate into nitrogen-free sucrose medium having a pH of 6.0, and incubation of the cultures at a decreased incubation temperature (10–12°C) favors the growth of *A. macrocytogenes.*

No selective enrichment method is available for *A. insignis.*

Isolation methods. The enrichment cultures are inspected daily. When macroscopic growth becomes apparent, the culture is examined by phase-contrast microscopy to detect the presence of *Azomonas* cells (ovoid cells 2.5 × 5.0 μm, or sometimes filamentous forms). Growth will usually occur within 2–5 days. Positive cultures are streak plated on nitrogen-free agar medium with glucose as the carbon source. For selective isolation, nitrogen-free agar media with appropriate selective

Table 4.48.

Characteristics differentiating the species of the genus **Azomonas**[a]

Characteristics	1. A. agilis	2. A. insignis	3. A. macro-cytogenes
Presence of enlarged cells in media with ethanol	−	−	+
Flagellar arrangement:			
Peritrichous	+	−	−
Lophotrichous	−	+	−
Monotrichous	−	−	+[b]
Gram-reaction	−	−	v
Formation of colony-retained homopolysaccharides	−	−	+
Diffusible pigments:			
Brown-black, on benzoate medium	−	d	−
Blue-white fluorescent, on iron-deficient medium[c]	+	−	d
Utilization as carbon sources:			
Mannitol	−	−	+
Maltose	d	−	+
Malonate	+	+	−

[a] Symbols: see standard definitions.
[b] Rarely, two flagella may occur at one pole.
[c] See Procedures for Testing Special Characters.

substances should also be used. Sometimes nitrogen-free iron-deficient agar can be used to advantage to encourage pigment production by some species.

Maintenance Procedures

For routine maintenance, *Azomonas* strains should be subcultured at monthly or bimonthly intervals on Winogradsky's agar medium* with glucose.

Paraffin oil preservation method (Tchan, unpublished data) and lyophilization method. These methods and their shortcomings are the same as those described for the genus *Azotobacter.*

Procedures for Testing Special Characters

Methods for determining cell morphology, Gram-reaction, flagellar arrangement, production of homopolysaccharides, pigment production and acid production from organic substrates are the same as those described for the genus *Azotobacter.*

Differentiation of the genus **Azomonas** from other genera

See Table 4.48 in the chapter on the genus *Azotobacter* for characteristics that can be used to differentiate *Azomonas* from morphologically or physiologically similar genera.

Taxonomic Comments

The genus *Azomonas* can be separated from *Azotobacter* by the lack of cyst formation. However the taxonomic position of *A. macrocytogenes* has presented difficulties since the initial description of the organism by H. Jensen (1955). The early statement regarding the formation of the microcyst was not precise. Jensen in his initial description indicated: "Formation of cysts could not be seen with certainty but took place possibly at 5–10°C where a number of smaller round cells with clearly visible cell walls were seen . . . ." However, electron microscopic investigation (Tchan, 1968) did not reveal the presence of typical *Azotobacter* microcysts, although vegetative cells escaping from capsu-

lar-like material were demonstrated. Baillie et al. (1962) moved *A. macrocytogenes* from *Azotobacter* and placed it in the genus *Azomonas.* In the eighth edition of *Bergey's Manual of Determinative Bacteriology* (1974), *A. macrocytogenes* was classified in *Azomonas*, a noncyst forming genus.

The existence of microcysts in *A. macrocytogenes* was again proposed by Thompson and Skerman (1979). Microcyst-like cells were produced at 28°C by four out of seven strains examined. In addition, the ability of *A. macrocytogenes* to survive desiccation was used by these authors as supporting evidence for the presence of microcysts. However, the production of unusually large cells, the relative tolerance to acid conditions, the monotrichous flagellation and the unusual tendency to Gram-positiveness differentiate this organism from *Azotobacter.* Numerical analysis (Thompson and Skerman) showed that *A. macrocytogenes* fused at a low hierarchical level with *A. paspali.* At a higher level the *A. macrocytogenes-A. paspali* group fused with the two other *Azo-*

---

* Defined in preceding article on *Azotobacter.*

Table 4.49.
Other characteristics of the species of the genus **Azomonas**[a]

Characteristics	1. A. agilis	2. A. insignis	3. A. macro-cytogenes
Nitrogen fixation occurs at pH:			
5.5	−	−	+
6.0	d	−	+
6.5–10.0	+	+	+
Acid production from carbohydrates in weakly buffered media	−	−	+
Titratable acidity = >10 microequivalents/ml	−	−	d
Growth at a temperature of			
9°C	d	d	d
14–28°C	+	+	+
32°C	+	+	d
37°C	+	−	−
Resistance to 50°C for 10 min	d	−	d
Colony-retained homopolysaccharides produced	−	−	+
Pigmentation:			
Diffusible red-violet	d	+	+
Diffusible red-purple	−	d	−
Diffusible yellow-green on iron-deficient media[b]	+	d	d
Diffusible blue-white fluorescent on iron-deficient media[b]	+	−	d
Utilization as sole carbon source:[c]			
D-fructose, D-glucose, ethanol, acetate, succinate, fumarate, lactate, pyruvate, gluconate or α-oxoglutarate	+	+	+
DL-Malate	+	+	d
D-Galactose, sorbitol	−	−	+
Trehalose	−	−	d
Raffinose	d	−	+
Propan-1-ol, butan-1-ol	d	d	d
Acetyl methyl carbinol, n-butyrate	+	d	+
Propionate, oxaloacetate	+	d	d
Glutarate, benzoate	−	−	−
Growth in presence of 1% NaCl	+	d	+
Habitat:			
Soil	−	−	+
Water	+	+	−
Susceptibility to antimicrobial agents:			
Streptomycin, 0.2 µg/ml	S	S	S
Chlortetracycline, 5.0 µg/ml	R	R	R
Oxytetracycline, 0.2 µg/ml	R	R	R
Chloramphenicol, 25 µg/ml	R	R	R
Sulfanilamide, 25 µg/ml	d	d	d
Penicillin G, 5 U/ml	d	S	d
Phenol, 0.05%	R	d	R
Sodium benzoate, 0.5%	R	d	R
Sodium benzoate, 1.0%	d	S	R
Sodium fluoride, 0.01 M	d	R	d
Mercuric chloride, 10 µg/ml	R	d	d
Iodoacetate, 1 mM	d	S	S

[a] For symbols see standard definitions; also S, susceptible, and R, resistant.
[b] See Procedures for Testing Special Characters.
[c] See also Table 4.48 for additional compounds.

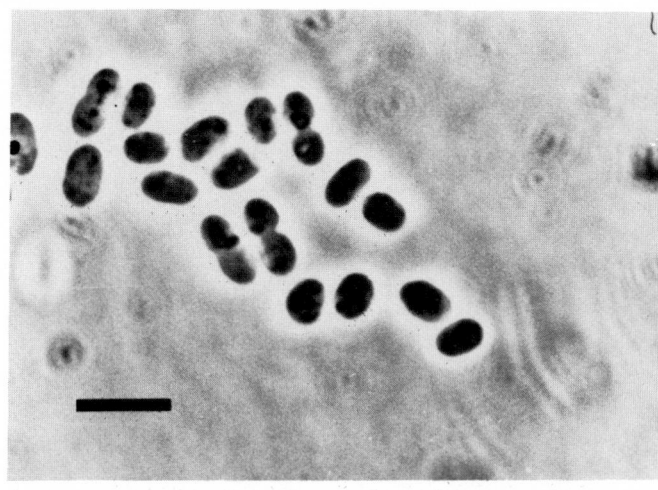

Figure 4.26. Cells of *Azomonas agilis* (phase-contrast microscopy). *Bar*, 10 µm.

monas species before fusing with *Azotobacter* (see Fig. 4.25 in the article on *Azotobacter*).

Thompson and Skerman (1979) proposed a new genus, *Azomonotrichon* Thompson and Skerman 1979, 297, for *A. macrocytogenes*. However, the rRNA cistrons of *Azomonas insignis* and *Azomonas agilis* differ as much from each other as they do from *A. macrocytogenes* and from *Azotobacter* (De Smedt et al., 1980). Therefore, if *A. macrocytogenes* should be placed in a new genus, another new genus should be created for *A. agilis*, to separate it from *A. insignis*. However, Thompson and Skerman have retained *A. agilis* and *A. insignis* in the same genus, *Azomonas*.

Further investigation on phenotypic analysis and on genome comparisons are needed to clarify the taxonomic position of members of the *Azotobacteraceae*, so it is desirable to temporarily delay the recognition of *Azomonotrichon*.

Azomonas can be distinguished from *Derxia* and *Beijerinckia* by its physiology and morphology. The similarity between *A. macrocytogenes* and *Beijerinckia* in their ability to produce polysaccharides and acid is only superficial. The very tenacious polysaccharide and the larger quantity of acid produced by *Beijerinckia* are absent in *A. macrocytogenes*.

Azomonas can be differentiated from *Azospirillum* by its morphology and the ability of some strains to utilize sucrose from nitrogen fixation. Furthermore *Azospirillum* is unable to fix nitrogen at an atmospheric partial pressure of oxygen.

There is no problem in differentiating *Azomonas* from *Rhizobium*; even though some rhizobia are capable of fixing nitrogen without their symbiont, none of the species of *Azomonas* is known to associate with the root system of plants. Also, immunoprecipitation methods have shown no cross-reaction with rhizobia (Tchan et al., 1980; Ann M. Simpson, Bachelor of Science (Honors) dissertation, University of Sydney, 1980). In addition, on the basis of nucleic acid comparisons, *Rhizobium* belongs to the fourth super family of De Smedt et al. (1980), whereas *Azomonas* is part of the second super family.

Some members of the *Enterobacteriaceae*, including *Klebsiella*, are capable of fixing nitrogen but not under a normal atmospheric partial pressure of oxygen. *Azomonas* cells also differ from *Klebsiella* in being motile and much larger, as well as in their inability to utilize lactose.

Pseudomonas lacks the ability to fix nitrogen and can thus be differentiated from *Azomonas*.

Acknowledgments

We acknowledge the technical help provided by Mrs. Z. Wyszomirska-Dreher in the immunological investigation and photomicrography. Our thanks are extended to Miss M. Hailwood for assistance in the preparation of the manuscript.

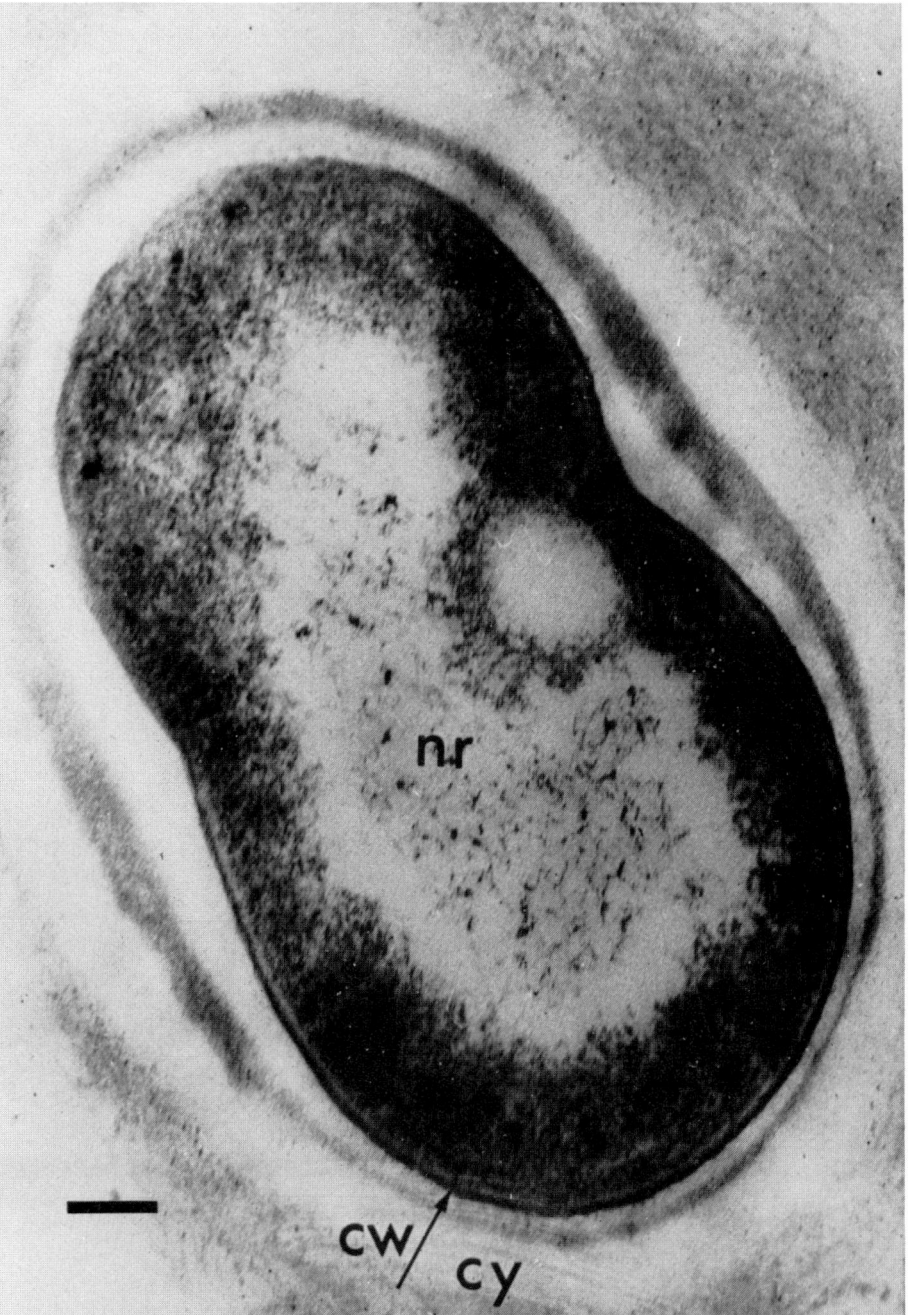

Figure 4.27. Electron micrograph of *Azomonas macrocytogenes*. The nuclear region (*nr*) and cell wall (*cw*) are readily visible. The structure *cy* probably consists of polysaccharide, but can be mistaken for the cyst coat of *Azotobacter*. *Bar, 0.2 μm.*

Differentiation and characteristics of the species of the genus **Azomonas**

The differential characteristics of the three species of *Azomonas* are presented in Table 4.48. Other characteristics of the species are indicated in Table 4.49.

List of the species of the genus **Azomonas**

1. **Azomonas agilis** (Beijerinck 1901) Winogradsky 1938, 400.[AL] (*Azotobacter agilis* Beijerinck 1901, 577.)

a'gi.lis. L. adj. *agilis* quick, agile.

The morphology is as depicted in Figure 4.26. The cells are motile by means of peritrichous flagella having a wavelength of 1.9–2.1 μm and an amplitude of 0.30–0.84 μm.

Physiological and nutritional characteristics are presented in Tables 4.48 and 4.49. Homopolysaccharides are not produced from sucrose or raffinose. Starch is not hydrolyzed. Ammonium ions can be used as a sole source of nitrogen for growth. Nitrate is used only rarely. Glutamate is not used.

The mol% G + C of the DNA is 52–53.2 (T_m).

Type strain: ATCC 7494 (WR 83). This strain was isolated by Kluyver and van den Bout (1936) probably from the same location as used by Beijerinck.

2. **Azomonas insignis** (Derx 1951) V. Jensen 1955, 156.[AL] (*Azotobacter insignis* Derx 1951, 344.)

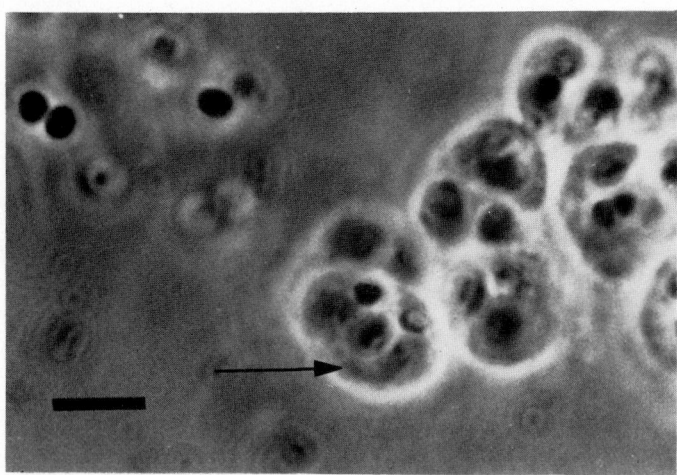

Figure 4.28. Cells of *Azomonas macrocytogenes* (phase-contrast microscopy). Note the large polysaccharide capsule (*arrow*) surrounding the vegetative cells. *Bar*, 10 μm.

in.sig'nis. L. adj. *insignis* distinguished by a mark.

The morphology is as described for the genus. Motile by lophotrichous flagella (full wavelength may exceed 4 μm). The numerous flagella tend to bind together to form thick, whiplike structures.

Colonies are translucent, low convex, butyrous and smooth. They are usually small, probably due to low production of extracellular polysaccharides.

Physiological and nutritional characteristics are presented in Tables 4.48 and 4.49. No homopolysaccharides are produced either with sucrose or raffinose. Starch is not hydrolyzed. Ammonium or nitrate, but not glutamate, can be utilized as a sole nitrogen source for growth.

The mol% G + C of the DNA is 55.1–58.3 (T_m).

Type strain: UQM 1966.

3. **Azomonas macrocytogenes** (ex Baillie, Hodgkiss and Norris 1962, 118) New and Tchan 1982, 381.[VP] (*Azotobacter macrocytogenes* Jensen 1955, 280[AL]; *Azomonotrichon macrocytogenes* (Jensen 1955) Thompson and Skerman 1981, 215.[VP]; *Azotobacter agilis* subsp. *jakutiae* Krasil'nikov 1949, 508.)

mac.ro.cy.to'gen.es. Gr. adj. *macrus* large; Gr. n. *kytos* cell; Gr. v. *gennaio* produce; M.L. part. adj. *macrocytogenes* large cell producing.

The morphology is depicted in Figures 4.27 and 4.28. Motile by polar flagella, usually one or, less frequently, two per cell. The flagellar wavelength is 1.3–1.7 μm and the amplitude is 0.25–0.52 μm. The cells are often surrounded by large capsules (Figs. 4.27 and 4.28) which differ from cysts by lacking the exocystorium (see the genus *Azotobacter*, Fig. 4.20).

Colonies on sucrose media are mucoid and large (15 mm in diameter in 7 days) due to the production of colony-retained homopolysaccharides. No diffusible homopolysaccharides are produced with sucrose or raffinose.

Physiological and nutritional characteristics are presented in Tables 4.48 and 4.49. Starch is not hydrolyzed. Ammonium ions can serve as a sole nitrogen source for growth. Nitrate and glutamate are not used.

The mol% G + C of the DNA is 58.2–58.6 (T_m).

Type strain: ATCC 12335 (NCIB 8700; WR-111). This strain is the original strain isolated by Jensen. Strain M (ATCC 12336) is a mutant also isolated by Jensen. Strain WR-140 is cotype of "*A. agilis* subsp. *jakutiae*" Krasil'nikov 1949.

FAMILY III. **RHIZOBIACEAE** CONN 1938, 321[AL]

D. Carlyle Jordan

Rhi.zo.bi.a'ce.ae. M.L. neut. n. *Rhizobium*, type genus of the family; *-aceae*, ending to denote a family; M.L. fem. pl. n. *Rhizobiaceae*, the *Rhizobium* family.

Cells **without endospores**, normally **rod-shaped. Motile**; one polar or subpolar flagellum, or 2–6 peritrichous flagella. Cells sometimes tend to form star-shaped clusters or rosettes in older broth cultures (*Rhizobium* and *Bradyrhizobium* weakly so). **Aerobic. Gram-negative. Many carbohydrates utilized.** Considerable extracellular slime usually produced during growth on carbohydrate-containing media. Some strains of rhizobia and agrobacteria show a close relationship in DNA base composition.

All species, with the exception of *Agrobacterium radiobacter*, **incite cortical hypertrophies on plants. Nodules are incited on roots** of leguminous species (family: *Leguminosae*) by strains of rhizobia (symbionts) **and on leaves of certain plants** in the families *Myrsinaceae* and *Rubiaceae* by strains of phyllobacteria (symbionts). Nodules are produced by rhizobia on both the roots and stems of the leguminous plant *Sesbania rostrata* (Dreyfus and Dommergues, 1980). **Gall hypertrophies are produced on roots and stems of diverse plant species** by strains of agrobacteria (tumorigenic phytopathogens). Can be isolated from nodules or galls; not easily identified when isolated from soil. Confirmation of isolates should be made by proper plant inoculation tests. The mol% G + C of the DNA is 57–65 (T_m).

Key to the genera of the family **Rhizobiaceae**

I. Cells stimulate nodule production on roots of leguminous plants (*Leguminosae*). Fix atmospheric nitrogen (dinitrogen) when in the symbiotic stage within root nodules, although ineffective nodules (unable to fix nitrogen) may be formed. Do not produce 3-ketolactose.
   A. Cause nodule production on roots of leguminous plants predominantly of the temperate zone. Fast growth on yeast extract-mannitol agar. Usually produce acid in mineral salts-mannitol medium. Mol% G + C of the DNA is 59–64 (T_m).
   Genus I. *Rhizobium*, p 235
   B. Cause nodule production on roots of tropical and some temperate zone leguminous plants. Slow growth on yeast extract-mannitol medium. Usually produce an alkaline reaction in mineral salts-mannitol medium. Mol% G + C of the DNA is 61–65 (T_m) (except for strains from *Lotononis*).
   Genus II. *Bradyrhizobium*, p 242
II. Cells do not cause root nodule production on leguminous plants but most species do produce other types of hypertrophies (uncontrolled and undifferentiated) on many plants. Do not fix dinitrogen. Mol% G + C of the DNA is 57–63 (T_m).

Genus I. **Rhizobium** *Frank 1889, 338*[AL]

D. CARLYLE JORDAN

Rhi.zo′bi.um. Gr. n. *rhiza* a root; Gr. n. *bios* life; M.L. neut. n. *Rhizobium* that which lives in a root.

Rods 0.5–0.9 × 1.2–3.0 μm. Commonly pleomorphic under adverse growth conditions. Usually contain granules of poly-β-hydroxybutyrate which are refractile by phase-contrast microscopy. Nonsporeforming. Gram-negative. **Motile** by one polar or subpolar flagellum or two to six peritrichous flagella. Fimbriae have been described on a few strains. **Aerobic, possessing a respiratory type of metabolism with oxygen as the terminal electron acceptor.** Often able to grow well under oxygen tensions less than 1.0 kPa (Wilson, 1940). Optimum temperature, 25–30°C. Optimum pH, 6–7. Colonies are circular, convex, semitranslucent, raised and mucilaginous, **usually 2–4 mm in diameter within 3–5 days on yeast-mannitol-mineral salts agar.** Pronounced turbidity develops after 2 or 3 days in agitated broth. Chemoorganotrophic, utilizing a wide range of carbohydrates and salts of organic acids as carbon sources, without gas formation. Cellulose and starch are not utilized. **Produce an acidic reaction in mineral-salts medium containing mannitol or other carbohydrates. Growth on carbohydrate media is usually accompanied by copious extracellular polysaccharide slime.** Ammonium salts, nitrate, nitrite and most amino acids can serve as nitrogen sources. Some strains will grow in a simple mineral salts medium with vitamin-free casein hydrolysate as the sole source of both carbon and nitrogen. Peptone is poorly utilized. Casein and agar are not hydrolyzed. Some strains require biotin or other water-soluble vitamins. **3-Ketoglycosides not produced** (Bernaerts and De Ley, 1963). The organisms are characteristically **able to invade the root hairs of temperate-zone and some tropical-zone leguminous plants** (family *Leguminosae*) **and incite production of root nodules** wherein the bacteria occur as intracellular symbionts. All strains exhibit host range affinities (host "specificity"). **The bacteria are present in root nodules as pleomorphic forms (bacteroids) which are normally involved in fixing atmospheric nitrogen** into a combined form (ammonia) utilizable by the host plant. The mol% G + C of the DNA is 59–64 (T_m).

Type species: *Rhizobium leguminosarum* (Frank 1879) Frank 1889, 338.

Further Descriptive Information

In young cultures the cells are short rods, but in old cultures or under adverse environmental conditions the cells are commonly pleomorphic (swollen and either globular, ellipsoidal, club shaped or branched). The inciting conditions vary somewhat with the strain but may include extremes of temperature and pH, low oxygen tension and growth in laboratory media containing low concentrations of calcium or magnesium or excessive amounts of various amino acids (such as glycine), or certain alkaloids, glycosides, dyes or antibiotics. Granules of poly-β-hydroxybutyrate are common in the cytoplasm of older cells, so that upon simple staining the rods appear banded. Strains of *R. leguminosarum* often contain metachromatic granules, demonstrated by staining with methylene blue, washing with dilute iodine and staining with neutral red (Graham and Parker, 1964). Within the root nodules, the pleomorphic bacteroids do not contain polyphosphate inclusions and usually no glycogen granules (Craig et al., 1973; J. P. Gourret, Ph.D. Thesis, Université de Rennes, France, 1981), but may have large concentrations of poly-β-hydroxybutyrate (Fig. 4.29).

Some strains are encapsulated. All produce abundant water-soluble extracellular polysaccharide, the principal constituent of which is acidic heteropolysaccharide (80–90%); the remainder, in most strains, being neutral, nonbranched, β-2-linked glucans (York et al., 1980). All strains within a species produce the same acidic heteropolysaccharide except for *R. leguminosarum* biovar *phaseoli* which possesses a unique heteropolysaccharide. Curdlan production has been reported in several isolates of *R. leguminosarum* biovar *trifolii* (Ghai et al., 1981).

Young cells are motile by several lateral or sometimes polar flagella (Fig. 4.30), except *R. loti* which possesses one flagellum, usually with subpolar insertion although a few cells may show a polar or lateral insertion (Abdel-Ghaffer and Jensen, 1966).

Growth on carbohydrate-containing solid media may be opaque, clear or translucent (Fig. 4.31) but also may exhibit small opaque areas within a clear slime. Mutation occurs to nongummy colonial forms, many of which may be small-colony variants which fix little or no nitrogen. Wrinkled or "rough" colonies have been described but this phenomenon is not related to ineffectiveness.

All strains grow rapidly on a mineral salts medium containing yeast extract and any one of a wide variety of carbohydrates, particularly mannitol, glucose, arabinose, fructose, galactose and sucrose (Vincent et al., 1979). Acid is usually produced from carbohydrate to a moderate degree, and is best estimated by the procedure of Norris (1965) with bromothymol blue as indicator. Dextrin is rarely utilized. Intermediates of the tricarboxylic acid cycle can be utilized as sole sources of carbon providing the basal medium has sufficient Ca^{2+} and Mg^{2+} to overcome the inhibitory chelating ability of these intermediates.

The nitrogen requirements can be satisfied by nitrate (and often nitrite), ammonium salts (providing the medium is well buffered) and by many amino acids and short-chain peptides. Certain amino acids, e.g. glycine, may be inhibitory. There have been several reports of possible nitrogen-fixation by free living strains but the situation in this respect is not yet clear. Some strains (*R meliloti* in particular) anaerobically produce N_2 from nitrate (Daniel et al., 1982).

Most strains lack the ability to absorb Congo red from a yeast extract-mannitol-mineral-salts medium containing a 0.0025% final concentration of this dye. This property, shared with species of *Bradyrhizobium*, results in colorless or faintly pink colonies, whereas contaminant colonies are often a deep red.

The temperature range, which is highly strain dependent, is 4–42.5°C; however, growth at 4°C is rare, and only *R. meliloti* can grow at 42.5°C. The temperature maximum for *R. leguminosarum* is 38°C. The pH range for the genus is 4.5–9.5, *R. meliloti* being the most alkali-tolerant species.

Rhizobium strains are only weakly proteolytic, but most strains produce a slow digestion in litmus milk, forming an upper clear "serum zone," usually with a slightly alkaline reaction or no change. *R. meliloti* strains tend to produce an acidic reaction.

The principal mechanisms of carbohydrate metabolism appear to be the Entner-Doudoroff pathway and the pentose cycle (Katznelson and Zagallo, 1957; Keele et al., 1969; Martinez-de Drets and Arias, 1972; Mulongoy and Elkan, 1977; and Ronson and Primrose, 1979). The Embden-Meyerhof-Parnas path is probably not physiologically important. Although the presence of fructose-1,6 biphosphate aldolase and 6-phosphofructokinase has been reported, subsequent studies have shown these enzymes to be very low in concentration or absent. Polyol entry into the carbohydrate degradation pathway is due to inducible dehydrogenases by which mannitol and sorbitol are converted to fructose, and arabitol to xylulose (Martinez-de Drets and Arias, 1970). L-Arabinose is metabolized to α-ketoglutarate (Duncan, 1979).

The tricarboxylic acid cycle is operative, and the enzymes of the glyoxylate bypass are present (Johnson et al., 1966). Pyruvate carboxylase is an important anaplerotic enzyme, at least in *R. leguminosarum* biovar *trifolii* (Ronson and Primrose, 1979).

Succinate, fumarate, and malate are taken up actively by the same porter system in both free living cells and bacteroids isolated from root nodules (T. Finan et al., 1981).

Transformation, transduction and conjugation have been demonstrated. Symbiotically defective mutants can be readily isolated indi-

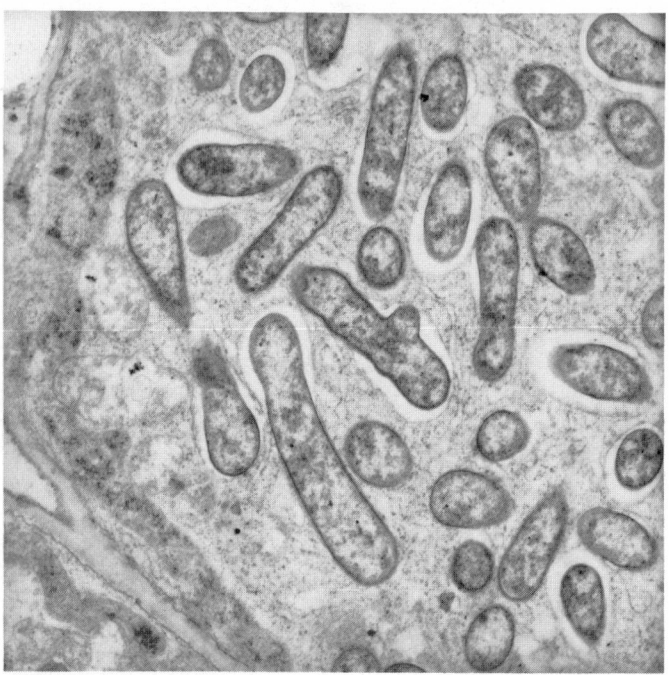

Figure 4.29. Transmission electron micrograph of bacteroids in a root nodule of *Medicago laciniata* (cut-leaf medic). Each bacteroid is enclosed by a plant-produced peribacteroidal membrane. (× 8444).

rectly by selection for auxotrophic or resistant mutants or directly by screening survivors of mutagenic agents. Auxotrophy toward adenine, uracil and leucine is often associated with symbiotic ineffectiveness in *R. meliloti* (Dénarié et al., 1976). Transposon mutagenesis is particularly effective, especially using Tn5 (which confers kanamycin resistance). This is on an R plasmid with a bacteriophage Mu insertion which prevents replication of the vector. Mutations conferring resistance to streptomycin and rifampicin are closely linked and provide a convenient screening method for demonstrating gene transfer.

Recombination in nodulating strains of *R. leguminosarum* and *R. meliloti* has been shown using P1 incompatibility group R plasmids (especially R68.45 which has superior chromosome-donor characteristics) as sex factors, and circular linkage maps have been constructed (Beringer and Hopwood, 1976; Meade and Singer, 1977; Kondorosi et al., 1977).

Cells contain plasmids, including large, naturally-occurring plasmids of $90-200 \times 10^6$ daltons. There is good evidence that nodulating ability is encoded by plasmid-located genes (Higashi, 1967; Dunican and Cannon, 1971). The transfer of a bacteriocinogenic plasmid, carrying Tn5, from *R. leguminosarum* biovar *viceae* to biovars *phaseoli* and *trifolii* confers the ability to nodulate *Pisum* (Johnston et al., 1978).

The nitrogen-fixation (*Nif*) genes also appear to be plasmid-borne (Dunican et al., 1976; Nuti et al., 1979) and are transmissible to *Agrobacterium* and *Klebsiella* (Stanley and Dunican, 1979), as well as to *Azotobacter* (Page, 1978).

The range of hosts susceptible to a particular *Rhizobium* bacteriophage is highly variable. In some cases it is limited to relatively few strains within a single host species; in others it may cross taxonomic boundaries. Bacteriophage-lysis is not sufficiently specific to be used to identify strains. Some strains of rhizobia are lysogenic. Bacteriocins have been reported (Roslycky, 1967), as well as a parasitic *Bdellovibrio*.

Most serological reactions (agglutination, gel diffusion, precipitin and fluorescent antibody) show strain specificity and can be of value in recognizing strains of rhizobia in nodules of field plants or in laboratory investigations. The agglutination reaction, using crushed nodule extracts, is the most widely used for field work because of its simplicity, although it is complicated by cross-reactions and autoagglutination. Surface antigens, although useful for strain recognition, are limited in their usefulness for species identification. Of particular taxonomic interest are those serological reactions related to the internal antigens found by Vincent and Humphrey (1970). Antiserum against a strain of *R. leguminosarum* biovar *trifolii* will detect several common

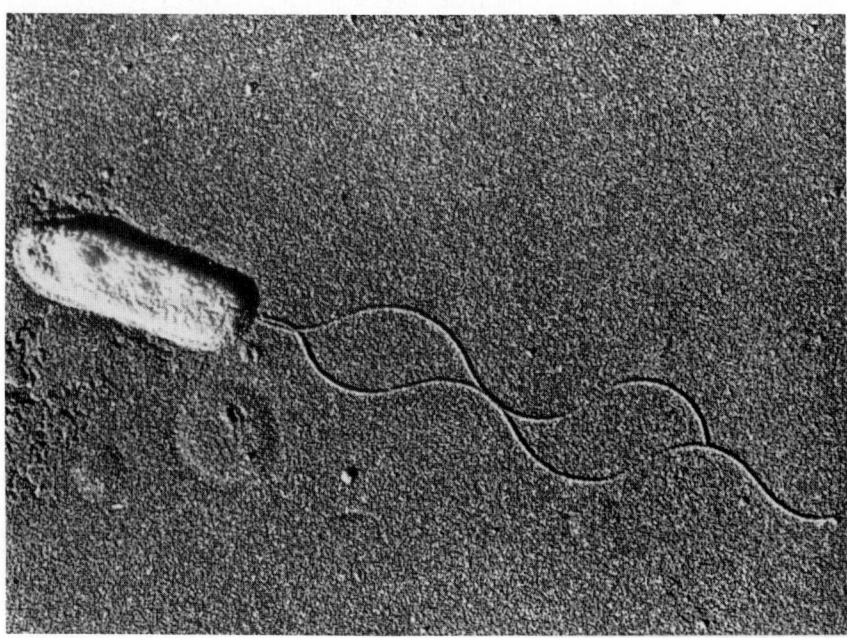

Figure 4.30. Cell of *Rhizobium leguminosarum* biovar *trifolii* showing two polar flagella. (× 14,000).

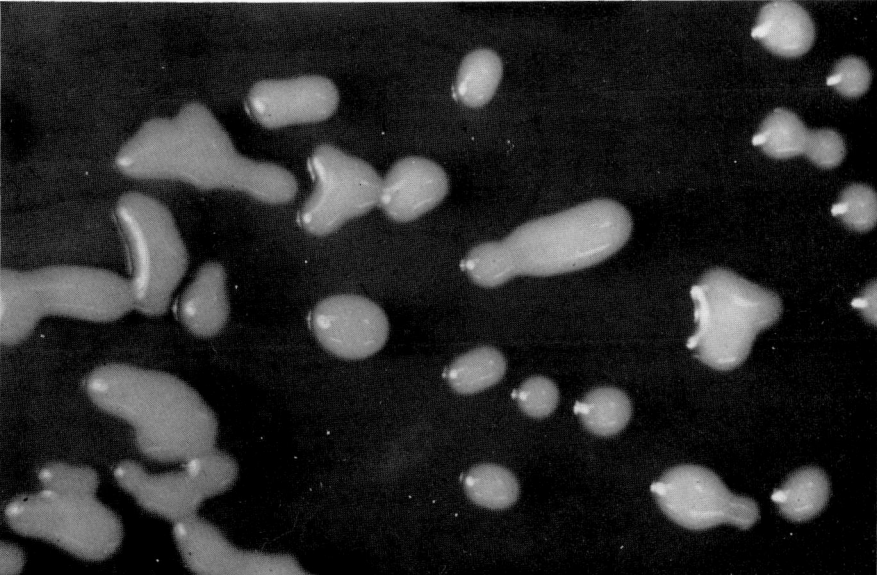

Figure 4.31. Colonies of *Rhizobium leguminosarum* on mineral-salts mannitol agar. (× 1.5).

internal group antigens among other strains of all three proposed biovars of this species. The same antiserum reveals fewer group antigens among *R. meliloti*, *R. loti* and *Agrobacterium* species, but none with any of the slow growing strains in the genus *Bradyrhizobium*. Cross-reaction occurs between *R. meliloti* and *Agrobacterium*.

The tetracycline antibiotics are generally the most active against the rhizobia. There is wide strain-to-strain variation in resistance but, in general, the fast growing rhizobia are intrinsically more sensitive than the slow-growing bradyrhizobia to the tetracyclines, penicillin G, viomycin, vancomycin and streptomycin. Spontaneous mutants for resistance to most antibiotics are a common component of all "wild type" strains. Resistance to streptomycin may be greater than 10 times that of the "wild type," whereas resistance to other antibiotics is generally of a lower order (2 or 3 times). Resistance to viomycin and neomycin often confers ineffectiveness (loss of the ability to fix nitrogen in symbiosis). Streptomycin-resistant mutants, which are usually effective, are important in ecological field studies on strain competition.

Controlled hyperplasia and hypertrophy of root tissue produce a differentiated root nodule (Figs. 4.32 and 4.33) on those plants of the family *Leguminosae* (see Table 4.54) which are commonly characteristic of the temperate zone.

Members of this genus are common soil inhabitants. Identification is relatively easy if isolated from host plant nodules, but difficult if isolated from the soil or if a noninfective mutant. The ability to cause nodule formation (infectiveness) is more stable than the ability to fix nitrogen in symbiosis (effectiveness). Strains tend to lose effectiveness after serial cultivation on media containing certain amino acids, especially D-forms, and after long term storage or continuous growth on laboratory media. Infectiveness and effectiveness represent discrete phenomena and vary within wide limits, depending upon genetic factors in both bacterial strain and host plant.

Enrichment and Isolation Procedures

Healthy root nodules, with a small portion of root attached, are placed in a small tube with a mesh bottom and washed for several minutes in running tap water. The nodules then are surface sterilized by a brief exposure to 95% alcohol followed by a 1- to 4-min exposure

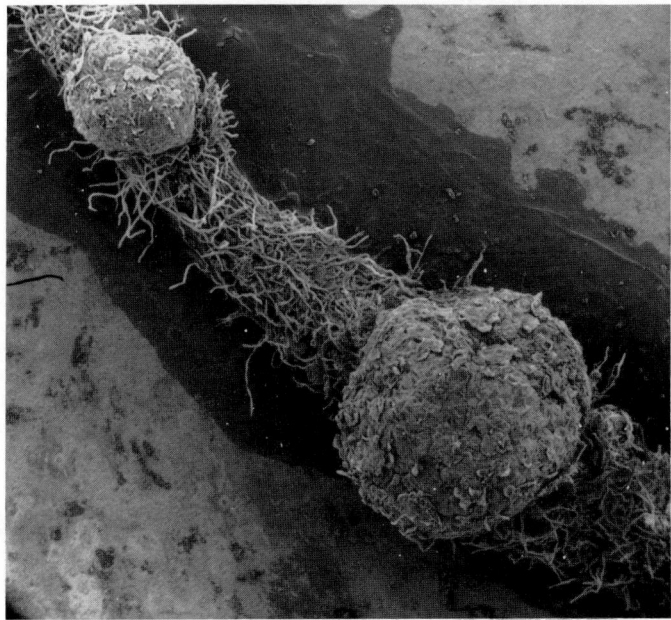

Figure 4.32. Scanning electron micrograph of immature root nodules of *Medicago sativa* (alfalfa). (× 36).

(depending on nodule size) in either 5% H_2O_2, a 0.1% acidified mercuric chloride solution ($HgCl_2$, 1 g; concentrated HCl, 5 ml; distilled water to 1 liter), or a solution of calcium hypochlorite (10 g/150 ml distilled water). This treatment is followed by 10 washes in sterile water, if the $HgCl_2$ solution is employed, or three changes if the H_2O_2 or hypochlorite solutions are employed. The nodules are crushed in a small drop of sterile water and a loopful transferred to 5 ml of sterile water. A 0.1-ml sample of this dilution is now spread on the surface of yeast extract-mannitol agar (YMA)* plates (lacking $CaCO_3$). Alternatively, a small

* Yeast extract-mannitol agar (YMA): mannitol, 10.0 g; KH_2PO_4, 0.5 g; $MgSO_4 \cdot 7H_2O$, 0.2 g; NaCl, 0.1 g; $CaCO_3$, 4.0 g; yeast water, 100 ml (or yeast extract (Difco), 0.4 g); agar, 15.0 g; and distilled water, 1 liter; pH 6.8–7.0. Sterilize at 121°C for 15 min. The $CaCO_3$ is omitted for the preparation of pour plates or for liquid medium (YMB) used for turbidimetric measurement of growth. The yeast water is prepared by mixing 100 g of baker's compressed yeast with 1 liter of cold water, allowing the mixture to stand at room temperature for 1 or 2 h, and then steaming for 40–60 min. After the mixture has been centrifuged or allowed to settle, the clear supernatant fluid is autoclaved at 121°C for 15 min (Kleczkowska et al., 1968).

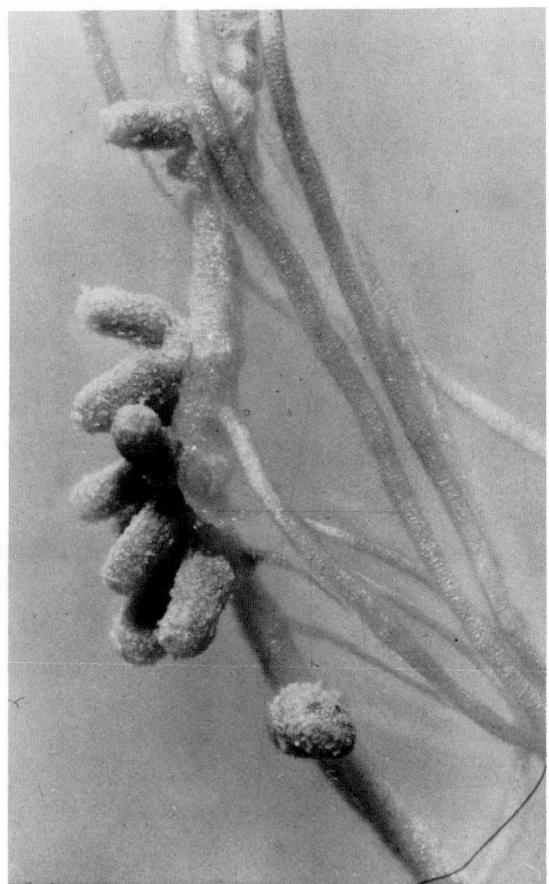

Figure 4.33. Mature, effective root nodules on *Medicago sativa*. (× 10).

loopful of the crushed nodule suspension can be streaked onto two successive plates of agar medium. Large nodules should be cut in half with a sterile blade and portions of the central area removed with a sterile wire flattened at one end. Incubation is at 28°C. Well isolated typical colonies are restreaked onto fresh plates for subsequent examination. If heavy fungal contamination is expected, the agar medium used for initial isolation should contain 0.002% actidione.

Isolation from soil requires the use of "trap hosts," which are leguminous plants grown in the soil and from which nodules are selected for subsequent isolation of rhizobia.

Nodules collected in the field can be briefly stored in small vials containing silica gel held under a cotton plug.

Rhizobia grow poorly on 0.04% peptone-1% glucose mineral salts agar and show little pH change. This medium can serve as a useful contamination check. Growth in 2 days at 28°C and marked pH change are contraindications for *Rhizobium*.

Maintenance Procedures

YMA-agar slant cultures in screw-capped containers can be stored for 2–3 years, or longer, at 2°C. At 15°C, survival is very good for at least 2 months. Lyophilization is best for long term storage. For lyophilization, the growth from a vigorous agar-slant culture is suspended in a sterile aqueous solution of 10% sucrose and 5% peptone and dispensed in 0.1-ml amounts in suitable containers, which are subsequently sealed under vacuum when freeze-drying is complete. Reconstitution is in YMB, followed by streaking over YMA plates and incubating at 28°C in an inverted position. Some strains may require incubation for up to 14 days before colonies develop to a suitable size. Norris (1963) has described a preservation method based on the use of small porcelain beads which, after inoculation and drying, can be removed one at a time and dropped into sterile YMB for subsequent recovery of the bacteria.

Procedures for Testing for Special Characters

Carbohydrate and organic acid utilization can be determined satisfactorily using inoculated plates of the medium of Elkan and Kwik (1968). On the dried surface of these plates are placed absorbent paper discs previously saturated with a 10% solution of the organic compound and slowly dried. During incubation at 28°C the plates are examined daily using indirect lighting against a black, nonreflecting background.

Estimation of carbohydrate utilization by the use of pH indicator dyes is not recommended because of effects brought about by differential uptake of ions in the absence of cellular growth.

A suitable defined medium for studies on the utilization of carbohydrates and organic acids, or of vitamin and amino acid requirements, is MOPS-salts (MS) medium* (T. Finan, personal communication).

Norris (1965) has published a suitable procedure for the estimation of acid production from carbohydrates.

Details on the methods used to assess nodulation response under greenhouse or growth room conditions are given by Vincent (1970).

Differentiation of the genus **Rhizobium** from other genera

Table 4.50 indicates the characteristics of *Rhizobium* that distinguish it from other genera which are morphologically or physiologically similar.

Taxonomic Comments

Previous classifications of the root nodule bacteria were based largely on the cross-inoculation group concept, with its assumption that those leguminous plants falling within a particular infection group were nodulated by one particular species of nodule bacteria (Fred et al., 1932). With repeated evidence of anomalous cross-infection among the different plant groups, this classification gradually lost credibility and now has been replaced by one which takes into account newer information based on techniques designed to examine larger portions of the bacterial genome. Nevertheless, it is highly likely that even this classification will be gradually modified when larger numbers of *Rhizobium* strains from a wide variety of leguminous plants have been studied. At present only ~8 or 9% of the 14,000 or so known species of leguminous plants have been examined for nodulation, and less than 0.5% have been studied relative to their symbiotic relationships with nodule bacteria. Most of the plants examined have been of agricultural importance and the huge reservoir of wild tropical leguminous plants is only now beginning to be investigated.

---

* MOPS-salts (MS) medium: morpholinopropane-sulfonic acid (MOPS buffer), 8.37 g; KOH, 1.12 g; NH$_4$Cl, 1.07 g; MgSO$_4$, 0.48 g; K$_2$HPO$_4$, 0.21 g; KH$_2$PO$_4$, 0.05 g; NaCl, 5 mg; CaCO$_3$, 20 mg; H$_3$BO$_3$, 1 mg; ZnSO$_4$·7H$_2$O, 1 mg; CuSO$_4$·5H$_2$O, 0.5 mg; MnCl$_2$·4H$_2$O, 0.5 mg; NaMoO$_4$·2H$_2$O, 1 mg; ethylenediamine tetraacetic acid (EDTA), 10 mg; Na-Fe-EDTA, 2 mg; and distilled water, 1 liter. Stock solutions of MOPS-KOH (sterilized by filtration) and of MgSO$_4$ (sterilized by autoclaving) are prepared and added aseptically to the autoclaved salts solution so as to provide the correct final concentrations. The final pH of the medium should be 7.2. All carbon sources or growth factors to be added are prepared as stock solutions (sterilized by filtration) and are added to the medium to give a final concentration of 15 mM (or, in the case of citrate, 5 mM).

Table 4.50.
Differential characteristics of the genus **Rhizobium** *and other morphologically or physiologically similar genera[a]*

Characteristics	Rhizobium	Bradyrhizobium	Agrobacterium	Phyllobacterium	Pseudomonas
Flagellar arrangement:					
Monotrichous	D	+	−	D	D[b]
Lophotrichous	−	−	−	D	D
Peritrichous	D	−	D	−	−[c]
Root nodules produced	+	+	−[d]	−	−
Leaf nodules produced	−	−	−	+	−
Nitrogenase activity occurs	+	+	−[d]	−[e]	−
Hypertrophy-initiating ability[f]	−	−	+	−	−
3-Ketolactose produced	−	−	D		
Fast growth on YMA[g]	+	−	+	+	+
Alkaline reaction in sugar media	−	+	−	−	−
H₂S production	D	−	−		
Response to biotin	+[h]	+	−		D
β-2-Linked glucans in extracellular gum	+	+	+		−
Mol% G + C of DNA	59–64	61–65	57–63	60–61	58–70

[a] Symbols: see standard definitions.
[b] Usually one polar flagellum, but some species have multiple polar flagella.
[c] *Pseudomonas stutzeri* and *Pseudomonas mendocina* possess a single polar flagellum when grown in liquid media; however, on agar media at 30°C numerous lateral flagella of shorter wavelength occur in addition to the polar flagellum.
[d] Not in naturally occurring strains. Genetically modified strains may produce nodules and may have nitrogenase activity.
[e] Controversial.
[f] By pinto bean leaf test (Lippincott and Heberlein, 1965) and carrot disc assay (Klein and Tenebaum, 1955, as used by Lippincott and Lippincott, 1969).
[g] Colonies 2–4 mm in diameter within 2 or 3 days at 28°C.
[h] But may differ among the strains of some species.

The fusion of the former species *R. phaseoli* and *R. trifolii* with *R. leguminosarum*, and the retention of *R. meliloti* as a separate species, is based on extensive evidence employing numerical taxonomy (Graham, 1964a; Moffet and Colwell, 1968), DNA/DNA homology (Heberlein et al., 1967; Jarvis et al., 1980), 2-dimensional polyacrylamide gel electrophoresis of cellular proteins (Roberts et al., 1980), serology (Graham, 1963; Vincent and Humphrey, 1970), composition of extracellular gum (Robertsen et al., 1981), and findings involving the transfer of infectivity via plasmids (Higashi, 1967; Dunican and Cannon, 1971; Johnson et al., 1978).

R. loti was proposed by Vincent (1974) on a variety of evidence as a species to contain those fast-growing polar and subpolar flagellated strains which nodulate a wide range of leguminous plants which, in part, also are susceptible to the slow growing polar and subpolar flagellated *Bradyrhizobium* strains (Abdel-Ghaffer and Jensen, 1966; Vincent and Humphrey, 1970; Humphrey et al., 1973; Roberts et al., 1980; Trinick, 1980). Fast and slow growing strains from *Lotus* do not share antigens (Pankhurst, 1979), lending support to the separation of these two physiological groups into two different genera. Additional data concerning the separation of such strains into two genera and criteria separating the fast-growing acid-producing *Lotus* bacteria from *R. leguminosarum* and *R. meliloti* are presented by Jarvis and Pankhurst, 1981.

The biovars indicated for *R. leguminosarum* are based largely, but not entirely, on host plant specificity (see Table 4.54) and there are reasons to suspect that biovar *phaseoli* is more distinctly related to *R. leguminosarum* than the other two biovars (Gibbins and Gregory, 1972; Jarvis et al., 1980; Roberts et al., 1980). Biovar *phaseoli* also has a tendency to produce a dark pigment on prolonged incubation, a characteristic not shared by other strains within the species, and it apparently has a unique acidic heteropolysaccharide in its extracellular gum (see Table 4.53).

Studies involving DNA base composition and DNA/DNA hybridization (Heberlein, De Ley and Tijtgat, 1967; De Ley, 1968), rRNA cistron similarities (De Smedt and De Ley, 1977) and numerical tax-

Table 4.51.
Characteristics differentiating **Rhizobium leguminosarum, R. meliloti,** *and* **R. loti**[a]

Characteristics	1. *R. leguminosarum*	2. *R. meliloti*	3. *R. loti*
Flagellar arrangement:			
Monotrichous	−	−	d
Peritrichous	+	+	−
Agglutination with antiserum against *R. meliloti*[b]	−	+	−
Growth in presence of 2% NaCl	−	d	−
Growth at 39–40°C	−	d	−
Thiamine required	d	−	
Pantothenate required	+	−	
H₂S produced	−	d	−
Precipitate on Ca glycerophosphate medium[c]	−	d	−
Acid reaction in litmus milk	−	d	−

[a] For symbols see standard definitions.
[b] Antisera for somatic and/or flagellar antigens.
[c] Hofer (1941).

onomy (Graham, 1964a; 't Mannetje, 1967) stress the close relationship between *Rhizobium* and *Agrobacterium* and have led to the proposal that these two genera should be amalgamated in part. However, distinctions between *Rhizobium*, which produce a differentiated root nodule, and *Agrobacterium*, which produce an uncontrolled, undifferentiated tumor, are possible on the basis of a variety of cultural and biochemical characteristics (Table 4.50).

Reasons for the separation of the slow growing root-nodule bacteria

Table 4.52.
Other characteristics of **Rhizobium leguminosarum, R. meliloti** *and* **R. loti**[a]

Characteristics	1. *R. leguminosarum*	2. *R. meliloti*	3. *R. loti*
Rapid growth on yeast extract-mannitol agar with acid production	+	+	+
Penicillinase production	d[b]	−	d
Response to biotin	d	d	
External requirement for niacin, pyridoxine, folic acid, p-aminobenzoic acid, inositol, B$_{12}$ and riboflavin	−	−	−
Utilization of glucose, galactose, fructose, arabinose, xylose, rhamnose, maltose, sucrose, lactose, trehalose, raffinose, mannitol, fumarate, malate succinate, citrate, pyruvate	+	+	d
Ability to grow on basal salts medium[c] containing sucrose and arabinose in presence of:			
20 mM Citrate	−	+	
20 mM Pyruvate	+	+	
20 mM Lactate	−	+	
20 mM Glyoxylate	d	+	
Utilization of dextrin	d	d	
Utilization of dulcitol	+	+	d
Growth at pH:			
3.5	−	−	−
4.0	−	−	d
4.5	d	−	+
5.0	+	d	+
8.0	+	+	+
9.0	d	+	d
9.5	−	+	d
Alkaline phosphatase activity	+	+	+

[a] For symbols see standard definitions.
[b] Biovars *trifolii* and *phaseoli* are negative.
[c] Skotnicki and Rolfe (1977).

from the fast growers are discussed in the description of the genus *Bradyrhizobium*.

Fast growing root-nodule bacteria, physiologically distinct from members of the genus *Bradyrhizobium*, recently have been obtained from soybean root nodules collected in China (Keyser et al., 1982). These bacteria form effective root nodules on wild soybean (*Glycine soja*) and on *G. max* cv Peking, a black seeded genetically unimproved

Table 4.53.
Composition of extracellular polysaccharides of **Rhizobium leguminosarum** *and* **R. meliloti**[a]

Composition	1. *R. leguminosarum*	2. *R. meliloti*
Acidic heteropolysaccharide:		
Structural Type 1[b]	−	+
Structural Type 2[b]	+	−
Structural Type 3[b]	d[c]	−
Glucuronic acid present	+	−
Methylated sugars present	−	−
Neutral Glucans		
β-1,2-[d]	+	+
β-1,3-	−	−
α-1,2-	−	−

[a] W. F. Dudman and P. Albersheim (personal communication). See also Janssen et al. (1977) and Robertsen et al. (1981). For examples see Table 4.50.
[b] See Figure 4.34 for the structures of the three types.
[c] Found in biovar *phaseoli* only.
[d] Also found in *Agrobacterium* and some strains of *Bradyrhizobium japonicum* (Dedonder and Hassid, 1964; Gorin et al., 1961; York et al., 1980).

line of soybean from China. However, ineffective nodules are produced on most commercial soybean cultivars so far examined. The taxonomic status of these bacteria is currently unknown but such strains may represent a transitional group falling between the genera *Rhizobium* and *Bradyrhizobium*.

Acknowledgments

The author is indebted to the members of the International Subcommittee on *Agrobacterium* and *Rhizobium*, who provided support and often contributed valuable advice: J. De Ley, P. H. Graham, J. Holding, B. D. W. Jarvis, E. Roslycky, B. W. Strijdom and J. M. Vincent. The author also wishes to thank J. E. Beringer, J. Burton, R. A. Date, W. F. Dudman, E. Schwinghamer and M. J. Trinick who took part in valuable discussions.

Further Reading

Bergersen, F.J. (Editor). 1980. *Methods for Evaluating Biological Nitrogen Fixation.* John Wiley, New York.
Gibson, A.H. and W.E. Newton (Editors). 1981. *Current Perspectives in Nitrogen Fixation.* Australian Academy of Science, Canberra.
Hardy, R.W.F. and W.S. Silver (Editors). 1977. *A Treatise on Dinitrogen Fixation.* Section III. Biology. John Wiley, New York.
Jarvis, B.D.W., A.G. Dick and R.M. Greenwood. 1980. Deoxyribonucleic acid homology among strains of *Rhizobium trifolii* and related species. Int. J. Syst. Bacteriol. *30:* 42–52.
Newton, W.E. and W.H. Orme-Johnson (Editors). 1980. *Nitrogen Fixation*, Vol. II. *Symbiotic Associations and Cyanobacteria.* University Park Press, Baltimore.
Vincent, J.M. 1970. *A Manual for the Practical Study of Root Nodule Bacteria.* International Biological Programme Handbook No. 12. Blackwell Scientific, Oxford.

Differential characteristics of the species of the genus **Rhizobium**

The differential characteristics of the species of *Rhizobium* are indicated in Table 4.51. Other characteristics are presented in Tables 4.52, 4.53 and 4.54, and Fig. 4.34.

List of the species of the genus **Rhizobium**

1. **Rhizobium leguminosarum** (Frank 1879) Frank 1889, 338.[AL] (*Schinzia leguminosarum* Frank 1879, 397.)

le.gu.mi.no.sa′rum. M.L. fem. n. *Leguminosae* old family name of the legumes; M.L. fem. gen. pl. n. *leguminosarum* of legumes.

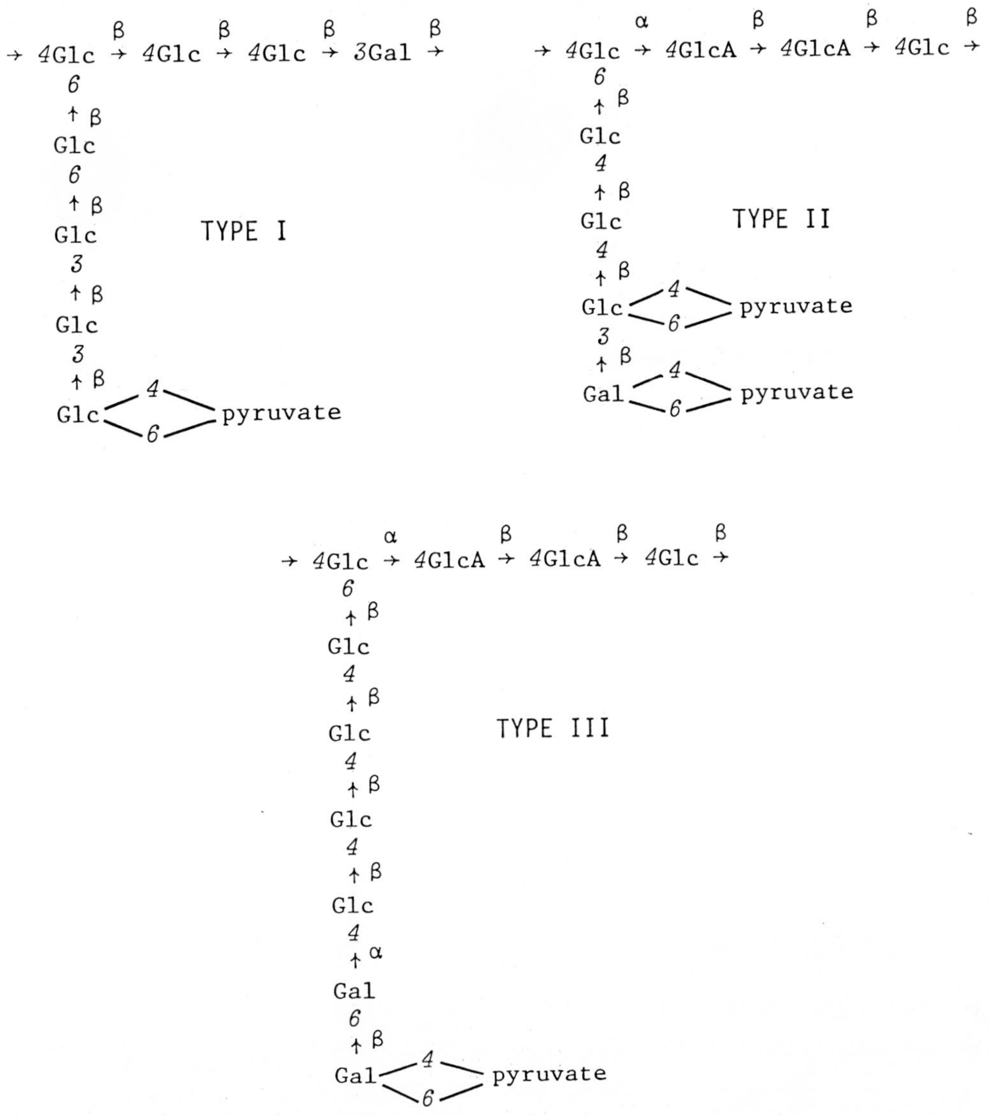

TYPE I

```
        β         β         β         β
→ 4Glc  →  4Glc  →  4Glc  →  3Gal  →
   6
   ↑ β
  Glc
   6
   ↑ β
  Glc
   3
   ↑ β
  Glc
   3
   ↑ β
          4
  Glc <        > pyruvate
          6
```

TYPE II

```
        α          β          β         β
→ 4Glc  →  4GlcA  →  4GlcA  →  4Glc  →
   6
   ↑ β
  Glc
   4
   ↑ β
  Glc
   4
   ↑ β
          4
  Glc <        > pyruvate
   3      6
   ↑ β
          4
  Gal <        > pyruvate
          6
```

TYPE III

```
        α          β          β         β
→ 4Glc  →  4GlcA  →  4GlcA  →  4Glc  →
   6
   ↑ β
  Glc
   4
   ↑ β
  Glc
   4
   ↑ β
  Glc
   4
   ↑ β
  Glc
   4
   ↑ α
  Gal
   6
   ↑ β
          4
  Gal <        > pyruvate
          6
```

Figure 4.34. Three types of acidic extracellular polysaccharides produced by *Rhizobium*. Type I is formed by *R. meliloti*, types II and III are formed by *R. leguminosarum*. *Glc*, glucose; *GlcA*, glucuronic acid; *Gal*, galactose. (Structural types according to P. Albersheim, personal communication.)

Table 4.54.
Nodulation characteristics of **Rhizobium** *species*[a,b]

Host Plant	1. *R. leguminosarum* Biovar *viceae*	Biovar *trifolii*	Biovar *phaseoli*	2. *R. meliloti*	3. *R. loti*
Pisum sativum, Vicia hirsuta and *Vicia sativa*	+[c]	±	±	−	−
Phaseolus vulgaris	(±)	(±)	+	−	±
Trifolium repens	±	+[d]	(±)	−	−
Lotus corniculatus	−	−	−	−	+
Medicago sativa	−	−	−	+[e]	(±)
Macroptilium atropurpureum	−	−	±	−	±

[a] Adapted from Vincent et al. (1979), with permission.
[b] Symbols: +, generally nodulates; ±, sometimes nodulates, nodules commonly ineffective; (±), rarely nodulates, nodules commonly ineffective; −, does not nodulate.
[c] Some negatives have been reported with *V. hirsuta* and *P. sativum*.
[d] Some negatives have been reported with isolates from *T. ambiguum* and African clovers.
[e] Other species of *Medicago* are likely to be more strain-specific.

The characteristics are as described for the genus and as indicated in Tables 4.51–4.54. Bacteroids in nodules are commonly irregular with X-, Y-, star- and club-shaped forms, except for *Phaseolus* bacteroids, which exhibit few branched forms in the nodules. Unevenly-stained forms predominate. The cells of *R. leguminosarum* have 2–6 peritrichous flagella.

Three biovars occur: *trifolii*, *phaseoli* and *viceae*.

The species normally causes formation of root nodules on some, but not necessarily all, species of *Pisum* (field pea), *Lathyrus* (pea), *Vicia* (vetch), *Lens* (lentil), temperate species of *Phaseolus* (*P. vulgaris* (kidney bean), *P. angustifolius* (bean), *P. multiflorus* (scarlet runner)) and *Trifolium* (clover).

The mol% G + C of the DNA is 59–63 (T_m).

Type strain: ATCC 10004.

2. Rhizobium meliloti Dangeard 1926, 194.[AL]

me.li.lo'ti. M.L. masc. n. *Melilotus* generic name of sweet clover; M.L. gen. n. *meliloti* of *Melilotus*.

The characteristics are as described for the genus and as indicated in Tables 4.51–4.54. Bacteroids in nodules are club shaped and branched. The cells of *R. meliloti* have 2–6 peritrichous flagella.

The species normally causes formation of root nodules on some, but not all, species of *Melilotus* (sweet clover), *Medicago* (alfalfa) and *Trigonella* (fenugreek).

The mol% G + C of the DNA is 62–63 (T_m).

Type strain: ATCC 9930.

3. Rhizobium loti Jarvis, Pankhurst and Patel 1982, 378.[VP]

lo'ti. M.L. masc. n. *Lotus* generic names of several kinds of plants; M.L. gen. n. *loti* of *Lotus*.

The characteristics are as described for the genus and as indicated in Tables 4.51–4.54. Bacteroids in nodules are club shaped and branched. The cells of *R. loti* have predominantly one polar or subpolar flagellum.

The species normally causes formation of root nodules on some, but not all, of the following hosts; *Lotus corniculatus* (birdsfoot trefoil), *Lotus tenuis* (slender birdsfoot trefoil), *Lupinus densiflorus* (lupine), *Anthyllis vulneraria* (kidney vetch), *Ornithopus sativus* (serradella), *Cicer arietinum* (chick pea), *Caragana arborescens* (siberian pea tree), *Leucaena leucocephala* (leadtree), and *Mimosa* (mimosa). Some of these plants are also nodulated by *Bradyrhizobium*. Information concerning additional plant hosts is presented by Jarvis et al. (1982).

The mol% G + C of the DNA is 59–64 (T_m).

Type strain: ATCC 33669.

Other Organisms

A small group of bacterial strains from three plant genera (*Coronilla* (crown vetch), *Onobrychis* (sainfoin) and *Sophora*) which are also nodulated by *R. loti*, appears to comprise a group distinct from the present *Rhizobium* species. Although all of these strains fix nitrogen with both *Leucaena leucocephala* and *Phaseolus vulgaris*, they are very different from typical strains isolated from either of these hosts, both with regard to DNA/DNA homology and plant specificity (Crow et al., 1981). The taxonomic significance of this group, however, is not known and must await results from investigations on a larger number of strains.

Genus II. Bradyrhizobium Jordan 1982, 137[VP]

D. CARLYLE JORDAN

Bra. dy. rhi. zo' bi. um. Gr. adj. *bradus* slow; M.L. neut. n. *Rhizobium* a bacterial generic name; M.L. neut. n. *Bradyrhizobium* the slow (growing) rhizobium.

Rods 0.5–0.9 × 1.2–3.0 μm. Commonly pleomorphic under adverse growth conditions. Usually contain granules of poly-β-hydroxybutyrate which are refractile by phase-contrast microscopy. Nonsporeforming. Gram-negative. **Motile** by one polar or subpolar flagellum. Fimbriae have not been described. **Aerobic**, possessing a respiratory type of metabolism with oxygen as the terminal electron acceptor. Optimum temperature, 25–30°C. Optimum pH, 6–7, although lower optima may be exhibited by strains from acid soils. Colonies are circular, opaque, rarely translucent, white and convex, and tend to be granular in texture; **they do not exceed 1 mm in diameter within 5–7 days incubation on yeast-mannitol-mineral salts agar.** Colonies produced by some strains isolated from *Lotononis bainesii* are red because of intracellular pigmentation. Only a moderate turbidity develops after 3–5 days or longer in agitated broth. Faster growing strains are uncommon. Chemoorganotrophic, utilizing a range of carbohydrates and salts of organic acids as carbon sources, without gas formation; pentoses are preferred as carbon sources. Cellulose and starch are not utilized. **Produce an alkaline reaction in mineral salts medium containing mannitol or many other carbohydrates. Growth on carbohydrate media is usually accompanied by extracellular polysaccharide slime. Some strains can grow chemolithotrophically in the presence of H_2, CO_2 and low levels of O_2.** Ammonium salts, usually nitrates, and some amino acids, can serve as nitrogen sources. Peptone is poorly utilized (except for strains isolated from *Lotononis*). Casein and agar are not hydrolyzed. There is usually no requirement for vitamins with the rare exception of biotin, which also may be inhibitory to some strains. **3-Ketoglycosides are not produced** (Bernaerts and De Ley, 1963). The organisms are characteristically able to invade the root hairs of tropical-zone and some temperate-zone leguminous plants (family *Leguminosae*) and incite the production of root nodules, wherein the bacteria occur as intracellular symbionts. All strains exhibit host range affinities (host "specificity"). **The bacteria are present in root nodules as swollen forms which are normally involved in fixing atmospheric nitrogen into combined forms utilizable by the host plant. Some strains fix nitrogen in the free living state** when examined under special conditions. The mol% G + C of the DNA is 61–65 (T_m).

Type species: *Bradyrhizobium japonicum* (Kirchner 1896) Jordan 1982, 137.

Further Descriptive Information

In young culture the cells are short rods, but in older cultures or under adverse growth conditions, including low concentrations of calcium or magnesium, the cells are commonly pleomorphic (swollen and elongated). Older cells stain to give a banded appearance because of large accumulations of poly-β-hydroxybutyrate. Within the root nodules the bacteroids are rod shaped and slightly swollen, but not branched or highly distorted, and they contain polyphosphate inclusions as well as poly-β-hydroxybutyrate accumulations.

All strains produce large amounts of water-soluble extracellular polysaccharide, the main constituent of which is acidic heteropolysaccharide (80–90%), the remainder being neutral β-1,2,-, β-1-3, and α-1,2- glucans (and perhaps others), with different degrees of branching (W. F. Dudman, The extracellular glucans of *Rhizobium japonicum* strain 3I1b135, Proceedings of the Fourth International Symposium on Nitrogen Fixation, Canberra, 1980; Dudman and Jones, 1981). About six different types of acidic heteropolysaccharides have been noted in an examination of 30 or more strains of *B. japonicum* and possess three distinctive features: (a) heterogeneity in composition and structure, (b) the frequent presence of methylated sugars (Dudman, 1976, 1978; Kennedy and Bailey, 1976), and (c) the presence of D-galacturonic acid (Dudman, 1976) in some strains (see the genus *Rhizobium*, Table 4.51).

Young cells are motile with one polar or subpolar flagellum, similar to *Rhizobium loti*. Growth on carbohydrate-containing solid medium is clear, translucent or opaque white with a tendency to exhibit a brownish pigmentation on prolonged incubation. Colonies of bacteria from *Lotononis bainesii* are pink to deep red, but bacteria from *Lotononis angolensis* give rise to colorless colonies. Small-colony variants occur.

Most strains grow on a mineral salts medium containing yeast and either glucose, galactose, gluconate, glycerol, fructose, arabinose, or mannitol. Maltose is utilized by about 10% of the strains but lactose, rhamnose, raffinose, trehalose, sucrose, dulcitol and dextrin are rarely utilized. However, tests carried out on basal medium devoid of yeast extract may give very different results, with a much greater range of carbon sources being utilized and with glucose apparently being a poor source of carbon (Meyer and Pueppke, 1980). Organic acids such as fumarate, malate, succinate, citrate, and pyruvate are utilized provided that the basal medium has sufficient Ca^{2+} and Mg^{2+} to overcome the inhibitory chelating effect of these acids.

Some strains can utilize ammonium salts or nitrate as sole source of nitrogen. Certain amino acids (glutamate, histidine, aspartate, proline) serve as sole nitrogen sources, but are inferior to vitamin-free casein hydrolysate. Peptone is not well utilized except for isolates from *Lotononis*. Anaerobic growth on nitrate medium and denitrification may occur (Daniel et al., 1982).

Most strains lack the ability to absorb Congo red from a yeast extract-mannitol-mineral-salts medium containing a 0.0025% final concentration of this dye. This results in colorless or faintly pink colonies, whereas contaminant colonies are often a deep red.

The maximum growth temperature ranges from 30–42°C, with most strains failing to grow above 39°C. Usually acid tolerant but not exclusively so (e.g. some *Lotus* bacteria; Cooper, 1982), most strains growing at pH 4.5. Over 30% of the strains will grow at pH 4.0 and a few as low as pH 3.5. Growth usually does not occur above pH 9.0.

An alkaline reaction is produced in litmus milk without the production of a clear, upper "serum zone."

Bradyrhizobia fail to grow in media containing 2% NaCl, do not produce H_2S, and do not form a precipitate in calcium glycerophosphate medium (Hofer, 1941). Penicillinase production is common.

The Entner-Doudoroff pathway is employed in carbohydrate degradation (Martinez-de Drets and Arias, 1972; Mulongoy and Elkan, 1977) perhaps with the simultaneous operation of the Embden-Meyerhof-Parnas pathway (Mulongoy and Elkan, 1977). However, the latter pathway is in doubt since adequate levels of fructose-1, 6-biphosphate aldolase may not be present. Since the key enzyme in the pentose phosphate route ($NADP^+$-dependent 6-phosphogluconate dehydrogenase) has not been found, the operation of this pathway also is debatable. The tricarboxylic acid cycle is operational in both free living cells (Keele et al., 1969; Mulongoy and Elkan, 1977) and bacteroids from root nodules (Storall and Cole, 1978). A pathway for the direct oxidation of gluconate by the tricarboxylic acid cycle, via 2-keto- and 2,5-diketogluconate and α-ketoglutarate, has been described by Keele et al. (1970).

Some strains possess an active uptake hydrogenase which enables them to grow chemolithotrophically in an atmosphere of 10% H_2, 5% CO_2 and 1% O_2 (the balance being N_2). Under such conditions, ribulose biphosphate carboxylase is primarily responsible for CO_2 fixation (Hanus et al., 1979; Lepo et al., 1980). Mutants unable to grow chemolithotrophically fall into several classes, including those impaired in H_2 uptake and those deficient in CO_2 uptake (Maier, 1981).

Nitrogenase activity by free living cells occurs in certain strains but

only in media containing selected carbon sources and under a low level of oxygen (Kurtz and LaRue, 1975; McComb et al., 1975; Pagan et al., 1975).

Studies on the genetics of the slow growing root nodule bacteria are few. Auxotrophic mutants are not easy to obtain and no linkage maps are available. However, genes required for nodulation have been transferred from *B. japonicum* to *Azotobacter vinelandii* (Maier et al., 1978) and P-1 group plasmids have been transferred within and between several serologically-distinct strains of *B. japonicum* (Pilacinski and Schmidt, 1981).

Little attention has been paid to bacteriophages active against the bradyrhizobia, but such agents have been isolated from rhizosphere soil and root nodules. Bacteriocins have been reported (Roslycky, 1967).

Although internal antigens are widely shared among the slow growing nodule bacteria, there is no sharing of somatic, flagellar or internal antigens between these bacteria and the fast growing *Rhizobium* spp. Also, immunodiffusion studies reveal that there is no sharing of antigens between fast or slow growing strains from *Lotus* (Pankhurst, 1979), adding support to the separation of these two bacterial groups from *Lotus* into two distinct genera. Bradyrhizobia are markedly more resistant to a number of antibiotics (tetracyclines, streptomycin, penicillin G, viomycin, vancomycin) than are strains of *Rhizobium*, but tend to be more sensitive to growth inhibitors such as D-alanine or ethidium bromide. However strain-to-strain variation in resistance is considerable. Spontaneous mutation to resistance to most antibiotics is a common occurrence in "wild type" strains.

Bradyrhizobia cause nodule production on *Glycine* (soybean), *Vigna* (cowpea), *Macroptilium* (siratro), certain species of *Lotus*, and a wide variety of leguminous plants which also are nodulated by *Rhizobium loti*, such as *Acacia*. In fact there appear to be three distinct groups of *Acacia* species: one group nodulated by *R. loti*, one group nodulated by *Bradyrhizobium* sp. (*Acacia*) and one group nodulated by both of these species (Dreyfus and Dommergue, 1981).

A highly specific nodulation of the nonleguminous plant *Parasponia* (*Trema*) by a strain of *Bradyrhizobium* occurs (Trinick, 1973, 1976). This is the only completely validated instance of nodulation by a *Rhizobium* or *Bradyrhizobium* species which occurs outside of the Leguminosae.

Norris (1956, 1965) considered the alkali-producing bradyrhizobia, which are particularly well adapted to acid tropical soils, to represent a survival of the ancestral type of root nodule bacteria, which ultimately gave rise to the acid-producing *Rhizobium* species associated with temperate zone leguminous plants. However this view is not universally accepted.

Enrichment and Isolation Procedures

Isolation from root nodules and soil is by the techniques described for the genus *Rhizobium*.

Maintenance Procedures

A suitable maintenance medium is yeast extract-mannitol agar (YMA) (see formulation under Enrichment and Isolation Procedures in the chapter on the genus *Rhizobium*). Storage recommendations are the same as given for *Rhizobium*.

Procedures for Testing for Special Characters

For studies of carbohydrate utilization, the sugar and $CaCO_3$ are omitted from YMA, and the yeast extract is decreased by 75%.

A defined basal medium for carbohydrate utilization studies has been published by Meyer and Pueppke (1980).* This medium can be modified to a general medium by omitting the indicator dye and adding $MgSO_4 \cdot 6H_2O$ (0.2 g/liter), yeast extract (Difco; 0.25 g/liter) and mannitol (5.0 g/liter).

---

* Defined basal medium (Meyer and Pueppke, 1980): $(NH_4)_2SO_4$, 0.5 g; 2-(N-morpholino)-ethanesulfonic acid (MES buffer), 1.1 g; 1,4-piperazinediethanesulfonic acid (PIPES buffer), 1.3 g; biotin, 0.01 g; bromthymol blue, 0.1 g; supplement B (Difco), 2.0 ml; carbon source, 5.0 g (except sodium glutamate and sodium citrate (2.0 g) or sodium malonate (3.0 g)); and distilled water, 1 liter. The final pH is 6.9.

Differentiation of the genus Bradyrhizobium from other genera

The characteristics of *Bradyrhizobium* that distinguish it from other genera which are morphologically or physiologically similar are given in table 4.50 in the chapter on the genus *Rhizobium*.

Taxonomic Comments

In the eighth edition of *Bergey's Manual* the genus *Rhizobium* was subdivided into two groups, distinguished by their growth rate in yeast extract-mannitol-mineral salts medium, flagellar arrangement, DNA base composition and the genera of host plants nodulated (Jordan and Allen, 1974). This subdivision was based on the proposal of 't Mannetje (1967) and on the desire to allow more information to accumulate which would permit the separation of the two groups into two different genera. Subsequently, a considerable amount of data has accumulated which has necessitated the placement of all of the slow growing, nonacid-producing root nodule bacteria in a genus (*Bradyrhizobium*) separate from that containing the fast growing, acid-producing nodule bacteria. These data have resulted from studies involving numerical taxonomy (Graham, 1964a; Moffet and Colwell, 1968; 't Mannetje, 1967), DNA base ratios (De Ley and Rassel, 1965; Wagenbreth, 1961), nucleic acid hybridization (Gibbins and Gregory, 1972; Heberlein *et al.*, 1967), cistron similarities (De Smedt and De Ley, 1977), serology (Graham, 1963; Humphrey et al., 1973; Vincent, 1977; Vincent and Humphrey, 1970), composition of extracellular gum (Dudman, 1976; Dudman, 1978; Kennedy, 1976; Kennedy and Bailey, 1976), carbohydrate utilization (Skotnicki and Rolfe, 1978) and metabolism (Martinez-de Drets and Arias, 1972), bacteriophage sensitivity (Napoli et al., 1980), antibiotic sensitivity (Strzelcowa, 1968), protein composition (Roberts et al., 1980), and type of bacteroid inclusion bodies (Craig et al., 1973).

The genus *Bradyrhizobium* represents an exceedingly heterogeneous group of nodule bacteria within which the taxonomic relationships are not well understood. Recent DNA/DNA hybridization studies by Hollis, Kloos and Elkan (1981) suggest that strains of what is now designated *B. japonicum* can be separated into at least three DNA homology groups. Therefore, it is quite possible that a series of species and/or biovars of *B. japonicum* will be generated within the genus as more information becomes available. Until such time as this occurs, the species formerly designated *R. lupini* (Jordan and Allen, 1974) is not being designated as such in the genus *Bradyrhizobium*, since its only major distinguishing characteristic was a high degree of nodulation affinity for *Ornithopus* and *Lupinus* spp. (see under Other Organisms Belonging to the Genus *Bradyrhizobium*).

The taxonomic position of nodule bacteria from the pasture legume *Lotononis* is uncertain (Norris, 1958). Isolates from *L. bainesii* are red because of an intracellular red carotenoid pigment, although isolates from *L. angolensis* are nonpigmented. As with other strains of *Bradyrhizobium*, these strains are monotrichous, grow slowly, and produce an alkaline reaction and no serum zone in litmus milk; however, cultured cells appear as enlarged, banded ovoids. Resistance of these strains to ultraviolet light is greater than that exhibited by other strains of *Bradyrhizobium* or of *Rhizobium* (Law, 1979). Peptone is utilized, the cells fail to react with antisera derived either from other slow growers or from species of *Rhizobium*, and the mol% G + C of the DNA is 68–69 (T_m) (Godfrey, 1972). The strains are extremely specialized, nodulating *Lotononis* spp. and *Macroptilium atropurpureum* effectively and selected species of *Aeschynomene* and *Crotolaria* ineffectively.

Comments concerning a group of fast growing root-nodule bacteria capable of producing effective nodules on *Glycine soja* and *G. max* cv Peking, but ineffective nodules on commerical lines of soybean, are presented under Taxonomic Comments for the genus *Rhizobium*.

Only one species of *Bradyrhizobium* is presently recognized.

Acknowledgments

The author is indebted to the members of the International Subcommittee on *Agrobacterium* and *Rhizobium*, who provided support and often contributed valuable advice: J. De Ley, P. H. Graham, J. Holding, B. D. W. Jarvis, E. B. Roslycky, B. W. Strijdom and J. M. Vincent. The author also wishes to thank J. E. Beringer, J. Burton, R. A. Date, W. F. Dudman, E. Schwinghamer and M. J. Trinick who took part in valuable discussions.

Further Reading

Additional reading is listed in the chapter on the genus *Rhizobium*.

List of the Species of the Genus Bradyrhizobium

1. **Bradyrhizobium japonicum** (Kirchner 1896) Jordan 1982, 137.[VP] (*Rhizobacterium japonicum* Kirchner 1896, 221; *Rhizobium japonicum* (Kirchner 1896) Buchanan 1926, 90[AL].)

ja.po′ni.cum. M.L. adj. *japonicum* pertaining to Japan.

The characteristics are as described for the genus. Bacteroids in root nodules are slightly swollen rods with rare branching, or coccus forms (in *Arachis* spp.). Cells of *B. japonicum* have one polar or subpolar flagellum. The species normally causes the formation of root nodules on species of *Glycine* (soybean) and on *Macroptilium atropurpureum* (siratro).

The mol% G + C of the DNA is 61–65 (T_m).

Type strain: ATCC 10324.

Other organisms belonging to the genus Bradyrhizobium

Other bradyrhizobia occur but have not yet been classified as species or biovars. These organisms cause nodule production on certain species of *Lotus* (*L. uliginosus* and *L. pedunculatus*), as well as on *Vigna* and species of *Lupinus, Ornithopus, Cicer, Sesbania, Leucaena, Mimosa, Lablab* and *Acacia*, which are also nodulated by the fast growing *Rhizobium loti*. Such strains also usually nodulate *Macroptilium* and, more rarely, *Glycine*. Some strains fix nitrogen in the free living state under special conditions. It is suggested that until such time as further species or biovars are created within the genus *Bradyrhizobium*, these organisms (other than *B. japonicum*) be designated as *Bradyrhizobium* sp. with the name of the appropriate host plant given in parenthesis immediately following; e.g. *Bradyrhizobium* sp. (*Vigna*) or *Bradyrhizobium* sp. (*Lupinus*).

Genus III. Agrobacterium Conn 1942, 359[AL]

KAREL KERSTERS AND JOZEF DE LEY

Ag.ro.bac.te′ri.um. Gr. n. *agros* a field; Gr. dim. neut. n. *bakterion* a small rod; M.L. neut. n. *Agrobacterium* a field rod.

Rods, 0.6–1.0 μm by 1.5–3.0 μm, occurring singly or in pairs. **Nonsporing. Gram-negative. Motile** by 1–6 **peritrichous flagella. Aerobic**, possessing a **respiratory type of metabolism** with oxygen as the terminal electron acceptor. Some strains are capable of anaerobic respiration in the presence of nitrate. Most strains are able to grow under reduced oxygen tensions in plant tissues. Optimum temperature:

25–28°C. Colonies are usually convex, circular, smooth, nonpigmented to light beige. **Growth on carbohydrate-containing media is usually accompanied by copious extracellular polysaccharide slime.** Catalase-positive, and usually oxidase- and urease-positive. **3-Ketoglycosides** are produced by the majority of strains belonging to *A. tumefaciens* biovar 1 and *A. radiobacter* biovar 1. **Chemoorganotrophs,** utilizing a wide range of carbohydrates, salts of organic acids and amino acids as carbon sources, but not cellulose, starch, agar or chitin. Produce an acid reaction in mineral salts media containing D-glucose, D-galactose and other carbohydrates. Ammonium salts and nitrates can serve as nitrogen sources for strains of some species and biovars; others require amino acids and additional growth factors. With the exception of *A. radiobacter*, members of this genus **invade the crown, roots and stems of a great variety of dicotyledonous and some gymnospermous plants, via wounds, causing the transformation of the plant cells into autonomously proliferating tumor cells.** The induced plant diseases are commonly known as **crown gall, hairy root and cane gall.** Some strains possess a wide host range, whereas others (e.g. grapevine isolates) possess a very limited host range. The tumors are self-proliferating and graftable. The tumor induction by *Agrobacterium* is correlated with the presence of a **large tumor-inducing plasmid (Ti-plasmid)** in the bacterial cells. Agrobacteria are **soil inhabitants.** Oncogenic strains occur mainly in soils previously contaminated with diseased plant material. Some nononcogenic *Agrobacterium* strains have been isolated from human clinical specimens. The mol% G + C of the DNA is 57–63 (T_m). The molecular weight of the *Agrobacterium* genome ranges from 3.0×10^9–3.6×10^9.

Type species: *Agrobacterium tumefaciens* (Smith and Townsend 1907, 672) Conn 1942, 359.

Further Descriptive Information

The phenotypic description of the genus *Agrobacterium* is mainly based on studies by De Ley et al. (1966), Lippincott and Lippincott (1969), Keane et al. (1970), White (1972), Kersters et al. (1973), Panagopoulos and Psallidas (1973), Kerr and Panagopoulos (1977), Süle (1978), Panagopoulos et al. (1978), and Holmes and Roberts (1981). The subdivision of several species of *Agrobacterium* into biovars is explained in the section Taxonomic Comments and later in Table 4.55.

In young cultures the cells occur mostly as short rods. The formation of star- or rosette-shaped aggregates of cells by several *Agrobacterium* strains has been described by Beijerinck and van Delden (1902), Stapp and Knösel (1956) and Knösel (1962).

The cell envelopes of *Agrobacterium* are generally similar to envelopes of other Gram-negative bacteria. The mucopeptide of *Agrobacterium* consists of glutamic acid, alanine, diaminopimelic acid and amino sugars. In addition leucine, phenylalanine, serine and aspartic acid have been detected in relatively large amounts in the mucopeptide layer of several agrobacteria. The enevelope lipopolysaccharides contain glucose, fucose and 2-keto-3-deoxyoctonic acid. Phosphatidylcholine, phosphatidylethanolamine, phosphatidyl-*N*-methylethanolamine and phosphatidylglycerol are the major phospholipids detected in the cell envelopes of both oncogenic and nononcogenic agrobacteria (Manasse and Corpe, 1967; Das et al., 1979). The fatty acid composition of the purified agrobacterial lipopolysaccharides is unique, because 3-hydroxytetradecanoic acid and 3-hydroxyhexadecanoic acid are the only major components (Salkinoja-Salonen and Boeck, 1978).

Neutral extracellular β-1,2-glucans are produced by most agrobacteria, as well as by *Rhizobium* and by some strains of *Bradyrhizobium japonicum* (York et al., 1980) (see also Table 4.53 of the genus *Rhizobium*). Water-soluble acidic heteropolysaccharides of various *Agrobacterium* strains have β-glucan as the main component of the polysaccharide, and have succinic, pyruvic and acetic acids as acidic components; they lack hexuronic acids. Their chemical structure is similar to the acidic heteropolysaccharides of *Rhizobium meliloti* (structural type 1 in Table 4.53 of the genus *Rhizobium*) (Zevenhuizen, 1971; Hisamatsu et al., 1980). The formation of water-insoluble β-1,3-glucans (curdlan-type polysaccharide) has been reported in some strains of *Agrobacter-*

ium (Nakanishi et al., 1976). The synthesis of glycogen by *A. tumefaciens* strain B6 is regulated at the level of ADP-glucose synthesis (Eidels et al., 1970).

Cellulose-containing fibrils are formed by oncogenic *Agrobacterium tumefaciens* strains during their attachment to plant cells in vitro and seem to anchor the bacteria to the plant cell surface (Matthysse et al., 1981). Lipopolysaccharides of the outer membrane of the *Agrobacterium* cell envelope play a role in the attachment of the bacteria to the wound site of the plant (Whatley et al., 1976).

Growth of *Agrobacterium* on nutrient agar is moderate, whereas abundant growth is obtained on media containing yeast-extract and a suitable carbohydrate such as glucose, sucrose or lactose (see Maintenance Procedures). On carbohydrate-containing solid media the majority of the strains produce circular, low-convex to convex, mucous, glistening, opaque, white to beige-colored colonies, with an entire edge and a diameter of 2–4 mm after 6 days of incubation. Nongummy and rough colonial forms are produced by some strains.

Intermediates of the tricarboxylic acid cycle and several amino acids can be utilized as sole sources of carbon. The majority of biovar-1 strains (see under Taxonomic Comments and later in Table 4.55) can grow on a minimal medium with nitrate or ammonium salts as the nitrogen source. No isolate of biovar 2 can utilize nitrate unless biotin is supplied; some biovar-2 strains require both L-glutamic acid and biotin. Strains belonging to *A. rubi* require L-glutamic acid and yeast extract (Starr, 1946; Lippincott and Lippincott, 1969; Keane et al., 1970).

Certain amino acids, such as glycine and various D-amino acids, may inhibit the growth of agrobacteria (Beardsley, 1962; Bopp, 1965). Denitrification occurs in some strains (Kersters et al., 1973; Pichinoty et al., 1977). Some strains strongly absorb Congo red or aniline blue from mannitol-containing media (Riker et al., 1930; Hendrickson et al., 1934); however, the absorption of dyes has no diagnostic value (Lippincott and Lippincott, 1969; Kersters et al., 1973). According to Skinner (1977), agrobacteria can be differentiated from rhizobia by a modified Nile blue test because only the former organisms reduce the dye.

The following physiological, biochemical and nutritional tests are positive for the whole genus: catalase; acid production from D-glucose, L-arabinose, D-xylose, D-fructose, adonitol, D-mannitol, L-rhamnose, lactose, maltose, cellobiose, sucrose, trehalose and salicin; and growth on D-glucose and β-hydroxybutyrate. The following physiological tests are negative for the whole genus: gas from D-glucose, gelatinase, production of indole, phenylalanine deamination, H_2S production in triple sugar iron agar, hydrolysis of starch and Tween 80, KCN tolerance, and fluorescence on King's B medium. Table 4.57 (shown later) should be consulted for further details because results obtained by different authors may be different for some of these tests. Agrobacteria are not known to fix dinitrogen or to grow chemolithotrophically with hydrogen. All the strains grow between 20°C and 28°C. Biovar-2 strains (see below) cannot grow above 30°C.

The changes in litmus milk are of diagnostic value. One group of strains (biovar 1) produces an alkaline reaction, usually accompanied by the formation of a brown serum zone. Strains of biovar 2 produce an acid reaction (pink color), sometimes accompanied by the formation of an "acid clot" (Lippincott and Lippincott, 1969; Kersters et al., 1973).

The principal mechanisms of glucose catabolism in *Agrobacterium* are the Entner-Doudoroff pathway and the pentose cycle (Vardanis and Hochster, 1961; Arthur et al., 1973, 1975). Glucose is scarcely metabolized via the glycolytic pathway because several strains show extremely low activities of phosphofructokinase and fructose-1,6-diphosphate aldolase (Arthur et al., 1975) and the activity of phosphoglyceromutase was found to be very low in another *A. tumefaciens* strain (Chern et al., 1976a). Glucuronic acid and glucaric acid are metabolized via 2-keto-3-deoxy-D-glucaric acid to α-ketoglutaric acid (Chang and Feingold, 1970). The initial step in the catabolism of L-sorbose by some biovar 1 strains is the reduction to sorbitol, followed by the oxidation of the latter compound to D-fructose (Van Keer et al., 1976). The majority of *Agrobacterium* strains belonging to biovar 1 characteristically oxidize a great number of carbohydrates (disaccharides, bionic acids and several monosaccharides) to the corresponding 3-uloses (Bernaerts and De Ley, 1960a, b; Fukui et al., 1963; De Ley et al., 1966). The vigorous and unusual oxidation of lactose to 3-ketolactose is the biochemical basis of the simple and specific diagnostic test of Bernaerts and De Ley (1963) for the rapid differentiation of biovar 1 from biovars 2 and 3 (see Table 4.56). These specific oxidations are catalyzed by an inducible hexopyranoside:cytochrome c oxidoreductase (D-glucoside 3-dehydrogenase), containing flavin adenine dinucleotide as cofactor (Hayano and Fukui, 1967; Van Beeumen and De Ley, 1968; Nakamura and Tyler, 1977). Although an α-3-ketoglucosidase was detected in a strain of Agrobacterium tumefaciens (Hayano and Fukui, 1970; Hayano et al., 1973), 3-ketosucrose and 3-ketolactose are probably not involved as essential intermediates in the metabolism of sucrose and lactose, respectively (Kurowski and Pirt, 1971; Janssens, Kersters and De Ley, unpublished results). Conditions have been worked out for increasing the yield of 3-ketoglycosides (Tyler and Nakamura, 1971; Fensom, Kurowski and Pirt, 1974; Kurowski et al., 1975).

The tricarboxylic acid cycle is operative in Agrobacterium (Arthur et al., 1973; Chern et al., 1976a) and the role of pyruvate carboxylase in CO_2 fixation has been demonstrated in at least one A. tumefaciens strain (Chern et al., 1976b).

Agrobacterium tumefaciens contains at least two soluble cytochromes c, a cytochrome c-552 and a cytochrome c-556. Cytochrome c-552 from A. tumefaciens strain IIChrys has been sequenced (Van Beeumen et al., 1980). It belongs to the cytochrome c sequence class IB (sensu Ambler, 1978) and, of all known procaryotic cytochromes c, shows the highest amino acid sequence homology with mitochondrial cytochrome c of tuna fish (Van Beeumen et al., 1980). Cytochrome c-556 from A. tumefaciens belongs to the cytochrome c sequence class II (sensu Ambler, 1978), because its single heme group is bound near the C-terminus (Van Beeumen et al., 1980) (see also Taxonomic Comments).

The uptake and metabolism of L-valine and L-proline by resting and growing cells of Agrobacterium tumefaciens has been studied by Behki and Hochster (1967) and Behki (1967).

Opines (octopine or nopaline) are unusual amino acid derivatives produced in tumor tissues induced by oncogenic strains of Agrobacterium. The arginine derivatives octopine and nopaline were the first to be identified (Petit et al., 1970). The synthesis of these "opines" by crown gall cells creates in the transformed plant tissue and in the rhizosphere a unique ecological niche for the oncogenic agrobacteria, because almost all oncogenic Agrobacterium strains can use either octopine or nopaline as a sole source of carbon and nitrogen, by way of Ti-plasmid encoded enzymes (Bomhoff et al., 1976; Montoya et al., 1977; Petit et al., 1978) (see below). However, in some strains (e.g., biovar-2 strains) chromosomal genes code for octopine degradation (Montoya et al., 1978). A catabolic pathway for octopine and octopinic acid has been proposed (Ellis et al., 1979a): scission of octopine into arginine and pyruvic acid, followed by the conversion of arginine via ornithine to glutamic acid.

Oncogenicity was first transferred from a donor A. tumefaciens strain to a recipient nononcogenic A. radiobacter strain by inoculating both strains together or in succession onto the same plant (Kerr, 1969, 1971). The cause of this phenomenon is the transfer of a tumor-inducing plasmid (pTi) from the oncogenic to the nononcogenic strain. Tumor-inducing Agrobacterium strains harbor one or more large extra-chromosomal DNA plasmids with a molecular weight of approximately 100×10^6 to 160×10^6 (Zaenen et al., 1974). Oncogenic Agrobacterium strains (A. tumefaciens) lose their ability to induce tumors when they are cured of their tumor-inducing plasmid. Conversely, introduction of the Ti-plasmid in avirulent Agrobacterium strains (A. radiobacter) by either DNA-mediated transformation or conjugation, yields oncogenic agrobacteria (Van Larebeke et al., 1974, 1975; Watson et al., 1975). During the course of infection of a wounded susceptible plant, the Ti-plasmid is somehow transferred to the plant cells, where part of it, the so-called T-DNA (molecular weight ~9–15×10^6), is integrated in the nuclear fraction of the transformed plant cells (Chilton et al., 1980;

Thomashow et al., 1980a; Lemmers et al., 1980; Zambryski et al., 1980; Willmitzer et al., 1980).

At least three genetic groups of Ti-plasmids can be recognized in wild type Agrobacterium strains according to the type of opine that is synthesized by the tumor and utilized by the bacterium. (a) The octopine-type Ti-plasmids are highly homologous to one another as indicated by DNA/DNA hybridization (Currier and Nester, 1976). (b) The nopaline-type Ti-plasmids form a more diversified group, sharing 50–100% DNA-homology among each other. (c) The agropine-type Ti-plasmids, formerly called null-type Ti-plasmids, induce neither nopaline nor octopine in the tumor but induce the production of agropine. The latter compound is a condensation product of the lactam form of glutamic acid (or glutamine) and a hexitol (Coxon et al., 1980; Guyon et al., 1980). Four areas of significant DNA homology were detected between a standard octopine and nopaline Ti-plasmid (Depicker et al., 1978; Van Montagu et al., 1980). Part of the T-DNA corresponds with one of these homologous areas. Physical maps of octopine and nopaline Ti-plasmids have been determined by restriction endonuclease analysis (Chilton et al., 1978; Depicker et al., 1980; De Vos et al., 1981). The functional organization of the nopaline plasmid pTi-C58 and octopine plasmids pTi-B6 and pTi-ACH5 has been determined by Holsters et al. (1980), Koekman et al. (1979) and De Greve et al. (1981).

The following functions have been found to be determined by the Ti-plasmids: (a) oncogenicity; (b) the nature of the opines synthesized in the transformed plant cells; (c) the utilization of opines and arginine (Ellis et al., 1979a); (d) the conjugative transfer of the Ti-plasmid; this transfer is promoted by the opines; (e) sensitivity to the bacteriocin agrocin 84 (see below); (f) the host range (Loper and Kado, 1979; Thomashow et al., 1980b); and (g) the exclusion of bacteriophage AP-1.

No correlation exists between DNA homologies of Ti-plasmids and chromosomal homology of the Agrobacterium tumefaciens strains investigated. Neither do plasmid homologies correlate with any numerical classification of the genus Agrobacterium (Currier and Nester, 1976) (see also Taxonomic Comments).

Another type of large plasmid is involved in the hairy root disease of plants caused by A. rhizogenes (Moore et al., 1979; White and Nester, 1980a). Little overall sequence homology with other Ti-plasmids was detected (White and Nester, 1980b). There is one small region of conserved homology between the hairy root plasmid and an octopine Ti-plasmid (pTi-B6806), but no homology with the region of the T-DNA of the latter plasmid. The A. rhizogenes plasmid is compatible with other Ti-plasmids and thus represents a new incompatibility class of Agrobacterium plasmids (White and Nester, 1980b). The investigated A. rhizogenes strains utilize octopine (Lippincott et al., 1973).

Large plasmids have been found in nononcogenic agrobacteria (A. radiobacter) (Merlo and Nester, 1977; Sheikholeslam et al., 1979) and, in several A. tumefaciens strains, multiple-size plasmids have been discovered besides the Ti-plasmid.

Further information concerning Ti-plasmids and their role in tumor formation can be found in the following review articles: Braun (1978), Drummond (1979), Van Montagu and Schell (1979), Schell et al. (1979), Van Montagu et al. (1980), Gordon (1981).

The best known bacteriocin from Agrobacterium is agrocin 84, synthesized by the nononcogenic A. radiobacter strain 84 (NCPPB 2407) (New and Kerr, 1972; Kerr and Htay, 1974). The production of agrocin 84 is encoded by a small plasmid (Ellis et al., 1979b). It is a toxic analogue of an adenine nucleotide (Roberts et al., 1977) selectively inhibiting Agrobacterium strains harboring a nopaline plasmid. Sensitivity towards agrocin 84 is determined by the Ti-plasmid (see above). Dipping seeds, roots or wounded plant surfaces (involved in grafting) in a suspension of A. radiobacter strain 84 has been used worldwide with success for the biological control of crown gall disease (for reviews see Moore and Warren, 1979 and Kerr, 1980).

Lysogeny for either active plaque-forming or defective bacteriophages is widespread in the genus Agrobacterium. Morphological, biological and physicochemical properties and genetic relationships of several of the isolated phages or phagelike particles have been determined (Beardsley, 1955; Zimmerer et al., 1966; Stonier et al., 1967; Manasse et al., 1972; De Ley et al., 1972; Vervliet et al., 1975). Virulent bacteriophages of *Agrobacterium* have been isolated from sewage and soil (Roslycky et al., 1963; Boyd et al., 1970a, b; J. P. Hernalsteens, Ph.D. thesis, V.U.B., Brussels, Belgium).

Most serological studies indicate that it is impossible to distinguish oncogenic from nononcogenic agrobacteria (Graham, 1971). However, strains belonging to biovars 1 and 2 can be distinguished from each other by serological reactions (Keane et al., 1970; Lopez, 1978). The fast-growing rhizobia show extensive cross-reaction with several *Agrobacterium* strains (Graham, 1971).

In general wild type agrobacteria are sensitive to aureomycin, gentamicin, neomycin, novobiocin, terramycin and tetracycline (Kersters et al., 1973). Growth of agrobacteria is inhibited by low concentrations (3–780 ng/ml of medium) of metacycline, doxycycline, sigmamycin (tetracycline + oleandomycin) and triacetyloleandomycin (Goedert, 1973).

Upon infection of wounded plant tissues, oncogenic *Agrobacterium* strains can transform plant cells into autonomously proliferating tumor cells. In nature the swellings mostly occur at the transition zone between the stem and the root system of the host plant, hence the name "crown gall disease." Small spherical growths or elongated ridges can occur on the stems of *Rubus* spp. such as raspberry and bramble bushes, hence the name "cane gall disease." One group of agrobacteria (*A. rhizogenes*) causes infectious hairy root or woolly knot on susceptible plants (such as apple trees and roses) but initiates crown gall on other plants. Some strains of *Agrobacterium* can induce on some plants (e.g., *Kalanchoë*) the formation of teratomata characterized by the development of aberrant shoots, leaves or roots developing from the tumor tissue.

A prerequisite for tumorigenesis is the wounding of the host. Infection can occur during various stages of the life of a plant via wounds caused by growth, germination (e.g. peaches and almond), subterranean insects or mechanical injuries (e.g., pruning, grafting and replanting of trees in nurseries). The host-range of *Agrobacterium tumefaciens* is very wide: at least 640 different plant species belonging to 93 different families of dicotyledonous and gymnospermous plants are susceptible to transformation by *Agrobacterium tumefaciens*. None of the 250 monocotyledonous species investigated was susceptible to the disease, except some members of the orders *Liliales* and *Arales* (De Cleene, unpublished results). Extensive host-range indexes for crown gall and infectious hairy root have been published (De Cleene and De Ley, 1976, 1981). The host range of *Agrobacterium* seems to be conferred by the virulence-specifying Ti-plasmid (see above). The type of disease produced (differentiated or undifferentiated tumors) is probably determined by both the bacterial Ti-plasmid and the host plant (Gresshoff et al., 1979).

Crown gall disease does not necessarily kill the plant, but its growth is often impaired and stunted. Significant damage and economic loss has been reported, e.g. on stone fruit in Australia and the United States (peach, almond, cherries) and on vineyards in Bulgaria, Greece and Hungary (De Cleene, 1979). Also greenhouse cultures (e.g. *Chrysanthemum*) can be attacked by the disease. Chemotherapeutic control of crown gall has been reported (Schroth et al., 1971), but nowadays biological control with the agrocin-producing *A. radiobacter* strain 84 has found wide application (see above). Unfortunately some *Agrobacterium* strains are resistant to agrocin 84. Further information concerning the infection process and mechanism of tumor formation can be found in Lippincott and Lippincott (1975), Braun (1978) and Gordon (1981).

Agrobacteria occur almost worldwide in soils and especially in the rhizosphere of plants. As many as 500 cells of *Agrobacterium*/g of soil have been reported. Agrobacteria can be isolated from young galls and tumorous outgrowths on different parts of diseased plants. Older tumor tissues are usually sterile. *Agrobacterium* strains have also been reported in a variety of human clinical specimens (CDC group Vd-3) (Lautrop, 1967; Riley and Weaver, 1977; Gilardi, 1978; Rubin et al., 1980). They are usually 3-ketolactose-positive and nononcogenic for tomato plants. It is believed that these clinical *Agrobacterium* isolates occur either as incidental inhabitants in the patient or as contaminants introduced during sample manipulation.

Enrichment and Isolation Procedures

Several selective media have been described for the isolation of agrobacteria from soil and crown gall tissues (Moore et al., 1980). The medium of Schroth et al. (1965) can be used for most biovar-1 strains.* The selective medium of New and Kerr (1971) is used for agrobacteria belonging to biovar2†; erythritol was selected as the sole carbon source in this medium because biovar-1 agrobacteria cannot utilize it (see Table 4.56). Agrobacteria can be isolated from soil and from young crown gall tissues by spreading 0.1 ml of the appropriate dilution of soil or extracts of gall tissue with an L-shaped glass rod over one of the media in Petri dishes, which are subsequently incubated at 27°C. Typical convex, glistening and circular colonies with an entire edge are transferred for storage and identification.

Other selective media were described by Clark (1969), Kado and Heskett (1970) and Moore et al. (1980), and recently a selective medium with tartrate as the carbon source has been developed for biovar-3 strains (Brisbane and Kerr, 1981).

Maintenance Procedures

Stock cultures of *Agrobacterium* may be maintained on agar slants in screw-capped vials at 4°C during 2 months on either one of the following media (in g/liter of tap water): (a) glucose, 20; yeast extract, 10; CaCO₃, 20; and agar, 20; or (b) glucose, 10; yeast extract, 10; (NH₄)₂SO₄, 1.0; KH₂PO₄, 0.25; and agar, 20. Agrobacteria remain viable for many years by preservation of the lyophilized cultures at 4°C.

Procedures for Testing Special Characters

Moore et al. (1980) give useful advice for performing inoculation tests for pathogenicity studies and summarize recipes for diagnostic media and tests.

Assay for virulence of Agrobacterium. It is difficult to recommend a single host plant for the assay of virulence of tumorigenic strains of *Agrobacterium*. The host range of some strains is indeed very restricted (Panagopoulos and Psallidas, 1973; Anderson and Moore, 1979). Sunflower seedlings (*Helianthus annuus*) provide one of the most rapid means. Usually 1–2 weeks after sowing the seeds, stems of the young sunflower plants can be used for inoculation with a sterile needle dipped in a heavy aqueous bacterial suspension prepared from a 2-day-

---

* The medium of Schroth et al. (1965) consists of (g/liter of distilled water): agar, 20.0; mannitol, 10.0; NaNO₃, 4.0; MgCl₂, 2.0; calcium propionate, 1.2; MgHPO₄·3H₂O, 0.2; MgSO₄·7H₂O, 0.1; NaHCO₃, 0.075; and magnesium carbonate, 0.075. The pH is adjusted to 7.1 with 1 N HCl. After the medium is autoclaved and cooled to 50–55°C, the following compounds are added aseptically to give final concentrations of (mg/liter): berberine, 275; sodium selenite, 100; penicillin G (1625 units/mg), 60; streptomycin sulfate, 30; cycloheximide, 250; tyrothricin, 1.0; and bacitracin (65 units/mg), 100.

† The medium of New and Kerr (1971) consists of (per liter of distilled water): agar, 18.0 g; *meso*-erythritol, 5.0 g; NaNO₃, 2.5 g; KH₂PO₄, 0.1 g; CaCl₂, 0.2 g; NaCl, 0.2 g; MgSO₄·7H₂O, 0.2 g; Fe-EDTA solution (0.65%, w/v), 2.0 ml; and biotin, 2.0 μg. The pH is adjusted to 7.0 with 1 N NaOH. After the medium is autoclaved and cooled to 50–55°C, the following compounds are added aseptically to give final concentrations of (mg/liter): cycloheximide, 250; bacitracin, 100; tyrothricin, 1.0; and sodium selenite, 100.

old culture from a nutrient agar slant or from an agar slant of the culture media mentioned in Maintenance Procedures. Sterile wooden prickers may also be used, the top of which have previously been dipped into a colony of the bacterium to be tested. Virulence tests on sunflower seedlings can usually be read within a week after inoculation. The young stems of various varieties of the following plants are also good indicators of virulence: tobacco (*Nicotiana tabacum*), tomato plants and *Kalanchoë daigremontiana* (Anderson and Moore, 1979). Young and well growing plants are recommended; tumor formation can be rather slow, taking from 2–4 weeks. As a control, wounded plants should be inoculated with known virulent and avirulent *Agrobacterium* strains. All the above mentioned plants should be kept in a greenhouse at 20–27°C and at a rather high relative humidity. When no greenhouse facilities are available disks of carrot roots (*Daucus carota*) are also useful (Klein and Tenenbaum, 1955; Lippincott and Lippincott, 1969). This is a convenient procedure, provided that at least 10 slices from different carrot roots are inoculated per strain and that proper controls are included, because false responses (cambial swellings) occasionally occur. *Agrobacterium tumefaciens* strains belonging to biovar 3 and isolated from grapevines usually display a restricted host specificity and virulence tests for these strains should therefore be performed on

green tender shoots of grapevines (Panagopoulos and Psallidas, 1973; Panagopoulos et al., 1978).

The pinto bean leaf bioassay of Lippincott and Heberlein (1965) has found wide application for studying quantitative aspects of tumor-initiating ability of *Agrobacterium*. A potato tuber bioassay for *Agrobacterium* tumors was originally described by Anand and Heberlein (1977) and modified by Pueppke and Benny (1981).

The root-inducing property of *A. rhizogenes* is usually tested by the carrot disk assay (Lippincott and Lippincott, 1969; Moore et al., 1979) or on *Kalanchoë daigremontiana* (White and Nester, 1980b). The interpretation of such experiments is sometimes difficult, because several *A. tumefaciens* strains are known to induce typical root teratomata on *Kalanchoë* plants (De Cleene and De Ley, 1981).

Sensitivity to bacteriocins. Methods for testing the sensitivity of bacterial strains to bacteriocins and methods for isolating bacteriocinogenic agrobacteria from soil have been described by Kerr and Panagopoulos (1977).

Test for biological control. Glasshouse tests using 4-week-old tomato seedlings have been described by Kerr and Panagopoulos (1977) for determining whether a virulent *Agrobacterium* strain is subject to biological control by a bacteriocinogenic strain (usually strain Kerr 84).

Differentiation of the genus **Agrobacterium** from other genera

The features that differentiate *Agrobacterium* from morphologically and physiologically similar genera *Rhizobium*, *Bradyrhizobium*, *Phyllobacterium* and *Pseudomonas* are given earlier in Table 4.50 for the genus *Rhizobium*. Features differentiating *Agrobacterium* from the genus *Alcaligenes* can be found later in Table 4.97 for the genus *Alcaligenes*.

Taxonomic Comments

The taxonomic structure of the genus *Agrobacterium* has been thoroughly studied. The following techniques have been used: (a) numerical analysis of phenotypic characteristics (White, 1972; Kersters et al., 1973; Holmes and Roberts, 1981); (b) biochemical and physiological tests (Keane et al., 1970; Kersters et al., 1973; Kerr and Panagopoulos, 1977; Süle, 1978; Holmes and Roberts, 1981); (c) DNA/DNA hybridizations (De Ley, 1972, 1974); (d) measurements of the thermal stability of DNA/DNA hybrids (De Ley et al., 1973); and (e) comparison of electrophoregrams of soluble proteins (Kersters and De Ley, 1975). In these studies at least 250 different *Agrobacterium* strains were involved, isolated from a great variety of habitats all over the world. The results obtained by the above mentioned methods corroborate each other (see below) and indicate that the genus *Agrobacterium* consists of at least three genetically and phenotypically different groups or clusters. Although most of these groups are large, one of them (*A. rubi*) contains only a few isolates.

These *Agrobacterium* clusters are listed below and Table 4.55 summarizes the correlations between various groupings and nomenclatures of *Agrobacterium* published after 1970. The term "biovar" is preferred here over "biotype," following the recommendations of Appendix 10 of the International Code of Nomenclature of Bacteria (Lapage et al., 1975). The nomenclature used by Holmes and Roberts (1981) is mentioned in quotation marks.

1. *Biovar 1* corresponds to biotype 1 of Keane et al. (1970), group I of White (1972), cluster 1 of Kersters et al. (1973) and to the "*A. tumefaciens*" group of Holmes and Roberts (1981). Biovar 1 encompasses more than 170, 3-ketolactose-positive, tumorigenic strains (named *A. tumefaciens* according to the Approved Lists of Bacterial Names, 1980), and avirulent strains (named *A. radiobacter* according to the Approved Lists), including the type strains of both species. According to Holmes and Roberts (1981), biovar 1 includes also some hairy root-forming strains (named *A. rhizogenes* according to the Approved Lists) isolated from cucumber;
2. *Biovar 2* corresponds to biotype 2 of Keane et al. (1970) (except for the two *A. rubi* strains ICPB TR2 (NCPPB 1856) and ICPB

TR3 (NCPPB 1854) studied by these authors), and also corresponds to group III of White (1972), cluster 2 of Kersters et al. (1973), and to the "*A. rhizogenes*" group of Holmes and Roberts (1981). It encompasses more than 60 3-ketolactose-negative tumorigenic (*A. tumefaciens*), rhizogenic (*A. rhizogenes*, including its type strain) and avirulent (*A. radiobacter*) strains;
3. Three strains constitute the *A. rubi* cluster (including its type strain);
4. A few 3-ketolactose-negative tumorigenic strains occupy a separate genotypic and phenotypic position (e.g. *A. tumefaciens* strains NCPPB 1650 and 1771). They are true members of the genus (De Smedt and De Ley, 1977).

A third biovar (i.e. biotype 3 *sensu* Kerr and Panagopoulos (1977)) has been described (Kerr and Panagopoulos, 1977; Süle, 1978; Panagopoulos et al., 1978). Biovar-3 strains were isolated from grapevines in Greece and Hungary and display a very limited host range (Panagopoulos et al., 1978). Holmes and Roberts (1981), on the basis of their numerical analysis, assigned two biovar-3 strains (NCPPB 2562 and 2611), together with the genetically separate strains NCPPB 1650 and 1771, to their *A. rubi* cluster. Future minor changes in the taxonomy of *Agrobacterium* can be anticipated when data on the genetic relationship between strains of biovar 3, *A. rubi* and strains NCPPB 1650 and 1771, become available. For the time being, we consider tumorigenic biovar-3 strains as a separate biovar of *A. tumefaciens* (see below).

DNA/DNA hybridizations (De Ley, 1972, 1974) show that biovar-1 strains, biovar-2 strains, the strains belonging to the small *A. rubi* cluster and the aberrant strains NCPPB 1650 and NCPPB 1771 hybridize only at about 15% DNA homology with each other (Fig. 4.35). The DNA homology group composed of biovar-1 strains is both genetically and phenotypically fairly heterogeneous and consists of at least seven genetically different subgroups hybridizing with each other at 45–50% DNA homology (Fig. 4.35). Within DNA homology group biovar 2, almost all the strains exhibit DNA homology values of 80–100% with each other. The genetic homogeneity of the biovar-2 strains is reflected in their great phenotypic and protein electrophoretic similarities (Kersters et al., 1973; Kersters and De Ley, 1975). The three strains of the *A. rubi* cluster share about 80% DNA relatedness. Strains named *A. tumefaciens* and *A. radiobacter* occur in both DNA homology groups biovars 1 and 2. Up to now the majority of strains named *A. rhizogenes* occur in one genetic group only (biovar 2), and *A. rubi* occurs in one small separate genetic cluster only.

The phenotypic and genotypic differences between biovars 1 and 2 are also confirmed by comparison of the physicochemical properties and the primary structures of the cytochromes *c*-556 of the *A. tumefaciens* strains IIChrys, B2a (belonging to 2 genetic subgroups of biovar 1) and strain Apple 185 (a biovar 2 strain). Cytochromes *c*-556 from both biovar 1 strains show 84% similarity in their amino acid sequences, whereas they share at most 67% similarity with cytochrome *c*-556 from the biovar-2 strain (Van Beeumen et al., 1980; Tempst and Van Beeumen, unpublished results).

The most obvious solution for the taxonomy and nomenclature of the genus *Agrobacterium* would be to give a separate species name or subspecies name for each of the three groups: biovar 1, biovar 2 and *A. rubi*. Unfortunately, such subdivision of the genus does not correspond to the four accepted species names on the Approved Lists of Bacterial Names. No correlation exists between the classical nomenclature (eighth edition of *Bergey's Manual* and the Approved Lists of Bacterial Names) of *Agrobacterium* on the one hand, and the real taxonomic structure of this genus as revealed by modern molecular and computer-assisted techniques on the other (see above).

In the previous edition of *Bergey's Manual* (Allen and Holding, 1974) the following four species belonged in the genus *Agrobacterium*; they were also repeated on the Approved Lists of Bacterial Names (see also Table 4.55):

1. *A. tumefaciens* (Smith and Townsend 1907) Conn 1942 (type species according to Opinion 33, Jud. Comm. 1970, 10),
2. *A. radiobacter* (Beijerinck and van Delden 1902) Conn 1942,
3. *A. rhizogenes* (Riker, Banfield, Wright, Keitt and Sagen 1930) Conn 1942,
4. *A. rubi* (Hildebrand 1940) Starr and Weiss 1943.

This classification is predominantly, and for some species solely, based on the phytopathogenic behavior of these bacteria and is of practical use for the phytopathologist, but is for three of the four species in complete disagreement with the real taxonomic structure of *Agrobacterium* (see above). No morphological, physiological or genotypical differentiation is possible between the *A. tumefaciens* and *A. radiobacter* strains within biovar 1, or between the *A. tumefaciens*, *A. radiobacter* and *A. rhizogenes* strains within biovar 2 (Keane et al., 1970; White, 1972; Kersters et al., 1973; De Ley, 1974; Holmes and Roberts, 1981). In addition, it is now firmly established that a clear-cut correlation exists in *Agrobacterium* between phytopathogenicity and presence or absence of Ti-plasmids (see Further Descriptive Information). Strains harboring Ti-plasmids provoke crown gall and should thus be named *A. tumefaciens*; strains without Ti-plasmids are not phytopathogenic and should be named *A. radiobacter*. However, Ti-plasmids can be lost, both naturally and in the laboratory, and can be transferred into *Agrobacterium* strains lacking them. One is thus faced with the absurd nomenclatural situation that a strain should change its species status according to the presence or absence of a plasmid only. On such a basis one cannot build a stable taxonomy. Moreover, it seems likely that a similar situation will be applicable to the specific status of *A. rhizogenes*; transfer of the large plasmid of a *A. rhizogenes* strain (biovar 2) to a plasmidless derivative of a *A. tumefaciens* strain (biovar 1) resulted in a rhizogenic biovar 1 strain (Moore et al., 1979; White and Nester, 1980a).

Several authors have already expressed their disagreement with the classification given in the eighth edition of *Bergey's Manual* and have proposed different classifications and nomenclatures (De Ley et al., 1966; Keane et al., 1970; White, 1972; Kerr et al., 1978; Holmes and Roberts, 1981). It has been suggested and it would be logical, for priority reasons, to consider *A. radiobacter* (Beijerinck and van Delden, 1902) Conn 1942 as the type species instead of *A. tumefaciens* (Smith and Townsend 1907) Conn 1942, which was described later. Tumorigenic and rhizogenic strains should not be considered as different species but as different pathovars of *A. radiobacter* (*A. radiobacter* pv. *tumefaciens* and *A. radiobacter* pv. *rhizogenes* (Keane et al., 1970; Kerr et al., 1978)). In those proposals the difference between biovars 1 and 2 (i.e. biotypes 1 and 2 *sensu* Keane et al., 1970) is kept at the infrasubspecific level. However, the percentage of DNA relatedness between biovars 1 and 2

is only about 20% (Fig. 4.35) and a clearcut phenotypic differentiation between both biovars is possible (Keane et al., 1970; White, 1972; Kersters et al., 1973; Holmes and Roberts, 1981) (see also Table 4.56). Therefore it seems better to separate both clusters at the species level as proposed by Holmes and Roberts (1981). We think that the best nomenclatural representation of *Agrobacterium* for the moment would be as follows:

1. *A. radiobacter* (type species) with *A. radiobacter* pv. *tumefaciens* for tumorigenic strains. *A. radiobacter* should include all the strains belonging to biovar 1 (i.e. biotype 1 (*sensu* Keane et al., 1970), cluster 1 (*sensu* Kersters et al., 1973; De Ley et al., 1973) and the "*A. tumefaciens*" group of Holmes and Roberts (1981)). Biovar 1 strains displaying clear-cut rhizogenic properties could be named *A. radiobacter* pv. *rhizogenes*;
2. A new *Agrobacterium* species (still to be named) including all the strains of cluster 2 (*sensu* Kersters et al., 1973; De Ley et al., 1973), the majority of the biotype-2 strains (*sensu* Keane et al., 1970) and the "*A. rhizogenes*" group of Holmes and Roberts (1981). This species contains two pathovars (pv. *tumefaciens* and pv. *rhizogenes*), according to the tumorigenic or rhizogenic properties of the bacteria;
3. *A. rubi* for the separate *A. rubi* cluster. This species is confined for the moment to three strains (ATCC 13335, ICPB TR2 and Braun EU6), displaying a high level of DNA homology among each other but hybridizing at only 15% with all the other agrobacteria investigated (De Ley, 1974). Two of these strains are authentic cultures isolated by E. M. Hildebrand from *Rubus* sp.

The nomenclatural status of the biovar-3 strains can be decided later when more information becomes available.

Unfortunately, although the above mentioned nomenclature seems to be an excellent solution and the species can be convincingly differentiated by genotypic and phenotypic criteria, this proposal can officially not be accepted for the time being, because we are mainly tied down by Opinion 33 issued by the Judicial Commission (1970) designating *A. tumefaciens* as the type species. All we can do for the moment is to reiterate the old nomenclature in the following adapted version (see also Table 4.55):

1. *A. tumefaciens* (Smith and Townsend 1907) Conn 1942 (type species), comprising tumorigenic strains belonging to biovars 1, 2 and 3 (i.e. respectively, biotypes 1, 2 (except *A. rubi* strains ICPB TR2 and TR3), *sensu* Keane et al., (1970), and biotype 3 *sensu* Kerr and Panagopoulos (1977));
2. *A. radiobacter* (Beijerinck and van Delden 1902) Conn 1942, comprising nontumorigenic strains belonging to biovars 1 and 2;
3. *A. rhizogenes* (Riker, Banfield, Wright, Keitt and Sagen 1930) Conn 1942 comprising rhizogenic strains belonging to biovar 2. Authentic rhizogenic strains of biovar 1 can be named *A. rhizogenes* biovar 1;
4. *A. rubi* (Hildebrand 1940) Starr and Weiss 1943.

This classification gives the freedom to add other biovars to the species when further work makes this necessary. However, the reader should keep in mind that three out of the four species do not reflect the true biological and taxonomic subdivisions of the genus *Agrobacterium*. The real taxonomic entities are represented by infrasubspecific subdivisions (biovars).

The taxonomic relatedness of *Agrobacterium* at the generic and suprageneric level was determined by hybridizing ^{14}C-labeled ribosomal RNA (rRNA) from either *A. tumefaciens* ICPB TT111 or *A. rhizogenes* ICPB TR7, with DNA from a number of *Agrobacterium* strains and a great variety of reference Gram-negative bacteria (De Smedt and De Ley, 1977). These studies show that the genus *Agrobacterium* belongs in the fourth rRNA superfamily (*sensu* De Ley, 1978) together with *Rhizobium, Phyllobacterium, Mycoplana, Bradyrhizobium, Acetobacter, Gluconobacter, Zymomonas, Beijerinckia, Rhodopseudomonas* (except *R. gelatinosa*), *Paracoccus denitrificans, Rhodomicrobium, Aquaspirillum, Rhodospirillum* (except *R. tenue*) and *Azospirillum* (De Smedt and De Ley, 1977; De Smedt et al., 1980; Gillis and De Ley, 1980; Gillis and

Table 4.55.
Correlations between various nomenclatures and classifications of the genus Agrobacterium

This *Manual*	Nomenclature According to			Phenotypic and Genetic Groups (or Clusters) According to		
	Allen and Holding (1974); (Bergey 8); Approved Lists (1980)	Keane et al. (1970); New and Kerr (1972); Kerr and Panagopoulos (1977); Panagopoulos et al. (1978)	Kerr et al. (1978)	White (1972)	Kersters et al. (1973); De Ley et al. (1973)	Holmes and Roberts (1981)
1. A. tumefaciens biovar 1[c]	A. tumefaciens	A. radiobacter var. tumefaciens biotype 1	A. radiobacter pv. tumefaciens	I	1	"A. tumefaciens"
A. tumefaciens biovar 2	A. tumefaciens	A. radiobacter var. tumefaciens biotype 2	A. radiobacter pv. tumefaciens	III	2	"A. rhizogenes"
A. tumefaciens biovar 3	A. tumefaciens	A. radiobacter var. tumefaciens biotype 3	A. radiobacter pv. tumefaciens	NE[a]	NE[a]	NE[a,b]
2. A. radiobacter biovar 1	A. radiobacter	A. radiobacter var. radiobacter biotype 1	A. radiobacter	I	1	"A. tumefaciens"
A. radiobacter biovar 2	A. radiobacter	A. radiobacter var. radiobacter biotype 2	A. radiobacter	III	2	"A. rhizogenes"
3. A. rhizogenes biovar 1[c]	A. rhizogenes	A. radiobacter var. rhizogenes biotype 1[c]	A. radiobacter pv. rhizogenes	I[c,d]	NE[a]	"A. tumefaciens"[c]
A. rhizogenes biovar 2[c]	A. rhizogenes	A. radiobacter var. rhizogenes biotype 2	A. radiobacter pv. rhizogenes	III	2	"A. rhizogenes"
4. A. rubi	A. rubi	A. radiobacter var. tumefaciens biotype 2	A. radiobacter pv. tumefaciens	Separate	A. rubi	A. rubi

[a] NE, not examined.
[b] Holmes and Roberts (1981) investigated two A. tumefaciens biovar-3 strains and allocated them in their A. rubi group.
[c] The majority of A. rhizogenes strains belong to biovar 2. Only a few strains of A. rhizogenes biovar 1 have been described up to now (Keane et al., 1970; Holmes and Roberts, 1981) and their allocation to A. rhizogenes is questionable as their rhizogenic properties are not thoroughly documented.
[d] One strain investigated.

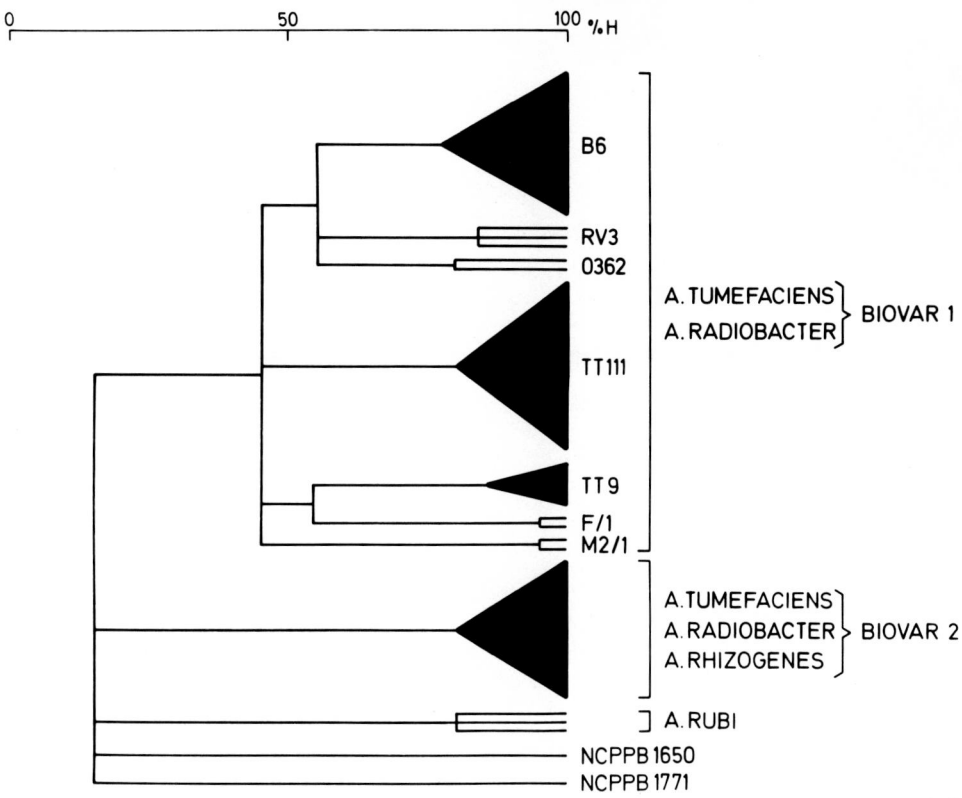

Figure 4.35. Genetic relationships within the genus *Agrobacterium* as determined by DNA/DNA hybridizations using the initial renaturation rate method (De Ley et al., 1970). The figure is modified from De Ley (1972, 1974). The %H represents percentage DNA relatedness. The species names are as used in this *Manual* (see also Table 4.55). The seven genetic subgroups of biovar 1 are labeled by representative strain numbers according to De Ley et al. (1973), Kersters et al. (1973) and De Ley (1974). The type strains of both *A. tumefaciens* and *A. radiobacter* are members of the B6 subgroup. NCPPB 1650 and 1771 represent single genetically and phenotypically aberrant *A. tumefaciens* strains, allocated by Holmes and Roberts (1981) in their *A. rubi* group. Biovar 3 strains were not yet included in DNA hybridization studies.

De Ley, unpublished results). The 16S rRNA cistrons of *Agrobacterium* show the highest similarity among each other ($T_{m(e)}$ of hybrids = 76–81°C) and exhibit a relatively high degree of similarity ($T_{m(e)}$ = 70.5–77°C) to those of the fast growing rhizobia, classified in this *Manual* as *Rhizobium, Mycoplana* ($T_{m(e)}$ = 75–77°C) and *Phyllobacterium* ($T_{m(e)}$ = 72–73°C) (De Smedt and De Ley, 1977). DNA/DNA hybridizations (Heberlein et al., 1967; Gibbins and Gregory, 1972; De Smedt and De Ley, 1977) and numerical analysis of phenotypic features (Graham, 1964; 't Mannetje, 1967; Moffett and Colwell, 1968; White, 1972; Holmes and Roberts, 1981) confirm that agrobacteria are indeed closely related to the fast growing rhizobia. However, the rRNA cistrons of *Agrobacterium* exhibit a lower degree of similarity ($T_{m(e)}$ = 65–70°C) to those of the slow growing rhizobia (De Smedt and De Ley, 1977), emphasizing the separate generic position of the latter organisms, now grouped in the genus *Bradyrhizobium* (this *Manual*). In the fourth rRNA superfamily, *Bradyrhizobium* constitutes a separate branch, which is as far removed from the branch composed of *Agrobacterium, Rhizobium,* and *Phyllobacterium* as, for instance, from the *Acetobacter* or *Rhodopseudomonas* branches (De Smedt and De Ley, 1977; Gillis and De Ley, 1980).

Ahrens (1968) and Ahrens and Rheinheimer (1967) classified seven new species of marine origin in the genus *Agrobacterium*, but withdrew this proposal later (Allen and Holding, 1974). Consequently these species names were not included on the Approved Lists of Bacterial Names in 1980. DNA/rRNA hybridizations showed that "*Agrobacterium ferrugineum*" Ahrens and Rheinheimer 1967 and "*A. kieliense*" Ahrens 1968 are not agrobacteria, although they belong with a $T_{m(e)}$ value of 73.5 and 72°C, respectively, on the *Agrobacterium-Rhizobium-*

Phyllobacterium branch (De Smedt and De Ley, 1977). Their exact taxonomic position remains unknown for the moment. The species "*A. luteum*" Ahrens and Rheinheimer 1967, "*A. gelatinovorum*" Ahrens 1968 and "*A. aggregatum*" Ahrens 1968 belong also in the fourth rRNA superfamily (*sensu* De Ley, 1978) but are much less related to *Agrobacterium* than the former two species. "*A. agile*" Ahrens 1968 belongs in the rRNA branch of *Pseudomonas* section I (*sensu* Bergey 8) (De Vos and De Ley, unpublished results). According to De Smedt and De Ley (1977) "*A. azotophilum*" Ulloa and Herrera 1972 (not on the Approved Lists) does not belong in *Agrobacterium*, and it is not even a member of the fourth rRNA superfamily.

Holmes and Roberts (1981) suggested that a small group of yellow-pigmented, 3-ketolactose-positive bacteria, including two "*Chromobacterium folium*" ("*Chromobacterium lividum*") strains NCTC 10590 and 10591, may constitute another distinct group of agrobacteria. Both above mentioned strains belong in the fourth rRNA superfamily (De Smedt and De Ley, 1977; De Ley et al., 1978). However, their rRNA cistrons are more closely related to the rRNA cistrons of the type strain of *Flavobacterium capsulatum* than to those of *Agrobacterium* (Bauwens and De Ley, 1981). It is unlikely that the so-called "*Agrobacterium* yellow group" of Holmes and Roberts (1981) belongs in *Agrobacterium*.

Acknowledgments

We thank M. De Cleene, M. Holsters, J. Van Beeumen and M. Van Montagu for helpful advice and critical reading of some parts of the manuscript. The aid of U. Torck is greatly appreciated. J. D. L. is indebted to the Fonds voor Kollektief Fundamenteel Onderzoek for research and personnel grants.

Further Reading

De Cleene, M. and J. De Ley. 1976. The host range of crown gall. Bot. Rev. *42:* 389–466.

De Cleene, M. and J. De Ley. 1981. The host range of infectious hairy-root. Bot. Rev. *47:* 147–194.

De Ley, J. 1974. Phylogeny of procaryotes. Taxon *23:* 291–300.

De Smedt, J. and J. De Ley. 1977. Intra- and intergeneric similarities of *Agrobacterium* ribosomal ribonucleic acid cistrons. Int. J. Syst. Bacteriol. *27:* 222–240.

Gordon, M. P. 1981. Tumor formation in plants. *In* Marcus (Editor), *Proteins and Nucleic Acids,* Vol. 6, Academic Press, New York, pp. 531–570.

Lippincott, J. A. and B. B. Lippincott. 1975. The genus *Agrobacterium* and plant tumorigenesis. Annu. Rev. Microbiol. *29:* 377–405.

Van Montagu, M., M. Holsters, P. Z. O'Farrell, J. P. Hernalsteens, A. Depicker, M. De Beuckeleer, G. Engler, M. Lemmers, L. Willmitzer and J. Schell. 1980. The interaction of *Agrobacterium* Ti-plasmid DNA and plant cells. Proc. R. Soc. Lond. B *210:* 351–365.

Differentiation and characteristics of species of **Agrobacterium**

The differential characteristics of the species and their biovars are shown in Table 4.56. Other characteristics are listed in Table 4.57.

List of the species of the genus **Agrobacterium**

Introductory note: The species status of *A. tumefaciens, A. radiobacter* and *A. rhizogenes* is highly controversial for reasons given in the section Taxonomic Comments. These epithets refer to phytopathogenic properties (tumorigenic, saprophytic and rhizogenic states, respectively).

Table 4.56.

Differential characteristics of the species and biovars of the genus **Agrobacterium**[a]

Characteristics	1. *A. tumefaciens*			2. *A. radiobacter*		3. *A. rhizogenes*[b]	4. *A. rubi*[c]
	Biovar 1	Biovar 2	Biovar 3[d]	Biovar 1	Biovar 2	Biovar 2	
Growth:							
At 35°C	+	−	d	+	−	−	d
At 28°C	+	+	+	+	+	+	+
On selective medium of Schroth et al. (1965)	+	−	−	+	−	−	
On selective medium of New and Kerr (1971)	−	+	−	−	+	+	
In the presence of 2% NaCl	+	−[e]	+	+	−[e]	−	
Production of 3-ketolactose	+	−	−[d]	+	−	−	−
Formation of acid from:							
meso-Erythritol	−	+	−[d]	−	+	+	+
Melezitose	+	−	−[d]				
Ethanol	+	−	−[d]	+	−	−	−
Formation of alkali from:							
Na malonate	−	+	+	−	+	+	+
Na L-tartrate	d	+	+				
Na propionate	d	−	−				
Simmons' citrate, supplemented with 0.0005% yeast extract	−	+		−	+	+	−
Reaction in litmus milk[e, f]:							
Alkaline	+	−	+	+	−	−	+
Acid	−	+	−	−	+	+	−
Formation of pellicle in ferric ammonium citrate solution	+	−	d	+	−	−	−
Growth factor requirements:							
biotin and/or L-glutamic acid[g]	−	+[e]		−	+[e]	+[e]	
L-glutamic acid and yeast extract[g, h]	−	−		−	−	−	+
Phytopathogenicity:							
Tumors produced on wounded stems of e.g. tomato plants, *Helianthus annuus, Nicotiana tabacum* and/or on discs of *Daucus carota*	+	+	d[j]	−	−	+	+
Roots produced on discs of *Daucus carota*	−	−	−	−	−	+[b]	−

[a] Data from Lippincott and Lippincott (1969), Keane et al. (1970), Kersters et al. (1973), Panagopoulos and Psallidas (1973), Kerr and Panagopoulos (1977), Süle (1978), Panagopoulos et al. (1978), Holmes and Roberts (1981). The methodology for the majority of the tests can be found in Keane et al. (1970), Kerr and Panagopoulos (1977), and Holmes and Roberts (1981). Symbols: see standard definitions.

[b] The majority of the investigated *A. rhizogenes* strains belong to biovar 2 (see Taxonomic Comments and footnote c of Table 4.55).

[c] Only the following three strains are considered to belong to *A. rubi:* ATCC 13334, 13335 and Braun EU6.

[d] According to Panagopoulos et al. (1978). Süle (1978) found positive or variable results for some other biovar-3 strains in the tests indicated by "d".

[e] For some strains Holmes and Roberts (1981) found reactions different from those reported here.

[f] An alkaline reaction in litmus milk is frequently accompanied by a brown discoloration; an acid reaction (pink color) is frequently accompanied by a clot formation.

[g] Keane et al. (1970).

[h] Starr (1946).

[i] See also under Procedures for Testing Special Characters.

[j] Biovar-3 strains were mainly isolated from grapevines. The majority of these isolates display a very limited host range. For such strains phytopathogenicity can only be demonstrated on young shoots of grapevine (Panagopoulos et al., 1978).

They are by no means correlated with the real taxonomic entities, which were determined by DNA homology and phenotypic studies. The real taxonomic groups are represented here as infrasubspecific subdivisions (biovars). This means that, e.g. *A. tumefaciens* biovar 1 is genetically and phenotypically much more related to *A. radiobacter* biovar 1 than to *A. tumefaciens* biovar 2.

1. **Agrobacterium tumefaciens** (Smith and Townsend 1907) Conn 1942, 359.[AL] (*Bacterium tumefaciens* Smith and Townsend 1907, 672; *Agrobacterium radiobacter* var. *tumefaciens* (Smith and Townsend 1907) Keane, Kerr and New 1970, 594; *Agrobacterium radiobacter* pv. *tumefaciens* (Smith and Townsend 1907) Kerr, Young and Panagopoulos 1978, 156.)

tu.me.fa'ci.ens. L. n. *tumor* a swelling, tumor; L. v. *facere* to make, to produce; M.L. part. adj. *tumefaciens* tumor producing.

The morphology of cells and colonies is as described for the genus.

The physiological, nutritional and phytopathological characteristics of three biovars of *A. tumefaciens* are summarized in Tables 4.56 and 4.57. The genetic relationships of *A. tumefaciens* biovar 3 (grapevine isolates) versus the other *Agrobacterium* species and biovars are unknown for the moment.

A large tumor-inducing plasmid has been detected in many strains.

Isolated from soil, plant rhizosphere and young gall tissues of different plants. Causes "crown gall disease" and oncogenic transformations (tumor formation) in a wide variety of higher plants after wounding. A host range index has been published (De Cleene and De Ley, 1976).

The mol% G + C of the DNA is 57–63 (T_m).

Type strain: ATCC 23308 (NCPPB 2437; DSM 30205; ICPB TT3; Braun B6). This strain belongs to biovar 1. Representative strains of

Table 4.57.

Other characteristics of the species and biovars of the genus **Agrobacterium**[a]

Characteristics	1. A. tumefaciens[b]			2. A. radiobacter		3. A. rhizogenes[c]	4. A. rubi[d]
	Biovar 1	Biovar 2	Biovar 3	Biovar 1	Biovar 2	Biovar 2	
Lysine and ornithine decarboxylase	−	−		−	−	−	−
Phenylalanine deaminase	−	−		−	−	−	−
Hydrolysis of casein, starch, Tween 80, and gelatin	−	−		−	−	−	−
Deoxyribonuclease	−	−		−	−	−	−
Urease	+	d		+		+	
β-Galactosidase	d	+		d	+	+	
Oxidase test:							
Method of Kovacs (1956)	+	−		+	−	−	−
Method of Holmes and Roberts (1981)	+	+		+	+	+	+
Indole formation	−	−		−	−	−	−
H₂S formation:							
Triple sugar iron agar	−	−		−	−	−	−
Lead acetate paper method	+	−		+	−	−	−
Fluorescence on King's B medium	−	−		−	−	−	−
Reduction of nitrate and nitrite	d	−		d	−	−	
Acid production from:							
D-Glucose, D-fructose, L-arabinose, D-xylose, adonitol, mannitol, lactose, maltose, sucrose, cellobiose, trehalose, salicin, L-rhamnose	+	+		+	+	+	+
Glycerol, dulcitol, sorbitol, inositol, raffinose	+	+		+	+	+	
Gas from glucose	−	−		−	−	−	−
Esculin hydrolysis	+	d		+	+	d	
Production of white precipitate on glycerophosphate agar	d	−		d	−	−	
Growth on:							
Sucrose + salts medium of Lippincott and Lippincott (1969)	+	−		+	−	−	−
L-Sorbose	d	+		d	+	+	
L-Ornithine, L-lysine, β-hydroxybutyrate	+	+		+	+	+	+
Na benzoate, Na oxalate, L-phenylalanine, L-tryptophan	−	−		−		−	−
Presence of virulence-specifying plasmids:							
Tumor-inducing type (pTi)	+	+	+	−	−		+
Hairy-root type (pAr)						+[e]	
Mol% G + C of DNA (T_m)	57–63	57–63		58–62.5	58–61	59–63	57.6–58.8

[a] Data compiled from Holmes and Roberts (1981), Keane et al. (1970), Kersters et al. (1973), and White (1972). Mol% G + C values are taken from De Ley et al. (1966), De Smedt and De Ley (1977), and De Ley (unpublished results). For symbols see standard definitions.

[b] Holmes and Roberts (1981) investigated two *A. tumefaciens* biovar-3 strains and allocated them to their *A. rubi* cluster (see also Taxonomic Comments). Extensive phenotypic data for other biovar-3 strains are not available.

[c] The majority of the investigated *A. rhizogenes* strains belong to biovar 2 (see Taxonomic Comments and also footnote c of Table 4.55).

[d] Only the following three strains are considered to belong to *A. rubi*: ATCC 13334, 13335 and Braun EU6.

[e] Detected in at least three *A. rhizogenes* strains (White and Nester, 1980b).

A. tumefaciens biovar 2 are: NCPPB 2303 (ICPB TT133 Volcani 156) and DSM 30200 (Kerr 38). A representative strain of biovar 3 is NCPPB 2562 (Panagopoulos Ag63).

2. **Agrobacterium radiobacter** (Beijerinck and van Delden 1902) Conn 1942, 359.[AL] (*Bacillus radiobacter* Beijerinck and van Delden, 1902, 3; *Agrobacterium radiobacter* var. *radiobacter* (Beijerinck and van Delden 1902) Keane, Kerr and New 1970, 594.)

ra.di.o.bac'ter. L. n. *radius* a ray, beam; M.L. *bacter* masc. equivalent of Gr. neut. n. *bakterion* a rod or staff; M.L. masc. n. *radiobacter* ray rod.

The morphology of cells and colonies is as described for the genus.

The physiological and nutritional characteristics of both biovars of *A. radiobacter* are summarized in Tables 4.56 and 4.57. No *A. radiobacter* biovar-3 strains have been isolated from nature up to now.

The absence of the tumor-inducing plasmid has been demonstrated in many strains.

Isolated from soil and plant rhizosphere. The majority of agrobacteria isolated from clinical specimens (CDC group Vd-3) belong to *A. radiobacter* biovar 1.

The mol% G + C of the DNA is 58–62.5 (T_m).

Type strain: ATCC 19358 (NCIB 9042; DSM 30147; IFO 13532; Delft EX 3.24.2). This strain belongs to biovar 1. Strain NCPPB 2407 (Kerr 84) is a representative strain of biovar 2. The latter strain produces agrocin and is used for the biological control of crown gall.

3. **Agrobacterium rhizogenes** (Riker, Banfield, Wright, Keitt and Sagen 1930) Conn 1942, 359.[AL] (*Bacterium rhizogenes* Riker, Banfield, Wright, Keitt and Sagen 1930, 536; *Agrobacterium radiobacter* var. *rhizogenes* (Riker, Banfield, Wright, Keitt and Sagen 1930) Keane, Kerr and New 1970, 594; *Agrobacterium radiobacter* pv. *rhizogenes* (Riker, Banfield, Wright, Keitt and Sagen 1930) Kerr, Young and Panagopoulos 1978, 156.)

rhi.zo'ge.nes. Gr. n. *rhiza* a root; Gr. v. *gennao* to make, to produce; M.L. adj. *rhizogenes* root-producing.

The morphology of cells and colonies is as described for the genus.

The physiological, nutritional and phytopathological characteristics are summarized in Tables 4.56 and 4.57. Most *A. rhizogenes* strains (including the type strain ATCC 11325) are genetically and phenotypically almost indistinguishable from *A. tumefaciens* biovar 2 and *A. radiobacter* biovar 2 (see Fig. 4.35 and Table 4.56). Keane et al. (1970) and Holmes and Roberts (1981) have reported on rhizogenic *Agrobac-*

terium strains belonging to biovar 1 (e.g. strains NCPPB 5 and NCPPB 2655). However, the rhizogenic properties of the latter strains are not thoroughly documented in the literature. When more evidence for their rhizogenic state becomes available such strains can later be designated as *A. rhizogenes* biovar 1.

Investigations on three *A. rhizogenes* strains indicated that their virulence plasmids represent a distinct plasmid type, based on DNA homology and compatibility (White and Nester, 1980b).

Isolated from plants attacked by infectious hairy root disease (e.g. apple trees, roses and *Spiraea* sp.). Causes hairy root disease or woolly knot on susceptible plants and tumorous outgrowths on the inoculated stems of other plants (*Vicia faba*, tomato plants) (Anderson and Moore, 1979; De Cleene and De Ley, 1981). In their host-range index of infectious hairy root, De Cleene and De Ley (1981) discuss the confusing situation concerning the specific phytopathogenicity of *A. rhizogenes* and *A. tumefaciens*.

The mol% G + C of the DNA is 59–63 (T_m).

Type strain: ATCC 11325 (DSM 30148; IFO 13257).

4. **Agrobacterium rubi** (Hildebrand 1940) Starr and Weiss 1943, 316.[AL] (*Phytomonas rubi* Hildebrand 1940, 694.)

ru'bi. L. n. *Rubus* generic name of blackberry; L. gen. n. *rubi* of *Rubus*.

The morphology of cells and colonies is as described for the genus.

The physiological, nutritional and phytopathological characteristics (three strains only) are listed in Tables 4.56 and 4.57. Growth rate on ordinary media is characteristically slower than for the other species. Requires L-glutamic acid and some vitamins (present in yeast extract) for growth (Starr, 1946; Keane et al., 1970).

The tumor-inducing plasmid of the type strain displays a high DNA homology with the Ti-plasmid of *A. tumefaciens* biovar 1 strain C-58 (Currier and Nester, 1976).

Isolated from cane galls on members of the genus *Rubus* (black raspberry, boysenberry). The host range is not limited to *Rubus* spp. The strains are tumorigenic for many other plants (Anderson and Moore, 1979).

The mol% G + C of the DNA is 57.6–58.8 (T_m).

Type strain: ATCC 13335 (NCPPB 1854; ICPB TR3). For the time being only two other strains are considered to belong to *A. rubi*: ATCC 13334 (NCPPB 1856; DSM 30149; IFO 13260; ICPB TR2) and strain Braun EU6 (see also *Taxonomic Comments*).

Genus IV. **Phyllobacterium** (ex Knösel 1962) nom. rev. (*Phyllobacterium* Knösel 1962, 96)

DIETER H. KNÖSEL

Phyl.lo.bac.te'ri.um. Gr. n. *phyllos* leaf; Gr. dim. neut. n. *bakterion* a small rod; M.L. neut. n. *Phyllobacterium* leaf bacterium.

Straight rods (in vitro), 0.4–0.8 μm in diameter and 0.8–2.0 μm in length. **Gram-negative. Motile** by a single polar flagellum or several polar or lateral flagella of long wavelength. **Aerobic**, having a strictly respiratory type of metabolism with oxygen as the terminal electron acceptor. Optimum temperature, 28–34°C. Colonies on glucose-yeast extract agar are translucent, colorless or beige colored, and slimy. Oxidase-positive. **Chemoorganotrophic**, using a variety of sugars or salts of organic acids as carbon sources. Do not hydrolyze starch, pectin or cellulose. **Occur in leaf nodules of higher plants** (species of *Myrsinaceae* (myrsine) and *Rubiaceae* (madder)). The mol% of the DNA is 59.6–61.3 (T_m) Gillis and De Ley, 1980).

Type species: *Phyllobacterium myrsinacearum* (ex Knösel 1962) nom. rev.

Further Descriptive Information

P. myrsinacearum and *P. rubiacearum* typically develop within leaf nodules into pleomorphic cells (rod shaped, ellipsoidal, or branched; see Fig. 4.36). In liquid media, especially in carrot juice medium,* the cells appear as motile straight rods which form characteristic star clusters (Fig. 4.37) after an initial phase of intensive swimming. The flagella have a long wavelength.

The maximum temperature at which growth occurs in 36°C; the minimum is 4–5°C. Cells heated in saline are killed within 10 min at 52°C.

Colonies are nonpigmented to beige colored, translucent to opaque in the center, slimy, circular and regular with an entire margin.

---

* Carrot juice medium has the following composition: fresh carrot juice, 500 ml; water, 500 ml; $FeSO_4 \cdot 7H_2O$, 0.1 g; $MnSO_4 \cdot H_2O$, 0.1 g; pH 7.2. The medium is sterilized by fractional sterilization. For a solid medium add 15.0 g of agar/liter.

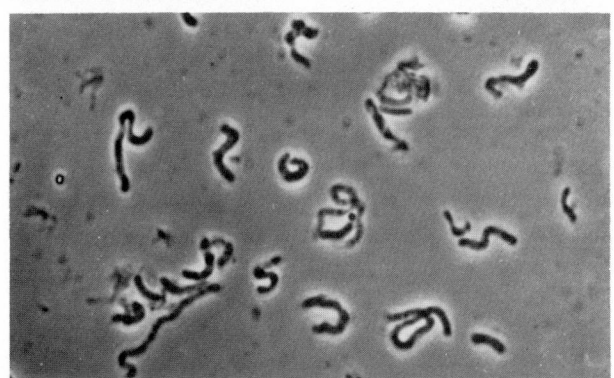

Figure 4.36. Phase-contrast photomicrograph showing bacteroids of *Phyllobacterium rubiacearum* from leaf nodules of *Pavetta zimmermanniana* (× 2000).

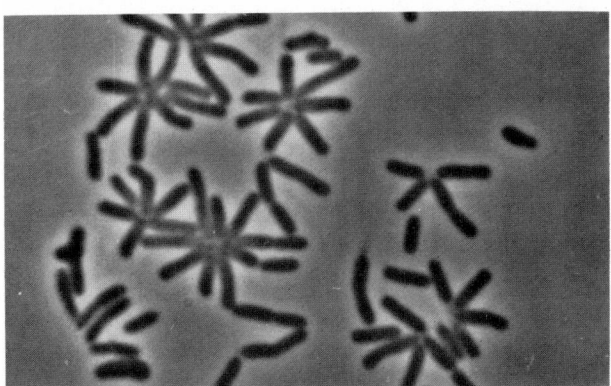

Figure 4.37. Phase-contrast photomicrograph showing star clusters of cells of *Phyllobacterium rubiacearum* in carrot juice medium after incubation at 28°C for 24 h (× 2000).

Acid is produced from a wide range of carbohydrates without gas formation: pentoses, hexoses, maltose, inulin, glycerol, adonitol, sorbitol, and dulcitol. Starch, pectin and cellulose are not hydrolyzed. Little H_2S is produced on bismuth sulfite agar. Ammonium salts, nitrates and most amino acids can serve as nitrogen sources. Strains may or may not reduce nitrate to nitrite.

It is not known whether the organisms are nitrogen fixers (van Hove, 1976).

Phyllobacteria seem to be limited to plants of tropical climates. Members of the genus are able to invade leaves and induce production of characteristic nodules (Figs. 4.38 and 4.39). According to Schwartz (1959), symbiotic stages exist in *Rubiaceae* (madder), *Myrsinaceae* (myrsine), *Myrtaceae* (myrtle), and *Dioscoreaceae* (yam), supposedly fixing nitrogen into combined forms utilizable by the host plant (but see van Hove, 1976). The bacteria described by Knösel (1962) have

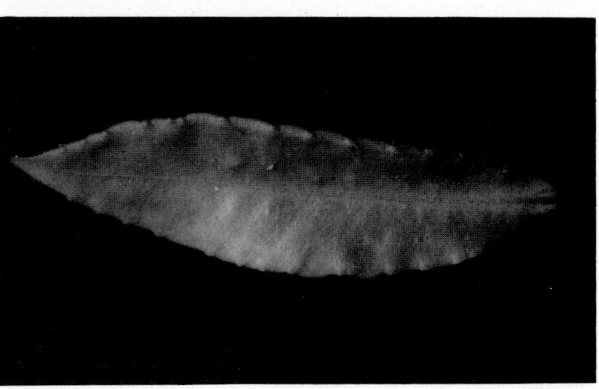

Figure 4.38. Photograph showing a leaf of *Ardisia crispa* with nodules localized at the leaf margin.

Figure 4.39. Photograph showing a leaf of *Pavetta zimmermanniana* with nodules distributed all over the leaf blade.

been isolated from plants growing in a tropical greenhouse in Stuttgart-Hohenheim, Germany. Earlier investigations were made by Zimmermann (1902), Miehe (1911), von Faber (1912), Nemeč (1922) and de Jongh (1938).

Isolation Procedures

Washed leaf pieces having nodules are macerated by rubbing and placed in saline. After shaking, dilutions are plated onto carrot juice agar (see preceding footnote) containing yeast extract. After incubation at 28°C a variety of colored and noncolored colonies develop. Typical colonies are transferred into liquid carrot juice medium for identification. The cultures should be observed after 24–48 h by phase-contrast microscopy to confirm for the presence of star clusters (Fig. 4.37).

Maintenance Procedures

Stock cultures on Tryptic soy agar (Difco) may be maintained at 4–5°C for at least 1 or 2 months. Cultures may also be preserved indefinitely by lyophilization.

Differentiation of the genus **Phyllobacterium** from other genera

The ability to induce hypertrophies on plants is considered to be a primary characteristic of species within the family *Rhizobiaceae*. Strains that stimulate nodule production on leaves of higher plants may be classified as members of the genus *Phyllobacterium*. See Table 4.50 in the article on the genus *Rhizobium* for other differential characteristics.

Taxonomic Comments

Although Miehe's first investigations were published in 1911, and although the generic name *Phyllobacterium* was published in 1962 by

Knösel, the various editions of *Bergey's Manual* failed to recognize the existence of symbiotic bacteria within leaves and the name of the genus was not included in the Approved Lists of Bacterial Names in 1980. Because the name *Phyllobacterium* presently has no standing in nomenclature, we propose it again in the present edition of the *Manual* as a nomen revictum, and we also again propose the species *P. myrsinacearum* and *P. rubiacearum* as nom. rev.

The placement of the genus in the family *Rhizobiaceae* is regarded as only tentative at present, although there is practical value in placing it near the other genera that induce plant hypertrophies. Future studies involving modern methods of taxonomic analysis may provide more details, especially with regard to the relationship of the genus to *Rhizobium* and *Agrobacterium*, to host range affinities, and to the fixation of N₂. One step in this direction has been done by Gillis and De Ley (1980), who reported, based on their DNA/rRNA studies, that

Phyllobacterium belongs to their fourth RNA superfamily, which consists mainly of taxa associated with the phytosphere: *Rhizobium*, *Agrobacterium*, *Zymomonas*, *Acetobacter*, *Gluconobacter*, and several other taxa. Their studies indicated that the relationship between *Phyllobacterium* and the acetic acid bacteria was not a very close one, but the relationship to other genera has not yet been precisely determined.

Differentiation of the species of the genus **Phyllobacterium**

The differential characteristics of the two species of the genus are summarized in Table 4.58.

Table 4.58.
Differential characteristics of the species of the genus **Phyllobacterium**[a]

Characteristics	1. *P. myrsinacearum*	2. *P. rubiacearum*
Flagellar arrangement:		
One to several flagella with polar or lateral attachment	+	−
Single polar flagellum only	−	+
Occurs in leaf nodules of:	*Myrsinaceae*	*Rubiaceae*
Nitrate reduced to nitrite	+	−

[a] Symbols: see standard definitions.

List of the species of the genus **Phyllobacterium**

1. **Phyllobacterium myrsinacearum** (ex Knösel 1962) nom. rev. (*Phyllobacterium myrsinacearum* Knösel 1962, 96.)

myr.si.na.ce.a′rum. M.L. fem. pl. n. *Myrsinaceae* family of plants; M.L. fem. gen. pl. n. *myrsinacearum* of the myrsine family.

When cultured in liquid media the cells are straight rods, 0.4–0.8 μm in diameter and 0.8–1.4 μm in length. The bacteroids in leaf nodules are rod shaped with some branched forms.

A pellicle is formed in liquid media.

Other characteristics are listed in Table 4.58.

Found in leaf nodules of *Ardisia crispa* (Fig. 4.38) and *A. crenata* (*Ardisia crenata* Sims; *A. crenulata* Lodd., not *A. crispa* (Thumb). A. DC.), *Myrsinaceae*.

The mol% G + C of the DNA is 59.8–60.4 (T_m) (Gillis and De Ley, 1980).

Type strain: LMG 2t2 (holotype).

2. **Phyllobacterium rubiacearum** (ex Knösel 1962) nom. rev.

(*Mycobacterium rubiacearum* von Faber 1912, 332; *Phyllobacterium rubiacearum* (von Faber 1912) Knösel 1962, 96.) (Not *Klebsiella rubiacearum* Centifano and Silver 1964, 780.)

ru.bi.a.ce.a′rum. M.L. fem. pl. n. *Rubiaceae* family of plants; M.L. fem. gen. pl. n. *rubiacearum* of the bedstraw family.

When cultured in liquid media the cells are straight rods, 0.4–0.8 μm diameter and 1.2–2.0 μm in length. The bacteroids in leaf nodules are rod shaped.

A pellicle is formed in liquid media.

Other characteristics are listed in Table 4.58.

Found in leaf nodules of *Pavetta zimmermanniana* Val. (Fig. 4.39), *Rubiaceae*.

The mol% G + C of the DNA is 60.3–61.3 (T_m) (Gillis and De Ley, 1980).

Type strain: LMG 1t1 (holotype).

Other organisms

An organism known as "*Phyllobacterium stappi*" was isolated by Knösel (1962) from dark spots on leaves of *Cienfuegosia gossypioides* Hochr. (syn: *Sturtia gossypioides* R. Br.; *Gossypium sturtii* F. Mn.), *Malvaceae*. The strain exhibits characteristics similar to those of phyl-

lobacteria, including production of star clusters in liquid media; however, there is no evidence that the organism induces hypertrophies on the leaves of plants.

FAMILY IV. **METHYLOCOCCACEAE** FAM. NOV.

ROGER WHITTENBURY AND NOEL R. KRIEG

Me.thyl.o.cocc.a′.ce.ae. M.L. masc. n. *Methylococcus* genus of methane-oxidizing bacteria; *-aceae* ending to denote a family; M.L. fem. pl. n. *Methylococcaceae* the *Methylococcus* family.

A diverse group of rods, vibrios and cocci having in common the ability to utilize methane (via methane monooxygenase) **as a sole carbon and energy source under aerobic or microaerobic conditions.** Bacteria incapable of utilizing methane are excluded from the family, even though they may be able to grow on other single-

carbon (C_1) compounds. Included in the family are not only **obligate methane oxidizers,** but also **facultative methane oxidizers** (i.e. organisms able to use complex organic compounds, such as glucose, as alternatives to methane for carbon and energy).

Gram-negative. All strains possess a **complex arrangement of**

intracytoplasmic membranes when grown on methane. These membrane arrangements may consist of bundles of vesicular disks or paired membranes aligned to the cell periphery. All strains have a **strictly respiratory type of metabolism** with oxygen as the terminal electron acceptor. Those strains tested are **catalase- and oxidase-positive and possess cytochromes of the** *c*, *a*, *o*, and *b* type. **All strains can use methanol** as a sole carbon and energy source; formaldehyde is probably used also by all strains. No strains appear able to grow on formate, although formate is oxidized. Occur in aerobic environments where methane is available, i.e. in soil, mud or water adjacent to or overlaying anaerobic environments.

Type genus: *Methylococcus* Foster and Davis 1966, 1929.

Further Descriptive Information

Some rod and coccal forms are motile, the rods possessing a polar flagellum or a polar tuft of flagella.

It is characteristic of methane-grown cells that they form a complex arrangement of intracellular membranes (Davies and Whittenbury, 1970). When facultative methane oxidizers are cultured on complex organic compounds, these membranes may be greatly decreased or absent (for example, see Lynch et al., 1980). Two major kinds of membrane arrangements have been defined in methane-grown cells (Davies and Whittenbury, 1970). In the "type I" arrangement there are vesicular disks of membrane organized into bundles which are distributed throughout the cell. The "type II" arrangement consists of pairs of membranes which either extend throughout the cell or run parallel to the cytoplasmic membrane. See Davies and Whittenbury (1970) for electron micrographs illustrating the two types.

Differences between the phospholipid composition of various obligate methane oxidizers have been found (e.g. see Andreev et al., 1977 and Makula, 1978). Some strains possess esterified fatty acids mainly of the $C_{16:0}$, $C_{16:1}$ type, whereas others have mainly $C_{18:1}$ fatty acids.

Resting stages may be formed by some strains. These can be: (a) *cysts* are indistinguishable from those formed by *Azotobacter* species, (b) *"lipid cysts"* characterized by large lipid inclusions (mainly composed of poly-β-hydroxybutyrate), absence of intracellular membranes, and formation of a thick layer on the outer surface of the cell; or (c) *exospores*, formed by budding at one cell pole. All three kinds are resistant to desiccation, and the exospores are also resistant to heat (85°C for 15 min). See Whittenbury et al. (1970) for representative photomicrographs.

Strains of methane oxidizers may be pigmented (yellow, ochre, pink or red). Some form a brown pigment when encystment occurs. Others release, irregularly, a water-soluble pigment (blue, yellow or brown).

Growth temperature ranges vary among strains. Some strains are thermophilic, failing to grow at 30°C but growing at 45–55°C.

Many strains are sensitive to the normal oxygen tension of an air atmosphere and require decreased oxygen levels (microaerobic conditions) for growth. Some strains are nitrogen fixers and are, therefore, extremely oxygen sensitive under nitrogen-fixing conditions.

Although all strains can use methanol as a sole carbon source, this substrate can be toxic when supplied above trace concentrations. Many strains, however, can give rise to methanol-tolerant mutants or can be "trained" to tolerate higher levels of methanol (Whittenbury, unpublished results).

Carbon dioxide, acetate, formate, some amino acids, and many co-oxidizable substrates (see below) serve as supplementary carbon sources for obligate methane oxidizers, provided formaldehyde is available via methane or methanol oxidation, or added separately. If metabolized, co-oxidizable substrates (e.g. primary alcohols, acetate, alkanes, dimethyl- and diethylether, alicyclic, aromatic and heterocyclic compounds and carbon monoxide) may provide supplementary carbon and energy to obligate methane oxidizers but cannot be used as sole carbon and energy sources. Facultative methane oxidizers can use glucose or other complex organic substrates as sole carbon and energy sources.

The pathway for methane oxidation is as follows: (a) oxidation to methanol, via methane monooxygenase; (b) methanol to formaldehyde,

via methanol dehydrogenase; (c) formaldehyde to formate, via formaldehyde dehydrogenase; and (d) formate to carbon dioxide, via formate dehydrogenase. Formaldehyde, formed in step (b), is assimilated by either the serine pathway (the initial step being the formation of serine from formaldehyde and glycine via serine transhydroxymethylase) or the ribose monophosphate (RMP) pathway (the initial steps being the formation of D-erythro-L-glycero-3-hexulose-6-phosphate from formaldehyde and ribulose-5-phosphate via 3-hexulose phosphate synthase, and subsequent conversion to fructose-6-phosphate via phospho-3-hexuloisomerase). See the reviews by Colby et al. (1979) and Higgins et al. (1981) for further details.

Differences in the tricarboxylic acid (TCA) cycle occur among methane oxidizers (Davey et al., 1972). Some strains have a complete cycle, but others lack the enzyme α-ketoglutarate dehydrogenase and therefore have an incomplete cycle that plays a mainly biosynthetic role rather than being used for both biosynthesis and energy production. Other differences among strains occur with respect to isocitrate dehydrogenase: in some strains the enzyme activity is dependent on either nicotinamide adenine dinucleotide (NAD) or NAD phosphate (NADP); in others it is dependent on NAD only and is unresponsive to NADP, and in still others it is NADP dependent and unresponsive to NAD (Davey et al., 1972).

At least one species (*Methylococcus capsulatus*) has the potential ability to fix carbon dioxide by the Benson-Calvin cycle (Taylor, 1977). To what extent carbon dioxide is actually fixed by this pathway is not yet clear.

Methane oxidizers are wide spread in nature and occur wherever both methane and oxygen are available, for example in the water or mud above the anaerobic sediment of a lake bottom. Much information on the ecology of the organisms has been presented by Whittenbury et al. (1976) and Hanson (1980).

Taxonomic Comments

An inevitable consequence of industrial interest in protein feedstuff and enzymes from the methane oxidizers has been a rapid accumulation of new information on their biology. This in turn has revealed the inadequacy or incorrectness of previous concepts concerning their nature and taxonomy, summarized in the eighth edition of *Bergey's Manual*. For instance, the definition of "*Methylomonadaceae*" as being a family of bacteria using *only* C_1 compounds, such as methane or methanol, is too rigorous because C_1 compound obligateness can no longer be reasonably maintained as a family feature. Moreover, the definition of "*Methylomonadaceae*" is not rigorous enough because it permits the inclusion of nonmethane-utilizing organisms that can grow on other C_1 compounds (i.e. methanol, formate, carbon monoxide and amines). For these reasons, we have defined a new family, *Methylococcaceae*, in place of "*Methylomonadaceae*" in this edition of the *Manual* to allow new approaches to the taxonomy of these diverse organisms. We also note that the family "*Methylomonadaceae*" was not included in the Approved Lists of Bacterial Names in 1980 and presently has no standing in nomenclature.

Only two of the species of methane oxidizers (and their respective genera) described in the eighth edition of the *Manual* appear, in our opinion, to be well authenticated and likely to survive future taxonomic revision: *Methylomonas methanica* (Söhngen 1906) Leadbetter 1974, 268 and *Methylococcus capsulatus* Foster and Davis 1966, 1929. The other two species described, "*Methylomonas methanooxidans*" (Brown and Strawinski 1958) Leadbetter 1974, 269 and "*Methylomonas methanitrificans*" (Davis, Coty and Stanley 1964) Leadbetter 1974, 268, may prove to be synonyms of each other and to be wrongly classified in the genus *Methylomonas* (see Table 4.59).

The situation is chaotic regarding the taxonomy and nomenclature of the methane oxidizers isolated and described more recently than those indicated above. This is partly due to a disregard of the Code governing the naming and defining of genera and species, but largely because of the lack of a generally agreed battery of effective tests and taxonomic guidelines. Although it is possible to categorize methane

Table 4.59.

Tentative scheme for the categorization of aerobic methane-oxidizing bacteria[a]

Characteristics	Group I	Group II
Membrane arrangement:		
Bundles of vesicular discs	+	−
Paired membranes aligned to cell periphery	−	+
Resting stages (if present):		
Cysts (*Azotobacter*-like)	+	−
Exospores or "lipid cysts"	−	+
Major carbon assimilation pathways:		
Ribulose monophosphate pathway (3-hexulose phosphate synthase-positive)	+	−
Serine pathway (hydroxypyruvate reductase-positive)	−	+
Tricarboxylic acid cycle:		
Complete	−	+
Incomplete (2-oxoglutarate dehydrogenase-negative)	+	−
Nitrogenase	d	+
Predominant fatty acid carbon-chain length	16	18

	Subgroup Ia	Subgroup Ib	Subgroup IIa (Obligate)[b]	Subgroup IIb (Facultative)[b]
Growth on glucose	−	−	−	+
Autotrophic CO_2 fixation (ribulose biphosphate carboxylase-positive)	−	+	−	−
Mol% G + C of DNA	50–64	62.5	62.5	62.5–63.0
Isocitrate dehydrogenase:				
NAD or NAD(P) dependent	+	−	−	−
NAD dependent only	−	+	−	−
NAD(P) dependent only	−	−	+	+
Cell shape:				
Cocci	d	+	−	−
Rods	d	−	d	+
Vibrios	−	−	d	−
Growth at 45°C	d	+	−	−
Named examples:	*Methylomonas methanica*	*Methylococcus capsulatus*	*"Methylosinus trichosporium"[c]*	*Methylobacterium organophilum*

[a] For extended descriptions of the determinants and for additional references, see Colby et al. (1979). Not all the strains classifiable into groups I and II have been shown to possess all the biochemical characteristics outlined in this scheme. Symbols: see standard definitions.

[b] Obligate: use methane, methanol and formaldehyde as carbon and energy source, but not 2-carbon or more compounds. Facultative: can use 2-carbon or more compounds in addition to methane, methanol and formaldehyde.

[c] Probably synonymous with *"Methylomonas methanooxidans"* and *"Methylomonas methanitrificans."*

oxidizers into groups and subgroups (Table 4.59), which may eventually prove to be further divisible into genera and species, the published attempts at separation into species cannot be regarded as being wholly successful. This becomes clear in the species descriptions themselves, including those in the most useful contribution in recent years to this subject, viz. the study by Romanovskaya et al. (1978). For example, the characteristics separating *Methylococcus luteus* Romanovskaya et al. 1981, 382, *Methylococcus bovis* Romanovskaya et al. 1981, 382, *Methylococcus chroococcus* Romanovskaya et al. 1981, 382, *Methylococcus whittenburyi* Romanovskaya et al. 1981, 382, and *Methylococcus vinelandii* Romanovskaya et al. 1981, 382, could be considered to be of a trivial and probably unstable nature. Certainly they are not ideal taxonomic criteria (e.g. colony pigmentation, segregation into R and S colony forms, growth temperature range, capsule formation, presence of gas vacuoles, cell size and shape, and ability to grow on media containing a specific concentration of methanol). In addition, it is unclear whether or not all strains have been rigorously tested for the expression of all characters.

One would have expected that a coccal morphology would be a feature common to all species of *Methylococcus*, but this is not so: some are described as being rodlike or oval in form. *Methylococcus mobilis* Hazeu et al. 1980, 676 is one of these and, although it appears to be a genuine new species of methane oxidizer, it is not certain that it should be classified as a species of *Methylococcus*. Hazeu et al. (1980a, b) clearly had reservations and noted that their organism was more similar to the *"Methylobacter"* strains described by Whittenbury et al. (1970) than to *Methylococcus capsulatus*. Hazeu et al. compared *M. mobilis* with the *Methylococcus* species described by Romanovskaya et al. (1978) and observed that it was most similar to *"Methylococcus ucrainicus."* But *"M. ucrainicus"* has properties (Romanovskaya et al., 1978) that place it more logically in the genus *Methylomonas* (rod-shaped cells with a mol% G + C of 52.1) rather than in the genus *Methylococcus* (cocci with a mol% G + C of 62.5) as defined in the eighth edition of *Bergey's Manual.* Examples such as these illustrate the uncertainty of the existing taxonomy of methane oxidizers.

We must conclude that there is insufficient information available for the design of a useful taxonomic system; this is a challenge to future taxonomists. A successful taxonomy will need to involve the use of nucleic acid studies (e.g. DNA/DNA and DNA/rRNA hybridization, 16S rRNA oligonucleotide cataloging), sequencing of selected protein molecules, and the identification of differentiating or determinative wall and membrane components, all in combination with the more conventional characters which have proved to be of limited value to date.

Despite this pessimism, there is a generally acceptable diagnostic scheme permitting a division of methane oxidizers into categories, but which does not involve the formal construction of genera and species. This scheme, initially proposed by Whittenbury et al. (1970) and subsequently updated (e.g. by Colby et al., 1979), seems to be a likely basis for a formal taxonomy in the future. A version is shown in Table 4.59 (not all categorizable strains have been subjected to all tests, it should be emphasized).

It is not possible or profitable now to discuss in detail the probable location within this scheme of all species, putative and tentative, mentioned in the literature. However, relationship of the facultative methane oxidizer, *Methylobacterium organophilum* Patt, Cole and Hanson 1976, 228 to the obligate methane oxidizers deserves comment. As can be seen in Table 4.59, this species appears to be a member of group II, differing from strains labeled as "*Methylosinus trichosporium*" Whittenbury et al. 1970, 210 in that it possesses the additional powers of using complex organic compounds. The question arises as to whether both the facultative and obligate varieties should be regarded as separate species within the one genus, a situation analogous to that seen within the genus *Thiobacillus*, or whether obligateness and facultativeness merits the devising of separate genera. Other problems involve the taxonomy of organisms not initially recognized as being methane oxidizers and which have been classified according to other criteria. For example, a recent report (Sampaio, da Silva, Döbereiner, Yates

and Pedrosa, 4th International Symposium N$_2$ Fixation, Canberra, 1981) provides evidence, not yet confirmed, that both *Azospirillum brasilense* and *Azospirillum lipoferum* can grow well on methane, methanol or formate as sole energy sources (see the genus *Azospirillum* in this *Manual*). If this report is confirmed, these organisms, presently classified mainly on their ability to fix N$_2$ in association with plant roots, may be eligible for inclusion in the family *Methylococcaceae*. Other similar cases may well occur.

As a final comment—or plea!—to research workers in this field, avoid the temptation to formally publish proposals of new genera and new species of methane oxidizers until a firm and widely accepted basis for a taxonomy is established.

Further Reading

Colby, J., H. Dalton and R. Whittenbury. 1979. Biological and biochemical aspects of microbial growth on C$_1$ compounds. Annu. Rev. Microbiol. *33:* 481–517.

Hanson, R.S. 1980. Ecology and diversity of methylotrophic organisms. Adv. Appl. Microbiol. *26:* 3–39.

Hazeu, W., W.H. Batenburg-van de Vegte and C. de Bruyn. 1980. Some characteristics of *Methylococcus mobilis* sp. nov. Arch. Microbiol. *124:* 211–220.

Higgins, I.J., D.J. Best, R.C. Hammond and D. Scott. 1981. Methane-oxidizing microorganisms. Microbiol. Rev. *45:* 556–590.

Romanovskaya, V.A., Y.R. Malashenko and V.N. Bogachenko. 1978. Corrected diagnoses of genera and species of methane-oxidizing bacteria. Mikrobiologiya *47:* 120–130.

Genus I. Methylococcus *Foster and Davis 1966, 1929[AL]*

ROGER WHITTENBURY AND NOEL R. KRIEG

Me.thyl.o.cocc′us. Fr. *méthyle* the methyl radical; Gr. n. *coccus* a grain, berry; M.L. masc. n. *Methylococcus* methyl coccus.

Cells **spherical,** usually occurring in pairs. **Nonmotile.** Resting stage is a **cyst.** Gram-negative. Aerobic, having a strictly respiratory type of metabolism with oxygen as the terminal electron acceptor. **Methane, methanol and formaldehyde are the only known compounds serving as sole carbon and energy sources.** No organic growth factors are required. The mol% G + C of the DNA is 62.5 (Bd).

Type species: *Methylococcus capsulatus* Foster and Davis 1966, 1929.

Further Descriptive Information

The intracellular membrane development that occurs in methane-grown cells persists when the cells are grown on methanol (Davies and Whittenbury, 1970; Linton and Vokes, 1978).

M. capsulatus possesses an esterified fatty acid composition of the C16:0, C16:1 type (Makula, 1978). Squalene is present in the cells (Bird et al., 1971).

M. capsulatus possesses phosphoribulokinase and ribulose biphosphate carboxylase and can fix CO$_2$ by the Benson-Calvin cycle. The importance of this for the overall carbon economy of the organisms is not yet known.

Other characteristics are as listed in Table 4.59 (subgroup 1b).

Enrichment and Isolation Procedures

Nitrate-mineral salts medium (NMS)* either as an agar or in liquid form will serve all purposes. Enrichment (e.g. soils, water and mud samples) can be made with the liquid medium in a variety of glass vessels in which 75% of the volume is taken up with a gas mixture of methane and air (30:70). The incubation temperature is 45°C for *M.*

capsulatus. Isolation is achieved best by spread-plating dilutions of the enrichment culture on NMS agar plates which are subsequently incubated at 45°C in plastic or glass containers to which a gas mixture similar to that used in enrichments is added. For details of these and other procedures, and the pitfalls, see Whittenbury et al. (1970).

Maintenance Procedures

Stock cultures, grown on NMS agar slopes in sealed containers, may be stored at 4°C; some strains may survive 3 months, others only 4 weeks. Lyophilization is not suitable for all strains; however, all can be preserved indefinitely in liquid nitrogen and subsequently stored at −100°C.

Procedures for Testing for Special Characters

The ability to use methane can be tested most simply as follows. Inoculate a liquid medium (NMS) contained in flasks of bottles sealable with an injectable rubber seal and over which a metal cap can be secured after injection to ensure gas-tightness. Inject methane, e.g. 15 ml in 100 ml of gas (air) space above 20 ml of medium, and incubate in a shaking incubator at 45°C. Subsequent growth should (after 1 week) be accompanied by a loss of most or all of the methane (assayed by gas chromatography). Uninoculated control flasks should show no detectable change of methane content in comparison with initial assay, and inoculated flasks lacking methane should show no growth.

Further Reading

See Further Reading listed for the family *Methylococcaceae*.

List of the species of the genus Methylococcus

1. **Methylococcus capsulatus** Foster and Davis 1966, 1929.[AL]

cap.su.la′tus. M.L. adj. *capsulatus* encapsulated.

---

* NMS medium (g/liter of distilled water): MgSO$_4$·7H$_2$O, 1.0; CaCl$_2$, 0.2; sequestrene iron complex (Geigy, Johnson of Hendon, Ltd), 0.004; KNO$_3$, 1.0; and agar (if added), 12.5. Trace elements may be necessary. The pH is adjusted to 6.8 before autoclaving. Finally, 20 ml of a sterile phosphate solution (NaH$_2$PO$_4$·12H$_2$O, 15 g in 300 ml of distilled water and adjusted to pH 6.8) is added to media cooled to 60°C.

Cocci 1.0 μm in diameter. Capsules are formed. Poly-β-hydroxybutyrate is not formed. The cells are not pigmented.

Nitrate is superior to ammonium salts or casein hydrolysate as a nitrogen source. No organic growth factors are required, although colony formation on solid media is enhanced by complex extracts such as casein hydrolysate.

Optimum temperature, 37°C. Growth occurs between 30 and 50°C but not at 55°C.

Other characteristics are as indicated in Table 4.59 (group Ia).

Type strain: ATCC 19069.

Species Incertae Sedis

For reasons discussed under Taxonomic Comments for the family *Methylococcaceae*, the following species cannot yet be definitely assigned to the genus *Methylococcus.*

a. *Methylococcus bovis* Romanovskaya, Malashenko and Bogachenko 1981, 382.^{VP} (Effective publication: Romanovskaya et al. 1978, 124.)

Type strain: strain CM of Whittenbury.

b. *Methylococcus chroococcus* Romanovskaya, Malashenko and Bogachenko 1981, 382.^{VP} (Effective publication: Romanovskaya et al. 1978, 124.)

Type strain: strain 9 of Whittenbury.

c. *"Methylococcus fulvus"* Malashenko, Romanovskaya and Kvasnikov 1972, 877.

d. *Methylococcus luteus* Romanovskaya, Malashenko and Bogachenko 1981, 382.^{VP} (Effective publication: Romanovskaya et al. 1978, 124.)

Type strain: UCM 53B.

e. *Methylococcus mobilis* Hazeu, Batenburg-van der Vegte and de Druyn 1980, 676.^{VP} (Effective publication: Hazeu et al. 1980, 211.)

Type strain: LMD 77.28.

f. *Methylococcus thermophilus* Malashenko, Romanovskaya, Bogachenko and Shved 1975, 859.^{AL}

Type strain: IMV-2Yu.

g. *"Methylococcus ucrainicus"* Malashenko, Romanovskaya and Kvasnikov 1972, 876.

h. *Methylococcus vinelandii* Romanovskaya, Malashenko and Bogachenko 1981, 382.^{VP} (Effective publication: Romanovskaya et al. 1978, 125.)

Type strain: strain Mexica of Whittenbury.

i. *Methylococcus whittenburyi* Romanovskaya, Malashenko and Bogachenko 1981, 382.^{VP} (Effective publication: Romanovskaya et al. 1978, 125.)

Type strain: strain 1521 of Whittenbury.

Genus II. **Methylomonas** (ex Leadbetter 1974) nom. rev.

ROGER WHITTENBURY AND NOEL R. KRIEG

Me.thyl.o.mo′nas. Fr. *méthyle* the methyl radical; Gr. *monas* a unit, monad; M.L. fem. n. *Methylomonas* methyl monad.

Straight, curved or branched rods, but not helical, 0.5–1.0 × 1.0–4.0 μm. **Motile by a single polar flagellum.** Sheaths or prosthecae are not known. **Resting stages are cysts.** Gram-negative. Aerobic, having a strictly respiratory type of metabolism with oxygen as the terminal electron acceptor. **Methane, methanol and formaldehyde are the only known sole sources of energy and carbon.** Organic growth factors are not required. Temperature range for growth, 20–35°C. The mol% G + C of the DNA is 52.1 (Bd).

Type species: *Methylomonas methanica* (ex Leadbetter 1974) nom. rev.

Further Descriptive Information

In static liquid culture (30°C) *M. methanica* gives rise to a surface pellicle (pink in color). When streaked on to agar plates, colony color may be pink or yellow, becoming consistently pink in continued subculture. A sapphire-colored soluble pigment is often found in iron-deficient media.

Other characteristics are as listed earlier in Table 4.59 (subgroup Ia).

Enrichment and Isolation Procedures

Media, gas atmospheres and procedures for enrichment and isolation are as described above for *Methylococcus capsulatus*. Static incubation of liquid cultures at 30°C often leads to enrichment of *M. methanica*.

Maintenance Procedures

These are as described for *Methylococcus capsulatus*.

Taxonomic Comments

The name *Methylomonas* was illegitimate when published by Leadbetter in 1974 because of the rule of priority of names; i.e. it was antedated by the name *"Methanomonas"* which had been published by Orla-Jensen in 1909. However, the prefix *"Methano-"* is presently in wide use for various genera of *anaerobic, methane-producing* bacteria, and its application to *aerobic, methane-oxidizing* bacteria would create considerable confusion. The problem of priority has now been eliminated by the fact that neither *"Methanomonas"* nor *"Methylomonas"* appeared on the Approved Lists of Bacterial Names in 1980; thus, neither name presently has any standing in nomenclature. We therefore propose that the name *Methylomonas* Leadbetter 1974, 267 be revived (nom. rev.), with the type species to be that proposed earlier by Leadbetter (1974), viz. *Methylomonas methanica*.

The species name *Methylomonas methanica* Leadbetter 1974, 268 similarly has no standing presently in nomenclature, and we propose that this name be revived as well. One factor complicating this is the fact that the type strain is not extant. Consequently, in the present chapter we propose NCIB strain 11130 as the neotype strain, since this strain is representative of the characteristics of the species.

For reasons discussed under Taxonomic Comments for the family *Methylococcaceae* we believe that other *Methylomonas* species should be considered as *species incertae sedis* for the present.

Further Reading

See the references listed under Further Reading for the family *Methylococcaceae*.

List of the species of the genus **Methylomonas**

1. **Methylomonas methanica** (ex Leadbetter 1974) nom. rev.
me.tha′ni.ca. M.L. n. *methanum* methane; M.L. fem. adj. *methanica* relating to methane.

Straight rods 0.6 × 1.0 μm. Extracellular slime is formed. Cells are pink, reflecting the presence of carotenoid pigments. Cysts are formed.

Catalase-positive. Oxidase-positive. Strict aerobes; grow best at oxygen concentrations of 20% or above. Optimum temperature, 30°C; no growth at 37°C. Optimum pH, ~7.0, with a range of 6.8–8.0 when sodium nitrate is used as the nitrogen source.

The mol% G + C of the DNA is 52.1 (Bd).

Type strain: not extant. NCIB strain 11130 is representative of the characteristics of the species, and we propose it as the *neotype strain.*

Species Incertae Sedis

For reasons discussed under Taxonomic Comments for the family *Methylococcaceae* the following species have an uncertain taxonomic status. None of the names presently has any standing in nomenclature.

a. "*Methylomonas flagellata*" Morinaga, Yamanaha, Otsuka and Hirose 1976, 1544.

b. "*Methylomonas gracilis*" Romanovskaya, Malashenko and Bogachenko 1978, 122.

c. "*Methylomonas maragarita*" Takeda, Motomatsu, Hachiya, Fukuoka and Takahara 1974, 795.

d. "*Methylomonas methanitrificans*" (Davis, Coty and Stanley 1964) Leadbetter 1974, 269. (*Pseudomonas methanitrificans* Davis, Coty and Stanley 1964, 471.)

e. "*Methylomonas methanooxidans*" (Brown and Strawinski 1958)

Leadbetter, 1974, 269. (*Methanomonas methanooxidans* Brown and Strawinski 1958, 122.)

Note: species d and e above may be synonyms of each other and may be wrongly classified in the genus *Methylomonas.*

Genera Incertae Sedis

For reasons given in Taxonomic Comments for the family *Methylococcaceae*, the following genera have an uncertain taxonomic status.

a. *Methylobacillus* Yordy and Weaver 1977, 254.[AL]
Type species: *Methylobacillus glycogenes* Yordy and Weaver 1977, 254.[AL]

b. *Methylobacterium* Patt, Cole and Hanson 1976, 228.[AL]
Type species: *Methylobacterium organophilum* Patt, Cole and Hanson 1976, 228.[AL]

c. "*Methylosinus*" Whittenbury, Phillips and Wilkinson 1970, 213.
Type species: none designated.

d. "*Methylovibrio*" Hazeu and Steenis 1970, 71.
Type species: "*Methylvibrio soehngenii*" Hazeu and Steenis 1970, 71.

FAMILY V. HALOBACTERIACEAE GIBBONS 1974, 269[AL]

HELGE LARSEN

Hal.o.bac.ter.i.a'ce.ae. M.L. n. *Halobacterium* type genus of the family; *-aceae* ending to denote a family; M.L. fem. pl. n. *Halobacteriaceae* the *Halobacterium* family.

Rods, cocci or discs; many strains display a multitude of involution forms. Reproduce by binary fission. Resting stages are not known. Give a Gram-negative staining reaction. Nonmotile, or motile by lophotrichous flagella. Mostly strict aerobes, but facultative strains, growing anaerobically with or without nitrate, have been reported. Require at least 1.5 M (~8%) NaCl, and in most cases 3–4 M (17–23%) NaCl in their environment for good growth. Colonies are various shades of red (pink, orange-red, vermilion, mauve-red) and are opaque to more or less translucent. Optimum temperature, 40–50°C. Chemoorganotrophic, using amino acids or carbohydrates as the main carbon source. Some strains require several amino acids for growth and are stimulated by vitamins. Occur ubiquitously in nature where the salt concentration is high, i.e. in salt lakes, salt ponds and marine salterns. Are common in crude solar salt and proteinaceous products (fish, hides) heavily salted with solar salt, and may spoil such products. Nonpathogenic. The DNA seems commonly composed of a major and a minor component, the latter making up ~10–35% of the total DNA. The mol% G + C of the major component is commonly 66–68, and of the minor component 57–60 (Bd).

Type genus: *Halobacterium* Elazari-Volcani 1957, 207.

Further Descriptive Information

The most striking feature of the *Halobacteriaceae* is their high requirement for sodium chloride. Although some strains may grow at a salt concentration as low as 1.5 M (~8%, w/v), most of the strains grow best at concentrations of 3.5–4.5 M (~20–26%, w/v) and do not grow well below 2.5 M (~15%, w/v). The *Halobacteriaceae* are often referred to as "the extremely halophilic bacteria." Besides the *Halobacteriaceae*, only some representatives of the genus *Ectothiorhodospira* are known to require such high salt concentrations (Trüper and Imhoff, 1981).

Another distinctive feature common to the natural isolates of the *Halobacteriaceae* is their red-to-orange pigmentation, due to the presence of carotenoids in the cells. Bacterioruberins (C_{50} carotenoids) are the predominant pigments. The common occurrence of small amounts of retinal is notable (Kushwaha et al., 1973). Colorless strains are only known as mutants produced in the laboratory. Isoprenoid quinones are of the menaquinone type, not the ubiquinone type (Collins et al., 1981).

The carotenoids seem to play a protective role against the strong sunlight in the habitats where these organisms are found, i.e. strongly

saline lakes (such as the Great Salt Lake, the Dead Sea, or the alkaline, saline lakes of East Africa) and the worldwide salterns where salt is commercially produced from seawater by evaporation (Dundas and Larsen, 1962). Representatives of the *Halobacteriaceae* may occur in these habitats in such large numbers that they impart a red color to the brine. They heavily contaminate the salt made from the brine. Without further processing, the solar salt may contain 10^5 to 10^6 viable cells/g, surviving for several years under practical storage conditions.

The *Halobacteriaceae* display a number of biochemical characteristics which clearly place them among the archaeobacteria. This pertains basically to the nucleotide sequence of the 16S rRNA (Magrum et al., 1978; Stackebrandt and Woese, 1981) and also to a number of other phenotypic properties which distinguish the organisms from the "eubacteria": (a) lack of a muramic acid-containing peptidoglycan in the cell envelope (Brown and Shorey, 1963; Kushner et al., 1964; Brown and Cho, 1970); (b) use of the mevalonate pathway instead of the malonate pathway in the formation of lipids, the membrane lipids being composed of ether-linked isoprenyl phosphoglycerides (Kates, 1978); (c) a pseudouridine or other nucleoside instead of the thymidine in the "common arm" of the transfer RNAs (tRNAs) (Magrum et al., 1978); (d) ADP-ribosylation of the peptide elongation factor EF-2 by diphtheria toxin (as yet shown only in *Halobacterium*) (Kessel and Klink, 1980), and (e) insensitivity towards a number of antibiotics, notably many cell wall and protein synthesis inhibitors (Pecher and Böck, 1981; Hilpert et al., 1981). Accordingly, the *Halobacteriaceae* can be grouped most closely with the methanogenic bacteria and the thermoacidophilic bacteria, representing with these a phylogenetic unit of ancient origin (Woese, 1981).

Besides the above-mentioned properties common to the archaeobacteria, the *Halobacteriaceae* possess some other rather special properties unique to them and pertaining to their life at high salt concentrations. Their requirement for NaCl for growth is specific. No other salt or chemical can replace NaCl, but sodium can be partially replaced by potassium (Brown and Gibbons, 1955). Attempts to adapt *Halobacteriaceae* to lower salt concentrations have failed (Kushner, 1978).

To compensate for the high salt concentration in the environment, the organisms accumulate salt within the cells in a concentration at least as high as that outside. It is remarkable that the internal salt is mainly KCl (up to 5 M), considering that the K^+ concentration in the environment, both in nature and under laboratory conditions, is normally very low (<0.05 M) (Christian and Waltho, 1962; Gochnauer and Kushner, 1971). The functional and structural units of the cells are adapted to these high salt concentrations. The enzymes function well in the presence of, and even require, salt in high concentrations (KCl and NaCl). The cell membrane maintains its integrity only in the presence of NaCl or KCl, and the ribosomes maintain their integrity only in the presence of KCl.

The molecular basis for the salt requirement of the enzymes and the structural units of the *Halobacteriaceae* is found partly in the acidic (Brown, 1963; Reistad, 1972), and partly in the hydrophobic (Lanyi, 1974), nature of their proteins. The salt neutralizes ionic forces and stabilizes hydrophobic bonds so that the proteins are kept in their proper conformational state. Upon removal of the salt, the proteins denature (Lanyi, 1974).

Further Reading

Bayley, S.T. and R.A. Morton. 1978. Recent developments in the molecular biology of extremely halophilic bacteria. Crit. Rev. Microbiol. *6:* 151–205.

Brown, A.D. 1976. Microbial water stress. Bacteriol. Rev. *40:* 832–846.
Dundas, I.E.D. 1977. Physiology of *Halobacteriaceae.* In Rose and Tempest (Editors), *Advances in Microbial Physiology,* Vol. 15, Academic Press, London, pp. 85–120.
Gibbons, N.E. 1969. Isolation, growth and requirements of halophilic bacteria. In Norris and Ribbons (Editors), *Methods in Microbiology,* Vol. 3B, Academic Press, London, pp. 169–183.
Kushner, D.J. 1978. Life in high salt and solute concentrations: halophilic bacteria. In Kushner (Editor), *Microbial Life in Extreme Environments,* Academic Press, London, pp. 317–368.
Lanyi, J.K. 1974. Salt-dependent properties of proteins from extremely halophilic bacteria. Bacteriol. Rev. *38:* 272–290.
Larsen, H. 1980. Ecology of hypersaline environments. *In* Nissenbaum (Editor), *Hypersaline Brines and Evaporitic Environments,* Elsevier, Amsterdam, pp. 28–39.
Larsen, H. 1981. The family *Halobacteriaceae.* In Starr, Stolp, Trüper, Balows and Schlegel (Editors), *The Prokaryotes: a Handbook on Habitats, Isolation and Identification of Bacteria,* Springer-Verlag, Berlin, pp. 985–994.

Key to the genera of the family **Halobacteriaceae**

I. Motile or nonmotile. Often regular rods or discs when grown under optimum conditions; often pleomorphic. Lyse upon dilution of the medium with water.

Genus I. *Halobacterium*, p. 262.

II. Nonmotile. Regular spheres, occurring in pairs, tetrads, sarcinae or irregular clusters. Do not lyse upon dilution of the medium with water.

Genus II. *Halococcus*, p. 266.

Genus I. **Halobacterium** *Elazari-Volcani 1957, 207*[AL]

HELGE LARSEN

Hal. o. bac. te′ ri. um. Gr. n. *hals, halos* the sea, salt; Gr. n. *bakterion* a small rod; M.L. neut. n. *Halobacterium* the salt-bacterium.

When grown under optimum conditions the cells may be **rod shaped** (0.5–1.2 μm in diameter and 1.0–6.0 or more μm in length) or **disk shaped** (1.0–3.0 × 2.0–3.0 μm, and 0.3–0.4 μm in thickness). At higher temperatures and in deficient media a variety of **involution forms** are seen (bent and swollen rods, clubs, ovoids, spheres, spindles, and other irregular forms). Some strains are highly pleomorphic even under optimum growth conditions. **The cells divide by constriction.** Resting stages are not known. The cells often contain gas vacuoles. Stain Gram-negative. **Nonmotile, or motile by lophotrichous flagella. Most strains are strict aerobes,** but facultative anaerobes, growing with or without nitrate, have been described. Optimum temperature, 40–50°C; no growth occurs below 10°C. **Colonies are pink, red, or red-to-orange,** and are opaque to translucent. Oxidase- and catalase-positive. Gelatin is usually liquefied. **Chemoorganotrophic,** using amino acids or sugars as the main carbon source. **Most isolates require at least 2.5 M (∼15%, w/v) NaCl and 0.1–0.5 M Mg²⁺ for growth; they grow best in 3.5–4.5 M (∼20–26%, w/v) NaCl,** and also grow well in saturated NaCl solutions (>5 M, or >29%, w/v). **Cell lysis occurs in hypotonic solutions. Occur ubiquitously in nature in strongly saline lakes and ponds.** These organisms are the most common cause of "pink" in salt fish, "red heat" in salted hides, and may cause spoilage of these and other salted proteinaceous products. The mol% G + C of the DNA ranges from 63–68 (Bd, T_m). The DNA is frequently composed of a major component (mol% G + C = 66–68) and a minor component (mol% G + C = 57–60); the latter component makes up from 10–36% of the total DNA (Bd).

Type species: *Halobacterium salinarium* (Harrison and Kennedy 1922) Elazari-Volcani 1957, 208.

Further Descriptive Information

Pleomorphism is a distinctive characteristic of the halobacteria. In some strains the cells are regular, slender rods when grown at moderate temperatures under optimum conditions, but display a multitude of involution forms, often rounded forms, when the cells are exposed to higher temperatures or media less suitable for growth, notably hypotonic media. Upon gradual dilution of the medium with water the cells change their shape, from rods through irregular transition forms to spheres, and the spheres undergo lysis (for most strains when the NaCl concentration reaches 5–10%). The cells are also transformed to spheres and lyse upon exposure to elevated temperatures (60–65°C).

The lysis phenomenon in hypotonic solutions is not primarily due to an osmotic effect, but rather the need for high concentrations of salt to maintain the cell envelope. When salt is removed, the envelope loses its rigidity and disintegrates. The cell envelope of the halobacteria is thus of a special kind. It is composed of a cell membrane closely bounded by an outer, proteinaceous layer and is special in that the proteins are strongly acidic and the lipids are ether-linked isoprenyl phosphoglycerides. A glycoprotein of a rather high molecular weight is shown to be a dominating component (50%) of the outer, proteinaceous layer. It may play a role in maintaining the shape of the cells (Mescher and Strominger, 1976 a, b). The saccharides of the glycoprotein contain covalently bound sulfate (Wieland et al., 1980). The proteinaceous subunits of the cell envelope are held together only in the presence of salt. The surface of the cell envelope has a hexagonal surface pattern. The carotenoids of the cell are embedded in the envelope.

Some strains of halobacteria, when grown at low oxygen tension, form patches in the cell membrane consisting of a special chromoprotein, bacteriorhodopsin. The coloring principle is retinal; the membrane patches are referred to as "the purple membrane" (Stoeckenius and Kunau, 1968; Oesterhelt and Stoeckenius, 1974). The purple membrane has been shown to possess important properties. It acts as a proton pump driven by light energy absorbed by the pigment. Important energy-requiring phenomena in the cell (ATP synthesis, amino acid uptake, K⁺ uptake, Na⁺ extrusion) utilize the energy made available in the light-driven proton translocation (Lanyi, 1978, 1980). Hartmann et al. (1980) demonstrated that halobacteria can utilize light for growth anaerobically, provided they contain bacteriorhodopsin.

Some strains of halobacteria contain gas vesicles, revealing themselves in the light microscope as strongly refractile bodies (gas vacuoles) which, in extreme cases, seem to fill the cell completely. Strongly gas vacuolated strains form pink, opaque colonies on agar plates; nonvacuolated strains form red, translucent colonies; moderately vacuolated strains give intermediate colony types. Gas vacuolation is most frequently seen in freshly isolated strains. Upon cultivation in the laboratory the halobacteria have a tendency to lose the ability to produce gas vacuoles, varying from strain to strain; in strains that lose the ability rapidly, translucent sectors may be seen in the opaque colonies. The function of the gas vacuoles seems to be to float the cells toward the surface, and thus toward oxygen, in the natural saline environments (Walsby, 1978).

Most halobacteria grow well in tryptone-yeast extract-salt media* when provided with good aeration. Many strains have a complex nutrition in the sense that they require several amino acids for growth. They also utilize amino acids as a source of energy. Purine and pyrimidines have been reported to be growth factors (Dundas et al., 1963; Onishi et al., 1965). Carbohydrates and vitamins stimulate growth of amino acid-requiring strains (Gochnauer and Kushner, 1969). Some strains utilize carbohydrates as the main source of energy (Tomlinson and Hochstein, 1976; Gonzalez et al., 1978), and may grow with glucose as the sole source of carbon (Rodriguez-Valera et al., 1980). Most strains are strict aerobes, but some strains are facultative anaerobes using nitrate as electron acceptor in anaerobic respiration (Volcani, 1940, 1957; Werber and Mevarech, 1978) and also having an apparent fermentative type of metabolism (Gonzalez et al., 1978). Hartmann et al. (1980) reported anaerobic growth at the expense of arginine being converted to ornithine. Growth is relatively slow; generation times of 3–6 h are the best that have been reported in laboratory experiments.

Biochemical reports on the basic metabolism of the halobacteria are limited. There are many reports on individual enzymes and how they are affected by salt (Kushner, 1978), but few reports on metabolic pathways exist. *H. saccharovorum* metabolizes glucose via the "modified Entner-Doudoroff pathway" in which glucose is oxidized via gluconate to 2-keto-3-deoxygluconate, and the latter split to 3-phosphoglyceraldehyde and pyruvate by a mechanism involving ATP (Tomlinson et al., 1974). Kawasaki et al. (1978) found the same pathway for glucose dissimilation in a halobacterium (strain R 113) isolated from salt.

Bayley considered the possibility that the high salt concentrations required by the halobacteria cause a misreading of the genetic code,

but experiments have shown that the codon assignments are the same as in other organisms. However, protein synthesis seems to be initiated with a nonformylated methionyl-tRNA$_{met}$ (Bayley and Morton, 1978); this corroborates the relation of the halobacteria to the archaeobacteria.

The high content of satellite DNA in the *Halobacteriaceae* is unique to this family. Evidence has been presented that the satellite DNA consists of plasmids, mostly of large size (molecular weight = 60–100 $\times$ 10^6) (Pfeifer et al., 1981). An alkalophilic halobacterium has been reported to contain no satellite DNA; its genome has a mol% G + C value lower than that reported for the other halobacteria (Tindall et al., 1980). The property of gas vacuole formation in the halobacteria is plasmid-mediated (Simon, 1978; Weidinger et al., 1979).

Bacteriophages infecting, and apparently specific for, halobacteria have been demonstrated by Torsvik and Dundas (1974) and Wais et al. (1975).

Enrichment and Isolation Procedures

Strong natural salines, crude solar salt, and proteinaceous products heavily salted with such salt, are potential sources of halobacteria. The material may be inoculated directly onto the surface of nutrient agar media containing 25% (w/v) NaCl. The plates are wrapped in plastic bags to prevent drying and are incubated at 35–38°C. Growth is indicated by development of a red color, and isolation can be achieved by suspending the growth in 25% NaCl and streaking suitable agar media.

Enrichments may be obtained from natural salines or from crude solar salt by inoculating the material into a suitable growth medium (see earlier footnote) in flasks and incubating at 35–38°C on a shaking machine that provides good aeration. Enrichments from salted proteinaceous products may be obtained in the same way, and also by incubating pieces of the product (fish, hides) above a shallow layer of 25% NaCl contained in loosely stoppered bottles. Growth is revealed by its red color and can be streaked for isolations (Gibbons, 1969; Larsen, 1981).

By these procedures members of the genus *Halobacterium* and/or the genus *Halococcus* may be obtained. Selective procedures that are specific for one or the other genus have not been devised.

Maintenance Procedures

Growth on agar slants in screw-capped tubes can be kept in the refrigerator for 4 or 5 months. Freeze-dried preparations, and preparations stored in liquid nitrogen, are viable for at least several years (Larsen, 1981).

Differentiation of the Genus **Halobacterium** from Other Genera

The key to the family *Halobacteriaceae* indicates those characteristics useful for distinguishing *Halobacterium* from *Halococcus*.

It should be noted that the salt-tolerant genus, *Halomonas*, does not belong to the archaeobacteria, does not have red pigmentation, and does not require high NaCl concentrations for growth, all in contrast to the genus *Halobacterium*. See the article on *Halomonas* for further information.

Taxonomic Comments

Klebahn (1919) in Germany was the first to describe distinctly the main morphological and physiological features of the *Halobacteriaceae*. He isolated from salt fish a salt-requiring, red-colored, pleomorphic, rod-shaped bacterium lysing in hypotonic solutions, which he named *Bacillus halobius ruber*. Harrison and Kennedy (1922) in Canada similarly isolated from salt fish a bacterium which from the published descriptions is well-nigh impossible to distinguish from Klebahn's

Bacillus halobius ruber. Harrison and Kennedy did not know about Klebahn's work, which was published in a serial of a rather limited circulation, so they named their isolate *Pseudomonas salinaria*. This was the beginning of a confusion that has lasted until now.

In Europe Petter (1932) in Holland extended Klebahn's work. She isolated several more strains of Klebahn's organism from salt fish, but found Klebahn's name improper because of the inability of the organism to form spores; therefore she changed the name to *Bacterium halobium*. This was later changed to *Halobacterium halobium* by Elazari-Volcani (1957) when proposing the genus *Halobacterium* in the seventh edition of *Bergey's Manual*. Strains of *H. halobium* were kept in Holland since the time of Petter and may be identical to strains kept in culture collections today (e.g. *H. halobium* NCMB 741).

In America, Lochhead (1934) in Canada, complying to a proposal of Bergey et al. (1923), listed the organism of Harrison and Kennedy under the name *Serratia salinaria*. He also described a new, closely

---

* An example of a suitable tryptone-yeast extract-salt medium consists of (g/liter of tap water): crude solar salt, 250.0; MgSO$_4$·7H$_2$O, 20.0; KCl, 5.0; CaCl$_2$·6H$_2$O, 0.2; yeast extract, 3.0–5.0; and tryptone (Difco), 5.0. The pH should be near neutrality except for the alkalophilic strains.

related species, *Serratia cutirubra*. Lochhead (1943) later pointed out that these organisms clearly differ from the *Serratia* type species *S. marcescens* Bizio both in their flagellation, pigmentation and ability to utilize carbohydrates, and that they should be assigned to the genus *Pseudomonas* as originally proposed by Harrison and Kennedy for *Pseudomonas salinaria*. Volcani (1957) later proposed that the two species described in Canada be named *Halobacterium salinarium* and *H. cutirubrum*, respectively. Strains are available from culture collections.

In the seventh edition of *Bergey's Manual* (Elazari-Volcani, 1957), two more species were included in the genus *Halobacterium: H. maris-mortui* Elazeri-Volcani 1940 and *H. trapanicum* (Peter 1931) Elazari-Volcani 1957. These species were distinguishable from the others by their ability to produce gas from nitrate.

In the eighth edition of *Bergey's Manual* (Gibbons, 1974) only two species were recognized: *H. salinarium* and *H. halobium. H. cutirubrum* was considered so similar to *H. salinarum* that ranking it as a separate species was not warranted. *H. halobium* was not clearly distinguished from *H. salinarium. H. maris-mortui* and *H. trapanicum* were listed as *species incertae sedis*, the former because no cultures were extant, the latter presumably because it was believed it had not been adequately described.

Colwell et al. (1979) subjected a large number of isolates of halobacteria to taxonomic analysis by numerical taxonomy. They concluded that the group might be divided into two clusters, one containing strains to be named *H. salinarium*, the other strains to be named *H. cutirubrum*. The difference between the clusters was small, however, and was based on rather special tests. Strains bearing the name *H. halobium* were suggested to be included among the strains of *H. salinarium*.

Thus, the literature reflects doubts and disagreements, if not confusion, as to whether *H. salinarium, H. cutirubrum* and *H. halobium* should be upheld as separate species. More recently, Fox et al. (1980) reported that strains of the three species display identical oligonucleotide catalogues for their 16S rRNA, a finding they believed to warrant considering the organisms to be strains of a single species. An organism previously known as *Amoeobobacter morrhuae*, which is halophilic but not phototrophic, exhibited a 16S rRNA catalogue that differed from that of *H. salinarium, H. cutirubrum* and *H. halobium* by a single base change; therefore, Fox et al. (1980) considered this organism to belong to the same species as the three halobacteria.

In the present edition of the *Manual*, the recommendation of Fox et al. (1980) has been adopted. It is proposed that *H. cutirubrum* and *H. halobium* be considered subjective synonyms of *H. salinarium*, the latter epithet having priority since 1922.

Since the eighth edition of *Bergey's Manual*, three more species of halobacteria have been recognized: *H. volcanii, H. saccharovorum*, and *H. vallismortis*. These names, as well as *H. trapanicum*, were included on the Approved Lists of Bacterial Names in 1980. The name "*H. maris-mortui*" is no longer accepted and did not appear on the Approved Lists. Soliman and Trüper (1981) have proposed a new species, *H. pharaonis*.

Further Reading

See references provided under Further Reading for the family *Halobacteriaceae*.

Differentiation of the species of the genus **Halobacterium**

Characteristics useful for differentiating the species of *Halobacterium* are listed in Table 4.60. Other characteristics of the species are presented in Table 4.61.

List of the species of the genus **Halobacterium**

1. **Halobacterium salinarium** (Harrison and Kennedy 1922) Elazari-Volcani 1957, 208.[AL] (*Pseudomonas salinaria* Harrison and Kennedy 1922, 120; *Halobacterium halobium* (Petter 1931) Elazari-Volcani 1957, 210; *Halobacterium cutirubrum* (Lochhead 1934) Elazari-Volcani 1957, 209.)

sal.in.ar′i.um. L. adj. *salinarium* belonging or pertaining to salt works.

Mainly rod shaped, 0.5–1.0 μm wide by 1.0–6.0 μm or more in length (see Fig. 4.40), but displays a multitude of involution forms, especially in deficient media and at elevated temperatures. Motile, especially in young cultures, by means of lophotrichous flagella.

Basically aerobic, but may grow anaerobically in the light when containing bacteriorhodopsin or in the dark when supplied with arginine.

Best growth occurs at 4–5 M NaCl; good growth occurs in saturated NaCl; no growth occurs below 3 M NaCl.

Has a complex nutrition; utilizes amino acids for growth. Growth may be stimulated by carbohydrates, but no acid production occurs. Other physiological and nutritional characteristics are listed in Tables 4.60 and 4.61.

Commonly found in proteinaceous products heavily salted with crude solar salt (fish, hides, viscera).

The mol% G + C of the major DNA component is 66–68, and of the minor component is 57–59.

Type strain: NRC 34002 (ATCC 33171)

2. **Halobacterium volcanii** Mullakhanbhai and Larsen* 1975, 213.[AL]

vol.can′i.i. M.L. gen. n. *volcanii* of Volcani; named after B. Elazari-Volcani, the Israeli microbiologist who discovered life in the Dead Sea.

Mainly disk shaped, 1.0–3.0 × 2.0–3.0 μm, and 0.3–0.4 μm in thickness, often slightly cupped (Fig. 4.41), but displays a multitude of involution forms, often ovoid, especially in deficient media and at elevated temperatures. In young cultures many elongated disks occur, some of which may rotate around their long axis. Flagella have not been demonstrated, and the movement is different from that of the lophotrichously flagellated strains of halobacteria.

Strictly aerobic. Grows best at 1.5–2.5 M (~9–15%, w/v) NaCl. Poor growth occurs above 4.5 M (26%) NaCl or below 1.0 M (~6%) NaCl. Very tolerant to 1.5 M MgCl₂. Grows on sugars as the sole source of carbon. Organic growth factors are not required. Other physiological and nutritional characteristics are listed in Table 4.61.

Found in the Dead Sea. A closely related organism, strain R-4 (CCM 3361), has been isolated from a marine saltern (Rodriguez-Valera et al., 1980).

The mol% G + C of the DNA is 63.4 ± 0.5 (T_m).

Type strain: NCMB 2012 (strain DS2).

3. **Halobacterium saccharovorum** Tomlinson and Hochstein 1977, 306.[AL] (Effective publication: Tomlinson and Hochstein 1976, 588.)

sacc.har.o′.vo.rum. L. n. *saccharum* sugar; L. v. *voro* to devour; M.L. neut. n. *saccharovorum* sugar devourer.

Rod-shaped cells, 0.6–1.2 × 2.5 μm. Motile.

Strictly aerobic. Optimum NaCl concentration for growth, 3.5–4.5 M (~20–26%, w/v).

A number of sugars (hexoses, pentoses, disaccharides) are readily utilized for growth, with acid production (mainly acetic acid). Growth

---

* Incorrectly cited as "Larson" in the Approved Lists of Names, 1980.

Table 4.60.

Differential characteristics of the species of the genus **Halobacterium**[a]

Characteristics	1. H. salinarium	2. H. volcanii	3. H. saccharovorum	4. H. vallismortis	5. H. pharaonis
Glucose used as a main carbon source	−	+	+	+	−
Facultatively anaerobic	−	−	−	+	−
Alkalophilic[b]	−	−	−	−	+
Cells disk shaped	−	+	−	−	−
Moderate salt requirement[c]	−	+	−	−	−

[a] Symbols: see standard definitions.
[b] Optimum pH is 8.5–9.5.
[c] Optimum NaCl concentration is less than 3.0 M (~17%, w/v).

Table 4.61.

Other characteristics of the species of the genus **Halobacterium**[a]

Characteristics	1. H. salinarium	2. H. volcanii	3. H. saccharovorum	4. H. vallismortis	5. H. pharaonis
Catalase test	+	+	+	+	+
Oxidase test	+	+	+	+	+
Gelatin hydrolysis	+		−	−	+
Starch hydrolysis	−		−	+	−
NO_3^- to NO_2^-	−		+	+	−
Gas from NO_3^-	−	−	−	+	−
H_2S production	+		+	+	+
Indole production	+		−	+	
Polymyxin, sensitive to	−	−	b	−	−
Bacitracin, sensitive to	+		−	+	+
Growth factors required	+	−	+		−
Optimum temperature, °C	50	45	50	40	45
Main carbon sources:					
Glucose	−[c]	+	+	+	−
Galactose	−[c]	+	+	+	
Mannose	−		+	−	
Fructose	−	+	+	+	−
Ribose	−	−	+		
Maltose		+	+	+	
Lactose	−	+	+		
Sucrose		+	+	+	
Glycerol	−[c]		+	+	
Acid production from glucose	−	−	+	+	−

[a] For symbols see standard definitions.
[b] Contradictory reports exist for polymyxin susceptibility.
[c] However, the compound stimulates growth in complex media.

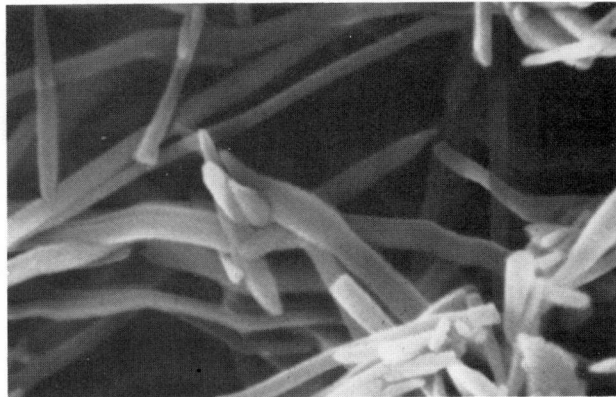

Figure 4.40. *Halobacterium salinarium*, strain 1, growing at 30°C in the medium given in the first footnote of the text. This is a young culture, in which this strain tends to form long rods (× 5000) (Preparation by Greta Bentzen; photography by the Laboratory of Clinical Electron Microscopy, University of Bergen.)

factors are required. Other physiological and nutritional characteristics are listed in Table 4.61.

Isolated from marine salterns.

The mol% G + C of the DNA is not known.

Type strain: ATCC 29252.

4. **Halobacterium vallismortis** Gonzalez, Gutierrez and Ramirez 1979, 436. (Effective publication: Gonzalez et al. 1978, 713.[AL])

val.lis.mor′tis. L. gen. n. *vallis* of the valley; L. gen. n. *mortis* of death; M.L. fem. n. *vallismortis* of the valley of death; named after Death Valley, California.

Rod-shaped cells, 0.6–1.0 × 3.0–5.0 μm. Extremely pleomorphic. Motile.

Facultatively anaerobic. Optimum NaCl concentration, 4.3 M (~25%, w/v). No growth below 2.5 M (~15%).

Sugars are metabolized, most with the production of acid. Gas production occurs from nitrate. Other physiological and nutritional characteristics are listed in Table 4.61.

Isolated from salt ponds in Death Valley.

The mol% G + C of the DNA is not known.

Type strain: ATCC 29715 (strain J.F.54).

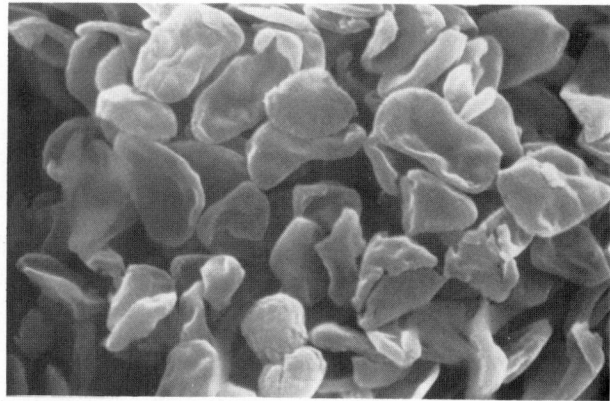

Figure 4.41. *Halobacterium volcanii*, strain DS2, growing at 30°C in a medium consisting of (g/liter of distilled water): NaCl, 125.0; MgCl$_2$·6H$_2$O, 50.0; K$_2$SO$_4$, 5.0; CaCl$_2$·6H$_2$O, 0.2; tryptone (Difco), 5.0; and yeast extract, 5.0; pH 6.8. The culture was photographed at the end of the exponential growth phase (× 7200) (Preparation by Greta Bentzen; photography by the Laboratory of Clinical Electron Microscopy, University of Bergen.)

5. **Halobacterium pharaonis** Soliman and Trüper 1983, 439.

pha.ra.o′nis. L. gen. n. *pharaonis* of pharao, title of the kings in ancient Egypt.

Rod-shaped cells, 0.8 × 2.0 to 3.0 μm. Motile by flagella.

Strictly aerobic. Optimum NaCl concentration, 3.5 M (~20%, w/v). No growth below 1.5 M (~9%).

Sugars are not utilized. Growth occurs on pyruvate, fumarate, formate and butyrate when glutamate is the source of nitrogen. Organic growth factors are not required. In contrast to the other halobacteria, *H. pharaonis* is alkalophilic (optimum pH is 8.5) and has a low magnesium requirement (below 0.01 M). Other physiological characteristics are listed in Table 4.61.

Isolated from alkaline, saline lakes of Wadi Natrun, Egypt. A closely related organism, strain SP-1, has been isolated from the alkaline, saline Lake Magadi, Kenya (Tindall et al., 1980).

The mol% G + C of the DNA is 64 (T_m).

Type strain: DSM 2160 (strain Gabara).

Species Incertae Sedis

The taxonomic status of the following species is uncertain because of lack of detailed descriptive information.

Halobacterium trapanicum (Petter 1931) Elazari-Volcani 1957, 211.[AL] (*Bacterium trapanicum* Petter 1932, 40.)

Description limited. Seems to have many characteristics in common with *H. vallismortis*, but is described as nonmotile, more orange in color, indole-negative, and not able to hydrolyze starch. Grows on asparagine as the sole source of carbon and nitrogen. Produces gas from nitrate. Isolated from solar salt (Trapani) and the Dead Sea.

The mol% G + C of the DNA is not known.

Type strain: NRC 34021.

Genus II. **Halococcus** *Schoop 1935, 817.*[AL]

HELGE LARSEN

Hal.o.coc′cus. Gr. n. *hals, halos* the sea, salt; Gr. n. *coccus* a berry; M.L. masc. n. *Halococcus* the salt-coccus.

Cocci, 0.8–1.5 μm in diameter, occurring in pairs, tetrads, sarcina packets, or irregular clusters. **Divide by septation. Nonsporeforming.** Stain Gram-negative. **Nonmotile. Strictly aerobic.** Optimum temperature, 30–37°C. **Colonies are red or red-to-orange** due to carotenoids in the cells. Oxidase- and catalase-positive. **Chemoorganotrophic. Require at least 2.5 M (~15%, w/v) NaCl for growth and 3.5–4.5 M (~20–26%, w/v) NaCl for best growth. No lysis occurs in hypotonic solutions. Occur ubiquitously in nature in strongly saline lakes and ponds,** in solar salt, and in proteinaceous materials heavily salted with solar salt. The mol% G + C of the DNA is 61–66 (T_M).

Type species: *Halococcus morrhuae* (Klebahn 1919) Kocur and Hodgkiss 1973, 154.

Further Descriptive Information

Much less information is extant on the red, extremely halophilic cocci than on the red, extremely halophilic rods. Sufficient is known, however, to place them unequivocally among the archaeobacteria. This pertains to the nucleotide sequence of their 16 S rRNA, as well as to the chemical composition of their cell wall and their lipids, and their sensitivity towards antibiotics.

In contrast to the halobacteria, the halococci have a thick (30–60 nm), rigid cell wall and form septa, not constrictions, during division (Steensland and Larsen, 1971; Kocur et al., 1972). The wall consists of a polysaccharide material which does not contain muramic acid, and is therefore not the typical peptidoglycan of eubacterial cell walls (Brown and Cho, 1970; Kandler et al., 1970). However, the wall material contains, in addition to simple sugars, amino sugars, uronic acids and glycine, and forms a complex heteroglycan which is highly sulfated, in some cases sulfonated, and apparently responsible for the rigid structure of the wall and the resistance against lysis in hypotonic solutions (Reistad, 1975, 1978; Steber and Schleifer, 1975).

The main lipids of the halococci are ether-linked isoprenyl phosphoglycerides, and are remarkably similar to those of the halobacteria

(Kates, 1978; Hunter et al. 1981). This pertains also to the carotenoids which are of the C$_{50}$ bacterioruberin type. The cells contain menaquinones, not ubiquinones, and retinal, as also reported for the halobacteria (Kushwaha et al., 1974; Collins et al., 1981).

As in the case of the halobacteria, the proteins of the halococci are acidic (Reistad, 1970), their enzymes are activated by salt (Larsen, 1967), and structural proteinaceous units such as the cell membrane disintegrate in hypotonic solutions (Reistad, 1978).

The mol% G + C of the DNA was determined for nine strains of halococci and found to vary from 60.5–65.8 (Kocur and Boháček, 1972). One strain has been reported to contain satellite DNA (31% of total DNA); the mol% G + C of the major DNA component was 67, of the minor component 59 (Moore and McCarthy, 1969). Appreciable amounts of satellite DNA may be a characteristic of the halococci, as well as of the halobacteria.

A tryptone-yeast extract-salt medium similar to that described for the halobacteria (see the first footnote in the previous genus, *Halobacterium*) is a suitable growth medium also for the halococci. Good aeration must be provided. Growth is slow even under optimum conditions, and generation times of ~14 h are the best that have been reported. The strains so far examined have a complex nutrition. A number of amino acids, purines and pyrimidines are required for growth (Onishi et al., 1965), as reported also for some strains of halobacteria.

The slow growth on laboratory media may be a reason why the halococci seem to be less frequently encountered in natural samples than the halobacteria. They are, however, found in the same places as the halobacteria, in strongly saline natural lakes and marine salterns, and have been shown to cause spoilage of salted fish and hides.

The halococci show their relatedness to the archaeobacteria also by their insensitivity towards a number of antibiotics. They are, however, very sensitive to bacitracin, as are some of the halobacteria and other archaeobacteria (Hunter et al., 1980; Hilpert et al., 1982). The presence of bacitracin in cultures of halococci lead to an accumulation of uridine nucleotides (Reistad, 1978).

Enrichment and Isolation Procedures

The methods are as given for the genus *Halobacterium*.

Maintenance Procedures

See the chapter on the genus *Halobacterium*.

Differentiation of the genus **Halococcus** from other genera

The key to the family *Halobacteriaceae* indicates those characteristics useful for distinguishing *Halococcus* from *Halobacterium*.

Taxonomic Comments

The striking relatedness of the red, extremely halophilic cocci to the archaeobacteria on the one hand, and to the members of the genus *Halobacterium* on the other hand, warrant their assignment to the family *Halobacteriaceae*. Their morphology, cell wall ultrastructure, mode of division and resistance towards lysis warrant their assignment to a separate genus, *Halococcus*.

These organisms are commonly said to have been described first by Farlow (1880), who proposed the name *Sarcina morrhuae* for some colony-forming matter on salt fish, appearing microscopically as tetrad-forming cocci. A re-evaluation of the view that Farlow dealt with the red, extremely halophilic cocci is in order, however. Farlow explicitly referred to his coccoid material as "always colorless"; moreover, the cocci measured 5–8 μm in diameter, the tetrads were "surrounded by a thin hyaline envelope," and Farlow never had this cryptically described material in culture. Farlow's description makes it very difficult, if not impossible, to accept him as the discoverer of the red, extremely halophilic cocci. Farlow (1886) later changed the name to *Sarcina litoralis* Poulsen 1879, allegedly for reasons of priority, but Poulsen's description shows rather convincingly that the material he studied was also not the red, extremely halophilic cocci. The name *S. litoralis* was occasionally used in the later literature for these organisms, however.

After Farlow's work, several authors described the occurrence of red-colored cocci from salty material, mostly salted fish. These might well have been of the kind we include today in the genus *Halococcus*, but they could also have been other kinds of cocci or even involution forms of representatives of the genus *Halobacterium*.

The first to describe the red, extremely halophilic cocci in a satisfactory manner was Klebahn (1919). He isolated and studied several strains from salt fish and proposed two species, *Sarcina morrhuae* Farlow and *Micrococcus* (*Diplococcus*) *morrhuae* sp. nov.

Petter (1932) argued that the red, extremely halophilic cocci are variable in that they tend to form packets or pairs, and that they vary in size depending on the growth conditions. She therefore proposed that only one species should be retained, *S. morrhuae* Klebahn, giving recognition to Klebahn as being the proper mentor to these organisms. In 1957, however, in the seventh edition of *Bergey's Manual*, two species, *S. litoralis* Poulsen and *M. morrhuae* Klebahn were listed, and both were described as red, extremely halophilic cocci.

The proposal by Schoop (1935b) to accord the red, extremely halophilic cocci generic rank was implemented by Kocur and Hodgkiss (1973). They investigated 22 strains of these organisms and found reasons to propose only one species, *Halococcus morrhuae* (Farlow 1880) comb. nov. This was adopted in the eighth edition of *Bergey's Manual* (Gibbons, 1974), and the genus and species were included on the Approved Lists of Bacterial Names in 1980. However, in view of the above discussion, it seems right that Klebahn's name should replace Farlow's to give the more proper designation *Halococcus morrhuae* (Klebahn 1919) Kocur and Hodgkiss 1973, 154.

In a numerical taxonomic analysis of a number of strains of red, extremely halophilic cocci, Colwell et al. (1979) found that the strains were less uniform in their characteristics than the strains of *Halobacterium* studied, yet there was no particular reason to distinguish between the strains at a species level. On the basis of present knowledge, it therefore seems best to recognize only a single species in the genus *Halococcus*.

Further Reading

See references provided under Further Reading for the family *Halobacteriaceae*.

List of the species of the genus **Halococcus**

Halococcus morrhuae (Klebahn 1919*) Kocur and Hodgkiss 1973, 154.^{AL} (Not *Sarcina litoralis* Poulsen 1879, 254; not *Sarcina morrhuae* Farlow 1880, 974.) (*Sarcina morrhuae* Klebahn 1919, 38; *Micrococcus* (*Diplococcus*) *morrhuae* Klebahn 1919, 42; *Halococcus litoralis* Schoop 1935, 817.)

morr.hu'ae. M.L. n. *morrhua* from the specific epithet of the codfish, *Gadus morhua* L. (often misspelled *morrhua*); M.L. gen. n. *morrhuae* of the codfish.

The morphology is as described for the genus.

Strictly aerobic. Amino acids are needed for growth and are also used as a source of energy. Glucose is not used as a main source of carbon for growth, and acid and gas are not produced from glucose when tested for by the usual methods. Nitrate is reduced to nitrite without production of gas. Catalase-positive. Oxidase-positive. H_2S is usually produced from thiosulfate and frequently from cysteine. Urease-negative. Some strains produce indole and hydrolyze gelatin, starch and esters.

Optimum pH, 7.2. Some growth occurs at pH 5.5; no growth occurs at pH 8.

Other characteristics are as described for the genus.

The mol% G + C of the DNA ranges from 61–66 (T_m). Appreciable amounts of satellite DNA may occur.

Type strain: ATCC 17082 (Approved Lists, 1980).

FAMILY VI. ACETOBACTERACEAE GILLIS AND DE LEY 1980, 23^{VP}

JOZEF DE LEY, MONIQUE GILLIS AND JEAN SWINGS

A.ce.to.bac.te.ra'ce. ae. M.L. masc. n. *Acetobacter* type genus of the family; *-aceae* ending to denote a family; M.L. fem. pl. n. *Acetobacteraceae* the *Acetobacter* family.

Gram-negative or Gram-variable, ellipsoidal to rod-shaped cells, occurring singly, in pairs or in chains. Involution forms may occur and are spherical, elongated, swollen, club shaped, curved or filamentous. **Motile by peritrichous flagella or by 1–8 polar flagella, or nonmotile.** Endospores are not formed. **Aerobic,** having a strictly respiratory type of metabolism with oxygen as the terminal electron acceptor. Chemoorganotrophic. Catalase-positive (except for *A. pasteurianus* subsp. *paradoxus* and *A. peroxydans*). Oxidase-negative. Optimum temperature, 25–30°C. Optimum pH, 5–6. Gelatin is not liquefied. Nitrates are not reduced. Indole is not formed. **Ethanol is oxidized to acetic acid in neutral and acid (pH 4.5) media.** For most strains, the final pH in D-glucose or D-xylose media is below 4.5.

---

*Incorrectly cited as Farlow 1880 on the Approved Lists of Bacterial Names (see Taxonomic Comments).

Occur in sugary and alcoholized, slightly acid niches, e.g. flowers, fruits, beer, wine, cider, vinegar, souring fruit juices, bees and honey. The mol% G + C of the DNA is 51–65.

Type genus: *Acetobacter* Beijerinck 1898, 215.

Further Comments

In the eighth edition of the *Manual*, the genera *Gluconobacter* and *Acetobacter* were separated on the basis of differences in flagellar arrangement. *Gluconobacter* was included in the polarly flagellated family *Pseudomonadaceae*, whereas *Acetobacter* was placed as a genus of uncertain affiliation among the aerobic Gram-negative rods and cocci. However, DNA/rRNA hybridization studies (Gillis and De Ley, 1980) have shown that *Acetobacter* and *Gluconobacter* are closely related (they link at an average $T_{m(e)}$ of 76.8°C) and that they are more closely related to each other than to any other genus studied. Moreover, the two genera have many phenotypic similarities. Consequently, the authors have placed both genera together in the family *Acetobacteraceae*. This family constitutes a separate branch of the fourth rRNA superfamily (sensu De Ley, 1978) consisting of such organisms as *Agrobacterium*, *Rhizobium*, *Beijerinckia*, *Paracoccus*, *Zymomonas*, most *Rhodopseudomonas* and *Rhodospirillum* species, *Azospirillum*, and some *Aquaspirillum* species.

Key to the genera of the family **Acetobacteraceae**

I. Lactate or acetic acid oxidized to CO_2. Motile by peritrichous flagella or nonmotile.
Genus I. *Acetobacter*, p. 268

II. No oxidation of lactate or acetic acid to CO_2. Motile by polar flagella or nonmotile.
Genus II. *Gluconobacter*, p. 275

Genus I. **Acetobacter** *Beijerinck 1898, 215*[AL]

JOZEF DE LEY, JEAN SWINGS AND FRANCIS GOSSELÉ

A.ce.to.bac′ter. L. n. *acetum* vinegar; M.L. n. *bacter* masc. equivalent of Gr. neut. n. *bakterion* rod, staff; M.L. masc. n. *Acetobacter* vinegar rod.

Cells **ellipsoidal** to **rod shaped**, straight or slightly curved, 0.6–0.8 μm by 1.0–4.0 μm, occurring singly, in pairs or in chains. Involution forms frequent in some strains; may be spherical, elongated, swollen, club shaped, curved or filamentous. Motile or nonmotile; if motile, the **flagella are peritrichous or lateral.** Endospores are not formed. Gram-negative (in a few cases Gram-variable). **Obligately aerobic;** metabolism is respiratory, never fermentative. Pale colonies; most strains produce no pigments. A minority of strains produce brown water-soluble pigments or show pink colonies due to porphyrins. Catalase-positive, oxidase-negative. Absence of gelatin liquefaction, indole or H_2S formation. **Oxidize ethanol to acetic acid. Acetate and lactate are oxidized to CO_2 and H_2O.** The best carbon sources for growth are ethanol, glycerol and lactate. Acid is formed from *n*-propanol, *n*-butanol and D-glucose. Neither lactose nor starch is hydrolyzed. Chemoorganotrophs. Optimum temperature, 25–30°C. The pH optimum for growth is 5.4–6.3. *Acetobacter* occurs in flowers, fruits, honey bees, **sake, tequila, palm wine, grape wine, cider, beer, South African Bantu Beer, kefir,** brewer's yeast, **vinegar, beechwood shavings of vinegar generators, vinegar acetifiers,** sugar cane juice, "tea fungus," vegetable tanning liquors, "**nata,**" garden soil and canal water. *Acetobacter* causes **pink disease** in pineapple fruit and **rot** in apples and pears. The mol% G + C of the DNA ranges from 51–65% (T_m).

Type species: *Acetobacter aceti* (Pasteur 1864) Beijerinck 1898, 215.

Further Descriptive Information

Many strains are pleomorphic and contain spherical, swollen, club-shaped, curved, or filamentous cells. These cell forms ("Abweichungs-formen," "Aberrationsformen," "Involutionsformen") retained the attention of the first students of acetic acid bacteria, e.g. Hansen (1894), who published accurate drawings of them. In a culture of *A. pasteurianus* he found filaments up to 200 μm long and swollen cells with a diameter of 11 μm. The question whether these involution forms are part of a life cycle, as postulated by Janke (1916), has never been answered. Several authors (Hansen, 1911; Frateur, 1950) stressed that involution forms are not immediately related to the aging process as thought before. The appearance or disappearance of involution forms seems to be correlated with changes of pH, temperature or medium composition (Loitsianskaya and Lebentrau, 1964). In many strains, two or more cell forms occur. Sometimes curved crescent-shaped pairs of cells are observed (Shimwell, 1959). In some strains, capsules were demonstrated by Kulka and Walker (1946) and were also clearly visible on many of Passmore's (1973) micrographs. Until 1954 it was believed that all acetic acid bacteria possessed polar flagella and they were therefore classified in the family *Pseudomonadaceae*. The flagella staining for light microscopy is rather difficult (Leifson, 1954; Passmore, 1973) and therefore very frustrating. It seems that the lateral or peritrichous flagella of *Acetobacter* are best observed by transmission electron microscopy (Passmore, 1973). According to Leifson (1954), the number of flagella per *Acetobacter* cell tends to be few, the arrangement uneven and their wavelength about 2.9 μm. Shimwell (1958) found numerous peritrichous flagella with a short wavelength of 1.4 μm in one *A. aceti* strain. Short wavelengths had been thought to be confined to *Gluconobacter* (Leifson, 1954). Many published micrographs (Loitsianskaya et al., 1977; Passmore, 1973) show lateral flagella. Concerning the motility of *Acetobacter*, Frateur (1950) warned: "La motilité est un critère qu'on ne peut passer sous silence, mais qu'il faut cependant manier avec grande prudence vu sa grande instabilité." Loitsianskaya et al. (1977) were still more radical in stating: "The type of flagellation cannot be considered to be reliable in taxonomy of acetic acid bacteria. It is also of little use for the identification and differentiation of *Acetobacter* and *Gluconobacter*."

In a cellulose-forming *Acetobacter* strain approximately 50 individual synthetic sites were demonstrated, organized in a row along the longitudinal axis, each generating cellulose microfibrils. These sites were visualized as a linear array of pores with a diameter of 120–150 Å on the lipopolysaccharide membrane (Brown et al., 1976; Zaar, 1979).

Most *Acetobacter* species and subspecies grow moderately to abundantly on standard medium (GYC).* The pale colonies are circular, raised or convex and in most cases not more than 3 mm in diameter. Some *A. pasteurianus* strains, i.e. those formerly classified as *A. peroxydans*, *A. pasteurianus* subsp. *paradoxus* and subsp. *ascendens* do not develop on GYC and should be cultivated on a modified Carr medium.† On mannitol medium (MYP),§ these strains develop scantily. All *Acetobacter* strains grow on Frateur's (1950) ethanol medium;¶ the acetic acid formed dissolves the calcium carbonate in a zone up to 12

---

* Standard medium (GYC) contains (g/liter distilled water): yeast extract, 10.0; D-glucose, 50.0; $CaCO_3$, 30.0; and agar, 25.0.
† Modified Carr medium (1968) contains (g/liter of distilled water): yeast extract, 30.0; ethanol, 20.0; and agar, 20.0.
§ Mannitol medium (MYP) contains (g/liter of distilled water): yeast extract, 5.0; mannitol, 25.0; peptone, 3.0; and agar, 25.0.
¶ Ethanol medium (Frateur, 1950) contains (g/liter of distilled water): yeast extract, 10.0; $CaCO_3$, 20.0; ethanol, 20.0; and agar, 20.0; pH 6–7.

mm from the edge of the growth streak. $CaCO_3$ is redeposited in this cleared zone within 3–10 days by overoxidation (nacreous lustre, "iris-ation"). The morphological development of the cellulose-synthesizing strains (formerly classified as *A. aceti* subsp. *xylinum*, in the present contribution in *A. pasteurianus* and *A. hansenii*) on agar is rather unique among bacteria because of the external sheath of cellulose microfibrils surrounding the dividing cell mass. Initially a smooth spheroid is observed which becomes after 8 days a rough, crinkled flattened colony merging afterwards in a coherent sheet (Sowden and Colvin, 1978). The authors state: "This morphology is the result of a repeated extrusion of cells from the confining sheath, followed by the regeneration of a new portion of the sheath on the extruded cell mass." The actual significance of the sheath of cellulose material for the organism is unknown. The cellulose-synthesizing *Acetobacter* strains have long been thought to be the only bacteria to produce extracellular cellulose but Deinema and Zevenhuizen (1971) showed that certain *Pseudomonas* species also do so. Cellulose biosynthesis by acetobacters has been a matter for active research during the last decade. Precursors, intermediates and final polymerization remain controversial topics in this field. In the first stage, glucose is transformed intracellularly from a free molecule to a polyglucosan, which is transported outside the cell. In the second stage, the polymer is progressively associated and crystallized to the rigid hard cellulose microfibril (for a short review see Colvin, 1977). A novel C_{35} terpene, a hopane derivative (bacteriohopane), was identified from a cellulose-synthesizing acetobacter (Förster et al., 1973). It is thought to be involved in the alignment of the extracellular cellulose microfibrils. The molecular mechanism by which this proceeds remains unknown (Haigh et al., 1973).

The aerobic *Acetobacter* develops at the surface of liquid media as a ring, film or pellicle, showing in some cases a uniform turbidity of the medium and a cell deposit. A firm cellulose pellicle is formed by a few strains. Schramm and Hestrin (1954) and Leisinger et al. (1966) found that less cellulose was formed in shake cultures than in static liquid cultures. Mutations of cellulose-synthesizing to cellulose-free cells were observed by Leisinger et al. (1966). In static cultures cellulose-synthesizing cells are favored whereas shake cultures favor cellulose-free mutant cells. Janke (1916, 1960) and Kulka et al. (1949) stressed that the capacity of *Acetobacter* spp. to form giant colonies on wort agar* could be useful for strain differentiation.

The growth factor requirements depend on the carbon source supplied (Rao and Stokes, 1953). In a synthetic medium containing nine vitamins and D-mannitol as a carbon source (Gosselé et al., 1980) some strains require *p*-aminobenzoic acid, niacin, thiamin or pantothenic acid as growth factors. Ameyama and Kondô (1966) found that in the presence of D-glucose, *Acetobacter* strains had no growth factor requirements. In media without a carbon source for growth, e.g. yeast extract broth, peptone broth or nutrient broth, no growth occurs. The best carbon sources for growth of *Acetobacter* are, in descending order: ethanol, glycerol and Na-DL-lactate. Single amino acids cannot be used as a sole source of nitrogen and carbon. Several strains are able to grow in a chemically defined medium† without amino acids, using NH_4^+ ions as a sole *N* source and ethanol as a carbon-source (Frateur's modified Hoyer medium). Only a few *Acetobacter* strains are able to use single amino acids as a sole source of nitrogen.§ Inhibition effects on the growth of *A. aceti* caused by individual addition of threonine, glycine, valine, serine, homoserine, cysteine and alanine have been reported by O'Sullivan (1974 a, b). No "essential" amino acids are known for *Acetobacter*.

In media containing 1%, 2%, 5% or 10% ethanol, 87%, 82%, 58% and 13% of the strains grow, respectively (Gosselé, Swings, Kersters,

Pauwels and De Ley, unpublished results). Acid is formed by more than 85% of the *Acetobacter* strains from ethanol, *n*-propanol, *n*-butanol and D-glucose. Less than 10% of the strains form acid from *meso*-erythritol, D-mannitol, sorbitol, *meso*-inositol, D-arabinose, L-sorbose, L-rhamnose, D-fructose, sucrose, cellobiose, lactose, raffinose, dextrin or starch.

For the breakdown of sugars, *Acetobacter* is equipped with the hexose monophosphate pathway and the tricarboxylic acid cycle (King et al., 1956; Gromet et al., 1957; Stouthamer, 1959, 1960; De Ley, 1961; Cooksey and Rainbow, 1962; Williams and Rainbow, 1964; White and Wang, 1964 a, b; Asai, 1968). Glycolysis is either absent or very weak as phosphofructokinase was found to be absent (Stouthamer, 1960; De Ley, 1961; White and Wang, 1964a; Leisinger, 1965). The Entner-Doudoroff pathway seems to occur only in cellulose-synthesizing *Acetobacter* strains, formerly classified as *A. aceti* subsp. *xylinum* (Leisinger, 1965; Kersters and De Ley, 1968) where it appears to be more active than the hexose monophosphate cycle (White and Wang, 1964 a, b). Strains which are able to grow on Hoyer's medium with ethanol as sole source of carbon and $(NH_4)_2SO_4$ as only source of nitrogen, utilize the enzymes of the glyoxylate bypass (Leisinger, 1965; Stouthamer et al., 1963). *Acetobacter*, like *Gluconobacter*, is remarkable for its direct oxidative capacities on sugars, alcohols and steroids (for reviews see De Ley and Kersters, 1964 and Asai, 1968). According to the Bertrand-Hudson rule, polyols with a *cis*-arrangement of two secondary hydroxyl groups in D-configuration to the adjacent primary alcohol group, are oxidized to the corresponding ketoses by most *A. aceti*, *A. liquefaciens* and *A. hansenii* strains but rarely by *A. pasteurianus* (for reviews see De Ley and Kersters, 1964 and Asai, 1968).

$$\begin{array}{ccc} CH_2OH & & CH_2OH \\ | & & | \\ HO\!-\!C\!-\!H & \rightarrow & C\!=\!O \\ | & & | \\ HO\!-\!C\!-\!H & & HO\!-\!C\!-\!H \\ | & & | \end{array}$$

Yamada et al. (1969) have shown the presence of ubiquinone-10 in *Gluconobacter* and ubiquinone-8 and -9 in the genus *Acetobacter*. The only exceptions were the *A. liquefaciens* and cellulose-synthesizing acetobacters which were found to possess ubiquinone-10. The following cytochromes have been found in *Acetobacter* but were not always present together in one strain: a_1, a_2, a_4, b, c, c_1 and d (Smith, 1961; Nakayama and De Ley, 1965; Bächi and Ettlinger, 1974). Cytochrome a_1 seems to be typical for *Acetobacter*; it has not been detected in *Gluconobacter*.

The following tests are always negative for all strains of the genus: oxidase, H_2S, indole formation, gelatin liquefaction, and reaction in litmus milk. Strains formerly classified as *A. peroxydans* and *A. pasteurianus* subsp. *paradoxus* do not show catalase activity. In some cases, it may be difficult to detect catalase with cellulose-synthesizing acetobacters due to the presence of polysaccharide material. Nitrate reduction was detected with a limited number of strains.

Acetobacter does not constitute a serologically homogeneous group, but common antigens have been found between certain strains (McIntosh, 1962).

Acetobacter is not known to have any pathogenic effect towards man or animals. One *Acetobacter* strain (AY) was demonstrated to possess a lethal activity towards yeasts from various genera (Gilliland and Lacey, 1964, 1966).

Acetobacter causes bacterial rot of apples and pears which is accompanied by different shades of browning. All the apple and pear varieties

---

* Wort agar was prepared from wort (specific gravity, 1.036), clarified with egg albumin and solidified with 0.7% agar.
† Frateur's modified Hoyer ethanol-vitamins medium contains (per liter of distilled water): ethanol, 38 ml; $(NH_4)_2SO_4$, 1.0 g; K_2HPO_4, 0.1 g; KH_2PO_4, 0.9 g; $MgSO_4 \cdot 7H_2O$, 0.25 g; $FeCl_3$, 0.005 g; D-biotin, calcium pantothenate, thiamine, folic acid, *p*-aminobenzoic acid, and vitamin B_{12}, 0.001 g each; pyridoxal-HCl, niacin and riboflavin, 0.0015 g each.
§ The medium for testing the utilization of L-amino acids as sole nitrogen sources contains (per liter of 0.2 M Tris-maleate buffer, pH 5.4): D-mannitol, 30.0 g; L-amino acid, 1.0 g; salt solution A (see below), 5.0 ml; salt solution B (see below), 5.0 ml; and the same levels of vitamins as in Frateur's Hoyer ethanol-vitamins medium (preceding footnote). *Salt solution A* contains (per liter of distilled water): KH_2PO_4, 100 g and K_2HPO_4, 100 g. *Salt solution B* contains (per liter of distilled water): $MgSO_4 \cdot 7H_2O$, 40.0 g; NaCl, 2.0 g; $FeSO_4 \cdot 7H_2O$, 0.2 g; and $MnSO_4 \cdot H_2O$, 1.5 g.

tested were susceptible but pears were found to be the most susceptible (Van Keer et al., 1981). Acetobacters are also capable of inducing pink disease of pineapple fruit: the diseased tissue turns pink to brown after heating, due to the presence of 2,5-diketogluconic acid formed by *Acetobacter* (Rohrbach and Pfeiffer, 1976; Kontaxis and Hayward, 1978; Cho et al., 1980).

Acetobacter clearly prefers alcohol-enriched niches in contrast to the sugar-loving *Gluconobacter*. From Pasteur's time on, its occurrence in wines, beer and vinegar has been extensively reported by many authors (see Swings and De Ley, 1981). The effects caused by *Acetobacter* in beer are: acetification, ropiness, turbidity, off-flavors, discoloration (Vaughn, 1942; Rainbow, 1971). Ploss et al. (1979) found *A. pasteurianus* in 70% of the brewery samples infected with acetic acid bacteria. The storage vessels and the filtration, filling and pressure equilibration devices were the items most frequently infected. Acetobacters present on grapes are responsible for rapid acetification of musts during the initial stages of fermentation (Vaughn, 1942; Peynaud and Domercq, 1961; Blackwood et al., 1969). Dupuy (1957) found *A. pasteurianus* and *A. hansenii* strains to be present in wines from southern France. The species *A. aceti*, *A. hansenii* and *A. pasteurianus*, but not *A. liquefaciens* have been isolated from vinegar by various authors (Frateur and Simonart, 1952; Turtura et al., 1973; Passmore and Carr, 1975). Whereas *Gluconobacter* is found on apples and at the beginning of cider manufacture, *A. aceti* and *A. pasteurianus* can be isolated in every following stage (Passmore and Carr, 1975).

Many pioneers of microbiology worked with acetic acid bacteria and published valuable studies on them: Pasteur, Persoon, Neuberg, Lafar, Janke, Henneberg, Hansen, Hermann, Hoyer, Frateur, Bernhauer, Beijerinck, Brown, Bertrand, Asai . . . At present, we can say that the genera *Acetobacter* and *Gluconobacter* are two well studied bacterial genera.

Enrichment and Isolation Procedures

In the past, many enrichment and isolation media have been described which mostly contain beer or yeast water (see Swings and De Ley, 1981). These ingredients are difficult to standardize. A simple and appropriate isolation medium for *Acetobacter* contains 0.5% Bacto yeast extract (Difco), 1.5% ethanol and 2.5% agar (Suomalainen et al., 1965).

Maintenance Procedures

Our working collection of *Acetobacter* strains is kept at 4°C on the standard (GYC) medium and is transferred monthly. Most strains remain viable for 3 months when kept on this medium in screw-capped vials. Some *A. pasteurianus* strains formerly classified as *A. peroxydans*, *A. pasteurianus* subsp. *paradoxus* and subsp. *ascendens*, which are unable to grow on GYC, are grown either on MYP or on Carr's modified ethanol medium (1968). The latter is not suitable for maintenance as the cells die rapidly. *Acetobacter* strains survive lyophilization for over 10 years. Another reliable method is preservation under liquid nitrogen.

Differentiation of the genus **Acetobacter** from other genera

The differentiation of *Acetobacter* from *Gluconobacter*, *Zymomonas*, *Vibrio*, *Aeromonas* and *Frateuria* is given in tables in the chapters on the genus *Zymomonas* and the genus *Frateuria*.

Taxonomic Comments

Acetobacter is a well defined genus. In the eighth edition of *Bergey's Manual* it contained three species: *A. aceti*, containing four subspecies (*aceti*, *orleanensis*, *xylinum* and *liquefaciens*); *A. pasteurianus*, containing five subspecies (*pasteurianus*, *lovaniensis*, *estunensis*, *ascendens*, and *paradoxus*); and *A. peroxydans*. The names of these species and subspecies were included in the Approved Lists of Bacterial Names in 1980. This system is not maintained in the present edition of *Bergey's Manual*.

In a numerical analysis of 177 phenotypic features, the present authors examined 98 *Acetobacter*, 98 *Gluconobacter* and 7 *Frateuria* strains. Four phenons were delineated: phenon 1 consisted of all *Frateuria* strains, phenon 2 consisted of all *A. aceti* subsp. *liquefaciens* strains, phenon 3 was *Gluconobacter*, and phenon 4 was *Acetobacter* minus *Acetobacter aceti* subsp. *liquefaciens*. Within phenon 4, three subphenons could be distinguished, two of which showed a fair internal homogeneity: *A. aceti* (Pasteur 1864) Beijerinck 1898, 215, emend. and *A. hansenii* sp. nov. The third subphenon, *A. pasteurianus* (Hansen 1879) Beijerinck 1916, 1199, emend., was heterogeneous. Its heterogeneity was reflected by its broad mol% G + C range from 52.8–62.5, and by protein electrophoregrams.

Of the four *Acetobacter* species, *A. liquefaciens* (phenon 2) most resembles *Gluconobacter* (Frateur, 1950; De Ley, 1961); it was formerly classified in *Gluconobacter* (Asai, 1935; Asai and Shoda, 1958; De Ley, 1961; Asai et al., 1964; Ameyama and Kondô, 1966; Asai, 1968). The presence of ubiquinone-10 and the numerical analysis illustrate this relatedness. Using DNA/rRNA hybridizations, Gillis and De Ley (1980) demonstrated that *A. liquefaciens* is a true *Acetobacter*.

Acknowledgment

J.D.L. is indebted to the Fonds voor Kollektief Fundamenteel Onderzoek (F.K.F.O.) for research and personnel grants; J.S. to the Nationaal Fonds voor Wetenschappelijk Onderzoek (N.F.W.O.); and F.G. to the Instituut tot Aanmoediging van het Wetenschappelijk Onderzoek in Nijverheid en Landbouw (I.W.O.N.L.) for research grants.

Further Reading

Asai, T. 1968. *Acetic acid bacteria. Classification and biochemical activities.* University of Tokyo Press, Tokyo, and University Park Press, Baltimore.
Swings, J. and J. De Ley. 1981. The genera *Gluconobacter* and *Acetobacter*. In Starr, Stolp, Trüper, Balows and Schlegel (Editors), The Prokaryotes: A Handbook on Habitats, Isolation, and Identification of Bacteria Springer-Verlag, Berlin, pp. 771–778.

Differentiation of the species of **Acetobacter**

The differentiation of the species is given in Table 4.62. Additional features are found in Table 4.63.

List of the species of the genus **Acetobacter**

1. **Acetobacter aceti** (Pasteur 1864) Beijerinck 1898, 215,[AL] emend. (*Mycoderma aceti* Pasteur 1864, 125.)

a.ce′ti. L. n. *acetum* vinegar, L. gen. n. *aceti* of vinegar.

Colonial and cell morphology are as described for the genus.

All strains are ketogenic towards glycerol and sorbitol and most strains also towards D-mannitol. Both 2-keto- and 5-ketogluconic acids

are synthesized from D-glucose. All strains acidify *n*-propanol, *n*-butanol, D-xylose, D-mannose and D-glucose. Ethanol, glycerol, D-mannitol, Na-acetate and Na-D,L-lactate are good carbon sources for growth. The majority of the strains utilize L-alanine and L-proline as a source of nitrogen in the presence of D-mannitol. With ethanol as a carbon source in the presence of growth factors, most *A. aceti* strains

Table 4.62.
Characteristics differentiating the species of genus **Acetobacter**[a]

Characteristics	1. *A. aceti*	2. *A. liquefaciens*	3. *A. pasteurianus*	4. *A. hansenii*
Formation of:				
Water-soluble brown pigments on GYC[b]	−	+	−	−
γ-Pyrones from D-glucose	−	d	−	−
γ-Pyrones from D-fructose	−	+	−	−
5-Ketogluconic acid from D-glucose	+	d	−	d
2,5-Diketogluconic acid from D-glucose	−	+	−	−
Ketogenesis from glycerol	+	+	−	+
Growth on carbon sources:				
Ethanol	+	+	d	−
Dulcitol	−	−	−	d
Na-Acetate	+	d	d	−
Growth on L-amino acids in the presence of D-mannitol as carbon source:				
L-Glycine, L-threonine, L-tryptophan	−	d	−	−
L-Asparagine, L-glutamine	d	+	−	+
Growth in the presence of 10% ethanol	−	−	d	−
Mol% G + C (T_m)[c]	55.9–59.5	62.3–64.6	52.8–62.5	58.1–62.6

[a] Symbols: see standard definitions.
[b] GYC, 5% D-glucose + 1% yeast extract + 3% $CaCO_3$ + 2.5% agar.
[c] Gillis and De Ley (1980).

assimilate ammonium. All strains require growth factors in the presence of D-mannitol. Other characteristics are listed in Table 4.63.

The mol% G + C of the DNA is 55.9–59.5 (T_m).

Type strain: NCIB 8621

2. **Acetobacter liquefaciens** sp. nov. (*Gluconoacetobacter liquifaciens* (sic) Asai 1935, 610; *Gluconobacter liquifaciens* (sic) Asai 1935, 679; *Acetobacter aceti* subsp. *liquefaciens* De Ley and Frateur 1974, 277.)

li.que.fa'ci.ens. L. v. *liquefacio* to liquefy; L. part. adj. *liquefaciens* liquefying.

Differentiation from the other species is indicated in Table 4.62.

Peritrichous flagellation; may show a mixed type of lateral and polar flagellation (Asai et al. 1964). Colonial and cell morphology are as described for the genus.

2-Keto-, 2,5-diketo- and sometimes also 5-ketogluconic acid are synthesized from D-glucose. The majority of the strains produce γ-pyrones from D-glucose and D-fructose; they are ketogenic towards glycerol, D-mannitol and sorbitol. Acid is produced from *n*-propanol, *n*-butanol, D-mannose and D-glucose. Most *A. liquefaciens* strains are able to grow on *n*-propanol, ethanediol, glycerol, *meso*-erythritol, D-mannitol, D-galactose, D-fructose, D-glucose, Ca-D-gluconate, Ca-D,L-glycerate and Na-D,L-lactate. In the presence of D-mannitol as a carbon source, each of the following amino acids can serve as a sole source of nitrogen for most strains: L-alanine, L-asparagine, L-glutamine, L-glutamic acid, L-glycine, L-lysine and L-proline. All strains are able to utilize ammonium as a sole source of nitrogen with ethanol as a carbon source, even without growth factors. Few strains do not require growth factors in the presence of D-mannitol. Other characteristics are listed in Table 4.63.

The mol% G + C of the DNA is 62.3–64.6 (T_m).

Type Strain: IAM 1834 (IFO 12388)

3. **Acetobacter pasteurianus** (Hansen 1879) Beijerinck 1916, 1199[AL] emend. (*Mycoderma pasteurianum* Hansen 1879, 230; *Acetobacter aceti* subsp. *xylinum* De Ley and Frateur 1974, 277; *A. aceti* subsp. *orleanensis* De Ley and Frateur 1974, 277; *A. pasteurianus* subsp. *estunensis* De Ley and Frateur 1974, 278; *A. pasteurianus* subsp. *ascendens* De Ley and Frateur 1974, 278; *A. pasteurianus* subsp. *paradoxus* De Ley and Frateur 1974, 278; *A. pasteurianus* subsp. *lovaniensis* De Ley and Frateur 1974, 278; *A. peroxydans* Visser 't Hooft 1925.)

pas.teur.i.a'nus. M.L. adj. *pasteurianus* of Pasteur; named after Louis Pasteur, French chemist and bacteriologist.

The colonial and cell morphology are as described for the genus.

Not all strains are able to grow on GYC. Ketogenesis from glycerol and formation of 5-ketogluconic acid from D-glucose occur seldom. Some strains lack catalase. Most strains produce acid from *n*-propanol and *n*-butanol, but not always from D-glucose. Ammonium is seldom utilized as sole source of nitrogen in the presence of ethanol or D-glucose. Only a few strains grow on L-alanine, L-glutamine, L-glutamic acid and L-proline as sole source of nitrogen in the presence of D-mannitol as a carbon source. All strains require growth factors in the presence of D-mannitol. Other characteristics are listed in Table 4.63.

The mol% G + C of the DNA is 52.8–62.5 (T_m).

Type strain: LMD 22.1t₁

4. **Acetobacter hansenii** sp. nov.

han.se'ni.i. M.L. n. *hansenii* of E. C. Hansen, Danish microbiologist, reknowned for his studies on the acetic acid bacteria.

Colonial and cell morphology are as described for the genus.

Ketogenic towards glycerol, D-mannitol and sorbitol; most strains produce 2-ketogluconic acid from D-glucose, sometimes also 5-ketogluconic acid. Acid from *n*-propanol, *n*-butanol, D-mannose and D-glucose. No growth on ethanol or Na-acetate as carbon sources. Ammonium, L-alanine, L-asparagine, L-glutamine and L-glutamic acid are assimilated as a sole source of nitrogen in the presence of D-mannitol as a carbon source. Some strains do not require growth factors. Other characteristics are listed in Table 4.63.

The mol% G + C of the DNA is 58.1–62.6 (T_m).

Type strain: NCIB 8746.

Table 4.63.

Other characteristics of the species of the genus **Acetobacter**[a]

Characteristics	1. *A. aceti* (7 strains)	2. *A. liquefaciens* (10 strains)	3. *A. pasteurianus* (66 strains)	4. *A. hansenii* (12 strains)
Colony morphology after 10 days on GYC medium:[b]				
Moderate to abundant growth	57	100	83	83
Translucent colony	0	10	2	0
Pale color	100	40	98	100
Pink color	0	0	2	25
Water-soluble brown pigment	0	90	0	0
Diameter <3 mm	71	70	90	75
Flat profile	14	10	21	8
Regular edge	43	20	57	25
Cell morphology of 2- to 3-day-old cultures:				
Gram-negative cells	100	100	79	100
Gram-variable cells	0	0	21	0
Rods	86	100	64	92
Ovoids-coccoids	71	30	73	33
Tapered cells	71	10	24	33
Filaments	0	70	32	50
Chains	0	20	34	33
Curved cells	57	40	39	75
Cells in pairs	57	90	83	92
Enlarged irregular involution forms	14	10	21	25
Diameter ≥0.7 μm	57	40	53	75
Motile cells	14	30	17	0
Biochemical reactions:				
Catalase	100	100	94	100
Nitrate reduction	0	0	9	8
Ferric chloride reaction on:				
D-Glucose	0	80	0	0
D-Fructose	0	90	2	0
D-Galactose	0	30	0	0
Ketogenesis from:				
Glycerol	100	100	9	92
D-Mannitol	86	100	47	92
Sorbitol	100	90	36	92
Oxidation of Ca-D,L-lactate	100	100	91	100
Formation of acetylmethylcarbinol from Ca-D,L-lactate	14	20	56	58
Formation of 2-ketogluconic acid from D-glucose	100	90	65	92
Formation of 5-ketogluconic acid from D-glucose	100	40	9	66
Formation of 2,5-diketogluconic acid from D-glucose	0	90	0	0
Final pH lower than 4.5 on:				
D-Xylose	100	80	36	66
D-Glucose	100	100	82	100
D-Galactose	14	0	8	8
Final pH lower than 5.9 on:				
Ethanol	100	100	91	83
n-Propanol	100	100	95	92
n-Butanol	100	100	89	92
n-Amyl alcohol	0	30	48	25
Glycerol	0	50	2	42
meso-Erythritol	0	60	2	17
D-Mannitol	0	0	2	0
Sorbitol	0	0	0	8
Meso-Inostiol	0	0	2	8
D-Arabinose	14	10	0	8
D-Ribose	57	80	9	50
D-Fructose	0	0	0	8
D-Mannose	100	100	59	92
Cellobiose	0	0	0	0
Lactose	0	0	0	0

Table 4.63.—*continued*

Characteristics	1. *A. aceti* (7 strains)	2. *A. liquefaciens* (10 strains)	3. *A. pasteurianus* (66 strains)	4. *A. hansenii* (12 strains)
Sucrose	0	10	2	33
Maltose	0	0	9	58
Dextrin	0	0	0	0
Starch	0	0	0	0
Growth on carbon sources:				
Methanol	0	0	11	0
Ethanol	100	100	79	0
n-Propanol	71	100	38	0
Ethanediol	14	80	3	0
meso-Erythritol	29	100	3	0
meso-Ribitol	14	60	2	0
L-Arabitol	14	30	0	0
meso-Xylitol	0	20	0	0
D-Mannitol	100	90	15	50
Sorbitol	43	60	9	0
meso-Inositol	14	20	8	17
Dulcitol	0	0	0	25
D-Ribose	14	60	9	0
D-Xylose	29	40	30	0
D-Lyxose	0	10	0	0
D-Fucose	0.	10	15	0
L-Arabinose	29	40	8	0
D-Galactose	14	100	23	0
D-Fructose	57	100	30	58
D-Glucose	57	100	30	50
Ca-D,L-Gluconate	14	100	17	58
D-Mannose	14	30	6	0
L-Sorbose	0	0	2	0
Cellobiose	14	20	0	0
Sucrose	14	70	0	17
Maltose	29	50	18	17
Raffinose	0	10	0	0
Dextrin	14	20	0	0
Na-Acetate	100	80	58	0
Ca-D,L-Glycerate	43	100	35	17
Na-D,L-Lactate	100	100	62	25
Na-Malonate	0	0	2	0
Na-L-Malate	0	40	33	8
Na-Citrate	0	20	11	0
Growth on single L-amino acids as sole source of nitrogen:				
L-Alanine	86	80	24	100
L-Arginine	0	60	0	17
L-Asparagine	14	90	5	100
L-Aspartic acid	29	70	2	75
L-Cysteine	0	70	2	17
L-Glutamine	29	90	9	100
L-Glutamic acid	57	90	15	100
L-Glycine	0	80	0	8
L-Histidine	0	30	0	8
L-Isoleucine	0	50	0	8
L-Leucine	0	40	0	17
L-Lysine	0	100	3	25
L-Methionine	0	60	0	8
L-Phenylalanine	0	50	0	17
L-Proline	86	100	27	42
L-Serine	0	0	0	8
L-Threonine	0	50	0	0
L-Tryptophan	0	70	0	0
L-Tyrosine	0	30	2	25
L-Valine	0	0	0	8
Growth on single L-amino acids as sole source of both carbon and nitrogen:				
L-Alanine	0	0	6	0

Table 4.63.—*continued*

Characteristics	1. *A. aceti* (7 strains)	2. *A. liquefaciens* (10 strains)	3. *A. pasteurianus* (66 strains)	4. *A. hansenii* (12 strains)
L-Glutamic acid	29	0	0	0
Growth on				
Frateur's modified Hoyer medium:				
+ Ethanol	71	100	12	0
+ Ethanol + growth factors	86	100	12	0
+ D-Glucose + growth factors	14	60	18	67
+ D-Mannitol + growth factors	43	80	32	100
SM medium[c]				
+ 0.5% NaCl	57	80	29	33
+ 1% Nacl	29	50	15	17
+ 2% NaCl	0	20	0	0
+ 1% Ethanol	100	90	83	92
+ 2% Ethanol	86	90	79	83
+ 5% Ethanol	71	20	59	75
+ 10% Ethanol	0	0	20	0
at 28°C	100	100	92	100
at 34°C	14	60	45	58
at 37°C	0	20	15	8
at pH 3.6	43	90	17	25
0.5% Yeast extract:				
+ 20% D-Glucose	14	10	27	33
+ 25% D-Glucose	14	0	17	8
+ 30% D-Glucose	0	0	6	8
Nutrient broth	0	0	3	0
Requirement for growth factors in the presence of D-mannitol as a carbon source:				
p-Aminobenzoic acid	14	11	2	25
Thiamine	14	11	2	17
Nicotinic acid	28	11	5	8
Pantothenic acid	14	11	2	17
Other growth factors from yeast	57	33	94	0
None	14	33	0	42
Growth in the presence of metals and dyes:				
0.001% Malachite green	29	100	41	17
0.0001% Crystal violet	71	100	79	83
0.001% Brilliant green	0	90	20	8
0.01% Cd(CH$_3$COO)$_2$·2H$_2$O	14	20	11	8
0.01% CoSO$_4$·7H$_2$O	86	90	12	0
0.001% HgCl$_2$	100	90	41	50
0.001% Tl-OOCCH$_3$	100	50	55	50
Resistance to antibiotics:[d]				
30 µg Kanamycin	86	10	33	33
10 µg Streptomycin	86	40	53	25
30 µg Tetracycline	57	30	38	42
30 µg Novobiocin	14	10	23	33
10 µg Ampicillin	100	100	74	42
30 µg Nalidixic acid	100	50	77	83
10 µg Gentamicin	100	100	97	100
30 µg Chloramphenicol	100	100	86	75
25 µg Cephaloridin	86	100	97	92
10 U Bacitracin	100	100	98	100
10 U Penicillin G	100	100	98	83
10 µg Erythromycin	100	100	92	92
10 µg Fusidic acid	100	100	98	100

[a] The results are expressed as the percentage of positive strains.
[b] For the composition of the GYC medium, see Table 4.62.
[c] SM contains 0.5% yeast extract (Oxoid) and 5% glucose.
[d] A strain was considered to be susceptible towards a given antibiotic when the inhibition area around the disk exceeded 12 mm after 2 days.

Genus II. Gluconobacter Asai 1935, 689, emend. mut. char. Asai, Iizuka and Komagata 1964, 100^{AL}

JOZEF DE LEY AND JEAN SWINGS

Glu.co.no.bac′ter. M.L. n. *acidum gluconicum* gluconic acid; M.L. n. *bacter* masc. equivalent of Gr. neut. n. *bacterion* rod or staff; M.L. masc. n. *Gluconobacter* gluconate rod.

Cells ellipsoidal to rod shaped, 0.5–0.8 × 0.9–4.2 μm, occurring singly and/or in pairs, rarely in chains. Enlarged, irregular cell forms (involution forms) may occur. Endospores are not formed. Gram-negative (in a few cases Gram-variable). **Motile or nonmotile; if motile, the cells have 3–8 polar flagella,** rarely a single flagellum. Obligately aerobic, having a strictly respiratory type of metabolism with oxygen as the terminal electron acceptor. Colonies are pale. Optimum temperature, 25–30°C; no growth at 37°C. Optimum pH, 5.5–6.0; most strains will grow at pH 3.6. Strongly catalase-positive. Oxidase-negative. Negative for nitrate reduction, gelatin liquefaction, indole production and H_2S formation. Chemoorganotrophic. **Oxidize ethanol to acetic acid.* Do not oxidize acetate or lactate to CO_2 and H_2O. Strong ketogenesis occurs from polyalcohols.** Acid formation from D-glucose and D-xylose is pronounced (pH $\leq$ 4.5). All strains produce **2-ketogluconic acid** from D-glucose, and the majority of strains also form 5-ketogluconic acid. No acid production or growth occurs on lactose or starch. *Gluconobacter* strains occur in **flowers,** garden soil, baker's yeast, **honey bees, fruits, cider,** beer, **wine, wine vinegar, South African Bantu beer, palmsap** and soft drinks. They cause **pink disease** in pineapples and **rot** in apples and pears. The mol% G + C of the DNA is 56–64 (T_m).

Type species: *Gluconobacter oxydans* (Henneberg 1897) De Ley 1961, 47.

Further Descriptive Information

Some strains may show irregular cells, swollen filaments, large sacs, or other aberrant cell forms. These involution forms are not found only in old cultures as thought before. Long filaments can start swelling at one point to produce a sac; the remaining part of the filament can segment, producing abnormally thickened cells. In Gram-stained preparations thin, cigar-shaped, poorly staining cells may occur. This phenomenon led Passmore (1973) to suggest a life cycle in some strains. Demonstration of the flagella by classical staining procedures for light microscopy is difficult, and transmission electron microscopy using a negative-staining technique seems preferable (Passmore, 1973). Early stationary-phase cells of *G. oxydans* show polar complexes of intracytoplasmic membranes accompanied by high concentrations of ribosomes (Fig. 4.42). The membrane-like material is either loosely coiled, organized in a lamellar arrangement, or as amorphous masses contained within a trilaminar membrane. It is probably formed by invagination of the plasma membrane at the end of active cell division (Fig. 4.43). Cells containing these intracytoplasmic membranes have demonstrated a 2-fold increase in their rate of glycerol oxidation and a 60% increase in free lipid compared to cells without them (Batzing and Claus, 1973; Claus et al., 1975; Heefner and Claus, 1976). During membrane formation, however, no major shift in extractable cellular phospholipids occurs, phosphatidylcholine being the most important fraction (30%) (Heefner and Claus, 1978; Tahara et al., 1976c). Hexadecanoic and octadecenoic acids account for more than 75% of the total fatty acids in both types of cells. Tahara et al. (1976a, b) isolated three ornithine-containing lipids from *G. oxydans* subsp. *suboxydans* IFO strain 3267,

one of which was a 2-hydroxy-fatty acid ester of N-α-3-hydroxypalmitoylornithinyltaurine.

Growth of *Gluconobacter* strains on standard medium (GYC†) is moderate to abundant. The colonies of the majority of strains are circular, raised or convex, regularly edged, milky white to yellowish, sometimes becoming brownish in the center, and not more than 3 mm in diameter. Some of the strains (25%) produce a pink, nondiffusible pigment (cytochromes), while 10% produce a soluble dark brown pigment. Crystals of 5-ketogluconate may be observed in this medium after 2–4 weeks. After 10 days on standard medium† or on mannitol medium,§ a large number of very small drops (microcolonies?) eventually appear on and at the edges of the colonies. Moderate to good growth occurs on Frateur's ethanol medium*; the acetic acid formed dissolves the calcium carbonate, which is not redeposited afterwards. In liquid media containing glucose, beer or wort, *Gluconobacter* grows in a pellicle or film at the surface accompanied eventually, but not always, by a uniform turbidity. Some strains produce viscous growth in beer, due to formation of dextrans or levans. Growth factor requirements depend on the carbon source supplied (Rao and Stokes, 1953). All *Gluconobacter* strains require growth factors in the presence of D-mannitol as the carbon source. Pantothenic acid, niacin, thiamine, and *p*-aminobenzoic acid are required by 96%, 40%, 8% and 4% of the strains, respectively (Gosselé et al., 1980). In media without a carbon source for growth (e.g. yeast extract broth, peptone broth or nutrient broth), no growth occurs. The best carbon sources for growth are (in descending order): D-mannitol, sorbitol, glycerol, D-fructose and D-glucose. A triphasic growth response is observed when *Gluconobacter* strains are grown in a glucose-containing defined medium at a constant pH (5.5) in batch culture (Olijve and Kok, 1979a, b). Single amino acids cannot be used as both a nitrogen and carbon source.

Gluconobacter strains are able to grow in a chemically defined medium¶ without amino acids, using ammonium ions as a sole nitrogen source. Single amino acids can be used as a sole nitrogen source ‖ by many *Gluconobacter* strains, and L-asparagine, L-aspartic acid, L-glutamine, L-glutamic acid and L-proline support growth of at least 80% of the strains. Valine and threonine may be inhibitory for some strains (Shamberger, 1960; Kerwar et al., 1964; O'Sullivan, 1974). No "essential" amino acids are known for *Gluconobacter* (Shamberger, 1960; Belly and Claus, 1972). The majority of strains are able to grow at pH 3.6. In media containing 1, 2 or 5% ethanol, 90%, 88% and 42% of strains exhibit growth, respectively (Gosselé, Swings, Kersters and De Ley, unpublished results). Acid is formed by more than 85% of *Gluconobacter* strains from *n*-propanol, *n*-butanol, glycerol, *m*-erythritol, D-mannitol, D-arabinose, D-ribose, D-fructose, D-galactose, D-mannose and maltose. No acid is formed from L-rhamnose, D-lactose, dextrin or starch.

Complete glycolytic and tricarboxylic acid cycle enzyme sequences are not present in *Gluconobacter*, and the hexose monophosphate pathway constitutes the most important route for the phosphorylative breakdown of sugars and polyols to CO_2 (Hauge et al., 1955b; Kitos, 1956; Kitos et al., 1958; Stouthamer, 1959, 1960; De Ley, 1961; Williams and Rainbow, 1964; Greenfield and Claus, 1972). The enzymes of the Entner-Doudoroff pathway have been demonstrated (Kovachevich and

---

* On Frateur's ethanol medium (see footnote ¶, on page 268, the genus *Acetobacter*) or on Carr's (1968) medium.

† See footnote *, p. 268, the genus *Acetobacter*.

§ See footnote §, p. 268, the genus *Acetobacter*.

¶ Frateur's Hoyer mannitol-vitamins medium contains (g/liter of distilled water): D-mannitol, 30.0; $(NH_4)_2SO_4$, 1.0; K_2HPO_4, 0.1; KH_2PO_4, 0.9; $MgSO_4 \cdot 7H_2O$, 0.25; $FeCl_3$, 0.005; D-biotin, calcium D-pantothenate, thiamine, folic acid, *p*-aminobenzoic acid, and vitamin B_{12}, 0.001 each; pyridoxal-HCl, niacin and riboflavin, 0.0015 each.

‖ See footnote §, p. 269, the genus *Acetobacter*.

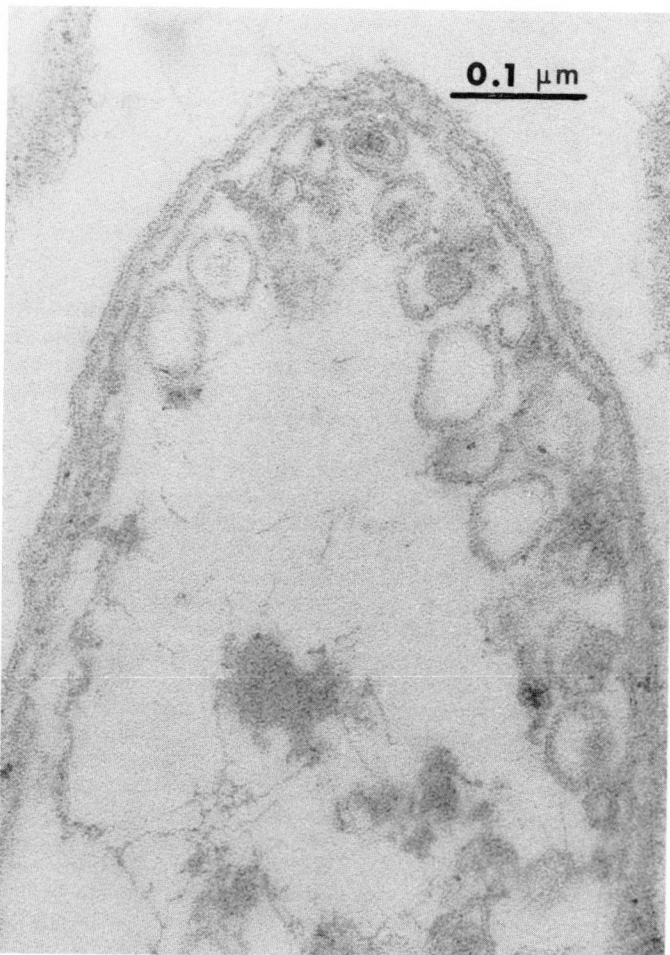

Figure 4.42. Longitudinal section of cell envelopes prepared from cultures of *Gluconobacter oxydans* ATCC 621, 6 h after obtaining maximum growth. Intracytoplasmic membranes are seen as polar accumulations of vesicles attached to the plasma membrane. (Reproduced with permission from G. W. Claus, B. L. Batzing, C. A. Baker and E. M. Goebel, Journal of Bacteriology *123:* 1169–1183, 1975, © American Society for Microbiology.)

Wood, 1955; Leisinger, 1965; Kersters and De Ley, 1968). Direct incomplete oxidations of sugars, aliphatic and cyclic alcohols and steroids proceed through one or two discrete steps and lead to nearly quantitative yields of the oxidation products. According to the Bertrand-Hudson rule, polyols, with a *cis-* arrangement of two secondary hydroxyl groups in D-configuration to the adjacent primary alcohol group, are oxidized to the corresponding ketoses by the majority of *Gluconobacter* strains. Several soluble and particulate polyol dehydrogenases have been described (Kersters and De Ley, 1963; for reviews see De Ley and Kersters, 1964 and Asai, 1968).

The following biochemical tests are uniformly negative for all strains of *Gluconobacter*: oxidase, reduction of nitrates, H_2S formation, indole production, oxidation of lactate, oxidation of acetate to CO_2 and H_2O, and gelatin liquefaction.

A double-stranded DNA bacteriophage (A-1) for *G. oxydans* has been isolated and described by Schocher et al. (1979).

Gluconobacter strains have been shown to share a common antigen (McIntosh, 1962.).

Gluconobacter strains are not known to have any pathogenic effect towards man or animals. They are capable of causing a bacterial rot of apples and pears which is accompanied by various shades of browning. The bacteria enter the apples through wounds in the cuticula and apple

tissue. All of the apple and pear varieties that have been tested are susceptible, but pears are the most susceptible (Van Keer et al., 1981). Strains of *G. oxydans* are also the causative agent of the pink disease of pineapple fruit; the diseased fruit turns pink or pink-brown to deep brown after heating (e.g. during canning), although the fruit possesses a normal appearance prior to heating (Kontaxis and Hayward, 1978); Rohrbach and Pfeiffer, 1975; Cho et al., 1980).

Gluconobacter strains flourish on sugars and are found in sugary niches such as flowers and fruits (e.g. ripe grapes (Blackwood et al., 1969; Passmore and Carr, 1975; Ameyama, 1975), apples, dates (Passmore and Carr, 1975), cherries, apricots, bananas, beets, guava, loquats, mandarins, mangoes, oranges, plums, ponkans', persimmons, sampô-kans, strawberries, tomatoes, and berries (*Zizyphus rotundifolia*) (Bhat and Rijshsinghani, 1955; Asai, 1968; Kahlon and Vygas, 1972). Some *Gluconobacter* strains cause off flavors in soft drinks (Sand, 1971, 1976), while others occur in high numbers in orange juice but leave the aroma and flavor unchanged (Franke, 1968). *Gluconobacter* strains are also found associated with palm trees, the sap of which is tapped in the tropics for the preparation of palm wine: they occur in the sap and also on the floret and in the tap hole (Faparusi, 1973, 1974). One would expect gluconobacters to be the true "working" strains in vinegar manufacture, since they do not overoxidize acetic acid to CO_2 and H_2O; however, the situation is more complex because the flora of vinegar

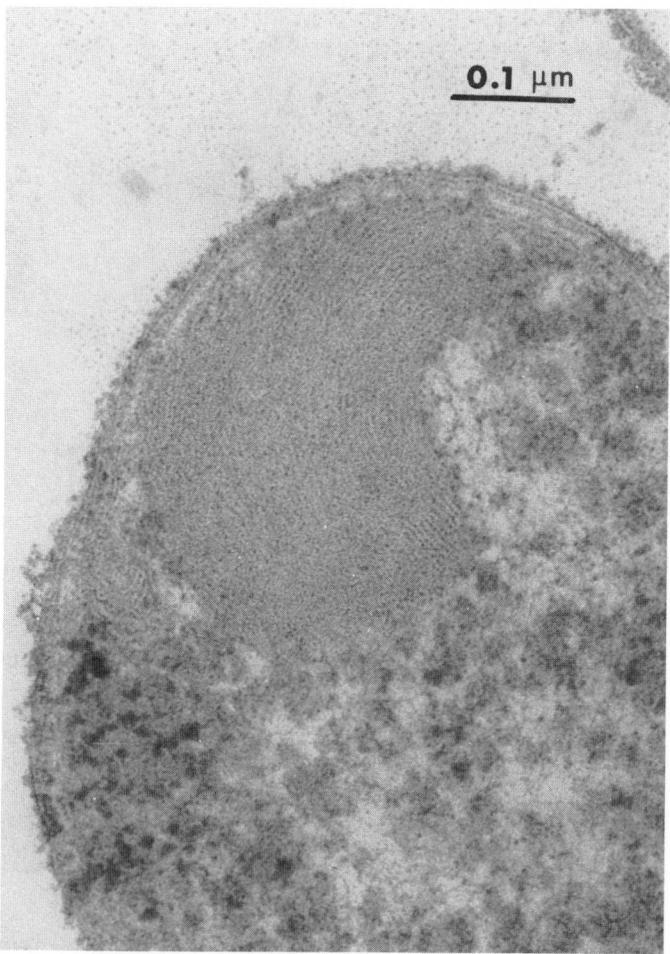

Figure 4.43. Thin section of late stationary-phase cells of *Gluconobacter oxydans* ATCC 621 showing the lamellar type of organization of membrane material. (Reproduced with permission from B. L. Batzing and G. W. Claus, Journal of Bacteriology *113:* 1455–1461, 1973 © American Society for Microbiology.)

plants, which is adapted to extreme conditions of acidity and ethanol concentration, is not well known and may be difficult to isolate. Moreover, most vinegar plants do not start from pure cultures; Hromatka and Leutner (1963a, b) are the only authors who have described the use of a pure culture of *G. oxydans* in submerged vinegar production.

Enrichment and Isolation Procedures

Enrichment of *Gluconobacter* strains present in flowers, fruits or bees can be set up using the following medium (g/liter of distilled water): yeast extract, 5.0; D-glucose, 50.0; actidione, 0.1; and bromophenol blue, 0.016. After incubation at 28°C for 2–4 days, those tubes showing

acidification are streaked onto plates of standard medium (GYC). Colonies which dissolve the $CaCO_3$ are further purified and characterized.

Similar enrichment and isolation procedures have been summarized by Swings and De Ley (1981).

Maintenance Procedures

Our working collection of *Gluconobacter* strains is maintained at 4°C on the standard (GYC) or mannitol (MYP) medium and is transferred monthly. Most strains remain viable for 3 months when kept on these media in screw-capped vials. *Gluconobacter* strains survive lyophilization for at least 10 years.

Table 4.64.
Characteristics of **Gluconobacter oxydans**[a]

Characteristics	Reaction or Result	Characteristics	Reaction or Result
Growth in Frateur's Hoyer medium with growth factors and:		Methanol, n-propanol, ethanediol, L-arabinose, D-xylose, D-galactose, D- or L-lyxose, raffinose, dextrin, formate, acetate, lactate, glycerate, malate, tartrate, citrate	−
Glucose	d	Single L-amino acids as sole nitrogen source:	
Mannitol	+	Asparagine, glutamine	+
Salt tolerance, growth in SM broth[b] containing 0.5–1.0% NaCl	d	Alanine, arginine, aspartic acid, cysteine, glutamic acid, glycine, histidine, isoleucine, leucine, lysine, methionine, phenylalanine, proline, serine, threonine, tryptophan, tyrosine, valine	d
pH tolerance, growth in SM broth at pH 3.6	d		
Temperature range, growth in SM broth at:		Antimicrobial agents (amount per disk)	
34°C	d	Ampicillin, 10 μg	d
37°C	−	Bacitracin, 10 U	R
Ethanol tolerance, growth in SM broth containing 1 to 5% ethanol	d	Cephaloridine, 25 μg	d
Glucose tolerance, growth in 0.5% yeast extract containing:		Chloramphenicol, 30 μg	R
		Colistin sulfate, 10 μg	R
20–25% Glucose	d	Erythromycin, 10 μg	R
30% Glucose	−	Fusidic acid, 10 μg	R
Tolerance to metals and dyes, growth in the presence of:		Gentamicin, 10 μg	R
Crystal violet (0.0001%), brilliant green (0.001%), malachite green (0.001%), $Cd(CH_3COO)_2 \cdot 2H_2O$ (0.01%), $CoSO_4 \cdot 7H_2O$ (0.01%), $HgCl_2$ (0.001%), or CH_3COOTl (0.001%)	d	Kanamycin, 30 μg	d
		Lincomycin, 2 μg	R
		Methicillin, 10 μg	R
		Nalidixic acid, 30 μg	R
Vitamin requirements:		Neomycin, 30 μg	R
Pantothenate	+	Nitrofurantoin, 200 μg	R
Niacin	d	Novobiocin, 30 μg	S
Thiamin, p-aminobenzoic acid	−	Penicillin G, 10 U	R
Final pH ≦ 4.5 in D-glucose or D-xylose media	+	Polymyxin B, 300 U	R
Acid also produced from[c]:		Streptomycin, 10 μg	d
n-Propanol, m-erythritol, D-arabinose, D-ribose, D-fructose, D-mannose	+	Sulfafurazole, 100 μg	R
		Tetracycline, 30 μg	S
Ethanol, n-amyl alcohol, glycerol, D-mannitol, sorbitol, m-inositol, L-sorbose, maltose, sucrose, n-butanol	d	Catalase	+
		Oxidase	−
L-Rhamnose, cellobiose, raffinose, dextrin	−	Nitrate reduction	−
Carbon sources[d]:		Indole production	−
Glycerol, D-mannitol, sorbitol	+	Gelatin hydrolysis	−
Ethanol, m-erythritol, m-ribitol, m-xylitol, L-arabitol, m-inositol, dulcitol, D-ribose, D-fructose, D-fucose, D-glucose, gluconate, L-sorbose, maltose, sucrose	d	Litmus milk, change	−
		Ferric chloride reaction[e] in media containing:	
		D-Glucose, D-fructose	d
		D-Galactose	−
		Acetyl methyl carbinol from lactate	d

[a] Symbols: see standard definitions.

[b] SM broth contains (g/liter): D-glucose, 50.0; yeast extract, 5.0.

[c] Acid production was tested in the following medium (g/liter): yeast extract, 5.0; bromocresol purple, 0.02; carbon source, 10.0. A final pH below 5.9 was considered as positive for acid formation.

[d] Growth was tested on the following medium (g/liter): carbon source, 3.0; yeast extract, 0.5; vitamin-free casamino acids, 3.0; agar (Oxoid No. 1), 25.0.

[e] Test for the formation of γ-pyrones.

Differentiation of the genus **Gluconobacter** from other genera

Characteristics useful for the differentiation of *Gluconbacter* from *Acetobacter*, *Zymomonas*, *Pseudomonas*, *Vibrio*, *Aeromonas*, *Frateuria* and *Xanthomonas* are listed in the chapters on *Zymomonas* and *Frateuria*. A rapid and accurate differentiation of *Gluconobacter* from *Acetobacter* can be based on ethanol and lactate oxidation: *Gluconobacter* does not overoxidize ethanol to CO_2 and H_2O via acetate and does not oxidize lactate to CO_2 and H_2O, whereas *Acetobacter* does. *Pseudomonas* species from sections I, II and III (Bergey 8) can be distinguished from *Gluconobacter* by their positive oxidase reaction, their capacity to oxidize lactate, the presence of a functional tricarboxylic acid cycle, and the absence of ketogenesis.

Taxonomic Comments

An extensive discussion of the classification and nomenclatural problems related to the genus *Gluconobacter* has been given by De Ley and Frateur (1970).

The genus *Gluconobacter* is a well-defined genus. In the previous edition of *Bergey's Manual* it contained a single species, *G. oxydans*, which was further subdivided into four subspecies: *oxydans*, *industrius*, *suboxydans* and *melanogenes*. Ameyama (1975) added a fifth subspecies, *G. oxydans* subsp. *sphaericus*. The names of the five subspecies are included on the Approved Lists of Bacterial Names. These subspecies were remnants of older species and no unequivocal diagnostic features existed to distinguish them, but the scheme presented in the eighth edition of *Bergey's Manual* seemed to be a first step towards a simplified and improved classification of the genus.

Recently, the existence of only a single species within the genus has been sustained based on a numerical analysis of 136 phenotypic features of 56 strains of acetic acid bacteria (Loitsyanskaya et al., 1979). The authors, however, rejected the subspecies differentiation as given in the eighth edition of *Bergey's Manual*. Another numerical analysis of 189 phenotypic features of 96 *Gluconobacter* strains showed that the strains

clustered together at 73% S_{SM}, supporting the concept of the single species *G. oxydans* within the genus (Gosselé, et al., 1983). It also became clear that the various subspecies cannot be regarded as biological entities. We propose to discontinue their use. In our opinion *Gluconobacter* strains form a broad spectrum within a single species (De Ley, 1961). This is also reflected by the wide mol% G + C span of 56–64 (Gillis and De Ley, 1980).

The genus *Gluconobacter* belongs to the fourth rRNA superfamily (sensu De Ley, 1978) in which it constitutes a separate branch together with the genus *Acetobacter* (Gillis and De Ley, 1980).

Miscellaneous Comments

Prescott and Dunn (1959) state that *G. oxydans* "is well adapted for industrial use, for it generally brings about the incomplete oxidation of sugars, alcohols and acids even when a liberal supply of oxygen is available, as is essential for a rapid dehydrogenation of the substrate." *Gluconobacter* strains can be used industrially to produce products such as L-sorbose from D-sorbitol; D-gluconic, 5-keto- and 2-ketogluconic acids, and D-tartaric acid from D-glucose; and dihydroxyacetone from glycerol.

Acknowledgment

J.D.L. is indebted to the Fonds voor Kollektief Fundamenteel Onderzoek (F.K.F.O.) for research and personnel grants, and J.S. to the Nationaal Fonds voor Wetenschappelijk Onderzoek (N.F.W.O.).

Further Reading

Asai, T. 1968. *Acetic acid bacteria.* University of Tokyo Press, Tokyo, University Park Press, Baltimore.

Gosselé, F. J. Swings, K. Kersters and J. De Ley. 1983. Numerical analysis of phenotypic features and protein gel electropherograms of *Gluconobacter* Asai 1935 emend. mut. char. Asai, Iizuka, and Komaqata 1964. Int. J. Syst. Bacteriol. *33:* 65–81.

List of the species of the genus **Gluconobacter**

1. **Gluconobacter oxydans** (Henneberg 1897) De Ley 1961, 47.[AL] (*Bacterium oxydans* Henneberg 1897, 224).

ox′y.dans. Gr. adj. *oxys* sharp, acid; L. part. adj. *dans* giving; M.L. part. adj. *oxydans* acid-giving, oxidizing.

The characteristics are as given for the genus. Additional characteristics are listed in Table 4.64.

The mol% G + C of the DNA is 56–64 (T_m).

Type strain: ATCC 19357.

Important Notes for Users of this Edition

1. Always read both generic and species descriptions because characters listed in the generic description are not usually listed in the species descriptions.

2. Unless otherwise indicated in footnotes to tables, the meanings of symbols are as follows:
   + 90% or more of strains are positive
   − 90% or more of strains are negative
   d 11–89% of strains are positive
   v strain instability (*not* equivalent to "d")
   D different reactions in different taxa (species of a genus or genera of a family)

3. All other symbols are defined in footnotes to tables.

FAMILY VII. **LEGIONELLACEAE** BRENNER, STEIGERWALT AND McDADE 1979, 658[AL]

DON J. BRENNER, JAMES C. FEELEY AND ROBERT E. WEAVER

Le.gi.on.el.la′ce.ae. M.L. n. *Legionella* type genus of the family; *-aceae* ending to denote family; M.L. fem. pl. n. *Legionellaceae* the *Legionella* family.

Rods 0.3–0.9 μm in width and 2–20 μm or more in length. Do not form endospores or microcysts. Not encapsulated. Not acid-fast. Gram-negative. Motile by one, two or more straight or curved polar or lateral flagella; nonmotile strains are occasionally seen. Aerobic. **L-cysteine-HCl and iron salts are required for growth.** The oxidase test is negative or weakly positive. Nitrates are not reduced. Urease-negative. Gelatin is liquefied. **Branched chain fatty acids predominate in the cell wall.** Chemoorganotrophic, using amino acids as carbon and energy sources. **Carbohydrates are neither fermented nor oxidized.** Isolated from surface water, mud, and from thermally polluted lakes and streams. There is no known soil or animal source. Pathogenic for man. The mol% G + C of the DNA is 39–43 (T_m, Bd).

Type genus: *Legionella* Brenner, Steigerwalt and McDade 1979, 656.

Further Descriptive Information

The family contains one genus, *Legionella*, with six species. It is circumscribed and readily distinguishable from other families on the basis of a unique set of phenotypic characteristics that include Gram-negativity, nonfermentative metabolism, requirement for L-cysteine and iron salts, and predominantly branched chain cellular fatty acids. Relatively large amounts of branched chain and hydroxy-substituted branched chain fatty acids are found in flavobacteria, but these organisms were shown by DNA/DNA hybridization to be unrelated to legionellae. DNA/DNA hybridization was also used to rule out relatedness between legionellae and members of other families that resemble *Legionellaceae* in some phenotypic characteristics (Brenner et al., 1978; Brenner, Steigerwalt, and McDade, 1979).

Legionellae were first isolated in guinea pigs and in embryonated hen's eggs (Tatlock, 1944; Jackson, Crocker, and Smadel, 1952; Bozeman, Humphries, and Campbell, 1968; McDade et al., 1977). Most environmental strains of all species were, and continue to be, isolated in this manner prior to cultivation on bacteriologic media. Legionellae do not grow on commonly employed media such as standard blood agar or nutrient agar. Mueller-Hinton agar, supplemented with hemoglobin and IsoVitaleX or GC base with similar supplements, was used to cultivate the first isolates of *L. pneumophila* (Center for Disease Control, 1977; Dumoff, 1979). Blood agar supplemented with cysteine and iron has been used to isolate and cultivate *L. pneumophila* in England (Greaves, 1980). A buffered charcoal yeast extract (BCYE) agar is now preferred for isolation of *L. pneumophila* in the United States (Pasculle et al., 1980). Several complex and semisynthetic broth media are now available for *L. pneumophila* (Pine et al., 1979; Warren and Miller, 1979; Ristroph, Hedlund, and Allen, 1980). Some species grow only on CYE agar (Feeley et al., 1979) and others will adapt to Feeley-Gorman (F-G) agar (Feeley et al., 1978) after isolation and multiple transfers on CYE agar (Hébert, personal communication).

Legionellae in paraffin-fixed tissue are not demonstrable with the staining techniques usually employed histologically. The Dieterle silver impregnation stain and the Giménez stain have been used successfully to demonstrate legionellae in paraffin-fixed tissue (Chandler et al., 1977; Greer et al., 1980). In nonparaffin-fixed, formalin-fixed tissue, legionellae are seen with Gram, Giménez, or Giemsa stains (Hernandez et al., 1980; Greer et al., 1980). Cells have vacuoles that stain with Sudan black B and may contain poly-β-hydroxybutyrate (Chandler et al., 1979). Diaminopimelic acid has been demonstrated in the cell wall (Guerrant, Lambert, and Moss, 1979). Some species show a blue-white autofluorescence under long wavelength (365 nm) ultraviolet light (Cordes et al., 1979; Lewallen et al., 1979), and many strains of most species produce a water-soluble, brown pigment on tyrosine-containing media (Baine et al., 1978; Thomason et al., 1979; Hébert et al., 1980a; Brenner, et al., 1980; Cordes et al., 1979; Morris et al., 1980; McKinney et al., 1981).

The term legionellosis is used by many to indicate all infections caused by legionellae. Strains of all species have been directly or, in one case, indirectly implicated in human pneumonia. Pneumonia caused by *L. pneumophila* is commonly referred to as Legionnaires' disease (Fraser et al., 1977) and pneumonia due to *L. micdadei* is commonly referred to as Pittsburgh pneumonia (Pasculle et al., 1980). At least one *Legionella* species, *L. pneumophila*, can also cause a mild, nonpneumonic, febrile disease termed Pontiac fever (Glick et al., 1978). A strain of *L. micdadei* was isolated from a guinea pig inoculated with the blood of a soldier (Tatlock, 1944; Hébert et al., 1980a), and several strains of *L. pneumophila* were isolated from the blood of patients with Legionnaires' disease (Edelstein et al., 1979; Macrae et al., 1979; Jackson et al., 1952).

Genus I. **Legionella** Brenner, Steigerwalt and McDade 1979, 658[AL]

DON J. BRENNER, JAMES C. FEELEY AND ROBERT E. WEAVER

Le.gi.on.el′la. M.L. n. *legio* legion or army; M.L. dim. ending *-ella*; M.L. fem. n. *Legionella* small legion or army.

As the only genus in the family, the definition of *Legionella* is identical to that of *Legionellaceae*.

Type species: *Legionella pneumophila* Brenner, Steigerwalt and McDade 1979, 658.

Further Descriptive Information

Legionellae are rod shaped or filamentous, 0.3–0.9 μm in width by 2–20 μm or more in length. The filaments are found after growth on agar media, less commonly in yolk sac material, and rarely in human lung or guinea pig tissue (Weaver and Feeley, 1979; Chandler et al., 1979). Electron microscopic examination reveals filamentous nucleoids, ribosomes, and Sudan black B-staining vacuoles thought to contain poly-β-hydroxybutyrate granules (Chandler et al., 1979). Cells are enclosed by a double envelope, composed of two three-layered unit membranes; a peptidoglycan layer has been demonstrated for *L. pneumophila*, and for *L. micdadei* (Flesher et al., 1979; Pasculle, Myerowitz and Rinaldo, 1979; Gress et al., 1980). Cell division occurs by a pinching process (Chandler et al., 1979; Keel, Finnerty and Feeley, 1979).

One, two, or occasionally more flagella occur per cell. The flagella are curved or straight and have a polar or lateral arrangement (Thomason et al. 1979; Chandler et al., 1980).

Legionellae are not acid-fast by the Ziehl-Neelsen procedure for mycobacteria (Hébert et al., 1980a), but *L. micdadei* may appear acid-fast in tissue preparations stained by other methods that use weak acids as decolorizing agents (Pasculle, Myerowitz and Rinaldo, 1979).

Pinpoint colonies appear on buffered charcoal yeast extract (BCYE) agar* in about 3 days after primary culture; the colony diameter reaches 3–4 mm after 5–7 days of incubation at 36 ± 1°C. Colonies are grey,

---

* BCYE agar. Add 10.0 g of ACES buffer (*N*-2-acetamido-2-aminoethane-sulfonic acid; pK_a = 6.9 at 20°C; available from Sigma Chemical Co.) to 500 ml of distilled water and dissolve by heating in a water bath at 45–50°C. Mix this solution with 440 ml of distilled water to which 40 ml of 1.0

glistening, convex, and circular with an entire edge (Weaver and Feeley, 1979; Feeley, Gorman and Gibson, 1979).

Legionellae do not grow on standard blood agar, nutrient broth or other commonly employed laboratory media. Mueller-Hinton agar supplemented with 1% hemoglobin and 1% IsoVitaleX (MH-IH) was first used by R. E. Weaver for cultivation of *L. pneumophila* (Feeley et al., 1978), but better media are now available. MH-IH was first replaced by Feeley-Gorman (F-G) agar* for routine culture of *L. pneumophila* (Feeley et al., 1978). Charcoal yeast extract (CYE) agar† was subsequently developed and found superior to F-G agar. BCYE was recently developed and is presently the most sensitive medium (Pasculle et al., 1980). CYE diphasic medium§ is used to culture blood. In all of these media, soluble ferric pyrophosphate or ferric nitrate replaces hemoglobin and L-cysteine-HCL replaces IsoVitaleX as growth factors for legionellae. All species grow on BCYE agar and on CYE agar. Most species will grow on F-G agar only after adaptation (Morris et al., 1980).

Several species exhibit a blue-white autofluorescence, and most species produce a diffusible, brown pigment on tyrosine-containing agar or F-G agar (Baine et al., 1978; Thomason et al., 1979; Lewallen et al., 1979; Cordes et al., 1979; Hébert et al., 1980a; Brenner et al., 1980; Morris et al., 1980; McKinney et al., 1981).

Growth of stock cultures obtained from humans occurs at temperatures between 25°C and 43°C, but not at 50°C. The optimum temperature is 36 ± 1°C (Weaver and Feeley, 1979; Thacker et al., 1981); however, in nature or in association with certain algae, the optimum growth temperature may be 45°C or higher (Fliermans et al., 1981; Orrison, Cherry and Milan, 1981).

The optimum pH for growth of legionellae is 6.8–7.0 (Feeley et al., 1978).

Carbon dioxide does not appear to stimulate the growth of legionallae, with the exception of *L. gormanii*. Because of this exception, and also the buffering effect of CO_2, incubation in an atmosphere of air + 2.5% CO_2, or in a candle jar, has been recommended for all legionellae; however, CO_2 may slightly inhibit growth in BCYE (Pasculle et al., 1980).

Carbohydrates are neither fermented nor oxidized by legionellae (Weaver and Feeley, 1979; Riley and Weaver, personal communication). Amino acids can presumably serve as the carbon source (Pine et al., 1979; Warren and Miller, 1979).

Legionellae are nutritionally fastidious. They require iron and L-cysteine for growth. As indicated previously, soluble ferric pyrophosphate, ferric nitrate or hemoglobin can provide the iron, and IsoVitaleX (BBL) can serve as a source of the L-cysteine. The following amino acids are essential for growth: cysteine, serine, methionine, arginine, valine, leucine, isoleucine and threonine; no vitamin or cofactor requirements have been demonstrated (George et al., 1980). Selenate was reported to stimulate the growth of *L. pneumophila* (Smalley, Jacques and Layne, 1980) but this observation has not been confirmed (Feeley, personal communication). Chemically defined and semisynthetic liquid media have been developed for legionellae (Pine et al., 1979; Warren and Miller, 1979; Ristroph, Hedlund and Allen, 1980).

One or more cryptic plasmids are present in many, but not all, strains of most species (Knudsen and Mikesell, 1980). Neither pathogenicity, antibiotic resistance, nor any other cell function was correlated with the presence of any plasmid (Aye, Wachsmuth and Feeley, personal communication; Mikesell, Ezzell and Knudson, 1981).

Motile strains of all *Legionella* species possess cross-reacting H antigens (Thomason, Chandler and Hollis, 1979; Lewallen et al., 1979; Hébert et al., 1980a). O antigens are largely unique for each species, although some cross-reactions have been demonstrated. More than one O-antigen group is present in *L. pneumophila* and *L. longbeachae* (Cherry and McKinney, 1979; England et al., 1980; McKinney et al., 1980; Bibb et al., 1981).

Legionella species are susceptible in vitro to a variety of antimicrobial agents using presently available methods to determine minimal inhibitory concentrations (MICS). Rifampin is the most active drug (lowest MIC) tested against these organisms, but they are susceptible in vitro to erythromycin, aminoglycosides, sulfamethoxazole-trimethoprim, chloramphenicol, cefoxitin, doxycycline, and minocycline (Thornsberry, Baker and Kirven, 1978). With the exception of *L. micdadei*, legionellae produce some level of β-lactamase activity when assayed with a chromogenic cephalosporin test (Hébert et al., 1980b). The β-lactamase is assumed to be chromosomal rather than of plasmid origin and is more active on cephalosporins than on penicillins (Thornsberry and Kirven, 1978). Erythromycin and rifampin were the most effective drugs in preventing mortality during in vivo egg and animal studies done with *L. pneumophila*. Aminoglycosides were effective in hen's egg studies, but not in animal studies; β-lactam antibiotics were not effective in any in vivo studies (Lewis et al., 1977; Fraser et al., 1978). The

---

N KOH has been added. Add the following ingredients: activated charcoal (Norit SG, available through Sigma Chemical Co., St. Louis, Missouri, Cat. No. C5510 (acid-washed with phosphoric and sulfuric acids)), 2.0 g; yeast extract, 10.0 g; and agar (Difco or Oxoid), 17.0 g. Dissolve by boiling. Autoclave 15 min at 121°C. Cool to 50°C. Aseptically add L-cysteine-HCl·H_2O solution (0.4 g in 10 ml of distilled water, sterilized by filtration), followed by ferric pyrophosphate solution (0.25 g in 10 ml of distilled water, sterilized by filtration). (Soluble ferric pyrophosphate is available on request from Dr. Morris Suggs, Director, Biological Products Div., Centers for Disease Control, Atlanta, Georgia 30333. It must be kept dry and stored in the dark until used. It is not usable if the color changes from green to yellow or brown. Do not heat over 60°C to dissolve the ferric pyrophosphate; a 50°C water bath is satisfactory.) The pH of the final solid medium should be 6.9 ± 0.05 at room temperature. Since reagents vary, each laboratory must determine the amount of KOH required. Do this by holding the bulk medium at 50°C while pouring one plate and checking its pH. When necessary, adjust the bulk medium with either 1.0 N KOH or 1.0 N HCl. Note that the pK_a of ACES buffer is influenced by temperature (0.02/°C); consequently, this must be considered with all pH determinations. Dispense 20-ml portions of the complete medium into 10 × 100-mm plastic Petri dishes; swirl the medium between pouring plates to keep charcoal particles suspended.

* F-G agar. Add the following ingredients to 980 ml of distilled water: casein (acid-hydrolyzed), 17.5 g; beef extractives, 3.0 g; starch, 1.5 g; and agar, 17.0 g. (These four ingredients can be replaced by 38.0 g of Mueller-Hinton agar (BBL)). Autoclave 15 min at 121°C. Cool to 50°C. Aseptically add L-cysteine-HCl·H_2O solution, followed by ferric pyrophosphate solution (see BCYE agar, preceding footnote, for details). Adjust the pH of the medium to 6.9 ± 0.05 as described for BCYE agar. Dispense 20-ml portions into 10 × 100-mm plastic Petri dishes.

† CYE agar is similar to BCYE agar (see first footnote) but lacks the ACES buffer. Add the yeast extract, charcoal and agar to 980 ml of distilled water. Dissolve by boiling. Autoclave 15 min at 121°C. Cool to 50°C. Aseptically add L-cysteine·HCl·H_2O solution, followed by ferric pyrophosphate solution (see first footnote). Adjust the pH to 6.9 ± 0.05 by adding 4.0–4.5 ml of 1.0 N KOH. Dispense as indicated for BCYE in first footnote.

§ CYE diphasic blood culture agar. For the agar phase, combine 2.0 g of activated charcoal (Norit SG), 17.0 g of agar (Difco or Oxoid), and 500 ml of distilled water. Boil and dispense 20-ml portions in 125-ml Wheaton serum bottles. Stopper loosely with a rubber stopper and a metal cap. Autoclave 20 min at 121°C. Cool to 50°C. Mix the warm charcoal and agar thoroughly and place the bottles at an angle to form an agar slant 6 cm in vertical height. This procedure should leave a portion of the agar protruding above the broth to be added subsequently. For the broth phase, combine 20.0 g of yeast extract and 500 ml of distilled water. Autoclave for 15 min at 121°C. Cool to 50°C. Aseptically add freshly prepared, filter-sterilized solutions of L-cysteine·HCl·H_2O (0.4 g in 10 ml distilled water) followed by ferric nitrate (0.1 g in 10 ml distilled water). Adjust the pH to 6.9 ± 0.05 by adding 6.0 ml of 1.0 N KOH. Finally, add sterile sodium polyanetholsulfonate (SPS) aseptically at a final concentration of 0.025% by volume to the complete medium (including the volume of the agar slant). SPS can be obtained from Hoffmann-LaRoche Inc. as Grobax. Dispense 20-ml portions into the Wheaton bottles of charcoal agar slants. Seal the bottles by crimping the metal caps over the rubber stoppers. Incubate at 36°C for 2 days to check sterility before using.

correlation between in vitro susceptibility and clinical efficacy of antibiotics for legionellae is poor (Kirby et al., 1978). Erythromycin is the drug of choice in treating human infection and should be given for a 3-week course (Kirby et al., 1978). Cotrimoxazole or doxycycline are the drugs of second choice. It has been suggested that rifampin should be used in combination with erythromycin when patients respond poorly to erythromycin alone (Cordes and Fraser, 1980).

All *Legionella* species are potentially pathogenic for humans. There is no information on whether they cause natural infections in animals, but to date legionellae have not been isolated from naturally infected animals even though more than 400 control guinea pigs have been cultured. Laboratory infections have been successfully induced in embryonated hen's eggs and in guinea pigs, gerbils, hamsters, rats, rabbits, and mice. Attempts to infect quail, pigeons, and chickens have not been successful (C. P. Patton and A. F. Kaufmann, unpublished data).

Disease caused by any *Legionella* species is often termed legionellosis. The two main types of legionellosis caused by *L. pneumophila* are Legionnaires' disease (LD) and Pontiac fever (PF). LD is a form of pneumonia with a 2- to 10-day incubation period after a presumed airborne exposure to the organism—there is no proof of person-to-person spread. The earliest symptoms include malaise, myalgia, headache, and often a nonproductive cough. Rapidly rising fever and chills usually appear within a day, at which time there are definite clinical and radiologic indications of pneumonia. Other common symptoms are chest pain, abdominal pain, vomiting, diarrhea, and mental confusion. Treatment with antibiotics is necessary; without the treatment the fatality rate is 15% or higher. In mild cases of LD, patients are febrile, have malaise, and may have respiratory tract symptoms. Hospitalization is usually not necessary and these patients usually go undiagnosed unless they are part of an LD outbreak. Smoking, preexisting respiratory illness, and conditions where the patient is immunocompromised are predisposing factors. LD occurs most often in men above the age of 50. Since the recognition of LD in 1976, more than 26 outbreaks and a total of almost 1000 sporadic cases have been detected, representing almost every state in the United States and more than 15 countries (A. L. Reingold, personal communication). Pontiac fever is an acute, nonpneumonic, febrile illness accompanied by chills, headache, and myalgia. It appears to have a high attack rate, is self-limiting and not fatal. Knowledge of PF is based largely on two outbreaks (Glick et al., 1978; Fraser et al., 1979), and it is likely that most sporadic cases would go undiagnosed. Legionellosis caused by *L. micdadei* (Pittsburgh pneumonia agent) is known as Pittsburgh pneumonia.

Most species of *Legionella* have been isolated from the environment. Natural environmental isolates are from streams or lakes, often thermally polluted water, or from moist soil adjacent to a body of water (Fliermans et al., 1981). There have been no dry soil isolates or isolates from insects (Fliermans et al., 1981). Urban environmental isolates were made from water collected from air conditioning cooling towers and evaporative condensers (Orrison et al., 1981). Isolates have recently been made from water collected from hospital showers (Tobin et al., 1981; Cordes et al., 1981) and from nebulizers (Gorman et al., 1980).

The primary hazard to laboratorians working with *Legionella* species is exposure to aerosols, although self-inflicted wounds are a potential hazard. The worker should adhere strictly to the general safety precautions of a bacteriology laboratory, particularly those that deal with minimizing and containing aerosols as set forth in the laboratory manual "*Legionnaires'*" *the Disease, the Bacterium and the Methodology* (Jones and Hébert, 1979). These include the use of a negative pressure class 1 or vertical laminar flow class 2 biological safety cabinet where possible, and the routine use of gloves, surgical masks and back-fastening gowns.

Enrichment and Isolation Procedures

Through 1980 there were more than 2000 diagnosed cases of legionellosis (A. L. Reingold, personal communication). The diagnoses were made largely on the basis of clinical symptoms and either direct fluorescent antibody (DFA) or indirect fluorescent antibody (IFA) tests.

Less than 10% of these cases have been culturally proven. Many of the early tissue samples were inappropriately handled before they reached the bacteriology laboratory. Since much better isolation methods are now available, the percentage of isolates should increase.

Human specimens. Lung tissue, pleural fluid, transtracheal aspirate or sputum, and blood are the specimens usually cultured when *Legionella* is suspected (McDade et al., 1977; Dumoff, 1979; Lattimer et al., 1978; Edelstein and Finegold, 1979; Edelstein, et al., 1979; Greaves, 1980; Feeley and Gorman, 1980). Samples that cannot be processed within 2 h of collection can be stored on wet ice or refrigerated for 2 days. Samples that will not be cultured within 2 days should be flash-frozen at $-70°C$ (Feeley and Gorman, 1980). Ten percent tissue suspensions in distilled water, body fluids, or homogenized sputum should be inoculated onto media designed to grow legionellae, such as BCYE, and onto a control medium such as blood agar or nutrient agar which will not support the growth of legionellae. Specimen samples should be placed on slides, air dried, fixed in acetone or 10% neutral Formalin, and tested for the presence of legionellae by DFA (Cherry et al., 1978; Cherry and McKinney, 1979). Samples of the specimen should also be flash-frozen in a dry ice-alcohol bath at $-70°C$ and stored for future diagnostic or reference use. When both culture and DFA results are negative, viable legionellae are either not present in the specimen or are present in numbers too small to detect. DFA results are presumptively positive for the presence of legionellae. Cultural results are definitively positive if confirmatory biochemical and serological tests and gas-liquid chromatographic (GLC) analysis of cellular fatty acids are done on clones of colonies as summarized below. A positive culture that is serologically negative by DFA indicates a possible new *Legionella* serogroup or species. If confirmatory tests indicate *Legionella* and serology remains negative, DNA relatedness studies should be done to determine whether the isolate represents a new serogroup in an existing species or a new species.

It is recommended that BCYE agar be used to isolate suspected legionellae, and that CYE agar diphasic blood culture medium be used specifically to culture blood for the possible presence of legionellae. F-G agar should be used only for examination of subcultures for colonial morphology and browning. Isolates that do not grow on F-G agar can be subcultured on tyrosine yeast extract agar (Baine et al., 1978).

Each batch of media must be tested for pH and sensitivity for primary isolation of legionellae. The pH of solid media is determined with a surface electrode or by emulsifying the agar from one plate in distilled water. If sensitivity of media for primary isolation is in doubt, prepare a control from a freshly isolated strain that has been minimally passed on CYE or BCYE agar. Pass through guinea pigs and use a homogenate of guinea pig spleen as inoculum. Laboratories without facilities for guinea pigs should suspend a freshly isolated strain, or other suitable reference strain, in sterile tap water to a turbidity of a McFarland No. 4 standard. Seed media with 0.05 ml of the standard inoculum; streak for isolated colonies; incubate at $36 ± 1°C$, in a moist environment, in air $+2.5\%$ CO_2. Agar plates usually show growth in heavily streaked areas after 2 days, and isolated, microscopic colonies, after 3–5 days. Blood bottles are held for 3 weeks and tilted daily to inoculate the slant from any growth in the broth. Positive blood bottles should be subcultured to BCYE agar. Media without antibiotics stored refrigerated, in the dark, in sealed, plastic containers remain stable for at least 2 months. Some antibiotic-containing media have long shelf lives, but to be safe, these media are freshly prepared at the Centers for Disease Control (CDC), Atlanta.

Environmental samples. Several methods have been used successfully for the isolation of legionellae from the environment (Morris et al., 1979; Tobin et al., 1981; Fliermans et al., 1981; Orrison et al., 1981). They differ mainly in the technique used for concentrating water samples. Water, soil or other samples are collected aseptically in chemically inert, sterile containers and held at ambient temperature or refrigerated until use. Morris et al. (1979) screen samples for the presence of legionellae by using the direct fluorescent antibody (DFA) test (Cherry et al., 1978; Cherry and McKinney, 1979). Water samples are inoculated undiluted and in 10-fold dilutions to 10^{-4} onto CYE agar in triplicate to obtain a total bacterial count. After 2–3 days of incubation, colony-forming units (CFU) are recorded as: sparse, $<10^4$/ml; moderate, 10^4–10^6/ml; or heavy, $>10^6$/ml. Negative plates are held for 14 days before being discarded. For samples with sparse CFU, 100 ml are sedimented at 2900 × g for 30 min and the sediment is suspended in 10 ml of sterile distilled water. Samples exhibiting high CFU are diluted in sterile distilled water to a concentration of 10^6 CFU/ml. Three milliliter samples of undiluted, moderate CFU samples, of diluted high CFU samples, or of concentrated, sparse CFU samples, are injected intraperitoneally into each of two guinea pigs. Soil samples (10 g) are inoculated into 100 ml of sterile distilled water containing 0.5% Tween 60 and are shaken at moderate speed for 30 min. Heavy particles are allowed to settle for 5–10 min. The sample is sedimented at 400 × g for 5 min, the supernatant fluid is sedimented at 2900 × g for 30 min, and the pellet is suspended in 10 ml of sterile distilled water. The CFU are determined, adjusted to 10^6 CFU, and injected into guinea pigs as for water samples. The guinea pigs are examined daily for a temperature response and for signs of illness. They are killed after a 0.6°C or greater rise in temperature on 2 consecutive days, or after any rise in temperature accompanied by one or more signs of illness, or at 7 days. The spleen, peritoneal exudate and swabs from the peritoneal wall are cultured on CYE agar.

Orrison et al. (1981) described a successful procedure for isolating legionellae from cooling towers. Fifty-milliliter to 4000-ml samples are filtered by nitrogen gas pressure though a 140-mm diameter filter (0.65 μm pore size) held in a steam-sterilized, stainless steel filter holder. After filtration, the membrane is aseptically transferred to a 250-ml sterile polycarbonate blending jar containing 50 ml of sample filtrate and blended in an Oster blender at the highest speed setting. The sample is then sedimented at 650 × g for 10 min to pellet membrane particles, and the supernatant is then sedimented at 5500 × g for 1.5 h. The resulting pellet is suspended in 5 ml of supernatant fluid. The number of legionellae is assessed by counting stained cells after DFA assay. If the number of *Legionella*-like cells is ~10^5/ml or greater, 1–3 ml are injected intraperitoneally into a guinea pig. Tissues from infected guinea pigs are cultured on CYE agar.

Fliermans et al. (1981) took 20-liter samples from lakes and rivers and concentrated them by continuous centrifugation at 15,000 rpm. The pellets were suspended in 40 ml of filter-sterilized water from the appropriate habitat. Samples were assayed by the DFA test and 3-ml samples were injected into guinea pigs. It must be emphasized that fluorescent antibody conjugates are not available for all species or all serogroups of *L. pneumophila*. Therefore, samples should be cultured or injected into guinea pigs even if they are negative by the fluorescent antibody test.

Selective procedures have recently been successfully used for the direct isolation of legionellae from clinical and environmental samples. Two recently described selective media for *L. pneumophila* have allowed differential recovery of legionellae from other bacteria commonly found as contaminants in lung specimens (Edelstein and Finegold, 1979) or in water (Thorpe and Miller, 1980). Greaves (1980) used an enriched blood agar supplemented with colistin, vancomycin, trimethoprim, and amphotericin B as a selective medium for legionellae from clinical sources. Bopp et al. (1981) pretreated aliquots of samples at pH 2 for 5 min and plated them directly to CYE and BCYE containing 0.5 mg/ml vancomycin, 4 μg/ml cephalothin, 16 μg/ml colistin, and 80 μg/ml cycloheximide. Feeley and Gorman (personal communication) have screened water samples by this method before injecting guinea pigs and found that the need to use guinea pigs was reduced by 80%.

Maintenance Procedures

Stock cultures on plates should be incubated in covered containers to prevent dehydration. BCYE slants are recommended for maintenance of stock cultures. Inoculated slants are incubated for 2 days at 36 ± 1°C and then stored in the dark at room temperature. Under these conditions, *L. pneumophila* cultures remain viable for 1–3 months

(Feeley, Gorman and Gibson, 1979; Hébert, 1980). Other species remain viable for variable periods from 4 months to 1 year (Hébert, 1980). Lyophilization or storage at −50°C or lower is recommended for long-term maintenance (Feeley and Weaver, personal communication); however, virulence and plasmids are often lost on subculture or lyophilization and, therefore, egg passage is recommended for maintenance of virulence (Aye, Wachsmuth and Feeley, personal communication).

Procedures for Testing Special Characters

Media. Legionellae grow on BCYE and CYE agars, and most species grow—either directly or after adaptation—on F-G agar. They do not grow on standard blood agar, nutrient agar, or other commonly employed laboratory media that do not contain both L-cysteine and iron salts.

Pigments. Brown pigment is formed on MH-IH agar or on other suitable media that contain tyrosine (Baine et al., 1978). Brown pigment cannot be observed on BCYE or CYE agar because these media are black (due to charcoal). Some species exhibit yellow fluorescence on F-G agar or in F-G broth (which has a formulation similar to that of F-G agar but without agar and with 100 mg/liter of soluble ferric pyrophosphate, and which is tubed in 2-ml portions) after 3–4 days of incubation. Yellow fluorescence is observed only in the dark by using a 365-nm ultraviolet light (Wood's light) (Weaver and Feeley, 1979). Some species exhibit blue-white autofluorescence on a variety of media when viewed under a Wood's light either in daylight or in darkness (Cordes et al., 1979; Lewallen et al., 1979).

Serology. The DFA test is highly specific and sensitive for the serological identification of legionellae at both the species and serogroup level. It is used both to determine whether legionellae are present in clinical and environmental samples, and as a rapid presumptive diagnostic test for the identification of legionellae (Cherry et al., 1978; Cherry and McKinney, 1979; Broome et al., 1979). Only ~15% of DFA-positive environmental samples prove culturally positive for legionellae, but false negative results have not been obtained (Fliermans et al., 1981; Cherry, personal communication). Detailed instructions for DFA are given by Cherry and McKinney (1979). A slide agglutination test can also be used for the rapid identification of legionellae (Wilkinson and Fikes, 1980). This test is especially useful for laboratories that do not have fluorescent antibody capability. Motile legionellae apparently possess cross-reacting flagellar antigens (Thomason et al., 1979). This property has not been used as a diagnostic tool because flagella are rarely seen in respiratory tract specimens and because fluorescent antibody conjugates prepared from purified flagella preparations are not yet available.

The indirect fluorescent antibody (IFA) test is the standard serological test used to diagnose legionellosis. Human test sera are reacted with heat-killed, whole *Legionella* cells. A 4-fold or greater rise in titer to ≥128 in paired acute- and convalescent-phase sera is considered definitive evidence of legionellosis. In a single serum, titers of 256 or higher are considered "presumptive" evidence of infection. The details of IFA are given by Wilkinson and colleagues (Wilkinson et al., 1979; Wilkinson, Fikes and Cruce, 1979). Most laboratories in the British Commonwealth use the method of Taylor et al. (1979) for IFA. More than 30,000 specimens have been tested for legionellosis at CDC since 1979 (H. W. Wilkinson, personal communication). Unfortunately, IFA reagents are currently available for only four of the six serogroups of *L. pneumophila*.

Gas-liquid chromatography (GLC). GLC analysis of cellular fatty acids is the method used, along with DFA, to test suspect isolates and to confirm the identification of known legionellae, new serogroups, and new species (Moss et al., 1977; Moss and Dees, 1979). The procedure is, thus far, totally specific in identifying species of *Legionella* at the genus level and can identify most, but not all, legionellae to the species level. The procedure involves saponification of whole cells, methylation of the fatty acids, extraction of the resultant methyl esters with diethyl ether and then hexane, followed by GLC analysis on a nonpolar stationary-phase column. The retention time of the resultant elution peaks is recorded, and tentative identification is made by comparison to reference standards. Confirmation and quantitation of the fatty acids are obtained by subsequent procedures, including mass spectrometry, use of polar columns, hydrogenation, and acetylation (Moss et al., 1977; Moss and Dees, 1979).

DNA relatedness. This is the only method whereby one can unequivocally speciate legionellae, confirm the existence of new serogroups in existing species, and identify new species (Brenner et al., 1978; Brenner, Steigerwalt and McDade, 1979; McDade, Brenner and Bozeman, 1979; Hébert, Steigerwalt and Brenner, 1980; Lewallen et al., 1979; Brenner et al., 1980; Morris et al., 1980; McKinney et al., 1981; Pasculle et al., 1980). DNA relatedness, however, is not adaptable to routine laboratory use and should be used only for isolates that cannot be characterized with certainty by biochemical, DFA and fatty acid analysis.

Differentiation of the genus **Legionella** *from other taxa*

Legionella pneumophila has no close relatives in other genera on the basis of DNA relatedness (Brenner et al., 1978; Brenner, Steigerwalt and McDade, 1979; Brenner et al., 1981). Although similar genetic studies have not been done on the other *Legionella* species, it is assumed that they have no close genetic relatives outside of the genus. Legionellae must be phenotypically differentiated from two groups of organisms that grow on some modified CYE media but not on common laboratory media. One is *Francisella tularensis* (K. McLeod, C. P. Patton and J. C. Feeley, unpublished data), and possibly *F. novicida*. These organisms are easily distinguished from legionellae because they produce acid from carbohydrates and do not contain branched chain fatty acids (Jantzen, Berdal and Omland, 1979). A number of thermophilic, spore-forming bacilli that stain Gram-negative mimic legionellae on CYE and do not grow on common laboratory media. They also have branched chain fatty acid profiles that superficially resemble those of one or more *Legionella* species (Thacker et al., 1981). These strains are distinguished from legionellae by their ability to grow at 50°C or higher, their ability to form spores, and their ability to grow on certain media which lack cysteine, such as tryptone broth (Thacker et al., 1981; Weaver, unpublished data).

Taxonomic Comments

Our knowledge of *Legionella* is in its infancy. Four strains that we now know to represent three separate *Legionella* species were isolated between 1943 and 1959 (Tatlock, 1944; Jackson, Crocker, and Smadel, 1962; Bozeman, Humphries, and Campbell, 1968). At the time of their discovery, however, none of these strains were cultivatable on bacteriologic media. They were treated as rickettsiae and not further studied until 1978 (McDade, Brenner, and Bozeman, 1979; Hébert et al., 1980a b). The "modern" study of what was to become the genus *Legionella* began in 1976 with the investigation of a large outbreak of respiratory disease at an American Legion Convention in Philadelphia, Pennsylvania (Fraser et al., 1977), resulting in the isolation of the causative agent, *L. pneumophila*, by Joseph McDade (McDade et al., 1977). Shortly thereafter the Flint 1 strain was isolated as an unknown on bacteriologic media by Dumoff (1979) and the Philadelphia strains were isolated on bacteriologic media by Weaver (Feeley et al., 1978). The first species, *L. pneumophila*, was named in 1979 and the remaining species followed in 1980 and 1981. The six named species of *Legionella* have extremely similar phenotypic characteristics (see Table 4.65 and 4.66). These include growth on media designed for legionellae, optimum pH and temperature for growth, common flagellar antigens, pathogenicity for guinea pigs and embryonated hen's eggs, nonfermentative metabolism, presence of catalase and gelatinase, absence of urease, and in vitro antibiotic susceptibility patterns. The cellular fatty acid compositions of all species are characterized by relatively large amounts of branched chain acids. All species have been directly or indirectly implicated as causative agents of human pneumonia. Available microbiological and biochemical tests are insufficient to separate all species with certainty. Once a species has been defined, it is quite simple and highly accurate to identify species by DFA.

New species and serovars have been confirmed by DNA relatedness. Relatedness between strains of the same species is 75% or higher for all species except *L. bozemanii* for which relatedness is 56%–77%. For all species, including *L. bozemanii*, the related sequences exhibit 3% or less divergence (unpaired bases within heteroduplex DNA), and there is only a small decrease in relatedness when reactions are done at a supraoptimal incubation temperature. A frequently used definition of a "genetic" species is "a group of strains whose DNAs are 70% or more related at optimal reassociation conditions; 55% or more related at less than optimal reassociation conditions, and have 6% or less divergence in their related sequences" (Brenner et al., 1972; Brenner, 1981). *Legionella* species fulfill this definition. There are no good genetic definitions for a genus or family. A priori, one would like a genus to contain species that are 40%–60% related; however, many well-established genera contain species that are from less than 5%–25% related (*Bacillus, Staphylococcus*, etc.). Relatedness between species of *Legionella* is generally low: 25% or less. The values obtained (Brenner et al., 1980; Table 4.68), although satisfactory qualitatively, show considerable variability in relatedness to different strains of the same species and in reciprocal relatedness values. Without guidance from a genus definition based on DNA relatedness, these data can be interpreted in at least four ways: (a) Create three or more families. Most species of *Enterobacteriaceae* share 20% relatedness. If a 20% cutoff is used, we can justify at least three separate families. (b) Place each species in a separate genus. Average relatedness of 25% or less would justify this. The only exception is the 46% relatedness of *L. dumoffii* to *L. gormanii*, but the reciprocal reaction using labeled *L. gormanii* DNA is only 24%. (c) Create a new genus for *L. micdadei* since it is less than 10% related to all other species. (d) Maintain all species in the same genus, by giving phenotypic characteristics priority over genetic relatedness in the formation of a genus.

A genus should ideally contain species that are both genetically and phenotypically similar. In their description and proposal of *L. bozemanii* and *L. dumoffii*, Brenner et al. (1980) argued that when both criteria could not be met, phenotypic relatedness should take precedence. In this way one ensures that genus designations are of practical use at the bench. For legionellae the phenotypic concept of a genus appears, at present, to be essential because few laboratories can separate the species of *Legionella*. Furthermore, little is known about *Legionellaceae*, and experience thus far indicates that many additional species will be encountered. As experience broadens, it may be possible to form genera that reflect both genetic and phenotypic similarity. For now, it is premature, impractical, and unnecessary to create additional genera.

Miscellaneous Comments

The use of trade names in this chapter is for identification only and does not constitute endorsement by the Public Health Service or by the U.S. Department of Health and Human Services.

Acknowledgments

The literature search for this chapter was concluded on December 1, 1980. We are extremely grateful to our colleagues for supplying preprints of manuscripts in press, for allowing access to their unpublished data, and for critically reviewing this manuscript. Special thanks in this regard are given to F. W. Chandler, W. B. Cherry, G. W. Gorman, G. A. Hébert, R. M. McKinney, G. K. Morris, C. W. Moss, A. G. Steigerwalt, B. M. Thomason, C. Thornsberry, and H. W. Wilkinson.

Further Reading

Jones, G.L. and G.A. Hébert (Editors). 1979. *"Legionnaires'" the disease, the bacterium and methodology.* Centers for Disease Control, Atlanta.
Jones, G.L. and G.A. Hébert (Editors). 1980. *Legionella* UPDATE. Centers for Disease Control, Atlanta, Georgia.

Table 4.65.

Characteristics of diagnostic value in identifying the genus **Legionella** *and its species*[a]

Characteristics	1. *L. pneumophila*	2. *L. bozemanii*	3. *L. micdadei*	4. *L. dumoffii*	5. *L. gormanii*	6. *L. long-beachae*
Growth on CYE agar	+	+	+	+	+	+
Growth on F-G agar	+	+[b]	+[b]	+[b]	−	−
Growth on blood agar and Trypticase soy agar base	−	−	−	−	−	−
Blue-white autofluorescence	−	+	−	+	+	−
Browning of F-G agar	+	+	−[c]	+	NG	NG
Browning of YE agar containing tyrosine	+	+	−	+	+	+
Oxidase test	+ or +/−	+/−	+	−	−	+
β-Lactamase	+	+/−	−	+	+	+/−
Acid from carbohydrates	−	−	−	−	−	−
Sodium hippurate hydrolysis	+	−	−	−	−	−

[a] Data from Weaver and Feeley (1979); Hébert et al. (1980a); Hébert, Steigerwalt and Brenner (1980); Brenner et al., (1980); Thornsberry and Kirven (1978); Morris et al. (1980); McKinney et al. (1981); Lewallen et al. (1979); Hébert, (1981). Symbols: see standard definitions; also NG, no growth; and +/−, weakly, or not always positive.

[b] Adapted for growth after passage on CYE agar.

[c] Browning was reported on F-G agar after the strain was well adapted to, and grew abundantly on, this medium (Hébert et al., 1980b).

Table 4.66.

Other characteristics of **Legionella** *species*[a]

Characteristics	1. *L. pneumophila*	2. *L. bozemanii*	3. *L. micdadei*	4. *L. dumoffii*	5. *L. gormanii*	6. *L. long-beachae*
Gram stain	−	−	−	−	−	−
Acid-fast stain (Ziehl-Neelsen)	−	−	−			
Occurrence of flagella (Leifson's flagella stain)	+	+	+	+	+	+
Motility	+	+	+	+	+	+
Sudanophilic granules (Sudan black B stain)	+					
Catalase test	+	+	+	+	+	+
Urease test	−	−	−	−	−	−
Gelatin liquefaction	+	+	+	+	+	+
Nitrate reduced to nitrite	−	−	−			−
Acid production from D-glucose, lactose, maltose, mannitol, sucrose or D-xylose	−	−	−	−	−	−

[a] Data from Weaver and Feeley (1979); Hébert et al. (1980a); Hébert, Steigerwalt and Brenner (1980); Brenner et al. (1980); Morris et al. (1980); McKinney et al. (1981); Thomason, Chandler and Hollis (1979); Lewallen et al. (1979); Hébert, (1981). For symbols see standard definitions.

Differentiation of the species of the genus **Legionella**

Characteristics useful for differentiating legionellae from other organisms and from one another are presented in Table 4.65. Other characteristics of *Legionella* species are given in Table 4.66. Cellular fatty acids of *Legionella* species are shown in Table 4.67, and Table 4.68 contains DNA relatedness data. Morphology is shown in Figs. 4.44–4.46.

List of the species of the genus **Legionella**

1. **Legionella pneumophila** Brenner, Steigerwalt and McDade 1979, 658.[AL]

pneu.mo′phi.la. Gr. n. *pneumo* lung; Gr. adj. *philos* loving; M.L. adj. *pneumophila* lung-loving.

The characteristics are as described for the genus and as listed in Tables 4.65–4.67.

The species contains six serogroups (Table 4.69).

Causative agent of legionellosis, presenting either as a pneumonia with a high fatality rate (Legionnaires' disease) or as an acute, febrile illness (Pontiac fever).

Isolated from human lung, sputum and blood, and from water.

The mol% G + C of the DNA is 39 (T_m, Bd). the DNA relatedness between strains of *L. pneumophila* is 75–100%. *L. pneumophila* is 3–20% related to other *Legionella* species (Table 4.68).

Table 4.67.
Cellular fatty acid composition of **Legionella** *species[a]*

Fatty Acid[b]	Approximate % of Total Acids[c]					
	1. *L. pneumophila*	2. *L. bozemanii*	3. *L. micdadei*	4. *L. dumoffii*	5. *L. gormanii*	6. *L. longbeachae[d]*
i-14:0	10	5	trace	5	5	
a-15:0	15	30	40	30	25	
i-16:0	35	15	10	15	15	
a-17:0	10	15	25	15	10	
a-17:1	0	0	4	0	0	

[a] Data adapted from Moss et al. (1977); Moss and Dees (1979); Hébert et al. (1980a); Lewallen et al. (1979); Cordes et al. (1979).

[b] i, branched chain with a methyl group located at the iso carbon atom; a, branched chain with a methyl group at the antepenultimate carbon. The number to the left of the colon is the number of carbon atoms. The number to the right of the colon is the number of double bonds.

[c] These are approximate values presented only for purposes of comparison. The original publications must be consulted for methodology, experimental parameters, range of values and interpretation of results.

[d] The cellular fatty acid composition of *L. longbeachae* is similar to that of *L. pneumophila*; however, the relative amounts of the major acids vary significantly among strains of *L. longbeachae*, but not among strains of *L. pneumophila* (C. W. Moss, personal communication).

Table 4.68.
DNA relatedness among **Legionella** *species[a]*

Source of Labeled DNA	% DNA Relatedness[b]					
	1. *L. pneumophila*	2. *L. bozemanii*	3. *L. micdadei*	4. *L dumoffii*	5. *L. gormanii*	6. *L. longbeachae*
L. pneumophila	75–100 (90)	8–25 (15)	5 (5)	15 (15)	20 (20)	2–4 (3)
L. bozemanii	2–13 (6)	56–77 (69)	0 (0)	15–32 (23)	12–22 (17)	
L. micdadei	0–5 (4)	0–6 (3)	90–100 (95)	1–3 (2)	6 (6)	
L. dumoffi	6–22 (14)	20–22 (21)	0–9 (5)	90 (90)	46 (46)	
L. gormanii	1 (1)	27 (27)	8 (8)	24 (24)	[c] [c]	
L. longbeachae	8 (8)	17 (17)	4 (4)	12 (12)	25 (25)	97–100 (99)

[a] Data from Brenner, Steigerwalt and McDade (1978); Hébert, Steigerwalt and Brenner (1980); Lewallen et al. (1979); Brenner et al. (1980); Morris et al. (1980).

[b] The range is given, followed by the average (in parentheses).

[c] Only one strain of *L. gormanii* is known.

Type strain: ATCC 33152 (CDC strain Philadelphia 1), isolated from human lung tissue by McDade and Weaver in 1977. The oldest isolate is strain OLDA isolated by E. Jackson in 1947 (Jackson et al., 1952).

2. **Legionella bozemanii** (sic) Brenner, Steigerwalt, Gorman, Weaver, Feeley, Cordes, Wilkinson, Patton, Thomason and Lewallen Sasseville 1980, 676.[VP] (*Effective publication*: Brenner et al. 1980, 111.)*

boze.man'i.i. M.L. gen. n. *bozemanii* named after F. Marilyn Bozeman, the microbiologist who isolated and first studied the organism.

The characteristics are as described for the genus and as listed in Tables 4.65–4.67.

Causative agent of human pneumonia.

Isolated from human lung.

The mol% G + C of the DNA is 43 (T_m). The DNA relatedness between strains of *L. bozemanii* is 56–77%. *L. bozemanii* is 6–23% related to other *Legionella* species (Table 4.68).

Type strain: ATCC 33217 (CDC strain WIGA; Bozeman strain WIGA), isolated in 1959 by Bozeman from human lung tissue.

3. **Legionella micdadei** Hébert, Steigerwalt and Brenner 1980, 676.[VP] (*Effective publication*: Hébert, Steigerwalt and Brenner 1980, 255.) (*Legionella pittsburghensis* Pasculle, Feeley, Gibson, Cordes, Myerowitz, Patton, Gorman, Carmack, Ezzell and Dowling 1980, 676.[VP])

mic.da'de.i. M.L. gen. n. *micdadei* of McDade; named after J. E. McDade, who isolated the etiologic agent of the 1976 Legionnaires' disease outbreak in Philadelphia.

The characteristics are as described for the genus and as listed in Tables 4.65–4.67.

---

* **Editorial Notes.** (1) The proper ending for the specific epithet is -*ae* as it is named for a woman. (2) The name *Legionella bozemanii* (sic) was effectively published in *Current Microbiology*, volume 4(2) (date of issue: October 13, 1980). The authors of the name submitted a notice to the *International Journal of Systematic Bacteriology* for validation of the name and it appeared on List No. 5 in volume 30(4) (date of issue: February 12, 1981). In that same issue a synonym, *Fluoribacter bozemanae* was validly published and was based on the same type strain (Garrity et al. 1980. Int. J. Syst. Bacteriol. *30:* 609–611). There is a question of priority of the specific epithet that cannot be settled by the *Bacteriological Code*, which does not address the matter of priority of names validly published in the same issue of the IJSB. The latter is to be discussed and decided by the Judicial Commission in the near future (Rules Revision Committee. 1982. Int. J. Syst. Bacteriol. *32:* 142–143).

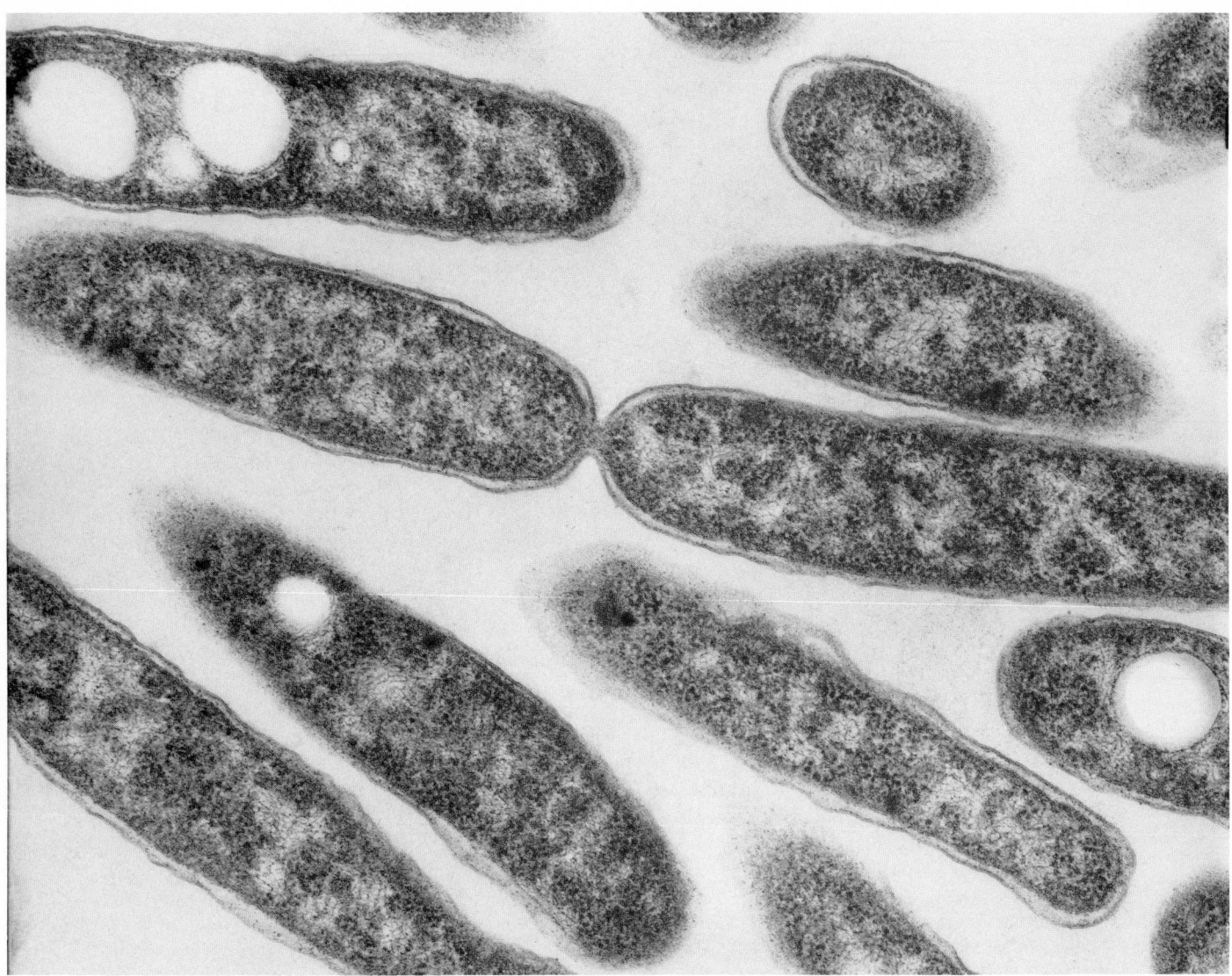

Figure 4.44.   Electron micrograph of *Legionella pneumophila* grown in broth (× 90,000).

Causative agent of Pittsburgh pneumonia. Commonly referred to as the Pittsburgh pneumonia agent.

Not acid-fast by the Ziehl-Neelsen method (Hébert et al., 1980b), but may appear acid-fast in tissue preparations stained by other methods (Pasculle, Myerowitz and Rinaldo, 1979).

Browning was demonstrated on F-G agar only after some strains were well adapted to and growing well on this medium. This took 12 or more subcultures (Hébert et al., 1980b). Without adaptation or in the absence of heavy growth, no browning occurs on F-G agar (Morris et al., 1980).

Isolated from human lung and blood, and from water.

The mol% G + C of the DNA is not known. The DNA relatedness between strains of *L. micdadei* is 90–100%. *L. micdadei* is 2–6% related to other *Legionella* species (Table 4.68).

Type strain: ATCC 33218 (CDC strain TATLOCK), isolated by H. Tatlock from a guinea pig inoculated with human blood in 1943.

4. **Legionella dumoffii** Brenner, Steigerwalt, Gorman, Weaver, Feeley, Cordes, Wilkinson, Patton, Thomason and Lewallen Sasseville 1980, 676.[VP] (*Effective publication*: Brenner et al. 1980, 111.)

du.mof′fi.i. M.L. gen. n. *dumoffii* of Dumoff; named after M. Dumoff, who first isolated *L. pneumophila* directly on bacteriologic media.

The characteristics are as described for the genus and as listed in Tables 4.65–4.67.

Causative agent of human pneumonia.

Isolated from human lung tissue and from water.

The mol% G + C of the DNA is not known. The DNA relatedness between strains of *L. dumoffii* is 90%. *L. dumoffii* is 5–46% related to other *Legionella* species (Table 4.68).

Type strain: ATCC 33279 (CDC strain NY-23), isolated from water by G. W. Gorman in 1978.

5. **Legionella gormanii** Morris, Steigerwalt, Feeley, Wong, Martin, Patton and Brenner 1980, 676.[VP] (*Effective publication*: Morris et al. 1980, 718.)

gor.man′i.i. M.L. gen. n. *gormanii* of Gorman; named after G. W. Gorman, who isolated and first studied the organism and pioneered in the isolation of legionellae from environmental and clinical sources.

The characteristics are as described for the genus and as listed in Tables 4.65–4.67.

Implicated serologically and by DFA staining of autopsy lung tissue, but not culturally proven, as a cause of human pneumonia.

The mol% G + C of the DNA is not known. The type strain and sole isolate is 1–27% related to other *Legionella* species.

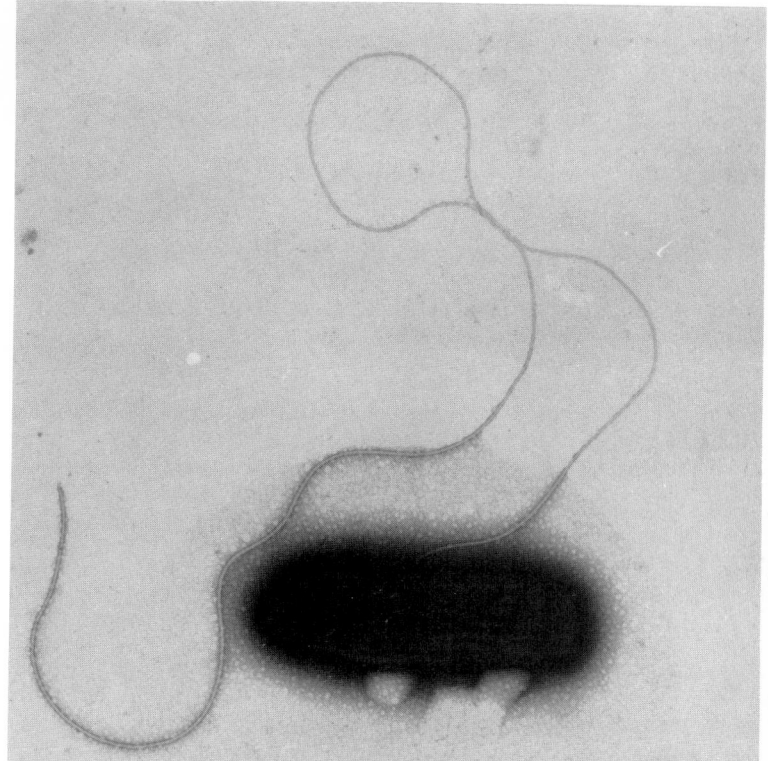

Figure 4.45. Transmission electron micrograph of *Legionella pneumophila* grown on CYE agar, showing a single, probably lateral, flagellum (× 61,700).

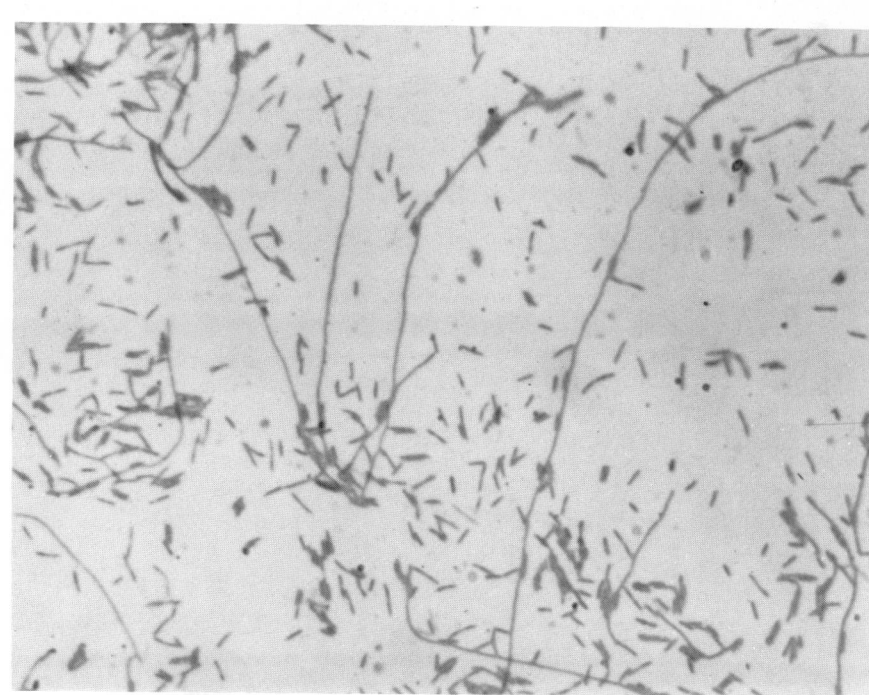

Figure 4.46. Carbol fuchsin-stained *Legionella pneumophila* grown on CYE agar, showing both short rods and filaments (× 1000).

Type strain: ATCC 33297 (CDC LS-13), isolated by G. W. Gorman from creek bank soil in 1978.

6. **Legionella longbeachae** McKinney, Porschen, Edelstein, Bissett, Harris, Bondell, Steigerwalt, Weaver, Ein, Lindquist, Kops and Brenner 1982, 266.[VP] (*Effective publication:* McKinney et al., 1981, 739)

long.beach'ae. M.L. gen. n. *longbeachae* of Long Beach (California) where the organism was first isolated.

The characteristics are as described for the genus and as listed in Tables 4.65–4.67.

Table 4.69.

Serogroups of **Legionella pneumophila**

Serogroup	Reference strain	Isolated by	Source
1	Philadelphia 1 (ATCC 33152)	J. E. McDade; R. E. Weaver	Human
1	Knoxville 1 (ATCC 33153)[a]	J. E. McDade	Human
2	Togus 1 (ATCC 33154)	J. E. McDade	Human
3	Bloomington 2 (ATCC 33155)	G. W. Gorman	Water
4	Los Angeles 1 (ATCC 33156)	P. H. Edelstein	Human
5	Dallas 1E (ATCC 33216)	P. Skaliy	Water
6	Chicago 2 (ATCC 33215)	H. M. Sommers	Human

[a] Knoxville 1 is included because it, as well as Philadelphia 1, is used to prepare antisera for serogroup 1.

The species contains two serogroups (Bibb et al., 1981). Causative agent of human pneumonia. Isolated from human lung tissue and transtracheal aspirates. The mol% G + C of the DNA is not known. The DNA relatedness between strains of *L. longbeachae* is 97–100% *L. longbeachae* is 4–25% related to other *Legionella* species.

Type strain: ATCC 33462 (CDC strain Long Beach 4), isolated from a transtracheal aspirate by R. Porschen in 1980.

Other Organisms

The research for this chapter was concluded on December 1, 1980. As of that date there were some six *Legionella*-like organisms (LLO) under study. These strains differed from all existing species by virtue of DNA relatedness (A. G. Steigerwalt, personal communication). It is therefore likely that several new *Legionella* species loom on the horizon.

Note Added in Proof

Substantial advances were made in our knowledge of legionellae in the period between the completion of this chapter and the publication of this volume. A seventh and an eighth serogroup of *L. pneumophila* were described (Bibb, W. F., P. M. Arnow, D. L. Dellinger and S. R. Perryman, 1983, Isolation and characterization of a seventh serogroup of *Legionella pneumophila*, J. Clin. Microbiol. *17:* 346–348; Bissett, M. L., J. O. Lee and D. S. Lindquist, 1983, A new serogroup of *Legionella pneumophila:* serogroup 8, J. Clin. Microbiol. *17:* 887–891) and strains that represent at least three additional serogroups of *L. pneumophila* are being studied (W. F. Bibb, A. G. Steigerwalt and H. W. Wilkinson, personal communication). Three additional *Legionella* species were described: (a) *Legionella jordanis* (Cherry, W. B., G. W. Gorman, L. H. Orrison, C. W. Moss, A. G. Steigerwalt, H. W. Wilkinson, S. E. Johnson, R. M. McKinney and D. J. Brenner, 1982, *Legionella jordanis:* a new species of *Legionella* isolated from water and sewage, J. Clin. Microbiol. *15:* 290–297); (b) *Legionella oakridgensis* (Orrison, L. H., W. B. Cherry, R. L. Tyndall, C. B. Fliermans, S. B. Gough, M. A. Lambert, W. F. Bibb, L. K. McDougal and D. J. Brenner, 1983, *Legionella oakridgensis:* unusual new species isolated from cooling tower water, Appl. Environ. Microbiol. *45:* 536–545); and (c) *Legionella wadsworthii*

(Edelstein, P. H., D. J. Brenner, C. W. Moss, A. G. Steigerwalt, E. M. Francis and W. L. George, 1982, *Legionella wadsworthii* species nova : a cause of human pneumonia, Ann, Intern. Med. *97:* 809–813). DNA hybridization studies in progress indicate the existence of 14 additional *Legionella* species (A. G. Steigerwalt, personal communication). A proposal, not subscribed to by the authors of this chapter, was made to divide the family *Legionellaceae* into three genera: *Legionella, Tatlockia,* and *Fluoribacter* (Brown, A., G. M. Garrity and R. M. Vickers, 1981, *Fluoribacter dumoffii* (Brenner et al.) comb. nov. and *Fluoribacter gormanii* (Morris et al.) comb. nov., Int. J. Syst. Bacteriol. *31:* 111–115; Garrity, G. M., A. Brown and R. M. Vickers, 1980, *Tatlockia* and *Fluoribacter:* two new genera of organisms resembling *Legionella pneumophila*, Int. J. Syst. Bacteriol. *30:* 609–614). New phenotypic tests, including the use of a dye-containing medium and the analysis of quinones, have been used to characterize legionellae (Vickers, R. M., A. Brown and G. M. Garrity, 1981, Dye-containing buffered charcoal yeast extract medium for differentiation of members of the family *Legionellaceae*, J. Clin. Microbiol. *13:* 380–382; Karr, D. E., W. F. Bibb and C. W. Moss, 1982, Isoprenoid quinones of the genus *Legionella*, J. Clin. Microbiol. *15:* 1044–1047). The reader is urged to search the literature for these and other advances, such as supplementation of BCYE agar with α-ketoglutarate and addition of improved selective agents (Edelstein, P. H., 1981, Improved semiselective medium for isolation of *Legionella pneumophila* from contaminated clinical and environmental specimens, J. Clin. Microbiol. *14:* 298–303; Wadowsky, R. M. and R. B. Yee, 1981, Glycine-containing medium for isolation of *Legionnellaceae* from environmental specimens, Appl. Environ. Microbiol. *42:* 768–772).

FAMILY VIII. **NEISSERIACEAE** PRÉVOT 1933, 119[AL]

KJELL BØVRE

Neis.se.ri.a′ce.ae. M.L. fem. n *Neisseria* type genus of the family; -aceae ending to denote family; M.L. fem. pl. n. *Neisseriaceae* the *Neisseria* family.

Organisms are either **coccal**, occurring singly, in pairs or in masses, often with adjacent sides flattened (different planes of division), or distinctly **rod shaped** or coccoid (one plane of division), frequently occurring in pairs or short chains. Endospores are not formed. May be capsulated. Gram-negative, but the cells often have a tendency to resist Gram decolorization. Flagella and swimming motility are absent.

Strains of several species are fimbriated (piliated) and may show surface-bound "twitching motility."

All species grow aerobically. Strains of some species may also grow weakly under anaerobic conditions. Strains of all recognized species usually have an optimum growth temperature of ~32–36°C, but psychophilic strains may be found in the genus *Acinetobacter* (see also

Table 4.70.
Differential characteristics of the genera of the family **Neisseriaceae**[a]

Characteristics	I. Neisseria	II. Moraxella		III. Acinetobacter	IV. Kingella	Moraxella urethralis (species incertae sedis)
		Subgenus Moraxella	Subgenus Branhamella[b]			
Cell morphology:						
Cocci	+	−	+	−	−	−
Rods	+[c]	+	−	+	+	+
Oxidase test	+	+	+	−	+[d]	+
Catalase test	+	+	+	+	−	+
Presence of carbonic anhydrase	+	−[e]	−			
Acid from glucose	[+]	−	−	D	+	−
Nitrite reduction	+	−	D	−	+	+
Presence of thymidine phosphorylase, nucleoside deoxyribosyltransferase and thymidine kinase	−	[−]	−[e]	−	+[e]	−
True waxes present in cell wall	−	[+]	+[e]	D	−[e]	−
Mol% G + C of DNA	46.5–53.5	40–47.5	40–47.5	38–47	47–55	46–47

[a] Symbols: +, positive for the majority of strains and some strains of each species; [+], positive for all strains of the majority of species (only one species uniformly negative); D, positive and negative species (or strains of *Acinetobacter*) about equally represented; [−], the only exception is one strain positive for thymidine kinase; −, all strains negative.

[b] In this table, the subgenus *Branhamella* includes the species *catarrhalis* and the "false neisseria" species *caviae*, *ovis*, and *cuniculi*. *Branhamella* has also been formally proposed as a separate genus to contain only one of the species, *Branhamella catarrhalis* (Catlin, 1970).

[c] Only one species, *N. elongata*, consists of rods.

[d] *Kingella indologenes* and *K. denitrificans* may be negative or weakly positive with the least sensitive test reagent, dimethyl-*p*-phenylenediamine, whereas they are distinctly positive when tetramethyl-*p*-phenylenediamine is used. For other organisms of this table (including *K. kingae*) the result is the same with both reagents.

[e] One or two species have not been examined.

Taxonomic Comments for the genus *Moraxella*). Several species have complex growth requirements while a few species grow regularly in simple defined media.

The oxidase test is positive for all genera except *Acinetobacter*. All presently recognized oxidase-positive species are parasitic in warm-blooded hosts, while strains of *Acinetobacter* are much less distinctly parasitic or are not parasitic, having water and soil as their main habitat (see also Taxonomic Comments for the genus *Moraxella*).

The mol% G + C of the DNA ranges from 38–55.

Type genus: *Neisseria* Trevisan 1885, 105.

Further Comments

Most species of *Neisseriaceae* contain strains which are highly competent in genetic transformation. This provides a tool for definite identification of numerous, although not all, species, and also forms an important basis for allocation of species within genera of the family (Bøvre, 1980; Bøvre and Hagen, 1981; Henriksen, 1976).

Genetic taxonomy in *Neisseriaceae* was pioneered by Catlin and Cunningham in 1961, showing close relationship between most *Neisseria* species except the species then named *Neisseria catarrhalis*.

The inclusion of rod-shaped organisms in the family was first formally proposed by Henriksen and Bøvre in 1968. In that proposal the genus *Moraxella* was transferred to the family *Neisseriaceae*, a transfer which has been generally accepted. An important basis for this taxonomic decision was the genetic relationship between the *catarrhalis* organism and *Moraxella*, as first reported by Bøvre in 1963 and shown independently by Catlin in 1964, and the further elaboration by genetic transformation and nucleic acid hybridization of the relationship between this species and other "false neisseriae" as a group and between this group and various *Moraxella* species. In fact, the transfer proposed by Henriksen and Bøvre in 1968 was combined with the proposed inclusion of these coccal species in the genus *Moraxella*. A different

arrangement was chosen by Catlin, who in 1970 proposed a new genus, *Branhamella*, for only one species, the coccal *Branhamella catarrhalis*.

In 1979, Bøvre proposed a division of the emended genus *Moraxella* into two subgenera, the subgenus *Moraxella* for the rod-shaped species, and the subgenus *Branhamella* for the group of genetically related coccal species (see Taxonomic Comments for the genus *Moraxella*). That a genus may contain both coccal and rod-shaped species from a genetical point of view has also been shown for the genus *Neisseria*, by the detection of the rod-shaped organism known as *Neisseria elongata* (Bøvre and Holten, 1970).

Genetic transformation and hybridization studies show some affinities between *Moraxella/Branhamella* and the genus *Acinetobacter*, which may be taken to support strongly the idea that these organisms belong to the same bacterial family (see reviews by Bøvre, 1980, and Bøvre and Hagen, 1981). Thus, rRNA/DNA hybridization studies have shown 66–69% homology between *Acinetobacter* on the one hand and *Moraxella osloensis/Moraxella (Branhamella) catarrhalis* on the other. The genetic indications that these organisms are related to the genus *Neisseria* are considerably weaker or absent.

The genus *Kingella*, proposed by Henriksen and Bøvre in 1976 for the organism known previously as *Moraxella kingae* (now *Kingella kingae*) and subsequently widened by the proposed inclusion of two more species (Snell and Lapage, 1976), apparently has no genetic affinity to *Moraxella/Branhamella* or to *Acinetobacter*. On the other hand, a slight genetic affinity between *K. kingae* and *N. elongata* has not been ruled out (Bøvre et al., 1977).

The species named *Moraxella urethralis* (see *Species Incertae Sedis* in the chapter on *Moraxella*) may be considered as a separate entity within the family, on the basis of little or no genetic affinity to the other genera (Bøvre and Hagen, 1981; see also Taxonomic Comments in the chapter on *Moraxella*).

It should be kept in mind that the impact of marginal or absent genetic affinity is uncertain in relation to the circumscription of families (and even the circumscription of genera).

The cellular lipid composition of *Neissericeae* is characterized by the absence of branched fatty acids. The quantitative and qualitative distributions of straight-chain fatty acids, including hydroxy acids, generally distinguish the genera as genetically circumscribed, and are also important in differentiation between several species (Jantzen et al., 1974, 1975; Bøvre et al., 1976). True waxes, i.e. simple esters of fatty alcohols and fatty acids, are present in the cell walls of most *Moraxella/Branhamella* species and often in *Acinetobacter* strains, but are absent in the genera *Neisseria* and *Kingella*, as well as in the species *Moraxella urethralis*.

While activities corresponding to the three enzymes thymidine phosphorylase, nucleoside deoxyribosyltransferase and thymidine kinase have been shown to be present in the genus *Kingella*, they are absent in the genera *Neisseria*, *Acinetobacter*, and (except for thymidine kinase in one strain) in *Moraxella/Branhamella*, as well as in *M. urethralis* (Jyssum, 1971, 1974; Jyssum and Bøvre, 1974).

Carbonic anhydrase has been detected in all *Neisseria* species, but not in the four *Branhamella* species, the rod-shaped *Moraxella* species so far examined, or in *M. urethralis* (Berger and Issi, 1971; Berger and Piotrowski, 1974). The genera *Acinetobacter* and *Kingella* have not been characterized in this manner.

Table 4.70 lists some differential characteristics of the genera of the family (for comparison, *M. urethralis* is included).

Editorial Note

Controversy exists concerning the taxonomic placement of three coccal species with respect to the *catarrhalis* organism. (These three species—*caviae, ovis*, and *cuniculi*—comprise a group often referred to as the "false neisseriae," because there is general agreement that they do not belong in the genus *Neisseria*.) According to one school of thought, the phenotypic similarities and the level of relatedness of these three species to the *catarrhalis* organism are sufficiently high to justify their inclusion in the same genus and subgenus. The other school maintains that the similarities and level of relatedness are not high enough to warrant such inclusion. Thus, in the chapter on the genus *Moraxella*, these species are listed as *Moraxella* (*Branhamella*) *caviae*, *M.* (*B.*) *ovis*, and *M.* (*B.*) *cuniculi*, whereas in the article on the genus *Neisseria* they are placed under *Species Incertae Sedis* and are designated by their original (but provisional) names *Neisseria caviae*, *N. ovis*, and *N. cuniculi*. See the Taxonomic Comments of these two articles for a more detailed exposition of these views. It is to be hoped that additional data, such as those derived from rRNA/DNA hybridization studies or rRNA oligonucleotide catalogs, can eventually resolve this controversy.

Further Reading

Bøvre, K. 1980. Progress in classification and identification of *Neisseriaceae* based on genetic affinity. In Goodfellow and Board (Editors), *Microbial Classification and Identification*, Society for Applied Bacteriology Symposium Series No. 8, Academic Press, London, pp. 55–72.

Bøvre, K. and N. Hagen. 1981. The family *Neisseriaceae*: rod-shaped species of the genera *Moraxella*, *Acinetobacter*, *Kingella*, and *Neisseria*, and the *Branhamella* group of cocci. In Starr, Stolp, Trüper, Balows and Schlegel (Editors), *The Prokaryotes: a Handbook on Habitats, Isolation and Identification of Bacteria*. Springer-Verlag, Berlin, pp. 1506–1529.

Henriksen, S.D. 1976. *Moraxella*, *Neisseria*, *Branhamella*, and *Acinetobacter*. Annu. Rev. Microbiol. *30:* 63–83.

Genus I. **Neisseria** *Trevisan 1885, 105*[AL]

NEYLAN A. VEDROS

Neis.se′ri.a. M.L. fem. n. *Neisseria* named after Dr. Albert Neisser, who discovered the etiological agent of gonorrhoea in the pus cells of patients in 1889.

Cocci 0.6–1.0 μm in diameter, occurring singly but more often in pairs with adjacent sides flattened; one species (*N. elongata*) is an exception and consists of **short rods** 0.5 μm wide, often arranged as diplobacilli or in short chains. Division of the coccal species is in two planes at right angles to each other, sometimes resulting in tetrads. Capsules and fimbriae (pili) may be present. Endospores are not present. Gram-negative, but there is a tendency to resist Gram decolorization. Swimming motility does not occur and flagella are absent. **Aerobic.** Some species produce a greenish yellow carotenoid pigment. Some species are nutritionally fastidious and hemolytic. Optimum temperature, 35–37°C. **Oxidase-positive. Catalase-positive except N. elongata.** Carbonic anhydrase is produced by all species except the "false neisseriae" *N. caviae*, *N. ovis*, and *N. cuniculi*.* All species reduce nitrite except *N. gonorrhoeae*, *N. canis*, and the "false neisseriae" *N. cuniculi* and *N. ovis*. Chemoorganotrophic. Some species are saccharolytic. Inhabitants of the mucous membranes of mammals. Some species are primary pathogens for man. The mol% G + C of the DNA is 46.5–53.5.

Type species: *Neisseria gonorrhoeae* (Zopf 1885) Trevisan 1885, 106.

Further Descriptive Information

Although the genus *Neisseria* has traditionally included only cocci, a rod-shaped species, *N. elongata*, is included in the genus in this edition of the *Manual* (see Taxonomic Comments).

All *Neisseria* species produce surface polysaccharides either as loosely associated envelopes or intact capsules. The latter have been observed on *N. subflava*, *N. mucosa*, some strains of *N. sicca* (Reyn, 1974), some strains of *N. meningitidis* (Vedros et al., 1968), and *N. gonorrhoeae*

(James and Swanson, 1977). Various factors affecting production of the capsules of *N. gonorrhoeae* have been investigated (De Marco de Hormaeche et al., 1978; Hendley et al., 1977; Richardson and Sadoff, 1977), but the actual presence of capsules in this species has not been confirmed by electron microscopy using India ink (Melly et al., 1979) or wheat germ agglutinin (Frasch, 1980). The chemical structure of the capsules and surface polysaccharides of *N. meningitidis* in relation to serogroups has been reviewed by Vedros (1978) and Robbins (1978).

Fimbriae have been observed on *N. meningitidis*, *N. gonorrhoeae*, *N. perflava*, and *N. elongata* (Devoe and Gilchrist, 1975; Frøholm et al., 1973; Wistreich and Baker, 1971; Bøvre et al., 1977). Fimbriation on *N. gonorrhoeae* has been associated with attachment of the bacteria to human mucosal cells (Swanson, 1973; Schlesinger, 1975; Buchanan, 1977; Pearce and Buchanan, 1978). The chemical composition (Morse et al., 1979) and antigenic differences among strains (Tramont, 1976; Buchanan, 1975; Novotny and Turner, 1975) have been studied. Strains with and without fimbriae appear to be equally susceptible to transformation (Sparling, 1966; Biswas et al., 1977; Baron and Saz, 1978). Loss of fimbriation can occur with high frequency in culture passage of *N. elongata* (Bøvre et al., 1977) and also *N. meningitidis* and *N. gonorrhoeae* unless special growth conditions are maintained (McGee et al., 1977). The presence or absence of fimbriae appears to be of little taxonomic value in the genus.

Although flagella and motility in liquid media are absent in *Neisseria* species. a "twitching motility" has been observed with *N. gonorrhoeae* and *N. meningitidis* (Henrichsen, 1975).

Colony morphology varies with the species and ranges from small, smooth, transparent, butyrous colonies to wrinkled, dry, adherent

---

* See Editorial Note for the family *Neisseriaceae*.

colonies (Catlin, 1978) Colony morphology has been particularly useful with *N. gonorrhoeae*. Five colony types have been described and labeled T₁-T₅ (Kellogg et al., 1963; Reyn et al., 1971). Types T₁ and T₂ were virulent for volunteers but not T₃ and T₄. T₁ colonies are small, raised dewdrop, slightly viscid, with entire edges; T₂ colonies may be friable and have defined or crenated edges or both; T₃-T₅ colonies are larger than T₁ (1.0 mm diameter), slightly convex, and light brown or colorless. The different types can be distinguished by oblique light as described by Juni and Heym (1977). Opaque and transparent variants of the colony types have been described (Swanson, 1978). The opaque colonies are less virulent in the chick embryo model, are typically isolated from the urethrae of males and the cervices of women near ovulation, are opaque to transmitted light, and the cells express an extra dominant, heat-modifiable protein in the outer membrane (see review by James and Swanson, 1978).

Colonies of *N. meningitidis* are larger than those of gonococci (1.0 mm or greater in diameter) and are smooth and moist with a glistening surface and entire edge (Morello and Bohnhoff, 1980). Mucoid colonies have been noted. Colonies of *N. lactamica* closely resemble those of meningococci but may be less moist and smaller (Morello and Bohnhoff, 1980).

Some strains of *N. lactamica* produce a yellow pigment (Hollis et al., 1969). A yellowish green pigment is also produced by *N. subflava*, *N. flavescens*, *N. perflava*, and *N. sicca*. *N. elongata* may also have a slight yellow tinge due to pigment production. Pigmentation of neisseriae varies with growth conditions and quantitative analysis of the extracted pigment indicates that it is of little taxonomic value but may be used to differentiate selected pigmented strains (Berger, 1961; Hoke and Vedros, 1982b).

Neisseria species other than *N. meningitidis* and *N. gonorrhoeae* will grow on plain nutrient agar at 35–37°C. *N. meningitidis* requires mineral salts, lactate, a few amino acids, and glutamic acid as a carbon source (Catlin, 1978; Reyn, 1974). Cystine is required by ~10% of the strains (Catlin, 1978), and some strains can be adapted to grow with ammonium salts as the sole nitrogen source (Jyssum, 1959). *N. gonorrhoeae* is more fastidious and requires glutamine for primary isolation of ~20% of strains and co-carboxylase for ~1% of strains (Reyn, 1974). Several defined media have been developed for growth of *N. gonorrhoeae* (La Scolea and Young, 1974; Catlin, 1973; Wong et al., 1980). The use of a biphasic growth system has resulted in a high density yield of cells (Gerhardt and Heden, 1960; Giard and Vedros, 1981). Iron is an essential growth factor for *N. gonorrhoeae* (Kellogg et al., 1963) and its availability in culture media influences the virulence of both meningococci and gonococci (Payne and Finkelstein, 1975, 1978). Iron (as iron dextran) in media also increases the colony size of both species (Payne and Finkelstein, 1977) and studies on iron assimilation and siderophore production have been reported (Yancey and Finkelstein, 1981a, b). The optimum temperature for growth of *Neisseria* species is 36–37°C with a range from 22–40°C, except for *N. meningitidis* and *N. gonorrhoeae* which grow poorly or not at all below 30°C. A high relative humidity (~50%) is beneficial to the growth of all species, and CO_2 (3–10%) is required for the growth of gonococci and enhances the growth of meningococci on solid media. After many passages, laboratory strains become less fastidious in their growth requirements than fresh isolates. See Catlin (1977) for a detailed review of gonococcal nutritional requirements.

A nontransducing bacteriophage has been isolated for *N. perflava* (*N. subflava*) (Stone et al., 1956; Phelps, 1967) and for *N. meningitidis* (Cary and Hunter, 1976). The latter has not been confirmed (Vedros, 1978). Bacteriocins have been reported for *N. meningitidis* (Kingsbury, 1966) and applied to a typing scheme (Counts et al., 1971). Bacteriocins have also been reported for *N. gonorrhoeae* (Flynn and McEntegart, 1972) but their presence was disputed (Senff et al., 1976).

Conjugal transfer of certain plasmids to appropriate gonococcal recipients has been reported (Eisenstein et al., 1977). Chromosomal gene transfer, first believed to be by means of conjugation (Roberts and Falkow, 1978), has been shown to be due to transformation (Sox

et al., 1979). Transformation has been the major procedure used in taxonomic studies of the genus (see Taxonomic Comments).

The antigenic make-up of the cell walls of gonococci has been studied in detail with particular emphasis on those components associated with virulence and on those antigens that may be used in rapid diagnosis, detection of disease in asymptomatic patients, and serological classification (Danielsson and Normark, 1980). Currently, immunofluorescence is routinely used in diagnostic laboratories for identification of *N. gonorrhoeae* (Deacon et al., 1959). Co-agglutination, using specific antibody-coated, Protein A-containing cells of *Staphylococcus aureus*, has also been useful in identification of gonococci (Danielsson and Maelund, 1978) and serological classification into the three antigenic groups W, J, and M (Danielsson and Sandstrom, 1980). The detection and use of the humoral antibody response in suspected but asymptomatic patients has not been successful.

N. meningitidis has been classified into serogroups A, B, C, D, X, Y, Z¹ (otherwise known as 29E), and W-135, based mainly on serological and chemical analysis of surface polysaccharides (see review by Vedros, 1978). Group D is rarely found in the human population. Further analysis of the complex lipoprotein-polysaccharide layer of the cell wall has resulted in further division of certain serological groups into serovars (see review by Frasch, 1979). One of the protein serotyping antigens (type 2) appears to correlate with virulence of group B and C strains (Frasch, 1977). The lipopolysaccharide (LPS) is devoid of O polysaccharide, but serotyping antigens have been observed by means of hemagglutination inhibition and solid-phase radioimmunoassay (Mandrell and Zollinger, 1977; Zollinger and Mandrell, 1977). The structural diversity of the LPS based on galactose and glucose ratios resulted in three distinct serological categories (Jennings et al., 1980). The cell envelope proteins and LPS of of most of the "nonpathogenic" neisseriae have been studied but not in sufficient detail to be of determinative value (Johnson et al., 1976; Russell et al., 1975; see also Taxonomic Comments).

The principal habitats of those *Neisseria* species isolated from humans are the mucous membrane surfaces. Similarly, for domestic and experimental animals the mucosal surfaces of the oropharynx are the principal habitats. Only *N. meningitidis* and *N. gonorrhoeae* are considered to be primary pathogens for man (Vedros, 1978). The other *Neisseria* species isolated from humans have been responsible for disease (e.g. *N. mucosa* (Berger et al., 1974), *N. flavescens* (Branham, 1930)), but it is generally considered that these species are opportunists and rarely cause infection. The species isolated from animals other than man have caused infections possibly as primary pathogens (e.g. *N. ovis* (Lindqvist, 1960) or as opportunists (e.g. *N. canis* (Hoke and Vedros, 1982c)). These species are generally part of the normal flora but may have a broader host range as indicate by the isolation of *N. mucosa* and *N. cuniculi* from marine mammals (Vedros et al., 1973, 1982).

Enrichment and Isolation Procedures

Details of the isolation and processing of *Neisseria* species from humans have been provided by Morello and Bohnhoff (1980) and Finegold et al. (1978). In brief, specimens considered to be pure cultures (blood, spinal fluid, urethral pus) are plated on chocolate agar (blood heated to 80–90°C to lake the blood cells) and incubated at 36–37°C for a minimum of 48 h in a 3–10% CO_2 atmosphere having high humidity. Specimens from body sites that may contain contaminants (cervix, oropharynx) are plated on modified Thayer-Martin medium (Martin et al., 1974) or New York City medium (Faur et al., 1973) and incubated as above. Generally only morphology, Gram-stain, oxidase reaction, and acidification of certain sugar media are used for routine identification. A modified rapid procedure for detection of acid production from sugars can be employed (Vedros, 1978) which gives results in 2–4 h. Immunofluorescence is used for the gonococci and serological grouping by means of agglutination tests for the meningococci. Other *Neisseria* species from humans or animals can be isolated on Mueller-Hinton agar with or without 3% defibrinated sheep blood; once obtained in pure culture they can be identified by the characteristics listed later in Table 4.71.

Maintenance Procedures

N. meningitidis and *N. gonorrhoeae* are particularly sensitive to cold temperatures and autolysis (see the review by Morse et al., 1979, for a discussion of autolysis of the gonococcus) and therefore need to be transferred frequently. Frequent subculturing may lead to the loss of important factors such as fimbriae. The best method of preservation is by lyophilization in a rich broth (e.g. trypticase soy broth) containing 6% lactose, with storage of the vials at 4°C (Heckly, 1961). Cultures may also be preserved in soft agar stabs tightly stoppered and stored at 37°C, on chocolate agar slants covered with sterile mineral oil (Cody, 1978), or suspended in broth containing 10% glycerol or 10–50% serum and stored at −70°C (Morello and Bohnhoff, 1980). To conserve space, thick suspensions of a culture can be drawn into sterile capillary tubes and stored at −70°C (T. Maier, personal communication).

Procedures for Testing for Special Characters

Initial identification of neisseriae is based on observing Gram-negative diplococci (except for *N. elongata* cells, which are short rods) taken from colonies that are oxidase-positive when tested with tetramethyl-*p*-phenylenediamine (Kovacs, 1956). Another routine test used in initial identification is the production of acid from various sugars, either by inoculation of cystine trypticase agar supplemented with 1% sugar or by measuring preformed enzymes in a rapid sugar fermentation test (see reviews by Vedros, 1978, 1981). All *Neisseria* species except *N. elongata* are catalase-positive by testing with 3% hydrogen peroxide and observing the prompt evolution of bubbles of gas. The biochemical reactions with the least variability among the species are the nitrate/nitrite reduction test (Cowan, 1974) and the synthesis of polysaccharide from 5% sucrose. The latter test employs bacteria grown on heart infusion agar containing 5% sucrose; after incubation for 2 days the colonies are tested with Lugol's iodine solution diluted 1:4 and the immediate development of a blue color indicates a positive reaction.

Extracellular enzymes of most *Neisseria* species have been measured but γ-glutamyl aminopeptidase appears to be the only substrate useful for differentiating *N. gonorrhoeae* (negative) from *N. meningitidis* (positive) (D'Amato et al., 1978; Hoke and Vedros, 1982b).

The presence of carbonic anhydrase is tested as described by Berger and Issi (1971) and Berger and Piotrowski (1974). The test is performed by determining the minimum inhibitory concentration (MIC) of acetazolamide for cultures grown on Heart infusion agar (Difco) containing 5% bovine serum. The cultures are incubated under an air atmosphere and also under air +10% CO_2. In general, if the MIC for an air-grown strain is ~32 μg/ml or lower, the strain produces carbonic anhydrase; this is confirmed by finding a much higher MIC for the strain when grown under 10% CO_2. The confirmation is especially important when the MIC is borderline (e.g. 16–62 μg/ml), as may occur, for example, with some strains of *N. elongata*.

Differentiation of the genus **Neisseria** from other genera

The genus *Neisseria* (including the "false neisseriae" *N. caviae*, *N. ovis*, and *N. cuniculi*) and *Moraxella* (*Branhamella*) *catarrhalis* are the only taxa of the family *Neisseriaceae* that contain cocci. Distinguishing cocci from short rods by microscopic observation alone may sometimes be difficult, and a reliable test that can be applied in doubtful instances is to culture the organisms in the presence of subinhibitory levels of penicillin: rod-shaped organisms form long, stringy cells, whereas cocci retain their coccal morphology (Kaffka, 1964; Catlin, 1975). It should be noted that *N. elongata*, the only rod-shaped species in the genus *Neisseria*, produces long cells by this method.

Table 4.70 of the family *Neisseriaceae* lists other characteristics that differentiate the genus *Neisseria* from other members of the family.

N. elongata can be distinguished from acinetobacters by its positive oxidase reaction and from the rod-shaped moraxellae by its ability to reduce nitrite. It can be distinguished from kingellae by its generally larger and more opaque colonies and frequent inability to produce acid from glucose; more specifically, it differs from *K. kingae* by being nonhemolytic, from *K. indologenes* by being indole-negative, and from *K. denitrificans* by failing to reduce nitrate (although some exceptional nitrate-positive *N. elongata* strains have been found).

Taxonomic Comments

Classification of the neisseriae into well-defined species has been the subject of intense study and has been in a state of continuous flux for the last two decades (see reviews by Henriksen, 1976; Bøvre and Hagen, 1981; and Vedros, 1981).

The degree of genetic relatedness between *N. gonorrhoeae* and *N. meningitidis* is extremely high. Kingsbury et al. (1969) found by thermal stability of hybrid DNA duplexes that the two species had at least 80% similarity in their nucleotide sequences, and the DNA/DNA hybridization studies by Hoke and Vedros (1982b) indicated a homology value of 93%. On purely genetic grounds, these two species could be considered as subspecies of a single species. Yet *meningitidis* and *gonorrhoeae* cause distinctly different kinds of clinical infections, and from a practical viewpoint it seems desirable to continue to consider the organisms as separate species.

Lactose-positive strains of organisms resembling *N. meningitidis* were recognized as early as 1934 (Jessen, 1934) but were largely ignored until the report by Hollis et al. (1969). The specific epithet *lactamicus* was subsequently changed to *lactamica* for grammatical reasons (Catlin, 1971). *N. lactamica* showed a close relationship to *N. meningitidis* by transformation studies and to the other "true neisseriae" by DNA/DNA hybridization, cellular fatty acid composition, and mol% G + C values (Hoke and Vedros, 1982a, b).

The close relationships between all of the "true neisseriae" have been demonstrated consistently by a variety of techniques (see Bøvre, 1980; Bøvre and Hagen, 1981; Hoke and Vedros, 1982a, b). Of particular interest have been the similarities between *N. flava*, *N. perflava*, and *N. subflava* (Henriksen and Bøvre, 1968). It is difficult to differentiate these three species by cultural and biochemical reactions, and in the eighth edition of *Bergey's Manual* they were incorporated into the single species *N. subflava* (Reyn, 1974). This close similarity is supported by DNA/DNA hybridization and genetic transformation studies (Hoke and Vedros, 1982a). Although a high level of DNA homology between *N. sicca* and the *N. subflava* group has been found (Hoke and Vedros, 1982a), biochemical distinction between *N. perflava* and *N. sicca* has also been reported (Berger and Catlin, 1975). Further confusion in the classification of these chromogenic neisseriae has been added by the finding of identical lipopolysaccharides in *N. subflava* and *N. canis* (Johnson et al., 1976).

Although the species *N. elongata* consists of rod-shaped cells, genetic transformation studies have shown that *N. elongata* has very high genetic affinities to the coccal species of *Neisseria* (Bøvre and Holten, 1970; Bøvre et al., 1977) and has no affinities to other genera of the family *Neisseriaceae* except possibly a very low affinity to *Kingella kingae*. *N. elongata* also possesses carbonic anhydrase, which is characteristic of the "true" neisseriae. It also possesses a similar fatty acid composition (Hoke and Vedros, 1982b). The evidence indicates that this rod-shaped microorganism properly belongs in the genus *Neisseria*.

Catlin (1960, 1961) first noted that two distinct genetic groups existed within the genus *Neisseria*: "*N. catarrhalis*" was distinct from the other species based on the mol% G + C of its DNA and on transformation using streptomycin resistance (Str^r) markers (Catlin and Cunningham, 1964). It was further noted that "*N. catarrhalis*" was distantly related to the genus *Moraxella*, but caution in making any definite conclusions on this taxonomic relationship was suggested (Catlin, 1964).

"*N. catarrhalis*" was formally proposed to be transferred to the new genus *Branhamella*, in honor of the late Sarah Branham, and this proposal has been generally accepted since subsequent data (e.g. DNA/DNA homology experiments (Kingsbury, 1967; Kingsbury et al., 1969)) have confirmed that the *catarrhalis* organism is indeed distinct from other *Neisseria* species.

It was later proposed that certain other *Neisseria* species (*N. caviae* and *N. ovis*) also be transferred to the genus *Branhamella* or be assigned to the genus *Moraxella* (Henriksen and Bøvre, 1968). This proposal has recently been expanded to include the additional species *N. cuniculi* (Bøvre, 1980). The central issue remains as to whether *N. caviae*, *N. ovis* and *N. cuniculi* should (a) be included in the genus *Branhamella*, (b) constitute a separate subgenus of *Neisseria*, or (c) be transferred to *Branhamella* as a subgenus in the genus *Moraxella*. The various alternatives have been reviewed by Bøvre (1980) and Bøvre and Hagen (1981), and alternative (c) has been implemented in the chapter on *Moraxella* in this edition of the *Manual*.

Certain phenotypic characteristics are shared by the four species *catarrhalis*, *caviae*, *ovis* and *cuniculi* and tend to support the idea of a separate group:

1. None of the four species forms acid from carbohydrates;
2. They all hydrolyze tributyrin;
3. The species *catarrhalis*, *caviae* and *ovis* contain true waxes; *cuniculi* has not yet been tested (Bryn et al., 1977);
4. An analysis of the fatty acids with chain lengths over 16 carbon reveals a similarity among the four species (Jantzen et al., 1974a, b; Lambert et al., 1971; Hoke and Vedros, 1982b);
5. They lack carbonic anhydrase (Berger and Issi, 1971);
6. They do not form extracellular polysaccharide.

On the other hand, certain dissimilarities occur:

1. Nitrate is reduced by *catarrhalis*, *caviae* and *ovis* but not by *cuniculi* (Riou, 1977; Vedros, unpublished results);
2. Nitrite is reduced by *caviae* and *catarrhalis* but not by *ovis* and *cuniculi*;
3. The core oligosaccharides from the LPS of *catarrhalis* are significantly different from those of *caviae* (Johnson et al., 1976);
4. Nutritional requirements have not been studied in detail, but one report has indicated that *catarrhalis* requires biotin whereas *caviae* requires nicotinic acid (McDonald and Johnson, 1975).

It is generally accepted that genetic relatedness as shown by DNA base composition, genetic transformation, and DNA/DNA hybridization have a greater significance in the taxonomy of bacteria than many phenotypic expressions. The mol% G + C of the DNA of the *catarrhalis* species is 41–42, whereas that for *caviae*, *ovis* and *cuniculi* is 44.5–50.4. With regard to transformation, the ratios of interspecific to intraspecific transformation between *catarrhalis* vs. *caviae* and *ovis* have been reported to be 10^{-4}–10^{-5} (Bøvre, 1980; Hoke and Vedros, 1982b). Jones and Sneath (1970) have considered such levels of transformation to be below those sufficient for classifying different species within the same genus. With regard to transformation between *catarrhalis* vs. *cuniculi*, a discrepancy exists in published data: Bøvre (1980) reported a value of 1–2×10^{-3} whereas Hoke and Vedros (1982b) found a value of 2.6×10^{-5}. The reasons for this discrepancy are not clear.

Nucleic acid hybridization studies have indicated only a low degree of relatedness between *catarrhalis* vs. the other "false neisseriae." By use of pulse RNA/DNA (i.e. messenger RNA/DNA) hybridization, Bøvre (1970) obtained 2.4–2.9% homology between *catarrhalis* and *ovis*, 2.5% between *catarrhalis* and *caviae*, and 5.5% between *ovis* and *caviae*. Messenger RNA homology values for bacteria are similar to those obtained by DNA homology (Johnson, 1981), and the values obtained by Bøvre represent a relatively low degree of relatedness. By use of DNA/DNA hybridization (optical method), Hoke and Vedros (1982b) reported homology values of 39% between *catarrhalis* and *ovis*, 33% between *catarrhalis* and *cuniculi*, and 58% between *ovis* and *cuniculi*. Since the homology between *catarrhalis* and some of the "true neisseriae" (i.e. *N. mucosa* and *N. denitrificans*) was 24–35%, and since *catarrhalis* is considered to have only a low degree of relatedness to the "true neisseriae," it seems reasonable to conclude that *catarrhalis* also has only a low degree of relatedness to *ovis* and *cuniculi*. It should be remembered, however, that DNA/DNA homology experiments are most helpful when they are used to detect similarities between closely related, rather than distantly related, organisms (Johnson, 1981). Similarities between distantly related organisms are more appropriately studied by means of rRNA/DNA hybridization experiments. To date, no rRNA/DNA homology data have been reported for the neisseriae.

It would seem premature at this time, therefore, to assign *caviae*, *ovis* and *cuniculi* to a genus or subgenus having *catarrhalis* as the type species. It appears preferable to place them in the genus *Neisseria* as *species incertae sedis*, as has been done in this present article of the *Manual*, until their proper placement can be ascertained more clearly.

Further Reading

Catlin, B.W. 1978. Characteristics and auxotyping of *Neisseria gonorrhoeae*. *In*: Bergan and Norris (Editors), *Methods in Microbiology*, Vol. 10, Academic Press, London, pp. 345–380.

Henricksen, S.D. 1976. *Moraxella*, *Neisseria*, *Branhamella* and *Acinetobacter*. Annu. Rev. Microbiol. *30*: 63–83.

Morello, J.A. and M. Bohnhoff. 1980. *Neisseria* and *Branhamella*. In: Lennette, Balows, Hausler and Truant (Editors), *Manual of Clinical Microbiology*, 3rd Ed., American Society for Microbiology, Washington, D.C., pp. 111–130.

Reyn, A. 1974. The genus *Neisseria*. In: Buchanan and Gibbons (Editors), *Bergey's Manual of Determinative Bacteriology*, 8th Ed., Williams & Wilkins, Baltimore, pp. 428–432.

Vedros, N.A. 1978. Serology of the meningococcus. In: Bergan and Norris (Editors), *Methods in Microbiology*, Vol. 10, Academic Press, London, pp. 293–314.

Differentiation of the species of the genus **Neisseria**

Neisseria species are relatively inert biochemically, but those activities and characteristics which are of determinative value are shown in Table 4.71.

List of the species of the genus **Neisseria**

1. **Neisseria gonorrhoeae** (Zopf 1885) Trevisan 1885, 106.[AL] (*Merismopedia gonorrhoeae* Zopf 1885, 54.)

go.nor.rhoe'ae. Gr. n. *gonorrhoeae* gonorrhoea; M.L. gen. n. *gonorrhoeae* of gonorrhoea.

Common name: gonococcus.

The biochemical characteristics are as described for the genus and as listed in Table 4.71.

Primary isolation is made on chocolate blood agar at temperatures of 35–36°C under an atmosphere containing 3–10% CO_2 and high relative humidity. (Minimum growth temperature, 30°C). At 48 h, colonies are 0.6–1.0 mm in diameter, opaque, greyish white, raised, finely granular, glistening and convex. They become mucoid with further incubation.

Primarily found in purulent venereal discharges. Also found in blood, the conjunctiva, petechiae, pharynx, and cerebrospinal fluid.

The mol% G + C of the DNA is 49.5–53.3 (T_m, chromatography).

Type strain: ATCC 19424.

Further comments: This species show very high genetic relatedness to *N. meningitidis* (Kingsbury et al., 1969; Hoke and Vedros, 1982b).

2. **Neisseria meningitidis** (Albrecht and Ghon 1901) Murray 1929, 8.[AL] (*Micrococcus meningitidis* Albrecht and Ghon 1903, 498.)

me.nin.gi'ti.dis. Gr. n. *meninx*, *meningis* the membrane enclosing the brain; M.L. fem. n. *meningitis*, *meningitidis* inflammation of the meninges.

Common name: meningococcus.

Table 4.71.
Differential characteristics of the species of the genus **Neisseria**[a]

Characteristics	1. N. gonorrhoeae	2. N. meningitidis	3. N. lactamica	4. N. sicca	5. N. subflava	6. N. flavescens	7. N. mucosa	8. N. cinerea	9. N. denitrificans	10. N. elongata	11. N. canis	a. N. caviae	b. N. ovis	c. N. cuniculi
Shape of cells:														
Cocci	+	+	+	+	+	+	+	+	+	−	+	+	+	+
Rods	−	−	−	−	−	−	−	−	−	+	−	−	−	−
Arrangement of cells														
Pairs	+	+	+	+	+	+	+	+	+	+	+	+	+	+
Tetrads	−	−	−	+	+	+	−	−	−	−	−	−	−	−
Short chains	−	−	−	−	−	−	−	−	−	+	−	−	−	−
Yellowish pigment	−	−	+	d	+	+	d	d	d	weak	+	−	−	−
Hemolysis on blood agar:														
Sheep	−	−	−	d	−	−	−	−	−	−	−	d	d	−
Horse	−	−	d	d	−	−	−	−	−	−	−	d	+	−
Rabbit	−	−	−	d	−	−	−	−	−	−	d	d	+	−
Human	−	−	−	d	−	−	−	−	−	−	−	d	+	−
Acid produced from:														
Glucose	+	+	+	+	+	−	+	−	+	−c	−	−	−	−
Maltose	−	+	+	+	+	−	+	−	−	−	−	−	−	−
Fructose	−	−	−	+	d	−	+	−	+	−	−	−	−	−
Sucrose	−	−	−	+	d	−	+	−	+	−	−	−	−	−
Mannose	−	−	−	−	−	−	−	−	+	−	−	−	−	−
Lactose	−	−	+	−	−	−	−	−	−	−	−	−	−	−
Nitrate reduction	−	−	−	−	−	−	+	−	−	−	+	+	+	+
Nitrite reduction	−d	−d	+	+	+	+	+	+	+	+	−d	+	−	−
Gas from nitrite	−	−	+	+	+	+	+	+	+	+	−	−	−	−
Synthesis of polysaccharide (iodine test)	−	−	−	+	de	+	+	−	+	−	−	−	−	−
Tributyrin hydrolysis	−	−	−	−	−	−	−	−	−	−	−	+	+	+
Mol% G + C of DNA	50–53	50–52	52	49–52	48–51	46–50	50–52	49–51	56	53	50	44–50	44–50	45–47

[a] Data compiled from: Bøvre and Hagen (1981), Catlin (1978), Hollis et al. (1969), Morello and Bohnhoff (1980), Reyn (1974), Riou (1977) and Hoke and Vedros (1982a). Symbols: see standard definitions.
[b] See Editorial Note for the family *Neisseriaceae*. See also Table 4.72 of the genus *Moraxella* for additional characteristics of these species.
[c] A few strains may form a small amount of acid from glucose, but most strains are negative.
[d] Nitrite in low concentrations can be reduced by *N. gonorrhoeae* and by serogroups A, D, and Y of *N. meningitidis* (Berger, 1970).
[e] Biovar *perflava* is positive; the reaction differs among strains of biovar *subflava*.

Cellular division occurs in two planes with the second division at a right angle to the first. Transient tetrads can therefore be observed in wet mounts of young, growing cultures (less than 8-h-old). This is important in differentiating the organisms from short rod forms which divide in a single plane.

Primary isolation is made on blood (or chocolate) agar or Mueller-Hinton agar. An atmosphere containing 5–8% CO_2 enhances growth. Optimum temperature, 36–37°C. Colonies vary in size depending on the medium, extent of crowding, and length of incubation; in 18–24 h they are ~1.0 mm in diameter. Colonies are round, smooth, glistening, and translucent on Mueller-Hinton agar and are often iridescent. Due to autolysis with age, colonies become more butyrous and rubbery to the touch of an inoculating needle.

Other characteristics are listed in Table 4.71. Some strains are erratic in producing acid from maltose and glucose (Jyssum and Jyssum, 1968). Rapid passage (5- to 6-h-intervals) may restore normal acid production and seroagglutination (Vedros, 1978).

Found in cerebrospinal fluid as the causative agent of cerebrospinal meningitis. Also found in blood, petechiae, joints, nasopharynx, and conjunctiva. Occasionally found in venereal discharges. Also occasionally found in sputum as a causative agent of pneumonia.

The mol% G + C of the DNA is 50–52 (T_m, chromatography).

Type strain: ATCC 13077.
Further comments: N. meningitidis shows a very high genetic relatedness to *N. gonorrhoeae* (Kingsbury et al., 1969; Hoke and Vedros, 1982b).

3. **Neisseria lactamica** Hollis, Wiggins and Weaver 1959, 72.[AL]
lact.a.mi'ca. L. n. *lac* milk, from whence *lactose* milk sugar; L. adj. *amicus* fond of; M.L. fem. adj. *lactamica* fond of lactose.

The characteristics are as described for the genus and as listed in Table 4.71.

Colonies are smooth, translucent, slightly butyrous, and often have a yellowish tinge. They are closely similar to those of meningococci but are smaller and less glistening.

N. lactamica is the only species to produce acid from lactose and to give a positive reaction for the hydrolysis of *o*-nitrophenyl-β-D-galactopyranoside (ONPG).

Rarely pathogenic. Commonly found in the nasopharynx of infants and children.

The mol% G + C of the DNA is 51.7 (T_m) (Hoke and Vedros, 1982a).
Type strain: ATCC 23970.
Further comments: Genetic transformation and DNA/DNA hybridization have indicated a close relationship between *N. lactamica, N.*

meningitidis and *N. gonorrhoeae*, but not as close as that between *N. meningitidis* and *N. gonorrhoeae* (Siddiqui and Goldberg, 1975; Hoke and Vedros, 1982a).

4. **Neisseria sicca** (von Lingelsheim 1908) Bergey, Harrison, Breed, Hammer and Huntoon 1923, 43.[AL] (*Diplococcus siccus* von Lingelsheim 1908, 476.)

sic'ca. L. fem. adj. *sicca* dry.

The characteristics are as described for the genus and as listed in Table 4.71.

Forms dry, wrinkled, adherent colonies but this may vary with some strains. Some strains may produce a xanthophyll pigment (Berger, 1961); when extracted, this pigment shows absorption peaks similar to the pattern exhibited by the pigments of *N. perflava*, *N. flava* and *N. mucosa* (Hoke and Vedros, 1982b).

Spontaneous agglutination occurs in saline. The strain used in the report on envelope proteins is questionable (Russell et al., 1975). *N. sicca* may be serologically distinct from *N. subflava* (*N. flava*, *N. perflava*) and *N. flavescens* (Berger and Wulf, 1961; Berger and Brunhoeber, 1961). The species is serologically related to *N. mucosa* (Véron et al., 1961).

Found in the nasopharynx, saliva and sputum of man.

The mol% G + C of the DNA is 49.0–51.5 (T_m, Bd, chromatography).

Type strain: NRL 30,016.

Further comments: Genetic transformation and DNA/DNA hybridization data indicate that *N. sicca* should be considered for inclusion with *N. perflava* and *N. flava* into the single species *N. subflava* (Hoke and Vedros, 1982a).

5. **Neisseria subflava** (Flügge 1886) Trevisan 1889, 32.[AL] (*Micrococcus subflavus* Flügge 1886, 159; *Neisseria flava* Bergey, Harrison, Breed, Hammer and Huntoon 1923, 43,[AL] *Neisseria perflava* Bergey, Harrison, Breed, Hammer and Huntoon 1923, 43.[AL]).

sub.fla'va. L. pref. *sub* less than; L. adj. *flavus* yellow; L. fem. adj. *subflava* yellowish.

The characteristics are as described for the genus and as listed in Table 4.71.

Colonies are smooth, transparent or opaque, often adherent. Some strains produce a yellowish pigment.

Often agglutinates spontaneously in saline (Reyn, 1974).

Found in the secretions from the human naspharynx and rarely in cerebrospinal fluid in cases of meningitis (Noguchi et al., 1963).

The mol% G + C of the DNA is 48.0–51.0 (T_m, Bd, chromatography).

Type strain: NRL 30,017.

6. **Neisseria flavescens** Branham 1930, 849.[AL]

fla.ves'cens. L. v. *flavesco* to become golden yellow; L. part. adj. *flavescens* becoming golden yellow.

The characteristics are as described for the genus and as listed in Table 4.71.

Colonies are smooth and opaque.

Found in cerebrospinal fluid from patients with meningitis and in blood in cases of septicemia. Rare (Branham, 1930; Wertlake and Williams, 1968).

The mol% G + C of the DNA is 46.5–50.1 (T_m, Bd, chromatography).

Type strain: ATCC 13120.

7. **Neisseria mucosa** (von Lingelsheim 1906) Véron, Thibault and Second 1959, 508.[AL] (*Diplococcus mucosus* von Lingelsheim 1906, 395.)

mu.co'sa. L. fem. adj. *mucosa* slimy.

The characteristics are as described for the genus and as listed in Table 4.71.

Colonies are mucoid and often adherent. Most strains are nonpigmented (Berger and Miersch, 1970) but one strain has been described as slightly yellow (Véron et al., 1959).

Found in the nasopharynx of man and, in one report, as part of the normal flora of the respiratory tissues in dolphins (Vedros et al., 1973). Occasionally pathogenic for man, causing pneumonia. Pathogenic for mice.

The mol% G + C of the DNA is 50.5–52.0 (T_m, Bd).

Type strain: ATCC 19606.

Further comments: An organism called "*N. mucosa* subsp. *heidelbergensis*" was described by Berger (1971). It differed from "*N. mucosa* subsp. *mucosa*" in that some strains were pigmented, possessed deoxyribonuclease, and produced more gas from nitrite. It is questionable as to whether a subspecies should be recognized.

8. **Neisseria cinerea** (von Lingelsheim 1906) Murray 1939, 283.[AL] (*Micrococcus cinereus* von Lingelsheim 1906, 396.)

ci.ne're.a. L. fem. adj. *cinerea* gray.

The cocci are plump and arranged in pairs or more often in scattered clusters.

Colonies are small (1.0–1.5 mm in diameter), grayish white with entire edges, and slightly granular.

The percentage of fatty acids having a chain length of over 16 carbon atoms is similar to that of the "true" neisseriae (Hoke and Vedros, 1982b). Carbonic anhydrase is produced (Berger and Issi, 1971).

Other characteristics are listed in Table 4.71.

Found in the nasopharynx of man.

The mol% G + C of the DNA is 49.0–50.9 (T_m, Bd) (Bøvre et al., 1969; Hoke and Vedros, 1982b).

Type strain: ATCC 14685.

9. **Neisseria denitrificans** Berger 1962, 455.[AL]

de.ni.tri'fi.cans. L. prep. *de* away from; L. n. *nitrum* soda; M.L. n. *nitrum* nitrate; M.L. v. *denitrifico* to denitrify; M.L. part. adj. *denitrificans* denitrifying.

The percentage of fatty acids with chain length over 16 carbon atoms and the fatty acid profile are similar to those of the "true" neisseriae (Hoke and Vedros, 1982b). Carbonic anhydrase is produced (Berger and Issi, 1971).

Other characteristics are as listed in Table 4.71.

Isolated from the throats of guinea pigs.

The mol% G + C of the DNA is 55.6 (T_m).

Type strain: ATCC 14686.

10. **Neisseria elongata** Bøvre and Holten 1970, 73.[AL]

e.lon'ga.ta. L. fem. part. adj. *elongata* elongated, stretched out.

Rods, short and slender, ~0.5 mm in diameter, often arranged as diplobacilli or in short chains. A marked elongation effect of sublethal concentrations of penicillin occurs during growth, with formation of very long filaments. Capsules are not formed. Nonmotile. May be fimbriated.

Colonies on blood agar are low convex or almost flat, ~2–3 mm in diameter after 48 h of incubation. Older colonies may attain a diameter of 4–5 mm and often show granular spreading zones around the periphery, or the colonies become irregular in outline with spreading projections. Agar corrosion with a peripheral groove and a central pit is often observed under the colony. The spreading and corrosion are related to the presence of fimbriae on the cells; nonfimbriated or less fimbriated variants give rise to colonies that are often smaller, nonspreading, and noncorroding. The colonies are semiopaque to grayish white with a yellowish tinge due to pigment production. The colony texture is usually claylike and coherent, and the growth mass when collected is lumpy and difficult or impossible to disperse. No hemolysis occurs.

Capable of growing weakly on blood agar under anaerobic conditions.

Grows on simple peptone media. The specific growth requirements are not known.

Catalase activity is usually not detectable but may be positive or weakly positive with some strains. Usually acid is not formed from glucose, but some strains may form small amounts of acid. No liquefaction of coagulated serum or gelatin occurs. Urease-negative. Phenylalanine is not produced except for weak reactions observed with some strains. Carbonic anhydrase is produced (Berger and Issi, 1971). Highly sensitive to penicillin. The species lacks true cellular waxes (Bryn et al., 1977) and contains heptose (Jantzen et al., 1976).

Other characteristics are listed in Table 4.71.

Isolated from the pharynx of healthy individuals and from cases of pharyngitis. Also isolated from bronchial aspirates, pus from perimandibular abscesses and from the urinary tract. So far recorded only from human sources. Considered as a largely harmless parasite.

The mol% G + C of the DNA is 53.0–53.5.

Type strain: ATCC 25295 (NCTC 10660; strain M2).

Further comments: The cells are frequently competent in genetic transformation and can be identified by transformation. The competence is apparently associated with the fimbriated state (Bøvre and Holten, 1970; Bøvre et al., 1977).

The species appears somewhat heterogeneous genetically (Bøvre et al., 1972). There is a distinct genetic affinity between *N. elongata* and the coccal species of the "true" neisseriae. The species also fits in with the "true" neisseriae with respect to cellular lipid and carbohydrate composition and the characteristics of glycolytic enzymes (see reviews by Bøvre, 1980, and Bøvre and Hagen, 1981).

10a. Neisseria elongata subspecies **elongata** Bøvre and Holten 1970, 73.[AL]

Differs from the subspecies *glycolytica* and "*intermedia*" by showing no acidification of glucose media and being catalase-negative.

10b. Neisseria elongata subspecies **glycolytica** Henriksen and Holten 1976, 480.[AL]

gly.co.ly'ti.ca. Gr. *glyko-*, from Gr. *glykys* sweet; Gr. adj. *lyticus* dissolving; M.L. fem. adj. *glycolytica* meant to indicate an ability to attack glucose.

Differs from the subspecies *elongata* by being catalase-positive, by causing a weak acidification of glucose media, and by forming colonies with a smooth texture. Quantitative genetic transformation data show identity reactions between the two subspecies, and also they are indistinguishable in terms of fatty acid composition (Bøvre et al., 1977).

Type strain: ATCC 29315.

Further comments. A third subspecies, "*N. elongata* subspecies *intermedia*," was proposed by Berger and Falsen (1976). This subspecies presently has no standing in nomenclature. It is catalase-positive and immunologically distinct from the type strain of the subspecies *elongata*. It may possibly be identical to the subspecies *glycolytica*.

11. Neisseria canis Berger 1962, 455.[AL]

ca'nis. L. gen. n. *canis* of the dog.

The cells are typical diplococci, rarely tetrads.

Colonies are smooth, butyrous, with a light yellowish tinge. Absorption peaks of the extracted pigment are similar to those of *N. lactamica* (Hoke and Vedros, 1982b).

The cellular fatty acids are similar to those of the "true" neisseriae. Carbonic anhydrase is produced (Berger and Issi, 1971).

Other characteristics are listed in Table 4.71.

Isolated from the throats of cats (Berger, 1962) and more recently as an opportunist in a cat-bite wound of a human (Hoke and Vedros, 1982c).

The mol% G + C of the DNA is 49.6 (T_m).

Type strain: ATCC 14678.

Further comments. DNA/DNA homology studies (Hoke and Vedros, 1982a) have indicated a higher degree of relatedness of *N. canis* to *N.*

mucosa and *N. perflava* (54–66%) than to *N. lactamica* (45%), *N. denitrificans* (29%) and to *Moraxella* (*Branhamella*) *catarrhalis* (16%).

Species Incertae Sedis

The following three species of "false neisseriae" are listed below under their original designations as *Neisseria* species for reasons discussed in the Taxonomic Comments section of this article. It is unlikely that they belong to the genus *Neisseria*, but in the author's opinion their proper taxonomic placement is not yet certain. It should be noted that the same species are also listed and described in the article on the genus *Moraxella* in this *Manual* as members of the subgenus *Branhamella*.

a. *Neisseria caviae* Pelczar 1953, 744.[AL] (*Moraxella caviae* (Pelczar 1953) Henriksen and Bøvre, 1968, 391[AL]; *Moraxella* (*Branhamella*) *caviae* (Pelczar 1953) Bøvre 1979, 404.[VP])

ca'vi.ae. M.L. fem. n. *Cavia* generic name of the guinea pig; M.L. gen. n. *caviae* of *Cavia*.

Typical cocci.

Colonies are nonpigmented but slightly brown and opaque on transparent media, and are low and conical. They average 2.0–2.5 mm in diameter after 48 h. Weakly hemolytic on sheep, horse, rabbit and human blood agar (similar to *N. sicca* and *N. ovis*).

Other characteristics are listed in Table 4.71.

Isolated from the throats of guinea pigs.

The mol% G + C of the DNA is 44.5–50.4 (T_m, Bd, chromatography).

Type strain: ATCC 14659 (NCTC 10293).

b. *Neisseria ovis* Lindqvist 1960, 165.[AL] (*Moraxella ovis* (Lindqvist 1960) Henriksen and Bøvre 1968, 391[AL]; *Moraxella* (*Branhamella*) *ovis* (Lindqvist 1960) Bøvre 1979, 404.[VP])

o'vis. L. gen. n. *ovis* of the sheep.

Typical cocci with frequent tetrads.

Colonies are grayish white, flat to slightly convex, friable and soft, and are ~2.5 mm in diameter after 48 h. They are hemolytic on human, rabbit and horse blood agar but variable on sheep blood agar.

Other characteristics are listed in Table 4.71.

Isolated from the conjunctiva of sheep and cattle.

The mol% G + C of the DNA is 44.5–50.4 (T_m, Bd, chromatography).

Type strain: ATCC 33078 (NCTC 11227; Lindqvist's strain 199/55).

c. *Neisseria cuniculi* Berger 1962, 455.[AL] (*Moraxella* (*Branhamella*) *cuniculi* Bøvre and Hagen 1981, 1522.[VP])

cun.i'cu.li. L. gen. N. *cuniculi* of the rabbit.

Typical cocci.

Colonies are conical, opaque, friable, and nonhemolytic.

Other characteristics are listed in Table 4.71.

Isolated from the mouth of healthy rabbits (Berger, 1962) and the nasopharynx of marine pinnipeds (Vedros et al., 1982).

The mol% G + C of the DNA is 44.6–46.8 (T_m).

Type strain: ATCC 14688 (CTC 10297; Berger strain K19).

The following species appears to be closely related to *N. denitrificans* but its taxonomic position is not yet certain.

d. *Neisseria animalis* Berger 1960, 160.[AL]

See Berger (1962) for description.

Type strain: ATCC 14678.

Genus II. **Moraxella** *Lwoff 1939, 173 emend. Henriksen and Bøvre 1968, 391*[AL]

KJELL BØVRE

(Includes *Branhamella* Catlin, 1970, 157.)

Mo.rax.el'la. M.L. dim. ending *-ella*; M.L. fem. n. *Moraxella* named after V. Morax, a Swiss ophthalmologist who pioneered the recognition of the type species.

Rods (subgenus *Moraxella*) or cocci (subgenus *Branhamella*).* The rods are often very short and plump, frequently approaching a coccus shape (1.0–1.5 μm wide by 1.5–2.5 μm in length; they usually occur in pairs and short chains (one plane of division).

---

* See Editorial Note for the family *Neisseriaceae*.

Variation in cell size, shape, and filament or chain formation is often seen in cultures, the pleomorphism being enhanced by lack of oxygen and by incubation temperatures above the optimum. **The cocci are usually smaller** (0.6–1.0 μm in diameter) and occur as single cells or in pairs with the adjacent sides flattened (differing planes of division); division in two planes at right angles to each other sometimes results in the formation of tetrads. May be capsulated. Gram-negative, but often with a tendency to resist Gram decolorization. Flagella are absent. Both rod shaped and coccal species may be fimbriated. Swimming motility is absent, but surface-bound "twitching motility" has been observed in some rod-shaped species. **Aerobic, but some strains may grow weakly under anaerobic conditions.** Most species (exception: M. (M.) osloensis) are nutritionally fastidious, but the specific growth requirements are unknown. Optimum temperature, 33–35°C. **Colonies are not pigmented. Oxidase-positive** (with either tetra- or dimethyl-p-phenylenediamine reagent). **Usually catalase-positive.** Chemoorganotrophic. **No acid is produced from carbohydrates.** Usually high sensitive to penicillin. Parasitic on the mucous membranes of man and other warm-blooded animals. The mol% G + C of the DNA ranges from 40.0–47.5 in each of the two subgenera.

Type species: *Moraxella (Moraxella) lacunata* (Eyre 1900) Lwoff 1939, 173.

Further Descriptive Information

Some species of the subgenus *Moraxella*, as well as *Moraxella (Branhamella) catarrhalis*, have been shown to possess fimbriae (Bøvre et al., 1970, 1976; Wistreich and Baker, 1971; Bøvre and Frøholm, 1972a). In the two species M. (M.) nonliquefaciens and M. (M.) bovis the fimbriated state is closely associated with the occurrence of agar-corroding and/or spreading colonies (Bøvre et al., 1970; Bøvre and Frøholm, 1972a) and surface-bound "twitching motility" of the cells Henrichsen et al., 1972). For M. (M.) bovis the fimbriated state has been shown to be of importance for infection of the bovine conjunctiva (Pedersen et al., 1972). The fimbriation is also associated with competence in genetic transformation in those species where the covariation of colony type permits such studies (Bøvre and Frøholm, 1970, 1971, 1972b). Loss of fimbriation and associated characters may occur with high frequency in cultures. Similar phenomena have been clearly demonstrated in *Kingella kingae* (Bøvre and Frøholm, 1971, 1972b; Frøholm and Bøvre, 1972; Henrichsen et al., 1972) and *Neisseria elongata* (Bøvre et al., 1977a). These characteristics are therefore of limited determinative value, except when used as a basis for selection of recipients in genetic identification by transformation (see under each species).

The genus *Moraxella* has been systematically studied and compared with other genera of the family with respect to cellular lipid composition. The genus is clearly distinguishable in these terms and some species of the subgenus *Moraxella* have specific fatty acid patterns of high identification value (Jantzen et al., 1974b, 1975). Both subgenera *Moraxella* and *Branhamella* contain waxes (except M. (M.) phenylpyruvica) (Bryn et al., 1977). They also lack cellular heptose (Jantzen et al., 1976).

Many physiological characteristics of the members of the genus are presented later in Table 4.72 (see also the Key to the Species of the Genus *Moraxella*). A comprehensive description of the rods of the subgenus *Moraxella* can be found in the review by Henriksen (1973). A corresponding description of the coccal species of the subgenus *Branhamella* has been published by Berger (1963), and both subgenera are covered in the reviews by Henriksen (1976) and Bøvre and Hagen (1981). Baumann et al. (1968) have described nutritional and other physiological properties of strains belonging to several rod-shaped and coccal species of the genus. Berger and Issi (1971) and Berger and Piotrowski (1974) have shown that the species of the genus *Moraxella* do not possess carbonic anhydrase (M. (M.) bovis and M. (M.) atlantae have not been examined). Other enzymatic characteristics are mentioned in the chapters on the family *Neisseriaceae* and the genus *Neisseria* in this *Manual*.

The sources of the various species of the genus are listed later in Table 4.72. The organisms are generally considered not to be of high pathogenicity and are usually harmless parasites of the mucous membranes of man and/or other animals, but most species may be opportunistic pathogens. M. (M.) lacunata was considered to be a significant causative agent of human conjunctivitis and keratitis in the past but is only rarely isolated at present. M. (M.) bovis is potentially pathogenic for cattle, being associated with infectious keratoconjunctivitis. M. (M.) phenylpyruvica (from man and other mammals) and M. (M.) osloensis (from man) may also be significant potential pathogens, and the frequently isolated human parasite M. (B.) catarrhalis may be of some clinical importance as an invader outside its normal habitat, the nasal cavity. A similar role of M. (M.) nonliquefaciens, having the same natural habitat, cannot be excluded at present.

Enrichment and Isolation Procedures

All species of the genus *Moraxella* will grow on ordinary blood agar media, but some small-colony strains of M. (M.) lacunata grow better on media containing heated blood (i.e. "chocolate" agar). Most strains, with a few exceptions among the aforementioned strains of M. (M.) lacunata, will grow on ordinary rich media without blood, such as Tryptose blood agar base (Oxoid) and heart infusion broth (Difco). Mueller-Hinton broth (Difco) supplemented with 5% yeast extract (Difco), with or without agar, supports growth generally, although weakly for some strains. The addition of serum may improve growth. A mineral salt medium with ammonium ions as the nitrogen source and acetate as the carbon source will be sufficient for most strains of M. (M.) osloensis and the species known as *Moraxella urethralis* (see Species Incertae Sedis), but is rarely or never used for isolation from clinical specimens; rather, blood agar media are generally used because of their differential value and broad coverage of the growth requirements of *Moraxella* and other genera of the family *Neisseriaceae*.

Selective procedures are not well developed for the genus *Moraxella*.

Incubation may generally be performed at 33–35°C in a humid atmosphere. Some strains of M. (M.) nonliquefaciens do not grow at 37°C in an ordinary dry atmosphere, and it is preferable to use a slightly lower temperature also at a high humidity. For coverage of the little-known group of psychrophilic, oxidase-positive rods mentioned under Taxonomic Comments, a parallel incubation at a lower temperature, e.g. 22°C, is necessary. A CO_2-enriched atmosphere during incubation is generally not required for moraxellae, but may improve the growth of some M. (M.) lacunata strains.

Maintenance Procedures

It has been found for some species of *Moraxella* that agar cultures may remain alive for several weeks at room or refrigerator temperatures under humid conditions; however, strains may frequently be lost when the interval between transfers exceeds a few days. The relatively frequent subculturing necessary may easily contribute to loss of variable factors that occur in recently isolated strains, such as fimbriation, and, therefore, maintenance of strains by lyophilization is important. Freezing at −70 to −80°C is a valuable supplement (see the genus *Neisseria* for suspending media).

Procedures for Testing for Special Characters

In the initial steps of identification of these bacteria, the oxidase reaction is of central importance. The sensitivity of the test when performed with dimethyl-p-phenylenediamine is considerably lower than when the tetramethyl reagent is used (Ellingworth et al., 1929). All species of *Moraxella* show a positive reaction with the dimethyl reagent, as do members of *Neisseria*, *Kingella kingae*, and the species named *Moraxella urethralis*; however, some other species of importance in the differential diagnosis of *Moraxella*, such as *Kingella indologenes*, *K. denitrificans*, *Eikenella corrodens* and *Cardiobacterium hominis*, are usually negative or only weakly positive with the dimethyl reagent, whereas they show a distinctly positive reaction with the tetramethyl reagent (Bøvre and Hagen, 1981).

Procedures for biochemical/cultural characterization and differentiation of *Moraxella* species are generally insufficiently developed, due to a large extent to the limited knowledge of the nutritional requirements of most species. Suboptimum growth during the tests may cause variable results and discrepancies between laboratories. The Minimal Standards for Description of New Taxa within the genera *Moraxella* and *Acinetobacter* (Bøvre and Henriksen, 1976) describe the methods of biochemical/cultural examination in current use, but the standards need improvement. Some refinements and additions to the standards may be found in the paper by Bøvre et al. (1976). For the nitrate and nitrite reduction tests, Mueller-Hinton broth (Difco) with 0.5% yeast extract (Difco) is recommended as the basal medium; this also takes into account the needs in differential diagnosis of *Moraxella* vs. *Kingella* and other genera (Bøvre and Hagen, 1981).

Cellular fatty acid analysis, when performed under well controlled standard conditions, may give valuable information for identification within the genus *Moraxella*, but some species are almost indistinguishable by this means (Jantzen et al., 1974a, b; Bøvre et al., 1976).

The procedures for genetic identification by transformation have been reviewed by Bøvre (1980) and Bøvre and Hagen (1981). These techniques may be superior to all other identification methods, particularly when strains of *Moraxella* are examined from a variety of habitats and hosts (see Further Comments following each species description).

Colony-form variation of the rod-shaped species, including corrosion of the agar and spreading of colonies, is studied upon prolonged incubation of blood agar cultures (Bøvre and Frøholm, 1972a). Fimbriation and twitching motility, often associated with the corroding/spreading colony forms, are studied by methods described by Bøvre and Frøholm (1972a) and Henrichsen et al. (1972).

Differentiation of the genus **Moraxella** from other genera

The subgenus *Branhamella* of the genus *Moraxella* may be differentiated from the coccal species of the genus *Neisseria* by its uniformly negative reactions in carbohydrate acidification tests, frequent reduction of nitrate, and absence of pigment production. When use is also made of the regularly friable, nonadherent texture of *M. (B.) catarrhalis* colonies, the distinction of *Branhamella* strains of human origin vs. the genus *Neisseria* is, from a practical viewpoint, easy and safe. In animal bacteriology, however, the distinction of *Branhamella* strains from *Neisseria* may be more difficult, and sometimes is dependent on tests for genetic affinity, composition of cellular lipids, and the presence or absence of carbonic anhydrase (Bøvre, 1980; Bøvre and Hagen, 1981) (see also Further Descriptive Information and Taxonomic Comments, and also the articles on the family *Neisseriaceae* and the genus *Neisseria* in this *Manual*).

The subgenus *Moraxella* is distinguished from the rod-shaped *Neisseria elongata* by its generally larger cellular dimensions, its inability to reduce nitrite, and by failing to exhibit the clay-like colony texture that is characteristic of *N. elongata*. The subgenus *Moraxella* is easily distinguished from the genus *Kingella* by its positive catalase reaction, inability to acidify glucose media, and inability to reduce nitrite. *Moraxella* is distinguished from *Acinetobacter* by its positive oxidase reaction. *M. (M.) osloensis* is distinguished from the species named *Moraxella urethralis* by its larger cells, failure to form white colonies, and by its failure to reduce nitrite. Fatty acid analysis may be helpful and genetic transformation decisive in the differentiation of the rod-shaped moraxellae from rods of other genera.

Some strains of the subgenus *Moraxella*, particularly of *M. (M.) osloensis*, may have a coccal appearance when recently isolated and may be difficult to differentiate from members of the subgenus *Branhamella* and from coccal *Neisseria* species. The effect of sublethal concentrations of penicillin in causing moraxellae to form elongated cells, in contrast to branhamellae and coccal neisseriae, may be helpful in this regard (see also the chapter on the genus *Neisseria* in this *Manual*). It should be noted that *M. (M.) osloensis*, which frequently exhibits coccus-like cells, forms fusiform cells with penicillin instead of the very long cells formed by other rod-shaped moraxellae and by *N. elongata* (Bøvre et al., 1977b; Bøvre and Hagen, 1981).

Taxonomic Comments

The unification of rod-shaped and coccal species in the genus *Moraxella* has as its main basis the genetic studies referred to in Further Comments for the family *Neisseriaceae*. The coccal species of "false neisseriae" employed in the genetic studies leading to their proposed transfer to the genus *Branhamella* (Henriksen and Bøvre, 1968) had been designated as *Neisseria catarrhalis*, *N. caviae* and *N. ovis*. Both quantitative genetic transformation data and results of DNA/mRNA hybridization showed that these "false neisseriae" and the rod-shaped *Moraxella* species had affinities to each other that were as high as or higher than those observed between the species of either set. It should also be noted that the stringent DNA/mRNA hybridization experiments performed have shown almost as high affinities between some rod-shaped moraxellae and the branhamellae (including the *catarrhalis* organism) as that found between *Neisseria elongata* and *Neisseria subflava* (Bøvre, 1970). Moreover, nutritional and physiological data (Baumann et al., 1968) and studies of cellular lipid composition (Jantzen et al., 1974b, 1975; Bryn et al., 1977) revealed a high similarity between the two sets of species and also a distinctness from other *Neisseriaceae* (see discussions by Bøvre (1979, 1980) and Bøvre and Hagen (1981)). Catlin's formal proposal in 1970 to transfer *N. catarrhalis* to the new genus *Branhamella* was mainly a reflection of the genetic incompatibility of *N. catarrhalis* with the genus *Neisseria*. Several workers in the field find it most natural to include other "false neisseriae" in the genus *Branhamella*, but some consider the possibility that these other coccal organisms represent a genus separate from *Branhamella*. Neither of these alternative views has been formally proposed. Agreeing with the need for a special group designation for the "false neisseriae" for determinative and communicative reasons, Bøvre in 1979 modified his earlier view by proposing to divide the emended genus *Moraxella* into two subgenera—the subgenus *Moraxella* for six rod-shaped species, and the subgenus *Branhamella* for the three coccal "false neisseriae" species. Bøvre (1980) and Bøvre and Hagen (1981) later showed that *N. cuniculi* should also be included in the subgenus *Branhamella* based on genetic affinities.

Among the presently recognized species of the genus *Moraxella*, the genetic interrelations are highest between three species: *M. (M.) lacunata*, *M. (M.) bovis* and *M. (M.) nonliquefaciens*, i.e. between the "classical moraxellae"; here, the interactions in quantitative streptomycin resistance transformation are 10^{-2}–10^{-3} of an intrastrain or intraspecies reaction. Mutual affinities that are nearly as high can be found between the species of the subgenus *Branhamella* and between this subgenus and the "classical moraxellae," while *M. (M.) osloensis*, *M. (M.) phenylpyruvica* and *M. (M.) atlantae* appear more genetically separated from each other and from the rest of the members of the genus *Moraxella* (see reviews by Bøvre, 1980, and Bøvre and Hagen, 1981).

A search for oxidase-positive rods and cocci occurring as parasites in animals indicates that there may be numerous undescribed species of the genus *Moraxella*. Thus, several new, genetically homogeneous entities—rod shaped as well as coccal—with typical affinities to the two subgenera have been preliminarily circumscribed by transformation (Bøvre, 1980; Bøvre and Hagen, 1981). In these reviews are also mentioned recent studies of this sort using collections of strains from poultry, fish, and human clinical material, which indicate the existence of partly psychrophilic and partly also saccharolytic oxidase-positive rods having a distinct genetic affinity to the genus *Moraxella*. Such organisms have been described phenotypically by Thornley (1967; "phenon 3"), Bøvre et al. (1974), and by others cited by Lautrop (1974).

The species named *Moraxella urethralis* is listed under *Species Incertae Sedis*. This organism has very little or no genetic affinity to the genus *Moraxella* and also differs from this genus in some other respects (Bøvre and Hagen, 1981). It is presently considered as a candidate for a new genus of the family *Neisseriaceae*.

The species named *Moraxella saccharolytica* Flamm 1956 (type strain,

ATCC 19245) and *Moraxella anatipestifer* (Hendrickson and Hilbert 1932) Bruner and Fabricant 1954 (type strain, ATCC 11845) are not considered to belong to the genus *Moraxella*, nor in the family *Neisseriaceae*. It has been shown by gas-chromatographic fatty acid analysis that these taxa contain large amounts of branched fatty acids, which is a feature incompatible with the family *Neisseriaceae*, as preliminarily reported by Bøvre and Hagen (1981).

Further Reading

Berger, U. 1963. Die anspruchlosen Neisserien. Ergebn. Mikrobiol. Immunol. Exp. Ther. *36:* 97–167.

Bøvre, K. 1980. Progress in classification and identification of *Neisseriaceae* based on genetic affinity. In Goodfellow and Board (Editors), *Microbial Classification and Identification*, Soc. for Appl. Bacteriol. Symp. Series No. 8, Academic Press, London, pp. 55–72.

Bøvre, K. and N. Hagen. 1981. *Neisseriaceae*: rod-shaped species of the genera *Moraxella, Acinetobacter, Kingella* and *Neisseria*, and the *Branhamella* groups of cocci. In Starr, Stolp, Trüper, Balows and Schlegel (Editors), *The Prokaryotes: A Handbook on Habitats, Isolation and Identification of Bacteria*, Springer-Verlag, Berlin, pp. 1506–1529.

Henriksen, S.D. 1973. *Moraxella, Acinetobacter*, and the *Mimeae*. Bacteriol. Rev. *37:* 522–561.

Henriksen, S.D. 1976. *Moraxella, Neisseria, Branhamella*, and *Acinetobacter*. Annu. Rev. Microbiol. *30:* 63–83.

Key to the species of the genus **Moraxella**

I. Subgenus *Moraxella*. Rod-shaped organisms.
    A. No growth in mineral media with acetate and ammonium salts.
        1. Phenylalanine deaminase- and urease-negative.
            a. Coagulated serum liquefied.
                b. No hemolysis on blood agar. Nitrate reduced.
                      1. *M. (M.) lacunata*
                bb. Usually hemolytic on blood agar. Nitrate usually not reduced. (Serum may occasionally not be liquefied.)
                      2. *M. (M.) bovis*
            aa. Coagulated serum not liquefied. Nonhemolytic.
                b. Nitrate reduced. Large colonies formed.
                      3. *M. (M.) nonliquefaciens*
                bb. Nitrate not reduced. Colonies are small.
                      4. *M. (M.) atlantae*
        2. Phenylalanine deaminase-positive and/or urease reaction distinctly positive.
                5. *M. (M.) phenylpyruvica*
    B. Growth occurs in mineral medium with acetate and ammonium salts.
                6. *M. (M.) osloensis*[*]
II. Subgenus *Branhamella*. Coccal organisms.
    A. Nitrate and nitrite usually reduced.[†]
        1. Colonies have a friable texture. Nonhemolytic.
                7. *M (B.) catarrhalis*
        2. Colonies have a butyrous consistency. Weak hemolysis may be observed.
                8. *M. (B.) caviae*
    B. Nitrate usually reduced. Nitrite not reduced. Usually hemolytic.
                9. *M. (B.) ovis*
    C. No reduction of nitrate or nitrite. Nonhemolytic.
                10. *M. (B.) cuniculi*

List of the Species of the Genus **Moraxella**

Subgenus **Moraxella** (Lwoff 1939) Bøvre 1979, 404[VP]

1. Moraxella (Moraxella) lacunata (Eyre 1900) Bøvre 1979, 404.[VP] (*Bacillus lacunatus* Eyre 1900, 5; *Moraxella lacunata* (Eyre 1900) Lwoff 1939, 173.[AL]) Type species of the subgenus *Moraxella*.

la.cu.na′ta. L. n. *lacuna* a shallow depression; M.L. fem. adj. *lacunata* pitted.

Medium thick to plump rods, 0.8–1.2 µm in diameter, coccoid to distinctly bacillary, occurring predominantly in pairs and short chains. Frequently pleomorphic. May form narrow capsules.

Colonies on blood agar may be small (0.1–0.3 mm in diameter) in 48 h as seen with the type strain, or may be larger (up to 3.0 mm) as seen with strains previously named "*M. liquefaciens*" (e.g. ATCC 17952). Guinea pig isolates form colonies of intermediate size. Colonies are translucent to semiopaque. Pitting of the agar may be observed. No

hemolysis occurs, but on heated blood (chocolate) agar, large dark zones usually occur around the colonies.

Strains which form small colonies may show improved growth on chocolate agar medium and some of them may not grow on certain rich media without serum or oleic acid. Strains which form large colonies, and also guinea pig isolates, are less fastidious.

Other characteristics of the species are given in Table 4.72.

Human strains have been derived mainly from inflamed as well as healthy conjunctiva and from sites in the upper respiratory tract. They have occasionally been isolated from the blood. The organism appears to have been a significant causative agent of human conjunctivitis (and keratitis) and was frequently isolated in the past. It is only rarely isolated at present.

---

[*] Some strains of *M. (M.) phenylpyruvica* may grow in the minimal medium; these strains are distinguished from *M. (M.) osloensis* by the phenylalanine deaminase reaction and/or a strong urease reaction. The ability of *M. (M.) phenylpyruvica* to grow weakly at 4–5°C will further help to distinguish it from *M. (M.) osloensis*. Another species known as *M. urethralis* (see *Species Incertae Sedis*) will regularly grow in the minimal medium; this species can be distinguished by its white colonies, small cells, and reduction of nitrite.

[†] Until recently it has been uncertain to what extent nitrite-negative strains of *M. (B.) catarrhalis* and *M. (B.) caviae* may occur (Snell and Lapage, 1976). Most other reports indicate that the two species usually reduce nitrite (Berger, 1962, 1963; Doern and Morse, 1980). Recent studies of 57 genetically and gas-chromatographically identified *M. (B.) catarrhalis* strains showed a uniformly positive nitrite reduction test (Bøvre and Hagen, unpublished results).

Table 4.72.
Characteristics of the species of the genus **Moraxella**[a]

Characteristics	Subgenus *Moraxella*						Subgenus *Branhamella*			
	1. *M. (M.) lacunata*	2. *M. (M.) bovis*	3. *M. (M.) nonliquefaciens*	4. *M. (M.) atlantae*	5. *M. (M.) phenylpyruvica*	6. *M. (M.) osloensis*	7. *M. (B.) catarrhalis*	8. *M. (B.) caviae*	9. *M. (B.) ovis*	10. *M. (B.) cuniculi*
Cell shape:										
Rods	+	+	+	+	+	+	−	−	−	−
Cocci	−	−	−	−	−	−	+	+	+	+
Oxidase test[b]	+	+	+	+	+	+	+	+	+	+
Catalase test	+	d	+	+	+	+	+	+	+	+
Carbonic anhydrase, presence of	−		−		−	−	−	−	−	−
Acid from glucose	−	−	−	−	−	−	−	−	−	−
Hemolysis (human blood)	−	[+]	−	−	−	−	−	W	[+]	−
Gelatin liquefaction		[+]						−	−	
Serum liquefaction	+	[+]						−	−	
Growth on mineral salts + NH₄⁺ + acetate	−	−	−	−	[−]	[+]	−	−	−	−
Growth at 5°C	−	[−]	−	−	[+]	−	−	−	−	−
Growth in presence of 6% NaCl	−	−	−	−	[+]	−	−	−	−	−
Growth stimulated by bile salts	−	−	−	+	+	−	−	−	−	−
Phenylalanine deaminase	−	−	−	−	[+]	−	−	−	−	−
Indole production	−	−	−	−	−	−	−	−	−	−
Urease	−	−	−	−	d	[−][c]	−	−	−	−
Nitrate reduction	+	[−]	+	−	[+]	d	[+]	+	[+]	−
Nitrite reduction	−	−	−	−	−	−	[+]	[+]	−	−
Sensitive to penicillin										
1.0 U/ml	+	+	+	+	[+][d]	[+][d]	[+][d]	+	+	+
0.1 U/ml (β-lactamase-positive strains omitted)	+	+	+	+	+	d	+	−	+	+
In complex media, utilization of:										
Butyrate	+	+	−		−	[+]	d	+	+	−
Caproate	+	+	−		−	[+]	−	−	+	
Ethanol	d	+	−		−	+	−		−	
Propionate	−	−	−		−	+	−			
Lactate	+	+	+		+	+	[+]	+	+	+
Acetate	+	+	[−]		+	+	d	d	d	
Presence of thymidine phosphorylase, nucleoside deoxyribosyltransferase and thymidine kinase	−	−	[−][e]		−	−	−	−	−	−
Mol% G + C of DNa	40.0–45.5	41.0–44.5	40–44	46.5–47.5	42.5–43.5	43–46	40–43	44.5–47.5	44.5–46.5	44.5
True waxes present in cell wall	+	+	+	+	−	+	+	+	+	
Has been isolated from:[f]										
Humans	+		+	+	+	+	+			
Cattle		+			+				+	
Sheep					+				+	
Goats					+					
Horses		+							+	
Pigs					+					
Rabbits										+
Guinea pigs	+							+		

[a] Symbols: +, all tested strains positive (but see footnote *f*); [+], most strains positive; d, result differs among strains; [−], most strains negative; −, all tested strains negative; W, weakly positive.

[b] By use of either the tetra- or dimethyl-*p*-phenylenediamine reagent.

[c] A few strains are weakly urease-positive.

[d] Some strains have been found to be penicillin resistant on the basis of β-lactamase production (no β-lactamase-negative strain grows in the presence of 1.0 U/ml pencillin).

[e] The type strain is positive for thymidine kinase only (Jyssum and Bøvre, 1974).

[f] Mainly based on strains with genetically or gas-chromatographically confirmed identity; at least one isolate from a host is indicated by the symbol +.

Genetically verified strains have also been isolated from conjunctiva of healthy guinea pigs.

The mol% G + C of the DNA is 40.0–44.5.

Type strain: ATCC 17967 (NCTC 11011).

Further comments. The occurrence of genetically competent strains permits identification by quantitative genetic transformation. The species is slightly heterogeneous, apparently having three genetic clusters corresponding to small- and large-colony type human isolates and guinea pig isolates, respectively. The entities are all within 10% of intrastrain affinities, which is compatible with their classification as a single species (Bøvre, 1965c). Although there is no sharp distinction in fastidiousness between the strains, they may be classified as *M. (M.) lacunata* subsp. *lacunata* for strains like the type strain, and "*M. (M.) lacunata* subsp. *liquefaciens*" for strains like ATCC 17952. As indicated by their intermediate phenotype (and their genetic affinities), guinea pig strains may not fit well in either of these subspecific entities.

2. **Moraxella (Moraxella) bovis** (Hauduroy, Ehringer, Urbain, Guillot and Magrou 1937) Bøvre 1979, 404.[VP] (*Haemophilus bovis* Hauduroy et al. 1937, 247; *Moraxella bovis* (Hauduroy et al. 1937) Murray 1948, 591.[AL]) (Includes *Moraxella equi* Hughes and Pugh 1970, 462.[AL])

bo'vis. L. gen. n. *bovis* of cattle.

Rods, showing variation in shape, size and arrangement as for the subgenus; may occur as more slender rods. The cells are often fimbriated and show "twitching motility."

A distinct variation in colony type occurs. Freshly isolated strains form hemispherical to flat colonies (~1 mm in diameter in 48 h) on blood agar; they corrode the agar and often show surface spreading. These characteristics are associated with fimbriation of the cells and "twitching motility" (Bøvre and Frøholm, 1972a, Henrichsen et al., 1972; Pedersen et al., 1972). The texture of such colonies is often friable. Nonfimbriated variants occur in cultures spontaneously and form noncorroding, nonspreading colonies which are convex and usually larger than their corroding counterparts (often 3.0 mm after 48 h) and have a butyrous consistency. Fimbriated cells of collection strains may also form large colonies, but these may be less corroding than the colonies of fresh isolates. With few exceptions, bovine strains produce distinct hemolytic zones around colonies. Dark zones are produced on chocolate agar. Equine isolates, named *M. equi*, presumably of the same species (see Further Comments), are nonhemolytic but produce dark zones on chocolate agar.

Other characteristics are listed in Table 4.72.

Nutritional fastidiousness is about the same as that for large-colony strains of *M. (M.) lacunata*.

Most frequently isolated from bovine eyes in cases of infectious keratoconjunctivitis, but has also been isolated from unaffected eyes and the nasal cavity of cattle. Strains originally named *M. equi* were isolated from equine eyes in cases of conjunctivitis.

Considered as potentially pathogenic, dependent on accessory factors to elicit disease.

The mol% G + C of the DNA is 41.0–44.5.

Type strain: ATCC 10900.

Further comments. The cells are frequently transformable, genetic competence being associated with fimbriation and corrosion (Bøvre and Frøholm, 1972b). Strains may be identified by quantitative genetic transformation (Bøvre, 1965a, c). The species is slightly genetically heterogeneous, with interstrain affinities as low as 10% of intrastrain transformation affinities (Bøvre, unpublished results).

M. equi Hughes and Pugh 1972, 462 (type strain, ATCC 25576) has affinities to bovine isolates of *M. (M.) bovis* comparable to those between some bovine strains, as shown by streptomycin resistance transformation (Bøvre, unpublished results). The equine isolates reported are therefore considered here as strains of *M. (M.) bovis*, probably with differing host predilection.

Similar strains from goats have been isolated by Pande and Sekariah (1960) and named "*Moraxella caprae*"; however, this name has no standing in nomenclature and the species cannot be studied further because representative strains are nonexistent.

3. **Moraxella (Moraxella) nonliquefaciens** (Scarlett 1916) Bøvre 1979, 404.[VP] (*Bacillus duplex non liquefaciens* Scarlett 1916, 107; *Moraxella nonliquefaciens* (Scarlett 1916) Lwoff 1939, 171.[AL])

non.li.que.fa'ci.ens. L. pref. *non* not; L. part. adj. *liquefaciens* dissolving; L. part. adj. *nonliquefaciens* not dissolving.

Rods, showing variation in shape, size and arrangement as for the subgenus.

The colony size and colony type variation in relation to fimbriation and "twitching motility" are similar to that described for *M. (M.) bovis*, but the corroding colonies of freshly isolated strains may often be larger and more spreading (Bøvre et al., 1970; Bøvre and Frøholm, 1972a; Henrichsen et al., 1972). Colonies are translucent to semiopaque. Some strains are very mucoid. No hemolysis occurs on blood agar.

Other characteristics are listed in Table 4.72.

M. (M.) nonliquefaciens is apparently more fastidious than other *Moraxella* species, except for some strains of *M. (M.) lacunata.* Thus, it differs from most other moraxellae by failing to grow on Hugh and Leifson's OF medium.

Most frequently isolated from the nasal cavity, which is probably the main natural habitat. Also found in other sites of the respiratory tract, including bronchial aspirates in chronic bronchitis. It is considered to be a well established parasite of man with good adaptation to the host, rarely if ever causing disease. A possible clinical role as an extranasal invader has not been ruled out.

The mol% G + C of the DNA is 40–44.

Type strain: ATCC 19975 (NCTC 10464; Bøvre and Henriksen strain 4663/62).

Further comments. The cells are frequently transformable, with competence in genetic transformation being associated with fimbriation of cells and with corroding, spreading colonies (Bøvre and Frøholm, 1972b). Strains can be identified by quantitative genetic transformation (Bøvre, 1964, 1965a, c). The species is genetically homogeneous after its revised definition in 1967 (Bøvre and Henriksen, 1967a). Isolates from the human genitourethral tract and from animals have not been genetically verified as yet.

4. **Moraxella (Moraxella) atlantae** (Bøvre, Fuglesang, Hagen, Jantzen and Frøholm 1976) Bøvre 1979, 404.[VP] (*Moraxella atlantae* Bøvre et al., 1976, 520.)

at.lan'tae. M.L. gen. n. *atlantae* of Atlanta, the American city where strains of the species were first recognized as a distinct group by Elizabeth O. King.

Rods, showing variation in shape and size as for the subgenus. There is little tendency to form chains. Cells are often fimbriated and may show "twitching motility."

Colonies are small (0.2–0.5 mm in diameter, occasionally up to 1 mm) in 48 h on either blood agar or chocolate agar. Agar corrosion and spreading of colonies may be seen. Permanent dissociation of such colonies formed by fimbriated cells into nonfimbriated, noncorroding variants has not been observed. Colonies are semiopaque and may appear slightly pink, probably due to blood pigment accumulation. Nonhemolytic. Bile salt (Oxoid) stimulation of growth occurs with levels up to 1%. Penicillin-sensitive, but usually resists 0.05 U/ml penicillin.

Other characteristics are listed in Table 4.72.

Only rarely isolated. Found in human blood, cerebrospinal fluid, and spleen. The natural habitats and pathogenicity are not yet defined.

The mol% G + C of the DNA is 46.5–47.5.

Type strain: ATCC 29525 (NCTC 11091; CDC 5118).

Further comments. The strains are often competent in genetic transformation and can be identified by transformation methods (Bøvre et al., 1976). The cellular lipid composition closely resembles that of *M. (M.) phenylpyruvica*; however, *M. (M.) atlantae* differs by containing true waxes (Bøvre et al., 1976; Bryn et al., 1977).

5. **Moraxella (Moraxella) phenylpyruvica** (Bøvre and Henriksen 1967) Bøvre 1979, 404.[VP] Epit. spec. cons. Opin. 42, Jud. Comm. 1971, 107. (*Moraxella phenylpyrouvica* (sic) Bøvre and Henriksen 1967, 344.)

phe.nyl.py.ru'vi.ca. M.L. n. *acidum phenylpyruvicum* phenylpyruvic acid; M.L. fem. adj. *phenylpyruvica* pertaining to phenylpyruvic acid, the product of deamination of phenylalanine by this organism.

Rods, showing variation in shape, size and arrangement as for the subgenus.

The colonies are relatively small (0.9–1.0 mm in diameter) in 48 h on blood agar. They often appear semiopaque with a very slight pink hue, probably due to blood pigment accumulation. Nonhemolytic.

Phenylalanine deaminase- and urease-positive, but strains negative for one or, very rarely, both activities may occur. Grows distinctly, but slowly, at 4–10°C. Usually grows at high salt or bile concentrations; the minimum inhibitory concentrations are usually 7.5–9.0% NaCl and 5% bile salts (Oxoid). Bile salt stimulation of growth occurs, with levels up to 4% (Snell et al., 1972; Bøvre et al., 1976).

Although usually highly sensitive to penicillin, several strains of human and animal origin have been found to be penicillin resistant due to β-lactamase production.

Other characteristics are listed in Table 4.72.

Isolated from human blood and cerebrospinal fluid, the genitourethral tract, and from other sites and specimens of man; also isolated from the genital tract and brain of sheep and cattle, the intestine of a goat, and the genital tract of pigs. The pathogenicity is unknown, but the organism may possibly be a significant potential pathogen.

The mol% G + C of the DNA is 42.5–43.5.

Type strain: ATCC 23333 (NCTC 10526; CDC 2863).

Further comments. Strains competent in genetic transformation have not yet been found, but genetic relationships to other taxa of the genus *Moraxella* have been detected by the use of strains as DNA donors in transformation. The cellular lipid composition is unique among recognized *Moraxella* species in that true waxes are absent; In other respects the lipid composition resembles that of *M. (M.) atlantae* (Jantzen et al., 1974b; Bøvre et al., 1976; Bryn et al., 1977).

Some organisms that resemble *M. (M.) phenylpyruvica* in certain respects have been isolated from poultry, fish, and human clinical material (see Taxonomic Comments for the genus). These strains grow either at low temperatures only, or at both low temperatures and ordinary incubation temperatures; they may be saccharolytic, are phenylalanine deaminase-positive, and have a high salt tolerance and sometimes a high bile tolerance which may be combined with growth

stimulation by bile. The strains produce much larger and more opaque (whitish) colonies than *M. (M.) phenylpyruvica*; also, the cellular lipid composition differs from that of *M. (M.) phenylpyruvica* (Bryn et al., 1977; Bøvre et al., unpublished results). The taxonomic status of these strains is not yet clear. The genetic homogeneity and taxonomic status of these strains are not yet clear. A representative strain that grows at both 4°C and 33°C and is nonsaccharolytic is ATCC 17955.

6. Moraxella (Moraxella) osloensis (Bøvre and Henriksen 1967) Bøvre 1979, 404.[VP] (*Moraxella osloensis* Bøvre and Henriksen 1967, 131.)

os.lo.en'sis. M.L. adj. *osloensis* pertaining to Oslo, Norway, where the species was first recognized.

Rods, showing variation in size, shape and arrangement as for the subgenus. The cells may be indistinguishable from cocci in specimens or when freshly isolated. A distinct rod shape occurs with pronounced formation of fusiform cells when cultured at sublethal concentrations of penicillin. Intracellular inclusions of poly-β-hydroxy butyrate are usually formed, especially under nitrogen-limiting conditions.

Colonies are 2.0–2.5 mm in diameter in 48 h and are semiopaque. They do not exhibit colony type variation.

Strains usually grow in mineral media with ammonium ions and acetate, but one genetically verified exception has been found.

Highly or, most often, moderately sensitive to penicillin, but resistant strains which possess β-lactamase have been found.

Other characteristics are listed in Table 4.72.

Isolated from the upper respiratory tract, genitourethral specimens, blood, cerebrospinal fluid, and pyogenic manifestations in joints, bursae and other sites from humans. Not yet isolated with certainty from nonhuman sources. Usually considered to be a harmless parasite, but a significant potential pathogenicity is possible.

The mol% G + C of the DNA is 43–46.

Type strain: ATCC 19976 (NCTC 10465; CDC A1920).

Further comments. Strains are frequently competent in genetic transformation and identification by transformation methods is simple (Juni, 1974; Bøvre, 1965d; Bøvre et al., 1977b). The species has a distinct cellular fatty acid composition (Jantzen et al., 1974b; Bøvre et al., 1977b).

Subgenus **Branhamella** (Catlin 1970) Bøvre 1979, 404.[VP]

7. Moraxella (Branhamella) catarrhalis (Frosch and Kolle 1896) Bøvre 1979, 404.[VP] (*Mikrokokkus catarrhalis* (sic) Frosch and Kolle, *in* Flügge 1896, 15; *Moraxella catarrhalis* (Frosch and Kolle 1896) Henriksen and Bøvre 1968, 391[AL]; *Branhamella catarrhalis* (Frosch and Kolle 1896) Catlin 1970, 157.[AL]) Type species of the subgenus *Branhamella*.

ca.tarrh.a'lis. Gr. adj. *catarrhus* downflowing, catarrh; M.L. adj. *catarrhalis* of catarrh.

Cocci, of a size, shape and division pattern as for the subgenus. May be fimbriated.

Colonies are ∼2.0 mm in diameter in 48 h, hemispherical, becoming considerably larger and convex, almost flat, on prolonged incubation. They usually have a friable texture, without adherence to the agar, and are opaque. Nonhemolytic.

Usually highly sensitive to penicillin, but strains resistant on the basis of β-lactamase production are now found frequently.

Very frequently isolated from the nasal cavity of man, which is considered to be the main natural habitat of the species. Also found less commonly in the pharynx. Has also been isolated from inflammatory secretions of the middle ear and maxillary sinus, from bronchial aspirate in bronchitis and pneumonia, and occasionally from systemic infections. Isolates resembling *M. (B.) catarrhalis* from nonhuman hosts have not yet been confirmed as such. The organism most often is considered to be a well adapted parasite of man, rarely causing disease. It may possibly play a significant role as an extranasal invader

in the respiratory tract, e.g. by causing or contributing to the pathology in low grade otitis media of infants (Coffey et al., 1967) and respiratory disease in a compromised host (Ninaue et al., 1978). Further studies on these points are warranted.

The mol% G + C of the DNA is 40–43.

Type strain: ATCC 25238 (NCTC 11020; Catlin's strain Ne 11).

Further comments. The cells are frequently competent in genetic transformation and are easily identified by a quantitative transformation procedure. The species appears genetically homogeneous, but one deviating strain, NCTC 4103, has been detected (Catlin and Cunningham, 1964; Bøvre, 1965b, 1980; Bøvre and Hagen, 1981). This strain may represent a separate entity of *Branhamella*. By use of sensitive techniques, the type strain can be distinctly transformed by the DNA from all other species of the genus *Moraxella* listed here and to a generally lesser extent by *Acinetobacter*, but not by *Neisseria* or *Kingella* donors.

8. Moraxella (Branhamella) caviae (Pelczar 1953) Bøvre 1979, 404.[VP] (*Neisseria caviae* Pelczar 1953, 744,[AL] *Moraxella caviae* (Pelczar 1953) Henriksen and Bøvre 1968, 391.[AL])

ca'vi.ae. M.L. fem. n. *Cavia* generic name of the guinea pig; M.L. gen. n. *caviae* of *Cavia*.

Cocci, of a size, shape and mode of division as for the subgenus.

The colonies are low, conical, and have a diameter of ∼2.5 mm in

48 h. They are semiopaque and have a butyrous consistency. Weak hemolysis may be observed, and in this respect the colonies may resemble those of *M. (B.) ovis*.

Sensitive to penicillin, but in some studies the organism has been shown to resist 0.1 U/ml penicillin.

Other characteristics are listed in Table 4.72.

Isolated from the pharynx and mouth of healthy guinea pigs.

The mol% G + C of the DNA is 44.5–47.5.

Type strain: ATCC 14659 (NCTC 10293).

9. **Moraxella (Branhamella) ovis** (Lindqvist 1960) Bøvre 1979, 404.[VP] (*Neisseria ovis* Lindqvist 1960, 165;[AL] *Moraxella ovis* (Lindqvist 1960) Henriksen and Bøvre 1968, 391.[AL])

o'vis. L. gen. n. *ovis* of the sheep.

Cocci, of a size, shape and mode of division as for the subgenus. Tetrad formation may be prominent.

Colonies are ~2.5 mm in diameter in 48 h, low-convex, becoming almost flat after longer incubation. They are greyish white and slightly smaller and more opaque than the largest colony forms of *M. (M.) bovis*. The colonies are friable or soft, and are usually surrounded by a narrow zone of clear hemolysis. Nonhemolytic variants may arise spontaneously in cultures, and primarily nonhemolytic strains (genetically confirmed as belonging to the species) have been isolated directly from sheep and cattle (and a horse).

Other characteristics are listed in Table 4.72.

Isolated from the conjunctiva of sheep and cattle and from upper respiratory sites in sheep (and a horse). Frequently found in cultures from eyes in infectious keratoconjunctivitis of sheep, and often found together with *M. (M.) bovis* in cases of such disease in cattle. Considered to be of low pathogenicity, most frequently occurring as a harmless parasite.

The mol% G + C of the DNA is 44.5–46.5.

Type strain: ATCC 33078 (NCTC 11227; Lindqvist's strain 199/55).

Further comments. The strains are frequently competent in genetic transformation and can be easily identified by quantitative transformation. Other closely-related coccal and rod-shaped moraxellae, not yet described, occur in sheep and cattle, and these must be taken into consideration when identifying *M. (B.) ovis*, as well as *M. (M.) bovis*, in these animals (Bøvre, 1980; Bøvre and Hagen, 1981).

10. **Moraxella (Branhamella) cuniculi** (Berger 1962) Bøvre and Hagen 1981, 1522. (*Neisseria cuniculi* Berger 1962, 455.[AL])

cun.i'cu.li. L. gen. n. *cuniculi* of the rabbit.

Cocci, of a size, shape, and mode of division as for the subgenus.

The colony size in 48 h is similar to or slightly smaller than that of *M. (B.) catarrhalis*. The colonies are conical, opaque, and often friable. Nonhemolytic.

Other characteristics are listed in Table 4.72.

Isolated from the mouths of healthy rabbits.

The mol% G + C of the DNA is 44.5.

Type strain: ATCC 14688 (NCTC 10297).

Further comments. The type strain is competent in genetic transformation, making genetic identification of new isolates possible.

Strains named "*Neisseria cuniculi* var. *gigantea*" (Berger, 1962, 1963) are no longer available.

Species Incertae Sedis

The taxonomic status of the following species is not yet certain.

Moraxella urethralis Lautrop, Bøvre and Frederiksen 1970, 255.[AL]

u.re.thra'lis. Gr. *ourethra* urethra; M.L. gen. n. *urethralis* of the urethra.

Rod-shaped cells, but much smaller than those of the recognized species of the subgenus *Moraxella*, being only ~0.6 μm in diameter; they also lack the typical plumpness of moraxellae. Intracellular poly-β-hydroxybutyrate inclusions are regularly present but may be difficult to observe in the small cells. Nonmotile; do not possess flagella.

Colonies on blood agar are 1.5–3.0 mm in 48 h. The organism grows rather slowly and often have a tendency to grow best in areas of heavy inoculation. The colonies are more overtly white than those of all recognized species of *Moraxella* and of *Acinetobacter*. Nonhemolytic.

Nitrate is usually not reduced. Nitrite is reduced, often with gas formation. Gelatin and coagulated serum are not liquefied. Phenylalanine deaminase- and urease-negative. Acid is not formed from carbohydrates. Catalase- and oxidase-positive.

Usually able to grow on simple mineral media with ammonium ions as the nitrogen source and acetate or butyrate as the carbon and energy source.

Usually highly sensitive to penicillin.

Habitat: probably the human genitourethral tract. Has been isolated from urine and from the female genital tract; the pathogenicity is unknown.

The mol% G + C of the DNA is 46–47.

Type strain: ATCC 17960.

Further comments. Competence of strains in genetic transformation is of only a low degree. Genetic identification is possible by use of nutritional marker transformation (Juni, 1977). Genetic studies performed by streptomycin resistance transformation and nucleic acid hybridization do not indicate any distinct genetic affinity to the genus *Moraxella* or to other genera of the family *Neisseriaceae*. The cellular lipid composition is unique compared to other taxa of *Neisseriaceae*, which makes identification by use of gas chromatography possible. The species may be considered as a candidate for a new genus in the family *Neisseriaceae*. For description and taxonomic discussion of *Moraxella urethralis* the chapter on the family *Neisseriaceae* in this *Manual* and the review by Bøvre and Hagen (1981) may be consulted.

Genus III. *Acinetobacter* Brisou and Prévot 1954, 727[AL]

ELLIOT JUNI

A.ci.ne'to.bac.ter. Gr. adj. *akinetos* unable to move; M.L. n. *bacter* the masculine form of the Gr. neut. n. *bactrum* a rod; M.L. masc. n. *Acinetobacter* nonmotile rod.

Rods 0.9–1.6 μm in diameter and 1.5–2.5 μm in length, becoming spherical in the stationary phase of growth. They commonly occur in pairs and also in chains of varible length. Do not form spores. Gram-negative but occasionally difficult to destain. Swimming motility does not occur but the cells display "twitching motility," presumably because of the presence of polar fimbriae. Aerobic, having a strictly respiratory type of metabolism with oxygen as the terminal electron acceptor. All strains grow between 20 and 30°C, with most strains having temperature optima of 33–35°C. Grow well on all common complex media. **Oxidase-negative. Catalase-positive.** Most strains grow in defined media containing a single carbon and energy source; they use ammonium or nitrate salts as the source of nitrogen

and display no growth factor requirements. D-glucose is the only hexose utilized by some strains. The pentoses D-ribose, D-xylose, and L-arabinose can also be utilized as carbon sources by some strains. Occur naturally in soil, water and sewage. Can cause nosocomial infections in humans. The mol% G + C of the DNA is 38–47 (T_m, Bd).

Types species: *Acinetobacter calcoaceticus* (Beijerinck 1911) Baumann, Doudoroff and Stanier 1968, 1538.

Further Descriptive Information

Rapidly growing cells tend to be plump rods, whereas cells in the stationary phase of growth are spherical and have a somewhat smaller diameter than the rods (Baumann et al., 1968b). Cells occur typically in pairs. Many strains are encapsulated and capsules may be seen readily in India ink wet mounts. Electron microscopy of thin sections of cells have revealed a cell wall ultrastructure that is typical of Gram-negative bacteria (Breuil et al., 1975; Scott et al., 1976). The peptidoglycan contains muramic acid, glucosamine, alanine, D-glutamic acid and *meso*-diaminopimelic acid (Martin et al., 1973; Horisberger, 1977). One strain has been shown to possess a cell wall lipopolysaccharide containing D-glucose, glucosamine, galactosamine, lipid A, ethanolamine, fatty acids, phosphate, and protein (Adams et al., 1970).

Colonies are generally nonpigmented and are mucoid when the cells are encapsulated.

Most strains of *Acinetobacter* can grow in a simple mineral medium containing a single carbon and energy source such as ethanol, acetate, lactate, pyruvate, malate, or α-ketoglutarate. Ammonium and nitrate salts serve as nitrogen sources. A wide variety of organic compounds can be used as carbon sources by particular strains and the ability to degrade members of a series of organic compounds has been used to cluster strains into a series of phenotypic groups (Baumann et al., 1968b). Among organic compounds degraded are hydrocarbons (*n*-hexadecane) (Stewart et al., 1959), aromatic compounds such as benzoate or quinate (Baumann et al., 1968b), and alicyclic compounds such as cyclohexanol (Donoghue and Trudgill, 1975).

Relatively few strains can use glucose as a carbon source for growth (Juni, 1972). One strain that can use it has been shown to catabolize it via the Entner-Doudoroff pathway (Taylor and Juni, 1961). Although unable to grow on glucose, many acinetobacters contain an aldose dehydrogenase and are able to acidify glucose media (gluconic acid being formed) as well as media containing other sugars such as D-xylose, L-arabinose, D-galactose, D-mannose, L-rhamnose, maltose, lactose, and cellobiose (Hauge, 1960). The aerobic acidification of glucose media has been considered to have taxonomic significance: acid producers were formerly considered to be strains of "*Bacterium anitratum*" or "*Herellea vaginicola*," whereas those acinetobacters that are unable to acidify glucose media were referred to as strains of "*Moraxella lwoffi*" or "*Mima polymorpha*."

Nitrite and nitrate can serve as nitrogen sources for acinetobacters and these organisms possess an assimilatory nitrate reductase (Jyssum and Joner, 1965). Although most acinetobacters are not able to reduce nitrate to nitrite in the conventional nitrate reduction assay, a few strains are able to do so (Riley and Weaver, 1974); these strains are unable to grow anaerobically with nitrate as the terminal electron acceptor (Juni, 1972).

Since all acinetobacters are oxidase-negative, they lack cytochrome *c* (Baumann et al., 1968a). However, they do contain cytochromes of the *a* and *b* variety (Whittaker, 1971). They also contain all the enzymes of the tricarboxylic acid cycle as well as those of the glyoxylate cycle (Juni, 1978). Many acinetobacters elaborate an extracellular lipase as determined by hydrolysis of Tween 80 (Baumann et al., 1968b) and other lipids (Breuil and Kushner, 1975). Starch and poly-β-hydroxybutyrate are not hydrolyzed but a few strains hydrolyze gelatin (Baumann et al., 1968b). Strains that show hemolysis on blood agar plates excrete phospholipase (Lehmann, 1971).

Although rare, competence for genetic transformation has been observed in a few strains. Nutritional as well as antibiotic resistance markers are transformed readily in a competent strain (Juni, 1972). Genes concerned with capsule biosynthesis of a competent strain have been modified by transformation (Juni and Janik, 1969). A partial mapping of genes directing synthesis of enzymes for the tryptophan biosynthetic pathway in *Acinetobacter* has been accomplished using transformation (Sawula and Crawford, 1972). A genetic analysis of the genes for proline biosynthesis has also been reported (Ginther, 1978).

Lytic phages for acinetobacters are isolated readily from sewage (Twarog and Blouse, 1968; Herman and Juni, 1974). A generalized transducing phage specific for one strain of *Acinetobacter* has been isolated and demonstrated to lysogenize the host strain (Herman and Juni, 1974). Plasmids are found in many acinetobacters (Murray and Moellering, 1979; Hinchliffe and Vivian, 1980) and some plasmids can

be transferred conjugally from *Pseudomonas* to *Acinetobacter* where they can reside stably (Olsen and Shipley, 1973). Using such plasmids, it has been possible to mobilize the *Acinetobacter* chromosome and conjugally transfer chromosomal genes to a recipient strain (Towner and Vivian, 1976a). In this manner it has been demonstrated, using a series of genetic markers, that the *Acinetobacter* chromosome is circular (Towner and Vivian, 1976b).

Evolution of acinetobacters has resulted in a large variety of surface antigens on different strains. A serologic system for identification of strains by immunofluorescent staining revealed 28 serovars among strains capable of forming acid aerobically from glucose as well as a series of other serovars among acinetobacters that do not form acid from glucose (Marcus et al., 1969). The capsular polysaccharide of one strain of *Acinetobacter* has been shown to interact with antisera prepared against group B and group G streptococci as well as with antipneumococcal type XX serum (Heidelberger et al., 1969). An antigen extracted from *Acinetobacter* fixes complement when reacted with sera containing chlamydial antibodies (Brade and Brunner, 1979).

Unlike most moraxellae, acinetobacters are resistant to penicillin. Growth of most strains is not inhibited by 1 U/ml of penicillin G and the majority of strains are resistant to 100 U/ml (Baumann et al., 1968a). Acinetobacters of hospital origin appear to be resistant to most antibiotics with the exception of kanamycin, tobramycin, carbenicillin, trimethoprim-sulfamethoxazole, minocycline, and doxycycline (Crues et al., 1979).

Although considered to be normally nonpathogenic, acinetobacters are causative agents of nosocomial infections, particularly in debilitated individuals (Henriksen, 1973; Glew et al., 1977). Infection with *Acinetobacter* has been shown to result in septicemia, meningitis, endocarditis, brain abscess, lung abscess, pneumonia, empyema, and urinary tract infection as well as other clinical manifestations (Glew et al., 1977). There appears to be a clear association between hospital instrumentation and subsequent infection with *Acinetobacter*. Evidence has been presented indicating that nosocomial infection with *Acinetobacter* has a distinct seasonal variation (Retailliau et al., 1979).

Acinetobacters occur naturally in soil and water (Baumann, 1968) and are also present in sewage (Warskow and Juni, 1972). It has been estimated that at least 0.001% of the total heterotrophic aerobic population in soil and water are acinetobacters (Baumann, 1968). Although isolated frequently from several areas of the human body, there is some uncertainty as to whether acinetobacters are present as contaminants rather than as commensals. The fact that acinetobacters are frequently isolated from the skins of hospital inpatients (Al-Khoja and Darrell, 1979) can account for the occurrence of nosocomial infections with strains of this organism. Although isolated readily from the hands of hospital personnel, there is no evidence for persistant colonization with *Acinetobacter* (Buxton et al., 1978).

Enrichment and Isolation Procedures

Isolation of acinetobacters can be accomplished using ordinary laboratory media such as heart infusion agar, brain heart infusion agar, or trypticase soy agar. The use of a selective medium that differentiates organisms able to grow on such a medium, such as MacConkey agar or eosin-methylene blue agar, has been useful in recognizing acinetobacters upon primary isolation. Although many acinetobacters grow well at 37°C the optimum growth temperature for most strains is 33–35°C. Acinetobacters having lower optimum temperatures and unable to grow at 37°C have been reported (Breuil et al., 1975). A general enrichment procedure for isolation of acinetobacters from soil and water involves inoculation of 20 ml of an acetate-mineral medium with 5 ml of a water sample or of a filtered 10% soil suspension followed by vigorous aeration during incubation at 30°C or at room temperature (Baumann, 1968). Vigorous aeration is especially favorable for enrichment of nonmotile acinetobacters since in a nonaerated culture motile, oxygen-consuming pseudomonads tend to move to the surface layer, thereby decreasing the amount of oxygen diffusing into the interior of the liquid volume. Acinetobacters have a slightly acid pH optimum for growth, and aeration at a pH from 5.5–6.0 favors enrichment of these organisms from soil and water samples.

Maintenance Procedures

Cultures of *Acinetobacter* can be maintained on heart infusion agar plates stored in plastic bags (to delay evaporation and drying) at room temperature for periods as long as 6 months before being transferred (Marcus et al., 1969). This procedure is not recommended, however, since it has been observed (Juni, unpublished) that the ability to use certain carbon sources is lost, probably as a result of spontaneous mutation, when cultures are transferred infrequently in this manner. Storage at low temperatures or by lyophilization appears to be best for preservation of initially observed properties of a particular strain. Cultures can be stored conveniently by preparing a heavy suspension of organisms in buffer or in heart infusion broth and adding 0.5 ml of this suspension to 1.0 ml of sterile glycerol in a screw capped test tube. After mixing, this suspension is stored at −20°C or below. Cultures stored at −20°C remain viable for at least 2 years, whereas cultures stored at lower temperatures (i.e. −40°C to −70°C) can be maintained indefinitely. Since 67% glycerol remains liquid at temperatures above −46°C a stored culture can be sampled conveniently by withdrawing a loopful, or 0.1 ml, of the glycerol stock and streaking this material directly on a heart infusion plate without the necessity of warming the glycerol suspension.

Procedures for Testing for Special Characters

Procedures for determination of phenotypic properties of strains of *Moraxella* and *Acinetobacter* have been proposed by the Subcommittee on *Moraxella* and Allied Bacteria of the International Committee on Systematic Bacteriology (ICSB) (Bøvre and Henriksen, 1976). Acid production from sugars is most conveniently determined by inoculating the surfaces of slants of phenol red agar base (PRAB) (Difco B98) containing 1% glucose and slants of PRAB containing 10% lactose (Samuels et al., 1969). Acid-producing strains result in yellow slants for both sugars whereas nonacid-producing strains either fail to change the color of the slant or act to intensify the initial red color.

Differentiation of the genus **Acinetobacter** *from other closely related genera*

Acinetobacters are distinguished readily from the enteric bacteria by the inability of strains of the former group to grow anaerobically in any medium. Species of *Moraxella* all give a positive oxidase reaction in contrast to the negative oxidase reaction characteristic of all strains of *Acinetobacter*. Some strains of *Acinetobacter* are pleomorphic and display long rod-shaped forms. Any doubts concerning identification of these, as well as other strains, can be dispelled by application of the genetic transformation assay where a tryptophan auxotroph (*trp E27*, ATCC 33308) of a highly competent strain is transformed to prototrophy by using a crude DNA preparation of the strain being tested (Juni, 1972). DNA preparations of strains of other genera do not react in this transformation assay.

Taxonomic Comments

Using particular nutritional markers in a genetic transformation assay it has been demonstrated that, of 265 strains tested, all acinetobacters are closely related (Juni, 1972). Analysis of DNA composition of various acinetobacters has revealed a range of 38–47 mol% G + C, thus indicating considerable evolution of different strains. Evidence that such evolution of strains has occurred is also provided by the observations that interstrain transformation of certain nutritional markers (Juni, 1972) and even of the high level streptomycin resistance marker (Bøvre et al., 1976) can be extremely poor in some cases. Further evidence for evolutionary divergence in acinetobacters comes from studies of DNA/DNA hybridization where 43 strains were classified into six homology groups based upon ability of DNA from strains within a given group to show 50% of more homology to DNA from the reference strain of a particular group (Johnson et al., 1970). No significant DNA homology could be demonstrated between DNA from acinetobacters and DNA from several strains of *Moraxella* (Johnson et al., 1970). By contrast, in rRNA/DNA hybridization experiments there was significant homology between 23 S rRNA from *Acinetobacter* and DNA from *Moraxella osloensis* or from *Branhamella catarrhalis* (Johnson et al., 1970). These results, as well as the demonstration of low level intergeneric transformation of the streptomycin resistance marker between *Acinetobacter* and some strains of *Moraxella* (Bøvre, 1967; Juni, 1972), are consistent with the conserved nature of ribosomal genes (Dubnau et al., 1965) and the fact that mutation to resistance to high concentrations of streptomycin is determined by a change in a ribosomal protein (Ozaki et al., 1969).

Speciation of acinetobacters has not yet been attempted since there do not appear to be clear criteria for establishing unique species. Baumann et al. (1968b) divided acinetobacters into several nutritional groups based chiefly upon ability of strains to use a given set of carbon sources. Johnson et al. (1970) found a correlation between DNA homology groups and the phenotypic groups of Baumann et al. (1968b). In a numerical analysis of 291 acinetobacters, Pagel and Seyfried (1976) made use of 89 characteristics based upon morphological, physiological, nutritional, and biochemical properties and showed that all strains could be divided into two major phenetic groups corresponding to the two major nutritional groups established by Baumann et al. (1968b). Although all strains of *Acinetobacter* are at the present time considered to be members of a single species, *A. calcoaceticus* (Henriksen, 1973), future studies may eventually provide a sound basis for establishing other species of *Acinetobacter*.

One characteristic that has been used widely to separate acinetobacters into two large groups is the ability or lack of ability to form acid from sugars. The names "*Herellea vaginicola*" and "*Bacterium anitratum*" have been used for acid-forming strains and the names "*Mima lwoffi*" and "*Moraxella lwoffi*" were formerly used for nonacid-forming strains. Henriksen (1973) has suggested the possibility that in place of the single species *A. calcoaceticus* the latter name might be used for acid-forming strains and that the name *Acinetobacter lwoffi* be used for the nonacid-forming strains. These two names were included on the Approved Lists of Bacterial Names in 1980. Unfortunately, there is no clear correlation betwen ability to form acid from sugars and ability to use various carbon sources, acid-forming strains being found in both of the large phenotypic groups proposed by Baumann et al. (1968b). Henriksen (1973) has pointed out that since it would not be considered a good taxonomic procedure to establish species based on only one characteristic it might be best, at the present time, to retain only the single species *A. calcoaceticus* for all acinetobacters.

Historical. The taxonomy of *Acinetobacter* involves a long history (Henriksen, 1973). Because acinetobacters are ubiquitous they have been isolated independently by many individuals from a variety of sources. It should be recognized that acinetobacters do not possess any unique characteristics that would make possible ready differentiation of these bacteria from other similar organisms. In fact, the acinetobacters are, at present, identified by a series of properties they do not possess, i.e. nonmotile, nonfermentative, oxidase-negative and, generally, inability to reduce nitrate to nitrite. It is, therefore, not surprising that different names were used to describe these organisms. The most common terms used were *Bacterium, Herellea, Moraxella, Mima, Achromobacter, Acinetobacter, Alcaligenes, Neisseria, Micrococcus, Diplococcus*, B5W, and *Cytophaga*.

Schaub and Hauber (1948) first proposed the name *Bacterium anitratum* for organisms now known to be acinetobacters because the term *Bacterium* was, at that time, used for bacteria not capable of being assigned to a known genus. The specific epithet *anitratum* was chosen since these bacteria failed to reduce nitrate. DeBord (1942) introduced

Table 4.73.

Differentiation of the phenotypic groups of **Acinetobacter calcoaceticus**[a]

Characteristics	Phenotypic Groups						
	A1	A2	A3	B1	B2	B3	B4
Gelatinase production	−	−	−	−	−	[+]	+
Acid from glucose	+	[−]	−	−	−	d	d
Growth on sole carbon sources:[b]							
D-Xylose and/or L-arabinose	[+]	−	−	−	−	−	−
Putrescine	[+]	−	d	−	−	−	−
Ornithine	[+]	−	d	−	[−]	−	−
Adipate	[+]	+	−	−	−	[−]	−
β-Alanine	+	d	[−]	−	−	[−]	−
Kynurenine	+	+	[−]	[−]	[−]	d	d
Anthranilate	[+]	+	d	−	[−]	d	[−]
L-Leucine	+	−	d	−	−	d	[+]
D-Malate	d	[+]	[+]	[−]	[−]	+	+
2,3-Butylene glycol	+	+	+	d	[−]	−	−
Quinate	[+]	+	+	[+]	[−]	d	[+]
L-Tyrosine	[+]	+	+	[+]	[−]	+	d
L-Aspartate	+	+	+	+	−	−	d
DL-Lactate	+	d	[−]	+	d	+	[−]

[a] From Baumann et al. (1968b). Symbols: +, all strains positive; [+], 80% or more of strains are positive; d, 21–79% of strains are positive; [−], 20% or less are positive; −, all strains negative. The type strain ATCC 23055 belongs to Group A but is unassigned to any subgroup. It more closely resembles subgroup A1 than any other group. See the species description for additional information about the type strain.

[b] Tested by the method of Stanier et al. (1966).

Table 4.74.

Comparison of the DNA homology groups with the phenotypic groups of **Acinetobacter calcoaceticus**

DNA Homology Group[a] (Johnson et al., 1970)	Phenotypic Group[b] (Baumann et al., 1968b)
1	A[c]
2	A[c] and B4
3	B3
4	B1
5	A[c] and B2
6	B1

[a] Defined as having 50% or more relative homology with the reference organism for the group, except group 6 which had 23–49% homology with the reference strains for the other five groups.

[b] Based on similarity coefficients in a numerical analysis. Group A is divided into subgroups A1, A2 and A3, except for a few strains (including the type strain of *A. calcoaceticus*) which are unassigned.

[c] Most of the group A strains are in homology group 1, but some are in homology groups 2 and 5.

the names *Herellea vaginicola* and *Mima polymorpha* to describe acinetobacters that, respectively, produced, or failed to produce, acid from sugars. In the first edition of *Bergey's Manual* (Bergey et al., 1923) the genus *Achromobacter* was introduced to include the colorless, Gram-negative saprophytes. Although the type strain of *Achromobacter* was motile, the genus was also considered to include nonmotile strains. In 1954 Brisou and Prévot proposed that nonmotile achromobacters be classified as strains of a new genus *Acinetobacter*. When the oxidase test became more widely used as a determinative key (Kovacs, 1956) it

became clear that some acinetobacters were oxidase-positive and some were oxidase-negative. In a study by Baumann et al. (1968b) of 106 oxidase-negative strains it was concluded that these were all members of one genus and it was proposed that the name *Acinetobacter* be used for members of this group. Baumann et al. (1968b) also suggested that the species epithet *calco-aceticus* be used since one of the first acinetobacters studied was isolated in Beijerinck's laboratory in 1909 and named *Micrococcus calco-aceticus* because of its ability to grow on a plate containing calcium acetate as the carbon source. DNA homology (Johnson et al., 1970) and genetic transformation studies (Juni, 1972) have served to verify the relatedness of all strains of *Acinetobacter* as well as to demonstrate the lack of close relatedness of acinetobacters to moraxellas and to the oxidase-positive, psychrotrophic achromobacters, the latter organisms constituting another distinct group of genetically interacting strains (Juni and Heym, 1980). Although the genus *Acinetobacter*, as originally conceived by Brisou and Prévot (1954), included oxidase-positive and oxidase-negative strains, the Subcommittee on the Taxonomy of *Moraxella* and Allied Bacteria proposed in 1971 that the genus *Acinetobacter* include only the oxidase-negative strains (Lessel, 1971).

Further Reading

Baumann, P., M. Doudoroff, and R.Y. Stanier. 1968. A study of the *Moraxella* group II. Oxidative-negative species (Genus *Acinetobacter*). J. Bacteriol. 95: 1520–1541.

Henriksen, S.D. 1973. *Moraxella, Acinetobacter*, and the *Mimeae*. Bacteriol. Rev. 37: 522–561.

Henriksen, S.D. 1976. *Moraxella, Neisseria, Branhamella* and *Acinetobacter*. Annu. Rev. Microbiol. 30: 63–83.

Juni, E. 1978. Genetics and Physiology of *Acinetobacter*. Annu. Rev. Microbiol. 32: 349–371.

Rosenthal, S.L. 1978. Clinical Role of *Acinetobacter* and *Moraxella*. In Gilardi (Editor), *Glucose Nonfermenting Gram-Negative Bacteria in Clinical Microbiology*, CRC Press, West Palm Beach, pp. 105–117.

List of the Species of the Genus **Acinetobacter**

1. **Acinetobacter calcoaceticus** (Beijerinck 1911) Baumann, Doudoroff and Stanier 1968, 1538.[AL] (*Micrococcus calco-aceticus* Beijerinck 1911, 1067.) Note: includes *Acinetobacter lwoffi* (Audureau 1940) Brisou

and Prévot 1954, 727.[AL] See the eighth edition of the *Manual* for additional synonymy.

cal.co.a.ce′ti.cus. L. n. *calx* chalk; L. n. *acetum* acetic acid; M.L. n.

calcoaceticus calcium acetate, which was used by Beijerinck in the enrichment medium from which he isolated the organism.

The characteristics are as described for the genus. Differentiation of the phenotypic groups within the species is indicated in Table 4.73, and comparison of the phenotypic groups with DNA homology groups is presented in Table 4.74.

The mol% G + C of the DNA is 38–47 (T_m, Bd).

Type strain: ATCC 23055 (Delft 1 of Beijerinck). This strain belongs to homology group 1. It also belongs to phenotypic group A, but was unassigned to any of the subgroups A1, A2 or A3. It more closely resembles subgroup A1 than any other subgroup. With reference to Table 4.73, it is gelatinase-negative, forms acid from glucose, and grows on putrescine, L-ornithine, adipate, L-leucine, 2,3-butylene glycol, quinate, L-tyrosine, L-aspartate and DL-lactate but not on D-xylose, L-arabinose, β-alanine, L-kynurenine, anthranilate, or D-malate.

Genus IV. **Kingella** *Henriksen and Bøvre 1976, 449^{AL}*

JEREMY J. S. SNELL

King.el′la. M.L. dim. ending *-ella*; M.L. fem. n. *Kingella* named after Elizabeth O. King, an American bacteriologist.

Straight rods, ∼1.0 μm in diameter and 2.0–3.0 μm in length with rounded or square ends. Occur in pairs and sometimes short chains. Endospores are not formed. **Gram-negative, but there is a tendency to resist Gram-decolorization.** Nonmotile by normal tests but may be fimbriated (piliated) and show "twitching motility." **Aerobic or facultatively anaerobic;** grow best aerobically but can grow weakly under anaerobic conditions on blood agar. Optimum temperature, 33–37°C. Two types of colonies occur on blood agar: (a) a spreading, corroding type associated with "twitching motility," fimbriation, and transformation competence, and (b) a smooth, convex type not showing twitching, fimbriation or competence. **Oxidase-positive** (when tested with tetramethyl-*p*-phenylene diamine; the dimethyl reagent may give weak or negative reactions). **Catalase-negative.** Coagulated serum is not liquefied. Urease-negative. Phenylalanine deaminase activity is negative or weak. Chemoorganotrophic. **Glucose and a limited number of other carbohydrates are fermented with acid production** but no gas. **Susceptible to penicillin.** Occur in human mucous membranes of the upper respiratory tract. The mol% G + C of the DNA is 47–55.

Type species: *Kingella kingae* (Henriksen and Bøvre 1968) Henriksen and Bøvre 1976, 449.

Further Descriptive Information

Cells of *Kingella* strains are characteristically plump, Gram-negative rods or coccobacilli occurring in pairs or chains. Films from 18-h-old cultures show a tendency to retain the crystal violet of the Gram stain. Both *K. kingae* and *K. denitrificans* show some pleomorphism with swollen, irregularly-stained cells. Cells of *K. indologenes* are more regular, longer and evenly stained. Cells of *K. kingae* may show "twitching motility" (Henrichsen et al., 1972). Twitching motility and competence in genetic transformation is associated with the presence of polar fimbriae (Frøholm and Bøvre, 1972; Henrichsen, 1972). Colonies formed by fimbriated cells are spreading and corroding, in contrast to the smooth, entire, convex colonies formed by nonfimbriated cells. Freshly isolated strains of *K. indologenes* and *K. denitrificans* also produced spreading corroding colonies and transformation has been demonstrated in both these species (Bøvre and coworkers, unpublished observations).

All species in the genus are nutritionally fastidious and little or no growth occurs on unsupplemented peptone media. Growth on nutrient agar is only marginally improved by addition of blood or serum and the colonies remain small, typically 0.5–1.0 mm in diameter after incubation for 48 h. There is no requirement for X or V factors or for a CO₂-enriched atmosphere. Growth occurs at 30 and 37°C but not at 5 or 45°C. Strains differ in their ability to grow at 22°C. There is some disagreement about the ability of *Kingella* species to grow anaerobically. Henriksen and Bøvre (1976) described *K. kingae* as aerobic, but Snell and Lapage (1976) found growth of all species of the genus on blood agar in an atmosphere of 95% H_2, 5% CO_2.

Cells of *K. kingae* contain the following fatty acids: *n*-dodecanoic, 3-hydroxydodecanoic, *n*-tetradecanoic, 3-hydroxytetradecanoic, hexadecanoic, *n*-hexadecanoic, octadecadienoic, octadecenoic, and *n*-octadecanoic. The cells do not contain *n*-pentadecanoic, 3-hydroxyhexadecanoic or heptadecenoic acids (Jantzen et al., 1974). Waxes have not been found in *K. kingae* (Bryn et al., 1977).

Strains of *Kingella* are normally sensitive to penicillin, sulfonamides, erythromycin, tetracycline, chloramphenicol and streptomycin.

The pathogenicity of members of the genus is probably low. *K. kingae* has been isolated from blood cultures (Tatum et al., 1974) and *K. indologenes* from angular conjunctivitis (Van Bijsterveld, 1970) and a corneal abscess (Sutton et al., 1972).

The natural habitat appears to be the upper respiratory tract of man, where the organisms are present on the mucous membranes as part of the normal flora. Strains of *K. denitrificans* have been recovered on Thayer-Martin medium from pharyngeal swabs taken for screening of *Neisseria meningitidis* (Hollis et al., 1972).

Enrichment and Isolation Procedures

No special procedures have been described for isolation of *Kingella* species. General purpose, enriched laboratory media such as blood agar serve to isolate these bacteria from pathological specimens. Incubation for 48 h may be required for development of colonies of reasonable size. Colonies of *K. kingae* are easily recognized on blood agar in mixed culture by a distinct zone of β-hemolysis surrounding the colonies. Thayer-Martin medium may be used selectively for the isolation of *K. denitrificans* (Hollis et al., 1972).

Maintenance Procedures

Kingella strains are difficult to maintain by serial transfer because blood agar cultures become sterile after 6–12 days at room temperature. Preservation is best achieved by freeze-drying. Horse serum containing 5% (w/v) inositol is a suitable suspending medium for freeze-drying (Redway and Lapage, 1974).

Procedures for Testing for Special Characters

The main difficulty in characterizing members of the genus is the poor growth obtained in peptone-based media. Growth is better on solid or semisolid media than in liquid media. Serum promotes a slight, although not a dramatic improvement of growth and addition of 5% horse serum to test media is worthwhile. Growth is slow and longer incubation to exclude negative results in characterization tests is needed than with faster growing organisms.

Acid from carbohydrates is best detected in media designed for use with *Neisseria* species. The "sugars for neisserias" medium described by Cowan (1974; medium A2.6.6., p. 147) has proved suitable.

Methods suitable for characterization of the genus have been described by Snell et al. (1972) and Snell and Lapage (1976).

Differentiation of the genus **Kingella** *from other genera*

Characteristics by which *Kingella* may be differentiated from phenotypically similar genera and species are shown in Table 4.75.

Taxonomic Comments

The genus *Kingella* is a grouping of convenience and further taxonomic study is needed to test its robustness. Although the constituent species share some common phenotypic characters, these bacteria are biochemically nonreactive and the number of characterization tests yielding positive reactions is insufficient to construct a sound taxonomy. There is no published information available on DNA/DNA hybridization between the species to support placing them in a single genus. Although all three species contain strains competent in genetic transformation and may be identified genetically (Henriksen and Bøvre, 1977b; Bøvre and coworkers, unpublished observations), no genetic affinity has yet been found between the species.

Historically, these bacteria are associated with the genus *Moraxella* and *K. kingae* was originally named as *Moraxella kingii* by Henriksen and Bøvre (1968). The epithet *kingii* was later corrected to *kingae* by Bøvre et al. (1974). The species was transferred to a newly created genus *Kingella* by Henriksen and Bøvre (1976). Characters contributing to the exclusion of this species from the genus *Moraxella* are saccharolytic activity, a negative catalase reaction, a distinctive pattern of fatty acid composition (Jantzen et al., 1974), absence of cellular waxes (Bryn et al., 1977), and absence of affinity as measured by genetic transformation (Henriksen, 1969) and DNA/mRNA hybridization (Bøvre, 1970).

There is slight evidence of a distant relationship to the genus *Neisseria*. The mol% G + C of the DNA of the two genera is in the same range. Cluster analysis based on data derived from analysis of fatty acids (Jantzen et al., 1975) has shown a closer similarity between *K. kingae* and *Neisseria* than with other members of the family *Neisseriaceae*. Very low, marginal levels of transformation have been demonstrated between *K. kingae* and *Neisseria elongata* subsp. *glycolytica* (Bøvre et al., 1977).

Further Reading

Bøvre, K. and N. Hagen. 1981. The family *Neisseriaceae*: rod-shaped species of the genera *Moraxella, Acinetobacter, Kingella,* and *Neisseria,* and the *Branhamella* group of cocci. In Starr, Stolp, Trüper, Balows and Schlegel (Editors), *The Prokaryotes: a Handbook on Habitats, Isolation and Identification of Bacteria,* Springer-Verlag, Berlin, pp. 1506–1529.

Henriksen, S.D. and K. Bøvre. 1976. Transfer of *Moraxella kingae* Henriksen and Bøvre to the genus *Kingella* gen. nov. in the family *Neisseriaceae*. Int. J. Syst. Bacteriol. *26:* 447–450.

Differentiation of the species of the genus **Kingella**

Characteristics differentiating the species of *Kingella* are listed in Table 4.76. Other characteristics of the species are listed in Table 4.77.

List of the species of the genus **Kingella**

1. **Kingella kingae** (Henriksen and Bøvre 1968) Henriksen and Bøvre 1976, 449.[AL] (*Moraxella kingii* (*sic*) Henriksen and Bøvre 1968, 383; *Moraxella kingae* Bøvre, Henriksen and Jonsson 1974, 307.)

king'ae. M.L. gen. n. *kingae* of King; named after Elizabeth O. King, an American bacteriologist.

The characteristics are as described for the genus and as listed in Tables 4.76 and 4.77.

The mol% G + C of the DNA ranges from 47.3–47.4 (T_m).

Type strain: ATCC 23330 (NCTC 10529 (Henriksen and Bøvre, 1976)).

2. **Kingella indologenes** Snell and Lapage 1976, 456.[AL] ("Bisterveld/Sutton" strains (van Bijsterveld, 1970; Sutton et al., 1972)).

in.dol.o'ge.nes. M.L. n. *indolum* indole; Gr. v. *gennaio* to produce; M.L. part adj. *indologenes* indole-producing.

Table 4.75.

Differentiation of the genus **Kingella** *from phenotypically similar genera and species*[a]

| | | Moraxella | | | | | | | |
Characteristics	*Kingella*	subgenus *Moraxella*	subgenus *Branhamella*	*Neisseria*	*Cardiobacterium hominis*	*Eikenella corrodens*	*Pasteurella*	*Actinobacillus actinomyetemcomitans*	*Haemophilus aphrophilus*
Cell shape:									
Cocci	−	−	+	+[b]	−	−	−	−	−
Rods	+	+	−	−[b]	+	+	+	+	+
Catalase	−	+	+	+[b]	−	−	+	+	+
Acid from glucose	+	−	−	+[b]	+	−	+	+	+
Acid from sorbitol	−	−	−	−	+	−	+	D	+
Ornithine decarboxylase	−	−	−	−	−	+	D	−	−
Nitrate reduction	D	D	D	D	−	+	+	+	+
Mol% G + C of DNA (T_m)	47.3–54.8[c]	42.8–45.0[c, d]	41.8–47.3[c]	49.4–54.5[c]	58.7–60.2[c]	55.6–58.2[c]	38.6–43.7[e]	46.4–47.1[e]	42.0–44.5[e]

[a] Symbols: see standard definitions.

[b] Exceptions: *Neisseria elongata* is rod shaped, usually catalase-negative, and usually does not form acid from glucose. *Neisseria flavescens* does not form acid from glucose.

[c] Values derived in the author's laboratory.

[d] *M. atlantae* was not tested in the author's laboratory. Published values are 46.0–47.5 (Bd) (Bøvre et al., 1976).

[e] Values from Höllander and Pohl (1980).

The characteristics are as described for the genus and as listed in Tables 4.76 and 4.77.

The mol% G + C of the DNA is 48.7 (T_m).

Type strain: ATCC 25869 (NCTC 10717 (Snell and Lapage, 1976)).

3. **Kingella denitrificans** Snell and Lapage 1976, 456.[AL] ("TMl group" (Hollis et al., 1972)).

de.ni.tri′fi.cans. L. prep. *de* away from; L. n. *nitrum* soda; M.L. noun *nitrum* nitrate; M.L. v. *denitrifico* to denitrify; M.L. part. adj. *denitrificans* denitrifying.

The characteristics are as described for the genus and as listed in Tables 4.76 and 4.77.

The mol% G + C of the DNA ranges from 54.1–54.8 (T_m).

Type strain: NCTC 10995 (Snell and Lapage, 1976).

Table 4.76.

Differential characteristics of the species of the genus **Kingella**[a]

Characteristics	1. *K. kingae*	2. *K. indologenes*	3. *K. denitrificans*
β-hemolytic	+	−	−
Growth in presence of 4% NaCl	−	+	−
Nitrate reduction	−	−	+
Nitrite reduction[b]	−	−	+
Gas produced from nitrite	−	−	+
Phosphatase activity	+	+	−
Casein digestion	+	+	−
Indole production	−	+	−
Tween 40 hydrolysis	−	+	−

[a] For symbols see standard definitions.

[b] Nitrite reduction is positive in all three species when a rich basal medium such as Mueller-Hinton broth with yeast extract is used. Using less rich media, *K. kingae* and *K. indologenes* give negative reactions.

Table 4.77.

Other characteristics of the species of the genus **Kingella**[a]

Characteristics	1. *K. kingae*	2. *K. indologenes*	3. *K. denitrificans*	Characteristics	1. *K. kingae*	2. *K. indologenes*	3. *K. denitrificans*
Catalase test	−	−	−	Lecithinase activity	−	−	−
Oxidase test[b]	+	+	+	Arginine dihydrolase, ornithine decarboxylase, lysine decarboxylase	−	−	−
Motility (swimming)	−	−	−				
Anaerobic growth (blood agar)	W	W	W	Deoxyribonuclease activity	−	−	d
Temperature tolerance, growth at:				H₂S production (detection by lead acetate strips)	d	+	−
5°C	−	−	−				
22°C	d	+	d	Susceptible to penicillin (1.0 U/ml)	+	+	+
30°C	+	+	+				
37°C	+	+	+	β-galactosidase activity (ONPG[c] test)	−	−	−
45°C	−	−	−				
Growth in presence of 6% NaCl	−	−	−	Starch hydrolysis	−	−	−
Bile tolerance, growth in:				Oxidation/Fermentation test	F	F	F
10% bile	−	d	d	Acid production from:			
40% bile	−	−	−	Glucose	+	+	+
Growth stimulation by bile	−	−	−	Sucrose	−	+	−
Liquefaction of gelatin and coagulated serum	−	−	−	Maltose	+	+	−
				Mannose	−	+	−
Citrate utilization	−	−	−	Fructose	−	+	−
Growth in mineral medium with β-hydroxybutyrate	−	−	−	Dextrin	−	+	d
Formation of intracellular poly-β-hydroxybutyrate (nutrient medium)	−	−	−	Adonitol, arabinose, cellobiose, dulcitol, ethanol, galactose, glycerol, inositol, lactose, mannitol, mannose, raffinose, rhamnose, salicin, sorbitol, trehalose, xylose	−	−	−
Urease activity	−	−	−				
Lipase activity: hydrolysis of							
Tween 20	−	+	d				
Tween 80	−	−	−	Mol% G + C of DNA[d]	47.3–47.4	48.7	54.1–54.8

[a] For symbols see standard definitions; also W, weak.

[b] Using tetramethyl-*p*-phenylenediamine reagent.

[c] ONPG, *o*-nitrophenyl-β-D-galactopyranoside.

[d] The mol% G + C values were derived from T_m values using the equation of Mandel et al. (1970): mol% G + C unknown = mol% G + C reference strain + [slope of equation × (T_m unknown − T_m reference strain)]. The slopes of the equations of Marmur and Doty (1962) and of De Ley (1972), which are both 2.44, were used, and the average T_m value from 11 determinations of the reference strain, *Escherichia coli* K-12 (mol% G + C of 51) was 91.51°C. Substituting in the equation gives: mol% G + C = 51 + [2.44 × (T_m − 91.51)].

OTHER GENERA

Table 4.78.

Some differential characteristics of genera of Section 4 that have not been assigned to any family[a]

Characteristics	Beijerinckia, p. 311	Derxia, p. 321	Xanthobacter, p. 325	Thermus, p. 333	Thermomicrobium, p. 338	Halomonas, p. 340	Alteromonas, p. 343	Flavobacterium, p. 353	Alcaligenes, p. 361	Serpens, p. 373	Janthinobacterium, p. 376	Brucella, p. 377	Bordetella, p. 388	Francisella, p. 394	Paracoccus, p. 399	Lampropedia, p. 402
Morphology:																
Rods or coccobacilli	+	+	+	+	+	+	+[b]	+	+	+	+	+	+	+	+[c]	−
Cocci	−	−	−	−	−	−	−	−	−	−	−	−	−	−	+	+
Nitrogenase activity present	+	+	+	−	−	−	−	−	−	−	−	−	−	−	−	−
N_2 fixed under atmospheric pO_2	+	+	−													
Cells stain Gram-positive to Gram-variable	−	−	+	−	−	−	−	−	−	−	−	−	−	−	−	−
Facultative hydrogen autotrophy	−	+	+	−	−	−	−	−	D	−	−	−	−	−	D	−
Grows at NaCl concentrations of 0.5% through 32.5%	−	−	−	−	−	+	−	−	−	−	−	−	−	−	−	−
Grows at NaCl concentrations of 3.0% through 20%	−	−	−	−	−	+	−	−	−	−	−	−	−	−	D	−
Optimum temperature 70–75°C	−	−	−	+	+	−	−	−	−	−	−	−	−	−	−	−
Marine organisms; require seawater-based media	−	−	−	−	−	−	+	−	D	−	−	−	−	−	−	−
Isolated from strong brines, solar salt, etc.	−	−	−	−	−	+	−	−	−	−	−	−	−	−	D	−
Oxidase test			+	+		+	+[d]	+	+	+	+	+	D	−	+	+
Catalase test	+	−	+	+	+	+		+	+	+	+	+	;	W	+	+
Motility (swimming)	D	+	D	−	−	+	+	−	+	+	+	−	D	−	−	−
Flagellar arrangement:																
Polar	−	+	−			+[e]	+		−	+[e]	+[e]	−				
Lateral	+	−	+			+[e]	−		+	+[e]	+[e]	−	+			
Extremely flexible and has vigorous serpentine motility in agar gels and highly viscous solutions	−	−	−	−	−	−	−	−	−	+	−	−	−	−	−	−
Occurs as sheets of rounded, almost cubical, cells arranged in square tablets of 16–64 cells	−	−	−	−	−	−	−	−	−	−	−	−	−	−	−	+
Colonies are yellow	−	−	+			+[f]	−[d]	+	D	−	−	−	−	−	−	−
Colonies are violet	−	−	−	−	−	−	−[d]	−	−	−	+	−	−	−	−	−
Glucose used as a carbon source	+	+	+[g]	+	−	+	+	D	D	−	+					−
Sugars are neither fermented nor oxidized	−	−	−	−		−		D	D	+	−					+
Parasites of warm-blooded hosts	−	−	−	−	−	−	−	−	−	−	−	−	+	+	+	−	−
Pathogenic for man	−	−	−	−	−	−	−	D	D	−	−	+	+	+	−	−
Mol% G + C of DNA	55–61	69–73	65–70	61–71	64	60–61	38–50	31–42	56–70	66	61–67	55–58	66–70	33–36	64–67	61

[a] Symbols: see standard definitions; also W, weak.

[b] Two species of *Alteromonas* consist of curved, rather than straight rods.

[c] *P. denitrificans* grows as either cocci or very short rods. *P. halodenitrificans* grows as a coccus at optimum NaCl concentrations (~6%), but the cells may become distorted, swollen and very sticky if the NaCl concentration becomes too low (e.g. ~3%).

[d] With one exception.

[e] Both polar and lateral occur on the same cell.

[f] Can also be white.

[g] Some strains of *X. autotrophicus* may not use glucose.

Genus Beijerinckia Derx 1950, 145^{AL}

JAN-HENDRIK BECKING

Beij.e.rinck'i.a. M.L. fem. n. Beijerinckia named after M. W. Beijerinck, the Dutch microbiologist (1851–1931).

Straight or slightly curved rods, ~0.5–1.5 μm in diameter and 1.7–4.5 μm in length, with rounded ends. Occur singly. Sometimes large, misshapen cells 3.0 × 5.0–6.0 μm occur; these are occasionally branched or forked. Large, highly-refractile, intracellular granules of **poly-β-hydroxybutyrate** occur, generally one at each pole. **Cysts** (enclosing one cell) and **capsules** (enclosing several cells) occur in some species. **Gram-negative. Motile by peritrichous flagella or nonmotile. Aerobic,** having a strictly respiratory type of metabolism with oxygen as the terminal electron acceptor. **Molecular nitrogen is fixed** under aerobic conditions and also under decreased oxygen pressures (microaerophilic conditions). Optimum temperature, 20–30°C; no growth occurs at 37°C. **Growth occurs between pH 3.0 and pH 9.5–10.0.** In liquid media **no surface pellicle is formed, but the whole medium becomes a homogeneous, highly viscous, semitransparent mass;** in some species the whole medium becomes opalescent and turbid, and adhering slime is not produced. On agar media, especially under N_2-fixing conditions, **copious tenacious and elastic slime is produced and giant colonies develop** with a smooth, folded, or plicated surface; some strains form slime having a more granular consistency similar to that formed by Azotobacter. Catalase-positive. **Glucose, fructose, and sucrose are utilized** by all strains and are oxidized to CO_2 and a small amount of acetic acid. **No growth occurs on peptone agar or in peptone broth. Glutamate is utilized poorly or not at all.** Occur in soils, particularly those of **tropical regions.** The mol% G + C ranges from 54.7–60.7 (T_M).

Type species: Beijerinckia indica (Starkey and De 1939) Derx 1950, 146.

The typical microscopic appearance of Beijerinckia cells is illustrated in Figures 4.47 and 4.48. Cells may be bicellular due to cross-wall formation in the middle of the longitudinal direction of the cell (Figs. 4.49 and 4.50). Under certain cultural conditions some Beijerinckia strains show coccoid cells without terminal lipoid globules (Fig. 4.51). Sometimes, especially in B. mobilis strains, more than two lipoid globules per cell occur (Fig. 4.52). The chemical nature of the lipoid material as being poly-β-hydroxybutyrate (PHB) has been determined by elemental analysis (C, 52.3%; H, 7.5%; O, 37.8%) and the crystalline structure by röntgen analysis (Becking, 1974b and unpublished results).

Cyst and capsule formation occur in some species (B. fluminesis and B. mobilis) and are illustrated in Figures 4.53 and 4.54.

The flagella of motile cells are of the peritrichous type, but flagellar arrangement has been studied in comparatively few strains (Hofer, 1944; Thompson and Skerman, 1979). Microscopic examination indicates that the flagella tend to originate from one-half of the often dumbbell-shaped cells. The wave pattern is normal or curly, the wavelength has an average value of ~1.1–1.3 μm, and the amplitude is 0.26–0.35 μm. The amplitude of the waves in B. fluminensis and B. derxii strains is usually somewhat larger (0.26–0.35 μm) than in strains of B. indica (0.26 μm) (Thompson and Skerman, 1979).

On agar media, especially under N_2-fixing conditions, Beijerinckia strains may produce giant colonies containing copious amounts of slime. This slime is often extremely tough, tenacious, or elastic, which makes it difficult to remove part of a colony with a loop. In B. mobilis and B. fluminensis the slime has a more granular consistency resembling that of Azotobacter and is therefore easier to remove. The polysaccharide slime of Beijerinckia species (Fig. 4.55) consists of glucose,

galactose, mannose, glucuronic acid and galacturonic acid, but no heptose. The main component (52% of the dry weight of the polysaccharide) is glucose (López and Becking, 1968).

On nitrogen-free, mineral agar media[*][†] Beijerinckia species exhibit various kinds of colonial characteristics and pigmentation, and in liquid cultures they show differences in viscosity and pellicle formation; see the individual species descriptions for details concerning these characteristics. The colony type of the individual species can be used for rapid screening of isolates prior to more precise identification.

The temperature range for growth of Beijerinckia species is from 10–35°C. Cells are resistant to freezing: no reduction of viability occurs when stored for 3–4 months at –4°C (Becking, 1961a).

Beijerinckia strains are aerobic, and the cells contain cytochromes with absorption peaks at 415–424 nm (Sorêt); 480, 518 nm (β); 525–527, 551–556 nm (αc); and 604, 630 nm ($a_1aa_3a_2$) (Moss and Tchan, 1958). Therefore, the cells contain cytochrome c (λ_{max} 524 and 552 nm) and cytochrome a (λ_{max} 590 and 630 nm).

Like other N_2 fixers, Beijerinckia species require molybdenum for optimum growth and N_2 fixation. The molybdenum requirement is notably higher than that of Azotobacteraceae, being ~0.004–0.034 ppm Mo (0.4–3.5 μg/100 ml) (Becking, 1962). Moreover, unlike Azotobacter, the molybdenum requirement cannot be replaced by vanadium (Becking, 1962).

The efficiency of N_2 fixation in Beijerinckia strains is usually 10–13 mg N/g glucose consumed in a nitrogen-free medium containing 1 or 2% carbohydrate. During the testing of 47 B. indica strains of various origins, it was observed that the efficiency of N_2 fixation was variable and related to the growth rate of the strains (Becking, unpublished results). Of the strains tested, 21% were fast growers and poor N_2 fixers (6.0–9.9 mg N/g glucose consumed), 53% were moderate growers and moderate N_2 fixers (10.0–13.9 mg N/g glucose), and 26% were slow growers and good N_2 fixers (14.0–16.9 mg N/g glucose). The efficiency of N_2 fixation is also dependent on the age of the culture and on the carbohydrate concentration. Low carbohydrate levels tend to increase efficiency. In a study of some fast and slow growing strains of B. indica, the average mg N fixed per g glucose consumed could be increased from 10 in the fast-growing strains to 20, and from 15 in the slow growing strains to 30, by decreasing the glucose level 10-fold (Becking, unpublished results). Decreasing the partial pressure of O_2 below that of a normal air atmosphere also distinctly increases the N_2-fixing efficiency (Becking, 1971, 1978; Spiff and Odu, 1973).

The mineral requirements of Beijerinckia species can generally be correlated with the mineral status of lateritic soils (Kluyver and Becking, 1955; Becking, 1961b, 1981). For example, such soils are rich in iron, and it is noteworthy that Beijerinckia species have a higher requirement for iron than do the Azotobacteraceae, and also they are able to tolerate extremely high iron levels, even at low pH values (Becking, 1961a, b); thus, these organisms can be used as an indicator of the degree of ferrallitization of these tropical soils (Dommergues, 1963).

Organisms of the genus Beijerinckia were originally isolated from a quartzite soil (pH 4.5) of Malaysia (Alston, 1936) and later from acid soils of Dacca, Bangladesh (pH 4.9) and Insein in Burma (pH 5.2) by Starkey and De (1939). They were later observed to be widely distributed in acidic tropical soils, especially in the eluvial, lateritic soils, and

---

[*] Nitrogen-free, mineral agar medium has the following composition (g/liter of distilled water): glucose, 20.0; K_2HPO_4, 0.8; KH_2PO_4, 0.2; $MgSO_4 \cdot 7H_2O$, 0.5; $FeCl_3 \cdot 6H_2O$, 0.025 or 0.05; $Na_2MoO_4 \cdot 2H_2O$, 0.005; $CaCl_2$, 0.05; and agar, 15.0. The pH is adjusted to 6.9. The $CaCl_2$ may be omitted to obtain a calcium-free medium.
[†] Nitrogen-free, mineral agar medium with $CaCO_3$; composition is similar to that given in footnote,[*] but the $CaCl_2$ is replaced by $CaCO_3$ (10–20 g/liter).

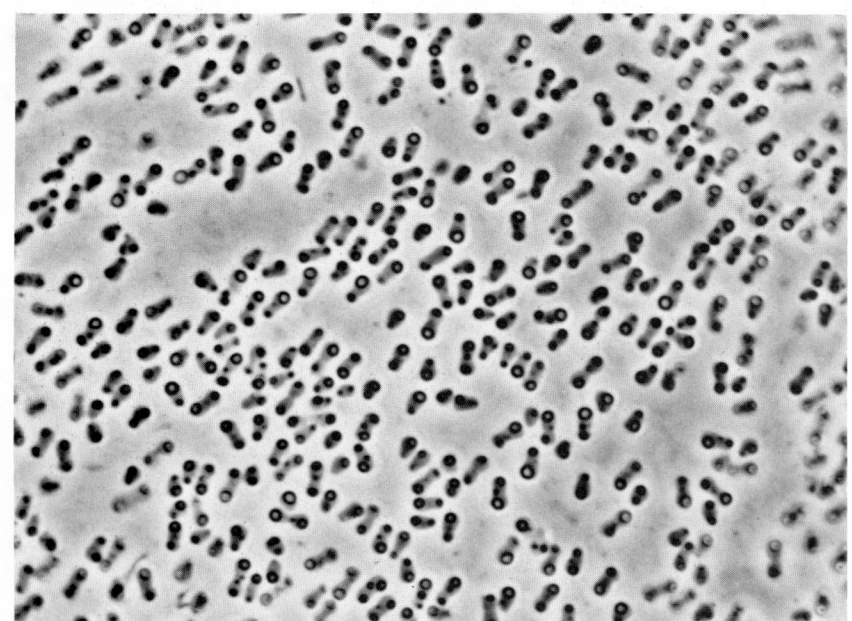

Figure 4.47. *Beijerinckia indica* cells cultured in nitrogen-free glucose mineral agar (pH 5.0). The typical appearance of the cells and their intracellular polar lipoid bodies is illustrated. Living preparation, phase-contrast microscopy (× 1500).

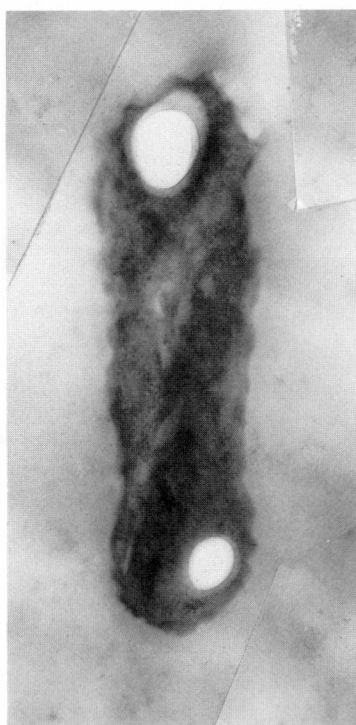

Figure 4.48. *Beijerinckia indica.* Electron micrograph of a thin section showing the two polar lipoid bodies, which are surrounded by a membrane (× 33,300).

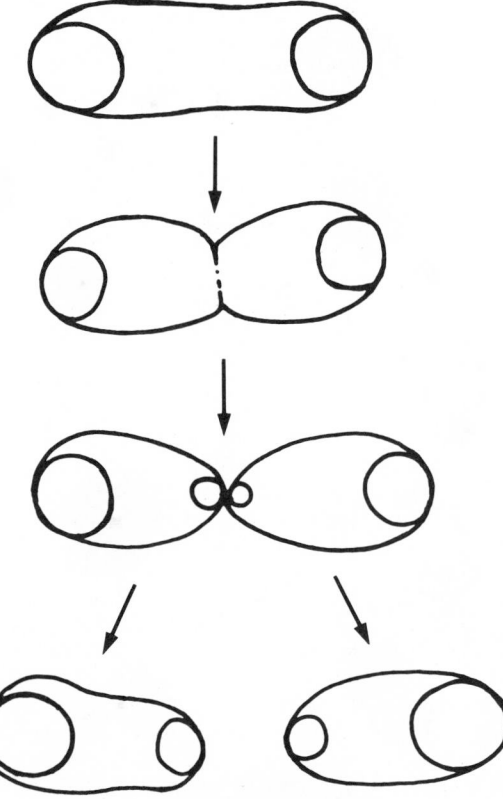

Figure 4.49. Diagram of the life cycle of a *Beijerinckia* cell. A cell in division forms a cross-wall in the middle of the longitudinal axis of the cell. In actively dividing cells intermediate stages can often be seen.

were normally absent in the illuvial alkaline soils (Kluyver and Becking, 1955; Becking, 1961a). They have been recovered only sporadically from soils outside the tropics.

In a survey of 392 soils of worldwide distribution, *Beijerinckia* was found only rarely in some temperate and subtropical soils, i.e. of Europe (Yugoslavia), South Africa, continental Asia (North India), Japan (Tokyo, Nikko National Park), and China (Hong Kong, Kwantung) (Becking, 1959, 1961a; see also the review by Becking, 1981).

Beijerinckia is commonly present of soils of tropical (equatorial) Africa (i.e. in soils of Ethiopia, Kenya, Uganda, Tanzania, etc.) and in

tropical soils of southeast Asia (Malaysia, Indonesia, Philippines) and South America (Trinidad, Surinam, Guyana, Venezuela, Bolivia, Brazil, etc.). The organism occurs primarily in soil, but it can also be recovered, although less readily, from water (e.g. irrigation water) of waterlogged soils of wet rice fields, the latter soil being a favorite habitat for *Beijerinckia* (Becking 1961a, 1978). *Beijerinckia* normally occurs in the biocoenose of leaves of tropical plants such as Coffea and

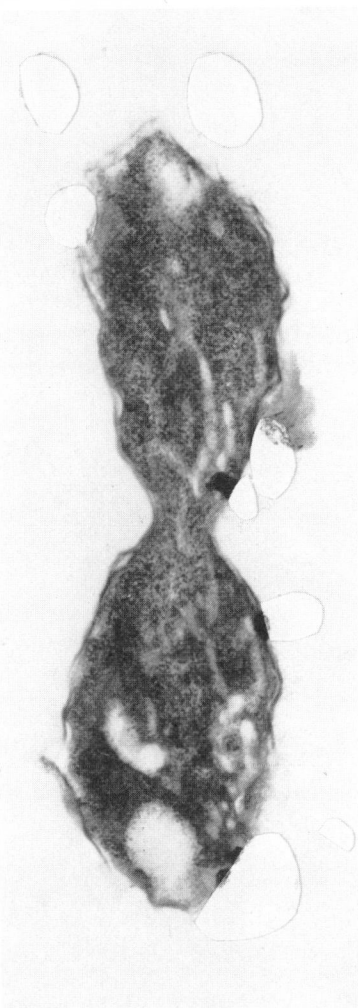

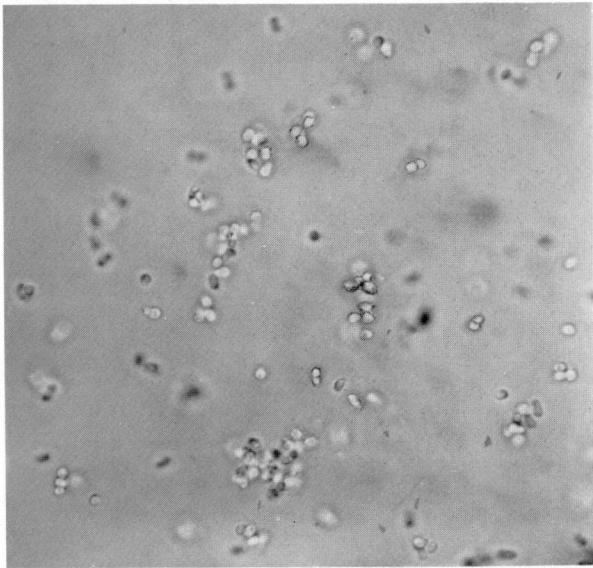

Figure 4.51. *Beijerinckia mobilis* from an aged culture. The individual cells often lack the characteristic polar lipoid bodies and are more rounded in form, resembling certain *Azotobacter* species (× 1000).

Figure 4.50. Electron micrograph of a *Beijerinckia* cell in the process of division. The constriction in the middle of the cell is clearly visible and the two terminal lipoid bodies of the original cell can also be seen (× 33,000).

Cocoa in some regions of southeast Asia (Indonesia) and South America (Surinam) (Ruinen, 1956, 1961). In the rhizosphere of soils cultivated with sugarcane under irrigated conditions in Brazil it is more numerous than in normal soil (Döbereiner and Alvahydo, 1959; Döbereiner, 1961).

B. indica is the most commonly encountered species and has been isolated from tropical soils of all continents and sometimes from nontropical regions. *B. mobilis* occurs mainly in very acid soils (pH 4.0–4.5) of tropical southeast Asia, Africa and South America and this is in accordance with its ability to fix nitrogen optimally at pH 3.9 (Becking, 1961a). The high degree of acid-tolerance might be useful for the specific enrichment and isolation of this species. *B. fluminensis* was originally isolated from a "Baixada Fluminensis" (pH 4.2–5.2) of Rio de Janeiro, Brazil, and also from some other Brazilian soils (Döbereiner and Ruschel, 1958). It was also isolated from many African soils (including South Africa) southeast Asian soils (Indonesia), and from soils of Hong Kong, China and India (Becking, 1961a). *B. derxii* was originally isolated and described by Tchan (1957) from an Australian soil (pH not mentioned); however, one isolate had already been obtained from an acid sandy soil (pH 4.5) from Tanzania, East Africa (Meiklejohn, 1954). Becking (unpublished results) found it rather common in neutral to alkaline soils (pH 6.5–7.0) from southeast Asia and particularly from South America. Matarassi et al. (1966) and Florenzano et al. (1968) also found it in the more alkaline soils of South

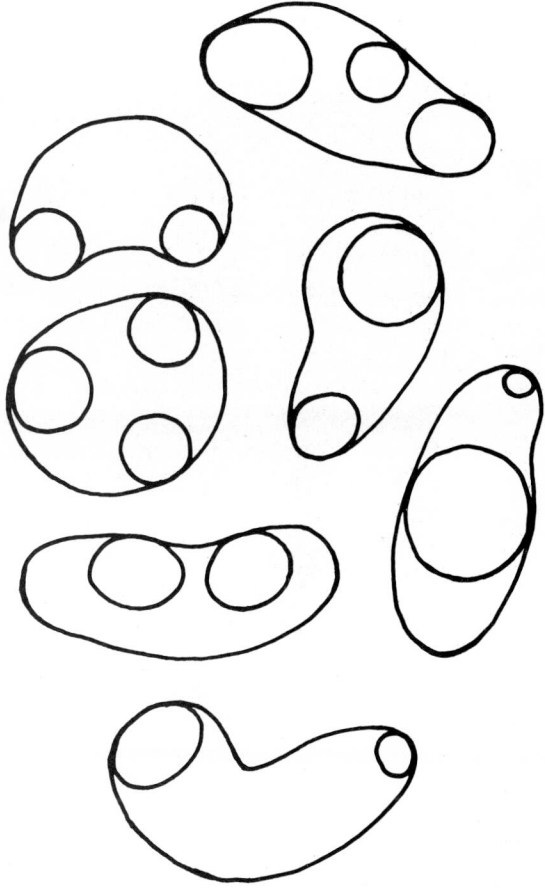

Figure 4.52. Deformed *Beijerinckia* cells showing round, dumbbell, pear-shaped or curved forms; sometimes even branched and forked forms may occur. More than two lipoid bodies per cell may be present. Such misshapen cells may occur in cultures of *B. mobilis* upon aging.

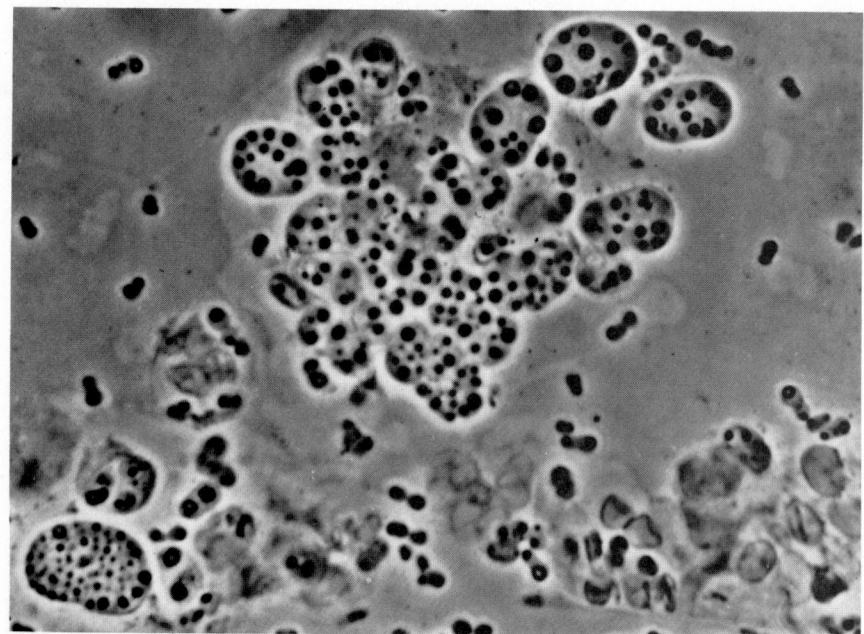

Figure 4.53. *Beijerinckia fluminensis* cultured in nitrogen-free glucose mineral agar (pH 5.0), showing distinct capsule formation. The capsules enclose a large number of individual cells. Living preparation, phase-contrast microscopy. (× 1500).

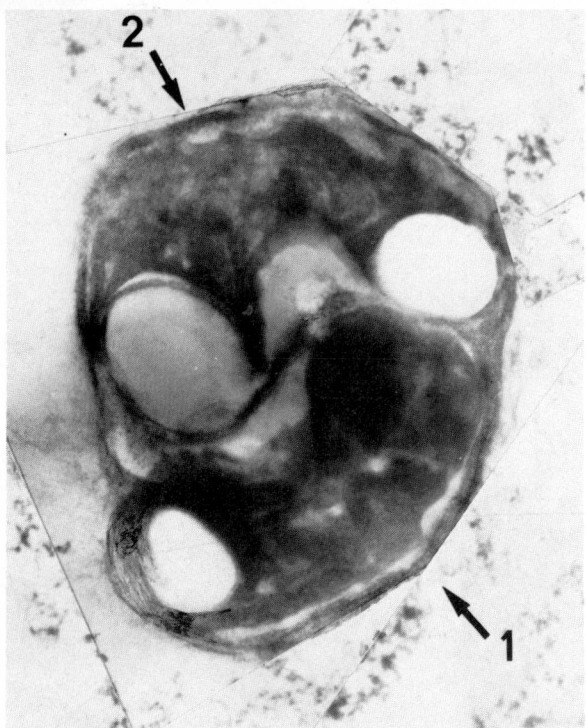

Figure 4.54. Electron micrograph of a capsule of *Beijerinckia mobilis* containing two cells (see *arrows 1* and *2*). The terminal lipoid bodies of each cell are visible and also the distinct capsular wall (see *arrow 1*) (× 33,000).

America (i.e. Venezuelan soils having a pH of 6.7–8.3). In an extensive survey of Australian nontropical soils, Thompson (1968; see also Thompson and Skerman, 1979) found *B. derxii* in several of these soils (krasnozems, red-brown earths, soloth) from Queensland (Toowoomba,

Oakey, Armidale, Hillgrove). The pH of these soils ranged from 5.1–6.7.

Enrichment and Isolation Procedures

An acidic, nitrogen-free medium* can be used for selective isolation of *Beijerinckia* from soil. The low pH of this medium favors development of beijerinckias, which are acid tolerant, and inhibits growth of other organisms, especially *Azotobacter* species. The requirement of trace elements in this medium (iron, molybdenum) is provided by the soil used as the inoculum. The medium is poured as thin layers (2–3 mm deep) into Petri dishes (~9–10 cm in diameter) to allow good aeration. This inhibits the development of anaerobic or facultative N_2 fixers, although the inhibition is not complete. Approximately 0.5 g of soil is used as the inoculum. The enrichment cultures are incubated at 30°C for 2 or more weeks (*Beijerinckia* strains grow slowly). The entire culture may eventually change into a viscous mass due to slime production. The cultures are examined microscopically at various times for the presence of characteristic *Beijerinckia*-like cells (Fig. 4.47). When such cells occur, the enrichment culture is plated onto a nitrogen-free, mineral agar medium (see first footnote). Although an acidic agar medium can be used for isolation, it is not recommended because the agar may be hydrolyzed or softened as a result of the heat sterilization.

The sieved-soil plate method of Winogradsky (1932) using nitrogen-free mineral agar (see first footnote) or silica-gel plates with glucose or sucrose (10 or 20 g/liter) can also be used. *Beijerinckia* colonies develop after 2–3 weeks around the soil particles on the plates. The medium should preferably be acidic (pH 4.5–5.0) to eliminate the growth of other N_2 fixers. The type strain of *B. indica* was obtained by this method. In general, however, it is less satisfactory than the use of liquid enrichment media because purification is more difficult (the slime is more tenacious) when one starts with solid media.

On agar media *Beijerinckia* species form characteristic, highly raised, glistening colonies containing a tough, elastic slime. For further purification a similar medium is used, but it is made neutral or alkaline by $CaCO_3$ (see second footnote). This is because the slime is more soluble under alkaline conditions and it is easier to suspend the cells in sterile tap water or liquid medium for further streaking. On the alkaline agar medium *B. indica* strains usually form highly raised colonies, whereas

---

* Enrichment medium (g/liter of distilled water): glucose, 20.0; KH_2PO_4, 1.0; and $MgSO_4 \cdot 7H_2O$, 0.5. The pH is adjusted to 5.0.

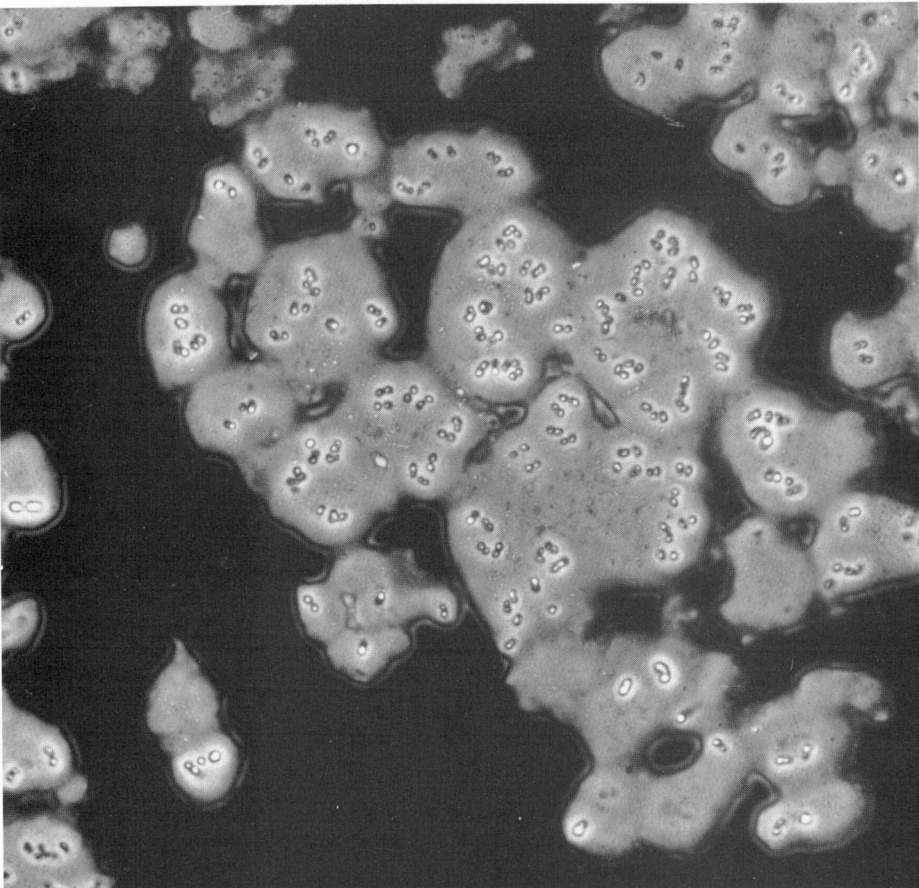

Figure 4.55. Cells of *Beijerinckia indica* suspending in India ink, showing the polysaccharide formation around the cells. Living preparation, phase-contrast microscopy. (× 1500).

B. mobilis colonies are flatter and produce a uniform reddish brown or amber-brown color on aging. The colonies are distinguished from those of *Azotobacter chroococcum* by being a more reddish brown or amber brown on aging.

Isolation from water sources (e.g. irrigation water of wet rice fields) is possible, but the inoculum size should be large (10 ml or more) because of the sparseness of the organisms. The sample is diluted with an equal volume of double-strength enrichment medium (see this footnote) and the pH of the medium is checked to make sure it is still 5.0 or lower. The medium is then dispensed into Petri dishes in thin layers as described above.

For isolation from phyllosphere habitats, the same liquid enrichment medium can be used; the detached leaves are partially submerged in shallow layers of medium in Petri dishes.

Specific enrichment procedures. No specific procedure is known to select for a particular *Beijerinckia* species, and all strains so far obtained are random isolates from soil using one of the general procedures outlined above. Certain carbon sources might, however, be useful for enrichment of one or another species because of their preferential use: e.g. citrate for enrichment of *B. indica*, formate or benzoate for *B. mobilis* (Becking, 1981). Thompson and Skerman (1979) have suggested that certain substrates or inhibitors might be useful for enrichment of various species, although this has not yet been experimentally tested. From the properties of the strains they studied by numerical analysis, they listed the following compounds as having potential value for enrichment or selection:

1. *B. indica*: L-arabinose, D-mannose, glycerol, caprylate, and *trans*-aconitate. Glycerol might be useful for *B. indica* and also for *Derxia gummosa* and some *B. fluminensis* strains. Nitrilotriacetate might also be useful for *B. indica* (unless its nitrogen content

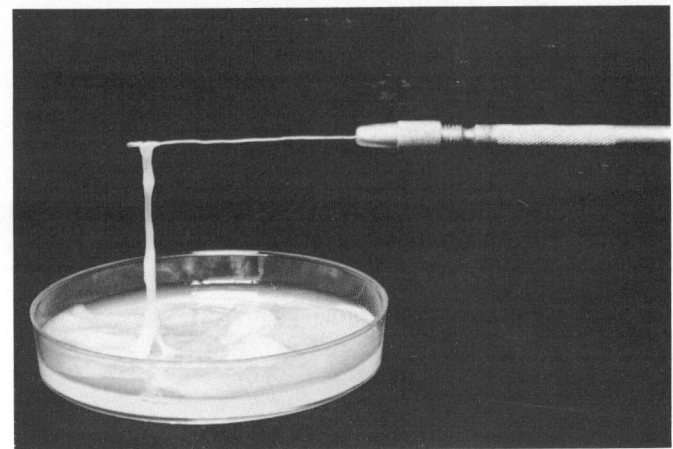

Figure 4.56. Enrichment culture of *Beijerinckia* inoculated with tropical soil, demonstrating the highly viscous consistency of the medium after 3 weeks (× 0.9).

decreases the selectivity of the medium for nitrogen-fixers);
2. *B. mobilis*: pentan-2-ol, 1,3-butylene glycol, and *n*-valerate might be useful. Phenol (0.05%) or erythromycin (125 μg/ml) might be used to inhibit other *Beijerinckia* species;
3. *B. indica* and *B. mobilis*: propan-1-ol, butan-1-ol, or 1,3-propylene glycol for enrichment;
4. *B. indica* subsp. *lacticogenes* and *B. mobilis*: caproate, *p*-hydroxybenzoate, or phenol for enrichment;
5. *B. derxii* and *B. fluminensis*: α-methyl-D-glucoside, maltose, mel-

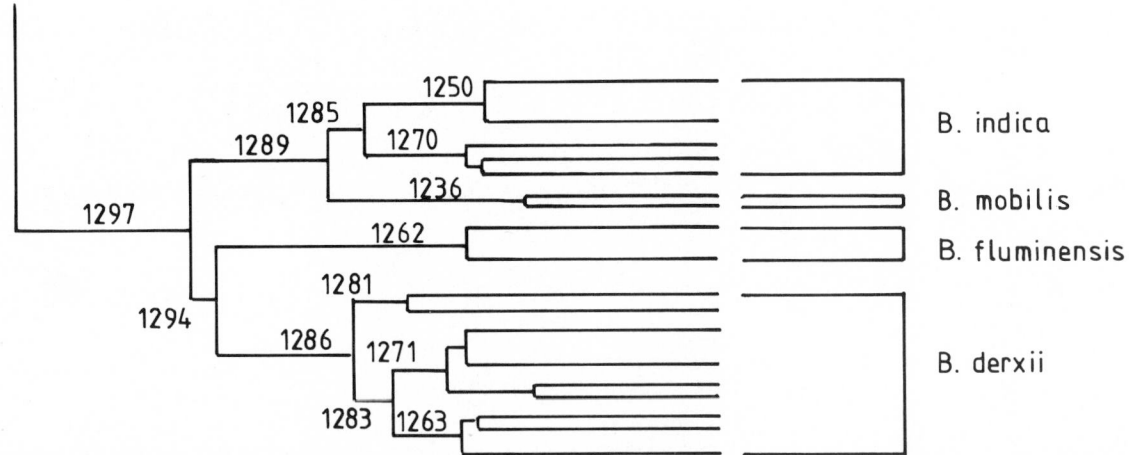

Figure 4.57. Simplified dendrogram of the hierarchical interrelations between the four *Beijerinckia* species according to Thompson and Skerman (1979). The numbers are related to the degree of similarity (or similarity attributes) within the species and strains. An increase in number denotes a decrease in similarity. (Reproduced with permission from J. P. Thompson and V. B. D. Skerman, *Azotobacteraceae*: the Taxonomy and Ecology of the Aerobic Nitrogen-fixing Bacteria, 1979, ©Academic Press, London.)

ibiose or melezitose for enrichment. Where both species are present, *B. derxii* would likely outgrow the extremely slow growing *B. fluminensis*;

6. *B. derxii* subsp. *venezuelae*: L-arabitol for enrichment.

Because the colony morphology and chromogenesis on plates differs among the various *Beijerinckia* species, the primary selection of colonies from the enrichment cultures is done mainly on the basis of these colonial characteristics. A more precise identification is done later by means of additional differential characteristics.

Maintenance Procedures

Beijerinckia strains are usually lyophilized in skim milk or dextran-sodium glutaminate solution on filter paper and stored in the dark at room temperature.

Storage has also been achieved on the usual agar media in tubes plugged with sterile rubber seals with storage in the dark at room temperature (Antheunisse, 1972, 1973); after 10 years 33% of the cultures retained viability. In the author's (Becking) laboratory, *Beijerinckia* cultures are stored under a seal of sterile liquid paraffin or mineral oil; such cultures generally survive for at least 3–5 years.

Strains may also be preserved indefinitely in liquid nitrogen. At the type culture collection in Delft, The Netherlands, dimethyl sulfoxide (10%, v/v) is added to cultures in the log phase or end of log phase,

and the cultures are frozen as rapidly as possible in liquid nitrogen. For recovery, the vials are thawed rapidly in a waterbath at 37°C.

Procedures for Testing Special Characters

Nitrogen fixation may be tested in nitrogen-free, mineral medium (e.g. the liquid versions of the media described in the first and second footnotes, or the medium described in footnote* below). It should be emphasized that for N_2-dependent growth *Beijerinckia* species do not require calcium, in contrast to most (but not all) *Azotobacteraceae*. Indeed, $CaCO_3$ is even slightly inhibitory because it tends to prolong the lag phase of growth. The fundamental nature of the inhibition is unknown, but it might involve a decrease in the trace elements in the medium due to precipitation by $CaCO_3$ during heat sterilization. Media supplemented with $CaCl_2$ rather than $CaCO_3$ invariably support good growth.

In testing *Beijerinckia* species for nitrate assimilation, precautions should be taken to avoid N_2 fixation. When tested under an air atmosphere, visible growth may be due to N_2 fixation and not to nitrate assimilation. Therefore, such tests must be conducted in closed systems (e.g. desiccator jars) in an atmosphere containing argon (78%, v/v), oxygen (20%, v/v) and carbon dioxide (2%, v/v). Moreover, nitrate disappearance (measured by chemical methods, e.g. the *o*-xylenol method) and nitrite appearance (measured by naphthylamine-sulfanilic acid-tartaric acid reagent) should be monitored simultaneously in the medium.

Differentiation of the genus **Beijerinckia** from other genera

Beijerinckia strains can be distinguished from other aerobic N_2 fixers by their great acid tolerance (which allows them to grow well at pH 4.0 or 5.0), their failure to form a pellicle on the surface of liquid media, and by their ability to make a liquid medium viscous by slime production (Fig. 4.56). Moreover, on solid media they produce characteristic large, slimy colonies having a tough, tenacious, and sometimes elastic slime. Because of this slime, it is often difficult to subculture portions of a colony for purification. The slime is semitransparent in liquid media but more opaque on solid media.

Beijerinckia cells can be distinguished from those of *Azotobacter* and *Azomonas* by their generally smaller size, their more rod-shaped or

sometimes pear- or dumbbell-shaped appearance, and by the characteristic presence of a lipoid body at each pole (Fig. 4.47).

Unlike *Azotobacter*, *Azomonas*, and many other bacteria, *Beijerinckia* strains do not grow on plain peptone agar. For this reason, plating on peptone agar can be used for a purity test of *Beijerinckia* isolates. Many strains of *Beijerinckia* utilize nitrate poorly or not at all and, in this respect, differ from strains of *Azotobacter*.

Although both *Beijerinckia* and *Derxia* produce slimy colonies on agar and viscosity in broth, *Beijerinckia* strains can be distinguished by (a) failure to produce dark mahogany-brown colonies with aging, (b) cells that contain bipolar lipoid bodies rather than numerous lipoid

---

* Nitrogen-free, mineral broth (g/liter of distilled water): glucose, 20.0; KH_2PO_4, 0.8; K_2HPO_4, 0.2; $MgSO_4 \cdot 7H_2O$, 0.5; $FeCl_3 \cdot 6H_2O$, 0.1; $CaCl_2 \cdot 2H_2O$, 0.05; and $Na_2MoO_4 \cdot 2H_2O$, 0.05. The pH is adjusted to 6.6. The medium is sterilized at 120°C for 20 min. Sucrose (20.0 g) can be substituted for glucose and the $CaCl_2$ can be omitted. Occasionally a more rapid initiation of growth is obtained in the presence of small amounts of $CaCl_2$.

bodies throughout the whole cell, (c) failure to form a pellicle at the surface of liquid media, and (d) a positive catalase reaction.

Taxonomic Comments

When numerical analysis methods are applied to *Beijerinckia* species and *Azotobacteraceae*, the *Beijerinckia* species fuse into a single, apparently coherent group (Thompson and Skerman, 1979). This group (the "1297" group; see Fig. 4.57) and the contrasting group containing the *Azotobacteraceae* (the "1300" group; see Fig. 4.25 of the article on the genus *Azotobacter* in this *Manual*) show maximal between-group variance, confirming the major division separating *Beijerinckia* strains from all other strains tested. By the use of a wide range of attributes, and considering all strains of named and unnamed *Beijerinckia* species, the numerical analysis supports the concept of a separate genus for these bacteria.

In experiments reported by De Smedt et al. (1980), ^{14}C-labeled rRNA from *Beijerinckia indica* was hybridized with filter-fixed DNA from a wide variety of Gram-negative bacteria. The hybrids were described by the temperature at which 50% of the hybrid was denatured ($T_{m(e)}$) and also by the percentage of rRNA binding. The data suggested that the genus *Beijerinckia* was a heterogenous group. From rRNA cistron similarities it was concluded that *Beijerinckia* and *Azotobacter/Azomonas* belong to different rRNA superfamilies. The constructed similarity maps of hybrids suggested that *Beijerinckia* is phylogenetically related to such organisms as *Xanthobacter*, "*Pseudomonas azotocolligans*," *Pseudomonas diminuta*, the authentic rhodopseudomonads, and other organisms such as *Azospirillum*, *Agrobacterium*, *Rhizobium*, *Acetobacter*, *Gluconobacter*, and *Zymomonas*. According to the present author, however, definite conclusions are difficult to draw. As already stressed by De Smedt et al., the percentage of rRNA binding is not a measure of rRNA homology, because the latter also depends on the number of rRNA cistrons per genome and on the size and state of replication of the genome.

In the opinion of the present author, representatives of the genus *Beijerinckia* show some affinity to members of the family *Azotobacteraceae*, since under certain environmental conditions *Beijerinckia* cells may lose their distinct bipolar lipoid bodies and assume a more globular diplococcus form (Fig. 4.51). In particular, the taxon *Azomonas agilis* may be related to *Beijerinckia* because of the following similarities: (a) poor or no assimilation of nitrate, (b) inability of vanadium to replace molybdenum in nitrogen fixation, (c) lack of a calcium requirement for nitrogen fixation, and (d) similarity in mol% G + C values (i.e. 53–59, which differs from the range for the genus *Azotobacter* (63–68)). Moreover, in both the hierarchical classification and in vector coordinate analyses (Thompson and Skerman, 1979), *Beijerinckia* appears to more closely resemble *Azomonas* and *Derxia gummosa* than *Azotobacter*.

Pending more data and further developments, the genus *Beijerinckia* is best listed under the noncommittal category "Other Genera" in this edition of *Bergey's Manual*.

Taxonomy of the species. Thompson and Skerman (1979) used many phenotypic characters of *Beijerinckia* strains for numerical analysis in a hierarchical classification and in a coordinate system. These authors confirmed the presence of four species within the genus (Fig. 4.57). It can be seen that the species fall within two main groups. Group 1294 (*B. fluminensis* + *B. derxii*) fuse with group 1289 (*B. indica* subsp. *indica* + *B. indica* subsp. *lacticogenes* + *B. mobilis*) to produce the group 1297. Diagnosis of this fusion has shown that group 1289 strains generally differed from group 1294 strains by having thinner cells and by utilizing caprylate, propan-1-ol, 1,3-propylene glycol, *trans*-aconitate, nitrilotriacetate, and L-arabinose, but not maltose and α-methyl-D-glycoside, as sole carbon sources. Moreover, strains of group 1289 produced acid from L-arabinose and glycerol and were resistant to 0.5% NaCl, 0.05% phenol, and 1.0% sodium benzoate.

In the hierarchical classification of the strains (Fig. 4.57), the number of strains should be noted (of which several are probably subcultures or variants of the same isolate): Group 1285 (including the type strain of "*Azotobacter indicus*" of Starkey and De (1939), 14 strains; group 1236, 2 strains (*B. mobilis*, apparently a rare species); group 1262, 10 strains (*B. fluminensis*), and group 1286, 21 strains (*B. derxii*). With regard to the latter species, it should be noted that this group contains the type strain of "*B. congensis*" (Hilger, 1965), the six co-types of "*B. venezuelae*" (Materassi et al., 1966), one co-type of *B. indica* (Starkey and De, 1939), and the type strain of "*B. indica* var. *alba*" (Derx, 1950b). Further, there are two main groups within *B. derxii*, group 1281 and 1283, which fuse at group 1286. Group 1281 contains only 4 strains, mainly of previously unclassified *Beijerinckia* species.

Further Reading

Becking, J. H. 1981. The family *Azotobacteraceae*. In Starr, Stolp, Trüper, Balows and Schlegel (Editors), *The Prokaryotes: a Handbook on Habitats, Isolation and Identification of Bacteria*, Springer-Verlag, Berlin, pp. 795–817.

Differentiation of the species of the genus **Beijerinckia**

Table 4.79 presents the main differential characteristics of the four species. Differences in utilization of various carbon sources are listed in Table 4.80.

List of the species of the genus **Beijerinckia**

1. **Beijerinckia indica** (Starkey and De 1939) Derx 1950, 146.^{AL} (*Azotobacter indicum* (*sic*) Starkey and De 1939, 337.)

in'di.ca. L. fem. adj. *indica* of India.

Straight or slightly curved rods 0.5–1.2 μm × 1.6–3.0 μm. Lipoid bodies persist in aged cultures. No resting stages occur; cyst or ascococcus formation is never observed.

Agar colonies are raised. At first they are semitransparent but soon become uniformly turbid or opaque white. On aging the colonies develop a light reddish pink, cinnamon or fawn color on neutral or alkaline media; on acid media they remain colorless. On acid media the slime is more tenacious, tough and elastic than on alkaline media. Giant colonies may develop, first with a smooth surface, but later with a folded, wrinkled or plicated surface (Figs. 4.58 to 4.60).

Liquid media become viscous. On aging color is produced, but is less prominent than on agar.

Grows between pH 3.0 and 10.0 (optimum is 4.0–10.0). Temperature range for growth, 10–35°C; no growth at 37°C.

Growth on, and utilization of nitrate is poor, and N$_2$ is fixed in preference to utilization of nitrate in the medium (Becking, 1962). Weak growth occurs on malt agar, no growth in plain broth or peptone agar.

Widely distributed in acid tropical soils.

The mol% G + C of the DNA is 54.7–58.5 (T_m) (De Ley and Park, 1966; De Smedt et al., 1980).

Type strain: ATCC 9039 (Delft E.III.12.1.1).

Further comments: Derx (1950b) described "*B. indica* var. *alba*," which was distinguished by its lack of pigmentation on aging; however, under extreme (alkaline) conditions it produced a pink pigment. In the hierarchical classification (Fig. 4.57) of Thompson and Skerman (1979) one cotype (strain WR-236) of "*B. indica* var. *alba*" is placed in *B. indica* subsp. *lacticogenes* (group 1270) and the other (strain WR-235) is placed in *B. derxii* subsp. *venezuelae* (group 1271).

1a. **Beijerinckia indica** subsp. **indica** (Starkey and De 1939) Derx 1950, 146.

Thompson and Skerman (1979) have distinguished the subspecies *indica* from the subspecies *lacticogenes* by differences in organic carbon utilization, nitrate reduction, and resistance to peptone-nitrogen. Based on their study of nine strains of subsp. *indica* and four or five strains of subsp. *lacticogenes*, the only absolute character that differentiated the two subspecies was the failure of *indica* to utilize *p*-hydroxybenzoate as a sole carbon source. Nitrate was reduced to nitrite by eight of nine

Table 4.79.

Differential characteristics of the species of the genus **Beijerinckia**[a]

Characteristics	1. B. indica	2. B. mobilis	3. B. fluminensis	4. B. derxii
Water-soluble, green fluorescent pigment	−	−	−	+
Colony color after aging	Fulvus or pink	Amber-brown	Fulvus or pink	Buff
Motility	[−][b]	+	[−][b]	−
Growth on casein agar	[−]	+	[−] (15)	+
Growth with nitrate as nitrogen source	d (40–50)	+	[−][c]	[−][c]
Growth on asparagine as the sole carbon and nitrogen source	[−] (6)	d (60)	[−][c]	[−][c]
Growth on carbon sources:				
Lactose	d (70)	−	[−] (20)	+
Erythritol	−	d (45)	−	−
Propanol	d (50)	+	−	−
Acetate, butyrate, fumarate, lactate, malate	d (50–95)	+	d (20–80)	[−][c]
Benzoate	[−] (6)	+	−	−

[a] Symbols: +, all strains positive; d, differs among strains; [−], most strains negative; −, all strains negative; ( ), % of strains giving positive reaction.
[b] If positive, motility occurs mostly in young stages and the cells are usually only weakly motile.
[c] If positive, the reactions are weak.

Table 4.80.

Utilization of carbon compounds by **Beijerinckia** *species*[a]

	1. B. indica	2. B. mobilis	3. B. fluminensis	4. B. derxii
Number of strains tested	48	7	5	3
Growth on carbon compounds:				
Arabinose	[+] (85)	+	+	[−][b]
Galactose	[+] (85)	d (45)	+	+
Inulin	d (65)	[−] (15)	−	[−][b]
Maltose	[+] (90)	d (70)	+	+
Rhamnose	d (35)	d (30)	−	[−][b]
Sorbose	d (70)	+	+	+
Starch	d (55)	d (30)	[−] (20)	[−][b]
Xylose	d (70)	d (45)	+	+
Butanol	d (80)	+	d (40)	−
Ethanol	[+] (90)	+	d (80)	+
Glycerol	d (70)	+	+	−
Inositol	d (55)	+	d (40)	−
Mannitol	[+] (90)	d (85)	d (40)	+
Acetate	d (50)	+	[−] (20)	[−][b]
Butyrate	d (85)	+	d (40)	[−][b]
Citrate	d (40)	−	−	[−][b]
Formate	−	d (70)[b]	−	[−][b]
Fumarate	d (80)	+	d (60)	[−][b]
Lactate	[+] (95)	+	d (80)	[−][b]
Malate	d (85)	+	[−] (20)	[−][b]
Malonate	d (50)	−	d (40)	[−][b]
Oxalate	[−] (2)	−	−	[−][b]
Propionate	−	−	−	[−][b]
Succinate	d (80)	+	d (40)	+
Tartrate	[−] (15)	−	−	[−][b]
Valerate	[−] (8)	d (70)[b]	−	[−][b]
Salicylate	[−] (6)	−	−	−

[a] Symbols: +, all strains positive; [+], most strains positive; d, differs among strains; [−], most strains negative; −, all strains negative; ( ), % of strains positive.
[b] If positive, the reactions are weak.

strains of subsp. *indica* but not by four of four strains of subsp. *lacticogenes*. The differences in utilization of sole carbon sources by subsp. *indica* vs. subsp. *lacticogenes* were as follows: propan-2-ol, 7/9 vs. 5/5; butan-2-ol, 0/9 vs. 3/5; D-arabitol, 4/9 vs. 1/5; phenol, 0/9 vs. 4/5; caproate, 0/9 vs. 4/5; adipate, pimelate, suberate, azelate, and sebacate, 0/9 vs. 1/5 or 1/4; α-oxybutyrate, 0/9 vs. 1/5; fumarate, DL-malate, tartrate (D, L, and *meso*), oxaloacetate, mucate, and *trans*-aconitate, 9/9 vs. 3-4/5.

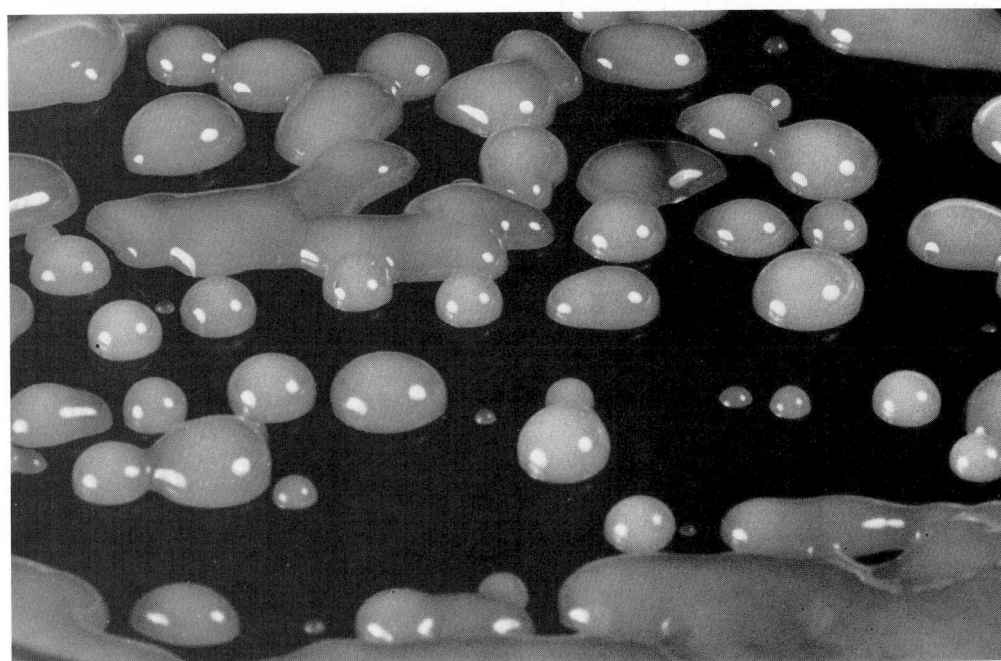

Figure 4.58. Typical colony type of *Beijerinckia indica* on nitrogen-free glucose mineral agar. The colonies are highly raised and have a very tough, elastic slime. In young cultures these colonies are colorless and transparent (× 2).

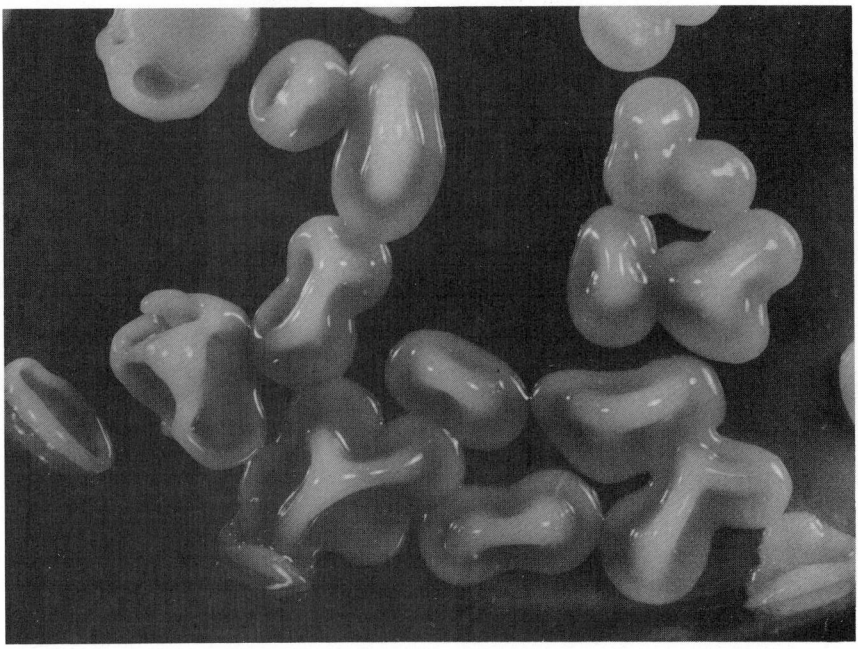

Figure 4.59. Typical colonies of an older culture of *Beijerinckia indica* on nitrogen-free glucose mineral agar. The colony turns opaque and its surface becomes plicated (× 2).

Resistance to 1% peptone in the medium was reported as 9/9 vs. 1/5. None of the strains of subsp. *indica* lacked flagella, whereas half the strains of subsp. *lacticogenes* did lack flagella.

Type strain: ATCC 9039.

1b. **Beijerinckia indica** subsp. **lacticogenes** (Kauffmann and Toussaint 1951) Thompson and Skerman 1981, 215.[VP] (Effective publication: Thompson and Skerman, 1979, 332.) (*Azotobacter lacticogenes* Kauffmann and Toussaint 1951, 710.)

lac.ti.co′ge.nes. M.L. n. *acidum lacticum* lactic acid; Gr. v. *gennaio* to produce; M.L. adj. *lacticogenes* lactic acid producing, which is an error (acetic acid is produced).

This subspecies is distinguished from the subspecies *indica* by the characteristics described above, and also by the consistency of the colonies (less elastic and rubbery (cartilagenous) and more butyrous and brittle than those of subsp. *indica*). Moreover, subsp. *lacticogenes* is somewhat less resistant to diamond fuchsin, brilliant green, sodium fluoride, and streptomycin. In contrast to subsp. *indica*, four of five strains of subsp. *lacticogenes* grown on agar containing *p*-hydroxybenzoate can metabolize protocatechuate via the *ortho*-cleavage pathway (Thompson and Skerman, 1979).

Type strain: ATCC WR-119 (ATCC 19361).

2. **Beijerinckia mobilis** Derx 1950, 10.[AL] (*Beijerinckia mobile* (*sic*) Derx 1950, 10.)

mo′bi.lis. L. fem. adj. *mobilis* movable, motile.

Straight, curved or pear-shaped rods, 0.6–1.0 μm × 1.6–3.0 μm. Sometimes misshaped or forked cells occur. Ascococcus-like clusters of

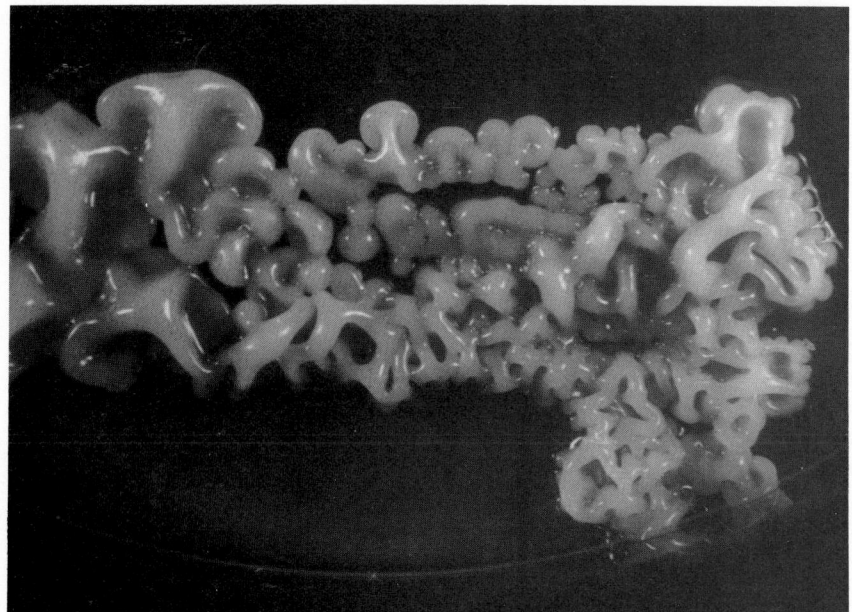

Figure 4.60. Typical colony of an aged culture of *Beijerinckia indica* on nitrogen-free glucose mineral agar. The colonies increase greatly in size due to copious slime production. The colonies become massive and opaque, with a plicated surface. In this stage they often attain a light reddish, pink or cinnamon color, especially on neutral or alkaline media (× 2).

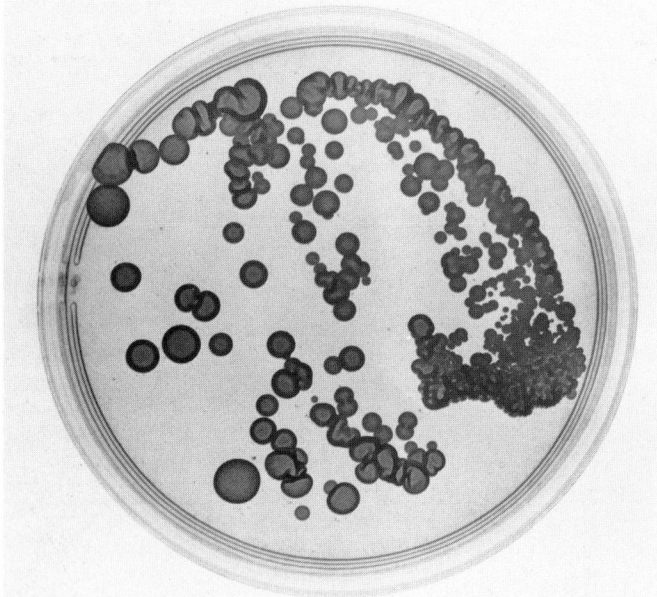

Figure 4.61. Colonies of *Beijerinckia mobilis* on a nitrogen-free glucose mineral agar containing $CaCl_2$. On this transparent medium, the species forms only small raised colonies having a typical amber-brown color on aging (× 0.8).

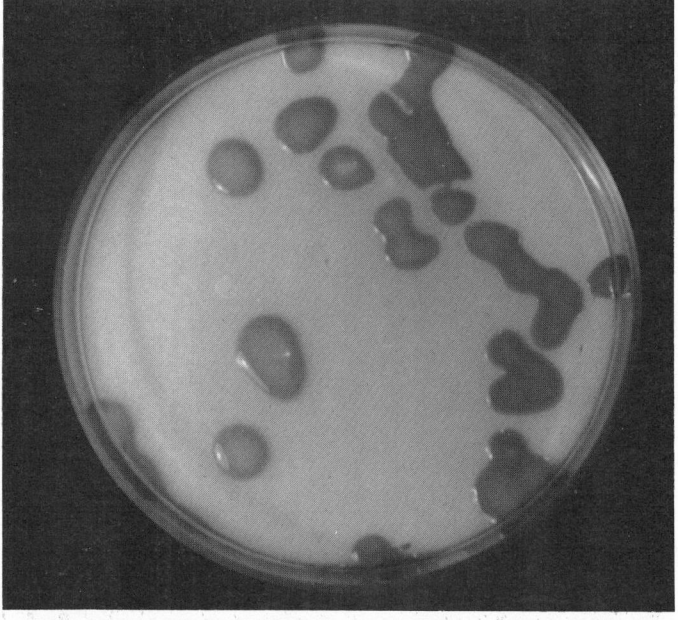

Figure 4.62. Colonies of *Beijerinckia mobilis* on nitrogen-free glucose mineral agar containing $CaCO_3$. On alkaline media this species forms opaque, relatively flat *Azotobacter*-like colonies with a characteristic amber-brown pigment. The slime is less sticky than that of the other *Beijerinckia* species (× 0.9).

cells are often visible in older cultures. The typical polar lipoid bodies may disappear in aging cells and the cells are then more rounded and resemble *Azotobacter* cells (Fig. 4.51).

Motility is conspicuous.

Agar colonies are not as raised as those of *B. indica*, and slime production is also less (Figs. 4.61 and 4.62). The slime is neither elastic nor sticky; its chemical composition has not yet been examined. Older cultures on neutral or alkaline agar media show a typcial dark amber or deep reddish brown color.

Broth cultures do not become viscous. There is a tendency to form a pellicle at the surface.

Grows between pH 3.0 and 10.0. Optimum growth and N_2 fixation occur at pH 4.0–5.0 and decrease sharply at the more alkaline values (Becking, 1961a). Temperature range for growth, 10–35°C; no growth at 37°C.

All strains tested have grown well on nitrate or ammonium salts as the nitrogen source (in contrast to *B. indica*). Weak growth or no growth occurs on urea, glycine, glutamate or tyrosine. All strains grow on leucine and casein agar. Moderate growth occurs on malt agar.

Common in Indonesian (Java) soils; also isolated from soils of South America (Surinam) and tropical Africa.

The mol% G + C of the DNA is 57.3 (T_m) (De Smedt et al., 1980).

Type strain: UQM 1969 (Delft E.III.12.2.1).

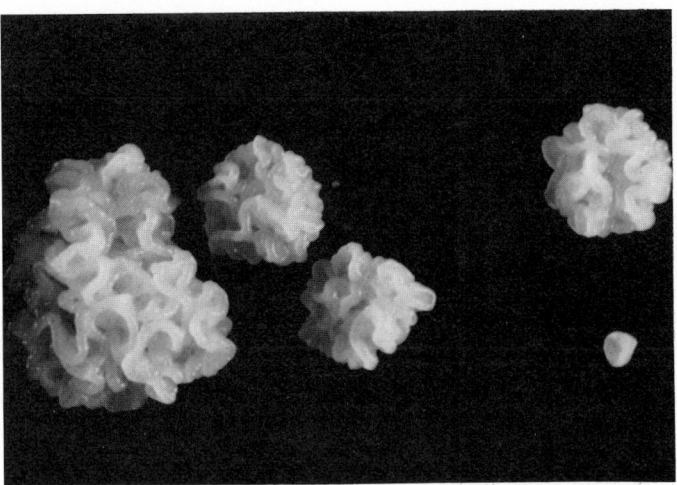

Figure 4.63. Typical colonies of *Beijerinckia fluminensis* on nitrogen-free glucose mineral agar. This species forms rather small, raised colonies with a highly plicated surface. In this species the slime has a granular consistency (× 4).

Further comments. The differences in levan production from sucrose observed by Derx (1950b) are variable and cannot be used for differentiation of this species.

3. **Beijerinckia fluminensis** Döbereiner and Ruschel 1958, 269.[AL]

flu.mi.nen′sis. M.L. adj. *fluminensis* named after the locality "Baixada Fluminense", State Rio de Janeiro, Brazil, from which soil it was first isolated.

Straight or slightly curved rods, 1.0–1.5 μm × 3.0–3.5 μm. Older cultures show characteristic large capsules enclosing from 2–10 or more individual cells. Division of the cells within the capsules has been observed.

Motility is slow or absent, especially in older cells.

Colonies are typically small and granular, moderately raised, with an irregular rough surface (Fig. 4.63). The slime is not liquid, tenacious or elastic, but more granular and stiff; its chemical composition has not yet been examined. Colonies are at first opaque white, becoming pink, reddish brown or fulvous (like *B. indica*) after 1–2 weeks on neutral or alkaline media.

Slime production in liquid media is reduced. No pellicle or viscosity occurs, but a bluish white turbidity develops.

Grows between pH 3.5 and 9.2. Temperature range for growth, 10–35°C (optimum is 26–33°C); no growth at 37°C.

Found in acidic soils of South America, Africa and Asia (China, Indonesia).

The mol% G + C of the DNA is 56.2 ± 1.8 (T_m) (type strain) (De Ley and Park, 1966; De Smedt et al., 1980).

Type strain: UQM 1685 (strain CD10 of Döbereiner and Ruschel).

4. **Beijerinckia derxii** Tchan 1957, 315.[AL]

derx′.i.i. M.L. gen. n. *derxii* of Derx; named after H. G. Derx, the Dutch microbiologist (1894–1953).

Single straight or curved rods, or rods with clavate extremities, 1.5–2.0 μm × 3.5–4.5 μm. Polar lipoid bodies are very large and conspicuous. No cyst or capsule formation occurs. Nonmotile.

Colonies are highly raised, slimy and smooth. The chemical composition of the slime has not yet been examined. Colonies are at first semitransparent or opaque white, but after 2–3 weeks a yellow-green, water-soluble fluorescent pigment is produced, particularly on iron-deficient media. When the pigment first appears it remains within the colony, but later it diffuses into the agar medium. Under certain conditions pigment production on agar media may be very poor or absent.

Liquid cultures become uniformly turbid and pigment production is usually less than on solid media.

Growth occurs between pH 4.0 and 9.0 (optimum is 6.0–7.0). There is no growth at pH 3.0 or 11.0. Temperature range for growth, 10–35°C; no growth at 37°C.

Isolated from soils from Queensland, Northern Australia, and neutral and alkaline soils of South Africa.

The mol% G + C of the DNA is 59.1 ± 1.6 (T_m) (De Ley and Park, 1966).

Type strain: UQM 1968 (strain Q13 of Tchan).

4a. **Beijerinckia derxii** subsp. **derxii** Tchan 1957, 315.

In the hierarchical classification (Fig. 4.57) of Thompson and Skerman (1979), within the species *B. derxii* the strains can be divided into two main groups: group 1263 and group 1271. Group 1263 contains a cotype of *B. derxii* Tchan 1957 and three cotype strains of "*B. congensis*" Hilger 1965 as well as three Australian isolates. This group is named by Thompson and Skerman (1979) as *B. derxii* subsp. *derxii*. Differences between this subspecies and the subspecies *venezuelae* (group 1271) are not markedly consistent apart from utilization of nitrate as outlined below in the description of the subspecies *venezuelae*.

Type strain: UQM 1968.

4b. **Beijerinckia derxii** subsp. **venezuelae** (Materassi et al. 1966) Thompson and Skerman 1981, 215.[VP] (Effective publication: Thompson and Skerman 1979, 343.) (*Beijerinckia venezuelae* Materassi, Florenzano, Balloni and Favilli 1966, 210.)

ven.e.zue′lae. M.L. gen. n. *venezuelae* of Venezuela, South America.

In the hierarchical classification of Thompson and Skerman (1979) group 1271 (Fig. 4.57) is classified as *B. derxii* subsp. *venezuelae*. This group consists of six cotype strains of *B. venezuelae* Materassi et al. 1966, one cotype of "*B. indica* var. *alba*" Derx 1950, and three Australian isolates.

Differences between group 1263 (subsp. *derxii*) and group 1271 (subsp. *venezuelae*) are not markedly consistent. In general, strains of group 1271 are distinguished by not utilizing nitrate as a sole source of nitrogen, being nonmotile, not hydrolyzing glycogen, growing over a slightly wider pH range (not more than 0.5 pH unit at each end of the range), and by differences in utilization of several organic compounds as sole sources of carbon. Of 14 stains of subsp. *venezuelae* and 7 strains of subsp. *derxii* tested by Thompson and Skerman (1979), the utilization of these substrates was as follows: melibiose, 5/14 vs. 7/7; propan-2-ol, 6/14 vs. 0/7; L-arabitol, 6/14 vs. 0/7; propionate, 7/14 vs. 0/7; fumarate, 8/14 vs. 1/7; and L-ascorbate, 6/14 vs. 5/7. Eleven of 14 strains of subsp. *venezuelae* used nitrate as a nitrogen source, whereas only 1 of 7 strains of subsp. *derxii* could do so.

Type strain: strain 2 of Materassi (WR-222).

Further comment: Of the six strains of subsp. *venezuelae* provided by G. Florenzano (i.e. strains WR-221 to WR-226), it is not clear how many independent isolates are represented.

Genus **Derxia** *Jensen, Petersen, De and Bhattacharya 1960, 193*[AL]

JAN-HENDRIK BECKING

Derx′i.a. M.L. fem. n. *Derxia* named after H. G. Derx, Dutch microbiologist (1894–1953).

Rod-shaped cells with rounded ends, 1.0–1.2 μm × 3.0–6.0 μm, occurring singly or in short chains. **Cells are rather pleomorphic,** depending on age and the medium. In aging cultures cells often remain together forming long filaments of sometimes locally swollen or distorted cells. Some cells may assume enormous sizes (up to 30 μm). **Young cells have a homogeneous cytoplasm; older cells show typical large refractile bodies throughout the whole cell. Resting stages are not known.** Gram-negative. **Motile by a short polar flagellum;** motile cells are numerous in liquid glucose media containing combined nitrogen, but rare on nitrogen-deficient solid media. **Aerobic,** having a strictly respiratory type of metabolism with oxygen as the terminal electron acceptor. **Molecular nitrogen is fixed** under aerobic conditions and also under decreased oxygen pressures (microaerophilic conditions). Optimum temperature, 25–35°C; growth is slow at 15°C, feeble at 40°C; no growth at 50°C. **Growth occurs between pH 5.5 and ~9.0;** no growth at pH 4.4. **Broth cultures turn into a gelatinous mass,** but growth near the surface is more luxuriant and forms a thick, tough pellicle. **Colonies on agar media are at first slimy and semitransparent, later massive and opaque, highly raised with a wrinkled surface. Older colonies develop a dark mahogany-brown color. Catalase-negative.** A wide range of sugars, alcohols and organic acids are oxidized mostly to CO_2 and a small amount of acid, probably acetic, when growing in an alkaline medium. **Can grow as a facultative hydrogen autotroph. Found in tropical soils** (Asia, Africa, South America). The mol% G + C of the DNA is 69.2–72.6 (T_m).

Type species: *Derxia gummosa* Jensen, Petersen, De and Bhattacharya 1960, 193.

Further Descriptive Information

The appearance of cells from young cultures is depicted in Fig. 4.64. Older cells on sugar-rich media contain large refractile bodies. On glucose-peptone agar especially, very elongated cells are produced containing many refractile bodies (Fig. 4.65). The refractile material is probably poly-β-hydroxybutyrate since it stains with Sudan III and Sudan black, but some vacuoles which do not stain may also be involved. On nitrogen-free glucose agar older cells undergo shrinkage (Fig. 4.66) and are finally enclosed by a slime envelope (Fig. 4.67).

Motile cells may become numerous in liquid glucose media containing ammonia or glutamate as the nitrogen source. The cells usually have a single polar flagellum, but some may have a flagellum at each pole. According to Thompson and Skerman (1979), the single polar flagellum is less than a full wave and is less than 3 μm in length.

Growth in liquid media usually starts as a ring at the glass-liquid interface and develops into a thick, wrinkled, tough pellicle. Shallow layers of medium change into a firm gelatinous mass after a couple of weeks. The color gradually becomes a dark red-brown.

Growth on nitrogen-deficient agar media begins as thin, whitish or semitransparent scattered colonies. Later, more massive, highly raised or dome-shaped colonies emerge which rapidly assume a diameter of 1 cm or more (giant colonies) (Fig. 4.68). These colonies are very reminiscent of those of *Beijerinckia* species. Colonies are at first whitish or dull yellow with a smooth surface, but the surface soon becomes coarse and wrinkled and the color deepens to a dark mahogany-brown. The slime of these colonies is very tenacious and gumlike, but in the other developmental stages it is more soft and smeary.

As noted by Jensen et al. (1960) in the original description of *D. gummosa*, aerobic growth on nitrogen-free, mineral glucose agar* consists of a mixture of a few "massive" colonies among many "thin," whitish colonies. It is the former that are associated with N_2 fixation; their slime affords some protection to the oxygen-sensitive nitrogenase system by decreasing the penetration of oxygen to the cells. The occurrence of two kinds of colonies under aerobic conditions is probably due to the occurrence of local areas of decreased oxygen concentration on the plates. It is only in these areas that N_2 fixation and extensive cell multiplication can occur that leads to the formation of the "massive" colonies. Under microaerophilic conditions (p_{O_2} less than 0.2

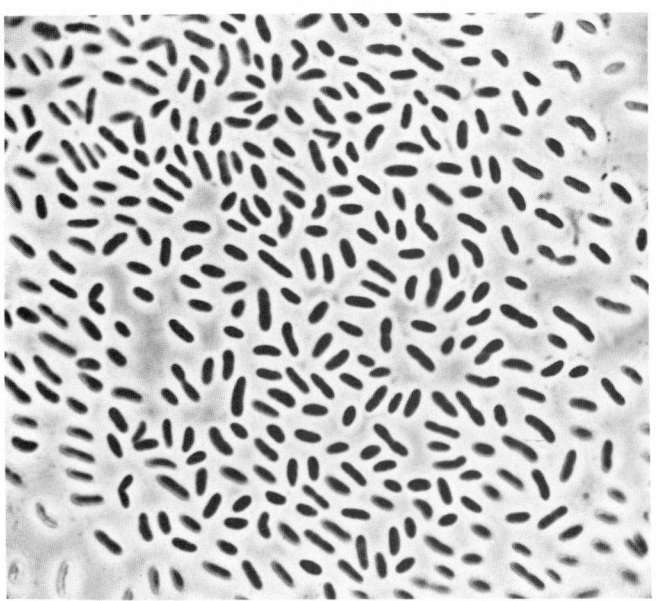

Figure 4.64. Seven-day-old cells of *Derxia gummosa* on nitrogen-free agar containing 2% glucose. Phase-contrast microscopy (× 950).

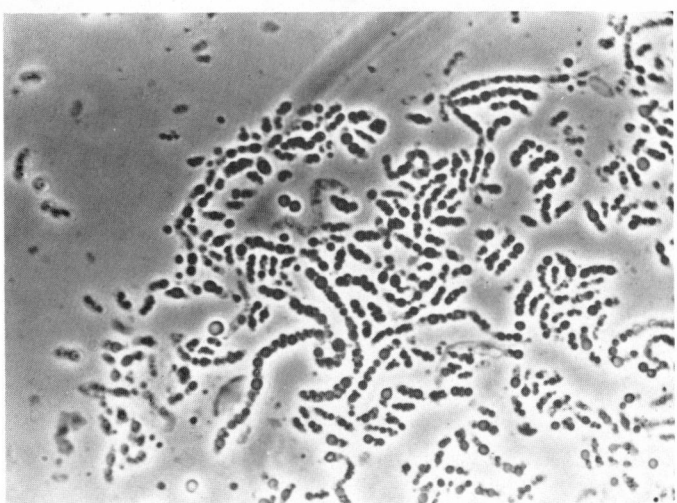

Figure 4.65. Ten-day-old cells of *Derxia gummosa* on peptone agar with 2% glucose. Phase-contrast microscopy (× 950).

atm), all of the colonies on a plate develop into the "massive" type (Hill, 1971).

The efficiency of N_2 fixation by *D. gummosa* varies between 9 and 25 mg N/g of glucose consumed, but in most strains it is distinctly lower than in *Azotobacter* or *Beijerinckia*. There is no requirement for amino acids, vitamins or growth factors, but trace elements, particularly molybdenum, are required. Vanadium cannot replace the molybdenum for N_2 fixation.

Growth with combined nitrogen sources is much faster than with N_2 and is completely uniform, in contrast to the uneven growth on nitrogen-free agar. Colonies change from a pale yellow through rust-brown to almost black (darkest if nitrate is present) and sometimes a light brown, water-soluble pigment is produced. Growth with glutamic acid,

---

* Nitrogen-free, glucose mineral medium (g/liter of distilled water): K_2HPO_4, 0.5; $MgSO_4 \cdot 7H_2O$, 0.25 (or 0.2); NaCl, 0.25; $FeSO_4 \cdot 7H_2O$, 0.1; $CaCl_2$ or $CaCO_3$, 0.1; $Na_2MoO_4 \cdot 2H_2O$, 0.005; and glucose, 10.0 (or 20.0); pH, 6.9. For a solid medium add 15.0 g agar/liter.

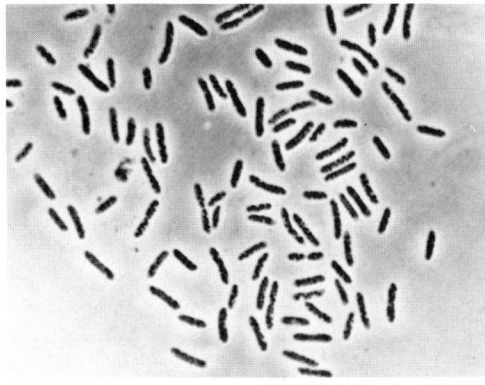

Figure 4.66. Three-week-old cells of *Derxia gummosa* on nitrogen-free glucose agar, showing shrinkage of the cells. Phase-contrast microscopy (× 950).

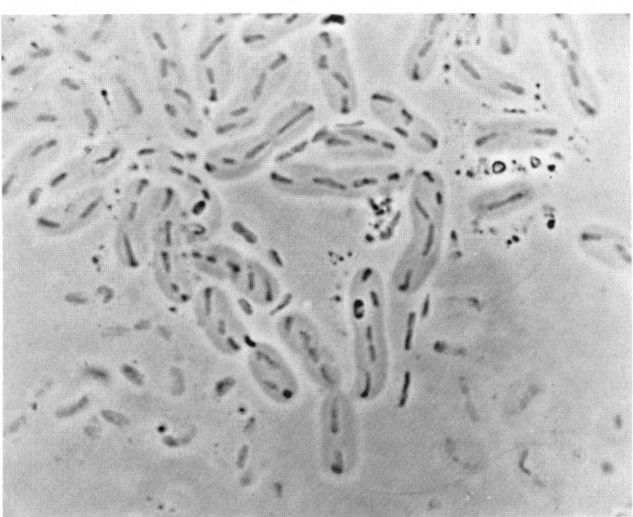

Figure 4.67. Three-month-old cells of *Derxia gummosa* on nitrogen-free glucose agar. The cells show shrinkage and are enclosed by a thick slime envelope. Phase-contrast microscopy (× 950).

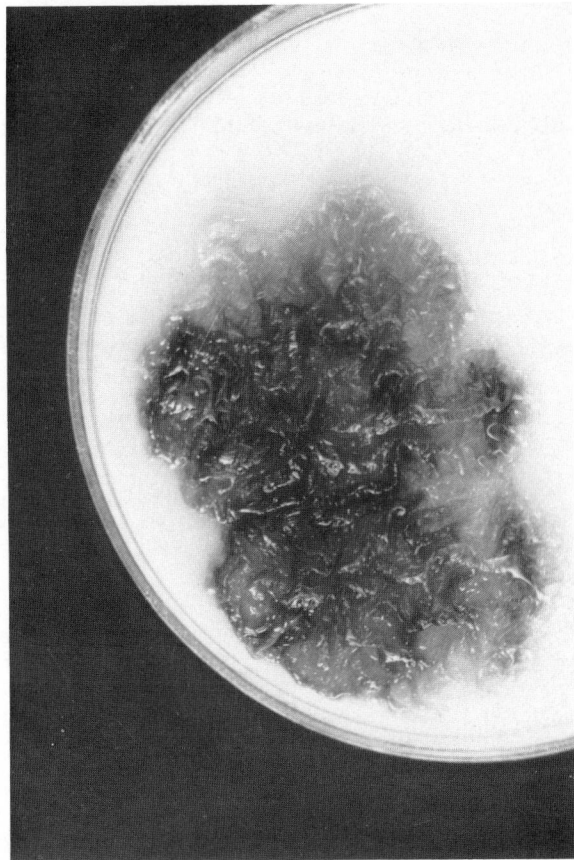

Figure 4.68. Colony type of *Derxia gummosa* on nitrogen-free glucose agar with calcium carbonate (× 1.0).

ammonium acetate, alanine, sodium nitrate and urea decreases from abundant to good in approximately the same sequence. Aspartic acid, asparagine and peptone give a much slower growth which is uneven and mostly confined to scattered colonies. Glycine seems to be toxic.

Growth on nitrogen-deficient media or under N_2-fixing conditions is stimulated by small amounts of combined nitrogen (nitrate, ammonia), particularly at the start (Tchan and Jensen, 1963). Small amounts of yeast extract are stimulatory for growth under N_2-fixing conditions (Becking, unpublished results), and although there is no apparent requirement for vitamins or growth factors, it is possible that small amounts of biotin are stimulatory for growth of some strains.

D. gummosa has been shown to grow as a hydrogen autotroph in an atmosphere containing $H_2 + CO_2 + O_2$, with either N_2 or NH_4^+ as the nitrogen source (Pedrosa et al., 1980). Indeed, it appears to grow nearly as well autotrophically as it does heterotrophically. Ribulose-1,5-biphosphate activity, which mediates the CO_2 fixation, occurs in autotrophically-grown cells but not in cells grown heterotrophically.

Growth on methane or methanol as the sole carbon source has been demonstrated for *D. gummosa* (Sampaio et al., 1981).

Growth on glucose, fructose, ethanol, glycerol, mannitol and sorbitol is good to excellent; growth on mannitol and lactate is scant. No growth or only a trace of growth occurs on lactose, galactose, maltose, sucrose, formate, acetate, propionate, pyruvate, succinate, malate, fumarate, dulcitol and starch. Butyrate, citrate, benzoate and xylose suppress growth.

Nitrate is not reduced to nitrite or N_2 in a glucose-nitrate medium. Indole is not produced from tryptophan.

Derxia occurs mainly in tropical soils. It was isolated originally by Jensen et al. (1960) from a West Bengal soil having a pH of 6.5. It has since been found to be widely distributed in Brazil (Campêlo and Döbereiner, 1970), Indonesia, China, and in southern Africa (Becking, 1981). In Brazil, Campêlo and Döbereiner (1970) found it to be most frequent in flooded soils, less frequent in humid soils, and least frequent in dry soils. In comparing isolations made on a nitrogen-free, mineral starch agar,* they found that *Derxia* could be more often isolated from root pieces of plants (mostly *Gramineae*) than from soil samples obtained from the same locality. In soil, the presence of the organisms seemed to be favored when the soil pH was between 5.1 and 5.5, although isolations were made from soils over a pH range of 4.5–6.5.

Enrichment and Isolation Procedures

The sieved soil method may be used for isolation. Soil particles are regularly distributed over the surface of nitrogen-free mineral media

---

* Nitrogen-free, mineral starch agar medium (per liter of distilled water): starch, 20.0 g; K_2HPO_4, 0.05 g; KH_2PO_4, 0.15 g; $MgSO_4 \cdot 7H_2O$, 0.2 g; $CaCl_2$, 0.02 g; $NaHCO_3$, 0.1 g; $FeCl_3$ solution (10%, w/v), 1 drop; $Na_2MoO_4 \cdot 2H_2O$, 0.002 g; brom thymol blue solution (0.5% alcoholic, w/v), 5.0 ml; and agar, 20.0 g. (Note: this medium is similar to that used by Lipman (1903) but has been modified by supplementation with $NaHCO_3$.)

Figure 4.69. Simplified dendrogram of the hierarchical interrelations between *Derxia*, *Azomonas* and "*Azo-tomonas*" species according to Thompson and Skerman (1979). The numbers are related to the degree of similarity (or similarity attributes) within the species and strains. An increase in number denotes a decrease in similarity. (Reproduced with permission from J. P. Thompson and V. B. D. Skerman, *Azotobacteraceae*: The taxonomy and ecology of the aerobic nitrogen-fixing bacteria, 1979 © Academic Press, London.)

containing mannitol† (Jensen et al., 1960), glucose§ (Becking, unpublished results), or starch* (Campêlo and Döbereiner, 1970). (It should be noted that some strains, including the type strain, produce only very scant growth on starch. Also, Thompson and Skerman (1979) reported that mannitol was not used by three of the six strains they tested.) One looks for the development of yellowish colonies around the soil particles. These colonies eventually become much larger and acquire a rust-brown color. Isolates should be further purified by repeated streaking.

Derxia strains are acid-tolerant and have been isolated from cultures designed to enrich for *Beijerinckia* strains by use of a liquid acidic nitrogen-free glucose medium¶ (Becking, unpublished results). This medium is dispensed into Petri dishes so as to form a thin layer 2–3 mm deep.

Maintenance Procedures

The maintenance procedures are the same as those described for the genus *Beijerinckia*.

Procedures for Testing Special Characters

Nitrogen fixation (heterotrophic growth) can best be tested in a liquid medium with glucose, fructose, mannitol, or other carbon source in a nitrogen-free mineral basal medium.* The addition of a low concentration of agar (0.1%, w/v) sometimes enhances growth by providing microaerophilic conditions in the manner described for the genus *Azospirillum*.

The ability to grow autotrophically can be tested on a solid or liquid medium ‖ in the presence or absence of $(NH_4)_2SO_4$ (1.0 g/liter). Cultures are incubated at 30°C in sealed vessels under the following atmospheres: (a) $O_2:N_2$, 1:99, v/v; (b) $O_2:CO_2:N_2$, 1:5:94; (c) $O_2:H_2:N_2$, 1:5:94; and (d) $O_2:H_2:CO_2:N_2$, 1:5:5:89. (Under atmospheres containing CO_2 the medium may also need to contain $NaHCO_3$ (1.0 g/liter) to prevent acidification.) Significant growth should occur only when both H_2 and CO_2 are provided, even when N_2 is the sole nitrogen source. Under conditions of $H_2 + CO_2$ dependent autotrophic growth under N_2-fixing or non-N_2-fixing conditions, the average doubling time (dt) of *D. gummosa* is 34.1 h (Pedrosa et al., 1980).

Differentiation of the genus **Derxia** from other genera

Derxia can be distinguished from other genera of N_2-fixing bacteria by its very slimy (gummy) growth, both on agar plates and in liquid media, combined with the very pleomorphic appearance of the cells (see Figs. 4.64–4.67) depending on age and type of medium. *Azotobacter* and *Azomonas* species never produce such slimy colonies. Confusion with *Beijerinckia* is unlikely since *Beijerinckia* cells usually show very characteristic cells with two polar lipoid bodies, whereas *Derxia* cells contain numerous lipoid bodies. Moreover, in contrast to *Beijerinckia* colonies, the colonies of *Derxia* acquire on aging a typical dark mahogany-brown color; also, *Derxia* is catalase-negative.

In contrast to the vibrioid cells of *Azospirillum*, *Derxia* cells are mainly straight rods and the lipoid bodies are more numerous. *Derxia* colonies are far more slimy than those of *Azospirillum*, and the mahogany-brown color they eventually acquire is in contrast to the pink or whitish color of *Azospirillum* colonies. Moreover, there are some minor physiological differences. *Derxia* strains can exhibit N_2-dependent growth on glucose, whereas only some strains of *Azospirillum* (i.e. only those belonging to the species *A. lipoferum*) can do so. On the other hand, *Derxia* strains generally do not grow well on malate, whereas all strains of *Azospirillum* can grow well on this carbon source. Moreover, *Derxia* strains are catalase-negative, whereas *Azospirillum* strains give reactions that range from strong to undetectable.

---

† Nitrogen-free, mineral mannitol agar (g/liter of distilled water): mannitol, 10.0; K_2HPO_4, 0.5; $MgSO_4·7H_2O$, 0.2; $CaCl_2$, 0.1; $CaCO_3$, 5.0; $Na_2WO_4·2H_2O$, 0.0005; $FeCl_3$ and $Na_2MoO_4·2H_2O$, traces; and agar, 15.0.

§ Nitrogen-free, mineral glucose agar for sieved-soil method (g/liter of distilled water): glucose, 20.0; K_2HPO_4, 0.8; KH_2PO_4, 0.2; $MgSO_4·7H_2O$, 0.5; $FeCl_3·6H_2O$, 0.025 (or 0.05); $Na_2MoO_4·2H_2O$, 0.005; $CaCl_2$, 0.05; and agar, 15.0. The pH is adjusted to 6.9.

¶ Liquid acidic nitrogen-free glucose medium (g/liter of distilled water): glucose, 20.0; KH_2PO_4, 1.0; and $MgSO_4·7H_2O$, 0.5. The pH is adjusted to 5.0.

‖ Medium for testing autotrophy (per liter of distilled water): KH_2PO_4, 1.2 g; K_2HPO_4, 0.8 g; $MgSO_4·7H_2O$, 0.2 g; NaCl, 0.2 g; $CaCl_2·2H_2O$, 0.02 g; $FeSO_4·7H_2O$, 0.002 g; trace element solution (*see below*), 2.0 ml; biotin, 10 μg; and agar, 15.0 g. *Trace element solution* (g/liter): $Na_2MoO_4·2H_2O$, 1.0; $MnSO_4·H_2O$, 1.75; H_3BO_3, 1.4; $CuSO_4·5H_2O$, 0.04; and $ZnSO_4·7H_2O$, 0.12. This medium was employed by Pedrosa et al. (1980) in their studies of *Derxia*. In these studies, a very low concentration of potassium malate (0.1 g/liter) was sometimes added to the medium. It should be noted that most strains of *Derxia* (including the type strain) give only scant growth with malate. Thompson and Skerman (1979) reported no growth on DL-malate in all of the six strains that they tested.

Unlike *Azotobacter*, *Azomonas*, and *Beijerinckia* species, *Derxia* can grow as a facultative hydrogen autotroph. In this ability it shows affinities to various other facultative hydrogen autotrophs that can also use H_2 to provide energy and reducing power for growth and CO_2 fixation: various *Pseudomonas* and *Alcaligenes* species ("*Hydrogenomonas*" species) (Buchanan and Gibbons, 1974), *Xanthobacter* species (Wiegel et al., 1978; Malik and Claus, 1979), nitrate-reducing *Paracoccus* species (Rittenberg, 1969), *Bradyrhizobium japonicum* (Hanus et al., 1979), and *Azospirillum lipoferum* (Malik and Schlegel, 1981; Sampaio et al., 1981).

Taxonomic Comments

In cell morphology and type of flagellar arrangement (one polar flagellum), the genus *Derxia* is like the *Pseudomonadaceae*, but studies on DNA base composition and DNA homology indicate that it is very different from *Pseudomonas* species (De Ley and Park, 1966).

By the use of rRNA cistron similarity as a criterion of genetic relatedness, De Smedt et al. (1980) found little relatedness between *Derxia* and the genera *Azotobacter*, *Azomonas*, and *Beijerinckia*. They placed *Derxia* in a different rRNA superfamily and indicated that the closest relatives of *Derxia* so far found were *Chromobacterium*, *Janthinobacterium*, *Pseudomonas acidovorans* and *Pseudomonas solanacearum*, and *Alcaligenes faecalis*.

Based on a numerical analysis of a large number of attributes, Thompson and Skerman (1979) showed a fusion of their five *Derxia* strains with "*Azotomonas*" ("*Azotomonas insolita*" Stapp 1940, sometimes regarded as "*Pseudomonas insolita*" (Stapp) Brisou 1961)) at group 1296 (Fig. 4.69). Moreover, there was a fusion with *Azomonas* (of the family *Azotobacteraceae*) at group 1299 (Fig. 4.69). In view of these relationships, the genus *Derxia* can be classified in the *Azobacteraceae* and the recognition of the group 1266 with *Derxia gummosa* as a single species within the genus is quite acceptable. It is evident from the dendrogram (Fig. 4.69) that *Derxia* shows closer affinities with the genus *Azomonas* rather than the genus *Azotobacter* (see also Fig. 4.25 of the article on the genus *Azotobacter* in this *Manual*).

It is the view of the present author that there must be some relation between *Derxia* and *Beijerinckia*, because both genera produce a viscous, tenacious slime, form colonies that are similar in several respects, and have a molydenum requirement for N_2 fixation that cannot be replaced by vanadium (in contrast to *Azotobacter* species, but like *Azomonas* species) (Becking, 1962; Jensen et al., 1960).

It is possible that other species of *Derxia* besides *D. gummosa* may exist. For instance, Roy and Sen (1962) described a new species of *Derxia* from a sample of partially retted jute plant (*Corchorus olitorius*) from Uttar Pradesh, India. The species description, however, does not present any cultural or physiological differences from *D. gummosa* to warrant the nomination of a new species.

Further Reading

Becking, J. H. 1981. The family *Azotobacteraceae*. In Starr, Stolp, Trüper, Balows and Schlegel (Editors), *The Prokaryotes: a Handbook on Habitats, Isolation and Identification of Bacteria*, Springer-Verlag, Berlin, pp. 795–817.

List of the Species of the Genus **Derxia**

1. **Derxia gummosa** Jensen, Petersen, De and Bhattacharya 1960, 193.[AL]

gum.mo′sa. L. fem. adj. *gummosa* slime (gum) producing.

The characteristics are as described for the genus.

Originally isolated from a soil of West Bengal of pH 6.5, but later also from slightly acidic or neutral soils of South America (Brazil, Surinam), South Africa and Java.

The mol% G + C of the DNA of the type strain is 70.4 ± 1.7 (T_m)

(De Ley and Park, 1966). In three strains later examined it ranged from 69.2–72.6 (T_m) (De Smedt et al., 1980).

Type strain: ATCC 15994 (Bhattacharya strain).

Further comments. The "*Derxia*-like" strains isolated by Petersen and Holmes (1964), i.e. strains N61 and N63, have been reported to show serological cross-reactions with *Xanthobacter flavus* (Postgate, 1981).

Genus **Xanthobacter** *Wiegel, Wilke, Baumgarten, Opitz and Schlegel 1978, 573.*[AL]

JÜRGEN K. W. WIEGEL AND HANS G. SCHLEGEL

Xan.tho.bac′ter. Gr. adj. *xanthos* yellow; M.L. masc. n. *bacter* the equivalent of Gr. neut. noun *bacterion* rod, staff; M.L. masc. noun *Xanthobacter* yellow rod.

Rod-shaped cells 0.4–1.0 μm in diameter and 0.8–6.0 μm in length. Pleomorphic cells are produced on succinate-containing media; coccoid cells as well as cells up to 10 μm long are produced on media containing an alcohol as the sole carbon source. Refractile (polyphosphate) and lipid (poly-β-hydroxybutyric acid) bodies are evenly distributed in the cells. Resting stages are unknown. **The Gram-reaction is positive or variable; however, the ultrastructure of the cell wall seems to be of the negative** Gram-type, the peptidoglycan is directly cross-linked by *meso*-diaminopimelic acid, and lipopolysaccharides are present. Nonmotile or motile (by peritrichous flagella). **Obligately aerobic,** having a strictly respiratory type of metabolism with oxygen as the terminal electron acceptor. Optimum temperature, 25–30°C. Optimum pH, 5.8–9.0. Colonies are opaque and slimy (although slime-free strains exist) and are **yellow** due to a water-insoluble carotenoid pigment (**zeaxanthin dirhamnoside**). **Catalase-positive.** All strains can grow as **chemolithoautotrophs** in mineral media under an atmosphere of H_2, O_2 and CO_2 (7:2:1, v/v) as well as **chemoorganotrophically** on methanol, ethanol, n-propanol, n-butanol, and various organic acids as sole carbon sources. Carbohydrate utilization is limited, and neither volatile/nonvolatile fatty acids nor gas are produced from some carbohydrates. Some strains require vitamins. **Atmospheric nitrogen is fixed** in nitrogen-deficient media, but by most strains only under a decreased oxygen pressure; the

efficiency is more than 10 mg of nitrogen fixed/g of sucrose consumed, and is usually 20 mg/g. The induction of root nodules on plants has so far not been observed. The organisms occur free-living in wet soil containing decaying organic material, and also in water. The mol% G + C of the DNA is 65–70 (T_m) and 66–68 (Bd).

Type species: *Xanthobacter autotrophicus* (Baumgarten, Reh and Schlegel 1974) Wiegel, Wilke, Baumgarten, Opitz and Schlegel 1978, 573.

Further Descriptive Information

Morphology: The cell shape of *Xanthobacter* depends strongly on the carbon source, whereas the nitrogen source has a minor effect (Fig. 4.70). Cells under N_2-fixing conditions are only slightly longer than in the presence of ammonia or other sources of combined nitrogen. Cells of both *X. autotrophicus* and *X. flavus* are coccoid when growing on n-propanol (Fig. 4.70D) and are long, filamentous rods when growing on n-butanol. One of the most significant morphological features is the formation of irregular and branched cells during growth on tricarboxylic acid cycle intermediates; e.g. succinate, which is a good substrate, leads to the most irregular cell forms during the late exponential growth phase (Fig. 4.70C; Wiegel and Schlegel, 1976). In contrast to nonmotile strains, the motile strains are only slightly pleomorphic and only after prolonged incubation (3–7 days) on succinate-containing agar media

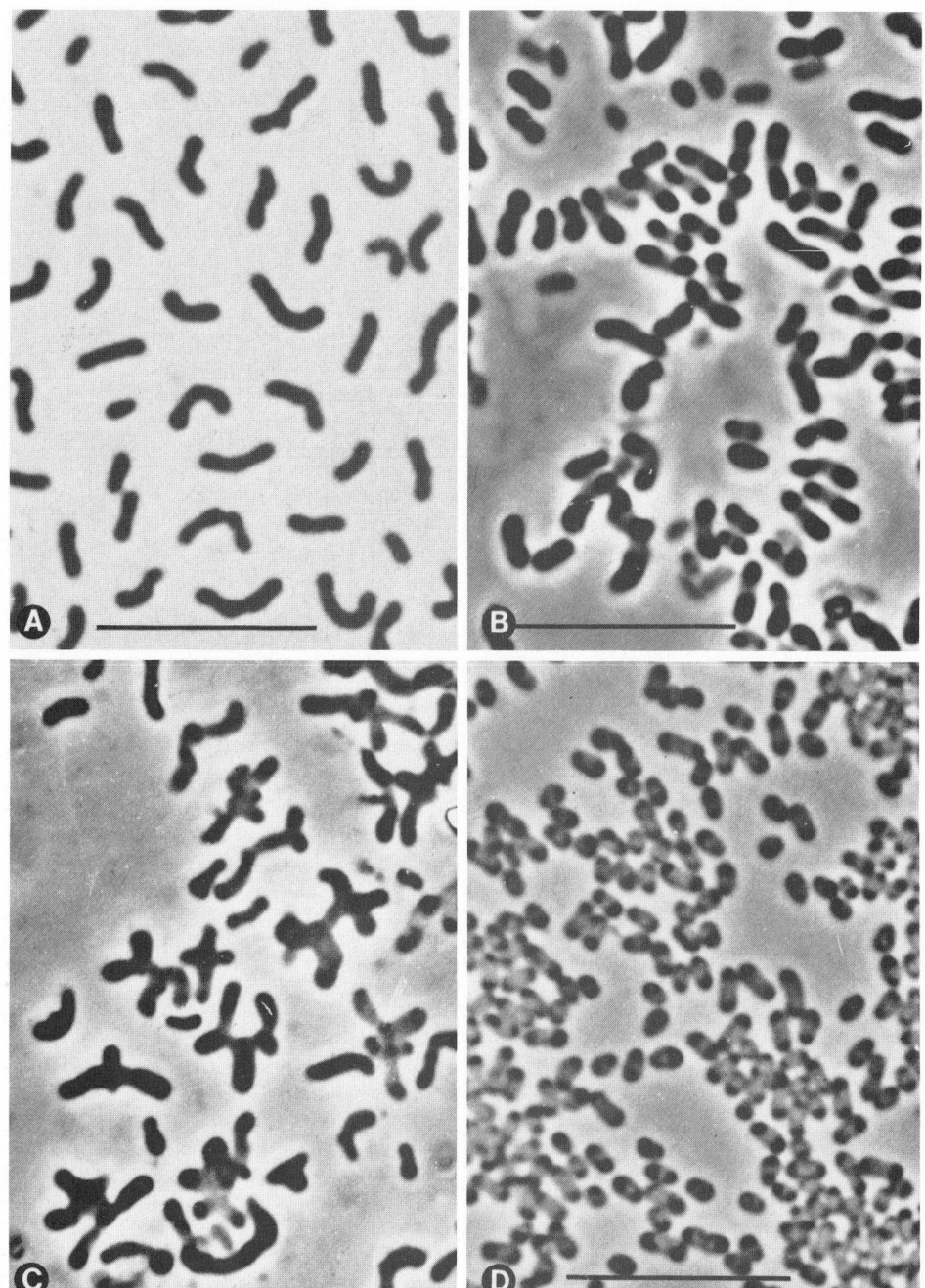

Figure 4.70.   Cells of *Xanthobacter autotrophicus*. *A*, strain 14g grown on malate in nonagitated liquid culture, showing irregular vegetative cells. *B–D*, strain JW 33 (reference strain). *B*, grown lithoautotrophically under conditions of nitrogen fixation, showing rod-shaped cells and palisade-like formations. *C*, grown in nutrient broth (0.7%) containing 0.1% succinate, showing the typical branched cell formation. *D*, grown on *n*-propanol in the presence of ammonium, showing coccoid cell formation. *E–F*, strain 14g, showing cell aggregation ("star-formation") on nutrient broth medium. This kind of aggregation is only observed after 2–5 h upon transfer into liquid nutrient broth media under nonagitated growth conditions. All *bars*, 10 μm. (Parts *A* and *E* reproduced with permission from K. Schneider et al., Archives of Microbiology *93*: 179–193, 1973, © Springer-Verlag, Heidelberg. Parts *B* and *C* reproduced with permission from J. Wiegel and H. G. Schlegel, Archives of Microbiology *107*: 139–142. 1976, © Springer-Verlag, Heidelberg, Part *F* reproduced by permission from M. Reh.)

(Aragno et al., 1977; J. Wiegel, unpublished data). The cells of the exponential growth phase and those grown under heterotrophic conditions are usually larger than those of the early stationary growth phase and those grown under chemolithoautotrophic conditions; the latter are mainly straight or only slightly-curved rods (Fig. 4.70*B*).

The occurrence of a snapping type of cell division, as was assumed

from light-microscopic inspection of *X. autotrophicus*, has not been confirmed by electron microscopic studies. The illusion of snapping cell division, as well as of "star formation" (Fig. 4.70 *E* and *F*) is presumably due to cell aggregation by copious amounts of slime. Surface patterns on the outer cell wall are not observed in electron micrographs.

Cell wall type and Gram-reaction. Many strains identified later as

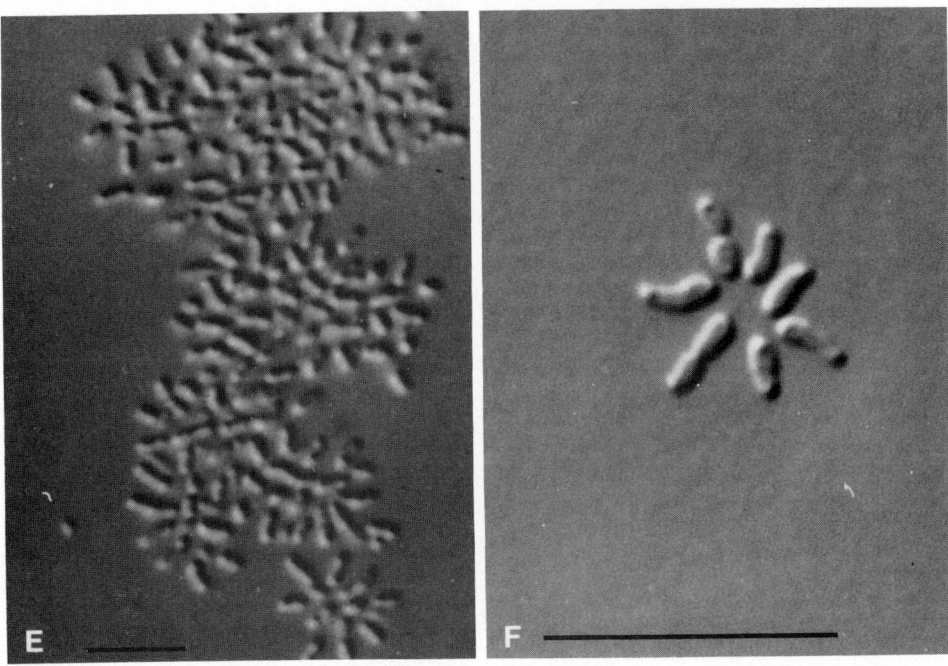

Figure 4.70 E and F.

strains of *Xanthòbacter*, were originally described to give a positive or variable Gram-reaction. However, *Xanthobacter* belongs to the negative Gram-type* bacterial group, since it contains lipopolysaccharides (Wiegel and Mayer, 1978; Wiegel, 1981). The impression of a positive Gram-reaction is feigned by the content of refractile bodies, which were identified as polyphosphate granules by electron microscopy as well as by volutin staining. Unfortunately, the polyphosphate content of the cells is high (15 mg of phosphate/g dry weight); even at a low phosphate concentration in the growth medium (0.01 mM) volutin granules do not disappear. By electron microscopy of thin sections of cells (Walther-Mauruschat et al., 1977) it was shown that the cell envelope of *X. autotrophicus* has a thin multilayered structure, resembling that of Gram-type negative cells (Walther-Mauruschat et al., 1977). The peptidoglycan content of the cell wall of the *X. autotrophicus* strains (15–25%) is intermediate compared to that of typical Gram-type positive* (30–70%) and Gram-type negative (10%) bacteria; teichoic acid and teichuronic acid are absent (Schleifer and Kandler, 1972; O. Kandler and F. Fiedler, personal communication).

Compared to other Gram-type negative bacteria the lipopolysaccharide content is relatively low for most strains tested. The lipopolysaccharide was either isolated or demonstrated by the lipopolysaccharide polymyxin B interaction technique using an electron microscope (Wiegel and Mayer, 1978, J. Wiegel and L. Quandt, unpublished data). The citrate synthase of both *Xanthobacter* species resembles that of Gram-type negative bacteria by having a molecular weight above 250,000 and by being inhibited by NADH (Weitzman, 1980; Berndt et al., 1976).

Slime formation. The majority of the strains produce copious amounts of slime (Fig. 4.71) which consists mainly of glucose, galactose, mannose and uronic acid (unidentified; 30% or more of the monomers contain a carboxyl group; Andreesen and Schlegel, 1974; R. Opitz, Ph.D. Thesis, Göttingen, 1977). At high C/N-ratios the slime formed can cause gelatinization and solidification of the medium without significant increase in cell mass. The amount of slime varies with the strain as well as growth conditions. Most strains produce copious amounts of slime during growth on carbohydrates or lactate, but minor

amounts during growth on O_2/H_2 or on alcohols. No correlation between slime formation and oxygen tolerance during nitrogen fixation is observed, although many of the slime producers grow better at high (2–3%) oxygen concentration in the gas atmosphere than the nonproducers; the majority of the strains produce more slime under nitrogen fixing conditions than in the presence of ammonium ions. Beside the motile strains and very few strains of *X. autotrophicus* (e.g. JW50) producing traces of slime, only one slimeless mutant (of strain 7c) could be isolated (Andreesen and Schlegel, 1974). The slime becomes soluble at pH values below 4.5 and above 10. By alkaline treatment the slimy cell aggregates can be separated into single cells as needed for isolation purposes (Wiegel and Schlegel, 1976). The slime has proven to be very recalcitrant and no degrading organism has been found under aerobic or anaerobic enrichment conditions.

Polyglutamine capsule. All *Xanthobacter* species (about 20 strains tested) produce an α-polyglutamine which is located between the cell wall and the slime. This polymer is not separated from the cells during the alkaline treatment. The glutamine polymer is not found on cells grown under N_2-fixing or chemolithoautotrophic conditions. Contrary to the slime, this polymer is apparently reutilized in the late stationary growth phase and under N-limitation (J. Wiegel, unpublished results). *Flexithrix* is the only other genus known which produces a similar α-polyglutamine capsule (H. König and O. Kandler, personal communication).

Storage material. Xanthobacter autotrophicus and *X. flavus* deposit poly-β-hydroxybutyric acid (PHB) as reserve material under heterotrophic as well as under chemolithoautotrophic conditions (Fig. 4.72). In the stationary growth phase, heterotrophically grown cells contain between 5 and 600 mg PHB/g dry weight of cells, depending on the strain and on the carbon source: fructose results normally in high yields whereas succinate results in very low values (less than 15 mg PHB/g dry weight). For example, reference strain JW33 has 668, 250 and 8 mg PHB/g dry weight when grown on fructose, sucrose and succinate, respectively (J. Wiegel, unpublished data; quantitative determination according to Jüttner et al., 1975).

---

* The terms "Gram-type negative" and "Gram-type positive" were proposed by Wiegel (1981) to describe bacteria according to ultrastructural and biochemical Gram characteristics (cell wall structure, presence or absence of indicator compounds such as lipopolysaccharide, etc. The terms are distinct from those referring to the results of the Gram-staining reaction: "Gram-reaction positive," "Gram-reaction negative" or "Gram-reaction variable."

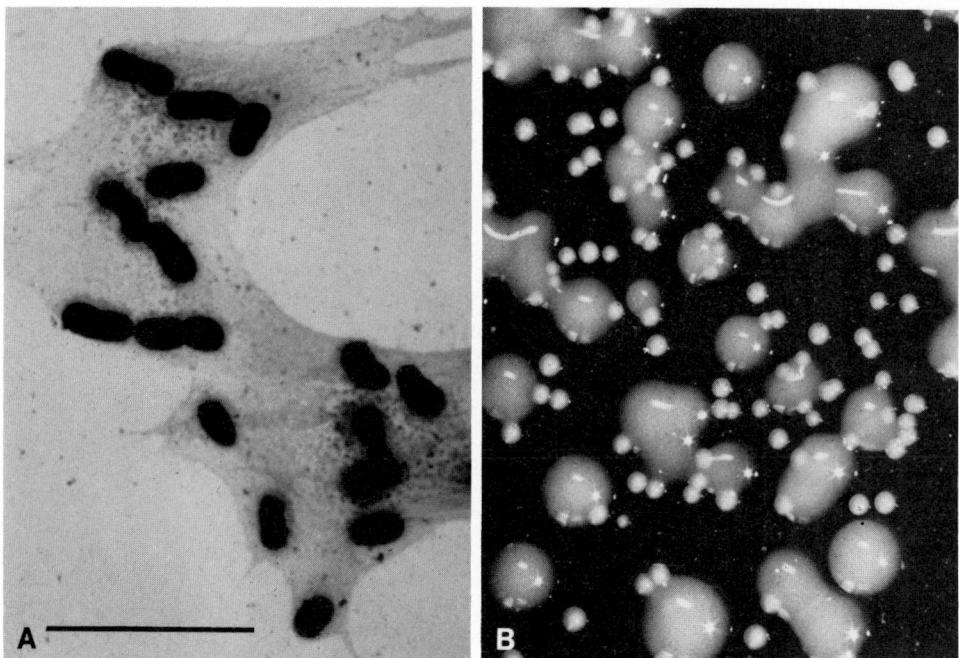

Figure 4.71. Slime production by *Xanthobacter autotrophicus*. *A*, Cells and slime of the type strain (DSM 432) stained negatively by uranyl acetate. *Bar*, 5 μm. (Reproduced with permission from N. Tunail and H. G. Schlegel, Archives of Microbiology *100:* 341–350, 1974, © Springer-Verlag, Heidelberg.) *B*, Colonies from the wild type (large, slimy colonies) and from the slimeless mutant (small, white colonies) of DSM 432. (Reproduced with permission from M. Andreesen and H. G. Schlegel, Archives of Microbiology *100:* 351–361, 1974, © Springer-Verlag, Heidelberg.)

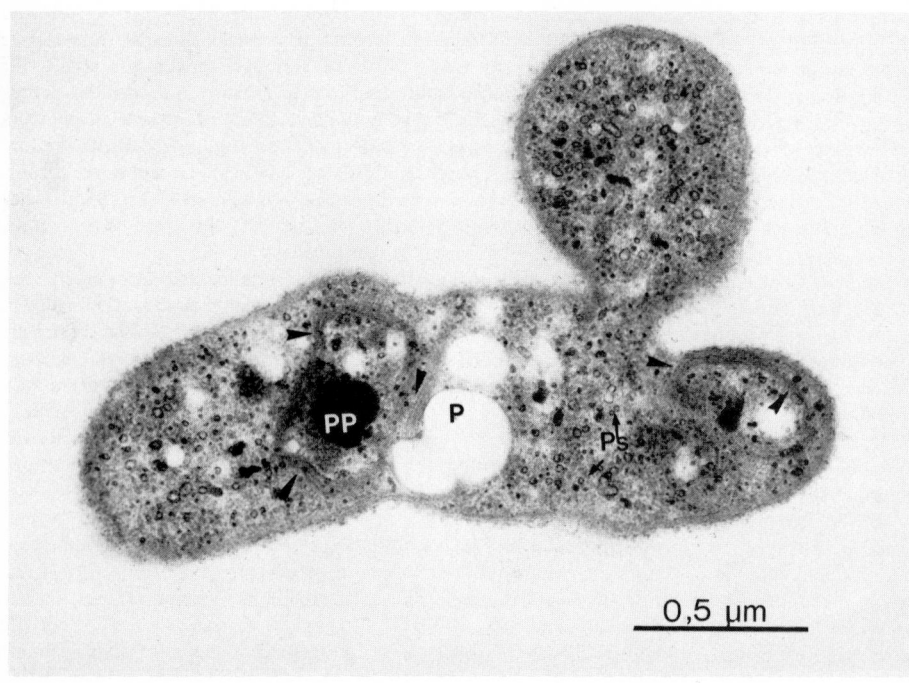

Figure 4.72. Ultrathin section of *Xanthobacter autotrophicus* strain 14g grown on a succinate-containing medium. The cells were prepared by glutaraldehyde-osmium tetraoxide fixation, uranyl acetate block staining, and lead citrate poststaining; preparation by A. Walther-Mauruschat. *P*, poly-β-hydroxybutyrate; *PP*, polyphosphate; *Ps*, small-type polyphosphate. (Reproduced with permission from J. Wiegel et al., International Journal of Systematic Bacteriology *28:* 573–581, 1978, © International Association of Microbiological Societies.)

Colonial characteristics. Both species of *Xanthobacter* form colonies that are smooth, convex, circular, filiform, opaque, and either white or yellow. The main component of the yellow pigment of all strains (the "white" strains contain just minor amounts) has been identified as zeaxanthin dirhamnoside (Herzberg et al., 1976; J. Wiegel and K. Schmidt, unpublished data). It is not water-soluble and is located in the cell wall (Eberhardt, 1971).

Nitrogen fixation and nitrogenase. The *Xanthobacter* species described so far, are microaerophilic nitrogen-fixing bacteria, as already described in the original reports on the discovery of nitrogen fixation in *X. flavus* (as *"Mycobacterium flavum"* 301) by Federow and Kalininskaya (1961) and in *X. autotrophicus* (as *"Corynebacterium autotrophicum"*) by Gogotov and Schlegel (1974) and Wiegel and Schlegel (1976). The nitrogen-fixing system, including its relationship to oxygen, has been studied mainly with *X. autotrophicus* strain GZ29 (Berndt et al., 1976, 1978) and *X. flavus* strain 301 (Biggins and Postgate, 1969, 1971). The composition of the nitrogenase components of heterotrophically grown cells of strain GZ29 and strain 301 are similar to that of other nitrogenases as far as metal and sulfur content as well as amino acid composition are concerned. As in other aerobic diazotrophs, the nitrogenase system seems to be loosely associated with membranes. The protection of the nitrogen-fixing system in *Xanthobacter* may occur through some conformational protection for the nitrogenase, through the respiration of associated aerobic microorganisms in the natural habitat, and through respiratory activity of *Xanthobacter* itself; the respiratory rate is, however, about one to two orders of magnitude lower than that in *Azotobacter* (Biggins and Postgate, 1971). *Xanthobacter* possesses an active ammonium (methylammonium) transport system. Methylammonium can serve as a sole carbon and nitrogen source (J. Wiegel and D. Kleiner, unpublished data). Additional information concerning the two well studied strains can be summarized as follows.

X. autotrophicus strain GZ29 (Berndt et al., 1976, 1978): the growth rates with ammonium as the nitrogen source and sucrose as the carbon source are highest at an oxygen partial pressure of 0.15 atm, whereas with N_2 as the sole nitrogen source maximum growth rates occur at 0.014 atm. In the whole cell assay for acetylene reduction the optimum oxygen partial pressure is 0.0036 atm. Even in the absence of any detectable oxygen, acetylene is reduced linearly for more than 1 h. The acetylene reduction activity has a very narrow pH optimum at 6.8. The nitrogenase is not cold-labile. The overall efficiency of nitrogen fixation is 22 mg nitrogen fixed/g of sucrose consumed, but in the early exponential growth phase values up to 65 mg/g have been found.

X. flavus strain 301 (Biggins and Postgate, 1969, 1971): heterotrophic growth in the presence of ammonium is not dependent on the oxygen partial pressure in the range of 0.01–0.20 atm, but with N_2 as the sole nitrogen source growth and nitrogen fixation have an optimum of ~0.1 atm. The optimum p_{O_2} for acetylene reduction is 0.05 atm (whole cells). The nitrogenase activity of cell-free extracts stirred under 0.2 atm O_2 decays exponentially, with a half life of ~5 min. ATP enhances the sensitivity towards O_2, presumably because of the formation of an active enzyme conformation. As with nitrogenases from other sources, the enzyme reduces H^+ to H_2, KCN to CH_4, and CH_3NC to CH_4, C_2H_4 and C_2H_6. The optimum concentration of ATP in the cell-free extract for acetylene reduction is 4–8 mM. As in other aerobic diazotrophs, but in contrast to the systems of *Clostridium pasteurianum*, pyruvate fails to promote acetylene reduction by cell-free extracts.

Hydrogenase. The strains of *X. autotrophicus* studied (e.g. 14g, 7c, GZ29) contain a membrane-bound hydrogenase serving the uptake of molecular hydrogen. Like the majority of membrane-bound hydrogenases, this enzyme is not able to reduce NAD directly, but channels electrons into the electron transport chain. A soluble, NAD-reducing hydrogenase is not present (Schneider and Schlegel, 1977). In strain 7c the enzyme is constitutive, and cells grown on various organic substrates and ammonium contain 10–100% of the hydrogenase activity compared to lithotrophically grown cells (Tunail and Schlegel, 1974). In strain 14g the hydrogenase is a strictly inducible enzyme (Schneider et al., 1973). In strain GZ29 hydrogenase is apparently a constitutive enzyme; its specific activity (up to 31 μmol H_2/h/mg protein in N_2-fixing cells) depends on the oxygen concentration during growth (Berndt and Wölfle, 1979; Pinkwart et al., 1979). Although the enzyme has been solubilized from the membranes and partially purified (Schink and Schlegel, 1980) a detailed description is still lacking. Immunological comparison of the membrane-bound hydrogenase of *Alcaligenes eutrophus* strain H16 reveals no relationship to that of *X. autotrophicus* GZ29 (Schink and Schlegel, 1980). Furthermore, an antiserum against the hydrogenase of *X. autotrophicus* strain GZ29 did not react with the membrane extracts of strains 7c, 14g and 12/60/x. This indicates major differences between the strains. The hydrogenase of the motile strain MA2 is apparently a loosely membrane-bound enzyme which tends to form aggregates. It is outstanding by its high specific activity (M. Pinkwart, unpublished data). The hydrogenases of *X. autotrophicus*

and related strains require detailed comparison. Furthermore, whether the hydrogenase involved in the uptake of H_2 for lithotrophic growth is identical with the hydrogenase which accompanies the nitrogenase will require further study.

Growth conditions. The spectrum of carbohydrates utilized is normally limited to fructose and/or sucrose and/or—in the presence of traces of yeast extract—to mannose. Some strains (*X. autotrophicus* 14g and the motile strains MA2 and SA35) do not utilize carbohydrates at all. *X. flavus* can utilize a few more sugars including glucose. Normally *X. autotrophicus* cannot use glucose, although the catabolic enzymes are induced during growth on fructose. However, glucose-utilizing cells appear spontaneously when cultures are incubated for a prolonged time in the presence of glucose. These adapted strains (transport mutants) grow very well on glucose; however, they may lose this ability again after growing for several transfers on other carbohydrates (J. Wiegel, unpublished data). Malik reported that the strain *X. autotrophicus* DSM 685 could use a wider range of carbohydrates than other strains of this species (Malik and Claus, 1979).

Catabolite repression is exerted by H_2 on the utilization of organic substrates; however, the extent of this effect varies with the organic substrate and the strains studied. Some strains need CO_2/CO_3^{2-} for induction of the enzymes for heterotrophic growth after transition from lithotrophic to organotrophic conditions (Schneider et al., 1973).

The doubling times reported for various strains vary between 1.5–5 h (heterotrophic conditions) and 3.0 to over 12 h (lithotrophic conditions). Under N_2-fixing conditions the doubling time is slightly longer (Wiegel and Schlegel, 1976; Berndt et al., 1978).

Traces of yeast extract shorten the lag phase after transition from heterotrophic to lithotrophic conditions. Three strains of *X. autotrophicus* are reported to require biotin (Aragno, 1975). High cell yields during mass culture under autotrophic conditions are obtained when oxygen levels are raised to match the increasing growth. For example, for autotrophic growth with ammonium as the nitrogen source an O_2 level of 5–8% (v/v) is usually optimum; at optical densities higher than 1.0 (at 600 nm) 10–15% O_2 is sufficient, and at optical densities higher than 2.5 (at 600 nm) 20% O_2 in the gas atmosphere is required for good growth. For strain *X. autotrophicus* GZ29, it has been shown that intact cells are able to reduce acetylene under anaerobic assay conditions, indicating the presence of a fermentative energy regenerating system. However, *growth* has not been observed under anaerobic conditions and no anaerobic *metabolic activity* has been ascertained, so far (Berndt et al., 1976). Since acid slime is produced and ammonium is consumed, the pH has to be adjusted by adding alkali.

Xanthobacter strains are sensitive to the following antibiotics at 100 μg/ml of growth medium: penicillin (minimum concentration, ~1 μg/ml), novobiocin and polymyxin B. They are resistant to erythromycin and bacitracin at 200 μg/ml. Violet red-bile medium (Oxoid), desoxycholate medium (Oxoid) and tellurite agar (selective for coryneform bacteria) support growth of the majority of the isolated strains. Both *Xanthobacter* species grow on mineral medium supplemented with an appropriate carbon source and 10^{-5} M crystal violet. However, the colonies formed are red instead of blue as colonies of other Gram-negative bacteria.

Metabolic pathways. Carbohydrates and gluconate are degraded via the Entner-Doudoroff pathway as has been shown for *X. autotrophicus* strains by determining enzyme activities after growth on various substrates (Tunail and Schlegel, 1972, 1974; Schneider et al., 1973). In addition, radiorespirometric studies showed that the pentose phosphate pathway is concomitantly used to a significant degree (R. Opitz, Ph.D. Thesis, Göttingen, 1977).

C_1 *carbon utilization.* CO_2 is mainly fixed via the ribulose bisphosphate pathway (Bowien and Schlegel, 1981). The key enzyme ribulose bisphosphate carboxylase is inducible. In addition, phosphoenolpyruvate carboxylase activity was found. Radiorespirometric experiments with $^{14}CO_2$ or $^{14}CH_3OH$ as substrates yielded radioactive malate. Other experiments also suggest that methanol is presumably utilized via CO_2 and the ribulose bisphosphate cycle (R. Opitz, Ph.D. Thesis, Göttingen, 1977). Methanol is oxidized to CO_2 via methanol dehydrogenase which is of the normal type (mol. wt. of about 120,000) stimulated by NH_4^+ and containing 2 mol of the coenzyme pyrrolo-quinoline quinone (PQQ) (J. A. Duine, personal communication). The glucose-6-phosphate dehydrogenase is allosterically inhibited by phosphoenolpyruvate (Tunail and Schlegel, 1972; Opitz and Schlegel, 1978).

Carbon monoxide is not oxidized to CO_2 and cannot serve as an energy or carbon source for lithotrophic growth (O. Meyer, personal communication). Thiosulfate can substitute for H_2 as a substrate for lithotrophic growth and provide energy for CO_2-fixation (Friedrich and Mitrenga, 1981; J. Wiegel, unpublished data, six strains tested).

Genetics. Intraspecific gene transfer has been detected in crosses involving the strains GZ29, GZ27 and JW50 of *X. autotrophicus* which produce only traces of slime. Strain GZ29 was used to study the transfer more closely. The involvement of a defective generalized transducing bacteriophage as well as conjugational gene transfer have been described (Wilke and Schlegel, 1979; Wilke, 198). The bacteriophage (CA3) was detectable only by its transducing activity and by electron microscopy; it did not form plaques. The genetic markers (resistance, auxotrophy, pigmentation) were transducible at frequencies of about 10^{-4}/marker and per phage particle. No cotransduction of markers was detected.

Agar-mating experiments have revealed a recombination system requiring direct cell contact. It allowed the transfer of large chromosomal segments at low frequency. All partners used functioned as donors as well as recipients. Two groups of closely linked markers have been found.

Cytochromes and ubiquinones. X. autotrophicus (only strain 14g investigated) contains cytochromes *a*, *b* (two different), *c* and *o*, irrespective of the growth conditions. The amounts of cytochromes *a*, *b*, and *c*-type were 0.03, 0.4–0.52 and 0.36 μmol/g of particle protein, respectively (Bernard et al., 1974). Ubiquinones Q10 (major), Q9 and Q8 are present in *X. autotrophicus* and *X. flavus* (Collins and Jones, 1981 and unpublished data). Menaquinones have not been found in these species (Wiegel et al., 1978). Both species contain the coenzyme pyrrolo-quinoline quinone after growth on methanol (J. A. Duine, personal communication).

Ferredoxin. X. flavus (as "*M. flavum*" 301) contains a [4Fe-4S]2- and a [4Fe-4S]-ferredoxin. It is likely that the latter one is a [3Fe-3S]-ferredoxin. *X. autotrophicus* GZ29 contains two different [4Fe-4S]2-ferredoxins (Bothe and Yates, 1976; Yates et al., 1978; Berndt et al., 1978; M. G. Yates, personal communication). The ferredoxin of *X. autotrophicus* GZ29 exhibits EPR features in the oxidized as well as in the reduced state. This is in contrast to the ferredoxins of *Azotobacter vinelandii* and *X. flavus*. Thus, it is possible that *X. autotrophicus*, at least strain GZ29, contains ferredoxins unique among the group of N2-fixing bacteria. Additional strains need to be analyzed before final conclusions regarding generic diversity can be drawn. There exists no evidence for the presence of constitutive flavodoxins in *X. flavus* 301 (Bothe and Yates, 1976) or in *X. autotrophicus* GZ29. In the latter

strain the ferredoxins probably serve as direct electron donors for the nitrogenase (Schrautemeier, 1981).

Enrichment and Isolation Procedures

The type strain of *Xanthobacter autotrophicus*, strain 7c, was isolated from black mud of a pond near Göttingen during enrichments for propane-oxidizing bacteria (D. Siebert, Ph.D. Thesis, Göttingen, 1969). However, strain 7c and all other strains tested do not utilize propane. De Bont and Leijten (1976) isolated a few strains which were able to grow on methane, butane, and/or hexane. These strains seem to be closely related to *Xanthobacter*, but DNA/DNA homology experiments have not been performed. Typical strains of *X. autotrophicus* can be specifically isolated from wet soil and mud containing organic material. The isolation procedure is based on the ability to fix nitrogen under chemolithotrophic growth conditions as described by Wiegel and Schlegel (1976). The soil sample is incubated in minimal media lacking any nitrogen source under a gas atmosphere consisting of 10% CO_2, 10% O_2 and 80% H_2 (v/v). After one or two transfers, the cultures are adjusted to a final pH of 10 or 11 by addition of NaOH and, after being whirled intensively for about 5 min, the cells are plated onto succinate-containing complex media. Yellow colonies containing branched cells (Fig. 4.70C) are probably strains of *Xanthobacter*. The purified strains should be tested for growth on methanol, N2 fixation under $H_2 + O_2 +$ CO_2, and for the presence of zeaxanthin dirhamnoside.

Other strains of *Xanthobacter* have been isolated as ordinary hydrogen oxidizers (e.g. Eberhardt, 1969; Schneider et al., 1973; D. Siebert, Ph.D. Thesis, Göttingen, 1969; Tunail and Schlegel, 1974) or as N2-fixing bacteria (e.g. *X. flavus* 301 as *Mycobacterium flavum* 301; Federov and Kalininskaya, 1961) and were later identified as strains of the genus *Xanthobacter* (Wiegel et al., 1978; Malik and Claus, 1979). In addition to ordinary strains, biotin-auxotrophic strains and two motile strains were isolated from water-samples of a small pond in Switzerland (Aragno, 1975) employing a membrane filter technique. Approximately 10 ml of pond water was passed through membrane filters which were then placed on agar and incubated under 80% H_2, 10% O_2 and 10% CO_2 at 28°C for 7–20 days. All these strains fixed nitrogen only under microaerobic conditions (at subatmospheric oxygen concentrations). Recently K. A. Malik (personal communication) isolated some aerobic N2-fixing strains of *Xanthobacter* (presumably a subspecies of *X. autotrophicus*) in course of enrichments for the isolation of the aerobic diazotroph *Azotobacter*.

Maintenance Procedures

All strains of *Xanthobacter* can be readily maintained as autotrophic cultures on minimal media containing vitamins and 0.02% yeast extract. The grown cultures can be kept at 2–5°C for at least 15 months or in the presence of 60% (v/v) glycerol at −20°C for at least 24 months. For long-term preservation, the cultures are lyophilized in the presence of skim milk and honey (10%); they have been kept in this form for

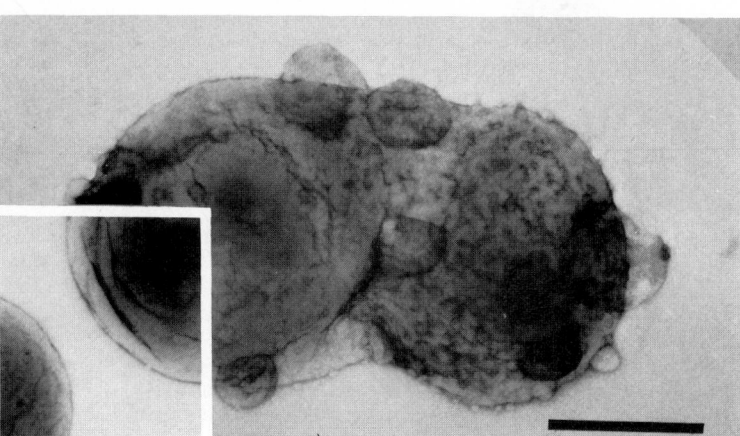

Figure 4.73. Demonstration of the presence of lipopolysaccharides in *Xanthobacter autotrophicus*. Negatively stained with 3% uranyl acetate, pH 5. Polymyxin B-treated cell of reference strain JW 33 showing lipopolysaccharide-polymyxin B interactions (bleb formation). The *inset* shows a control without polymyxin B treatment. *Bar*, 0.5 μm. (Reproduced with permission from L. Quandt.)

more than 10 years without significant loss of viability (Malik, 1976; Aragno and Schlegel, 1981).

Procedures for Testing Special Characters

The cultivation of hydrogen-oxidizing bacteria requires the handling of explosive gas mixtures. Therefore, before starting experimental work the relevant safety precautions should be studied. For a comprehensive outline on handling gases see Aragno and Schlegel (1981).

Chemolithoautotrophic growth on $H_2 + CO_2 + O_2$. The ability to grow lithoautotrophically should be tested in liquid minimal media containing no, or less than 0.05%, yeast extract and kept under a gas atmosphere of 80% H_2 + 10% CO_2 + 10% O_2. For testing the ability to fix nitrogen under lithotrophic conditions, ammonium is omitted and the culture is shaken under a gas atmosphere of 10% H_2 + 3–5% air + 10% CO_2 + 75–77% N_2. The growth under N_2-fixing conditions has to be tested at least for three subcultures with about a 2% (v/v) inoculation volume. Since slime production occurs, special attention has to be paid to the purity of the strains. For isolating pure cultures it is recommended to include an alkaline treatment step prior to streaking the suspension onto nutrient broth agar containing 0.3% succinate (Wiegel and Schlegel, 1976). If pure cultures have been achieved, the presence of nitrogenase is checked first using the acetylene reduction assay as described by Berndt et al. (1976) for microaerophilic diazotrophs and then by $^{15}N_2$-incorporation (Wiegel and Schlegel, 1976; Biggins and Postgate, 1971). Hydrogenase is tested according to Aragno and Schlegel (1981) to demonstrate the ability to oxidize H_2.

Gram-type. Since the Gram reaction is often doubtful with these species, other methods should be used to determine the Gram-type (Wiegel, 1981), e.g. the polymyxin B-lipopolysaccharide interaction as described by Wiegel and Mayer (1978; Fig. 4.73) and Wiegel and Quandt, unpublished data).

Qualitative determination of zeaxanthin dirhamnoside (K. Schmidt, personal communication). To identify zeaxanthin dirhamnoside the peracetylated pigment is used. It is prepared by the following procedure. Approximately 1 g of freeze-dried cells are extracted with acetone-methanol (1:1, v/v) overnight. After the organic solvent has been removed the residue is dissolved in ~1 ml of pyridine, the solution is flushed with N_2, and 0.1 ml of acetic anhydride is added per milliliter of pyridine. This reaction mixture is allowed to stand for 20 h in the dark at ~25°C. The acetylated products are driven from the pyridine solution into diethyl ether by adding a 3% (w/v) aqueous NaCl solution. The pyridine is quantitatively removed from the ether fraction by repeatedly washing with the aqueous NaCl solution. The ether extract is then evaporated to dryness and the dry residue is redissolved in a few drops of diethyl ether. This solution is used for thin layer chromatography on silica gel (Merck, Darmstadt FRG, No. 5721) using acetone:benzine ether (40–60°C fraction) 3:7 (v/v). R_F values: zeaxanthin, 0.8; zeaxanthin monorhamnoside-peracetate, ~0.63; and zeaxanthin dirhamnoside-peracetate, 0.4. An authentic sample derived from the type strain or from a reference strain should be used as reference. Spectral data (acetone as solvent): maxima (*cis*) at 342, (425), 452 and 478 nm; maxima (*trans*) (428), 453, and 480 nm.

Differentiation of the genus Xanthobacter from other genera

Originally two main characters were considered to be typical for *Xanthobacter*: (a) lithoautotrophic growth and (b) nitrogen fixation under heterotrophic as well as autotrophic conditions. At that time, each of these characters had been found only singly in different bacteria. However, there are now indications that these properties can be cross-transferred between various bacteria (Dalton, 1980). Furthermore, several lithotrophic strains of nitrogen-fixing bacteria, such as *Rhizobium*, *Azotobacter* and *Azospirillum* have been isolated and partially described, (e.g. Malik and Schlegel, 1980; Lepo et al., 1980). Therefore, the ability to fix nitrogen under lithotrophic conditions does no longer indicate that an isolate belongs to the *Xanthobacter* group.

So far, all isolated strains of *Xanthobacter autotrophicus* and *X. flavus* contained as their main carotenoid zeaxanthin dirhamnoside. This pigment is also present in recently isolated strains which are motile or which can fix N_2 under aerobic (atmospheric oxygen pressure) conditions. This pigment separates, so far, *Xanthobacter* from other yellow-pigmented lithotrophic or N_2-fixing bacteria of other genera. The carotenoid zeaxanthin has also been found in some strains of *Flavobacterium* with a high mol% G + C value; however, members of this group neither fix N_2 nor grow chemolithoautotrophically, and flavobacteria do not show features of pleomorphism similar to those of *Xanthobacter* (Oyaizu and Komagata, 1981).

The high mol% G + C of *Xanthobacter*, its pleomorphism (on succinate) or the ability to utilize short aliphatic alcohols are further properties which separate this genus from many other known hydrogen-oxidizing bacteria.

The differentiation from other free-living N_2-fixing bacteria is given in Table 2.7 in the article on the genus *Azospirillum*. It should be noted that some strains of *Xanthobacter* sp. can fix nitrogen not only under microaerobic conditions but also under the oxygen partial pressure of air.

Taxonomic Comments

Due to the pleomorphic cell shape and the tendency to exhibit a positive Gram-reaction, the high mol% G + C of the DNA, and the impression of a "snapping-type" or "palisade" cell formation, *Xanthobacter* strains were originally described as *Mycobacterium* and as *Corynebacterium* (coryneform bacteria). However, there is no relation on the DNA or RNA level to this group. In contrast to *Arthrobacter* and coryneform bacteria, snapping occurs due to the presence of adhesive slime and not to the rupture of the connective cell walls (Figs. 4.70B, 4.71A). Moreover, *Xanthobacter* is fundamentally a Gram-type negative organism; thus, a proper assay for the presence or absence of lipopolysaccharides has to be made to confirm the results of the Gram-reaction.

On the basis of the DNA/RNA homologies (De Smedt et al., 1980) it is not justified to assign *Xanthobacter* to the families *Azotobacteraceae* or *Rhizobiaceae*. The phototroph *Rhodopseudomonas* was the closest

Table 4.81.
Differential characteristics of the species of the genus **Xanthobacter**[a]

Characteristics	1. X. autotrophicus	2. X. flavus[b]
Vitamins required for growth (biotin, vitamin B_{12} and riboflavin)	–[c]	+
N_2-fixing system similar to that of Azotobacter[d]	–	+
Utilization of:[e]		
Propionate	+	–
Malonate	–	+
Maltose	–	+
Ribose	–	+
Phenylalanine	–	+
Histidine	–	+

[a] Symbols: see standard definitions.

[b] Data from Malik and Claus (1979).

[c] Some biotin-requiring strains exist and presumably need to be assigned as a subgroup of *X. autotrophicus*.

[d] Berndt et al., 1978.

[e] Since substrate utilization is very diverse among the strains of *X. autotrophicus*, DNA/DNA homology studies are required to substantiate any assignment to one of the species, except when a strain shows the exact properties of the reference strain JW33 (DSM 1618) of *X. autotrophicus*.

Table 4.82.

Other characteristics of the species of the genus **Xanthobacter**[a]

Characteristics	1. X. autotrophicus	2. X. flavus	Characteristics	1. X. autotrophicus	2. X. flavus
Cell diameter, μm	0.4–0.8	0.5–0.7	Membrane-bound hydrogenase (uptake) that does not reduce NAD	+	+
Cell length, μm	0.8–4.0[b]	1.0–2.5[c]	Hydrogenase activity:		
Motility	–	–	Inducible only	–	+
Water-insoluble zeaxanthin pigment produced[d]	+	+	Inducible or constitutive	+	–
Slime produced[e]	+	+	Oxygen-labile	+	
α-Polyglutamine capsule-like material produced	+	+	Carbohydrates catabolized via the Entner-Doudoroff and pentose phosphate pathways	+	+
Polyphosphate granules formed	+	+	Nitrate reduced to nitrite	+	+
Intracellular poly-β-hydroxybutyrate formed	+	+	Tetrathionate to thiosulfate	+	+
Growth at:			Tetrazolium salts reduced	+	
15°C	Weak	Weak	Lecithinase, deoxyribonuclease activities	–	–
28 – 32°C	+	+	Catalase, oxidase, phosphatase activities	+	+
37°C	Weak	Weak	Urease activity	d	–
45°C	d and Weak	–	Tyrosinase activity	+	
pH range for growth	5.0–9.0		Indole production	–	–
Growth in the presence of:			Litmus milk (alkaline reaction only change)	+	+
2.5% (w/v) NaCl	d	+	Voges-Proskauer test	–	
5.0% (w/v) NaCl	d	–	H_2S production	–	–
7.5% (w/v) NaCl	–	–	Gelatin liquefaction	–	–
1.0% (w/v) dodecylsulfate	–	–	Hydrolysis of Tween, starch, casein, or cellulose	–	–
10 μM crystal violet	d	+	Main component (60–90%) of fatty acid is 11-octadecenoic acid (cis)	+	+
Growth on tellurite agar	d	+	Sole carbon sources:		
Growth on H_2 (with O_2) as energy source	+	+	Succinate, malate, glutamate, methanol, ethanol, n-propanol, n-butanol, CO_2	+	+
Growth on thiosulfate as energy source	+	+			
Utilization of CO_2 as sole carbon source	+	+			
Growth in submerged liquid culture	+	+	Acetate, citrate, lactate, fumarate, pyruvate, gluconate, glutarate, glucose, fructose, sucrose	d	+
Strictly respiratory type of metabolism	+	+			
Nitrate as terminal electron acceptor	–	–			
Heterotrophic growth under air atmosphere	+	+	Salicin, β-hydroxybutyrate, tartrate	–	–
Autotrophic growth occurs:			Mol% G + C of DNA	69–70; (65–68)[f]	69; (68)[f]
Up to 20% O_2 atmosphere	+	–			
Only below 5% O_2 atmosphere	–	+			
Nitrogen fixation (microaerobic conditions)	+	+			

[a] Symbols: see standard definitions.

[b] Occasionally as long as 8.0 μm.

[c] Occasionally as long as 5.0 μm.

[d] Some "white" strains produce only minor amounts.

[e] Amounts range from copious to small. The slime contains sugar acids.

[f] Values in parentheses are from De Smedt et al. (1980).

related organism. Therefore, classification on the family level is so far not possible. *Xanthobacter* belongs to the fourth RNA superfamily (De Smedt et al., 1980) of the Gram-type negative bacteria studied.

The motile and aerobic N$_2$-fixing strains as well as the biotin-requiring strains probably belong to new species and subspecies, respectively (unpublished data, further studies are in progress). Further isolations and a more detailed taxonomic study may result in revealing more diverse features of this genus in the future. Although *X. autotrophicus* strain 7c is designated historically as the type strain of the type species, the reference strain JW33 (DSM 1618) should be used for comparative studies (Wiegel et al., 1978). Strain 7c is a relatively atypical strain.

Differentiation of the species of the genus **Xanthobacter**

Characteristics useful for the differentiation of the two species presently recognized in the genus *Xanthobacter* are listed in Table 4.81.

List of the species of the genus **Xanthobacter**

1. **Xanthobacter autotrophicus** (Baumgarten, Reh and Schlegel 1974) Wiegel, Wilke, Baumgarten, Opitz and Schlegel 1978, 580.[AL]

(*Corynebacterium autotrophicum* Baumgarten, Reh and Schlegel 1974, 214).

au.to.tro'phi.cus. Gr. pref. *auto-* self; Gr. n. *trophos* one who feeds; M.L. masc. adj. *autotrophicus* self-feeding, referring to the ability of the organism to use CO_2 as a sole carbon source.

The characteristics are as described for the genus and as listed in Tables 4.81 and 4.82. The morphology is depicted in Fig. 4.70.

Habitat: soil, mud, water. Widely distributed in nature.

The mol% G + C of the DNA is 69–70 (T_m; Wiegel et al., 1978), 65–68 (T_m; De Smedt et al., 1980), or 66–68 (Bd; M. Aragno, unpublished data).

Type strain: DSM 432 (strain 7c (Tunail and Schlegel, 1974; Baumgarten et al., 1974)). This strain is atypical of the species, and the reference strain DSM 1618 (strain JW33 (Wiegel and Schlegel, 1976; Wiegel et al., 1978)) should be used for comparative purposes.

2. **Xanthobacter flavus** Malik and Claus 1979, 286.[AL] (strain 301, misclassified by Federov and Kalininskaya (1961) as *Mycobacterium flavum* (Orla-Jensen 1919) strain 301.)

fla'vus. L. adj. *flavus* yellow.

The characteristics are as described for the genus and as listed in Tables 4.81 and 4.82.

The properties of *X. flavus* are very similar to those of *X. autotrophicus*, except the nitrogenase system of *X. flavus* 301 differs considerably from *X. autotrophicus* strain GZ29 (Berndt et al., 1978). However, only these two strains have been tested. The DNA/DNA homology (T_d-method) of the two type strains of *X. flavus* and *X. autotrophicus* is about 25%. Whether the vitamin requirement, the higher sensitivity to oxygen under autotrophic conditions of *X. flavus* 301 and the substrate range of carbohydrates are valuable properties to distinguish between these two species (Malik and Claus, 1979) has to be examined with additional isolates in the future.

Isolated from soil in the U.S.S.R.

The mol% G + C of the DNA is 68–69 (T_m).

Type strain: DSM 338 (NCIB 10071; strain 301 (Federov and Kalininskaya, 1961)), isolated from turf podozol soil (U.S.S.R.). No other strains have yet been identified.

Further comments. Several strains with properties other than described above have been isolated and are under investigation. It is expected that more species or subspecies will be described in the near future.

Other organisms

Several isolates have been described which were not assigned to known species and resemble the *Xanthobacter* strains in various properties. However, either the strains were not available on request or are still under investigation; thus they will be listed here only.

1. Sixteen strains isolated and described by De Bont and Leijten

(1976)
2. "*Mycobacterium butanitrificans*" (Coty, 1967)
3. Group IV, 10 strains (Kuono and Ozaki, 1975)
4. Strains N61 and N63 (Jensen and Holm, 1975)
5. Several strains among 40 strains (Ooyama, 1971, 1976).

Genus **Thermus** *Brock and Freeze 1969, 295*[AL]

THOMAS D. BROCK

Ther'mus. Gr. adj. *thermus* hot; M.L. masc n. *Thermus* to indicate an organism living in hot places.

Straight rods, 0.5–0.8 μm in diameter and 5.0–10.0 μm in length. Filaments from 20 to more than 200 μm may occur under some cultural conditions. Most strains form **rotund bodies**—large spheres 10–20 μm in diameter —derived from the association of individual cells; such bodies are usually seen in old cultures. **Nonmotile**; do not possess flagella. Endospores absent. Gram-negative. Most strains form **yellow, orange or reddish colonies**, with pigmentation due to carotenoid pigments. **Aerobic,** having a strictly respiratory type of metabolism with oxygen as the terminal electron acceptor. Oxidase- and catalase-positive. Gelatin is usually hydrolyzed. Starch is usually weakly digested. Nitrates are usually reduced to nitrites. **Thermophilic, with an optimum temperature of 70–75°C.** The optimum pH for growth is around neutrality. Found in **hot springs** of neutral to alkaline pH, as well as in **hot water heaters.** Also found in **natural waters subject to thermal pollution.** The mol% G + C of the DNA of the strains so far examined is 61–71.

Type species: *Thermus aquaticus* Brock and Freeze 1969, 295.

Further Descriptive Information

A detailed study on the fine structure of *Thermus aquaticus* was carried out by Brock and Edwards (1970) and subsequent studies have confirmed most of their findings. The general appearance of a *T. aquaticus* rod is shown in Figure 4.74 and it can be seen that the cell envelope structure has characteristics of Gram-negative bacteria, although the outer wall layer shows characteristic scallops and appears to be more defined than the diffuse outer layer of most other Gram-negative bacteria. The scallops arise because the outer layer is connected in a regular manner with the underlying peptidoglycan layer. The cell division mechanism resembles that seen in other Gram-negative bacteria, in which an invagination of the whole cell envelope

occurs by a furrowing process which progresses to the center of the cell.

A characteristic structure in most strains of *Thermus* is a large sphere which Brock and Edwards (1970) termed "rotund body." These are not conventional spheroplasts, but structures formed by the aggregation and association of a number of separate cells, as shown in Figure 4.75A. As seen in Figure 4.75B, the structure holding the rotund body together is the outer wall layer, which has peeled back from part of the cell and has become associated with the outer wall layer of adjacent cells. Rods and filaments of *T. aquaticus* form linear arrays or aggregates, perhaps brought together by outer slime material, and fusion may occur at this time. The average number of cells per rotund body is around 14. No studies have been done on the viability or function of rotund bodies. In their study of a large number of strains isolated from hot water heaters, Brock and Boylen (1973) found that about 70% of the isolates produced rotund bodies. See also the paper by Kraepilin and Gravenstein (1980) on induction of rotund bodies.

The sensitivity of *T. aquaticus* to penicillin noted by Brock and Freeze (1969) and by subsequent workers suggested that the cell wall contained peptidoglycan, and Pask-Hughes and Williams (1978) proved this chemically. A detailed study of the cell-wall chemistry of a number of newly-isolated thermophilic bacteria has been carried out by Merkel et al. (1978). All isolates of *Thermus* studied contain muramic acid but lack diaminopimelic acid. In place of diaminopimelic acid, these organisms contain ornithine as the major diamino acid. Merkel et al. (1978) note that the peptidoglycan of *Thermus* resembles that of the Gram-positive bacteria by the presence of ornithine, the high concentration of glycine, and glucosamine occurring in a high concentration nearly equimolar with alanine.

None of the workers reporting the isolation of *Thermus* have ever

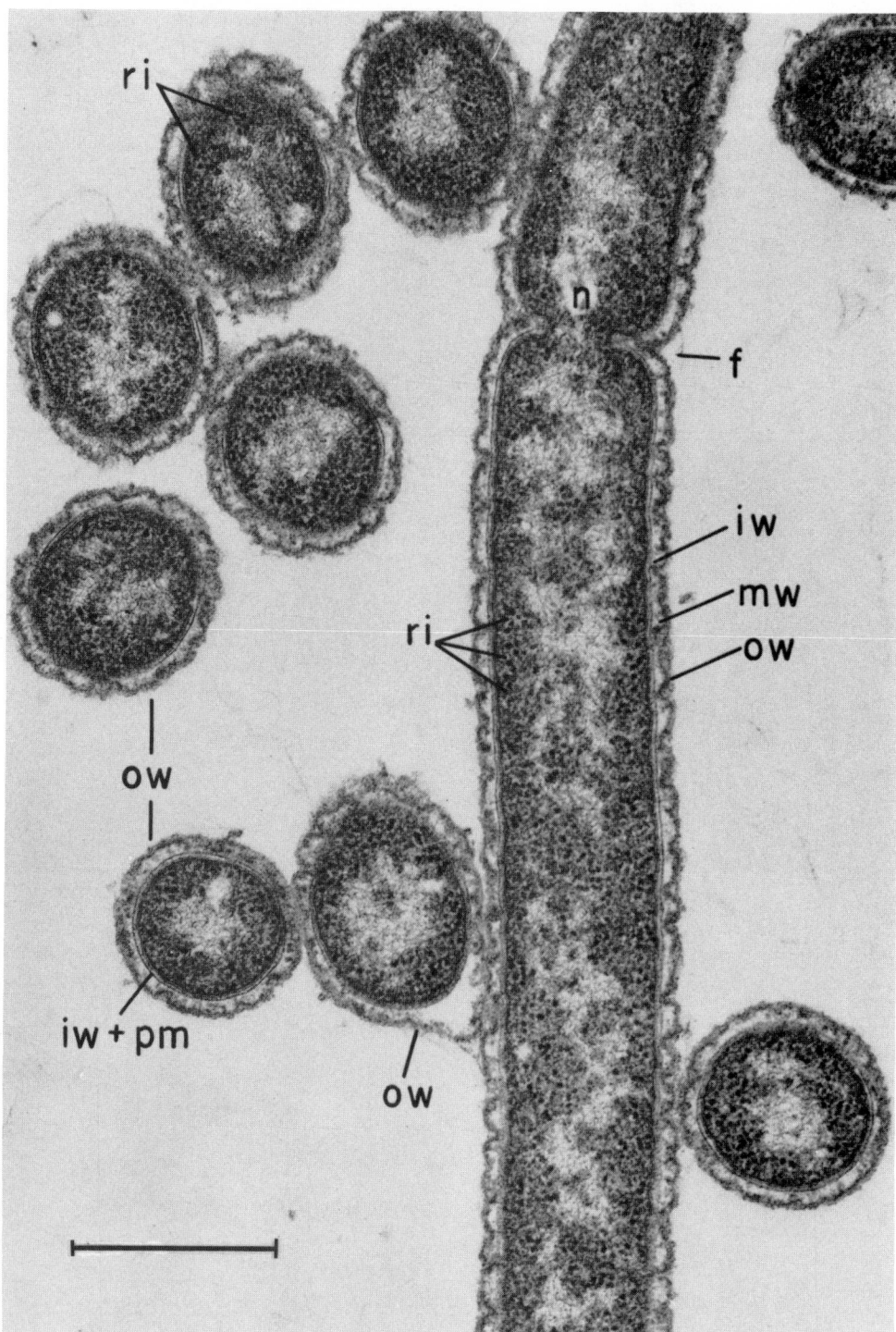

Figure 4.74.   Longitudinal, transversal and oblique views of cells of *Thermus aquaticus* (type strain), revealing nucleoplasm (*n*) with dense thin DNA fibrils, surrounded by cytoplasm containing numerous ribosomes (*ri*). The cell envelope comprises plasma membrane (*pm*) and wall exhibiting outer dense layers (*ow*), middle light zone (*mw*), and inner dense layer (*iw*). Note cell division furrowing (*f*). Where two cells are in contact, the external wall layer (*ow*) has separated from the inner wall, which remains adherent to the plasma membrane (*bar*, 0.5 μm). (Reproduced with permission from T. D. Brock and M. Edwards, Journal of Bacteriology *104:* 509–517, 1970, © American Society for Microbiology.)

observed evidence of sporulation. However, since spore-formation often is influenced by culture conditions, and since many thermophiles are spore-formers, it is important to realize that the characteristic "non-sporulating" is a negative property and hence has less taxonomic meaning than positive characteristics. Brock and Boylen (1973), in their study of isolates from hot water tanks, attempted to induce sporulation by culturing on starch agar, since this medium often favors spore formation of bacilli. No sign of spore formation was seen on this medium.

No evidence of motility can be detected in wet mounts or by the

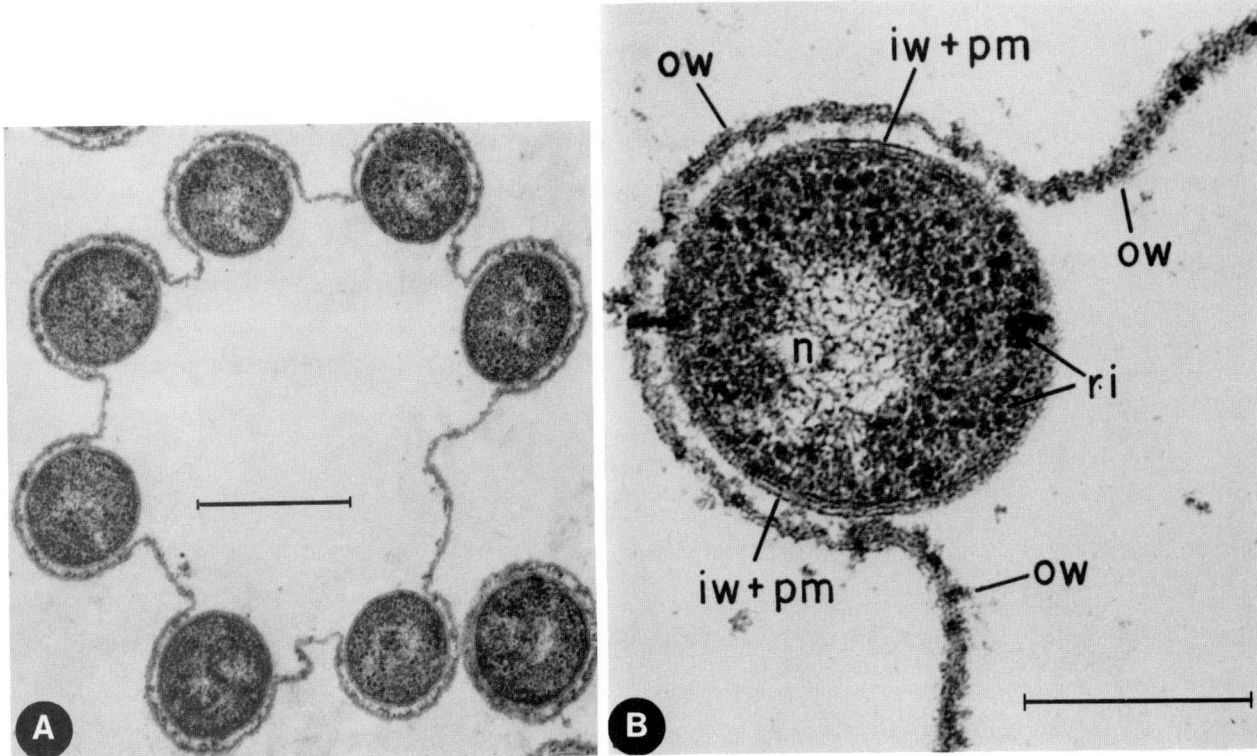

Figure 4.75. *A*, cross-section through a "rotund body" containing seven cells, showing the manner by which the cells are connected (*bar*, 0.5 μm). *B*, portion of a rotund body showing cross-section through a single rod. The outer wall (*ow*) is separated by a light zone from the inner wall (*iw*) which remains apposed to the plasma membrane (*pm*); *n*, nucleoplasm; and *ri*, ribosomes. *Bar*, 0.25 μm. (Reproduced by permission from T. D. Brock and M. Edwards, Journal of Bacteriology *104:* 509–517, 1970, © American Society for Microbiology.

appearance of colonies on agar plates. Brock and Freeze (1969) speculated that *Thermus aquaticus* had some resemblance to a gliding bacterium (overall morphology, unusual sensitivity to actinomycin) and suggested that lack of detection of gliding motility could be a technical problem related to inability to observe living cells under the microscope for extended periods of time at high temperature. However, no slime tracks or other signs of gliding can be detected on agar plates. Flagella are absent on electron micrographs of shadow-cast material (Brock and Freeze, 1969; Saiki et al. 1972; Oshima and Imahori, 1974).

Two culture media have been widely used for cultivating *Thermus*: medium D of Brock and Freeze (1969) and medium 162 of Degryse et al. (1978). The formulations of these media are given in Table 4.83.

Because *Thermus* is an obligate aerobe and the concentration of oxygen is low in high temperature conditions, careful attention to aeration is important. Brock and Rose (1969) have discussed some of the general principles of cultivation at high temperature. Specifically, attention must be taken to avoid evaporation of culture media during aeration, and the preferred procedure is to use covered water bath shakers. In unshaken culture tubes, *Thermus* isolates often grow in a pellicle at the top of the tube. When pure culture isolation on agar plates is to be carried out, the agar concentration should be at 3%, and the plates should be wrapped in Saran plastic to avoid drying out during the incubation period. The colonies of *Thermus* are slowly spreading, relatively compact, and frequently yellow to bright orange. For virtually all work, conventional bacteriological procedures can be used with only minor modifications for the higher temperatures involved.

With regard to temperature relations, the word *thermophile* has a number of meanings, depending on the group of organisms under study. The definition of obligate thermophily is somewhat arbitrary, but is generally considered to include those organisms unable to grow at temperatures below about 45–50°C. Extreme thermophiles have been considered to be those organisms which grow at temperatures above 70°C. In a reconsideration of the terminology for thermophilic bacteria,

Williams (1975) made the following proposals:

1. *Caldoactive* bacteria, maximum growth temperature above 70°C, optimum above 65°C, minimum above 40°C;
2. *Thermophilic* bacteria, maximum growth temperature above 60°C, optimum above 50°C, minimum above 30°C.

However, this classification, as is true of earlier classifications, is arbitrary and subject to misinterpretation. It seems preferable to use merely the general word "thermophilic," with the proviso that the specific temperature range of an organism be given.

Brock and Freeze (1969) found the growth limits of the type strain of *T. aquaticus* to be: maximum, 79°C; minimum, 40°C; and optimum, 70–72°C. A number of subsequent studies on temperature limits of various isolates have been carried out, the most detailed of which is that of Pask-Hughes and Williams (1977). In this study the temperature optimum ranged from 66.0–74.5, with the mean generation time at the optimum temperature ranging from 31–67 min. The temperature maximum for the strains ranged from 77.6–81.0. In such studies the temperature parameters of any organism are somewhat flexible, depending upon culture media and other conditions of growth. One strain isolated by Oshima and Imahori (1974) and called by them "*Thermus thermophilus*" has been reported to have a maximum temperature for growth of 85°C, while still having an optimum temperature the same as that of the type species (70–72°C).

Although little work has been done on thermal death of *Thermus*, Saiki et al. (1972) showed that their isolate survived 120 min at 90°C with only a slight loss of viability, but was killed rapidly in a boiling water bath (one log killing in about 30 min).

Although *Thermus* isolates will not grow at room temperature or lower, the organism does not seem to be cold sensitive, and samples can be allowed to cool for a reasonable length of time before enrichment cultures are set up.

Table 4.83.

Media used for the growth of thermophilic bacteria[a]

Medium D (×10 Concentrated)	Medium 162 (×10 Concentrated)
1000 ml bidistilled water	Idem
1 g nitrilotriacetic acid	Idem
10 ml micronutrient solution[b]	Idem
1 ml FeCl$_3$ solution (0.3 g/1000 ml)	5 ml Fe citrate solution (0.01 M)
0.6 g CaSO$_4$·2H$_2$O	0.4 g CaSO$_4$·2H$_2$O
1 g MgSO$_4$·7H$_2$O	2 g MgCl$_2$·6H$_2$O
0.8 g NaCl	Autoclave at 120°C for 20 min
1 g KNO$_3$	
3 g NaNO$_3$	Before use, the following additions are made aseptically per liter of 10-fold diluted medium:
1.1 g NaHPO$_4$	
Adjust to pH 7.2 with 1 N NaOH	60 ml NaHPO$_4$·12H$_2$O solution (0.2 M)
After sterilization by autoclaving, the pH is 7.2–7.1.	20 ml KH$_2$PO$_4$ solution (0.2 M) The pH is 7.2

[a] From Degryse et al. (1978), reproduced with permission. If not mentioned above, NH$_4$Cl (0.01 final concentration) was used as a nitrogen source.

[b] Micronutrient solution (per liter): H$_2$SO$_4$ (concentrated), 0.5 ml; MnSO$_4$·H$_2$O, 2.28 g; ZnSO$_4$·7H$_2$O, 0.50 g; H$_3$BO$_3$, 0.50 g; CuSO$_4$·5H$_2$O, 0.025 g; Na$_2$MoO$_4$·2H$_2$O, 0.025 g; and CoCl$_2$·6H$_2$O, 0.045 g.

Brock and Freeze (1969) reported that the type strain could use either amino acids or ammonium salts as the nitrogen source, and sugars and organic acids as the carbon source. No growth factor requirement was evident; however, growth was best in complex media such as 0.1–0.3% tryptone plus yeast extract (when added to the basal salts medium). Degryse et al. (1978) carried out nutritional studies on several strains, using a synthetic medium which had better buffering capacity. A wide variety of substances were suitable as carbon on nitrogen sources (see Tables 4.84 and 4.85). One strain, HB8, required lysine as a growth factor, but all the other strains were devoid of growth factor requirements. However, Saiki et al. (1972) isolated a strain (AT-62) that appeared to require amino acids as the nitrogen source (viz., a mixture of glutamate and aspartate, or aspartate, isoleucine and proline), as well as the vitamins biotin, folic acid and *p*-aminobenzoic acid.

Many isolates of *Thermus* are pigmented. However, nonpigmented isolates are not uncommon (Ramaley and Hixson, 1970) and Brock and Boylen (1973) showed that most isolates from hot water heaters were nonpigmented. If the pigment plays a photoprotective role, as is thought to occur in other bacteria, than it would be reasonable that isolates from hot springs would be pigmented and isolates from hot water heaters would be nonpigmented. There seems to be no compelling reason to classify nonpigmented strains in a different species than *T. aquaticus*. The pigments of *T. aquaticus* are carotenoids associated with the cell membrane (Ray et al., 1971b). The major carotenoids, identified chromatographically, are phytoene and deltacarotene, and there are several polar carotenoids that constitute a small fraction of the total. Absorption spectra of solvent extracts could be used to characterize the pigments of various strains, because differences in absorption spectra can be detected even in crude extracts. However, for any detailed taxonomic characterization using pigments, chromatographic procedures would be essential. Absorption spectra of various strains have been published by Pask-Hughes and Williams (1977), Saiki et al. (1972) and Oshima and Imahori (1974).

Thermus isolates are obligate aerobes. Growth is inhibited by sodium azide, a known inhibitor of cytochromes. Pask-Hughes and Williams (1975) carried out a specific study on the cytochromes of strains NH and DI, both nonpigmented strains isolated from London hot water

heaters. Both strains had characteristic *c*-type cytochromes, and there was some spectroscopic evidence for *b*-type and *a*-type cytochromes as well. A detailed study on the cytochrome *c*-552 of strain HB8 has been carried out by Hon-Nami and Oshima (1977). This cytochrome was highly thermoresistant and reacted with several cytochrome-dependent enzymes from mesophiles and mammalian sources.

A large number of studies have been carried out on enzymes of *Thermus* isolates to determine their thermostability. Many studies have also been carried out on RNA, ribosomes, and the protein-synthesizing machinery. Space does not permit review of this work, but in general it can be stated that all enzymes and macromolecules that have been looked at show unusual thermostability in comparison to mesophiles. Reviews of some of this work can be found in Zuber (1975), Shilo (1979) and Zeikus (1979). A new polyamine, designated *thermine*, has been isolated from strain HB8 (Oshima, 1975). Also, a thermostable bacteriophage (phi-YS40) has been isolated for this strain (Sakaki and Oshima, 1975). The bacteriophage was larger than phage T4 of *Escherichia coli* and had a similar complex morphology. It withstood heating at 80°C for an hour, but was inactivated rapidly at 90°C. Tests of a number of *Thermus* isolates revealed 5 out of 11 isolates were sensitive to this phage but the thermophile *Bacillus stearothermophilus* was resistant. A novel glycolipid has also been isolated from strain HB8 (Oshima and Yamakawa, 1972). Details of fatty acids and lipids of strain YT-1 can be found in Ray et al. (1971a, b). The organism is very low in unsaturated fatty acids and has a relatively high proportion of branched chain fatty acids. The fatty acid composition is markedly influenced by the temperature at which the organism is grown.

It was reported initially by Brock and Freeze (1969), and has been confirmed by subsequent workers, that *Thermus* isolates are unusually sensitive to several antibiotics that most Gram-negative bacteria are resistant to: penicillin, actinomycin, novobiocin. Sensitivity to actinomycin D was unusually high, with complete inhibition by a concentration as low as 0.8 μg/ml (Brock and Boylen, 1973). Antibiotic sensitivities of various strains have been reported by Williams (1975), Degryse et al. (1978), Pask-Hughes and Williams (1977), Oshima and Imahori (1974), and Saiki et al. (1972).

Although the hot spring habitat of *Thermus* has been clearly defined, virtually nothing is known about the ecological relationships of this genus. Growth rates in nature, actual nutrient sources used, dispersal mechanisms and relationships with other organisms have not been studied.

Enrichment and Isolation Procedures

Members of the genus *Thermus* were first isolated from effluents of hot springs of neutral to alkaline pH (Brock and Freeze, 1969) and were subsequently also isolated from man-made thermal habitats such as hotwater heaters (Brock and Boylen, 1973). Previous to the 1960s, most of the thermophilic bacteria that had been isolated were members of the genus *Bacillus*, which had temperature optima around 55–60°C and were unable to grow above 70°C. The key factors in permitting the isolation of this new group of thermophiles were several: (a) use of a higher temperature for enrichment (70°C instead of 55°C); (b) use of media more dilute in organic consituents; and (c) use of inocula from hot springs or high temperature hot water heaters (since *Thermus* is not found commonly in conventional soils, such as are the thermophilic bacilli). In their original work, Brock and Freeze (1969) indicated that rich culture media (1% tryptone plus yeast extract) inhibited the growth of *Thermus*, and that culture media of 0.1–0.3% tryptone plus yeast extract must be used. Subsequently, Saiki et al. (1972) isolated a strain (called by them "*Thermus flavus*") that grew at 1% tryptone plus yeast extract, and indicated that the problem with rich culture media was a decrease in the pH at higher temperature. The original culture medium D of Brock and Freeze (1969), which had been developed for culture of hot-spring cyanobacteria, was only weakly buffered (Table 4.83). Degryse et al. (1978) also pointed out the importance of pH control for the successful cultivation of *Thermus*, especially because of the loss of carbon dioxide at high temperatures, leading to depletion of inorganic

Table 4.84.
Carbon sources utilized by **Thermus aquaticus**[a]

Carbon Source	Growth Response
Glucose, glutamate, acetate, β-hydroxybutyrate,[b] pyruvate, phenyl acetate[b]	+
Glycerol[b]	+ or ±
Starch	±
Butyrate,[b] malate	d
Propionate, citrate, octanol, benzoate, p-hydroxybenzoate, glycine, acetamide, hexane, pentane, heptane, petroleum ether, methylamine, formate, methanol, hippurate, creatine, itaconate, mesaconate	−

[a] Based on a study of six strains by Degryse et al. (1978), reproduced with permission. One strain (HB8) had a lysine requirement and was tested with 20 μg L-lysine/ml; the other strains had no growth factor requirements. Symbols: +, good growth; −, no growth; ±, variable or slight growth; d, differs among strains.
[b] Not all six strains were tested.

Table 4.85.
Nitrogen sources utilized by **Thermus aquaticus**[a]

Nitrogen Source	Growth Response
Gelatin, glutamate, urea,[b,c] ammonium salts[b]	+
Nitrate[b]	d
Dinitrogen,[b] glycine,[b] lysine[b,d]	−

[a] Based on a study of seven strains by Degryse et al. (1978), reproduced with permission. One strain (HB8) had a lysine requirement and was tested with 20 μg lysine/ml; the other strains had no growth factor requirements. Symbols: +, good growth; −, no growth; d, differs among strains.
[b] With acetate as a carbon source.
[c] One strain not tested.
[d] Only one strain tested.

carbon from the culture medium. They developed a culture medium which was more buffered than the original medium D of Brock and Freeze, and reported improved enrichment culture procedures.

By use of the improved basal salts medium and a knowledge of the nutrition of known isolates, Degryse et al. (1978) developed an enrichment culture procedure in which *Thermus* strains can be isolated by aerobic incubation at 65°C in medium 162 (Table 4.83) supplemented with a single organic compound such as malate or acetate and a low nitrogen concentration (0.0054% ammonium chloride or 1.0% glutamate). In the same medium after anaerobic incubation in the presence of 1% fermentable carbohydrate, sporeforming organisms can be selectively recovered. However, it has been the author's general experience that if inocula from aquatic habitats are used at an enrichment temperature of 70°C or higher, members of the genus *Thermus* are almost always isolated. This is one of the easiest organisms to enrich for, because the microbial diversity in habitats where it is found is so low.

Maintenance Procedures

There are no special difficulties in maintaining cultures of *Thermus*. Agar media should be prepared with 3% agar in order to maintain a good surface at the high incubation temperature, and attention should be taken to avoid evaporation during incubation, either by sealing cultures or by incubating in moist chambers. After suitable incubation at proper temperature, stock cultures can be maintained on agar slants in the refrigerator. Cultures can also be frozen in liquid nitrogen or freeze-dried using conventional bacteriological procedures. Freeze-dried cultures have maintained viability for up to 10 years in the author's laboratory.

Differentiation of **Thermus** from other genera

There are few other organisms which can be confused with *Thermus*. Thermophilic bacilli are generally Gram-positive, and exhibit spore formation. Thermophilic clostridia are anaerobic, whereas *Thermus* is aerobic. Merkel et al. (1978) have described a group of hydrocarbon-oxidizing thermophiles that have some similarities to *Thermus*, but *Thermus* does not utilize hydrocarbons. Another extreme thermophile, *Thermomicrobium roseum*, has similar physiological conditions to *Thermus* but differs morphologically. *Thermomicrobium roseum* is a pleomorphic rod and lacks a peptidoglycan cell wall (Merkel et al, 1980; Merkel et al., 1978).

Taxonomic Comments

As noted, several species of the genus *Thermus* have been described. However, in a detailed taxonomic analysis, Degryse et al. (1978) found that the differences between the various named species were really very minor. They concluded that only a single species should be maintained, *Thermus aquaticus*, until a more detailed study had been done on a larger number of strains. They believed that the species "*Thermus thermophilis*," based as it was on a single strain, HB8, did not merit species rank. Other described species, already mentioned in the present work, "*T. ruber*," "*T. flavus*," probably also do not merit recognition in the absence of further taxonomic work.

One characteristic that prompted the initial definition of *Thermus* as a separate genus from *Flavobacterium* was the high mol% G + C content of its DNA. Brock and Freeze (1969) reported DNA base compositions of four isolates to range from 65.4–67.4 (Bd). Subsequently, the DNA base compositions of a number of other isolates have been determined (Aleksandrushkina and Egorova, 1978; Oshima and Imahori, 1974; Pask-Hughes and Williams, 1977; Ramaley and Hixson, 1970; Saiki et al., 1972). All strains are in the high mol% G + C range, with a spread from 60.5–70.8%. It is not known whether the differences among the strains are real, or the result of methodological variations. No clear picture of species separations is revealed by examination of DNA base compositions of various isolates.

List of the species of the genus **Thermus**

1. **Thermus aquaticus** Brock and Freeze 1969, 295.[AL]
a.qua'ti.cus. L. adj. *aquaticus* living in water.
The characteristics are as described for the genus. Additional characteristics are listed in Tables 4.84 and 4.85.

Isolated from hot springs, hot water heaters and thermally polluted natural waters.
The mol% G + C of the DNA is 61–71 (T_m, Bd).
Type strain: ATCC 25104 (strain YT1 of Brock and Freeze).

Other organisms

Other organisms which should be considered in relation to *Thermus* are *Thermomicrobium* and the hydrocarbon-oxidizing thermophiles described by Merkel et al. (1978). There is also a large number of extreme thermophiles living in boiling hot springs that have not been cultured (Brock, 1978), and some of them may be related to *Thermus*. The diversity of extremely thermophilic bacteria living in neutral to alkaline environments is much higher than would be concluded from the range of organisms that have been described taxonomically.

Genus **Thermomicrobium** *Jackson, Ramaley and Meinschein 1973, 34*[AL]

JEROME J. PERRY

Ther.mo.mi.cro'bi.um. Gr. n. *therme* heat; Gr. adj. *micrus* small; Gr. n. *bius* life, M.L. neut. n. *Thermomicrobium* indicates a small organism living in hot environments.

Short, irregularly-shaped rods, 1.3–1.8 μm in diameter and 3.0–6.0 μm in length. The pleomorphic forms are dumbbell shaped or appear irregular in diameter and occur singly or in pairs. **Neither resting stages nor endospores are formed.** Gram-negative. **No peptidoglycan diamino acid occurs in the cell walls in significant amounts.** Nonmotile. Obligately **aerobic. Optimum temperature for growth, 70–75°C;** maximum, 80°C, minimum, 45°C. Optimal pH, 8.2–8.5, but good growth occurs between 7.5 and 8.7. Chemoorganotrophic, having a strictly respiratory type of metabolism with oxygen as the terminal electron acceptor. Catalase-positive. Colonies have a rose-pink color. Maximum growth occurs on a medium consisting of yeast extract and peptone (0.5% each). **Growth does not occur on glucose.** *n*-Alkanes are not utilized. Generation time, 5.5 h. Isolated from a hot spring in Yellowstone National Park, U.S.A. The mol% G + C of the DNA is 64 (T_m).

Type species: *Thermomicrobium roseum* Jackson, Ramaley and Meinschein 1973, 34.

Further Descriptive Information

Although the original description of the species noted the appearance of possibly-motile cells, there is no evidence of flagella in electron micrographs and motility is not presently included in the description.

Electron micrographs of longitudinal sections of the organism indicate that it has a layered cell wall similar to that of Gram-negative bacteria. The outermost layer appears to be a repeating structure covering the cell surface in a regular mosaic pattern (Ramaley et al., 1978). Attempts to isolate a peptidoglycan from *T. roseum* by various techniques (Merkel et al., 1980) revealed that the cell wall material obtained was unlike that from other Gram-negative thermophilic bacteria (Merkel et al., 1978). The purified wall fraction from *T. roseum* is composed mainly of a protein with a monomeric molecular weight of 75,000. The amino acid composition of the cell wall fraction is shown in Table 4.86, column A, and more closely resembles the amino acid composition of subunit cell wall polymers (Mescher and Strominger, 1976; Thornley, 1975) than the typical cell wall of Gram-negative thermophilic organisms (Merkel et al., 1978). The major cell wall

protein has been purified electrophoretically and the amino acid composition (Table 4.86, column B) indicates high concentrations of proline, glutamic acid, glycine and alanine. There are some minor differences in composition and in the molar ratios of the amino acids when compared to the purified cell wall. The obvious similarities do indicate that the monomeric protein occurs as a major component of the outer envelope. The role of this protein in the stability of the organism is not known at the present time. The atypical nature of the cell wall of *T. roseum* is a distinct feature and of probable major taxonomic significance. Other Gram-negative, obligately thermophilic bacteria studied appear to have a more uniform peptidoglycan composition.

Nutritional studies have been accomplished with Allen's salts medium* (Allen, 1959) and in a salts medium devised by Castenholz (1969) with nitrate as a source of inorganic nitrogen.† *T. roseum* grew better with increasing concentrations of yeast extract from 0.1–0.7% with an equal amount of added tryptone. Total growth decreased at concentrations above 0.7%. In media with defined carbon sources, e.g. glycerol or succinate, the organism did not grow unless glutamate was present, and then solely in Castenholz's medium. When complex media were employed, better growth was attained with Allen's salts. A possible vitamin requirement has not been completely assessed.

The pink pigment of *T. roseum* is a carotenoid with absorbance maxima similar to torulene and 3,4-dehydrolycopene (Jackson et al., 1973).

Enrichment and Isolation Procedures

Only one strain of *T. roseum* has been isolated. Isolation was accomplished from Toadstool Spring in Yellowstone National Park: a sample of mat and water taken near the source of the spring (pH 8.9) at 74°C yielded compact pink colonies after 1 week incubation on plates at 70°C, using a medium consisting of 0.1% yeast extract, 0.1% tryptone and the mineral salts mixture described by Allen (1959). The other colonies that grew on primary isolation were of the genus *Thermus*. A pure culture was obtained by continued restreaking and incubation for 5 days.

Differentiation of the genus **Thermomicrobium** from other genera

Under phase-contrast microscopy, *Thermomicrobium* cells are quite small and pleomorphic, whereas *Thermus* cells appear as long, thin, regular rods. *Thermomicrobium* can also be distinguished from the genus *Thermus* on the basis of generation time (5–6 h for *Thermomicrobium*, 1 h for *Thermus*), nutrition and cell wall composition. *Thermus* strains have few growth factor requirements and grow on a wide array of sugars and organic acids as carbon sources. The genus *Thermomicrobium* requires glutamate (and possibly other factors) and is more limited in substrate range. For details, see Tables 4.87 and 4.88.

Taxonomic Comments

The only species currently included in this genus is *T. roseum*, and this species is represented by only a single strain. Another species, *T. fosteri* represented by ATCC strain 29033, was initially placed in the genus (Phillips and Perry, 1976), but subsequent study of *T. roseum* (Merkel and Perry, 1980) has indicated an atypical cell wall composition which *T. fosteri* does not have. *T. fosteri* is more closely related to other hydrocarbon-utilizing thermophiles (Merkel et al., 1978) than it is to *T. roseum*.

List of the species of the genus **Thermomicrobium**

1. **Thermomicrobium roseum** Jackson, Ramaley and Meinschein 1973, 34.[AL]

ro'se.um M.L. adj *roseus* rose colored.

The description is as given for the genus. The cell wall composition

---

* Allen's basal salts (mg/liter deionized distilled water): $(NH_4)_2SO_4$, 1,300; KH_2PO_4, 280; $MgSO_4 \cdot 7H_2O$, 247; $CaCl_2 \cdot 2H_2O$, 74; $FeCl_3 \cdot 6H_2O$, 19; $MnCl_2 \cdot 4H_2O$, 1.8; $Na_2B_4O_7 \cdot 10H_2O$, 4.4; $ZnSO_4 \cdot 7H_2O$, 0.22; $CuCl_2 \cdot H_2O$, 0.05; $Na_2MoO_4 \cdot 2H_2O$, 0.03; and VCl_2, 0.03; pH adjusted to 2.0 with H_2SO_4 during storage to prevent precipitation.

† Castenholz's basal salts, × 10 stock solution (per liter of distilled water): nitrilotriacetic acid, 1.0 g; $CaSO_4 \cdot 2H_2O$, 0.6 g; $MgSO_4 \cdot 7H_2O$, 1.0 g; NaCl, 0.08 g; KNO_3, 1.03 g; $NaNO_3$, 6.89 g; Na_2HPO_4, 1.11 g; $FeCl_3$ solution (0.28 g/liter of distilled water), 10.0 ml; and micronutrient solution, 10.0 ml. The micronutrient solution contains (per liter of distilled water): H_2SO_4, 0.5 ml; $MnSO_4 \cdot H_2O$, 2.2 g; $ZnSO_4 \cdot 7H_2O$, 0.5 g; H_3BO_3, 0.5 g; $CuSO_4$, 0.016 g; $Na_2MoO_4 \cdot 2H_2O$, 0.025 g; and $CoCl_2 \cdot 6H_2O$, 0.046. The Castenholz × 10 basal salts stock is adjusted to pH 8.2 with 1 N NaOH and autoclaved.

is given in Table 4.86.

Isolated from a hot spring in Yellowstone National Park.

The mol% G + C of the type strain is 64 (T_m).

Type strain: ATCC 27502.

Other organisms

Several hydrocarbon-utilizing, obligately thermophilic, Gram-negative nonsporulating rods have been isolated, including ATCC Strain 29033 which was initially named *Thermomicrobium fosteri*. All of these organisms possess a typical peptidoglycan and therefore differ from *Thermomicrobium roseum*. They are presently unassigned to any established genus.

All of the strains are capable of growth on normal alkanes with a carbon chain length from C_{13} to C_{20}. One group of strains contains diaminopimelic acid (DAP) as the major diamino acid in the peptidoglycan, has a mol% G + C from 52–58, forms nonpigmented colonies, has a generation time of 1.8–3.7 h on *n*-heptadecane or 0.7–1.8 h on glucose, grows on complex media, and has an optimum growth temperature of 55–65°C. A second group, including the strain previously named *T. fosteri*, contains DAP plus lysine or ornithine as a peptidoglycan constituent, has a generation time of 4–6 h on *n*-heptadecane and an optimum temperature for growth at 60°C. Members of this group cannot utilize sugars, fatty acids or any of the carbon sources tested except the C_{13} to C_{20} normal alkanes. Complex media also fail to support growth. The mol% G + C ranges from 68–72 and only one strain (*T. fosteri*) forms pigmented colonies (light pink).

For further descriptive information, see the articles by Merkel, Underwood and Perry (1978) and Merkel, Stapleton and Perry (1978).

Table 4.86.

Amino acid analysis of the cell wall and cell wall protein of **Thermomicrobium roseum**[a]

Amino Acid	Cell Wall (A)		Cell Wall Protein (B)	
Threonine	8.1[b]	0.5[c]	3.6[b]	0.3[c]
Serine	6.2	0.4	4.0	0.3
Proline	8.1	0.5	14.0	1.2
Muramic acid	3.1	0.2	0	0
Glucosamine	0	0	0	0
Glutamic acid	12.8	0.8	11.9	1.0
Glycine	19.3	1.2	33.8	2.9
Alanine	16.8	1.0	11.5	1.0
Valine	4.0	0.2	1.8	0.2
Diaminopimelic acid	1.2	0.1	0	0
Leucine	6.2	0.4	3.2	0.3
Isoleucine	0	0	1.8	0.2
Tyrosine	4.7	0.3	2.2	0.2
Galactosamine	3.1	0.2	0	0
Histidine	1.2	0.1	3.2	0.3
Lysine	0	0	Tr	Tr
Arginine	3.7	0.2	Tr	Tr
Phenylalanine	0	0	1.8	0.2
Ornithine	0	0	7.2	0.6
Tryptophan	ND	ND	ND	ND

[a] Symbols: Tr, trace; and ND, not determined. (Reproduced with permission from G. J. Merkel, D. R. Durham and J. J. Perry, Canadian Journal of Microbiology *26*: 556–559, 1980, National Research Council of Canada.)

[b] Percentage of total micromoles of amino acid.

[c] Molar ratio with alanine equal to 1.

Table 4.88.

Physiological characteristics of **Thermomicrobium roseum**[a]

Characteristics	Reaction or Result
Pink pigmentation	+
Growth on *n*-heptadecane	−
Substrate utilization[b]:	
In Castenholz's salts medium	
D-Fructose, D-glucose, glycerol, sodium succinate, mannitol, sucrose, sodium acetate, sodium citrate, peptone, brain heart infusion, Trypticase soy broth, or tryptone	−
In Castenholz's salts with 0.2% glutamate	
Glycerol, sucrose, nutrient broth or yeast extract	+
Sodium glutamate or casein hydrolysate	Weak
In Allen's salts	
Peptone, casein hydrolysate, brain heart infusion, nutrient broth, Trypticase soy broth, tryptone, yeast extract	+
Susceptible to the following antibotics (amt/disk):	
Chloramphenicol, 30 µg; erythromycin, 15 µg; kanamycin, 5 µg; neomycin, 5 µg; novobiocin, 5 µg; penicillin, 2 U; streptomycin, 2 µg; tetracycline, 5 µg.	+

[a] For symbols, see standard definitions.

[b] Substrate added at 0.2%.

Table 4.87.

Differential characteristics of **Thermus** sp., **Thermomicrobium roseum** *and unclassified, Gram-negative, nonsporulating thermophilic rods*[a]

Characteristics	Thermus Strains	Thermomicrobium roseum	Unclassified Strains[b]	
			Group A	Group B
Peptidoglycan diamino acid present in significant amounts	Ornithine	None	Diaminopimelic acid (DAP)	DAP plus lysine or ornithine
n-Alkane utilization	−	−	+	+
Growth on glucose	+	−	+	−
Growth on complex media	+	+	+	−
Generation time	20–60 min	5.5 h	1–2h[c]	5–6 h
Optimum growth temperature (°C)	60–70	70–75	55–60[c]	60
Mol% G + C of DNA	61–71	64	52–58	68–72

[a] Symbols: see standard definitions.

[b] Merkel, Stapleton and Perry, 1978. Also see Other Organisms at end of article.

[c] Glucose as substrate.

Genus **Halomonas** Vreeland, Litchfield, Martin and Elliot 1980, 494[VP]

RUSSELL H. VREELAND

Ha.lo.mo'nas. Gr. n. *hals, halos* salt, the sea; Gr. n. *monas* a unit, monad; M.L. fem. n. *Halomonas* salt(-tolerant) monad.

Rod shaped or pleomorphic. Rods are generally 0.6–0.8 μm wide and 1.6–1.9 μm long. Elongated, flexuous filaments may be formed under certain conditions. Spores are not formed. Gram-negative. **Motile by 4–7 unsheathed lateral or polar flagella. Possess mainly a respiratory type of metabolism with oxygen as the terminal electron acceptor; growth can occur anaerobically with nitrate. Growth on glucose can occur under anaerobic conditions in the absence of nitrate,** indicating some fermentative ability; however, **anaerobic growth with other carbohydrates or with amino acids requires the presence of nitrate.** Colonies are **white to yellow, never red.** Catalase- and oxidase-positive. Nitrate is reduced to nitrite. Chemoorganotrophic. Carbohydrates, amino acids and some polyols can serve as sole carbon sources in mineral media. Ammonium sulfate can serve as a sole nitrogen source. **Halotolerant, able to grow in NaCl concentrations ranging from 0.1–32.5% (w/v).** Isolated from a solar salt facility, but may inhabit intertidal areas and saline lakes. The mol% G + C of the DNA is 60.5 ± 0.5 (T_m, Bd).

Type species: *Halomonas elongata* Vreeland, Litchfield, Martin and Elliot 1980, 495.

Further Descriptive Information

During exponential growth, *Halomonas* strains consist of a mixture of straight and curved rods which lack intracellular granules (Fig. 4.76). Upon entry into stationary phase *Halomonas* may form elongated flexuous filaments of indeterminate length (Fig. 4.76). When formed, the percentage and length of elongated cells depends upon the type of growth medium. In complex casamino acids medium (CAS)* with 8% NaCl all of the cells may be elongated with many filaments containing irregular loops and bends. In a chemically defined mineral salts medium (MS)† containing 8% NaCl as few as 25% of the cells may be elongated. In this case filaments seldom produce irregular looping or extensive bending. Electron microscopy of thin sections from optimum growth conditions reveal a typical Gram-negative appearance (Vreeland and Martin, 1980).

The number and arrangement of flagella on short cells depends upon the strain examined. Cells generally possess from 4–7 flagella. In some strains the flagella are arranged laterally, while others possess only polar flagella (Fig. 4.77 *A* and *B*). One biovar of this genus has both lateral and polar flagellation. Motile *Halomonas* cells describe a helix when viewed by light microscopy. Motility may be rapidly lost under low oxygen concentrations. Flagella have not been observed on elongated cells or cells grown in low NaCl concentrations (0.2%).

On solid CAS or MS medium containing 8% NaCl, *Halomonas* colonies are white to cream colored, smooth, glistening, opaque, and ~2 mm in diameter after 24 h at 30°C. The colonies generally become yellow and spread during prolonged incubation.

Halomonas strains require Na⁺ for growth (at least 0.1% NaCl in

CAS medium or 0.3% NaCl in MS medium). Members of this genus are tolerant to NaCl in that they will grow in a wide range of NaCl concentrations (0.1%–32% NaCl w/v). The optimum NaCl concentration is 2.2–8.0% at 30°C. The Na⁺ requirement can be satisfied by NaCl, NaNO₃, NaBr (Vreeland and Martin, 1980) or Na glutamate. Tolerance is affected by growth temperature (Vreeland and Martin, 1980). In terms of ability to promote salt tolerance, temperatures can be arranged as follows: 30°C > 23°C, 37°C > 15°C > 45°C > 4°C (Vreeland et al., 1980). The carbon source used in MS medium has not been found to affect NaCl tolerance (Vreeland and Martin, 1980).

In CAS medium containing 8% NaCl, *Halomonas* strains grow at pH values from 5.0–9.0. The pH tolerance has not been tested on MS medium and has not been tested at different temperatures.

The physiological characteristics of *Halomonas* strains are listed later in Tables 4.89 and 4.90. Although considered to be mainly aerobic, the organisms grow well on glucose under anaerobic conditions in the absence of nitrate; if the glucose is replaced by other carbohydrates or an amino acid, nitrate is required for anaerobic growth.

Members of the genus are nutritionally versatile, and the following compounds can serve as sole carbon sources for growth: glucose, gluconate, glycerol, fructose, mannose, sucrose, cellobiose, succinate, mannitol, alanine, glutamine, glutamate, asparagine, aspartate, lysine, histidine, phenylalanine, tyrosine, tryptophan, proline, arginine, leucine, isoleucine, valine, methionine, cysteine, serine and threonine.

At this writing *Halomonas* strains have been isolated only from a solar salt facility on the island of Bonaire, Netherlands Antilles (Vreeland et al., 1980). Representatives of the genus may, however, be more widespread in nature.

Enrichment and Isolation Procedures

Halmonas strains may be isolated from salterns by the use of either CAS or the high-salt casein (HSC) medium§ described by Colwell et al. (1979). Brine samples are spread onto the surface of the medium and incubated close to environmental temperature and under high humidity for 2–7 days. *Halomonas* colonies will be white to cream colored and easily distinguished from red-pigmented halophiles. Alternatively, the brine samples can be added directly to filter pads saturated with medium. This has the advantage of allowing the use of small (9 mm) Petri plates, which facilitate shipment to distant field areas. A disadvantage of the technique is that the white *Halomonas* colonies can be difficult to see on white filter pads; also, the pad fiber makes colonial isolation tedious.

A more direct isolation technique might be to take advantage of the organism's ability to grow at either low or high salt concentrations. High salt brines could be mixed with low salt water, then spread onto a low salt medium. Alternatively, low salt samples could be added to high salt media to select for salt-tolerant organisms.

---

* CAS medium is a modification of the medium of Abram and Gibbons as described by Gibbons (1969). The medium contains (g/liter of distilled water): yeast extract, 1.0; casamino acids (Difco) (not "vitamin-free"), 7.5; Proteose peptone No. 3 (Difco), 5.0; sodium citrate, 3.0; MgSO₄·7H₂O, 20.0; Fe(NH₄)₂(SO₄)₂, 0.005; K₂HPO₄, 7.5; and NaCl (or solar salt), 80.0. The pH is adjusted to 8.0 ± 0.1 with NaOH prior to sterilization and is 7.5 ± 0.1 after autoclaving at 121°C for 20 min (Vreeland et al., 1980). CAS medium is stable for several weeks when stored in the dark. It should be discarded if any crystal formation is seen.

† MS medium is similar to that described by Vreeland and Martin (1980). It contains (g/liter): MgCl₂·6H₂O, 5.3; KCl, 0.75; and (NH₄)₂SO₄, 4.1. It is supplemented with a carbon source (10–50 mM) and NaCl (2.92–198.7 g/liter). Phosphate (K₂HPO₄·3H₂O) is added as a sterile 10 × concentrate after the medium has been autoclaved, to give a final concentration of 0.5 g/liter. The medium may also be supplemented with CaCl₂ (0.11 g/liter) to enhance growth (R. H. Vreeland and R. G. E. Murray, unpublished data). The pH of the medium is adjusted to 7.2 with 1 N KOH prior to autoclaving and is 7.0 after sterilization.

§ HSC medium is prepared as two separate solutions (A and B). Solution A contains NaCl or solar salt (25%, w/v). The pH of this solution is adjusted to 11 with 1 N NaOH and "vitamin-free" casein (Difco) is added to a concentration of 15 g/liter. This mixture is incubated at 30°C overnight to allow protein acidic groups to be exposed. The pH is then adjusted to 7.9 ± 0.1 with sterile NaOH or HCl. Solution B also contains 25% (w/v) solar salt or NaCl and is supplemented with yeast extract (0.2%), sodium citrate (0.6%), MgSO₄·7H₂O (4.0%) and ferric ammonium sulfate (0.01%). This solution is adjusted to 7.9 with 1 N NaOH and sterilized by autoclaving. Just before use solutions A and B are mixed 1:1 (v/v). If a solid medium is to be used, agar (4.0%) is added to solution B prior to autoclaving.

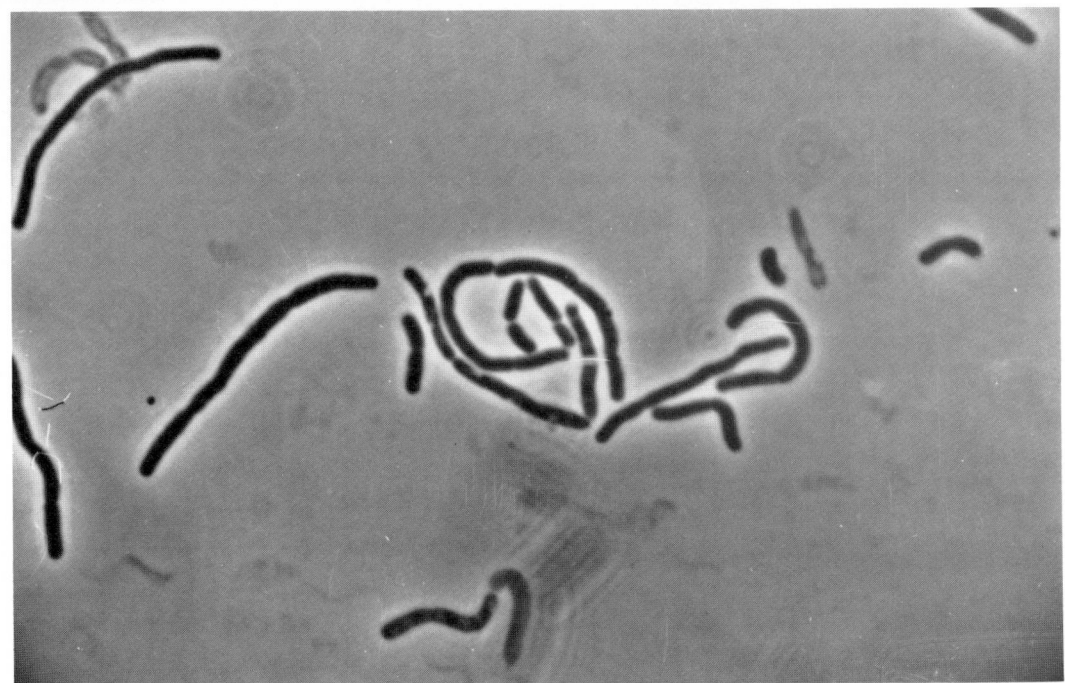

Figure 4.76. Phase contrast micrograph of *Halomonas elongata* strain 1H9 (ATCC 33173) showing both long and short cell forms (× 2200).

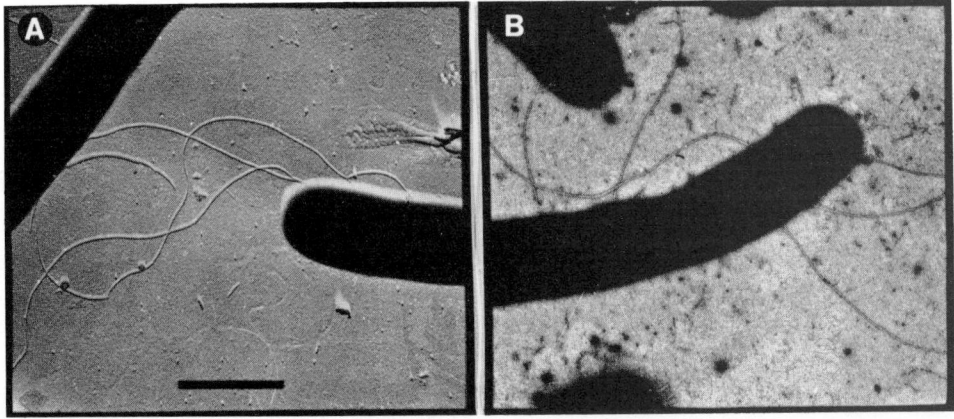

Figure 4.77. Electron micrographs of *Halomonas elongata* showing flagellar arrangement. *A.* strain 1H9 (ATCC 33173); carbon shadowed. *B.* strain 1H15 (ATCC 33174); negatively stained (*Bar*, 1.0 μm).

Maintenance Procedures

Halomonas strains can be maintained on CAS slants containing 8% NaCl. Stock cultures are transferred every 6 months, allowed to grow at 30°C for 1–7 days, and then stored in the dark at 4°C (Vreeland *et al.*, 1980).

All strains of *Halomonas* have been lyophilized, but survival rates vary with the strain. Strains ATCC 33173 and 33174 survive for up to 2 years. The growth from a fresh CAS slant or broth is suspended in 0.5 ml of a solution containing Proteose Peptone No. 3 (Difco) (0.5%), yeast extract (1.0%) and NaCl (0.2%). A small amount of the suspension (0.1–0.2 ml) is transferred to a lyophilization vial, dipped into liquid N_2 for 5–10 min, and lyophilized. Preserved cultures are reconstituted by suspending the cells in a small amount of CAS lacking salt. After 15–30 min, the cells are transferred to CAS broth containing 8% NaCl and incubated at 30°C.

Halomonas may also be stored frozen under liquid N_2. Due to the variation in survival for lyophilized cells, liquid N_2 storage is the method of choice for long term preservation.

Procedures for Testing Special Characters

An unusual aspect of *Halomonas* is its ability to grow at either low or high salt concentrations. This characteristic is best tested by use of CAS medium containing 0, 3.5, 8.0, 15.0, 20.0 and 32.0% (w/v) NaCl or solar salt. Temperature tolerance and qualitative effects of temperature on salt tolerance can be tested by incubating tubes of each salt concentration at temperatures from 4–45°C. After 2 or 3 weeks all tubes without visible growth should be incubated for an additional 2 or 3 weeks at 30°C to determine whether any apparent inhibitory conditions have actually been lethal.

DNA may be extracted from *Halomonas* by the method of Marmur (1961) with some modification. *Halomonas* suspensions and the sodium dodecyl sulfate (SDS) used for lysis must be heated to 50°C. Following SDS addition (2% final concentration) the lysate is held at 50°C for 15 min. Pronase is not added to the suspension. Water-saturated phenol (also at 50°C) is then added to fully denature the protein. The preparation may then be cooled to room temperature and the rest of the Marmur procedure followed. This modification is necessary since *Hal-*

omonas strains possess a nuclease which, unless inactivated by heating, causes rapid breakdown of DNA following cell lysis.

All other characterization tests may be performed using conventional

techniques (Holding and Collee, 1971). All media are supplemented with 8% NaCl, and uninoculated controls are necessary to ensure that any reactions detected are not artifacts caused by the NaCl.

Differentiation of **Halomonas** from other genera

The characteristics which differentiate *Halomonas* from other halophilic genera are listed in Table 4.89.

Despite their helical motility, flexibility and high mol % G + C value, *Halomonas* is easily differentiated from such genera as *Aquaspirillum*, *Oceanospirillum* or *Serpens* on the basis of salt tolerance, fermentative ability, and ability to catabolize several carbohydrates. *Halomonas* can be quickly differentiated from these other genera by the ability to survive exposure to, and to grow in, very high NaCl (≤ 20%) concentrations.

Taxonomic Comments

A comparison of *Halomonas* strains with *Halobacterium* isolated from the same source was conducted by Colwell et al., (1979). This comparison, based on 250 characteristics, indicated that the two groups possessed less than 40% similarity. This value is much lower than would be expected from closely related bacterial groups (Colwell et al., 1979). Unlike the red halophiles, *Halomonas* has a cell wall ultrastructure typical of Gram-negative bacteria and has membrane lipis that consist of fatty acids esterified to glycerol rather than of phytanols that are ether-linked to glycerol. When subjected to osmotic shock from rapid dilution or concentration of NaCl *Halomonas* respond as osmometers swelling or plasmolyzing as necessary. Osmotically shocked *Halomonas* cells do not lose viability and soon reestablish osmotic balance and resume growth. Although these considerations make it unlikely that *Halomonas* belongs to the family *Halobacteriaceae*, the precise taxonomic placement of the genus is uncertain. Nucleic acid studies, such as rRNA/DNA hybridization or rRNA oligonucleotide cataloging,

would be helpful in determining the phylogenetic relationships of *Halomonas* to other genera.

A numerical taxonomic study of the nine available strains of *Halomonas elongata* indicates some heterogeneity in the species, with S_J coefficients between strains ranging from 100–72% (Vreeland et al., 1980). Whether the strains are closely related by DNA/DNA hybridization has not yet been determined.

Table 4.89.

Characteristics differentiating **Halomonas** *from other halophilic genera of Gram-negative bacteria[a]*

Characteristics	*Halomonas*	*Halobacterium*	*Halococcus*
Morphology:			
Rods	+	+[b]	−
Cocci	−	−	+
Anaerobic growth with glucose (in the absence of nitrate)	+[c]	−	−
Red pigment formed	−	+	+
NaCl concentration (%) for optimum growth rate	3–8	25–30	25–30
Range of NaCl concentration (%) at which growth occurs	0.1–32.5	12–32	15–32
Cells lyse when shifted to low (<1% NaCl) concentrations	−	+	−
Typical Gram-negative cell wall profile	+	−	−
Lipid type:			
Fatty acids ester-linked to glycerol	+	−	−
Phytanols ether-linked to glycerol	−	+	+
Mol % G + C of DNA	60–61	57–68	61–66

[a] Symbols: see standard definitions.

[b] Disk-shaped cells also occur.

[c] With other carbohydrates, anaerobic growth occurs only in the presence of nitrate.

Table 4.90.

Other characteristics of **Halomonas elongata**[a]

Characteristics	Reaction or Result
Nitrate reduced to nitrite	+
Growth occurs anaerobically in the presence of nitrate	+
Growth occurs anaerobically in the absence of nitrate	−[b]
Surface growth occurs in thioglycolate broth	d (90)
Catalase test	+
Oxidase test (Kovacs')	+
Glucose fermented	+
Glycerol, sucrose, mannose and cellobiose oxidized	+
Ornithine decarboxylase	+
Lysine decarboxylase	+
Malonate utilized	+
Gluconate oxidized	d (78)
Lactose oxidized	d (67)
Urease activity	d (67)
Esculin hydrolysis	d (56)
Gelatin liquefaction	d (44)
β-Galactosidase (ONPG[c] hydrolysis)	d (33)
Indole production	d (22)
Methyl red test	d (11)
Voges-Proskauer test	d (11)
H$_2$S produced from cysteine	−
Casein hydrolysis	−
Starch hydrolysis	−
Agar liquefaction	−
Phenylalanine deaminase	−
Susceptible to:	
HgCl$_2$, 1:5000	+
Chloramphenicol, 30 U/disk	d (89)
Ampicillin, 30 U/disk	d (50)
Penicillin G, 10 U/disk	d (25)
Bacitracin, 10 U/disk	d (25)
Novobiocin, 30 U/disk	d (25)
Neomycin, 30 U/disk	d (13)
Tetracycline, 30 U/disk	d (13)
Nalidixic acid, 30 U/disk	−
Streptomycin, 10 μg/disk	−
0/129[d]	−

[a] From R. H. Vreeland et al. (1980), based on results for nine strains. (Reproduced with permission from the American Society for Microbiology and International Association of Microbiology Society); Symbols: +, all strains positive; −, all strains negative; and d, differs among strains. Numbers in parentheses indicate the percentage of strains that are positive.

[c] Except with glucose; see Table 4.89.

[c] ONPG, o-nitrophenyl-β-D-galactopyranoside.

[d] Vibriostatic agent 0/129, 2,4-diamino-6,7-diisopropylpteridine phosphate. Crystals of the compound were applied to an inoculated plate.

The type strain of *H. elongata* was chosen on the basis of its close similarity to the hypothetical median organism (Vreeland et al., 1980).

Acknowledgments

The author wishes to express his appreciation to Drs. C. D. Litchfield and E. L. Martin for their suggestions during the preparation of this manuscript; Mrs. M. Hall for providing the electron micrographs; and Dr. C. F. Robinow for the phase-contrast micrography.

Further Reading

Colwell, R.R., C.D. Litchfield, R.H. Vreeland, L.A. Kiefer and N.E. Gibbons. 1979. Taxonomic studies of red halophilic bacteria. Int. J. Syst. Bacteriol. *29:* 379–399.

Vreeland, R.H. and E.L. Martin. 1980. Growth characteristics, effects of temperature and ion specificity of the halotolerant bacterium *Halomonas elongata.* Can. J. Microbiol. *26:* 746–752.

Vreeland, R.H., C.D. Litchfield, E.L. Martin and E. Elliot. 1980. *Halomonas elongata,* A new genus and species of extremely salt-tolerant bacteria. Int. J. Syst. Bacteriol. *30:* 485–495.

List of the species of the genus **Halomonas**

1. **Halomonas elongata** Vreeland, Litchfield, Martin and Elliot 1980, 495.[VP]

e.lon'.ga.ta. L. fem. part. adj. *elongata* elongated, stretched out.

The characteristics are as described for the genus. Morphological features are depicted in Figures 4.76 and 4.77. Other characteristics are listed in Tables 4.89 and 4.90.

The species contains one biovar which is differentiated from the type strain on the basis of possessing both lateral and polar flagella and of failing to survive in 32% NaCl.

The mol% G + C of the DNA of the type strain is 60.5 ± 0.5 (T_m, Bd). The reference strain for the biovar has not yet been analyzed.

Type strain; ATCC 33173 (strain 1H9 of Vreeland et al., 1980). *Reference strain for biovar*: ATCC 33174 (strain 1H15 of Vreeland et al., 1980).

Other organisms

Two other strains of halotolerant bacteria have been isolated, but neither strain has been fully characterized. Some characteristics of these strains are given below.

a. *NRCC strain 41227.* This organism appears to be closely related to *Halomonas.* The following description has been supplied by R. K. Latta, National Research Council of Canada, Ottawa, Ontario K1A-OR6.

Straight or curved rods, occurring singly, in pairs or short chains. The age of the culture and the NaCl concentration affect the cell size and the chain length. Spores are not formed. Gram-negative. Motile by a single polar flagellum.

Colonies formed on the medium of Seghal and Gibbons* containing 7.0% NaCl are white, translucent and smooth in 24 h at 30°C. Colonies occasionally spread upon longer incubation.

Requires Na+. KCl will not substitute for NaCl in the medium. The optimum NaCl concentration is 7.0%. Grows well from 3.0–25.0% NaCl. Growth is poor in NaCl concentrations below 1.0%.

Optimum temperature, 30°C. Temperature range, 5–37°C. The effect of temperature on NaCl tolerance has not been tested.

Grows from pH 5.0 to 8.0.

Oxygen or nitrate will serve as the terminal electron acceptor. Grows anaerobically in the presence of nitrate. Nitrate is reduced to nitrite.

Oxidizes carbohydrates; produces acid but no gas from glucose and xylose. Catalase- and urease-positive. Oxidase-positive (Kovacs'). Indole and H_2S are not produced. Methyl red and Voges-Proskauer tests are negative. Gelatin, casein, starch and agar are not hydrolyzed.

The following compounds serve as carbon and energy sources: acetate, citrate, DL-malate, succinate, lactate, glucose, xylose, L-arabinose, D-arabinose, D-fructose, D-mannitol, *meso*-inositol and ethanol. The following compounds are not utilized: butyrate, isobutyrate, propionate, tartrate, benzoate, lactose, sucrose, cellobiose, maltose, melibiose, L-sorbose, D-ribose, D-sorbitol, *n*-propanol, *n*-butanol, ethylene glycol, *n*-hexadecane, pyridine-1-oxide, formaldehyde, formamide, *N,N*-dimethylacetamide, and ethylenediamine.

Sensitive to penicillin (1.5 U/disc), chloramphenicol (10 μg/disk) and oxytetracycline (30 μg/disk). Not sensitive to vibriostatic agent 0/129 or to novobiocin (10 μg/disk).

Isolated as a contaminant from Sehgal and Gibbons' medium containing 25% NaCl.

The mol% G + C of the DNA is 57 ± 1 (Bd) and 54–55 (T_m).

b. *Strain Ba₁.* Gram-negative rods. Obligately aerobic. Grow at NaCl concentrations from 0.0–3.0 M NaCl. Isolated from solar evaporation ponds on the Dead Sea (Rafaeli-Eshkol, 1968).

Genus **Alteromonas** Baumann, Baumann, Mandel and Allen 1972, 418.[AL]

PAUL BAUMANN, MICHEL J. GAUTHIER AND LINDA BAUMANN

Al.te.ro.mo'nas. L. *alter* another; Gr. *monas* a unit, monad; M.L. fem.n. *Alteromonas* another monad.

Straight or curved rods, 0.7–1.5 μm in diameter and 1.8–3.0 μm in length. **Do not accumulate poly-β-hydroxybutyrate** (PHB) as an intracellular reserve product. Microcysts or endospores are not formed. **Gram-negative.** Motile by means of **single polar flagella. Chemoorganotrophs** capable of **respiratory but not fermentative metabolism.** Molecular oxygen is a universal electron acceptor; **do not denitrify.** None of the strains has a constitutive arginine dihydrolase system. All **require a seawater base for growth;** many strains require organic growth factors. All grow at 20°C. **Common inhabitants of coastal waters and the open oceans.** The mol% G + C content of the DNA is 38–50 mol% (T_m, Bd).

Type species: *Alteromonas macleodii* Baumann, Baumann, Mandel and Allen 1972, 418.

Further Descriptive Information

Species of *Alteromonas* are composed of either straight or curved rods; phase contrast micrographs of two representative species, *A. macleodii* and *A. communis*, are shown in Figures 4.78 and 4.79, respectively. Ultrathin sections of *A. macleodii* (Fig. 4.80) as well as *A. haloplanktis* (Forsberg et al., 1970) and *A. espejiana* (Dahlberg and Franklin, 1970; Diedrich and Cota-Robles, 1974) indicate a morphology typical of many Gram-negative bacteria. Cells of all species are motile by means of polar flagella and do not possess the sheath characteristic of the flagella of other common marine species belonging to the genus *Vibrio* (Figs. 4.81–4.84). *A. communis* and *A. vaga* have bipolar single flagella; the other species have single flagella at only one pole. Involu-

---

* Medium of Seghal and Gibbons (1960), g/liter: Casamino acids (Difco) (not "vitamin free"), 7.5; yeast extract, 1.0; sodium citrate, 3.0; KCl, 2.0; $MgSO_4 \cdot 7H_2O$, 20.0; Fecl₂, 0.023; pH 7.4.

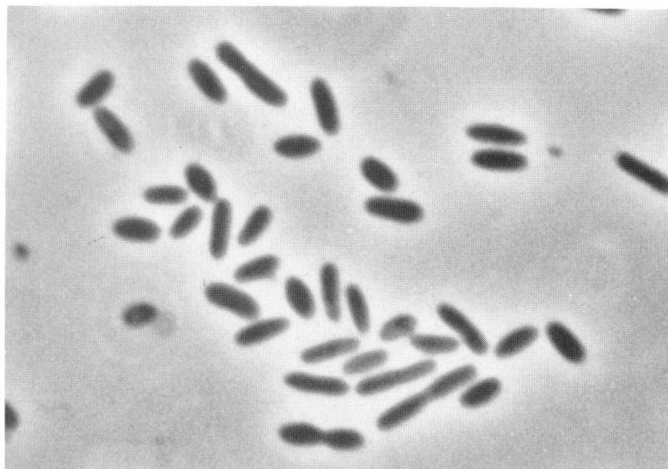

Figure 4.78. Phase-contrast micrograph of *Alteromonas macleodii* in exponential phase of growth in YEB (*bar*, 5 μm).

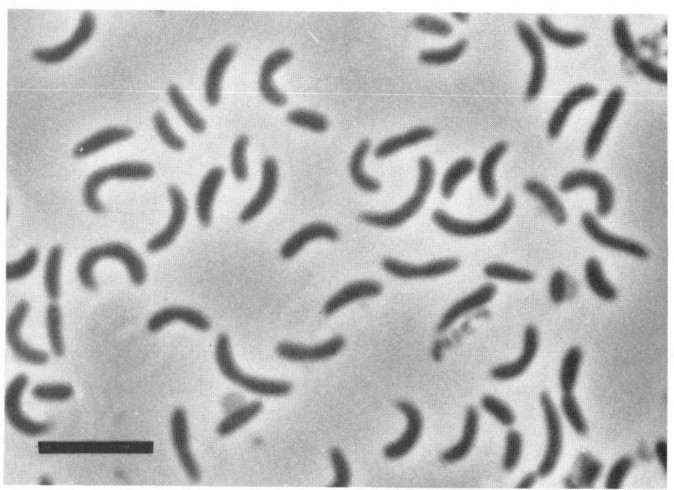

Figure 4.79. Phase contrast micrograph of *Alteromonas communis* in exponential phase of growth in YEB (*bar*, 5 μm).

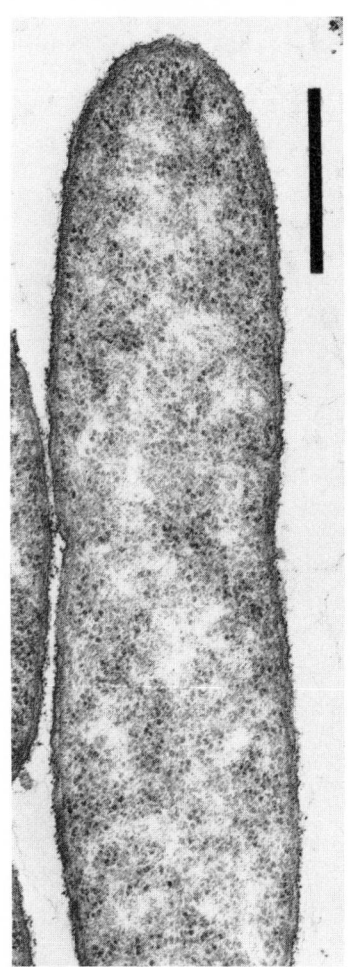

Figure 4.80. Ultrathin section of *Alteromonas macleodii* in exponential phase of growth in YEB (*bar*, 1.0 μm). (Reproduced with permission from R. D. Allen.)

tion forms are frequently observed in old cultures or under adverse conditions of cultivation.

When streaked on media such as Yeast extract agar* (YEA), Difco Marine agar (MA) (Zobell, 1941), or Basal medium agar* (BMA) containing 0.2% (w/v) D-glucose, the colonies of most species are not distinctive, being smooth and convex, with entire edges and a creamy color. Several species produce insoluble pigments: *A. rubra*, prodigiosin (Gauthier, 1976a; Gerber and Gauthier, 1979); *A. luteoviolacea*, violacein (Gauthier, 1976b); *A. citrea*, a lemon-yellow noncarotenoid pigment (Gauthier, 1977); and *A. aurantia*, an orange noncarotenoid pigment (Gauthier and Breittmayer, 1979). Many strains of *A. hanedai* produce a soluble brown pigment on complex media (Jensen et al., 1980).

All of the species grow in media containing a seawater base but fail to grow when all of the Na⁺ is replaced by equimolar amounts of K⁺, indicating a requirement for the former cation (MacLeod, 1968). The concentration of Na⁺ required for optimal growth ranges from 125–600 mM (Chan et al., 1978; Gauthier, 1976a, b, 1977; Gauthier and Breitt-

mayer, 1979). In some species the requirement for Na⁺ is considerably reduced by seawater levels of Mg⁺⁺ and Ca⁺⁺ (Reichelt and Baumann, 1974). With the exception of a few strains of *A. hanedai*, all of the species utilize D-glucose. No single organic compound serves as a universal sole or principal source of carbon and energy for species of *Alteromonas* so that a combination of substrates such as acetate and D-glucose must be used for a medium to support the growth of all strains. There is considerable variation in the requirement for organic growth factors. Strains of some species are able to grow in media containing a single organic carbon and energy source and NH₄Cl as a nitrogen source. Other strains require amino acids which can be provided in the growth medium either as a mixture of the individual amino acids, at a concentration of 1 mg/liter each, or as casein hydrolysate (0.5% w/v). A number of isolates of *A. hanedai* appear to require vitamins since they will only grow in the presence of yeast extract (0.5% w/v). *A. hanedai*, *A. aurantia*, and some strains of *A. haloplanktis* and *A. undina* grow at 4°C. With the exception of *A. hanedai*, all the species will grow at 30°C; none is able to grow at 45°C.

Species of *Alteromonas* are able to utilize from 20–58 organic compounds as sole or principal sources of carbon and energy, including pentoses, hexoses, disaccharides, sugar acids, sugar alcohols, monocar-

---

* All media contain an artificial sea water base (ASW; MacLeod, 1968) having the following composition (g/liter): NaCl, 23.4; MgSO₄·7H₂O, 24.6; KCl, 1.5; and CaCl₂·2H₂O, 2.9. (The salts are dissolved separately and then combined.) Basal medium (BM) consists of (g/liter): Tris-(hydroxymethyl)-aminomethane hydrochloride (Tris-HCl), 6.1 or 12.1, with HCl added to give a pH value of 7.5; NH₄Cl, 1.0; K₂HPO₄·3H₂O, 0.075; FeSO₄·7H₂O, 0.028; and ½ strength ASW. Yeast extract broth (YEB) consists of BM with yeast extract (5.0 g/liter). YEA and BMA are obtained by adding agar (20 g/liter) to YEB and BM, respectively.

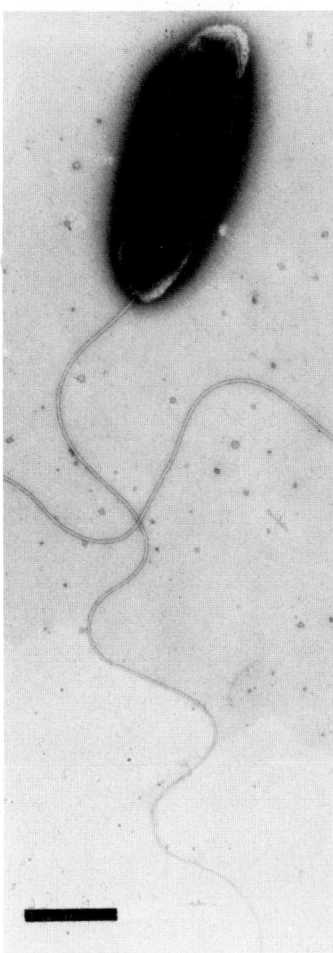

Figure 4.81. Electron micrograph of *Alteromonas macleodii*. Negatively stained (*bar*, 1.0 μm). (Reproduced with permission from R. D. Allen.)

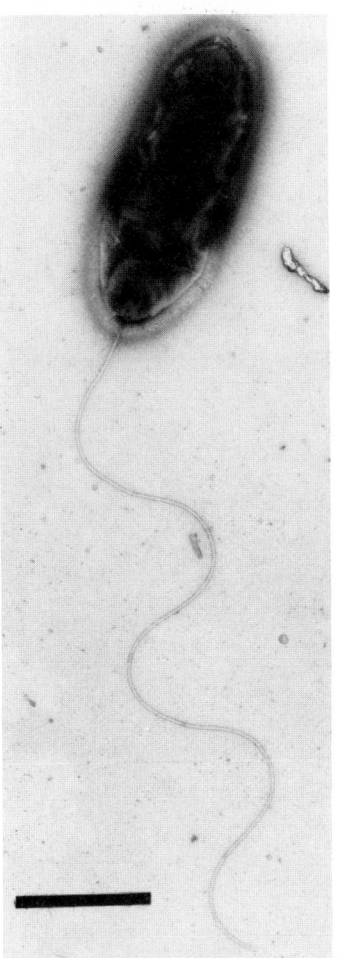

Figure 4.82. Electron micrograph of *Alteromonas haloplanktis*. Negatively stained (*bar*, 1.0 μm). (Reproduced with permission from: J. L. Reichelt and P. Baumann, International Journal Systematic Bacteriology 23: 438–441, 1973, © American Society for Microbiology.)

boxylic fatty acids, tricarboxylic acid cycle intermediates, amino acids, and aromatic compounds. Most species have a number of extracellular enzymes which include an amylase, gelatinase, and lipase; some species also have an extracellular alginase and/or chitinase.

A. macleodii, A. haloplanktis, A. espejiana, A. undina, A. communis, and *A. vaga* utilize D-glucose and D-fructose via an inducible Entner-Doudoroff pathway (Baumann and Baumann, 1973; Chan et al., 1978; Sawyer et al., 1977). When grown on D-fructose none of these species contains a 1-phosphofructokinase activity, indicating that in these organisms, unlike some marine species of *Pseudomonas* and *Alcaligenes* (Baumann and Baumann, 1975; Sawyer et al., 1977), fructose-1-phosphate and fructose-1,6-diphosphate are not intermediates of D-fructose catabolism. Enzymes of the tricarboxylic acid cycle and the glyoxylate cycle have been detected in cell-free extracts of *A. haloplanktis* (MacLeod and Hori, 1960; MacLeod et al., 1960a, b).

A. communis and *A. vaga*, the only species of *Alteromonas* which are able to utilize aromatic compounds, degrade *m*- and *p*-hydroxybenzoate via the α-ketoacid pathway (Ornston, 1971) as is indicated by the *meta* cleavage of the intermediate protocatechuate (Baumann et al., 1972). Both species also appear to have a single aspartokinase, the first enzyme in the biosynthetic pathway leading to the synthesis of amino acids of the aspartate family. The activity of this enzyme is inhibited by L-threonine and L-lysine and exhibits concerted feedback inhibition in the presence of low concentrations of these effectors (Baumann and Baumann, 1974). Aspartokinase activity has not been detected in cell-free extracts of *A. macleodii* and *A. haloplanktis*, the only other species tested.

A. hanedai is the only species of this genus which is able to luminesce (Jensen et al., 1980). The presence of this property in an obligately respiratory organism is unusual since all of the previously described luminous species are facultative anaerobes capable of a fermentative as well as a respiratory metabolism. The reaction sequence leading to light emission by *A. hanedai* is similar to that of other procaryotes (Jensen et al., 1980; Nealson and Hastings, 1979).

With the exception of *A. vaga*, all species of *Alteromonas* are oxidase positive. As is the case with some species of *Photobacterium*, occasional strains of *A. vaga* will give a positive reaction if toluene is used to make the cells more permeable to the oxidase reagent (Baumann and Baumann, 1981). Differential oxidized/reduced cytochrome spectra have indicated that these strains, as well as those which remain oxidase negative, have low levels of cytochrome *c* (which accepts electrons from the oxidase reagent (West et al., 1978)). The low levels of this cytochrome might explain the negative reaction obtained with many strains of *A. vaga* even after treatment with toluene. Franklin et al. (1971) have obtained evidence for the presence of cytochromes *b*, *c*, *a*, and *o* in *A. espejiana*, using whole cells as well as cell membrane preparations.

Mutants of *A. haloplanktis* and *A. espejiana* appear to be readily obtained by exposure to nitrosoguanidine or ultraviolet light (Fein and MacLeod, 1975; Tsukagoshi et al., 1975). In *A. espejiana*, the penicillin counter-selection technique has been successfully employed in the isolation of an auxotrophic mutant (Tsukagoshi et al., 1975).

Bacteriophages have been isolated for *A. espejiana* and *A. haloplanktis* (Johnson, 1968; Espejo and Canelo, 1968a, b; Franklin et al., 1976).

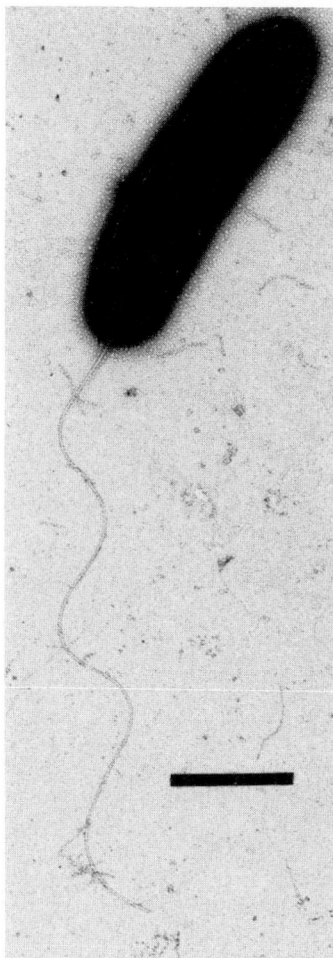

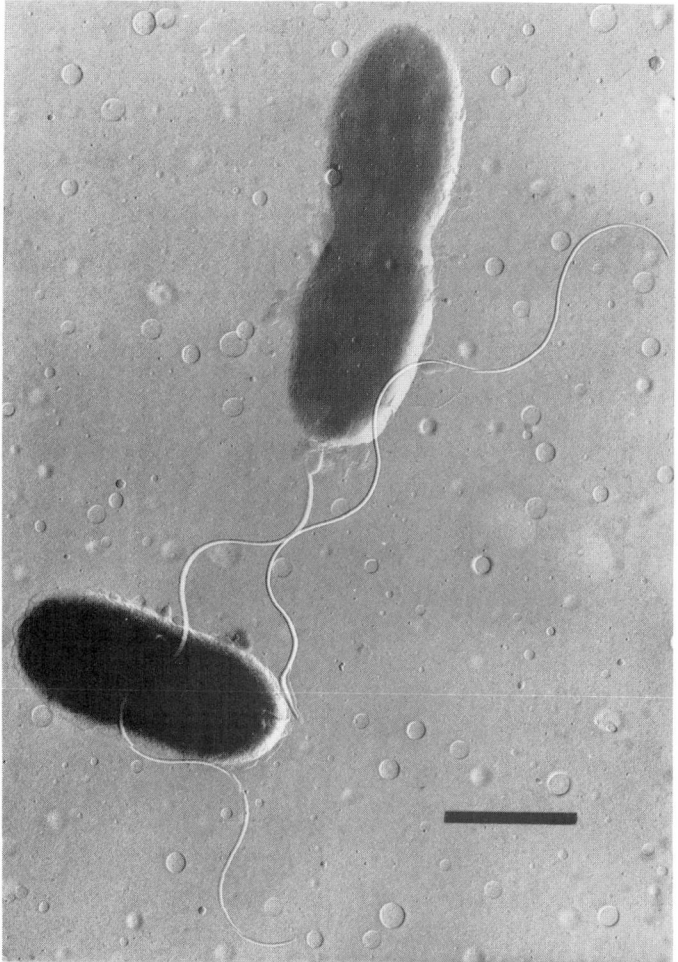

Figure 4.83. Electron micrograph of *Alteromonas hanedai*. Negatively stained (*bar*, 1.0 μm). (Reproduced with permission from: M. J. Jensen et al., Current Microbiology, *3:* 311–315, 1980, © Springer-Verlag.)

Figure 4.84. Electron micrograph of *Alteromonas aurantia.* Shadowed preparation (*bar*, 1.0 μm). (Reproduced with permission from M. J. Gauthier and V. A. Breittmayer, International Journal of Systematic Bacteriology *29:* 366–372, 1979, © American Society for Microbiology.)

A. communis, A. vaga, and *A. haloplanktis* but not *A. macleodii* were found to be suitable hosts for bdellovibrios, initially isolated using species of *Vibrio* (Taylor et al., 1974).

A. rubra, A. luteoviolacea, A. citrea, and *A. aurantia* produce high molecular weight antibiotic substances which besides being autotoxic are also active against both Gram-negative and Gram-positive bacteria. These antibiotics are produced on a variety of complex media containing a seawater base. The antibiotic from *A. rubra* has been partially purified and shown to be a complex glycoprotein which is bound to the outer layer of the cell wall (Ballester et al., 1977). *A. citrea* produces two polyanionic antibiotics with marked differences in electrical charge (Gauthier, 1977). The mode of action of these antibiotics appears to involve a stimulation of the oxygen consumption by sensitive bacteria with a concomitant lethal production of hydrogen peroxide (Gauthier, 1976c). In addition, *A. luteoviolacea* produces one, or possibly two, brominated bactericidal compounds (Gauthier and Flatau, 1976).

Strains of *A. haloplanktis* may be pathogenic for oysters (Colwell and Sparks, 1967).

Little information is currently available on the ecology of *Alteromonas*. Most species appear to be common in the marine environment. *A. haloplanktis* has been isolated from the coastal waters (both Atlantic and Pacific) of North America, as well as from the Indian Ocean. *A. undina* and *A. espejiana* have been obtained off the Pacific coast of North America and, also, in the case of the latter species, off the coast of Chile. *A. rubra, A. luteoviolacea, A. citrea,* and *A. aurantia* have been isolated from the Mediterranean Sea, off the coast of France. The first

two of these four species, as well as *A. communis* and *A. vaga*, have been isolated from seawater off the Hawaiian Archipelago. Strains of *A. hanedai* have been obtained from the Arctic and the Antarctic as well as from Sannich Inlet on the coast of British Columbia, Canada. The seasonal distribution of some of the pigmented species of *Alteromonas* in the coastal waters of Nice, France, has been studied by Gauthier et al. (1975).

Two species of *Alteromonas* have been the objects of extensive investigation. *A. haloplanktis*, strain B-16 (ATCC 19855), has been used in the studies by MacLeod and his collaborators which were designed to elucidate the biochemical basis of the Na^+ requirement of this organism. Their findings indicated that Na^+ is essential for the function of all the permease systems tested, including the uptake systems for amino acids, tricarboxylic acid cycle intermediates, galactose, orthophosphate, and K^+ (Fein and MacLeod, 1975; Thompson and MacLeod, 1973, 1974; Thompson et al., 1970; Wong, 1969), as well as for the maintenance of the integrity of the cell wall (Forsberg et al., 1970). *A. espejiana* BAL-31 is the host of bacteriophage PM2 which is unusual in that it contains lipid. This phage, which has double-stranded circular DNA, has been the subject of numerous investigations (Espejo and Canelo, 1968a, b; Franklin et al., 1976). Studies of *A. espejiana* have been primarily concerned with its role as a host and have involved a biochemical and structural analysis of the changes which occur during phage replication, with considerable emphasis on lipid metabolism (Franklin et al., 1976).

Enrichment and Isolation Procedures

A relatively specific enrichment of *A. communis* and *A. vaga* can be obtained by adding a 500-ml seawater sample to a sterile 2-liter Erlenmeyer flask containing 25 ml 1 M Tris-HCl (pH 7.5), 0.5 g NH₄Cl, 38 mg K₂HPO₄·7H₂O, 14 mg FeSO₄·7H₂O, and 0.5 g *m*-hydroxybenzoate. The culture is incubated at 20–25°C, observed for signs of growth for up to 10 days, then streaked onto BMA containing 0.1% (w/v) *m*-hydroxybenzoate. Many of the strains of *A. macleodii* have been obtained by direct isolation from seawater. In this procedure, samples are filtered through 0.22 or 0.45 μm nitrocellulose filters which are subsequently placed onto BMA plates containing 0.2% (w/v) lactose. After 2–10 days incubation at 25°C, colonies are picked and streaked onto homologous medium.

There are no selective procedures for the remaining species which have been obtained by plating seawater samples onto complex media such as MA, YEA, or a defined medium such as BMA containing 0.2% (w/v) D-glucose and a mixture of amino acids, each at a concentration of 1 mg/liter (Baumann and Baumann, 1981; Chan et al., 1978). *A. rubra*, *A. luteoviolacea*, *A. citrea*, and *A. aurantia* were obtained from seawater samples plated on complex media and were chosen for study because of their distinctive pigmentation and antibiotic production. *A. hanedai*, the sole luminous species of *Alteromonas*, was obtained by direct isolation during ecological studies of luminous bacteria. As with other luminous organisms, light production by this species is best on complex media such as Luminous medium (LM) which is composed of BMA supplemented with 0.3% (v/v) glycerol and (in g/liter) 5 g yeast extract, 5 g tryptone, and 1 g CaCO₃. Since there is strain variation in the intensity and the duration of light emission, it is best to streak *A. hanedai* onto LM plates, incubate at 15°C for 12–36 h, and periodically examine the cultures with dark-adapted eyes (Baumann and Baumann, 1981).

Maintenance Procedures

A. macleodii, *A. haloplanktis*, *A. undina*, *A. espejiana*, *A. communis*, and *A. vaga* have been maintained at 15–18°C on MA slants with monthly transfers. After each transfer, the cultures are allowed to grow at 25°C for 1–2 days and then returned to 15–18°C for storage. Most of these species do not survive well at 4°C. *A. hanedai* has been maintained at 4°C on LM slants, transferred every 2 weeks, and allowed to grow at 15°C for 2–3 days before being stored at 4°C. A number of strains which had no organic growth factor requirements when first isolated acquired amino acid requirements when maintained for several years on MA slants. This problem can be overcome by culturing the strains on BMA slants containing 0.2% D-glucose rather than on a complex medium. In most cases, all of the above species which have been lyophilized and kept at 4°C have yielded viable cells after 5–7 years of storage. The viability of *A. hanedai*, a more recent acquisition, has only been tested after 2 years of lyophilization. For the preparation of lyophils the growth from a fresh slant is suspended in about 0.5 ml of a sterile solution containing ¼ ASW, 5 g/liter yeast extract, and 5 g/liter peptone (adjusted to pH 7.5), and transferred into a lyophil tube. The cell suspension is quickly frozen in a mixture of dry ice and acetone and placed under vacuum for 10–12 h (Baumann and Baumann, 1981). Lyophils are reconstituted by suspending the powder in about 0.5 ml Yeast extract broth (YEB) or, for *A. hanedai*, LM (without agar); a portion of the liquid is streaked onto YEA or LM and the remainder is inoculated into a tube containing 4 ml of the same medium. Growth is generally observed after 1–2 days incubation at 25°C; for *A. hanedai* the incubation temperature should be reduced to 15°C.

In the case of the species which produce antibiotics (*A. rubra*, *A. luteoviolacea*, *A. citrea*, and *A. aurantia*), the accumulation of these autotoxic substances leads to a loss of viability which is not retarded by storage at reduced temperatures (4–15°C). Consequently, maintenance of these strains requires that they be subcultured at weekly intervals. Preservation by lyophilization has not been successful with these organisms. The only adequate method of long-term preservation has been storage in liquid nitrogen in Difco marine broth (ZoBell, 1941) or in ¼ ASW containing 10 g/liter glycerol. Under these conditions viable cells of all the antibiotic-producing species have been recovered after 3 years of storage.

Procedures for Testing Special Characters

Species of *Alteromonas* are not able to ferment sugars. A suitable test medium for fermentation is YEB with 100 mM Tris-HCl (pH 7.5), 1% (w/v) D-glucose and 2 g/liter agar. After boiling to melt the agar, 10-ml aliquots of the medium are dispensed into test tubes, autoclaved, and inoculated by means of a stab. About 5 ml of 2% (w/v) agar (autoclaved and cooled to 41°C) are carefully layered over the medium (which is prechilled for *A. hanedai*) to make an agar plug. The test tubes are incubated at 25°C (15°C for *A. hanedai*) and observed for turbidity and gas production for a period of 4 days. Strict aerobes will not grow under these conditions while facultative anaerobes, which may or may not produce gas, grow to a high density and lower the pH of the medium.

None of the species of *Alteromonas* accumulates PHB as a reserve product and several species are unable to utilize the monomer, β-hydroxybutyrate. The simplest way to determine PHB accumulation is to grow the cells in BM which is limited for nitrogen [0.02% (w/v) (NH₄)₂SO₄] and contains an excess of the carbon and energy source [0.2–0.4% (w/v) β-hydroxybutyrate or acetate]; in the case of strains with organic growth factor requirements, (NH₄)₂SO₄ is reduced to 0.1% and 0.01–0.02% (w/v) yeast extract is added. The culture is incubated at 25°C (15°C for *A. hanedai*) and examined daily by means of the phase-contrast microscope for a period of 4 days for the presence of the bright intracellular granules characteristic of PHB. It is best to include as a control an organism such as *Vibrio nereis* or *V. natriegens* which is known to accumulate PHB. In some cases visual determination of PHB accumulation is complicated by the fact that the granules may be small and difficult to detect or that some strains, especially in old cultures, have involution forms with inclusions which might be confused with PHB. A relatively easy chemical means of identifying this reserve product involves alkaline hypochlorite digestion of the cells followed by solubilization and precipitation of PHB as described by Williamson and Wilkinson (1958). A quantitative estimation of the polymer can be performed as described by Slepecky and Law (1960).

The differentiation of species of *Alteromonas* is primarily based on differences in nutritional spectra and on the ability to grow at different temperatures. For determining the latter property, YEB and, for *A. hanedai*, LM (without agar) are suitable media. Growth at 4°C is observed periodically for a period of 7 days, while at 35° and 40°C the results are read within 2 days. The ability to utilize different organic compounds as sole sources of carbon and energy is determined by replica plating strains onto BMA containing 0.2% (w/v) of the tested sugar or 0.1% (w/v for solids and v/v for liquids) of most of the other compounds. For strains having requirements for organic growth factors, the media can be supplemented with amino acids (1 mg/liter) or with low levels of yeast extract, tryptone, or casein hydrolysate, either alone or in combination and generally not exceeding 0.1 g/liter (Baumann and Baumann, 1981; Chan et al., 1978). The plates are incubated at 25°C for 6 days and examined for growth every other day; the temperature is reduced to 15°C for *A. hanedai*. For a detailed description of the methods used for the nutritional screening, the determination of the presence of extracellular hydrolases, and the mechanism of aromatic ring cleavage consult Baumann and Baumann (1981) and Palleroni and Doudoroff (1972).

Differentiation of the genus **Alteromonas** from other genera

Species of *Alteromonas* are common in the marine environment and appear to be rare or absent in most terrestrial habitats. The requirement for over 100 mM Na⁺ for optimal growth restricts the potential habitats of the present members of this genus and distinguishes these organisms from terrestrial *Pseudomonas*. In practice, therefore, the major concern is with the differentiation of *Alteromonas* from physiologically and morphologically similar bacteria indigenous to the marine environment. Table 4.91 lists these organisms and includes the characteristics which differentiate them from *Alteromonas*. It should be stressed that the information concerning marine species of *Pseudomonas* and *Alcaligenes* is limited and that there are a number of groups of polarly flagellated aerobes which have been characterized but not given species designations. For the differentiation of *Alteromonas* from these groups consult Baumann and Baumann (1981).

Studies dealing with the ecology of luminous bacteria frequently are initiated by an examination of the decay kinetics of light emission in the presence of dodecanal (Hastings et al., 1978; Nealson and Hastings, 1979). Luminous bacteria can be subdivided into two major categories according to those having "slow" or "fast" decay kinetics. The latter category includes *A. hanedai*, *Photobacterium phosphoreum*, *P. leiognathi*, *Vibrio fischeri*, and *V. logei*. The differentiation of *A. hanedai* from these other species is readily accomplished since it is the only luminous species lacking the ability to ferment D-glucose.

Taxonomic Comments

In a study of nonfermentative marine bacteria, Baumann et al. (1972) found that four phenotypically distinct clusters of polarly flagellated rods had G + C contents in their DNAs of 38–50 mol%. Rather than increase the diversity of the already heterogenous genus *Pseudomonas* (Palleroni et al., 1973) (G + C range of 58–70 mol%), it was decided that these four clusters should become species in a new genus, *Alteromonas*. It was recognized that the separation of these two genera was somewhat arbitrary, but it was hoped that future studies would provide other properties distinguishing these genera, thereby permitting a better definition of *Alteromonas*. Seven subsequently described species of *Alteromonas* have been found to share certain distinctive properties present in *A. macleodii* and *A. haloplanktis* but not *A. communis* and *A. vaga*. They all have a number of extracellular hydrolases, are not able to utilize aromatic compounds and, in many cases, are not able to utilize several tricarboxylic acid cycle intermediates as well as such common carbon sources as glycerol and lactate. Preliminary studies of the amino acid sequence divergence of glutamine synthetase (Baumann et al., 1980) and nucleotide divergence of ribosomal RNA (De Smedt et al., 1980) have also revealed that *A. communis* and *A. vaga* are

Table 4.91.
Differential characteristics of the genus **Alteromonas** and other aerobic Gram-negative marine bacteria.[a]

Characteristics	Altero-monas	Pseudo-monas[b]	Alcali-genes[c]	Oceano-spirillum
Flagellar arrangement				
Polar	+	+	−	+
Peritrichous	−	−[d]	+	−
Mol % G + C of DNA	38–50	55–64	52–68	42–51
Gelatinase	D[e]	−	−	D
Utilization of:				
DL-Malate	D[f]	+	+	D
D-sorbitol, *m*-hydroxybenzoate	D[f]	−	D	

[a] Symbols: see standard definitions
[b] The species considered are *P. nautica*, *P. marina* and *P. doudoroffii* (Baumann et al., 1972).
[c] The species considered are *A. pacificus*, *A. cupidus*, *A. venustus* and *A. aestus* (Baumann et al., 1972).
[d] Some strains of *P. nautica* may have a few lateral flagella in addition to the polar flagellum.
[e] Species 1 through 9 are +, species 10 and 11 are −.
[f] Species 1 through 9 are −, species 10 and 11 are +.

different from *A. macleodii* and *A. haloplanktis*. The latter approach suggests that *A. macleodii* and *A. haloplanktis* are related to such genera as *Escherichia* and *Vibrio* indicating that these organisms are part of the evolutionary branch represented by the 16S ribosomal RNA group III of Gibson et al., 1979. More extensive investigations involving the evolutionary relationships between additional species of *Alteromonas* are necessary before considering the subdivision of this genus. Pending such studies *Alteromonas* should be regarded as a temporary refuge for a number of common marine species whose evolutionary relationship is, for the most part, unknown.

Further Reading

Baumann, L., P. Baumann, M. Mandel and R.D. Allen. 1972. Taxonomy of aerobic marine eubacteria. J. Bacteriol. *110:* 402–429.
Baumann, P. and L. Baumann. 1981. The marine Gram-negative eubacteria. In Starr, Stolp, Trüper, Balows and Schlegel (Editors), *The Prokaryotes: A Handbook on Habitats, Isolation and Identification of Bacteria.* Springer-Verlag, New York, pp. 1302–1331.
Hendrie, M.S. and J.M. Shewan. 1979. The identification of pseudomonads. In Skinner and Lovelock (Editors), *Identification Methods for Microbiologists*, 2nd Ed., Academic Press Inc., London, pp. 1–14.

Differentiation and characteristics of species of **Alteromonas**

The differential characteristics of species of *Alteromonas* are presented in Table 4.92. Other characteristics of the species are provided in Table 4.93. Since different sets of traits are diagnostic for each of the species of *Alteromonas*, a table which includes selected diagnostic properties for all 11 species will contain a considerable number of traits. Therefore, in order to facilitate the identification of unknown isolates of *Alteromonas* we have selected a total of 13 readily determinable diagnostic traits (designated by boldface type in Table 4.92) which are useful for a preliminary species identification. Of the total 55 possible comparisons between different pairs of species, only five pairs are distinguishable by as few as two traits (*A. espejiana* and *A. macleodii*; *A. espejiana* and *A. haloplanktis*; *A. citrea* and *A. rubra*; *A. citrea* and *A. aurantia*; *A. communis* and *A. vaga*). The remaining pairs are distinguishable by three to nine traits.

List of the species of the genus **Alteromonas**

1. **Alteromonas macleodii** Baumann, Baumann, Mandel and Allen 1972, 420.[AL]

mac.leo′di.i. M.L. gen. n. *macleodii* of MacLeod; named after R. A. MacLeod, a Canadian microbiologist who pioneered studies on the biochemical bases of the Na+ requirement of marine bacteria.

Morphology is as depicted in Figures 4.78–4.81.

Physiological and nutritional characteristics of the species are presented in Tables 4.92 and 4.93.

The following combination of properties distinguishes *A. macleodii* from the other species of *Alteromonas* by five to seven traits: positive for sucrose, cellobiose, melibiose, lactose, salicin, D-gluconate, DL-glycerate, and glycerol; negative for citrate and pigmentation.

The mol% G + C content of the DNA is 44–47 (Bd).

Type strain: ATCC 27126 (strain 107 of Baumann, Baumann, Mandel and Allen 1972).

2. **Alteromonas haloplanktis** (ZoBell and Upham 1944) Reichelt and Baumann 1973, 438.[AL] (*Vibrio haloplanktis* Zobell and Upham 1944, 261; *Pseudomonas enalia* (ZoBell and Upham) Colwell and Sparks 1967, 985; *Alteromonas marinopraesens* (ZoBell and Upham) Baumann, Baumann, Mandel and Allen 1972, 422.)

ha.lo.plank′tis. Gr. n. *halos* sea; Gr. adj. *planktos* wandering; M.L. adj. *haloplanktis* sea wandering.

Morphology is as depicted in Figure 4.82.

Physiological and nutritional characteristics of the species are presented in Tables 4.92 and 4.93.

The following combination of properties distinguishes *A. haloplanktis* from the other species of *Alteromonas* by 3–8 traits: straight rods positive for D-mannose, sucrose, maltose, succinate, and citrate; negative for alginase, melibiose, lactose, glycerol, L-threonine, putrescine, and pigmentation.

The mol% G + C content of the DNA is 41–45 (T_m, Bd).

Type strain: ATCC 14393 (strain 215 of Baumann, Baumann, Mandel and Allen 1972).

3. **Alteromonas espejiana** Chan, Baumann, Garza and Baumann 1978, 220.[AL]

es.pe.ji.a′na M.L. fem. adj. *espejiana* of Espejo; named after R. T. Espejo, a Chilean microbiologist who isolated one of the first lipid-containing bacteriophages.

Physiological and nutritional characteristics of the species are presented in Tables 4.92 and 4.93.

The following combination of properties distinguishes *A. espejiana* from the other species of *Alteromonas* by four to eight traits: positive for alginase, sucrose, melibiose, lactose, citrate, and D-mannitol; negative for salicin, D-gluconate, N-acetylglucosamine, succinate, glycerol, and pigmentation.

The mol% G + C content of the DNA is 43–44 (T_m).

Type strain: ATCC 29659 (strain 261 of Chan, Baumann, Garza and Baumann 1978).

Table 4.92
Differential characteristics of the species of the genus **Alteromonas**[a]

Characteristics	1. A. macleodii	2. A. haloplanktis	3. A. espejiana	4. A. undina	5. A. hanedai	6. A. rubra	7. A. luteoviolacea[l]	8. A. citrea[l]	9. A. aurantia	10. A. communis	11. A. vaga
Cell shape	St	St	St	Cu	St	St	St	St	St	Cu	St
Luminescence	−	−	−	−	+	−	−	−	−	−	−
Growth at:											
4°C	−	−	−	d	+	−	d	−	+	−	−
35°C	+	d	d	−	−	+	+	d	−	+	+
40°C	d	−	−	−	−	−	−	d	−	+	−
Reduction of NO_3^- to NO_2^-	−	d	−	−	+	−	−	−	−	−	−
Requirement for organic growth factors	−	d	+	+	+	+	+	+	+	−	−
Production of:											
Amylase	+	d	+	d	−	+	+	+	+	−	−
Gelatinase, lipase	+	+	+	+	+	+	+	+	+	−	−
Alginase	d	−	+	−	−				−	−	−
Chitinase	−	d	−	+	+	−	−	d	−	−	−
Utilization of:											
D-Mannose	−	+	d	−	−	+	−	+	+	+	+
D-Galactose	+	d	+	−	d	−	−	−	−	d	+
D-Fructose	+	d	d	−	−	−	−	+	+	+	+
Sucrose	+	+	+	+	−	−	−	−			
Maltose	+	+	+	+	−	−	+	−	d	d	d
Cellobiose	+	−	d	−	−	−	−	−	−	−	d
Melibiose	+	−	+	−	−	−	−	−	−	−	−
Lactose	+	−	+	−	−	−	−	−	−	−	−
Salicin	+	−	−	−	−	−	−	−	−	−	−
D-Gluconate	+	−	−	−	d	−	−	−	−	+	+
N-Acetylglucosamine	d	+	−	+	+	−	+	−	+	−	+
Succinate, fumarate	−	+	−	+	−	−	−	−	−	+	+
DL-Lactate	d	−	−	−	−	−	d	−	−	+	+
DL-Glycerate	+	−	−	−	+	−	−	−	−	d	−
Citrate	−	+	+	−	−	−	−	−	−	+	+
Aconitate	−	+	+	−	−				−	+	+
Erythritol	−	−	−	−	−	−	−	−	−	−	+
D-Mannitol	d	d	+	−	−	−	−	−	−	+	+
Glycerol	+	−	−	−	−	−	−	−	−	+	+
γ-Aminobutyrate, sarcosine	−	−	−	−	−				−	+	+
L-Threonine	−	−	+	d	−		+	−	d	−	−
L-Tyrosine	+	+	+	+	d		d	−	+	−	−
Putrescine	−	−	−	−	+				−	+	+
D-Sorbitol, DL-malate, α-ketoglutarate, m-hydroxybenzoate	−	−	−	−	−	−	−	−	−	+	+
Pigmentation	−	−	−	−	d	+	+	+	+	−	−

[a] Boldface type indicates traits useful for the preliminary identification of species. For symbols see standard definitions; also St, straight rods; and Cu, curved rods.

Table 4.93.
Additional characteristics of species of **Alteromonas**[a]

Characteristics	1. *A. macleodii*[b]	2. *A. haloplanktis*[b]	3. *A. espejiana*[b]	4. *A. undina*[b]	5. *A. hanedai*[x]	6. *A. rubra*[d]	7. *A. luteoviolacea*[e]	8. *A. citrea*	9. *A. aurantia*[g]	10. *A. communis*[b]	11. *A. vaga*[b]
Production of deoxyribonuclease						+	+	+	+		
Utilization of:											
D-Ribose	−	−	−	−	d					−	d
D-Xylose	d	−	−	−	−					d	d
D-Arabinose	−	−	−	d	−					−	d
L-Arabinose	−	d	−	−	−					d	+
L-Rhamnose	−	−	−	−	−					d	+
D-Glucose	+	+	+	+	d	+	+	+	+	+	+
Trehalose	+	d	+	+	−	+	+	+	+	d	d
Saccharate	−	−	−	−	−					+	+
Glucuronate	−	−	−	−	−	−			−	d	d
Galacturonate	+	−	−	−	−					d	d
Acetate	+	+	+	+	+	−	−	−	−	+	+
Propionate	+	+	+	+	+					d	d
Butyrate	+	d	+	d	+					−	−
Isobutyrate	+	d	+	+	−					−	−
Valerate	+	d	d	+	−					−	−
Isovalerate	d	−	+	−	−					−	−
Caproate	+	−	−	−	d					−	d
Heptanoate	d	d	−	−	d					−	+
Caprylate	+	d	−	−	−					−	d
Pelargonate	+	d	+	+	d					−	d
Caprate	+	+	+	+	+					d	d
L-Tartrate	−	−	−	−	−	d				d	d
DL-β-Hydroxybutyrate	d	d	d	+	−		−			+	d
Pyruvate	+	+	+	−	+	−	d	−	−	+	+
Aconitate	−	+	+	−	−				−	+	+
meso-Inositol	−	−	−	−	−	−	−	−	−	+	+
Adonitol	−	−	−	−	−	−	−	−	−	−	d
Propyleneglycol	d	−	−	−	−					−	−
Ethanol	d	+	−	+	−					+	−
n-Propanol	d	d	−	+	−					+	−
n-Butanol	−	−	d	+	−					−	−
p-Hydroxybenzoate	−	−	−	−	−	−			−	+	+
Phenylacetate	−	−	−	−	−					d	d
Quinate	−	−	−	−	−					+	+
Glycine	d	d	+	+	+				+	−	−
L-α-Alanine	+	+	+	+	+		+	−	−	+	+
D-α-Alanine	−	−	−	−	d			−	−	+	+
β-Alanine	−	−	−	−	−					+	d
L-Serine	d	+	+	+	+				−	−	d
L-Leucine	d	d	d	+	d		−	−	d	−	−

4. Alteromonas undina Chan, Baumann, Garza and Baumann 1978, 220.[AL]

un.di′na. L. fem. n. *undina* undine, water nymph.

Physiological and nutritional characteristics of the species are presented in Tables 4.92 and 4.93.

The following combination of properties distinguishes *A. undina* from the other species of *Alteromonas* by 3–8 traits: curved rods positive for chitinase, sucrose, maltose, and succinate; negative for 35°C, NO₃⁻ to NO₂⁻, D-mannose, D-fructose, citrate, and pigmentation.

The mol% G + C content of the DNA is 43–44 (T_m).

Type strain: ATCC 29660 (strain 272 of Chan, Baumann, Garza and Baumann 1978).

5. Alteromonas hanedai Jensen, Tebo, Baumann, Mandel and Nealson 1981, 382.[VP] (*Effective publication:* Jensen, Tebo, Baumann, Mandel and Nealson 1980, 311.)

ha.ne′dai. M.L. gen. n. *hanedai* of Haneda; named after Y. Haneda, a Japanese biologist who pioneered studies of bioluminescence.

Morphology is as depicted in Figure 4.83.

Physiological and nutritional characteristics of the species are presented in Tables 4.92 and 4.93.

The following combination of properties distinguishes *A. hanedai* from the other species of *Alteromonas* by five to nine traits: positive for luminescence, 4°C, NO₃⁻ to NO₂⁻, chitinase, and putrescine; negative for amylase, D-mannose, D-fructose, sucrose, maltose, and succinate.

The mol% G + C content of the DNA is 44–47 (Bd).

Type strain: ATCC 33224 (strain 281 of Jensen, Tebo, Baumann, Mandel and Nealson 1980).

6. Alteromonas rubra Gauthier 1976, 464.[AL]

rub′ra. L. fem. adj. *rubra* red.

Table 4.93.—*continued*

Characteristics	1. *A. macleodii*[b]	2. *A. haloplanktis*[b]	3. *A. espejiana*[b]	4. *A. undina*[b]	5. *A. haneda*[c]	6. *A. rubra*[d]	7. *A. luteoviolacea*[e]	8. *A. citrea*[f]	9. *A. aurantia*[g]	10. *A. communis*[b]	11. *A. vaga*[b]
L-Isoleucine	d	−	d	d	+				−	−	−
L-Valine	d	−	−	−	−				−	−	−
L-Aspartate	d	d	+	−	+				−	d	d
L-Glutamate	d	d	−	−	+		−	−	−	+	+
L-Lysine	−	−	−	−	−		d	−	−	+	d
L-Arginine	d	d	d	d	−		d	−	+	+	d
L-Ornithine	−	−	d	d	−		−	−	−	+	+
L-Citrulline	−	−	+	d	−				d	d	d
L-Histidine	−	d	−	−	−		d	−	d	+	+
L-Proline	−	d	+	d	−		d	−	−	+	+
L-Phenylalanine	−	−	d	d	−		d	−	−	d	d
Ethanolamine	−	−	−	−	−				−	d	−
Spermine	−	−	−	−	+				−	+	d
Betaine	−	−	−	−	−				−	+	+
Trigonelline	−	−	−	−	−				−	−	d

[a] For symbols see standard definitions.

[b] These species are not able to utilize the following organic compounds as sole or principal sources of carbon and energy: D-fucose, inulin, cellulose, mucate, formate, oxalate, malonate, maleate, glutarate, adipate, pimelate, suberate, azelate, sebacate, D-tartrate, *meso*-tartrate, glycolate, levulinate, citraconate, itaconate, mesaconate, ethyleneglycol, 2,3-butanediol, methanol, isopropanol, isobutanol, D-mandelate, L-mandelate, benzoylformate, benzoate, *o*-hydroxybenzoate, phenylacetate, phenylethanediol, phenol, naphthalene, L-norleucine, DL-α-aminobutyrate, DL-α-aminovalerate, δ-aminovalerate, L-tryptophan, D-tryptophan, DL-kynurenine, kynurenate, anthranilate, *m*-aminobenzoate, *p*-aminobenzoate, methylamine, benzylamine, histamine, tryptamine, butylamine, α-amylamine, 2-amylamine, pentylamine, creatine, hippurate, pantothenate, acetamide, nicotinate, nicotinamide, allantoin, adenine, guanine, cytosine, thymine, uracil, and *n*-dodecane.

[c] The following organic compounds could not be utilized as principal sources of carbon and energy: malonate, glutarate, benzoate, δ-aminovalerate, and hippurate.

[d] The following organic compounds could not be utilized as principal sources of carbon and energy: raffinose, malonate, D-tartrate, galactitol, and *o*-hydroxybenzoate.

[e] The following organic compounds could not be utilized as principal sources of carbon and energy: D-2-deoxyribose, raffinose, inulin, esculin, malonate, galactitol, *o*-hydroxybenzoate, L-glutamate, L-tryptophan, creatine, deoxycholate, cholesterol, tributyrin, and lecithin.

[f] The following organic compounds could not be utilized as principal sources of carbon and energy: 2-D-deoxyribose, D-glucosamine, D-galactosamine, raffinose, inulin, oxalate, malonate, D-tartrate, galactitol, *o*-hydroxybenzoate, L-glutamate, L-cysteine, glycylglycine, L-ascorbate, and L-tryptophan.

[g] The following compounds could not be utilized as principal sources of carbon and energy: 2-D-deoxyribose, D-fucose, D-tagatose, D-galactosamine, L-sorbose, gentiobiose, turanose, amygdalin, arbutin, esculin, methyl-D-glucoside, methyl-D-mannoside, raffinose, inulin, 2-ketogluconate, 5-ketogluconate, oxalate, malonate, D-tartrate, D-arabitol, L-arabitol, xylitol, galactitol, *o*-hydroxybenzoate, L-norleucine, DL-norvaline, L-aspartate, L-glutamate, L-cysteine, L-methionine, glycylglycine, L-ascorbate, L-tryptophan, DL-kynurenine, anthranilate, *m*-aminobenzoate, *p*-aminobenzoate, benzylamine, histamine, tryptamine, α-amylamine, diaminobutane, creatine, and acetamide.

Physiological and nutritional characteristics of the species are presented in Tables 4.92 and 4.93.

The following combination of properties distinguishes *A. rubra* from the other species of *Alteromonas* by two to six traits: positive for 35°C, amylase, D-mannose, and prodigiosin; negative for 4°C, D-fructose, sucrose, maltose, succinate, citrate, and D-mannitol.

Produces a polyanionic autotoxic antibiotic on complex media.

The mol% G + C content of the DNA is 46–48 (T_m).

Type strain: ATCC 29570 (strain 18 of Gauthier 1976).

7. **Alteromonas luteoviolacea** Gauthier 1982, 85[VP] (*Effective publication:* Gauthier 1976, 147.) (*Alteromonas luteoviolaceus*. (sic) Gauthier 1976, 147)

lu.te.o.vi.o.la′ce.a. L. adj. *luteus* yellow; L. adj. *viola* violet; M.L. adj. *luteoviolacea* yellow-violet.

Physiological and nutritional characteristics of the species are presented in Tables 4.92 and 4.93.

The following combination of properties distinguishes *A. luteoviolacea* from the other species of *Alteromonas* by three to six traits: positive for 35°C, amylase, gelatinase, maltose, L-threonine and violacein; negative for chitinase, D-mannose, D-fructose, sucrose, and lactose.

All strains produce a polyanionic, autotoxic antibiotic and some may also produce bactericidal brominated compounds.

The mol% G + C content of the DNA is 40–43 (T_m).

Type strain: NCMB 1893 (strain CH130 of Gauthier 1976).

8. **Alteromonas citrea** Gauthier 1977, 354.[AL]

ci′tre.a. L. adj. *citreus* lemon-yellow; L. fem. adj. *citrea*.

Physiological and nutritional characteristics of the species are presented in Tables 4.92 and 4.93.

The following combination of properties distinguishes *A. citrea* from the other species of *Alteromonas* by two to six traits: positive for amylase, D-mannose, D-fructose, and a lemon-yellow pigment; negative for 4°C, sucrose, maltose, succinate, citrate, D-mannitol, and glycerol.

Produces two polyanionic, autotoxic antibiotics on complex media.
The mol% G + C content of the DNA is 41–45 (T_m).

Type strain: ATCC 29719 (strain 10 of Gauthier 1977).

9. **Alteromonas aurantia** Gauthier 1979, 371.[AL]

au.ran'tia. M.L. n. *aurantium* orange; M.L. fem. adj. *aurantia* orange colored.

Morphology is as depicted in Figure 4.84.

Physiological and nutritional characteristics of the species are presented in Tables 4.92 and 4.93.

The following combination of properties distinguishes *A. aurantia* from the other species of *Alteromonas* by two to seven traits: positive for 4°C, amylase, D-mannose, D-fructose, and an orange pigment; negative for 35°C, chitinase, sucrose, succinate, citrate, and D-mannitol.

Produces a polyanionic, autotoxic antibiotic on complex media.

The mol% G + C content of the DNA is 38–43 (T_m).

Type strain: ATCC 33046 (strain 208 of Gauthier 1979).

10. **Alteromonas communis** Baumann, Baumann, Mandel and Allen 1972, 420.[AL]

com.mu'nis. L. adj. *communis* common.

Morphology is as depicted in Figure 4.79.

Physiological and nutritional characteristics of the species are presented in Tables 4.92 and 4.93.

A. communis is distinguished from *A. vaga* by its cell curvature, growth at 40°C, and inability to utilize *N*-acetylglucosamine and erythritol. Unlike the remaining species of *Alteromonas*, *A. communis* is able to utilize D-sorbitol, DL-malate, α-ketoglutarate, and *m*-hydroxybenzoate and does not have an extracellular gelatinase and lipase.

The mol% G + C content of the DNA is 45–48 (Bd).

Type strain: ATCC 27118 (strain 8 of Baumann, Baumann, Mandel and Allen 1972).

11. **Alteromonas vaga** Baumann, Baumann, Mandel and Allen 1972, 420.[AL]

va'ga. L. adj. *vaga* wandering.

Physiological and nutritional characteristics of the species are presented in Tables 4.92 and 4.93.

A. vaga, a straight rod, is distinguished from *A. communis* by its cell shape as well as by its ability to utilize *N*-acetylglucosamine and erythritol and its inability to grow at 40°C. Unlike the remaining species of *Alteromonas*, it is able to utilize D-sorbitol, DL-malate, α-ketoglutarate, and *m*-hydroxybenzoate and does not have an extracellular gelatinase and lipase.

The mol% G + C content of the DNA is 46–50 (Bd).

Type strain: ATCC 27119 (strain 40 of Baumann, Baumann, Mandel and Allen 1972).

Species Incertae Sedis

A number of species assigned to *Pseudomonas* on the basis of their nonfermentative metabolism and flagellar arrangement appear to belong to the genus *Alteromonas* and phenotypically resemble the majority of the species of this genus (excluding *A. communis* and *A. vaga*). All require either a description of additional strains or more extensive characterization before their formal assignment to *Alteromonas* can be performed.

a. "*Pseudomonas piscicida*" (Bein 1954) Buck, Meyers and Leifson 1963, 1125. (*Flavobacterium piscicida* Bein 1954, 115.)

This species has a DNA base composition of 43–46 mol% G + C, produces a yellow-orange pigment, and is pathogenic for marine fish and crabs (Buck et al., 1963; Hansen et al., 1963, 1965; Mandel et al., 1965; Meyers et al., 1959; Weeks et al., 1962).

Type strain: ATCC 15251.

b. "*Pseudomonas atlantica*" Humm 1946, 58.

ATCC strain 19262, isolated by Yaphe (1957) is a marine agar decomposer; the DNA base composition is 43.5 mol% G + C (Mandel, 1966).

Type strain: although ATCC strain 19262 (NCMB 301) has been considered to be the type strain (Sneath and Skerman, 1966), it is not one of the original strains on which Humm based his description of the species.

c. "*Pseudomonas nigrifaciens*" White 1940, 640.

ATCC strain 19375 causes discoloration of butter (White, 1940). The DNA base composition is 42.9 mol% G + C (Baumann et al., 1972).

Type strain: none. *Co-type strain:* ATCC 19375.

d. *Alteromonas putrefaciens* Lee, Gibson and Shewan 1981, 215.[AL] (Effective publication: Lee, Gibson and Shewan 1977, 449). (*Pseudomonas putrefaciens* Derby and Hammer 1931) Long and Hammer 1941, 176; *Achromobacter putrefaciens* Derby and Hammer 1931, 414.)

This species contains a group of strains of importance in fish spoilage and also isolates from clinical samples. The strains have DNA base compositions of 43 to 55 mol% G + C and constitute at least two phenotypically distinguishable groups and possibly four DNA homology groups (Riley et al., 1971; Owen et al., 1978; Levin, 1972). Since their DNA base composition excludes these organisms from *Pseudomonas*, yet is somewhat higher than the published range for *Alteromonas*, Lee et al. (1977) redefined the latter genus to include these strains. However, since "*Pseudomonas putrefaciens*" is clearly heterogeneous, a generic assignment and other nomenclatural changes should await an adequate speciation within this collection of strains.

The mol% G + C of the DNA is 44.7–54.7.

Type strain: NCIB 10471 (ATCC 8071; Hammer 95).

Important Notes for Users of this Edition

1. Always read both generic and species descriptions because characters listed in the generic description are not usually listed in the species descriptions.

2. Unless otherwise indicated in footnotes to tables, the meanings of symbols are as follows:

    +   90% or more of strains are positive

    −   90% or more of strains are negative

    d   11–89% of strains are positive

    v   strain instability (*not* equivalent to "d")

    D   different reactions in different taxa (species of a genus or genera of a family)

3. All other symbols are defined in footnotes to tables.

Genus Flavobacterium Bergey, Harrison, Breed, Hammer and Huntoon 1923, 97[AL]

BARRY HOLMES, ROBERT J. OWEN AND THOMAS A. McMEEKIN

Fla.vo.bac.te′ri.um. L. adj. *flavus* yellow; Gr. neut. n. *bakterion* a small rod; M.L. neut. n. *Flavobacterium* a yellow bacterium.

Rods with parallel sides and rounded ends, typically 0.5 μm wide and 1.0–3.0 μm long. Intracellular granules of poly-β-hydroxybutyrate are absent. Endospores not formed. Gram-negative. **Nonmotile. Do not glide or spread. Aerobic,** having a strictly respiratory type of metabolism. Environmental isolates grow at temperatures from 5–30°C; most clinical isolates also grow at 37°C. Growth on solid media is typically pigmented (yellow to orange) but nonpigmented strains occur. Colonies are translucent (occasionally opaque), circular (diameter 1–2 mm), convex or low convex, smooth and shiny with entire edges. **Catalase-, oxidase- and phosphatase-positive.** Agar is not digested. Chemoorganotrophic. Acid but no gas is produced from carbohydrates in media having a low peptone concentration. Widely distributed in soil and water; also found in raw meats, milk and other foods, and in the hospital environment and in human clinical material. The mol% G + C of the DNA is 31–42 (T_m).

Type species: *Flavobacterium aquatile* (Frankland and Frankland 1889) Bergey, Harrison, Breed, Hammer and Huntoon 1923, 100.

Further Descriptive Information

The cells of *Flavobacterium* strains are generally rod shaped. Some unusually long rods (≧5 μm) may be formed in liquid cultures of *F. meningosepticum* and *F. odoratum*.

The main cellular fatty acid components in *F. aquatile* and *F. breve* are iso C15:0 and 3-OH iso C17:0 acids (Fautz et al., 1981; Oyaizu and Komagata 1981). *F. meningosepticum* was reported by Moss and Dees (1978) and by Oyaizu and Komagata (1981) to contain in addition significant amounts of 2-OH iso C15:0, although Fautz et al. (1981) found the main fatty acids in this species were C16:1 and C16:0 and that hydroxy acids were totally absent. Group IIb was reported by Moss and Dees (1978) to have a similar fatty acid composition to *F. meningosepticum*. Group IIk, biovar (biotype) 2, herein referred to as *F. multivorum* (Holmes et al., 1981), was reported by Yabuuchi et al. (1981) to contain iso C15:0 and 2-OH iso C15:0 as the major fatty acid components and the main cellular lipids were identified as sphingophospholipids.

Cells are nonmotile in soft agar and in hanging drop preparations and are normally regarded as lacking flagella. However, Weeks (1955) by means of electron microscopy found structures on *F. aquatile* F36 that he thought were similar to pseudoflagella. This strain also shows surface translocation characteristics according to Perry (1973), but not unequivocal gliding motility as defined by Henrichsen (1972). Webster and Hugh (1979), using light microscopy, likewise found evidence of nonfunctional flagella on *F. aquatile* (ATCC 11947, F36) as well as on *F. meningosepticum* (ATCC 13253, NCTC 10016). Thomson et al. (1981) were unable to confirm the presence of flagella on these two species using electron microscopy. However, nonflagellar appendages, morphologically similar to those described on gliding bacteria, were observed on *F. aquatile*. Levine et al. (1980) described the presence of 'floppy' nonfunctional polar flagella on strains of CDC Group IIk, biovar 2 (*F. multivorum*).

The hue and intensity of pigmentation varies considerably and may be affected by the growth medium, the temperature and the incubation period. The degree of pigmentation may be more pronounced at lower temperatures (15–20°C) and daylight may be required for maximum pigmentation to develop. Growth on casein, milk and starch agar may also enhance pigmentation. The pigments are nonfluorescent in ultraviolet light and are insoluble in growth media. Preliminary studies by Weeks (1981) show the principal pigments of *F. breve* and *F. odoratum* are not carotenoid but are probably of the flexirubin type described by Reichenbach et al. (1981).

All strains are chemoorganotrophic and may prove difficult to maintain following primary isolation. They are generally not fastidious, but

amino acids and supplementary growth factors (particularly aneurine or biotin) may be required for growth, and the majority will not grow in a simple salts/glucose medium. Strains are strictly aerobic and unable to grow in an atmosphere of pure hydrogen although they may grow anaerobically in the presence of 7% (v/v) carbon dioxide and, consequently, may be mistaken for facultative anaerobes. The temperatures for growth of environmental isolates are usually between 5 and 30°C, but clinical isolates will grow at 37°C. Some strains of *F. meningosepticum* and of group IIb will also grow at 42°C.

The metabolism is strictly respiratory. Menaquinones are the sole respiratory quinones (see also Taxonomic Comments). Gas is not produced from carbohydrates and acid production is only observed in media of low peptone concentration. The carbohydrates most frequently attacked are glucose, fructose, glycerol, maltose and trehalose. Acid is not produced from adonitol, dulcitol, inositol and sorbitol. Cellulose, agar and other high molecular weight carbohydrates are not attacked. Most strains are proteolytic, hydrolyzing casein and gelatin. Extracellular deoxyribonuclease (DNase) is normally produced and tributyrin is hydrolyzed. Nitrate is generally not reduced. Some strains of *F. meningosepticum* and of group IIb utilize citrate. Most strains grow on MacConkey's agar and urease is frequently produced. Although most strains give an acid (oxidative) reaction in glucose O-F medium, strains of *F. odoratum* characteristically yield an alkaline reaction in that medium. β-D-Galactosidase (ONPG test) is commonly produced but results are invariably negative in tests for KCN tolerance, H_2S production (unless the medium contains added cysteine), gluconate oxidation, malonate utilization, phenylalanine deamination, and production of arginine dihydrolase and lysine and ornithine decarboxylases. Esculin is generally hydrolyzed. Lipid inclusions are invariably absent after growth on β-hydroxybutyrate. Strains in several species produce indole.

There are not as yet established systems for the investigation of classical genetic phenomena in species of *Flavobacterium*. Strains representing the various species were screened for the presence of plasmids but none were detected (Owen and Holmes, 1981). However, Kono et al. (1980) have reported plasmid DNA conferring resistance to ampicillin, carbenicillin and erythromycin in a strain of *F. odoratum*.

Six serological groups (A–F) of *F. meningosepticum* are recognized (King, 1959; Owen and Lapage, 1974). Most of the typed strains belong to serovar C. In France, Richard et al. (1979a, b) described an additional six serovars (G–L) of *F. meningosepticum* of which serovar G was the commonest. However, several of the strains representing these new serovars (notably serovars I, J and L) appear to be strains of *F. breve* (Holmes and Owen, unpublished results). Serological studies have not been carried out on other *Flavobacterium* species, although cross-reactions have been observed between some *F. meningosepticum* antisera and strains of group IIb (King, 1959; Owen and Lapage, 1974; Price and Pickett, 1981). Price and Pickett (1981) also reported cross-reactions between *F. meningosepticum* antisera and strains of group IIf (see Other Organisms).

Flavobacterium strains, in particular those from clinical sources, are resistant to many antimicrobials, which include streptomycin, ampicillin, amikacin, carbenicillin, gentamicin, kanamycin, polymyxin B and tobramycin. Strains are either resistant or moderately susceptible to sulfamethoxazole, co-trimoxazole, trimethoprim, cephaloridine, erythromycin, chloramphenicol, tetracycline and nalidixic acid. A characteristic of the strains is their susceptibility to antimicrobial agents such as novobiocin which are normally used in the treatment of infections caused by Gram-positive organisms (Altmann and Bogokovsky, 1971; von Graevenitz and Grehn, 1977). Von Graevenitz (1981) reported that considerable discrepancies could arise between the results of disk and tube dilution tests and considered it important to perform dilution susceptibility tests in cases of drugs that display sensitivity in the disk test. It is possible that chromosomal genes play a significant role in the antimicrobial response of *Flavobacterium* strains, as indicated by the nature of their resistance patterns and by the rarity of detectable plasmids.

F. meningosepticum is well recognized as a cause of neonatal meningitis and such infections have a poor prognosis. Since the disease was first described in 1944, about 100 cases have been published (von Graevenitz, 1981). The most frequent serovar found was C (which was also involved in several epidemics) followed by B, F, A and E. The overall case-fatality rate was about 55%; and in most of the survivors hydrocephalus developed. Meningitis due to *F. meningosepticum* in adults is rare, as is septicemia, and all the patients in whom it has been described had underlying diseases predisposing to Gram-negative infection. This species may also cause pneumonia in both infants and adults. Colonization of the respiratory tract in compromised hosts may occur under epidemic situations. The pathogenicity for man of other *Flavobacterium* species is less well documented but cases of meningitis (Bagley et al., 1976), bacteremia (Stamm et al., 1975) and upper respiratory tract colonization of seriously ill patients (Du Moulin, 1979) have been attributed to group IIb. A case of bacterial peritonitis (Dhawan et al., 1980) has been attributed to group IIk, biovar 2 (*F. multivorum*). *F. odoratum* is most commonly recovered from urine specimens sometimes in significant numbers; infection of amputation sites has also been reported (Davis et al., 1979; Holmes et al., 1979) but no infections have been definitely ascribed to this organism. *F. meningosepticum* is not pathogenic for mice or rabbits (King, 1959; Grehn, 1976), but a case of meningitis in a cat has been reported (Sims, 1974). *F. odoratum* shows no animal pathogenicity (Stutzer, 1923).

Strains closely allied to *F. breve* and group IIb are widely distributed in soil and water and are commonly found in food such as raw meats and poultry, vegetables during commercial processing, and in dairy products (McMeekin et al., 1971; Hayes, 1977). Strains are not part of the normal human flora, but isolates are obtained from the upper respiratory tract, the gastrointestinal tract, urogenital tract, various wounds or ulcers, and blood (see review by von Graevenitz, 1981). Although in early reports *F. meningosepticum* was predominantly associated with meningitis in infants it has subsequently been identified in a wide range of clinical specimens (Holmes and Owen, unpublished results; Richard et al., 1979b).

Enrichment and Isolation Procedures

Special procedures are not normally required for the isolation of *Flavobacterium* strains; however, growth of environmental strains may be slow on nutrient media and 5- to 7-days incubation may be required for full development of colonies. Unlike all other *Flavobacterium* species, *F. aquatile* (ATCC 11947) grows poorly in nutrient broth and on nutrient agar containing meat extract and peptone. However, this strain grows well on a medium containing an enzymatic digest of casein or gelatin, supplemented with glucose and yeast extract (Weeks, 1974).

Maintenance Procedures

Stock cultures of most species may be maintained for several months on a slant of Dorset egg medium (Cowan, 1974) in a metal screw-capped bijou bottle stored at 4°C. Cultures can be maintained for longer periods by suspending growth from an 18-h-old agar slant culture in defibrinated rabbit blood, transferring to a small tube (plugged or capped) and freezing in a mixture of dry ice and alcohol prior to storage at −50°C. Cultures may also be freeze-dried without loss of viability.

Procedures for Testing for Special Characters

Morphology and Gram-stain. Microscopic examinations should be carried out on cultures grown on solid media and in liquid media. Because some Gram-positive bacteria are easily decolorized by usual Gram-staining procedures and may appear Gram-negative, Weeks (1974) recommended the use of the Kopeloff-Beerman modification of the Gram-strain. Staining may be complemented by testing for growth on crystal violet agar (concentrations of 2.5×10^{-5} and 1×10^{-6}, w/v) or on sodium dodecyl sulfate (SDS) (1%, w/v) agar. Treatment of a thick cell suspension with a few drops of a 10% (w/v) solution of SDS or of a 3% (w/v) solution of potassium hydroxide is also a simple confirmatory test, because cells of Gram-negative bacteria are rapidly lysed with a marked increase in viscosity.

Gliding motility and spreading growth. The demonstration of these two characteristics, which are important in differentiating between *Flavobacterium* and *Cytophaga* or *Flexibacter*, can present difficulties. It is essential to culture the organism on agar with a low nutrient concentration such as the medium of Anacker and Ordal (1959) which contains (g/liter): tryptone (Difco), 0.5; yeast extract, 0.5; sodium acetate, 0.2; beef extract, 0.2, and agar (Difco), 9.0; adjusted to pH 7.2–7.4. Gliding motility is also affected by the amount of surface moisture on the medium and best results are obtained with freshly poured media incubated in a humid atmosphere (Henrichsen, 1972). Gliding may be studied by direct microscopy of the colony edge on thinly poured agar in a Petri dish following overnight incubation. Henrichsen (1972) recommends the use of a high power dry lens, arguing that the addition of a coverslip to allow observation under oil introduces additional surface forces which may lead to the formation of artefacts. Hayes (1977) observes under oil, and Perry (1973) additionally recommends the use of small glass beads placed on the agar to provide pools of liquid of varying depths.

The interpretation, and hence the definition, of gliding varies. Henrichsen (1972) defines gliding as movement which "is continuous and regularly follows the long axis of the cells which are predominantly aggregated in bundles during the movement." Perry's definition appears to be somewhat wider and apparently includes strains showing unrestricted movement on a solid substratum without necessarily continuous directional movement. Thus, Perry (1973) using his glass bead technique recorded *F. aquatile* (NCIB 8694, ATCC 11947) as a glider, a view not shared by other workers such as Henrichsen (personal communication) and Hayes (1977). For the present, Henrichsen's definition of gliding motility is recommended as this includes only those isolates which show unequivocal gliding motility.

Isolates exhibiting obvious gliding motility might also be expected to produce spreading colonies on low nutrient concentration media, but these two characteristics are not always correlated. Henrichsen (1972) reported that spreading may result from a variety of surface translocation mechanisms. Several workers recorded an apparently anomalous result for NCIB 9059, "*Flavobacterium pectinovorum*," now considered a junior synonym of *Cytophaga johnsonae* (Christensen, 1977), which shows true gliding but fails to spread (Lund, 1969; McMeekin et al., 1971; Perry, 1973; Hayes, 1977). Gliding motility, therefore, cannot be inferred with certainty from a spreading habit, and must be confirmed by direct microscopic observation.

DNA base composition. The accurate estimation of the mol% G + C of DNA requires special equipment. The fact that ultraviolet (UV) light sensitivity appears to be correlated with the mol% G + C (McMeekin, 1977) provides a simple alternative method to allow presumptive separation of the low G + C ($\leq$ 42%) strains corresponding to *Flavobacterium* from other yellow-pigmented strains with higher G + C contents. Samples (5 μl) of a dilution series prepared from 24- to 48-h-old nutrient broth cultures are applied in duplicate on nutrient agar and one of each pair is exposed to UV radiation. McMeekin (1977) used an exposure time of 90 s, with the culture placed 55 cm from a 15-watt germicidal lamp in a laminar flow cabinet. This resulted in at least a 10^5-fold decrease in survivors for strains with low G + C content, and a <10^3-fold decrease in survivors for strains with high G + C content. As different UV outputs may be emitted from different lamps, it is necessary to determine the optimum exposure time for each system with control organisms of known G + C contents. A simple qualitative test may be performed by streaking the organism on nutrient agar and protecting half the agar from UV light. Low G + C organisms grow only on the shaded portion.

Differentiation of the genus **Flavobacterium** from other taxa

Table 4.94 displays characteristics useful for differentiation of *Flavobacterium* from morphologically or physiologically similar taxa.

Taxonomic Comments

The main criteria for including a species in *Flavobacterium* as originally described by Bergey et al. (1923) were the formation of yellow- or orange-pigmented colonies on culture media and the ability to produce acid weakly from carbohydrates. Because the genus was established on color and so few other characteristics, it was taxonomically heterogeneous from the start and contained Gram-negative species that were either motile by peritrichous or polar flagella or nonmotile, as well as some Gram-positive species. The degree of heterogeneity in the genus was not reduced until the fifth edition of the *Manual* (Bergey et al., 1939), from which species containing polarly flagellated strains were excluded. Gram-positive species were excluded from the seventh edition of the *Manual* (Weeks and Breed, 1957). In the eighth edition of the *Manual* (Weeks, 1974), the genus was further restricted to species not showing gliding or spreading, and was divided into two sections. Section I contained the nonmotile species with mol% G + C contents of 30–42 (low G + C strains), and section II contained species that were either nonmotile or motile with peritrichous flagella and which had mol% G + C contents of 63–70 (high G + C strains). The definition given in the present edition of the *Manual* further restricts *Flavobacterium* to low G + C content strains.

F. aquatile is the present type species. The problems of its acceptability as such are discussed fully elsewhere (McMeekin and Shewan, 1978; Holmes and Owen, 1979), and a request has been made for the name to be rejected as a *nomen dubium* (Holmes and Owen, 1979). It has been suggested that *F. breve* is a logical choice to replace *F. aquatile* as type species. *F. breve* was included in the original description of *Flavobacterium* (Bergey et al., 1923) and, apart from *F. aquatile*, was the only original species retained by Weeks (1974). The description of *F. breve* has recently been revised (Holmes et al., 1978), and the species is represented by several reference strains and a neotype strain (NCTC 11099). A proposal has been made to effect revival of the name *F. breve* (Holmes and Owen, 1982, Int. J. Syst. Bacteriol., *32*: 233–234). Moreover, there is no evidence of a spreading or gliding motility that would suggest an affinity to typical cytophagas.

The main taxonomic argument for replacing *F. aquatile* rests on the debatable question of its motility and flagella. Bergey et al. (1923) described *F. aquatile* as being motile by peritrichous flagella, which was contrary to the original description of the species by Frankland and Frankland (1889). There are various conflicting reports about the motility and flagella of strain F36 (ATCC 11947) the type strain of *F. aquatile*, as indicated in Further Descriptive Information and in Procedures for Testing Special Characters. This strain was reisolated from purportedly the same source from which Frankland and Frankland (1889) isolated the species. The evidence from chemotaxonomic studies of rRNA cistrons (see below) does indicate, however, that *F. aquatile* is much more closely related to some *Cytophaga* strains than are *F. breve* and the various other species herein placed in *Flavobacterium*. This is supported to some extent by numerical taxonomic studies of phenotypic characteristics (Hayes, 1977) and cellular fatty acids (Oyaizu and Komagata, 1981). Callies and Mannheim (1978) originally reported *F. aquatile* to contain ubiquinones, a feature at variance with other low G + C strains; however, the species has been shown subsequently to produce menaquinones (Oyaizu and Komagata, 1981; Mannheim, personal communication). However, despite the arguments presented, the Judicial Commission (L. G. Wayne, Int J Syst. Bacteriol. *32*: 464–465, 1982) has declined to issue the opinion requested by Holmes and Owen (1979).

At the suprageneric level, *Flavobacterium* as herein defined appears to have closer similarities to members of the family *Cytophagaceae* than to any other group. There is considerable biochemical and chemotaxonomic evidence of the similarity between *Flavobacterium* and *Cyto-* *phaga*, particularly that derived from mol% G + C contents, respiratory quinones and cellular fatty acid compositions. These similarities lead to problems when one attempts to distinguish between the two genera. Investigation of rRNA cistrons has, however, provided a valuable insight into the suprageneric affinities of the members of the so-called *Flavobacterium-Cytophaga* complex. Bauwens and De Ley (1981) showed by DNA/rRNA hybridization experiments that the low G + C flavobacteria are very heterogeneous among themselves and comprise at least five subgroups. One subgroup contains *F. aquatile*, "*F. pectinovorum*," "*F. tirrenicum*" and *C. johnsonae*. A second subgroup contains "*C. marinoflava*," *C. lytica*, *C. salmonicolor*, *F. uliginosum* and *Flexibacter aurantiacus* subsp. *excathedrus*. *F. odoratum*, *F. breve* and *F. meningosepticum* each belong to separate subgroups. Three of the *Flexibacter* species were considered to be taxonomically quite different from the *Flavobacterium-Cytophaga* complex. The exact taxonomic position of *Flavobacterium heparinum*, *F. ferrugineum* and *F. okeanokoites* is uncertain. On the basis of these results, Bauwens and De Ley (1981) stated that "the authentic genus *Flavobacterium* will have to consist of all or some of the above named flavobacteria," and concluded that it was not possible to distinguish unequivocally between *Cytophaga* and *Flavobacterium* solely on the basis of rRNA cistron similarity because the strains are so heterogeneous. All the other named *Flavobacterium* species examined were considered to belong in different taxonomic areas.

Fourteen species that appear on the Approved Lists of Bacterial Names (Skerman et al., 1980) are now excluded from the genus *Flavobacterium*; these are listed under *Species Incertae Sedis*; three other species listed by Weeks (1974) which were not placed on the Approved Lists and therefore have no standing in nomenclature, are also listed in this section.

Of the six species placed in *Flavobacterium* section I by Weeks (1974), only three (*F. aquatile*, *F. meningosepticum* and *F. breve*) have been retained as acceptable members of the genus. Six species were placed by Weeks (1974) in section II, but none of these are now retained in *Flavobacterium* although recent data indicate that "*F. tirrenicum*" could be revived if necessary. The only approved species with an uncertain place in the present arrangement is *F. indoltheticum*. Recent studies using numerical taxonomic analysis and DNA base compositions indicate that this species is closely similar to *F. breve* and would be indistinguishable from it by the range of tests given here (Table 4.95). On taxonomic grounds, its exclusion may be warranted because it could be considered a later synonym of *F. breve*.

The six named species now placed in *Flavobacterium* and the new species *F. spiritivorum* are, with the exception of *F. aquatile* and *F. balustinum*, all phenotypically well described and are each represented by a number of strains which have been deposited in major culture collections. The DNA results show that group IIb, which phenotypically is virtually indistinguishable from *F. balustinum* (Tables 4.95 and 4.96) is a relatively heterogeneous taxon which may comprise several species (Owen and Snell, 1976; Owen and Holmes, 1981). Studies on the chromosomal DNA characteristics, in particular mol% G + C contents and DNA/DNA hybridization experiments, have provided valuable information about the relationships within and between the various species of *Flavobacterium*. The species *F. breve* and *F. spiritivorum* represent respectively the lower and upper limits of the mol% G + C values for the genus and, with a mol% G + C difference of about 10%, these two species are unlikely to share significant amounts of common G + C regions. The spread in G + C content is acceptable for a bacterial genus.

In general, the intraspecific DNA/DNA hybridization results on *F. breve*, *F. meningosepticum* and *F. odoratum* show that these species comprise a "core" of highly related strains (Owen and Holmes, 1978, 1980, 1981). The majority of strains are typical in that they belong to this core, but there are a number of "outliers" or atypical strains, with varying degrees of similarity to the typical strains. For example, the

Table 4.94.

Differential characteristics of the genus **Flavobacterium** *and other morphologically or physiologically similar taxa[a]*

Characteristics	Gliding Motility	Functional Flagella	Fermentative in Glucose O-F Medium	Catalase Production	Oxidase Production	Penicillin Susceptibility	Gram Reaction
Flavobacterium	−	−	−	+	+	−	−
Cytophaga	+						
Flexibacter	+						
Agrobacterium		+					
Alcaligenes		+					
Alteromonas		+					
Bordetella bronchiseptica		+					
Pseudomonas		+					
Xanthomonas		+					
Enterobacteriaceae			+				
Pasteurellaceae			+				
Vibrionaceae			+				
Kingella				−			
Acinetobacter					−		
Bordetella parapertussis					−		
Branhamella catarrhelis[b]						+	
Moraxella[c]						+	
Neisseria						+	
Coryneforms							+

[a] For symbols see standard definitions.
[b] *Moraxella (Branhamella) catarrhalis.*
[c] *Moraxella* subgenus *Moraxella.*

Table 4.95.

Differential characteristics of the species of the genus **Flavobacterium**[a]

Characteristics	1. F. aquatile	2. F. breve	3. F. balustinum	4. F. meningosepticum	5. F. odoratum	6. F. multivorum	7. F. spiritivorum
Acid produced aerobically from carbohydrates:[b]							
Glucose	+[c]	d	+(+)	d	−	+	+
Arabinose	−	−	−(d)	−	−	+	d
Cellobiose	−	−	−(−)	−	−	+	+
Ethanol	−	−	+(d)	d	−	−	+
Lactose	+[c]	−	−(−)	d	−	+	+
Mannitol	−	−	−(−)	d	−	−	+
Raffinose	−	−	−(−)	−	−	+	+
Salicin	−	−	−(−)	−	−	+	+
Sucrose	+[c]	−	−(d)	−	−	+	+
Xylose	−	−	−(d)	−	−	+	+
Casein digestion	+	+	+(+)	+	+	−	−
Esculin hydrolysis	−	−	+(+)	+	−	+	+
Indole production	−	+	+(+)	d	−	−	−
Nitrite reduction	NG	−	−(d)	d	+	−	−
Starch hydrolysis	NG	−	−(d)	−	−	−	−
Urease production	−	−	−(d)	d	+	+	+
β-Galactosidase production (ONPG)	−	−	−(d)	+	−	+	+

[a] Symbols: see standard definitions; and NG, no growth on test medium. Results in parentheses are for the phenotypically similar group IIb.
[b] Tested in ammonium salt medium.
[c] Delayed reaction.

type strain of *F. meningosepticum* (NCTC 10016) is atypical of the species and shares only 30% of its DNA sequences with most other *F. meningosepticum* strains, including strains of the same serovar (Sottile et al., 1973; Owen and Snell, 1976; Callies and Mannheim, 1980). Two distinct DNA-relatedness groups are apparent within *F. odoratum* and this species may warrant division into two or three species (Owen and Holmes, 1978). However, there is considerable genome heterogeneity within group IIb (Owen and Snell, 1976; Owen and Holmes, 1981); eventually this group may warrant separation into several new species. Levine et al. (1980) noted some DNA heterogeneity within their strains of group IIk, biovar 2 (*F. multivorum*).

Interspecies DNA/DNA hybridizations indicate that each of the

Table 4.96.
Other characteristics of the species of the genus **Flavobacterium**[a]

Characteristics	1. F. aquatile	2. F. breve	3. F. balustinum	4. F. meningo-septicum	5. F. odoratum	6. F. multi-vorum	7. F. spiriti-vorum
Catalase production	+	+	+(+)	+	+	+	+
Good growth on nutrient agar	−	+	+(+)	+	+	+	+
Growth at room temperature (18–22°C)	+	+	+(+)	+	+	+	+
Growth on:							
MacConkey agar	−	+	+(+)	+	+	+	+
β-Hydroxybutyrate	−	+	+(+)	+	+	+	+
Hydrolysis of:							
Tributyrin	NG	+	+(+)	+	+	+	+
Tween 20	+[b]	+	+(+)	+	+	+	+
Oxidase production	+	+	+(+)	+	+	+	+
Phosphatase production	+[b]	+	+(+)	+	+	+	+
Acid produced aerobically[c] from:							
Adonitol, dulcitol, inositol, sorbitol	−	−	−(−)	−	−	−	−
Arginine desimidase	−	−	−(−)	−	−	−	−
Arginine dihydrolase	−	−	−(−)	−	−	−	−
Fluorescent pigment produced on King's medium B	NG	−	−(−)	−	−	−	−
Gas from glucose[d]	NG	−	−(−)	−	−	−	−
Gluconate oxidation	NG	−	−(−)	−	−	−	−
H₂S production	NG	−	−(−)	−	−	−	−
KCN tolerance	−	−	−(−)	−	−	−	−
Lysine or ornithine decarboxylase	−	−	−(−)	−	−	−	−
Malonate utilization	−	−	−(−)	−	−	−	−
Motility	−	−	−(−)	−	−	−	−
Phenylalanine deaminase	−	−	−(−)	−	−	−	−
Poly-β-hydroxybutyrate inclusion granules	NG	−	−(−)	−	−	−	−
Reduction of 0.4% (w/v) selenite	−	−	−(−)	−	−	−	−
3-Ketolactose production	−	−	−(−)	−	−	−	−
Acid produced aerobically[c] from:							
Glycerol, trehalose	−	−	−(d)	d	−	+	+
Fructose	−	−	+(d)	d	−	+	+
Maltose	+[b]	d	−(+)	+	−	+	+
Rhamnose	−	−	−(−)	−	−	d	−
Acid produced from glucose[d]	NG	−	−(−)	−	−	d	d
Acid from 10% (w/v) glucose	−	−	−(d)	d	−	+	+
Acid from 10% (w/v) lactose	−	−	−(−)	−	−	+	d
Alkali production on Christensen's citrate	NG	−	−(d)	d	−	−	−
Deoxyribonuclease production	NG	+	+(d)	+	+	d	+
Gelatin hydrolysis	−	+	+(+)	+	+	d	+
Growth at:							
5°C	−	−	−(d)	−	−	−	−
37°C	−	d	+(+)	+	+	+	+
42°C	−	−	−(d)	d	−	−	−
Growth on:							
Cetrimide agar	−	−	−(−)	d	−	−	−
Simmons' citrate agar	−	−	−(−)	d	−	−	−
Nitrate reduction	NG	−	+(d)	−	−	−	−
Opalescence on lecithovitellin agar	−	−	+(d)	d	−	−	−
Oxidative in glucose O-F medium	−	d	+(d)	+	−[e]	+	+
Pigment production on tyrosine agar	−	−	+(d)	d	d	−	−
Production of yellow pigment	+	+	+(+)	d	+	+	d
Tween 80 hydrolysis	−	d	+(+)	d	d	+	+
Tyrosine hydrolysis	−	−	−(d)	d	d	−	−

[a] Results in parentheses are for the phenotypically similar group IIb. For symbols, see Table 4.95.
[b] Delayed reaction.
[c] Tested in an ammonium salt medium.
[d] Tested in a peptone water medium.
[e] Strains produce an alkaline reaction.

species of *Flavobacterium* constitutes a separate and distinctive DNA group, although there is a background level of hybridization of about 20% which might represent a common DNA complement (Callies and Mannheim, 1980; Owen and Holmes, 1981). The evidence of DNA characteristics is that several *Flavobacterium* species contain distinct subgroups which might eventually justify separation into new species. However, virtually no phenotypic characteristics, and these include new approaches such as the use of commercial enzyme tests, have been found yet that correlate with the genomic differences. Subdivision of the species at present would serve no useful purpose.

The species herein placed in *Flavobacterium*, except *F. aquatile* and *F. balustinum*, are described mainly from isolates from the clinical environment where selection pressures tend to be constant, thus providing more discrete species. Because of changing environmental conditions nonclinical isolates do not form such discrete species and their allocation to the species herein described is less certain. As defined here, the genus *Flavobacterium* has been modified extensively so it is difficult to anticipate what changes there may be in the future. The present arrangement is somewhat tentative because of the acknowledged heterogeneity of the *Flavobacterium-Cytophaga* complex. As more studies are made, the need for revision and new genera within the complex will undoubtedly arise. The decision of the Judicial Commission not to replace *F. aquatile* as the type species of *Flavobacterium* together with the DNA/rRNA data of Bauwens and De Ley (1981) means that *Flavobacterium* should strictly be comprised only of the single species *F. aquatile* which is, itself, represented by a single strain. Only *Cytophaga johnsonae* and "*Flavobacterium tirrenicum*" are perhaps closely enough related to *F. aquatile* to be in the same genus. In Table 4.95 the named species are arranged to illustrate how they might be reclassified along the lines discussed previously (Holmes and Owen,

1981). In such a scheme, one prospective new genus would be constituted by the species *F. breve*, *F. balustinum*, *F. meningosepticum*, and group IIb. Another prospective new genus could contain *F. odoratum*, strains of which differ from those of the majority of *Flavobacterium* species in being nonsaccharolytic and in failing to produce indole. The DNA hybridization data already indicate the likelihood of several species allied to *F. odoratum*. This species also differs from other *Flavobacterium* species in its cellular fatty acid composition (Oyaizu and Komagata, 1981). The third prospective genus could contain *F. multivorum* and *F. spiritivorum*, strains of which are saccharolytic but differ from those of the majority of *Flavobacterium* species in failing to produce indole and in being less actively proteolytic. There is additional evidence that group IIk, biovar 2 (*F. multivorum*) is unique in that it contains sphingophospholipids, the novel nature of which is a feature that Yabuuchi et al. (1981) believe would be evidence for a new genus. Chemotaxonomic studies, in particular rRNA cistron similarities, have played an important role in providing a rational basis for the revision of *Flavobacterium* and it may well be that the future genera within the *Flavobacterium-Cytophaga* complex will be based on the rRNA subgroups.

Further Reading

Holmes, B. and R.J. Owen, 1979. Proposal that *Flavobacterium breve* be substituted as the type species of the genus in place of *Flavobacterium aquatile* and emended description of the genus *Flavobacterium*: status of the named species of *Flavobacterium*. Request for an opinion. Int. J. Syst. Bacteriol. 29: 416–426.

McMeekin, T. A. and J. M. Shewan, 1978. Taxonomic strategies for *Flavobacterium* and related genera. J. Appl. Bacteriol. 45: 321–332.

Reichenbach, H. and O.B. Weeks (Editors). 1981. The *Flavobacterium-Cytophaga* Group (Proceedings of the International Symposium on Yellow-Pigmented Gram-Negative Bacteria of the *Flavobacterium-Cytophaga* Group, Braunschweig July 8 to 11, 1980). Verlag Chemie, Weinheim, pp. 1–217.

Differentiation and characteristics of the species of **Flavobacterium**

The differential characteristics of the named species of *Flavobacterium* are given in Table 4.95. Other characteristics of the species appear in Table 4.96. The methods used are described by Holmes et al. (1975, 1981).

List of the species of the genus **Flavobacterium**

1. **Flavobacterium aquatile** (Frankland and Frankland 1889) Bergey, Harrison, Breed, Hammer and Huntoon 1923, 100.[AL] (*Bacillus aquatilis* Frankland and Frankland 1889, 381.)

a.qua'ti.le. L. neut. adj. *aquatile* living in water.

Rods 0.5–0.7 × 1.0–3.0 µm, approaching a coccobacillary form in young cultures. Longer rods and filamentous forms occur in liquid and on solid media.

Discrete colonies form on solid media at 25–30°C. Swarming of colonies has been reported for cultures maintained at 15°C (Mitchell et al., 1969) and gliding movement of cells has been demonstrated using a special microscopic technique (Perry, 1973). Pigmentation of colonies is light yellow-brown at 30°C and bright orange at 15–20°C. Colonies are 1–3 mm in diameter, smooth, entire, glistening and transparent, but more mucoid and spreading at 15°C.

Unlike other *Flavobacterium* species, *F. aquatile* will not grow well on nutrient agar and it requires special growth media (see Enrichment and Isolation Procedures).

Growth does not occur under anaerobic conditions, at 37°C, or in media containing more than 1% added NaCl.

The physiological characteristics as given in Tables 4.95 and 4.96 were derived from Holmes and Owen (unpublished results).

Isolated from a deep well in the chalk region of Kent, England.

The mol% G + C of the DNA is 32.0 (T_m).

Type strain: ATCC 11947 (NCIB 8694). *Reference strains:* there are no other authentic strains available.

2. **Flavobacterium breve** (Lustig 1890) Holmes and Owen 1982, 233.[VP] (*Bacillus brevis* Lustig 1890, 52).

bre've. L. neut. adj. *breve* short.

Rods 0.5 × 1.0–2.0 µm, but some longer rods may be present.

Nonmotile. Swarming of colonies has not been reported.

Pigmentation is light yellow and the hue does not change with variation of medium and temperature.

Colonies on nutrient agar at 30°C have an entire edge and are low convex, circular, smooth and shining. The colony size is pinpoint to 2.0 mm on nutrient agar; colonies on blood agar are pinpoint to 2.5 mm and nonhemolytic.

Although most strains grow at 37°C, the majority will only produce acid from carbohydrates when incubated at 30°C and prolonged incubation may be necessary.

The physiological characteristics of the species as given in Tables 4.95 and 4.96 were derived from Holmes et al. (1978).

The mol% G + C of the DNA is 32.4 ± 0.6 (T_m).

Type strain: NCTC 11099 (strain CL88/76 of Holmes et al., 1978).

Reference strains: NCTC 11162 (strain CL626/75 of Holmes et al., 1978); NCTC 11163 (strain CL669/76 of Holmes et al., 1978); ATCC 14234.

3. **Flavobacterium balustinum** Harrison 1929, 233.[AL]

ba.lus.ti'num. Etymology uncertain but possibly derived from the L. n. *balux* gold ore or gold dust.

Rods 0.5 × 1.0–1.8 µm. Neither cellular gliding movement nor swarming growth is evident.

Always brightly yellow pigmented.

Colonies on nutrient agar at 30°C are typically circular, 2 mm in diameter, entire, and viscid or butyrous in consistency. Colonies become mucoid after 2–3 days and are nonhemolytic on blood agar.

The physiological characteristics of the species as given in Tables

4.95 and 4.96 were derived from Holmes and Owen (unpublished results).

The mol% G + C of the DNA is 33.1 (T_m).

Type strain: NCTC 11212 (La 724).

Reference strains: there are no other authentic strains available. Since strains of group IIb (mol% G + C of the DNA is 35.0–38.5 [T_m]) are phenotypically similar to the type strain of *F. balustinum* the following reference strains should be included in future studies of *F. balustinum*: NCTC 11390 (A140/68), NCTC 11409 (CL42/78); these strains are particularly typical of group IIb; NCTC 10795 (group IIb strain 3531); NCTC 10796 (group IIb strain 3716).

4. Flavobacterium meningosepticum King 1959, 247.[AL]

me.nin.go.sep'ti.cum. Gr. n. *meninx, meningos* meninges, membrane covering the brain; Gr. adj. *septikos* putrefactive; M.L. adj. *meningosepticum* apparently referring to association of the bacterium with both meningitis and septicemia and not to septic meningitis as the name implies.

Rods 0.5 × 1.0–2.0 μm, slender and slightly curved or short with rounded ends. Filaments are common. Some strains are encapsulated, especially following animal passage (mice). Neither cellular gliding movement nor swarming has been reported.

Strains are frequently nonpigmented and, when yellow-pigmented strains are encountered, the degree of pigmentation is generally slight.

Colonies on blood agar at 37°C vary from punctiform to 2 mm in diameter (usually 1.0–1.5 mm) and are smooth, entire, glistening, translucent, butyrous and gray-white. Blood agar is not hemolyzed, but the medium may show a green discoloration.

The physiological characteristics of the species as given in Tables 4.95 and 4.96 were derived from Holmes and Owen (unpublished results).

The mol% G + C of the DNA is 37.0 ± 0.5 (T_m).

Type strain: NCTC 10016 (strain 14 of King, 1959; serovar A). *Reference strains:* NCTC 10585 (strain 422 of King, 1959; serovar B); NCTC 10586 (strain 3375 of King, 1959; serovar C); NCTC 10587 (serovar D); NCTC 10588 (serovar E); NCTC 10589 (serovar F); NCTC 11305 (CIP 78.30; Richard et al., 1979a, b; serovar G); NCTC 11306 (CIP 79.05; Richard et al., 1979b; serovar H); NCTC 11309 (CIP 79.29; Richard et al., 1979b; serovar K).

The following strains deposited in the National Collection of Type Cultures (NCTC) as representing, respectively, the new serovars I, J and L of *F. meningosepticum* (Richard et al., 1979b) appear phenotypically to be strains of *F. breve*: NCTC 11307, NCTC 11308 and NCTC 11310.

5. Flavobacterium odoratum Stutzer *in* Stutzer and Kwaschnina 1929, 221.[AL]

o.do.ra'tum. L. adj. *odoratus* perfumed.

Rods 0.5 × 1.0–2.0 μm, but longer rods and long chains (4–10 cells) may occur in broth medium. Neither cellular gliding movement nor swarming growth has been reported.

Only rarely not yellow pigmented.

Various colony types occur. Colony type 1 is effuse with spreading edges, 3–4 mm in diameter, dull and matt, but on further incubation the colonies become smooth and shiny. Colony type 2 is similar to type 1 at 24 h but is smaller (~1.0–1.5 mm in diameter). Type 3 colonies are smooth, shiny and convex with no spreading edge and with a diameter of ≤1.0 mm at 24 h; on further incubation the colonies resemble those of type 1. Colony variation correlates with genomic differences (Owen and Holmes, 1978). Nonhemolytic on blood agar.

A characteristic fruity odor is produced. Nitrite is characteristically reduced, but not nitrate. Strains of the species differ from those of the majority of *Flavobacterium* species in being nonsaccharolytic.

The physiological characteristics of the species as given in Tables 4.95 and 4.96 are derived from Holmes et al. (1977, 1979).

The mol% G + C of the DNA is 31.4–36.1 (T_m).

Type strain: NCTC 11036 (ATCC 4651; an original strain of Stutzer). *Reference strains:* NCTC 11179 (strain CL41/66 of Holmes et al., 1977);

NCTC 11180 (strain CL229/67 of Holmes et al., 1977).

6. Flavobacterium multivorum Holmes, Owen and Weaver 1981, 25.[VP]

mul.ti.vo'rum. L. adj. *multus* many; L. trans. v. *vorare* to swallow; M.L. adj. *multivorus* produces acid from many carbohydrates.

Rods 0.5 μm in diameter and 1.0–2.0 μm in length. Cellular gliding movement or swarming growth has not been reported.

A pale yellow pigment is produced that becomes clearly yellow at 48-h incubation.

Colonies on nutrient agar at 30°C are circular, 1 mm in diameter, low convex, smooth and opaque. Nonhemolytic on blood agar.

Strains of the species differ from those of the majority of *Flavobacterium* species in failing to produce indole and in not being actively proteolytic.

The physiological characteristics of the species are given in Tables 4.95 and 4.96 and were derived from Holmes et al. (1981).

The mol% G + C of the DNA is 39.6 ± 0.5 (T_m).

Type strain: NCTC 11343 (Group IIk, biovar 2 strain B5533). *Reference strains:* NCTC 11033 (Group IIk, biovar 2 strain A8895); NCTC 11034 (Group IIk, biovar 2 strain B3159).

7. Flavobacterium spiritivorum Holmes, Owen and Hollis 1982, 159.[VP]

spi.ri.ti'vo.rum. L. n. *spiritus* spirit; L. trans. v. *vorare* to swallow; M.L. adj. *spiritivorus* spirit-devouring, referring to its ability to produce acid from spirits or alcohols.

Rods 0.5 μm in diameter and 1.0–1.4 μm in length. Cellular gliding movement or swarming growth is not known to occur.

Occasional strains are nonpigmented; in other strains a pale yellow pigment becomes clearly yellow at 48 h of incubation.

Colonies on nutrient agar at 30°C are typically circular, 1–3 mm in diameter, low convex, smooth and opaque.

Strains of the species differ from those of the majority of *Flavobacterium* species in failing to produce indole and in not being actively proteolytic.

The physiological characteristics of the species as given in Tables 4.95 and 4.96 are derived from Holmes, Owen and Hollis (1982).

The mol% G + C of the DNA is 41.4 ± 0.4 (T_m).

Type strain: NCTC 11386 (strain E7288). *Reference strains:* NCTC 11387 (strain A14/65); NCTC 11388 (strain CL48/80).

Species Incertae Sedis

The following species appear on the Approved Lists of Bacterial Names (Skerman et al., 1980) but are excluded from the genus *Flavobacterium* as presently defined for the reasons given below.

a. *Flavobacterium acidificum* Steinhaus 1941, 772.[AL]

ac.i.di'fi.cum. L. adj. *acidus* acid, sour; L. v. *facere* to make; M.L. adj. *acidificum* making acid.

Motile. Although described as Gram-positive (Weeks, 1974) the species is in fact Gram-negative and the type strain is considered to be a strain of *Erwinia herbicola* subsp. *ananas*.

The mol% G + C of the DNA is 52.7 (Callies and Mannheim, 1978).

Type strain: ATCC 8366 (NCIB 9891).

b. *Flavobacterium acidurans* Millar 1973, 147.[AL]

a.ci.du'rans. L. adj. *acidus* sour, acid; L. part. adj. *durans* resisting; M.L. adj. *acidurans* acid-resisting.

The mol% G + C of the DNA is 66.3.

Type strain: ATCC 27383.

c. *Flavobacterium capsulatum* Leifson 1962, 163.[AL]

cap.su.la'tum. L. n. *capsula* a small chest, capsule; M.L. neut. adj. *capsulatum* encapsulated.

Placed in section II of *Flavobacterium* by Weeks (1974). Ribosomal RNA cistron analysis places it in a separate rRNA superfamily that includes *Pseudomonas paucimobilis* as its nearest taxon (Bauwens and De Ley, 1981).

The mol% G + C of the DNA is 65.1 (T_m).

Type strain: ATCC 14666 (NCIB 9890).

d. *Flavobacterium devorans* (Zimmermann 1890) Bergey, Harrison, Breed, Hammer and Huntoon 1923, 102.[AL]

de'vo.rans. L. part. adj. *devorans* consuming, devouring.

Motile. Placed in section II of *Flavobacterium* by Weeks (1974). The type strain was identified as *Pseudomonas paucimobilis* by Yabuuchi et al. (1979). This is supported by rRNA cistron analysis which places it in the same rRNA superfamily as *F. capsulatum* (Bauwens and De Ley, 1981).

The mol% G + C of the DNA is 65.2 (T_m).

Type strain: ATCC 10829 (NCIB 8195).

e. *Flavobacterium esteraromaticum* (Omelianski 1923) Bergey, Harrison, Breed, Hammer and Huntoon 1930, 150.[AL]

es.ter.a.ro.ma'ti.cum. German, *ester* probably synthesis of *essig* vinegar, *äther* ether and *säure* acid; Gr. adj. *aromatikos* sweet-smelling; M.L. neut. adj. *esteraromaticum* smelling sweet due to esters.

Gram-positive and probably a coryneform.

The mol% G + C of the DNA is 69.

Type strain: ATCC 8091.

f. *Flavobacterium ferrugineum* Sickles and Shaw 1934, 429.[AL]

fer.ru.gi'ne.um. L. neut. adj. *ferrugineum* resembling iron rust, dark red.

Placed in section I of *Flavobacterium* by Weeks (1974). Contains menaquinones (Callies and Mannheim, 1978). The taxonomic position based on rRNA cistron similarity is uncertain (Bauwens and De Ley, 1981) but is not allied to the main rRNA subgroups in the *Flavobacterium-Cytophaga* complex.

The mol% G + C of the DNA is 48.6 (T_m).

Type strain: ATCC 13524.

g. *Flavobacterium halmophilum* Elazari-Volcani 1940, 85.[AL]

halm.o.phi'lum. Gr. n. *halme* brine, seawater; Gr. adj. *philos* loving; M.L. neut. adj. *halmophilum* seawater loving.

Placed in section I of *Flavobacterium* by Weeks (1974). Respiratory quinones are ubiquinones (Callies and Mannheim, 1978). Belongs in a separate rRNA superfamily with closest affinities to "*Alcaligenes aquamarinus*" (Bauwens and De Ley, 1981).

The mol% G + C of the DNA is 49.7 (Callies and Mannheim, 1978).

Type strain: ATCC 19717.

Further comments. The name of this species is given incorrectly as *F. halmephilium* on the Approved Lists of Bacterial Names.

h. *Flavobacterium heparinum* Payza and Korn 1956, 854.[AL]

he.pa.ri'num. Gr. n. *hepar* liver; M.L. neut. adj. *heparinum* pertaining to the liver.

Gliding motility has been reported (Perry, 1973). Transferred to *Cytophaga* as *C. heparina* by Christensen (1980), but the taxonomic position is uncertain from rRNA cistron analysis (Bauwens and De Ley, 1981).

The mol% G + C of the DNA is 45.6 (T_m).

Type strain: ATCC 13125.

i. *Flavobacterium indoltheticum* Campbell and Williams 1951, 903.[AL]

in.dol.the'ti.cum. M.L. n. *indolum* indole; Gr. adj. *theticus* positive; M.L. neut. adj. *indoltheticum* indole-positive.

Originally described as motile. Placed in section II of *Flavobacterium* by Weeks (1974). The type strain is nonmotile, however (see Taxonomic Comments).

The mol% G + C of the DNA is 33.8 (T_m).

Type strain: ATCC 27950.

j. *Flavobacterium marinotypicum* ZoBell and Upham 1944, 268.[AL]

ma.ri.no.ty'pi.cum. L. adj. *marinus* of the sea; Gr. adj. *typicus* conformable, typical; M.L. adj. *marinotypicum* probably intended to mean typical of the sea.

Possesses peritrichous flagella. Gram-positive.

Type strain: ATCC 19260.

k. *Flavobacterium oceanosedimentum* Carty and Litchfield 1978, 563.[AL]

o.ce.an.o.sed.i.men'tum. Gr. n. *okeanos* the ocean; L. n. *sedimentum* a settling, subsidence; M.L. neut. adj. *oceanosedimentum* intended to mean from marine sediments.

The mol% G + C of the DNA is 67.5.

Type strain: ATCC 31317.

l. *Flavobacterium okeanokoites* ZoBell and Upham 1944, 270.[AL]

o.ke.a.no.ko.i'tes. Gr. n. *okeanos* the ocean; Gr. fem. n. *coite, coites* bed; M.L. fem. gen. n. *okeanokoites* of the ocean bed.

Possesses peritrichous flagella. The taxonomic position is uncertain on the basis of rRNA cistron analysis (Bauwens and De Ley, 1981).

Type strain: CCM 320.

m. *Flavobacterium resinovorum* Delaporte and Daste 1956, 831.[AL]

re.si.no.vo'rum. Gr. n. *rhetine* resin or gum of trees; L. n. *resina* resin; L. trans. v. *vorare* to swallow; M.L. adj. *resinovorum* resin swallowing.

Possesses peritrichous flagella.

The mol% G + C of the DNA is 66.4.

Type strain: NCIB 8767.

n. *Flavobacterium uliginosum* ZoBell and Upham 1944, 263.[AL]

u.li.gi.no'sum. L. neut. adj. *uliginosum* wet, damp.

Agarolytic. Placed in section I of *Flavobacterium* by Weeks (1974). The affinities to the *Cytophaga lytica* subgroup (Bauwens and De Ley, 1981) that are evident from rRNA cistron analysis agree with the suggestion of Holmes and Owen (1981) that this species should be included in the marine agarolytic cytophagas.

The mol% G + C of the DNA is 32.

Type strain: ATCC 14397.

Several other species included in *Flavobacterium* by Weeks (1974) were not placed on the Approved Lists of Bacterial Names and therefore currently have no standing in nomenclature. These species are as follows.

o. "*Flavobacterium tirrenicum*" Marini and Spalla 1964. 37.

tir.re'ni.cum. L. adj. *tyrrhenus* of Tyrrhen, Tirrene; M.L. adj. *tirrenicum* of or pertaining to Tuscany.

Placed in section II of *Flavobacterium* by Weeks (1974) and reported as motile; however, the type strain is nonmotile. The organism is a member of the *Flavobacterium-Cytophaga* complex and has rRNA cistron similarities to the *F. aquatile* sub-group (Bauwens and De Ley, 1981).

The mol% G + C of the type strain is 34.4 (Callies and Mannheim, 1978).

Type strain: ATCC 15997.

p. "*Flavobacterium rigense*" Bergey, Harrison, Breed, Hammer and Huntoon 1923, 100.

ri.gen'se. M.L. neut. adj. *rigense* pertaining to Riga, the city where this species was isolated.

This species was placed in section II of *Flavobacterium* by Weeks (1974).

Type strain: none designated. *Reference strain*: F18 (O. B. Weeks, isolated by Fred Mindach, Indianapolis, Indiana.)

q. "*Flavobacterium lutescens*" (Migula 1900) Bergey, Harrison, Breed, Hammer and Huntoon 1923, 114.

lu.tes'cens. L. part. adj. *lutescens* becoming muddy.

This species was placed in section II of *Flavobacterium* by Weeks (1974). It belongs to another rRNA superfamily with members of the genus *Pseudomonas* (Bauwens and De Ley, 1981).

Type strain: none designated. *Reference strain*: Brisou 2611.

Other Organisms

Strains ascribed to groups IIf and IIj (Tatum et al., 1974) have a number of characteristics in common with *Flavobacterium*. They produce indole and are proteolytic. They are excluded at present from *Flavobacterium* because of their greater susceptibility to antimicrobial agents, their nonsaccharolytic nature, their failure to produce obvious yellow pigment, and the fact that they tend to be strict parasites rather than free living. Owen and Snell (1973) compared group IIf with *Flavobacterium* and *Moraxella* and failed to reach any definite conclu-

sion as to which of these two genera group IIf might be ascribed. In DNA base composition, ability to produce indole, ability to grow well on nutrient agar and ability to produce a brown melanin-like pigment on tyrosine agar, group IIf shows closer affinity to *Flavobacterium*. However, the nonsaccharolytic nature of group IIf, its susceptibility to low concentrations of penicillin and small genome size suggest affinity to *Moraxella*. Price and Pickett (1981) recommend that group IIf should be recognized as a species in *Flavobacterium*.

Genus **Alcaligenes** Castellani and Chalmers 1919, 936[AL]

KAREL KERSTERS AND JOZEF DE LEY

Al.ca.li′ge.nes. Arabic *al* the; Arabic noun *galīy* the ash of saltwort; French noun *alcali* alkali; Gr. v. gennaio to produce; M.L. masc. n. *Alcaligenes* alkali-producing (bacteria).

Rods, coccal rods or cocci, 0.5–1.0 μm in diameter and 0.5–2.6 μm in length, usually occurring singly. Resting stages not known. **Gram-negative. Motile** with 1–8 (occasionally up to 12) **peritrichous flagella. Obligately aerobic**, possessing a strictly respiratory type of metabolism with oxygen as the terminal electron acceptor. Some strains are capable of anaerobic respiration in the presence of nitrate or nitrite. Optimum temperature: 20–37°C. Colonies on nutrient agar are **non-pigmented. Oxidase-positive.** Catalase-positive. Indole not produced. Cellulose, esculin, gelatin and DNA usually not hydrolyzed. **Chemoorganotrophic, using a variety of organic acids and amino acids as carbon sources. Alkali produced** from several organic salts and amides. Carbohydrates usually not utilized. Some strains produce acid from D-glucose and D-xylose and utilize both carbohydrates as carbon source. **Occur in water and soil.** Some are common, apparently saprophytic, inhabitants of the intestinal tract of vertebrates. Numerous strains have been **isolated from clinical material such as blood, urine, feces, purulent ear discharges, spinal fluid, wounds**, etc. Occasionally causing opportunistic infections in man. The mol% G + C of the DNA is 56–70 (T_m, Bd).

Type species: *Alcaligenes faecalis* Castellani and Chalmers 1919, 936.

Further Descriptive Information

The information given here is limited to the authentic *Alcaligenes* strains comprising *A. faecalis* and *A. denitrificans* and including the formerly named "*A. odorans*," "*A. odorans* var. *viridans*," *A. ruhlandii* and "*Achromobacter xylosoxidans*" (see Taxonomic Comments). For information on the hydrogen-oxidizing and marine organisms known as *A. eutrophus*, *A. paradoxus*, *A. latus*, *A. aestus*, *A. aquamarinus*, *A. cupidus*, *A. pacificus* and *A. venustus*, consult the individual descriptions given under *Species Incertae Sedis* at the end of the chapter.

A fairly large number of physiological and biochemical studies have been published on so-called *Alcaligenes* and *Achromobacter* strains which were poorly characterized and sometimes not deposited in culture collections. Data on such strains will obviously not be used here, because some of these strains were later shown to belong to other genera such as *Acinetobacter*, *Agrobacterium* and *Pseudomonas*. The information under this heading (see also Tables 4.98 and 4.99) concerns only strains which were proven to belong either to *A. faecalis* or to *A. denitrificans*. The descriptions are mainly based on studies by De Ley et al. (1970), Gilardi (1973), Hendrie et al. (1974), Pintér and Kántor (1974), Tatum et al. (1974), Yabuuchi et al. (1974), Gilardi (1978), Pichinoty et al. (1978), Rarick et al. (1978), Rubin et al. (1980), Kiredjian et al. (1981), Yamasato et al. (1982), and Kersters, Popoff and De Ley (unpublished results). The earlier literature has been summarized by Hendrie et al. (1974). Our knowledge on the fine structure, genetics and biochemistry of the genus *Alcaligenes* is limited, probably because these bacteria lack pronounced biochemical activities and have not been thoroughly investigated.

Data on the micromorphology are available for the hydrogen-oxidizing strain ATCC 15749 (DSM 653), classified here in *A. denitrificans* subsp. *xylosoxydans* and previously named *A. ruhlandii* (see Taxonomic Comments). Its cells contain poly-β-hydroxybutyrate, polyphosphates, glycogen and mesosomes (Walther-Mauruschat et al., 1977) The peritrichous flagella of this strain are sheathed, similar to the flagellar structure of the type strain of *A. denitrificans* subsp. *xylosoxydans* (previously named *Achromobacter xylosoxidans*; see Taxonomic Comments) (Yabuuchi et al., 1974; Aragno et al., 1977).

The majority of the *Alcaligenes* strains produce a uniform turbidity in nutrient broth. After growth for 24 h on blood agar, colonies of *A. faecalis* are usually 1.0–1.5 mm in diameter, often with a characteristic flat, spreading periphery. *A. denitrificans* produces smaller colonies about 0.5 mm in diameter with an entire edge.

Most *Alcaligenes* strains will grow with ammonium salts or nitrate as the sole nitrogen source. Some strains require organic nitrogen compounds (amino acids and/or vitamins) (Hendrie et al., 1974; Pichinoty et al., 1978). Good growth is usually obtained on blood agar and peptone-containing media.

The following biochemical and nutritional tests are positive for the whole genus: oxidase; catalase; growth on acetate, succinate, fumarate, D,L-lactate, D,L-β-hydroxybutyrate, L-malate, citrate, L-alanine, L-aspartate, L-glutamate, L-proline and L-phenylalanine. The following tests are negative for the whole genus: L-arginine dihydrolase; L-lysine and L-ornithine decarboxylase; formation of pigments, indole, acetyl methyl carbinol, H_2S and 3-ketolactose; hydolysis of esculin, gelatin, lecithin, starch and DNA.

Tests for nitrate reduction and denitrification have always been considered of great importance for distinguishing *A. faecalis* from *A. denitrificans*. The majority of the *A. denitrificans* strains indeed reduce nitrate and nitrite to nitrogen gas and can grow anaerobically with nitrate or nitrite as electron acceptor. Most *A. faecalis* strains (including the previously named "*A. odorans*") reduce nitrite but not nitrate and cannot grow anaerobically in the presence of nitrate. The determination of DNA base compositions, ribosomal ribonucleic acid (rRNA) cistron similarities and carbon substrate utilization profiles indicate, however, that a number of formerly named *A. faecalis* strains (e.g., CIP 60.75) should be classified as *A. denitrificans* (De Ley et al., 1970; Pichinoty et al., 1978; Kersters, Popoff, Segers and De Ley, unpublished results). It should be recalled here that tests for nitrate reduction and denitrification showed fairly low reproducibility levels when performed with the same strains in different laboratories (Sneath and Collins, 1974). Moreover, Hendrie et al. (1974) reported that the property of anaerobic respiration in the presence of nitrate may be lost in *A. denitrificans* on subculturing.

The distribution of assimilatory and dissimilatory nitrate reductases in several *Alcaligenes* strains has been studied by Pichinoty and Chatelain (1973).

Some strains of *A. denitrificans* can also grow anaerobically in the presence of tetrathionate (Pichinoty et al., 1978).

Alkalinization of organic salts and amides has been used to differentiate *Alcaligenes* strains from other weakly saccharolytic and non-saccharolytic Gram-negative bacteria (Pickett and Pedersen, 1970; Otto and Pickett, 1976; Oberhofer et al., 1977).

Carbohydrates are not oxidized by resting cells of *A. faecalis* NCIB 8156 (type strain), whereas intermediates of the tricarboxylic acid cycle, L-glutamate, L-aspartate, L-histidine, glyoxylate, glycolate and formate are readily attacked (D. M. Gibson, Ph.D. dissertation, University of Aberdeen, Scotland, 1968). Ragland et al. (1966) reported an active NADP-linked isocitrate dehydrogenase, but no glucose-6-phosphate or 6-phosphogluconate dehydrogenases in citrate-grown cells of this strain; however, in addition to enzymes of the tricarboxylic acid cycle, Gibson (Ph.D. dissertation, University of Aberdeen, Scotland, 1968) also found active NAD- and NADP-linked glucose-6-phosphate dehydrogenase activity in extracts of cells of this strain grown on a medium containing peptone and beef extract.

De Ley et al. (1970) detected a correlation between the DNA base composition and the distribution of the key enzymes of a modified Entner-Doudoroff pathway in a large number of *A. faecalis*, "*A. odorans*" and *A. denitrificans* strains. All "*A. odorans*" and the majority of the *A. faecalis* strains had a mol% G + C of their DNA of 55.9–59.4 and lacked D-gluconate-6-phosphate dehydratase, gluconate dehydratase and 2-keto-3-deoxy-D-gluconate-6-phosphate (KDPG) aldolase. All the *A. denitrificans* strains, some misnamed *A. faecalis* strains and some unspecified *Alcaligenes* and "*Achromobacter*" strains with mol% G + C of their DNA in the range of 63.9–69.8 possessed an active D-gluconate dehydratase and KDPG-aldolase and metabolized D-gluconate via a modified Entner-Doudoroff pathway, where dehydratation precedes phosphorylation.

Some *Alcaligenes* strains oxidize arsenite to arsenate (Turner, 1954; Hendrie et al., 1974; Osborne and Ehrlich, 1976; Phillips and Taylor, 1976). One of these strains, previously designated as "*Achromobacter arsenoxydans-tres*" (NCIB 8687), is a normal member of *A. faecalis* (Hendrie et al., 1974; Kersters, Segers and De Ley, unpublished results).

None of the 32 strains of *A. denitrificans*, *A. faecalis* and "*A. odorans*" investigated by Pichinoty et al. (1978) was able to grow chemolithotrophically in a mixture of H_2, CO_2, and N_2; however, the only available strain of the hydrogen-oxidizing species *A. ruhlandii* belongs genetically in *A. denitrificans* (Fig. 4.85; see Taxonomic Comments). *A. ruhlandii* possesses a soluble and a particulate hydrogenase (Vishniac and Trudinger, 1962; Schink and Schlegel, 1980).

The morphology of several temperate bacteriophages from lysogenic strains of *A. faecalis* has been described by Maré et al. (1966). Several *A. faecalis* and *A. denitrificans* strains produce bacteriocins active against other strains of *Alcaligenes*, *Proteus*, *Escherichia coli* and *Staphylococcus* (Maré and Coetzee, 1964; Djambazov et al., 1971).

Gilardi (1971a) reported that the determination of sensitivities towards antibiotics is of limited diagnostic value for clinical isolates of the genus *Alcaligenes*. The following authors have published summaries of antimicrobial susceptibility patterns: Yabuuchi et al. (1974), Oberhofer et al. (1977) and von Graevenitz (1978).

The clinical significance of these bacteria is difficult to assess; however, they can cause opportunistic infections in patients with underlying illnesses (Igra-Siegman et al., 1980). Cases of purulent meningitis and cerebral ventriculitis have been attributed to "*Achromobacter xylosoxidans*" (Yabuuchi et al., 1974; Shigeta et al., 1978), which is classified in this *Manual* as *Alcaligenes denitrificans* subsp. *xylosoxydans*. In the hospital environment *Alcaligenes* strains have been isolated from moist items such as respirators, hemodialysis systems, intravenous solutions and even disinfectants (Holmes et al., 1977a; Shigeta et al., 1978; Rubin et al., 1980). They have also been found in

a great variety of clinical specimens such as blood, sputum, urine, feces, spinal fluid, pleural fluid, bronchial washings, wounds, burns, throat swabs, purulent ear discharges and swabs from eye and pharynx (Tatum et al., 1974; von Graevenitz, 1978; Rubin et al., 1980).

Bacteria belonging to *A. faecalis* and *A. denitrificans* are usually considered to be difficult to identify because they do not possess specific physiological or biochemical characteristics. Since the development of commercial multitest systems more attention has been paid during the last years to the identification of clinical isolates belonging to this group of glucose-nonfermenting bacteria (Gilardi, 1971b, 1973; Tatum et al., 1974; Otto and Pickett, 1976; Isenberg and Sampson-Scherer, 1977; Oberhofer et al., 1977; Hofherr et al., 1978; Rarick et al., 1978; Otto and Blachman, 1979; Appelbaum et al., 1980; Igra-Siegman et al., 1980; Rubin et al., 1980).

Little ecological information is available on bacteria belonging to the genus *Alcaligenes*. They seem to be ubiquitous as they have been isolated from soil, water, dairy products, rotten eggs (Board, 1965), and the intestinal tract of vertebrates, nematodes and insects (Málek et al., 1963). *A. denitrificans* may play a role in the denitrification process in soil (Gamble et al., 1977).

Enrichment and Isolation Procedures

A. denitrificans has been isolated from soil by enrichment procedures described by Pichinoty et al. (1978) using the following liquid minimal medium: $Na_2HPO_4 \cdot 12H_2O$, 3.575 g; KH_2PO_4, 0.98 g; $MgSO_4 \cdot 7H_2O$, 0.03 g; NH_4Cl, 0.5 g; 0.2 ml of a trace element solution*; and 4.0 g of a carbon source in 1 liter of distilled water; pH 7.0. Organic acids such as L-malate, succinate, adipate, *meso*-tartrate and itaconate have been used successfully as the carbon source. Soil samples are suspended in 50 ml of medium in 250-ml flasks and the air is replaced by pure N_2O. Primary cultures are kept at 32°C for several days without shaking and then serially transferred in the minimal medium. Colonies can be isolated by streaking plates of minimal medium solidified by the addition of 1.4% agar. These plates are incubated aerobically at 32°C for 3 days. The isolates should be checked for a strong oxidase reaction and anaerobic growth with N_2O. It should be noted that this enrichment procedure will allow the growth of other denitrifying bacteria, such as e.g., *Pseudomonas stutzeri*.

A. faecalis and *A. denitrificans* can be isolated from sources such as clinical specimens by the use of an ordinary medium (e.g. blood agar) and a selective enteric medium (preferably MacConkey agar). Conventional microbiological procedures (Gilardi, 1978; Rubin et al., 1980) can be used for the identification of oxidase-positive, peritrichously flagellated glucose-nonfermentative rods or coccoids belonging to the genus *Alcaligenes*.

Facultatively chemolithotrophic, peritrichously flagellated, hydrogen-oxidizing bacteria, such as *A. eutrophus*, *A. paradoxus* and *A. latus* (here classified as *species incertae sedis*) can be isolated from soil, mud and water samples by enrichment in a liquid basal mineral medium (D. H. Davis, Ph.D. dissertation, University of California, Berkeley, 1967; Davis, et al., 1970; Palleroni and Doudoroff, 1972; Palleroni and Palleroni, 1978). Incubation at 30°C should be carried out in a sealed vessel in which the air is replaced by a mixture of 50% H_2, 5% CO_2, 4–20% O_2, 25–41% N_2. The nonautotrophic strains of *A. paradoxus* have been isolated from soil by enrichment with pantothenate (Davis et al. 1970) or poly-β-hydroxybutyrate (Delafield et al., 1965). The marine *A. aquamarinus*, *A. aestus*, *A. cupidus*, *A. pacificus* and *A. venustus* (here classified as *species incertae sedis*) have been isolated from sea water by direct isolation procedures using membrane filters or by enrichment methods described by Baumann et al. (1971) and Baumann et al., (1972). L-Lysine, L-histidine, L-glutamate, L-valine, glycolate, adipate, *o*-hydroxybenzoate, nicotinamide and *meso*-inositol have been used as carbon sources for the isolation (Baumann et al., 1972).

---

* Trace element solution (g/liter of distilled water): ethylenediaminetetraacetic acid, disodium salt dihydrate, 50.0; $ZnSO_4 \cdot 7H_2O$, 2.2; $CaCl_2$, 5.54; $MnCl_2 \cdot 4H_2O$, 5.06; $FeSO_4 \cdot 7H_2O$, 4.79; $NH_4Mo_7O_{24} \cdot 4H_2O$, 1.1; $CuSO_4 \cdot 5H_2O$, 1.57; $CoCl_2 \cdot 6H_2O$, 1.6; H_3BO_3, 0.05. The pH is adjusted to 6.0 with KOH. (Pichinoty et al., 1977).

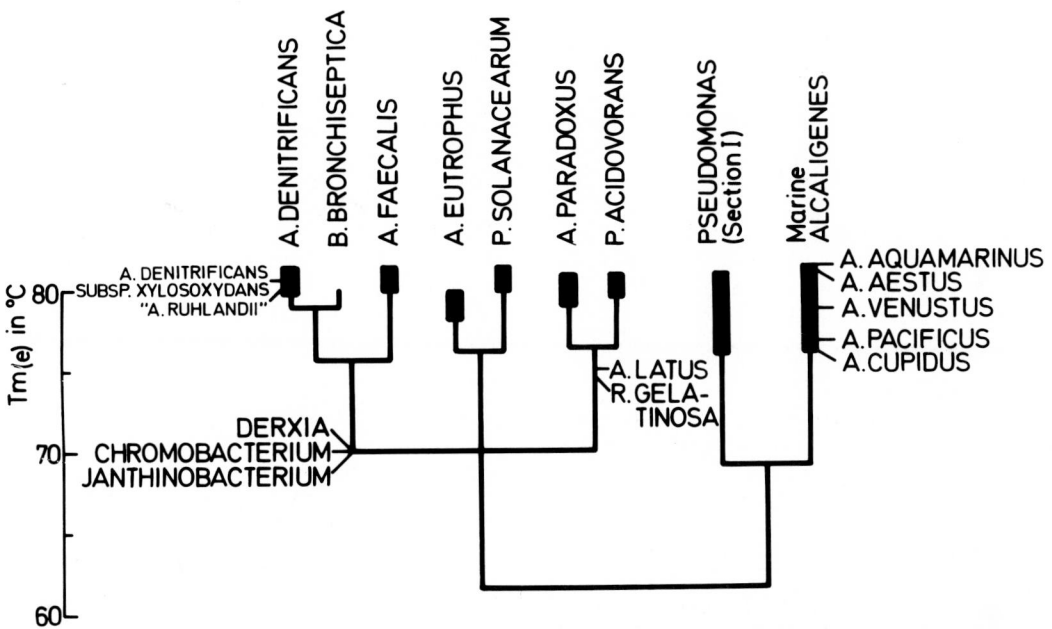

Figure 4.85. Dendrogram of the similarities between rRNA cistrons (expressed as $T_{m(e)}$ in °C) from the *Alcaligenes* species and *species incertae sedis* and some other Gram-negative genera and species. $T_{m(e)}$ is the midpoint temperature of the thermal denaturation curves of the DNA/rRNA hybrids. The black rectangles on top of most branches represent the range of $T_{m(e)}$ values of the reference taxon. For the source of the ^{14}C-rRNAs see text. *A. faecalis*, *A. denitrificans*, *A. eutrophus*, *A. paradoxus* and *A. latus* belong in the same rRNA superfamily (sensu De Ley, 1978) as *Pseudomonas* (Sections II and III, Bergey 8), *Chromobacterium*, *Janthinobacterium*, *Derxia*, *Bordetella bronchiseptica* and *Rhodopseudomonas gelatinosa*. The closest rRNA superfamily consists of *Pseudomonas* (Section I, Bergey 8) and the marine *Alcaligenes* species (*A. aquamarinus*, *A. aestus*, *A. cupidus*, *A. pacificus* and *A. venustus*). (The data for the graph are from De Ley, 1978; De Ley et al., 1978; De Smedt et al., 1980; De Vos, 1981, and Segers, De Vos, Gillis, De Smedt, Van Landschoot, Tytgat and De Ley, unpublished results.)

Maintenance Procedures

Stock cultures of *Alcaligenes* strains may be maintained on media such as nutrient agar, preferably in screw-capped vials, at 4°C for 1–3 months. The media for the marine bacteria should be supplemented with 2 or 3% NaCl. *Alcaligenes* strains can be preserved for many years by lyophilization.

Differentiation of the genus **Alcaligenes** from other genera

One of the most obvious characteristics of *Alcaligenes* is its extremely limited action on carbohydrates (see Tables 4.98 and 4.99). *Alcaligenes* strains can therefore be confused with *Bordetella bronchiseptica*, *Pseudomonas acidovorans*, *P. testosteroni*, *P. alcaligenes*, *P. pseudoalcaligenes* and similar organisms. *B. bronchiseptica* is distinguished by its strong urease reaction. In the past some *Agrobacterium* strains were mistaken for *Alcaligenes* strains, even though *Agrobacterium* can grow on a great variety of carbohydrates. Table 4.97 indicates the most salient features that differentiate *Alcaligenes* from some of these taxa. It should be noted that *B. bronchiseptica*, *P. acidovorans*, *P. testosteroni* and *Alcaligenes* belong in the same rRNA superfamily (*sensu* De Ley, 1978; see Taxonomic Comments).

Taxonomic Comments

In the past the genera *Alcaligenes* and "*Achromobacter*" have all too often been a dumping ground for very different types of bacteria, most of which were poorly characterized. This was due to the lack of an adequate description of both genera and to the inactivity of these bacteria in the commonly used biochemical tests. Because great overall similarities existed between the descriptions of both genera, and because no extant culture of the type species of the genus "*Achromobacter*" existed, Hendrie et al. (1974) proposed to abandon the genus name "*Achromobacter*" Bergey, Harrison, Breed, Hammer and Huntoon 1923, 132. Consequently, the genus "*Achromobacter*" was not described in the eighth edition of the *Manual* (Buchanan and Gibbons, 1974); moreover, it was not mentioned on the Approved Lists of Bacterial Names (Skerman et al., 1980). In 1981 the generic name *Acromobacter* was revived by Yabuuchi and Yano with a different type species, *Achr. xylosoxidans*.

According to the previous edition of the *Manual* the following four species belonged in the genus *Alcaligenes*: *A. faecalis*, *A. aquamarinus*, *A. eutrophus* and *A. paradoxus* (Holding and Shewan, 1974). The DNA base composition of the genus ranges from 57.9–70 mol% G + C. Later, the taxonomic structure of the genus became more heterogeneous. As a result of an extensive phenotypic analysis, Baumann et al. (1972) accommodated by a process of elimination four new groups of aerobic marine bacteria in the genus *Alcaligenes* as *A. aestus*, *A. cupidus*, *A. pacificus* and *A. venustus*. The authors stated that these organisms were very similar to the genus *Pseudomonas* and were distinguishable from this genus solely on the basis of their flagellation. Aragno and Schlegel (1977) transferred the hydrogen-oxidizing "*Pseudomonas ruhlandii*" (Packer and Vishniac 1955) Davis 1969 (in Davis et al., 1969) to the genus *Alcaligenes* as *A. ruhlandii* because of its flagellar arrangement. Finally, Palleroni and Palleroni (1978) created the new species

Table 4.97.

*Differential characteristics of the genus **Alcaligenes** and other taxa[a]*

Characteristics	Alcaligenes	Bordetella bronchiseptica	Pseudomonas acidovorans	Pseudomonas testosteroni	Agrobacterium
Polar flagella	−	−	+	+	−
Peritrichous flagella	+	+	−	−	+
Urease[b]	−[c]	+[c]	−	−	+[c]
Carbon sources for growth:					
D-Glucose	D	−	−	−	+
L-Arabinose	−	−	−	−	+
D-Fructose	−	−	+	−	d
D-Mannitol	−	−	+	−	+
D-Gluconate	D	−	+	+	+
Sucrose	−	−	−	−	+
p-Hydroxybenzoate	−	−	+	+	
Mol% G + C of DNA	56–70	68–69.5[d]	67	62	57–63

[a] See also Tables 4.103 and 4.104 of the genus *Bordetella*. Symbols: see standard definitions.

[b] Gilardi (1978); Rubin et al. (1980).

[c] The urease reaction is rapid in *B. bronchiseptica* and *Agrobacterium* and sometimes slow in *A. denitrificans*.

[d] De Ley and Segers, unpublished results.

A. latus for three strains of hydrogen-utilizing bacteria possessing fairly large and thick cells. Accordingly the following *Alcaligenes* species were accepted on the Approved Lists (Skerman et al., 1980):

1. *A. aestus* Baumann, Baumann, Mandel and Allen 1972
2. *A. aquamarinus* (ZoBell and Upham 1944) Hendrie, Holding and Shewan 1974
3. *A. cupidus* Baumann, Baumann, Mandel and Allen 1972
4. *A. eutrophus* Davis 1969 in Davis, Doudoroff, Stanier and Mandel 1969
5. *A. faecalis* Castellani and Chalmers 1919
6. *A. latus* Palleroni and Palleroni 1978
7. *A. pacificus* Baumann, Baumann, Mandel and Allen 1972 (not "*A. pacifus*" as erroneously printed in the Approved Lists)
8. *A. paradoxus* Davis 1969 in Davis, Doudoroff, Stanier and Mandel, 1969
9. *A. ruhlandii* (Packer and Vishniac 1955) Aragno and Schlegel 1977
10. *A. venustus* Baumann, Baumann, Mandel and Allen 1972

The genus *Alcaligenes* thus became an agglomerate of bacteria occurring in widely different ecological niches and possessing widely different phenotypic characteristics (see Tables 4.98 and 4.99). The span of its DNA base composition became so wide (52.9–70 mol% G + C) that it casts serious doubts on the presumably close genetic relatedness of the 10 *Alcaligenes* species on the Approved Lists. (It should be noted that *A. denitrificans* was not on the Approved Lists because Hendrie et al. (1974) considered *A. denitrificans* Leifson and Hugh 1954 as a subjective synonym of *A. faecalis* Castellani and Chalmers 1919; however, the great difference in DNA base composition (approximately 10 mol% G + C, see below) between the majority of *A. denitrificans* and *A. faecalis* strains (including the type strains) is clearly not in agreement with the proposed synonymy.) In 1983 *A. denitrificans* was revived by Rüger and Tan.

This puzzling taxonomic situation of and within the genus *Alcaligenes* was recently clarified to a great extent by the determination of the similarities of rRNA cistrons of a great number of *Alcaligenes* strains by De Ley and co-workers. Their studies revealed a great heterogeneity within the genus *Alcaligenes* as it was defined in previous editions of the *Manual* and on the Approved Lists. Figure 4.85 summarizes the results of hybridizations between DNA from a great variety of named Gram-negative bacteria and ^{14}C-rRNA from the type strains of *A. faecalis* (NCIB 8156, ATCC 8750), *A. denitrificans* (ATCC 15173), *A. eutrophus* (ATCC 17697), *A. paradoxus* (ATCC 17713), *A. aquamarinus* (NCMB 557, ATCC 14400), *P. fluorescens* (ATCC 13525), *P. acidovorans* (ATCC 15668) and *P. solanacearum* (NCPPB 325, ATCC 11696) (De Ley, 1978; De Ley et al. 1978; De Smedt et al., 1980; De Vos, 1981;

De Ley, Segers, De Vos and Gillis, unpublished results). These studies show that *A. faecalis*, *A. denitrificans*, *A. ruhlandii*, *A. eutrophus*, *A. paradoxus* and *A. latus* belong in the third rRNA superfamily (*sensu* De Ley, 1978) together with *Pseudomonas* sections II and III (except *P. pseudoalcaligenes*) (Bergey 8), *Chromobacterium*, *Janthinobacterium*, *Derxia*, *Bordetella bronchiseptica* and *Rhodopseudomonas gelatinosa* (Fig. 4.85). The comparison of oligonucleotide catalogs obtained from 16 S rRNAs shows also that *A. faecalis* belongs in the subline defined by *Rhodopseudomonas gelatinosa* and *Rhodospirillum tenue* (Fox et al., 1980). Also the comparison of the known primary structures of the class II cytochromes c reveals that cytochrome c′ from *Alcaligenes* strain NCIB 11015 (a normal member of *A. denitrificans* subsp. *xylosoxydans* (Kersters, Segers and De Ley, unpublished results)) displays its greatest sequence similarity (48%) with cytochrome c′ from *Rhodopseudomonas gelatinosa* (Ambler et al., 1979).

The five marine *Alcaligenes* species (*A. aquamarinus*, *A. aestus*, *A. cupidus*, *A. pacificus* and *A. venustus*) are related to one another but belong in the second rRNA superfamily (*sensu* De Ley, 1978), as they are more closely related to *Pseudomonas* section I (Bergey 8) than to *A. faecalis* (Fig. 4.85). These five marine species are, besides their morphology and flagellation, phenotypically very different from the type species *A. faecalis* (see Tables 4.98 and 4.99). They are therefore considered here as *species incertae sedis*, until their exact taxonomic and nomenclatural status will be settled.

With regard to rRNA cistrons, *A. eutrophus* and *A. paradoxus* belong to different branches in rRNA superfamily III; they differ almost as much from each other and from *A. faecalis* (Fig. 4.85), as e.g. the genera *Agrobacterium*, *Acetobacter* and *Zymonomas* differ from each other (De Smedt and De Ley, 1977; De Ley et al., 1978; Gillis and De Ley, 1980). Auling et al. (1980) could not detect any relatedness between the type strains of *A. faecalis* and *A. paradoxus* by DNA/DNA reassociation kinetics. However, the rRNA cistrons of *A. paradoxus* closely resemble those of *P. acidovorans* ($T_{m(e)}$ is 76–76.5°C), whereas the rRNA cistrons of *A. eutrophus* show great similarities with those of *P. solanacearum* ($T_{m(e)}$ is 75.5–77°C) (Fig. 4.85) (De Vos, 1981; De Ley, De Vos and Segers, unpublished results). Also Baumann and Baumann (1978) concluded that *A. eutrophus* is related to *P. solanacearum* and *P. cepacia*, and that *A. paradoxus* is related to *P. acidovorans*, *P. testosteroni*, *P delafieldii* and *P. facilis* by immunological comparisons of glutamine synthetase. Studies on comparative enzymology and allostery of pathways of aromatic amino acid biosynthesis indicate likewise that *A. eutrophus* and *A. paradoxus* are related to *P. solanacearum* and *P. acidovorans*, respectively (Byng et al., 1980; Whitaker et al., 1981). These combined results indicate that the peritrichously flagellated *A. paradoxus* strains belong to the same rRNA branch as the following polarly flagellated species of *Pseudomonas* section III (Bergey 8): *P. acidovorans*, *P. testosteroni*, *P. delafieldii* and the three hydrogen-oxidizing species *P. facilis*, *P. flava* and *P. palleronii*. The peritrichously-flagellated *A. eutrophus* belongs to the same rRNA branch as the polarly flagellated *P. solanacearum (Pseudomonas* section III), *P. cepacia*, *P. gladioli* and *P. caryophylli (Pseudomonas* section II) (De Vos, 1981; De Ley, De Vos and Segers, unpublished results). These results strengthen the opinion (Gillis and De Ley, 1980) that the importance of the type of flagellation in establishing the suprageneric ranks in bacterial taxonomy should be decreased. The rRNA cistrons of the three strains of *A. latus* exhibit a relatively high degree of similarity to those of *A. paradoxus* and *P. acidovorans* (Fig. 4.85) ($T_{m(e)}$ is 73.5–76°C; De Ley and Segers, unpublished results). The peritrichously flagellated hydrogen-oxidizing species *A. eutrophus*, *A. paradoxus* and *A. latus* are considered here as *species incertae sedis* until their exact taxonomic position will be settled, because these species are phenotypically (see Tables 4.98 and 4.99) and phylogenetically (Fig. 4.85) different from each other and from the type species *A. faecalis*.

There is strong genetic and phenotypic evidence that *A. faecalis* and *A. denitrificans* are not subjective synonyms but must be regarded as separate species (De Ley et al., 1970; Pichinoty et al., 1978; Kersters and De Ley (1980); Kiredjian et al. (1981); Yamasato et al. (1982); Rüger and Tan (1983): (a) considerable differences exist between the DNA base compositions of *A. faecalis* and *A. denitrificans*: 55.9–59.4 and 63.9–69.8 mol% G + C (T_m and Bd), respectively; (b) with regard to rRNA cistrons, *A. faecalis* and *A. denitrificans* differ as much from each other ($T_{m(e)}$ is 75.5°C, Fig. 4.85) as *A. eutrophus* differs from *P. solanacearum* (Fig. 4.85); (c) both taxa can be differentiated from each other by a great number of phenotypic characteristics (see Tables 4.98 and 4.99) and by their gel electrophoretic protein patterns (Kersters, 1978; Kersters and De Ley, 1980); (d) *A. denitrificans* strains possess enzymes of a modified Entner-Doudoroff pathway for the metabolism of D-gluconate, whereas *A. faecalis* strains lack the enzymes of both the normal and modified Entner-Doudoroff pathway (De Ley et al., 1970).

"*A. odorans*" (Málek and Kazdová-Koviškóvá 1946) Málek, Radochová and Lysenko 1963 (not on the Approved Lists) is genotypically and phenotypically very similar to *A. faecalis*. The type strains of both species belong in the same genetic, protein electrophoretic and phenotypic groups (Popoff, Kersters and De Ley, unpublished results). The subjective synonymy of *A. faecalis* Castellani and Chalmers 1919, "*A. odorans*" (Málek and Kazdová-Kožiškóvá 1946) Málek, Radochová and Lysenko 1963, "*A. odorans* var. *viridans*" Mitchell and Clarke 1965 and "*Achromobacter arsenoxydans-tres*" Turner 1954 is retained here as proposed by Hendrie et al. (1974). The latter taxa form, indeed, a very homogeneous group of organisms both on genetic and on phenotypic grounds (Kersters and De Ley, 1980; Kiredjian et al., 1981).

The rRNA cistrons of *A. denitrificans* are indistinguishable from those of *Achromobacter xylosoxidans* Yabuuchi and Yano 1981 by hybridization techniques ($T_{m(e)}$ is 80–81.5°C; Fig. 4.85) (De Ley and Segers, unpublished results). Yabuuchi and Ohyama (1971) proposed strain ATCC 27061 as type strain for *Achr. xylosoxidans* and gave a detailed description of the species (Yabuuchi et al., 1974). Strains of *A. xylosoxidans* were mainly isolated from clinical specimens and correspond to King's groups IIIa and IIIb (King, 1964; Yabuuchi et al., 1974; Rubin et al., 1980). Strains assigned to *A. xylosoxidans* can be clearly differentiated from other *Alcaligenes* strains by their growth and acid formation on D-glucose and D-xylose (Table 4.98). DNA/DNA hybridizations indicated that these xylose-oxidizing strains form a homogeneous group of bacteria related by 30–40% DNA homology to *A. denitrificans* (Kiredjian et al., 1981). Therefore the following two subspecies are proposed here for *A. denitrificans*: *A. denitrificans* subsp. *denitrificans* subsp. nov. (proposed type strain ATCC 15173) and *A. denitrificans* subsp. *xylosoxydans* subsp. nov. (proposed type strain ATCC 27061).

A comparison of carbon substrate utilization profiles and protein electrophoregrams indicates that some previously named *A. faecalis* strains, with a DNA base composition in the range 64–66 mol% G + C should be reclassified in *A. denitrificans* subsp. *denitrificans*. Also

Rarick et al. (1978) found that one group of so-called *A. faecalis* strains was morphologically and biochemically more similar to *A. denitrificans*.

The rRNA cistrons of the only extant strain (ATCC 15749, DSM 653) of the hydrogen-oxidizing *A. ruhlandii* (Packer and Vishniac 1955) Aragno and Schlegel 1977 (on the Approved Lists) are indistinguishable by hybridization techniques from those of *A. denitrificans* (Fig. 4.85). Moreover the protein electrophoregrams and phenotypic features of *A. ruhlandii* and *A. denitrificans* subsp. *xylosoxydans* are indistinguishable from each other (Kersters and Popoff, unpublished results). The type strains of both *A. ruhlandii* and *A. denitrificans* subsp. *xylosoxydans* have been reported to possess sheathed flagella (Yabuuchi et al., 1974; Aragno and Schlegel, 1977; Aragno et al., 1977). The great genotypic (Fig. 4.85) and phenotypic similarities between *A. denitrificans* subsp. *denitrificans*, *A. denitrificans* subsp. *xylosoxydans* and *A. ruhlandii* allow but one species name for all these organisms. Although *A. ruhlandii* is the only species name on the Approved Lists, there are several serious arguments not to use it for this entire large group of organisms: (a) the names of both *A. denitrificans* and *Achr. xylosoxidans* are widely used by clinical bacteriologists; to change these names to the unfamiliar *A. ruhlandii* would destabilize the nomenclature; (b) the species name *A. ruhlandii* was specifically created for the hydrogen-oxidizing capacities of the organism, being represented by only one extant strain (ATCC 15749); (c) the oxidation of hydrogen gas could not be demonstrated in over 20 *A. denitrificans* strains (Pichinoty et al., 1978); (d) on the analogy of *A. eutrophus* and *A. paradoxus* (see below) it may well be that the hydrogenases responsible for the oxidation of hydrogen by *A. ruhlandii* are encoded by plasmid-linked genes; therefore *A. ruhlandii* cannot even be a subspecies of *A. denitrificans*, because a loss of the eventual plasmid, would mean a switch of taxa, an undesirable situation; (e) according to Recommendation 12c of the International Code of Nomenclature of Bacteria (Lapage et al., 1975) the epithet *xylosoxydans* is to be preferred as it points to the characteristic property to oxidize D-xylose (Table 4.98). As long as but one strain of *A. ruhlandii* is known and the nature of the genetic background of the H_2-oxidizing system is unknown, it seems to us that it is best to keep strain ATCC 15749 at an infrasubspecific level, e.g., as biovar *ruhlandii*. Consequently only *A. faecalis* and *A. denitrificans* are considered here as true species of the genus *Alcaligenes*.

The rRNA cistrons of *Bordetella bronchiseptica* are very similar to those of *A. denitrificans* ($T_{m(e)}$ is 78.5–79.5°C, Fig. 4.85). Fifteen *B. bronchiseptica* strains were investigated, including the type strain ATCC 19395 (Kersters and De Ley, 1980; De Ley and Segers, unpublished results). *B. bronchiseptica* is a pathogenic organism causing diseases of the respiratory tract in mammals (Goodnow, 1980). *B. bronchiseptica* shares common antigens with *B. pertussis* and *B. parapertussis* (Goodnow, 1980), but possesses also many phenotypic properties in common with *A. faecalis* and *A. denitrificans* (Johnson and Sneath, 1973). Azurins (blue copper-containing proteins) occur in *A. faecalis*, *A. denitrificans* and also in *Bordetella bronchiseptica* (Sutherland, 1966); the primary structure of some of them is known (Dayhoff, 1976).

The correct taxonomic position of an unnamed group of bacteria, isolated from clinical specimens and assigned to "*Achromobacter*" as CDC group Vd (Tatum et al., 1974; Chester and Cooper, 1979; Rubin et al., 1980) remains unknown for the moment.

Recently, Austin et al. (1981) described a new group of bacteria isolated from moribund lobsters and named them *A. faecalis* subsp. *homari* (type strain, ATCC 33127). The taxonomic status of these organisms should be verified by DNA/rRNA hybridization techniques, because the organisms were reported to possess several growth characteristics uncommon for *A. faecalis*.

Acknowledgments

We thank Dr. M. Popoff for the communication of unpublished results and critical reading of the manuscript. The aid of U. Torck is greatly appreciated. J. D. L. is indebted to the Fonds voor Kollektief Fundamenteel Onderzoek for research and personnel grants.

Table 4.98.

Differential characteristics of the species, subspecies and species incertae sedis of the genus **Alcaligenes**[a]

Characteristics	Authentic *Alcaligenes* Species			Species incertae sedis							
	1. A. faecalis[b]	2. A. denitrificans		a. A. eutrophus[d]	b. A. paradoxus[d]	c. A. latus[e]	d. A. aestus[f]	e. A. aquamarinus[g]	f. A. cupidus[f]	g. A. pacificus[f]	h. A. venustus[f]
		2a. subsp. denitrificans[b]	2b. subsp. xylosoxydans[b,c]								
Number of strains investigated	>50	>30	>50	14	16	2	6	1	5	6	14
Width of cells >1.2 μm	−	−	−	−	−	+					
Yellow carotenoid cellular pigments	−	−	−	−	+	−[h]	−	−	−	−	−
Oxidase reaction	+	+	+	+	+	+	+		−	+	+
Nitrate reduced to nitrite	−	[+]	+	+	d[i]	+	[+]	+	[+]	+	+
Nitrite reduction[j]	+	[+]	+	+	−	−					
Anaerobic growth with nitrate	−	[+]	+	+	−	−			−	−	−
Anaerobic growth with nitrite[j]	+	[+]	+	−	−	−					
Chemolithotrophic growth with H₂	−[k]	−[k]	−[l]	+	d[i]	+	−		−	−	−
Hydrolysis of gelatin	−	−	−	−	d	+			−	−	−
Acid from D-xylose and D-glucose in O-F medium	−	−	+								
Carbon sources for growth:											
D-Glucose	−	−	+	M	+	+	+	+	+	+	+
L-Arabinose	−	−	−	−	+	−	−	−	+	−	−
D-Xylose	−	−	+	−	+	−	−	−	[+]	d	−
D-Fructose	−	−	d	+	+	+	+	−	+	+	d
D-Mannitol	−	−	−	−	+	−	+	+	+	−	[+]
D-Mannose	−	−	d	−	+	−	−	−	+	−	−
D-Gluconate	−	[+]	+	+	+	+	[+]	+	+	+	+
Acetate	+	+	+	+	+	−	+	+	+	+	+
Adipate	−	+	+	+	[+]	−	[+]	−	−		d
Pimelate	−	+	+	+	[+]	−	+	+	−		
Sebacate	−	+	+	+	+	d	+	+	−	−	−
Suberate	−	+	+	+	d	+	+	+	−		−
meso-Tartrate	−	+	+	+	d	−	d	+	[+]	−	−
Itaconate	−	+	+	+	d	+	d	−			
Isolated from:											
Marine environment	−	−	−	−	−	−	+	+	+	+	+
Soil	+	+	+	+	+	+	−	−	−	−	−
Clinical material	+	+	+	−	−	−	−	−	−	−	−
Mol% G + C of DNA	55.9–59.4	63.9–68.9	66.0–69.8	66.3–67.5	66.8–69.4	69.1–71.1	56.9–57.9	57.9	59.9–62.1	67.3–68.3	52.9–54.5

[a] Symbols: +, 90% or more of the strains are positive; [+], 80% or more of the strains are positive; d, 11–79% of the strains are positive; −, 10% or less of the strains are positive; M, mutant growth.

[b] Gilardi, 1968; Pichinoty et al., 1978; Rarick et al., 1978; Rubin et al., 1980; Kiredjian et al., 1981; Yamasato et al., 1982; Kersters and De Ley, unpublished results.

[c] Yabuuchi et al. (1974).

[d] Davis et al. (1969, 1970); D. H. Davis (Ph.D. dissertation, University of California, Berkeley, 1967).

[e] Palleroni and Palleroni (1978).

[f] Baumann et al. (1972).

[g] ZoBell and Upham (1944); Holding and Shewan (1974); Kersters and De Ley, unpublished results.

[h] The colonies of *A. latus* are greyish pink or yellowish.

Further Reading

Aragno, M. and H.G. Schlegel. 1981. The hydrogen-oxidizing bacteria. In Starr, Stolp, Trüper, Balows and Schlegel (Editors), *The Prokaryotes: a Handbook on Habitats, Isolation, and Identification of Bacteria*, Springer-Verlag, Berlin, pp. 865–893.

Baumann, P. and L. Baumann. 1981. The marine Gram-negative Eubacteria: genera *Photobacterium, Beneckea, Alteromonas, Pseudomonas*, and *Alcalignes*. In Starr, Stolp, Trüper, Balows and Schlegel (Editors), *The Prokaryotes: a*

Handbook on Habitats, Isolation, and Identification of Bacteria, Springer-Verlag, Berlin, pp. 1302–1331.

Kiredjian, M., M. Popoff, C. Coynault, M. Lefèvre and M. Lemelin. 1981. Taxonomie du genre *Alcaligenes*. Ann. Microbiol. (Inst. Pasteur) *132B*: 337–374.

Yamasato, K., M. Akagawa, N. Oishi and H. Kuraishi. 1982. Carbon substrate assimilation profiles and other taxonomic features of *Alcaligenes faecalis, Alcaligenes ruhlandii* and *Achromobacter xylosoxidans*. J. Gen. Appl. Microbiol. *28*: 195–213.

Differentiation and characteristics of species of **Alcaligenes**

The differential characteristics of the species and the *species incertae sedis* are shown in Table 4.98. Other characteristics are listed in Table 4.99.

List of the species of the genus **Alcaligenes**

1. **Alcalizenes faecalis** Castellani and Chalmers 1919, 936.[AL] (*Achromobacter arsenoxydans-tres* Turner 1954, 475; *Alcaligenes odorans* (Málek and Kazdová-Kožiškova (1946) Málek, Radochová and Lygenko 1963, 353; *A. odorans* var. *viridans* Mitchell and Clarke 1965, 347.)

fae.ca′lis. L. n. *faex, faecis* dregs; M.L. adj. *faecalis* fecal.

The morphological characteristics are as described for the genus. Colonies on nutrient agar are nonpigmented to grayish-white, translucent to opaque, flat to low convex, margin usually entire, usually smooth, sometimes dull or rough. Most strains form colonies with a thin spreading irregular edge. Some strains (previously named "*A. odorans*" (Málek and Kazdová-Kožiškova 1946) Málek, Radochová and Lysenko 1963) produce a characteristic aromatic fruity odor and/or a green discoloration on blood agar (those strains producing the latter characteristic were previously named "*A. odorans* var. *viridans*" Mitchell and Clarke 1965).

Physiological and nutritional characteristics of *A. faecalis* are presented in Tables 4.98 and 4.99. Carbohydrates are not utilized as sole carbon sources. Good growth is obtained on several organic acids and amino acids.

Chemolithotrophic growth using hydrogen gas has not been demonstrated.

Nitrite, but not nitrate, is reduced. Anaerobic respiration with nitrite, but not with nitrate, as electron acceptor is possible for most strains. Some strains oxidize arsenite.

Isolated from soil, water, feces, urine, blood, sputum, wounds, pleural fluid, nematodes and insects.

The mol% G + C of the DNA is 55.9–59.4 (T_m and Bd) (De Ley et al., 1970; Pichinoty et al., 1978).

Type strain: ATCC 8750 (NCIB 8156; CIP 60.80; DSM 30030; CCM 1052; and Conn 16).

Other representative culture collection strains belonging to this taxon: ATCC 15554, CCEB 554, NCTC 10416, CIP 71.8 (formerly "*A. odorans*"); ATCC 19209, NCTC 10388, Burchill 1 (formerly "*A. odorans* var. *viridans*"); NCIB 8687 (formerly "*Achromobacter arsenoxydans*").

Further comments. The taxonomic position of the recently proposed new subspecies *A. faecalis* subsp. *homari* Austin, Rodgers, Forns and Colwell 1981, 75, should be verified by DNA/DNA hybridization techniques (see Taxonomic Comments).

2. **Alcaligenes denitrificans** Rüger and Tan 1983, 88.[VP]

de.ni.tri′fi.cans. L. prep. *de* away, from; L. n. *nitrum* soda; M.L. n. *nitrum* nitrate; M.L. v. *denitrificare* to denitrify; M.L. pres. part. *denitrificans* denitrifying.

The morphological characteristics are as described for the genus. Colonies on nutrient agar are circular, nonpigmented to grayish-white, translucent to opaque, flat to convex, usually smooth, sometimes dull or rough, margin usually entire.

Physiological and nutritional characteristics of the species and its two subspecies are presented in Tables 4.98 and 4.99. Most strains characteristically use *meso*-tartrate, itaconate, adipate, pimelate and other dicarboxylic acids as sole carbon sources. Some strains are auxotrophic and require organic nitrogenous compounds for growth.

Nitrates and nitrites are usually reduced. Most strains carry out an anaerobic respiration in the presence of nitrate and nitrite.

The mol% G + C of the DNA ranges from 63.9–69.8 (T_m and Bd) (De Ley et al., 1970; Yabuuchi et al., 1974; Pichinoty et al., 1978).

Type strain: ATCC 15173 (NCTC 8582; CIP 77.15; DSM 30026; Hugh 12).

2a. **Alcaligenes denitrificans** subsp. **denitrificans** Rüger and Tan 1983, 88.

See Table 4.98 for differentiation of this subspecies from the subspecies *xylosoxydans*.

Isolated from soil, and a variety of clinical specimens such as feces, urine, blood, pleural fluid, purulent ear discharges, prostatic secretions and throat swabs.

The mol% G + C of the DNA is 63.9–68.9 (T_m and Bd) (De Ley et al., 1970; Pichinoty et al., 1978).

Suggested type strain: ATCC 15173 (see above).

Other representative culture collection strains belonging to this taxon: ATCC 13138, CIP 60.81 (formerly *A. denitrificans*).

2b. **Alcaligenes denitrificans** subsp. **xylosoxydans** (Yabuuchi and Yano 1981) subsp. nov. (*Achromobacter xylosoxidans* (Yabuuchi and Ohyama 1971) Yabuuchi and Yano 1981, 477; *Alcaligenes ruhlandii* (Packer and Vishniac 1955) Aragno and Schlegel 1977, 280.)

xy.los.ox′.y.dans. Gr. n. *xylon* wood, xylose, wood sugar; Gr. adj. *oxys* sharp, acid; L. part. *dans* giving; M.L. pres. part. *oxydans* acid-giving, oxidizing; M.L. *xylosoxydans*, oxidizing xylose.

See Table 4.98 for differentiation of this subspecies from the subspecies *denitrificans*.

Nearly all strains utilize D-glucose, D-xylose, D-gluconate, adipate and pimelate as sole carbon source and characteristically form acid from D-xylose in the O/F medium of Hugh and Leifson (1953).

One strain (previously named *A. ruhlandii* ATCC 15749 (DSM 653)) can grow autotrophically with hydrogen.

Mostly isolated from clinical specimens such as blood, sputum,

---

[i] Two biovars are recognized: biovar I can grow chemolithotrophically with H_2 and reduces nitrate to nitrite; biovar II does not possess these properties.

[j] Kiredjian et al. (1981).

[k] Tested for a limited number of strains (Pichinoty et al., 1978).

[l] Strain ATCC 15749 (previously named *A. ruhlandii*) can grow chemolithotrophically with H_2 (see Taxonomic Comments).

Table 4.99.

Other characteristics of the species, subspecies and species incertae sedis of the genus **Alcaligenes***

Characteristics	Authentic *Alcaligenes* species			Species incertae sedis							
	1. *A. faecalis*	2a. *A. denitrificans* subsp. *denitrificans*	2b. *A. denitrificans* subsp. *xylosoxydans*	a. *A. eutrophus*	b. *A. paradoxus*	c. *A. latus*	d. *A. aestus*	e. *A. aquamarinus*	f. *A. cupidus*	g. *A. pacificus*	h. *A. venustus*
Gram reaction, presence of spores	−	−	−	−	−	−	−	−	−	−	−
Peritrichous flagella	+	+	+	+	+[a]	+	+	+	+	+	+
Accumulation of poly-β-hydroxybutyrate		+[b]		+[c]	+[c]	+[d]	+[e]		+[e]	+[e]	+[e]
Catalase	+	+	+	+	+	+		+			
Urease[f]	−	−[g]	−[h]	−	+						
Hydrolysis of:											
Starch	−	−	−	−	−	+[i]	d	−	−	−	−
Tween 80	−[f]	−[f]	−[h,j]	+[f,k]	+[f,k]	+[f]					
Enzymes of Entner-Doudoroff (ED) pathway present	−[l]	−[l]	−[l]	+[m]			+[n]	+[l]	+[n]	+[n]	+[n]
Enzymes of modified ED-pathway[l] present	−	+	+					−			
m-Hydroxybenzoate metabolized via:											
Gentisate				+	−	+					
Protocatechuate				−	+	−					
ortho-Cleavage of protocatechuate				+	+	−	−		+	+	+
meta-Cleavage of protocatechuate				−	−	+	−		−	−	−
Carbon sources for growth[o]:											
Number of strains investigated and references	>50[b,p,q,r]	>30[b,p,r]	>50[r]	14[c,k,r]	16[c,k,r]	2[d]	6[e]	1[r]	5[e]	6[e]	14[e]
D-Arabinose	−	−	d	−	d	−	−	−	d	−	−
D-Ribose	−	−	−	−	d	−	−	+	d	−	−
D-Galactose	−	−	−	−	+	−	d	−	+	−	−
D-Lyxose, L-xylose, D-tagatose, L-fucose, gentiobiose, amygdalin, arbutin, esculin, glycogen, L-arabitol, 5-ketogluconate	−	−	−					−			
L-Sorbose, raffinose, dulcitol	−	−	−			−		−			
D-Fucose	−	−	−	−	d	−	−		−		−
L-Rhamnose, cellobiose	−	−	−	−	d	−	−		d	−	−
Lactose, salicin	−	−	−	−	−	−	−		d		−
Maltose, sucrose	−	−	−	−	−	+	d	+	d	−	d
Melezitose, turanose	−	−	−					+			
Melibiose	−	−	−			−	−		d	−	−
Trehalose	−	−	−	−	−	d	d	+	d	−	d
N-Acetylglucosamine	−	−	−			−	−	+	+		d
Inulin	−	−	−	−	−	−	−	−	−		−
Adonitol	−	−	−	−	d	−	−	−	d	d	−
D-Arabitol	−	−	−			d	+				

* Symbols: see standard definitions.

[a] Degenerate peritrichous flagella (Davis et al., 1969).

[b] Pichinoty et al. (1978), Kiredjian et al. (1981), Yamasato et al. (1982).

[c] Davis et al. (1970).

[d] Palleroni and Palleroni (1978).

[e] Baumann et al. (1972).

[f] Popoff, unpublished results.

[g] Weakly positive for some strains according to Gilardi (1973, 1978) and Rubin et al. (1980).

[h] Yabuuchi et al. (1974).

[i] Weak reaction (Palleroni and Palleroni, 1978).

[j] Rubin et al. (1980).

[k] D. H. Davis (Ph.D. Dissertation, University of California, Berkeley, 1967).

[l] De Ley et al. (1970).

[m] Gottschalk et al. (1964).

[n] Sawyer et al. (1977).

[o] Auxanographic techniques were used. Small differences exist between the methods used by different authors; the original references should be consulted for details. Significant discrepancies between results of different authors are indicated by footnotes.

[p] Gilardi (1973); Rarick et al. (1978).

[q] Pintér and Kántor (1974).

[r] Kersters and De Ley (unpublished results), using a standardized auxanographic micromethod (API-strips, API, Montalieu-Vercieu, France).

[s,t,u] −, +, d, respectively, according to Kersters and De Ley (unpublished results).

Table 4.99.—*continued*

Characteristics	Authentic *Alcaligenes* species			Species *incertae sedis*							
	1. *A. faecalis*	2a. *A. denitrificans* subsp. *denitrificans*	2b. *A. denitrificans* subsp. *xylosoxydans*	a. *A. eutrophus*	b. *A. paradoxus*	c. *A. latus*	d. *A. aestus*	e. *A. aquamarinus*	f. *A. cupidus*	g. *A. pacificus*	h. *A. venustus*
meso-Erythritol	−	−	−	−	−	−	d	+	d	−	−
meso-Inositol	−	−	−	−	d	−	d	−	d	−	+
Sorbitol	−	−	−	−	+	−	+	+	d	−	d
2-Ketogluconate	−	−	−	+	d	+					
Mucate				+	d	+	−		−	−	−
Glycerol	−	d	d	−	+	+	+	+	+	+	+
Propionate	+	+	+	+	d	d	+	+	+	+	+
Butyrate	+	+	+	+	d	+	+	+	+	+	+
Isobutyrate	d	d	+	+	d	+	+	+	+	+	+
Valerate	d	d	+	−	d	−	+	+	+	d	+
Isovalerate	d	d	+	−	d	−	+	+	+	+	+
Caproate	d	d	d	+	d	−	+	+	+	+	+
Heptanoate	d	d	d	−	−	−	−	+	+	+	d
Caprylate	d	d	d	−	−	−	−	−	+	d	+
Pelargonate	d	d	d	−	−	−	d	−	+	d	+
Caprate	+	d	+	−	−	−	d	−	+	d	+
Oxalate	−	−	−	−	−	−	−	−	−	−	−
Malonate	+	d^s	−	−	d	+	d	−	d	−	d
Succinate, fumarate, D, L-lactate, D, L-β-hydroxybutyrate	+	+	+	+	+	+	+	+	+	+	+
Maleate	d	d	d	−	d	−	−	−	−	−	−
Glutarate	d	+	+	+	d	−	d	−	d	+	+
Azelate	−	+	+	+	d		+	+	−	−	d
Glycolate	+	d^s	−	+	d	d	−	−	+	−	−
D, L-Glycerate	d^t	d^t	d^t	−	+	+	+	+	+	d	d
D-Malate	−	+	+	d	d	−	+				
L-Malate	+	+	+	+	+	+	+				
D, L-Malate						+		+	d	+	
D-Tartrate	−	d	d	−	d	+	−	−	d	−	−
L-Tartrate	−	d	−	−	d	d	d	+	d	−	−
Pyruvate	d	+^{u}	d	+	+	−	+	−	+	+	+
Levulinate	−	−	−	+	d	−	d	−	d	−	−
α-Ketoglutarate	d	+^{u}	+^{u}	+	+	−	d	−	+	+	+
Aconitate	d^s	+^{u}	+	+	d	+	−	−	+	d	+
Citraconate	d	d	+	+	d	−	d	−	−	−	−
Mesaconate	−	d^t	+	+	+	−	d	−	−	−	−
Citrate	+	+	+	+	+	d	d	−	+	+	+
Benzoate	d	d	d	+	−	−	−	−	+	+	d
o-Hydroxybenzoate	−	−	−	−	−	−	−	−	d	−	d
m-Hydroxybenzoate	−	d^s	−	+	+	+	−	−	d	d	−
p-Hydroxybenzoate	−	−	−	+	+	+	−	−	+	d	d
D-Mandelate	−	−	−	−	−	−	−	−	−	−	−
L-Mandelate	d	d	−	−	d	+	−	−	d	−	d
Phthalate	−	−	−	−	d	−	−				
Phenylacetate	+	d^t	+	+	d	+	d	+	−	+	+
Quinate		−		−	+	+	−		d	−	d
Glycine	d^t	d^s	d	−	d	−	−	−	d	+	−
L-α-Alanine, L-glutamate, L-proline	+	+	+	+	+	+	+	+	+	+	+
D-α-Alanine	+	+	+	+	d	+	+	+	+	+	+
β-Alanine	−	d	d	+	d	+	d	−	d	d	d
L-Serine	−	d^t	+	−	d	+	−	+	+	+	d
L-Threonine	d	+	+	−	d	+	−	d	−	−	+
L-Leucine	+	d^t	+	d	+	+	d	+	d	d	+
L-Isoleucine	+	d^t	+	d	d	d	+	+	d	−	d
L-Valine	d	d^t	+	+	−		d		−	−	d
L-Aspartate	+	+	+	+	+	+	d	+	+	+	+
L-Lysine	d	d	d	−	−	−	+	−	+	+	+

Table 4.99.—*continued*

	Authentic *Alcaligenes* species			*Species incertae sedis*							
Characteristics	1. *A. faecalis*	2a. *A. denitrificans* subsp. *denitrificans*	2b. *A. denitrificans* subsp. *xylosoxydans*	a. *A. eutrophus*	b. *A. paradoxus*	c. *A. latus*	d. *A. aestus*	e. *A. aquamarinus*	f. *A. cupidus*	g. *A. pacificus*	h. *A. venustus*
L-Arginine	−	−	−			−	+	−	+	+	d
L-Ornithine	d	d	d			+	d	−	d	+	d
L-Citrulline	−	d	d			+	−	−	−	−	−
DL-Norleucine	+	d	d	+		−	−	−	−	−	−
γ-Aminobutyrate	−	d	d	+	+	+	d	−	d	d	d
δ-Aminovalerate	−	d	d	−	−	−	−	∸	d	+	+
L-Histidine	d	d^t	+	+	+	−	−		d	+	d
L-Tyrosine	d	d	d	d	d	d	−		+	+	+
L-Phenylalanine	+	+	+	+	+	−	−		d	+	+
L-Tryptophan	+	d^t	d	+	d	−	−	−	−	−	d
L-Cysteine	d	d	d			−					
L-Methionine	d	−	−			−					
D, L-Kynurenine	d	d	d			−	−	−	+	−	
Kynurenate				−	d	−	−		d	+	−
Anthranilate				−	−	−			d	+	d
Ethanolamine	−	−	−	−	d	−	−	−	−	−	+
Benzylamine	−	−	−	−	−	−	−	−		d	−
Putrescine				−	−	−	−		+	+	+
Spermine, histamine	−	−	−	−	−	−	−		d	+	−
Tryptamine	d	d	d	−	−	−	−		−	−	−
Butylamine	−	−	−	−	d	+	−		−	+	d
Pentylamine	−	−	−	−	d	−	−		d	d	−
Betaine, sarcosine	−	−	−	−	−	+	−		+	+	+
Creatine	−	−	−	−	−	+	d		+	+	+
Acetamide	d	d	−	−	−	+	d	−	d	−	−
m-Aminobenzoate	−	−	−	−	−		−	−	−	−	d
p-Aminobenozoate	−	−	−	−	−		−	−	−	−	−

wounds, purulent ear discharge, spinal fluid, cerebral tissue, urine, feces and in a few cases also from disinfectant solutions.

The mol% G + C of the DNA is 66–69.8 (T_m)(De Ley et al., 1970; Yabuuchi et al., 1974; Holmes et al., 1977).

Suggested type strain: ATCC 27061, CIP 71.32, Yabuuchi KM 543, NCTC 10807, Hugh 2838. This strain was proposed as type strain for "*Achromobacter xylosoxidans*" by Yabuuchi and Ohyama (1971).

Other representative culture collection strains belonging to this taxon: CIP 58.72 and CIP 61.20 (formerly *A. denitrificans*); ATCC 15749, DSM 653, NCIB 11475 (formerly *A. ruhlandii*).

Species Incertae Sedis

The peritrichously, flagellated, aerobic, facultatively chemolithotrophic hydrogen-oxidizing bacteria previously classified as *Alcaligenes eutrophus*, *A. paradoxus* and *A. latus*, as well as the nonfermentative, peritrichously flagellated marine bacteria *A. aestus*, *A. aquamarinus*, *A. cupidus*, *A. pacificus* and *A. venustus* (all on the Approved Lists, 1980), previously assigned to the genus *Alcaligenes*, are considered here as *species incertae sedis*. They are all validly published species names with known type strains (Approved Lists, 1980), but do not belong to the genus *Alcaligenes* for the reasons given in the section Taxonomic Comments. They should be reclassified and generically renamed.

The morphological, physiological and nutritional characteristics of these eight *species incertae sedis* are presented in Tables 4.98 and 4.99. Some additional information pertaining to *A. aestus*, *A. aquamarinus*, *A. cupidus*, *A. pacificus* and *A. venustus* can be found after the description of *A. venustus*.

a. **Alcaligenes eutrophus** Davis 1969 in Davis, Doudoroff, Stanier and Mandel 1969, 386.[AL]

eu.troph′us. Gr. prep. *eu* good, beneficial; Gr. n. *trophus* one who feeds; M.L. n. *eutrophus* good nutrition, well nourished.

Straight rods, 0.7 μm in diameter and 1.8–2.6 μm in length. Gram-negative. One to four (rarely 5 or more) peritrichous flagella are present. Colonies on nutrient agar are circular, opaque and white or cream colored. Fructose is the only sugar used by freshly isolated strains, but mutants capable of growth on D-glucose occur.

Physiological and nutritional characteristics of *A. eutrophus* are given in Tables 4.98 and 4.99.

Capable of denitrification in mineral media with lactate, but not in complex media or under autotrophic conditions; this ability may be lost on prolonged cultivation on organic media (Davis et al., 1969).

p-Hydroxybenzoate is metabolized via the *ortho*-cleavage of protocatechuate; *m*-hydroxybenzoate is metabolized via the gentisate pathway (Davis et al., 1969).

Facultatively chemolithotrophic in an atmosphere containing hydrogen, oxygen and carbon dioxide. Optimum temperature of growth: about 30°C.

Isolated from soil and water by enrichment with hydrogen gas.

The molecular weight of the genome DNA is 4.1–5.1 × 10⁹ (Auling et al., 1980).

The mol% G + C of the DNA is 66.3–67.5 (Bd and T_m) (Davis et al., 1969; De Ley, Segers and Tytgat, unpublished results).

Type strain: ATCC 17697 (DSM 531; Stanier 335).

Further Descriptive Information

Mesosomal structures and inclusions of polyphosphates and poly-β-hydroxybutyrate have been detected in cells of *A. eutrophus* (Walther-Mauruschat et al., 1977). Electron microscopic investigations revealed a type of cell wall similar to the walls of the majority of Gram-negative bacteria and of the following Gram-negative hydrogen oxidizing bacteria: *A. paradoxus, Pseudomonas facilis, P. flava, P. palleronii, P. pseudoflava* and *Aquaspirillum autotrophicum* (Walther-Mauruschat et al., 1977). *A. eutrophus* and *A. paradoxus* possess a similar type of flagellar fine structure, which differs from that of other hydrogen-oxidizing bacteria such as *P. facilis, P. flava, P. palleronii* and *A. ruhlandii* (the latter organism is classified here as *A. denitrificans* subsp. *xylosoxydans* (see Taxonomic Comments)).

Considerable knowledge of the biochemistry and physiology of *A. eutrophus* was obtained by Schlegel and co-workers, Göttingen, West Germany. The majority of these biochemical studies were performed with strain H 16 (ATCC 17699, NCIB 10442, DSM 428) and mutants therefrom. From the numerous metabolic and enzymological features studied in *A. eutrophus* only a few are mentioned here: growth on and metabolism of D-fructose and D-glucose via the Entner-Doudoroff pathway (Gottschalk et al., 1964; Gottschalk, 1965; Schlegel and Gottschalk, 1965); utilization of 2-ketogluconate (Nandadasa et al., 1974); regulatory mechanisms of carbohydrate metabolism (Blackkolb and Schlegel, 1968; König, et al., 1969; Abdelal and Schlegel, 1974; Bowien, et al., 1974); formation and utilization of poly-β-hydroxybutyric acid (Schindler, 1964; Hippe, 1967); fixation of carbon dioxide (Hirsch, 1963; Hirsch et al., 1963; Hirsch and Schlegel, 1963); the biosynthesis of isoleucine and valine (Hill and Schlegel, 1969; Reh and Schlegel, 1969; Wiegel and Schlegel, 1977a, b); the biosynthesis of aromatic amino acids (Friedrich et al., 1976); autotrophic growth on formate (Friedrich, et al., 1979); dissimilation of aromatic compounds and the regulation of the β-ketoadipate pathway (Johnson and Stanier, 1971); and excretion of glycolate (Codd, et al., 1976), ethanol, lactic acid and butanediol (Schlegel and Volbrecht, 1980).

In contrast to other hydrogen-oxidizing bacteria, *A. eutrophus* contains two distinctly different and independent hydrogenases: a soluble NAD-reducing enzyme which generates reducing power for CO_2-fixation, and a membrane-bound enzyme which is unable to react with NAD or NADP and functions as part of the respiratory chain (Wittenberger and Repaske, 1961; Eberhard, 1966; Schneider and Schlegel, 1977). The soluble NAD-reducing enzyme has been purified and characterized as a conjugated iron-sulfur protein containing iron-sulfur centers and FMN as an enzyme-bound electron carrier (Schneider and Schlegel, 1976, 1978; Schneider, et al., 1979). The molecular weight, subunit composition and isoelectric point of the soluble and membrane-bound hydrogenases are remarkably different (Schink and Schlegel, 1979). The latter enzyme contains no FMN and was found to be immunologically related to the membrane-bound hydrogenase of *P. pseudoflava* (strain DSM 1034) (Schink and Schlegel, 1980).

The key enzyme for CO_2 fixation, D-ribulose-1,5-diphosphate carboxylase, belongs to the high molecular weight multimeric enzymes that consist of eight large and eight small subunits (Bowien, et al., 1976; Bowien and Mayer, 1978; Bowien, et al., 1980). Such enzymes are characteristic for higher plants, unicellular green algae, cyanobacteria and some *Chromatiaceae*.

The autotrophic growth of *A. eutrophus* has received special attention as a potential source of single-cell protein and as a bioregenerative life support system for space travel. The conditions for obtaining dense autotrophic cultures (25 g dry weight/liter medium) of *A. eutrophus* have been described (Repaske and Mayer, 1976).

Widespread defective lysogeny has been reported in strains of *A. eutrophus* (Auling et al., 1977). A large plasmid pAE1 was detected in three *A. eutrophus* strains (Lim, et al., 1980). Plasmid-less derivatives of *A. eutrophus* have been found to completely lack hydrogenase activity but still possess D-ribulose-1,5-diphosphate carboxylase activity. Friedrich et al. (1981) found that hydrogen metabolism by *A. eutrophus* is genetically linked to a self-transmissible plasmid.

"*Alcaligenes hydrogenophilus*" (a species proposed by Ohi, Takada, Komemushi, Okazaki and Miura, 1979; not on Approved Lists, 1980) does not merit a separate species rank, since DNA/rRNA hybridizations, protein electrophoregrams and nutritional properties show that it is indistinguishable from *A. eutrophus* (Kersters, Segers and De Ley, unpublished results).

Immunological comparisons of glutamine synthetase (Baumann and Baumann, 1978) and DNA/rRNA hybridizations (De Ley, Segers and De Vos, unpublished results; see Fig. 4.85 and Taxonomic Comments) suggest that *A. eutrophus* is genetically more related to *P. solanacearum* than to *A. faecalis* or *A. paradoxus*.

b. **Alcaligenes paradoxus** Davis 1969 in Davis, Doudoroff, Stanier and Mandel 1969, 387.[AL]

pa.ra.dox′us. Gr. prep. *para* amiss, contrary to; Gr. n. *doxus* an opinion; M.L. n. *paradoxus* contrary to expectation, in reference to the chemolithotrophic and/or organotrophic metabolism of the organism.

Straight rods, 0.5 μm in diameter and 1.5–2.6 μm in length, occurring singly or in pairs. Gram-negative. Motile by one to two (rarely three to four) "degenerately peritrichous" flagella, which have predominantly a subpolar or lateral insertion. The flagella are fragile and may be 4–6 times as long as the cell. Pili have been observed at the polar caps of *A. paradoxus* (Aragno et al., 1977).

Colonies on nutrient agar are yellow and usually glistening and slimy. Nonmucoid variants can be selected from mucoid strains (Davis, 1967). The yellow pigments are carotenoids with absorption maxima at ~405 and/or 425 nm in acetone. *A. paradoxus* strains possess a characteristic cytochrome a_2 with a peak at 625 nm in the difference spectrum.

Many, but not all, strains are chemolithotropic in an atmosphere of hydrogen, oxygen and carbon dioxide.

Physiological and nutritional characteristics of *A. paradoxus* are presented in Tables 4.98 and 4.99.

Denitrification does not occur.

Both *meta*- and *para*-hydroxybenzoate are metabolized via the *ortho*-cleavage of protocatechuate.

Two biovars were recognized by Davis et al. (1969). Strains of biovar I can use the aerobic oxidation of hydrogen gas as a source of energy for autotrophic growth, and are able to reduce nitrate to nitrite in organic media; strains belonging to biovar II cannot grow autotrophically and rarely reduce nitrate to nitrite in organic media.

Optimum temperature for growth: about 30°C.

A. paradoxus is a common soil inhabitant.

The molecular weight of the genome DNA is 4.7–4.8 × 10⁹ (Auling et al., 1980).

The mol% G + C of the DNA is 66.8–69.4 (Bd and T_m) (D.H. Davis, Ph. D. dissertation, University of California, Berkeley, 1967; Davis et al., 1969; De Ley, Segers and Tytgat, unpublished results).

Type strain: ATCC 17713 (DSM 30034; Stanier 351). The type strain belongs to biovar I; *A. paradoxus* ATCC 17549 (DSM 30162; Stanier 180) is considered as a representative strain of biovar II (Davis et al., 1969).

Further Descriptive Information

Mesosomal structures, inclusions of polyphosphates and poly-β-hydroxybutyrate have been demonstrated in two strains (including the type strain) of *A. paradoxus* (Walther-Mauruschat et al., 1977). Micromorphological studies indicated that the fine structure of the cell wall of *A. paradoxus* is similar to the cell walls of e.g. *A. eutrophus, P. flava, P. palleronii, Aquaspirillum autotrophicum* and other Gram-negative bacteria (Walther-Mauruschat et al., 1977). The flagella of *A. paradoxus* and *A. eutrophus* possess the same type of fine structure (Aragno et al., 1977).

Pesticide-degrading plasmids have been isolated from *A. paradoxus* (Don and Pemberton, 1981). There are indications that a plasmid may play a role in the autotrophic metabolism of *A. paradoxus* biovar I (Lim et al., 1980).

In contrast to *A. eutrophus*, only a membrane-bound hydrogenase and no soluble hydrogenase is present in *A. paradoxus* (Schneider and Schlegel, 1977).

Immunological comparisons of glutamine synthetase (Baumann and Baumann, 1978) and DNA/rRNA hybridizations (De Ley, Segers and De Vos, unpublished results; see Fig. 4.85 and Taxonomic Comments) suggest that *A. paradoxus* is genetically more related to *P. acidovorans* than to *A. faecalis* or *A. eutrophus*.

c. **Alcaligenes latus** Palleroni and Palleroni 1978, 423.[AL]
la′tus. L. adj. *latus* broad.

The cells are short to coccoid rods, 1.2–1.4 μm in diameter and 1.6–2.4 μm in length, occurring singly, in pairs or in short chains. Gram-negative. Motile by means of 5–10 peritrichous flagella. The cells are frequently heavily granulated.

Under autotrophic growth conditions (Palleroni and Palleroni, 1978) on a solid mineral medium colonies are round, greyish pink and opaque. Colonies are wrinkled in fresh isolates but can become smooth upon subcultivation.

Facultatively chemolithotrophic in an atmosphere containing hydrogen, oxygen and carbon dioxide. Can grow with dinitrogen as sole nitrogen source (Schlegel, personal communication).

Physiological and nutritional characteristics are given in Tables 4.98 and 4.99.

Optimum temperature of growth: about 35°C.

A membrane-bound hydrogenase, but no soluble, NAD-reducing hydrogenase has been found in three strains of *A. latus* (Palleroni and Palleroni, 1978).

meta-Hydroxybenzoate is metabolized via the gentisate pathway and protocatechuate is degraded by *meta*-cleavage by cells grown on *p*-hydroxybenzoate or on quinate.

Three strains were isolated from soil (Palleroni and Palleroni, 1978).

A. latus belongs in the rRNA branch of *P. acidovorans*, *P. testosteroni* and *A. paradoxus* (Fig. 4.85; see Taxonomic Comments).

The mol% G + C of the DNA is 69.1–71.1 (Bd and T_m) (Palleroni and Palleroni, 1978; De Ley and Segers, unpublished results).

Type strain: ATCC 29712 (DSM 1122; Palleroni H-4).

d. **Alcaligenes aestus** Baumann, Baumann, Mandel and Allen 1972, 426.[AL]
aes′tus. L. gen. n. *aestus* of the tide.

Straight rods. The cell dimensions are not given in the original description, but cells of the type strain (ATCC 27128) are 0.8 μm × 2.0–3.0 μm. Gram-negative. Motile by means of peritrichous flagella.

The colony morphology is not given in the original description. Grows on Marine agar (Difco) and on nutrient agar supplemented with 2% NaCl.

Physiological and nutritional characteristics are presented in Tables 4.98 and 4.99.

All the strains grow on C_7–C_{10} dicarboxylic acids but no strain utilizes aromatic compounds as carbon source.

Isolated from seawater near Oahu, Hawaii at depths of 100–600 m.

The mol% G + C of the DNA is 56.9–57.9 (Bd) Baumann *et al.*, 1972).

Type strain: ATCC 27128 (Baumann 134).

e. **Alcaligenes aquamarinus** (ZoBell and Upham 1944) Hendrie, Holding and Shewan 1974, 537.)[AL] (*Achromobacter aquamarinus* ZoBell and Upham 1944, 264.)
a.qua.ma.ri′nus. L. n. *aqua* water; L. adj. *marinus* of the sea; M.L. adj. *aquamarinus* pertaining to seawater.

Straight rods, 0.8 × 2.0–4.0 μm. Gram-negative. Motile by one to eight peritrichous flagella.

Grows on Marine agar (Difco) and on nutrient agar supplemented with 2% NaCl.

Physiological and nutritional characteristics of *A. aquamarinus* are presented in Tables 4.98 and 4.99.

Can use various dicarboxylic acids as carbon source. Utilization of ammonium or nitrate salts as sole nitrogen source has not been demonstrated. Nitrate is reduced to nitrite but not to nitrogen gas.

Isolated from sea water.

The mol% G + C of the DNA is 57.9 (T_m) (De Ley *et al.*, 1970).

Type strain: ATCC 14400 (NCMB 557; DSM 30161; ZoBell 558).

f. **Alcaligenes cupidus** Baumann, Baumann, Mandel and Allen 1972, 426.[AL]
cu′pi.dus. L. adj. *cupidus* desiring.

Straight rods. The cell dimensions are not given in the original description, but cells of the type strain (ATCC 27124) are 0.7–1.0 μm × 2.0–3.0 μm. Gram-negative. Motile by peritrichous flagella.

The colony morphology is not given in the original description. Grows on Marine agar (Difco) and on nutrient agar supplemented with 2% NaCl.

Physiological and nutritional characteristics of *A. cupidus* are given in Tables 4.98 and 4.99.

Oxidase-negative. *A. cupidus* is a nutritionally versatile organism being able to utilize at least 70 different carbon compounds as sole sources of carbon and energy.

Isolated from surface water off the immediate coast of Oahu, Hawaii.

The mol% G + C of the DNA ranges from 59.9–62.1 (Bd) (Baumann et al., 1972).

Type strain: ATCC 27124 (Baumann 79).

g. **Alcaligenes pacificus** Baumann, Baumann, Mandel and Allen 1972, 426.[AL]
pa.ci′fic.us. M.L. adj. *pacificus* pertaining to the Pacific ocean.

Straight rods. The cell dimensions are not given in the original description, but cells of the type strain (ATCC 27122) are 0.7 μm × 1.4–2.0 μm. Gram-negative. Motile by peritrichous flagella.

The colony morphology is not given in the original description. Grows on Marine agar (Difco) and on nutrient agar supplemented with 2% NaCl.

Physiological and nutritional characteristics of *A. pacificus* are presented in Tables 4.98 and 4.99.

Utilizes D-glucose and D-fructose but no other hexoses as carbon and energy source.

Isolated from surface water off the immediate coast of Oahu, Hawaii.

The mol% G + C of the DNA is 67.3–68.3 (Bd) (Baumann et al., 1972).

Type strain: ATCC 27122 (Baumann 62).

h. **Alcaligenes venustus** Baumann, Baumann, Mandel and Allen, 1972, 426.[AL]
ve.nus′tus. L. adj. *venustus* lovely, beautiful.

Straight rods. The cell dimensions are not given in the original description, but cells of the type strain (ATCC 27125) are 0.8–1.0 μm × 1.4–1.6 μm.

The colony morphology is not given in the original description. Grows on Marine agar (Difco) and on nutrient agar supplemented with 2% NaCl.

Physiological and nutritional characteristics are given in Tables 4.98 and 4.99. Is able to grow at 4°C.

Isolated from surface water off the immediate coast of Oahu, Hawaii.

The mol% G + C of the DNA is 52.9–54.5 (Baumann et al., 1972).

Type strain: ATCC 27125 (Baumann 86).

Further Comments on the Species Incertae Sedis d–h

All the strains belonging to *A. aestus*, *A. aquamarinus*, *A. cupidus*, *A. pacificus* and *A. venustus* are from marine origin. They do not form fluorescent or yellow cell-associated pigments. They are not luminescent and do not grow chemolithotrophically with hydrogen gas (Baumann et al., 1972). *A. aestus*, *A. cupidus*, *A. pacificus* and *A. venustus* accumulate poly-β-hydroxybutyrate as intracellular reserve product and metabolize D-glucose and D-fructose primarily via the Entner-Doudoroff pathway (Sawyer et al., 1977). These four marine species display also a similar type of regulation of their aspartokinase activity (Baumann and Baumann, 1974).

All these marine bacteria belong in the second rRNA superfamily (*sensu* De Ley, 1978) and are more related to *Pseudomonas fluorescens* than to *A. faecalis* (Fig. 4.85). Baumann et al. (1972) stated in their original description that these peritrichously flagellated marine bacteria were physiologically more related to the polarly flagellated pseudomonads. The degree of similarity of the rRNA cistrons of the five marine species is represented in Fig. 4.85. *A. aestus* and *A. aquamarinus* possess similar physiological and nutritional characteristics (Tables 4.98 and 4.99), similar protein electrophoregrams, similar DNA base composition and similar rRNA cistrons (Fig. 4.85). Both species may therefore be highly interrelated.

Genus **Serpens** Hespell 1977, 380[AL]

ROBERT B. HESPELL

Ser′pens. L. fem. n. *serpens* snake, serpent.

Rod-shaped cells, 0.3–0.4 µm wide by 8–12 µm long. Occur singly or in pairs. Cysts or coccoid bodies not formed, but cells in the stationary phase of growth are longer (16–25 µm) and often possess blebs or spherical protuberances. Gram-negative. **Extremely flexible and capable of serpentine-like motility in agar gels. Possess bipolar tufts of 4–10 flagella and also a few lateral flagella.** Poly-β-hydroxybutyrate or other internal granules not formed. Have a strictly respiratory type of metabolism with oxygen as the sole electron acceptor. **Grow aerobically but prefer oxygen concentrations less than that of an air atmosphere.** Catalase- and **oxidase-positive**. Chemoorganotrophic. **Lactate is the only effective carbon and energy source**, although very slight growth occurs with acetate or α-ketoglutarate. Carbohydrates, fatty acids and sugar alcohols are not catabolized. Casein hydrolysate, peptone, yeast extract and, for most strains, ammonium chloride can serve as nitrogen sources; nitrates and nitrites are not used. Vitamins are stimulatory but not required. Optimum temperature, 28–30°C. On media containing 1.8–2.0% agar, colonies are cream colored, round, 3–6 mm in diameter, and have a filamentous edge. **On media with less than 1.5% agar, only subsurface spreading colonies occur.** Found in the sediments of eutrophic freshwater ponds. The mol% G + C of the DNA is 66 (Bd).

Type species: *Serpens flexibilis* Hespell 1977, 381.

Further Descriptive Information

The most striking morphological feature is the complex, serpentine-like movement of the cells when observed in slides of agar pieces removed from the edge of a spreading colony (Fig. 4.86). The organisms move very rapidly and display a furious lashing and bending of the cells; coiling into a knotlike configuration is also common. In liquids, the movement is less violent and the cells move rapidly with reversals along a straight line axis with a gentle flexing of the body, and the distal or trailing tip of the cell vibrates intensely (Fig. 4.86 *E–H*). No tumbling type of movement is evident. Motility occurs in buffer solutions of polyvinylpyrrolidine having a viscosity greater than 1000 centipoise (Greenberg and Canale-Parola, 1977). In comparison, viscosities of 60 and 1000 centipoise prevent the motility of *Escherichia coli* and *Spirochaeta halophila*, respectively. The structural basis permitting such cell flexibility is unknown, but the organisms possess a thin peptidoglycan layer and a morphologically typical Gram-negative outer membrane.

Although growth occurs aerobically, *S. flexibilis* prefers oxygen tensions less than that of an air atmosphere, growing subsurface (1–3 mm deep) in stationary liquid media or in media with agar concentrations of less than 1.5%. The type strain and all other strains examined can use only lactate as an effective oxidizable carbon and energy source. Cell extracts contain only trace levels of the key enzymes of the Embden-Meyerhof-Parnas or hexose monophosphate pathways, but the tricarboxylic acid cycle enzymes are in high levels. Vitamins are not required for growth, but yeast extract is often stimulatory. Most strains can use ammonium chloride as a sole nitrogen source, but increased growth rates and cell yields occur with media containing peptone or casein hydrolysate.

Isolation and Enrichment Procedures

S. flexibilis can be readily isolated from the upper few centimeters of the sediments of eutrophic freshwater ponds. Traditional enrichment culture techniques employing lactate as the carbon source are not useful for isolation as the organisms often become overgrown by other heterotrophic bacteria. However, selective isolation can be accomplished by the use of membrane filters overlayed on a Petri dish of appropriate media, a technique which has also been employed for the isolation of thin spirilla and spirochaetes (Canale-Parola et al., 1966). A small inoculum of pond water/mud slurry is deposited in the center of a sterile cellulose filter disk (0.3–0.45 µm pore diameter) which has been placed on the surface of a plate of isolation medium.* After incubation for 6–12 h at 30°C, the disk is removed and the plate is incubated for 2–4 days. The organisms grow as a subsurface, whitish veil that diffuses from the center of the plate. By picking from the edge of the veil and streaking onto a second plate, cloned cultures may be obtained. Although aerobic spirilla and spirochaetes can also be obtained in this manner, these organisms can usually be distinguished from *S. flexibilis* by light microscopic observations, but definitive confirmation should be done by electron microscopy.

Maintenance Procedures

S. flexibilis is grown routinely in LYPP broth† and can be maintained by biweekly transfers on slants of LYPP agar. The organism can be preserved indefinitely by lyophilization or by storage in liquid nitrogen.

Taxonomic Comments

At present, the relationship of the genus *Serpens* to other genera of Gram-negative aerobic rods is unclear. However, preliminary evidence from an analysis of the oligonucleotide catalog obtained from the 16S rRNA suggests that the genus may be phylogenetically related to the family *Pseudomonadaceae* (Hespell and Woese, unpublished data).

The genus presently consists of a single species, *S. flexibilis*, as all strains thus far isolated appear to be quite similar. *Editorial Note*: Woese et al (1983) recently reported that *A. flexibilis* is closely related to *Pseudomonas pseudoalcaligenes* (S_{AB} = 0.9) and might be considered as a variant pseudomonad that had initially developed a defect in its system for synthesizing septa.

---

* Isolation medium (per 90 ml of water): yeast extract, 0.2 g; peptone, 0.1 g; hay extract, 10 ml; and agar, 1.0 g. The hay extract is prepared by boiling 10 g of hay in 100 ml of water for 15 min and clarifying the mixture by centrifugation. The pH of the isolation medium is adjusted to 7.0 with KOH before sterilization.

† LYPP broth (per 100 ml water): 60% sodium lactate syrup, 1.0 ml; yeast extract, 0.3 g; peptone, 0.2 g; and K_2HPO_4, 0.35 g. The pH of the medium is adjusted to 7.2 prior to autoclaving. LYPP agar is prepared by adding 1.5 g of agar.

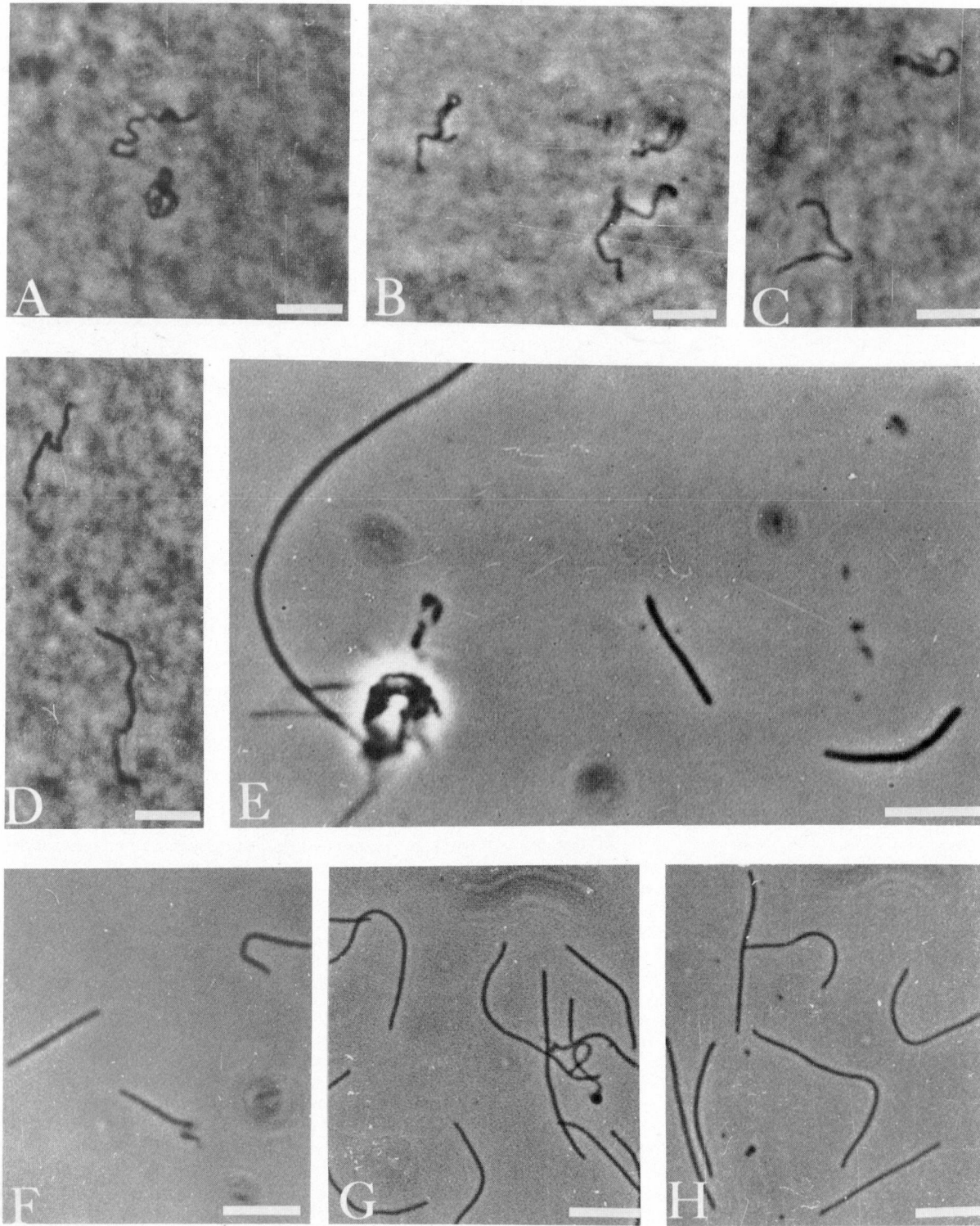

Figure 4.86. Living cells of *Serpens flexibilis* PFR-1. Wet-mount preparations, phase contrast. Cells in agar pieces excised from subsurface, spreading growth (*A–D*) displayed rapid motion with bending, lashing, and serpentine-like movements. Cells rapidly reversed themselves between coiled and uncoiled forms (*A*). In liquid menstrua (*E–H*), logarithmic-phase cells (*F* and *H*) showed straight-line movements with flexing of the cell body. Stationary-phase cells often clumped (*E*) and had protuberances (*G*). *Marker bars* represent 3.0 μm. (Reproduced with permission from R. B. Hespell, International Journal of Systematic Bacteriology *27:* 371–381, 1977, © International Association of Microbiological Societies.)

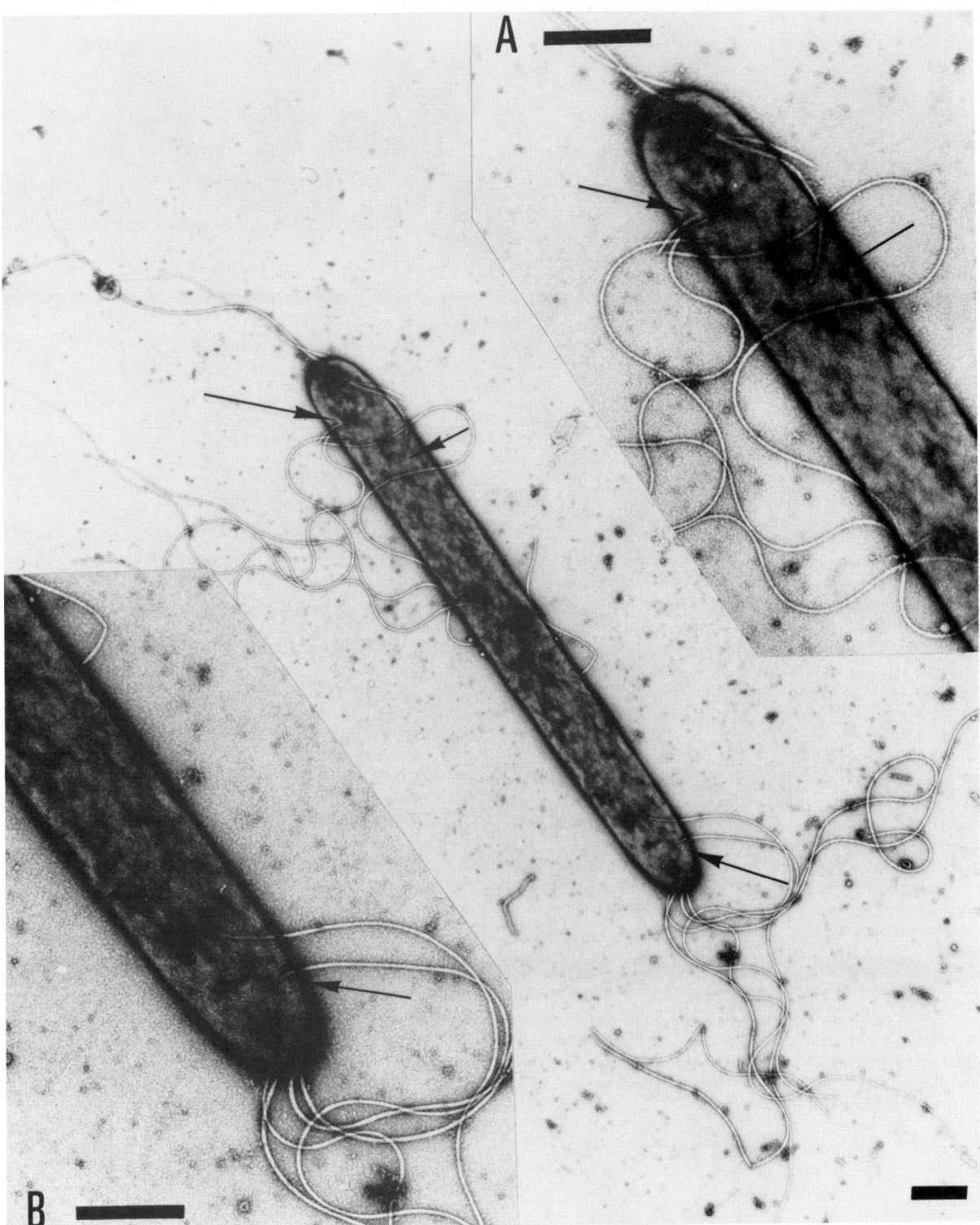

Figure 4.87. Transmission electron micrograph of *S. flexibilis* PFR-1 cells. The *insets* (*A* and *B*) are higher magnifications of the ends of the same cell stained with phosphotungstic acid. Although the overall cell shape has been preserved, some flagella have detached, and an aggregate of four flagella remains at one end of the cell. Several laterally inserted flagella are present, and their insertion points (*arrows*) are more clearly evident in the insets. The *marker bars* represent 0.25 μm. (Reproduced with permission from R. B. Hespell, International Journal of Systematic Bacteriology *27*: 371–381, c 1977, © International Association of Microbiological Societies.

List of the species of the genus **Serpens**

1. **Serpens flexibilis** Hespell 1977, 381.[AL]

flex.i.bi'lis. L. adj. *flexibilis* flexible, pliant.

The description of the species is the same as that for the genus. The morphological features of the type strain are depicted in Figs. 4.86 and 4.87.

Type strain: ATCC 29606.

Genus **Janthinobacterium** De Ley, Segers and Gillis 1978, 164[AL]

PETER H. A. SNEATH

Janth.in.o.bac.ter′i.um. L. adj. *Janthinus* violet-colored; Gr. n. *bakterion* a small rod; M.L. neut. n. *Janthinobacterium* a small violet-colored rod.

Rods 0.8–1.2 × 2.5–6.0 μm with rounded ends, sometimes slightly curved. Occur singly, occasionally with some pairs or short chains. Definite capsules are not evident, although sometimes intercellular slime is formed. **No resting stages are known**. Gram-negative, occasionally with barred or bipolar staining and lipid inclusions. **Motile by means of both a single polar flagellum and usually one to four subpolar or lateral flagella. Strict aerobes**. Produce low convex, round, **violet colonies** on solid media; in nutrient broth, a **violet ring** is formed at the junction of the liquid surface and the container wall. Optimum temperature, 25°C; minimum, 2°C; maximum, 32°C. Optimum pH, 7–8; no growth below pH 5. No growth occurs in media containing 6% or more of NaCl. Chemoorganotrophs, having a strictly **respiratory type of metabolism with oxygen as the terminal electron acceptor**. Acid but no gas is produced from glucose and certain other carbohydrates. Lactate is oxidized to CO_2. Usually oxidase-positive by the method of Kovacs (1956), although the violet pigment may interfere with the reading. Catalase-positive. Indole-negative. Voges-Proskauer-negative. Nitrate and nitrite are reduced, sometimes with visible gas production. Ammonia is formed from peptone. Phosphatase-positive. Arylsulfatase-negative. Grow on ordinary peptone media. Utilize citrate and ammonia as sole carbon and nitrogen sources for growth and grow rapidly. Growth factors are not required. **Resistant to benzylpenicillin** (10 μg/ml) **and to vibriostatic agent 0/129** (2,4-diamino-6,7-diisopropylpteridine, 30 μg/disc). Soil and water organisms, common in temperate climates. Occasionally cause food spoilage. The mol% G + C of the DNA is 61–67 (T_m).

Type species: *Janthinobacterium lividum* (Eisenberg 1891) De Ley, Segers and Gillis 1978, 164.

Further Descriptive Information

The growth of *Janthinobacterium* is often gelatinous or rubbery (70% of strains) and a tough pellicle is then seen on broth cultures. The violet ring in broth is usually viscous and frondlike. Violet pigmentation is often less intense and is produced more slowly in colonies, sometimes as concentric rings in the colonies. A few strains show a pale yellow diffusible fluorescent pigment in young cultures. The cultures do not smell of ammonium cyanide.

Properties of the pigment violacein are described under the genus *Chromobacterium*, together with a description of the characteristic flagellar arrangement.

Strains of *Janthinobacterium* are common in soil and water in temperate regions and are occasionally involved in food spoilage (strains from food sources may produce active metalloproteinases (Dainty et al., 1978)). Strains from leaf nodules of plants (Bettelheim et al., 1968) seem to belong to some other genus (De Ley et al., 1978).

Enrichment and Isolation Procedures

Isolation from soil may be made by the method of Corpe (1951). A few crumbs of soil are placed in a Petri dish and 10–25 ml of sterile water added. About 5 or 6 grains of heat-sterilized, polished rice are added, and the plates are closed and incubated at 20°C for several weeks. Incubation at 4–10°C may be advantageous. Strains can be isolated by plating from violet patches on the rice onto nutrient agar and incubating at 20°C. In place of nutrient agar one can also use 0.025% yeast extract in watery agar, or a medium consisting of 1% mashed and strained, boiled rice in 1.5% water agar supplemented with L-tryptophan (25 mg/liter) to improve pigmentation. *Janthinobacterium* can also be isolated on the selective medium of Ryall and Moss (1975) or of Keeble and Cross (1977) as described for the genus *Chromobacterium*.

Maintenance Procedures

The organisms survive for several years in dilute peptone water (0.1% peptone) at 4°C or frozen at −76°C in nutrient broth containing 15% glycerol. They also survive freeze-drying, but viability may be low.

Procedures for Testing Special Characters

The testing methods for HCN production, carbohydrate reactions and hydrolysis of casein and esculin, together with the characteristics of violacein, are described by Sneath (1979) and are also given in the chapter on *Chromobacterium* in this *Manual*.

Table 4.100. *Physiological characteristics of* **Janthinobacterium lividum**[a]

Characteristic	Reaction or Result
Acid produced from:[b]	
L-Arabinose, fructose, galactose, glucose, *m*-inositol, maltose, mannose, sorbitol, xylose	+ (95)
Lactose, glycerol, mannitol, sucrose	[+] (80)
cellobiose, inulin	d (50)
Salicin, starch, trehalose	[−] (10)
Dulcitol	−
Esculin hydrolysis	+ (95)
Chitin	[−] (5)
Reducing substance (e.g. 2,5-diketogluconate) produced from gluconate	[−] (30)
Urease	− or W
Phenylalanine deaminase	−
β-Galactosidase (ONPG method)	[−] (10)
Gelatin liquefaction, 7 days	[−]
Casein hydrolysis	[−] (10)
Hemolysis on horse blood agar	d (20)
Amino acid decarboxylases:	
Lysine decarboxylase	−
Ornithine decarboxylase	−
Arginine dihydrolase	−
Arginine decarboxylase	[−] (20)
Melanin produced from phenylalanine	[−]
Nitrate reduction	+ (95)
Nitrite reduction	[+] (80)
Visible gas production from nitrate	d (40)
Turbid zone on egg yolk agar (lecithinase)	−
HCN produced	−
Methylene blue reduction	[−]
Selenite reduction	−
Growth in KCN medium	[−]
Growth on MacConkey agar	d (50)
Sole carbon sources:	
Citrate	+
L-Leucine, acetate	[+] (80)
Malonate	[−] (10)
L-Ornithine, L-phenylalanine	−

[a] Symbols: +, positive for 95% or more strains; [+], positive for 80% or more strains; d, positive for 31–79% of strains; [−], positive for 30% or fewer strains; −, negative for all strains; and W, weak reaction. Numbers in parentheses indicate the % of strains giving a positive reaction.
[b] Tested in the medium of Hugh and Leifson (1953).

Differentiation of **Janthinobacterium** from other genera

See Table 5.84, the genus *Chromobacterium*, for characteristics that differentiate *Janthinobacterium* from *Chromobacterium*.

Taxonomic Comments

The nomenclature of the genus has been confused, and the name "*Chromobacterium lividum*" has been widely used in the past for the organism now known as *Janthinobacterium lividum* (see *C. lividum* in the eighth edition of the *Manual*). "*C. lividum*" and *C. violaceum* strains were found by De Ley et al. (1978) to differ phenotypically and genotypically (by DNA/DNA hybridization) and to form tight, separate clusters around their type strains on RNA similarity maps; consequently, the "*C. lividum*" cluster was elevated to genus rank as *Janthinobacterium lividum*.

Membrane-forming (gelatinous) strains earlier called "*Chromobacterium amethystinum*" and "*C. membranaceum*" cross-react serologically with other strains of *J. lividum* and do not differ notably in other ways.

The genus *Janthinobacterium* has been grouped with *Rhizobium* and *Agrobacterium*; however, the findings of De Ley, Segers and Gillis (1978) suggest that it is more closely related to *Alcaligenes, Bordetella, Chromobacterium*, and some nonfluorescent pseudomonads. Unpigmented strains could be mistaken for bacteria such as these.

The description of *J. lividum* is based on the studies by Leifson (1956), Sneath (1956, 1960) and De Ley, Segers and Gillis (1978).

List of the species of the genus **Janthinobacterium**

1. **Janthinobacterium lividum** (Eisenberg 1891) De Ley, Segers and Gillis 1978, 164.[AL] (*Bacillus lividus* Eisenberg 1891, 81; *Chromobacterium lividum* Bergey et al. 1923, 119).

li'vi.dum. L. adj. *lividum* leaden-colored, dark-blue.

The description is the same as that given for the genus. Other characteristics are given in Table 4.100.

Common in soil and water in temperate regions.

The mol% G + C of the DNA is 61–67 (T_m).

Type strain: NCTC 9796 (HB; ATCC 12473; NCIB 9130; D 303).

Other organisms

A number of violet-pigmented Gram-negative rods of marine origin have been described which produce violacein, but their relation to *Janthinobacterium* and *Chromobacterium* are still uncertain. An example is "*Chromobacterium marinum*" (Hamilton and Austin, 1967). They require further study but appear to share the following features: they have a requirement for at least 1% NaCl, are strictly aerobic, show oxidative attack on glucose, have restricted growth temperature from about 10–30°C, do not hydrolyze esculin or arginine, usually hydrolyze casein and give turbidity from egg yolk, do not produce HCN, and may acidify trehalose but not xylose.

Genus **Brucella** Meyer and Shaw 1920, 173[AL]

M. J. CORBEL AND W. J. BRINLEY-MORGAN

Bru.cel'la. L. dim. ending-*ella*; M.L. fem. n. *Brucella* named after Sir David Bruce, who first recognized the organism causing undulant (Malta) fever.

Cocci, coccobacilli or short rods, 0.5–0.7 μm in diameter and 0.6–1.5 μm in length. Arranged singly and, less frequently, in pairs, short chains or small groups. True capsules are not produced. Do not usually show true bipolar staining. Resting stages are not known. Gram-negative. **Nonmotile;** do not produce flagella. **Aerobic,** possessing a respiratory type of metabolism and having a **cytochrome-based electron transport system** with oxygen or nitrate as the terminal electron acceptor. Nitrate reductase is produced. **Many strains require supplementary CO$_2$ for growth,** especially on primary isolation. Colonies on serum-dextrose agar or other clear medium are transparent, raised, convex, with an entire edge and a smooth, shiny surface. They appear a **pale honey color** by transmitted light. Nonsmooth variants of the smooth species occur but there are also stable nonsmooth species with a distinctive host range. Optimum temperature 37°C. Growth occurs between 20 and 40°C. Optimum pH, 6.6–7.4. **Catalase-positive.** Usually oxidase-positive, but negative strains occur. Chemoorganotrophic. Most strains require complex media containing several amino acids, thiamin, nicotinamide and magnesium ions; some strains may be induced to grow on minimal media containing an ammonium salt as the sole nitrogen source. **Growth is improved by serum or blood, but hemin** (X-factor) **and nicotinamide adenine dinucleotide** (NAD: V-factor) **are not essential. Acid production does not occur from carbohydrates in conventional media,** except for *B. neotomae*. Do not produce indole. **Do not liquefy gelatin** or inspissated serum. Do not lyse erythrocytes. Do not produce **acetyl methyl carbinol** (Voges-Proskauer test.) **The methyl red test is negative. Possess characteristic intracellular antigens specific for the genus.** Intracellular parasites, transmissible to a wide range of animal species including man. The mol% G + C of the DNA is 55–58.

Type species: *Brucella melitensis* (Hughes 1893) Meyer and Shaw 1920, 173.

Further Descriptive Information

Cell morphology. When grown in nutritionally-adequate liquid or solid media such as serum-dextrose broth (SDB)* or serum-dextrose agar (SDA†, *Brucella* cells are coccoid, coccobacilli or short rods with slightly convex sides and rounded ends. Freshly-isolated strains tend to be more coccoid than laboratory-adapted cultures; this is also true

---

* Serum-dextrose broth (SDB): tryptone-soya broth (Oxoid), 30 g; distilled water, 1000 ml; sterile horse serum (inactivated at 56°C for 30 min), 50 or 100 ml; D-glucose (25% (w/v) solution autoclaved at 105°C for 20 min), 40 ml. The medium is prepared by dissolving the tryptone soya broth powder in the water and autoclaving at 115°C for 15 min. After cooling, the sterile horse serum and glucose are added aseptically. Note: Trypticase soy broth (BBL), brucella broth (BBL) or brucella broth (Gibco) may be used as alternatives to tryptone soya broth.

† Serum-dextrose agar (SDA): blood agar base No. 2 (Oxoid), 40 g; distilled water, 1000 ml; sterile horse serum (inactivated at 56°C for 30 min), 50 or 100 ml; D-glucose ((25% w/v) solution autoclaved at 105°C for 20 min), 40 ml. The blood agar base is dissolved in the water with the aid of gentle heating and then autoclaved at 121°C for 15 min. After cooling, the horse serum and glucose are added aseptically and plates or slopes poured immediately. Note: Trypticase soy agar (BBL) brucella agar (BBL) or brucella agar (Gibco) are satisfactory alternative basal media.

of organisms growing in vivo. In general, the morphology of *Brucella* strains is fairly constant and pleomorphic forms are rare except in old cultures growing under adverse conditions.

Brucella cells stain readily by conventional methods. Although not truly acid fast, they do tend to resist decolorization by weak acids and thus stain red by Macchiavello's stain or by the modified Ziehl-Neelsen technique used by Stamp et al (1950). They are usually stained red by the modified Köster method (Christofferson and Ottosen, 1941), but *B. ovis* is an exception. True capsules do not occur although capsule-like structures have been reported on occasion (Huddleson 1940).

Fine structure. The ultrastructure of the *Brucella* cell is broadly similar to that of other Gram-negative bacteria but shows a number of significant differences from that of cells of the *Enterobacteriaceae* as typified by *Escherichia coli* (De Petris et al. 1964; Dubray, 1972; 1976; Dubray and Plommet, 1976).

The *Brucella* cell wall is composed of an outermost layer of lipopolysaccharide protein about 9 nm thick with the polysaccharide chains exposed to the external surface (Dubray, 1976). This is the location of the major surface antigens in both smooth and nonsmooth cells. The lipopolysaccharide-protein layer is linked to an electron-dense inner layer 3–5 nm thick corresponding to cross-linked muramic acid-containing peptidoglycan. This layer is much more prominent in *Brucella* than in *E. coli* (Dubray 1976; Dubray and Plommet 1976).

Beneath this is located the periplasmic space which is apparent as a zone of low electron density 3–6 nm thick in smooth-phase cells but up to 30 nm thick in nonsmooth cells (Dubray and Plommet, 1976). The function of this is not known but in other bacteria it is known to be the location of the periplasmic enzymes involved in cell wall synthesis and degradation during cell division. It has been suggested as the location of cell wall component precursors such as native hapten and polysaccharide B (Moreno et al., 1981).

The cytoplasmic membrane, which has a typical triple layered lipoprotein structure, is located beneath the periplasmic space (De Petris et al., 1964; Dubray and Plommet, 1976). Granular aggregations adjacent to the cytoplasmic membrane mark the location of polyribosomal complexes (Dubray, 1972; 1976).

The *Brucella* cytoplasm is fairly homogeneous and is interspersed with small vacuoles and polysaccharide-containing granules (Dubray, 1972). The position of the nuclear vacuole is difficult to identify in some cells but can usually be located as an osmiophobic mass intersected by osmiophilic filamentous structures (De Petris et al., 1964; Peschkov and Feodorov, 1978).

Cell wall composition. The detailed chemical structure of the *Brucella* cell wall has not been fully determined. Gross analyses have indicated that the cell wall accounts for about 21% of the total bacterial dry weight in smooth cultures and 14% in nonsmooth strains. The walls of smooth *Brucella* cells contain approximately 37% protein, 14% carbohydrate, 18% lipid, 0.46% muramic acid and 0.1% 2-keto-3-deoxyoctulosonic acid (KDO). For nonsmooth *Brucella* cell walls the corresponding values are approximately 47.5% protein, 13% carbohydrate, 17% lipid, 0.4% muramic acid and 0.1% KDO (Kreutzer et al., 1977).

The amino acids alanine, arginine, aspartic, acid, glutamic acid, glycine, histidine, leucine/iso-leucine, lysine, phenylalanine, proline, serine, threonine and tyrosine have been identified as components of purified cell walls of smooth *B. abortus B. melitensis* and *B. suis.* Phenylalanine occurs in the N-terminal position (Kellerman et al., 1970). In addition α-ϵ-diaminopimelic acid has been identified in the peptidoglycan fraction (Dubray and Plommet, 1976).

The carbohydrate components of *Brucella* cell walls include glucose, mannose, rhamnose, galactose, heptose, hexosamine, dideoxyhexose and KDO (Kellerman et al., 1970). Other carbohydrates have been identified in extracts containing the lipopolysaccharide fraction (Bowser et al., 1964; Lacave et al., 1969; Renoux, Renoux and Tinelli, 1973; Kreutzer et al., 1979) but their role in the cell wall structure is unknown.

The lipid components of the *Brucella* cell wall have taxonomic significance and comprise about 18% of the total dry weight. This consists of 11.5% of free or loosely bound lipid and ~6.5% of firmly bound lipid. At least in the case of *B. abortus*, the loosely bound lipids are composed of 22.1% phospholipids and 76.1% free fatty acids and neutral lipids. The phospholipid fraction contains diphosphatidylglycerol compounds of the cardiolipin type and minor proportions of phosphatidylethanolamine, lysophosphatidyl ethanolamine, phosphatidylcholine and several still uncharacterized compounds (Bobo and Eagon, 1968; Wober et al., 1964). An unusual "ornithine-lipid" is also present (Kreutzer and Robertson, 1979).

The bulk of the free fatty acids are C_{18} saturated, monounsaturated and polyunsaturated acids with minor proportions of C_{12}, C_{14}, C_{16}, and C_{19} acids. A C_{19} cyclopropane fatty acid has also been detected in association with the phosphatide fraction. Hydroxy fatty acids are present, particularly in the series β-OH C_{21}, but β-hydroxymyristic acid is absent (Bobo and Eagon, 1968; Thiele and Kehr, 1969; Wober et al., 1964; Kreutzer et al., 1979).

The overall fatty acid composition of the *Brucella* cell is sufficiently distinctive to be of value for identification and classification at both the genus and subgenus level (Tanaka et al., 1977).

Using gas liquid chromatography of the fatty acid methyl esters, all *Brucella* species are differentiated from other Gram-negative bacteria, particularly by their long-chain fatty acid composition. The species *B. abortus, B. melitensis, B. neotomae, B. ovis,* and to a lesser extent *B. suis,* show a high degree of similarity although they can be distinguished from each other. *B. canis* is the most distinctive species and shows marked differences from the others in its fatty acid elution profile (Tanaka et al., 1977).

B. canis differs from the other species, including nonsmooth strains of *B. suis,* in containing a substantial proportion of a C16:1 fatty acid but lacking the major C19 cyclopropane fatty acid (Dees et al., 1981).

As in the case of the other Gram-negative bacteria, the *Brucella* cell wall is composed of distinct lipopolysaccharide-protein and mucopeptide-lipoprotein layers. The composition of both of these varies with the colonial phase of the culture. Thus the mucopeptide-lipoprotein fraction comprises about 25% of the total dry weight of the smooth *Brucella* cell wall and contains ~5.3% lipoprotein, 11.2% hexosamine, 4.4% carbohydrate (as hexose) and 2% muramic acid (Kreutzer and Robinson, 1979). In nonsmooth *Brucella* cell walls, the mucopeptide-lipoprotein fraction accounts for about 46% of the total dry weight and contains 29.2% lipoprotein, 8% hexosamine, 1% carbohydrate (as hexose) and 1.6% muramic acid. KDO is absent in both cases (Kreutzer and Robertson, 1979).

The lipopolysaccharide-protein complexes of smooth *Brucella* cell walls are unusual in that they partition predominantly into the phenol layer on phenol-water extraction (Baker and Wilson, 1965). Those of the nonsmooth strains are extremely hydrophobic and cannot be extracted with phenol-water but are soluble in phenol-petroleum ether-chloroform (Moreno et al., 1979).

The lipopolysaccharide-protein complexes of smooth *Brucella* cell walls contain ~6.3% protein, 11.6% carbohydrate (as hexose), 27% lipid, 0.62% KDO and 0.4% hexosamine, whereas those of nonsmooth *Brucella* cell walls contain 1.5% protein, 6.5% carbohydrate (as hexose), 27% lipid, 0.74% KDO and 0.04% hexosamine (Moreno et al., 1979).

The carbohydrate fraction of smooth *Brucella* lipopolysaccharide-protein is reported to contain 13.8% glucose, 10.4% mannose, 3.5% quinovosamine, 3.4% glucosamine and 0.86% KDO (Moreno et al., 1981). Additional carbohydrate components, including galactose, heptose, rhamnose, dideoxyaldose and 4-aminoarabinose, have been reported in studies of less extensively purified preparations (Bowser et al., 1964; Lacave et al., 1969; Dranovskaya and Vershilova, 1977; Kreutzer et al., 1979). So far, no qualitative differences have been found in the carbohydrate composition of the lipopolysaccharide complexes from smooth and nonsmooth *Brucella* strains (Kreutzer et al., 1979; Moreno et al., 1979).

The failure to find heptose in highly purified *Brucella* lipopolysaccharide-protein complex (Hurvell and Lindberg, 1973; Moreno et al., 1979; 1981) is significant, as this class of sugar is an integral component of the lipopolysaccharides of the *Enterobacteriaceae* (Westphal, 1974).

The lipid composition of *Brucella* lipopolysaccharide-protein complexes has not been completely elucidated. Stearic and palmitic acid are present as major components, together with at least six other saturated or unsaturated fatty acids (Kreutzer et al., 1979; Moreno et al., 1979). Hydroxy fatty acids in the range C_{16}–C_{21} are also present (Kreutzer et al., 1979) but β-hydroxymyristic acid, an essential component of the lipid A fraction of the lipopolysaccharides of *Enterobacteriaceae*, is absent (Lacave et al., 1969; Renoux et al., 1973; Kreutzer et al., 1979; Moreno et al., 1979).

Cultural characteristics. Most *Brucella* strains behave as slow growing, fastidious organisms on primary isolation. Although laboratory-adapted strains may be induced to grow in synthetic media containing an ammonium salt as the sole nitrogen source, the majority of fresh isolates have complex nutritional requirements and grow poorly on ordinary nutrient media unless these are supplemented with blood, serum or tissue extracts. Liver infusion agar was at one time widely employed for the cultivation of *Brucella* (Huddleson, 1943) but better-defined media of more consistent composition are now preferred.

For most purposes, SDA is the medium recommended and will support the growth of all species and most strains (Jones and Morgan, 1958). Trypticase soy agar (BBL), brucella agar (BBL), Albimi brucella agar* and tryptone soya agar (Oxoid) will support the growth of most strains without serum supplementation and the growth of nearly all if heated equine serum is added to a final concentration of 5–10% (v/v). The function of the serum is not simply nutritional but serves also to neutralize inhibitors present in the peptone component of ordinary culture media. Apart from serum or blood, other colloids, including Tween 40, may be used to supplement the medium (Huddleson, 1954; I. F. Huddleson, Bacteriol. Proc., Soc. Amer. Bacteriologists, p. 53, G-58, 1956). Strains of *B. abortus* biovar 2 and some of biovar 4, and *B. ovis* are the most sensitive to inhibitors and grow best on media containing 10% (v/v) serum.

Potato infusion agar† supports the growth of many *Brucella* strains and is often employed as the medium of choice for antigen or vaccine production as it does not favor dissociation (Alton et al., 1975). For the most fastidious strains, it may be necessary to supplement this medium with horse serum to achieve satisfactory growth.

Although wide variations occur between strains within species, in general the most rapid growth and largest colony size are achieved by isolates of *B. suis* biovars 1 and 3 and *B. canis*. Growth is least vigorous for *B. ovis*, followed by most strains of *B. melitensis* biovar 1 and *B. suis* biovar 2. Strains of the other species and biovars usually occupy an intermediate position.

On primary isolation, colonies of any *Brucella* strain are rarely visible before 48 h. At this stage colonies on SDA are usually from 0.5–1.0 mm in diameter, raised, convex with a circular outline and an entire edge. In transmitted light, the colonies of smooth strains have a shiny surface and appear a clear pale yellow. In reflected light, the colonies have a smooth glistening surface but appear slightly opalescent and bluish gray. The colonies of nonsmooth strains are of similar size and shape to smooth colonies but vary considerably in color, consistency and surface texture. They range from smooth-intermediate (SI) variants which are morphologically indistinguishable from smooth (S) colonies but may differ in antigenic properties and phage susceptibility, through intermediate (I) forms to rough (R) and mucoid (M) variants. R colonies are usually much less transparent than S colonies, with a more granular, dull surface and range in color from matt white, yellowish white or buff, to brown. Unlike S colonies, which are soft and easily emulsifiable to form stable suspensions in saline solutions, R colonies are often friable or viscous and difficult to detach cleanly from the agar surface.

They will not form uniform suspensions in saline solutions but produce granular aggregates, threads or clumps. M colonies are similar to R colonies in color and opacity but have a sticky glutinous texture.

The colonial variants of *Brucella* are best studied after 4 days growth on glycerol-dextrose agar (GDA),§ under oblique illumination as described by Henry (1933). Differentiation of the various colonial types is greatly facilitated by staining with ammonium oxalate-crystal violet before examination in reflected light (White and Wilson, 1951). Under these conditions S colonies appear pale yellow, R colonies are stained red with a coarse granular appearance and other dissociated colonies are stained various shades of pink, purple or blue.

Apart from S, R and M colonies, numerous phases transitional between these may occur in cultures undergoing dissociation to the nonsmooth state.

The genetic and physiological mechanisms responsible for the changes observed in dissociation from smooth to nonsmooth colonial phases have not been fully determined. The process is not simply a result of a genetic deletion producing synthesis of an incomplete lipopolysaccharide structure as in the case of some rough (Re) mutants of *Salmonella* (Westphal, 1974); but also involves deeper structures within the cell wall (Dubray and Plommet, 1976; Kreutzer and Robertson, 1979).

Most *Brucella* strains grow moderately well on sheep blood agar but the colonial appearance is not distinctive. The organisms are nonhemolytic, but a greenish brown discoloration may develop around the colonies. This is most apparent in old cultures and is probably attributable to alkali production.

The more vigorous strains of *B. abortus, B. melitensis* and *B. suis* will grow on MacConkey agar producing small lactose-negative colonies. In general, the growth of most *Brucella* strains is inhibited on media containing bile salts, tellurite or selenite. Tolerance to synthetic dyes varies considerably between strains and is employed as the basis for differentiation of biovars (see below).

Nutrition and growth conditions. Growth in simple nutrient liquid media is usually poor unless these are supplemented with blood, serum or tissue extracts. Most strains will grow fairly well on unsupplemented high quality enriched peptone-based media such as brucella broth (Gibco or BBL) or trypticase soy broth (BBL).

Good growth is obtained in SDB or other media supplemented with serum. In all cases it is essential to maintain adequate aeration if satisfactory growth is to be obtained.

After static incubation for 7 days at 37°C in SDB, smooth strains produce a slight to moderate uniform turbidity with a light powdery deposit. Nonsmooth strains may produce a granular or slimy deposit, variable turbidity and pellicle formation sometimes accompanied by a "stalactite" appearance. Growth in static liquid media favors dissociation of S-phase cultures to nonsmooth forms. Vigorous aeration will prevent this provided that the medium remains adequately buffered near neutral pH.

In semisolid media, CO_2-requiring cultures of *B. abortus* and *B. ovis* produce a disk of growth a few millimeter below the surface of the medium. CO_2-independent *Brucella* species produce a uniform turbidity from the surface down to a depth of a few millimeters. There is no growth under the anaerobic conditions prevailing in static deep cultures.

The optimum growth temperature for all *Brucella* strains is 36–38°C, but growth of most will occur in the range 20°C–40°C. All strains lose viability at 56°C although temperatures as high as 85°C may be necessary to ensure sterilization (Swann et al., 1981).

---

* Albimi brucella agar: the original formulation is no longer available but an equivalent product is obtainable as Brucella agar from Gibco Laboratories, Grand Island, New York, NY 14072, U.S.A.

† Potato infusion agar: Bacto potato infusion agar (Difco), 49.0 g; glycerol, 20.0 g; distilled water, 1000 ml. The glycerol is dissolved in the water and the dehydrated medium is suspended in this solution and heated to the boiling point until dissolved. The medium is sterilized at 121°C for 15 min. It should be prepared immediately before use.

§ Glycerol dextrose agar: to blood agar base No. 2 (Oxoid) or Trypticase soy agar (BBL) that has been autoclaved and cooled to 56°C, sterile solutions of glycerol and D-glucose are added to give final concentrations of 2% (w/v) and 1% (w/v), respectively.

Metabolism and metabolic pathways. The electron transport of *B. abortus* (the only species closely studied in this regard) consists of a branched system involving cytochromes $a + a_3$, b, c and o, flavoproteins and ubiquinone. It is unusually resistant to respiratory inhibitors (Rest and Robertson, 1975).

Energy-yielding processes are essentially oxidative, and *Brucella* cultures show little ability to acidify carbohydrate media in conventional tests. They have been reported to lack phosphofructokinase (Robertson and McCullough, 1968b), although this enzyme was detected in extracts of *B. suis* by Roessler et al., (1952). The inability to acidify carbohydrate media has also been attributed to inhibition by peptone constituents, and acidic reactions have been demonstrated in peptone-free media (Pickett and Nelson, 1955). *B. neotomae* is exceptional in that it produces acid (but not gas) from glucose, galactose, arabinose and xylose in conventional peptone-water sugar media (Stoenner and Lackman, 1957).

Glucose catabolism occurs via the hexose monophosphate pathway in conjunction with the tricarboxylic acid cycle (Rest and Robertson, 1974; Robertson and McCullough, 1968a, b). *meso*-Erythritol is metabolized by many *Brucella* strains in preference to glucose (Anderson and Smith, 1965) and D-erythritol 1-phosphate and other intermediates in the erythritol pathway will reduce the entire electron transport system (Rest and Robertson, 1975).

Although some laboratory-adapted strains will grow in minimal medium with an ammonium salt as the sole nitrogen source (McCullough and Dick, 1943), the nutritional requirements of *Brucella* cultures in general are complex. Multiple amino acids, thiamin, biotin, nicotinamide and magnesium ions are essential for growth and iron and manganese exert a regulatory action. The growth of some strains is stimulated by calcium pantothenate and *meso*-erythritol. Very few strains will grow with citrate as the sole carbon source.

Sulfur-containing amino acids and proteins are degraded and may be reduced to H_2S, but the production of this varies with species and biovar. Indole is never produced from tryptophan or its proteins and acetyl methyl carbinol is not produced from glucose (Voges-Proskauer reaction).

Hydrolytic activity towards proteins in general is very limited and gelatin, inspissated serum and litmus milk are not digested. *Brucella* strains either render the latter alkaline or produce no visible change.

The supplementary CO_2 required by strains of *B. abortus* and *B. ovis* for growth is used as a nutritional factor and not simply to lower oxygen tension or pH. It is incorporated directly into pyrimidines, glycine and alanine (Newton et al., 1956).

The optimum pH for growth is between 6.6 and 7.4. Cultures die rapidly at pH 3.5 or below. Most *Brucella* strains produce alkali on protein or peptone-containing media and this may act as a growth-limiting factor. Culture media should be adequately buffered near pH 7 for optimum growth. The optimum osmotic pressure is between 2 and 6 atmospheres, equivalent to between 0.05 and 0.15 M NaCl.

A wide range of carbohydrate and amino acid substrates is oxidized. Manometric measurement of oxidation rates with selected substrates produces metabolic patterns which are characteristic of each species and some biovars (Meyer, 1961; 1969; Meyer and Cameron, 1961a, b; Wundt, 1963). The oxidative metabolic patterns show a close correlation with the phage-sensitivity pattern and the preferred natural hosts of the species (Meyer, 1962; Meyer and Morgan, 1962; Philippon, 1968). They are of primary importance in defining the species of *Brucella* and are of additional value in classifying the biovars of *B. suis* (Stableforth and Jones, 1963; Jones, 1967; Jones and Wundt, 1971).

Procedures for the determination of oxidative metabolic patterns by manometric methods are described by Morgan and Gower (1966), Alton et al., (1975) and Corbel et al., (1979). A nonquantitative technique, using thin layer chromatography to detect substrate utilization may also be employed (Balke et al., 1977; Corbel et al., 1978). The range of substrates used includes L-arabinose, D-glucose, D-galactose, D-ribose, D-xylose, *meso*-erythritol, L-alanine, L-asparagine, L-glutamic acid, L-

arginine, L-lysine, DL-citrulline and DL-ornithine. For some purposes, other substrates including adonitol, L-histidine, L-serine and D-amino acids may also be used.

The species *B. abortus*, *B. melitensis*, *B. suis* and *B. canis* are usually oxidase positive in tests with tetramethyl-*p*-phenylenediamine but some strains are oxidase negative and this can be a useful epidemiological marker (Verger, et al., 1979).

Most *Brucella* strains produce nitrate reductase and reduce nitrates to nitrites. Nitrites may be further reduced (Zobell and Meyer, 1932). *B. ovis* and some *B. canis* strains do not reduce nitrate or nitrite.

B. canis, *B. neotomae* and *B. suis* strains normally show very strong urease activity. Most strains of *B. abortus* and *B. melitensis* also produce urease, but a few do not and this may be of epidemiological value. *B. ovis* does not usually hydrolyze urea, but some strains may do so on prolonged incubation.

Genetics and variation. Spontaneous variation in some of the properties of *Brucella* cultures is not uncommon. This usually results from mutations involving the modification of individual characteristics. Probably the most frequent mutation is the production of nonsmooth variants from S-phase organisms. The reverse mutation also occurs but much less frequently (Jones and Berman, 1951). Other characteristics are also subject to variation. CO_2-independent variants are rapidly selected from CO_2-dependent cultures of *B. abortus* and much less frequently from *B. ovis*. Loss of H_2S production and urease production occurs occasionally. Changes in resistance to various dyes, including basic fuchsin, thionine, thionine blue, safranin O and to *meso*-erythritol and various antibiotics, occur with varying frequency. Modification of such characteristics may result in alteration of biovar. For variation of *B. abortus* biovar 2 to biovar 1, the mutation rate for dye resistance is 6.9×10^{-10}/cell division for basic fuchsin (Shibata et al., 1962).

Variations in surface antigens or phage sensitivity in S-phase cultures occur infrequently; nevertheless, smooth phage-resistant variants of *B. abortus* have been isolated under both laboratory and field conditions (Corbel and Morris, 1974; 1975; Harrington et al., 1977).

The production of variants can be stimulated by extraneous factors including physiological concentrations of progesterone, testosterone or diethyl stilbestrol (Meyer, 1976b, c). Mutations resulting in change from one species to another have been reported to have occurred under such conditions (Meyer, 1976c). In general, *Brucella* strains maintain their species identity which is closely related to their preferred host specificity.

Cell wall-defective variants of *Brucella* may be induced by penicillin or glycine (Hines et al., 1964; Roux and Sassine, 1971; Hatten, 1973), hormones (Meyer, 1976c) or cell cultures of immune macrophages (McGhee and Freeman, 1970a). They may also be recovered from the blood or tissues of animals or man infected with *Brucella* strains (Nelson and Pickett, 1951; Ross and Corbel, 1980).

Artificially induced spheroplasts are osmotically unstable and require hypertonic media for survival and growth. Naturally occurring *Brucella* L-forms are less sensitive osmotically but have more exacting growth requirements than the parent *Brucella* strains. Thus they may require specially enriched media containing reducing agents (Nelson and Pickett, 1951) or high concentrations of horse serum (Corbel et al. 1980).

Cell wall-defective *Brucella* variants are highly pleomorphic and, if induced to grow on solid media, they produce bizarre colonies. These may vary from tenacious, granular "fried egg" colonies of the *Mycoplasma* type which may or may not be surrounded by a lipid film, to colonies resembling those of normal smooth or intermediate *Brucella* cultures, but very much smaller (Nelson and Picket 1951; Corbel et al., 1980). *Brucella* L-forms and spheroplasts normally show partial or complete loss of surface antigens but may also express surface antigens characteristic of nonsmooth cultures. They usually show a reduced sensitivity to brucella-phages but may be susceptible to growth inhibition or lysis. Their pathogenicity towards experimental animals is slight or absent, unless they revert to the parent form.

Genetic transformation, conjugation and transduction have not been demonstrated in the genus *Brucella*. Examination of a limited number of strains has also failed to demonstrate the presence of plasmids or other forms of extra-chromosomal DNA (F. Simon, Contribution a l'étude des bacteriophages du genre *Brucella*; phage Tb, BM29 et Bk, Thesis, University of Paris, 1979). Indirect evidence for the occurrence of plasmids, such as infectious antibiotic resistance, is also lacking.

The existence of bacteriocins specific for the genus *Brucella* has not been confirmed, although bacteriocin-like effects have been occasionally reported (Pickett and Nelson, 1950; Todorov and Koleva-Todorova, 1971).

Phages and phage typing. A large number of bacteriophages active upon members of the genus *Brucella* has been described. These phages have not been shown to lyse bacteria of other genera and thus are of taxonomic value for identification at both genus and species level. On the basis of their host range, the phages may be classified into five distinct groups.

Those in group 1, typified by the Tbilisi (Tb) strain (Popkhadze and Abashidze, 1957), are capable of efficient replication only in cells of *B. abortus* in the S, SI or I phases. Limited replication also occurs in S, SI or I cultures of *B. neotomae*, but the efficiency of plating is low. At high concentrations, these phages produce lysis of S, SI or I cultures of *B. suis* by a bacteriocin-like effect. Cultures of *B. melitensis*, *B. canis* and *B. ovis* are not lysed nor are cultures of *B. abortus*, *B. neotomae* or *B. suis* in the M or R phases.

Group 2 comprises those phages typified by the Firenze (Fi) strain 75/13. (Corbel and Thomas, 1976). These replicate in and form plaques on S, SI and I cultures of *B. abortus*, *B. neotomae* and *B. suis* but not on those of other *Brucella* species. They are also inactive on M and R strains of *Brucella*. The efficiency of plating is somewhat higher on *B. abortus* strains than on those of *B. neotomae* or *B. suis* biovar 4 and very much higher on these than on *B. suis* biovars 1, 2 and 3.

Group 3 includes those phages typified by the Weybridge (Wb) strain (Morris and Corbel, 1973). All of these replicate and form plaques on S, SI and I cultures of *B. abortus*, *B. neotomae* and *B. suis*. The efficiency of plating on the three species varies between the phages of this group within a range of about 2 $\log_{10}$ units. Some of these phages have been reported to form plaques on cultures of *B. melitensis*, but the efficiency of plating is very low (Moreira-Jacob, 1968; Douglas and Elberg, 1976). None of the phages of group 3 will lyse M or R strains of any *Brucella* species, including *B. canis* and *B. ovis*.

Group 4 comprises the Berkeley phages Bk_0, Bk_1 and Bk_2 (Douglas and Elberg, 1976; 1978). The most useful of these, Bk_2, replicates and causes lysis in S-phase cultures of *B. abortus*, *B. melitensis*, *B. neotomae* and *B. suis*. It shows no lytic activity towards *B. canis*, *B. ovis* or other nonsmooth *Brucella* strains, even at high concentrations. Its lytic activity towards the smooth strains is drastically reduced as these undergo dissociation to nonsmooth forms. This is particularly the case with *B. melitensis* cultures and the efficiency of plating of Bk_2 phage on this species may vary considerably between strains.

Group 5 includes those phages lytic for nonsmooth *Brucella* strains. All are derived from phage R which was developed as a mutant selected from a mixture of phages active on smooth *Brucella* strains (Corbel, 1977a; 1979). Phage R is lytic for nonsmooth cultures of *B. abortus* but not other species. The strain is genetically unstable, however, and produces smooth-specific phage mutants at high frequency during replication. Phage R/O is lytic for *B. ovis* and some S-phase *B. abortus* and *B. suis* strains, but not for *B. melitensis* or nonsmooth cultures of *B. abortus*, *B. melitensis*, *B. suis*, *B. neotomae* or *B. canis*. It is genetically unstable and its host range is subject to variation between successive batches.

Phage R/M is lytic for some R phase cultures of *B. melitensis* and, like phages R and R/O, is genetically unstable. Phage R/C was developed by passage of phage R/O on cultures of *B. canis* RM 6/66. Unlike the other phages in this group it is relatively stable in host range and is therefore the most useful for taxonomic purposes. It is lytic for *B. canis*, *B. ovis* and nonsmooth strains of *B. abortus*. Some nonsmooth

strains of *B. suis* and *B. melitensis* are inhibited by it at high concentration. It is not lytic for completely smooth strains of *B. abortus*, *B. melitensis*, *B. suis* or *B. neotomae*, although it will produce plaques on strains of *B. abortus* in the SI, I, M or R phases.

All of the brucella-phages appear to belong to the same phage family. They are morphologically similar with a hexagonal head from 55–80 nm in width and with a tubular, apparently noncontractile tail from 14–33 nm long. Most of the phage strains have a mean head diameter of about 60 nm and a tail length of about 25 nm. Attachment of the phages is mediated via short fibrous structures linked to the distal end of the tail which interact with protein or glycoprotein components forming part of the lipopolysaccharide-protein complex of the outer envelope of the bacteria (Corbel, 1977b).

All of the brucella phages are relatively stable to organic solvents, nonionic and anionic detergents. They are inactivated by heat, cationic detergents and oxidizing agents. They show a variable stability to proteolytic enzymes and reducing agents, and do not require divalent cations for their interaction with *Brucella* cells. All contain DNA as their genetic material and, for the phages of group 1, this has a mol% G + C of 45.3–46.7 (T_m).

The phage-susceptibility pattern is of major importance in definition of the species of *Brucella*. For culture identification it is convenient to use phage preparations standardized at the routine test dilution (RTD). This is defined as the minimum concentration which produces complete lysis of the propagating strain for that phage.

Antigenic structure. Smooth species of *Brucella* show complete cross-reaction in agglutination tests with unabsorbed antisera to smooth brucella organisms. This cross-reaction does not extend to nonsmooth variants in the M or R phases. Cross-reactions between nonsmooth strains can be demonstrated by agglutination tests with unabsorbed antiserum to a rough *Brucella* strain or to the R antigen of *B. ovis*. By using cross-absorbed antisera, different quantitative distribution of the two major surface antigens, A and M, can be demonstrated in smooth strains (Wilson and Miles, 1932) and is of value in differentiating biovars of the major species (see Table 4.102). These antigenic determinants form part of the high molecular weight lipopolysaccharide-protein complex which constitutes the major agglutinogen of smooth *Brucella* species (Diaz et al., 1968).

Minor surface or subsurface antigens common to smooth and certain nonsmooth *Brucella* strains have been described. These include the "native hapten", also referred to as component 1 or second polysaccharide, present in phenol-water, ether-water, or trichloroacetic acid extracts of smooth strains (Diaz et al., 1968; Moreno et al., 1981) and the related but not identical polysaccharide B present in the cytoplasmic fraction of rough *B. melitensis* strain B115 (Diaz and Dorronsoro, 1971; Diaz et al., 1979; Moreno et al., 1981). These components are nontoxic, nonantigenic polysaccharides containing glucose, mannose and a trace of quinovosamine (Moreno et al., 1981). They precipitate with antisera from *Brucella*-infected animals in the presence of high concentrations of sodium chloride and are useful in diagnostic serological tests (Diaz et al., 1979).

Other surface or subsurface antigens reported in smooth and some nonsmooth *Brucella* strains include the f antigen (Freeman et al., 1970; McGhee and Freeman, 1970b) and the χ, β and γ antigens (Raybould and Chantler 1980; Raybould et al., 1981). A total of 4 surface antigens has been detected in studies with spheroplast preparations (Baughn and Freeman, 1966).

Cross-agglutination reactions have been reported between *Brucella* and various other genera including *Pasteurella*, *Proteus*, "*Pfeifferella*", *Francisella*, *Vibrio*, *Campylobacter*, *Leptospira*, *Pseudomonas*, *Salmonella* and *Yersinia* (Mallman, 1930; Morse et al., 1953). Some of these cross-reactions have not been substantiated and Wilson (1934) attributed the apparent cross-agglutination between *Brucella*, *Pasteurella*, *Proteus* and "*Pfeifferella*" either to the use of incompletely smooth antigen preparations or to the coincidental presence in the same antiserum of antibodies to several antigenically unrelated organisms.

An antigenic relationship has been confirmed between the surface antigens of smooth *Brucella* species and *Yersinia enterocolitica* 0:9 (Ahvonen et al., 1969; Corbel and Cullen, 1970), *Francisella tularensis* (Francis and Evans, 1926; Ohara et al., 1974), *Salmonella* 0:30 serotypes (Cioglia, 1950; Wundt, 1959; Corbel, 1975), *Vibrio cholerae* (Wong and Chow, 1937; Feeley, 1969) and *Escherichia coli* 0:157 (Corbel and Stuart, unpublished results).

The most important of these cross-reactions is that with the O antigen of *Y. enterocolitica* 0:9. This can present serious problems in differential diagnosis by serological methods and has been the subject of a recent review (Mittal and Tizard, 1981).

Serological cross-reactions between nonsmooth *Brucella* strains and organisms of other genera have received relatively little attention. Evidence for cross-reactions between *B. canis* and *Actinobacillus equuli*, mucoid strains of *Pseudomonas aeruginosa* and some serotypes of *Pasteurella multocida* has been presented (Carmichael et al., 1980; A. Weber, Untersuchungen zur mikrobiologischen Diagnose und Epidemiologie der *Brucella canis*-Infektion des Hundes, Thesis, Giessen, 1976).

The soluble internal antigens released on disruption of *Brucella* cells are, in many cases, common to both smooth and nonsmooth strains. Most also appear to be unique to the *Brucella* genus (Diaz et al., 1967; 1968). They can be demonstrated by immunodiffusion or immunoelectrophoresis and details of these methods for identification at genus level are described by Corbel et al., (1979).

The major antigens of the smooth *Brucella* species have been classified by, *inter alia*, Freeman et al., (1970). All species evidently have a very similar antigenic composition although occasional strain differences have been reported (Stemshorn and Nielsen, 1977; Hatten and Brodeur, 1978; Raybould et al., 1981).

Nonsmooth strains contain the R antigen and possibly at least one other surface antigen in place of the A and M agglutinogen of smooth strains (Diaz and Bosseray, 1973; Zoha and Carmichael, 1981).

The R antigens of *B. ovis* and other nonsmooth *Brucella* strains have been characterized as lipopolysaccharide complexes with a low protein content (Moreno et al., 1979). They contain antigenic determinants common to most, if not all, nonsmooth *Brucella* strains (Diaz and Bosseray, 1973). In addition, species-specific antigenic determinants may also be present (Diaz and Bosseray, 1973). At least two other surface antigens may be present in some nonsmooth strains, one of which is associated with the mucoid colonial phase and may be strain-specific (Zoha and Carmichael, 1981).

Antibiotic sensitivity. On primary isolation, nearly all *Brucella* strains are sensitive in vitro to gentamicin, tetracycline and its derivatives and rifampicin.

Most strains are also susceptible to ampicillin, chloramphenicol, erythromycin, kanamycin, novobiocin, spectinomycin, streptomycin and sulfamethoxisole and trimethoprim in combination. Variations in susceptibility occur between species and even between biovars and strains within species. Most *Brucella* strains are resistant to penicillins, cephalosporins, polymyxin and nalidixic acid, and nearly all are resistant to amphotericin B, bacitracin, cycloheximide, clindamycin, lincomycin, nystatin and vancomycin at concentrations that are therapeutically attainable. Differences in antibiotic sensitivity have been used for identifying strains (Fritzsch and Abadjieff, 1966) but the procedure is of limited value.

Pathogenicity. Brucella species are pathogenic for a wide variety of animals, frequently producing generalized infections with a bacteremic

phase followed by localization in the reproductive organs and the reticuloendothelial system. Infection in the pregnant animal often results in placental and fetal infection and this frequently causes abortion. The organisms may localize in mammary tissue and can be excreted in the milk. Typically, growth in vivo is intracellular and the organisms can survive within both granulocytes and monocytes. Infections in the natural host are rarely lethal and often are mild, with clinical manifestations occurring mainly in the pregnant animal. Nevertheless localization can occur in a wide range of organs with production of a variety of lesions.

All of the *Brucella* species may produce infections in laboratory animals including guinea-pigs, mice and rabbits but the severity of the infection varies considerably with the virulence of the infecting strain. For the guinea-pig the order of pathogenicity is *B. melitensis* $\geq$ *B. suis* $\geq$ *B. abortus* > *B. neotomae* $\geq$ *B. canis* > *B. ovis*. (Braude, 1951; Isayama et al., 1977).

The more pathogenic strains usually produce a local abscess at the site of inoculation followed by bacteremia of varying duration. The regional lymph nodes may become enlarged and granulomatous changes develop. Similar changes occur in the liver and spleen and frequently in other organs, particularly the testes and epididymides. *B. melitensis* and *B. suis* biovars sometimes produce fatal infections (Braude, 1951). The other species rarely produce severe disease and infection is usually self limiting within a period varying from a few weeks to more than 6 months. The mouse is more susceptible to persistent infection with *B. neotomae*, *B. canis* and *B. ovis* than the guinea-pig (Isayama et al., 1977). Pathogenic effects of these species are limited to slight to moderate splenic enlargement.

Infection in most species is accompanied by production of specific antibodies and by the development of a cell-mediated immune response which may be demonstrable by tests for dermal hypersensitivity (Bhonghbhibat et al. (1970).

Ecology. Although easily cultivated in vitro, under natural conditions *Brucella* species behave as obligate parasites and do not pursue an existence independent of their animal hosts. Their distribution is world wide, apart from the few countries from which they have been successfully eradicated. Considerable local variations occur in the prevalence of particular species and biovars.

Enrichment and Isolation Procedures

The isolation of *Brucella* may be attempted from any tissue or secretion. Those most likely to yield positive cultures include abortion material (placental cotyledon, amniotic fluid, vaginal discharge, fetal gastric contents, fetal lung and fetal liver), lymph nodes, bone marrow, mammary gland, uterus, seminal vesicles and accessory glands, testes and epididymides or other organs with local lesions, milk, colostrum, semen and blood.

Uncontaminated materials may be inoculated directly onto SDA plates. Where contaminating organisms are likely to be present, a selective medium should be used. SDA supplemented with the antibiotic formulation of Farrell (1974)* is satisfactory for the isolation of most smooth strains of *Brucella* but may be too inhibitory for *B. canis* and *B. ovis*. For the isolation of these from contaminated material, modified Thayer-Martin medium (Brown et al., 1971) or SDA containing 10% (v/v) heated horse serum and VCN-F inhibitor (BBL)† may be used. These media are less selective than Farrell's SDA and it is advisable to dilute heavily contaminated material, such as semen, in 5–10 volumes of sterile isotonic saline and filter it through a membrane filter (0.8 µm

---

* Farrell's selective medium: prepared from SDA (see footnote † on page 377). After addition of the horse serum and glucose to the molten medium cooled to 56°C, antibiotics are added to give the following final concentrations: bacitracin, 25 U/ml; polymyxin B, 5 U/ml; actidione, 100 µg/ml; vancomycin, 20 µg/ml; nalidixic acid, 5 µg/ml; and nystatin, 100 U/ml.

† Modified SDA antibiotic medium: SDA is prepared as described in footnote † page 377, but with horse serum added to a final concentration of 10% (v/v) and No. 1 agar (Oxoid) added to give a final concentration of 3% (w/v). To each liter of molten medium cooled to 56°C, one vial of VCN inhibitor (BBL; reconstituted in 10 ml of sterile distilled water) is added. This is followed by 1 ml of furadantin solution (10 mg/ml of 0.1 N NaOH) and the plates are poured immediately. The medium contains vancomycin (3 µg/ml), sodium colistimethate (7.5 µg/ml), nystatin (12.5 U/ml) and furadantin (10 µg/ml) (final concentrations).

pore size) before plating it out. All cultures should be incubated at 37°C under air supplemented with 5–10% (v/v) CO_2 unless isolation of a CO_2-independent species of *Brucella* is being attempted.

Direct inoculation of selective media is usually satisfactory for isolation of *Brucella* from heavily infected materials even in the presence of contaminating organisms. However, when the *Brucella* concentration is likely to be low, as in blood, milk or semen samples, enrichment procedures should be used. Enrichment may be achieved either by intramuscular inoculation of the sample into guinea pigs, followed by culture of the spleen tissue 4 weeks later, or by growth in a liquid enrichment medium. Guinea pig inoculation is only likely to succeed with virulent strains of *B. melitensis*, *B. abortus* and *B. suis*.

Enrichment cultures may be performed by the two phase system of Castañeda (1947). As liquid phase, SDB supplemented with 2% (w/v) trisodium citrate is used. The solid phase is SDA containing 2.5% (w/v) agar. The medium is made selective by adding the antibiotic formulation of Farrell (1974) to the liquid phase. For cultures of *B. ovis* and *B. canis* this may be replaced by VCN-F inhibitor. The sample is mixed with the liquid phase and the bottles incubated at 37°C under air (with supplementary CO_2 if required). At intervals of 2 or 3 days the liquid phase is tipped over the solid phase and the bottles reincubated. Incubation is continued for up to 6 weeks or until colonies appear on the solid phase, whichever is sooner.

Blood or tissue samples may be enriched for *B. canis* by inoculation of the yolk sacs of 6- to 8-day-old chick embryos. Yolk from eggs with dead embryos is plated onto SDA or other suitable medium (A. Weber, Untersuchungen zur mikrobiologischen Diagnose und Epidemiologie der *Brucella canis*-Infektion des Hundes, Thesis, Giessen, 1976).

Culture of cell wall-defective forms of *Brucella* may be attempted by culturing samples of blood, synovial fluid and solid tissue on Farrell's selective medium enriched with 20% (v/v) horse serum. Incubation should be continued for at least 14 days before discarding the plates. The plates should be examined for microcolonies under a stereomicroscope.

Maintenance Procedures

Cultures may be maintained for short periods by streaking onto SDA slopes, incubating for 72 h under air (+ 10% CO_2 (v/v) if required) and then sealing the slope and storing at 4°C. This procedure needs to be repeated every 6–8 weeks. It is unsatisfactory for long-term maintenance of strains as these are liable to change their characteristics on repeated subculturing.

Brucella cultures may be preserved satisfactorily by vacuum drying. The strains are grown on SDA slopes incubated at 37°C under air (+ 10% CO_2 if required) for 72 h. The growth is washed off the slopes and suspended in sterile rabbit serum to give a suspension of about 10^{10} organisms/ml. This is distributed in 0.1-ml volumes into sterile tubes sealed with cotton-wool plugs. The tubes are stored over phosphorus pentoxide in a desiccator kept at 4°C. The desiccator is evacuated daily until a pressure of 0.05 mm Hg can be obtained. The tubes are then sealed under vacuum in glass ampoules containing a small quantity of silica gel. Under these conditions *Brucella* cultures may remain viable for many years.

Brucella strains may also be preserved satisfactorily by freeze-drying. General directions for this have been given by Lapage et al., (1970) and details of the technique by Boyce and Edgar (1966).

Storage in liquid nitrogen will maintain *Brucella* cultures with a smaller decrease in viability than is produced by either vacuum drying or freeze-drying (Davies, et al., 1973). Cultures are grown for 72 h on SDA slopes at 37°C in air (+10% CO_2 if required) and the growth suspended in single strength Bacto-glutamate medium* to form a dense suspension. This is left undisturbed at 4°C for 7 days, after which volumes of up to 1 ml are placed in sterile glass screw-capped vials.

The vials are allowed to equilibrate in the vapor phase or immersed in the liquid.

Procedure for Testing Special Characters

Safety precautions. All *Brucella* isolates should be regarded as potentially pathogenic for man. Adequate precautions against accidental infection should be taken at all stages of work involving live cultures. It is strongly recommended that all such procedures should be performed by suitably trained staff using efficient exhaust protective cabinets. Particularly hazardous techniques, such as the determination of oxidative metabolism rates by respirometric methods, are best left to reference laboratories experienced in their use. Specific indications of safety precautions to be adopted during work with *Brucella* cultures have been given by Alton et al. (1975) and Corbel et al. (1979).

Oxidative metabolism rates. For respirometric methods, cells are grown in Roux flasks containing SDA for 48 h. They are harvested in Sorensen's phosphate buffer (pH 7.0), washed 3 times by centrifugation (10,000 × g for 15 min), and standardized turbidimetrically to a cell concentration equivalent to 0.9 mg N/ml as determined by micro-Kjeldahl analysis. Measurements of oxygen uptake are performed using either a Warburg constant volume respirometer or a Gilson differential respirometer (see *Manometric Techniques* by Umbreit et al., 1972). Oxygen uptake with various substrates is expressed as microliters per milligram of cell nitrogen per hour ($Q_{O_2}N$). The following substrates are normally used for *Brucella* studies. *Group 1*: L-alanine, L-asparagine and L-glutamic acid. *Group 2*: L-arginine, DL-citrulline, DL-ornithine and L-lysine. *Group 3*: L-arabinose, D-galactose, D-glucose, D-ribose, D-xylose and *meso*-erythritol. Other substrates are sometimes used. The substrates are prepared as 10% (w/v) solutions in Sorensen's phosphate buffer (pH 7.0), sterilized by filtration and stored at −20°C until needed. The DL-citrulline solution should be sterilized by autoclaving at 121°C for 15 min.

The $Q_{O_2}N$ values obtained by either the Gilson or the Warburg method are corrected for endogenous O_2 uptake (by substracting the $Q_{O_2}N$ value for controls lacking substrate) and are compared with expected values for each species (Meyer, 1969; Meyer and Cameron, 1961a, b; Verger and Grayon, 1977). The results may conveniently be expressed in the form of a three-level metabolic profile (Verger and Grayon, 1977) or as a pattern of + and − signs (see Table 4.101).

Thin-layer chromatography. The oxidative metabolic pattern may be determined quantitatively by thin-layer chromatography. The results do not always coincide precisely with those obtained by respirometry, possibly because nonoxidative degradation of substrates may occur; however, the overall pattern is generally satisfactory for culture identification.

The method used is a modification of that described by Balke et al., (1977). The amino acid and carbohydrate substrates are incubated with the *Brucella* suspension and the supernatant fluids are then tested for the presence of the substrate. It is essential to include *Brucella* reference strains as standards during such testing.

The cultures are grown on SDA layers in Roux flasks for 48 h at 37°C. The cells are harvested in Sorensen phosphate buffer (pH 7.2), washed twice by centrifugation and suspended in the buffer to 0.9 mg N/ml. The suspension is then divided into two portions. One portion is heated for 30 min at 100°C; this constitutes the negative control.

Volumes of 0.5 ml of the negative control and unheated cell suspension are transferred to separate centrifuge tubes. To each is added 25 μl of a 0.5% solution of the amino acid or carbohydrate substrate and the mixtures are incubated in a shaking water bath at 37°C for 2.5 h. The tubes are then centrifuged at 9000 × g for 15 min. Drops (20 μl) of the supernatants from the control and tests are placed on a base line marked ∼2 cm from the edge of a 20 × 20-cm silica gel G 1500 thin-

---

* Bacto-glutamate medium: Bacto-casitone (Difco) (25 g) is dissolved in 1000 ml of distilled water and autoclaved at 115°C for 20 min. To this solution, 50 g of sucrose and 10 g of monosodium glutamate are added and dissolved by steaming for 10 min. The medium is filtered under positive pressure through Seitz EKS filters into sterile containers and finally autoclaved at 106°C for 15 min.

layer plate (Schleicher and Schüll, West Germany). Separate plates must be used for the amino acids and carbohydrates. The plates are also spotted with 20-μl drops of reference solutions of the various substrates (0.5%). The spots should be at least 1–2 cm apart.

After the spots have dried, the plates are developed in n-butanol:glacial acetic acid: water (4:1:1, v/v) until the solvent front has migrated 8–10 cm. The plates are dried at room temperature. Amino acid plates are sprayed with ninhydrin reagent (0.25 g ninhydrin + 25 ml acetone + 1.25 ml glacial acetic acid) and heated for 15 min at 100°C. Carbohydrate plates are sprayed first with periodic acid reagent (2.28 g periodic acid/100 ml of water, diluted immediately before use in 19 volumes of acetone) and, after drying, with 3% (w/v) p-anisidine hydrochloride in n-butanol; the plates are then heated at 110°C for 10 min. The sugars are located as yellow-brown spots and the sugar alcohols as white spots, on a blue background.

By comparison with the control, a reaction can be assigned to one of the following three categories: (a) the substrate is completely metabolized (no test sample spot); (b) the substrate is partially metabolized, and (c) the substrate is not metabolized. If the negative control is absent, this indicates that the preparation will require a longer heating to inactivate the enzyme. The amino acid and carbohydrate reference preparations should give give clear, well defined spots; if they give weakly-stained or multiple spots, the substrate should be replaced by a fresh batch.

Determination of phage sensitivity. The colonial phase of the test culture must first be determined by streaking GDA plates and incubating under a suitable gas atmosphere at 37°C for up to 5 days. The colonial morphology is then examined in obliquely-transmitted light according to Henry (1933) or after ammonium oxalate-crystal violet staining according to White and Wilson (1951). If the culture has colonies of only a single phase, representatives of these are selected for phage typing. If a mixture of smooth and nonsmooth colonial forms is present, then smooth colonies are selected for subculturing and phage typing.

For routine phage typing, a suspension of cells is prepared as described for the slide agglutination test (see below). Plates of SDA or TSA are inoculated with this so as to produce confluent growth. Typing may be performed using the Tb phage standardized at the routine test dilution (RTD) and at 10,000 × RTD, or by using phages representing groups 1–5 of Corbel and Thomas (1980) standardized at RTD. For typing smooth cultures, Tb phage alone when used at both concentrations is usually adequate, but for nonsmooth or atypical cultures the additional phages are useful. Discrete drops of phage preparation (~25 μl) are applied to the surface of the inoculated plate and allowed to soak into the agar. The plates are then inverted and incubated for 37°C under the appropriate gas atmosphere. They are inspected for growth and phage lysis after 24-h incubation and at further 24-h intervals if required. The results of the phage sensitivity tests can only be interpreted accurately if the colonial phase of the culture is known. In addition to the initial examination of the colonial morphology of the culture, a further check should be done by testing a sample of growth from each page plate for agglutinability in 0.1% aqueous acriflavine (clumping indicates nonsmooth cells).

CO_2 requirement. The culture is streaked onto duplicate SDA slopes. One of these is incubated at 37°C under an air atmosphere, the other under air supplemented with 5–10% CO_2. Growth on both slopes indicates an absence of CO_2-dependence, whereas growth only with CO_2 supplementation indicates a CO_2 requirement. The test should be performed on freshly-isolated cultures as CO_2-dependence can rapidly be lost by *Brucella.*

H_2S production. The culture is inoculated onto an SDA or TSA slope and a strip of lead acetate-impregnated paper is placed in the mouth of the tube. This must not come into contact with the medium or with condensation on the wall of the tube. The culture is incubated under the appropriate atmosphere for 4 or 5 days. The paper strip is examined and changed daily. Blackening of the paper indicates H_2S production. Slight blackening of the tip of the paper for the first day only is not considered positive.

Agglutination with monospecific antisera. The growth from an SDA slope incubated for 48 h is suspended in ~0.5 ml of sterile saline to give an opacity equivalent to ~10^{10} cells/ml. One drop of this suspension is mixed with an equal volume of 0.1% aqueous acriflavine on a glass slide and examined for agglutination. The absence of agglutination indicates a smooth culture. In this case, one loopful of the suspension is added to each of a series of loopfuls of A and M monospecific antisera and to negative control serum on a glass slide. After agitation for 1 min, each drop is examined for agglutination. Smooth *Brucella* cultures will produce agglutination with either A and/or M sera but not with the negative control serum.

If the culture is nonsmooth (agglutination occurs with acriflavine), one loopful of culture suspension is added to a loopful of R antiserum and to a loopful of negative control serum, with subsequent agitation for 1 min. The absence of agglutination in either serum will exclude a nonsmooth *Brucella* strain. Agglutination with only the R antiserum is strongly suggestive of a nonsmooth *Brucella.* Agglutination with both sera may also indicate a nonsmooth *Brucella,* although rough organisms of other genera occasionally react in this way.

Urease activity. Christensen's agar slopes are inoculated with single loopfuls of culture suspension prepared as described for the slide agglutination tests. The slopes are examined immediately and after 15-min, 1-h, 2-h and 24-h incubation at 37°C. Cultures of *B. suis, B. canis* and *B. neotomae* almost invariably produce an immediate positive reaction, turning the medium magenta within 15 min. Most strains of *B. abortus* and *B. melitensis* will give positive reactions after 1 or 2 h and nearly all after 18–24 h. The reference strain *B. abortus* 544 is an exception to this and is urease-negative even at 24 h. *B. ovis* cultures are also urease-negative under these conditions, although some strains will give positive reactions if incubated for up to 7 days.

Growth in the presence of dyes. Cell suspensions prepared as for the slide agglutination tests are used to inoculate plates of SDA or other basal media containing the dyes basic fuchsin or thionine.* The plates are divided into four quadrants and one loopful of suspension is applied to each quadrant and streaked five times in succession without recharging the loop. The plates are incubated at 37°C under the appropriate gas atmosphere for up to 4 days. Growth on three or more streaks is considered to indicate resistance to the dye. Growth on only one or two streaks is not considered significant. Other dyes, such as safranin O, methyl violet, pyronine Y, thionine blue or malachite green, are occasionally used in addition to basic fuchsin and thionine. In each case, it is essential that plates inoculated with the biovar 1 reference strains for each *Brucella* species are incubated and examined in parallel with the test cultures.

Differentiation of the genus **Brucella** from other genera

Most of the bacterial species known to cross-react serologically with *Brucella* are unlikely to be misidentified as members of this genus, as they are easily distinguishable by their morphological, cultural and biochemical characteristics. A possible exception is *Francisella tularensis;* however, this organism will not grow on ordinary brucella culture media, is not susceptible to any known brucella phages, and is usually

---

* Dye sensitivity test media: a 0.1% solution of basic fuchsin or thionine (National Aniline Division, Allied Chemical and Dye Co., New York, N.Y. U.S.A.) is prepared in distilled water and heated at 100°C for 1 h. Portions of this stock solution are then added to molten basal media such as SDA or TSA so as to produce a final dye concentration between 10 and 40 μg/ml. The exact concentration required to produce satisfactory differentiation of biovars must be determined for each batch of dye medium by use of *Brucella* reference strains. The plates should be incubated overnight at 37°C before use, to reveal any contaminants.

very much smaller than *Brucella* cells; it also shows fermentative activity towards glucose and a variety of other carbohydrates, is catalase-negative, has a low cytochrome *c* content, produces rapidly lethal infections in mice, and does not share internal antigens with *Brucella* (although unabsorbed antisera will cross-react in agglutination and immunofluorescence tests). Moreover, its DNA base composition (33–36 mol% G + C) is quite distinct from that of *Brucella* and there is no homology of polynucleotide sequences (Hoyer and McCullough, 1968a).

Confusion is most likely to occur with other small Gram-negative nonfermentative bacteria. These include *Bordetella* species, particularly *B. bronchiseptica* which is often isolated from animal sources, and species of *Achromobacter*, *Acinetobacter*, *Branhamella*, *Kingella*, *Moraxella* and *Neisseria*. Some of the less frequently identified *Pseudomonas* species have also been confused with *Brucella* as have occasional strains of *Haemophilus* and *Pasteurella*.

Differentiation of these is achieved in the first instance by careful examination of Gram-stained smears. The morphology of many of these organisms is usually sufficiently distinctive to differentiate them from *Brucella*. Examination of growth and colonial morphology on brucella culture media incubated at 20°C and 37°C and motility tests at both these temperatures will enable many other isolates to be eliminated.

At this stage the colonial phase of a suspected *Brucella* should be determined. Slide agglutination tests with unabsorbed antiserum to smooth and nonsmooth *Brucella* strains will then permit recognition of true members of the genus in nearly every case. Occasionally, nonsmooth organisms other than *Brucella* are agglutinated by antisera to nonsmooth *Brucella* strains. Subsequent tests with phage R/C will permit *B. canis*, *B. ovis* and nonsmooth *B. abortus* strains to be definitely identified as members of the genus. The identification of nonsmooth strains of the other species can occasionally present difficulties, particularly if they give atypical results in one or more of the routine typing procedures. In such cases it is necessary to resort to further tests for genus identification. These will include examination of the electrophoretic pattern of phenol-acetic acid-soluble proteins (Morris, 1973), examination of intracellular antigens by immunodiffusion or immunoelectrophoresis (Corbel et al., 1979), determination of DNA base composition (Hoyer and McCullough, 1968a), examination of the cytochrome absorption spectrum (Dranovskaya and Kushnarev, 1968) and gas-liquid chromatography of the fatty acid methyl esters (Tanaka et al., 1977; Dees et al., 1981).

Taxonomic Comments

The genus *Brucella* forms a discrete homogeneous group which is not closely related to any other genus. Although a number of bacterial genera share similar gross DNA base compositions with *Brucella*, none of these shows any other resemblance. DNA/DNA homology studies within the genus have shown all species to have similar polynucleotide sequences. Even the species producing the least efficient DNA hybridization, namely *B. ovis*, shows >90% homology with the *B. suis* reference strain (Hoyer and McCullough, 1968a, b).

This DNA homology is indicative of a very close genetic similarity between all members of the genus and is consistent with the concept of a single species. Nevertheless consistent differences do exist in the biochemical, cultural and pathogenic properties of the strains forming the recognized subdivisions within the genus and these are correlated with the host specificity. This strengthens the case for a division into species which has proved to be of epidemiological value.

The present system of classification was devised by the Subcommittee on Taxonomy of the genus *Brucella* at its first meeting (Stableforth and Jones, 1963), with amendments introduced at its subsequent meetings (Jones, 1967; Jones and Wundt, 1971; Wundt and Morgan, 1975). Although a number of studies, including those based on numerical taxonomy (Šaferšteijn, 1973; Gargani, 1977), have produced suggestions for modifications of this scheme it is generally accepted as satisfactory for most practical purposes. Most of the modifications suggested have been in relation to the subdivision of the major species and the status of the minor species *B. canis*, *B. ovis* and *B. neotomae* (Meyer, 1969; Šaferšteijn, 1973; Gargani, 1977).

Overall, the current evidence points to a high degree of species stability although it has long been recognized that the differences between *Brucella* species are quantitative rather than qualitative and that indeterminate forms are encountered. The evidence for species-to-species transformation has been reviewed by Meyer (1976a) and must be considered inconclusive at present. The evidence for biovar to biovar variation within species is better substantiated and interchange clearly occurs (Shibata et al 1962).

The conventional biotyping tests do not always indicate the full extent of differences between biovars. For example, *B. abortus* biovar 2 nominally differs from biovar 1 only in sensitivity to basic fuchsin. Nevertheless, strains of *B. abortus* biovar 2 are distinctive in being sensitive to a wide range of antibiotics, dyes and other inhibitors and in having more exacting nutritional requirements than strains of biovar 1. Some strains of *B. abortus* biovar 4 resemble biovar 2 in these properties and possibly should be distinguished as a separate group.

On the other hand, the differences between some biovars may be very small and dependent upon a single characteristic. For example, strains of *B. abortus* biovar 3 are consistently distinguishable from those of biovar 6 solely on the basis of their ability to grow in the presence of thionine at 1:25,000. This difference is not clear-cut. Some strains show partial sensitivity to thionine and cultures may cover a complete spectrum, with those of typical biovar 3 at one end and those of typical biovar 6 at the other, while many strains occupy an intermediate position (Tolari, Thomas and Corbel, unpublished results).

The existing system of classification does not permit the identification of all *Brucella* isolates. Thus strains have been described whose properties do not coincide with the descriptions of the recognized species and biovars. These include the dye-resistant variants of *B. suis* biovar 1 isolated in South America and the South Pacific (Corbel, Thomas and Garcia-Carillo, unpublished results) and strains cultured from rodents in Africa and Australia (Heisch et al., 1963; Cook et al., 1966) which behave as variants of *B. suis* biovar 3. Other *Brucella suis* strains isolated from rodents in the U.S.S.R. (Taran et al., 1966) and those designated as "*Brucella murium*" (Korol and Parnas, 1967), are probably sufficiently distinctive to be considered as a new biovar. Indeed it seems likely that the existing biovar system will have to be modified or extended to accommodate new variants as they are discovered.

Differentiation of the species and biovars of the genus **Brucella**

Tables 4.101 and 4.102 list characteristics useful for differentiation of the species and biovars of *Brucella*.

List of the Species of the Genus **Brucella**

1. **Brucella melitensis** (Hughes 1893) Meyer and Shaw 1920, 179.[AL] (*Streptococcus miletensis* (sic) Hughes 1893, 235.)

me.li.ten'sis. L. adj. *melitensis* of or pertaining to the Island of Malta (Melita).

The morphological and cultural characteristics are as described for the genus. Other characteristics of the species and its biovars are indicated in Tables 4.101 and 4.102.

The usual natural hosts are sheep and goats, but other species may be infected including cattle, pigs and man. Smooth cultures are usually pathogenic for the guinea pig and the mouse. In general, smooth

Table 4.101.

Differentiation of the species of the genus **Brucella**[a]

	Lysis by Phage at RTD						Oxidation of Substrates													
Species[b]	Tb	Wb	Fi	Bk$_2$	R/O	R/C	L-Alanine	L-Asparagine	L-Glutamic acid	L-Arabinose	D-Galactose	D-Ribose	D-Glucose	D-Xylose	L-Arginine	DL-Citrulline	DL-Ornithine	L-Lysine	*meso*-Erythritol	Preferred Host
1. *B. melitensis*	NL	NL	NL	L	NL	NL	+	+	+	−	−	−	+	−	−	−	−	−	+	Sheep, goats
2. *B. abortus*	L	L	L	L	PL	NL	+	+	+	+	+	+	+	d	+	+	(+)	+	+	Cattle
3. *B. suis*																				
Biovar 1	NL	L	PL	L	NL	NL	d	−	−	+	d	+	+	+	+	+	+	+	+	Swine
Biovar 2	NL	L	PL	L	NL	NL	−	d	d	+	d	+	+	+	+	+	+	−	+	Swine, hares
Biovar 3	NL	L	PL	L	NL	NL	d	−	d	−	−	+	+	+	+	+	+	+	+	Swine
Biovar 4	NL	L	L	L	NL	NL	−	−	d	−	−	+	+	+	+	+	+	+	+	Reindeer
4. *B. ovis*	NL	NL	NL	NL	L	L	d	+	+	−	−	−	−	−	−	−	−	−	−	Sheep
5. *B. neotomae*	PL	L	L	L	NL	NL	d	+	+	+	+	d	+	−	−	−	−	−	+	Desert wood rat
6. *B. canis*	NL	NL	NL	NL	NL	L	d	−	+	d	d	+	+	−	+	+	+	+	d	Dogs

[a] Symbols: NL, no lysis; L, lysis; PL = partial lysis; +, Q$_{O_2}$N is 50 or greater; −, Q$_{O_2}$N is less than 50; d, Q$_{O_2}$N values differ among strains (may be higher or lower than 50).

[b] Results apply to smooth strains only, except for *B. ovis* and *B. canis* (which occur only in the non-smooth phase).

Table 4.102.

Differentiation of the species and biovars of the genus **Brucella**[a]

Species	Biovar	CO$_2$ requirement	H$_2$S produced	Growth on Media Containing[b]		Agglutination with Monospecific Antisera		
				Thionine	Basic Fuchsin	A	M	R
1. *B. melitensis*	1	−	−	+	+	−	+	−
	2	−	−	+	+	+	−	−
	3	−	−	+	+	+	+	−
2. *B. abortus*	1	[+]	+	−	+	+	−	−
	2	[+]	+	−	−	+	−	−
	3^c	[+]	+	+	+	+	−	−
	4	[+]	+	−	+	−	+	−
	5	−	−	+	+	−	+	−
	6^c	−	[−]	+	+	+	−	−
	7	−	[+]	+	+	+	+	−
	9	−	+	+	+	−	+	−
3. *B. suis*	1	−	+	+	[−]	+	−	−
	2	−	−	+	−	+	−	−
	3	−	−	+	+	+	−	−
	4	−	−	+	[−]	+	+	−
4. *B. ovis*	None	+	−	+	[−]	−	−	+
5. *B. neotomae*	None	−	+	−^{d}	−	+	−	−
6. *B. canis*	None	−	−	+	[−]	−	−	+

[a] Symbols: +, positive for all strains; [+], positive for most strains; [−], negative for most strains; −, negative for all strains.

[b] Dye concentration, 1:50,000 (w/v).

[c] For more certain differentiation of biovar 3 and 6, thionine at 1:25,000 (w/v) is used; biovar 3 gives a positive growth response, biovar 6 is negative.

[d] Growth will occur in the presence of thionine at a concentration of 1:150,000 (w/v).

cultures of *B. melitensis* tend to be more virulent for laboratory animal than *B. abortus* and may cause fatal infections. Nonsmooth cultures are usually avirulent for both laboratory animals and the natural hosts.

The mol% G + C of the DNA is 58 (T_m).

Type strain: 16M (ATCC 23456; NCTC 10094). This is also the reference strain for biovar 1. The reference strain for biovar 2 is 63/9 (ATCC 23457; NCTC 10508), and that for biovar 3 is Ether (ATCC 23458; NCTC 10509).

2. **Brucella abortus** (Schmidt 1901) Meyer and Shaw 1920, 176.[AL] (*Bacterium abortus* Schmidt in Schmidt and Weiss, 1901, 266).

a.bor′tus. L. gen. n. *abortus* of abortion, miscarriage.

The morphological and cultural characteristics are as described for the genus. Other characteristics of the species and its biovars are indicated in Tables 4.101 and 4.102.

The usual natural hosts are cattle and other bovidae. Horses, camels, sheep, deer, dogs, man and other species may also be infected. Placentitis and abortion are usually produced in the pregnant animal.

Pathogenic for laboratory animals including rabbits, guinea pigs and mice; the guinea pig is probably the most susceptible. Rough strains are usually avirulent, but mice retain the organisms in the spleen for some time after inoculation.

The mol% G + C of the DNA is 56 (Bd) or 57 (T_m).

Type strain: 544 (ATCC 23448; NCTC 10093). This is also the reference strain for biovar 1. For biovar 2 (referred to as "dye-sensitive *B. abortus*" by Wilson (1933)) the reference strain is 86/8/59 (ATCC 23449; NCTC 10501). For biovar 3 (referred to as "Rhodesian abortus", "non-CO_2-requiring, thionine-resistant *B. abortus*" (Bevan, 1930) or "CO_2-requiring, thionine-resistant *B. abortus*" (Van der Schaaf and Rosa, 1940), the reference strain is Tulya (ATCC 23450; NCTC 10502). The reference strain for biovar 4 is 292 (ATCC 23451; NCTC 10503). For biovar 5 (referred to as "British *melitensis*" (Stableforth, 1959)) the reference strain is B3196 (ATCC 23452; NCTC 10504). The reference strain for biovar 6 is 870 (ATCC 23453; NCTC 10505), and for biovar 7 is 63/75 (ATCC 23454; NCTC 10506). For biovar 9 (referred to as "H_2S-producing *B. melitensis*" by Taylor et al., (1932)) the reference strain is C68 (ATCC 23455; NCTC 10507). (Biovar 8 was originally described as "CO_2-requiring *B. melitensis*" by Taylor et al., (1932), but as no cultures of this biovar are known to exist the status of this biovar was suspended by the Subcommittee on the Taxonomy of the genus *Brucella* in 1978. (Corbel, 1982)).

3. Brucella suis Huddleson 1929, 12.[AL]

su'is. L. gen. n. *suis* of the pig.

The morphological and cultural characteristics are as described for the genus. Other characteristics of the species and its biovars are indicated in Tables 4.101 and 4.102.

Biovars 1, 2 and 3 are naturally pathogenic for pigs. Biovar 2 also naturally infects hares. Strains resembling biovar 3 have been isolated from various species of rodents (Heisch et al., 1963; Cook et al., 1966). Biovar 4 is naturally pathogenic for reindeer (Davȳdov, 1961). All biovars, with the possible exception of biovar 2, are pathogenic for man. Other species may also be infected, including dogs, horses and many species of rodents. In the natural hosts, generalized infections are produced with localizing lesions, particularly in the genitalia. The testes, epididymides and seminal vesicles are usually severely affected in the male. Metritis, placentitis and abortion are produced in the pregnant female.

Pathogenic for laboratory animals including rabbits, guinea pigs and mice. Splenomegaly and wide-spread granulomatous and suppurative lesions are produced. Biovars 1 and 3 are usually the most virulent, and heavy inocula may produce fatal infections in guinea pigs. The organisms are located intracellularly in vivo.

The mol% G + C of the DNA is 56–57 (T_m).

Type strain: 1330 (ATCC 23444; NCTC 10316). This is also the reference strain for biovar 1 (referred to as "American *suis*"). For biovar 2 (referred to as "Danish or European *B. suis*" by Thomsen (1929)) the reference strain is Thomsen (ATCC 23445: NCTC 10510). For biovar 3 (referred to as "American *melitensis*" or "dye-resistant *B. suis* (Huddleson, 1957)) the reference strain is 686 (ATCC 23446; NCTC 10511). For biovar 4 (formerly *Brucella rangiferi tarandi* (Davȳdov, 1961)) the reference strain is 40 (ATCC 23447; NCTC 11364).

4. Brucella ovis Buddle 1956, 351.[AL]

o'vis. L. gen. n. *ovis* of the sheep.

The morphological and cultural characteristics are as described for the genus. Althogh other species of *Brucella* stain red by modified Köster's stain, *B. ovis* stains blue.

Other characteristics are listed in Tables 4.101 and 4.102. No biovars are recognized.

Adonitol and DL-serine are oxidized.

Cultures are known to exist only in the non-smooth colonial phase. They do not agglutinate with antisera monospecific for A and M surface antigens but are agglutinated by antisera to the R surface antigen of *B. ovis.* Cross-reactions occur with the surface antigens of non-smooth strains of other *Brucella* species. Many of the internal antigens are shared with other *Brucella* species irrespective of colonial phase.

Pathogenic for sheep, producing epididymo-orchitis in the male and placentitis and abortion in the pregnant female. Goats may be infected experimentally and sub-clinical infections may be produced in cattle, guinea pigs, rabbits, mice and gerbils.

The mol% G + C of the DNA is 57–58 (T_m).

Type strain: 63/290 (ATCC 25840; NCTC 10512).

Further comments. B. ovis is the only *Brucella* species showing less than 100% homology with *B. suis* reference DNA.

5. Brucella neotomae Stoenner and Lackman 1957, 947.[AL]

ne.o.to'mae. M.L. fem. n. *Neotoma* generic name of the desert wood rat of the Western U.S.A., *Neotoma lepida* Thomas; M.L. fem. gen. n. *neotomae* of the desert wood rat, the host from which the organism was first isolated.

On morphological and cultural characteristics are as described for the genus. Other characteristics are indicated in Tables 4.101 and 4.102.

Nonpathogenic for cattle, sheep, goats and pigs. Not known to be pathogenic for man. Does not apparently produce disease in its natural host (the desert wood rat) and shows minimal pathogenicity for laboratory animals. Guinea pigs develop slight splenomegaly and sometimes epididymo-orchitis or testicular abscesses following intraperitoneal inoculation. Small granulomatous lesions develop in the liver and spleen. Mice are more susceptible to infection and lesions may be produced by frequently passaged strains.

The mol% G + C of the DNA is 56–57 (T_m).

Type strain: 5K33 (ATCC 23459; NCTC 10084).

6. Brucella canis Carmichael and Bruner 1968, 579.[AL]

ca'nis. L. gen. n. *canis* of the dog.

The morphological and cultural characteristics are as described for the genus. Other characteristics are indicated in Tables 4.101 and 4.102. No biovars are recognized.

On incubation for more than a few days, the colonies become very tenacious and viscous. The growth is almost impossible to emulsify and forms a ropy agglutinate in physiological saline. In Albimi brucella broth at 37°C for 7 days a moderate turbidity is produced with a ropy or viscous sediment which cannot be uniformly resuspended. Older cultures may form a fine surface pellicle which is easily disrupted. In serum-dextrose broth or well buffered media, the ropy sediment or pellicle is often not produced.

Cultures are always in the rough or mucoid phase on primary isolation. A smooth phase is not known. Cultures are not agglutinated by antisera monospecific for the A and M antigens but do agglutinate with antiserum to the R antigen of *B. ovis.* Cross-reactions with the surface antigens of nonsmooth strains of other *Brucella* species also occurs. Internal antigens are shared with smooth and nonsmooth strains of all *Brucella* species.

Pathogenic for the dog, producing chronic bacteremia and localizing granulomatous lesions. Epididymo-orchitis and prostatitis are produced in the male, and metritis, placentitis and abortion are produced in the pregnant female. The infection is occasionally transmitted to man, but natural infections in other species have not been authenticated. Infection in guinea pigs and mice may be established by inoculation of large doses of organisms. The mouse retains the infection longer than the guinea pig.

The mol% G + C of the DNA is 56 (T_m).

Type strain: RM6/66 (ATCC 23365; NCTC 10854).

Other organisms

Several *Brucella*-like isolates have been described, the properties of which do not coincide with the descriptions for recognized species. Some of these isolates may be regarded as variants of existing species or biovars. These include the dye-resistant variants of *B. suis* biovar 1 isolated in South America and the South Pacific (Corbel et al., unpublished results) and the strains isolated from rodents in Australia and Africa (Cook, Campbell and Barrow, 1966; Heisch et al., 1965) which behave as variants of *B. suis* biovar 3. The taxonomic position of the cultures designated as "*Brucella murium*" (Korol and Parnas, 1967) has still to be decided, but probably insufficient evidence exists to justify their nomination as a new species. Similar cultures from rodents in the U.S.S.R. (Taran et al., 1966) resemble *B. suis* biovar 3 but oxidize asparagine, are agglutinated by M but not A monospecific serum and show a relatively high pathogenicity for murine and cricetine rodents. These cultures should probably be regarded as representing a separate biovar of *B. suis*.

Genus **Bordetella** Moreno-López 1952, 178[AL]

MARGARET PITTMAN

Bor.de.tel′la. M.L. dim. ending-*ella*; M.L. fem. n. *Bordetella* named after Jules Bordet, who with O. Gengou first isolated the organism causing pertussis.

Minute coccobacillus, 0.2–0.5 μm in diameter and 0.5–2.0 μm in length, often bipolar stained, and arranged singly or in pairs, more rarely in chains. **Gram-negative. Nonmotile,** two species; **motile,** one species, by peritrichous flagella. **Strictly aerobic. Optimum temperature, 35–37°C.** Colonies on Bordet-Gengou medium are smooth, convex, pearly, glistening, nearly transparent and surrounded by a zone of hemolysis without definite periphery. **Metabolism respiratory, never fermentative.** Chemoorganotrophic, **require nicotinamide, organic sulfur** (e.g. cysteine) and **organic nitrogen** (amino acids). **Utilize oxidatively** glutamic acid, proline, alanine, aspartic acid and serine with production of ammonia and CO_2. Litmus milk is made alkaline. **Mammalian parasite** and **pathogen. Localize** and **multiply among the epithelial cilia of the respiratory tract.** The mol% G + C of the DNA is 66–70 (T_m).

Type species: *Bordella pertussis* (Bergey et al. 1923) Moreno-López 1952, 178.

Further Descriptive Information

Cells of freshly isolated strains (phase I) of all bordetellae appear as minute coccobacilli arranged singly or in pairs. Cells of degraded strains (phase IV) are slightly larger and may appear in short chains. *B. pertussis* is enveloped by a capsule (Lawson, 1940) or a slime sheath (Klieneberger-Nobel, 1948). A capsule (Lawson strain) on *B. bronchiseptica* has been reported (see Nakase, 1957). The envelope, which is not enhanced by antiserum as in the "quellung reaction," might consist of the extruded particles that have been demonstrated by electron microscopy (Morse and Morse, 1970; Novotny and Cownley, 1979). Pili-like filaments extend from the surface of all bordetellae (Morse and Morse, 1970). Cell walls of *Bordetella bronchiseptica* are highly convoluted with one or two less electron-dense spots on the surface (Bemis et al., 1977b). The periodic structure of the few short peritrichous flagella of *B. bronchiseptica* is 190 by 139 Å (Labaw and Mosley, 1955).

Cell walls of *B. pertussis*, in common with other Gram-negative cells, are composed of five layers (Morse and Morse, 1970). A proposed scheme of the location of the biologically active antigens in the cell wall and the cytoplasm has been reported by Zakharova (1979).

Culture supernatant fluids of *B. pertussis* contain filamentous particles 20 Å in diameter and 40–70 μm in length (Morse and Morse, 1970) and ring-shaped (spheroid) particles, 75–80 Å in diameter, with molecular weight mass of approximately 75,000 (Morse and Morse, 1976). Similar filaments are present in supernatant fluids of *B. parapertussis* and *B. bronchiseptica* cultures (Morse and Morse, 1970) but the ring-shaped particles are absent. The latter particles are pertussis toxin that cause several kinds of pathophysiological reactions. Prior to the evidence that a single entity induces the reactions, different names that reflect the nature of the reactions were used: histamine-sensitizing factor (HSF) (Parfentjev and Goodline, 1948), lymphocytosis-promoting factor (LPF) (Iwasa et al., 1966) and islet-activating protein (IAP), (Ui, Katada and Yajima, 1979). Since this substance has the characteristic of an exotoxin ("true toxin") it was designated pertussis toxin (HSF-LPF-IAP) to conform with the custom to name a toxin after the disease in which it causes the harmful effects (Pittman, 1979). Its molecular structure and intracellular catalytic activity are in accord with the A-B model of toxins (Tamura et al., 1983).

The time of development and size of the colonies differ between the species. On Bordet-Gengou (B-G) medium* the colonies of *B. pertussis*, pinpoint in size, appear in 3–6 days; colonies of *Bordetella parapertussis*, slightly larger, develop in 2 or 3 days; and colonies of *B. bronchiseptica*, the largest, develop the most rapidly. On peptone agar, only the *B. parapertussis* colonies are surrounded by a zone of diffused brown pigment. *B. bronchiseptica* colonies on MacConkey agar are reddish and surrounded by a small red zone with amber discoloration of the underlying medium (I. A. P. McCandlish, 1977, Ph.D. Thesis, University of Glasgow, Glasgow, Scotland).

A temperature of 35–37°C favors maintenance of characteristics including surface antigens. Lower temperatures favor reversible phenotypic changes (Lacy, 1960).

In vivo, growth of the genus occurs on and among cilia of the respiratory epithelial cells. Injected intracerebrally in the mouse, *B. pertussis* grows only among the cilia of the ependymal cells and does not invade the parenchyma, whereas *B. bronchiseptica* grows initially among the cilia, then in the ventricular cavity and the brain parenchyma (Iida et al., 1962). Cilia seem to be essential for in vivo growth of *B. pertussis* (Gallavan and Goodpasture, 1937). Iida and Ajika (1975) postulated that the bacteria are dependent on a cellular source of

---

* Modified Bordet-Gengou medium for isolation of *Bordetella pertussis* (Kendrick and Eldering, 1969; Blair, 1970): *Base agar:* potato infusion, 500 ml; NaCl, 11.25 g; agar, 50 g; and distilled water, 1500 ml. Potato infusion is made by boiling 500 g of peeled, sliced potatoes in 1000 ml distilled water and 40 ml of glycerol. When potatoes are soft, the infusion is poured into tall cylinders for sedimentation. The clear supernatant is used in the base agar. For the base agar, the salt is dissolved and the agar is thoroughly wetted with a portion of the water. The remainder of the water is added, the agar is dissolved by heating, and the potato infusion is added. No adjustment of the pH is necessary. The medium is dispensed as desired and autoclaved at 121°C for 15 min. *Completed medium:* Defibrinated blood, 15–20 ml (the authors use sheep blood less than 72-h-old) is added per 100 ml of melted agar cooled to 45°C. If penicillin is desired, 0.25–0.50 U/ml are added. The final mixture is dispensed into plates or tubes as desired and incubated overnight at 35°C for a preliminary sterility check. Each lot should be checked for growth-promoting properties, colony appearance, and characteristic hemolytic zone, with a control culture of *B. pertussis*. The medium is stored at 4–6°C in containers such as plastic bags to preserve moisture. Satisfactory plates should have a moist surface and be cherry red at the time of use. To promote mass growth, 1% peptone may be added.

energy. In vitro, the major metabolic activity of bordetellae is oxidation of amino acids. Ammonia, CO_2 and small quantities of other materials are produced. The main amino acid used is glutamate. *B. pertussis* and *B. parapertussis* deaminate also proline, alanine, aspartic acid, serine and glycine and *B. bronchiseptica* utilizes all of the amino acids except arginine, lysine and histidine (Meyer and Cameron, 1957; Rowatt, 1955, 1957). Essential additional requirements are nicotinomide and organic sulfur (cystine, cysteine or glutathione).

B. pertussis is more fastidious relative to growth conditions than are the other two species. Growth, particularly growth initiated by small inocula, is inhibited by unsaturated fatty acids, colloidal sulfur or sulfides formed during autoclaving, and a metabolic substance(s), probably organic peroxides (Rowatt, 1957). These substances may be absorbed with starch, charcoal, albumin or anion-exchange resins. Adjustment of the nutritional, physical and adsorbent factors is very complex. No absorbent is required in the Stainer-Scholte (1970) defined liquid medium* for mass growth of *B. pertussis* from high levels of seeding. Aeration of liquid medium is obtained by vortex stirring with sterile air blown over the surface, shaking or use of shallow layers of medium.

The following physiological tests are invariably negative for the genus: gelatin, esculin, casein and starch hydrolysis, oxidative fermentation of carbohydrates, ability to grow anaerobically, indole production, and phosphatase. Tests are positive for peroxidase, adenylate cyclase in culture supernatant and ammonia from amino acids. Tests that differ between species are nitrate reduction, urease and oxidase. The catalase test is positive for *B. parapertussis* and *B. bronchiseptica* but it varies between strains of *B. pertussis* (Portwood, 1946; Lautrop, 1960).

B. pertussis has a marked propensity to undergo two distinct types of variation. (a) Serial mutation changes, phase I to phase IV, occur with repeated passage on culture medium and independently (Leslie and Gardner, 1931). Phase I has the characteristics of freshly isolated cells and phase IV represents the final stage of spontaneous degradation—growth on nutrient agar, avirulence, nonprotective, nontoxic, nonhemagglutinable, agglutinogen changes, etc. (see Dobrogosz et al., 1979). Properties of intermediate phases represent mutations that occur in any order (Parker, 1979) and are not constant between the degraded phases. (b) Phenotypic variants, freely reversible, occur within a few generations of growth in an altered environment (mode X to mode C) (Lacy, 1960). Phase I cells grown in high magnesium, low sodium medium show a loss in certain envelope proteins, intracerebral protective activity, pertussis toxin (HSF-LPF-IAP), heat-labile (56°C) toxin (HLT), adenylate cyclase activity and adjuvanticity (Wardlaw and Parton, 1979). The concomitant losses suggest (a) that a single site of action is controlled by one enzyme, (b) that the enzyme affects the fastidiousness of the growth of phase I *B. pertussis*, and (c) that the enzyme is lost during mutation to phase IV. The gene that controls the enzyme remains to be identified.

B. parapertussis and *B. bronchiseptica* show the usual mutational changes associated with transition from smooth to rough colonies but not modulation (Lacy, 1951). They do not produce pertussis toxin (HSF-LPF-IAP).

A close genetic relationship between the species of the genus is shown by (a) DNA/DNA reassociation reactions between *B. pertussis*, *B. parapertussis* and *B. bronchiseptica*, (b) transformation of leucine and tryptophan auxotrophs of *B. pertussis* by DNA of *B. pertussis* and *B. bronchiseptica* close to homologous values, and (c) confirmation of genetic exchange by treatment of *B. pertussis* with *N*-methyl-*N'*-nitro-*N*-nitrosoguanidine (Kloos et al., 1979). In a numerical study, Johnson and Sneath (1973) found, for the strains of the genus that they studied, a cophenetic correlation of 0.96.

With DNA from a streptomycin-resistant strain of *B. pertussis*, Branefors (1964) transferred resistance to a streptomycin-sensitive strain. Terakado and Mitsuhashi (1974) transferred resistance among *B. bronchiseptica* strains and other host bacteria. It has been suggested that the antibiotic resistance gene is carried on plasmids (Hedges et al., 1974).

Kloos et al. (1979) reported that most strains of *B. pertussis* and *B. parapertussis* contained a presumed, small cryptic plasmid of approximately 3 megadaltons and that four of seven strains of *B. bronchiseptica* carried one or more larger plasmids (8–28 megadaltons) in addition to a smaller plasmid similar in size to the *B. pertussis* plasmid.

Bacteriophages have shown a close relationship between *B. bronchiseptica* and *B. parapertussis* but not with *B. pertussis*, *Alcaligenes faecalis* or seven species of *Brucella* (Rauch and Pickett, 1961).

A bacteriocin typing scheme for the species has not been reported. Litkenhaus and Liu (1967) reported that two rough strains of *B. pertussis*, ATCC 190 and 6627, inhibited the growth of 22 of 24 smooth strains of *B. pertussis* but not the growth of *B. parapertussis* or *B. bronchiseptica* strains. Subsequently, Kawai and Yabuuchi (1975) found that the inhibitor strains were misidentified and classified them as *Pseudomonas pertucinogena*.

Information on the agglutinogens (Kauffmann, 1947) of the genus is deficient. Eldering et al. (1957) identified 14 heat-labile (120°C for 1 h) agglutinogens among the three species. Antigen 7 was common between the species and each species had a species-specific antigen. The other 10 antigens occurred in various combinations within and between strains of a species. All *B. parapertussis* strains, that had been studied, had the same pattern of four antigens (Kendrick and Eldering, 1969). Strains of *B. bronchiseptica* differ in both K- and O-antigen patterns. There are indications of a greater similarity in the patterns among strains from a single animal species than between strains from heterologous animal species (Nakase, 1957; Pedersen, 1975). Unfortunately, the recently designated K antigens have not been identified with the scheme of Eldering et al. (1957) which has been very valuable in the serotyping of *B. pertussis*. The relations of the designated O antigens likewise are not known. There are many pitfalls in the serotyping of agglutinogens (Dolby and Bronne-Shanbury, 1975) and K antigens differ in thermal resistance to lower than 100°C up to 120°C (Kendrick and Eldering, 1969). It appears that one O antigen is common between the species.

The common 56°C-labile, dermonecrotic toxin (HLT) is located in the cytoplasm or periplasm (Morse, 1976; Zakharova, 1979). Since it is demonstrable only after the cell is ruptured, it may be present as a precursor or zymogen that must be activated before toxicity occurs (Anderson and North, 1943; Morse, 1976). There is no supporting evidence that HLT plays a role in immunity (Pittman, 1979).

Pertussis toxin (HSF-LPF-IAP), labile at 80°C, is produced only by *B. pertussis*. The suggestion of Goodnow (1980) that a histamine-sensitizing factor is produced by *B. bronchiseptica* relates to a nonspecific increase in histamine sensitivity of dogs infected with *B. bronchiseptica* (Dixon et al., 1979). Dixon et al. noted that this mechanism of bronchial hyperactivity was similar to the response of subjects with "colds" (presumably viral respiratory infections) following inhalation of histamine or citric acid aerosols. The HSF of *B. pertussis* is a specific antigen which renders the mouse highly sensitive to histamine (see references in Munoz et al. (1981) and in Pittman (1979)). Since pertussis toxin is a potent adjuvant, it appears that the early clearance of *B. pertussis* and the prolonged immunity after whooping cough, in contrast to the prolonged infection and short duration of immunity after canine and swine bordetellosis (Bemis et al., 1977a), is due to antitoxin (see Pittman, 1979).

The roles of pertussis toxin, (also, designated pertussigen (Munoz et

---

* Stainer-Scholte defined medium (g/liter of distilled water): sodium glutamate, 10.72; proline, 0.24; L-cystine, 0.04; ascorbic acid, 0.02; niacin, 0.004; glutathione, 0.1; NaCl, 2.5; KH$_2$PO$_4$, 0.5; KCl, 0.2; MgCl$_2$·6H$_2$O, 0.1; CaCl$_2$, 0.02; FeSO$_4$·7H$_2$O, 0.01; Tris buffer "121," 6.075. The cystine, glutathione, ascorbic acid, niacin and FeSO$_4$ are sterilized by filtration and added aseptically to the autoclaved medium. The pH is adjusted to 7.6 with 2.5 N HCl before autoclaving.

al., 1981) and LPF-HA (Sato et al., 1981)) and filamentous antigen (F-HA) in pertussis immunity are under investigation (Munoz et al., 1981; Sato et al., 1981). It is definite that purified pertussis toxin or the toxoid is capable of protecting mice against lethal challenge of *B. pertussis*; but there may not be total clearance of the bacteria from the respiratory tract. Human protective activity remains to be explored, as well as whether F-HA antibody can effectively prevent attachment of *B. pertussis* to cilia. F-HA is present also in *B. bronchiseptica* and *B. parapertussis* (Irons and MacLennan, 1979; Munoz et al., 1981); pertussis toxin is absent.

A comparison of the chemical structures of the lipopolysaccharide (LPS) of the individual species has not been reported. LPS isolated from *B. pertussis* differs from enterobacterial LPS in gross chemical structure and in significantly lower pyrogenicity (Chaby et al., 1979).

Antibiotic sensitivity of bordetellae may vary between the species and concentrations tested (Johnson and Sneath, 1973). In general, bordetellae are sensitive to erythromycin, chloramphenicol, streptomycin and oxytetracycline and are resistant to penicillin and bacitracin. Erythromycin has been favored for prophylaxis and early treatment of *B. pertussis* infection (Bass et al., 1969; Altemeir and Ayoub, 1977). However, failures have been reported (Halsey et al., 1980; Grob, 1981). Antibiotics do not alleviate the clinical symptoms induced by pertussis toxin.

A number of antibacterial agents, including sulfamethazine and sulfaquinoxalin trimethoprim, reduce the severity of *B. bronchiseptica* and "*Bordetella bronchiseptica*-like bacterium" (Hinz et al., 1978) infections but not carriage (Appel and Bemis, 1977; Ganaway et al., 1965; Glünder et al., 1979; Switzer, 1963). High resistance developed in swine *B. bronchiseptica* to sulfamethazine in rations (Harris et al., 1969). Alexander et al. (1980) reported that trimethoprim-sulfonamide combinations were prophylactic for piglets against *B. bronchiseptica*.

Kloos et al. (1979) indicated that phase IV strains of *B. pertussis* could be differentiated from phase I strains by their antibiotic resistance.

Bordetellae are highly communicable pathogenic, obligatory parasites of man and animals and are worldwide in distribution. They are transmitted by intimate exposure to expired droplets. The bacteria localize and multiply among the cilia of the epithelial cells of the respiratory tract and do not invade the underlying tissue. Each species induces bronchial and pulmonary pathology (Pittman, 1970; Bemis et al., 1977a; Duncan et al., 1966). Only *B. pertussis* induces prolonged paroxysmal coughing, lymphocytosis-leucocytosis, altered glucose metabolism and other symptoms which are induced by pertussis toxin (Pittman et al., 1980). The acute stage of *B. bronchiseptica* infection is much shorter in duration. In piglets, there is osseous resorption in the nasal turbinate that causes atrophy of the snout. In the infant of man and animal, susceptibility to infection is highest, the disease is most severe and mortality is the highest.

Man is the only natural host of *B. pertussis* and *B. parapertussis*. The carriage of *B. pertussis* is low or nil (Linnemann et al., 1968). *B. bronchiseptica*, primarily a pathogen of laboratory, domestic and wild animals, infects man only occasionally (Lautrop, 1960). It infects (often in epidemic form) rabbits, guinea pigs, rats, nonhuman primates, dogs, swine, cats, horses, foxes, opossums, raccoons, skunks and voles (Bemis et al., 1977b; Switzer et al., 1966; Jensen and Duncan, 1980). The reported carrier rates for dogs, swine and rabbits are high. The "*Bordetella bronchiseptica*-like bacterium" infects turkey poults (Hinz et al., 1978). This organism has been isolated, also, from a hen and a duck (Hinz et al., 1979). Farrington and Jorgenson (1976) isolated a bordetella strain from one of 47 house sparrows. This strain may have been a transience from swine. The reported slow urease and negative nitrate reduction activities suggest that the strains isolated from turkeys by Filion et al. (1967) may have been "*B. bronchiseptica*-like bacteria." Information is deficient on avian distribution of "bordetellae."

Isolation and Enrichment Procedures

Bordetellae in human specimens are isolated on modified B-G medium (16–20% blood, Kendrick and Eldering, 1969) or B-G medium (50% blood, Lautrop, 1960). The lesser amount of blood favors detection of the hemolytic zone around a colony. Fresh blood is used. Penicillin, 0.25–0.5 U/ml, is effective in reducing growth of Gram-positive organisms. Plates incubated at $36 \pm 1.0°C$ are examined daily until positive but not longer than 7 days. Lacy's medium (Lautrop, 1960), especially designed for isolation of *B. pertussis*, has not gained general acceptance. Peptone should not be present in isolation media as it inhibits growth of small numbers of *B. pertussis*. Charcoal agar, containing penicillin, serves as a transport medium for swab specimens (Holwerda, 1971).

For the isolation of *B. bronchiseptica* from animals in relatively clean areas, Bemis et al. (1977b) recommended Levine eosin-methylene blue agar (Difco) and blood agar (brucella agar base plus 5% sheep blood and isoVitaleX enrichment (BBL), final concentration 1:50). Mac-Conkey agar (1% glucose) with added antimicrobial agents has had wide application (see Smith and Baskerville, 1979). The G20G medium* of Smith and Baskerville (1979) takes into account that growth of *B. bronchiseptica* is inhibited by acid produced by more rapidly growing associated bacteria.

Maintenance Procedures

Strains of *B. pertussis* and *B. parapertussis* are grown on B-G slopes. After incubation at 35–37°C to allow abundant growth, the cultures are stored at 4–6°C in closed containers for 2–3 weeks. Strains of *B. bronchiseptica* are grown on a variety of nutrient slopes and incubated for 1–2 days. No more than a few passages of strains of either species are made after recovery from the preserved state.

For preservation, heavy suspensions of cells, harvested from an agar medium suitable for the species (not liquid medium), are prepared (a) in sterile skim milk or 10% powdered milk, dispensed in vials and freeze-dried, or (b) in 0.05 M Tris-HCl buffer (pH 7.6) containing glycerol, dispensed in vials and stored in the gas phase of liquid nitrogen (Novotny and Brookes, 1975).

Procedures for Testing for Special Characters

With the "*B. bronchiseptica*-like bacterium," the nitrate test was regularly positive only when the medium was supplemented with nicotinamide adenine dinucleotide (NAD) (100 μg/ml) and 1% serum; and the oxidase reaction was not positive until 48–72 h, using the sensitive test of Kovacs (1956) (Hinz et al., 1979). Johnson and Sneath (1973) used a basic medium that contained NAD and serum to test for biochemical activities of bordetellae and related bacteria. They also used the Kovacs method for oxidase activity.

Further Reading

Iida, T., N. Kusano, A. Yamoto and M. Konosu. 1966. An immunofluorescence study of the action of antibody in experimental intracerebral infection in mice with *Bordetella pertussis*. J. Pathol. Bacteriol. *92:* 359–367.

Lautrop, H. 1971. Epidemics of parapertussis in Denmark. Lancet *1:* 1195–1198.

Linnemann, C.C. Jr. and E.B. Perry. 1977. *Bordetella parapertussis* Recent experiences and a review of the literature. Am. J. Dis. Child. *131:* 560–563.

Manclark, C.B. and J.C. Hill. 1979. *International Symposium on Pertussis*, Department of Health, Education and Welfare, Publication No. 79-1830, U.S. Government Printing Office, Washington, D.C.

Munoz, J.J. and R.K. Bergman. 1977. *Bordetella pertussis* immunological and other biological activities. Marcel Dekkar, New York, pp. 1–235.

Parker, C. 1976. Role of the genetics and physiology of *Bordetella pertussis* in the production of vaccine and the study of host-parasite relationships in pertussis. In Perlman (Editor), *Advances in Applied Microbiology*, Vol 20. Academic Press, New York, pp. 27–42.

Pittman, M. and A.C. Wardlaw. 1981. The genus *Bordetella*. In Starr, Stolp, Trüper, Balows and Schlegel (Editors), *The Prokaryotes: a Handbook on Habitats, Isolation and Identification of Bacteria*. Springer-Verlag, New York, pp. 1075–1085.

Tamura, M., K. Nogimori, M. Yajima, K. Ase and M. Ui. 1983. A role of the B-oligomer moiety of islet-activating protein, pertussis toxin, in the development of the biological effects of intact cells. J. Biol. Chem. *258:* 6756–6761.

---

* G20G medium (per milliliter): Bacto peptone (Difco), 20 mg; glucose, 10 mg; lactose, 10 mg; NaCl, 5 mg; bromothymol blue, 80 μg; agar (Oxoid L13), 15 mg; penicillin (Crystapen: Glaxo), 20 μg; furaltadone, 20 μg; and gentamicin (Genticin: Nicolas, Slough), 0.5 μg.

Differentiation of the genus **Bordetella** from other genera

Table 4.103 provides the primary characteristics that differentiate this genus from the three genera, *Alcaligenes*, *Brucella*, and *Haemophilus*, in which one or more of the species of *Bordetella* has been classified in the past. Another genus, *Acinetobacter*, may be differentiated by nonproduction of a diffusible pigment from tyrosine and nonsensitivity to M and B 938 (4:4'-diamidinodiphenylamine dihydrochloride) (Johnson and Sneath, 1973), by a lower mol% G + C of the DNA (39–47) and by its saprophytic ecology (Reyn, 1974).

The greatest generic confusion has been between *Alcaligenes* and *Bordetella bronchiseptica*. Johnson and Sneath (1973) highlighted a close relationship in a numerical taxonomic study. The genus *Alcaligenes* and particularly the species *Alcaligenes faecalis*, as defined by Holding and Shewan (1974), are heterogeneous. Table 4.104 differentiates *B. bronchiseptica* from *A. faecalis* and *Alcaligenes denitrificans*. Kersters and De Ley (1980) provided evidence that strains of *B. bronchiseptica* form a tight cluster at the border of *A. denitrificans* and that *A. faecalis* and *A. denitrificans* are distinct.

Taxonomic Comments

Recent information on the close genetic relationship (Kloos et al., 1979) and on the similarity of physiological activities (Dobrogosz et al., 1979) confirms the classification of *B. pertussis*, *B. parapertussis* and *B. bronchisepticus* in one genus, *Bordetella* (Moreno-López, 1952). Furthermore, the mol% G + C of DNA of the three species appears to be similar, falling in the range of 66–70 (67.4–67.8 (Bacon et al., 1967; 67.0–69.8 (J. Johnson, cited by Johnson and Sneath, 1973); and 66–70

(Kawai and Yabuuchi, 1975)). Reported lower value may not be reliable. The value of 57.5 (Bacon et al., 1967) appears to be anomalous, and the value of 61 (H. B. Hoyer, cited by Pittman, 1974) has not been published.

The mol% G + C of DNA of *B. bronchiseptica* is similar to the value for *Alcaligenes denitrificans* (Table 4.104) and the numerical phenetic correlation between the two species is greater than 0.90. Some differences are shown in Table 4.104. Both species are definitely different from *Alcaligenes faecalis* (Kersters and De Ley, 1980).

The name *Bordetella bronchiseptica* was placed on the Approved Lists of Bacterial Names (Skerman et al., 1980). Ferry (1911) named the organism, isolated from dogs, *Bacillus bronchicanis*. After isolation of the organism from other animal species, he changed the epithet to *bronchiseptica* (Ferry, 1912). The latter epithet has been used almost universally except by Haupt (1935).

The "*Bordetella bronchiseptica*-like bacterium" isolated from turkey poults (Hinz, Glünder, and Lüders, 1978) resembles *B. bronchiseptica* in respiratory pathogenesis, pathological lesions, epidemiology, and common O antigen. Reported differences are (a) urease-negative, (b) oxidase-positive only after 48–72 h incubation, (c) nitrate reduction when the medium is supplemented with 100 µg NAD/ml and 1% serum (Hinz et al., 1979), and (d) lower mol% G + C of DNA (61.6 (K. Kersters and J. De Ley, personal communication)). The serotypes of the K-antigens were not reported. For the time being, it seems advisable to include the "*B. bronchiseptica*-like bacterium" in *Bordetella* until its taxonomic position can be more thoroughly assessed.

Differential characteristics of species of **Bordetella**

The differential characteristics of the species of *Bordetella* are listed in Table 4.105. Other characteristics of the species are presented in Tables 4.106 and 4.107.

List of the species of the genus **Bordetella**

1. **Bordetella pertussis** (Bergey, Harrison, Breed, Hammer and Huntoon 1923) Moreno-López 1952, 178.[AL] (Microbe de coqueluche Bordet and Gengou 1906, 731; *Hemophilus pertussis* Bergey, Harrison, Breed, Hammer and Huntoon 1923, 269.)

Table 4.103.
Differential characteristics of the genus **Bordetella** *and other morphologically and physiologically similar genera*[a]

Characteristics	Bordetella	Alcaligenes	Brucella	Haemophilus
Strictly parasitic	+	−	+	+
Saprophytic	−	+	−	−
Localize on respiratory cilia	+	−	−	−
Strictly aerobic	+	+	+	
Growth requirement:				
Thiamine	−	−	+	−
Nicotinamide	+	−	−	−
X and/or V factor	−	−	−	+
Ferment carbohydrates	−	−	−	+
Nitrate reduction	D	D	+	+
Litmus milk, alkaline	+	+	−	−
Oxidation of amino acids	+	+	+	−
Tetrazolium reduction	+	−		+
Growth on 320 mg potassium tellurite per liter	−	+		−
Citrate utilized	D	+	−	−
PAGE resemblance[b]	−		−	−
Mol% G + C of DNA	66–70 (T_m)	56–70 (T_m)	55–58 (Bd)	38–44 (T_m)

[a] Symbols: see standard definitions.
[b] Polyacrylamide gel electrophoresic (PAGE) patterns are distinct for each genus (Nicolet et al. 1980).

Table 4.104.

Differentiation between **Bordetella bronchiseptica, Alcaligenes faecalis ("Alcaligenes odorans"),** *and* **Alcaligenes denitrificans ("Achromobacter xylosoxidans")**[a]

Characteristics	3. *B. bronchiseptica*	*A. faecalis*	*A. denitrificans*
Pathogen	+	−	−
PAGE resemblance[a,b]	+	−	+
rRNA cistron resemblance[c]	+	−	+
D-Gluconate dehydratase[a]		−	+
Tetrazolium reduction[d]	+	−	−
Growth on 320 mg potassium tellurite per liter[d]	−	+	+
Urease	+	−	−
Nitrate reduction to nitrite	+	d	+
Denitrification[e]	−	−	+
Alkalization of[e]			
Amides:			
Acetamide	−	+	+
Allantoin	+	−	+
Asparagine	+	+	−
Nicotinamide	+	+	
Organic salts:			
Citrate	+	+	+
Mucate	+	−	−
Tartrate	+	+	−
Mol% G + C of DNA (T_m)[a]	67–69	56–59	64–70

[a] Kersters and De Ley (1980). (For symbols see standard definitions.)

[b] Polyacrylamide gel electrophoregram.

[c] K. Kersters and J. De Ley (personal communication).

[d] Johnson and Sneath (1973).

[e] Pickett and Pedersen (1970).

per.tus'sis. L. prefix *per* very, severe; L. n. *tussis* cough; M.L. gen. n. *pertussis* of a severe cough, of whooping cough.

Minute coccobacillus, 0.2–0.5 μm in diameter and 0.5–1.0 μm in length, encapsulated or surrounded by a slime sheath composed of extruded filaments or secreted blebs. Potato-glycerol-blood agar (Bordet-Gengou) has been preferred for primary isolation. Colonies are minute in size and are surrounded by a zone of hemolysis on media containing about 20% blood. Uniqueness in slow rate of growth and susceptibility to growth inhibitors (Table 4.106) are associated with the production of pertussis toxin (Table 4.107). A defined broth medium may be used for mass growth. No pellicle is formed.

Physiological and nutritional characteristics are presented in Tables 4.105 and 4.106. Heat-labile (120°C for 1 h) agglutinogen factors are presented in Table 4.107.

There is a marked propensity to mutate from phase I to phase IV (degradation) with acquired ability to grow rapidly on peptone agar, change with regard to agglutinogens and biological activities associated with pertussis toxin, etc. Reversible phenotypic modulation occurs with altered growth conditions (Wardlaw and Parton, 1979).

Parasitic, pathogenic, found only in the respiratory tract of man.

The mol% G + C of the DNA is 66–70 (T_m). Reported lower values are questionable.

Type strain: ATCC 9797 (strain 18–232 of Kendrick et al., 1947).

2. **Bordetella parapertussis** (Eldering and Kendrick 1938) Moreno-López 1952, 178.[AL] (*Bacillus parapertussis* Eldering and Kendrick 1938, 571.)

pa.ra.per.tus'sis. Gr. prep. *para* resembling; M.L. n. *pertussis* the specific epithet of *Bordetella pertussis*; M.L. adj. *parapertussis* resembling *Bordetella pertussis*.

The morphology is similar to that of *B. pertussis*. Colonies are slightly larger and appear earlier on Bordet-Gengou medium than *B. pertussis*.

Growth on peptone agar is accompanied by a brown discoloration of the medium that is attributed to the action of tyrosinase on the tyrosine in the medium (Ensminger, 1953). A pellicle forms on liquid media. Growth is more rapid than with *B. pertussis*.

The physiological and nutritional characteristics are presented in Tables 4.105 and 4.106. The differential antigens are presented in Table 4.107.

Parasitic, pathogenic, found only in the respiratory tract of man.

The mol% G + C of the DNA is 66–70 (T_m).

Type strain: ATCC 15311 (strain 522 of Eldering and Kendrick, 1938).

3. **Bordetella bronchiseptica** (Ferry 1912) Moreno-López 1952, 178.[AL]

(*Bacillus bronchicanis* Ferry 1911, 404; *Bacillus bronchisepticus* Ferry 1912, 377.)

bron.chi.sep'ti.ca. Gr. n. *bronchus* the trachea; Gr. adj. *septicus* putrefactive, septic; M.L. fem. adj. *bronchiseptica* intended to mean with an infected bronchus.

The morphology is as described for the genus. Colonies appear earlier on Bordet-Gengou medium and are larger than those of *B. pertussis* or *B. parapertussis*. Colonies on MacConkey agar are reddish and surrounded by a small red zone with amber discoloration of the

Table 4.105.

Differentiation between **Bordetella pertussis, Bordetella parapertussis, Bordetella bronchiseptica,** *and* **"Bordetella bronchiseptica-***like bacteria"*[a]

Characteristics	1. *B. pertussis*	2. *B. parapertussis*	3. *B. bronchiseptica*	Unclassified: "*B. bronchiseptica*-like"
Motility	−	−	+	+
Growth on Bordet-Gengou medium:				
1–2 Days	−	+	+	+
3–6 Days	+			
Growth on peptone agar:				
Phase I	−	+	+	+
Phase IV	+	+	+	+
Browning	−	+	−	−
Growth on MacConkey agar	−	+	+	+
Citrate utilization	−	+	+	+
Nitrate reduction	−	−	+[b]	+[c]
Urease	−	+	+[d]	−
Oxidase	+	−	+	+[e]
Mol% of G + C of DNA (T_m)	67–70[f]	66–70[f]	68.9[g]	61.6[h]

[a] Symbols: +, typically positive; −, typically negative.

[b] Regularly positive in nitrate medium supplemented with nicotinamide adenine dinucleotide (NAD) and serum. Exceptions occur in conventional nitrate-test medium.

[c] Only in medium supplemented with NAD and serum (Hinz et al., 1979).

[d] Positive within 4 h.

[e] Kovacs' test positive after 48–72 h incubation (Hinz et al., 1979).

[f] J. Johnson (cited by Johnson and Sneath, 1973); Kawai and Yabuuchi, 1975.

[g] Type strain (K. Kersters and J. De Ley, personal communication).

[h] Strain 591/77 of Hinz (K. Kersters and J. De Ley, personal communication).

Table 4.106.

Other characteristics of **Bordetella pertussis, Bordetella parapertussis** *and* **Bordetella bronchiseptica**[a]

Characteristics	1. B. pertussis	2. B. para-pertussis	3. B. bronchi-septica
Strictly aerobic	+	+	+
Catalase	d	+	+
Peroxidase	+	+	+
Phosphatase	−	−	−
Ability to catabolize carbohydrates	−	−	−
Adenylate cyclase activity:[b]			
Extracellular	+	+	+
Intracellular	+	−	−
Production by phase IV cells	−	−	−
Glutamic decarboxylase activity	−	−	+
Gelatin hydrolysis	−	−	−
Litmus milk (alkaline reaction)	+	+	+
Growth requirements:			
Nicotinamide	+	+	+
Organic sulfur (e.g. cysteine)	+	+	+
Organic nitrogen (amino acids)	+	+	+
Utilization of amino acids:[c,d]			
L-glutamic acid, L-proline, DL-alanine, DL-aspartic acid, L-serine, glycine	+	+	+
All amino acids except arginine, lysine and histidine[d]	−	−	+
Growth inhibited by unsaturated fatty acids or colloidal sulfur[e]	+	−	−
Sensitive to:[f]			
Bacitracin, thionine, pyronin	−	−	−
Erythromycin, chloramphenicol, streptomycin	+	+	+
Penicillin	d	−	−
Sulfonamides	−	+	d
Safranine	+	+	−
Nitrofurantoin	+	+	d
Basic fuchsin	d	+	−
Degree of tolerance to NaCl, bile, phenol and temperature[g]	Lowest	Inter-mediate	Highest

[a] For symbols see standard definitions.
[b] Endoh et al. (1980); Hewlett et al. (1979).
[c] Meyer and Cameron (1957).
[d] Rowatt (1955).
[e] Rowatt (1957).
[f] Johnson and Sneath (1973). Strains of *B. pertussis* vary in sensitivity to penicillin (Kendrick and Eldering, 1969).
[g] *B. pertussis* is less tolerant to NaCl, phenol and bile and has a narrower range in growth temperature than *B. parapertussis* and *B. bronchiseptica*. *B. bronchiseptica* is the most tolerant (Johnson and Sneath, 1973).

Table 4.107.

Differential and common antigens of phase I strains of the species of the genus **Bordetella**[a]

Antigens	1. B. pertussis	2. B. para-pertussis	3. B. bronchi-septica
K-antigens[b]			
Common to genus:			
Factor 7	+	+	+
Species specific:			
Factor 1	+	−	−
Factor 14	−	+	−
Factor 12	−	−	+
Other factors occurring in one or more strains:			
Factors 2, 3, 4, 5, 6, 13	+		
Factors 8, 9, 10		+[c]	
Factors 8, 9, 10, 11, 13			+
O-antigen, common[b]	+	+	+
Filamentous hemagglutinin[d]	+		+
Toxins:			
HLT[e]	+	+	+
Pertussis[f]	+	−	−
LPS[g]	+	+	+

[a] For symbols see standard definitions.
[b] Eldering et al. (1957).
[c] Found in all strains of *B. parapertussis*.
[d] Irons and McLennan (1979); Munoz et al. (1981).
[e] Heat-labile (56°C) dermonecrotic toxin, demonstrable after rupture of the cell wall (see Pittman, 1979).
[f] Exotoxin ("true toxin"), (HSF-LPF-IAP-pertussigen), heat-labile (80°C) (Pittman 1979).
[g] Lipopolysaccharide. The occurrence of distinct or common chemical structures between the species has not been reported.

underlying medium (I. A. P. McCandlish, 1977, Ph.D. Thesis, University of Glasgow, Glasgow, Scotland). A pellicle forms on liquid media.

The physiological and nutritional characteristics are presented in Tables 4.105 and 4.106. Differential antigens are presented in Table 4.107. Characteristics that differentiate this species from two related species of *Alcaligenes* are given in Table 4.104.

Parasitic, pathogenic, found in the respiratory tract of domestic and wild mammalian animals (dogs, swine, guinea pigs, rabbits, raccoons, et al.). It may be transmitted from animals to man. Chronic carriage occurs and is in contrast to noncarriage of *B. pertussis*.

The mol% G + C of the DNA is 67–69 (T_m).

Type strain: ATCC 19395 (strain Dog 71 of Ferry, 1912). Reference strain ATCC 4617 is used in the assay of polymyxin and sodium colistimate.

Other organisms

The taxonomic status of the following organism has not yet been determined.

"*Bordetella bronchiseptica*-like" (Hinz, Glünder and Lüders, 1978).

The organism was isolated from turkey poults, also from one chicken and one duck. It resembles *B. bronchiseptica* in morphology, motility, growth conditions, epidemiology with predilection for young birds, clinical symptoms, histopathology, etc., and it has an O antigen in common with a strain of *B. bronchiseptica* isolated from a hog (Hinz et al., 1979). The differences, listed in Table 4.105, include a negative urease test, a delayed oxidase reaction and a low mol% G + C of the DNA.

The organism is not *Alcaligenes faecalis* (K. Kersters and J. De Ley, personal communication). Although Goodnow (1980) considered the organism to be a *B. bronchiseptica*, its affiliation with the genus *Bordetella* is as yet uncertain.

The mol% G + C of the DNA is 61.6 (T_m).

Reference strain: 591-77 of Hinz et al. (1979).

Genus **Francisella** Dorofe'ev 1947, 176[AL]

HENRY T. EIGELSBACH AND VIRGINIA G. MCGANN

Fran.cis.el'la. M. L. dim. ending -*ella*; M. L. fem. n. *Francisella* named after Edward Francis, an American bacteriologist who extensively studied the etiologic agent and pathogenesis of tularemia and is credited with naming the disease.

Rod-shaped cells, 0.2 × 0.2–0.7 μm (*Francisella tularensis*) or 0.7 × 1.7 μm (*Francisella novicida*), when cultured in appropriate media and examined during active growth; pleomorphism occurs subsequently. Gram-negative, faintly-staining. **Nonmotile. Obligately aerobic.** On glucose-cysteine-blood agar smooth gray colonies are formed which reach a maximum size in 2–4 days and are surrounded by a characteristic green zone of discoloration. Weakly catalase-positive. **Oxidase-negative.** Catabolism of carbohydrates is characteristically slow with the production of acid but no gas. **Cysteine (or cystine) is either required for growth (*F. tularensis*) or is greatly stimulatory for growth (*F. novicida*). H₂S is produced.** Unlike other bacteria, the type species contains relatively large amounts of long-chain saturated and monoenoic C_{20} to C_{26} fatty acids as well as 3-hydroxy-hexadecanoate, 2-hydroxy-decanoate and 3-hydroxy-octadecanoate. ***F. tularensis* is the causative agent of tularemia in man and animals;** *F. novicida* causes experimental infections in laboratory animals. The mol% G + C of the DNA is 33–36 (T_m, Bd).

Type species: *Francisella tularensis* (McCoy and Chapin 1912) Dorofe'ev 1947, 178.

Further Descriptive Information

When cultured in an appropriate liquid medium and examined during active growth, both *F. tularensis* and *F. novicida* are small, singly occurring, nonmotile, nonsporulating, rod-shaped organisms. *F. tularensis* is smaller (0.2 μm wide × 0.2–0.7 μm in length) than *F. novicida* (0.7 μm in diameter × 1.7 μm in length). Somewhat shorter forms of both organisms occur in infected tissues. Generally, dye-stained clinical material is unsatisfactory for organism identification; application of direct or indirect fluorescent antibody techniques is recommended for rapid, specific recognition (Eigelsbach and McGann, 1981). In actively growing cultures *Francisella* cells appear as minute, faintly-staining, Gram-negative coccoidal organisms; on closer inspection, however, these forms are visualized as bipolar components of a rod-shaped organism separated by an even more faintly stained central area enclosed by a delicate cell wall. This characteristic morphology is more readily demonstrable with a polychrome (Giemsa) stain. During the logarithmic growth phase in a liquid medium shaken at 37°C, cells are relatively uniform; soon thereafter, pleomorphism is observed. During the phase of decline (after 24-h incubation), the cell wall elongates, filamentous cells form and filament fragmentation frequently occurs; also, at this time, a small number of rods not larger than 300 nm in diameter can be visualized by electron microscopy (Ribi and Shepard, 1955). Minute forms, infectious for mice, pass through membranes with 600-nm average pore diameter and are estimated to be in the range of 300–350 nm in diameter (Foshay and Hesselbrock, 1945). These authors found no morphological feature to differentiate a virulent from a nonvirulent strain.

Divergent views exist concerning the reproduction of *F. tularensis*. Using light microscopy, Hesselbrock and Foshay (1945) described multiple modes of reproduction in gelatin-hydrolysate liquid medium. Budding was observed frequently and appeared to be the chief mode of reproduction; binary fission, though not seen, was still considered a possibility. Many morphological units including "minimal reproductive units" were described. Electron microscopic studies (Ribi and Shepard, 1955) performed with organisms taken from actively growing cultures revealed cells with constrictions suggesting multiplication by binary fission. During the death phase, filamented cells easily fractured into pieces corresponding to the forms reported by Hesselbrock and Foshay (1945). Ribi and Shepard (1955) concluded that the possession of a complex life cycle is unlikely with regard to *F. tularensis*. Excellent electron micrographs of *F. tularensis* (Hood, 1977) provide additional evidence of reproduction by binary fission.

Virulent organisms of *F. tularensis* are surrounded by a relatively thick (0.02–0.04 μm), electron-transparent capsule that is removed quite easily in solutions containing sodium chloride; loss of capsule is accompanied by loss of virulence but viability is unaffected (Hood, 1977). With loss of capsule the organisms become less resistant to acid pH levels that simulate the lysosomal environment of infected macrophages (P. G. Canonico, 1981, unpublished results). In material from infected animals, stained organisms are found to be surrounded by a clear area presumed to be capsule, and if virulent organisms from culture are mixed with serum, capsules may be demonstrated. Underlying the capsule is the delicate, double-layered cell wall, which is easily distorted during cytological processing. The lipid concentration in capsule and cell wall, 50% and 70%, respectively, is unusually high for Gram-negative bacteria; the lipid composition is also characteristic: saturated straight-chain 16:0 (48%) and α-OH 14:0 (31%) fatty acids in the capsule as compared with the cell wall, which has only a trace of α-OH 14:0 but contains principally β-OH 10:0 (30%) with six other fatty acids (60%) in approximately equal proportions. Capsule and cell wall differ quantitatively and qualitatively in sugar composition, but amino acid analysis suggests similarity in structure of their protein moiety with a difference in the amount present: 35% total amino acids in the capsule and 8% in the cell wall (Hood, 1977).

On glucose-cysteine-blood agar (GCBA) and peptone-cysteine agar (PCA) viscous, readily emulsifiable colonies reach maximum size after 2–4 days at 37°C (1–4 mm for *F. tularensis* and 6–8 mm for *F. novicida*). Growth of *F. novicida* is usually more rapid and luxuriant than that of *F. tularensis*. On GCBA medium colonies are surrounded by a characteristic green discoloration not associated with true hemolysis; on "chocolatized" agar the discoloration is brownish in appearance. Colony-type variants, including those associated with virulence and immunogenicity, are best detected when observed on transparent agar illuminated by obliquely transmitted light (Fig. 4.88) (Eigelsbach et al., 1951, 1952; Eigelsbach and Downs, 1961).

Francisella species are strictly aerobic and grow optimally at 37°C. *F. tularensis* grows much more slowly and less abundantly than *F. novicida* (Larson et al., 1955; Owen et al., 1964). *F. novicida* is less fastidious than is *F. tularensis* and does not require addition of a sulfhydryl compound for cultivation but growth is markedly enhanced by addition of cysteine or cystine to media. Unlike *F. tularensis*, *F. novicida* grows moderately well with uniform turbidity in proteose peptone broth without cystine and somewhat less abundantly in nutrient gelatin without causing liquefaction (Owen et al., 1964).

F. tularensis cannot be cultured routinely on conventional laboratory media, but two commercially produced solid media, glucose cysteine agar (GCA) with thiamine* or cystine heart agar† (when either is

---

* GCA agar with thiamine (BBL). When used with added blood the medium is commonly referred to as GBCA and can be substituted for the original, noncommercial medium described by Downs et al. (1947). Suspend 58 g of the dry material in a liter of distilled or demineralized water. Mix thoroughly. Heat with frequent agitation and boil for 1 min. Dispense into tubes and sterilize by autoclaving at 118–121°C for 15 min. For liter quantities, autoclave at the same temperatures for 30 min. Cool to 45–48°C. Aseptically add 25 ml of packed, human blood cells or 50 ml of defibrinated rabbit or sheep blood. Mix thoroughly and pour into plates. Incubate at 37°C for 24 h before use to decrease surface moisture and to test for sterility.

† Cystine heart Agar (Difco). Suspend 102 g of the dry material in a liter of distilled or demineralized water, then follow the procedure indicated in the preceding footnote.

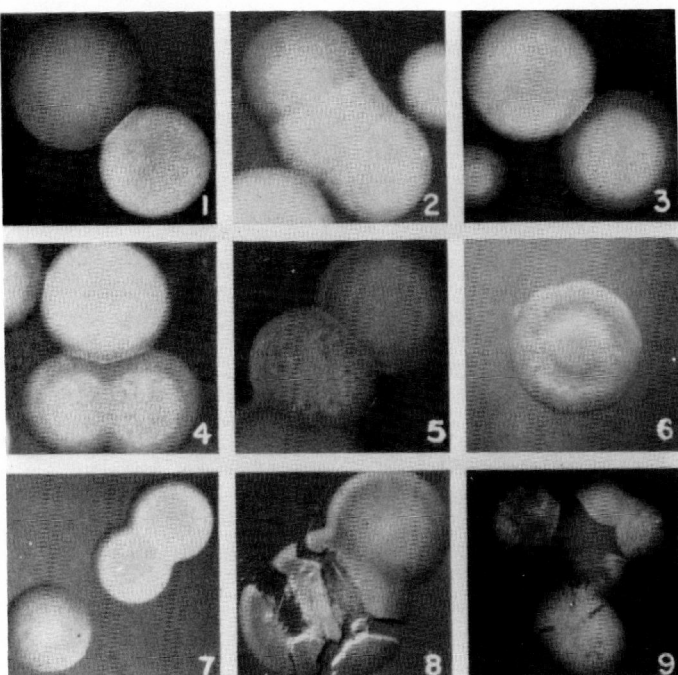

Figure 4.88. Colony type variants of *Francisella tularensis* on peptone-cysteine agar (× 100). *1*, smooth, blue colony (*upper left*) and smooth, dense buff colony (*lower right*). *2*, three smooth colony types: one less dense, blue, watery in consistency (*upper left*); one more dense, buff, and buttery in consistency (*middle*); and one dense, buff, and buttery in consistency (*lower right*). *3*, dense, buff colony (*upper left*) next to a less dense colony having a blue peripheral zone (*lower right*). *4*, gray, viscous colonies next to a single, dense, buff colony. *5*, daughter colonies arising from a dense buff colony after prolonged incubation. *6*, rough-appearing colony type possessing a central plateau surrounded by an irregular pitted area that gradually blends with a smooth peripheral zone. *7*, gray colony (*lower left*); dense iridescent colony (*middle*); and a dense, buff colony (*upper right*). *8*, dense, firm, dry, dull, flat, light gray colonies which, when touched with a wire, break into several pieces with sharp, irregular edges (*lower left*). *9*, sectored colonies, typical of unstable variants. (Reproduced with permission from H. T. Eigelsbach, W. Braun and R. D. Herring, Journal of Bacteriology *61*: 557–569, 1951, © American Society for Microbiology).

enriched with defibrinated rabbit blood or outdated, settled, human blood cells) provide excellent growth from small inocula. Another solid medium§ containing blood, equally satisfactory but not produced commercially, has been described (Gaspar et al., 1961). A transparent medium such as PCA¶ (Snyder et al., 1946) is required for the detection of colony type variants described in the literature.

Each lot of commercial medium and the protein-derived ingredients in the other media described should be tested before use for their ability to support satisfactory growth of avirulent *F. tularensis* ATCC 6223 or live vaccine strain LVS (Eigelsbach and Downs, 1961). Occasionally a marginal lot of medium will yield more rapid and abundant growth upon addition of 0.5 g ferrous sulfate/liter of medium. Incorporation of antibiotics (penicillin, polymyxin B and cycloheximide) is required when clinical specimens containing normal flora are cultured for *F. tularensis*.

Satisfactory liquid media for the cultivation of *F. tularensis* include: peptone cysteine broth (Snyder et al., 1946); casein partial hydrolysate (Mills et al., 1949); and several chemically defined media (Mager et al., 1954; Traub et al., 1955; Nagle et al., 1960; and Chamberlain, 1965). Continuous shaking for 12–18 h during incubation at 37°C is required for maximal growth (10×10^9–40×10^9 viable organisms/ml).

Under suitable conditions, anaerogenic dissimilation of a limited number of carbohydrates occurs with slight acid production; slow development of minimal reactions is characteristic of *F. tularensis*. Both species form acid from glucose, fructose and mannose; acid from maltose is commonly seen with *F. tularensis* and acid from sucrose is a distinguishing feature of *F. novicida* (Girard and Gallut, 1957). Inability of *F. tularensis* biovar *palaearctica* to utilize glycerol assists differentiation of these organisms from biovar *tularensis* (Olsufiev et al., 1959). A third proposed biovar, *mediaasiatica*, presently included in biovar *palaearctica*, is reported to ferment glycerol but not maltose (Kunitsa et al., 1972). Growth in litmus milk is scant; slight acidity may develop in *F. tularensis* cultures in 2 weeks and an acid reaction with soft coagulation in *F. novicida* cultures (Owen et al., 1964). Possession of a citrulline ureidase system has been reported for *F. tularensis* biovar *tularensis* (Marchette and Nicholes, 1961). Catalase activity is weak but strongest in strains of *F. tularensis* biovar *tularensis*; attenuation of virulence is accompanied by a marked reduction in activity (Rodionova, 1976). Strains of *F. tularensis* biovar *tularensis* contain cytochromes a_1, a_2 and b_1 but no a_3 or c-type cytochromes or cytochrome oxidase (Fellman and Mills, 1960); the amount of cytochrome b_1 in the whole cell and cell-free extract is reported to be greater for virulent than for avirulent organisms (Mizuhara and Yamanaka, 1961). An active succinoxidase system in *F. tularensis* biovar *tularensis* is located in the particulate fraction of the cell and interacts with cytochrome b_1 (Fellman and Mills, 1960).

Transformation and recombination studies within the genus *Francisella* are in accord with the close phylogenetic relationship between the two species of the genus (Tyeryar and Lawton, 1969, 1970; see also Taxonomic Comments). Koliaditskaia et al. (1959) described lytic activity in *F. tularensis* cultures consistent with that expected of a phage; however, no phage for either *F. tularensis* or *F. novicida* has been characterized or is available for use.

The antigenic composition of *F. tularensis* has been examined primarily for detection of infection and identification of the infectious organism or of nonviable immunogenic substances. Conventional serologic procedures, including direct and indirect immunofluorescent techniques, indicate that all *F. tularensis* strains are qualitatively similar in antigenic composition, and convalescent sera or immune sera produced with whole organisms react with and can be absorbed by all strains. Live vaccine is a highly effective immunogen against infection of susceptible hosts with fully virulent strains (Eigelsbach and Downs, 1961; Saslaw et al., 1961; McCrumb, 1961; Hornick et al., 1966; Hornick and Eigelsbach, 1966); nonviable vaccines induce similar protection only in resistant hosts or against strains of moderate virulence in the susceptible host.

Antigenic material has been extracted from *F. tularensis* with phenol, acetone, water saturated with ether, or trichloroacetic acid (Nicholes, 1946; Ormsbee and Larson, 1955; Nutter, 1971; Hambleton et al., 1974; Holm et al., 1980). Chemical analysis and electron microscopy suggest that these preparations lack nucleic acid and are derived from the cell wall. All products contain multiple antigenic factors, induce agglutinins and/or precipitins, are nontoxic and have immunogenic activity similar to that of killed whole cells. Other extracts, rich in RNA, have similar immunogenic properties but their activity is destroyed by RNase (Andron and Eigelsbach, 1975). Capsular material is neither immunogenic

---

§ *Francisella tularensis* isolation medium (g/liter of distilled or demineralized water): Tryptose broth with thiamine (Difco), 20.0; cysteine-HCL, 5.0; sodium thioglycolate, 2.0; glucose, 10.0; and agar, 10.0. Combine all ingredients except the agar and adjust the pH to 7.2. Add the agar and heat in flowing steam for 5 min. Autoclave at 121°C for 20 min. Cool to 45–48°C. Aseptically add 50 ml of defibrinated rabbit blood. Mix thoroughly and pour into plates. Incubate at 37°C for 24 h before use to decrease surface moisture and to test for sterility.
¶ Peptone-cysteine agar (PCA) (g/liter of distilled or demineralized water): Bacto peptone (Difco), 20.0; NaCl, 10.0; glucose, 1.0; cysteine-HCl, 1.0; and agar, 20.0. Combine all ingredients except the agar and adjust the pH to 6.8. Add the agar, heat with agitation and boil for 1 min. Cool to 45–48°C and pour into plates. Incubate at 37°C for 24 h before use to decrease surface moisture and to test for sterility.

nor toxic (Hood, 1977). One type of nonviable preparation, a suspension of cells killed by gamma-irradiation, protects mice against infection with fully virulent *F. tularensis* but the immunogenic activity disappears within 4 days after irradiation (Gordon et al., 1964). This preparation has two toxic components lethal for mice, guinea pigs and rabbits: the activity of one is labile at 4°C, declining with the decrease in immunogenicity, whereas that of the other is stable and has endotoxin-like properties (Landay et al., 1968). Immune or convalescent sera can protect mice against the toxic activities but are ineffective against infection.

F. tularensis has a minor serological relationship to *Brucella* (Saslaw and Carlisle, 1961) and *Y. pestis* (Larson et al., 1951). A closer relationship is seen with *F. novicida* (Owen et al., 1964). Live *F. tularensis* vaccine protects against both species, while live *F. novicida* vaccine protects against homologous infection and against *F. tularensis* strains of moderate virulence. Nonliving vaccines protect only against homologous organisms. Cross-reactions between *Francisella* species occur in complement fixation tests with guinea pig antisera, but agglutination, hemagglutination, and passive cutaneous anaphylaxis reactions are highly specific. Nonspecific induction of protection against *F. novicida* but not against *F. tularensis* has followed immunization with BCG vaccine (Claflin and Larson, 1972).

The aminoglycosides, streptomycin, gentamicin, and kanamycin are bactericidal for *F. tularensis*, whereas the tetracylines and chloramphenicol are bacteriostatic. Streptomycin is preferable for the treatment of severe disease including pneumonic tularemia. Although gentamicin and kanamycin are highly effective in these illnesses, experience with these antibiotics is meager and greater risk of oto- and nephrotoxicity is incurred (Woodward, 1979; Hoover, 1979). Because of their bactericidal nature, these aminoglycosides are not associated with disease relapse but the drugs must be administered parenterally.

Less severe tularemia (ulceroglandular without secondary pneumonia) is usually treated by oral administration of tetracycline. Relapses with this bacteriostatic drug are more frequent when therapy is initiated during the first week of the disease than 10–12 days after onset of illness. Retreatment with the same antibiotic is followed by a prompt response. Choramphenicol is as effective as tetracycline for treatment of less severe tularemia but is not recommended because of potential hematologic toxicity. Antibiotic-resistant strains have been induced in vitro but have never been isolated in cases of naturally occurring tularemia.

Routine susceptibility testing is accomplished with the disk agar diffusion method. Readings are made after overnight incubation at 37°C.

F. tularensis is widely distributed in nature and is found on all continents throughout the world except Australia and Antarctica. The two main regions in which numerous infections of humans have occurred over a relatively large area are the U.S.A. (all states except Hawaii) and southern U.S.S.R. Two biovars are recognized and differ in virulence. One appears only in North America, where it is predominant and highly virulent for numerous hosts. The other is found in Europe and Asia as well as in North America but causes a milder disease in humans and domestic rabbits. The principal reservoirs for this biovar are voles and water rats in the U.S.S.R. and beavers, muskrats and voles in North America. The organism can be isolated from natural waters (streams, rivers, lakes or ponds) in areas frequented by these animals. *F. tularensis* has been isolated from approximately 100 types of wildlife, about one-half of which have transmitted tularemia to man. They include wild rabbits, muskrats, water rats, beavers, squirrels, woodchucks, sheep, mice, voles and game birds as well as biting insects (usually ticks or deer flies). Infection follows handling of infected animal carcasses, insect bites, ingestion of improperly cooked meat or contaminated water, or inhalation of airborne organisms, especially during processing of agricultural products contaminated by infected rodent excreta. Bites or scratches by resistant wild or domestic

carnivores (dog, cat, skunk, coyote, fox, hog or bull snake) whose mouth parts have been contaminated by eating infected animals may also result in human infection. *F. tularensis* has great invasive ability and is able to penetrate the unbroken skin. Although humans of all ages, sexes and races are susceptible, man-to-man transmission is extremely rare.

Human tularemia is an acute, febrile, granulomatous, infectious, zoonotic disease. During evaluation of vaccine prophylaxis in volunteers it was determined that nonvaccinated individuals could be infected subcutaneously with 10 organisms or by inhalation of 10–50 cells (Saslaw et al., 1961). The clinical picture and severity vary appreciably according to the route of infection and the virulence of the organism. The incubation period in humans is usually 3–4 days but ranges from 2–10 days dependent primarily on dose.

Before the advent of antibiotic therapy, tularemia in untreated patients in North America resulted in a case mortality rate of approximately 5–30%. In Eurasia, mortality rates were lower for all untreated clinical types (averaging less than 1%) because of inherently lower virulence of the variety common to that area.

In general, *F. novicida* is less virulent than *F. tularensis*. It is experimentally pathogenic for white mice, guinea pigs and hamsters producing lesions similar to those of tularemia, while rabbits, white rats and pigeons are resistant (Owen, 1974). Human infections have not been reported.

Enrichment and Isolation Procedures

Detailed procedures have been described by Eigelsbach (1974) for the collection, transport and storage of clinical specimens and by Eigelsbach and McGann (1981) for testing water samples from suspected streams or wells or collection fluid from impinged air samples. Successful isolation of *F. tularensis* from the collected materials will depend on the precautions taken to maintain the viability of *F. tularensis* organisms and to prevent their overgrowth by normal flora. All specimens or samples should be kept at a temperature of 10°C or lower and procedures for isolation should be performed immediately (at the bedside or in the field) or soon after samples arrive in the laboratory.

Direct culture on any of the recommended agar media has the advantage of being more rapid and less hazardous than inoculation of susceptible laboratory animals. When enriched with blood, either commercially prepared GCA with thiamine or cystine heart agar is satisfatory for isolation from most environmental samples or clinical specimens. Incorporation of antibiotics into the medium (penicillin, 100,000 U/ml; polymyxin B sulfate, 100,000 U/ml and cycloheximide, 0.1 mg/ml) is recommended for suppression of normal flora. Occasionally ordinary media inoculated with infected tissue will initially support growth of *F. tularensis*. The addition of colistin, nystatin, lincomycin and trimethoprim to chocolate agar routinely used for the isolation of gonococci has been reported to allow the isolation of *F. tularensis* from some experimental mixtures containing flora prepared from human throat cultures (Berdal and Søderlund, 1977).

Media inoculated with clinical material that potentially contains a high concentration of *F. tularensis* organisms in a relatively small volume (0.1 ml of fluid expressed from a primary lesion or draining lymph node, or a swab impregnated with conjunctival scrapings) should be streaked with a wire loop to ensure development of isolated colonies. Other clinical specimens and environmental samples contain a relatively low concentration of *F. tularensis* cells and a considerably larger volume of material is required for culture. The volume of inoculum, however, should not exceed 0.2–0.3 ml/plate of recommended media. A total volume of at least 3 ml should be cultured from samples of blood (without anticoagulant), gastric aspirates or pharyngeal washes and 2 ml from samples of pleural fluid, lymph node perfusate or suspensions of excised tissue (processed in Ten Broeck grinders with 2 ml of sterile gelatin-saline*). Similar volumes of inocula are required for culture of processed environmental materials; impinged air samples or water from

---

* Gelatin-saline: 0.85% NaCl, 0.1% gelatin, pH 6.8.

suspected streams or wells are passed through 0.45-μm cellulose acetate filters and the trapped organisms are freed by vigorous agitation of the filter in sterile gelatin-saline. These fluid inocula of clinical or environmental origin should be deposited on the surface of the medium in the center of the plate and thoroughly spread with a sterile U-shaped glass rod. Smeared or streaked plates are inverted and incubated at 35–37°C for 48–96 h. Increased CO_2 is not required, but is not harmful. These procedures for isolation of *F. tularensis* are equally satisfactory for *F. novicida*.

A portion of all specimens and samples should be held at −30 to −70°C for further reference. If adequate facilities are available for inoculating and housing highly infectious animals (restricted entry of personnel, a vented hood for animal inoculation and necropsy, and clear plastic animal holding cages fitted with filter tops to be opened only in the vented hood) and isolation by direct culture was unsuccessful, materials that fail to yield isolates can be inoculated intraperitoneally into guinea pigs. Inoculation with one to five viable cells of *F. tularensis* usually results in death within 5–10 days. Percutaneous inoculation is sometimes successful if high numbers of *F. tularensis* are present. Moribund animals should be examined immediately. Spleen pathology is pathognomonic of tularemia (enlargement 4–5 times normal size and numerous, minute, gray foci of necrosis) and *F. tularensis* can be isolated readily from heart-blood, spleen and liver.

Maintenance Procedures

Repeated serial transfer of *F. tularensis* in liquid or on solid media routinely leads to a loss in virulence of the culture. Solid media recommended for isolation and cultivation are satisfactory for limited subculturing. Viability and virulence remain constant for years if 12- to 18-h growth is harvested from liquid or solid medium and frozen or lyophilized with equal volumes of skim milk or 20% sucrose, 2.6% gelatin and 0.2% agar (Faibich, 1959; Faibich and Tamarkina, 1946). Frozen cultures are stored at −30 to −70°C and lyophilized preparations are held at 5 to −70°C. *F. novicida* remains viable and virulent when lyophilized in skim milk and stored under vacuum at 4°C.

Differentiation of the genus **Francisella** from other genera

Table 4.108 indicates the characteristics of *Francisella* that can be used to distinguish this genus from other genera of small, Gram-negative, nonsporing coccobacillary bacteria that are pathogenic for man and other mammals.

Taxonomic Comments

The taxonomic position of the genus *Francisella* remains uncertain as indicated in the previous edition of this *Manual*. The highly infectious nature of these bacteria is one of the reasons responsible for slow development of critical evidence for taxonomic classification. The following information is derived from studies with a limited number of representative strains. DNA hybridization studies indicate that the genus is not closely related to *Pasteurella*, *Yersinia* or the coliforms (Ritter and Gerloff, 1966), and the mol % G + C of the DNA is significantly lower than that of *Brucella* or *Yersinia*. In addition, the cellular fatty acid composition of *Francisella* is distinctly different from that of other Gram-negative bacteria, including *Brucella*, *Pasteurella* and *Yersinia* (Jantzen et al., 1979), and the unusually high lipid content of the cell wall is unlike that of other Gram-negative bacteria (Hood, 1977).

The DNA/DNA hybridization experiments of Ritter and Gerloff (1966) reveal a distinct reciprocal relationship between the DNA of *F. tularensis* and *F. novicida* (more than 78% of ^{32}P-labeled DNA bound by the heterologous type) and essentially no hybridization between DNA from *Francisella* and that of *Yersinia* or *Pasteurella*. The relationship between the two species of *Francisella* is also indicated by the

Table 4.108.
Differential characteristics of the genus **Francisella** *and other genera of small, Gram-negative, nonsporing, coccobacillary bacteria*[a]

Characteristics	Francisella	Brucella	Pasteurella	Yersinia
Cells: ≤ 0.25 μm × ≤ 0.5 μm[b]	+	−	D	−
Capsule easily demonstrated	−	−	+	+
Gram-stain: weak counterstain	+	+	−	−
Strictly aerobic	+	+	−	−
Optimum temperature is 37°C	+	+	+	−
CO_2 enhances growth	−	+	D	−
Cysteine/cystine requirement	D	−	−	−
Chains occur in liquid media	−	+	D	D
Carbohydrates are dissimilated[c]	+[d]	−	+	+
Acid produced from sucrose	−	−	+	D
Catalase	+[d]	+	+	+
Oxidase	−	D	+	−
NH_3 produced in liquid media	−	+	+[d]	+
Sodium ricinoleate solubility	+	−	+	+
Penicillin-sensitive in vitro	−	−	D	D
Arthropod vectors can occur	+	−	−	+
Francisella antisera reactions				
Agglutinin	+	D	−	D
Fluorescent antibody	+	−	−	−
Francisella antibodies absorbed	+	−	−	−
Mol% G + C of DNA	33–36	55–58	40–45	46–50

[a] Symbols: see standard definitions.
[b] Cell size in infected tissue.
[c] Acid but no gas produced.
[d] Weak reaction.

demonstration of high transformation frequencies in *F. novicida* treated with homologous DNA or with *F. tularensis* DNA (Tyeryar and Lawton, 1970).

The species *F. novicida*, presently consisting of a single known isolate (ATCC 15482) which is apparently avirulent for man, has less fastidious metabolic requirements than *F. tularensis*; these metabolic as well as serologic differences support designation of a separate species because loss of virulence in *F. tularensis* generally is accompanied by a requirement for more exacting growth conditions.

Olsufiev et al. (1959) recognized two major tularemia pathogens and recommended the designation *Francisella tularensis* var. (biovar) *tularensis* for the organism prevalent in North America and responsible for the higher incidence and greater severity of human illness there, and *Francisella tularensis* var. (biovar) *palaearctica* for the organism encountered in Europe, Asia and the Americas. *F. tularensis* biovar *tularensis* is associated with tick-borne tularemia in rabbits, produces the classical illness described in most medical texts and is highly virulent for man; *F. tularensis* biovar *palaearctica* is frequently linked with water-borne disease of rodents in North America and Eurasia and causes a milder form of illness. In North America the two kinds of tularemia are characterized by epidemiological pattern and virulence of the etiologic agent; tentative designations of type "A" and type "B" have been proposed for the more virulent and less virulent types (Jellison, 1974).

Since 1970, investigators in the U.S.S.R. have used the designations *F. tularensis nearctica* Olsufiev for *F. tularensis* biovar *tularensis* and *F. tularensis holarctica* Olsufiev for *F. tularensis* biovar *palaearctica*. Additional subdivisions of the latter group are designated *F. tularensis holarctica* var. *japonica* Rodionova for Japanese-type strains and *F. tularensis mediaasiatica* Aikimbaev for strains from Central Asia. Additional information may be found in a review of tularemia in North America from 1930–1974 (Jellison, 1974) and in a review of investigations in the U.S.S.R. (Pollitzer, 1967).

Differentiation and characteristics of the species of the genus **Francisella**

Differential characteristics of the species of *Francisella* are presented in Table 4.109 and other characteristics of the species in Table 4.110.

List of the species of the genus **Francisella**

1. **Francisella tularensis** (McCoy and Chapin 1912) Dorofe'ev 1947, 176.[AL] (*Bacterium tularense* McCoy and Chapin 1912, 61.)

Table 4.109.

Differential characteristics of **Francisella tularensis** *and* **Francisella novicida**[a]

Characteristics	1. *F. tularensis*		2. *F. novicida*
	Biovar *tularensis*	Biovar *palaearctica*	
Capsule present	v[b]	v[b]	−
Growth requirement:			
Cystine or cysteine	+	+	−
Growth on ordinary media (blood agar, gelatin, peptone broth)	−	−	+
Colonies on GCBA[c] > 5 mm in diameter	−	−	+
Acid produced from:			
Maltose	+	+[d]	−
Sucrose	−	−	+
Glycerol	+	−[e]	+
Citrulline ureidase	+	−[e]	+
Serum agglutination:			
F. tularensis	+	+	−
F. novicida	−	−	+
Vaccine efficacy:			
Killed vaccine			
F. tularensis	+[f]	+[f]	−
F. novicida	−	−	+
Live vaccine			
F. tularensis	+	+	+
F. novicida	+[f]	+[f]	+

[a] For symbols see standard definitions.
[b] Capsule associated with virulence of the strain.
[c] GCBA, glucose-cystine-blood agar.
[d] Central Asian strains (proposed biovar *mediaasiatica*) reported to be negative.
[e] Central Asian strains (proposed biovar *mediaasiatica*) reported to be positive.
[f] Protects fully susceptible hosts only against strains of partially reduced virulence.

tu.la.ren'sis. M.L. adj. *tularensis* pertaining to Tulare County, California, where the disease tularemia was first observed.

The characteristics are as given for the genus and as listed in Tables 4.108–4.110.

Two biovars are recognized, *tularensis* and *palaearctica*; these are differentiated by the characteristics listed in Table 4.109. Other differences are related to severity of infections and to geographical distri-

Table 4.110.

Other Characteristics of **Francisella tularensis** *and* **Francisella novicida**[a]

Characteristics	1. *F. tularensis*		2. *F. novicida*
	Biovar *tularensis*	Biovar *palaearctica*	
Pleomorphic cells occur following logarithmic growth	+	+	+
Bipolar staining (Giemsa)	+	+	+
H₂S from cysteine/cystine	+	+	+
Motility	−	−	−
NH₃ produced in liquid media	−	−	−
Gelatin hydrolysis	−	−	−
Indole production	−	−	−
Litmus milk after 2-week incubation:			
Slight acidity	+	+	+
Soft coagulation	−	−	+
Acid from glucose, maltose and mannose[b]	+	+[c]	+
Median subcutaneous infectious dose, < 10³ organisms:			
Mice, guinea pigs	+[d]	+[d]	+
Domestic rabbits	+[d]	−	−
White rats	−	−	−
Humans	+[d]	+	−

[a] For symbols see standard definitions.
[b] Slight acidity develops slowly in media with cysteine or cystine.
[c] Central Asian strains (proposed biovar *mediaasiatica*) reported to be negative.
[d] Median infectious dose, 1–10 organisms; this dose is lethal for mice, guinea pigs and domestic rabbits.

bution. Wild strains of biovar *tularensis* are highly infectious and virulent for man; they are as virulent for the domestic rabbit as for the white mouse or guinea pig. Spontaneous loss of virulence can occur when strains are maintained on artificial media. Wild strains of biovar *palaearctica* cause milder human illness than biovar *tularensis*, and at least 10^6 more organisms are required to kill the domestic rabbit than to kill the white mouse or guinea pig. Strains of biovar *tularensis* are found in nature only in North America, particularly in lagomorphs and wild rodents. The organisms may be transmitted by the bite of ticks or deer flies or by contact with, or ingestion of, contaminated animal carcasses. Strains of the biovar *palaearctica* occur in Europe, Asia and the Americas and are associated with a variety of wild rodents. They are also present in natural waters or in agricultural products contaminated by these animals.

Type strain: ATCC 6223 (strain B38, an avirulent strain with more fastidious growth requirements and a greater tendency toward variation in colony type than virulent strains). This strain is also the reference strain for the biovar *tularensis*. Virulent strains, such as SCHU, are available only from individual investigators. No reference strain for biovar *palaearctica* is available from the ATCC; the avirulent live vaccine strain can be obtained by consultation with the Centers for Disease Control, Attention: Immunobiologics Activity, Atlanta, Geor-gia 30333, U.S.A., and virulent strains are available only from individual investigators.

2. **Francisella novicida** (Larson, Wicht and Jellison 1955) Olsufiev, Emelyanova and Dunaeva 1959, 146.[AL] (*Pasteurella novicida* Larson, Wicht and Jellison 1955, 253.)

no.vi'ci.da. L. adj. *novus* new; L. v. suff. *-cida* from L. v. *caedo* to cut, kill; M.L. n. *novicida* new killer.

The characteristics are as described for the genus and as listed in Tables 4.108–4.110. In tissues the organisms appear as coccoid to ovoid or short rod-shaped cells 0.2–0.3 μm $\times$ 0.3 μm; on solid media they are 0.5 $\times$ 0.5–0.9 μm; in liquid media they are 0.7 $\times$ 1.7 μm. Capsules are not produced.

The methyl red test and Voges-Proskauer tests are negative. Nitrate is not reduced to nitrite. Methylene blue is reduced.

The organisms produce lesions similar to tularemia in white mice, guinea pigs and hamsters, but virulence is lower than that of most strains of *F. tularensis*. Rabbits, white rats and pigeons are resistant. Not known to infect man.

Isolated from a water sample taken from Ogden Bay, Utah in 1951.
Type strain: ATCC 15482.

Genus **Paracoccus** Davis 1969, 384[AL]

MILOSLAV KOCUR

Pa.ra.coc'cus. Gr. prep. *para* like, alongside of; Gr. n. *coccus* a grain, berry; M.L. masc. n. *Paracoccus* like a coccus.

Spherical cells (0.5–0.9 μm in diameter) or short rods (0.9–1.2 μm long). Occur singly, in pairs or in clusters. **Intracellular granules of poly-β-hydroxybutyrate present.** No resting stages are known. Gram-negative. **Nonmotile. Aerobic,** having a strictly respiratory type of metabolism; anaerobic growth can occur if nitrate, nitrite or nitrous oxide are available as terminal electron acceptors. **Nitrate is reduced to nitrous oxide and molecular nitrogen under anaerobic conditions.** One species can grow either **autotrophically with H_2 and CO_2** or heterotrophically with a wide variety of organic compounds as sole carbon sources; this species is not halophilic. A second species is not capable of autotrophic growth but is **halophilic.** Optimum temperature, 25–30°C. **Oxidase- and catalase-positive.** Occur in soil and presumably in natural and artificial brines. The mol% G + C of the DNA is 64–67 (T_m, Bd).

Type species: *Paracoccus denitrificans* (Beijerinck 1910) Davis 1969, 384.

Further Descriptive Information

In usual complex media such as nutrient agar, *P. denitrificans* grows as cocci or very short rods. The cell wall consists of several layers with a total thickness of 25–55 nm, and the ultrastructure is similar to that of other Gram-negative bacteria. Special cellular structures related to the hydrogen autotrophy of the organisms have not been observed (Kocur et al., 1968; Sleytr and Kocur, 1973; Walther-Mauruschat et al., 1977; Nokhal and Mayer, 1979).

P. halodenitrificans grows as a normal coccus in media containing 1.0 M (5.8%) NaCl, but cells grown in 0.6 M (3.5%) or 0.55 M (3.2%) NaCl without additional ions are generally distorted, swollen and very sticky. Cells grown at low salt concentrations synthesize less cell wall material than normal, and the observed morphological changes are considered to result, at least partly, from failure of the cells to synthesize new cell wall during division (Takahashi and Gibbons, 1959).

The ultrastructure of *P. halodenitrificans* is similar to that of *P. denitrificans* (Kocur et al., 1968).

Both species have intracellular granules of poly-β-hydroxybutyrate (PHB). Cells of *P. halodenitrificans* have a high content of this polymer, which accounts for the high endogenous respiration of this organism; the amount of polymer in resting cells decreases with time, and viability can be correlated with the polymer content of the cells. PHB-rich cells of *P. halodenitrificans* have been found to maintain their viability for longer periods than PHB-poor cells. Cells of *P. halodenitrificans* oxidize PHB in the presence of NaCl but not in water; optimum oxidation occurs in 0.33–0.75 M (1.9–4.4%) NaCl (Sierra and Gibbons, 1962a, b; 1963).

The lipid composition of *P. denitrificans* has been studied extensively. The cells contain ~15% of their dry weight as lipid. Of the total fatty acids present, 78% (Girard, 1971) or 86% (Wilkinson et al., 1972) are of the C_{18} monoenoic acid type; this has been identified as oleic acid (Wilkinson et al., 1972). Phosphatidylcholine comprises ~31% of the total phospholipids, with the remainder consisting of phosphatidylglycerol (52%), phosphatidylethanolamine (5.8%), an unknown phospholipid (5.3%), cardiolipin (3.2%), and lesser amounts of other phospholipids. The occurrence of phosphatidylcholine as a major phospholipid is unusual among eubacteria, but is characteristic of mammalian mitochondria (Wilkinson et al., 1972).

The total lipid content of *P. halodenitrificans* grown at various salt concentrations is nearly constant (11 $\pm$ 2% of the dry weight of the cells), but the proportion of phosphatide, unsaponifiable material and PHB are influenced by the salt concentration of the growth medium (Kates et al., 1961). *P. halodenitrificans* contains 62% C_{18} monoenoic fatty acids (Girard, 1971).

Unlike *P. denitrificans*, *P. halodenitrificans* does not grow at NaCl concentrations of 1.75% or less. It grows optimally in 4.5% NaCl, and can grow at NaCl concentrations up to 20% (but at a reduced rate). It will not grow in other salts such as NaBr, $NaNO_3$, LiCl and KCl, although it will survive in their presence for varying periods depending on the salt used and its concentration (Robinson and Gibbons, 1952). *P. halodenitrificans* grows on ordinary laboratory media if salt is added. One suitable medium consists of (g/liter): Proteose peptone (Difco), 5.0; tryptose, 5.0; and NaCl, 35–40 (Gibbons, 1969); another medium is nutrient agar containing 40.0–60.0 g NaCl/liter. No synthetic medium for *P. halodenitrificans* has been devised.

P. denitrificans is a versatile organism able to adapt its metabolism to the prevailing environmental conditions (Stouthamer, 1980). Under heterotrophic conditions it is capable of utilizing a wide variety of compounds as sources of carbon and energy (Vogt, 1965; Pichinoty et al., 1977; Davis et al., 1969). In the absence of an organic source of carbon it can use hydrogen as a reductant and it can fix CO_2 by the operation of the reductive pentose phosphate cycle (Kornberg et al., 1960). *P. denitrificans* possesses a complete tricarboxylic acid cycle (Forget and Pichinoty, 1965). Glucose is metabolized via the Entner-Doudoroff pathway or by the hexose monophosphate pathway, or by a combination of these two pathways. The glycolytic pathway is absent (Slabas and Whatley, 1977).

P. denitrificans can grow aerobically on methanol, methylamine or formate as the sole source of carbon and energy (Cox and Quayle, 1975; Pichinoty et al., 1977; van Verseveld and Stouthamer, 1978). Methanol or formate is oxidized to CO_2, which is again fixed via the reductive pentose phosphate cycle. Adaptation of *P. denitrificans* to growth on methanol involves the synthesis of an ammonia-dependent, dye-linked methanol dehydrogenase (Cox and Quayle, 1975). The organism can also grow anaerobically on methanol using nitrate or nitrite as the terminal electron acceptor (Bamforth and Quayle, 1978). It is not known how widespread this property is among different strains of *P. denitrificans*; Pichinoty et al. (1977) found no growth of six strains under these conditions. Because of its ability to use 1-C compounds, *P. denitrificans* can be classified as a facultative methylotroph; it can also be classified as a facultative autotroph on the basis of its hydrogen autotrophy (Bamforth and Quayle, 1978).

P. denitrificans cannot use organic compounds as terminal electron acceptors; therefore, anaerobic growth is strictly dependent upon the presence of nitrate, nitrite or nitrous oxide (Kluyver, 1956); however, cells grown anaerobically with nitrate are still able to use oxygen since the oxidase is a constitutive feature of the cells (Kluyver, 1956; Fewson and Nicholas, 1961). *P. denitrificans* produces two different nitrate reductases—A and B. When grown anaerobically at the expense of nitrate, it produces a particle-bound reductase ("nitrate reductase A") which has the ability to reduce chlorate as well as nitrate. A second, soluble nitrate reductase ("nitrate reductase B") is distinguished by its inability to reduce chlorate (Pichinoty, 1964, 1970). The optimum pH for denitrification by *P. denitrificans* has been found to be between 7.0 and 8.5, and the optimum temperature between 30 and 40°C (Shimizu et al., 1978). The rate of denitrification can be enhanced by adding a small amount of yeast extract (0.05–0.1 g/liter of medium) (Verhoeven et al., 1954; Banerjee and Schlegel, 1966; Bovell, 1967; Shimizu et al., 1978).

P. halodenitrificans is not capable of autotrophic growth, in contrast to *P. denitrificans*. It requires thiamin for growth and is stimulated by purines and pyrimidines (Katznelson and Lochhead, 1952). Its lactic dehydrogenase and cytochrome oxidase are more active at high NaCl concentrations, but the isocitric, succinic, malic, α-ketoglutaric and glutamic dehydrogenases are most active at much lower concentrations (Baxter and Gibbons, 1956).

P. halodenitrificans synthesizes nitrate reductase A only (Pichinoty, 1971). This enzyme has been found to be a non-heme iron protein containing molybdenum (Rosso et al., 1973). In *P. halodenitrificans*, denitrification proceeds optimally in cultures at an NaCl concentration

of 2.2% and in cell-free preparations at 0.9%. This indicates that the salt content of the cell is less than that of its environment and that at least one enzyme system is sensitive to "salting out" (Robinson et al., 1952).

The genetic aspects of *P. denitrificans* have been studied by only a few authors. Paraskeva (1979) found that plasmid R68.45 mediates the transfer of kanamycin resistance from *Pseudomonas aeruginosa* to *P. denitrificans*. Kanamycin resistance could be transferred from one strain of *P. denitrificans* to another, which opens up the possibility of using R68.45 as a sex factor in *P. denitrificans*. Four mutants of *P. denitrificans* requiring methionine or leucine were isolated by Banerjee (1966), and mutants deficient in c-type cytochrome were isolated by Willison and John (1979). The growth requirements of 40 cysteine-requiring mutants were consistent with sulfite and sulfide being intermediates in the reduction of sulfate to cysteine in *P. denitrificans* (Paraskeva and Whatley, 1980).

Enrichment and Isolation Procedures

For *P. denitrificans*, an enrichment procedure called "molecular hydrogen-nitrate system" may be used (Kluyver, 1956; Schlegel et al., 1961; Vogt, 1965; Nokhal and Mayer, 1979). A liquid minimal medium containing tartrate or succinate incubated under a nitrous oxide atmosphere at 32°C may also be used (Pichinoty et al., 1977). Only a few strains have been isolated: Verhoeven et al. (1954), Vogt (1965) and Bollag and Russel (1976) each isolated one strain; Pichinoty et al. (1977) isolated two strains, and Nokhal and Mayer (1979) isolated 11 strains.

Gamble et al. (1977) isolated 146 cultures of denitrifying bacteria from 90 soil samples; however, none of these was *P. denitrificans*. This indicates that *P. denitrificans* is not one of the major denitrifiers found in nature.

As for *P. halodenitrificans*, only a single strain has been isolated (from Wiltshire bacon-curing brine), although it is presumed that the organism may be widely distributed in natural and artificial brines (Robinson and Gibbons, 1952).

Maintenance Procedures

Strains of *P. denitrificans* are initially cultured on nutrient agar; *P. halodenitrificans* is cultured on media containing 4% NaCl. After incubation at 25–30°C to allow abundant growth, the cultures may be maintained in a refrigerator (4°C) for 2 or 3 months. They may also be stored on the above media under liquid paraffin in a refrigerator (5°C) for 1–2 years, or they can be preserved indefinitely by lyophilization.

Procedures for Testing Special Characters

For the study of autotrophic or heterotrophic growth of *P. denitrificans*, the mineral base of Schlegel et al. (1961) is recommended (Vogt, 1964; Banerjee, 1966; Aragno et al., 1977; Nokhal and Mayer, 1979). For the study of chemolithotrophic growth of *P. denitrificans*, the medium of Doudoroff (1940) as described in Stanier et al. (1966) is recommended (Pichinoty et al., 1977).

Differentiation of the genus **Paracoccus** from other genera

Table 4.111 indicates those characteristics of *Paracoccus* that differentiate it from other morphologically or physiologically similar organisms.

Taxonomic Comments

Until recently the taxonomic position of the genus *Paracoccus* has not been well defined because of lack of experimental data; therefore, it was placed among "genera of uncertain affiliation" (Doudoroff, 1974). Recent analyses of the oligonucleotide catalogs of 23 S rRNA (McKay et al., 1979) and 16 S rRNA (Fox et al., 1980) indicate a close degree of relatedness of *P. denitrificans* to certain purple photosynthetic bacteria, particularly *Rhodopseudomonas capsulata* and *R. sphaeroides*.

In the case of 16 S rRNA, the S_{AB} values involved were ~0.7 (Fox et al., 1980). These results are consistent with the idea that *P. denitrificans* may be a nonphotosynthetic species of purple photosynthetic bacteria, particularly of the genus *Rhodopseudomonas* (McKay et al., 1979). A number of phenotypic similarities between *Paracoccus* and *Rhodopseudomonas* are indeed evident (Table 4.111), including the ability of some strains of *R. sphaeroides* to grow anaerobically in the dark by denitrification (Satol et al., 1976); however, several differences are also evident such as those concerning motility or acid production from glucose.

Other studies have indicated that *P. denitrificans* and *Rhodopseudomonas* species exhibit a mitochondria-like respiration, and that the cytochrome c of some species of *Rhodopseudomonas*, the cytochrome

Table 4.111.
Differential characteristics of the genus **Paracoccus** *and other morphologically or physiologically similar organisms*[a]

Characteristics	Paracoccus	Rhodopseudomonas sphaeroides	Rhodopseudomonas capsulata	Xanthobacter	Alcaligenes paradoxus	Alcaligenes denitrificans	Pseudomonas stutzeri	Branhamella catarrhalis	Acinetobacter
Have mainly a spherical shape	+[b]	+	+[c]	−	−	−	−	+	−[d]
Intracellular poly-β-hydroxybutyrate formed	+	+	+	+	+	+	−	−	−
Motile	−	+	+	−	+	+	+	−	−
Phototrophic	−	+	+	−	−	−	−	−	−
Nitrogenase activity	−	−	−	+	−	−	−	−	−
NO_3^- reduced to NO_2^-	+	d[e]	−	−	d	+	+	+	−
NO_2^- reduced to N_2O	+	d[e]	−	−	−	+	+	+	−
N_2O reduced to N_2	+	d[e]	−	−	−	+	+	+	−
Chemolithotrophic growth with H_2	D	+	+	+	d	−	−	−	−
Hydrolysis of Tween 80	−	−		+	−	+	−	+
Acid from glucose	−	+	+		+	d	+	−	+
3-Ketolactose produced from lactose oxidation	−	−			+	−	−	−
Pigment produced	−	Br-R	Br-R	Y	Y	−	−	−	−
Mol% G + C of DNA	64–67	68–70	65–67	65–70	67–69	64–70	61–66	40–43	38–47

[a] Symbols: +, typically positive; −, typically negative; d, differs among strains; D, differs among species; Y, yellow or greenish yellow; Br-R, brown or red.
[b] Rod-shaped in exponential phase.
[c] Spherical below pH 7.0, ovoid or rod-shaped above pH 7.0.
[d] Rod-shaped in exponential phase but nearly spherical in stationary phase.
[e] Satoh et al. (1976).

Table 4.112.
Differential characteristics of the species of the genus **Paracoccus**[a]

Characteristics	1. *P. denitrificans*	2. *P. halo-denitrificans*
Halophilic, requiring at least 3% NaCl for growth	−	+
Facultative hydrogen autotroph	+	−
Thiamin required for growth	−	+

[a] For symbols see standard definitions.

c_{550} of *P. denitrificans*, and the cytochromes *c* of mitochondria, all show strong structural homology (John and Whatley, 1975a, b; Dickerson, et al., 1976). *P. denitrificans* occupies a unique position among bacteria, in that it collects in a single prokaryotic organism those features of a mitochondrion which are otherwise randomly distributed among a wide variety of aerobic bacteria. The currently available evidence suggests that *P. denitrificans* might be viewed as a free-living, highly-adaptable mitochondrion (John and Whatley, 1977a, b).

The degree of relatedness between *P. denitrificans* and *P. halodenitrificans* has not yet been assessed by nucleic acid homology methods.

Further Reading

Dickerson, R.E. 1980. Cytochrome *c* and the evolution of energy metabolism. Sci. Amer. *242:* 136–153.
John, P. and F.R. Whatley. 1975. *Paracoccus denitrificans* and the evolutionary origin of the mitochondrion. Nature *254:* 495–498.
John, P. and F.R. Whatley. 1975b. *Paracoccus denitrificans:* a present-day bacterium resembling the hypothetical free-living ancestor of the mitochondrion. Symp. Soc. Exp. Biol. *29:* 39–40.
John, P. and F.R. Whatley. 1977a. The bioenergetics of *Paracoccus denitrificans.* Biochim. Biophys. Acta *463:* 129–153.
John, P. and F. R. Whatley. 1977b. *Paracoccus denitrificans* Davis (*Micrococcus denitrificans* Beijerinck) as a mitochondrion. Adv. Bot. Res. *4:* 51–115.
Haddock, B.A. and C.W. Jones. 1977. Bacterial respiration. Bacteriol. Rev. *41:* 47–99.
Payne, W.J. 1973. Reduction of nitrogenous oxides by microorganisms. Bacteriol. Rev. *37:* 409–452.
Stouthamer, A.H. 1980. Bioenergetic studies on *Paracoccus denitrificans.* Trends Biochem. Sci. *5:* 164–166.

Differentiation of the species of the genus **Paracoccus**

Table 4.112 presents those features which differentiate the two species of the genus. Table 4.113 indicates other characteristics of the species.

List of the species of the genus **Paracoccus**

1. **Paracoccus denitrificans** (Beijerinck 1910) Davis 1969, 384.[AL]
(*Micrococcus denitrificans* Beijerinck 1910, 53.)
de.ni.tri′fi.cans. M.L. part. adj. *denitrificans* denitrifying.
The morphology is as described for the genus.

Colonies on nutrient agar are 2 to 3 mm in diameter, usually circular, entire, smooth, glistening, white and opaque.
Physiological and nutritional characteristics are presented in Tables 4.112 and 4.113. Organic growth factors are not required for aerobic

Table 4.113.

Other characteristics of **Paracoccus denitrificans** *and* **Paracoccus halodenitrificans**[a]

Characteristics	1. *P. denitrificans*	2. *P. halo-denitrificans*
Oxidase and catalase activity	+	+
Anaerobic growth with nitrate	+	+
NO_3^- reduced to NO_2^-, N_2O and N_2	+	+
Facultative methylotroph (growth occurs on methanol, methylamine or formate)	+	
Growth in 6% NaCl	−	+
Pigment produced on nutrient agar	−	−
Hydrolysis of esculin, starch, gelatin or Tween 80	−	−
Phosphatase	−	−
Urease	−	−
Indole test	−	−
H_2S production	−	−
Voges-Proskauer test	−	−
Simmons' citrate	−	−
Hemolysis	−	−
Phenylalanine deaminase		−
Arginine dihydrolase, lysine decarboxylase, ornithine decaboxylase		−
Acid from glucose[b]	−	−
β-Galactosidase	−	−
Extracellular hydrolysis of poly-β-hydroxybutyrate	−	
Growth with sarcosine, creatine, *m*-hydroxybenzoate[c] *p*-hydroxybenzoate, sucrose, trehalose, maltose, glycerate, histidine and glucose	+	
Growth with testosterone, pantothenate or hydroxymethylglutarate	−	

[a] For symbols see standard definitions.

[b] Tested in the medium of Hugh and Leifson (1953). *P. denitrificans* can oxidize glucose by the Entner-Doudoroff or hexose-monophosphate pathways (Slabas and Whatley, 1977).

[c] *m*-Hydroxybenzoate is metabolized by the protocatechuate pathway (Davis et al., 1969).

and anaerobic growth with organic substrates or for autotrophic growth under H_2 and CO_2 (Doudoroff, 1974). Both ammonium salts and nitrate can serve as nitrogen sources.

Grows readily on nutrient agar containing 0–5% NaCl; growth becomes scanty above 5% NaCl. Optimum temperature, 30°C; range, 5–37°C.

Isolated from soil by enrichment for denitrifying bacteria and by enrichment for autotrophic hydrogen bacteria.

The mol% G + C of the DNA is 64–67 (T_m, Bd).

Type strain: ATCC 17741.

Further comments. The citation given for the species on the Approved Lists of Names (*P. denitrificans* (Beijerinck and Minkman 1910) Davis 1969)) appears to be incorrect with respect to the authors of the specific epithet. Minkman was not a co-author of the original article; rather, he was acknowledged as having given technical assistance in the work.

2. **Paracoccus halodenitrificans** (Robinson and Gibbons 1952) Davis 1969, 386.[AL] (*Micrococcus halodenitrificans* Robinson and Gibbons 1952, 154.)

ha.lo.de.ni.tri′fi.cans. Gr. n. *hals, halis* salt; M.L. v. *denitrifico* to denitrify; M.L. part. ajd. *halodenitrificans* salt (-requiring) denitrifying.

The morphology is as described for the genus. The cells may stick together due to a viscous surface slime which is either excreted or absorbed by the cells and which can be digested with deoxyribonuclease (Smithies and Gibbons, 1955).

Colonies on nutrient agar + 6% NaCl are circular, entire, convex, butyrous, glistening, cream colored and opaque.

Physiological and nutritional characteristics are presented in Tables 4.112 and 4.113.

Optimum temperature, 25–30°C; range, 5–32°C.

No growth occurs in media containing less than 3% NaCl. Optimum growth occurs with 4.4–8.8% NaCl, slow growth occurs with up to 20% NaCl.

Isolated from meat-curing brines. Presumably widely distributed in natural and artificial brines.

The mol% G + C of the DNA is 64–66 (T_m, Bd).

Type strain: ATCC 13511.

Genus **Lampropedia** Schroeter 1886, 151[AL]

R. G. E. MURRAY

Lam·pro·pe′di·a. Gr. adj. *lampros* bright, radiant; Gr. n. *pedia* a plain, flat country; M.L. fem. n. *Lampropedia* a shining flat sheet (of cells).

Sheets of rounded, almost cubical cells, arranged in square tablets of 16–64 cells, occasionally separated into pairs or tetrads. **Divide synchronously in a sheet and alternately in two planes.** The cells of a tablet are enclosed within a **complex, structured envelope.** Each cell is enclosed in a **Gram-negative type of cell wall.** Intracellular granules of **poly-β-hydroxybutyrate** are prominent. No flagella occur. **Twitching movements** of small groups of cells occur during active growth. **Obligately aerobic,** having a strictly respiratory type of metabolism with oxygen serving as the terminal electron acceptor. **Growth occurs as a thin, hydrophobic, extending pellicle** on the surface of both liquid and solid media. Nonpigmented. Optimum temperature, 30°C. Optimum pH, 7.0. Oxidase- and catalase-positive. Chemoorganotrophic. **Energy sources are limited to intermediates of the tricarboxylic acid cycle.** Carbohydrates, alcohols, glucosides and fatty acids are not utilized. Ammonium salts or certain amino acids can serve as sole nitrogen sources. **Vitamins may be required for growth.** The ecological niche is unknown, but

observations and isolations indicate an environment rich in organic matter. The mol% G + C of the DNA is 61 (Bd).

Type species: Lampropedia hyalina Schroeter 1886, 151.

Further Descriptive Information

This is a genus based mainly on morphological characters and it consists of a single species, *L. hyalina.* The organism is cultivable and strains are available for study.

The characteristic appearance of growth is a hydrophobic pellicle consisting of square tablets of Gram-negative cells, apparently dividing synchronously in two planes (Fig. 4.89), forming continuous but rumpled sheets on the surface of media (Kuhn and Starr, 1965; Puttlitz and Seeley, 1968; Seeley, 1974). Three other species have been named in the distant past (de Toni and Trevisan, 1889), including pigmented species, but they have not been reisolated in the intervening years and have no validity today.

The distinctive morphology (Fig. 4.89) allows recognition in its

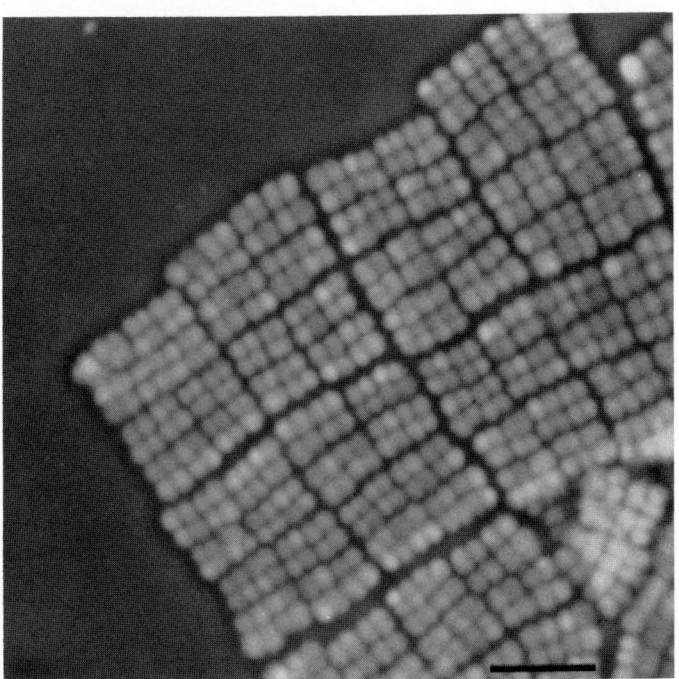

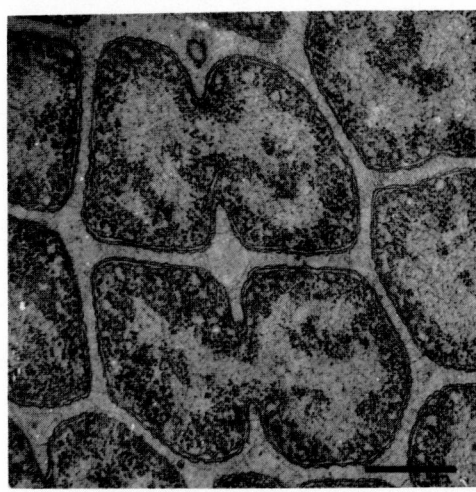

Figure 4.90. Electron micrograph of a section in the plane of a tablet to show division and the relatively large volume of nucleoplasm A matrix substance separates the cell walls (of usual Gram-negative character; see Figure 4.92) of adjacent cells. Divisions always show a "constrictive" form (*bar*, 0.5 μm).

Figure 4.89. Light micrograph of a nigrosin preparation of *L. hyalina* showing a corner of a sheet of actively growing tablets of cells. Adjacent tablets are almost synchronized in division (*bar*, 5 μm).

natural surroundings. In rich organic environments, and especially in laboratory media loaded with sodium acetate, the cells are full of poly-β-hydroxybutyrate (PHB) (see Fig. 4.91) as several small or one large granules (Kuhn and Starr, 1965).

L. hyalina exhibits a peculiar form of motility involving sudden, irregular shifts in the position of small *groups* of cells. Pringsheim (1955) and Puttlitz and Seeley (1968) describe this phenomenon and it has been compared (Pringsheim, 1966) to movements seen in the *Chroococcales* as an argument for relationship to cyanobacteria. However, such a "twitching" motility is also seen in other nonflagellated bacteria and most of them have fimbriae (Henrichsen, 1972). *Lampropedia* strains have not been shown to have either flagella or fimbriae. Twitching, generally (Henrichsen, 1972) and in this case (Puttlitz and Seeley, 1968), is a function of living, metabolizing cultures.

The cell wall is of the Gram-negative type with an outer membrane and a thin, underlying peptidoglycan layer. It has not been isolated and characterized in chemical terms. All layers of this wall intrude at once to separate the sister cells (Fig. 4.90).

The structure of the cells is not remarkable, but the envelope that encloses each tablet of cells has unusual features (Murray, 1963; Chapman et al., 1963; Pangborn and Starr, 1966). The separate tablets of 16, 32 or 64 cells are surrounded by a hexagonal array of complex spindle-shaped units (spacing is 23–26 nm) on a thin continuous but perforated layer (7.5-nm holes; spacing is 13.5–14.5 nm). This highly structured integument bridges over the spaces between the cells in the tablet (Fig. 4.91). The space between cells, between the walls and the structured envelope, and within the intruding septa (Figs. 4.91–4.93) contains fibrous materials and is called the intercalated layer. None of these layers have been isolated for biochemical studies. The enveloping layers together accomplish the division and separation of tablets of cells.

Colonial characteristics are determined by the growth habit and the physical nature of the surface of the enveloping layer. The sheets of cells, one cell thick, extend as they grow over the surface of liquid or solid media and tend to wrinkle as they grow to large size and meet some obstacle to spreading. Many microcolonies are square. An irregularly shaped piece transferred to a liquid medium floats on the surface

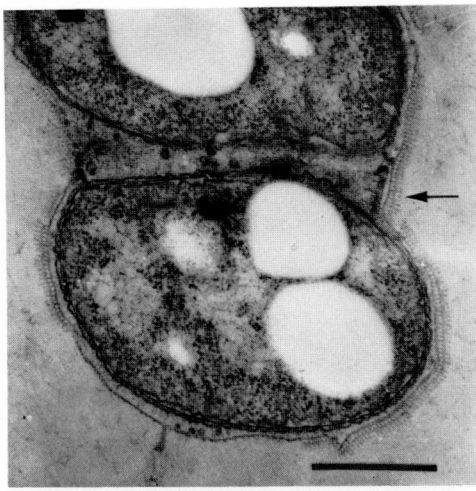

Figure 4.91. Section transverse to the edge of a tablet of cells showing a bi-partite external structured envelope. The inner layer is obvious in an area exposed by stripping. The envelope encloses the tablet and bridges over the matrix separating adjacent cells (*arrow*). The low density vesicles represent poly-β-hydroxybutyrate granules (*bar*, 0.5 μm).

and maintains that odd shape as it grows; old cultures provide a rain of cells and debris that sink to the bottom. The capacity to float and grow as a pellicle suggests a hydrophobic surface, which is believed to be mediated by fine fibrous material (? proteinaceous) external to the envelope array (S. G. Lanys, M. Sc. thesis "Morphological studies of cell envelope differences among colony variants of *Lampropedia hyalina*," University of Western Ontario, 1972). There is no life cycle.

The description of the genus and its growth characteristics assumes morphological stability. However, mutants are easily derived by isolation of natural variants or stimulated by a mutagen such as nitrosoguanidine (unpublished results and S. G. Lanys, thesis, *cited above*) producing different smooth or rough colonies rather than sheets. Many of these have lost the ability to form tablets or even tetrads, and most have lost the envelope layers (Figs. 4.93 and 4.95). Rough variants usually have retained the envelope arrays (Fig. 4.94) and these most commonly have been derived with the help of nitrosoguanidine. There-

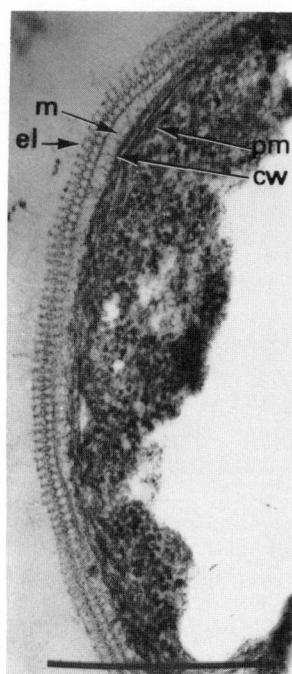

Figure 4.92. A high magnification of a section showing the complex structured envelope (*el*), the envelope matrix (*m*), the cell wall (*cw*) and plasma membrane (*pm*) (*bar*, 0.5 μm).

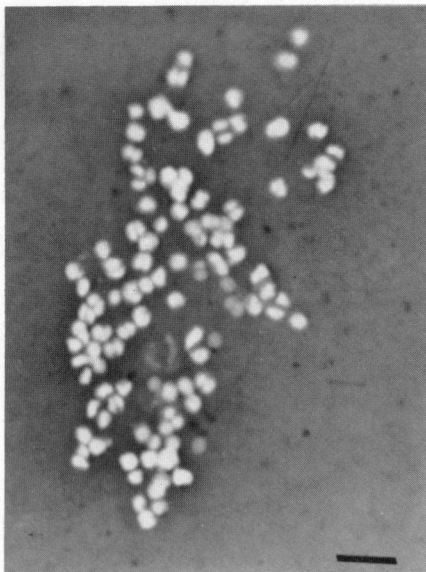

Figure 4.93. Light micrograph of a nigrosin preparation of a non-sheeting strain derived from the original isolation of *L. hyalina*. ATCC 11041 (Murray, 1963) (*bar*, 5.0 μm).

fore, it is possible—even probable—that clones exist in nature that are phylogenetically related to *Lampropedia* but show none of the unique morphological characters that we use to identify *L. hyalina* in the absence of clear metabolic distinctions. The cultural, physiological and biochemical characters of the available strains have been studied by

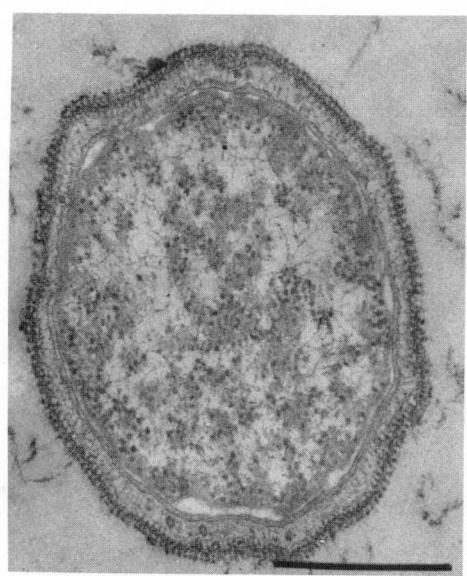

Figure 4.94. Electron micrograph of a section of a nonsheeting derivative of *L. hyalina* ATCC 13691 which possesses both the structured and the intercalated enveloping layers outside the cell wall (*bar*, 0.5 μm).

Puttlitz and Seeley (1968). It is evident that this respiratory, catalase-positive chemoorganotroph requires Krebs cycle intermediates as energy sources and utilizes ammonium salts and a limited palette of amino acids as nitrogen sources. It is slightly fastidious in that it requires biotin and thiamine for growth, but can be cultured in a simple defined medium* if these vitamins are supplied. It is a mesophile that grows best at neutral pH and does not tolerate either 0.5% bile or 1.0% salt. It does not produce exoenzymes that degrade proteins, lipids or the major carbohydrate polymers. So that without its peculiar morphology it is not a distinguished aerobe and would be hard to recognize (see Figs. 4.93–4.95).

Nothing is known about the genetics, antigenic structure or antibiotic sensitivity of this organism. The mol% G + C of the DNA (M. Mandel, personal communication) ranges from 60.7–61.2 for all available strains.

The ecological niche is unknown. The microbe is sufficiently distinctive (Starr and Skerman, 1965) to be recognized in its tablet or sheeting form by simple microscopy of natural specimens and undoubtedly it has been seen many times (Starr, 1981) even if there have been no more than three isolations (Pringsheim, 1955; Julius Kirchner, noted by Hungate, 1966; Frank Kovacses strain "Mac 583," described by Schad et al., 1964). The isolations and sightings (sometimes called "window-pane sarcinas") involved waters infused with quantities of organic material, probably well digested and populated with many other microbes. A partial listing includes: waters polluted with sugar refinery wastes, swamp water (Schroeter, 1886), stagnant water including aquatic plants (de Toni and Trevisan, 1889), liquid manure from a barnyard (Pringsheim, 1955), rumen fluid (see Hungate, 1966; Eadie, 1962; Smiles and Dobson, 1956; Clarke, 1979), intestinal content of herbivorous reptiles and their nematodes (Schad et al., 1964), and sewage-polluted, muddy water (R. Kolkwitz, 1909, cited by Starr, 1981). Considering the need for organic acids (Krebs cycle intermediates) and for vitamins, as well as the lack of exoenzymes and growth as a pellicle at the air/water interface, it is not surprising that it thrives in many of these situations. Its real habitat remains to be discovered and could

---

* Chemically defined medium for growth of *L. hyalina* (Puttlitz and Seeley, 1968) consists of the following (per liter): basal salts solution (CaCl$_2$·2H$_2$O, 0.1 g; FeSO$_4$·7H$_2$O, 0.2 g; MgSO$_4$·7H$_2$O, 0.2 g; MnCl$_2$·4H$_2$O, 0.25 g; distilled water, 1000 ml), 100 ml; sodium pyruvate, 3.0 g; NH$_4$Cl, 3.0 g; phosphate buffer solution (KH$_2$PO$_4$, 7.0 g; K$_2$HPO$_4$, 13.0 g; distilled water, 1000 ml), 50 ml; thiamine-HCl, 1.0 mg; biotin, 1.0 μg; and NaOH, 70 mg. The vitamins are added aseptically to the sterile basal medium from stock solutions sterilized by filtration.

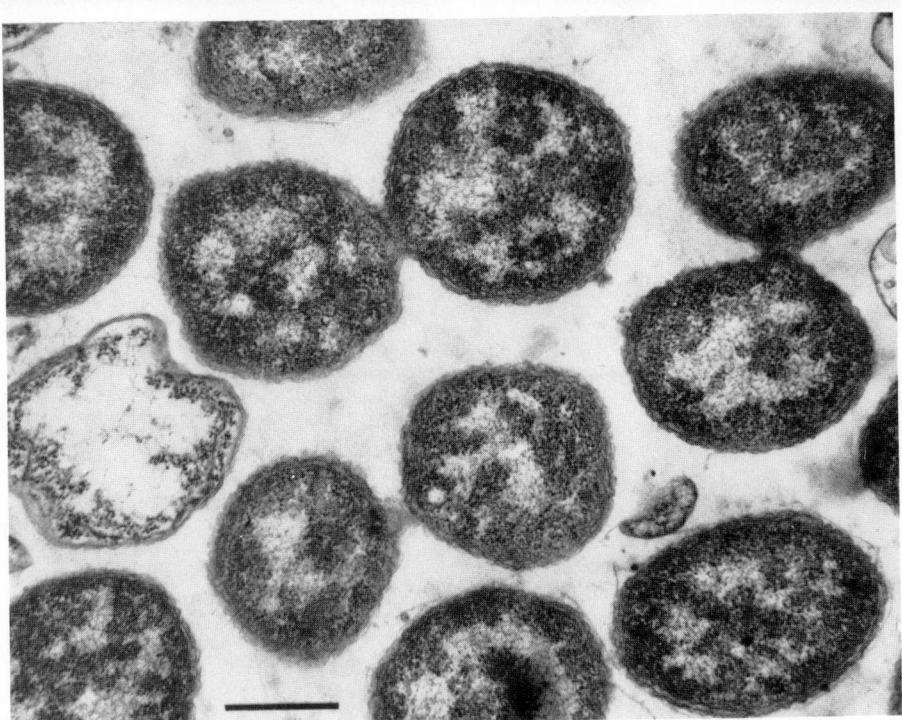

Figure 4.95. A derivative of *L. hyalina* ATCC 11041 in section showing that it is almost, if not completely, devoid of the enveloping layers (*bar*, 0.5 µm).

be many other places accessible to these sources, perhaps associated with plants or the soil around them. It is hard to believe that this strict aerobe can be even an irregular inhabitant of cattle or sheep rumen unless (R. E. Hungate, personal communication) the lampropedias found in this site can use some hydrogen acceptor other than oxygen. However, there is no doubt of the occasional presence of enough window-pane sarcinas to be detected by microscopy.

Enrichment and Isolation Procedures

L. hyalina will grow on many rich media (both liquid and solidifed with agar) but isolation is made difficult by enormous numbers of other bacteria in the source material. Pringsheim (1955) used a capillary to transfer individual tablets of cells to dishes with water over a layer of soil in which starch particles or a grain of wheat was embedded; the lampropedias multiplied as a pellicle in the surface film and, after a day or two at room temperature, could be streaked on an agar medium (0.1% sodium acetate, 0.2% yeast extract and 0.1% peptone) to give a few isolated microcolonies. Microcolonies can be identified readily by

low power microscopy and subcultured. Kirschner (cited by Hungate, 1966) left rumen fluid standing for several days in an open flask and isolated *Lampropedia* by streaking the pellicle that formed onto rumen fluid agar, which was then incubated aerobically at room temperature. There are a number of remarks on pellicle formation in the literature, so it seems likely that simple retention at room temperature is the preliminary approach to enrichment.

Maintenance Procedures

The common laboratory practice is maintenance by monthly subculture at room temperature on a neutral pH yeast extract-acetate-peptone agar (0.1–0.5% of each). It survives lyophilization to a fair degree. Undoubtedly, constant subculture can lead to selection of mutants. Murray (1963) observed that ATCC strain 11041, derived from Pringsheim's isolation, had lost its ability to form sheets (Figs. 4.93 and 4.95). Nowadays, safe stocks can be kept in liquid N_2 suspended in the above fluid medium with 20% glycerol added.

Differentiation of the genus **Lampropedia** from other genera

The ability of *Lampropedia* to grow as sheets of cells on liquid or solid media distinguishes it from other genera of nonphotosynthetic bacteria. An organism known as *Thiopedia rosea* has a morphological resemblance to *Lampropedia*, but is photosynthetic (see Taxonomic Comments below).

Taxonomic Comments

The phylogenetic relatives of *L. hyalina* are not yet identified and so it must remain a Gram-negative coccus of unknown affiliations. However, some aspects of its unique tablet-forming morphology are repeated in *Thiopedia rosea*, a nonsulfur purple, photosynthetic bacterium (Hirsch, 1977). A similar type of photosynthetic organism was studied in the electron microscope (J. A. Lauritis, Ph.D. thesis "Fine structure of an unusual photosynthetic bacterium," Iowa State University, 1967) and identified as "*Rhodothece*" (now *Amoebobacter*), which did not get beyond the diplococcal form. The extraordinary feature of this organism was the possession of a superficial wall array virtually identical

in fine structure to the complex envelope array of *L. hyalina*. In fact, apart from the chromatophore membranes it was similar to some of the enveloped but nonsheeting variants of *L. hyalina* (Fig. 4.94; S. G. Lanys and R. G. E. Murray, unpublished observations). Natural variants isolated from cultures often lose the enveloping structures and the ability to form tablets (Murray, 1963). Associations dependent upon such similarities are fraught with danger because we have no understanding of the genetics of such structures and whether or not the determinants are genetically transferable to any receptive organisms. It is probable that at least one of the old named species, "*L. violacea*" (see Seeley, 1974), may have been one of the *Chromatiaceae* (such as *Lamprocystis roseopersicina*) because the color of cell suspensions is purple to purple-violet (Pfennig and Trüper, 1974).

No less a problem is presented by Pringheim's (1966) proposal that *L. hyalina* should be regarded as an apochlorotic species of *Merismopedia*, a large coccoid cyanobacterium which forms squared sheets of cells embedded in a glutinous matrix but without evidence of an enveloping structured array. Somewhat similar associations are evident in *Thiocapsa* and *Lamprocystis* among the Chromatiaceae.

Few, if any, of these taxonomic hypotheses can be given any credence without strong supportive evidence based, in all probability, on several accepted molecular markers for phylogenetic relationship. Furthermore, comparative studies of fine structure of a wider range of genera as well as studies, if possible, of heterotrophically grown *Thiopedia* have yet to be undertaken. Uncertainty about taxonomic associations is exemplified by association with sulfur bacteria in the sixth edition of *Bergey's Manual*, complete omission from the seventh edition, and relegation to a genus of uncertain affiliation among Gram-negative cocci in the eighth edition. Obviously we are no further ahead now, but the tools to solve these problems may be available.

Further Reading

Starr, M.P. 1981. The Genus *Lampropedia*. In Starr, Stolp, Trüper, Balows and Schlegel (Editors) The Prokaryotes: a Handbook on Habitats, Isolation, and Identification of Bacteria, Springer-Verlag, Berlin, pp. 1530–1536.

List of the species of the genus **Lampropedia**

1. **Lampropedia hyalina** (Ehrenberg 1832) Schroeter 1886, 151.[AL] (*Gonium hyalinum* Ehrenberg 1832, 63.)

hy·a·li′na. Gr. adj. *hyalinos* glassy, shiny; M.L. fem. adj. *hyalina* hyaline.

Cells 1.0–1.5 × 1.0–2.5 μm. Morphological characteristics are as described for the genus and as depicted in Figs. 4.89–4.92.

Physiological and biochemical characteristics are as described for the genus and as listed in Table 4.114.

Utilizes pyruvate, lactate, butyrate, fumarate, malate, succinate (and acetate in the presence of catalytic levels of pyruvate) as sole energy sources.

Utilizes NH_2Cl, alanine, arginine and tyrosine as sole nitrogen sources.

Biotin and thiamine are required for growth.

Temperature range for growth, 10–35°C; optimum, 30°C. pH range for growth, 6.0–8.6; optimum, 7.0.

The mol% G + C of the DNA is 60.7–61.2 (Bd; M. Mandel, personal communication).

Type strain: ATCC 11041. *However, this strain is inappropriate* because it has lost the characteristic of forming tablets and sheets and cells, and has lost the ability to cover its surface with a structured envelope (Murray, 1963). *A neotype strain must be designated.* A possible candidate is reference strain ATCC 13871.

Table 4.114.

Physiological and biochemical characteristics of **Lampropedia hyalin**[a]

Characteristics	Reaction or Result
Growth under anaerobic conditions	−
Intracellular poly-β-hydroxybutyrate formed	+
Oxidase test	+
Catalase test	+
Indole production	−
Acetyl methyl carbinol production (Voges-Proskauer test)	−
Litmus milk	−[b]
Benzidine test for heme groups	+
Hemolysis on blood agar	−
Arginine deaminase	+
Hydrolysis of gelatin, casein, fats and fatty acids, starch, hippurate, DNA and urea	−
Growth in the presence of:	
1.0% NaCl	+
1.5% NaCl	−
2.0% sucrose	+
4.0% sucrose	−
0.5% bile	−
Final pH in culture media	8.4–8.6

[a] Data from Puttlitz and Seeley (1968). Symbols: see standard definitions.
[b] No change.

Other organisms

It is possible that the "window-pane sarcina" seen in rumen contents (which usually do not show poly-β-hydroxybutyrate granules; R. E. Hungate, personal communication) is separable from *L. hyalina*, but recognition would require cultivation and metabolic studies. An early sighting gave rise to the name "*Bacterium merismopedioides*" (Zopf, 1883), which might be an appropriate name if a rumen strain is ever differentiated from *L. hyalina*.

Many other synonyms probably have been created. Among them the generic name "*Pedioplana*" Wolf 1907, 10 was considered to be synonymous with *Lampropedia* by Seeley (1974) but, as Starr (1981) points out, it was described as motile and as possessing flagella.

The *species incertae sedis* included by Seeley (1974) are no longer valid but memory of them should not be erased because they may represent other physiological variations for rediscovery: "*Lampropedia reitenbachii*" (Caspary) de Toni and Trevisan 1889, 1048; "*Lampropedia violacea*" (Brébisson) de Toni and Trevisan 1889, 1048, and "*Lampropedia ochracea*" (Mattenheimer) de Toni and Trevisan 1889, 1049.

Important Notes for Users of this Edition

1. Always read both generic and species descriptions because characters listed in the generic description are not usually listed in the species descriptions.

2. Unless otherwise indicated in footnotes to tables, the meanings of symbols are as follows:

+   90% or more of strains are positive

−   90% or more of strains are negative

d   11–89% of strains are positive

v   strain instability (*not* equivalent to "d")

D   different reactions in different taxa (species of a genus or genera of a family)

3. All other symbols are defined in footnotes to tables.

Facultatively Anaerobic Gram-Negative Rods

Table 5.1.

Some differential characteristics of the families of Section 5[a]

Characteristics	Entero-bacteriaceae (p. 408)	Vibrio-naceae (p. 516)	Pasteurel-laceae (p. 550)
Cell diameter, μm	0.3–1.5	0.3–1.3	0.2–0.3
Straight rods	+	D	+
Curved rods	–	D	–
Motility	D	+[b]	–
Flagellar arrangement (liquid media):			
Polar	–	+	
Lateral	+[c]	–	
Oxidase test	–	+[b]	+[b]
Na$^+$ required or stimulatory for growth	–	D	–
Contain enterobacterial common antigen	+[d]	–[e]	–
Cells contain menaquinones[f]	D	D	–
Parasitic on mammals and birds	D	–[b]	+
Heme and/or nicotinamide adenine dinucleotide required for growth	–	–	D
Plant pathogenicity	D	–	–
Organic nitrogen sources required	–[b]	–[b]	+

[a] Symbols: see standard definitions.

[b] A few exceptions may occur.

[c] Except *Tatumella*, which may have polar, subpolar or lateral flagella.

[d] *Erwina chrysanthemi* does not contain the antigen.

[e] *Pleisomonas shigelloides* contains the antigen.

[f] *Pasteurellaceae* do contain demethylmenaquinones but not menaquinones; ubiquinones may or may not be produced. *Enterobacteriaceae* and *Vibrionaceae* may contain menaquinones, demethylmenaquinones and ubiquinones.

FAMILY I. ENTEROBACTERIACEAE RAHN 1937, Nom. fam. cons. Opin. 15, Jud. Comm. 1958, 73; Ewing, Farmer, and Brenner 1980, 674; Judicial Commission 1981, 104.

Don J. Brenner

En.te.ro.bac.te.ri.a′ce.ae. M.L. n. *enterobacterium* an intestinal bacterium; -*aceae* ending to denote a family; M.L. fem. pl. n. *Enterobacteriaceae* the family of the enterobacteria. Rahn's original derivation is not certain. It may have come from his genus *Enterobacter*, or may have come from the root enterobacterium.

Gram-negative straight rods, 0.3–1.0 × 1.0–6.0 μm; motile by peritrichous flagella, except for *Tatumella*, or nonmotile. Do not form endospores or microcysts; not acid-fast. Grow in the presence and absence of oxygen. Grow well on peptone, meat extract, and usually MacConkey's media. Some grow on D-glucose as the sole source of carbon, others require vitamins and/or amino acids. Chemoorganotrophic; respiratory and fermentative metabolism. Not halophilic. Acid and often visible gas is produced during fermentation of D-glucose, other carbohydrates and polyhydroxyl alcohols. Catalase-positive except for *Shigella dysenteriae* 0 group 1 and *Xenorhabdus nematophilus*; oxidase

408

negative. Nitrate reduced to nitrite except by some strains of *Erwinia* and *Yersinia*. G + C content of DNA is 38–60 mol% (T_m, Bd).

DNAs from species within most genera are at least 20% related to one another and to *Escherichia coli*, the type species of the family. Notable exceptions are species of *Yersinia, Proteus, Providencia, Hafnia* and *Edwardsiella*, whose DNAs are 10–20% related to those of species from other genera.

Except for *Erwinia chrysanthemi* (Le Minor et al., 1972) all species tested contain the enterobacterial common antigen (Kunin, 1963; Kunin et al., 1962; Whang and Neter, 1962; Vosti et al., 1964; Le Minor et al., 1972).

Type genus: Escherichia Castellani and Chalmers 1919, 941. Designated type genus Opin. 15, Jud. Comm. 1958, 73.

Further Comments

Circumscription. The definition circumscribes a large biochemically and genetically related group that shows substantial heterogeneity in its ecology, host range and pathogenic potential for man, animals, insects and plants. The delimitation of *Enterobacteriaceae* from members of other families seems complete, except as mentioned below; however, systematic studies have rarely been done.

The genera *Vibrio, Photobacterium, Aeromonas* and *Plesiomonas* are oxidase-positive and have polar flagella when grown in liquid media—characteristics which distinguish them from *Enterobacteriaceae*. However, at least two *Vibrio* species (*V. metschnikovii* and *V. gazogenes*) are oxidase-negative; strains of other species are oxidase-negative or weakly positive; and, under certain conditions (often on solid media), members of these genera produce peritrichous flagella. *P. shigelloides* is the only member of *Vibrionaceae* to contain the enterobacterial common antigen (Le Minor et al., 1972). Some *Aeromonas* strains show higher DNA relatedness to *E. coli*, the type species of *Enterobacteriaceae*, than that seen with several genera within the family. In fact, *Enterobacteriaceae* and *Vibrionaceae* have been treated as a superfamily. Nonetheless, the functional distinction between *Enterobacteriaceae* and *Vibrionaceae* is extremely useful and essentially exclusive. If one imagines an evolutionary continuum from a common ancestor, it is not surprising to find gray areas or areas of overlap between families.

Subdivision of the family. In previous editions of *Bergey's Manual* the family was divided into tribes largely on the basis of fermentation of D-glucose by the mixed acid pathway (positive methyl red reaction) or by the 2,3-butanediol pathway (positive Voges-Proskauer reaction), urease and KCN. The use of tribes is of no diagnostic significance and of questionable taxonomic significance. The latter contention is supported by the fact that the tribes listed and the genera included in various tribes changed markedly between Bergey's seventh and eighth editions. In the seventh edition, the tribe *Escherichieae* contained the genera *Escherichia, Enterobacter,* and *Klebsiella,* but the genera *Salmonella* and *Shigella* were in the tribe *Salmonelleae.* In the eighth edition the tribe *Salmonelleae* was deleted, *Klebsiella* was placed in the tribe *Klebsielleae,* and *Salmonella* and *Shigella* were transferred to *Escherichieae.* A further problem with the tribe concept is that the only tribes appearing on the Approved Lists of Bacterial Names (Skerman et al., 1980) are *Erwinieae, Escherichieae, Proteeae, Salmonelleae* and *Serratieae.* The tribe concept, therefore, is not used in the present edition of the *Manual.* Although far from perfect, arbitrary DNA relatedness groupings (Figs. 5.1 and 5.2) approximate evolutionary divergence within the genera.

Further notes. Enterobacteriaceae are distributed worldwide. They are found in soil, water, fruits, vegetables, grains, flowering plants and trees, and in animals from insects to man. Their medical and economic importance, as well as their rapid generation time, ability to grow on defined media, and ease of genetic manipulation have made them the objects of intense laboratory study.

Many species are of considerable economic importance. Erwiniae cause blight, wilt and soft-rot disease in corn, potatoes, pineapples and many other crops, often destroying substantial amounts of the crops (Starr and Chatterjee, 1972). The commercial and tropical fish industries are severely affected by the diseases caused by *Yersinia ruckeri* and species of *Edwardsiella* (Ewing et al., 1978; Shotts and Snieszo, 1976).

Salmonellosis in poultry is a worldwide problem, both for poultry farmers and as a vehicle for human disease (Williams, 1965; Von Rockel, 1965; Hall, 1965). Stillbirths and wool damage in sheep are usually caused by salmonellae (Jensen, 1974). *Escherichia coli* strains that have the K99 colonization factor and produce enterotoxin are primarily responsible for diarrhea in lambs. Enterotoxigenic strains of *E. coli* containing specific colonization factors are also responsible for highly fatal diarrhea in piglets and calves (Bruner and Gillespie, 1973). Klebsiellae and *Citrobacter freundii* cause bovine mastitis. Salmonellosis is also common in pigs, cows, horses, dogs and cats (Barnes and Sorensen, 1975; Ewing, 1969).

Numerous other animal infections are caused by *Enterobacteriaceae*. A few examples are sexually transmitted uterine infections in horses caused by a limited number of capsular types of *Klebiella pneumoniae*; infections in snakes, turtles and lizards caused by salmonellae; diarrheal and septicemic infection in rabbits, other rodents and minks caused by yersiniae; and shigellosis in monkeys. Salmonellae remain the most frequently encountered etiologic agents of food-borne disease.

Until the 1940s only *Salmonella* (including *Arizona*) and *Shigella* were considered as gastrointestinal pathogens. It is now well established that *E. coli* is a significant cause of diarrheal illness both in infants and adults in many areas of the world. Invasive and enterotoxigenic strains of *Yersinia enterocolitica*, apparently restricted to certain serovars ("serotypes"; see preface to the *Manual*, p. xiii), cause diarrhea and mesentery lymphadenitis. Enterotoxigenic strains of *Klebsiella pneumoniae* have frequently been isolated from patients with tropical sprue (Klipstein et al., 1973). Enterotoxin production has also been reported for an occasional strain of *Enterobacter*. Since the enterotoxin genes in *E. coli* are on transmittable plasmids (Smith and Halls, 1968; Gyles et al., 1974), it would not be surprising to find enterotoxin-producing strains in other species of *Enterobacteriaceae*.

Species of *Enterobacteriaceae* not normally associated with diarrheal disease are often referred to as opportunistic pathogens. Most of these species can cause a variety of extraintestinal infections. The compromised host (for example, the malnourished, diabetic, immunosuppressed, catheterized, burn, cancer, respiratory or elderly patient) is vulnerable to nosocomial infections caused by opportunistic pathogens. *Enterobacteriaceae* have been responsible for about 50% of nosocomial infections in the United States (Center for Disease Control, 1977). These infections were most frequently caused by *E. coli, Klebsiella, Enterobacter, Proteus, Providencia,* and *Serratia marcescens*.

Compared with the eighth edition of *Bergey's Manual,* the present volume contains many nomenclatural changes, several new genera, and many new species. Also, in contrast to the eighth edition, type strains have been designated for all species. The main reasons for these changes are: (a) a conservative approach to *Enterobacteriaceae* was taken in the eighth edition, and therefore, descriptions of several known species and several nomenclatural proposals were omitted or only mentioned informally in the text; (b) contributions to the chapter on *Enterobacteriaceae* in the eighth edition were completed before 1970, so more than a decade has elapsed since the last edition; (c) environmental and animal studies have uncovered new species from water, insects, nematodes, plants, fish and small animals; and (d) data from DNA relatedness studies have provided criteria for a species definition that can be used to determine whether any new or biochemically atypical group represents a new species or a biogroup within an existing species.

In Table 5.2 the current classification is compared with that in the eighth edition. There are a number of species for which nomenclatural synonyms exist (Table 5.2). Some of these, *S. paratyphi* B, *S. paratyphi* C, *S. daressalaam, A. hinshawii,* are not on the Approved Lists of Bacterial Names (Approved Lists) (Skerman et al., 1980). The *Salmonella* serovar names have no standing in nomenclature, but continue to be used as an extremely useful form of communication. *Arizona hinshawii* was included on the final list of species sent from the

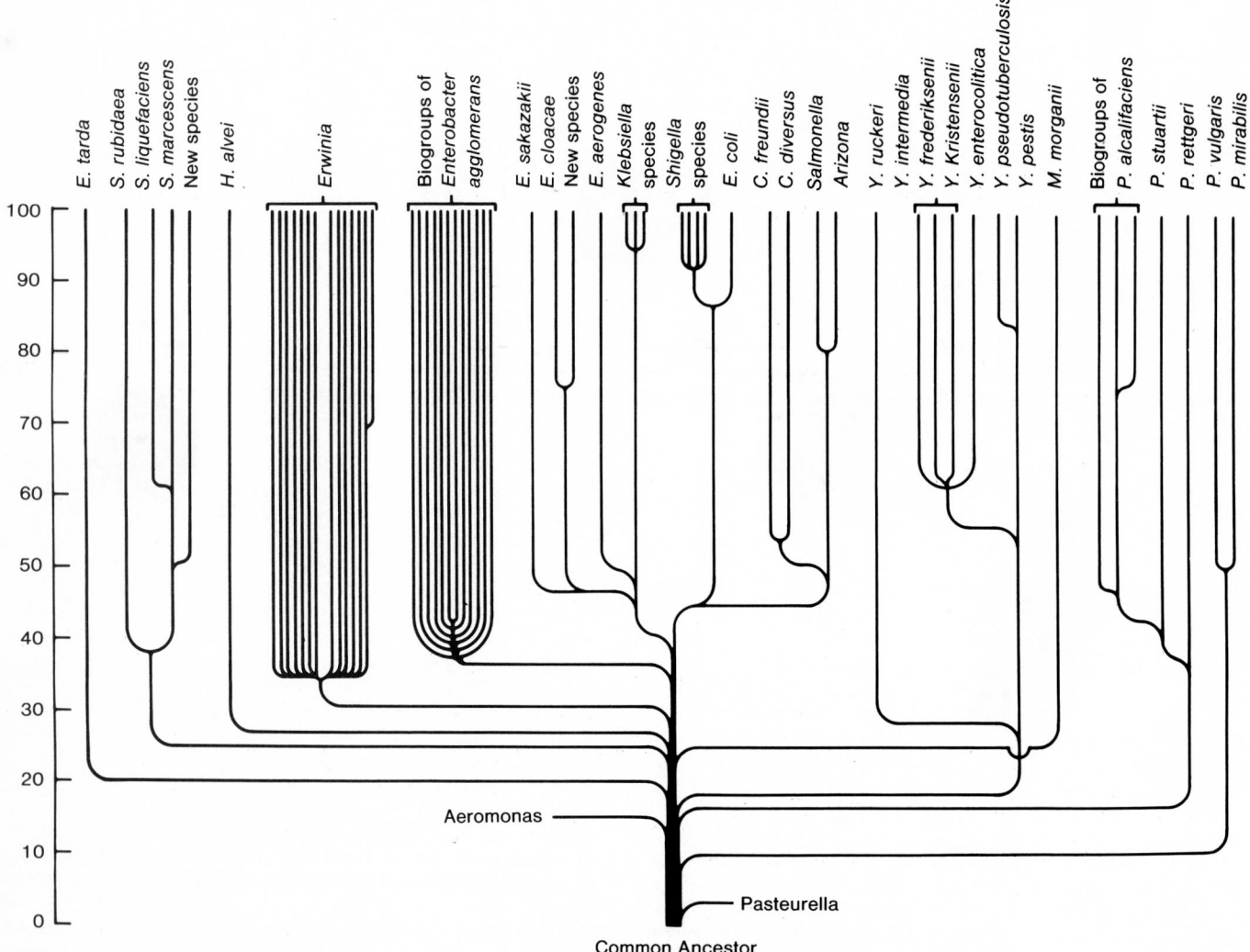

Figure 5.1. Divergence of *Enterobacteriaceae*. The *ordinate* is percentage of relatedness. This figure is a simplified attempt to depict relatedness of each species of enterobacteria to all other species. It assumes a common ancestor from which all of the organisms have diverged. The horizontal branches depict the degree of relatedness of the group of organisms to all organisms that have not yet branched. For example, *E. tarda* is ~20% related to all organisms except *Aeromonas*, *Proteus*, *Providencia* and *Pasteurella*; *Citrobacter* species are ~45% related to all species above them and *C. diversus* and *C. freundii* were speciated at a point in time such that they are now 50% related.

International Committee on Systematic Bacteriology Subcommittee on the Taxonomy of *Enterobacteriaceae* (*Enterobacteriaceae* Subcommittee), but was omitted from the Approved Lists. Therefore, *Arizona hinshawii* no longer has standing in the literature, although it is widely used in the United States and elsewhere. It remains to be seen whether *Arizona hinshawii* will be reproposed (see section on "Taxonomic and Nomenclatural Problems" below).

Taxonomic and Nomenclatural Problems

Escherichia and **Shigella.** The organism named *Escherichia adecarboxylata* (Leclerc, 1962) appears on the Approved Lists. It is negative in decarboxylase reactions and positive in reactions for KCN, malonate, cellobiose, and often urea. It may belong in the *Erwinia herbicola-Erwinia stewartii-Erwinia uredovora-Enterobacter agglomerans* (*Erwinia-E. agglomerans*) complex rather than in the genus *Escherchia* (Bascomb et al., 1971; Ewing and Fife, 1972). DNA relatedness studies will be necessary to resolve the status of *E. adecarboxylata*.

The four species of *Shigella* and *E. coli* are a single species on the basis of DNA relatedness (Brenner et al., 1972, 1973). *Shigella* and *E. coli* strains are often extremely difficult to separate biochemically

because there are aerogenic (gas-producing) shigellae and lactose-negative, anaerogenic, nonmotile *E. coli*. *E. coli* strains can cause a dysentery-like diarrhea, so pathogenicity does not provide definitive separation. Shigellae are actually metabolically inactive biogroups of *E. coli*. It is taxonomically difficult to justify separate genera or even separate species status for these organisms. They remain separate species because of the ease of communication these names provide in medical microbiology and because of the resistance and confusion that would be caused by reclassification. (However, the original usage implied that shigellae were pathogenic and that *E. coli* was not; this is certainly not true). Nonetheless, *Vibrio cholerae* was similarly reclassified, and it is now proposed that *Yersinia pestis* be taxonomically, but not practically, considered as a subspecies of *Y. pseudotuberculosis* (see *Yersinia* below). Perhaps a similar future recommendation will be made for *E. coli* and *Shigella*—namely, that they be a single species with five subgroups for taxonomic purposes, but that they continue to be treated and written as separate genera or species.

Edwardsiella. The name *Edwardsiella anguillimortifera* was proposed as a senior synonym for *Edwardsiella tarda* (Sakazaki and Tamura, 1975). No available strains correspond to the description of

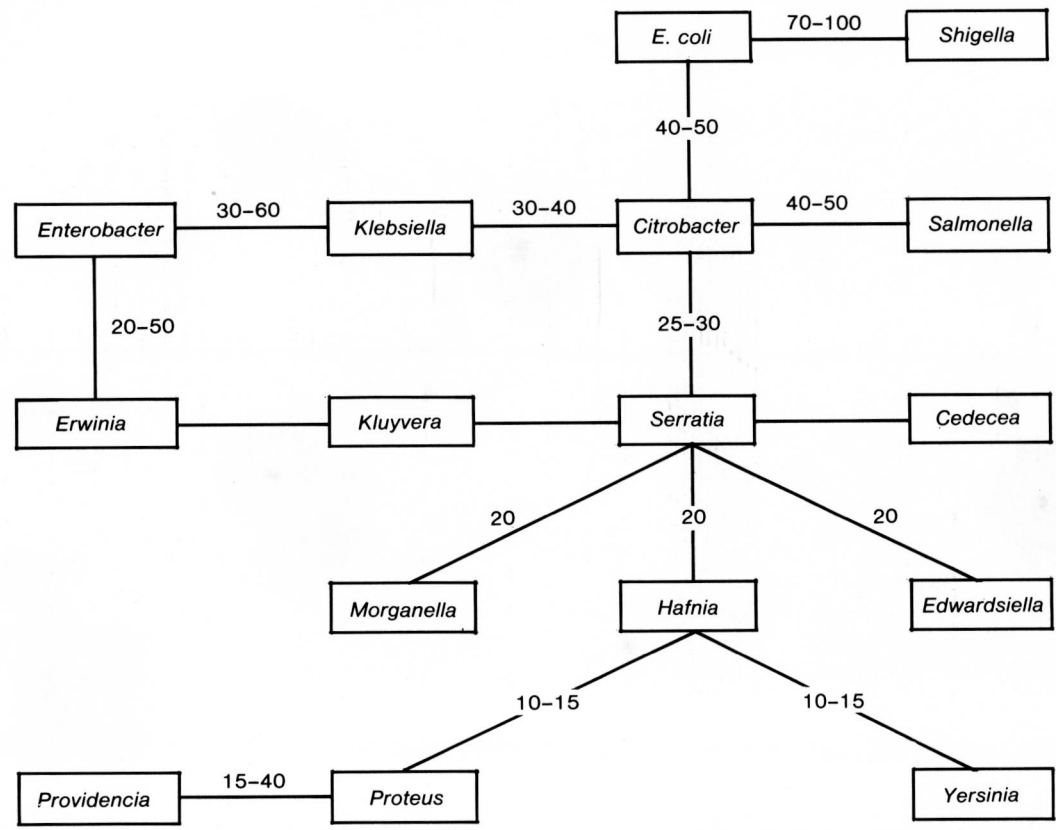

Figure 5.2. DNA relatedness among *Enterobacteriaceae*. The numbers represent the approximate percentage of relatedness.

E. anguillimortifera (which differs from the description of *E. tarda*). This problem is discussed in the chapter on *Edwardsiella*.

Citrobacter. DNA relatedness studies leave little doubt that *Levinea malonatica (Citrobacter diversus)* and *Levinea amalonatica* belong with *Citrobacter freundii* in the genus *Citrobacter* (Crosa et al., 1974), rather than in the genus *Levinea* as proposed by Young et al. (1971). Both *Levinea* species were contained as biogroups of "*Citrobacter intermedius*" in the eighth edition of *Bergey's Manual. C. intermedius* did not have a type strain and does not appear on the Approved Lists. The specific epithets *diversus, koseri* and *malonatica* are admittedly synonyms, are all validly published, and all appear on the Approved Lists. Both *C. diversus* and *C. koseri* (Frederiksen, 1970) have priority over the subjective synonym *L. malonatica. C. diversus* dates back to 1928, whereas *C. koseri* was named in 1970; however, questions have been raised about the correspondence of *C. diversus* to the original description of this organism (Holmes et al., 1974). *Levinea amalonatica* became *Citrobacter amalonaticus* when it was transferred to the genus *Citrobacter* (Brenner and Farmer, 1981; Brenner et al., 1977). This three-species concept is available for those who do not accept *Levinea* as a separate genus (Farmer, 1981).

Salmonella. In the eighth edition of *Bergey's Manual*, Le Minor and Rohde (1974) stated that "scientifically none of the present methods of nomenclature of *Salmonella* is satisfactory" and that "the International *Enterobacteriaceae* Subcommittee has not given clear guidance on the naming of the different types." Unfortunately, these statements remain just as true today as they were more than a decade ago. The use of "species" names for *Salmonella* serovars is extremely useful in many fields. As long as these serovar names are not *taxonomically* equated with species, this practice should be encouraged as stated in the chapter on *Salmonella*. It is the taxonomic treatment of salmonellae that is untenable. DNA relatedness data have shown that representative strains of biotypically typical *Salmonella* serovars (subgenus I),

biochemically atypical *Salmonella* serovars (subgenera II and IV) and *S. arizonae* (subgenus III) belong to a single genetic species (Crosa et al., 1973; Stoleru et al., 1976). Five subgroups were distinguishable within this single genetic *Salmonella* species. They corresponded to *Salmonella* subgenera I; II; III (*S. arizonae* with monophasic flagellar antigens); III (*S. arizonae* with diphasic flagellar antigens); and IV.

The logical classification of salmonellae should therefore be as a single diverse species with five subspecies. An acceptable name would have to be proposed for the single species, and names, as well as type strains, would have to be proposed for each subspecies. Since *S. houtenae* was the type for subgenus IV and *S. salamae* was the type for subgenus II and type strains were designated for each of these (they are not on the Approved Lists), they could be reproposed and serve for two of the subspecies. *S. arizonae* has a type strain with monphasic flagellar antigens which could serve for the monophasic subspecies of what is now subgenus III. One possibility for the subspecies with diphasic flagella would be subspecies "*hinshawii*." A type strain would have to be designated. The remaining problem would be to designate a name and a type species for the subspecies corresponding to subgenus I. This would also be the species name. The oldest serovar appears to be *S. typhi*, which entered the literature in 1886; however, naming the single species after any existing serovar, especially *S. typhi*, would cause massive confusion and would be unwise. Perhaps this dilemma could be solved by following the suggestion of Kauffmann and Edwards (1952) to designate "*S. enterica*" as the single species name. The type strain might then be a strain of serovar *S. typhi*, or, since *S. typhi* is not biochemically typical, a strain of *S. typhimurium*, the most frequently occurring serovar, could serve. Another alternative for the type strain could be a rough strain.

In medial bacteriology Latin binomials would be used for serovars from subgenus I and for the names serovars in subgenera II and IV. Unnamed serovars in subgenera II, III and IV could be listed by

Table 5.2.

Comparison of current classification with that in the eighth edition of Bergey's Manual

Current Classification	Synonyms[a]	Bergey's 8th
Escherichia coli		*E. coli*
E. blattae		NL[b]
Shigella dysenteriae		*S. dysenteriae*
S. flexneri		*S. flexneri*
S. boydii		*S. boydii*
S. sonnei		*S. sonnei*
Edwardsiella tarda	*E. anguillimortifera*	*E. tarda*
E. ictaluri		NL
E. hoshinae		NL
Citrobacter freundii		*C. freundii*
C. diversus	*Levinea malonatica, C. koseri*	*C. intermedius* biogroup b
C. amalonaticus	*Levinea amalonatica*	*C. intermedius* biogroup a
Salmonella choleraesuis		*S. cholerae-suis*
S. hirschfeldii[c]	*S. paratyphi C[c]*	*S. hirschfeldii*
S. typhi		*S. typhi*
S. paratyphi A[c]		*S. paratyphi* A
S. schottmuelleri	*S. paratyphi B[c]*	*S. schottmuelleri*
S. typhimurium		*S. typhimurium*
S. enteritidis		*S. enteritidis*
S. gallinarum[c]		*S. gallinarum*
S. salamae[c]	*S. daressalaam[c]*	*S. salamae*
S. arizonae	*Arizona hinshawii[c]*	*S. arizonae*
S. houtenae[c]		*S. houtenae*
Klebsiella pneumoniae subsp. *pneumoniae*		*K. pneumoniae*
K. pneumoniae subsp. *ozaenae[c]*		*K. ozaenae*
K. pneumoniae subsp. *rhinoscleromatis[c]*		*K. rhinoscleromatis*
K. oxytoca		*K. pneumoniae*, indole-positive biogroup
K. planticola		NL
K. terrigena		NL
Enterobacter cloacae		*E. cloacae*
E. aerogenes	*Klebsiella mobilis*	*E. aerogenes*
E. agglomerans		*Erwinia herbicola,* *Erwinia stewartii,* and *Erwinia uredovora*
E. gergoviae		NL
E. amnigenus		NL
E. sakazakii		yellow-pigmented *E. cloacae*
E. intermedium		NL
Hafnia alvei		*H. alvei*
Serratia marcescens		*S. marcescens*
S. liquefaciens		NL
S. rubidaea	*S. marinorubra*	NL
S. plymuthica		biogroup of *S. marcescens*
S. proteamaculans		NL
S. odorifera		NL
S. fonticola		NL
S. ficaria		NL
Proteus vulgaris		*P. vulgaris*
P. mirabilis		*P. mirabilis*
P. myxofaciens		NL
Providencia alcalifaciens		*Proteus inconstans* biogroup A
P. stuartii		*Proteus inconstans* biogroup B
P. rettgeri		*Proteus rettgeri*
Morganella morganii		*Proteus morganii*
Yersinia pseudotuberculosis		*Y. pseudotuberculosis*
Y. pestis		*Y. pestis*
Y. enterocolitica		*Y. enterocolitica*
Y. ruckeri		NL
Y. intermedia		biogroup of *Y. enterocolitica*
Y. frederiksenii		biogroup of *Y. enterocolitica*
Y. kristensenii		biogroup of *Y. enterocolitica*
Erwinia amylovora		*E. amylovora*
E. salicis		*E. salicis*

Table 5.2—*continued*

Current Classification	Synonyms[a]	Bergey's 8th
E. tracheiphila		*E. tracheiphila*
E. nigrifluens		*E. nigrifluens*
E. quercina		*E. quercina*
E. rubrifaciens		*E. rubrifaciens*
E. herbicola	*Enterobacter agglomerans*	*E. herbicola*
E. stewartii	*Enterobacter agglomerans*	*E. stewartii*
E. uredovora	*Enterobacter agglomerans*	*E. uredovora*
E. carotovora	*Pectobacterium carotovorum*	*E. carotovora*
E. chrysanthemi	*Pectobacterium chrysanthemi*	*E. chrysanthemi*
E. cypripedii	*Pectobacterium cypripedii*	*E. crypripedii*
E. rhapontici	*Pectobacterium rhapontici*	*E. rhapontici*
E. carnegieana	*Pectobacterium carnegieana*	NL
E. mallotivora		NL
Obesumbacterium proteus		NL
Kluyvera ascorbata		NL
K. cryocrescens		NL
Cedecea lapagei		NL
C. davisae		NL
Tatumella ptyseos		NL
Xenorhabdus nematophilus		NL
X. luminescens		NL
Rahnella aquatilis		NL

[a] Synonyms are of several types: objective, subjective, challenged, those that are on the Approved Lists of Bacterial Names (Skerman et al., 1980), and those which no longer have standing in the literature. See text for explanation.

[b] NL, not listed.

[c] Not on Approved Lists of Bacterial Names (Skerman et al., 1980) and have no current standing in nomenclature.

subspecies names followed by the antigenic formula (e.g., *S. salamae* 40:b:-; *S. houtenae* 43:z_{29}:-; *S. arizonae* 5:1,6,7:-; "*S. hinshawii*" 1,4:33:31).

The *Enterobacteriaceae* Subcommittee must reconsider the problems in classification of salmonellae and recommend a solution that is both consistent with the taxonomic data and that will serve the needs of all microbiologists. The suggestions given above are neither original nor formal proposals. They are meant to reiterate the problem and to exemplify a possible solution.

Klebsiella. *K. pneumoniae, K. ozaenae* and *K. rhinoscleromatis* are bio-sero-pathogroups of one genetic species (Brenner et al., 1972). As with shigellae, they have traditionally been separated because of their medical interest, but it would be more accurate taxonomically to treat them as subspecies of *K. pneumoniae*, and this has been done in the present edition of the *Manual*. For medical purposes, however, one would expect them to continue to be reported as if they were separate species. *Klebsiella mobilis* appears on the Approved Lists. It is an objective synonym for *Enterobacter aerogenes* and will be discussed with *Enterobacter*. "*Klebsiella aerogenes*," "*Klebsiella edwardsii*," and "*Klebsiella atlantae*" are usually considered as biogroups of *K. pneumoniae*. They do not have standing in the literature, but are used clinically in Great Britain and in other parts of the former British Commonwealth. Their descriptions are given by Cowan (1974). Strains of *Klebsiella pneumoniae* that are motile and Voges-Proskauer-negative were recently reported by Ferragut and Leclerc (1978). This finding cannot be considered definitive as these strains are no longer motile (D. Izard, personal communication).

Enterobacter. *Erwinia dissolvens* and *Erwinia nimipressuralis*, two species formerly thought to be atypical erwiniae, are closely related to the genus *Enterobacter* both biochemically and by DNA hybridization (Dye, 1969; Steigerwalt et al., 1975). Both are most closely related to *Enterobacter cloacae* and belong in the genus *Enterobacter* either as biogroups of *E. cloacae* or as separate species. Very few strains have been studied; therefore, no formal proposals have been made for their classification (Steigerwalt et al., 1975). A third species, *Erwinia cancerogena*, is also thought to belong to *Enterobacter* (Lelliott, 1974).

Enterobacter aerogenes is biochemically and by DNA hybridization (Brenner et al., 1972; Steigerwalt et al., 1975) equally or more related to klebsiellae than to other *Enterobacter* species. For this reason, there have been attempts to place it in the genus *Klebsiella*. *Klebsiella mobilis* was proposed by Bascomb et al. (1971). *K. mobilis* and *E. aerogenes* have the same type strain and are, by definition, objective synonyms. If *E. aerogenes* is transferred to the genus *Klebsiella*, it would become the new combination "*K. aerogenes*." This name would be most confusing, however, because it formerly was used for some strains of *Klebsiella pneumoniae* (Cowan et al., 1960). At present "*K. aerogenes*" has no standing in the literature (it is not on the Approved Lists), and to repropose it for a different group would only cause confusion. *K. mobilis* is, in a sense, confusing, since there are nonmotile strains of *E. aerogenes*. The use of *K. mobilis* would also cause confusion due to the loss of the epithet "*aerogenes*," which is now well accepted. A decision must be made as to whether the classification of *Enterobacter aerogenes* should remain status quo, whether it should be transferred to *Klebsiella* as *K. mobilis*, or whether it should be transferred to *Klebsiella* with a new species name.

Enterobacter agglomerans was proposed by Ewing and Fife (1971, 1972) for an organism(s) that had been in the literature previously in *Erwinia*, under a large number of species names. They argued that all of the previous names, including *Erwinia herbicola, Erwinia stewartii* and *Erwinia uredovora*, which were listed in the eighth edition and the present volume of *Bergey's Manual*, were synonyms and junior to the description of *Bacillus agglomerans*. They felt that biochemically the organism belonged in the genus *Enterobacter* rather than in the genus *Erwinia*. They further stated that this species, represented by seven anaerogenic and four aerogenic biogroups, might subsequently be transferred to a new genus which, if necessary, could be subdivided into more than one species. DNA hybridization studies (D.J. Brenner, J. Leete, G.R. Fanning, R.G. Steigerwalt and M. Krichevsky, unpublished data) indicate that neither the three-species *Erwinia* scheme nor the 11-biogroup scheme agree with DNA relatedness data. There appear to be 10 or more DNA relatedness groups within the *Erwinia-E. agglomerans* complex. The type strain for *E. agglomerans* is different from the

Table 5.3.

Biochemical identification of **Enterobacteriaceae**[a]

Characteristics	Cedecea davisae	Cedecea lapagei	Citrobacter amalonaticus	Citrobacter diversus	Citrobacter freundii	Edwardsiella hoshinae	Edwardsiella ictaluri	Edwardsiella tarda	Edwardsiella tarda biogroup 1	Enterobacter aerogenes	Enterobacter agglomerans	Enterobacter cloacae	Enterobacter gergoviae	Enterobacter intermedium	Enterobacter sakazakii	Escherichia adecarboxylata	Escherichia blattae	Escherichia coli	Escherichia coli, inactive	Hafnia alvei
Indole production		−	+	+	−	[−]	−	+	+	−	[−]	−	−	−	[−]	+	−	(+)	[+]	−
Methyl red	+	d	+	+	+	+	−	+	+	−	d	−	d	d	[−]	+	+	+	+	d
Voges-Proskauer	[+]	+	−	−	−	−	−	−	−	+	d	+	+	+	+	−	−	−	−	d
Citrate, Simmons'	+	[+]	[+]	+	+	−	−	−	−	+	d	+	+	+	+	−	d	−	−	−
Hydrogen sulfide on TSI	−	−	−	−	[+]	−	−	+	+	−	−	−	−	−	−	−	−	−	−	−
Urease, Christensen's	−	−	[+]	[+]	d	−	−	−	−	−	[−]	d	+	d	−	d	−	−	−	−
Phenylalanine deaminase	−	−	−	−	−	−	−	−	−	−	[−]	−	−	−	d	−	−	−	−	−
Lysine decarboxylase	−	−	−	−	−	+	+	+	+	+	−	−	+	−	−	−	+	[+]	d	+
Arginine dihydrolase	d	[+]	[+]	d	d	−	−	−	−	−	−	+	−	d	+	−	[−]	−	−	−
Ornithine decarboxylase	+	−	+	+	[−]	+	d	+	+	+	−	+	+	+	+	−	+	d	[−]	+
Motility	+	[+]	+	+	+	+	−	+	+	+	[+]	+	+	+	+	+	−	[+]	−	+
Gelatin liquefaction at 22°C	−	−	−	−	−	−	−	−	−	−	−	−	−	−	d	−	−	−	−	−
KCN, growth in	[+]	d	+	−	+	−	−	−	+	d	+	−	+	+	+	−	−	−	−	+
Malonate utilization	[+]	+	[−]	+	[−]	+	−	−	+	d	[+]	+	+	[−]	d	+	−	−	−	d
D-Glucose, acid production	+	+	+	+	+	+	+	+	+	+	+	+	+	+	+	+	+	+	+	+
D-Glucose, gas production	[+]	+	+	+	+	d	d	+	d	+	[−]	+	+	+	+	+	+	+	−	+
Lactose	−	d	d	d	d	−	−	−	+	d	+	d	+	+	+	+	−	+	[−]	−
Sucrose	+	−	[−]	[−]	d	+	−	−	−	+	[+]	+	+	d	+	+	−	d	[−]	−
D-Mannitol	+	+	+	+	+	+	+	+	+	+	+	+	+	+	+	+	+	+	+	+
Dulcitol	−	−	−	d	d	−	−	−	−	[−]	[−]	−	d	−	+	−	d	d	−	−
Salicin	+	+	d	[−]	−	d	−	−	−	d	[+]	+	+	+	+	−	−	d	−	[−]
D-Adonitol	−	−	−	+	−	−	−	−	−	+	−	[−]	−	−	−	+	−	−	−	−
myo-Inositol	−	−	−	−	−	−	−	−	−	+	[−]	[−]	−	−	[+]	−	−	−	−	−
D-Sorbitol	−	−	+	+	+	−	−	−	−	+	d	+	−	+	−	−	−	+	[+]	−
L-Arabinose	−	−	+	+	+	[−]	−	+	+	+	+	+	+	+	+	+	−	+	[+]	+
Raffinose	−	−	−	−	d	−	−	−	−	+	d	+	+	+	+	+	−	d	[−]	−
L-Rhamnose	−	−	+	+	+	−	−	−	−	+	[+]	+	+	+	+	+	+	[+]	d	+
Maltose	+	+	+	+	+	+	+	+	+	+	[+]	+	+	+	+	+	+	+	[+]	+
D-Xylose	+	−	+	+	+	−	−	−	−	+	+	+	+	+	+	+	+	+	d	+
Trehalose	+	+	+	+	+	+	−	−	+	+	+	+	+	+	+	+	[+]	+	+	+
Cellobiose	+	+	+	+	d	−	−	−	−	d	+	+	+	+	+	−	−	−	−	[−]
α-Methyl-D-glucoside	−	−	−	d	−	−	−	−	−	−	[+]	−	d	+	−	−	−	−	−	−
Esculin hydrolysis	d	+	−	−	−	−	−	−	−	d	d	+	d	+	+	−	d	−	−	−
Melibiose	−	−	[−]	−	d	−	−	−	−	+	d	+	d	+	+	−	−	[+]	d	−
D-Arabitol	+	+	−	+	−	−	−	−	−	d	[−]	+	−	−	+	−	−	−	−	−
Mucate	−	−	+	+	+	−	−	−	−	d	[+]	d	−	+	d	+	d	+	d	−
Lipase, corn oil	[+]	+	−	−	−	−	−	−	−	−	−	−	−	−	−	−	−	−	−	−
Deoxyribonuclease at 25°C	−	−	−	−	−	−	−	−	−	−	−	−	−	−	−	−	−	−	−	−
NO₃⁻ → NO₂⁻	+	+	+	+	+	+	+	+	+	+	[+]	+	+	+	+	+	+	+	+	+
Oxidase, Kovacs'	−	−	−	−	−	−	−	−	−	−	−	−	−	−	−	−	−	−	−	−
ONPG (β-galactosidase)	[+]	+	+	+	+	−	−	−	+	+	+	+	+	+	+	+	+	+	d	+
Yellow pigment	−	−	−	−	−	−	−	−	−	−	[+]	−	−	−	+	+	−	−	−	−
D-Mannose	+	+	+	+	+	+	+	+	+	+	+	+	+	+	+	+	+	+	+	+

[a]Symbols: +, 90–100% of strains are positive; [+], 76–89% positive; d, 26–75% positive; [+], 11–25% positive; −, 0–10% positive. Data are calculated for a 48-h incubation period unless otherwise indicated (gelatin liquefaction and deoxyribonuclease). The incubation temperature was $36 \pm 1°C$ for all species except *Yersinia ruckeri* and *Xenorhabdus* species, which were incubated at $25 \pm 1°C$.

Table 5.3—*continued*

Characteristics	Klebsiella oxytoca	Klebsiella pneumoniae	subsp. ozaenae	subsp. pneumoniae	subsp. rhinoscleromatis	Kluyvera ascorbata	Kluyvera cryocrescens	Morganella morganii	Obesumbacterium proteus biogroup 1	Obesumbacterium proteus biogroup 2	Proteus mirabilis	Proteus myxofaciens	Proteus vulgaris	Providencia alcalifaciens	Providencia rettgeri	Providencia stuartii	Rahnella aquatilis	Salmonella I	Salmonella II	Salmonella III = Arizona
Indole production	+	−	−	−		[+]	[+]	+	−	−	−	−	+	+	+	+	−	−	−	−
Methyl red	d	+	[−]	+	+	+	+	+	+	(−)	+	+	+	+	+	+	[+]	+	+	+
Voges-Proskauer	+	−	+	−	−	−	−	−	d	−	[−]	+	−	−	−	−	+	−	−	−
Citrate, Simmons'	+	d	+		−	+	[+]	−	−	−	d	d	[−]	+	+	+	+	+	+	+
Hydrogen sulfide on TSI	−	−	−		−	−	−	−	−	−	+	−	+	−	−	−	−	+	+	+
Urease, Christensen's	+	−	+	−	(−)	(−)	−	+	−	−	+	+	+	(−)	+	d	−	−	−	−
Phenylalanine deaminase	−	−	−	−	−	−	−	+	−	−	+	+	+	(+)	+	+	−	−	−	−
Lysine decarboxylase	+	d	+		+	d		−	+	+	−	−	−	−	−	−	−	+	+	+
Arginine dihydrolase	−	−	−	−	−	−	−	−	−	−	−	−	−	−	−	−	−	d	+	[+]
Ornithine decarboxylase	−	−	−		−	+	+	+	−	+	+	−	−	−	−	−	−	+	+	+
Motility	−	−	−	−	−	+	[+]	+	−	−	+	+	+	(+)	+	[+]	−	+	+	+
Gelatin liquefaction at 22°C	−	−	−	−	−	−	−	−	−	−	+	[+]	+	(−)	−	−	−	−	−	−
KCN, growth in	+	[+]	+	+	+	[+]	+	+	−	−	+	+	+	+	+	+	+	−	−	−
Malonate utilization	+	−	+	+	+	[+]	+	−	−	−	−	−	−	−	−	−	+	−	+	+
D-Glucose, acid production	+	+	+	+	+	+	+	+	+	+	+	+	+	+	+	+	+	+	+	+
D-Glucose, gas production	+	d	+	−	+	+	+	[+]	−	−	+	+	[+]	[+]	−	−	[+]	+	+	+
Lactose	+	d	+	−	+	+	+	−	−	−	−	−	−	−	−	−	+	−	−	d
Sucrose	+	[−]	+	[+]	+	[+]	+	−	−	−	[−]	+	+	[−]	[−]	d	+	−	−	−
D-Mannitol	+	+	+	+	+	+	+	−	−	−	−	−	−	+	+	[−]	+	+	+	+
Dulcitol	d	−	d		[−]	−	−	−	−	−	−	−	−	−	−	−	[+]	+	+	−
Salicin	+	+	+	+	+	+	+	−	−	−	−	[+]	−	d	−	+	−	−	−	−
D-Adonitol	+	+	+	+	−	−	−	−	−	−	−	−	−	+	+	−	−	−	−	−
myo-Inositol	+	d	+	+	−	−	−	−	−	−	−	−	−	−	+	+	−	d	[−]	−
D-Sorbitol	+	[+]	+	+	d	d		−	−	−	−	−	−	−	−	−	+	+	+	+
L-Arabinose	+	+	+	+	+	+		−	−	−	−	−	−	−	+	−	+	+	+	+
Raffinose	+	+	+	[+]	+	+		−	−	−	−	−	−	−	−	−	−	−	−	−
L-Rhamnose	+	d	+	+	+	+		−	d		−	−	−	d	−	−	+	+	+	+
Maltose	+	+	+	+	+	+		−	d		−	+	+	−	−	−	+	+	+	+
D-Xylose	+	+	+	+	+	+		−	d		+	−	+	−	−	−	+	+	+	+
Trehalose	+	+	+	+	+	+		[−]	+	+	+	+	d	−	−	+	+	+	+	+
Cellobiose	+	+	+	+	+	+		−	−	−	−	−	−	−	−	+	+	−	−	−
α-Methyl-D-glucoside	+	d	[+]	−	+	+		−	−	−	−	+	[+]	−	−	−	+	−	−	−
Esculin hydrolysis	+	[+]	+	d	+	+		−	−	−	−	−	[+]	−	d	−	+	−	−	−
Melibiose	+	+	+	+	+	+		−	−	−	−	−	−	−	−	−	+	+	−	+
D-Arabitol	+	+	+	+	−	−	−	−	−	−	−	−	−	−	+	−	−	−	−	−
Mucate	+	[−]	[+]	−	+	[+]		−	−	−	−	−	−	−	−	−	−	d	+	d
Lipase, corn oil	−	−	−	−	−	−	−	−	−	−	+	+	[+]	−	−	−	−	−	−	−
Deoxyribonuclease at 25°C	−	−	−	−	−	−	−	−	−	−	−	−	d	−	−	−	−	−	−	−
NO3− → NO2−	+	[+]	+	+	+	+	[+]	+	+		+	+	+	(+)	+	+	+	+	+	+
Oxidase, Kovacs'	−	−	−	−	−	−	−	−	−	−	−	−	−	(−)	−	−	−	−	−	−
ONPG (β-galactosidase)	+	[+]	+		−	+	+	−	−	−	−	−	−	−	−	−	+	−	d	+
Yellow pigment	−	−	−	−	−	−	−	−	−	−	−	−	−	−	−	−	+	−	−	−
D-Mannose	+	+	+	+	+	+	+	+	[+]	−	−	−	−	(+)	+	+	+	+	+	+

[a]Symbols: +, 90–100% of strains are positive; [+], 76–89% positive; d, 26–75% positive; [−], 11–25% positive; −, 0–10% positive. Data are calculated for a 48-h incubation period unless otherwise indicated (gelatin liquefaction and deoxyribonuclease). The incubation temperature was 36 ± 1°C for all species except *Yersinia ruckeri* and *Xenorhabdus* species, which were incubated at 25 ± 1°C.

Table 5.3—*continued*

Characteristics	*Salmonella IV* 41	*Salmonella choleraesuis* 42	*Salmonella gallinarum* 43	*Salmonella paratyphi A* 44	*Salmonella pullorum* 45	*Salmonella typhi* 46	*Serratia ficaria* 47	*Serratia fonticola* 48	*Serratia liquefaciens* 49	*Serratia marcescens* 50	*Serratia odorifera* 51	*Serratia plymuthica* 52	*Serratia rubidaea* 53	*Shigella boydii* 54	*Shigella dysenteriae* 55	*Shigella flexneri* 56	*Shigella sonnei* 57	*Tatumella ptyseos* 58	*Yersinia enterocolitica* 59
Indole production	−	−	−	−	−	−				d	−	−	−	[−]	d	d	−	−	d
Methyl red	+	+	+	+	+	+	[+]	+	[+]	[−]	[+]	+	[−]	+	+	+	+	−	+
Voges-Proskauer	−	−	−	−	−	−	[+]	−	[+]	+	[+]	d	+	−	−	−	−	[−]	−
Citrate, Simmons'	+	[−]	−	−	−	−	+	+	+	+	+	d	+	−	−	−	−	−	−
Hydrogen sulfide on TSI	+	d	+	−	+	+	−	−	−	−	−	−	−	−	−	−	−	−	−
Urease, Christensen's	−	−	−	−	−	−	[−]	−	[−]	−	−	−	−	−	−	−	−	−	[+]
Phenylalanine deaminase	−	−	−	−	−	−	−	−	−	−	−	−	−	−	−	−	−	[+]	−
Lysine decarboxylase	+	+	+	−	+	+	−	+	+	+	+	−	d	−	−	−	−	−	−
Arginine dihydrolase	d	d	−	[−]	d	−	−	−	−	−	−	−	[−]	−	−	−	−	−	−
Ornithine decarboxylase	+	+	−	+	+	−	−	+	+	+	d	−	−	−	−	−	+	−	+
Motility	+	+	−	+	−	+	+	+	+	+	+	d	[+]	−	−	−	−	−	−
Gelatin liquefaction at 22°C	−	−	−	−	−	−	+	−	+	+	+	d	+	−	−	−	−	−	−
KCN, growth in	+	−	−	−	−	−	d	d	+	+	d	d	[−]	−	−	−	−	−	−
Malonate utilization	−	−	−	−	−	−	−	+	−	−	−	−	[+]	−	−	−	−	−	−
D-Glucose, acid production	+	+	+	+	+	+	+	+	+	+	+	+	+	+	+	+	+	+	+
D-Glucose, gas production	+	+	−	+	[+]	−	−	[+]	d	d	−	d	d	−	−	−	−	−	−
Lactose	−	−	−	−	−	−	[−]	+	−	−	+	[+]	+	−	−	−	−	−	−
Sucrose	−	−	−	−	−	−	+	[−]	+	+	d	+	+	−	−	−	−	+	+
D-Mannitol	+	+	+	+	+	+	+	+	+	+	+	+	+	+	−	+	+	−	+
Dulcitol	−	−	+	+	−	−	−	+	−	−	−	−	−	−	−	−	−	−	−
Salicin	d	−	−	−	−	+	+	+	+	d	+	+	−	−	−	−	−	d	d
D-Adonitol	−	−	−	−	−	−	+	−	d	d	−	+	−	−	−	−	−	−	−
myo-Inositol	d	−	−	−	−	−	d	d	d	[+]	+	d	[−]	−	−	−	−	−	d
D-Sorbitol	+	[+]	−	+	[−]	+	+	+	[+]	+	+	d	−	d	d	d	−	−	+
L-Arabinose	+	−	[+]	+	+	−	+	+	+	−	+	+	+	+	d	d	+	−	+
Raffinose	−	−	−	−	−	−	d	+	+	−	d	+	+	−	−	d	−	[−]	−
L-Rhamnose	+	+	−	+	+	−	d	d	[−]	−	+	−	−	−	d	−	[+]	−	−
Maltose	+	+	+	+	−	+	+	+	+	+	+	+	+	[−]	[−]	d	+	−	d
D-Xylose	+	+	d	−	[+]	[+]	+	[+]	+	−	+	+	+	[−]	−	−	−	−	d
Trehalose	+	−	d	+	[+]	+	+	+	+	+	+	+	+	[+]	[+]	d	+	+	+
Cellobiose	[−]	−	−	[−]	−	−	+	−	[−]	[−]	+	[+]	+	−	−	−	−	−	[+]
α-Methyl-D-glucoside	−	−	−	−	−	−	−	+	−	−	−	d	+	−	−	−	−	−	−
Esculin hydrolysis	−	−	−	−	−	−	+	+	+	+	d	[+]	+	−	−	−	−	−	[−]
Melibiose	+	d	−	+	−	+	d	+	[+]	−	+	+	+	[−]	−	d	[−]	d	−
D-Arabitol	−	−	−	−	−	−	+	+	−	−	−	−	+	−	−	−	−	−	d
Mucate	−	−	d	−	−	−	−	−	−	−	−	−	−	−	−	−	−	−	−
Lipase, corn oil	−	−	−	−	−	−	[+]	−	[+]	+	d	d	+	−	−	−	−	−	d
Deoxyribonuclease at 25°C	−	−	−	−	−	−	+	−	+	+	+	+	+	−	−	−	−	−	−
NO$_3^-$ → NO$_2^-$	+	+	+	+	+	+	+	+	+	+	+	+	+	+	+	+	+	+	+
Oxidase, Kovacs'	−	−	−	−	−	−	−	−	−	−	−	−	−	−	−	−	−	−	−
ONPG (β-galactosidase)	−	−	−	−	−	−	+	+	+	+	+	[+]	+	−	d	−	[+]	−	+
Yellow pigment	−	−	−	−	−	−	−	−	−	−	−	−	−	−	−	−	−	−	−
D-Mannose	+	[+]	+	+	+	+	+	+	+	+	+	+	+	+	+	+	+	+	+

[a]Symbols: +, 90–100% of strains are positive; [+], 76–89% positive; d, 26–75% positive; [−], 11–25% positive; −, 0–10% positive. Data are calculated for a 48-h incubation period unless otherwise indicated (gelatin liquefaction and deoxyribonuclease). The incubation temperature was 36 ± 1°C for all species except *Yersinia ruckeri* and *Xenorhabdus* species, which were incubated at 25 ± 1°C.

Table 5.3—*continued*

Characteristics	*Yersinia frederiksenii* 60	*Yersinia intermedia* 61	*Yersinia kristensenii* 62	*Yersinia pestis* 63	*Yersinia pseudotuberculosis* 64	*Yersinia ruckeri* 65	*Xenorhabdus luminescens* 66	*Xenorhabdus nematophilus* 67
Indole production	+	+	d	−	−	−	d	d
Methyl red	+	+	+	[+]	+	+	−	−
Voges-Proskauer	−	−	−	−	−	−	−	−
Citrate, Simmons'	[−]	−	−	−	−	−	d	−
Hydrogen sulfide on TSI	−	−	−	−	−	−	−	−
Urease, Christensen's	[+]	[+]	+	−	+	−	[−]	−
Phenylalanine deaminase	−	−	−	−	−	−	−	−
Lysine decarboxylase	−	−	−	−	−	[+]	−	−
Arginine dihydrolase	−	−	−	−	−	−	−	−
Ornithine decarboxylase	[+]	+	+	−	−	+	−	−
Motility	−	−	−	−	−	[+]	+	+
Gelatin liquefaction at 22°C	−	−	−	−	−	d	d	[+]
KCN, growth in	−	−	−	−	−	d	−	−
Malonate utilization	−	−	−	−	−	−	−	−
D-Glucose, acid production	+	+	+	+	+	+	[+]	[+]
D-Glucose, gas production	d	[−]	d	−	−	[−]	−	−
Lactose	d	d	d	−	−	−	−	−
Sucrose	+	+	−	−	−	−	−	−
D-Mannitol	+	+	+	+	+	+	−	−
Dulcitol	−	−	−	−	−	−	−	−
Salicin	[+]	+	d	d	[−]	−	−	−
D-Adonitol	−	−	−	−	−	−	−	−
myo-Inositol	[−]	[−]	d	−	−	−	−	−
D-Sorbitol	+	+	+	−	−	−	−	−
L-Arabinose	+	+	+	+	d	−	−	−
Raffinose	d	d	−	−	[−]	−	−	−
L-Rhamnose	+	+	+	−	+	−	−	−
Maltose	+	+	+	[+]	+	+	[−]	−
D-Xylose	+	+	+	+	+	−	−	−
Trehalose	−	+	+	+	+	+	−	−
Cellobiose	+	+	+	−	−	−	−	−
α-Methyl-D-glucoside	−	[+]	−	−	−	−	−	−
Esculin hydrolysis	[+]	+	−	[+]	+	−	−	−
Melibiose	−	[+]	−	d	+	−	−	−
D-Arabitol	+	d	d	−	−	−	−	−
Mucate	−	−	−	−	−	−	−	−
Lipase, corn oil	d	[−]	−	−	−	d	−	−
Deoxyribonuclease at 25°C	−	−	−	−	−	−	−	[−]
NO₃⁻ → NO₂⁻	+	+	+	[+]	+	[+]	−	[−]
Oxidase, Kovacs'	−	−	−	−	−	−	−	−
ONPG (β-galactosidase)	+	+	+	[+]	d	d	−	−
Yellow pigment	−	−	−	−	−	−	d	d
D-Mannose	+	+	+	+	+	+	+	[+]

[a]Symbols: +, 90–100% of strains are positive; [+], 76–89% positive; d, 26–75% positive; [−], 11–25% positive; −, 0–10% positive. Data are calculated for a 48-h incubation period unless otherwise indicated (gelatin liquefaction and deoxyribonuclease). The incubation temperature was 36 ± 1°C for all species except *Yersinia ruckeri* and *Xenorhabdus* species, which were incubated at 25 ± 1°C.

type strain for *E. herbicola*, *E. stewartii* and *E. uredovora*; therefore, there is a scientific as well as a nomenclatural problem. For this reason, the *Enterobacteriaceae* Subcommittee, without prejudice, put all four names on the Approved Lists. Three additional names on the Approved Lists, *Escherichia adecarboxylata*, *Erwinia ananas* (now a variety of *Erwinia herbicola*) and *Erwinia milletiae*, may also represent different species now included in this very heterogeneous group. Further studies

are required before a final determination can be made with respect to classification and nomenclature. At present, the phytopathologists are comfortable with the three-species concept and many medical bacteriologists use *E. agglomerans*—thus, the use of both systems in this volume.

Hafnia. DNA hybridization studies on *H. alvei* strains revealed two separate relatedness groups (Steigerwalt et al., 1975). A second species

Table 5.4.
Additional biochemical reactions of **Enterobacteriaceae**[a]

Species	Nitrate Reductase	Tetrathionate Reductase	Galacturonate	2-Keto-gluconate	γ-Glutamyl-transferase
Citrobacter amalonaticus (*Levinea amalonatica*)		+		+	+
Citrobacter diversus (*C. koseri, L. malonatica*)		−		+	+
Citrobacter freundii	+A	+	+	+	+
Edwardsiella hoshinae		−			
Edwardsiella tarda	+B	+		−	−
Erwinia carotovora	+A	−	+	−	+
Enterobacter aerogenes	+A	−	+	+	+
Enterobacter agglomerans	+A	−	+	+	+
Enterobacter cloacae	+A	−	d	+	+
Enterobacter gergoviae				+	+
Enterobacter sakazakii				+	
Escherichia coli	+A	−		−	+
Hafnia alvei	dA or B	d	+	+	+
Klebsiella oxytoca	dA	d		+	+
Klebsiella ozaenae		−		+	+
Klebsiella pneumoniae	+A	−		+	+
Klebsiella rhinoscleromatis		−		d	−
Morganella morganii	dA	+		−	+
Proteus mirabilis	+A	+		−	+
Proteus vulgaris	+A	+		−	+
Providencia alcalifaciens	+B	+		−	+
Providencia rettgeri	+A	+		d	+
Providencia stuartii	+A	+		−	+
Salmonella subgenus I	+A	+	−	−	+
Salmonella subgenus II	+A	+	+	−	+
Salmonella subgenus III monophasic	+A	+	−		
Salmonella subgenus III diphasic	+A	+	+	−	+
Salmonella subgenus IV	+A	+	+	−	+
Serratia ficaria				+	+
Serratia liquefaciens	+A	+	+	+	+
Serratia marcescens	+A	d	+	+	+
Serratia odorifera	+A	−		+	+
Serratia plymuthica	+A	−	+	+	+
Serratia rubidaea (*S. marinorubra*)	+A	−	+	+	+
Shigella boydii	+A	−		−	d
Shigella dysenteriae		−		−	d
Shigella flexneri	+A	−		−	d
Shigella sonnei	+A	−		−	−
Yersinia enterocolitica	+B	d		d	+
Yersinia pestis	dB	d		−	−
Yersinia pseudotuberculosis	+B	−		−	+
Yersinia frederiksenii	+B	+			
Yersinia intermedia	+A or B	+			
Yersinia kristensenii	+B	d			

[a] Symbols: +, 90% or more of strains are positive; d, 10.0%–89.9% positive; −, 0–9.9% positive; blank space, data not reported or not available; A, type A; B, type B. The nitrate reductase test was incubated at 32°C; all other tests were incubated at 35–37°C. The γ-glutamyltransferase test was read at 24 h; all other tests results were read at 48 h. Data compiled from the following references: Pichinoty et al. (1969); Le Minor et al. (1979); Richard (1977); Giammanco et al. (1980); Buissière et al. (1981); Grimont et al. (1978); Grimont (1977); Bercovier et al. (1980); Bercovier et al. (1980); Bercovier et al. (1980); Brenner et al. (1980); and Ursing et al. (1980).

was not designated because no single biochemical test or series of tests served to unequivocally separate the two DNA relatedness groups (F.W. Hickman and J.J. Farmer, III, personal communication). *Obesumbacterium proteus* ("*Hafnia protea*") is considered in the section on "Other Genera."

Serratia. *Serratia rubidaea* and *Serratia marinorubra* are subjective synonyms; both appear on the Approved Lists, but have different type strains (the type for *S. marinorubra* is ATCC 27614, not ATCC 27593 as incorrectly shown on the Approved Lists because of an error by the *Enterobacteriaceae* Subcommittee). The *Enterobacteriaceae* Subcommittee must address this problem. *Serratia proteamaculans* was proposed by Grimont and Starr (1978) as a senior synonym for *Serratia liquefaciens*. Holmes (1980) then requested that the epithet *liquefaciens* be conserved over *proteamaculans* because of its worldwide acceptance. This controversy may resolve itself because Grimont et al. (1981) have studied additional strains and have concluded that *S. proteamaculans* is a species distinct from *S. liquefaciens* (see *Serratia* chapter). *Serratia plymuthica* had species status in the seventh edition of *Bergey's Manual* (Breed and Murray, 1957), but was considered a biogroup of *S. marcescens* in the eighth edition (Sakazaki, 1974). DNA relatedness studies have now shown *S. plymuthica* to be a separate species (Grimont et al., 1978).

Proteus, Providencia, and Morganella. Biochemical, serologic, guanine plus cytosine content, and DNA relatedness data precipitated several nomenclatural and taxonomic changes (Brenner et al., 1978) in the genus *Proteus* as constituted in the eighth edition of *Bergey's Manual* (Lautrop, 1974). *Proteus morganii* was moved to a new genus, *Morganella*, which on the basis of mol% G + C content and DNA relatedness showed a closer relationship to other genera of *Enterobacteriaceae* than to *Proteus* or *Providencia*. *Proteus inconstans* was moved to the genus *Providencia* as two species, *Providencia alcalifaciens* (*P. inconstans* biogroup A) and *Providencia stuartii* (*P. inconstans* biogroup B). *Proteus rettgeri* was transferred to the genus *Providencia*. The names *Proteus morganii* and *Proteus rettgeri* appear on the Approved Lists. *Proteus myxofaciens*, thought not to be a *Proteus* species in the eighth edition (Lautrop, 1974), has been shown to be a valid and separate species in the genus *Proteus* (Brenner et al., 1978). There is only one available strain of *P. myxofaciens* and, therefore, most people have no familiarity with this species.

Yersinia. Three groups of *Yersinia* previously considered as either biochemically atypical *Y. enterocolitica* or *Y. enterocolitica*-like strains have now been speciated. The new species, *Yersinia intermedia*, *Yersinia frederiksenii* and *Yersinia kristensenii* are separable from *Y. enterocolitica* and from each other by their fermentation reactions for L-rhamnose, raffinose, melibiose and sucrose (Brenner et al., 1980; Ursing et al., 1980; Bercovier et al., 1980). *Y. pestis* and *Y. pseudotuberculosis* were shown to be a single species (Bercovier et al., 1980). It was proposed that they be referred to as *Y. pseudotuberculosis* subsp. *pseudotuberculosis* and *Y. pseudotuberculosis* subsp. *pestis* for taxonic purposes and be written as separate species for medical purposes. This presents a problem because *Y. pestis* is the type species of *Yersinia*. *Yersinia ruckeri*, a fish pathogen, was included in the genus *Yersinia* as an alternative to creating a new genus for this organism (Ewing et al., 1978). "*Yersinia*" *philomiragia* appears on the Approved Lists. This species, first proposed in 1969 (Jensen et al., 1969), was not mentioned in the eighth edition of *Bergey's Manual*. Recent work indicates that it is not a member of *Yersinia* or of the family *Enterobacteriaceae* (Ursing et al., 1980).

Erwinia. The taxonomic problems with *Erwinia herbicola*, *Erwinia stewartii*, *Erwinia uredovora* and *Enterobacter agglomerans* have already been discussed under *Enterobacter*. Also discussed under *Enterobacter* were "*Erwinia*" *dissolvens*, "*Erwinia*" *nimipressuralis* and "*Erwinia*" *cancerogena*. *Erwinia paradisiaca* appears on the Approved Lists. Very little is known about this organism, and strains have not been readily available. All of the so-called soft-rot or Carotovora group

appear on the Approved Lists as both *Erwinia* species and as *Pectobacterium* species. Plant pathologists prefer not to split erwiniae at the genus level. The *Enterobacteriaceae* Subcommittee is expected to come to a similar conclusion, and, thus the genus *Pectobacterium* will not be used.

Erwiniae have been mainly studied by phytomicrobiologists and phytopathologists. The media, biochemical and other phenotypic tests used for their isolation, enrichment cultivation and identification are quite different from those used for other *Enterobacteriaceae*. The 37°C incubation temperature used for other *Enterobacteriaceae* is near, at, or above the maximum growth temperatures of erwiniae (excluding the Herbicola group). The isolation of erwiniae from humans or animals is rarely reported. It is not known, however, if they are actually rarely seen or whether their seeming lack of occurrence reflects improper isolation and enrichment procedures, or, if they are isolated, failure to identify them. A study using optimum isolation procedures and an optimum incubation temperature would help resolve this problem. Also needed is a large-scale characterization study of all *Erwinia* species by tests and methods used for other *Enterobacteriaceae*.

DNA relatedness data (Gardner and Kado, 1972; Brenner et al., 1973; Brenner et al., 1974) and phenotypic characteristics indicate the existence of extreme heterogeneity among *Erwinia* species; several species are more closely related to members of *Enterobacter* than to other erwiniae. Starr and Chatterjee (1972), in considering this heterogeneity, reviewed and espoused the possibility of reclassifying erwiniae into one or more existing genera in *Enterobacteriaceae*.

***Enterobacteriaceae** and its type genus.* Rule 21a of the *International Code of Nomenclature of Bacteria* requires that family names be formed by adding "aceae" to the stem of the type genus (Lapage et al., 1975). Rahn proposed the name *Enterobacteriaceae* before there was an international code. When the first code was written in 1948, *Enterobacteriaceae* Rahn became illegitimate, not because of its ending, but because it was not in accord with several provisions of the Code. *Enterobacteriaceae* had become so widely used and accepted that in 1958 the Judicial Commission voted to conserve the name and to designate *Escherichia* as the type genus of the family (Judicial Commission, 1958). This Judicial Commission ruling (Opinion 15) was incorporated into the 1958 version of the code and remains in the current 1975 version (Lapage et al., 1975).

In 1978 (International Committee on Systematic Bacteriology, 1979) the Judicial Commission voted to change the family name to "*Enterobacteraceae*" (no "i") and to change the type genus of the family to *Enterobacter*. Several arguments have been raised in opposition to changing the family name. These are based upon the principle of nomenclatural stability, an apparent conflict with the letter and the spirit of several rules in the *International Code of Nomenclature of Bacteria*, and upon the question of whether the Judicial Commission action was procedurally legitimate. These objections have been presented in detail in the *International Journal of Systematic Bacteriology* (Farmer et al., 1980).

A further, more acute problem arose when neither *Enterobacteriaceae* nor "*Enterobacteraceae*" appeared in the body of the Approved Lists (*Enterobacteriaceae* was mentioned in a footnote). Some interpreted this omission to mean that *Enterobacteriaceae* Rahn 1937 as conserved by Judicial Commission Opinion 15 in 1958 had no standing in the literature and that the family was without a name. The name *Enterobacteriaceae* was therefore reproposed (Ewing et al., 1980). The Judicial Commission recently reviewed the present status of the name and concluded that *Enterobacteriaceae* Rahn 1937 is presently valid (Judicial Commission of the International Committee on Systematic Bacteriology, 1981). This dispute will no doubt have to be settled by a decision from the Judicial Commission.

Biochemical Identification of **Enterobacteriaceae**

Enteric Section,* Centers for Disease Control. People tend to accept without question the reactions given in biochemical charts for various

* Mary Alyce Asbury, Don J. Brenner, Geraldine H. Carter, Betty R. Davis, C. Elais, J.J. Farmer, III, F.W. Hickman, Alma M. McWhorter, Conradine Riddle, and H.G. Wathen. Data compiled and tabulated by J.J. Farmer, III, C. Elais, and W.H. Ewing; section written by Don J. Brenner.

species. To do so, especially in regard to *Enterobacteriaceae*, is a dangerous practice. Results vary according to a number of parameters, some of which are size of inoculum, incubation temperature, duration of incubation, composition and volume of media, test method, criteria for judging a test positive, and the environment from which the tested strains were obtained. Many of these parameters are often not given. Biochemical reactions for *Enterobacteriaceae* are presented in Table 5.3. In an effort to be at least semiquantitative, symbols are given for five (rather than three) percentage ranges. All species were studied in a single laboratory (the Enteric Section at the Centers for Disease Control) by a single set of described methods (Edwards and Ewing, 1972; Hickman and Farmer, 1978). For some species, where the Enteric Section data were biased due to a large number of biochemically atypical strains, the percentages were adjusted to more accurately reflect the percentages expected from a more representative strain sample. Because different methods and tests are often used, the percentage of positive reactions obtained may differ somewhat from those presented in the chapters on specific genera. For example, reactions on *Yersinia* are often done at 28°C. The purpose of Table 5.3 is not to advocate any given test or to put undue emphasis on the percentages obtained, but to present a comprehensive comparison derived from a single set of data obtained by tests commonly done in a diagnostic laboratory. *Erwinia* species and certain of the newly described species are not included because data on these species are insufficient. It must be emphasized that the data in Table 5.3 were obtained at 36 ± 1°C after *48 h of incubation*. For example, *Yersinia enterocolitica* strains are more than 99% positive for urea, but, in our hands, *at 48 h*, slightly less than 90% are positive.

From 50 to more than 200 biochemical tests have been used in phenetic or numerical taxonomic studies of *Enterobacteriaceae* (Bascomb et al., 1971; Johnson et al., 1975; Véron, 1975; Véron and Le Minor, 1975a; Véron and Le Minor, 1975b). These include tests for the fermentation of a large number of carbohydrates and polyhydroxyl alcohols; tests for the ability to use a wide variety of organic substrates as the sole source of carbon and energy, and tests for the presence of specific enzymes. A number of these tests are useful for the differentiation of species or biogroups within *Enterobacteriaceae* (Véron and Le Minor, 1975a; Véron and Le Minor, 1975b). Some tests of particular

diagnostic value are for nitrate reductase type A or type B (Pichinoty and Piéchaud, 1968; Pichinoty et al., 1969), tetrathionate reductase (Richard, 1977), fermentation or growth on sodium galacturonate (Le Minor, Buissière and Brault, 1979), presence of α-glutamyltransferase (Giammanco et al., 1980), and fermentation or growth on 2-ketogluconate (Buissiere et al., 1981). A summary of data obtained by the Institut Pasteur group for these tests is given in Table 5.4.

Enterobacteriaceae will soon contain some 20 genera with more than 100 species. To identify the new species it will often be necessary to use tests that are not now used routinely. Furthermore, it is increasingly difficult and risky to identify a strain on the basis of a small number of biochemical characteristics. Each laboratory, depending on its area of specialization, must make several basic decisions. The first is which "nonroutine" tests to add for routine use or for use in special cases. A second decision is whether to speciate in all cases, a third is which species to ignore, and an important corollary of these is what percentage of incorrect identification a laboratory is willing to tolerate. For example, current knowledge indicates that a brewery laboratory must be concerned with *Obesumbacterium*, but probably not *Xenorhabdus*; that fishery laboratories must cope with all *Edwardsiella* species and *Yersinia ruckeri*, but not *Erwinia*; that agricultural laboratories must be aware of *Serratia ficaria* and *Klebsiella*, but not *Edwardsiella*; etc. These are rules of thumb, not absolutes, that allow a laboratory to decrease its work load without significantly decreasing its efficiency.

If there is a certainty with respect to *Entérobactériaceae* it is that the family will continue to be dynamic and will continue to pose a challenge to microbiologists in all specialties.

Acknowledgments

I am extremely grateful to my colleagues in the Enteric Section: Mary Alyce Asbury, Geraldine A. Carter, Betty R. Davis, J. J. Farmer, III, Frances W. Hickman, Alma C. Murlin, Conradine F. Riddle, Arnold G. Steigerwalt and H. Gail Wathen; and to Leon Le Minor, Service des Entérobactéries, Institut Pasteur, Paris, for providing unpublished data. I am indebted to J. J. Farmer, III, and Leon Le Minor for critically reviewing this manuscript. The literature search for this manuscript was concluded on December 30, 1980.

Genus I. **Escherichia** Castellani and Chalmers 1919, 941[AL]*

Frits Ørskov

Esch.er.i′chi.a. M.L. fem. n. *Escherichia* named after Theodor Escherich, who isolated the type species of the genus.

Straight rods, 1.1–1.5 μm × 2.0–6.0 μm, occurring singly or in pairs. Capsules or microcapsules occur in many strains. Gram-negative. **Motile by peritrichous flagella or nonmotile.** Facultatively anaerobic, having both a respiratory and a fermentative type of metabolism. The remainder of the description is restricted to *E. coli* because *E. blattae* is not well studied and only a few strains exist. Optimum temperature, 37°C. Colonies on nutrient agar may be smooth (S), low convex, moist, gray, with a shiny surface and entire edge and easily dispersible in saline, or they may be rough (R), dry and difficult to disperse well in saline. There are intermediate forms between these extremes. Mucoid and slime-producing forms occur. Chemoorganotrophic. Oxidase-negative. Acetate can usually be used as a sole carbon source, **but citrate cannot be used. Glucose and other carbohydrates are fermented with the production of pyruvate, which is further converted into lactic, acetic and formic acids. Part of the formic acid is split by a complex hydrogenlyase system into equal amounts of CO_2 and H_2.** Some strains are anaerogenic. **Lactose is fermented by most strains but fermentation may be delayed or absent.** Occur in the lower part of the intestine of warm-blooded animals and, in the case of *E. blattae*, of cockroaches. The mol% G + C of the DNA is 48–52 (T_m).

Type species: *Escherichia coli* (Migula 1895) Castellani and Chalmers 1919, 941.

Further Descriptive Information

Many strains, especially those isolated from extraintestinal sites, have polysaccharide capsules or microcapsules (Ørskov et al., 1977).

According to the state of the lipopolysaccharide (LPS) of the outer membrane, strains can be described as smooth (S) or rough (R). S forms, which usually grow as glistening colonies on ordinary agar media and show turbid growth in fluid media, have developed polysaccharide side chains whereas R forms, which usually will show dry and wrinkled colony forms on agar and which will agglutinate spontaneously in fluid media, have lost their polysaccharide side chains by mutation (Lüderitz et al., 1966).

In addition to the proteinaceous flagella, most strains have fimbriae (pili) or fibrillar proteins often extending in great numbers from the bacterial surface and far out into the surrounding medium. They have a width of 5–9 nm (Duguid, 1964; Brinton, 1965). Some fimbriae have specific functions as adhesive organs.

Two main varieties of fimbriae have been described based on their hemagglutinating ability. One is made up of the so-called type 1

---

* "*AL*" denotes the inclusion of this name on the Approved Lists of Bacterial Names (1980).

fimbriae (pili) characterized by hemagglutination (HA) that is inhibited by mannosides, the HA being mannose-sensitive (MS). Fimbriae of the other variety also cause HA, but the reaction is not inhibited by mannose, the HA being mannose-resistant (MR). Type 1 fimbriae are found in a great majority of *E. coli* strains and constitute antigenically, as far as they have been examined, a group of more or less related antigens (Gillies and Duguid, 1958), while there are many antigenically different MR fimbriae (Ørskov et al., 1980b; Ørskov et al., 1982). It has been shown by Ørskov et al. (1980a) that an important function of type 1 fimbriae is to bind to mucous material (slime) on mucous surfaces. They suggested that the binding of type 1 fimbriae to Tamm-Horsfall protein (urinary slime) is an important part of unspecified host defense. The many serologically diverse MR fimbriae which often function as virulence factors can be both species-specific and organ-specific in their adhesive characters.

Some strains of *E. coli* produce enterotoxins. Two enterotoxins have been well studied: the thermolabile toxin (LT), which is closely related to choleratoxin, and the thermostable toxin (ST). Both are found alone or together in enterotoxigenic *E. coli* (ETEC) strains, and are often associated with a limited number of O:K:H serovars and O groups. ETEC strains often have MR fimbriae.

LT and ST are plasmid-determined. LT can be demonstrated by several techniques. Some are based on LT's ability to stimulate hormone-producing tissue cultures and thereby changing their morphology, others on its immunological properties. ST is demonstrated by the infant mouse assay. For a review, see Rowe (1979).

Subdivision of *E. coli* can be carried out in many ways, but serology is one of the most useful ways to subdivide the species on a global basis. This method is based on the many antigenic differences found in structures on the bacterial surface. The main aspect of this analysis is the O antigen determination based on antigenicity of the LPS; 171 O antigens are presently listed, many of which cross-react.

The K antigens, which originally were defined exclusively according to their agglutinating abilities, have been redefined (Ørskov et al., 1977) and now the definition is also based on their chemical nature. The K antigens are the polysaccharide capsular antigens. Nearly 80 different K antigens are known. A description of the serology, chemistry and genetics of *E. coli* O and K antigens is given by Ørskov et al. (1977).

Flagellar or H antigens make up the third main group of serotyping antigens. A total of 56 H antigens are established. A serovar is recorded in the following way: 018acK1:H7 or 0111:H2 (the latter antigenic formula indicates that K antigens are not present in the strain). MR fimbriae, which are present only in some, often pathogenic, serovars, can also be used for the serological characterization (Ørskov et al., 1977, 1980b). Thus, enterotoxigenic strains from newborn piglets will usually belong to a limited number of serovars and in addition carry fimbrial antigens which are responsible for the necessary adhesion of the strain to the epithelium of the small intestine. The most common antigen, which was originally described as K88 at a time when its chemical character was unknown, is now termed F4 (Ørskov and Ørskov, in Bergan and Norris (eds.): *Methods in Microbiology*, London, Academic Press, in press). Similarly, enterotoxigenic strains isolated from newborn calves may carry an adhesive virulence factor originally named K99, now F5. Since the proteinaceous nature of these MR fimbrial antigens was recognized and more and more similar antigens were found in strains isolated not only from diarrheal diseases but also from extraintestinal diseases, a special new category was proposed for such fimbrial and fibrillar antigens: the F antigens. Thus, K88 is now F4, and K99 is F5. The labels proposed for the CF1 and CF2 antigens found in human enterotoxigenic strains will be F2 and F3, respectively

(Ørskov and Ørskov, in Bergan and Norris (eds.): *Methods in Microbiology*, London, Academic Press, in press). Some of the MR fimbriae are plasmid-determined. The MS fimbriae, which for many reasons make up the type 1 group with a separate position from the MR fimbriae, are designated as F1, but this antigen number covers a large group of antigens probably sharing common factors.

From the above description of the many known surface antigens in *E. coli*, it is easy to understand that the number of possible serovars is extremely high and, even though complete serotyping involving O, K and H anigens has been carried out in only a very few laboratories, it is well known that the existing number of serovars is very high.

For a description of other methods for subdivision of *E. coli*, i.e., phage typing, colicin typing, biotyping, typing by outer membrane protein (OMP) pattern, typing by antibiotic resistance patterns and typing by direct hemagglutination, see Ørskov and Ørskov (in Bergan and Norris (eds.): *Methods in Microbiology*, London, Academic Press, in press). Very useful is phage typing of the K1 antigen because K1 antisera are difficult to produce (Gross et al., 1977).

E. coli can be looked upon as primarily an opportunistic pathogen, but investigations in recent years have shown that a rather limited number of serovars or clones also play important and more specific roles in intestinal and extraintestinal diseases. Such clones often possess plasmids which provide them with special virulence traits (Ørskov and Ørskov, 1977).

E. blattae has not been associated with pathogenicity either in humans or in cockroaches.

Extraintestinal diseases. Neonatal meningitis is frequently associated with serovars that have the K1 antigen (Sarff et al., 1975). A limited number of O:K:H serovars, usually with MR fimbriae (F antigens) and often hemolytic, are associated with invasive urinary tract infections (UTI). Other extraintestinal diseases such as urinary tract injections and septicemia may be associated with similar sets of serovars (Ørskov and Ørskov, 1975).

Intestinal diseases. The letters EPEC (enteropathogenic *E. coli*) cover the few serovars associated with infantile diarrhea, mostly occurring in infant institutions. The pathophysiological role of most EPEC serovars has yet to be established. ETEC (enterotoxigenic *E. coli*) consist of a rather limited number of strains which produce enterotoxins (mostly plasmid-determined) causing diarrhea in animals and man. Many ETEC strains carry adhesive F antigens. A high degree of species specificity is characteristic of these clones. The term EIEC (enteroinvasive *E. coli*) covers those serovars that may cause dysentery-like disease. For a recent review, see Rowe (1979).

Enrichment and Isolation Procedures

Many simple agar media can be used for isolation. Media used for selective isolation from feces usually contain substances that partly or completely inhibit growth of bacteria other than *Enterobacteriaceae* (tetrathionate, deoxycholate, bile salts, etc.). The addition of Maranil (dodecylbenzolsulfonate) at a concentration of 0.005% will inhibit swarming of *Proteus* organisms. For details, see Edwards and Ewing (1972) or Kauffmann (1966) or any catalogue from one of the medium-producing companies. At Statens Seruminstitut, Copenhagen, we use a medium developed in the Media Department of this institute: brom-othymol blue (BTB) agar.*

Maintenance Procedures

E. coli strains can be kept alive for many years in beef extract agar stabs (tightly closed, e.g. by corks soaked in melted paraffin wax) or on

---

* Bromothymol blue agar (selective for *Enterobacteriaceae*). Combine the following ingredients: peptone (Orthana Ltd., Copenhagen), 10.0 g; NaCl, 5.0 g; yeast extract (Oxoid), 5.0 g; and distilled water, 1000 ml. The pH is adjusted to 8.0, agar powder is added, and the preparation is autoclaved at 120°C for 20 min. The following components are then added aseptically from sterile stock solutions: Maranil solution (Paste A75 (dodecylbenzolsulfonate), Henkel, West Germany), 1.0 ml; sodium thiosulfate (50% solution), 2.0 ml; bromothymol blue (Riedel de Haen, West Germany; 1.0% solution), 10.0 ml; lactose (33% solution), 27 ml; and glucose (33% solution), 1.2 ml. The pH is adjusted to 7.7–7.8. In order to obtain optimum results, the amount of glucose must be adjusted for every new batch of yeast extract, peptone and agar. This medium is very useful for differentiation of lactose-fermenting colonies based on their color.

Dorset egg medium. Cultures are initially incubated at 37°C followed by storage in the dark at room temperature (20–22°C). After a few weeks or months such cultures often contain many mutational forms such as R forms and acapsular forms; consequently, we prefer to store important cultures in beef broth containing 10% glycerol at −80°C. Screw-capped vials are used for easy access.

Differentiation of the genus **Escherichia** from other genera

See Table 5.3 of the family *Enterobacteriaceae* for characteristics that can be used to differentiate this genus from other genera of the family.

Taxonomic Comments

The identification of *Escherichia* strains seldom causes problems; however, many studies have shown that "*Escherichia* is a genus (or species) made up of phenotypically variable strains" (Farmer and Brenner, 1977). DNA/DNA hybridization studies have been an invaluable tool for solving problems in this field. The genus *Shigella* is closely related to *Escherichia* and only historical reasons make it acceptable that these two genera are not united. Several typical *Escherichia* types, the above mentioned EIEC types, have been found in recent years which have pathogenic traits that are similar to those of *Shigella*. The Sereny test (Sereny, 1967), which demonstrates the capacity to cause keratoconjunctivitis in the guinea pig, typical of *Shigella* strains, is also found in these special *Escherichia* strains. Day et al. (1981) described a tissue culture technique which can be used as a substitute for the Sereny test. Typically, such dysentery-provoking *Escherichia* strains have O antigens that are closely related or identical to *Shigella* O antigens. Brenner et al. (1972) by DNA reassociation studies found high homology between *Shigella* strains and these special *Escherichia* strains. Not unexpectedly, many strains are phenotypically intermediate between *Escherichia* and *Shigella*, but for obvious reasons a special taxonomic status for such strains is not warranted. In the older literature the name Alkalescens-Dispar can be found, but, as stated by Brenner (1978), this group is virtually indistinguishable from *E. coli* strains and is, in fact, a biogroup of *E. coli* that is anaerogenic, lactose-negative (or delayed) and nonmotile.

While most or all characters which classically have been used for

Procedures for Testing Special Characters

Kilian and Bülow (1976) have found that a very high percentage of *Escherichia-Shigella* strains, exclusively among the *Enterobacteriaceae*, produce β-glucuronidase (PGUA test). This test therefore holds promise as a screening test for bacteria belonging to this group.

definition of the genus *Escherichia* are chromosomally determined, several traits which are not characteristic of *Escherichia* have in recent years been found in otherwise typical *Escherichia* strains. Lautrop et al. (1971) and Layne et al. (1971) described H_2S-positive strains of *Escherichia*; the H_2S character was plasmid-determined. It is not known which selective forces account for the simultaneous isolation of H_2S-positive *Escherichia* strains in different parts of the world.

Other "forbidden" phenotypic traits have similarly been described in *Escherichia*, many of which are undoubtedly plasmid-determined. Ørskov et al. (1961) found many urease-producing strains among typical serovars from piglet diarrhea. Wachsmuth et al. (1979) demonstrated the plasmid-determined nature of a similar urease-positive phenotype in human *E. coli* strains. Citrate-utilizing *E. coli* strains were described by Washington and Timm (1978) and the plasmid background of similar strains was demonstrated by Sato et al. (1978). Carbon dioxide-dependent cultures can be found (Eykyn and Phillips, 1978). *Escherichia blattae* was isolated from the hindgut of healthy cockroaches in England (Burgess et al., 1973) and on Easter Island (Nogrady and Aubert, personal communication). A citrate-positive, malonate-positive biogroup and a biogroup negative in these reactions were described (Burgess et al., 1973).

Further Reading

Burgess, N.R.H., S.N. McDermott and J. Whiting. 1973. Aerobic bacteria occurring in the hind-gut of the cockroach *Blatta orientalis*. J. Hyg. (Lond.) *71:* 1–7.

Edwards, P.R. and W.H. Ewing. 1972. Identification of *Enterobacteriaceae*, 3rd Ed., Burgess Publishing, Minneapolis, Minn.

Ewing, W.H. and W.J. Martin. 1967. The biochemical reactions of the genus *Escherichia*. Monograph, National Communicable Disease Center, Atlanta, Ga.

Kauffmann, F. 1954. *Enterobacteriaceae*, 2nd Ed., Munksgaard, Copenhagen.

Differentiation of the species of the genus **Escherichia**

Characteristics useful in distinguishing the two species of *Escherichia* are given in Table 5.3 of the family *Enterobacteriaceae*.

List of the species of the genus **Escherichia**

1. **Escherichia coli** (Migula 1895) Castellani and Chalmers 1919, 941.[AL] (*Bacillus coli* Migula 1895, 27.)

co′li. Gr. n. *colon* large intestine, colon; M.L. gen. n. *coli* of the colon.

The characteristics are as described for the genus and as listed in Table 5.3 of the family *Enterobacteriaceae*.

Some O groups, O:H and O:K:H serovars from human *E. coli* enter-

Table 5.5.
Some O Groups, O:H and O:K:H serovars from human **Escherichia coli** *enteropathies*

Infantile Diarrhea EPEC[a]	Diarrhea in Adults and Children		
	ETEC[b]		EIEC[c]
026, 044, 055, 086, 0111, 0114, 0119, 0125, 0126, 0127, 0128, 0142, 0158	O6:K15:H16, O8:K40:H9, O8:K47:H-, O8:K25:H9, O11:H27, O15:H11, O20:H-, O25:K7:H42, O25:K98: H-; O27:H7, O27:H20, O63:H12, O73:H45, O85:H7; O78:H11, O78:H12, O114:H21; O115:[H51],[d] O128:H7, O128:H12, O128:H21, O139:H28, O148:H28, O149:H4, O159:H4, O159:H20, O159:H34; O166:H27, O169:H-		028ac, 0112, 0124, 0136, 0143, 0144, 0152, 0164

[a] EPEC, enteropathogenic *E. coli*. The O groups are listed; however, only a limited number of O:H types have been shown to have an association with infantile diarrhea.

[b] ETEC, enterotoxigenic *E. coli*. The data presented are primarily from Ørskov and Ørskov (1980). 0166 = OX8, and 0169 = OX2.

[c] EIEC, enteroinvasive *E. coli*.

[d] [ ], nonmotile variants exist.

opathies are indicated in Table 5.5 in this chapter.

Occurs in the lower part of the intestine of warm-blooded animals. The mol% G + C of the DNA is 48–52 (T_m).

Type strain: ATCC 11775.

2. **Escherichia blattae** Burgess, McDermott and Whiting 1973, 4.[AL]

blat'tae. L. fem. n. *blatta* cockroach; L. gen. n. *blattae* of the cockroach.

The characteristics are as described for the genus and as listed in Table 5.3 of the family *Enterobacteriaceae*.

Isolated from the hindgut of the cockroach *Blatta orientalis*.

Type strain: CDC 9005-74.

Species Incertae Sedis

Escherichia adecarboxylata Leclerc 1962, 736.[AL]

a.de.car.box'y.la.ta. Gr. pref. *a* not; M.L. adj. *adecarboxylata* not decarboxylating.

Little additional information concerning this organism has come forward since its mention in the eighth edition of the *Manual*, but unpublished studies indicate that it probably belongs to the *Erwinia herbicola-Enterobacter agglomerans* complex (Bascomb et al., 1971).

Type strain: ATCC 23216.

Genus II. **Shigella** *Castellani and Chalmers 1919, 936* [AL]

BERNARD ROWE AND ROGER J. GROSS

Shi.gel'la. M.L. dim. ending-*ella*; M.L. fem. n. *Shigella* named after K. Shiga, the Japanese bacteriologist who first discovered the dysentery bacillus.

Straight rods similar in morphology to other *Enterobacteriaceae*. Gram-negative. **Nonmotile.** Facultatively anaerobic, having both a respiratory and a fermentative type of metabolism. Catalase-positive (with exceptions in one species). Oxidase-negative. Chemoorganotrophic. **Ferment sugars without gas production** (a few exceptions produce gas). **Do not use citrate or malonate as a sole carbon source. Do not grow in KCN or produce H₂S.** Intestinal pathogens of man and other primates, causing **bacillary dysentery.** The mol% G + C of the DNA is 49–53 (Normore, 1973).

Type species: Shigella dysenteriae (Shiga 1898) Castellani and Chalmers 1919, 935.

Further Descriptive Information

The genus consists of four species, *S. dysenteriae, S. flexneri, S. boydii* and *S. sonnei*. These are often referred to as subgroups A, B, C and D, respectively.

The biochemical characteristics of the genus are listed in Table 5.6.

The species have been well characterized antigenically. *S. dysenteriae* contains 10 serovars, each with a distinctive antigen by which it can be recognized; there are few cross-reactions, either within the species or with other species.

S. flexneri contains eight serovars and nine subserovars. The serovars are antigenically related, but each has a qualitatively distinct major (type) antigen; the group antigens are shared by other members of the species. Because of the important intragroup relations, highly absorbed sera are needed for the detailed serotyping of *S. flexneri*. The immunochemical and genetic basis of the complex antigenic structure of the species has been summarized by Petrovskaya and Bondarenko (1977). The lipopolysaccharide O antigen of all serovars except *S. flexneri* 6 contains group antigens 3, 4 as a main primary structure. The type-specific antigens I, II, IV and V and the group antigens 7, 8 are all the result of phage conversion of the 3, 4 antigens resulting in the incorporation of α-glycosyl secondary side chains. Type-specific antigen III and group antigen 6 differ from the above antigens in that they contain acetyl groups. Nevertheless, these antigens are also formed as a result of phage conversion of the 3, 4 antigens. The lipopolysaccharide O antigen of *S. flexneri* serovar 6 differs from that of other *S. flexneri* serovars and does not contain the immunochemical determinants of the 3, 4 antigens. Strains of serovar 6 therefore resemble strains of *S. boydii* immunochemically, and Petrovskaya and Bondarenko have proposed that they be reclassified as such.

S. boydii contains 15 serovars and each has a qualitatively distinct antigen; there may be some cross-reactions with antisera to other *Shigella* species, but these seldom interfere with diagnosis. Serovars 10 and 11 share a major antigen, although each possesses a specific antigen.

S. sonnei contains only one serovar, which exists in two "phases," I and II; each has a distinctive antigen. Phase II is regarded as a loss

variation, but organisms in that phase may be isolated from patients, usually during convalescence and toward the end of an outbreak. An antiserum containing agglutinins for both phases should be used for identification

In addition to the recognized serovars of shigellae, Ewing et al. (1958) have described a number of provisional *Shigella* serovars. These may be added to the serotyping scheme in the future, but in the meantime they remain *sub judice* and antisera for their identification is usually available only at reference laboratories. Provisional serovars under consideration at present include *S. dysenteriae* 3873-50, 2000-53 and 3341-55 and *S. boydii* 3615-53, 2710-54 and 1621-54.

Colicin typing is of value in epidemiological studies of *S. sonnei*. The scheme is based on that described by Abbott and Shannon (1958) and distinguishes 14 types using 15 indicator strains (see *Procedures for Testing Special Characters*). Phage-typing schemes have also been described. Only a few reports have appeared for *S. dysenteriae* and *S. boydii* but a number of schemes have been described for *S. flexneri* and *S. sonnei* (Bergan, 1979).

Shigellae are pathogens of man and other primates and although there have been occasional reports of infections in dogs, other animals are resistant to infection. Laboratory animals such as mice, rabbits and guinea pigs may be infected orally but only following starvation and treatment with gastric antacids and antiperistaltic agents.

In humans, the lesions of bacillary dysentery are usually restricted to the rectum and large intestine, but in severe cases part of the terminal ileum may be affected. Typically there is acute inflammation with ulceration of the epithelium; the organisms rarely spread deeper than the lamina propria, and bloodstream involvement is uncommon. Infections due to *S. sonnei* rarely extend beyond the epithelial inflammatory stage, but infections with *S. dysenteriae* serovar 1 (Shiga's bacillus) or *S. flexneri* strains often cause ulceration.

The invasive properties of *Shigella* have been demonstrated using tests for the ability to produce keratoconjunctivitis in the guinea pig eye (Sérèny test), and to invade HeLa cells in tissue culture (Ogawa et al., 1967; Day et al., 1981). The rabbit ileal loop test has also been used as an experimental model. It has been shown that *S. dysenteriae* serovar 1 and *S. flexneri* serovar 2a produce toxins which are lethal to mice, enterotoxic in rabbit ileal loops, and cytotoxic for Hela cells (O'Brien et al., 1977). The demonstration of related toxins from both *S. dysenteriae* serovar 1 and *S. flexneri* might suggest that the enterotoxin has a role in the pathogenesis of bacillary dysentery. It was first thought that the enterotoxin of *S. dysenteriae* serovar 1 did not stimulate adenyl cyclase, unlike the cholera enterotoxin and the heat-labile enterotoxin of *Escherichia coli*. However, it has now been shown that under optimum assay conditions adenyl cyclase is stimulated by *S. dysenteriae* serovar 1 enterotoxin (Charney et al., 1976). Further work is needed, and in any case there is little doubt that epithelial invasion and multiplication are the main virulence factors.

Table 5.6.

Characteristics of the genus **Shigella**[a]

Test or Substrate	Result
β-Galactosidase	D[b]
Simmons' citrate	−
Christensen's citrate	−
Sodium acetate	D[c]
Arginine decarboxylase	−
Lysine decarboxylase	−
Ornithine decarboxylase	D[d]
Gelatin liquefaction	−
Gluconate	−
H₂S (triple sugar iron agar)	−
Indole production	D[e]
KCN, growth in	−
Malonate utilization	−
Methyl red test	+
Voges-Proskauer test	−
Phenylalanine deaminase	−
Urease	−
Motility	−
Glucose:	
Acid	+
Gas	D[f]
Acid from:	
Adonitol	−
Cellobiose	−
Dulcitol	−
Inositol	−
Lactose	D[g]
Mannitol	D[h]
Raffinose	D
Salicin	−
Sucrose	D[i]
Xylose	−

[a] Symbols: see standard definitions.

[b] Strains of *S. dysenteriae* 1 and *S. sonnei* are positive; positive strains of *S. flexneri* 2a and *S. boydii* 9 have been described.

[c] Some biovars of *S. flexneri* 4a are positive; all other biovars are negative.

[d] Strains of *S. boydii* 13 and *S. sonnei* are positive.

[e] Some strains of some serovars of *S. dysenteriae*, *S. flexneri* and *S. boydii* produce indole while strains of other serovars are always negative. *S. sonnei* is always negative.

[f] Some biovars of *S. flexneri* 6 are positive; positive strains of *S. boydii* 13 and 14 have been described.

[g] Strains of *S. sonnei* are usually positive after several days of incubation; positive strains of *S. flexneri* 2a and *S. boydii* 9 have been described.

[h] Strains of *S. dysenteriae* are negative; negative biovars of *S. flexneri* 4a ("*S. rabaulensis*," "*S. rio*") and *S. flexneri* 6 (Newcastle biovar) occur; negative biovars of *S. sonnei* occur rarely.

[i] Strains of *S. sonnei* are usually positive after several days of incubation.

Although infections are frequently mild and self-limiting, antibiotic treatment may be required in severe cases. Treatment is complicated by the increasing incidence of multiple drug resistance among *Shigella* strains. Indeed, the first observation of multiple, transferable drug resistance was in *Shigella* in Japan (Ochiai et al., 1959). Subsequent surveys in the United States (Neu et al., 1975) and in England (Thomas and Tillett, 1973) showed that the majority of *S. sonnei* strains were multiply resistant. Furthermore, a recent survey in England and Wales (Gross et al., 1981) showed that almost 50% of *Shigella* strains belonging to subgroups A, B and C were resistant to three or more drugs.

Enrichment and Isolation Procedures

Food and water. The minimum infecting dose of shigellae is small and occurrence of the organisms in food, milk and water may be significant even when only a small number of organisms are present. There are no reliable and effective enrichment methods, however, and the true incidence of *Shigella* contamination of foodstuffs cannot be accurately determined. The GN (Gram-negative) broth of Hajna (1955) may be useful for enrichment of *Shigella* and it is recommended that the investigation of foodstuffs should include an enrichment step using this medium. Subsequent steps in the isolation of *Shigella* from foods should follow the procedure recommended for fecal specimens.

Fecal specimens. Freshly passed stools should be examined, although if this is not possible fecal swabs showing marked fecal staining may be used. The specimens should be collected during the acute stage of the disease and before any chemotherapy is started. Specimens should be examined as soon after collection as possible. Enrichment with GN broth may be of value, but isolation is usually effected by direct plating. If the specimen includes blood and mucus, these should be included in the portion examined.

Some strains grow poorly on inhibitory media, and both a relatively noninhibitory medium such as MacConkey or eosin methylene blue (EMB) agar, and an inhibitory medium such as deoxycholate citrate agar (DCA) or shigella-salmonella (SS) agar should be used. Instructions for preparation of these media are given by Edwards and Ewing (1972). Specimens are streaked onto the chosen media and after overnight incubation at 37°C non-lactose-fermenting colonies are selected for further examination. Even when stool specimens from acute dysentery are examined, there may be only a scanty growth of *Shigella*.

Maintenance Procedures

Cultures of *Shigella* may be maintained on Dorset egg medium at room temperature, but rough and degraded variants frequently arise. Important cultures are best maintained lyophilized or in liquid nitrogen.

Procedures for Testing Special Characters

For colicin typing of *S. sonnei*, the organism under investigation is inoculated heavily in a broad streak across a blood agar plate and incubated at 37°C for 24 h. The bacterial growth is then removed from the agar by scraping with a glass slide and the organisms remaining are killed with chloroform. The 15 indicator strains are streaked onto

Table 5.7.

Differential characteristics of the species of the genus **Shigella**[a]

	S. dysenteriae	S. flexneri	S. boydii	S. sonnei
β-Galactosidase	D[b]	−	D	+
Ornithine decarboxylase	−	−	−[c]	+
Gas from glucose[d]	−	−	−	−
Acid from:				
Dulcitol[e]	−	−	−	−
Lactose	−	−	−	(+)[f]
Mannitol	−	+	+	+
Raffinose	−	D	−	(+)[f]
Sucrose	−	−	−	(+)[f]
Xylose	−	−	D	−
Indole production[g]	D	D	D	−

[a] For symbols see standard definitions.

[b] *S. dysenteriae* 1 strains are positive; some other serovars are sometimes positive.

[c] *S. boydii* 13 strains are positive.

[d] Gas production from glucose: only certain biovars of *S. flexneri* 6, and of *S. boydii* 13 (Rowe et al., 1975) and *S. boydii* 14 (Carpenter, 1961) are aerogenic.

[e] *S. dysenteriae* 5 and *S. flexneri* 6 may ferment dulcitol.

[f] (+), positive reaction delayed (more than 24 h).

[g] *S. dysenteriae* 1, *S. flexneri* 6 and *S. sonnei* never produce indole, while strains of *S. dysenteriae* 2 always produce indole.

the plate at right angles to the original line of growth. After further incubation for 8–12 h the patterns of inhibition of growth of the indicator strains can be examined and compared with a key. It is important that controls be included in every batch of tests.

Differentiation of the genus **Shigella** from other closely related taxa

The biochemical identification of *Shigella* is complicated by the similarity of some strains of other genera. In particular, strains of *Hafnia alvei*, *Providencia sp.*, *Aeromonas sp.* and atypical *Escherichia coli* frequently cause difficulties.

Nonlactose-fermenting or anaerogenic strains of *E. coli* are a common problem. Of particular interest are members of the Alkalescens Dispar (A-D) group which are now defined as nonmotile, anaerogenic biotypes of *E. coli*. These are best differentiated from *Shigella* by means of the Christensen's citrate and lysine decarboxylase tests in which *Shigella* is always negative. The members of the A-D group were divided into eight serogroups on the basis of their O antigens (Frantzen, 1950), although most of these are identical with or closely related to *E. coli* antigens. Now that these organisms are regarded as *E. coli*, no further serogroups will be added to the A-D scheme.

Taxonomic Comments

The occurrence of biochemically atypical strains of *E. coli* has prompted Shmilovitz et al. (1974) to suggest the recognition of an intermediate group to be known as Intermediate Shigella Coli Alkalescens Dispar (ISCAD). Stenzel (1978) proposed the inclusion of such

Table 5.8.
Earlier designations and antigenic formulae of **Shigella** *species*

Subgroup and Species	Serovar	Sub-serovar	Antigenic Formula	Main Earlier Designations or Synonyms
Subgroup A				
S. dysenteriae	1			*S. shigae*
	2			*S. schmitzii, S. ambigua*
	3			*S. largei* Q771, *S. arabinotarda* A
	4			*S. largei* Q1167, *S. arabinotarda* B
	5			*S. largei* Q1030
	6			*S. largei* Q454
	7			*S. largei* Q902
	8			Serotype 599-52 (Ewing et al.)
	9			Serotype 58 (Cox and Wallace)
	10			Serotype 2050 (Ewing)
Subgroup B				
S. flexneri	1	1a	I:2,4	V (Andrewes and Inman)
		1b	I:'S':6:2,4	VZ (Andrewes and Inman)
	2	2a	II:3,4	W (Andrewes and Inman)
		2b	II:7,8	WX (Andrewes and Inman)
	3	3a	III:6,7,8	Z (Andrewes and Inman)
		3b	III:6,3,4	
		3c	III:6:	
	4	4a[a]	IV:'B':3,4	103 (Boyd)
		4b	(IV):'B':6:3,4	103Z (Rewell and Bridges)
	5		V:7,8	P119 and P119X (Boyd), (Bridges)
	6		VI:(2),4	*S. newcastle*; Manchester bacillus; Boyd 88 (Newcastle and Manchester-aerogenic; Newcastle-mannitol-negative)
	X		−:7,8	X (Andrewes and Inman)
	Y		−:3,4	Y (Andrewes and Inman)
Subgroup C				
S. boydii	1			170 (Boyd)
	2			P288 (Boyd)
	3			D1 (Boyd)
	4			P274 (Boyd)
	5			P143 (Boyd)
	6			D19 (Boyd)
	7			Lavington I; *S. etousae*
	8			Serotype 112 (Cox and Wallace)
	9			Serotype 1296/7 and 1320 (Francis)
	10			Serotype 430 (Ewing); D15 (Szturm et al.)
	11			Serotype 34 and 732 (Ewing)
	12			Serotype 123 (Ewing and Hucks)
	13			Serotype 425 (Ewing and Hucks)
	14			Serotype 2770-51 (Ewing and Hucks)
	15			Serotype 703 (Ewing et al.)
Subgroup D				
S. sonnei				Duval's bacillus; *B. ceylanesis* A

[a] The group phase of this subserovar, corresponding to Boyd's 103B organism, has the formula −:'B':3,4.

strains in *Shigella* subgroup D and suggested that this subgroup should be renamed "*S. metadysenteriae.*" The situation is further complicated by the fact that some strains of *E. coli* share with *Shigella* the ability to cause bacillary dysentery and to cause keratoconjunctivitis of the guinea pig eye in the Sérèny test (Sakazaki et al., 1974). However, the Enterobacteriaceae Sub-Committee of the International Committee on Bacteriological Nomenclature (Carpenter, 1963) has advised that pathogenicity should not be considered in the classification of *Enterobacteriaceae* and strains with biochemical reactions which do not conform strictly to those of *Shigella* should be classified as atypical *E. coli.* Nevertheless, it should be realized that *E. coli* and *Shigella* strains (except *S. boydii* serovar 13) are indistinguishable on the basis of DNA hybridization studies (Brenner et al., 1973) and it may be largely for historical reasons that the two genera remain separate.

Further Readings

Edwards, P.R. and W.H. Ewing. 1972. Identification of Enterobacteriaceae, Burgess Publishing, Minneapolis, Minn.
Keusch, G.T. 1979. In Lambert (Editor), Clinics in Gastroenterology, Vol. 8, No. 3, W. B. Saunders, London, pp. 645–662.
DuPont, H.L. and L.K. Pickering (Editors). 1980. Infections of the Gastrointestinal Tract, Plenum Medical Book, New York, pp. 61–82.
Lapage, S.P., Rowe, B., Holmes, B. and R.J. Gross. 1979. In Skinner and Lovelock (Editors), Identification Methods for Microbiologists, 2nd Ed., Academic Press, London, pp. 123–141.

Differentiation of the species of the genus **Shigella**

Biochemical characteristics useful for differentiating the species of *Shigella* are listed in Table 5.7.

List of the species of the genus **Shigella**

1. **Shigella dysenteriae** (Shiga 1898) Castellani and Chalmers 1919, 935, epit. spec. cons. Opin. 11, Jud. Comm. 1954, 149.[AL] (*Bacillus dysenteriae* Shiga 1898, 817.)

dys.en.te′ri.ae. Gr. n. *dysenteria* dysentery; M.L. gen. n. *dysenteriae* of dysentery.

Also known as subgroup A.

Colonies of serovar 1 often have a pinkish tinge on Leifson's deoxycholate citrate agar. Catalase is not produced by serovar 1, but is usually produced by strains of other serovars.

Mannitol is not fermented. Dulcitol is fermented by strains of serovar 5. Indole is not produced by serovar 1 but is always produced by strains of serovar 2; strains of other serovars vary in indole production.

All the serovars have, at one time or another, been known by other designations, and these are shown in Table 5.8.

Type strain: ATCC 13313 (NCTC 4837; Newcastle 1934) (Jud. Comm. 1963, Opin. 26).

2. **Shigella flexneri** Castellani and Chalmers 1919, 937, epit. spec. cons. Opin. 11, Jud. Comm. 1954, 149.[AL]

flex′ner.i. M.L. gen. n. *flexneri* of Flexner; named after Simon Flexner, an American bacteriologist.

Also known as subgroup B.

Catalase is produced.

Mannitol is fermented, except by biovar Newcastle, serovar 6 and a mannitol-negative, xylose-positive biovar of serovar 4a (sometimes known as "*S. rabaulensis*"). Dulcitol is fermented by certain biovars of serovar 6 (see Table 5.7), some of which produce gas from fermentable sugars.

Indole is not produced by serovar 6; in other serovars indole production is variable.

The reactions of *S. flexneri* strains in diagnostic absorbed antisera are shown in Table 5.9.

Type strain: ATCC 29903.

3. **Shigella boydii** Ewing 1949, 634, epit. spec. cons. Opin. 11, Jud. Comm. 1954, 149.[AL]

boy′di.i. M.L. gen. n. *boydii* of Boyd; named after Sir John Boyd, a British bacteriologist.

Also known as subgroup C.

Catalase is produced.

Mannitol is fermented. Dulcitol is usually fermented by serovars 2, 3, 4, 6 and 10, but this may be delayed. Xylose fermentation is variable.

Indole may or may not be produced. Gas-producing biovars of *S. boydii* serovar 13 (Rowe et al., 1975) and serovar 14 (Carpenter, 1961) have been described.

Type strain: ATCC 8700.

4. **Shigella sonnei** (Levine 1920) Weldin 1927, 182, epit. spec. cons. Opin. 11, Jud. Comm. 1954, 149.[AL] (*Bacterium sonnei* Levine 1920, 31.)

son′ne.i. M.L. gen. n. *sonnei* of Sonne; named after Carl Sonne.

Also known as subgroup D.

On deoxycholate citrate agar colonies are at first colorless, but after

Table 5.9.

Reactions of **S. flexneri** *serovars in diagnostic absorbed slide-agglutinating serums*[a]

Serum		Serovar with Simplified Antigenic Formula[b]												
Type	Agglu-tinins	1a I:2,4	1b I:S:6:2,4	2a II:3,4	2b II:7,8	3a[c] III:6:7,8	3b III:6:3,4	3c III:6...	4a IV:B:3,4	4b IV:B:6:3,4	5[d] V:7,8	5 V:(3,4)	6 VI:2,4	X −:7,8	Y −:3,4
1	I	++	++	−	−	−	−	−	−	−	−	−	−	−	−
2	II	−	−	++	++	−	−	−	−	−	−	−	−	−	−
3	III:6	−	+	−	−	++	++	++	−	+	−	−	−	−	−
4	IV:B	−	−	−	−	−	−	−	++	++	−	−	−	−	−
5	V	−	−	−	−	−	−	−	−	−	++	++	−	−	−
6	VI	−	−	−	−	−	−	−	−	−	−	−	++	−	−
X	7,8	−	−	−	++	++	−	−	−	−	+	−	−	++	−
Y	3,4	−/±	−/±	−/+	−	−	++	−	−/+	−/±	−	−/±	−	−	++

[a] Symbols: ++, strong reaction; +, moderate reaction; ±, weak reaction; −, no reaction.
[b] Arabic numerals are used to designate serovars, but it is customary to use Roman numerals to express **type-specific** antigens or agglutinins, and arabic numerals for **group antigens or agglutinins.**
[c] Occasional variants may also react in absorbed Y serum.
[d] Subserovars of *S. flexneri* 5 have not yet been designated.

a few days show bright pink papillae consisting of lactose-fermenting cells. On MacConkey's taurocholate lactose agar, phase I colonies are indistinguishable from colonies of other shigellas, but phase II colonies are larger, flatter and more translucent and have an irregular edge. On subculture, phase I colonies produce both phase I and phase II colonies, but phase II colonies give rise to phase II colonies only.

Mannitol is fermented rapidly, lactose and sucrose more slowly. Some strains may ferment xylose.

Catalase is produced. Indole is not produced.

Ornithine is decarboxylated; arginine may be decarboxylated.

Type strain: ATCC 29930.

Genus III. **Salmonella** *Lignières 1900, 389*[AL]

L. Le Minor

Sal.mon.el′la. M.L. dim. ending -ella; M.L. fem. n. *Salmonella* named after D. E. Salmon, an American bacteriologist.

Straight rods, $0.7–1.5 \times 2.0–5.0$ μm, conforming to the general definition of the family *Enterobacteriaceae*. Gram-negative. **Usually motile** (peritrichous flagella). Facultatively anaerobic. Colonies are generally 2–4 mm in diameter. Nitrates are reduced to nitrites. **Gas is usually produced from glucose.** Hydrogen sulfide is usually produced on triple-sugar iron agar. Indole-negative. **Citrate is usually utilized as a sole carbon source.** Lysine and ornithine decarboxylase (Møller's) reactions are usually positive. Urease-negative. Phenylalanine and tryptophan are not oxidatively deaminated. Sucrose, salicin, inositol and amygdalin are usually not fermented. Lipase and deoxyribonuclease are not produced. Pathogenic for humans, causing enteric fevers, gastroenteritis and septicemia; may also infect many animal species besides humans. Some serovars are strictly host-adapted. The mol% G + C of the DNA is 50–53 (Ch, T_m, Bd) (Hill, 1966).

Type species: *Salmonella choleraesuis* (Smith 1894) Weldin 1927, 155.

Further Descriptive Information

Although most salmonellae are motile, nonmotile mutants may occur, and one type ("*S. gallinarum*" or "*S. pullorum*") is always nonmotile.

Certain *Salmonella* types may form unusually small colonies (~1 mm diameter), whereas most types form larger colonies (2–4 mm).

Most salmonellae are aerogenic; however, *S. typhi*, an important exception, never produces gas. Anaerogenic variants of normally gas-producing *Salmonella* serovars may occur; this is particularly common with *S. dublin*.

Hydrogen sulfide is produced by most salmonellae, but a few types do not form it (e.g., some strains of *S. choleraesuis*, and most strains of "*S. paratyphi A*.")

Citrate is generally utilized by salmonellae, but some types do not use it (particularly *S. typhi* and "*S. paratyphi-A*"). Most salmonellae do not utilize malonate, but *S. arizonae* does use it.

The lysine decarboxylase reaction (Møller's) is positive for most salmonellae; an important exception is "*S. paratyphi-A*." Most salmonellae are also positive for ornithine decarboxylase (Møller's), but *S. typhi* is negative.

Lactose is generally not fermented by salmonellae, but many strains of *S. arizonae* ferment it rapidly or slowly, and nearly all strains of *S. arizonae* have β-galactosidase activity (by the ONPG test).

Other biochemical characteristics of the genus are indicated in Table 5.3 in the article on the family *Enterobacteriaceae*. Subdivision of the genus *Salmonella* into the so-called "subgenera" of Kauffman (1960, 1963a, b, 1964) on the basis of biochemical characteristics is shown in Table 5.10 of the present chapter. These subdivisions correspond more closely to species or subspecies in other groups of bacteria, but whatever rank is assigned to them, the worthiness of these subdivisions was confirmed by Rohde (1965, 1966, 1967). A new "subgenus," V, is added in the present chapter. Salmonellae belonging to this "subgenus" grow in the presence of KCN (as those of "subgenus" IV); they are lactose-, malonate- and gelatin-negative and dulcitol- and mucate-positive (as those of "subgenus" I); and they are negative for *d*-, *l*- and *i*-tartrates and positive for the ONPG test (as those of "subgenus" III).

Division into serovars. The Kauffmann-White scheme, in which

Table 5.10.

Differential characteristics of the "subgenera" of the genus **Salmonella**[a]

	"Subgenus"				
	I	II	III	IV	V[b]
β-galactosidase (ONPG test)	−	− or x	+	−	+
Acid production from:					
Lactose	−	−	+ or x	−	−
Dulcitol	+	+	−	−	+
Mucate	+	+	d	−	+
Galacturonate[c]	−	+	d	+	+
Utilization of:					
Malonate	−	+	+	−	−
d-Tartrate	+	− or x	− or x	− or x	−
Gelatin hydrolysis (film method)	−	+	+	+	−
Growth in presence of KCN	−	−	−	+	+
Habitat of the majority of strains:					
Warm-blooded animals	+	−	−	−	−
Cold-blooded animals and environment	−	+	+	+	+

[a] Symbols: +, positive for 90% or more of strains in 1–2 days; d, positive for 11–89% of strains in 1–2 days; −, positive for 0–10% of strains in 1–2 days; x, late and irregularly positive (3–7 days). The temperature for all reactions is 37°C.

[b] L. Le Minor, M. Véron and M. Popoff, 1982, Ann. Microbiol. (Inst. Pasteur): 133B: 223–243.

[c] From Le Minor et al. (1979). Monophasic serovars of "Subgenus" III are galacturonate-negative; diphasic serovars are positive.

organisms are represented by the numbers and letters given to the different O (somatic), Vi (capsular) and H (flagellar) antigens, indicates only those antigens of primary diagnostic importance and is not a complete record of the antigenic complement or its complexity (Kauffmann, 1966). The scheme, expanded to include all five "subgenera," is given in Table 5.11. The original "*Arizona*" antigens (given in brackets) have been converted to the presently used *Salmonella* designations.

Antigenic formulae (for example, 6,7:r:1,7) represent the O antigens: the phase 1 H antigen(s): the phase 2 H antigen(s), respectively. Those formulae with particular O antigens in common are collected into an O group and arranged alphabetically by H antigens within the group.

Lysogenization by certain converting phages may produce changes in the O antigenic formulae of salmonellae. In antigenic groups A, B and D the presence of O antigen 1 (factor 1) is associated with lysogenization (Iseki and Kashiwagi, 1955, 1957; Stocker, 1958; Zinder, 1957), but the presence or absence of this factor in strains of these groups does not change the name of the organism (for example, the name *S. typhimurium* applies to both the "1$^+$" and "1$^-$" strains). On the other hand, in group E the name is changed: phage ϵ_{15} (Iseki and Sakai, 1953) alters the O antigen 3, 10 to 3, *15*, thereby making "*S. anatum*" become "*S. newington*," and in a similar way phage ϵ_{34} changes "*S. newington*" to "*S. minneapolis*." The same applies to "*S. cerro*" and "*S. siegburg*," the latter being merely the lysogenic variant of the former (Le Minor, 1965), and also to all of the strains of group C$_4$ (O antigen 6,7,<u>14</u>) which are lysogenic variants of group C$_1$ (O antigen 6,7) although they bear different names. For this reason, all the factors associated with phage conversion are <u>underlined</u> in the joint Kauffmann-White scheme (Table 5.11) which includes all "*Arizona*" serovars with their corresponding *Salmonella* formulae. The converting phages of *Salmonella* are identical in morphology (Vieu et al., 1965), but their action is limited to certain O groups and they are serologically different from one another (Le Minor, 1968).

The specificities of the O factors in *Salmonella* is determined by the composition and structure of the polysaccharides. Specificity is modified during S → R mutation and by bacteriophage conversions (see reviews by Stocker and Mäkelä, 1971, 1978; Lüderitz et al., 1971). Thus, the only difference between the 4,12 and the 9,12 O-specific repeating units is in the di-deoxyhexose branch unit attached to the mannose, which is abequose in 4,12 and tyvelose in 9,12. In the conversion of 3,10 → 3,15, the terminal acetyl radical of the chain is suppressed and the α-linkage between galactose and mannose is transformed into a β-linkage.

Other modifications of the specificity of somatic (O) antigens may occur after a mutation, resulting in new specificities called T$_1$ and T$_2$ by Kauffmann (1956) and in different R types (reviewed by Stocker and Mäkelä, 1971, 1978, and by Lüderitz et al., 1971).

Subdivision of serovars. Biovars are different sugar fermentation patterns shown by strains of the same serovar. They are determined by the presence or absence of enzymes and hence are genetically determined. Biovars may serve as markers and be of interest epidemiologically (for example, the xylose$^+$ and xylose$^-$ character of *S. typhi*).

Phagovars are determined by the sensitivity of cultures to a series of bacteriophages at appropriate dilutions. Phage typing of *S. typhi* and other salmonellae which possess the Vi antigen ("*S. hirschfeldii*" and rarely "*S. dublin*") is based on a series of adapted phages from phage Vi-II of Craigie and Yen (1938). Phage typing of "*S. schottmuelleri*" (Felix and Callow, 1943) and *S. typhimurium* (Anderson, 1964) uses a different series of phages. Analogous methods have been proposed for other serovars of *Salmonella*, some of them making use of the lysogenicity of the strains.

Other subdivisions of the serovars may be made on the basis of the production of, or the sensitivity to, bacteriocins and on the basis of the resistance to antibiotics.

Genetics. The genetic map of *S. typhimurium* (Sanderson and Hartman, 1978) is not very different from that of *E. coli* K12 (Bachmann and Low, 1980). Hfr strains of *Salmonella* may be selected after F plasmid transfer. Conjugative chromosomal transfer may occur from

Salmonella to *E. coli*, from *E. coli* to *Salmonella*, and from one serovar of *Salmonella* to another. Chromosomal genes responsible for O, Vi and H antigens can be transferred from one genus to the other (Iino and Lederberg, 1964). Crosses may be used to localize the regions of the bacterial chromosome which specify avirulence for mice (Krishnapillai and Baron, 1964) or to study the role of O antigen factors in the virulence of *Salmonella* (Mäkelä et al., 1973).

As for other *Enterobacteriaceae*, salmonellae may harbor "foreign" replicons—temperate phages or plasmids that may code for antibiotic resistance or for metabolic characters commonly used in diagnostic identification, e.g. lactose or sucrose fermentation (Le Minor et al., 1973, 1974). Thus it is unwise to exclude *Salmonella* solely on the basis of a positive lactose or sucrose reaction. It is also more difficult to identify salmonellae when a pleiotrophic mutation occurs, such as one that simultaneously affects nitrate, tetrathionate, and thiosulfate reductase as well as hydrogenlyase (Le Minor, et al., 1969).

About 5% of *Salmonella* strains produce bacteriocins active against *Escherichia coli*, *Shigella* and/or *Salmonella* (Fredericq, 1948). Most of these bacteriocins adsorb to the same receptor as that for colicins B, E$_1$, E$_2$ or I. *Salmonella* bacteriocins differ from colicins *sensu stricto* by their activity spectra on colicin indicator strains. Some of these *Salmonella* bacteriocins are not even active against colicin indicator strains but are active against *Salmonella* strains only (Hamon and Péron, 1966).

Susceptibility to the 01 phage. Most strains of the genus *Salmonella* are susceptible to the 01 phage of Felix and Callow (1943); this phage is highly specific for *Salmonella*, lysing more than 98% of the strains studied in routine *Salmonella* diagnosis (Kallings, 1967). Whereas the majority of strains of *Salmonella* "subgenus" I and II of diphasic "subgenus" III are lysed (some strains, chiefly of the E group, are resistant), monophasic cultures of "subgenus" III and strains of "subgenus" IV are generally resistant (Bockemühl, 1972). Mutations conferring resistance to 01 phage have been studied by Lindberg (1969), McPhee et al. (1975) and Hudson (1978).

A *Salmonella* phage which attacks only flagellated bacteria was isolated by Sertic and Boulgakov (1936). Sensitivity to this phage depends on the H antigen. For example, bacteria with antigens of the "g" complex are resistant (Meynell, 1961).

Pathogenicity. *Salmonella* serovars may be strictly adapted to one particular host (these serovars are auxotrophic), may be ubiquitous (found in a large number of animal species), or may be of still unknown pathogenicity.

Serovars adapted to man (e.g., *S. typhi*, "*S. paratyphi A*," "*S. sendai*") usually cause grave diseases with septicemia-typhoidic syndrome. They are not pathogenic for other animal species. Salmonellosis is transmitted from man to man, without an intermediate host, through fecal contamination of water and food. The incidence is higher in developing countries with poor hygiene. Other serovars are adapted to one animal species; e.g., "*S. abortusovis*" is adapted to sheep and is a major cause of abortion in ewes, whereas "*S. typhisuis*" and "*S. gallinarum*" ("*S. pullorum*") are adapted to swine and poultry, respectively.

Ubiquitous *Salmonella* serovars (e.g., *S. typhimurium*) are mostly responsible for food-borne infections. It is necessary to ingest a sufficiently high number of bacteria (10^8 to 10^9) to express clinical symptoms. Salmonellosis of newborns and infants (who are more susceptible to infections than adults) presents diverse clinical symptoms, from a grave typhoid-like illness with septicemia to a mild or asymptomatic infection. In pediatric wards the infection is transmitted by the hands of personnel.

The entrance of a serovar into a food chain may be the origin of its implantation in a country. For example, many countries have become infected with "*S. hadar*" introduced by imported turkeys, or by "*S. agona*" introduced by fish meal imported from South America.

After recovery from a clinical case of salmonellosis, some patients—although asymptomatic—remain carriers for weeks, months, or years (i.e. continue to eliminate salmonellae in feces). Carriage contributes to the dissemination of salmonellosis, especially if the diagnosis of the

(Text continues on p. 445)

Table 5.11.

Antigenic formulae of the serovars of the genus **Salmonella**[a]

Serovar	Somatic (O) antigens	Flagellar (H) Antigens	
		Phase 1	Phase 2
	Group 02 (A)		
S. paratyphi A	1,2,12	a	[1,5]
S. nitra	2,12	g,m	−
S. kiel	1,2,12	g,p	−
	Group 04 (B)		
S. kisangani	1,4,[5],12	a	1,2
S. hessarek	4,12,27	a	1,5
S. fulica	4,[5],12	a	1,5
S. arechavaleta	4,[5],12	a	[1,7]
S. bispebjerg	1,4,[5],12	a	e,n,x
S. tinda	1,4,12,27	a	e,n,z_{15}
S. II makoma	4,[5],12	a	−
S. nakura	1,4,12,27	a	z_6
S. paratyphi B[b]	1,4,[5],12	b	1,2
S. limete	1,4,12,27	b	1,5
S. canada	4,12	b	1,6
S. uppsala	4,12,27	b	1,7
S. abony	1,4,[5],12,27	b	e,n,x
S. abortusbovis	1,4,12,27	b	e,n,x
S. II sofia	1,4,12,27	b	[e,n,x]
S. wagenia	1,4,12,27	b	e,n,z_{15}
S. wien	1,4,12,27	b	1,w
S. schleissheim	4,12,27	b	−
S. legon	1,4,12,27	c	1,5
S. abortusovis	4,12	c	1,6
S. altendorf	4,12,27	c	1,7

Serovar	Somatic (O) antigens	Flagellar (H) Antigens	
		Phase 1	Phase 2
S. jericho	1,4,12,27	c	e,n,z_{15}
S. hallfold	1,4,12,27	c	l,w
S. bury	4,12,27	c	z_6
S. stanley	1,4,[5],12,27	d	1,2
S. eppendorf	1,4,12,27	d	1,5
S. brezany	1,4,12,27	d	1,6
S. schwarzengrund	1,4,12,27	d	1,7
S. II kluetjenfelde	4,12	d	e,n,x
S. sarajane	4,[5],12,27	d	e,n,x
S. duisburg	1,4,12,27	d	e,n,z_{15}
S. salinatis	4,12	d,e,h	d,e,n,z_{15}
S. mons	1,4,12,27	d	l,w
S. ayinde	1,4,12,27	d	z_6
S. saintpaul	1,4,[5],12	e,h	1,2
S. reading	1,4,[5],12	e,h	1,5
S. eko	4,12	e,h	1,6
S. kaapstad	4,12	e,h	1,7
S. chester	1,4,[5],12	e,h	e,n,x
S. sandiego	4,[5],12	e,h	e,n,z_{15}
S. II makumira	1,4,12,27	e,n,x	1,[5],7
S. derby	1,4,[5],12	f,g	[1,2]
S. agona	1,4,12	f,g,s	−
S. II	1,4,[5],12	f,g,t	$z_6:z_{42}$
S. essen	4,12	g,m	−
S. hato	4,[5],12	g,m,s	−

[a] Supplemented by the formulae approved up to the end of 1980 and including those for *S. arizonae* ("*Arizona*"). A supplement to the Kauffmann-White scheme, describing the formulae and biochemical characteristics of new *Salmonella* serovars, is published annually in the Annales de Microbiologie (Institut Pasteur), Paris.

Symbols: [ ], may be absent; ( ), not well developed (weakly agglutinable). The symbols for somatic factors whose presence is connected with phage conversion are underlined (e.g. 6,14,18). They are present only if the culture is lysogenized by the corresponding converting phage. These factors are mentioned in the table for serovars in which they were found. It is probable that most, if not all, serovars in a group could be converted by these bacteriophages.

All the "subgenus" I serovars bear a name (e.g. *S. paratyphi A*). "Subgenus" II serovars have the designation "*S. II*," and atypical members of "subgenus" II are designated "*S. (II)*." The serovars belong to this "subgenus" and those which were described before the Moscow International Congress (1966) bear a name (e.g. S. II *sofia*). Those described subsequently are designated solely by their antigenic formula (e.g. S. II 1,4,12,27:z:1,5). Members of "subgenus" III are designated "*S. III.*" The serovars of this "subgenus" appear in the table with the name *S. arizonae*, followed by the formula according to the symbols used in the Kauffmann-White scheme and, in parentheses, the formula according to Edwards, Fife and Ewing. The extent to which these two formulae correspond has been established by Dr. R. Rohde. Members of "subgenus" IV are designated "*S. IV.*" Members of "subgenus" V are designated "*S. V*" and a name (e.g. *S. V bongor*). This is provisional because initially they were considered as atypical strains of "subgenus" I.

Groups C_4, E_2 and E_3 are retained in this table, although it has been shown that the serovars belonging to them are, respectively, those of groups C_1 lysogenized by phage 14 (6, 7), and E_1 lysogenized by ϵ_{15} or $\epsilon_{15} + \epsilon_{34}$. No further serovars have been added to these groups, which are retained provisionally.

[b] Biovar *d*-tartrate positive is often called var. *java*.

[c] May possess an R-phase H antigen: z_{43}.

[d] May possess an R-phase H antigen: z_{40}.

[e] May possess an R-phase H antigen: 1,11; z_{37}, z_{49}.

[f] May possess an R-phase H antigen: z_{50}.

[g] May possess an R-phase H antigen: z_{47}; z_{50}.

[h] May possess an R-phase H antigen: z_{50}.

[i] May possess an R-phase H antigen: z_{45}.

[j] May possess an R-phase H antigen: j; z_{66}.

[k] May possess an R-phase H antigen: z_{40}.

[l] May possess an R-phase H antigen: 1,13.

[m] The serovars of this group also contain the factors 0:3 and (10), the latter not very well developed. They can be lysogenized by phages ϵ_{15} and ϵ_{34} and in the case

of double lysogenization become strongly agglutinable, like strains of group E_3, by antisera against 0:34 and 0:12₂.

[n] May possess an R-phase H antigen: z_{45}.

[o] May possess an R-phase H antigen: z_{48}.

[p] May possess an R-phase H antigen: z_{37}.

[q] May possess an R-phase H antigen: z_{45}.

[r] May possess an R-phase H antigen: z_{45}

[s] May possess an R-phase H antigen: z_{49}.

[t] May possess an R-phase H antigen: z_{37}.

[u] May possess an R-phase H antigen: z_{27}; z_{34}; z_{43}; z_{45}; z_{46}.

[v] May possess an R-phase H antigen: z_{59}.

[w] May possess an R-phase H antigen: z_{37}.

[x] May possess an R-phase H antigen: z_{37}; z_{43}.

[y] May possess an R-phase H antigen: z_{45}.

[z] May possess an R-phase H antigen: z_{33}; z_{49}.

[aa] The antigenic factor described for this strain as *Ar.* 32a,32c is very different from other factors H_{32} of *Arizona* 32a,32b. Factor 32b is strongly related to *Salmonella* H factor c.

[bb] May possess an R-phase H antigen: z_{59}.

[cc] May possess an R-phase H antigen: z_{58}.

[dd] May possess an R-phase H antigen: z_{27}.

[ee] May possess an R-phase H antigen: z_{50}.

[ff] May possess an R-phase H antigen: z_{45}.

[gg] May possess an R-phase H antigen: z_{50}.

[hh] May possess an R-phase H antigen: z_{50}.

[ii] May possess an R-phase H antigen: z_{58}.

[jj] May possess an R-phase H antigen: z_{50}.

[kk] May possess an R-phase H antigen: 2_{50}

[ll] May possess an R-phase H antigen: z_{45}.

[mm] May possess an R-phase H antigen: z_{47}; z_{50}.

[nn] May possess an R-phase H antigen: z_{58}.

[oo] May possess an R-phase H antigen: z_{50}.

[pp] This group is not homogeneous and certain serovars possess factors other than 54. Moreover, factor 054 (which has some antigenic resemblance to 042) can be lost by certain serovars: *S. tonev*, which possess factor 21, then becomes similar to *S. minnesota*, *S. uccle* retains factor 3 on this segregation. *S. poeseldorf*, which posessses factors 8,20, becomes similar to *S. kentucky*, and *S. ochsenwerder*, which possesses factors 6₁,6₂,7, becomes similar to *S. thompson*. *S. steinwerder* can, moreover, be converted by phage ϵ_{34} and acquire factors 34 and 12₂.

[qq] May possess an R-phase H antigen: z_{47}.

[rr] The group 064 is combined with the group 048 (Winkle, I. 1976, Ann. Microbiol. (Inst. Pasteur) *127B:* 463–472.)

Table 5.11—*continued*

Serovar	Somatic (O) antigens	Phase 1	Phase 2	Serovar	Somatic (O) antigens	Phase 1	Phase 2
		Flagellar (H) Antigens				Flagellar (H) Antigens	
S. II *caledon*	1,4,12,27	g,m,[s],t	e,n,x	*S. fortune*	1,4,12,27	z_{10}	z_6
S. II *bechuana*	1,4,12,27	g,[m],t	[1,5]	*S. vellore*	1,4,12,27	z_{10}	z_{35}
S. II	4,12	g,m,t	z_{39}	*S. brancaster*	1,4,12,27	z_{29}	—
S. california	4,12	g,m,t	—	*S.* II *helsinki*	1,4,12	z_{29}	[e,n,x]
S. kingston[c]	1,4,[5],12,27	g,s,t	[1,2]	*S. pasing*	4,12	z_{35}	1,5
S. budapest	1,4,12,27	g,t	—	*S. tafo*	1,4,12,27	z_{35}	1,7
S. travis	4,[5],12	g,z_{51}	1,7	*S. sloterdijk*	1,4,12,27	z_{35}	z_6
S. tennyson	4,5,12	g,z_{51}	e,n,z_{15}	*S. yaounde*	1,4,12,27	z_{35}	e,n,z_{15}
S. II	4,12	g,z_{62}	—	*S. tejas*	4,12	z_{36}	—
S. banana	4,[5],12	m,t	1,5	*S. wilhelmsburg*	1,4,[5],12,27	z_{38}	—
S. typhimurium	1,4,[5],12	i	1,2	*S.* II *durbanville*	1,4,12,27	z_{39}	1,[5],7
S. lagos	1,4,[5],12	i	1,5	*S. thayngen*	1,4,12,27	z_{41}	1,(2),5
S. agama	4,12	i	1,6	*S. abortusequi*	4,12	—	e,n,x
S. tsevie	4,12	i	e,n,z_{15}		Group 06,7 (C$_1$)		
S. gloucester	1,4,12,27	i	l,w	(The strains of this group may be lysogenized by phage 14 → 6,7,14)			
S. massenya	1,4,12,27	k	1,5	*S. sanjuan*	6,7	a	1,5
S. neumuenster	1,4,12,27	k	1,6	*S. umhlali*	6,7	a	1,6
S. II	1,4,12,27	k	1,6	*S. austin*	6,7	a	1,7
S. ljubljana	4,12,27	k	e,n,x	*S. oslo*	6,7	a	e,n,x
S. texas	4,[5],12	k	e,n,z_{15}	*S. denver*	6,7	a	e,n,z_{15}
S. fyris	4[5],12	l,v	1,2	*S. coleypark*	6,7	a	l,w
S. azteca	4,[5],12,27	l,v	1,5	*S.* II	6,7	a	z_6
S. clackamas	4,12	l,v	1,6	*S.* II *calvinia*	6,7	a	z_{42}
S. bredeney[d]	1,4,12,27	l,v	1,7	*S. brazzaville*	6,7	b	1,2
S. kimuenza	1,4,12,27	l,v	e,n,x	*S. edinburg*	6,7	b	1,5
S. II	1,4,12,27	l,v	e,n,x	*S. adime*	6,7	b	1,6
S. brandenburg	1,4,12	l,v	e,n,z_{15}	*S. koumra*	6,7	b	1,7
S. II	1,4,12,27	l,v	z_{39}	*S. georgia*	6,7	b	e,n,z_{15}
S. mono	4,12	l,w	1,5	*S.* II *bloemfontein*	6,7	b	[e,n,x]:z_{42}
S. togo	4,12	l,w	1,6	*S. ohio*	6,7	b	l,w
S. II *kilwa*	4,12	l,w	e,n,x	*S. leopoldville*	6,7	b	z_6
S. ayton	1,4,12,27	l,w	z_6	*S. kotte*	6,7	b	z_{35}
S. kunduchi	1,4,[5],12,27	l,[z_{13}], z_{28}	1,2	*S.* II	6,7	b	z_{39}
S. tyresoe	4,12	l,[z_{13}],z_{28}	1,5	*S. paratyphi C*	6,7,[Vi]	c	1,5
S. kubacha	1,4,12,27	l,z_{13},z_{28}	1,7	*S. choleraesuis*	6,7	[c]	1,5
S. kano	1,4,12,27	l,z_{13},z_{28}	e,n,x	*S. typhisuis*	6,7	c	1,5
S. vom	1,4,12,27	l,z_{13},z_{28}	e,n,z_{15}	*S. birkenhead*	6,7	c	1,6
S. reinickendorf	4,12	l,z_{28}	e,n,x	*S. kisii*	6,7	d	1,2
S. II	4,12	l,z_{28}	—	*S. isangi*	6,7	d	1,5
S. heidelberg	1,4,[5],12	r	1,2	*S. kivu*	6,7	d	1,6
S. bradford	4,12,27	r	1,5	*S. kambole*	6,7	d	1,7
S. remo	1,4,12,27	r	1,7	*S.* II	6,7	d	1,7
S. bochum	4,[5],12	r	l,w	*S. amersfoort*	6,7	d	e,n,x
S. southampton	1,4,12,27	r	z_6	*S. gombe*	6,7	d	e,n,z_{15}
S. drogana	1,4,12,27	r,i	e,n,z_{15}	*S. livingstone*	6,7	d	l,w
S. africana	4,12	r,i	l,w	*S. wil*	6,7	d	l,z_{13},z_{28}
S. coeln	4,[5],12	y	1,2	*S. larochelle*	6,7	e,h	1,2
S. trachau	4,12,27	y	1,5	*S. lomita*	6,7	e,h	1,5
S. teddington	1,4,12,27	y	1,7	*S. norwich*	6,7	e,h	1,6
S. ball	1,4,[5],12,27	y	e,n,x	*S. braenderup*	6,7	e,h	e,n,z_{15}
S. jos	1,4,12,27	y	e,n,z_{15}	*S. rissen*	6,7	f,g	—
S. kamoru	4,12,27	y	z_6	*S. eingedi*	6,7	f,g,t	1,2,7
S. shubra	4,[5],12	z	1,2	*S. afula*	6,7	f,g,t	e,n,x
S. kiambu	4,12	z	1,5	*S. montevideo*	6,7	g,m,[p],s	[1,2,7]
S. II	1,4,12,27	z	1,5	*S.* II	6,7	g,m,[s],t	e,n,x
S. indiana	1,4,12	z	1,7	*S.* II	6,7	(g),m,[s],t	1,5
S. neftenbach	4,12	z	e,n,x	*S.* II	6,7	g,m,s,t	z_{42}
S. II *nordenham*	1,4,12,27	z	e,n,x	*S. othmarschen*	6,7	g,m,[t]	—
S. koenigstuhl	1,4,12	z	e,n,z_{15}	*S. menston*	6,7	g,s,t	[1,6]
S. preston	1,4,12,27	z	l,w	*S.* II	6,7	g,t	e,n,x:z_{42}
S. entebbe	1,4,12,27	z	z_6	*S. riggil*	6,7	g,t	—
S. stanleyville	1,4,[5],12,27	z_4,z_{23}	[1,2]	*S. alamo*	6,7	g,z_{51}	1,5
S. kalamu	4,[5],12	z_4,z_{24}	[1,5]	*S. haelsingborg*	6,7	m,p,t,[u]	—
S. haifa	1,4,[5],12	z_{10}	1,2	*S. oranienburg*	6,7	m,t	—
S. ituri	1,4,12	z_{10}	1,5	*S. augustenborg*	6,7	i	1,2
S. tudu	4,12	z_{10}	1,6	*S. oritamerin*	6,7	i	1,5
S. albert	4,12	z_{10}	e,n,x	*S. garoli*	6,7	i	1,6
S. tokoin	4,12	z_{10}	e,n,z_{15}	*S. lika*	6,7	i	1,7
S. mura	1,4,12	z_{10}	l,w	*S. athinai*	6,7	i	e,n,z_{15}

Table 5.11—*continued*

Serovar	Somatic (O) antigens	Flagellar (H) Antigens Phase 1	Phase 2	Serovar	Somatic (O) antigens	Flagellar (H) Antigens Phase 1	Phase 2
S. norton	6,7	i	l,w	S. II	6,7	z_{41}	1,7
S. galiema	6,7	k	1,2	S. hillsborough	6,7	z_{41}	l,w
S. thompson	6,7	k	1,5	S. tamilnadu	6,7	z_{41}	z_{35}
S. daytona	6,7	k	1,6	S. II sullivan	6,7	z_{42}	1,7
S. baiboukoum	6,7	k	1,7	S. II	6,7	z_{42}	e,n,x:1,6
S. singapore	6,7	k	e,n,x	S. III arizonae	6,7	—	1,6
S. escanaba	6,7	k	e,n,z_{15}	(Ar.27:–:30)			
S. III arizonae (Ar. 27:22:31:37)	6,7	(k)	z:[z_{55}]		Group 06,8 (C$_2$)		
S. II	6,7	k	[z_6]	S. doncaster	6,8	a	1,5
S. concord	6,7	l,v	1,2	S. curacao	6,8	a	1,6
S. irumu	6,7	l,v	1,5	S. nordufer	6,8	a	1,7
S. mkamba	6,7	l,v	1,6	S. narashino	6,8	a	e,n,x
S. kortrijk	6,7	l,v	1,7	S. II	6,8	a	e,n,x
S. bonn	6,7	l,v	e,n,x	S. leith	6,8	a	e,n,z_{15}
S. potsdam	6,7	l,v	e,n,z_{15}	S. II tulear	6,8	a	z_{52}
S. gdansk	6,7	l,v	z_6	S. skansen	6,8	b	1,2
S. III arizonae (Ar. 27:23:25)	6,7	l,v	z_{53}	S. nagoya	6,8	b	1,5
S. gabon	6,7	l,w	1,2	S. stourbridge	6,8	b	1,6
S. colorado	6,7	l,w	1,5	S. eboko	6,8	b	1,7
S. II	6,7	l,w	1,5,7	S. gatuni	6,8	b	e,n,x
S. nessziona	6,7	l,z_{13}	1,5	S. presov	6,8	b	e,n,z_{15}
S. kenya	6,7	l,z_{13}	e,n,x	S. bukuru	6,8	b	l,w
S. neukoelln	6,7	$l,z_{13},[z_{28}]$	e,n,z_{15}	S. banalia	6,8	b	z_6
S. makiso	6,7	l,z_{13},z_{28}	z_6	S. wingrove	6,8	c	1,2
S. II heilbron	6,7	l,z_{28}	1,5:[z_{42}]	S. utah	6,8	c	1,5
S. virchow	6,7	r	1,2	S. bronx	6,8	c	1,6
S. infantis[e]	6,7	r	1,5	S. belfast	6,8	c	1,7
S. nigeria	6,7	r	1,6	S. belem	6,8	c	e,n,x
S. colindale	6,7	r	1,7	S. quiniela	6,8	c	e,n,z_{15}
S. papuana	6,7	r	e,n,z_{15}	S. muenchen	6,8	d	1,2
S. grampian	6,7	r	l,w	S. manhattan	6,8	d	1,5
S. richmond	6,7	y	1,2	S. sterrenbos	6,8	d	e,n,x
S. bareilly	6,7	y	1,5	S. herston	6,8	d	e,n,z_{15}
S. oyonnax	6,7	y	1,6	S. II	6,8	d	z_6:z_{42}
S. gatow	6,7	y	1,7	S. newport[h]	6,8	e,h	1,2
S. hartford[f]	6,7	y	e,n,x	S. kottbus	6,8	e,h	1,5
S. mikawasima[g]	6,7	y	e,n,z_{15}	S. cremieu	6,8	e,h	1,6
S. II tosamanga	6,7	z	1,5	S. tshiongwe	6,8	e,h	e,n,z_{15}
S. oakland	6,7	z	1,6[7]	S. sandow	6,8	f,g	e,n,z_{15}
S. cayar	6,7	z	e,n,x	S. chincol	6,8	g,m,[s]	[e,n,x]
S. businga	6,7	z	e,n,z_{15}	S. II	6,8	g,m,t	[e,n,x]
S. bruck	6,7	z	l,w	S. nanergou	6,8	g,s,t	—
S. II	6,7	z	z_6	S. II baragwanath	6,8	m,t	1,5
S. II	6,7	z	z_{39}	S. II germiston	6,8	m,t	e,n,x
S. II oysterbeds	6,7	z	z_{42}	S. bassa	6,8	m,t	—
S. obogu	6,7	z_4,z_{23}	1,5	S. lindenburg	6,8	i	1,2
S. aequatoria	6,7	z_4,z_{23}	e,n,z_{15}	S. takoradi	6,8	i	1,5
S. goma	6,7	z_4,z_{23}	z_6	S. warnow	6,8	i	1,6
S. IV roterberg	6,7	z_4,z_{23}	—	S. malmoe	6,8	i	1,7
S. somone	6,7	z_4,z_{24}	—	S. bonariensis	6,8	i	e,n,x
S. IV kralendyk	6,7	z_4,z_{24}	—	S. aba	6,8	i	e,n,z_{15}
S. II cape	6,7	z_6	1,7	S. cyprus	6,8	i	l,w
S. menden	6,7	z_{10}	1,2	S. blockley	6,8	k	1,5
S. inganda	6,7	z_{10}	1,5	S. schwerin	6,8	k	e,n,x
S. eschweiler	6,7	z_{10}	1,6	S. charlottenburg	6,8	k	e,n,z_{15}
S. ngili	6,7	z_{10}	1,7	S. litchfield	6,8	l,v	1,2
S. djugu	6,7	z_{10}	e,n,x	S. loanda	6,8	l,v	1,5
S. mbandaka	6,7	z_{10}	e,n,z_{15}	S. manchester	6,8	l,v	1,7
S. redba	6,7	z_{10}	z_6	S. holcomb	6,8	l,v	e,n,x
S. II	6,7	z_{10}	z_{35}	S. II	6,8	l,v	e,n,x
S. tennessee	6,7	z_{29}	[1,2,7]	S. edmonton	6,8	l,v	e,n,z_{15}
S. II	6,7	z_{29}	—	S. fayed	6,8	l,w	1,2
S. palime	6,7	z_{35}	e,n,z_{15}	S. hiduddify	6,8	l,z_{13},z_{28}	1,5
S. II bacongo	6,7	z_{36}	z_{42}	S. breukelen	6,8	$l,z_{13},[z_{28}]$	e,n,z_{15}
S. IV argentina	6,7	z_{36}	—	S. bovismorbificans	6,8	r	1,5
S. rumford	6,7	z_{38}	1,2	S. akanji	6,8	r	1,7
S. lille	6,7	z_{38}	—	S. hidalgo	6,8	r	e,n,z_{15}
S. II gilbert	6,7	z_{39}	1,5,7	S. goldcoast	6,8	r	l,w
				S. tananarive	6,8	y	1,5

Table 5.11—*continued*

Serovar	Somatic (O) antigens	Flagellar (H) Antigens Phase 1	Phase 2
S. bulgaria	6,8	y	1,6
S. II	6,8	y	1,6:z_{42}
S. inchpark	6,8	y	1,7
S. praha	6,8	y	e,n,z_{15}
S. mowanjum	6,8	z	1,5
S. II	6,8	z	1,5
S. kalumburu	6,8	z	e,n,z_{15}
S. kuru	6,8	z	l,w
S. lezennes	6,8	z_4,z_{23}	1,7
S. chailey	6,8	z_4,z_{23}	e,n,z_{15}
S. duesseldorf	6,8	z_4,z_{24}	—
S. tallahassee	6,8	z_4,z_{32}	—
S. zerifin	6,8	z_{10}	1,2
S. mapo	6,8	z_{10}	1,5
S. cleveland	6,8	z_{10}	1,7
S. hadar	6,8	z_{10}	e,n,x
S. glostrup	6,8	z_{10}	e,n,z_{15}
S. wippra	6,8	z_{10}	z_6
S. II	6,8	z_{29}	1,5
S. uno	6,8	z_{29}	[e,n,z_{15}]
S. yarm	6,8	z_{35}	1,2
S. aesch	6,8	z_{60}	1,2
Group 08 (C_3)			
S. be	8,<u>20</u>	a	—
S. djelfa	8	b	1,2
S. korbol	8,<u>20</u>	b	1,5
S. sanga	8	b	1,7
S. konstanz	8	b	e,n,x
S. shipley	8,<u>20</u>	b	e,n,z_{15}
S. tounouma	8,<u>20</u>	b	z_6
S. alexanderpolder	8	c	l,w
S. santiago	8,<u>20</u>	c	e,n,x
S. tado	8,<u>20</u>	c	z_6
S. virginia	8	d	1,2
S. yovokome	8	d	1,5
S. labadi	8,<u>20</u>	d	z_6
S. bardo	8	e,h	1,2
S. ferruch	8	e,h	1,5
S. atakpame	8,<u>20</u>	e,h	1,7
S. rechovot	8,<u>20</u>	e,h	z_6
S. emek	8,<u>20</u>	g,m,s	—
S. reubeuss	8,<u>20</u>	g,m,t	—
S. alminko	8,<u>20</u>	g,s,t	—
S. yokoe	8	m,t	—
S. bargny	8,<u>20</u>	i	1,5
S. kentucky	8,<u>20</u>	i	z_6
S. haardt	8	k	1,5
S. pakistan	8	l,v	1,2
S. amherstiana	8	l,v	1,6
S. hindmarsh	8,<u>20</u>	r	1,5
S. cocody	8,<u>20</u>	r,i	e,n,z_{15}
S. brikama	8,<u>20</u>	r,i	l,w
S. altona	8,<u>20</u>	r,[i]	z_6
S. giza	8,<u>20</u>	y	1,2
S. brunei	8,<u>20</u>	y	1,5
S. alagbon	8	y	1,7
S. sunnycove	8	y	e,n,x
S. kralingen	8,<u>20</u>	y	z_6
S. bellevue	8	z_4,z_{23}	1,7
S. dabou	8,<u>20</u>	z_4,z_{23}	l,w
S. corvallis	8,<u>20</u>	z_4,z_{23}	[z_6]
S. albany[i]	8,<u>20</u>	z_4,z_{24}	—
S. bazenheid	8,<u>20</u>	z_{10}	1,2
S. paris	8,<u>20</u>	z_{10}	1,5
S. istanbul	8	z_{10}	e,n,x
S. chomedey	8	z_{10}	e,n,z_{15}
S. molade	8,<u>20</u>	z_{10}	z_6
S. II	8	z_{29}	e,n,x:z_{42}
S. tamale	8,<u>20</u>	z_{29}	[e,n,z_{15}]
S. angers	8,<u>20</u>	z_{35}	z_6

Serovar	Somatic (O) antigens	Flagellar (H) Antigens Phase 1	Phase 2
S. apeyeme	8,<u>20</u>	z_{38}	—
S. diogoye	8,<u>20</u>	z_{41}	z_6
Group 06,7,<u>14</u> (C_4)			
(*Salmonella* serovars of group C_1 lysogenized by "phage 14")			
S. lockleaze	6,7,<u>14</u>	b	e,n,x
S. nienstedten	6,7,<u>14</u>	b	[l,w]
S. hissar	6,7,<u>14</u>	c	1,2
S. kaduna	6,7,<u>14</u>	c	e,n,z_{15}
S. omderman	6,7,<u>14</u>	d	e,n,x
S. eimsbuettel	6,7,<u>14</u>	d	l,w
S. nieukerk	6,7,<u>14</u>	d	z_6
S. ardwick	6,7,<u>14</u>	f,g	—
S. thielallee	6,7,<u>14</u>	m,t	—
S. gelsenkirchen	6,7,<u>14</u>	l,v	z_6
S. jerusalem	6,7,<u>14</u>	z_{10}	l,w
S. bornum	6,7,<u>14</u>	z_{38}	—
S. III *arizonae* (Ar. 27:45:30)	6,7,<u>14</u>	z_{39}	1,2
Group 09, 12 (D_1)			
S. sendai	1,9,<u>12</u>	a	1,5
S. miami	1,9,<u>12</u>	a	1,5
S. II	9,12	a	1,5
S. os	9,12	a	1,6
S. saarbruecken	1,9,<u>12</u>	a	1,7
S. lomalinda	1,9,<u>12</u>	a	e,n,x
S. II	1,9,<u>12</u>	a	e,n,x
S. durban	9,12	a	e,n,z_{15}
S. II	9,12	a	z_{39}
S. onarimon	1,9,<u>12</u>	b	1,2
S. frintrop	1,9,<u>12</u>	b	1,5
S. II *mjimwema*	1,9,<u>12</u>	b	e,n,x
S. II *blankenese*	1,9,<u>12</u>	b	z_6
S. II *suederelbe*	1,9,<u>12</u>	b	z_{39}
S. goeteborg	9,12	c	1,5
S. ipeko	9,12	c	1,6
S. elokate	9,12	c	1,7
S. alabama	9,12	c	e,n,z_{15}
S. ridge	9,12	c	z_6
S. ndolo	1,9,<u>12</u>	d	1,5
S. tarshyne	9,12	d	1,6
S. II *rhodesiense*	9,12	d	e,n,x
S. zega	9,12	d	z_6
S. jaffna	1,9,<u>12</u>	d	z_{35}
S. typhi[j]	9,12,[Vi]	d	—
S. bournemouth	9,12	e,h	1,2
S. eastbourne	1,9,<u>12</u>	e,h	1,5
S. israel	9,12	e,h	e,n,z_{15}
S. II *lindrick*	9,12	e,n,x	1,[5],7
S. II	9,12	e,n,x	1,6
S. berta	1,9,<u>12</u>	f,g,t	—
S. enteritidis	1,9,<u>12</u>	g,m	[1,7]
S. blegdam	9,12	g,m,q	—
S. II	1,9,<u>12</u>	g,m,[s],t	[1,5]:[z_{42}]
S. II *kuilsrivier*	1,9,<u>12</u>	g,m,s,t	e,n,x
S. dublin	1,9,12[Vi]	g,p	—
S. naestved	1,9,<u>12</u>	g,p,s	—
S. rostock	1,9,<u>12</u>	g,p,u	—
S. moscow	9,12	g,q	—
S. II *neasden*	9,12	g,s,t	e,n,x
S. newmexico	9,12	g,z_{51}	1,5
S. II	1,9,<u>12</u>	g,z_{62}	—
S. antarctica	9,12	g,z_{63}	—
S. II	9,12	m,t	e,n,x
S. pensacola	1,9,<u>12</u>	m,t	—
S. seremban	9,12	i	1,5
S. claibornei	1,9,<u>12</u>	k	1,5
S. goverdhan	9,12	k	1,6
S. mendoza	9,12	l,v	1,2
S. panama	1,9,<u>12</u>	l,v	1,5
S. kapemba[k]	9,12	l,v	1,7

Table 5.11 (*Continued*)

Serovar	Somatic (O) antigens	Phase 1	Phase 2	Serovar	Somatic (O) antigens	Phase 1	Phase 2
		Flagellar (H) Antigens				Flagellar (H) Antigens	
S. II	9,12	l,v	e,n,x	*S.* II	9,46	z_{10}	z_6
S. goettingen	9,12	l,v	e,n,z_{15}	*S.* II	9,46	z_{10}	z_{39}
S. II	9,12	l,v	z_{39}	*S. ouakam*[n]	9,46	z_{29}	—
S. victoria	1,9,12	l,w	1,5	*S. hillegersberg*	9,46	z_{35}	1,5
S. II *daressalaam*	1,9,12	l,w	e,n,x	*S. basingstoke*	9,46	z_{35}	e,n,z_{15}
S. itami	9,12	l,z_{13}	1,5	*S. trimdon*	9,46	z_{35}	z_6
S. miyazaki	9,12	l,z_{13}	1,7	*S. fresno*	9,46	z_{38}	—
S. napoli	1,9,12	l,z_{13}	e,n,x	*S.* II	9,46	z_{39}	1,7
S. javiana[l]	1,9,12	l,z_{28}	1,5	*S. wuppertal*	9,46	z_{41}	—
S. II	9,12	l,z_{28}	e,n,x	**Group 01,9,12,(46),27 (D₃)**			
S. jamaica	9,12	r	1,5	*S.* II *zuerich*	1,9,12,(46),27	c	z_{39}
S. camberwell	9,12	r	1,7	*S.* II	9,12,(46),27	g,t	e,n,x
S. campinense	9,12	r	e,n,z_{15}	*S.* II	1,9,12,(46),27	l,z_{13},z_{28}	z_{39}
S. lome	9,12	r	z_6	*S.* II	1,9,12,(46),27	y	z_{39}
S. lawndale	1,9,12	z	1,5	*S.* II	1,9,12,(46),27	z_4,z_{24}	1,5
S. kimpese	9,12	z	1,6	*S.* II	1,9,12,(46),27	z_{10}	e,n,x
S. II *stellenbosch*	1,9,12	z	1,7	*S.* II	1,9,12,(46),27	z_{10}	z_{39}
S. II *angola*	1,9,12	z	z_6	**Group 03,10 (E₁)**			
S. II *hueningen*	9,12	z	z_{39}	*S. aminatu*	3,10	a	1,2
S. wangata	1,9,12	z_4,z_{23}	[1,7]	*S. goelzau*	3,10	a	1,5
S. portland	9,12	z_{10}	1,5	*S. oxford*	3,10	a	1,7
S. II *canastel*	9,12	z_{29}	1,5	*S. masembe*	3,10	a	e,n,x
S. II	1,9,12	z_{29}	e,n,x	*S.* II *matroosfontein*	3,10	a	e,n,x
S. penarth	9,12	z_{35}	z_6	*S. galil*	3,10	a	e,n,z_{15}
S. elomrane	1,9,12	z_{38}	—	*S.* II	3,10	a	z_{39}
S. II *wynberg*	1,9,12	z_{39}	1,7	*S. kalina*	3,10	b	1,2
S. ottawa	1,9,12	z_{41}	1,5	*S. butantan*	3,10	b	1,5
S. gallinarum-pullorum	1,9,12	—	—	*S. allerton*	3,10	b	1,6
Group 09,46 (D₂)[m]				*S. huvudsta*	3,10	b	1,7
S. baildon	9,46	a	e,n,x	*S. benfica*	3,10	b	e,n,x
S. doba	9,46	a	e,n,z_{15}	*S.* II	3,10	b	e,n,x
S. zadar	9,46	b	1,6	*S. yaba*	3,10	b	e,n,z_{15}
S. worb	9,46	b	e,n,x	*S. epicrates*	3,10	b	l,w
S. II *lundby*	9,46	b	e,n,x	*S.* II	3,10	b	z_{39}
S. bamboye	9,46	b	l,w	*S. gbadago*	3,10	c	1,5
S. linguere	9,46	b	z_6	*S. ikayi*	3,10	c	1,6
S. itutaba	9,46	c	z_6	*S. pramiso*	3,10	c	1,7
S. ontario	9,46	d	1,5	*S. agege*	3,10	c	e,n,z_{15}
S. quentin	9,46	d	1,6	*S. anderlecht*	3,10	c	l,w
S. strasbourg	9,46	d	1,7	*S. okefoko*	3,10	c	z_6
S. olten	9,46	d	e,n,z_{15}	*S. stormont*	3,10	d	1,2
S. plymouth	9,46	d	z_6	*S. shangani*	3,10	d	1,5
S. bergedorf	9,46	e,h	1,2	*S. lekke*	3,10	d	1,6
S. guerin	9,46	e,h	z_6	*S. onireke*	3,10	d	1,7
S. II	9,46	e,n,x	1,5,7	*S. souza*	3,10	d	e,n,x
S. wernigerode	9,46	f,g	—	*S.* II	3,10	d	e,n,x
S. hillingdon	9,46	g,m	—	*S. madjorio*	3,10	d	e,n,z_{15}
S. II *duivenhoks*	9,46	g,m,s,t	e,n,x	*S. birmingham*	3,10	d	l,w
S. gateshead	9,46	g,s,t	—	*S. weybridge*	3,10	d	z_6
S. II	9,46	m,t	e,n,x	*S. maron*	3,10	d	z_{35}
S. sangalkam	9,46	m,t	—	*S. vejle*	3,10	e,h	1,2
S. mathura	9,46	i	e,n,z_{15}	*S. muenster*[o]	3,10	e,h	1,5
S. potto	9,46	i	z_6	*S. anatum*	3,10	e,h	1,6
S. marylebone	9,46	k	1,2	*S. nyborg*	3,10	e,h	1,7
S. cochin	9,46	k	1,5	*S. newlands*	3,10	e,h	e,n,x
S. ceyco	9,46	k	z_{35}	*S. meleagridis*	3,10	e,h	l,w
S. india	9,46	l,v	1,5	*S. sekondi*	3,10	e,h	z_6
S. geraldton	9,46	l,v	1,6	*S.* II *chudleigh*	3,10	e,n,x	1,7
S. toronto	9,46	l,v	e,n,x	*S. regent*	3,10	f,g,[s]	[1,6]
S. shoreditch	9,46	r	e,n,z_{15}	*S. alfort*	3,10	f,g	e,n,x
S. sokode	9,46	r	z_6	*S. suberu*	3,10	g,m	—
S. benin	9,46	y	1,7	*S. amsterdam*	3,10	g,m,s	—
S. mayday	9,46	y	z_6	*S.* II	3,10	g,m,s,t	—
S. II *haarlem*	9,46	z	e,n,x	*S. westhampton*[p]	3,10	g,s,t	—
S. bambylor	9,46	z	e,n,z_{15}	*S.* II *islington*	3,10	g,t	—
S. ekotedo	9,46	z_4,z_{23}	—	*S. southbank*	3,10	m,t	[1,6]
S. II *maarssen*	9,46	z_4,z_{24}	z_{39}:z_{42}	*S.* II *stikland*	3,10	m,t	e,n,x
S. lishabi	9,46	z_{10}	1,7	*S. cukmere*	3,10	i	1,2
S. inglis	9,46	z_{10}	e,n,x	*S. amounderness*	3,10	i	1,5
S. louisiana	9,46	z_{10}	z_6	*S. truro*	3,10	i	1,7

Table 5.11—*continued*

Serovar	Somatic (O) antigens	Flagellar (H) Antigens Phase 1	Phase 2	Serovar	Somatic (O) antigens	Flagellar (H) Antigens Phase 1	Phase 2
S. bessi	3,10	i	e,n,x	S. goerlitz	3,15	e,h	1,2
S. falkensee	3,10	i	e,n,z_{15}	S. newhaw	3,15	e,h	1,5
S. yeerongpilly	3,10	i	z_6	S. newington	3,15	e,h	1,6
S. wimborne	3,10	k	1,2	S. selandia	3,15	e,h	1,7
S. zanzibar	3,10	k	1,5	S. cambridge	3,15	e,h	l,w
S. yundum	3,10	k	e,n,x	S. drypool	3,15	g,m,s	—
S. marienthal	3,10	k	e,n,z_{15}	S. II parow	3,15	g,m,s,t	—
S. newrochelle	3,10	k	l,w	S. halmstad	3,15	g,s,t	—
S. nchanga	3,10	l,v	1,2	S. nancy	3,15	l,v	1,2
S. sinstorf	3,10	l,v	1,5	S. portsmouth	3,15	l,v	1,6
S. london	3,10	l,v	1,6	S. newbrunswick	3,15	l,v	1,7
S. give	3,10	[d]:l,v	1,7	S. kinshasa	3,15	l,z_{13}	1,5
S. II	3,10	l,v	e,n,x	S. lanka	3,15	r	z_6
S. ruzizi	3,10	l,v	e,n,z_{15}	S. tuebingen	3,15	y	1,2
S. II fuhlsbuettel	3,10	l,v	z_6	S. binza	3,15	y	1,5
S. sinchew	3,10	l,v	z_{35}	S. tournai	3,15	y	z_6
S. assinie[q]	3,10	l,w	z_6:	S. manila	3,15	z_{10}	1,5
S. freiburg	3,10	l,z_{13}	1,2	Group 03,15,34 (E₃)			
S. uganda	3,10	l,z_{13}	1,5	(*Salmonella* serovars of group E₁ lysogenized by phages ε₁₅ and ε₃₄)			
S. fallowfield	3,10	l,z_{13},z_{28}	e,n,z_{15}	S. khartoum	3,15,34	a	1,7
S. hoghton	3,10	l,z_{13},z_{28}	z_6	S. arkansas	3,15,34	e,h	1,5
S. II	3,10	l,z_{28}	1,5	S. minneapolis	3,15,34	e,h	1,6
S. joal	3,10	l,z_{28}	1,7	S. wildwood	3,15,34	e,h	l,w
S. lamin	3,10	l,z_{28}	e,n,x	S. canoga	3,15,34	g,s,t	—
S. II westpark	3,10	l,z_{28}	e,n,x	S. menhaden	3,15,34	l,v	1,7
S. II	3,10	l,z_{28}	z_{39}	S. thomasville	3,15,34	y	1,5
S. ughelli	3,10	r	1,5	S. illinois	3,15,34	z_{10}	1,5
S. elisabethville	3,10	r	1,7	S. harrisonburg	3,15,34	z_{10}	1,6
S. simi	3,10	r	e,n,z_{15}	Group 01,3,19 (E₄)			
S. weltevreden	3,10	r	z_6	S. juba	1,3,19	a	1,7
S. seegefeld	3,10	r,i	1,2	S. gwoza	1,3,19	a	e,n,z_{15}
S. dumfries	3,10	r,i	1,6	S. gnesta[t]	1,3,19	b	1,5
S. amager[r]	3,10	y	1,2	S. visby	1,3,19	b	1,6
S. orion	3,10	y	1,5	S. tambacounda	1,3,19	b	e,n,x
S. mokola	3,10	y	1,7	S. kande	1,3,19	b	e,n,z_{15}
S. ohlstedt	3,10	y	e,n,x	S. broughton	1,3,19	b	l,w
S. bolton	3,10	y	e,n,z_{15}	S. accra	1,3,19	b	z_6
S. langensalza	3,10	y	l,w	S. madiago	1,3,19	c	1,7
S. stockholm	3,10	y	z_6	S. ahmadi	1,3,19	d	1,5
S. fufu	3,10	z	1,5	S. liverpool	1,3,19	d	e,n,z_{15}
S. II alexander	3,10	z	1,5	S. tilburg	1,3,19	d	l,w
S. huddinge	3,10	z	1,7	S. niloese	1,3,19	d	z_6
S. II finchley	3,10	z	e,n,x	S. vilvoorde	1,3,19	e,h	1,5
S. clerkenwell	3,10	z	l,w	S. sanktmarx	1,3,19	e,h	1,7
S. landwasser	3,10	z	z_6	S. sao	1,3,19	e,h	e,n,z_{15}
S. II tafelbaai	3,10	z	z_{39}	S. calabar	1,3,19	e,h	l,w
S. adabraka	3,10	z_4,z_{23}	[1,7]	S. rideau	1,3,19	f,g	—
S. florian	3,10	z_4,z_{24}	—	S. maiduguri	1,3,19	f,g,t	e,n,z_{15}
S. II	3,10	z_4,z_{24}	—	S. kouka	1,3,19	g,m,[t]	—
S. okerara	3,10	z_{10}	1,2	S. senftenberg[u]	1,3,19	g,[s],t	—
S. lexington[s]	3,10	z_{10}	1,5	S. cannstatt	1,3,19	m,t	—
S. coquilhatville	3,10	z_{10}	1,7	S. stratford	1,3,19	i	1,2
S. kristianstad	3,10	z_{10}	e,n,z_{15}	S. machaga	1,3,19	i	e,n,x
S. biafra	3,10	z_{10}	z_6	S. avonmouth	1,3,19	i	e,n,z_{15}
S. II	3,10	z_{29}	e,n,x	S. zuilen	1,3,19	i	l,w
S. jedburgh	3,10	z_{29}	—	S. taksony	1,3,19	i	z_6
S. zongo	3,10	z_{35}	1,7	S. ngor	1,3,19	l,v	1,5
S. shannon	3,10	z_{35}	l,w	S. parkroyal	1,3,19	l,v	1,7
S. cairina	3,10	z_{35}	z_6	S. westerstede	1,3,19	l,z_{13}	[1,2]
S. macallen	3,10	z_{36}	—	S. winterthur	1,3,19	l,z_{13}	1,6
S. bolombo	3,10	z_{38}	[z_6]	S. lokstedt	1,3,19	l,z_{13},z_{28}	1,2
S. II mpila	3,10	z_{38}	z_{42}	S. stuivenberg	1,3,19	l,z_{13},z_{28}	1,5
S. II winchester	3,10	z_{39}	1,7	S. bedford	1,3,19	l,z_{13},z_{28}	e,n,z_{15}
Group 03,15 (E₂)				S. tomelilla	1,3,19	l,z_{28}	1,7
(*Salmonella* serovars of group E₁ lysogenized by phage ε₁₅)				S. yalding	1,3,19	r	e,n,z_{15}
S. clichy	3,15	a	1,5	S. fareham	1,3,19	r,i	l,w
S. rosenthal	3,15	b	1,5	S. gatineau	1,3,19	y	1,5
S. westminster	3,15	b	z_{35}	S. krefeld	1,3,19	y	l,w
S. pankow	3,15	d	1,5	S. korlebu	1,3,19	z	1,5
S. eschersheim	3,15	d	e,n,x	S. lerum	1,3,19	z	1,7

Table 5.11—*continued*

Serovar	Somatic (O) antigens	Phase 1	Phase 2
S. schoeneberg	1,3,19	z	e,n,z_{15}
S. carno	1,3,19	z	l,w
S. sambre	1,3,19	z_4,z_{24}	—
S. dallgow	1,3,19	z_{10}	e,n,z_{15}
S. llandoff	1,3,19	z_{29}	$[z_6]$
S. chittagong	1,3,10,19	b	z_{35}
S. bilu	1,3,10,19	f,g,t	1,(2),7
S. ilugun	1,3,10,19	z_4,z_{23}	z_6
S. dessau	1,3,15,19	g,s,t	—
S. cannonhill	1,3,15,19	y	e,n,x
	Group 011 (F)		
S. gallen	11	a	1,2
S. marseille	11	a	1,5
S. toowong	11	a	1,7
S. luciana	11	a	e,n,z_{15}
S. epinay	11	a	l,z_{13},z_{28}
S. II glencairn	11	a	$z_6:z_{42}$
S. atento	11	b	1,2
S. leeuwarden	11	b	1,5
S. wohlen	11	b	1,6
S. II	11	b	1,7
S. II srinagar	11	b	e,n,x
S. pharr	11	b	e,n,z_{15}
S. chiredzi	11	c	1,5
S. gustavia	11	d	1,5
S. chandans	11	d	e,n,x
S. II montgomery	11	d,(a)	d,e,n,z_{15}
S. findorff	11	d	z_6
S. chingola	11	e,h	1,2
S. adamstua	11	e,h	1,6
S. redhill	11	e,h	l,z_{13},z_{28}
S. II grabouw	11	g,m,s,t	z_{39}
S. IV mundsburg	11	g,z_{51}	—
S. II lincoln	11	m,t	e,n,x
S. aberdeen	11	i	1,2
S. brijbhumi	11	i	1,5
S. heerlen	11	i	1,6
S. veneziana	11	i	e,n,x
S. pretoria	11	k	1,2
S. abaetetuba	11	k	1,5
S. sharon	11	k	1,6
S. colobane	11	k	1,7
S. kisarawe	11	k	$e,n,x[z_{15}]$
S. amba	11	k	l,z_{13},z_{28}
S. III arizonae (Ar. 17:29:25)	11	k	z_{53}
S. stendal	11	l,v	1,2
S. maracaibo	11	l,v	1,5
S. fann	11	l,v	e,n,x
S. bullbay	11	l,v	e,n,z_{15}
S. III arizonae (Ar. 17:23:31)	11	l,v	z
S. III arizonae (Ar. 17:23:25)	11	l,v	z_{53}
S. glidji	11	l,w	1,5
S. osnabrueck	11	l,z_{13},z_{28}	e,n,x
S. II huila	11	l,z_{28}	e,n,x
S. senegal	11	r	1,5
S. rubislaw	11	r	e,n,x
S. volta	11	r	l,z_{13},z_{28}
S. solt	11	y	1,5
S. jalisco	11	y	1,7
S. herzliya	11	y	e,n,x
S. nyanza	11	z	z_6
S. II soutpan	11	z	z_{39}
S. remete	11	z_4,z_{23}	1,6
S. etterbeek	11	z_4,z_{23}	e,n,z_{15}
S. III arizonae (Ar. 17:1,2,5:–)	11	z_4,z_{23}	—
S. IV parera	11	z_4,z_{23}	—
S. yehuda	11	z_4,z_{24}	—
S. IV	11	z_4,z_{32}	—
S. wentworth	11	z_{10}	1,2
S. straengnaes	11	z_{10}	1,5
S. telhashomer	11	z_{10}	e,n,x
S. lene	11	z_{38}	—
S. maastricht	11	z_{41}	1,2
S. II	11	—	1,5
	Group 013,22 (G_1)		
S. mim	13,22	a	1,6
S. marshall	13,22	a	l,z_{13},z_{28}
S. ibadan	13,22	b	1,5
S. oudwijk	13,22	b	1,6
S. rottnest	1,13,22	b	1,7
S. vaertan	13,22	b	e,n,x
S. bahati	13,22	b	e,n,z_{15}
S. II	1,13,22	b	z_{42}
S. haouaria	13,22	c	e,n,x,z_{15}
S. friedenau	13,22	d	1,6
S. diguel	1,13,22	d	e,n,z_{15}
S. willemstad	1,13,22	e,h	1,6
S. raus	13,22	f,g	e,n,x
S. II	13,22	(f),g,t	—
S. bron	13,22	g,m	$[e,n,z_{15}]$
S. II limbe	1,13,22	g,m,t	[1,5]
S. newyork	13,22	g,s,t	—
S. II rotterdam	1,13,22	g,t	1,5
S. washington	13,22	m,t	—
S. II	13,22	k	$1,5:z_{42}$
S. lovelace	13,22	l,v	1,5
S. borbeck	13,22	l,v	1,6
S. II	13,22	l,z_{28}	1,5
S. tanger	1,13,22	y	1,6
S. poona[v]	1,13,22	z	1,6
S. bristol	13,22	z	1,7
S. tanzania	1,13,22	z	e,n,z_{15}
S. ried	1,13,22	z_4,z_{23}	$[e,n,z_{15}]$
S. III arizonae (Ar. 18:1,2,5)	13,22	z_4,z_3	—
S. roodepoort	1,13,22	z_{10}	1,5
S. II clifton	13,22	z_{29}	1,5
S. II goodwood	13,22	z_{29}	e,n,x
S. agoueve	13,22	z_{29}	—
S. mampong	13,22	z_{35}	1,6
S. nimes	13,22	z_{35}	e,n,z_{15}
S. leiden	13,22	z_{38}	—
S. II	13,22	z_{39}	1,5,(7)
S. III arizonae (Ar. 18:–:–)	13,22	—	—
	Group 013,23 (G_2)		
S. chagoua	1,13,23	a	1,5
S. wyldegreen	13,23	a	l,w
S. II tygerberg	1,13,23	a	z_{42}
S. mississippi	1,13,23	b	1,5
S. II acres	1,13,23	b	$[1,5]:z_4$
S. bracknell	13,23	b	1,6
S. ullevi	1,13,23	b	e,n,x
S. durham	13,23	b	e,n,z_{15}
S. handen	1,13,23	d	1,2
S. mishmarhaemek	1,13,23	d	1,5
S. wichita[w]	1,13,23	d	[1,6]
S. grumpensis	13,23	d	1,7
S. II	13,23	d	e,n,x
S. telelkebir	13,23	d	e,n,z_{15}
S. putten	13,23	d	l,w
S. isuge	13,23	d	z_6
S. tschangu	1,13,23	e,h	1,5
S. II epping	1,13,23	e,n,x	1,7
S. havana	1,13,23	f,g,[s]	—
S. agbeni	13,23	g,m	—
S. II	13,23	g,m,s,t	1,5
S. II luanshya	1,13,23	g,m,[s],t	[e,n,x]
S. congo	13,23	g,m,s,t	—
S. okatie	13,23	g,s,t	—
S. II gojenberg	1,13,23	g,t	1,5
S. II	1,13,23	g,t	z_{42}

Table 5.11—*continued*

Serovar	Somatic (O) antigens	Flagellar (H) Antigens Phase 1	Phase 2
S. III *arizonae* (Ar. 18:13,14:–)	1,13,23	g,z_{51}	—
S. II *katesgrove*	1,13,23	m,t	1,5
S. II *worcester*	1,13,23	m,t	e,n,x
S. II *boulders*	1,13,23	m,t	z_{42}
S. *kintambo*	13,23	m,t	—
S. *idikan*	1,13,23	i	1,5
S. *jukestown*	13,23	i	e,n,z_{15}
S. *kedougou*	1,13,23	i	l,w
S. II	13,23	k	z_{41}
S. *nanga*	1,13,23	l,v	e,n,z_{15}
S. II	13,23	l,z_{28}	1,5
S. II	13,23	l,z_{28}	z_6
S. II *vredelust*	1,13,23	l,z_{28}	z_{42}
S. *adjame*	13,23	r	1,6
S. *linton*	13,23	r	e,n,z_{15}
S. *yarrabah*	13,23	y	1,7
S. *ordonez*	1,13,23	y	l,w
S. *tunis*	1,13,23	y	z_6
S. II *nachshonim*	1,13,23	z	1,5
S. *farmsen*	13,23	z	1,6
S. *worthington*	1,13,23	z	l,w
S. *ajiobo*	13,23	z_4,z_{23}	—
S. III *arizonae* (Ar. 18:1,6,7:–)	13,23	z_4,z_{23},z_{32}	—
S. *romanby*	13,23	z_4,z_{24}	—
S. III *arizonae* (Ar. 18:1,3,11:–)	1,13,23	z_4,z_{24}	—
S. *demerara*	13,23	z_{10}	l,w
S. II	1,13,23	z_{29}	e,n,x
S. *cubana*ᕽ	1,13,23	z_{29}	
S. *anna*	13,23	z_{35}	e,n,z_{15}
S. *fanti*	13,23	z_{38}	—
S. II *stevenage*	1,13,23	[z_{42}]	1,[5],7
S. II	13,23	—	1,6
Group 06,14 (H)			
S. *garba*	1,6,14,25	a	1,5
S. *ferlac*	1,6,14,25	a	e,n,x
S. *banjul*	1,6,14,25	a	e,n,z_{15}
S. *ndjamena*	1,6,14,25	b	1,2
S. *tucson*	[1],6,14,[25]	b	[1,7]
S. III *arizonae* (Ar. 7a,7c:43:28)	(6),14	b	e,n,x
S. *blijdorp*	1,6,14,25	c	1,5
S. *kassberg*	1,6,14,25	c	1,6
S. *runby*	1,6,14,25	c	e,n,x
S. *minna*	1,6,14,25	c	l,w
S. *heves*	6,14,24	d	1,5
S. *finkenwerder*	[1],6,14,[25]	d	1,5
S. *midway*	6,14,24	d	1,7
S. *florida*	[1],6,14,[25]	d	1,7
S. *lindern*	6,14,25	d	e,n,x
S. *charity*	1,6,14,25	d	e,n,x
S. *teko*	1,6,14,25	d	e,n,z_{15}
S. *encino*	1,6,14,25	d	l,z_{13},z_{28}
S. *albuquerque*	1,6,14,24	d	z_6
S. *bahrenfeld*	6,14,24	e,h	1,5
S. *onderstepoort*	1,6,14,[25]	e,h	1,5
S. *magumeri*	1,6,14,25	e,h	1,6
S. *beaudesert*	[1],6,14,[25]	e,h	1,7
S. *warragul*	1,6,14,25	g,m	—
S. *caracas*	[1],6,14,[25]	g,m,s	—
S. *catanzaro*	6,14	g,s,t	—
S. II *rooikrantz*	1,6,14	m,t	1,5
S. II *emmerich*	6,14	[m,t]	e,n,x
S. *kaitaan*	1,6,14,25	m,t	—
S. *mampeza*	1,6,14,25	i	1,5
S. *buzu*	1,6,14,25	i	1,7
S. *schalkwijk*	6,14,24	i	e,n,..
S. *moussoro*	1,6,14,25	i	e,n,z_{15}
S. *harburg*	1,6,14,25	k	1,5
S. II	6,14	k	[e,n,x]
S. III *arizonae* (Ar. 7a,7c:29:31)	(6),14	k	z
S. II	1,6,14	k	z_6:z_{42}
S. III *arizonae* (Ar. 7a,7c:29:25)	(6),14	k	z_{53}
S. *boecker*	[1],Ƅ,14,[25]	l,v	1,7
S. *horsham*	[1],6,14,[25]	l,v	e,n,x
S. III *arizonae* (Ar. 7a,7c:23:31)	(6),14	l,v	z
S. III *arizonae* (Ar. 7a,7c:23:21)	(6),14	l,v	z_{35}
S. *aflao*	1,6,14,25	l,z_{28}	e,n,x
S. III *arizonae* (Ar. 7a,7c:24:31)	(6),14	r	z
S. *surat*	[1],6,14,[25]	r,[i]	e,n,z_{15}
S. *carrau*	6,14[24]	y	1,7
S. *madelia*	1,6,14,25	y	1,7
S. *fischerkietz*	1,6,14,25	y	e,n,x
S. *mornington*	1,6,14,25	y	e,n,z_{15}
S. *homosassa*	1,6,14,25	z	1,5
S. *soahanina*	6,14,24	z	e,n,x
S. *sundsvall*	1,6,14,25	z	e,n,x
S. *poano*	1,6,14,25	z	1,z_{13},z_{28}
S. *bousso*	1,6,14,25	z_4,z_{23}	[e,n,z_{15}]
S. IV	6,14	z_4,z_{23}	—
S. *chichiri*	6,14,24	z_4,z_{24}	—
S. *uzaramo*	1,6,14,25	z_4,z_{24}	—
S. *nessa*	1,6,14,25	z_{10}	1,2
S. II *bornheim*	1,6,14,25	z_{10}	1,(2),7
S. II *simonstown*	1,6,14	z_{10}	1,5
S. III *arizonae* (Ar. 7a,7c:27:28)	(6),14	z_{10}	e,n,x,z_{15}
S. III *arizonae* (Ar. 7a,7c:27:[31]:[38])	(6),14	z_{10}	[z]:[z_{56}]
S. II *slangkop*	1,6,14	z_{10}	z_6:z_{42}
S. *potosi*	6,14	z_{36}	1,5
S. *sara*	1,6,14,25	z_{38}	[e,n,x]
S. II	1,6,14	z_{42}	1,6
S. III *arizonae* (Ar. 7a,7c...:26:21)	1,6,14,25	z_{52}	z_{35}
Group 016 (I)			
S. *hannover*	16	a	1,2
S. *brazil*	16	a	1,5
S. *amunigun*	16	a	1,6
S. *nyeko*	16	a	1,7
S. *togba*	16	a	e,n,x
S. *fischerhuette*	16	a	e,n,z_{15}
S. *heron*	16	a	z_6
S. *hull*	16	b	1,2
S. *wa*	16	b	1,5
S. *glasgow*	16	b	1,6
S. *hvittingfoss*	16	b	e,n,x
S. II	16	b.	e,n,x
S. *sangera*	16	b	e,n,z_{15}
S. *malstatt*	16	b	z_6
S. II	16	b	z_{39}
S. II	16	b	z_{42}
S. *vancouver*	16	c	1,5
S. *gafsa*	16	c	1,6
S. *shamba*	16	c	e,n,x
S. *hithergreen*	16	c	e,n,z_{15}
S. *oldenburg*	16	d	1,2
S. II	16	d	1,5
S. *sherbrooke*	16	d	1,6
S. *gaminara*	16	d	1,7
S. *barranquilla*	16	d	e,n,x
S. *nottingham*	16	d	e,n,z_{15}
S. *caen*	16	d	l,w
S. *barmbek*	16	d	z_6
S. *malakal*	16	e,h	1,2
S. *saboya*	16	e,h	1,5
S. *rhydyfelin*	16	e,h	e,n,x
S. *weston*	16	e,h	z_6
S. II *bellville*	16	e,n,x	1,(5),7
S. *tees*	16	f,g	—
S. *adeoyo*	16	g,m	—
S. *nikolaifleet*	16	g,m,s	—
S. II *mobeni*	16	g,[m],[s],t	e,n,x

Table 5.11—*continued*

Serovar	Somatic (O) antigens	Flagellar (H) Antigens Phase 1	Phase 2
S. II *merseyside*	16	g,t	[1,5]
S. II	16	m,t	e,n,x
S. II *rowbarton*	16	m,t	[z_{42}]
S. mpouto	16	m,t	—
S. amina	16	i	1,5
S. wisbech	16	i	1,7
S. frankfurt	16	i	e,n,z_{15}
S. pisa	16	i	l,w
S. abobo	16	i	z_6
S. III *arizonae* (Ar. 25:33:21)	16	i	z_{35}
S. szentes	16	k	1,2
S. nuatja	16	k	e,n,x
S. orientalis	16	k	e,n,z_{15}
S. III *arizonae* (Ar. 25:29:31)	16	k	z
S. III *arizonae* (Ar. 25:22:21)	16	(k)	z_{35}
S. III *arizonae* (Ar. 25:29:25)	16	k	z_{53}
S. III *arizonae* (Ar. 25:23:30)	16	l,v	1,5,7
S. shanghai[y]	16	l,v	1,6
S. welikade	16	l,v	1,7
S. salford	16	l,v	e,n,x
S. burgas	16	l,v	e,n,z_{15}
S. III *arizonae* (Ar. 25:23:31:[41])	16	l,v	z:[z_{61}]
S. losangeles	16	l,v,	z_6
S. III *arizonae* (Ar. 25:23:21)	16	l,v	z_{35}
S. III *arizonae* (Ar. 25:23:25)	16	l,v	z_{53}
S. westeinde	16	l,w	1,6
S. lomnava	16	l,w	e,n,z_{15}
S. II *noordhoek*	16	l,w	z_6
S. mandera	16	l,z_{13}	e,n,z_{15}
S. enugu	16	l,[z_{13}],z_{28}	[1,5]
S. battle	16	l,z_{13},z_{28}	1,6
S. ablogame	16	l,z_{13},z_{28}	z_6
S. II *sarepta*	16	l,z_{28}	z_{42}
S. rovaniemi	16	r,i	1,5
S. annedal	16	r,i	e,n,x
S. zwickau	16	r,i	e,n,z_{15}
S. saphra	16	y	1,5
S. akuafo	16	y	1,6
S. kikoma	16	y	e,n,x
S. avignon	16	y	e,n,z_{15}
S. fortlamy	16	z	1,6
S. lingwala	16	z	1,7
S. II *louwbester*	16	z	e,n,x
S. brevik	16	z	e,n,z_{15}
S. II	16	z	z_{42}
S. kibi	16	z_4,z_{23}	—
S. II *haddon*	16	z_4,z_{23}	—
S. IV *ochsenzoll*	16	z_4,z_{23}	—
S. IV *chameleon*	16	z_4,z_{32}	—
S. II	16	z_6	1,6
S. III *arizonae* (Ar. 25:27:30)	16	z_{10}	1,5,7
S. lisboa	16	z_{10}	1,6
S. III *arizonae* (Ar. 25:27:28)	16	z_{10}	e,n,x,z_{15}
S. redlands	16	z_{10}	e,n,z_{15}
S. angouleme	16	z_{10}	z_6
S. saloniki	16	z_{29}	—
S. II *jacksonville*	16	z_{29}	—
S. dakota	16	z_{35}	e,n,z_{15}
S. naware	16	z_{38}	—
S. II *woodstock*	16	z_{42}	1,(5),7
S. II *elsiesrivier*	16	z_{42}	1,6
S. III *arizonae* (Ar. 25:26:21)	16	z_{52}	z_{35}
Group 017 (J)			
S. bonames	17	a	1,2
S. jangwani	17	a	1,5
S. kinondoni	17	a	e,n,x
S. kirkee	17	b	1,2
S. II *hillbrow*	17	b	e,n,x,z_{15}
S. bignona	17	b	e,n,z_{15}
S. II	17	b	z_6
S. victoriaborg	17	c	1,6
S. II *woerden*	17	c	z_{39}
S. berlin	17	d	1,5
S. niamey	17	d	l,w
S. jubilee	17	e,h	1,2
S. II *verity*	17	e,n,x,z_{15}	1,6
S. II	17	e,n,x,z_{15}	1,7
S. II *bleadon*	17	(f),g,t	[e,n,x,z_{15}]
S. II	17	g,t	z_{39}
S. bama	17	m,t	—
S. II	17	m,t	—
S. ahanou	17	i	1,7
S. III *arizonae* (Ar. 12:33:21)	17	i	z_{35}
S. irenea	17	k	1,5
S. matadi	17	k	e,n,x
S. II	17	k	—
S. morotai	17	l,v	1,2
S. michigan	17	l,v	1,5
S. carmel	17	l,v	e,n,x
S. III *arizonae* (Ar. 12:23:28)	17	l,v	e,n,x,z_{15}
S. III *arizonae* (Ar. 12:23:21)	17	l,v	z_{35}
S. granlo	17	l,z_{28}	e,n,x
S. lode	17	r	1,2
S. III *arizonae* (Ar. 12:24:31)	17	r	z
S. II	17	y	—
S. gori	17	z	1,2
S. warengo	17	z	1,5
S. tchamba	17	z	e,n,z_{15}
S. II *constantia*	17	z	l,w:z_{42}
S. III *arizonae* (Ar. 12:1,2,5:-) (Ar.12:1,2,6:-)	17	z_4,z_{23}	—
S. III *arizonae* (Ar. 12:1,6,7,9:-)	17	z_4,z_{23},z_{32}	—
S. III *arizonae* (Ar. 12:1,3,11:-)	17	z_4,z_{24}	—
S. III *arizonae* (Ar. 12:1,6,7:-) (Ar.12:1,7,8:-)	17	z_4,z_{32}	—
S. djibouti	17	z_{10}	e,n,x
S. III *arizonae* (Ar. 12:27:28:[38])	17	z_{10}	e,n,x,z_{15}:[z_{56}]
S. III *arizonae* (Ar. 12:27:31)	17	z_{10}	z
S. kandla	17	z_{29}	—
S. III *arizonae* (Ar. 12:16,17,18:-)	17	z_{29}	—
S. III *arizonae* (Ar. 12:17,20:-)	17	z_{36}	—
Group 018 (K)			
S. brazos	6,14,18	a	e,n,z_{15}
S. fluntern	6,14,18	b	1,5
S. rawash	6,14,18	c	e,n,x
S. groenekan	18	d	1,5
S. usumbura	18	d	1,7
S. pontypridd	18	g,m	—
S. III *arizonae* (Ar. 7a,7b:13,14:-)	18	g,z_{51}	—
S. II	18	m,t	1,5
S. langenhorn	18	m,t	—
S. memphis	18	k	1,5
S. III *arizonae* (Ar. 7a,7b:22:25)	18	(k)	z_{53}
S. III *arizonae* (Ar. 7a,7b:22:34)	18	(k)	z_{54}
S. III *arizonae* (Ar. 7a,7b:23:28)	18	l,v	e,n,x,z_{15}
S. orlando	18	l,v	e,n,z_{15}
S. III *arizonae* (Ar. 7a,7b:23:31)	18	l,v	z
S. toulon	18	l,w	e,n,z_{15}
S. III *arizonae* (Ar. 7a,7b:24:31)	18	r	z

Table 5.11—*continued*

Serovar	Somatic (O) antigens	Phase 1	Phase 2	Serovar	Somatic (O) antigens	Phase 1	Phase 2
		Flagellar (H) Antigens				**Flagellar (H) Antigens**	
S. II	18	y	e,n,x,z_{15};	*S. mundonobo*	28	d	1,7
S. cerro	6,14,18	z_4,z_{23}	[1,5]	*S. mocamedes*	28	d	e,n,x
S. aarhus	18	z_4,z_{23}	z_{64}	*S. patience*	28	d	e,n,z_{15}
S. II	18	z_4,z_{23}	—	*S. cullingworth*	28	d	l,w
S. III *arizonae*	18	z_4,z_{23}	—	*S. kpeme*	28	e,h	1,7
(Ar. 7a,7b:1,2,5:–)				*S.* II	28	e,n,x	1,7
(Ar.7a,7b:1,2,6:–)				*S. friedrichsfelde*	28	f,g	—
S. blukwa	18	z_4,z_{24}	—	*S. abadina*	28	g,m	$[e,n,z_{15}]$
S. III *arizonae*	18	z_4,z_{32}	—	*S.* II *llandudno*	28	g,(m),[s],t	1,5
(Ar. 7a,7b:1,7,8:–)				*S. croft*	28	g,m,s	—
S. carnac	18	z_{10}	z_6	*S.* II	28	g,m,t	e,n,x
S. II *zeist*	18	z_{10}	z_6	*S.* II	28	g,s,t	e,n,x
S. II *beloha*	18	z_{36}	—	*S. ona*	28	g,s,t	—
S. IV	18	z_{36},z_{38}	—	*S.* II	28	m,t	[e,n,x]
S. sinthia	18	z_{38}	—	*S. vinohrady*	28	m,t	—
S. cotia	18	—	1,6	*S. doorn*	28	i	1,2
	Group 021 (L)			*S. cotham*	28	i	1,5
S. assen	21	a	[1,5]	*S. volkmarsdorf*	28	i	1,6
S. ghana	21	b	1,6	*S. dieuppeul*	28	i	1,7
S. minnesota^z	21	b	e,n,x	*S. warnemuende*	28	i	e,n,x
S. hydra	21	c	1,6	*S. kuessel*	28	i	e,n,z_{15}
S. rhone	21	c	e,n,x	*S. guildford*	28	k	1,2
S. II	21	c	e,n,x	*S. ilala*	28	k	1,5
S. spartel	21	d	1,5	*S. adamstown*	28	k	1,6
S. magwa	21	d	e,n,x	*S. ikeja*	28	k	1,7
S. madison	21	d	z_6	*S. taunton*	28	k	e,n,x
S. good	21	f,g	e,n,x	*S. ank*	28	k	e,n,z_{15}
S. III *arizonae* (Ar. 22:13,14:–)	21	g,z_{51}	—	*S. leoben*	28	l,v	1,5
S. diourbel	21	i	1,2	*S. vitkin*	28	l,v	e,n,x
S. III *arizonae* (Ar. 22:33:30)	21	i	1,5,7	*S. nashua*	28	l,v	e,n,z_{15}
S. III *arizonae* (Ar. 22:33:28)	21	i	e,n,x,z_{15}	*S. ramsey*	28	l,w	1,6
S. III *arizonae* (Ar. 22:29:28)	21	k	e,n,x,z_{15}	*S. fajara*	28	l,z_{28}	e,n,x
S. III *arizonae* (Ar. 22:29:31)	21	k	z	*S. bassadji*	28	r	1,6
S. III *arizonae* (Ar. 22:23:31)	21	l,v	z	*S. kibusi*	28	r	e,n,x
S. III *arizonae* (Ar. 22:23:40$_a$,40$_c$)	21	l,v	z_{57}	*S.* II *oevelgoenne*	28	r	e,n,z_{15}
S. keve	21	l,w	—	*S. chicago*	28	r,[i]	1,5
S. ruiru	21	y	e,n,x	*S. banco*	28	r,i	1,7
S. II	21	z	—	*S. sanktgeorg*	28	r,[i]	e,n,z_{15}
S. baguida	21	z_4,z_{23}	—	*S. oskarshamn*	28	y	1,2
S. III *arizonae* (Ar. 22:1,2,6:–)	21	z_4,z_{23}	—	*S. nima*	28	y	1,5
S. IV *soesterberg*	21	z_4,z_{23}	—	*S. pomona*	28	y	1,7
S. II *gwaai*	21	z_4,z_{24}	—	*S. kitenge*	28	y	e,n,x
S. III *arizonae* (Ar. 22:1,3,11:–)	21	z_4,z_{24}	—	*S. telaviv*	28	y	e,n,z_{15}
S. III *arizonae* (Ar. 22:27:28)	21	z_{10}	e,n,x,z_{15}	*S. shomolu*	28	y	l,w
S. III *arizonae* (Ar. 22:27:31)	21	z_{10}	z	*S. selby*	28	y	z_6
S. II *wandsbek*	21	z_{10}	z_6	*S. ezra*	28	z	1,7
S. III *arizonae* (Ar. 22:16,17,18:–)	21	z_{29}	—	*S. brisbane*	28	z	e,n,z_{15}
S. gambaga	21	z_{35}	e,n,z_{15}	*S.* II *ceres*	28	z	z_{39}
S. III *arizonae* (Ar. 22:32^{aa}:28)	21	z_{65}	e,n,x,z_{15}	*S. teltow*	28	z_4,z_{23}	1,6
	Group 028 (M)			*S. babelsberg*	28	z_4,z_{23}	$[e,n,z_{15}]$
S. solna	28	a	1,5	*S. rogy*	28	z_{10}	1,2
S. dakar	28	a	1,6	*S. farakan*	28	z_{10}	1,5
S. bakau	28	a	1,7	*S. malaysia*	28	z_{10}	1,7
S. seattle	28	a	e,n,x	*S. umbilo*	28	z_{10}	e,n,x
S. honelis	28	a	e,n,z_{15}	*S. luckenwalde*	28	z_{10}	e,n,z_{15}
S. moero	28	b	1,5	*S. moroto*	28	z_{10}	l,w
S. ashanti	28	b	1,6	*S.* III *arizonae* (Ar. 35:27:[40a,40c])	28	z_{10}	$[z_{57}]$
S. bokanjac	28	b	1,7	*S. djermaia*	28	z_{29}	—
S. langford	28	b	e,n,z_{15}	*S. babili*	28	z_{35}	1,7
S. II *kaltenhausen*	28	b	z_6	*S. aderike*	28	z_{38}	e,n,z_{15}
S. hermannswerder	28	c	1,5		**Group 030 (N)**		
S. eberswalde	28	c	1,6	*S. overvecht*	30	a	1,2
S. halle	28	c	1,7	*S. zehlendorf*	30	a	1,5
S. dresden	28	c	e,n,x	*S. guarapiranga*	30	a	e,n,x
S. wedding	28	c	e,n,z_{15}	*S. doulassame*	30	a	e,n,z_{15}
S. techimani	28	c	z_6	*S.* II *odijk*	30	a	z_{39}
S. amoutive	28	d	1,5	*S. louga*	30	b	1,2
S. hatfield	28	d	1,6	*S. aschersleben*	30	b	1,5
				S. urbana	30	b	e,n,x

Table 5.11 (*Continued*)

Serovar	Somatic (O) antigens	Phase 1	Phase 2	Serovar	Somatic (O) antigens	Phase 1	Phase 2
S. neudorf	30	b	e,n,z_{15}	*S.* III *arizonae* (Ar. 20:24:21)	35	r	z_{35}
S. II	30	b	z_6	*S.* III *arizonae* (Ar. 20:24:41)	35	r	z_{61}
S. zaire	30	c	1,7	*S. alachua*[ff]	35	z_4,z_{23}	—
S. morningside	30	c	e,n,z_{15}	*S.* III *arizonae* (Ar. 20:1,2,6:−)	35	z_4,z_{23}	—
S. II	30	c	z_{39}	*S. westphalia*	35	z_4,z_{24}	—
S. messina	30	d	1,5	*S.* III *arizonae* (Ar. 20:1,7,8:−)	35	z_4,z_{32}	—
S. livulu	30	e,h	1,2	*S. camberene*	35	z_{10}	1,5
S. II *slatograd*	30	f,g,t	—	*S. enschede*	35	z_{10}	l,w
S. godesberg	30	g,m	—	*S. ligna*	35	z_{10}	z_6
S. II	30	g,m,s	e,n,x	*S.* III *arizonae* (Ar. 20:27:21)	35	z_{10}	z_{35}
S. giessen	30	g,m,s	—	*S.* II *utbremen*	35	z_{29}	e,n,x
S. sternschanze[bb]	30	g,s,t	—	*S. widemarsh*	35	z_{29}	—
S. wayne	30	g,z_{51}	—	*S.* III *arizonae*	35	z_{29}	—
S. landau	30	i	1,2	(Ar. 20:16,17,18:−)			
S. morehead	30	i	1,5	*S.* III *arizonae* (Ar. 20:17,20:−)	35	z_{36}	—
S. soerenga	30	i	l,w	*S. haga*	35	z_{38}	—
S. hilversum	30	k	1,2	*S.* III *arizonae* (Ar. 20:26:30)	35	z_{52}	1,5,7
S. ramatgan	30	k	1,5	*S.* III *arizonae* (Ar. 20:26:28)	35	z_{52}	e,n,x,z_{15}
S. aqua	30	k	1,6	*S.* III *arizonae* (Ar. 20:26:31)	35	z_{52}	z
S. angoda	30	k	e,n,x	*S.* III *arizonae* (Ar. 20:26:21)	35	z_{52}	z_{35}
S. odozi	30	k	$e,n,[x],z_{15}$			Group 038 (P)	
S. II	30	k	e,n,x,z_{15}	*S.* II	38	b	1,2
S. ligeo	30	l,v	1,2	*S. rittersbach*	38	b	e,n,z_{15}
S. donna	30	l,v	1,5	*S. sheffield*	38	c	1,5
S. morocco	30	l,z_{13},z_{28}	e,n,z_{15}	*S. kidderminster*	38	c	1,6
S. gege	30	r	1,5	*S.* II *carletonville*	38	d	[1,5]
S. matopeni	30	y	1,2	*S. thiaroye*	38	e,h	1,2
S. bietri	30	y	1,5	*S. kasenyi*	38	e,h	1,5
S. steinplatz	30	y	1,6	*S. korovi*	38	g,m,[s]	—
S. baguirimi	30	y	e,n,x	*S.* II *foulpointe*	38	g,t	—
S. nijmegen	30	y	e,n,z_{15}	*S.* III *arizonae* (Ar. 16:13,14:−)	38	g,z_{51}	—
S. bodjonegoro	30	z_4,z_{24}	—	*S.* IV	38	g,z_{51}	—
S. II	30	z_6	1,6	*S. mgulani*	38	i	1,2
S. sada	30	z_{10}	1,2	*S. lansing*	38	i	1,5
S. kumasi	30	z_{10}	e,n,z_{15}	*S.* III *arizonae* (Ar. 16:33:25)	38	i	z_{53}
S. aragua	30	z_{29}	—	*S. echa*	38	k	1,2
S. kokoli	30	z_{35}	1,6	*S. mango*	38	k	1,5
S. wuiti	30	z_{35}	e,n,z_{15}	*S. inverness*	38	k	1,6
S. ago	30	z_{38}	—	*S. njala*	38	k	e,n,x
S. II	30	z_{39}	1,7	*S.* III *arizonae* (Ar. 16:29:31)	38	k	z
		Group 035 (O)		*S.* III *arizonae* (Ar. 16:29:25)	38	k	z_{53}
S. umhlatazana	35	a	e,n,z_{15}	*S.* III *arizonae* (Ar. 16:22:30)	38	(k)	1,5,7
S. tchad	35	b	—	*S.* III *arizonae* (Ar. 16:22:31)	38	(k)	z
S. yolo	35	c	—	*S.* III *arizonae*	38	(k)	$z_{35}:[z_{56}]$
S. dembe[cc]	35	d	l,w	(Ar. 16:22:21:[38])			
S. gassi	35	e,h	z_6	*S.* III *arizonae* (Ar. 16:22:34)	38	(k)	z_{54}
S. adelaide[dd]	35	f,g	—	*S.* III *arizonae* (Ar. 16:22:37)	38	(k)	z_{55}
S. II	35	f,g,t	1,5	*S. alger*	38	l,v	1,2
S. ealing	35	g,m,s	—	*S. kimberley*	38	l,v	1,5
S. II	35	g,m,s,t	—	*S. roan*	38	l,v	e,n,x
S. ebrie	35	g,m,t	—	*S.* III *arizonae* (Ar. 16:23:31)	38	l,v	z
S. anecho	35	g,s,t	—	*S.* III *arizonae* (Ar. 16:23:21)	38	l,v	z_{35}
S. II	35	g,t	z_{42}	*S.* III *arizonae*	38	l,v	$z_{35}:[z_{54}]$
S. agodi	35	g,t	—	(Ar. 16:23:25:[34])			
S. III *arizonae* (Ar. 20:13,14:−)	35	g,z_{51}	—	*S. lindi*	38	r	1,5
S. monschaui	35	m,t	—	*S.* III *arizonae* (Ar. 16:24:30)	38	r	1,5,7
S. III *arizonae* (Ar. 20:33:28)	35	i	e,n,x,z_{15}	*S. emmastad*	38	r	1,6
S. gambia	35	i	e,n,z_{15}	*S.* III *arizonae*	38	r	$z:[z_{57}]$
S. bandia	35	i	l,w	(Ar. 16:24:31:[$40_a,40_b$])			
S. III *arizonae* (Ar. 20:33:31)	35	i	z	*S.* III *arizonae* (Ar. 16:24:21)	38	r	z_{35}
S. III *arizonae* (Ar. 20:33:21)	35	i	z_{35}	*S. freetown*	38	y	1,5
S. III *arizonae* (Ar. 20:29:31)	35	k	z	*S. colombo*	38	y	1,6
S. III *arizonae* (Ar. 20:22:31)	35	(k)	z	*S. perth*	38	y	e,n,x
S. III *arizonae* (Ar. 20:22:21)	35	(k)	z_{35}	*S. yoff*	38	z_4,z_{23}	1,2
S. III *arizonae*[ee] (Ar. 20:29:25)	35	k	z_{53}	*S.* IV	38	z_4,z_{23}	—
S. III *arizonae* (Ar. 20:23:30)	35	l,v	1,5,7	*S. bangkok*	38	z_4,z_{24}	—
S. III *arizonae* (Ar. 20:23:21)	35	l,v	z_{35}	*S.* III *arizonae* (Ar. 16:27:31)	38	z_{10}	z
S. II	35	l,z_{28}	—	*S.* III *arizonae* (Ar. 16:27:25)	38	z_{10}	z_{53}
S. III *arizonae* (Ar. 20:24:28)	35	r	e,n,x,z_{15}	*S. klouto*	38	z_{38}	—
S. massakory	35	r	l,w	*S.* III *arizonae* (Ar. 16:39:25)	38	z_{47}	z_{53}

Table 5.11—*continued*

Serovar	Somatic (O) antigens	Flagellar (H) Antigens Phase 1	Phase 2
S. III arizonae (Ar. 16:26:21)	38	z_{52}	z_{35}
S. III arizonae (Ar. 16:26:25)	38	z_{52}	z_{53}
Group 039 (Q)			
S. II	39	a	z_{39}
S. wandsworth	39	b	1,2
S. abidjan	39	b	l,w
S. II	39	c	e,n,x
S. logone	39	d	1,5
S. mara	39	e,h	1,5
S. hofit	39	i	1,5
S. champaign	39	k	1,5
S. kokomlemle	39	l,v	e,n,x
S. oerlikon	39	l,v	e,n,z_{15}
S. II mondeor	39	l,z_{28}	e,n,x
S. anfo	39	y	1,2
S. windermere	39		1,5
Group 040 (R)			
S. shikomah	40	a	1,5
S. greiz	40	a	z_6
S. II	1,40	a	z_6
S. II springs	40	a	z_{39}
S. riogrande	40	b	1,5
S. saugus	40	b	1,7
S. johannesburg	1,40	b	e,n,x
S. duval	1,40	b	e,n,z_{15}
S. benguella	40	b	z_6
S. II	40	b	—
S. II suarez	1,40	c	e,n,x,z_{15}
S. II	1,40	c	z_{39}
S. driffield	1,40	d	1,5
S. II ottershaw	40	d	—
S. tilene	1,40	e,h	1,2
S. II	1,40	(f),g	e,n,x,z_{15}
S. bijlmer	1,40	g,m	—
S. II boksburg	40	g,m,s,t	e,n,x
S. II alsterdorf	1,40	g,m,t	1,5
S. II	1,40	g,t	1,5
S. II	1,40	g,t	e,n,x
S. II	1,40	g,t	z_{42}
S. III arizonae (Ar. 10a,10b:13,14:28)	40	g,z_{51}	e,n,x,z_{15}
S. IV seminole	1,40	g,z_{51}	—
S. II	40	m,t	z_{39}
S. II	1,40	m,t	z_{42}
S. IV	40	m,t	—
S. III arizonae (Ar. 10a,10b:33:30)	40	i	1,5,7
S. goulfey	1,40	k	1,5
S. allandale	1,40	k	1,6
S. hann	40	k	e,n,x
S. II sunnydale	1,40	k	e,n,x,z_{15}
S. III arizonae (Ar. 10a,10b:29:31:40a,40c)	40	k	$z:z_{57}$
S. III arizonae (Ar. 10a,10b:29:25)	40	k	z_{53}
S. millesi	1,40	l,v	1,2
S. III arizonae (Ar. 10a,10b,(10c):23:31)	40	l,v	z
S. III arizonae (Ar. 10a,10b:23:25)	40	l,v	z_{53}
S. overchurch	40	l,w	—
S. bukavu	1,40	l,z_{28}	1,5
S. santhiaba	40	l,z_{28}	1,6
S. II bulawayo	1,40	z	1,5
S. casamance	40	z	e,n,x
S. nowawes	40	z	z_6
S. II	1,40	z	z_6
S. II	40	z	z_{39}
S. III arizonae (Ar. 10a,10b:1,2,5:-) (Ar.10a,10b:1,2,6:-)	40	z_4,z_{23}	—
S. IV sachsenwald	1,40	z_4,z_{23}	—
S. II degania	40	z_4,z_{24}	z_{39}
S. III arizonae (Ar. 10a,10b:1,3,11:-)	40	z_4,z_{24}	—
S. IV	40	z_4,z_{24}	—
S. III arizonae (Ar.10a,10b:1,7,8:-)	40	z_4,z_{32}	—
S. IV	40	z_4,z_{32}	—
S. II	1,40	z_6	1,5
S. trotha	40	z_{10}	z_6
S. III arizonae (Ar. 10a,10b:27:21)	40	z_{10}	z_{35}
S. omifisan	40	z_{29}	—
S. III arizonae (Ar. 10a,10b:16,17,18:-)	40	z_{29}	—
S. II fandran	1,40	z_{35}	e,n,x,z_{15}
S. III arizonae (Ar. 10a,10b:17,20:-)	40	z_{36}	—
S. II grunty	1,40	z_{39}	1,6
S. karamoja	1,40	z_{41}	1,2
S. II	1,40	$[z_{42}]$	1,(5),7
Group 041 (S)			
S. II	41	b	[1,5]
S. II	41	b	1,7
S. vietnam	41	b	$[z_6]$
S. III arizonae (Ar. 13:32a,32b:28)	41	c	e,n,x,z_{15}
S. II	41	c	z_6
S. egusi	41	d	[1,5]
S. II hennepin	41	d	z_6
S. II lethe	41	g,t	—
S. III arizonae (Ar. 13:13,14:-)	41	g,z_{51}	—
S. leatherhead	41	m,t	1,6
S. II	41	k	—
S. III arizonae (Ar. 13:22:21)	41	(k)	z_{35}
S. II	41	l,z_{13},z_{28}	e,n,x,z_{15}
S. lumbumbashi	41	r	1,5
S. II dubrovnik	41	z	1,5
S. waycross	41	z_4,z_{23}	—
S. III arizonae (Ar. 13:1,2,5:-) (Ar.13:1,2,6:-)	41	z_4,z_{23}	—
S. IV	41	z_4,z_{23}	—
S. III arizonae (Ar. 13:1,6,7:-)	41	z_4,z_{23},z_{32}	—
S. ipswich	41	z_4,z_{24}	[1,5]
S. III arizonae (Ar. 13:1,3,11:-)	41	z_4,z_{24}	—
S. III arizonae (Ar. 13:1,7,8:-)	41	z_4,z_{32}	—
S. II negev	41	z_{10}	1,2
S. leipzig	41	z_{10}	1,5
S. landala	41	z_{10}	1,6
S. inpraw	41	z_{10}	e,n,x
S. II lurup	41	z_{10}	$[e,n,x,z_{15}]$
S. II lichtenberg	41	z_{10}	z_6
S. lodz	41	z_{29}	—
S. III arizonae (Ar. 13:16,17,18:-)	41	z_{29}	—
S. III arizonae (Ar. 13:17,20:-)	41	z_{36}	—
S. offa	41	z_{38}	—
S. II	41	—	1,6
Group 042 (T)			
S. faji	1,42	a	e,n,z_{15}
S. II chinovum	42	b	1,5
S. II uphill	42	b	e,n,x,z_{15}
S. tomegbe	1,42	b	e,n,z_{15}
S. egusitoo	1,42	b	z_6
S. antwerpen	1,42	c	e,n,z_{15}
S. kampala	1,42	c	z_6
S. II fremantle	42	(f),g,t	—
S. maricopa	1,42	g,z_{51}	1,5
S. III arizonae (Ar. 15:13,14:-)	42	g,z_{51}	—
S. II	42	m,t	$[e,n,x,z_{15}]$

Table 5.11—*continued*

Serovar	Somatic (O) antigens	Flagellar (H) Antigens Phase 1	Phase 2
S. waral	1,42	m,t	—
S. kaneshie	1,42	i	l,w
S. middlesbrough	1,42	i	z_6
S. haferbreite	42	k	1,6
S. III *arizonae* (Ar. 15:29:31)	42	k	z
S. gwale	1,42	k	z_6
S. III *arizonae* (Ar. 15:22:21)	42	(k)	z_{35}
S. III *arizonae* (Ar. 15:23:30)	42	l,v	1,5,7
S. II *portbech*	42	l,v	e,n,x,z_{15}
S. III *arizonae* (Ar. 15:23:28)	42	l,v	e,n,x,z_{15}
S. coogee	42	l,v	e,n,z_{15}
S. III *arizonae* (Ar. 15:23:31)	42	l,v	z
S. III *arizonae* (Ar. 15:23:25)	42	l,v	z_{53}
S. II	42	l,z_{13},z_{28}	z_6
S. II	42	l,z_{28}	—
S. sipane	1,42	r	e,n,z_{15}
S. brive	1,42	r	l,w
S. III *arizonae* (Ar. 15:24:31)	42	r	z
S. III *arizonae* (Ar. 15:24:25)	42	r	z_{53}
S. II *nairobi*	42	r	—
S. III *arizonae*[gg] (Ar. 15:24:–)	42	r	—
S. harvestehude	1,42	y	z_6
S. II *detroit*	42	z	1,5
S. ursenbach	42	z	1,6
S. II *rand*	42	z	e,n,x,z_{15}
S. II *nuernberg*	42	z	z_6
S. gera	1,42	z_4,z_{23}	1,6
S. III *arizonae* (Ar. 15:1,2,5:–) (Ar.15:1,2,6:–)	42	z_4,z_{23}	—
S. toricada	1,42	z_4,z_{24}	—
S. III *arizonae* (Ar. 15:1,3,11:–)	42	z_4,z_{24}	—
S. II	42	z_6	1,6
S. II	42	z_{10}	e,n,x,z_{15}
S. III *arizonae* (Ar. 15:27:28)	42	z_{10}	e,n,x,z_{15}
S. III *arizonae* (Ar. 15:27:31)	42	z_{10}	z
S. loenga	1,42	z_{10}	z_6
S. II	42	z_{10}	z_6
S. III *arizonae* (Ar. 15:27:21)	42	z_{10}	z_{35}
S. III *arizonae* (Ar. 15:27:38)	42	z_{10}	z_{56}
S. djama	1,42	z_{29}	—
S. kahla	1,42	z_{35}	1,6
S. weslaco	42	z_{36}	—
S. IV	42	z_{36}	—
S. vogan	1,42	z_{38}	z_6
S. taset	1,42	z_{41}	—
S. III *arizonae* (Ar. 15:26:31)	42	z_{52}	z
S. II	42		1,6
Group 043 (U)			
S. graz	43	a	1,2
S. berkeley	43	a	1,5
S. II	43	a	z_6
S. II *kommetje*	43	b	z_{42}
S. montreal	43	c	1,5
S. II	43	d	e,n,x,z_{15}
S. II	43	d	z_{39}
S. II	43	d	z_{42}
S. II	43	e,n,x,z_{15}	1,(5),7
S. II	43	e,n,x,z_{15}	1,6
S. milwaukee	43	f,g	—
S. II	43	f,g,t	1,5
S. II *mosselbay*	43	g,m,[s],t	$[z_{42}]$
S. veddel	43	g,t	—
S. IV	43	g,z_{51}	—
S. II	43	g,z_{62}	e,n,x
S. mbao	43	i	1,2
S. thetford	43	k	1,2
S. ahuza	43	k	1,5
S. III *arizonae* (Ar.21:29:31)	43	k	z
S. III *arizonae* (Ar. 21:23:25)	43	l,v	z_{53}
S. III *arizonae* (Ar. 21:23:38)	43	l,v	z_{56}
S. III *arizonae* (Ar. 21:24:28)	43	r	e,n,x,z_{15}
S. III *arizonae* (Ar. 21:24:31)	43	r	z
S. III *arizonae* (Ar. 21:24:25)	43	r	z_{53}
S. farcha	43	y	1,2
S. kingabwa	43	y	1,5
S. ogbete	43	z	1,5
S. II	43	z	1,5
S. III *arizonae* (Ar. 21:1,2,5:–)	43	z_4,z_{23}	—
S. IV *houten*	43	z_4,z_{23}	—
S. III *arizonae* (Ar. 21:1,3,11:–)	43	z_4,z_{24}	—
S. IV	43	z_4,z_{24}	—
S. IV *tuindorp*	43	z_4,z_{32}	—
S. adana	43	z_{10}	1,5
S. II	43	z_{29}	e,n,x
S. II	43	z_{29}	z_{42}
S. IV	43	z_{29}	—
S. ahepe	43	z_{35}	1,6
S. III *arizonae* (Ar. 21:17,20:–)	43	z_{36}	—
S. IV *volksdorf*	43	z_{36},z_{38}	—
S. irigny	43	z_{38}	—
S. II *bunnik*	43	z_{42}	[1,5,7]
S. III *arizonae* (Ar. 21:26:25)	43	z_{52}	z_{53}
Group 044 (V)			
S. niakhar	44	a	1,5
S. tiergarten	44	a	e,n,x
S. niarembe	44	a	l,w
S. sedgwick	44	b	e,n,z_{15}
S. madigan	44	c	1,5
S. quebec	44	c	e,n,z_{15}
S. bobo	44	d	1,5
S. kermel	44	d	e,n,x
S. fischerstrasse	44	d	e,n,z_{15}
S. II	1,44	e,n,x	1,6
S. vleuten	44	f,g	—
S. gamaba	44	g,m,s	—
S. II	44	g,t	z_{42}
S. carswell	44	g,z_{51}	—
S. IV	44	g,z_{51}	—
S. muguga	44	m,t	—
S. lawra	44	k	e,n,z_{15}
S. malika	44	l,z_{28}	1,5
S. brefet	44	r	e,n,z_{15}
S. V *camdeni*	44	r	—
S. uhlenhorst	44	z	l,w
S. kua	44	z_4,z_{23}	—
S. II	44	z_4,z_{23}	—
S. III *arizonae* (Ar. 1,3:1,2,5:–) (Ar.1,3:1,2,6:–)	44	z_4,z_{23}	—
S. IV	44	z_4,z_{23}	—
S. III *arizonae* (Ar. 1,3:1,6,7,9:–) (Ar. 1,3:1,2,10:–)	44	z_4,z_{23},z_{32}	—
S. christiansborg	44	z_4,z_{24}	—
S. III *arizonae* (Ar. 1,3:1,3,11:–)	44	z_4,z_{24}	—
S. IV	44	z_4,z_{24}	—
S. III *arizonae* (Ar. 1,3:1,7,8:–)	44	z_4,z_{32}	—
S. IV *lohbruegge*	44	z_4,z_{32}	—
S. guinea	44	z_{10}	[1,7]
S. IV	44	$z_{36},[z_{38}]$	—
S. koketime	44	z_{38}	—
S. II *clovelly*	1,44	z_{39}	$[e,n,x,z_{15}]$
Group 045 (W)			
S. II *vrindaban*	45	a	e,n,x
S. meekatharra	45	a	e,n,z_{15}
S. II *ejeda*	45	a	z_{10}
S. riverside	45	b	1,5
S. fomeco	45	b	e,n,z_{15}
S. deversoir	45	c	e,n,x
S. dugbe	45	d	1,6
S. karachi	45	d	e,n,x
S. suelldorf	45	f,g	—
S. tornow	45	g,m,[s]	—

Table 5.11—*continued*

Serovar	Somatic (O) antigens	Phase 1	Phase 2
S. II *windhoek*	45	g,m,s,t	1,5
S. II *bremen*	45	g,m,s,t	e,n,x
S. II *perinet*	45	g,m,t	e,n,x,z_{15}
S. binningen	45	g,s,t	–
S. III *arizonae* (Ar. 11:13,14:–)	45	g,z_{51}	–
S. IV	45	g,z_{51}	–
S. II	45	m,t	1,5
S. apapa	45	m,t	–
S. casablanca	45	k	1,7
S. cairns	45	k	e,n,z_{15}
S. II *klapmuts*	45	z	z_{39}
S. IV	45	z_4,z_{23}	–
S. III *arizonae* (Ar. 11:1,3,11:–)	45	z_4,z_{24}	–
S. III *arizonae* (Ar. 11:1,7,8:–)	45	z_4,z_{32}	–
S. II	45	z_{29}	1,5
S. II	45	z_{29}	z_{42}
S. jodhpur	45	z_{29}	–
S. III *arizonae* (Ar. 11:16,17,18:–)	45	z_{29}	–
S. lattenkamp	45	z_{35}	1,5
S. balcones	45	z_{36}	–
Group 047 (X)			
S. II *bilthoven*	47	a	[1,5]
S. II	47	a	e,n,x,z_{15}
S. II *phoenix*	47	b	1,5
S. II *khami*	47	b	[e,n,x,z_{15}]
S. sya	47	b	z_6
S. saka	47	b	–
S. III *arizonae* (Ar. 28:43:–)	47	b	–
S. III *arizonae* (Ar. 28:32:30)	47	c	1,5,7
S. III *arizonae* (Ar.23:32:28) (Ar. 28:32:28:[$40_a,40_c$])	47	c	e,n,x,z_{15}:[z_{57}]
S. III *arizonae* (Ar. 28:32:31)	47	c	z
S. III *arizonae* (Ar. 28:32:21)	47	c	z_{35}
S. kodjovi	47	c	–
S. stellingen	47	d	e,n,x
S. II *quimbamba*	47	d	z_{39}
S. sljeme	1,47	f,g	–
S. luke	1,47	g,m	–
S. anie	47	(g),m,t	–
S. II	47	g,t	e,n,x
S mesbit	47	m,t	e,n,z_{15}
S. III *arizonae*[hh] (Ar. 23:33:28)	47	i	e,n,x,z_{15}
S. bergen	47	i	e,n,z_{15}
S. III *arizonae* (Ar. 28:33:31)	47	i	z
S. III *arizonae* (Ar. 23:33:21) (Ar.28:33:21)	47	i	z_{35}
S. III *arizonae* (Ar. 23:33:25) (Ar.28:33:25:[$40_a,40_c$])	47	i	z_{53}:[z_{57}]
S. staoueli	47	k	1,2
S. bootle	47	k	1,5
S. III *arizonae* (Ar. 28:29:30)	47	k	1,5,7
S. dahomey[ii]	47	k	1,6
S. III *arizonae* (Ar. 28:29:28)	47	k	e,n,x,z_{15}
S. lyon	47	k	e,n,z_{15}
S. III *arizonae* (Ar. 28:29:31)	47	k	z
S. III *arizonae* (Ar. 23:29:21)	47	k	z_{35}
S. III *arizonae* (Ar. 23:29:25)	47	k	z_{53}
S. III *arizonae*[jj] (Ar. 23:23:30)	47	l,v	1,5,(7)
S. III *arizonae* (Ar. 28:23:28)	47	l,v	e,n,x,z_{15}
S. III *arizonae* (Ar. 28:23:21)	47	l,v	z_{35}
S. III *arizonae* (Ar. 28:23:25)	47	l,v	z_{53}
S. III *arizonae* (Ar. 28:23:$40a,40c$)	47	l,v	z_{57}
S. teshie	1,47	l,z_{13},z_{28}	e,n,z_{15}
S. dapango	47	r	1,2
S. III *arizonae* (Ar. 23:24:30)	47	r	1,5,7
S. III *arizonae* (Ar. 23:24:31)	47	r	z
S. III *arizonae* (Ar. 23:24:21)	47	r	z_{35}
S. III *arizonae* (Ar. 23:24:25:[44])	47	r	z_{53}:[z_{60}]
S. III *arizonae*[kk] (Ar. 23:24:–)	47	r	
S. moualine	47	y	1,6
S. blitta	47	y	e,n,x
S. mountpleasant	47	z	1,5
S. kaolack	47	z	1,6
S. II	47	z	e,n,x,z_{15}
S. II *chersina*	47	z	z_6
S. tabligbo	47	z_4,z_{23}	e,n,z_{15}
S. bere[ll]	47	z_4,z_{23}	z_6
S. tamberma	47	z_4,z_{24}	–
S. II	47	z_6	1,6
S. III *arizonae* (Ar. 28:27:30)	47	z_{10}	1,5,7
S. III *arizonae* (Ar. 28:27:31)	47	z_{10}	z
S. III *arizonae* (Ar. 28:27:21)	47	z_{10}	z_{35}
S. ekpoui	47	z_{29}	–
S. III *arizonae* (Ar. 28:16,17,18:–)	47	z_{29}	–
S. bingerville	47	z_{35}	e,n,z_{15}
S. alexanderplatz	47	z_{38}	–
S. quinhon	47	z_{44}	–
S. III *arizonae* (Ar. 28:26:30)	47	z_{52}	1,5,7
S. III *arizonae* (Ar. 28:26:28)	47	z_{52}	e,n,x,z_{15}
S. III *arizonae* (Ar. 28:26:31)	47	z_{52}	z
S. III *arizonae* (Ar. 28:26:21)	47	z_{52}	z_{35}
Group 048 (Y)			
S. hisingen	48	a	1,5,7
S. II	48	a	z_6
S. III *arizonae* (Ar. 5:35:[21])	48	a	[z_{35}]
S. II	48	a	z_{39}
S. II	48	b	z_6
S. III *arizonae* (Ar. 5,29:32:31)	48	c	z
S. II *hagenbeck*	48	d	z_6
S. fitzroy	48	e,h	1,5
S. II *hammonia*	48	e,n,x,z_{15}	z_6
S. II *erlangen*	48	g,m,t	–
S. III *arizonae* (Ar. 5:13,14:–)	48	g,z_{51}	–
S. IV *marina*	48	g,z_{51}	–
S. III *arizonae* (Ar. 5,29:33:31)	48	i	z
S. III *arizonae* (Ar. 29:33:21:[40])	48	i	z_{35}:[z_{57}]
S. III *arizonae* (Ar. 5:33:25)	48	i	z_{53}
S. III *arizonae* (Ar. 5:29:30)	48	k	1,5,(7)
S. II	48	k	e,n,x,z_{15}
S. III *arizonae* (Ar. 5:29:28)	48	k	e,n,x,z_{15}
S. dahlem	48	k	e,n,z_{15}
S. III *arizonae* (Ar. 5,29:31)	48	k	z
S. III *arizonae* (Ar.5:29:21)	48	k	z_{35}
S. II *sakaraha*	48	k	z_{39}
S. III *arizonae* (Ar. 5,29:29:25)	48	k	z_{53}
S. III *arizonae* (Ar. 5:22:25)	48	(k)	z_{53}
S. III *arizonae*[mm] (Ar. 5:23:30)	48	l,v	1,5,(7)
S. III *arizonae* (Ar. 5,29:23:31)	48	l,v	z
S. III *arizonae* (Ar. 5:24:28)	48	r	e,n,x,z_{15}
S. III *arizonae* (Ar. 5,29:24:31)	48	r	z
S. toucra[nn]	48	z	1,5
S. III *arizonae* (Ar. 5:1,2,5:–) (Ar.5:1,2,5,6:–) (Ar.5:1,6:–)	48	z_4,z_{23}	–
S. III *arizonae* (Ar. 5:1,6,7:–)	48	z_4,z_{23},z_{32}	–
S. djakarta	48	z_4,z_{24}	–
S. III *arizonae* (Ar. 5:1,3,11:–)	48	z_4,z_{24}	–
S. III *arizonae* (Ar. 5:1,7,8:–)	48	z_4,z_{32}	–
S. IV	48	z_4,z_{32}	–
S. II *ngozi*	48	z_{10}	[1,5]
S. isaszeg	48	z_{10}	e,n,x
S. III *arizonae* (Ar. 5:27:28)	48	z_{10}	e,n,x,z_{15}
S. III *arizonae* (Ar. 5,29:27:31)	48	z_{10}	z
S. II	48	z_{29}	–

Table 5.11—*continued*

Serovar	Somatic (O) antigens	Flagellar (H) Antigens Phase 1	Phase 2
S. V *bongor*	48	z_{35}	—
S. III *arizonae* (Ar. 5,29:17,20:–)	48	z_{36}	—
S. IV	48	z_{36},z_{38}	—
S. V *balboa*	48	z_{41}	—
S. III *arizonae* (Ar. 5,29:26:28)	48	z_{52}	e,n,x,z_{15}
S. III *arizonae* (Ar. 5:26:31)	48	z_{52}	z
Group 050 (Z)			
S. rochdale	50	b	e,n,x
S. II	50	b	z_6
S. II *krugersdorp*	50	e,n,x	1,7
S. II *namib*	50	g,m,s,t	1,5
S. IV *wassenaar*	50	g,z_{51}	—
S. II *atra*	50	m,t	$z_6{:}z_{42}$
S. III *arizonae* (Ar. 9a,9c:33:30)	50	i	1,5,7
S. III *arizonae* (Ar. 9a,9c:33:28)	50	i	e,n,x,z_{15}
S. III *arizonae* (Ar. 9a,9c:33:31)	50	i	z
S. III *arizonae* (Ar. 9a,9c:29:30)	50	k	1,5,7
S. III *arizonae* (Ar. 9a,9c:29:28)	50	k	e,n,x,z_{15}
S. III *arizonae*°° (Ar. 9a,9b:29:31) (Ar. 9a,9c:29:31)	50	k	z
S. III *arizonae* (Ar. 9a,9b:22:31)	50	(k)	z
S. II *seaforth*	50	k	z_6
S. III *arizonae* (Ar. 9a,9b:29:21)	50	k	z_{35}
S. III *arizonae* (Ar. 9a,9b:22:21)	50	(k)	z_{35}
S. III *arizonae* (Ar. 9a,9c:29:25)	50	k	z_{53}
S. fass	50	l,v	1,2
S. III *arizonae* (Ar. 9a,9b:23:28)	50	l,v	e,n,x,z_{15}
S. III *arizonae* (Ar. 9a,9c:23:31)	50	l,v	z
S. III *arizonae* (Ar. 9a,9c:23:21)	50	l,v	z_{35}
S. II	50	l,w	$e,n,x,z_{15}{:}z_{42}$
S. II	50	l,z_{28}	z_{42}
S. III *arizonae* (Ar. 9a,9b:24:30)	50	r	1,5,(7)
S. III *arizonae* (Ar. 9a,9c:24:28)	50	r	e,n,x,z_{15}
S. III *arizonae* (Ar.9a,9b:24:31) (Ar. 9a,9c:24:31)	50	r	z
S. III *arizonae* (Ar. 9a,9b:24:21)	50	r	z_{35}
S. III *arizonae* (Ar. 9a,9b:24:25)	50	r	z_{53}
S. dougi	50	y	1,6
S. II *greenside*	50	z	e,n,x
S. III *arizonae* (Ar. 9a,9b:1,2,5:–) (Ar.9a,9b:1,2,6:–)	50	z_4,z_{23}	—
S. IV *flint*	50	z_4,z_{23}	—
S. III *arizonae* (Ar. 9a,9b:1,6,7:–)	50	z_4,z_{23},z_{32}	—
S. III *arizonae* (Ar. 9a,9b:1,3,11:–)	50	z_4,z_{24}	—
S. IV	50	z_4,z_{24}	—
S. III *arizonae* (Ar. 9a,9b:1,2,10:–) (Ar. 9a,9b:1,7,8:–)	50		—
S. IV *bonaire*	50	z_4,z_{32}	—
S. III *arizonae* (Ar. 9a,9c:27:31:[38])	50	z_{10}	$z{:}[z_{56}]$
S. II *hooggraven*	50	z_{10}	$z_6{:}z_{42}$
S. III *arizonae* (Ar. 9a,9c:27:25)	50	z_{10}	z_{53}
S. III *arizonae* (Ar. 9a,9b:16,17,18:–)	50	z_{29}	—
S. III *arizonae*	50	z_{36}	

Serovar	Somatic (O) antigens	Flagellar (H) Antigens Phase 1	Phase 2
(Ar. 9a,9b:17,20:–)			
S. II *faure*	50	z_{42}	1,7
S. III *arizonae*	50	z_{42}	1,7
(Ar. 9a,9b:26:30) (Ar. 9a,9c:26:30)	50	z_{52}	1,5,7
S. III *arizonae* (Ar. 9a,9b:26:31) (Ar. 9a,9c:26:31)	50	z_{52}	z
S. III *arizonae* (Ar. 9a,9b:26:21) (Ar. 9a,9c:26:21)	50	z_{52}	z_{35}
S. III *arizonae* (Ar. 9a,9b:26:25) (Ar. 9a,9c:26:25)	50	z_{52}	z_{53}
Group 051			
S. tione	51	a	e,n,x
S. II	51	c	—
S. gokul	1,51	d	[1,5]
S. meskin	51	e,h	1,2
S. III *arizonae* (Ar. 1,2:13,14:–)	51	g,z_{51}	—
S. kabete	51	i	1,5
S. dan	51	k	e,n,z_{15}
S. III *arizonae* (Ar. 1,2:29:21)	51	k	z_{35}
S. overschie	51	l,v	1,5
S. dadzie	51	l,v	e,n,x
S. III *arizonae* (Ar. 1,2:23:31)	51	l,v	z
S. II *askraal*	51	l,z_{28}	$[z_6]$
S. antsalova	51	z	1,5
S. treforest	1,51	z	1,6
S. lechler	51	z	e,n,z_{15}
S. III *arizonae* (Ar. 1,2:1,2,5:–) (Ar. 1,2:1,2,6:–)	51	z_4,z_{23}	—
S. IV *harmelen*	51	z_4,z_{23}	—
S. III *arizonae* (Ar. 1,2:1,3,11:–)	51	z_4,z_{24}	—
S. II	51	z_{29}	e,n,x,z_{15}
S. II *roggeveld*	51	—	1,7
Group 052			
S. uithof	52	a	1,5
S. ord	52	a	e,n,x,z_{15}
S. molesey	52	b	1,5
S. flottbek	52	b	[e,n,x]
S. II	52	c	k
S. utrecht	52	d	1,5
S. II	52	d	e,n,x,z_{15}
S. butare	52	e,h	1,6
S. derkle	52	e,h	1,7
S. saintemarie	52	g,t	—
S. II	52	g,t	—
S. III *arizonae* (Ar.31:29:21)	52	k	z_{35}
S. III *arizonae* (Ar. 31:29:25)	52	k	z_{53}
S. III *arizonae* (Ar. 31:23:25)	52	l,v	z_{53}
S. II *lobatsi*	52	z_{44}	1,5,7
S. III *arizonae* (Ar. 31:26:31)	52	z_{52}	z
Group 053			
S. II	53	d	1,5
S. II	1,53	d	z_{39}
S. II	53	d	z_{42}
S. III *arizonae* (Ar. 1,4:13,14:–)	53	g,z_{51}	—
S. IV	1,53	g,z_{51}	—
S. III *arizonae* (Ar. 1,4:33:31)	53	i	z
S. III *arizonae* (Ar. 1,4:29:28)	53	k	e,n,x,z_{15}
S. III *arizonae* (Ar. 1,4:29:31)	53	k	z
S. III *arizonae* (Ar. 1,4:22:31)	53	(k)	z
S. III *arizonae* (Ar.1,4:22:21)	53	(k)	z_{35}
S. III *arizonae* (Ar. 1,4:23:28)	53	l,v	e,n,x,z_{15}
S. III *arizonae* (Ar. 1,4:23:21)	53	l,v	z_{35}
S. II *midhurst*	53	l,z_{28}	z_{39}
S. III *arizonae* (Ar. 1,4:24:31)	53	r	z

Table 5.11—*continued*

Serovar	Somatic (O) antigens	Flagellar (H) Antigens		Serovar	Somatic (O) antigens	Flagellar (H) Antigens	
		Phase 1	Phase 2			Phase 1	Phase 2
S. III *arizonae* (Ar. 1,4:24:21)	53	r	z_{35}	*S.* III *arizonae*[qq]	58	r	z_{53}:[z_{57}]
S. III *arizonae* (Ar. 1,4:24:38)	53	r	z_{56}	(Ar. 1,33:24:25[40a,40c])			
S. II	53	z	1,5	*S.* II	58	z_{10}	1,6
S. III *arizonae* (Ar. 1,4:31:30)	53	z	1,5,(7)	*S.* II	58	z_{10}	z_6
S. II	53	z	z_6	*S.* III *arizonae* (Ar. 1,33:26:31)	58	z_{52}	z
S. III *arizonae* (Ar.1,4:1,2,5:–)	53	z_4,z_{23}	—	*S.* III *arizonae* (Ar. 1,33:26:21)	58	z_{52}	z_{35}
(Ar.1,4:1,2,6:–)					**Group 059**		
S. IV	53	z_4,z_{23}	—	*S.* III *arizonae* (Ar. 19:32:28)	59	c	e,n,x,z_{55}
S. III *arizonae* (Ar.1,4:1,6,7:–)	53	z_4,z_{23},z_{32}	—	*S.* III *arizonae* (Ar. 19:33:31)	59	i	z
(Ar.1,4:1,6,7,9:–)				*S.* III *arizonae* (Ar. 19:33:21)	59	i	z_{35}
S. II *humber*	53	z_4,z_{24}	—	*S.* III *arizonae* (Ar. 19:22:28)	59	(k)	e,n,x,z_{15}
S. III *arizonae* (Ar.1,4:1,3,11:–)	53	z_4,z_{24}	—	*S.* II *betioky*	59	k	(z)
S. III *arizonae* (Ar. 1,4:27:21)	53	z_{10}	z_{35}	*S.* III *arizonae* (Ar. 19:22:31)	59	(k)	z
S. III *arizonae* (Ar. 1,4:16,17,18:–)	53	z_{29}	—	*S.* III *arizonae* (Ar. 19:22:21)	59	(k)	z_{35}
S. IV *bockenheim*	1,53	z_{36},z_{38}	—	*S.* III *arizonae* (Ar. 19:29:25)	59	k	z_{53}
S. III *arizonae* (Ar. 1,4:26:21)	53	z_{52}	z_{35}	*S.* III *arizonae* (Ar. 19:23:31)	59	l,v	z
S. III *arizonae* (Ar. 1,4:26:25)	53	z_{52}	z_{53}	*S.* III *arizonae* (Ar. 19:23:25)	59	l,v	z_{53}
	Group 054[pp]			*S.* III *arizonae* (Ar.19:1,2,5:–)	59	z_4,z_{23}	
S. tonev	21,54	b	e,n,x	(Ar. 19:1,2,6:–)			
S. winnipeg	54	e,h	1,5	*S.* III *arizonae* (Ar. 19:27:25)	59	z_{10}	z_{53}
S. rossleben	54	e,h	1,6	*S.* III *arizonae* (Ar.19:27:40a,40c)	59	z_{10}	z_{57}
S. borreze	54	f,g,s	—				
S. uccle	3,54	g,s,t	—	*S.* III *arizonae* (Ar. 19:16,17,18:–)	59	z_{29}	—
S. poeseldorf	8,20,54	i	z_6				
S. ochsenwerder	6,7,54	k	1,5	*S.* III *arizonae* (Ar. 19:17,20:–)	59	z_{36}	—
S. czernyring	54	r	1,5	*S.* III *arizonae* (Ar. 19:26:–)	59	z_{52}	—
S. steinwerder	3,15,54	y	1,5		**Group 060**		
S. yerba	54	z_4,z_{23}	—	*S.* II *setubal*	60	g,m,t	z_6
S. canton	54	z_{10}	e,n,x	*S.* III *arizonae* (Ar. 24:33:28)	60	i	e,n,x,z_{15}
	Group 055			*S.* III *arizonae* (Ar. 24:33:21)	60	i	z_{35}
S. II *tranoroa*	55	k	z_{39}	*S.* III *arizonae* (Ar.24:29:31)	60	k	z
	Group 056			*S.* III *arizonae* (Ar. 24:29:21)	60	k	z_{35}
S. II *artis*	56	b	—	*S.* III *arizonae* (Ar. 24:22:25)	60	(k)	z_{53}
S. II	56	d	—	*S.* III *arizonae* (Ar. 24:23:31)	60	l,v	z
S. II	56	e,n,x	1,7	*S.* III *arizonae* (Ar. 24:24:28)	60	r	e,n,x,z_{15}
S. II	56	l,z_{28}	—	*S.* III *arizonae* (Ar. 24:24:31)	60	r	z
S. III *arizonae* (Ar. 14:1.2.5:–)	56	z_4,z_{23}	—	*S.* III *arizonae* (Ar. 24:24:21)	60	r	z_{35}
(Ar. 14:1,2,6:–)				*S.* III *arizonae* (Ar. 24:24:25)	60	r	z_{53}
S. III *arizonae* (Ar. 14:1,6,7,9:–)	56	z_4,z_{23},z_{32}	—	*S.* II *luton*	60	z	e,n,x
S. II	56	z_{10}	e,n,x	*S.* III *arizonae* (Ar. 24:27:31)	60	z_{10}	z
S. III *arizonae* (Ar. 1,14:16,18:–)	56	z_{29}	—	*S.* III *arizonae* (Ar. 24:27:21)	60	z_{10}	z_{35}
				S. III *arizonae* (Ar. 24:26:30)	60	z_{52}	1,5,7
	Group 057			*S.* III *arizonae* (Ar.24:26:31)	60	z_{52}	z
S. antonio	57	a	z_6	*S.* III *arizonae* (Ar.24:26:21)	60	z_{52}	z_{35}
S. maryland	57	b	1,7	*S.* III *arizonae* (Ar. 24:26:25)	60	z_{52}	z_{53}
S. III *arizonae* (Ar. 34:32:31:44)	57	c	$z:z_{60}$		**Group 061**		
S. II	57	d	1,5	*S.* III *arizonae* (Ar. 26:32:30)	61	c	1,5,(7)
S. II	57	g,m,s,t	z_{42}	*S.* III *arizonae* (Ar. 26:32:21)	61	c	z_{35}
S. II	57	g,t	—	*S.* III *arizonae* (Ar. 26:33:28)	61	i	e,n,x,z_{15}
S. III *arizonae* (Ar. 34:33:28)	57	i	e,n,x,z_{15}	*S.* III *arizonae* (Ar. 26:33:31)	61	i	z
S. III *arizonae* (Ar. 34:33:31)	57	i	z	*S.* III *arizonae* (Ar. 26:33:21)	61	i	z_{35}
S. IV	57	z_4,z_{23}	—	*S.* III *arizonae* (Ar. 26:33:25)	61	i	z_{53}
S. II *locarno*	57	z_{29}	z_{42}	*S.* III *arizonae* (Ar. 26:29:30)	61	k	1,5,(7)
S. II *manombo*	57	z_{39}	e,n,x,z_{15}	*S.* III *arizonae* (Ar. 26:22:25)	61	(k)	z_{53}
S. II *tokai*	57	z_{42}	$1,6:z_{53}$	*S.* III *arizonae* Ar. 26:23:30:[40a,40b])	61	l,v	$1,5,7:[z_{57}]$
	Group 058			*S.* III *arizonae* (Ar. 26:23:31)	61	l,v	z
S. II	58	a	[z_6]	*S.* III *arizonae* (Ar. 26:23:21)	61	l,v	z_{35}
S. II	58	b	1,5	*S.* III *arizonae* (Ar. 26:24:30)	61	r	1,5,7
S. II	58	c	z_6	*S.* III *arizonae* (Ar. 26:24:21)	61	r	z_{35}
S. II	58	d	z_6	*S.* III *arizonae* (Ar. 26:24:25)	61	r	z_{53}
S. III *arizonae* (Ar. 1,33:33:28)	58	i	e,n,x,z_{15}	*S.* III *arizonae* (Ar. 26:27:21)	61	z_{10}	z_{35}
S. III *arizonae* (Ar. 1,33:23:28)	58	l,v	e,n,x,z_{15}	*S.* III *arizonae* (Ar. 26:26:30)	61	z_{52}	1,5,7
S. III *arizonae* (Ar. 1,33:23:21)	58	l,v	z_{35}	*S.* III *arizonae* (Ar. 26:26:31)	61	z_{52}	z
S. II *basel*	58	l,z_{13},z_{28}	1,5	*S.* III *arizonae* (Ar. 26:26:21)	61	z_{52}	z_{35}
S. III *arizonae* (Ar. 1,33:24:28)	58	r	e,n,x,z_{15}	*S.* III *arizonae* (Ar. 26:26:25)	61	z_{52}	z_{53}
S. III *arizonae* (Ar. 1,33:24:31)	58	r	z		**Group 062**		
				S. III *arizonae* (Ar. 6:13,14:–)	62	g,z_{51}	—
				S. III *arizonae* (Ar. 6:1,2,5:–)	62	z_4,z_{23}	—

Table 5.11—*continued*

Serovar	Somatic (O) antigens	Flagellar (H) Antigens	
		Phase 1	Phase 2
S. III *arizonae* (Ar. 6:1,7,8:–)	62	z_4,z_{32}	–
Group 063			
S. III *arizonae* (Ar. 8:13,14:–)	63	g,z_{51}	–
S. III *arizonae* (Ar. 8:1,2,5)	63	z_4,z_{23}	–
S. III *arizonae* (Ar. 8:1,7,8:–)	63	z_4,z_{32}	–
S. III *arizonae* (Ar. 8:17,20:–)	63	z_{36}	–
Group 065″			
S. III *arizonae* (Ar. 30:32:30)	65	c	1,5,7
S. III *arizonae* (Ar. 30:32:31)	65	c	z
S. III *arizonae* (Ar. 30:32:25)	65	c	z_{53}
S. II	65	(f),g,t	–
S. III *arizonae* (Ar. 30:33:28)	65	i	e,n,x,z_{15}
S. III *arizonae* (Ar. 30:22:31)	65	(k)	z
S. III *arizonae* (Ar. 30:22:21)	65	(k)	z_{35}
S. III *arizonae* (Ar. 30:22:25)	65	(k)	z_{53}
S. III *arizonae* (Ar. 30:23:28)	65	l,v	e,n,x,z_{15}
S. III *arizonae* (Ar. 30:23:31)	65	l,v	z
S. III *arizonae* (Ar. 30:23:21)	65	l,v	z_{35}
S. III *arizonae* (Ar. 30:23:25)	65	l,v	z_{53}
S. III *arizonae* (Ar. 30:27:28)	65	z_{10}	e,n,x,z_{15}
S. III *arizonae* (Ar. 30:27:31)	65	z_{10}	z
S. III *arizonae* (Ar. 30:26:31)	65	z_{52}	z
S. III *arizonae* (Ar. 30:26:21)	65	z_{52}	z_{35}
S. III *arizonae* (Ar. 30:26:25)	65	z_{52}	z_{53}
S. II	65		1,6
Group 066			
S. V *maregrosso*	66	z_{35}	–
S. V *brookfield*	66	z_{41}	–
S. V *malawi*	66	z_{65}	–
Group 067			
S. crossness	67	r	1,2

carrier state is not monitored by periodic stool cultures. Antibiotics that are active in curing the disease (e.g., chloramphenicol or thiophenicol for typhoid fever) are ineffective in the treatment of the carrier state.

Strains of *Salmonella* from urine are often of the R form. Bilharziosis has to be controlled in *Salmonella* carriers (LoVerde, 1980). Sickle-cell anemia must be suspected in cases of osteomyelitis due to *Salmonella* in black children (Vandepitte, 1953).

Antibiotic and drug sensitivity. *Salmonella* strains such as *E. coli* can readily acquire plasmids that contain genes that confer resistance to antibiotics. Multiple resistance is selected for when antibiotics are used extensively in hospitals or added to feed. The same plasmids may be found in strains of human or animal origin (Anderson et al., 1975). Serovars strictly adapted to man, such as *S. typhi*, may acquire resistance to chloramphenicol as the result of the long term, indiscriminate use of this drug or other antibiotics (Anderson, 1975).

Ecology. Although some *Salmonella* serovars are strictly host-adapted, the majority have a wide host range (e.g. *S. typhimurium*). Some are localized in a particular region of the globe (e.g. "*S. sendai*" in the Far East, "*S. berta*" in North America), but others are ubiquitous (e.g. *S. typhimurium*). Strains belonging to "subgenera" II and III are frequently isolated from the intestinal contents of cold-blooded animals and only rarely from warm-blooded animals. Strains of "subgenera" IV and V are isolated chiefly from the environment and are rarely pathogenic for man.

Isolation and Enrichment Procedures

Isolation from blood is done according to the classical method for hemoculture. A biphasic culture bottle containing a vertical agar layer along one side and a broth medium at the bottom (Castaneda, 1957; Hall et al., 1979; Krieg and Gerhardt, 1981) prepared with tryptic soy agar/broth containing 2% sodium citrate is convenient. Isolated colonies grow on the agar layer. Identification is usually done by (a)

diagnosis of the family *Enterobacteriaceae*, (b) diagnosis of the genus *Salmonella* (diagnosis of the "subgenus" for strains isolated from blood cultures is not routinely necessary, because almost all blood isolates belong to "subgenus" I), (c) diagnosis of the serovar, (d) determination of the antibiotic susceptibility pattern, and (e) further study of the biovar and phagovar if indicated.

Selective procedures are needed for the isolation of *Salmonella* from specimens containing mixed bacterial flora (fecal samples, autopsy samples, food, environmental samples, etc. Enrichment (i.e. an increased ratio of *Salmonella* cells to other bacterial cells during incubation) is obtained using liquid nutrient media containing selective agents that inhibit or retard the growth of bacteria other than *Salmonella*. Use of enrichment media is essential when the number of salmonellae in a sample is very low, i.e., when the probability of finding colonies by direct isolation is low. Three media may be recommended for general use: (a) the tetrathionate medium of Muller (1923); (b) Muller's medium modified by Kauffman (1935) by addition of bile and brilliant green; and (c) selenite broth devised by Leifson (1936). Tetrathionate and selenite broth are suitable for all *Salmonella* serovars. Tetrathionate-bile-brillant green medium is suitable for all except host-adapted serovars such as *S. typhi*. Enrichment media should be heavily inoculated, e.g. 0.5 ml of fecal suspension per 10 ml of medium. After incubation for 18 h at 37°C, a loopful of enrichment culture is streaked onto agar plating medium.

The same enrichment media may be used for detection of salmonellae in water. The simplest method is to add one volume of the water sample to an equal volume of double-strength medium. For detecting salmonellae in food, a generally suitable procedure is to inoculate 25 g of the suspected food into 225 ml of selenite F broth, incubate for 24 h and isolated on selective agar media. In the case of a dehydrated food, nutrient broth containing the sample is incubated overnight before inoculation of enrichment media.

Agar media are used for isolation of salmonellae. Streaking a loopful of enrichment culture or a suspension of the sample (e.g., stool) should be done carefully in order to obtain the greatest number of perfectly isolated colonies. Because the most discriminating character is lactose fermentation, the majority of media for isolation contain lactose and a pH indicator. In addition, the media contain selective agents to inhibit the growth of non-*Salmonella* organisms and the swarming of *Proteus mirabilis* and *P. vulgaris*. Some media also contain ferrous citrate for the detection of H_2S-producing bacteria.

Examples of media of moderate selectivity are: (a) *MacConkey agar*, which contains lactose, neutral red, and the selective inhibitors crystal violet and bile salts. Lactose-positive colonies are red, lactose-negative colonies are colorless. (b) *desoxycholate citrate agar*, which contains lactose, neutral red, and the selective agent desoxycholate. Ferric ammonium citrate is included as an indicator of H_2S production. Lactose-positive colonies are red, lactose-negative colonies are colorless. If H_2S is produced, the inner part of the colony is black.

Examples of media of higher selectivity are: (a) *SS agar*, which contains lactose, neutral red, and the selective agents brilliant green and bile salts. Ferric citrate is an indicator of H_2S production. The appearance of colonies is the same as on desoxycholate citrate agar. (b) *brilliant green agar*, which contains lactose, phenol red, and the selective agent brilliant green. This medium is easy to prepare and is suitable for all salmonellae except host-adapted serovars. It is not suitable for shigellae. Lactose-positive colonies are green, lactose-negative colonies are pink. All of the above-mentioned media are reviewed in the books by Kauffmann (1966) and Edwards and Ewing (1972). (c) *Hektoen medium* (King and Metzer, 1968), which contains lactose, sucrose and salicin, a mixture of bromothymol blue and Andrade's pH indicator, ferric citrate to detect H_2S production, and sodium desoxycholate as a selective inhibitor. Colonies that do not ferment any of the three sugars (e.g. *Salmonella*) are blue-green, with a black center if H_2S is produced. Colonies fermenting one or more of the sugars (e.g., *Escherichia coli*, *Enterobacter cloacae*) are salmon-colored. This medium is suitable for all *Salmonella* serovars and for shigellae.

A general procedure for the detection of salmonellae in feces or food is as follows. A suspension of the sample in saline is streaked onto the chosen isolation medium and also inoculated into an enrichment broth. After overnight incubation, the plating medium is examined for suspect colonies (lactose-negative, H₂S-positive or negative); also, a loopful of the enrichment culture is streaked onto another plate of selective agar medium. After overnight incubation, this plate is also examined for suspect colonies. A quick screening of several suspect colonies is done by inoculating each into a few drops of urea medium and incubating at 37°C for 2 h. Biochemical characterization is continued only for urease-negative colonies (urease-positive colonies growing at 18 h are likely to be *Proteus*). *Salmonella* must be differentiated mainly from *Citrobacter freundii*, *Proteus mirabilis*, *Hafnia alvei*, and, in food bacteriology,

Alteromonas putrefaciens. To detect *Salmonella arizonae*, attention should be given to *lactose-positive*, H₂S-positive colonies on plating media.

Maintenance Procedures

Salmonella cultures remain viable for many years when stored on peptone agar) meat extract, 5.0 g; peptone, 10.0 g; NaCl, 3.0; Na₂HPO₄·12H₂O, 2.0 g; agar, 10.0 g; distilled water, 1,000 ml; pH 7.4) distributed into small, tightly stoppered, screw-capped tubes. This medium is stab-inoculated and kept in the dark at room temperature. Lyophilization also gives good results. For lyophilization, it is necessary to isolate each subculture and to select a colony with the desired serologic characteristics.

Differentiation from other closely related genera

Characteristics useful for differentiating the genus *Salmonella* from other *Enterobacteriaceae* are given in Table 5.3 of the chapter on the family *Enterobacteriaceae*.

Taxonomic Comments

If one accepts the principle that bacteria which are related by 70% or more on the basis of DNA/DNA hybridization experiments belong to the same "genospecies," the so-called "genus" *Salmonella* is, in fact, one species (Crosa, 1973). In other words, all salmonellae and arizonae form one species composed of five subgroups: typical *Salmonella*, atypical *Salmonella* "subgenus" II, atypical *Salmonella* "subgenus" IV, monophasic "subgenus" III (*S. arizonae*), and diphasic "subgenus" III (*S. arizonae*) (Brenner, 1978; Stoleru et al., 1976). Genetically, the level of Kauffman's four "subgenera," including the discrimination between the monophasic and diphasic strains of "subgenus" III and the new "subgenus" V (Le Minor, unpublished results; see Table 5.10) is that of subspecies. "Nevertheless the schemes now in use will continue to be used because people are familiar with them and are very slow to adjust to a new system" (Brenner, 1978).

The names given to salmonellae do not follow the usual rules of nomenclature. Because of their importance in pathology, the first salmonellae were given names which indicated the disease and/or the animal from which the organism was isolated, and names of this kind (such as *S. typhi*, "*S. paratyphi-A*," *S. choleraesuis*, *S. typhimurium* and "*S. abortusovis*") continue to be used in clinical bacteriology. This nomenclature was abandoned by the more systematically minded, for these names implied that pathogenicity was limited to definite host species, whereas this is not generally true. For example, *S. typhimurium* and "*S. bovismorbificans*" are frequently isolated from human infections. New types are now given the name of the town, region or country in which the first strain was isolated, e.g., "*S. london*," "*S. panama*," "*S. stanleyville*," etc. New types of "subgenera" II, III and IV described since 1966 are designated simply by antigenic formula; this allows the "Arizona" group ("subgenus" III, or *S. arizonae*) of Edwards, Fife and Ramsey (1959) to be included in the Kauffmann-White scheme, simplifies the terminology of the antigenic factors, and allows the same antisera to be used to establish antigenic formulae (Kauffmann and Rohde, 1962; Kauffmann, 1965; Rohde, 1967). With few exceptions, the formulae of "*Arizona*" serovars published by Edwards, Fife and Ewing (1965) may be translated into *Salmonella* formulae and included in the Kauffmann-White scheme.

The International Subcommittee on *Enterobacteriaceae* has not given clear guidance on the naming of the differing serovars. It is paradoxical that serovars of "subgenus" I bear species-like epithets, while those of *Escherichia coli* and *Salmonella* "subgenus" III (i.e., *S. arizonae*) do not. *S. typhi* owes its name to the importance of the bacterium in human pathology, but when these infection syndrome names were first applied no one could have imagined that by 1981 there would be more than

2,000 closely related serovars. Borman, Stuart and Wheeler (1944) proposed the subdivision of the genus into three species, *S. choleraesuis* (the type species), "*S. typhosa*" (*S. typhi*) and "*S. kauffmannii*", the last to serve as a species for all the serological types. Kauffman and Edwards (1952) made a similar proposal, but designated the all-embracing species "*Salmonella enterica*." Ewing (1966) proposed a three-species concept, with *S. enteritidis* representing all serovars other than *S. typhi* and *S. choleraesuis*. Another proposal (Le Minor, Rohde and Taylor, 1970) was to consider Kauffmann's "subgenera" as species: "*S. kauffmannii*" ("subgenus" I), *S. salamae* ("subgenus" II), *S. arizonae* ("subgenus" III) and "*S. houtenae*" ("subgenus" IV). Serovars of "*S. kauffmannii*" would be designated by their species names followed by that of their serovar (e.g., "*S. kauffmannii*" serovar *typhi*), and serovars of "*S. salamae*," *S. arizonae* and "*S. houtenae*" would be designated by their species names followed by their antigenic formulae. Kauffmann (1971, 1973) disagreed with all of the preceding propositions and considered a species as a "group of related sero-fermentative phage-types" in his "Realität Theorie" (reviewed 1978).

Scientifically, none of the present methods of nomenclature of salmonellae is satisfactory. Without prejudice as to what constitutes a species, the *Enterobacteriaceae* subcommittee considers the diagnostic use of the Kauffmann-White scheme to be overridingly important and that the practice of giving names to the serovars of "subgenus" I should continue, but that new serovars of the other subgenera should be designated only by their antigenic formulae.

Editorial Note

On the basis of numerical taxonomy and DNA relatedness studies, Le Minor, Véron and Popoff (Ann. Microbiol. (Inst. Pasteur) 133B: 245–254, 1982) recently proposed nomenclatural changes for salmonellae, as follows. The genus should consist of a single species, *S. choleraesuis*, having six subspecies: (a) the subspecies *choleraesuis*, corresponding to the former subgenus I; (b) the subspecies *salamae*, corresponding to the former subgenus II; (c) the subspecies *arizonae*, corresponding to the monophasic serovars of the former subgenus III; (d) the subspecies *diarizonae*, corresponding to the diphasic serovars of the former subgenus III; (e) the subspecies *houtenae*, corresponding to the former subgenus IV, and (f) the subspecies *bongori*, composed of strains that are positive for dulcitol, ONPG and KCN. Type strains were proposed for each subspecies.

Further Reading

Edwards, P.R. and W. H. Ewing. 1972. Identification of *Enterobacteriaceae*, 3rd Ed, Burgess Publishing, Minneapolis, Minn.

Kaufmann, F. 1966. The Bacteriology of *Enterobacteriaceae*, Munksgaard, Copenhagen.

Kauffmann, F. 1978. *Das Fundament*, Munksgaard, Copenhagen.

Kelterborn, E. 1967. Salmonella-Species, Hirzel, Leipzig.

Van Oye, E. 1964. *The World Problem of Salmonellosis*, Junk, The Hague.

Differentiation of the "subgenera" of the genus **Salmonella**

The biochemical characteristics which differentiate the five "subgenera" of the genus *Salmonella* are presented in Table 5.10.

Differentiation of the serovars of the genus **Salmonella**

The antigenic formulae of the salmonellae (i.e. the Kauffmann-White scheme) are given in Table 5.11. An alphabetical listing of *Salmonella* serovars, indicating their "subgenus" and O group, is presented in Table 5.12.

List of selected serovars of the genus **Salmonella**

"Subgenus" I

a. *Salmonella choleraesuis* (Smith 1894) Weldin 1927, 155.[AL] (*Bacillus cholerae suis* Smith 1894, 9.) *Editorial Note*: although the specific epithet *cholerae-suis* is listed in the Approved Lists of Bacterial Names (1980), the hyphen should not be used (J. J. Farmer III, Int. J. Syst. Bacteriol. *33*: 425, 1983).

chol.er.ae.su′is. Gr. n. *cholera* cholera; L. n. *sus* hog; M.L. gen. n. *suis* of a hog; M.L. gen. n. *choleraesuis* of hog cholera.

Antigenic formula: 6,7,c:1,5. The detailed O antigen formula is normally 6_2,7, but this may be transformed by lysogenization into 6_1,7 or 6_2,7,14.

Arabinose and trehalose are not fermented; dulcitol is slowly and irregularly fermented.

Those strains which produce H_2S are designated as *S. choleraesuis* biovar *kunzendorf*.

Pathogenic for man and other animals.

Type strain: ATCC 13313 (NCTC 5735).

b. "*Salmonella hirschfeldii*" Weldin 1927, 161. (Paratyphoid C bacillus, Hirschfeld 1919, 296; *Salmonella paratyphi-C* Salmonella Subcommittee 1934.)

hirsch.fel.di.i. M.L. gen. n. *hirschfeldii* of Hirschfeld; named after Hirschfeld, who first called the organism the paratyphoid C bacillus, a name still in common use today.

Antigenic formula; 6,7,[Vi]:c:1,5.

Ferments dulcitol and trehalose; produces H_2S. Arabinose fermentation is variable.

c. *Salmonella typhi* (Schroeter 1886) Warren and Scott 1930, 416.[AL] (*Bacillus typhi* Schroeter 1886, 165.)

ty′phi. Gr. n. *typhus* a stupor; M.L. gen. n. *typhi* of typhoid.

Antigenic formula: 9,12,[Vi]:d:−. Wild strains may possess H antigen z_{66} instead of H antigen d (Guinée et al., 1981).

Does not grow on Simmons' citrate medium or on a minimal defined medium; requires tryptophan as a growth factor.

Does not produce gas from glucose or other sugars. Fermentation of xylose is variable.

Many strains are agglutinated by anti-Vi serum and are inagglutinable by O9 serum; their colonies are opaque and have an iridescent appearance when examined by transmitted light. Colonies of intermediate appearance agglutinable by both Vi and O antisera, may occur (VW colonies).

Pathogenic only for man, causing typhoid (enteric) fever; transmitted by water or food contaminated by human excreta.

Type strain: ATCC 19430.

d. "*Salmonella paratyphi-A*" (Brion and Kayser 1902) Castellani and Chalmers 1919, 939. (*Bacterium paratyphi* Kayser 1902, 426; *Bacterium paratyphi* typus A Brion and Kayser 1902, 613.)

pa.ra.ty′phi. Gr. prep. *para* alongside of; Gr. n. *typhus* a stupor; M.L. gen. n. *paratyphi-A* of type A typhoid-like infection.

Antigenic formula: 1,2,12:a:−. As with other strains of O antigen groups A, B and D, the presence of factor 1 is connected with lysogenization.

Aerogenic. Ferments arabinose but no xylose.

The majority of strains do not produce H_2S, and in this respect "*S. paratyphi-A*" is unlike most other salmonellae.

Lysine decarboxylase is weak or negative.

Pathogenic only for man.

e. "*Salmonella schottmuelleri*" (Winslow et al., 1919) Bergey et al. 1923, 213. (*Bacterium paratyphi* typus B Brion and Kayser 1902, 613; *Bacillus schottmuelleri* Winslow, Kligler and Rothberg 1919, 479.)

schott.muel′ler.i. M.L. gen. n. *schottmuelleri* of Schottmüller; named after Prof. R. Schottmüller, who isolated the organism in 1899.

Antigenic formula; 1,4,[5],12:b:1,2.

Produces a slime layer when grown on a medium containing 0.5% glucose and 0.2 M sodium phosphate, pH 7 (Anderson, 1961).

Negative for *d*-tartrate.

Causes enteric fever in man and very rarely infects animals.

A variant known as *S. java* is positive for *d*-tartrate, fails to produce a slime layer, and usually causes enteritis in man and not uncommonly in animals as well (Kauffmann, 1941).

Some strains are intermediate between these two extremes.

f. *Salmonella typhimurium* (Loeffler 1892) Castellani and Chalmers 1919, 939.[AL] (*Bacillus typhimurium* Loeffler 1892, 134.)

ty.phi.mu′ri.um. Gr. n. *typhus* a stupor; L. n. *mus* mouse; l. gen. pl. n. *murium* of mice; M.L. gen. pl. n. *typhimurium* typhoid of mice.

Antigenic formula: 1,4,[5],12:i:1,2. The presence of factor 1 follows lysogenization by a converting phage named *iota* or PLT_{22}.

Ubiquitous and frequently the cause of infections in man and animals; also the most frequent agent of *Salmonella* gastroenteritis in man.

The well-known chromosome map of *Salmonella* is that of *S. typhimurium* strain LT_2 (for a review see Sanderson and Hartman, 1978).

Type strain: ATCC 13311.

g. *Salmonella enteritidis* (Gaertner 1888) Castellani and Chalmers 1919, 939.[AL] (*Bacillus enteritidis* Gaertner 1888, 573.)

en.te.ri′ti.dis. Gr. n. *enteron* gut, intestine; M.L. n. *enteritis* enteritis, inflammation of the intestine; M.L. gen. n. *enteritidis* of enteritis.

Antigenic formula: 1,9,12:g,m:−.

Frequently occurs in man and animals.

Type strain: ATCC 13076.

h. "*Salmonella gallinarum*" (Klein 1889) Bergey et al., 1925, 236. (*Bacillus gallinarum* Klein 1889, 689; *Bacterium pullorum* Rettger 1909, 123; *Salmonella gallinarum-pullorum* Taylor et al. 1952, 140.)

gal.li.na′rum. L. n. *gallina* hen; L. gen. pl. n. *gallinarum* of hens.

Antigenic formula: 1,9,12:−:−.

Always nonmotile. Maybe subdivided into biovars on the basis of fermentation characteristics, production of gas and production of H_2S.

Does not grow on a minimal defined medium.

Isolated chiefly from chickens and other birds. Causative agent of fowl typhoid.

"Subgenus" II

4. "*Salmonella salamae*" Le Minor, Rohde and Taylor 1970, 209. (*Salmonella dar-es-salaam* Salmonella Subcommittee 1934, 346.)

sa.la′mae. M.L. gen. n. *salamae* of (Dar-es) salaam.

Antigenic formula: 1,9,12:1,w:e,n,x.

Mucate and malonate positive; gelatin liquefaction slow.

Isolated in 1922 from the urine of a patient in Dar-es-Salaam (Tanzania) and the antigenic structure was determined by White (1926). Biochemical characteristics differ from previously identified salmonellae (Table 5.10) and the organism became the type of species of "subgenus" II.

Type strain: NCTC 5773 (ATCC 6959).

"Subgenus" III

j. *Salmonella arizonae* (Borman 1957) Kauffmann in van Oye, 1964.[AL] (*Paracolobactrum arizonae* Borman 1957, 347.)

(Text continues on p. 458)

Table 5.12.

Alphabetical list of names of **Salmonella** *serovars classified by "subgenus" and indicating the O group*

Serovar	O Group	Serovar	O Group
"Subgenus" I		S. anecho	O
S. aarhus	K	S. anfo	Q
S. aba	C_2	S. angers	C_3
S. abadina	M	S. angoda	N
S. abaetetuba	F	S. angouleme	I
S. aberdeen	F	S. anie	X
S. abidjan	Q	S. ank	M
S. ablogame	I	S. anna	G_2
S. abobo	I	S. annedal	I
S. abony	B	S. antarctica	D_1
S. abortusbovis	B	S. antonio	57
"S. abortuscanis" 4,5,12:b:z_5 (phase R)	B	S. antsalova	51
S. abortusequi	B	S. antwerpen	T
S. abortusovis	B	S. apapa	W
S. accra	E_4	S. apeyeme	C_3
S. adabraka	E_1	S. aqua	N
S. adamstown	M	S. aragua	N
S. adamstua	F	S. ardwick	C_4
S. adana	U	S. arechavaleta	B
S. adelaide	O	S. arkansas	E_3
S. adeoyo	I	S. aschersleben	N
S. aderike	M	S. ashanti	M
S. adime	C_1	S. assen	L
S. adjame	G_2	S. assinie	E_1
S. aesch	C_2	S. atakpame	C_3
S. aequatoria	C_1	S. atento	F
S. aflao	H	"S. atherton" = S. waycross	S
S. africana	B	S. athinai	C_1
S. afula	C_1	S. atlanta (combined with S. mississippi)	G_2
S. agama	B	S. augustenborg	C_1
S. agbeni	G_2	S. austin	C_1
S. agege	E_1	S. avignon	I
S. ago	N	S. avonmouth	E_4
S. agodi	O	S. ayinde	B
S. agona	B	S. ayton	B
S. agoueve	G_1	S. azteca	B
S. ahanou	J	S. babelsberg	M
S. ahepe	U	S. babili	M
S. ahmadi	E_4	S. baguida	L
S. ahuza	U	S. baguirmi	N
S. ajiobo	G_2	S. bahati	G_1
S. akanji	C_2	S. bahrenfeld	H
S. akuafo	I	S. baiboukoum	C_1
S. alabama	D_1	S. baildon	D_2
S. alachua	O	S. bakau	M
S. alagbon	C_3	S. balcones	W
S. alamo	C_1	S. ball	B
S. albany	C_3	S. bama	J
S. albert	B	S. bambesa (combined with S. miami)	D_1
S. albuquerque	H	S. bamboye	D_2
S. alexanderplatz	X	S. bambylor	D_2
S. alexanderpolder	C_3	S. banalia	C_2
S. alfort	E_1	S. banana	B
S. alger	P	S. banco	M
S. allandale	R	S. bandia	O
S. allerton	E_1	S. bangkok	P
S. alminko	C_3	S. banjul	H
S. altendorf	B	"S. bantam" = S. meleagridis	E_1
S. altona	C_3	S. bardo	C_3
S. amager	E_1	S. bareilly	C_1
S. amba	F	S. bargny	C_3
S. amersfoort	C_1	S. barmbek	I
S. amherstiana	C_3	S. barranquilla	I
S. amina	I	S. basingstoke	D_2
S. aminatu	E_1	S. bassa	C_2
S. amounderness	E_1	S. bassadji	M
S. amoutive	M	"S. batavia" = S. lexington	E_1
S. amsterdam	E_1	S. battle	I
S. amunigun	I	S. bazenheid	C_3
S. anatum	E_1	S. be	C_3
S. anderlecht	E_1	S. beaudesert	H

Table 5.12.—*continued*

Serovar	O Group	Serovar	O Group
S. bedford	E$_4$	S. bukavu	R
S. belem	C$_2$	S. bukuru	C$_2$
S. belfast	C$_2$	S. bulgaria	C$_2$
S. benfica	E$_1$	S. bullbay	F
S. benguella	R	S. burgas	I
S. benin	D$_2$	S. bury	B
S. bere	X	S. businga	C$_1$
S. bergedorf	D$_2$	S. butantan	E$_1$
S. bergen	X	S. butare	52
S. berkeley	U	S. buzu	H
S. berlin	J	S. caen	I
S. berta	D$_1$	S. cairina	E$_1$
S. bessi	E$_1$	S. cairns	W
S. biafra	E$_1$	S. cairo (combined with S. stanley)	B
S. bietri	N	S. calabar	E$_4$
S. bignona	J	S. california	B
S. bijlmer	R	S. camberene	O
S. bilu	E$_4$	S. camberwell	D$_1$
S. bingerville	X	S. cambridge	E$_2$
S. binningen	W	S. campinense	D$_1$
S. binza	E$_2$	S. canada	B
S. birkenhead	C$_1$	S. cannonhill	E$_4$
S. birmingham	E$_1$	S. cannstatt	E$_4$
S. bispebjerg	B	S. canoga	E$_3$
S. blegdam	D$_1$	S. canton	54
S. blijdorp	H	S. caracas	H
S. blitta	X	"S. cardiff" 6,7:k:1,10 (phase R)	C$_1$
S. blockley	C$_2$	S. carmel	J
S. blukwa	K	S. carnac	K
S. bobo	V	S. carno	E$_4$
S. bochum	B	S. carrau	H
S. bodjonegoro	N	S. carswell	V
S. boecker	H	S. casablanca	W
S. bokanjac	M	S. casamance	R
S. bolombo	E$_1$	S. catanzaro	H
S. bolton	E$_1$	S. cayar	C$_1$
S. bonames	J	S. cerro	K
S. bonariensis	C$_2$	S. ceyco	D$_2$
S. bonn	C$_1$	S. chagoua	G$_2$
S. bootle	X	S. chailey	C$_2$
S. borbeck	G$_1$	S. champaign	Q
S. bornum	C$_4$	S. chandans	F
S. borreze	54	S. charity	H
S. bournemouth	D$_1$	S. charlottenburg	C$_2$
S. bousso	H	S. chester	B
S. bovismorbificans	C$_2$	S. chicago	M
S. bracknell	G$_2$	S. chichiri	H
S. bradford	B	S. chincol	C$_2$
S. braenderup	C$_1$	S. chingola	F
S. brancaster	B	S. chiredzi	F
S. brandenburg	B	S. chittagong	E$_4$
S. brazil	I	S. choleraesuis	C$_1$
S. brazos	K	S. chomedey	C$_3$
S. brazzaville	C$_1$	S. christiansborg	V
S. bredeney	B	S. clackamas	B
S. brefet	V	S. claibornei	D$_1$
S. breukelen	C$_2$	S. clerkenwell	E$_1$
S. brevik	I	S. cleveland	C$_2$
S. brezany	B	S. clichy	E$_2$
S. brijbhumi	F	S. cochin	D$_2$
S. brikama	C$_3$	S. cocody	C$_3$
S. brisbane	M	S. coeln	B
S. bristol	G$_1$	S. coleypark	C$_1$
S. brive	T	S. colindale	C$_1$
S. bron	G$_1$	S. colobane	F
S. bronx	C$_2$	S. colombo	P
S. broughton	E$_4$	S. colorado	C$_1$
"S. broxbourne" = S. wien	B	S. concord	C$_1$
S. bruck	C$_1$	S. congo	G$_2$
S. brunei	C$_3$	S. coogee	T
S. budapest	B	"S. cook" 39:z$_{48}$:1,5 (phase R)	Q
"S. buenosaires" = S. bonariensis	C$_2$	S. coquilhatville	E$_1$

Table 5.12.—*continued*

Serovar	O Group	Serovar	O Group
S. corvallis	C_3	S. ekpoui	X
S. cotham	M	S. elisabethville	E_1
S. cotia	K	S. elokate	D_1
S. cremieu	C_2	S. elomrane	D_1
S. croft	M	S. emek	C_3
S. crossness	67	S. emmastad	P
S. cubana	G_2	S. encino	H
S. cuckmere	E_1	S. enschede	O
S. cullingworth	M	S. entebbe	B
S. curacao	C_2	S. enteritidis	D_1
S. cyprus	C_2	S. enugu	I
S. dabou	C_3	S. epicrates	E_1
S. dadzie	51	S. epinay	F
S. dahlem	Y	S. eppendorf	B
S. dahomey	X	S. escanaba	C_1
S. dakar	M	S. eschersheim	E_2
S. dakota	I	S. eschweiler	C_1
S. dalat (combined with S. ball)	B	S. essen	B
S. dallgow	E_4	S. etterbeek	F
S. dan	51	S. ezra	M
S. dapango	X	S. fajara	M
S. daytona	C_1	S. faji	T
S. decatur (combined with S. choleraesuis)	C_1	S. falkensee	E_1
S. dembe	O	S. fallowfield	E_1
S. demerara	G_2	S. fann	F
S. denver	C_1	S. fanti	G_2
S. derby	B	S. farakan	M
S. derkle	52	S. farcha	U
S. dessau	E_4	S. fareham	E_4
S. deversoir	W	S. farmsen	G_2
S. dieuppeul	M	S. fass	Z
S. diguel	G_1	S. fayed	C_2
S. diogoye	C_3	S. ferlac	H
S. diourbel	L	S. ferruch	C_3
S. djakarta	Y	S. findorff	F
S. djama	T	S. finkenwerder	H
S. djelfa	C_3	S. fischerhuette	I
S. djermaia	M	S. fischerkietz	H
S. djibouti	J	S. fischerstrasse	V
S. djugu	C_1	S. fitzroy	Y
S. doba	D_2	S. florian	E_1
S. doncaster	C_2	S. florida	H
S. donna	N	S. flottbek	52
S. doorn	M	S. fluntern	K
S. dougi	Z	S. fomeco	W
S. doulassame	N	S. fortlamy	I
S. dresden	M	S. fortune	B
S. driffield	R	S. frankfurt	I
S. drogana	B	S. freetown	P
S. drypool	E_2	S. freiburg	E_1
S. dublin	D_1	S. fresno	D_2
S. duesseldorf	C_2	S. friedenau	G_1
S. dugbe	W	S. friendrichsfelde	M
S. duisburg	B	S. frintrop	D_1
S. dumfries	E_1	S. fufu	E_1
S. durban	D_1	S. fulica	B
S. durham	G_2	S. fyris	B
S. duval	R	S. gabon	C_1
S. ealing	O	S. gafsa	I
S. eastbourne	D_1	S. galiema	C_1
S. eberswalde	D_1	S. galil	E_1
S. eboko	C_2	S. gallen	F
S. ebrie	O	S. gallinarum	D_1
S. echa	P	S. gamaba	V
S. edinburg	C_1	S. gambaga	L
S. edmonton	C_2	S. gambia	O
S. egusi	S	S. gaminara	I
S. egusitoo	T	S. garba	H
S. eimsbuettel	C_4	S. garoli	C_1
S. eingedi	C_1	S. gassi	O
S. eko	B	S. gateshead	D_2
S. ekotedo	D_2	S. gatineau	E_4

Table 5.12.—*continued*

Serovar	O Group	Serovar	O Group
S. gatow	C_1	*S. hessarek*	B
S. gatuni	C_2	*S. heves*	H
S. gbadago	E_1	*S. hidalgo*	C_2
S. gdansk	C_1	*S. hiduddify*	C_2
S. gege	N	*S. hillegersberg*	D_2
S. gelsenkirchen	C_4	*S. hillingdon*	D_2
S. georgia	C_1	*S. hillsborough*	C_1
S. gera	T	*S. hilversum*	N
S. geraldton	D_2	*S. hindmarsh*	C_3
S. ghana	L	*S. hisingen*	Y
S. giessen	N	*S. hissar*	C_4
S. give	E_1	*S. hithergreen*	I
S. giza	C_3	*S. hofit*	Q
S. glasgow	I	*S. hoghton*	E_1
S. glidji	F	*S. holcomb*	C_2
S. glostrup	C_2	*S. homosassa*	H
S. gloucester	B	*S. honelis*	M
S. gnesta	E_4	*S. horsham*	H
S. godesberg	N	*S. huddinge*	E_1
S. goelzau	E_1	*S. hull*	I
S. goerlitz	E_2	*S. huvudsta*	E_1
S. goeteborg	D_1	*S. hvittingfoss*	I
S. goettingen	D_1	*S. hydra*	L
S. gokul	51	*S. ibadan*	G_1
S. goldcoast	C_2	*S. idikan*	G_2
S. goma	C_1	*S. ikayi*	E_1
S. gombe	C_1	*S. ikeja*	M
S. good	L	*S. ilala*	M
S. gori	J	*S. illinois*	E_3
S. goulfey	R	*S. ilugun*	E_4
S. goverdhan	D	*S. inchpark*	C_2
S. grampian	C_1	*S. india*	D_2
S. granlo	J	*S. indiana*	B
S. graz	U	*S. infantis*	C_1
S. greiz	R	*S. inganda*	C_1
S. groenekan	K	*S. inglis*	D_2
S. grumpensis	G_2	*S. inpraw*	S
S. guarapiranga	N	*S. inverness*	P
S. guerin	D_2	*S. ipeko*	D_1
S. guildford	M	*S. ipswich*	S
S. guinea	V	*S. irenea*	J
S. gustavia	F	*S. irigny*	U
S. gwale	T	*S. irumu*	C_1
S. gwoza	E_4	*S. isangi*	C_1
S. haardt	C_3	*S. isaszeg*	Y
S. hadar	C_2	*S. israel*	D_1
S. haelsingborg	C_1	*S. istanbul*	C_3
S. haferbreite	T	*S. isuge*	G_2
S. haga	O	"*S. italiana*" 9,12:1,v:1,11 (phase R)	D_1
S. haifa	B	*S. itami*	D_1
S. halle	M	*S. ituri*	B
S. hallfold	B	*S. itutaba*	D_2
S. halmstad	E_2	"*S. iwo-jima*" = *S. kentucky*	C_3
"*S. hamilton*" 3,15:e,h:1,2:z_{27} (phase R) (combined with	E_2	*S. jaffna*	D_1
S. goerlitz)		*S. jaja* (combined with *S. stanleyville*)	B
S. handen	G_2	*S. jalisco*	F
S. hann	R	*S. jamaica*	D_1
S. hannover	I	*S. jangwani*	J
S. haouaria	G_1	*S. java* (combined with *S. paratyphi B*)	B
S. harburg	H	*S. javiana*	D_1
S. harrisonburg	E_3	*S. jedburgh*	E_1
S. hartford	C_1	*S. jericho*	B
S. harvestehude	T	*S. jerusalem*	C_4
S. hatfield	M	*S. joal*	E_1
S. hato	B	*S. jodhpur*	W
S. havana	G_2	*S. joenkoeping* (combined with *S. kingston*)	B
S. heerlen	F	*S. johannesburg*	R
S. heidelberg	B	*S. jos*	B
S. hermannswerder	M	*S. juba*	E_4
S. heron	I	*S. jubilee*	J
S. herston	C_2	*S. jukestown*	G_2
S. herzliya	F	*S. kaapstad*	B

Table 5.12.—*continued*

Serovar	O Group	Serovar	O Group
S. kabete	51	*S. kuru*	C_2
S. kaduna	C_4	*S. labadi*	C_3
S. kahla	T	*S. lagos*	B
S. kaitaan	H	*S. lamin*	E_1
S. kalamu	B	*S. landala*	S
S. kalina	E_1	*S. landau*	N
S. kalumburu	C_2	*S. landwasser*	E_1
S. kambole	C_1	*S. langenhorn*	K
S. kamoru	B	*S. langensalza*	E_1
S. kampala	T	*S. langford*	M
"*S. kanda*" = *S. meleagridis*	E_1	*S. lanka*	E_2
S. kande	E_4	*S. lansing*	P
S. kandla	J	*S. larochelle*	C_1
S. kaneshie	T	*S. lattenkamp*	W
S. kano	B	*S. lawndale*	D_1
S. kaolack	X	*S. lawra*	V
S. kapemba	D_1	*S. leatherhead*	S
S. kaposvar (combined with *S. reading*)	B	*S. lechler*	51
S. karachi	W	*S. leeuwarden*	F
S. karamoja	R	*S. legon*	B
S. kasenyi	P	*S. leiden*	G_1
S. kassberg	H	*S. leipzig*	S
S. kedougou	G_2	*S. leith*	C_2
S. kentucky	C_3	*S. lekke*	E_1
S. kenya	C_1	*S. lene*	F
S. kermel	V	*S. leoben*	M
S. keve	L	*S. leopoldville*	C_1
S. khartoum	E_3	*S. lerum*	E_4
S. kiambu	B	*S. lexington*	E_1
S. kibi	I	*S. lezennes*	C_2
S. kibusi	M	*S. ligeo*	N
S. kidderminster	P	*S. ligna*	O
S. kiel	A	*S. lika*	C_1
S. kikoma	I	*S. lille*	C_1
S. kimberley	P	*S. limete*	B
S. kimpese	D_2	*S. lindenburg*	C_2
S. kimuenza	B	*S. lindern*	H
S. kingabwa	U	*S. lindi*	P
S. kingston	B	*S. linguere*	D_2
S. kinondoni	J	*S. lingwala*	I
S. kinshasa	E_2	*S. linton*	G_2
S. kintambo	G_2	*S. lisboa*	I
S. kirkee	J	*S. lishabi*	D_2
S. kisangani	B	*S. litchfield*	C_2
S. kisarawe	F	*S. liverpool*	E_4
S. kisii	C_1	*S. livingstone*	C_1
S. kitenge	M	*S. livulu*	N
S. kivu	C_1	*S. ljubljana*	B
S. klouto	P	*S. llandoff*	E_4
S. kodjovi	X	*S. loanda*	C_2
S. koenigstuhl	B	*S. lockleaze*	C_4
S. koketime	V	*S. lode*	J
S. kokoli	N	*S. lodz*	S
S. kokomlemle	Q	*S. loenga*	T
S. konstanz	C_3	*S. logone*	Q
S. korbol	C_3	*S. lokstedt*	E_4
S. korlebu	E_4	*S. lomalinda*	D_1
S. korovi	P	*S. lome*	D_1
S. kortrijk	C_1	*S. lomita*	C_1
S. kottbus	C_2	*S. lomnava*	I
S. kotte	C_1	*S. london*	E_1
S. kouka	E_4	*S. losangeles*	I
S. koumra	C_1	*S. louga*	N
S. kpeme	M	*S. louisiana*	D_2
S. kralingen	C_3	*S. lovelace*	G_1
S. krefeld	E_4	*S. lubumbashi*	S
S. kristianstad	E_1	*S. luciana*	F
S. kua	V	*S. luckenwalde*	M
S. kubacha	B	*S. luke*	X
S. kuessel	M	*S. lyon*	X
S. kumasi	N	*S. maastricht*	F
S. kunduchi	B	*S. macallen*	E_1

Table 5.12.—*continued*

Serovar	O Group	Serovar	O Group
S. machaga	E$_4$	*S. mokola*	E$_1$
S. madelia	H	*S. molade*	C$_3$
S. madiago	E$_4$	*S. molesey*	52
S. madigan	V	*S. mono*	B
S. madison	L	*S. mons*	B
S. madjorio	E$_1$	*S. monschaui*	O
S. magumeri	H	*S. montevideo*	C$_1$
S. magwa	L	*S. montreal*	U
S. maiduguri	E$_4$	*S. morehead*	N
S. makiso	C$_1$	*S. morningside*	N
S. malakal	I	*S. mornington*	H
S. malaysia	M	*S. morocco*	N
S. malika	V	*S. morotai*	J
S. malmoe	C$_2$	*S. moroto*	M
S. malstatt	I	*S. moscow*	D$_1$
S. mampeza	H	*S. moualine*	X
S. mampong	G$_1$	*S. mountpleasant*	X
S. manchester	C$_2$	*S. moussoro*	H
S. mandera	I	*S. mowanjum*	C$_2$
S. mango	P	*S. mpouto*	I
S. manhattan	C$_2$	*S. muenchen*	C$_2$
S. manila	E$_2$	*S. muenster*	E$_1$
S. mapo	C$_2$	*S. muguga*	V
S. mara	Q	*S. mundonobo*	M
S. maracaibo	F	*S. mura*	B
S. maregrosso	66	*S. naestved*	D$_1$
S. maricopa	T	*S. nagoya*	C$_2$
S. marienthal	E$_1$	*S. nakuru*	B
S. maron	E$_1$	*S. nancy*	E$_2$
S. marseille	F	*S. nanergou*	C$_2$
S. marshall	G$_1$	*S. nanga*	G$_2$
S. maryland	57	*S. napoli*	D$_1$
S. marylebone	D$_2$	*S. narashino*	C$_2$
S. masembe	E$_1$	*S. nashua*	M
S. massakory	O	*S. naware*	I
S. massenya	B	*S. nchanga*	E$_1$
S. matadi	J	*S. ndjamena*	H
S. mathura	D$_2$	*S. ndolo*	D$_1$
S. matopeni	N	*S. neftenbach*	B
S. mayday	D$_2$	*S. nessa*	H
S. mbandaka	C$_1$	*S. nessziona*	C$_1$
S. mbao	U	*S. neudorf*	N
S. meekatharra	W	*S. neukoelln*	C$_1$
S. meleagridis	E$_1$	*S. neumuenster*	B
S. memphis	K	*S. newbrunswick*	E$_2$
S. menden	C$_1$	*S. newhaw*	E$_2$
S. mendoza	D$_1$	*S. newington*	E$_2$
S. menhaden	E$_3$	*S. newlands*	E$_1$
S. menston	C$_1$	*S. newmexico*	D$_1$
S. mesbit	X	*S. newport*	C$_2$
S. meskin	51	*S. newrochelle*	E$_1$
S. messina	N	*S. newyork*	G$_1$
S. mexicana (combined with *S. muenchen*)	C$_2$	*S. ngili*	C$_1$
S. mgulani	P	*S. ngor*	E$_4$
S. miami	D$_1$	*S. niakhar*	V
S. michigan	J	*S. niamey*	J
S. middlesbrough	T	*S. niarembe*	V
S. midway	H	*S. nienstedten*	C$_4$
S. mikawasima	C$_1$	*S. nieukerk*	C$_4$
S. millesi	R	*S. nigeria*	C$_1$
S. milwaukee	U	*S. nijmegen*	N
S. mim	G$_1$	*S. nikolaifleet*	I
S. minna	H	*S. niloese*	E$_4$
S. minneapolis	E$_3$	*S. nima*	M
S. minnesota	L	*S. nimes*	G$_1$
S. mishmarhaemek	G$_2$	*S. nissii* combined with *S. nienstedten*)	C$_4$
S. mission (combined with *S. isangi*)	C$_1$	*S. nitra*	A
S. mississippi	G$_2$	*S. njala*	P
S. miyazaki	D$_1$	*S. nordufer*	C$_2$
S. mkamba	C$_1$	*S. norton*	C$_1$
S. mocamedes	M	*S. norwich*	C$_1$
S. moero	M	*S. nottingham*	I

Table 5.12.—*continued*

Serovar	O Group	Serovar	O Group
S. nowawes	R	*S. portland*	D_1
S. nuatja	I	*S. portsmouth*	E_2
S. nyanza	F	*S. potosi*	H
S. nyborg	E_1	*S. potsdam*	C_1
S. nyeko	I	*S. potto*	D_2
S. oakland	C_1	*S. pramiso*	E_1
S. obogu	C_1	*S. praha*	C_2
S. ochsenwerder	54	*S. presov*	C_2
S. odozi	N	*S. preston*	B
S. oerlikon	Q	*S. pretoria*	F
S. oevelgoenne	M	*S. pueris* (combined with *S. newport*)	C_2
S. offa	S	*S. pullorum*	D_1
S. ogbete	U	*S. putten*	G_2
S. ohio	C_1	*S. quebec*	V
S. ohlstedt	E_1	*S. quentin*	D_2
S. okatie	G_2	*S. quinhon*	X
S. okefoko	E_1	*S. quiniela*	C_2
S. okerara	E_1	*S. ramatgan*	N
S. oldenburg	I	*S. ramsey*	M
S. olten	D_2	*S. raus*	G_1
S. omderman	C_4	*S. rawash*	K
S. omifisan	R	*S. reading*	B
S. ona	M	*S. rechovot*	C_3
S. onarimon	D_1	*S. redba*	C_1
S. onderstepoort	H	*S. redhill*	F
S. onireke	E_1	*S. redlands*	I
S. ontario	D_2	*S. regent*	E_1
S. oranienburg	C_1	*S. reinickendorf*	B
S. ord	52	*S. remete*	F
S. ordonez	G_2	*S. remo*	B
S. oregon (combined with *S. muenchen*)	C_2	*S. reubeuss*	C_3
S. orientalis	I	*S. rhone*	L
S. orion	E_1	*S. rhydyfelin*	I
S. oritamerin	C_1	*S. richmond*	C_1
S. orlando	K	*S. rideau*	E_4
S. os	D_1	*S. ridge*	D_1
S. oskarshamn	M	*S. ried*	G_1
S. oslo	C_1	*S. riggil*	C_1
S. osnabrueck	F	*S. riogrande*	R
S. othmarschen	C_1	*S. rissen*	C_1
S. ottawa	D_1	*S. rittersbach*	P
S. ouakam	D_2	*S. riverside*	W
S. oudwijk	G_1	*S. roan*	P
S. overchurch	R	*S. rochdale*	Z
S. overschie	51	*S. rogy*	M
S. overvecht	N	*S. romanby*	G_2
S. oxford	E_1	*S. roodepoort*	G_1
S. oyonnax	C_1	*S. rosenthal*	E_2
S. pakistan	C_3	*S. rossleben*	54
S. palime	C_1	*S. rostock*	D_1
S. panama	D_1	*S. rottnest*	G_1
S. pankow	E_2	*S. rovaniemi*	I
S. papuana	C_1	*S. rubislaw*	F
S. paratyphi A	A	*S. ruiru*	L
S. paratyphi B = *S. schottmuelleri*	B	*S. ruki* (combined with *S. ball*)	B
S. paratyphi C = *S. hirschfeldii*	C_1	*S. rumford*	C_1
S. paris	C_3	*S. runby*	H
S. parkroyal	E_4	"*S. rutgers*" 3,10:l,z_{40}:1,7 (phase R)	E_1
S. pasing	B	*S. rruzizi*	E_1
S. patience	M	*S. saarbruecken*	D_1
S. penarth	D_1	*S. saboya*	I
S. pensacola	D_1	*S. sada*	N
S. perth	P	*S. sainte-marie*	52
S. pharr	F	*S. saint-paul*	B
S. pikine (combined with *S. altona*)	C_3	*S. saka*	X
S. pisa	I	"*S. sakai*" = *S. potsdam*	C_1
S. plymouth	D_2	*S. salford*	I
S. poano	H	*S. salinatis*	B
S. poeseldorf	54	*S. saloniki*	I
S. pomona	M	*S. sambre*	E_4
S. pontypridd	K	*S. sandiego*	B
S. poona	G_1	*S. sandow*	C_2

Table 5.12.—*continued*

Serovar	O Group	Serovar	O Group
S. sanga	C_3	*S. stormont*	E_1
S. sangalkam	D_2	*S. stourbridge*	C_2
S. sangera	I	*S. straengnaes*	F
S. sanjuan	C_1	*S. strasbourg*	D_2
S. sanktgeorg	M	*S. stratford*	E_4
S. sanktmarx	E_4	*S. stuivenberg*	E_4
S. santhiaba	R	*S. suberu*	E_1
S. santiago	C_3	*S. suelldorf*	W
S. sao	E_4	*"S. suez"* = *S. shubra*	B
S. saphra	I	*"S. suipestifer"* = *S. cholerae-suis*	C_1
S. sara	H	*S. sundsvall*	H
S. sarajane	B	*S. sunnycove*	C_3
S. saugus	R	*S. surat*	H
S. schalkwijk	H	*S. sya*	X
S. schleissheim	B	*S. szentes*	I
S. schoeneberg	E_4	*S. tabligbo*	X
"S. schottmuelleri" = *S. paratyphi B*	B	*S. tado*	C_3
S. schwarzengrund	B	*S. tafo*	B
S. schwerin	C_2	*"S. taihoku"* = *S. meleagridis*	E_1
S. seattle	M	*S. takoradi*	C_2
S. sedgwick	V	*S. taksony*	E_4
S. seegefeld	E_1	*S. tallahassee*	C_2
S. sekondi	E_1	*S. tamale*	C_3
S. selandia	E_2	*S. tambacounda*	E_4
S. selby	M	*S. tamberma*	X
S. sendai	D_1	*S. tamilnadu*	C_1
S. senegal	F	*S. tananarive*	C_2
S. senftenberg	E_4	*S. tanger*	G_1
S. seremban	D_1	*S. tanzania*	G_1
S. shamba	I	*S. tarshyne*	D_1
S. shangai	I	*S. taset*	T
S. shangani	E_1	*S. taunton*	M
S. shannon	E_1	*S. tchad*	O
S. sharon	F	*S. tchamba*	J
S. sherbrooke	I	*S. techimani*	M
S. sheffield	P	*S. teddington*	B
S. shikmonah	R	*S. tees*	I
S. shipley	C_3	*S. tejas*	B
S. shomolu	M	*S. teko*	H
S. shoreditch	D_2	*S. telaviv*	M
S. shubra	B	*S. telelkebir*	G_2
S. simi	E_1	*S. telhashomer*	F
"S. simsbury" 1,3,19:z_{27}:- (phase R)	E_4	*S. teltow*	M
S. sinchew	E_1	*S. tennessee*	C_1
S. singapore	C_1	*S. tennyson*	B
S. sinstorf	E_1	*S. teshie*	X
S. sinthia	K	*S. texas*	B
S. sipane	T	*S. thaygen*	B
S. skansen	C_2	*S. thetford*	U
S. sladun (combined with *S. abony*)	B	*S. thiaroye*	P
S. sljeme	X	*S. thielallee*	C_4
S. sloterdijk	B	*S. thomasville*	E_3
S. soahanina	H	*S. thompson*	C_1
S. soerenga	N	*S. tiergarten*	V
S. sokode	D_2	*S. tilburg*	E_4
S. solna	M	*S. tilene*	R
S. solt	F	*S. tim* (combined with *S. newington*)	E_2
S. somone	C_1	*S. tinda*	B
S. southampton	B	*S. tione*	51
S. southbank	E_1	*S. togba*	I
S. souza	E_1	*S. togo*	B
S. spartel	L	*S. tokoin*	B
S. stanley	B	*S. tomegbe*	T
S. stanleyville	B	*S. tomelilla*	E_4
S. staoueli	X	*S. tonev*	54
S. steinplatz	N	*S. toowong*	F
S. steinwerder	54	*S. toricada*	T
S. stellingen	X	*S. tornow*	W
S. stendal	F	*S. toronto*	D_2
S. sternschanze	N	*S. toucra*	Y
S. sterrenbos	C_2	*S. toulon*	K
S. stockholm	E_1	*S. tounouma*	C_3

Table 5.12.—*continued*

Serovar	O Group	Serovar	O Group
S. tournai	E_2	*S. westeinde*	I
S. trachau	B	*S. westerstede*	E_4
S. travis	B	*S. westhampton*	E_1
S. treforest	51	*S. westminster*	E_2
S. trimdon	D_2	*S. weston*	I
S. trotha	R	*S. westphalia*	O
S. truro	E_1	*S. weybridge*	E_1
S. tschangu	G_2	*S. wichita*	G_2
S. tsevie	B	*S. widemarsh*	O
S. tshiongwe	C_2	*S. wien*	B
S. tucson	H	*S. wil*	C_1
S. tudu	B	*S. wildwood*	E_3
S. tuebingen	E_2	*S. wilhelmsburg*	B
S. tunis	G_2	*S. willemstad*	G_1
S. typhi	D_1	*S. wimborne*	E_1
S. typhimurium	B	*S. windermere*	Q
S. typhisuis	C_1	*S. wingrove*	C_2
S. tyresoe	B	*S. winnipeg*	54
S. uccle	54	*S. winterthur*	E_4
S. uganda	E_1	*S. wippra*	C_2
S. ughelli	E_1	*S. wisbech*	I
S. uhlenhorst	V	*S. wohlen*	F
S. uithof	52	*S. womba* (combined with *S. altendorf*)	B
S. ullevi	G_2	*S. worb*	D_2
S. umbilo	M	*S. worthington*	G_2
S. umhlali	C_1	*S. wuerzburg* (combined with *S. miami*)	D_1
S. umhlatazana	O	*S. wuiti*	N
S. uno	C_2	*S. wuppertal*	D_2
S. uppsala	B	*S. wyldegreen*	G_2
S. urbana	N	*S. yaba*	E_1
S. ursenbach	T	*S. yalding*	E_4
S. usumbura	K	*S. yaounde*	B
S. utah	C_2	*S. yarm*	C_2
S. utrecht	52	*S. yarrabah*	G_2
S. uzaramo	H	*S. yeerongpilly*	E_1
S. vaertan	G_1	*S. yehuda*	F
S. vancouver	I	*S. yerba*	54
S. vejle	E_1	*S. yoff*	P
S. vellore	B	*S. yokoe*	C_3
S. veneziana	F	*S. yolo*	O
S. venusberg (combined with *S. nchanga*)	E_1	*S. yovokome*	C_3
S. victoria	D_1	*S. yundum*	E_1
S. victoriaborg	J	*S. zadar*	D_2
S. vietnam	S	*S. zagreb* (combined with *S. saintpaul*)	B
S. vilvoorde	E_4	*S. zaire*	N
S. vinohrady	M	*S. zanzibar*	E_1
S. virchow	C_1	*S. zega*	D_1
S. virginia	C_3	*S. zehlendorf*	N
S. visby	E_4	*S. zerifin*	C_2
S. vitkin	M	*S. zongo*	E_1
S. vleuten	V	*S. zuilen*	E_4
S. vogan	T	*S. zwickau*	I
S. volksmarsdorf	M	Subgenus" II	
S. volta	F	*S. II acres*	G_2
S. vom	B	*S. II alexander*	E_1
S. wagenia	B	*S. II alsterdorf*	R
S. wandsworth	Q	*S. II angola*	D_1
S. wangata	D_1	*S. II artis*	56
S. waral	T	*S. II askraal*	51
S. warengo	J	*S. II atra*	Z
S. warnemuende	M	*S. II bacongo*	C_1
S. warnow	C_2	*S. II baragwanath*	C_2
S. warragul	H	*S. II basel*	58
S. washington	G_1	*S. II bechuana*	B
S. waycross	S	*S. II bellville*	I
S. wayne	N	*S. II beloha*	K
S. wedding	M	*S. II betioky*	59
S. welikade	I	*S. II bilthoven*	X
S. weltevreden	E_1	*S. II blankenese*	D_1
S. wentworth	F	*S. II bleadon*	J
S. wernigerode	D_2	*S. II bloemfontein*	C_1
S. weslaco	T	*S. II boksburg*	R

Table 5.12.—*continued*

Serovar	O Group	Serovar	O Group
S. II *bornheim*	H	S. II *lincoln*	F
S. II *boulders*	G_2	S. II *lindrick*	D_1
S. II *bremen*	W	S. II *llandudno*	M
S. II *bulawayo*	R	S. II *lobatsi*	52
S. II *bunnik*	U	S. II *locarno*	57
S. II *caledon*	B	S. II *louwbester*	I
S. II *calvinia*	C_1	S. II *luanshya*	G_2
S. II *canastel*	D_1	S. II *lundby*	D_2
S. II *cape*	C_1	S. II *lurup*	S
S. II *carletonville*	P	S. II *luton*	60
S. II *ceres*	M	S. II *maarssen*	D_2
S. II *chersina*	X	S. II *makoma*	B
S. II *chinovum*	T	S. II *makumira*	B
S. II *chudleigh*	E_1	S. II *manica* (combined with S. II 1,9,12:g,m,[s],t:[1, 5]:[z_{42}])	D_1
S. II *clifton*	G_1	S. II *manombo*	57
S. II *clovelly*	V	S. II *matroosfontein*	E_1
S. II *constantia*	J	S. II *merseyside*	I
S. II *daressalaam*	D_1	S. II *midhurst*	53
S. II *degania*	R	S. II *mjimwema*	D_1
S. II *detroit*	T	S. II *mobeni*	I
S. II *dubrovnik*	S	S. II *mondeor*	Q
S. II *duivenhoks*	D_2	S. II *montgomery*	F
S. II *durbanville*	B	S. II *mosselbay*	U
S. II *eilbek* (combined with S. III *arizonae* 61:i:z)	61	S. II *mpila*	E_1
S. II *ejeda*	W	S. II *muizenberg* (combined with S. II 1,9,12:g,m,[s],t:[1, 5]:[z_{42}])	D_1
S. II *elsiesrivier*	I	S. II *nachshonim*	G_2
S. II *emmerich*	H	S. II *nairobi*	T
S. II *epping*	G_2	S. II *namib*	Z
S. II *erlangen*	Y	S. II *neasden*	D_1
S. II *fandran*	R	S. II *negev*	S
S. II *faure*	Z	S. II *ngozi*	Y
S. II *finchley*	E_1	S. II *noordhoek*	I
S. II *foulpointe*	P	S. II *nordenham*	B
S. II *fremantle*	T	S. II *neurnberg*	T
S. II *fuhlsbuettel*	E_1	S. II *odijk*	N
S. II *germiston*	C_2	S. II *ottershaw*	R
S. II *gilbert*	C_1	S. II *oysterbeds*	C_1
S. II *glencairn*	F	S. II *parow*	E_2
S. II *gojenberg*	G_2	S. II *perinet*	W
S. II *goodwood*	G_1	S. II *phoenix*	X
S. II *grabouw*	F	S. II *portbech*	T
S. II *greenside*	Z	S. II *quimbamba*	X
S. II *grunty*	R	S. II *rand*	T
S. II *gwaai*	L	S. II *rhodesiense*	D_1
S. II *haarlem*	D_2	S. II *roggeveld*	51
S. II *haddon*	I	S. II *rooikrantz*	H
S. II *hagenbeck*	Y	S. II *rotterdam*	G_1
S. II *hamburg* (combined with S. II 1,9,12:g,m[s],t:[1,5]:[z_{42}])	D_1	S. II *rowbarton*	I
S. II *hammonia*	Y	S. II *sakaraha*	Y
S. II *heilbron*	C_1	S. II *sarepta*	I
S. II *helsinki*	B	S. II *seaforth*	Z
S. II *hennepin*	S	S. II *setubal*	60
S. II *hillbrow*	J	S. II *shomron* (combined with S. III *arizonae* 18:z_4,z_{32}:-)	K
S. II *hooggraven*	Z	S. II *simonstown*	H
S. II *hueningen*	D_1	S. II *slangkop*	H
S. II *huila*	F	S. II *slatograd*	N
S. II *humber*	53	S. II *sofia*	B
S. II *islington*	E_1	S. II *soutpan*	F
S. II *jacksonville*	I	S. II *springs*	R
S. II *kaltenhausen*	M	S. II *srinagar*	F
S. II *katesgrove*	G_2	S. II *stellenbosch*	D_1
S. II *khami*	X	S. II *stevenage*	G_2
S. II *kilwa*	B	S. II *stikland*	E_1
S. II *klapmuts*	W	S. II *suarez*	R
S. II *kluetjenfelde*	B	S. II *suederelbe*	D_1
S. II *kommetje*	U	S. II *sullivan*	C_1
S. II *kraaifontein* (combined with S. II *luanshya*)	G_2	S. II *sunnydale*	R
S. II *krugersdorp*	Z	S. II *sydney* (combined with S. III *arizonae* 48:i:z)	Y
S. II *kuilsrivier*	D_1	S. II *tafelbaai*	E_1
S. II *lethe*	S	S. II *tokai*	57
S. II *lichtenberg*	S	S. II *tosamanga*	C_1
S. II *limbe*	G_1	S. II *tranoroa*	55

Table 5.12.—*continued*

Serovar	O Group	Serovar	O Group
S. II *tulear*	C$_2$	S. IV *chameleon*	I
S. II *tygerberg*	G$_2$	S. IV *flint*	Z
S. II *uphill*	T	S. IV *harmelen*	51
S. II *utbremen*	O	S. IV *houten*	U
S. II *veddel*	U	S. IV *kralendyk*	C$_1$
S. II *verity*	J	S. IV *lohbruegge*	V
S. II *vredelust*	G$_2$	S. IV *marina*	Y
S. II *vrindaban*	W	S. IV *mundsburg*	F
S. II *wandsbek*	L	S. IV *ochsenzoll*	I
S. II *westpark*	E$_1$	S. IV *parera*	F
S. II *wilhemstrasse* (combined with S. II *lobatsi*)	52	S. IV *roterberg*	C$_1$
S. II *winchester*	E$_1$	S. IV *sachsenwald*	R
S. II *windhoek*	W	S. IV *seminole*	R
S. II *woerden*	J	S. IV *soesterberg*	L
S. II *woodstock*	I	S. IV *tuindorp*	U
S. II *worcester*	G$_2$	S. IV *volksdorf*	U
S. II *wynberg*	D$_1$	S. IV *wassenaar*	Z
S. II *zeist*	K	"Subgenus" V	
S. II *zuerich*	D$_3$	S. V *balboa*	Y
"Subgenus" IV		S. V *bongor*	Y
S. IV *argentina*	C$_1$	S. V *brookfield*	66
S. (IV) *bern* (combined with S. IV 40:z$_4$, z$_{32}$)	R	S. V *camdeni*	V
S. IV *bockenheim*	53	S. V *malawi*	66
S. IV *bonaire*	Z	S. V *maregrosso*	66

a.ri.zo′nae. M.L. gen. n. *arizonae* of Arizona, a state in the United States.

Antigenic formula: 51:z$_4$,z$_{23}$:–. (The corresponding "*Arizona*" formula is 1,2,:1,2,5:–.)

The original strains isolated from reptiles were designated dar-es-Salaam type var. from Arizona (Caldwell and Ryerson, 1939). The antigenic formula was determined by Kauffman (1941) as 33:z$_4$,z$_{23,36}$:–, and he gave it the name *Salmonella* sp. (serotype) *arizona*. After Edwards et al. (1947) established *Arizona* as an independent group, the O antigen 33 was deleted from the Kauffmann-White scheme. O antigen 51 is identical with the old O antigen 33 and with the *Arizona* antigen designated 1,2 by Edwards et al. The H antigens z$_4$, z$_{23}$, z$_{36}$ (simplified to z$_4$, z$_{23}$) correspond to H antigens 1, 2, 5 of Edwards et al.

Type strain: ATCC 13314 (NCTC 9297).

"Subgenus" IV

k. "*Salmonella houtenae*" Le Minor, Rohde and Taylor 1970, 209. (*Salmonella houten* Kauffmann 1962, 353.)

hou′te.nae. M.L. gen. n. *houtenae* of Houten, a town in Holland.

Antigenic formula: 43:z$_4$,z$_{23}$:–.

The type species of *Salmonella* "subgenus" IV. It is the oldest known member of the "subgenus" (see discussion by Kauffmann, 1966, p. 244 on *S. delplata*, a mixed culture from which the serovar *S. houtenae* was obtained).

Type strain: NCTC 10401.

"Subgenus" V

l. "*Salmonella bongor*" Le Minor, Chamoiseau, Chairé-Marsaines and Egrou 1969, 775.

bon′gor. M.L. n. *bongor* Bongor, a town in Chad.

Antigenic formula: 48:z$_{35}$:–.

It is the oldest known member of the "subgenus." Isolated from a lizard in Chad.

Genus IV. *Citrobacter* Werkman and Gillen 1932, 173[AL]

RIICHI SAKAZAKI

Cit.ro.bac′ter. L. n. *citrus* lemon; M.L. n. *bacter* a small rod; M.L. masc, n. *Citrobacter* a citrate-utilizing rod.

Straight rods, ~1.0 μm in diameter and 2.0–6.0 μm in length. Occur singly and in pairs. Conform to the general definition of the family *Enterobacteriaceae*. Usually not encapsulated. Gram-negative. **Usually motile by peritrichous flagella.** Facultatively anaerobic, having both a respiratory and a fermentative type of metabolism. Grow readily on ordinary media. Colonies on nutrient agar are generally 2–4 mm in diameter, smooth, low convex, moist, translucent or opaque and gray with a shiny surface and entire edge. Mucoid or rough forms may occur occasionally. Oxidase-negative. Catalase-positive. Chemoorganotrophic. **Citrate can be utilized as a sole carbon source.** Nitrate is reduced to nitrite. **Lysine is not decarboxylated.** Phenylalanine deaminase, gelatinase, lipase and deoxyribonuclease are not produced. Alginate and pectate are not decomposed. Glucose is fermented with the production of acid and gas. The methyl red test is positive; **the Voges-Proskauer test is negative.** Occur in the feces of man and other animals; probably normal intestinal inhabitants. Often isolated from clinical specimens as opportunistic pathogens. Also found in soil, water, sewage and food. The mol% G + C of the DNA is 50–52 (T_m).

Type species: *Citrobacter freundii* Werkman and Gillen 1932, 173.

Further Descriptive Information

Members of *Citrobacter* may or may not ferment lactose promptly but nearly always produce β-galactosidase. L-Arabinose, cellobiose, maltose, L-rhamnose, trehalose, D-xylose, D-mannitol, D-sorbitol, and glycerol are fermented rapidly by the majority of strains. Raffinose and *myo*-inositol are rarely attacked.

Ornithine is decarboxylated by almost all strains of *C. diversus* and *C. amalonaticus*, but less than 20% of strains of *C. freundii* produce this enzyme. Strains of *C. freundii* and *C. amalonaticus* in contrast to *C. diversus* can grow in media containing potassium cyanide.

Strains of *C. diversus* ferment D-adonitol, but nearly all strains of *C. freundii* and *C. amalonaticus* fail to ferment this substrate. Malonate is utilized as a sole carbon source by most strains of *C. diversus*, but can be used by less than 15% of the strains of *C. freundii* and not by *C. amalonaticus*.

The majority of strains of *C. freundii* produce abundant H$_2$S in the butt of Kligler iron agar and triple-sugar iron agar. Lactose is fermented by many strains of *C. freundii*, but the reactions are frequently delayed.

Indole is not produced by *C. freundii* with few exceptions, but all strains of *C. diversus* and *C. amalonaticus* give a positive indole test.

Nitrogen fixation under anaerobic condition has been reported in some strains of *C. freundii* isolated from the hindgut of Australian termites and from paper mill process water (Bergensen, 1980).

West and Edwards (1954) first established an antigenic schema of the Bethesda-Ballerup group of bacteria, which is now called *C. freundii*, based on their early studies (Edwards et al., 1948; Bruner et al., 1949; Moran and Bruner, 1949). The antigenic schema included 32 O groups and 87 H antigens. Sedlák and Slajsová (1966, 1967) and Sedlák (1974) expanded the antigenic schema by adding further O and H antigens, increasing the total number to 42 O and more than 90 H antigens. The antigens of many serovars of *C. freundii* relate to those of many *Salmonella* and *Escherichia* cultures (West and Edwards, 1954; Sakazaki and Namioka, 1957; Davis and Ewing, 1963; Sedlák and Slajsová, 1966). O antigenic relationships between *C. freundii* and *Hafnia alvei* were reported by Sakazaki (1971) and Sedlák and Slajsová (1966). The H antigens of *C. freundii* are monophasic. Some strains of O groups 5 and 29 of *C. freundii* may possess an antigen serologically identical with the Vi antigen of *Salmonella typhi* (Kauffmann and Møller, 1940; Monteverde, 1944). In contrast to *S. typhi*, however, quantitative variation of the Vi antigen in *C. freundii* cultures is reversible, and the presence of the Vi antigen is not related to the virulence of the cultures.

Serological studies of *C. diversus* were first reported by Gross et al. (1973) using four isolates from infantile meningitis. Later, Gross and Rowe (1974, 1975) and Gross et al. (1981) designated 17 O groups without any account of the H antigens. Popoff and Richard (1975), who studied the serology of *C. diversus* independently of Gross and Rowe, established an antigenic schema which contained 6 O groups and 7 H antigens. Sourek and Aldová (1976) also studied O antigens of *C. diversus* and independently proposed 9 O antigens.

Although no antigenic schema was proposed, van Oye et al. (1975) reported that the O antigens of 35 of 38 strains of *C. amalonaticus* were closely related to those of several serovars of *Shigella dysenteriae* and *Shigella boydii*. Sourek and Aldová (1976) presented an O grouping system in which 13 O groups were designated.

Members of *C. freundii* are usually susceptible to the aminoglycosides, chloramphenicol and colistin. Susceptibility of *C. freundii* to ampicillin, tetracycline and the cephalosporins differs among the strains. *C. diversus* and *C. amalonaticus* are susceptible to the amino-glycosides, cephalosporins, colistin, chloramphenicol and tetracycline. *C. diversus* and *C. amalonaticus* generally appear to be resistant to ampicillin and carbenicillin (Lund et al. 1974).

Members of the genus *Citrobacter* occur not only in feces of man and other animals with no disorder but also in water, sewage, soil and food. They are also found in clinical bacteriology not only in stools but also in urine, sputum and specimens from bacteremia, meningitis, otitis media, wounds, abscesses, the throat and autopsies; their role seems to be that of an opportunistic pathogen. Recently cases of neonatal meningitis caused by *C. diversus* have often been reported (Gross et al., 1973; Gwynn and George, 1973; Puentes et al., 1975; Tamborlane and Soto, 1975; Ribeiro et al., 1976). Although *C. freundii* was once considered to be an enteropathogen, it seems rather to be a normal inhabitant of the intestine (Sakazaki et al., 1960). Some investigators, however, have suggested a possible role of certain strains of *C. freundii* and *C. diversus* in causing diarrhea (Kleinmeier and Schafer, 1956; Sakazaki and Namioka, 1957; Sedlák, 1957; Nestorescu et al., 1964; Popovic et al., 1964; Guerrant et al., 1976; Wadström et al., 1976; Finn, 1978).

Enrichment and Isolation Procedures

The majority of *C. freundii* strains can grow in liquid enrichment media such as selenite broth and tetrathionate broth and on selective isolation media such as salmonella-shigella agar, deoxycholate-citrate agar, brilliant green agar and bismuth sulfite agar. Colonies which ferment lactose slowly can resemble *Salmonella* colonies in many instances.

Although *C. diversus* and *C. amalonaticus* strains are usually able to grow on the selective media indicated above, many strains are inhibited to some extent; therefore, less inhibitory media such as MacConkey agar and xylose-lysine-deoxycholate agar may be preferable.

Maintenance Procedures

Stock cultures of *Citrobacter* strains may be maintained at room temperature in a semisolid medium containing 1.0% Bacto-casitone (Difco), 0.3% yeast extract, 0.5% NaCl and 0.3% agar, pH 7.0. The cultures remain viable up to a year without subculturing if they are sealed with a rubber stopper or a cork which has been soaked in hot paraffin wax. Strains may also be preserved indefinitely by lyophilization.

Differentiation of the genus **Citrobacter** from other genera

Table 5.13 indicates the characteristics of *Citrobacter* that differentiate it from biochemically similar genera.

Taxonomic Comments

The genus *Citrobacter* was proposed by Werkman and Gillen (1932) for the citrate-utilizing "coli-aerogenes intermediates." Until recent years, however, the name did not gain acceptance and the organisms have been described under a variety of designations. *C. freundii* was described as "*Escherichia freundii*" by Yale (1939), and as "*Colobactrum freundii*" (for rapid lactose fermenters) and *Paracolobactrum intermedium* (for slow lactose fermenters) by Borman et al. (1944). The role of citrobacters as possible pathogens was first noticed by Kauffmann and Møller (1940), who described an organism called "*Salmonella ballerup*" which is presently classified in *C. freundii*. Monteverde (1944) reported an organism similar to *S. ballerup* under the name "*Salmonella hormaechei*." Later, this biogroup of organisms was removed from the genus *Salmonella* and was called the Ballerup group (Harhoff, 1949; Bruner et al., 1949). Independently of the Ballerup group of organisms, Edwards et al. (1948) and Moran and Bruner (1949) studied a group of bacteria characterized by Barnes and Cherry (1946) and referred to it as the Bethesda group of bacteria. West and Edwards (1954) found that organisms of both the Bethesda and Ballerup groups were biochemically and serologically indistinguishable and combined the two groups into the Bethesda-Ballerup group. Moreover, West and Edwards (1954) and Møller (1954) called attention to the close biochemical relationship between members of the Bethesda-Ballerup group and strains of *E. freundii*. Accordingly, Kauffmann (1954) reclassified the Bethesda-Ballerup group into *E. freundii*, and later revived the genus *Citrobacter* for *E. freundii* (Kauffmann, 1956).

More recently, Young et al. (1971) described a new genus, *Levinea*, which contained two species, *L. malonatica* and *L. amalonatica*. Ewing and Davis (1972) noted, however, that *L. malonatica* was a later synonym of "*Citrobacter diversum*" which was designated by Werkman and Gillen (1932); consequently, they revived the name *C. diversus* for this species (with a grammatical modification of the ending of the specific epithet). Prior to the work of Young et al. (1971) and Ewing and Davis (1972), Frederiksen (1970) had described a new species, *Citrobacter koseri*. It was confirmed by numerical taxonomy (Sakazaki et al., 1976) and by DNA relatedness (Crosa et al., 1974) that *C. koseri* was also a synonym of *C. diversus*. Although the name *C. diversus* is accepted in the United States, there are many workers in Europe who believe that the original description of "*C. diversum*" by Werkman and Gillen was based on strains which were different from the strains of *C. diversus* described by Ewing and Davis. These workers therefore believe that the name *C. koseri* has priority (Holmes et al., 1974). Because no original strains of "*C. diversum*" exist, it is difficult to judge the dispute

Table 5.13.

Differential characteristics of the genus **Citrobacter** *and biochemically similar genera[a]*

Characteristics	Citro-bacter	Sal-monella	Escher-ichia	Entero-bacter
Lysine decarboxylase	−	+	+	D
Citrate (Simmons')	+	+	−	+
Voges-Proskauer test	−	−	−	+
Growth in KCN medium	D[b]	−	−	+
Indole production	D[c]	−	+	−
Ornithine decarboxylase	D[d]	+	+	+
ONPG hydrolysis[e]	+	D	+	+
Mol% G + C of DNA	50–52	50–53	48–52	52–59

[a] Symbols: +, 90–100% of strains are positive; −, 90–100% of strains are negative; D, different reactions given by different species of a genus.
[b] Only *C. diversus* is negative.
[c] Only *C. freundii* is negative.
[d] Less than 20% of *C. freundii* are negative.
[e] ONPG, *o*-nitrophenyl-β-D-galactopyranoside.

immediately. Thus, the names *C. koseri* and *L. malonatica*, in addition to *C. diversus*, have been included in the Approved Lists of Bacterial Names in 1980.

Ewing and Davis (1972) regarded *L. amalonatica* as a biovar of *C. freundii*. On the basis of results of numerical taxonomy (Sakazaki et al., 1976) and DNA relatedness (Crosa et al., 1974), however, it was obvious that *L. amalonatica* should be placed in a species separate from *C. freundii*, although this organism was more closely related to *C. freundii* and *C. diversus* than to other genera of the family *Enterobacteriaceae*. Thus, Brenner et al. (1977) suggested moving *L. amalonatica* to *Citrobacter*, and the name *L. amalonaticus* has been formally proposed (Brenner and Farmer, 1981, 1982). Macierevicz (1966) studied a group of organisms which were H$_2$S-negative and ornithine decarboxylase-positive and proposed an illegitimate generic name "*Padlewskia*" without a designation of any specific epithet for the organisms of this genus. From the biochemical characteristics described by Macierewicz, it is clear that *Padlewskia* organisms are identical to *C. amalonaticus*.

"*Citrobacter intermedium*" was proposed by Werkman and Gillen (1932) for H$_2$S-negative strains of *Citrobacter*. Vaughn and Levine (1942) transferred this species to the genus *Escherichia* as "*E. intermedia*." Sedlák (1974) revived this species in the eighth edition of *Bergey's Manual* as "*C. intermedius*." He described two biovars in *C. intermedius*: biovar "a," corresponding to *L. amalonatica*, and biovar "b," corresponding to *L. malonatica*. It was found, however, that one of Werkman's original strains of *C. intermedius*, ATCC 6750, was a typical *C. freundii* (Frederiksen, 1970). The name *C. intermedius* was, therefore, not included on the Approved Lists of Bacterial Names in 1980 and has no nomenclatural standing.

Differentiation and characteristics of species of **Citrobacter**

The differential characteristics of the species of *Citrobacter* are indicated in Table 5.14. Table 5.15 lists other characteristics of the species.

List of the species of the genus **Citrobacter**

1. **Citrobacter freun'dii** (Braak 1928) Werkman and Gillen 1932, 173,[AL] (*Bacterium freundii* Braak 1928, 140.)

freun'di.i. M.L. gen. n. *freundii* of Freund; named after A. Freund, the bacteriologist who first observed that trimethylene glycol was a product of fermentation.

The morphology is as given for the genus. Usually motile. Usually not encapsulated, although encapsulated strains may occur in some strains belonging to certain O antigen groups.

The colony morphology is similar to that of *Escherichia coli*, but growth may occur on some selective inhibitory media for the isolation of *Salmonella* on which *E. coli* is inhibited.

Physiological and biochemical characteristics are presented in Tables 5.14 and 5.15. Less than 20% of the strains produce ornithine decarboxylase.

Found in man and other animals including mammals, birds, reptiles and amphibians. Also found in soil, water, sewage and food. Often

Table 5.14

Characteristics differentiating **Citrobacter freundii**, **Citrobacter diversus** *and* **Citrobacter amalonaticus**[a]

Characteristics	1. *C. freundii*	2. *C. diversus*	3. *C. amalonaticus*
Indole production	−	+	+
H$_2$S production[b]	+	−	−
Arginine dihydrolase	d	+	+
Ornithine decarboxylase	d	+	+
Growth in KCN medium	+	−	+
Malonate utilization	−	+	−
D-Adonitol, acid from	−	+	−
Mol% G + C of DNA	50–51	51–52	51–52

[a] Symbols: +, 90–100% of strains are positive; −, 90–100% of strains are negative; d, different reactions given by different strains of a species.
[b] In Kligler iron agar and triple-sugar iron agar.

Table 5.15.

Other characteristics of **Citrobacter freundii, Citrobacter diversus** *and* **Citrobacter amalonaticus**[a]

Characteristics	1. *C. freundii*	2. *C. diversus*	3. *C. amalonaticus*
Voges-Proskauer test	−	−	−
H$_2$S production	+	−	−
Urease (Christensen)	d	d	d
Gelatin hydrolysis	−	−	−
Phenylalanine deaminase	−	−	−
d-Tartrate (Kauffmann-Petersen)	(+)	d	−
Mucate, acid from	+	+	+
Esculin hydrolysis	−	d	+
Lipase (Tween 80)	−	−	−
Deoxyribonuclease	−	−	−
Acid from carbohydrates:			
D-Glucose, L-arabinose, cellobiose, maltose, L-rhamnose, trehalose, D-xylose, D-mannitol, D-sorbitol, glycerol	+	+	+
Lactose	d	+	+
Sucrose	d	−	−
Dulcitol	d	d	−
Salicin	−	+	+
Raffinose, erythritol, myo-inositol	−	−	−
Gas from D-glucose	+	+	+
ONPG hydrolysis[b]	+	+	+

[a] Symbols: +, 90–100% of strains are positive; (+), 90–100% of strains are positive after 3 days or more of incubation; −, 90–100% of stains are negative; d, different reactions given by different strains of a species.
[b] ONPG = *o*-nitrophenyl-β-D-galactopyranoside.

found in clinical specimens such as urine, throat, sputum, blood and wound swabs as an opportunistic or secondary pathogen.

The mol% G + C of the DNA is 50–51 (T_m).

Type strain: ATCC 8090.

2. **Citrobacter diversus** (Burkey 1928) Werkman and Gillen 1932, 180.[AL] (*Aerobacter diversum* Burkey 1928, 77; *Citrobacter koseri* Frederiksen, 1970, 93.)

di.ver′sus. L. v. *divertere* to turn in different directions; L. part. adj. *diversus* differing.

The morphology is as given for the genus. Motile. Not encapsulated.

Colonies on nutrient agar are translucent to opaque, resembling those of *Escherichia coli*.

Physiological and biochemical characteristics are presented in Tables 5.14 and 5.15.

Found in the feces of man and other animals and in soil, water, sewage and food. Also isolated from human clinical specimens such as urine, throat, nose and sputum and wound swabs. Occasionally causes neonatal meningitis.

The mol% G + C of the DNA is 51–52 (T_m).

Type strain: ATCC 27156.

3. **Citrobacter amalonaticus** (Young Kenton, Hobbs and Moody 1971) Brenner and Farmer 1982, 266. (*Levinea amalonatica* Young et al. 1971, 58.)

a.ma.lo.na′ti.cus. Gr. prefix *a* not; M.L. adj. *malonaticus* pertaining to malonate; M.L. adj. *amalonaticus* not pertaining to malonate (i.e. not able to utilize malonate).

The morphology is as given for the genus. Motile. Not encapsulated.

Colonies on nutrient agar are translucent to opaque, resembling those of *Escherichia coli*.

Physiological and biochemical characteristics are indicated in Tables 5.14 and 5.15.

Found in the feces of man and other animals and in soil, water and sewage. Also found in a variety of human clinical specimens as an opportunistic pathogen.

The mol% G + C of the DNA is 51–52 (T_m).

Type strain: ATCC 25405.

Genus V. Klebsiella Trevisan 1885, 105^{AL}

IDA ØRSKOV

Kleb.si.el′la. M.L. dim. ending *-ella*; M.L. fem. n. *Klebsiella* named after Edwin Klebs (1834–1913), a German bacteriologist.

Straight rods, 0.3–1.0 μm in diameter and 0.6–6.0 μm in length, arranged singly, in pairs or short chains. Conform to the general definition of the family *Enterobacteriaceae*. **Capsulated.** Gram-negative. **Nonmotile.** Facultatively anaerobic, having both a respiratory and a fermentative type of metabolism. Grow on meat extract media, producing more or less dome shaped, glistening colonies of varying degrees of stickiness depending on the strain and the composition of the medium. There are no special growth factor requirements. **Oxidase-negative. Most strains can use citrate and glucose as a sole carbon source.** Glucose is fermented with the production of acid and gas (more CO_2 is produced than H_2), but anaerogenic strains occur. Most strains produce 2,3-butanediol as a major end product of glucose fermentation and the **Voges-Proskauer test is usually positive**; lactic, acetic and formic acids are formed in smaller amounts and ethanol in larger amounts than in a mixed acid fermentation. **Fermentation of inositol, hydrolysis of urea, and lack of production of ornithine decarboxylase or H_2S are further distinctive characters.** Some strains fix nitrogen. Occur in intestinal contents, clinical specimens, soil, water, grain, etc. The mol% G + C of the DNA is 53–58 (T_m).

Type species: *Klebsiella pneumoniae* (Schroeter 1886) Trevisan 1887, 94.

Further Descriptive Information

The outermost layer of *Klebsiella* bacteria consists of a large polysaccharide capsule, a character which distinguishes members of this genus from most other bacteria in the family (*Enterobacter aerogenes* and *Escherichia coli* strains with a heat-stable K antigen (A type) may form similar capules). The cell wall itself, however, is structured as that of other *Enterobacteriaceae*, i.e. when going from within: (a) the cytoplasmic membrane, (b) the peptidoglycan layer and (c) the outer membrane containing the lipopolysaccharide (LPS). In addition, *Klebsiella* strains may possess fimbriae (pili), some with a mannose-sensitive adhesin (type 1) and others with a mannose-resistant adhesin (type 3) or with both types (Duguid, 1959).

The production of the large capsules gives rise to large mucoid colonies of a viscid consistency. The capsular material also diffuses freely into the surrounding liquid medium as extracellular capsular material.

Klebsiella strains grow readily on all kinds of media since they have no particular growth requirements. A carbohydrate-rich medium gives a better development of the capsule than a carbohydrate-poor medium. In the author's laboratory a bromothymol blue lactose medium* is most often used.

In general the methyl red test is negative and the Voges-Proskauer (VP) test is positive in *Klebsiella*,, meaning that acetoin and 2,3-butanediol are formed from pyruvic acid and that these neutral end products predominate over the acidic end products as a result of the sugar fermentation. Some strains, e.g. *K. rhinoscleromatis*, do not form acetoin and 2,3-butanediol. Other strains produce acetoin and 2,3-butanediol in such small amounts that the methyl red reaction remains positive. In some strains the acetoin will disappear before the VP reaction is tested. Seemingly paradoxical methyl red and VP reactions may therefore occur (i.e. both tests positive or both tests negative).

Some strains of *Klebsiella* have the ability to fix molecular nitrogen. No particular correlation of this property with the source of a strain seems evident; according to Postgate (1978), "A *Klebsiella* from the gut is as likely—or as unlikely—to fix nitrogen as one from a soil or water sample." Since the nitrogenase is rapidly inactivated in the cells in the presence of oxygen, the nitrogen-fixing ability of *Klebsiella* strains is generally expressed only under anaerobic conditions; however, several reports indicate that low levels of dissolved oxygen (<10 mm Hg) can be tolerated (Klucas, 1972; Hill, 1975; Neilson and Sparell, 1976) or even used to support nitrogen fixation (Hill, 1976). The genetics and regulation of nitrogen fixation in *Klebsiella* have recently been reviewed by Brill (1980).

Klebsiella strains may be lysogenic, but phages used by some workers for phage typing have been isolated from stools or sewage (Slopek et al., 1967; Slopek, 1978).

Many *Klebsiella* strains produce bacteriocin (klebecin) and typing sets of such producers can be selected (Slopek and Maresz-Babczyszyn, 1967; Edmondson and Cook, 1979).

Successful genetic recombinations have been reported in *Klebsiella* (Matsumoto and Tazaki, 1970), and *K. pneumoniae* has been used by several workers for detailed genetic analysis of the genes involved in N_2 fixation (*Nif* genes). These genes are clustered near the *His* region on the chromosome but can be mobilized and transferred to other organisms.

---

* For composition of bromothymol blue lactose medium, see the genus *Escherichia*.

A high percentage of *Klebsiella* strains from clinical isolates and particularly those from nosocomial infections contain R factors that determine resistance to a variety of drugs, such as β-lactams, cephalosporins, aminoglycosides, tetracyclines, chloramphenicols, sulfonamides and trimethoprim. All *Klebsiella* strains are resistant to ampicillin and this resistance may reside in chromosomal genes or be mediated by genes present on the chromosome and on a plasmid.

In general klebsiellae are good recipients for R factors, a fact that may have made *Klebsiella* a culprit in serious nosocomial epidemic diseases (Falkow, 1975).

Reeve and Braithwaiter (1975) demonstrated two classes of *Klebsiella* strains, one with a strong and the other with a weak lactose-positive phenotype. This was shown to be due to the presence of a *Lac* plasmid in the strongly fermenting strains.

Klebsiella possesses both O (lipopolysaccharide, LPS) and K (polysaccharide) antigens, but serological typing is based on examination of the K antigens. This is because the number of O antigen types is lower than that of the K antigen types and because O antigen determination is hampered by the heat-stable K antigens.

Capsular types A to C of Julianelle (1926) and C to F of W.R.O. Goslings (Onderzoekungen over de bacteriologie en de epidemiologie van het scleroma respiratorium, Thesis, Amsterdam, 1933, pp. 199–201) and of Goslings and Snijders (1936) were redesignated 1 to 6 by Kauffmann (1949) who also established eight new types. Other workers have brought the total number of K types up to 82 (for a review see Ørskov and Ørskov: Serotyping of *Klebsiella*. In Bergan and Norris (Editors), *Methods in Microbiology*, Academic Press, London, in press. The capsular polysaccharides have been analyzed qualitatively (Nimmich, 1968, 1971) and the structures of a majority of them have been determined (for reviews see Heidelberger and Nimmich, 1976; Sutherland, 1977; Rieger-Hug and Stirm, 1981).

The majority of K antigens contain only one charged monosaccharide constituent, most often glucuronic acid, and two to four of the following sugars: galactose, D-glucose, mannose, fucose and L-rhamnose. Other noncarbohydrate constituents, such as acetate or pyruvate, may also be present. For a review of the O antigen structures, see Jann and Jann (1977).

Klebsiellae are opportunistic pathogens that can give rise to bacteremia, pneumonia, urinary tract and several other types of human infection. In recent years there has been an increase in *Klebsiella* infections, particularly in hospitals, due to strains with multiple antibiotic resistance (for a review see Montgomerie, 1979). The gastrointestinal tract is considered to be the main reservoir and the hands of the personnel the main factor for transmission. These outbreaks particularly occur in urological patients and in neonatal and intensive care units. Enterotoxin-producing *Klebsiella* strains have been described (Klipstein et al., 1977).

Klebsiellae are also widely distributed in nature, occurring in soil, water, grain, etc. Many of these environmental strains, however, probably belong to the two newly proposed species, *K. terrigena* and *K. planticola*.

Enrichment and Isolation Procedures

Although klebsiellae are normal inhabitants of the intestinal tract, they are usually present in such low numbers, compared with *E. coli*, that they may be difficult to select after growth for only 24 h; however, they usually will appear as characteristic, elevated, mucoid colonies after incubation for 48 h. The detection and isolation from sources such as feces or water can be facilitated by use of a selective medium. Since *Klebsiella* strains can utilize citrate as a sole carbon source, citrate-containing media have long been used to facilitate their isolation. Methyl violet and double violet agar have been proposed as selective media (Campbell and Roth, 1975; Campbell et al., 1976). A synthetic medium containing *myo*-inositol as the sole carbon source was used successfully for selection of *Klebsiella* (and *Serratia*) by Legakis et al. (1976), and a MacConkey-inositol-carbenicillin agar medium was devised by Bagley and Seidler (1978); the selectivity of the later medium is based upon the high resistance of *Klebsiella* to carbenicillin, in contrast to that of other *Enterobacteriaceae*.

Maintenance Procedures

Klebsiella strains can be easily maintained in meat extract agar stabs or on egg medium when kept at room temperature in the dark. They can be preserved either by storage in broth containing 10% glycerol at −80°C or by lyophilization.

Procedures for Testing Special Characters

Von Riesen (1976) reported that indole-positive strains of *Klebsiella* were able to digest polypectate, and this ability was later shown to be a distinctive character of *K. oxytoca*. The pectate test is negative in the medium of Martin and Ewing (Edwards and Ewing, 1972) but positive in that of Starr (1947). The procedure used by Starr et al. (1967) is as follows. The following ingredients are added to 100 ml of distilled water while stirring: $CaCl_2 \cdot 2H_2O$ (10% solution), 0.6 ml; bromthymol blue (0.1% solution in 6.4×10^{-4} N NaOH), 1.0 ml; yeast extract (Difco), 0.6 g; and sodium polygalacturonate (P-1879, Sigma Chemical, St. Louis, Mo.; or 102921, ICN, Cleveland, Ohio), 3.0 g (added very slowly so that each particle is wetted). After the polygalacturonate is uniformly swelled it is dissolved by bringing the temperature almost to the boiling point with continuous stirring. The pH is adjusted to 7.3 with 1 N NaOH by monitoring the color of the indicator. The medium is sterilized at 121°C for 15 m and dispensed into Petri dishes or tubes. The cultures are either spotted onto or stabbed into the medium, which is then incubated at 30°C and inspected daily for up to 6 days for evidence of liquefaction and/or sinking of the colonies.

The test for liquefaction of gelatin should preferably be the rapid method of Kohn (1953) as described by Lautrop (1956a) and Edwards and Ewing (1972). Most *Klebsiella* strains which liquefy gelatin will do so within 4 days of this method.

Differentiation of the genus **Klebsiella** from other genera

See Table 5.3 of the family *Enterobacteriaceae* for characteristics that can be used to distinguish this genus from other genera of the family. The greatest problem is to distinguish *Klebsiella pneumoniae* strains from nonmotile *Enterobacter aerogenes* strains which liquefy gelatin very slowly. The urease test may be of decisive importance in such cases (*K. pneumoniae* is urease-positive).

Taxonomic Comments

In the eighth edition of *Bergey's Manual* three species were described in the genus *Klebsiella*: *K. pneumoniae*, *K. ozaenae* and *K. rhinoscleromatis*. Because DNA reassociation studies have shown that these three species belong to the same DNA relatedness group (Brenner et al., 1972), *K. ozaenae* and *K. rhinoscleromatis* are considered as subspecies of *K. pneumoniae* in the present edition of the *Manual*. Both subspecies may be considered as metabolically inactive biogroups of *K. pneumoniae*: *K. rhinoscleromatis* is the most metabolically inactive, while the metabolic activity in *K. ozaenae* strains is variable. In contrast to *K. pneumoniae* strains (Ewing and Martin, 1974), some *K. ozaenae* strains are arginine dihydrolase-positive. Traditionally, *K. ozaenae* belongs to capsule type 4 (3, 5, 6 or 1/5 have also been described), and if no serotyping is done it is very difficult to distinguish a metabolically active strain of *K. ozaenae* from a strain of *K. pneumoniae*. *K. rhinoscleromatis* may have an as yet undetected phenotypic property, since it is found constantly and exclusively in patients with rhinoscleroma as well as in their contacts. Three subspecies of *K. pneumoniae* are proposed: *K. pneumoniae* subsp. *pneumoniae* (subsp. nov.), *K. pneumoniae* subsp. *ozaenae* (subsp. nov.), and *K. pneumoniae* subsp. *rhinoscleromatis* (subsp. nov.). The description of each of the three subspecies corresponds to that of the former species. It is recommended that clinical laboratories omit these subspecies designations for routine reporting.

Indole-positive and gelatin-liquefying strains of *Klebsiella* have been a taxonomic problem for years. Some authors have considered them as biogroups of *K. pneumoniae* (Edwards and Ewing, 1975; Ørskov, 1974); others as a separate group (Lautrop, 1956b; Stenzel et al., 1972), and still others have excluded such strains from their studies of *Klebsiella*. The name *Bacterium oxytocus* Flügge was revived by Lautrop (1956b) for these strains, and the name *Aerobacter oxytocum* was recognized in earlier editions of *Bergey's Manual*. Korth et al. (1960) showed that the "oxytoca" variants of *Klebsiella* produce a dark brown pigment when grown on a defined medium containing gluconate and ferric citrate. On the basis of DNA/DNA hybridization studies, it has been proposed that indole- and gelatin-positive strains be removed from the genus *Klebsiella* (Jain et al., 1974) or that they be considered as a separate *Klebsiella* species designated *K. oxytoca* (Brenner et al., 1977). The latter course has been followed in the present edition of the *Manual*.

A third proposed species of *Klebsiella* is *K. terrigena* (Izard et al., 1981), a name recently coined for strains which are derived mainly from aquatic and soil environments (Izard et al., 1981). According to numerical taxonomic analysis (Gavani et al., 1977; Naemura et al., 1979) and DNA/DNA hybridization studies (Woodward et al., 1979; Izard et al., 1981), *K. terrigena* forms a species distinct from both *K. pneumoniae* and *K. oxytoca*. Phenotypically, *K. terrigena* is closely related to *K. pneumoniae*, but three tests (growth at 10°C (positive for *K. terrigena*), gas production from lactose when incubated at 44.5°C (negative for *K. terrigena*), and fermentation of melizitose (positive for *K. terrigena*)) can differentiate the two species.

A fourth proposed species of *Klebsiella* is *K. planticola* (Bagley et al., 1981), which contains strains isolated primarily from botanical and soil environments (Bagley et al., 1981). *K. planticola* is distinct from other *Klebsiella* species on the basis of numerical taxonomy (Gavini et al., 1977; Naemura et al., 1979) and by DNA relatedness (Woodward et al., 1979; Izard et al., 1981). Like *K. terrigena*, *K. planticola* can be separated from *K. pneumoniae* by growth at 10°C (positive for *K. planticola*) and by gas production from lactose when incubated at 44.5°C (negative for *K. planticola*). Melizitose is fermented by *K. terrigena* but not by *K. pneumoniae* or *K. planticola*.

In this edition of the *Manual*, the genus *Klebsiella* is confined to nonmotile strains. Proposals have been made to transfer *Enterobacter*

aerogenes to the genus *Klebsiella* as *K. mobilis* (Bascomb et al., 1971; Izard et al., 1980). *E. aerogenes* is biochemically and genetically as related or more related to klebsiellae than to most other *Enterobacter* species (Bascomb et al., 1971; Brenner et al., 1972; Steigerwalt et al., 1975; Izard et al., 1980). If transferred to the genus *Klebsiella*, *E. aerogenes* would normally become the new combination "*K. aerogenes*," a name that has no standing in nomenclature. It had been used to designate certain strains of *K. pneumoniae* and, therefore, should not be reproposed for a different group. The proposal of *K. mobilis* poses several problems: (a) a Judicial Commission decision would be required to change the specific epithet from *aerogenes* to *mobilis*; (b) the well accepted epithet *aerogenes* would be lost; (c) the epithet *mobilis* is misleading because not all strains of *E. aerogenes* are motile; (d) the important genus characteristic, lack of motility, would no longer be definitive for the genus *Klebsiella*.

Further Comments

Cowan et al. (1960) recognized five species in the *Klebsiella* group: *K. aerogenes*, *K. pneumoniae* (sensu stricto), *K. ozaenae*, *K. rhinoscleromatis*, and *K. edwardsii* with two varieties: *K. edwardsii* var. *edwardsii* and *K. edwardsii* var. *atlantae*. Durlakowa et al. (1967) and Slopek and Durlakowa (1967) divided *Klebsiella* into the six taxa of Cowan et al.; the names and the rank were, however, somewhat changed. Bascomb et al. (1971) divided *Klebsiella* into six taxa, one of which was *K. pneumoniae* (sensu stricto) and another composed of *K. aerogenes*, *K. edwardsii* and indole-forming *Klebsiella* strains. Brenner et al. (1972) found 80–90% DNA relatedness between *K. pneumoniae* (sensu lato), *K. ozaenae*, *K. rhinoscleromatis* and *K. edwardsii*. No *K. pneumoniae* (sensu stricto, according to Cowan et al., 1960, or Bascomb et al., 1971) was included in that study. However, the neotype strain of *K. pneumoniae*, ATCC 13883 (Ørskov, 1974), which is a *K. pneumoniae* (sensu stricto) strain (VP-negative, KCN-negative), has been shown to be genetically indistinguishable from other *Klebsiella pneumoniae* (sensu lato) strains (Seidler et al., 1975; Woodward et al., 1979). The classification of Cowan et al. (1960) is used in the United Kingdom and at other places, but never in the United States. This means that the same organism will be classified either as *K. pneumoniae* or *K. aerogenes*, depending on the country.

The existence of two additional *Klebsiella* species that contain strains of environmental origin is suggested in the studies by Naemura et al. (1979) and Woodward et al. (1979).

Differentiation and characteristics of the species of the genus **Klebsiella**

Table 5.16 presents the characteristics differentiating the four species of *Klebsiella*, and Table 5.17 lists additional characteristics of the species. Table 5.18 lists those characteristics that differentiate the three subspecies of *K. pneumoniae*.

List of the species of the genus **Klebsiella**

1. **Klebsiella pneumoniae** (Schroeter 1886) Trevisan 1887, 94.[AL] (Includes *Aerobacter aerogenes* as described in the seventh edition of *Bergey's Manual* (Breed, 1957). (*Hyalococcus pneumoniae* Schroeter 1886, 1952.)

pneu.mo′ni.ae. Gr. n. *pneumonia* pneumonia, inflammation of the lungs; M.L. gen. n. *pneumoniae* of pneumonia.

The characteristics are as described for the genus and as listed in Tables 5.16 to 5.18.

K. pneumoniae can be divided into many biovars (Ørskov, 1957; Rennie and Duncan, 1974).

K. pneumoniae is normally found in the intestinal tract of man and animals, but in low numbers compared with *E. coli*. It may be isolated in association with several pathological processes in man, e.g. infection of the urinary and respiratory tracts. Capsule types 1, 2 and 3 may be the causative agent of pneumonia. In animals, *K. pneumoniae* may be isolated from metritis in mares and bovine mastitis.

The mol% G + C of the DNA is 56–58 (T_m) (Seidler et al., 1975). The intraspecies DNA relative reassociation values is ~80–90% (Brenner et al., 1972) or 73–100% (Woodward et al., 1979).

Type strain: ATCC 13883 (NCTC 9633; CDC 298-56).

1a. **Klebsiella pneumoniae** subspecies **pneumoniae** (Schroeter 1886) Trevisan 1887, 94.[AL]

Distinguished from the subspecies *ozaenae* and *rhinoscleromatis* by the characteristics listed in Table 5.18.

Type strain: ATCC 13883.

1b. **Klebsiella pneumoniae** subspecies **ozaenae** subsp. nov. (*Klebsiella ozaenae* (Abel 1893) Bergey, Harrison, Breed, Hammer and Huntoon 1925, 266; *Bacillus mucosus ozaenae* Abel 1893, 167; *Bacillus ozaenae* (Abel 1893) Lehmann and Neumann 1896, 204.)

o.zae′nae. L. fem. n. *ozaena* ozena; L. gen. n. *ozaenae* of ozena.

Table 5.16.
Differential characteristics of the species of the genus **Klebsiella**[a]

Characteristics	1. K. pneumoniae	2. K. oxytoca	3. K. terrigena	4. K. planticola
Indole production	−	+	−	d
Pectate degradation	−	+	−	−
Fecal coliform test (gas production from lactose at 44.5°C)	+	−	−	−
Growth at 10°C	−	+	+	+
Fermentation of:				
Inulin	−	+	d	d
D-Melizitose	−	d	+	−
L-Sorbose	d	+	+	+
Utilization of:				
Gentisate or *m*-hydroxybenzoate	−	+	+	−
Hydroxy-L-proline	d	d	d	+

[a] Symbols: see standard definitions.

Table 5.17.
Other characteristics of the species of the genus **Klebsiella**[a]

Characteristics	1. K. pneumoniae	2. K. oxytoca	3. K. terrigena	4. K. planticola
Methyl red test	−	−	+	d
Voges-Proskauer test	+	+	+	+
Fermentation of:				
L-Arabinose, myo-inositol, lactose, D-mannitol, L-rhamnose, sucrose, D-glucose, raffinose, D-sorbitol	+	+	+	+
Adonitol	d	+	+	d
Dulcitol	d	d	−	d
Utilization of:				
Citrate (Simmons')	+	+	+	+
Malonate	+	d	d	+
Utilization of organic acids:				
Sodium citrate	d	+		
d-Tartrate	d	+	+	d
Arginine dihydrolase (Møller)	−	−	−	−
Lysine decarboxylase (Møller)	+	+	+	+
Ornithine decarboxylase (Møller)	−	−	−	−
Gelatin hydrolysis	−	d	−	−
H₂S production (triple-sugar iron agar)	−	−	−	−
Urease	+	+	+	+
Formation of 2-ketogluconate from gluconate	−	d	−	−

[a] For symbols see standard definitions.

Distinguished from the subspecies *pneumoniae* and *rhinoscleromatis* by the characteristics listed in Table 5.18.

Occurs in ozena and other chronic diseases of the respiratory tract.

Type strain: ATCC 11296 (NCTC 5050).

Table 5.18.
Differential characteristics of the subspecies of **Klebsiella pneumoniae**[a]

Characteristics	1a. pneumoniae	1b. ozaenae	1c. rhinoscleromatis
Gas from glucose	+	d	+
Acid from:			
Lactose	+	(+)	−
Dulcitol	d	−	−
Methyl red test	−	+	+
Voges-Proskauer test	+	−	−
Utilization of:			
Citrate (Simmons')	+	d	−
Malonate	+	−	+
Urease	+	d	−
Utilization of organic acids (Kauffmann-Petersen):			
Citrate	d	d	−
d-Tartrate	d	d	−
Mucate	+	d	−
Lysine decarboxylase (Møller)	+	d	−
Arginine dihydrolase (Møller)	−	d	−

[a] For symbols see standard defintions; also (+), slow fermentation.

1c. **Klebsiella pneumoniae** subspecies **rhinoscleromatis** subsp. nov. (*Klebsiella rhinoscleromatis* Trevisan 1887, 95; *Bacterium rhinoscleromatis* (Trevisan 1887) Migula 1900, 352.)

rhi.no.scle.ro′ma.tis. M.L. adj. *rhinoscleromatis* pertaining to rhinoscleroma.

Distinguished from the subspecies *pneumoniae* and *ozaenae* by the characteristics listed in Table 5.18.

Found in patients with rhinoscleroma.

Type strain: ATCC 13884 (NCTC 5046).

2. **Klebsiella oxytoca** (Flügge 1886) Lautrop 1956, 375.[AL] (*Bacillus oxytocus perniciosus* Flügge 1886, 268.)

ox.y.to′ca. Gr. *oxys* sour, acid; Gr. suffix *-tokos* bearer, producer; M.L. n. *oxytocus* acid-producer; spurious M.L. adj. *oxytoca* (sic) acid-producing.

The characteristics are as described for the genus and as listed in Tables 5.16 and 5.17.

Present in the intestinal tract of man and animals. Can be isolated from various pathological processes and also from botanical and aquatic environments.

K. oxytoca strains are encapsulated. Some of the K antigen test strains are *K. oxytoca*; however, in very few, if any, cases has a particular kind of K antigen been found only in *K. oxytoca* strains.

The mol% G + C of the DNA ranges from 55–58 (T_m). The intraspecies DNA relative reassociation values was 75% in the study by Brenner et al. (1975) and 95% (average value) in the study by Woodward et al. (1979).

Type strain: ATCC 13182.

3. **Klebsiella terrigena** Izard, Ferragut, Gavini, Kersters, De Ley and Leclerc 1981, 116.[VP]

ter.ri.ge′na. L. n. *terra* soil; L. suffix *gena* origin; M.L. n. *terrigena* from soil.

The characteristics are as described for the genus and as listed in Tables 5.16 and 5.17.

Isolated mainly from aquatic and soil environments.

Phenotypically, *K. terrigena* resembles *K. pneumoniae*; however, it can be distinguished by its ability to grow at 10°C, its inability to produce gas from lactose at 44.5°C, and by its ability to ferment melizitose.

The mol% G + C of the type strain was 56.7 (T_m) (Izard et al., 1981). The average intraspecies DNA relative reassociation value is above 86% (Izard et al., 1981).

Type strain: CIP 80-07 (CUETM 77-176; Gavini et al. L 84).

4. Klebsiella planticola Bagley, Seidler and Brenner 1982, 266.[VP*] (Effective publication: Bagley et al. 1981, 105.)

plan.ti′co.la. L. fem. n. *planta* a plant; L. suff. *-cola* dweller; M.L. fem. n. *planticola* plant-dweller.

The characteristics are as described for the genus and as listed in Tables 5.16 and 5.17.

Isolated mainly from botanical, aquatic and soil environments. Three biovars have been described (Naemura et al., 1979).

K. planticola can be distinguished from *K. pneumoniae* by its ability to grow at 10°C and by its inability to produce gas from lactose at 44.5°C. Its inability to ferment melizitose distinguishes *K. planticola* from *K. terrigena*.

Encapsulated; typable with *Klebsiella* K antisera.

The mol% G + C of the two strains tested was 53.9 and 55.4 (T_m) (Seidler et al., 1975). The average intraspecies DNA relative reassociation value is above 75% (Woodward et al., 1979).

Type strain: ATCC 33531 (V-236; CDC 4245-72).

Genus VI. Enterobacter Hormaeche and Edwards 1960, 72[AL]; Nom. Cons. Opin. 28, Jud. Comm. 1963, 38

C. RICHARD

En.te.ro.bac′ter. Gr. neut. n. *enteron* intestine; M.L. masc. n. *bacter* equivalent of bacterium, a small rod; M.L. masc. n. *Enterobacter* intestinal small rod.

Straight rods, 0.6–1.0 μm wide × 1.2–3.0 μm long, conforming to the general definition of the family *Enterobacteriaceae*. Gram-negative. **Motile by peritrichous flagella** (generally 4–6). Facultatively anaerobic. Grow readily on ordinary media. Ferment glucose with production of acid and gas (generally $CO_2:H_2 = 2:1$). Gas is not produced from glucose at 44.5°C. Most strains give a **positive Voges-Proskauer reaction** and a negative methyl red test. Citrate and malonate are usually utilized as sole sources of carbon and energy. Hydrogen sulfide is not produced from thiosulfate. Gelatin is liquefied slowly by most strains. **Deoxyribonuclease (DNase), Tween 80 esterase and lipase are not produced.** Optimum temperature for growth, 30°C. Most clinical strains grow at 37°C; some environmental strains give erratic biochemical reactions at 37°C. Widely distributed in nature; common in man and animals. The mol% G + C of the DNA is 52–60 (Bd).

Type species: *Enterobacter cloacae* (Jordan 1890) Hormaeche and Edwards 1960, 72.

Further Descriptive Information

The genus *Enterobacter* belongs to group II of the family *Enterobacteriaceae* as indicated in the eighth edition of the *Manual* and is therefore characterized by a positive Voges-Proskauer (VP) reaction and β-galactosidase (ONPG test). Unlike the genus *Klebsiella*, *Enterobacter* is motile, and unlike the genus *Serratia*, *Enterobacter* is negative for lipase, Tween 80 esterase and DNase.

Enterobacter species grow rapidly on the usual enteric media. In general, strains from environmental sources grow better at 20–30°C rather than 37°C, whereas strains from clinical sources grow better at 37°C. On Drigalski lactose agar, *E. cloacae* forms colonies that are lactose-positive or negative, round (2–3 mm in diameter), and slightly iridescent or flat with irregular edges. On Hektoen medium, colonies have a similar diameter and are salmon-pink colored. On eosin methylene blue agar the colonies are pinkish, mucoid and convex, 3–4 mm in diameter. *E. sakazakii* grows rapidly on nutrient agar or tryptic soy agar, forming bright yellow colonies at 25°C or pale yellow colonies at 37°C, 1–3 mm in diameter. Various colony types of *E. sakazakii* are observed: typical smooth colonies, mucoid rubbery colonies, and occasionally dry colonies (Farmer et al., 1980). Aerogenic strains of *E. agglomerans* form colonies resembling those of *E. cloacae*, whereas anaerogenic strains, especially those of biogroup 1, may present different morphologies: (a) rough and wrinkled colonies that are rather difficult to remove with a platinum wire, (b) smooth, irregularly round colonies, (c) "cauliflower" rough colonies, and (d) convex mucoid colonies (particularly on media containing carbohydrates). Anaerogenic strains often elaborate a yellow pigment (75% of all strains, 85% of biogroup 1 strains), whereas this is less common with aerogenic bio-

group strains (less than 50%) (Richard, 1978). Low temperatures (20–30°C) are better for pigment production than 37°C. The carotenoid-like yellow pigment is soluble in ethanol and acetone but is insoluble in water and chloroform. Colonies of *E. aerogenes* resemble those of *E. cloacae*. The colonial morphology occurring on media containing methyl violet as a selective agent can be used to differentiate *E. aerogenes* and other Gram-negative organisms from *Klebsiella pneumoniae* (Campbell and Roth, 1975). *E. gergoviae* colonies resemble those of *E. cloacae* and *E. aerogenes*.

Biochemical reactions differ widely among the species and biogroups of *Enterobacter*, and carbohydrate fermentation tests and amino acid decarboxylase tests are useful for differentiation. *E. cloacae* produces acid and gas rapidly from cellobiose and produces acid slowly from glycerol. Although lactose may be fermented slowly, *E. cloacae* is always positive for β-galactosidase (ONPG test). Some strains of *E. cloacae* utilize malonate and ferment adonitol. The enzyme β-xylosidase is present. *E. cloacae* is positive for arginine dihydrolase (ADH) and ornithine decarboxylase (ODC) but is negative for lysine decarboxylase (LDC). Most strains liquefy gelatin slowly. Some strains of *E. cloacae* (and *E. aerogenes*) have the ability to fix molecular nitrogen under anaerobic conditions (e.g., Neilson and Sparell, 1976; Nelson et al., 1976).

E. sakazakii shows biochemical characters similar to those of *E. cloacae*, but does not ferment D-sorbitol or mucate and has a delayed DNase reaction.

E. agglomerans is negative for ADH, ODC and LDC.

E. aerogenes is motile, positive for ODC, negative for urease, and can utilize *m*-hydroxybenzoate as a sole carbon and energy source. These are useful tests for distinguishing this species from *K. pneumoniae*.

E. gergoviae does not ferment D-sorbitol or mucate, is negative for β-xylosidase and gelatinase, and is positive for ODC and LDC but negative for ADH. *E. gergoviae* is urease-positive, whereas other *Enterobacter* species are urease-negative.

E. intermedium (Izard et al., 1980a) and *E. amnigenus* (Izard et al., 1981) are new species on the basis of DNA relatedness to one another and to other *Enterobacter* species. They can be separated from other *Enterobacter* species by their inability to grow at 41°C and by the reactions given in Table 5.20.

Biotyping, sometimes serotyping (by use of O and H antigens and occasionally capsular antigens) and antibiotic susceptibility may be used as epidemiological markers for *Enterobacter* strains. With regard to antigenic characters, 53 O antigens, 56 H antigens and 79 different serovars have been described for *E. cloacae* (Sakazaki and Namioka, 1960). The fermentation of various carbohydrates (adonitol, lactose, mucate, L-rhamnose, dulcitol, salicin, sucrose, α-methylglucoside, glycerol), malonate utilization, and the presence of β-galactosidase, β-

---

* *VP* denotes that this name has been validly published in the official publication, International Journal of Systematic Bacteriology.

xylosidase, gelatinase, ODC and ADH may be used as markers for epidemiological studies of *E. cloacae*. With regard to *E. aerogenes*, ~80% of the strains possess a thin capsule that is antigenically related to the capsular antigens of *Klebsiella* (chiefly antigens K68 and K26, and occasionally antigens K4, K11, K42 and K59). *E. aerogenes* and *Klebsiella* antigens are not identical but do have common fractions which are responsible for the cross-reactions (Richard, 1977).

Concerning antibiotic susceptibility, most *Enterobacter* strains are resistant to ampicillin and cephalosporins (Toala et al., 1970), but are generally sensitive to carbenicillin and the newer cephalosporins, such as cefotaxime (Sirot et al., 1980). Some strains of *E. cloacae* and *E. aerogenes* found in hospitals are resistant to tetracycline, aminoglycosides and sulfonamides.

E. cloacae is the most frequently isolated *Enterobacter* species from man and animals. It is found in human and animal feces, but is not known to be an enteric pathogen. It is, however, an opportunistic pathogen isolated from urine, sputum and the respiratory tract, pus, and occasionally from blood or spinal fluid. It has an increasing importance in hospitals, especially in intensive care units, emergency units and urology.

E. sakazakii is often a commensal without clinical significance and is occasionally a pathogen causing neonatal meningitis and bacteremia.

E. agglomerans can behave as an opportunistic pathogen in immunologically compromised patients such as neonates, premature infants, burned or multiply traumatized patients, and patients with leukemia or who are undergoing immunosuppressive therapy. Strains of *E. agglomerans* are frequently isolated by blood culture because they are generally introduced by such invasive procedures as catheterization, intubation, and surgical or medical acts. Such contaminations result in a transitory bacteremia and occasionally septicemia (Richard, 1978).

E. aerogenes is found in human and animal feces, but is not known to be an enteric pathogen. It is an opportunistic pathogen and is isolated from the respiratory tract, genitourinary tract, pus, and occasionally from blood and spinal fluid. Like *Klebsiella pneumoniae*, it appears to be a normal constituent of the preputial flora of healthy stallions and, therefore, may be an etiologic agent of epidemic metritis in mares (Plate and Atherton, 1976).

E. gergoviae sometimes appears to be an opportunistic pathogen and has been isolated from urine, pus, sputum, blood and other clinical specimens. The species has been implicated in a long term nosocomial outbreak of urinary tract infections (Richard et al., 1976).

E. amnigenus and *E. intermedium* have not been isolated from human infection.

All *Enterobacter* species are found in the natural environment (water, sewage, soil, vegetables), especially *E. agglomerans*—called *Erwinia herbicola* by phytopathologists. *E. agglomerans* is a saprophytic microorganism frequently isolated from plants, flowers, seeds and vegetables (it is probably not phytopathogenic) and from a wide variety of environmental sources such as water, soil and foodstuffs. *E. cloacae* is found in water, sewage, soil and meat. *E. sakazakii* is rarely encountered in clinical specimens and is more prevalent in the environment and in food. *E. aerogenes* is found in water, sewage, soil and dairy products. *E. gergoviae* has been isolated from various environmental sources (cosmetics, water, etc.). *E. amnigenus* and *E. intermedium* are found in drinking and surface water and in unpolluted soil.

Enrichment and Isolation Procedures

All media designed for the isolation of *Enterobacteriaceae* can be used for the isolation of *Enterobacter* species: MacConkey agar, Drigalski lactose agar, Hektoen agar, deoxycholate lactose citrate agar, etc. *Enterobacter* can also grow on media for general use, such as blood agar, nutrient agar, tryptic soy agar, bromocresol purple lactose agar, etc.

Media specifically selective for *Enterobacter* are not available.

Maintenance Procedures

Strains are initially grown on tryptic soy agar at their optimum temperature. They are then inoculated by stabbing a maintenance medium* designed for maintenance of *Enterobacteriaceae* and related organisms. The cultures are then stored at room temperature in a dark, dry place.

Cultures may be also preserved by freeze-drying. Freeze-drying is the best procedure for preservation of pigmented strains.

Differentiation of the genus **Enterobacter** from other genera

Table 5.19 provides the main characteristics that can be used to differentiate the genus *Enterobacter* from the genera *Klebsiella*, *Hafnia* and *Serratia*.

Taxonomic Comments

Enterobacter cloacae is the type species of the genus *Enterobacter*. Strains of *E. sakazakii* were previously called yellow-pigmented *E. cloacae*; however, DNA/DNA hybridization studies have shown that *E. cloacae* strains form one DNA relatedness group different from that containing the yellow strains (which are now named *E. sakazakii*) (Steigerwalt et al., 1976). The type strain of *E. sakazakii* is 83–89% related to other *E. sakazakii* strains and only 31–54% related to nonpigmented *E. cloacae* strains (Farmer et al., 1980).

E. agglomerans is a heterogeneous species that is synonymous with *Erwinia herbicola*, *Erwinia uredovora* and *Erwinia stewartii*. Ewing and Fife (1972) proposed that the strains from clinical sources be designated as *Enterobacter agglomerans* because the characteristics of the organisms were in conformity with the genus *Enterobacter*. On the other hand, strains of interest to phytopathologists have been placed in the genus *Erwinia* (see that article on *Erwinia* in this *Manual*). It currently is difficult, if not impossible, to distinguish strains from different sources due to the diversity in this group of organisms. Further phenotypic and genotypic studies must be done to define the groups now

referred to as *E. agglomerans* and *Erwinia* species. See the article on the family *Enterobacteriaceae* in this *Manual* for further information concerning this problem.

E. aerogenes and *Klebsiella pneumoniae* have a number of characteristics in common; however, it has been shown that they represent two distinct DNA/DNA homology groups. Only 56% relatedness is observed between *Klebsiella* and *E. aerogenes*, whereas *E. cloacae* exhibits 40% relatedness with *K. pneumoniae* and *E. aerogenes* (Brenner et al., 1972). Because *E. aerogenes* does exhibit some phenotypic and genetic similarity to *K. pneumoniae*, some bacteriologists have proposed the transfer of *E. aerogenes* into the genus *Klebsiella* as *K. mobilis* (Bascomb et al., 1971; Izard et al., 1980).

E. gergoviae is a urease-positive *Enterobacter* species which shares many characters with *E. aerogenes*. The biochemical homogeneity within *E. gergoviae* is reflected by a high level of genetic relatedness among strains from France, the United States and Africa (relative binding ratio at 60°C, 76–97%) (Brenner et al., 1980).

E. hafniae has been transferred previously to the genus *Hafnia* as *Hafnia alvei* because it has few phenotypic or genetic similarities with other *Enterobacter* species.

E. liquefaciens has been transferred to the genus *Serratia* as *S. liquefaciens* because it is closely related to *S. marcescens* by biochemical and genetic properties.

---

* Maintenance medium (g/liter): Bacto-peptone (Difco), 10.0; NaCl, 5.0; Bacto-agar (Difco), 10.0; pH 7.4. The medium should be dispensed into small (9.5–10 × 90 mm) screw-capped tubes.

Enterobacter intermedium and *Enterobacter amnigenus* (see "Other Organisms Belonging to the Genus Enterobacter") are phenotypically closest to *E. cloacae* but are distinct by DNA/DNA hybridization (Izard et al., 1980a; Izard et al., 1981).

Atypical strains that are difficult to assign to the described species of the genus *Enterobacter* are occasionally encountered.

Differential characteristics of **Enterobacter** species

Tables 5.20, 5.21 and 5.22 give characteristics useful for the differentiation of the various species of *Enterobacter*.

List of the species of the genus **Enterobacter**

1. **Enterobacter cloacae** (Jordan 1890) Hormaeche and Edwards 1960, 72;[AL] Nom. Cons., Opinion 28, Jud. Comm. 1963, 38. (*Bacillus cloacae* Jordan 1890, 836.)

clo.a′cae. L. n. *cloaca* a sewer; L. gen. n. *cloacae* of sewer.

The characteristics are described in Tables 5.20, 5.21 and 5.22. The following tests should be emphasized for identification of the species: LDC-negative, ODC- and ADH-positive.

E. cloacae has a natural resistance to ampicillin. Many strains are resistant to cephalosporins, chloramphenicol, tetracycline and sulfonamides. Most strains are sensitive to aminoglycosides (except streptomycin), colistin, nalidixic acid and nitrofuranes.

E. cloacae is less susceptible to chlorination than *Escherichia coli*.

Occurs in water, sewage, soil, meat, hospital environments and on the skin and in the intestinal tracts of man and animals as a commensal.

The mol% G + C of the DNA is 52–54 (T_m).

Type strain: ATCC 13047 (NCTC 10005, CDC 279-56).

2. **Enterobacter sakazakii** Farmer, Asbury, Hickman and Brenner 1980, 575.[VP]

sa.ka.za′ki.i. M.L. gen. n. *sakazakii* of Sakazaki; named after the Japanese bacteriologist Riichi Sakazaki.

Previously known as "yellow-pigmented *E. cloacae*."

The characteristics are as described in Tables 5.20 and 5.22. The biochemical characteristics are similar to those of *E. cloacae*, but *E. sakazakii* does not ferment D-sorbitol and mucate and gives a delayed positive DNase test. The nondiffusible yellow pigment (best formed at 25°C) is useful for identification; this pigment may be lost upon subculturing. Approximately 10% of the strains produce indole. Table 5.22 presents the main characters useful for differentiating *E. sakazakii*, *E. agglomerans* and *E. cloacae*.

Generally susceptible to ampicillin, carbenicillin, aminoglycosides, chloramphenicol, tetracycline and nalidixic acid; 87% of the strains are resistant to cephalothin.

Occurs in the environment and in foods, rarely in clinical specimens.

The mol% G + C of the DNA is 57 (T_m).

Type strain: ATCC 29544 (CDC 4562-70).

Table 5.19

Differentiation between **Enterobacter** *and related genera*[a]

Characteristics	Entero-bacter	Klebsiella	Hafnia	Serratia
Motility	+	−	+	+
Ornithine decarboxylase	+	−	+	[+]
Arginine dihydrolase	D	−	−	−
Deoxyribonuclease	−	−	−	+
Gelatinase	D	D	−	[+]
Citrate utilization	[+]	[+]	−	+
Susceptible to *Hafnia* phage[b]	−	−	+	−
D-Sorbitol (acid)	[+]	+	−	[+]

[a] Symbols: +, all strains positive in 24–48 h; [+], majority of strains positive (generally more than 89%); −, all strains negative after 7 days D, differs among species.

[b] Guinée and Valkenburg, 1968.

Table 5.20.

Differential characteristics of the species of the genus **Enterobacter** *and of* **Hafnia alvei**[a]

Characteristics	1. E. cloacae	2. E. sakazakii	3. E. agglomerans	4. E. aerogenes	5. E. gergoviae	a. E. intermedium	b. E. amnigenus	Hafnia alvei
KCN	+	+	d	+	−	d	d	+
Urease	−	−	−	−	+	−	−	−
Gelatinase	(+)	[+]	[+]	d	−	−	−	−
Decarboxylases:								
Lysine	−	−	−	+	+/(+)	−	−	+
Ornithine	+	+	−	+	+	+	+	+
Arginine	+	+	−	−	−	−	+	−
β-Xylosidase[b]	+	+	d	+	−	+	+	−
Acid from:								
Sorbitol	+	−	d	+	−	+	d	−
Sucrose	+	+	d	+	+	d	d	−
Raffinose	[+]	+	d	+	+	+	+	−
α-Methylglucoside	+	+	d	+	−	+	d	−
Mucate	d	−	d	+	−	+	+	−
Citrate (Simmons')	+	+	[+]	+	+	+	+	−
Indole	−	d	d	−	−	−	−	−
Yellow pigment formed	−	+	d	−	−	−	−	−

[a] Symbols: +, all strains positive in 24–48 h; [+], majority of strains positive (generally more than 89%); (+), delayed positive (positive between 3 and 7 days); −, all strains negative after 7 days; +/(+), some strains positive in 24–48 h, some strains positive between 3 and 7 days; d, differs among strains (generally between 11 and 80% positive).

[b] From Brisou et al. (1972).

Table 5.21.

Biochemical characteristics of **Enterobacter agglomerans** *and* **Enterobacter cloacae**[a]

Characteristics	1. *E. cloacae*	3. *E. agglomerans*	
		Anaerogenic strains	Aerogenic strains
Yellow pigment	−	[+]	d
Gas from D-glucose	+	−	+
Nitrate reductase	+	[+]	+
Indole	−	[−]	[−]
Voges-Proskauer	+	[+]	d
Decarboxylases:			
Ornithine	+	−	[−]
Arginine	+	−	−
Lysine	−	−	−
KCN	+	[−]	[+]
Methyl red	−	[+]	[+]
D-Sorbitol, acid	+	[−]	[+]
Raffinose, acid	+	[−]	d
Mucate, acid	[+]	[−]	d
β-Xylosidase	+	[−]	[+]
Gelatinase	+	[+]	d
Motility	+	[+]	[+]
Acid from:			
D-Xylose, L-arabinose, D-mannitol	+	+	+
D-Adonitol	d	[−]	[−]
L-Rhamnose, maltose	+	[+]	+
Sorbose, D-tartrate	−	−	−
Lactose	[+]	[−]	d
Sucrose	+	d	d
myo-Inositol	[−]	[−]	[−]
Salicin	[+]	d	d
Cellobiose	+	d	+
Glycerol	d	d	d
Melibiose	[+]	[−]	[+]
β-Galactosidase (ONPG test)	+	+	+
Citrate:			
Simmons'	+	[+]	[+]
Christensen's	+	+	+
Tetrathionate reductase (TTR)	−	−	−
Malonate	[+]	d	[+]
Urease, H$_2$S	−	−	−
Phenylalanine and tryptophan deaminase	−	−	−

[a] Symbols: +, all strains positive in 24–48 h; [+], majority of strains positive (generally more than 89%); [−], majority of strains negative (generally more than 89%) after 7 days; −, all strains negative after 7 days; d, differs among strains (generally from 11–89% positive).

3. **Enterobacter agglomerans** (Beijerinck 1888) Ewing and Fife 1972, 10.[AL] (*Bacillus agglomerans* Beijerinck 1888, 749; *Erwinia herbicola* (Geilinger 1921) Dye 1964, 268; *Bacterium herbicola* Geilinger 1921, 105; *Erwinia uredovora* (Pon, Townsend, Wessman, Schmitt and Kingsolver 1954) Dye 1963, 149; *Xanthomonas uredovorus* Pon, Townsend, Wessman, Schmitt and Kingsolver 1954, 710: *Erwinia stewartii* (Smith 1898) Dye 1963, 504; *Pseudomonas stewartii* Smith 1898, 422; *Escherichia adecarboxylata* Leclerc 1962, 736.)

ag.glo'mer.ans. L. v. *Agglomerare* to form into a ball; L. part. adj. *agglomerans* forming into a ball (referring to the occurrence of the bacteria in aggregates surrounded by a translucent sheath (symplasmata) in anaerogenic strains).

The biochemical characteristics of *E. agglomerans* are as described in Tables 5.20, 5.21 and 5.22. Tables 5.23 and 5.24 present the biochemical characters of the 11 biogroups (7 anaerogenic and 4 aerogenic).

The majority of strains are anaerogenic (80%: and Ewing and Fife, 1972; 62%: Richard, 1975, 1978).

The biochemical characters of *E. cloacae* and *E. agglomerans* (aerogenic and anaerogenic groups) are compared in Table 5.21. the aerogenic biogroups are closely related to *E. cloacae*, particularly biogroup G1. *E. agglomerans* is chiefly characterized by the *absence of LDC, ODC and ADH* and by the synthesis of a nondiffusible yellow pigment. Biogroup 1 is usually yellow pigmented, has a strongly active gelatinase, lacks β-xylosidase, does not ferment D-sorbitol or mucate, and is inhibited by KCN. Pectinolytic, lipolytic and alginolytic activities have not been detected in *E. agglomerans*.

Most strains are sensitive to antibiotics, except for possible resistance to ampicillin and cephalothin and sometimes to carbenicillin and nitrofuranes.

Isolated from plants, flowers, seeds, vegetables, water, soil and foodstuffs. Some strains are of human and animal origin.

Table 5.22.

Main characteristics differentiating **Enterobacter cloacae**, **Enterobacter sakazakii** *and* **Enterobacter agglomerans**[a]

Characteristics	1. *E. cloacae*	2. *E. sakazakii*	3. *E. agglomerans*	
			Aerogenic Biogroups	Anaerogenic Biogroups
Gas from D-glucose	+	+	+	−
Yellow pigment	−	+	d	[+]
Decarboxylases:				
Lysine	−	−	−	−
Ornithine	+	+	[−]	−
Arginine	+	+	−	−
Acid from:				
Sorbitol	+	−	[+]	[−]
Mucate	d	−	d	d
Indole	−	d	d	d

[a] For symbols see Table 5.21.

Table 5.23.

Differentiation of the biogroups of anaerogenic strains of **Enterobacter agglomerans**[a]

Biogroup	Nitrate Reduction	Indole	Voges-Proskauer	No. of Strains Examined
1	+	−	+	157
2	+	−	−	52
3	−	−	−	21
4	−	+	+	19
5	+	+	−	19
6	−	−	+	12
7	+	+	+	8

[a] From Fife and Ewing (1972). Symbols: +, all strains positive in 24–48 h; −, all strains negative after 7 days.

Table 5.24.

Differentiation of the biogroups of aerogenic strains of **Enterobacter agglomerans**[a]

Biogroup	Indole	Voges-Proskauer	No. of Strains Examined
G1	−	+	33
G2	−	−	15
G3	+	−	15
G4	+	+	6

[a] From Fife and Ewing (1972). Symbols: +, all strains positive in 24–48 h; −, all strains negative after 7 days.

The mol% G + C of the DNA is 53–58 (Bd).
Type strain: ATCC 27155 (NCTC 9381, CDC 1461-67).

Further Comments

The synonymy of *E. agglomerans* with *Erwinia* species has already been discussed (see Taxonomic Comments). Also, a species called *Escherichia adecarboxylata* (see the chapter on the genus *Escherichia* in this *Manual*) probably belongs to the *Enterobacter agglomerans* complex (Bascomb et al., 1971).

4. **Enterobacter aerogenes** (Kruse 1896) Hormaeche and Edwards 1960, 72.[AL] (*Bacillus aerogenes* Kruse 1896, 340.)

a.e.ro′ge.nes. Gr. masc. n. *aer* air; Gr. v. *gennanio* to produce; M.L. adj. *aerogenes* gas-producing.

The characteristics are described in Table 5.20. *E. aerogenes* shares many biochemical characters with *Klebsiella pneumoniae*, such as acidification of many carbohydrates with gas, utilization malonate, and a positive LDC reaction. *Motility, ODC and urease* are the major characteristics to differentiate these two species (Table 5.25).

Most strains of *E. aerogenes* are resistant to ampicillin and cephalosporins and sensitive to carbenicillin.

Occur in water, sewage, soil, dairy products and the feces of man and animals.

The mol% G + C of the DNA is 53–54 (Bd).
Type strain: ATCC 13048 (NCTC 10006, CDC 819-56).

5. **Enterobacter gergoviae** Brenner, Richard, Steigerwalt, Asbury and Mandel 1980, 1.[VP]

ger.go′vi.ae. M.L. gen. n. *gergoviae* of Gergovie Highland; intended to pertain to the fact that the type strain was isolated from samples

Table 5.25.

Main Characteristics differentiating **Enterobacter aerogenes** *from* **Klebsiella pneumoniae**[a]

Characteristics	E. aerogenes	K. pneumoniae
Motility	+	−
Ornithine decarboxylase	+	−
Urease	−	+
Sorbose, acid	−	d
Lactose, acid	(+) or +	+
m-Hydroxybenzoate	+	−
Gelatinase	(+) or +	−
Carbenicillin	S	R
Cephalothin	R	S

[a] Symbols: +, all strains positive in 24–48 h; (+), delayed positive (positive between 3 and 7 days); −, all strains negative after 7 days; S, susceptible; R, resistant; d, differs among strains.

taken during a urinary infection outbreak in Clermont-Ferrand University Hospital near Gergovie Highland in France.

The characteristics are as described in Table 5.20. *E. gergoviae* is closest to *E. aerogenes* phenotypically (Richard et al., 1976) but is *urease-positive.* Table 5.20 indicates other characteristics that distinguish between these two species.

Most strains are susceptible to antibiotics, but strains isolated from a urinary infection outbreak in France were multiresistant.

Occur in various environmental sources such as cosmetics, water, etc. Have also been recovered from clinical specimens.

The mol% G + C of the DNA is 60 (Bd).
Type strain: CIP 76.01 (ATCC 33028; CDC 604-77).

Other organisms belonging to the genus **Enterobacter**

Two newly described species of *Enterobacter* have been validly published and are distinct from each other and from other *Enterobacter* species on the basis of DNA relatedness. They can be differentiated phenotypically from other *Enterobacter* by their inability to grow at 41°C and by the characteristics listed in Table 5.20.

a. *Enterobacter intermedium* Izard, Gavani and Leclerc 1980, 601.[VP] (*Effective publication:* Izard, Gavini and Leclerc 1980, 51.)

in.ter.me′di.um. L. adj. *intermedium* intermediate.
Type strain: CIP 79-27 (CUETM 77-130; strain E86 of Gavini).

b. *Enterobacter amnigenus* Izard, Gavini, Trinel and Leclerc 1981, 37.[VP]

am.ni′ge.nus. L. adj. *amnigenus* coming from water.
The mol% G + C of the DNA is 60 (Bd).
Type strain: ATCC 33072 (CUETM 77-118).

Two other organisms presently listed under the genus *Erwinia* belong to the genus *Enterobacter* on the basis of DNA relatedness.

c. *Erwinia dissolvens* (Rosen 1922) Burkholder 1948, 472.[AL] (*Pseudomonas dissolvens* Rosen 1922, 497.)

dis.sol′vens. L. part. adj. *dissolvens* dissolving.
This organism belongs to the genus *Enterobacter* (Waldee, 1945; Dye, 1969; Steigerwalt et al., 1976) as a new species or as a biogroup of *E. cloacae.* Its DNA is 60–80% related to DNA from *E. cloacae* and the two organisms are very similar biochemically (Steigerwalt et al., 1976).
Type strain: ATCC 23373.

d. *Erwinia nimipressuralis* Carter 1945, 423.[AL]

ni.mi.pres.su.ra′lis. L. adv. *nimis* overmuch; L. n. *pressura* pressure; M.L. adj. *nimipressuralis* with excessive pressure.
This organism belongs to the genus *Enterobacter* (Graham, 1964; Dye, 1969; Steigerwalt et al., 1976) as a new species or as a biogroup of *E. cloacae.* Its DNA is 55–65% related to that of *E. cloacae* and *E. dissolvens* (Steigerwalt et al., 1976). *E. nimipressuralis* is negative in sucrose and raffinose reactions—characteristics which separate it from *E. cloacae* (Steigerwalt et al., 1976). It was reported as the causative agent of "wetwood" disease in elm trees (Carter, 1945).
Type strain: ATCC 9912.

Genus VII. **Erwinia** Winslow, Broadhurst, Buchanan, Krumwiede, Rogers and Smith 1920, 209[AL]

R. A. LELLIOTT AND ROBERT S. DICKEY

(Includes *Pectobacterium* Waldee 1945, 469.[AL])

Er.wi′ni.a. M.L. fem. n. *Erwinia*; named after Erwin F. Smith.

Straight rods, 0.5–1.0 × 1.0–3.0 μm; occur singly, in pairs and sometimes in short chains. Gram-negative. **Motile** (one exception) by peritrichous flagella. **Facultatively anaerobic,** but anaerobic growth by some species is weak. Optimum temperature, 27–30°C; maximum varies between 32°C and at least 40°C. **Oxidase-negative. Catalase-positive. Acid is produced from** fructose, galactose, D-glucose, β-metylglucoside and sucrose. **Utilize** acetate, fumarate, gluconate, malate and succinate, but not benzoate, oxalate, or propionate as carbon- and energy-yielding sources. **Associated with plants as pathogens, saprophytes, or as constituents of the epiphytic flora.** At least one species has also been isolated from human and animal hosts. The mol% G + C of the DNA is 50–58 (T_m, Bd).

Type species: *Erwinia amylovora* (Burrill 1882) Winslow, Broadhurst, Buchanan, Krumwiede, Rogers and Smith 1920, 209.

Further Descriptive Information

Acid is usually produced from mannitol, mannose, ribose and sorbitol, but rarely from adonitol, dextrin, dulcitol or melezitose. Gas production is comparatively weak or absent. Decarboxylases for arginine, lysine or ornithine cannot be detected by Møller's method (Møller, 1955) except in a few (usually 5% or less) strains of *E. carotovora* and *E. chrysanthemi*. Formation of putrescine occurs when the amino acids are decarboxylated under aerobic conditions (Zherebilo and Gvozdyak, 1976). Glutamic acid is not decarboxylated. Urease and lipases rarely are produced. Additional characters of the species and subspecies of the genus are given in Tables 5.26 to 5.28 with data for *E. cypripedii*, *E. nigrifluens*, *E. quercina*, *E. rubrifaciens*, *E. salicis*, *E. tracheiphila* and *E. uredovora* based on relatively small numbers of strains for each species.

Pectate lyases are produced by strains of *E. carotovora* (Mount et al., 1970), *E. carotovora* subsp. *atroseptica* (Hall and Wood, 1970), *E. chrysanthemi* (Garibaldi and Bateman, 1971), and *E. rubrifaciens* (Gardner and Kado, 1976). Cellulase (Cx) also is produced by strains of *E. carotovora*, *E. carotovora* subsp. *atroseptica*, and *E. chrysanthemi* in the presence of carboxymethyl cellulose (El-Helaly et al., 1979).

Fermentation end products from glucose are CO_2 and different combinations of succinate, lactate, formate and acetate; some form 2,3-butanediol and some ethanol (White and Starr, 1971). Starch is not hydrolyzed beyond dextrins.

Naturally occurring plasmids have been detected in strains of *E. amylovora*, *E. carotovora*, *E. chrysanthemi* and *E. herbicola*, and plasmids from bacteria other than *Erwinia* have been introduced into strains of the foregoing *Erwinia* species and strains of *E. carotovora* subsp. *atroseptica*, *E. nigrifluens*, and *E. uredovora* (Lacy and Leary, 1979; Chatterjee and Starr, 1980). Plasmid-mediated transfer of chromosome genes by conjugation also has been reported for strains of *E. amylovora*, *E. carotovora*, *E. chrysanthemi* and *E. herbicola*.

Virulent or temperate phages have been isolated, characterized and reported to be active against strains of *E. amylovora* (Ritchie and Klos, 1979), *E. carotovora* (Chapman et al., 1951; Faltus and Kishko, 1980), *E. chrysanthemi* (Paulin and Nassan, 1978), *E. herbicola* (Harrison and Gibbins, 1975), *E. nigrifluens* and *E. rubrifaciens* (Zeitoun and Wilson, 1969). Bacteriocinogeny or production of bacteriocin-like substances has been noted for strains of *E. carotovora* (Itoh et al., 1978), *E. chrysanthemi* (Echandi and Moyer, 1979), *E. herbicola* (Beer and Vidaver, 1978), and *Erwinia* species from sugar beet (Stanghellini et al., 1977).

Antisera prepared against live or heat-killed cells, nonpurified or purified immunogens have been used for the differentiation or identification of all *Erwinia* species except *E. ananas*, *E. cypripedii*, *E. mallotivora*, *E. rhapontici* and *E. uredovora* (Elrod, 1946; DeKam, 1976; Schaad, 1979). Serogroups have been determined for *E. carotovora* (De Boer et al., 1979) and *E. chrysanthemi* (Samsun and Nassan-Agha, 1978; Yakrus and Schaad, 1979).

Erwinia species cause plant diseases which include blights, cankers, die back, leaf spots, wilts, discoloration of plant tissues, and soft rots variously described as stalk rot, crown rot, stem rot, or fruit collapse. Ingress by the pathogen generally occurs through natural openings and wounds. *Erwinia uredovora* is a parasite of rust fungi and multiplies in the plant tissue infected by the rust organism (Hevesi and Mashaal, 1975). *Erwinia tracheiphila* overwinters in the bodies of cucumber beetles (*Diabrotica vittata* Fabr. and *D. duodecimpunctata* Oliv.) (Leach, 1964), whereas *E. stewartii* overwinters primarily in a flea beetle (*Chaetocnema pulicaria* Melsh.) (Pepper, 1967). Strains of *E. herbicola* are common in the epiphytic microflora of plants; instances have been reported in which *E. herbicola* has produced symptoms on plants, sometimes possibly in association with other phytopathogenic bacteria (Gibbins, 1978).

Enrichment and Isolation Procedures

The pathogens generally can be easily isolated. The affected plant material should be washed in tap water, followed by sterile water, and dried with paper toweling. Surface sterilization (3 min in 1:10 dilution of 5.25% active sodium hypochlorite) sometimes is detrimental for isolation. Affected tissue is removed from a young lesion or the edge of older necrotic areas by a sterile scalpel; the tissue is comminuted in sterile water, saline, or buffer solution and is streaked onto a solid medium, such as nutrient agar or YDC (Dye, 1968). The isolation of *E. tracheiphila* is more easily accomplished by aseptically cutting the affected stem, placing the two cut stem surfaces together, and gently pulling apart, removing a portion of the threads of bacteria and placing the bacteria in nutrient broth or onto a solid medium (Burkholder, 1960). The delicate growth of *E. tracheiphila* will appear in 3 or 4 days; frequent transfer is necessary, but virulence may be reduced or lost with repeated transfers.

The isolation of some *Erwinia* species can be facilitated by use of selective-differential media, but such media are usually not necessary. *Erwinia amylovora*, *E. herbicola*, *E. nigrifluens*, *E. quercina* and *E. rubrifaciens* will grow on MS medium (Miller and Schroth, 1972) and produce characteristic colonies. Sorbitol is substituted for mannitol in the MS medium for the isolation of *E. amylovora* (Schroth and Hildebrand, 1980). The medium of Crosse and Goodman (Crosse and Goodman, 1973) also can be used for *E. amylovora*. Selective media have been developed for the isolation of pectolytic erwinias (Kelman and Dickey, 1980). The CVP medium containing crystal violet and sodium polypectate (Cuppels and Kelman, 1974) is commonly used. Although pectolytic pseudomonads also will grow on CVP, they can be eliminated by adding manganese. A soluble pink pigment is produced by *E. rubrifaciens* and *E. rhapontici* grown on YDC.

Maintenance Procedures

Stock cultures of *Erwinia* species should be grown on standard media of choice at 25–30°C until good growth occurs. The cultures can be maintained for short term storage in a refrigerator (4–5°C); some strains of *E. chrysanthemi* are nonviable after 3 or 4 weeks at 4°C, but remain viable for longer periods when stored at 12°C.

For long term preservation, erwinias can be successfully stored as lyophilized cultures usually suspended in equal amounts of 10% glucose and 10% peptone (Ferguson and Nuttall, 1964; Lelliott, 1965). Strains also have been stored in distilled water at 10°C by the method of DeVay and Schnathorst (1963), in soil or under mineral oil (Lelliott, 1965), in liquid nitrogen (cells suspended in 10% skim milk) (Moore and Carlson, 1975), and in glycerol at −70°C or on silica gel at −20°C (Sleesman and Leben, 1978).

Taxonomic Comments

The taxonomy of the genus *Erwinia* and designation of species in the genus has been complicated by the heterogeneity of the strains included in the taxon. It has been suggested that members of the genus be placed into new groupings with other members of the *Enterobacteriaceae* (Starr and Mandel, 1969; White and Starr, 1971). This concept also is supported by studies of selected strains by DNA/DNA homology (Gardner and Kado, 1972), DNA relatedness (Brenner, Fanning and Steigerwalt, 1974) and DNA/DNA segmental homology (Murata and Starr, 1974). The data for the successful implementation of this proposal currently are however, not available. Therefore, the order of the species and subspecies used herein reflects relatedness to the type species, *E. amylovora*, based on cluster analysis using 54 phenotypic characteristics (Dickey, unpublished observations). The results of four numerical analyses have shown that a different relationship between the various nomenspecies was indicated by each method of analysis (Dye, 1981).

The heterogeneity within the genus also is reflected in the genetic clusters which have been proposed. Waldee (1945) suggested that *Erwinia* should be limited to pathogens (*E. amylovora*, *E. salicis* and

E. tracheiphila) that cause necrotic or wilt diseases, utilize a restricted range of carbon compounds and usually require organic nitrogen compounds for growth; and that the biochemically more active soft rotting pathogens (*E. carotovora* and *E. chrysanthemi*) should be placed in a separate genus *Pectobacterium*. Although some workers have supported this suggestion (Brenner et al., 1973, 1974), it has not been generally accepted because there are species taxonomically intermediate between these two groups, and there are pathogens that resemble *E. carotovora* in most of their characteristics but do not cause rots. A proposal also has been made whereby one genus is retained and the organisms are separated into three groups, namely, the Amylovora, Herbicola, and Carotovora groups (Dye, 1968, 1969); however, these groupings are subject to the same inconsistencies mentioned above. A core of relatedness and genetic clusters have been demonstrated for strains and nomenspecies of *Erwinia* by molecular hybridization and segmental homology, although the affinities between most members of the genus are no greater than for other enterobacteria (Gardner and Kado, 1972; Brenner et al., 1973; Brenner et al., 1974; Murata and Starr, 1974; Azad and Kado, 1980).

Erwinia herbicola includes an assortment of yellow and nonpigmented strains from plant lesions, plant surfaces, man and animals, and occasionally from soil, water and air. These organisms previously have been assigned to various genera and species (Dye, 1969; Gibbins, 1978). Strains of interest to phytopathologists have been placed in the genus *Erwinia*. Ewing and Fife (1972) proposed that strains from clinical sources be designated as *Enterobacter agglomerans* because the characteristics of the organisms were in conformity with the genus *Enterobacter* (see the article on *Enterobacter* in this *Manual*). It currently is difficult, if not impossible, to distinguish strains from different sources due to the diversity in this group of organisms. Further phenotypic and genotypic studies must be done to define the groups now referred to as *E. agglomerans* and *Erwinia* species. See the family *Enterobacteriaceae* for additional discussion of this problem.

Strains of *E. chrysanthemi* have been isolated from numerous plant species and cultivars (Dickey, 1981). Six pathovars (pv. *chrysanthemi*, pv. *dianthicola*, pv. *dieffenbachiae*, pv. *paradisiaca*, pv. *parthenii* and pv. *zeae*) have been designated for *E. chrysanthemi* (Dye et al., 1980). The relationship between pathogenicity, phenotypic properties and serological reactions of strains of the pathovars is not entirely clear (Samson and Nassan-Agha, 1978; Yakrus and Schaad, 1979; Dickey, 1981).

Differentiation and characteristics of the species of the genus **Erwinia**

The differential characteristics of the species of *Erwinia* are given in Tables 5.26 to 5.28. Only small numbers of strains of *E. tracheiphila*, *E. rubrifaciens*, *E. quercina*, *E. salicis*, *E. cypripedii*, *E. nigrifluens* and *E. uredovora* have been studied, and data for these species should be treated with reserve.

List of the species and subspecies of the genus **Erwinia**

1. **Erwinia amylovora** (Burrill 1882) Winslow, Broadhurst, Buchanan, Krumwiede, Rogers and Smith 1920, 209[AL] (*Micrococcus amylovorus* Burrill 1882, 134.)

a.my.lo'.vo.ra. Gr. n. *amylum* starch; L. v. *voro* to devour; M.L. fem. adj. *amylovora* starch-destroying.

The characteristics are as given for the genus and as listed in Tables 5.26–5.28.

Colonies on 5% sucrose nutrient agar are typically white, domed, shining, mucoid (levan type) with radial striations and a dense flocculent center or central ring after 2 or 3 days at 27°C. Non-levan forms are isolated rarely. (See Enrichment and Isolation Procedures for selective media.)

Agglutination with *E. amylovora* antiserum is the most rapid and accurate method of determination (Lelliott, 1968); the species is serologically homogeneous and has few agglutinogens in common with related species or with the saprophytes found in diseased material.

Causes a necrotic disease (fireblight) of most species of the *Pomoideae* and of some species in other subfamilies of the *Rosaceae*. A *forma specialis* has been described from raspberry (*Rubus idaeus*) by Starr and Folsom (1951).

The mol% G + C of the DNA of seven strains ranges from 53.6–54.1 (Bd).

Type strain: ATCC 15580 (strain BS1114 of Martinec and Kocur, 1964).

2. **Erwinia tracheiphila** (Smith 1895) Bergey, Harrison, Breed, Hammer and Huntoon 1923, 173.[AL] (*Bacillus tracheiphilus* Smith 1895, 364.)

tra.che.i'phi.la. L. n. *trachia* the windpipe; Gr. adj. *philus* loving; M.L. adj. *tracheiphila* trachea-loving, i.e. growing in the tracheiphila of the vascular bundles.

The characteristics are as given for the genus and as listed in Tables 5.26–5.28.

Grows very poorly on nutrient agar but moderately well on yeast extract glucose chalk agar (YDC) or glucose nutrient agar.

Causes a vascular wilt of *Cucurbita* species.

The mol% G + C of the DNA of three strains ranges from 50–52 (Bd).

Type strain: NCPPB 2452 (Approved Lists, 1980).

3. **Erwinia mallotivora** Goto 1976, 472.[AL]

mal.lo.ti'vo.ra. M.L. n. *Mallotus* a genus of trees; L. v. *voro* to devour; M.L. adj. *mallotivora Mallotus*-destroying.

The characteristics are as described for the genus and as listed in Tables 5.26–5.28.

Colonies on nutrient agar without sucrose are white, raised, transparent, and circular with smooth surfaces and entire margins after 2 days. Colonies on nutrient agar with 5% sucrose are flat, white, circular with entire margins and smooth surfaces, butyrous, and transparent after 1 day; after 4 days colonies are domed, circular, white, mucoid, and translucent, and sometimes possess radial striations.

Causes a leaf spot of Akamegashiwa (*Mallotus japonicus*).

The mol% G + C of the DNA of two strains is 49.8 and 51.0 (Bd).

Type strain: ATCC 29573 (strain AM1 of Goto, 1976).

4. **Erwinia rubrifaciens** Wilson, Zeitoun and Fredrickson 1967, 621.[AL] (*Erwinia amylovora* var. *rubrifaciens* (Dye 1968, 605.)

rub.ri.fac'i.ens. L. adj. *ruber* red; L. v. *facio* make; M.L. part. adj. *rubrifaciens* red-producing.

The characteristics are as given for the genus and as listed in Tables 5.26–5.28.

Grows poorly on nutrient agar, but well on yeast exract glucose chalk agar (YDC) on which colonies are cream to yellow, low convex, smooth, shining with entire margins. Craters form around colonies on the polypectate gel B and C of Hildebrand (1971).

Causes a phloem necrosis of Persian walnut trees (*Juglans regia*).

The mol% G + C of the DNA of three strains ranges from 52.0–52.6 (Bd).

Type strain: ATCC 29291 (Dye, 1968).

5. **Erwinia quercina** Hildebrand and Schroth 1967, 253.[AL] (*Erwinia amylovora* var. *quercina* (Hildebrand and Schroth 1967) Dye 1968, 605.)

Table 5.26.

Cultural, physiological and biochemical characteristics of the species of the genus **Erwinia**[a,b]

Characteristics	1. E. amylovora	2. E. tracheiphila	3. E. mallotivora	4. E. rubrifaciens	5. E. quercina	6. E. salicis	7. E. herbicola	8. E. ananas	9. E. rhapontici	10. E. carotovora	11. E. chrysanthemi	12. E. cypripedii	13. E. nigrifluens	14. E. stewartii	15. E. uredovora
Motility	+	+	+	+	+	+	+	+	+	+	+	+	+	−	+
Anaerobic growth	W	W	+	+	+	W	+	+	+	+	+	+	+	+	+
Growth factors required[c]	+	+	+	−	+	−	−	−	−	−	−	−	−	−	−
Pink diffusible pigment[d]	−	−	−	+	−	−	−	−	+	−	−	−	−	−	−
Blue pigment[e]	−	−	−	−	−	−	−	−	−	−	d	−	−	−	−
Yellow pigment[f]	−	−	−	−	−	−	+	+	−	−	−	−	−	+	+
Mucoid growth[g]	+	−	+	+	+	+	d	+	+	d	d	d	−	+	−
Symplasmata[h]							d	−						−	d
Growth at 36°C	−	−	−	+	+	−	+	+	d	d	+	+	+	d	+
H₂S from cysteine[i]	−	+	−	+	+	+	+	d	+	+	+	+	+	−	−
Reducing substances from sucrose[j]	+	d	+	−	+	+	d	+	d	d	−	−	−	d	+
Acetoin[i]	+	d	+	−	+	+	+	+	+	+	+	−	+	−	+
Urease[i]	−	−	−	−	−	−	−	−	−	−	−	−	+	+	−
Pectate degradation[k]	−	−	−	+	−	−	−	−	+	+	−	−	−	−	−
Gluconate oxidation[l]	−	−	−	−	−	−	−	−	d	−	−	+	−	−	−
Gas from D-glucose[m]	−	−	−	−	−	−	−	−	−	d	+	+	−	−	−
Casein hydrolysis[i]	−	−	−	−	−	−	−	−	d	d	−	−	−	−	−
Growth in KCN broth	−	−	−	−	−	−	−	+	d	d	+	−	−	−	−
Cotton seed oil hydrolysis[i]	−	−	−	−	−	−	−	−	d	d	d	+	−	−	−
Gelatin liquefaction[i]	+	−	−	−	−	−	+	+	−	+	−	−	−	−	+
Phenylalanine deaminase[n]	−	−	−	−	−	−	+	−	−	−	−	+	−	−	+
Indole test[o]	−	−	−	−	−	−	−	+	−	+	−	−	−	−	+
Nitrate reduction[i]	−	−	−	−	−	−	+	−	+	+	+	+	−	−	+
Growth in 5% NaCl		−	−				+	+	+	+	d	+		+	+
Deoxyribonuclease (DNase)[p]	−	−	−	−	−	−	−	−	−	−	−	−	−	−	+
Phosphatase[q]		−								d	−	+	d		
Lecithinase[r]										−	−	+	−		
Sensitivity to erythromycin(15 μg/disk)										+	−	+	+		

[a] Data mostly from Dye (1968, 1969) with supplemental data from Graham (1972), Goto (1976), Sellwood and Lelliott (1978), Dickey (1979) and Dickey and Victoria (1980). For invariant characters see generic description.

[b] Symbols: +, 80% or more of strains positive; −, 20% or less of strains positive; d, 21–79% of strains positive; W, weak growth; blank space, insufficient or no data.

[c] *E. amylovora* requires nicotinic acid. Other growth factor-requiring species will grow in an inorganic salts medium with utilizable C source and yeast extract; their exact requirements are not known.

[d] On 1% yeast extract, 1% D-glucose, 2% ppt. chalk, 2% agar (YDC). Pigment production by *E. rhapontici* is more consistent on media containing 2% sucrose, 0.5% peptone, 0.05% K_2HPO_4, 0.025 $MgSO_4$, 2% agar (pH 7.2–7.4), or 5% sucrose nutrient agar.

[e] On YDC (see footnote *d* above) after 5–10 days at 27°C.

[f] On nutrient agar. Nonpigmented strains of *E. herbicola* occur (Billing and Baker, 1963) but their frequency in relation to pigmented strains is not known.

[g] On 5% sucrose nutrient agar.

[h] See Graham and Hodgkiss (1967).

[i] By the methods of Dye (1968).

[j] After 2 days shake culture at 27°C in 4% sucrose, 1% peptone, 0.5 beef extract broth. The production of an orange or brown color (with or without precipitate) with an equal volume of Benedict's quantitative reagent after 10 min in a boiling water bath constitutes a positive reaction.

[k] In 3 days at 27°C on Paton's medium (Paton, 1959).

[l] After 4 days shake culture at 27°C in the medium of Shaw and Clarke (1955) and tested and read as footnote *j* above.

[m] In the sealed tube of Hugh and Leifson's (1953) O/F medium. *E. quercina* and *E. rubrifaciens* produce small amounts of gas (possibly from peptone) on some other media.

[n] On phenylalanine agar test No. 18, 2–3 days at 27°C (Report, 1958). The reaction is weaker than that given by *Proteus* species.

[o] After 2 and 5 days at 27°C in 1% tryptone, 0.1% tryptophan broth and tested with Kovacs' reagent. *E. chrysanthemi* probably converts tryptophan to α-methyl indole and not indole (Lelliott, 1956).

[p] On DNase test agar after 2 days at 27°C (Graham and Hodgkiss, 1967).

[q] As described by Cowan and Steel (1965), using 0.05% sodium phenolphthalein diphosphate agar after 2 days at 27°C.

[r] On egg-yolk agar after 7 days at 27°C.

Table 5.27.
Acid production from organic compounds by **Erwinia** *species*[a,b]

Compound	1. E. amylovora	2. E. tracheiphila	3. E. mallotivora	4. E. rubrifaciens	5. E. quercina	6. E. salicis	7. E. herbicola	8. E. ananas	9. E. rhapontici	10. E. carotovora	11. E. chrysanthemi	12. E. cypripedii	13. E. nigrifluens	14. E. stewartii	15. E. uredovora
Melibiose	−	−	−	−	−	+	−	+	+	+	+	+	+	+	+
Inositol	−	−	−	−	−	+	−	+	+	d	d	+	+	−	+
Raffinose	−	−	−	−	−	+	d	+	+	+	+	−	+	+	+
Inulin	−	−	−	−	−	−	+	d	+	−	d	−	−	d	+
Starch	−	−	−	−	−	+	+	+	−	−	−	−	−	−	+
Maltose	−	−	−	−	−	+	+	+	d	−	+	−	−	−	+
L-Arabinose	d	−	−	+	−	−	+	+	+	+	+	+	+	+	+
Sorbitol	d	−	−	+	+	+	+	+	+	+	+	+	+	+	+
Ribose	+	−	+	+	+	+	+	+	+	+	+	+	+	+	+
Mannose	−	−	+	+	+	+	+	+	+	+	+	+	+	+	+
Mannitol	−	−	+	+	+	+	+	+	+	+	+	+	+	+	+
Cellobiose	−	−	(+)	−	−	−	−	+	+	+	+	+	−	−	+
Lactose	−	−	−	−	−	−	d	+	+	+	d	−	−	+	+
Rhamnose	−	−	−	−	−	−	+	d	+	+	+	+	+	+	−
Esculin	−	−	−	−	+	+	d	d	+	+	+	+	+	−	d
Salicin	−	−	−	+	+	d	+	+	+	+	+	+	+	−	d
Xylose	−	−	+	−	−	+	+	d	+	+	+	+	+	+	+
Trehalose	+	−	+	−	−	+	+	+	+	−	+	+	+	+	+
Dulcitol	−	−	−	−	−	+	−	d	−	−	−	−	−	−	−
Glycerol	−	−	(+)	d	+	d	−	+	+	d	+	d	+	−	+
Adonitol	−	−	−	−	−	−	−	−	−	−	−	−	−	−	+
Dextrin	−	−	−	−	−	−	−	−	−	−	−	−	−	−	+
Melezitose	−	−	−	−	−	−	−	−	−	d	−	−	−	−	+
α-methyl glucoside	−	−	−	+	+	−	−	−	d	d	−	−	−	−	−

[a] After 7-days growth at 27°C in unshaken aqueous solution of 1% organic compound, 1% peptone with bromcresol purple as an indicator. *E. tracheiphila* grows very slowly in the medium.

[b] Symbols: +, 80% or more of strains positive; −, 20% or less of strains positive; (+), delayed positive reaction; d, 21–79% of strains positive. For invariant characters, see generic description. Data mostly from Dye (1968, 1969) with supplemental data from Graham (1972), Goto (1976), Sellwood and Lelliott (1978), Dickey (1979) and Dickey and Victoria (1980).

quer.ci'na. L. n. *quercus* oak; L. suff. *-ina* belonging to; M.L. part. adj. *quercina* oak-belonging.

The characteristics are as given for the genus and as listed in Tables 5.26–5.28.

Growth on potato glucose peptone calcium carbonate (PGPC) agar is luxuriant and after 24 h colonies are white, circular and raised with entire margins. Craters form around colonies on polypectate gel of Hildebrand (1971).

Small amounts of gas are produced (possibly from peptone) in a glucose peptone medium and in PGPC.

Superficially rots onion (but not potato) slices and induces profuse lateral root development in 3 or 4 days on slices of carrot, turnip or beet.

Causes copious oozing of sap from acorns and, by artificial inoculation, shoot blight of *Quercus agrifolia* and *Q. wislizeni.*

The mol% G + C of the DNA of two strains is 54.6 and 55.1 (Bd).
Type strain: ATCC 29281 (Dye, 1968).

6 **Erwinia salicis** (Day 1924) Chester 1939, 406.[AL] (*Bacterium salicis* Day 1924, 14.)

sa'li.cis. L. n. *salix* the willow; L. gen. n. *salicis* of the willow.

The characteristics are as given for the genus and as listed in Tables 5.26–5.28.

Grows poorly on nutrient agar but moderately well on yeast extract glucose chalk agar (YDC) or on glucose nutrient agar.

Colonies on 0.5% starch potato agar (pH 6.5) are yellowish in 2–3 days. A bright yellow pigment is produced on autoclaved potato tissue.

Craters form around colonies on the pectate gel of Paton (1959).

Causes a vascular wilt of *Salix* species.

The mol% G + C of the DNA of two strains is 51.3 and 51.5 (Bd).
Type strain: ATCC 15712 (Martinec and Kocur, 1963).

7. **Erwinia herbicola** (Löhnis 1911) Dye 1964, 268.[AL] (*Bacterium herbicola* Löhnis 1911, 141; *Enterobacter agglomerans* (Beijerinck 1888) Ewing and Fife 1972, 10.)

her.bi'co.la. L. n. *herba* grass, green plants; L. suff. *-cola* dweller; M.L. n. *herbicola* grass-dweller.

The characteristics are as given for the genus and as listed in Tables 5.26–5.28.

The yellow (YC) and nonpigmented (DC) *Erwinia*-like organisms from plant sources described by Billing and Baker (1963) are included in this species.

Colonies of most strains are yellow; nonpigmented forms have been isolated and may be common. Small craters form around colonies on polypectate gel C of Hildebrand (1971).

Exists on plant surfaces and as secondary organisms in lesions caused by many plant pathogens. Some strains (syn. *Erwinia milletiae*) are reported to cause galls on *Milletia japonica*, some on *Wistaria floribunda* and *W. brachybotrys* (Goto et al., 1980) and some (syn. *Agrobacterium gypsophilae*) to cause galls on *Gypsophila paniculata*. Has been isolated from water (syn. *Flavobacterium rhenanum*), the enteric tract of man (syn. *Bacterium typhi flavum, Enterobacter* pigmentées anaérogènes Le Clerc, 1962, and see Gilardi et al., 1970), from septic tonsils of man, from the spleen and liver of symptomless deer (Muraschi et al., 1965)

Table 5.28.

Utilization of some organic compounds as a source of carbon and energy for **Erwinia** *species[a,b]*

Species	Citrate	Formate	Lactate	Tartrate	Galacturonate	Malonate
1. *E. amylovora*	+	+	+	−	−	−
2. *E. tracheiphila*	d	d	−	−	−	−
3. *E. mallotivora*	+	−	−	−	−	−
4. *E. rubrifaciens*	+	+	+	+	−	−
5. *E. quercinia*	+	+	+	−	−	−
6. *E. salicis*	−	−	−	−	−	−
7. *E. herbicola*	+	+	+	d	−	d
8. *E. ananas*	+	+	+	+	d	−
9. *E. rhapontici*	+	+	+	d	d	+
10. *E. carotovora*	+	+	+	−	d	−
11. *E. chrysanthemi*	+	+	+	d	d	+
12. *E. cypripedii*	+	+	+	+	+	d
13. *E. nigrifluens*	−	+	+	+	−	−
14. *E. stewartii*	+	+	+	+	−	−
15. *E. uredovora*	+	+	+	+	−	−

[a] In 21 days at 27°C on OY medium (Dye, 1968).

[b] Symbols: +, 80% or more of strains positive; −, 20% or less of strains positive; d, 21–79% of strains positive. Data mostly from Dye (1968, 1969) with supplemental data from Graham (1972), Goto (1976), Sellwood and Lelliott (1978), Dickey (1979) and Dickey and Victoria (1980). For invariant characteristics see generic description.

Table 5.29.

Characteristics differentiating the subspecies of **Erwinia carotovora**[a]

Characteristics	10a. *E. carotovora* subsp. *carotovora*	10b. *E. carotovora* subsp. *atroseptica*
Mucoid growth	d	−
Growth at 36°C	+	−
Reducing substances from sucrose	−	+
Casein hydrolysis	+	d
Cotton seed oil hydrolysis	d	−
Acid production from:		
Inositol	d	−
Maltose	−	+
Glycerol	+	d
α-Methyl glucoside	−	+
Utilization of galacturonate as a carbon and energy source	+	d

[a] For symbols and conditions see footnotes to Tables 5.26 to 5.28.

and from man and animals in the role of opportunistic pathogens (Gibbins, 1978; von Graevenitz, 1977).

The mol% G + C of the DNA of 30 strains ranges from 52.6–57.7 (Bd).

Type strain: NCPPB 2971 (Approved Lists, 1980).

8. **Erwinia ananas** Serrano 1928, 271.[AL]

a'na.nas. M.L. n. *Ananas* generic name of the pineapple.

The characteristics are as described for the genus and as listed in Tables 5.26–5.28.

The original description by Serrano (1928) is indistinguishable from *E. herbicola*, but studies of recent isolates of this pineapple pathogen indicate that it should be regarded as a distinct species.

Causes rot of pineapple (*Ananas sativus*) fruitlets.

The mol% G + C of the DNA of four strains is 53.1–54.1 (Bd).

Type strain: NCPPB 1846 (Approved Lists, 1980).

9. **Erwinia rhapontici** (Millard 1924) Burkholder 1948, 475.[AL] (*Phytomonas rhapontica* (*sic*) Millard 1924, 11; *Pectobacterium rhapontici* (Millard 1924) Patel and Kulkarni 1951, 80[AL]; *Erwinia carotovora* var. *rhapontici* (Millard 1924) Dye 1969, 93.)

rha.pon'ti.ci. M.L. n. *rhaponticum* specific epithet of *Rheum rhaponticum*, rhubarb; M.L. gen. n. *rhapontici* of rhubarb.

The characteristics are as described for the genus and as listed in Tables 5.26–5.28.

Rots potato, onion and cucumber slices slowly, weakly and erratically (Sellwood and Lelliott, 1978).

Causes a crown rot of rhubarb (*Rheum rhaponticum*), pink grain of wheat (Roberts, 1974), internal browning of hyacinth and occurs epiphytically and saprophytically in lesions caused by other bacteria (Sellwood and Lelliott, 1978).

The mol% G + C of the DNA of three strains ranges from 51.0–53.1 (Bd).

Type strain: ATCC 29283 (Approved Lists, 1980).

10. **Erwinia carotovora** (Jones 1901) Bergey, Harrison, Breed, Hammer and Huntoon 1923, 171.[AL] (*Bacillus carotovorus* Jones 1901, 12; *Pectobacterium carotovorum* (Jones 1901) Waldee 1945, 469[AL]; *E. carotovora* var. *carotovora* Dye 1969, 81.)

ca.ro.to'vo.ra. L. n. *carota* carrot; L. v. *voro* to devour; M.L. adj. *carotovora* carrot-devouring.

The characteristics are as described for the genus and as listed in Tables 5.26–5.28.

Causes rotting, particularly of storage tissues, of a wide variety of plants and causes a vascular and parenchymatal disease (blackleg) of potato plants.

The species is divided into two subspecies.

The mol% G + C of the DNA ranges from 50.5–53.1 (Bd).

Type strain: ATCC 15713 (Martinec and Kocur, 1963).

10a. **Erwinia carotovora** subspecies **carotovora** (Jones 1901) Bergey, Harrison, Breed, Hammer and Huntoon 1923, 171.[AL]

Characteristics distinguishing this subspecies from the subspecies *atroseptica* are indicated in Table 5.29. Gas production from carbohydrates is erratic; some strains (syn. *E. aroideae*) are anaerogenic when isolated, others produce moderate or small amounts of gas and often become anaerogenic after prolonged culture.

Causes rotting, particularly of storage tissues, of a wide variety of plants.

The mol% G + C of the DNA of 11 strains ranges from 50.5–53.1 (Bd).

Type strain: ATCC 15713.

10b. **Erwinia carotovora** subspecies **atroseptica** (van Hall 1902) Dye 1969, 81.[AL] (*Bacillus atrosepticus* van Hall 1902, 134.)

at.ro.sep'ti.ca. L. adj. *ater* black; Gr. adj. *septicus* producing a putrefaction; M.L. adj. *atrospetica* producing a black rot.

Characteristics distinguishing this subspecies from the subspecies *carotovora* are indicated in Table 5.29.

Causes a vascular and parenchymatal disease (blackleg) of potato (*Solanum tuberosum*) plants and a storage rot of potato tubers.

The mol% G + C of the DNA of two strains is 51.3 and 53.1 (Bd).

Type strain: NCPPB 549 (Approved Lists, 1980).

Further Comments

A third subspecies of *E. carotovora* has recently been described: "*Erwinia carotovora* subspecies *betavasculorum*" Thomson, Hildebrand and Schroth 1981, 1040.

Characteristics distinguishing this subspecies from subspecies *carotovora* and *atroseptica* are growth at 36°C, reducing substances formed from sucrose, utilization of inositol, maltose, glycerol, and α-methyl glucoside, but not galacturonate. Other nutritional and physiological characteristics useful for distinguishing the subspecies include: utilization of D-lactate, ethanol, L-lysine, palatinose and D-asparagine, but not cellobiose, melibiose, malonate and raffinose; no production of indole, phosphatase or gas from glucose; and resistance to erythromycin.

Causes soft rot of sugar beet.

The mol% G + C of the DNA of three strains is 54.4–54.7 (T_m).

Designated type strain: NCPPB 2795 (Thomson et al., 1981).

11. **Erwinia chrysanthemi** Burkholder, McFadden and Dimock 1953, 526.[AL] [*Pectobacterium chrysanthemi* (Burkholder, McFadden and Dimock 1953) Brenner, Steigerwalt, Miklos and Fanning 1973, 205.[AL]) (Subj. syns.: *Erwinia carotovora* var. *paradisiaca* Victoria and Barros 1969, 189; *Erwinia paradisiaca* Fernández-Borrero and López-Duque 1970, 22.)

chrys.an'.the.mi. M.L. n. *Chrysanthemum* generic name; M.L. gen. n. *chrysanthemi* of chrysanthemums.

The characteristics are as described for the genus and as listed in Tables 5.26–5.28.

Colonies on potato-glucose-agar (pH 6.5) are characteristically umbonate with undulate to coralloid margins ("fried egg") at 3–6 days of growth.

Causes vascular wilts or parenchymatal necroses of a wide range of plant species and cultivars (Dickey, 1981). There is evidence for differentiation into pathovars (see Taxonomic Comments).

The mol% G + C of the DNA of six strains is 55.1–57.1 (Bd).

Type strain: ATCC 11663 (Approved Lists, 1980).

12. **Erwinia cypripedii** (Hori 1911) Bergy, Harrison, Breed, Hammer and Huntoon 1923, 171.[AL] (*Bacillus cypripedii* Hori 1911, 91; *Erwinia carotovora* var. *cypripedii* (Hori 1911) Dye 1969, 93: *Pectobacterium cypripedii* (Hori 1911) Brenner, Steigerwalt, Miklos and Fanning 1973, 205.[AL])

cyp.ri.ped'i.i. M.L. n. *Cypripedium* generic name; M.L. gen. n. *cypripedii* of cypripedium orchids.

The characteristics are as described for the genus and as listed in Tables 5.26–5.28.

Causes a brown rot of cypripedium orchids (*Cypripedium* spp.).

The mol% G + C of the DNA of two strains is 54.1 and 54.6 (Bd).

Type strain: PDDCC 1591 (Approved Lists, 1980).

13. **Erwinia nigrifluens** Wilson, Starr and Berger 1957, 673.[AL] (*Erwinia amylovora* var. *nigrifluens* Dye 1968, 605.)

ni.gri.flu'ens. L. adj. *niger, nigra* black; L. v. *fluo* flow; M.L. part. adj. *nigrifluens* black flowing.

The characteristics are as described for the genus and as listed in Tables 5.26–5.28.

Colonies on Bacto-EMB (Difco) agar are dark violet with a green metallic sheen. Craters form around colonies on the polypectate medium of Hildebrand (1971).

Growth media should contain yeast extract and should be at pH 7–8.

Causes a bark necrosis of the Persian walnut (*Juglans regia*).

The mol% G + C of the DNA of one strain is 56.1 (Bd).

Type strain: ATCC 13028 (Dye, 1968).

14. **Erwinia stewartii** (Smith 1898) Dye 1963, 504.[AL] (*Pseudomonas stewarti* Smith 1898, 422.)

stew.ar'ti.i. M.L. gen. n. *stewartii* of Stewart; named after F. C. Stewart.

The characteristics are as described for the genus and as listed in Tables 5.26–5.28.

Growth slow, but better on nutrient media with a utilizable carbohydrate such as glucose or sucrose than without.

Causes a vascular wilt of corn (*Zea mays*) and some related plants, and exists in its insect vector, *Chaetocnema pulicaria*.

The mol% G + C of the DNA of two strains is 54.6 and 55.1 (Bd).

Type strain: ATCC 8199 (Approved Lists, 1980).

15. **Erwinia uredovora** (Pon, Townsend, Wessman, Schmitt and Kingsolver 1954) Dye 1963, 149.[AL] (*Xanthomonas uredovorus* Pon, Townsend, Wessman, Schmitt and Kingsolver 1954, 710).

ur.e.do'vo.ra. L. n. *uredo* blight; L. v. *voro* to devour; M.L. adj. *uredovora* blight-devouring (i.e., eats uredospores and uredia).

Craters form around colonies on polypectate gel A of Hildebrand (1971).

Attacks uredia of *Puccinia graminis* and can exist in soil.

The mol% G + C of the DNA of five strains ranges from 53.0–54.5 (Bd, T_m).

Type strain: ATCC 19321 (Sneath and Skerman, 1966).

Species Incertae Sedis

The taxonomic position of the following species is doubtful.

a. *Erwinia cancerogena* Urosevic 1966, 500.[AL]

can.cer.o'ge.na. L. n. *cancer* crab, the disease cancer; L. v. *gigno* to produce; M.L. fem. adj. *cancerogena* cancer-inducing.

Causes a canker disease of poplar (*Populus* species).

This species produces positive reactions for arginine and ornithine decarboxylase. It is probably a species of *Enterobacter*.

Type strain: NCPPB 2176 (Approved Lists, 1980).

b. *Erwinia carnegieana* Standring 1942, 310.[AL] (*Pectobacterium carnegieana* (Standring 1942) Brenner, Steigerwalt, Miklos and Fanning 1973, 205.[AL])

car.ne.gie.a'na. M.L. adj. *carnegieana* pertaining to *Carnegiea*, the name of a cactus.

In the original description *E. carnegieana* is described as, *inter alia*, a Gram-positive organism which does not ferment lactose, produces a necrotic disease of *Carnegiea gigantea* and does not attack *Opuntia* species or rot carrots. Later, Boyle (1949) with other isolates showed that the Gram reaction became nearly negative with continued culture, confirmed the lactose reaction and showed that they were not agglutinated by *E. carotovora* antiserum. Burkholder (1957) emended the description to, *inter alia*, Gram-negative with Gram-positive granules in the cells of old cultures and lactose-positive. Alcorn (1961) obtained isolates from *C. gigantea* and *Opuntia* species that would cross-infect.

Strain NCPPB 439 is Gram-negative, lactose-positive, does not rot carrots or liquefy pectate gel and produces lysine decarboxylase; two of Alcorn's isolates (NCPPB 671 and 672) are typical of *E. carotovora* (Lelliott and Graham, unpublished observations). There may therefore be two pathogens of *C. gigantea*: *E. carnegieana* and *E. carotovora*, both of which cause a similar disease.

Type strain: NCPPB 439 (ATCC 33259) (Sneath and Skerman, 1966).

Editorial Note

The type strain of *E. carnegieana* (ATCC 33259) has been identified as a typical *Klebsiella pneumoniae* (R. L. Gherna, American Type Culture Collection, personal communication).

c. *Erwinia dissolvens* (Rosen 1922) Burkholder 1948, 472.[AL] (*Pseudomonas dissolvens* Rosen 1922, 497.)

dis.sol'vens. L. part. adj. *dissolvens* dissolving.

Nonmotile. Produces large amounts of gas from many carbohydrates and decarboxylates arginine and/or lysine (Dye, 1969).

Isolated from rotting cornstalks (*Zea mays*).

This organism belongs to the genus *Enterobacter* (Waldee, 1945; Dye, 1969; Steigerwalt et al., 1976) as a new species or as a biogroup of *E. cloacae*. Its DNA is 60–80% related to DNA from *E. cloacae* and the two organisms are very similar biochemically (Steigerwalt et al., 1976).

Type strain: ATCC 23373 (Approved Lists, 1980).

d. *Erwinia nimipressuralis* Carter 1945, 423.[AL]

ni.mi.pres.su.ra'lis. L. adv. *nimis* overmuch; L. n. *pressura* pressure; M.L. adj. *nimipressuralis* with excessive pressure.

Isolated from wet wood of elms (*Ulmus* species) but its pathogenicity is doubtful.

Strains produce large amounts of gas from many carbohydrates (including lactose), decarboxylate arginine and produce lipase.

This organism belongs to the genus *Enterobacter* (Graham, 1964; Dye, 1969b; Steigerwalt et al., 1976) as a new species or as a biogroup of *E. cloacae*. Its DNA is 55–65% related to the DNA of *E. cloacae* and *Erwinia dissolvens* (see paragraph c above) (Steigerwalt et al., 1976). It is sucrose- and raffinose-negative, characteristics which separate it from *E. cloacae* (Steigerwalt et al., 1976).

Type strain: ATCC 9912 (Sneath and Skerman, 1966).

Important Notes for Users of this Edition

1. Always read both generic and species descriptions because characters listed in the generic description are not usually listed in the species descriptions.

2. Unless otherwise indicated in footnotes to tables, the meanings of symbols are as follows:

+ 90% or more of strains are positive

− 90% or more of strains are negative

d 11–89% of strains are positive

v strain instability (*not* equivalent to "d")

D different reactions in different taxa (species of a genus or genera of a family)

3. All other symbols are defined in footnotes to tables.

Genus VIII. *Serratia* Bizio 1823, 288[AL]

PATRICK A. D. GRIMONT AND FRANCINE GRIMONT

Ser.ra′ti.a. M.L. fem noun *Serratia*, named after Serafino Serrati, an Italian physicist.

Straight rods, 0.5–0.8 μm in diameter and 0.9–2.0 μm in length, with rounded ends. Conform to the general definition of the family *Enterobacteriaceae*. Generally **motile**, by means of peritrichous flagella. Facultatively anaerobic. Colonies are most often opaque, somewhat iridescent, and either **white, pink or red in color.** Almost all strains can grow at temperatures between 10 and 36°C, at pH 5–9, and in the presence of 0–4% (w/v) NaCl. The catalase reation is strongly positive. **Acetoin is produced from pyruvate.** Reducing compound(s) are produced from gluconate. D-Glucose is fermented in the presence (and in the absence) of 0.001 M iodoacetate. Maltose, mannitol and trehalose are fermented and utilized as sole carbon sources. D-Alanine, L-alanine, 4-aminobutyrate, **caprylate**, citrate, **L-fucose**, D-glucosamine, kynurenate, L-proline, putrescine and tyrosine are utilized as sole carbon sources. Dulcitol and tagatose are neither fermented nor utilized as sole carbon sources. Butyrate and 5-amino-valerate are not utilized as sole carbon sources. **Extracellular enzymes hydrolyze DNA, lipids** (tributyrin, corn oil) **and proteins** (gelatin, casein), but not starch (in 4 days), polygalacturonic acid or pectin. Phenylalanine (and tryptophan) deaminase and thiosulfate reductase (H$_2$S from thiosulfate) are not produced. *o*-**Nitrophenyl-β-D-galactopyranoside (ONPG) is hydrolyzed by most strains. Chlorate is reduced** anaerobically by *Serratia* nitrate reductase (anaerobic growth does not occur with chlorate). **Growth factors are generally not required** by *Serratia* strains. The organisms **occur in the natural environment** (soil, water, plant surfaces) **or as opportunistic human pathogens.** The mol% G + C of the DNA is 52–60 (T_m, Bd).

Type species: *Serratia marcescens* Bizio 1823, 288.

Further Descriptive Information

Cells of *Serratia* rarely show a visible capsule in India ink mounts, although mucoid colonies can be observed in *S. plymuthica* and occasionally in other *Serratia* species; however, cells of *S. odorifera* possess a microcapsule which can be evidenced by the quellung reaction (capsular swelling) using *Klebsiella* anticapsule K4 or K68 sera (Richard, 1979). Polysaccharides, excreted by cells of *S. marcescens*, can be extracted from the cell surface layer or from the culture medium. These polysaccharides contain chiefly D-glucose and glucuronic acid and lower proportions of D-mannose, heptose, L-fucose and L-rhamnose (Adams and Martin, 1964; Adams and Young, 1965).

Colony diameters are ~1.5–2.0 mm after overnight growth on nutrient agar. Swarming does not occur.

Two different pigments can be produced by various *Serratia* strains: prodigiosin and pyrimine (Williams and Qadri, 1980). Prodigiosin, a nondiffusible, water-insoluble pigment bound to the cell envelope, is produced by two biogroups (A1 and A2) of *S. marcescens* and by most strains of *S. plymuthica* and *S. rubidaea*. Prodigiosin-producing colonies are totally red or show either a red center, a red margin or red sectors. The exact color given by the pigment depends upon cultural conditions (e.g. amino acids, carbohydrates, pH, inorganic ions, temperature) and may include orange, pink, red, or magenta. Prodigiosin is best produced on peptone-glycerol agar* at 20–35°C. The temperature range for pigment production is 12–36°C. Prodigiosin is not produced anaerobically. Chemically, prodigiosin is 2-methyl-3-amyl-6-methoxyprodigiosene (prodigiosene is 5-(2-pyrryl)-2,2′-dipyrrylmethene). In the cell, prodigiosin is formed by condensation of a volatile 2-methyl-3-amyl-pyrrol (MAP) and a nonvolatile 4-methoxy-2-2′-bipyrrole-5-carboxaldehyde (MBC). Several classes of nonpigmented mutants have been isolated that are either blocked on the MAP pathway or the MBC

pathway. Syntropic pigmentation may occur when two different class mutants are grown side by side (Williams and Qadri, 1980).

Pyrimine, a water-soluble, diffusible pink pigment (Williams and Qadri, 1980), is produced by some strains of *S. marcescens* biogroup A4. Ferrous iron is required for the production of pyrimine. Pyrimine is L-2(2-pyridyl)-Δ'-pyrroline-5-carboxylic acid. When pyrimine is produced, the agar medium turns pink while the colonies are white to pinkish.

Cultures can produce two kind of odors, a fishy to urinary odor attributed to trimethylamine (mixed with some NH$_3$), or a musty, potato-like odor resembling that of 2-methoxy-3-isopropyl-pyrazine. The musty odor is produced by *S. odorifera*, *S. ficaria*, and a few strains of *S. rubidaea*. All other strains and species produce the fishy-urinary odor.

Several species can grow readily at 4–5°C (*S. liquefaciens*, *S. plymuthica*, *S. odorifera* and *S. ficaria*) or at 40°C (*S. marcescens* and several strains of *S. rubidaea* and *S. odorifera*); however, the temperature of 37°C is not favorable for the isolation of *S. plymuthica*. When *S. liquefaciens* and *S. plymuthica* are studied, many tests that are positive at 28–35°C give negative results at 37°C (e.g. Voges-Proskauer, decarboxylases, tetrathionate reductase tests).

A strong catalase activity, which can be evidenced with 3% (or less) H$_2$O$_2$, is produced by *Serratia* strains (Taylor and Achanzar, 1972).

There is no sodium ion requirement for growth in the genus *Serratia*; however, the optimum concentration of NaCl for growth is ~0.5% (w/v) for *S. marcescens* or 1% (w/v) for *S. rubidaea* (unpublished results). Tolerance to NaCl ranges from 5–6% (w/v) for *S. plymuthica* to 10% (w/v) for *S. rubidaea*.

In a minimal medium containing ammonium sulfate as the nitrogen source, the following compounds serve universally as sole carbon sources for all *Serratia* strains: *N*-acetylglucosamine, D-alanine, L-alanine, 4-aminobutyrate, aspartate, caproate, caprylate, citrate, D-fructose, L-fucose, fumarate, D-galactose, D-galacturonate, D-glucose, D-glucuronate, L-glutamate, L-histidine, inositol, kynurenate, L-malate, maltose, D-mannitol, D-mannose, L-proline, putrescine, pyruvate, D-ribose, L-serine, succinate, trehalose and L-tyrosine. Most strains of all species can utilize acetate, caprate, D-glucosamine, glycerate, glycerol, lactate, phenylacetate, salicin and L-tryptophan. The following compounds are never utilized as sole carbon sources: acetamide, adipate, 4-aminobenzoate, DL-5-aminovalerate, α-amylamine, azelate, benzylamine, butanol, butylamine, butyrate, citraconate, DL-citrulline, creatine, dulcitol, ethylene glycol, D-fucose, glutarate, glycolate, L-isoleucine, isophthalate, isopropanol, isovalerate, levulinate, L-leucine, D-mandelate, L-mandelate, mesaconate, methanol, methylamine, DL-norleucine, oxalate, pantothenate, phenol, phthalate, pimelate, propanol, propylene glycol, sebacate, sorbose, spermine, suberate, terephthalate, testosterone, tryptamine, D-tryptophan, turanose, uracil, urea, L-valine, and L-xylose (Grimont et al., 1977b, 1978a, 1979b).

Characteristic extracellular enzymes are produced. All species recognized herein can hydrolyze DNA, gelatin, soluble casein, tributyrin, and corn oil. Only rare strains fail to produce one or more of these extracellular enzymes. All species, except *S. odorifera*, can hydrolyse Tween 80. Chitin is hydrolyzed by all species except *S. rubidaea* and *S. odorifera*. Lecithin is also hydrolyzed by many strains. Spot-inoculated starch agar,† incubated for 4 days and then flooded with Lugol's iodine, shows no zone of clearing (Grimont et al., 1977b); however, longer incubation (6–14 days) may allow detection of some amylase-producing strains (M. Popoff, personal communication).

---

* Peptone glycerol agar: Bacto-peptone (Difco), 5.0 g; glycerol, 10.0 ml; Bacto-agar (Difco), 20.0 g; distilled water, 1000 ml.

† Starch agar: nutrient agar containing 0.5% (w/v) soluble starch.

A red-pigmented *S. marcescens* has been found to produce a carboxymethyl cellulase (Thayer, 1978). Depolymerization of a carboxymethyl cellulose gel is faster with *S. marcescens*, *S. rubidaea*, and *S. liquefaciens* than with *S. odorifera*, *S. ficaria* and *S. plymuthica* (unpublished results).

Up to 11 proteinases have been revealed by agar gel electrophoresis. Each strain produces one to four different proteinases. Different species have different proteinase patterns (Grimont et al., 1977a). Isoelectric points of the 11 proteinases are between pH 3.6 and pH 6.0 (Grimont and Grimont, 1978a).

Fructose, maltose, D-mannitol, D-mannose, ribose, and trehalose are fermented by all strains. Most strains ferment glycerol and *myo*-inositol. Fermentation of D-glucose is not prevented by 0.001 M iodoacetate (Grimont et al., 1977b, 1978a, 1979b), an inhibitor of the Embden-Meyerhof-Parnas glycolytic pathway and other enzymic reactions. *Serratia* species can produce gluconate-6-phosphate dehydrase and 2-keto-3-deoxygluconate-6-phosphate aldolase (Kersters and De Ley, 1968), which are the characteristic enzymes of the Entner-Doudoroff pathway. Under aerobic conditions, 2-ketogluconic acid is produced from D-glucose (Misenheimer et al., 1965). A reducing compound (probably 2-ketogluconate) is also produced from gluconate by all species (Grimont et al., 1977b, 1978a, 1979b).

The Voges-Proskauer (VP) test, when done on a 3-day-old culture in Clark-Lubs medium, is negative for 40% of the strains of *S. plymuthica*, although acetoin can be detected after incubation for 18 h by use of a sensitive method (Richard, 1972). These strains which are VP-negative after 3 days incubation, can utilize 2,3-butanediol as a sole carbon source (Grimont et al., 1977b). *Serratia* strains that cannot produce acetoin from pyruvate (under any experimental conditions) are very rare. A tiny gas bubble is commonly produced by *S. marcescens* in a peptone-water-glucose medium with Durham tube. *S. plymuthica* and *S. liquefaciens* produce a larger amount of gas. The end products of glucose fermentation by *S. marcescens* are 2,3-butanediol, ethanol, formate, lactate, succinate and CO_2, with small amounts of acetate, acetoin and glycerol and very little or no H_2 (Neish et al., 1948; White and Starr, 1971). The end products yielded by *S. plymuthica* are 2,3-butanediol, ethanol, lactate, succinate, CO_2, H_2 and small amounts of formate, acetate, acetoin, and glycerol (Neish et al., 1948). The 2,3-butanediol produced by *S. marcescens* is mostly a *meso*-isomer, whereas *S. plymuthica* is unique in producing a *levo*-rotatory 2,3-butanediol (Neish et al., 1948).

Transduction systems have been described in *S. marcescens* (Kaplan and Brendel, 1969; Matsumoto et al., 1973). The earliest genetic transfer described in *S. marcescens* (Belser and Bunting, 1956) is also suggestive of a transduction mechanism.

Lactose plasmids have been demonstrated in *S. liquefaciens* (Le Minor et al., 1974) and in *S. marcescens* (C. Coynault, personal communication). Antibiotic resistance plasmids of incompatibility groups H_2, C, M, P, W, and F_{II} have been identified in *S. marcescens*. Plasmids of groups M and N have been found in *S. liquefaciens* (Hedges, 1980).

Bacteriophages active on *Serratia* are easily found in river water or sewage. Phages that are active on one species of *Serratia* are usually active on strains of other species of that genus but rarely on strains of other genera (Grimont and Grimont, 1978a). Lyosgeny is very common in *Serratia* species (Prinsloo, 1966). Several phage typing systems have been studied (Pillich et al., 1964; Hamilton and Brown, 1972; Farmer, 1975; F. Grimont, Doctorate in Pharmacy thesis, University of Bordeaux II, 1977).

Bacteriocins produced by *Serratia* are of two kinds: (a) a trypsin-resistant, acid-sensitive (pH 2) structure (Hamon and Péron, 1961) called "group A bacteriocin" by Prinsloo (1966) and later found by electron microscopy to resemble phage tails (Traub, 1972); and (b) a trypsin-sensitive, acid-resistant protein (Hamon and Péron, 1961) called "group B bacteriocin" by Prinsloo (1966). Bacteriocins produced by one species of *Serratia* frequently cross-react with other species of this same genus. *Serratia* bacteriocins are also frequently active on *Escherichia coli* K12. *S. marcescens* strains produce group A and/or group B bacteriocins. *S. rubidaea* strains produce only group A bacteriocins. *S. liquefaciens* and *S. ficaria* produce only group B bacteriocins. *S. odorifera* produce neither group A or group B bacteriocins (Hamon and Péron, 1979; Y. Hamon, personal communication). Bacteriocin typing can be used for epidemiological purposes (Traub, 1980).

The antigenic structure of only one species (*S. marcescens*) has been detailed. The present scheme consists of 21 somatic antigens (01 to 021) and 25 flagellar antigens (H1 to H25) (Edwards and Ewing, 1972; Le Minor and Pigache, 1978; Traub and Fukushima, 1979; Le Minor and Sauvageot-Pigache, 1981).

Subdivision of antigens 05 (into 05a, 05b, 05c), 010 (into 010a, 010b), and 016 (into 016a, 016b, 016c, 016d) has been proposed (Le Minor and Sauvageot-Pigache, 1981). Cross-reactions between factors 06 and 014 are very extensive and the distinction between these two factors does not seem worthwhile. H antigens are monophasic in *S. marcescens*.

Resistance to cephalothin, colistin, and polymyxin (with respect to achievable serum levels of antibiotics) is very frequent in the genus and almost constant in *S. marcescens*. With the antibiotic disk method, a zone phenomenon develops around disks impregnated with colistin and polymyxin: the inhibition zone contains colonies close to the disk. However, this zone phenomenon is not restricted only to *Serratia*. Resistance to tetracycline and ampicillin is very frequent in *S. marcescens* and rare in other *Serratia* species. Plasmid-determined resistance to aminoglycoside antibiotics, carbenicillin, chloramphenicol, trimethoprim, sulfonamides and mercury ions can be found in clinical strains of *S. marcescens*. Resistance to cetyltrimethylammonium chloride (1.5 mg/ml) and thallous acetate (0.8 mg/ml) is very frequent (Grimont et al., 1977b). Of all the *Serratia* species, *S. marcescens* is the most resistant to antibiotics, antiseptics, and metal ions; *S. plymuthica* is the least resistant to these antimicrobials.

A typical hypersensitivity reaction is produced by inoculation of plants such as tobacco and king protea with *Serratia* (Lakso and Starr, 1970; Grimont et al. 1978b). *S. proteamaculans* was isolated from a leaf spot disease of *Protea cynaroides* (Paine and Stanfield, 1919) and *S. marcescens* (under the name *Erwinia amylovora* var. *alfalfae*) was isolated from a root disease of alfalfa (Shinde and Lukezic, 1974).

S. marcescens and *S. liquefaciens* are potential pathogens for insects. Pathogenicity is correlated with the production of lecithinase, proteinase and chitinase (Lysenko, 1974; Lysenko, 1976; Kaska, 1976).

Mastitis in cows and other animal infections have been associated with *Serratia* species (Grimont and Grimont, 1978a). Pathogenicity in experimental animals is of the type expected of a Gram-negative bacterium. Experimental depression of phagocytic cell number or function in animals enhances susceptibility to *Serratia* infections (Simberkoff, 1980).

S. marcescens is a prominent opportunistic pathogen for hospitalized human patients. Other *Serratia* species can be involved in bacteremia, especially when accidentally injected into the body (contaminated perfusion or irrigation liquid). They can also be isolated from sputum without having clinical significance (Grimont and Grimont, 1978a).

Serratia species occur on plants, in the digestive tract of rodents (unpublished data), and in soil and water. *S. ficaria* is especially associated with the fig/fig-wasp ecosystem (Grimont et al., 1979b).

Enrichment and Isolation Procedures

Fecal samples (diluted with distilled water) or plant material washings are inoculated onto caprylate-thallous (CT) agar* (Starr et al.,

---

* Caprylate-thallous (CT) agar: autoclaved solutions A and B are mixed aseptically in equal volumes and the mixture is poured into Petri dishes. *Solution A*: $CaCl_2 \cdot 2H_2O$, 0.0147 g; $MgSO_4 \cdot 7H_2O$, 0.123 g; KH_2PO_4, 0.68 g; K_2HPO_4, 2.61 g; trace element solution (see below), 10 ml; caprylic acid, 1.1 ml; yeast extract (Difco), 5% (w/v) solution, 2.0 ml; thallous sulfate, 0.25 g; distilled water to 500 ml. Adjust the pH to 7.2. *Trace element solution*: distilled water, 1000 ml; H_3PO_4, 1.96 g; $FeSO_4 \cdot 7H_2O$, 0.0556 g; $ZnSO_4 \cdot 4H_2O$, 0.0287 g; $MnSO_4 \cdot 4H_2O$, 0.0223 g; $CuSO_4 \cdot 5H_2O$, 0.025 g; $Co(NO_3)_2 \cdot 6H_2O$, 0.003 g; H_3BO_3, 0.0062 g. This solution keeps well at 4°C for at least 2 years. *Solution B*: NaCl, 7.0 g; $(NH_4)_2SO_4$, 1.0 g; agar

1976). After 2–5 days, the growth is removed by scraping and tested for deoxyribonuclease (DNase) activity. DNase-positive cultures are then purified by streaking a nonselective medium (e.g. tryptic soy agar). Different colonial types are tested for DNase, and DNase-positive isolates are then thoroughly characterized and identified. This procedure allows isolation of all *Serratia* species as defined in this chapter. *Providencia, Acinetobacter* and fluorescent *Pseudomonas* strains can grow on CT agar when samples contain large numbers of these organisms. Other selective media based on DNase production and antibiotic resistance have been proposed (Farmer et al., 1973; Cate, 1972; Berkowitz and Lee, 1973). These antibiotic-containing media are efficient for the isolation of *S. marcescens* but may not be as reliable for more sensitive species (e.g. *S. plymuthica*).

Maintenance Procedures

For short term preservation (several months), heavy suspension of bacteria in sterile distilled water are made from bacterial growth scraped with a platinium loop from a nutrient agar slant. The suspensions are stored at room temperature.

For longer preservation (several years), screw-capped tubes containing semisolid nutrient agar† are stab-inoculated. After overnight growth at 30°C, the tubes are tightly closed and kept at room temperature in the dark. Maintenance failure may occur if the tube is not protected from desiccation by a rubber seal in the screw cap. Rubber corks dipped in melted paraffin wax may be preferred in place of screw caps.

For long term preservation (over 5 years), freeze-drying is preferred.

Procedures for Testing Special Characters

Carbon source utilization test. The defined medium M70 of Véron (1975) without yeast extract is made by mixing equal volumes of freshly autoclaved solutions A and B. *Solution A* contains $CaCl_2 \cdot 2H_2O$, 0.0147 g; $MgSO_4 \cdot 7H_2O$, 0.123 g; KH_2PO_4, 0.680 g; K_2HPO_4, 2.610 g; trace element solution (same as for CT agar), 10 ml; and distilled water to 500 ml. Adjust to pH 7.2 with NaOH. *Solution B* is the same as in CT agar. The carbon source solution (1 g of carbon source in ~5 ml of distilled water; salts or hydrated molecules are weighted so as to have 1 g of the organic ion) is adjusted to pH 7, sterilized by filtration, and added to the complete medium while the latter is still hot and molten. The medium is then dispensed into divided Petri dishes ("Replidish" Sterilin, Teddington, England) with 3 ml per well. After drying over-

night at 37°C, the plates are inoculated with a multipoint inoculator (Denley). Inoculated plates are incubated at 30°C and examined for growth every other day for 12 days. If the room temperature is about 20–25°C, the plates can be removed from the incubator after 4 days and kept at room temperature; this can be useful if incubators are crowded with cultures or materials that can exhale volatile carbon sources. Only unequivocal growth should be recorded as positive.

Voges-Proskauer test (Richard's modification). Clark and Lubs medium (BBL) is dispensed in large 22 × 215 mm tubes (0.5 ml per tube) and inoculated with 0.05 ml of a heavy bacterial suspension in distilled water. After incubation at 30°C for 18 h, 0.5 ml of α-naphthol solution (6% w/v alcoholic solution) and 0.5 ml of 4 M NaOH are added. The tubes are shaken, heated briefly in a Bunsen flame, and examined for a red color (Richard, 1972).

Tetrathionate reduction. The medium of Le Minor et al. (1970) contains: peptone (Difco), 10.0 g; NaCl, 5.0 g; $K_2S_4O_6$, 5.0 g; bromthymol blue (0.2% aqueous solution), 25 ml; and distilled water to 1 liter. Adjust the pH to 7.4, sterilize by filtration, and dispense into 12 × 120 mm tubes (4 ml per tube). The size of the tubes (for a rather limited aeration) is critical. Inoculated tubes are incubated at 30°C for 24 h and examined for a yellow color (tetrathionate reduction).

β-Xylosidase. Paper disks (0.5 cm) are loaded with 0.1 ml of a 2% (w/v) aqueous solution of p-nitrophenyl-β-D-xylopyranoside and kept dry in a tightly-capped flask at 4°C. the test is performed exactly like the β-galactosidase test, but with p-nitrophenyl-β-D-xylopyranoside disks in place of ONPG disks (Brisou et al., 1972).

H-Immobilization test. The motility of each isolate to be typed must be enhanced by passage through a 0.3% semisolid agar U-tube.

The following autoclaved semisolid medium is dispensed in 2.0-ml volumes into small (92 × 13 mm) screw-capped tubes: tryptic peptone, 20.0 g; D-mannitol, 2.0 g; KNO_3, 1.5 g; phenol red solution (1%), 4 ml; agar, 4.5 g; distilled water, 1000 ml; pH 7.4. The tubes of semisolid medium are melted (boiling water bath), cooled to 50°C in a water bath, supplemented with 0.05 ml of each serum dilution under sterile conditions, and allowed to gel.

Tubes with serum dilutions (and control tubes without serum) are stab-inoculated with a highly motile culture. After overnight incubation, tubes are examined for immobilization. This H-immobilization test is very specific and much easier to perform than the classical H-agglutination (Le Minor and Pigache, 1977).

Differentiation of the genus **Serratia** from other genera

Table 5.30 provides the primary characteristics that can be used to differentiate the genus *Serratia* (as defined in this chapter) from biochemically similar taxa.

Taxonomic Comments

A number of changes have been made since the eighth edition of the *Manual* in which it was indicated that the genus *Serratia* was composed of only one species, *S. marcescens* (the type species).

Transfer of *Enterobacter liquefaciens* to the genus *Serratia* was first proposed by Barbe (Doctor in Pharmacy thesis, University of Marseille, France, 1969) and supported by studies on bacteriocin cross-reactions between *S. marcescens* and *E. liquefaciens* (Hamon et al., 1970). Valid publication of the new combination *S. liquefaciens* followed a numerical taxonomy study (Bascomb et al., 1971).

A phenon named "biotype 2" (Bascomb et al., 1971) and "phenon B" (Grimont and Dulong de Rosnay, 1972) was thought identical to *Bacterium rubidaeum* Stapp 1940 and named *S. rubidaea* (Ewing et al., 1973). The same phenon was also identified as *S. marinorubra* Zobell and Upham 1944 (Grimont et al., 1977, 39). *S. rubidaea* and *S. mari-*

norubra were based on different type strains (ATCC 27593 and ATCC 27614, respectively). The Approved Lists of Bacterial Names, however, give both names *S. rubidaea* and *S. marinorubra* with the same type strain (viz. ATCC 27614, the type strain of *S. rubidaea*). Hence, both names, which were subjective synonyms, are now objective synonyms and redundant. To avoid further confusion, the name *S. rubidaea* (Stapp) Ewing et al. should now be used exclusively to designate the same (*S. rubidaea-S. marinorubra*) taxon.

The ancient species *S. plymuthica* (Lehman and Neumann 1896) Breed, Murray, and Hitchens 1948 was shown to be a valid species by numerical taxonomy (Grimont et al., 1977b) and by DNA/DNA hybridization (Grimont et al., 1978a).

Recently, two new species, *S. odorifera* Grimont et al., 1978, 453 and *S. ficaria* Grimont, Grimont, and Starr 1981 were defined by DNA relatedness, carbon source utilization tests, and by standard biochemical tests.

DNA relatedness studies have shown that *S. marcescens, S. plymuthica, S. rubidaea, S. odorifera,* and *S. ficaria* are homogeneous and discrete genospecies (Steigerwalt et al., 1976; Grimont et al., 1978a;

(Difco), 15.0 g; distilled water to 500 ml. Adjust the pH to 7.2. The complete medium keeps well at 4°C. It should not be remelted once it has solidified.

† Semisolid nutrient agar:meat extract (Liebig), 3.0 g; yeast extract (Difco), 10.0 g; agar (BBL or Difco), 7.5 g; distilled water to 1000 ml. Adjust pH to 7.4.

Table 5.30.
Differential characteristics of the genus **Serratia** *and other biochemically similar taxa*[a]

Characteristics	*Serratia*	*"Serratia" fonticola*	*Erwinia herbicola-Enterobacter agglomerans* group	*Enterobacter cloacae*	Pectinolytic *Erwinia*[b]	*Klebsiella*[c]
Carbon source utilization test:						
4-Aminobutyrate	+	−	+	d	−	D
5-Aminovalerate	−	−	−	−	−	D
Arginine	−	−	−	−	−	+
Caprate	+	−	−	−	−	−
Caproate	+	−	−	−	−	−
Caprylate	+	−	−	−	−	−
D-Dulcitol	−	+	−	−	−	−
L-Fucose	+	−	−	−	−	+
Pelargonate	D	−	−	−	−	−
Tagatose	−	+	−	−	−	D
Tyrosine	+		−	−	−	D
Voges-Proskauer test	+	−	+	+	+	+
Gelatin, hydrolyzed	+	−	D	d	D	D
Tributyrin, hydrolyzed	+	−	−	−	−	−
Deoxyribonuclease	+	−	−	−	D	−
Gluconate test[d]	+	+	−	+	−	+
Iodoacetate test[e]	+	+	+	+	−	+
Mol% G + C of DNA	52–60	48.8–52.5	53.5–56	53	51–54	53.8–57

[a] Symbols: see standard definitions.

[b] Including *Erwinia carotovora*, *E. atroseptica* and *E. chrysanthemi*.

[c] Including *Klebsiella pneumoniae* and *K. mobilis* (*Enterobacter aerogenes*).

[d] Production of reducing compound(s) from gluconate.

[e] Production of acid from glucose in the presence of 0.001 M iodoacetate.

1979b). *S. liquefaciens* is heterogeneous (Steigerwalt et al., 1976) and is probably composed of several genospecies. One biovar (C1c) of *S. liquefaciens* was identified as *Erwinia proteamaculans* (Paine and Stansfield, 1919) Dye 1966 and renamed *S. proteamaculans* (Grimont et al., 1978, 503). Reexamination of DNA relatedness in *S. liquefaciens* disclosed at least three genospecies: *S. liquefaciens* sensu stricto, *S. proteamaculans* (Grimont et al., 1981), and a third group containing strain ATCC 14460 and named *S. grimesii* (Grimont et al., 1982 a, b).

A group of strains called *"Citrobacter* lysine⁺" or *"Citrobacter*-like" was found to be related significantly to the genus *Serratia* in DNA/DNA hybridization studies (Crosa et al., 1974). This genospecies has recently been named *Serratia fonticola* Gavini et al., 1979; however, a difficulty is that *S. fonticola* does not have the key characteristics of the genus *Serratia* (Table 5.31). Furthermore, *Serratia* phages which are active on strains of any *Serratia* species (as defined herein) have been found to be inactive on all *S. fonticola* strains tested (unpublished data). Bacteriocins from *Serratia* are also inactive on *S. fonticola* (Hamon, personal communication). In this chapter, *S. fonticola* is considered to be a *species incertae sedis* pending further study.

All molecular approaches to taxonomy (e.g. genome size, DNA relatedness, immunologic cross-reactions between iso-functional enzymes, physical properties and regulation of enzymes, amino acid sequences of enzymes) support the distinction of the genus *Serratia* from the other members of the family *Enterobacteriaceae* (reviewed by Grimont and Grimont, 1978a).

Further Reading

Edwards, P.R. and W.H. Ewing. 1972. Identification of *Enterobacteriaceae*, 3rd Ed., Burgess Publishing, Minneapolis, Minn.

Grimont, P.A.D. and F. Grimont. 1978. The genus *Serratia*. Annu. Rev. Microbiol. *32:* 221–248.

Grimont, P.A.D. and F. Grimont. 1981. The genus *Serratia*. In Starr, Stolp, Trüper, Balows, and Schlegel (Editors), *The Prokaryotes: A Handbook on Habitats, Isolation, and Identification of Bacteria*, Springer-Verlag, New York, pp. 1187–1203.

Grimont, P.A.D., F. Grimont, H.L.C. Dulong de Rosnay, and P.H.A. Sneath. 1977. Taxonomy of the genus *Serratia*. J. Gen. Microbiol. *98:* 39–66.

Von Graevenitz, A. and S.J. Rubin. 1980. The genus *Serratia*, CRC Press, Boca Raton, Fla.

Differentiation and characteristics of the species of the genus **Serratia**

The differential characteristics of the species of *Serratia* are indicated in Table 5.31. Other characteristics of the species are listed in Table 5.32.

List of the species of the genus **Serratia**

1. Serratia marcescens Bizio 1823, 288.[AL]

mar.ces'cens. M.L. v. *marcesco* to fade; L. part. adj. *marcescens* fading away.

The cell morphology and colonial morphology are as given for the genus. Prodigiosin or pyrimine can be produced.

Physiological and nutritional characteristics are presented in Tables 5.31 and 5.32.

A biotyping system based on pigment production, tetrathionate reduction, and utilization of *meso*-erythritol, trigonelline, quinate, benzoate, 3-hydroxybenzoate, 4-hydroxybenzoate, and DL-carnitine as sole

Table 5.31.
Characteristics differentiating the species of the genus **Serratia**[a]

Characteristics	1. S. marcescens	2. S. liquefaciens	3. S. plymuthica	4. S. rubidaea	5. S. odorifera	6. S. ficaria
Growth on[b] and acid production from:						
L-Arabinose, D-melibiose, D-xylose	−	+	+	+	+	+
Xylitol	+	−	−	−	+	+
D-Melezitose	−	+	+	db	−	+
L-Rhamnose	−	db	−	−	+	+
D-Sorbitol	+	+	db	−	+	+
D-Arabitol	−	−	−	+	−	+
Growth on:						
Betaine	−	−	db	+	−	−
Nicotinate	+	+	+	−	+	+
D-Tartrate	−	−	−	db	+	−
Trigonelline	db	−	−	+	+	+
Prodigiosin production	db	−	d	+	−	−
Musty odor	−	−	−	d	+	+
Good growth at 4°C	−	+	+	−	+	+
Indole production	−	−	−	−	+	−
Tetrathionate reduction	db	+	−	−	−	−
Lysine decarboxylase (Møller)	+	+	−	db	+	−
Ornithine decarboxylase (Møller)	+	+	−	−	db	−
β-Xylosidase	−	−	db	+	+	v
Tween 80 hydrolysis	+	+	+	+	−	+
Chitin hydrolysis	+	d	+	−	−	+
Gas from glucose	−	+	d	−	−	−

[a] Data from Grimont et al. (1977, 1978, 1979). For symbols see standard definitions; also db, test differentiates biovars.
[b] Utilization as sole carbon source.

Table 5.32.
Other characteristics of the species of the genus **Serratia**[a]

Characteristic	1. S. marcescens	2. S. liquefaciens	3. S. plymuthica	4. S. rubidaea	5. S. odorifera	6. S. ficaria
Growth on[b]:						
Acetate	+	d	d	d		+
trans-Aconitate	+	db[c]	d	+	+	+
Adonitol	+	−	−	+	+	+
β-Alanine	+	d	d	−		
Anthranilate	d	−	−	d		
DL-Arginine	−	−	−	d		
Benzoate	db	db	d	+	−	+
Benzylformate	−	−	d	d		
2,3-Butanediol	−	−	d	−		
Caprate	+	+	d	+		
Caproate	+	+	d	+		
Caprylate	+	+	+	+	+	+
DL-Carnitine	d	d	−	−	−	−
D-Cellobiose	−	−	+	+	+	+
meso-Erythritol	db	db	−	+	db	+
Ethanol	d	d	d	d		
D-Glucosamine	+	+	d	+	+	+
Glycerate	+	+	d	+		
Glycerol	+	+	d	+		+
Heptanoate	+	−	d	−		
Hippurate	d	−	−	d		
Histamine	−	−	−	d	−	−
3-Hydroxybenzoate	db	−	−	−	−	−
4-Hydroxybenzoate	db	d	d	−	−	d
3-Hydroxybutyrate	d	−	−	−		
Inulin	−	−	d	−		
α-Ketoglutarate	d	+	d	+		
Lactate	+	+	d	+		
D-Lactose	−	d	+	+	+	d
D-Malate	d	db	d	d		

Table 5.32.—*continued*

Characteristics	1. S. marcescens	2. S. liquefaciens	3. S. plymuthica	4. S. rubidaea	5. S. odorifera	6. S. ficaria
α-Methylglucoside	−	+	d	+	−	+
Mucate	−	−	d	+	+	+
L-Ornithine	d	d	−	−	db	−
Pelargonate	+	d	d	+		
Phenylacetate	+	+	d	+		
L-Phenylanine	d	d	d	+		
Propionate	+	−	d	−		
Quinate	db	db	+	+	−	+
Raffinose	−	+	+	+	db	+
Salicin	+	db	+	+		
Sarcosine	−	−	d	+		
Sucrose	+	+	+	+	db	+
L-Tartrate	−	−	−	d		
meso-Tartrate	d	db	−	−		
L-Tryptophan	d	d	d	+		
Valerate	d	−	−	−		
Growth at 37°C	+	+	d	+	+	+
Growth at 40°C	+	−	−	d	+	
Growth in NaCl:						
7% (w/v)	+	d	d	+		
8.5% (w/v)	d	d	−	+		
10% (w/v)	−	−	−	d		
Tetrathionate reduced	db	+	−	−	−	−
Methyl red test	−	d	d	−	+	−
H₂S from cysteine	+	+	+	d		+
Arginine decarboxylase (Møller)	−	db	−	−	−	−
Malonate test	−	−	−	db	−	−
Tween 40 hydrolysis	+	+	+	+	+	+
Tween 60 hydrolysis	+	+	+	+	d	+
Lecithinase (turbidity)	+	d	d	d		
Esculin hydrolysis	+	db	+	+	+	+
Growth on colistimethate (10 and 100 μg/ml)	+	+	d	d		
Acid produced from:						
Adonitol	v	−	−	+	v	+
myo-Inositol	d	+	d	d	+	+
Lactose	−	−	d	+	+	d
Raffinose	−	+	+	+	db	+
Salicin	+	db	+	+	+	+
Sucrose	+	+	+	+	db	+

[a] Data from Grimont et al. (1977b, 1978a, 1979b). For symbols see standard definitions; also db, test differentiates biovars.

[b] Carbon source utilization tests.

carbon sources, has been described (Grimont and Grimont, 1978b). Groups of biovars (called biogroups) (Table 5.33) correspond to definite, non-overlapping sets of serovars (Table 5.34) (Grimont et al., 1979a).

Nonpigmented biogroups A3 and A4 are ubiquitous. Nonpigmented biogroups A5/8 and TCT are almost confined to hospitalized patients. Pigmented biogroups A1 and A2/6 are found in the natural environment and occasionally in human patients.

The mol% G + C of the DNA is 57.5 to 60 (T_m, Bd).

Type strain: ATCC 13880.

2. Serratia liquefaciens (Grimes and Hennerty 1931) Bascomb, Lapage, Willcox and Curtis 1971, 293.[AL] (*Aerobacter liquefaciens* Grimes and Hennerty 1931, 93).

li.que.fa′ciens. M.L. part. adj. *liquefaciens* dissolving.

The cell morphology and colonial morphology are as given for the genus. Prodigiosin is not produced.

Physiological and nutritional characteristics are presented in Tables 5.31 and 5.32.

Several biovars can be recognized (Table 5.35) Grimont et al., 1977b: and unpublished data). The present species *S. liquefaciens* is formed of at least three genospecies: one corresponds to biovar Clab (including the type strain of *S. liquefaciens*); another corresponds to biovars Clc (including the type strain of *S. proteamaculans*), EB, RB and RQ; and a third one corresponds to biovars Cld and Adc.

S. liquefaciens is the most prevalent *Serratia* species in the natural environment (plants, digestive tract of rodents). Occasionally encountered as an opportunistic pathogen.

The mol% G + C of the DNA is 53 to 54 (T_m, Bd).

Type strain: ATCC 27592.

The present species is biochemically and genetically heterogeneous. Splitting the present species into three species, *S. liquefaciens* sensu stricto, *S. proteamaculans*, and a new species, *S. grimesii*, can be anticipated.

3. Serratia plymuthica (Lehmann and Neumann 1896) Breed, Murray and Hitchens 1948, 481. (*Bacterium plymuthicum* (sic) Leh-

Table 5.33.

Identification of **S. marcescens** *biogroups and biovars[a]*

Characteristics	Biogroup						
	A1	A2/6	A3	A4	A5/8	TCT	TC
Growth on[b]:		•					
meso-Erythritol	+	+	+	+	−	−	−
Benzoate and hippurate	+	−	−	−	−	−	−
Quinate and 4-hydroxy-benzoate	−	db[c,d]	−	db[e]	+	−	−
3-Hydroxybenzoate	−	−	db[f]	−	db[g]	−	−
Trigonelline	−	db[h]	db[i]	−	+	+	−
DL-Carnitine	db[j]	d	d	+	db[k]	db[l]	+
Tetrathionate reduction	+	+	+	−	+	+	+
Prodigiosin production	+	+	−	−	−	−	−

[a] For symbols see standard definitions.
[b] Carbon source utilization test.
[c] db, test differentiates biovars.
[d] Positive for biovars A6, negative for A2a and A2b.
[e] Positive for biovar. A4a, negative for A4b.
[f] Positive for biovars A3a and A3b, negative for A3c and A3d.
[g] Positive for biovar A8b, negative for A5 and A8a.
[h] Positive for biovar A2b, negative for A2a and A6.
[i] Positive for biovars A3b and A3d, negative for A3a and A3c.
[j] Positive for biovar A1a, negative for A1b.
[k] Positive for biovar A5, negative for A8a and A8b.
[l] Positive for biovar TCT, negative for biovar TT.

Table 5.34.

Correspondence between serovars and biogroups in **S. marcescens**[a]

Biogroup	O:H Serovars[b]
A1	5:2, 5:3, 5:13, 5:23, 10:6, 10:13
A2/6	6,14:2, 6,14:3, 6,14:8, 6,14:9, 6,14:10, 6,14:13, 8:3, 13:5
A3	3:5, 3:11, 4:9, 4:18, 5:6, 5:15, 6,14:5, 6,14:6, 6,14:20, 9:11, 9:17, 12:5, 12:9, 12:11, 12:17, 12:20, 13:11, 13:17, 15:3, 15:5, 15:8, 15:9, 17:4, 18:21
A4	1:1, 1:4, 2:1, 2:8, 3:1, 4:1, 4:4, 5:1, 5:6, 5:8, 5:24, 9:1, 13:1, 13:13
A5/8	3:12, 3,21:12, 4:12, 5:4, 6,14:4, 6,14:12, 8:12, 15:12, 21:12
TCT	1:7, 2:7, 4:7, 5:7, 5:19, 7:23, 11:4, 13:7, 13:12, 16:19, 18:9, 18:16, 19:14
TC	10:8, 20:12

[a] Data from Grimont et al. (1979a), and unpublished data.
[b] Serovars for which exceptions to the correspondence occur are in italics.

mann and Neumann 1896, 264,) *Note: S. plymuthica* is cited on the Approved Lists of Names as *Serratia plymuthica* (Dyar 1895) Bergey, Harrison, Breed, Hammer and Huntoon 1923, 88. This is incorrect, for reasons discussed by Grimont et al. (1977b).

ply. mu'thi.ca. M.L. adj. *plymuthica* pertaining to Plymouth.

The cell morphology and colonial morphology are as given for the genus. Prodigiosin is produced by most strains.

Physiological and nutritional characteristics are presented in Tables 5.31 and 5.32.

Most *S. plymuthica* strains studied were isolated from fresh water. Very rarely found in human sputum. No human infection reported.

The mol% G + C of the DNA is 53.5–56.5 (T_m).

Type strain: ATCC 183.

4. **Serratia rubidaea** (Stapp 1940) Ewing, Davis, Fife and Lessel 1973, 224.[AL] (*Bacterium rubidaeum* Stapp 1940, 259; *Serratia marinorubra* Zobell and Upham 1944, 255.)

Table 5.35.

Identification of **S. liquefaciens** *biovars[a]*

Characteristics	S. liquefaciens biovars						
	Clab	Clc	Cld	EB	RB	RQ	Adc
Growth on[b]:							
trans-Aconitate	−	+	−	+	−	+	+
Adonitol	−	−	−	+	−	−	−
Benzoate	−	−	+	+	+	−	−
meso-Erythritol	−	−	−	+	−	−	−
D-Malate	+	−	d	−	−	d	d
Quinate	−	−	−	−	−	+	−
L-Rhamnose	−	−	−	−	+	d	−
meso-Tartrate	+	−	−	−	−	d	−
Arginine decarboxylase (Møller)	−	−	+	−	−	−	+
Tetrathionate reduction	+	+	+	+	+	d	+
Esculin hydrolysis	+	+	+	+	+	−	+

[a] For symbols see standard definitions.
[b] Carbon source utilization tests.

ru.bi'dae.a. from the Latin name *Rubus idaeus* (raspberry), contracted and made to agree in gender with *Serratia*.

The cell morphology and colonial morphology are as given for the genus. Prodigiosin is produced by most strains.

Physiological and nutritional characteristics are presented in Tables 5.32 and 5.33.

S. rubidaea strains are rarely isolated, both in the natural environment and in human patients. May be found in ripe coconuts (P. A. D. Grimont, F. Grimont and M. P. Starr, unpublished data).

The mol% G + C of the DNA is 53.5–58.3 (T_m).

Type strain: ATCC 27593.

5. **Serratia odorifera** Grimont, Grimont, Richard, Davis, Steigerwalt and Brenner 1978, 461.[AL]

o.do.ri.fe'.ra. M.L. fem. adj. *odorifera*, bringing odors, fragrant.

The cell morphology and colonial morphology are as given for the genus. Prodigiosin is not produced. Cultures give off a musty, potato-like odor.

Physiological and nutritional characteristics are presented in Tables 5.31 and 5.32.

The capsular antigen reacts with *Klebsiella* antisera K4 or K68.

Rare opportunistic pathogen. Occasionally isolated from plants or food.

The mol% G + C of the DNA is 54.6 (T_m).

Type strain: ATCC 33077.

6. **Serratia ficaria** Grimont, Grimont and Starr 1981, 216.[VP] (Effective publication: Grimont, Grimont and Starr 1979, 282.)

fi.ca'ri.a. M.L. fem. adj. *ficaria* of figs.

The cell morphology and colonial morphology are as given for the genus. Prodigiosin is not produced. Cultures give off a musty, potato-like odor.

Physiological and nutritional characteristics are presented in Tables 5.31 and 5.32.

Associated with the fig/fig-wasp biological cycle. Occasionally found on plants other than fig trees.

The mol% G + C of the DNA is 59.6 (T_m).

Type strain: ATCC 33105.

Species Incertae Sedis

a. **Serratia fonticola** Gavini, Ferragut, Izard, Trinel, Leclerc, Lefebvre and Mossel 1979, 98.[AL]

fon.ti'co.la. M.L. n. *fons, fontis* spring, fountain; L. suffix *-cola* dweller; M. L. noun *fonticola* spring-dweller.

Rod-shaped cells, described as being 0.5 × 30 μm; the latter value, however, is probably a misprint. Gram-negative. Motile by peritrichous flagella.

Conform to the definition of the family *Enterobacteriaceae*, do not conform to the present definition of the genus *Serratia*.

Growth occurs between 4°C and 37°C. No growth at 41°C.

The following biochemical tests are positive: Simmons' citrate, malonate test, lysine and ornithine decarboxylase, tetrathionate reduction, Tween 80 esterase, β-galactosidase, β-xylosidase, and esculin hydrolysis; fermentation of adonitol, L-arabinose, D-dulcitol, D-fructose, D-galactose, D-glucose, D-glycerol, *myo*-inositol, lactose, maltose, D-mannitol, D-mannose, melibiose, α-methylglucoside, raffinose, L-rhamnose, D-ribose, salicine, D-sucrose, D-trehalose.

The following biochemical tests are negative: arginine decarboxylase, indole, H₂S, DNase, gelatinase, Voges-Proskauer, phenylalanine, deaminase, urease, fermentation of inulin, D-melezitose, and L-sorbose.

The following compounds can serve as sole carbon source: adonitol, D-alanine, L-alanine, L-arabinose, L-aspartate, citrate, D-dulcitol, *meso*-erythritol, D-fructose, D-galactose, gluconate, D-glucose, L-glutamate, DL-glycerate, D-glycerol, L-histidine, *myo*-inositol, DL-lactate, D-maltose, D-mannitol, D-mannose, L-proline, putrescine, pyruvate, L-rhamnose, D-ribose, salicin, D-sorbitol, succinate, tagatose (upublished data), and D-trehalose.

The following compounds cannot serve as sole carbon source: adipate, β-alanine, 2-aminobenzoate, 4-aminobenzoate, DL-2-aminobutyrate, DL-3-aminobutyrate, DL-4-aminobutyrate, 5-aminovalerate, amylamine, D-arabinose, L-arginine, azelate, benzoate, benzylamine, benzylformate, betaine, 2,3-butanediol, butylamine, butyrate, caprate, *n*-caproate, D-cellobiose, citraconate, L-citrulline, creatine, diphenylamine, dodecane, ethanol, ethanolamine, ethylene glycol, D-fucose, L-fucose (unpublished data), geraniol, glutarate, glycine, glycolate, hexadecane, heptanoate, hippurate, histamine, 4-hydroxybenzoate, DL-3-hydroxybutyrate, inulin, isobutyrate, L-isoleucine, isophthalate, isopropanol, isovalerate, itaconate, 2-ketoglutarate, L-leucine, levulinate, L-lysine, D-mandelate, L-mandelate, mesaconate, methanol, L-methionine, naphthalene, nicotinate, oxalate, pantothenate, pelargonate, phenol, phthalate, pimelate, *n*-propanol, proponiate, propylene glycol, salicylate, sarcosine, sebacate, spermine, suberate, D-sucrose, D-tartrate, L-tartrate, *meso*-tartrate, terephthalate, L-threonine, trigonelline, tryptamine, D-tryptophan, urate, urea, *n*-valerate, and L-valine (data from Gavini et al., 1979).

Occur in fresh water.

The mol% G + C of the DNA is 48.8–52.5 (T_m).

Type strain: ATCC 29844.

Genus IX. *Hafnia* Møller 1954, 272^{AL}

RIICHI SAKAZAKI

Haf'ni.a. O.L. fem. n. *Hafnia* the old name for Copenhagen.

Straight rods, ~1.0 μm in diameter and 2.0–5.0 μm in length. Conform to the general definition of the family *Enterobacteriaceae*. Not encapsulated. Gram-negative. **Motile by peritrichous flagella at 30°C**, but nonmotile strains may occur. Facultatively anaerobic, having both a respiratory and a fermentative type of metabolism. Grow readily on ordinary media. Colonies on nutrient agar are generally 2–4 mm in diameter, smooth, moist, translucent, and gray with a shiny surface and entire edge. Oxidase-negative. Catalase-positive. Chemoorganotrophic. **The majority of strains utilize citrate, acetate and malonate as a sole carbon source** after 3–4 days of incubation. Nitrate is reduced to nitrite. H₂S is not produced in the butt of Kligler iron agar. Gelatinase, lipase, and deoxyribonuclease are not produced. Alginate is not utilized. Pectate is not decomposed. Phenylalanine deaminase is not produced. **Lysine and ornithine decarboxylase tests are positive**, but the arginine dihydrolase test is negative. Glucose is fermented with the production of acid and gas. **Acid is not produced from D-sorbitol, raffinose, melibiose, D-adonitol and *myo*-inositol.** The methyl red test is usually positive at 35°C and negative at 22°C. Acetylmethylcarbinol is usually produced from glucose at 22–28°C but may not be produced at 35°C. Occur in the feces of man and other animals including birds; also occur in sewage, soil, water and dairy products. The mol% G + C of the DNA is 48–49 (T_m).

Type species: *Hafnia alvei* Møller 1954, 272.

Further Descriptive Information

Members of *Hafnia* are able to grow at 35°C, but many of their physiological and biochemical activities at this temperature are irregular. Many strains are nonmotile at 35°C, but the majority are motile at 25–30°C. Although most strains do not produce acetylmethylcarbinol from glucose at 35°C, they give a positive Voges-Proskauer reaction when incubated at 22–28°C. At 25°C, they produce gas from glucose and about 10% of them grow on Simmons' citrate agar within 24 h, but all of these reactions may be negative at 35°C.

Lactose is not fermented, but plasmid-mediated lactose-positive strains may occur (Le Minor and Coynault, 1976).

Hafnia is defined as an H₂S-negative organism. Møller (1954) and Kauffmann (1954) reported *Hafnia* as producing H₂S since most strains of *Hafnia alvei* slightly darken ferric chloride-gelatin medium (Kauff-

mann, 1951) and SIM medium (Difco), as well as peptone iron agar (Difco). They fail, however, to blacken the butt of Kligler iron agar and of triple-sugar iron agar. Ewing (1960) suggested that either Kligler iron agar or triple-sugar iron agar must be a standard medium for the H₂S test of the family *Enterobacteriaceae*, because each permits easy differentiation of genera or species within the family.

The maximum temperature for growth is usually 40–42°C. No growth occurs at 5°C.

The serology of *Hafnia* was first studied by Stuart and Rustigian (1943) who divided their cultures of biotype 32011, the majority of which are now classified into *Hafnia*, into eight serovars. Eveland and Faber (1953) studied 58 strains of biotype 32011 serologically and reported 21 somatic and 22 flagellar antigens. Deacon (1952) also carried out a serological study on 17 cultures of "*Aerobacter cloacae*" including biotype 32011 and recognized 12 somatic and 6 flagellar antigens among the cultures. However, Sakazaki and Namioka (1957) and Sakazaki (1961) found that cultures of biotype 32011 studied by those authors mentioned above included not only *Hafnia* but also *Enterobacter cloacae*. Serological studies on 294 biochemically well-defined *Hafnia* cultures were performed by Sakazaki (1961) who established an antigenic schema of *Hafnia* consisting of 29 O groups and 23 H antigens. Later, Matsumoto (1963, 1964) expanded this schema to 68 O groups and 34 H antigens. Deacon (1952) reported the diaphasic variation in the H antigens of the strains he studied, but Sakazaki (1961) and Matsumoto (1963) failed to observe such variation. Some *Hafnia* strains may be O-inagglutinable with their homologous O antisera in unheated cultures. Sakazaki (1961) suggested that the antigen that inhibited the O-agglutination was a slime antigen. The alpha antigen (Stamp and Stone, 1944) may be recognized in some strains (Sakazaki, 1961; Emslie-Smith, 1961). In addition to this, intergeneric relationships of O antigens were recognized between *H. alvei* and other genera of the family *Enterobacteriaceae* (Sakazaki, 1961; Matsumoto, 1963, 1964; Sedlák and Slajsová, 1966). Eveland and Faber (1953) reported O antigenic relationships between *Hafnia* (biotype 32011) and *Salmonella*.

Baturo and Raginskaya (1978) have recently published an antigenic schema including 39 O and 35 H antigens of *H. alvei*, independent of that of previous investigators.

Table 5.36.

Differential characteristics of the genus **Hafnia** *and biochemically similar genera*[a]

Characteristics	Hafnia	Enterobacter	Serratia
Citrate (Simmons')	−[b]	+	+
Gelatin hydrolysis	−	D	+
Lysine decarboxylase	+	D	D
Arginine dihydrolase	−	D	−
Lipase (Tween 80)	−	−	+
Deoxyribonuclease	−	D	+
Acid from carbohydrates:			
Raffinose, sucrose	−	+	D
Lactose, D-adonitol, myo-inositol, D-sorbitol	−	D	D
Hafnia specific bacteriophage lysis[c]	+	−	−
Mol% G + C of DNA	48–49	52–60	52–60

[a] Symbols: +, 90–100% of strains are positive; −, 90–100% of strains are negative; D, different reaction given by different species of a genus.
[b] Late positive reactions are given by ~50% of the strains of *Hafnia*.
[c] Guinée and Valkenburg (1968).

The majority of strains of *H. alvei* are susceptible to carbenicillin, streptomycin, tetracycline, polymyxin B, and nalidixic acid, but resistant to cephalosporins and ampicillin.

A *Hafnia*-specific bacteriophage that provides a reliable tool for the identification of *Hafnia* strains was described by Guinée and Valkenburg (1968).

H. alvei occurs not only in man and animals and birds, but also in natural environments such as soil, sewage and water. In medical bacteriology, *H. alvei* is found in clinical specimens, especially from feces in healthy humans, occasionally from blood, sputum, urine, and from wounds, abscesses, the throat, abdominal cavity and autopsies. In most cases, however, they are found in mixed culture and seem to be opportunistic pathogens which produce infections in patients with some underlying illness or predisposing factors.

H. alvei has been reported as a possible causative agent of intestinal disorders by some investigators. However, no conclusive evidence has been obtained on its enteropathogenicity. Matsumoto (1963) reported the isolation of this organism from 13% of stool specimens from apparently healthy individuals. Sakazaki (1966, unpublished data) found *H. alvei* in 42% of fecal samples of healthy persons.

Enrichment and Isolation Procedures

Hafnia can grow on less-selective isolation media for enterobacteria such as eosin-methylene blue, deoxycholate-lactose, MacConkey, xylose-lysine-deoxycholate and Hektoen enteric agars. The majority of *Hafnia* strains may also grow on salmonella-shigella and deoxycholate-citrate agars. Colonies of *H. alvei* on these plating agar media are of

Table 5.37.

Characteristics of **Hafnia alvei**[a]

Characteristics	H. alvei
Indole production	−
Voges-Proskauer test (22°C)	+
Voges-Proskauer test (35°C)	d
Citrate (Simmons') (22°C)	d
Citrate (Simmons') (35°C)	−
H₂S (triple-sugar iron agar)	−
Urease (Christensen)	−
Gelatin hydrolysis	−
Phenylalanine deaminase	−
Lysine decarboxylase	+
Arginine dihydrolase	−
Ornithine decarboxylase	+
Growth in KCN medium	+
Malonate utilization	d
Esculin hydrolysis	−
Lipase (Tween 80)	−
Deoxyribonuclease	−
ONPG hydrolysis[b]	d
Gas from glucose	+
Acid from carbohydrates:	
D-Glucose, L-arabinose, maltose, L-rhamnose, trehalose, D-xylose, D-mannitol, glycerol	+
Lactose, melibiose, raffinose, sucrose,[c] D-adonitol, dulcitol, D-sorbitol, myo-inositol, mucate	−
Salicin	d
d-Tartrate (Kauffmann-Petersen)	−

[a] For symbols see standard definitions.
[b] ONPG, o-nitrophenyl-β-D-galactopyranoside. This test is generally positive especially if it is carried out from a culture incubated at 22°C.
[c] Late positive reactions are given by ~50% of the strains of *Hafnia*.

colorless and translucent and resemble those of *Salmonella* (*Hafnia* strains are sometimes misidentified as *Salmonella* H₂S-negative) but rare strains may produce red or pink colonies on media which contain sucrose. Sakazaki (1966, unpublished data) devised a differential isolation medium, deoxycholate-lactose-sucrose-sorbitol agar.*

There are no selective enrichment broth media for the isolation of *H. alvei*. Some strains fail to grow in selenite and tetrathionate broths.

Maintenance Procedures

Stock cultures may be maintained at room temperature in a semisolid medium consisting of 1.0% Bacto-casitone (Difco), 0.3% yeast extract, 0.5% NaCl and 0.3% agar, pH 7.0. *Hafnia* strains remain viable up to a year without subculture if the culture is sealed with a rubber stopper or a cork which has been soaked in hot paraffin wax. Strains may also be preserved indefinitely by lyophilization.

Differentiation of the genus **Hafnia** from other genera

Table 5.36 indicates the characteristics of *Hafnia* that differentiate it from biochemically similar genera.

Taxonomic Comments

The bacteria of the genus *Hafnia* have been described under several names. Møller (1954) found a new group of organisms, in which a supposedly authentic strain of *Bacillus paratyphi-alvei* of Bahr (1919) was included. He proposed the name *Hafnia alvei* for this bacterial

group, because he considered that Bahr's strain ought to be regarded as the type of this group. Sakazaki (1961) suggested a new combination *Enterobacter alvei* for *H. alvei*, because of its biochemical similarity to *Enterobacter*. Ewing and Fife (1968) pointed out that Bahr's strain, which had been designated as the type strain of *H. alvei* by Møller (1954), was not an authentic strain of this species, since biochemical reactions of the strain were not the same as those described by Bahr (1919). They considered therefore that the specific epithet *alvei* was

---

* Deoxycholate-lactose-sucrose-sorbitol agar (per liter of distilled water): yeast extract, 5.0 g; trypticase (BBL), 5.0 g; lactose, 10.0 g; sucrose, 5.0 g; D-sorbitol, 10.0 g; sodium deoxycholate, 2.5 g; sodium citrate, 20.0 g; ferric citrate, 1.0 g; neutral red, 0.02 g; agar, 15.0 g. The medium is adjusted to pH 7.4.

illegitimate, and proposed the name *Enterobacter hafniae* for *H. alvei.* However, *Hafnia alvei* Møller 1954 is the only correct name for this group of bacteria, because there is no doubt that the Bahr's strain studied by Møller (1954) was a new bacterium at that time. In addition, numerical taxonomy studies by Johnson et al. (1975) and Gavini et al. (1976) indicated that *Hafnia* strains occupy a position separate from *Enterobacter.* In DNA/DNA hybridization studies, Steigerwalt et al. (1976) reported only 11–26% homology between *H. alvei* and *Enterobacter.*

Only a single species, *Hafnia alvei,* has been designated. Steigerwalt et al. (1976) indicated that *H. alvei* consists of two DNA relatedness

groups, but these two groups have not been defined biochemically.

Priest et al. (1973) proposed that *Obesumbacterium proteus* Shimwell 1964, a common brewery contaminant, should be placed in the genus *Hafnia* as *H. protea.* They described two groups in this species by numerical analysis of biochemical and physiological characteristics. Brenner (1979, personal communication) determined DNA relatedness in both groups and found that one group appears to be a biovar of *H. alvei,* whereas the other group is a new species that does not belong to the genus *Hafnia* (see article on "Other Genera of the Family *Enterobacteriaceae*").

List of the species of the genus **Hafnia**

1. **Hafnia alvei** Møller 1954, 272.[AL]

al've.i. L. n. *alveus* a beehive; L. gen. n. *alvei* of a beehive.

The morphology is as given for the genus. Motility is most pronounced at 30°C and often absent at 37°C. Nonmotile strains may be encountered occasionally. Capsules are usually not present.

Grows readily on ordinary media. Colonies are translucent. Rare strains may produce mucoid colonies. The majority of strains grow on

salmonella-shigella agar.

Physiological and biochemical characteristics are presented in Tables 5.36 and 5.37.

Found in the feces of man and other animals, including birds. Also found in sewage, soil, water and daily products.

The mol% G + C of the DNA is 48.0–48.7 (T_m).

Type strain: NCTC 8106 (ATCC 13337).

Genus X. **Edwardsiella** *Ewing and McWhorter 1965, 37*[AL]

JOHN J. FARMER III AND ALMA C. MCWHORTER

("Asakusa group" Sakazaki and Murata 1962, 616; "Bartholomew group" King and Adler 1964, 230; "Bacterium 1483-59" Ewing et al., 1965, 33.)

Ed.ward.si.el'la. M.L. dim. ending *ella*; M.L. fem. n. *Edwardsiella*; named after P. R. Edwards (1901–1966), the American bacteriologist who was chief of the Enteric Laboratories, Centers for Disease Control, U.S.A., from 1948–1962 and made many contributions to our knowledge of the *Enterobacteriaceae* (Cherry and Ewing, 1966).

Small straight rods, about 1 μm in diameter × 2–3 μm conforming to the general definition of the family *Enterobacteriaceae.* Gram-negative. **Motile** by peritrichous flagella. Facultatively anaerobic. Catalase-positive. Oxidase-negative. Reduce nitrate to nitrite. Optimum temperature, 37°C, except for *E. ictaluri* which prefers a lower temperature. Growth occurs on peptone and similar agar media with small colonies (~0.5–1 mm in diameter) after 24 h incubation. **Vitamins and amino acids are required for growth.** Ferment D-glucose with the production of acid and often visible gas. Also ferment a few other compounds **but are inactive compared to many taxa in the family *Enterobacteriaceae.* Usually resistant to colistin but have large zones around most other antibiotic disks, including penicillin.** Frequently isolated from cold-blooded animals and their environment, particularly fresh water. Pathogenic for eels, catfish, and other animals, sometimes causing economic losses; also a rare opportunistic pathogen for humans. The mol% G + C of the DNA is 53–59 (T_m, Bd).

Type species: *Edwardsiella tarda* Ewing and McWhorter 1965, 37.

Further Descriptive Information

Most of the available information concerns *E. tarda* since the other two species of *Edwardsiella* have been described only recently. *Edwardsiella* strains grow less luxuriantly than many other *Enterobacteriaceae* and form smaller colonies in 24 h at 36°C. This may be related to their growth requirements. d'Émpaire (1969) reported that *E. tarda* requires cysteine, methionine and nicotinamide. Hoshina (1962) stated that "*Paracolobactrum anguillimortiferum*," an organism now thought to have been a *Edwardsiella (E. anguillimortifera),* required niacin, phenylalanine, threonine and valine, and that aspartic acid, glutamic acid, isoleucine and cysteine were "important for growth." Other groups of *Edwardsiella* probably have similar nutritional requirements.

E. ictaluri is the most fastidious species of the genus. Growth is very slow on plating media and 2–3 days of incubation are often required for colonies to reach 1 mm in diameter. Although characteristic biochemical reactions are apparent at 36°C (see Tables 5.38 and 5.39), a lower temperature seems to be preferred (Hawke, 1979). Biochemically, *E. ictaluri* is also the least active of the *Edwardsiella* species.

Two independent serotyping schemes have been described for *E. tarda.* Sakazaki (1967) recognized 17 O antigens, 11 H antigens, and 18 O-H combinations. Edwards and Ewing (1972) described a scheme with 49 O antigens, 37 H antigens, and 148 O-H combinations among 394 cultures studied. Currently, efforts are being made to standardize the schema for serotyping *Edwardsiella* (R. Sakazaki and D. J. Brenner, personal communication). This schema will be a combination of the O and H antigens described by Ewing and McWhorter (Edwards and Ewing, 1972, p. 145) and those of Sakazaki (1967). Other typing techniques such as bacteriocin production or susceptibility, bacteriophage typing, and biotyping have seldom been used for *Edwardsiella,* although Hamon et al. (1969) did demonstrate bacteriocin production and sensitivity.

Many strains of *Edwardsiella* have high-level intrinsic resistance to colistin (Muyembe et al. 1973), but some strains have small zones of inhibition around colistin-impregnated disks. All three species have large zones around penicillin-impregnated disks, an unusual finding for members of *Enterobacteriaceae.* They also have large zones of inhibition around most other antibiotics. Occasionally resistance to sulfonamides or other drugs has been observed in *E. tarda.* Antibiotic resistance that is mediated by R plasmids (R factors) is very rare in *Edwardsiella* (no examples were encountered in our survey of the literature), which suggests that human contact with this genus of organisms is rare.

When *E. tarda* was first described, it was thought to be a possible cause of diarrhea (Ewing et al., 1965). Some intriguing evidence later came from a study of the Orang Asli, a group of jungle-dwelling natives of West Malaysia (Gilman et al., 1971). There were 29 isolates of *E. tarda* from stool cultures of 208 patients hospitalized with blood diarrhea but only one isolate from 120 stool cultures of control individuals (hospital patients without diarrhea). An interesting relationship between *E. tarda* and the protozoan *Entamoeba histolytica* was also shown. Twenty-five of the patients with bloody diarrhea had both organisms, and 4 had *E. tarda* only. Twenty-four of the 25 patients with both organisms had significant antibody titers to a whole-cell antigen of *E. tarda,* whereas a control group of 15 patients was negative. All of the patients who were culture-negative for *E. tarda* but positive for *Entamoeba histolytica* also had antibodies to *E. tarda*. These data indicate that *E. tarda* may be involved in the pathogenesis of amoebic dysentery, although an alternate explanation is that the presence of *E. tarda* is due to a change in the gut micro-environment and that the organism plays no role in diarrhea. Makulu et al. (1973) also found an association between *E. tarda* and *Entamoeba histolytica* in patients from Zaire with bloody diarrhea, but the correlation was lower than in the previous study. They postulated a possible triggering role of *E. tarda* in initiating invasive amoebic infection.

E. tarda is rarely present in the feces of healthy people. Onogawa et al. (1976) in Japan found only one positive culture from 97,704 food handlers and only 25 positive cultures from 255,896 school children. Makulu et al. (1973) found no positive cultures among 841 healthy subjects in Zaire. Several studies indicate that the number of *E. tarda* isolations depends upon the methods used in processing stool cultures, the geographic area of the study, and the season in which the survey is done (Iveson, 1973). These variables have not always been considered by those trying to determine the relative incidence of *E. tarda* in patients with diarrhea and in controls. A higher isolation rate has invariably been found among the diarrhea patients (Bhat et al., 1967; Ewing et al., 1965; Gilman et al., 1971; Makulu et al., 1973; Nguyen-Van-Ai et al., 1975). Some strains of *E. tarda* may be able to cause diarrhea, particularly in underdeveloped countries, but *E. tarda* should not be considered as an "inherent pathogen," a status given to *Salmonella* and *Shigella*. The role of *Edwardsiella* in diarrhea needs further study. One promising technique is to test a patient's acute-phase and convalescent-phase sera against the particular strain of *Edwardsiella* isolated from feces. Chatty and Gavan (1968) reported a case in which *E. tarda* was isolated from a patient with nutritional cirrhosis of the liver, diarrhea, and low-grade fever. The person had lived in Central and South America. A convalescent-phase serum from the patient had an antibody titer of 1:160 to both somatic and flagella antigens of the *E. tarda* strain isolated from feces. In this case *E. tarda* was incriminated as the probable cause of the diarrhea. Similar studies are needed for all isolates of *Edwardsiella* from stools of people with diarrhea and from healthy controls.

E. tarda is now well documented as an opportunistic pathogen, but it is rarely found in most industrialized countries. It seldom causes meningitis, endocarditis, bacteremia, or urinary tract infections but is often isolated from wounds (Jordan and Hadley, 1969). A typical example is the report of Chatty and Gavan (1968) of a boy who struck a submerged log while swimming in a lake. A splinter entered his right thigh and eventually led to gas gangrene, a diagnosis confirmed by the isolation of *Clostridium perfringens*. *E. tarda* was also isolated but probably only colonized the wound. Wound cultures have often yielded other bacteria in addition to *E. tarda*, so its role is difficult to assess. Antibody responses to the particular strain would be very useful in defining the role of *E. tarda* in these infections.

E. tarda has been isolated from many animals including pets (Nguyen-Van-Ai et al., 1975), domestic animals (Owens et al., 1974), animals in zoos (Otis and Behler, 1973), rats (Nguyen-Van-Ai et al., 1975), aquatic animals and birds (White et al., 1973), fish (Nguyen-Van-Ai et al., 1975), frogs (Bartlett et al., 1977), turtles (Otis and Behler, 1973), and marine animals (Nguyen-Van-Ai et al., 1975). It is also frequently found in the environment, particularly where these animals live (White et al., 1973). Most *E. tarda* isolates have come from stools or other specimens from healthy animals, but *E. tarda* can cause outbreaks of "red disease" in pond-cultured eels (Wakabayashi and Egusa, 1973) or of "emphysematous putrefactive disease" (gas-filled lesions in the muscles) of channel catfish (Meyer and Bullock, 1973). Isolated cases of septicemia have been reported in other animals (Chamoiseau, 1967).

Another *Edwardsiella* species, *E. hoshinae* is also associated with animals, but only eight isolates were originally reported (Grimont et al, 1980). Three were from monitor lizards (*Varanus* sp.) in Chad, two

from puffins (*Fratercula arctica*) in Brittany, France, one from a lizard in Senegal, one from a flamingo (*Phoenicopterus ruber*) in France and one from water. Two recent isolates were from feces of patients without diarrhea (R. Sakazaki, personal communication); thus there is no evidence that *E. hoshinae* can cause human disease. Another distinct group of *Edwardsiella* strains originally called *Edwardsiella* group "GA 7752" (Hawke, 1979) has caused many outbreaks of enteric septicemia of catfish. This new *Edwardsiella*, which was recently named *E. ictaluri* (Hawke et al., 1981), has been isolated from pond-raised catfish in the southeast, particularly the Mississippi delta area where catfish farming is most intense. The disease is seasonal, occurring almost exclusively in the spring and again in the fall when water temperatures are about 25°C, which seems to be the optimum growth temperature of *E. ictaluri* in the laboratory (Hawke, 1979). *E. ictaluri* has also been isolated from white catfish (*Ictalurus catus*) and the brown bullhead (*Ictalurus nebulosus*). No human isolates have been reported.

The natural resorvoir of *Edwardsiella* appears to be the intestine of animals, from which feces disseminate the organism into the environment. Most human infections caused by *E. tarda* probably result from contact with the organism in the environment. Endogenous human infections, although probably rare, may occur if gut carriage has been established.

Enrichment and Isolation Procedures

Very little information is available on the selective isolation of *Edwardsiella*, and most of what exists concerns *E. tarda*, the most common species. Little has been written about the other species because they were described only recently.

Most data on the isolation of *E. tarda* have come from culture surveys to detect *Salmonella* and *Shigella*. Unfortunately, there has been no systematic study to evaluate growth and survival of *E. tarda* in enrichments and on plating media commonly used in enteric bacteriology. *E. tarda* strains usually grow on plating media commonly used, including the following agar media: sheep blood, chocolate, MacConkey, SS (salmonella-shigella) and deoxycholate citrate. However, strains of *E. tarda* do grow more slowly than most other species of *Enterobacteriaceae*. Pure cultures grow on brilliant green and bismuth sulfite agar (Sakazaki, 1967), but Iveson (1973) found these two media useless in isolating *E. tarda* from feces.

Strains of *E. tarda* are often isolated from liquid enrichments such as tetrathionate and selenite F (media used to isolate *Salmonella*) and occasionally these enrichments have resulted in a higher yield than direct plating (Makulu et al., 1973). Iveson (1973) described an efficient method for isolating *E. tarda* from stool cultures. Specimens were first enriched (either at 37°C or 43°C) with strontium chloride B medium* (Iveson, 1971). After 24 h of incubation, plates of deoxycholate citrate agar were streaked. This method was excellent for isolating *Samonella*, *Arizona*, *Shigella*, and *E. tarda* from stool cultures and could presumably be adapted to all types of specimens including those from the environment.

E. tarda and often other *Edwardsiella* species (Farmer and Mc-Whorter, unpublished data) have intrinsic resistance to the polypeptide antibiotic colistin, and advantage can be taken of this for isolation procedures. Muyembe et al. (1973) showed that all *E. tarda* strains grew in the presence of 10 μg/ml of colistin and that more than 80% grew in 100 μg/ml. Other metabolic properties of *E. tarda* could be used in designing a differential and selective medium. Colistin could be added to peptone iron agar so that strains that grow and produce black colonies (because of H_2S production) would probably be *E. tarda* or H_2S-positive species of *Proteus*. A different approach would be to incorporate several carbohydrates (or related compounds) not fermented by *E. tarda* into a fermentation base such as MacConkey agar base without lactose (Difco), but with added colistin. Most *Enterobacteriaceae* would either be inhibited (only *Serratia*, *Proteus*, *Providencia*, *Morganella*, *Cedecea* and some *Yersinia* strains are colistin-resistant)

---

* Strontium chloride B medium (g/liter): Bacto-tryptone (Difco), 5.0; NaCl, 8.0; KH_2PO_4, 1.0; and $SrCl_2$, 34.0. "Sterilization" is done by heating at 100°C for 30 min (final pH, 5.0–5.5).

or form red colonies because they ferment one or more of the sugars. Colonies of *E. tarda* would be colorless. Various combinations of the carbohydrates used to differentiate the four *Edwardsiella* groups (Table 5.38) could be used to make a differential medium for one of the groups. Many other approaches are feasible which would combine colistin enrichment with a differential biochemical reaction. None of the above methods has actally been tried and only represent theoretical possibilities.

Maintenance Procedures

Edwardsiella strains survive well in the laboratory without transfer. Cultures are inoculated into 100 × 13 mm tubes containing a "peptone" medium (Trypticase soy agar slants, Trypticase soy semisolid (0.4% agar), or blood agar base slants with 0.3% added yeast extract) and are incubated overnight. The tubes are then tightly sealed with a stopper (white rubber, No. 000, for a 13 × 100 mm screw cap tube) or with paraffin-coated corks. It is essential that the seal be airtight so that the water in the medium does not evaporate because drying may kill the strain. Almost all cultures of *E. tarda* have remained viable for over 10 years without transfer with this storage method, but there has been little experience with the other *Edwardsiella* species.

In addition to those "working stocks," important cultures should also be preserved as "freezer stocks." Growth from a Trypticase soy agar plate is removed with cotton swab and a heavy suspension is made in sterile 10% w/w skim milk in water (or in sterile sheep or rabbit blood). This suspension is quick frozen in 95% alcohol (which is kept in −70°C freezer) or in a dry ice-acetone bath. Other workers prefer to put the skim milk suspension into the freezer directly so it is frozen slowly rather than quickly. The freezer stocks should be kept in the freezer at the lowest temperature available. Freeze-drying presumably can also be used for long term preservation.

Procedures for Special Testing

Indole production, method 1. This is the method described by Edwards and Ewing (1972) and is the "standard method" used by the Enteric Section for testing all cultures. A tube of peptone water (20 g of Bacto-peptone (Difco), 5 g of NaCl and 1000 ml of distilled water) is inoculated and incubated at 36°C for 48 h. About 0.6 ml of Kovacs' reagent (10 g of *p*-dimethylaminobenzaldehye, 50 ml of 12 N HCl, 150 ml of isoamyl alcohol) is then added. A positive test is the presence of a pink or red color in the upper layer.

Indole production, method 2. This is a more sensitive method which detects indole production by some strains which are indole-negative by method 1. Heart infusion broth (Difco) is inoculated, incubated for 48 h, and then tested with Kovacs' reagent as described above.

Differentiation of the genus **Edwardsiella** from other genera

There is no single test to differentiate *Edwardsiella*. The best method is to do a complete set of biochemical reactions. This will indicate that the culture is a member of *Enterobacteriaceae* and that it belongs to the genus *Edwardsiella*. *Edwardsiella* is more fastidious than many other *Enterobacteriaceae* and forms colonies in 24 h at 36°C which are smaller than those of most other *Enterobacteriaceae*. *Edwardsiella* is apparently more susceptible to 2,4-diamino-6,7-diisopropyl pteridine (vibriostatic compound "O/129", Sigma Chemical) than other *Enterobacteriaceae* (Chatelain et al., 1979, Grimont et al., 1980). Many groups of *Enterobacteriaceae* have zones of inhibition around disks impregnated with the antibiotic colistin but have no zone around penicillin. *Edwardsiella* strains usually have the opposite pattern. *Edwardsiella* is biochemically somewhat similar to *Escherichia coli*, the *Salmonella-Arizona* group and the *Proteus-Providencia-Morganella* group but is easily differentiated on the basis of a complete set of biochemical test results or on the basis of antibiotic susceptibility patterns. These phenotypic differences correlate with the phylogenetic divergence of *Edwardsiella* from these other groups.

Taxonomic Comments

Edwardsiella was discovered independently in 1959 by two research groups. It was called "Bacterium 1483-59" by Ewing and his coworkers at the Centers for Disease Control (Ewing et al. 1965). Almost all of their isolates were from human clinical specimens and most were from feces. In 1959 Sakazaki and coworkers (Sakazaki, 1967) independently discovered the same group of organisms isolated mainly from snakes. The name "Asakusa group" was coined by the Japanese workers (Sakazaki and Murata, 1962). King and Adler (1964) proposed the name "Bartholomew group" in 1964, but it was Ewing and colleagues who in 1965 coined the scientific name *Edwardsiella tarda*, which has standing in nomenclature.

When *Edwardsiella* was proposed, there was some doubt whether it deserved status as a separate genus, since there was only one species. It was even questioned whether *E. tarda* was a separate species. Cowan (and Steel, 1974, p. 105) makes the following statement: "*Edwardsiella* has much in common with some shigellae which, within themselves have differences comparable with those between *Escherichia* and *Edwardsiella*. In short, *Edwardsiella* is a good example of the excessive splitting at 'generic' level that has taken place within the enterobacteria. In our opinion it is better regarded as a biotype of *Escherichia coli*; less satisfactorily as a species, *Escherichia tarda*." However, Bren-

ner and coworkers (1974) showed that 20 strains of *E. tarda* from diverse sources and different countries were highly related by DNA/DNA hybridization (82–96% related at 60°C with small values for percent divergence, 81–93% related at 75°C; done by the hydroxyapatite method with ^{32}P). *Edwardsiella tarda* was only 8–29% related to other genera in the family *Enterobacteriaceae*, and was 17–25% related to *Escherichia coli*, the type species of the type genus for the family. These data argue convincingly that *Edwardsiella* should be maintained as a separate genus in the family *Enterobacteriaceae*. All of the *Edwardsiella* strains studied by Brenner et al. (1974) were highly related to each other and formed a single species. However, two new *Edwardsiella* species have been described in the last two years. *E. hoshinae* is distinct from *E. tarda* by DNA/DNA hybridization (S1 nuclease method; Grimont et al., 1980) and is phenotypically distinct from *E. tarda* and *E. ictaluri*. *E. ictaluri* is distinct from, but closely related to, *E. tarda* by DNA/DNA hybridization (hydroxyapatite method with ^{32}P; Hawke et al., 1981). These additional species now make *Edwardsiella* a much better "phylogenetic genus" and nicely counter the previous argument that there had been excessive splitting in establishing *Edwardsiella*.

A nomenclatural problem in *Edwardsiella* concerns the name *Edwardsiella anguillimortifera* (Hoshina 1962, Sakazaki and Tamura 1975). This name appears on the *Approved Lists of Bacterial Names* (Skerman et al., 1980, p. 292) with the type strain listed as ATCC 15947. This strain was proposed as the neotype strain (Sakazaki and Tamura, 1975) but has been challenged (J. J. Farmer III, 1976–1977, unpublished letters to P.H.A. Sneath, Chariman of the Judicial Commission) under rule 18e of the 1975 Bacteriological Code because it was considered a doubtful name (*nomen dubium*) and because the properties of the proposed neotype strain differed from those given in the original description of "*Paracolobactrum anguillimortiferum*" Hoshina 1962 (see Table 5.40). Furthermore, no other strains have been isolated which fit the original description of "*P. anguillimortiferum*." It could be argued that ATCC 15947 became established as the type strain of "*P. anguillimortiferum*" with the implementation of the *Approved Lists of Bacterial Names* (Skerman et al., 1980, page 229, paragraph 4). An extension of this argument might be that since "*anguillimortifera*" is the senior synonym, it must replace "*tarda*." Although there is possible nomenclatural validity to this argument, we believe that *E. tarda* is the name that should be used to avoid unnecessary confusion in the literature and provide stability in nomenclature. The controversy is quite complex (involving two different versions of the *Bacteriological*

Code, as well as the *Approved Lists of Bacterial Names*), and will eventually require a ruling of the Judicial Commission. In the meantime, we consider "*Paracolobactrum anguillimortiferum*" as only a possible subjective synonym of *Edwardsiella tarda* and consider the *E. anguillimortifera* as being under judicial consideration (*sub judice*) and will not use it. Since its proposal in 1975, few workers have used *E. anguillimortifera* but instead have used the well known and accepted name *E. tarda*. We will follow this convention.

Acknowledgments

We thank W. H. Ewing for helpful discussions and for his recollections of the early days of *Edwardsiella* research, Riichi Sakazaki for interesting discussion about his experience with *Edwardsiella* and about the problem of "*Paracolobactrum anguillimortiferum*," P. A. D. Grimont who furnished strains of *E. hoshinae* and *E. tarda* biogroup 1, J. P. Hawke who furnished strains of *Edwardsiella ictaluri*, and Kathleen

Ann Kelley of Emory University for assistance in writing the Greek and Latin origins of scientific names.

Further Reading

Ewing, W. H., A. C. McWhorter, M. R. Escobar, and A. H. Lubin. 1965. *Edwardsiella*, a new genus of *Enterobacteriaceae* based on a new species, *E. tarda*. Int. Bull. Bacteriol. Nomen. Taxon. *15:* 33–38.

Farmer, J. J. III. 1981. The genus *Edwardsiella*. In Starr, Stolp, Trüper, Balows and Schlegel (Editors), *The Prokaryotes*. Springer-Verlag, New York, pp. 1135–1139.

Grimont, P. A. D., F. Grimont, C. Richard and R. Sakazaki. 1980. *Edwardsiella hoshinae*, a new species of *Enterobacteriaceae*. Curr. Microbiol. *4:* 347–351.

Hawke, J. P. 1979. A bacterium associated with disease of pond cultured channel catfish, *Ictalurus punctatus*. J. Fisheries Res. Board Canada *36:* 1508–1512.

Hawke, J. P., A. C. McWhorter, A. G. Steigerwalt, and D. J. Brenner. 1981. *Edwardsiella ictaluri* sp. nov., the causative agent of enteric septicemia of catfish. Int. J. Syst. Bacteriol. *31:* 396–400.

Jordon, G. W. and W. K. Hadley. 1969. Human infection with *Edwardsiella tarda*. Ann. Intern. Med. *70:* 283–288

Differentiation and characteristics of the species and biogroups of the genus **Edwardsiella**

Table 5.38 presents the differential characteristics of the species and biogroups of *Edwardsiella*. Table 5.40 presents additional biochemical features of the organisms.

List of the species of the genus **Edwardsiella**

1. **Edwardsiella tarda** Ewing and McWhorter 1965, 37[AL] in Ewing, McWhorter, Escobar and Lubin (1965).

tar'da. L. fem. adj. *tarda* slow (intended meaning was "inactive," referring to the fermentation on only a few carbohydrates compared to many other *Enterobacteriaceae*).

The characteristics are as given for the genus and listed in Tables 5.38 and 5.39. Grimont et al. (1980) recently described a group of D-mannitol-positive, sucrose-positive, L-arabinose-positive strains which were closely related to "biochemically" typical strains of *E. tarda* by DNA/DNA hybridization. In this chapter this phenotypically distinct group will be referred to as "*E. tarda* biogroup 1" (see Tables 5.38 and 5.39). Strains of *E. tarda* which are negative for D-mannitol, sucrose and L-arabinose are much more common (perhaps a thousand times so), and are designated as "*E. tarda* wild type."

As indicated previously under Taxonomic Comments, the name *Edwardsiella anguillimortifera* (Hoshina 1962) Sakazaki and Tamura 1975, might be considered to be an objective synonym of *E. tarda* because it has the same type strain (ATCC 15947) on the *Approved*

Lists of Bacterial Names (Skerman et al., 1980). The status of this strain, however, has been challenged because the characteristics of the strain differ in several important respects from Hoshina's description of "*Paracolobactrum anguillimortiferum*" (see Table 5.40), and the matter will need clarification by the Judicial Commission.

Occurs in a wide variety of animals, rarely in the feces of healthy people. It is an opportunistic human pathogen, which may cause wound infections and probably also some cases of diarrhea.

The mol% G + C of the DNA of *E. tarda* is 55–58 (T_m, Bd).

Type strain (holotype): ATCC 15947.

2. **Edwardsiella hoshinae** Grimont, Grimont, Richard and Sakazaki, 1981, 216.[VP] (Effective publication: Grimont, Grimont, Richard and Sakazaki, 1980, 349.)

ho.shi'nae. M.L. gen. n. *hoshinae* of Hoshina; named after the late Toshikazu Hoshina, the Japanese bacteriologist who was one of the first to describe an organism which was probably an *Edwardsiella*.

The characteristics are as described for the genus and indicated in Tables 5.38 and 5.39.

Most isolates have come from animals. However, two human isolates were from human feces but there is no evidence that this species causes diarrhea.

Four of the available isolates of *E. hoshinae* have no zones of inhibition around disks impregnated with the antibiotic colistin; however, the four other strains have zones of 10–17 mm. *E. hoshinae* has large zone around the following antibiotics: nalidixic acid, sulfadiazine, gentamicin, streptomycin, kanamycin, tetracycline, chloramphenicol, ampicillin, carbinicillin, cephalothin and penicillin G (range of zones of inhibition for penicilin G, 28–31 mm).

The mol% G + C of the DNA is 56–57 (T_m).

Type strain: (holotype): ATCC 33379 (CIP 78-56, Grimont 2-78).

3. **Edwardsiella ictaluri** Hawke, McWhorter, Steigerwalt and Brenner 1981, 400.[VP] (*Edwardsiella* GA 7752 Hawke, 1979, 1509.)

ic.ta.lu'ri. *Ictalurus* the genus name for catfish; M.L. fem. adj. *ictaluri* pertaining to catfish.

The chacteristics are as described for the genus and as indicated in Tables 5.38 and 5.39.

E. ictaluri is the most fastidious of the three *Edwardsiella* species. Growth is very slow on plating media, often requiring 2 or 3 days of incubation for colonies to become 1 mm in diameter. It seems to prefer a lower temperature, although characteristic biochemical reactions are apparent at 36°C (Tables 5.38 and 5.39). Biochemically, it is also the least active of the three *Edwardsiella* species.

Table 5.38.

Differentiation of the species and biogroups of the genus **Edwardsiella**[a]

Characteristics	1. *E. tarda* Wild Type	1. *E. tarda* Biogroup 1	2. *E. hoshinae*	3. *E. ictaluri*
Acid production from:				
D-Mannitol	−	+	+	−
Sucrose	−	+	+	−
Trehalose	−	−	+	−
L-Arabinose	−	+	[−]	−
Tetrathionate reduction[b]	+	−	+	
Malonate utilization	−	−	+	−
Indole production (method 1)	+	+	[−]	−
H₂S production on triple-sugar iron agar	+	−	−	−
Motility	+	+	+	−
Citrate (Christensen's)	+	+	[+]	−

[a] Symbols: +, positive for 90–100% of strains (at 36°C in 48 h); [+], positive for 75–89% of strains; [−], positive for 11–25% of strains; −, positive for 0–10% of strains.

[b] Based on the data of Grimont et al. (1981).

Table 5.39.

Other characteristics of the species and biogroups of the genus **Edwardsiella**[a]

Characteristics	1. E. tarda Wild Type	1. E. tarda Biogroup 1	2. E. hoshinae	3. E. ictaluri	Characteristics	1. E. tarda Wild Type	1. E. tarda Biogroup 1	2. E. hoshinae	3. E. ictaluri
Indole production[b]:					D-Mannitol, sucrose	−	+	+	−
Method 1	+	+	[−]	−	Trehalose	−	−	+	−
Method 2	+	+	d^w		L-Arabinose	−	+	[−]	−
Methyl red	+	+	+	−	Glycerol	d	−	d	−
Voges-Proskauer	−	−	−	−	Salicin	−	−	d	−
Citrate:					Adonitol, D-arabitol, cel-	−	−	−	−
Simmons'	−	−	−	−	lobiose, dulcitol, eryth-				
Christensen's	+	+	[+]	−	ritol, lactose, i-(myo)-				
H$_2$S production:					inositol, melibiose, α-				
Triple-sugar iron agar	+	−	−	−	methyl-D-glucoside,				
Peptone iron agar	+	[+]^w	+^{w}	−	raffinose, L-rhamnose,				
Urea (Christensen's)	−	−	−	−	D-xylose				
Phenylalanine deaminase	−	−	−	−	Acid production from mu-	−	−	−	−
Amino acid decarboxylases					cate				
(Møller's)					Tartrate (Jordan's)	[−]	−	−	−
Lysine decarboxylase	+	+	+	+	Esculin hydrolysis	−	−	−	−
Arginine dihydrolase	−	−	−	−	Acetate utilization	−	−	−	−
Ornithine decarboxylase	+	+	+	+^{w}	Nitrate reduced to nitrite	+	+	+	+
Motility:					Deoxyribonuclease	−	−	−	−
36°C	+	+	+	−	Lipase (corn oil)	−	−	−	−
25°C (within 3 d)	+	d	+	+^{w}	β-Galactosidase (ONPG[c]	−	−	−	−
Gelatin hydrolysis (22°C)	−	−	−	−	test)				
KCN, growth in	−	−	−	−	Pectate hydrolysis	−	−	−	−
Malonate utilization	−	−	+	−	Pigment production	−	−	−	−
Gas production, D-glucose	+	−	[−]	d	Tyrosine clearing	−	−	−	−
Gas production, any sugar	+	+	d	+	Oxidase test (Kovacs')	−	−	−	−
Acid production from:					Tetrathionate reductase[d]	+	−	+	
D-Mannose, maltose	+	+	+	+	Mol% G + C of DNA	55–58		56–57	53

[a] Symbols: +, positive for 90–100% of strains (at 36°C in 48 h unless otherwise indicated); [+], positive for 75–89% of strains; [−], positive for 11–25% of strains; −, positive for 0–10% of strains; d, positive for 26 to 74% of strains; superscript "w", weak reaction.

[b] Grimont et al. (1981) reported that all eight strains of *E. hoshinae* were indole-positive. In our hands two strains were indole-negative and the others produced small amounts.

[c] ONPG, *o*-nitrophenyl-β-D-galactopyranoside.

[d] Based on the data of Grimont et al. (1981).

Table 5.40.

Differences in the reported phenotypic properties of **Edwardsiella tarda** *and* **"Paracolobactrum anguillimortiferum"**[a]

Property[b]	E. tarda	"P. anguillimortiferum"
Methyl red test	+	−
Phenylalanine required	−	+
Threonine required	−	+
Valine required	−	+
Cysteine required	+	−
Methionine required	+	−
Pathogenic for trout[c]	−	+

[a] Symbols: see standard definitions.

[b] Results for the methyl red test for *E. tarda* are based on data obtained in the Enteric Section. The nutritional requirements of *E. tarda* are as given by d'Émpaire (1969). Properties of *"P. anguillimortiferum"* are based on the original description by Hoshina (1962). Two different workers (unpublished results) have not been able to duplicate the work of d'Émpaire on the nutritional requirements of *E. tarda*, so this point needs clarification. The properties of *"P. anguillimortiferum"* cannot be verified because no authentic strains are available.

[c] Meyer and Bullock (1973) found no pathogenicity for "fingerling brown trout" (*Salmo trutta*); Hoshina (1962) found pathogenicity for rainbow trout.

The antibiotic susceptibility by disk diffusion is difficult to determine because the strains grow so poorly on Mueller-Hinton agar at 37°C. They must be incubated at 25°C instead. If 12 antibiotic disks are placed on a plate, it cannot be read after incubation because the large zones of inhibition overlap and no growth is visible. Instead, it is preferable to place only four disks on each of three 150 × 20 mm Mueller-Hinton agar plates. After 24 h at 25°C the growth is too faint to read, but at 48 h zones of inhibition are clear. No zone of inhibition occurs around colistin (10 μg disk) but very large (20–50 mm) zones were around the other agents tested, including a large zone (range, 29–35 mm) around penicillin G (10 U/disk). Thus, *E. ictaluri* appears to be susceptible to nalidixic acid, sulfadiazine, streptomycin, kanamycin, gentamicin, tetracycline, chloramphenicol, penicillin, ampicillin, carbenicillin and cephalothin but resistant to colistin.

Occurs as a pathogen of catfish.

The mol% G + C of the DNA is 53 (Bd).

Type strain (holotype): ATCC 33202 (CDC 1976-78, GA 7752).

Genus XI. *Proteus* Hauser 1885, 12.[AL]

JOHN L. PENNER

Pro'te.us. Gr. n. *Proteus*: an ocean god able to change himself into different shapes.

Straight rods, 0.4–0.8 μm in diameter × 1.0–3.0 μm in length. Gram-negative. **Motile** by peritrichous flagella. **Most strains swarm with periodic cycles of migration producing concentric zones, or spread in a uniform film, over moist surfaces** of nutrient media solidified with agar or gelatin. The organisms in this genus conform to the definition of the family *Enterobacteriaceae*. They **oxidatively deaminate phenylalanine and tryptophan. Urea is hydrolyzed.** They produce acid from several mono- and disaccharides. They do not produce acid from inositol or from straight chain tetra-, penta- or hexahydroxy alcohols, but generally do produce acid from glycerol. **Hydrogen sulfide is produced.** Pathogenic, causing urinary tract infections; also are secondary invaders, causing septic lesions at other sites of the body. Occur in the intestines of humans and a wide variety of animals; also occur in manure, soil and polluted waters. One species has been isolated only from gypsy moth larvae. The mol% G + C of the DNA is 38–41 (T_m) (Falkow et al., 1962).

Type species: *Proteus vulgaris* Hauser 1885, 12.

Further Descriptive Information

In broth cultures, the cells are short rods about 0.6 μm wide and 1.2 μm long. On solid media, cells are 0.8 μm wide and 1.2 μm long (Williams, 1978). Swarming (the movement of cells in periodic cycles of migration and consolidation) occurs on media solidified with agar or gelatin to produce concentric rings on the plate around the point of inoculation. During migration, the cells (swarm cells) are 20–80 μm long and possess many flagella. During consolidation, swarm cells divide for a period of time before producing another generation of swarm cells (Williams, 1978). Some strains (or variants) produce a single uniform film without periodic cycles (C variant of Belyavin, 1951, and the Z variant of Coetzee and Sacks, 1960). Some strains neither swarm nor spread and merely form distinct colonies.

New insights into the phenomenon of swarming have been gained in the last decade but the mechanisms basic to the induction of swarming remain a mystery. Factors critical to the initiation of swarming appear to be the development of the elongated swarm cells, the increased manufacture of flagella and the production of extracellular slime.

The swarming of *Proteus* makes it difficult to isolate bacteria of other species from pathological specimens plated on agar media and therefore methods have been contrived to prevent swarming. In the enteric laboratory, media have been formulated to inhibit swarming by incorporating in the media bile salts or detergents, by reducing the sodium chloride concentration or by increasing the concentration of the agar to 4% (New Zealand agar) or 7% (Japanese agar). The incorporation of 0.1–0.3 mM *p*-nitrophenyl glycerol in solid media also inhibits swarming without affecting flagellation or motility and, because it is of low toxicity to *Proteus* and other bacteria, an evaluation for its use in the clinical laboratory has been advocated (Kopp et al., 1966; Williams, 1973).

Swarms of different strains may fail to penetrate into each other and a sharp line of demarcation is produced between the two swarms (Dienes phenomenon) (Dienes, 1946). In other cases, the swarms may merge into each other without the production of such a line. The occurrence of the line was interpreted to reflect differences in the strains and the absence of the line to signify that the strains were the same. These observations have been exploited for differentiating strains, mostly *P. mirabilis*, in epidemiological studies (Story, 1954). However, strains of different biochemical types may swarm together (Kippax, 1957). Thus, a negative Dienes test (absence of the demarcation line) is less reliable for indicating that the strains are the same than is a positive test (production of the demarcation line) for indicating that the strains are different (Fránce and Markham, 1968; De Louvois, 1969). Results obtained with Dienes tests may fail to correlate with results obtained by bacteriophage typing (Hickman and Farmer, 1976). The production of the line of demarcation appears to be unrelated to the flagellar (H) antigens of the strains (Sourek, 1968; Skirrow, 1969), but appears to depend both on the bacteriocins produced by the swarming strains and on the bacteriocins to which they are sensitive (Senior, 1977). The Dienes test can be usefully employed in epidemiological studies when used in combination with other typing schemes and when its limitations are recognized. The Dienes test has not been tested on *P. myxofaciens* strains.

Another important distinguishing feature of *Proteus* and the other *Proteeae* (*Providencia* and *Morganella*) is their ability to oxidatively deaminate a variety of amino acids, producing keto acids and ammonia (Bernheim et al., 1935; Stumpf and Green, 1944; Singer and Volcani, 1955). Addition of ferric chloride solution to keto acids in aqueous solution produces different colors dependent upon the amino acid from which the keto acid was produced (Singer and Volcani, 1955), and the same colors are produced when ferric chloride solution is added to bacteria grown on nutrient media supplemented with the amino acids. Tests for the differentiation of *Proteus*, *Providencia* and *Morganella* from other *Enterobacteriaceae* that do not produce the deaminases have been developed. Tests for phenylalanine deaminase and for tryptophan deaminase are widely used (Henriksen, 1950; Thibault and Le Minor, 1957).

Bacteriophages lytic for *P. mirabilis* and *P. vulgaris* may be obtained from lysogenic strains or from sewage (Vieu, 1963; Coetzee, 1972). Strains of both *Proteus* species may be differentiated by bacteriophage and, although several schemes, mostly for *P. mirabilis*, have been described, no one scheme has been widely adopted (France and Markham, 1968; Pavlatou et al., 1965; Hickman and Farmer, 1976; Izdebska-Szymona et al., 1971; Schmidt and Jeffries, 1974; Vieu and Capponi, 1965).

Bacteriocins (proticins) may be produced spontaneously or sometimes only after induction with mitomycin C, and bacteriocin typing of *Proteus* strains has been advocated (Cradock-Watson, 1965; Al-Jumaili, 1975; Senior, 1977; Kusek and Herman, 1980). Agreement has not been reached on whether differentiation of the strains should be accomplished on the basis of the inhibitory activity of the bacteriocins produced by the strains under examination, on the basis of sensitivity to a selected set of bacteriocins, or by the use of both methods in combination.

Serotyping of *P. vulgaris* and *P. mirabilis* may be accomplished on the basis of 49 somatic (O) antigens using the simplified scheme of Kauffmann and Perch (Kauffmann, 1966). This scheme includes strains of both species for preparing O antisera. Seventeen of the O antigens are present on *P. vulgaris* strains, 27 on *P. mirabilis* strains, and 5 occur on strains of both species. Three O antigens designated A, B and C, and 11 others designated 100–104 and 200–205, have been defined in other studies but have not been systematically included in an expanded Kauffmann-Perch scheme (Larsson and Olling, 1977; Penner and Hennessy, 1980). Isolates generally agglutinate in antisera against strains of the same species and, therefore, separation of the serovars to provide individual schemes for each species facilitates serotyping (Penner and Hennessy, 1980). The most frequently isolated strains are *P. mirabilis* with O antigens 3, 6 or 10 (Lanyi, 1956; de Louvois, 1969; Larsson and Olling, 1977; Kauffmann, 1966; Penner and Hennessy, 1980).

The number of flagellar (H) antigens in the Kauffmann-Perch scheme is 19. The most common are H antigens 1, 2 and 3. Crossreactions among the H antigens are numerous and complex, and the use of the H antigens in differentiating Proteus strains has been limited essentially to the initial studies of Kauffmann and Perch (Perch, 1948). Capsular (K) antigens (designated C antigens) have been demonstrated for some strains of *P. vulgaris* and *P. mirabilis* (Namioka and Sakazaki, 1959). The antigenic structure of *P. myxofaciens* has not been examined.

Antibodies formed in humans during the course of certain rickettsial infections may react with O antigens of three *Proteus* strains designated X19, X2 and XK. These three strains are used for preparing antisera against *P. vulgaris* 01 and 02 and *P. mirabilis* 03 antigens, respectively. The diagnostic test for antibodies in human sera against these specificities is called the Weil-Felix reaction (see the family *Rickettsiaceae*). Results of this test as an indication of rickettsial infection should be interpreted with caution because of the fact that *Proteus* infections may also evoke antibodies against these antigens and that *P. mirabilis* strains with the 03 antigen are the most frequently isolated *Proteus* strains from human infections.

P. vulgaris and *P. mirabilis* have intrinsic resistance to bacitracin, polymyxin and colistin but are generally susceptible to nalidixic acid. Both species have strains resistant and susceptible to nitrofurantoin. Strains of both species may be either susceptible or resistant to tetracyclines but the proportion of resistant strains is on the increase. The majority of *P. mirabilis* and over 50% of *P. vulgaris* strains are susceptible to chloramphenicol. *P. mirabilis* strains are generally susceptible to penicillins and cephalosporins whereas *P. vulgaris* strains are generally resistant. Most strains of both species are susceptible to the aminoglycosides. Strains of both species may acquire plasmids coding for antibiotic resistances giving rise to marked increase in the resistance to aminoglycosides and/or other antibiotics to which the species is generally susceptible.

Antibiotic susceptibility studies on *P. myxofaciens* have not been reported.

P. mirabilis and *P. vulgaris* may cause primary and secondary infections in man. *P. mirabilis* is much more frequently isolated from clinical specimens than is *P. vulgaris* and is one of the leading pathogens of the human urinary tract. *P. mirabilis* urinary tract infections acquired outside the hospital are often associated with an underlying condition such as diabetes or structural abnormalities of the tract (Wallace and Petersdorf, 1971; Grossberg et al., 1962). *Proteus* urinary tract infections occur more commonly in infection-susceptible hospital patients with predisposing conditions such as catheterization, surgery or urological instrumentation of the tract. Approximately one-quarter of the population are intestinal carriers of *Proteus* (Rustigian and Stuart, 1945) and the patient may become infected with his own flora (autoinfection). Infections may also be contracted through transmission of the bacteria from other patients or from a common reservoir (Dutton and Ralston, 1957; Kippax, 1957). An often-mentioned factor contributing to the pathogenicity of *Proteus* in the urinary tract is the activity of

the urease enzyme in producing ammonia and raising the pH (Braude and Siemienski, 1960; MacLaren, 1968; Musher et al., 1975; Griffith et al., 1973; Phillips, 1955). *Proteus* urinary tract infections may give rise to bacteremias that are difficult to treat and often fatal.

Under suitable conditions *Proteus* bacteria may be opportunistic invaders and cause septic lesions at other sites of the body. They have been isolated from infections of wounds, burns, respiratory tract, eyes, ears and throat.

Circumstantial evidence has been cited to implicate *P. mirabilis* as the etiological agents of outbreaks of gastroenteritis resulting from the consumption of contaminated food (Cooper et al., 1941; Cherry et al., 1946) and as the agents causing infantile enteritis (Lanyi, 1956), but their roles as the principal pathogens have been difficult to assess in light of the high carriage rate of *Proteus* in healthy individuals (Carpenter, 1964).

Neonatal umbilical stumps contaminated with *Proteus* bacteria may lead to highly fatal bacteremias and meningitis (Becker, 1962; Burke et al., 1971; Levy and Ingall, 1967; Librach, 1968; Shortland-Webb, 1968).

P. myxofaciens has been isolated only from living and dead gypsy moth larvae (*Porthetria dispar*) but its role as a pathogen of the larvae has not been critically examined.

Proteus strains are widely distributed in nature. *P. mirabilis* is the more common of the two species (Levine and Hoyt, 1945). Both species occur in the intestines of mice, rats, monkeys, raccoons, dogs, cats, cattle, pigs, birds, reptiles and in a large proportion of the human population (Cantu, 1911; Phillips, 1955; Muller, 1972; Wilson and Miles, 1975; Rustigian and Stuart, 1945). The role of *Proteus* in the intestine is not well understood. The bacteria may assist in the hydrolysis of urea although their contribution must be minor in comparison to the large populations of urease-producing anaerobes (Brown et al., 1971; Sabbaj et al., 1970). More important may be their role in the oxidative deamination of amino acids producing keto acids and ammonia (Drasar et al., 1974).

Bacteria of the two species are found in manure, soil and polluted waters where they are thought to have an important function in the decomposition of organic materials (Wilson and Miles, 1975).

Enrichment and Isolation Procedures

Growth of *Proteus* from stool samples is regarded as a nuisance because these bacteria are not generally considered to be infectious agents of the intestine and because their tendency to swarm interferes with the isolation of other bacterial species. Isolation media in the enteric laboratory are therefore designed to inhibit swarming and to preferentially select known pathogens such as *Salmonella* and *Shigella*. Most such media are also suitable for the direct isolation of *Proteus* and are routinely used to isolate *Proteus* from urines and other clinical specimens. However, primary isolation media specific for *Proteus*, *Providencia* and *Morganella* have been designed (Malinowski, 1966; Xilinas et al., 1975; Zarett and Doetsch, 1949).

Tetrathionate or selenite broth are suitable liquid enrichment media when feces are to be examined for *Proteus*. The rate of isolation is increased from 8.2 to 23.6% for *P. mirabilis* and from 0 to 2.7% for *P. vulgaris* when primary plating is preceded by enrichment with tetrathionate (Hynes, 1942; Rustigian and Stuart, 1945).

Maintenance Procedures

Proteus may be maintained on Trypticase soy agar at 4°C with monthly transfers or may be preserved indefinitely by lyophilization. The Enteric Section, Centers for Disease Control, stores cultures at room temperature in tubes of blood agar base or Trypticase soy agar. These tubes are sealed with a cork or rubber stopper and the cultures have remained viable for many years without transfer (J. Farmer and F. Hickman, personal communication).

Differentiation of **Proteus** from **Providencia** and **Morganella**

Key characteristics for differentiating these three closely related genera are shown in Table 5.41.

Taxonomic Comments

A number of changes have been made since the last edition of the manual in which it was indicated that the genus *Proteus* was composed of five species, namely *P. vulgaris*, *P. mirabilis*, *P. rettgeri*, *P. morganii* and *P. inconstans*. *P. myxofaciens*, on the other hand, was excluded from the genus *Proteus* because it was said to be *Erwinia herbicola* (*Enterobacter agglomerans*). Convincing evidence for major changes in the classification was derived from deoxyribonucleic acid (DNA) relatedness studies (Brenner et al., 1978). Two species were recognized in *Providencia*: cultures previously in biochemical subgroup A were placed in *Providencia alcalifaciens* and those in subgroup B in *Providencia stuartii*. *Proteus rettgeri* was found to be more closely related to the latter two species than to *Proteus vulgaris* or *P. mirabilis* and was, therefore, assigned to the genus *Providencia*. *Proteus morganii* was found to be related to *Proteus* and *Providencia* at levels no greater than to other *Enterobacteriaceae* and was placed in *Morganella*, the genus proposed earlier by Fulton (Fulton, 1943). *Proteus myxofaciens* was included in the genus *Proteus* because of its phenotypic similarity and because of its relatedness by DNA/DNA hybridization to *P. vulgaris* and *P. mirabilis*. Its DNA was only 10% related to DNA from a strain of *Erwinia herbicola*

Further Reading

Brenner, D. J., J. J. Farmer III, G. R. Fanning, A. G. Steigerwalt, P. Klykken, H. G. Wathen, F. W. Hickman and W. H. Ewing. 1978. Deoxyribonucleic acid relatedness of *Proteus* and *Providencia* species. Int. J. Syst. Bacteriol. *28:* 269–282.

Kauffmann, F. 1966. The Bacteriology of *Enterobacteriaceae*, Williams & Wilkins, Baltimore, pp 333–360.

Rustigian, R. and C. A. Stuart. 1943. Taxonomic relationships in the genus *Proteus*. Proc. Soc. Exp. Biol. Med. *53:* 241–243.

Rustigian, R. and C. A. Stuart. 1945. The biochemical and serological relationships of the organisms of the genus *Proteus*. J. Bacteriol. *49:* 419–436.

Williams, F. D. 1978. Nature of the swarming phenomenon in *Proteus*. Annu. Rev. Microbiol. *32:* 101–122.

Table 5.41.
Characteristics differentiating **Proteus, Providencia** *and* **Morganella**[a,b]

Characteristics	Proteus	Providencia	Morganella
Swarming	+	−	−
H₂S production	+	−	−
Gelatin hydrolysis	+	−	−
Lipase (corn oil)	+	−	−
Utilization of citrate (Simmons')	D	+	−
Ornithine decarboxylase	D	−	+
Acid production from:			
Mannose	−	+	+
Maltose	D	−	−
Acid from one or more of the following polyhydric alcohols:			
Inositol, D-mannitol, adonitol, D-arabitol, erythritol	−	+	−

[a] Symbols: see standard definitions.

[b] Temperature of reactions, 36 ± 1°C. All reactions are for 48 h.

List of the species of the genus **Proteus**

1. Proteus vulgaris Hauser 1885, 12.[AL]

vul.ga′ris. L. adj. *vulgaris* common.

Morphological characteristics are as described for the genus. Other characteristics are listed in Tables 5.41 to 5.43.

Some strains are hemolytic on blood agar.

Less frequently found in clinical specimens than *P. mirabilis*.

Generally resistant to penicillins and cephalosporins.

The mol% of the DNA is 39.3 ± 1.2% (T_m) (Falkow et al., 1962).

Type strain: ATCC 13315.

2. Proteus mirabilis Hauser 1885, 34.[AL]

mi.ra′bi.lis. L. adj. *mirabilis* wonderful.

Morphological characteristics are as described for the genus. Other characteristics are listed in Tables 5.41 to 5.43.

Some strains are hemolytic on blood agar.

More frequently found in clinical specimens than *P. vulgaris*.

Most common site of infection is the urinary tract. Generally susceptible to ampicillin and cephalosporins.

The mol% of the DNA is 39.3 ± 1.4% (T_m) (Falkow et al., 1962).

Type strain: ATCC 29906.

3. Proteus myxofaciens Cosenza and Podgwaite 1966, 188.[AL]

myx.o.fac′i.ens. Gr. fem. n. *myxa*, slime; M.L. masc. n. *faciens* producing; *myxofaciens* slime-producing (bacteria).

Morphological characteristics are as described for the genus. Other characteristics are listed in Tables 5.41 to 5.43. Only one strain studied

Table 5.42.
Differential characteristics of the species of the genus **Proteus**[a]

Characteristics	1. P. vulgaris	2. P. mirabilis	3. P. myxofaciens[b]
Indole production	+	−	−
Ornithine decarboxylase	−	+	−
Acid from:			
Maltose	+	−	+
α-Methylglucoside	d	−	+
D-Xylose	d	+	−
Tyrosine clearing	+	+	−
Slime production, 25°C in TSB[c]	−	−	+

[a] Temperature of reactions, 36 ± 1°C. All reactions are for 48 h. For symbols see standard definitions

[b] Reactions based on study of only one strain (ATCC 19692).

[c] TSB, Trypticase soy broth.

in detail. Thin film of growth on solid media. Produces highly viscous slime. Hemolytic on blood agar.

Isolated from living and dead gypsy moth larvae (*Porthetria dispar* L.).

Type strain: ATCC 19692.

Table 5.43.
Other characteristics of the species of the genus **Proteus**[a]

Characteristics	1. P. vulgaris	2. P. mirabilis	3. P. myxofaciens[b]	Characteristics	1. P. vulgaris	2. P. mirabilis	3. P. myxofaciens[b]
Phenylalanine deaminase	+	+	+	Oxidase test	−	−	−
Urease	+	+	+	ONPG hydrolysis[c]	−	−	−
NO$_3^-$ reduced to NO$_2^-$	+	+	+	Pectate liquefaction	−	−	−
Motility	+	+	+	Malonate utilization	−	−	−
Swarming	+	+	+	Amino acid decarboxylases (Møller):			
Gelatin liquefaction (22°C)	+	+	+	Lysine decarboxylase	−	−	−
H$_2$S production (triple-sugar iron agar)	+	+	+ (3–4 days)	Arginine dihydrolase	−	−	−
Growth in KCN	+	+	+	Acid production from:			
Acid from glucose	+	+	+	Sucrose	+	d	+
Gas from glucose	+	+	+	Trehalose, glycerol	d	+	+
Methyl red test	+	+	+	Salicin, esculin	d	−	−
Voges-Proskauer test	−	d	+	Lactose, L-arabinose, raffinose, L-rhamnose, cellobiose, mannose, melibiose, mucate, inositol, D-mannitol, adonitol, D-arabitol, s-sorbitol, dulcitol, erythritol	−	−	−
Citrate utilization (Simmons')	d	d	+				
Tartrate utilization (Jordan)	+	d	+				
Acetate utilization	d	d	−				
Lipase activity (corn oil)	d	+	+				
Deoxyribonuclease (25°C)	d	d	−				

[a] Temperature of reactions, 36 ± 1°C unless otherwise noted. All reactions are for 48 h except where otherwise noted. For symbols see standard definitions.

[b] Based on study of only one strain (ATCC 19692).

[c] ONPG, *o*-nitrophenyl-β-D-galactopyranoside.

Genus XII. **Providencia** Ewing 1962, 96[AL]

JOHN L. PENNER

Pro.vi.den′ci.a. M.L. fem. n. *Providencia* named after the city of Providence, Rhode Island, U.S.A.

Straight rods, 0.6–0.8 × 1.5–2.5 μm, conforming to the general definition of the family *Enterobacteriaceae*. Gram-negative. **Motile by peritrichous flagella. Swarming does not occur.** Facultatively anaerobic. **Oxidatively deaminate phenylalanine and tryptophan. Produce acid from one or more of the following polyhydric alcohols:** inositol, D-mannitol, adonitol, D-arabitol, erythritol. **Acid is produced from mannose.** Indole-positive. **Citrate** (Simmons') and tartrate (Jordan) are utilized. Isolated from diarrhetic stools, urinary tract infections, wounds, burns and bacteremias. The mol% G + C of the DNA is 39–42% (Falkow et al., 1962).

Type species: *Providencia alcalifaciens* Ewing 1962, 96.

Further Descriptive Information

Providencia strains, like those of *Proteus* and *Morganella*, deaminate phenylalanine, and at least some strains of the genus deaminate other amino acids (Singer and Volcani, 1955). Like other *Proteeae*, *Providencia* strains decompose tyrosine to produce a clearing on the agar media in which the insoluble amino acid is incorporated (Sheth and Kurup, 1975), and produce a reddish-brown pigment when cultured on nutrient agar containing 5% tryptophan (Polster and Svobodova, 1964). *Providencia* differs from other *Proteeae* by being able to produce acid from inositol and straight-chain tetra-, penta- or hexahydroxy alcohols, and the species of *Providencia* are differentiated on the basis of their reactions on these substrates. Yellow-orange-centered colonies are produced by *Providencia* on deoxycholate citrate agar (Cook, 1948; Buttiaux et al., 1954). The color is apparently caused by the precipitation

of ferric hydroxide as a result of the alkalinity produced by the growth of the bacteria on the medium (Catsaras et al., 1965).

Urease is produced characteristically by strains of only one species, *P. rettgeri*. The proportion of urease-positive strains of *P. stuartii* has been estimated to be 15% (Brenner et al., 1978), although subsequent calculations based on a larger number of strains indicate 6–10% (Penner et al., 1979). The urease enzyme of at least some *P. stuartii* strains is encoded on a transferable plasmid (Grant et al., 1981). The presence of the plasmid in endemic strains of some hospitals could be expected to cause variations among the hospitals in the frequency of isolation of urease-positive *P. stuartii*.

Providencia bacteriophages have lytic activity on *Providencia* and *Proteus* strains but not on *Morganella* strains (Coetzee, 1963). The phages may be isolated from sewage and from lysogenic strains. A scheme consisting of 12 selected bacteriocins may be used to differentiate strains of *P. alcalifaciens* and *P. stuartii* (Al-Jumaili and Fenwick, 1978).

Thermostable somatic (O) antigens, thermolabile flagellar (H) antigens and capsular (K) antigens occur in *Providencia*. The original antigenic scheme was for two species (*P. alcalifaciens* and *P. stuartii*) and consisted of 56 O antigens, 28 H antigens and 2 K antigens (Ewing et al., 1954). For differentiation of strains on the basis of the O antigens the schemes have been extended and separated according to species, so that, currently, 46 O antigens for *P. alcalifaciens* and 17 O antigens for *P. stuartii* may be identified (Penner et al., 1979a, b). The original schemes for *P. rettgeri* listed 34 O antigens and 26 H antigens (Namioka and Sakazaki, 1958). New serovars have been isolated and the number of O antigens now recognized is 93 (Penner and Hennessey, 1979).

The practice of listing *P. rettgeri* with *Proteus vulgaris* and *Morganella morganii* in the "indole-positive *Proteus* group," and *P. alcalifaciens* and *P. stuartii* in the "Providence" group, in antimicrobial agent-susceptibility studies has tended to obscure significant species differences in susceptibility. Generally, strains of *P. alcalifaciens* are more susceptible than are *P. stuartii* and *P. rettgeri* to penicillins, cephalosporins and aminoglycosides (Overturf et al., 1974; Penner and Preston, 1980). The most resistant *Providencia* strains are found in the species *P. stuartii*. Amikacin is often effective against *P. stuartii* strains that are resistant to other antibiotics.

The urinary tract of the catheterized or compromised patient is the most common site of *P. stuartii* and *P. rettgeri* infections. Strains of the two species may also produce wound and burn infections and bacteremias. The rise in medical importance of these organisms is associated with their tendency to cause nosocomial infections and with their marked resistance to numerous antibiotics.

P. alcalifaciens strains are generally isolated from stool specimens taken from patients with diarrhea. The most common serotype isolated is 0:3. Whether these bacteria, particularly of this serotype, are indeed the causative agents of the diarrheas as claimed (Carpenter, 1964) or whether they are commensals that flourish during infections caused by viral or other bacterial agents remains to be determined.

Providencia isolates recovered in studies on *Proteus* indicate that there is some overlapping of habitats between the two genera. *Providencia* strains are rarely isolated from intestines of healthy individuals by methods routinely employed in examining fecal specimens (Singer and Bar-Chay, 1954). Rigorous examinations have not been conducted to determine if this reflects a genuinely low incidence or if it reflects small bacterial populations that are detectable only with special media.

Enrichment and Isolation Procedures

Media used in the clinical laboratory for isolation of *Enterobacteriaceae* may be used to isolate *Providencia*. Tetrathionate or selenite broths may be used for enrichment. Media for the specific isolation of *Providencia* have not been reported, but the medium of Malinowski (1966) should be considered because differentiation from other *Enterobacteriaceae* does not depend upon hydrolysis of urea.

Maintenance Procedures

Providencia strains may be maintained on trypticase soy agar at 4°C with monthly transfers or may be preserved indefinitely by lyophilization. The Enteric Section, Centers for Disease Control, stores cultures at room temperature in tubes of blood agar base or trypticase soy agar. These tubes are sealed with a cork or rubber stopper and the cultures have remained viable for many years without transfer (J. Farmer and F. Hickman, personal communication).

Taxonomic Comments

Major changes in the classification of members of the tribe *Proteeae* since the last edition of the *Manual* have led to the emergence of the genus *Providencia*. Two species of the genus, *P. alcalifaciens* and *P. stuartii*, were previously included in one species of the genus *Proteus* (*Proteus inconstans*) or were often grouped together and called the "Providence" strains. Bacteria that are now known to be urease-positive strains of *P. stuartii* were included along with typical *P. rettgeri* in *Proteus rettgeri*. The new classification in the present *Manual* was introduced because it was confirmed through DNA/DNA hybridization studies that these bacteria were a group distinct from other *Proteeae* (Brenner et al., 1978). In *Proteus inconstans* two distinct groups that corresponded to the biochemical types (subgroups A and B) were recognized. Since there is doubt about the validity of the epithet *inconstans* introduced by Ornstein (1921) to indicate variability in the fermentation of glucose by a bacterium for which no subculture of the original strain exists, the validity published epithets *alcalifaciens* and *stuartii* were selected, the former for subgroup A strains and the latter for subgroup B strains. This was in accordance with proposals previously published (Ewing, 1962).

Proteus rettgeri was also found to consist of two groups on the basis of DNA/DNA hybridization studies. One group consisted of typical *P. rettgeri*. The other group consisted of urease-positive strains of *P. stuartii* and were reassigned to that species.

Interpreting data from earlier studies may cause problems because sometimes it is not clear, in light of the new classification, to which species the bacteria under study actually belonged. A case in point concerns the data on the mol% G + C content of DNA. It is not certain whether the value reported for *Proteus rettgeri* (39 ± 1.5) is for a typical strain or for a urease-positive *P. stuartii* strain, or if the value reported for *Proteus inconstans* (41.5 ± 0.6) is for *P. alcalifaciens* or *P. stuartii* (Lautrop, 1974; Falkow et al., 1962). Similar problems may arise in other studies in which the earlier classifications were used.

Further Reading

Brenner, D. J., J. J. Farmer III, G. R. Fanning, A. G. Steigerwalt, P. Klykken, H. G. Wathen, F. W. Hickman and W. H. Ewing. 1978. Deoxyribonucleic acid relatedness of *Proteus* and *Providencia* species. Int. J. Syst. Bacteriol. *28:* 269–282.

Ewing, W. H. 1962. The Tribe *Proteeae:* its nomenclature and taxonomy. Int. Bull. Bacteriol. Nomencl. Taxon. *12:* 93–102.

Differentiation of the species of the genus **Providencia**

Table 5.44 presents characteristics for differentiation of the three species of *Providencia*.

Table 5.44.

Differential characteristics of the species of the genus **Providencia**[a]

Characteristics	1. *P. alcalifaciens*	2. *P. stuartii*	3. *P. rettgeri*
Urease production	−	d	+
Acid production from:			
Inositol	−	+	+
D-Mannitol	−	d	+
Adonitol	+	−	+
D-Arabitol	−	−	+
Erythritol	−	−	d
Trehalose	−	+	−

[a] Temperature of reactions, 36 ± 1°C. All reactions are for 48 h. For symbols see standard definitions.

List of the species of the genus **Providencia**

1. Providencia alcalifaciens (De Dalles Gomes 1944) Ewing 1962, 96.[AL] (*Eberthella alcalifaciens* de Salles Gomes 1944, 183; *Proteus inconstans* (Ornstein 1921) Shaw and Clarke 1955, 155.)

al.cal.i.fac′i.ens. Fr. n. *alcali* alkali; L. v. *facere* to do, make; L. part. adj. *faciens* making; M.L. part. adj. *alcalifaciens* alkali-producing.

The characteristics are as described for the genus and as listed in Tables 5.44 and 5.45.

Most strains are susceptible to penicillins and cephalosporins.

Generally isolated from diarrhetic stools, particularly from children, but the role in disease production is not known. The most frequently isolated strains are serovar 0:3.

Type strain: ATCC 9886.

2. Providencia stuartii (Buttiaux et al., 1954) Ewing 1962, 96.[AL] (*Proteus stuartii* Buttiaux, Osteux, Fresnoy and Moriamez 1954, 385; *Proteus inconstans* (Ornstein 1921) Shaw and Clarke 1955, 155.)

stu.ar′ti.i. M.L. gen. n. *stuartii* of Stuart; named after C. A. Stuart, bacteriologist at Providence, Rhode Island, U.S.A.

The characteristics are as described for the genus and as listed in Tables 5.44 and 5.45.

Many strains are resistant to penicillins and cephalosporins. Some strains are resistant to gentamicin and kanamycin. Some exceptional strains are resistant to most antibiotics in current use.

Isolated most often from urine specimens of hospitalized and catheterized patients. Less frequently isolated from wounds, burns and bacteremias. May cause nosocomial infections. Rarely isolated from stool specimens.

Type strain: ATCC 29914.

3. Providencia rettgeri (Hadley, Elkins and Caldwell 1918) Brenner, Farmer, Fanning, Steigerwalt, Klykken, Wathen, Hickman and Ewing 1978, 269.[AL] (*Bacterium rettgeri* Hadley, Elkins and Caldwell 1918, 180; *Proteus rettgeri* (Hadley et al. 1918) Rustigian and Stuart 1943, 242.)

rett′ge.ri. M.L. gen. n. *rettgeri* of Rettger; named after L. F. Rettger, the American bacteriologist who first isolated the organism in 1904.

The characteristics are as described for the genus and as listed in Tables 5.44 and 5.45.

Many strains are resistant to penicillins and cephalosporins, but the strains are generally not as resistant as *P. stuartii* strains.

Generally isolated from urine specimens of hospitalized and catheterized patients. Less frequently isolated from other sites. May cause nosocomial infections. Rarely isolated from stool specimens.

Type strain: ATCC 29944.

Table 5.45.
Other characteristics of the species of the genus **Providencia**[a]

Characteristics	1. *P. alcalifaciens*	2. *P. stuartii*	3. *P. rettgeri*
Phenylalanine deaminase	+	+	+
Indole production	+	+	+
Nitrates reduced to nitrites	+	+	+
Motility, 36°C	+	d	+
Growth in KCN	+	+	+
Methyl red test	+	+	+
Voges-Proskauer test	−	−	−
Citrate utilization (Simmons')	+	+	+
Tartrate utilization (Jordan)	+	+	+
Acetate utilization	d	d	d
Lipase activity (corn oil)	−	−	−
Oxidase test	−	−	−
β-Galactosidase (ONPG test)	−	−	−
Pectate liquefaction	−	−	−
Tyrosine clearing	+	+	+
Malonate utilization	−	−	−
Amino acid decarboxylases (Møller)			
Lysine decarboxylase	−	−	−
Arginine dihydrolase	−	−	−
Ornithine decarboxylase	−	−	−
Gelatin liquefaction (22°C)	−	−	−
H₂S production (triple-sugar iron agar)	−	−	−
Acid production from:			
Glucose, mannose	+	+	+
Sucrose, glycerol	d	d	d
Esculin	−	d	−
D-Xylose, salicin, L-rhamnose	−	−	d
Lactose, L-arabinose, raffinose, maltose, cellobiose, α-methylglucoside, melibiose, mucate, dulcitol, D-sorbitol	−	−	−
Gas from glucose	d	−	d

[a] Temperature of reactions, 36 ± 1°C unless otherwise noted. All reactions are for 48 h. For symbols see standard definitions.

Genus XIII. *Morganella* Fulton 1943, 81[AL]

JOHN L. PENNER

Mor.ga.nel'la. M.L. dim. ending *-ella*; M.L. fem. n. *Morganella* named after H. de R. Morgan, who first studied the organism.

Straight rods, 0.6–0.7 μm in diameter and 1.0–1.7 μm in length, conforming to the general definition of the family *Enterobacteriaceae*. Gram-negative. **Motile** by means of peritrichous flagella, but some strains do not form flagella above 30°C. After 48 h on 1% agar media at 22°C growth may spread to form a surface film. **Swarming does not occur.** Facultatively anaerobic. **Deaminate phenylalanine and tryptophan oxidatively. Urease-positive. Indole-positive. Ornithine is decarboxylated.** A few carbohydrates can be fermented. **Produce acid from mannose.** Utilize Jordan tartrate but not Simmons' citrate. Occur in the feces of humans, dogs, other mammals and reptiles. Opportunistic secondary invaders, isolated from bacteremias, respiratory tract, wound and urinary tract infections. The mol% G + C of the DNA is 50 (T_m).

Type species: Morganella morganii (Winslow et al. 1919) Brenner et al. 1978, 269.

Further Descriptive Information

Until recently, the members of *Morganella* were classified as *Proteus* and were thus considered in the light of their membership in that genus rather than as a separate group. Like *Proteus*, *Morganella* strains can be cultured on laboratory media used for enteric bacteria but some strains may not form flagella above 30°C (Coetzee and De Klerk, 1964). After 48 h on 1% agar media at 22°C, growth may spread to form a film (Rauss, 1936; Coetzee and De Klerk, 1964) and the culture may consist of semifilamentous forms resembling those of *Proteus* (Rauss, 1936); however, swarming on 1.5% agar (with cycles of migration and consolidation typical of *Proteus*) has not been demonstrated (Sevin and Buttiaux, 1939). Some *Morganella* strains are hemolytic on blood agar.

Like *Proteus* and *Providencia*, *Morganella* strains produce urease and phenylalanine deaminase; however, the *Morganella* enzymes are serologically unrelated to those of the other two genera (Guo and Liu, 1965; Smit and Coetzee, 1967), and the urease has other properties which differ markedly from those of the ureases of *Proteus* and *Providencia* (Richard, 1965; Rosenstein et al., 1981). *Morganella* strains also decompose tyrosine to produce a clearing on media containing the insoluble amino acid (Sheth and Kurup, 1975). *Morganella* strains also produce a reddish-brown pigment when cultured on nutrient media supplemented with 5% tryptophan (Polster and Svobodova, 1964). Unlike *Proteus* and *Providencia*, *Morganella* strains are noted for their inability to ferment carbohydrates. Glucose and mannose are the only sugars from which *Morganella* strains typically produce acid. Trehalose and glycerol are fermented by some strains (Hickman et al., 1980) and lactose-positive strains have been isolated occasionally (Sutter and Foecking, 1962; Tierno and Steinberg, 1975); the ability to ferment lactose is plasmid-encoded (Le Minor and Coynault, 1976). Unlike *Proteus* and *Providencia*, *Morganella* strains do not produce a red color on lysine iron agar (Edwards and Ewing, 1972). Typically, *Morganella* strains decarboxylate only ornithine, but a few strains decarboxylate both ornithine and lysine, and a few decarboxylate neither (Hickman

et al., 1980); lysine decarboxylase in *Morganella* is plasmid-encoded (Cornelis et al., 1981). *Morganella* strains require niacin and pantothenate for growth (Pelczar and Porter, 1940).

Morganella bacteriophages do not generally attack *Proteus* and *Providencia* strains (Coetzee, 1963). Twelve lytic patterns have been found among 26 *Morganella* strains using seven bacteriophages (Schmidt and Jeffries, 1974). The activities of 12 *Morganella* bacteriocins (morganocins) are detectable on MacConkey agar but not on nutrient agar (Coetzee, 1967).

The original antigenic scheme based on somatic (O) and flagellar (H) antigens (Rauss and Vörös, 1959) has been extended to 42 serogroups and 75 serovars (Rauss et al., 1975). The O antigens can be determined by passive hemagglutination (Penner and Hennessy, 1979).

Morganella strains are generally resistant to colistin, erythromycin, penicillin, ampicillin and cephalothin, and are generally susceptible to nalidixic acid, carbenicillin, the aminoglycosides and chloramphenicol. There is much variation among the strains in susceptibility to tetracyclines and sulfonamides.

Morganella was once considered to be a cause of diarrhea (Morgan, 1906; Tribondeau and Fichet, 1916; Magheru, 1923; Thjøtta, 1920; Rauss, 1936) because it was found as the predominant species in diarrhetic stools and because other known pathogens (*Salmonella*, *Shigella*) were not present. Reports of this type have been lacking in recent years, however, and firm evidence for an etiological role in enteritis has not been forthcoming. There is considerably more evidence for a pathogenic role in urinary tract infections, particularly for those of nosocomial origin (Sevin and Buttiaux, 1939; Lanyi, 1957; Von Graevenitz and Spector, 1969; McMillan, 1972). It is an opportunistic, secondary invader rather than a primary pathogen at other sites and has been isolated from blood, sputa and pus from patients with bacteremias, respiratory tract and wound infections.

The habitat of *Morganella* has not been examined systematically, but it has been isolated from the intestines of humans, dogs, other mammals and reptiles (Phillips, 1955; Müller, 1972).

Enrichment and Isolation Procedures

Media for primary isolation of *Enterobacteriaceae* are usually used for isolating *Morganella*. The culturing of feces in tetrathionate or selenite broth prior to plating on enteric media increases the rate of *Morganella* isolations from 1.8 to 10% in studies on human intestinal carriage (Rustigian and Stuart, 1945).

Maintenance Procedures

Morganella strains may be maintained on trypticase soy agar with monthly transfers or may be preserved indefinitely by lyophilization. The Enteric Section, Centers for Disease Control, stores cultures at room temperature in tubes of blood agar base or trypticase soy agar. These tubes are sealed with a cork or rubber stopper and the cultures have remained viable for many years without transfer (J. Farmer and F. Hickman, personal communication).

Differentiation of the genus **Morganella** from other genera

See the genus *Proteus*, Table 5.41, for characteristics that can be used to differentiate *Morganella* from other related genera of *Enterobacteriaceae*.

Taxonomic Comments

A major change since the last edition of the *Manual* has been the transfer of this group from the genus *Proteus* to the genus *Morganella*. Rauss (1936) concluded that the organism "belongs taxonomically to

the genus *Proteus*," and Yale (1939) attributed the name *Proteus morganii* to him. Yale, however, not Rauss, was cited as the author in the eighth edition of the *Manual* because it was pointed out that Rauss had not actually published the name (Lessel, 1971). The inclusion in the genus *Proteus* received further support because the organisms, like *P. vulgaris* and *P. mirabilis*, hydrolyzed urea (Rustigian and Stuart, 1943). The major criteria supporting the elevation of *P. morganii* to generic rank are that the DNA contains 50 mol% G + C, which is similar to the DNA base composition of *Escherichia coli* and *Salmonella* rather than *Proteus*, and that DNA/DNA hybridization studies show that the organisms are related at only a 20% level to most enteric bacteria and at not more than 20% to *Proteus* (Brenner et al., 1978).

Further Reading

Brenner, D. J., J. J. Farmer III, G. R. Fanning, A. F. Steigerwalt, P. Klykken, H. G. Wathen, F. W. Hickman and W. H. Ewing. 1978. Deoxyribonucleic acid relatedness of *Proteus* and *Providencia* species. Int. J. Syst. Bacteriol 28: 269–282.

Hickman, F. W., J. J. Farmer III, A. E. Steigerwalt and D. J. Brenner. 1980. Unusual groups of *Morganella* ("Proteus") *morganii* isolated from clinical specimens: lysine-positive and ornithine-negative biogroups. J. Clin. Microbiol. 12: 88–94.

Rauss, K. and S. Vörös. 1959. The biochemical and serological properties of *Proteus morganii*. Acta Microbiol. Acad. Sci. Hung. 6: 233–246.

Rustigian, R. and C. A. Stuart. 1943. Taxonomic relationships in the genus *Proteus*. Proc. Soc. Exper. Biol. Med. 53: 241–243.

List of the species of the genus **Morganella**

1. **Morganella morganii** (Winslow, Kligler and Rothberg 1919) Brenner, Farmer, Fanning, Steigerwalt, Klykken, Wathen, Hickman and Ewing 1978, 269.[AL] (*Bacillus morgani* (sic) Winslow, Kligler and Rothberg 1919, 481; *Proteus morganii* (Winslow et al.) Yale 1939, 435.)

mor.ga′ni.i. M.L. gen. n. *morganii* of Morgan; named after H. de R. Morgan, a British bacteriologist who first studied the organism.

The description is the same as that for the genus. See Table 5.46 for other characteristics.

Occur in the feces of humans, dogs, other mammals and reptiles. Opportunistic human pathogens.

The mol% G + C of the DNA is 50 (Falkow et al., 1962).

Type strain: ATCC 25830.

Table 5.46.
Characteristics of **Morganella morganii**[a]

Test	Reaction or Result
Phenylalanine deaminase	+
Urease	+
Indole	+
Growth in KCN	+
Amino acid decarboxylases (Møller):	
Ornithine decarboxylase	+
Lysine decarboxylase	−
Arginine dihydrolase	−
Methyl red test	+
Voges-Proskauer test	−
NO_3^- reduced to NO_2^-	d
Tyrosine clearing	+
Oxidase test	−
ONPG hydrolysis[b]	−
Deoxyribonuclease	−
Lipase	−
Tartrate utilization (Jordan)	+
H_2S production (triple-sugar iron agar)	−
Motility	+
Gelatin liquefaction	+
Utilization of citrate (Simmons')	−
Utilization of acetate or malonate	−
Acid production from:	
Glucose, mannose	+
Trehalose	d
Lactose, sucrose, L-arabinose, raffinose, L-rhamnose, D-xylose, cellobiose, α-methyl-glucoside, melibiose, salicin, esculin, mucate	−
Gas from glucose	d

[a] Temperature of reactions, 36 ± 1°C. All reactions are for 48 h. Symbols: see standard definitions.
[b] ONPG, *o*-nitrophenyl-β-galactopyranoside.

Genus XIV. **Yersinia** *Van Loghem 1944, 15.*[AL]

HERVE BERCOVIER AND HENRI H. MOLLARET

Yer.si′ni.a. M.L. fem. n. *Yersinia* named for the French bacteriologist A. J. E Yersin, who first isolated the causal organisms of plague in 1894.

Straight rods to coccobacilli, 0.5–0.8 μm in diameter and 1–3 μm in length. Endospores are not formed. Capsules are not present, but an envelope occurs in *Y. pestis* strains grown at 37°C or in cells from in vivo samples. Gram-negative. **Nonmotile at 37°C, but motile with** **peritrichous flagella when grown below 30°C, except for** *Y.* *pestis* **which is always nonmotile.** Growth occurs on ordinary nutrient media. Colonies on nutrient agar are translucent to opaque, 0.1–1.0 mm in diameter after 24 h. Optimum temperature, 28–29°C.

Facultatively anaerobic, having both a respiratory and a fermentative type of metabolism. Oxidase-negative. Catalase-positive. Nitrate is reduced to nitrite with a few exceptions in specific biovars. Glucose and other carbohydrates are fermented with acid production **but little or no gas. Phenotype characteristics are often temperature-dependent**, and usually more characteristics are expressed by cultures incubated at 25–29°C than at 35–37°C. The enterobacterial common antigen is expressed by all species investigated. Occur in a broad spectrum of habitats (live and inanimate), with some species adapted to specific hosts. The mol% G + C of the DNA is 46–50 (T_m, Bd).

Type species: *Yersinia pestis* (Lehmann and Neumann 1896) Van Loghem 1944, 15.

Further Descriptive Information

Cells of *Yersinia* species are small, coccoid-shaped Gram-negative bacilli that resemble cells of *Pasteurellaceae* rather than of *Enterobacteriaceae*. Pleomorphism occurs depending on the type of medium used and the temperature of incubation. Rods, coccobacilli, and small chains of 4 or 5 elements (especially in liquid media) can be seen in a Gram stain, which reveals a more pronounced tendency to bipolar staining in *Y. pestis* than in the other species. No spores or specific inclusions are formed. No definite capsules occur, but *Y. pestis* displays an envelope that might be taken for a capsule when cultured in proper media (Burrows, 1963) incubated at 37°C, or when stained in samples taken from live hosts (mice, guinea pigs, humans). L forms have been described for *Y. enterocolitica* (Pease, 1979).

All *Yersinia* species are nonmotile when incubated at 37°C but motile at 22–29°C, except *Y. pestis*, which is never motile. Fresh isolates of *Y. enterocolitica* and *Y. pseudotuberculosis* may require a few subcultures to express their motility. Motile cells have 2–15 peritrichous flagella characterized by a long wavelength (Nilehn, 1969).

Yersinias do not differ from other *Enterobacteriaceae* in their fine structure and overall cell wall composition. Lipopolysaccharides (O antigens) have been isolated and characterized (Davies, 1958; Rische et al., 1973). The whole-cell lipid composition of all *Yersinia* species investigated exhibits a pattern shared with other *Enterobacteriaceae* (Tornabene, 1973, Jantzen and Lassen, 1980).

Yersinia species grow on nutrient agar without enrichment. A small colony diameter differentiates yersinias from all other *Enterobacteriaceae*. After incubation for 24–30 h at 30 or 37°C, *Y. pestis* forms minute colonies (0.1 mm) that can be discerned only with difficulty by the naked eye. After 48 h their diameter increases to 1.0–1.5 mm. The colonies are slightly opaque, butyrous, smooth, round, and have somewhat irregular edges. The use of enriched media (serum, blood, yeast extract) does not dramatically improve the growth, and after 48 h the colony sizes are similar to those found on nutrient agar. All other *Yersinia* species grown on nutrient agar at 25–37°C produce visible colonies in 24 h. The colonies reach a diameter of 1.0–1.5 mm after 24–30 h, and 2.0–3.0 mm after 48 h. After 18 h they are translucent, smooth and round with irregular edges, but after 48 h the centers become elevated and the edges become more regular, producing a "chinese hat" shape. When cultured for 48 h, all *Yersinia* species dissociate into small (0.5 mm) and large colonies (2 mm). This phenomenon appears to depend on the medium used (Bercovier et al., 1979).

Growth is moderate in liquid media: incubation of yersinias for 48 h will yield the same turbidity that occurs in 18 h with other *Enterobacteriaceae*. When grown in nutrient broth *Y. pestis* forms a deposit at the bottom of the tube and the supernatant remains relatively clear; this is followed by the appearance of a pellicle, which in turn disintegrates to form flocculent masses and a larger deposit. This phenomenon is attenuated in peptone water. *Y. pseudotuberculosis* occasionally grows in a manner similar to that of *Y. pestis*. All other *Yersinia* species give uniform turbidity in nutrient broth and in peptone water.

Y. pestis and *Y. pseudotuberculosis* give variable growth responses on MacConkey agar. All the other species grow well on this medium, with colonies reaching a size similar to that observed on nutrient agar. On salmonella-shigella agar incubated at 25°C *Y. pestis* hardly grows at all, whereas all the other species produce pin-point colonies in 24–30

h. When incubated on this medium at 37°C, *Y. enterocolitica* is only partially inhibited, whereas all other species are severely inhibited (Bottone, 1977; Nilehn, 1969; Bercovier et al., 1979).

All *Yersinia* species except *Y. pestis* can grow at 25°C on synthetic mineral-salt media with various carbohydrates as the energy source (Burrows and Gillet, 1966; Bercovier et al., 1979). *Y. pestis* requires L-methionine and L-phenylalanine. When incubated at 37°C on synthetic mineral-salt media all *Yersinia* species become auxotrophic, and the addition of at least biotin and thiamine is necessary to promote growth (Burrows and Gillet, 1966). The growth of *Y. pestis* on such media is enhanced by the addition of L-isoleucine, L-valine, glycine, L-threonine, and reducing agent, and by incubation in a CO_2-enriched atmosphere (Brubaker, 1972). Virulent strains of *Y. pestis* require Ca^{2+} or ATP for growth at 37°C but not at 25°C (Zahorchak et al., 1979). This temperature-dependent requirement for Ca^{2+} has also been described for some virulent strains of *Y. pseudotuberculosis* and *Y. enterocolitica*.

All *Yersinia* species grow at temperatures of 4–42°C, with an optimum temperature of 28–29°C. *Y. pestis* and *Y. pseudotuberculosis* tolerate a pH range of 5.0–9.6; other *Yersinia* species can grow in a pH range of 4.0–10.0. The optimum pH for all species is 7.2–7.4.

Yersinia species can grow in peptone water without the addition of NaCl. *Y. pestis* and *Y. pseudotuberculosis* tolerate up to 3.5% NaCl, and the other species can tolerate up to 5% NaCl. *Y. pseudotuberculosis* is the only species which grows well on media containing 0.06% tellurite (Brzin, 1968).

Yersinias do not differ significantly from other *Enterobacteriaceae* in their general metabolism (Brubaker, 1972). They produce acid during fermentation of glucose. *Y. enterocolitica*, *Y. frederiksenii* and *Y. intermedia* produce acetoin when incubated at 28°C, whereas this characteristic is variable for *Y. ruckeri* and is always absent in *Y. pestis* and *Y. pseudotuberculosis*. No species produces acetoin at 37°C.

The main physiological and biochemical characteristics of the various *Yersinia* species are given in Tables 5.48 and 5.49. Yersinias ferment carbohydrates without gas production; this characteristic is constant for *Y. pestis* and *Y. pseudotuberculosis*, but other species may produce a few bubbles after 2 or 3 days at 28°C. Because the optimum growth temperature of yersinias is 28–29°C, some biochemical activities are often temperature-dependent (cellobiose and raffinose fermentation, ornithine decarboxylase, ONPG (o-nitrophenyl-β-D-galactopyranoside) hydrolysis, indole production, and the Voges-Proskauer reaction) and are more constantly expressed at 28°C rather than at 37°C. all species except *Y. intermedia* reduce nitrate to nitrite by a type B nitrate reductase; *Y. intermedia* strains have either a type A nitrate reductase, like most *Enterobacteriaceae*, or a type B reductase. The ONPG activity of yersinias does not correspond to a true β-galactosidase, but only to an ONPG-ase (Le Minor et al., 1977). In addition to the characteristics given in Tables 5.48 and 5.49, *Yersinia* species are able to attack polypectate in 5–7 days and starch in 3–7 days. Yersinias are neither hemolytic nor proteolytic, except *Y. ruckeri*, which liquefies gelatin, and some strains of *Y. pestis* which have fibrinolytic and coagulase activity linked to the production of Pesticin I. Lecithinase activity in *Y. enterocolitica* is strain-dependent. *Y. pseudotuberculosis*, *Y. enterocolitica* and *Y. ruckeri* strains have a lipase that is active on corn oil, but only *Y. intermedia*, *Y. frederiksenii* and *Y. enterocolitica* biovar 1 express a lipase-esterase that is active on Tween 80.

Transformation of auxotrophic strains of *Y. enterocolitica* by prototrophic strains using the Juni-Janik technique has been reported (Callahan and Koroma, 1979). F lac$^+$ episomes from *E. coli* have been transferred to *Y. pestis* (Martin and Jacob, 1962), to *Y. pseudotuberculosis* (Lawton et al., 1963b) and to *Y. enterocolitica* (Cornelis and Colson, 1975), but usually with a low frequency (10^{-4}–10^{-6}). This has allowed chromosomal mapping of *Y. pseudotuberculosis* (Lawton and Stull, 1971; McMahon, 1973). Gene transfer by conjugation between *Y. pseudotuberculosis* and *Y. pestis* has also been demonstrated (Lawton et al., 1968a).

R factors have been transferred to *Y. pestis* and *Y. pseudotuberculosis* (Ginoza and Matney, 1963) and to *Y. enterocolitica* (Knapp and Lebeck, 1967). Wild strains of *Yersinia* carrying R plasmids (Cornelis et al.,

1973; Kanazawa et al., 1979) appear to be rare. This could be explained, at least for *Y. enterocolitica*, by the presence of a retriction-modification system (Cornelis and Colson, 1975). Metabolic plasmids coding for lactose and raffinose fermentation have been described in *Y. enterocolitica* (Cornelis et al., 1976).

Other plasmids related to various virulence tests (Ca^{2+} dependency, autoagglutination, lethality for mice and gerbils, Sereny test) have been demonstrated in *Y. pestis* (Ferber and Brubaker, 1981), *Y. pseudotuberculosis* (Gemski et al., 1980b) and *Y. enterocolitica* (Zink et al., 1980; Gemski et al., 1980a). These plasmids of 40–48 megadalton molecular weight constitute a family of related plasmids (Portnoy et al., 1981; Ben Gurion and Shafferman, 1981). *Y. pestis* and *Y. pseudotuberculosis* have never been found to be lysogenic, whereas of 1252 strains of *Y. enterocolitica* studied, 86.4% were lysogenic when grown at 25°C but not at 37°C (Nicolle et al., 1973). Phages active on *Y. pestis* and *Y. pseudotuberculosis* have been described (Gunnison et al., 1951; Girard, 1953), but they are not host-specific and are used only for presumptive bacteriological diagnosis. Coliphages T$_2$, T$_3$ and T$_7$ are also active on *Y. pseudotuberculosis* and *Y. pestis* (Hertman, 1964; Ackerman and Poty, 1969). A phage typing system, useful in epidemiology, has been developed for *Y. enterocolitica* (Nicolle et al., 1973): strains of *Y. enterocolitica* serogroup 03 are associated with phagovar VIII in Europe, IXa in the Republic of South Africa, and IXb in Canada.

Strains of *Y. pestis* produce a bacteriocin active on *Y. pseudotuberculosis* Ben Gurion and Hertman, 1958). This was named Pesticin I by Brubaker and Surgalla (1962) after they detected a second bacteriocin (Pesticin II) which was produced by *Y. pestis* and *Y. pseudotuberculosis*. Pesticin I is also active on certain strains of *E. coli*. *Y. pestis* strains that produce Pesticin I also elaborate a fibrinolytic factor and a coagulase (Brubaker, 1972). A bacteriocin-like activity associated with the presence of phage tails has been ascribed in *Y. enterocolitica* (Nicolle et al., 1973). *Y. intermedia* produces a bacteriocin-like substance at 25°C but not at 37°C that is active on certain strains of *Y. enterocolitica*, *Y. intermedia*, *Y. frederiksenii* and *Y. kristensenii* (Botonne et al., 1979).

The antigenic structure of *Yersinia* species is complex, but some antigens are shared by *Y. pestis*, *Y. pseudotuberculosis* and *Y. enterocolica*. The common enterobacterial antigen has been found in all species investigated (Le Minor et al., 1972a; Maeland and Digranes, 1975). The Fraction 1 envelope antigen (F1) of *Y. pestis* is best produced when cultures are incubated at 37°C on protein-rich media (Fox and Higuchi, 1958). This antigen is heat-labile (10 min at 100°C), water-soluble, and contains a carbohydrate protein (F1A) and a carbohydrate-free protein (F1B). Passive hemagglutination with F1 antigen is used for serologic surveys in plague foci. The presence of this antigen has also been demonstrated in *Y. pseudotuberculosis* (Quan et al., 1965). V and W antigens expressed by virulent strains of *Y. pestis* cultivated at 37°C appear to be related to the presence of a 45 megadalton plasmid (Ferber and Brubaker, 1981; Ben Gurion and Shafferman, 1981). Production of plasmid-mediated V and W antigens has also been described in *Y. pseudotuberculosis* Gemski et al., 1981b) and in *Y. enterocolitica* (Gemski et al., 1981a). The somatic antigen of *Y. pestis* is rough (R antigen) and therefore no serogroups have been described in this species. This R antigen is also present in *Y. pseudotuberculosis* (Thal and Knapp, 1971). In addition, *Y. pestis* and *Y. pseudotuberculosis* share at least 11 out of 18 antigens studied by Lawton et al. (1960). *Y. pestis* and *Y. enterocolitica* express common protein antigens (Barber and Eylan, 1976). The antigenic scheme for *Y. pseudotuberculosis* (Thal and Knapp, 1971) comprises 6 main thermostable serogroups (I to VI) with subgroups (A, B, 2 to 15), and 5 thermolabile flagellar H antigens (a to e). Antigenic relationships have been demonstrated between *Y. pseudotuberculosis* (serogroups II, IV, IVA and VI) and the following organisms: *Salmonella* serogroups B and D, *E. coli* serogroups 017, 055 and 077, and *Enterobacter cloacae* (Knapp, 1968; Mair and Fox, 1973).

Wauters et al. (1972) described 34 different O antigen and 20 H antigen serogroups in *Y. enterocolitica*. This classification included some serogroups defined by strains belonging to *Y. intermedia* (017) and *Y.*

kristensenii (011, 012, 028). Nevertheless, these serogroups are useful epidemiological markers. Crossreactions occur between *Y. enterocolitica* serogroup 09 and *Brucella* species (Hurvell and Lindberg, 1973), and between *Y. enterocolitica* serogroup 012 and *Salmonella* factor 047 (Le Minor et al., 1972b).

Yersinia species are susceptible in vitro to the following antimicrobial agents: tetracycline, chloramphenicol, aminoglycosides (streptomycin, gentamicin, kanamycin and neomycin), sulfonamides (alone or in combination with trimethoprim), and nalidixic acid. They are susceptible to some degree towards colistin and are resistant to erythromycin and novobiocin. *Y. pestis* and *Y. pseudotuberculosis* are usually susceptible to β-lactam antibiotics but their susceptibility to penicillin is in the range of sensitive to intermediate. Resistance to ampicillin (Borowski and Zaremba, 1973) and to streptomycin (Kanazawa and Ikemura, 1979) has been described for *Y. pseudotuberculosis*, *Y. enterocolitica*, *Y. intermedia* (Botonne, 1977). *Y. frederiksenii* and *Y. kristensenii* probably are resistant to penicillin and slightly susceptible or resistant to other β-lactam antibiotics (ampicillin, carbenicillin, cephalothin) (Bercovier et al., 1979). The level of resistance is strain-dependent (Zaremba and Aldova, 1979) and temperature-dependent (Chester and Stotzky, 1976). *Y. enterocolitica* strains produce both a constitutive β-lactamase (active on ampicillin, carbenicillin, penicillin and cephalosporins) and an inducible β-lactamase (active only on cephalosporins and penicillin) (Cornelis and Abraham, 1975). *Y. enterocolitica* strains that are resistant to tetracycline, chloramphenicol, streptomycin and kanamycin have been reported (Zaremba and Aldova, 1979).

Y. pestis is the causative agent of plague. Plague is primarily a disease of wild rodents. *Y. pestis* is transmitted among wild rodents by fleas, in which the bacteria multiply and block the esophagus and the pharynx. The fleas regurgitate the bacteria when they take their next blood meal and transmit the disease to man if no other hosts are available. Infective flea bites produce the typical bubonic form of plague in humans. *Y. pestis* multiplies intracellularly in the host and proceeds through the lymphatic system. The lymph nodes near the flea bite are the first to become inflamed and enlarged, constituting the bubo. The evolution of the infection is usually so rapid that no characteristic lesions are found in the spleen or liver at autopsy. If not treated, the disease evolves in 5–10 days to septicemia and sometimes to a secondary pneumonia. From the latter situation, primary pneumonic plague can spread by means of droplets from man to man. In this clinical form death generally occurs in less than 4 days. *Pestis minor* cases, in which the bacteria remain self-limited in buboes followed by self-cures have been described in endemic plague areas (Pollitzer, 1954).

The virulence of *Y. pestis* is associated with the presence of 6 factors (Surgalla et al., 1968): (a) the ability to produce the F1 antigen, (b) the V,W antigens (associated with Ca^{2+} dependency, growth inhibition on oxalate medium, and autoagglutination when cultures are incubated at 37°C), (c) a pigment (incorporation of Congo red dye or hemin), (d) Pesticin I, (e) a toxin (the "murine toxin," whose activity is not clearly established), and (f) the ability to synthesize purines. The LD$_{50}$ dose for mice inoculated with strains expressing the aforementioned virulence factors is 1–10 organisms. Avirulent strains of *Y. pestis* never produce V,W antigens except in the case of the vaccine strain EV76, whose attenuated virulence has resulted from a mutation in its iron metabolism. Virulent strains and the EV76 strain harbor a 45 megadalton plasmid. In contrast to the V,W antigens, the lack of any of the other virulence factors does not completely abolish the virulence of *Y. pestis* strains.

Y. pseudotuberculosis is responsible for epizootics in nearly all animal species, especially in rodents. Animals are usually contaminated by the oral route and, after 1 or 2 weeks of incubation, the bacteria are found in the mesenteric lymph node. The main symptoms are mesenteric adenitis and chronic diarrhea. The infection evolves either to a self-cure or to a fatal septicemia. *Y. pseudotuberculosis* is an intracellular parasite and, like *Y. pestis*, reaches the lymphatic system. At autopsy, caseous lesions are found in the Peyer's patches, the mesenteric lymph node, the spleen and the liver. Humans orally contaminated by *Y.*

pseudotuberculosis develop either a mesenteric adenitis which simulates an acute appendicitis, or, in the compromised host, a severe septicemia. *Y. pestis* and *Y. pseudotuberculosis* appear to share at least two virulence factors: the F1 antigen and the V,W antigens.

Y. enterocolitica has been recognized as pathogenic for chinchillas, hares, monkeys and humans. The pathogenicity for animals is similar to that of *Y. pseudotuberculosis*. In children, *Y. enterocolitica* is responsible for acute adenitis simulating appendicitis, and also for terminal ileitis with diarrhea. In adults the main clinical forms of infection are arthritis, septicemia and erythema nodosum. The infection is probably acquired orally, and the bacteria multiply first in the Peyer's patches of the host. Then, depending on its serogroup and on the presence of a plasmid associated with virulence (V,W antigens expressed or positive autoagglutination test), the bacteria remain localized in the gut (ileitis) or invade lymphatic organs (mesenteric adenitis) and eventually reach the blood circulation (septicemia). Arthritis is caused mainly by serogroup 09, which has antigens in common with *Brucella*. This symptom is closely associated with the presence of the histocompatibility antigen HLA-27 in man (Bottone, 1977; Mollaret et al., 1979). Production of a heat-stable enterotoxin (ST) resembling *E. coli* ST (Okamoto et al., 1981) has been demonstrated in vitro (Pai and Mors, 1978), but its role in pathogenicity is not clear: *Y. enterocolitica* strains do not produce ST when incubated in vitro at temperature above 30°C, and no direct proof of production of ST in vivo has been reported.

Virulent strains of *Y. pestis*, *Y. pseudotuberculosis* and *Y. enterocolitica* rapidly become avirulent when subcultured on nutrient media incubated at 37°C. This is a result of the loss of the virulence plasmid associated with Ca²⁺ dependency and production of the V,W antigens. Cross-immunity among these three species has been demonstrated (Thal, 1973; Alonso et al., 1978). Human chemoprophylaxis with sulfonamides, vaccination, and the spreading of insecticides and rodenticides are the suggested measures for controlling plague. The drugs of choice for treatment for all *Yersinia* infections are streptomycin, the sulfonamides, chloramphenicol and the tetracyclines.

The pathogenicity of *Y. intermedia*, *Y. kristensenii* and *Y. frederiksenii* in man and animals is not clearly established. They all behave more like opportunistic pathogens than true pathogens (Bottone, 1977; Bercovier et al., 1978). ST-producing strains of these three species have been described (Kapperud, 1980), but their clinical significance is still unknown. *Y. ruckeri* is a fish pathogen responsible for red mouth disease, especially in rainbow trout. An inflammation of the mouth and the throat is the main characteristic of the disease which is enzootic (Rucker, 1966). The bacterium is usually isolated from the kidneys of fish undergoing a systemic infection.

The geographical distribution of *Y. pestis* is widespread, and the organism has been isolated from all the continents. Plague is enzootic in Africa (Central, East and South Africa), in North and South America, and in Asia (Southeast Asia, U.S.S.R., Iran). Between epidemics, *Y. pestis* remains localized in definite foci (Balthazard, 1964). It has been isolated from more than a hundred different naturally infected species of rodents, but rarely from predatory animals (carnivores and birds, the latter being resistant to the infection). The spread of plague is usually accomplished by the cycle of rodents to fleas, fleas to rodents. The reservoir of the bacteria is the soil contaminated by infected dead fleas and rodents. The bacteria survive for months in deep burrows. Rodents coming from noninfected areas become infected when they dig burrows in previously contaminated areas (Mollaret et al., 1963). This cycle constitutes the "sylvatic plague." When urban rodents are in contact with rural rodents, the bacteria can spread to humans through flea bites. The epidemiology of plague is linked to the ecology of both fleas and rodents.

Y. pseudotuberculosis is distributed worldwide. It has been found in numerous animal species, especially rodents and birds, in soil, and in man (Wetzler, 1970). Wild animals, which are often asymptomatic carriers, are considered the reservoir of the bacteria. Man and animals are contaminated orally either by direct contact with sick or asymptomatic animals or through food contaminated by the excretions of these

animals. The incidence of this infectious disease varies with the seasons and is highest during the cold seasons. *Yersinia* species multiply even at 4°C and therefore have a selective advantage over other bacteria at low temperatures; this explains why *Y. pseudotuberculosis*, *Y. enterocolitica*, *Y. frederiksensii* and *Y. kristensenii* are more frequently isolated from the environment during the cold seasons than during the hot seasons. Human and animal infections follow this seasonal distribution as well.

Y. enterocolitica has been isolated from a wide variety of sources (live and inanimate) in every country in which it has been looked for and probably has a worldwide distribution (Mollaret et al., 1979). Biovar 1 strains are ubiquitous, having been found in a wide range of animals and environmental sources (including foods), whereas other biovars or serogroups are frequently associated with a specific host (Bercovier et al., 1980a): biovar 5 strains have been isolated mainly from hares in Europe; biovar 4, serogroup 03 strains and biovar 3, serogroup 05,27 strains are responsible for most human gastrointestinal infections in Europe, Canada and the Republic of South Africa; serogroup 08 strains are frequently isolated from various syndromes in the United States; serogroup 09 strains are closely associated with human arthritis in Europe.

Y. intermedia and *Y. frederiksenii* have been identified in Europe, America, Australia and New Zealand, Israel and Japan. These two species have been isolated mainly from fresh water and foods and only rarely from nonirrigated soil, man or animals other than fish (Bercovier et al., 1978; Brenner et al., 1980a; Kapperud 1977; Ursing et al., 1980a). *Y. kristensenii* has been found in Europe, America, Japan and Australia. Strains of this species have been isolated mainly from soil, foods, and asymptomatic animals; isolates from other environmental sources and from human infection are rare (Bercovier et al., 1980c).

Y. ruckeri has been encountered only in the United States and Canada. It seems to be a natural component of the fresh water ecosystem. The red mouth disease appears only when fish are exposed to large number of bacteria, as has been shown experimentally (Ross et al., 1966). The disease is usually enzootic and occasionally epizootic in fish hatcheries.

Enrichment and Isolation Procedures

Isolation of *Yersinia* strains from noncontaminated samples (blood, lymph nodes) can be performed by using blood agar or nutrient agar incubated for 48 h at 28°C, or 24 h at 37°C followed by 24 h at room temperature. The isolation of *Y. pestis* from contaminated samples requires inoculation (subcutaneously or percutaneously) of animals (guinea pigs, mice or rats). The organism can be cultured from the spleen, liver or lymphatic nodes of the inoculated animals. All other *Yersinia* species will usually be isolated from stools or food samples by inoculating standard or special selective bile-salt media such as MacConkey agar (Lee, 1977), DCL agar, salmonella-shigella agar, SS-D agar (Wauters, 1973), CAL medium (Dudley and Shotts, 1979), CIN medium (Schiemann, 1979), oxalate medium (Soltesz et al., 1980) and BABY 4 medium (Bercovier, unpublished results). All these media should preferably be incubated for 48 h at 28–29°C or for 24 h at 37°C followed by 48–72 h at room temperature. Recovery of *Yersinia* strains from contaminated samples can be improved by various cold enrichment techniques (Lee et al., 1980; van Pee and Straiger, 1979).

Maintenance Procedures

Stab inoculations of *Yersinia* strains in conventional stock culture media stored in the dark at room temperature or at 4°C provide living cultures for 10 years or more, if the tubes are tightly sealed. Lyophilization and deepfreeze storage in 10% glycerol are suitable preservation techniques. To keep a strain fully virulent, it should never be subcultured at 37°C, but always at 25–28°C.

Procedures for Testing Special Characteristics

Methods to test tetrathionate reduction, tellurite reduction and the type of nitrate reductase have been described or referenced by Bercovier et al. (1979). The Ca^{2+}-dependency of virulent *Yersinia* strains is evaluated on magnesium oxalate medium* (Higuchi and Smith, 1961), as follows. Inoculate 0.1 ml of a bacterial suspension (10^5 bacterial/ml) onto two plates: one is incubated at 37°C, the other at 26°C. Check colony numbers on the two plates after 2 or 3 days. Colonies growing at 26°C but not at 37°C are Ca^{2+}-dependent. A fully virulent strain should give confluent growth at 26°C, whereas only 10–100 colonies should appear at 37°C. The *autoagglutination test* (Laird and Cavanaugh, 1980) to detect virulent *Yersinia* strains is done by inoculating

10 or more isolated colonies, each one into a pair of tubes (13 × 100 mm) containing 2 ml of RPMI-1640 medium containing 10% fetal calf serum and 25 mM HEPES buffer (N-2-hydroxyethylpiperazine-N′-2-ethanesulfonic acid). One tube is incubated at 37°C, the other at 26°C. After incubation at 26°C for 18 h, virulent colonies give a uniform turbidity; at 37°C a layer of agglutinated bacteria appears at the bottom of the tube and the supernatant remains clear. Avirulent strains give uniform turbidity at both 26°C and 37°C, and rough strains show spontaneous agglutination at both temperatures.

Differentiation of the genus **Yersinia** from other genera

Characteristics useful for differentiating *Yersinia* from other physiologically similar genera are listed in Table 5.47.

Taxonomic Comments

The genus *Yersinia* was proposed by van Loghem (1944) in order to separate *Y. pestis* and *Y. pseudotuberculosis* (formerly in the genus *Pasteurella*) from *Pasteurella* species sensu stricto (i.e. *P. multocida*, etc.), from which they differ in their oxidase reaction and in their DNA base composition (Mollaret, 1965). The genus *Yersinia* belongs to the family *Enterobacteriaceae*. *E. coli* tDNA (i.e. the genes coding for transfer RNA) and *Y. pestis* DNA are 63% related (Brenner et al., 1977), a value similar to that found for *E. coli* tDNA and *Hafnia alvei* DNA. All *Yersinia* species express the common enterobacterial antigen. Their physiological characteristics and their fatty acid contents are similar to those of all *Enterobacteriaceae* species. The mol% G + C of *Yersinia* species ranges from 46–50 and is consistent with that for *Enterobacteriaceae* species.

The genus *Yersinia* presently consists of seven different species. On the basis of DNA/DNA hybridization studies, all of these species are more closely related to each other than to any other *Enterobacteriaceae* species (Brenner et al., 1978; Brenner et al., 1980b; see also Table 5.51. The genus *Yersinia* can be considered a very homogeneous taxon.

DNA relatedness among *Yersinia* species is 40% or higher except for *Y. ruckeri* which is at most 38% related to other *Yersinia* species. DNAs of *Y. ruckeri* strains have been shown to be 30% related to

Serratia species (Ewing et al., 1978). *Y. ruckeri* was included in *Yersinia* because its mol% G + C of 48 is closer to that of *Yersinia* species than to that of *Serratia* species. Because the phenotypic characteristics of *Y. ruckeri* are very different from those of other *Yersinia* species (see Tables 5.48 and 5.49), it might constitute a new genus by itself. Phylogenetic studies would be helpful in clarifying this problem.

Strains of *Y. enterocolitica* belonging to the five different biovars (see Table 5.50), including the metabolically inactive biovar 5 strains, constitute a homogeneous genospecies (Bercovier et al., 1980a). The strains described as *Y. enterocolitica*-like organisms or atypical *Y. enterocolitica* are separated into three different species: *Y. intermedia* (Brenner et al., 1980a), *Y. frederiksenii* (Ursing et al., 1980a), and *Y. kristensenii* (Bercovier et al., 1980c). *Y. frederiksenii* consists of three genetic groups on the basis of DNA/DNA hybridization (Ursing et al., 1980a). For practical reasons, because there are no phenotypic differences among the three genetic groups, only one species has been proposed for the rhamnose-positive strains. More study on phenotypic characteristics is needed in order to separate the three genetic groups.

The DNAs of *Y. pestis* strains, regardless of biovar, and of *Y. pseudotuberculosis* are 90% or more interrelated. This explains the antigenic and biochemical similarities of the two species (Mollaret, 1965). On the basis of DNA data, Bercovier et al. (1980b) proposed that the two species constitute a single species, divided into two subspecies: *Y. pseudotuberculosis* subspecies *pseudotuberculosis* and *Y. pseudotuberculosis* subspecies *pestis*. This proposal was made in order

Table 5.47.

Differential characteristics of the genus **Yersinia** *and other physiologically similar genera[a]*

Characteristics	Yersinia	Hafnia	Citrobacter	Escherichia	Enterobacter	Klebsiella	Salmonella	Proteus	Pasteurella
Oxidase test (tetramethylphenylenediamine)	−	−	−	−	−	−	−	−	+
Colony size greater than 1.0 mm on nutrient agar, 24 h, 37°C	−	+	+	+	+	+	+	+	−
Motility at:									
37°C	−	+	+	D	+	−	+	+	−
25°C	D	+	+	+	+	−	+	+	−
Gas from glucose fermentation	− or W	+	+	+	+	D	+	+	−
Citrate (Simmons'), 37°C	−	−	+	−	+	D	+	D	−
Voges-Proskauer test, 25°C	D	+	−	−	D	D	−	D	−
Lysine decarboxylase	D	+	−	+	D	D	+	−	−
H$_2$S production (Kligler)	−	−	D	−	−	−	+	D	−
Phenylalanine deaminase	−	−	−	−	−	−	−	+	−
Mol% G + C of DNA	46–50	48–49	50–52	48–52	52–60	53–58	50–53	38–41	40–45

[a] Symbols: see standard definitions; also, W, weak reaction.

* Magnesium oxalate medium: blood agar base (BBL, or any other manufacturer if the Ca^{2+} content of the base is low), 40.0 g; distilled water, 830 ml. Sterilize at 121°C for 15 min and cool to 45°C. From stock solution sterilized by filtration, aseptically add the following ingredients: MgCl$_2$ solution (23.8 g/liter),80 ml; sodium oxalate solution (33.5 g/liter), 80 ml; and glucose solution (180.2 g/liter), 10 ml.

to embrace the available scientific knowledge and to comply with public health requirements.

Ursing et al. (1980b) have shown, on the basis of DNA and physiological data, that *Y. philomiragia* (Jensen et al., 1969) is not related to the genus *Yersinia* and, furthermore, that it is not a member of the family *Enterobacteriaceae*. These authors stated that until a proper assignment is made for this species it should be referred to as the "Philomiragia bacterium."

Acknowledgments

We wish to acknowledge our indebtedness to D. J. Brenner and L. Levy for reviewing this chapter and to R. Siegel for her help in typing, patience and skillful assistance.

Further Reading

Bercovier, H., D. J. Brenner, J. Ursing, A. G. Steigerwalt, G. R. Fanning, J. M. Alonso, G. P. Carter and H. H. Mollaret. 1980. Characterization of *Yersinia enterocolitica sensu stricto*. Curr. Microbiol. *4:* 201–206.

Bercovier, H., H. H. Mollaret, J. M. Alonso, J. Brault, G. R. Fanning, A. G. Steigerwalt and D. J. Brenner. 1980. Intra- and interspecies relatedness of *Yersinia pestis* by deoxyribonucleic acid hybridization and its relationship to *Yersinia pseudotuberculosis*. Curr. Microbiol. *4:* 225–230.

Bercovier, H., J. Ursing, D. J. Brenner, A. G. Steigerwalt, G. R. Fanning, G. P. Carter and H. H. Mollaret. 1980. *Yersinia kristensenii*: a new species of *Enterobacteriaceae* composed of sucrose-negative strains (formerly called atypical *Yersinia enterocolitica* or *Yersinia enterocolitica*-like). Curr. Microbiol. *4:* 219–224.

Brenner, D. J., H. Bercovier, J. Ursing, J. M. Alonso, A. G. Steigerwalt, G. P. Carter and H. H. Mollaret. 1980. *Yersinia intermedia*: a new species of *Enterobacteriaceae* composed of rhamnose-positive, melibiose-positive, raffinose-positive strains (formerly called atypical *Yersinia enterocolitica* or *Yersinia enterocolitica*-like). Curr. Microbiol. *4:* 207–212.

Brenner, D. J., J. Ursing, H. Bercovier, A. G. Steigerwalt, G. R. Fanning, J. M. Alonso and H. H. Mollaret. 1980. Deoxyribonucleic acid relatedness in *Yersinia enterocolitica* and *Yersinia enterocolitica*-like organisms. Curr. Microbiol. *4:* 195–200.

Carter, P. B., L. Lafleur and S. Toma (Editors). 1979. Third International Symposium on *Yersinia*, Contr. Microbiol. Immunol. Vol. 2, Karger, Basel.

Pollitzer, R. 1954. Plague. W. H. O. Monograph Series No. 22, World Health Organization, Geneva.

Ursing, J., D. J. Brenner, H. Bercovier, G. R. Fanning, A. G. Steigerwalt, J. Brault and H. H. Mollaret. 1980. *Yersinia frederiksenii*: a new species of *Enterobacteriaceae* composed of rhamnose-positive strains (formerly called atypical *Yersinia enterocolitica* or *Yersinia enterocolitica*-like). Curr. Microbiol. *4:* 213–218.

Differentiation of the species of the genus **Yersinia**

Characteristics useful in differentiating the various species of *Yersinia* are listed in Table 5.48.

List of the species of the genus **Yersinia**

1. **Yersinia pestis** (Lehmann and Neumann 1896) Van Loghem 1944, 15.[AL] (*Bacterium pestis* Lehmann and Neumann 1896, 194; *Yersinia pseudotuberculosis* subsp. *pestis* Bercovier, Mollaret, Alonso, Brault, Fanning, Steigerwalt and Brenner 1981, 383.[VP])

pes'tis. L. noun *pestis* plague, pestilence.

The characteristics are as described for the genus and as listed in Tables 5.48 and 5.49.

Three biovars have been described in relation to the geographical distribution of the organism: (a) biovar *antiqua* produces acid aerobically from glycerol, reduces nitrate to nitrite, does not ferment melibiose, and is found in Central Asia and Central Africa; (b) biovar *medievalis* produces acid from both glycerol and melibiose but does not reduce nitrate to nitrite; it is found in Iran and the U.S.S.R.; and (c) biovar *orientalis* (synonym: *oceanic*) does not produce acid from either glycerol or melibiose but reduces nitrate to nitrite and is distributed worldwide.

Some rare atypical strains positive in their reactions for urease and rhamnose have been reported.

Table 5.48.
Characteristics differentiating the species of the genus **Yersinia**[a]

Characteristics	1. *Y. pestis*	2. *Y. pseudo-tuberculosis*	3. *Y. enterocolitica*	4. *Y. intermedia*	5. *Y. frederiksenii*	6. *Y. kristensenii*	7. *Y. ruckeri*
Motility (25°C)	−	+	+	+	+	+	d
Lysine decarboxylase (Møller)	−	−	−	−	−	−	+
Ornithine decarboxylase (Møller)	−	−	+	+	+	+	+
Urease	−	+	+	+	+	+	−
β-Xylosidase[b]	+	+	−	−	d	−	−
Gelatinase	−	−	−	−	−	−	+
Citrate (Simmons'), 25°C	−	−[c]	−	+	d	−	+
Voges-Proskauer test, 25°C	−	−	+	+	+	−	d
Indole production	−	−	d	+	+	d	−
γ-Glutamyl transferase	−	d	+	+	+	+	+
Acid production from:							
Rhamnose	−	+	−	+	+	−	−
Sucrose	−	−	+	+	+	−	−
Cellobiose	−	−	+	+	+	+	−
Melibiose	d	+	−	−	−	−	−
α-Methyl-D-glucoside	−	−	−	+	−	−	−
Sorbose	−	−	+	+	+	+	−
Sorbitol	−	−	+	+	+	+	−
Raffinose	−	d	−	+	−	−	−

[a] For symbols see standard definitions.
[b] Using *p*-nitrophenyl-β-D-xylopyranoside as substrate.
[c] Strains belonging to serogroup IV are citrate-positive.

SECTION 5. FACULTATIVELY ANAEROBIC GRAM-NEGATIVE RODS

Table 5.49.

Other characteristics of the species of the genus **Yersinia**[a]

Characteristics	1. Y. pestis	2. Y. pseudo-tuberculosis	3. Y. enterocolitica[b]	4. Y. intermedia	5. Y. frederiksenii	6. Y. kristensenii	7. Y. ruckeri
Catalase	+	+	+	+	+	+	+
Oxidase	−	−	−	−	;	−	−
Pigment formed	−	−	−	−	−	−	−
Motility, 37°C	−	−	−	−	−	−	−
Methyl red test, 37°C	+	+	+	+	+	+	+
Voges-Proskauer test, 37°C	−	−	−	−	−	−	−
Citrate (Simmons'), 37°C	−	−	−	−	−	−	−
KCN, growth in, 37°C	−	−	−	−	−	−	d
Malonate utilization	−	−[c]	−	−	−	−	−
D-Tartrate utilization	−	−	−				−
Mucate utilization	−	−	−	d	−	−	−
Citrate (Christensen)	−	−	d	d	d	d	+
Nitrate reduced to nitrite	d	+	+	+	+	+	+
Oxidation-fermentation test (Hugh-Leifson)	O/F	O/F	O/F	O/F	O/F	O/F	O/F
D-glucose, gas production	−	−	v and W[d]	v and W	v and W	v and W	v and W
H₂S production (Kligler)	−	−	−	−	−	−	−
Tetrathionate reductase	−	d	d	+	+	d	
Phenylalanine or tryptophan deaminase	−						
Arginine dihydrolase (Møller)	−	−	−	−	−	−	−
β-Galactosidase[e]	+	+	+	+	+	+	+
Lipase (Tween 80)	−	−	d	d	d	d	
Deoxyribonuclease	+	d	d	−	−	−	−
Acid production from:							
Glucose, fructose, galactose, ribose, mannose, maltose, trehalose, N-acetylglucosamine, mannitol	+	+	+	+	+	+	+
L-Arabinose	+	+	+	+	+	+	−
Glycerol	d	+	+	+	+	+	−
i-Inositol	−	−	+	+	+	+	−
D-Xylose	+	+	d	+	+	+	−
Esculin	+	+	d	+	+	d	−
Amygdalin	−	−	v	+	+	v	−
Arbutin	+	d	v	+	+	v	
Salicin, dextrin	d	d	v	+	+	v	
Lactose	−	−	d	−	d	d	−
Adonitol, erythritol, dulcitol, D-arabinose, L-xylose, methyl-D-mannoside, methyl-xyloside, melezitose, inulin	−	−	−	−	−	−	−

[a] Tests were incubated at 28°C except where indicated, and were read during 3 days. For symbols see standard definitions.
[b] Tests are given for biovars 1 to 4.
[c] Strains belonging to serogroup IV are malonate-positive.
[d] W, weak reaction.
[e] Using o-nitrophenyl-β-D-galactopyranoside as substrate.

Y. pestis is the causative agent of plague. The disease can be reproduced experimentally in mice, rats, guinea-pigs and monkeys.

The mol% G + C of the DNA is 46 (T_m).

Type strain: ATCC 19428 (NCTC 5923).

2. **Yersinia pseudotuberculosis** (Pfeiffer 1889) Smith and Thal 1965, 220.[AL] (*Bacillus pseudotuberculosis* Pfeiffer 1889, 5; *Yersinia pseudotuberculosis* subsp. *pseudotuberculosis* (Pfeiffer 1889) Smith and Thal 1965, 220; see Bercovier et al. 1983, 383.)

pseu.do.tu.ber.cu.lo'sis. Gr. adj. *pseudes* false; M.L. fem. n. *tuberculosis* tuberculosis; M.L. gen. n. *pseudotuberculosis* of false tuberculosis.

The characteristics are as described for the genus and as listed in Tables 5.48 and 5.49.

Table 5.50.

Differentiation of the biovars of **Yersinia enterocolitica**

Characteristics	Biovar				
	1	2	3	4	5
Lipase (Tween 80)	+	−	−	−	−
Deoxyribonuclease	−	−	−	+	+
Indole production	+	+	−	−	−
Nitrate reduced to nitrite	+	+	+	+	−
Acid production from:					
D-Xylose	+	+	+	−	−
Sucrose	+	+	+	+	d
D-Trehalose	+	+	+	+	−

Some freshly isolated strains may require subculturing before expressing their motility.

Strains belonging to serogroup IV are citrate-positive (Simmons') and malonate-positive.

Up to 5% of *Y. pseudotuberculosis* strains have been reported to produce acid from adonitol.

Some strains, mostly of serogroup III, produce an exotoxin that differs from the *Y. pestis* toxin. The biological activity is not well defined.

Y. pseudotuberculosis is a human and animal pathogen responsible for mesenteric lymphadenitis, diarrhea and septicemia. The disease can be reproduced experimentally in guinea-pigs challenged *per os* and in mice. Aureomycin given orally to guinea-pigs induces the disease in healthy carriers.

The mol% G + C of the DNA is 46.5 (T_m).

Type strain: ATCC 29833 (NCTC 10275). This strain belongs to serogroup I.

3. Yersinia enterocolitica (Schleifstein and Coleman 1943) Frederiksen 1964, 104.[AL] (*Bacterium enterocoliticum* Schleifstein and Coleman 1943, 56.)

en.ter.o.co.li'ti.ca. Gr. n. *enteron* intestine; Gr. n. *colon* the colon; Gr. suff. *-iticos* pertaining to; M.L. fem. adj. *enterocolitica* pertaining to the intestine and colon.

The characteristics are as described for the genus and listed in Tables 5.48 and 5.49.

The Voges-Proskauer test is usually positive at 22–28°C and negative at 37°C.

Biovars of *Yersinia enterocolitica* are listed in Table 5.50 and, like phagovars and serogroups, are useful epidemiological tools.

Rare atypical strains that are either positive for their reactions on Simmons' citrate, for acid production from lactose and raffinose (due to a metabolic plasmid), or negative for urease activity have been reported.

When incubated at 20°C, *Y. enterocolitica* strains produce a broad spectrum mannose-resistant hemagglutinin which is lost at 37°C (MacLaglen and Old, 1980).

Y. enterocolitica is responsible for diarrhea, terminal ileitis, mesenteric lymphadenitis, arthritis and septicemia in man and animals. The disease can be reproduced experimentally in mice, gerbils and monkeys.

The species has been isolated from a wide variety of sources in the environment (live and inanimate) including foods and from healthy man and animals.

The mol% G + C of the DNA is 48.5 ± 1.5 (T_m), Bd.

Type strain: ATCC 9610 (strain 161; CIP 80-27). This strain belongs to biovar 1, serogroup 08 and phagover X_z.

4. Yersinia intermedia Brenner, Bercovier, Ursing, Alonso, Steig-

erwalt, Fanning, Carter and Mollaret 1981, 217.[VP] (Effective publication: Brenner et al. 1980, 207.)

in.ter.me'di.a. L. fem. adj. *intermedia* intermediate; here it implies that biochemical reactions of this species seem midway between *Y. enterocolitica* and *Y. pseudotuberculosis*.

The characteristics are as described for the genus and as listed in Tables 5.48 and 5.49.

Media with a high bile salt content (0.8%) are inhibitory, especially when incubated at 37°C.

Some biochemial characteristics (citrate utilization; cellobiose, rhamnose and raffinose fermentation) are always expressed at 25–28°C but are inconstant at 37°C.

Either a type A or a type B nitrate reductase is present.

Eight biovars have been described (Brenner et al., 1980a) based on the fermentation of melibiose, rhamnose, α-methyl-D-glucoside, raffinose and on the utilization of citrate (Simmons'). Of the strains studied, 96% are positive for at least four of these five tests.

Y. intermedia has been isolated mainly from fresh water sources, fish, foods, and occasionally from sick and healthy humans.

The mol% G + C of the DNA is 48.5 ± 0.5 (T_m, Bd).

Type strain: ATCC 29909 (strain 3953; Bottone 48; Chester 48; CIP 80-28).

5. Yersinia frederiksenii Ursing, Brenner, Bercovier, Fanning, Steigerwalt, Brault and Mollaret 1981, 217.[VP] (Effective publication: Ursing et al. 1980, 213.)

fred.er.ik.sen'i.i. M.L. gen. n. *frederiksenii* of Frederiksen; named after the Danish microbiologist Wilhelm Frederiksen, who made a substantial contribution to the study of the genus *Yersinia*.

The characteristics are as described for the genus and as listed in Tables 5.48 and 5.49.

This species is composed of three different genetic groups. One group is positive for β-xylosidase and citrate (Simmons'), and the type strain belongs to this group. The other two groups are variable or negative for these tests. More phenotypic studies are needed to differentiate the three groups.

Some strains are able to ferment raffinose and lactose when they harbor a metabolic plasmid.

Y. frederiksenii has been isolated mainly from fresh water sources, fish, foods, and occasionally from healthy or sick man and animals.

The mol% G + C of the DNA is 48 (T_m).

Type strain: CIP 80-29 (strain 6175).

6. Yersinia kristensenii Bercovier, Ursing, Brenner, Steigerwalt, Fanning, Carter and Mollaret 1981, 217.[VP] (Effective publication: Bercovier et al. 1980, 219.)

kris.ten.se'ni.i. M.L. gen. n. *kristensenii* of Kristensen, named after

Table 5.51.

DNA relatedness and divergence in related sequences in the genus **Yersinia**[a]

Source of Unlabeled DNA	1. Y. pestis	2. Y. pseudo-tuberculosis	3. Y. enterocolitica	4. Y. intermedia	5. Y. frederiksenii	6. Y. kristensenii	7. Y. ruckeri
Y. pestis	97(0)	88(0.1)	43				
Y. pseudotuberculosis		92(1)	59(12)	54(12)		50	30(15)
Y. enterocolitica		48(11.5)	96(2.5)	58(11)	60(12.5)	69(9)	30(15)
Y. intermedia		44(12)	59(12)	95(1.5)	61(11)	62(12)	
Y. frederiksenii		44(11)	67(11)	58(12.5)	81(5)	59(10)	
Y. kristensenii		44(11)	70(9.5)	62(12.5)	55(12)	84(4)	
Y. ruckeri		33(13)	30(15)	38			95(0.1)

[a] Data are from Bercovier et al. (1980), Brenner et al. (1976), Brenner et al. (1980), Ewing et al. (1978), and Ursing et al. (1980). Hybridizations were carried out at 60°C. The first number is the average relatedness in per cent of all unlabeled strains with the specific labeled DNA. Homologous reactions are not included in the average. The second number, in parentheses, is the percentage of divergence calculated on the basis of 1% unpaired bases per 1°C decrease in duplex stability.

the Danish microbiologist Martin Kristensen, who first isolated this organism.

The characteristics are as described for the genus and as listed in Tables 5.48 and 5.49.

Growth is delayed (7 days) when cultures are incubated at 41°C and even at 37°C for some isolates.

Some strains utilize citrate (Simmons') after 7 days incubation at 25°C.

Most strains produce a "musty" or "cabbage-like" odor when grown on nutrient agar.

Some strains produce an enterotoxin (ST) when incubated at 22°C and also at 37°C (Kapperud, 1980).

Y. kristensenii strains have been isolated mainly from soil, from various environmental sources (fresh water, foods) and rarely from healthy or sick man and animals.

The mol% G + C of the DNA is 48.5 ± 0.5 (T_m, Bd).

Type strain: CIP 80-30 (strain 105).

7. **Yersinia ruckeri** Ewing, Ross, Brenner and Fanning 1978, 37.[AL]
ruck′er.i. M.L. gen. n. *ruckeri* of Rucker; name after R. R. Rucker, who studied the red mouth disease and its etiological agents.

The characteristics are as described for the genus and as listed in Tables 5.48 and 5.49.

The cells are 1 μm in width and 2–3 μm in length. Filaments can be seen in old cultures (48 h at 22°C).

Colonies on nutrient agar are smooth, circular and slightly raised. Growth is delayed or inhibited on salmonella-shigella agar incubated at 37°C but not at 22°C.

Corn oil is hydrolyzed when the test is performed at 22°C but not at 37°C.

Y. ruckeri is one of the agents responsible for red mouth disease in rainbow trout. The disease can be transmitted experimentally from fish to fish. The organism has been isolated only in North America.

The mol% G + C of the DNA is 48 ± 0.5 (Bd).

Type strain: ATCC 29473.

Species Incertae Sedis

Yersinia philomiragia Jensen, Owen and Jellison 1969, 1237.[AL]
phi.lo.mi.ra′gi.a. Gr. adj. *philos* loving; M.L. n. *miragia* plural of Latinized English word *mirage*; *philomiragia* loving mirages, because of the mirages that are seen in the area where the isolations of this species were made.

A fastidious organism pathogenic for muskrats. Because of a morphological resemblance to *Y. pestis* and some degree of DNA relatedness to *Y. pestis* (Ritter and Gerloff, 1966), the organism was assigned to the genus *Yersinia*. Later studies indicated, however, that no significant DNA relatedness occurred between *Y. philomiragia* and other *Yersinia* species, other *Enterobacteriaceae* or *Pasteurella multocida* (Ursing et al., 1980b). The strains that have been investigated are phenotypically similar to one another and form a homogenous DNA relatedness group (Ursing et al., 1980b). Until an appropriate genus assignment can be made, it is recommended that the organism be referred to as the "Philomiragia bacterium."

Motility is negative at both room temperature and 36°C.

Strains of the species give positive tests for the following reactions (36°C): catalase, indole production, gelatin hydrolysis, Voges-Proskauer test, and acid production from D-glucose, maltose and sucrose. Reactions for the following tests differ among strains: β-galactosidase and acid production from D-fructose and galactose. The following reactions are negative: oxidase, nitrate reduction to nitrite, H2S production, urease, arginine dihydrolase, lysine decarboxylase, ornithine decarboxylase, phenylalanine deaminase, citrate utilization (Simmons'), esculin hydrolysis, and acid production from L-arabinose, L-rhamnose, D-raffinose, lactose, D-mannitol, D-sorbitol, i-inositol and salicin. Gas is not produced from carbohydrate fermentation.

First isolated in 1959 from a dead muskrat found in a marshy area at the Bear River Migratory Bird Refuge in northern Utah. Other strains have been isolated from water in the same area.

Type strain: ATCC 25015.

Other genera of the family **Enterobacteriaceae**

JOHN J. FARMER III

The previous sections describe the genera of *Enterobacteriaceae* which have been known and thoroughly studied for many years and whose generic names are familiar to most microbiologists. A number of other genera in the family, however, have received little attention: *Obesumbacterium* and *Xenorhabdus* because of their limited ecological niche, *Kluyvera* because of its poor recognition by the scientific community, and *Rahnella*, *Tatumella* and *Cedecea* because of their newness.

The purpose of this section is to acquaint the reader with the limited information about these new genera. For obvious reasons, the material presented cannot be as complete as that given for the other genera in the family. Over the last few years, the Enteric Laboratories at the Centers for Disease Control (CDC) has characterized isolates of all these new genera. This section will summarize the material in the literature about the new groups as well as our own findings.

Unless otherwise stated, all data are based on cultures studied by the Enteric Laboratories, CDC. Biochemical testing was by the method of Edwards and Ewing (1972), which has been updated (Farmer et al., 1980). Incubation was at 36 ± 1°C except for cultures of *Xenorhabdus*, which grew poorly or not at all at 36°C and which were tested at 25°C

(Table 5.52). All enzyme names should be understood to be in quotation marks since actual enzyme assays with cell-free extracts were not done.

Stocks cultures of the six genera were prepared and stored in the same way as other *Enterobacteriaceae*. Cultures were preserved by two methods. In method 1, growth was taken from a trypticase soy agar plate (or any agar medium that allows optimum growth) and a heavy suspension was made in 10% w/v skim milk. This was "quick frozen" in a beaker of 95% ethanol (kept in the −70°C freezer) and then stored at −70°C. In method 2, cultures were inoculated into a solid or semisolid medium (100-× 13-mm screw-cap tubes) such as trypticase soy agar, trypticase soy semisolid (0.4% agar) or blood agar base, incubated 1–2 days until growth was obvious, sealed with a number "000" white rubber stopper and stored in the dark at room temperature. This latter stock culture is called the "working stock" and the −70°C culture is called the "freezer stock." Almost all *Enterobacteriaceae* survive well with both methods, except *Tatumella* cultures which may die when only method 2 is used. Important cultures of *Enterobacteriaceae* are preserved by both methods, but routine cultures are stocked only by method 2. *Tatumella* stocks are done both ways.

Genus **Obesumbacterium** Shimwell 1963, 759[AL]

O.be′sum.bac.te′ri.um. L. neut. adj. *obesum* fat; L. neut. n. *bacterium* rod; M.L. neut. n. *Obesumbacterium* a fat, rod-shaped bacterium.

Pleomorphic rods 0.8–2.0 μm in diameter, 1.5–100 μm in length (short, "fat" rods predominate when grown in beer wort with live yeasts, long pleomorphic rods usually predominate when grown in

most bacteriological media), conforming to the general definition of the family *Enterobacteriaceae*. **Nonmotile.** Facultatively anaerobic. Very slow growing, forming colonies less than 0.5 mm in diameter on

Table 5.52.

Characteristics of the species of the genera **Obesumbacterium, Xenorhabdus, Kluyvera, Rahnella, Cedecea** *and* **Tatumella**[a]

Characteristics	Obesumbacterium proteus biogroup 1	Obesumbacterium proteus biogroup 2	Xenorhabdus nematophilus	Xenorhabdus luminescens	Kluyvera ascorbata	Kluyvera cryocrescens	Rahnella aquatilis	Cedecea davisae	Cedecea lapagei	Tatumella ptyseos
% Hybridization (DNA/DNA) with *E. coli*	27	43	4	4	34	34	19	10	14	27
Mol% G + C of DNA	48–49	48–49	43–44	43–44	56–57	55	51–56	49–50	48–52	53–54
Catalase	+	+	−	+	+	+	+	+	+	+
Nitrate reduced to nitrite	+	+	[−]	−	+	+	+	+	+	+
Cell length >4 μm	d	d	+	+	−	−	−	−	−	−
Colonies >1 mm on Trypticase soy agar, 24 h, 36°C	−	−	−	−	+	+	+	+	+	−
Bioluminescent	−	−	−	+	−	−	−	−	−	−
Survive for 1 month on agar	+	+	+	+	+	+	+	+	+	−
Inhibited by penicillin (10 U/disk)	−	−	−	−	−	−	−	−	−	+
Incubation temperature (°C) for the tests listed below:	36	36	25	25	36	36	36	36	36	36
Indole production	−	−	d	d	+	[+]	−	−	−	−
Methyl red	[+]	[−]	−	−	+	+	[+]	+	d	−
Voges-Proskauer	d	−	−	−	−	−	+	[+]^w	+	[−]
Citrate utilization (Simmons')	−	−	−	d	+	[+]	+	+	+	−
H$_2$S on triple-sugar iron agar	−	−	−	−	−	−	−	−	−	−
Urea hydrolysis	−	−	−	[−]	−	−	−	−	−	−
Phenylalanine deaminase	−	−	−	−	−	−	+^{w}	−	−	[+]^w
Amino acid decarboxylases (Møller's):										
Lysine decarboxylase	+	+	−	−	+	[−]	−	−	−	−
Arginine dihydrolase	−	−	−	−	−	−	−	d	[+]	−
Ornithine decarboxylase	d	+	−	−	+	+	−	+	−	−
Motility	−	−	+	+	+	[+]	−	+	[+]	−
Gelatin hydrolysis, 22°C	−	−	[+]	d	−	−	−	−	−	−
KCN, growth in	−	−	−	−	+	[+]	−	[+]	+	−
Malonate utilization	d	−	−	−	+	[+]	+	+	+	−
Gas from D-glucose	−	−	−	−	+	+	[+]	[+]	+	−
Acid production from:										
D-Glucose	+	+	[+]^w	[+]^w	+	+	+	+	+	+
Adonitol	−	−	−	−	−	−	−	−	−	−
L-Arabinose	−	−	−	−	+	+	+	−	−	−
D-Arabitol	−	−	−	−	−	−	−	+	+	−
Cellobiose	−	−	−	−	+	+	+	+	+	−
Dulcitol	−	−	−	−	[−]	−	[+]	−	−	−
Erythritol	−	−	−	−	−	−	−	−	−	−
Glycerol	−	−	−	−	d	−	[−]	−	−	−
i(myo)-inositol	−	−	−	−	−	−	−	−	−	−
Lactose	−	−	−	−	+	+	+	[−]	d	−
Maltose	−	d	−	[−]^w	+	+	+	+	+	−
D-Mannitol	d	−	−	−	+	+	+	+	+	−
D-Mannose	+	+	[+]^w	+	+	+	+	+	+	+
Melibiose	−	−	−	−	+	+	+	−	−	d
α-Methyl-D-glucoside	−	−	−	−	+	+	−	−	−	−
Mucate	−	−	−	−	+	[+]	d	−	−	−
Raffinose	−	−	−	−	+	+	+	−	−	[−]
L-Rhamnose	−	[−]	−	−	+	+	+	−	−	−
Salicin	d	−	−	−	+	+	+	+	+	d
D-Sorbitol	−	−	−	−	d	d	+	−	−	−
Sucrose	−	−	−	−	+	[+]	+	+	−	+
Trehalose	d	d	−	−	+	+	+	+	+	+
D-Xylose	−	[−]	−	−	+	+	+	+	−	−
Esculin hydrolysis	−	−	−	−	+	+	+	d	+	−
Tartrate, Jordan's	d	[−]	d	d	d	[−]	−	−	−	−
Acetate utilization	−	−	−	−	d	[+]	−	−	d	−
Lipase (corn oil)	−	−	−	−	−	−	−	+	+	−
Deoxyribonuclease (DN'ase), 25°C	−	−	[−]	−	−	−	−	−	−	−
Oxidase (Kovacs')	−	−	−	−	−	−	−	−	−	−
β-Galactosidase (ONPG test)	d	−	−	−	+	+	+	+	+	−
Yellow pigment on trypticase soy agar	−	−	d	d	−	−	−	−	−	−

[a] Symbols: +, positive for 90–100% of strains; [+], positive for 75–89% of strains; d, positive for 26–74% of strains; [−], positive for 11–25% of strains; −, positive for 0–10% of strains. All reactions are for 36°C unless otherwise indicated and 48 h incubation. The superscript "w" indicates a weak reaction.

ordinary plating media at 24 h. Optimum temperature, ~32°C. **Acid formed from D-glucose and D-mannose; very few other carbohydrates are fermented.** Gas formation during fermentation is variable (original description says gas is produced, but none of the strains studied produced gas). **Lysine decarboxylase is positive.** Nitrate is reduced to nitrite. **Many biochemical tests normally used for differentiation of *Enterobacteriaceae* are negative or delayed. Occurs as a brewery contaminant** which can survive and grow in the presence of live yeasts during beer production. The mol% G + C of the DNA is 48–49 (Bd). The genus has a single species, *O. proteus*, with two defined biogroups (1 and 2). Howwever, these two biogroups are really distinct species which are phenotypically different and only distantly related by DNA/DNA hybridiation.

Type species: *Obesumbacterium proteus* (Shimwell and Grimes 1936) Shimwell 1963, 759.

Further Descriptive Information

Shimwell's original description of an organism he called *Flavobacterium proteum* centered on its cellular morphology (Shimwell, 1936; Shimwell and Grimes, 1936). He noted that it appeared as plump rods 0.8–1.2 × 1.5–4 μm when grown in wort media or when taken directly from breweries during fermentation. This morphology had no doubt led to the term "short fat rod of pitching yeasts" which had been used in breweries for many years. Shimwell (1936) also noted much pleomorphism when the organism was grown in laboratory media which were alkaline or neutral. Chains of cells up to 100 μm were observed under these conditions.

A number of reports discuss the ecology of *Obesumbacterium proteus* in breweries (Case, 1965; Shimwell, 1936, 1948, 1963, 1964; Shimwell and Grimes, 1936; Strandskov et al., 1953). Unfortunately, no distinction has been made between biogroups 1 and 2, so it is usually impossible to determine whether these two biogroups, which are really distinct species, are different in their ecology, distribution, and other factors. Future studies in breweries should resolve this problem, but only if the two distinct biogroups of *O. proteus* are differentiated. This approach was recently used by van Vuuren (1978) in South African breweries.

Taxonomic Comments

Obesumbacterium was first proposed as a genus by Shimwell in 1963 (Shimwell, 1963, 1964) to accomodate the organism known as "*Flavobacterium proteus*" (Shimwell and Grimes, 1936), which was called "*Flavobacterium proteum*" in the original proposal (Shimwell and Grimes, 1936). This specific epithet "proteus" was chosen because the organism has a very pleomorphic cell morphology depending on the particular growth conditions (Shimwell 1936, 1948, 1964). The organism was first named in 1936 when Shimwell was doing his studies on the "short fat rods of pitching yeasts." He gave an adequate description (based on the techniques of that time) of the organism, but much of his description is not helpful today in identifying the organism. A pure culture of "*Flavobacterium proteus*" (isolated from yeast of Beamish and Crawford's brewery, Cork, Ireland) was deposited by Shimwell in the National Collection of Type Cultures (NCTC), England. However, when Shimwell's own culture was "lost" he wrote the NCTC and found they had also "lost" this culture (unpublished letter of 27 February 1964 from J. L. Shimwell to E. F. Lessel of the ATCC). Thus it appears that no culture has survived from those originally studied by Shimwell in writing his description of "*F. proteus.*" The strain most studied is apparently ATCC 12841 (NCIB 8771; strain 42 of Strandskov and Bockelmann), and until 1980 it was only a reference strain (Bergey 8, p. 363). This strain was isolated from lager and ale yeasts and deposited by the Shaefer Brewing Co. (Strandskov and Bockelmann, 1955). This strain was given status without comment as the type strain (neotype) of *Obesumbacterium proteus* when the *Approved Lists of Bacterial Names* were issued in 1980.

There is confusion whether the current type of *Obesumbacterium proteus* is the same organism that Shimwell studied and named "*Fla-*

vobacterium proteus." In 1956, the NCTC sent three cultures (No. 42, No. 2, No. 41) to Shimwell to examine and determine whether he thought they were "*F. proteus.*" These cultures had been isolated and described by Strandskov and Bockelmann (1955). For No. 42 (the current type strain of *O. proteus*) Shimwell concluded: "This is almost certainly an authentic strain. Its morphology is almost exactly that of my original isolation, namely thick (up to two or more μ), long (up to 100 μ or more) filaments etc. etc. together with the usual short fat rods in fair numbers. . . . I have not studied the biochemical properties of these strains. I have assumed that those published by Strandskov and Bockelmann of the F. & M. Schaefer Brewing Co., Brooklyn, New York, are correct i.e. Indol, acetylmethylcarbonal, H₂S and starch, all negative; nitrite from nitrate positive. As you know, however, probably hundreds of Gram-negative bacteria would answer to this description. Indeed, the species is very poorly characterized biochemically, its main characteristic being its extremely large cell size, and its almost incredible pleomorphism, by means of which, (taken in conjunction with its presence in a brewery fermentation) it can readily be identified. In cataloguing any of the strains I suggest that No. 42 could be safely named *F. proteus*; No. 2 a little doubtfully, and No. 41 very doubtful indeed." (Letter of 7 June, 1956 from J. H. Shimwell to W. S. Greaves of the NCTC.) Thus the current concept of *Obesumbacterium proteus* based on its type strain ATCC 12841 (strain No. 42, above) is not incompatible with "*Flavobacterium proteus*" defined by Shimwell based on strains which no longer exist. Unfortunately, Shimwell's original description of "*Flavobacterium proteum*" fits *Obesumbacterium proteus* biogroup 1 and biogroup 2, so it is uncertain which of these latter two organisms (or perhaps both) was originally studied by Shimwell (1936). It is even possible that it was neither since Shimwell described "*Flavobacterium proteum*" as a producer of gas during fermentation of carbohydrates, but neither biogroup of *O. proteus* produces gas (Priest et al., 1973; see also Table 5.52). However, Shimwell's concept of *O. proteus* was based mainly on its cellular morphology and its ecological niche in beer wort fermentations rather than its biochemical reactions.

"*Flavobacterium proteus*" has been known for many years as a brewery contaminant which can survive and grow in the presence of live yeasts during beer production. Because it fermented D-glucose and other carbohydrates, it was incompatible with a redefined genus *Flavobacterium* which was limited to oxidative rather than fermentative bacteria. The removal from *Flavobacterium* was later confirmed by Bauwens and De Ley (1981), who showed that by DNA/rRNA hybridization experiments that "*F. proteus*" was not closely related to *Flavobacterium*. Shimwell formed the new genus *Obesumbacterium* for "*F. proteus*," because its phenotypic properties and ecological niche differed from those of other genera. Shimwell (1963, 1964) did not assign *Obesumbacterium* to a particular family, but its properties (see Table 5.52) are compatible with the family *Enterobacteriaceae*.

Priest and co-workers (1973) determined the phenotypic properties and did DNA/DNA hybridization experiments on 19 strains of *O. proteus*, including 16 brewery isolates. From their overall results, they defined two biogroups, which had the same guanine plus cytosine content of the DNA, 48.0–48.5 mol%. They also proposed that *O. proteus* be transferred to the genus *Hafnia* where its citation would be "*Hafnia protea*" (Shimwell and Grimes, 1936) Priest et al., 1973. (The new combination could have been proposed as "*Hafnia proteus*," since "proteus" is a substantive which need not agree in gender with its genus.) This change in classification was accepted to some extent, but in reality both *O. proteus* and "*H. protea*" have been rarely mentioned in the literature. Most of the existing citations have been in journals related to brewing. The name "*Hafnia protea*" lost standing in nomenclature on January 1, 1980, because it did not appear on the *Approved Lists of Bacterial names*; however, the names *Obesumbacterium* and *O. proteus* have had standing in nomenclature since they appear on the lists.

The classification of *Obesumbacterium* was clarified by Brenner and co-workers (Brenner, 1981), who examined the taxonomic position of *O. proteus* biogroups 1 and 2 (of Priest et al., 1973) by DNA/DNA hybridization techniques. *O. proteus* biogroup 1 was very highly related to *Hafnia alvei*, and it was concluded that this biogroup is a synonym of *Hafnia alvei*. It can best be thought of as the pleomorphic, KCN-negative, nonmotile, nongas-producing, salicin-positive, L-arabinose-negative, L-rhamnose-negative, maltose-negative, D-xylose-negative, β-galactosidase-negative biogroup of *Hafnia alvei* which has become adapted to the brewery environment. This adaptation to the brewery environment was noted by Shimwell and Grimes (1936) in the original description of "*Flavobacterium proteus*": "The organism sometimes failed to grow in dilute media, probably owing to its having become accustomed to the more concentrated nature of beer-wort in the brewery." This adaptation has presumably made the organism very "sluggish" in its metabolic activities, as is reflected in its slow growth rate and diminished activity in the tests normally done for identification of *Enterobacteriaceae*. The classification of *O. proteus* biogroup 1 as *H. alvei* is strengthened by the fact that strains of *O. proteus* biogroup 1 are lysed by the *Hafnia*-specific bacteriophage 1672 described by Guinee and Valkenburg (1968) (Enteric Laboratories, CDC, unpublished data, summarized later in Table 5.54).

Van Vuuren and co-workers (1981) studied 10 cultures of "*Hafnia alvei-Obesumbacterium proteus*" isolated from South African lager-beer breweries. One of their five distinct biogroups (based on API 20E Profiles) was very active biochemically and more like typical cultures of *Hafnia alvei*. The other four biogroups were progressively less biochemically active with the least active group more like *Obesumbacterium proteus*. Based on these data, it appears that biogroups intermediate between typical *Hafnia alvei* and *Obesumbacterium proteus* biogroup 1 may occur in breweries. A complete set of biochemical tests (see Table 5.52) and lysis by the *Hafnia* specific bacteriophage are needed to characterize isolates from breweries that resemble *Hafnia/Obesumbacterium*.

O. proteus biogroup 2 is different biochemically from *O. proteus* biogroup 1 and from *H. alvei* (see Table 5.53). In DNA/DNA hybridization studies *O. proteus* biogroup 2 was only 25–30% related to *O. proteus* biogroup 1 (Brenner, 1981). Interestingly, among the *Enterobacteriaceae* the closest relative to *O. proteus* biogroup 2 was *Escherichia blattae* which was 60–65% related. Thus *O. proteus* biogroup 2 should not be included in the same species as *O. proteus* biogroup 1. In fact, the two taxa are so distantly related in a phylogenetic sense that they probably should be classified in different genera.

The type strain of *Obesumbacterium proteus* is ATCC 12841 (NCIB 8771). The 14th edition of the ATCC catalog has it listed as "*Hafnia protea*"; the 15th edition has it listed as *Obesumbacterium proteus*. This culture is lysed by the *Hafnia*-specific bacteriophage 1672, is methyl red-negative, D-mannitol-negative, salicin-positive (6 days), maltose-negative, D-xylose-negative, esculin-positive (7 days) and ONPG-negative. Thus from Table 5.53 (below) it clearly belongs to *Obesumbacterium proteus* biogroup 1, although it is somewhat more inactive or slower in its biochemical reactions than the other strains.

At present *O. proteus* biogroup 1 can be logically considered as a subjective synonym of *Hafnia alvei*. The best eventual solution is probably to name it as a distinct subspecies or biogroup of *H. alvei*. However, its classification as a named subspecies could cause some serious nonmenclatural problems in *Hafnia* because of priorities. Since the name *Obesumbacterium* has traditionally been well known in the brewing industry, which seems to be the ecological niche for this organism, the most practical solution at present is to continue use of the names *Obesumbacterium proteus* biogroup 1 and *Obesumbacterium proteus* biogroup 2 until all the taxonomic problems are resolved and a "final" nonmenclature and classification can be proposed after careful analysis. The one disadvantage to this proposal is that the name "*Hafnia protea*" is now being used by brewery bacteriologists (H. J. J. van Vuuren, personal communication); however, this name has no standing in nomenclature. Since *O. proteus* is a heterogeneous species, it is essential to append "biogroup 1" or "biogroup 2" to the name which will correspond to the two distinct species shown by DNA/DNA hybridization and biochemical tests. Unless this is done, the intended meaning of *Obesumbacterium proteus* is unclear.

Further Reading

Case, A.C. 1965. Conditions controlling *Flavobacterium proteus* in brewery fermentations. J. Inst. Brew. *71:* 250–256.
Priest, F.G., H.J. Somerville, J.A. Cole and J.S. Hough. 1973. The taxonomic position of *Obesumbacterium proteus*, a common brewery contaminant. J. Gen. Microbiol. *75:* 295–307.
Shimwell, J.L. 1936. A study of the common rod bacteria of brewers' yeast. J. Inst. Brew. *42:* 119–127.
Shimwell, J.L. 1963. *Obesumbacterium* gen. nov. Brewers' J. *99:* 759–760.
Shimwell, J.L. and M. Grimes, 1936. The distinguishing characters of *Flavobacterium proteum* (sp. nov.), the common rod bacterium of brewers' yeast. J. Inst. Brew. *42:* 348–350.

List of the species of the genus **Obesumbacterium**

1. **Obseumbacterium proteus** (Shimwell and Grimes 1936) Shimwell 1963, 759.* (*Flavobacterium proteum* (*sic*) Shimwell and Grimes 1936, 348).

pro'te.us. Gr. masc. n. *Proteus*, the ancient Greek sea-god noted for being able to change his form at will; Gr. masc. noun *proteus* pleomorphic.

Obesumbacterium proteus is really two different species, which can be differentiated by phenotypic tests (Table 5.53) and DNA/DNA hybridization. The type strain has the properties of *Obesumbacterium proteus* biogroup 1. The description is as given for the genus and as listed in Table 5.52.

Occurs in breweries where it grows in beer wort along with yeasts early in the fermentation.

The mol% G + C of the DNA is 48 to 49 (Bd).

Type strain (neotype): ATCC 12841 (NCIB 8771; strain No. 42 of Strandskov and Bockelmann (1955)).

Table 5.53.

Differentiation of **Obesumbacterium proteus** *biogroup 1,* **Obesumbacterium proteus** *biogroup 2 and* **Hafnia alvei**[a]

Characteristics	Incubation time, (days)	O. proteus Biogroup 1	O. proteus Biogroup 2	H. alvei
Lysis by the *Hafnia*-specific bacteriophage of Guinée and Valkenburg (1968)	1	+	−	+
Voges-Proskauer (22°C)	4	+	−	+
Acid production from:				
D-Mannitol	10	+	−	+
Salicin	7	+	−	[−]
D-Xylose	7	−	+	+
Esculin hydrolysis	7	+	−	[−]

[a] Symbols: same as indicated for Table 5.52, but they apply to the incubation times given in Table 5.53.

---

* This citation of authors for the name *O. proteus* is different from the one that appears on the 1980 *Approved Lists of Bacterial Names*, which omits "Shimwell and Grimes 1936." Presumably this was an inadvertent omission.

Genus **Xenorhabdus** Thomas and Poinar 1979, 354[AL]

Xe.no.rhab′dus. Gr. n. *xenos* unwanted guest (less literally, "pathogen"); Gr. fem. n. *rhabdus* rod; M.L. masc. n. *Xenorhabdus* pathogenic rod-shaped bacterium.

Rod-shaped cells 0.8–2 μm × 4–10 μm. In older cultures the cells contain crystalline inclusions (not poly-β-hydroxybutyrate). **Coccoid bodies** (spherical cells), resulting from the disintegration of the cell wall, are **also formed in older cultures** and average 2.6 μm in size. **Motile** by means of peritrichous flagella. Facultatively anaerobic, having both a respiratory and a fermentative type of metabolism. One species is catalase-negative; the other is bioluminescent. **Optimum temperature, ~25°C; grow poorly or not at all at 36°C. Acid production from glucose is weak or delayed, even at 25°C; most older carbohydrates are not fermented or the amount of acid produced is very small.** Nitrate is not reduced to nitrite. **Most biochemical tests used for differentiation of the *Enterobacteriaceae* are negative. Isolated only from nematodes** of the genera *Neoaplectana* and *Heterorhabditis* and from the insect larvae they parasitize. The mol% G + C of the DNA is 43–44 (Bd)

Type species: *Xenorhabdus nematophilus* (Poinar and Thomas 1965) Thomas and Poinar 1979, 355.

Further Descriptive Information

Xenorhabdus strains grow poorly or not at all at 36°C. They grow very slowly on plating media and are sluggish in their biochemical reactions, even at 25°C (see Table 5.52). These properties should make them easy to differentiate from other named species in *Enterobacteriaceae*. Acid production from D-glucose and other carbohydrates is very weak (to the point that a different pathway of glucose fermentation may be involved). In enteric fermentation base with Andrade's indicator, there was no evidence that most compounds were fermented (Table 5.52). These data are different from those of Thomas and Poinar (1979), who found strains of *Xenorhabdus* much more active in their fermentation reactions, although they also state that these reactions were weak. These differences must be considered by those trying to identify strains suspected to be *Xenorhabdus*. *Xenorhabdus* strains grow poorly on MacConkey agar; only a fraction of the population forms visible colonies (low plating efficiency). *Xenorhabdus* produces various pigments on different media (Thomas and Poinar, 1979). Antibiograms (Bauer et al., 1966) must be done at 25°C and incubated for 3 days to observe clear zones. Both species have large zones of inhibition around disks impregnated with nalidixic acid, gentamicin, streptomycin, kanamycin, tetracycline, and chloramphenicol, but there is no zone around penicillin. Resistance is variable from strain to strain with colistin, ampicillin, carbenicillin and cephalothin.

X. nematophilus has been isolated only from nematodes of the genus *Neoaplectana* and from the insect larvae they parasitize. *X. luminescens* has been isolated only from nematodes of the genus *Heterorhabditis* and the insect larvae they parasitize. None of the several thousand *Enterobacteriaceae* cultures reported as "unidentified" by CDC's Enteric Laboratories resembles *Xenorhabdus*, so there is no evidence that

it causes human infection or occurs in clinical specimens; however, *Xenorhabdus* probably would not be isolated if it did occur in a clinical specimen.

Taxonomic Comments

This genus was proposed by Thomas and Poinar (1979) in the *International Journal of Systematic Bacteriology* and was included on the *Approved Lists of Bacterial Names* with its two species, *X. nematophilus* and *X. luminescens*.

In 1965, Poinar and Thomas noted the association of an unusual bacterium in the intestinal lumen of the nematode species *Neoaplectana carpocapsae* (in the original article, this nematode was referred to as an undescribed species (DD-156) of the nematode genus *Neoaplectana*). After the bacterium was studied by several reference laboratories, who reported they could not identify it, Poinar and Thomas (1965) named it "*Achromobacter nematophilus*," with its holotype strain designated as ATCC 19061. The etymology of "nematophilus" was not given in the original proposal. About 10 years later, Hendrie and coworkers (1974) proposed that the genus *Achromobacter* be rejected and that the species of *Achromobacter* be transferred to *Alcaligenes* and to other genera. Although this proposal was satisfactory for many of the nonfermentative species of *Achromobacter*, it left "*A. nematophilus*" without a genus because it is a fermenter (although very weak).

In 1975, Poinar and Thomas received a bacterium isolated from an Australian nematode which was somewhat similar to their strain isolated in 1965, but the Australian isolate was bioluminescent. They were confronted with a problem in classification, since *Achromobacter* had been "abolished" and they now had two similar species without a genus. Their solution was to propose the new genus *Xenorhabdus* in the family *Enterobacteriaceae* to include "*A. nematophilus*" and the luminescent species which they named *X. luminescens*. Table 5.52 compares the properties of *Xenorhabdus* with the other new genera. Strains of *Xenorhabdus* contain the *Enterobacteriaceae* common antigen (Enteric Laboratories, upublished data). However by DNA/DNA hybridization (hydroxyapatite, 60°C) *Xenorhabdus* is only 4% related to *Escherichia coli*, the type species on the type genus for the family (Enteric Section, unpublished data). These data indicate that at present *Xenorhabdus* should be retained in the family *Enterobacteriaceae* although it is only distantly related to the genera that comprise the "core" of the family.

Further Reading

Poinar, G.O., Jr. and G.M. Thomas. 1965. A new bacterium, *Achromobacter nematophilus* sp. nov. (*Achromobacteriaceae: Eubacteriales*) associated with a nematode. Int. Bull. Bacteriol. Nomencl. Taxon. *15:* 249–252.

Poinar, G.O., Jr. and G.M. Thomas. 1967. The nature of *Achromobacter nematophilus* as an insect pathogen. J. Invertebr. Pathol. *9:* 510–514.

Thomas G.M. and G.O. Poinar Jr. 1979. *Xenorhabdus* gen. nov., a genus of entomopathogenic nematophilic bacteria of the family *Enterobacteriaceae*. Int. J. Syst. Bacteriol. *29:* 352–360.

List of the species of the genus **Xenorhabdus**

1. **Xenorhabdus nematophilus** (Poinar and Thomas 1965) Thomas and Poinar 1979, 355.[AL] (*Achromobacter nematophilus* Poinar and Thomas 1965, 249.)

ne.ma.to′phi.lus. Modern entomological term *nematode*; Gr. adj. *philus* loving, or having affinity for; M.L. adj. *nematophilus* nematode-loving.

The characteristics are as described for the genus and as listed in Tables 5.52 and 5.54.

Isolated from nematodes of the genus *Neoaplectana*.

The mol% G + C of the DNA is 43–44 (Bd).

Type strain (holotype): ATCC 19061.

2. **Xenorhabdus luminescens** Thomas and Poinar 1979, 355.[AL]

lu.mi.nes′cens. M.L. pres. part. *luminescens* luminescing (for its bioluminescence).

The characteristics are as described for the genus and as listed in Tables 5.52 and 5.54.

Isolated from nematodes of the genus *Heterorhabditis*.

The mol% G + C of the DNA is 43–44 (Bd).

Type strain (holotype): ATCC 29999 (strain Hb).

Table 5.54.
Differential characteristics of the species of the genus **Xenorhabdus**[a]

Characteristics	1. X. nematophilus	2. X. luminescens
Catalase	−	+
Bioluminscence[b]	−	+[w]
Association with nematode genus:		
Neoaplectana	+	−
Heterorhabditis	−	+
Pigment on Loeffler's blood serum		
Brown	+	−
Orange	−	+

[a] For symbols see Table 5.52.

[b] Based on cultures grown for 2–3 days at 25°C on nutrient agar or trypticase soy agar. It takes 10–20 min adaptation of the eyes in a totally dark room to observe this weak bioluminscence. This is in contrast to the strong bioluminescence of many marine bacteria, which take only 5–10 sec to 2–3 min of dark adaptation.

Genus Kluyvera Farmer, Fanning, Huntley-Carter, Holmes, Hickman, Richard and Brenner 1981b, 382.[VP] (Effective publication: Farmer, Fanning, Huntley-Carter, Holmes, Hickman, Richard and Brenner 1981a, 927.). ("Enteric Group 8," Fanning, Farmer, Parker, Huntley-Carter and Brenner, Abst. Annu. Mtg. Amer. Soc. Microbiol., 1979, p. 100, I-30; similar to but not identical with Kluyvera Asai et al., 1956, 489)

Kluy′ ve.ra. M.L. fem. n. *Kluyvera*, named for the Dutch microbiologist. A.J. Kluyver, who made many contributions to microbial physiology and taxonomy.

Small rod-shaped cells, 0.5–0.7 × 2–3 μm, conforming to the general definition of the family *Enterobacteriaceae.* **Motile** by scant peritrichous flagella. Facultatively anaerobic. Catalase-positive. Nitrate is reduced to nitrite. Acid and gas are produced from glucose. **Large amounts of α-ketoglutaric acid are formed during glucose fermentation. Most other carbohydrates are fermented**, but polyhydroxyl alcohols generally are not. The majority of strains are **indolepositive. Citrate is usually used as a sole carbon source.** Methyl red-positive, Voges-Proskauer-negative. Occur in food, soil and sewage. Probably also an infrequent opportunistic pathogen of man. The mol% G + C is 55–57 (Bd).

Type species: *Kluyvera ascorbata* Farmer, Fanning, Huntley-Carter, Holmes, Hickman, Richard and Brenner 1981b, 382.

Further Descriptive Information

Kluyvera occurs in human clinical specimens, and may be an opportunistic pathogen. Schwach (1979) reported three isolates of Enteric Group 8, all from upper respiratory tract specimens, which the Enteric Laboratories of the CDC subsequently received and identified (phenotypically) as *K. ascorbata*. Since the strains were in mixed culture and were not detected in subsequent specimens, she concluded that they were probably not clinically significant. Braunstein and co-workers (1980) reported on two cases which yielded cultures that we had identified as Enteric Group 8 (both are *K. ascorbata*). One of these was from the sputum of a 6-year-old boy with pulmonary tuberculosis. This isolate was not considered clinically significant. A second isolate was from gall bladder drainage fluid of a 63-year-old woman with acute pancreatitis. On the basis of chart review, this isolate was considered clinically significant. These are the only published reports assessing the clinical significance of Enteric Group 8 or *Kluyvera*. Of our 144 *Kluyvera* strains, none has been from spinal fluid, but five strains have been from blood: three strains (two from France) of *K. ascorbata*, one strain of *K. cryocrescens* (a 3-month-old, at autopsy), and one of *Kluyvera* species group 3. No other information was included to allow evaluation of the cultures' clinical significance. These five blood isolates and the report of Braunstein and co-workers (1980) suggest that *Kluyvera* is more than a benign saprophyte. Most new species of

Enterobacteriaceae have at least attained the status of "infrequent opportunistic pathogen." On the basis of present knowledge, this status also seems appropriate for *Kluyvera*. The respiratory tract was the most common source for *Kluyvera*, but there is no strong evidence that it is clinically significant here (one isolate of *K. ascorbata* was, however, from a lung at autopsy). The respiratory tract (particularly sputum) is notoriously difficult to evaluate for clinical significance except in carefully designed prospective studies. The urinary tract was the next most common source, but there was no mention of more than 100,000 organisms/ml of urine or the presence of white blood cells or of red blood cells. Feces were a common site of isolation, and the presence of *Kluyvera* in food is an obvious source of these isolates.

Taxonomic Comments

Kluyvera is a genus with a turbulent history. Although the name has been used in the literature, the genus name, and the two original species names ("*K. citrophila*" and "*K. noncitrophila*") did not appear on the *Approved Lists of Bacterial Names*. This nomenclatural problem needed resolution and resulted in the proposal of a redefined genus *Kluyvera* (Fanning et al., Abstr. Annu. Mtg. Amer. Soc. Microbiol., 1979, I-30; Farmer, et al., 1981a).

In 1956 and 1957, Asai and coworkers in Japan proposed the genus *Kluyvera* for a group of polarly flagellated bacteria which produced large amounts of α-ketoglutaric acid during the fermentation of glucose. Five strains were originally studied, one from soil and four from sewage. Two species names were proposed, *K. citrophila* and *K. noncitrophila*, which were based on the difference among the strains in utilizing citric acid as the sole source of carbon and energy (Asai et al., 1956). Asai and co-workers (1955, 1956) state (on the basis of their interpretation of the Kluyver and van Niel paper of 1936) that the genus was named to honor Professor A. J. Kluyver who, with C. B. van Niel, in 1936 postulated that there may be a group of polarly flagellated organisms in the tribe *Pseudomonadeae* which have a mixed acid type of fermentation similar to *Escherichia* (called *Bacterium* in the paper). If such a group were to be discovered it could be a separate genus, which would differentiate it from the genus *Aeromonas*, which has a butylene glycol fermentative pathway rather than a mixed acid pathway. Asai and co-workers (1955, 1956) thought they had discovered this postulated group of polarly flagellated organisms and named the group *Kluyvera* in honor of A. J. Kluyver for his many contributions to microbial metabolism and physiology. *Kluyvera* was classified in the tribe *Pseudomonadeae*, which at that time included nonfermentative genera, but also included the fermenters of the genus *Aeromonas*. Today, the family *Pseudomonadaceae* is restricted to bacteria which do not ferment glucose.

In 1962, Asai and co-workers confirmed the observations of J. M. Shewan and Rudolph Hugh that all five of their *Kluyvera* strains actually had peritrichous rather than polar flagella (Asai et al., 1962). Accordingly, they proposed that the two species *K. citrophila* and *K. noncitrophila* be transferred to the genus *Escherichia* in the family *Enterobacteriaceae*. Thus these organisms presumably became "*Escherichia citrophila*" comb. nov. and "*Escherichia noncitrophila*" comb. nov., since the genus *Kluyvera* was abolished by its original proposers.

Our interest in the group of organisms which eventually became the redefined genus *Kluyvera* began in 1977 when one of the Enteric Laboratories' computer programs listed over a dozen strains with almost identical biochemical reactions which had been reported as "unidentified." At the time, groups of unidentified *Enterobacteriaceae* were being formed, and this was the eighth group. Thus, the vernacular name "Enteric Group 8" was given to the group, and strains were then reported with this designation and with a request to send similar ones. Soon after Enteric Group 8 was defined, Barry Holmes of the Computer Identification Laboratory, National Collection of Type Cultures, England pointed out that they had strains labeled *Kluyvera* which were almost identical to our Enteric Group 8. From then on, Enteric Group 8 was thought of as a synonym of *Kluyvera*. The DNA/DNA hybridization studies of Fanning and co-workers (Abstr. Annu. Mtg. Amer. Soc. Microbiol., 1979, I-30) defined two large species groups in *Kluyvera*

and defined one group of strains which probably represents additional *Kluyvera* species. The latter group is being referred to as *Kluyvera* species group 3. Although *Kluyvera* was abolished in 1962 (Asai et al., 1962) reports in the literature have continued to use *Kluyvera* (Beck et al., 1980; Kleeberger, 1979; Nakazawa et al., 1972; Nara et al., 1971; Serizawa et al., 1979; Thanbichler and Beck, 1974; Thurner and Busse, 1978). For this reason, the name *Kluyvera* was reproposed (Farmer et al., 1981a). This alternative was considered better than proposing a new genus name, which could cause confusion because of the continuous use of *Kluyvera*.

Kluyvera is a distinct group in the family *Enterobacteriaceae* both phenotypically and by DNA/DNA hybridization (Fanning et al., Abstr. Annu. Mtg. Amer. Soc. Microbiol., 1979, I-30; Farmer et al., 1981a). *Kluyvera* is reported to accumulate extremely large amounts of α-ketoglutaric acid during glucose fermentation (Asai et al., 1956, 1957), but few other *Enterobacteriaceae* have been investigated for this property. Table 5.52 indicates that *Kluyvera* is otherwise a typical genus of *Enterobacteriaceae* which can be differentiated from the others by a complete set of biochemical tests.

Further Reading

Asai, T., H. Iizuka, and K. Komagata. 1962. The flagellation of genus *Kluyvera*. J. Gen. Appl. Microbiol. (Japan) 8: 187–191.
Asai, T., S. Okumura, and T. Tsunoda. 1956. On a new genus, *Kluyvera*. Proc. Japan Acad. 32: 488–493.
Fanning, G. R., J. J. Farmer III, J. N. Parker, G. P. Huntley-Carter and D. J. Brenner. 1979. *Kluyvera*: A new genus in *Enterobacteriaceae*. Abst. Annu. Mtg. Amer. Soc. Microbiol. p.100, I-30.
Farmer, J. J. III, G. R. Fanning, G. P. Huntley-Carter, B. Holmes, F. W. Hickman, C. Richard and D. J. Brenner. 1981a. *Kluyvera*: A new (redefined) genus in the family *Enterobacteriaceae*: identification of *Kluyvera ascorbata* sp. nov. and *Kluyvera cryocrescens* sp. nov. in clinical specimens. J. Clin. Microbiol. 13: 919–933.

Differential characteristics of the species of the genus **Kluyvera**

The two named species of *Kluyvera*, *K. ascorbata* and *K. cryocrescens* are very close phenotypically, but can be differentiated by simple tests (Table 5.55) and by differences in the zones of inhibition around the

antibiotic disks carbenicillin (100 μg/disk) and cephalothin (30 μg/disk). *K. cryocrescens* usually has big zones (sometimes with resistant colonies in the zones), but *K. ascorbata* has much smaller zones.

Table 5.55.
Differential characteristics of the species of the genus **Kluyvera**[a]

Characteristics	1. *K. ascorbata*	2. *K. cryocrescens*
Ascorbate fermentation[b]	+	−
Growth at 5°C within 21 days	−	+

[a] For symbols see Table 5.52.
[b] Acid production from L-ascorbate (or from a "break down" product of L-ascorbate formed during autoclaving) (Farmer et al. 1981a).

List of the species of the genus **Kluyvera**

1. **Kluyvera ascorbata** Farmer, Fanning, Huntley-Carter, Holmes, Hickman, Richard and Brenner 1981b, 382.[VP] (Effective publication: Farmer, Fanning, Huntley-Carter, Holmes, Hickman, Richard and Brenner, 1981a, 927.)

a.scor.ba'ta. Modern chemical term ascorbate, salt of ascorbic acid; M.L. fem. adj. *ascorbata* pertaining to ascorbate (referring to the positive ascorbate test).

The characteristics are as described for the genus and as listed in Tables 5.52 and 5.55.

Occur in water, sewage and food. Probably an infrequent opportunistic pathogen of man.

The mol% G + C of the DNA is 56–57 (Bd).

Type strain (holotype): ATCC 33433 (CDC 0648-74).

2. **Kluyvera cryocrescens** Farmer, Fanning, Huntley-Carter, Holmes, Hickman, Richard and Brenner 1981b, 382.[VP] (Effective publication: Farmer, Fanning, Huntley-Carter, Holmes, Hickman, Richard and Brenner, 1981a, 927.)

cry.o.cres'cens. Gr. noun *kryos* cold; L. fem. pres. part. *crescens* growing; M.L. fem. adj. *cryocrescens* growing in the cold.

The characteristics are as described for the genus and as listed in Tables 5.52 and 5.55.

Occurs in soil, water, sewage, the hospital environment, and human clinical specimens where it is probably an infrequent opportunistic pathogen of man.

The mol% G + C of the DNA is 55 (Bd).

Type strains (holotype): ATCC 33435 (CDC 2065-78).

Other organisms

a. *Kluyvera* species group 3.

This group has five strains that probably represent at least one additional species of *Kluyvera*, based on the DNA/DNA hybridization studies of Fanning et al. (Abst. Annu. Mtg. Amer. Soc. Microbiol., 1979, I-30). Phenotypically, the group is presently indistinguishable from the other two named *Kluyvera* species.

b. *Buttiauxella agrestis* Ferragut, Izard, Gavini, Lefebvre and Leclerc, 1981, 40.

The genus *Buttiauxella* was recently described by Ferragut et al. (1981) and is phenotypically very similar to *Kluyvera*. *Kluyvera* strains ferment sucrose within 1 or 2 days, but strains of *Buttiauxella agrestis* are negative at 1 week. Comparisons of the two new genera by DNA/DNA hybridization and phenotype are in progress which should result in a logical proposal on the nomenclature and classification of the two taxa.

Type strain: ATCC 33320 (CIP 80-31; CUETM 77-167; Gavini strain F-44).

Note Added in Proof

The following new genus name and species name have gained standing in nomenclature since preparation of this manuscript:

Genus *Buttiauxella* Ferragut, Izard, Gavini, Lefebvre and Leclerc 1982, 266^{VP} (Effective publication: Ferragut et al. 1981, 40.) ("Group F," Gavini et al., 1976.)

Type species: *Buttiauxella agrestis* Ferragut, Izard, Gavini and Leclerc 1982, 266^{VP} (Effective publication: Ferragut et al. 1981, 40.)

Type strain (holotype): ATCC 33320 (CIP 80-31; CUETM 77-167).

Reference

Ferragut, C.D., D. Izard, F. Gavini, B. Lebvre and H. Leclerc. 1982. Validation of the publication of new names and new combinations previously effectively published outside the IJSB. List No. 8. Int. J. Syst. Bacteriol. *32:* 266–268.

Genus **Rahnella** Izard, Gavini, Trinel and Leclerc 1981, 382.^{VP} (Effective publication: Izard, Gavini, Trinel and Leclerc 1979, 174.). (Group H2, Gavini et al., 1976.)

Rahn'el.la. M.L. dim. ending *-ella*; M.L. fem n. *Rahnella* named after Otto Rahn, the German-American microbiologist who made many contributions (Pederson, 1957) to bacterial systematics and who coined the name *Enterobacteriaceae* in 1937.

Small rod-shaped cells 0.5–0.7 × 2–3 μm, conforming to the general definition of the family *Enterobacteriaceae*. **Nonmotile at 36°C; motile when grown at 25° C.** Nitrate is reduced to nitrite. D-Glucose is fermented with the production of acid and, for the majority of strains, gas. **Negative for lysine and ornithine decarboxylases and for arginine dihydrolase. Weakly positive for phenylalanine deaminase.** The majority of strains are methyl red-positive; **all strains are Voges-Proskauer-positive. Many carbohydrates are fermented** including lactose, maltose, L-rhamnose, raffinose and salicin. Occurs in fresh water. May occasionally be isolated from human clinical specimens, but clinical significance is not known.

The mol% G + C of the DNA is 51–56 (Tm).

Type species: *Rahnella aquatilis* Izard, Gavini, Trinel and Leclerc 1981, 382.

Further Descriptive Information

The natural habitat of *Rahnella* is water, and all of the isolates of Gavini and co-workers (1976) were from waters in France. The Enteric Laboratories at CDC has identified several American water isolates as *R. aquatilis*. One puddle of water standing over red-clay soil in Gwinnett County, Georgia, U.S.A., had two different strains (distinct biogroups) of *R. aquatilis*. *R. aquatilis* may also occasionally occur in human clinical specimens. One of our strains was from a burn wound. Only a few *Rahnella* strains have been described, so much more study is needed to define its ecology and possible role in human disease.

Taxonomic Comments

In 1976, Gavini and co-workers defined a new group of *Enterobacteriaceae* and gave it the vernacular name "group H." The original definition was based on clustering by numerical taxonomy and the phenotypic differences between group H2 and other *Enterobacteriaceae*. In 1979, Izard and co-workers used DNA/DNA hybridization to compare strains of group H2 to each other and to named species of *Enterobacteriaceae*. On the basis of the close relatedness within group H2 and the low relatedness to other *Enterobacteriaceae*, they proposed the new genus *Rahnella* with one species *Rahnella aquatilis*, and designated a holotype strain. *Rahnella* was named to honor Otto Rahn, the German-American microbiologist who made many contributions to systematic bacteriology (Pederson, 1957) and who coined the family name *Enterobacteriaceae* in 1937. The species name *aquatilis* was taken from the fact that the strains of *R. aquatilis* were isolated from water. The names *Rahnella* and *R. aquatilis* were validly published, but were not validated in the International Journal of Systematic Bacteriology before January 1, 1980. They did not appear on the *Approved Lists of Bacterial Names* (Skerman et al., 1980); however, both names have now been validly published (Izard et al., 1981) and have standing in nomenclature.

Rahnella has no single distinguishing feature to differentiate it from other *Enterobacteriaceae* (Table 5.52). Phenotypically, *Rahnella* is nonmotile at 36°C but motile at 25°C, negative for lysine and ornithine decarboxylases and for arginine dihydrolase, is weakly positive (less so than the *Proteus* group) for phenylalanine deaminase, and does not produce a yellow pigment. These properties differentiate *Rahnella* from the heterogeneous group of bacteria classified in the *Enterobacter agglomerans*/*Erwinia herbicola* complex. Previously, strains now identified as *Rahnella* would have represented a small part of this complex group.

Further Reading

Gavini, F., C. Ferragut, B. Lefebvre and H. Leclerc. 1976. Étude taxonomique d'entérobactéries appartenant ou apparentées au genre *Enterobacter*. Ann. Microbiol. (Inst. Pasteur) *127B:* 317–335.

Izard, D., F. Gavini, P. A. Trinel and H. Leclerc. 1979. *Rahnella aquatilis*, nouveau membre de la famille des *Enterobacteriaceae*. Ann. Microbiol. (Inst. Pasteur) *130A:* 163–177.

List of the species of the genus **Rahnella**

1. **Rahnella aquatilis** Izard, Gavini, Trinel and Leclerc 1981, 382.^{VP} (Effective publication: Izard, Gavini, Trinel and Leclerc 1979, 174.) (Group H2, Gavini et al., 1976.)

a.qua'ti.lis. L. adj. *aquatilis* living in water.

The characteristics are as described for the genus and as listed in Table 5.52.

Occurs in freshwater. May occasionally be isolated from human clinical specimens, but clinical significance is unknown.

The mol% G + C of the DNA is 51–56 (T_m).

Type strain (holotype): ATCC 33071 (CIP 78-65 (strain 133)].

Genus **Cedecea** *Grimont, Grimont, Farmer and Asbury 1981, 325.*[VP] (Enteric Group 15, Farmer, Grimont, Grimont and Asbury, Abstr. Annu. Mtg. Amer. Soc. Microbiol., 1980, C-123).

Ce.de′ ce.a. M.L. fem n. *Cedecea*, formed from the letters CDC; named by P. A. D. Grimont and F. Grimont after the Centers for Disease Control, Atlanta, Georgia, where the organisms were originally recognized as a new group.

Rod-shaped cells 0.5–0.6 × 1–2 μm, conforming to the general definition of the family *Enterobacteriaceae.* **Most strains are motile. Facultatively anaerobic.** D-Glucose is fermented with the production of acid and usually gas. Nitrate is reduced to nitrite. **The majority of strains are lipase-positive (for corn oil). Negative for deoxyribonuclease and gelatinase. Resistant to the antibiotics colistin and cephalothin.** Isolated from human clinical specimens, usually from the respiratory tract, but clinical significance is unknown.

The mol% G + C of the DNA is 48–52 (T_m).

Type species: *Cedecea davisae* Grimont, Grimont, Farmer and Asbury 1981, 325.

Further Descriptive Information

All the *Cedecea* strains have been isolated from human clinical specimens, and over 50% have been from the respiratory tract. None of the original isolates were from blood or spinal fluid, but recently an isolate of "*Cedecea* species 3" and one of "*Cedecea* species 4" were from blood cultures. *Cedecea* will probably gain status as an infrequent colonizer or as an infrequent opportunistic pathogen. Very little is currently known about its ecology, epidemiology, or role in human disease.

Taxonomic Comments

Cecedea has been proposed as a new genus in *Enterobacteriaceae* (Farmer et al., Abst. Annu. Mtg. Amer. Soc. Microbiol., 1980, C-123; Grimont et al., 1981). This group was designated Enteric Group 15 in 1977 as a diagnostic culture was being reported. There was a group of over a dozen other cultures which were very similar to it and all had been reported "unidentified." The strains were lipase-positive (corn

oil) and resistant to the antibiotics colistin and cephalothin. Among *Enterobacteriaceae*, these properties are unique to the genus *Serratia*, but the new group differed from *Serratia* in that it was negative for deoxyribonuclease and gelatin hydrolysis. Enteric Group 15 was thought to be a uniform group of strains which would probably be a new species of *Serratia*, intermediate between the "typical" *Serratia* (deoxyribonuclease-positive, lipase-positive, gelatinase-positive) and *Serratia fonticola*, which is negative for all three of these tests (Farmer et al., Abst. Annu. Mtg. Amer. Soc. Microbiol., 1980, C-123). Enteric Group 15 was studied by the Grimonts in France (Farmer et al., 1980, Abst. C-123; Grimont et al., 1981), who used phenotypic characterization and DNA/DNA hybridization (S1 nuclease method). Fifteen strains of Enteric Group 15 were more closely related to each other (32–100%) than to strains of the six named *Serratia* species (6–10%) or to other named taxa in *Enterobacteriaceae* (1–23%). Five groups of strains within Enteric Group 15 were phenotypically distinct (Table 5.56) and distinct by DNA hybridization. Two of these have been named as species, but the vernacular names "*Cedecea* species 3", "*Cedecea* species 4" and "*Cedecea* species 5" will be used for the other three until more strains are available. Table 5.56 gives the biochemical tests useful in differentiating the 5 species of *Cedecea*.

Further Reading

Bae, B.H.C., S.B. Sureka and J.A. Ajamy. 1981. Enteric Group 15 (*Enterobacteriaceae*) associated with pneumonia. J. Clin. Microbiol. *14:* 596–597.

Farmer, J.J. III, P.A.D. Grimont, F. Grimont and M.A. Asbury. 1980b. *Cedecea:* A new genus in *Enterobacteriaceae*. Abstr. Annu. Mtg. Amer. Soc. Microbiol. p. 295, Abstract C 123.

Grimont, P.A.D., F. Grimont, J.J. Farmer III and M.A. Asbury. 1981. *Cedecea davisae* gen. nov., sp. nov. and *Cedecea lapagei* sp. nov., new *Enterobacteriaceae* from clinical specimens. Int. J. Syst. Bacteriol. *31:* 317–326.

Table 5.56.

Differential characteristics of the named species and unnamed species of the genus **Cedecea**[a]

Characteristics	1. *C davisae*	2. *C. lapagei*	Unnamed Species		
			"Species 3"	"Species 4"	"Species 5"
Voges-Proskauer test:					
Method 1[b]	−	[+]	−	−	−
Method 2[c]	+	+	+[w]	+	+
Ornithine "decarboxylase" (Møller's)	+	−	−	−	d
Acid production from:					
Sucrose, D-xylose	+	−	+	+	+
Raffinose	−	−	+	−	+
D-Sorbitol	−	−	−	+	+
Melibiose	−	−	+	−	+
Malonate utilization	+	+	−	+	−
Growth in media without thiamine	−	+	+	+	−

[a] For symbols see Table 5.52.

[b] Method 1: cultures were grown in 1 ml of MR-VP broth for 48 h at 36°C, then 1 ml of modified O'Meara test reagent (Edwards and Ewing, 1972, p. 353) was added. A pink-red color within 5 min was scored as positive.

[c] Method 2: cultures were grown in 1 ml of MR-VP broth for 5 days at 25°C, then 0.5 ml of 40% KOH and 0.5 ml of 6% α-naphthol in ethanol (absolute) was added. A pink-red color within 5 min was scored as positive.

List of the species of the genus **Cedecea**

1. Cedecea davisae Grimont, Grimont, Farmer and Asbury 1981, 325.[VP]

da′vi.sae. M.L. gen. n. *davisae* of Davis; named after Betty Davis, the American bacteriologist of the Enteric Section, Centers for Disease

Control, Atlanta, Georgia, who has made many contributions to the biochemical and serological identification of *Enterobacteriaceae* and *Vibrionaceae*.

The characteristics are as given for the genus and as listed in Tables 5.52 and 5.56.

Isolated from human clinical specimens; clinical significance unknown.

The mol% G + C of the DNA is 49–50 (T_m).

Type strain (holotype): ATCC 33431 (CDC 3278-77).

2. **Cedecea lapagei** Grimont, Grimont, Farmer and Asbury 1981, 325.[VP]

la·pa′ge·i. M.L. gen. n. *lapagei* of Lapage; named after Stephen Lapage, the British bacteriologist who has made many contributions to bacterial systematics, particularly as an author of the *Bacteriological Code*.

The characteristics are as given for the genus and as listed in Tables 5.52 and 5.56.

Isolated from human clinical specimens.

The mol% G + C of the DNA is 48–52 (T_m).

Type strain (holotype): ATCC 33432 (CDC 0485-76).

Other organisms

Three unnamed groups of *Cedecea*, believed to be additional species of the genus on the basis of DNA/DNA hybridization experiments and phenotypic differences, are listed in Table 5.56 and are differentiated from the two named species of *Cedecea*.

Note Added in Proof

The following new name for "*Cedecea* species 4" has gained standing in nomenclature since preparation of this manuscript:

Cedecea neteri Farmer, Sheth, Hudzinski, Rose and Asbury 1983, 438.[VP] (Effective publication: Farmer et al. 1982, 777.)

Type strain (holotype): ATCC 33855 (CDC 0621-75; strain 002).

References

Farmer, J.J., III, N.K. Sheth, J.A. Hudzinski, H.D. Rose and M.F. Asbury. 1982. A case of bacteremia due to *Cedecea neteri* sp. nov. J. Clin. Microbiol. *16:* 775–778.

Farmer, J.J., III, N.K. Sheth, H.A. Hudzinski, H.D. Rose and M.F. Asbury. 1983. Validation of the publication of new names and new combinations previously effectively published outside the IJSB. List No. 10. Int. J. Syst. Bacteriol. *33:* 438–440.

Genus **Tatumella** Hollis, Hickman, Fanning 1982b, 267[VP] (Effective publication: Hollis, Hickman, Fanning, Farmer, Weaver and Brenner 1981, 86.) (Group EF-9 Hollis, Hickman, Fanning, Brenner and Weaver, Abstr. Annu. Mtg. Amer. Soc. Microbiol., 1980, p. 295, C-122.)

Ta.tum.el′ la. M.L. dim. neut, ending *-ella*; M.L. fem. n. *Tatumella* named to honor Harvey Tatum, the American bacteriologist, who has made many contributions to our understanding of the classification and identification of fermentative and nonfermentative bacteria of medical importance.

Small rod-shaped cells 0.6–0.8 × 0.9–3 μm, conforming to the general definition of the family *Enterobacteriaceae*. **Nonmotile at 36°C; over half the strains are motile** by means of polar, subpolar or lateral flagella **when grown at 25°C.** Facultatively anaerobic. **Biochemically more active at 25°C than at 36°C.** D-Glucose is fermented with the production of acid but no gas. Few other sugars are fermented at 36°C. Stock cultures often die within a few weeks on laboratory media. **Large zones of inhibition are formed around disks containing penicillin (10 U),** in contrast to most other *Enterobacteriaceae*. Isolated from human clinical specimens, mainly from the respiratory tract. Probably an infrequent opportunistic pathogen (three isolates were from blood) or colonizer.

The mol% G + C of the DNA is 53–54 (Bd).

Type species: *Tatumella ptyseos* Hollis, Hickman, and Fanning 1982, 267.

Further Descriptive Information

Tatumella shares many of the properties of *Enterobacteriaceae* (Table 5.52) but is unusual in several ways. Stock cultures may die within a few weeks on agar or in semisolid stock culture media, unlike most *Enterobacteriaceae*, which can be kept almost indefinitely in sealed tubes kept at room temperature. However, *Tatumella* cultures frozen in 5% rabbit blood and stored in −40°C have remained viable after storage for up to 14 years. This latter method (or perhaps freeze-drying) should be used for long-term preservation. In contrast to most other *Enterobacteriaceae*, *Tatumella* has large zones of inhibition around 10-U penicillin G disks (range of 15–36 mm, mean of 24 mm, standard deviation of 4.6 mm; method of Bauer et al., 1966). The flagellation of *Tatumella* is also unusual. Strains are non-motile at 36°C, but 66% are motile at 25°C. No flagella are seen on most cells, but those seen are polar, subpolar or lateral (Leifson, 1960) rather than peritrichous. Biochemically, *T. ptyseos* is more active at 25°C than 36°C. The biochemical reactions are summarized in Table 5.52. All the strains of *T. ptyseos* were isolated from human clinical specimens, and 86% were from the respiratory tract. Three cultures were from blood, so it appears that *T. ptyseos* should at least be considered as a rare opportunistic pathogen until more data are available.

Taxonomic Comments

The name *Tatumella* has been proposed (Hollis et al. 1981b) for the group of organisms which had previously been known as Group EF-9 by the Special Bacteriology Section at the Centers for Disease Control (Hollis et al., Abst. Annu. Mtg. Amer. Soc. Microbiol., 1980, p. 295, C-122). This group had been known for many years, but its taxonomic position was only recently investigated. Twenty-seven "suspect" strains of Group EF-9 were studied by DNA/DNA hybridization, biochemical reactions and antibiotic susceptibility. Twenty-five of these strains were related by 85% or more by DNA/DNA hybridization (60°C, hydroxyapatite method with ³²P) to the type strain ATCC 33301 but one strain was not related. By DNA hybridization, other taxa in the family *Enterobacteriaceae* were related by 7–38%, including 25–30% relatedness of *Escherichia*, the type genus of the family. On the basis of these data, Group EF-9 was proposed as a new genus, *Tatumella*, with *T. ptyseos* as the only species. The citation following the genus and species names was proposed by the original authors as "Hollis, Hickman and Fanning, 1981" to recognize the greater contribution of the first three authors (Hollis et al., 1981b).

Further Reading

Hollis, D.G., F.W. Hickman, G.R. Fanning, D.J. Brenner, and R.E. Weaver. 1980. EF-9: A newly described group of *Enterobacteriaceae*. Abstr. Annu. Mtg. Amer. Soc. Microbiol. p. 295, C-122.

Hollis, D.G., F.W. Hickman, G.R. Fanning, J.J. Farmer III, R.E. Weaver and D.J. Brenner. 1981. *Tatumella ptyseos* gen. nov., sp. nov., a member of the family *Enterobacteriaceae* found in clinical specimens. J. Clin. Microbiol. *14:* 79–88.

List of the species of the genus **Tatumella**

1. **Tatumella ptyseos** Hollis, Hickman and Fanning 1982, 267.[VP] (Effective publication: Hollis, Hickman, Fanning, Farmer, Weaver and Brenner 1981, 86.)

pty′se.os. Gr. n. *ptyseos* a spitting (or less literally from sputum, the most common source of clinical isolates).

The characteristics are as given for the genus and as listed in Table 5.52.

Isolated from human clinical specimens, particularly from the respiratory tract.

The mol% G + C of the DNA is 53–54 (Bd).

Type strain (holotype): ATCC 33301 (CDC 9591-78 (D6168)).

Acknowledgments

I thank F. G. Priest for cultures of *Obesumbacterium proteus* biogroups 1 and 2 (sent to W. H. Ewing); G. M. Thomas and G. P. Poinar for cultures of *Xenorhabdus nematophilus* and *X. luminescens*; H. Leclerc, F. Gavini, F. Izard and P. A. Trinel for cultures of *Rahnella* and *Buttiauxella*; Martin Busse, Karla Tomfohrde, T. Schwach, Robert H. Gherna, Barry Holmes, and C. Richard for cultures of *Kluyvera*; F. W. Hickman, M. A. Asbury, G. P. Huntley-Carter, and the other members of the Enteric Laboratories for much of the original characterization used in writing the descriptions of the six genera; Professor Kathleen Ann Kelley, Emory University for helping with the Latin and Greek derivations not given by the original authors; Robert H. Gherna and Margaret S. Hendrie for furnishing copies of letters from the ATCC and NCIB file concerning the history of important strains of *Obesumbacterium proteus*; and the following who read the whole chapter—Don J. Brenner, Barry Holmes, P. A. D. Grimont, Frances W. Hickman, R. Sakazaki; or the part concerning their area of specialization; H. J. J. van Vuuren (*Obesumbacterium*), G. O. Poinar and G. M. Thomas (*Xenorhabdus*), D. Izard, F. Gavini, P. A. Trinel and H. Leclerc (*Rahnella*), and G. P. Huntley-Carter (*Kluyvera*).

FAMILY II. **VIBRIONACEAE** VERON 1965, 5245[AL]

PAUL BAUMANN AND RALPH H. W. SCHUBERT

Vib.ri.o.na′ce.ae. M.L. masc. n. *Vibrio* type genus of the family; -*aceae* ending to denote family; M.L. fem. pl. n. *Vibrionaceae* the *Vibrio* family.

Gram-negative straight or curved rods. Motile by means of **polar flagella.** Under certain conditions of cultivation (e.g. on solid media) additional lateral flagella may be synthesized which differ in wavelength from the polar flagellum and may number from a few to over 100 flagella/cell. Do not form endospores or microcysts. **Chemoorganotrophs and facultative anaerobes** capable of respiratory and fermentative metabolism. Oxygen is a universal electron acceptor. Do not denitrify. **Most are oxidase-positive. All utilize D-glucose** as a sole or principal source of carbon and energy. Most utilize ammonium salts as sole sources of nitrogen. A few have relatively simple organic growth factor requirements. Most species require 2–3% NaCl or a seawater base for optimum growth. Primarily **aquatic inhabitants** found in sea- and freshwater and in association with aquatic animals. Several species are pathogenic for man, fish, eels, and frogs, as well as other vertebrates and invertebrates. The mol% G + C of the DNA ranges from 38–63.

Type genus: *Vibrio* Pacini 1854, 411.

Further Comments

When the family *Vibrionaceae* was initially proposed by Véron in 1965, his primary intent was to group a number of genera comprised of species which were, for the most part, oxidase-positive and motile by means of polar flagella. This grouping was not necessarily meant to imply an evolutionary relationship among these species but, rather, was intended as a convenience for the purpose of differentiating these organisms from the *Enterobacteriaceae*, comprised of oxidase-negative species having peritrichous flagella. Subsequent to this proposal, studies dealing with the comparative physiology and genetics of members of the *Vibrionaceae* and *Enterobacteriaceae* have established that these organisms share a number of distinctive attributes which have a degree of complexity suggesting a common evolutionary origin. The most extensive comparative studies have dealt with the regulation of the activity of aspartokinase (Baumann and Baumann, 1973; Cohen et al., 1969) and 3-deoxy-D-*arabino*-heptulosonate-7-phosphate (DAHP) synthetase (Chludzinski et al., 1972; Jensen and Rebello, 1970; Jensen and Stenmark, 1970) as well as with the genetic organization and regulation of enzymes of tryptophan biosynthesis (Bieger and Crawford, 1978; Crawford, 1975). The reactions catalyzed by aspartokinase and DAHP synthetase initiate complex branched pathways leading to the biosynthesis of amino acids of the aspartate family and the aromatic amino acids, respectively. Members of the *Vibrionaceae* and *Enterobacteriaceae* have three isofunctional aspartokinases, one which is feedback inhibited by L-threonine, a second by L-lysine, and a third which is

unaffected by any of the amino acids of the aspartate family. Similarly, members of the *Vibrionaceae* and the *Enterobacteriaceae* have three isofunctional DAHP synthetases, each feedback inhibited by a different amino acid, L-phenylalanine, L-tyrosine, or L-tryptophan. Detailed studies of the regulation and organization of genes coding for enzymes of tryptophan biosynthesis have indicated that in both families the genes are part of a single operon and have the same unique organizational and regulatory pattern (Bieger and Crawford, 1978; Crawford, 1975, personal communication). Evidence of a common evolutionary origin for the *Vibrionaceae* and *Enterobacteriaceae* has also been obtained from 5S ribosomal RNA (rRNA) sequencing (Hori and Osawa, 1979), 16S RNA oligonucleotide cataloging (Fox et al., 1980; Gibson et al., 1979), rRNA/DNA hybridization (Baumann and Baumann, 1976; 1981), and an immunological comparison of glutamine synthetase (GS) (Baumann and Baumann, 1978; Baumann, et al., 1980a, b). According to the results of 16S RNA oligonucleotide cataloging, members of the

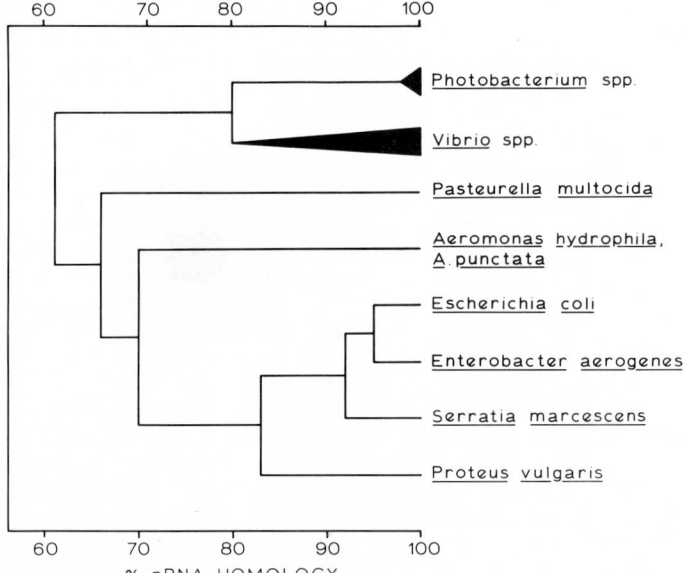

Figure 5.3. Single linkage analysis of the results of rRNA/DNA homology studies (Baumann and Baumann, 1976, 1981). Data for *Pasteurella multocida* are from Fox et al. (1980).

Vibrionaceae and the *Enterobacteriaceae* belong to a major evolutionary branch which includes *Chromatium vinosum*, *Pseudomonas aeruginosa*, and *Acinetobacter calcoaceticus* and which probably evolved from a sulfur purple bacterium.

Recent studies of a number of informational molecules allow tentative conclusions as to the evolutionary relationships among genera of the *Vibrionaceae* and *Enterobacteriaceae*. The most extensive studies have involved rRNA/DNA hybridization experiments (Baumann and Baumann, 1976) and 16S oligonucleotide cataloging (Fox et al., 1980; Gibson et al., 1979); both approaches have provided virtually identical results and are summarized in Figure 5.3. Many or most of these interrelationships have been confirmed by immunological studies of sequence similarities of GS (Baumann and Baumann, 1978; Baumann et al., 1980a) and alkaline phosphatase (Cocks and Wilson, 1972) which have also indicated that *Erwinia* (for which there are no rRNA data) is about as distantly related to *E. coli* as is *Proteus*. One conclusion which is evident from the data summarized in Figure 5.3 is that the members of the family *Vibrionaceae* represent two distinct lines of descent, the first comprised of *Photobacterium* and *Vibrio* species and the second of *Aeromonas*; a third line of descent is represented by *Pasteurella multocida*. The latter genus, along with *Actinobacillus* and *Haemophilus*, has been assigned to the family *Pasteurellaceae*. In addition, it appears that members of the *Enterobacteriaceae* are an offshoot of the *Pasteurella-Aeromonas* branch. This is supported by differences in the superoxide dismutases (SODs) of the *Vibrionaceae* and the *Enterobacteriaceae* (Bang et al., 1978). All members of the *Vibrionaceae* (*Vibrio*, *Photobacterium*, *Aeromonas*, *Plesiomonas*) have a single Fe-SOD, while a more limited investigation of the *Enterobacteriaceae* suggests the presence of either a single Mn-SOD or both an Fe-SOD and an Mn-SOD. The extensive amino acid sequence similarity between the latter two enzymes indicates a gene duplication of the ancestral Fe-SOD (Harris and Steinman, 1977), possibly in the immediate precursor of the *Enterobacteriaceae*. Subsequently, some of the members of this family appear to have lost the Fe-SOD and retained only the Mn-containing enzyme. It should be stressed that the conclusions drawn from Figure 5.3 are tentative since the following key genera have not been included in the studies of rRNA homology: *Plesiomonas*, additional genera of *Pasteurellaceae* (*Actinobacillus* and *Haemophilus*), and *Xenorhabdus*. The latter genus, which is oxidase-negative and contains an Mn-SOD (Bang et al., 1978), has been found by immunological studies of GS to be more distant from *Escherichia coli* than *Proteus* or *Erwinia* (Baumann et al., 1980a), thereby potentially extending the diversity within the *Enterobacteriaceae*. *Plesiomonas* (for

which only GS data is available) appears to be somewhat more closely related to *E. coli* than *Vibrio* and might, therefore, be affiliated with *Pasteurella* or *Aeromonas*. From the presently available data (Fig. 5.3) it is already apparent that the families *Vibrionaceae*, *Enterobacteriaceae*, and *Pasteurellaceae* (assuming the relationships of *P. multocida* are representative of this family) are not equivalent and that, for the sake of consistency, *Aeromonas* may have to be removed from the *Vibrionaceae* and given a new family designation. Any new proposals should, however, await further studies of relationship of additional genera within this group of organisms by either rRNA/DNA hybridization or 16S oligonucleotide cataloging.

See Table 5.57, genus *Vibrio*, for the differentiation of the genera in this family.

Table 5.57.

Differential characteristics of the genus **Vibrio** *and other morphologically or physiologically similar genera*[a]

Characteristics	Vibrio	Photo-bacterium	Aeromonas	Plesiomonas
Sheathed polar flagella	+	−	−	−
Accumulation of poly-β-hydroxybutyrate coupled with the inability to utilize β-hydroxybutyrate	−	+		
Na$^+$ is required for growth or stimulates growth	+	+	−	−
Production of:				
Lipase	[+][b]	D	+	−
Utilization of:				
D-Mannitol	[+][c]	−	[+]	−
Mol% G + C of DNA	38–51	40–44	57–63	51

[a] Symbols: +, all positive; [+], most positive; −, all negative; D, some species positive, some species negative.

[b] *V. nereis*, *V. anguillarum* biovar II, and *V. costicola* are negative for this trait.

[c] *V. nereis*, *V. anguillarum* biovar II, and *V. marinus* are negative for this trait.

Important Notes for Users of this Edition

1. Always read both generic and species descriptions because characters listed in the generic description are not usually listed in the species descriptions.

2. Unless otherwise indicated in footnotes to tables, the meanings of symbols are as follows:

    +   90% or more of strains are positive

    −   90% or more of strains are negative

    d   11–89% of strains are positive

    v   strain instability (*not* equivalent to "d")

    D   different reactions in different taxa (species of a genus or genera of a family)

3. All other symbols are defined in footnotes to tables.

Genus I. **Vibrio** Pacini 1854, 411[AL]

PAUL BAUMANN, A. L. FURNISS AND JOHN V. LEE

(*Beneckea* Campbell 1957, 328)

Vib′ri·o. L. v. *vibrio* move rapidly to and fro, vibrate; M.L. masc. n. *Vibrio* that which vibrates.

Straight or curved rods, 0.5–0.8 μm in width and 1.4–2.6 μm in length. Involution forms usually present in old cultures or under adverse conditions of cultivation. Do not form endospores or microcysts. **Gram-negative.** In liquid media, motile by monotrichous or multitrichous **polar flagella which are enclosed in a sheath** continuous with the outer membrane of the cell wall. **On solid media, may synthesize numerous lateral flagella** with a wavelength shorter than that of the sheathed polar flagellum. **Facultative anaerobes** capable of both **fermentative and respiratory metabolism.** Molecular oxygen is a universal electron acceptor. Do not denitrify or fix molecular nitrogen.* All are **chemoorganotrophs**; most are able to **grow in a mineral medium containing D-glucose** and NH$_4$Cl. A few strains have organic growth factor requirements. **Sodium ions stimulate the growth of all species and are an absolute requirement for most**; the minimal concentration necessary for optimal growth ranges from 5–700 mM. Most species grow well in media containing a seawater base. Fermentation of D-glucose results in the production of acidic end products but, generally, no gas. All utilize D-glucose, D-fructose, maltose, and glycerol. **Most species are oxidase-positive.** All grow at 20°C; most grow at 30°C. **Found in aquatic habitats with a wide range of salinities.** Very common in marine and estuarine environments and on the surfaces and in the intestinal contents of marine animals. Some species are also found in freshwater habitats. Several species are pathogenic for man as well as for marine vertebrates and invertebrates. The mol% G + C of the DNA is 38–51 (Bd; T_m).

Type species: *Vibrio cholerae* Pacini 1854, 411.

Further Descriptive Information

Cell morphology. Species of *Vibrio* consist of straight or curved rods (Figs. 5.4 and 5.5) with an ultrastructure typical of most other Gram-negative bacteria (Figs. 5.6 and 5.7; Colwell, 1970; Felter et al., 1969; Glauert et al., 1963). In some strains, cell curvature is more pronounced in early stationary phase in liquid media than during exponential growth. In late stationary phase or under adverse conditions, involution forms generally predominate in the culture. While the morphology of most cells becomes very irregular, some assume a uniform spherical shape (Baker and Park, 1975; Das and Chatterjee, 1969; Felter et al., 1969, 1970; Kennedy et al., 1970). In V. cholerae NCTC 4716, Baker and Park (1975) found that the conversion of rods to spheres during stationary phase was accompanied by a loss of viability, a decrease in peptidoglycan, and by leakage of intracellular constituents. When grown in liquid media, some of the marine strains possess numerous tubular appendages of unknown function (Fig. 5.8) which are continuous with the outer membrane of the cell wall (Allen and Baumann, 1971). A number of species accumulate intracellular refractile granules of the reserve product poly-β-hydroxybutyrate (Figs. 5.9 and 5.10) when grown in a medium containing excess β-hydroxybutyrate and limiting nitrogen. Rhapidosomes have been found in V. harveyi (Yamamoto, 1967).

Flagella. In liquid media, most species of *Vibrio* have a single, sheathed, polar flagellum with a wavelength of 1.4–1.8 μm (Fig. 5.11) occasional cells have been observed with up to three polar flagella (Baumann et al., 1980b; de Boer et al., 1975a). The flagella of two species, V. fischeri and V. logei, are somewhat different in that they occur in tufts of 3–12 and have a wavelength of about 3.6 μm (Fig. 5.12) (Allen and Baumann, 1971; Bang et al., 1978a). The polar flagella of all species of *Vibrio* are 24–30 nm thick (Figs. 5.11 and 5.12) and are composed of a 14- to 16-nm core surrounded by a sheath (Fig. 5.13)

which is continuous with the outer membrane of the cell wall (Allen and Baumann, 1971; Baumann et al., 1980b; Das and Chatterjee, 1966; de Boer et al., 1975a, b; Follet and Gordon, 1963; Glauert et al., 1963; Hendrie et al., 1970; Hodgkiss and Shewan, 1968; Johnson et al., 1943; McCarthy, 1975; Miwatani et al., 1970). Studies with *V. cholerae* have shown that the sheath and the outer membrane contain a common protein, but antibody to the lipopolysaccharide of this species reacts with the outer membrane and not with the sheath (Hranitzky et al., 1980). When grown on solid media, many species of *Vibrio* synthesize additional unsheathed lateral flagella (Fig. 5.14) (Allen and Baumann, 1971; Buttiaux and Voisin, 1958/1959; de Boer et al., 1975a, b; Ulitzer, 1974; Ulitzer and Kessel, 1973; Yabuuchi et al., 1974); there is an immediate cessation of synthesis upon transfer from solid to liquid media (Shinoda and Okamoto, 1977). These flagella have a diameter of 14–15 nm and a wavelength of 0.9 μm, and in some strains may number over 100/cell. The flagellins of the polar flagella differ from the flagellins of the lateral flagella in their amino acid composition,

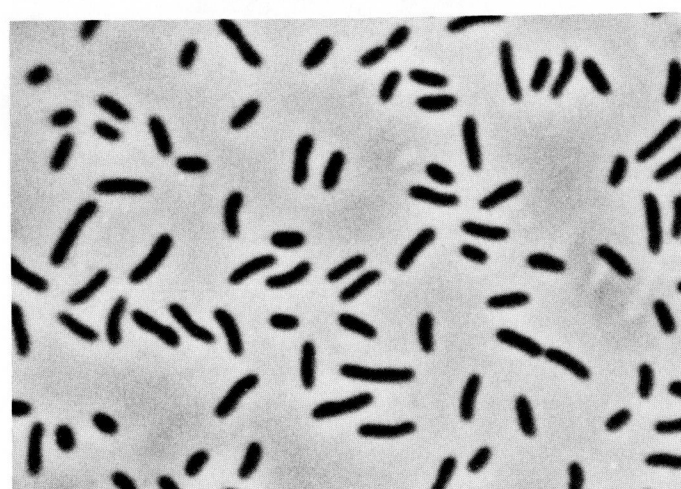

Figure 5.4. Phase contrast micrograph of *Vibrio nereis* in exponential phase of growth in YEB (× 3000).

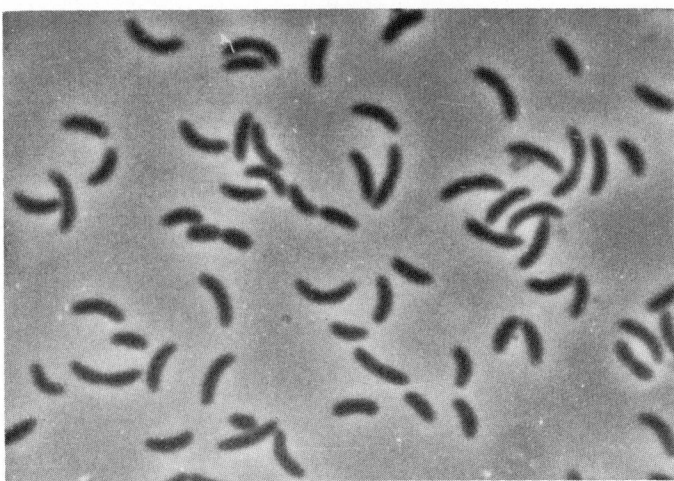

Figure 5.5. Phase contrast micrograph of *Vibrio splendidus* biovar II in exponential phase of growth in YEB (× 3000).

---

* See Other Organisms at the end of this chapter.

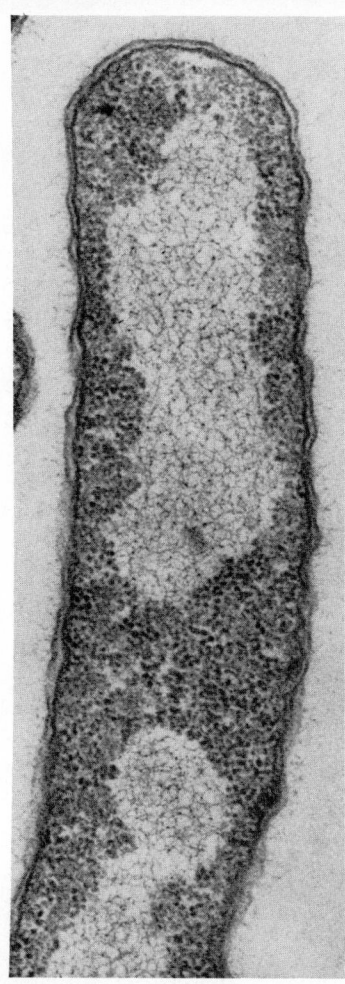

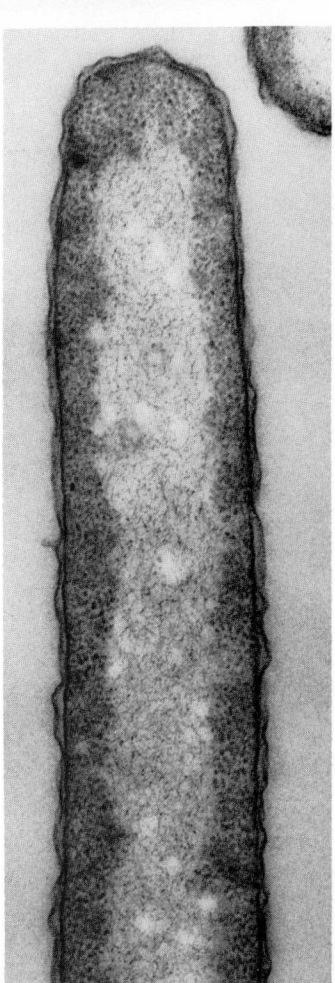

Figure 5.6. Ultrathin section of *Vibrio harveyi* in exponential phase of growth in YEB (× 43,000). (Reproduced with permission of S. W. Watson.)

Figure 5.7. Ultrathin section of *Vibrio fischeri* in exponential phase of growth in YEB (× 40,000). (Reproduced with permission of S. W. Watson.)

Figure 5.8. Electron micrograph of *Vibrio campbellii* illustrating the appearance of the tubular appendages observed in some marine vibrios. Negatively stained (× 36,000). (Reproduced with permission from R. D. Allen and P. Baumann, Journal of Bacteriology *107:* 295–302, 1971, © American Society for Microbiology.)

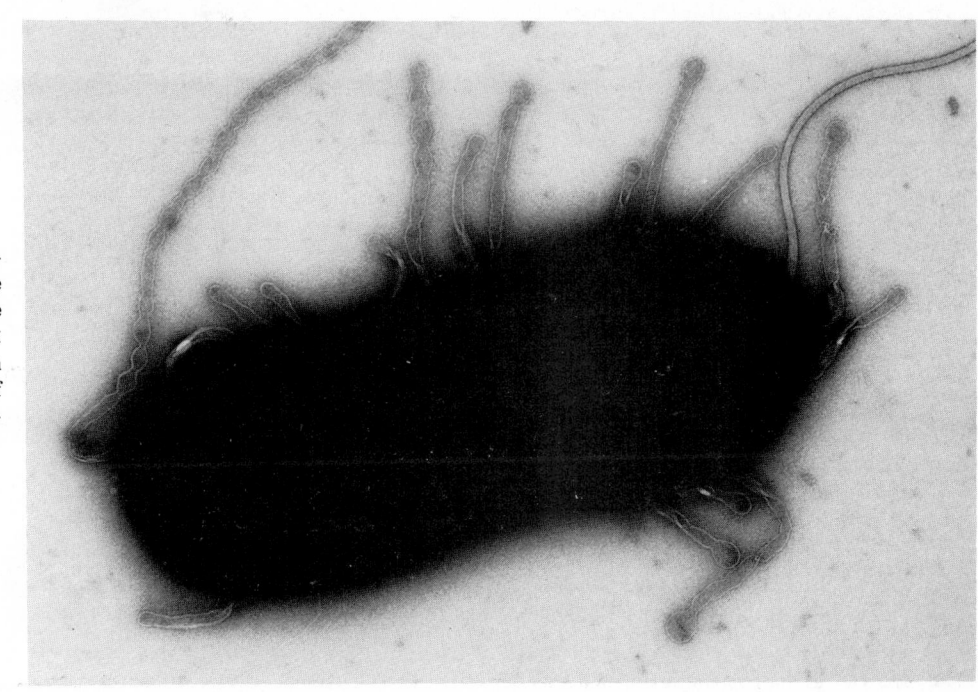

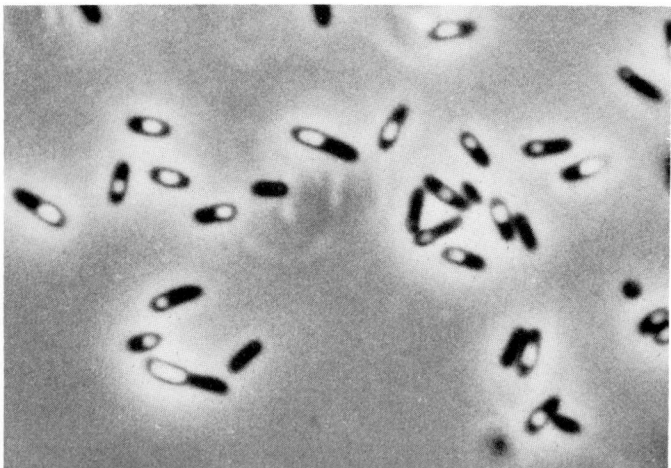

Figure 5.9. Phase contrast micrograph of *Vibrio nereis* containing refractile granules of PHB accumulated during stationary phase; grown in BM containing limiting nitrogen and excess β-hydroxybutyrate (× 3000).

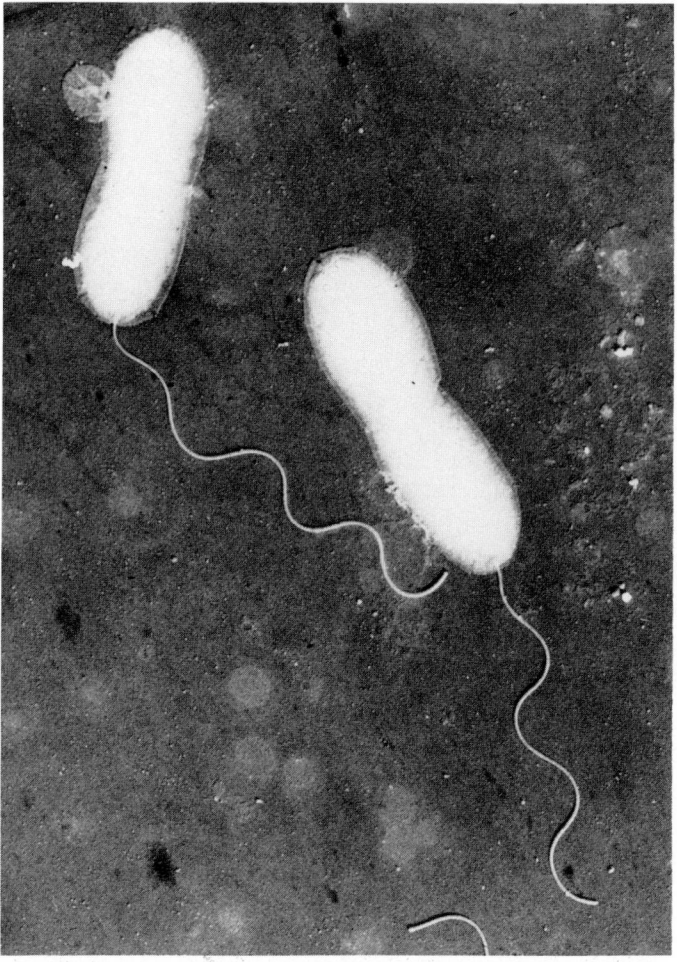

Figure 5.11. Electron micrograph of *Vibrio alginolyticus* grown in liquid medium showing the sheathed polar flagellum. Shadowed preparation (× 13,000). (Reproduced with permission from C. Golten and W. A. Scheffers, Netherland Journal of Sea Research 9: 351–364, 1975, © E. J. Brill.)

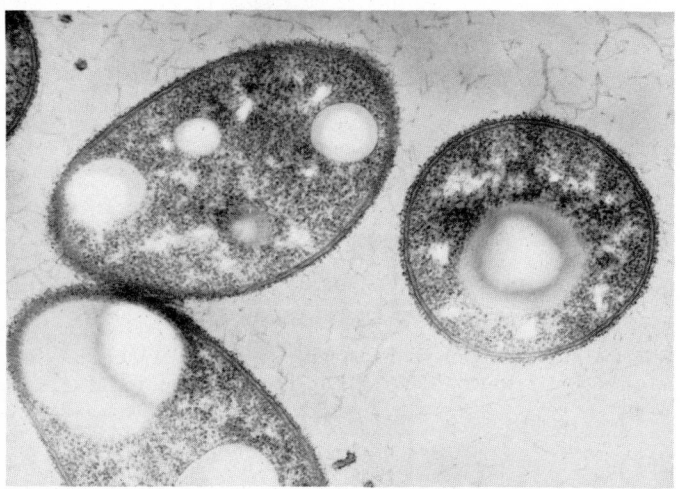

Figure 5.10. Ultrathin section of *Vibrio natriegens* containing granules of PHB accumulated during stationary phase of growth in BM with limiting nitrogen and excess β-hydroxybutyrate (× 32,000). (Reproduced with permission of R. D. Allen.)

antigenic properties, and behavior on hydroxyapatite columns (Shinoda et al., 1974a, b; 1976).

Many strains of marine vibrios are able to swarm on solid media. Swarming appears to be associated with the formation of long cells with many lateral flagella and is affected by a number of chemical and physical parameters including concentration of the agar, complexity of the medium, and temperature (Baumann and Baumann, 1977; de Boer et al., 1975a, b; Ulitzer, 1975a, b). The lateral (but not the polar) flagella are essential for swarming on solid media (Shinoda and Okamoto, 1977).

Cultures of some *Vibrio* strains grown on solid media may occasionally contain large bundles of detached flagella (Fig. 5.15) which are readily detected by phase microscopy (Ulitzer and Kessel, 1973). Fimbriae have been observed in a number of species (Hodgkiss and Shewan, 1968; Tweedy et al., 1968).

Colonial morphology and pigmentation. Vibrios will grow readily on a variety of media.* Most species give rise to convex, smooth, creamy white colonies with entire edges. Variants in colonial morphology may be detected in some species, particularly after repeated culture and storage on more complex media. Colonies may be rough or sometimes rugose; such colonies appear firmly attached to the medium and cannot be emulsified.

V. nigripulchritudo produces an insoluble blue-black pigment which accumulates in crystalline form within the colony (Fig. 5.16); pigment production is enhanced by growth on a minimal medium (Baumann et

---

* Many of the species of fresh water, estuarine or marine origin, particularly those pathogenic for humans, will grow on a wide range of the commonly used nutrient agars, blood agar and other media, provided they contain NaCl (normally 0.5–1.0% (w/v)). Most species of marine and estuarine origin will grow better on seawater agar (Lee et al., 1979b), Difco marine agar (MA) (Zobell, 1941), medium M of leifson (1970), or media containing the artificial seawater base (ASW) of MacLeod (1968) consisting of 400 mM (23.4 g/liter) NaCl, 100 mM (24.6 g/liter) $MgSO_4 \cdot 7H_2O$, 20 mM (1.5 g/liter) KCl, and 20 mM (2.9 g/liter) $CaCl_2 \cdot 2H_2O$ (the salts are dissolved separately and combined). A *basal medium* (*BM*) satisfactory for the cultivation of these isolates contains 50 or 100 mM (6.1 or 12.1 g/liter) Tris(hydroxymethyl)aminomethane (adjusted to pH 7.5 with HCl), 19 mM (1.0 g/liter) NH_4Cl, 0.33 mM (75 mg/liter) $K_2HPO_4 \cdot 3H_2O$, 0.1 mM (28 mg/liter) $FeSO_4 \cdot 7H_2O$, and ½ strength ASW. *Yeast extract broth* (*YEB*) is obtained by supplementing BM with yeast extract (5.0 g/liter); *yeast extract agar* (*YEA*) and *basal medium agar* (*BMA*) are made by

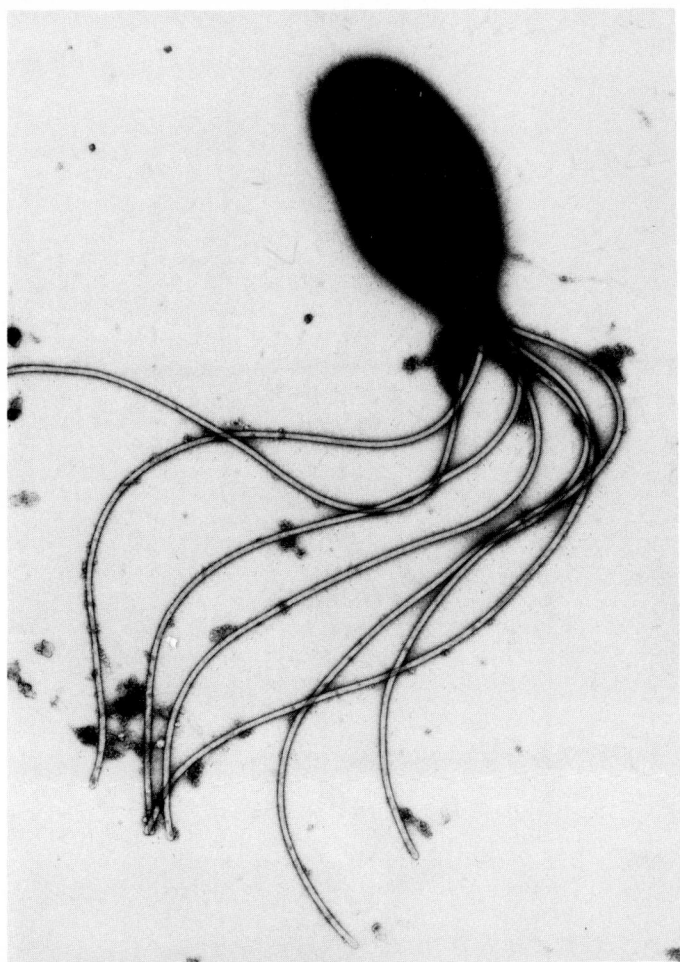

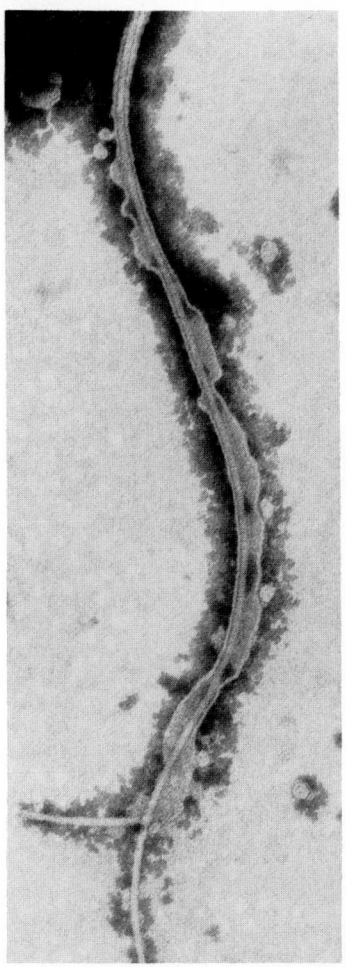

Figure 5.12. Electron micrograph of *Vibrio fischeri* showing tufts of sheathed polar flagella. Negatively stained (× 23,000). (Reproduced with permission from: J. L. Reichelt and P. Baumann, Arch. Mikrobiol. *94:* 283–330, 1973, © Springer-Verlag.)

Figure 5.13. Electron micrograph of a polar flagellum of *Vibrio alginolyticus* in which the sheath has partially disintegrated exposing the inner core. Negatively stained (× 30,000). (Reproduced with permission of R. D. Allen.)

al., 1971b). *V. gazogenes* makes a red pigment (probably similar or identical to prodigiosin) (Harwood, 1978), and *V. fischeri* and *V. logei* produce a yellow-orange cell-associated pigment which is readily evident after incubation for 3–4 days on solid complex media (Bang et al., 1978a; Reichelt and Baumann, 1973).

Growth conditions and nutrition. Growth of all species of *Vibrio* is stimulated by Na⁺. The minimal concentration required for optimal growth ranges from 5–15 mM for *V. cholerae* and *V. metschnikovii* to 600–700 for *V. costicola* (Baumann et al., 1980b; Kushner, 1978; Reichelt and Baumann, 1974). A screening of a relatively small number of strains of *V. cholerae* has indicated that this species differs from the remaining species of *Vibrio* in its ability to grow, and be serially transferred, in TBM containing no added Na⁺ (contaminating levels below 0.1 mM). The growth rate and cell yield is, however, reduced to 50–80% of that observed at the optimal Na⁺ concentration. Unlike *V. cholerae*, however, most strains of *V. metschnikovii* are not able to grow in 1% (w/v) tryptone broth with no added NaCl, a medium used in the

salt tolerance test for identification. All of the remaining species have an absolute requirement for Na⁺. In some species of *Vibrio*, the requirement for Na⁺ is reduced by levels of Mg⁺⁺ (50 mM) and Ca⁺⁺ (10 mM) such as are in seawater (Reichelt and Baumann, 1974). Since there is considerable variation in the ionic requirements of the different species of *Vibrio*, no single medium will allow optimal growth of all strains.

Most species of *Vibrio* do not require organic growth-factors. Some strains, however, may develop requirements for amino acids, particularly after prolonged storage and subculture (Lee et al., 1978; P. Baumann, unpublished observations). More complex growth-factor requirements are found in *V. anguillarum* biovar II, *V. marinus*, and some strains of *V. logei* which will grow only in the presence of yeast extract (Bang et al., 1978a; Baumann et al., 1978; Reichelt and Baumann, 1973).

Species of *Vibrio* vary with respect to the temperatures at which growth will occur. All grow at 20°C and most at 30°C; some grow at 4 and 45°C; none grows at 50°C.

adding agar (20 g/liter) to YEB and BM, respectively. *V. cholerae* and *V. metschnikovii* may be grown in the *terrestrial basal medium* (*TBM*) of Palleroni and Doudoroff (1972), modified to contain 50 or 100 mM (6.1 or 12.1 g/liter) Tris(hydroxymethyl)aminomethane (adjusted to pH 7.5 with HCl), 30 mM (1.75 g/liter) NaCl, 19 mM (1.0 g/liter) NH₄Cl, 10 mM (0.75 g/liter) KCl, 2 mM (0.5 g/liter) MgSO₄·7H₂O, 1 mM (0.23 g/liter) K₂HPO₄·3H₂O, 0.55 mM (81 mg/liter) CaCl₂·2H₂O, and 50 mg/liter ferric ammonium citrate. *Terrestrial yeast extract broth* (*TYEB*) is TBM with yeast extract (5.0 g/liter). *Terrestrial basal medium agar* (*TBMA*) and *terrestrial yeast extract agar* (*TYEA*) are made by adding agar (20 g/liter) to TBM and TYEB, respectively. Either the marine or terrestrial media can be made suitable for the cultivation of *V. costicola* by increasing the concentration of NaCl to 800 mM (46.8 g/liter).

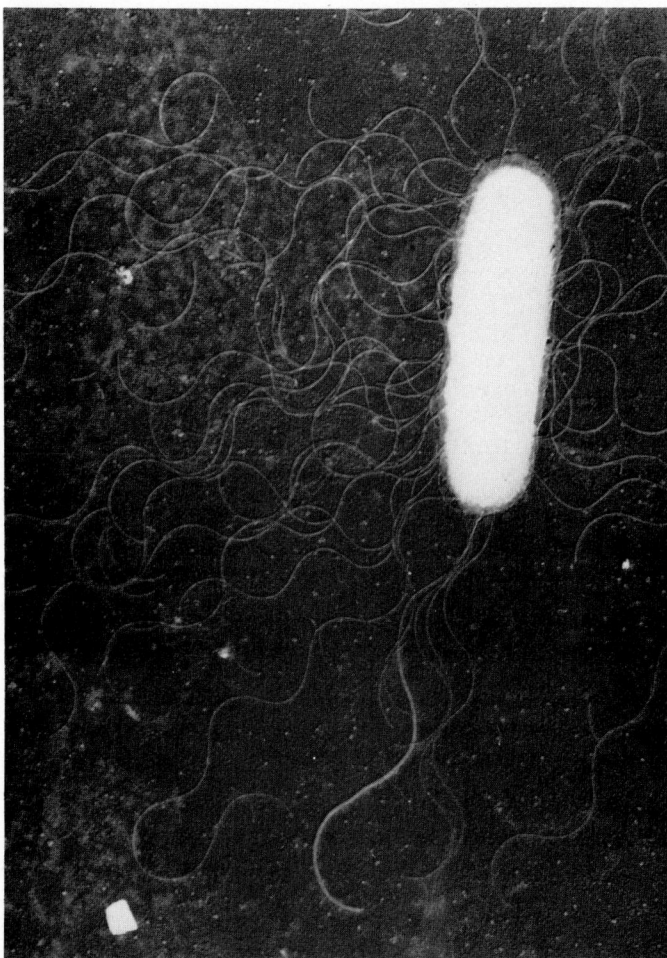

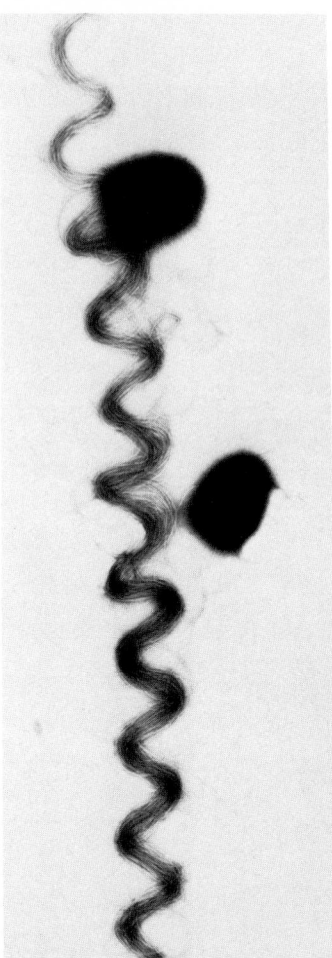

Figure 5.14. Electron micrograph of *Vibrio alginolyticus* grown on solid medium showing the thick, sheathed polar flagellum and the numerous unsheathed lateral flagella. Shadowed preparation (× 18,000). (Reproduced with permission from W. E. de Boer, C. Golten and W. A. Scheffers, Netherland Journal of Sea Research 9: 197–213, 1975, © E. J. Brill.)

Figure 5.15. Large bundle of flagella in a culture of *Vibrio harveyi.* Such bundles are frequently observed in solid grown cultures. Negatively stained (× 13,000). (Reproduced with permission of R. D. Allen.)

Many species of *Vibrio* will tolerate moderately alkaline conditions and will grow at pH 9, and some, notably *V. cholerae* and *V. metschnikovii,* will grow at pH 10.

The species vary in their nutritional versatility; some grow on as few as 12 and others on as many as 67 of the 150 organic compounds tested as sole or principal sources of carbon and energy. These compounds include pentoses, hexoses, disaccharides, sugar acids, sugar alcohols, C_2–C_{10} monocarboxylic fatty acids, tricarboxylic acid cycle intermediates, amino acids and monocyclic aromatic compounds. Most species have a number of extracellular hydrolases which may include amylase, gelatinase, lipase, chitinase, alginase and deoxyribonuclease.

Physiology and metabolism. Under anaerobic conditions, species of *Vibrio* ferment D-glucose by means of a mixed acid fermentation. The principal end-products are formic, acetic, lactic, succinic and pyruvic acids as well as ethyl alcohol (Doudoroff, 1942; Harwood, 1978; Payne et al., 1961; Unger et al., 1961). Upon completion of the fermentation, the pH of the medium ranges from 4.6–5.8. Some species produce acetoin and/or diacetyl as well as 2,3-butanediol. Two species, *V. fluvialis* biovar II and *V. gazogenes,* produce gas which in the latter species has been shown to be a mixture of H_2 and CO_2 (Harwood, 1978). D-Glucose is catabolized via a phosphoenolpyruvate:D-glucose phosphotransferase system and a constitutive Embden-Meyerhof pathway while D-gluconate is degraded by means of an inducible Entner-Doudoroff pathway (Bag, 1973; Baumann et al., 1973; Eagon and Wang,

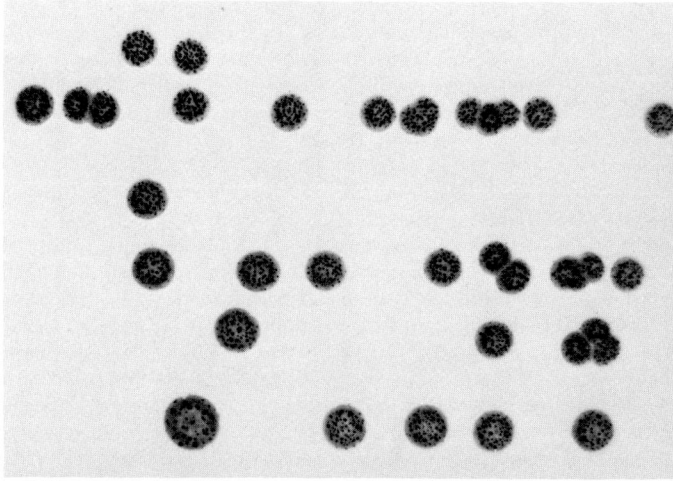

Figure 5.16. Colonies of *Vibrio nigripulchritudo* containing crystals of a blue-black pigment after 2 days incubation on BMA with 0.2% (v/v) glycerol (× 4.5).

1962; Kubota et al., 1979; Matsumoto et al., 1974; Reichelt and Baumann, 1973; Unger et al., 1961). D-Fructose is utilized via an inducible phosphoenolpyruvate:D-fructose phosphotransferase system and 1-phosphofructokinase which convert this sugar to fructose-1-phosphate and fructose-1,6-diphosphate, respectively; the subsequent catabolism of this sugar-phosphate is performed by enzymes of the Embden-Meyerhof pathway (Bag, 1974; Gee et al., 1975).

Several species of *Vibrio* are able to utilize benzoate, *p*-hydroxybenzoate or quinate. These compounds are degraded by means of the α-ketoacid pathway as is indicated by the *m*-cleavage of the intermediate protocatechuate (Ornston, 1971).

Most species of *Vibrio* are oxidase-positive, a property which correlates with the presence of cytochromes of the *c* type (West et al., 1978). Cytochromes of the *b* and *c* types have been found in most species (Baumann et al., 1973; P. Baumann, unpublished observations). Several species have an apparently soluble *c* type cytochrome with the capacity to bind carbon monoxide (West et al., 1978). Cytochromes *b*, *c*, *d*, *o* and a_1 occur in *V. cholerae*, *V. anguillarum* biovar I, *V. natriegens* and *V. alginolyticus* (Unemoto and Hayashi, 1979; West et al., 1978; Weston and Knowles, 1974). *V. metschnikovii*, which is oxidase-negative, contains cytochromes *b*, *d*, *o* and a_1 but lacks cytochromes of the *c* type (West et al., 1978).

Several species of *Vibrio* have a constitutive arginine-dihydrolase system as determined by assaying for ornithine produced from arginine under anaerobic conditions (Baumann and Baumann, 1981; Baumann et al., 1971a).

With the exception of one isolate, strains of all species of *Vibrio* have a single iron-containing superoxide dismutase (SOD) as is indicated by one activity band in polyacrylamide gels (Bang et al., 1978b, and unpublished observations). Strain ATCC 27519, which has been identified as *V. parahaemolyticus* by both the immunological properties of its alkaline phosphatase and its phenotype (M. J. Woolkalis, unpublished observations), has two activity bands in polyacrylamide gels (Bang et al., 1978b; Daily et al., 1978). The major band corresponds to the iron-containing SOD; the nature of the SOD in the minor band has not been established.

The description of the morphological, physiological, and nutritional properties of *Vibrio* is primarily based on material considered in publications by Baumann and Baumann (1977, 1981), Baumann et al. (1980b), Barua and Burrows (1974), Chumakova et al. (1973), Colwell (1970), Desmarchelier and Reichelt (1981), Hendrie et al. (1970), Lee et al. (1978, 1981a), Sakazaki (1979), and Véron (1965, 1966).

Bioluminescence. Some strains of *V. cholerae*, *V. harveyi*, *V. splendidus*, *V. fischeri* and *V. logei* are able to emit light of a blue-green color. The reaction leading to light emission, catalyzed by the enzyme luciferase, has been shown to be similar in all procaryotes (Nealson and Hastings, 1979). The substrates are reduced flavin mononucleotide ($FMNH_2$), a long chain aldehyde (RCHO; probably tetradecanal), and molecular oxygen which react according to the following overall stoichiometry:

$$FMNH_2 + O_2 + RCHO \xrightarrow{\text{luciferase}} FMN + H_2O + RCOOH + \text{light}$$

(Hastings and Nealson, 1977; Ziegler and Baldwin, 1981). Bacterial luciferase is a heterodimer having a molecular weight (MW) of about 80,000 and consisting of α and β subunits with MWs of about 42,000 and 38,000, respectively. The subunits are functionally distinct with the active center of the enzyme located on the α subunit. Amino-terminal-sequence-determinations of the α and β subunits from *V. harveyi* and *V. fischeri* (Baldwin et al., 1979) have indicated considerable sequence similarity between homologous subunits (57% identity between α subunits and 71% between β subunits). A comparison of the heterologous subunits has shown 36–46% sequence identity; these values are sufficiently high to indicate that the α and β subunits originated via duplication of a common ancestral sequence.

A strain of *V. logei* has been found which emits yellow light at temperatures below 18°C and blue-green light at temperatures above 23°C. This unusual emission of yellow, as opposed to blue-green, light at lower temperatures appears to be due to the transfer of energy from luciferase to another molecular species (Ruby and Nealson, 1977; Ziegler and Baldwin, 1981). A blue-fluorescence protein, differing in molecular weight from that of *Photobacterium phosphoreum* but having

properties suggesting a common prosthetic group, has been detected in *V. fischeri* by Lee et al. (1979a).

In species of *Vibrio*, luciferase synthesis is dependent on the cell density of the culture, a finding which is consistent with the synthesis and excretion into the medium of a substance which induces luciferase (Nealson and Hastings, 1979). Such a compound (designated an autoinducer) has been purified from *V. fischeri* and its structure determined to be *N*-(3-oxohexanoyl)-3-aminodihydro-2(3*H*)-furanone (*N*-β-ketohexanoyl-homoserine lactone).

Genetics and plasmids. Bhaskaran (1958, 1959) discovered the sex factor P which mediates conjugal mating in *V. cholerae*. This system was used to produce the first linear maps of the chromosome of *V. cholerae* (Barua and Burrows, 1974; Parker et al., 1979). Chromosomal transfer by P occurs at a low frequency but may be increased by the transposon-facilitated recombination system of Johnson and Romig (1979). Their system has been used to construct circular maps of the chromosomes of the eltor and classical biovars of *V. cholerae* serovar O:1 and demonstrate that the two maps are very similar (Johnson et al., 1980). The structural genes for cholera toxin have not yet been located but they appear to be chromosomal (Mekalanos et al., 1979). Various regulatory genes governing toxin production have been detected and are located on the chromosome (Holmes et al., 1979; Mekalanos and Murphy, 1980; Mekalanos et al., 1979; Vasil et al., 1975). Transduction has also been demonstrated in *V. cholerae* (Ogg et al., 1981).

In contrast to *V. cholerae*, little or no information is available on the genetics of the remaining *Vibrio* species. Keynan et al. (1974) isolated a bacteriophage for *V. harveyi* which transduces only the genes of the tryptophan operon (Crawford and Nealson, 1976). To date, this is the sole genetic system which has been clearly established in a marine vibrio. Treatment with *N*-methyl-*N'*-nitro-*N*-nitrosoguanidine has been used to obtain mutants of the luminous systems of *V. harveyi* (Cline, 1978) and *V. fischeri* (Ulitzer and Hastings, 1978), and mutants of *V. parahaemolyticus* and *V. alginolyticus* altered in flagellation and their ability to swarm on solid media (Ulitzer, 1975a; Shinoda and Okamoto, 1977). Ethylmethane sulfonate has been used to isolate mutants of *V. parahaemolyticus* deficient in the binding of adenosine-3',5'-monophosphate (Iuchi et al., 1975) as well as mutants of the phosphoenolpyruvate:D-glucose phosphotransferase system (Kubota et al., 1979). Genes coding for ribosomal RNA (rRNA) from *V. harveyi* have been cloned in *Escherichia coli* (Lamfrom et al., 1978).

Resistance transfer (R) factors are generally rare in *V. cholerae* O:1 but have been reported for a few strains isolated in Algeria, India, the Philippines and the U.S.S.R. (O'Grady et al., 1976) and for many strains isolated in an outbreak in Tanzania (Mhalu et al., 1979) and another in Bangladesh (Huq et al., 1980b). In all cases, the plasmid was of compatability group C (Threlfall et al., 1980). Aoki et al. (1975) have shown that R-factors from *V. anguillarum* and other marine vibrios have unique incompatibility properties distinguishing them from previously established groups. R-factors have also been detected in *V. fluvialis* (McNicol et al., 1980). In *V. anguillarum* a plasmid encoding an iron-sequestering system has been found to be associated with high virulence (Crosa, 1980). A bacteriocinogenic plasmid in *V. harveyi* has been characterized by McCall and Sizemore (1979), and cryptic plasmids have been detected in *V. parahaemolyticus* (Guerry and Colwell, 1977).

Phage, phage-typing and bdellovibrios. The extensive studies on the bacteriophages active against *V. cholerae* and the typing schemes based on them have been reviewed by Mukerjee (1978). Only the scheme of Mukerjee et al. (1957) has been used widely for the typing of the classical biovar of *V. cholerae*. Similarly, the scheme of Basu and Mukerjee (1968) is the only one to have been used extensively for typing the eltor biovar but this is of limited epidemiological value since it divides strains into only six types. Two new schemes for typing all strains of *V. cholerae* O:1 have recently been developed. The scheme of Drozhevkina and Arutyunov (1979) divides the eltor strains into six types. The scheme of Lee and Furniss (1981) is based on nine phages derived from the earlier schemes and five new isolates. This scheme appears to be potentially useful epidemiologically since over 1000 recent isolates of the eltor biovar fell into 24 types, although 3 types still predominated. Vibriocin typing schemes for *V. cholerae* O:1 and non-O:1 *V. cholerae* strains have been developed but not widely applied (Chakrabarty et al., 1970, 1971). Vibriocins and vibriocin typing have been reviewed by Brandis (1978).

Bacteriophages of different morphological types have been isolated for *V. parahaemolyticus* (Nakanishi et al., 1966; Sklarow et al., 1973) and *V. natriegens* (Zachary, 1974). An enumeration of bacteriophages occurring in a variety of marine samples and active against several species of marine vibrios has been performed by Baross et al. (1978a).

Twelve out of 13 species of *Vibrio* tested have been found to serve as hosts for strains of marine *Bdellovibrio* (Marbach et al., 1976; Miyamoto and Kuroda, 1975; Taylor et al., 1974).

Antigenic structure. Strains of *V. cholerae*, both O:1 and non-O:1, share a common flagellar (H) antigen (Bhattacharyya, 1975; Gardner and Venkatraman, 1935; Sakazaki et al., 1970) which is demonstrated most easily by the method of Sil and Bhattacharyya (1979); the nature of this flagellar antigen is not known. Flagellins from the polar flagella of many species of *Vibrio* do share a common antigenic determinant(s); these species include *V. cholerae*, *V. metschnikovii*, *V. harveyi*, *V. campbellii*, *V. parahaemolyticus*, *V. alginolyticus*, *V. vulnificus*, *V. nereis*, *V. fluvialis*, *V. splendidus*, *V. pelagius*, *V. nigripulchritudo*, *V. anguillarum*, *V. fischeri*, and *V. costicola* (Shinoda et al., 1976; D. A. Barnes and S. Shinoda, unpublished observations). The lateral flagella of those species that form them are antigenically distinct from their polar flagella. In *V. parahaemolyticus* and *V. alginolyticus* the lateral flagella are antigenically similar to one another but distinct from those of *V. harveyi* and *V. campbellii* (Shinoda et al., 1976).

V. cholerae may be subdivided on the basis of its O antigens (Gardner and Venkatraman, 1935). Two O-typing schemes are used, one with 60 serovars (Sakazaki et al., 1970; Shimada and Sakazaki, 1977) and the other with 72 serovars (Smith, 1979). Undoubtedly there are many more serovars as only about 60% of strains are typable at present. The O:1 serovar is the same in both schemes and into this type fall all of the strains responsible for epidemic or pandemic cholera. Using absorbed antisera the O:1 serovar may be divided into the Ogawa and Inaba subtypes; a Hikojima subtype has been proposed, but further work is necessary to establish its reality. An antigen that has been termed R has also been described but its exact nature has not been determined (Shimada and Sakazaki, 1973).

V. parahaemolyticus may be serotyped on the basis of its O and K antigens (Sakazaki et al., 1968; Committee on the serological typing of *V. parahaemolyticus*, 1970). The qualitative chemical nature of the antigens of *V. parahaemolyticus* is reviewed by Miwatani and Takeda (1976). The quantitative chemical analysis of the sugar composition of the O antigenic lipopolysaccharides has recently been described by Hisatsune et al. (1980a). Studies of the chemical composition of the lipopolysaccharides of *V. cholerae*, *V. metschnikovii*, and *V. fluvialis* have shown that they do not have 2-keto-3-deoxy-octonic acid in the polysaccharide core of their lipopolysaccharide. This is similar to *Aeromonas* and *V. parahaemolyticus* but unlike the majority of other Gram-negative bacteria (Hisatsune et al., 1980b). Little is known about the antigenic structure of the other species of *Vibrio*.

Antibiotic sensitivity. Only the antibiotic sensitivities of the clinically significant species have been studied in detail (Hollis et al., 1976; Huq et al., 1980a; Lee et al., 1981a; Miwatani and Takeda, 1976; O'Grady et al., 1976). With very rare exceptions they are sensitive to tetracycline which is usually the antibiotic of choice for the treatment of infections. In addition, they are sensitive to a relatively wide range of antibiotics including chloramphenicol, gentamicin, kanamycin, streptomycin and

sulfonamides. *V. parahaemolyticus* and *V. alginolyticus* are relatively more resistant to polymyxin than other *Vibrio* species. Sensitivity to penicillin and novobiocin varies from species to species, *V. cholerae* generally being fully sensitive while strains of *V. parahaemolyticus* and *V. fluvialis* are relatively resistant.

Pathogenicity. *V. cholerae* serovar O:1 is the causative agent of epidemic or asiatic cholera; other serovars may produce diarrhea, and outbreaks have occurred (Aldová et al., 1968; Blake et al., 1980b; Dakin et al., 1974; Kamal and Zinnaka, 1971; McIntyre and Feeley, 1965). The pathogenic strains of *V. cholerae* O:1 produce cholera toxin whose action on the mucosal cells of the small intestine is responsible for the characteristic diarrhea of the disease cholera. The toxin is an oligomeric protein (MW 84,000) composed of an A_1 subunit (MW 21,000), an A_2 subunit (MW 7,000) and five B subunits (MW 10,000). The B subunit is responsible for binding to the receptor ganglioside G_{M1} on the cell membrane and the A_1 subunit activates adenylate cyclase leading to increased intracellular levels of cyclic AMP and hypersecretion of salt and water. Cholera toxin has been extensively studied and there are several reviews (Collier and Mekalanos, 1980; Enomoto and Gill, 1980; Field, 1979; Holmes et al., 1979; Moss and Vaughn, 1979; Richards and Douglas, 1978; Van Heyningen, 1977). Other factors important for the pathogenesis of *V. cholerae* O:1 have not been so extensively studied but they include motility (Yancey et al., 1978) and adhesion (Srivastava et al., 1980).

The pathogenic mechanisms of the other serovars of *V. cholerae* have not been extensively examined, but many strains produce a toxin identical or very similar to cholera toxin. Different pathogenic mechanisms, as yet unidentified, operate in other strains which do not produce cholera toxin but which are enteropathogenic in animal models and apparently cause diarrhea in humans. A few strains of environmental origin produce a toxin similar to the heat-stable toxin of *E. coli* while many others, particularly of environmental origin, do not produce detectable enterotoxins and appear to be nonpathogenic (Spira and Daniel, 1980; Spira et al., 1979).

V. parahaemolyticus causes gastroenteritis in humans which is usually contracted by eating contaminated seafoods (Fujino et al., 1974; Miwatani and Takeda, 1976). The mechanisms whereby *V. parahaemolyticus* causes gastrointestinal disease have not been elucidated despite the extensive research on this subject which is reviewed by Miwatani and Takeda (1976) and Blake et al. (1980b). More than 95% of strains isolated from patients with diarrhea are Kanagawa-positive,[*] i.e. they hemolyze human erythrocytes on Wagatsuma agar (Miyamoto et al., 1969). The hemolysin is a heat-stable direct hemolysin (MW 42,000) which is cytotoxic and cardiotoxic, but its clinical significance is not known. Kanagawa-positive strains differ from Kanagawa-negative strains in a number of ways: they produce reactions in rabbit ileal loops more frequently, cause bacteremia in orally challenged infant rabbits, and they adhere to HeLa cells and are more rapidly cytotoxic to them.

V. vulnificus may invade the human body from the gastrointestinal tract leading to a septicemia particularly in patients with hepatic disease (Blake et al., 1979). *V. vulnificus*, *V. cholerae*, *V. parahaemolyticus*, and *V. alginolyticus* have all been isolated from superficial lesions on humans where they may simply be colonizers or opportunistic pathogens (Baumann et al., 1973; Blake et al., 1979, 1980b; Pezzlo et al., 1979; Spark et al., 1979).

V. fluvialis biovar I has been isolated from humans with diarrhea suggesting that it may also be an enteropathogen but this remains to be proven (Huq et al., 1980a; Lee et al., 1981a).

V. anguillarum is a pathogen of marine fish and eels and a major cause of disease in fish culture (Anderson and Conroy, 1970; Sinder-

---

[*] The Kanagawa phenomenon is determined on Wagatsuma agar, which contains (g/liter): yeast extract, 3.0; Bacto-peptone (Difco), 10.0; NaCl 70.0; D-mannitol, 10.0; K_2HPO_4, 5.0; crystal violet 0.001; and agar, 15.0. The ingredients are dissolved by heating at 100°C, the medium is cooled to 50°, and the pH adjusted to 8.0 and a saline suspension of washed human erythrocytes added to a final concentration of 5% (v/v). Plates are inoculated with small drops of fresh broth cultures of the test organisms and incubated for 20–24 h at 37°C. Strains producing a zone of β-hemolysis are Kanagawa-positive. Known positive and negative controls should be tested in parallel. In the original method it was recommended that the basal medium was not sterilized but this is not always practicable and autoclaving the basal medium is usually satisfactory.

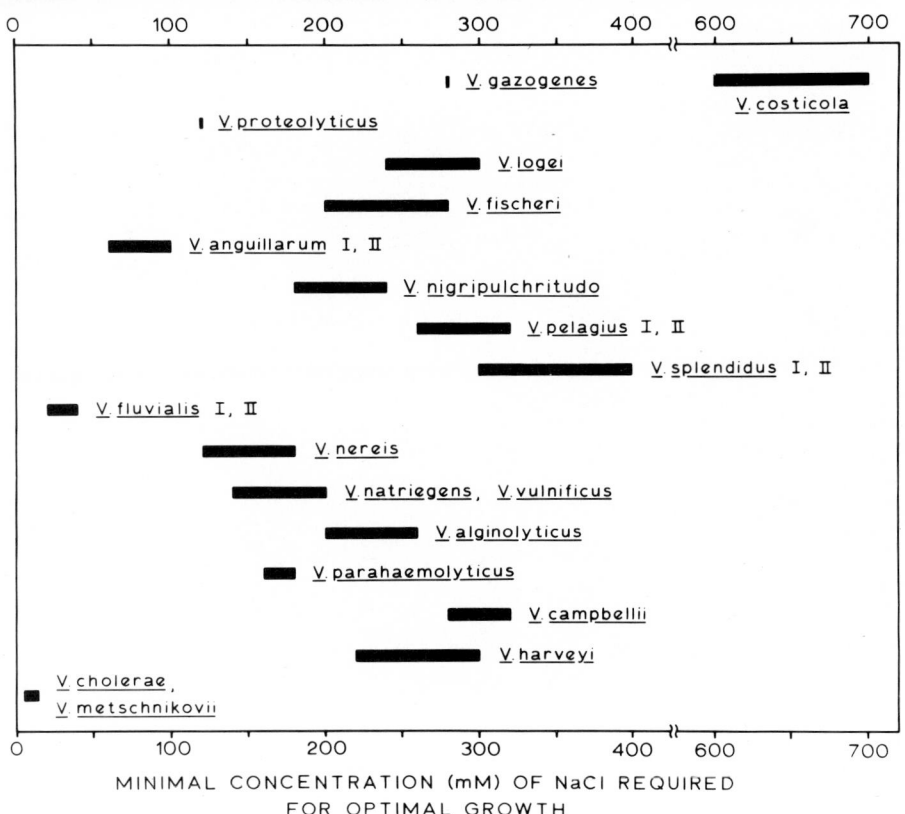

Figure 5.17. Range of the minimal NaCl concentrations required for optimal growth of species of *Vibrio*. Data from Baumann et al. (1980b).

man, 1970). A number of vibrios have been isolated from diseased or dead shellfish; these include *V. alginolyticus*, *V. parahaemolyticus* and *V. logei* (Baross et al., 1978c; Bowser et al., 1980; Vanderzant et al., 1971).

Ecology. The wide range of Na⁺ concentrations necessary for the optimal growth of species of *Vibrio* (Fig. 5.17) suggests that the different species may vary greatly with respect to their potential to inhabit environments of different salinities. At one extreme are *V. cholerae* and *V. metschnikovii* which require 5–15 mM Na⁺ for optimal growth; both species have been isolated from freshwater as well as estuarine habitats (Desmarchelier and Reichelt, 1981; Kaper et al., 1979; Lee et al., 1978; Szita et al., 1979). At the other extreme are *V. splendidus* and *V. costicola* which require 300–400 and 600–700 mM Na⁺, respectively, and would not be expected to do well in environments with considerably lower salt concentrations. On the other hand, *V. fluvialis* and *V. anguillarum*, which have respective requirements of 20–40 and 60–100 mM Na⁺, could potentially coexist with *V. cholerae* and *V. metschnikovii* in a number of estuarine and brackish habitats with salinities considerably below that of seawater. Evidence supporting this has been provided by the studies of Lee et al. (1978, 1981a) and West et al. (1980).

Transmission of *V. cholerae* O:1 to man is by water or food and has been extensively studied (Barua and Burrows, 1974). The natural reservoir of toxigenic *V. cholerae* O:1, the cause of epidemic cholera, is still not known. It has long been assumed to be in humans but recent evidence has suggested it may be in the aquatic environment (Blake et al., 1980a; Colwell et al., 1980; Rogers et al., 1980). These two theories are not mutually exclusive. Nontoxigenic *V. cholerae* O:1 also occurs naturally in brackish water in regions where indigenous cases of cholera are not known to occur (Bashford et al., 1979).

Non-O:1 *V. cholerae* have been isolated from freshwater and estuarine environments throughout the world as well as from birds, frogs and freshwater fish (Bashford et al., 1979; Bisgaard et al., 1978; Colwell

et al., 1980; Desmarchelier and Reichelt, 1981; Kaper et al., 1979; Müller, 1977; Năcescu and Ciufecu, 1978; Szeness et al., 1979; Szita et al., 1979; West et al., 1980). Many of these organisms were identified by a limited range of characters, but some were extensively characterized (Colwell et al., 1980; Desmarchelier and Reichelt, 1981; P. A. West, 1980 (Ecology and taxonomy of the genus *Vibrio*, Ph.D. dissertation, University of Kent, Canterbury, England)). In a brackish water drainage ditch in England there was a marked seasonal incidence of non-O:1 *V. cholerae* (West et al., 1980, and unpublished observations). The majority of the non-O:1 strains of *V. cholerae* isolated from the environment do not appear to be enteropathogenic (Spira and Daniel, 1980) and probably are indigenous there. The reservoir of the enteropathogenic strains of the various serovars of *V. cholerae* remains uncertain.

The epidemiology and ecology of *V. parahaemolyticus* have been extensively studied but, unfortunately, these organisms have not always been adequately distinguished from other marine vibrios (Anderson and Ordal, 1972; Baumann and Baumann, 1981; Baumann et al., 1973; Kaneko and Colwell, 1978). For example, two of the best clinically established diagnostic traits for *V. parahaemolyticus*, namely the ability to grow at 40–43°C and the inability to utilize sucrose, are shared by *V. vulnificus*, *V. proteolyticus* and some luminous and nonluminous strains of *V. harveyi*. In addition, the utilization of L-arabinose (a diagnostic trait present in some strains of *V. parahaemolyticus*) has also been observed in nonluminous isolates of *V. harveyi* which are able to grow at 43°C and are unable to utilize sucrose. Nevertheless, *V. parahaemolyticus* appears to be a common inhabitant of coastal waters and estuaries in tropical and temperate regions (Ayres and Barrow, 1978; Fujino et al., 1974; Kaneko and Colwell, 1978; Miwatani and Takeda, 1976; Sakazaki, 1979). In temperate regions there is a seasonal cycle; the organism is present only in sediment in the winter but increases in numbers and appears in the water column in warmer months when the water temperature is above 14°C (Kaneko and Colwell, 1978). It also appears to be associated with zooplankton and

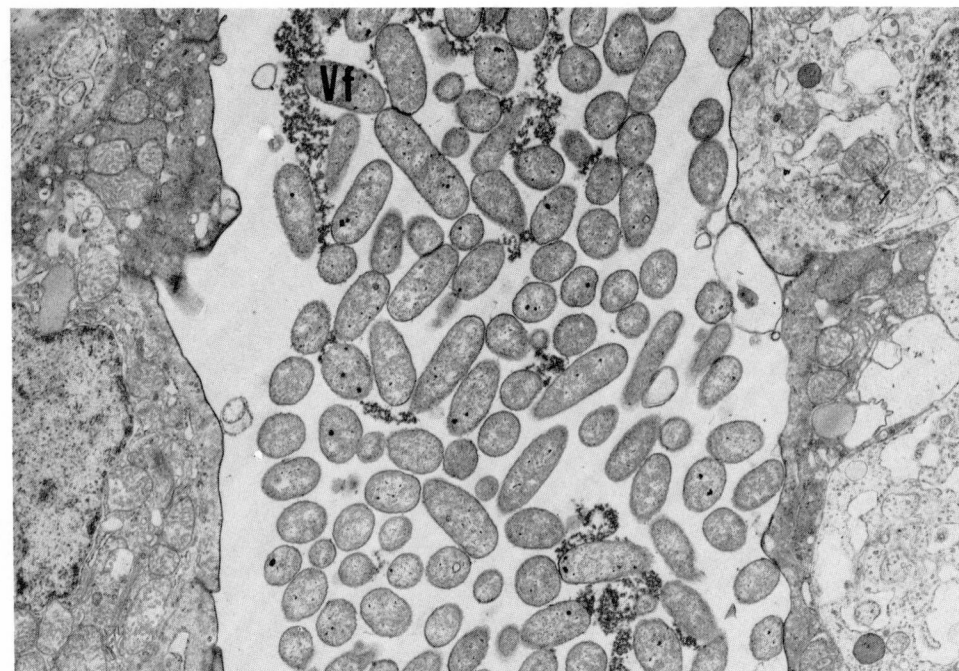

Figure 5.18. Ultrathin section through the luminous organ of *Monocentris japonicus*. *Vf*, *Vibrio fischeri* (× 9000). (Reproduced with permission of K. H. Nealson and B. M. Tebo.)

may be important in the breakdown of zooplanktonic chitin (Kaneko and Colwell, 1978). The presence of *V. parahaemolyticus* in the open ocean is uncertain although it may be present there in association with pelagic fish (Aoki et al., 1967; Yasunaga, 1965).

The older literature contains references to the isolation of luminous organisms resembling *V. cholerae* from the stools of cholera patients (Jermoljewa, 1926) and from freshwater environments (Yasaki, 1927). Until recently only one such strain was available in culture collections. This isolate, designated *V. albensis* (ATCC 14547), has been shown to be indistinguishable from *V. cholerae* by DNA homology as well as by the immunological and electrophoretic properties of its SOD (Bang et al., 1978b; Reichelt et al., 1976; S. S. Bang, unpublished observations). Recently additional luminous isolates, indistinguishable from *V. cholerae* on the basis of an extensive nutritional characterization, have been isolated from freshwater (Desmarchelier and Reichelt, 1981) and brackish water (West, 1980, Ph.D. dissertation, University of Kent, Canterbury, England).

In most ecological studies of marine luminous bacteria, *V. harveyi* has not been distinguished from *V. splendidus* biovar I, and *V. fischeri* has not been differentiated from the recently described *V. logei*. Studies of the distribution of luminous species in water columns from the open ocean in the North Atlantic indicated that *V. harveyi* was the predominant organism in the upper 150 meters during the spring, decreased in the fall, and was absent in the winter. *V. fischeri* was present in low numbers throughout most of the water column (Ruby et al., 1980). Ruby and Nealson (1978) found a seasonal variation in the relative numbers of *V. harveyi* and *V. fischeri* in seawater off the coast of San Diego, California, with a major peak of the latter species occurring during the spring prior to an increase in *V. harveyi*. An excellent correlation was observed between the water temperature and the numbers of *V. harveyi* present. Different and more complex seasonal variations were observed at several stations off the coast of Israel and the Gulf of Aqaba-Elat (Yetinson and Shilo, 1979); a correlation was observed between the distribution patterns and a number of physiological parameters (Shilo and Yetinson, 1979). *V. harveyi* was the only luminous species detected during a year-long study of a lagoon off Galveston Island, Texas (O'Brien and Sizemore, 1979).

Some luminous bacteria have the capacity to enter into a symbiotic association with marine animals. Of the five species of *Vibrio* contain-

ing luminous strains, only *V. fischeri* have been found in the specialized luminous organs of teleost fishes (Fig. 5.18) and squid (Fitzgerald, 1977; Herring and Morin, 1978; Nealson and Hastings, 1979; Ruby and Nealson, 1976; Tebo et al., 1979). Both *V. fischeri* and *V. harveyi* have been found on the surfaces and in the intestinal contents of marine animals (O'Brien and Sizemore, 1979; Reichelt and Baumann, 1973; Ruby and Morin, 1979).

There have been few studies of the ecology of the other *Vibrio* species. *V. alginolyticus* is very common in coastal waters of temperate and tropical regions (Baumann and Baumann, 1981; Golten and Scheffers, 1975). *V. campbellii*, *V. natriegens*, *V. pelagius*, *V. nigripulchritudo*, *V. splendidus* biovar II and *V. nereis* have been isolated in the vicinity of the Hawaiian Archipelago, either from offshore samples or from the open ocean at depths of up to 1300 meters (Baumann et al., 1971a, b). *V. nereis* has also been isolated from the coastal waters of England (Lee et al., 1981a). *V. fluvialis* biovar II has been obtained from estuarine habitats in England (Lee et al., 1981a) as well as from coastal waters off the Hawaiian Islands (P. Baumann, unpublished observations). *V. costicola* has been isolated from brines (Kushner, 1978). *V. metschnikovii* has been isolated from sewage, as well as from freshwater and estuarine environments in England (Lee et al., 1978). The remaining species, *V. gazogenes*, *V. proteolyticus* and *V. marinus*, are represented by single strains isolated from a saltwater marsh, a wood-boring crustacean and a seawater sample collected at 1200 meters in the North Pacific, respectively (Colwell and Morita, 1964; Harwood, 1978; Merkel et al., 1964).

Miscellaneous physiological studies. Several species of *Vibrio* have been the subjects of extensive physiological and biochemical studies. Investigations of *V. natriegens*, a species which has one of the shortest recorded doubling times (9.8 min at 37°C in a complex medium (Eagon, 1962)), have focused on its respiratory chain as well as on its growth physiology in batch and chemostat cultures (recent references: Linton et al., 1977; Nazly et al., 1980). In *V. alginolyticus* phosphohydrolases have been studied in considerable detail (reviewed by Unemoto et al., 1974). The chemical and physical properties of a proteinase from *V. proteolyticus* have been determined by Bayliss et al. (1980). *V. costicola*, usually described as a "moderate halophile," is the most extensively studied organism in this category (reviewed by Kushner, 1978). The two *Vibrio* species which are able to grow at 4°C, *V. marinus* (ATCC

15381, MP-1) and *V. logei* (ATCC 15382, PS-207), have been the subjects of investigations dealing with the physiological effects of temperature and pressure (reviewed by Morita, 1975).

Enrichment and Isolation Procedures

The simplest and most widely used enrichment for *V. cholerae* is alkaline peptone water (APW) which contains 1% (w/v) peptone and 1% (w/v) NaCl, at pH 8.6. About 1 g of fecal material is inoculated into 20 ml APW and incubated for 6–8 h at 37°C before subculturing to a suitable selective medium (see below). A secondary enrichment may be carried out by inoculating a second APW with 1 or 2 ml of the primary enrichment. This APW may also be used in a similar manner for the enrichment of *V. parahaemolyticus* and *V. fluvialis* from feces. Many other species of *Vibrio* will be enriched in APW, particularly if the temperature of incubation is reduced. If food, water or other specimens are to be examined for vibrios it may be necessary to vary the salt concentration, temperature and period of incubation according to the particular vibrio being sought or the nature of the specimen. This is discussed in more detail by Furniss et al. (1978). Alkaline peptone water tellurite (Pal et al., 1967) and trypticase-tellurite-taurocholate-peptone water (Monsur, 1963) are also suitable for the enrichment of *V. cholerae* and possibly other species.

Specific and widely used enrichment broths more suitable for *V. parahaemolyticus* are glucose-salt-teepol broth (Beuchat, 1977) and salt-colistin broth (Miwatani and Takeda, 1976). Neither is suitable for the isolation of *V. cholerae* or other species.

Many selective media have been described for the isolation of *V. cholerae* and *V. parahaemolyticus* from clinical material. Thiosulfate-citrate-bile salt-sucrose agar (TCBS) (Kobayashi et al., 1963), which was originally formulated for the isolation of *V. parahaemolyticus*, is the most suitable selective agar being readily available commercially, easy to prepare and use, and highly selective. There may be batch-to-batch and brand-to-brand variation, and each lot should be tested for its suitability (McCormack et al., 1974; Nicholls et al., 1976). After incubation at 37°C for 18–24 h *V. cholerae* normally appears as yellow, sucrose-fermenting, flattish colonies of about 2 mm diameter whereas *V. parahaemolyticus* colonies are green or blue (nonsucrose-fermenting) and 2–5 mm in diameter. Many other species of *Vibrio* grow on TCBS, but bacteria of most other genera grow slowly or not at all.

Other selective media frequently used for the isolation of *V. cholerae* are alkaline taurocholate-tellurite-gelatin agar (Monsur, 1963) and bile salt agar (Lankford and Burrows, 1965). Being less selective than TCBS, both are only suitable for use by experienced workers in laboratories handling large numbers of *V. cholerae*. Neither is available commercially. Vibrio agar (Tamura et al., 1972), which is available commercially, is also used for the isolation of *V. cholerae*. *V. parahaemolyticus* may also be isolated from clinical material on the modified bromothymol blue teepol agar of Sakazaki (1965).

Isolation from other pathological material (wound swabs, urine, blood) presents no problem and selective procedures are unnecessary. The problem here is recognition rather than isolation, as clinically significant vibrios will grow well on commonly used media such as blood agar and nutrient agar.

Many of the marine species have been obtained from enrichments involving the addition of 500-ml seawater samples to sterile 2-liter Erlenmeyer flasks containing 25 ml 1 M Tris-HCl (pH 7.5), 0.5 g NH$_4$Cl, 0.38 g K$_2$HPO$_4$·7H$_2$O, 14 mg FeSO$_4$·7H$_2$O, and 0.5 g or 0.5 ml of one of a variety of different organic carbon and energy sources. Enrichments exhibiting visible turbidity (generally within 10 days at 20–25°C) are streaked onto BMA containing 0.1–0.2% (w/v for solids and v/v for liquids) of the homologous carbon and energy source. Suitable organic compounds include sugars, sugar alcohols, sugar acids, tricarboxylic acid cycle intermediates, fatty acids and amino acids. Strains of marine vibrios have also been obtained by direct isolation from seawater. Samples are filtered through 0.22 or 0.45 μm pore size nitrocellulose filters which are subsequently placed onto BMA contain-

ing 0.1 or 0.2% of the carbon and energy source or onto YEA or MA. After 2–10 days incubation at 20–25°C, colonies are picked and streaked onto homologous media. Vibrios have also been isolated by plating onto nonselective seawater-based media. For additional information pertaining to the isolation of marine vibrios see Baumann and Baumann (1981).

Some strains of *Vibrio* swarm on complex solid media, consequently isolated colonies cannot be obtained under these conditions of cultivation. Swarming can be abolished, or greatly reduced, by increasing the concentration of agar in complex media to 4% (w/v) (Reichelt and Baumann, 1973) or by streaking the organism onto a minimal medium such as BMA containing 0.2% (v/v) glycerol (Baumann et al., 1971a).

The isolation of luminous species of *Vibrio* does not involve enrichment procedures. Instead, samples are placed directly onto a suitable complex medium and luminescence is detected by the dark-adapted eyes of the investigator. The selection and purification of luminous colonies is facilitated by the use of sterile toothpicks and a red light connected to a rheostat (Cosenza and Buck, 1966; Hastings and Nealson, 1981). A medium promoting the luminescence of marine isolates (luminous medium (LM) is BMA containing 0.3% (v/v) glycerol and (in g/liter) 5.0 g yeast extract, 5.0 g tryptone, 1.0 g CaCO$_3$ and 20.0 g agar). For the observation of luminescence in terrestrial isolates, terrestrial luminous medium (TLM) is made by adding the same supplements to TBMA. Since there is strain variation with respect to the intensity and duration of light emission and since some strains will only luminesce as isolated colonies, it is best to examine, periodically, cultures streaked onto LM or TLM within 12–36 h of inoculation. A suitable incubation temperature for all species is 15°C; only some will luminesce well at 20° or 25°C. Sources of luminous isolates include seawater, the surfaces and intestinal contents of marine animals, luminous organs and, more rarely, freshwater. Additional details for the isolation of luminous bacteria are given by Baumann and Baumann (1981), Hastings and Nealson (1981), Ruby et al. (1980) and Desmarchelier and Reichelt (1981).

Maintenance Procedures

Strains of *V. cholerae* and *V. metschnikovii* can be maintained on TYEA or any nutrient agar not containing a fermentable carbohydrate, marine luminous isolates and *V. marinus* on LM, and the remaining species (including *V. costicola*) on MA. In the case of most species, strains are transferred monthly, allowed to grow at 20–25°C for 1–2 days and then stored at 15–18°C. *V. anguillarum* biovar II and *V. logei* are transferred every 2 weeks and are grown and stored at 15°C. *V. marinus* is also transferred every 2 weeks, allowed to grow for 2 days at 15°C and subsequently maintained at 4°C. With the exception of this species, most of the remaining species of *Vibrio* do not survive well at 4°C. Most species have been successfully stored at 15–18°C on agar slants or stabs completely covered with sterile mineral oil (USP) and transferred once or, preferably twice, a year, although under these conditions they may remain viable for 3 years. Oil-covered stab cultures of most species may also be stored at −70°C. Before adopting a maintenance procedure for a new isolate the survival time at 4°C and 15–18°C should be determined by checking viability at weekly intervals for up to 1 or 1-½ months. Strains which are able to grow on minimal media when first isolated occasionally acquire organic growth factor requirements upon prolonged maintenance on complex media. To overcome this problem, organisms may be cultivated and stored on BM or TBM containing 0.2% glycerol. Cultures of luminous bacteria often acquire nonluminous mutants which may predominate with time. In order to maintain light-producing strains, it is advisable to renew the stock culture periodically from several brightly luminous colonies. The ability to synthesize lateral flagella is another trait which may be lost upon prolonged cultivation.

All species of *Vibrio* have been lyophilized and recovered after more than 8 years of storage at 4°C. For the preparation of lyophils of *V. cholerae* and *V. metschnikovii*, an 18- to 24-h-old culture grown on a

TYEA slant or other appropriate media is suspended in about 0.5 ml of 10% (w/v) skimmed milk and transferred to a lyophil tube which is subsequently dipped into a mixture of dry ice and acetone and placed under vacuum for 10–12 h. For marine species, the organisms are grown on YEA or LM and suspended in ¼ strength ASW, 5.0 g/liter yeast extract and 5.0 g/liter peptone (adjusted to pH 7.5). A suspending medium containing one part nutrient broth, two parts horse serum and 5% (w/v) inositol has also been successfully used for lyophilizing some vibrio species (Furniss et al., 1978). All lyophils are reconstituted by suspending the powder in about 0.5 ml of liquid medium of the same composition as the medium in which the organism was grown prior to lyophilization. A portion of the suspension is streaked onto the appropriate solid medium and the remainder inoculated into a tube containing 4 ml of the corresponding liquid medium. Growth is generally observed after 1–2 days at 25°C. For *V. logei*, *V. anguillarum* biovar II and *V. marinus*, the cultures should be kept at 15°C and incubated for at least 4 days.

Procedures For Testing Special Characters

Since species of *Vibrio* vary in their ionic requirements, most of the tests should be performed in one of three appropriate simple or complex media based on TBM (for *V. cholerae* and *V. metschnikovii*), BM (for strains from marine and estuarine sources) and TBM or BM containing 800 mM NaCl for *V. costicola*. Incubation temperatures of 25° and 30°C are adequate for most species, except *V. logei*, *V. anguillarum* biovar II and *V. marinus*, which should be incubated at 15°C.

Fermentation and gas production can be detected in a semisolid medium containing 1% (w/v) D-glucose and 2% (w/v) agar. These constituents are added to TYEB or YEB containing 100 mM Tris-HCl (pH 7.5). After melting the agar, 10-ml aliquots of the medium are dispensed into test tubes, then autoclaved and inoculated by means of a stab. About 5 ml of 2% (w/v) agar (autoclaved and cooled to 41°C) are layered over the medium (prechilled in the case of *V. logei*, *V. anguillarum* biovar II and *V. marinus*) to make an agar plug. The cultures are incubated and observed for turbidity and gas production for 4 days at 25° or 30°C or for 6 days at 15°C. In this medium, facultative anaerobes grow to a high turbidity and gas, if produced, is trapped in the semisolid medium; strict aerobes will not grow or will give only slight turbidity under these conditions.

The Voges-Proskauer test is important for the identification of several species. The medium used is YEB containing 100 mM Tris-HCl (pH 7.5) and 2% (w/v) D-glucose. Since the amount of acetoin and/or diacetyl produced by many of the strains of *Vibrio* is relatively low and since its accumulation may be transient, the test should be performed in triplicate and the culture fluid tested at 2, 4, and 6 days after inoculation using the method described by Skerman (1967). If only a slight color is detected, a quantitative determination of acetoin and/or diacetyl can be performed as described by Eggleton et al. (1943).

To test for a Na$^+$ requirement, a strain is inoculated into BM containing 0.2% (v/v) glycerol as well as into a medium identical to BM except for the substitution of equimolar amounts of K$^+$ for the Na$^+$ (MacLeod, 1968). If a strain requires Na$^+$ but not organic growth factors, good growth will be observed in the first but not in the second medium. For most growth factor-requiring strains, these two media can be suitably modified by the addition of 0.02–0.05% (w/v) yeast extract. For *V. costicola*, the final concentration of Na$^+$ is increased to 800 mM and is replaced by equimolar amounts of K$^+$ in the Na$^+$-free medium. With *V. cholerae*, which is able to grow in both the Na$^+$-containing and Na$^+$-free media, visible turbidity may appear somewhat earlier in the former than in the latter medium. For some terrestrial strains it may be necessary to perform the test in TBM, since the high levels of Na$^+$ or K$^+$ in BM may inhibit growth of the organisms (J. V. Lee and P. Baumann, unpublished observations).

The presence of an arginine-dihydrolase system is diagnostic for a number of species of *Vibrio*. A convenient means of determining whether an organism may contain this enzyme system is the Thornley method (1960). (The slight modification described by Baumann and Baumann (1981) must be used for the marine strains.) Since this procedure is not specific for the arginine-dihydrolase system (in that it only detects an arginine-dependent increase in pH which could occur by means of several different reactions), it is necessary to confirm positive results by chemically testing for the production of ornithine from arginine under anaerobic conditions (Baumann and Baumann, 1981; Baumann et al., 1971a). *V. metschnikovii* is the only species of *Vibrio* containing strains which are positive by the Thornley method although they do not produce detectable ornithine.

The assignment of strains into species is, to a considerable extent, based on a nutritional analysis which tests the ability of isolates to utilize different organic compounds as sole or principal sources of carbon and energy. The nutritional spectra of all the species of *Vibrio* have been determined by replica plating the organisms onto TBMA or BMA containing 0.2% (w/v) of the tested sugars and 0.1% (w/v for solids and v/v for liquids) of most of the other organic compounds. For strains requiring organic growth factors the media can be supplemented with amino acids (1 mg/liter) or with low levels of yeast extract, tryptone or casein hydrolysate, either alone or in combination and generally not exceeding 0.1 g/liter (Bang et al., 1978a; Baumann and Baumann, 1981; Desmarchelier and Reichelt, 1981). Different basal media may give good growth of *Vibrio* species and have been used in nutritional studies. In general the results will correlate well, but there may be exceptions for some substrates (P. Baumann and J. V. Lee, unpublished observations). The plates are incubated at the appropriate temperature and examined for growth, every other day, for a total of 6 days. Further details as well as other procedures used in the characterization of strains are given in Baumann and Baumann (1981) and Palleroni and Doudoroff (1972). These references include the compounds used in the nutritional screening, the concentrations of the carbon and energy sources and their methods of sterilization as well as procedures for determining growth at different temperatures, the production of extracellular enzymes and the mechanism of aromatic ring cleavage. Additional methodologies are considered by Lee et al. (1978, 1981a).

The lateral flagella synthesized by many species of *Vibrio* on solid media are readily detached. Special care is, therefore, essential in the preparation of specimens for staining by the Leifson method. For a detailed description of the precautions to be taken during the harvesting, fixation and washing of the cells, consult Baumann and Baumann (1981).

Swarming on solid media is a property which is affected by a number of different parameters (see section on *Flagella*). Nevertheless, under fairly standard conditions this property is of use for the differentiation of some species of *Vibrio*. Freshly-poured media such as MA, YEA or other comparable complex media containing 1.5–2.0% (w/v) agar are dried overnight at 37°C and the center of the plate is inoculated with a small drop or loopful of a fresh culture. After an incubation of 18–48 h at 25 or 30°C, swarming organisms have generally spread over the entire surface of the agar (Baumann et al., 1973; Sakazaki, 1968).

Cell-morphology should be examined in the appropriate complex liquid medium (TYEB or YEB) during exponential as well as early stationary phase of growth. In most cases, the decision as to whether the strain is a straight or curved rod does not present difficulties since the majority of the cells are of one or the other type.

Luminous strains of *V. harveyi*, *V. splendidus* biovar I and *V. cholerae* can be differentiated from *V. fischeri* and *V. logei* by the decay kinetics of light emission by luciferase in the presence of the aldehyde dodecanal. The former group of species has "slow" decay kinetics while the latter has "fast" decay kinetics (Nealson and Hastings, 1979; Hastings and Nealson, 1981). A description of the method used in the determination of the decay kinetics is given by Hastings et al. (1978).

The tests that are routinely in use in clinical laboratories may be readily adapted for the identification of the species likely to be encountered by medical microbiologists. All media should have the concentration of NaCl increased to 1% (w/v) which is suitable for the routine identification of both the *Vibrionaceae* and *Enterobacteriaceae*. The temperature of incubation may be 30° or 37°C and tests are incubated for up to 7 days, although 1 or 2 will normally be sufficient. Acid and gas production from sugars may be determined in peptone water sugars containing Durham tubes and lysine and ornithine decarboxylase by the method of Møller (1955) using Difco decarboxylase base. The media for the decarboxylase tests should be heavily inoculated with several drops of a broth culture or a colony picked off with a straight wire. The Voges-Proskauer test may be done after incubation for 18–24 h by one of the many methods available, but method 3 of Cowan (1974) is the most reliable.

Esculin hydrolysis is determined in a liquid medium containing:

Tryptone 1% (w/v), NaCl 1% (w/v), KCl 0.1% (w/v), $MgCl_2 \cdot 6H_2O$ 0.4% (w/v), esculin 0.1% (w/v) and ferric citrate 0.005% (w/v).

Sensitivity to the water-soluble vibriostatic agent 2,4-diamino-6,7-diisopropyl pteridine (0/129) phosphate is determined by a disk sensitivity testing method on nutrient agar containing 0.5% (w/v) NaCl (Furniss et al., 1978). The discs contain 10 μg and 150 μg of 0/129 phosphate. Any zone of inhibition around the 150 μg disk is read as sensitive. The test is particularly useful for differentiating *V. fluvialis* from *V. anguillarum* and *Aeromonas*.

Growth at different concentrations of NaCl is determined in tryptone (1% w/v) broth containing the appropriate amount of NaCl. The broths are inoculated with 25 μl of a fresh turbid broth culture (3- to 6-h-old) of the tested organism and incubated for 2 days. Growth at 43°C after 24 h is determined in tryptone broth containing 2% (w/v) NaCl.

Differentiation of the genus **Vibrio** from other genera

Table 5.57 presents the characteristics which distinguish *Vibrio* from morphologically and physiologically similar genera. While the number of readily determinable diagnostic traits between these genera is limited, there is little difficulty in their differentiation since the constituent species are generally well defined. Furthermore, a difference in their ecology simplifies the task since only some species have the potential to coexist in the same habitat (see Ecology, above).

Taxonomic Comments

Speciation within the genus *Vibrio* is based on an extensive phenotypic characterization as well as on DNA/DNA hybridization and an immunological comparison of SOD and alkaline phosphatase (AP) of representative strains. The first three approaches were applied to all or most of the 20 species of *Vibrio*, while the latter included only the

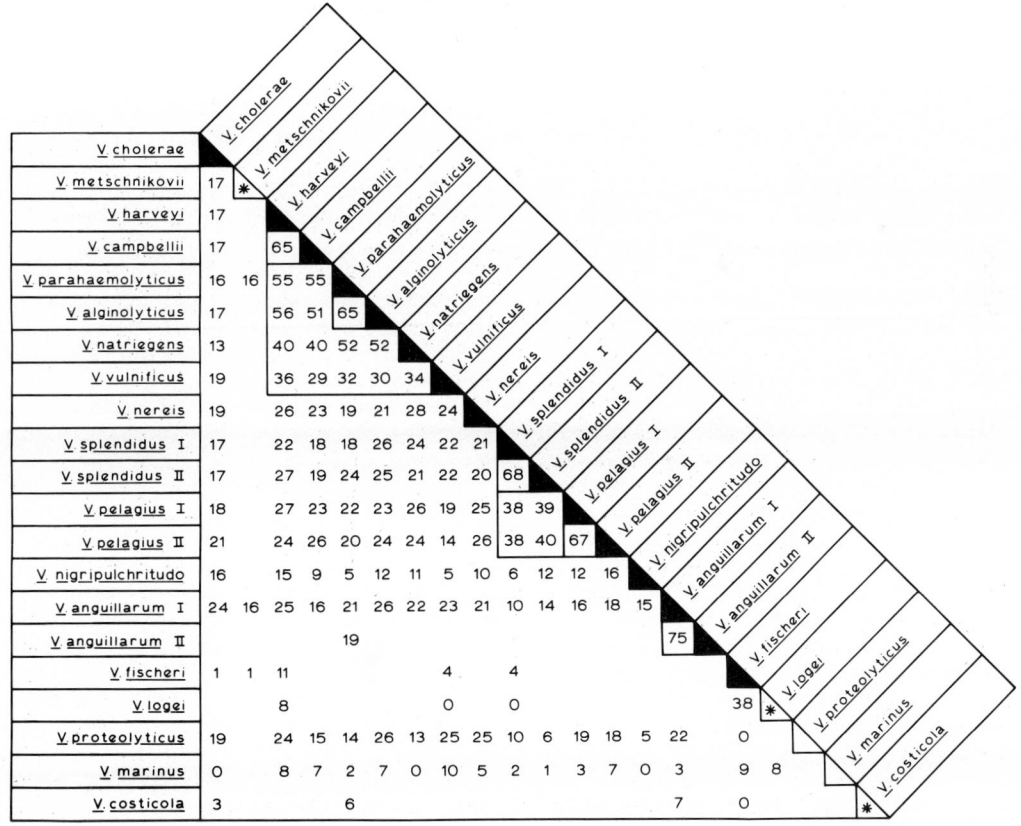

Figure 5.19. Summary of the results of DNA homology studies among species of *Vibrio*. The homology values represent an average for all the strains tested. Data from Reichelt et al. (1976); data for *V. anguillarum* biovar II are normalized averages from Schiewe et al. (1977). The species not included in the DNA homology studies are *V. fluvialis* and *V. gazogenes*. *Black triangle*, the DNA homology between strains of the species is over 80%; *asterisk*, the DNA homology between strains of the species has not been determined; *white triangle*, species consisting of a single strain.

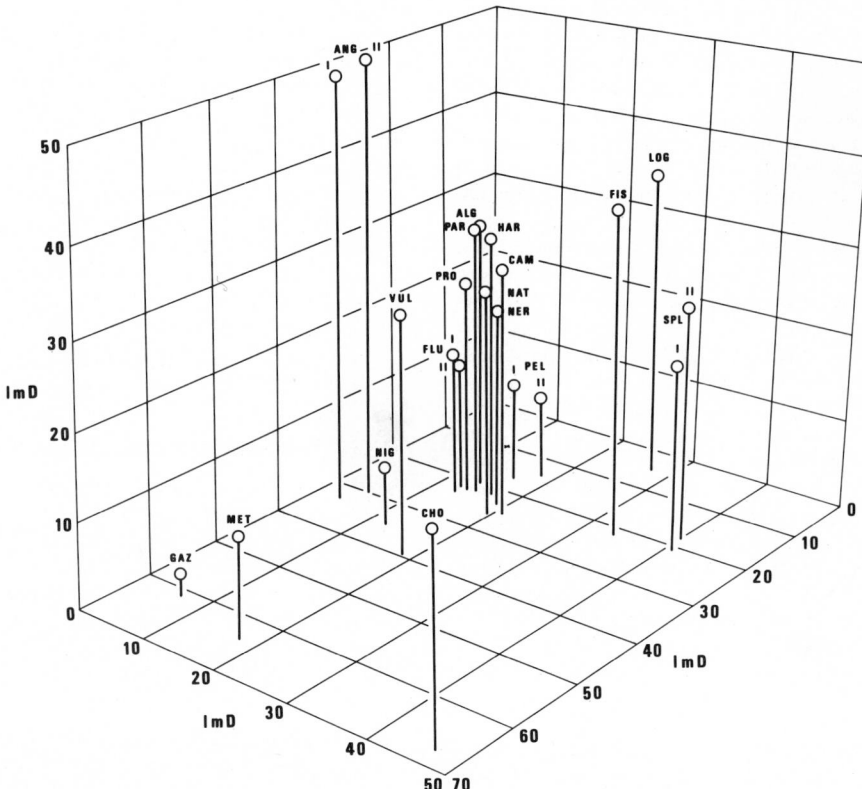

Figure 5.20. Schematic three-dimensional representation of the immunological relationships of superoxide dismutases from species and biovars of *Vibrio*. The spheres represent the relative positions of the species and biovars. *ImD*, immunological distance, a parameter related to percentage of amino acid sequence difference. All species are designated by the first three letters of their species names. A total of 3–4 strains from each species or biovar was analyzed and the maximal standard deviation was ±1.2 ImD. *V. costicola* was a considerable distance from these species and its position is not shown in this figure. (Data from Baumann et al. (1980a) and unpublished observations of S. S. Bang, L. Baumann, and P. Baumann.)

14 species which had an AP. The results of these four studies are in complete agreement with respect to the species or biovar assignments of strains. In the genus *Vibrio* a species or biovar comprises strains which are (a) phenotypically similar and, with a few exceptions, readily distinguishable from other species or biovars, (b) genotypically similar in having DNA homologies of over 80% and (c) closely related on the basis of similarities in the amino acid sequences of their SODs and APs. Although the biovars of most of the species of *Vibrio* are sufficiently distinct, phenotypically and genotypically, to warrant separate species rank, biovar status has been retained whenever one of the clusters had fewer than five strains so that the diagnostic traits used for their identification could be validated by the characterization of additional isolates. In the case of *V. fluvialis*, the two biovars are well represented but their DNA homology has not been determined.

References for the phenotypic studies are found in appropriate places in the text. The principal investigations of DNA homology are those of Anderson and Ordal (1972), Citarella and Colwell (1970), Clark and Steigerwalt (1977), Reichelt et al. (1976), Schiewe et al. (1977) and Staley and Colwell (1973a, b). References to the immunological studies are Baumann et al. (1980a) as well as the unpublished observations of S. S. Bang, L. Baumann, M. J. Woolkalis and P. Baumann. A summary of the relationships established by DNA homology and the immunological studies of SOD are presented in Figs. 5.19 and 5.20, respectively.

A major impetus for undertaking studies of the evolutionary relationships among species of *Vibrio* and related organisms has been the goal of establishing genera on the basis of the natural relationships of the constituent species. In the family *Enterobacteriaceae*, the present tendency is to make generic assignments using data obtained from DNA/DNA hybridization studies (which measures the average rela-

tionship of total genomes). Since this method is far too specific to permit the assessment of relationships to more distant species, its application to generic assignments can lead to excessive fragmentation and create genera composed of a few closely related species without necessarily adding significant informational content to the generic definitions. More distant relationships can be determined by analyzing the divergence of informational molecules with sequences which are conserved to a greater extent than most of the genome. This had been the rationale for studies of relationships among species of *Vibrio* and related genera by rRNA/DNA hybridization as well as by an immunological comparison of the amino acid sequences of glutamine synthetase (GS) and SOD (Baumann and Baumann, 1976, 1980; Baumann et al., 1980a; S. S. Bang, L. Baumann, M. J. Woolkalis and P. Baumann, unpublished observations). With a few exceptions, the results of these different approaches are consistent in indicating the existence of a group of closely related, primarily marine species of *Vibrio* with a more distant affinity to *V. cholerae*, *V. metschnikovii*, *V. gazogenes* and *V. costicola* (Fig. 5.20). The species in the last category do not comprise a homogeneous group since they differ in their relationships to one another as well as to the group of marine vibrios. The decision to place all these organisms in the genus *Vibrio* and to abolish *Beneckea* and *Lucibacterium* (genera previously containing *Vibrio* species) is based on the fact that no simple subdivisions could be made among the species which would be consistent with all the results of the evolutionary studies (for a discussion see Baumann et al., 1980b). The genetic diversity within this genus, as measured by rRNA homology, GS and AP, is greater than that between *Escherichia* and *Serratia* and somewhat less than that between *Escherichia* and *Proteus*.

The results of oligonucleotide cataloging of 16S rRNA (Fox et al.,

1980) as well as rRNA/DNA hybridization have shown that *Vibrio* and *Photobacterium* are closely related. (The organism designated *Photobacterium fischeri* in this reference is actually *P. phosphoreum* (G. E. Fox, personal communication)). These two genera constitute a single line of descent from a common ancestor, the other major lines of divergence being represented by *Aeromonas*, *Pasteurella*, and the *Enterobacteriaceae*. A common ancestor for all of these organisms is also indicated from the pattern of regulation of aspartokinase activity and the organization of the genes coding for the enzymes of the tryptophan biosynthetic pathway. In the case of aspartokinase, the activity which initiates the pathway leading to the biosynthesis of amino acids of the aspartate family, species of *Vibrio*, *Aeromonas*, and the *Enterobacteriaceae* share a complex and unique pattern of regulation involving three isofunctional enzymes, each of which is subject to a different mode of feedback inhibition and/or repression (Baumann and Baumann, 1973; Cohen et al., 1969). Similarly, *Vibrio* resembles *Aeromonas* and the *Enterobacteriaceae* in having the genes encoding the enzymes of tryptophan biosynthesis arranged in the same order within a single operon (Bieger and Crawford, 1978; Crawford, 1975). An additional conclusion derived from 16S oligonucleotide cataloging is that *Vibrio* and related genera are part of a major evolutionary line of descent which includes such species as *Pseudomonas aeruginosa*, *Acinetobacter calcoaceticus* and *Chromatium vinosum* and which probably originated from a purple sulfur bacterium (Fox et al., 1980; Gibson et al., 1979).

Further Reading

Barua, D. and W. Burrows (Editors). 1974. *Cholera*. W. B. Saunders, Philadelphia.

Baumann, P. and L. Baumann. 1977. Biology of the marine enterobacteria: genera *Beneckea* and *Photobacterium*. Annu. Rev. Microbiol. *31:* 39–61.

Baumann, P. and L. Baumann. 1981. The marine Gram-negative eubacteria. In Starr, Stolp, Trüper, Balows and Schlegel (Editors), *The Prokaryotes*, Springer-Verlag, New York, pp. 1302–1331.

Baumann, P., L. Baumann, M.J. Woolkalis and S.S. Bang. 1983. Evolutionary relationships in *Vibrio* and *Photobacterium*: a basis for a natural classification Annu. Rev. Microbiol. *37:* 369–398.

Blake, P.A., R.E. Weaver and D.G. Hollis. 1980b. Diseases of humans (other than cholera) caused by vibrios. Annu. Rev. Microbiol. *34:* 341–367.

Collier, R.J. and J.J. Mekalanos. 1980. ADP-Ribosylating exotoxins. In Bisswanger and Schmincke-Ott (Editors), *Multifunctional Proteins*, Wiley & Sons, New York, pp. 261–291.

Fujino, T., G. Sakaguchi, R. Sakazaki and Y. Takeda (Editors). 1974. International Symposium on *Vibrio parahaemolyticus*. Saikon Publishing, Tokyo.

Furniss, A.L., J.V. Lee and T.J. Donovan. 1978. *The Vibrios*. Public Health Laboratory Service Monograph Series, Her Majesty's Stationery Office, London.

Miwatani, T. and Y. Takeda. 1976. *Vibrio parahaemolyticus* a Causative Bacterium of Food Poisoning. Saikon Publishing, Tokyo.

Nealson, K.H. and J.W. Hastings. 1979. Bacterial bioluminescence: its control and ecological significance. Microbiol. Rev. *43:* 496–518.

Sakazaki, R. 1979. *Vibrio* infections. In Riemann and Bryan (Editors), *Foodborne Infections and Intoxications* 2nd Ed., Academic Press, New York, pp. 173–209.

Ziegler, M.M. and T.O. Baldwin. 1981. Biochemistry of bacterial bioluminescence. Curr. Top. Bioener. *12:* 65–113.

Differentiation and characteristics of species of **Vibrio**

The differential characteristics of species of *Vibrio* are presented in Table 5.58. Other characteristics of the species are presented in Table 5.59. These two tables summarize the results of a phenotypic characterization of over 750 strains of *Vibrio* using a uniform methodology consisting primarily of an extensive nutritional characterization (Baumann and Baumann, 1977, 1981; Jensen et al., 1980; Desmarchelier and Reichelt, 1981).

Table 5.60 gives the differential characteristics of *Vibrio* species that may be encountered by clinical bacteriologists. This table is based on methods employed in clinical laboratories. The nonpathogenic species in Table 5.60 are not common in specimens from humans but are in certain marine or estuarine water and food samples that may be examined for epidemiological purposes. *V. harveyi* and *V. campbellii* will often need further characterization to distinguish them adequately from *V. parahaemolyticus* and *V. vulnificus* using some of the tests in Table 5.58.

In the selection of diagnostic traits for Table 5.58, preference was given to traits which are easily scored and, consequently, least subject to misinterpretation. The inclusion in one case of the somewhat subjective trait, cell curvature, and in several other cases of a few carbon compounds (e.g. propionate, valerate, heptanoate, ethanol, L-glutamate) supporting relatively light growth was, nevertheless, unavoidable. In order to minimize the possibility of misinterpreting the light growth with these carbon compounds, it is advisable to include known strains as positive and negative controls. Since usually different sets of traits are diagnostic for different species, a table which includes selected diagnostic properties for all 24 species and biovars of *Vibrio* will contain a large number of traits. This is the major drawback in the use of Table 5.58 for the routine identification of strains. In order to simplify this procedure, 13 readily determinable traits (designated by boldface type in Table 5.58) have been selected for the initial screening of isolates. Of the total 276 possible comparisons between different pairs of species and biovars of *Vibrio*, only 3 cannot be distinguished by this set of diagnostic traits (*V. harveyi* and *V. vulnificus*; *V. campbellii* and *V. splendidus* biovar II; *V. nereis* and *V. fluvialis* biovar I). One trait differentiates 27 pairs; all the remaining pairs differ by 2–9 traits. The tentative identification provided by this initial screening should be confirmed by testing for the additional traits given in the species description and can be supplemented by others chosen from Tables 5.58 and 5.59 by the investigator.

Although most *Vibrio* species and biovars are differentiated by a considerable number of independent phenotypic properties, a few are distinguishable by only two or three diagnostic traits; these include (a) *V. harveyi* and *V. splendidus* biovar I, (b) *V. campbellii* and *V. splendidus* biovar II, (c) *V. fluvialis* biovars I and II and (d) *V. fischeri* and *V. logei*. In some cases their identification can be aided by testing for diagnostic traits present in fewer than 90% of the strains (consult Baumann and Baumann, 1981, or original publications).

It should be noted that species 17–19 consist of single strains and species 20 of two strains. Future characterizations of additional isolates will undoubtedly change some of the traits which are now diagnostic for these species.

List of the species of the genus **Vibrio**

1. **Vibrio cholerae** Pacini 1854, 411.[AL]

chol′er.ae. Gr. n. *cholera* cholera; M.L. gen. n. *cholerae* of cholera, an intestinal disease.

Physiological and nutritional characteristics of the species are presented in Tables 5.58–5.60.

The following combination of properties distinguishes *V. cholerae* from other species of *Vibrio* (or their biovars) by four to six traits: positive for oxidase, NO_3^- to NO_2^- reduction, acetoin and/or diacetyl production, growth at 40°C, as well as utilization of sucrose, α-ketoglutarate and growth in the absence of Na^+; negative for lateral flagella on solid medium, arginine dihydrolase and utilization of D-sorbitol, L-α-alanine, L-tyrosine and putrescine. Epidemic strains of *V. cholerae* serovar 0:1 may be divided into the classical and eltor biovars. The classical biovar was called biotype cholerae by Shewan and Véron (1974), but the name "classical" is used more commonly. The biovars are differentiated from one another as shown in Table 5.61.

The present pandemic of cholera is caused by the eltor biovar; the classical biovar has almost disappeared except for rare isolations on the Indian subcontinent. The original eltor strains were very hemolytic to sheep erythrocytes, but those isolated in recent years have been very

Table 5.58.

Differential characteristics of the species and biovars of the genus **Vibrio**[a]

Characteristics	1. V. cholerae	2. V. metschnikovii	3. V. harveyi	4. V. campbellii	5. V. parahaemolyticus	6. V. alginolyticus	7. V. natriegens	8. V. vulnificus	9. V. nereis	10. V. fluvialis I	10. V. fluvialis II	11. V. splendidus I	11. V. splendidus II	12. V. pelagius I	12. V. pelagius II	13. V. nigripulchritudo	14. V. anguillarum I	14. V. anguillarum II	15. V. fischeri	16. V. logei	17. V. proteolyticus	18. V. gazogenes	19. V. marinus	20. V. costicola
Number of strains tested	161	6	91	44	134	38	8	15	6	15	7	4	15	7	4	14	20	5	12	11	1	1	1	2
Flagellation:																								
3–12 Polar flagella	−	−	−	−	−	−	−	−	−	−	−	−	−	−	−	−	−	−	+	+	−	−	−	−
Lateral flagella when grown on solid media	−	−	+	+	+	+	−	−	−	d	d	−	d	−	−	−	−	−	−	−	+	−	−	−
Swarming on solid complex media	−	−	−	−	−	+	−	−	−	−	−	−	−	−	−	−	−	−	−	−	+	−	−	−
Straight rods[b]	d	d	+	+	+	+	+	−	+	d	−	d	+	−	+	+[c]	−	d	+	+	+	−	+	−
PHB-accumulation	−	−	−	−	−	−	+	−	+	−	d	−	−	−	−	d	−	−	−	−	−	−	−	−
Pigmentation:																								
Yellow-orange	−	−	−	−	−	−	−	−	−	−	−	−	−	−	−	−	−	−	+	+	−	−	−	−
Blue-black	−	−	−	−	−	−	−	−	−	−	−	−	−	−	−	+	−	−	−	−	−	−	−	−
Red	−	−	−	−	−	−	−	−	−	−	−	−	−	−	−	−	−	−	−	−	−	−	+	−
Arginine dihydrolase[d]	−	−	−	−	−	−	−	+	+	+	+	−	−	−	−	−	+	−	−	−	+	−	−	+
Oxidase	+	−	+	+	+	+	+	+	+	+	+	+	+	+	+	+	+	+	+	+	+	−	+	+
Reduction of NO_3^- to NO_2^-	+	−	+	+	+	+	+	+	+	+	+	+	+	+	+	+	+	−	d	+	+	−	+	d
Luminescence	−[e]	−	d	−	−	−	−	−	−	−	−	−	+	−	−	−	−	−	+	+	−	−	−	−
Gas from D-glucose	−	−	−	−	−	−	−	−	−	+	−	−	−	−	−	−	−	−	−	−	−	+	+	−
Production of acetoin and/or diacetyl	+	+	−	−	−	+	−	−	−	−	−	−	−	−	−	−	+	−	−	−	+	−	−	−
Na⁺ required for growth	−	d	+	+	+	+	+	+	+	+	+	+	+	+	+	+	+	+	+	+	+	+	+	+
Requirement for organic growth factors	d	d	−	−	−	−	−	−	−	−	−	−	−	−	−	−	+	−	d	−	−	−	+	d
Growth at:																								
4°C	−	−	−	−	−	−	−	d	−	−	d	−	d	d	−	−	−	−	−	+	−	−	+	−
30°C	+	+	+	+	+	+	+	+	+	+	+	+	+	+	+	+	+	−	+	+	+	+	−	+
35°C	+	+	+	+	+	+	+	+	+	+	+	d	−	+	+	−	+	−	d	−	+	+	+	+
40°C	+	+	d	−	+	+	+	+	d	+	+	−	−	−	−	−	−	−	−	−	+	+	−	−
Production of:																								
Amylase	+	+	+	+	+	+	d	+	−	+	d	+	+	−	+	+	+	−	−	+	+	+	−	−
Gelatinase	+	+	+	+	+	+	d	+	d	+	d	+	+	−	+	+	+	−	−	+	+	+	+	−
Lipase	+	+	+	+	+	+	+	+	−	+	+	+	+	+	+	+	+	−	+	d	+	+	+	−
Alginase	−	−	d	−	−	−	−	−	−	−	−	d	−	+	+	−	−	−	−	−	−	−	−	−
Chitinase	+	+	+	+	+	+	−	+	d	+	+	+	+	d	+	+	+	d	d	+	+	−	+	−
Utilization of:																								
D-Xylose	−	−	−	−	−	−	−	−	−	−	−	−	−	−	−	−	−	−	−	−	−	+	+	−
L-Arabinose	−	−	d	−	d	−	+	−	−	+	+	−	−	−	−	−	+	−	−	−	+	+	−	−
D-Mannose	d	d	+	d	+	d	d	+	−	+	d	+	−	d	+	+	+	−	+	d	+	+	−	−
D-Galactose	+	d	d	−	+	d	+	+	−	+	+	+	−	+	+	+	d	−	+	+	−	+	+	−
Sucrose	+	+	d	−	−	+	+	−	+	+	+	d	−	+	d	−	+	+	−	−	+	−	−	+
Trehalose	+	+	+	+	+	+	+	+	+	+	+	+	+	+	+	+	+	−	d	−	+	−	−	+
Cellobiose	−	−	+	d	−	−	d	+	−	d	d	+	d	−	−	+	+	−	+	+	−	+	−	−
Melibiose	−	−	−	−	−	−	d	−	−	−	−	−	−	−	−	+	−	−	−	−	−	−	−	−
Lactose	−	d	−[f]	−	−	−	−	d[f]	−	−	−	−	−	d[f]	d[f]	+	−	−	−	−	d	−	−	−
Salicin	−	−	d	−	−	−	+	−	−	+	−	−	−	−	−	−	−	−	d	−	−	−	+	−
D-Gluconate	+	+	+	−	+	+	+	+	+	+	+	d	−	+	+	d	+	−	+	+	−	+	+	+
D-Glucuronate	−	−	+	−	d	−	d	+	−	+	−	+	−	−	−	−	+	−	−	−	−	−	−	−
D-Galacturonate	−	−	−	−	−	−	−	−	−	+	+	−	−	−	−	−	−	−	−	−	−	−	−	−

[a] Traits useful for the preliminary identification of species (see text) are designated by boldface type. For symbols see standard definitions (except for footnote b below).

[b] +, straight rods; −, curved rods.

[c] Straight rods in exponential phase of growth becoming curved in stationary phase.

[e] Determined by the anaerobic production of ornithine from arginine. See text for a discussion of problems associated with the Thornley method which measures alkali production from arginine.

[d] Luminous strains of this species have been found by Desmarchelier and Reichelt (1981) and West (1980, Ph.D. thesis).

[f] Wild-type strains are unable to utilize lactose; many strains may readily acquire this property by mutation.

Table 5.58—*continued*

Characteristics	1. *V. cholerae*	2. *V. metschnikovii*	3. *V. harveyi*	4. *V. campbellii*	5. *V. parahaemolyti*	6. *V. alginolyticus*	7. *V. natriegens*	8. *V. vulnificus*	9. *V. nereis*	10. *V. fluvialis I*	10. *V. fluvialis II*	11. *V. splendidus I*	11. *V. splendidus II*	12. *V. pelagius I*	12. *V. pelagius II*	13. *V. nigripulchritudo*	14. *V. anguillarum I*	14. *V. anguillarum II*	15. *V. fischeri*	16. *V. logei*	17. *V. proteolyticus*	18. *V. gazogenes*	19. *V. marinus*	20. *V. costicola*
Propionate	+	−	+	+	+	+	+	+	+	+	+	+	+	+	+	+	+	−	−	−	−	+	−	+
Valerate	−	−	−	−	+	+	+	−	+	−	−	−	−	−	−	−	+	−	−	−	−	−	−	−
Heptanoate	−	d	+	−	+	+	+	−	+	−	+	−	−	−	d	−	−	−	−	−	−	−	−	−
Glutarate	−	−	−	−	−	+	−	−	+	−	+	−	−	−	−	−	d	−	−	−	−	−	−	−
DL-Malate	+	+	d	d	+	+	+	d	+	+	+	+	d	d	−	+	+	−	−	−	+	+	+	−
β-Hydroxybutyrate	−	−	−	−	−	+	−	+	+	d	−	−	−	−	−	+	−	−	−	−	−	−	−	−
DL-Lactate	+	+	+	+	+	+	+	+	+	+	+	+	+	+	+	+	+	−	−	+	+	+	+	+
Citrate	d	d	+	d	+	+	+	+	+	+	+	+	+	+	+	+	+	+	d	−	+	−	+	−
α-Ketoglutarate	+	−	+	+	+	+	+	+	+	+	+	+	+	−	−	+	+	−	−	−	+	+	+	−
Pyruvate	+	+	+	+	+	+	+	+	+	+	+	+	+	+	+	+	+	+	−	−	+	+	+	+
D-Mannitol	+	+	+	d	+	+	+	d	−	+	+	+	+	+	+	+	+	+	−	−	+	+	+	+
D-Sorbitol	−	d	−	−	−	−	−	−	−	−	−	−	−	−	z	+	+	−	−	−	−	−	−	−
meso-Inositol	−	d	−	−	−	−	d	−	−	−	−	−	−	−	−	+	d	−	−	−	−	−	−	−
Ethanol	−	−	−	−	+	d	+	−	+	+	+	−	−	d	−	d	−	−	−	−	−	−	−	−
p-Hydroxybenzoate	−	−	−	−	−	−	+	−	−	+	d	−	−	−	−	−	−	−	−	−	−	−	−	+
L-α-Alanine	−	+	d	d	d	+	+	+	+	+	+	+	d	+	−	+	+	d	−	−	+	−	−	+
D-α-Alanine	d	d	+	+	+	+	+	+	+	+	+	+	+	+	+	+	−	−	−	+	−	+	+	+
β-Alanine	−	−	−	−	−	d	−	−	−	−	−	−	−	d	d	−	−	−	−	d	+	+	+	+
L-Serine	d	+	+	d	+	+	+	−	d	+	+	+	d	+	+	+	+	+	−	−	d	+	+	−
L-Leucine	−	−	−	−	+	+	d	−	+	−	−	−	−	−	−	−	−	−	−	d	+	+	+	−
L-Glutamate	+	+	+	−	+	+	+	+	+	+	+	+	d	+	+	+	+	+	d	+	+	+	+	−
γ-Aminobutyrate	−	−	−	−	−	−	+	−	+	+	+	−	−	+	−	−	−	−	−	−	−	−	−	−
δ-Aminovalerate	−	−	−	−	−	−	+	−	+	−	d	−	−	+	−	−	−	−	−	−	−	−	−	−
L-Histidine	d	d	d	−	+	+	+	d	d	+	d	d	d	d	d	+	+	d	−	−	+	−	−	+
L-Proline	+	+	+	+	+	+	+	+	+	+	+	+	+	+	+	+	+	+	+	+	+	+	+	−
L-Tyrosine	−	−	+	d	+	+	d	+	d	+	+	d	d	d	−	−	−	−	−	−	−	−	−	+
Putrescine	−	−	−	−	+	d	+	−	+	d	+	−	−	+	+	−	−	−	−	−	+	−	−	−
L-Rhamnose, malonate, benzoate, spermine, betaine, sarcosine, hippurate	−	−	−	−	−	+	−	−	−	−	−	−	−	−	−	−	−	−	−	−	−	−	−	−

poorly hemolytic or even nonhemolytic. The nontoxigenic strains of *V. cholerae* 0:1 that may be isolated from the environment even in areas where the disease is absent are frequently strongly hemolytic but usually do not belong clearly to either biovar.

It should be noted that neither an extensive phenotypic characterization nor DNA/DNA hybridization studies support the subdivision of *V. cholerae* into the classical and eltor biovars (Citarella and Colwell, 1970; Colwell, 1970; Desmarchelier and Reichelt, 1981).

The mol% G + C of the DNA is 47–49 (T_m, Bd).

Type strain: ATCC 14035 (NCTC 8021 of Desmarchelier and Reichelt 1981, 127).

2. **Vibrio metschnikovii** Gaméia 1888, 485.[AL] (*Vibrio cholerae* biovar *proteus* Shewan and Véron 1974, 344.)

metsch.ni.ko′vi.i. M.L. masc. gen. n. *metschnikovii* of Metschnikoff; named after E. Metschnikoff, a Russian biologist.

Physiological and nutritional characteristics of the species are presented in Tables 5.58–5.60.

The following combination of properties distinguishes *V. metschnikovii* from other species of *Vibrio* (or their biovars) by four to six traits: positive for growth at 40°C, acetoin and/or diacetyl production, utilization of sucrose and D-gluconate and growth in the absence of Na⁺; negative for oxidase, NO_3^- to NO_2^- reduction and utilization of cellobiose, α-ketoglutarate, L-tyrosine and putrescine.

The mol% G + C of the DNA is 44–46 (T_m, Bd).

Type strain: NCTC 8443 (Lee, Donovan and Furniss 1978, 110).

3. **Vibrio harveyi** (Johnson and Shunk 1936) Baumann, Baumann, Bang and Woolkalis 1981a, 217.[VP] (Effective publication: Baumann, Baumann, Bang and Woolkalis 1980b, 128.) (*Achromobacter harveyi* Johnson and Shunk 1936, 587; *Lucibacterium harveyi* (Johnson and Shunk 1936) Hendrie, Hodgkiss and Shewan 1970, 166; *Beneckea harveyi* (Johnson and Shunk 1936) Reichelt and Baumann 1973, 320.)

har′vey.i. M.L. gen. n. *harveyi* of Harvey; named after E. N. Harvey, a pioneer in studies of bioluminescence.

Morphology is as depicted in Figures 5.6 and 5.15.

Physiological and nutritional characteristics of the species are presented in Tables 5.58–5.60.

The following combination of properties distinguishes *V. harveyi* from other species of *Vibrio* (or their biovars) by three to eleven traits: straight rods synthesizing lateral flagella on solid medium, positive for growth at 35°C and utilization of D-mannose, cellobiose, D-gluconate, D-glucuronate, heptanoate, α-ketoglutarate, L-serine, L-glutamate and L-tyrosine; negative for arginine dihydrolase, acetoin and/or diacetyl production and utilization of β-hydroxybutyrate, D-sorbitol, ethanol, L-leucine, γ-aminobutyrate and putrescine.

The mol% G + C of the DNA is 46–48 (T_m).

Type strain: ATCC 14126 (strain 384 of Reichelt and Baumann 1973, 320).

4. **Vibrio campbellii** (Baumann, Baumann and Mandel 1971) Baumann, Baumann, Bang and Woolkalis 1981a, 217.[VP] (Effective publi-

Table 5.59.

Additional characteristics of species and biovars of **Vibrio**[a]

Characteristics	1. V. cholerae[b]	2. V. metschnikovii	3. V. harveyi[b]	4. V. campbellii[b]	5. V. parahaemolyticus[b]	6. V. alginolyticus[b]	7. V. natriegens[b]	8. V. vulnificus[b]	9. V. nereis[b]	10. V. fluvialis I[c]	10. V. fluvialis II[c]	11. V. splendidus I[b]	11. V. splendidus II[b]	12. V. pelagius I[b]	12. V. pelagius II[b]	13. V. nigripulchritudo[b]	14. V. anguillarum I	14. V. anguillarum II	15. V. fischeri[b]	16. V. logei	17. V. proteolyticus[b]	18. V. gazogenes	19. V. marinus	20. V. costicola
Growth at:																								
20°C	+	+	+	+	+	+	+	+	+	+	+	+	+	+	+	+	+	+	+	+	+	+	+	+
45°C	−	d	−	−	−	−	−	−	d	−	−	−	−	−	−	−	−	−	−	−	−	−	−	−
50°C	−	−	−	−	−	−	−	−	−	−	−	−	−	−	−	−	−	−	−	−	−	−	−	−
Utilization of:																								
D-Ribose	+	+	+	+	+	+	+	+	+	+	+	+	+	+	+	−	d	+	+	d	+	+	+	+
D-Glucose	+	+	+	+	+	+	+	+	+	+	+	+	+	+	+	+	+	+	+	+	+	+	+	+
D-Fructose	+	+	+	+	+	+	+	+	+	+	+	+	+	+	+	+	+	+	+	+	+	+	+	+
Maltose	+	+	+	+	+	+	+	+	+	+	+	+	+	+	+	+	+	+	+	+	+	+	+	d
N-Acetylglucosamine	+	+	+	+	+	+	+	+	+	+	+	+	+	+	+	+	+	+	+	+	+		+	+
Acetate	+	+	+	d	+	+	+	+	d	+	+	d	+	+	+	+	+	+	−	−	−	+	+	+
Butyrate	−	−	−	−	d	+	+	−	+	−	−	−	−	−	−	−	−	−	−	−	−	+	−	+
Isobutyrate	−	−	−	−	−	d	+	−	+	−	−	−	−	−	−	−	−	−	−	−	−	−	−	−
Isovalerate	−	−	−	−	d	d	−	−	d	−	−	−	−	−	−	−	−	−	−	−	−	−	−	−
Caproate	+	+	d	−	d	−	d	−	d	+	+	−	−	−	−	+	+	−	−	−	−	−	−	−
Caprylate	+	d	d	d	+	+	+	+	d	+	+	−	−	−	−	d	+	−	−	−	+	−	−	−
Pelargonate	d	−	d	d	+	+	−	d	+	+	−	−	−	−	−	−	−	−	−	−	+	−	−	−
Caprate	d	d	d	d	+	+	+	+	+	+	+	d	+	+	+	−	d	−	−	−	+	−	+	−
Succinate	+	+	+	+	+	+	+	+	+	+	+	+	+	+	+	−	+	+	+	+	+	+	+	+
Fumarate	+	+	+	+	+	+	+	+	+	+	+	+	+	+	+	−	+	+	+	+	+	+	+	+
L-Tartrate	−	−	−	−	−	−	d	−	−	−	−	−	−	−	−	−	−	−	−	−	−	−	−	−
DL-Glycerate	d	−	+	d	+	+	+	+	−	+	+	d	d	−	−	+	+	d	d	−	+	−	−	−
Aconitate	+	d	+	+	+	+	+	+	+	+	+	+	+	+	+	+	+	d	d	−	+	−	−	−
Erythritol	−	−	−	−	−	−	−	−	−	−	−	−	−	−	−	−	d	−	−	−	−	−	−	−
Glycerol	+	+	+	+	+	+	+	+	+	+	+	d	+	+	+	+	+	+	+	+	+	+	+	+
Propanol	−	−	d	−	+	d	+	−	+	+	+	−	−	d	d	d	−	−	−	−	−	−	−	−
Phenylacetate	−	−	−	−	−	−	−	d	−	−	−	−	−	−	−	−	−	−	−	−	−	−	−	−
Quinate	−	−	−	−	−	+	−	−	+	d	d	−	−	d	−	−	−	−	−	−	−	−	−	−
Glycine	−	−	d	−	d	+	+	−	+	d	d	+	d	+	+	+	−	−	−	−	+	−	−	−
L-Threonine	d	+	+	+	+	+	+	+	+	+	+	+	d	+	+	+	+	+	d		+		+	d
L-Aspartate	d	+	d	−	+	d	d	+	d	+	+	d	d	+	+	+	+	d			+	+	+	−
L-Arginine	+	+	d	−	+	+	+	+	+	+	+	−	+	+	−	−	−	−	−	−	+	−	−	−
L-Ornithine	+	+	−	−	−	+	−	+	+	+	−	−	d	+	−	−	−	−	−	−	+	−	−	−
L-Citrulline	−	−	d	−	−	−	+	−	+	−	−	+	−	+	+	−	−	−	−	−	−	−	−	−

[a] For symbols see standard definitions.

[b] These species or biovars are not able to utilize the following organic compounds as sole or principal sources of carbon and energy: D-arabinose, D-fucose, inulin, cellulose, saccharate, mucate, formate, oxalate, maleate, adipate, pimelate, suberate, azelate, sebacate, D-tartrate, *meso*-tartrate, glycolate, laevulinate, citraconate, itaconate, mesaconate, adonitol, ethylene glycol, propylene glycol, 2,3-butanediol, methanol, isopropanol, *n*-butanol, isobutanol, D-mandelate, L-mandelate, benzoyl formate, *o*-hydroxybenzoate, *m*-hydroxybenzoate, phenylethanediol, phenol, naphthalene, L-isoleucine, L-norleucine, L-valine, L-lysine, L-phenylalanine, L-tryptophan, D-tryptophan, anthranilate, *m*-aminobenzoate, *p*-aminobenzoate, methylamine, ethanolamine, benzylamine, histamine, tryptamine, butylamine, α-amylamine, 2-amylamine, pentylamine, creatine, pantothenate, acetamide, nicotinate, nicotinamide, trigonelline, allantoin, adenine, guanine, cytosine, thymine, uracil and *n*-dodecane.

[c] This biovar is not able to utilize the following organic compounds as sole sources of carbon and energy: adipate, pimelate, suberate, azelate, sebacate, adonitol, *m*-hydroxybenzoate, phenylacetate, L-isoleucine, L-valine and L-lysine.

cation: Baumann, Baumann, Bang and Woolkalis 1980b, 128.) (*Beneckea campbellii* Baumann, Baumann and Mandel 1971a, 288.)

camp.bel′li.i. M.L. gen. n. *campbellii* of Campbell; named after L. L. Campbell, an American bacteriologist.

Morphology is as depicted in Figure 5.8.

Physiological and nutritional characteristics of the species are presented in Tables 5.58–5.60.

The following combination of properties distinguishes *V. campbellii* from other species of *Vibrio* (or their biovars) by two to nine traits: positive for lateral flagella on solid medium, growth at 35°C, amylase and utilization of propionate; negative for arginine dihydrolase, growth at 40°C and utilization of D-galactose, sucrose, D-gluconate, D-glucuronate, heptanoate, L-glutamate, L-histidine and putrescine.

The mol% G + C of the DNA is 46–48 (T_m, Bd).

Type strain: ATCC 25920 (Strain 40 of Baumann, Baumann and Mandel 1971a, 289).

5. **Vibrio parahaemolyticus** (Fujino, Okuno, Nakada, Aoyama, Fukai, Mukai and Ueho 1951) Sakazaki, Iwanami and Fukumi 1963, 181.[AL] (*Pasteurella parahaemolytica* Fujino, Okuno, Nakada, Aoyama, Fukai, Mukai and Ueho 1951, 11; *Beneckea parahaemolytica* (Fujino et al. 1951) Baumann, Baumann and Mandel 1971a, 291.)

Table 5.60.

Identification of **Vibrio** *species and allied genera likely to be encountered in clinical laboratories[a]*

Characteristics	1. V. cholerae	2. V. metschnikovii	3. V. harveyi [b]	4. V. campbellii [b]	5. V. parahaemolyticus [b]	6. V. alginolyticus [b]	8. V. vulnificus [b]	10. V. fluvialis I [c]	10. V. fluvialis II [c]	14. V. anguillarum I	Aeromonas	Plesiomonas
Thornley arginine[d]	−	+	−	−	−	−	−	+	+	+	+	+
Lysine decarboxylase	+	d	+	+	+	+	+	−	−	−	D	+
Ornithine decarboxylase	+	−	+	−	+	+	+	−	−	−	−	+
Glucose gas	−	−	−	−	−	−	−	+	−		D	−
L-Arabinose acid	−	−	d	−	d	−	−	+	+	d	D	−
Inositol acid	−	d	−	−	−	−	−	−	−	−	−	+
Salicin acid	−	−	d	−	−	−	+	+	−	−	D	−
Sucrose acid	+	+	d	−	−	+	−	+	+	+	+	−
Voges-Proskauer (24 h)	d	+	−	d	−	+	−	−	−	+	D	−
ONPG hydrolysis (24 h)	+	d	d	−	−	−	+	+	+	+	+	+
Growth at 43°C	+	d	d	−	+	+	+	d	−	−	D	+
Inhibition by 0/129 phosphate[e]:												
10 µg	S	S	d	S	R	R	S	R	R	S	R	d
150 µg	S	S	S	S	S	S	S	S	S	S	R	S
Growth in % NaCl:												
0	+	d	−	−	−	−	−	d	d	d	+	+
3	+	+	+	+	+	+	+	+	+	+	+	+
6	−	+	+	+	+	+	+	+	+	d	−	−
8	−	d	d	−	+	+	−	d	d	−	−	−
10	−	−	d	−	−	+	−	d	−	−	−	−
Esculin hydrolysis	−	d	−	−	−	−	−	d	−	−	D	−
Oxidase (Kovacs')	+	−	+	+	+	+	+	+	+	+	+	+
NO3− to NO2−	+	−	+	+	+	+	+	+	+	+	+	+
Growth on TCBS	Y	G	Y/G	G	G	Y	G	Y	Y	Y	−	−

[a] For symbols see standard definitions; also S, sensitive; R, resistant; Y, yellow; and G, green.

[b] To differentiate *V. parahaemolyticus*, *V. alginolyticus*, *V. harveyi*, *V. campbellii* and *V. vulnificus* adequately it may be necessary to test for growth on cellobiose, gluconate, ethanol, L-serine, L-leucine, L-glutamate, and putrescine (see Table 5.58).

[c] To differentiate *V. fluvialis* biovars I and II adequately it is necessary to test for growth on δ-aminovalerate, putrescine, glutarate and glucuronate (see Table 5.58).

[d] See text for discussion of the problems associated with the Thornley method.

[e] 0/129 phosphate: 2,4-diamino-6,7-diisopropylpteridine phosphate.

Table 5.61.

Characteristics that distinguish the biovars of **V. cholerae**[a]

Characteristics	Classical	Eltor
Hemolysis[b]	−	d[c]
Voges-Proskauer[d]	−	+
Hemagglutination[e]	−	+
Sensitivity to:		
Polymyxin B 50 IU[f]	+	−
Classical phage IV[g]	+	−
Eltor phage V[h]	−	+

[a] For symbols see standard definitions.

[b] Sheep erythrocytes, brain-heart-thioglycolate-cystine agar of Sakazaki et al. (1971).

[c] The first strains of the current pandemic were strongly hemolytic but later strains have been nonhemolytic.

[d] Method 3 of Cowan (1974).

[e] Chicken erythrocytes, slide test of Finkelstein and Mukerjee (1963).

[f] Disk test of Gan and Tija (1963).

[g] Mukerjee et al. (1957).

[h] Basu and Mukerjee (1968).

para.hae.mo.ly'ti.cus. Gr. prep. *para* by the side of, beside; Gr. n. *haema* blood; Gr. adj. *lyticus* dissolving; M.L. adj. *parahaemolyticus* dissolving blood.

Physiological and nutritional characteristics of the species are presented in Tables 5.58–5.60.

The following combination of properties distinguishes *V. parahaemolyticus* from other species of *Vibrio* (or their biovars) by four to eight traits: positive for lateral flagella on solid medium, growth at 40°C and utilization of ethanol, L-leucine and putrescine; negative for arginine dihydrolase, acetoin and/or diacetyl production, swarming on complex solid media and utilization of sucrose, cellobiose, valerate, β-hydroxybutyrate and γ-aminobutyrate.

The mol% G + C of the DNA is 46–47 (T_m, Bd).

Type strain: ATCC 17802 (strain 113 of Baumann, Baumann and Mandel 1971a, 291).

6. **Vibrio alginolyticus** (Miyamoto, Nakamura and Takizawa 1961) Sakazaki 1968, 360.[AL] (*Oceanomonas alginolytica* Miyamoto, Nakamura and Takizawa 1961, 481; *Beneckea alginolytica* (Miyamoto et al.) Baumann, Baumann and Mandel 1971a, 289.)

al.gi.no.ly'ti.cus. L. fem. n. *alga* seaweed; M.L. adj. *alginicus* pertaining to alginic acid from seaweed; Gr. adj. *lyticus* dissolving; M.L. adj. *alginolyticus* alginic acid-dissolving.

Morphology is as depicted in Figures 5.11, 5.13 and 5.14.

Physiological and nutritional characteristics of the species are presented in Tables 5.58–5.60.

The following combination of properties distinguishes *V. alginolyticus* from other species of *Vibrio* (or their biovars) by four to ten traits: positive for lateral flagella and swarming on solid complex media, acetoin and/or diacetyl production, growth at 40°C and utilization of sucrose, valerate, L-leucine and L-tyrosine; negative for arginine dihydrolase and utilization of cellobiose, β-hydroxybutyrate and γ-aminobutyrate.

The mol% G + C of the DNA is 45–47 (T_m, Bd).

Type strain: ATCC 17749 (strain 118 of Baumann, Baumann and Mandel 1971a, 290).

7. **Vibrio natriegens** (Payne, Eagon and Williams 1961) Baumann, Baumann, Bang and Woolkalis 1981a, 217.[VP] (Effective publication: Baumann, Baumann, Bang and Woolkalis 1980b, 128.) (*Pseudomonas natriegens* Payne, Eagon and Williams 1961, 125; *Beneckea natriegens* Baumann, Baumann and Mandel 1971a, 291.)

na.tri.e′gens. M.L. *natrium* sodium; L. *egens* pres. part. *egere*, to be in need; M.L. adj. *natriegens* sodium-needing.

Morphology is as depicted in Figure 5.10.

Physiological and nutritional characteristics of the species are presented in Tables 5.58 and 5.59.

V. natriegens is distinguished from all the remaining species of *Vibrio* by its ability to utilize L-rhamnose, malonate, benzoate, spermine, betaine, sarcosine and hippurate.

The mol% G + C of the DNA is 46–47 (T_m, Bd).

Type strain: ATCC 14048 (strain 111 of Baumann, Baumann and Mandel 1971a, 291).

8. **Vibrio vulnificus** (Reichelt, Baumann and Baumann 1979) Farmer 1980, 656.[VP] (*Beneckea vulnifica* Reichelt, Baumann and Baumann 1979, 80.)

vul.ni′fi.cus. L. adj. *vulnificus* inflicting wounds.

Physiological and nutritional characteristics of the species are presented in Tables 5.58–5.60.

The following combination of properties distinguishes *V. vulnificus* from other species of *Vibrio* (or their biovars) by four to ten traits: curved rods, positive for growth at 40°C and utilization of D-galactose, cellobiose, D-gluconate, D-glucuronate and L-tyrosine; negative for lateral flagella on solid medium, arginine dihydrolase, acetoin and/or diacetyl production and utilization of sucrose, heptanoate, L-serine, γ-aminobutyrate and putrescine.

In the initial phenotypic characterization of *V. vulnificus* (Baumann et al., 1973; Group C-2), all of the strains were found to be unable to utilize lactose as a sole source of carbon and energy. Hollis et al. (1976) subsequently reported that strains of this species fermented lactose. When the cause of this discrepancy was analyzed (Reichelt et al., 1976; P. Baumann, unpublished observations), it was found that wild type strains of this species were, in fact, unable to utilize lactose, however some strains readily acquired the ability to utilize this sugar by mutation. This would explain the positive fermentation tests of Hollis et al. (1976) since they were performed in complex media where sufficient substrate would have been available to support the growth of a population of cells large enough to potentially contain spontaneous lactose-utilizing mutants. In view of the problems associated with the utilization of lactose by *V. vulnificus*, its application as a trait diagnostic for this species should be regarded with considerable caution.

The mol% G + C of the DNA is 46–48 (T_m, Bd).

Type strain: ATCC 27562 (strain 324 of Reichelt, Baumann and Baumann 1976, 114).

9. **Vibrio nereis** (Harwood, Bang, Baumann and Nealson 1980) Baumann, Baumann, Bang and Woolkalis 1981a, 217.[VP] (Effective publication: Baumann, Baumann, Bang and Woolkalis 1980b, 128.)

(*Beneckea nereida* (sic) Harwood, Bang, Baumann and Nealson 1980, 655.)

ne′re.is. L. n. *nereis* a sea nymph.

Morphology is as depicted in Figures 5.4 and 5.9.

Physiological and nutritional characteristics of the species are presented in Tables 5.58 and 5.59.

The following combination of properties distinguishes *V. nereis* from other species of *Vibrio* (or their biovars) by four to seven traits: positive for arginine dihydrolase and utilization of glutarate, β-hydroxybutyrate, L-leucine and δ-aminovalerate; negative for lipase and utilization of L-arabinose and D-galactose.

The mol% G + C of the DNA is 46–47 (T_m, Bd).

Type strain: ATCC 25917 (strain 80 of Baumann, Baumann and Mandel 1971a, 289).

10. **Vibrio fluvialis** Lee, Shread, Furniss and Bryant 1981b, 217.[VP] (Effective publication: Lee, Shread, Furniss and Bryant 1981a, 92.)

flu.vi.a′lis. L. adj. *fluvialis* of or belonging to a river.

Physiological and nutritional characteristics of the species are presented in Tables 5.58–5.60.

The following combination of properties distinguishes *V. fluvialis* from other species of *Vibrio* (or their biovars) by three to eight traits: positive for arginine dihydrolase, growth at 40°C and utilization of L-arabinose, D-galactose, D-galacturonate and γ-aminobutyrate; negative for acetoin and/or diacetyl production and utilization of L-rhamnose, malonate and L-leucine. Biovar I, unlike biovar II of this species, is able to utilize D-glucuronate and is unable to utilize glutarate or produce gas from D-glucose.

The mol% G + C of the DNA is 49–51 (T_m, Bd).

Type strain: NCTC 11327 (strain VL 5125 of Lee, Shread, Furniss and Bryant 1981a).

11. **Vibrio splendidus** (Beijerinck 1900) Baumann, Baumann, Bang and Woolkalis 1981a, 217.[VP] (Effective publication: Baumann, Baumann, Bang and Woolkalis 1980b, 128.) (*Photobacter splendidum* Beijerinck 1900, 362; *Beneckea splendida* (Beijerinck 1900) Reichelt, Baumann and Baumann 1979, 80.)

splen′di.dus. L. adj. *splendidus* brilliant.

Morphology is as depicted in Figure 5.5.

Physiological and nutritional characteristics of the species are presented in Tables 5.58 and 5.59.

The following combination of properties distinguishes *V. splendidus* biovar I from other species of *Vibrio* (or their biovars) by three to ten traits: curved rods, positive for arginine dihydrolase, luminescence, utilization of D-galactose, cellobiose, D-glucuronate, α-ketoglutarate and L-serine; negative for lateral flagella on solid medium, acetoin and/or diacetyl production, growth at 40°C and utilization of L-arabinose, β-hydroxybutyrate, D-sorbitol, γ-aminobutyrate and putrescine.

The following combination of properties distinguishes *V. splendidus* biovar II from all the remaining species of *Vibrio* (or their biovars) by two to eight traits: reduces NO_3^- to NO_2^-, positive for amylase and utilization of citrate, α-ketoglutarate and pyruvate; negative for arginine dihydrolase, acetoin and/or diacetyl production, growth at 35°C and utilization of D-mannose, D-galactose, sucrose, D-gluconate, D-glucuronate and propanol.

A total of nine traits (Tables 5.58 and 5.59) distinguish *V. splendidus* biovar I from biovar II, which are related by a DNA homology of 68% (Fig. 5.19). Although these two biovars have DNA homologies of 19–22% to *V. harveyi* and *V. campbellii* they differ from the former and latter species by only three and two traits, respectively.

The mol% G + C of the DNA is 45–46 (T_m, Bd).

Type strain: ATCC 33125 (NCMB 1 of Reichelt, Baumann and Baumann 1976, 114).

12. **Vibrio pelagius** (Baumann, Baumann and Mandel 1971) Baumann, Baumann, Bang and Woolkalis 1981a, 217.[VP] (Effective publication: Baumann, Baumann, Bang and Woolkalis 1980b, 128.) (*Beneckea pelagia* Baumann, Baumann and Mandel 1971a, 291.)

pe.la'gius. L. adj. *pelagius* of the sea.

Physiological and nutritional characteristics of the species are presented in Tables 5.58 and 5.59.

V. pelagius is positive for alginase and utilization of D-gluconate, citrate and putrescine; negative for arginine dihydrolase, growth at 40°C and utilization of L-rhamnose, cellobiose and α-ketoglutarate. Biovar I, unlike biovar II of this species, is negative for amylase, gelatinase and utilization of γ-aminobutyrate and δ-aminovalerate. This combination of properties distinguishes biovars I and II of *V. pelagius* from the remaining species of *Vibrio* (or their biovars) by four to eight traits.

The mol% G + C of the DNA is 45–47 (T_m, Bd).

Type strain: ATCC 25916 (strain 99 of Baumann, Baumann and Mandel 1971a, 291).

13. Vibrio nigripulchritudo (Baumann, Baumann, Mandel and Allen 1971) Baumann, Baumann, Bang and Woolkalis 1981a, 217.[VP]

(Effective publication: Baumann, Baumann, Bang and Woolkalis 1980b, 128.) (*Beneckea nigrapulchrituda* (sic) Baumann, Baumann, Mandel and Allen 1971b, 1383.)

ni.gri.pul.chri.tu'do. L. *niger* black; L. *pulchritudo* beauty; M.L. adj. *nigripulchritudo* black beauty.

Physiological and nutritional characteristics of the species are presented in Tables 5.58 and 5.59.

The following combination of properties distinguishes *V. nigripulchritudo* from other species of *Vibrio* (or their biovars) by five to eight traits: produces crystals of blue-black pigment (Fig. 5.16), positive for utilization of melibiose, lactose, β-hydroxybutyrate, D-sorbitol and meso-inositol; no growth at 40°C and negative for utilization of sucrose.

The mol% G + C of the DNA is 46–47 (T_m, Bd).

Type strain: ATCC 27043 (strain 164 of Baumann, Baumann, Mandel and Allen 1971b, 1383).

14. Vibrio anguillarum Bergeman 1909, 28.[AL]

an.guil.la'rum. L. n. *anguilla* eel; L. gen. pl. n. *anguillarum* of eels.

Physiological and nutritional characteristics of the species are presented in Tables 5.58–5.60.

The following combination of properties distinguishes *V. anguillarum* biovar I from *V. anguillarum* biovar II as well as the remaining species of *Vibrio* (or their biovars) by four to seven traits: positive for arginine dihydrolase, NO$_3^-$ to NO$_2^-$ reduction, acetoin and/or diacetyl production, amylase and utilization of L-arabinose, sucrose and D-sorbitol; negative for growth at 40°C and utilization of β-alanine, L-tyrosine and putrescine.

The following combination of properties distinguishes *V. anguillarum* biovar I from *V. anguillarum* biovar II and all the remaining species of *Vibrio* (or their biovars) by four to eight traits: requires growth-factors and is positive for gelatinase and utilization of sucrose; negative for amylase, lipase and utilization of trehalose, D-gluconate, DL-lactate and D-mannitol.

Although biovars I and II of *V. anguillarum* are closely related by a high DNA homology (Fig. 5.19), they differ by a total of 23 phenotypic properties which include the requirement for organic growth factors by biovar II (Tables 5.58 and 5.59). Biovar I is nutritionally more versatile than biovar II, and there are no nutritional or physiological properties which are present in the latter but not the former biovar. Since the strains of biovar II were isolated from diseased fish, its loss of functional attributes might reflect an adaptation to a parasitic mode of existence.

The mol% G + C of the DNA is 44–46 (T_m, Bd).

Type strain: ATCC 19264 (strain 144 of Baumann, Bang and Baumann 1978, 86).

15. Vibrio fischeri (Beijerinck 1889) Lehmann and Neumann 1896, 342.[AL] (*Photobacterium fischeri* Beijerinck 1889, 402.)

fisch'er.i. M.L. gen. n. *fischeri* of Fischer; named after Bernhard Fischer, one of the earliest students of luminescent bacteria.

Morphology is as depicted in Figures 5.7, 5.12 and 5.18.

Physiological and nutritional characteristics of the species are presented in Tables 5.58 and 5.59.

The following combination of properties distinguishes *V. fischeri* from other species of *Vibrio* (or their biovars) by three to six traits: positive for polar tufts of two to eight flagella, yellow-orange cell-associated pigment and growth at 30°C; negative for growth at 4°C, gelatinase and utilization of D-gluconate, DL-lactate and pyruvate.

The mol% of G + C of the DNA is 39–41 (T_m, Bd).

Type strain: ATCC 7744 (strain 398 of Reichelt and Baumann 1973, 322).

16. Vibrio logei (Harwood, Bang, Baumann and Nealson 1980) Baumann, Baumann, Bang and Woolkalis 1981a, 217.[VP] (Effective publication: Baumann, Baumann, Bang and Woolkalis 1980b, 128.) (*Photobacterium logei* Harwood, Bang, Baumann and Nealson 1980, 655.)

log'e.i. M.L. gen. n. *logei* of Loge; from German Loge, Norse god of fire and mischief.

Physiological and nutritional characteristics of the species are presented in Tables 5.58 and 5.59.

The following combination of properties distinguishes *V. logei* from the remaining species of *Vibrio* (or their biovars) by three to eight traits: positive for polar tufts of two to eight flagella, yellow-orange cell-associated pigment, growth at 4°C and utilization of D-gluconate; negative for growth at 30°C, gelatinase and utilization of DL-lactate and pyruvate.

The mol% G + C of the DNA is 40–42 (T_m, Bd).

Type strain: ATCC 29985 (strain 584 of Bang, Baumann and Nealson 1978a, 287).

17. Vibrio proteolyticus (Merkel, Traganza, Mukherjee, Griffin and Prescott 1964) Baumann, Baumann, Bang and Woolkalis 1982, 267. (Effective publication: Baumann, Baumann, Bang and Woolkalis 1980b, 128.) (*Aeromonas proteolytica* Merkel, Traganza, Mukherjee, Griffin and Prescott 1964, 1230; *Aeromonas hydrophila* subspecies *proteolytica* (Merkel et al. 1964) Schubert 1969, 412.)

pro.te.o.ly'ti.cus. Ger. protein from Gr. *protos* first; Gr. adj. *lyticus* dissolving; M.L. adj. *proteolyticus* protein-dissolving.

Physiological and nutritional characteristics of the species are presented in Tables 5.58 and 5.59.

The following combination of properties distinguishes *V. proteolyticus* from the remaining species of *Vibrio* (or their biovars) by four to eight traits: positive for swarming on solid complex media, arginine dihydrolase, acetoin and/or diacetyl production, growth at 40°C and utilization of D-sorbitol, β-alanine and putrescine; negative for utilization of sucrose.

The mol% G + C of the DNA is 50.5 (Bd).

Type strain: ATCC 15338 (strain 145 of Baumann, Baumann and Mandel 1971a, 290).

18. Vibrio gazogenes (Harwood, Bang, Baumann and Nealson 1980) Baumann, Baumann, Bang and Woolkalis 1981a, 217.[VP] (Effective publication: Baumann, Baumann, Bang and Woolkalis 1980b, 128.) (*Beneckea gazogenes* Harwood, Bang, Baumann and Nealson 1980, 655.)

ga.zo'ge.nes. Fr. n. *gaz* gas; L. n. *genesis* birth; M.L. adj. *gazogenes* gas-producing.

Physiological and nutritional characteristics of the species are presented in Tables 5.58 and 5.59.

The following combination of properties distinguishes *V. gazogenes* from the remaining species of *Vibrio* (or their biovars) by four to six traits: positive for red pigment, gas from D-glucose and utilization of D-xylose, salicin and D-sorbitol; negative for NO$_3^-$ to NO$_2^-$ reduction. *V. gazogenes* is oxidase-negative. The determination of this property of *V. gazogenes* is complicated by its red pigmentation, a problem which can be overcome by using the spontaneous nonpigmented mutants which are relatively common in cultures of this species.

The mol% G + C of the DNA is 47.1 (T_m).
Type strain: ATCC 29988 (Harwood 1978, 238).

19. Vibrio marinus (ex Ford 1927, 347) nom. rev.

ma.ri′nus. L. adj. *marinus* marine, of the sea.

Physiological and nutritional characteristics of the species are presented in Tables 5.58 and 5.59.

The following combination of properties distinguishes *V. marinus* from all the remaining species of *Vibrio* (or their biovars) by four to seven traits: positive for growth at 4°C and utilization of DL-malate, α-ketoglutarate and D-α-alanine; negative for growth at 30°C, amylase and utilization of sucrose, trehalose, D-mannitol and putrescine.

The mol% G + C of the DNA is 42.2 (T_m).
Type strain: ATCC 15381 (Reichelt and Baumann 1973, 322).

20. Vibrio costicola Smith 1938, 29.[AL]

cos.ti′co.la. L. n. *costa* rib; L. subst. *cola* dweller; M.L. n. *costicola* rib dweller.

Physiological and nutritional characteristics of the species are presented in Tables 5.58 and 5.59.

The following combination of properties distinguishes *V. costicola* from the remaining species of *Vibrio* (or their biovars) by four to eight traits: positive for arginine dihydrolase and utilization of sucrose and L-tyrosine; negative for gelatinase, lipase and utilization of D-galactose, DL-malate, γ-aminobutyrate, L-proline and putrescine. Strains of this species will not grow in media containing less than approximately 200–250 mM NaCl. The minimal concentrations required for optimal growth are presented in Figure 5.17.

The mol% G + C of the DNA is 50.0 (Bd).
Type strain: ATCC 33508.

Other organisms

Although the number of well characterized species of *Vibrio* has increased considerably during the last 10 years, many undescribed species undoubtedly remain. Four groups, with an inadequate number of strains, are already known to be sufficiently distinct, phenotypically and genotypically, to represent new species of *Vibrio*. These four groups consist of (a) strain 77 (ATCC 33506), (b) strains 84 (ATCC 33507), 85, (c) strains 76, 142 (ATCC 33505) and (d) strains 94, 95 (ATCC 33523) (Baumann and Baumann, 1977; Reichelt et al., 1976). The latter two strains have mol% G + C contents in their DNAs of 54%; the formal species assignment of these two isolates will necessitate an extension of the G + C range of *Vibrio*. Baross et al. (1978b) have found a number of strains of agar-decomposing vibrios with a DNA homology of 23–58% to *V. parahaemolyticus*. It is not known whether these organisms can be equated with any of the currently recognized species. The group of marine luminous isolates having a G + C content of 46–48 mol% which has been designated as "*Photobacterium belozerskii*" by Chumakova et al. (1973) has properties of the genus *Vibrio*. The reported phenotypic traits are, however, too few to allow an assignment to any of the described species.

Organisms that are unable to ferment or grow on sucrose but are otherwise phenotypically similar to *V. cholerae* have been isolated from shellfish, brackish water and the stools of humans with diarrhea in many parts of the world. They have usually been identified as *V. cholerae*. Unlike *V. cholerae*, they do not produce amylase and can utilize glucuronate and often ethanol (J. V. Lee et al., unpublished results). They are also Voges-Proskauer-negative and sensitive to 50 International Units (IU) of Polymyxin B (Furniss et al., 1978). Recent studies on more than 50 such organisms show that they have species level DNA relatedness to one another but are related by only 20–50% DNA homology to *V. cholerae* (Fanning et al., 1980, 1981; D. J. Brenner, personal communication). Thus they appear to be a distinct species for which the name *V. mimicus* has been proposed (Davis et al., 1981 and 1982).

Recently two vibrio-like organisms that fix nitrogen have been described (Guerinot and Patriquin, 1981). Strains YL 1491 and VLb 110 (Lee et al., 1981a) and nine other recently isolated strains also fix nitrogen (M. L. Guerinot, R. R. Colwell, P. A. West and J. V. Lee, unpublished observations). The mol% G + C contents and the phenotypic properties of these strains suggest their assignment to *Vibrio*. Their inclusion, if warranted, would require a modification of the definition of this genus.

A group of isolates from freshwater reservoirs, sewage and fecal samples has been characterized by Kalina et al. (1980) and assigned to a newly created genus, "*Allomonas*." Although it is not possible to equate these isolates definitively with any of the species of *Vibrio*, their general properties suggest their possible inclusion in this genus. A similar conclusion appears to be applicable to the isolates of Turova and Levanova (1980). Finally, it should be noted that *Vibrio succinogenes* is a strict anaerobe (Smibert and Holdeman, 1976; Wolin et al., 1961) and, therefore, clearly excluded from the genus *Vibrio* as defined in this chapter; it has been reclassified in the genus *Wolinella*.

Note Added in Proof

Since the writing of this article the following new species of *Vibrio* have been described: *V. aestuarianus* (D. L. Tison and R. J. Seidler, personal communication), *V. damsela* (Love et al., 1981), *V. diazotrophicus* (Guerinot et al. 1982), *V. hollisae* (Hickman et al., 1982), *V. mimicus* (Davis et al., 1981), *V. ordalii* for *V. anguillarum* biovar II (Schiewe et al., 1981), and *V. orientalis* (Yang et al., 1983). In addition, it has been shown that the G + C content of *Allomonas* is 57 mol% (Turova et al., 1983) thereby excluding this proposed genus from *Vibrio*.

References

Davis, B.R., G.R. Fanning, J.M. Madden, A.G. Steigerwalt, H.B. Bradford, H.L. Smith and D.J. Brenner. 1981. Characterization of biochemically atypical *Vibrio cholerae* strains and designation of new pathogenic species, *Vibrio mimicus*. J. Clin. Microbiol. *14:* 631–639.

Guerinot, M.L., P.A. West, J.V. Lee and R.R. Colwell, 1982. *Vibrio diazotrophicus* sp. nov., a marine nitrogen-fixing bacterium. Int. J. Syst. Bacteriol. *32:* 350–357.

Hickman, F.W., J. J. Farmer, D.G. Hollis, G.R. Fanning, A.G. Steigerwalt, R.E. Weaver and D.J. Brenner. 1982. Identification of *Vibrio hollisae* sp. nov. from patients with diarrhea. J. Clin. Microbiol. *15:* 395–401.

Love, M., D. Teebken-Fischer, J.E. Hose, J.J. Farmer, F.W. Hickman and G.R. Fanning. 1981. *Vibrio damsela*, a marine bacterium, causes skin ulcers on the damselfish *Chromis punctipinnis*. Science *214:* 1139–1140.

Schiewe, M.H., T.J. Trust and J.H. Crosa. 1981. *Vibrio ordalii* sp. nov.: a causative agent of vibriosis in fish. Curr. Microbiol. *6:* 343–348.

Turova, T.P., T.I. Grafova and I.M. Badalova. 1983. *Allomonas*, a new group of microorganisms of the family *Vibrioanceae*. Taxonomic status of *Allomonas*, determined on the basis of the study of their DNA (Russian). Microbiol. Epidemiol. Immunolobiol. (Eng. Transl.) *1:* 22–23.

Yang, Y., L. Yeh, Y. Cao, L. Baumann, P. Baumann, J.S. Tang and B. Beaman. 1983. Characterization of marine luminous bacteria isolated off the coast of China and description of *Vibrio orientalis* sp. nov. Curr. Microbiol. *8:* 95–100.

Genus II. *Photobacterium* Beijerinck 1889, 401^AL

PAUL BAUMANN AND LINDA BAUMANN

Pho.to.bac.te′ri.um. Gr. n. *phos* light; Gr. neut. dim. n. *bakterion* a small rod; M.L. neut. n. *Photobacterium* light (-producing) bacterium.

Plump, straight rods, 0.8–1.3 μm in diameter and 1.8–2.4 μm in length. **Accumulate poly-β-hydroxybutyrate (PHB) under certain** conditions of cultivation; **do not utilize the exogenous monomer β-hydroxybutyrate.** Involution forms usually seen in old cultures or under adverse conditions of cultivation. Do not form endospores or microcysts. **Gram-negative.** Motile by **one to three unsheathed polar flagella;** some nonmotile. **Chemoorganotrophs** capable of **respiratory and fermentative metabolism.** Grow in presence and absence of oxygen which is a universal electron acceptor. Do not denitrify. Do not fix molecular nitrogen. Fermentation of D-glucose results in production of acidic end products. Sodium ions are required for growth. Most strains **grow in a mineral medium containing a seawater base,** D-glucose, and NH₄Cl; other strains require L-methionine in addition. In addition to D-glucose, all utilize D-mannose, D-fructose, and glycerol and grow at 20°C. Two species are bioluminescent. **Common in the marine environment** and on the surfaces and in the intestinal contents of marine animals; some found as symbionts in specialized luminous organs of marine fish. The G + C content of the DNA is 40–44 mol% (T_m, Bd).

Type species: *Photobacterium phosphoreum* (Cohn 1878) Beijerinck 1889, 401.

Further Descriptive Information

During exponential phase of growth in a complex medium such as yeast extract broth* (YEB), cells of all species are rod shaped and lack visible intracellular granules when observed by phase contrast microscopy (Fig. 5.21). In early stationary phase of growth in basal medium*

volatilize under the electron beam. As is seen from this figure, cells of *Photobacterium* have a cell wall structure which is typical of many Gram-negative bacteria.

Motile cells harvested from liquid or solid medium usually have one to three polar flagella (Figs. 5.24 and 5.25), although occasional cells may have up to five. The flagella of *Photobacterium*, unlike those of *Vibrio* species which are common in the marine environment, are not enclosed in a sheath (Allen and Baumann, 1971; Baumann et al., 1980; Reichelt and Baumann, 1973).

The colonies of *Photobacterium* are not distinctive, being convex, smooth, with entire edges and a white color when grown on a complex or mineral medium such as yeast extract agar* (YEA), marine agar (MA) (ZoBell, 1941), or basal medium agar* (BMA) containing 0.2% (w/v) D-glucose. The color of the colonies is somewhat whiter than that of many Gram-negative marine bacteria, probably due to a relatively low content of cytochromes.

Species of *Photobacterium* require Na⁺ for growth and are unable to grow in media containing a seawater base in which the Na⁺ has been replaced by an equimolar amount of K⁺ (MacLeod, 1968). For optimal growth, 160–280 mM Na⁺ is required; some species may also require seawater levels of Mg⁺⁺ (50 mM) and Ca⁺⁺ (10 mM) (Reichelt and Baumann, 1974, 1975). Most strains of *P. leiognathi* and *P. angustum* have no organic growth factor requirements; some strains of *P. phosphoreum* require L-methionine, either alone or in combination with other amino acids (Reichelt and Baumann, 1973; Ruby et al., 1980). All species grow at 20°C, *P. phosphoreum* and some strains of *P. angustum* grow at 4°C, while none grows at 40°C.

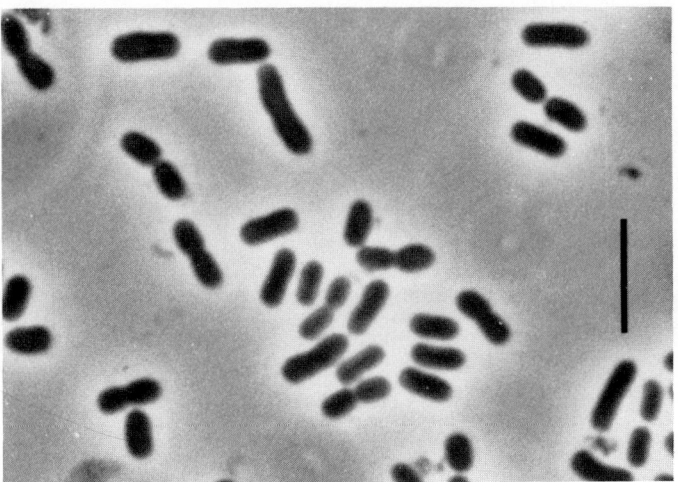

Figure 5.21. Phase-contrast micrograph of *Photobacterium leiognathi* in exponential phase of growth in YEB (*bar*, 5 μm).

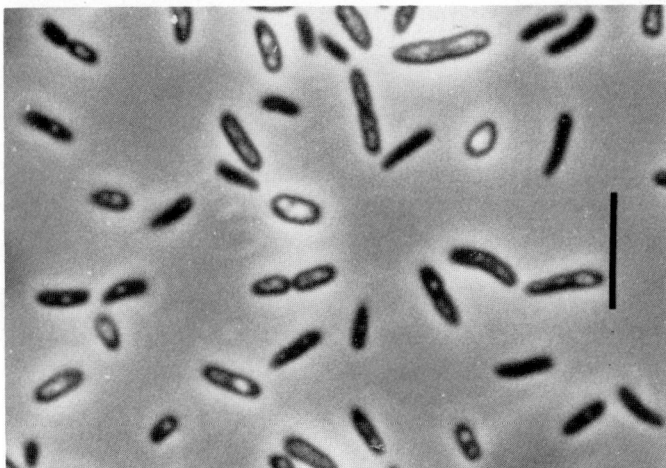

Figure 5.22. Phase contrast micrograph of *Photobacterium leiognathi* showing accumulation of PHB in early stationary phase cells grown in BM containing 0.2% (w/v) D-glucose (*bar*, 5 μm).

(BM) containing 0.2% (w/v) D-glucose, PHB, which is stored by the cells as a reserve product, is detectable as discrete, refractile, intracellular granules (Fig. 5.22) (Reichelt and Baumann, 1973). When a substantial amount of PHB is accumulated, the entire cell appears refractile. This reserve material can account for as much as 40% of the dry weight of the cell. Figure 5.23 is an electron micrograph of thin sections of cells containing granules of PHB which have a tendency to

The nutritional versatility of species of *Photobacterium* is relatively limited; only 7–22 carbon compounds can be utilized as sole or principal sources of carbon energy. These compounds include hexoses and a few pentoses, disaccharides, sugar acids, tricarboxylic acid cycle intermediates, and amino acids. None of the species has an extracellular

* All media contain an artificial seawater base (ASW; MacLeod, 1968) having the following composition (g/liter): NaCl, 23.4; MgSO₄·7H₂O, 24.6; KCl, 1.5; and CaCl₂·2H₂O, 2.9. (The salts are dissolved separately and then combined.) Basal medium (BM) consists of (g/liter): tris (hydroxymethyl) aminomethane hydrochloride (Tris-HCl), 6.1 or 12.1, with HCl added to give a pH value of 7.5; NH₄Cl, 1.0; K₂HPO₄·3H₂O, 0.075; FeSO₄·7H₂O, 0.028; and 1/2 strength ASW. YEB consists of BM with yeast extract (5.0 g/liter). YEA and BMA are obtained by adding agar (20 g/liter) to YEB and BM, respectively.

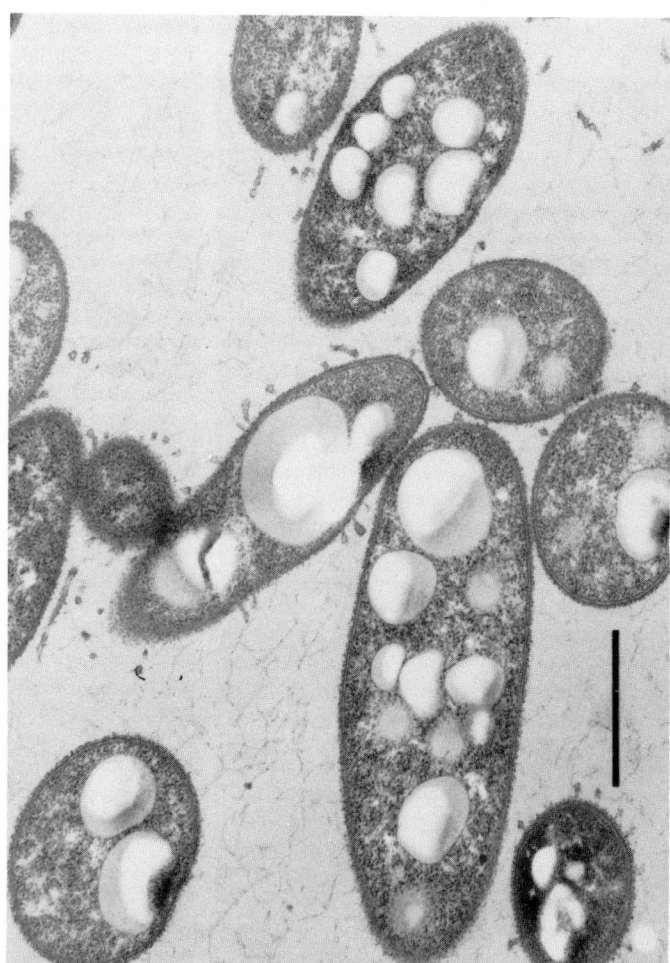

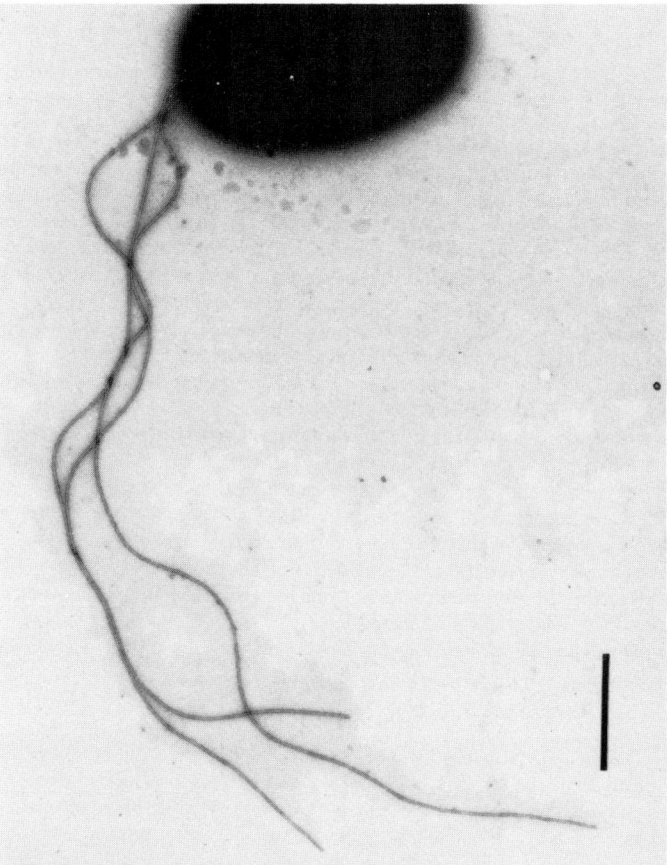

Figure 5.24. Electron micrograph of *Photobacterium phosphoreum*. Negatively stained (*bar*, 1 μm). (Reproduced with permission from: Reichelt and Baumann, Arch. Microbiol. *94:* 283–330, 1973, © Springer-Verlag.)

Figure 5.23. Ultrathin section of *Photobacterium phosphoreum* showing granules of PHB in early stationary phase cells grown in BM containing 0.2% (w/v) D-glucose (*bar*, 1 μm). (Reproduced with permission of R. D. Allen.)

amylase or alginase; some strains may have an extracellular chitinase, lipase, or gelatinase.

Under anaerobic conditions, D-glucose is catabolized via a mixed acid fermentation, the principal end products being formic, acetic, lactic, and succinic acids as well as ethyl alcohol (Doudoroff, 1942). Upon completion of the fermentation, the final pH of the medium ranges from 4.8–5.2. A mixture of CO_2 and H_2 is produced by most strains of *P. phoshoreum* and a few strains of *P. leiognathi* but none of the strains of *P. angustum*. Acetoin and/or diacetyl and 2,3-butanediol are produced by the majority of strains of *P. phosphoreum*, a few strains of *P. leiognathi*, and none of the strains of *P. angustum*. D-Glucose is dissimilated via a constitutive Embden-Meyerhof pathway while D-gluconate is degraded via an inducible Entner-Doudoroff pathway (Reichelt and Baumann, 1973). The utilization of D-fructose is initiated by an inducible phosphoenolpyruvate:D-fructose phosphotransferase system which converts D-fructose to fructose-1-phosphate. Following conversion of this sugar phosphate to fructose-1,6-diphosphate by an inducible 1-phosphofructokinase, dissimilation proceeds via the Embden-Meyerhof pathway (Gee et al., 1975).

A negative oxidase test (Kovacs, 1956; Stanier, 1966) is often obtained with strains of *Photobacterium*; some strains give a positive reaction when treated with toluene prior to the addition of the oxidase reagent (Baumann and Baumann, 1981; Reichelt and Baumann, 1973; Reichelt et al., 1976). A positive oxidase test correlates well with the presence of a cytochrome of the c type (West et al., 1978). Differential oxidized/reduced cytochrome spectra have indicated that low levels of

cytochrome c are present in representative strains of *Photobacterium*, probably accounting for the absence of a positive result with the oxidase reagent in some strains. Cytochrome c, in addition to an unusual cytochrome of the *bd* type, has been isolated from *P. phosphoreum* (Watanabe et al., 1979).

A constitutive arginine dihydrolase system (as determined by the anaerobic conversion of arginine to ornithine) has not been detected in species of *Photobacterium* (Baumann et al., 1971; Reichelt and Baumann, 1973). Some strains form alkaline products from arginine when tested by a modification of the method of Thornley or Møller (Baumann and Baumann, 1981; Hendrie et al., 1970).

The description of the morphological, physiological, and nutritional properties of *Photobacterium* is based on the work of Baumann and Baumann, 1977, 1981; Baumann et al., 1980a; Chumakova et al., 1973; Hendrie et al., 1970; Reichelt and Baumann, 1973, 1975.

P. phosphoreum and *P. angustum*, like species of *Vibrio* and *Aeromonas*, have a single iron-containing superoxide dismutase (Bang et al., 1978). *P. leiognathi* is distinctive in that it has an additional superoxide dismutase which resembles the eucaryotic enzyme in containing copper and zinc but differs from it in having two nonidentical subunits and a different isoelectric point (Bang et al., 1978; Puget et al., 1977).

Strains of *P. phosphoreum* and *P. leiognathi* are able to emit light of a blue-green color, a property not found in strains of *P. angustum*. The reaction sequence resulting in light emission and the enzyme catalyzing this process, luciferase, are similar in all procaryotes (Nealson and Hastings, 1979). Luciferase is a heterodimer which, in *P. phosphoreum*, has a molecular weight of about 82,000 and is composed of α and β subunits having molecular weights of 42,000 and 38,000, respectively

(Watanabe et al., 1976; Yoshida and Nakamura, 1973). One strain of *P. leiognathi* is unusual in that it has two luciferases, one soluble and similar to that previously reported for other luminous bacteria and, the other, a cell envelope-associated glycoprotein which is inactivated by treatment with lysozyme (Balakrishnan and Langerman, 1977). The reaction catalyzed by luciferase involves the luminescent oxidation of reduced flavin mononucleotide (FMNH₂) and a long chain aliphatic aldehyde (probably tetradecanal) by molecular oxygen. The reaction can be expressed as

$$FMNH_2 + O_2 + RCHO \xrightarrow{\text{luciferase}} h\upsilon + FMN + RCOOH$$

and constitutes a bypass of the electron transport chain since the electrons are shunted from flavin to molecular oxygen without involvement of cytochromes (Hastings and Nealson, 1977). Riendeau and Meighen (1979) have detected an enzyme activity in *P. phosphoreum* which involves the recycling of the long chain fatty acid to the aldehyde and is dependent on adenosine triphosphate and reduced nicotinamide adenine dinucleotide phosphate. Ulitzer and Hastings (1980) have found a similar activity in *P. leiognathi*. A blue fluorescent protein with a lumazine prosthetic group has been isolated from *P. phosphoreum* (Koka and Lee, 1979).

Experiments with batch cultures as well as chemostats have shown a cell density-dependent luciferase synthesis in *P. phosphoreum* and *P. leiognathi* which is believed to occur as the result of an accumulation of an extracellular autoinducer. Such an autoinducer has been purified from the luminescent species *Vibrio fischeri* and its chemical structure

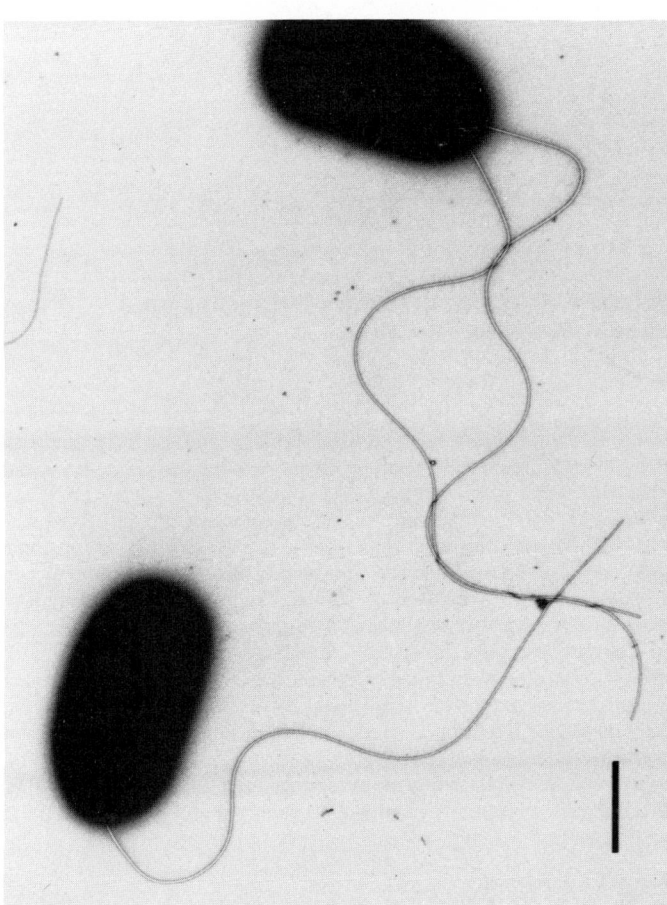

Figure 5.25. Electron micrograph of *Photobacterium angustum*. Negatively stained (*bar*, 1 μm). (Reproduced with permission from: Allen and Baumann, J. Bacteriol. *107*: 295–302, 1971, © American Society for Microbiology.)

determined (Nealson and Hastings, 1979). Autoinduction does not appear to occur in some strains of *P. phosphoreum* and *P. leiognathi* which synthesize luciferase constitutively throughout the growth cycle (Katznelson and Ulitzer, 1977; Watanabe et al., 1975).

Bacteriophages active against *P. phosphoreum* have been isolated from the marine environment (Spencer, 1960, 1963). All three species of *Photobacterium* have been found to be suitable hosts for marine bdellovibrios (Taylor et al., 1974; Yetinson and Shilo, 1979). Spontaneous and induced mutants involving amino acid requirements and changes in luminescence of *P. leiognathi* have been studied by Popova and Shenderov (1979a, b).

Strains of *P. phosphoreum* and *P. leiognathi* are widespread in the marine environment and have been isolated from seawater, the surfaces and intestinal contents of marine animals, and from the specialized luminous organs of marine fish (Herring and Morin, 1978; Nealson and Hastings, 1979; Reichelt and Baumann, 1973; Ruby and Morin, 1979). No information is available on the distribution of *P. angustum*; all strains of this species have been isolated from the open ocean off the Hawaiian Archipelago. In a study of the distribution of luminous species in water columns at stations in the open ocean of the North Atlantic and over the Puerto Rico Trench, Ruby et al. (1980) noted an increase in the numbers of *P. phosphoreum* at depths ranging from 200–1000 meters, with maximal concentrations ranging from 3–8 colony-forming units/100 ml of seawater. *P. leiognathi* was found in considerably fewer numbers throughout the water columns. These two species differ in their growth-temperature profiles, the minimal and maximal temperature permitting growth being lower for *P. phosphoreum* than for *P. leiognathi*. This property correlates with the distribution of *P. phosphoreum* in colder waters and with the distribution of both species in the luminous organs of teleost fishes. *P. leiognathi* is found in fish inhabiting shallow tropical waters (0–15 meters) (Reichelt et al., 1977) while *P. phosphoreum* is associated with the luminous organs of fish from midwater (200–600 meters) and bathyal (600–1200 meters) habitats (Ruby and Morin, 1978). Figure 5.26 is an electron micrograph of a thin section through the luminous organ of a leiognathid fish (*Equulites novaehollandiae*) harboring cells of *P. leiognathi*.

Ecological studies of the distribution of luminous bacteria in the inshore waters of the Gulf of Elat have revealed a seasonal variation in numbers of *P. leiognathi* and *Vibrio harveyi* which correlated both with changes in environmental parameters and with the physiological properties of these two species (Shilo and Yetinson, 1979; Yetinson and Shilo, 1979). Studies with samples, ranging to depths of 600 meters, from an open water station in the Gulf of Elat indicated that *P. leiognathi* was the sole species present, in low numbers, throughout the year. The presence of only one species and the lack of seasonal variation was attributed to the relatively constant environmental parameters in the water column throughout the year.

Enrichment and Isolation Procedures

No enrichment methods are available for the isolation of luminous marine bacteria. All of the methods used are based on allowing the growth of luminous organisms, as well as any other chemoorganotrophs present, to reach a cell density sufficient to permit detection of light by the dark-adapted eyes of the investigator. Since luminescence is favored by complex media, a suitable luminescence medium (LM) is BM containing 0.3% (v/v) glycerol and (in g/liter) 5 g yeast extract, 5 g tryptone, 1 g CaCO₃, and 20 g agar. The duration of light emission and its intensity is variable in different strains of the same species and is occasionally detected only in isolated colonies. Therefore, it is best to periodically examine cultures streaked onto solid medium for luminescent colonies within 12–36 h of inoculation. A suitable incubation temperature for both luminous species of *Photobacterium* is 15°C. Sources of luminous bacteria include the surfaces and intestinal contents of marine animals, luminous organs, as well as seawater. A method which is relatively specific for *P. phosphoreum* involves the incubation of fresh fish, squid, or octopus at 10–15°C, half-submerged in either natural or artificial seawater, for 10–18 h. Periodic examination usually reveals luminous sites which can be transferred onto LM plates with sterile toothpicks. The removal of luminous bacteria from specimens or Petri plates having both luminous and nonluminous organisms is facilitated by the use of a red light connected to a rheostat (Cosenza and Buck, 1966; Hastings and Nealson, 1981). Additional details of use in the isolation of luminous species of *Photobacterium* from different marine sources (including luminous organs) are given by Baumann and Baumann (1981), Hastings and Nealson (1981), and Ruby et al. (1980). No specific methods are available for the isolation of *P. angustum*; all strains of this species were among a large collection of organisms obtained by direct isolation from seawater.

Maintenance Procedures

Strains of *P. phosphoreum* and *P. leiognathi* can be maintained on LM and transferred monthly. After each transfer, the cultures are allowed to grow at 15–18°C (*P. phosphoreum*) or 25°C (*P. leiognathi*) for 1–2 days and stored at 4°C and 15–18°C, respectively. *P. angustum* has been maintained on YEA or MA in the same manner as *P. leiognathi*; neither of these species survives well at 4°C (Baumann and Baumann, 1981).

Strains of all luminous species develop dark mutants which, with time, may predominate in the culture. For the continued maintenance of light-producing strains, it is advisable to periodically begin a new stock culture from a single, brightly-luminous colony.

Strains of all three species have been lyophilized and kept at 4°C; with a few exceptions, viable cells have been recovered after 5–7 years of storage. To prepare lyophils, the growth from a fresh slant is suspended in about 0.5 ml of a sterile solution consisting of 1/4 strength ASW, 5 g/liter yeast extract and 5 g/liter peptone (adjusted to pH 7.5) and transferred into a lyophil tube which is dipped into a mixture of dry ice and acetone prior to being placed under vacuum for 10–12 h. The lyophils are reconstituted by suspending the powder in about 0.5 ml LM (without agar) or YEB. A portion of the suspension is streaked onto LM or YEA and the remainder is inoculated into a tube containing 4 ml of the same medium. Growth is generally observed after 1–2 days incubation at 25°C; for *P. phosphoreum* a lower temperature (15–18°C) should be used. For a description of the methods used for the preservation of strains in agar stabs as well as at low temperatures see Nealson (1978) and Hastings and Nealson (1981).

Procedures for Testing Special Characters

A somewhat unusual combination of properties characteristic of *Photobacterium* is the ability to accumulate PHB as an intracellular reserve product coupled with the inability to utilize the exogenous monomer, β-hydroxybutyrate, as a sole or principal source of carbon and energy. Accumulation of refractile PHB granules can be readily detected by phase microscopy in early stationary phase cultures grown in BM containing 0.2% (w/v) D-glucose (Fig. 5.23). In the case of organisms which require growth factors, the medium is supplemented with 0.05–0.1% (w/v) yeast extract. A chemical method for the identification of PHB is described by Williamson and Wilkinson (1958) and Slepecky and Law (1960).

Fermentation and gas production are best detected in YEB containing 100 mM Tris-HCl (pH 7.5), 1% (w/v) D-glucose, and 2 g/liter agar. After melting the agar, 10 ml aliquots of the medium are dispensed into test tubes, autoclaved, and inoculated by means of a stab. About 5 ml of 2% (w/v) agar (autoclaved and cooled to 41°C) are carefully layered over the medium (prechilled in the case of *P. phosphoreum*) to make an agar plug. The test tubes are incubated at 18°C and observed

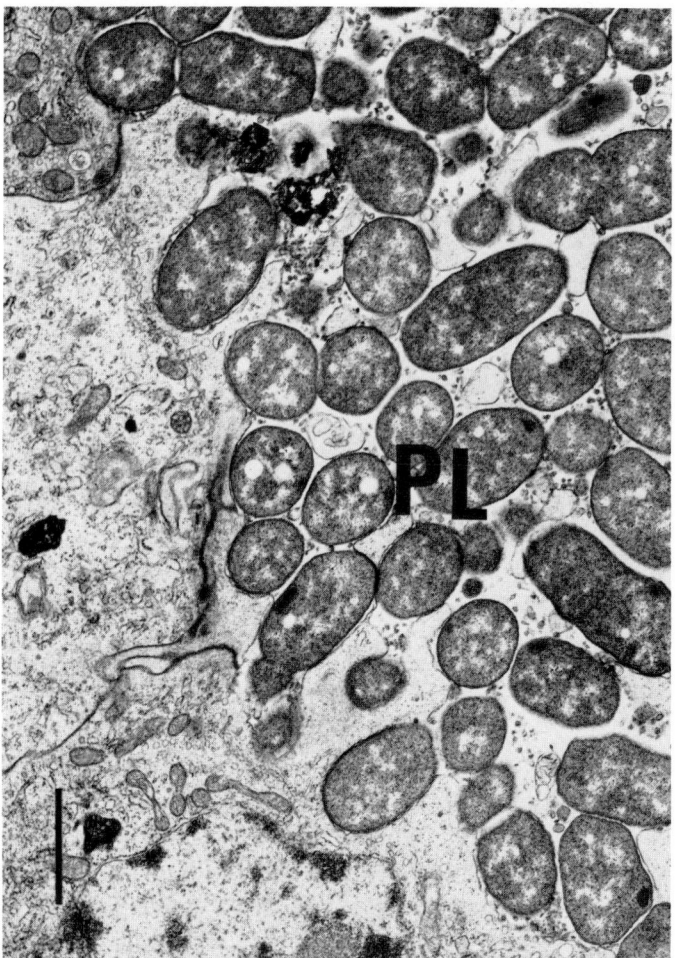

Figure 5.26. Ultrathin section through the luminous organ of a leiognathid fish *Equulities novaehollandiae (PL, Photobacterium leiognathi; bar,* 2 μm). (Reproduced with permission from: J. M. Bassot, Archives de Zoologie Experimentale et Generale, *116:* 359–373, 1975, © Centre National de la Recherche Scientifique.)

for turbidity and gas production for a period of 6 days. In this medium, facultative anaerobes grow to a high turbidity and gas, if produced, is trapped in the semisolid medium; strict aerobes do not grow or give only very slight turbidity under these conditions.

The ability to grow at 4°C and 35°C is determined in either YEB or LM (without agar). Incubation at the former temperature is continued for a period of 1 week with periodic examination; at the latter temperature, incubation is for 2 days. The ability of carbon compounds to serve as sole or principal sources of carbon and energy is determined by replica plating the organisms onto BMA containing 0.2% (w/v) of the tested sugars or 0.1% (w/v for solids and v/v for liquids) of most other organic compounds. The plates are incubated at 15–20°C and examined every other day for a total of 6 days. For a detailed description of the procedures used for the characterization of strains, including the nutritional screening and tests for extracellular enzymes, consult Baumann and Baumann (1981) as well as Palleroni and Doudoroff (1972).

Differentiation of the genus **Photobacterium** from other genera

Table 5.62 indicates the characteristics of *Photobacterium* which distinguish it from other morphologically or physiologically similar genera. Since a high Na⁺ requirement appears to restrict *Photobacterium* to the marine environment, the differentiation of *Photobacterium* from marine species of *Vibrio* has the greatest practical application. Although there are very few readily determinable traits distinguishing

these two genera, their differentation, in practice, is not difficult since the individual species differ by many unrelated phenotypic properties.

The decay kinetics of light emission by luciferase in the presence of dodecanal subdivides luminous bacteria into two major groups having either "slow" or "fast" kinetics (Hastings and Nealson, 1977; Hastings et al., 1978; Jensen et al., 1980; Hastings and Nealson, 1981). "Slow"

Table 5.62.

Differential characteristics of the genus **Photobacterium** *and other morphologically or physiologically similar genera[a]*

Characteristics	Photo-bacterium	Vibrio	Aero-monas	Plesio-monas
Accumulation of PHB coupled with the inability to utilize β-hydroxybutyrate	+	−		
Sheathed polar flagella	−	+	−	−
Requirement for over 100 mM Na⁺ for optimal growth	+	D	−	−
Utilization of D-mannitol	−	D[b]	D	−
Mol% G + C in DNA	40–44	38–51	57–63	51

[a] Symbols: all positive; D, most positive; −, all negative.
[b] *V. nereis*, *V. anguillarum* biovar II, and *V. marinus* are negative for this trait. Most strains of the remaining species utilize D-mannitol.

decay kinetics are characteristic of the enzyme from *Vibrio harveyi*, *V. splendidus* biovar I, *V. cholerae*, and *Xenorhabdus luminescens*, while "fast" decay kinetics occur with the luciferases of *Photobacterium*, *V. fischeri*, *V. logei*, and *Alteromonas hanedai*. Although *Photobacterium* is not unique in having "fast" decay kinetics, this property still has diagnostic value since the luminous species of *Vibrio* and *Alteromonas* are readily distinguished from *Photobacterium* by a few easily determined properties; *V. fischeri* and *V. logei* have a yellow-orange, cell-associated pigment while *Photobacterium* does not, and *A. hanedai* is a strict aerobe whereas *Photobacterium* is a facultative anaerobe.

Taxonomic Comments

In vitro DNA/DNA hybridization studies (Reichelt et al., 1976) have shown that within each species of the genus *Photobacterium* the strains are related by >84% DNA homology. With regard to the percentage of DNA homology between species, *P. leiognathi* and *P. angustum* are related by 53–61% and both have 19–36% DNA homology to *P. phosphoreum*. There is only 0–13% homology between species of *Photobacterium* and *Vibrio*. Major groups of bacteria having a common evolutionary origin have been found to share similar regulatory patterns of enzymes involved in key biosynthetic reactions. This is particularly well illustrated by genera of the *Enterobacteriaceae* as well as *Aero-*

monas, *Vibrio*, and *Photobacterium* which possess a distinctive mode of regulating aspartokinase activity (Baumann and Baumann, 1973; Cohen et al., 1969). These genera contain three isofunctional aspartokinases; the activity of the first is inhibited by L-threonine and the second by L-lysine, while the third is unaffected by any of the amino acids of the aspartate family (Baumann and Baumann, 1973).

Studies of ribosomal RNA (rRNA)/DNA homology (Baumann and Baumann, 1976, 1981) and a comparison of the similarities in the amino acid sequences of glutamine synthetase and superoxide dismutase (Baumann et al., 1980a, b) indicate considerable evolutionary divergence between *Photobacterium* and the genera *Aeromonas*, *Escherichia*, *Enterobacter*, *Serratia*, *Proteus*, and *Erwinia*; the genus most closely related to *Photobacterium* is *Vibrio*. These results, as well as those from studies of aspartokinase regulation, are confirmed by 5S rRNA nucleotide sequencing (Hori and Osawa, 1979; Woese et al., 1975) and 16S oligonucleotide cataloging (Fox et al., 1980; Gibson et al., 1979). (The organism designated *P. fischeri* in Fox et al. (1980) is actually *P. phosphoreum* (G. E. Fox, personal communication)). The latter studies indicate that all of these genera are part of an evolutionary branch which includes *Pseudomonas aeruginosa*, *Acinetobacter*, and *Chromatium* and which probably originated from a purple sulfur bacterium.

The genus *Photobacterium* as presently defined does not include bioluminescence as a diagnostic trait since only two of the species are able to emit light. Additional luminous species are found in the genera *Vibrio*, *Alteromonas*, and *Xenorhabdus*. One species (*V. fischeri*) is also listed as *Photobacterium fischeri* in the Approved List of Bacterial Names (Skerman et al., 1980). Recent studies on the evolution of glutamine synthetase and superoxide dismutase clearly assign this species to *Vibrio* (Baumann and Baumann, 1980a, b).

Further Reading

Baumann, P and L. Baumann. 1977. Biology of the marine enterobacteria: genera *Beneckea* and *Photobacterium*. Annu. Rev. Microbiol. *31:* 39–61.

Baumann, P. and L. Baumann. 1981. The marine Gram-negative eubacteria. In Starr, Stolp, Trüper, Balows, Schlegel (Editors), *The Prokaryotes*, Springer-Verlag, New York, pp. 1302–1331.

Hastings, J.W. and K.H. Nealson. 1977. Bacterial bioluminescence. Annu. Rev. Microbiol. *31:* 549–595.

Nealson, K.H. and J.W. Hastings. 1979. Bacterial bioluminescence: its control and ecological significance. Microbiol. Rev. *43:* 496–518.

Reichelt, J.L. and P. Baumann. 1973. Taxonomy of the marine, luminous bacteria. Arch. Mikrobiol. *94:* 283–330.

Differentiation and Characteristics of species of **Photobacterium**

The differential characteristics of species of *Photobacterium* are presented in Table 5.63. Other characteristics of the species are provided in Table 5.64.

Table 5.63.

Characteristics differentiating species of **Photobacterium**.[a]

Characteristics	1. *P. phosphoreum*	2. *P. leiognathi*	3. *P. angustum*
Gas from D-glucose	+	−	−
Luminescence	+	+	−
Growth at 4°C	+	−	d[b]
Growth at 35°C	−	+	+
Production of:			
Gelatinase	−	−	d[b]
Lipase	−	+	d
Utilization of:			
D-Xylose	−	−	+
Maltose	+	−	d
Acetate	−	+	+
DL-Glycerate	+	d	−
Pyruvate	−	+	+
L-Proline	−	+	−

[a] Unless otherwise indicated the incubation temperatures are 15–20°C. For symbols see standard definitions.
[b] Positive in 80% of the strains.

Table 5.64.

Other characteristics of species of **Photobacterium**[a]

Characteristics	1. *P. phosphoreum*	2. *P. leiognathi*	3. *P. angustum*	Characteristics	1. *P. phosphoreum*	2. *P. leiognathi*	3. *P. angustum*
Reduction of NO_3^- to NO_2^-	+	+	d	D-galactose, D-fructose, N-acetylglucosamine, glycerol			
Growth at:				D-Ribose, D-gluconate, DL-lactate, succinate, fumarate	d	+	+
20°C	+	+	+				
30°C	d	+	+	L-Aspartate	d	+	d
40°C	–	–	–	L-α-Alanine, L-serine	d	d	+
Requirement for L-methionine	d	–	–	DL-Glycerate, DL-malate, L-threonine, L-glutamate	d	d	–
Production of:				D-Glucuronate	d	–	–
Chitinase	+	+	d	Caprate	–	d	–
Amylase, alginase	–	–	–	Sucrose, trehalose	–	–	d
2,3-Butanediol	d	–	–				
Utilization of:							
D-Glucose, D-mannose,	+	+	+				

[a] The following compounds cannot be utilized as sole or principal sources of carbon and energy by any of the species of *Photobacterium*: D-arabinose, L-arabinose, D-fucose, L-rhamnose, cellobiose, melibiose, lactose, inulin, salicin, cellulose, saccharate, mucate, D-galacturonate, formate, propionate, butyrate, isobutyrate, valerate, isovalerate, caproate, heptanoate, caprylate, pelargonate, oxalate, malonate, maleate, glutarate, adipate, pimelate, suberate, azelate, sebacate, D-tartrate, L-tartrate, *meso*-tartrate, DL-β-hydroxybutyrate, glycolate, citrate, α-ketoglutarate, aconitate, levulinate, citraconate, itaconate, mesaconate, erythritol, D-mannitol, D-sorbitol, *meso*-inositol, adonitol, ethylene glycol, propylene glycol, 2,3-butanediol, methanol, ethanol, *n*-propanol, isopropanol, *n*-butanol, isobutanol, D-mandelate, L-mandelate, benzoyl formate, benzoate, *o*-hydroxybenzoate, *m*-hydroxybenzoate, *p*-hydroxybenzoate, phenylacetate, quinate, glycine, D-α-alanine, β-alanine, L-leucine, L-isoleucine, L-norleucine, L-valine, L-lysine, L-arginine, L-ornithine, L-citrulline, γ-aminobutyrate, δ-aminovalerate, L-histidine, L-tyrosine, L-phenylalanine, L-tryptophan, D-tryptophan, anthranilate, *m*-aminobenzoate, *p*-aminobenzoate, methylamine, ethanolamine, benzylamine, putrescine, spermine, histamine, tryptamine, butylamine, α-amylamine, 2-amylamine, pentylamine, betaine, sarcosine, creatine, hippurate, pantothenate, acetamide, nicotinate, nicotinamide, trigonelline, allantoin, adenine, guanine, cytosine, thymine, uracil. For Symbols see standard definitions.

List of species of the genus **Photobacterium**

1. **Photobacterium phosphoreum** (Cohn 1878) Beijerinck 1889, 401.[AL] (*Micrococcus phosphoreus* Cohn 1878, 126.)

phos.pho′re.um. Gr. v. *phosphoreo* bring light; M.L. neut. adj. *phosphoreum* light-bearing.

Morphology is as depicted in Figures 5.23 and 5.24.

Physiological and nutritional characteristics of the species are presented in Tables 5.63 and 5.64.

The mol% G + C of the DNA is 41–42.

Type strain: ATCC 11040 (strain 439 of Reichelt and Baumann, 1973).

2. **Photobacterium leiognathi** Boisvert, Chatelain and Bassot 1967, 521.[AL] (*Photobacterium mandapamensis* Hendrie, Hodgkiss and Shewan 1970, 165.)

lei.o.gna′thi. M.L. gen. n. *leiognathi* named after fish of the family *Leiognathidae*.

Important Notes for Users of this Edition

1. Always read both generic and species descriptions because characters listed in the generic description are not usually listed in the species descriptions.

2. Unless otherwise indicated in footnotes to tables, the meanings of symbols are as follows:

   +   90% or more of strains are positive

   –   90% or more of strains are negative

   d   11–89% of strains are positive

   v   strain instability (*not* equivalent to "d")

   D   different reactions in different taxa (species of a genus or genera of a family)

3. All other symbols are defined in footnotes to tables.

Morphology is as depicted in Figures 5.21, 5.22 and 5.26.

Physiological and nutritional characteristics of the species are presented in Tables 5.63 and 5.64. The mol% G + C of the DNA is 42–44.

Type strain: ATCC 25521.

3. **Photobacterium angustum** Reichelt, Baumann and Baumann, 1979, 79.[AL] (*Effective publication:* Reichelt, Baumann and Baumann 1976, 112.)

an.gus'tum. L. neut. adj. *angustum* limited, with respect to nutritional versatility.

Morphology is as depicted in Figure 5.25.

Physiological and nutritional charactristics of the species are presented in Tables 5.63 and 5.64.

The mol% G + C of the DNA is 40–42.

Type strain: ATCC 25915 (strain 68 of Baumann, Baumann and Mandel, 1971, 268).

Species Incertae Sedis

"*Vibrio psychroerythrus*" D'Aoust and Kushner 1972, 342.

This species consists of a single prodigiosin-producing strain (D'Aoust and Kushner, 1971; D'Aoust and Gerber, 1974) which has some of the properties of the genus *Photobacterium*: it ferments D-glucose, is motile by means of a single unsheathed polar flagellum, and has a DNA base composition of 40 mol% G + C. However, DNA homology studies have indicated little or no nucleotide complementarity (0–5%) to species of *Photobacterium* or *Vibrio* (Reichelt et al., 1976). This strain has not been included in any of the studies involving evolutionary relationships within *Photobacterium*, *Vibrio* and *Aeromonas*.

Type stain: ATCC 27364.

Genus III. Aeromonas *Kluyver and Van Niel 1936, 398*[AL]

MICHEL POPOFF

Ae.ro.mo'nas. Gr. n. *aer* air, gas; Gr. n. *monas* unit, monad; M.L. fem. n. *Aeromonas* gas (-producing) monad.

Cells straight, rod shaped with rounded ends to coccoid, 0.3–1.0 μm in diameter and 1.0–3.5 μm in length. Occur singly, in pairs or short chains. **Resting stages not known.** Gram-negative. Generally **motile by a single polar flagellum;** peritrichous flagella may be formed on solid media in young cultures. One species is nonmotile. Facultative anaerobes. **Metabolism of glucose is both respiratory and fermentative.** Carbohydrates are broken down to acid or acid and gas (CO_2 and H_2). Nitrate is reduced to nitrite. **Oxidase-positive.** Catalase-positive. Optimum temperature, 22–28°C; some strains do not grow at 35°C. **Resistant to the vibriostatic agent** 2,4-diamino-6, 7-diisopropylpteridine **(0/129).** Chemoorganotrophic, using a variety of sugars and organic acids as carbon sources. Occur in freshwater and sewage. Some species are pathogenic to frogs and fish. The mol% G + C of the DNA is 57–63 (Bd, T_m).

Type species: *Aeromonas hydrophila* (Chester 1901) Stanier 1943, 213.

Further Descriptive Information

Two well separated groups are included in the genus *Aeromonas*. Psychrophilic and nonmotile aeromonads are clustered in the first group, named *Aeromonas salmonicida*. The second group consists of mesophilic and motile bacteria; this group can be divided into three species: *Aeromonas hydrophila*, *Aeromonas caviae* and *Aeromonas sobria* (Popoff and Véron, 1976; Popoff et al., 1981). These three species will be referred to in this chapter as motile *Aeromonas* species.

On complex media, cells of *A. salmonicida* appear as coccobacilli; the length is less than twice the width. This species commonly develops short chains and clumps (Smith, 1963). Cells of motile *Aeromonas* species show considerable variation in shape and size. Some strains appear as short rods whereas others produce thin and filamentous forms. Rare cells may exhibit a somatic curvature, but this is by no means as marked as that shown by some *Vibrio* species (McCarthy, 1975). Cells are arranged in singles and pairs.

A. salmonicida cells are nonmotile and atrichous. Cells of motile *Aeromonas* species possess a single polar flagellum generally with a wavelength of 1.7 μm. Most of these flagellated strains form lateral flagella in young cultures. These lateral flagella have a shorter wavelength than the polar flagellum. When incubated past the logarithmic phase of growth, cells exhibit only polar flagellation (Ewing et al., 1961).

The optimum growth temperature for *A. salmonicida* is 22–25°C. Most strains grow at 5°C. The maximum temperature at which growth occurs is usually 35°C. When cultured on nutrient agar at 22°C for 48 h, colonies of *A. salmonicida* are round, raised, entire, translucent and friable (Smith, 1963). Most strains of *A. salmonicida* produce a brown water-soluble pigment on media containing 0.1% tyrosine or phenylalanine. Pigment production does not occur anaerobically (Williamson, 1928). Some nonpigmented strains may be isolated (Smith, 1963). On blood agar, hemolysis occurs rapidly, the colonies becoming greenish after 7 days.

The optimum growth temperature for motile *Aeromonas* species is 28°C. Some strains can grow at 5°C. The maximum temperature at which growth occurs is usually 38–41°C. On nutrient agar, colonies of motile aeromonads are round, raised, with an entire edge and a smooth surface. They are translucent and white to buff in color. The culture odor varies from extremely strong to absent (McCarthy, 1975). Motile *Aeromonas* species do not produce pigment. However, Ross (1962) reported the isolation of one motile *Aeromonas* strain which produced a dark red-brown pigment indistinguishable in appearance from that produced by *A. salmonicida*.

The biochemical characteristics of *Aeromonas* have been studied by Eddy (1960, 1962), Ewing et al. (1961), Smith (1963), Popoff (1969), McCarthy (1975) and Popoff and Véron (1976). Acid is produced by all strains of *Aeromonas* from glucose and maltose, but not from xylose, dulcitol, inositol, adonitol, malonate and mucate. All strains of *Aeromonas* possess gelatinase, deoxyribonuclease, ribonuclease and Tween 80 esterase. Hydrogen sulfide is not produced from thiosulfate.

The following carbohydrates are usually fermented by *A. salmonicida*: arabinose, trehalose, galactose, mannose and dextrin. The following biochemical tests are universally negative for *A. salmonicida*: growth in KCN broth, growth in nutrient broth containing 7.5% NaCl, urease, ornithine decarboxylase (ODC), tetrathionate reductase and acidification of media containing rhamnose, sorbose, sorbitol, lactose, raffinose and cellobiose. Arginine is catabolized by *A. salmonicida* via an arginine dihydrolase (ADH) system. Arginine desimidase, ornithine transcarbamylase, carbamyl phosphokinase, carbamyl phosphatase, and adenosine tri- di-, and monophosphatases have been demonstrated in cell extracts of *A. salmonicida* (Shieh and Reddy, 1972). Strains of *A. salmonicida* may be grown in the chemically defined medium developed by O'Leary et al. (1956).

The following physiological tests are universally positive for motile *Aeromonas* species: catalase, starch hydrolysis, lecithinase, phosphatase, ADH, hydrolysis of *o*-nitrophenyl-β-D-galactopyranoside (ONPG), growth in nutrient broth without NaCl, and fermentation of mannitol, trehalose, fructose, galactose and dextrin. The following tests are universally negative: pectinase, ODC, tryptophan and phenylalanine deaminase, growth on cetrimide agar, growth in nutrient broth containing 5% NaCl, and acid production from sorbose, erythritol and raffinose.

By an auxanographic method using inorganic M-70 medium (Véron,

1975) containing ammonium sulfate as the nitrogen source, the following compounds serve universally as sole carbon sources for motile *Aeromonas* species: D-ribose, D-fructose, D-galactose, D-glucose, D-maltose, D-trehalose, D-gluconate, caprylate, pelargonate, caprate, succinate, fumarate, DL-glycerate, L-malate, glycerol, D-mannitol, L-aspartate and L-glutamate.

A. salmonicida is a serologically homogeneous species (Karlsson, 1964; Spence et al., 1965; Popoff, 1969). The possibility of serogrouping motile *Aeromonas* species was demonstrated by Kulp and Borden (1942), Miles and Miles (1951), Kjems (1955), Ewing et al. (1961), Page (1962), De Meuron and Peduzzi (1979) and Leblanc (1981). It is possible to delineate 12 O antigens and 9 H antigens within this group of bacteria. De Meuron and Peduzzi (1979) have also reported the presence of K antigens which partly inhibit the O agglutination.

Transferable drug resistance factors (R factors) have been detected in *Aeromonas*. In naturally occurring *A. salmonicida* strains, plasmids conferring resistance to streptomycin, chloramphenicol, tetracycline and sulfathiazole are transferable to *Escherichia coli* (Aoki et al., 1971; Popoff and Davaine, 1971). These R factors belong to the fi⁻ type or to R(I) or I-like R factors (Aoki et al., 1971). In motile *Aeromonas* species, the presence of R factors has also been detected (Aoki and Egusa, 1971). These R factors have markers of resistance to sulfonamide, tetracycline, streptomycin and chloramphenicol. All these R factors belong to the fi⁻ type and are assigned to compatibility group A (Datta and Hedges, 1973).

A. salmonicida is a strict parasite under natural conditions. Although these bacteria were isolated from natural water, their existence in river water is very short-lived (McCraw, 1952). *A. salmonicida* is the causal agent of fish furunculosis traditionally associated with salmon and trout (Emmerich and Weibel, 1894). Furunculosis appears to be a specific infection of fish. The pathogenic action of *A. salmonicida* may be due to its abundant growth in fish blood and tissues, and to the production of a leucocidin (Klontz et al., 1966).

Motile *Aeromonas* species occur widely in water, sludge and sewage (Leclerc and Buttiaux, 1962; Schubert et al., 1972; Hazen et al., 1978). These organisms have been isolated occasionally from apparently healthy people (Lautrop, 1961; Catsaras and Buttiaux, 1965). However a fecal origin cannot explain the presence of these bacteria in surface water or sewage. Motile aeromonads have long been recognized as the causal agent of "red-leg" disease in amphibians (Russell, 1898; Emerson and Norris, 1905; Shotts et al., 1972). They are also considered to be responsible for diseases in reptiles (Camin, 1948; Page, 1961; Marcus, 1971), fishes (Haley et al., 1967), snails (Mead, 1959), cows (Wohlege-

muth et al., 1972) and humans (Davis et al., 1978). They may be a secondary invader in virus-infected fish (Heuschmann-Brunner, 1965). Mice may be infected experimentally (Schubert, 1964). Several cases of fatal human septicemia have been reported, but in all instances the patient was debilitated by some other disease (Davis et al., 1978). Recently, motile *Aeromonas* species were reported to be pathogenic for humans when wounds were exposed to polluted water (Davis et al., 1978). Some strains may also act as primary agents of acute diarrheal diseases (Bhat et al., 1974; Chatterjee and Neogy, 1972; Sanyal et al., 1975). A possibile explanation of the enteropathogenic potential of motile *Aeromonas* species has come with the finding that some strains produced a heat-labile enterotoxin (Sanyal et al., 1975; Wadström et al., 1976).

Enrichment and Isolation Procedures

Strains of *A. salmonicida* can be isolated on Trypticase soy agar. Plates are incubated at 22–25°C for 48 h and checked for the development of small colonies producing oxidase and brown pigmentation. *A. salmonicida* grows sparsely on selective media for *Enterobacteriaceae*. No specific selective medium has been thus far developed for the isolation of *A. salmonicida*. In both diseased fishes and healthy carriers, the kidney is the organ from which *A. salmonicida* is most readily isolated (usually in pure culture).

Strains of motile *Aeromonas* species can be isolated on nutrient agar or on Trypticase soy agar. Motile aeromonads also grow well on some selective media for *Enterobacteriaceae* (Drigalski agar, McConkey agar): The colonies are usually lactose-negative on these media, but a few strains may develop lactose-fermenting colonies. The growth of these bacteria is usually inhibited on thiosulfate-citrate-bile salt-sucrose (TCBS) agar, a selective medium for the isolation of *Vibrio cholerae*. Specific selective media have been proposed for the isolation of motile aeromonads from stools. Von Graevenitz and Zinterhofer (1970) described a "supplemented DNase medium" based on deoxyribonuclease production by *Aeromonas* strains. "Shotts-Rimler agar" (Shotts and Rimler, 1973) containing novobiocin, and "pril-xylose-ampicillin agar" (Rogol et al., 1979) were also devised for the isolation of motile *Aeromonas* species.

Maintenance Procedures

Strains of *Aeromonas* can be maintained on Trypticase soy agar: after incubation to allow good growth, the cultures may be kept in a refrigerator (4–8°C) for at least 1 month. They may also be preserved by freeze-drying.

Differentiation of the genus **Aeromonas** from other genera

The characteristics that distinguish *Aeromonas* from other morphologically or physiologically similar genera are presented in Table 5.57 of the article on the genus *Vibrio* in this *Manual*. Characteristics that differentiate *Aeromonas* species from *Plesiomonas shigelloides* are indicated in Table 5.65.

Taxonomic Comments

Members of the genus *Aeromonas* are now clearly differentiated from members of the *Enterobacteriaceae* (Report of the *Enterobacteriaceae* Subcommittee, 1958) and from members of the genera *Pseudomonas* and *Vibrio* (Eddy, 1960 and 1962; Schubert, 1960, 1964 and 1969); however, *Aeromonas* classification at a specific level is not yet unequivocally established, even though *A. salmonicida* and motile *Aeromonas* species form two clearly distinguished clusters.

On the basis of morphological, cultural, physiological and biochemical characteristics, Smith (1963) proposed the new genus "*Necromonas*" to accomodate *A. salmonicida* strains. The name "*Necromonas salmonicida*" was suggested as an alternative to "brown-pigmented strains of *A. salmonicida*," and "*Necromonas achromogenes*" as an alternative to "non-pigmented strains of *A. salmonicida*". DNA homology studies do not support this proposition, however, since pigmented and nonpig-

mented *A. salmonicida* strains show a fairly high degree of homology with motile *Aeromonas* species (MacInnes et al., 1979). Moreover the generic name "*Necromonas*" was not retained in the last edition of the *Manual* and did not appear in the Approved Lists of Bacterial Names. For the time being, "*N. salmonicida*" and "*N. achromogenes*" are synonymous with *A. salmonicida* and its nonpigmented variants.

Schubert (1967, 1969) proposed the differentiation of *A. salmonicida* into three subspecies: *A. salmonicida* subsp. *salmonicida*, *A. salmonicida* subsp. *achromogenes* and *A. salmonicida* subsp. *masoucida*. These three subspecies can be distinguished by biochemical characters (Table 5.65). But *A. salmonicida* subsp. *salmonicida* and *A. salmonicida* subsp. *masoucida* appear to be a genetically homogenous group by DNA homology studies (MacInnes et al., 1979). Although there is no evidence that *A. salmonicida* subsp. *masoucida* and possibly other biochemically atypical isolates of *A. salmonicida* warrant their present subspecies status, the name of these three subspecies was retained in the Approved Lists of Bacterial Names.

Phage typing can be used for epidemiological typing of *A. salmonicida*. Phages have been isolated from water and sewage, or from lysogenic isolates (Christison et al., 1938; Paterson et al., 1969; Popoff and Vieu, 1970). *A. salmonicida* phages can be divided into three

Table 5.65.
Differentiation between **Aeromonas hydrophila, Aeromonas caviae, Aeromonas sobria, Aeromonas salmonicida** *and* **Plesiomonas shigelloides**[a]

Characteristics	1. A. hydrophila	2. A. caviae	3. A. sobria	4. A. salmonicida subsp.			P. shigelloides
				salmonicida	achromogenes	masoucida	
Motility	+	+	+	−	−	−	+
Monotrichous flagellation in liquid medium	+	+	+	−	−	−	−
Lophotrichous flagellation in liquid medium	−	−	−	−	−	−	+
Coccobacilli in pairs, chains and clumps	−	−	−	+	+	+	−
Rods in singles and pairs	+	+	+	−	−	−	+
Brown water-soluble pigment	−	−	−	+	−	−	−
Growth in nutrient broth at 37°C	+	+	+	−	−	−	+
Indole production in 1% peptone water	+	+	+	−	+	+	+
Esculin hydrolysis	+	+	−	+	−	+	
Growth in KCN broth (Møller technique)	+	+	−	−	−	−	−
L-Histidine and L-arginine utilization	+	+	−	−	−	−	
L-Arabinose utilization	+	+	−	+	−	+	−
Fermentation of salicin	+	+	−	d	d	d	−
Fermentation of sucrose	+	+	+	−	+	+	−
Fermentation of mannitol	+	+	+	+	+	+	−
Breakdown of inositol	−	−	−	−	−	−	+
Acetoin from glucose (Voges-Proskauer)	+	−	d	−	−	+	−
Gas from glucose	+	−	+	+	−	+	−
H₂S from cysteine	+	−	+	−	−	+	

[a] Symbols: +, typically positive; −, typically negative; d, differs among strains.

morphological groups and 10 serological types. The phage typing set is based upon studies with eight phages. Fourteen phage types have been defined by this set of phages (Popoff, 1971).

The classification of motile aeromonads into species is complex. The controversy arises from the sharp discordance which exists between the views of several workers. The opinion of Ewing et al. (1961), Eddy and Carpenter (1964) and McCarthy (1975) favors a single species for all motile aeromonads. Schubert (1967, 1969) recognizes two separate species with various subspecies and biotypes. Popoff et al. (1981) divide motile aeromonads into three species.

There is phenotypic and genetic evidence for division of motile aeromonads into three species, namely *A. hydrophila, A. caviae* and *A. sobria.* The differential phenotypic characteristics (Popoff and Véron, 1976) are indicated in Table 5.65. The composition of core oligosaccharides from motile aeromonads was analyzed by Shaw and Hodder (1978); the strains were examined with respect to the hexose and heptose monosaccharide residues present in the core region of the cell wall lipopolysaccharides. On the basis of the various combinations of hexose and heptose residues, motile aeromonads are divided into three distinctly separate groups which correspond respectively to *A. hydrophila, A. caviae* and *A. sobria.* From DNA hybridization studies using the S1-nuclease method (Popoff et al., 1981), *A. hydrophila, A. caviae* and *A. sobria* exhibit interspecies DNA/DNA homology values of 35–50% with 8–12% divergence. These DNA hybridization values are consistent with the criteria required to define groups at a species level (Brenner et al., 1972) and they support the differentiation between *A. hydrophila, A. caviae* and *A. sobria* at a species level rather than at a subspecies level. Each of these three species contains more than one

DNA hybridization group: three hybridization groups can be delineated in *A. hydrophila,* two groups in *A. caviae,* and at least two groups in *A. sobria.* Within each of these hybridization groups, strains exhibit homology values of 70–97% (with 1–4% divergence) to the reference strain. Unfortunately, these DNA hybridization groups within a species are so far biochemically undistinguishable from one another. Pending further phenotypic studies, it seems desirable to split motile aeromonads into three nomenspecies (*A. hydrophila, A. caviae* and *A. sobria*) until the taxonomic significance of the DNA hybridization groups within these species can be stated precisely.

A Gram-negative halophilic bacterium was isolated from the intestine of a marine isopod (*Limnoria tripunctata*) by Merkel and Traganza (1958). On the basis of its polar flagellation, fermentative action on carbohydrates and positive reaction in the oxidase test, it was classified in the genus *Aeromonas* as a new species, "*Aeromonas proteolytica*" (Merkel et al., 1964). This classification was supported by Schubert (1969) who renamed the organism *A. hydrophila* subsp. *proteolytica.* However this strain differs from motile *Aeromonas* species as follows: (a) phenotypically, by its sodium ion requirement; and (b) genotypically, by a mol% G + C content of its DNA of 50% (Baumann et al., 1971; McCarthy, 1975, Popoff and Véron, 1976) which is outside the reported range for motile aeromonads (57–62%). These differences, together with a failure to react with six *A. hydrophila* antisera (McCarthy, 1975), are convincing reasons for excluding *A. hydrophila* subsp. *proteolytica* from the genus *Aeromonas. Editorial note:* In this edition of the *Manual* the organism is classified in the genus *Vibrio,* i.e. *V. proteolyticus.*

Differentiation and characteristics of species of **Aeromonas**

The differential characteristics of the species of *Aeromonas* are presented in Table 5.65. Other characteristics of the species are presented in Table 5.66.

List of the species of the genus **Aeromonas**

1. **Aeromonas hydrophila** (Chester 1901) Stanier 1943, 213.[AL] (*Bacillus hydrophilus* Chester 1901, 235.)

hy.dro'phi.la. Gr. n. *hydro* water; Gr. adj. *philos* loving; M.L. adj. *hydrophila* water-loving.

Table 5.66.

Other characteristics of motile **Aeromonas** *species (species 1–3) and* **Aeromonas salmonicida** *(species 4)[a]*

Characteristics	Motile Aeromonas Species	A. salmonicida
Oxidase	+	+
NO_3^- reduced to NO_2^-	+	+
Lysine decarboxylase, Møller's medium	d	d
Ornithine decarboxylase, Møller's medium	−	−
Arginine dihydrolase, Møller's medium	+	+
Tryptophan and phenylalanine deaminases	−	−
Urease	−[b]	−
Starch, gelatin, DNA and RNA hydrolysis	+	+
Tween 80 esterase	+	+
Citrate (Simmons')	d	−
Citrate (Christensen's)	d	−
Growth in peptone water without NaCl	+	+
ONPG test	+	d
Fermentation of maltose, galactose and trehalose	+[b]	+[b]
Fermentation of cellobiose, lactose and sorbitol	d	−
Fermentation of dulcitol, rhamnose, inositol, xylose, raffinose and adonitol	−	−
Breakdown of malonate, mucate and D-tartrate	−	−
Fermentation of glycerol	d	d
Tetrathionate reductase	d	−

[a] For symbols see Table 5.65.
[b] Aberrant strains occur.

Straight rods, 0.3–1.0 μm in diameter and 1.0–3.5 μm in length. Motile by a single polar flagellum in liquid medium; peritrichous flagella may occur on solid media in young culture. Not encapsulated.

Optimum growth temperature, 28°C. On nutrient agar, colonies are white to buff, circular and convex with an entire margin.

Physiological and nutritional characteristics are presented in Tables 5.65 and 5.66. The strains can use L-histidine, L-arabinose, L-arginine and salicin as sole carbon sources. They hydrolyze esculin, grow in KCN medium, ferment salicin, produce gas and acetoin from glucose, and produce H_2S from cysteine.

May be pathogenic for frogs, fish and mammals (including man). Found in fresh water and sewage.

The mol % G + C of the DNA ranges from 58–62 (Bd, T_m).

Type strain: ATCC 7966.

2. **Aeromonas caviae** (ex Eddy 1962, 145.) nom. rev.

ca'vi.ae. M.L. fem. n. *Cavia* generic name of guinea pig; M.L. gen. n. *caviae* of guinea pig.

Morphological and cultural characteristics are as indicated for *A. hydrophila.*

Physiological and nutritional characteristics are presented in Tables 5.65 and 5.66. The strains use L-histidine, L-arabinose, L-arginine and salicin as sole carbon sources. They hydrolyze esculin, grow in KCN

medium, ferment salicin, and do not produce gas or acetoin from glucose or H_2S from cysteine.

Found in fresh water and sewage, and on fishes.

The mol % G + C of the DNA ranges from 61–63 (Bd, T_m).

Type strain: ATCC 15468.

3. **Aeromonas sobria** Popoff and Véron 1981, 215.[VP] (*Effective publication*: Popoff and Véron 1976, 20.)

so.bri'a. M.L. fem. adj. *sobria* moderate.

Morphological and cultural characteristics are as indicated for *A. hydrophila.*

Physiological and nutritional characteristics are presented in Tables 5.65 and 5.66. Strains do not use L-histidine, L-arabinose, L-arginine and salicin as sole carbon sources. They do not hydrolyze esculin, do not grow in KCN medium, and do not ferment salicin. They do produce gas from glucose and H_2S from cysteine.

Found in fresh water and sewage, and on fishes.

The mol % G + C of the DNA ranges from 58–60 (T_m).

Type strain: CIP 7433.

4. **Aeromonas salmonicida** (Lehmann and Neumann 1896) Griffin, Snieszko and Friddle 1953, 138.[AL] (*Bacterium salmonicida* Lehmann and Neumann 1896, 240.)

sal.mon.ic'i.da. L. n. *salmo, salmonis* salmon; L. suff. *-cida* from L. v. *caedo* cut or kill; M.L. n. *salmonicida* salmon-killer.

Coccobacilli; length less than twice the width. In nutrient broth, pairs, chains and clumps are usually seen in phase-contrast preparations. Nonmotile. Not encapsulated.

Optimum growth temperature, 22–25°C. Colonies are circular, raised, translucent and friable.

Physiological and nutritional characteristics are presented in Tables 5.65 and 5.66.

Found on, and pathogenic to, salmonid fishes, causing furunculosis. May also cause serious infections in other fish. Not found in surface waters.

The mol % G + C of the DNA ranges from 57–59 (Bd, T_m).

4a. **Aeromonas salmonicida** subsp. **salmonicida** (Lehmann and Neumann 1896) Griffin, Snieszko and Friddle 1953, 138.[AL]

Strains produce a brown water-soluble pigment on media containing 0.1% tyrosine or phenylalanine. They do not produce indole. They do hydrolyze esculin and ferment mannitol.

Type strain: NCMB 1102.

4b. **Aeromonas salmonicida** subsp. **achromogenes** (Smith 1963) Schubert 1967, 278.[AL] (*Necromonas achromogenes* Smith 1963, 273.)

a.chro.mo.gen'es. Gr. adj. *achromos* colorless; Gr. v. *gennaio* produce; M.L. adj. *achromogenes* not producing color.

Strains do not produce brown water-soluble pigment. They may produce indole. They do not hydrolyze esculin and do not ferment mannitol.

Type strain: NCMB 1110.

4c. **Aeromonas salmonicida** subsp. **masoucida** Kimura 1969, 52.[AL]

ma.sou.ci'da. Japanese noun *masou* specific epithet of *Oncorhynchus masou*; L. v. suff. *-cida* from L. v. *caedo* cut or kill; M.L. fem. n. *masoucida Oncorhynchus masou*-killer.

Strains do not produce brown water-soluble pigment. They do produce indole, hydrolyze esculin, and ferment manitol.

Type strain: ATCC 27013.

Genus IV. **Plesiomonas** Habs and Schubert 1962, 324[AL]

RALPH H. W. SCHUBERT

Ple.si.o.mo'nas. Gr. masc. n. *plesios* neighbor; Gr. fem. n. *monas* unit, monad; M.L. fem. n. *Plesiomonas* neighbor monad (to *Aeromonas*).

Cells **round-ended, straight, rod-shaped**, 0.8–1.0 × 3.0 μm. Resting stages not known. Gram negative. **Motile by polar flagella,** generally lophotrichous. **Facultatively anaerobic.** Chemoorganotrophic, having **both a respiratory and a fermentative type of**

metabolism. Carbohydrates are catabolized with production of acid but no gas. Most strains grow on mineral media containing ammonium salts as a sole nitrogen source and glucose as a sole source of carbon. **Oxidase* and catalase reactions are positive. Negative for diastase, lipase, proteinases. Positive for lysine, ornithine and arginine decarboxylases** (Møller technique). Most strains are **sensitive to vibriostatic agent 0/129** (2,4-diamino-6,7-diisopropyl pteridine). Occur in fish and other aquatic animals and in a variety of mammals; probably does not belong to the normal intestinal flora of man, but can cause diarrhea in man. The mol% G + C of the DNA is 51 (Ch) (Sebald and Véron, 1963).

Type species: *Plesiomonas shigelloides* (Bader 1954) Habs and Schubert 1962, 324.

Further Descriptive Information

Cells of *Plesiomonas shigelloides* typically appear as straight, sometimes rather long, rods or even filaments. No microcysts occur. No granules of intracellular poly-β-hydroxybutyrate are present. The organisms are usually motile, but nonmotile flagellated and nonmotile atrichous strains are known to occur (Ewing et al., 1961). The flagella are long and polar (Habs and Schubert, and have an undulating wavelength which averages 3.5–4.0 μm in stained preparations (Ewing et al., 1961); in electron micrographs a short wavelength of 2.0–3.0 μm and a long wavelength of 2.3 and 9.0 μm occurs (Schubert, 1963). In young cultures (2- to 4-h-old) lateral flagella occur in addition to polar flagella; these lateral flagella definitely have a shorter wavelength (less than 1.7 μm in stained preparations) than the polar flagella (Ewing et al., 1961).

Plesiomonas shigelloides grows in peptone broth with uniform turbidity; no sediment or pellicle is formed. On nutrient agar or blood agar at 24 h, colonies are 1.0–1.5 mm in diameter, grayish, shiny and opaque, with a slightly raised center and a smooth surface and entire edge. In older cultures, complex variations are observed in colony morphology (Sakazaki et al., 1959; Habs and Schubert, 1962). No water-soluble fluorescent or brown pigment is produced. Optimum growth occurs between 37 and 38°C; the maximum growth temperature is between 40 and 44°C, while the minimum temperature is 8°C. No growth occurs in nutrient broth containing 7.5% NaCl. The pH range for growth is 5.0–7.7; some strains can grow at pH 8.0. No growth occurs at pH 3.0.

Glucose, maltose, trehalose, inositol and glycerol are fermented with acid but no gas. The following carbohydrates are not fermented; starch, dextrin, glycogen, mannitol, fructose, sucrose, arabinose, esculin, raffinose, cellobiose, salicin, sorbitol, inulin, melezitose, rhamnose, xylose, dulcitol and adonitol. Catabolism of lactose, galactose, mannose, salicin and chitin is variable.

The methyl red reaction is variable. The Voges-Proskauer test is negative and the butanediol dehydrogenase reaction is negative (Schubert and Kexel, 1964). *Other negative tests:* gluconate, malonate, citrate utilization; growth with KCN; production of H_2S; gelatin liquefaction; casein digestion; fibrinolysin activity; elastase activity; lecithinase activity; urease; and phenylalanine deaminase. No filterable hemolysin is produced.

Indole is formed. Phosphatase activity occurs.

Only a few *P. shigelloides* strains share a common O antigen with *Shigella sonnei* (Sakazaki et al., 1959); 57 strains studied serologically were distributed in 16 O groups by Quincke (1957), and only 1 of these groups showed antigenic relationships with *S. sonnei*. Shimada and Sakazaki (1978) defined 30 O antigen groups, including some somatic antigén groups also found in the genus *Shigella*, and 11 H antigens. Whang et al. (1972) showed that strains of *P. shigelloides* possess the common antigen of *Enterobacteriaceae* (CA).

Antibiotic sensitivity testing of 12 *P. shigelloides* strains indicated sensitivity to all compounds tested: ampicillin, tetracycline, chloramphenicol, cephalothin, cefotaxim, mezocilin and sulfamethoxazole (Schubert, unpublished results).

With few exceptions (e.g. Ellner and McCarthy, 1973), *P. shigelloides* has not been isolated from wounds or inflammatory processes. The species does, however, play a role as a pathogen of the human intestine. Cases of diarrhea in various degrees of severity are increasingly reported. Symptoms range from loose stools to watery, choleriform excrements. A number of epidemic outbreaks of diarrhea attributed to *P. shigelloides* as the causative agent have been reported from Africa, India and Japan (Vandepitte et al., 1974; Vandepitte et al., 1980; Bhat et al., 1974; Sanyal et al., 1975; Tsukamoto et al., 1978). Descriptions of well documented isolated cases exist equally from temperate climates (e.g. Jandl and Linke, 1976; Davis et al., 1978; Cooper and Brown, 1968; Zajc-Satler et al., 1972). Apparently *P. shigelloides* does not belong to the normal intestinal flora of man (Nakanishi et al., 1969; Catsaras and Buttiaux, 1965); in only a few instances has man been found to be a symptomless carrier of the organism (e.g. Wuthe, 1972; Gravenitz and Mensch, 1968). In various animals a carrier state is more frequent: *Plesiomonas* has thus far been isolated from fish, other aquatic animals, and from mammals such as swine, dogs, cats, goats, sheep and monkeys (Arai et al., 1980; Vandepitte et al., 1980). Some animals, however, develop symptoms of disease (e.g. Davis et al., 1978; Bader, 1954). Our present knowledge seems to indicate that the reported infections are attributable to contact—direct or indirect—with contaminated surface water.

Research on the ecology and on the epidemiology of *P. shigelloides* has been strongly stimulated by incidental findings of the organism in aquatic animals in warm climates (see Vandepitte et al., 1980). These findings have led to a systematic search for the organism in the intestine of fish in various parts of the world (Arai et al., 1980; Vandepitte et al., 1980; Schubert, 1981). After the presence of the organism in the large bowel had been established in a high percentage of the fish examined, subsequent detection of the organism in surface water followed (Arai et al., 1980; Schubert, 1981). Seasonal variations in the numbers of organisms detectable in surface water samples, as observed in Japan (Arai et al., 1980) and in Europe (Schubert, 1981) where positive findings were obtained only in the warm season, are explained by the fact that the bacteria do not multiply at temperatures below 8°C (Schubert, 1981). *P. shigelloides* can still be isolated in declining numbers for some time after the onset of the cold season (at temperatures between 3 and 7°C) when very sensitive methods are used (Schubert, unpublished results); after prolonged spells of cold weather, however, even these methods fail to produce positive results (Schubert, unpublished results).

Enrichment and Isolation Procedures

In most countries, diagnostic laboratory procedures used in diarrhea concentrate on the detection of *Salmonella* and *Shigella*; consequently, most observations of *P. shigelloides* have been made on media developed for the primary culture of *Salmonella* and *Shigella*. However, most of these media are not ideally suited to the culture of *P. shigelloides* because the selective reagents incorporated are toxic for most strains (Schubert, 1977).

Good results may be obtained on SS agar and IBB† agar. The latter medium offers the additional advantage of higher electiveness. IBB agar favors the growth of *P. shigelloides* by providing inositol as a carbon source which can be utilized by a few competing bacteria. After 48 h *P. shigelloides* colonies are easily distinguished from others in mixed cultures: they become generally about 1 mm larger than colonies on SS agar, whitish in appearance or with various degrees of red coloration depending on their distance from other colonies (due to acid formation from inositol in the presence of neutral red indicator). Acid

---

* As, the Nadi-reaction (α-naphthol and dimethyl-*p*-phenylenediamine reagents) (Schultze, 1910).

† Inositol-brilliant green-bile salts (IBB) agar has the following composition (g/liter): Proteose peptone (Difco), 10.0; meat extract (Lab Lemco (Oxoid)), 5.0; NaCl, 5.0; Bile salts No. 3 (Difco), 8.5; brilliant green (Merck), 0.00033; neutral red (Merck), 0.025; *meso*-inositol (Merck), 10.0; and agar (Difco), 15.0; pH adjusted to 7.2.

formation tends to be so weak that even single colonies will give a somewhat belated cytochrome oxidase reaction after receiving a drop of Nadi's reagent (Schubert, 1977).

Maintenance Procedures

Plesiomonas strains are easily maintained for several years in par-

affin-sealed stab cultures on trypticase soy agar. Cultures may also be preserved by lyophilization.

Procedures for Testing for Special Characters

For the demonstration of the properties of *Plesiomonas* strains, the same methods are used as those employed for the characterization of *Aeromonas*, *Vibrio* and members of the family *Enterobacteriaceae*.

Differentiation of the genus **Plesiomonas** from other genera

See Table 5.57, the genus *Vibrio*, for characteristics that differentiate *Plesiomonas* from the other genera of the family *Vibrionaceae* (*Vibrio*, *Photobacterium* and *Aeromonas*).

Taxonomic Comments

Ferguson and Henderson (1947) assigned their strain "C 27" (now a reference strain of *P. shigelloides* designated ATCC 14030) to "Paracolon" of the family *Enterobacteriaceae*, but did not assign a specific epithet. The present specific epithet *shigelloides* is due to Baker (1954) who, because of the polar flagellation of the organism, classified the species in the genus *Pseudomonas*. Unlike *Pseudomonas*, however, the species was capable of fermentation and was therefore transferred to the genus *Aeromonas* by Ewing et al. (1961). This was followed by its

transfer to the newly created genus *Plesiomonas* (Habs and Schubert, 1962). The creation of this genus was inevitable because *P. shigelloides* strains do not exhibit some essential features of either *Aeromonas* or *Vibrio*, the hitherto known genera of oxidase positive, fermentative, polarly flagellated, Gram-negative rods. *Plesiomonas* is characterized by two to five lophotrichous polar flagella, whereas *Vibrio* and *Aeromonas* species are usually monotrichous. Moreover, *Plesiomonas* lacks exoenzymes whereas *Vibrio* and *Aeromonas* generally produce lipase, deoxyribonuclease (DNase) and various proteinases such as gelatinase and caseinase. In addition, *Plesiomonas* ferments inositol, whereas this trait is rare in *Aeromonas* and *Vibrio*. *Plesiomonas* also differs from members of *Aeromonas* and *Vibrio* by the restricted range of carbohydrates it ferments (Eddy and Carpenter, 1964) and by its DNA base composition.

List of the species of the genus **Plesiomonas**

1. **Plesiomonas shigelloides** (Bader 1954) Habs and Schubert 1962, 324.[AL] (*Pseudomonas shigelloides* Bader 1954, 455.)

shi.gel.loi′des. M.L. fem. n. *Shigella* a generic name; Gr. suffix *eides* similar; M.L. adj. *shigelloides* Shigella-like.

The description is as given for the genus.

Occur in fish and other aquatic animals and in a variety of mammals; probably do not belong to the normal intestinal flora of man but can cause diarrhea in man.

The mol% G + C of the DNA is 51 (Ch).

Type strain: ATCC 14029; NCIB 9242.

FAMILY III. **Pasteurellaceae** POHL 1981a, 382.[VP] (*Effective publication*: Pohl 1979, 81.)

WALTER MANNHEIM

Pas.teu.rel.la′ce.ae. M.L. fem. n. *Pasteurella* type genus of the family; -*aceae* suffix to denote a family; M.L. fem. pl. n. *Pasteurellaceae* the *Pasteurella* family.

Straight rigid, **coccoid to rod-shaped cells, usually 0.2–0.3 × 0.3–2.0 µm. Pleomorphism** with cell swelling and formation of filaments may occur. Nonsporeforming. **Gram-negative. Nonmotile. The cells contain demethylmenaquinones;** ubiquinones may or may not be produced. **Aerobic with varying degrees of microaerophilia, facultatively anaerobic.** Mesophilic. **Chemoorganotrophic, with both respiratory and fermentative types of metabolism.** Acid is produced by fermentation of glucose, other carbohydrates, sugar alcohols or glycosides; however, conventional fermentation test media may fail to detect the accumulation of acid fermentation products with some of the most fastidious members of the family. Usually anaerogenic, but gas-producing species do occur. As a rule, fumarate is used as a terminal electron acceptor in demethylmenaquinone-mediated anaerobic respiration. **Nitrites are formed from nitrates. Oxidase, catalase and alkaline phosphatase reactions are characteristically positive,** but negative reactions occur in some species. Complex media supplemented with yeast extract and serum or whole blood lysate are used for primary isolation. Require organic nitrogen sources. **Varying patterns of nutritional requirements may include several amino acids, B vitamins, β-nicotinamide adenine nucleotides, and haematin or protoporphyrin.** Usually sensitive to benzylpenicillin and other β-lactam antibiotics, inhibitors of 70S ribosomal protein biosynthesis, sulfonamides, trimethoprim, erythromycin and colistin. **Parasitic in vertebrates,** particularly

mammals and/or birds. The mol% G + C of the DNA is 38–47 (T_m). Genome size is 1.2–2.2×10^9 daltons.

Type genus: *Pasteurella* Trevisan 1887, 88.

Circumscription, position and rank. The description given above comprises the recognized species of the genera *Actinobacillus*, *Haemophilus* and *Pasteurella*. On the basis of a numerical phenetic analysis of a limited number of species, Sneath and Johnson (1973) indicated that the three genera could form a family. This has been confirmed by recent phenotypic comparisons (Mannheim et al., 1980) and by DNA/DNA reassociation studies (Pohl, 1979; Pohl and Mannheim, unpublished results) using the optical method for determining quantitative renaturation rates (De Ley et al., 1970). Single-linkage clustering of the DNA hybridization data reveals complex intra- and intergeneric relatedness patterns at or above the 30% DNA/DNA reassociation level.

The family rank of the group seems justified mainly by the following reasons. (a) The taxonomic structure of the group emerging from the genetic relatedness patterns allows discrimination of specific, generic and tribus-like structures that are generally in accordance with the traditional species concepts. (b) Classification of the group as a family will produce structures and ranks of subgroups closely resembling those of the related family, *Enterobacteriaceae*. (c) The creation of a family permits one to conserve the well-known generic names *Actinobacillus*, *Haemophilus* and *Pasteurella* and to use the oldest recognized genus name within the group, *Pasteurella*, as a basis for the name of the family, thereby following the principle of nomenclatural stability.

In the 16S rRNA oligonucleotide classification by Woese and co-workers (Fox et al., 1980), *Pasteurella* clusters together with other Gram-negative, chemoorganotrophic derivatives of the purple photosynthetic bacteria branch that are capable of both fermentative glycolysis and respiration (i.e. members of the families *Enterobacteriaceae* and *Vibrionaceae*). In accordance with these data, DNA/rRNA hybridization experiments have located representative members of *Pasteurellaceae* in a particular branch of the first Gram-negative superfamily as defined by De Ley and co-workers (De Ley et al., unpublished data). Undoubtedly, the delineation of the family *Pasteurellaceae* from the neighboring families deserves further investigation. Although measurable DNA/DNA reassociation at permissive conditions has been obtained between strains of *Pasteurella multocida* and *Yersinia pseudotuberculosis* (Brenner et al., 1976), no evidence for closer relationships detectable by this method has been found in a comparison of the DNA obtained from representative members of *Enterobacteriaceae* and *Pasteurella* (Ursing, 1981); in particular, no measurable DNA/DNA reassociation has been obtained between members of *Enterobacteriaceae* and some gas-producing pasteurellas that have been considered as taxonomic intermediates (McAllister and Carter, 1974).

The description of the family does not exclude organisms that are phenotypically related and exhibit similar DNA base compositions, but have not been associated so far with the family at the genotypic level (e.g. so-called "*Actinobacillus seminis*" (Baynes and Simmons, 1960)). The pasteurella-like fish pathogen described by Janssen and Surgalla (1968) deviates from the above description of *Pasteurellaceae* by its larger genome, its cryophilia, its lack of nitrate reductase and its unusual host range (Pohl, 1979; Mannheim et al., 1980; nevertheless, it does display some DNA base sequence relatedness to several members of *Pasteurellaceae*; this remains to be studied in detail. The relationships of pasteurella-like organisms with a mol% G + C of 50 or greater, e.g. the "SP" group (Frederiksen, 1982) and the EF-4 organisms (Holmes, 1981), remain to be explored.

Identification and differentiation. Members of the family *Pasteurellaceae* are easily isolated from mucous membranes, tissues or secretions of their animal hosts using a suitable chocolate agar* with 1–3 days of incubation at 36°C in a candle jar. Unless typical strains of well-known, established species are found, the isolate should be carefully characterized by an appropriate set of tests (see Frederiksen, 1973; Kilian, 1976; Mannheim et al., 1980). Nonmotile and nutritionally exacting variant strains of *Enterobacteriaceae* or *Vibrionaceae*, Gram-labile fastidious *Bacillus*, or aerotolerant *Clostridium* strains are sometimes mistaken for one or another of the species of *Pasteurellaceae*. Even nonfermentative organisms may be confused with *Pasteurellaceae* due to the technical limitations of conventional fermentation tests (cf. Holländer, 1976a). Oxidase and phosphatase reactions and sensitivity to benzylpenicillin are useful differential criteria in many instances (Mannheim et al., 1980); however, there is no general agreement about the optimal methods for the detection of oxidase and phosphatase activities (cf. Henriksen and Jyssum, 1961; Kilian, 1976; Gadberry et al., 1980; Mannheim et al., 1980; Bercovier et al., 1981). Among the newer cytochemical criteria, the respiratory quinones have proven to be useful in defining the family (Holländer and Mannheim, 1975; Mannheim et al., 1978; Mannheim et al., 1980; Holländer et al., 1981). Wild-type strains of *Pasteurellaceae* use demethylmenaquinone as their main respiratory quinone in anaerobic and aerobic electron transport (Holländer, 1976b), although ubiquinones are also produced under aerobic conditions by most species. Members of the family *Pasteurellaceae* do not produce menaquinones, whereas *Enterobacteriaceae* and *Vibrionaceae* may contain menaquinones, demethylmenaquinones and ubiquinones. Further lipoquinone patterns that exclude an organism from *Pasteurellaceae* are: (a) no quinones (in the micromolar order of mag-

nitude per gram of cell protein), as with *Clostridium* and other anaerobic Gram-labile rods, and as with *Gemella* (Hess et al., 1979); (b) menaquinones as sole respiratory quinones, as in aerobic Gram-positive and anaerobic Gram-negative rods, *Cytophagaceae*, and so-called "*Pasteurella anatipestifer*", and (c) ubiquinones as sole respiratory quinones, e.g. as in most of the strictly aerobic and some facultative Gram-negative bacteria including *Cardiobacterium*, *Eikenella*, some *Erwinia* species, *Haemophilus* (*Gardnerella*) *vaginalis*, and so-called *Haemophilus piscium* (Mannheim et al., 1980; Holländer, 1980; and others). Other cytochemical traits such as cellular fatty acids, proteins, or pyrolysis mass spectral patterns may facilitate the differentiation of the family in the future.

Major subdivisions. Although genetic relationships among most of the nomenclatural type strains of recognized *Actinobacillus*, *Haemophilus* and *Pasteurella* species, and a limited number of additional strains, have been determined (Pohl, 1979; Pohl, 1981b; Mannheim and co-workers, unpublished data), the present data matrix does not allow a description of the entire taxonomic structure of the family. However, the molecular data support the view that the three traditional genera are heterogenous and should be redefined in terms of genetic relatedness. If genera are confined to about 50% DNA base sequence homology clusters containing the respective type species, several additional clusters of equal rank become candidates for further genera. Some of the genus-like clusters are interrelated at about the 40% level and may form tribus-like groups. One tribe (which should receive the name *Pasteurelleae*) unites (a) the redefined genus *Pasteurella*, consisting of *P. multocida* and an unnamed species (so-called *Pasteurella* "gas", or type Henriksen of *P. pneumotropica*); *P. gallinarum* and *Haemophilus avium*; (b) genus *Haemophilus* sensu stricto, so far containing only *H. influenzae* and *H. aegyptius*, and (c) *Haemophilus parasuis*. Another tribe ("*Actinobacilleae*") would contain three generic groups, one being *Actinobacillus* sensu stricto and consisting of a large DNA homology group that comprises, in addition to *A. lignieresii* and *A. equuli*, *P. ureae* (as suggested by Jones, 1962), *Haemophilus pleuropneumoniae* (Pohl et al., unpublished data), *A. suis*, *A. capsulatus*, the unnamed porcine *Actinobacillus* species of Ross et al. (1972), and several yet unrecognized species. *P. haemolytica* (all biovars and serovars) is situated in the close vicinity of *Actinobacillus* spp. (in accordance with the proposal of Mráz, 1969). *A. actinomycetemcomitans* is less closely related to *A. lignieresii* and probably represents a separate genus. Further candidates for separate genera are the *Haemophilus aphrophilus*/*H paraphrophilus*/*H. paraphrohaemolyticus* group and the *H. parainfluenzae* group. Several other species have not yet been affiliated to genus-like DNA homology groups. The genetic relatedness patterns show clearly that some of the outstanding nutritional requirements, such as that for the V-factor, are not necessarily generic features; the groups can, however, be redefined by careful analysis of catabolic reactions. The majority of recognized *Pasteurella* species so far studied have clustered at or above the 70% DNA homology level; in some cases, lumping or splitting of species, recognition of new species or subspecies, and nomination of new type strains on the basis of both molecular and phenotypic data will be necessary.

Further Remarks

The molecular data mentioned above allow an understanding of the family *Pasteurellaceae* as a group of organisms that have lost much of the genome information of their free-living ancestors during phylogenetic adaptation to parasitic life, as had been postulated by Lwoff (1944). The extent and patterns of the resulting defects in their biosynthetic capabilities have been only partly studied, but have impaired the detection, phenotypic characterization and classification of those organisms in the past. Considering the present taxonomic situation of the family *Pasteurellaceae*, much of the descriptive work remains

---

* For example, Tryptic soy agar (Difco) supplemented with 5–10% (v/v) of sheep blood heated for 15 min at 80°C.

to be done. During the past two decades, a variety of new isolates from man and domestic animals has been described. Many of these isolates are awaiting recognition as species of *Pasteurellaceae*; others have been only poorly described, or misclassified. In wild mammals and birds, many hitherto undescribed members of *Pasteurellaceae* are also likely to exist.

Table 5.67.
Differentiation of members of the family **Pasteurellaceae**[a]

Characteristics	Pasteurella species						Haemophilus species								Actinobacillus species	
	1	2	3	4	5	6	1–3	4	5	6–10	11–16	"a"	"b"	"c"	1–4	5
Symbiotic growth with *Staphylococcus aureus* on blood agar	−	−	−	−	−	−	+	−	−	+	d	+[b]	−	−	−	−
V-factor requirement	−	−	−	−	−	−	+	−	−	+	d	−	−	−	−	−
δ-Aminolevulinic acid → porphyrins	+	+	+	+	+	+	−	−	−	+	+[c]	+	+	−[d]	+	+
Indole production	+	+	−	−	−	−	d	+	−	−[e]	−	+	−	−	−	−
Urease test	−	+	−	+	+	−	+[e]	−	−	d	−	−	+	−	+	−
Ornithine decarboxylase	+	+	−[e]	−	+	d	d	−	−	d	−	−	+	−	−	−
Hydrolysis of *o*-nitrophenyl-β-D-galactopyranoside (ONPG test)	d	+	d	−	+	d	−	d	−	d	d	d		−	+	−

[a] Based on Kilian and Frederiksen (1981). Symbols: see standard definitions.
[b] Satellitic growth occurs in an atmosphere of increased CO_2, but there is no response to either X or V factors.
[c] 90% or more of strains of *H. aprophilus* give a delayed positive reaction.
[d] Some growth stimulation occurs with X and XV factor disks, but not by V factor, on nutrient agar.
[e] Occasional strains may differ from this reaction.

Genus I. **Pasteurella** Trevisan 1887, 94,[AL] Nom. cons. Opin. 13, Jud. Comm. 1954, 153.

G. R. CARTER

Pas.teu.rel′la. M.L. dim. fem. n. *Pasteurella* named after Louis Pasteur.

Cells spherical, ovoid or rod-shaped, 0.3–1.0 μm in diameter and 1.0–2.0 μm in length. Occur singly or less frequently in pairs or short chains. **Bipolar staining is common,** especially in preparations made from infected animal tissues. Resting stages are not formed. Not acid-fast. Gram-negative. **Nonmotile.** Facultatively anaerobic, having a fermentative type of metabolism. Temperature range for growth, 22–44°C; optimum, 37°C. **Catalase-positive; almost always oxidase-positive. Nitrates are reduced to nitrites.** Gelatinase-negative. Methyl red- and Voges-Proskauer-negative. Lysine and arginine decarboxylase are not produced. Glucose and other fermentable compounds are fermented with the production of acid but usually no gas. **Parasitic on the mucous membranes of the upper respiratory and digestive tracts of mammals (rarely man) and birds.** The mol% G + C of the DNA is 40–45 (T_m) (Pohl, 1979).

Type species: *Pasteurella multocida* (sic) (Lehmann and Neumann 1899) Rosenbusch and Merchant 1939, 85.

Further Descriptive Information

Morphology. Members of the genus occur most frequently as coccobacilli or short rods. Filamentous forms of varying length are seen after strains have been subcultured several times. Some strains of *P. multocida* are strikingly pleomorphic on primary culture from clinical materials.

Most of the virulent strains of *P. multocida* and *P. haemolytica* produce capsules of varying size. These capsules, which are carbohydrate in nature, are frequently lost after several subcultures. The noncapsulated cells of canine and feline strains display varying degrees of acid-agglutination, whereas strains from cattle and swine do not (Smith, 1958).

The various pasteurellae form colonies that are similar in appearance. They are round and grayish and range from 1–3 mm in diameter after 24 h. One variety of *P. multocida* has large capsules consisting mainly of hyaluronic acid which imparts a moist, mucoid character to the colonies (Carter, 1958). Both *P. multocida* and *P. haemolytica* have been shown to dissociate readily, producing several different colonial

variants (Carter, 1957; Biberstein et al., 1958). In *P. multocida* mucoid, smooth and blue variants occur most commonly (Carter, 1957).

Only *P. haemolytica* produces a distinct zone of β-hemolysis. Hemolysis varies with different species of blood: bovine blood is preferred for optimum hemolysis with *P. haemolytica*. Organisms from poultry called *P. haemolytica* have larger zones of β hemolysis than *P. haemolytica* from mammals. They differ in other respects from the latter and probably constitute a distinct species. Yellow-pigmented strains are frequent among the *Pasteurella* "gas"/Henriksen-type of *P. pneumotropica* organisms, do occur in strains of the types Heyl and Jawetz of *P. pneumotropica*, and are occasionally observed in the "dog-type" strains of *P. multocida* (Schulz et al., 1977; Mannheim et al., 1980).

Certain physiological features. Although pasteurellae usually do not produce gas from carbohydrates, *P. aerogenes* and an organism referred to as *Pasteurella* n. sp. I "gas" (King, 1972) do produce gas. The latter organism is urease-positive and probably is a biovar of *P. pneumotropica*. *P. ureae*, *P. aerogenes*, and the "gas" organism are the only members of the genus that are urease-positive, and *P. haemolytic* and *P. aerogenes* are the only ones that are able to grow on MacConkey agar.

Scharmann et al. (1970) observed that 102 of 104 strains of *P. multocida* and 3 of 5 strains of *P. haemolytica* produced neuraminidase. Drzeniek et al. (1972) reported that the neuraminidase of *P. multocida* was bound to the cell and that it was inducible by *N*-acetyl-D-mannosamine and by free and bound sialic acid.

Antigenic properties. *P. multocida* and *P. haemolytica* have been the subject of numerous serologic and antigenic studies. Four different capsular types of *P. multocida*, designated A, B, C and D, were identified by a passive hemagglutination procedure (Carter, 1955). Type C was subsequently discarded as not being a valid type and an additional capsular variety, type E, was added (Carter, 1961). In later studies Namioka and Murata (1961) showed that strains of *P. multocida* possessed several different O or somatic types, based on the use of acid-treated cells and agglutinin-absorption procedures. They designated serovars (serotypes) by a number which indicated the O type, followed by a letter indicating the capsular type. For example, the important hemorrhagic septicemia serovar was designated 6:B. A total of 11 somatic or O groups were identified by Namioka and Bruner (1963). Heddleston et al. (1972) identified different serovars by means of a gel diffusion precipitin test which employed as antigen an extract obtained by heating cells to 100°C. They ultimately identified 16 somatic groups, and their system of identifying serovars was an extension of the earlier classification devised by Little and Lyon (1943).

Neither of the systems of identifying somatic antigens is altogether satisfactory, no doubt because of the number and variety of antigenic determinants associated with the endotoxin. It is clear that *P. multocida* consists of more than 16 serovars, some of which are associated with certain animal species and particular disease manifestations. Carter and Chengappa (1981a) have recommended that the systems developed by Carter (1955, 1961) and Heddleston et al. (1972) be combined and used as a standard way of designating serovars. A serovar in this system would be designated by its capsular type (A, B, D, or E) followed by an arabic number indicating the somatic type. The serovars B:2 and E:2 are the causative agents of hemorrhagic septicemia in cattle and water buffaloes. Serovars A:1 and A:3 are important causes of fowl cholera.

P. haemolytica has been classified serologically in much the same way as *P. multocida*. Carter (1956) employed indirect hemagglutination (IHA) to identify the principal capsular type from cattle with pneumonia, which was later designated type 1 (Biberstein et al., 1960). In a comprehensive study, passive hemagglutination and agglutination procedures were used to identify capsular and somatic types, respectively (Biberstein et al., 1960). The capsular types were designated by arabic numbers and the somatic types by capital letters. Capsular types were later identified by a simple slide agglutination test (Frank and Wessman, 1978). The distribution of serovars vis-à-vis diseases, hosts and biovars has been summarized by Biberstein (1978).

Despite many studies, the precise nature of the antigens of *P. multocida* has not been fully described. The capsular antigens or substances of capsular types B and E are acidic polysaccharides (Knox and Bain, 1962; Pennard and Nagy, 1976). The capsules of capsular type A organisms consist largely of hyaluronic acid (Carter and Annau, 1953; Carter, 1958). Several observations indicate that the serologic activity of the type A and D capsular substances is due to haptenes (Carter and Chengappa, 1981b). Based on immunoelectrophoretic antigenic analysis of capsular types B and E and other varieties of *P. multocida*, Prince and Smith (1966) identified three principal components associated with the capsules. These components and some of their characteristics are summarized as follows:

1. β-Antigen: type-specific polysaccharide; adsorbed to red cells in the IHA procedure; found in organisms that give rise to iridescent and mucoid variants.
2. α-Complex: probably a polysaccharide-protein complex; closely adherent to the cell wall; immunogenic; probably somewhat labile.
3. γ-Antigen: lipopolysaccharide found in organisms from all variants; makes up cell wall; each has one or more antigenic determinants responsible for different O or somatic serological varieties.

In contrast to *P. multocida*, the antigenic structure of *P. haemolytica* and other species in the genus has received little attention.

Pathogenicity in natural hosts. *P. multocida* has been recovered from many animal species, both wild and domesticated, normal and diseased. The major diseases of domestic animals caused by this species are the following (Carter, 1967):

Primary disease
1. Hemorrhagic septicemia of cattle and buffaloes. This important economical disease occurs mainly in tropical and subtropical countries and is caused by two serovars of *P. multocida*, neither of which has been reported to cause human infections.
2. Fowl cholera, an acute to chronic disease of chickens, turkeys, ducks, geese and wild fowl. The disease is widespread in many countries, resulting in great losses particularly in turkeys in North America.
3. Occasional primary pasteurellosis may occur in many animal species.

Secondary disease
P. multocida is a common and important secondary invader of pneumonic lesions in cattle, sheep, swine and goats particularly, causing many cases of chronic to severe pneumonias. Shipment, transport and crowding of animals, particularly during inclement weather, predispose to this disease.

In addition to these principal diseases, *P. multocida* is recovered from a wide range of sporadic infections in many species. Dogs, cats, and probably other species frequently harbor *P. multocida* in their mouths as commensals, and consequently bites inflicted on man and other animals are often infected with *P. multocida* (Hubbert and Rosen, 1970a, b). *P. multocida* is mainly an opportunist or secondary invader in internal infections in humans; a wide variety of sporadic infections have been reported, including meningitis, encephalitis, otitis, septicemia, sinusitis, peritonitis, bronchiectasis and arthritis (Carter, 1981).

P. haemolytica has a primary or secondary role in pneumonia of cattle, sheep and goats particularly. It is a principal component of the etiology of pneumonic pasteurellosis or "shipping fever" of cattle. Other important diseases it causes are mastitis of ewes and septicemia of lambs. Only a few infections have been attributed to this species in human beings. Avian *P. haemolytica* is recovered frequently from the intestine and upper respiratory tract of apparently normal chickens and turkeys. It is occasionally associated with respiratory infections and salpingitis (Heddleston, 1975) in poultry, but in general it has a low potential for causing disease.

P. pneumotropica can be recovered from the nasopharynx of some apparently normal guinea pigs, rats, hamsters, mice, dogs and cats. When associated with disease it is usually a secondary invader, as in pneumonic disease in rats and mice. Various other sporadic infections occur infrequently in mice, dogs, rats and hamsters (Brennan et al., 1965). Only a few infections have been reported in humans (Carter, 1981).

P. ureae is a rather uncommon commensal in the upper respiratory tract of humans. In one report it was found in 1% of the sputa examined (Jones and O'Conner, 1962), and in another study it was recovered from the upper respiratory tract of only 10 individuals (Hendriksen and Jyssum, 1960, 1961). It has been associated infrequently with bronchiectasis and chronic bronchitis (Jones, 1962). Other diseases with which it has been occasionally associated are pneumonia, meningitis, septicemia, sinusitis and ozaena (Carter, 1981). Like *P. pneumotropica* it has a weak potential for causing disease.

P. gallinarum occurs as a commensal in the upper respiratory tract of poultry and occasionally of cattle and sheep. It has a weak capacity for producing disease and is most often associated with chronic respiratory infections of poultry (Heddleston, 1975).

The habitat of *P. aerogenes* is the intestine of swine, where it appears to be a harmless member of the intestinal flora (Carter and McAllister, 1974). Isolations have been made from a swine-bite and leg ulcer involving humans (Frederiksen and Kilian, 1981).

Experimental infections in laboratory animals. The heterogeneity of strains of *P. multocida* is reflected in their varying pathogenicity for experimental animals. Many of the strains from disease processes in domestic mammals and poultry are remarkably pathogenic for mice and rabbits. On the other hand, cultures from dogs and cats have little pathogenicity for mice and rabbits even when large numbers of organisms are inoculated. The capsular type B and E strains from cases of hemorrhagic septicemia are the only ones that have a consistent ability to fatally infect cattle, water buffaloes, sheep, goats and swine experimentally.

Although experimental infections can be established in calves and lambs with *P. haemolytica*, members of this species have a low pathogenicity experimentally for rabbits, mice and other laboratory animals.

The remaining species in the genus are practically nonpathogenic for laboratory animals.

Enrichment and Isolation Procedures

Pasteurellae grow well on blood agar, and this medium is routinely used for their isolation from clinical specimens. It is difficult to initiate cultures of *P. multocida* from small inocula without media containing blood or hematin. Increased CO_2 tension does not appreciably increase the growth or number of colonies. Selective media have been described by Morris (1958) for the isolation of *P. multocida* and *P. haemolytica* from heavily contaminated materials; however, their use is seldom required in the clinical laboratory. Many strains of *P. multocida* are moderately to highly pathogenic for mice and rabbits, and these animals can be inoculated with heavily contaminated clinical materials. Cultures of *P. multocida* can then be readily recovered with few or no contaminants from the blood and tissues of dead or moribund mice or rabbits. The recent observation that *P. multocida* is resistant to nafcillin suggests that this antibiotic might be useful in selective media (Neter, 1981).

Maintenance Procedures

Surface cultures of most of the pasteurellae will survive on solid media or in the commonly-used liquid media for at least a week. Many cultures of *P. multocida*, *P. pneumotropica* and *P. gallinarum* can be maintained in Stock Culture Agar (Difco) butts at room temperature for weeks or even months if the tubes are tightly sealed. However, some cultures will not survive for more than a few days, and those that do remain viable may undergo some dissociation. Biovar A of *P. haemolytica* loses its viability more rapidly than biovar T in Stock Culture Agar and other media (Smith, 1959). *P. multocida* loses its viability in and on media when stored for more than 2–3 weeks in the refrigerator.

All of the pasteurellae can be preserved for years in the lyophilized state by use of conventional suspending media for freeze-drying. Cells harvested from culture media maintain viability for long periods if suspended in defibrinated blood and stored at −40 to −70°C. Viability can be maintained indefinitely if the organisms are stored in liquid nitrogen.

Differentiation of the genus **Pasteurella** from other genera

The genera *Pasteurella* and *Actinobacillus* are closely related and differentiation of one from the other sometimes poses problems. The principal differential characteristics are listed in Table 5.77 of the article on *Actinobacillus* in this *Manual*.

Several characteristics distinguish *Pasteurella* from *Yersinia* and *Haemophilus*. In contrast to *Pasteurella* and *Actinobacillus*, *Yersinia* species are motile at 22°C, are oxidase-negative, and grow in media containing 4.5% NaCl. Members of *Haemophilus* require either (or both) the X and V factors, whereas *Pasteurella* and *Actinobacillus* do not. The sticky character of the colonies of *Actinobacillus* and their growth on MacConkey agar help to differentiate them from all of the *Pasteurella* species except *P. haemolytica* and *P. aerogenes*.

Taxonomic Comments

The similarities between the two genera *Pasteurella* and *Actinobacillus* have led to uncertainty in the generic placement of some organisms. In a numerical taxonomic study of strains of *Haemophilus*, *Actinobacillus* and *Pasteurella*, Sneath and Johnson (1973) found that one cluster contained strains of *Pasteurella* and *Actinobacillus*, which were associated at a similarity index of 75%. They concluded that this constituted strong evidence against two distinct genera. Frederiksen (1973) examined species of *Pasteurella* and *Actinobacillus* and concluded that arrangement into clear-cut genera and species would require genetic studies.

Mannheim et al. (1978) examined strains of *Actinobacillus*, *Haemophilus* and *Pasteurella* for the presence of respiratory quinones and on the basis of finding demethylmenaquinones in all three taxa concluded that the group should rank as a family. According to Pohl (1979), the three genera are interrelated at or above the 30% DNA homology level and could be divided into six tribes linking at the 40% level. On the basis of DNA/DNA hybridization studies, several organisms which have been included in the genus *Pasteurella* should probably be placed in the genus *Actinobacillus* instead. For example, the observations by Mráz (1969) and Pohl (1979) indicate that *P. haemolytica* biovar A is closer to the genus *Actinobacillus* than to *Pasteurella*. After examining strains of *P. ureae*, Frederiksen (1973) concluded that they formed a homogeneous group that differs from *P. haemolytica* and other *Pasteurella* species by five or more characteristics and from *A. lignieresii* by only three characteristics; based on these data and also on the results of DNA/DNA hybridization studies, it seems that *P. ureae* and *P. haemolytica* biovar A should eventually be included in the genus *Actinobacillus* instead of the genus *Pasteurella* (Pohl, 1981). Hacking and

Sileo (1977) have also noted the similarities between *P. ureae*, *A. lignieresii* and *A. suis*.

Two new species, *P. gallinarum* and *P. aerogenes*, which were not listed in the eighth edition of *Bergey's Manual*, were included in the Approved Lists of Bacterial Names in 1980 and have been added to the genus *Pasteurella* in the present edition of the *Manual*. *P. aerogenes* was considered by Ursing (1981) to be sufficiently distinctive on the basis of DNA/DNA hybridization studies to constitute a new species. It has been suggested that the gas-producing organisms called *Pasteurella* n. sp. I "*gas*" (King, 1972) are a biovar of *P. pneumotropica* (Frederiksen, 1982). Another gas-producing pasteurella-like group provisionally termed *Pasteurella* SP is usually isolated from guinea pigs (Frederiksen, 1982); the DNA contains ca. 50% mol% G + C, which is outside the range of the classical pasteurellas.

The species "*Pasteurella anatipestifer*," which had been listed in *species incertae sedis* as "*Pfeifferella anatipestifer*" in the eighth edition of *Bergey's Manual*, was included on the Approved Lists of Bacterial Names as *Moraxella anatipestifer* in 1980. However, the species contains large amounts of branched fatty acids, which is a feature incompatible with the family *Neisseriaceae* (see the chapter on the genus *Moraxella* in this edition of the *Manual*); consequently, that taxonomic placement of this species continues to remain uncertain.

In addition to conventional methods and also the use of nucleic acid hybridization for taxonomic studies of pasteurellae, other methods are presently being used. For example, Jantzen et al. (1981) studied the cellular fatty acid composition of strains of *Pasteurella*, *Haemophilus* and *Actinobacillus*; however, the fatty acids appeared to be of little value for intragroup differentiation. On the other hand, differences in the polyacrylamide gel electrophoresis (PAGE) patterns of cellular proteins seemed able to differentiate *P. multocida*, *P. gallinarum*, and "*P. anatipestifer*" (see above) from *P. haemolytica* (Harry and Brown, 1981), and PAGE protein patterns might therefore be useful for taxonomic studies of pasteurellae.

Further Reading

Carter, G.R. 1967. Pasteurellosis: *Pasteurella multocida* and *Pasteurella haemolytica*. Adv. Vet. Sci. *11*: 321–379.

Carter, G.R. 1981. *Pasteurella*. In Blobel and Schlieber (Editors), *Handbuch der bakteriellen Infektionen bei Tieren*. Gustav Fischer, Jena, pp. 557–593.

Carter, G.R. 1981. The genus *Pasteurella*. In Starr, Stolp. Trüper, Balows and Schlegel (Editors), *The Prokaryotes: a Handbook on Habitats, Isolation and Identification of Bacteria*, Springer-Verlag, Berlin, pp. 1383–1391.

Differentiation of the species of the genus **Pasteurella**

Characteristics useful for the differentiation of the species of *Pasteurella* are presented in Table 5.68.

Table 5.68.
Differential characteristics of the species of the genus **Pasteurella**[a]

Characteristics	1. *P. multocida*	2. *P. pneumotropica*	3. *P. haemolytica*	4. *P. ureae*	5. *P. aerogenes*	6. *P. gallinarum*
Hemolysis (β)	−	−	+	−	−	−
Growth on MacConkey's agar	−	−	+	−	+	−
Indole production	+	+	−	−	−	−
Urease activity	−	+	−	+	+	−
Gas from carbohydrates	−	−	−	−	+	
Acid production from:						
Lactose	−	d	d	−	−	−
Mannitol	+[b]	−	+	+	−	−

[a] Data from Carter (1981). For symbols see standard definitions.
[b] Strains from dogs and cats may be negative for mannitol.

List of the species of the genus **Pasteurella**

1. **Pasteurella multocida** (*sic*) (Lehmann and Neumann 1899) Rosenbusch and Merchant 1939, 85.[AL] (*Bacterium multocidum* (*sic*) *multocidum* Lehmann and Neumann 1899, 196; *Pasteurella gallicida* (Burill 1883) Buchanan 1925, 414.[AL])

mul.to.ci'da. L. adj. *multus* many; L. adj. suf. *-cidus* from L. v. *caedo* to kill; M.L. fem. adj. *multocida* many-killing; i.e. pathogenic for many (species of animals). *Note:* since *multus* is Latin, not Greek, the grammatically correct form of the specific epithet should be *multicida*.

Cells usually coccobacillary or short rods in cultures from diseased tissues; strains from healthy animals are often pleomorphic with longer bacillary forms and occasional short filaments. Bipolar staining is usual in the coccobacillary and rod forms. Capsules, when present, are best demonstrated in cultures by negative staining methods; in infected tissues they are better seen in Giemsa-stained preparations.

Growth from small inocula is poor except on media containing blood or hematin. Nonhemolytic, but most strains produce a brownish discolorization of blood media in regions of confluent growth. Mucoid, smooth and rough variants occur (Carter and Bain, 1960); colonies of capsulated smooth strains are iridescent in obliquely transmitted light (Smith, 1958). Cultures on blood agar have a faint but distinctive smell of value in recognition.

Growth occurs between 25 and 40°C. Most avian strains and a few others grow at 42°C. Optimum temperature, 37°C.

Gluconate oxidation is variable. Other physiological characteristics are listed in Tables 5.68 and 5.69.

Most strains are susceptible to penicillin, tetracyclines, chloramphenicol, novobiocin, erythromycin, neomycin, polymyxin, gentamicin, cephalothin and sulfonamides. Multiple drug resistance has been reported (Chang and Carter, 1976).

Highly pathogenic for mice and rabbits; injection of 1–10 cells of a virulent strain often produces a rapidly fatal septicemia. Strains from birds are also pathogenic for pigeons.

The mol% G + C of the DNA is 40.8–43.2 (T_m) (Pohl, 1979).

Type strain: NCTC 10322.

Further comments. *P. multocida* includes strains showing considerable variation in fermentative ability. Many strains from dogs and cats differ in their fermentative properties from strains recovered from cattle, sheep, swine and domestic poultry (Smith, 1958; Carter, 1976; Ghoniem et al., 1973). It has been suggested that the species includes at least five biovars based upon hyaluronidase decapsulation, flocculation by acriflavine, colonial iridescence, fermentation pattern, mouse

pathogenicity, host predilection, and serologic and immunologic characteristics (Carter, 1976). The names suggested for these proposed biovars were: (a) the "mucoid" biovar, (b) the "hemorrhagic septicemia" biovar, (c) the "porcine" biovar, (d) the "canine" biovar, and (e) the "feline" biovar. Frederiksen (1973) described a biovar which he referred to as the "dog-type." It was negative for sorbitol and mannitol fermentation, whereas his other strains were usually positive.

Burrill gave the specific epithet *gallicida* to an organism he considered to be the cause of fowl cholera. Buchanan (1925) placed Burrill's organism in the genus *Pasteurella* but provided an inadequate description of the species. A type strain, NCTC 10322, was designated in the Approved Lists of Bacterial Names in 1980, but the basis for such a designation is not known by this author. The strain, also known as W-9217, is a porcine strain isolated by this author in 1962 (Carter, 1963) and is a typical *P. multocida*. A reference strain of *P. gallicida*, ATCC 29997, is also a typical *P. multocida* (avian). It thus appears that *P. gallicida* is synonymous with *P. multocida*. Both names appear on the Approved Lists of Bacterial Names and therefore have the same date of valid publication (i.e. January 1, 1980). According to Rule 24b of the Bacteriological Code (Lapage et al., 1975), "If two names compete for priority and if both names date from 1 January 1980 on an approved list, the priority shall be determined by the date of the original publication of the name before 1 January 1980." Because the specific epithet *gallicida* was published in 1883 and the epithet *multocida* in 1899, the former would have priority. However, the name *multocida* has been used for many years and a change to *gallicida* would introduce much confusion. A Request for an Opinion needs to be submitted to the Judicial Commission to see if the name *multocida* can be conserved against *gallicida*. *Editorial Note*: A request for an opinion to confirm *P. multocida* as the type species of the genus has been published recently by P. H. A. Sneath (Int. J. Syst. Bacteriol. 32: 459–460, 1983).

2. **Pasteurella pneumotropica** Jawetz 1950, 179.[AL]

pneu.mo.tro'pi.ca. Gr. n. *pneumon* lung; Gr. n. *tropicus* tropic, circle; M.L. fem. adj. *pneumotropica* having an affinity for the lungs.

Cells rod-shaped, ~0.5 × 1.2 μm with occasional longer forms. Bipolar staining is not common.

Colonies on blood agar are 1.6–2.0 mm in diameter after 48 h at 37°C and are smooth, grayish translucent, and butyrous, with a characteristic odor. Nonhemolytic.

Growth occurs between 22 and 40°C; optimum temperature, 37°C.

Table 5.69.

Other characteristics of the species of the genus **Pasteurella**[a]

Characteristics	1. P. multocida	2. P. pneumotropica	3. P. haemolytica	4. P. ureae	5. P. aerogenes	6. P. gallinarum
Catalase test	+	+	+	+	+	+
Oxidase test (Kovacs')	[+]	+	+	+	+	+
Ornithine decarboxylase	[+]	+			+	[−]
Lysine decarboxylase	−	−			−	−
Arginine dihydrolase	−	−			−	−
H$_2$S production (lead acetate strip)	+	[+]			+	[+]
Nitrate reduced to nitrite	[+]	+	+		+	+
Growth in KCN	[+]	d				
Methyl red and Voges-Proskauer tests	−	−	−	−	−	−
Acid production from:						
Adonitol	[−]	[−]	−			−
Amygdalin	[−]	d				
Arabinose	d[b]	[−]			+	−
Cellobiose		[−]				
Dextrin	d		+			+
Dulcitol	d[b]	[−]	d	[−]	−	−
Erythritol	[−]	[−]	−			
Esculin	[−]					
Fructose				+		+
Galactose	+	[+]				+
Glucose	+	[+]	+	+	+	+
Glycerol	d	[+][b]	d	[−]	+	−
Glycogen	[−]	[−]	+			
Inositol	[−]	d	+			−
Inulin	[−]		−			−
Maltose	d[c]	[+]	+		+	+
Mannose	+	[+]				+
Melizitose	[−]					
Melibiose	d		+			
Raffinose	d	d	+		−	d
Rhamnose	[−]	[−]	d		d	−
Salicin	[−]	[−]		[−]	−	
Sorbitol	[+][d]		+	+	−	d
Sorbose	[−]					
Starch		[+]	+			
Trehalose		[+]			−	+
Xylose	[+][d]	[−]		[−]	+	d

[a] Data compiled from Carter (1981), Ghoniem et al. (1973) and Smith (1958, 1974). Symbols: +, all strains positive; [+], most strains positive; d, differs among strains; [−], most strains negative; −, all strains negative.

[b] Arabinose and dulcitol fermentation is most common in strains from birds.

[c] Maltose fermentation is characteristic of strains from dogs and cats.

[d] Strains from cats and dogs may be negative for sorbitol and xylose.

Physiological characteristics are listed in Tables 5.68 and 5.69.

Susceptible to penicillin, tetracyclines, chloramphenicol, colistin, cephalothin, gentamicin, kanamycin, polymyxin and sulfonamides.

The mol% G + C of the DNA is 40.3–42.8 (T_m) (Pohl, 1979).

Type strain: NCTC 8141.

3. **Pasteurella haemolytica** Newsom and Cross 1932, 715.[AL]

hae.mo.ly'ti.ca. Gr. n. *haema* blood; Gr. adj. *lyticus* dissolving M.L. fem. adj. *haemolytica* blood-didsolving.

Cells rod-shaped, somewhat larger than those of *P. multocida*. Slight pleomorphism and bipolar staining are seen. Capsules are present in cultures from diseased tissues.

A zone of hemolysis surrounds colonies of freshly isolated strains but may be reduced or lost after a few subcultures. A double zone of hemolysis on lamb's blood agar is characteristic. There is no effect on ovine erythrocytes suspended in broth. Growth occurs on MacConkey's agar.

Growth occurs between 25 and 40°C; optimum temperature, 37°C.

Physiological characteristics are listed in Tables 5.68–5.69.

The susceptibility to antimicrobial agents is similar to that of *P. multocida*. Multiple drug resistance has been reported (Chang and Carter, 1976).

The mol% G + C of the DNA is 42.3–43.6 (T_m) (Pohl, 1979).

Type strain: NCTC 9380.

Further comments. Two different biovars of *P. haemolytica* have been described based upon differences in fermentative activity, serologic characteristics, and pathogenicity (Smith, 1959; Biberstein and Francis, 1968). The differential characteristics of the two biovars are listed in Table 5.70. As was mentioned earlier, biovar A should ultimately be transferred to the genus *Actinobacillus* on the basis of DNA relatedness. *P. haemolytica* serovar 11 may represent a third biovar or taxon (Frederiksen, personal communication).

4. **Pasteurella ureae** Jones 1962, 150.[AL] (*Pasteurella haemolytica* var. *ureae* Henriksen and Jyssum 1960, 443.)

u.re'ae. Gr. n. *urum* urine; M.L. gen. n. *ureae* of urine.

Cell rod-shaped, pleomorphic, depending on the growth medium. Bipolar staining is occasionally seen.

Growth is best on media containing blood or serum.

Growth occurs from 25–40°C; optimum temperature, 37°C.

Physiological characteristics are listed in Tables 5.68 and 5.69.

Occurs infrequently in the noses of healthy humans and in occasional cases of ozaena and other infections of the respiratory tract.

The mol% G + C of the DNA is 41.2 (T_m) (Pohl, 1979).

Type strain: ATCC 25976 (NCTC 10219; Henriksen strain 3520/59).

5. **Pasteurella aerogenes** McAllister and Carter, 1974, 920.[AL]

a.e.ro′gen.es. Gr. masc. n. *aer* air; Gr. v. *gennaio* to produce; M.L. adj. *aerogenes* gas-producing.

The cells are coccobacilli, 0.5–1.0 μm in diameter and 1.1–2.0 μm in length. Filaments 6–15 μm long are seen in older cultures. Noncapsulated.

Colonies on blood agar after 48 h at 37°C are circular, smooth, entire, convex and translucent, and range in diameter from 0.5–1.0 mm. No hemolysis occurs. The organisms grow equally well on tryptose and MacConkey agars. On the latter medium *Salmonella*-like colonies develop in 24 h; these acquire a faint pink color with additional incubation. No growth occurs on SS agar. Broth cultures are uniformly turbid.

Optimum temperature, 37°C. No growth occurs at 25–28°C in 24 h.

Phenylalanine deaminase is not produced. Reactions in triple-sugar iron agar include an acid slant, acid butt, gas, but no blackening (H_2S negative). β-Galactosidase is produced. Citrate is not utilized. Other physiological characteristics are listed in Tables 5.68 and 5.69.

The mol% G + C of the DNA is 41.8 (T_m) (Pohl, 1979).

Type strain: ATCC 27883.

6. **Pasteurella gallinarum** Hall, Heddleston, Legenhausen and Hughes 1955, 604.[AL]

gal.li.na′rum. L. n. *gallina* hen; L. gen. pl. n. *gallinarum* of hens.

Cells are coccobacilli 0.5 ± 0.1 μm in diameter and 1.5 ± 0.5 μm in length; some forms are 10 μm or longer after subculturing. Bipolar staining occurs. Freshly isolated cultures are encapsulated.

Colonies on serum or dextrose-starch agar are iridescent, circular, smooth, entire, convex and translucent. After incubation for 24 h on blood or serum agar colonies may be as large as 1.5 mm in diameter. No growth occurs on MacConkey's agar. Nonhemolytic.

Growth occurs between 30 and 42°C; optimum temperature, 37°C.

Table 5.70.

Differential characteristics of the two biovars of **Pasteurella haemolytica**[a]

Characteristics	Biovar A	Biovar T
Acid production from:		
L-Arabinose, D-xylose	+	−
Trehalose, salicin	−	+
Susceptibility to penicillin	High	Low
Serovars	1, 2, 5, 6, 7, 8, 9, 11, 12	3, 4, 10
Principal location in natural host	Nasopharynx	Tonsils
Principal disease association	Pneumonia of cattle and sheep; septicemia of nursing lambs	Septicemia of feeder lambs

[a] Data from Biberstein (1978). For symbols see standard definitions.

Physiological characteristics are listed in Tables 5.68 and 5.69.

The mol% G + C of the DNA is 41.2–44.8 (T_m) (Mráz et al., 1977).

Type strain: ATCC 13361.

Species Incertae Sedis

Moraxella anatipestifer (Hendrickson and Hilbert 1932) Bruner and Fabricant 1954, 461.[AL] (*Pfeifferella anatipestifer* Hendrickson and Hilbert 1932, 249; *Pasteurella anatipestifer* (Hendrickson and Hilbert 1932) Breed 1957, 397.)

a.na.ti.pes′ti.fer. L. fem. n. *anas, anatis* duck; L. fem. n. *pestis* plague; L. adj. *pestifer* pestilence-carrying; M.L. adj. *anatipestifer* duck-plague-carrying.

See the eighth edition of *Bergey's Manual* and also Henriksen (1973) for descriptions of the organism.

The nonfermentative nature of the organism and other characteristics exclude it from the genus *Pasteurella*. It also does not appear to belong to the genus *Moraxella* (see Taxonomic Comments for the genus *Moraxella* in this *Manual*) and its proper taxonomic placement is uncertain.

Type strain: ATCC 11845.

Addendum to the genus **Pasteurella**

REINIER MUTTERS AND WALTER MANNHEIM

Recently the taxonomy of the genus *Pasteurella* was studied by DNA/DNA hybridization using the initial renaturation rate method of De Ley et al. (1970), i.e. stringent reassociation conditions (R. Mutters, University of Marburg, F. R. G., unpublished data). By single-linkage clustering of the DNA binding values a genus-like group (*Pasteurella sensu stricto*) was obtained consisting of organisms that clustered at or above the 50% DNA binding level. Within the group four subgroups of organisms clustering at or above the 70% DNA binding level could be discriminated: (i) *Pasteurella multocida*, except biovars 1 and 6 of Frederiksen (1973); (ii) *Pasteurella* "gas" or type Henriksen of *Pasteurella pneumotropica*, *Pasteurella gallinarum*, and *Haemophilus avium* (in part); (iii) biovar 6, or "dog-type" strains of *Pasteurella multocida*; and (iv) a new group of organisms isolated from cats, dogs, and human dog-bite lesions and *Haemophilus avium* (in part).

Phenotypically, the group consists of small Gram-negative, coccobacillary, nonmotile and nonsporeforming rods that grow as facultative anaerobes on chocolate or sheep blood agar without hemolysis and are usually catalase- and oxidase-positive. Faint yellowish pigments are formed by a minority of the strains. Glucose is fermented and nitrites are produced from nitrates. Demethylmenaquinones are used, particularly in anaerobic fumarate respiration; ubiquinones are synthesized by most strains. Indole (except for *P. gallinarum* and *H. avium*) and phosphatase reactions are usually positive. Acid is produced (usually within 48 h) from glucose, galactose, fructose, mannose and sucrose.

The following tests are negative: lysine decarboxylase, arginine dihydrolase, gelatinase, and catabolism of sorbose, rhamnose, starch, inositol, adonitol, dulcitol, salicin and esculin. The four subgroups can be separated by the characteristics given in Table 5.71. The three species

Table 5.71.

Differential characteristics of four DNA homology subgroups within the genus-like group **Pasteurella** *sensu stricto*[a]

Characteristics	Subgroup (i)	(ii)	(iii)	(iv)
Mol% G + C of DNA	41–44	39–42	38–39	40–42
Acid from mannitol	+	d	−	−
Ornithine decarboxylase	+	−	+	−
Acid from maltose	−[b]	+	−	−
Urease test	−	d	−	−
Indole production	+	d	+	d

[a] Data from R. Mutters (unpublished results). The genus-like group *Pasteurella sensu stricto* is composed of strains that cluster at or above the 50% DNA binding level; the four subgroups consist of strains clustering at or above the 70% DNA binding level. See Addendum to the Genus *Pasteurella* for further details. For symbols see standard definitions.

[b] Occasional strains may differ from this reaction.

contained in subgroup (ii) are easily identified using the tables for *Pasteurella* and *Haemophilus*.

Although these data implicate the transfer of *Haemophilus avium* to the genus *Pasteurella*, the type or reference strains so far tested of *P. haemolytica* biovar A, *P. pneumotropica* type Jawetz and type Heyl, *P. ureae*, *P. aerogenes*, and the pasteurella-like group "SP," remain outside

the genotypic cluster *Pasteurella sensu stricto*, as had already been indicated by Pohl (1979, and unpublished data). On the other hand, there do exist strains of organisms that appear to be phenotypic intermediates between *P. ureae* and *P. "gas,"* and also several pasteurella-like groups of avian isolates, but their genetic relationships have not yet been studied.

Genus II. **Haemophilus** Winslow, Broadhurst, Buchanan, Krumwiede, Rogers and Smith 1917, 561.[AL]

MOGENS KILIAN AND ERNST L. BIBERSTEIN

Hae.mo′phi.lus or Haem.oph′il.us. Gr. n. *haima* blood; Gr. n. *philos* lover; M.L. masc. *Haemophilus* blood-lover.

Minute to medium-sized **coccobacilli or rods**, generally less than 1 μm in width and variable in length, sometimes forming threads or filaments and showing marked **pleomorphism**. Gram-negative. **Nonmotile. Aerobic or facultatively anaerobic. Require preformed growth factors present in blood**, particularly X factor (protoporphyrin IX or protoheme) and/or V factor (nicotinamide adenine dinucleotide (NAD) or NAD phosphate (NADP)). Even after specific growth factors have been provided, growth is best on complex media. Optimum temperature, 35–37°C. **Nitrates are reduced to, or beyond, nitrites.** Oxidase and catalase reactions vary among strains. **Chemoorganotrophic.** All species can **attack carbohydrates fermentatively**, yielding acetic, lactic, and succinic acids as end products in glucose broth. Occur as **obligate parasites on the mucous membranes** of man and a variety of animals species. The mol% G + C of the DNA is 37–44 (T_m).

Type species: *Haemophilus influenzae* (Lehmann and Neumann 1896) Winslow, Broadhurst, Buchanan, Krumwiede, Rogers, and Smith 1917, 561.

Further Descriptive Information

Cell morphology. Cells of *Haemophilus* species tend to occur as individual short rods or coccobacilli. Filament formation is environmentally influenced and develops as cultures age and are under less than optimum conditions of growth. Capsules are present in a number of species and are of particular interest in *H. influenzae*, *H. paragallinarum*, and *H. pleuropneumoniae*, where they play a part in pathogenesis, determination of type specificity, and production of antiinfective immunity.

Cell wall composition. The cell walls resemble those of other Gram-negative bacteria in structure, composition and endotoxic activity. Analysis of the lipopolysaccharide (LPS) obtained from *H. influenzae* yields glucose, galactose, glucosamine, heptose, and a 2-keto-3-deoxy-octonate-like (KDO) molecule. The latter component is present in much lesser amounts than KDO of *Enterobacteriaceae*: 0.4–1.5% of LPS in *H. influenzae* compared to 5–8% in *Salmonella* spp. (Flesher and Insel, 1978).

As in other members of the family *Pasteurellaceae*, the number of cell wall fatty acids found in *Haemophilus* species is comparatively low. The general pattern is characterized by relatively large amounts of *n*-tetradecanoate (14:0), 3-hydroxy-tetradecanoate (3-OH-14:0), hexadecanoate (16:1), and *n*-hexadecanoate (16:0). The three C_{18} fatty acids, octadecadienoate (18:2), octadecenoate (18:1), and *n*-octadecanoate (18:0) are also present but in low concentrations. Small species-to-species variations in the relative amounts of these fatty acids have been demonstrated (Jantzen et al., 1981).

Stable wall-deficient variants (L-forms) have been produced experimentally from *H. pleuropneumoniae* and *H. influenzae* (Neil et al., 1970; Roberts et al., 1974).

Fine structure. The cell wall of *Haemophilus* organisms is typical of Gram-negative bacteria, having an ultrastructure composed of multiple wavy outer membranes and a poorly defined plasma membrane, with an intervening electron-transparent space (Sherwin and Wilkins, 1973; Kilian and Theilade, 1975; Doern and Buckmire, 1976; Kilian and Theilade, 1978; Holt et al., 1980). The entire cell wall, including cell membrane, averages 20 nm in thickness. Vesicular structures ("blebs")

have been demonstrated on the outer wall of several *Haemophilus* species (Holt et al., 1980). These are morphologically identical to lipopolysaccharide vesicles and are excreted into the surroundings. Methods for the isolation of the inner and outer membranes from *H. influenzae* serovar b cells have been devised by Loeb et al. (1981). In overall composition, the outer membrane and inner membrane of *H. influenzae* are similar to those of other Gram-negative bacteria (Loeb et al., 1981).

Surface-associated thin hairlike structures of an unidentified nature have been demonstrated in *H. aphrophilus* and *H. paraphrophilus* (Holt et al., 1980). Fimbriae have been detected in hemagglutinating strains of *H. influenzae* and *H. aegyptius* (Scott and Old, 1981).

Colonial and cultural characters. Surface colonies of *Haemophilus* species on sufficiently rich media are usually nonpigmented or slightly yellowish, flat, convex, and attain a diameter of 0.5–2.0 mm within 48 h at 37°C. Most species produce smooth colonies but some variation is seen particularly in *H. parainfluenzae*. Some species show β hemolysis on blood agar (see Table 5.73). Growth in broth media usually shows even turbidity but strains of *H. aphrophilus*, *H. paraphrophilus* and some strains of *H. parainfluenzae* show granular growth with heavy deposits.

Nutrition and growth conditions. Apart from the specific growth factor requirements, the various species of *Haemophilus* exhibit some variation in their nutritional needs and preferred growth conditions. The most universally satisfactory propagative media are chocolate agar and Levinthal media (agar and broth). The former has the virtue of relative ease of preparation, the latter that of transparency, facilitating the recognition of colonial phases and dissociation phenomena. Colonial iridescence, a property highly correlated with encapsulation, is most readily recognized on Levinthal's agar.

The most critical ingredients of any media for *Haemophilus* spp., whether for propagation or characterization, are the growth factors X and/or V. X factor is usually protoporphyrin IX but in some instances the iron-containing protoheme (Lwoff and Lwoff, 1937; White and Granick, 1963). Blood or blood derivatives including hemin are the traditional and adequate sources of X factor. The customary 5% of blood used in blood and chocolate agar are ample. When crystalline hemin is used, required amounts have been shown to vary between 0.1 and 10 μg/ml for *H. influenzae* (Gilder and Granick, 1947; Brumfitt, 1959; Biberstein and Spencer, 1962; Evans et al., 1974), 80 and 100 for "*H. influenzaemurium*"-like organisms (Csúkas, 1976), and found to be 200 for optimal growth of *H. ducreyi* (Hammond et al., 1978). The V factor is minimally nicotinamide mononucleoside (Gingrich and Schlenck, 1944) but is usually described as nicotinamide adenine dinucleotide (NAD) or NAD phosphate (NADP). Although present in blood, it is unavailable owing in part to its intracellular location and partly to the presence of NADase in the blood of many species (Krumwiede and Kuttner, 1938). In chocolate agar the NAD has been liberated from the cells and the enzyme destroyed. Traditional sources of V factor, apart from blood, have been yeast derivatives (Thjøtta and Avery, 1921). When met by crystalline NAD, V factor requirements of *H. influenzae* were found to range from 0.2–1 μg/ml; those of *H. parainfluenzae*, from 1–5 μg/ml, with some strains requiring as much as 25 μg/ml for optimum growth (Evans et al., 1974).

When inoculated on media deficient in one or both of the growth factors, *Haemophilus* colonies will cluster around contaminant colonies of other bacteria which produce the critical factors in excess (Grassberger, 1897). This phenomenon is called satellitism and is sometimes utilized for propagation and characterization of *Haemophilus* spp. To demonstrate it, inoculate a medium such as blood agar with a strain requiring the V factor. Cross the inoculated area with a single streak of *Staphylococcus aureus*. Early growth of the *Haemophilus* strain will be confined to the area immediately adjacent to the line of staphylococcal growth. It will gradually spread peripherally but remain heaviest in the area nearest the "feeder" streak, reflecting the diffusion gradient of the limiting growth factor. Satellitism on blood agar constitutes strongly suggestive evidence of a V-factor requirement.

Other growth factors, apart from X and V, have been described for *H. influenzae* and include pantothenic acid, thiamine, and uracil. Some strains also require a purine, and cysteine has been shown to stimulate luxuriant growth of the species (Holt, 1962).

Haemophilus paragallinarum is reported to have a sodium chloride requirement of 1.0–1.5% for maximal growth (Rimler et al., 1977). Raised carbon dioxide tension is beneficial, and may be required, for surface growth of a number of species including *H. paragallinarum*, *H. aphrophilus*, *H. paraphrophilus*, and *H. paraphrohaemolyticus*. In the case of *H. paragallinarum* propagated in liquid media, CO_2 has been found to replace a 5% serum requirement observed under fully aerobic conditions (Rimler et al., 1976). A serum requirement exists also for *H. parasuis* and *H. ducreyi*. The precise role of serum in these situations is not known. Detoxification of media constituents has been suggested (Page, 1962). Completely synthetic media have been devised repeatedly, particularly for *H. influenzae* and *H. parainfluenzae* (Herbst and Snell, 1949; Talmadge and Herriott, 1960; Butler, 1962; Wolin, 1963; Herriott et al., 1970; Klein and Luginbuhl, 1979). In all instances, the adequacy of the medium was tested either with a very limited number of cultures, or was found to support the growth of only a portion of the strains tested. The simplest (Klein and Luginbuhl, 1979) and one of the most complex (Herriott et al., 1970), when tested extensively in parallel, performed exactly alike with regard to their growth-supporting properties.

Metabolism. Enzymes of the Embden-Meyerhof-Parnas, the hexose monophosphate, the Entner-Doudoroff, and the tricarboxylic acid cycle pathways have been demonstrated in several *Haemophilus* species (Klein, 1940; White, 1966; Holländer, 1976). Glycolytic growth in the "classical" (White and Sinclair, 1971) sense, however (i.e. in the absence of a functional electron transport chain), has been observed only in X factor-requiring members of the genus, e.g. *H. influenzae*, *H. aegyptius*, *H. haemoglobinophilus* (White, 1963). In the *H. parainfluenzae* strain studied, glycolysis (as well as oxidative catabolism of glucose) occurred only in the presence of a functional electron transport system and a terminal electron acceptor, viz. oxygen, nitrate or fumarate (White, 1966; White and Sinclair, 1971). The electron transport chain in the species examined proceeds from flavoproteins (dehydrogenases) via cytochromes *b*, *c*, *d*, and *a* to demethylmenaquinone (DMK) to cytochrome *o* and/or oxygen or nitrate (White and Sinclair, 1971; Holländer, 1976). The presence of DMK to the exclusion of other respiratory quinones has been described as typical of the genus and proposed as a taxonomic criterion (Holländer and Mannheim, 1975; Holländer et al., 1981). By the present definition of the genus, taxa are included that contain ubiquinone in addition to DMK (see Table 5.74). It has been suggested that these taxa resemble more closely the genera *Pasteurella* and *Actinobacillus* than *Haemophilus* (Holländer and Mannheim, 1975).

Genetics. Several forms of genetic transfer mechanisms have been detected in *Haemophilus* species. Alexander and Leidy (1951) utilized capsulation as a genetic marker to demonstrate the occurrence of transformation in *H. influenzae*. Noncapsulated R variants derived from each of the six capsular serotypes a–f were examined for the ability to be transformed by DNA extracted from capsulated donor strains. Strain Rd (selected from a capsular type d) proved to be the most dependable recipient, being transformed to capsular types a, b, c,

d, e or f (or combinations of these), according to the capsular type of the DNA donor (Alexander et al., 1954). Catlin and Tartagni (1969) have detected type b antigen by immunofluorescence microscopy 40 min after adding transforming DNA to a population of strain Rd. The size of the DNA segment required for transformation of capsular type b polysaccharide has been detected to be about 33 megadaltons (Catlin et al., 1972).

Antibiotic resistance genes located in the chromosome as well as on small plasmids have been transferred between *H. influenzae* strains by genetic transformation (Stuy, 1979).

Both intra- and interspecific transformation has been demonstrated in *Haemophilus* (Schaeffer, 1958; Leidy et al., 1956, 1959; White et al., 1964; Leidy et al., 1965; Steinhardt and Herriott, 1968; Beattie and Setlow, 1970).

Genetic transformation in *H. influenzae* occurs by different mechanisms and is more efficient than in enteric bacteria. *H. influenzae* undergoes transformation with high efficiency (>1%) under shift-down nutritional conditions (Herriott et al., 1970). The increased competence under such conditions is paralleled by an altered envelope composition (Zoon and Scocca, 1975). A specific binding protein for homologous double-stranded DNA has been detected in the cell envelope of *H. influenzae* (strain Rd) (Deich and Smith, 1980). This membrane protein recognizes a specific sequence of 11 DNA nucleotides that appears with much higher frequency in *Haemophilus* than in other genera (Danner et al., 1980).

Plasmids coding for resistance to a number of antibiotics have been demonstrated in *H. influenzae*, *H. parainfluenzae*, *H. parahaemolyticus*, *H. ducreyi* and *H. pleuropneumoniae* (see below). Two types of R plasmids occur in *H. influenzae*: 30- to 38-megadalton plasmids and 2.5- to 4.4-megadalton plasmids (De Graff et al., 1976; Elwell et al., 1977; van Klingeren et al., 1977; Kaulfers et al., 1978; Laufs et al., 1979). The large *Haemophilus* R plasmids that have been described have most of their base sequences in common independent of geographic origin and their antibiotic resistance markers (Elwell et al., 1977; Laufs and Kaulfers, 1977) and are closely related to plasmids coding for β-lactamase in gonococci and enterobacteria (Laufs et al., 1978; Laufs et al., 1979).

Small cryptic plasmids have been detected in several *Haemophilus* species (Elwell et al., 1977; Stuy, 1979).

Large plasmids coding for resistance to one to three antibiotics have been shown to be transferable between strains of *H. influenzae* and between *H. influenzae* and *Escherichia coli* by a mechanism which requires cell-to-cell contact (Thorne and Farrar, 1975; van Klingeren et al., 1977; Stuy, 1979). The genetic transfer is likely to be mediated by conjugation although sex pili are not known in haemophili. The same transfer mechanism has been demonstrated in *H. ducreyi* (Brunton et al., 1979; Deneer et al., 1982).

Four different bacteriophages of *H. influenzae* have been reported: HP1 (and the mutants c1 and c2), HP3, S2, and N3 (Boling et al., 1973; Stuy, 1978). The latter is morphologically distinct by having a longer tail and a contractile sheath. Differences in phage sensitivity and lysogeny have been demonstrated among *H. influenzae* serotypes and a number of *Haemophilus* species (Stuy, 1978).

A variety of phage modification and restriction properties have been demonstrated in *Haemophilus* species (Setlow et al., 1968; Stuy, 1976). With regard to restriction endonucleases, the genus *Haemophilus* has been the most productive of all genera so far examined. More than 20 different enzymes have been isolated from a variety of *Haemophilus* species (for review, see Roberts, 1976).

Strains of *H. influenzae* serovar b have been shown to produce a bacteriocin to which other serovars and nontypable *H. influenzae*, *H. parahaemolyticus*, and some strains of *H. parainfluenzae* and *H. haemolyticus* are sensitive (Venezia and Robertson, 1975; Stuy, 1978).

Antigenic structure. There is a great deal of antigenic diversity in the genus and even within several species: *H. influenzae* (Pittman, 1931), *H. parasuis* (Bakos, 1955). *H. paragallinarum* (Page, 1962; Kume et al., 1980), *H. pleuropneumoniae* (Nicolet, 1971; Gunnarsson et al., 1977),

H. paraphrophilus, H. haemoglobinophilus (Frazer et al., 1975). In capsulated species, the capsular antigens furnish the basis for type specificity. All the capsular antigens studied in sufficient detail have proven to be polysaccharides. They are serologically quite specific although more than one antigen may be present in a capsule (Williamson and Zinnemann, 1951, 1954; Branefors-Helander, 1972). Capsular structure and composition has been studied most thoroughly for *H. influenzae* and shown to be polysaccharide in all six types (teichoic acids, in the case of serovars a, b, c, and f (Crisel et al., 1975; Branefors-Helander, 1977; Branefors-Helander et al., 1979; Egan et al., 1980; Branefors-Helander et al., 1980). The structure of the serovar a–f capsules is shown later in Table 5.76. The molecular weight of the serovar b capsular teichoic acid (ribosyl ribitol phosphate) has been demonstrated to be 152,000 by ultracentrifugation or >200,000 by gel filtration (Rodrigues et al., 1971). The serovar d and e capsules do not contain phosphate. Little is known of the physicochemical properties of the capsules of other *Haemophilus* species.

Among somatic antigens, obtained by various extraction procedures and sonic disruption, cross-reactions occur between serovars within species (Branefors-Helander, 1979; Gunnarsson et al., 1978) and between species within the genus (Tunevall, 1953; Omland, 1964; Branefors-Helander, 1979; Schiøtz et al., 1979). The cell wall LPS of an *H. influenzae* serovar b strain was shown to carry at least three antigenic specificities (Flesher and Insel, 1978).

Of particular interest is the occurrence of bacterial antigens cross-reactive with the *H. influenzae* serovar b capsule across a wide and unrelated variety of bacteria including *Staphyloccus aureus, Staphylococcus epidermidis, Streptococcus pyogenes, Streptococcus pneumoniae* (serovars 6, 15a, 29, 35a), *E. coli* (K 100), *Lactobacillus plantarum, Streptococcus faecium, Bacillus alvei,* and *Bacillus pumilis* (Alexander, 1958; Bradshaw et al., 1971, Argaman et al., 1974).

Antibiotic sensitivity. Most strains, particularly of *H. influenzae,* are inhibited by the following concentrations or less of the common antimicrobials (μg/ml): penicillin G, 1.0; ampicillin, 2.0; cephalosporins, 6.0 (except for cephaloridine and especially cephalexin which usually have higher MICs); cloxacillin, 25; dicloxacillin, 25; nafcillin, 25; tetracycline, 4.0; chloramphenicol, 4.0; gentamicin, 6.0; kanamycin, 6.0; tobramycin, 4.0; streptomycin, 2.0; rifampin, 1.0; and sulfamethoxazole/trimethoprim, 2.5/0.12 (data compiled from Emerson et al., 1975; Green et al., 1979; Phillips et al., 1977; Sabbath et al., 1970).

The macrolide (erythromycin, oleandomycin) and lincosamide antibiotics (lincomycin, clindamycin) are generally less active, and bacitracin has so little effect on *H. influenzae* and other *Haemophilus* species that it is used, in a concentration of 5–19 U/ml, in media for the selective isolation of haemophili (Ederer and Schurr, 1971; see below).

Since 1973 ampicillin resistance, usually though not invariably due to a TEM-type β-lactamase, which is coded for by a transmissible plasmid, has appeared especially in *H. influenzae* (Elwell et al., 1975; Bell and Plowman, 1980; Markowitz, 1980). In some large surveys 3–5% of *H. influenzae* isolates carrying plasmid-mediated β-lactamase activity have been detected (Ward et al., 1978; Green et al., 1979). Plasmid-coded resistance to other antimicrobics has been observed. These include chloramphenicol (van Klingeren et al., 1977; Bryan, 1978), tetracycline (Dang Van et al., 1975a; Goldstein et al., 1977; Bryan, 1978), kanamycin (Dang Van et al., 1975b; Goldstein et al., 1977). Plasmids transferring simultaneously ampicillin, chloramphenicol, and tetracycline resistance have been found in *H. influenzae* (Bryan, 1978). While tetracycline resistance appears to occur at rates comparable to penicillin-ampicillin resistance in some populations (Piot et al., 1977; Green et al., 1979), chloramphenicol resistance is still extremely rare (Brotherton et al., 1976; Ward et al., 1978). Resistance to trimethoprim has been observed (Green et al., 1979), although the extent is under some dispute (May and Davies, 1972; Bushby, 1973; May, 1973).

Haemophilus parainfluenzae generally shows higher resistance levels than nonresistant *H. influenzae* to the antimicrobial agents commonly used on *Haemophilus* species (Kamme, 1969; Mayo and McCarthy,

1977). Plasmids coding for β-lactamase in *H. parainfluenzae* have been detected at rates of 5.8 and 14% in a United States and New Zealand study, respectively. The corresponding values for *H. influenzae* were 2.7 and 3.5% (Kauffman et al., 1979; Greene et al., 1979). Ampicillin resistance independent of β-lactamase has also been reported (Walker & Smith, 1980). Transmissible chloramphenicol resistance has been described in *H. parainfluenzae* and credited to an acetyl transferase (Cavanaugh et al., 1975; Shaw et al., 1978), while plasmid-coded aminoglycoside resistance in the same species was found to be mediated by a phosphotransferase (Le Goffic et al., 1977).

β-Lactamase activity has been demonstrated also in *H. parahaemolyticus* and *H. paraphrophilus* but not in *H. haemolyticus* (Kauffman et al., 1979; Jones et al., 1976; Green et al., 1979).

The following minimum inhibitory concentrations (μg/ml) were observed on 19 isolates of *H. ducreyi* (Hammond et al., 1978): vancomycin, 8–128; polymyxin, 32–128; penicillin G or ampicillin, 4.0 (unless β-lactamase-positive); cloxacillin, 32–64; cephalothin, 4.0–8.0; tetracycline, 0.4–32; doxycycline, 0.25–8.0; chloramphenicol, ≦4.0; rifampin, ≦4.0; sulfisoxazole, ≦8.0; and nalidixic acid, ≦8.0.

Plasmid-mediated β-lactamase and sulfonamide resistance has been demonstrated in *H. ducreyi* (Brunton et al., 1979; Albritton et al., 1982) and *H. pleuropneumoniae* (Hirsh et al., 1981). The latter was also shown to carry a plasmid coding for streptomycin resistance; tetracycline resistance, based on unknown genetic mechanisms, also occurred (Hirsh et al., 1981). For nonresistant strains of *H. pleuropneumoniae* the following minimum inhibitory concentrations (μg/ml) have been reported (Derijcke et al., 1978): ampicillin, 0.06; tetracycline, 0.5–2.0; and chloramphenicol, 0.25–1.0.

Ecology. Haemophilus organisms form part of the indigenous flora of the mucous membranes of the human upper respiratory tract and mouth, and may be isolated from the vagina and intestinal canal. The same applies to pigs, sheep, monkeys, and various fowl species. Haemophili have also been isolated from healthy dogs, cats, cattle, guanacos, rabbits, rats, mice, guinea pigs, and ferrets, and there is reason to believe that most mammalian and avian species can be included in the list of carriers (for review see Kilian and Frederiksen, 1981). With the exception of humans, pigs, and sheep, detailed information on the carrier rates is lacking, and many of the organisms isolated from other carriers have been only partly characterized. However, it appears that, with the possible exception of *H. parainfluenzae,* natural carriage of the individual species is strictly related to specific hosts.

Man is the natural host of the following species: *H. influenzae, H. aegyptius, H. haemolyticus, H. ducreyi, H. parainfluenzae, H. parahaemolyticus, H. paraphrohaemolyticus, H. aphrophilus, H. paraphrophilus,* and *H. segnis.* Occurrence of the two species *H. aegyptius* and *H. ducreyi* in healthy individuals has not been documented. The natural habitats of the remaining species is shown in Table 5.72. Special points are discussed under the description of the individual species.

H. parasuis and *H. pleuropneumoniae* have their natural habitats in pigs. The former species is part of the normal flora of the porcine upper respiratory tract, whereas the latter is rarely encountered in healthy animals.

The two species *H. paragallinarum* and *H. avium* are found in the respiratory tracts of poultry where, in particular, *H. avium* appears to be a member of the normal microflora.

H. haemoglobinophilus is a frequent commensal inhabitant of the lower genital tract of dogs.

Pathogenicity. Among the *Haemophilus* species that colonize man, *H. influenzae* is clearly the most important from a clinical point of view. Although not responsible for epidemic influenza, as the name suggests, it is involved in a variety of numerically important and severe infections (for review see Turk and May, 1967). These infections can be divided into two groups: (a) acute, pyogenic and usually invasive infections in which *H. influenzae* is the primary pathogen, and (b) infections (usually chronic) in which *H. influenzae* seems to play a secondary part.

The acute infections include meningitis in children, of which *H.*

Table 5.72.

Percentage distribution into species of **Haemophilus** *strains isolated from infectious diseases and normal floras of man*

Sources	No. of Strains	H. influenzae Biovars						H. haemo- lyticus	H. parain- fluenzae[a]	H. segnis	H. para- phrophilus	H. aphro- philus
		I	II	III	IV	V	VI					
		%	%	%	%	%	%	%	%	%	%	%
Meningitis and epiglot- titis	157	94	5	0	1	0	0	0	0	0	0	0
Ear infections	53	26	30	11	2	21	0	0	0	0	0	0
Conjunctivitis	104	9	41	46[b]	0	0	0	0	4	0	0	0
Lower respiratory tract, CF-patients	56	38	21	21	2	8	0	0	11	0	0	0
Healthy upper respira- tory tract	496	<1	4	2	1	2	1	1	73	11	5	0
Oral cavity	649	0	0	0	0	0	0	0	74	19	2	5

[a] Includes hemolytic strains (*H. parahaemolyticus* and *H. paraphrohaemolyticus*).

[b] Includes strains with the characteristics of *H. aegyptius*.

influenzae is the leading cause, and other septicemic conditions with local implications such as epiglottitis, cellulitis, arthritis and osteomyelitis. The *H. influenzae* organisms that can be isolated from these conditions virtually always possess a serovar b capsule, and the majority belong to biovar I (Table 5.72). Capsulated strains of several serovars may also cause pneumonia, usually in adults.

Noncapsulated strains of *H. influenzae* are often implicated in chronic bronchitis, sinusitis, conjunctivitis, and otitis media, and occur frequently in the lower respiratory tract of patients with cystic fibrosis during acute exacerbations. Isolates from such conditions usually belong to the biovars II and III, the same biovars that colonize the nasopharynx of most healthy individuals (Table 5.72).

H. aegyptius is a frequent cause of acute and contagious conjunctivitis mainly in hot climates. Due to difficulties in differentiating *H. aegyptius* from *H. influenzae*, which also can cause conjuctivitis, the natural history of *H. aegyptius* infections is poorly understood. There are indications, however, that *H. aegyptius* is associated with a more acute form of conjunctivitis and that the organism, in contrast to *H. influenzae*, can colonize eyes without predisposing conditions being present.

H. ducreyi appears to be the causative agent of the venereal disease soft chancre or chancroid, although a variety of microorganisms can be isolated from such lesions.

The species *H. parainfluenzae, H. parahaemolyticus, H. paraphrohaemolyticus, H. segnis, H. aphrophilus*, and *H. paraphrophilus*, which occur in the mouth and nasopharynx of healthy persons, are only opportunistic pathogens. The oral cavity is usually a likely source of the etiologic agents in infections caused by these organisms. Such infections include endocarditis, brain abscesses, dental abscesses, jaw infections and infections following human bites or finger sucking (for review, see Frederiksen and Kilian, 1981).

The clinical condition of swine in which *H. parasuis* is most frequently encountered is Glässer's disease (polyserositis), a systemic infection producing fibrinous inflammation of the membranes lining the large body cavities, joints, and meninges. While no antecedent virus infection has been invoked as a precipitant of this condition, stressful events like weaning, weather changes, or movement to new quarters commonly, but not invariably, precede it. In contrast, *H. pleuropneumoniae* is a consistent pathogen of swine capable of causing highly contagious and often fatal infections in previously unstressed animals. The disease is primarily an infection of the respiratory tract causing fibrinous pneumonia and pleuritis (for review see Biberstein, 1981).

H. paragallinarum causes an infection of the upper respiratory tract of chickens called fowl coryza. It begins in the nasal passages and sinuses and spreads to the conjunctivae, and can extend to the air sacs and lungs.

H. haemoglobinophilus is of low pathogenicity. On rare occasions it has been implicated in urogenital inflammatory disease of dogs.

Enrichment and Isolation Procedures

The traditional way of demonstrating *Haemophilus* organism in samples from mucosal membranes is to inoculate a blood agar plate and cross-inoculate it with a staphylococcus strain to provide the V-factor. Most *Haemophilus* species will grow to sizeable colonies in the vicinity of the feeder strain after overnight incubation at 35–37°C in air supplemented with 5–10% extra CO_2. However, in mixed cultures haemophili may easily be overgrown. Thus, respiratory samples often give rise to heavy growth of haemophili on selective media in cases where no apparent haemophilus colonies are detectable by routine cultivation methods.

Chocolate agar supplemented with 300 μg/ml (18.9 U/ml) of bacitracin is a very satisfactory medium for the selective isolation of haemophili (Hovig and Aandahl, 1969). Some pharyngeal and oral *Neisseria* species will grow on this medium, and so will *Actinobacillus actinomycetemcomitans* and *Eikenella corrodens*. An alternative selective medium which gives equally satisfactory results is chocolate agar supplemented with bacitracin (5 U/ml) and cloxacillin (5 μg/ml) (Sims, 1970).

For the isolation of the more fastidious haemophili (e.g. *H. ducreyi* and *H. aegyptius*) higher isolation rates are obtained with chocolate agar enriched with 1% IsoVitaleX (BBL) (Hammond et al., 1978; Vastine, et al., 1974).

Maintenance Procedures

Plate and tube cultures of haemophili will usually survive for no longer than 1 week without subcultivation. Survival is often better at room temperature than at 4°C. The most satisfactory way of conserving cultures is by lyophilization, e.g. in skim milk. Levinthal broth cultures will remain viable for at least 2 years when frozen at −70°C in sealed glass ampoules.

Procedures for Testing Special Characters

Test for porphyrin synthesis (X-factor requirement). Like many other bacteria, hemin-independent haemophilus strains excrete porphobilinogen (PBG) and porphyrins—intermediates in the hemin biosynthetic pathway (Fig. 5.27)— when supplied with δ-aminolevulinic acid (ALA). In contrast, hemin-requiring strains do not excrete these compounds because they lack the enzymes responsible for their synthesis (Biberstein et al., 1963). This is the background for the porphyrin test, which is the easiest and most reliable means of determining X-factor requirement in *Haemophilus* organisms (Kilian, 1974):

A substrate of the following composition is used: δ-aminolevulinic acid, (Sigma), 2 mM, and $MgSO_4$, 0.08 mM, in 0.1 M phosphate buffer, pH 6.9. The substrate is distributed in 0.5-ml quantities in small glass tubes, and is inoculated with a heavy loopful of bacteria from an agar plate culture. After incubation for 4 h at 37°C, the result is read under

PATHWAY OF PORPHYRIN BIOSYNTHESIS   PORPHYRIN TEST

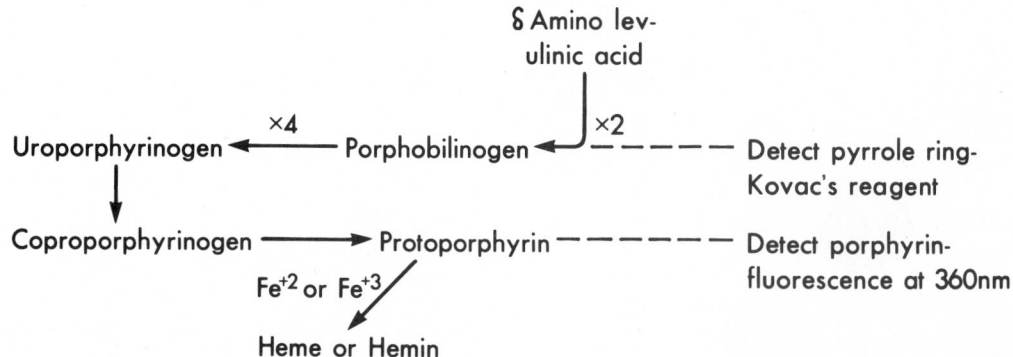

Figure 5.27. Pathway of porphyrin biosynthesis in hemin-independent *Haemophilus* species.

Wood's light (360 nm), preferably in a dark room. A red fluorescence from the bacterial cells and/or from the fluid is indicative of porphyrins (Fig. 5.27); i.e. the strain is independent of the X factor.

An alternative method of reading is: add 0.5 ml of Kovacs' reagent, shake vigorously, and allow the phases to separate. A red color in the lower water phase, indicative of porphobilinogen (Fig. 5.27), means that the strain is independent of X factor. With this method of reading, an inoculated "substrate" without ALA must be included to avoid false positive reactions due to indole, which also gives a red color with Kovacs' reagent.

The traditional use of paper disks impregnated with X factor on agar media cannot be recommended for the demonstration of X-factor requirement. Even with care, use of this method results in misidentifications approaching 20% (Kilian and Eriksen, unpublished results).

V-factor requirement. The satellite phenomenon mentioned above is widely used as a means of determining V-factor requirement. This phenomenon may be demonstrated on an agar medium lacking the V factor by cross-inoculating the plate with an appropriate feeder strain (e.g. *Staphylococcus* or *Pseudomonas*). More unequivocal results may be obtained by the application of a V factor-impregnated paper disk. Media in which all ingredients have been autoclaved are consistently free of V factor, whereas ordinary blood agar contains varying amounts of available V factor. A convincing satellite phenomenon is, therefore, sometimes difficult to achieve on ordinary blood agar plates. The best results are obtained on a blood agar medium in which the blood (5–10%) has been added before autoclaving. This medium is completely devoid of V factor but otherwise satisfies all growth requirements of the *Haemophilus* species, including the X factor.

Fermentation of carbohydrates can be studied in phenol red broth base (Difco) supplemented with 1% of the respective carbohydrates and 10 μg/ml each of NAD and hemin added from solutions sterilized by filtration (Kilian, 1976). In this medium most strains of *H. parainfluenzae*, *H. parahaemolyticus*, *H. aphrophilus*, *H. paraphrophilus*, and *H. haemolyticus* release hydrogen and carbon dioxide during fermentation of glucose (Kilian, 1976). This process may be demonstrated by an inverted Durham tube in the glucose broth. *H. paracuniculus* is reported to be unable to grow in phenol red broth base supplemented

with V factor. Phenol red sorbitol broth is recommended for the examination of fermentation reactions in this organism (Targowski and Targowski, 1979).

Micromethods for Examination of **Haemophilus** *Strains*

Micromethods that do not require growth are recommended for the detection of indole production, urease, ornithine and lysine decarboxylases and arginine dihydrolase, and for the demonstration of glycosidase activities. All these tests use 0.5-ml quantities of the substrate and are inoculated by suspension of a heavy loopful of bacteria in the substrate. The results can usually be read after incubation for 4 h at 37°C, but the incubation may be extended to 24 h.

Indole test (Clarke and Cowan, 1952). The substrate is 0.1% L-tryptophan in 0.067 M phosphate buffer at pH 6.8. After incubation, add 0.5 ml of Kovacs' reagent and shake. A red color in the upper alcohol phase indicates the presence of indole.

Urease test (Lautrop, 1960). The basal medium consists of (per 100 ml of distilled water): KH_2PO_4, 0.1 g; K_2HPO_4, 0.1 g; NaCl, 0.5 g; and phenol red solution (1:500, prepared by dissolving 0.2 g in 92 ml of distilled water + 8 ml of 1 N NaOH), 0.5 ml. Adjust the pH to 7.0 with 5 N NaOH, autoclave, and add 10.4 ml of urea solution (20% aqueous, sterilized by filtration). A red color that develops within 4 h after inoculation indicates urease activity.

Amino acid decarboxylases. Ornithine and lysine decarboxylases and arginine dihydrolase can be demonstrated in Møller's medium (Møller, 1955). When this is heavily inoculated a purple color develops within 4 h with positive strains.

Glycosidases. Glycosidase activities can be demonstrated in 0.1% buffered solutions (w/v) of the respective chromogenic nitrophenol derivatives (Kilian, 1978). A yellow color developing within 4 h indicates glycosidase activity.

Hemagglutination. The hemagglutinating activity as found in some isolates from conjunctivitis (mainly *H. aegyptius*), and some isolates from fowl and swine is demonstrated as described by Davis et al. (1950).

Several of the biochemical tests mentioned above can be performed by the use of the Minitek System (BBL) (Oberhofer and Back, 1979) or the strips of the PathoTec System (Warner-Lambert Co.).

Differentiation of the genus **Haemophilus** *from other closely related taxa*

According to the present definition of the genus, the demonstrable need of X or V factor on the part of a Gram-negative rod or coccobacillus would qualify that organism as a *Haemophilus* strain. Conversely, the absence of such need would exclude a bacterium from the genus. Of all the secondary criteria usually cited as characteristic of the genus, such as carbohydrate fermentation, alkaline phosphatase activity, and

nitrate reduction, only the last named is without exception among the currently recognized species (Kilian, 1976). A G + C percentage conspicuously outside that encompassing the genus, at present, would certainly raise doubts as to the appropriate classification of a bacterium as *Haemophilus* regardless of growth factor needs. There are no motile members of the genus, and the question has never arisen whether motility would exclude an organism meeting the other standards of the genus.

Among all the present members but one, the growth factor requirements are stable, i.e. they do not disappear in the course of numerous subcultures on artificial media. The one exception is *H. aphrophilus*. This organism has an apparent X-factor requirement when freshly isolated despite the fact that it contains all the enzymes of the heme biosynthetic pathway (White and Granick, 1963). On subculture, variant colonies consisting of X factor-independent cells are said to arise, which apparently supplant the original population (Boyce et al., 1969). Strains carried by type culture collections usually do not have the X requirement. In practice, *H. aphrophilus* is identified by criteria other than the growth factor requirement (see below).

Organisms with growth factor requirements comparable to those of *Haemophilus* spp. occur sporadically among other, unrelated taxa (Jensen and Thofern, 1953; Caldwell et al., 1965; Beljanski, 1955). The basis for excluding them from the genus *Haemophilus* would be morphological, tinctorial, and cultural characteristics identifying them with their parent taxa. If these characteristics are similar to those prevailing in the genus *Haemophilus*, as they would be in case of, for example, a *Pasteurella* or *Actinobacillus* species, it is likely that the bacterium in question would indeed be classified as a *Haemophilus* species.

The validity of growth factors as primary generic criteria is currently being questioned (see below).

Taxonomic Comments

As indicated above, the entire basis for circumscribing the genus by its growth factor requirements has recently been challenged in the light of DNA homology studies. These have revealed that the members of the family *Pasteurellaceae* which have such requirements do not form a homogeneous group suggestive of a common genus. Instead they are widely distributed among the members of the family that have no such requirements. For example, *H. pleuropneumoniae* has been found to be related to a *Pasteurella*-like bacterium at a binding level of more than 90%, which suggests species identity. It is also highly related to *Actinobacillus lignieresii* and has been proposed as a V factor-requiring *Actinobacillus* rather than a *Haemophilus* species (Pohl, 1981). Its DNA homology with *H. influenzae*, on the other hand, was about 30% (Pohl, 1981; Mannheim, 1981).

There are at present no phenotypic traits known that correlate with the DNA homology data and would thus provide a workable basis for reconciling the genetic information with the practical needs for laboratory identification of genera and species.

It can be expected that much of the taxonomic effort in the immediate future will be directed at resolving the apparent conflict between the historically established and currently accepted criteria for classifying the genus *Haemophilus* on the one hand and the taxonomic implications generated by the nucleic homology investigations on the other.

Further Reading

Kilian, M., W. Frederiksen and E.L. Biberstein. 1981. *Haemophilus, Pasteurella* and *Actinobacillus*. Academic Press, London.

Sell, S.H. and P.F. Wright, 1982. *Haemophilus influenzae*. Epidemiology, immunology and prevention of disease. Elsevier Biomedical, New York.

Turk, D.C. and J.R. May. 1967. *Haemophilus influenzae*. Its clinical importance. The English Universities Press, London.

Differentiation of the species of the genus **Haemophilus**

The differential characteristics of the species of *Haemophilus* are indicated in Table 5.73. Other characteristics of the species are presented in Table 5.74.

List of the species of the genus **Haemophilus**

1. **Haemophilus influenzae** (Lehmann and Neumann 1896) Winslow, Broadhurst, Buchanan, Krumwiede, Rogers and Smith 1917, 561.[AL] (*Bacterium influenzae* Lehmann and Neumann 1896, 187.)

in.flu.en′zae. Italian noun *influenza* influenza; M.L. gen. n. *influenzae* of influenza.

Coccobacilli or small regular rods $0.3–0.5 \times 0.5–3.0$ μm. Colonies on chocolate agar are smooth, low, convex, grayish, translucent, and attain a diameter of 0.5–1.0 mm in 24 h. Capsulated strains usually produce larger and more mucoid colonies (1–3 mm) which show a tendency to coalesce with no visible line of demarcation. On transparent agar media, colonies of capsulated strains show iridescence when examined under obliquely transmitted light.

The majority of *H. influenzae* strains can be assigned to any of six biovars (I–VI) on the basis of a few biochemical characteristics (see Tables 5.73 and 5.74) (Kilian, 1976; Oberhofer and Back, 1979). Six serovars (a–f) have been identified on the basis of capsular polysaccharides (Pittman, 1931; Pittman, unpublished results). The relationship between biovars and the production of capsules is shown in Table 5.75. The structure of the serovar a–f capsules is shown in Table 5.76. A total of 41 antigenic determinants has been detected in a strain of H. influenzae serovar b by crossed immunoelectrophoresis. A considerable number of these determinants are related to antigenic determinants of strains of *H. haemolyticus*, *H. parainfluenzae*, and various enterobacteria (Schiøtz et al., 1979).

Like other *Haemophilus* species that have been examined, *H. influenzae* possesses neuraminidase activity (Müller and Hinz, 1977). Virtually all strains produce an extracellular endopeptidase (IgA1 protease) capable of inducing specific cleavage of a proline-serine or a proline-threonine peptide bond in the hinge region of human immunoglobulin A1 (Kilian et al. 1979; Male, 1979; Kilian et al., 1980; Mulks et al., 1982).

H. influenzae is present in the nasopharynx of 75% of healthy children and the carriage rate in adults is somewhat lower (Kilian and Frederiksen, 1981). It is rarely encountered in the human oral cavity and has not been detected in any animal species. Capsulated strains are harbored intermittently in the nasopharynx of a minority of healthy individuals (3–7%). Serovars b and f are the serovars most frequently encountered, whereas serovar c strains are notoriously rare.

H. influenzae was isolated originally from cases of endemic influenza and regarded as its causative agent at the time. It is frequently isolated from chronic infections of the upper and lower human respiratory tract, paranasal sinuses, middle ears, and conjunctivae, in which conditions *H. influenzae* plays an important, though probably secondary, part. Strains implicated in such conditions are usually noncapsulated and belong to the biovars II or III like the majority of isolates from healthy upper respiratory tracts (Kilian and Frederiksen, 1981). Encapsulated strains of serovar b are the commonest causes of bacterial meningitis in children and occasionally cause acute epiglottitis (obstructive laryngitis) cellulitis, osteomyelitis and joint infections (Turk and May, 1967). More than 90% of *H. influenzae* isolates from meningitis and epiglottitis belong to biovar I (Kilian and Frederiksen, 1981).

The mol% G + C of the DNA is 39 (T_m).

Type strain: NCTC 8143 (strain 680 of Pittman; noncapsulated, biovar II). *Reference strains:* NCTC 8466 (serovar a, biovar I); NCTC 7279 (serovar b, biovar I); NCTC 8469 (serovar c, biovar II); NCTC 8470 (serovar d, biovar IV); NCTC 10479 (serovar e, biovar IV); NCTC 8473 (serovar f, biovar I); NCTC 4560 (noncapsulated, biovar III); NCTC 11394 (noncapsulated, biovar V).

2. **Haemophilus aegyptius** (Trevisan 1889) Pittman and Davis 1950, 413.[AL] (*Bacillus aegyptius* Trevisan 1889, 13.)

ae.gyp′ti.us. L. adj. *aegyptius* Egyptian.

Table 5.73.
Differential characteristics of the species of the genus **Haemophilus**[a]

Characteristics	1. H. influenzae Biovar I	Biovar II	Biovar III	Biovar IV	Biovar V	Biovar VI	2. H. aegyptius	3. H. haemolyticus	4. H. haemoglobinophilus	5. H. ducreyi	6. H. parainfluenzae Biovar I	Biovar II	Biovar III	7. H. parahaemolyticus	8. H. paraphrohaemolyticus	9. H. pleuropneumoniae	10. H. paracuniculus	11. H. aphrophilus	12. H. paraphrophilus	13. H. segnis	14. H. parasuis	15. H. paragallinarum	16. H. avium	Species Incertae Sedis a. "H. somnus"	b. "H. agni"	c. "H. equigenitalis"
V-factor requirement	+	+	+	+	+	+	+	+	−	−	+	+	+	+	+	+	+	−	+	+	+	+	+	−	−	−
ALA → porphyrins	−	−	−	−	−	−	−	−	−	−	+	+	+	+	+	+	+	W	+	+	+	+	+	+	+	+
Indole production	+	+	−	−	+[a]	−	−	d	+[c]	−	−	−	−	−	−	−	+	−	−	−	−	−	−	+[b]	−	−
Urease	+	+	+	+	−	−	+	+	−	−	−	+	+	+	+	+	+	−	−	−	−	−	−	−	d	−
Ornithine decarboxylase	+	−	−	+	+	+	−	−	−	−	+	+	−	d	−	−	+	−	−	−	−	−	−	−	+	−
Arginine dihydrolase	−	−	−	−	−	−	−	−	−	−	−	−	−	−	−	−	+	−	−	−	−	−	−	−	−	−
Hemolysis	−	−	−	−	−	−	−	+	−	[d]	−	−	−	+	+	+	−	−	−	−	−	−	−	−	−	−
CAMP reaction	−	−	−	−	−	−	−	−	−	−	−	−	−	−	+	−	−	−	−	−	−	−	−	−	−	−
D-Glucose, acid	+	+	+	+	+	+	[+]	+	+	[d]	+	+	+	+	+	+	+	+	+	W	+	+	+	+	+	+
D-Glucose, gas	−	−	−	−	−	−	−	[d]	−	−	[d]	[d]	−	[d]	−	−	−	+	+	−	−	−	−	−	−	−
Acid from:																										
D-Fructose	−	−	−	−	−	−	−	W	−	−	+	+	+	+	+	+	+	+	+	W	+	+	+	+	+	+
Sucrose	−	−	−	−	−	−	−	−	+	−	+	+	+	+	+	+	+	+	+	W	+	+	−	−	−	−
Lactose	−	−	−	−	−	−	−	−	−	−	−	−	−	−	−	d	−	+	−	d	−	d	−	−	−	−
D-Xylose	+	+	+	+	+	+	−	d	+	−	−	−	−	−	−	+	−	−	−	−	−	d	d	+	W	−
D-Ribose	+	+	+	+	+	+	[+]	+	d	−	−	−	−	−	−	+	−	+	+	−	+	+	−	−	−	−
D-Mannose	−	−	−	−	−	−	−	−	+	−	+	+	+	−	−	+	−	+	+	−	+	+	+	+	−	−
D-Manitol	−	−	−	−	−	−	−	−	+	−	−	−	−	−	−	+	−	−	−	−	−	+	d	+	W	−
D-Sorbitol	−	−	−	−	−	−	−	−	−	−	−	−	−	−	−	−	−	−	−	−	−	+	d	+	−	−
β-Galactosidase (ONPG test)[b]	−	−	−	−	−	−	−	d	−	+	d	d	d	d	+	+	+	+	d	d	+	d	d	d		
α-Fucosidase	−	−	−	−	−	−	−	−	−	−	−	−	−	−	d	−	−	−	−	+	−	−	−	−	−	−
Catalase	+	+	+	+	+	+	+	+	+	−	d	d	+	d	+	d	+	−	−	d	+	−	+	−	−	+
CO₂ enhances growth	−	−	−	−	−	−	−	−	−	−	−	−	−	−	+	−	+	+	+	−	d	+	−	+	+	+
Alkaline phosphatase	+	+	+	+	+	+	+	+	−	+	+	+	+	+	+	+	+	+	+	+	+	+	+	−	+	+

[a] Data compiled from: Kennedy et al. (1960), Bailie et al. (1973), Garcia-Delgado et al. (1974), Corboz and Wild (1981), Kilian (1976), Kilian et al. (1979), Kilian and Theilade (1978), Kilian (unpublished results), Taylor et al. (1978), Mannheim et al. (1980), Holländer et al. (1981), Oberhofer and Back (1979), Hinz and Kunjara (1977), and Targowski and Targowski (1979). Symbols: +, 90% or more of strains are positive; d, 11–89% of strains are positive; −, 10% or less of strains are positive; [d], 11–89% of strains give a delayed positive reaction; [+], 90% or more of strains give a delayed positive reaction; blank space, not determined; and W, weakly positive reaction.
[b] ONPG, o-nitrophenyl-β-D-galactopyranoside.

Common name: Koch-Weeks bacillus.

Long slender rods 0.2–0.3 × 2.0–3.0 μm. Growth of freshly isolated strains on chocolate agar is slow. After 48 h colonies attain a diameter of about 0.5 mm and are smooth, low, convex, grayish, and translucent. The species does not grow on Tryptic soy agar (Difco) with X and V factors added, in contrast to *H. influenzae*. Forms comet-like colonies in semisolid agar media (Pittman and Davis, 1950).

Capsules have not been demonstrated. Bacteria of this species produce an extracellular IgA1-cleaving endopeptidase (Kilian et al., 1979) and have neuraminidase activity (Müller and Hinz, 1977).

Features which may be of use in distinguishing *H. aegyptius* from *H. influenzae* include poorer growth on most media, a lack of indole production and xylose fermentation, the distinct bacillary morphology, a hemagglutinating activity, the comet-like growth in semi solid media, and the susceptibility to troleandomycin. Apart from the latter, none of these features will unequivocally differentiate the two species.

Some strains lack the enzyme ferrochelatase which inserts iron into protoporphyrin IX and thus require protoheme as X-factor (White and Granick, 1963; Kilian et al., 1976; Ruckelshausen and Holländer, 1978).

The organism causes acute or subacute infectious conjunctivitis in hot climates. It has not been demonstrated in healthy individuals.

The high frequency of transformation (Leidy et al., 1959; Leidy et al., 1965) and an observed 78% DNA homology (Pohl, 1981) between *H. influenzae* and *H. aegyptius* may indicate that it is unjustified to maintain the separation between the two species. However, until further information is available, the fact that *H. aegyptius* appears to be associated with a more acute and contagious form of conjunctivitis is to us a reasonable cause for maintaining *H. aegyptius* as a separate taxon.

The mol% G + C of the DNA is 39 (T_m).

Type strain: ATCC 11116 (strain 180a of Pittman).

3. **Haemophilus haemolyticus** Bergey, Harrison, Breed, Hammer and Huntoon 1923, 269.[AL]

hae.mo.ly'ti.cus. Gr. n. *haema* blood; Gr. adj. *lyticus* loosening, dissolving; M.L. adj. *haemolyticus* blood dissolving.

Small, regular, noncapsulated coccobacilli or short rods with occasional filamentous forms. Colonies on chocolate agar are smooth, convex, grayish, translucent, and reach a diameter of 0.5–1.5 mm within 24 h.

Produce clear hemolytic zones on bovine or sheep blood agar. However, the hemolytic activity may be lost upon subcultivations.

Is found in the nasopharynx of a minority of the healthy human population. A pathogenic potential has never been demonstrated.

The mol% of G + C of the DNA is 39 (T_m).

Type strain: NCTC 10659 (strain AQ/3273 of McIves). Strain NCTC

Table 5.74.

Other characteristics of the species of the genus **Haemophilus**[a]

Characteristics	H. influenzae Biovar I	Biovar II	Biovar III	Biovar IV	Biovar V	Biovar VI	H. aegyptius	H. haemolyticus	H. haemoglobinophilus	H. ducreyi	H. parainfluenzae Biovar I	Biovar II	Biovar III	H. parahaemolyticus	H. paraphrohaemolyticus	H. pleuropneumoniae	H. paracuniculus	H. aphrophilus	H. paraphrophilus	H. segnis	H. parasuis	H. paragallinarum	H. avium	"H. somnus"	"H. agni"	"H. equigenitalis"
Lysine decarboxylase	d	d	—	d	d	—	—	—	—	—	d	—	—	—	—	—	—	—	+	—	—	+	—	+	+	+
Oxidase	+	+	+	d	+	+	+	+	+	+	+	+	+	+	+	d	+	—	+	—	—	—	$+^{7}$	+	+	+
Ubiquinone[b]	—	—	—	—	—	—	—	$+^{8}$	—	—	—	—	—	—	—	$+^{8}$	—	—	—	—	—	$+^{7}$	+	+	—	+
Naphthoquinone[b,c]	+	+	$+^{7}$	+	+	+	$+^{7}$	+	$+^{8}$	+	+	+	$+^{7}$	+	$+^{7}$	+	$+^{6}$	$+^{6}$	$+^{6}$	+	+	+	+	+	—
H$_2$S (lead acetate)	—	—	—	d	—	—	—	+	d	—	+	+	+	+	+	+	—	—	+	—	d	—	+	+	—	—
Hemagglutination	—	—	—	—	—	—	d	—	—	—	—	—	—	—	—	—	—	—	—	—	—	+	+	—	—	—
Acid production from:																										
L-Arabinose	—	—	—	—	—	—	—	—	—	—	—	—	—	—	—	—	—	—	—	—	—	—	—	d	d	W
L-Rhamnose	—	—	—	—	—	—	—	—	—	—	—	—	—	—	—	—	—	—	—	—	—	—	—	—	—	—
D-Galactose	+	+	+	+	+	+	[+]	+	—	—	+	+	+	d	d	+	—	+	—	W	+	—	+	d	W	—
D-Mannose	—	—	—	—	—	—	—	+	—	+	+	+	—	—	+	+	+	+	W	+	+	+	+	—	—	—
Sorbose	—	—	—	—	—	—	—	—	—	—	—	—	—	—	—	—	—	+	—	—	—	—	—	d	—	—
Cellobiose	—	—	—	—	—	—	—	—	—	—	—	—	—	—	—	—	—	—	—	—	—	—	—	d	—	—
Maltose	+	+	+	+	+	+	[+]	+	+	—	+	+	+	+	+	+	+	+	+	W	+	+	+	+	W	—
Melibiose	—	—	—	—	—	—	—	—	—	—	—	—	—	—	—	—	—	+	—	—	—	—	—	+	+	—
Trehalose	—	—	—	—	—	—	—	—	—	—	—	—	—	—	—	—	—	+	—	—	—	—	—	+	+	—
Melizitose	—	—	—	—	—	—	—	—	—	—	—	—	—	—	—	—	—	+	—	—	—	—	—	—	—	—
Raffinose	—	—	—	—	—	—	—	—	—	—	—	—	—	—	—	d	—	+	—	—	—	—	—	d	—	—
Inulin	—	—	—	—	—	—	—	—	—	—	—	—	—	—	—	—	—	—	—	—	+	—	—	—	—	—
Dulcitol	—	—	—	—	—	—	—	—	—	—	—	—	—	—	—	—	—	—	—	—	—	—	—	d	—	—
Glycerol	—	—	—	—	—	—	—	—	—	—	—	—	—	—	—	—	—	—	d	—	W	—	—	—	—	—
meso-Erythritol	—	—	—	—	—	—	—	—	—	—	—	—	—	—	—	—	—	—	d	—	—	d	—	d	—	—
Inositol	—	—	—	—	—	—	—	—	—	—	—	—	—	—	—	—	—	—	—	—	—	—	—	—	—	—
Xylitol	—	—	—	—	—	—	—	—	—	—	—	—	—	—	—	—	—	—	—	—	—	—	—	—	—	—
Esculin, salicin, adonitol	—	—	—	—	—	—	—	—	—	—	—	—	—	—	—	—	+	—	—	—	—	—	—	—	—	—
α-Galactosidase	—	—	—	—	—	—	—	—	—	—	—	—	—	—	—	—	—	d	d	—	d	d	+	—		—
α-Glucosidase	—	—	—	—	—	—	—	—	—	—	—	—	—	—	—	—	—	—	—	—	d	—	—	—		—
β-Glucosidase	—	—	—	—	—	—	—	—	—	—	—	—	—	—	—	—	—	—	—	—	—	—	—	—		—
α-Mannosidase	—	—	—	—	—	—	—	—	—	—	—	—	—	—	—	—	—	—	—	—	—	—	—	—		—
β-Xylosidase	—	—	—	—	—	—	—	—	—	—	—	—	—	d	—	—	—	—	—	—	—	d	—	—		—
β-Glucuronidase	d	—	d	—	—	—	—	—	—	—	—	—	—	—	—	—	—	—	—	—	—	—	—	—	—	—
Nitrate reduction	+	+	+	+	+	+	+	+	+	+	+	+	+	+	+	+	+	+	+	+	+	+	+	+	+	—
Nitrite reduction	—	—	—	—	—	—	—	d	d	—	+	+	+	+	+	+	+	+	+	d	—	—	—	—		—

[a] Data compiled from: Kennedy et al. (1960), Bailie et al. (1973), Garcia-Delgado et al. (1974), Corboz and Wild (1981), Kilian (1976), Kilian et al. (1979), Kilian and Theilade (1978), Kilian (unpublished results), Taylor et al. (1978), Mannheim et al. (1978), Holländer et al. (1981), Oberhofer and Back (1979), Hinz and Kunjara (1977), and Targowski and Targowski (1979). For symbols see Table 5.73.

[b] Superscript numbers indicate the number of isoprenoid units.

[c] Where "+" is indicated, demethylmenaquinone is the naphthoquinone.

8479 (ATCC 10014) which is listed by Sneath and Skerman (1966) as suggested working type of the species, is a strain of *H. parahaemolyticus*.

4. Haemophilus haemoglobinophilus (Lehmann and Neumann 1907) Murray 1939, 309.[AL] (*Bacterium haemoglobinophilus* Lehmann and Neumann 1907, 270.)

hae.mo.glo.bi.no'phi.lus. M.L. n. *haemoglobinum* hemoglobin; Gr. adj. *philos* loving; M.L. adj. *haemoglobinophilus* hemoglobin-loving.

Small, slightly pleomorphic, noncapsulated rods. Colonies on chocolate agar are smooth, convex and translucent with a small granular area on top, and reach a diameter of 1–2 mm within 24 h. In contrast to other members of the genus, they grow well on blood agar and show no satellite growth around a streak of staphylococcus.

Differs from all other *Haemophilus* species in lacking alkaline phosphatase.

Belongs to the normal flora of the preputial sac of dogs, and is probably of low pathogenicity.

The mol% G + C of the DNA is 38 (T_m).

Type strain: NCTC 1659 (strain XIII of Kristensen).

5. Haemophilus ducreyi (Neveu-Lemaire 1921) Bergey, Harrison, Breed, Hammer and Huntoon 1923, 271.[AL] (*Coccobacillus ducreyi* Neveu-Lemaire 1921, 20.)

du.crey'i. M.L. gen. n. *ducreyi* of Ducrey; named after Ducrey, the bacteriologist who first isolated this organism.

Slender rods in pairs or chains, measuring 0.5 × 1.5–2.0 μm. Capsules have not been detected. Growth is poor on most laboratory media. Colonies on chocolate agar after 72 h are predominantly small (~0.5 mm in diameter), flat, smooth, grayish and translucent but often with a few interspersed larger colonies having an otherwise identical appearance. Growth on blood agar is very sparse, and there is no satellite growth around a staphylococcus. Some strains may show weak β hemolysis.

Clinical isolates usually appear to be asaccharolytic. However, under favorable growth conditions, some strains show a late positive reaction for glucose fermentation. The species is inert in most of the traditional biochemical tests (Table 5.74) but produces a wide array of peptidases.

Causes the human venereal disease known as soft chancre or chancroid. A carrier state in healthy individuals has not been detected.

The mol% G + C of the DNA is 38 (T_m).

Type strain: CIP 542.

6. Haemophilus parainfluenzae Rivers 1922, 431.[AL]

pa.ra.in.flu.en′zae. Gr. prep. *para* alongside of, resembling; M.L. n. *influenzae* specific epithet; M.L. gen. n. *parainfluenzae* intended to mean like the species of *H. influenzae*.

Small, pleomorphic rods usually with long filamentous forms. Occasional strains possessing a capsule have been described (Sims, 1970). Colonies on chocolate agar are grayish white or yellowish opaque and reach a diameter of 1–2 mm after 24 h. Some strains produce flat, smooth, colonies with an entire edge, others show a serrated edge, and yet others grow as very rough wrinkled colonies. The irregular forms of colonies are usually coherent in texture and can be slid intact across

the surface of the agar plate. Strains growing as rough type colonies often convert into the smooth type after some in vitro transfers. Growth in broth media may or may not be granulated.

H. parainfluenzae is ubiquitous in the human oral cavity and pharynx, and may be present in the normal vaginal flora (Sims, 1970; Kilian and Schiøtt, 1975, Tuyau and Sims, 1975). Organisms closely resembling *H. parainfluenzae* have been isolated from monkeys, swine, rabbits, and rats (Kilian and Frederiksen, 1981). Has neuraminidase activity (Tuyau and Sims, 1974). The majority of human isolates can be assigned to any of three biovars (I–III) defined on the basis of some differences in biochemical reactions (Tables 5.73 and 5.74).

Is of low pathogenicity, but is occasionally implicated in endocarditis in humans.

The mol% G + C of the DNA is 40–41 (T_m).

Type strain: NCTC 7857 (biovar I). *Reference strains:* NCTC 10665 (biovar II), NCTC 11607 (biovar III).

7. Haemophilus parahaemolyticus Pittman 1953, 750.[AL]

pa.ra.hae.mo.ly′ti.cus. Gr. prep. *para* alongside of, resembling; M.L. n. *haemolyticus* specific epithet; M.L. adj. *parahaemolyticus* (*Haemophilus*) *haemolyticus*-like.

Small, pleomorphic rods usually with long filamentous forms. Growth on chocolate agar is similar to that of *H. parainfluenzae*. A more or less distinct β hemolytic zone is produced on blood agar plates, most clearly detectable on bovine or sheep blood agar. The hemolytic activity is often lost upon subcultivations.

Regular member of the microflora of the human oral cavity and pharynx. Has been associated with acute pharyngitis, purulent oral infections, and occasional cases of endocarditis.

Some strains of this species (including the type strain) produce an extracellular endopeptidase capable of cleaving human IgA1 (Male, 1979). All strains have neuraminidase activity (Tuyau and Sims, 1974).

Hemolytic V-requiring strains from swine previously assigned to this species (Zinnemann and Biberstein, 1974) have been shown to belong to a separate species, *H. pleuropneumoniae* (Kilian et al., 1979).

The mol% G + C of the DNA is 40–41 (T_m).

Table 5.75.

Relationship between capsular serovar and biovar of 423 clinical isolates of **Haemophilus influenzae**

Biovar	No. of Strains	Serovar						Non-capsulated Strains
		a	b	c	d	e	f	
I	227	3	182	0	0	0	10	32 (14%)
II	87	0	8	3	0	0	1	75 (86%)
III	59	0	0	0	0	0	0	59 (100%)
IV	34	0	4	0	8	10	0	12 (35%)
V	16	0	0	0	0	0	0	16 (100%)

Table 5.76.

Structures of *H. influenzae* capsular polysaccharides[a]

Type	Structure
a	4)-β-D-Glc-(1 → 4)-D-ribitol-5-(PO$_4$ →
b	3)-β-D-Rib-(1 → 1)-D-ribitol-5-(PO$_4$ →
c	4)-β-D-GlcNAc-(1 → 3)-α-D-Gal-1-(PO$_4$ →
	3
	↑ R = OAc (0.8)
	│ H (0.8)
	R
d	4)-β-D-GlcNAc-(1 → 3)-β-D-ManANAc-(1 →
	6
	↑
	│ R = L-serine (0.41)
	│ L-threonine (0.14)
	R L-alanine (0.41)
e	3)-β-D-GlcNAc-(1 → 4)-β-D-ManANAc-(1 →
e′	3)-β-D-GlcNAc-(1 → 4)-β-D-ManANAc-(1 →
	3
	↑
	2
	β-D-fructose
f	3)-β-D-GalNAc-(1 → 4)-α-D-GalNAc-1-(PO$_4$ →
	3
	↑
	OAc

[a] Ribose and fructose are in the furanose ring form; Glc, Gal, GlcNAc, and ManANAc are in the pyranose ring form (Egan et al., 1980a, b; Tsui et al., 1981 a, b; Sutton et al., 1982).

Type strain: NCTC 8479 (strain 536 of Pittman; has been mislabeled *H. haemolyticus*).

8. Haemophilus paraphrohaemolyticus Zinnemann, Rogers, Frazer and Deveraj, 1971, 143.[AL]

par.aph.ro.hae.mo.ly′ti.cus. M.L. adj. *paraphro*-resembling *H. aphrophilus*; M.L. adj. *haemolyticus* blood dissolving; M.L. adj. *paraphrohaemolyticus* like *H. aphrophilus* but hemolytic.

Growing in air with 10% extra CO_2, short to medium length rods 0.75–2.5 μm and 0.4–0.5 μm in width with occasional short filaments; in air without added CO_2 short to long, coarse rods with involution forms and twisted filaments.

Requires increased CO_2 tension in incubation atmosphere for satisfactory growth and good β hemolysis on blood agar.

Apart from the CO_2 requirement, the biochemical, physiological and ecological characteristics are identical to those of *H. parahaemolyticus*.

Has been isolated from human sore throats, ulcers of the mouth, in sputum and in urethral discharge of adult males, but pathogenic significance is unknown.

The mol% G + C of the DNA is 40–41 (T_m).

Type strain: NCTC 10670 (strain L1 of Zinnemann, Rogers, Frazer and Devaraj, 1971).

9. Haemophilus pleuropneumoniae* Shope 1964, 362.[AL]

pleu.ro.pneu.mo′ni.ae. Gr. n. *pleura* lung sac; Gr. n. *pneumon* the lungs; M.L. fem. gen. n. *pleuropneumoniae* of pleuropneumonia.

Coccobacilli to pleomorphic rods with occasional filamentous forms. Pathogenic strains are capsulated.

Colonies on chocolate agar are grayish opaque and reach a diameter of 1–2 mm in 48 h. Two types of colonies can be seen: a rounded, hard "waxy" type, which adheres to a platinum loop, or a flatter, soft, glistening type. Capsulated strains produce iridescent colonies on clear agar media (e.g. Levinthal's agar). Colonies on bovine or sheep blood agar usually give rise to a β hemolytic zone. The hemolysin acts synergistically with *Staphylococcus aureus* β toxin on bovine or sheep blood cells resulting in a positive CAMP reaction on blood agar.

Six serovars (1–6) have been recognized on the basis of surface antigens, which, in the case of serovars 1–3, have been identified as capsular polysaccharides (Nicolet, 1971; Gunnarsson et al., 1977; Nielsen, 1982).

H. pleuropneumoniae has been isolated from pneumonic lesions, blood, and various other tissues of swine, a case of arthritis in a lamb, and a brain abscess in a steer. A carrier state in healthy animals has not been demonstrated.

Produces extensive lobar pneumonia with accompanying fibrinous pleuritis in swine; when bacteremic, also causes meningitis and arthritis.

The mol% G + C of the DNA is 42 (T_m).

Type strain: ATCC 27088 (strain 4074 of Shope, 1964; serovar 1). *Reference strains:* serovar 2: ATCC 27089; serovar 3: ATCC 27090; serovar 4: ATCC 33378 (strain M62 of Biberstein et al., 1963); serovar 5: ATCC 33377 (strain K17 of Biberstein et al., 1963); serovar 6: ATCC 33590 (NCTC 11407 (strain Femö of Nielsen, 1979)).

10. Haemophilus paracuniculus Targowski and Targowski 1979, 33.[VP]

pa.ra.cu.ni′cu.lus Gr. prep. *para* alongside of, resembling; L. n. *cuniculus* rabbit, or possibly the specific epithet of "*Haemophilus cuniculus*," a species which has never been described; M.L. adj. *paracuniculus* like "*H. cuniculus*."

Small coccobacilli or pleomorphic rods with occasional filamentous forms. In broth media supplemented with V factor, the organism forms chains of rods. Colonies on chocolate agar are smooth, high convex, grayish and opaque. Capsules have not been described.

Does not grow in phenol red broth base supplemented with V factor.

Has been isolated from the gastrointestinal tract of rabbits with mucoid enteritis. The pathogenic significance of the organism has not been established.

The mol% G + C of the DNA is 40 (Bd).

Type strain: ATCC 29986.

11. Haemophilus aphrophilus Khairat 1940, 505.[AL]

a.phro′phi.lus. Gr. n. *aphros* foam; Gr. adj. *philos* loving; M.L. adj. *aphrophilus* foam loving.

Short regular rods 0.45–0.55 μm × 1.5–1.7 μm with only occasional filamentous forms. Colonies on chocolate agar incubated in air supplemented with 10% extra CO_2 are high convex, granular, yellowish, opaque, and reach a diameter of 1.0–1.5 mm within 24 h. When incubated without extra CO_2, the growth is characteristically stunted, with very small colonies interspersed with a few larger colonies. Grows on blood agar incubated in air plus CO_2 without feeder strain. Growth in broth media is granular, with a heavy sediment on the bottom of the tube and adhering colonies on the walls, which are difficult to remove.

H. aphrophilus possesses all the enzymes of the hemin biosynthetic pathway characteristic for hemin-independent species (White and Granick, 1963). Accordingly, it gives a positive though usually weak reaction for both porphobilinogen and prophyrins in the porphyrin test. The reason for an apparent need for hemin-containing media at primary isolation remains to be elucidated.

H. aphrophilus is a frequent member of the microflora of human dental plaque particularly between the teeth and in the gingival pockets (Kraut et al., 1972; Kilian and Schiøtt, 1975). Occasionally causes endocarditis and brain abscesses in humans. Has been isolated from spinal fluid, wounds and jaw infections (King and Tatum, 1962).

The mol% G + C of the DNA is 42 (T_m).

Type strain: NCTC 5906 (strain PM1 of Khairat, 1940).

12. Haemophilus paraphrophilus Zinnemann, Rogers, Frazer and Boyce 1968, 418.[AL]

pa.ra.phro′phi.lus. Gr. prep. *para* alongside of, resembling; M.L. adj. *aphrophilus* specific epithet; M.L. adj. *paraphrophilus* (*Haemophilus*) *aphrophilus*-like.

Short regular rods with occasional filamentous forms. Involution forms occur under fully aerobic incubation. Growth characteristics are identical to those described for *H. aphrophilus*. In contrast to *H. aphrophilus*, *H. paraphrophilus* requires V factor and is not dependent on X factor on primary isolation. The two species have otherwise almost identical properties (Table 5.73) and are genetically closely related (Pohl, 1981).

Found as a member of the normal flora of the human oral cavity and pharynx (Kilian and Frederiksen, 1981). May cause subacute endocarditis, paronychia, brain abscesses and has been isolated from osteomyelitis of the jaw, an inflamed appendix, from urine of children with congenital malformation of urogenital tract and from the vagina of mature women.

The mol% G + C of the DNA is 42 (T_m).

Type strain: NCTC 10557 (strain Reece of Zinneman et al., 1968).

13. Haemophilus segnis Kilian 1977, 306.[AL] (Effective publication: Kilian 1976, 35.)

seg′nis. L. adj. *segnis*, slow, sluggish.

Pleomorphic rods, often showing a predominance of irregular filamentous forms. Colonies on chocolate agar are smooth or granular, convex, grayish white, opaque and reach a diameter of about 0.5 mm after incubation for 48 h. Growth in fermentation media is slow and reactions are negative or weakly positive.

Regular member of the human oral flora particularly in dental plaque and can be isolated from the pharynx. Has been isolated in pure culture from pancreas abscess.

---

* Although Matthews and Pattison (1961) are indicated as the authors of the specific epithet in the Approved Lists of Bacterial Names, and although they were the first to describe the organism in the literature, they called it *H. parainfluenzae* and did not propose it as a new species.

The mol% G + C of the DNA is 43–44 (T_m).

Type strain: NCTC 10977 (strain HK 316 of Kilian, 1976).

14. Haemophilus parasuis Biberstein and White 1969, 77.[AL]

pa.ra.su'is. Gr. prep. *para* alongside of, resembling, M.L. n. *suis* specific epithet; M.L. adj. *parasuis* (*Haemophilus*) *suis*-like.

Thin, pleomorphic rods of varying length. Growth on chocolate agar is very feeble after 48–72 h. The colonies are smooth, grayish, translucent and reach a diameter of about 0.5 mm. A marked enhancement of growth on chocolate agar with colonies attaining a diameter of 1–2 mm, is seen around a streak of a staphylococcus.

Four serovars (A–D) based on capsular polysaccharides (Bakos et al. 1952). Noncapsulated strains are antigenically heterogeneous. Two types (I and II) have been identified on the basis of differences in patterns of cellular proteins (Nicolet and Krawinkler, 1981).

Member of the normal flora of the upper respiratory tract of swine. Causes respiratory tract infections and polyserositis (Glässer's disease).

Most strains in culture collections labeled *H. suis* are only V-factor requiring and belong to *H. parasuis* according to the proposal of Biberstein and White (1969).

The mol% G + C of the DNA is 41–42 (T_m).

Type strain: NCTC 4557 (strain 1374 of Shope).

15. Haemophilus paragallinarum Biberstein and White 1969, 77.[AL]

pa.ra.gal.li.na'rum. Gr. prep. *para* alongside of, resembling; M.L. gen. pl. n. *gallinarum* specific epithet; M.L. adj. *paragallinarum* (*Haemophilus*)-*gallinarum*-like.

Coccobacilli to pleomorphic rods with occasional filamentous forms. Most strains are capsulated. Colonies on chocolate agar are smooth, convex, grayish, semiopaque, and attain a diameter of 0.5–1.0 mm within 48 h, in air supplemented with 10% CO_2. Growth is feeble in air without extra CO_2. Young cultures (8- to 24-h-old) of capsulated strains produce iridescent colonies on transparent agar media. Growth is enhanced by serum added to cultivation media (Hinz, 1973; Rimler et al., 1976).

Three serovars (A–C) have been defined on the basis of heat-labile surface antigens (Page, 1962). Kato and Tsubahara (1962) have used the designations I, II and III for three serovars of which serovar III may be a noncapsulated variant of serovar I (Kume et al., 1980). Hinz (1980) has established six serovars on the basis of heat-stable somatic antigens.

Produces neuraminidase and *N*-acetylneuraminate pyruvate lyase (Hinz and Müller, 1977).

Present in the respiratory tract of poultry. Causes respiratory tract disease in chickens known as infectious coryza.

The species includes organisms previously labelled *H. gallinarum* since no such strains require both X and V factors.

The mol% G + C of the DNA is 42 (T_m).

Type strain: ATCC 29545 (strain IPDH 2403 of Hinz and Kunjara, 1977; serovar A).

16. Haemophilus avium Hinz and Kunjara 1977, 324.[AL] (*Haemophilus paravium* Hinz and Müller, 1977, 72).

a'vi.um. L. n. *avis* a bird; L. gen. pl. n. *avium* of birds.

Coccoid to pleomorphic rods which occur singly, in pairs, and in filamentous forms. Colonies on chocolate agar are usually smooth, convex, grayish white or yellowish opaque, but may be wrinkled or granular. Most strains are capsulated and give rise to iridescent colonies on transparent agar media. CO_2 is usually not required for optimum growth (Hinz and Kunjara, 1977).

Produces neuraminidase and *N*-acetylneuraminate pyruvate lyase (Hinz and Müller, 1977).

Is a regular member of the respiratory tract flora of poultry. A pathogenic potential has not been demonstrated.

The mol% G + C of the DNA is 42 (T_m).

Type strain: ATCC 29546 (strain IPDH 2654 of Hinz and Kunjara, 1977; serovar 2).

Species Incertae Sedis

The following three species have been described as members of the genus *Haemophilus*. By current standards they do not qualify for inclusion in the genus.

a. "*Haemophilus somnus*" Bailie 1969, 64. (*Haemophilus*-like organism, Kennedy, Biberstein, Howarth, Frazier and Dungworth 1960, 403; *Actinobacillus actinoides*-like organism, Bailie, Anthony and Weide 1966, 165; *Actinobacillus* sp., Gossling 1966, 18; *Haemophilus somnifer* Miles, Anthony and Dennis 1972, 431.)

som'nus. L. masc. n. *somnus* sleep.

Gram-negative short rods, coccobacilli, or filaments. Most strains noncapsulated. Nonmotile. Not acid-fast. Colonial characteristics: 0.2–0.6 mm in diameter, raised circular, smooth, entire colonies on beef blood agar after 24 h (Kennedy et al., 1960); 0.5–1.5 mm on PPLO-Haemophilus agar (Nicolet, 1971; Corboz and Nicolet, 1975); 1–2 mm in 3 days on brain-heart infusion calf blood-yeast extract plates (Garcia-Delgado et al., 1976). Prolonged incubation on blood agar produces colonies with slightly granular appearance, papillate centers, and flattened peripheries. Most authors report no or weak hemolytic activity.

Nutrition and growth conditions: Usually no growth of fresh isolates on infusion, tryptose, trypticase soy, serum, or hemoglobin agar under any atmospheric condition. Growth occurs on blood and chocolate agars, and on serum-yeast extract PPLO-agar under 5–20% carbon dioxide. Poor or no growth in ambient air, or anaerobically on any solid medium (Kennedy et al., 1960; Garcia-Delgado et al., 1976). Satellitic growth on *Staphylococcus aureus* occurs in an atmosphere of increased carbon dioxide (Kennedy et al., 1960). Occasional strains lack the special atmospheric and nutritional requirements described, and all cultures apparently adapt gradually to fully aerobic growth. No growth response to X and V factors. Metabolism and biochemical activities are shown in Table 5.73 and 5.74.

Antigenic structure: all isolates examined have antigens in common regardless of geographic and anatomic origin (Shigidi and Hoerlein, 1970; Corboz and Nicolet, 1975; Garcia-Delgado et al., 1976). Cross-reactions at low titer with a number of other mostly Gram-negative bacteria have been reported (Miller et al., 1975).

Antimicrobial susceptibility: only rough, qualitative data exist, based on nonstandardized disk diffusion tests. These suggest susceptibility to ampicillin, bacitracin, cephaloridine, chloramphenicol, dihydrostreptomycin, erythromycin, novobiocin, penicillin G, polymyxin B, and tetracycline; and resistance to lincomycin, neomycin, and sulfonamides (Garcia-Delgado et al., 1976).

Pathogenicity: "*H. somnus*" causes septicemia and meningoencephalomyelitis in cattle (Kennedy et al., 1960) and is involved in respiratory (Brown et al., 1970) and genital infections (Waldhalm et al., 1974), including abortions (Chladek, 1975) in that species. Experimental infections of mice, guinea pigs, rabbits, and sheep have been reported (Kennedy et al., 1960; Panciera et al., 1968; Miles et al., 1972).

Ecology: the reservoir of "*H. somnus*" appears to be the mucous membranes of the normal bovine respiratory and genital tracts (Corstvet et al., 1973; Waldhalm et al., 1974).

The mol% G + C of the DNA is 37.3 ± 0.2 (T_m).

Type strain: none designated.

b. "*Haemophilus agni*" Kennedy, Frazier, Theilen and Biberstein 1958, 645.

ag'ni. L. gen. n. *agni* of the lamb.

Morphology: Gram-negative, nonmotile, non acid-fast pleomorphic rods or coccobacilli 0.3–0.7 × 0.5–2.0 μm. Long, thread-like forms containing spherical and thickened fusiform bodies occur. Pleomorphism diminishes on passage. Presence of capsules suggested.

Colonial characteristics: in 5–10% CO_2 the colonies are convex and translucent and range from <0.5 mm–1.5 mm. Become flattened peripherally and acquire more sharply contoured edges but no significantly larger size on further incubation.

Nutrition and growth conditions: very slight growth on blood agar in air. Colonial growth on blood agar under 5–10% CO_2 as described above. Satisfactory growth occurs on hemoglobin cystine agar and chocolate agar. No growth response to X and V factors. No satellitic growth with staphylococci. No growth on MacConkey, tryptose, 10% equine serum (with and without 1% yeast hydrolysate), mycoplasma agar, coagulated blood serum, or gelatin.

Metabolism and biochemical activities are shown in Table 5.73 and 5.74.

Antigenic structure: cross-react consistently with "*H. somnus*".

Antimicrobial susceptibility: not reported.

Pathogenicity: associated with septicemia, meningitis, polyarthritis myositis, pneumonia and mastitis of sheep. Of experimental animals only suckling mice were susceptible.

Ecology: the reservoir of *H. agni* has not been determined.

The mol% G + C of the DNA is 36.8 ± 0.3.

Type strain: none designated.

Further comments: an organism occupying a position close to both "*H. agni*" and "*H. somnus*" has been encountered in a number of inflammatory diseases of sheep in Australia (Claxton and Everett 1966, Rahaley and White 1977, Roberts 1956) and New Zealand (Kater et al., 1962) and named "*Histophilus ovis*" (Roberts 1956).

c. "*Haemophilus equigenitalis*" Taylor, Rosenthal, Brown, Lapage, Hill and Legros 1978, 136.

e.qui.ge.ni.ta'lis. L. gen. n. *equi* of the horse; L. adj. *genitalis* genital; L. adj. genital of the horse.

Morphology: Gram-negative, nonacid-fast, nonmotile, short rods, 0.7 μm wide × 0.7–1.8 μm in length, with occasional filaments 5–6 μm long.

Ultrastructure: typical of Gram-negative bacteria. Triple-layered outer membrane, 7 nm thick; dense intermediate layer, 4 nm thick; triple-layered cell membrane, 7 nm thick. Presence of thread-like capsule, 20–30 nm thick, has been demonstrated in ruthenium red-stained preparations under transmission electron microscopy (Swaney and Breese, 1980).

Colonial characteristics: raised pinpoint colonies, smooth butyrous, gray, at 48–72 h on horse blood chocolate agar (Taylor et al., 1978). On chocolated Eugon agar (Swaney and Sahu, 1978), colonies are visible in 24 h under ×15 magnification; they are 1 mm in diameter at 48 h and 1.5 mm in diameter at 72 h. Nonhemolytic on Columbia horse blood agar (Taylor et al., 1978).

Nutrition and growth conditions: good growth on Columbia base chocolate agar (Taylor et al., 1978) or chocolated Eugon agar (Swaney and Sahu, 1978). Appreciable growth also on Columbia base blood agar and chocolate agar with other base. Little growth on plain blood agar, serum agar, starch agar, casein agar, egg yolk agar, nutrient agar. No growth on MacConkey, poly-β-hydroxybutyrate, Tween, 3.5% NaCl, or 1% glycine agars. Some growth stimulation occurs with X and XV factor (disks) but not by V factor, on nutrient agar. Optimum growth occurs under 5–10% carbon dioxide. Very little growth occurs in air or anaerobically. Grows over a temperature range from 30–41°C; optimum temperature, 37°C.

Metabolism and biochemical activities: in addition to the information given in Table 5.73 and 5.74, the following characterize "*H. equigenitalis*." No hydrolysis of starch, gelatin, lecithin, coagulated serum (Loeffler), and DNA. Negative for phenylalanine deaminase and desimidase. No utilization of citrate or malonate. Gluconate is not oxidized.

Antigenic structure: Antiserum agglutinates at low titers (1:20) *B. abortus, P. multocida, P. pneumotropica, H. influenzae,* "*H. influenzaemurium*," *Y. pseudotuberculosis, Y. enterocolitica* (serovar not stated), *Actinobacillus* sp., *Moraxella osloensis, Neisseria elongata* subsp. *glycolytica.*

Antimicrobial agents: minimum inhibitory concentrations (μg/ml) for the type strain have been reported as follows: penicillin G, <0.25; ampicillin, 0.5; tetracycline, 1.0; erythromycin, <0.06; clindamycin, 16; lincomycin, 32; gentamicin, 0.25; kanamycin, 1.0; neomycin, 1.0; streptomycin, >512 (*note:* streptomycin-susceptible strains exist); polymyxin B, 0.25; nalidixic acid, 4.0; nitrofurantoin, 1.0; sulfamethoxazole, 32; and trimethoprim, 4.0. Disk diffusion tests have also suggested susceptibility to cephaloridine, carbenicillin, amikacin, tobramycin and chloramphenicol, and resistance to metronidazole.

Pathogenicity: "*H. equigenitalis*" causes endometritis and cervicitis in mares as a result of venereal exposure to carrier stallions.

Ecology: the agent appears to be a strict parasite leading a commensal existence on the external genitalia of healthy stallions but may survive for prolonged periods in the clitoridal region and vulva of mares following infection.

The mol% G + C of the DNA is 36–37 (T_m).

Type strain: NCTC 11184 (strain 61717/77).

Other organisms

Two species, previously listed as *incertae sedis*, have been definitively removed from the genus. One, *H. piscium*, has been excluded on the basis of its DNA base composition, growth factor requirements (Kilian, 1976), its cell wall structure (Kilian and Theilade, 1975), respiratory quinones (Holländer and Mannheim, 1975), antigenic constitution, and phage susceptibility (Paterson et al., 1980). It has been concluded that it represents an atypical form of *Aeromonas salmonicida* (Paterson et al., 1980). The other *species incertae sedis*, *H. vaginalis*, has been placed into a distinct genus, *Gardnerella* (q.v.).

Some ambiguity continues to surround the species "*H. suis*" and "*H. gallinarum*." Both of these were originally described as requiring both X and V factors. In recent years, and by current testing procedures, isolates obtained under comparable conditions lack the X requirement. An organism having both X and V requirements is occasionally recovered from the respiratory tract of swine, but its identity with the original "*H. suis*," particularly with regard to pathogenic activity, remains unproven. The names *H. parasuis* and *H. paragallinarum* were

introduced to designate the more common X factor-independent types (Biberstein and White, 1969), but there is real doubt as to the existence of anything resembling "*H. gallinarum*" since isolation of an X-dependent strain of "*H. gallinarum*" has not been documented since 1936.

Previous editions of *Bergey's Manual* have listed several *Haemophilus* species that were omitted from the Approved Lists of Bacterial Names, because no representative cultures existed. These species included "*H. putoriorum*," "*H. citreus*," "*H. influenzaemurium*" and "*H. ovis*." Recently, the isolation of organisms with characteristics closely following the original descriptions of the two last species have been reported (Czukás, 1976; Little et al., 1980). However, a revival of the two names has not yet been proposed.

Unclassified *Haemophilus* organisms, which may represent new species, have been isolated from man (Ryan, 1968; Kilian, 1976), swine (Biberstein et al., 1977; Kilian et al., 1978), sheep (Little et al., 1980), cats (Nicolet, unpublished results), and fowl (Grebe and Hinz, 1975).

Genus III. *Actinobacillus* Brumpt 1910, 849.[AL]

J. E. PHILLIPS

Ac.ti.no.ba.cil'lus. Gr. n. *actis, actinis* a ray; L. dim. masc. n. *bacillus* a small staff or rod; M.L. masc. n. *Actinobacillus* ray bacillus or rod.

Cells spherical, oval or rod-shaped, $0.4\pm0.1 \times 1.0\pm0.4$ μm. **The cells are mostly bacillary but are interspersed with coccal elements which often lie at the pole of a bacillus giving a characteristic "Morse code" form.** Occasionally longer forms occur up to 6 μm, especially on media containing glucose or maltose. The cells are arranged singly, in pairs or, more rarely, in chains. Do not form endospores. Gram-negative. Not acid-fast. Exhibit irregular staining. Small amounts of extracellular slime may be demonstrated in wet India ink preparations. **Nonmotile.** Facultative anaerobes. Optimum temperature, 37°C; temperature range for growth, 20–42°C. Cultures are very sticky on primary isolation; colonies may be difficult to remove completely from the agar surface. Surface cultures have low viability and die in 5–7 days. Chemoorganotrophs, having a fermentative type of metabolism. Acid but no gas is produced within 24 h from glucose and fructose. Dulcitol, inositol and inulin are not fermented. Other carbohydrates may be fermented with acid but no gas. **Positive for β-galactosidase.** Methyl red test is negative. **Nitrates are reduced to nitrites. Indole is not produced.** All species except one are **positive for urease.** Growth has been reported only on complex media; minimal nutritional requirements are unknown. All species except one can grow on MacConkey agar. Parasitic upon mammals (including man) and birds. The mol% G + C of the DNA is 40–43 (T_m).

Type species: *Actinobacillus lignieresii* Brumpt 1910, 849.

Further Descriptive Information

All actinobacilli show a marked pleomorphism usually with bacillary forms predominating, but also with coccobacilli and longer filamentous forms. The filamentous forms are seen more frequently in cultures growing on media containing glucose or maltose and it is not unusual to find the filaments breaking up into short bacillary forms or granules, giving an appearance of streptobacilli or even streptococcal chains.

Most species do not form capsules, but evidence of extracellular slime is often apparent especially with *A. equuli* and *A. suis*. *A. capsulatus* does produce capsules.

Colonies of actinobacilli growing on nutrient agar or blood agar tend to be sticky, especially on primary isolation from tissues. This property is much more marked with some species than with others and may or may not be lost on repeated subcultivation. Strains producing very sticky colonies also usually produce highly viscous cultures in fluid media, especially those containing glucose. This increased viscosity is seen especially with *A. equuli* and to a lesser extent with *A. suis* and may be demonstrated by the strand of sticky material which is drawn out between the surface of the broth and an inoculating loop.

Most species do not exhibit hemolytic activity on blood agar, but colonies of *A. suis* on sheep blood agar are always surrounded by a clearly marked zone of complete hemolysis. Some strains of *A. equuli* also exhibit hemolytic activity on sheep blood agar. *A. suis* on horse blood agar has some hemolytic action, but there may be only partial lysis of the cells.

Members of the genus do not produce pigmented colonies, but pigment production can be demonstrated in *A. suis* by centrifuging broth-grown cultures and resuspending the deposit in saline (Kim, 1976). A creamy yellow color is apparent, contrasting with a white suspension with other species of actinobacilli.

There are differences in hydrogen sulfide production (tested by lead acetate paper) between species of actinobacilli; most strains of *A. lignieresii* produce hydrogen sulfide, whereas the majority of *A. equuli* strains do not do so and only occasional strains of *A. suis* give a positive result.

Gelatinase activity is not characteristic of the genus, but a small number of strains of *A. equuli* are found which are positive when tested

by the gelatin agar plate method of Frazier (1926) (Frederiksen, 1973; Vallée et al., 1974).

Fermentation of carbohydrates and alcohols with the production of acid but no gas usually occurs within 24 hrs, but there may be strain or species variations with late fermentations occurring up to 7 days. With some substrates late fermentation is the more usual.

The antigenic structure of actinobacilli is complex. In *A. lignieresii*, *A. equuli* and *A. suis*, both heat-stable (somatic) and heat-labile (surface) antigens have been demonstrated (Phillips, 1967; Mráz, 1969a; Kim, 1976) and these may allow the division of species into antigenic types. Six antigenic types of *A. lignieresii* have been described (Phillips, 1967) which can be distinguished by differences in heat-stable antigens. There is evidence of geographical variations in the frequencies of these types. Of the strains examined in Britain none of those falling into type 3 was recovered from cattle (Phillips, 1967), in which host the majority of isolates were of type 1; however, in Japan type 3 was found in cattle and the predominant type was 5 (Nakazawa et al., 1979). In *A. lignieresii*, heat-labile antigens associated with extracellular slime can be demonstrated.

The existence of antigenic groups in *A. equuli* was noted by Mráz (1968) and Kim (1976) showed 28 groups on the basis of heat-stable O antigens, but no evidence of host specificity of groups was found.

No major examination of the antigens of *A. suis* has been made, but Kim (1976) found that 16 strains of this taxon drawn from Britain, Denmark, Germany and the United States showed remarkable uniformity in their antigenic structure.

A. actinomycetemcomitans is reported as being serologically uniform (Heinrich and Pulverer, 1959a).

There is considerable evidence of antigenic cross-relationships between *A. lignieresii*, *A. equuli* and *A. suis* (Haupt, 1934; Valée et al., 1963, Wetmore et al., 1963; Bouley, 1966; Vallée et al., 1974; Kim, 1976). Kim (1976) showed that with immunodiffusion there is at least one antigenic determinant in heat-stable antigenic complexes which is shared by *A. lignieresii*, *A. equuli* and *A. suis* and also by *Pasteurella haemolytica* but not *P. multocida*. Sharing of antigens between actinobacilli and pasteurellae has been described by others (Mráz, 1969a; Ross et al., 1972; Mráz, 1977). Haupt (1934) reported cross-agglutination reactions between *A. equuli* and *Pseudomonas mallei*.

Most members of the genus are found both as pathogens and as commensal organisms in domestic animals. Occasionally they may be found associated with disease in man, and one species (*A. actinomycetemcomitans*) occurs, with only isolated exceptions, in man alone. As commensal organisms actinobacilli are to be found in the alimentary, respiratory and genital tracts of normal animals. The translation of the organism from the commensal to the pathogenic status usually involves some factor which assists in the entry of the organism into and its establishment within the tissues; i.e. actinobacilli are opportunistic pathogens. The diseases caused by actinobacilli are usually of a sporadic nature, but occasionally a group of animals may be affected when a common "trigger factor" is present.

A. lignieresii is pathogenic for both cattle and sheep. In cattle, chronic granulomatous lesions are found most frequently in the tongue ("wooden tongue") and other soft tissues of the head and upper alimentary tract and, less often, in lungs, liver, pleura, heart muscle and skin. The lymph nodes of the affected region are almost invariably involved. In sheep, *A. lignieresii* was first described by Christiansen (1917) as *Bacterium purifaciens*, but Tunnicliff (1941), comparing it with the organism from cattle, concluded that the two were identical. Lesions are seen in sheep involving the skin (especially in the head region), lungs, testes and mammary glands. The commensal role of *A. lignieresii* has been demonstrated in the mouth and rumen of healthy cattle and sheep (Phillips, 1961, 1964).

Other animal species are not commonly infected with *A. lignieresii* but lesions have been reported in the tongue of dogs (Kemenes and Markói, 1959), in the brain of a horse (Chladeck and Ruth, 1976), in suppurative lymphadenitis in laboratory rats (Vallée et al., 1959) and in salpingitis in ducks (Bisgaard, 1975). It has also been reported as an etiological agent of disease in man (Thompson and Willius, 1932; Pathak and Ristic, 1962).

A. equuli is pathogenic for horses and pigs, causing purulent nephritis and arthritis especially in foals and piglets. In foals it may also give rise to an acute septicemic condition (sleepy foal disease). Adult horses and pigs may show lesions of endocarditis, meningitis and metritis. Pregnant animals may abort. *A. equuli* has been reported as a commensal organism in healthy horses in the intestinal tract (Laudien, 1923), in the mouth (Cottew and Francis, 1954), in the tonsillar region (Dimock et al., 1947) and in tracheal mucus (Kim et al., 1976), but its recovery from normal swine has not yet been reported.

A. equuli does not commonly occur as a pathogen in other animal species, but it has been isolated from monkeys with septic embolism (Moon et al., 1969), from calves associated with enteritis (du Plessis et al., 1967; Osbaldiston and Walker, 1972), from a dog with skin lesions (Vallée et al., 1960) and from a rabbit with pneumonia (Vallée, 1959).

A. suis is mainly a pathogen of pigs of all ages, being found associated with septicemic disease, pneumonia and arthritis (Zimmermann, 1964; Mair et al., 1974). It has also been reported having a pathogenic role in horses. Its occurrence as a commensal organism in pigs has not been clearly recognized, although an actinobacillus (probably *A. suis*) isolated from cases of septic embolism in pigs was also recovered from the tonsils of normal animals in the same herd (Cutlip et al., 1972). *A. suis* has also been found in the upper respiratory tract of normal horses (Kim et al., 1976).

Although the occurrence of *A. suis* as a pathogen of other animal species has not been recorded, actinobacilli isolated from ducks and swans (Hacking and Sileo, 1977; Onderka and Kierstead, 1979) have characteristics most closely related to *A. suis*.

A. capsulatus has been described only as the etiological agent of arthritis in rabbits. Its occurrence as a commensal organism has not been recorded and it is not known to infect other hosts.

A. actinomycetemcomitans has been regarded as of doubtful pathogenicity for man since it occurs in conjunction with actinomycetes in lesions in over 30% of cases of actinomycosis (Heinrich and Pulverer, 1959b). However, infections due solely to *A. actinomycetemcomitans* have been described in the jaw (Thjøtta and Sydnes, 1951) and in endocarditis (Page and King, 1966). The occurrence of *A. actinomycetemcomitans* as part of the normal flora of the human mouth has been recorded (Heinrich and Pulverer, 1959b).

Experimental pathogenicity of the actinobacilli is low. *A. lignieresii*, *A. equuli* and *A. suis* are not pathogenic for rabbits and guinea pigs. In mice also *A. lignieresii* and *A. equuli* are not pathogenic, but *A. suis* infects mice by the intraperitoneal route. *A. capsulatus* is pathogenic for rabbits, but guinea pigs are less susceptible and mice develop only transient infections after subcutaneous injection. *A. actinomycetemcomitans* is not pathogenic for laboratory animals, although it has been isolated from abscesses in a naturally occurring infection in laboratory mice (Vallée and Gaillard, 1953).

Enrichment and Isolation Procedures

Primary isolation of actinobacilli from tissues can be effected by cultivation on the enriched media (blood agar or serum agar) usually employed for the isolation of pathogens from animal tissues. Increased CO_2 tension (5–10%) may improve the growth of *A. capsulatus* (Arseculeratne, 1962) and *A. actinomycetemcomitans* (Holm, 1954). Cultures prepared by aseptic techniques from unopened lesions will usually yield actinobacilli as the sole organisms.

Cultivation of actinobacilli from situations in which mixed bacterial populations may exist lead to difficulties in isolation because of possible overgrowth by other bacteria. Selective media have been used for the recovery of *A. lignieresii* from the mouths of normal cattle (Till and Palmer, 1960)* and from the rumens of normal cattle and sheep (Phillips, 1961, 1964).†

Maintenance Procedures

Surface cultures on blood agar or serum agar seldom survive more than 10 days and most strains die out within 4 days. Cultures grown in Robertson's cooked meat medium may be stored for up to 4 weeks. Cultures may be preserved by lyophilization in sterile rabbit serum or sterile 20% peptone solution and will remain viable for many years (up to 20) in sealed ampoules. Heavy suspensions of cultures grown on nutrient or blood agar and washed off in sterile rabbit serum or sterile 20% peptone solution may be stored at −70°C for up to 2 years without undue loss of viability.

Differentiation of the genus **Actinobacillus** from other genera

The close similarities between actinobacilli and the genera *Haemophilus* and *Pasteurella* pose problems of differentiation which may necessitate consideration of many characteristics. The main features of difference and similarity between these three taxa and *Yersinia* are set out in Table 5.77. (See also Table 5.67 for the family *Pasteurellaceae*.)

MacConkey agar containing crystal violet may be of value in differentiating between *Actinobacillus* and *Pasteurella* (Mráz, 1975), but the distinction is not clearcut. The vibriostatic agent 0/129 (2,4-diamino-6,7-diisopropylpteridine) inhibits *Actinobacillus* and *Pasteurella* but not *Yersinia* (Chatelain et al., 1979).

Taxonomic Comments

Three species which were listed in the last edition of the Manual as *species incertae sedis* and which are now included in the Approved Lists of Bacterial Names have been added to the genus in the present edition.

Similarities between the two genera *Actinobacillus* and *Pasteurella* have led to difficulties and indecisions in the placement of organisms in these two groups, especially those organisms isolated from sources other than the well recognized pathological conditions associated with the classical species. Sneath and Johnson (1973), in a numerical taxonomy study of *Haemophilus* and related bacteria based on 134 characteristics, found that in the *Actinobacillus-Haemophilus-Pasteurella* group one cluster contained the strains of *Actinobacillus* and *Pasteurella* associated at a similarity index of 75%, and concluded that there was strong evidence against the separation of the two as distinct genera. Frederiksen (1973) also pointed out the difficulties of arranging these two genera into well delineated species, but suggested that genetic studies would be needed to arrive at a more meaningful classification. Mannheim et al. (1980), examining collection cultures representing the *Actinobacillus-Haemophilus-Pasteurella* group, reported that the group should rank as a family and that a common feature was the occurrence of demethylmenaquinones as the respiratory quinones produced by members of the group.

---

* Till and Palmer's medium: Hartley's broth (Cruickshank, 1965), 900 ml; agar, 10.0 g; Filde's peptic digest (Cruickshank, 1965), 100 ml; oleandomycin phosphate, 20 mg; neomycin sulfate, 1.5 mg. The peptic digest and the oleandomycin are added after the basal nutrient agar has been sterilized and cooled to 50°C.

† Phillips' medium: Hartley's broth (Cruickshank, 1965), 930 ml; agar, 10.0 g; horse blood (oxalated), 50 ml; oleandomycin phosphate stock solution containing 100 μg/ml (stored at −20°C prior to use), 10 ml; nystatin stock suspension containing 20,000 U/ml (stored at −20°C prior to use), 10 ml. The horse blood, oleandomycin and nystatin are added after the basal nutrient agar has been sterilized and cooled to 50°C.

Table 5.77.

Differential characteristics of the genus **Actinobacillus** *and other genera.* [a]

Characteristics	Actino-bacillus	Haemophilus	Pasteurella	Yersinia
Colonies sticky	+	−	−	−
Motile at 22°C	−	−	−	D
Catalase	D	D	+	+
Oxidase	D	D	D	−
Phosphatase	+	+[b]	+	−
Growth on MacConkey agar	+	−	D	+
Growth on 4.5% NaCl	−	−	−	+
Methyl red test (37°C)	−	D	−	+
Voges-Proskauer test (37°C)	D	−	−	−
Fermentation of				
Glucose	+	+	+	+
Fructose	+	D	+	+
Xylose	+	D	D	D
Dulcitol	−	−	−[c]	−
Inositol	−	−[d]	D	D
Inulin	−	−[e]	−	−
Hydrolysis of Tween 80	−	−	−	D
Sensitivity to 0/129[f]	+	+	D	−
Mol% G + C of DNA	40–43	38–44	40–45	46–50

[a] Symbols: see standard definitions.
[b] Except *H. haemoglobinophilus.*
[c] Except some strains of *P. multocida* and *P. haemolytica.*
[d] Except some strains of *H. aphrophilus* and *H. parasuis.*
[e] Except *H. parasuis.*
[f] 2,4-diamino-6,7-diisopropylpteridine.

DNA hybridization techniques have shown that the three genera are interrelated at or above the 30% DNA binding level (Pohl, 1981) and can be divided into six tribes linking at the 40% level, for one of which the name "Actinobacilleae" is proposed (Mannheim, 1981). Into this tribe fall *A. lignieresii, A. equuli, A. suis* and *A. capsulatus,* while *A. actinomycetemcomitans* shows a more remote relatedness which would support its removal, along with biovar Heyl of *Pasteurella pneumotropica* (Frederiksen, 1973), into another (new) genus of the tribe "Actinobacilleae."

In addition to the four species of *Actinobacillus* names, DNA hybridization studies suggest that other organisms of the *Actinobacillus-Haemophilus-Pasteurella* group should be included in the genus as four possible new species, viz:

1. *Pasteurella haemolytica* biovar A together with *P. mastitidis,*
2. *Actinobacillus*-like organisms isolated from the sow vagina (Ross et al., 1972),
3. *P. ureae,*
4. *Haemophilus pleuropneumoniae* together with the "*Pasteurella haemolytica*-like organisms" isolated from porcine necrotic pleuropneumonia (Bertschinger and Seifert, 1978).

The inclusion of *P. haemolytica* biovar A in the genus *Actinobacillus* supports, at least in part, the earlier proposal by Mráz (1969b) that *P. haemolytica* should be renamed "*Actinobacillus haemolyticus.*" Mráz included both biovars A and T in his proposal.

The results of the DNA hybridization studies with *P. ureae* lend support to the view that this organism recovered from the human respiratory tract resembles the actinobacilli as much as it does *P. haemolytica* (Jones, 1962) and to the opinion that it can be regarded as a human *Actinobacillus* (Frederiksen, 1973). Strong similarities between *P. ureae, A. suis* and *A. lignieresii* have been noted by Hacking and Sileo (1977).

Haemophilus pleuropneumoniae Shope 1964, 362 has been isolated from pleuropneumonia in swine as also has the "*Pasteurella haemolytica*-like organism" of Bertschinger and Seifert (1978) and both are phenotypically very similar except for the V factor requirement of *H. pleuropneumoniae.* Moreover, both, besides being linked together in DNA hybridization at the 90% level, are linked to *A. lignieresii* at the 80% level.

The organism first described by Kohlert (1968) from the reproductive tract of hens as "*Pasteurella salpingitidis*" and subsequently considered by Mráz et al. (1976) as "*A. salpingitidis*" has been shown by DNA hybridization studies (Pohl, 1981) not to be closely related to the actinobacilli.

"*Actinobacillus seminis*" Baynes and Simmons 1960, 459, which was excluded from the genus in the last edition of the *Manual,* has been shown to have low relatedness to other members of the *Actinobacillus-Haemophilus-Pasteurella* group (Pohl, 1981) thus supporting its continued exclusion.

Further Reading

Kilian, M, W. Frederiksen and E.L. Biberstein. 1981. *Haemophilus, Pasteurella* and *Actinobacillus.* Academic Press, London.

Differentiation and characteristics of species of **Actinobacillus**

The differential characteristics of the species of *Actinobacillus* are given in Table 5.78. Other characteristics of the species are listed in Table 5.79.

Table 5.78.
Differential characteristics of **Actinobacillus** *species[a]*

Characteristics	1. A. lignieresii	2. A. equuli	3. A. suis	4. A. capsulatus	5. A. actinomycetemcomitans
Hemolysis on sheep Blood agar	−	d	+	−	−
Hydrolysis of:					
Sodium hippurate	−	+	+		
Esculin	−	−	+	(+)[b]	−
Fermentation of:					
Cellobiose	−	−	+		−
Lactose	(+)	+	+	+	−
Mannitol	+	+	−	+	d
Melibiose	−	+	+		−
Salicin	−	−	+	+	−
Trehalose	−	+	+	+	−

[a] For symbols see standard definitions; (+), delayed reaction.
[b] Result with one strain only.

Table 5.79.
Other characteristics of **Actinobacillus** *species[a]*

Characteristics	1. A. lignieresii	2. A. equuli	3. A. suis	4. A. capsulatus	5. A. actinomycetemcomitans
Capsules formed	−	−	−	+	−
Extracellular slime present in:					
Wet preparations	+	+	+	−	
Stained smears	−	+	+	−	
Growth on MacConkey agar	+	+	+	+	−
Catalase	d	d	+	+	+
Oxidase	+	d	d	+	+
Phosphatase	+	+	+	+	+
Gelatinase	−	d	−	−	−
Hydrogen sulfide produced	+	d	−	−	−
Methyl red test	−	−	−		
Voges-Proskauer test	d	−	−		
Methylene blue test	+	d	d		
Sodium gluconate oxidized	−	−	−		
Starch[a] synthesized from:					
Glucose	+	d	−		
Maltose	+	d	−		
Lysine decarboxylase	−	−	−	−	−
Ornithine decarboxylase	−	−	−	−	−
Arginine dihydrolase	−	−	−	−	−
Fermentation (acid no gas) of:					
Adonitol	−	−	−	−	−
Arabinose	d	d	+	−	−
Dextrin	+	+	+		v
Dulcitol	−	−	−	−	−
Fructose	+	+	+	+	+
Galactose	+	+	+	+	d
Glucose	+	+	+	+	+
Glycerol	d	d	+		−
Inositol	−	−	−		−
Inulin	−	−	−		−
Maltose	+	+	+	+	+
Mannose	+	+	+	+	+
Raffinose	d	+	+	+	
Rhamnose	−	−	−	−	−
Sorbitol	−	d	−	+	
Sorbose	−	−	−	−	−
Starch	−	d	d	−	d
Sucrose	+	+	+	+	−
Xylose	+	+	+	+	d

[a] For symbols see standard definitions.
[b] Blue-black color with 1 in 10 Gram's iodine.

List of the species of the genus **Actinobacillus**

1. **Actinobacillus lignieresii** Brumpt 1910, 849.[AL]

lig.ni.e.re′si.i. M.L. gen. n. *lignieresii* of Lignières; named for J. Lignières, one of the bacteriologists who first isolated this organism.

Cells are usually rod shaped, but marked variability occurs depending upon the growth medium. Long bacillary or filamentous forms are most frequently seen on media containing glucose and maltose; shorter bacillary and coccobacillary forms are more usual on media containing blood or serum.

Colonies on primary isolation are slightly sticky but this property is lost on repeated subcultivation. Fluorescent, granular and dwarf colonial variants have been described (Ristic et al., 1956). Broth cultures are uniformly turbid with little deposit.

Fermentation of arabinose, glycerol and lactose is usually slow (3–7 days); a small number of strains may be slow in giving a positive reaction with galactose and sucrose. While some strains ferment raffinose promptly or slowly, the majority fail to do so.

Growth occurs between 20 and 39°C. Optimum temperature 37°C. Does not grow at 44°C.

The mol% G + C of the DNA is 41.8–42.6 (T_m) (Boháček and Mráz, 1967).

Type strain: NCTC 4189.

2. **Actinobacillus equuli** (van Straaten 1918) Haupt 1934, 513.[AL] (*Bacillus equuli* van Straaten 1918, 75.)

e.quu'li. L. n. *equulus* a foal; L. gen. n. *equuli* of a foal.

Cells usually rod shaped, but showing marked variability depending upon the growth medium. Long bacillary forms similar to those seen with *Actinobacillus lignieresii* occur on media containing glucose or maltose.

Stickiness of colonies is not lost on repeated subculturing. Colonial variation occurs. When first isolated from pathological material colonies are usually rough but may become smooth on repeated subculture. The change from rough to smooth may be accompanied by a reduction in the stickiness of the colonies, but the viscous nature of the growth is never lost. Dwarf colonial variants have been reported (Edwards, 1931). Broth cultures exhibit an extreme viscosity. There is a low viability of fluid cultures in media containing even small amounts of fermentable substrates.

Glycerol and mannose are usually fermented slowly. With dextrin, fructose, maltose, melibiose, raffinose, sucrose, trehalose and xylose fermentation may be delayed. The majority of strains fail to ferment arabinose, cellobiose, salicin and sorbitol but occasional strains are found which give fermentation either promptly or slowly.

Growth occurs between 20 and 39°C. Some strains will grow at 44°C.

The mol% G + C of the DNA is 40.0–41.8 (T_m) (Boháček and Mráz, 1967).

Type strain: ATCC 19392.

3. **Actinobacillus suis** van Dorssen and Jaartsveld 1962, 456.[AL]

su'is. L. n. *sus* the pig, swine; L. gen. n. *suis* of the pig.

Cells usually rod shaped but considerable variability may be seen with long rods and filaments from media both with and without glucose or maltose.

Colonies are usually sticky and adherent to the medium, but not so markedly as *Actinobacillus equuli*. Stickiness increases with prolonged incubation up to 72 h. Older colonies develop a transparent border zone giving the appearance of a fried egg. With repeated subculturing, colonies may lose their marked adherence to the medium. Viscous growth occurs in nutrient broth, but less marked than in *Actinobacillus equuli*. Broth-grown cells sedimented by centrifugation are pigmented creamy yellow. Strains have low viability: nutrient agar and nutrient broth cultures die out within 15 days at 4°C.

Arabinose, dextrin and galactose are usually fermented promptly but a few strains may give late fermentation with these substrates. Glycerol and mannose are usually slow to be fermented.

Sensitive to ampicillin, cephaloridine, chloramphenicol, colistin methane sulfonate, streptomycin, tetracycline, penicillin and gentamicin.

Optimum temperature, 37°C.

The mol% G + C of the DNA is 40.5 (T_m) (Mráz, 1968).

Type strain: CCM 5586.

4. **Actinobacillus capsulatus** Arseculeratne 1962, 38.[AL]

cap.su.la'tus. L. n. *capsula* a small chest, capsule; M.L. masc. adj. *capsulatus* encapsulated.

Cells rod shaped. Old cultures show filamentous bacilli with fragmentation into minute coccoid bodies. Moniliform bodies are produced in 5-day-old cultures on Loeffler's serum. Capsules are present.

Primary cultures will not grow on nutrient agar or in nutrient broth, but subcultures grow as pinpoint colonies or a faint turbidity.

Colonies on sheep blood agar are very sticky. "Flower Head" colonies are produced on rabbit blood agar.

Growth on Loeffler's serum is scanty, but profuse on Dorset egg.

Small discrete mural colonies occur in serum broth.

Growth is favored on primary isolation only by the addition of 10% CO_2.

Viability is best on Dorset egg on which it survives no longer than 10 days.

Resistant to penicillin; sensitive to streptomycin, tetracycline and chloramphenicol.

Optimum temperature 37°C. No growth at 22°C. Killed by heat at 60°C for 10 min.

The mol% G + C of the DNA is 42.4 (T_m) (Mannheim et al., 1980).

Type strain: Frederiksen P243 (Dr. W. Frederiksen, Statens Seruminstitut, Copenhagen).

5. **Actinobacillus actinomycetemcomitans** (Klinger) Topley and Wilson 1936, 279.[AL] (*Bacterium actinomycetem comitans* Klinger 1912, 198.)

ac.ti.no.my.ce.tem.co'mi.tans. Gr. n. *actis* a ray; Gr. n. *myces, mycetis* a fungus; M.L. n. *actinomyces* ray fungus; L. part. adj. *comitans* accompanying; M.L. part. adj. *actinomycetemcomitans* accompanying an actinomycete.

Cells spherical or rod shaped, the latter being seen more frequently in agar cultures than in broth or gelatin cultures.

Agar colonies are small (1 mm diameter after 2–3 days), adherent to the medium and difficult to break up. Colonies are described as star-like (Colebrook, 1920) or like "crossed cigars" (Heinrich and Pulverer, 1959a).

Growth in broth is in the form of granules at the bottom and up the sides of the tube. This characteristic growth is usually maintained on subculturing (King and Tatum, 1962) although some strains may produce a uniformly turbid growth in broth on repeated subculture.

Fermentation reactions with galactose, mannitol and xylose permit definition of 8 biotypes (Pulverer and Ko, 1970).

Grows slightly better under anaerobic conditions (Thjøtta and Sydnes, 1951). Microaerophilic (Heinrich and Pulverer, 1959a). Growth is improved by increased CO_2 tension of not less than 0.5% (Holm, 1954).

Sensitive to chlortetracycline, chloramphenicol, streptomycin, erythromycin, polymyxin B, oxytetracycline and novobiocin. Some strains are sensitive to penicillin.

Optimum growth at 37°C (Heinrich and Pulverer, 1959a). No growth at 22°C.

The mol% G + C of the DNA is 42.7 (T_m) (Kilian, 1976).

Type strain: NCTC 9710.

Species Incertae Sedis

The taxonomic placement of the following species is not yet known, but it should not be included in the genus *Actinobacillus*.

"*Actinobacillus seminis*" Baynes and Simmons 1960, 459.

sem.in'is. L. n. *semen* seed; L. gen. n. *seminis* of semen.

Cells are Gram-negative, pleomorphic, ranging from coccobacilli to rods 1 µm × 4–5 µm, arranged singly, in pairs and in short chains. Capsules are not formed. Nonmotile at both 22 and 37°C.

On primary isolation, the organism grows under aerobic and microaerophilic conditions, but best results are obtained with an atmosphere containing 10–20% CO_2. The colonies increase in size with prolonged incubation and range from pinpoint at 24 h to 3 or 4 mm at 4–5 days when they are umbonate with grayish white centers and a transparent periphery. Some strains grow less well and produce smaller colonies which are more transparent, low convex and undifferentiated (van Tonder, 1979). There is no hemolysis of sheep or horse blood. Strains producing the smaller colonies show a lemon-yellow pigmentation of the cells packed by centrifugation, whereas strains producing the larger colonies have grayish white packed cells (van Tonder, 1979). Growth does not occur on MacConkey agar.

No degradation of carbohydrates occurs even after incubations for periods of up to 14 days. The following carbohydrates have been reported to give such negative results: glucose, adonitol, arabinose, dextrin, dulcitol, fructose, galactose, glycerol, inositol, inulin, lactose, maltose, mannitol, mannose, raffinose, salicin, sorbitol, starch, sucrose, trehalose and xylose. Some strains may show slight acid production after incubation for 28 days with arabinose, fructose, mannose and trehalose (Baynes and Simmons, 1960), and more rapid breakdown of glucose, arabinose, fructose, maltose, mannitol and xylose has been reported (Mannheim et al., 1980).

Most strains do not reduce nitrate to nitrite, but occasional positive results may occur. The catalase test is usually positive, the oxidase test is usually negative. H_2S (detected by lead acetate paper) is usually not produced, but weak reactions may occur. The methyl red and Voges-Proskauer tests are negative. Other negative reactions include: indole production, phosphatase, urease, gelatinase and citrate utilization.

Antigenic differences between strains have been reported (van Tonder, 1979).

"*A. seminis*" occurs as a pathogen of sheep, causing epididymitis and polyarthritis. It is not pathogenic for mice or guinea pigs injected intraperitoneally or intramuscularly.

The mol% G + C of the DNA is 43.7 (T_m) (Mannheim et al., 1980), but a wider range of values—37.8–48.8 (T_m)—has been reported (Gumbrell and Smith, 1974).

Holotype strain: K3844-C, G. C. Simmons; ATCC 15768 (Mannheim et al., 1980).

Other organisms

The following organisms have certain characteristics in common with actinobacilli but their taxonomic placement is not yet certain.

(i) *Organisms from the vagina of postparturient sows.*

Organisms described as actinobacilli have been isolated from the vagina of postparturient sows (Ross et al., 1972). They are Gram-negative, pleomorphic, nonmotile, rod shaped to coccoid. Colonies on horse blood agar are 1–2 mm in diameter, grayish white, convex, and have a narrow zone of complete hemolysis. Growth is best under anaerobic conditions. The presence of 5% CO_2 stimulates growth. Growth can occur on MacConkey agar.

The organisms produce acid but no gas from the following carbohydrates: glucose, arabinose, galactose, glycerol, inositol, mannitol, mannose, rhamnose, sorbitol and xylose. The following carbohydrates are not fermented: adonitol, dulcitol, inulin, lactose, maltose, raffinose, salicin, sucrose and trehalose. Esculin is not hydrolyzed.

Nitrates are reduced to nitrites. Catalase and urease are produced. Hydrogen sulfide is produced (detected by lead acetate paper). The methyl red and Voges-Proskauer tests are negative. Other negative tests: indole production, gelatinase, citrate utilization.

All strains are antigenically similar. Some agglutination cross-reactions occur with *A. suis*, *A. lignieresii* and "*A. seminis.*" Precipitation reactions also show cross-relationships with *A. equuli* and *Pasteurella haemolytica.*

Not pathogenic for young pigs by intraperitoneal and intravenous routes. Kills 7-day-old chick embryos in 24–48 h after yolk-sac inoculation.

The mol% G + C of the DNA is 41.9 (T_m) (Mannheim et al., 1980).

Representative strain: strain 192 (Ross et al., 1972); ATCC 27072; NCTC 10801.

(ii) *Organisms from porcine necrotic pleuropneumonia.*

The *Pasteurella haemolytica*-like organism isolated from porcine necrotic pleuropneumonia (Bertschinger and Seifert, 1978) may be regarded as a porcine actinobacillus closely related to *Haemophilus pleuropneumoniae* (Mannheim, 1981).

The morphological and cultural characters resemble those of *Pasteurella haemolytica*. Growth can occur on MacConkey agar.

The organisms produce acid but no gas from glucose, dextrin, fructose, galactose, maltose, mannitol, mannose, sucrose and xylose. There is a late fermentation of lactose. The following carbohydrates are not fermented: arabinose, dulcitol, inositol, raffinose, rhamnose, salicin, sorbitol and trehalose. Esculin is not hydrolyzed.

Nitrates are reduced to nitrites. Catalase may or may not be produced. Urease, phosphatase and oxidase tests are positive. Indole is not produced. Gelatinase is not produced.

The mol% G + C of the DNA is 42.2–42.5 (T_m) (Mannheim et al., 1980).

Holotype strain: 2008/76 (Bertschinger) (Dr. H. U. Bertschinger, Veterinär-bakteriologisches Institut der Universität Zürich).

Important Notes for Users of this Edition

1. Always read both generic and species descriptions because characters listed in the generic description are not usually listed in the species descriptions.

2. Unless otherwise indicated in footnotes to tables, the meanings of symbols are as follows:

   +   90% or more of strains are positive

   −   90% or more of strains are negative

   d   11–89% of strains are positive

   v   strain instability (*not* equivalent to "d")

   D   different reactions in different taxa (species of a genus or genera of a family)

3. All other symbols are defined in footnotes to tables.

OTHER GENERA

Table 5.80.

Some differential features of the genera of Section 5 not assigned to any family

Characteristics	Zymomonas[a] (p. 576)	Chromo-bacterium (p. 580)	Cardio-bacterium (p. 583)	Calymmato-bacterium (p. 585)	Gardner-ella (p. 587)	Eikenella (p. 591)	Strepto-bacillus (p. 598)
May stain Gram-variable	−	−	+	−	+	−	−
Acid from glucose	+	+[b]	+		+	−	+
Major product of sugar fermentation:							
Ethanol	+		−		−		
Lactic acid	−		+		−		
Acetic acid	−		−		+		
Parasitic on warm-blooded animals and/or humans	−	−	+	+	+	+	+
Pathogenic for humans	−	+	+	+	+	+	+
Causes Donovanosis (granuloma inguinale) in man	−	−	−	+	−	−	−
Causes one form of rat-bite fever in humans	−	−	−	−	−	−	+
Causes vaginitis in humans	−	−	−	−	+	−	−
Colonies are violet	−	+[c]	−	−	−	−	−
Motility (swimming)	−[d]	+[e]	−	−	−	−	−
Oxidase test	−	+	+		−	+	−
Catalase test	+	+	−		−	−	−
Nitrate reduced to nitrite	−	+[f]	−		−	+	−
Indole produced	−	−[g]	+[h]		−	−	−
Hemin usually required for growth under aerobic conditions	−	−	+	−	−	−	−
Many strains appear to corrode the surface of the agar medium	−	−	−	−	−	+	−
Mol% G + C of DNA	47–50	50–68	59–60		42–44	56–58	24–26

[a] Some strains are anaerobic.

[b] Most strains show a fermentative attack on glucose, but ~20% show an oxidative attack.

[c] White variants are difficult to identify and may be mistaken for *Aeromonas* or *Vibrio* species.

[d] Most strains are not motile, but a few are motile by means of 1 to 4 polar flagella.

[e] Cells have both a single polar flagellum plus 1–4 subpolar or lateral flagella.

[f] Most strains also reduce nitrite.

[g] Negative by usual testing methods, but under some conditions compounds that give positive reactions with indole test reagents may accumulate.

[h] Only small amounts of indole are formed.

Genus **Zymomonas** *Kluyver and van Niel, 1936, 399.*[AL]

JEAN SWINGS AND JOZEF DE LEY

Zy.mo′mo.nas or Zy.mo.mo′nas. Gr. n. *zyme* leaven, ferment; Gr. n. *monas* a unit, monad; M.L. fem.n. *Zymomonas* fermenting monad.

Rod-shaped cells with rounded ends, occasionally ellipsoidal, usually in pairs, 2–6 μm long and 1.0–1.4 μm wide. Gram-negative. Usually nonmotile; if motile, they possess **one to four polar flagella.** Motility may be lost spontaneously. Facultative anaerobic; some strains are obligately anaerobic. Chemoorganotrophic, **growing on and fermenting 1 mol of glucose or fructose to almost 2 mol of ethanol, 2 mol of CO₂ and some lactic acid.** Some strains may also utilize sucrose, but other carbon sources are not used. Optimum temperature 25–30°C. Colonies on the standard medium* are glistening, regularly edged, white to cream colored, 1–2 mm in diameter after 2 days at 30°C. Oxidase-negative. Gelatinase-negative. Nitrates are not reduced and indole is not produced. *Zymomonas* tolerates 5% ethanol and is acid tolerant, growing at pH 3.5–7.5. Good growth is obtained only when a mixture of amino acids is present in the medium, but no one amino acid is essential. All strains **require biotin and pantothenate.** *Zymononas* occurs as a spoiler in **beers, ciders** and **perries;** as

---

* The standard medium (SM) has the following composition (per liter of distilled water): D-glucose, 20 g; and yeast extract, 5 g.

fermenting agents in **Agave sap**, **palm sap** and **sugarcane juice**; and on **honey bees** and in **ripening honey**. The mol% G + C of the DNA is 47.5–49.5 (T_m).

Type species: *Zymononas mobilis* (Lindner 1928) Kluyver and van Niel 1936, 399.

Further Descriptive Information

Zymomonas cells are mostly straight rods with rounded or ovoid ends, occurring singly or in paris. They form neither spores nor capsules, and contain no detectable intracellular lipids, glycogen, polyphosphates or poly-β-hydroxybutyrate. Some individual strains form either rosette-like cell aggregations, cell chains, curved or U-shaped cells, or filamentous cells. Most strains are nonmotile.

Deep colonies in solid standard medium are lenticular, regular, entire edged, butyrous, white or cream colored, and 1–2 mm in diameter after 2–4 days at 30°C. Anaerobic surface colonies are spreading, entire-edged, convex or umbonate, and 1–4 mm in diameter after 2–7 days at 30°C (Swings and De Ley, 1977). When incubated aerobically, colonies reach a maximum diameter of 1.5 mm or appear as microcolonies (Swings et al., 1977).

Although *Zymomonas* has a fermentative type of metabolism it is able to grow aerobically, and should therefore be qualified a facultative anaerobe.

Zymomonas grows easily in liquid media containing either D-glucose or D-fructose: a dense turbidity accompanied by abundant CO_2 formation develops after 1–2 days at 30°C. The final pH in the standard medium after 3 days at 30°C is 4.8–5.2. The acidification of the medium is more pronouced upon incubation at higher temperatures. Strain-specific flocculent or compact cell deposits are formed. Half of the strains grow in glucose concentrations up to 40% (Swings and De Ley, 1977). In continuous cultures containing 15 and 25% glucose, the glucose is not fully metabolized (Lee et al., 1979). Sucrose is fermented and used for growth by many *Zymomonas* strains. This property is lost occasionally upon subculturing on D-glucose (Shimwell, 1950). Sucrose fermentation is inducible (Dadds et al., 1973; Kluyver and Hoppenbrouwers, 1931; Richards and Corbey, 1974). Less than 2% of the sucrose provided is converted into levan (Dawes et al., 1966).

The nitrogen source for growth can be supplied either as peptone, yeast extract, nutrient broth, beer, palm juice or apple juice or a mixture of 20 amino acids. (Groups of amino-acids, individual amino acids or NH_4Cl can also serve as nitrogen sources, but this has not been verified for every *Zymomonas* strain.) In synthetic media, the growth yield is lower and the generation time longer than in complex media, but the ethanol yield remains constant (Belaïch and Senez, 1965). A synthetic medium containing 20 amino acids, 10 vitamins, 5 purine and pyrimidine bases, and D-glucose sustains good growth of 38 *Zymomonas* strains through five serial transfers (Van Pee et al., 1974). The withdrawal of any one of the amino acids from this medium does not depress growth. Amino acids serve both as nitrogen and as carbon source (Swings and De Ley, 1977). In the absence of glucose, however, amino acids are not used as sole carbon sources for growth or fermentation (Belaïch, 1963).

Zymomonas strains require biotin and pantothenate as growth factors. No strain requires nictotinic acid. Only six strains need additional growth factors. The most exacting strain (VP3) requires additional vitamin B_{12}, lipoic acid, riboflavin and folic acid (Van Pee et al., 1974).

Most (90%) of the *Zymomonas* strains are able to grow between pH 3.85 and pH 7.55. At pH 3.5, 43% of the strains develop, illustrating a high acid tolerance (Swings and De Ley, 1977). This feature is not at all surprising as the natural niche of the genus is in acid palm wines, ciders and beers at pH 4 or below. *Zymomonas* cannot grow in liquid standard medium at pH 3.05.

Zymomonas grows best between 25 and 30°C. At 38°C, 74% of the strains grow, but at 40°C growth is rare (Swings and De Ley, 1977; De Ley and Swings, 1976). *Zymomonas* slowly develops at 15°C (Millis, 1951; Dadds et al., 1973) but not at 4°C. Growth at 36°C is the best phenotypic test to differentiate the subspecies *mobilis* (+) from *pomacii*

(−) (Swings et al., 1977). *Zymomonas* is killed by exposure to 60°C for 5 m. Forrest (1967) observed that above 33°C the specific growth rate of *Zymomonas* dropped, but the rate of glucose consumption continued to increase, suggesting that above this temperature the coupling between anabolism and catabolism is not very efficient.

With regard to ethanol tolerance, *Zymomonas* grows in the presence of 5% ethanol, and many strains grow at even higher concentrations.

Zymomonas is an unusual bacterium in that it ferments glucose anaerobically by the Entner-Doudoroff mechanism, followed by a pyruvate decarboxylation, according to the following general fermentation balance:

$$1 \text{ glucose} \rightarrow (1.58\text{–}1.93) \text{ ethanol} + (1.7\text{–}1.9) \text{ } CO_2 + (0.02\text{–}0.2) \text{ lactate} + (0.011) \text{ cell material } [CH_2O]$$

Small amounts of acetaldehyde, acetyl methyl carbinol and glycerol are also formed. During the dissimilation of glucose under aerobic conditions, ethanol and acetate are formed. The aerobic mechanism seems to be limited to the oxidation of ethanol to acetic acid. Fragments of the tricarboxylic acid cycle are present in *Zymomonas* (Dawes et al., 1970). Only 2% of the glucose is incorporated in the cells, producing 48% of the cellular carbon (Belaïch and Senez, 1965); the rest of the carbon is derived from yeast extract components. The growth yield coefficients ($Y_{glucose}$ = 3.88–9.32, as determined by several authors) indicate that the growth of these organisms is not very efficient.

The following physiological tests are positive for the genus: catalase; reduction of methylene blue, thionin and 2,3,5-triphenyltetrazolium chloride; formation of traces of acetyl methyl carbinol and the production of a characteristic fruity odor. The following tests are negative: growth in 0.5% yeast extract, nutrient broth or 1% peptone broth, in liquid standard medium + 2% NaCl; indole production, nitrate reduction, hydrolysis of gelatin, hydrolysis of Tween 60 and Tween 80, and oxidase.

Plasmids of the incompatibility groups P1 (pRD1, pJB4JI and R68.45) and FII (R1drd19) have been transferred into *Zymomonas mobilis* by conjugation and stably maintained; the third plasmid was from a *Pseudomonas aeruginosa* and the other three from a *Escherichia coli* host (Skotnicki et al., 1980).

Dadds et al. (1973) suggested the possible existence of two serovars within the genus *Zymomonas*.

Zymomonas is not known to be pathogenic for man, animals or plants. Lindner (1929, 1931) recommended the use of *Zymomonas* in human nutrition as a kind of yogurt. Antagonistic effects of *Zymomonas* against bacteria and fungi in vitro (Gonçalves et al., 1970, 1968, 1972) and the therapeutic use of *Zymomonas* in cases of chronic enteric and gynecological infections (Wanick et al., 1971, 1970; De Paula Gomes, 1959; De Souza and De Souza, 1973) have been reported.

In sweet English ciders *Zymomonas* is the causative agent of a secondary fermentation, known as "cider sickness." The first description of this phenomenon was given by Barker and Hillier (1912). Cider sickness is recognized by frothing and abundant gas formation, a typical change in the aroma and flavor, reduction of sweetness and marked turbidity forming a heavy deposit afterwards (Millis, 1951; Barker, 1948; Carr and Passmore, 1971). The cause of a cider disorder known as framboisé, or "framboisement" in France, is also attributed to *Zymomonas*. Whether *Zymomonas* is involved in any case of framboisé is uncertain, as lactic and acetic acid bacteria seem also to play a role in its development (Guitonneau et al., 1939; Bidan, 1959; Pollard, 1959). Millis (1951) also isolated *Zymomonas* from "sick" perries. Lindner (1928a) discovered that *Zymomonas* is the fermentative agent that transforms the sugary *Agave* sap (aguamiel) to pulque in Mexico. *Zymomonas* is a serious beer contaminant, particularly in English cask beers, producing a heavy turbidity and an unpleasant odor due to acetaldehyde and H_2S. It has not been reported in lager beers. Palm wines are prepared in the far East and in Africa from the sap of *Arenga*, *Raphia* and *Elaeis* palms and are known to harbor *Zymomonas* as a fermentative agent. *Zymomonas* is also present in fermenting sugarcane juice in Brazil, and on bees and ripening honey in Spain.

Enrichment and Isolation Procedures

The following medium, originally designed as a detection medium for *Zymomonas* in breweries (Dadds, 1971), can be recommended for enrichment and has the following composition (g/liter): malt extract, 3 g; yeast extract, 3 g; D-glucose, 20 g; peptone, 5 g; and actidione, 0.02 g. The pH is adjusted to 4.0. Ethanol is added to 3% (v/v). The presence of *Zymomonas* is indicated by abundant gas production after 2–6 days of 30°C. Isolated colonies of *Zymomonas* are obtained from enrichments or samples by streaking onto WL differential medium (Difco), standard medium or standard medium + 2% CaCO₃, with incubation in a Gas-Pak anaerobic system (BBL). Another isolation method consists of mixing dilutions of the enrichments or samples with the WL differential medium (at 50°C) and pouring into Petri dishes. Samples can also be streaked onto WL or standard medium and then covered by a second layer of medium. In WL differential medium, the colonies are lenticular, 1–4 mm in diameter after 4 days at 30°C, and deep green.

Maintenance Procedures

Zymomonas cultures held in the standard medium at room temperature are transferred every 2–3 weeks. *Zymomonas* survives the ordinary lyophilization procedure for many years.

Differentiation of the genus Zymomonas from other genera

Table 5.81 indicates the most salient features that differentiate *Zymomonas* from other genera. *Zymomonas* is phenotypically and genotypically well defined and is easily recognized. Its most outstanding feature is the quantitative fermentation of glucose, fructose or sucrose—but no other sugars—to equimolar amounts of ethanol and CO₂. This feature makes *Zymomonas* a unique ethanol-producing bacterium.

Zymomonas is excluded from the *Enterobacteraceae* on the basis of its polar flagellation, its inability to reduce nitrates, its growth at pH 4 and its growth in the presence of 5% ethanol. *Zymomonas* has some phenotypic resemblance to the Gram-negative, polarly flagellated, facultatively anaerobic, fermentative genera *Vibrio* and *Aeromonas*; however, the latter genera are oxidase-positive and reduce nitrates. *Vibrio* and *Zymomonas*, which have partly overlapping mol% G + C values, can be further differentiated by the resistance of *Zymomonas* towards the vibriostatic compound 0/129 (2,4-diamino,6,7-diisopropyl pteridine). Unlike *Zymomonas*, *Aeromonas* is sensitive to novobiocin, cannot grow at pH 4 and has a mol% G + C range of 57–62.

Genetically, phenotypically and ecologically, *Zymomonas* is related to the acetic acid bacteria: they both occur in acid, sugary and alcoholized niches such as tropical plant juices and beer. They are ecologically complementary in that *Zymomonas* produces ethanol which is further oxidized by the acetic acid and bacteria. *Zymomonas* more closely resembles *Gluconobacter* than *Acetobacter* because of its polar flagella, its incomplete tricarboxylic acid cycle and the occurrence of the Entner-Doudoroff pathway. It has been suggested (Swings and De Ley, 1977) that *Zymomonas* and the acetic acid bacteria might be derived from a common aerobic ancestor. The acetic acid bacteria are differentiated from *Zymomonas* by their strictly aerobic growth requirements and their mol% G + C.

Pseudomonas species of sections I, II, III (see this *Manual*) are characterized by a higher mol% G + C value (58–70) than *Zymomonas* (47.5–49.5); also, they are strictly aerobic, generally reduce nitrates, do not grow at pH 4 and have no growth factor requirements.

Taxonomic Comments

The genus *Zymomonas* belongs in the fourth rRNA superfamily (sensu De Ley, 1978) in which it constitutes a separate branch comparable with, for example, the *Acetobacter-Gluconobacter*, *Azospirillum-Spirillum*, *Agrobacterium* and *Rhizobium* branches (Gillis and De Ley, 1980; De Smedt et al., 1980).

Table 5.81.

Differential characteristics of the genus **Zymomonas** *and other genera[a]*

Characteristics	Zymomonas	Acetobacter	Glucono-bacter	Pseudo-monas[b]	Vibrio	Aero-monas
Gram variability occurs	−	+	+	−	−	−
Flagellar arrangement:						
Polar only	+	−	+	+	+	+
Peritrichous	−	+	−	−	−	−
Oxygen tolerance:						
Growth under both aerobic and anaerobic conditions	+	−	−	−	+	+
Growth under aerobic conditions only	−	+	+	+	−	−
Oxidase	−	−	−	+[c]	+[c]	+
Sensitive to vibriostatic compound 0/129	−			−	+	−
Carbohydrate metabolism:						
Fermentative and respiratory	+	−	−	−	+	+
Respiratory only	−	+	+	+	−	−
Gas from D-glucose	+	−	−	−	−	D
1 mol of glucose fermented to 2 mol of ethanol and 2 mol of CO₂	+					
Nitrate reduction	−	−	−	D	+	+
Growth factors required	+[d]		+[e]	−	−	
Growth at pH 4.0	+	+	+	−	−	−
Inhibited by novobiocin	+	D	+		d	−
Mol% G + C of DNA	47.5–49.5	51–65	56–64	58–70	38–51	57–62

[a] Symbols: +, typically positive; −, typically negative; D, differs among species.

[b] *Pseudomonas* sections I, II and III (this *Manual*).

[c] Some species exhibit a negative or weak oxidase reaction.

[d] Pantothenate and biotin.

[e] Pantothenate and/or niacin.

All the *Zymomonas* strains have mol% G + C values within the narrow range of 47.5–49.5. The tightness of the genus *Zymomonas* is also reflected in the high phenotypic similarity ($S_{SM} = > 88\%$) between the strains (De Ley and Swings, 1976). The genus *Zymomonas* contains only one species: *Zymomonas mobilis*. DNA/DNA hybridizations show a nucleotide sequence similarity of >76%; only strain ATCC 29192 is aberrant with less than 32% DNA duplexing (Swings and De Ley, 1975). The homogeneity of the genus *Zymomonas* is further demonstrated by the computer-assisted comparison of electrophoregrams of the soluble cell proteins. All strains except three (ATCC 29192, NCIB 8777 and NCIB 15565) cluster together at or above a correlation coefficient $r = 0.88$, confirming the visual inspection of the gels (Swings et al., 1976). Infrared spectra of intact *Zymomonas* cells reveal that strain ATCC 29192 is characterized by a shoulder at 960 cm^{-1} whereas all the other strains have a distinct peak (Swings and Van Pee, 1977).

Strains ATCC 29192, NCIB 8777 and 10565 are almost identical and are united in *Zymomonas mobilis* subsp. *pomacii* (Swings et al., 1977). All the other strains belong in *Zymomonas mobilis* subsp. *mobilis*.

Miscellaneous Comments

Zymomonas may be important as industrial ethanol producer and it offers advantages over traditional yeast fermentation: it has higher specific rates of glucose uptake and ethanol production; it gives higher ethanol yield and lower biomass; it grows anaerobically and it has a high ethanol tolerance. Some authors studied extensively the kinetics of ethanol fermentation both in batch and continuous cultures at high glucose concentrations (Lee et al., 1979; Rogers et al. 1979, 1980).

Further Reading

Swings, J. and J. De Ley. 1977. The biology of *Zymomonas*. Bacteriol. Rev. *41:* 1–46.

Characteristics of the species **Zymomonas mobilis** and differentiation of its subspecies

The characteristics of *Zymomonas mobilis* are shown in Tables 5.81 and 5.82. The differentiation of its two subspecies is indicated in Table 5.83.

Table 5.82.

Other characteristics of **Zymomonas mobilis**[a]

Characteristics	Reaction or Result	Characteristics	Reaction or Result
Occurrence of "fruity" odor when cultured in standard medium	+	D-Glucose, D-fructose	+
		Sucrose	d
Growth in 0.5% yeast extract broth, in 0.5% peptone broth, or in beer	−	D-Mannose, L-sorbose, D- and L-arabinose, L-rhamnose, D-xylose, D-ribose, D-sorbitol, salicin, dulcitol, D-mannitol, adonitol, erythritol, glycerol, ethanol, D-galacturonate, D,L-malate, succinate, pyruvate, D,L-lactate, tartrate, citrate, starch, dextrin, raffinose, D-trehalose, maltose, lactose, D-cellobiose	−
Growth in beer containing 2% glucose or on malt agar	+		
Salt tolerance: growth in SM in presence of:			
0.5% NaCl	+		
1.0% NaCl	d		
2.0% NaCl	−		
pH range: growth in SM at:		Urease	d
pH 3.05	−	Methylene blue reduction	+
pH 3.50	d	Thionin reduction	+
pH 4.0–7.0	+	Triphenyltetrazolium reduction	+
pH 7.5	d	Indole	−
pH 8.0	−	Hydrolysis of gelatin, Tween 60, Tween 80	−
Temperature range, growth in SM at:		Decarboxylases:	
30–36°C	+	L-Ornithine	d
38°C	d	L-Arginine	d
40°C	−	L-Lysine	d
Ethanol tolerance, growth in SM containing:		Antimicrobial agents (amount per disk)	
5.5% Ethanol	+	Ampicillin, 10 μg	d
7.7% Ethanol	d	Bacitracin, 5 U	R
Glucose tolerance, growth in SM containing:		Cephaloridine, 10 μg	d
20% Glucose	+	Chloramphenicol, 30 μg	S
40% Glucose	d	Erythromycin, 10 μg	d
Neutral red:		Fusidic acid, 10 μg	S
growth in SM containing 0.1% dye	+	Gentamicin, 10 μg	R
Vitamin requirements:		Kanamycin, 10 μg	R
Pantothenate and biotin	+	Lincomycin, 10 μg	R
Lipoic acid, folic acid, niacin, p-aminobenzoic acid, riboflavin, cyanocobalamin	−	Methicillin, 10 μg	R
		Nalidixic acid, 30 μg	R
Catalase	+	Neomycin, 10 μg	R
Oxidase	−	Novobiocin, 30 μg	S
Acetyl methyl carbinol formed (Voges-Proskauer)	W	Penicillin, 5 U	R
		Polymyxin, 300 U	R
Reduction of nitrate	−	Streptomycin, 10 μg	R
H₂S produced	d	Sulfafurazole, 500 μg	S
Survival at 60°C for 5 min	−	Tetracycline, 10 μg	S
Final pH in SM at 30°C	4.8–5.2	Vancomycin, 10 μg	d
Carbon sources:		Actidione, 0.01%	R

[a] For symbols see standard definitions; also R, resistant; S, susceptible; and W, weak.

List of the species and subspecies of the genus **Zymomonas**

1. **Zymomonas mobilis** (Lindner 1928) Kluyver and van Niel 1936, 399.[AL] (*Termobacterium mobile* Lindner 1928, 253, *Zymomonas anaerobia* (Shimwell 1937) Kluyver 1957, 199.)

mo′bi.lis. L. adj. *mobilis* movable, motile.

The description of the species is as for the genus. See also Tables 5.81 and 5.82.

Table 5.83.
Differentiation between subspecies of **Zymomonas mobilis**[a]

Characteristics	1a. Z. mobilis subsp. mobilis	1b. Z. mobilis subsp. pomacii
Colony diameter after aerobic growth on SM for 7 days at 30°C	1.5 mm	<1.0 mm
Growth in SM at 36°C	+	−
Percent DNA/DNA homology with strain 5.3[b]	76–100	<32
Clustering level of protein electrophoregrams[c]	Cluster together above $r = 0.88$	Cluster at $r = 0.75$ with subsp. *mobilis*
Infrared spectra of intact cells[d]:		
Distinct peak at 960 cm⁻¹	+	−
Shoulder only, 960 cm⁻¹	−	+

[a] For symbols see Table 5.81.
[b] Swings and De Ley, 1975.
[c] Swings et al., 1975.
[d] Swings and Van Pee, 1977.

The mol% G + C of the DNA is 47.5–49.5 (T_m).
Type strain: ATCC 10988 (NCIB 8938; NRRL B-806; Queensland 410; L192 Delft; 1TH Delft; DSM 424; IMG 1655).

1a. **Zymomonas mobilis** subsp. **mobilis** (Lindner 1928) De Ley and Swings 1976, 156.[AL] (*Zymomonas mobilis* var. *anaerobia* Richards and Corbey 1974, 243; *Zymomonas mobilis* var. *recifensis* Gonçalves de Lima, De Araújo, Schumacher and Cavalcanti Da Silva 1970, 3; *Zymomonas anaerobia* var. *anaerobia* (Shimwell) Carr 1974, 353; *Saccharomonas anaerobia* var. *immobilis* Shimwell 1950, 182; *Zymomonas anaerobia* var. *immobilis* (Shimwell) Carr 1974, 353.)

See Table 5.83 for differentiation of this subspecies from the subspecies *pomacii*.

Isolated from bees, from ripening honey in Spain, from the fermenting sap of *Agave americana* in Mexico, from fermenting palm juice (*Arenga pinnata*) in Java, Indonesia, and *Elaeis guineensis* and *Raphia vinifera* in Zaire and Nigeria, and from fermenting sugarcane juice in Brazil. It has also been isolated in England from beer, from the surface of brewery yards and from the brushes of cask-washing machines.
Type strain: ATCC 10988. Phenotypic centrotype: ATCC 29191.

1b. **Zymomonas mobilis** subsp. **pomacii** (Millis 1956) De Ley and Swings 1976, 156.[AL] (*Zymomonas anaerobia* subsp. *pomaceae* (sic) Millis 1956, 527; *Zymomonas mobilis* subsp. *pomaceae* (sic) (Millis 1956) De Ley and Swings 1976, 156.)

pom.a′ci.i. V.L. n. *pomacium* cider; M.V. L. gen. n. *pomacii* of cider.

See Table 5.83 for differentiation of this subspecies from the subspecies *mobilis*.

Isolated in England from sick cider and from apple pulp.
Type strain: ATCC 29192 (NCIB 11200). Reference strains: NCIB 8777 and NCIB 10565.

Genus **Chromobacterium** Bergonzini 1881, 153[AL]

PETER H. A. SNEATH

Chro.mo.bac.te′ri.um. Gr. n. *chroma* color; Gr. n. *bakterion* a small rod; M.L. neut. n. *Chromobacterium* a small colored rod.

Rods 0.6–0.9 × 1.5–3.5 μm with rounded ends, sometimes slightly curved. Occur singly; occasionally pairs, elongated forms or short chains occur. Definite capsules are not evident. **No resting stages known.** Gram-negative, often with barred or bipolar staining and lipid inclusions. **Motile by means of both a single polar flagellum and usually one to four subpolar or lateral flagella. Facultative anaerobes.** Produce butyrous, **violet colonies** on solid media; in nutrient broth, a **violet ring** is formed at the junction of the liquid surface and the container wall. Growth occurs at 25°C, but species differ in their optimum, maximum and minimum temperatures. Optimum pH, 7–8; no growth below pH 5. No growth occurs in media containing 6% or more of NaCl. Chemoorganotrophs, having mainly a **fermentative attack on carbohydrates.** Acid but no gas is produced from glucose and certain other carbohydrates. Lactate is oxidized to CO_2. Usually oxidase-positive by the method of Kovacs (1956), although the violet pigment may interfere with the reading. Catalase-positive. Indole-negative by usual testing methods, but under some conditions compounds may accumulate that give positive reactions with indole test reagents (Corpe, 1963). Voges-Proskauer-negative. Nitrate and usually nitrite are reduced, often with visible gas production. Ammonia is formed from peptone. Phosphatase-positive. Arylsulfatase-negative. Grow on ordinary media. May utilize citrate and ammonia as sole carbon and nitrogen sources, but growth occurs slowly. Growth factors are not required. **Resistant to benzylpenicillin** (10 μg/ml) **and to vibriostatic agent 0/129** (2,4-diamino-6,7-diisopropylpteridine, 30 μg/disc); sensitive to tetracycline (30 μg/ml). Soil and water organisms, occasionally causing infections of mammals, including man. The mol% G + C of the DNA is 50–68 (T_m).

Type species: *Chromobacterium violaceum* Bergonzini 1881, 153.

Further Descriptive Information

The characteristic flagellar arrangement is best seen in young cultures on solid media. The single polar flagellum is inserted at the tip of the cell, shows long, shallow waves and often stains faintly. The lateral flagella are usually long and one to four in number, although up to eight may occur. They may be inserted subpolarly or laterally, usually show deep short waves and stain readily. The two types of flagella are antigenically distinct. Old cultures and cultures in liquid media show few lateral flagella. Occasional strains lack the lateral flagella.

The pigment violacein is perhaps the most distinctive aspect of the genus and occurs in both species. It is readily identified by spectrophotometric means and shows an absorption maximum in ethanolic solution at 579 nm and a minimum at 430 nm. In 10% (v/v) H_2SO_4 in ethanol the pigment gives a green solution with an absorption maximum at 700 nm. If NaOH is added to an ethanolic solution it becomes green, then reddish brown. Pigment is only freely produced on media containing tryptophan, and may be suppressed by certain brands of peptone. Violacein has antibiotic properties (DeMoss, 1967). Violacein production has also been used as an assay for L-tryptophan (Sebek, 1965).

C. violaceum occurs mainly in soil and water and is common in tropical countries. *C. fluviatile* has been isolated from river water in England.

C. violaceum occasionally causes serious pyogenic or septicemic infections of mammals, including man (reviewed by Sneath, 1960). Pathogenic but nonpigmented variants may not be easily identified in the clinical laboratory (see Differentiation of the Genus *Chromobacterium* from Other Genera).

Enrichment and Isolation Procedures

A selective medium for *C. fluviatile* is that of Keeble and Cross (1977), which contains (g/liter): yeast extract (Difco), 1.0; beef extract (Lab Lemco Oxoid), 1.0; Casitone (Difco), 2.0; glucose, 10.0; and agar (Lab M No. 2), 18.0. To the molten, cooled medium add aseptically solutions (sterilized by filtration) of neomycin hydrochloride, cycloheximide and nystatin to give a final concentration each of 50 mg/liter. Pigmentation is good on this medium. An alternative medium is that of Ryall and Moss (1975), or a citrate ammonium salts agar (see Moss and Ryall, 1981).

Selective media for *C. violaceum* have not been developed.

For screening isolates, Sivendra and Tan (1977) recommended Kligler iron agar (KIA) and triple sugar iron agar (TSI). Acid but no blackening is produced in the butt of both media. Acid is not produced on the slant of KIA, and by only some strains on TSI.

Maintenance Procedures

The organisms survive for several years in dilute peptone water (0.1% peptone) at room temperature (for *C. violaceum*) or 4°C (for *C. fluviatile*). They also can be preserved indefinitely by lyophilization or by freezing in nutrient broth containing 15% glycerol.

Procedures for Testing for Special Characters

Production of HCN is best tested by stab inoculation into a tube of semisolid medium (nutrient agar diluted with an equal volume of water). An indicator paper (prepared by dipping filter paper into saturated aqueous picric acid, drying, dipping into 10% aqueous sodium carbonate and drying again) is placed between the tube and the plug. Production of HCN is shown by the paper turning from yellow to brick-red in 1 or 2 days at 25°C.

Casein hydrolysis is tested by streak inoculation of 50% skim milk with 1.5% agar added. A clear zone occurring around the growth after incubation for 4 days at 25°C indicates casein hydrolysis.

Turbidity from egg yolk is tested by streak inoculation of plates containing 1 part of egg-yolk emulsion (egg yolk removed aseptically and suspended in 4 volumes of aqueous 5% NaCl) to 9 parts of melted nutrient agar after cooling to 55°C. Incubate for 4 days at 25°C.

Esculin hydrolysis is tested in a medium containing (g/liter): peptone, 10.0; sodium citrate 1.0; esculin, 1.0; ferric citrate, 0.05; pH, 7.0. The medium is tubed in 5-ml quantities and sterilized by autoclaving. After inoculation and incubation at 25°C for 4 days, hydrolysis of the esculin is indicated by a brown coloration.

Arginine hydrolysis is demonstrated using the following medium (g/liter): peptone, 1.0; NaCl, 5.0; K$_2$HPO$_4$, 0.3; agar, 3.0; phenol red, 0.01; and L-arginine hydrochloride, 10.0; pH, 7.2. The medium is dispensed into tubes to a depth of 2 cm and sterilized by autoclaving. The medium is inoculated by stabbing, covered with sterile melted petrolatum and incubated at 25°C for 4 days. A red coloration indicates a positive reaction.

Differentiation of **Chromobacterium** from other genera

Table 5.84 presents characteristics differentiating the genus *Chromobacterium* from the genus *Janthinobacterium*.

Unpigmented strains of *C. violaceum* may be difficult to identify and may be mistaken for *Aeromonas* or *Vibrio* species. Distinction from *Aeromonas* is aided by the following: *C. violaceum* is indole-negative, methyl red-negative, and produces HCN, whereas *Aeromonas* shows the reverse reactions. Also, unlike *C. violaceum*, most strains of *Vibrio* and *Aeromonas* are able to utilize D-mannitol as a carbon source (see article on the genus *Vibrio*).

Taxonomic Comments

Some organisms previously classified in the genus *Chromobacterium* in the eighth edition of the *Manual* are now classified in the genus *Janthinobacterium* (i.e. *J. lividum*). The previous classification should be borne in mind when consulting the older literature of the field.

The present description of the genus *Chromobacterium* is based mainly on the following references: Leifson (1956), Sneath (1956, 1960), Steel and Midgley (1962), Moffett and Colwell (1968), De Ley et al. (1978), and Moss et al. (1978). Strains of *C. violaceum* that do not acidify carbohydrates anaerobically (e.g. *C. laurentium* of Leifson, 1956) do not differ in most other respects from the fermentative strains, and they do extensively cross-react serologically with them; they are here considered to belong to *C. violaceum*.

The genus *Chromobacterium* historically has been grouped with the genera *Rhizobium* and *Agrobacterium*, but evidence from DNA/rRNA studies suggests that it is closest to *Alcaligenes*, *Bordetella*, *Janthinobacterium*, and certain nonfluorescent pseudomonads (De Ley et al., 1978).

Table 5.84.

Characteristics differentiating the genus **Chromobacterium** *from the genus* **Janthinobacterium**[a]

Characteristics	Chromo-bacterium	Janthino-bacterium
Glucose is catabolized by:		
Fermentation	D (80)	−
Oxidation	D (20)	+ (95)
Turbid zone on egg yolk agar (lecithinase activity)	+	−
Acid from:		
Trehalose	+	−
L-Arabinose	−	+ (95)
Xylose	−	+ (95)
Casein hydrolysis	+	− or weak (10)
Esculin hydrolysis	−	+ (95)

[a] For symbols see standard definitions. Numbers in parentheses indicate the % of strains giving a positive reaction.

Miscellaneous Comments

The genus *Chromobacterium* is of interest to biochemists in several areas: indole metabolism and the biosynthesis of violacein; production of HCN; occurrence of unusual sugar compounds; and the production of extracellular polysaccharides (DeMoss, 1967; Brysk et al., 1969; Stevens et al., 1963; Niven et al., 1975; Corpe, 1964).

Differentiation between the species of the genus **Chromobacterium**

Table 5.85 presents the main features that differentiate the two species of the genus.

List of the species of the genus **Chromobacterium**

1. **Chromobacterium violaceum** Bergonzini 1881, 153^{AL}

vi.o.la′ce.um. L. adj. *violaceum* violet colored.

Rods 0.6–0.9 × 1.5–3.0 μm, often coccobacillary. Rarely contain metachromatic granules. Usually (80%) contain poly-β-hydroxybutyrate inclusions.

Colonies are low convex, violet, smooth (although rough and nonpigmented variants may occur), and not gelatinous. A violet ring is formed in nutrient broth at the surface with a fragile pellicle.

Attack on carbohydrates is usually fermentative (80% of strains, rarely with gas), sometimes oxidative (20%). Acidity from carbohydrates is detectable in ordinary peptone water, but the medium of Hugh and Leifson (1953) is preferable.

Grows on ordinary peptone media. Usually grows on MacConkey agar, giving pale violet colonies; grows slowly on eosin methylene blue agar.

Facultatively anaerobic, although the oxidative strains grow slowly anaerobically. Optimum temperature, 30–35°C; minimum, 10–15°C; maximum, 40°C (20% of the strains can grow at 44°C).

Other characteristics are given in Tables 5.85 and 5.86.

Soil and water organisms, common in tropical countries. Occasionally cause infections in mammals, including humans.

The mol% G + C of the DNA is 65–68 (T_m).

Type strain: ATCC 12472 (MK, NCTC 9757, NCIB 9131, D 252).

2. **Chromobacterium fluviatile** Moss, Ryall and Logan 1981, 216.[VP] (*Effective publication:* Moss, Ryall and Logan 1978, 18.)

flu.vi.a.ti′le. L. adj. *fluviatile* of rivers.

Rods 0.7 × 3.0–3.5 μm, occurring singly or in short chains with occasional elongated forms.

Colonies are flat, very thin, irregular in outline, spreading, pale violet with a copper-beaten, slightly rough surface, and are not gelatinous. A uniform turbidity is produced in nutrient broth with a violet ring at the surface but usually no pellicle.

Grows on ordinary peptone media. Grows on MacConkey agar, giving violet colonies.

Facultatively anaerobic. Optimum temperature, 25°C; minimum, 4°C; maximum, ~30°C.

Some strains produce reducing substances from gluconate.

Weakly hemolytic on goat blood agar.

Other characteristics are given in Tables 5.85 and 5.86.

Isolated from river water in England.

The mol% G + C of the DNA is 50–52 (T_m).

Type strain: NCTC 11159 (strain 165).

Table 5.85.

Characteristics differentiating between the species of the genus **Chromobacterium**[a]

Characteristics	1. C. violaceum	2. C. fluviatile
Growth at 4°C, 7 days	−	+
Growth at 37°C, 7 days	+	−
HCN production	+	−
Turbid zone on egg-yolk agar	+	Weak, only under colony
Arginine hydrolysis	+	−
Colonies flat, spreading	−	+

[a] For symbols see standard definitions.

Table 5.86.

Other characteristics of the species of the genus **Chromobacterium**[a]

Characteristics	1. C. violaceum	2. C. fluviatile
Acid but no gas from:		
Glucose, fructose, trehalose	+	+
Maltose	d (50)	+
Mannose	[+] (80)	[+] (90)
Sorbitol	d (60)	−
Rhamnose	d (50)	
Sucrose	[−] (25)	[−] (10)
Starch, dextrin, glycogen	[−] (20)	
Glycerol	[−] (10)	[−] (30)
Salicin	[−] (10)	−
L-Arabinose, galactose	−	[−] (10)
Cellobiose	−	[−] (30)
Dulcitol, m-inositol, inulin, lactose, mannitol, xylose	−	−
Adonitol, melibiose, melezitose, raffinose	−	
Gelatin liquefaction (infundibuliform)	+	+
Casein hydrolysis	+	+
Hemolytic on horse blood	[+] (95)	W or −
Chitin digestion	[+] (95)	+
Starch digestion	W or −	[−] (10)
Agar digestion	−	
Esculin hydrolysis	−	−
Acetic acid produced from ethanol	−	
Urease	W or −	−
Phenylalanine deaminase	−	−
β-Galactosidase (ONPG method)	−	
Acetamide hydrolysis	−	
Pectate digestion	−	
Nitrate reduction	[+] (95)	+
Nitrite reduced	[+] (80)	−
H₂S production	− or W	−
Amino acid decarboxylases:		
Lysine decarboxylase	−	−
Arginine dihydrolase	d (50)	−
Ornithine decarboxylase	−	−
Melanin produced from phenylalanine	[+] (80)	
HCN production; cultures smell of ammonium cyanide	[+] (95)	
Penicillinase production	[−] (30)	
Methylene blue reduction	+ or W	+
Deoxyribonuclease (DNase)	−	
Selenite reduction	−	
3-Ketolactose produced from lactose	−	
2-Ketogluconate produced from gluconate	−	
Sole carbon sources:		
Citrate	+[b]	+[b]
L-Alanine, L-glutamate, L-histidine, L-lysine, L-ornithine, L-phenylalanine, L-tyrosine	+	
L-Arginine, L-serine, L-proline	d (60)	
L-Cystine, L-leucine	−	
Acetate	d (50)	−
Malonate	−	−
Formate	[−] (30)	
Growth in KCN medium	+	

[a] Symbols: +, all strains positive; [+], positive for 80% or more strains; d, positive for 31–79% of strains; [−], positive for 30% or fewer strains; −, negative for all strains; W, weak. Numbers in parentheses indicate the % of strains giving a positive reaction.

[b] Growth occurs slowly.

Genus **Cardiobacterium** Slotnick and Dougherty 1964, 271[AL]

ROBERT E. WEAVER

Car.di.o.bac.te'ri.um. Gr. n. *cardia* heart; Gr. n. *bakterion* small rod; M.L. neut. n. *Cardiobacterium* bacterium of the heart.

Straight rods 0.5–0.75 μm **in diameter** and 1.0–3.0 μm in length, with rounded ends. Occasional long filaments, 7.0–35.0 μm may occur. **Pleomorphic.** Cells are arranged singly, in pairs, in short chains and in rosette clusters. **Gram-negative, but retention of crystal violet may occur in the swollen ends or central portions of cells. Nonmotile. Facultatively anaerobic.** Carbon dioxide is required by some strains on isolation. Aerobic growth is scant unless humidity is elevated. Growth in candle jars or under anaerobic conditions is not dependent on an elevated humidity. Optimum temperature, 30–37°C. Colonies on blood agar are smooth, convex and opaque. Chemoorganotrophic, having a strictly **fermentative** type of metabolism. Acid but not gas is produced from fructose, glucose, mannose, sorbitol and sucrose. **Lactic acid is the major product of glucose fermentation;** smaller amounts of pyruvate, formate and propionate are formed. **Oxidase-positive. Catalase-negative. Small amounts of indole are formed. Nitrates are not reduced.** No growth occurs on MacConkey agar. Urease-negative. Ornithine decarboxylase (ODC) negative. Occur in nasal flora of humans; isolated from the blood of humans with bacterial endocarditis. The mol% G + C of the DNA is 59–60 (T_m).

Type species: *Cardiobacterium hominis* Slotnick and Dougherty 1964, 271.

Further Descriptive Information

Cells of *Cardiobacterium* grown on tryptose blood agar (containing 5% human blood) are pleomorphic and have rounded ends. The usual cell dimensions are ~0.5–0.6 μm in diameter and from 1.0–2.2 μm in length. Occasionally filaments are formed which vary in length from 7.0–35.0 μm. Tear-drop shaped cells may occur. One or both ends of *Cardiobacterium* cells are frequently enlarged, and the crystal violet of the Gram stain tends to be retained in these areas. Other cells may show crystal violet retention in the central portions (Slotnick and Dougherty, 1964).

The degree of pleomorphism is apparently influenced by the medium on which *Cardiobacterium* is grown. Pleomorphism has been reported in cultures grown on agar media which do not contain yeast extract. On yeast extract-containing media, the cells have been found to be generally uniform Gram-negative rods 0.5 μm and 2.0 μm long (Savage et al., 1977).

Sudanophilic bodies and metachromatic granules have been reported with the use of the Sudan Black B and Albert's stain, respectively (Slotnick and Dougherty, 1964). However, Midgley et al. (1970) were unable to confirm these observations.

The cell wall is of the Gram-negative type with coherent layers: a dense outer layer, a unit membrane, and a dense inner layer. The substructure of the surface layer consists of closely packed units which are nearly spherical and have an average diameter of 3.5 nm. Cells studied by electron microscopy have been found to have 20- to 40-nm-thick polar caps consisting of a dense, tufty material. In the periphery of the cytoplasm, numerous unit-membrane profiles have been observed and are thought to be intrusions of the plasma membrane (Reyn et al., 1971).

After incubation for 24 h at 37°C on tryptose blood agar (containing 5% human blood), colonies are punctiform. After 48 h the maximum size is 1–2 mm. The colonies are circular, convex, smooth, moist, glistening, opaque and butyrous, and have an entire edge. No clear zones of hemolysis are formed around the colonies (Slotnick and Dougherty, 1964).

Strains of *C. hominis* may be grown in a defined medium* described by Slotnick and Dougherty (1965). Essential components of the medium are pantothenate, niacinamide, thiamine, threonine, proline, leucine, histidine, glycine and arginine. Biotin, pyridoxine, tyrosine, valine and glutamate are not required but must be included for optimum growth. Glucose is not an essential carbon source or the sole energy source, but it is fermented and does increase the growth yield. Maltose, sucrose, fructose and mannose are also fermented and can be substituted effectively for glucose in the medium. Lactose, arabinose and galactose are not fermented or utilized for growth. Low concentrations (~1.5–2.0 μg/ml) of riboflavin or flavin mononucleotide inhibit the growth of *C. hominis*. This inhibitory effect is neutralized by the presence of excess leucine in the medium.

The major end product of glucose fermentation is lactic acid (Slotnick and Dougherty, 1964).

Growth at 30°C and 37°C is equally good. Growth at 25°C is sporadic and light. No growth occurs above 42°C or at 22°C (Midgley et al., 1970; Slotnick and Dougherty, 1964).

The optimal pH range for growth is 7.0–7.2. *C. hominis* grows poorly in air unless the humidity is increased by a method such as placing filter paper strips, saturated with water, in a closed container. In a candle jar or under anaerobic conditions it is not necessary to increase the humidity. The optimal atmosphere contains 3–5% CO_2 (Slotnick and Dougherty, 1964).

Midgley et al. (1970) found that three of four strains grew equally well in a candle jar and in a sealed jar without increased CO_2. The fourth strain grew less well in the sealed jar than in the candle jar.

Slotnick and Dougherty (1964) produced antisera to four strains. All of the strains reacted to a high titer in each of the unabsorbed sera; however, by absorption studies the four strains were shown not to be antigenically identical. No antigenic relationship was demonstrated to members of the genera: *Brucella, Pasteurella, Bordetella, Moraxella, Haemophilus, Streptobacillus, Corynebacterium, Bacteroides, Neisseria, Escherichia,* and *Lactobacillus.*

Savage et al. (1977) tested five strains and found the minimum inhibitory concentrations for all strains to be less than 2 μg/ml for ampicillin, carbenicillin, cephalothin, chloramphenicol, penicillin, tetracycline, streptomycin, kanamycin, gentamicin, and colistin.

In humans, *C. hominis* is a cause of bacterial endocarditis. Tucker et al. (1962) found no evidence of infection in mice, guinea pigs, rabbits, hamsters or pigeons after injection of 10^7–10^8 viable organisms intravenously, intraperitoneally or subcutaneously.

C. hominis is apparently part of the normal nasal and pharyngeal flora of humans. It was isolated from the nose or throat of 68 of 100 individuals examined by Slotnick et al. (1964). The organism was also isolated from cervical and vaginal cultures obtained from 2 of 159 subjects examined (Slotnick, 1968).

Isolation and Enrichment Procedures

Cardiobacterium hominis has been isolated from blood in several different media under both aerobic and anaerobic conditions. Media which have been used successfully are brain heart infusion broth

---

* Defined medium has the following composition (per liter of distilled water): $MgSO_4 \cdot 7H_2O$, 0.4 g; $MnSO_4 \cdot H_2O$, 3 mg; NaCl, 50 mg; Na_2HPO_4, 2.84 g; KH_2PO_4, 2.72 g; $ZnSO_4 \cdot 7H_2O$, 4.43 mg; $FeSO_4 \cdot 7H_2O$, 4 mg; $CuSO_4 \cdot 5H_2O$, 0.5 mg; glucose, 5.0 g; arginine, 160 mg; glutamic acid, 200 mg; glycine, 176 mg; histidine, 126 mg; leucine, 426 mg; proline, 100 mg; threonine, 276 mg; valine, 185 mg; tyrosine, 35 mg; calcium pantothenate, 10 μg; niacinamide, 10 μg; pyridoxine hydrochloride, 20 μg; thiamine hydrochloride, 10 μg; and biotin, 1 μg. Final pH is 7.0.

containing sodium polyanetholsulfonate (SPS) and *p*-aminobenzoic acid, casein soy broth with SPS, thioglycolate broth with SPS, and glucose broth (Midgley et al., 1970; Savage et al., 1977).

Slotnick et al. (1964) isolated *C. hominis* from the nose and throat of humans on trypticase soy agar plates containing 5% human blood. The plates were incubated at 37°C for 48–72 h in candle jars. Because of difficulty in recognizing the colonies of *C. hominis*, smears of selected areas of the plates were stained with fluorescent antibody. Isolation of the organisms was accomplished by subculturing from the areas which showed positive fluorescent staining reactions.

Maintenance Procedures

Strains of *C. hominis* may be preserved by lyophilization. They also may be preserved by freezing. Cells from 18- to 24-h-old agar cultures suspended in defibrinated rabbit blood have survived for more than 15 years at −50°C.

Procedures for Testing for Special Characters

Indole formation is an important characteristic for the identification of *C. hominis*. Indole production, however, may be weak and may not be detected by test procedures which do not concentrate the indole by xylene extraction.

Tryptone broth and heart infusion broth are suitable media for the indole test. Remove a portion of an 18- to 24-h-old broth culture and test. If the reaction is negative, the remaining portion of the broth culture should be tested after incubation for an additional 24 h. To perform the test add approximately 0.5–1 ml of xylene to 3 or 4 ml of broth culture. Shake vigorously to extract the indole. After the xylene layer has formed at the surface, add either Ehrlich's or Kovacs' reagent, allowing the reagent to flow down the side of the slightly tilted tube.

Differentiation of **Cardiobacterium** from other genera

Characteristics by which *C. hominis* can be distinguished from other Gram-negative rod-shaped organisms which have similar physiological characteristics are listed in Table 5.87.

Taxonomic Comments

The genus contains a single species. Its characteristics are not sufficiently similar to those of other existing genera to justify classifying *C. hominis* within any of them. Reasons for not placing it in any of the genera which, at the time, were considered to be in the family *Brucellaceae* have been discussed by Slotnick and Dougherty (1964).

Further Reading

Snell, J.J.S. and S.P. Lapage. 1976. Transfer of some saccharolytic *Moraxella* species to *Kingella* Henriksen and Bøvre 1976, with descriptions of *Kingella indologenes* sp. nov. and *Kingella denitrificans* sp. nov. Int. J. Syst. Bacteriol. *26:* 451–458.

Table 5.87.

Differentiation of **Cardiobacterium hominis** *and phenotypically similar genera and species[a]*

Characteristics	*Cardiobacterium hominis*	*Haemophilus aphrophilus*	*Kingella*	*Actinobacillus actinomycetemcomitans*	*Pasteurella*	*Eikenella corrodens*	*Capnocytophaga*
Oxidase	+	d	+	d	+	+	−
Catalase	−	−	−	+	+	−	
Indole	+	−	D[b]	−	D	−	−
Fermentative ability present	+[c]	+	+[c]	+	+	−	+
Nitrate reduction	−	+	D[d]	+	+	+	D
Mol% G + C of DNA	59–60	42	47.3–54.8	42.7	40–45	56.2–58.2	33–41

[a] Symbols: see standard definitions.

[b] Produced only by *K. indologenes*

[c] *C. hominis* acidifies sorbitol and usually acidifies mannitol; *K. indologenes* does not acidify mannitol or sorbitol.

[d] Not reduced by *K. indologenes*

List of the species of the genus **Cardiobacterium**

1. **Cardiobacterium hominis** Slotnick and Dougherty 1964, 271. (Group II-D, Tucker et al., 1962, 2–3.)

ho′mi.nis. L. gen. n. *hominis* of man.

The characteristics are as described for the genus. Additional characteristics are listed in Table 5.88.

Occurs in nasal and oral flora of humans; isolated from the blood of humans with endocarditis.

The mol% G + C of the DNA is 59–60 (T_m).

Type strain: ATCC 15826 (strain 6573 of Slotnick and Dougherty, 1964).

Table 5.88.
Characteristics of **Cardiobacterium hominis**[a]

Characteristics	1. *Cardiobacterium hominis*
Oxidase	+
Catalase	−
Motility	−
Urease	−
Esculin hydrolysis	−
Nitrate reduction	−
Indole	Weak +
H_2S (triple-sugar iron agar)	−
Litmus milk acidification	d
Growth on MacConkey agar	−
Gelatin liquefaction	−
Fermentative ability present	+
Tween 20 hydrolysis	−
Tween 40 hydrolysis	−
Acidification of carbohydrate media[b]	
Glucose, sucrose, fructose, mannose, sorbitol	+
Maltose and mannitol	d[c]
Xylose, lactose, salicin, arabinose, adonitol, dulcitol, galactose, rhamnose, trehalose, inositol, cellobiose, erythritol, melibiose, melezitose	−

[a] For symbols see standard definitions.

[b] Incubated for 7 days.

[c] Maltose and mannitol reactions reported to be variable by Slotnick and Dougherty (1964) and Midgely et al. (1970). Weaver and Hollis found mannitol to be acidified by 92% and maltose by 99% of 74 strains (unpublished data).

Genus **Calymmatobacterium** Aragao and Vianna 1913, 221[AL]

ROBERT B. DIENST AND GEORGE H. BROWNELL

Ca.lym.ma.to.bac.te′ri.um. Gr. n. *calymma* mantle, sheath; Gr. dim. neut. n. *bakterion* a small rod; M.L. neut. n. *Calymmatobacterium* the sheathed rodlet.

Pleomorphic rods, 0.5–1.5 μm wide by 1.0–2.0 μm in length, with rounded ends. Occur singly or in clusters. **The cells exhibit single or bipolar condensation of chromatin. Capsules are present.** Gram-negative. Nonmotile. The exudate from infected tissues, when stained by Wright's stain or by Giemsa stain, demonstrates **characteristic intracellular organisms in the cytoplasm of large mononuclear phagocytes.** Can be cultivated in vivo in the yolk sac of embryonated chicken eggs or in vitro on special egg yolk-containing media. Optimum temperature, 37°C. Pathogenic for humans, causing **Donovanosis (granuloma inguinale).** The mol% G + C of the DNA is not known.

Type species: *Calymmatobacterium granulomatis* Aragao and Vianna 1913, 221.

Further Descriptive Information

In diseased tissue smears stained by the Wright's method, *C. granulomatis* occurs within the cytoplasm of large mononuclear monocytes as blue to purple pleomorphic rods surrounded by pink capsules (Fig. 5.28). The organism may occasionally be observed free in extracellular spaces. The single or bipolar condensation of chromatin gives rise to characteristic "safety pin" forms. The ultrastructure of the intracellular organisms has been described by Davis and Collins (1969), Dodson et al. (1973) and Kuberski et al. (1980). Electron micrographs reveal encapsulated bacilliforms with characteristic Gram-negative cell walls (Fig. 5.29).

Calymmatobacterium granulomatis has been clinically proven to be the causal agent of granuloma inguinale (Dienst et al., 1938). Presently, the disease is referred to as Donovanosis because initial lesions have been diagnosed in skin areas other than the genital region. The organism is pathogenic only for man and infection cannot be produced in laboratory animals.

There are no protective antibodies produced by a patient infected with *C. granulomatis*. Once the infection occurs the disease persists chronically and may spread through the lymphatics to all tissue unless treated with antibiotics. The patient does produce specific sensitizing antibodies as shown by skin testing (Chen et al., 1949). Antibodies can be detected in the serum of patients by complement-fixation procedures (Dulaney and Packer, 1947). The test antigens used in these techniques have included pus from granulomatous lesions, whole or ruptured *C. granulomatis* or boiled or extracted egg yolk medium following growth of the organism.

Donovanosis is seen throughout the world and is endemic in many areas including the United States. This infection is encountered primarily in the dark skinned races and occurs mostly in regions where the climate is warm and humid for several months of the year. Most researchers support the contention that Donovanosis is not primarily a venereal disease, as such, but is an infection resulting from intimate contamination and poor hygiene.

Enrichment and Isolation Procedures

At present, the only sources for *C. granulomatis* are the lesions of Donovanosis. An organism has reportedly been isolated from human feces (Goldberg, 1962) which has antigenic similarities to *C. granulomatis*.

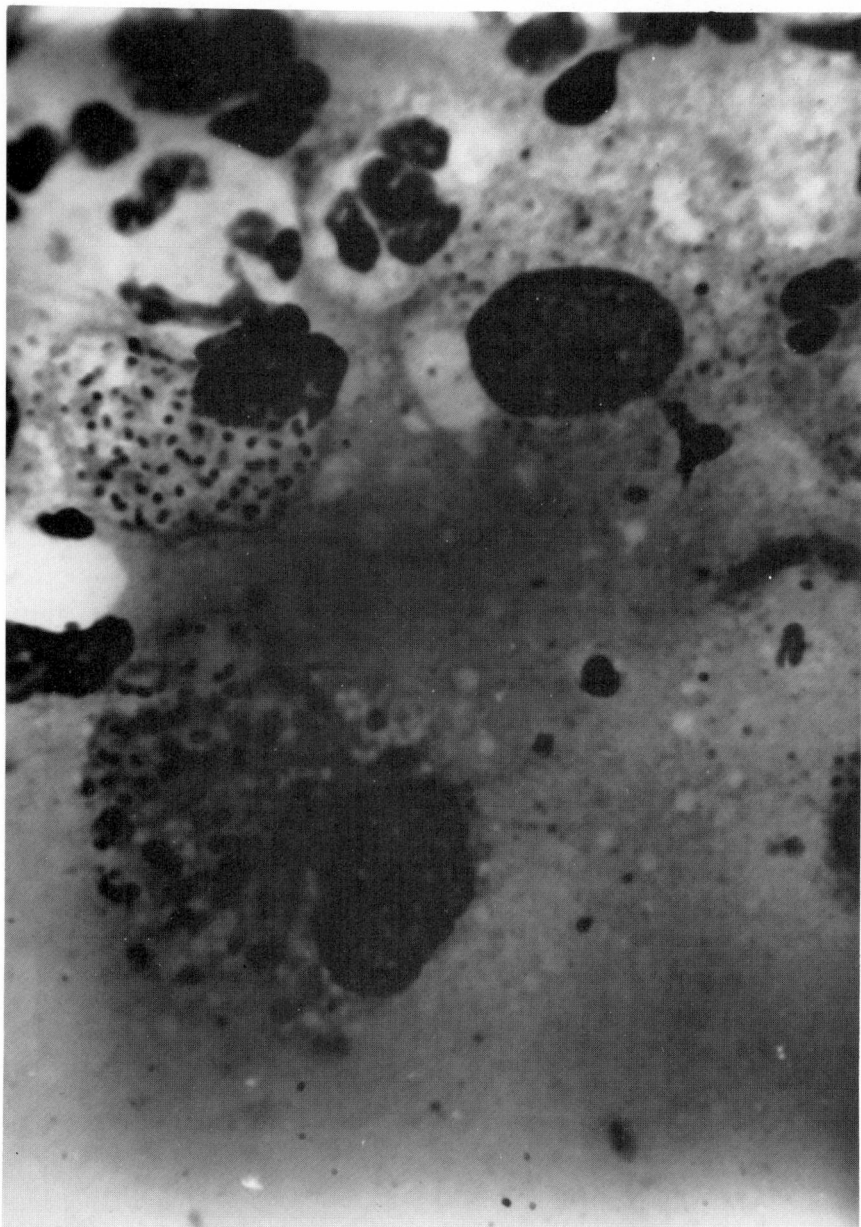

Figure 5.28. Large mononuclear phagocytes filled with *C. granulomatis*. Wright's stain (× 900). (Reproduced with permission from R. B. Dienst and G. H. Brownell *in* M. P. Starr et al. (editors) *The Prokaryotes: a Handbook on Habitats, Isolation and Identification of Bacteria.* p. 1410, 1981, © Springer-Verlag, New York.)

The following procedure is given by Morse (1980). The ulcerative lesions are cleansed with sterile, saline-soaked gauze before obtaining samples, in order to decrease contamination and remove tissue debris. Samples of tissue are removed by scraping or by means of a biopsy punch from beneath the border of the lesion, and small cleansed pieces of tissue are minced into small particles. Inoculation is made into the yolk sacs of 5-day-old embryonated eggs. After incubation for 72 h, the organisms can be detected in the yolk sac fluid.

C. granulomatis can also be isolated and grown in vitro. For example, a pure culture was isolated by Dienst (1948) by inoculating fresh egg medium* exudate aspirated from a pseudobubo of a patient. This isolate could be subcultured and maintained in the same culture medium, and examination of the subcultures revealed large numbers of encapsulated organisms consistent with the morphology of *C. granulomatis* as seen in stained smears of exudates from patients with Donovanosis. Dienst et al. (1948) indicated that several factors were important for isolation and cultivation: (a) maintenance of a low oxidation-reduction potential, (b) the requirement for a growth factor found in egg yolk, and (c) use of semisolid media containing 0.12% agar.

Dulaney slants† have also been used for isolation and cultivation of *C. granulomatis* (Dulaney et al., 1948). After inoculation of lesion material onto a Dulaney slant, Locke's fluid† is added to cover three-

---

* Dienst's egg yolk medium. The basal medium contains (g/liter): peptone, 10.0; tryptone, 3.0; glucose, 3.0; commercial sea salts, 2.0; and agar, 1.2; pH 7.3. The basal medium is sterilized by autoclaving and cooled to 45°C. Egg yolk which has been removed aseptically from fresh eggs is diluted with an equal volume of sterile saline (0.85% NaCl) and then mixed in equal portions with the basal medium.

† Dulaney slants (Dulaney et al., 1948). The yolks are removed aseptically from 5- to 8-day-old embryonated chicken eggs and placed in an equal volume of sterile Locke's solution (see below) containing sterile glass beads. The yolks are homogenized and the homogenate is dispersed into tubes and coagulated in a slanted position with steam at 80°C for 15 min. *Locke's solution* (g/liter): NaCl, 9.0; CaCl$_2$, 0.24; KCl, 0.42; Na$_2$CO$_3$, 0.20; and dextrose, 2.5.

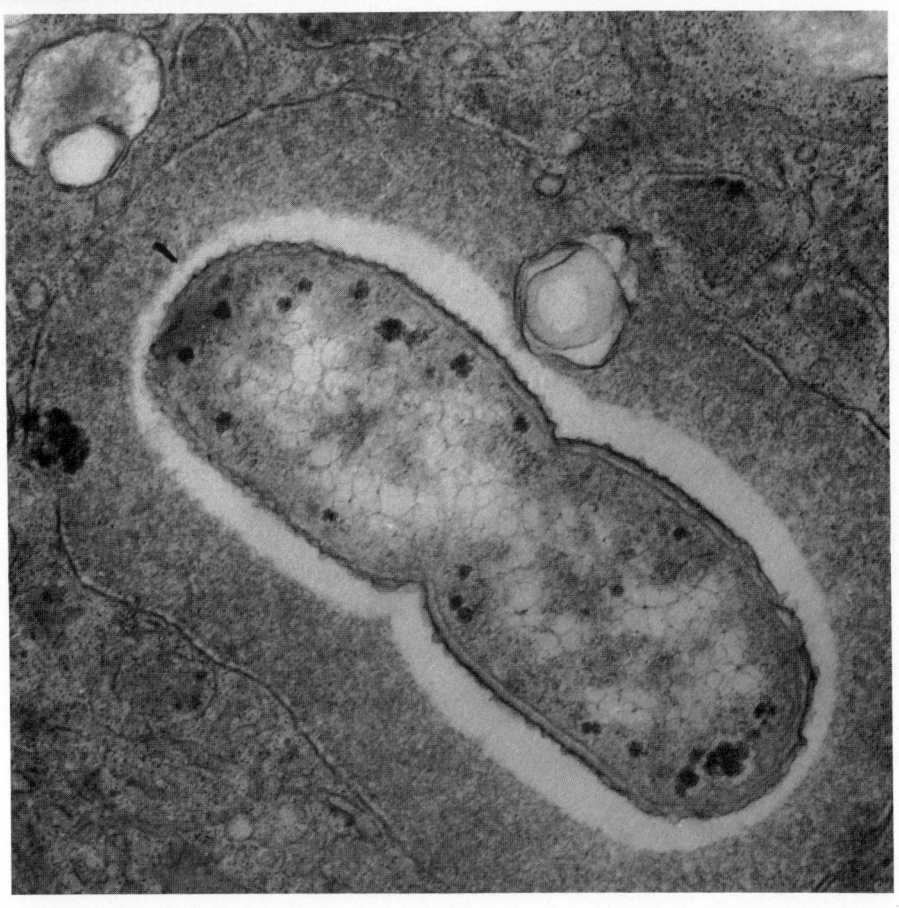

Figure 5.29. A dividing *C. granulomatis* cell within a phagosome (× 68,500). (Reproduced with permission from T. Kuberski, J. M. Papadimitriou and P. Phillips, Journal of Infectious Diseases *142:* 744–749, 1980, © The University of Chicago Press.)

quarters of the slant, and the tubes are then incubated in a vertical position for 48–72 h (Morse, 1980).

A semisynthetic medium has been devised by Goldberg (1959) for cultivation of laboratory strains of *C. granulomatis.* In this medium, the requirement for egg yolk is replaced by lactalbumin hydrolysate or by papaic digest of soy meal USP.

Differentiation of the genus **Calymmatobacterium** *from other genera*

The examination of diseased tissue smears stained by Wright's blood stain or Giemsa stain is the simplest procedure for identification for *C. granulomatis.* The characteristic appearance of the intracellular organisms (Fig. 5.28) is specific for the diagnosis of Donovanosis.

Taxonomic Comments

The coccobacillary microorganisms first observed by Donovan (1905) were frequently referred to as "Donovan bodies" when seen in tissue smears from patients with granulomatis lesions in the inguinal region. The Donovan bodies were later called *Calymmatobacterium granulom-* *atis* by Aragao and Vianna (1913). When Anderson et al. (1945) first isolated the organisms by yolk sac inoculation, they termed the etiologic agent for granuloma inguinale as *Donovania granulomatis*; however, the name *Donovania* did not have priority over *Calymmatobacterium.*

In the eighth edition of the *Manual,* the genus was classified in the family *Brucellaceae.* Others have suggested that it should be placed in the family *Klebsiella* (Rake, 1948). The taxonomic relationships of *Calymmatobacterium* to other bacterial genera, however, are not yet understood, and in the present edition of the *Manual* it seems desirable not to ally the genus with any established family.

List of the species of the genus **Calymmatobacterium**

1. **Calymmatobacterium granulomatis** Aragao and Vianna 1913, 221.[AL]

gran.u.lo′ma.tis. L. dim. n. *granulum* a small grain; Gr. suff. *-oma* a swelling or tumor; M.L. n. *granuloma* a granuloma; M.L. gen. n. *granulomatis* of a granuloma.

The characteristics are as described for the genus and as depicted in Figures 5.28 and 5.29.

The mol% G + C of the DNA is not known.

Type strain: no strain extant.

Genus **Gardnerella** *Greenwood and Pickett 1980, 170*[VP]

J. R. GREENWOOD AND M. J. PICKETT

Gard.ne.rel′la. M.L. dim. ending *-ella*; M.L. fem. n. *Gardnerella* named after H. L. Gardner.

Pleomorphic rods ~**0.5** μm **in diameter and 1.5–2.5** μm **in length.** Filaments do not occur. No capsules or endospores formed. **Stain Gram-negative to Gram-variable;** the cell walls are laminated. **Nonmotile.** Facultatively anaerobic. Fastidious in growth re-

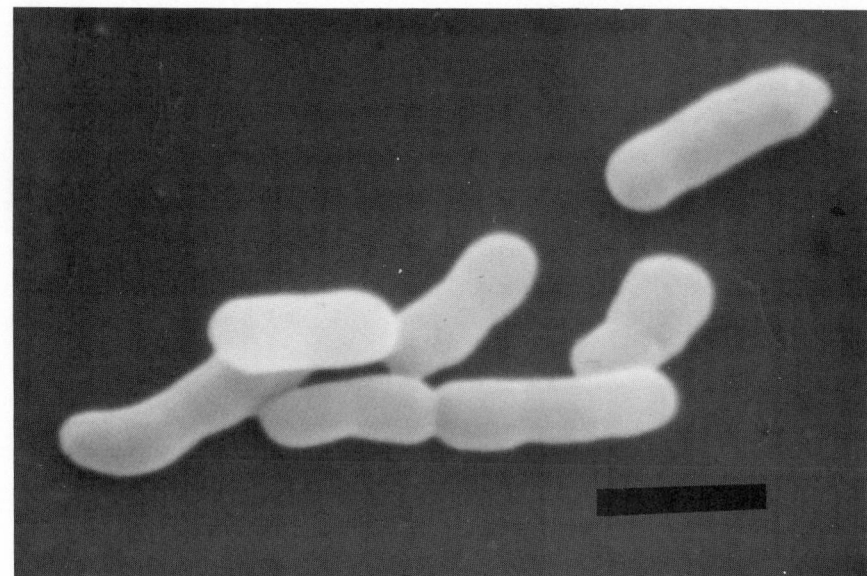

Figure 5.30. Scanning electron micrograph showing *Gardnerella vaginalis* ATCC 14018. Note pleomorphic morphology (*bar*, 1 μm).

quirements. **Catalase- and oxidase-negative.** Chemoorganotrophic, having a fermentative type of metabolism. **Acid but no gas** is produced from a variety of carbohydrates including **maltose and starch.** Acetic acid is the major product of fermentation. **Hippurate is hydrolyzed. Human blood, but not sheep blood, is hemolyzed.** Found in the human genital/urinary tract. **Considered to be a major cause of bacterial "nonspecific" vaginitis.** The mol% G + C of the DNA is 42–44 (Bd).

Type species: *Gardnerella vaginalis* (Gardner and Dukes 1955) Greenwood and Pickett 1980, 170.

Further Descriptive Information

The cells appear as small pleomorphic bacilli and coccobacilli (Fig. 5.30) which stain Gram-negative to Gram-variable. They have been described as staining Gram-positive when grown on inspissated serum (Zinnemann and Turner, 1963). They do not give an acid-fast reaction. Sudanophilic inclusions and metachromatic granules occur in the cells.

The cell walls contain the amino acids *N*-acetylglucosamine, alanine, aspartic acid, glutamic acid, glycine, histidine, lysine, methionine, proline, serine, threonine and tryptophan, but diaminopimelic and teichoic acids have not been detected (Criswell et al., 1971). Cellular fatty acid analysis has shown laurate, myristate, stearate and oleate (Greenwood and Pickett, 1980) and palmitic, stearic and C_{18} monoenoic acids (Moss and Dunkelberg, 1969). Carbohydrate analysis has indicated 6-deoxytalose and no arabinose in the cell walls of the type strain (Vickerstaff and Cole, 1969). A lipopolysaccharide-like fraction is associated with the cell walls.

Electron microscopy of the cell walls has presented conflicting results. Reyn et al. (1966) reported a microscopic morphology, particularly of the cell walls and septa, closely resembling that of a Gram-positive organism. On the other hand, Criswell et al. (1971, 1972) reported that the fine structure of the walls was more typical of a Gram-negative organism.

Although most strains of *Gardnerella* are facultatively anaerobic, obligately anaerobic strains have been described.

No growth or only slight growth occurs on nutrient agar (Piot et al., 1980). No growth occurs on most common selective media (see Table 5.90). On Vaginalis agar* colonies are pinpoint after incubation for 24

h and are 0.4–0.5 mm in diameter after 48 h. The colonies are round, opaque and smooth. They become larger than 0.5 mm after incubation beyond 48 h, but their viability decreases rapidly.

Colonies are nonhemolytic on sheep blood agar but the majority of strains exhibit diffuse β hemolysis on human or rabbit blood (Fig. 5.31). Little or no hemolysis occurs on horse blood.

The optimum growth temperature is 35–37°C. Growth may also occur at 25 and 42°C. The optimum pH range is 6.0–6.5 No growth occurs at pH 4.0 and only slight growth occurs at pH 4.5 (Greenwood and Pickett, 1979). Obligately anaerobic strains have been described (Malone et al., 1975).

Gardnerella strains are fastidious in their nutritional requirements but do not need nicotinamide adenine dinucleotide (V factor), hemin (X factor) or coenzyme-like substances (Dunkelberg and McVeigh, 1969; Edmunds, 1962). They have been reported to require biotin, folic acid, niacin, thiamine, riboflavin and two or more purines/pyrimidines (Dunkelberg and McVeigh, 1969). Growth is improved with fermentable carbohydrates and certain peptones.

The major product of sugar fermentation is acetic acid (Moss and Dunkelberg, 1969), but some strains also produce one or more of the following organic acids: lactic, formic or succinic (Malone et al., 1975). Gas is not formed from sugar fermentation. Tables 5.89 and 5.90 (below) indicate the variety of carbohydrates that can be fermented.

A common antigen has been shown by tube agglutination and by fluorescent antiserum (Redmond and Kotcher, 1963).

All strains are susceptible to ampicillin, carbenicillin, oxacillin, penicillin and vancomycin, and all are uniformly resistant to nalidixic acid, neomycin, colistin and sulfadiazine at usual therapeutic levels. Strains differ in susceptibility to kanamycin, tetracycline, gentamicin and tobramycin (McCarthy et al., 1979; Greenwood, unpublished results). Metronidazole is effective in vivo but gives variable results in in vitro susceptibility tests (Balsdon et al., 1980).

G. vaginalis is believed by many to be the cause of bacterial "nonspecific" vaginitis. It has also occasionally been reported to cause bacteremia in postpartum women and in patients following septic abortion and transurethral resection of the prostate.

G. vaginalis is isolated from the human genital/urinary tract and appears to have world-wide distribution.

---

* Vaginalis agar consists of Columbia agar base (BBL) containing 1% Proteose Peptone No. 3 (Difco). After the medium has been autoclaved and cooled to 45–50°C, human blood (5%, v/v) preserved with citrate, phosphate and dextrose is added aseptically to the medium.

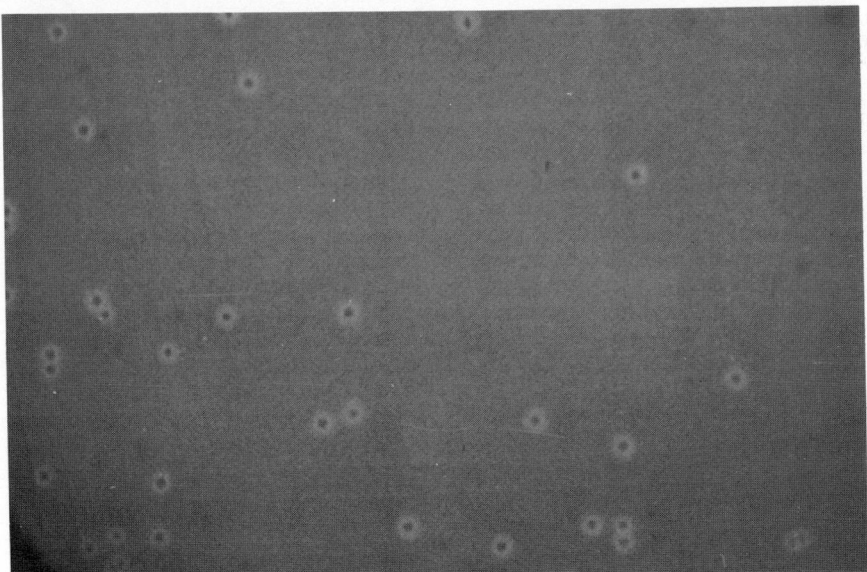

Figure 5.31. Diffuse β hemolysis produced by *G. vaginalis* colonies on V agar after 48 h incubation ($\times 2$).

Enrichment and Isolation Procedures

Swabs of clinical material are plated on Vaginalis agar within 4–6 h of collection. One quadrant is inoculated with the swab and the plate is streaked for isolation of colonies. Inoculated plates are incubated for 48 h at 35°C in either a candle extinction jar lined with water-saturated absorbent paper or in a CO_2 incubator. *G. vaginalis* forms opaque, domed, entire colonies ~ 0.5 mm in diameter that are surrounded by a small zone of diffuse β hemolysis. For final purification, a single colony is streaked onto a chocolate agar plate.

Another medium for isolation is peptone-starch-dextrose medium* (Dunkelberg et al., 1970). This medium is inoculated and incubated as outlined above for Vaginalis agar. On this medium, colonies of *G. vaginalis* are 0.5–2.0 mm in diameter, dull white, convex, domed, somewhat conical in shape and entire.

Other methods for isolation of *G. vaginalis* include those described by Smith (1975), Golberg and Washington (1976), Gardner and Dukes (1955) and Mickelson et al. (1977).

Maintenance Procedures

Working stock cultures should be transferred every 48 h to ensure viability. This requirement for frequent transfer appears to be independent of the type of maintenance medium or temperature of storage.

Stock strains may be preserved indefinitely by lyophilization in either rabbit serum or 10% skim milk. Strains may also be preserved for at least 5 years by storage at −70°C in glycerol-brucella broth (brucella broth (Difco) supplemented with 15% glycerol). This is most easily accomplished by autoclaving 0.5 ml of broth in a half-dram vial. A dense suspension of organisms harvested from 24- to 48-h-old cultures grown on Vaginalis agar is prepared directly in the vial. After incubation for 20–30 min at room temperture the vial is transferred to a −70°C freezer.

Procedures for Testing Special Characters

Acid production from carbohydrates. The following basal medium is used (g/liter): proteose peptone No. 3 (Difco), 20.0; phenol red, 0.02; agar (Difco), 5.0; pH 7.3. The carbohydrate (10.0 g/liter) is added aseptically to the autoclaved, cooled basal medium from a 10% stock solution which has been sterilized by filtration. The complete medium is dispensed in 3-ml volumes into screw-capped test tubes (13 $\times$ 100 mm). The tubes are inoculated with 48-h-old growth by stabbing the medium. The cultures are incubated under an air atmosphere at 35°C for up to 5 days. Acidification is indicated by a yellow color.

Oxidase and catalase tests. For the oxidase test, growth from a 48-h-old chocolate agar plate is smeared with a platinum loop onto a filter paper strip saturated with a solution of 1% tetramethyl-*p*-phenylene-diamine dihydrochloride in 0.2% ascorbic acid. For the catalase test, growth from a chocolate agar plate is placed on a glass slide with a wooden applicator stick, overlayed with a drop of 3% H_2O_2 and observed for evolution of bubbles. It is important to note that false positive catalase reactions have been observed when growth for this test has been taken from media containing human blood (e.g. Vaginalis agar).

Hemolysis. Hemolytic activity of cultures is tested on Vaginalis agar. Blood (e.g. human, sheep, rabbit) is added to the basal medium to provide a 5% (v/v) final concentration.

Hippurate hydrolysis. This test is by the rapid method of Hwang and Ederer (1975).

Differentiation of the genus **Gardnerella** from other genera

Table 5.89 provides the primary characteristics that can be used to differentiate this genus from other morphologically or physiologically similar genera.

Taxonomic Comments

Based on either superficial growth characteristics or morphology, previous studies classified *G. vaginalis* as a member of the genus *Haemophilus* or the genus *Corynebacterium*. Other genera have also been suggested (Lapage, 1974). DNA/DNA hybridization studies by Greenwood and Pickett (1980) and by Piot et al. (1980) have shown that *G. vaginalis* is not closely related to members of the following genera: *Actinobacillus, Bifidobacterium, Branhamella, Brevibacterium, Corynebacterium, Capnocytophaga, Haemophilus, Lactobacillus, Pasteurella, Propionibacterium* and *Streptococcus*. Precise knowledge of broader phylogenetic relationships to other genera or higher taxa will depend on future studies of the ribosomal RNA, i.e. DNA/rRNA hybridization or rRNA oligonucleotide cataloging.

---

* Peptone-starch-dextrose agar (g/liter): proteose peptone No. 3 (Difco), 20.0; soluble starch, 10.0; dextrose, 2.0; Na_2HPO_4, 1.0; $NaH_2PO_4 \cdot H_2O$, 1.0; agar, 15.0. A more recent formulation of this medium omits the phosphate buffer (W. E. Dunkelberg, personal communication).

Table 5.89.
Differential characteristics of the genus **Gardnerella** *and other morphologically or physiologically similar genera*[a]

Characteristics	Gardnerella	Capnocytophaga	Actinobacillus	Cardiobacterium	Haemophilus
Oxidase test	−	−	−[b]	+	D
Catalase test	−	+	+	−	D
Cell morphology:					
Pleomorphic	+	−	−	+	−
Short to coccoid	−	−	+	−	+
Fusiform	−	+	−	−	−
Yellow-orange pigmented colonies	−	+	−	−	−
Nitrate reduced to nitrite	−	D	+	−	+
Indole production	−	−	−	weak	D
Growth on:					
Blood agar, 35°C, in air	+	−	−[b]	d	−
Blood agar, 35°C, in air + 5% CO_2	+	+	+	+	−[c]
Kligler iron agar reactions:					
Slant	NG	K or A	K or A	K or NG	NG
Butt	NG	N or A	A	A or NG	NG
β Hemolysis, 5% human blood agar	+	−	−	−	−
Acid production from:					
Lactose	d	D	D	−	D
Maltose	+	+	+	+	D
Mannitol	−	−	D	+	D
Mol% G + C of DNA	42–44	33–41	40–43	59–60	38–44

[a] Symbols: +, typically positive; D, differs among species; d, differs among strains of a genus containing only a single species; −, typically negative; A, acid; K, alkaline; N, neutral; NG, no growth.
[b] An occasional strain may be weakly positive.
[c] One species, *H. aphrophilus*, can grow on blood agar.

Because of the unusual cell wall of *Gardnerella* and the apparent lack of a genetic relationship to other genera with comparable mol% G + C values, the genus is not presently assignable to any existing family.

Further Reading

Dunkelberg, W.E. 1974. Monograph. A bibliographic review of *Corynebacterium vaginale (H. vaginalis)*. Printing Office, Fort McPherson, Georgia.
Dunkelberg, W.E. 1977. *Corynebacterium vaginale*. Sex. Trans. Dis. *4:* 69–75.
Gardner, H. L. 1980. *Haemophilus vaginalis* vaginitis after twenty-five years. Am. J. Obstet. Gynecol. *137:* 385–391.
Greenwood, J.R., and M.J. Pickett. 1979. Salient features of *Haemophilus vaginalis*. J. Clin. Microbiol. *9:* 200–204.
Greenwood, J.R., and M.J. Pickett. 1980. Transfer of *Haemophilus vaginalis* Gardner and Dukes to a new genus, *Gardnerella: G. vaginalis* (Gardner and Dukes) comb. nov. Int. J. Syst. Bacteriol. *30:* 170–178.
Pheifer, T.A., P.S. Forsyth, M.A. Durfee, H.M. Pollock, and K.K. Holmes. 1978. Nonspecific vaginitis: role of *Haemophilus vaginalis* and treatment with metronidazole. N. Engl. J. Med. *298:* 1429–1434.
Piot, P., E. Van Dyck, M. Goodfellow, and S. Falkow. 1980. A taxonomic study of *Gardnerella vaginalis (Haemophilus vaginalis)* Gardner and Dukes 1955. J. Gen. Microbiol. *119:* 373–396.

List of the species of the genus **Gardnerella**

1. **Gardnerella vaginalis** (Gardner and Dukes 1955) Greenwood and Pickett 1980, 170.[VP] (*Haemophilus vaginalis* Gardner and Dukes 1955, 963.)

va.gi.na′lis. L. adj. *vaginalis* of the vagina.

The morphology and cultural characteristics are as described for the genus and as depicted in Figures 5.30 and 5.31.

Physiological and nutritional characteristics are as described for the genus and as listed in Tables 5.89 and 5.90.

Considered to be a major cause of bacterial "nonspecific" vaginitis. Isolated from the human genital/urinary tract.

The mol% G + C of the DNA is 42–44 (Bd).

Type strain: ATCC 14018 (strain 594 of Gardner and Dukes, 1955).

Table 5.90.
Physiological characteristics of **Gardnerella vaginalis**[a]

Test	Reaction or Result
Indole production	−
Urease	−
ONPG hydrolysis	d
Voges-Proskauer test	−
Methyl red test	+
Phenylalanine deaminase	−
H$_2$S production	−
Lipase	d
Hydrolysis of:	
Hippurate	+
Tributyrin	−
Tween 80	−
Casein	d
Starch	+
Esculin	−
Gelatin	−
H$_2$O$_2$ inhibition	+
Benzidine test for cytochromes	−
Nitrite from nitrate	−
Amino acid decarboxylases (Møller):	
Arginine decarboxylase	−
Lysine dihydrolase	−
Ornithine decarboxylase	−
Gluconate oxidized to 2-ketogluconate	−
Growth on selective media:	
Tellurite (0.01%) agar	−
Sodium chloride (3%) agar	−
Bile (1%) agar	−
Rogosa agar	−
Thayer-Martin agar	−
Growth at:	
pH 4	−
pH 8	d
25 C	d
30 C	d
Acid produced from:	
Dextrose, dextrin, maltose, ribose, starch	+
L-Arabinose, fructose, galactose, inulin, lactose, mannose, sucrose, xylose	d
Arbutin, cellobiose, glycerol, inositol, mannitol, melibiose, raffinose, rhamnose, salicin	−

[a] For symbols see standard definitions; also ONPG, *o*-nitrophenyl-β-D-galactopyranoside.

Genus **Eikenella** Jackson and Goodman 1972, 74.[AL]

F. L. JACKSON AND Y. GOODMAN

Ei.ke.nel'la. M.L. dim. ending -*ella*; M.L. fem. n. *Eikenella* named after M. Eiken, who first named the type species of the genus.

Straight rods, 0.3–0.4 × 1.5–4.0 µm, unbranched, with rounded ends and a regular morphology. Short filaments are occasionally formed. Nonsporeforming. Gram-negative. **Nonmotile,** possessing no flagella; however a "twitching motility" may occur on agar surfaces. **Facultatively anaerobic.** Optimum temperature, 35–37°C. **Colonies may appear to corrode the surface of the agar;** noncorroding strains may also occur. Nonhemolytic; a slight greening of blood media around colonies may occur. **Oxidase-positive** (Kovacs' method). **Negative for catalase, urease, arginine dihydrolase and indole. Lysine decarboxylase-positive.** Nitrates are reduced to nitrites. **No acid is formed from glucose or other carbohydrates. Hemin is usually required for growth under aerobic conditions.** Occur in

the human mouth and intestine; can be opportunistic pathogens. The mol% G + C of the DNA is 56–58 (T_m).

Type species: *Eikenella corrodens* (Eiken 1958) Jackson and Goodman 1972, 75.

Further Descriptive Information

Electron micrographs of *E. corrodens* negatively stained with phosphotungstate show a finely convoluted (cerebral) cell surface (Fig. 5.32). Sections stained with ruthenium red-OsO$_4$ show a cytoplasmic membrane and outer membrane, characteristic of Gram-negative bacteria, and a slime layer, loosely organized and fibrous, was present on strains examined by Progulske and Holt, 1980 (Fig. 5.33A). The slime layer is

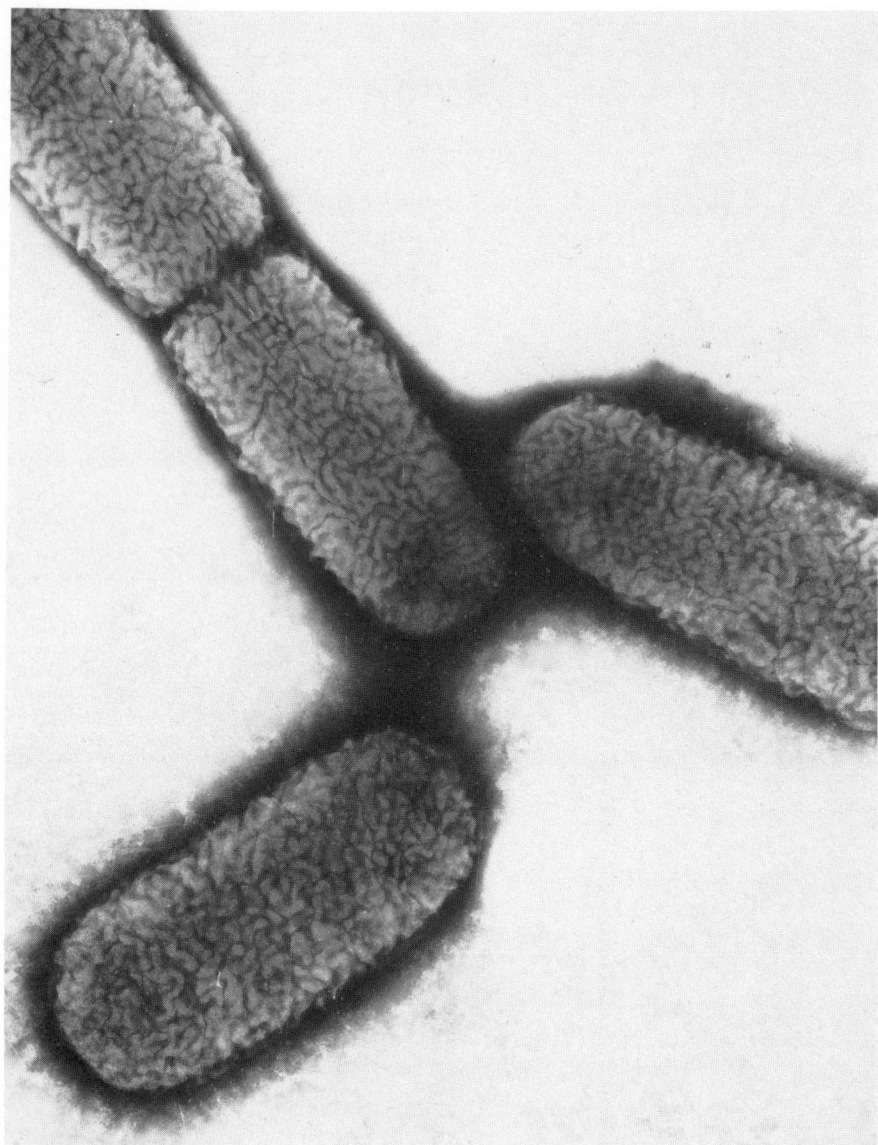

Figure 5.32. *Eikenella corrodens*. Phosphotungstate negative staining ($\sim \times$ 74,000).

associated with the outer surface of the outer membrane, and it is possible that the pilus-like structures demonstrable by negative staining represent components of this layer modified by preparation techniques (see Jackson et al., 1971; Progulske and Holt, 1980). Scanning electron microscopic adherences are shown in Figure 5.33*B*.

Lipopolysaccharide from the organisms has been reported to have endotoxin activity, whereas the slime layer has little endotoxin activity, but may be immunosuppressive (Behling et al., 1979). A component with endotoxin activity (*Limulus* amoebocyte lysate test), probably lipopolysaccharide, containing 0.5% ketodeoxyoctonate, has been obtained from cells by phenol-water extraction followed by differential centrifugation and gel filtration. This may represent a group antigen, and is distinct from type-specific protein antigens which may also be present (Malizewski and Badger, 1980).

The organisms appear nonmotile by conventional tests. Corroding colonies (see below) may show spreading edges, and microscopic observation of corroding strains growing on agar surfaces has shown that a form of surface translocation termed "twitching motility" occurs, and is correlated with the presence of asymetrically arranged pilus-like structures of uncertain nature (Henrichsen, 1975a, b; Henrichsen and Blom, 1975; Schröter, 1975). This form of translocation involves small,

intermittent jerks, leading to displacement over only short distances, not regularly related to the long axis of the cell, at speeds of 1–2 to 2–5 μm/min. The use of coverslips may prevent this movement, as organisms adhere to the glass, and the phenomenon is dependent on presence of a thin film of water at the surface of the medium, as found in cultures incubated in a humidified atmosphere.

Two types of colony may occur on media solidified with agar. Typical "corroding" strains are so named because the colonies appear to corrode the surface of the agar. The organisms penetrate into the surface of the medium, and under humid conditions tend to spread by twitching motility (see above). The colonies appear as if in small pits in the surface of the medium, probably because of the combination of growth into the surface, localized physical alterations of the medium and optical properties of the colony. The appearance of pitting is not produced on inspissated serum medium. Problems of interpretation of these features have been discussed by Jackson et al. (1971) and by Khairat (1967). Colonies of corroding strains growing on Columbia agar base with 5% sheep blood at 36°C under 15% CO_2 and 100% humidity for 72 h (Dorff et al., 1974) show an opaque, yellowish, moist center, a clear, glistening central zone (granular, pearly and refractile) and a rough, nonrefractile opaque perimeter. Colonies are small (0.2–

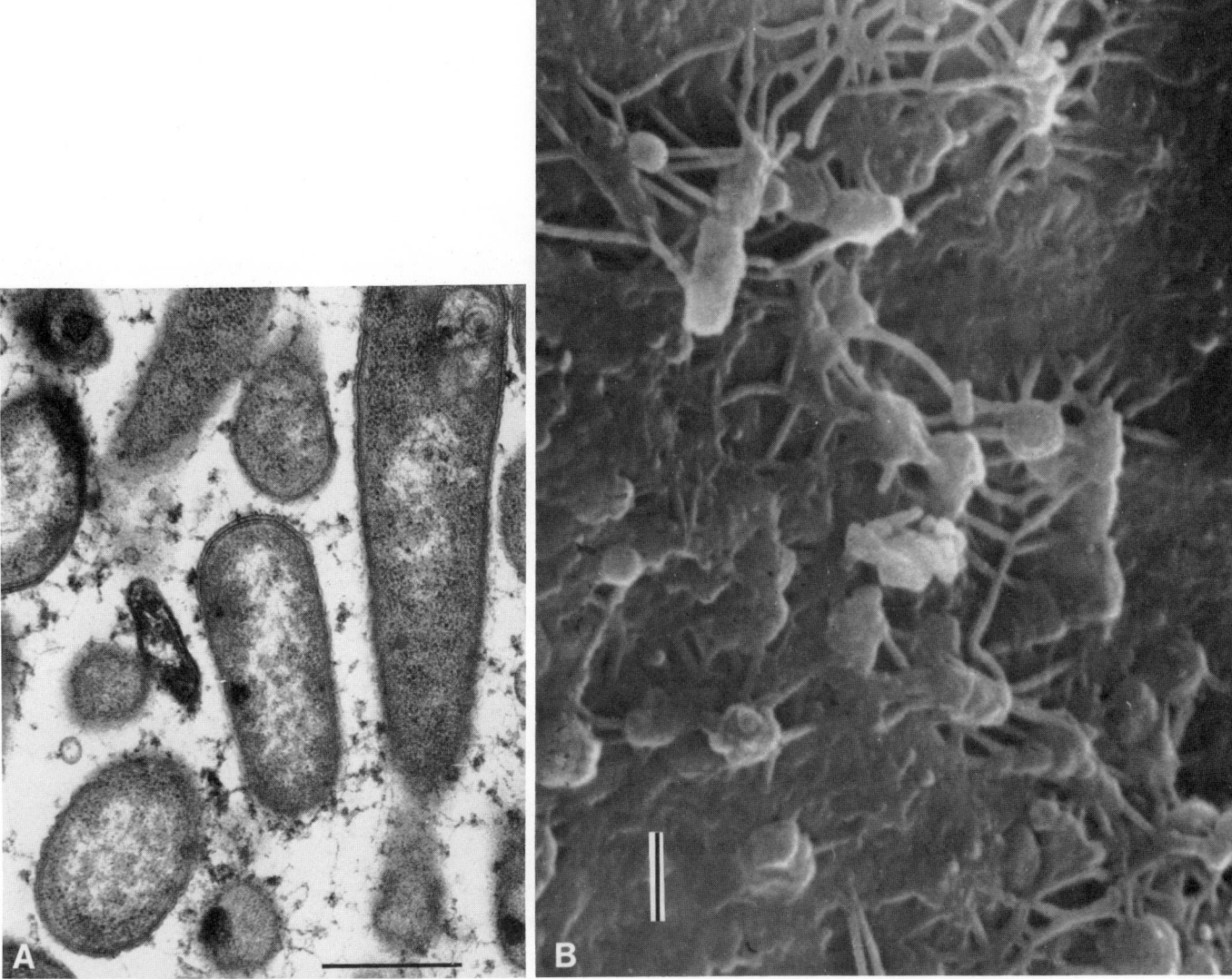

Figure 5.33. *A*, sections of *Eikenella corrodens* stained by ruthenium red-OsO₄, showing cell envelope layers and ruthenium red-positive fibrils (*bar*, 0.5 μm). *B*, scanning electron micrograph of a 7-day-old culture of *E. corrodens* showing fibrillar material connecting cells. The colonies were covered with a thick, slime-like material (*bar*, 1 μm). (Reproduced with permission from A. Progulske and S. C. Holt, Journal of Bacteriology *143:* 1003–1018, 1980,© American Society for Microbiology).

0.5 mm at 24 h; 0.5–1.0 mm at 48 h). A spreading edge may develop to give a final diameter of ∼3.0 mm (Jackson et al., 1971). Colony sizes are similar on the cystine-hemin agar used in addition to blood agar by these authors (Fig. 5.34).

Noncorroding strains, forming colonies 0.5–1.0 mm in diameter which are translucent and dome-shaped, may be isolated by selection from corroding strains and may also be encountered on primary isolation. These variants do not produce colonies with spreading edges, do not exhibit twitching motility and lack pilus-like surface appendages (Henrichsen, 1975b).

Plate cultures have an odor described as "bleach-like" or as resembling that of *Haemophilus* and *Pasteurella* species.

Growth in fluid media is usually described as poor, but may be improved by addition of cholesterol (10 μg/ml; Henricksen, 1969b) or 3% blood serum. The addition of 0.2% agar improves growth (Jackson et al., 1971). Whether the organisms are able to degrade agar is unknown.

The optimum temperature for growth is 35–37°C. At 25°C, minute colonies are visible in 5–7 days. Growth at 40°C is good, but is poor at 42°C. No growth occurs at 44°C.

The optimum pH for growth is 7.3.

Hemin (5 to 25 μg/ml) is required for aerobic growth of freshly-isolated strains, but not for anaerobic growth. The hemin requirement may be influenced by other constituents of the medium, as indicated by the demonstration that addition of cystine to a concentration of 0.005% raises the minimum hemin requirement to 20–25 μg/ml. Under these conditions the colony size is increased. Cystine at 0.1% is strongly inhibitory to growth at all hemin concentrations that support growth in the absence of added cystine (Jackson et al., 1971).

Under aerobic or anaerobic conditions, growth of freshly isolated strains is enhanced by 5–10% CO_2. Repeated subculture in air leads to loss of this response (Hill et al., 1970), but it is retained if strains are repeatedly transferred in a CO_2-enriched atmosphere (Jackson et al., 1971).

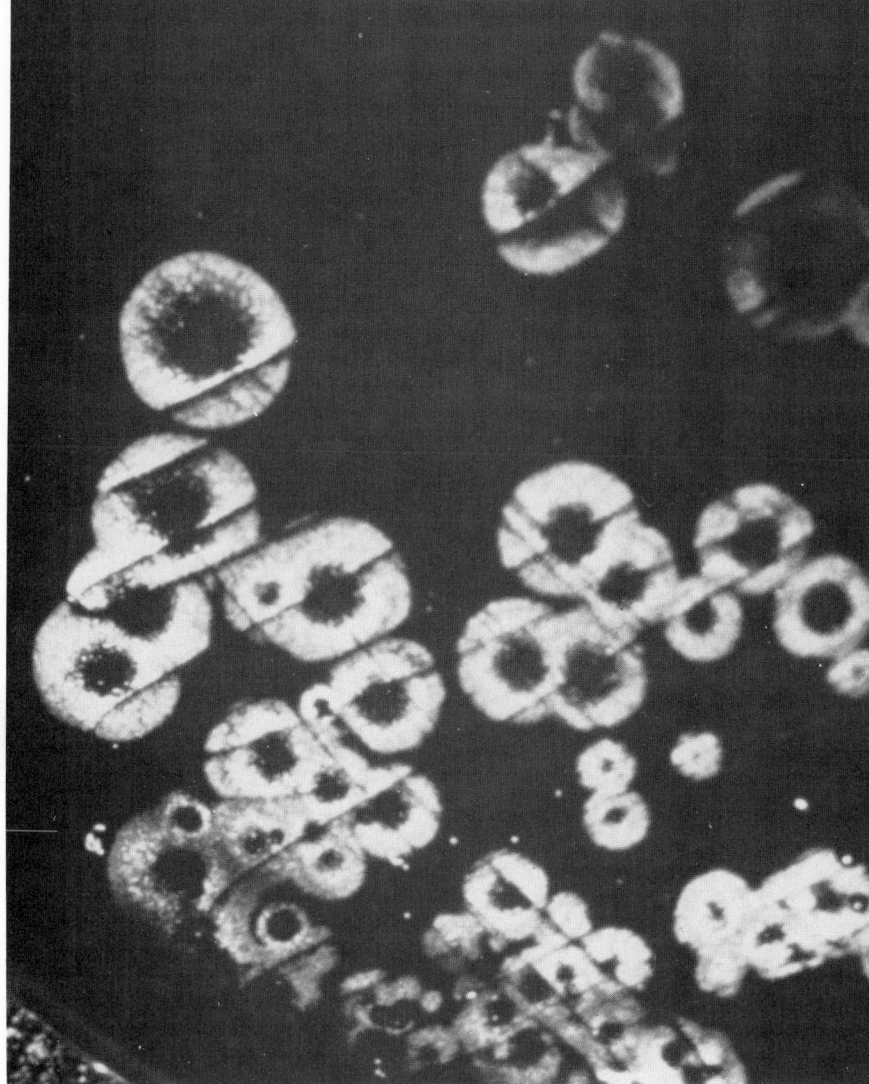

Figure 5.34. Colonies of *Eikenella corrodens* on sheep blood agar at 48 h. Colonies are 0.5–1.0 mm. (Reproduced with permission from E. J. Bottone, J. Kittick and S. S. Schneierson, American Journal of Clinical Pathology *59:* 560–566, 1973, © American Society of Clinical Pathologists.)

The strains that have been tested for bile tolerance have been inhibited under aerobic conditions by 5 or 10% bile; under anaerobic conditions growth has been reported to occur in the presence of 10% bile, and there is evidence that this resistance may be higher under microaerobic conditions in the presence of nitrate (Jackson et al., 1971).

Organisms grown anaerobically or aerobically, with or without added CO_2, are strongly oxidase-positive by the method of Kovacs (1956) using either dimethyl-*p*-phenylenediamine or the tetramethyl reagent. The standardized tube test (Jackson: see Jackson and Goodman, 1978) can also be used and shows that the reaction is azide-sensitive.

Spectroscopic examination by the method of Shobe (1974) has shown the presence of cytochromes of the *b* and *c* groups (Jackson, unpublished results). Ubiquinone is present, but not desmethylmenaquinone (Holländer and Mannheim, 1975). Growth yields of oxygen-limited cultures are not increased by fumarate.

An examination of six strains of *E. corrodens* by gas-liquid chromatography/mass spectrophotometry has shown that the organisms contain hexadecanoic and octadecenoic acids as major fatty acid constituents (Prefontaine and Jackson, 1972). This is in contrast to *Bacteroides ureolyticus* which has a high content of octadecenoic acid but a low content of hexadecanoic acid.

Antigenic differences between strains have been demonstrated by agglutination reaction and also by immunodiffusion studies in which up to four antigens were detected (although some strains lacked one or two components (Jackson et al., 1971). Badger has divided 46 strains into four groups by microagglutination tests (see Cokyendall and Kaczmarek, 1980). Maliszewski and Badger (1980) have isolated a group antigen common to three serotypes. It appears to be a lipopolysaccharide, contains 0.5% ketodeoxyoctonate, and shows endotoxin activity by the *Limulus* amoebocyte lysate test. A type-specific antigen (protein, sensitive to trypsin and heat) was prepared from one strain. According to Johnson et al. (1978), endotoxin prepared from a corroding strain had the unusual effect of causing a decrease in microviscosity of cell membranes. Anaerobic rods may show some cross-antigenicity with *E. corrodens* (Robinson and James, 1973), but as pointed out by Jackson and Goodman (1978) the difference in mol% G + C content of the DNA between *E. corrodens* and these anaerobes precludes a close genetic relationship, and the designation by Robinson and James of strain NCL-20 (now known to be a strain of *Bacteroides ureolyticus*) as a "link" strain between the facultative and anaerobic species is not justifiable (see also Table 5.92).

For determination of antibiotic sensitivities, best results are obtained by standardized-inoculum plate-dilution methods, but other techniques

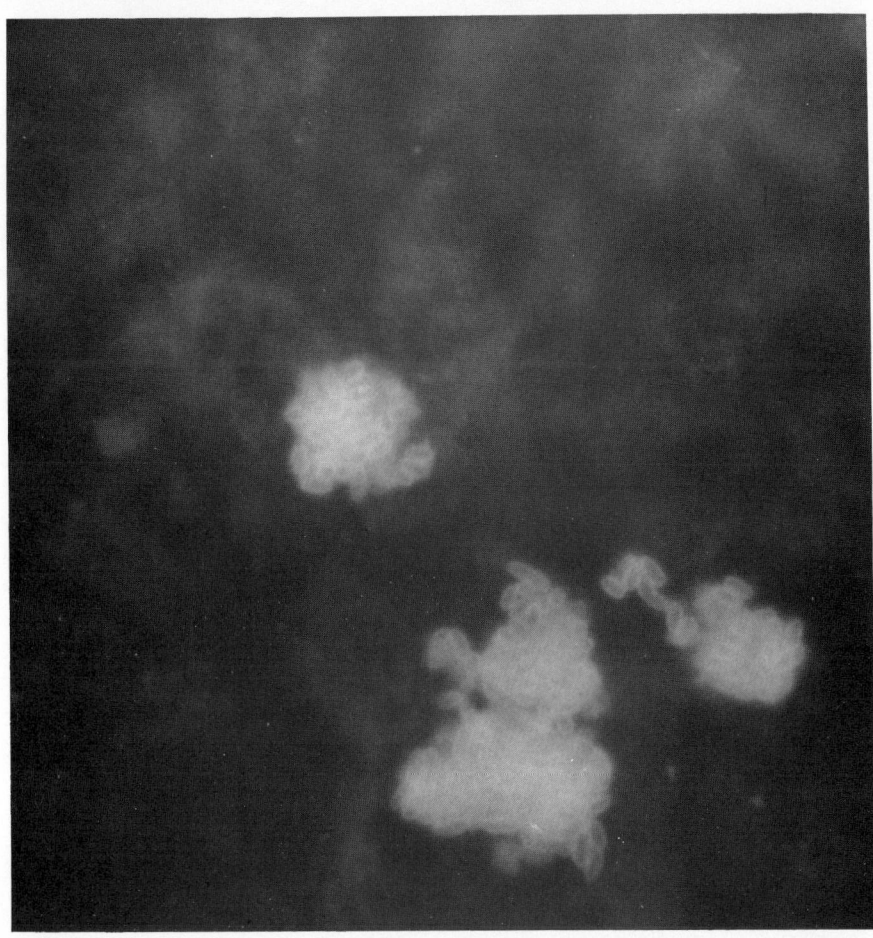

Figure 5.35. Section of aortic vegetation from a rabbit with experimental *Eikenella corrodens* endocarditis. Microcolonies stained with specific rat antiserum and fluorescent goat anti-rat serum. (Reproduced with permission from S. J. Badger et al., Infection and Immunity *23:* 751–757, 1979, © American Society for Microbiology.)

have been used with variable results (Hill et al., 1970; Jackson et al., 1971; Zinner et al., 1973; Brooks et al., 1974; Robinson and James, 1974; Labbé et al., 1977; Goldstein et al., 1978; Slee and Tanzer, 1978). The reports presently available indicate the strains are usually sensitive to penicillin G, ampicillin and cefoxitin, but resistant to penicillinase-resistant penicillins and moderately sensitive or resistant to cephalothin, cephapirin and cephaloridine. Resistance to aminoglycosides is variable, but usually sufficient to preclude clinical effectiveness. Strains are often sensitive to chloramphenicol, tetracycline, rifampicin and colistin. Resistance to lincomycin, clindamycin and metronidazole is a constant feature.

There is good evidence that *E. corrodens* is sometimes pathogenic in humans. When isolated from lesions, it is usually present in mixed culture with other facultative bacteria or with anaerobic bacteria; however, in 10–15% of positive specimens it is present in pure culture, and may cause serious diseases. In some cases, it may be the sole survivor in antibiotic-treated mixed infections. Holm (1950) noted the presence of "corroding bacilli" in actinomycotic lesions, and Reinhold (1966) studied strains from human sources. Marsden and Hyde (1971) and Kaplan et al. (1973) reported infections in children. Khairat (1967) recovered corroding bacilli from 16% of blood cultures drawn 1 min after dental extraction, but the true identity of some of the strains remains uncertain. King (1964) included *E. corrodens* (termed "HB-1") in a discussion of unusual pathogenic Gram-negative bacteria. Brooks et al. (1974) produced infection by injecting mixtures of *E. corrodens* and streptococci, with further potentiation by addition of methylphenidate, thus simulating lesions found in drug addicts. Experimental endocarditis, rarely fatal and seldom bacteremic, has been produced in catheterized rabbits (Badger et al., 1979); the organism

could be shown in vegetations by fluorescent-antibody staining (Fig. 5.35).

The possible role of *E. corrodens* in the production of *periodontal disease* with bone destruction has been discussed and investigated (Socransky, 1977; Johnson et al., 1978; Listgarten et al., 1978; Behling et al., 1979; Progulske and Holt, 1981). Infection related to the mouth and upper and lower respiratory tract sinuses, lips and face is not uncommonly reported (Schröter and Stawru, 1970; Marsden and Hyde, 1971; Jackson et al., 1971; Carruthers and Sommers, 1973; Kaplan et al., 1973; Bottone et al., 1973; Zinner et al., 1973; Brooks et al., 1974; see also Rubenstein et al., 1976; Goodman, 1977; Piéron and Mafart, 1977; Colloc et al., 1980; DeMello and Leonard, 1979; Dudley et al., 1978; Jones and Romig, 1979; Megraud et al., 1981). Serious infections including *brain abscesses* have sometimes been encountered. Transfer from the mouth may cause infection of human bites, and more than 60 cases of this sort have been reported (Bilos et al., 1978; Goldstein et al., 1978).

The ability of *E. corrodens* to survive in the intestine leads to its presence in abdominal infections, including wound infection, *liver and other abscesses* and peritonitis (Jackson et al., 1971; Lutwick, 1976; Nagesh et al., 1977; Maia et al., 1980). Meningitis and fatal endocarditis were reported by Dorff et al. (1974), endocarditis responding to therapy by Geraci et al. (1974), and *polymicrobial endocarditis* by Marcus and Phelps (1978) and Shinhar et al. (1980).

Osteomyelitis has been mentioned by several authors, including Jackson et al. (1971), Carruthers and Sommers (1973), Litwack and Borchardt (1973), Johnson and Pankey (1976), and Maia et al. (1980). Subcutaneous abscess formation in a splenectomized patient was found by Peloux et al. (1977).

Several reports indicate that *pus* formed in response to infection is greenish and has a *foul odor*, even when other bacteria are not present.

E. corrodens is regarded by several investigators as an opportunistic pathogen, particularly likely to produce infection in compromised hosts.

E. corrodens is probably a normal inhabitant of the human mouth and intestine, and has also been isolated from the genitourinary tract.

Enrichment and Isolation Procedures

Specimens should be plated on blood agar medium (5% sheep or horse blood, Columbia or Oxoid No. 2 base, or similar media) and incubated anaerobically and aerobically (in each case with 5–15% CO_2) at 35–37°C. About half of the strains isolated have been recovered from the anaerobic plates but will grow aerobically on first subculture. This may be partly accounted for by suppression of associated organisms and by faster initial growth anaerobically on transfer from relatively anaerobic body sites to artificial media. In addition, the typical corroding appearance of colonies may develop better under the humid conditions that are usual in anaerobic cultures and so lead to easier recognition of the organism. For further information on media and atmospheres of incubation, see Goldstein et al. (1981).

Isolation from mixed infections is facilitated by the use of the medium described by Slee and Tanzer (1976), which contains clindamycin (0.5 µg/ml), KNO_3 (2.0 mg/ml) and hemin (5.0 µg/ml) in agar-solidified Todd-Hewitt medium.

Aerobic plates should be maintained in a humid atmosphere during incubation to encourage production of typical colonies, e.g. in properly humidified CO_2 incubators, or in plastic bags gassed with a CO_2-air mixture, and containing wet absorbent paper. Plate cultures should be examined daily for up to 5 days.

Maintenance Procedures

Cultures on blood agar plates should be subcultured at weekly intervals to avoid loss of viability. They may be stored in plastic bags at 4°C after 3 days of incubation.

For preservation in liquid nitrogen, a portion of the growth from a plate is transferred to 1 ml of horse serum, placed in a sterile screw-capped plastic vial and stored in a liquid-nitrogen refrigerator. Survival for at least 2 years has been reported (Labbé et al., 1977).

For freeze-drying, growth from a single plate is washed off with a suspending medium (either bovine serum containing 7.5% glucose or double-strength reconstituted dried skim milk), placed in 0.2-ml volumes in freeze-drying tubes and lyophilized. Freeze-dried cultures in vacuum-sealed glass ampoules should be stored in a refrigerator (4°C). Some strains may be serum-sensitive and for these the milk medium may be preferable. The milk should be free from antibiotics. Samples of cultures preserved by any method should be tested for viability before storage of batches.

Differentiation of the genus **Eikenella** from superficially similar organisms

Table 5.91 indicates those characteristics that distinguish *Eikenella* from a number of other genera and species. Additional features that distinguish *E. corrodens* from *Bacteroides ureolyticus* are presented in Table 5.92.

Taxonomic Comments

The genus was defined to include facultative organisms formerly grouped under the name *Bacteroides corrodens* (Eiken, 1958). Exami-

Table 5.91.

Differentiation of **Eikenella** *from superficially similar Gram-negative bacteria*[a]

Characteristics	Eikenella	Actinobacillus	Bordetella (excluding B. bronchiseptica)	Brucella	Cardiobacterium	Haemophilus	Moraxella	Kingella	Pasteurella[b]	Yersinia	Bacteroides 3936 and 4482[c]	Bacteroides ureolyticus	"Vibrio-like anaerobes"[d]
Oxidase test (using dimethyl-*p*-phenylenediamine	+	−					+	+	+			v[e]	
Catalase	−	+	+	+			+	−	+	+		−	
Indole	−				+								
Acid from glucose	−	+			+	+		+	+	+	+		
Urease	−			+						−		+	−
Grow only anaerobically	−										+	+	+
Proteolytic	−											+	
Motile, flagellated	−									+[f]			+
Mol% G + C of DNA	56–58	40–43	66–70	55–58	59–60	38–44	40–45	47–55	40–45	46–50	38–39	28–30	43–45

[a] Symbols: see standard definitions.

[b] A corroding strain of *Pasteurella multocida* was described by Henriksen and Frøholm (1975).

[c] Jackson, Goodman and Rhodes (1971); Prefontaine and Jackson (1972).

[d] Motile bacteria of the 10278 group (straight rods) described by Smibert and Holdeman (1976) are stimulated by formate-fumarate and form "pitting" or corroding colonies. Two motile, somewhat curved, polar-flagellated corroding bacteria we have examined were stimulated by formate-fumarate and had G + C contents of 43.0 and 44.7 mol%. These organisms probably belong to the genus *Wolinella* Tanner, Badger, Listgarten, Visconti and Socransky, 1981.

[e] Some strains of *B. ureolyticus* may give a positive Kovacs' reaction with the dimethyl-reagent (F. L. Jackson, unpublished)

[f] Not *Y. pestis*.

Table 5.92.

Differentiation of **Eikenella corrodens** *from* **Bacteroides ureolyticus**[a]

Characteristics	E. corrodens	B. ureolyticus
Aerobic growth, with hemin	+	−
Urease	−	+
Oxidase test (dimethyl-p-phenylenediamine)	+	v
Lysine decarboxylase[b]	+[c]	−
Ornithine decarboxylase[b]	d	−
Gelatin hydrolysis[d]	−	+
Anaerobic growth enhanced by formate-fumarate; succinate is major end product	−	+
Susceptible to clindamycin (5 μg/ml)	−	+
Susceptible to metronidazole (5 μg/ml)	−	+
Anaerobic growth in presence of 10% bile	+	−
Growth with 0.02% sodium azide	−	+
Odor of plate cultures: "bleach-like"	+	−
Mol% G + C of DNA	56–58	27–29

[a] For symbols see standard definitions.

[b] Möller's method (Möller, 1955).

[c] Negative reactions reported in the literature are probably method-related.

[d] Frazier (1926) method preferred. 30% trichloroacetic acid may be substituted for mercuric chloride. If conventional 12% gelatin is used, formate-fumarate supplementation is necessary to ensure growth of *B. ureolyticus* (Smibert and Holdeman, 1976).

The use of different methods for any test may affect the results. For suitable methods, see Lapage et al. (1968); Midgley et al. (1970); Hill et al., 1970; Jackson et al. (1971); Brooks et al. (1974); and Cowan and Steel (1974, appendix C).

nation of strains to which this name had been applied revealed that they were heterogeneous. Most were not strict anaerobes, but would grow in the presence of oxygen on hemin-containing media (see Hen-

riksen, 1969a; Hill et al., 1970; Jackson et al., 1971). Certain less oxygen-tolerant strains which were urease-positive have been assigned to the species *Bacteroides ureolyticus* (Jackson and Goodman, 1978). It has been found that some other anaerobic "corroding bacteria" are poorly characterized flagellated organisms (Henrichsen, 1975b; Smibert and Holdeman, 1975; Jackson and Goodman, 1978). For more recent information on these organisms the paper by Tanner et al. (1981) should be consulted. The statement made by Smibert and Holdeman (1976) that the anaerobic urease-positive strain VPI 7814 is flagellated is incorrect, as is the report by Brooks et al. (1974) that this strain has a DNA G + C content of 55.0 mol%. The true value is 28.4%, and this strain is *B. ureolyticus* and not an atypical *Eikenella* (see Jackson and Goodman, 1978).

The organisms termed HB-1 by King (1964) are the same as *Eikenella corrodens* (Jackson and Goodman, 1972; Riley et al., 1973). Similar organisms were included among those referred to as "corroding bacilli" by Holm (1950). The Gram-negative anaerobes forming spreading colonies described by Henriksen (1948) were heterogeneous.

Coykendall and Kaczmarek (1980) found that 22 *E. corrodens* strains examined had an overall mean G + C content of 56.3 mol%, and showed at least 70% DNA/DNA homology. Variations in homology findings (70–100%) did not correlate with antigenic differences. It was concluded that *E. corrodens* can be regarded as a "molecularly homogeneous species," taking into account DNA homology, reported G + C ratios and fatty acid profiles (see Further Descriptive Information).

Further Reading

Cokyendall, A.L. and F.S. Kaczmarek. 1980. DNA homologies among *Eikenella corrodens* strains. J. Periodontal Res. *15*: 615–620.

Jackson, F.L., Y.E. Goodman, F.R. Bel, P.C. Wong and R.L.S. Whitehouse. 1971. Taxonomic status of facultative and strictly anaerobic "corroding bacilli" that have been classified as *Bacteroides corrodens*. J. Med. Microbiol. *4*: 171–184.

Jackson, F.L. and Y.E. Goodman. 1972. Transfer of the facultatively anaerobic organism *Bacteroides corrodens* to a new genus, *Eikenella*. Int. J. Syst. Bacteriol. *22*: 73–77.

Progulske, A. and S.C. Holt. 1980. Transmission-scanning electron microscope observations of selected *Eikenella corrodens* strains. J. Bacteriol. *143*: 1003–1018.

Tanner, A.C.R., S. Badger, C.-H. Lai, M.A. Listgarten, R.A. Visconti and S.S. Socransky. 1981. *Wolinella* gen. nov., *Wolinella succinogenes* (*Vibrio succinogenes* Wolin et al.) comb. nov., and description of *Bacteroides gracilis* sp. nov., *Wolinella recta* sp. nov., *Campylobacter concisus* sp. nov., and *Eikenella corrodens* from humans with periodontal disease. Int. J. Syst. Bacteriol. *31*: 432–445.

List of the species of the genus **Eikenella**

1. **Eikenella corrodens** (Eiken 1958) Jackson and Goodman 1972, 75.[AL] (*Bacteroides corrodens* Eiken 1958, 415.)

cor.ro'dens. M.L. part. adj. *corrodens* gnawing.

The characteristics are as described for the genus and as listed in Tables 5.91–5.93.

Probably normal inhabitants of the human mouth and intestine. Can be opportunistic pathogens.

The mol% G + C of the DNA is 56–58 (T_m).

Type strain: ATCC 23834 (NCTC 10596; Henriksen's strain 33/54–55).

Table 5.93.

Other characteristics of **Eikenella corrodens**[a]

Test	Reaction or Result
Anaerobic growth can occur	+
Catalase	−
Oxidase test (di- or tetramethyl-p-phenylenediamine)	+
Cytochromes present of b and c groups	+
Nitrates reduced to nitrites[b]	+
Anaerobic growth enhanced by nitrate	+
Arginine dihydrolase[c]	−
Indole	−
Acid from glucose and other carbohydrates	−
Flagella present	−
Corroding and noncorroding colonies occur	+

[a] For symbols see standard definitions.

[b] The method of Cook (1950) is convenient for this test (Jackson et al., 1971), but strains giving a negative reaction by this method have been encountered (Hill et al., 1970). We have not found negative reactions if plates are incubated anaerobically and then exposed to air.

[c] Method of Möller (1955).

Genus **Streptobacillus** *Levaditi, Nicolau and Poincloux 1925, 1188*[AL]

NORMAN SAVAGE

Strep.to.ba.cil′lus. Gr. adj. *streptos* twisted, curved; L. dim. n. *bacillum* a small rod; M.L. n. *bacillus* a small rod; M.L. masc. n. *Streptobacillus* a twisted or curved small rod.

Rods, 0.1–0.7 × 1–5 μm long, with rounded or pointed ends. Occur singly or form long, wavy chains or filaments 10–150 μm long. **May be highly pleomorphic,** depending on cultural conditions. Single rods may show central swelling; chains or filaments may have a series of swellings resulting in a "string of beads" appearance. Gram-negative. Nonmotile. Capable of growth anaerobically or aerobically although the metabolism is fermentative. Optimum temperature, 35–37°C. A moist environment or soft agar may enhance growth. **Conversion to L-phase or transitional-phase variants may occur spontaneously during cultivation. Catalase- and oxidase-negative. Indole not produced. Nitrate not reduced to nitrite.** Chemoorganotrophic, having a fermentative type of metabolism. Acid but no gas is produced from glucose. **Serum, ascitic fluid or blood is required for growth. Inhabitants of the throat and nasopharynx of wild and laboratory rats. Causes one form of rat-bite fever in man.** The mol% G + C of the DNA is 24–25 for the bacillary form and 24–26 for the L-phase variant (T_m) (Williams et al., 1969; and unpublished data of Williams and Wittler, 1970, as reported by Wittler and Cary, 1974).

Type species: *Streptobacillus moniliformis* Levaditi, Nicolau and Poincloux 1925, 1188.

Further Descriptive Information

The cell morphology of *S. moniliformis* is extremely variable (Fig. 5.36) and is dependent on culture age and conditions (Heilman, 1941). Smears made from colonies grown on tryptose-phosphate agar supplemented with 20% horse serum show single rods, chains or filaments uniformly shaped or with fusiform or bulbar swellings. Smears made from cultures grown in tryptose-phosphate broth with 20% horse serum show clumps of cells, but the cells are more uniformly rods and filaments with fewer and smaller fusiform and bulbar swellings. The swellings stain more intensely than the rods or regular filaments. Old cultures are difficult to stain by the Gram method. Giemsa or Wayson stains may be better than the Gram stain for all age cultures (Rogosa, 1970).

The cell envelope of the type strain of *S. moniliformis* contains 64% protein, 28% lipid and 6% carbohydrate (Knipp and Sokatch, 1969). Low levels of glucosamine and muramic acid are possibly related to the spontaneous development of the transitional or L-phase variants (Knipp and Sokatch, 1969).

Membranes of a stable L-phase variant of *S. moniliformis* contain 53% protein and 40% lipid. Two-thirds of the lipids are polar and the remaining one-third is cholesterol (Razin and Boschwitz, 1968).

The swollen elements of *Streptobacillus* appear in thin sections to be bounded by a continuous cell wall but compartmentalized by cytoplasmic divisions. The swollen elements seem to have the same structure as the uniform rods and filaments (Bisset and Hale-McCaughey, 1967).

S. moniliformis requires media supplemented with serum, ascitic fluid or blood. The L-phase variant will grow in Edward broth (Razin, 1963) in which the serum component has been replaced by 0.01% (w/v) Tween 80 + 1% (w/v) bovine serum albumin (Razin and Shafer, 1969). When the organism is grown in serum, cholesterol is concentrated into the membranes; however, cholesterol is not essential for growth (Razin and Shafer, 1969; Razin and Tully, 1970). No growth occurs in milk. Poor growth occurs on Loeffler's serum agar slants. No growth occurs on media supplemented with hemoglobin or yeast extract (Van Rooyen, 1936).

Colonies of *S. moniliformis* on serum agar are 1–2 mm in diameter after incubation for 3 days and are circular, convex, grayish, smooth and glistening with a butyrous consistency. Colonies of the L-phase variant are much smaller (300–500 nm after 24 h) and exhibit a typical "fried-egg" appearance with a dense center that penetrates the agar to a depth of 30–50 μm and a lacy, peripheral portion (Wittler and Cary, 1974). Other colonies that are intermediate in size between typical bacterial and L-phase colonies, and which exhibit only a granular and course appearance, may be observed. These colonies are presumably transitional (Brown and Nunemaker, 1942; Freundt, 1956a).

Colonies on horse or sheep blood agar are not hemolytic.

Serum broth cultures are characterized by a white, flocculent sediment at the bottom and on the sides of tubes, or by whitish granules; the supernatant medium is usually clear.

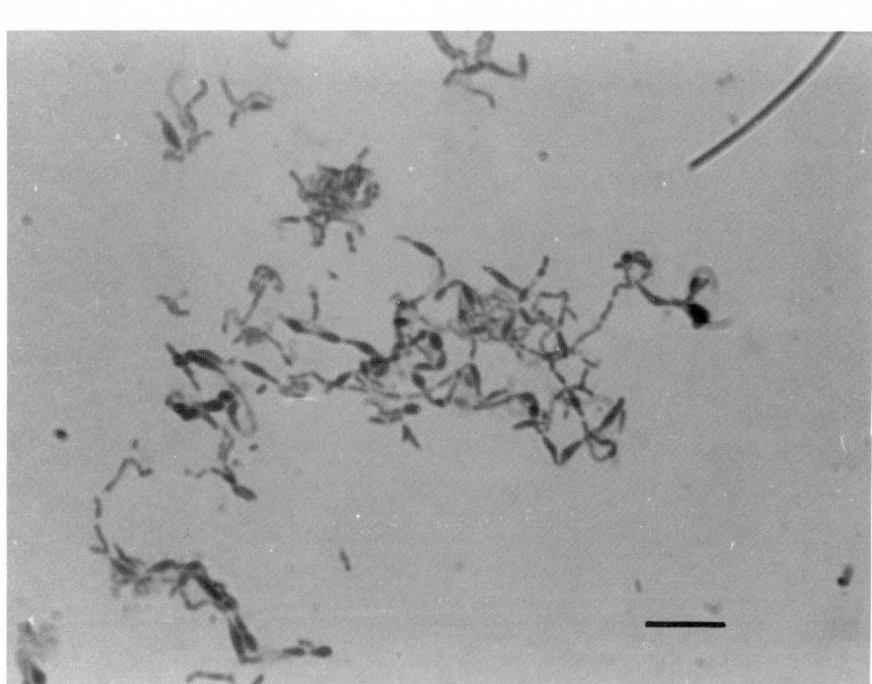

Figure 5.36. Gram stain of *Streptobacillus moniliformis* cultured on serum-supplemented agar. The typical chains, filaments and bulbar swellings can be seen (*bar*, 10 μm).

The biochemical reactions of *S. moniliformis* are presented later in Table 5.94.

The antigenic structure of *S. moniliformis* is not well defined. The bacillary phase has an antigen that is not present in the L-phase (Klieneberger, 1942). Immunization of mice with a vaccine prepared from the bacillary phase provides protection against subsequent challenge with either bacillary or L-phase organisms, whereas immunization with a vaccine prepared from the L-phase provides no protection against challenge with a culture in the bacillary phase (Freundt, 1956b).

The susceptibility of *S. moniliformis* to various antimicrobial agents *in vitro* is indicated later in Table 5.94. Antibiotic therapy of infected mice has been partially successful with either penicillin G or streptomycin (Levey and Levey, 1948). Antibiotic therapy in humans has also been effective with either penicillin or streptomycin. Arsenicals have been effective in 50% of human cases and sulfonamides have been ineffective (Roughgarden, 1965).

Human infection with *S. moniliformis* following rat bite causes one form of rat-bite fever (Brown and Nunemaker, 1942). Human infection following consumption of *S. moniliformis* contaminated milk is presumed to have been the cause of a disease known as Haverhill fever (Place et al., 1926). Complications of streptobacillary rat-bite fever may include endocarditis (McDermott et al., 1945; Stokes et al., 1951; Hamburger and Knowles, 1953), brain abscesses (Oeding and Pedersen, 1950), amnionitis (Faro et al., 1980), as well as bronchitis and pneumonia, abscesses and persistent severe arthritis (Roughgarden, 1965).

Diseases produced in animals include fatal epizootics in laboratory mice (Levaditi et al., 1932; Mackie et al., 1933; Freundt, 1956a), tendonsheath infection in turkeys (Boyer et al., 1958; Mohamed et al., 1969), arrested pregnancy and abortion in laboratory mice (Sawicki et al., 1962), cervical abscesses in guinea pigs (Aldred et al., 1974), and pleuritis in a koala (Russell and Straube, 1979).

The normal habitat of *S. moniliformis* is the nasopharynx of laboratory rats and possibly wild rats (Tunnicliff, 1916; Strangeways, 1933). The organism is frequently isolated from lesions in laboratory rats having bronchopneumonia (Tunnicliff, 1916; Smith, 1921; Bell and Elmes, 1969; Gay et al., 1972), middle ear infections (Nelson, 1940; Olson and McCune, 1968) and conjunctivitis (Young and Hill, 1974; Hill, 1974). Reproduction of these diseases in rats by inoculation with *S. moniliformis* has been unsuccessful (Bell and Elmes, 1969; Gay et al., 1972; Hill, 1974).

The L-phase variant of *S. moniliformis* is generally considered to be nonpathogenic (Freundt, 1956b); however, it has been isolated from pathologic states (Klieneberger, 1938; Dolman et al., 1951).

Enrichment and Isolation Procedures

Isolation of *S. moniliformis* from most clinical lesions is difficult because of the requirement for serum or a similar supplement, the requirement for a humid environment, the slow growth of the organism, and the probability of overgrowth by contaminating organisms. Isolation is enhanced by use of large amounts of specimen. Packed blood cells from 10 ml of a citrated blood sample, citrated joint fluid or material from a wound, cutaneous eruption or abscess should be inoculated into serum-supplemented broth and on serum agar plates (Rogosa, 1970). Good growth is obtained with a meat infusion or tryptose basal medium supplemented with 20% horse serum. Neopeptone is not recommended as a base for optimum growth (Brown and Nunemaker, 1942). Plates should be incubated in a humid environment and examined daily for development of colonies. Colony development may require a minimum of 3 days incubation. Broth cultures should show typical fluff balls within 3 days. The viability of cultures, especially broth cultures, is brief, and the organism will die rapidly unless transferred or preserved. Transfers should be made soon after detection of growth. An entire "fluff ball" should be transferred from broth to fresh broth or agar (Rogosa, 1970). Swabs may be used to transfer bacterial phase agar cultures. For transfer of L-phase colonies, small agar blocks containing heavy growth should be streaked on fresh agar, leaving the inverted block in place on the fresh agar (Rogosa, 1970). Penicillin (2 U/ml) and thallium acetate (1:2000) may be useful for enrichment and isolation of the L-phase variant from mixed flora (Freundt, 1956a).

Maintenance Procedures

Cultures can be maintained by frequent transfer to fresh media supplemented with 20% serum. Broth cultures may require transfer at 24-h intervals. Agar cultures can be transferred less frequently, usually at 3-day intervals. Refrigeration may prolong survival (7–15 days) of either broth or agar cultures (Wittler and Cary, 1974), but survival under these conditions is not dependable. The organisms can be lyophilized and will remain viable indefinitely. Organisms in infected tissues, body fluids and cultures remain viable for several years when stored at −25 to −70°C (Wittler and Cary, 1974). Subculturing of stored cultures requires serum-supplemented infusion media and a moist environment.

Procedures for Testing Special Characters

All media for biochemical tests require serum, blood or ascites supplementation to allow growth and determination of characteristics. Biochemical reactions, particularly for the L-phase variant, may best be determined using agar plates containing the basal medium supplemented with horse serum and filter-sterilized substrates and/or indicators that are added prior to pouring the plates (Cohen et al., 1968). All tests should include uninoculated controls containing substrate and inoculated controls in which distilled water is substituted for the substrate. For carbohydrate breakdown, cystine-tryptic agar base may be the medium of choice (Lambe et al., 1973).

Differentiation of **Streptobacillus** from Other Genera

Genera likely to be confused with *Streptobacillus*, particularly in blood cultures from patients with clinical endocarditis, include *Cardiobacterium, Actinobacillus* and *Haemophilus* (Midgley et al., 1970). *Streptobacillus* can be differentiated from these related genera on the basis of its serum requirement, flocculent growth in broth, small butyrous colony on agar, characteristic microscopic appearance, absence of catalase and oxidase activity and failure to reduce nitrate to nitrite or produce indole. Genera related to *Streptobacillus* will be positive for one or more of the catalase, oxidase, nitrate reduction or indole production characteristics while *Streptobacillus* is negative for all (Midgley et al., 1970; Lapage, 1974).

Taxonomic Comments

The generic name *Streptobacillus*, although previously considered to be illegitimate (Buchanan et al., 1966), has been in common use for many years and has now been conserved by virtue of its appearance on the Approved Lists of Names. The name *Streptobacillus* was first used by Ucke (1898) for an aerobic sporeformer (type species, *S. terrae*), but this is clearly not synonymous with *Streptobacillus* as described by Levaditi et al. (1925), who named the nonsporeformer *S. moniliformis* as type species. *Streptobacillus moniliformis* may be synonymous with an organism called *Haverhillia multiformis* (Parker and Hudson, 1926). This relationship is based on comparisons of stained smears (Van Rooyen, 1936) and the clinical courses of Haverhill fever (Place et al., 1926; Parker and Hudson, 1926) and rat-bite fever (Levaditi et al., 1925; Brown and Nunemaker, 1942). The Haverhill epidemic was presumed to have been transmitted by ingestion of contaminated milk instead of by rat bite. Infection by *S. moniliformis* via the digestive tract has been accomplished in laboratory animals by feeding large doses of broth cultures (Freundt, 1956a), but since *S. moniliformis* does not grow in milk to allow development of an infective inoculum, transmission via milk seems doubtful. Since no direct comparison of cultures has ever been made, a definitive relationship between *S. moniliformis* and *H. multiformis* is not certain.

The taxonomic position of *S. moniliformis* has not been clarified since the eighth edition of the *Manual*. With the exception of the cell wall, which seems to be rudimentary and lost spontaneously when the organism converts to the L-phase, the characteristics of *S. moniliformis* seem more similar to some of the *Mycoplasmatales* than to other bacteria. The similarity includes growth requirements (e.g. serum), DNA base composition, membrane composition and cholesterol incorporation, pleomorphism, animal parasitism and disease production.

The sixth edition of the *Manual* (Breed et al., 1948) included *Streptobacillus* in the family *Parvobacteriaceae* (tribe *Haemophileae*), and the seventh edition (Breed et al., 1957) transferred it to the family *Bacteroidaceae*. The eighth edition (Buchanan and Gibbons, 1974) listed *Streptobacillus* as a genus of "uncertain affiliation." It seems desirable to continue not to ally it with any particular family until more definitive taxonomic studies, such as nucleic acid hybridizations, are done.

List of the species of the genus **Streptobacillus**

1. **Streptobacillus moniliformis** Levaditi, Nicolau and Poincloux 1925, 1188.[AL]

mo.ni.li.for′mis. L. n. *monile* necklace; L. n. *forma* shape or form; M.L. adj. *moniliformis* necklace-shaped.

The characteristics are as described for the genus and as listed in Table 5.94. A photomicrograph of the organism is given in Figure 5.36.

Inhabitants of the throat and nasopharynx of wild and laboratory rats.

The mol% G + C of the DNA is 24–25 for the bacillary form and 24–26 for the L-phase variant (T_m).

Type strain: ATCC 14647.

Table 5.94.

Physiological characteristics of **Streptobacillus moniliformis**[a]

Test	Reaction or Result
Serum, blood or ascitic fluid required for growth	+
Gas produced from carbohydrate fermentation	−
Acid produced from:	
Dextrin, fructose, galactose, glucose, glycogen, maltose, mannose, salicin, inulin or starch	+
Adonitol, arabinose, cellobiose, dulcitol, glycerol, inositol, lactose, mannitol, melizitose, melibiose, raffinose, rhamnose, sorbitol, sorbose, sucrose, trehalose or xylose	−
Gelatin liquefaction	−
Indole produced	−
Catalase, oxidase or urease activity	−
Phenylalanine deaminase	−
Nitrate reduced to nitrite	−
Esculin hydrolysis	d
Benzidine test for cytochrome enzymes[b,c]	−
Oxidation of gluconate to 2-ketogluconate[c]	−
Methyl red and Voges-Proskauer tests	−
H_2S production	+
Arginine dihydrolase	+
Anaerobic reduction of methylene blue	+
Anaerobic and aerobic reduction of potassium tellurite and 2,3,5-triphenyltetrazolium chloride	+
Phosphatase activity[d]	+
End products of glucose fermentation include acetic, formic, lactic and succinic acids[e]	+
In vitro susceptibility to antimicrobial agents[f]:	
Ampicillin, chloramphenicol, chlortetracycline, erythromycin, nitrofurantoin, nitrofurazone, novobiocin, penicillin, oxytetracycline, streptomycin	S
Bacitracin, dihydrostreptomycin, tetracycline	WS
Neomycin, polymyxin B, sulfonamides	R

[a] Symbols: see standard definitions; also S, sensitive; WS, weakly sensitive; and R, resistant.
[b] Deibel and Evans, 1960.
[c] Wittler and Cary, 1974.
[d] Cohen et al., 1968.
[e] Lambe et al., 1973.
[f] Roughgarden, 1965; Hamburger and Knowles, 1953.

Important Notes for Users of this Edition

1. Always read both generic and species descriptions because characters listed in the generic description are not usually listed in the species descriptions.

2. Unless otherwise indicated in footnotes to tables, the meanings of symbols are as follows:

+   90% or more of strains are positive

−   90% or more of strains are negative

d   11–89% of strains are positive

v   strain instability (*not* equivalent to "d")

D   different reactions in different taxa (species of a genus or genera of a family)

3. All other symbols are defined in footnotes to tables.

SECTION 6

Anaerobic Gram-Negative Straight, Curved and Helical Rods

FAMILY I. BACTEROIDACEAE PRIBRAM 1933, 10^{AL}*

LILLIAN V. HOLDEMAN, ROGER W. KELLEY AND W. E. C. MOORE

Bac.te.ro.i.da'ce.ae or Bac.te.roi.da'ce.ae. M.L. n. *Bacteroides* type genus of the family; *-aceae* ending to denote a family; M.L. fem. pl. n. *Bacteroidaceae* the *Bacteroides* family.

Obligately anaerobic, Gram-negative, nonsporeforming, straight, curved or helical **rods,** that are either motile or nonmotile. **Chemoorganotrophic,** metabolizing carbohydrates, peptones, or metabolic intermediates. Most species produce detectable amounts of organic acids.

Further Comments

Members of the genera *Selenomonas* and *Anaerovibrio* are very similar in many major characteristics (e.g. propionate and acetate are major fermentation products and cells are curved and motile). According to the classic definitions, *Selenomonas* species have tufts of flagella on the concave side of crescent-shaped cells and *Anaerovibrio* have polar or subpolar flagella. Although the position of flagella in *Selenomonas* has been established by Kingsley and Hoeniger (1973) on the basis of electron micrographs of cell sections, the tufts of flagella are not easy to demonstrate by routinely used flagella stains. We (W.E.C.M., L.V.H.) have isolated many oral strains of Gram-negative rods with the biochemical characteristics of *S. sputigena* and with polyacrylamide gel electrophoresis (PAGE) patterns identical with the type strain of *S. sputigena*. Most of these isolates, by conventional flagella stain methods, appear to have 1 or 2 subpolar flagella. Although this may be an artifact of the staining method, it causes no problem in differentiating among described species of *Selenomonas* and *Anaerovibrio* because the species differ in many of their other characteristics. Routine flagella stains may not be sufficient to differentiate between these two genera if an isolate has properties of the genera but characteristics unlike described species. Although *A. lipolytica* is distinct from any described species of *Selenomonas*, whether the difference is sufficient to represent a separate genus is not certain to us on the basis of available information.

Syntrophomonas (McInerney et al., 1981, 1982) and "*Syntrophobacter*" (Boone and Bryant, 1980) are two genera of anaerobic rods that derive their energy from oxidation of short chain fatty acids. *Syntrophomonas* consists of Gram-negative, slightly helical rods with round ends, having 2–8 lateral flagella along the concave side of the cell. They metabolize saturated fatty acids (butyrate through octanoate) by beta-oxidation to acetate or acetate and propionate, using protons as electron acceptors. "*Syntrophobacter*" consists of nonmotile rods that oxidize only propionate. The products are acetate, hydrogen and carbon dioxide. Members of these two genera have not yet been isolated in pure culture. Both require co-culture with hydrogen-utilizing organisms such as *Desulfovibrio* or methanogens. "*Syntrophobacter*" grows only (a) in co-culture with *Desulfovibrio* with sulfate serving as the hydrogen sink or (b) in low sulfate medium in co-culture with both *Methanospirillum hungateii* (corrig.) and *Desulfovibrio* with methane serving as the main electron sink product. Unlike *Syntrophomonas*, "*Syntrophobacter*" has not yet been grown with the methanogen in the complete absence of *Desulfovibrio*.

The correct taxonomic position of the genus *Butyrivibrio* is uncertain. Although members of the genus stain Gram-negative, electron micrographs show that the cells have a cell wall structure resembling that found in Gram-positive organisms (Cheng and Costerton, 1977). However, on the basis of the Gram stain reaction, the genus is retained in *Bacteroidaceae* in this edition of the *Manual*. A similar "Gram-positive type" cell wall structure is reported for *Acetivibrio cellulolyticus* (Patel, genus *Acetivibrio*) and *Lachnospira multiparus* (Cheng et al., 1979). Such discrepancies between Gram stain reactions and cell wall structure pose unsolved problems for determinative bacteriology and identification, as opposed to taxonomic and hierarchical relationships. For practical aspects of determinative bacteriology, the classical stain reaction is most useful. However, wall structure may indicate close genetic or taxonomic relationships to genera in other families and might provide important insight concerning ecological significance of these taxa. Buffered Gram stain reagents (e.g. Kopeloff's modification) and, if possible, media with little or no carbohydrate and young cultures are recommended for determinations of Gram stain reactions.

The sulfur- or sulfate-reducing bacteria, many of which are both chemoorganotrophic and chemoauxotrophic, are included separately in a group of sulfate-reducing bacteria in this edition of the *Manual*. Although most of these organisms are not thought to be associated with the normal or pathogenic flora of man or other animals, *Desulfomonas pigra* is part of the colon and fecal flora of humans and strains of *Desulfovibrio* occasionally have been reported from human clinical specimens (Porschen and Chan, 1977).

---

* *AL* denotes the inclusion of this name on the Approved Lists of Bacterial Names (1980).

Other characteristics of members of the family **Bacteroidaceae**

Although not all species of the genera *Bacteroides* and *Fusobacterium* have been tested, there may be differences in cell wall and lipopolysaccharide composition that are genus specific. None of the 14 *Bacteroides* species tested contain heptose or 2-keto-3-deoxyoctonate (KDO) in their lipopolysaccharide, while all of the 8 *Fusobacterium* species tested contain heptose and KDO. *Leptotrichia buccalis* also contains heptose and KDO (Hofstad, 1974). Sphingolipids are common in many plants and animals; however, among procaryotes, they have only been demonstrated in the genus *Bacteroides*. Not all *Bacteroides* species contain sphingolipids (Fritsche, 1975; Miyagawa et al., 1978).

Pathogenicity

Many significant human and animal pathogens are found in this family. An excellent review of the literature concerning anaerobic bacteria in human disease is that of Finegold (1977).

Further Reading

Balows, A., R.M. DeHaan, V.R. Dowell and L.B. Guze (Editors). 1974. *Anaerobic Bacteria: Role in Disease*, Charles C Thomas, Springfield, Ill.

Clarke, R.T.J. and T. Bauchop (Editors). 1977. *Microbial Ecology of the Gut*, Academic Press, London.

Drasar, B.S. and M.J. Hill. 1974. *Human Intestinal Flora*, Academic Press, London.

Finegold, S.M. 1977. *Anaerobic Bacteria in Human Disease*, Academic Press, New York.

Hungate, R.E. 1966. *The Rumen and Its Microbes*, Academic Press, New York.

International Symposium on Anaerobes. 1980. Nippon Merck-Banyu Co., Ltd., Tokyo.

Lambe, D.W., Jr., R.J. Genco and K.J. Mayberry-Carson (Editors). 1980. *Anaerobic Bacteria. Selected Topics*, Plenum Press, New York.

Philips, I. and M. Sussman (Editors). 1974. *Infection with Non-Sporing Anaerobic Bacteria*, Churchill Livingston, Edinburgh.

Smith, L. DS. 1975. *The Pathogenic Anaerobic Bacteria*, 2nd Ed., Charles C Thomas, Springfield, Ill.

Willis, A.T. 1977. *Anaerobic Bacteriology: Clinical and Laboratory Practice*, 3rd Ed., Butterworths, London.

Key to the genera of the family **Bacteroidaceae**

I. Nonmotile or peritrichous; straight rods.
  A. Produce mixture of fermentation products including succinate, acetate, formate, lactate and propionate from carbohydrates or peptone; butyrate usually not a major product (if present, produced in combination with isobutyrate and isovalerate).
       Genus I. *Bacteroides*, p. 604.
  B. Produce butyrate as a major product; smaller amounts of acetate and sometimes propionate, formate, or lactate also may be produced; all described species are nonmotile.
       Genus II. *Fusobacterium*, p. 631.
  C. Lactate is the sole major fermentation product; small amounts of formate, acetate, or succinate also may be produced. All described species are nonmotile.
       Genus III. *Leptotrichia*, p. 637.

II. Motile, not peritrichous.
  A. Curved rods with monotrichous or lophotrichous polar or subpolar flagella; produce butyric acid as a major fermentation product.
       Genus IV. *Butyrivibrio**, p. 641.
  B. Produce succinate and acetate as major fermentation products.
    1. Short rods or coccobacilli with a single polar flagellum.
         Genus V. *Succinimonas*, p. 643.
    2. Helical or spiral-shaped cells.
      a. Single polar flagellum.
           Genus VI. *Succinivibrio*, p. 644.
      b. Bipolar tufts of flagella.
           Genus VII. *Anaerobiospirillum*, p. 645.
  C. Single polar flagellum; helical, curved, or straight rods; either hydrogen or formate are electron donors for reduction of fumarate to succinate. Carbohydrates not fermented.
       Genus VIII. *Wolinella*, p. 646.
  D. Produce propionate and acetate as fermentation products.
    1. Tufts of flagella on concave side of crescent-shaped cell (only central or subpolar flagella may be seen with conventional flagella stains, see Further Comments).
         Genus IX. *Selenomonas*, p. 650.
    2. Single polar flagellum, lipolytic, curved cells.
         Genus X. *Anaerovibrio*, p. 653.
    3. Lateral flagella aligned on concave side of curved cell.
         Genus XI. *Pectinatus*, p. 655.
  E. Produce acetic acid, ethanol, hydrogen, and carbon dioxide as major fermentation products, straight to slightly curved rods with nonpolar (single or tufts) or lateral flagella (see Further Comments).
       Genus XII. *Acetivibrio*, p. 658.
  F. Produce ethanol, formate, lactate, acetate, carbon dioxide, and hydrogen (from glucose); straight to slightly curved rods with a single lateral to subpolar flagellum; may stain weakly Gram-positive in young cultures (see Further Comments).
       Genus XIII. *Lachnospira*, p. 661.

---

* See Further Comments.

Genus I. **Bacteroides** *Castellani and Chalmers 1919, 959*[AL]

LILLIAN V. HOLDEMAN, ROGER W. KELLEY AND W. E. C. MOORE

Bac.te.ro.i′des or Bac.te.roi′des. M.L. n. *bacter* the masc. equivalent of Gr. neut. n. *bacterum* a staff or rod; Gr. n. *idus* form, shape; M.L. masc. n. *Bacteroides* rodlike.

Nonsporeforming rods that are **Gram-negative, nonmotile** or motile by peritrichous flagella, and **obligately anaerobic**. Chemoorganotrophs. Metabolize carbohydrates, peptone, or metabolic intermediates. Fermentation products of saccharoclastic species include combinations of **succinate, acetate, lactate, formate or propionate**, sometimes with short-chained alcohols; butyrate usually is not a major product. Trace to moderate amounts of isobutyrate and isovalerate may be produced (from peptone). From peptone, nonsaccharoclastic species produce either (a) combinations of trace to moderate amounts of succinate, formate, acetate and lactate or (b) combinations of moderate to major amounts of acetate, butyrate, succinate, isovalerate, propionate, isobutyrate and alcohols. **When *n*-butyrate is produced, isobutyrate and isovalerate also are present.** The mol% G + C of the DNA (in the species examined) ranges from 28 to 61.

Type species: *Bacteroides fragilis* (Veillon and Zuber 1898) Castellani and Chalmers 1919, 959.

Further Descriptive Information

Morphology. Terminal or central swellings, vacuoles, or filaments are common in many species. Cells of many anaerobic rods are more pleomorphic when cultural conditions are less than ideal; i.e. when grown in media that are not highly reduced (Doetsch et al., 1957), in medium that is not nutritionally optimal, or after acid products have accumulated. Cells of many species are longer when stained from broth cultures than from growth on solid medium. Among fermentative species, cells often are larger in the presence of fermentable carbohydrate than in basal peptone media.

Cellular composition. Information on the fatty acid content and cell wall composition of various *Bacteroides* species can be found in Hofstad (1979), Miyagawa et al. (1979a, b; 1981) Miyagawa and Suto (1980), Mayberry (1980a, b) and Shah and Collins (1980). None of 14 species tested contains heptose or 2-keto-3-deoxyoctonate (KDO) in the lipopolysaccharide (Hofstad, 1974).

Nutrition and growth conditions. Carbon dioxide is utilized or required by many saccharoclastic species and is incorporated into succinic acid (Caldwell et al., 1969). Hemin is required or highly stimulatory for growth of many species, particularly the saccharoclastic species that produce succinate. Vitamin K is required or highly stimulatory for growth of some species. Hemin and vitamin K routinely are added to media for culture of bacteroides.

Growth of described species is most rapid at 37°C and pH near 7.0. Some species do not grow well in medium with a pH below 7.

The maximum E_h that will allow initiation of growth varies among species and, within species, differs with the size of inoculum used. E_h sensitivity of species ranges from those that can initiate growth in partially oxidized media to those that have not yet been cultured in media with an initial oxidation-reduction potential above −100 mV, even when large inocula are used. Addition of serum (10% v/v) or ascitic fluid to oxidized media may enable them to support growth. Serum also provides growth-stimulatory factors for some species.

Natural habitat. Members of the genus have been isolated from the cavities of man, other animals, and insects; infections of soft tissue; and sewage. Some species may be pathogenic.

Miscellaneous. All of 14 fermentative species of *Bacteroides* were positive for phosphatase (pH range 6.0–7.2) and were able to ferment glucose-1-phosphate, whereas five nonfermentative species tested were negative for both (Wilkins and Walker, 1976).

Several *Bacteroides* species, including *B. fragilis*, *B. vulgatus*, *B. distasonis*, and *B. thetaiotaomicron*, are able to deconjugate and transform various bile acids. They also can dehydroxylate, dehydrogenate, and cleave side chains of these bile acids (e.g. cholate, chenodeoxycholate and deoxycholate). For further information, see Midtvedt and

Norman (1967), Hill and Drasar (1968), Edenharder (1976), Edenharder et al. (1976), Hylemon and Stellwag (1976), and Owen et al. (1977).

Reductive carboxylation of succinate to α-ketoglutarate occurs in *B. fragilis*, *B. vulgatus*, *B. distasonis*, *B. multiacidus* and *B. uniformis* (Allison et al., 1979).

Isolation Procedures

A complex medium containing peptone, yeast extract, vitamin K and hemin is recommended for isolation of most species from body sites. For determination of pigmentation of *B. melaninogenicus* and related species, rabbit blood (hemolyzed) is preferable to blood from other animals. Some species require a medium containing volatile fatty acids or serum. Formulas for basal media are given in Holdeman et al. (1977), Sutter et al. (1980) and Dowell and Hawkins (1974). Formulas for selective media for certain groups of bacteroides are given in Sutter et al. (1980).

Inoculated media should be incubated in an oxygen-free gas containing at least 5% carbon dioxide. Carbon dioxide is required for optimal growth of most of the saccharoclastic species. If anaerobe jars or chambers (glove boxes) are used, the gas also must contain hydrogen (to enable the palladium catalyst to reduce oxygen to water). Hydrogen is not required to maintain anaerobic conditions in the anaerobic tube culture method because oxygen is excluded from the tube during medium preparation and the neck of the tube is placed under a stream of oxygen-free gas (to exclude oxygen) each time the tube is opened.

Maintenance Procedures

Freezing at −80°C or lyophilization of young cultures grown in a well-buffered medium containing no fermentable carbohydrate is satisfactory for storage of most species. Storage of lyophilized cultures at about 4°C is recommended. Even with the best storage conditions, only a portion of the original cell population survives. Therefore, large inocula in the best medium (nutritionally) with least exposure to oxygen are recommended for recovery of a viable culture from stored material.

To maintain stock strains in the laboratory, it is advisable to transfer them weekly in chopped meat (or other suitable) medium that does not contain a fermentable carbohydrate.

It is very difficult to maintain cultures of *B. penumosintes*, but freezing generally is more satisfactory than lyophilization.

Procedures and Methods for Characterization Tests

Use of a heavy inoculum (~5% v/v of broth culture) of a young culture (late log or early stationary phase of growth) gives the most reliable results in biochemical tests. Unless otherwise cited, the characteristics listed for the species were determined with prereduced media by the methods described in Holdeman et al. (1977), and susceptibility to antimicrobial agents by the broth-disk method of Wilkins and Thiel (1973). The basal prereduced peptone-yeast extract (PY) medium contained (per 100 ml): 0.5 g trypticase, 0.5 g peptone, 1 g yeast extract, 0.1 μl vitamin K_1, 0.5 mg hemin, cysteine hydrochloride and salts solution.

Additives that may enhance growth. If a strain is not growing well, it can be tested in media with various additives to see if any enhances growth. The "growth stimulation media" that we normally use are (amounts listed are final concentrations in the medium) serum or ascitic fluid (5–10%), rumen fluid (10–30%), Tween 80 or oleate (0.02%), ammonium formate (0.2%) with either sodium fumarate (0.2%) or nitrate (0.1%, usual nitrate medium) and pyruvate (0.9%). These additives, except for pyruvate or nitrate, normally are tested in peptone-yeast extract-glucose (PYG) medium and the amount of growth and pH compared with PYG medium without any additive.

Serum, ascitic fluid, or rumen fluid can be added to all differential media. However, because serum normally contains an amylase that hydrolyzes starch, maltose and glycogen, positive reactions in these media may be due to the serum enzyme(s) rather than to the metabolism of the organism. An occasional lot of rumen fluid also may hydrolyze starch or contain sufficient indole to produce a positive reaction in the absence of growth of organisms. When these substances are added to all biochemical media, an uninoculated control of starch plus additive and indole medium plus additive should be tested for starch hydrolysis and indole, respectively. Hemin and a mixture of volatile fatty acids are the major growth factors in rumen fluid (Caldwell and Bryant, 1966) and usually can be substituted for rumen fluid. Tween 80 will enhance growth of an occasional strain of Gram-negative anaerobe, but generally is thought to be used more often by Gram-positive organisms. Therefore, if Tween 80 markedly stimulates growth, one should check carefully to confirm the Gram reaction of the strain.

Metabolic intermediates (e.g. fumarate, pyruvate) should not be added to carbohydrate-containing media to be inoculated with fermentative bacteria.

Agar or agar-absorbed broth are required for growth of an unnamed "fusiform bacteroides" isolated from gingival sulci of persons with periodontitis [A. C. R. Tanner: *Journal of Dental Research* (Special Issue A), *60:* 485 (Abstr. 703), 1981].

Although most species grow equally well in CO_2 concentrations ranging from 5 to 100% as long as no oxygen is present in the gas, some species grow better (in a poorly buffered medium) in an anaerobic gaseous atmosphere containing no more than 10% CO_2; others grow better in 100% CO_2. One can suspect that less CO_2 will be beneficial if the growth is better in chopped meat (well buffered) than in PYG (poorly buffered) medium. Alternatively, the organism may require the additional peptone present in chopped meat medium.

Some bacteroides have a requirement for sodium (~0.5% NaCl). The PY-based medium described in the VPI Anaerobe Laboratory Manual (Holdeman et al., 1977) does not contain sufficient sodium to satisfy this requirement.

Morphology and Gram reaction. For determination of Gram reaction, a buffered Gram's stain procedure (such as Kopeloff's modification) of cells from young (log phase) cultures is recommended. Cells of many Gram-positive species stain Gram-negative in older cultures or when acid has been produced in the culture medium.

Hydrogen production. Hydrogen production was determined by gas chromatography of the headspace gas (Holdeman et al., 1977). Production of hydrogen may not cause disruption of the agar in PYG agar deep cultures.

Gas production. Gas production was determined by disruption of agar in loosely covered PYG agar deep cultures. Accumulation of small (lenticular) bubbles near the colonies in the agar deep tube was recorded as "1" gas, a small split of the agar column across the tubes as "2" gas, agar separated and displaced in one or more places as "3" gas and agar forced to the top of the tube as "4" gas.

Esculin hydrolysis and esculin fermentation. These are independent

reactions. Hydrolytic products are determined by the appearance of a black color upon addition of ferric ammonium citrate reagent to esculin-PY broth cultures. Acid production is determined by pH measurement.

Indole. Cultures in media containing sufficient tryptophan (chopped meat, tryptone, etc.) and no fermentable carbohydrate are required. The culture medium should be extracted with xylene and the reagents (either Ehrlich's or Kovac's) poured down the side of the tube to layer next to the xylene. Do not shake the tube after the reagent has been added. Alternatively, a loopful of growth from a *pure culture* can be smeared on filter paper saturated with 1% paradimethylamino-cinnamaldehyde in 10% (v/v) hydrochloric acid (Sutter et al., 1980). Development of a blue color indicates indole. This method may not detect weak indole production.

Gelatin digestion. Cultures of 12% gelatin-PY medium and uninoculated tubes of gelatin medium are chilled until the originally liquid controls are solid (about 15 min at 4°C). Failure of the gelatin culture to solidify at 4°C indicates complete digestion (+). Liquefaction of cultures at room temperature within 30 min or in less than half the time of the control tube indicates partial liquefaction (w).

Milk proteolysis. Acid production (a) in milk often causes protein coagulation (c). The curd may or may not show evidence of gas streaks. Shrinking of the curd may leave a clear whey. This liquid is sometimes mistaken for digestion (proteolysis) of the milk. Digestion of curd is evidenced by dissolution of the curd (first) and increasing turbidity of the whey and often takes many days. Digestion may occur without previous curd formation and results (usually slowly) in decreased opacity. Digestion with or without curd formation results in a yellowish clear liquid after extended incubation (up to 3 weeks). *Clostridium sporogenes* is a good positive control culture for milk proteolysis.

Acid production. Positive reactions listed for carbohydrate fermentation in *Bacteroides* represent a pH below 5.5 and a decrease in pH of at least 0.5 pH unit below the control PY basal medium culture. Weak (w) acid production represents a pH of 5.5–5.7 and at least 0.3 pH unit below the PY culture control. A pH of 5.7 or above, or within 0.3 unit of the PY control culture pH, is considered negative (−). When in doubt, examination of the amount of growth in the sugar-containing medium vs. that in medium without sugar can help in interpretation of a weak reaction; i.e. with slight pH decrease, if growth is much better in the sugar-containing medium the carbohydrate probably was fermented. For rapid-growing saccharoclastic species, the final pH is reached after incubation for 18–24 h in nutritionally adequate media.

Uninoculated xylose and arabinose, under CO_2, often are pH 5.9. For cultures in these media, a final pH of 5.4 or below is considered strong acid production.

Neutral red reduction. Positive neutral red reduction is indicated by disappearance of the red color within 5 days of incubation in fructose broth containing neutral red [add 0.1 ml of sterile stock solution (0.1% w/v of neutral red dissolved in 60% absolute ethanol) per 3 ml of PY-fructose].

Differentiation of the genus **Bacteroides** *from other genera*

Characteristics useful for differentiating the genus from the other genera of the family *Bacteroidaceae* are given in the key to the family.

Taxonomic Comments

Several species described in the eighth edition of *Bergey's Manual* have been abolished or reassigned to other genera. Because there are no extant strains of *B. constellatus*, *B. niger* or *B. serpens*, these species were not included on the Approved Lists of Bacterial Names in 1980. *Bacteroides ochraceus* was reassigned to the genus *Capnocytophaga* as *C. ochracea* (Leadbetter et al., 1979).* *Bacteroides clostridiiformis* subsp.

clostridiiformis and subsp. *girans* were shown to produce spores. They were combined and named *Clostridium clostridiiforme* (Kaneuchi et al., 1976; Cato and Salmon, 1976). Strains of *B. biacutus* were also shown to produce spores. Some of these strains were assigned to species *C. clostridiiforme* and others were allocated to an unnamed *Clostridium* sp. (Kaneuchi et al., 1976).

Descriptions of many unnamed *Bacteroides* species isolated from human feces can be found in Moore and Holdeman (1974b) and Holdeman et al. (1976). Phenotypic characteristics of various homology groups of saccharoclastic *Bacteroides* species are described in Johnson and Ault (1978).

---

* *Editorial note:* The genus *Capnocytophaga* is described in Volume III of the *Manual.*

Acknowledgments

The DNA homology work was supported by Public Health Service Grant DE-05218 from the National Institute of Dental Research. Phenotypic characterization of these species was supported by Public Health Service Grants DE-05054 and DE-05139 from the National Institute of Dental Research and AI-15244 from the National Institute of Allergy and Infectious Diseases. Much of the basic information retained from the eighth edition of the *Manual* was supported by Public Health Service Grant GM-14604 from the National Institute of Medical Sciences and by project 2022820 from the Commonwealth of Virginia.

We thank Thomas O. MacAdoo, Department of Foreign Languages, Virginia Polytechnic Institute and State University, for providing the etymology of some of the specific epithets.

We gratefully acknowledge the microbiological or technical assistance of Loretta P. Albert, Pauletta C. Atkins, Ella W. Beaver, Ruth Z. Beyer, Maeve N. Crowgey, Ann P. Donnelly, Jackie G. Eudaly, Luba S. Fabrycky, Barbara A. Harich, Donald E. Hash, Linda K. Hoffman, Carolyn L. Hubbard, Jane L. Hungate, Daniel M. Linn, June M. McElwee, Ann C. Ridpath, Carolyn W. Salmon, Gail S. Selph, Debra B. Sinsabaugh, Sue C. Smith, Christina J. Spittle, Susan E. Stevens, Barbara C. Thompson, Dianne M. Wall, Catherine A. Waters, and Bonnie M. Williams; the support assistance of Eleanor A. Johnston, Julia J. Hylton, Ruth E. McCoy, Claudine P. Saville, Phyllis V. Sparks and Margaret L. Vaught; and the secretarial assistance of Mary P. Harvey and Donya B. Stephens, whose work has contributed to our taxonomic studies since 1974.

We are especially grateful to Elizabeth P. Cato for collaborative assistance in the taxonomic studies.

Further Reading

See Further Reading for the family *Bacteroidaceae* (p. 603).

Differentiation of the species of the genus **Bacteroides**

Characteristics by which species in the genus can be differentiated are given in the key to species in the genus (see below) and in Tables 6.1 (for saccharoclastic species) and 6.2 (for nonsaccharoclastic species). Acid production in the key refers to both strong and weak fermentations; i.e. a species is positive if the terminal pH is below 5.7 and 0.3 pH unit below the control basal medium culture. Additional characteristics are given in the text concerning each species. In the text and tables, the species are arranged according to the general similarity of their phenotypic reactions. The order of species in the key is dependent only upon the most reliable distinguishing properties and therefore does not follow the same sequence.

Only the basionym and selected synonyms are given for the species. The reader should consult the *Index Bergeyana* (Buchanan et al., 1966) and the *Supplement to Index Bergeyana* (Gibbons et al., 1981) for other synonyms.

Colony descriptions are from blood agar plates or streak tube cultures incubated anaerobically for 2 days, unless otherwise indicated.

Reported G + C contents of the DNAs are expressed as the nearest whole number.

The drawings of each species, given in the text, are composites of the type strain and, where available, other strains showing DNA homology with the type. Individual cultures may show somewhat less variation in cellular morphology than is depicted. The scale for the drawings is given in the drawing of *B. fragilis* (species 1). Excellent color photographs of colonies and strains of cells of most species are depicted in Mitsuoka, 1980 (text in Japanese, species names and pictures in English).

Unless specified otherwise in the text, cells are not motile. Catalase is produced by *B. fragilis* and a few other saccharoclastic species, but not by most species. Hippurate usually is not hydrolyzed. Lecithinase and lipase are not produced (on McClung-Toabe egg yolk agar) by described species. Adonitol, erythritol, glycerol, sorbitol and sorbose are not fermented by most species (positive exceptions are noted in the text concerning the species). Galactose is fermented by species that ferment lactose. Strains that produce acid from starch usually completely hydrolyze starch after extended incubation (as indicated by no dark color upon addition of iodine).

Fermentation products listed in the tables do not include alcohols. If alcohols are produced, they usually will correspond, in carbon-chain length, to the major volatile acids. Not all species have been tested for production of phenylacetic acid. If the species is known to produce phenylacetic acid, the product is listed in the tables. Phenylacetic acid has not been detected in cultures of *Bacteroides vulgatus* (Mayrand, 1979) and *B. asaccharolyticus* (Mayrand et al. 1980; Kaczmarek and Coykendall, 1980).

Key to species in the genus **Bacteroides**

I. Lactose acid.
  A. Sucrose acid.
    1. Zoogleal mass not formed.
      a. Indole not produced.
        b. No giant cells (≥ 2 μm diameter).
          c. Good growth in 20% bile.
            d. Inositol not acid.
              e. Trehalose not acid.
                f. Rhamnose not acid.
                  1. *B. fragilis*
                ff. Rhamnose acid.
                  2. *B. vulgatus*
              ee. Trehalose acid.
                3. *B. distasonis*
            dd. Inositol acid.
              4. *B. multiacidus*
          cc. Growth partially or completely inhibited by bile.
            d. Cellobiose acid.
              e. Xylose acid (pH < 5.4).
                f. Growth at 45°C, human origin.

    g. Do not produce β-glucosidase.
     11. *B. oris*
    gg. Produce β-glucosidase.
     12. *B. buccae*
   ff. No growth at 45°C, isolated from ruminants.
     21a. *B. ruminicola* subsp. *ruminicola*
  ee. Xylose not acid.
   f. Arabinose acid.
     21b. *B. ruminicola* subsp. *brevis*
   ff. Arabinose not acid.
    g. Salicin not acid, pigment produced on blood agar.
     13. *B. loescheii*
    gg. Salicin acid, no black pigment on blood agar.
     14. *B. oralis*
  dd. Cellobiose not acid.
   e. Esculin acid.
     15. *B. denticola*
   ee. Esculin not acid.
     16. *B. melaninogenicus*
 bb. Giant cells, >2 μm diameter.
     22. *B. hypermegas*
aa. Indole produced.
 b. Trehalose acid.
     5. *B. ovatus* (salicin acid)
     6. *B. thetaiotaomicron* (salicin usually not acid)
 bb. Trehalose not acid.
     7. *B. uniformis*
2. Zoogleal mass produced.
     10. *B. zoogleoformans*
B. Sucrose not acid.
 1. Indole produced.
  a. Esculin hydrolyzed.
   b. Maltose acid.
     8. *B. eggerthii*
   bb. Maltose not acid.
     9. *B. splanchnicus*
  aa. Esculin not hydrolyzed.
     24. *B. macacae*
 2. Indole not produced.
  a. Melibiose not acid.
   b. Cellobiose not acid.
    c. Maltose acid.
     17. *B. bivius*
    cc. Maltose not acid.
     23. *B. levii*
   bb. Cellobiose acid.
     25. *B. succinogenes*
  aa. Melibiose acid.
     26. *B. microfusus*
II. Lactose not acid.
 A. Starch acid.
  1. Indole not produced.
   a. Mannose not acid.
    b. Glucose acid.
     18. *B. disiens*
    bb. Glucose not acid.
     29. *B. amylophilus*
   aa. Mannose acid.
     19. *B. corporis*
  2. Indole produced.
     20. *B. intermedius*
 B. Starch not acid.
  1. Maltose acid.
     27. *B. termitidis*
  2. Maltose not acid.
   a. Indole not produced.
    b. Esculin hydrolyzed.

(Key continues on p. 610)

Table 6.1.
Biochemical reactions of the fermentative species of the genus **Bacteroides**[a]

Characteristics	1. B. fragilis	2. B. vulgatus	3. B. distasonis	4. B. multiacidus	5. B. ovatus	6. B. thetaiotaomicron	7. B. uniformis	8. B. eggerthii	9. B. splanchnicus	10. B. zoogleoformans	11. B. oris	12. B. buccae	13. B. loescheii	14. B. oralis	15. B. denticola	16. B. melaninogenicus
Products from PYG	SAppa (ibivl)	SAp (ivlbl)	SAppa (ivibl)	LSA (f)	SAppa (ibivl)	SAppa (ivibl)	Sapl (ivib)	SAp (ivibl)	SAP bivib(l)	SAP (ivib)	SA (pibiv)	SA(fpiv lfuibb)	Sa (flpy)	SAf (l)	SA(lf ivib)	SA(f ivibl)
Hydrogen produced	2,3	1,3			2,1	1,3	±2	4	4							
Growth in 20% bile	4S	4S	4S	4	4S	4	2,4	4	2S	−	−	−	−	−	−	−
Esculin hydrolyzed	+	+	+	+	+	+	+	+	+	+	+	+	+	+	+	+
Indole produced	−	−	−	−	+	+	+	+	+	+−	+	+	+	+	+	+
Nitrate reduced	−	−	−	+	−	−	−	−	−	−	−	−	−	−	−	−
Gelatin digested	−w	+	−w	−	d	−w	−	−	+−	d	+−	+w	+	+	+	+
Milk reaction	c	c	c	c	c	c	c	c	c	c	c	c	c	c	c	c
Meat digested	−	−	−	−	−	−	−	−	−	−	−	−	−	−	−	−
Acid produced from:																
Amygdalin	−+	−	+w	+−	+−	+w	+	−	−	−w	+−	+−	−	+	−	−
Arabinose	−	+	−	+	+	+	+w	+	+	d	+−	+	−	−	−	−
Cellobiose	−w	−	+	+	+	+	+w	−w	−	+w	+w	+	+	+	+	−
Dextrin	+w	+	+	+	+	+	+	+	−	+	+	+	+−	nt	+w	+
Esculin	+w	−	w+	+−	w+	w+	+−	−w	d	−	+−	d	−+	+	+	−
Fructose	+	+	+	+	+	+	+	+	+	+w	+	+	d	+	+	+
Glucose	+	+	+	+	+	+	+	+	−	+w	+	+	+	+	+	+
Glycogen	+	+	−	−	+−	+	+−	+	−	+w	−	−	+	−	−	−
Inositol	−	−	+	+	−	−	−	−	−	−	+	+	−	+	−	+
Inulin	d	+−	+	−	d	d	+w	+	+	+w	+	+−	−+	+	+	+
Lactose	+	+	+	+	+	+	+	+	+	+w	+	+	+	+	+	+
Maltose	+	+	+	+	+	+	+w	+	−	+w	+	+	+	+	+	+
Mannitol	−	−	−	+−	−	−	−	−	−	−	−	−	−	−	−	−
Mannose	+	+	+	+w	+	+	+w	+	+w	+w	+	+	+w	+	+w	+
Melezitose	−	−	+	w−	+−	d	−	−	−	d	−	d	d	−	−	−
Melibiose	+w	+w	+	+	+	+	+w	+	−+	d	d	+	d	+	+	+w
Raffinose	+	+	+	+	+	+	+	+	−	+	+	+	+	+	+	+
Rhamnose	−	−+	+−	−+	+w	+w	−+	w−	−	−	d	+−	d	d	d	−
Ribose	−	+	+	+−	+	+	−+	−	−	d	d	d	d	−w	d	−
Salicin	−	−	+	+	+	+	−+	+w	−	d	+	+	+−	+	+	+
Starch	+	+	+	+	+	+	+	−	−	+w	+	+	+−	+	+	+
Sucrose	+	−	+	+	+	+	+	+	−	+	+	+	+	+	+	+
Tehalose	−	+	+	+	+	+	−	+	−	d	+	+	−	−	−	−
Xylose	+	+	+	+	+	+	+	+	−	−	−	+	−	−	−	−
Gas (PYG agar)	d	1,3	±−	1−	1,2	1,2	1,2	−2	1,4	−1	−	−2	−	±	−	−1

Table 6.1—*continued*

Characteristics	17. *B. bivius* SAiV (ibf)	18. *B. disiens* SA(ib ivfp)	19. *B. corporis* SAivib (b)	20. *B. intermedius* SAiv (ibpf)	21. *B. ruminicola* a. subsp. ruminicola SAF (pivib)	21. *B. ruminicola* b. subsp. brevis SAfp (ibivbl)	22. *B. hypermegas* PAl s(py)	23. *B. levii* ABp ivibs	24. *B. macacae* PSa ivbib	25. *B. succinogenes* AS (f)	26. *B. microfusus* ASp (lf)	27. *B. termitidis* AL (f)	28. *B. capillosus* sa (lfp)	29. *B. amylophilus* SAF (l)	30. *B. furcosus* La (fs)
Hydrogen produced	-	-	-	-	-	-	-	-	-	-	4	-	-	-	-
Growth in 20% bile	-1	-	-1	-	-	-	1,4	1,3	1	1-	S2	-	-1	-	-2
Esculin hydrolyzed	-	-	-	-	+	+	+	-	-	-	-	-	+	-	+
Indole produced	-	-	-	+	-	-	-	-	+	-	-	-	-	-	-
Nitrate reduced	-	-	+	+	+	+	-	+	-	-	-	-	-	-	-
Gelatin digested	+	+	+	p	c	c	c	p	cp	a	w	-	-w	-	-w
Milk reaction	p	p	pc	p	c	c	c	-	cp	a	c-	-	c	-	-
Meat digested	+	+	-+	+-	-	-	-	+	+	-	-	-	-	-	-
Acid produced from:															
Amygdalin	-	-	-	-	+	w	w-	-	-	-	-	-	-	-	-
Arabinose	-	-	-	-	+	+w	+	-	-	-	-	-	-	-	-
Cellobiose	-	-	-	-	r	+w	+-	-	-	+w	-	+	-	-	-
Dextrin	+	+	+	+	r	+w	w-	-	-	-	-	-	-	+	-
Esculin	-	-	-	-	+w	r	+	-	-w	-	-	-	-	-	-
Fructose	w	-w	d	+-	+	+	+	+w	+	+w	-w	+	-	-	w-
Glucose	+	+	+	+	d	+	+	+	+	+w	w	+	w-	+	w+
Glycogen	+	+	+	+-	-	w+	-	-	-	-	-	-	-	+	-
Inositol	-	-	-	-	-	-	+-	-	-	-	-	-	-	-	-
Inulin	-	-	-	+-	+	+w	+-	+w	-	-	-	-	-	-	-
Lactose	+	-	-	-	+	+w	+	-	+	w	+w	-	-	+	-
Maltose	+	+	-	+	+	+	+	+w	+	r	-	+	-	+	-
Mannitol	-	-	-	-	-	-	+	-	-	-	-	-	-	-	-
Mannose	+	-	+w	d	+	+	+	+w	+	-	w	-	-	-	-
Melezitose	-	-	-	-	-	-w	-w	-	-	-	-	-	-	-	-
Melibiose	-	-	-	d	+	+	+	-	-	-	+w	-	-	-	-
Raffinose	-	-	-	-	-	+w	+	-	-	-	-	-	-	-	-
Rhamnose	-	-	-	-	-	-	-+	-	-	-	-	-	-	-	-
Ribose	-	-	-	-	+	r	+w	-	-	-	-	-	-	-	-
Salicin	-	-	-	-	+	r	d	-	-	r	-	+	-	-	-w
Starch	+	+	+	+w	+	d	-	-	-	-	-	-	-	+	-
Sucrose	-	+	-	+	+	+	+	-	-	-	-	+	-	-	+
Trehalose	-	-	-	-	-	-	+	-	-	-	-	+	-	-	-
Xylose	-	-	-1	-	+	-	+	-	-	-	-	+	-	-	-
Gas (PYG agar)	-1	-1	-1	-±	-	-	2	d	-	-	±1	-	-3	-	-2

ᵃ Products from 1% peptone-1% yeast extract-1% glucose (PYG) broth cultures: Capital letters indicate an average (from multiple cultures) of >1 meq of acid/100 ml broth; small letters, <1 meq/100 ml. A, acetic; B, butyric; F, formic; Fu, fumaric; iB, isobutyric; iV, isovaleric; L, lactic; P, propionic; Pa, phenylacetic; Py, pyruvic; S, succinic. Except for *B. macacae* and *B. gingivalis*, the cited production of phenylacetic acid is from Mayrand (1979). Products in parentheses (approximate concentrations in headspace gas): ±, trace; 1, 0.5%; 2, 1%; 3, 2%; 4, 3% or more. Growth in 20% bile, 1, poor; 2, moderate; 4, excellent; S, stimulated (more growth than in the PYG control tube). + (sugars), pH below 5.5; - (sugars), pH 5.7 or above; w, weak reaction or (sugars) pH 5.5 to 5.7; a (milk), acid; c (milk), curd; p (milk), proteolyzed; r, reported result varies between laboratories; where two reactions are given, the first is the more common.

Table 6.2.
Biochemical reactions of nonfermentative species of the genus **Bacteroides**[a]

Characteristics	31. *B. ureolyticus*	32. *B. praeacutus*	33. *B. gracilis*	34. *B. nodosus*	35. *B. pneumosintes*	36. *B. putredinis*	37. *B. coagulans*	38. *B. gingivalis*	39. *B. asaccharolyticus*
Products from PYG	SA (lf)	ABi*V* p*ibs*(lf)	Sa	asp (f*iv*b*ibl*)	a (sl)	S*iV* Pa*ibb*(l)	a (fpls)	B*iV* ap*ibspa*	AB *iV*Sp*ib*
Hydrogen produced	1, 2	–	1	–	–	2, 4	–	–	d
Growth in 20% bile	–	d	–	–±	–	–3	d	–	–
Indole produced	–	–	–	–	–	+	+	+	+
Nitrate reduced	+	+–	+	–	–	–	–	–	–
Urease produced	+	–	–	–	–				
Gelatin digested	+–	+w	–	+	–	+	+	+	+
Milk reaction	p–	–	–	p	–	p	cp	p	p
Meat digested	–+	–	–	+	–	d	–	+	+
Gas (PYG agar)	–1	–2	–	–	–	d	–2	–	–

[a] No acid is produced from the carbohydrates listed in Table 6.1. For the purposes of this key it is sufficient to determine that esculin is not hydrolyzed and glucose and maltose are not fermented. See footnote to Table 6.1 for explanation of the symbols.

    c. Lactic acid not a major product.
        28. *B. capillosus*
   cc. Major product is lactic acid.
        30. *B. furcosus*
 bb. Esculin not hydrolyzed.
    c. Urease positive.
        31. *B. ureolyticus*
   cc. Urease negative.
     d. Milk and meat not digested.
       e. Produces major butyric and isovaleric acids.
          32. *B. praeacutus*
      ee. Does not produce major butyric and isovaleric acids.
        f. Nitrate reduced.
          33. *B. gracilis*
        ff. Nitrate not reduced, tiny cells.
          35. *B. pneumosintes*
    dd. Milk and meat digested.
        34. *B. nodosus*
aa. Indole produced.
  b. No black colonies on blood agar.
    c. Produces isovaleric and isobutyric acids.
        36. *B. putredinis*
   cc. Does not produce isovaleric and isobutyric acids.
        37. *B. coagulans*
 bb. Black colonies on blood agar.
    c. Phenylacetic acid produced.
        38. *B. gingivalis*
   cc. Phenylacetic acid not produced.
        39. *B. asaccharolyticus*

Description of species of the genus **Bacteroides**

1. **Bacteroides fragilis** (Veillon and Zuber 1898) Castellani and Chalmers 1919, 959.[AL] [*Bacillus fragilis* Veillon and Zuber 1898, 536; *Fusiformis fragilis* (Veillon and Zuber 1898) Topley and Wilson 1929, 393; *Ristella fragilis* (Veillon and Zuber 1898) Prévot 1938, 2980; *Bacteroides fragilis* subsp. *fragilis* (Veillon and Zuber 1898) Castellani and Chalmers 1919, 959.]

fra′gi.lis. L. adj. *fragilis* fragile (relating to the brittle colonies that may form under some culture conditions).

Bacillus fragilis was first described by Veillon and Zuber (1898). In 1919, Castellani and Chalmers transferred the species to the genus *Bacteroides*. Holdeman and Moore (1970, 1974a) studied 326 strains of bacteria that fit the general description of *B. fragilis* and included type or reference strains for the species described by Eggerth and Gagnon (1933) and of other saccharoclastic bacteroides that grow well in bile. They found a continuum of variants, but some clusters were discernible. Due to the overall morphologic and phenotypic similarities of the clusters, they were designated as subspecies of *B. fragilis* (Holdeman and Moore, 1970, 1974a).

Subsequent DNA homology studies (Johnson, 1973) demonstrated that the *B. fragilis* subspecies were all genetically distinct. Therefore, their species rank was reinstated by Cato and Johnson (1976). Because *B. distasonis*, *B. ovatus*, *B. thetaiotaomicron*, and *B. vulgatus* were considered subspecies of *B. fragilis* from about 1970 until 1978, literature references to "*B. fragilis*" during this era include all of the subspecies unless the subspecies designation was used. Current literature may still refer to the saccharoclastic bacteroides that grow well in bile as the "*B. fragilis* group," which will include, in general, *B. fragilis, B. distasonis, B. ovatus, B. thetaiotaomicron* and *B. vulgatus.*

(See Further Comments at the end of this species description.)

The description of *B. fragilis* presented below, unless otherwise indicated, is based on our study of the type strain, 20 strains with high DNA homology with the type (Johnson, 1978), and other phenotypically similar strains.

Cells from glucose broth cultures are 0.8–1.3 by 1.6–8.0 μm; they occur singly or in pairs and have rounded ends. Vacuoles often are present, particularly in broth media containing a fermentable carbohydrate. Kaspar (1976a), Kaspar et al. (1977) and Babb and Cummins (1978) concur that many strains of *B. fragilis* are encapsulated but the two research groups define the term "capsule" differently. The presence of a capsule may play a role in the pathogenicity of *B. fragilis* and other *Bacteroides* species, but its relative importance is not yet clear (Walker and Wilkins, 1976; Onderdonk et al., 1977, 1978; Babb and Cummins, 1978; Maskell, 1981).

PY   **PYG**

Figure 6.1. *Bacteroides fragilis.* (*Bar* = 10 μm.)

Surface colonies on horse blood agar plates are 1–3 mm, circular, entire, low convex, and translucent to semiopaque. They often have an internal structure of concentric rings when viewed by obliquely transmitted light. In general, strains produce no hemolysis on horse blood or rabbit blood agar; a few strains may be slightly hemolytic, particularly in the area of confluent growth . A very few strains (less than 1%) are β-hemolytic.

Glucose broth cultures are turbid with sediment. The sediment usually is smooth, but may be stringy or ropy. The terminal pH in glucose (1%) broth is between 5.0 and 5.5. Typical positive reactions usually are not obtained unless heme is added to the differential media. Growth is often enhanced by 20% bile. Growth may occur at 25 or 45°C. Upper and lower temperatures permitting growth have not been determined for most strains, but they vary with growth conditions and size of inoculum. Most strains grow at pH 8.5, but more slowly and less luxuriantly than at pH 7.0. Cells can survive exposure to air for at least 6–8 h (Loesche, 1969). Over 10% of the cells survive exposure to 3 ATA (atmospheres absolute pressure) of 100% oxygen for 18 h on agar medium containing blood, hemin and menadione, but only 0.14% survive when blood, hemin and menadione are omitted from the agar plating medium (Hill and Osterhout, 1974).

B. fragilis grows well in a minimal medium containing hemin, glucose, minerals, NH₄Cl, sulfide, bicarbonate/CO₂ buffer and vitamin B₁₂. The vitamin B₁₂ requirement can be replaced by methionine. Although cultures grow well in PY medium, Varel and Bryant (1974) report that there is little or no ability to use organic nitrogen compounds as a nitrogen source.

Hemin is either required for growth or markedly stimulates growth of strains of *B. fragilis*. The generation time in glucose-minimal medium (without hemin) is 8 h, but decreases to 2 h (with an increased cell yield) when hemin is present (Macy et al., 1975; Frantz and McCallum, 1979). In a glucose-enriched medium, the maximum generation time is 60 min (Frantz and McCallum, 1979).

No cytochromes are detected in cells grown in medium without hemin; with hemin, cytochromes *b* and *o* are present (Macy et al., 1975; Reddy and Bryant, 1977). The metabolic end products also vary depending on the presence of hemin. In its absence or at suboptimal concentrations, fumarate (or malate) and lactate are formed; with

hemin, the major products are succinic and acetic acids with lesser amounts of propionic acid (Macy et al., 1978). The formation of fumarate or malate is strain dependent (Chen and Wolin, 1981). If hemin is present, fumarate is reduced to succinate via NADH, H⁺ and various cytochromes and quinones, which results in additional ATP production. The hemin is probably necessary for the formation of cytochrome *b* and NADH:fumarate oxidoreductase (Macy et al., 1978). Both formate and hydrogen can serve as electron donors to reduce fumarate (Harris and Reddy, 1977). Succinate is then further decarboxylated to propionate using a pathway similar to that used by *Veillonella* species (Macy et al., 1978). Vitamin B₁₂ is required for the production of propionate from succinate (Chen and Wolin, 1981).

No heptose or KDO is detected in the lipopolysaccharide (Hofstad, 1974; Kaspar, 1976b). No lysine decarboxylase activity is detected (Werner, 1974). Catalase is produced (see Further Comments). Indoleacetic acid is produced from tryptophan (Chung et al., 1975).

B. fragilis is relatively resistant to penicillin. Many strains produce β-lactamases, including penicillinases and cephalosporinases, although usually at low concentrations (Del Bene and Farrar, 1973; Olsson et al., 1976; Weinrich and Del Bene, 1976). The production of penicillinase by *B. fragilis* is sufficient to protect penicillin sensitive species in mixed culture infections in vivo (Hackman and Wilkins, 1975).

Resistance to tetracycline is widespread among clinical isolates and probably is plasmid mediated (Privitera et al., 1979). Regulation of tetracycline resistance and its transferability is inducible in the majority of the strains. Resistance to erythromycin and clindamycin often is cotransferred with tetracycline resistance (Privitera et al., 1981).

Susceptibility to additional antimicrobial agents is discussed in Further Comments.

B. fragilis produces a bacteriocin that inhibits synthesis of RNA but not of DNA (Mossie et al., 1979).

Other characteristics of the species are given in Table 6.1 (species 1) and in Further Comments.

Isolated, either with other organisms or as the only organism present, from various types of human clinical specimens including appendicitis, peritonitis, heart valve infections, blood, rectal abscesses, pilonidal cysts, postsurgical wounds and lesions of the urogenital tract; occasionally isolated from the mouth and vagina. Although *B. fragilis* is the most common species of anaerobic bacteria isolated from human soft tissue infections and anaerobic bacteremia, it accounts for less than 1% of the normal human intestinal flora (Moore and Holdeman, 1974b; Holdeman et al., 1976).

The mol% G + C content of the DNA is 41 to 44 (Johnson, 1978).

Type strain: ATCC 25285 (NCTC 9343).

Further comments. Many comparative studies have been performed on the former subspecies of *B. fragilis*, which currently are classified as *B. fragilis, B. vulgatus, B. distasonis, B. ovatus* and *B. thetaiotaomicron.* The results of some of these studies are summarized below and in Table 6.3.

All five species produce glutamic acid decarboxylase (Werner, 1970b), deoxyribonuclease and phosphatase (Porschen and Sonntag, 1974; Porschen and Spaulding, 1974); acid from glucuronate and glucosamine (Salyers et al., 1977a); sphingolipids (Miyagawa et al., 1978); superoxide dismutase (Carlsson et al., 1977; Gregory et al., 1978) and glycogen storage granules (Lindner et al., 1979). The superoxide dismutase of *B. fragilis* is an iron-containing enzyme (Gregory and Dapper, 1980). *B. fragilis* has a much higher level of neuraminidase activity than do the other four species (Müller and Werner, 1970; Hammann et al., 1981). None of the five species produces acid from alginate, guar gum, locust bean gum or porcine gastric mucin (Salyers et al., 1977a).

B. fragilis, B. distasonis, B. thetaiotaomicron and *B. vulgatus* (but not *B. ovatus*) are capable of good growth in a minimal medium containing hemin, glucose, minerals, NH₄Cl, sulfide, bicarbonate-carbon dioxide buffer and vitamin B₁₂ or methionine (Varel and Bryant, 1974). Hemin either is required or markedly stimulates growth.

Growth (as estimated by colony diameters on agar media) may be inhibited by blood. This inhibition can be reversed by addition of hemin

Table 6.3.

Additional characteristics of **Bacteroides fragilis** *and some other saccharoclastic bacteroides[a]*

Characteristics	1. B. fragilis	2. B. vulgatus	3. B. distasonis	5. B. ovatus	6. B. thetaiotaomicron
Phenylacetic acid produced[b]	+	−	+	+	+
Dextran hydrolyzed[c]	−	−	−	+	d
Hyaluronidase[c]	+	−	+	+	+
Chondroitin sulfatase[d]	+	−	+	+	+
Acid from:[e]					
Amylopectin	+	+	−	d	+
Amylose	+	+	−	d	+
Chondroitin sulfate	−	−	−	+	+
Dextran	−	−	−	+	+
Fucose	+	+	−	+	+
Galacturonate	−	+	−+	+	+
Gum tragacanth	−	−	−	+−	−
Heparin	−	−	−	+	+
Hyaluronate	−	−	−	+	+
Laminarin	−	−	+	−	d
Larch arabinogalactan	−	d	−	d	+
Ovomucoid	−	−	−	+−	+
Pectin	−	d	−	+	+
Polygalacturonate	−	d	−	+−	+
Xylan	−	−+	−	+	−

[a] Where two reactions are given, the first is the more common. See footnote to Table 6.1 for explanation of symbols.
[b] Van Assche (1978); Mayrand (1979).
[c] Holbrook and McMillan (1977).
[d] Rudek and Haque (1976).
[e] Salyers et al. (1977a).

(Wilkins et al., 1976). Growth of strains of *B. fragilis* was inhibited less than growth of *B. distasonis, B. ovatus, B. thetaiotaomicron* and *B. vulgatus.*

Because of the potential significance of isolation of *B. fragilis* from clinical specimens, various selective isolation media or screening methods for this species have been devised. Most of the selective media are based on the ability of *B. fragilis* to grow in the presence of 20% bile and a high concentration of one or more antibiotics (Vargo et al., 1974; Sutter et al., 1980). Many of these media probably permit growth of saccharoclastic intestinal bacteroides other than *B. fragilis, B. distasonis, B. ovatus, B. thetaiotaomicron* and *B. vulgatus,* but strains of most of these other species have not been tested for growth on these media. Surface growth of *B. splanchnicus,* occasionally isolated from clinical specimens, is not inhibited by a disk containing 1 mg kanamycin (Holdeman *et al.,* 1977). Although *B. fragilis, B. distasonis, B. ovatus, B. thetaiotaomicron* and *B. vulgatus* are almost always resistant to 2 units of penicillin, this is not a satisfactory selective method specifically for these bacteria because many other bacteroides also are resistant to this concentration of penicillin.

In vitro susceptibility of *B. fragilis* (probably including tested isolates of *B. distasonis, B. ovatus, B. thetaiotaomicron* and *B. vulgatus*) is summarized by Finegold (1977): 5% are susceptible to penicillin G (2 U/ml), 10% to ampicillin (3 μg/ml), 26% to amoxicillin (4 μg/ml), 80% to carbenicillin (64 μg/ml), 6% to cephalothin (8 μg/ml), 13% to cefazolin (12.5 μg/ml), 67% to cefoxitin (8 μg/ml), 40% to tetracycline (2 μg/ml), 50% to doxycycline (2 μg/ml), 70% to minocycline (2 μg/ml), 99% to chloramphenicol (16 μg/ml), 20% to erythromycin (1 μg/ml), 13% to lincomycin (3 μg/ml), 96% to clindamycin (4 μg/ml) and 99% to metronidazole (8 μg/ml).

The use of catalase production as a distinctive diagnostic test for *B. fragilis* and similar saccharoclastic species of bacteroides has been studied by Hansen and Stewart (1978) and Wilkins et al. (1978). All tested strains of *B. fragilis, B. ovatus, B. thetaiotaomicron* and *B. distasonis* and 4 of 5 strains of *B. vulgatus* were catalase-positive with 15% H_2O_2-Tween 80 (Hansen and Stewart, 1978). Wilkins et al., using cultures in chopped meat broth (containing 1 μg hemin/ml and no

fermentable carbohydrate) and testing with an equal volume of 3% H_2O_2, report that all tested strains of *B. fragilis* and *B. distasonis* produced catalase; catalase production was variable among strains of *B. thetaiotaomicron, B. ovatus* and *B. eggerthii;* and strains of *B. vulgatus, B. uniformis* and DNA homology groups "3452A" and "subsp. a" (Johnson, 1978) were catalase negative.

B. fragilis and phenotypically similar *Bacteroides* species have been shown to consist of multiple serovars. Both direct and indirect fluorescent antibody procedures have been developed and used successfully to identify clinical isolates of *B. fragilis* (Stauffer et al., 1975; Abshire et al., 1977; Kaspar et al., 1977; Holland et al., 1979; Lambe, 1979; Labbé et al., 1980). Agglutination tests also have been used with varying degrees of success to identify and classify *B. fragilis* strains. No correlation between the agglutination serogroups and the clinical source of the strains has yet been demonstrated (Lambe and Moroz, 1976; Elhag and Tabaqchali, 1978). Babb and Cummins (1981) used agglutination tests and were able to differentiate *B. fragilis* from phenotypically similar *Bacteroides* species and from DNA homology groups which were phenotypically indistinguishable.

2. **Bacteroides vulgatus** Eggerth and Gagnon 1933, 401.[AL] [*Pasteurella vulgata* (Eggerth and Gagnon 1933) Prévot 1938, 292; *B. fragilis* subsp. *vulgatus* (Eggerth and Gagnon 1933) Holdeman and Moore 1970, 35.]

vul.ga′tus. L. adj. *vulgatus* common (referring to the predominance of the species in fecal flora).

This description, unless otherwise indicated, is based on our study of the type, 30 strains having high DNA homology with the type (Johnson, 1978), and other phenotypically similar strains.

Cells in glucose broth are 0.5–0.8 by 1.5–8.0 μm. They may be pleomorphic with swellings or vacuoles, but less so than strains of *B. fragilis.* In broth cultures, cells usually occur singly, occasionally in pairs or short chains. Capsules are detected in some strains (Babb and Cummins, 1978).

Surface colonies on blood agar are 1–2 mm, circular, entire, convex, grayish, and semiopaque. There is no hemolysis on sheep blood agar.

Figure 6.2. *Bacteroides vulgatus.*

Glucose broth cultures are turbid with a smooth sediment and a final pH of 5.0–5.5.

No heptose or KDO is detected (Hofstad, 1974). Hemin is required or highly stimulatory for growth. No catalase is produced (Wilkins et al., 1978); catalase is produced by 4 of 5 strains tested (Hansen and Stewart, 1978).

Other characteristics of the species are given in Table 6.1 (species 2) and Further Comments for *B. fragilis.*

Isolated primarily from human feces where it is 1 of the 10 most common species (Moore and Holdeman, 1974b; Holdeman et al., 1976). Occasionally isolated from human infections.

The mol% G + C content of the DNA is 40–42 (Johnson, 1978).

Type strain: ATCC 8482 (NCTC 11154).

3. **Bacteroides distasonis** Eggerth and Gagnon 1933, 403.[AL] [*Ristella distasonis* (Eggerth and Gagnon 1933) Prévot 1938, 291; *Bacteroides fragilis* subsp. *distasonis* (Eggerth and Gagnon 1933) Holdeman and Moore 1970, 35.]

dis.ta.so′nis. M.L. gen. n. *distasonis* of Distaso; named after A. Distaso, a Romanian bacteriologist.

This description, unless otherwise indicated, is based on our study of the type strain, 10 strains having high DNA homology with the type (Johnson, 1978), and other phenotypically similar strains.

Cells in glucose broth are 0.6–1.0 by 1.6–11.0 μm with rounded ends. In broth cultures they occur singly, occasionally in pairs. No capsules are detected (Babb and Cummins, 1978).

Figure 6.3. *Bacteroides distasonis.*

Surface colonies on blood agar are pinpoint to 0.5 mm, circular, entire, convex, translucent to opaque, gray-white, and smooth. Some strains are α-hemolytic (sheep blood).

Glucose broth cultures are turbid with smooth sediment and a final pH of 5.0–5.5.

No heptose or KDO is detected (Hofstad, 1974). Hemin is required or highly stimulatory for growth. Catalase is produced.

Other characteristics of the species are given in Table 6.1 (species 3) and Further Comments for *B. fragilis.*

Isolated primarily from human feces, where it is one of the most common species (Moore and Holdeman, 1974b; Holdeman et al., 1976); occasionally isolated from human clinical specimens.

The mol% G + C content of the DNA is 43–45 (Johnson, 1978).

Type strain: ATCC 8503 (NCTC 11152).

Further comments. Some strains that we (L.V.H., W.E.C.M.) had identified as *B. distasonis*, using phenotypic characteristics given in the eighth edition of the *Manual* were found to have high DNA homology with reference DNA from VPI strains 3452A or T4-1 but no homology with reference DNA from the type strain of *B. distasonis* (Johnson, 1978).

The mol% G + C of the DNA of strain 3452A is 41. Strains of the 3452A group are indole-negative, but are more closely related (by DNA hybridization studies) to *B. thetaiotaomicron* and *B. ovatus* than to

Table 6.4.

Differential phenotypic characteristics of **Bacteroides distasonis** *and DNA homology groups T4-1 and 3452A*[a]

Characteristics	*B. distasonis*	Homology Group	
		T4-1	3452A
pH below 5.5 from			
Arabinose	14	10	100
Cellobiose	100	0	32
Melezitose	100	10	91
Salicin	100	0	73
Catalase produced	100	0	5
Hydrogen produced	0	0	100

[a] From Johnson and Ault (1978). The figures represent percent of strains positive.

indole-negative species (Johnson, 1978). This genetically distinct group of saccharoclastic bacteroides has not yet been named.

The mol% G + C of the DNA of strains in the T4-1 homology group ranges from 43 to 46. By DNA homology studies, the T4-1 group is not closely related to any of the other bacteroides tested (Johnson, 1978).

Phenotypic reactions of the 3452A and T4-1 groups are very similar to those of *B. distasonis.* Phenotypic characteristics helpful in differentiating among them (summarized from Johnson and Ault, 1978) are given in Table 6.4.

Bacteriocins have been found in several strains of the 3452A homology group (Booth et al., 1977).

Strains of the 3452A group have been isolated from human clinical specimens and human feces. Nine of 10 strains in the T4-1 homology group were isolated from human feces; the other was from hog cecal contents (Johnson, 1978).

4. **Bacteroides multiacidus** Mitsuoka, Terada, Watanabe and Uchida 1974, 40.[AL]

mul.ti.a′ci.dus. L. adj. *multus* many, much; L. adj. *acidus* sour; M.L. neut. n. *acidum* acid; M.L. adj. *multiacidus* producing much acid (referring to the large number of carbohydrates fermented).

The following description is from Mitsuoka et al. (1974) and our study of the type and other phenotypically similar strains.

Cells from glucose broth cultures are 0.8–1.5 by 3–20 μm with rounded ends and occur singly, in short chains, or in irregular groups.

Figure 6.4. *Bacteroides multiacidus.*

Colonies on glucose blood agar are grayish white, 3–8 mm in diameter, generally circular, and convex with an irregular surface. Some strains are slightly hemolytic; no brown pigment is produced.

There is little or no growth in media without a fermentable carbohydrate. The final pH in glucose broth is 4.1–4.3.

The optimum temperature for growth is 37°C; usually no growth at 45°C. Bile neither stimulates nor inhibits growth. Strains are resistant to 1.5% sodium propionate, 0.001% brilliant green, and rifampin (10 μg/ml) and are inhibited by 0.005% crystal violet, kanamycin (100 μg/ml), polymyxin B (10 μg/ml) and colistin (10 μg/ml). Sensitivity to penicillin (10 μg/ml), erythromycin (50 μg/ml) and bacitracin (3 U/ml) is variable among strains.

Glutamic acid decarboxylase is not produced; α-methylmannoside is not fermented, α-methylglucoside usually is not fermented. No sphingolipids are detected (Miyagawa et al., 1978).

Other characteristics of the species are given in Table 6.1 (species 4).

Mitsuoka et al. separated *B. multiacidus* into two groups on the basis of fermentation of mannitol and sorbitol. The type strain belongs to group 1 (mannitol positive, sorbitol negative). ATCC 27724 (NCTC 10935) was designated as a reference strain for group 2 (mannitol negative, sorbitol positive).

Isolated from feces of man and pigs (10^6–10^8/g feces) and occasionally from human clinical specimens.

The mol% G + C of the DNA is 56–58.

Type strain: ATCC 27723 (NCTC 10934).

5. **Bacteroides ovatus** Eggerth and Gagnon 1933, 405.[AL] [*Pasteurella ovata* (Eggerth and Gagnon 1933) Prévot 1938, 292; *Bacteroides fragilis* subsp. *ovatus* (Eggerth and Gagnon 1933) Holdeman and Moore 1970, 35.]

o.va′tus. L. adj. *ovatus* ovate, egg-shaped (relating to the cellular shape).

This description, unless otherwise indicated, is based on our study of the type strain, 25 strains having high DNA homology with the type (Johnson, 1978), and other phenotypically similar strains.

Cells in glucose broth are 0.6–0.8 by 1.6–5.0 μm with rounded ends. Oval forms predominate. In broth cultures, cells occur singly, occasionally in pairs. Capsules are detected in some strains (Babb and Cummins, 1978).

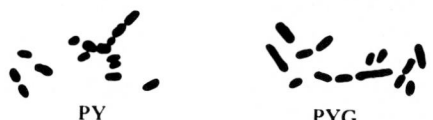

Figure 6.5. *Bacteroides ovatus.*

Surface colonies on blood agar are 0.5–1.0 mm, circular, entire, convex, pale buff colored, semiopaque, and may have a mottled appearance. There is no hemolysis of sheep blood.

Glucose broth cultures are turbid with a smooth or granular sediment and a final pH of 5.0–5.4.

Hemin is required or highly stimulatory for growth. Catalase is produced (Hansen and Stewart, 1978); catalase production is variable among strains (Wilkins et al., 1978). Sorbitol is weakly fermented by a few strains.

Other characteristics of the species are given in Table 6.1 (species 5) and Further Comments for *B. fragilis.*

Cellular morphology may be helpful in differentiating *B. ovatus* from salicin fermenting strains of *B. thetaiotaomicron* (species 6).

Isolated from human feces and occasionally from human clinical specimens.

The mol% G + C content of the DNA is 39–43 (Johnson, 1978).

Type strain: ATCC 8483 (NCTC 11153).

6. **Bacteroides thetaiotaomicron** (Distaso 1912) Castellani and Chalmers 1919, 960.[AL] [*Bacillus thetaiotaomicron* Distaso 1912, 444; *Sphaerocillus thetaiotaomicron* (Distaso) Prévot 1938, 300 (*Note*: Prévot placed this organism in the genus *Sphaerocillus* because Distaso thought the strains he examined were motile, but Eggerth and Gagnon (1933) believe that Distaso "observed Brownian movement, which is very active"); *Bacteroides fragilis* subsp. *thetaiotaomicron* (Distaso 1912) Holdeman and Moore 1970, 35.]

the.ta.i.o.ta.o′mi.cron. M.L. n. *thetaiotaomicron* a combination of the Greek letters *theta, iota* and *omicron* (relating to the morphology of vacuolated forms).

This description, unless otherwise indicated, is based on our study of the type strain, 25 strains having high DNA homology with the type (Johnson, 1978), and other phenotypically similar strains.

Cells in glucose broth are 0.7–1.1 by 1.3–8.0 μm and are pleomorphic. They occur singly or in pairs.

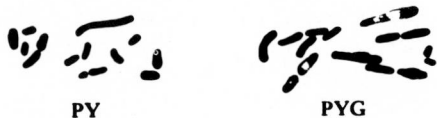

Figure 6.6. *Bacteroides thetaiotaomicron.*

Cells from blood agar plates or peptone-yeast extract broth are smaller and more homogeneous in size and shape than cells from media with a fermentable carbohydrate. Capsules are present in some strains (Babb and Cummins, 1978) and may be correlated with colonial morphology and bacteriophage resistance (Burt et al., 1978).

Surface colonies on blood agar are punctiform, circular, entire, convex, semiopaque, whitish, soft and shiny. Sheep blood is not hemolyzed. Colonies on laked blood agar do not fluoresce or have pigment.

Glucose broth cultures are turbid with smooth sediment and a final pH of 5.0–5.5.

Hemin is required or highly stimulatory for growth. Catalase is produced (Hansen and Stewart, 1978); catalase production is variable among strains (Wilkins et al., 1978).

No lysine decarboxylase is detected (Werner, 1974). No heptose or KDO is detected (Hofstad, 1974). Indoleacetic acid is formed from tryptophan (Chung et al., 1975).

Amino acids are not utilized as nitrogen sources but are incorporated into cellular material (Smith and Salyers, 1981).

Other characteristics of the species are given in Table 6.1 (species 6) and Further Comments for *B. fragilis* (species 1) and *B. uniformis* (species 7).

Cellular morphology may be helpful in differentiating salicin fermenting strains of *B. thetaiotaomicron* from *B. ovatus* (species 5).

Isolated from human feces. Frequently found in human clinical specimens.

The mol% G + C of the DNA is 40–43 (Johnson, 1978).

Type strain: ATCC 29148 (NCTC 10582).

7. **Bacteroides uniformis** Eggerth and Gagnon 1933, 400.[AL]

u.ni.form′is. L. masc. adj. *uniformis* having only one form, uniform.

This description, unless otherwise indicated, is based on our study of the type strain, 19 strains having high DNA homology with the type (Johnson, 1978), and other phenotypically similar strains.

Cells are 0.6–1.0 by 1.5–11 μm. In stains from broth cultures, cells occur singly and in pairs; an occasional filament may be seen. Vacuoles that do not greatly swell the cell often are present in cells grown in media containing a fermentable carbohydrate.

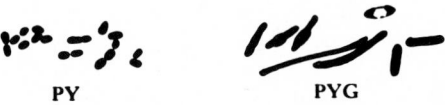

Figure 6.7. *Bacteroides uniformis.*

Surface colonies on blood agar plates are 0.5–2.0 mm in diameter, circular, entire, low convex, gray-to-white, and translucent to slightly opaque. There usually is no hemolysis of blood agar; some strains may produce a slight greening of the agar.

Glucose broth cultures are turbid with a smooth sediment and a final pH of 5.0–5.5. The optimum temperature for growth is 35–37°C. Growth may occur at 25 or 45°C. Catalase is not produced (Wilkins et al., 1978).

Other characteristics are given in Table 6.1 (species 7).

Strains of *B. uniformis* and *B. thetaiotaomicron* are very similar phenotypically, but they differ in their G + C content and are not related by DNA/DNA homology (Johnson, 1978). In general, strains of *B. uniformis* do not grow as well in 20% bile as do strains of *B.*

thetaiotaomicron. Fermentation of salicin (usually positive for *B. uniformis*) and trehalose (positive for *B. thetaiotaomicron*) also are helpful in differentiating members of these two species (Table 6.1). Also see Further Comments, for *B. eggerthii* (species 8).

Isolated as part of normal human and swine fecal flora; also isolated from various human clinical specimens. There is evidence for a positive correlation between the concentration of *B. uniformis* in fecal flora and stress or anger experienced by the individual (Holdeman et al., 1976).

The mol% G + C of the DNA is 45–48 (Johnson, 1978).

Type strain: ATCC 8492.

Further comments. Many strains referred to as *B. thetaiotaomicron* in previous publications (Moore and Holdeman, 1974b; Holdeman et al., 1976) are actually *B. uniformis* (Moore et al., 1981).

Bacteriocins have been found in several strains (including VPI 0061-1 = ATCC 8492) studied by Booth et al. (1977) that later were identified as *B. uniformis*.

8. Bacteroides eggerthii Holdeman and Moore 1974b, 270.[AL]

eg.gerth′i.i. M.L. gen. n. *eggerthii* of Eggerth, named after Arnold H. Eggerth, an American bacteriologist.

This description, unless otherwise indicated, is based on our study of the type strain, 12 strains having high DNA homology with the type (Johnson, 1978), and other phenotypically similar strains.

Cells are pleomorphic rods, ranging from coccoid to large rods with vacuoles or swellings, occurring singly or in pairs. Dimensions range from 0.4–1.0 by 1.0–6.0 μm in glucose broth.

Figure 6.8. *Bacteroides eggerthii.*

Colonies on anaerobic blood agar plates are punctiform, circular, entire, convex, translucent, gray-white, shiny, smooth and nonhemolytic. No black pigment is observed on laked blood agar roll tubes within 8 days.

Cultures in PY broth are moderately turbid. Glucose broth cultures are turbid with a smooth (occasionally stringy) sediment and a final pH of 4.9–5.6.

Hemin markedly stimulates growth, but is not required. Without hemin, malate and lactate are produced; with hemin, succinate and acetate are produced. Vitamin B_{12} is required for production of propionate from succinate (Chen and Wolin, 1981). Growth is not affected by 10% rumen fluid or 0.1% Tween 80.

The optimum temperature for growth is 37°C. There is good growth at 30 and 45°C, moderate growth at 25°C.

Neutral red is reduced. Strains are negative for acetylmethylcarbinol (Voges-Proskauer reaction) and growth in 6.5% NaCl. Catalase production is variable among strains (Wilkins et al., 1978).

Acid is produced from amylopectin, amylose, galacturonate, glucosamine, glucuronate, heparin and xylan. No acid from alginate, chondroitin sulfate, dextran, fucose, guar gum, gum tragacanth, hyaluronate, laminarin, larch arabinogalactan, locust bean gum, ovomucoid, pectin, polygalacturonate and porcine gastric mucin (Salyers et al., 1977a).

Strains usually are susceptible (in vitro) to chloramphenicol (12 μg/ml), clindamycin (1.6 μg/ml) and penicillin G (2 U/ml). Some strains are resistant to 6 μg/ml of tetracycline.

Other characteristics are given in Table 6.1 (species 8).

Isolated from human feces and occasionally from clinical specimens.

The mol% G + C content of DNA is 44–46.

Type strain: ATCC 27754 (NCTC 11155).

Further comments. *B. eggerthii* and *B. uniformis* are 30 to 35% interrelated by DNA/DNA hybridization studies. *B. eggerthii* is related

at ~50%, to another group of indole-positive intestinal bacteroides designated "subsp. a" (Johnson, 1978). Fermentation of sucrose (positive for *B. uniformis* and "subsp. a," negative for *B. eggerthii*) and salicin and melibiose (positive for *B. uniformis*, negative for *B. eggerthii* and "subsp. a") are helpful in differentiating these genetically distinct entities (Johnson and Ault, 1978).

9. Bacteroides splanchnicus Werner, Rintelen and Kunštek-Santos 1975, 133.[AL]

splanch′ni.cus. Gr. pl. n. *splanchna* the "innards"; M.L. masc. adj. *splanchnicus* pertaining to the internal organs (referring to the source of isolation).

The description is from Werner et al. and study of the type and 11 other strains.

Cells from glucose broth cultures are 0.7 by 1–5.0 μm.

Figure 6.9. *Bacteroides splanchnicus.*

Colonies on anaerobic blood agar plates are punctiform to 1 mm, circular to slightly irregular, entire, convex, translucent, opalescent to yellowish, smooth and shiny. No internal structure is visible in obliquely transmitted light. Surface colonies in roll-streak tubes are similar but larger (1–2 mm) and exhibit a mosaic internal structure in obliquely transmitted light.

Glucose broth cultures are turbid with a smooth sediment and a final pH of 4.6–5.0.

Hemin is highly stimulatory, but not required. Without hemin, malate, acetate and hydrogen are produced; with hemin, succinate, acetate, butyrate and decreased amounts of hydrogen are produced. Presence of vitamin B_{12} results in propionate production from succinate and decreased butyrate production (Chen and Wolin, 1981).

Produces glutamic acid decarboxylase. Butyrate is produced from lysine.

Cells contain sphingolipids (Miyagawa et al., 1978).

All strains tested are resistant to aminoglycosides and polymyxins [minimum inhibitory concentration (MIC) values greater than 60 μg/ml] and susceptible to tetracyclines, lincomycin, clindamycin, rifampicin and erythromycin (MIC values less than 0.5 μg/ml). Chloramphenicol, penicillins, and cephalosporins show bacteriostatic activity at 5–40 μg/ml.

Other characteristics are given in Table 6.1 (species 9).

Isolated from human feces and vagina and occasionally from abdominal infections.

The mol% G + C of the DNA has not been reported.

Type strain: ATCC 29572 (NCTC 10825).

10. Bacteroides zoogleoformans *corrig.* (Weinberg, Nativelle and Prévot 1937) Cato, Kelley, Moore and Holdeman 1982, 271.[VP]* [*Bacterium zoogleiformans* (sic) Weinberg, Nativelle and Prévot 1937, 725; *Capsularis zoogleiformans* (sic) Prévot 1938, 293.]

zo.o.gle′o.for′mans. Gr. adj. *zoos* alive, living; Gr. masc. noun *gloios* gum, glue; M.L. fem. n. *zoogleoea* living glue; L. part. adj. *formans* forming; M.L. part. adj. *zoogleoformans* forming zoogloea (pertaining to the glutinous mass produced in broth cultures).

This description is from our study of the type and nine other isolates.

Cells in glucose broth are 0.6–1.0 by 0.8–8.0 μm, have rounded ends, and may stain irregularly. They usually occur singly, occasionally in pairs. No capsules are detected. Cells form a viscous, glutinous mass in broth cultures and produce distinctive towers of "zooglial" slime on meat particles.

---

*VP denotes that this name has been validly published in the official publication, International Journal of Systematic Bacteriology.

Figure 6.10. *Bacteroides zoogleoformans.*

Surface colonies on blood agar plates are 0.5–2 mm, circular, entire, pulvinate, opaque, buff-colored, shiny, smooth and butyrous.

Hemin is required for growth. No growth at 25°C, variable growth at 45°C. In prereduced media, the most reliable fermentation reactions are obtained when 10% (v/v) sterile serum is added and cultures are inoculated and incubated in a gaseous atmosphere containing 90% N₂ and only 10% CO₂ (i.e. pH of medium is 6.9–7.0 until fermentation occurs).

Other characteristics are given in Table 6.1 (species 10).

Isolated from the gingival sulcus of man.

The mol% G + C of DNA from the type strain is 47.

Type strain: ATCC 33285.

Further comments. All strains tested, except the type strain, produce indole. The typical slimy sediment, particularly in chopped meat medium, is the most distinctive characteristic of this species. Some strains that were isolated from bovine rumen contents and were identified as *B. ruminicola* also produced a viscous mass in broth cultures (Bryant et al., 1958). Although the biochemical reactions of *B. zoogleoformans* are very similar to those of several other *Bacteroides* species, the polyacrylamide gel electrophoresis patterns of soluble cellular proteins are quite distinctive (Cato et al., 1982).

11. **Bacteroides oris** Holdeman, Moore, Churn, and Johnson 1982, 126.*VP*

or′is. L. gen. n. *oris* of the mouth (referring to a major natural habitat of the species).

The description is based on the study of the type and 19 other strains with high DNA homology with the type strain.

Cells in glucose broth are 0.5–1.2 by 1.0–8.0 μm and are arranged singly, in pairs and in short chains.

Figure 6.11. *Bacteroides oris.*

Surface colonies on blood agar plates are 0.5 to 1 mm in diameter, circular with an entire edge, low convex, translucent to semiopaque, white-to-buff, shiny and smooth. There is no hemolysis of rabbit blood.

Glucose broth cultures are uniformly turbid with a smooth sediment. The terminal pH in media containing a fermentable carbohydrate ranges from 4.5 to 4.9 with carbohydrates fermented by all strains and from 5.0 to 5.6 in media with carbohydrates that were not fermented by all strains (Table 6.1). Most strains grow well at 45°C; growth at 25 and 30°C is variable among strains.

Hemin is highly stimulatory for growth of the type strain and is required for growth of some strains.

Produces phosphatase and α-glucosidase but not β-glucosidase (unpublished data).

Half the strains were resistant to 2 U/ml of penicillin G and 20% of the strains were resistant to 6 μg/ml of tetracycline. All strains were susceptible to chloramphenicol (12 μg/ml), clindamycin (1.6 μg/ml) and erythromycin (3 μg/ml).

Other characteristics of the species are given in Table 6.1 (species 11).

Isolated from the gingival crevice, systemic human infections, and

the large intestine of chickens. Nonoral human clinical isolates have been from face, neck and chest abscesses and drainages; abdominal wound drainages and peritoneal fluid; blood; and spinal fluid.

The mol% G + C of the DNA is 42–46.

Type strain: ATCC 33573.

Further Comments. [Also see Further Comments for *B. ruminicola* (species 21).] Production of β-glucosidase by strains of *B. buccae* and not by strains of *B. oris* help differentiate these two species. Also, 8 of 9 strains of *B. oris* reduced neutral red, whereas all 6 strains of *B. buccae* tested were negative.

12. **Bacteroides buccae** Holdeman, Moore, Churn, and Johnson 1982, 128.*VP*

buc′cae. L. gen. n. *buccae* of the mouth (referring to a major natural habitat of the species).

The description is based on the study of the type and 12 other strains that have high DNA homology with the type strain.

Cells in glucose broth are 0.5–0.7 by 0.8–8.0 μm, occurring as single cells, in pairs or occasionally in short chains.

Figure 6.12. *Bacteroides buccae.*

Surface colonies on blood agar plates are 0.5–1.0 mm in diameter, circular with an entire edge, low convex, translucent to semiopaque, gray-to-white or buff, shiny and smooth. There is no hemolysis of rabbit blood.

Glucose broth cultures are uniformly turbid with a smooth sediment. The terminal pH in media containing a fermentable carbohydrate generally ranges from 4.5 to 5.0 in media with carbohydrates fermented by all strains and from 5.1 to 5.6 in media with carbohydrates that were not fermented by all strains (Table 6.1). Most strains grow well at 30 and 45°C; growth at 25°C is variable among strains.

Hemin is required for growth of the type strain.

Produces phosphatase, α-glucosidase, β-glucosidase and leucyl-glycine deaminase (unpublished data).

The type strain and most of the other strains are susceptible to chloramphenicol (12 μg/ml), clindamycin (1.6 μg/ml), erythromycin (3 μg/ml), penicillin G (2 U/ml) and tetracycline (6 μg/ml). One strain was resistant to 2 U/ml of penicillin G and another strain was resistant to 6 μg/ml of tetracycline.

Other characteristics of the species are given in Table 6.1 (species 12). Also see Further Comments for *B. oris* (species 11) and *B. ruminicola* (species 21).

Isolated from the gingival crevice and from human clinical specimens, including chest drainages, blood, sinus aspirate (sinusitis), peritoneal fluid and a mandibular cyst.

The mol% G + C of the DNA is 50 to 52.

Type strain: ATCC 33574.

Further comments. Production of β-glucosidase and leucyl-glycine deaminase by *B. buccae* and not by *B. oris* help differentiate these two species.

The name *Bacteroides capillus* was proposed (Kornman and Holt, 1981, 1982) for a group of saccharoclastic bacteroides with characteristics identical with those of *B. buccae*. Johnson found that Kornman strains 938.11 and 1363.01 have 87% DNA homology with the type strain of *B. buccae*. Because no type strain was designated for *B. capillus* by Kornman and Holt and the species had not been validated in the *International Journal of Systematic Bacteriology* before the description of *B. buccae* was published (Holdeman et al., 1982), the name *B. buccae* has precedence. Kornman and Holt (1981) report that the cells of this species are encapsulated (as observed in India ink preparations) and that they have a surface layer external to the outer membrane. They observed hairlike projections emerging from the cell surface. The thick and copious extracellular slime appears to be a potent B-cell mitogen. Kornman and Holt also observed that the strains are extremely sensitive to exposure to oxygen (i.e. air) and to sonication for 15 sec. Collagenolytic activity and in vitro resorbing activity have been detected [C. Trummel, K. Kornman, S. Holt, and P. Robertson: *Journal of Dental Research* (Special Issue A), *60:* 641 (Abstr. 1326), 1981, as cited in Kornman and Holt, 1981].

13. Bacteroides loescheii Holdeman and Johnson 1982, 406.[VP]

loesche'i.i. L. gen. n. *loescheii* of Loesche, named after Walter J. Loesche, an American dental microbiologist.

This description is based on our study of the type strain and five strains having DNA homology with the type (*B. melaninogenicus* subsp. *melaninogenicus* homology group C of Johnson (1980)).

Cells in glucose broth are 0.4–0.6 by 0.8–15 μm, occurring singly, in pairs, and in short chains.

PY   PYG

Figure 6.13. *Bacteroides loescheii.*

Surface colonies on anaerobic blood agar plates are 1.0–2.0 mm, circular, entire, low convex, translucent, shiny, and smooth. After incubation for 48 h on plates containing whole blood, colonies are white or buff. Upon continued anaerobic incubation (up to 14 days), colonies may become brown (often only light brown). Although pigment develops more rapidly on media containing hemolyzed blood, definite dark brown or black colonies may not develop with some strains, even if cultures are incubated for 21 days. Most strains produce slight clearing or β-hemolysis on rabbit blood agar.

Glucose broth cultures are turbid with a smooth sediment and a final pH of 4.9–5.4.

Growth is completely or markedly inhibited by 20% bile. Hemin is required or highly stimulatory for growth. Even in the presence of hemin, serum (10% v/v) often enhances growth and fermentation.

No H_2S production is detected in SIM (sulfide-indole-motility medium, BBL). Dextrin is fermented by 4 of 6 strains, inulin by 2 of 6. Fermentation of gum arabic, larch arabinogalactan, and xylan is variable among strains. The 6 strains tested are susceptible to chloramphenicol (12 μg/ml), clindamycin (1.6 μg/ml), penicillin G (2 U/ml) and tetracycline (6 μg/ml).

Sphingolipids are produced by the type strain (referred to as "*B. oralis* ATCC 15930") (Miyagawa et al., 1978). Extracellular phospholipase A is produced (Bulkacz et al., 1979).

Other characteristics of the species are given in Table 6.1 (species 13). Also see Further Comments for *B. melaninogenicus* (species 16).

Hydrolysis of esculin and fermentation of cellobiose are helpful in differentiating *B. loescheii*, *B. melaninogenicus*, and *B. denticola*.

Isolated from the gingival crevice of humans.

The mol% G + C of the DNA is 46 [Williams et al., 1975 (for VPI 9085 = ATCC 15930)] and 47 to 48 [Johnson, 1980 (for *B. melaninogenicus* subsp. *melaninogenicus* homology group C)].

Type strain: ATCC 15930 (NCTC 11321, Loesche 8B, VPI 0037, VPI 9085).

Further comments. The type strain (ATCC 15930) is Loesche's strain 8B, originally deposited in ATCC as a brown (B) variant of *B. oralis*. Johnson found (Holdeman and Johnson, 1982) that this strain has no homology with either VPI 2381 (*B. melaninogenicus*, ATCC 25845, type strain) or VPI 10043 (*B. denticola*, ATCC 33185). The current classification of some commonly used reference strains of pigmenting

bacteroides is given in Table 6.5. Also, see Further Comments for *B. oralis* (species 14).

14. Bacteroides oralis Loesche, Socransky and Gibbons 1964, 1334.[AL] [*Ristella oralis* (Loesche et al. 1964) Prevot et al. 1967, 264.]

o.ra'lis. M.L. adj. *oralis* of the mouth.

The description is based on that of Loesche et al. and study of the type strain and two other strains [ATCC 33321 (VPI 5832) and ATCC 33322 (VPI 9958)] that have high DNA homology with the type strain (T. Mitsuoka, personal communication).

Cells from glucose broth are 0.5–0.8 by 1.0–5.0 μm, arranged in pairs and chains.

PY   PYG

Figure 6.14. *Bacteroides oralis.*

Surface colonies on hemolyzed rabbit blood agar are 1.0 mm, circular, entire, convex, shiny, smooth, semiopaque and buff. Strains are nonhemolytic on horse blood infusion agar (Loesche et al., 1964).

In 24 h, glucose broth cultures are moderately turbid with a smooth sediment and pH of 4.5–4.9. Neither hemin nor vitamin K is required for growth of the type strain. Strains grow in up to 3% oxygen and survive exposure to air for up to 80 min (Loesche, 1969).

Gelatin usually is completely liquefied after incubation for 3 weeks. No decarboxylase for L-lysine, L-ornithine and L-arginine. No H_2S is detected in peptone iron agar or in semisolid agar with 0.02% ferrous sulfate and 0.03% sodium hyposulfite, but there is slight blackening of lead acetate paper suspended over cultures in trypticase broth (Loesche et al., 1964).

No heptose or KDO is detected (Hofstad, 1974).

Other characteristics of the species are given in Table 6.1 (species 14).

Isolated from the gingival crevice area of man and from infections, usually of the oral cavity and upper respiratory and genital tracts.

The mol% G + C of the DNA is 43 (Watabe et al., 1983).

Type strain: ATCC 33269 (VPI D27B-24)

Further comments. Since its description, numerous types of organisms (generally saccharoclastic with growth inhibited by bile) have been identified as *B. oralis*. Not all investigators determined that strains they identified as *B. oralis* fermented cellobiose and did not ferment arabinose and xylose (characteristics consistent with those of the original description of the species). Some strains previously referred to as *B. oralis* in the literature are *B. oris* or *B. buccae* (unpublished results). The fact that the type strain originally deposited by Loesche et al. was lost was unfortunate, particularly since another strain that they deposited (ATCC 15930, strain 8B of Loesche et al.) was available from ATCC. Concurrent with the interest in bacteroides in the last two decades and with improved methods for culture of the organisms, many investigators independently observed that ATCC 15930 produced colonies with brown or black pigment on blood agar, particularly if the blood was hemolyzed (Shah et al., 1976; Holbrook et al., 1977). Because pigment was observed in this saccharoclastic organism, ATCC 15930 was reidentified (by W.E.C.M., L.V.H.) as a strain of *B. melaninogenicus* subsp. *melaninogenicus*. It has been referred to as both *B. oralis* and *B. melaninogenicus* in the literature. Recent studies have shown that ATCC 15930 has no DNA homology with the type strain of either *B. melaninogenicus* or *B. intermedius* and this strain has been designated the type strain of *B. loescheii* [Holdeman and Johnson, 1982]. Also see "Further Comments" for *B. loescheii* (species 13).

Indeed, there was so much confusion concerning the identification of *B. oralis* that the International Taxonomic Subcommittee on Anaerobic Gram-negative Rods was unable to select a type strain to represent

Table 6.5.

Current classification of some commonly used reference strains of pigment-forming bacteroides previously classified as **Bacteroides melaninogenicus**[a]

Published Strain Numbers[b]	Identification or Classification		Cited in[d]
	Present[c]	Previous	
ATCC 25845[e] (VPI 2381, WAL B282)	*B. melaninogenicus*	*B. melaninogenicus* subsp. *melaninogenicus*	1, 5, 9, 12, 13, 15, 18, 19, 20
DWL 138-71, serogr. A (VPI 9847)	*B. melaninogenicus*	*B. melaninogenicus* subsp. *melaninogenicus*	6, 15
VPI 9343 (Lambe 58-71A, serogr. A)	*B. melaninogenicus* "9343 subgroup"	*B. melaninogenicus* subsp. *melaninogenicus*	15, 18, 19, 20
WAL 2728 (VPI 9343)	*B. melaninogenicus* "9343 subgroup"	*B. melaninogenicus* subsp. *melaninogenicus*	15, 16, 18, 19, 20
ATCC 33185 (VPI 10043, DWL 1282-74B)	*B. denticola*	*B. melaninogenicus* subsp. *melaninogenicus*	5
VPI 7570A	*B. denticola*	*B. melaninogenicus* subsp. *melaninogenicus, B. oralis*	2, 3
Socransky 287 (VPI 12530)	*B. denticola*	*B. melaninogenicus*	5
ATCC 15930[e] (VPI 9085, Loesche 8B)	*B. loescheii*	*B. melaninogenicus* subsp. *melaninogenicus, B. oralis*	1, 2, 9, 13, 15, 16, 18, 20, 22, 24
Socransky 379	*B. loescheii*	*B. melaninogenicus* subsp. *melaninogenicus*	9
ATCC 25611[e] (VPI 4197, WAL B422)	*B. intermedius*	*B. melaninogenicus* subsp. *intermedius*	1, 5, 9, 12, 13, 20
VPI 4200 (WAL B485)	*B. intermedius*	*B. melaninogenicus* subsp. *intermedius*	1
NCTC 9336 (VPI 8944; DWL 729-74, serogr. C)	*B. intermedius* "8944 subgroup"	*B. melaninogenicus*	1, 3, 5, 7, 15, 18, 19, 20, 22
NCTC 9338 (DWL 206-74)	*B. intermedius* "8944 subgroup"	*B. melaninogenicus* subsp. *intermedius*	1, 3, 6, 15, 18, 22
ATCC 25261 (VPI 4203; WAL B515; DWL 724-74, serogr. C)	*B. intermedius* "8944 subgroup"	*B. melaninogenicus* subsp. *intermedius*	7, 12, 15, 20, 23
Williams T584 (DWL 731-74, serogr. C; VPI 9145)	*B. intermedius* "8944 subgroup"	*B. melaninogenicus* subsp. *intermedius*	7, 12, 15
T588 (Lambe 733-74, serogr. C)	*B. intermedius* "8944 subgroup"	*B. melaninogenicus* subsp. subsp. *intermedius*	7, 15, 16, 18, 19, 20, 22
DWL 297-74G, serogr. C	*B. intermedius* "8944 subgroup"	*B. melaninogenicus* subsp. *intermedius*	7
G11a-d2 (Carlsson)	*B. intermedius* "8944 subgroup"	*B. melaninogenicus* subsp. *intermedius*	13, 15
LH-107	*B. intermedius* "8944 subgroup"	*B. melaninogenicus* subsp. *intermedius*	15
JP1	*B. intermedius* "8944 subgroup"	*B. melaninogenicus* subsp. *intermedius*	15
ATCC 33547[e] (VPI 9342; DWL 532-70A, serogr. C-1)	*B. corporis*	*B. melaninogenicus* subsp. *intermedius*	5, 7, 13, 18, 20
DWL 728-74, serogr. C-1 (VPI 8667)	*B. corporis*	*B. melaninogenicus* subsp. *intermedius*	7
ATCC 29147[e] (LEV strain, VPI 3300, VPI 10450)	*B. levii*	*B. melaninogenicus, B. melaninogenicus* subsp. *levii*	4, 8. 12, 13, 14, 18, 19
JP-2	*B. levii*	*B. melaninogenicus* subsp. *levii*	12, 18, 19
ATCC 25260[e] (VPI 4198; WAL B440; DWL 208-74, serogr. B)	*B. asaccharolyticus*	*B. melaninogenicus* subsp. *asaccharolyticus*	1, 3, 6, 9, 12, 13, 15, 18, 19, 20
NCTC 9337 (VPI 8945)	*B. asaccharolyticus*	*B. melaninogenicus* subsp. *asaccharolyticus*	1, 3, 15, 18, 19, 20, 22
B536	*B. asaccharolyticus*	*B. melaninogenicus* subsp. *asaccharolyticus*	1, 13, 18, 19, 20, 22, 24
CR₂a (VPI 3280)	*B. asaccharolyticus*	*B. melaninogenicus* subsp. *asaccharolyticus*	14, 21
WAL B477 (VPI 4199)	*B. asaccharolyticus*	*B. melaninogenicus* subsp. *asaccharolyticus*	14, 18, 19, 20, 22

Table 6.5—*continued*

Published Strain Numbers[b]	Identification or Classification		Cited in[d]
	Present[c]	Previous	
ATCC 27067 (VPI 5834)	*B. asaccharolyticus*	*B. melaninogenicus* subsp. *asaccharolyticus*	9, 12, 17, 20
DWL 407-71I, serogr. B	*B. asaccharolyticus*	*B. melaninogenicus* subsp. *asaccharolyticus*	6
Socransky 376 (VPI 12520)	*B. gingivalis*	*B. asaccharolyticus*	5, 10, 11, 20
Socransky 381	*B. gingivalis*	*B. asaccharolyticus*	9, 10, 12, 13, 15, 20
Socransky 382	*B. gingivalis*	*B. asaccharolyticus*	9, 10, 11, 12, 20
Werner W50	*B. gingivalis*	*B. asaccharolyticus*	12, 13, 15, 18, 19, 20
Werner W83	*B. gingivalis*	*B. asaccharolyticus*	11, 13, 15, 16, 18, 19, 20

[a] Based on Johnson (1980), Van Steenbergen (1981), Holdeman and Johnson (1982) and Johnson and Holdeman (1983).

[b] ATCC, American Type Culture Collection; Carlsson strain from J. Carlsson, University of Umeå, Umeå, Sweden; DWL, Dwight W. Lambe, Jr., East Tennessee State University; NCTC, National Collection of Type Cultures; VPI, Virginia Polytechnic Institute and State University Anaerobe Laboratory; WAL, Wadsworth VA Hospital Anaerobe Laboratory (usually received from S. M. Finegold or Vera Sutter); Loesche strain is from Walter J. Loesche, University of Michigan; Socransky strains are from S. S. Socransky, Forsyth Dental Center, Boston.

[c] Strains designated *B. melaninogenicus* have high DNA homology with the type strain (ATCC 25845, VPI 2381) and ~45% DNA homology with strains in the "9343 subgroup" [see *Further Comments* for *B. melaninogenicus* (species 16)]. Strains designated *B. intermedius* have high DNA homology with the type strain (ATCC 25611, VPI 4197) and 35 to 45% DNA homology with strains in the "8944 subgroup" [see *Further Comments* for *B. intermedius* (species 20)].

[d] 1, Finegold and Barnes (1977); 2, Hofstad (1974); 3, Holbrook et al. (1977); 4, Holdeman et al. (1977); 5, Johnson (1980); 6, Lambe (1974); 7, Lambe and Jerris (1976); 8, Lev (1980); 9, Listgarten and Lai (1979); 10, Mansheim and Coleman (1980); 11, Mansheim and Kasper (1977); 12, Mayberry et al. (1982); 13, Mouton et al. (1980); 14, Reddy and Bryant (1977); 15, Shah et al. (1976); 16, Shah et al. (1979); 17, Slots and Genco (1979); 18, Swindlehurst et al. (1977); 19, Van Steenbergen et al. (1979); 20, Van Steenbergen (1981); 21, Wahren and Gibbons (1970); 22, Williams et al. (1975); 23, Wong et al. (1977); 24, Woo et al. (1979).

[e] Type strain of the species.

the species when they met in Munich in 1978. When we (W.E.C.M., L.V.H.) were contributing our suggestions for species names and strains to be included in the Approved Lists of Bacterial Names we selected an oral isolate (VPI D27B-24, later deposited as ATCC 33269) with characteristics consistent with those originally described by Loesche et al. to represent the species and submitted that strain number as a "reference" strain of *B. oralis*. All suggested reference strains were designated type strains in the 1980 list. As a result, this strain automatically became the type of *B. oralis*. Current classification of some commonly used reference strains originally identified as *B. oralis* is given in Table 6.5.

Because of the unfortunate confusion concerning the identification of *B. oralis*, much of the previous literature concerning this species must be read very critically to determine if the authors used identification tests that would differentiate this species from taxa that we now know to be genetically distinct. The characteristics listed below have been reported for organisms previously identified as *B. oralis* and may pertain to *B. oralis* or to species now recognized as *B. oris* or *B. buccae*.

No glutamic acid is produced (Werner, 1970b). Urea is not hydrolyzed. No superoxide dismutase activity is evident (Gregory et al., 1978). Hydrolysis of dextran varies among strains (Holbrook and McMillan, 1977). A cell-bound β-lactamase is produced (Salyers et al., 1977b).

15. Bacteroides denticola Shah and Collins 1982, 266[VP] (Effective publication: Shah and Collins 1981, 235).

den.ti′co.la. L. masc. n. *dens, dentis* tooth; L. v. suff. *cola* from L. v. *colo* to dwell; M.L. noun *denticola* tooth dweller.

Unless otherwise designated, this description is based on our study of 21 strains [including strain 1221 (Shah and Collins, 1981)] that have homology with ATCC 33185 (Holdeman and Johnson, 1982).

Cells from glucose broth cultures are 0.5–0.7 by 0.7–6.0 μm and usually occur in pairs and short chains.

Figure 6.15. *Bacteroides denticola.*

Surface colonies on blood agar are 1.0–2.0 mm in diameter, circular, entire, low convex, semiopaque to opaque, shiny, smooth, and often appearing to have white or buff "concentric rings." Pigmented colonies develop more rapidly, and to a darker color, on rabbit than on sheep blood. Most strains have dark brown or black colonies within 7 days on medium containing hemolyzed rabbit blood. Two of 20 strains never developed colonies with definite pigment.

Glucose broth cultures are turbid with a smooth sediment and pH of 4.2–4.8.

Hemin is required or highly stimulatory for growth of most strains; vitamin K generally is not required. Growth is completely or markedly inhibited in 20% bile.

The type and strain 1221 have malate dehydrogenase, glutamate dehydrogenase, β-galactosidase and β-glucosaminidase. Glucose-6-phosphate and 6-phosphogluconate dehydrogenases were absent.

Sphingolipids were present. The long chain fatty acid composition is mainly *anteiso*- and *iso*-methyl branched acids with small amounts of straight chain acids. The major fatty acid is 12-methyltetradecanoic acid. The principle respiratory quinones are unsaturated menaquinones with 11 and 12 isoprene units (Shah and Collins, 1981).

Fifteen strains tested were susceptible to chloramphenicol (12 μg/ml), clindamycin (1.6 μg/ml), penicillin G (2 U/ml) and tetracycline (6 μg/ml). The type and four other strains were resistant to 2 U/ml of penicillin G and one of these strains also was resistant to 6 μg/ml of tetracycline.

No heptose or KDO was detected in VPI 7570A (referred to as "*B. oralis*") (Hofstad, 1974).

Although Shah and Collins (1981) report that gelatin is not liquefied by the type strain or strain 1221, all strains we tested (including 1221) liquefied gelatin.

Other characteristics of the species are given in Table 6.1 (species 15). Also see "Further Comments" for *B. melaninogenicus* (species 16) and Table 6.5.

Tests for hydrolysis of esculin and fermentation of cellobiose are helpful in differentiating *B. denticola* from *B. melaninogenicus* and *B. loescheii*.

Isolated from the gingival crevice and human clinical specimens.

The mol% G + C of the DNA is 49 to 51.

Type strain: NCDO 2352, (Socransky 1210).

16. **Bacteroides melaninogenicus** (Oliver and Wherry 1921) Roy and Kelly 1939, 569.[AL] [*Bacterium melaninogenicum* Oliver and Wherry 1921, 341; *Bacteroides melaninogenicus* subsp. *melaninogenicus* (Oliver and Wherry 1921) Roy and Kelly 1939, 569.]

me.la.ni.no.ge'ni.cus. Gr. adj. *melas* black; M.L. n. *melaninum* melanin; M.L. adj. *genicus* producing, probably derived from Gr. n. *genetes* a producer; M.L. adj. *melaninogenicus* melanin producing. [*Note:* The black pigment produced by this organism is due to protohemin (Shah et al., 1979) and not to melanin, as originally thought.]

This description is based on our study of the type and 11 other strains having DNA homology with the type strain [*B. melaninogenicus* subsp. *melaninogenicus* homology group A of Johnson (1980)], supplemented with information from Shah et al. (1976) and Holbrook et al. (1977).

Cells from glucose broth cultures are 0.5–0.8 by 0.9–2.5 μm with occasional cells of 10 μm or longer.

Figure 6.16. *Bacteroides melaninogenicus.*

Surface colonies on blood agar are 0.5–2.0 mm, circular, entire, convex, and shiny. They are usually darker in the center of the colony; edges are gray to light brown. Colonies become darker upon continued incubation (5–14 days). Dark pigment usually develops more rapidly when laked blood, rather than blood containing whole red blood cells, is used. Some strains do not produce colonies with black pigment on horse blood agar but pigment well on agar with rabbit blood. A few strains are β-hemolytic on rabbit blood agar.

Glucose broth cultures are usually turbid with smooth or stringy sediment and a pH of 4.6–5.0.

Hemin (1 μg/ml), menadione (0.1 μg/ml), or both are required for or enhance growth of most strains. Most strains grow at pH 8.5 and 25°C; some grow at 45°C.

Cell wall peptidoglycan may contain either lysine or diaminopimelic acid. No heptose or KDO is detected (Hofstad, 1974). Cells contain sphingolipids (Miyagawa et al., 1978) and have superoxide dismutase activity (Gregory et al., 1978).

Other characteristics of the species are given in Table 6.1 (species 16).

Isolated from the gingival crevice and from human clinical specimens.

The mol% G + C of the DNA is 36–40 (Shah et al., 1976; Johnson, 1980).

Type strain: ATCC 25845 (Finegold B282).

Further comments. Historically, the predominant differentiating characteristic of *B. melaninogenicus* was its production of pigmented colonies on a blood-containing medium. Sawyer et al. (1962) grouped 31 strains of *B. melaninogenicus* as (a) strongly fermentative, (b) weakly fermentative and (c) nonfermentative. Holdeman and Moore (1970) divided this phenotypically heterogeneous species into three subspecies that corresponded, in general, to the groups of Sawyer et al. On the basis of studies of the LEV strain (ATCC 29147) of pigmenting bacteroides isolated from the bovine rumen by M. Lev (1958), Holdeman et al. (1977) proposed another taxon, *B. melaninogenicus* subsp. *levii*. This is recognized as *B. levii* in this edition of the *Manual. B. melaninogenicus* subsp. *asaccharolyticus* was elevated to species rank by Finegold and Barnes (1977) because it differed from the other subspecies of *B. melaninogenicus* in many phenotypic characteristics and in the G + C content of the DNA (51–52 mol%). A second species, *B. gingivalis*, was proposed by Coykendall et al. (1980) for nonfermentative pigmenting bacteroides isolated principally from the human gingival sulcus and differing from *B. asaccharolyticus* by the G + C content of the DNA (46.5–48.5 mol%), production of phenylacetic acid, and hemagglutination of sheep erythrocytes (Slots and Genco, 1979). *B. macacae* (Slots and Genco) Coykendall 1980, 563 was proposed to include weakly fermentative, catalase-positive, pigmenting bacteroides isolated from periodontal pockets of the monkey, *Macaca arctoides*. The G + C content of the DNA is 42–43 mol%. There is little or no DNA homology among strains of *B. asaccharolyticus*, *B. gingivalis*, and *B. macacae* (Coykendall et al., 1980).

Recently, DNA homology studies by J. L. Johnson have shown that *B. melaninogenicus* subsp. *melaninogenicus* and subsp. *intermedius* also are genetically distinct and that there are three DNA homology groups distinct at the species level within each subspecies (Johnson, 1980). Of the three homology groups in *B. melaninogenicus* subsp. *melaninogenicus*, Johnson's homology group A is *B. melaninogenicus* and the names "*B. socranskii*" and *B. loescheii* were proposed for homology groups B and C, respectively, represented by ATCC 33185 and 15930 [L. V. Holdeman, J. L. Johnson, and W. E. C. Moore: *Journal of Dental Research* (Special Issue A), *60:* 414 (Abstr. 415), 1981]. Barnes strain 1221 was one of the strains in the "*B. socranskii*" homology group and was one of the two strains used by Shah and Collins (1981) in the description of *B. denticola*, which was validated in 1982 (Skerman et al.). *B. denticola* therefore has precedence over "*B. socranskii*."

Johnson (Holdeman and Johnson, 1982) also has recognized two related DNA homology subgroups within *B. melaninogenicus* and designated them "2381 (ATCC 25845) subgroup" and "9343 subgroup." The mol% intergroup homology is 30–50. Strains in these two subgroups are not distinguishable by usual phenotypic tests. Similarly, Johnson has recognized two subgroups ("ATCC 15930" and "DIC-20") with 40–45 mol% intergroup homology among strains of *B. loescheii*.

The status of the homology groups (A, B and C) of *B. melaninogenicus* subsp. *intermedius* (Johnson, 1980) is discussed in Further Comments for *B. intermedius* (species 20).

Because many previous studies of organisms called *B. melaninogenicus* included strains that might have been members of any of 9 presently recognized species, the earlier literature is difficult to correlate with present species designations. The present classification of some commonly studied strains referred to as *B. melaninogenicus* is given in Table 6.5.

Strains of "pigmenting bacteroides" vary in the rapidity with which they form dark (often brown, rather than black) colonies on blood agar medium. The type and lot of blood also can affect the pigmentation. Most reliable development of pigment occurs on rabbit or human blood, and pigmentation occurs more rapidly on hemolyzed (laked) blood rather than whole blood (Shah et al., 1976). With optimum conditions for pigment production, most strains will develop dark colonies within two weeks. Some strains may require incubation for three weeks, and an occasional strain will not show definite pigmentation with some lots of blood.

Meyers et al. (1969) noted that colonies of *B. melaninogenicus* fluoresced under long-wave ultraviolet light (e.g. Wood's light). The colonies fluoresce before they become dark and do not fluoresce once the dark color develops. However, when the black colonies were emulsified in methanol, red fluorescence was again apparent. The compound producing this red fluorescence has been identified as protoporphyrin (Shah et al., 1979). The fluorescence often has been used for early identification of "*B. melaninogenicus*" and is reliable only if the "brick red" fluorescence is considered as indicative of "pigmenting bacteroides." Many strains of saccharoclastic, pigmenting bacteroides give an orange or yellow fluorescence, as will some other species (e.g. *B. bivius, B. disiens*) of bacteroides (unpublished data).

On the basis of fluorescent antibody studies, Lambe (1974, 1980) and Lambe and Jerris (1976) described four serogroups of pigmenting bacteroides and designated them A, B, C and C-1. Strains of *B. melaninogenicus* subsp. *melaninogenicus* reacted with group A conjugate, strains of *B. asaccharolyticus* reacted with group B conjugate and strains of *B. melaninogenicus* subsp. *intermedius* reacted with group C or C-1 conjugate. Lambe (1974) reported that ATCC 25847 (no longer maintained in ATCC because it is the same strain as ATCC 25845, the type strain of *B. melaninogenicus*) and Lambe strain 221-72E reacted with his group A conjugate. Lambe 221-74E now is reclassified as *B. loescheii*. Mouton et al. (1980) report that VPI 9085 (ATCC 15930, now the type strain of *B. loescheii*) reacted with *B. melaninogenicus* subsp. *melaninogenicus* conjugate (Fluoretec reagent, Pfizer), but with less intense staining than observed with ATCC 25845 and other strains of *B. melaninogenicus* subsp. *melaninogenicus* tested. The serological reaction of *B. denticola* with fluorescent antibody conjugates has not been reported. Zambon et al. (1981) reported that their strains of *B. melaninogenicus* subsp. *melaninogenicus* (pigmented bacteroides that hydrolyzed esculin and were indole-negative, indicating either *B. denticola* or *B. loescheii*) reacted with specific antisera. They did not report the source of the antiserum or test the antisera with the type strain or with any other strain now known to be *B. melaninogenicus*.

B. asaccharolyticus and *B. gingivalis* are serologically distinct (Mouton et al., 1980).

17. **Bacteroides bivius** Holdeman and Johnson 1977, 341.[AL]

bi'vi.us. L. adj. *bivius* having two ways, pertaining to the saccharoclastic and proteolytic activities of the species.

This description is from our study of the type strain and of 13 other strains that show DNA homology with the type.

Cells are 0.7–1.0 by 1.0–4.5 μm and occur in pairs or short chains. In some cultures, cells may be mistaken for Gram-negative cocci.

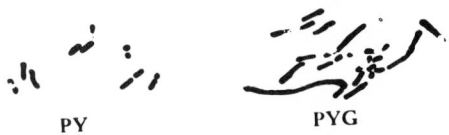

PY   PYG

Figure 6.17. *Bacteroides bivius.*

Colonies 0.5–2.0 mm, circular, entire to slightly erose, convex, translucent to semiopaque, smooth and glistening. Colonies of some strains produce slight greening of blood agar and show light orange to pink fluorescence under longwave ultraviolet light.

Cell walls contain *meso*-diaminopimelic acid (Hammann and Werner, 1981).

PY broth cultures are turbid with smooth sediment; growth is increased slightly with a fermentable carbohydrate. Final pH in glucose broth is 4.7–5.0. Final pH in PY broth frequently is 5.8–6.0.

Growth is inhibited by 6.5% NaCl. Hemin is required for growth. Casein is digested within 48 h. Most strains ferment glycerol.

No superoxide dismutase activity is detected (Gregory et al., 1978).

All strains tested were susceptible to chloramphenicol (12 μg/ml), clindamycin (1.6 μg/ml) and erythromycin (3 μg/ml). Ten of 14 strains were susceptible to tetracycline (6 μg/ml); 4 of 14 were sensitive to penicillin (2 U/ml).

All strains were inhibited by carbenicillin (32 μg/ml) and clindamycin (0.25 μg/ml). Variable resistance to cephalothin, cefazolin, cefaclor, penicillin, ampicillin, and moxalactam due to the production of a β-lactamase by some strains (Kirby et al., 1980; Bourgault and Rosenblatt, 1979).

Other characteristics of the species are given in Table 6.1 (species 17).

Usually isolated from infections in the urogenital or abdominal region. Found as normal flora of the vagina in 33% of women tested (G. B. Hill, S. A. Gall, A. P. Kohan and O. M. Ayers: Abstr. Annu. Mtg. Amer. Soc. Microbiol., 1978, C32). Occasionally isolated from the mouth, blood, chest fluid, and breast abscesses.

The mol% G + C of DNA is 40.

Type strain: ATCC 29303 (NCTC 11156).

18. **Bacteroides disiens** Holdeman and Johnson 1977, 340.[AL]

di'si.ens. L. part. adj. *disiens* going in two directions; intended to refer to the fact that the organism is both saccharoclastic and proteolytic.

This description is based on our study of the type and 10 other strains having DNA homology with the type.

Cells are 0.6–0.9 by 2.0–7.0 μm and occur in pairs, or occasionally in short chains. Longer rods occur both singly and in chains with the short rods.

PY   PYG

Figure 6.18. *Bacteroides disiens.*

Colonies are minute to 2 mm in diameter, circular, entire, convex, translucent to opaque, smooth, shiny and white. There is no hemolysis on blood agar; occasionally a slight greening is observed in the area of confluent growth. Colonies of some strains show light orange to pink fluorescence on blood agar plates.

Cell walls contain *meso*-diaminopimelic acid (Hammann and Werner, 1981).

Peptone-yeast extract broth cultures are moderately turbid, sometimes with a slight, smooth sediment; the pH of basal medium frequently is 5.9–6.1. Growth is enhanced by fermentable carbohydrates, with smooth (sometimes granular or flocculent) sediment. The final pH in glucose broth is 4.9–5.2.

Optimum growth occurs at 37°C. Some strains grow at 25 and 45°C. Growth is inhibited by 6.5% NaCl. Hemin is required for growth.

Casein is digested within 48 h. Milk and meat are digested in 10–14 days with freshly isolated strains, in 1–3 days after several transfers. Hippurate is hydrolyzed. Produces superoxide dismutase (Gregory et al., 1978).

All strains tested were susceptible to chloramphenicol (12 μg/ml), clindamycin (1.6 μg/ml) and erythromycin (3 μg/ml). Resistance to penicillin (2 U/ml) and tetracycline (6 μg/ml) is variable.

Inhibited by clindamycin (0.125 μg/ml) and metronidazole (4 μg/ml). Relatively resistant to penicillins and cephalosporins, presumably because of β-lactamase production (Wüst and Wilkins, 1978; Kirby et al., 1980).

Other characteristics of the species are given in Table 6.1 (species 18).

Strains are phenotypically similar to *B. corporis* except that no black or brown pigmentation is observed. The species are distinct by DNA homology.

Isolated from abdominal and urogenital infections. Found as normal vaginal flora in 20% of women tested (G. B. Hill, S. A. Gall, A.P.

Kohan and O. M. Ayers: Abst. Annu. Mtg. Amer. Soc. Microbiol., 1978, C32) and in the mouth.

The mol% G + C of DNA is 40–42.

Type strain: ATCC 29426 (NCTC 11157).

19. **Bacteroides corporis** Johnson and Holdeman 1983, 19.[VP]

cor′po.ris. L. gen. n. *corporis* of the body (pertaining to its isolation from human clinical specimens).

This description is based on our study of the type and five other strains having high DNA homology with the type.

Cells in glucose broth cultures are 0.9–1.6 by 1.6–4.0 μm, occurring singly, in pairs, and in short chains. In many cultures cells are coccoid. Occasionally, long filaments (11 μm) may be observed.

Figure 6.19. *Bacteroides corporis.*

Surface colonies on anaerobic blood agar plates are minute to 1.0 mm, circular, entire, convex and (in 48–72 h) buff with brown edges. Dark brown colonies develop by 4–7 days.

Glucose broth cultures are turbid, often with a smooth or ropy sediment that tends to adhere to the bottom of the tube, and have a final pH of 4.8–5.1. Addition of 10% (v/v) serum enhances growth and fermentation of some strains. Hemin and vitamin K are required for growth.

Of six strains tested, two were resistant to penicillin G (2 U/ml), one was resistant to tetracycline (6 μg/ml) and two were resistant to both penicillin G and tetracycline. All six strains were susceptible to chloramphenicol (12 μg/ml) and clindamycin (1.6 μg/ml).

Other characteristics of the species are given in Table 6.1 (species 19). Also see Table 6.5 and Further Comments for *B. melaninogenicus* (species 16) and *B. intermedius* (species 20).

Isolated from various types of human clinical specimens.

The mol% G + C of the DNA is 43–46 (Johnson, 1980; Van Steenbergen, 1981; Johnson, and Holdeman, 1982).

Type strain: ATCC 33547 (VPI 9342, Lambe 532-70A).

20. **Bacteroides intermedius** Johnson and Holdeman 1983, 18.[VP]
(*Bacteroides melaninogenicus* subsp. *intermedius* Holdeman and Moore 1970, 33.)

in.ter.me′di.us. L. adj. *intermedius* intermediate.

This description is based on our study of the type and 38 other strains having high DNA homology with the type.

Most cells are 0.4–0.7 μm by 1.5–2.0 μm with some cells up to 12 μm long.

Figure 6.20. *Bacteroides intermedius.*

Surface colonies on blood agar are circular, entire, low convex, 0.5–2.0 mm in diameter, translucent, smooth and hemolytic. Older or larger colonies may be opaque. After anaerobic incubation for 48 h, colonies may be tan, gray, reddish brown or black. Pigmentation of colonies occurs more rapidly on agar containing hemolyzed blood than on agar with whole blood. On hemolyzed rabbit blood agar, about one-third of the strains have dark brown to black colonies within 2 days, another third within 7 days and most of the rest within 14 days. An occasional

strain requires 18–21 days for definite pigmentation to be noted. Almost all strains fluoresce under shortwave ultraviolet light within 2–4 days, but only a brick-red fluorescence should be interpreted as indicative of probable pigmentation (see Further Comments for *B. melaninogenicus*, species 16).

Cell walls contain *meso*-diaminopimelic acid (Hammann and Werner, 1981). Diaminopimelic acid is the only dibasic amino acid in the peptidoglycan. No heptose or KDO is detected (Hofstad, 1974).

Glucose broth cultures are turbid with smooth (sometimes ropy or slightly mucoid) sediment and final pH of 4.9–5.4.

Hemin is required for growth of most strains. Vitamin K is often highly stimulatory or required for fermentation. Growth is inhibited by 6.5% NaCl. The type strain grows well at temperatures between 25 and 45°C. In poorly buffered media (e.g., prereduced PY-sugar media; Holdeman et al., 1977), some strains grow better in an atmosphere containing 10% (rather than 100%) carbon dioxide, probably due to the lower pH resulting from the higher concentration of carbon dioxide.

Gum arabic, larch arabinogalactan and xylan are not fermented.

Produces superoxide dismutase (Gregory et al., 1978). Dextran is not hydrolyzed (Holbrook and McMillan, 1977).

Other characteristics are given in Table 6.1 (species 20).

Common inhabitant of the gingival crevice. Also isolated from human clinical specimens from head, neck, and pleural infections; occasionally isolated from blood, abdominal, and pelvic sites.

The mol% G + C of the DNA is 41–44 mol% (Williams et al., 1975; ohnson, 1980).

Type strain: ATCC 25611 (Finegold B422).

Further comments. Johnson has determined that there are two DNA homology groups, distinct at the species level, with the phenotypic characteristics of *B. intermedius* (Johnson and Holdeman, 1982). One group has high DNA homology with the type strain (ATCC 25611 = VPI 4197) and has been referred to as the "4197 subgroup" or "*B. intermedius* I." The other has high homology to ATCC 33563 (VPI 8944 = NCTC 9336) and has been referred to as the "8944 subgroup" or "*B. intermedius* II." At present, these have not been separated into different species. The mol% intergroup homology is 36 ± 4 to 42 ± 5 (Johnson, and Holdeman, 1982).

No differences were detected between the two homology subgroups using the API ZYM system (Analytab Products, Plainview, N.Y.). The type and five homologous strains from each group were strongly positive for acid and alkaline phosphatase, phosphoamidase, and α-glucosidase. Very weak reactions occurred for C4 esterase and C8 esterase lipase (unpublished data).

Aspartate, asparagine, cysteine, or serine stimulate growth and are metabolized by ATCC 25261 (8944 subgroup, Table 6.5) (Miles et al., 1976; Wong et al., 1977).

Strains of the "8944 homology subgroup" have been isolated from the gingival crevice and various types of human clinical specimens. Other collection strains known to belong to the "8944 subgroup" include ATCC 25261 (Finegold B515) and NCTC 9338. Reference strains are listed in Table 6.5.

A third homology group among strains formerly identified as *B. melaninogenicus* subsp. *intermedius* is now designated *B. corporis*.

21. **Bacteroides ruminicola** Bryant, Small, Bouma and Chu 1958, 18.[AL] [*Ruminobacter ruminicola* (Bryant et al. 1958) Prévot 1966, 121.]

ru.mi.ni′co.la. M.L. adj. *rumin-* of or relating to the rumen; L. substantive ending *-cola* inhabitant; M.L. n. *ruminicola* inhabitant of the rumen.

This description is based on that of Bryant et al. and study of Bryant strain 23 (ATCC 19189) and GA33 (ATCC 19188) unless otherwise indicated.

Cells are 0.8–1.0 by 0.8–8.0 μm. They have slightly tapered, rounded ends and often are encapsulated. Most cells are 1.2–6.0 μm long. Cells become swollen with large round inclusion bodies after 2–3 days of incubation.

Subsurface colonies in agar roll tubes are 2–3 mm in diameter and

Table 6.6.
Characteristics of subspecies and biovars of **Bacteroides ruminicola**[a]

Characteristics	21a. *B. ruminicola* subsp. *ruminicola*								21b. *B. ruminicola* subsp. *brevis*		
Biovar	1	2	3	4	5	6	7	8	1	2	3
Gelatin digestion	+	+	−	+	+	−	−	−	+	+	+
Starch hydrolysis	+	+	+	+	−	−	−	−	+	+	+
Acid production from:											
Arabinose	+	+	+	+	+	−	+	−	+	+	+
Dextrin	+	+	+	+	−	−	−	−	+	+	+
Esculin	+	+	+	+	+	+	+	−	+	+	+
Gum arabic	−	−	−	−	−	−	−	−	+	+	−
Inulin	+	+	+	+	+	−	−	−	+	+	+
Maltose	+	+	+	+	−	−	+	−	+	+	+
Salicin	+	+	+	+	+	+	+	+	+	−	+
Sucrose	+	+	+	+	+	+	−	−	+	+	+
Xylan	+	+	+	+	+	+	+	+	−	−	+
Xylose	+	+	+	+	+	+	+	+	−	−	+
Rumen fluid required[b]	+	+	+	−	+	+	+	+	−	−	−
H₂S production[c]	−	weak	−	weak	−	−	−	−	+	−	+

[a] From Bryant et al. (1958).
[b] Hemin replaces the rumen fluid requirement of the majority of strains (Caldwell et al., 1965).
[c] Medium of Bryant and Small (1956).

lenticular. Surface colonies in roll tubes are 1–2 mm, entire, smooth, convex, translucent and light tan in color.

Glucose broth cultures are turbid with slime or a slimy sediment. The final pH of glucose broth cultures is 4.6–5.7.

Growth occurs at 30°C but not at 22 or 45°C. The rumen strains tested could not be grown in anaerobe jars on the surface of horse blood agar plates with rumen fluid or egg yolk agar plates with rumen fluid (Holdeman and Moore, 1974a).

Most strains grow well in a defined medium containing glucose, CO₂, minerals, heme, B vitamins, certain volatile fatty acids, methionine and cysteine (Bryant and Robinson, 1962; Pittman and Bryant, 1964). Ammonia or peptides, but not free amino acids, serve as the main source of nitrogen. Amino acids are transported and incorporated (Stevenson, 1979). Methionine and cysteine are essential, stimulatory, or not stimulatory depending on the strain (Pittman and Bryant, 1964; Pittman et al., 1967). Heme or a related tetrapyrrole is essential to growth of most strains (Caldwell et al., 1965) but is synthesized by others. Acetate and 2-methylbutyric or isobutyric acids are highly stimulatory to growth of most strains, especially when grown in media without peptides, and the latter acids are essential for some strains (Dehority, 1966).

Propionate is formed via the acrylate pathway (Wallnöfer and Baldwin, 1967).

Contains b-type and o-type cytochromes (White et al., 1962; Reddy and Bryant, 1977). Able to use extracellular hydrogen to reduce fumarate to succinate via cytochrome b (Henderson, 1980). Ammonia is produced from peptides (Bladen et al., 1961). Glutamine decarboxylase negative (Terada et al., 1976). Contain sphingolipids (Kunsman, 1973; Kunsman and Caldwell, 1974; Miyagawa et al., 1978).

Isolated from the reticulo-rumen of cattle, sheep and elk and presumed to be among the more numerous bacteria in the rumen of most ruminants. Also isolated from the intestinal contents of chickens.

Bryant et al. divided *B. ruminicola* into two subspecies that differ in their ability to ferment xylose and xylan, their requirement for heme, and their cellular morphology (Table 6.6). *B. ruminicola* subsp. *brevis* grows slightly better than *B. ruminicola* subsp. *ruminicola* in a peptone-yeast extract medium.

21a. **Bacteroides ruminicola** subsp. **ruminicola** Bryant, Small, Bouma and Chu 1958, 18.[AL]

Cells tend to be longer than those of *B. ruminicola* subsp. *brevis*.

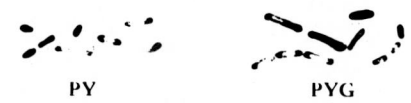

Figure 6.21a. *Bacteroides ruminicola* subsp. *ruminicola*.

Heme or a related tetrapyrrole is essential for growth. Xylan and pectin are strongly fermented. No acid from gum arabic.

Other characteristics of the subspecies are given in Table 6.1 (subspecies 21a) and Table 6.6 (*B. ruminicola* subsp. *ruminicola* biovar 1). Also see Further Comments (below).

The mol% G + C of the DNA is 49 (Reddy and Bryant, 1977).

Type strain: ATCC 19189 (Bryant 23).

21b. **Bacteroides ruminicola** subsp. **brevis** Bryant, Small, Bouma and Chu 1958, 18.[AL]

bre'vis. L. adj. *brevis* short.

Although this subspecies does not require heme or a related tetrapyrrole in the growth medium [i.e. it synthesizes heme (White et al., 1962)], heme may enhance growth of some strains. Most cells are much shorter (mainly coccoid to oval) than those of *B. ruminicola* subsp. *ruminicola*.

Figure 6.21b. *Bacteroides ruminicola* subsp. *brevis*.

Negative for fibrinolysin, elastase, hyaluronidase, and chondroitin sulfatase. Produces deoxyribonuclease and phosphatase (Rudek and Haque, 1976).

Other characteristics of the subspecies are given in Table 6.1 (subspecies 21b) and Table 6.6 (*B. ruminicola* subsp. *brevis* biovar 1). Also see Further Comments (below).

The mol% G + C of the DNA is 50 (Reddy and Bryant, 1977).

Type strain: ATCC 19188, (Bryant GA33).

Further comments. Bryant et al. (1958) described the two subspecies of *B. ruminicola.* They further differentiated *B. ruminicola* subsp. *ruminicola* into eight biovars (biotypes) and *B. ruminicola* subsp. *brevis* into three. The characters differentiating these biovars are listed in Table 6.6.

During studies of anaerobic bacteria isolated from human clinical infections and normal flora, many strains were isolated with characteristics similar to those of *B. ruminicola* subsp. *brevis* biovar 3. The description of *B. ruminicola* subsp. *brevis* biovar 3 given in the fourth edition of the VPI Anaerobe Laboratory Manual (Holdeman et al., 1977) was based, in large part, on results obtained with the human strains.

Johnson and colleagues have studied the DNA relationships among rumen, porcine, and human strains labeled *B. ruminicola.* Churn and Johnson (Annu. Mtg. Amer. Soc. Microbiol., 1980, Abstr. C40) reported no DNA homology between the type strain of *B. ruminicola* subsp. *ruminicola* and the type strain of *B. ruminicola* subsp. *brevis.* It is reasonable to assume that the two subspecies probably will be elevated to species rank. Churn and Johnson also found that other rumen strains of *B. ruminicola* had only 0–50% homology with the type strains of the two subspecies. There were not enough bovine strains representing the original biotypes to determine if the biotypes corresponded to any of the homology groups.

Human strains, most of which previously were identified as *B. ruminicola* subsp. *brevis* biovar 3, did not have any homology with the bovine strains, nor did pig strains show any homology to the human strains. About 70% of the human isolates belonged to one of two homology groups. The two main homology groups of human isolates later were designated *B. oris* and *B. buccae* (Holdeman et al., 1982). They can be differentiated from *B. ruminicola* by their ability to grow at 45°C and by the more rapid growth and much higher turbidity they produce in peptone-yeast extract broth.

Strain ATCC 27518 (VPI 7649) originally was deposited as representative of a human isolate of *B. ruminicola* subsp. *brevis* biovar 3. DNA homology studies show that it is a strain of *B. oris.*

22. **Bacteroides hypermegas** Harrison and Hansen 1963, 28.[AL] [*Sphaerophorus hypermegas* (Harrison and Hansen 1963) Prévot 1966, 145.]

hy.per.meg′as. Gr. pref. *hyper* excessive; Gr. adj. *megas* great; M.L. adj. *hypermegas* excessively great, referring to the large cells.

This description is from Harrison and Hansen (1963), Goldberg et al. (1964), Mitsuoka et al. (1974), Cato and Barnes (1976) and study of the type and 5 other strains.

Cells are 2.0–3.0 by 5.0–11.0 μm with rounded ends and granular appearance. They contain volutin but no fat or glycogen (Harrison and Hansen). They may be encapsulated on initial isolation.

Figure 6.22.   *Bacteroides hypermegas.*

Surface colonies on horse blood agar are 1–2 mm in diameter, circular, entire, flat to convex, semiopaque, granular or mottled, gray-white and not hemolytic. Surface colonies on Reinforced Clostridial Medium Agar (BBL) are 3–5 mm in diameter, convex, gray, opaque and smooth.

Glucose broth cultures are uniformly turbid, sometimes with flocculent sediment; the final pH is 4.8.

Growth is stimulated by fermentable carbohydrates. Optimum temperature for growth is 37°C. Strains grow poorly, if at all, at temperatures below 25°C or above 45°C. Strains do not grow in media with pH below 4.8 or above 8.6.

Trace to moderate amounts of gas are produced in glucose agar deep

cultures (our laboratory); no gas produced (Harrison and Hansen, 1963). Lactate is utilized (Cato and Barnes, 1976). Growth in media with 1:100,000 brilliant green is variable. Growth is inhibited in media with NaCl concentration much above 1.5%. Glutamic acid is not decarboxylated (Werner, 1970b). Cellulose not digested. Acid produced from sorbitol and α-methylglucoside. No acid from α-methylmannoside (Mitsuoka et al., 1974).

No sphingolipids (Miyagawa et al., 1978), heptose, or KDO are detected (Hofstad, 1974).

Growth is inhibited by neomycin (25 μg/ml), kanamycin (100 μg/ml), polymyxin B (10 μg/ml), colistin (10 μg/ml) and erythromycin (50 μg/ml), but not by penicillin (10 μg/ml), bacitracin (3 U/ml) and rifampin (10 μg/ml) (Mitsuoka et al., 1974).

Other characteristics of the species are given in Table 6.1 (species 22).

Isolated from the intestinal tracts of poultry, humans, and dogs (Mitsuoka et al., 1974).

The mol% G + C of the DNA of EBF/61/42 is 35 (Barnes and Impey, 1968).

Type strain: ATCC 25560 (NCTC 10570; Barnes EBF/61/42).

Further comments. Strains phenotypically identical to *B. hypermegas,* except with much smaller cells, have been isolated from human feces (Moore and Holdeman, 1974b). These strains are probably not members of the species *B. hypermegas.*

23. **Bacteroides levii** Johnson and Holdeman 1983, 21.[VP] [*Bacteroides melaninogenicus* subsp. *levii* Holdeman, Cato, and Moore 1977, 32.)

lev′i.i. L. gen. n. *levii* of Lev, named after Meir Lev, the American-English microbiologist who first isolated this organism.

Except where otherwise noted, this description is based on our study of the LEV strain (ATCC 29147) and two phenotypically similar strains [including Garcia strain JP-2 (Swindlehurst et al., 1977)] isolated from cattle.

In glucose broth cultures, cells are 0.6–1.2 by 2.0–7.0 μm, occurring in pairs and short chains.

Figure 6.23.   *Bacteroides levii.*

Surface colonies on anaerobic blood agar plates are minute, circular, entire, and low convex. After incubation for 2–3 days colonies are buff to light brown; dark brown colonies develop after incubation for 5–7 days.

Glucose broth cultures are turbid with a smooth sediment and a final pH of 5.5.

Cell walls of the type strain contain *meso*-diaminopimelic acid (Hammann and Werner, 1981).

Vitamin K and hemin are required for growth (Lev, 1958, 1968). Succinate stimulates growth and can replace the requirement for heme. Cytochrome *b* is synthesized from the heme and is not detected in its absence (Lev et al., 1971). Glutamine stimulates growth and results in increased uptake of many amino acids and dipeptides (Lev, 1980). The cells possess superoxide dismutase activity (Gregory et al., 1978).

Further information on the biochemistry of *B. levii* can be found in Lev (1959, 1977) and Lev and Milford (1972, 1973, 1977, 1978).

Other characteristics of the species are given in Table 6.1 (species 23).

Isolated from bovine rumen, cattle horn abscess, and bovine summer mastitis.

The mol% G + C of the DNA is 48 (Reddy and Bryant, 1977), 45–47 (Van Steenbergen et al., 1979).

Type strain: ATCC 29147 (LEV strain).

Further comments. Strains with similar phenotypic characteristics have been isolated from human clinical specimens. The genetic relatedness between the human and bovine isolates has not been investigated.

24. **Bacteroides macacae** (Slots and Genco 1980) Coykendall, Kaczmarek and Slots 1980, 563.[VP] (*Bacteroides melaninogenicus* subsp. *macacae* Slots and Genco 1980, 83.)

ma.ca'cae. M.L. fem. n. *Macaca* genus name of the macaque; M.L. gen. n. *macacae* of the macaque.

This description is based on Slots and Genco (1980), Coykendall et al. (1980), and study of the type strain.

From glucose broth, cells are coccoid to rod-shaped and 0.6–0.8 by 1.0–2.5 μm.

Figure 6.24. *Bacteroides macacae.*

Colonies on blood agar (5 days) are 0.5–1.5 mm in diameter. They become grayish green and then black after prolonged incubation. Nonhemolytic. Black pigment develops faster on rabbit blood than sheep blood.

Glucose broth cultures are turbid with smooth sediment and a final pH of 5.5–5.8.

Growth is enhanced by menadione (1 μg/ml) but not by 0.0005% hemin, 0.3% formate-0.3% fumarate or 0.02% Tween 80.

Strains produce catalase and hydrolyze arginine. Acetylmethylcarbinol, oxidase and urease are not produced.

Other characteristics are given in Table 6.1 (species 24).

Isolated from the oral cavity of macaque monkeys.

The mol% G + C of DNA is 43–44 (Coykendall et al., 1980).

Type strain: ATCC 33141.

25. **Bacteroides succinogenes** Hungate 1950, 13.[AL] [*Ruminobacter succinogenes* (Hungate 1950) Prévot 1966, 122.]

suc.ci.no'ge.nes. M.L. n. *acidum succinicum* succinic acid; Gr. v. *gennaio* produce; M.L. adj. *succinogenes* succinic acid producing.

This description is from Hungate (1950, 1966), Bryant and Doetsch (1954) and Cato et al. (1978).

After 24-h incubation in glucose-rumen fluid-carbonate broth, the cells of the type strain are coccoid to oval, 0.8–1.6 by 0.9–1.6 μm and occur singly, in pairs, and occasionally in short chains.

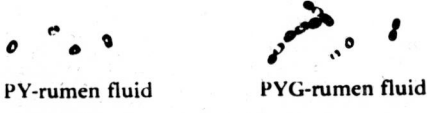

PY-rumen fluid   PYG-rumen fluid

Figure 6.25. *Bacteroides succinogenes.*

Surface colonies in rumen fluid-glucose-cellobiose agar (RGCA) in roll tubes are entire, slightly convex, translucent to opaque, nonpigmented or sometimes yellow, often with "frosted glass" appearance. One strain tested did not grow on the surface of agar plates incubated in an anaerobe jar. Deep colonies (RGCA) are 1–3 mm in diameter and lenticular. Colonies of some strains are not visible macroscopically in cellulose agar roll tubes, but are surrounded by definite clear zones of cellulose digestion; rods may be observed microscopically at the periphery of the area cleared of cellulose. Cells migrate through agar. Strains may degrade cellulose in liquid media but not produce clear zones in agar medium. The amount and rate of degradation is very dependent on the source of cellulose (Stewart et al., 1981).

Glucose broth cultures are turbid with smooth sediment. Broth clears in older cultures as cells lyse. Final pH in glucose broth is about 5.5.

Branched chain fatty acids [isobutyric or isovaleric (DL-2-methylbutyric)] and a straight-chained acid (5-carbon or longer), sulfide or cysteine, sodium, biotin and sometimes *p*-aminobenzoic acid, NH_3 and CO_2 are essential for growth. Amino acid nitrogen is not used (Bryant and Robinson, 1962; Bryant, 1973; Bryant and Doetsch, 1955; Bryant and Robinson 1963; Bryant et al., 1959). Good growth and cellulose digestion occur in media containing only cellulose, *p*-aminobenzoic acid, biotin, cysteine, alanine, phenylalanine, valerate, isobutyrate, a carbonic acid-bicarbonate buffer, resazurin and minerals (Bryant et al., 1959).

The optimum temperature for growth is ∼ 40°C; strains grow well in glucose medium at 30–38°C but do not grow at 25 or 45°C. Grows in glucose medium with an initial pH of 6–7.7 but not at pH 5.5 (Bryant and Doetsch, 1954). Final pH of cellobiose broth cultures of the type strain is 5.4 after incubation for 4 days. Very sensitive to inhibition by heavy metals and trace elements (Forsberg, 1978).

Cultures show CO_2 uptake in fermentation of cellulose or cellobiose. No acid is produced from gum arabic or xylan.

Capable of adhering to and degrading plant cell walls (Latham *et al.*, 1978).

No sphingolipids are detected (Miyagawa et al., 1978).

Produces succinic and acetic acid as major endproducts. In co-culture with other organisms such as *Selenomonas ruminantium*, the succinate is decarboxylated to propionate and CO_2, which presumably explains why succinate is not found as a major product in the rumen (Scheifinger and Wolin, 1973).

B. succinogenes, like *B. fragilis*, is capable of reducing fumarate to succinate in order to derive additional ATP; however, the fumarate reductase is $FMNH_2$ specific and cannot use NADH, H^+ (Miller, 1978).

Other characteristics of the species are given in Table 6.1 (species 25).

Isolated from rumen contents of cattle, sheep, and several wild African antelopes and from the cecum of rats.

The mol% G + C of the DNA is 47–49.

Type strain: ATCC 19169 (Bryant S85).

26. **Bacteroides microfusus** Kaneuchi and Mitsuoka 1978, 478.[AL] mic.ro.fus'us. Gr. adj. *micro* small; L. n. *fusus* a spindle; M.L. adj. *microfusus* pertaining to a small spindle (referring to cellular morphology).

The description is from Kaneuchi and Mitsuoka (1978) and study of the type strain.

Cells from glucose broth cultures are spindle-shaped and 0.5–0.9 by 1.0–5.0 μm in size and occur singly, in pairs, and sometimes in short chains. Short, filamentous, or swollen forms sometimes occur in media with fermentable carbohydrate.

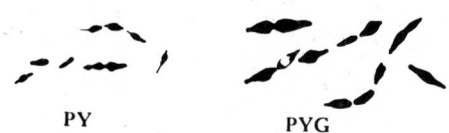

PY   PYG

Figure 6.26. *Bacteroides microfusus.*

Surface colonies on Eggerth-Gagnon agar are 0.5–1 mm in diameter, circular, convex, entire, smooth, translucent, grayish and β-hemolytic. Some strains form raised, opaque, gray colonies that are 1.0–1.5 mm in diameter.

Glucose broth cultures are turbid with a smooth sediment and final pH of 5.5–5.8.

Growth is enhanced by 20% bile or 0.0005% hemin plus 0.00005% menadione. Growth is not affected by 10% rumen fluid or 0.1% Tween 80. Growth is inhibited by 0.001% brilliant green, 0.005% crystal violet, neomycin (1600 μg/ml) and rifampin (15 μg/ml). Strains are resistant to bacitracin (3 U/ml), cefalotin (300 μg/ml), colistin (10 μg/ml), polymyxin B (10 μg/ml) and vancomycin (10 μg/ml). There is slight to moderate growth in the presence of kanamycin (1000 μg/ml), penicillin (15 μg/ml) or erythromycin (60 μg/ml).

Optimum growth occurs at 37°C. There is poor growth at 45°C and usually no growth below 25°C. No strains survived 70°C for 10 min.

Produces β-glucuronidase and glutamic acid decarboxylase. No acid from α-methylglucoside or α-methylmannoside. Casein not digested. No hydrolysis of orotic acid.

Other characteristics of the species are given in Table 6.1 (species 26).

Isolated from the feces of calves, Japanese quails, and the ceca of chickens. The type strain was present in the feces of Japanese quail at a concentration of 10^9/g (wet weight).

The mol% G + C of DNA is 60–61 for the type strain and two other strains tested.

Type strain: ATCC 29728 (NCTC 11190).

27. Bacteroides termitidis (Sebald 1962) Holdeman and Moore 1970, 33.[AL] (*Sphaerophorus siccus* var. *termitidis* Sebald 1962, 124.)

ter.mi′ti.dis. L. n. *tarmes, tarmit-* (L.L. var. *termes, termit-*) worm that eats wood; M.L. adj. *termitidis* pertaining to the termite.

The description is from Sebald (1962) and Potrikus and Breznak (1980a, b).

Cells are 0.3 by 2.0–12.0 μm with central swellings and occur singly and in pairs and filaments.

after Sebald, 1962

Figure 6.27. *Bacteroides termitidis.*

Surface colonies are 1–2 mm in diameter, circular and transparent to opaque. Colonies in deep agar are lenticular and not pigmented.

Glucose broth cultures are slightly turbid with sediment and gas (H_2).

No growth occurs at 56°C. Growth is not inhibited by 1:100,000 gentian violet or by brilliant green-N₃Na medium (Beerens and Tahon-Castel, 1965).

Coagulated proteins are not attacked. Neither urease, chitinase, nor toxin is produced. Uric acid degraded to CO_2, acetate, and ammonia (Potrikus and Breznak, 1980a). Not pathogenic for mice or guinea pigs.

Other characteristics are given in Table 6.1 (species 27).

Isolated from posterior intestinal contents of termites, where they are part of the predominant bacterial flora.

The mol% G + C of the DNA is 34–36.

Type strain: NCTC 11300 (ATCC 33386).

28. Bacteroides capillosus (Tissier, 1908) Kelly 1957, 433.[AL] [*Bacillus capillosus* Tissier 1908, 193; *Ristella capillosa* (Tissier 1908) Prévot 1938, 292.]

ca.pil.lo′sus. L. adj. *capillosus* very hairy.

This description is from Sebald (1962) and Cato et al. (1979).

After 24-h incubation in glucose broth, cells of the type strain are straight or curved rods, 0.7–1.1 by 1.6–7.0 μm, occurring singly, in pairs, or in short chains. Vacuoles, swellings, and filaments with tapered ends are observed.

PY   PYG

Figure 6.28. *Bacteroides capillosus.*

Surface colonies are minute to 1 mm, circular, entire, convex, translucent and smooth.

No diaminopimelic acid detected in cell walls of the type strain (Hammann and Werner, 1982).

Broth cultures are lightly turbid, sometimes with sediment (smooth to slightly stringy).

Growth of most strains is enhanced by hemin, rumen fluid, or Tween 80. Strains are generally nonfermentative unless Tween 80 is added to the medium, in which case they may be slightly fermentative. Good growth occurs at 37 and 45°C, slight growth at 30°C, and no growth at 25°C.

Deoxyribonuclease and phosphatase are produced. No elastase, hyaluronidase, or chondroitin sulfatase is detected (Rudek and Haque, 1976). About half of 43 strains tested are resistant to penicillin G (Cato et al., 1979).

Of eight strains tested, all were inhibited by concentrations of chloramphenicol, metronidazole, and doxycycline that are attainable in blood. Resistance to penicillins and cephalosporins was variable (Kirby et al., 1980).

Other characteristics are given in Table 6.1 (species 28).

Isolated from cysts and wounds, human mouth, human infant and adult feces, intestinal tracts of hogs, mice and *R. lucifugus* (termite) and from sludge.

The mol% G + C of the DNA from the type strain is 60 (Cato et al., 1979).

Type strain: ATCC 29799.

29. Bacteroides amylophilus Hamlin and Hungate 1956, 552.[AL] [*Ruminobacter amylophilum* (Hamlin and Hungate 1956) Prévot 1966, 121.]

am.y.lo′phi.lus. Gr. n. *amylo* starch; Gr. part. *philo* loving; M.L. adj. *amylophilus* starch loving.

Description from Hamlin and Hungate, Cato et al. (1978), and from study of two strains.

Cells of the type strain, after incubation for 2 days in PY-maltose broth, are 0.9–1.2 by 1–3 μm, pleomorphic, oval-to-long rods with round or tapered ends. Swollen forms and irregularly curved cells may occur in other media.

PY   PY-maltose

Figure 6.29. *Bacteroides amylophilus.*

Surface colonies on rumen fluid-glucose-cellobiose agar roll tubes are 1 mm in diameter, circular, entire, slightly convex, translucent, smooth, glistening, white to tan. No growth on the surface of plates incubated in an anaerobic jar. Colonies in deep agar are 0.8–1 mm in diameter, lenticular, entire or irregular, white and soft butyrous.

Starch broth cultures are turbid and have a final pH of 5.3–5.5. Amylase production is stimulated by Tween 80 (McWethy and Hartman, 1977).

Carbon dioxide, ammonia and a fermentable carbohydrate are required for growth. Carbon dioxide is fixed and ammonia is assimilated. Amino acids and fatty acids are not required for growth (Blackburn, 1968; Miura et al., 1980). Optimum temperature for growth is near 37°C. Cellulose is not digested. Strains are very sensitive to inhibition by heavy metals and trace elements (Forsberg, 1978).

No cytochromes (Reddy and Bryant, 1977) or sphingolipids were detected (Kunsman and Caldwell, 1974; Miyagawa et al., 1978).

Other characteristics are given in Table 6.1 (species 29).

Occurs sporadically in rumen contents of cattle but, when present, may be the predominant starch digester and constitutes 10% of the bacterial population of the rumen. Also occurs in the ovine rumen (Bryant and Robinson, 1962; Blackburn and Hobson, 1962).

The mol% G + C of the DNA is 40–42 (Reddy and Bryant, 1977).

Type strain: ATCC 29744 (Hobson H 18).

30. **Bacteroides furcosus** (Veillon and Zuber 1898) Hauduroy, Ehringer, Urbain, Guillot and Magrou 1937, 61.*AL* [*Bacillus furcosus* Veillon and Zuber 1898, 541; *Fusiformis furcosus* (Veillon and Zuber 1898) Topley and Wilson 1929, 302; *Ristella furcosa* (Veillon and Zuber 1898) Prévot 1938, 291.]

fur.co′sus. L. adj. *furcosus* forked (pertaining to cell shape).

This description is based on Prévot et al. (1967), Cato et al. (1979), and study of the type and 3 other strains.

The cells are short, nonmotile, pleomorphic rods, 0.3–1.5 by 1–3 μm and occur singly, in pairs, and in short chains. Some cells appear to be forked or Y-shaped.

Figure 6.30. *Bacteroides furcosus.*

Surface colonies are 0.5 mm in diameter, circular, entire, convex, translucent to semiopaque, grayish white, shiny and smooth.

No diaminopimelic acid detected in cell walls of the type strain (Hammann and Werner, 1981).

Glucose broth cultures are turbid with smooth sediment and have a final pH of 5.7–5.9. Growth is stimulated by rumen fluid.

Cultures grow at pH 8.0. There is good growth at 30 and 37°C, slight growth at 25°C, and none at 45°C.

Although this species has been reported to ferment maltose and mannitol (Prévot et al., 1967), none of the strains tested in our laboratory ferment these substrates. Produces phosphatase. No fibrinolysin, elastase, hyaluronidase, chondroitin sulfatase, or deoxyribonuclease is detected (Rudek and Haque, 1976).

No sphingolipids (Miyagawa et al., 1978), heptose, or KDO are detected (Hofstad, 1974).

Additional characteristics are given in Table 6.1 (species 30).

Isolated from infected appendix, lung abscesses, abdominal abscesses, and mud snails. Infrequently isolated from human and pig feces.

The mol% G + C of the DNA from the type strain is 34 (Cato et al., 1979).

Type strain: ATCC 25662 (Suzuki T-301-A2).

31. **Bacteroides ureolyticus** Jackson and Goodman 1978, 199.*AL* [*Bacteroides corrodens* Eiken 1958, 415 (in part); *Ristella corrodens* (Eiken 1958) Prévot 1966, 118 (in part).]

ur′e.o.ly.ti.cus. M.L. n. *urea* urea; Gr. adj. *lyticus* dissolving; M.L. adj. *ureolyticus* urea dissolving.

This description is from Jackson and Goodman (1978).

Cells are 0.5 by 1.5–4.0 μm. Filaments exceeding 20 μm in length may occur. Cells of some strains have polar tufts of long pili in electron micrographs (Jackson et al., 1971) and exhibit "twitching" motility. The pili sometime form a bundle and may be mistaken for flagella with light microscopy (Jackson and Goodman, 1978).

Figure 6.31. *Bacteroides ureolyticus.*

Surface colonies on blood agar may be either (a) 1 mm in diameter, circular with entire to slightly undulating margins, convex to slightly

umbonate and gray-white, or (b) spreading or "swarming" growth extending for a few millimeters from the slightly raised center. The flat spreading growth appears to occupy a slight depression in the agar, so that the edge of the spreading growth is not detectably raised above the surface of the medium. The "pitting" usually can be seen best if the plate surface is observed at about a 30–45° angle.

Broth cultures are lightly turbid, sometimes with granular or slightly stringy sediment.

Grows in 1% oxygen, but not in 5% oxygen. Hemin (25 μg/ml), 0.1% nitrate and 10% CO_2 may enhance growth. Resistant to 0.02% sodium azide and 0.1% KCN. Growth enhanced by formate or hydrogen as an electron donor and fumarate or nitrate as an electron acceptor. The fumarate is reduced to succinate. Aspartate, malate, and asparagine can also serve as electron acceptors (Smibert and Holdeman, 1976).

Contains a c-type cytochrome, and probably a b-type cytochrome (Jackson et al., 1970). Octadecenoic acid comprises ∼70% of C_{12}–C_{20} fatty acids. Hexadecanoic is 10% of the total (Prefontaine and Jackson, 1972).

Oxidase-positive with 0.3% tetramethyl-*p*-phenylenediamine. Urease positive (Christensen urea agar medium). Lysine decarboxylase negative. Most strains show gelatinase activity (Frazier method) and H_2S is produced in SIM medium. No DNase or phosphatase detected (Porschen and Sonntag, 1974; Porschen and Spaulding, 1974).

Other characteristics are given in Table 6.2 (species 31).

Isolated from infections of the respiratory and intestinal tracts and from the buccal cavity, intestinal tract, urogenital tract; also from blood drawn after dental extraction.

The mol% G + C of the DNA is 28–30 (Jackson et al., 1971).

Type strain: NCTC 10941 (ATCC 33387).

Further comments. Bacteroides corrodens as originally described by Eiken (1958) included both obligately anaerobic and facultative strains. ATCC 23834 [NCTC 10596, Henriksen 333/54-55 (Eiken)], deposited by Henriksen (1969) to represent *B. corrodens*, is a facultative strain and the type strain of *Eikenella corrodens* (Eiken) Jackson and Goodman 1972, 73. The DNA G + C content of *E. corrodens* is 56–58 mol% compared to 28–30 mol% for the anaerobic, urease-positive strains which now are classified as *Bacteroides ureolyticus* (Jackson and Goodman, 1978).

The urease activity of *B. ureolyticus* separates this species from other nitrate-positive, nonfermentative species (e.g. *B. gracilis, Campylobacter concisus* and *Wolinella recta*) that use hydrogen (or formate) and fumarate.

32. **Bacteroides praeacutus** (Tissier 1908) Holdeman and Moore 1970, 33.*AL* [*Coccobacillus praeacutus* Tissier 1908, 193; *Zuberella praeacuta* (Tissier 1908) Prévot 1938, 293.]

prae.a.cu′tus. L. pref. *prae* very, quite; L. adj. *acutus* sharp; M.L. adj. *praeacutus* quite sharp.

Description from Prévot et al. (1967) and Cato et al. (1979).

Cells are motile, peritrichous, and occur singly or in pairs. They have rounded or occasionally sharply pointed ends. Swellings near the ends of cells are common. Cells from glucose broth are 0.6–0.9 by 2–8 μm. In some cultures, cells may be filamentous, up to 20 μm in length.

Figure 6.32. *Bacteroides praeacutus.*

Surface colonies on horse blood agar are 0.5 mm in diameter, circular, flat with a scalloped or diffuse edge, grayish, dull, smooth, translucent with mosaic appearance when viewed by obliquely transmitted light.

Cell walls of the type strain contain *meso*-diaminopimelic acid (Hammann and Werner, 1981).

Glucose broth cultures are moderately turbid with a smooth sediment. Optimum growth temperature is 37°C. Moderate turbidity at 25, 30 and 45°C. Nitrate was completely reduced by one of three strains. Hippurate was hydrolyzed by two of three strains.

No heptose or KDO detected (Hofstad, 1974).

Other characteristics are given in Table 6.2 (species 32).

Isolated from intestinal tracts of infants and adults, from gangrenous lesions, lung abscesses, and from blood.

The mol% G + C of the DNA from the type strain is 28 (Cato et al., 1979).

Type strain: ATCC 25539 (NCTC 11158).

33. **Bacteroides gracilis** Tanner, Badger, Lai, Listgarten, Visconti, and Socransky 1981, 442.[VP]

gra′cil.is. L. adj. *gracilis* slim, slender, thin.

This description is from Tanner et al. (1981).

Cells are slim, straight rods, 0.4 by 4–6 µm, with both tapered and rounded ends. Cells may demonstrate "twitching" movements. No flagella.

Figure 6.33. *Bacteroides gracilis.*

Three colony types are observed on blood agar: (a) translucent, 1 mm, convex; (b) agar-pitting or corroding, diameter up to 5 mm and (c) spreading with diameter up to 5 mm.

Some strains grow in the presence of 5% oxygen, none grow in air with 10% CO$_2$. Growth is stimulated by formate and fumarate; hydrogen can also be used as an electron donor. Growth is inhibited by 4% NaCl.

Oxidase-negative. Lysine, ornithine, and arginine are not decarboxylated; dextran is not hydrolyzed. Urease, ammonia and acetylmethylcarbinol are not produced.

Other characteristics are given in Table 6.2 (species 33).

B. gracilis is very similar to *Wolinella recta* phenotypically but the two species are distinct by DNA homology. *W. recta* has a single polar flagellum and *B. gracilis* has no flagella.

Isolated from the gingival crevices of humans.

The mol% G + C of the DNA is 44–46.

Type strain: ATCC 33236 (Forsyth strain 1084).

34. **Bacteroides nodosus** (Beveridge 1941) Mráz 1963, 85.[AL] ["Organism K," Beveridge 1938, 1; *Fusiformis nodosus* Beveridge 1941, 23; *Ristella nodosa* (Beveridge 1941) Prévot 1948, 82.]

no.do′sus. L. adj. *nodosus* knotty or swollen, pertaining to the shape of the cells.

Description from Beveridge (1941) and Cato et al. (1979).

Cells are fairly large, nonmotile, straight or slightly curved rods, 1–1.7 by 3–6 µm. They often have swollen ends and occur singly and occasionally in paris. Terminal enlargements of cells are more pronounced in lesions than in cultures. Cells have large numbers of pili. The number of pili is correlated with virulence and a change in colonial morphology (Short et al., 1976). There are conflicting reports regarding the presence of a capsule (Cooper, 1977; Stewart and Egerton, 1979; Every and Skerman, 1980).

Surface colonies of the type strain are 0.5–2 mm, smooth, convex, translucent or semiopaque. Skerman et al. (1981) described three basic colony types: (a) papillate or beaded (B)-type (most pathogenic) from ovine foot rot, (b) mucoid (M)-type (less pathogenic) from noninvasive infections of the interdigital skin of sheep and cattle and (c) circular (C)-type (nonpathogenic) resulting from repeated passage in liquid cultures. Changes in colony type within a strain correlates with changes in pathogenicity, elastase activity, and immunoprotective properties. Colonies often etch into the surface of the medium immediately under the colony, producing a sunken appearance.

Broth cultures are lightly turbid, sometimes granular. Growth is enhanced by at least 10% CO$_2$ and 10% horse serum. Growth also is stimulated by Tween 80. Growth is most rapid at pH 6.4–7.6. This species does not grow at pH 4–6 but grows at pH 8–9. Good growth occurs at 37 and 45°C; poorer growth at 30°C. Growth is best in media containing trypticase; 0.02–0.05 M arginine required for maximum growth (Skerman, 1975).

Inhibited by 1.6 µg/ml metronidazole (Chow et al., 1977). A selective medium using lincomycin has been devised by Gradin and Schmitz (1977).

In addition to the characteristics given in Table 6.2 (species 34), elastin particles, hoof powder, and hide powder are digested. Tyrosine crystals develop in chopped meat after incubation for 4–6 weeks.

Causative agent of foot rot in sheep (and possibly in goats); infected hoofs apparently are the only natural habitat.

The mol% G + C of the DNA of Skerman strain 10 is 45 (L. R. Hill, personal communication).

Type strain: ATCC 25549 (VPI 5731-1; McMaster 198A).

Further comments. Ovine foot rot is a mixed infection requiring the presence of both *Bacteroides nodosus* and *Fusobacterium necrophorum.* The transmitting agent is *B. nodosus; F. necrophorum* is considered part of the normal flora (Roberts and Egerton, 1969).

There are at least 14 different serovars of *B. nodosus.* The distribution of serovars varies around the world, which probably explains the lack of efficacy of Australian-produced vaccines in the United States (Schmitz and Gradin, 1980). Antibodies to pili preparations produce typical K-type agglutination reactions and are capable of protecting sheep against experimental challenge (Stewart, 1978).

Strains that are phenotypically similar to animal isolates of *B. nodosus* have been isolated from human clinical specimens (pilonidal cyst, rectal fistula, decubitus ulcer, and leg wound). These strains probably are not *B. nodosus.* They did not react with fluorescent antibody conjugates to *B. nodosus,* and their colony morphology differed from that of the animal isolates (Joseph Gradin, Oregon State University, Corvallis, personal communication). Also, the colonies did not pit the agar beneath the colonies and the cells were shorter (Cato et al., 1979).

35. **Bacteroides pneumosintes** (Olitsky and Gates 1921) Holdeman and Moore 1970, 33.[AL] [*Bacterium pneumosintes* Olitsky and Gates 1921, 727; *Dialister pneumosintes* (Olitsky and Gates 1921) Bergey et al. 1923, 271.]

pneu.mo.sin′tes. Gr. n. *pneuma* air; Gr. n. *sintes* a spoiler, thief; M.L. adj. *pneumosintes* breath destroying.

This description is from Hitchens (1957), Prévot et al. (1967), and Cato et al. (1979).

Cells of the type strain are 0.2–0.4 by 0.3–0.6 µm, arranged singly, in pairs, or in very short chains. Gram stains usually give the impression of very small cocci. Cells from glucose broth may be 0.5 to 1.0 µm long after several transfers. Cells from nasopharynx are 0.15–0.30 µm long and one-half to one-third as wide; filterable through Berkefeld V and N filters. Many nonreactive Gram-negative rods may appear biochemically similar to *B. pneumosintes,* but they can be distinguished easily by cell morphology.

Deep agar colonies are punctiform, granular, and white with no evidence of gas production. Surface colonies on horse blood agar are

Figure 6.35. *Bacteroides pneumosintes.*

punctiform, circular, entire, convex, clear, transparent, shiny and smooth.

Broth cultures usually are not turbid, but numerous very small cells can be seen in stains from the cultures. Because the cells are so small, stains have to be studied carefully to distinguish cells from "background" stain.

Equivalent growth at 30 and 37°C; no growth at 25 or 45°C.

No DNase or phosphatase detected (Porschen and Sonntag, 1974; Porschen and Spaulding, 1974).

Pathogenic for rabbits when injected intratracheally (Prévot et al., 1967).

Isolated from nasopharyngeal washings from normal individuals (may be involved in secondary infections of the upper respiratory tract), gingival crevice and periodontal pockets, blood, respiratory tract, various head and neck infections, and brain abscesses (in association with *Streptococcus intermedius* (*S. anginosus*)).

The mol% G + C of the DNA is not known.

Type strain: ATCC 33048.

36. **Bacteroides putredinis** (Weinberg, Nativelle and Prévot 1937) Kelly 1957, 420.[AL] [*Bacillus putredinis* Weinberg, Nativelle, and Prévot 1937, 755; *Ristella putredinis* (Weinberg et al.) Prévot 1938, 291.]

put.re′di.nis. L. N. *putredo* putridity; M.L. gen. n. *putredinis* of putridity.

The description is based on previous literature descriptions, study of the type and 15 other strains, and Cato et al. (1979).

Cells of the type strain after 24-h incubation in glucose broth are straight or curved with rounded ends, 0.3–0.5 by 0.9–3.0 μm, occurring singly or in pairs. Occasional swellings at one end are seen, but spores have not been detected and the organisms do not survive heating at 70 or 80°C for 10 min.

Figure 6.36. *Bacteroides putredinis.*

Surface colonies are pinpoint to 0.5 mm, circular to slightly irregular with entire to erose margins, low convex, translucent, gray, dull, smooth.

Broth cultures are lightly turbid with smooth or stringy sediment. Growth is slightly stimulated by Tween 80.

Optimum growth is at 37°C. The type strain produces little or no growth at 30 and 45°C. Other strains grow between 25 and 45°C.

Some strains digest chopped meat, most strains produce trace amounts of catalase. Glutamic acid is decarboxylated (Werner, 1970b). Produces fibrinolysin. No elastase, hyaluronidase, chondroitin sulfatase, or phosphatase detected (Rudek and Haque, 1976). Rudek and Haque reported no deoxyribonuclease from a strain of *B. putredinis*, but Werner (1970a) found that the type strain was positive for DNase. Hydrogen sulfide was not detected with SIM medium, but was positive using lead acetate paper (Cato et al., 1979). Catalase positive.

No heptose or KDO detected (Hofstad, 1974).

Sensitive to clindamycin, cefoxitin, chloramphenicol, erythromycin

and metronidazole. Moderately resistant to tetracycline and doxycycline (Kirby et al., 1980).

Other characteristics are given in Table 6.2 (species 36).

Isolated from feces, abdominal and rectal abscesses, from cases of acute appendicitis, from foot rot in sheep, farm soil, and, rarely, from the human mouth.

The mol% G + C of the DNA is not known.

Type strain: ATCC 29800 (Suzuki strain Gifu Ando).

37. **Bacteroides coagulans** Eggerth and Gagnon 1933, 409.[AL] [*Pasteurella coagulans* Eggerth and Gagnon 1933) Prévot 1938, 292.]

co.a′gu.lans. L. part. adj. *coagulans* curdling, coagulating.

This description is from Eggerth and Gagnon, and Cato et al. (1979).

Cells of the type strain are small ovoid rods, 0.6 μm wide by 0.8–2.5 μm long, occurring singly, in pairs, or in short chains.

Figure 6.37. *Bacteroides coagulans.*

Surface colonies on horse blood agar are punctate, circular, entire, slightly raised, translucent; nonhemolytic.

Glucose broth cultures are only slightly turbid; no acid is produced. Growth is inhibited by Tween 80. Grows better in a 90% N_2-10% CO_2 atmosphere than in 100% CO_2 (Cato et al., 1979). Grows poorly, if at all, at 25 and 45°C; does not grow at pH 8.5.

Produces elastase. No fibrinolysin, hyaluronidase, chondroitin sulfatase, deoxyribonuclease, or phosphatase detected (Rudek and Haque, 1976). Hydrogen sulfide detectable with lead acetate paper but not SIM medium (Cato et al., 1979). Hippurate is hydrolyzed.

Inhibited by a 1000 μg kanamycin disk (Porschen and Stalons, 1976).

No heptose or KDO detected (Hofstad, 1974).

Other characteristics are given in Table 6.2 (species 37).

Isolated from human feces, urogenital tract, and occasionally from human clinical specimens.

The mol% G + C of the DNA from the type strain is 37 (Cato et al., 1979).

Type strain: ATCC 29798.

38. **Bacteroides gingivalis** Coykendall, Kaczmarek and Slots 1980, 563.[VP]

gin.gi.val′is. L. n. *gingiva* gum; L. gen. n. *gingivalis* of the gums.

Most cells in broth are coccobacillary, 0.5 μm wide by 1–2 μm long. Rod-shaped cells up to 5 μm may be observed.

Figure 6.38. *Bacteroides gingivalis.*

Colonies on blood agar are 1–2 mm, convex, and form black pigment in 7–10 days.

The peptidoglycan contains lysine. No KDO is detected (Shah et al., 1976; Mansheim and Kaspar, 1977).

Resistant to 10 μg/ml colistin (Shah et al., 1976) and 1000 μg kanamycin disk. Inhibited by 5 μg metronidazole disk, 0.1–0.5% bile salts, ethyl violet (1/80,000), gentian violet (1/100,000) and brilliant green (1/80,000) (Holbrook et al., 1977).

Produces a cell-bound, oxygen-sensitive collagenase. Symptoms from subcutaneous injection of guinea pigs with whole cells plus a mixed culture of organisms ranged from localized abscesses to a rapidly spreading necrotic infection resulting in death in 18–24 h. The mixed culture alone produced no symptoms. Other *B. melaninogenicus* strains were not pathogenic with the above test (Mayrand et al., 1980).

B. gingivalis has a strong reaction for trypsin-like activity (as tested in the API ZYM system or with similar substrate); strains of *B. asaccharolyticus* and of saccharolytic pigmented bacteroides are negative [B. E. Laughon, S. A. Syed and W. J. Loesche: *Journal of Dental Research* (Special Issue A), *60:* 332 (Abstr. 88), 1981; Slots, 1981].

Produces extracellular phospholipase A (Bulkacz et al., 1979).

Other characteristics are given in Table 6.2 (species 38).

Isolated from the human mouth.

The mol% G + C of the DNA is 46–48 (Coykendall et al., 1980).

Type strain: ATCC 33277 (Slots 2561).

Further comments. *B. gingivalis* is almost identical biochemically to *B. asaccharolyticus*. It differs in the source of isolation and G + C content (*B. asaccharolyticus* is 49–53 mol%). The species has little or no DNA homology with *B. asaccharolyticus* (Coykendall et al., 1980; Van Steenbergen et al., 1981). *B. gingivalis* strains can agglutinate sheep erythrocytes and produce phenylacetic acid whereas *B. asaccharolyticus* cannot (Slots and Genco, 1979; Mayrand, 1979; Kaczmarek and Coykendall, 1980). Also, *B. gingivalis* does not react with a commercial fluorescent antibody staining system (Fluoretec-M; Mouton et al., 1980).

39. Bacteroides asaccharolyticus (Holdeman and Moore 1970) Finegold and Barnes, 1977, 388.[AL] (*Bacteroides melaninogenicus* subsp. *asaccharolyticus* Holdeman and Moore 1970.)

a.sac.cha.ro.ly'ti.cus. Gr. pref. *a* not; Gr. n. *sacchar* sugar; Gr. adj. *lyticus* able to loosen; M.L. adj. *asaccharolyticus* not digesting sugar.

This description is based on our study of ATCC 25260 and numerous similar strains.

Cells from broth cultures are 0.8–1.5 by 1.0–3.5 μm. Longer cells occasionally are seen. From solid medium the cells may be much shorter (almost spherical).

PY   **PYG**

Figure 6.39. *Bacteroides asaccharolyticus.*

Colonies are 0.5–1.0 mm in diameter, round, convex, opaque and light gray after incubation for 48 h; 6–14 days are required for black color formation. Hemolysis of rabbit blood is variable.

The peptidoglycan contains lysine. No KDO is detected (Hofstad, 1974; Shah et al., 1976). No diaminopimelic acid detected in cell walls of NCTC 9337 (Hammann and Werner, 1981).

Addition of 0.5% NaCl to PY extract medium stimulates growth (F. Jimenez, personal communication).

Does not produce phenylacetic acid (Kaczmarek and Coykendall, 1980). Oxidase-negative (Holbrook et al., 1977). Does not decarboxylate glutamate (Werner, 1970b; Werner et al., 1971). Cells have fibrinolytic activity (Nitzan *et al.*, 1978).

Resistant to 10 μg/ml colistin (Shah et al., 1976), and 1000 μg kanamycin disk. Inhibited by 5 μg metronidazole disk, 0.1 to 0.5% bile salts, ethyl violet (1/80,000), gentian violet (1/100,000) and brilliant green (1/80,000) (Holbrook et al., 1977).

Six strains were sensitive to penicillins, cephalosporins, bacitracin, chlortetracycline, chloramphenicol, erythromycin and rifampicin. Six strains were resistant to streptomycin, colistin, polymyxin B, and neomycin, having MIC values of 10–100 μg/ml (Werner et al., 1971).

Other characteristics are given in Table 6.2 (species 39). Also see Further Comments for *B. melaninogenicus* (species 16) and *B. gingivalis* (species 38).

Isolated from clinical specimens.

The mol% G + C of the DNA is 50–51 (Williams et al., 1975), 51–52 (Coykendall et al., 1980).

Type strain: ATCC 25260 (Finegold B440).

Addendum

After this manuscript for the genus *Bacteroides* was completed, Shah and Collins (1981) published descriptions of three new species of bacteroides from the oral cavity: *B. buccalis* (type strain is National Collection of Dairy Organisms (NCDO) 2354, strain HS4), *B. denticola* (type strain is NCDO 2352, strain 1210), and *B. pentosaceus* (type strain is NCDO 2353, strain NP 333). These species are distinct from each other and from *B. oralis* by DNA reassociation studies (van Steenbergen et al., 1980).

The two strains of *B. buccalis* differ from the type strain of *B. oralis* by not hydrolyzing starch [API 20A system (Analytab Products, Inc., Montalieu-Varcieu, France)], not producing leucine aminopeptidase, and having α-fucosidase acitivity (API ZYM system).

From the characteristics given in the description, *B. pentosaceus* appears to be a synonym of either *B. oris* or *B. buccae*.

The reader is referred to Shah and Collins (1981) for a complete description of these organisms.

Also after this manuscript was completed, Patel and Breuil (1981, 1982) described *B. polypragmatus*, an obligately anaerobic, encapsulated, motile (peritrichous), Gram-negative bacillus isolated from sewage sludge. The species is saccharoclastic and grows between pH 5.6 and 9.2 and at temperatures ranging from 10 to 43°C. Optimum growth is at pH 7.0–7.8 and 30–35°C. Substrates utilized are amygdalin, arabinose, cellobiose, dulcitol, esculin, fructose, galactose, glucose, glycerol, glycogen, lactose, maltose, mannitol, mannose, melezitose, melibiose, pyruvate, raffinose, rhamnose, ribose, salicin, sorbose, soluble starch, trehalose and xylose. Not utilized are arginine, erythritol, hippurate, inositol, inulin, lactate, sorbitol, sucrose, threonine and urea. Indole and H_2S are produced. Esculin is hydrolyzed. Tests for acetylmethylcarbinol, ammonia, nitrate reduction, and digestion of gelatin, milk and meat are negative. Major fermentation products are CO_2, H_2, ethanol and acetic acid. The mol% of the G + C of the DNA is 61. The type strain is GP4, National Research Council (Canada) 2288.

Note Added in Proof

The readers's attention is directed to the following names published while this manuscript was in press.

Bacteriodes veroralis Watabe, Benno and Mitsuoka 1983, 62.[VP]

This species is phenotypically similar to *B. oralis* and *B. buccalis* but has little or no DNA homology with the type strains of either of these species. Helpful phenotypic characteristics to differentiate the species are fermentation of esculin and salicin by *B. oralis* but not by *B. veroralis* or *B. buccalis* and fermentation of xylan by *B. veroralis* but not by *B. buccalis* (Watabe et al., 1983).

Isolated from the human buccal cavity.

The mol% G + C of the DNA is 42.

Type strain: ATCC 33779 (VPI D22A-7).

Megamonas hypermegas (Harrison and Hensen 1963) Shah and Collins 1983, 439.[VP] (Effective publication: Shah and Collins 1982, 395). Basonym: *Bacteroides hypermegas.*

Mitsuokella multiacidus (Mitsuoka et al. 1974) Shah and Collins 1983, 439.[VP] (Effective publication: Shah and Collins 1982, 493.) Basonym: *Bacteroides multiacidus.*

References:

Shah, H.N. and M.D. Collins. 1982. Reclassification of *Bacteroides hypermegas* (Harrison and Hansen) in a new genus *Megamonas*, as *Megamonas hypermegas* comb. nov. Zentralbl. Bakteriol. Parasitenkd. Infektionskr. Hyg. Abt. Orig. *C3:* 394–39.

Shah, H.N. and M.D. Collins. 1982. Reclassification of *Bacteroides multiacidus* (Mitsuoka, Terada, Watanabe and Uchida) in a new genus *Mitsuokella*, as *Mitsuokella multiacidus* comb. nov. Zentralbl. Bakteriol. Parasitenkd. Infektionskr. Hyg. Abt. 1 Orig. *C3:* 491–494.

Shah, H.N. and M.D. Collins. 1983. Validation of the publication of new names and new combinations previously effectively published outside the IJSB. List No. 10. Int. J. Syst. Bacteriol. *33:* 438–440.

Watabe, J., Y. Benno and T. Mitsuoka. 1983. Taxonomic study of *Bacteroides oralis* and related organisms and proposal of *Bacteroides veroralis* sp. nov. Int. J. Syst. Bacteriol. *33:* 57–64.

Genus II. **Fusobacterium** *Knorr 1922, 4* [AL]

W. E. C. MOORE, LILLIAN V. HOLDEMAN AND ROGER W. KELLEY

Fu.so.bac.te′ri.um. L. n. *fusus* a spindle; Gr. dim. n. *bakterion* a small rod; M.L. neut. n. *Fusobacterium* a small spindle-shaped rod.

Nonsporeforming rods that are **Gram-negative** and **obligately anaerobic.** All described species are nonmotile. **Chemoorganotrophs.** Metabolize peptone or carbohydrates. Major products from peptone or carbohydrate include **butyrate,** often with acetate and lactate and lesser amounts of propionate, succinate, and formate and short-chained alcohols. Isobutyrate and isovalerate are not produced. The mol% G + C of the DNA (in the species examined) ranges from 26 to 34 for the type and 5 other species and is from 52 to 57 mol% for *F. prausnitzii.*

Type species: *Fusobacterium nucleatum* Knorr 1922, 17.

Further Descriptive Information

Cell morphology. Although cells of the type species are fusiform (spindle-shaped), not all species in the genus are fusiform. Neither are all Gram-negative fusiform rods members of the genus *Fusobacterium*; they may be either bacteroides or clostridia, depending upon the major acids produced or the presence of spores.

Cell wall composition. Information on the fatty acid content and cell wall composition of various *Fusobacterium* species can be found in Hofstad (1979), Kato et al. (1979, 1981), Miyagawa et al. (1979a), Hofstad and Skaug (1980), Jantzen and Hofstad (1981) and Vasstrand (1981). All eight species so far tested contain heptose and 2-keto-3-deoxyoctonate (KDO) in the lipopolysaccharide (Hofstad, 1974).

Nutrition and growth conditions. Fusobacteria do not grow on the surface of agar plates incubated aerobically or in air enriched to 5–10% with CO_2. The maximum E_h permitting growth varies, depending upon the species, the size of inoculum, and the medium. Oxidized media are inhibitory for some species; addition of serum (5–10%, v/v) or ascitic fluid to such media may enable them to support growth.

Most strains grow in a complex medium containing peptone (about 1%) and yeast extract (0.5–1.0%). Rumen fluid (heme and/or volatile fatty acids) is stimulatory for some species.

Growth usually is most rapid at 37°C and at a pH near 7.

Pathogenicity. Some species are pathogenic and occur in various purulent or gangrenous infections and in organ infarcts.

Ecology. Found in cavities of man and other animals.

Enrichment and Isolation Procedures

See the genus *Bacteroides* for general methods and pertinent references for selective media.

Maintenance Procedures

See Further Comments in the genus *Bacteroides*.

Procedures for Testing Special Characters

Conversion of threonine to propionate. Cultures grown in peptone-yeast extract (PY) medium with and without added DL-threonine (Holdeman et al., 1977) are acidified and extracted with ether. The amount of propionic acid detected by gas-liquid chromatography of the PY-threonine culture is compared with that from the PY culture. The organism converts threonine if more propionic acid is present in the PY-threonine culture than in the PY culture. (Propionic acid in the PY culture is from metabolism of peptone.) In general, those strains that convert threonine to propionate also make a small amount of propionate from PY extract or chopped meat medium.

Conversion of lactate to propionate. This test is particularly useful for confirmation of the identity of *F. necrophorum.* Cultures grown in PY medium with and without added lactate are acidified and extracted with ether. The amount of propionic acid detected by gas-liquid chromatography of the PY-lactate culture is compared with that from the PY culture. The organism converts lactate to propionate if more propionic acid is present in the PY-lactate culture than in the PY culture.

Conversion of lactate to propionate can be detected from other types of lactate-containing media. If fresh meat is used in preparation of chopped meat medium, the medium contains large amounts of lactate (which accumulates in the muscle of the animal as the tissue cells metabolize anaerobically after death). Thus, a large amount of propionate is present in these chopped meat cultures of *F. necrophorum.* The presence of lactate in chopped meat medium can be determined by chromatographing a chloroform extract of a methylated sample of uninoculated chopped meat medium.

The described species of fusobacteria are nonfermentative or only weakly fermentative (pH seldom lower than 5.5). All described species of fusobacteria make ammonia from peptone, which sometimes can mask weak acid production (the ammonia raises the pH of the poorly buffered media used to determine fermentation of carbohydrates).

If pH indicators are used to determine presence or absence of acid, some of the species listed here as weak fermenters may appear to make strong acid because the color of the indicator(s) will change at a pH higher than 5.5. Species that make large amounts of hydrogen (see Table 6.7) may reduce the pH indicator. Sterile indicator can be added to the tubes of media at the time the pH is read to confirm an acid pH.

Other procedures and methods are given in the genus description of the genus *Bacteroides.*

Further Reading

See Further Reading for the family *Bacteroidaceae.*

Taxonomic Comments

Several species described in the eighth edition of *Bergey's Manual* have been transferred to other genera or abolished. *F. symbiosum* was shown to contain spores and was reassigned to the genus *Clostridium* as *C. symbiosum* (Kaneuchi et al., 1976). Most strains of *F. bullosum* also contained spores (unpublished data).

Heat-resistant cells have been found in the organism described as *Fusobacterium polysaccharolyticum* (van Gylswyk, 1980), so this species is now considered to belong to the genus *Clostridium* (van Gylswyk et al., 1980).

No strains of *F. glutinosum* or *F. stabile* are extant so these species were not included on the Approved Lists of Bacterial Names in 1980. The strain on which our previous description of *F. aquatile* was based was a mixed culture (unpublished data), so no strains of this species are known to be extant.

F. plauti has been shown to have a Gram-positive type cell wall structure and its transfer to the genus *Eubacterium* has been proposed (Hofstad, 1982).

Descriptions of several unnamed species of *Fusobacterium* isolated from human feces can be found in Moore and Holdeman (1974b) and Holdeman et al. (1976).

Acknowledgements

Much of the basic information retained from the eighth edition of the *Manual* was supported by Public Health Service Grant GM-14604 from the National Institute of General Medical Sciences and by project 2022820 from the Commonwealth of Virginia. New information that we have obtained since the eighth edition of the *Manual* was supported by the Commonwealth of Virginia and grants DE-05054 and DE-05139 from the National Institute of Dental Research.

We gratefully acknowledge the microbiological, technical, support, and secretarial assistance as listed in the acknowledgements, genus *Bacteroides*.

Differentiation of the species of the genus **Fusobacterium**

Characteristics by which species in the genus can be differentiated are given in the key to species (p. 633) in the genus and in Table 6.7. Additional characteristics are given in the text concerning each species.

Only the basionym and selected synonyms are given for the species. The reader should consult the *Index Bergeyana* (Buchanan et al., 1966) and the *Supplement to Index Bergeyana* (Gibbons et. al., 1981) for other synonyms.

Colony descriptions are from blood agar plates or streak tube cultures incubated anaerobically for two days, unless otherwise indicated.

Reported G + C contents of the DNAs are expressed as the nearest whole number.

The drawings of each species given in the text are composites of the type strain and, where available, other strains showing DNA homology with the type. Individual cultures may show somewhat less variation in cellular morphology than is depicted. The scale for the drawings is given in Figure 6.1 of *Bacteroides* (*B. fragilis*). Excellent color photographs of colonies and strains of cells of most species are depicted in Mitsuoka (1980b) (text in Japanese, species names in English).

None of the described species is motile.

Described species do not ferment adonitol, arabinose, dulcitol, glycerol, glycogen, inositol, inulin, mannitol, melezitose, rhamnose, ribose, sorbitol or sorbose; reduce nitrate; or produce catalase, lecithinase, or acetylmethylcarbinol. In addition to butyric, propionic and acetic acids, species (except *F. prausnitzii*) produce variable amounts of butanol from PY medium. Small amounts of formate, lactate, succinate, and ethanol also may be produced. Some species convert threonine or

Table 6.7.

Characteristics of the species of the genus **Fusobacterium**[a]

Characteristics	1. *F. nucleatum*	2. *F. gonidiaformans*	3. *F. varium*	4. *F. necrophorum*	5. *F. perfoetens*	6. *F. naviforme*	7. *F. russii*	8. *F. mortiferum*	9. *F. necrogenes*	10. *F. prausnitzii*
Products from PYG	Bap (FLs)	BAp (lfs)	BLap (s)	Bpa (ls)	Bap (L)[b]	BLa (fps)	BLa (f)	Bap (sflv)	Bap (fsl)	BLF (s)
Propionate from:										
threonine	+	+	+	+	+	−	−	+	+	−
lactate	−	−	−	+	−	−	−	−	−	−
Hydrogen produced	−	4	4	4	4	−	−	4	4	−
Esculin hydrolyzed		−	−	−	−	−	−	+	+	+
Indole produced	+	+	d	+	−	+	−	−	−	−
Growth in 20% bile	±	−1	4,1	−4	−	−1	−1	4S	4,1	1−
Gelatin digested	−	−w	−	+−	−	−	−w	−	−	−
Milk reaction	−	−	−	cp−	−	−	−	−c	−	−a
Meat digestion	−	−	−	+−	−	−	−	−	−	−
Acid from:										
amygdalin	−	−	−	−	−	−	−	−w	−	−
cellobiose	−	−	−	−	−	−	−	w−	−w	−w
esculin	−	−	−	−	−	−	−	−	−	−w
fructose	−w	−	w+	−w	w	−	−	+w	w+	w−
glucose	−w	−	w+	−w	w	w−	−	+w	w+	w−
lactose	−	−	−	−	−	−	−	+w	−	w−
maltose	−	−	−	−	−	−	−	w−	−	w−
mannose	−	−	w+	−	−	−	−	+w	w+	−w
melibiose	−	−	−	−	−	−	−	w−	−w	−
raffinose	−	−	−	−	−	−	−	+w	−w	−
salicin	−	−	−	−	−	−	−	−w	−w	−
starch	−	−	−	−	−	−	−	−w	−	−w
sucrose	−	−	−	−	w	−	−	+w	−w	−w
trehalose	−	−	−	−	−	−	−	−+	w−	−w
xylose	−	−	−	−	−	−	−	−w	−	−
Gas (PYG agar)	−2	4,2	4	4,2	3	−3	2−	4	4	−2

[a] Products from 1% peptone-1% yeast extract-1% glucose (PYG) broth cultures: capital letters indicate an average (from multiple cultures) of >1 meq of acid/100 ml broth; small letters, <1 meq/100 ml. A, acetic; B, butyric; F, formic; L, lactic; P, propionic; S, succinic; V, valeric. Products in parentheses ( ) may or may not be detected. Hydrogen production (approximate concentrations in headspace gas), −, none detected; 4, 3% or more. Growth in 20% bile, −, no growth; ±, very poor; 1, poor; 2, moderate; 4, excellent; S, stimulated (more growth than in the PYG control tube). + (sugars), pH below 5.5; − (sugars), pH 5.7 or above; w, weak reaction or (sugars) pH 5.5 to 5.7; a (milk), acid; c (milk), curd; p (milk), proteolyzed. Gas (PYG agar deeps), −, no gas detected; 2, splits across the agar; 3, agar displaced; 4, agar displaced to the top of the tube. Where two reactions are given, the first is the more common.

[b] See comments in text.

lactate to propionate. Pyruvate is converted to acetate and butyrate, sometimes also to formate, succinate, and lactate. H_2S is produced [SIM (sulfide-indole-motility medium), BBL] unless otherwise noted.

Further data on antimicrobial susceptibilities of species in the genus can be found in George et al. (1981).

Key for species in the genus **Fusobacterium**

I. Esculin not hydrolyzed.
  A. Propionate formed from threonine.
    1. Indole produced.
      a. Propionate not formed from lactate.
        b. Mannose not acid.
          c. No hydrogen produced, thin cells with tapered ends.
            1. *F. nucleatum*
          cc. Abundant hydrogen produced, pleomorphic cells.
            2. *F. gonidiaformans*
        bb. Weak acid from mannose.
          3. *F. varium*
      aa. Propionate formed from lactate.
        4. *F. necrophorum*
    2. Indole not produced.
      a. Weak acid from mannose.
        3. *F. varium*
      aa. Mannose not acid, coccoid cells.
        5. *F. perfoetens*
  B. Propionate not formed from threonine.
    1. Indole produced.
      6. *F. naviforme*
    2. Indole not produced.
      7. *F. russii*
II. Esculin hydrolyzed.
  A. Propionate formed from threonine.
    1. Lactose acid.
      8. *F. mortiferum*
    2. Lactose not acid.
      9. *F. necrogenes*
  B. Propionate not formed from threonine.
    10. *F. prausnitzii*

List of the Species of the Genus **Fusobacterium**

1. **Fusobacterium nucleatum** Knorr 1922, 17.[AL] [*Bacillus fusiformis* Veillon and Zuber 1898, 540 and many other combinations using "*Fusiformis*" except the organism described as *F. fusiforme* by Hoffman in the seventh edition of the *Manual*; Group I, Spaulding and Rettger 1937, 535; Group III (and probably *Fusobacterium polymorphum*) Baird-Parker 1960, 458; NOT *Fusobacterium plauti-vincentii* Knorr 1922, 5.]

nu.cle.a'tum. L. neut. adj. *nucleatum* having a kernel, nucleated.

This description is based on Knorr (1922) and study of the type strain and numerous other strains having 70–98% DNA homology with the type strain (J. L. Johnson, unpublished data).

Cells from glucose broth cultures are 0.4–0.7 by 3–10 μm, have tapered to pointed ends, and often have central swellings and intracellular granules. Cell length is variable but is usually fairly uniform within actively growing cultures. Cells do not possess pili or flagella (Falkler and Hawley, 1977; Dahlen *et al.*, 1978).

Surface colonies on blood agar are 1–2 mm in diameter, circular to slightly irregular, convex to pulvinate, translucent, often with a "flecked" appearance when viewed by transmitted light; usually non-

hemolytic (horse or rabbit blood), but may be slightly hemolytic under the area of confluent growth or may produce greenish discoloration of the blood agar upon exposure to oxygen.

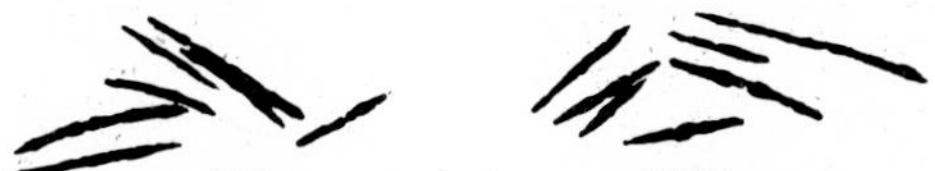

Glucose broth cultures have a flocculent or granular sediment, with or without turbidity, a final pH of 5.6–6.2, and a foul "bad breath" odor.

Produces DNase (Porschen and Sonntag, 1974). No phosphatase detected (Porschen and Spaulding, 1974). Most strains produce H_2S.

Capable of hemagglutinating human and animal erythrocytes.

Contains the diamino acid lanthionine in cell wall peptidoglycan as a major component; no lysine, diaminopimelic acid, or ornithine (Kato et al., 1979; Vasstrand, 1981). Heptose and KDO are present in the lipopolysaccharide (Hofstad, 1974).

Grows in the presence of up to 6% oxygen; survives exposure to air for 100 min (Loesche, 1969).

Resistant to 3 μg/ml of erythromycin (broth disk test). All strains tested (8–18) susceptible to less than 0.8 μg/ml ampicillin and lincomycin; less than 1 μg/ml cephalothin, tetracycline, doxycycline, minocycline and clindamycin; 4 μg/ml amoxicillin, cefoxitin, chloramphenicol and metronidazole; and 8 μg/ml carbenicillin. Ninety-four percent of the strains tested were susceptible to 2 U/ml penicillin G and 11% of the strains to 1 μg/ml erythromycin (Finegold, 1977).

Other characteristics of the species are given in Table 6.7 (species 1).

The inability of *F. nucleatum* to convert lactate to propionate is useful in differentiating it from *F. necrophorum*.

Figure 6.40. *Fusobacterium nucleatum.* (See Figure 6.1 for scale.)

Isolated from the gingival margin and sulcus and from infections of the upper respiratory tract and pleural cavity, occasionally from wounds and other kinds of infections.

The mol% G + C of the DNA is 27–28.

Type strain: ATCC 25586.

Further comments. From DNA/DNA homology studies of 50 strains isolated from human gingival sulci and adjacent tooth surfaces, homology values (hydroxylapatite method) ranged from 60 to over 90% with DNA from the type and 3 other reference strains [Y. Selin and J. L. Johnson: *Journal of Dental Research* (Special Issue A), *60*: 415 (Abstr. 420), 1981], indicating a fair amount of genetic diversity within this species. This diversity was seen among multiple strains isolated from different sites within the same mouth.

There still appears to be some confusion and concern about the relationship between *F. nucleatum* and "*F. fusiforme*" (Guillermet, 1981). Most certainly *Fusobacterium nucleatum* (type strain = ATCC 25586) is the same organism as "*Fusiformis fusiformis*" (Veillon and Zuber) Topley and Wilson 1938, 357 and "*Fusiformis fusiformis*" (Vincent) Topley and Wilson as cited in Prévot (1938, 1948, 1966) and as recognized by Prévot. Rules for bacteriological nomenclature were patterned after rules for botanical, rather than zoological, nomenclature. In bacteriology and botany it is the rule that a specific epithet is illegitimate if it is a tautonym [if it exactly repeats the name of the genus (Rule 25c, *International Code of Nomenclature*, 1958)]. Therefore, the name "*Fusiformis fusiformis*" is illegitimate because it is a tautonym. "*Fusobacterium fusiforme*" does not appear on the Approved Lists of Bacterial Names (Skerman et al., 1980). Since publication of the Lists (Skerman, 1980), neither "*Fusobacterium fusiforme*" nor "*Fusiformis fusiformis*" has any legitimate taxonomic standing.

"*Fusobacterium fusiformis*" does not appear on the Lists because of another taxonomic rule [9c(3b), *International Code of Nomenclature*, 1958], which states that the type species of a genus must be one of those included when the genus originally was described. When Knorr described the genus *Fusobacterium* he proposed only three species: "*F. plauti-vincentii*," *F. nucleatum*, and "*F. polymorphum*." "*F. plauti-vincentii*" is considered a later synonym of *Leptotrichia buccalis*. Therefore, if the genus *Fusobacterium* were to maintain legitimate taxonomic standing, the type species had to be either *F. nucleatum* or "*F. polymorphum*." These two species were considered synonymous by later workers (Spaulding and Rettger, 1937; Omata and Braunberg, 1960) and the name *F. nucleatum* usually was used. The species recognized by others as "*Fusiformis fusiformis*" was identical with *F. nucleatum*. "*Fusiformis fusiformis*" therefore is a synonym of *F. nucleatum*.

Had "*Fusiformis*" been used for the name of this genus, "*Fusiformis fusiformis*" still would have no taxonomic standing (because it is a tautonym). Therefore, due to an unfortunate set of circumstances, a specific epithet to which many persons were attached (one of these authors included) could not be used legitimately in either of these genera.

2. **Fusobacterium gonidiaformans** (Tunnicliff and Jackson 1925) Moore and Holdeman 1970, 45.*AL* [*Bacillus gonidiaformans* Tunnicliff and Jackson 1925, 430; *Sphaerophorus gonidiaformans* (Tunnicliff and Jackson 1925) Prévot 1938, 299.]

go.ni.di.a.for′mans. Gr. n. *gone* offspring, seed; M.L. n. *gonidium* gonidium; L. part. adj. *formans* forming; M.L. part adj. *gonidiaformans* gonidia forming.

This description is based on our study of the type strain and 20 phenotypically similar other strains.

Cells from glucose broth cultures are pleomorphic and vacuolated, 0.4–0.7 by 0.7–3.0 μm, often with degenerate filaments or long strands. The spheroid or gonidial forms implied by the name of this organism are seen most often in old cultures or in media that are not highly reduced.

Surface colonies on horse blood agar plates are punctiform to 1 mm in diameter, circular, entire, low convex, translucent and smooth.

Glucose broth cultures are turbid with smooth sediment and a final pH of 5.6–6.2

PY   PYG

Figure 6.41.   *Fusobacterium gonidiaformans.*

Produces DNase (Porschen and Sonntag, 1974). No phosphatase is detected (Porschen and Spaulding, 1974). Hippurate is hydrolyzed.

Resistant to 3 μg/ml erythromycin; sensitive to 2 U/ml penicillin G, 1.6 μg/ml clindamycin, 12 μg/ml chloramphenicol and 6 μg/ml tetracycline.

Other characteristics of the species are given in Table 6.7 (species 2).

Differences in cellular morphology help differentiate *F. gonidiaformans* and *F. nucleatum*.

Isolated from the intestinal and urogenital tracts of humans, various types of human infections and from a lamb with pneumonia.

The mol% G + C of the DNA is not known.

Type strain: ATCC 25563.

3. **Fusobacterium varium** (Eggerth and Gagnon 1933) Moore and Holdeman 1969, 12.*AL* [*Bacteroides varius* Eggerth and Gagnon 1933, 409; *Sphaerophorus varius* (Eggerth and Gagnon 1933) Prévot 1938, 299.]

va′ri.um. L. neut. adj. *varium* diverse, varied.

This description is based on our study of the type strain and 7 strains that have DNA homology with the type (J. L. Johnson, unpublished data).

Cells from glucose broth cultures are pleomorphic, coccoid and rod-shaped, and stain unevenly. Cells are 0.3–0.7 by 0.7–2.0 μm and occur singly and in pairs.

PY   PYG

Figure 6.42.   *Fusobacterium varium.*

Surface colonies on blood agar are punctiform to 1 mm in diameter, circular with entire edges, flat to low convex, translucent, usually with gray-white centers and colorless edges.

Glucose broth cultures are turbid with smooth sediment and a final pH of 5.3 to 5.7.

Dextran is not hydrolyzed (Holbrook and McMillan, 1977). Produces lysine decarboxylase (Werner, 1974). Produces DNase (Porschen and Sonntag, 1974). No phosphatase detected (Porschen and Spaulding, 1974).

Heptose and KDO are present in the lipopolysaccharide (Hofstad, 1974).

A bacteriophage active against four strains was isolated from filtrate of feline feces (Huet and Thouvenot, 1964).

Two strains tested were resistant to 3.1 μg/ml ampicillin but not to 6.2 μg/ml. Susceptible to 8 μg/ml cephalothin (8/8 strains), 12.5 μg/ml cefazolin (10/10 strains), less than 1 μg/ml metronidazole (10/10 strains). Resistance to lincomycin (2/8 strains susceptible to 3.1 μg/ml) and clindamycin (4/8 strains susceptible to 4 μg/ml) was variable among strains (Finegold, 1977). Resistance to 6 μg/ml tetracycline and 2 U/ml penicillin G also is variable among strains (broth disk test). All strains are resistant to 3 μg/ml erythromycin.

Other characteristics of the species are given in Table 6.7 (species 3).

Isolated from human feces, purulent infections of man (upper respiratory tract, surgical wounds, peritonitis), cecal contents of mice, intestinal contents of *Blatta orientalis* (roach), posterior intestinal tract of *R. lucifugus* (termite) and vaginal swab of chinchilla.

The mol% G + C of the DNA is 29 [chromatographic separation (Sebald, 1962)]; 26–28 [T_m (Johnson, unpublished data)].

Type strain: ATCC 8501 (NCTC 10560).

Further comments. Strains that have a high degree of nucleic acid similarity with the type strain all weakly ferment mannose. Several strains that do not ferment mannose, but are similar to *F. varium* in other phenotypic characteristics show no DNA homology with the type strain (J. L. Johnson, unpublished data).

4. **Fusobacterium necrophorum** (Flügge 1886) Moore and Holdeman 1969, 12.[AL] [*Bacillus necrophorus* Flügge 1886, 273; *Fusiformis necrophorus* (Flügge) Topley and Wilson 1929, 299; *Sphaerophorus necrophorus* (Flügge) Prévot 1938, 298.]

ne.cro′pho.rum. Gr. adj. *necros* dead; Gr. adj. *phorum* bearing; M.L. neut. adj. *necrophorum* necrosis producing.

This description is based on our study of the type strain and numerous other strains with similar characteristics.

Cells in glucose broth cultures are 0.5–0.7 μm in diameter with swellings up to 1.8 μm. The ends of the cells may be round or tapered. Cell length ranges from coccoid bodies to filaments over 1.00 μm. Filamentous forms with granular inclusions are more common in broth, while bacilli are more common in older cultures and growth on agar.

PY   **PYG**

Figure 6.43. *Fusobacterium necrophorum.*

Surface colonies on blood agar are 1–2 mm in diameter; circular with scalloped to erose edges, convex to umbonate, often with bumpy, ridged, or uneven surface; translucent to opaque, often with mosaic internal structure when viewed by transmitted light. Most strains produce either α- or β-hemolysis on rabbit blood agar. In general, the β-hemolytic strains are lipase-positive (on egg yolk agar) and the α-hemolytic or nonhemolytic strains are lipase-negative. No lecithinase is produced.

Glucose broth cultures have a smooth, flocculent, granular, or stringy sediment and usually are turbid. The final pH of fructose and glucose cultures is 5.6–6.3. A few strains produce a pH of 5.8–5.9 in maltose medium.

Human, rabbit and guinea pig red blood cells are agglutinated, bovine and ovine red blood cells are not (Simon, 1975).

Dextran is not hydrolyzed (Holbrook and McMillan, 1977). No superoxide dismutase (Gregory et al., 1978) or lysine decarboxylase (Werner, 1974) is detected.

Produces DNase (Porschen and Sonntag, 1974). No phosphatase is detected (Porschen and Spaulding, 1974).

Heptose and KDO are present in the lipopolysaccharide (Hofstad, 1974).

Resistance to 3 μg/ml erythromycin is variable among strains. Susceptible to 12 μg/ml chloramphenicol, 1.6 μg/ml clindamycin, 2 U/ml penicillin and 6 μg/ml tetracycline. Four of 4 strains tested were susceptible to less than 1 μg/ml cephalothin (Finegold, 1977).

Other characteristics of the species are given in Table 6.7 (species 4).

Isolated from the natural cavities of man and other animals and from clinical specimens (necrotic lesions, abscesses, and blood) of man and other animals, particularly liver abscesses and foot rot of cattle.

For a review of natural and experimental pathogenicity, see Prévot et al. (1967, pp. 307–321 and 335–343) and Langworth (1977).

The mol% G + C of the DNA is 31–34 [chromatographic separation (Sebald 1962)].

Type strain: ATCC 25286 (from bovine liver abscess; Fievez strain 2358).

Further comments. Beerens et al. (1971) suggest that this species may exist in three phases. Phase A corresponds to organisms previously recognized as "*S. necrophorus*" and is hemolytic, has a hemagglutinin, and is pathogenic for mice. Phase B corresponds to organisms previously recognized as "*S. funduliformis*" and is hemolytic, has no hemagglutinin, and demonstrates little, if any, pathogenicity for mice. Phase C corresponds to organisms previously recognized as "*S. pseudonecrophorus*" and is not hemolytic, has no hemagglutinin, and is not pathogenic for mice. A tendency toward mutation to penicillin resistance (500 units/ml) also is reported to accompany Phase A to C mutation. Phase A organisms are isolated more often from cattle than from man and Phase B and C organisms from man more often than from other animals.

Strains of *F. necrophorum* that produces α-hemolysis or are nonhemolytic may be differentiated from *F. nucleatum* by their conversion of lactate to propionate and by their production of large amounts of gas in peptone yeast extract glucose agar deeps. The gas splits the agar and often raises parts of the agar to the top of the tube.

There is considerable doubt concerning the validity of "*F. necrophorum*" or "*Sphaerophorus necrophorus*" as identified in many publications prior to 1970.

5. **Fusobacterium perfoetens** (Tissier 1905) Moore and Holdeman 1973, 72.[AL] [*Coccobacillus perfoetens* Tissier 1905, 110; *Ristella perfoetens* (Tissier 1905) Prévot 1938, 291; *Sphaerophorus perfoetens* (Tissier 1905) Sebald 1962, 149.]

per.foe′tens. L. perf. *per* very; L. part. adj. *foetens* stinking; M.L. part. adj. *perfoetens* very stinking.

Description from Prévot et al. (1967, 381), Weinberg et al. (1937, 790), van Assche and Wilssens (1977), and our study of the type strain.

Cells from glucose cultures are 0.6–0.8 by 0.8–1.0 μm, oval, never elongated, occurring singly, in pairs, in chains of no more than three cells, or in masses. No flagella or capsule.

PY   **PYG**

Figure 6.44. *Fusobacterium perfoetens.*

Colonies in deep agar (2 days) are 1 mm in diameter and lenticular. Surface colonies on blood agar are 1–2 mm in diameter, circular with an entire edge, convex to raised, grayish white, translucent, smooth, and nonhemolytic on horse blood. Colonies of some strains are slightly umbonate with diffuse edges and a slightly mottled or granular appearance.

Growth in glucose broth is rapid. Cultures are turbid with a fine to ropy sediment and a pH of 5.6. Gas and a fetid odor are produced. Galactose is weakly fermented (van Assche and Wilssens, 1977). Growth is enhanced in media containing fructose, glucose, mannose, sucrose and trehalose. The pH of cultures in these media ranges from 5.6 to 5.95.

Produces CO_2 and NH_3. Major amounts of lactic acid produced from PYG cultures (van Assche and Wilssens); no lactic acid detected in PYG cultures of the type strain (unpublished data).

Inhibited by 0.001% polymyxin. Resistant to 0.001% brilliant green. Optimum temperature is 37°C; good growth occurs at 45°C, poor growth at 25–30°C. Survives up to 24 h of exposure to air.

Isolated by Tissier in 1900 from an infant with diarrhea and in 1905 from nursing infants. Strain CC1 isolated in 1947 by Prévot from the cecum of a horse and studied by Sebald (1962) has been lost. Van Assche and Wilssens (1977) studied six isolates from the feces of a 2-week-old pig, one of which was designated as the neotype strain.

The mol% G + C of the DNA is 28–30 (Sebald, 1962; van Assche and Wilssens, 1977).

Type strain: ATCC 29250.

6. **Fusobacterium naviforme** (Jungano 1909) Moore and Holde-

man 1970, 45.[AL] [*Bacillus naviformis* Jungano 1909, 123; *Ristella naviformis* (Jungano 1909) Prévot 1938, 291.]

na.vi.for′me. L. n. *navis* ship; L. n. *forma* shape; M.L. neut. adj. *naviforme* in the shape of a ship.

This description is based on our study of the type strain and 30 other phenotypically similar strains.

Cells from glucose broth cultures are 0.5–0.7 by 3–12 μm, usually with pointed ends, and occurring as pairs and in chains. Cells in old cultures have a beaded appearance.

PY   **PYG**

Figure 6.45. *Fusobacterium naviforme.*

Surface colonies are punctiform to 2.0 mm in diameter, circular, entire, low convex, gray-white, translucent with mottled appearance when viewed by obliquely transmitted light.

Glucose broth cultures are lightly turbid with smooth to clumpy sediment and final pH of 5.5–6.4.

Produces DNase (Porschen and Sonntag, 1974). No phosphatase is detected (Porschen and Spaulding, 1974).

Heptose and KDO are present in the lipopolysaccharide (Hofstad, 1974).

About one-half of the strains are resistant to 3 μg/ml erythromycin. Susceptible to 12 μg/ml chloramphenicol, 1.6 μg/ml clindamycin, 2 U/ml penicillin G and 6 μg/ml tetracycline.

Other characteristics of the species are given in Table 6.7 (species 6).

Isolated by Jungano from the large intestine of a laboratory rat. Other strains have been isolated from the human gingival sulcus, from various human clinical specimens, and the bovine rumen.

The mol% G + C of the DNA is not known.

Type strain: ATCC 25832.

7. **Fusobacterium russii** (Hauduroy, Ehringer, Urbain, Guillot and Magrou 1937) Moore and Holdeman 1970, 45.[AL] (*Bacteroides russii* Hauduroy, Ehringer, Urbain, Guillot and Magrou 1937, 73.)

rus′si.i. M.L. gen. noun *russii* of V. Russ, the bacteriologist who first cultured this organism.

This description is based on our study of the type and 27 phenotypically similar human clinical and intestinal strains.

Cells in glucose broth are 0.3–0.7 by 1.5–4.0 μm with thin filaments 10–15 μm in length. Pallisade cellular arrangement often is seen in smears. In more oxidized media (such as thioglycolate), thin filamentous beaded forms with pointed ends are common.

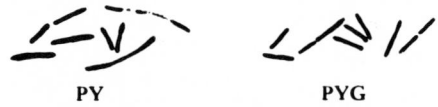

PY   **PYG**

Figure 6.46. *Fusobacterium russii.*

Surface colonies on horse blood agar are 0.5–1 mm in diameter, circular, smooth, shiny, entire, convex and translucent. The type strain is β-hemolytic on horse blood agar.

Glucose broth cultures are turbid, often with stringy sediment and have a final pH of 5.9–6.1.

Produces DNase and phosphatase (Porschen and Sonntag, 1974; Porschen and Spaulding, 1974). The type strain produces H₂S in SIM.

Heptose and KDO are present in the lipopolysaccharide (Hofstad, 1974).

Susceptible to 12 μg/ml chloramphenicol, 1.6 μg/ml clindamycin and 2 U/ml penicillin G. Some strains are resistant to 6 μg/ml tetracycline.

Other characteristics of the species are given in Table 6.7 (species 7).

Isolated by Russ (1905) from perianal abscess. Also isolated from infections of cats, including actinomycosis of cats, and from human and animal feces.

The mol% G + C of the DNA is 31 (Sebald, 1962).

Type strain: ATCC 25533 (isolated from a cat lesion).

Further comments. Polyacrylamide gel electrophoresis (PAGE) patterns of soluble cellular proteins of the type strain and of intestinal isolates are exceedingly similar. The PAGE patterns of oral strains with similar phenotypic characteristics are different (unpublished data).

8. **Fusobacterium mortiferum** (Harris 1901) Moore and Holdeman 1970, 45.[AL] [*Bacillus mortiferus* Harris 1901, 546; *Sphaerophorus mortiferus* (sic) (Harris 1901) Prévot 1938, 299.]

mor.ti′fer.um. L. n. *mors, mortis* death; L. v. *fero* bear; M.L. neut. adj. *mortiferum* death bearing.

This description is from Prévot et al. (1967), Prévot (1966) and our study of the type and 35 other phenotypically similar strains.

Cells from glucose broth cultures are 0.8–1.0 by 1.5–10 μm, occurring singly and in pairs and short chains. Cells stain irregularly and may be extremely pleomorphic with globular forms, swellings, and threads.

PY   **PYG**

Figure 6.47. *Fusobacterium mortiferum.*

Surface colonies on horse blood agar are 1–2 mm in diameter, circular with entire, diffuse, or slightly scalloped edge; convex or slightly umbonate; translucent; smooth. Colonies in glucose agar deeps are lenticular or irregular and may be surrounded by smaller colonies.

Glucose broth cultures are uniformly turbid with smooth or semiviscous sediment.

No superoxide dismutase is detected (Gregory et al., 1978). Lysine decarboxylase is not produced (Werner, 1974). H₂S is produced in SIM.

Produces DNase and phosphatase (Porschen and Sonntag, 1974; Porschen and Spaulding, 1974).

Heptose and KDO are present in the lipopolysaccharide (Hofstad, 1974).

Eight of 11 strains were susceptible to 8 μg/ml cephalothin, all of 15 strains were susceptible to 3.1 μg/ml cefazolin, 11/11 strains were susceptible to 4 μg/ml cefoxitin, 10/10 were susceptible to less than 0.8 μg/ml lincomycin and to less than 1 μg/ml clindamycin; and 17/17 were susceptible to less than 1 μg/ml metronidazole (Finegold, 1977). Resistant to 3 μg/ml erythromycin; susceptible to 12 μg/ml chloramphenicol, 2 U/ml penicillin G and 6 μg/ml tetracycline (broth disk method).

Other characteristics of the species are given in Table 6.7 (species 8).

Isolated from blood and various human clinical specimens, intestinal tract and feces; and once from irradiated mice.

The mol% G + C of the DNA is 26–28 (chromatographic separation, Sebald, 1962; T_m, J. L. Johnson, personal communication).

Type strain: ATCC 25557.

Further comments. Strains similar to *F. mortiferum* have been isolated from human feces. However, these isolates did not grow on blood agar plates incubated in anaerobic jars, had a uniform cellular morphology, and produced little or no hydrogen from glucose (Holdeman et al., 1976).

9. **Fusobacterium necrogenes** (Weinberg, Nativelle and Prévot 1937) Moore and Holdeman 1970, 45.[AL] [*Bacillus necrogenes* Weinberg,

Nativelle and Prévot 1937, 681; *Spherophorus necrogenes* (*sic*) (Weinberg et al. 1937) Prévot 1937, 298.]

ne.cro'ge.nes. Gr. adj. *necros* dead; Gr. v. *gennaio* produce; M.L. adj. *necrogenes* necrosis producing.

This description is based on our study of the type and three phenotypically similar strains.

Cells from 1-day-old glucose broth cultures are extremely pleomorphic with coccoid cells about 0.3–0.8 μm and thin filamentous forms 0.2–0.8 μm in diameter and up to 20 μm in length. Cells in older cultures are somewhat more uniform irregularly staining rods 0.7–0.8 by 1.5–4.0 μm.

PY   **PYG**

Figure 6.48. *Fusobacterium necrogenes.*

Surface colonies on horse blood agar are minute to 0.5 mm in diameter, circular, flat to low convex, entire, translucent, white, smooth, shiny. Colonies are surrounded by zones of β-hemolysis on VL agar containing horse blood (Barnes and Impey, 1968).

Glucose broth cultures are moderately turbid with pH of 5.7 to 6.0.

Glutamine decarboxylase-negative (Terada et al., 1976).

Heptose and KDO present in the lipopolysaccharide (Hofstad, 1974).

Other characteristics of the species are given in Table 6.7 (species 9).

Originally isolated by Kawamura (1926) from necrotic abscess of a chicken. Barnes strain EB/D/1/4a was isolated from cecal contents of a duck (Barnes and Impey, 1968). Other strains have been isolated from human feces.

The mol% G + C of the DNA of EB/D/1/4a (parent strain of the type strain) is 28 (Barnes and Impey, 1968).

Type strain: ATCC 25556 (NCTC 10723).

10. **Fusobacterium prausnitzii** (Hauduroy, Ehringer, Urbain, Guillot and Magrou 1937) Moore and Holdeman 1970, 45.[AL] [*Bacteroides praussnitzii* (*sic*) Hauduroy, Ehringer, Urbain, Guillot and Magrou 1937, 68.]

praus.nit'zi.i. M.L. gen. n. *prausnitzii* of Prausnitz; named for C. Prausnitz, the bacteriologist who first isolated this organism.

Literature description supplemented with results from study of the type and 40 strains (Cato et al., 1974).

Cells from glucose broth cultures are 0.5–0.9 by 2.5–14 μm; short forms are straight with tapered or rounded ends, longer cells are curved. Occur singly, in pairs, or in short chains; may stain irregularly with colorless vacuoles. A thick capsule is present.

PY   **PYG**

Figure 6.49. *Fusobacterium prausnitzii.*

Colonies on blood agar streak tubes are pinpoint to 1 mm in diameter, circular, entire to slightly erose, convex, translucent to transparent, colorless to milky white; occasionally have a granular or mottled appearance. Difficult to cultivate on anaerobic agar plates.

Glucose broth cultures are turbid with viscous or ropy sediment and final pH of 5.5–6.0. Most strains are stimulated by 5% rumen fluid.

Converts galacturonate to acetate, butyrate, and formate. Neutral red is reduced, hippurate is not hydrolyzed, and deoxyribonuclease is variable. May be urease-positive on initial isolation (Wozny et al., 1977).

Originally isolated from purulent pleurisy. Usually isolated from human or animal feces. One of the 10 most predominant species in the human intestine (Moore and Holdeman, 1974b; Holdeman et al., 1976).

The mol% G + C of the DNA is 52–57.

Type strain: ATCC 27768.

Further comments. Although this species predominates in human intestinal contents and feces, we have not encountered any strains in clinical specimens.

Note Added in Proof

The reader's attention is directed to the following species published while this manuscript was in press:

Fusobacterium simiae Slots and Potts 1982, 193.[VP]

sim'i.ae. L. fem. n. *simia* monkey; L. gen. n. *simiae* of the monkey.

Cellular morphology resembles that of *F. nucleatum* (*q.v.*, species 1).

Fructose and glucose are fermented (pH of 5.5–5.6). Twenty-seven other substrates tested are not fermented. Indole and lipase are produced; hippurate is hydrolyzed. Grows in media containing 2% oxgall.

From glucose, butyrate is the major fermentation product; major amounts of acetate and small amounts of propionate, lactate, and succinate also are produced. Neither hydrogen nor gas is detected. Lactate and threonine are converted to propionate.

The type strain has 48% DNA homology with the type and one other strain of *F. nucleatum* and 9% homology with the type strain of *F. necrophorum*.

Isolated from the mouth of the stump-tailed macaque (*Macaca arctoides*).

The mol% G + C of the DNA is 27–28.

Type strain: ATCC 33568 (Slots and Potts 7511 R2-13).

Reference:

Slots, J. and T.V. Potts. 1982 *Fusobacterium simiae*, a new species from monkey dental plaque. Int. J. Syst. Bacteriol. *32:* 191–194; ibid. **33:** 442.

Genus III. **Leptotrichia** Trevisan 1879, 138[AL]

TOR HOFSTAD

Lep.to.trich'i.a. Gr. adj. *leptus* fine, small; Gr. n. *thrix*, *thricis* hair; M.L. fem. n. *Leptotrichia* fine hair.

Straight or slightly curved rods, 0.8–1.5 μm wide and 5–15 μm long, **with one or both ends pointed or rounded**. Frequently arranged in pairs, chains or septate filaments. No club formation or branching. Nonmotile. Gram-negative, often with Gram-positive granules distributed evenly along the long axis. May be Gram-positive in very young cultures. **Anaerobic on first isolation; many strains subsequently grow aerobically in the presence of CO₂.** Optimum temperature, 35–37°C; little or no growth occurs at 25°C. Good growth occurs at pH 7.0–7.4. Chemoorganotrophic. **Metabolize carbohydrates with formation of acid without gas. The major product of glucose fermentation is lactic acid.** Acetic and succinic acids may be produced in trace amounts. Catalase, hydrogen sulfide and

indole are not produced. Nitrate is not reduced. **Habitat: oral cavity of man.** Also found in the female periurethral region and is present in the oral cavity of guinea pigs. The mol% G + C of the DNA is 25 (T_m, Bd).

Type species: *Leptotrichia buccalis* (Robin) Trevisan 1897, 147.

Further Descriptive Information

In Gram-stained smears of 48-h-old cultures, *Leptotrichia* generally appears as Gram-negative single cells or in pairs end to end (with the adjacent ends flattened), and in a "strawlike" arrangement. Young cultures show a varying proportion of Gram-positive cells. Thread forms and chains up to 100 μm long are particularly evident in smears made from colonies with pronounced filamentous margins and from old fluid cultures. Bizarre irregular forms, such as coccoid bodies, bulbous swellings and "ropelike" twisted filaments have been observed (Kasai, 1961).

Electron microscopy of *Leptotrichia* reveals an ultrastructure characteristic of a Gram-negative bacterium (Hofstad and Selvig, 1969). An interesting feature is the occurrence of scalelike membranous folds (Fig. 6.50) which protrude from the surface of the outer membrane (Listgarten and Lai, 1975). The diamino acid of the peptidoglycan is diaminopimelic acid (Baboolal, 1969; Davis and Baird-Parker, 1959). The purified peptidoglycan of *L. buccalis* strain L11 contains muramic acid, *N*-acetylglucosamine, glutamic acid, alanine and diaminopimelic acid in a probable molar ratio of 1.0 : 1.0 : 1.0 : 1.8 : 1.2 (Vasstrand et al., 1982).

Although the primary structure of the peptidoglycan has not been examined, the amino acid composition of strain L11 indicates that the peptidoglycan of *L. buccalis* belongs to the A1γ type characteristic for Gram-negative bacteria (Schleifer and Kandler, 1972). Crude cell walls

of *L. buccalis* contain 10–12% lipid, as measured by the amount of material that could be extracted by diethyl ether from walls treated with hydrochloric acid (Baboolal, 1969). The cellular fatty acid composition of whole cells includes hexadecanoic, octadecanoic and 3-hydroxytetradecanoic acids as major components (Hofstad and Jantzen, 1982) (see Table 6.10).

Endotoxic lipopolysaccharide (LPS) has been isolated from strains of *L. buccalis* extracted with 45% aqueous phenol (de Araujo et al., 1966; Gustafson et al., 1966; Knox and Parker, 1973). Little is known about the chemical structure of the LPS. Knox and Parker (1973) detected heptose, but no deoxyoctulosonic acid in their preparations. Both components, viz., L- (occasionally D-) *glycero*-D-*manno*-heptose and 3-deoxy-D-*manno*-octulosonic acid, are characteristic constituents of the polysaccharide core backbone of a bacterial LPS. LPS from three *L. buccalis* strains (ATCC 19616, ATCC 14201, L11) is currently being studied at our laboratory. All three LPS have a lipid A component containing 3-hydroxytetradecanoic and dodecanoic acids, and a polysaccharide moiety. D-*glycero*-D-*manno*-heptose is present in all LPS,

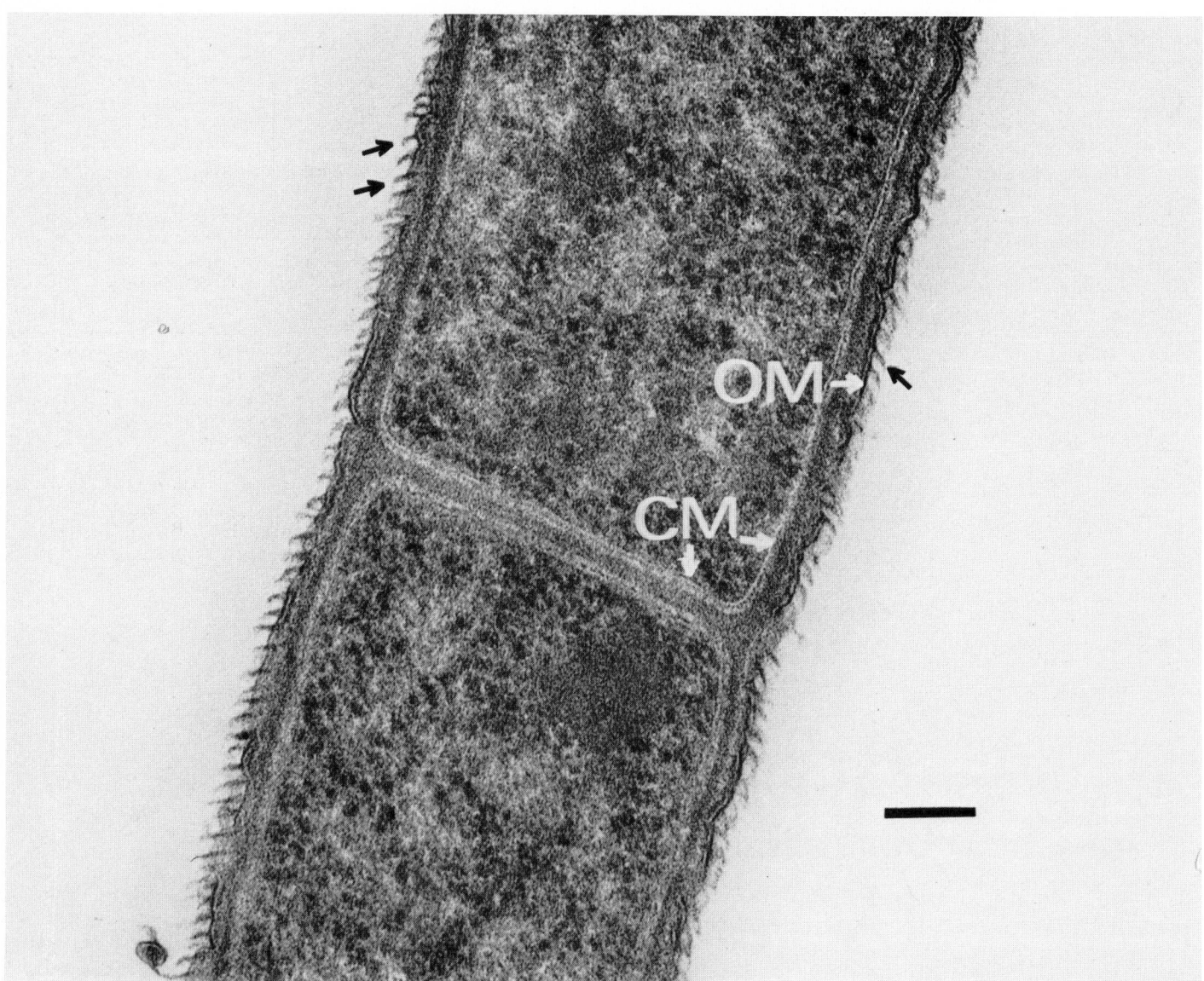

Figure 6.50. Longitudinal section of *Leptotrichia buccalis* ATCC 19616, which shows a typical Gram-negative cell wall structure and associated membraneous folds (*arrows*). *OM*, outer membrane; *CM*, cytoplasmic membrane. *Bar* = 0.1 μm. (Reproduced by permission from M. A. Listgarten and C.-H. Lai: *Journal of Bacteriology, 123:* 747–749, 1975, © American Society for Microbiology.

but 3-deoxy-D-*manno*-octulosonic acid has been detected only in the LPS of strain L11 (Birkeland and Hofstad, 1982).

Colonies of *Leptotrichia* grown anaerobically on blood agar or the tryptone-yeast extract medium described by Kasai (1961)* are distinctive. After 2 or 3 days of incubation they are smooth, colorless, 2 to 3 mm in diameter, convex and with a convoluted surface resembling that of a human brain. The colonies are sometimes raised with a filamentous edge. The last colony type is especially seen after incubation for 24 h or less (Hamilton and Zahler, 1957; Kasai, 1961). The colonies are nonhemolytic and nonadherent to the medium. Pleomorphism in colony morhology may be seen (Kasai, 1961). Occasionally both brainlike colonies and colonies with wide filamentous edges are seen in subcultures made from single colonies.

The nutritional requirements of *Leptotrichia* and the metabolic pathways utilized for the generation of energy have not been examined in detail. Like many other microorganisms parasitizing the mucous membranes of man and animals, *Leptotrichia* possibly requires several growth factors.

Leptotrichia is highly saccharoclastic. Several mono- and disaccharides are fermented with the production of D- and L-lactic acid alone, or accompanied by trace amounts of acetic and succinic acids. The terminal pH is 5.4 or less. CO_2 is not produced. A few strains hydrolyze starch within 2–4 days. Hamilton and Zahler (1957) found that strains which were unable to hydrolyze starch initially did so upon reinoculation of the culture with the same strain. Pyruvate is fermented with production of CO_2, acetic and formic acids (Jackins and Barker, 1951).

Some strains have been tested for production of α- and β-galactosidase, α- and β-glucosidase, glucuronidase, N-acetyl-β-glucosaminidase, α-mannosidase and α-fucosidase, using the API ZYM test kit (API System S.A., La Balme Les Grottes, Montalieu Vercieu, France) (Hofstad, unpublished results). A positive reaction was obtained for α-glucosidase. Although the strains examined fermented lactose, which is a natural substrate for β-galactosidase, production of this enzyme could not be demonstrated.

L. buccalis does not form ammonia, indicating that nitrogeneous compounds are not metabolized. Kasai (1965), however, observed some growth of *L. buccalis* in a medium based on tryptone and yeast extract without carbohydrate added.

L. buccalis grows in media based on peptones and yeast extract supplemented with a fermentable carbohydrate. Growth is enhanced by addition of serum to the medium. Prereduced, anaerobically sterilized (PRAS) media are preferable for consistent growth, but not indispensable. The organism grows well in ordinary semifluid media, and in fluid media containing a reducing agent and dispensed in narrow-necked bottles or tubes. An anaerobic atmosphere with 5–10% CO_2 is essential for isolation of *Leptotrichia* and for good growth on solid media. Many strains become aerotolerant upon transfer, and grow slowly on the surface of solid media in an ordinary CO_2 incubator.

Page and Krywolap (1976) examined the DNA base composition of *L. buccalis* strains ATCC 19616 and ATCC 23471 by thermal denaturation of DNA and buoyant-density determination. The mol% G + C was similar in both strains: 24.5 and 24.8, respectively. The polynucleotide homology between the same strains and two strains of *Fusobacterium nucleatum* was less than 30%. Hofstad (1970), using thermal denaturation, found the mol% G + C of the DNA of *L. buccalis* strain L11 to be 34.

In early years, serological studies were performed in order to clarify the taxonomic relationships between anaerobic fusiform rods that differed with respect to biochemical properties (Varney, 1927; Slanetz and Rettger, 1933; Spaulding and Rettger, 1937). These studies showed that there was no serological relationship between nonsaccharoclastic and saccharoclastic fusiform bacteria, viz. between *Fusobacterium* and *Leptotrichia*. Gustafson and Kroeger (1962) examined 39 strains of *L. buccalis* by agglutination, precipitation and indirect immunofluores-

cence. They found both group-reactive and type-specific antigens, and postulated that the type specific antigens were located in the outermost layer of the bacterial cell. Their results were corroborated by de Araujo et al. (1963), who presented evidence for the presence of a trypsin resistant group antigen in *L. buccalis* and for type specificity harbored by the LPS. The LPS of *L. baccalis* ATCC 14201 and ATCC 19616 show serological cross-reactivity. Both are R-form LPS, the oligosaccharides of which are in acid-labile linkage to the lipid A moiety (Birkeland and Hofstad, unpublished observations).

The sensitivity of *L. buccalis* to antimicrobial agents has not been examined systematically. Observations by Kunz (1974) indicate that the organism is sensitive to penicillin G, cephalosporins, tetracyclines and chloramphenicol, but resistant to gentamicin, kanamycin and erythromycin. *L. buccalis* is sensitive to metronidazole (Hofstad, unpublished observations).

The habitat of *Leptotrichia* is the oral cavity of man. The principal niche is the dental plaque, i.e. the bacterial deposit which forms on the tooth surface and at the gingival margin. The number of *L. buccalis* in plaque material is uncertain, possibly less than 1% of the total viable count (Slack and Bowden, 1965). The occurrence of *L. buccalis* is not solely dependent on tooth eruption since the organism has been isolated from the mouth of pre-dentate infants (McCarthy et al., 1965). *L. buccalis* has been found also in the oral cavity of guinea pigs fed commercial pellets (Hofstad, unpublished observations).

A Gram-negative organism identified as *Leptotrichia* is present as a member of the normal microflora of the periurethral region of healthy girls (Bollgren et al., 1979). The organism has also been isolated from the cervix of a pregnant woman with premature rupture of the membranes (Evaldson et al., 1980).

L. buccalis is considered apathogenic for man or animals. However, a case of cavitary pulmonary disease with associated septicemia due to *L. buccalis* in an immunocompromised patient has recently been reported (Morgenstein et al., 1980). Being a saccharoclastic organism, *L. buccalis* may participate in the development of tooth decay. It has also been suggested that *L. buccalis* may play a role in periodontal disease (Krywolap and Page, 1977). However, the organism is present in gingival crevices and pockets at less than 0.2 or 2.7% of the total viable counts of fusiform bacteria in normal individuals or patients with advanced chronic periodontal disease, respectively (Hadi and Russell, 1969). On the other hand, the cell wall LPS is a potent endotoxin (Gustafson et al., 1966). Antibodies reacting with LPS are present in normal human serum (Mergenhagen et al., 1965; Falkler and Hawley, 1976). These are IgM antibodies (Hawley and Falkler, 1976) and may be included among the so-called "natural" antibodies. Mashimo et al. (1976) found that with age there were increasing titers of antibodies reactive with a dialyzed and concentrated culture supernatant of *L. buccalis*. Antibodies were also present in subjects who were edentulous for 10–20 years. Their findings suggest heteroimmunization, i.e. that the antigenic stimulus is a cross-reactive antigen.

L. buccalis has been observed to cause hemagglutination of red blood cells from man and warm-blooded animals (Kondo et al., 1976). Human buccal epithelial cells, HeLa and embryonic kidney cells were able to bind fragments of bacterial cells causing hemagglutination (Falkler et al., 1978). A salivary glycoprotein is able to aggregate cells of *L. buccalis* (Kondo et al., 1978). This glycoprotein may be identical with the human and animal cell receptor for the hemagglutinating substance.

Isolation Procedures

L. buccalis is best isolated from plaque between adjacent teeth and gingival crevices. Plaque samples are taken by a sterile metal instrument which is used to scrape the surface of the tooth. Sampling from the gingival crevice is best performed by the use of sterile filter paper points which are gently inserted into the crevice. Plaque material can

---

* Kasai's medium (g/liter): tryptone, 10.0; yeast extract, 2.0; NaCl, 5.0; $K_2HPO_4 \cdot 3H_2O$, 5.0; soluble starch, 20.0; cysteine hydrochloride, 0.5; pH adjusted to 7.2–7.4.

be inoculated either directly to the medium or after being suspended or diluted. Suitable diluents are the serum-containing diluent of Bowden and Hardie (1971) and the WAL diluent of Sutter et al. (1971). The infected tapering end of the paper point is streaked on a small area of the surface of a solid medium, with further spreading of the deposited material being carried out by a wire loop.

Isolation under selective conditions is preferred. We have found the CVE agar medium (Walker et al., 1979)* prepared for selective isolation of oral fusobacteria most useful. This medium takes advantage of the resistance of *L. buccalis* to erythromycin and low concentrations of certain dyes. Large brainlike or rough colonies of *L. buccalis* are easily differentiated from the smaller and smooth or slightly irregular colonies of *F. nucleatum*. The selective media of Omata and Disraely (1956) and Baird-Parker (1957) may also be used. It is advisable to check individual batches of the dye for their suitability before preparing a dye-containing medium. It is not yet known whether the different selective media are usable for isolation of *Leptotrichia* from the periurethral region.

The characteristic colony morphology of *L. buccalis* makes nonselective isolation on blood agar or the starch-containing basal medium of Kasai (1961) fully possible.

Maintenance Procedure

L. buccalis can be maintained by weekly serial subculture on blood agar. The viability of some strains may be lost after 4–5 days of incubation in a fluid or semifluid medium. Viable cells may be stored in the lyophilized state. Lyophilization is, however, not always successful. Storage of live cultures in liquid nitrogen may be a better alternative for maintaining viability.

Procedures for Testing Special Characters

Fermentation of carbohydrates and testing for other biochemical properties may be performed as outlined in the VPI Anaerobe Laboratory Manual (Holdeman et al., 1977). PRAS media are preferable, but good growth of *L. buccalis* may be obtained in ordinary rich semifluid media dispensed in narrow tubes.

Large amounts of microbial cells are necessary for isolation of LPS or other cell components. Mass cultivation is performed in fluid medium based on tryptone and yeast extract and supplemented with glucose (0.5%), cysteine (0.1%) and animal or human serum or plasma (2 to 5%). The employment of well-filled narrow-necked screw-cap bottles or containers makes the use of PRAS media unnecessary.

Differentiation of **Leptotrichia** from morphologically similar organisms

Table 6.8 indicates the characteristics that distinguish *L. buccalis* from organisms with a similar microscopic or colony morphology.

Taxonomic Comments

L. buccalis is a well circumscribed species. The strains examined are all isolated from the oral cavity. Their microscopical and colony morphology is similar, and they vary only slightly with respect to biochemical properties. The only reported difference, which might be of importance, is that some strains grow in air containing CO_2, whereas others remain obligately anaerobic. Whether the *Leptotrichia* strains isolated from the human female periurethral region (Bollgren et al., 1979) are representatives of *L. buccalis* or another species of *Leptotrichia* remains to be examined.

Thjötta et al. (1939), who gave the first adequate description of *L. buccalis*, were of the opinion that the organism was a Gram-negative bacterium related to *Fusobacterium nucleatum* (Böe and Thjötta, 1944). Hamilton and Zahler (1957), Gilmour et al. (1961) and Kasai (1965) concluded that *L. buccalis* was a Gram-positive organism related to

Table 6.8.
*Differentiation of **Leptotrichia buccalis** from **Fusobacterium nucleatum** and **Eubacterium saburreum**[a]*

Characteristics	L. buccalis	F. nucleatum	E. saburreum
Gram reaction	−[b]	−	+[c]
Morphology:			
Cells generally thick	+	−	−
Length of cells, μm	5.0–15.0	3.0–10.0[d]	5.0–20.0
Cells are fusiform	+	+	−
Filaments common	−	−	+
Bulbous swellings common	−	−	+
Colony morphology (blood agar)	Smooth, convoluted or raised with filamentous edge	Smooth, convex, with "flecked" appearance	Rough, rhizoid, adherent or Smooth, rhizoid or convoluted, nonadherent
Terminal pH in glucose media	<5.4	>5.8	<5.4
Indole production	−	+	+
Gas production	−	−	+
Predominant fatty acid end products from peptone-yeast extract-glucose broth:			
Lactic acid	+	−	−
Butyric acid	−	+	+
Acetic acid	−[e]	−	+

[a] Modified from Hofstad (1981).
[b] Often Gram-positive in young cultures.
[c] Often Gram-negative in old cultures.
[d] Occasionally up to 15 μm.
[e] *L. buccalis* may produce a small amount of acetic acid.

---

* CVE agar (g/liter): trypticase, 10.0; yeast extract, 5.0; NaCl, 5.0; glucose, 2.0; tryptophan, 0.2; agar, 15.0; crystal violet, 0.005; erythromycin, 0.004; defibrinated sheep blood, 50.0 ml; pH adjusted to 7.0–7.2 before autoclaving. Crystal violet is added before autoclaving, erythromycin and blood after autoclaving and cooling to 50–55°C.

Table 6.9.
Biochemical characteristics of **Leptotrichia buccalis**[a]

Acid[b] from:	
Amygdalin, cellobiose, fructose, glucose, lactose, maltose, mannose, melezitose, salicin, sucrose, trehalose	+
Galactose, starch	d
Arabinose, dulcitol, glycerol, inositol, inulin, mannitol, melibiose, raffinose, rhamnose, ribose, sorbitol, sorbose, xylose	−
Hydrolysis of esculin	+
Liquefaction of gelatin	−
Coagulation of milk	d
Production of ammonia, gas, indole, H$_2$S, lecthinase, urease	−
Reduction of nitrate	−
Propionate from threonine	−
Growth inhibition by 20% bile	−

[a] Based on data from Thjøtta et al. (1939); Hamilton and Zahler (1957); Kasai (1965); Holdeman et al. (1977); and examination of 15 own strains. Symbols: see standard definitions.
[b] pH ≤ 5.4.

Lactobacillus. The electron microscopical study by Hofstad and Selvig (1969) and the isolation of a potent endotoxin from the organism definitely established that *L. buccalis* is a Gram-negative bacterium.

Table 6.10.
Fatty acid composition of **Leptotrichia buccalis**[a]

Fatty Acid	Mol% of Total	
	Range	Mean
12:0	tr[b]–1.6	0.8
14:0	4.1–6.5	5.2
15:0	0.4–0.9	0.6
16:1	0.9–2.9	2.3
16:0	35.2–42.5	39.9
18:2	1.0–2.1	1.4
18:1	26.8–35.2	30.8
18:0	2.2–3.9	3.2
3-OH-14:0	12.7–18.8	15.8

[a] From an examination of 5 strains (Hofstad and Jantzen, unpublished results).
[b] tr = <0.1%.

Based on similarity in DNA base composition of *L. buccalis* and *F. nucleatum* Page and Krywolap (1976, 1977a) suggested a phylogenetic relationship between the two morphologically similar organisms. Nucleic acid hybridization experiments were inconclusive, however.

List of the species of the genus **Leptotrichia**

1. **Leptotrichia buccalis** (Robin) Trevisan 1879, 147.[AL] (*Leptothrix buccalis* Robin 1853, 345.)

buc.ca′lis. L. adj. *buccalis* buccal, pertaining to the mouth.

The description is the same as that given for the genus. Additional characteristics are listed in Tables 6.8, 6.9 and 6.10.

Habitat: the oral cavity of man.

The mol% G + C of the DNA is 25 (T_m, Bd).

Type strain: ATCC 14201.

Genus IV. **Butyrivibrio** Bryant and Small 1956, 18, emend. Moore, Johnson and Holdeman 1976, 241[AL]

MARVIN P. BRYANT

Bu.ty.ri.vib′ri.o. M.L. adj. *butyricus* butyric; L. v. *vibro* to vibrate; M.L. n. *vibrio* that which vibrates; a generic name; M.L. masc. n. *Butyrivibrio* a butyric vibrio.

Curved rods, 0.3–0.8 μm by 1.0–5.0 μm, single or in chains or filaments which may or may not be helical. No resting stages known. Stain Gram-negative but the type species is structurally Gram-positive. **Motile by polar or subpolar flagella,** monotrichous or lophotrichous; some strains may be nonmotile. **Strictly anaerobic.** Chemoorganotrophic, having a **fermentative type of metabolism** with carbohydrates being the main fermentable substrates. **Glucose or maltose is fermented with butyrate as one of the important products.** Under some conditions large amounts of lactic acid and little butyric acid may be produced. Occur in the rumen of ruminants and sometimes in human, rabbit and horse feces. The mol% G + C of the DNA is 36–41 (T_m).

Type species: *Butyrivibrio fibrisolvens* Bryant and Small 1956, 19.

Further Descriptive Information

In wet mounts prepared from the water of syneresis at the base of RGCA slants,* stab-inoculated into the base and incubated overnight at 37°C, the organisms appear as curved rods with tapered and rounded ends. Cells occur as singles and in short or long chains and sometimes as filaments. Chains may or may not show a helical arrangement and pairs may be in an "S" arrangement. Motility is rapid and vibrating and often progressive, but often only a few cells in a culture show motility. Flagellation is polar or subpolar and a few nonmotile, aflagellated strains have been reported. Flagellation is monotrichous (*Butyrivibrio fibrisolvens*) or lophotrichous (*Butyrivibrio crossotus*). Cells of the latter tend to be the larger in diameter. Cell walls of *B. fibrisolvens*, while staining Gram-negative, contain teichoic acid (Sharpe et al., 1975) and have a Gram-positive fine structure, though the wall is much thinner than is usual for Gram-positive bacteria (Cheng and Costerton, 1977). The fine structure of cells of *B. crossotus* has not been determined.

The genus is fermentative and little growth occurs in media lacking in carbohydrate energy source. Formic, lactic and butyric acid are produced from glucose (*B. fibrisolvens*) or maltose (*B. crossotus*). CO$_2$ is a product of *B. fibrisolvens* but has not yet been determined in *B. crossotus*. *B. fibrisolvens* produces H$_2$ but *B. crossotus* does not. Acetic acid may be produced or used. Under some conditions of culture, lactic acid production may be increased and only a small amount of butyric acid is formed (Gill and King, 1958). Small amounts of ethanol, propionic, pyruvic and succinic acids may be formed.

Nothing is known concerning the antigenic structure of *B. crossotus*, but Margherita and Hungate (1963) found *B. fibrisolvens* strains to be

---

* RGCA slants (g/liter): K$_2$HPO$_4$, 0.23; KH$_2$PO$_4$, 0.23; (NH$_4$)$_2$SO$_4$, 0.45; NaCl, 0.45; MgSO$_4$, 0.023; CaCl$_2$, 0.023; resazurin, 0.001; glucose, 0.5; cellobiose, 0.5; soluble starch, 0.5; rumen fluid (centrifuged), 400; cysteine·HCl, 0.25; Na$_2$S·9H$_2$O, 0.25; Na$_2$CO$_3$, 4.0; agar (Difco), 10.0; pH 6.7; gas phase, 100% CO$_2$. Prepared and used with the Hungate technique (Bryant, 1972).

antigenically very diverse. Sharpe et al. (1975) and Hewett et al. (1976) found that a number of strains contained lipoteichoic acids, but this was not found in other strains.

Pathogenicity has not been detected.

The members of the genus are among the most numerous bacteria in the rumen of ruminants under a wide variety of dietary conditions, and are sometimes found among the predominant bacteria in human, rabbit and horse feces (Brown and Moore, 1960; Moore and Holdeman, 1974). They are highly versatile in energy sources utilized for growth and some strains ferment cellulose and many ferment starch, pectin, xylan and othe polysaccharides.

Enrichment and Isolation Procedures

Butyrivibrio strains are commonly isolated by nonselective procedures from among the predominant bacteria grown as colonies in RGCA roll-tubes inoculated with high dilutions of rumen fluid (Bryant and Small, 1956) or less often from human feces (Moore et al., 1976). *B.*

fibrisolvens is often isolated from high dilutions of rumen fluid using rumen fluid agar roll-tube media with finely-ground cellulose as the energy source. In this case zones of cellulose digestion around colonies are seen, although the zones are often less distinct and the colonies less numerous than those of other cellulolytic bacteria such as *Ruminococcus albus* and *R. flavefaciens* (Hungate, 1966; Shane et al., 1969). *B. fibrisolvens* is the main rumen organism fermenting the bioflavonoid rutin, and can be isolated from high dilutions of rumen fluid as colonies in RGCA roll-tube medium containing 0.2% rutin in place of other energy sources. Colonies form clear zones in the suspended rutin and a yellow precipitate of quercetin forms around some (Cheng et al., 1969; Leedle and Hespell, 1980).

Maintenance Procedures

Strains can be maintained in stabbed-slants of RGCA medium held at −70°C for periods of a year or more. They can also be preserved by lyophilization.

Differentiation from other taxa

Differentiation of the genus from other taxa with which it might be confused is indicated in Table 6.11.

Taxonomic Comments

The genus constitutes one of the most numerous and biochemically versatile groups of bacteria in the rumen, and there is great variation among the features of strains. For example, great variation occurs in energy sources, fermentation products (Bryant and Small, 1956; Shane et al., 1969) and nutrition (Bryant and Robinson, 1962; Roche et al., 1973). Shane et al. (1969) placed most cellulolytic strains in two groups. One group produced appreciable lactate and low levels of formate, and removed acetate from the medium during cellobiose fermentation; a second group produced acetate, more formate but little or no lactate. The latter group was more exacting in nutritional requirements (Roche et al, 1973); however, both groups varied greatly in other features.

Under somewhat acidic conditions, the rumen may contain a species similar to *B. fibrisolvens* except that it is somewhat larger and exhibits lophotrichous flagella. This group has been referred to as the "B-385-like strains" (Bryant, 1956; Bryant et al., 1961).

Moore and Holdeman (1974) observed that human fecal isolates of *Eubacterium rectale* may stain very weakly Gram-positive or Gram-negative and their differentiation from *B. fibrisolvens* is in doubt.

The genus *Butyrivibrio* and apparently related organisms needs further detailed study with modern taxonomic methods including studies of ultrastructure and ribosomal RNA analyses to determine if further species and genera need to be differentiated and the relationship of the genus to higher taxa.

Further Reading

Bryant, M.P. and N. Small. 1956. The anaerobic, monotrichous, butyric acid-producing, curved rod-shaped bacteria of the rumen. J. Bacteriol. 72: 16–21.

Hungate, R.E. 1966. *The Rumen and Its Microbes*, Academic Press, New York.

Moore, W.E.C., J.L. Johnson and L.V. Holdeman. 1976. Emendation of *Bacteroidaceae* and *Butyrivibrio* gen. nov. and ten new species in the genera *Desulfomonas, Butyrivibrio, Eubacterium, Clostridium* and *Ruminococcus*. Int. J. Syst. Bacteriol. 26: 238–252.

Shane, B.S., L. Gouws and A. Kistner. 1969. Cellulolytic bacteria occurring in the rumen of sheep conditioned to low-protein teff hay. J. Gen. Microbiol. 55: 445–457.

Table 6.11.

Differential characteristics of the genus **Butyrivibrio** *and other morphologically similar, nonsporing anaerobic genera[a]*

Characteristics	Butyri-vibrio	Eubac-terium	Seleno-monas	Lachnospira	Succini-vibrio	Pectinatus
Curved rods	+	D	+	−	+	+
Stain Gram-positive	−	+[b]	−	+[b]	−	−
Flagellar arrangement:						
Polar to subpolar	+	D	−	−	+	−
Laterally attached	−	D	+	+	−	+
Monotrichous	D	D	−	+	+	−
Tufts	D	D	+	−	−	−
Major products of glucose or maltose fermentation:						
Butyrate	+	D	−	−	−	−
Succinate	−	−	v	−	+	+
Propionate	−	−	+	−	−	+

[a] Symbols: see standard definitions.
[b] Cells of some strains may be very weakly Gram-positive or Gram-negative.

Differentiation of the species of the genus **Butyrivibrio**

Some differential characteristics of the species of *Butyrivibrio* are indicated in Table 6.12.

List of the species of the genus **Butyrivibrio**

1. **Butyrivibrio fibrisolvens** Bryant and Small 1956, 19.[AL]

fi.bri.sol'vens. L. n. *fibra* fiber; L. part. adj. *solvens* dissolving; M.L. part. adj. *fibrisolvens* fiber-dissolving.

The morphology is indicated in the generic description. While strains stain uniformly Gram-negative, they have a Gram-positive ultrastructure with a very thin wall.

In RGCA roll-tubes, surface colonies are usually smooth, entire, slightly convex, translucent, light tan in color and 2–4 mm in diameter. Some strains have rough colonies that are more flat, lighter in color and have filamentous margins. Deep colonies are usually lenticular or Y-shaped but some form compound lenticular colonies.

In rumen fluid-cellulose agar, colonies of cellulolytic strains vary from lens-shaped to triangular to compound lenticular or rhizoidal colonies. Zones of cellulose digestion around colonies vary from very narrow with slow and indistinct digestion to broad zones with rapid and complete digestion.

The type of zones formed in rumen fluid-rutin agar is indicated under Enrichment and Isolation Procedures.

Growth in liquid glucose medium varies from a uniform turbidity to a flocculant or granular sediment, some of which may adhere to the walls of tubes.

Good growth occurs at 37°C and usually at 45°C, but more slowly at 30°C. No growth occurs at 22 or 50°C.

The final pH in poorly buffered glucose medium is usually 5.0–5.6.

Most strains grow in chemically defined culture media containing glucose or cellobiose as the energy source, amino acid mixtures and ammonium salts added as nitrogen sources, minerals, B-vitamins and cysteine. Many strains grow with an ammonium salt as the nitrogen source, but amino acid mixtures are usually stimulatory. Acetate is often stimulatory to growth; propionate or branched-chain volatile acids are stimulatory to some strains but not to others (Bryant and Robinson, 1962; Bryant, 1973; Roche et al., 1973).

The energy-yielding metabolism is indicated in the generic description and in Tables 6.12 and 6.13. Some strains ferment polysaccharides such as cellulose, starch, pectin and/or hemicelluloses and xylan and other materials such as the flavone glycoside, rutin.

Indole and catalase are not produced. Nitrate is usually not reduced. Hydrogen sulfide is usually not produced.

Gelatin liquefaction is variable.

The species is nonpathogenic.

Ecological information is indicated above and in the generic description.

Type strain: ATCC 19171 [strain D1 of Bryant and Small (1956)].

2. **Butyrivibrio crossotus** Moore, Johnson and Holdeman, 1976, 241.[AL]

cros.so'tus. Gr. adj. *crossotus*, tasseled.

The morphology is indicated in the generic description. The fine structure has not yet been elucidated.

After incubation at 37°C for 5 days in RGCA roll-tubes, subsurface colonies are 0.5–1.0 mm in diameter, lenticular and translucent to transparent. Surface colonies on brain heart infusion agar (supplemented) roll-streaks are 0.2–1.0 mm in diameter, circular, entire, convex, translucent to semiopaque and smooth (Moore et al., 1976). Only a few strains grow as surface colonies on blood agar plates incubated

anaerobically, and no hemolytic activity is observed. On egg yolk agar few strains grow and no lecithinase or lipase reactions are seen.

There is poor growth in peptone-yeast extract broth unless a fermentable carbohydrate is provided. In maltose broth, cultures show abundant growth with a smooth, flocculent or ropy sediment and usually some uniform turbidity.

Optimum growth temperature is 37°C. Some strains grow at 45°C. Growth is slow at 30°C.

The energy-yielding metabolism is indicated in the generic description and in Tables 6.12 and 6.13. Materials fermented are largely limited to starch, glycogen, dextrin and maltose.

No ammonia is produced from peptone.

All strains have been isolated from human feces or rectal contents.

The mol% G + C of the DNA is 36–37 (T_m).

Type strain: ATCC 29175 [T9-40A; Moore et al. (1976)].

Table 6.12.
Characteristics differentiating **Butyrivibrio fibrisolvens** *and* **Butyrivibrio crossotus**[a]

Characteristics	1. *B. fibrisolvens*	2. *B. crossotus*
Flagella monotrichous	+	−
Flagella lophotrichous	−	+
Produce H_2 from glucose or maltose	+	−
Ferment glucose and fructose	+	Weak
Ferment sucrose, cellobiose and xylose	+	−

[a] Symbols: see standard definitions.

Table 6.13.
Other characteristics of **Butyrivibrio fibrisolvens** *and* **Butyrivibrio crossotus**[a]

Characteristics	1. *B. fibrisolvens*	2. *B. crossotus*
Energy sources:		
Maltose	+	+
Galactose, inulin, salicin	+	−
Lactose	+	d
Starch	d	+
Esculin, trehalose	d	−
Glycerol, inositol, mannitol	−	−
Esculin hydrolysis	d	−
Acids produced from fermentation:[b]		
Butyrate, formate	+	+
Lactate, acetate	d	+

[a] Symbols: see standard definitions.
[b] From glucose or cellobiose in *B. fibrisolvens*, and from maltose in *B. crossotus*.

Genus V. **Succinimonas** Bryant, Small, Bouma and Chu 1958, 21[AL]

MARVIN P. BRYANT

Suc.ci.ni.mo'nas. M.L. n. *acidum succinicum* succinic acid; Gr. n. *monas* a unit, monad; M.L. fem. n. *Succinimonas* succinic acid monad.

Short, straight rods to coccobacilli, 1.0–1.5 μm wide and 1.0–3.0 μm long, with rounded ends. No spores or resting stages are produced. Gram-negative. **Motile by a single polar flagellum.**

Strictly anaerobic. Chemoorganotrophic, having a fermentative type of metabolism. **Glucose, maltose, dextrin or starch can serve as the energy source. Succinate and acetate are the main products of**

glucose fermentation. No butyrate or gas is formed. Catalase-negative. Occur in the bovine rumen. The mol% G + C of the DNA is not known.

Type species: *Succinimonas amylolytica* Bryant, Small, Bouma and Chu 1958, 21.

Further Descriptive Information

In the water of syneresis of RGCA slants (see the genus *Butyrivibrio* for composition of medium and anaerobic techniques) incubated overnight at 37°C the cells are arranged as singles, pairs and clumps. No capsules are evident. Motility is relatively slow and progressive and is rapidly lost upon exposure to air.

Surface colonies on RGCA agar in roll tubes are smooth, convex, translucent, light tan and 0.7–1.5 mm in diameter after 3 days incubation at 37°C. Deep colonies are lenticular and 0.7–1.0 mm in diameter. In liquid media with glucose as the energy source growth occurs as light, uniform turbidity.

The type strain of *S. amylolytica* grows well in a glucose medium with rumen fluid replaced by trypticase and yeast extract. It also grows in a chemically defined medium containing glucose, CO_2/HCO_3^- buffer, minerals, B-vitamins and acetate. Ammonium ions serve as the nitrogen source and cannot be replaced by amino acids or peptides. Sulfide serves as both the reducing agent and the sulfur source. Acetate (30 mM) is a highly stimulatory supplement even though the organisms produce acetate from glucose (Bryant and Robinson, 1962; Roberton and Bryant, unpublished data).

Growth occurs at 30 and 37°C but not at 22 or 45°C. The final pH

in poorly buffered, liquid glucose medium is 5.2–5.8. Good growth occurs at pH 6.5–7.0. The upper pH limit has not been determined.

During fermentation of glucose to succinate and acetate, a net uptake of CO_2 occurs. Of the substrates tested, only glucose, maltose, dextrin and starch are fermented. Arabinose, xylose, fructose, cellobiose, lactose, sucrose, cellulose, inulin, xylan, glycerol, esculin, mannitol, lactate, amino acids and peptides are not fermented.

H_2S and indole are not produced. Gelatin is not liquefied. The Voges-Proskauer test is variable. Nitrate is not reduced.

S. amylolytica appears to be nonpathogenic for humans or animals. It occurs in the rumen of cattle fed diets containing roughage and some grain, where it is involved in fermentation of starch and its hydrolytic products. It is usually present as only a small proportion of the total viable bacteria in the bovine rumen (less that 6% of total). Whether *S. amylolytica* occurs in the rumen of ruminants other than cattle or in nonruminal ecosystems is not known.

Enrichment and Isolation Procedures

Succinimonas is isolated nonselectively in anaerobic roll tubes of RGCA medium from the rumen of cattle fed hay-grain diets. It constitutes only a small proportion of the colonies which develop after 3 days or more of incubation at 37°C.

Maintenance Procedures

S. amylolytica strains can be maintained on stab-inoculated RGCA slants at −70°C for a year or more. They can also be preserved indefinitely by lyophilization.

Differentiation of the genus **Succinimonas** from other genera

See the key to the family *Bacteroidaceae* for primary characteristics that differentiate *Succinimonas* from other genera of anaerobic bacteria.

Taxonomic Comments

The phylogenetic relationships of *Succinimonas* to other genera and higher taxa are not known. Studies of the ribosomal RNA, such as DNA/rRNA hybridization or rRNA oligonucleotide cataloging, would

be useful in this regard.

Only one species is recognized presently in the genus.

Further Reading

Bryant, M.P., N. Small, C. Bouma and H. Chu. 1958. *Bacteroides ruminicola*, sp. nov. and *Succinimonas amylolytica*, gen. nov., species of succinic acid-producing anaerobic bacteria of the bovine rumen. J. Bacteriol. *76:* 15–23.
Bryant, M.P. and I.M. Robinson. 1962. Some nutritional characteristics of predominant culturable ruminal bacteria. J. Bacteriol. *84:* 605–614.

List of the species of the genus **Succinimonas**

1. **Succinimonas amylolytica** Bryant, Small, Bouma and Chu 1958, 21.[AL]

am.y.lo.ly′ti.ca. Gr. n. *amylum* fine meal, starch; Gr. adj. *lyticus* loosening, dissolving; M.L. fem. adj. *amylolytica* starch dissolving.

The characteristics are as described for the genus.
Occur in the bovine rumen.
The mol% G + C of the DNA is unknown.
Type strain: ATCC 19206 [strain $B_2$4 of Bryant et al. (1958)].

Genus VI. **Succinivibrio** Bryant and Small 1956, 22[AL]

MARVIN P. BRYANT

Suc.ci.ni.vib′ri.o. M.L. n. *acidum succinicum* succinic acid; M.L. masc. n. *Vibrio* that which vibrates, a generic name; M.L. masc. n. *Succinivibrio* the succinic acid vibrio.

Curved rods, 0.4–0.6 by 1.0–7.0 μm, with pointed ends. The cells are helically twisted with less than 1 coil to 3 or more coils per cell. The cells may become straight or only slightly curved after maintenance on artificial media. Gram-negative. **Possess a progressive vibrating type of motility by means of a single polar flagellum.** Strictly anaerobic. Chemoorganotrophic, having a fermentative type of metabolism with carbohydrates being the main fermentable substrates. **The major products of glucose fermentation are succinate, acetate, formate and sometimes lactate.** Butyrate and H_2 are not formed. A large net uptake of CO_2 may occur. Occur in the rumen of cattle and sheep. The mol% G + C of the DNA is not known.

Type species: *Succinivibrio dextrinosolvens* Bryant and Small 1956, 22.

Further Descriptive Information

Cells from young cultures on RGCA slants (see the genus *Butyrivibrio* for composition of medium and anaerobic techniques) are short with 1 or less than 1 complete coil, but longer cells containing 2 or 3 coils are commonly present also. Cells are mainly single but a few short chains may occur. In aging cultures swollen, giant spirillar forms and round bodies may be present.

Surface colonies on RGCA agar in roll tubes are 1–2 mm in diameter, entire, translucent, slightly convex and light tan in color after 3 days incubation at 37°C. Deep colonies are lenticular. In liquid media with glucose as the energy source growth occurs as a heavy, flocculent sediment which is easily dispersed; light turbidity also occurs.

Good growth of *S. dextrinosolvens* occurs in a chemically-defined medium containing glucose, minerals, *p*-aminobenzoic acid, 1,4-naphthoquinone, ammonium ions, cysteine, methionine, leucine and serine. The use of a CO_2/HCO_3^- buffer system (pH 6.7) is necessary for good growth (Gomez-Alarcon and Bryant, 1982, manuscript in preparation).

The final pH in lightly buffered glucose medium is 4.8–5.2.

Growth occurs at 30–39°C but not at 22 or 45°C.

S. dextrinosolvens requires a fermentable carbohydrate for growth and does not ferment amino acids. Succinic and acetic acids, and often formic and lactic acids, are produced from fermentation of glucose. Gas is not produced. Xylose, galactose, glucose, maltose and dextrin can serve as fermentable substrates. Arabinose, fructose, cellobiose, sucrose, esculin, salicin and mannitol are fermented by some strains. No fermentation of lactose, trehalose, cellulose, xylan, inulin, glycerol or inositol occurs. Starch is partially fermented but not completely hydrolyzed.

Nitrate is not reduced. Gelatin is not hydrolyzed. Acetoin, indole and H_2S are not produced. The catalase test is negative.

Urease is produced by many strains but is strongly repressed in media containing large amounts of ammonia or other utilizable nitrogen sources (Wozny et al., 1977).

S. dextrinosolvens is not pathogenic for animals as far as is known. Isolates corresponding to the description of the organism have been isolated from a few cases of human bacteremia (Porschen and Chan, 1977; Southern, 1975).

S. dextrinosolvens is found in the rumen of cattle and sheep, especially when high grain diets containing large amounts of starch are fed (Bryant and Small, 1956; Bryant et al., 1961; Wozny et al., 1977). The organism appears to be a major fermenter of dextrins under these conditions.

Enrichment and Isolation Procedures

S. dextrinosolvens is sometimes a predominant organism from the rumen of cattle fed high grain diets, and can be isolated nonselectively in anaerobic roll tubes of RGCA medium incubated for 3 days or more at 37°C.

Maintenance Procedures

S. dextrinosolvens can be maintained on stab-inoculated RGCA slants at −70°C for 1 year or more. The organisms can also be preserved indefinitely by lyophilization.

Differentiation of the genus **Succinivibrio** from other genera

See Table 6.11 in the article on the genus *Butyrivibrio* for those characteristics useful in differentiating *Succinivibrio* from other morphologically or physiologically similar genera of anaerobic bacteria.

Succinivibrio differs from *Anaerobiospirillum* by having monotrichous rather than lophotrichous flagella. It differs from the genus *Anaerovibrio* by not producing a large amount of propionate, by not fermenting glycerol and by its ability to ferment dextrin.

Taxonomic Comments

The phylogenetic relationships of *Succinivibrio* to other genera or higher taxa are not known. Studies of the ribosomal RNA, such as DNA/rRNA hybridization or rRNA oligonucleotide cataloging, would be useful in this regard.

Further Reading

Bryant, M.P. and N. Small. 1956. Characteristics of two new genera of anaerobic curved rods isolated from the rumen of cattle. J. Bacteriol. *72:* 22–26.
Porschen, R.K. and P. Chan. 1977. Anaerobic vibrio-like organisms cultured from blood: *Desulfovibrio desulfuricans* and *Succinivibrio* species. J. Clin. Microbiol. *5:* 444–447.

List of the species of the genus **Succinivibrio**

1. **Succinivibrio dextrinosolvens** Bryant and Small 1956, 22[AL]
dex.tri.no.sol'vens. M.L. n. *dextrinosum* dextrin; L. part. adj. *solvens* dissolving; M.L. part. adj. *dextrinosolvens* dextrin-dissolving.
The characteristics are as described for the genus.

Occur in the bovine and ovine rumen.
The mol% G + C of the DNA is unknown.
Type strain: ATCC 19716 [strain 24 of Bryant and Small (1956)].

Genus VII. **Anaerobiospirillum** Davis, Cleven, Brown and Balish 1976, 503[AL]

MARVIN P. BRYANT

An.ae.ro.bi.o.spi.ril'lum. Gr. pref. *an* not; Gr. n. *aer* air; Gr. n. *bios* life; M.L. dim. neut. n. *spirillum* a small spiral; M.L. neut. n. *Anaerobiospirillum* anaerobic small spiral.

Helical rods, having a cell width of 0.6–0.8 μm and a cell length of 3.0–8.0 μm. The ends of the cells are rounded. Spores are not formed. Gram-negative. **Motile by means of bipolar tufts of flagella.** Anaerobic. Chemoorganotrophic, having a fermentative type of metabolism. **The major products of glucose fermentation are succinic and acetic acids**; smaller amounts of lactic and formic acids may be formed. Found in the throats and colons of dogs. The mol% G + C of the DNA is 44 (T_m).

Type species: *Anaerobiospirillum succiniciproducens* Davis, Cleven, Brown and Balish 1976, 503.

Further Descriptive Information

The diameter of the cell helix is 0.9–1.1 μm. Some helical cells may reach 20 μm in length. The cells usually occur singly. Straight rods and spherical forms may occur, especially in liquid cultures.

The cells have a corkscrew-like motility. The flagella are 14 nm in diameter and are arranged as bipolar tufts of ~16 flagella.

Surface colonies on A II agar* (Aranki and Freter, 1972) or blood agar are 0.5–1.0 mm in diameter, circular, convex and translucent after 3 days incubation at 37°C. Some of the colonies may have a raised center.

---

* Medium AII (Aranki and Freter, 1972) has the following composition (g/liter): Tryptic soy agar (Difco), 40.0; yeast extract (Difco), 5.0; K_2HPO_4, 2.5; hemin, 0.001; agar (Difco), 5.0; glucose, 0.5; palladium chloride, 0.332 mg; Na_2CO_3, 0.63; cysteine-HCl, 0.5; placenta powder (Nutritional Biochemicals Corp., Cleveland, Ohio), 2.0; and menadione, 0.005. The placenta powder and cysteine are autoclaved separately in aqueous solution and added aseptically to the autoclaved and cooled (56°C) basal medium. The menadione is kept in an alcoholic stock solution (sterilized by filtration) stored in the cold, and is added aseptically to the medium.

Growth is most rapid at 37–40°C and does not occur below 18°C or above 50°C. Cells do not survive 80°C for 10 min.

All strains ferment glucose, lactose and sucrose. Some also ferment fructose, maltose and raffinose. Strains are variable or negative for glycerol, inositol, melibiose or trehalose. The following compounds are not fermented: adonitol, arabinose, cellobiose, dulcitol, esculin, galactose, mannitol, rhamnose, salicin, sorbitol and xylose. Starch hydrolysis is variable.

The catalase test is negative. Lipase activity does not occur. Indole is not produced. Gelatin is not hydrolyzed and meat is not digested. Nitrate is not reduced.

Of the three strains so far isolated, one came from the throat of a dog and two from a cecal homogenate of a dog. Although the organisms seem to be part of the normal microbiota of the throat and bowel of dogs, no definite ecological function is known.

Enrichment and Isolation Procedures

The following procedure was described by Davis et al. (1976). Throat samples from dogs are taken with cotton swabs and cecal samples are obtained by surgery. The samples are placed in tubes of an anaerobic transport broth for transport to the laboratory, where they are placed in an anaerobic glove box (Aranki and Freter, 1972). The cecal tissue samples are homogenized in a blender. Both the throat and cecal samples are diluted by 10-fold serial dilutions and plated onto prere-duced A II agar. The plates are incubated at 37°C in a specially designed incubator within the glove box containing a gas mixture of 80% N_2: 10% H_2: 10% CO_2. The plates are incubated for 3 days or more. The colonies of *Anaerobiospirillum* are not easily differentiated from many others on the plates, and it is necessary to examine the cells from many colonies microscopically in order to find a few colonies of *Anaerobiospirillum*. The organisms may be undetectable in some samples, whereas they may occur in others at a level of 10^5–10^7 organisms per ml or per g (dry weight). This should be contrasted with a total colony count for anaerobes in the range of 10^9–10^{10} per g of dry tissue (Davis et al., 1976). Colonies of *Anaerobiospirillum* selected from plates can be cultured in peptone-yeast-glucose (PYG) broth (see Holdeman and Moore, 1972, for formulation of medium).

Maintenance Procedures

The organisms may be cultured in prereduced, screw-capped tubes of enriched milk medium [52 g of litmus milk (Difco) and 375 ml of water]. After incubation for 18–48 h in an anaerobic glove box the tubes are frozen in liquid nitrogen and stored at −70°C (Davis et al., 1976).

A colony on a small piece of blood agar can be added to a vial containing 1 ml of enriched milk medium and lyophilized (Davis et al., 1976).

Differentiation of the genus **Anaerobiospirillum** from other genera

See the key to the genera of the family *Bacteroidaceae* for primary characteristics that distinguish *Anaerobiospirillum* from other genera of anaerobes. The genus is distinguished from *Succinivibrio* by its lophotrichous flagella and by its ability to ferment lactose but not galactose or xylose.

Taxonomic Comments

The phylogenetic relationships of *Anaerobiospirillum* to other genera or higher taxa are not known. Studies of the ribosomal RNA, such as DNA/rRNA hybridization or rRNA oligonucleotide cataloging, would be helpful in this regard.

Further Reading

Davis, C.P., D. Cleven, J. Brown and E. Balish. 1976. *Anaerobiospirillum*, a new genus of spiral-shaped bacteria. Int. J. Syst. Bacteriol. 26: 498–504.

List of the species of the genus **Anaerobiospirillum**

1. **Anaerobiospirillum succiniciproducens** Davis, Cleven, Brown and Balish 1976, 503.[AL]

suc.ci.ni.ci.pro.du′cens. M.L. n. *acidum succinicum* succinic acid; L. pres. part. *producens* producing; M.L. part. adj. *succiniciproducens* producing succinic acid.

The characteristics are as described for the genus.
Found in the throats and colons of dogs.
The mol% G + C of the DNA of the type strain is 44 (T_m).
Type strain: ATCC 29305 [strain S411 of Davis et al. (1976)].

Genus VIII. **Wolinella** Tanner, Badger, Listgarten, Visconti and Socransky, 1981, 439[VP]

ANNE C. R. TANNER AND SIGMUND S. SOCRANSKY

Wo.li.nel′la. M. L. ending *-ella*; M.L. fem. n. *Wolinella* named after M. J. Wolin, American bacteriologist who first isolated the type species.

Helical, curved, or straight, unbranched cells, 0.5–1.0 μm in diameter and 2–6 μm in length, with rounded or tapered ends. Endospores are not produced. **Gram-negative.** Rapid, darting motility is by means of a **single polar flagellum.** Colonies are pale yellow-opaque to gray translucent with convex, pitting, and spreading variants. **Anaerobic. Hydrogen and formate are electron donors and are used as energy sources. Fumarate and NO_3 are used as electron acceptors.** The formate is oxidized to CO_2, while fumarate is reduced to succinate. Carbohydrates are not fermented and do not support growth. Hydrogen sulfide is produced. **Catalase-negative.** Strains have been isolated from the **bovine rumen,** from the **human gingival sulcus,** and from **dental root canal** infections. The mol% G + C of the DNA is 42–48 (T_m).

Type species: *Wolinella succinogenes* (Wolin, Wolin and Jacobs 1961) Tanner, Badger, Listgarten, Visconti and Socransky 1981, 439.

Further Descriptive Information

Cells of *Wolinella succinogenes* typically appear as curved or helical rods and can form spiral chains. In contrast, cells of *Wolinella recta* are usually straight, but curved and helical cells can occasionally be observed. Cells have a single polar flagellum. The ultrastructure of *W. succinogenes* has not been described. Ultrastructural examination of *W. recta* strains reveals that the peripheral layers of the cell consist of distinct inner and outer membranes which are straight and parallel to one another, each with a typical trilaminar structure (Lai et al., 1981). The membranes are separated by a distance of approximately 20 nm and the intervening space does not contain a distinct peptidoglycan layer. The outer surface of the outer membrane is covered by an array of bullet-shaped macromolecular subunits with a circular base measuring approximately 17 nm in diameter. These are packed in hexagonal arrays consisting of a central subunit surrounded by 6 subunits with a center-to-center distance of about 20 nm, which is best visualized in negatively stained specimens. The subunits are attached to the outer surface of the outer membrane by their flat surfaces, with the convex portion facing outward. When cells are disrupted, the subunits can be detected on both sides of the outer membrane. No subunits are observed on either surface of the inner membrane. There are electron-dense cytoplasmic inclusions ranging in diameter from 5 to 45 nm, with a mean diameter of 30 nm.

Wolinella strains, in common with strains of *Eikenella corrodens*, *Bacteroides ureolyticus* and *B. gracilis*, can exhibit a range of colony types similar to that described for *E. corrodens* (Jackson and Goodman, 1972, 1978; Smibert and Holdeman, 1976; Tanner et al., 1981). One colonial variant is agar-corroding or pitting, a second has a spreading, transparent, matte appearance, and a third consists of small, transparent, yellow or gray convex colonies. The ability of colonies to corrode and spread is dependent on the medium. Corroding colonies commonly appear on media made in the laboratory but appear as the surface-spreading type when transferred to commercially prepared Trypticase soy agar (BBL) supplemented with 5% sheep blood. Small convex colonies appear with either of the other types or alone. High DNA/DNA homology between colonial variants of *W. recta* isolates indicate that they are pure cultures. There can be differences in susceptibilities to certain inhibitory agents by different colonial variants of *W. recta*. In general, the cells from spreading colonies are more susceptible to inhibitors than cells from the non-spreading variants. Similar cell-wall ultrastructures are observed in cells derived from different colonial types of a pure culture. However, slight antigenic differences between the two forms can be noted (Badger and Tanner, 1981). Each colonial variant can be maintained in pure culture on agar plates, but serial transfer in broth or revival from the freeze-dried state frequently leads to reversion to multiple colony types.

Wolinella strains grow at 37°C in an anaerobic atmosphere consisting of 80% N_2, 10% CO_2 and 10% H_2. Cells will also grow on the surface of blood agar plates in an atmosphere including 5% O_2 when a heavy cell suspension is inoculated using a Steers' replicator. *Wolinella* strains grow in complex media (for example Mycoplasma or Todd Hewitt broth) supplemented with hemin, sodium formate (0.2%) and sodium fumarate (0.3%). Strains can also be grown on a similarly supplemented synthetic medium. Culture media for *W. succinogenes* have been described by Wolin et al. (1961) and Kafkewitz (1975). All *Wolinella* strains tested grow in the simple medium described by Kafkewitz.

Wolinella strains oxidize formate or hydrogen as an energy source, formate being oxidized to produce carbon dioxide. The oxidation can be coupled with any of several electron acceptors including fumarate and nitrate. Ammonium formate was found to yield better growth than sodium formate of *W. succinogenes* cells when fumarate, but not nitrate, was used as electron acceptor (Kafkewitz, 1975). Alternate electron acceptors have been described for *W. succinogenes* and for human isolates which resemble *W. succinogenes*. Fumarate could be replaced by L-malate, L-aspartate or by L-asparagine as electron acceptors (Wolin et al., 1961; Kafkewitz, 1975). Oxidation of formate or hydrogen can also be coupled with reduction of nitrate or, for *W. succinogenes*, with nitrous oxide (Wolin et al., 1961; Yoshinari, 1980). Nitrate is reduced to nitrite or to ammonia, but not nitrogen gas; nitrous oxide is reduced to nitrogen.

A growth requirement for succinate was described for *W. succinogenes* when nitrate replaced fumarate as the electron acceptor (Niederman and Wolin, 1972) but not nitrous oxide (Yoshinari, 1980). Added succinate was not needed in the presence of substances which could be converted to succinate, including malate, fumarate, and L-asparagine. The succinate requirement could not be replaced by oxaloacaetate, pyruvate and bicarbonate, acetate, propionate, butyrate, L-glutamate, D-aspartate or δ-aminolevulinic acid (Niederman and Wolin, 1972). L-aspartate and L-asparagine could substitute for fumarate, malate and nitrate as electron acceptors when *W. succinogenes* was grown in the presence of formate.

Formate dehydrogenase was isolated from *W. succinogenes* as a dimer of two identical subunits of M.W. 110,000, with a *b*-cytochrome (−200mV) which mediated electron transfer to menaquinone (Kröger et al., 1979). Fumarate reductase was isolated in two forms, one of which contained a *b*-cytochrome (−20mV), which mediated electron transfer from menaquinone to fumarate reductase (Unden et al., 1980). The reactive sites of formate dehydrogenase appear to face the outside, while those of fumarate reductase face the inside of the cytoplasmic membrane (Kroger et al., 1980).

L-Asparaginase has been isolated from *W. succinogenes* (Kafkewitz and Goodman, 1974) and from similar organisms isolated from humans (Albanese and Kafkewitz, 1978; Abuchowski et al., 1979; Radcliffe et al., 1979). This enzyme has been of particular interest for its potential antitumor activity. *W. succinogenes* asparaginase was used to inhibit growth of an in vitro culture of human pancreatic carcinoma (Wu et al., 1978).

W. recta strains are extremely sensitive to antibiotics, dyes and indicators when tested by an in vitro dilution technique. *W. recta* strains are also sensitive to sonic oscillation, a method routinely used to disperse gingival dental plaque samples.

The pathogenic potential of *Wolinella* strains is unknown although *W. recta* strains have been isolated in higher proportions from periodontal sites of adults with alveolar bone loss than from healthy sites. *W. recta* strains have been isolated from periodontal pockets of adults with periodontal disease; from necrotic dental root canals; and from periapical granulomas associated with nonvital teeth (Sundqvist, 1976; Van Palenstein Helderman and Rosman, 1976; Tanner et al., 1979; Jacobs, 1980).

Additional *Wolinella* strains have been isolated from other human clinical sites (Smibert and Holdeman, 1976) which, although phenotypically similar to the two named species, show low DNA homology and represent additional distinct species (Tanner et al., 1982). An isolate resembling *W. succinogenes* has been isolated from sewage (Yoshinari, 1980).

Enrichment and Isolation Procedures

The rumen strain of *W. succinogenes* was isolated from an inoculum of bovine rumen fluid after serial transfer in an anaerobic methanogenic enrichment medium containing formate, sulfide, and inorganic salts. Secondary enrichment was made in a broth medium containing formate and fumarate (Wolin et al., 1961).

Oral strains are isolated using continuous anaerobic techniques for sampling, dispersal, and dilution. The diluted samples are incubated on Trypticase soy agar plates supplemented with 5% sheep blood, in an anaerobic atmosphere consisting of 80% N_2, 10% CO_2 and 10% H_2 (Tanner et al., 1979).

Maintenance Procedures

The organisms can be maintained in the laboratory by weekly transfer on commercially prepared Trypticase soy agar (BBL) supplemented with 5% sheep blood. Several broth media have been described to cultivate *W. succinogenes* (Wolin et al., 1961; Kafkewitz, 1975). *Wolinella* strains, including *W. succinogenes*, will grow in a broth medium (MFF broth) consisting of *Mycoplasma* broth supplemented with hemin (5 mg/liter), sodium formate (2 g/liter) and sodium fumarate (3 g/liter). *Wolinella* strains grow at 35°C in an atmosphere of 80% N_2, 10% H_2, and 10% CO_2.

For preservation by lyophilization, a dense suspension of cells harvested from surface growth on blood agar media is prepared in broth containing 5% serum and 1% glucose. Organisms can also be preserved in liquid nitrogen by placing young colonies from blood agar plates into broth supplemented with 5% dimethyl sulfoxide, for slow freezing before final storage.

Procedures for Testing Special Characters

Cell morphology and active motility can be determined by darkfield microscopy of young broth cultures. Alternatively, MFF broth partially solidified with 0.3% agar (Difco) can be used to test for motility.

Differentiation of the genus **Wolinella** from other taxa

Table 6.14 indicates the characteristics of *Wolinella* that distinguish it from other genera of morphologically or physiologically similar motile genera.

Campylobacter sputorum and *C. concisus* have a metabolism similar to that of *Wolinella* species. It is difficult to distinguish these oral *Campylobacter* species from *Wolinella* species using routine physiological tests. They can be distinguished by their lower mol% G + C (29–38) and by their relative resistance to antibiotics, dyes and indicators (Tanner et al., 1981). Oral *Wolinella* and *Campylobacter* species may also be distinguished using serologic techniques (Badger and Tanner, 1981).

Wolinella strains isolated from the human oral cavity frequently exhibit agar-corroding colonies. *Bacteroides ureolyticus* and *B. gracilis* are also agar-corroding species which are stimulated by formate and fumarate in broth and have biochemical features similar to those of *Wolinella* species; however, unlike *Wolinella* species, they are nonmotile. *B. ureolyticus* also can be distinguished by its lower mol% G + C value of 28 and by its positive urease reaction. *B. gracilis* cells, unlike those of *Wolinella*, are slender with tapered ends and although they have a mol% G + C value of 42–44, they show no DNA homology with any *Wolinella* isolates tested (Tanner et al., 1981). A third species of agar-corroding organisms, *Eikenella corrodens*, isolated from the human oral cavity and human infections, differs from *Wolinella* by being nonmotile, facultatively anaerobic, lysine and ornithine decarboxylase-positive, and by having a higher mol% G + C value (56–58). Moreover, *E. corrodens*, although stimulated by nitrate, is not stimulated by formate and fumarate in broth culture.

Taxonomic Comments

The species previously known as *Vibrio succinogenes* (Wolin et al., 1961), and certain strains of the organism previously known as *Vibrio sputorum* or *Campylobacter sputorum* (Prévot, 1940; Véron and Chatelain, 1973) have now been classified in the genus *Wolinella*. The genus *Vibrio* contains saccharolytic, facultatively anaerobic organisms and therefore is inappropriate for anaerobic, nonfermentative organisms. Of the strains previously classified as *V. sputorum*, those having mol% G + C value of 30–38 would now be recognized as members of the genus *Campylobacter*, and those having a mol% G + C value of 42–48 are now included in the genus *Wolinella*.

The results of DNA/DNA hybridizations between *W. recta* strains and helical strains of *W. succinogenes* are summarized in Table 6.18. Eight strains of *W. recta* were clearly closely related to each other on the basis of DNA homology, indicating a distinct species. The type strain of *W. recta* showed no significant homology to the type strain of *W. succinogenes*, substantiating a distinction between those two species.

The occurrence of other, as yet unnamed, species of *Wolinella* is discussed under Other Organisms Belonging to the Genus *Wolinella*.

Further Comments

The identification of a number of laboratory strains, currently labeled *V. succinogenes* and *C. sputorum* subsp. *sputorum* may have to be reevaluated to determine whether they belong in the genus *Wolinella* or the genus *Campylobacter*.

Acknowledgments

We would like to acknowledge the kindness of Drs. M. J. Wolin, R. M. Smibert, and L. V. Holdeman for providing reference strains, and also for their assistance in assessing taxonomic status of isolates involved in the creation of the genus *Wolinella*.

Further Reading

Badger, S.J. and A.C.R. Tanner. 1981. Serological studies of *Bacteroides gracilis*, *Campylobacter concisus*, *Wolinella recta*, and *Eikenella corrodens*, all from humans with periodontal disease. Int. J. Syst. Bacteriol. *31:* 446–451.

Kafkewitz, D. 1975. Improved growth media for *Vibrio succinogenes*. Appl. Microbiol. *29:* 121–122.

Lai, C.-H., M.A. Listgarten, A.C.R. Tanner and S.S. Socransky. 1981. Ultrastructures of *Bacteroides gracilis*, *Campylobacter concisus*, *Wolinella recta*, and *Eikenella corrodens*, all from humans with periodontal disease. Int. J. Syst. Bacteriol. *31:* 465–475.

Smibert, R.M. and L.V. Holdeman. 1976. Clinical isolates of anaerobic gram-negative rods with a formate-fumarate energy metabolism: *Bacteroides corrodens*, *Vibrio succinogenes*, and unidentified strains. J. Clin. Microbiol. *3:* 432–437.

Tanner, A.C.R., S. Badger, C.-H. Lai, M.A. Listgarten, R.A. Visconti and S.S. Socransky. 1981. *Wolinella* gen. nov., *Wolinella succinogenes* (*Vibrio succinogenes* Wolin et al.) comb. nov., and description of *Bacteroides gracilis* sp. nov., *Wolinella recta* sp. nov., *Campylobacter concisus* sp. nov., and *Eikenella corrodens* all from humans with periodontal disease. Int. J. Syst. Bacteriol. *31:* 432–445.

Wolin, M.J., E.A. Wolin and N.J. Jacobs. 1961 Cytochrome-producing anaerobic vibrio, *Vibrio succinogens*, sp. n. J. Bacteriol. *81:* 911–917.

Table 6.14.
Characteristics distinguishing **Wolinella** *from other genera of motile organisms having polar flagella[a]*

Characteristics	Wolinella	Vibrio	Succinivibrio	Succinimonas	Butyrivibrio	Desulfovibrio	Anaerobvibrio	Campylobacter
Growth occurs under an air atmosphere (21% O₂)	−	+	−	−	−	−	−	−
Fermentative (acid or gas produced from carbohydates)	−	+	+	+	+	−	+	−
Oxidase reaction	+	+						+
H₂S produced from SO₄²⁻	−	−	−	−	−	+	−	−
Growth in broth stimulated by a combination of formate and fumarate	+	−	−	−	−	−	−	D
Mol% G + C of DNA	42–48	38–51				46–61		30–38

[a] Data derived from Bryant (1974a, b, c), Hungate (1966), Postgate (1974), Shewan and Véron (1974), Smibert (1978), and Tanner et al. (1981). Symbols: see standard definitions.

Differentiation and characterization of species of **Wolinella**

The differentiating characteristics of the species of *Wolinella* are indicated in Table 6.15. Other characteristics of the species are presented in Table 6.16 and antibiotic minimum inhibitory concentrations (MICs) in Table 6.17.

Table 6.15.
Characteristics differentiating **Wolinella succinogenes** *and* **Wolinella recta**[a]

Characteristics	1. W. succinogenes	2. W. recta
Helical cells dominant	+	−
Straight cells dominant	−	+
Hexagonal units in cell wall	−	+
Growth in the presence of (g/liter): Janus green (0.1), basic fuchsin (0.05), safranine (0.1), sodium fluoride (0.5), sodium deoxycholate (1.0), crystal violet (0.005), alizarine red (0.32), azure II (0.05), methyl orange (0.25), or rifampin (0.064)	+	−

[a] Symbols: see standard definitions.

Table 6.16.
Other characteristics of **Wolinella succinogenes** *and* **Wolinella recta**[a]

Characteristics	1. W. succinogenes	2. W. recta
Mol % G + C content	47	42–47
Active motility	+	+
Branching, or endospore formation	−	−
Growth anaerobically	+	+
Growth in air + 10% CO_2	−	−
Growth stimulated by formate and fumarate	+	+
Benzidine and oxidase reactions	+	+
Reduction of nitrate, nitrite	+	+
Reduction of benzyl viologen or neutral red	+	+
H_2O_2 decomposition, urease	−	−
Lysine, ornithine or arginine decarboxylase	−	−
End products:		
H_2S, succinate, H_2, CO_2	+	+
Formate, acetate, lactate, pyruvate, propionate, isobutyrate, butyrate, valerate, isovalerate, caproate, isocaproate or nitrous oxide	−	−
Growth in the presence of (g/l):		
Malachite green (0.00002); brilliant green (0.0125); Evans blue (0.05)	+	+
Oxgall (10); phenol (1); sodium chloride (40); indulin scarlet (0.5); potassium cyanide (1)	−	−
Hydrolysis of: starch, dextran, esculin, casein, DNA, gelatin	−	−
Production of: hydrogen peroxide, lecithinase, lipase, indole, ammonia, acetylmethylcarbinol, gas in glucose broth	−	−
Acid from: adonitol, amygdalin, arabinose, cellobiose, dextran, dulcitol, fructose, esculin, galactose, glucose, glycerol, inositol, inulin, lactose, maltose, mannitol, mannose, melezitose, melibiose, raffinose, rhamnose, ribose, salicin, sorbitol, starch, sucrose, trehalose or xylose	−	−

[a] Symbols: see standard definitions

Table 6.17.
Minimum inhibitory concentrations (MIC) of antibiotics for **Wolinella succinogenes** *and* **Wolinella recta**

Antibiotic	MIC, μg/ml[a]	
	1. W. succinogenes	2. W. recta
Bacitracin	128	64–>128
Chloramphenicol	2–8	1–2
Clindamycin	1–2	≦0.25–0.5
Colistin	2–4	<0.25–1.0
Erythromycin	2–8	0.5–2.0
Gentamicin	2–8	1–2
Kanamycin	8	≦0.25–1.0
Metronidazole	1–2	0.5–2.0
Minocycline	0.05–4.0	≦0.25–2.0
Naladixic acid	32–128	16–128
Neomycin	16–32	16
Penicillin	8–>128	≦0.25–4.0
Polymyxin B	1–8	≦0.25–1.0
Rifampin	≧128	2
Streptomycin	2–8	0.5–4.0
Tetracycline	0.5–2.0	≦0.25–0.5
Vancomycin	>128	>128

[a] Tested using an agar dilution technique on a basal medium containing (per liter): Trypticase (BBL), 15.0 g; yeast extract (Difco), 5.0 g; sodium pyruvate, 2.0 g; sodium formate, 1.0 g; sodium fumarate, 1.5 g; sodium succinate, 100 mg; NaCl, 5.0 g; hemin, 5.0 mg; and agar, 20.0 g; pH 7.4.

List of species of the genus **Wolinella**

1. **Wolinella succinogenes** (Wolin, Wolin and Jacobs, 1961) Tanner, Badger, Lai, Listgarten, Visconti and Socransky, 1981, 439.[VP] (*Vibrio succinogenes* Wolin, Wolin and Jacobs 1961, 917.[AL])

suc.ci.no′ge.nes. M.L. n. *acidum succinicum* succinic acid; Gr. v. *gennaio* to produce; M.L. adj. *succinogenes* succinic acid producing.

Mainly helical cells.

Surface colonies can appear either yellowish convex 1–2 mm in diameter or highly translucent resembling droplets of water.

Anaerobic; can grow in the presence of up to 2% oxygen but not in air enriched with 10% CO_2.

Physiological and nutritional characteristics are presented in Tables 6.15 and 6.16. Antibiotic MICs are in Table 6.17.

Found in the bovine rumen.

The mol% G + C of the DNA is 47 (T_m).

Type strain: ATCC 29543.

2. **Wolinella recta** Tanner, Badger, Lai, Listgarten, Visconti and Socransky, 1981, 441.[VP] (Includes some of the strains previously assigned to *Vibrio sputorum* Prévot 1940, 85.)

rec′ta. L. adj. *recta* straight, direct, in a straight line.

Mainly straight cells.

The inner and outer cell membranes are generally straight (or smooth) and parallel, about 20 nm apart. There is no distinct peptidoglycan cell-wall layer. The outer surface of the outer membrane is covered with a distinctive array of hexagonal, packed, macromolecular subunits, each about 17 nm in diameter.

Gray, translucent colonies are produced on blood agar plates. There are three colony types: a 1-mm diameter, convex variant; an agar-pitting or corroding variant up to 5 mm in diameter; and a variant which spreads on agar, up to a diameter of 5 mm.

Anaerobic; some strains will grow in the presence of 2% oxygen but not in air enriched with 10% CO_2.

Physiological and nutritional characteristics are presented in Tables 6.15 and 6.16. Antibiotic MICs are in Table 6.17.

The *W. recta* strains differ from those of *W. succinogenes* in cellular and ultrastructural morphology (Lai et al., 1981), in their serologic reactions (Badger and Tanner, 1981), and by their greater susceptibility to dyes and antibiotics (Tables 6.15–6.17).

Found in the gingival crevice of humans; pathogenicity is unknown.

The mol% G + C of the DNA is 42–46 (T_m).

Type strain: ATCC 33238.

Other organisms belonging to the genus **Wolinella**

Additional fresh human isolates have characteristics of the genus *Wolinella* but do not show high DNA homology with the currently named species in this genus (Tanner et al., 1982) (also see Table 6.18). Such isolates include ATCC 33567 (VPI 10659), FDC 286, and VPI strains 9584, 10279 and 10296. Since some of these strains show a high degree of DNA homology with each other, additional species should be created to include these isolates.

Table 6.18.
Percent DNA homology between **W. succinogenes, W. recta** *and* **Wolinella** *strains FDC 286 and VPI 9584*[a]

| | % Homology with DNA from Strain: | | | |
| | *W. recta* | | *Wolinella* sp. | |
Strain	ATCC 33238	FDC 285	FDC 286	VPI 9584
1. *Wolinella succinogenes* ATCC 29543	1	16	20	20
2. *Wolinella recta*				
ATCC 33238	100	75	25	24
372	95	85		17
303	90	73		24
302	79	86		23
1087	78	92		28
267	80	100		30
285	75	100	43	30
1219	68	100	37	21
Wolinella sp.				
FDC 286	25	44	100	43
VPI 9584	24	30	43	100

[a] Data derived using initial renaturation rate method, De Ley et al. (1970).

Genus IX. **Selenomonas** Von Prowazek 1913, 36[AL]

Marvin P. Bryant

Se.le.no.mo′nas. Gr. n. *selene* the moon; Gr. n. *monas* a unit, monad; M.L. fem. n. *Selenomonas* moon(-shaped) monad.

Curved to helical rods, usually 0.9–1.1 by 3.0–6.0 μm. The ends are usually tapered and rounded to give short kidney- to crescent-shaped or **vibrioid cells.** Long cells and chains of cells are often helical. Capsules are not formed. Resting stages are not known. Gram-negative. **Motile with active tumbling; flagella (up to 16) are arranged linearly as a tuft near the center of the concave side in the area of cell fission. Strictly anaerobic.** Optimum temperature, 35–40°C; maximum 45°C; minimum, 20–30°C. Chemoorganotrophic, having a **fermentative type of metabolism. Carbohydrates and sometimes amino acids and lactate can serve as fermentable substrates. Fermentation of glucose yields chiefly acetic acid, propionic acid, CO_2 and/or lactate.** Small amounts of H_2 and succinate may be produced. Catalase-negative. The mol% G + C of the DNA is 54–61 (Bd).

Type species: *Selenomonas sputigena* (Flügge 1886) Boskamp 1922, 70.

Further Descriptive Information

Cells are rigid, vibrioid to helical to crescentic in shape. They occur singly, in pairs and short chains and may occur in clumps, especially in *S. sputigena* (Kingsley and Hoeniger, 1973). Some strains, especially those of *Selenomonas ruminantium*, may be smaller (0.7 by 2.0–4.0 μm) (John et al., 1974); some strains, such as those of *S. ruminantium* subsp. *bryanti* (see under Other Organisms below) may be larger (2.0–3.0 by 5–10 μm) (Prins, 1971). Large cells may also be seen during microscopic examination, especially of sheep rumen fluid.

The flagella are 17 nm or more in diameter and their insertion into the cell is linear and ~68 nm apart, center to center (Kingsley and Hoeniger, 1973). The earlier confusion about the points of attachment of the flagella seems to have been clarified by the detailed studies of Kingsley and Hoeniger (1973). Some workers believed that the flagella were randomly attached or peritrichous (see MacDonald et al., 1959). This may have been due to the use of media with nutritional deficiencies or with too large an amount of the energy source, e.g. 1% glucose (Kingsley and Hoeniger, 1973). Thus, long cells which have not undergone fission may exhibit flagella typical of two or more cells, and, since fission occurs through the region where the flagella are located, the flagella may be near one end of the individual daughter cells immediately post fission.

Much information has been published on the cell envelope, membrane structure and lipid chemistry of *S. ruminantium* (see Kamio and Takahashi (1980) for further references).

The antigenic structure has received some attention but is not yet very definitive (Kingsley and Hoeniger, 1973; Hobson et al., 1962).

S. sputigena has been found only in the buccal cavity of man, especially in the gingival crevice. Its function in this habitat is not readily apparent, although it is a strong fermenter of certain carbohydrates. *S. ruminantium* is often numerous in the rumen of various ruminants, and similar members of the genus have been found in large numbers in the cecum of ground squirrels (Barnes and Burton, 1970), the rat (Ogimoto, 1972) and the pig (Robinson et al., 1981). Leifson (1960) observed morphologically similar organisms in highly contaminated river water.

Enrichment and Isolation Procedures

S. sputigena can be among the predominant colonies obtained from human gingival debris inoculated into anaerobic blood agar (Loesche and Gibbons, 1965; Gibbons et al., 1963). It can also be isolated selectively, by streaking onto anaerobic plates of agar medium containing veal heart infusion, sodium lauryl sulfate and sheep's serum. The plates are incubated 5–8 days at 37°C in a Brewer jar under H_2. *S. sputigena* grows as a surface film, and the sodium lauryl sulfate inhibits part of the oral flora which tends to overgrow the organism (MacDonald and Madlener, 1957). The sheep's serum permits an essential rapid drop in E_h of the surface layers of the medium in the Brewer jar. It seems probable that the organism could be selectively isolated using a sodium lauryl sulfate agar medium with mannitol as the main energy source. Many possible contaminants would not utilize mannitol. Moreover, an anaerobic glovebox might be a useful alternative to the Brewer jar.

Strains of *S. ruminantium* have often been isolated nonselectively from the rumen using RGCA medium roll tubes (Bryant and Small, 1956); see also the section on the genus *Butyrivibrio* in this *Manual*). High dilutions (10^{-7}–10^{-8}) of rumen fluid are inoculated into roll tubes and incubated for 3 days at 37–39°C. When a large number of colonies are studied, a few may be *S. ruminantium*. Under some animal dietary conditions such as high grain feeding where the rumen pH is lowered, *S. ruminantium* is sometimes a major species cultured in nonselective media (Caldwell and Bryant, 1966).

Strains can also be isolated selectively, using the agar medium of Tiwari et al. (1969) which contains mannitol, trypticase (BBL), yeast extract, *n*-valerate, acetate, cysteine, minerals and a CO_2 phase. The medium is selective because few other rumen bacteria grow with mannitol as an energy source, and only a small amount of bicarbonate is added so that the pH will be ~6.0, thereby inhibiting some rumen bacteria. Moreover, some rumen bacteria require branched-chain acids such as isobutyrate and 2-methylbutyrate which are not added to the medium. Cysteine and ferrous sulfate are added at levels higher than usual so that organisms producing sulfide from cysteine will form black colonies containing ferrous sulfide (as does *S. ruminantium* subsp. *ruminantium*). The only other bacterium producing large colonies on this medium inoculated from some rumen content samples is *Megasphaera elsdenii*.

S. ruminantium subsp. *lactilytica* is isolated with some degree of selectivity in roll tubes agar media containing glycerol as the main energy source (Hobson and Mann, 1961).

Maintenance Procedures

Selenomonas strains can be lyophilized by common procedures used for anaerobes.

Differentiation of the genus **Selenomonas** from other genera

The genus *Selenomonas* is separated from many other genera of anaerobic, curved, motile Gram-negative nonsporing rods by its type of flagellar arrangement, i.e. a tuft of flagella arranged in a closely spaced linear fashion near the middle of the concave side of the cell, and also by its production of propionate as a major product of glucose fermentation. It differs from the genus *Pectinatus* in that the latter has linearly arranged flagella that may be placed all along one side of the cell. *Selenomonas* has an optimum temperature of ~37°C or higher and does not grow below 20–30°C, whereas *Pectinatus* grows between 15 and 40°C and has an optimum temperature of ~32°C. *Selenomonas* species usually ferment esculin, lactose, raffinose and sucrose, whereas *Pectinatus* strains do not. Also, the mol% G + C of the DNA of *Selenomonas* is 54–61, in contrast to a value of ~40 for *Pectinatus*. Characteristics differential for the genus *Selenomonas* are listed in Table 6.11 in the article on the genus *Butyrivibrio*.

Taxonomic Comments

Precise knowledge of the phylogenetic relationships of the genus *Selenomonas* to other genera or higher taxa will depend on future studies of its nucleic acids, such as the degree of similarity observed in studies on rRNA oligonucleotide sequences.

The detailed comparative studies of *S. sputigena* and *S. ruminantium* by Kingsley and Hoeniger (1973) indicate that these bacteria should not be placed in *Spirillum* or *Aquaspirillum* (MacDonald et al., 1959).

Further Reading

Bryant, M.P. 1956. The characteristics of strains of *Selenomonas* isolated from bovine rumen contents. J. Bacteriol. *72:* 162–167.

Judicial Commission. 1958. Opinion 21. Conservation of the generic name *Selenomonas* von Prowazek. Int. Bull. Bacteriol. Nomen. Taxon. *8:* 163–165.

Kingsley, V.V., and J.F.M. Hoeniger. 1973. Growth, structure, and classification of *Selenomonas*. Bacteriol. Rev. *37:* 479–521.

Lessel, E.F. and R.S. Breed. 1954. *Selenomonas* Boskamp, 1922—a genus that includes species showing an unusual type of flagellation. Bacteriol. Rev. *18:* 165–169.

MacDonald, J.B., E.M. Madlener and S.S. Socransky. 1959. Observations on *Spirillum sputigenum* and its relationship to *Selenomonas* species with special reference to flagellation. J. Bacteriol. *77:* 559–565.

Differentiation of the species of the genus **Selenomonas**

Some differential features of the species *S. sputigena* and *S. ruminantium* are indicated in Table 6.19.

List of the species of the genus **Selenomonas**

1. **Selenomonas sputigena** (Flügge 1886) Boskamp 1922, 58[AL] (*Spirillum sputigenum* Flügge 1886, 387).

spu.ti′ge.na. L. n. *sputum* spit, sputum; L. v. *gigno* to produce; M.L. fem. adj. *sputigena* sputum-produced.

See Tables 6.19 and 6.20 and the generic description for many features.

Colonies on anaerobic blood agar are generally smooth, convex, grayish yellow, sometimes mottled and sometimes with a more opaque center. Usually the diameter is less than 0.5 mm. Sometimes the colonies are smooth, granular, flat, grayish and translucent with an irregular border, and up to 2 mm in diameter.

In Brewer's thioglycolate broth (Difco) growth occurs as heavy floccules and coarse granules (MacDonald, 1953). In MPB broth (Kingsley and Hoeniger, 1973) growth is turbid, often with a granular sediment.

No definitive studies have been done on the nutrition, but *S. sputigena* is a strong fermenter of certain carbohydrates. Its ability to utilize amino acids or many other potential energy sources has not been determined. It does not usually produce H_2S from cysteine, suggesting that it does not degrade cysteine.

Glucose is fermented with production of propionate and acetate (Loesche and Gibbons, 1965) and may produce some lactate and small amounts of succinate (Holdeman et al., 1977). The nature of the fermentation products suggests that the organisms would produce some CO_2. The ability to produce at least small amounts of H_2 has not been adequately studied.

S. sputigena appears not to be pathogenic for mice or guinea pigs. Intravenous injection into rabbits has caused some symptoms and even death in some cases but not in others (MacDonald et al., 1959). Although *S. sputigena* is found in the human gingival crevice, its relationship to the etiology of periodontal disease has not been established.

The mol% G + C of the DNA is 61 (Bd).

Type strain: ATCC 33150 (VPI 10068).

2. **Selenomonas ruminantium** (Certes 1889) Wenyon 1926, 311[AL] (*Ancyromonas ruminantium* Certes 1889, 70).

ru.mi.nan′ti.um. M.L. pl. n. *ruminantia* ruminants; M.L. pl. gen. n. *ruminantium* of ruminants.

See Tables 6.19 and 6.20 and the generic description for many features.

On RGCA-medium roll tubes (see the genus *Butyrivibrio* for medium composition) colonies are entire, slightly convex, translucent, light tan and 2–4 mm in diameter after incubation for 3 days at 37°C; deep colonies are thin and lenticular. Growth in liquid media is heavily turbid, often with some lightly flocculent sediments (Bryant, 1956).

All strains of the species so far studied grow in a chemically defined medium containing glucose as the energy source, B-vitamins, CO_2, ammonia as the nitrogen source, sulfide as the sulfur source, *n*-valerate and minerals (Bryant and Robinson, 1962; Kanegasaki and Takahashi, 1967; John et al., 1974; Tiwari et al., 1969; Linehan et al., 1978). Some strains require only biotin among the B-vitamins, and some strains do not require *n*-valerate or similar acid. Some strains can utilize amino acids such as aspartate, histidine, serine or cysteine as sole nitrogen sources. Cysteine can also serve as a sole sulfur source. A few strains contain urease and can utilize urea as a sole nitrogen source (John et al., 1974; see also Robinson et al., 1981). *S. ruminantium* subsp. *lactilytica* strains, which require *n*-valerate when grown on glucose, do not require it for growth on lactate but have added requirements for aspartate and *p*-aminobenzoate (Linehan et al., 1978; Kanegasaki and Takahashi, 1967). Nitrate is reduced to ammonia in a few strains and thus serves as a sole nitrogen source (John et al., 1974), but other strains reduce nitrate only to nitrite or do not reduce nitrate at all.

S. ruminantium ferments glucose, producing propionate, acetate, CO_2 and/or lactate as major products. Strains form L-lactate and sometimes a mixture of L- and D-lactate (Scheifinger et al., 1975). The amount of lactate produced as compared to propionate and acetate depends on the strain, the amount of glucose in the medium and, in chemostat cultures, the growth rate (Scheifinger et al., 1975). Strains may also produce some succinate. All strains so far studied produce some H_2, although the amount produced is usually very small except in the presence of methanogens which use the H_2 to produce methane (Scheifinger et al., 1975).

The mol% G + C of the DNA is 54 (Bd).

Type strain: ATCC 12561.

2a. **Selenomonas ruminantium** subsp. **ruminantium** Bryant 1974, 425.[AL]

Table 6.19.
Characteristics differentiating **Selenomonas sputigena** *and* **Selenomonas ruminantium**[a]

Characteristics	1. *S. sputigena*	2. *S. ruminantium*
Habitat:		
Human buccal cavity	+	−
Rumen	−	+
Ferment cellobiose, dulcitol and salicin	−	+
Produce H_2S from cysteine	−	+
Mol% G + C of DNA	61	54

[a] Symbols: see standard definitions.

Table 6.20.
Other physiological features of **Selenomonas sputigena** *and* **Selenomonas ruminantium**[a]

Characteristics	1. *S. sputigena*	2. *S. ruminantium*
Fermentation of:		
Arabinose, esculin, galactose, glucose, lactose, maltose, sucrose and mannitol	+	+
Dextrin, inulin, glycerol and starch	+	d
Cellulose and xylan	−	−
Trehalose	d	d
Raffinose	d	+
Sorbitol and inositol	−	d
Nitrate reduced to nitrite	+	d
Indole produced	−	−
Gelatin liquefied	−	−

[a] Symbols: see standard definitions.

The description is as for the species. Differs from the subspecies *lactilytica* by being unable to ferment lactate and glycerol.

Type strain: ATCC 12561.

2b. **Selenomonas ruminantium** subsp. **lactilytica** (Hungate 1966) Bryant 1956, 165.[AL] (*Selenomonas lactilytica* Hungate 1966, 68.)

lac.ti.ly'ti.ca. L. n. *lac, lactis* milk; Gr. adj. *lyticus* dissolving; M.L. adj. *lactilytica* milk (lactate) dissolving.

This subspecies possesses the same characteristics as the species except that lactate and glycerol are fermented. Lactate is fermented with production of propionate, acetate and CO_2 (Bryant, 1956). Glycerol is fermented mainly to propionate with small amounts of lactate, succinate and acetate (Hobson and Mann, 1961).

"O" antisera from sheep strains have been found to react with a bovine strain of the same subspecies but not with a bovine strain of the subspecies *ruminantium*, suggesting that the two subspecies may differ in "O" antigens (Hobson et al., 1962).

Type strain: ATCC 19205.

Other organisms

The following organisms either have no available strains or have not been obtained in pure culture.

i. "*Selenomonas ruminantium* subspecies *bryanti*" Prins 1971, 825.

bry.an'ti. M.L. gen. n. *bryanti* of Bryant.

This subspecies possess characteristics similar to the subspecies *ruminantium* except that the cells in pure culture are larger, measuring 2.0–3.0 by 5.0–10.0 μm. H_2S is not produced from cysteine, and arabinose, xylose, galactose, lactose and dulcitol are not fermented.

This subspecies was isolated using anaerobic rumen-fluid agar medium containing minerals similar to those of RGCA medium (see *Butyrivibrio*) and deep agar tubes rather than roll tubes. The rumen fluid source material was differentially centrifuged to enrich the large selenomonads before culture (Prins, 1971).

No type strain has been designated and none of the originally described strains are now available.

ii. "*Selenomonas palpitans*" Simons 1921, 50.

pal'pi.tans. L. part. adj. *palpitans* trembling.

Simons (1920) first observed an organism similar to *Selenomonas* in the cecum of guinea pigs and it was later named *S. palpitans* (Simons, 1921, 1922; Boskamp, 1922). It has not been obtained in pure culture, however, and in a study of the cytology of the organism from guinea pig cecal contents, Kingsley and Hoeniger (1973) found that the flagella of the tuft were bunched together in a single area rather than forming a line of insertions as is the case in *Selenomonas*. The further taxonomic status of this organism must be delayed until it is isolated in pure culture.

Genus X. **Anaerovibrio** Hungate 1966, 80[AL]

RUDOLF A. PRINS

An.ae.ro.vib'ri.o. Gr. pref. *an* not; Gr. n. *aer* air; *anaero* not (living) in air; L. v. *vibrio* to move rapidly to and fro, vibrate; M.L. masc. n. *vibrio* that which vibrates; M.L. n. *Anaerovibrio* vibrio not living in air.

Slightly curved rods, ~0.5 μm thick and 1.2–3.6 μm in length. Gram-negative. **Motile in liquid media by a single polar flagellum.** Strictly anaerobic; no scavenging system for oxygen is present. Growth occurs at 38°C but not at 20, 30 or 50°C. Optimum pH for growth, 6.3; no growth occurs below pH 5.9 or above pH 7.0. Minute brownish colonies develop within 5–7 days in linseed oil agar and produce zones of clearing in the agar where the triacylglycerols have been hydrolyzed. **Lipolytic.** In glycerol agar the colonies are lens-shaped, mucoid, whitish discs. Glycerol, fructose, ribose and DL-lactate can be used as carbon sources. **Glycerol is fermented mainly to propionate. Fermentation products from DL-lactate, ribose and fructose are acetate, propionate, CO_2,** and traces of H_2 and succinate. **Peptides, triacylglycerols and phospholipids also sustain growth.** An organic nitrogen source and vitamins (folic acid, pantothenate and pyridoxal-HCl) are required. Occur free-living in the rumen of sheep and cattle. The mol% G + C of the DNA is not known.

Type species: *Anaerovibrio lipolytica* Hungate 1966, 80.

Further Descriptive Information

Cells of *Anaerovibrio lipolytica* are slightly curved Gram-negative rods, variable in size but generally 0.5 μm wide and 1.2–1.8 μm long in young cultures in liquid medium.* Organisms in older cultures become more difficult to stain and develop a central granule. These older organisms are generally longer, up to 3.0–3.6 μm. Still later, the organisms seem to disintegrate, leaving only a granular mass (Hobson and Mann, 1961).

Bacteria harvested from cultures in the early logarithmic phase, or bacteria in continuous cultures at a dilution rate of 0.12 h^{-1}, show the presence of "blebs" on the surface. These structures are believed to contain lipase activity, which can be found associated with membranous material having a phospholipid composition similar to that of the whole bacteria (Henderson and Hodgkiss, 1973). Freeze-etching electron microscope studies have indicated that *A. lipolytica* is likely to show phase separations of lipids and proteins in the lateral plane of the membranes similar to those found in membranes of *Escherichia coli* and other microbes (Verkley et al., 1975).

The cells are actively motile with a single polar flagellum. No iodine-staining polysaccharide or definite capsules are evident, but some strains produce slime.

Colonies in linseed oil-rumen fluid-agar roll tubes† are minute or invisible to the naked eye unless incubation is continued for a long time, and growth is usually evident only by the circular clear areas which develop in the agar due to lipolytic activity. After 3–5 days the colonies are very small and grayish with narrow zones of clearing. These zones may require magnification to be seen (Hobson, 1969). After initial isolation, subsequent subculturing results in larger colo-

---

* Medium 12 of Hobson and Mann (1961) contains (per liter): mineral solution A [KH_2PO_4, 3.0 g; $(NH_4)_2SO_4$, 6.0 g; NaCl, 6.0 g; $MgSO_4$, 0.6 g; $CaCl_2$, 0.6 g; distilled wate, 1000 ml], 150 ml; mineral solution B (K_2HPO_4, 3.0 g; distilled water, 1000 ml), 150 ml; rumen fluid (strained through gauze and centrifuged at 62,000 × *g* for 10 min), 400 ml; distilled water, 290 ml; resazurin [0.1% (w/v) aqueous solution], 1.0 ml; $NaHCO_3$, 4.0 g; cysteine-HCl, 0.5 g; and linseed oil [50% (v/v) emulsion in sterile rumen fluid], 20.0 ml. The minerals, rumen fluid, water and resazurin are autoclaved together at 120°C for 15 min. The $NaHCO_3$ and cysteine-HCl are added aseptically from a stock solution sterilized by filtration, and the linseed oil is added aseptically from the sterile emulsion. The medium is prepared and dispensed (with shaking to emulsify the oil) under a 100% CO_2 atmosphere.

† Medium 13 of Hobson and Mann (1961) is similar to Medium 12 (see footnote *) except for the addition of 20 g of agar/liter. The linseed oil emulsion is added to the medium at a temperature of ~70°C and the medium is shaken immediately so that the agar does not set while the oil is being emulsified. It is dispensed into roll tubes and kept at 50°C until inoculated, and the tubes are subsequently rolled under cold water. For the roll-tube technique, see Hungate (1969).

nies. In glycerol-rumen fluid-agar (containing 2 g of glycerol/liter) colonies are lens-shaped, mucoid whitish discs (Prins et al., 1975).

The organism is strictly anaerobic and does not possess an oxygen-removing system (de Vries et al., 1974).

Despite the fact that the *A. lipolytica* strains isolated by Hobson and Mann (1961) grew somewhat better in a medium containing 30–40% (v/v) rumen fluid than in media containing 10–20% rumen fluid, volatile fatty acids (including branched-chain acids), hemin and menadione were not required for the growth of later isolates (Prins et al., 1975). Yeast extract and trypticase (BBL) were required in a medium containing glycerol as the energy source. The yeast extract could be replaced by a combination of the vitamins folic acid, pantothenate and pyridoxal-HCl. No requirement could be demonstrated for thiamine, niacin, riboflavin, *p*-aminobenzoic acid, biotin or cyanocobalamin. The trypticase could be replaced by certain combinations of amino acids (e.g. serine + aspartate, glutamine + leucine, but not by glycine + alanine). In a medium lacking glycerol, trypticase sustained some growth in the presence of yeast extract and higher levels of acetate and propionate were produced, indicating that amino acid fermentation had occurred (Prins et al., 1975).

A. lipolytica does not utilize the long chain fatty acids resulting from its hydrolysis of linseed oil, and no growth occurs in a medium containing the acids from hydrolyzed linseed oil (Hobson and Mann, 1961). Henderson (1973a) reported that growth of *A. lipolytica* strain 55 was unaffected by oleic acid up to 100 μg/ml.

The following substrates can be fermented by *A. lipolytica*: glycerol, fructose, ribose, and DL-lactate. The following substances will not sustain growth: arabinose, xylose, rhamnose, glucose, galactose, mannose, sorbose, glucose-1-phosphate, glucuronic acid, galacturonic acid, glucosamine, cellobiose, esculin, salicin, maltose, lactose, sucrose, melibiose, trehalose, melezitose, raffinose, dextrin, starch, glycogen, amylose, pectin, inulin, cellulose, xylan, mannitol, adonitol, sorbitol, dulcitol, inositol and propylene glycol (Hobson and Mann, 1961; Prins et al., 1975; Counotte, 1981).

In addition to glycerol, fructose, ribose, DL-lactate and trypticase, triglycerides and phospholipids can also sustain growth of *A. lipolytica* (Prins et al., 1975).

The final pH in Medium 12* containing glycerol is 5.6, and in Medium 14‡ containing glycerol it is 5.0 (Hobson and Mann, 1961). The final pH in the medium of Prins et al. (1975) with glycerol as the energy source is 5.9 and no growth occurs below pH 5.8. The pH range for growth of strain 5S of Hobson and Summers (1967) is 5.7–7.0. The optimum pH for growth on glycerol of strain A22 (received from C. Henderson, Aberdeen, Scotland, U.K.) is 6.3 (Counotte, 1981). The maximum growth rate (μ_{max}) of this strain (0.21 h^{-1}) corresponds closely to the value of 0.2 h^{-1} found for strain 5S on glycerol by Hobson and Summers (1967) in continuous culture experiments. From the data of Hobson and Summers, a μ_{max} of 0.16 h^{-1} and K_S of 3.33 mM can be calculated for the growth of strain 5S on fructose in continuous culture.

Growth of *A. lipolytica* occurs at 38°C but not at 20, 30 or 50°C.

Hydrogen sulfide is produced, weak urease activity is present, and esterase and lipase activity are present. The indole test is negative, nitrate is not reduced, and neither gelatin nor esculin is hydrolyzed (Counotte, 1981; Hobson and Mann, 1961).

Fermentation products from glycerol are: propionate, succinate, and traces of H$_2$ and L-lactate (Prins et al., 1975). Less succinate is formed from 60 mM glycerol than from 20 mM glycerol. Fermentation products from DL-lactate, ribose and fructose are acetate, propionate and CO$_2$, and traces of succinate and H$_2$ (and L-lactate from the sugars) (Prins et al., 1975; Hobson and Summers, 1967).

Propionate is formed by *A. lipolytica* for DL-lactate by the dicarboxylic acid pathway via succinate, similar to the pathway occurring in propionibacteria, *Veillonella* and *Selenomonas ruminantium* (Prins et al., 1975). Like these organisms, *A. lipolytica* contains cytochrome *b*

(0.34 μmol/g bacterial dry weight after growth on glycerol in batch culture); absorption peaks corresponding to cytochrome *a* have also been observed, as well as a CO-binding pigment (de Vries et al., 1974). Glycerol appears to be fermented via glycerol kinase, since NADH or glycerol-1-phosphate—but not glycerol—can serve as a hydrogen donor for reduction of cytochrome *b*. Reoxidation of cytochrome *b* requires fumarate. HOQNO (2-*n*-heptyl-4-hydroxyquinolin-*N*-oxide) retards the reduction of cytochrome by NADH. The generation of 1 mol of ATP during electron transport from glycerol-1-phosphate to fumarate via cytochrome *b* would explain the molar growth yield of *A. lipolytica* on glycerol (22 g of bacteria/mol of glycerol), since Hobson (1965) concluded that 2 mol of ATP were formed per mole of glycerol fermented. The other mole of ATP must be formed by substrate-level phosphorylation in the conversion of glycerol to pyruvate.

Membrane suspensions of *A. lipolytica* have shown only fumarate reductase activity when prepared under strictly anaerobic conditions, indicating that no scavenging system for oxygen exists in this organism. Moreover, strain 5S did not take up oxygen in continuous culture experiments by Hobson and Summers (1967).

A. lipolytica is able to reduce fumarate to succinate with extracellular H$_2$. The apparent K_m value for dissolved H$_2$ of the membrane-bound hydrogenase of strain 5S is 1.4×10^{-5} M, and the apparent K_m value for fumarate of the membrane-bound fumarate reductase is 1.7×10^{-4} M. In view of this and the fact that cytochrome *b* is present in fructose-grown cells at a level of 0.24 μmol/g bacterial dry weight, it has been suggested that, when methanogenesis in the rumen is inhibited, accumulated H$_2$ could serve to reduce fumarate to succinate and that cytochrome *b* could play a role in the oxidation of the H$_2$ (Henderson, 1980).

The lipase of *A. lipolytica* is constitutive, and cell-free extracts of strain L1741 grown in batch culture on linseed oil, glycerol, DL-lactate, ribose or fructose liberate 22–24 nmol of glycerol per min per mg protein from a 25% (v/v) linseed oil emulsion (Prins et al., 1975). Cells of strains L1741 and L1342 grown on glycerol hydrolyze trilinolein at a rate of 0.4 μmol of linoleate liberated/h/0.5 $\times$ 10^7 cells. Diglycerides are attacked more rapidly by the lipase than are triglycerides, and there is no activity against water-soluble esters such as tributyrin. Olive oil is more rapidly hydrolyzed than the triglycerides isolated from it, which may be due to the presence of diglycerides in the olive oil. For other properties and purification of the lipase, see Hobson and Summers (1966, 1967) and Henderson (1968, 1970, 1971).

As is the case with other anaerobic bacteria capable of lactate fermentation, *A. lipolytica* contains large amounts of the unusual phospholipid phosphatidylserine [19.8% of the total lipid phosphate (Van Golde et al., 1975)]. The fatty acids C$_{15:0}$, C$_{17:0}$, C$_{15:1}$ and C$_{17:1}$ are the principal fatty acids of both phosphatidylserine and phosphatidylethanolamine.

In sheep fed a basal ration of linseed cake meal, casein and a concentrate mixture (ground corn, crushed oats and bran), with hay fed separately, *A. lipolytica* was found in the rumen in concentrations of at least 10^8/ml, when linseed oil-agar was used as the plating medium (Hobson and Mann, 1961). The bacteria seemed to be regular inhabitants of the rumen of these sheep, since feeding of linseed oil or glycerol did not increase the number of lipolytic bacteria; however, it should be noted that the feed already contained some lipids.

Slyter et al. (1976) identified 18 strains as *A. lipolytica* from among 81 strains isolated from a mixed ruminal population being changed from forage to concentrates. The bacteria were isolated by means of a lactate medium and were present at levels of 10^8 per ml of rumen contents. Lower concentrations (0.5 $\times$ 10^7 and 1.1 $\times$ 10^7/ml) of *A. lipolytica* were found by Prins et al. (1975) in two sheep fed at maintenance on hay only, and levels of 1.6 $\times$ 10^7/ml occurred in a cow fed hay at maintenance. Counotte (1981) found 5 strains of *A. lipolytica* among 21 strains isolated from the rumen of cattle changed from a hay diet to

‡ Medium 14 of Hobson and Mann (1961) is similar to Medium 12 (see footnote *) except that the rumen fluid is replaced by Bacto-casitone (Difco) and yeast extract (1.5 g and 0.25 g/liter, respectively).

a hay + concentrates diet; the organisms were mainly isolated while the animals were still on the hay diet.

Because of its relative sensitivity to low pH values, *A. lipolytica* can play only a modest role in the fermentation of lactate in the rumen. When the organism is present, however, it probably has an important function in the hydrolysis of triacylglycerols and phospholipids and in the fermentation of glycerol.

Enrichment and Isolation Procedures

Rumen contents are serially diluted in a linseed oil-agar medium such as Medium 13 of Hobson and Mann (1961) (see Footnote †) in roll tubes, using the Hungate technique (Hungate, 1969). After 5–7 days of incubation at 40°C, clearing zones develop in the agar where the oil has been hydrolyzed. Small brownish colonies can be seen in the center of these cleared areas. Single colonies are removed under a dissecting microscope and subcultured in linseed oil agar. For final purification cultures are transferred to an agar medium in which the linseed oil is replaced by glycerol (0.2%) as the substrate.

Henderson (1973b) advocates the use of trilaurin as the substrate for isolation, since growth of lipolytic bacteria is detectable after 2 days with trilaurin and since clearing zones are more clearly defined than with linseed oil emulsions. Emulsions can be prepared by dispersing 600 mg of trilaurin (or 1 ml of linseed oil) in 20 ml of water using an ultrasonic disintegrator. When trilaurin emulsions are made, the water is heated slightly to melt the trilaurin prior to sonic oscillation. Emulsions are added (10 ml/100 ml of medium) to the other medium constituents with the temeprature being maintained at 45°C. The complete media are autoclaved and dispersed under CO_2 using anaerobic techniques.

Maintenance Procedures

Stock cultures can be maintained in the medium of Henderson (1971)§ with biweekly transfers.

Differentiation of **Anaerovibrio** from other genera

The key to the genera of the family *Bacteroidaceae* indicates characteristics which distinguish the genus *Anaerovibrio* from other genera of anaerobic, Gram-negative bacteria. The outstanding property of *Anaerovibrio* is, of course, its lipolytic ability. In addition, the limited variety of sugars fermented distinguishes the organism from *Butyrivibrio* and the morphology distinguishes it from *Selenomonas*. Although the fermentation products and the few sugars fermented suggest a similarity to *Succinimonas*, different sugars are fermented and the cell morphology is different.

Taxonomic Comments

The relationships of *Anaerovibrio* to other genera of anaerobic,

Gram-negative bacteria are not known. Nucleic acid studies, such as rRNA/DNA hybridization or rRNA oligonucleotide cataloging, would be helpful in this regard.

Further Reading

Hobson, P.N. and S.O. Mann. 1961. The isolation of glycerol-fermenting and lipolytic bacteria from the rumen of sheep. J. Gen. Microbiol. 25:227–240.

Prins, R.A., A. Lankhorst, P. can der Meer and C. J. van Nevel. 1975. Some characteristics of *Anaerovibrio lipolytica*, a rumen lipolytic organism. Antonie van Leeuwenhoek J. Microbiol. Serol. 41:1–11.

List of the species of the genus **Anaerovibrio**

1. **Anaerovibrio lipolytica** Hungate 1966, 80.[AL]

   li.po.ly′ti.ca. Gr. n. *lipos* fat; Gr. adj. *lytikos* dissolving; M.L. adj. *lipolytica* fat-dissolving.

   The description is as given for the genus.

Found in the rumen of sheep and cattle.

The mol% G + C of the DNA is not known.

Type strain: VPI 7553.

Genus XI. **Pectinatus** Lee, Mabee and Jangaard 1978, 582[AL]

SUN Y. LEE

Pec.ti.na′tus. L. part. adj. *pectinatus* combed; M.L. masc. n. *Pectinatus* combed (bacteria).

Slightly curved rods, 0.7–0.8 μm in diameter and 2.0–32 μm or more in length, with rounded ends; they occur singly, in pairs, and only rarely in short chains. Shorter, younger cells do not show a helical shape, but the elongated older cells tend to form helices. After cultures have passed the stationary and declining phases, round cell forms are often observed. Gram-negative. **Young cells move very actively, giving the appearance of an "X" shape as they swim.** Older and longer cells have a snakelike motion as they move. **Flagella emanate from only one side of a cell. Flagella are not limited to the center portion of the concave side of the cell as they are in the genus *Selenomonas*.** The number of flagella per cell depends upon the cell size and its condition, but generally ranges from 1 to 23 or more. The organisms are **obligately anaerobic, nonsporeforming mesophiles.** They grow between 15 and 40°C, with optimum growth at ~32°C. Originally isolated from **spoiled, packaged beer.** The mol% G + C of the DNA is 39.8 (T_m).

Type species: *Pectinatus cerevisiiphilus* Lee, Mabee and Jangaard 1978, 582.

Further Descriptive Information

Cells of *Pectinatus cerevisiiphilus* typically appear as curved rods. In the decline phase, cultures often exhibit round cell forms and these increase with the time of incubation (Fig. 6.51). These globular forms are believed to be similar to the spheroplasts of selenomonads (Kingsley and Hoeniger, 1973). The rods are motile by means of flagella which emanate from only one side of the cell body (Fig. 6.52).

No growth is obtained on agar plates under aerobic conditions or under a CO_2 environment. Pour plates of thioglycolate agar (Difco) incubated in a GasPak jar (BBL) occasionally support growth and the streak plate completely fails to demonstrate surface growth. Using the Hungate roll-tube technique (Hungate, 1970) or the Lee tube system (Ogg et al., 1979) with thioglycollate agar, colonies develop after 3–4

---

§ Maintenance medium (per liter): agar, 20.0 g; yeast extract, 6.0 g; casein hydrolysate (or trypticase), 7.5 g; K_2HPO_4 (0.3% solution), 150 ml; and resazurin (0.1% aqueous solution), 1.0 ml. After autoclaving the basal medium, 100 ml of a solution containing 5% glycerol, 0.5% cysteine-HCl and 6% $NaHCO_3$, sterilized by filtration, is added per liter.

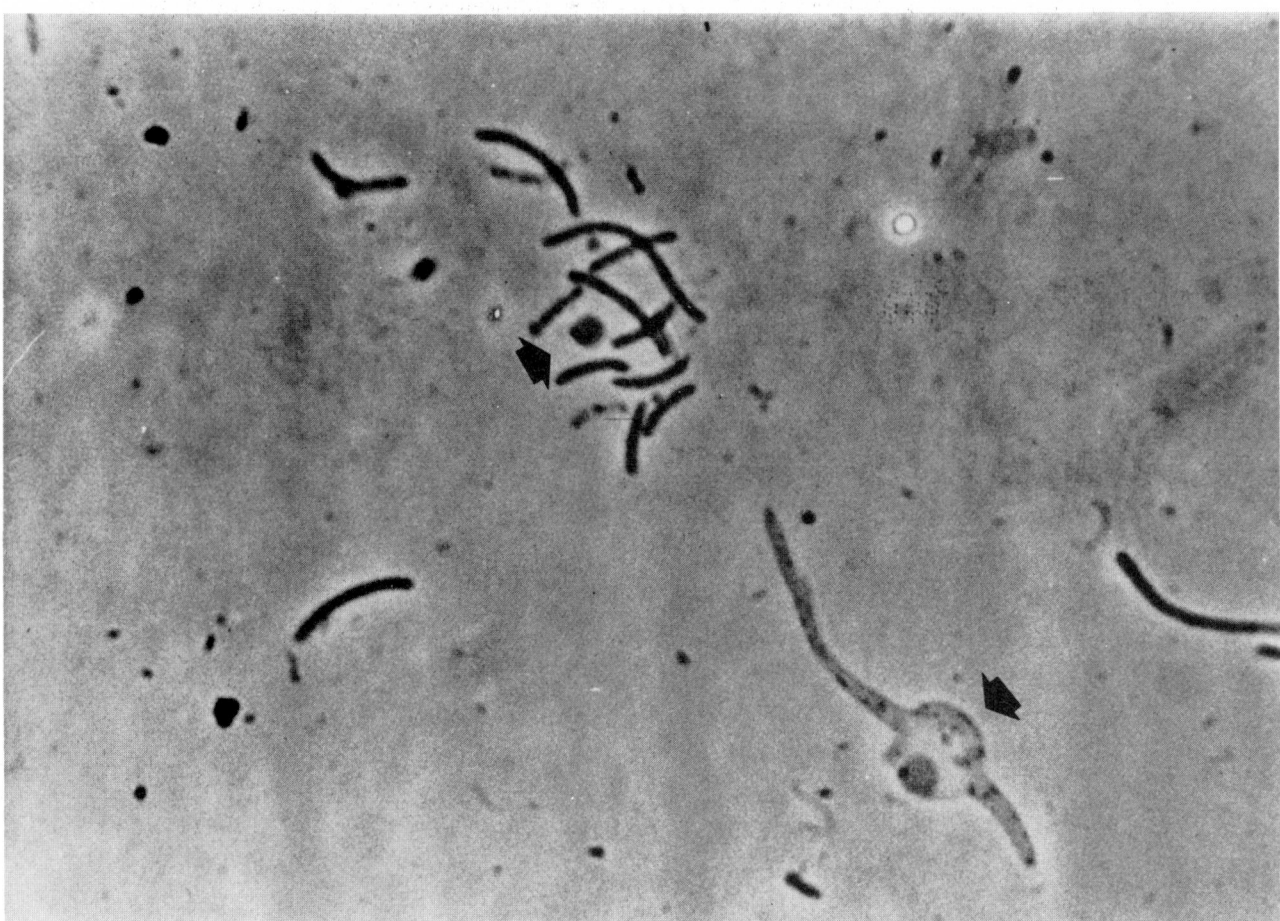

Figure 6.51. Globular forms (*arrows*) of *Pectinatus cerevisiiphilus* as seen by phase-contrast microscopy (×1,200). (Reproduced with permission from S. Y. Lee et al., *International Journal of Systematic Bacteriology, 28:* 582–594, 1978, © American Society for Microbiology, Washington, D.C.)

days at 30°C. Colonies are circular, entire, low convex to pulvinate, white, glistening and opaque.

Acid is produced from adonitol, arabinose, cellobiose, dulcitol, erythritol, fructose, galactose, glucose, glycerol, lactate, maltose, mannitol, mannose, rhamnose and ribose. Amygdalin, lactose, raffinose, salicin, sorbitol, sorbose, sucrose, trehalose and xylose are not fermented. Nitrate is not reduced to nitrite, and starch is not hydrolyzed. Acetic, propionic, succinic and lactic acids, and CO_2, are produced from lactate. Catalase is not produced.

Enrichment and Isolation Procedures

A glove box atmosphere containing 80% N_2, 10% CO_2 and 10% H_2 will support growth. Many agar media having fermentable sugars can be used for cultivation and isolation of this bacterium. For example, thioglycolate medium (Difco) with 1% glucose and 1.5% agar, and MRS lactobacilli broth (Difco) fortified with 1.5% agar, are excellent media when used in a glove box. Hungate's roll-tube technique (Hungate, 1970) is another alternative for *cultivation*, but is inadequate for *isolation* purposes where the pH of the inoculum is relatively low, as in beer (pH 4.3–4.5); this low pH can soften the agar and allow it to fall down from the wall of the roll tube.

A selective, differential agar medium (LL-agar*; Lee et al., 1981) has been used for the isolation of strains of *Pectinatus*. In this medium

strains of *Pectinatus* produce black colonies, whereas other brewery organisms are either inhibited or form white colonies. In brewing applications, a suspected beer sample (0.1–1.0 ml) is directly introduced into the Lee Tube containing LL-agar that has been boiled (autoclaving is not recommended) and cooled to 47°C in a water bath. After tightening the screw cap, the tube is inverted slowly several times for mixing. The oxidation-reduction indicator will exhibit an oxidized color (blue) but soon will return to a reduced (colorless) condition except for the top portion of the tube. The tube is then incubated at 30–32°C for 5–10 days. After colonies have developed, a white paper roll is inserted into the opening in the bottom of the tube for contrast. An isolated black colony may then be removed using a modified long needle with a syringe (Ogg et al., 1979).

Maintenance Procedures

Cultures can be maintained in semisolid thioglycolate medium (Difco) with 1.0% glucose. However, the organism also grows well in freshly prepared MRS lactobacilli broth (Difco) with a 2% (v/v) inoculum and minimum head space. The culture tends to lose its viability with repeated transfer. Between transfers, cultures can be stored at 2–5°C for 4–5 days. For preservation, active cells grown in either thioglycolate medium or MRS lactobacilli broth can be placed in vials and stored in liquid nitrogen.

---

* LL-agar (lactate-lead acetate agar), g/liter: yeast extract, 5.0; beef extract, 3.0; sodium lactate (60% syrup), 17 ml; ascorbic acid, 1.0; $Na_2S_2O_3 \cdot 5 H_2O$, 0.1; lead acetate, 0.2; methylene blue, 0.002; phenethyl alcohol, 2.0 ml and agar, 15.0

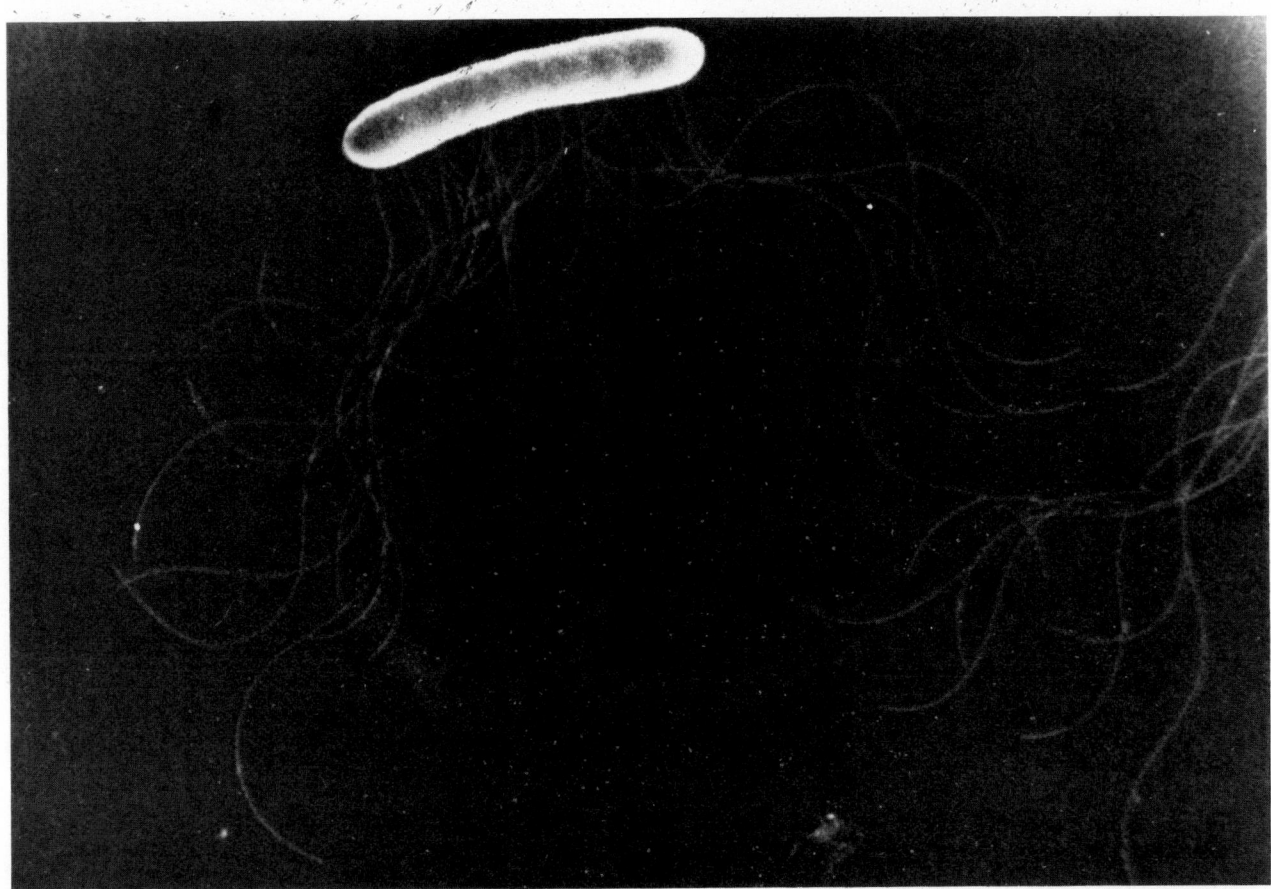

Figure 6.52. Scanning electron micrograph illustrates flagella of *Pectinatus cerevisiiphilus* (×13,680). (Reproduced with permission from S. Y. Lee et al., *International Journal of Systematic Bacteriology, 28:* 582–594, 1978, © American Society for Microbiology, Washington, D.C.)

Differentiation of the genus **Pectinatus** from other genera

See the key to the genera of the family *Bacteroidaceae* for characteristics distinguishing the genus *Pectinatus* from other anaerobic bacteria.

Both *Selenomonas* and *Pectinatus* have lateral flagella emanating from the concave side of the cell; however, the flagella of *Selenomonas* occur as a tuft at the center of the concave side of the cell, whereas the flagella of *Pectinatus* are not limited to the central region. Table 6.21 indicates other differences between *Pectinatus* and the selenomonads.

Taxonomic Comments

The phylogenetic relationships of *Pectinatus* to other anaerobic bacteria have not yet been determined by nucleic acid hybridization or by studies of the oligonucleotide sequences of rRNA.

Lists of the species of the genus **Pectinatus**

Pectinatus cerevisiiphilus Lee, Mabee and Jangaard 1978, 582.[AL]

ce.re.vi.si.i'phi.lus. L. n. *cerevisia* beer; Gr. adj. *philus* loving M.L. adj. *cerevisiiphilus* beer-loving (bacteria).

Because only a single strain of this new species has been isolated to date, the generic description given above for this strain also serves as the species description.

Isolated from spoiled beer.

The mol% G + C of the DNA is 39.8 (T_m).

Type strain: ATCC 29359.

Other organisms

As of this writing, there have been reports describing the isolation of *Pectinatus* outside of the United States. Back *et al.* (1979) have recovered *Pectinatus* from spoiled German beer (DSM strains 20465 and 20466), and Haikara *et al.* (1981) isolated this bacterium in Scandinavian (Finnish) beer. In both cases, complete biochemical and electron microscopic studies were performed along with the type strain of *P. cerevisiiphilus*. All morphological and biochemical characteristics of these reported isolates were very similar to those of *P. cerevisiiphilus* with minor differences in sugar utilization, mol% G + C and serological reactions. However, the taxonomic positions of these isolates within the genus *Pectinatus* have not yet been established.

Table 6.21.

Characteristics useful in differentiating **Pectinatus cerevisiiphilus** *from* **Selenomonas** *species*[a]

Characteristics	P. cerevisii-phimus	S. rumin-antium	S. sputi-gena
Cell shape:			
Crescent moon	−	+	+
Slightly curved rods with rounded ends	+	−	−
Cell diameter, μm	0.7–0.8	0.9–1.1	0.9–1.1
Cell length, μm	2.0–32.0 or longer	3.5–6.0	3.0–5.5
Flagellar arrangement:			
Tuft in center of concave side	−	+	+
Multiple flagella on one side, but not as a tuft	+	−	−
Glucose fermentation products:			
Acetic acid	+	+	+
Lactic acid	+	+	−[b]
Propionic acid	+	+	+
Succinic acid	+	−[c]	−[b]
Fermentation of:			
Erythritol	+	−	−
Esculin	−	+	+
Lactose	−	+	+
Melibiose	−	+	+
Raffinose	−	+	d
Sucrose	−	+	+
Xylose	−	+	+
Adonitol	Weak	−	−
Curd produced in milk	−	+	+

[a] Symbols: see standard definitions.

[b] Some lactate and small amounts of succinate may be produced (see the genus *Selenomonas*).

[c] Some succinate may be produced (see the genus *Selenomonas*)

Genus XII. **Acetivibrio** *Patel, Khan, Agnew and Colvin 1980, 184,*[VP] *emend. Robinson and Ritchie 1981, 335*[VP]

G. B. PATEL

A.ce′ti.vib.ri.o. L. n. *acetum* vinegar; M.L. masc. n. *Vibrio* genus of bacteria; M.L. masc. n. *Acetivibrio* vinegar (acetic) vibrio.

Straight to slightly curved rods, 0.5–0.9 by 1.5–10.0 μm occurring singly, in pairs and in chains of cells up to 40 μm long. **Gram-negative, nonsporeforming. Motile by means of a single flagellum** located about one-third of the distance from pole to pole **or by multiple flagella** which are arranged linearly and often fascicled, **emanating from the concave side of the cells in both cases. Obligately anaerobic.** Optimum temperature 35–37°C. Chemoorganotrophic. May or may not metabolize peptone. **Acetic acid is the sole or major organic acid end product of carbohydrate fermentation.** Formic acid is produced by some strains. **Propionic, butyric, succinic and lactic acids are not formed.** The other products of carbohydrate fermentation are H_2, CO_2, and ethanol. The mol% G + C of the DNA ranges from 37–40 (UV spectroscopy and T_m).

Type species: *Acetivibrio cellulolyticus* Patel, Khan, Agnew and Colvin 1980, 184.

Further Descriptive Information

Single cells are straight to slightly curved rods and are present singly, in pairs, and occasionally in chains (up to 40 μm). In indole-nitrite broth supplemented with 0.5% cellobiose, *A. cellulolyticus* grows predominantly as very long filaments with round swellings at various points; in PY-cellobiose broth (Holdeman and Moore, 1975) long cells without septa are observed occasionally, but the swellings are not as numerous. On CCA medium (Allison et al., 1979). *A. ethanolgignens* grows as helical curved rods with pointed ends in the early growth phase (12–24 h), but filamentous and swollen forms are present in older cultures.

Endospores are not formed. *A. cellulolyticus* forms capsules.

The cell wall of *A. cellulolyticus* has five distinct layers and resembles in morphology the well-differentiated walls of Gram-positive organisms rather than those of Gram-negative cells (Salton, 1964), although the organisms stain Gram-negative. The outermost layer of the wall is fuzzy, amorphous, and soft enough to adsorb and sometimes bury portions of cellulose microfibrils with which the cells come into contact. The cell surface has a net positive charge. The cytoplasm contains electron-dense, vaguely striated bodies and small globules without membranes, all embedded in a finely granular matrix.

The cell wall of *A. ethanolgignens* has not been examined in thin section. Negative staining indicates the presence of intracellular inclusions consisting of dense, spherical to oblate granules (30–80 nm in diameter) with close clustering of subunits, and also rhapidosome-like rodlets 20 by 100 nm.

The flagellum of *A. cellulolyticus* is unusual because of its location (about one-third the distance from pole to pole). This location seems to be responsible for the "tumbling motility" of the cells. The flagellum has a diameter of ~20 nm and a length of 6.0–7.0 μm. The cells of *A. ethanolgignens* have 10–15 flagella, often fasciculated, arranged linearly on the concave surface. The flagella are 10–12 nm in diameter.

The optimum temperature for growth is ~35°C.

Surface colonies of *A. cellulolyticus* on cellulose agar are clearly visible after 3–4 weeks of incubation and are surrounded by clear zones of cellulose digestion which extend into the agar below. The colonies are round, raised, cream-colored and have an undulated margin. Surface colonies of *A. ethanolgignens* on CCA medium (Allison et al., 1979) are circular, convex, smooth, translucent, 0.5–1.5 mm in diameter and nonhemolytic. In contrast to *A. cellulolyticus*, *A. ethanolgignens* does not digest cellulose.

In cellulose broth, *A. cellulolyticus* produces a yellow pigment in the medium as well as on the particles of cellulose. The intensity of the yellow pigmentation varies with the growth conditions. The pigment is also formed in cellobiose broth cultures.

Growth of *Acetivibrio* strains fails to occur in media whose E_h is about −115 mV or above. Growth of *A. cellulolyticus* fails to occur in peptone-yeast extract broth (PY broth; Holdeman and Moore, 1975) in the absence of a fermentable carbohydrate. In modified PY broth (Robinson et al., 1981) *A. ethanolgignens* shows slight to moderate growth in the absence of a fermentable carbohydrate.

Only cellulose, cellobiose and salicin have been found to be capable of supporting the growth of *A. cellulolyticus*, although cells can be adapted to grow on glucose by transferring a cellobiose-grown culture to 1% glucose medium (10% inoculum, v/v) and subsequently transferring the slight growth that occurs after ~1 week to fresh glucose medium. Glucose-adapted cells can be maintained by transferring to fresh medium at 24–48-h intervals. Glucose-adapted cells grow readily when transferred to cellulose broth. Growth occurs in PY-cellobiose broth under an atmosphere of either 100% N_2 or 80% N_2:20% CO_2, but it is completely inhibited under 80% H_2:20% CO_2. There appears to be no requirement for vitamin K or heme; however, the lag period of cultures in PY-cellobiose broth is decreased by addition of these components. *A. cellulolyticus* can grow and be maintained in a defined medium* containing a fermentable carbohydrate, and, indeed, growth is more vigorous in this medium supplemented with 1% (w/v) cellobiose than in PY-cellobiose broth. Cells grown under nitrogen-limited conditions in 1% cellobiose broth accumulate an iodophilic polysaccharide (up to 37% of cell dry weight). No accumulation occurs during cultivation in cellulose (1% w/v) broth. The sulfur requirement for growth can be supplied by either Na_2S or by cysteine. Inorganic nitrogen [NH_4Cl or $(NH_4)_2SO_4$] can serve as a nitrogen source.

Growth of *A. ethanolgignens* is slight to moderate in modified peptone-yeast extract broth (MPY medium of Robinson et al., 1981) in the absence of a fermentable carbohydrate. Growth is enhanced, and a thick sediment accumulates, in MPY medium containing a fermentable carbohydrate. It is not known if the organism can grow in a defined medium.

The main products of fermentation by *Acetivibrio* strains are acetic acid, ethanol, H_2 and CO_2. In the case of *A. cellulolyticus* grown in PY-cellobiose broth, the amount of ethanol increases with the partial pressure of H_2 produced, yet is at least 2–3 times *lower* (on a molar

Table 6.22.

Differential characteristics of the species of the genus **Acetivibrio**[a]

Characteristics	1. A. cellulo-lyticus	2. A. ethanol-gignens
Substrates utilized:		
Cellobiose, cellulose	+	−
Fructose, galactose, lactose, maltose, mannitol, mannose, pyruvate	−	+
Glucose	−[b]	+
Reduction of nitrate to nitrite	−	+
H_2S producton	−	+
Voges-Proskauer test for acetyl-methylcarbinol	−	+
Ammonia production	−	+
Flagellar arrangement:		
Single, subpolar, from concave side of cell	+	−
Multiple, often fasciculated, from concave side of cell	−	+

[a] Data for *A. cellulolyticus* taken mainly from Patel et al. (1980) and for *A. ethanolgignens* from Robinson and Ritchie (1981). Symbols: see standard definitions.

[b] Can be adapted to grow on glucose.

basis) than acetic acid; in contrast, *A. ethanolgignens* grown on glucose, lactose, maltose or mannitol produces at least 3 times *more* ethanol than acetic acid.

Other physiological characteristics of *Acetivibrio* species are indicated in Table 6.22.

The distribution of *Acetivibrio* in nature is not yet known, since the two presently known species have only been reported recently. The type strain of *A. cellulolyticus* was isolated by Patel et al. (1980) from a methanogenic enrichment culture (Khan et al., 1979), whereas strains of *A. ethanolgignens* were isolated by Robinson and Ritchie (1981) from the colons of pigs experimentally infected with *Treponema hyodysenteriae*.

The pathogenicity of *Acetivibrio* species is not known. Although *A. ethanolgignens* is consistently present in high numbers on the colonic mucosa of pigs showing clinical signs of swine dysentery, and although it satisfies the requirement that at least one other organism is necessary for *T. hyodysenteriae* to colonize the colons of gnotobiotic pigs, pathogenicity was not observed by Robinson and Ritchie (1981) when gnotobiotic pigs were inoculated with both organisms.

Enrichment and Isolation Procedures

The enrichment culture which was used for isolation of the type strain of *A. cellulolyticus* (Patel et al., 1980) was started by inoculating a mineral salts medium containing cellulose with effluent from a municipal sewage sludge digester. The anaerobic enrichment fermenter was maintained using batch culture techniques with repeated feeding of homogenized tissue paper (1 g/liter/week), and the fermenter contents were withdrawn to achieve a retention time of 24–30 weeks (Khan et al., 1979). A sample of the established mixed culture was diluted

---

* Basal broth (mg/liter): NaHCO₃, 2060; NH₄Cl, 680; K₂HPO₄, 296; KH₂PO₄, 180; (NH₄)₂SO₄, 150; MgSO₄·7H₂O, 120; CaCl₂·2H₂O, 61; FeSO₄·7H₂O, 21; N(CH₂COOH)₃, 15; NaCl, 10; MnSO₄·H₂O, 5; CoCl₂·6H₂O, 1; ZnSO₄·7H₂O, 1; CuSO₄·5H₂O, 0.1; AlK(SO₄)₂·12H₂O, 0.1; H₃BO₃, 0.1; Na₂MoO₄·2H₂O, 0.1; pyridoxine-HCl, 0.1; thiamine-HCl, 0.05; riboflavin, 0.05; niacin, 0.05; *p*-aminobenzoic acid, 0.05; lipoic acid, 0.05; biotin, 0.02; folic acid, 0.02; vitamin B₁₂, 0.005; and resazurin, 1. This basal medium is supplemented with 1 g/liter of cellulose prepared from absorbent cotton as described by Hungate (1950), 250 mg of cysteine-HCl/liter and 250 mg of Na₂S·9H₂O/liter. The medium is prereduced by the Hungate (1950) technique and is sterilized by autoclaving under an atmosphere of 80% N₂:20% CO₂. For cellulose agar, the basal medium is supplemented with (per liter) 18 g of agar, 2 g of cellulose and 0.25 g of cysteine-HCl, and the pH is adjusted to 7.5. The medium (lacking the cysteine) is sterilized aerobically by autoclaving and cooled to 50°C. The cysteine-HCl is then added aseptically from a sterile stock solution and the medium is dispensed into Petri dishes, followed by storage in an anaerobic chamber (5% CO₂:10% H₂:85% N₂) for a 24-h period to reduce the medium. The final pH of all media is 7.0 ± 0.2.

serially into cellulose broth (see footnote *) and 0.1 ml portions were spread on the surface of cellulose agar plates (see footnote *) inside the anaerobic chamber. The plates were transferred into Brewer anaerobic jars which were flushed with 80% N$_2$:20% CO$_2$ (by means of two valves fitted into the lid) and the jars were incubated at 35°C. Colonies showing cellulose digestion were transferred (inside the anaerobic chamber) to cellulose broth. These cultures were in turn serially diluted and streaked several times to obtain a pure culture. Purity was confirmed by examination of colony type and by Gram-staining.

Methods for isolation of *A. ethanolgignens* from the colons of pigs have been referred to by Robinson and Ritchie (1981).

Maintenance Procedures

Stock cultures of *A. cellulolyticus* are maintained by weekly transfer [using a 2% (v/v) inoculum from each previous culture] into cellulose broth (3 g of cellulose/liter) with incubation at 35°C under an atmosphere of 80% N$_2$:20% CO$_2$. Procedures for long term preservation have not been evaluated in detail but preliminary data indicate that both

freeze-dried and liquid nitrogen-stored stock cultures (previously grown for 72 h) can be maintained for periods of at least 1 year.

Procedures for Testing Special Characters

Biochemical tests are done using either MPY broth (Robinson et al., 1981) which is a modification of peptone yeast extract broth (PY medium of Holdeman and Moore, 1975) or PY broth supplemented with 0.5% (w/v) of carbohydrate or other substrate (Holdeman and Moore, 1975). The inoculum for *A. cellulolyticus* consists of 0.04 ml of a 72-h-old cellulose broth grown culture/10 ml of test medium. The inoculum for *A. ethanolgignens* consists of 0.1 ml of culture grown in MPY broth supplemented with 0.5% (w/v) glucose. The characterization methods described by Holdeman and Moore (1975) and Robinson et al. (1981) are employed. Cultures showing turbid growth are tested at 1 week of incubation or earlier, and all negative tests are incubated up to 4 weeks. Media containing an insoluble substrate, such as cellulose, are incubated on a rotary shaker (120 rpm).

Determination of H$_2$ and CO$_2$ in the headspace of culture vessels is done according to the method of van Huyssteen (1967).

Differentiation of the genus **Acetivibrio** from other genera

See the key to the family *Bacteroidaceae* for the primary characteristics that distinguish *Acetivibrio* from other genera of anaerobic bacteria. The formation of acetic acid as the sole or the major organic acid end product of carbohydrate fermentation distinguishes *Acetivibrio* from other genera of motile, curved, anaerobic rods (*Selenomonas, Anaerobiospirillum, Pectinatus, Anaerovibrio, Lachnospira, Butyrivibrio* and *Succinivibrio*) which form other or additional acids. The flagellar arrangement of *Acetivibrio* differs from that of *Selenomonas* (tuft of flagella near the center of the concave side of the cell), *Pectinatus* (comblike lateral flagella along the concave side) and the genera having polar flagella (*Butyrivibrio, Succinivibrio, Anaerobiospirillum* and *Anaerovibrio*).

Taxonomic Comments

The phylogenetic relationships of *Acetivibrio* to other genera or higher taxa are not known. Application of the techniques of DNA/

DNA hybridization, DNA/rRNA hybridization or rRNA oligonucleotide cataloging would be helpful in this regard.

Further Reading

Patel, G.B. and C. Breuil. 1981. Accumulation of an iodophilic polysaccharide during growth of *Acetivibrio cellulolyticus* on cellobiose. Arch. Microbiol. *129:* 265–267.
Patel, G.B., A.W. Khan, B.J. Agnew and J.R. Colvin. 1980. Isolation and characterization of an anaerobic, cellulolytic microorganism, *Acetivibrio cellulolyticus* gen. nov., sp. nov. Int. J. Syst. Bacteriol. *30:* 179–185.
Robinson, I.M. and A.E. Ritchie. 1981. Emendation of *Acetivibrio* and description of *Acetivibrio ethanolgignens*, a new species from the colons of pigs with dysentery. Int. J. Syst. Bacteriol. *31:* 333–338.

Differentiation of the species of the genus **Acetivibrio**

Characteristics useful in differentiating the two species of the genus are listed in Table 6.22.

List of the species of the genus **Acetivibrio**

1. **Acetivibrio cellulolyticus** Patel, Khan, Agnew and Colvin 1980, 184.[VP]

cel.lu.lo.ly'ti.cus. M.L. n. *cellulosum* cellulose; Gr. adj. *lyticus* dissolving; M.L. adj. *cellulolyticus* cellulose dissolving.

The characteristics are as described for the genus and as listed in Tables 6.22 and 6.23.

Optimum temperature, ~35°C; range for growth, 20–40°C. The optimum pH is ~7.0; range for growth is ~6.0–7.7.

The mole products formed per mole of cellobiose used after 24 h of incubation under 80% N$_2$:20% CO$_2$ in defined basal broth containing 1% cellobiose and 14.8 mM ammonia nitrogen are: H$_2$, 6.10; CO$_2$, 4.27; acetic acid, 1.61; ethanol, 0.30, and glucose, 0.08. A large portion of the carbon from the cellobiose is tied up in biomass, partly due to the accumulation of an iodophilic polysaccharide. Traces of propanol and butanol occur in cultures incubated for 8 days under either 100% N$_2$ or 80% N$_2$:20% CO$_2$.

Isolated from a methanogenic enrichment culture. Only one strain has been described.

The mol% G + C of the type strain is 38 ± 1.2 (ultraviolet spectroscopy).

Type strain: NRC 2248 (ATCC 33288; strain CD2 of Patel et al., 1980).

2. **Acetivibrio ethanolgignens** Robinson and Ritchie 1981, 335.[VP]

eth.a.nol.gig'nens. M.L. n. *ethanol* ethanol; L. part. adj. *gignens* giving birth to, producing; M.L. part. adj. *ethanolgignens* producing ethanol.

The characteristics are as described for the genus and as listed in Tables 6.22 and 6.23.

Optimum temperature, ~37°C; slight growth occurs at 45°C but none at 15°C.

In addition to ethanol, acetic acid, H$_2$ and CO$_2$, growth on glucose yields trace amounts of formic acid. During fructose fermentation the amounts of acetic acid and formic acid produced are about equal. Propionic, lactic and succinic acids are not formed.

The type strain is susceptible to the following antimicrobial agents when tested in MPY-glucose medium: cephalothin (6 µg/ml), chloramphenicol, 5 µg/ml), clindamycin (2 µg/ml), penicillin (400 U/ml) and vancomycin (6 µg/ml). The strain is resistant to rifampin (5 µg/ml), kanamycin (12 µg/ml) and neomycin (12 µg/ml).

Forty-six isolates have been obtained from the colons of pigs experimentally infected with *T. hyodysenteriae*.

The mol% G + C of the DNA of the type strain is 40 (*T$_m$*).

Type strain: ATCC 33324 (strain 77-6 of Robinson and Ritchie, 1981).

Table 6.23.
Other characteristics of the species of the genus **Acetivibrio**[a]

Characteristics	1. A. cellulo-lyticus	2. A. ethanol-gignens
Cell diameter, μm	0.5–0.8	0.5–0.9
Cell length, μm	4.0–10.0	1.5–2.5
Products of carbohydrate fermentation:[b]		
Acetic acid, ethanol, H_2, CO_2	+	+
Formic acid	–	Trace
Propionic, butyric, succinic and lactic acids	–	–
Indole production	–	–
Catalase	–	–
Urease	–	–
Growth in presence of 20% bile	–	–
Esculin hydrolysis	–	–
Gelatin liquefaction	–	–
Litmus milk: curd and digestion		
Utilization of:		
Arabinose, esculin, inositol, inulin, lactate, raffinose, rhamnose, ribose, sorbitol, sucrose, trehalose, xylose	–	–
Adonitol		–
Amygdalin, arginine, dulcitol, erythritol, glycogen, hippurate, melezitose, melibiose, sorbose, threonine	–	

[a] Data for *A. cellulolyticus* taken mainly from Patel et al. (1980) and for *A. ethanolgignens* from Robinson and Ritchie (1981). Symbols: see standard definitions.
[b] *A. cellulolyticus* end products from cellobiose fermentation, and *A. ethanolgignens* end products from glucose fermentation.

Genus XIII. Lachnospira Bryant and Small 1956, 24[AL]

MARVIN P. BRYANT

Lach.no.spi′ra. Gr. n. *lachnos* woolly hair, down; L. n. *spira* a coil; M.L. fem. n. *Lachnospira* woolly (colony producing) spiral.

Curved rods, 0.4–0.6 μm wide and 2.0–4.0 μm long, with bluntly pointed ends. Weakly Gram-positive; the Gram stain may be negative except in very young cultures. Motile by monotrichous lateral to subpolar flagella. Strictly anaerobic. **The colonies are filamentous, woolly in appearance.** Chemoorganotrophic, having a fermentative type of metabolism. **The products of glucose fermentation are formate, lactate, acetate, ethanol, CO_2 and H_2. Pectin is also fermented** with the same products plus **methanol** being produced. Succinate, butyrate and propionate are not produced. Occur in the bovine rumen. The mol% G + C of the DNA is not known.

Type species: Lachnospira multiparus Bryant and Small 1956, 24.

Further Descriptive Information

Single cells appear as curved or helical rods containing one or more turns. Some very long chains of cells which are only slightly curved and which have more bluntly rounded ends, and also filamentous cells, are often observed, especially in liquid media lacking rumen fluid.

The Gram stain may be only weakly positive or seem negative; however, the cell wall is structurally of the Gram-positive type (Cheng et al., 1979).

Surface colonies on RGCA agar in roll tubes (see the genus *Butyrivibrio* for composition of medium and anaerobic methods) incubated at 37°C for 3 days are 2–5 mm in diameter and characteristically flat and filamentous; some spreading may occur. Deep agar colonies appear as woolly balls which penetrate the agar to some extent. In liquid rumen-fluid glucose medium growth occurs as a large flocculent sediment. The floc may be difficult to disperse; this is especially true in media in which yeast extract and Trypticase or chemically defined ingredients replace the rumen fluid.

Growth of *L. multiparus* occurs at 30–45°C but not at 22 or 50°C.

The final pH in lightly buffered glucose medium is 4.8–5.2. Abundant growth occurs at pH 6.0 to 7.0. The upper pH limit for growth is not known.

L. multiparus grows well in a chemically defined medium containing B-vitamins, cysteine, glucose, ammonium ions, minerals, acetate and CO_2/HCO_3^- buffer (Bryant and Robinson, 1962). Acetate is highly stimulatory, especially in defined media lacking amino acids or peptides. One strain of *L. multiparus* has been found to require only *p*-aminobenzoic acid and biotin among the B-vitamins (Emery et al., 1957). Either ammonium ions or mixtures of amino acids or peptides can serve as the nitrogen source.

Glucose, fructose, cellobiose, esculin, pectin, salicin and sucrose are fermented. Xylose fermentation is variable. Arabinose, cellulose, dextrin, galactose, glycerol, gum arabic, inositol, inulin, lactose, maltose, mannitol, trehalose and xylan are not fermented. Starch is not hydrolyzed. The organism contains pectinmethylesterase, as indicated by its production of methanol during pectin fermentation (Rode et al., 1981), but does not ferment galacturonic acid (Dehority, 1969). It presumably ferments polygalacturonic acid, since pectin is an excellent energy source. Amino acids are not fermented.

Indole and H_2S are not produced. Nitrate is not reduced. Gelatin is not liquefied. The Voges-Proskauer test differs among strains.

L. multiparus is very important in pectin fermentation in the rumen, as indicated by its greatly increased numbers in the rumen when lush legume forages such as alfalfa or ladino clover are the chief dietary components (Bryant et al., 1960). Its importance is also indicated by its ability to macerate clover and grass leaves (Cheng et al., 1979).

Enrichment and Isolation Procedures

L. multiparus can be isolated from rumen contents using nonselective procedures similar to those described for the genus *Butyrivibrio*, i.e. anaerobic roll tubes containing RGCA medium. The distinctive woolly and filamentous appearance of the colonies makes isolation of *L.*

multiparus quite easy (Bryant and Small, 1956). Isolation from the rumen contents of cattle fed lush alfalfa or clover diets is especially easy because of the large numbers of *L. multiparus* present (Bryant et al., 1960).

RGCA medium can be made somewhat selective by replacing the usual carbohydrates by 0.3% pectin (Dehority, 1969); however, colonies of other pectin-fermenting bacteria such as *Bacteroides ruminicola* or *Butyrivibrio fibrisolvens* may also be quite numerous.

Maintenance Procedures

L. multiparus can be maintained on stab-inoculated RGCA slants at $-70°C$ for 1 year or more. The organisms can also be preserved indefinitely by lyophilization.

Differentiation of the genus **Lachnospira** from other genera

See Table 6.11 in the section on the genus *Butyrivibrio* for those characteristics useful in differentiating *Lachnospira* from other morphologically or physiologically similar genera of anaerobic bacteria. The key to the genera of the family *Bacteroidaceae* also provides useful distinguishing characteristics.

Taxonomic Comments

In this edition of the *Manual* the genus *Lachnospira* is included in the family *Bacteroidaceae* solely for convenience, since the cells generally stain Gram-negative. The cell wall, however, is of the Gram-

positive type. The relationship of *Lachnospira* to genera or higher taxa of Gram-positive bacteria needs to be elucidated, and nucleic acid studies such as DNA/rRNA hybridization or rRNA oligonucleotide cataloging would be quite useful.

Further Reading

Bryant, M.P., and N. Small. 1956. Characteristics of two new genera of anaerobic curved rods isolated from the rumen of cattle. J. Bacteriol. 72: 22–26.

Cheng, K.-J., D. Dinsdale and C.S. Stewart. 1979. Maceration of clover and grass leaves by *Lachnospira multiparus*. Appl. Environ. Microbiol. 38: 723–729.

Dehority, B.A. 1969. Pectin-fermenting bacteria isolated from the bovine rumen. J. Bacteriol. 99: 189–196.

List of the species of the genus **Lachnospira**

1. **Lachnospira multiparus** Bryant and Small 1956, 24.[AL]

mul·ti·par'us. L. adj. *multus* much, many; L. v. suff. *parus* from L. v. *pario* to produce; M.L. adj. *multiparus* many (products) produced.

The characteristics are as described for the genus.

Occurs in the rumen of the bovine and probably other ruminants. The mol% G + C of the DNA is not known.

Type strain: ATCC 19207 (strain D32 of Bryant and Small, 1956).

Important Notes for Users of this Edition

1. Always read both generic and species descriptions because characters listed in the generic description are not usually listed in the species descriptions.

2. Unless otherwise indicated in footnotes to tables, the meanings of symbols are as follows:

+ 90% or more of strains are positive

− 90% or more of strains are negative

d 11–89% of strains are positive

v strain instability (*not* equivalent to "d")

D different reactions in different taxa (species of a genus or genera of a family)

3. All other symbols are defined in footnotes to tables.

SECTION 7

Dissimilatory Sulfate- or Sulfur-reducing Bacteria

Friedrich Widdel and Norbert Pfennig

This group is a physiological assemblage of morphologically diverse, strictly anaerobic, Gram-negative eubacteria that utilize as electron acceptors either (i) sulfate and other oxidized sulfur compounds, or (ii) elemental sulfur; these are reduced to H_2S. In some species the metabolism may also be fermentative; however, carbohydrates are rarely degraded. Cells may contain c-type cytochromes and/or b-type cytochromes. Many species of the sulfate-reducing bacteria contain, as sulfite reductase, desulfoviridin, desulforubidin or P582. Hydrogenase is present in many species. Nitrate reduction to ammonia is rare but has been reported in a few strains. The ability to fix nitrogen has been demonstrated for some species. Genera and species differ with respect to their utilization of organic compounds: many species carry out an incomplete oxidation of substrates such as lactate to CO_2 and acetate; the latter cannot be oxidized further. Other species are capable of oxidizing acetate and other organic compounds completely to CO_2.

Habitats: anaerobic mud and sediments of freshwater, brackish water and marine environments, and the gastrointestinal tract of man or animals.

The mol% G + C of the DNA ranges from 37 to 67.

Key to the genera of the dissimilatory sulfate- or sulfur-reducing bacteria

I. Bacteria able to oxidize acetate completely to CO_2 using elemental sulfur as the electron acceptor. Sulfate is never reduced. **Dissimilatory sulfur-reducing bacteria.**
   Genus *Desulfuromonas*, p. 664.

II. Bacteria capable of dissimilatory sulfate reduction. Sulfite, thiosulfate, or other oxidized sulfur compounds may also serve as electron acceptors. **Dissimilatory sulfate-reducing bacteria.**
   A. Cells are free-living and occur singly, in pairs or in short chains.
      1. Cells are rod shaped, helical, vibrioid or spirilloid, occasionally straight and are motile. Incomplete oxidation of lactate to acetate and CO_2 is common. Some strains utilize fatty acids that may be oxidized completely.
         Genus *Desulfovibrio*, p. 666.
      2. Cells are rod shaped, straight and nonmotile. No endospores are formed. Incomplete oxidation of pyruvate to acetate and CO_2.
         Genus *Desulfomonas*, p. 672.
      3. Cells are spherical under all conditions and are capable of complete oxidation of fatty acids or benzoate.
         Genus *Desulfococcus*, p. 673.
      4. Cells are ellipsoidal to rod shaped with rounded ends, sometimes appearing coccoid. Acetate is completely oxidized to CO_2. Growth occurs preferentially in brackish water or sea water media.
         Genus *Desulfobacter*, p. 674.
      5. Cells are ellipsoidal, often lemon or onion shaped with pointed ends. Incomplete oxidation of propionate or lactate to acetate and CO_2 occurs.
         Genus *Desulfobulbus*, p. 676.
      6. Cells are rod shaped, straight, or slightly curved. The ends of the cells may be pointed. All species are capable of endospore formation.
         Genus *Desulfotomaculum*
      The genus description is given among the anaerobic sporeforming bacteria treated in Volume 2 of the Manual.
   B. Cells are arranged in sarcina-like packets of irregular or distorted appearance; coccoid or ellipsoidal single cells may occur that are occasionally motile. Complete oxidation of fatty acids or benzoate occurs.
      Genus *Desulfosarcina*, p. 677.
   C. Cells are arranged in uniseriately multicellular, flexible filaments with gliding motility. Complete oxidation of fatty acids occurs.
      Genus *Desulfonema*
      The genus description is given among the gliding bacteria treated in Volume 3 of the Manual.

Genus **Desulfuromonas** Pfennig and Biebl 1977, 306[AL]*

NORBERT PFENNIG

De.sul.fu.ro.mo'nas. L. pref. *de* from; L. *sulfur* sulfur; Gr. n. *monas* a unit, monad; M.L. fem. n. *Desulfuromonas* a monad that reduces sulfur.

Straight or slightly curved rods and elongated ovoid rods, 0.4–0.9 μm in diameter and 1.0–4.0 μm in length. Resting stages are not known to occur. Gram-negative. **Motile, generally by means of a single flagellum located at a lateral or subpolar position**; the cells exhibit a characteristic propeller-like movement. Some of the strains have polar flagella. Strictly anaerobic. **Possess mainly a respiratory type of metabolism with elemental sulfur as the terminal electron acceptor**, being reduced to H_2S (dissimilatory sulfur reduction). L-Malate or fumarate may be fermented to acetate and succinate. Optimum growth temperature, 30°C. Colonies on anaerobic agar media are translucent to opaque and either **peach-colored or pink**. Chemoorganotrophic, using acetate and other simple organic compounds as carbon sources and electron donors in the presence of bicarbonate; **the acetate and other substrates are completely oxidized to CO_2.** Occur regularly in anaerobic sediments of salt lakes and in marine or brackish habitats; have also been enriched from freshwater sediments. The mol% G + C of the DNA is 50–63 (T_m).

Type species: *Desulfuromonas acetoxidans* Pfennig and Biebl 1977, 306.

Further Descriptive Information

Strains of *Desulfuromonas* isolated from anaerobic sediments of saline lakes or marine habitats are generally straight or slightly curved slender rods (Fig. 7.1) which are highly motile and have a single lateral to subpolar flagellum (Fig. 7.2). Most strains isolated from anaerobic freshwater sediments are elongated ovoid rods which are motile by means of a subpolar or polar flagellum. The cells of freshwater strains have a tendency to form nonmotile clumps which stick to the bottom of the culture vessel.

Most marine and salt lake strains require 2.0% NaCl and 0.3% $MgCl_2 \cdot 6H_2O$ for growth; some strains can grow in both seawater or freshwater media. Biotin may be required as a growth factor. All strains of *Desulfuromonas* can be cultured in a defined anaerobic medium† with 0.03% $Na_2S \cdot 9H_2O$ as a reductant. The electron donor and carbon source (usually acetate) is added aseptically from a sterile stock solution (final concentration = 0.05%). The electron acceptor, elemental sulfur, is added aseptically from a sterile slurry of sulfur flower in distilled water‡; a pea-size amount is used for 50 ml of medium. *Desulfuromonas* strains that can utilize malate or fumarate are grown with 0.2% of one of these substrates. With the strains originally isolated, no growth was observed within 3 weeks on malate or fumarate when acetate was omitted; however, later studies showed that malate or fumarate could be fermented in bicarbonate-free medium supplemented with yeast extract in the absence of additional electron donors (Pfennig and Widdel, 1981). Nevertheless, strains do differ in their ability to ferment malate or fumarate. The ability to use the following electron donors and carbon sources also varies among *Desulfuromonas* strains: ethanol, propanol, pyruvate, lactate, propionate and glutamate. The following substrates generally cannot be used as electron donors and carbon

sources: polysaccharides, sugars, sugar alcohols, most amino acids, higher fatty acids, and $H_2 + CO_2$.

Besides elemental sulfur, poly-or disulfide bond-containing compounds such as polysulfide, cystine or oxidized glutathione can serve as electron acceptors. Sulfate, sulfite, thiosulfate, nitrate or oxygen cannot act as electron acceptors.

The pH range for growth is 6.5–8.5, with the optimum being 7.2–7.5.

Colonies of *Desulfuromonas* in agar media, as well as accumulations of cells in sediments or centrifuged pellets, are pink- to peach-colored, due to c-type cytochromes with absorption maxima at 419, 523 and 553 nm. In redox difference spectra the β and α peaks shift to 522 and 551.5 nm, respectively. A major component of the cytochromes of *D. acetoxidans* strain 5071 is the low potential, 3-heme cytochrome $c_{551.5}$ (c_7); its amino acid sequence is identical with that of cytochrome c_7 from "Chloropseudomonas" cultures (Probst et al., 1977).

It is characteristic of all *Desulfuromonas* strains that they grow well in syntrophic cultures containing phototrophic green sulfur bacteria and acetate (Biebl and Pfennig, 1978). Marine or salt lake isolates show particularly efficient growth in combined cultures containing the green bacterium *Prosthecochloris aestuarii* at all light intensities between 5 and 1000 lux. This is readily attributed to the fact that the illuminated *Prosthecochloris* oxidizes H_2S almost exclusively to extracellular elemental sulfur, which in turn is immediately reduced back to H_2S by *Desulfuromonas* as long as acetate is present.

Dissimilatory sulfur-reducing bacteria resembling *Desulfuromonas* occur regularly in the anaerobic sulfide-containing sediments of marine and brackish water habitats and salt lakes. They have also been isolated, although much less successfully, from sulfide-containing anaerobic freshwater sediments.

Enrichment and Isolation Procedures

For the selective enrichment of *Desulfuromonas* strains the basal mineral medium supplemented with 0.05% sodium acetate and a spatula of ground, wetted sulfur flower is used. Bottles or tubes (100-, 50- or 20-ml capacity, kept anaerobic by tight screw caps having autoclavable rubber seals) serve as culture vessels. Depending on the salinity of the anaerobic mud sample that is used as the inoculum, sea- or freshwater medium is used. When the sulfur flower is kept in suspension by incubating the bottles containing glass beads on a rotary shaker, sulfur-reducers may become enriched after 1 or 2 weeks of incubation at 28°C. Successful enrichments are recognized by their strong odor of H_2S and by their lowered pH (6.5–6.8); microscopically, slender motile rods with a propeller-like movement are seen. Second and third transfers in media supplemented with vitamins usually grow within a few days. Although the tolerance of *Desulfuromonas* to H_2S is considerable, growth is limited by the inhibitory effect of the H_2S formed from sulfur, and not more than 0.1% H_2S can be tolerated.

The growth inhibition due to H_2S can be eliminated if the enrichment cultures for *Desulfuromonas* are inoculated with a phototrophic green

---

* *AL* denotes the inclusion of this name on the Approved Lists of Bacterial Names (1980).

† Defined basal medium has the following composition (per liter of distilled water): KH_2PO_4, 1.0 g; NH_4Cl, 0.5 g; $CaCl_2 \cdot 2H_2O$, 0.1 g; NaCl (only for seawater medium), 20.0 g; $MgCl_2 \cdot 6H_2O$, 0.4 g (for seawater medium, 3.0 g); trace element solution (see below), 1.0 ml. After the medium has been autoclaved and cooled under anaerobic conditions, the following components are added aseptically from sterile stock solutions: 20 ml of a 20% (w/v) solution of $NaHCO_3$ (saturated with CO_2 and autoclaved under a CO_2 atmosphere), 3 ml of a 10% (w/v) solution of $Na_2S \cdot 9H_2O$ (autoclaved under an N_2 atmosphere), and 1 ml of a vitamin solution containing 4 mg of p-aminobenzoic acid and 1 mg of biotin per 100 ml. The pH of the basal medium is adjusted to 7.1–7.3 with sterile dilute HCl or Na_2CO_3 solution. The medium is then distributed aseptically into sterile screw-capped bottles or test tubes and sealed against air. *Trace element solution* (per liter): 10 ml of 25% (w/w) HCl; $FeCl_2 \cdot 4H_2O$, 1.5 g; $CoCl_2 \cdot 6H_2O$, 190 mg; $MnCl_2 \cdot 4H_2O$, 100 mg; $ZnCl_2$, 70 mg; H_3BO_3, 6 mg; $Na_2MoO_4 \cdot 2H_2O$, 36 mg; $NiCl_2 \cdot 6H_2O$, 24 mg; $CuCl_2 \cdot 2H_2O$, 2 mg. Initially the $FeCl_2$ is dissolved in the 10 ml of HCl; distilled water is then added followed by the other components.

‡ Highly purified sulfur flower is ground thoroughly in a mortar together with distilled water. The slurry is autoclaved for 30 m at 112°C in a screw-capped bottle. The excess water is then decanted.

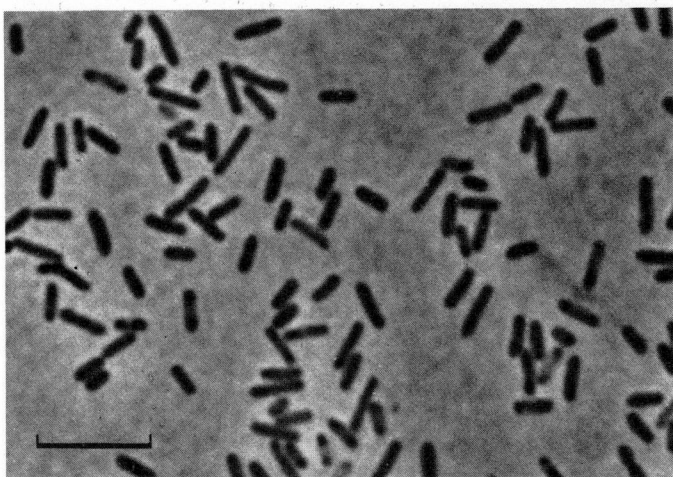

Figure 7.1. Phase contrast photomicrograph of *Desulfuromonas acetoxidans* DSM strain 684 grown in ethanol-malate medium. Bar, 5 μm.

sulfur bacterium and incubated in dim light from a tungsten lamp. In this case no sulfur flower is added to the medium because the green sulfur bacterium continuously forms elemental sulfur and at the same time consumes the H_2S formed by the sulfur reducers being enriched. Repeated transfers may yield fast-growing syntrophic cultures which can be used directly for isolation of a pure culture.

Pure cultures are obtained by repeated application of the agar shake dilution method. Repeatedly washed agar is used to prepare a sterile 3.3% (w/v) solution in water, which is then dispensed in 3-ml amounts into test tubes. The molten agar is then combined with 6-ml amounts of prewarmed (42°C) complete culture medium, with 6–8 tubes being prepared for each dilution series. [The complete culture medium is prepared with polysulfide solution instead of sulfur flower in order to obtain a fine and homogeneous distribution of sulfur in the agar deeps. Three drops of an autoclaved polysulfide solution (10 g of $Na_2S \cdot 9H_2O$ and 3 g of sulfur flower dissolved in 15 ml of distilled water) are added per 50 ml of medium.] The inoculated tubes are placed in a cold water bath and immediately sealed against air by overlaying with a mixture of 1 part paraffin wax and 3 parts mineral oil. The sealed tubes are incubated at 28–30°C. Clearing of the opaque, yellowish polysulfide and appearance of pink to peach-colored colonies indicate growth of *Desulfuromonas*.

In addition to microscopic examination, newly isolated strains are checked for purity by inoculating them into AC medium (Difco) and

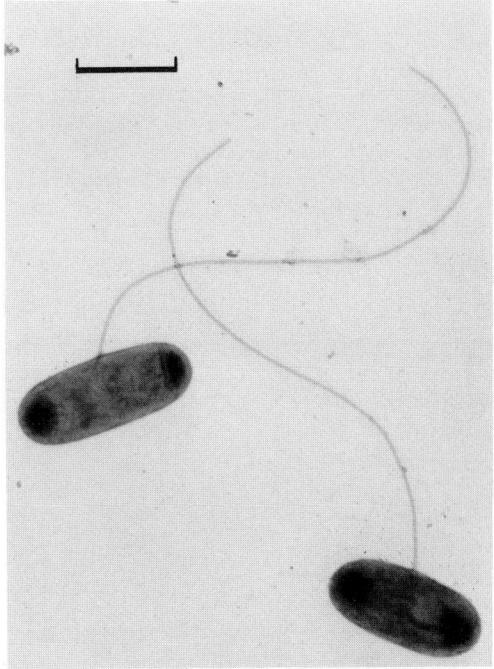

Figure 7.2. Negatively stained cells of *Desulfuromonas acetoxidans* strain 5071 grown in ethanol-malate medium. Bar, 1 μm. (Courtesy of F. Mayer, Göttingen.) (Reproduced with permission from N. Pfennig and H. Biebl, *Archives of Microbiology, 110:* 3–12, 1976, © Springer-Verlag, Heidelberg.)

ethanol-lactate medium for sulfate-reducing bacteria. Absence of turbidity in both media and failure to form iron sulfide in the latter medium indicate lack of contamination.

Maintenance Procedures

Stock cultures may be maintained in liquid medium in tightly closed screw-capped bottles at 4–8°C, with transfers every 3 months. Strains capable of using malate or fumarate are maintained on these substrates; strains lacking this ability are maintained with elemental sulfur as the electron acceptor.

For long term preservation, vials with cell suspensions in anaerobic culture medium containing 5% (v/v) dimethyl sulfoxide are stored in liquid nitrogen.

Differentiation of the genus **Desulfuromonas** from other genera

The genus *Desulfuromonas* is distinguished from other anaerobic, dissimilatory sulfur-reducing bacteria primarily by physiological characteristics. The special feature of the genus is the capacity to grow on acetate as the sole organic carbon source by an anaerobic respiration in which the oxidation of acetate to CO_2 is stoichiometrically linked to the reduction of elemental sulfur to sulfide. Sulfate, sulfite or thiosulfate cannot be utilized as electron acceptors. This inability to reduce oxidized sulfur compounds distinguishes *Desulfuromonas* from acetate-oxidizing, sulfate-reducing bacteria of the genera *Desulfobacter* and *Desulfotomaculum*. The ability to reduce elemental sulfur to H_2S in a dissimilatory metabolism also occurs in *Wolinella succinogenes*, saprophytic *Campylobacter* strains (Laanbroek et al., 1978) and rod-shaped sulfate-reducing bacteria of the genus *Desulfovibrio* (Biebl and Pfennig, 1977); however, these organisms are unable to oxidize acetate to CO_2 and require H_2 or formate (or lactate in the case of *Desulfovibrio*) as electron donors for reduction of sulfur.

In contrast to other rod-shaped anaerobic bacteria exhibiting a dissimilatory reduction of sulfur compounds, most *Desulfuromonas* strains are motile by means of a single flagellum located at a lateral or subpolar position.

Taxonomic Comments

The phylogenetic relationships of *Desulfuromonas* to other genera or higher taxa are not known. Nucleic acid studies, such as DNA/rRNA hybridization or rRNA oligonucleotide cataloging, would be helpful in this regard.

Two morphologically different groups occur within the genus *Desulfuromonas*: (i) strains having slender, rod-shaped or slightly curved cells which are motile by a single lateral to subpolar flagellum, and (ii) strains having elongated, ovoid cells with a single polar or subpolar flagellum. The latter group also differs physiologically from the first group by being unable to utilize alcohols as electron donors; some grow on acetate exclusively, others use pyruvate in addition. The first group is presently classified in the only established species of the genus, *D. acetoxidans*. The second group will eventually be classified in a new species, "*D. acetexigens*" (unpublished results).

List of the species of the genus **Desulfuromonas**

1. **Desulfuromonas acetoxidans** Pfennig and Biebl 1977, 306.[AL]

a.cet.o′xi.dans. L. n. *acetum* vinegar; M.L. n. *acidum aceticum* acetic acid; M.L. v. *oxido* make acid, oxidize; M.L. part. adj. *acetoxidans* oxidizing acetate.

Straight and slightly curved slender rods, 0.4–0.8 μm in diameter and 1–4 μm in length. Motile by a single lateral to subpolar flagellum.

Acetate, ethanol, propanol and other simple organic compounds are oxidized to CO_2 and elemental sulfur is reduced to H_2S. L-Malate or fumarate may be used as an electron acceptor instead of sulfur and are reduced to succinate; malate or fumarate may also be fermented.

Colonies or densely packed cells are peach-colored or pink due to the presence of c-type cytochromes.

Occur regularly in anaerobic sediments from marine or brackish water habitats or salt lakes.

The mol% G + C of the DNA is 50–52 (T_m).

Type strain: DSM 684 (strain 11070 "South Orkney Islands").

Other organisms belonging to the genus **Desulfuromonas**

Certain strains of sulfur-reducing bacteria appear to belong to the genus *Desulfuromonas* but differ from *D. acetoxidans* in several respects. The cells are elongated ovoid rods, 0.8–1.2 μm in diameter and 1.2–2.5 μm in length. They are motile by a single polar or subpolar flagellum. Some of the strains use only acetate as an electron donor; others can use pyruvate in addition. None of the strains can use alcohols as electron donors. They are isolated from anaerobic freshwater sedi-ments and have a tendency to form nonmotile clumps which stick to the bottom of the culture vessel. The mol% G + C of strain 4970 is 62.5.

These strains appear to constitute a species distinct from *D. acetoxidans*. The name "*Desulfuromonas acetexigens*" will be proposed for them at a later time.

Genus **Desulfovibrio** Kluyver and van Niel 1936, 397[AL]

JOHN R. POSTGATE

De.sul.fo.vi′bri.o. L. pref. *de* from; L. n. *sulfur* sulfur; L. v. *vibrio* to vibrate; M.L. masc. n. *Vibrio* that which vibrates, a generic name; M.L. masc. n. *Desulfovibrio* a vibrio that reduces sulfur compounds.

Curved or occasionally straight rods, sometimes sigmoid or spirilloid, 0.5–1.5 by 2.5–10.0 μm. The morphology is influenced by age and environment; descriptions refer to freshly grown cultures in anaerobic sulfate-enriched media. **Endospores not formed.** Gram-negative. **Motile** by means of a single or lophotrichous polar flagella. Obligately anaerobic. **Possess mainly a respiratory type of metabolism with sulfate or other sulfur compounds as the terminal electron acceptors, being reduced to H_2S;** however, the metabolism is sometimes fermentative. Media containing a reducing agent are required for growth. In a few cases a vitamin requirement has been reported. Some species and subspecies are moderately halophilic. Optimum temperature, usually 25–35°C; upper limit normally ~44°C; some strains can grow at temperatures at or below 0°C. A thermophilic species has been reported. Chemoorganotrophic. **Most species oxidize organic compounds such as lactate incompletely to acetate, which cannot be utilized further.** In a few cases acetate and other substrates are oxidized completely to CO_2. Carbohydrates are rarely utilized. Gas is never formed from carbohydrates. Cells contain c-type cytochromes (c_3) and usually b-type cytochromes. Many species contain desulfoviridin. Hydrogenase is usually present. Strains of some species may show mixotrophic growth using H_2 as the energy source and assimilating acetate + CO_2, or yeast extract, as carbon sources. Gelatin is not liquefied. Nitrates are sometimes reduced. Molecular nitrogen is sometimes fixed. Species generally show some degree of antigenic cross-reaction. Habitats: anaerobic mud of fresh and brackish water and marine environments; intestines of animals; manure and feces. The mol% G + C of the DNA is 46.1–61.2 (Bd).

Type species: *Desulfovibrio desulfuricans* (Beijerinck 1895) Kluyver and van Niel 1936, 397.

Further Descriptive Information

Desulfovibrios had a reputation for pleomorphism and it is true that old, stressed or very sulfide-rich cultures show many aberrant forms. However, young (i.e. early stationary phase) cultures are morphologically self-consistent, though the morphology is somewhat influenced by the composition of the culture medium. In medium B (Table 7.1), phase-contrast microscopy generally discloses small curved rods. Stationary cells show bean-shaped, single-wave curvature; sigmoid or spirilloid forms are rare, but the species *D. africanus* has a tendency to an elegant sigmoid form and *D. salexigens* tends to be stubby. Rod-shaped strains have been reported but are a minority. Strains of consistent semilunar form exist. Internal structure is rarely visible except in the large species *D. gigas*, which often shows refractile zones.

Old cultures, or cultures stressed with antibiotics, Mg^{2+} deprivation or Na_2SO_3, show sigmoid and spirilloid forms, the latter sometimes of considerable length (up to 20 or 30 waves); filamentous forms such as those in stressed cultures of desulfotomacula or enterobacteria are not seen. Forms which have been described as coccoid also occur in old cultures, particularly of *D. salexigens*; they have a plasmolyzed appearance with much of the cytoplasm concentrated in one sector of the sphere. Such aberrants are probably viable and cultures containing them yield normal forms on subculture.

Desulfovibrios exhibit a random, rapid, progressive motility, readily distinguishable from the nonprogressive, tumbling motility of desulfotomacula. Motility is suppressed at high sulfide concentrations. One or two truly nonmotile strains exist (see Postgate, 1979). Flagella are rarely visible by light microscopy, probably because of their small size and rapid movement; they can be demonstrated by electron microscopy or traditional flagellar staining, but they are readily detached during the necessary manipulations. They are always polar and are lophotrichous in certain species.

The cultivation of *Desulfovibrio* species has been discussed by Postgate (1979) and Widdel (1980). Strains of *Desulfovibrio* are recognized primarily by sulfate reduction, i.e. by their ability to produce substantial amounts of sulfide in anaerobic media (e.g. those of Baars, 1930) based on a simple carbon source (such as lactate) and a chemically equivalent amount of sulfate. Often a ferrous salt is incorporated into the medium so that blackening due to H_2S is diagnostic for growth and sulfate reduction. Single colony isolation from deep agar is facilitated by inclusion of a ferrous salt into the medium, since black colonies consist of presumptive sulfate reducers; in lactate-sulfate agar the colonies of *Desulfovibrio* are wholly black, but in peptone-glucose-sulfate agar (often used to check for contaminant anaerobes) they show a golden sheen when young. Media should be of a salinity appropriate to the natural environment or the salinity habitat of the strain. Freshly isolated strains often show tactophily, so that the sediment or wall of an enrichment culture may have most of the population adhering to it.

Although desulfovibrios are not usually considered to be pathogenic, one instance of transient pathogenicity to man, a bacteremia which was probably opportunist, has been recorded (see Postgate, 1979).

Enrichment and Isolation Procedures

Enrichment procedures for desulfovibrios using anaerobic cultivation methods are well established (Postgate, 1965, 1979; Widdel, 1980). Media for enrichment are simple. A bacteriological salt mixture containing a suitable carbon source and sulfates will yield these bacteria if it is incubated anaerobically with an appropriate inoculum at a pH value around 7.5. The E_h of the medium must be low; although with many samples of natural origin the nonsulfate-reducing bacteria grow first and lower the E_h as they grow, it is usually wise to add chemical reducing agents to bring the E_h to below -150 mV. Ferrous salts are usually present in excess to indicate sulfate reduction by blackening; salts of many other metals that form characteristic insoluble sulfides may be used, but copper, even as the "insoluble" carbonate, is toxic. Medium B (Table 7.1) is recommended for general use.

It is common experience that media containing a precipitate, such as medium B, are more satisfactory than those without a precipitate, and growth is often seen to begin as a zone of blackening in the precipitate. In all cases, success is aided by artificially lowering the E_h value of the medium chemically. The most satisfactory general reducing agent is Na_2S which, at 1 mM, does not blacken the medium seriously. Thioglycolate (thiolacetate) at 1 mM may also be used, but in the writer's experience is best when combined with ascorbate at ~1 mM. Fresh sodium dithionite has also been used but cannot readily be sterilized. Ascorbate alone may be used in many instances, but cysteine, though ideal for pure cultures of D. desulfuricans (Grossman and Postgate, 1953), should never be used for enrichment cultures because cysteine-decomposing bacteria can produce H_2S and cause false positives. A reduced iron nail (Abd-el-Malek and Rizk, 1958) can provide a satisfactory reducing environment for enrichment cultures. The ini-

tial pH value should be ~7.5; the salinity of the enrichment medium should be adjusted to that of the material used as inoculum.

Glass-stoppered bottles (30–60-ml capacity) are satisfactory containers for enrichment cultures. The stoppers should be lightly coated with silicone grease (which resists autoclaving). The bottles should be filled to the brim with medium B plus inoculum and the stopper pressed home to eliminate all air. The stopper should be pressed home again after a few hours incubation because expansion of the culture fluid on warming tends to loosen it.

Unequivocal blackening after incubation indicates successful enrichment of sulfate-reducing bacteria. A gray color, or blackening restricted to the precipitate, is more probably due to release of H_2S from organic sulfur compounds in the inoculum. Sulfide exceeding 30 mg of sulfur/liter suggests that dissimilatory sulfate-reducing bacteria are present; values lower than this may be obtained as a result of assimilatory or putrefactive microbial reactions.

The oxygen sensitivity and slow growth of Desulfovibrio makes isolation inconvenient, and sequential dilution of dispersed colonies from deep agar, or of sediment from a liquid enrichment culture, is often preferred. Populations are tolerant of air and elaborate precautions to handle them anaerobically need not normally be taken. The use of soft agar facilitates removal of colonies from the depths of agar media.

A general procedure for isolation of Desulfovibrio species is as follows. Prepare a stock of medium B (Table 7.1) of appropriate salinity. Take ~25 ml and supplement with agar (~1%), sodium ascorbate (0.01%) and sodium thioglycolate (0.01%). Autoclave briefly (5 min at 121°C). This medium does not store and should be used at once. After autoclaving, check that the pH is ~7.5 and distribute 4-ml portions into 6 long sterile tubes (10 × 150 mm external; sometimes called "vanilla" tubes) and hold molten at 40°C. The levels of the medium in the 6 tubes should be similar. Dip a closed, flamed Pasteur pipette into the enrichment culture and then successively into tubes 1–6; allow the medium to gel. The agar "dilutions" may then be incubated at 30°C (or other appropriate temperature) in air and the anaerobes will grow in the depths of the agar. After 4 or 5 days well-separated black colonies should be visible in the 4th or 5th tubes. These are colonies of sulfate-reducing bacteria; other colonies should be ignored.

Break the tube at a convenient point and withdraw 2 or 3 colonies with a fine Pasteur pipette. Break up each colony in a few drops of sterile saline and inspect the cell suspension under the microscope. If one of the suspensions is apparently pure, inoculate part of it into liquid medium B of E_h poised with Na_2S. Check the culture for purity as described below. Use the remainder of the suspension to repeat "Pasteur pipette dilution" as above in case it is necessary to isolate colonies again. If the first series is successful, these tubes may be discarded.

For isolation of the fatty acid-utilizing sulfate reducers D. sapovorans and D. baarsii, the agar shake culture method (as described for Desulfobacter in this edition of the Manual) is recommended. In the agar shakes the insoluble fatty acids palmitate and stearate should be replaced by mixture of butyrate, caproate and caprylate (see Table 7.1). For selective growth of colonies of D. baarsii in agar shakes, formate may also be used as the only organic substrate.

Purity of cultures is checked as follows. Contaminant aerobes are excluded by plating out on any nutrient agar containing glucose and peptone; no colonies should appear. Contaminant anaerobes can be a more serious problem. Prepare 25 ml of a sterile medium consisting of peptone (0.4%), glucose (1%), Na_2SO_4 (0.2%), $MgSO_4$ (0.1%) and agar (1.5%), and add sterile $Fe(NH_4)_2(SO_4)_2 \cdot 6H_2O$ to a final concentration of 0.05%. Adjust the pH to between 7.0 and 7.6 and distribute into 5 or more sterile plugged test tubes (10 × 150 mm). Cool to 45°C, make "Pasteur pipette dilutions" from the test culture, allow the medium to gel, and incubate at 30°C. This medium is highly selective for nonsulfate-reducing anaerobes; if the culture is pure, a relatively small number of exclusively black colonies will appear after 3 or 4 days, although because of an effect of peptone on the solubility of FeS they may appear golden brown when they start to form. Colonies which are not black,

Table 7.1.
*Composition of media for the cultivation of **Desulfovibrio** species[a,b]*

Component	Concentration, g/liter of distilled water		
	Medium B	Medium C	Medium D
KH_2PO_4	0.5	0.5	0.5
NH_4Cl	1.0	1.0	1.0
$CaSO_4$	1.0	—	—
Na_2SO_4	—	4.5	—
$MgSO_4 \cdot 7H_2O$	2.0	2.0	—
Sodium lactate	3.5	3.5	—
Sodium pyruvate	—	—	3.5
$CaCl_2 \cdot 2H_2O$	—	0.06	0.1
$MgCl_2$	—	—	1.6
$FeSO_4 \cdot 7H_2O$	0.5	0.004	0.004
Yeast extract	1.0	1.0	1.0
Sodium citrate	—	0.3	—

[a] The reducing agents sodium thioglycolate and sodium ascorbate, when used, should be at ~0.1 g/liter each. Agar may be used up to 10 g/liter for gelling. See Postgate (1979) for a further discussion of media.
[b] For cultivation of the fatty acid-utilizing sulfate reducers D. sapovorans and D. baarsii, the mineral medium described by Widdel (see footnote * in the section on Desulfobacter in this Manual) is first prepared without NaCl and $MgCl_2$. Sodium salts of fatty acids are added from stock solutions (26 g of palmitic acid or 28 g of stearic acid plus 4 g NaOH/liter, heated in a boiling water bath until clear, then sterilized by autoclaving; the solutions must be melted by reheating before use). For D. sapovorans, the following are added per liter of medium: 10–20 ml of palmitate solution followed by 1.0 ml of $MgCl_2$ solution (400 g $MgCl_2 \cdot 6H_2O$/liter) and 4.0 ml of NaCl solution (300 g/liter). For D. baarsii, the following are added per liter of medium: 10 ml of stearate solution followed by 3.0 ml of $MgCl_2$ solution and 25 ml of NaCl solution. Alternatively, for pure cultures of D. sapovorans and D. baarsii, one can add 5–10 ml/liter of medium of a neutralized solution containing (per liter): butyric acid, 6.0 g; caproic acid, 2.5 g; and caprylic acid, 1.5 g.

which form gas or which have pronounced halos are due to contaminant anaerobes.

The isolation procedure outlined above will also yield pure cultures of *Desulfotomaculum* species, though it will probably not yield cultures of other genera of sulfate-reducing bacteria.

Maintenance Procedures

Pure cultures are best maintained for routine study in medium B. For populations free of precipitate, medium C (Table 7.1) may be used. Chemostat culture is recommended as providing physiologically stable populations for metabolic research. Medium D (Table 7.1) supports growth of some species and can then be used to obtain populations free of sulfide, as well as for diagnostic purposes. Modifications of the media

described in Table 7.1 with various carbon or nitrogen sources have been used in research (see Postgate, 1979; Widdel, 1980); their composition is largely self-evident. Populations sealed anaerobically in media containing sediment, such as medium B, survive many years at room temperature.

The fatty acid-utilizing sulfate reducers *D. sapovorans* and *D. baarsii* should be kept at 2–5°C and transferred to fresh media every 2–4 months.

Lyophilization is used by culture collections for preservation of desulfovibrios. Cultures may be grown in medium C, harvested by aseptic centrifugation and lyophilized in *mist. desiccans* or in a cysteine-peptonized milk preparation (see Postgate, 1979).

Differentiation of **Desulfovibrio** from other sulfate-reducing genera

A curved morphology and progressive motility often provide a provisional identification of *Desulfovibrio* species. Members of the genus do not have heat-resistant forms and sensitivity to heating for 5 min at 100°C distinguishes them unequivocally from *Desulfotomaculum*. Most members of the genus *Desulfovibrio* show the desulfoviridin test: a characteristic red fluorescence (due to the sirohydrochlorin chromophore of that pigment) appears when a cell suspension is inspected in light of 365 nm immediately after addition of a few drops of 2.0 N NaOH. This test, and morphological characters such as motility and vibrioid form distinguish most strains from *Desulfobacter, Desulfococcus, Desulfosarcina* and *Desulfobulbus*. Although the writer regards the acceptance of the genus *Desulfomonas* as uncertain (Postgate, 1979), the principal feature that presently distinguishes this genus from *Desulfovibrio* is a mol% G + C of 66–67; moreover, *Desulfomonas* strains, unlike most strains of *Desulfovibrio*, are nonmotile and have a straight rod shape.

Taxonomic Comments

The classification of desulfovibrios given in the present edition of the *Manual* has some tentative features but is a useful working system.

It is an extension of that of Postgate and Campbell (1966) which had some official status (Postgate, 1967). Comments on some of the criteria were made by Postgate (1979), who recorded criticisms by other authors. In particular, the hibitane resistance test seems to be of limited value and some strains of *D. salexigens* do not have the high resistance quoted. Strains of *D. desulfuricans* lacking desulfoviridin have been reported (e.g. see Miller and Saleh (1964)). Serology can be ambiguous within the genus. DNA/rRNA hybridization studies by Pace and Campbell (1971) show close similarity to DNA from *D. vulgaris* in mRNA from *D. desulfuricans* (97%), but less with mRNA from *D. africanus* (59%) and *D. salexigens* (67%). Riederer-Henderson and Wilson (1970) have shown that nitrogen fixation is widespread within the genus, so the earlier subspecies epithet *azotovorans* must lapse. A commentary on individual strains held by the NICB is available as Appendix 1 of Postgate (1979).

Further Reading

Postgate, J.R. 1979. *The sulphate-reducing bacteria.* Cambridge University Press, Cambridge-London-New York-Melbourne.

Differentiation of the species of the genus **Desulfovibrio**

Characteristics useful in differentiating the species of *Desulfovibrio* are listed in Table 7.2.

List of the species of the genus **Desulfovibrio**

1. **Desulfovibrio desulfuricans** (Beijerinck 1895) Kluyver and van Niel 1936, 397.[AL] (*Spirillum desulfuricans* Beijerinck 1895, 113).

de.sul.fu'ri.cans. L. pref. *de* from; L. n. *sulfur* sulfur; M.L. part. adj. *desulfuricans* reducing sulfur compounds.

The morphology is as described in Table 7.2 and as depicted in Figure 7.3. Sigmoid forms may occur.

Other characteristics are listed in Table 7.2.

Requires media containing sulfates which blacken if iron salts are present. Cultures have a pronounced smell of H_2S, which can reach 200–1000 mg/liter in stoppered vessels. Sulfite or thiosulfate can replace sulfate.

Energy sources are restricted to lactate, pyruvate, formate, choline, or certain simple primary alcohols including methanol, ethanol, propanol and butanol. Choline and pyruvate supports fermentative growth without sulfate.

Usually inhibited by 10–25 mg of hibitane/liter, but not half of these concentrations (Saleh, 1964).

Found in fresh water, particularly polluted waters showing blackening and sulfide formation; also found in soils, particularly anaerobic or water-logged soils rich in organic materials, and in marine or brackish waters.

The mol% G + C of the DNA is 55 ± 1 (Bd).
Type strain: NCIB 8307 (Essex 6).

1a. **Desulfovibrio desulfuricans** subsp. **desulfuricans** (Beijerinck 1895) Kluyver and van Niel 1936, 397.[AL]
Differs from the subspecies *aestuarii* by not having a requirement for NaCl.
Type strain: NCIB 8307 (Essex 6).

1b. **Desulfovibrio desulfuricans** subsp. **aestuarii** Postgate and Campbell 1966, 734.[AL]
aes.tu.a'ri.i. L. n. *aestuarium* low ground covered by the sea at high water; L. gen. n. *aestuarii* of an estuary.
Differs from the subspecies *desulfuricans* in that strains have a requirement for NaCl, usually 2.5%. The strains, of marine or brackish origin, are incapable of adapting to fresh water environments.
Type strain: NCIB 9335 (Sylt 3).

2. **Desulfovibrio vulgaris** Postgate and Campbell 1966, 734.[AL] (Includes strains earlier accepted as *D. desulfuricans* as defined by Zobell, 1957.)
vul.ga'ris. L. adj. *vulgaris* common.

Table 7.2.

Differential characteristics of the species of the genus **Desulfovibrio**[a]

Characteristics	1. *D. desulfuricans*	2. *D. vulgaris*	3. *D. salexigens*	4. *D. africanus*	5. *D. gigas*	6. *D. baculatus*	7. *D. sapovorans*	8. *D. baarsii*	9. *D. thermophilus*
Typical shape	V	V	V	Sg	Sp	R	V	V	R
Cell diameter, μm	0.5–1.0	0.5–1.0	0.5–1.0	0.5	1.2–1.5	0.6	1.5	0.5–1.0	0.5
Cell length, μm	3.0–5.0	3.0–5.0	3.0–5.0	5.0–10.0	5.0–10.0	1.3	3.5–5.5	2.0–4.0	2.0
Flagellar arrangement:									
Single polar	+	+	+	−	−	+	+	+	+
Lophotrichous	−	−	−	+	+	−	−	−	−
Thickness of flagellum (nm)	20–25	20–25[b]	20–25	12	9	21			
Growth in:									
Lactate + sulfate	+	+	+	+	+	+	+	−	+
Pyruvate + sulfate	+	+	+	+	+	+	+	−	+
Pyruvate, no sulfate	+	d[c]	−	−	−	−	+	−	−
Malate + sulfate	+	−	+	+	−	+	−	−	−
Malate, no sulfate	−	−	−	−	+	−	−	−	−
Choline + sulfate	+	d[c]	−	−	−				−
Choline, no sulfate	+	d[c]	−	−	−				−
Palmitate + sulfate							+	+	
Acetate + sulfate	−	−	−	−	−	−	−	+	−
Butyrate + sulfate	−	−	−	−	−	−	+	+	+
Desulfoviridin present	+	+	+	+	+	−	−	−	+
NaCl requirement	d[d]	−	+	−	−	−	−	−	−
Hibitane resistance, mg/liter	10–25	2.5	1000	2.5	2.5				
Optimum temperature, °C	34–37	34–37	34–37	34–37	34–37	28–37	34	35	65
Temperature maximum, °C	42–45	40–45[e]	42–45	ca. 40	ca. 40	41	38	39	85
Mol% G + C of DNA	55[f]	61[b]	46	61	60	57	53		

[a] Symbols: see standard definitions; V, vibrioid; Sg, sigmoid; Sp, spirilloid; R, straight rod.

[b] Not determined for subspecies *oxamicus*.

[c] Positive for subspecies *oxamicus*; negative for subspecies *vulgaris*.

[d] Positive for subspecies *aestuarii*; negative for subspecies *desulfuricans*.

[e] About 40°C for subspecies *oxamicus*; 42–45°C for subspecies *vulgaris*.

[f] Not determined for subspecies *aestuarii*.

The morphology is as described in Table 7.2 and as depicted in Figure 7.4.

Similar to *D. desulfuricans* except for the characteristics indicated in Table 7.2; notably, it does not show sulfate-free growth and has a higher mol% G + C value of 61 ± 1 (Bd).

Type strain: DSM 644 (NCIB 8303; Hildenborough).

2a. Desulfovibrio vulgaris subsp. **vulgaris** Postgate and Campbell 1966, 734.[AL]

Differs from the subspecies *oxamicus* by being incapable of metabolizing oxamate, oxalate or choline and by failing to grow on pyruvate unless sulfate is present.

Type strain: DSM 644 (NCIB 8803; Hildenborough).

2b. Desulfovibrio vulgaris subsp. **oxamicus** Postgate and Campbell 1966, 734.[AL]

ox.am′i.cus. M.L. masc. adj. *oxamicus* pertaining to oxamic acid.

Differs from the subspecies *vulgaris* in being capable of metabolizing oxamate, oxalate or choline and of growing on pyruvate in the absence of sulfate.

Type strain: NCIB 9442 (Monticello 2).

3. Desulfovibrio salexigens Postgate and Campbell 1966, 735.[AL]

sal.ex′i.gens. L. n. *sal* salt; L. v. *exigo* to demand; M.L. part. adj. *salexigens* salt demanding.

The morphology is as described in Table 7.2 and is similar to that of *D. desulfuricans*. Fat, semilunar forms are seen in old cultures.

Other characteristics are as listed in Table 7.2.

Requires Cl⁻ for growth, supplied as NaCl (>0.6%, usually 2.5–5.0%).

Hibitane resistance varies and is often more than 1 g/liter. Some strains with a resistance of 250 mg/liter have been reported (Miller et al., 1968).

Found in sea water, marine and estuarine muds, and pickling brines.

The mol% G + C of the DNA is 46 ± 1 (Bd).

Type strain: ATCC 14822 (NCIB 8403; British Guiana).

4. Desulfovibrio africanus Campbell, Kasprzycki and Postgate 1966, 1127.[AL]

af.ri.ca′nus. L. adj. *africanus* pertaining to Africa.

The morphology is as described in Table 7.2 and as depicted in Figure 7.5.

The characteristics are similar to those for *D. vulgaris*, except as noted in Table 7.2.

Isolated from salt and fresh waters from Africa. Has a wide salt tolerance.

The mol% G + C of the DNA is 61.2 ± 1 (Bd).

Type strain: NCIB 8401 (Benghazi).

5. Desulfovibrio gigas Le Gall 1963, 1120.[AL]

gi′gas. L. n. *gigas* giant.

The morphology is as described in Table 7.2 and depicted in Figure 7.6. The cells are often in chains appearing as spirilla. Young organisms show areas of low contrast when examined by phase-contrast microscopy.

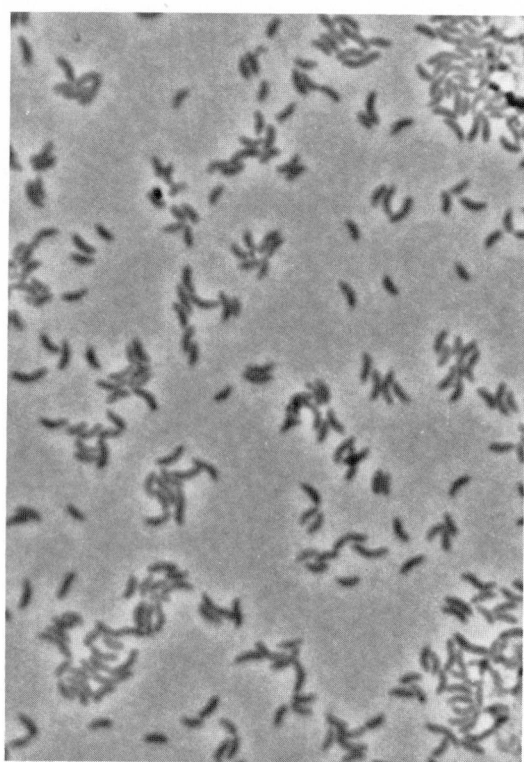

Figure 7.3. Phase-contrast photomicrograph of *Desulfovibrio desulfuricans* NCIB 8307 (×2000). (Reproduced by permission from Dr. N. Pfennig.)

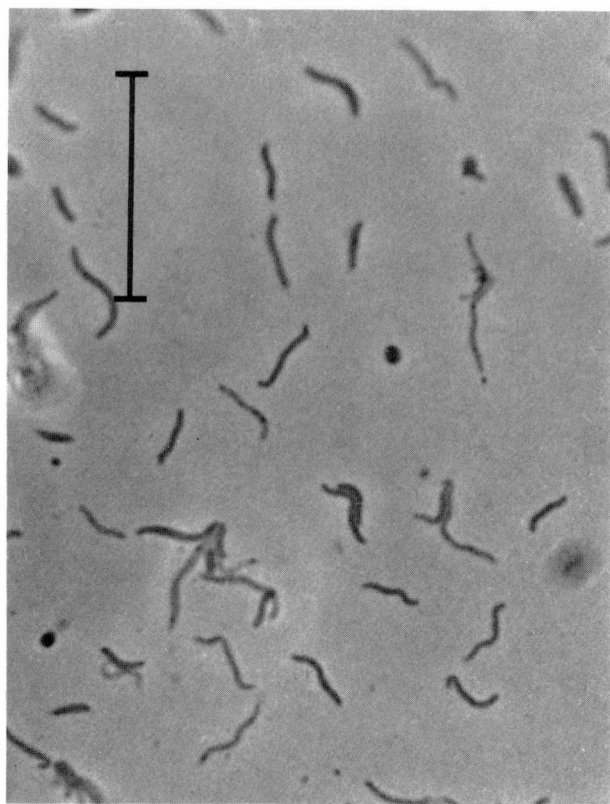

Figure 7.5. Phase-contrast photomicrograph of *Desulfovibrio africanus* strain "Benghazi" (NCIB 8401). *Bar*, 10 μm. (Reproduced by permission from Dr. Crawford Dow.)

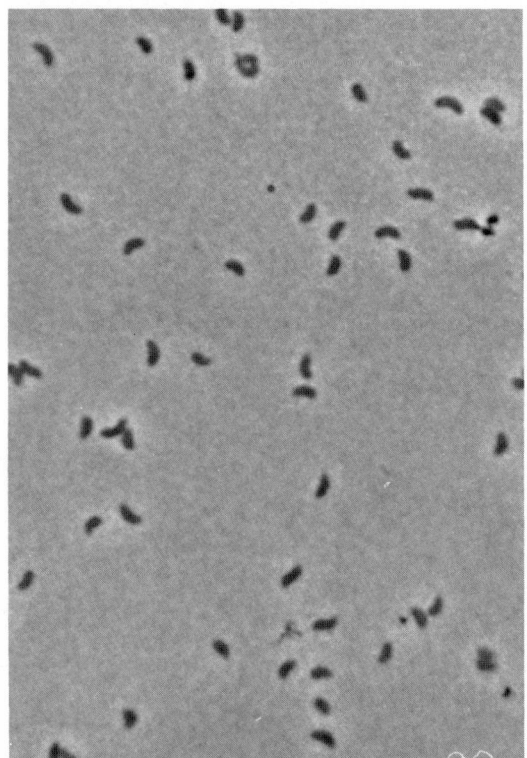

Figure 7.4. Phase-contrast photomicrograph of *Desulfovibrio vulgaris* NCIB 8303 (×2000). (Reproduced by permission from Dr. N. Pfennig.)

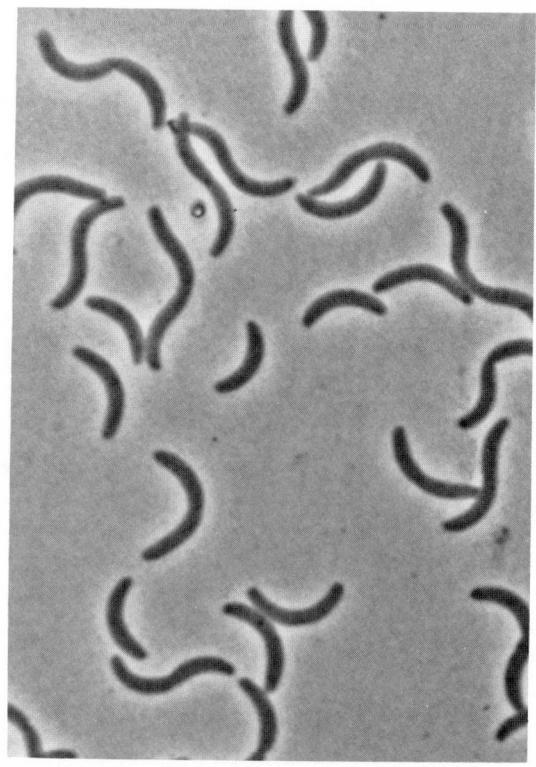

Figure 7.6. Phase-contrast photomicrograph of *Desulfovibrio gigas* (×2000). (Reproduced by permission from Dr. N. Pfennig.)

Other characteristics are listed in Table 7.2.

An E_h of ~80 mv (ascorbate at pH 7) seems most suitable for growth, which is slower than that of other species.

Isolated from Etang de Berre, near Marseilles, France. Despite its salt water origin, saline media were not used for its cultivation.

The mol% G + C of the DNA is 60.2 (Bd).

Type strain: NCIB 9332.

6. Desulfovibrio baculatus Rozanova and Nazina 1976, 825.[VP*]

ba.cu.la′tus. L. adj. *baculatus* rod shaped.

The morphology is as described in Table 7.2: short straight rods with rounded ends.

Differs from *D. vulgaris* mainly by its straight morphology and negative desulfoviridin test.

Grows between 2 and 41°C. Optimum temperature, 28–37°C.

Growth occurs with lactate, pyruvate or malate if sulfate is provided. Acetate, alcohols, oxalate and sugars are not utilized. Formate or H_2 supports detectable growth if yeast autolysate is present. Sulfite or thiosulfate can replace sulfate as a terminal electron acceptor. Cytochromes are present.

Isolated from manganese ore in the U.S.S.R.

The mol% G + C of the DNA is 56.8 (chemical analysis).

Type strain: strain X, Institute of Microbiology, U.S.S.R.

7. Desulfovibrio sapovorans Widdel 1981, 382.[VP]

(*Effective publication*: Widdel 1980, 388.)

sa.po′vo.rans. L. n. *sapo* soap; L. v. *voro* to devour; M.L. part adj. *sapovorans* devouring soap (i.e. higher fatty acids).

The morphology is as described in Table 7.2 and as depicted in Figure 7.7. Sigmoid or spirilloid cells sometimes occur. Storage granules of poly-β-hydroxybutyrate occur. The flagellum is sheathed.

Other characteristics are listed in Table 7.2.

Butyrate, 2-methylbutyrate, higher fatty acids up to 18 carbon atoms, lactate and pyruvate serve as electron donors and are incompetely oxidized to acetate; fatty acids with odd numbers of carbon atoms are oxidized to acetate and propionate. Growth on palmitate is good, but is rather slow on stearate. The following substrates are not utilized: H_2, formate, acetate, propionate, isobutyrate, 3-methylbutyrate, alcohols, succinate, fumarate, benzoate, cyclohexancarboxylate and sugars. Sulfate and sulfite serve as electron acceptors and are completely reduced to H_2S. Thiosulfate, elemental sulfur, fumarate, nitrate and oxygen are not utilized. Grows on pyruvate without sulfate, but does not grow on lactate or fumarate.

Growth requires mineral media with sulfide as reductant. Vitamins are not required.

Optimum temperature, 34°C; range for growth, 15–38°C. Optimum pH, 7.7; range for growth, 6.5–9.3.

The cell membrane and cytoplasmic fraction contain b- and c-type cytochromes, the former being extractable with acetone plus HCl. Desulfoviridin is not present.

Habitat: anaerobic mud of fresh water environments. Specific enrichment can be accomplished with palmitate.

The mol% G + C of the DNA is 52.7 (T_m).

Type strain: DSM 2055 (Lindhorst; 1pa3).

8. Desulfovibrio baarsii Widdel 1981, 382.[VP]

Effective publication: Widdel 1980, 389.)

baar′si.i. M.L. gen. n. *baarsii* of Baars; named after J. K. Baars, a Dutch microbiologist who did the first comprehensive studies on nutrition of sulfate-reducing bacteria.

The morphology is as indicated in Table 7.2 and as depicted in Figure 7.8. Sigmoid or spirilloid cells sometimes occur.

Formate, acetate, propionate, butyrate, 2-methylbutyrate, 3-methylbutyrate, and higher fatty acids up to 18 carbon atoms are utilized as electron donors and are completely oxidized to CO_2. Growth occurs on formate without the presence of other carbon sources. Growth on

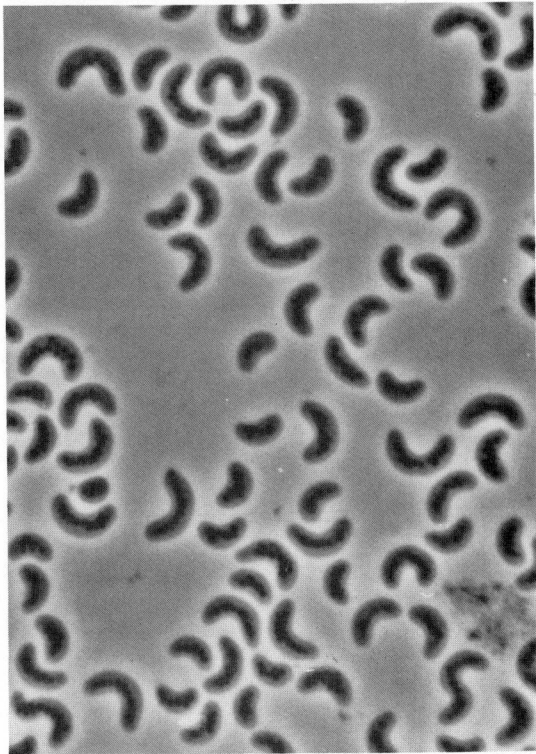

Figure 7.7. Phase-contrast photomicrograph of *Desulfovibrio sapovorans* (×2000). (Reproduced by permission from Dr. N. Pfennig.)

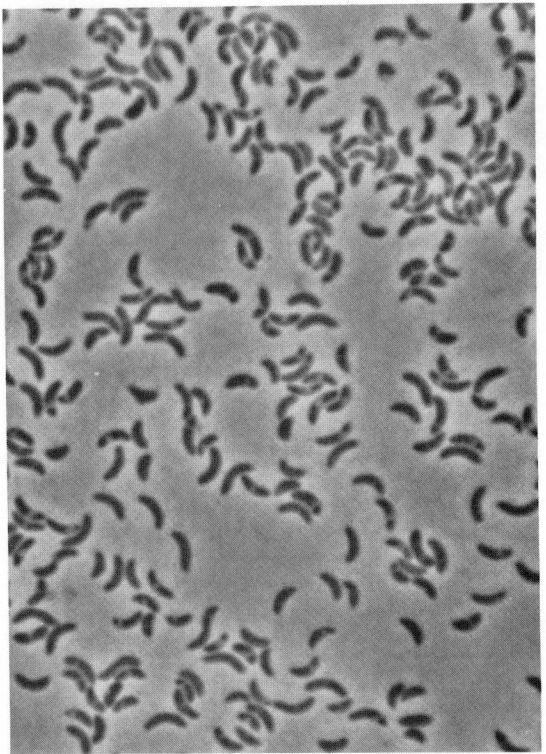

Figure 7.8. Phase-contrast photomicrograph of *Desulfovibrio baarsii* (×2000). (Reproduced by permission from Dr. N. Pfennig.)

---

* *VP* denotes that this name has been validly published in the official publication, *International Journal of Systematic Bacteriology.*

acetate or propionate alone is rather slow. The following substrates are not utilized: H_2, isobutyrate, alcohols, lactate, pyruvate, succinate, fumarate, malate, benzoate, cyclohexancarboxylate and sugars. Sulfate, sulfite and thiosulfate serve as electron acceptors and are reduced to H_2S. Elemental sulfur, fumarate, nitrate and oxygen are not reduced. No fermentation of organic substrates occurs.

Growth requires mineral media with sulfide as reductant. Vitamins are not required. Optimum growth occurs in freshwater media with up to 7 g NaCl and 1 g $MgCl_2 \cdot 6H_2O$/liter; other strains grow well at higher salt concentrations.

Optimum temperature, 35–39°C; range for growth, 20–43°C. Optimum pH, 7.3; range for growth, 6.5–8.2.

Cell membranes and cytoplasm contain b- and c-type cytochromes, but no desulfoviridin.

Habitat: anaerobic mud of fresh and brackish water environments. Specific enrichment can be accomplished with stearate.

The mol% G + C of the DNA of the type strain is 65.9 (T_m).

Type strain: DSM 2075 (2st14).

9. **Desulfovibrio thermophilus** Rozanova and Khudakova 1974, 1069.[AL]

ther.mo′phil.us. Gr. n. *therme* heat; Gr. adj. *philus* loving; M.L. adj. *thermophilus* heat loving.

The morphology is as described in Table 7.2: straight rods with rounded ends, often paired or in chains.

Lactate or pyruvate support growth in the presence of sulfate; methanol, ethanol, butanol, isobutanol, acetate, malate, oxalate, glucose and lactose do not. Lactate, pyruvate or choline do not support growth without sulfate. Sulfite or thiosulfate can replace sulfate.

NaCl not required.

Optimum temperature, 65°C; range for growth, 45–85°C.

The cells contain desulfoviridin.

Habitat: stratal eater at 84°C in a petroleum deposit near the Caspian Sea.

The mol% G + C of the DNA is not known.

Type strain: VKM V-1128, Institute of Microbiology, U.S.S.R.

Species Incertae Sedis

The status of the following species has been discussed by Postgate and Campbell (1966), and the Subcommittee on Sulfate-reducing Bacteria (ICSB) has not accepted them.

a. "*Desulfovibrio rubentschikii*" (Baars 1930) Zobell 1948, 248. [*Vibrio rubentschickii* (sic) Baars 1930, 89; *Sporovibrio rubentschickii* (sic) (Baars 1930) Brisou 1955, 227.]

Baars described a strain similar to *D. desulfuricans* but which was distinguished by its ability to oxidize acetate to CO_2 and water. No strains are extant and there is some doubt as to the purity of Baars' culture.

b. "*Desulforistella hydrocarbonoblastica*" Hvid-Hansen 1951, 332.

Short, pointed, nonmotile, nonsporulating rods (in a formate-sulfate medium) from subterranean water in Sjaelland, Denmark. Described as an obligate anaerobe and facultative autotroph. Grew best at 30°C; did not grow at 37°C. Utilized formate, lactate, propionate and acetate, and also other organic compounds. Old cultures sometimes contained an ether-soluble bituminous material.

The original culture has been lost and the organism has not been reisolated in Denmark or elsewhere.

Genus **Desulfomonas** Moore, Johnson and Holdeman 1976, 238[AL]

W. E. C. MOORE AND LILLIAN V. HOLDEMAN

De.sul.fo.mo′nas. L. pref. *de* from: L. n. *sulfur* sulfur; Gr. fem. n. *monas* unit; M.L. fem. n. *Desulfomonas* a cell that reduces sulfur compounds.

Nonsporeforming rods that are straight, Gram-negative, non-motile, aflagellate, obligately anaerobic, and reduce sulfate to hydrogen sulfide. Nonsaccharoclastic, nonproteolytic. Utilize metabolic intermediates to produce organic acids and hydrogen. Cells contain cytochromes and desulfoviridin. The mol% G + C of the DNA in the described species is 66–67.

Type species: *Desulfomonas pigra* Moore, Johnson and Holdeman 1976, 238.

Enrichment and Isolation Procedures

Complex media usually used for growth of anaerobic bacteria will support growth of members of this genus. Biebl and Pfennig (1977) report that VPI 11112 (ATCC 29098, the type strain of the type species) grows well with 0.1% ethanol in mineral media supplemented with 0.02% yeast extract. This medium is used for many other species of sulfate-reducing bacteria.

Maintenance Procedures

Cultures produce moderate growth through serial transfers in prereduced anaerobically sterilized peptone-yeast extract (PY) broth

(Holdeman et al., 1977). PY broth cultures survive lyophilization or freezing at −80°C.

Procedures for Testing for Special Characters

Sulfate reduction can be determined by comparing growth in PY-pyruvate (or other suitable metabolic intermediates) with that in PY-pyruvate supplemented with 20 mM $MgSO_4$. In the presence of 20 mM $MgSO_4$ the evidence for sulfate reduction is indicated by the increase in substrate utilization, increase in organic acid production, and decrease or elimination of hydrogen in the headspace gas.

Desulfoviridin is detected by the method of Postgate (see the genus *Desulfovibrio*).

Spores are not observed and cultures do not survive heating at 80°C for 10 min.

Cells are not motile when actively growing cultures are observed microscopically. No flagella are detected by King's modification of the Hugh-Leifson's flagella stain (Holdeman et al., 1977).

Differentiation from other sulfate-reducing anaerobes

Desulfomonas is differentiated from other rod-shaped sulfate-reducing anaerobes by absence of spore production, inability to use mono-carboxylic fatty acids as electron donors, absence of flagella or motility, and DNA with a G + C content of 66–67 mol%.

Description of species of the genus **Desulfomonas**

1. **Desulfomonas pigra** Moore, Johnson and Holdeman 1976, 238.[AL]

pig′ra. L. adj. lazy (referring to the limited number of substrates utilized by the species).

This description is based on Moore et al. (1976) and our study of the type and 10 phenotypically similar strains.

Cells from PY-glucose broth are 0.8–1.0 μm by 2.5–10 μm with blunt rounded ends and occur singly and in pairs (Fig. 7.9; scale: 1.4 mm = 1.0 μm).

Figure 7.9. *Desulfomonas pigra.*

Colonies on anaerobic blood agar are translucent, 1–2 mm in diameter, circular to slightly irregular, low-convex or umbonate, and non-hemolytic on rabbit blood agar. Subsurface colonies in rumen fluid-glucose-cellobiose agar after 5 d of incubation at 37°C are 0.5 mm in diameter, lenticular, translucent to opaque and may produce a dark halo indicating copious H_2S production.

Moderate growth with turbidity and smooth sediment is produced in prereduced anaerobically sterilized peptone-yeast extract (PY) broth. Abundant growth is produced in PY-pyruvate broth.

The optimum temperature for growth is 37°C. Some strains grow at 30 and 45°C.

Adonitol, amygdalin, arabinose, cellobiose, dextrin, esculin, fructose, galactose, glucose, glycerol, inositol, inulin, lactose, maltose, mannitol, mannose, melibiose, raffinose, rhamnose, ribose, salicin, sorbitol, sorbose, starch, sucrose, trehalose and xylose are not fermented. Esculin and starch are not hydrolized.

Gelatin, milk and meat are not proteolyzed. Indole, lecithinase, lipase, and oxidase are not produced. Nitrate is reduced by one of the 11 strains.

Growth is not affected by 20% bile.

Acetic acid (0.3–1.5 meq/100 ml of medium), H_2S, and traces of hydrogen are produced in PY or PY-glucose cultures; 2.6–10 or more meq of acetic acid/100 ml of medium and greater than 3% hydrogen in headspace gas are produced in PY-pyruvate cultures. Addition of 20 mM $MgSO_4$ to PY-pyruvate (which originally has 0.07 mM $MgSO_4$) increases pyruvate utilization and acetate and H_2S production and decreases or eliminates the accumulation of hydrogen in the headspace gas. Lactate utilization varies among the strains tested. Ethanol is utilized by VPI strain 11112 (ATCC 29098). Formate and fumarate are not utilized. Elemental sulfur is not utilized as an electron acceptor by VPI strain 11112 (Biebl and Pfennig 1977).

Cells contain cytochrome c and desulfoviridin (Sperry and Wilkins, 1977).

Isolated from human feces at 10^8/g dry weight (Holdeman, Good and Moore, 1976), from colon contents, peritoneal fluid, pylonidal cyst abscess and ruptured sigmoid colon.

The mol% G + C of the DNA is 66–67.

Type strain: ATCC 29098.

Genus **Desulfococcus** Widdel 1981, 382.^{VP} (Effective publication: Widdel 1980, 376.)

FRIEDRICH WIDDEL AND NORBERT PFENNIG

De.sul.fo.coc′cus. L. pref. *de* from; L. n. *sulfur* sulfur; M.L. masc. n. *coccus* equivalent of Gr. masc. n. *coccos* grain, berry; M.L. masc. n. *Desulfococcus* a berry-shaped (spherical) sulfate-reducer.

Spherical cells, 1.5–2.2 μm in diameter. Occur singly or in pairs. Spore formation is not observed. Cells often contain granules of poly-β-hydroxybutryate. Gram-negative. Generally nonmotile, but slowly motile cells have been observed in some of the strains and enrichment cultures. Strictly anaerobic, having both a respiratory and a fermentative type of metabolism. Optimum temperature, 30–36°C. Media containing a reductant and vitamins are necessary for growth. Colonies in anaerobic agar media are whitish or greyish to yellowish and tend to be slimy. Chemoorganotrophic, using formate, acetate, propionate, butyrate, higher fatty acids, lactate, pyruvate, alcohols, **benzoate or similar aromatic compounds** as carbon sources and also as electron donors for anaerobic respiration; these compounds are completely oxidized to CO_2. **Sulfate and other oxidized sulfur compounds serve as terminal electron acceptors and are reduced to H_2S. In the absence of an external electron acceptor, growth occurs by fermentation of lactate or pyruvate to acetate and propionate.** Occur in anaerobic mud from freshwater, brackish water and marine habitats; also occur in sludge from anaerobic sewage digestors. The mol% G + C of the DNA of the type strain is 57.4 (T_m).

Type species: *Desulfococcus multivorans* Widdel 1981, 382.

Further Descriptive Information

The type strain of *D. multivorans* has spherical cells under all growth conditions (Fig. 7.10). Cells of other strains may sometimes be irregular. The type strain stains Gram-negative, but one strain that was morphologically and nutritionally similar stained Gram-positive; this strain, however, exhibited an ultrastructure characteristic of Gram-negative bacteria.

The type strain of *D. multivorans* is nonmotile, but cells from other strains and from enrichment cultures may exhibit a slow motility.

The optimum growth temperature of the type strain of *D. multivorans* is 35°C. Growth can occur at temperatures as low as 15°C.

D. multivorans reduces sulfate, sulfite or thiosulfate to H_2S. Elemental sulfur, fumarate, malate and nitrate cannot serve as electron acceptors. The following componds are utilized as electron donors and organic carbon sources: formate, acetate, propionate, butyrate, isobutyrate, valerate, 2-methylbutyrate, 3-methylbutyrate, higher fatty acids up to 14 carbon atoms, lactate, pyruvate, ethanol, propanol and butanol. Growth also occurs with aromatic compounds such as benzoate, phenylacetate, 3-phenylpropionate, 2-hydroxybenzoate, and with cyclohexanecarboxylate. All organic substrates are completely oxidized to CO_2. Growth on formate requires no carbon sources besides the formate.

In the absence of an external electron acceptor, growth is possible with pyruvate or lactate which are fermented to acetate and propionate. Sugars are not fermented.

Strains of *Desulfococcus* may be cultured in a defined medium* with sulfate as the electron acceptor and benzoate as the electron donor and carbon source. The best growth is obtained with NaCl concentrations between 7 and 20 g/liter and $MgCl_2 \cdot 6H_2O$ concentrations between 1 and 3 g/liter, although growth does occur with lower concentrations of these salts. Ammonium ions are used as the nitrogen source. When *D.*

---

* The defined medium is prepared as described in footnote * of the article on the genus *Desulfobacter*, with the following additions (g/liter): NaCl, 7.0 and $MgCl_2 \cdot 6H_2O$, 1.2. After autoclaving the medium the following additional components are added aseptically from sterile stock solutions (per liter of medium): sodium benzoate solution (150 g/liter), 5.0 ml; vitamin solution (*p*-aminobenzoic acid, 40 mg/liter; biotin, 10 mg/liter; thiamine hydrochloride, 100 mg/liter; solution sterilized by filtration), 1.0 ml; sodium selenite solution ($Na_2SeO_3 \cdot 5H_2O$, 3 mg/liter; NaOH, 0.5 g/liter), 1.0 ml. For stimulation of growth on benzoate, 1.0 ml of a solution of fatty acids is added per liter of medium; this solution contains (g/liter): iso-butyrate, 5.0; *n*-valerate, 5.0; 2-methylbutyrate, 5.0; 3-methylbutyrate, 5.0; *n*-caproate, 2.5; *n*-heptanoate, 2.5; and *n*-octanoate, 2.5. The pH of the medium is adjusted to 7.2–7.4.

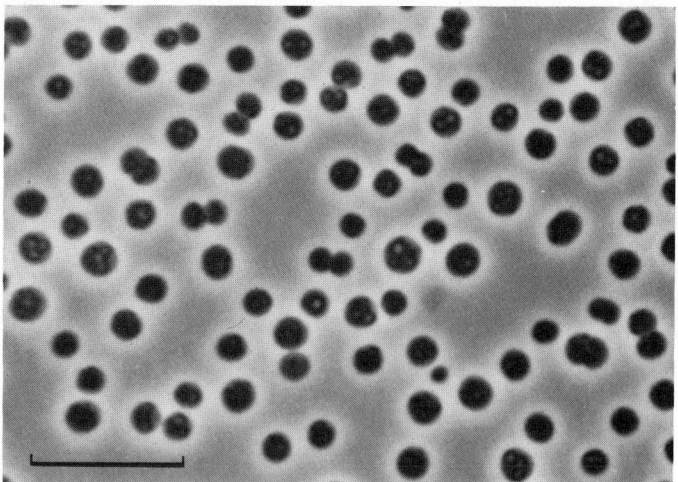

Figure 7.10. Phase-contrast photomicrograph showing cells of *Desulfococcus multivorans* isolated from a sewage digestor. Bar, 10 μm.

multivorans is cultured on benzoate or phenyl-substituted fatty acids, molybdate and selenite are required as trace elements. Growth on benzoate is stimulated by low levels of fatty acids, especially those having an odd number of carbon atoms; with such supplementation the doubling time for the organisms is ~24 h. *p*-Aminobenzoic acid, biotin and thiamine are required as growth factors. Sulfide serves as a reductant. When inoculation is done from old cultures, initiation of growth is favored by the addition of 10–30 mg of sodium dithionite per liter of medium as a further strong reductant.

Cytochromes of the *b* and *c* type were shown to be present mainly in the membrane fraction. The sulfite reductase desulfoviridin has been identified in the cytoplasm of *D. multivorans*.

Desulfococcus strains occur in anaerobic mud from ditches, brackish water and marine habitats. They also occur in the sludge from anaerobic sewage digestors.

Enrichment and Isolation Procedures

For the selective enrichment of *Desulfococcus* strains, the benzoate-sulfate medium is used. Bottles or tubes (100-, 50- or 20-ml capacity) can serve as culture vessels and are kept anaerobic with tight screw caps or by flushing with a mixture of 90% N_2: 10% CO_2 and sealing with butyl rubber stoppers. The medium is inoculated with black anaerobic mud from ditches, brackish water or marine habitats or with sludge from anaerobic sewage digestors; 2–5% of the total culture volume should be added. Enrichments are incubated at 30–35°C and mixed by shaking every second day. After intense formation of H_2S has occurred (generally after 2–4 weeks), transfers to fresh medium are made. The enrichments should be mixed well before subculturing.

Pure cultures of *Desulfococcus* strains are obtained by repeated application of the agar shake dilution method. The method is similar to that described for the genus *Desulfobacter*, except that benzoate medium is added to the molten concentrated agar solution.

Maintenance Procedures

Pure cultures are maintained in liquid medium in screw-capped bottles or in anaerobically gassed, butyl rubber-sealed bottles, with storage at 2–5°C. Stock cultures are transferred every 3–5 months.

Desulfococcus strains may be preserved indefinitely by suspending the cells in anaerobic medium containing 5% dimethyl sulfoxide and storing in liquid nitrogen.

Differentiation of the genus Desulfococcus from other genera

Desulfococcus can be differentiated by its coccoid cell form, which is obvious even during cell division, and by its ability to grow well with benzoate as the carbon source. *Desulfosarcina variabilis* resembles *Desulfococcus* by its benzoate-oxidizing, sulfate-reducing ability and by further nutritional characteristics (Pfennig and Widdel, 1981); moreover, free-living cells of *Desulfosarcina* often appear coccoid. It differs from *Desulfococcus*, however, by its formation of sarcina-like aggregates, the development of which is favored in agar media.

Taxonomic Comments

The phylogenetic relationships of *Desulfococcus* to other genera or higher taxa are not known. Nucleic acid studies, such as DNA/rRNA hybridization or rRNA oligonucleotide cataloging, would be helpful in elucidating these relationships.

List of the species of the genus Desulfococcus

1. **Desulfococcus multivorans** Widdel 1981, 382.[VP] (Effective publication: Widdel 1980, 377.)

mul.ti.vo'rans. L. adj. *multus* many, numerous; L. v. *voro* to devour, swallow; M.L. part. adj. *multivorans* devouring numerous kinds of substrates.

The characteristics are as described for the genus.

Occurs in anaerobic mud from ditches, brackish water and marine habitats. Also occurs in the sludge from anaerobic digestors.

The mol% G + C of the DNA of the type strain is 57.4 (T_m).

Type strain: DSM 2059 (strain 1be1 of Widdel, 1980).

Genus Desulfobacter Widdel 1981, 382.[VP] (Effective publication: Widdel 1980, 373.)

FRIEDRICH WIDDEL AND NORBERT PFENNIG

De.sul.fo.bac'ter. L. pref. *de* from; L. n. *sulfur* sulfur; M.L. masc. n. *bacter* equivalent of Gr. neut. n. *bakterion* rod or staff; M.L. masc. n. *Desulfobacter* a rod-shaped sulfate reducer.

Rod-shaped to ellipsoidal cells, 1.0–2.0 μm in diameter and 1.7–3.5 μm in length, with rounded ends. Occur singly or in pairs, or may stick together and form clumps. Spore formation is not observed. Gram-negative. Nonmotile, or motile by a single polar flagellum. Strictly anaerobic, having a respiratory type of metabolism **with sulfate or other oxidized sulfur compounds serving as terminal electron acceptors and being reduced to H_2S.** Optimum growth temperature, 28–32°C. Media containing a reductant and vitamins are necessary for growth. Most strains require at least 0.5% NaCl and 0.1% $MgCl_2 \cdot$

$6H_2O$. Colonies in anaerobic agar media are whitish to greyish and smooth. Chemoorganotrophic, **using acetate** or other simple organic compounds as carbon sources and also as electron donors for anaerobic respiration; **these compounds are completely oxidized to CO_2. Occur in the anaerobic parts of brackish water and marine habitats;** have also been enriched from freshwater mud by use of salt water media. The mol% G + C of the type strain is 45.9 (T_m).

Type species: *Desulfobacter postgatei* Widdel 1981, 382.

Further Descriptive Information

D. postgatei has ellipsoidal to short rod-shaped cells with rounded ends (Fig. 7.11); some strains form longer rods. The cells are often in pairs. Cells of marine strains have the tendency to stick together in clumps which settle to the bottom. In the first few enrichment passages on acetate and sulfate, motile cells can often be seen; after isolation most strains become nonmotile.

The temperature maximum for growth is 37°C; the minimum is 10°C. The pH range for growth is 6.2–8.5, with optimum growth at 7.3.

Growth occurs only in the presence of sulfate, sulfite or thiosulfate as electron acceptors. Nitrate, fumarate and malate are not reduced. Elemental sulfur inhibits growth. The type strain of *D. postgatei* uses only acetate as the electron donor and carbon source; however, other strains can also use ethanol and/or lactate. Sugars, pyruvate and fumarate are not fermented.

Strains of *Desulfobacter* may be cultured in a defined medium* with sulfate as the electron acceptor and acetate as the electron donor and carbon source. Ammonium ions are used as the nitrogen source. Sulfide is added as a reductant. Marine strains require 20.0 g of NaCl and 3.0 g of $MgCl_2 \cdot 6H_2O$/liter; brackish water strains tolerate lower levels of these salts. *p*-Aminobenzoic acid and biotin are required as growth factors. When inoculation is done from old cultures, initiation of growth is favored by the addition of sodium dithionite as a further strong reductant.

D. postgatei contains membrane-bound and soluble cytochromes of the *b* and *c* type. Desulfoviridin is not present.

Desulfobacter strains occur in the black anaerobic sediments from brackish water and marine habitats. Occurrence in anaerobic freshwater habits (ditches) has also been shown by use of enrichment cultures prepared with brackish water media.

Enrichment and Isolation Procedures

For the selective enrichment of *Desulfobacter* strains, the acetate-sulfate medium is used. Bottles or tubes (100-, 50- or 20-ml capacity) can serve as culture vessels and are kept anaerobic with tight screw caps or by flushing with a mixture of 90% N_2: 10% CO_2 and sealing with butyl rubber stoppers. The medium is inoculated with black anaerobic mud from brackish water or marine habitats; 2–5% of the total culture volume should be added. The NaCl concentration of the medium should correspond approximately to the salinity of the natural source. The addition of $MgCl_2 \cdot 6H_2O$ need not be higher than 0.3%. Brackish water medium should be used for enrichment of *Desulfobacter* strains from anaerobic freshwater habitats; media containing only ~1 g of NaCl and 0.5 g of $MgCl_2 \cdot 6H_2O$/liter are not suitable for such enrichment, but once strains grow well in brackwish water medium they may subsequently develop at low salinities. Addition of dithionite to enrichment cultures is recommended. Enrichments are incubated at 28–30°C and are mixed every second day by shaking. After intense formation of H_2S has occurred (generally in 10–20 days), transfers to fresh medium are made. Because cells of *Desulfobacter* often attach to the bottom layers of enrichment cultures, the cultures should be mixed well before subculturing.

Pure cultures are obtained by repeated application of the agar shake dilution method. A 3.3% (w/v) of agar is prepared in distilled water; the agar should be washed several times in distilled water before

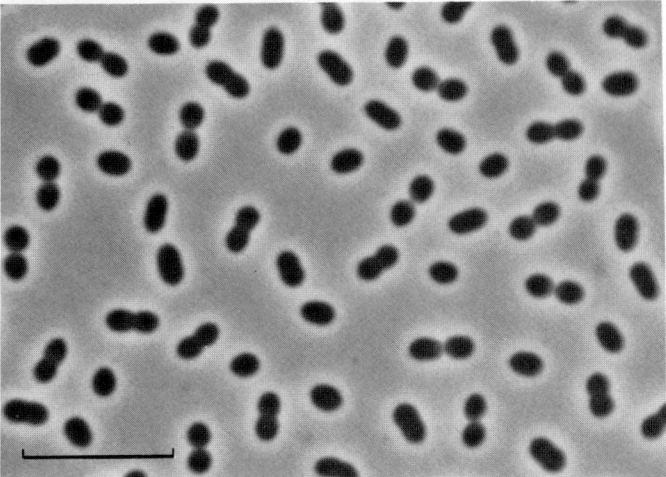

Figure 7.11. Phase-contrast photomicrograph of *Desulfobacter postgatei* DSM 2034 isolated from brackish water mud. *Bar*, 10 μm.

preparing the solution. Depending on the salinity of the culture medium, 7.0 or 20.0 g of NaCl and 1.2 or 3.0 g of $MgCl_2 \cdot 6H_2O$ are added per liter of agar solution. The agar solution is dispensed in 3-ml amounts into test tubes, which are stoppered with cotton and autoclaved. The agar tubes are kept molten in a water bath at 55°C. Defined culture medium is prewarmed to 41°C and 6-ml amounts are added to the tubes of liquefied agar; exposure to air is minimized by dipping the tip of the pipette into the agar medium. Starting with a few drops of an enrichment culture as inoculum, serial dilutions are made using 6–8 tubes. Before the agar medium solidifies, dithionite solution is added aseptically to each tube from an 0.1-ml pipette which is at the same time used for gently mixing starting at the highest dilution. All tubes are then hardened in cold water and immediately sealed with an overlay consisting of 1 part of paraffin wax and 3 parts of mineral oil; the overlay should be ~2 cm thick. The tubes are finally closed with a Wright-Burry seal, or they can be flushed with 90% N_2 + 10% CO_2 and sealed with butyl rubber stoppers. During the first 2 days of incubation the paraffin overlay is reheated to achieve a better sealing effect. The tubes are incubated in the dark for 2–4 weeks at 28–30°C. Well-separated colonies that occur in the higher dilutions can be removed with sterile Pasteur pipettes; the cells are suspended in 0.5–1.0 ml of anaerobic medium and used as the inoculum for subsequent agar shake cultures. The process is repeated until the cultures are purified.

Maintenance Procedures

Pure cultures are maintained in liquid medium in screw-capped bottles or in anaerobically gassed bottles sealed with butyl rubber stoppers, with storage at 2–5°C. Stock cultures are transferred every 3–5 months. Strains may also be preserved indefinitely by suspending the cells in anaerobic medium containing 5% (v/v) dimethyl sulfoxide and storing in liquid nitrogen.

---

* The defined medium has the following composition (g/liter of distilled water): Na_2SO_4, 3.0; KH_2PO_4, 0.2; NH_4Cl, 0.3; NaCl, 7.0 (for brackish water strains) or 20.0 (for marine strains); $MgCl_2 \cdot 6H_2O$, 1.2 (for brackish water strains) or 3.0 (for marine strains); KCl, 0.5; $CaCl_2 \cdot 2H_2O$, 0.15. After the medium is autoclaved and cooled under an atmosphere of 90% N_2 + 10% CO_2, the following components are added per liter of medium from sterile stock solutions, while access of air is prevented by continuous flushing with a mixture of 90% N_2 + 10% CO_2: acetate solution ($CH_3COONa \cdot 3H_2O$, 280 g/liter), 10.0 ml; trace element solution (see below), 1.0 ml; bicarbonate solution ($NaHCO_3$, 84 g/liter, saturated with CO_2 and autoclaved under a CO_2 atmosphere), 30.0 ml; sulfide solution ($Na_2S \cdot 9H_2O$, 120 g/liter, autoclaved under an N_2 atmosphere), 3.0 ml; vitamin solution (*p*-aminobenzoic acid, 40 mg/liter, and biotin, 10 mg/liter), 1.0 ml. *Trace element solution* (without complexing agent) contains (per liter): HCl (25% solution), 10 ml; $FeCl_2 \cdot 4H_2O$, 1.5 g; $CoCl_2 \cdot 6H_2O$, 190 mg; $MnCl_2 \cdot 4H_2O$, 100 mg; $ZnCl_2$, 70 mg; H_3BO_3, 6 mg; $Na_2MoO_4 \cdot 2H_2O$, 36 mg; $NiCl_2 \cdot 6H_2O$, 24 mg; $CuCl_2 \cdot 2H_2O$, 2 mg. The $FeCl_2$ is initially dissolved in the HCl solution and distilled water is added, followed by the other components. The pH of the complete defined medium is adjusted to 7.0–7.3. After the medium is inoculated, sodium dithionite (10–30 mg/liter) is added from a freshly prepared 5% solution sterilized by filtration under anaerobic conditions; alternatively, dry crystals can be added with a sterile spatula.

Differentiation of the genus **Desulfobacter** from other genera

The differentiation of *Desulfobacter* from other morphologically similar sulfate-reducing bacteria can be done primarily on the basis of physiological characteristics (Pfennig and Widdel, 1981). *Desulfobacter* strains grow well in sea water or brackish water media containing acetate and sulfate (with a 15–20-h doubling time). Sulfide concentrations of more than 20 mM may be produced. In contrast to *Desulfotomaculum acetoxidans*, which also grows on acetate and sulfate, *Desulfobacter* cells have rounded ends and do not form spores; moreover, they have a lower optimum growth temperature and they utilize only a few organic substrates besides acetate, such as ethanol or lactate, all of which are completely oxidized to CO_2.

Morphologically, *Desulfobacter* cells exhibit some resemblance to the cells of *Desulfococcus multivorans* and *Desulfosarcina variabilis*; however, strains of the latter two species develop only slowly on acetate.

They are also capable of utilizing a larger variety of organic compounds, such as propanol, butanol, higher fatty acids and benzoate.

Taxonomic Comments

The phylogenetic relationships of *Desulfobacter* to other genera or higher taxa are not known. Nucleic acid studies, such as DNA/rRNA hybridization or rRNA oligonucleotide cataloging, would help to elucidate such relationships.

Further Reading

Widdel, F. and N. Pfennig. 1981. Studies on dissimilatory sulfate-reducing bacteria that decompose fatty acids. I. Isolation of new sulfate-reducing bacteria enriched with acetate from saline environments. Description of *Desulfobacter postgatei* gen. nov., sp. nov. Arch. Microbiol. *129:* 395–400.

List of the species of the genus **Desulfobacter**

1. **Desulfobacter postgatei** Widdel 1981, 382.^{VP} (Effective publication: Widdel 1980, 373.)

post.ga′te.i. M.L. gen. n. *postgatei* of Postgate; named after J. R. Postgate, an English microbiologist who has made extensive studies of sulfate-reducing bacteria.

The characteristics are as described for the genus.

Occur in anaerobic parts of brackish water and marine habitats; may also be enriched from freshwater mud by use of salt water media.

Type strain: DSM 2034 (strain 2ac9 of Widdel (1980)).

Genus **Desulfobulbus** Widdel 1981,382^{VP} (Effective publication: Widdel 1980, 374.)

FRIEDRICH WIDDEL AND NORBERT PFENNIG

De.sul.fo.bul′bus. L. pref. *de* from; L. noun *sulfur* sulfur; L. n. *bulbus* onion; M.L. masc. n. *Desulfobulbus* onion-shaped sulfate reducer.

Ellipsoidal cells, 1.0–1.3 μm in diameter and 1.5–2.0 μm in length, often lemon or onion shaped with pointed ends. Occur singly, in pairs or in chains. Spore formation is not observed. Gram-negative. **Many strains are motile by a single polar flagellum.** Strictly anaerobic, having both a respiratory and a fermentative type of metabolism. Optimum growth temperature, 28–39°C. Media containing a reductant and vitamins are necessary for growth. Marine strains require higher NaCl and $MgCl_2$ concentrations than do freshwater strains. Colonies in anaerobic agar media are whitish to greyish and smooth. Chemoorganotrophic, using **propionate**, lactate, pyruvate, ethanol or propanol as carbon sources and also as electron donors for anaerobic respiration; these compounds are oxidized incompletely to acetate. **Sulfate and other oxidized sulfur compounds serve as terminal electron acceptors and are reduced to H_2S. In the absence of an external electron acceptor, growth occurs by fermentation of pyruvate or lactate to propionate and acetate.** Occur in the anaerobic parts of freshwater, brackish water and marine habitats; also isolated from rumen contents, animal dung and sewage sludge. The mol% G + C of the DNA of the type strain is 59.9 (T_m).

Type species: *Desulfobulbus propionicus* Widdel 1981, 382.

Further Descriptive Information

Cells of *D. propionicus* are generally lemon or onion shaped (Fig. 7.12), although some of the strains may have a more ovoid cell shape and less pointed ends. The cells are often in pairs and chains. Many strains are motile by a single polar flagellum; however, the type strain of *D. propionicus* is nonmotile, and electron microscopy of the cells has shown the presence of pili but no flagella.

The temperature range for growth is 10–43°C. The type strain has

an optimum growth temperature of 39°C; other strains may have lower temperature optima of ~28–30°C. The pH range for growth is 6.0–8.6, with an optimum pH of 7.2.

D. propionicus reduces sulfate, sulfite or thiosulfate to H_2S. Nitrate can also be used as a terminal electron acceptor and is reduced to ammonia. Fumarate, malate and elemental sulfur cannot serve as electron acceptors. The following compounds are utilized as electron donors and organic carbon sources: propionate, lactate, pyruvate, ethanol and propanol. Oxidation of these compounds is incomplete and leads to the formation of acetate as a regular end product. *Desulfobulbus* can grow with H_2 as an electron donor, but growth does not occur chemoautotrophically: acetate is necessary as a carbon source in the presence of bicarbonate.

In the absence of a terminal electron acceptor, growth is possible with pyruvate or lactate which are fermented to propionate and acetate. Fumarate and sugars are not fermented.

Strains of *Desulfobulbus* may be cultured in a defined medium* with sulfate as the electron acceptor and propionate, lactate or propanol as the electron donor and carbon source. Ammonium ions serve as the nitrogen source. Sulfide serves as a reductant in the medium. When inoculation is done from old cultures, initiation of growth is favored by the addition of 10–30 mg of sodium dithionite as a further strong reductant.

Membrane-bound and soluble cytochromes of the *b* and *c* type have been identified in *D. propionicus*. Desulfoviridin is not present.

Desulfobulbus occurs in the anaerobic black mud of freshwater, brackish water and marine habitats. It also occurs in bovine rumen fluid, in the intestinal tract of animals and in the anaerobic sludge of animal manure deposits and sewage plants.

---

* The defined medium is prepared as described in footnote * of the section on the genus *Desulfobacter*, with the following additions (g/liter): NaCl, 1.0 (for freshwater strains) or 20.0 (for marine strains); $MgCl_2 \cdot 2H_2O$, 0.5 (for freshwater strains) or 3.0 (for marine strains). Sodium propionate (1.5 g/liter) is added aseptically from a sterile stock solution to provide an organic substrate; however, the best growth occurs with sodium lactate (2.0–4.0 g/liter) in the presence of sulfate. The only growth factor required by *Desulfobulbus* is *p*-aminobenzoic acid. The pH of the complete medium is adjusted to 7.3–7.4.

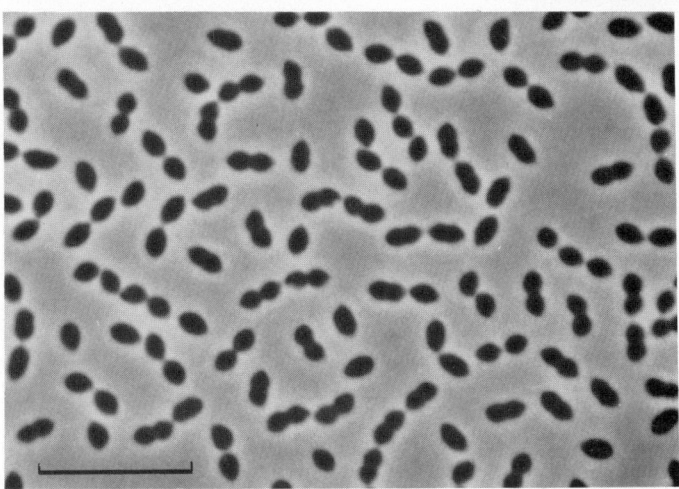

Figure 7.12. Phase-contrast photomicrograph showing cells of *Desulfobulbus propionicus* DSM 2032 isolated from freshwater mud. *Bar*, 10 μm.

Enrichment and Isolation Procedures

For the selective enrichment of *Desulfobulbus* strains, the propionate-sulfate medium is used. Bottles or tubes (100-, 50- or 20-ml capacity) can serve as culture vessels and are kept anaerobic with tight screw caps or by flushing with a mixture of 90% N_2: 10% CO_2 and sealing with butyl rubber stoppers. The medium is inoculated with black anaerobic mud from ditches, ponds, brackish water or marine habitats, manure deposits, sludge from sewage digestors, or with rumen fluid; 3–5% of the total culture volume should be added. The NaCl concentration of the medium should correspond approximately to the salinity of the natural source, but the concentration of $MgCl_2 \cdot 6H_2O$ need not be higher than 0.3%. Addition of dithionite to enrichment cultures is recommended. Enrichments are incubated at 28–36°C and mixed by shaking every second day. When intense formation of H_2S has occurred (normally after ~10 days), transfers to fresh medium are made. The enrichments should be mixed well before transfer.

Pure cultures of *Desulfobulbus* strains are obtained by repeated application of the agar shake dilution method. The method is similar to that described for the genus *Desulfobacter* except that propionate medium is added to the molten concentrated agar solution.

In enrichments with sea water or brackish water medium, cells of *Desulfobacter postgatei* may develop that grow on the acetate excreted by *Desulfobulbus*. In the highest dilutions of agar shake cultures, however, only *Desulfobulbus* develops into large colonies.

Maintenance Procedures

Pure cultures are maintained in liquid medium in screw-capped bottles or in anaerobically gassed, butyl rubber-sealed bottles, with storage at 2–5°C. Stock cultures are transferred every 1–3 months.

Desulfobulbus strains may be preserved indefinitely by suspending the cells in anaerobic medium containing 5% dimethyl sulfoxide and storing in liquid nitrogen.

Differentiation of the genus **Desulfobulbus** from other genera

The differentiation of *Desulfobulbus* from morphologically similar, anaerobic, sulfate-reducing bacteria is based primarily on physiological characteristics (Pfennig and Widdel, 1981). *Desulfobulbus* strains grow well (with doubling times of ~10 h) with sulfate and propionate, and the latter substrate is a characteristic organic substrate for the genus. *Desulfococcus multivorans* and *Desulfosarcina variabilis* are other sulfate-reducing bacteria that can also utilize propionate; however, these species carry out complete oxidation of the propionate and also grow more slowly than does *Desulfobulbus*.

Taxonomic Comments

The phylogenetic relationships of *Desulfobulbus* to other genera and higher taxa are not known. Nucleic acid studies, such as DNA/rRNA hybridization or rRNA oligonucleotide cataloging, would be helpful in elucidating these relationships.

Further Reading

Widdel, F. and N. Pfennig. 1982. Studies on dissimilatory sulfate-reducing bacteria that decompose fatty acids. II. Incomplete oxidation of propionate by *Desulfobulbus propionicus* gen. nov. sp. nov. Arch. Microbiol. *131*: 360–3655.

List of the species of the genus **Desulfobulbus**

1. **Desulfobulbus propionicus** Widdel 1981, 382.^{VP} (Effective publication: Widdel 1980, 375.)

pro.pi.o'ni.cus. M.L. n. *acidum propionicum* propionic acid; M.L. adj. *propionicus* pertaining to propionic acid.

The characteristics are as described for the genus.

Occur in the anaerobic part of freshwater, brackish water or marine habitats. Also occur in bovine rumen fluid and in the anaerobic sludge of animal manure deposits and sewage digestors.

The mol% G + C of the DNA of the type strain is 59.9 (T_m).

Type strain: DSM 2032 (strain 1pr3 of Widdel (1980)).

Genus **Desulfosarcina** Widdel 1981, 382^{VP} (Effective publication: Widdel 1980, 382.)

FRIEDRICH WIDDEL AND NORBERT PFENNIG

Irregularly shaped cells occurring in large, sarcina-like packets which form sediments in liquid media. Repeated transfer from the culture supernatant fluid favors the development of coccoidal to ellipsoidal cells 1.0–1.5 μm by 1.5–2.5 μm which occur singly or in pairs. Spore formation is not observed. Granules of poly-β-hydroxybutyrate frequently occur within the cells. Gram-negative. Usually nonmotile, but cells motile by means of a single polar flagellum may occur. Strictly anaerobic, having both a respiratory and a fermentative type of metabolism. Optimum growth temperature, 28–33°C. Media containing a reductant and not less than 1.0% NaCl and 0.2% $MgCl_2 \cdot 6H_2O$ are necessary for growth. Colonies in anaerobic agar media are greyish to yellowish, compact and irregular in shape. **Chemoorganotrophic or chemoautotrophic**, using formate, acetate, propionate, butyrate, higher fatty acids, other organic acids, alcohols and **benzoate or similar aromatic compounds as electron donors for anaerobic respiration** and also as carbon sources; these compounds are oxidized completely to CO_2. For chemoautotrophic growth, **H_2 can serve as the electron donor and CO_2 as the carbon source. Sulfate and other oxidized sulfur compounds serve as terminal electron acceptors and are reduced to H_2S. In the absence of an external electron acceptor, growth occurs by fermentation of lactate or pyruvate to acetate and propionate.** Occur in the anaerobic parts of brackish water and marine habitats. The mol% G + C of the DNA of the type strain is 51.2 (T_m).

Type species: *Desulfosarcina variabilis* Widdel 1981, 382.

Further Descriptive Information

D. variabilis forms large, sarcina-like packets having an irregular or distorted appearance (Fig. 7.13). These packets form sediments in liquid media. Cells liberated from the packets (e.g. by squeezing) have flattened shapes and are irregularly arranged. Repeated subculturing from culture supernatant fluid leads to development of coccoidal or ellipsoidal cells which occur singly or in pairs (Fig. 7.14). Single cells of the type strain are nonmotile, although some possess a single polar flagellum. In enrichment cultures of *Desulfosarcina* motile cells have been observed.

The temperature range for growth is 15–38°C, with optimum growth occurring at 33°C. The pH range for growth is 6.7–9.0, with pH 7.4 being the optimum.

D. variabilis reduces sulfate, sulfite or thiosulfate to H_2S. Elemental sulfur and nitrate cannot serve as electron acceptors. In the presence of sulfate, chemoautotrophic growth occurs with H_2 as the electron donor and CO_2 as the carbon source. The following compounds serve as both electron donors and carbon sources for chemoorganotrophic growth: formate, propionate, butyrate, valerate, 2-methylbutyrate, 3-methylbutyrate, higher fatty acids up to 14 carbon atoms, lactate, pyruvate, succinate, fumarate, ethanol, propanol, butanol, benzoate, phenylacetate, 3-phenylpropionate, 3-hydroxybenzoate, 4-hydroxybenzoate, hippurate and cyclohexanecarboxylate. Growth on acetate alone is very slow. All organic electron donors are completely oxidized to CO_2.

In the absence of an external electron acceptor, growth can occur with lactate or pyruvate, which are fermented to acetate and propionate. Fumarate can be slowly fermented to acetate, propionate and succinate; it does not act as a terminal electron acceptor for anaerobic respiration. Sugars are not fermented.

Strains of *Desulfosarcina* may be cultured in a defined medium* with sulfate as the electron acceptor and benzoate as the electron donor and carbon source. Ammonium ions are used as the nitrogen source. For growth on benzoate, addition of molybdate as a trace element is necessary. It is not yet known if selenite is required, but selenite is added routinely to the medium. The type strain of *D. variabilis* does not require vitamins; other strains have not yet been studied in this regard. Sulfide acts as a reductant in the medium. When inoculation is

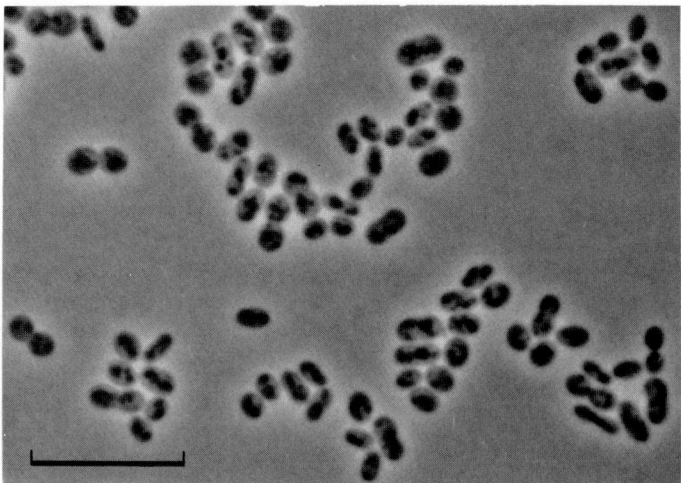

Figure 7.14. Phase-contrast photomicrograph showing *Desulfosarcina* cells occurring singly and in pairs after repeated transfer from the supernatant fluid of cultures initially showing mainly packet formation. *Bar*, 10 μm.

done from old cultures, initiation of growth is favored by the addition of 10–30 mg of dithionite/liter of medium as a further strong reductant.

Cell membranes and cytoplasm contain *b*- and *c*-type cytochromes, but no desulfoviridin.

D. variabilis is sensitive toward light and has to be incubated in the dark. Diffuse daylight in the laboratory inhibits growth completely.

Desulfosarcina occurs in anaerobic black mud of brackish water and marine habitats.

Enrichment and Isolation Procedures

For the selective enrichment of *Desulfosarcina* strains, the benzoate-sulfate medium is used. Bottles or tubes (100-, 50- or 20-ml capacity) can serve as culture vessels and are kept anaerobic with tight screw caps or by flushing with a mixture of 90% N_2: 10% CO_2 and sealing with butyl rubber stoppers. The medium is inoculated with black anaerobic mud from brackish or sea water sediments; 2–5% of the total culture volume should be added. Enrichments are incubated in the dark at 25–30°C and mixed by shaking every second day. After intense formation of H_2S has occurred (generally after 10–20 days), transfers to fresh medium are made. The enrichments should be mixed well before subculturing.

Pure cultures of *Desulfosarcina* are obtained by repeated application of the agar shake dilution method. The method is similar to that described for the genus *Desulfobacter*, except that benzoate medium is added to the molten concentrated agar solution. Before inoculation of the agar medium, cell packets of *Desulfosarcina* should be broken carefully with a sterile glass rod in anaerobic medium. In agar media *Desulfosarcina* forms irregularly shaped, compact colonies which can be removed with a sterile Pasteur pipette.

Maintenance Procedures

Pure cultures are maintained in liquid medium in screw-capped bottles or in anaerobically gassed, butyl rubber-sealed bottles, with storage in the dark at 2–5°C. Stock cultures are transferred every 2 to 3 months.

Desulfosarcina strains may be preserved indefinitely by suspending the cells in anaerobic medium containing 5% dimethyl sulfoxide and storing in liquid nitrogen.

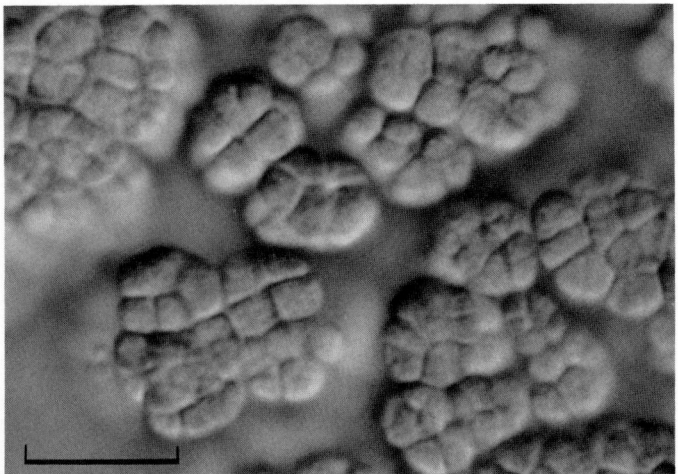

Figure 7.13. Phase-contrast photomicrograph showing cell packets of *Desulfosarcina variabilis* isolated with benzoate from brackish water mud. *Bar*, 10 μm.

---

* The defined medium is prepared as described in footnote * in the chapter on the genus *Desulfobacter*. The following additions are recommended (g/liter): NaCl, 13.0 and $MgCl_2 \cdot 6H_2O$, 2.0. After autoclaving the medium the following additional components are added from sterile stock solutions (per liter of medium): benzoate solution (sodium benzoate, 150 g/liter), 5.0 ml; and selenite solution ($Na_2SeO_3 \cdot 5H_2O$, 3 mg/liter and NaOH, 0.5 g/liter), 1.0 ml. The pH of the complete medium is adjusted to 7.3–7.5.

Differentiation of the genus **Desulfosarcina** from other genera

Desulfosarcina differs from other anaerobic sulfate-reducing bacteria by its formation of sarcina-like cell packets. A characteristic physiological property of *Desulfosarcina* is its growth with benzoate and phenyl-substituted fatty acids. This property is also shared by the genus *Desulfococcus*; however, strains of the latter genus are coccoid under all cultural conditions and never form cell packets.

Taxonomic Comments

The phylogenetic relationships of *Desulfosarcina* to other genera or higher taxa are not known. Nucleic acid studies, such as DNA/rRNA hybridization or rRNA oligonucleotide cataloging, would be helpful in elucidating these relationships.

List of the species of the genus **Desulfosarcina**

1. **Desulfosarcina variabilis** Widdel 1981, 382.[VP] (Effective publication: Widdel 1980, 383).

va.ri.a′bi.lis. L. adj. *variabilis* changeable, variable.

The characteristics are as described for the genus.

Occur in the anaerobic mud of brackish water and marine habitats. The mol% G + C of the DNA of the type strain is 51.2 (T_m).

Type strain: DSM 2060 (strain 3be13 of Widdel (1980)).

Important Notes for Users of this Edition

1. Always read both generic and species descriptions because characters listed in the generic description are not usually listed in the species descriptions.

2. Unless otherwise indicated in footnotes to tables, the meanings of symbols are as follows:
   + 90% or more of strains are positive
   − 90% or more of strains are negative
   d 11–89% of strains are positive
   v strain instability (*not* equivalent to "d")
   D different reactions in different taxa (species of a genus or genera of a family)

3. All other symbols are defined in footnotes to tables.

SECTION 8

Anaerobic Gram-Negative Cocci

FAMILY I. **VEILLONELLACEAE** ROGOSA 1971, 232[AL]*

MORRISON ROGOSA

Veil.lo.nel.la′ce.ae. M.L. fem. n. *Veillonella* type genus of the family; *-aceae* ending to denote a family; M.L. fem. pl. n. *Veillonaceae* the *Veillonella* family.

Cocci, varying in diameter from ~0.3–0.5 μm to ~2.5 μm. Occur characteristically in pairs. Single cells, masses or chains may also occur, although the chains may show gaps, illustrating the basic diplococcal arrangement. Adjacent sides of cell pairs may be flattened. Endospores are not formed. Nonmotile; flagella do not occur. Gram-negative, but tend to resist decolorization. Anaerobic. Oxidase-negative. Catalase-negative, but some strains decompose peroxide by a pseudocatalase (non-heme-containing). Chemoorganotrophic. Possess complex nutritional requirements. Gas is produced, often abundantly. Carbohydrates may or may not be fermented. Lactic acid may not be produced; if present, it is not a major product of fermentation. Lactate is fermented by some genera with the production of CO_2, H_2, and various volatile fatty acids containing 2–6 carbon atoms. Parasites of homothermic animals such as man, ruminants, rodents and pigs; particularly found in the alimentary tract.

Type genus: *Veillonella* Prévot 1933, 118.

The major differentiating characteristics of the genera of *Veillonellaceae* are presented in Table 8.1.

Further Comments

Fifty-one human clinical isolates of *Veillonellaceae* were examined by Chow et al. (1975), who classified them according to Holdeman and Moore (1973). All 31 strains of *Veillonellaceae* exhibited a pink to red fluorescence of colonies when illuminated with a long-wave (360 nm) ultraviolet lamp in a dark room. Nine strains of *Acidaminococcus fermentans* and 12 strains of *Megasphaera elsdenii* did not fluoresce. Although *Bacteroides melaninogenicus* fluoresced similarly to *Veillonellaceae*, the fluorescence of *Veillonella* faded rapidly in air (15 min to 2 h), whereas that of *B. melaninogenicus* was stable. *B. melaninogenicus* may be coccobacillary and thus confused with *Veillonella*, but the normal presence of black colony pigments on blood agar at 2–5 day in *B. melaninogenicus* and their absence in *Veillonella* and other members

Table 8.1.
Differential characteristics of the genera of the family **Veillonellaceae**[a]

Characteristics	Veillonella	Acidaminococcus	Megasphaera
Cell diameter, μm	0.3–0.5	0.6–1.0	1.7–1.9; 2.4–2.6
Red fluorescence of colonies under ultraviolet light (360 nm)	+	−	−
Ability to ferment carbohydrates	−[b]	−[b]	+
Lactate fermented	+	−	+
Amino acids are the main energy source	−	+	−
Pyruvate utilized	+	−	+
Succinate decarboxylated	+	−	−
Products in growth media:			
Gases:			
CO_2	+	+	+
H_2	+	−	w
Volatile fatty acids:			
2-carbon	+	+	+
3-carbon	+	−	+
4-carbon	−	+	+
5-carbon	−	−	+
6-carbon	−	−	+
Mol% G + C of DNA	40.3–44.4	56.6	53.1–54.1

[a] Symbols: +, positive reaction; −, negative reaction; w, slight reaction.
[b] Generally negative; slight or variable reactions may occur.

*AL denotes the inclusion of this name on the Approved Lists of Bacterial Names (1980).

of *Veillonellaceae* should aid in their differentiation. Definitive identification requires biochemical testing, gas-liquid chromatography of metabolic end products, nucleic acid studies, etc.

The anaerobic, hourglass-shaped organisms belonging to the genus *Gemmiger* (Gossling and Moore, 1975) might possibly be classifiable in the family *Veillonellaceae*. The effect of sublethal concentrations of penicillin on the cellular morphology, however, suggests that they are probably rods; moreover, the cells appear to reproduce by constriction, giving the appearance of buds. The genus is presently classified with other budding bacteria in Volume 3 of the *Manual*.

Further Reading

Chow, A.W., V. Patten and L.B. Guze. 1975. Rapid screening of *Veillonella* by ultraviolet fluorescence. J. Clin. Microbiol. *2:* 546–548.

Ellner, P.D., P.A. Granato and C.B. May. 1973. Recovery and identification of anaerobes: a system suitable for the routine clinical laboratory. Appl. Microbiol. *26:* 904–913.

Holdeman, L.V. and W.E.C. Moore (Editors). 1973. *Anaerobe Laboratory Manual*, 2nd Ed. Virginia Polytechnic Institute and State University, Blacksburg, Va.

Holdeman, L.V., E.P. Cato and W.E.C. Moore (Editors). 1977. *Anaerobe Laboratory Manual*, 4th Ed. Virginia Polytechnic Institute and State University, Blacksburg, Va.

Genus I. **Veillonella** *Prévot 1933, 118, emend. mut. char. Rogosa 1965, 706*[AL]

MORRISON ROGOSA

Veil.lo.nel′la. M.L. dim. ending -*ella*; M.L. fem. n. *Veillonella* named after A. Veillon, the French bacteriologist who isolated the type species.

Cocci, 0.3–0.5 μm in diameter, appearing by light microscopy as diplococci, masses and short chains. Nonsporulating. Gram-negative. Nonmotile. **Anaerobic.** Optimum temperature, 30–37°C. Optimum pH, 6.5–8.0. Oxidase-negative. Catalase-negative, but some species produce an atypical catalase lacking porphyrin. Chemoorganotrophic. **Pyruvate, lactate,** malate, fumarate and oxaloacetate **are fermented; carbohydrates and polyols are not fermented,** except for one species where fructose fermentation has been detected. **Acetate, propionate, CO_2 and H_2 are produced from lactate.** Nutritional requirements are complex. CO_2 is required for growth. Parasitic in the mouths and in the intestinal and respiratory tracts of man and other animals. The mol% G + C of the DNA is 36–43 (T_m) or 40 to 44 (Bd).

Type species: *Veillonella parvula* (Veillon and Zuber 1898) Prévot 1933, 119.

Further Descriptive Information

By electron microscopy, single cells in the logarithmic phase of growth appear spherical, whereas diplococci have a flattening at the cell junction.

Colonies in lactate agar media are lens, diamond or heart shaped, 1–3 mm in their greatest dimension, smooth entire, opaque, grayish white and butyrous. Nonhemolytic in blood agar.

No growth occurs in amino acid-containing media containing necessary vitamins and other cofactors unless supplemented with pyruvate (which supports growth best) or with lactate, malate, fumarate or oxaloacetate. No growth occurs with added succinate, carbohydrates, polyols, phosphorylated hexoses or trioses. Riboflavin and folic acid are not required; niacin and calcium pantothenate are often stimulatory but dispensable; biotin and *p*-aminobenzoic acid are frequently stimulatory and sometimes indispensable; pyridoxal and thiamine are required. Some organisms require putrescine or cadaverine.

Growth is poor at 40°C, slow at 24°C and absent at 18 and 45°C. Cells are killed at 60°C for 30 min.

The benzidine test for porphyrin is negative, but some species produce an atypical catalase lacking porphyrin.

Resting cells respire on lactate; oxaloacetate can be attacked aerobically with the production of CO_2 and H_2. Although there is no growth with succinate as an energy source, resting cells produce CO_2 and propionate from succinate.

D-Ribose is not fermented but is incorporated into nucleic acids. Glucokinase, fructokinase and glucose permease are not detectable. Formate, citrate, isocitrate and malonate are not attacked. Hypoxanthine may be fermented.

H_2S is produced from reduced glutathione, cysteine, cystine, thiosulfate, thiocyanate and thioglycolate. Gelatin is not liquefied. Indole is not produced. Nitrate is reduced to nitrite.

Growth occurs in 1% Tween 80, but is variable or inhibited by 10% bile, 0.0002% crystal violet or brilliant green, and 1% NaCl. No growth occurs in the presence of 4% NaCl, 0.25% phenethyl alcohol or 0.001% potassium tellurite. A variety of inorganic compounds, dyes and indicators are reduced, but 2,3,5-triphenyltetrazolium chloride is not reduced.

The organisms are sensitive to penicillin G (0.4–3.1 U/ml); chloramphenicol, chlortetracycline, oxytetracycline, polymyxin B (<1.0 μg/ml); and erythromycin (1.3 to 5.0 μg/ml). They are resistant to streptomycin (>25 μg/ml) and vancomycin (500 μg/ml).

Serologically specific endotoxins (lipopolysaccharides) induce pyrogenicity and the Shwartzman reaction in rabbits.

The majority of strains have mol% G + C values clustering around 39% when tested by the thermal denaturation method (Mays et al., 1982). Buoyant density values of identical strains tend to be slightly higher (Rogosa, 1974b).

Enrichment and Isolation Procedures

Lactate agar media containing vancomycin (7.5 μg/ml) favor isolation.

Maintenance Procedures

Cells suspended in skim milk can be successfully lyophilized.

Differentiation of the genus **Veillonella** from other closely related genera

Characteristics useful for differentiation of the genus *Veillonella* from other genera of anaerobic Gram-negative cocci are listed in Table 8.1.

Taxonomic Comments

The eighth edition of *Bergey's Manual* recognized *Veillonella parvula*, *V. parvula* subsp. *parvula*, *V. parvula* subsp. *rodentium*, *V. parvula* subsp. *atypica*, *V. alcalescens*, *V. alcalescens* subsp. *alcalescens*, *V. alcalescens* subsp. *ratti*, *V. alcalescans* subsp. *criceti*, and *V. alcalescens* subsp. *dispar*. These taxa fell into 7 serological groups (Rogosa, 1965, 1974b). An additional 7 strains from the guinea pig were isolated by

Rogosa and comprised an eighth antigenic group (unpublished data). All serological groups were proved to be distinct by reciprocal absorption procedures during which minor shared antigenic reactions were eliminated. Because serovars do not necessarily designate different species, the serovars were named as subspecies of the genus (Rogosa, 1965, 1974a).

Initial studies by Rogosa (1964, 1965) indicated clear differentiations in group requirements for cadaverine or putrescine, and these were reported in the eighth edition of the *Manual* (Rogosa, 1974b). In many subsequent trials over a decade or more, this clear nutritional difference disappeared and the requirement for cadaverine or putrescine varied among strains in some of the subspecies. There may have been unknown variations in different lots of nutritional test media or in experimental conditions, or genetic factors such as plasmids may have been involved. Mays et al. (1982) have had a similar experience and have also reported some variations in the decomposition of hydrogen peroxide among subcultures of a given strain. Thus, the separation of the species of *Veillonella* has become more difficult.

DNA/DNA hybridization experiments have clarified the speciation within the genus. In a recent study (Mays et al., 1982), seven DNA homology groups were distinct at the species level. Because the type strains of *V. parvula* subsp. *parvula* and *V. alcalescens* subsp. *alcalescens* had high homology, *V. alcalescens* was considered a subjective synonym of *V. parvula* and is so treated here. In keeping with the proposals of Mays et al., the following subspecies are elevated to species rank: *V. parvula* subsp. *rodentium* is recognized as *V. rodentium*, *V. parvula* subsp. *atypica* as *V. atypica*, *V. alcalescens* subsp. *ratti* as *V. ratti*, *V. alcalescens* subsp. *criceti* as *V. criceti*, and *V. alcalescens* subsp *dispar* as *V. dispar*. In addition, a new species, *Veillonella caviae*, was recognized by Mays et al. and is included herein.

Because most strains of *V. criceti* ferment fructose, the genus description has been emended. Other species of *Veillonella* do not ferment fructose.

Most human oral isolates belong to *V. parvula*, *V. atypica* or *V. dispar*. Of the human strains examined serologically, most are in serogroup VI or IV and a very small number are in serogroup V or VII. Human isolates have not been found in any other serogroups.

Two human isolates, VPI strains 3312A and 6788D (from the abdomen and blood, respectively), were found by Mays et al. to form a separate DNA homology group and a name for them has been deferred until additional strains are available. DNA from 3 other strains (VPI 8638 from the vagina, VPI 7944 from urine and VPI 1184B from feces), and also the DNA from strains Rogosa MV3 (VPI 12095, from the mouth of the mouse), had no homology with any reference DNA (Mays et al., 1982). The mouse strain is also of interest because it is the only strain ever isolated by Rogosa from a very large number of mice; evidently, the genus *Veillonella* is absent or poorly represented in the oral ecosystem of the mouse. A similar situation appears to exist in fowls: only one strain has ever been isolated from chickens despite multiple oral samplings.

Further Reading

Rogosa, M. 1964. The genus *Veillonella*. I. General cultural, ecological and biochemical considerations. J. Bacteriol. *87:* 162–170.
Rogosa, M. 1965. The genus *Veillonella*. IV. Serological groupings and genus and species emendations. J. Bacteriol. *90:* 704–709.
Rogosa, M. 1974. Genus I. *Veillonella*. In Buchanan and Gibbons (Editors), *Bergey's Manual of Determinative Bacteriology*, 8th Ed., Williams & Wilkins, Baltimore, pp. 446–447.
Mays, T.D., L.V. Holdeman, W.E.C. Moore, M. Rogosa and J.L. Johnson. 1982. Taxonomy of the genus *Veillonella*. Int. J. Syst. Bacteriol. *32:* 28–36.

Differentiation of the species of the genus **Veillonella**

Although DNA/DNA hybridization can clearly distinguish the various species of *Veillonella*, differentiation by phenotypic characters that are correlated with the genetic groups is in an unsatisfactory state.

Further work is required in order to arrive at a satisfactory phenotypic diffentiation scheme. The few phenotypic characters that can presently be correlated with some of the species are listed in Table 8.2.

Table 8.2.
Characteristics associated with the various species of **Veillonella**

Species	Serological Group	Catalase	Putrescine or Cadaverine Requirement	Fructose Fermented	Habitat
1. *V. parvula*	Human strains: IV or VI	−	Most −	−	Buccal or intestinal, from man, the rat, and the rabbit
2. *V. rodentium*	II	−	~33%+	−	Buccal or intestinal, from hamster, rat, rabbit
3. *V. atypica*	V or VI	−	−	−	Buccal cavity of man, rarely rodents
4. *V. ratti*	III	+	−	−	Mouth and intestine of rats
5. *V. criceti*	I	+	Most +	+	Mouth of hamster
6. *V. dispar*	VII	+	+	−	Mouth and respiratory tract of man
7. *V. caviae*	VIII	−	78%+	−	Mouth of guinea pig

List of the species of genus **Veillonella**

1. **Veillonella parvula** (Veillon and Zuber 1898) Prévot 1933, 119.[AL] (*Staphylococcus parvulus* Veillon and Zuber 1898, 542.) (Includes *Veillonella alcalescens* subsp. *alcalescens* Prévot 1933, 127.)

par′vu.la. L. fem. dim. adj. *parvula* very small.

The characteristics are described for the genus and as listed in Table 8.2.

Prévot (1933) and the seventh edition of *Bergey's Manual* describe the species as fermenting glucose and producing indole. Study of the type strain demonstrated that glucose is not fermented and indole is not produced.

The mol% G + C values of the DNA cluster around 38 (T_m) and 41 (Bd). DNA homology values within 44 strains of the species range from

53 to 100% (average = 70%). Homologies with other reference DNAs range from 1 to 44 (see Table 8.3).

Type strain: ATCC 10790 (Prévot Te 3).

2. Veillonella rodentium (Rogosa 1965) Mays, Holdeman, Moore, Rogosa and Johnson 1982, 34.[VP*] (*Veillonella parvula* subsp. *rodentium* Rogosa 1965, 707.)

ro.den′ti.um. L. v. *rodere* to gnaw; L. part. adj. *rodens* gnawing.

The characteristics are as described for the genus and as listed in Table 8.2.

The mol% G + C of the DNA averages 43 (T_m) and 44.4 (Bd). DNA homology with other *Veillonella* species is indicated in Table 8.3 and ranges from 4 to 21% (average = 15%).

Type strain: ATCC 17743 [strain HV19 (Rogosa, 1965)].

3. Veillonella atypica (Rogosa 1965) Mays, Holdeman, Moore, Rogosa and Johnson 1982, 34.[VP] (*Veillonella parvula* subsp. *atypica* Rogosa 1965, 707.)

a.typ′i.ca. L. prep. *a* from, away from; M.L. adj. *typicalis* typical, of a type; M.L. adj. *atypicalis* not typical.

The characteristics are as described for the genus and as listed in Table 8.2.

The type strain and one other strain belong to serogroup V. Seventeen other strains belong to serogroup VI.

The mol% G + C of the DNA averages 39. DNA homology with other *Veillonella* species is indicated in Table 8.3. Thirty-one strains have only 3 to 24% homology with any other reference DNA.

Type strain: ATCC 17744 [strain KON (Langford et al., 1950; Rogosa, 1965)].

4. Veillonella ratti (Rogosa 1965) Mays, Holdeman, Moore, Rogosa and Johnson 1982, 34[VP] (*Veillonella alcalescens* subsp. *ratti* Rogosa 1965, 708.)

rat′ti. M.L. n. *Rattus* genus of rats; M.L. gen. n. *ratti* of the rat genus.

The characteristics are as described for the genus and as listed in Table 8.2.

The mol% G + C of the DNA averages 42 (T_m) and 44 (Bd). The DNA homology with reference DNA from other *Veillonella* species is very low (Table 8.3).

Type strain: ATCC 17746 [strain RV-12X (Rogosa, 1965)].

5. Veillonella criceti (Rogosa 1965) Mays, Holdeman, Moore, Rogosa and Johnson 1982, 34.[VP] (*Veillonella alcalescens* subsp. *criceti* Rogosa 1965, 708.)

cri.ce′ti. M.L. n. *Cricetus* genus of hamsters; M.L. gen. n. *criceti* of the hamster genus.

The characteristics are as described for the genus and as listed in Table 8.2.

Nine out of 10 strains lowered the pH of fructose media by 0.4–1.0 pH unit (i.e. to pH 5.6 or lower) and produced increased amounts of acetic and propionic acids as compared to appropriate controls.

The mol% G + C of the DNA averages 39 (T_m) and 40.3 (Bd). DNA homologies within 15 strains of the species range from 72 to 100% (average = 84%). Homologies with other reference DNAs are very low (3–9%; see Table 8.3).

Type strain: ATCC 17747 [strain HV1 (Rogosa, 1965)].

6. Veillonella dispar (Rogosa 1965) Mays, Holdeman, Moore, Rogosa and Johnson 1981, 34.[VP] (*Veillonella alcalescens* subsp. *dispar* Rogosa 1965, 708.[AL])

dis′par. L. adj. *dispar* dissimilar, different.

The characteristics are as described for the genus and as listed in Table 8.2.

The mol% G + C of the DNA averages 39 (T_m) and 42 (Bd). DNA homologies within 8 strains of the species range from 67 to 100% (average = 76%). Homologies with other reference DNAs range from 5 to 38% (average = 15%) (see Table 8.3).

Type strain: ATCC 17748 [strain ERN (Langford et al., 1950)].

7. Veillonella caviae Mays, Holdeman, Moore, Rogosa and Johnson, 1982, 34[VP].

ca′vi.ae. L. gen. n. *caviae* of the guinea pig (genus *Cavia*).

The characteristics are as described for the genus and as listed in Table 8.2. Serogroup VIII; this serogroup was not included with the other seven in the original publication by Rogosa (1965).

The mol% G + C of the DNA averages 39 (T_m). DNA homologies within 7 strains of the species range from 71 to 100% (average = 87%). Homologies with other reference DNAs are low, ranging from 4 to 20% (average = 13%).

Type strain: ATCC 33540 (Rogosa strain PVI; VPI 12140).

Table 8.3.

DNA/DNA homology values within and among **Veillonella** *species[a]*

Unlabeled DNA from	Serogroup and no. of strains	Mol% G + C of DNA	Percent homology with reference labeled DNA from						
			1. *V. parvula*	2. *V. rodentium*	3. *V. atypica*	4. *V. ratti*	5. *V. criceti*	6. *V. dispar*	7. *V. caviae*
1. *V. parvula*	II, 40 VI, 4	37–40 (38)	**53–100 (70)**	5–24 (11)	8–29 (22)	1–5 (3)	0–9 (5)	29–44 (28)	9–20 (15)
2. *V. rodentium*	II, 4	42–43 (43)	18–22 (20)	**97–100 (99)**	12–21 (17)	3–6 (4)	4–10 (7)	15–24 (21)	16–19 (18)
3. *V. atypica*	VI, 29 V, 2	36–40 (39)	7–40 (20)	2–19 (7)	**56–100 (78)**	2–4 (3)	1–8 (6)	17–34 (24)	10–17 (15)
4. *V. ratti*	III, 2	41–43 (42)	5–10 (8)	1–2 (2)	2 (2)	**88–100 (94)**	14–16(15)	1–5 (3)	1 (1)
5. *V. criceti*	I, 15	38–40 (39)	3–8 (6)	1–9 (3)	3–10 (6)	3–12 (9)	**72–100 (84)**	1–9 (5)	3–13 (8)
6. *V. dispar*	VII, 8	38–40 (39)	29–54 (38)	4–12 (7)	20–30 (7)	3–6 (5)	4–9 (7)	**67–100 (76)**	13–18 (15)
7. *V. caviae*	VIII, 7	37–39 (39)	13–17 (15)	6–21 (14)	17–23 (20)	2–7 (4)	2–10 (7)	15–24 (17)	**71–100 (87)**

[a] Data from T. D. Mays et al., 1982. Values given in parentheses are the average percent homology values. Values given in **boldface type** are intraspecies homology values.

---

* *VP* denotes that this name has been validly published in the official publication, International Journal of Systematic Bacteriology.

Genus II. **Acidaminococcus** Rogosa 1969, 765[AL]

MORRISON ROGOSA

A.cid.a.min.o.coc′cus. M.L. n. *acidum* acid; M.L. adj. *amino* amino; *Gr. n. coccus* a grain, berry; M.L. masc. n. *Acidaminococcus* the amino acid coccus.

Cocci, 0.6–1.0 μm in diameter, often occurring as oval or kidney-shaped diplococci. Nonsporulating. Gram-negative. Nonmotile; flagella are not present. **Anaerobic;** no growth on the surface of agar media incubated in the air. Optimum temperature, 30–37°C. Optimum pH, 7.0. Oxidase- and catalase-negative. Chemoorganotrophic; **amino acids,** especially glutamate, **are the main energy sources. Pyruvate, lactate,** fumarate, malate, succinate and citrate **are not used as energy sources.** Only ~40% of strains catabolize glucose and the reaction is weak. In amino acid-containing media, **acetic and butyric acids** accumulate in a molar ratio of 2:1; CO_2 is also formed, **but H_2 and propionate are not detectable.** Nutritional requirements are complex. Isolated from the intestinal tract of the pig and man. The mol% G + C of the DNA is 56 (Bd).

Type species: *Acidaminococcus fermentans* Rogosa 1969, 765.

Further Descriptive Information

An outer cell wall membrane is demonstrable in thin sections by electron microscopy. Lipopolysaccharide (endotoxin) is present, and a Shwartzman reaction occurs in rabbits.

Surface colonies on complex media incubated in 95% H_2 + 5% CO_2 are 0.1 to 0.2 mm in diameter in 48 h, round, entire, slightly raised, whitish gray or nearly transparent. In peptone-yeast extract broth growth starts at the bottom of the tube and the broth becomes evenly turbid. Supplements of sodium glutamate (0.4–0.5%) enhance growth and gas (CO_2) production. Derivative products from glucose autoclaved in amino acid media are necessary or highly stimulatory for growth.

Growth is poor or absent at 25 and 45°C. Cells do not survive 60°C for 30 min. Growth occurs at initial pH values between 6.2 and 7.5, although best growth occurs at a neutral reaction. Final pH values in media initially at pH 7.5 range from ~6.1 to 6.7.

Nutritional requirements are multiple. Tryptophan, glutamate, valine and arginine are required by all strains. Cysteine and histidine are required by 93% of strains, tyrosine by 75%, phenylalanine and serine by 50%. Glycine is sometimes stimulatory. Alanine, leucine, isoleucine, proline, threonine, methionine, lysine and aspartate are not required for growth. In amino acid-containing media vitamin B_{12}, pyridoxal, pantothenate and biotin are indispensable for growth. *p*-Aminobenzoic

acid is essential or highly stimulatory. Exogenous putrescine, folic acid, folinic acid, thiamine, niacin and riboflavin are not required. No growth occurs in lactate or pyruvate media which support the growth of *Veillonella;* indeed, pyruvate suppresses growth completely.

Polyols including adonitol, dulcitol, erythritol, glycerol, inositol, mannitol and sorbitol are not attacked. Amygdalin, arabinose, fructose, galactose, inulin, maltose, mannose, melezitose, α-methyl-D-glucoside, α-methyl-D-mannoside, raffinose, salicin, sorbose, sucrose, trehalose, xylose, erythrose and esculin are not attacked. Ambiguous, extremely weak, or negative reactions occur with cellobiose, fucose, lactose, melibiose, rhamnose and ribose.

Ammonia is produced. Gelatin is generally not liquefied, although slow and partial liquefaction may sometimes occur. H_2S is not produced. Indole is generally not produced. Oxidase- and catalase-negative. The benzidine test for porphyrin is negative. Nitrate is not reduced. Sulfonthalein indicators are not reduced.

There is no serological cross-reaction between strains of *Acidaminococcus* and either *Veillonella* serovars or *Peptococcus aerogenes.*

Resistant to vancomycin (7.5 μg/ml). Sensitive to colistin (10 μg/disk).

Acidaminococcus strains have been isolated from the intestinal tract of the pig and man. They have been isolated in relatively large numbers from 25% of normal human feces; they have also been isolated from a closed abdominal abscess and from a putrid lung abscess as part of mixed anaerobic and facultative flora (Sugihara et al., 1974). They may be widespread in the intestinal tracts of various homothermic animals.

Enrichment and Isolation Procedures

In a study of the isolation of *Acidaminococcus* from human feces, Sugihara et al. (1974) found that the highest recovery was obtained from dilutions of the fecal samples when kanamycin-vancomycin blood agar, neomycin blood agar, or non-antibiotic-containing blood agar were used.

Maintenance Procedures

Cells suspended in skim milk may be lyophilized successfully.

Differentiation of the genus **Acidaminococcus** from other closely related genera

Characteristics useful for differentiating *Acidaminococcus* from other genera of anaerobic Gram-negative cocci are indicated in Table 8.1.

Although both *Acidaminococcus* and *Peptococcus aerogenes* ferment glutamate to very similar products, the latter genus is differentiated nutritionally, serologically, by Gram reaction, and by a widely different mol% G + C content of the DNA.

Taxonomic Comments

The phylogenetic relationship of *Acidaminococcus* to other anaerobic genera of bacteria is not known. Nucleic acid studies, such as DNA/

rRNA hybridization or rRNA oligonucleotide cataloging, would be useful in this regard.

Further Reading

Rogosa, M. 1969. *Acidaminococcus fermentans* sp. nov., anaerobic Gram-negative diplococci using amino acids as the sole energy source for growth. J. Bacteriol. *98:* 756–766.
Sugihara, P.T., V.L. Sutter, H.R. Attebery, K.S. Bricknell and S.M. Finegold. 1974. Isolation of *Acidaminococcus fermentans* and *Megasphaera elsdenii* from normal human feces. Appl. Microbiol. *27:* 274–275.

List of the species of the genus **Acidaminococcus**

1. **Acidaminococcus fermentans** Rogosa 1969, 765.[AL]
fer.men′tans. M.L. part. adj. *fermentans* fermenting.
The characteristics are as described for the genus.
The mol% G + C of the DNA from 15 strains was 56 ± 0.9 (Bd).

Type strain: ATCC 25085 (strain VR4). This strain does not ferment glucose. Reference strains fermenting glucose weakly are: ATCC 25086 (strain VR7) and ATCC 25087 (strain VR11).

Genus III. **Megasphaera** Rogosa 1971, 187^AL

MORRISON ROGOSA

Me.ga.sphae′ra. Gr. adj. *megas* big; Gr. n. *sphaera* a sphere; M.L. fem. n. *Megasphaera* big sphere.

Cocci, 2.0 μm or more in diameter, in pairs or occasionally in chains. Nonsporulating. Gram-negative. Nonmotile. **Anaerobic.** Growth occurs from 25 to 40°C but generally not at 45°C. Catalase-negative. Chemoorganotrophic. Gas is produced. **Lactate is fermented** with the production of acetate, propionate, **4-carbon straight- and branched-chain fatty acids, valerate,** little or no caproic acid, a large quantity of CO_2, and small amounts of H_2. **Glucose is fermented** with different products: some formate is produced, less acetate, propionate, butyrate and valerate, and **caproate is the most copious product** (60% or more of the total). **Pyruvate is utilized,** but succinate, fumarate and malate are not attacked. Nutritional requirements are complex. Found in the rumen of cattle and sheep and in the feces and intestine of man. The mol% G + C of the DNA is 53.6 (Bd).

Type species: *Megasphaera elsdenii* (Gutierrez, Davis, Lindahl and Warwick 1959) Rogosa 1971, 189.

Further Descriptive Information

In wet mounts the cells are spherical, 2.4–2.6 μm in diameter. In stained or fixed preparations, adjacent sides of diplococci are flattened and the diameter ranges from 1.2 to 1.9 μm. Occasionally 8–10 diplococci are arranged in a chain. Thin smears are Gram-negative even in 4-h-old cultures.

Surface colonies after 4 days are 0.5–2.0 mm in diameter, slightly raised, circular and entire, with a glistening to slightly rough surface, and are adherent to butyrous in consistency. Deep colonies are thin and disk shaped, up to 4 mm in diameter, and are greenish yellow or honey colored.

Although most strains do not grow at 45°C, a few do grow at this temperature. Such strains do not grow at 50°C.

Strains grow well in 0.4% yeast extract medium containing KH_2PO_4 (0.05%), NH_4Cl (0.05%), $MgCl_2 \cdot 6H_2O$ (0.03%), thioglycolic acid (0.03%), soluble starch (2.0%) and sodium DL-lactate (1.3%), pH = 7.4.

The use of the starch, and also incubation in 5% or more CO_2, facilitates isolation.

In a medium containing 0.4% yeast extract, 0.03% thioglycolic acid, and 1% substrate, at pH 7.4 and in an atmosphere of 95% H_2 + 5% CO_2 at 35–38°C, there is good growth and gas production with lactate, glucose and fructose. Variable growth and fermentation occurs with glycerol, maltose, mannitol, sorbitol and sucrose, and no growth occurs with arabinose, cellobiose, dextrin, esculin, galactose, inulin, lactose, mannose, raffinose, rhamnose, salicin, starch, trehalose or xylose. The final pH on glucose or fructose is 4.0–5.0; on lactate it is 7.8–8.0.

Megasphaera differs from *Veillonella*, which produces propionate from succinate, by producing propionate by an acrylic pathway as in *Clostridium propionicum.*

H_2S is produced. Gelatin is not liquefied. Indole is not produced. Nitrate is not reduced.

Megasphaera occurs in the rumen of cattle and sheep fed large amounts of starch, and also in the cecum of pigs. It is isolated in relatively large numbers from the feces and intestine of man (10% or more of samples) (see Werner (1973) and Sugihara et al. (1974)). It has been isolated from a case of endocarditis (Brancaccio and Legendre, 1979).

Enrichment and Isolation Procedures

Sugihara et al. (1974) obtained high recovery of *Megasphaera* on neomycin blood agar, *Veillonella*-neomycin agar, Eugonagar-maltose, and *Bifidobacterium* selective agar. No recovery occurred on blood agar, kanamycin-vancomycin blood agar, Fusobacterium agar, china blue agar or rifampin-vancomycin blood agar.

Maintenance Procedures

Cultures stored at 4°C must be transferred at least every 2 weeks. The organisms may not survive freeze-drying procedures, but can be preserved in liquid nitrogen.

Differentiation of the genus **Megasphaera** from other closely related genera

Characteristics useful for the differentiation of *Megasphaera* from other genera of anaerobic Gram-negative cocci are indicated in Table 8.1.

Taxonomic Comments

The phylogenetic relationship of *Megasphaera* to other anaerobic genera of bacteria is not known. Nucleic acid studies, such as DNA/rRNA hybridization of rRNA nucleotide cataloging, would be helpful in this regard.

Strains from the cecum of the pig appear to be different from rumen strains in having smaller cells, in fermenting raffinose and variably fermenting xylose. The mol% G + C of pig strains has not yet been

determined. The taxonomic relationship of these strains to the rumen strains needs to be clarified.

Further Reading

Rogosa, M. 1971. Transfer of *Peptostreptococcus elsdenii* Gutierrez et al. to a new genus, *Megasphaera* [M. elsdenii (Gutierrez et al.) comb. nov.]. Int. J. Syst. Bacteriol. *21:* 187–189.
Sugihara, P.T., V.L. Sutter, H.R. Attebery, K.S. Bricknell and S.M. Finegold. 1974. Isolation of *Acidaminococcus fermentans* and *Megasphaera elsdenii* from normal human feces. Appl. Microbiol. *27:* 274–275.
Werner, H. 1973. *Megasphaera elsdenii*—a normal inhabitant of the human intestines? Zentralbl. Bakteriol. Parasitenkd. Infektionskr. Hyg. Abt. I Orig. Reihe A *223:* 343–347.

List of the species of the genus **Megasphaera**

1. **Megasphaera elsdenii** (Gutierrez, Davis, Lindahl and Warwick 1959) Rogosa 1971, 189.^AL (*Peptostreptococcus elsdenii* Gutierrez, Davis, Lindahl and Warwick 1959, 20; organism LC (Elsden and Lewis, 1953, 183; rumen organism LC (Elsden et al., 1956, 686.)

els.de′ni.i. M.L. gen. n. *elsdenii* of Elsden; named after S.R. Elsden who first isolated the organism.

The characteristics are as described for the genus.

The mol% G + C of the DNA of the type strain and two similar strains was 53.6 ± 0.5 (Bd).

Type strain: ATCC 25940 [NCIB 8927; strain LC1 (Elsden et al., 1956)]; obtained from the rumen of a sheep.

Important Notes for Users of this Edition

1. Always read both generic and species descriptions because characters listed in the generic description are not usually listed in the species descriptions.

2. Unless otherwise indicated in footnotes to tables, the meanings of symbols are as follows:
  + 90% or more of strains are positive
  − 90% or more of strains are negative
  d 11–89% of strains are positive
  v strain instability (*not* equivalent to "d")
  D different reactions in different taxa (species of a genus or genera of a family)

3. All other symbols are defined in footnotes to tables.

SECTION 9

The Rickettsias and Chlamydias

ORDER I. **RICKETTSIALES** GIESZCZKIEWICZ 1939, 25[AL]*

Emilio Weiss and James W. Moulder

Rick.ett.si.a′les. M.L. fem. n. *Rickettsia* type genus of order; *-ales* ending to denote order; M.L. fem. pl. n. *Rickettsiales* the *Rickettsia* order.

Mainly rod-shaped, coccoid and often pleomorphic Gram-negative microorganisms with typical bacterial cell walls and no flagella. Multiply only inside host cells. They may be cultivated in living tissues such as embryonated chicken eggs or metazoan cell cultures. Except for binary fission, which is common to all members of this order, there are notable exceptions to any one of the characteristics listed above. For example, microorganisms are included that appear ring-shaped in stained preparations, or have a flagellum, or are Gram-positive, or multiply on bacteriological media of moderate complexity. All are regarded as parasitic or mutualistic. The parasitic forms are associated with the reticuloendothelial and vascular endothelial cells or erythrocytes of vertebrates and often with various organs of arthropods which may act as vectors or primary hosts. May cause disease in man or in other vertebrate and invertebrate hosts. The mutualistic forms in insects are regarded as essential for development and reproduction of the host.

In the eighth edition of *Bergey's Manual*, Part 18, "The Rickettsias," was divided into two orders, *Rickettsiales* and *Chlamydiales*. This division, retained here, although contributing to the classification of the *Chlamydiales*, does not reduce the complexity of the order *Rickettsiales*. The removal of additional taxa from *Rickettsiales* to provide for a more homogeneous characterization of the order cannot rationally be done at this time with the information available. This edition recognizes, however, that a number of microorganisms have been described that have been incompletely classified or not classified at all. These microorganisms, sometimes regarded as possible members of the *Rickettsiales*, are discussed in Section 11 of this *Manual*.

Three families are accepted in the order *Rickettsiales*, viz., *Rickettsiaceae*, *Bartonellaceae* and *Anaplasmataceae*.

Table 9.1.
Differential characteristics of the families of the order **Rickettsiales**[a]

Characteristics	I *Rickettsiaceae*	II *Bartonellaceae*	III *Anaplasmataceae*
Trilaminar cell wall	+	+	−
Axenic cultivation	D[b]	+	−
Association with vertebrate cells:			
Nucleated cells	+	+	−
Erythrocytes	−	+	+

[a] Symbols: see standard definitions.
[b] See Table 9.2.

Key to the families of order **Rickettsiales**

The differential characteristics of the families of the order *Rickettsiales* are presented in Table 9.1. Since information on some of the taxa is limited, the criteria shown may not apply in every case.

FAMILY I. **RICKETTSIACEAE** PINKERTON 1936, 186[AL]

Rick.ett.si.a′ce.ae. M.L. fem. n. *Rickettsia* type genus of the family; *-aceae* ending to denote family; M.L. fem. pl. n. *Rickettsiaceae* the *Rickettsia* family.

Small, rod-shaped, coccoid and diplococcus-shaped, often pleomorphic organisms which **are often intimately associated with arthropod tissues, usually in an intracellular position**. Gram-negative. **With one exception** the genera pathogenic for vertebrates

---

*AL denotes the inclusion of this name in the Approved Lists of Bacterial Names (1980).

Table 9.2.

Differential characteristics of the genera of the tribe **Rickettsieae**

Characteristics	I *Rickettsia*	II *Rochalimaea*	III *Coxiella*	Characteristics	I *Rickettsia*	II *Rochalimaea*	III *Coxiella*
Axenic cultivation	−	+	−	Metabolism[a]			
Growth in association with eucaryotic cells:				Optimal pH	7.0	7.0	4.5
In cytoplasm or nucleus	+	−	−	CO$_2$ produced from:			
In phagolysosomes	−	−	+	Glucose	−	−	Weak[b]
Epicellularly	−	+	−	Glutamate	+	Weak[b]	+
Presence of endospore-like forms	−	−	+	Succinate	Weak	+	+
				Mol% G + C of DNA	29–33	39	43

[a] When separated from host cells.

[b] Activity enhanced when certain other substrates are present.

have not yet been cultivated in cell-free media. May be parasitic in man and other vertebrates, causing diseases (e.g. typhus and related ills in man, or tropical canine pancytopenia or other diseases of domestic animals) that are transmitted by arthropods (lice, fleas, ticks and mites) and rarely by other invertebrates. Some are confined to the invertebrate host as pathogens or symbiotes.

The family *Rickettsiaceae* includes three tribes: *Rickettsieae*, *Ehrlichieae* and *Wolbachieae*.

Key to the tribes of family **Rickettsiaceae**

Variation in intrinsic properties of the organisms within each tribe is so great that the most useful key is based on the host.

I. Pathogenic for man or related to species that are pathogenic for man, although in most cases man is an accidental host.

Tribe I. *Rickettsieae*, p. 688

II. Pathogenic for one or more vertebrate hosts, usually domestic animals, but not known to be pathogenic for man. The one exception, a possible human pathogen, is closely related to animal pathogens.

Tribe II. *Ehrlichieae*, p. 704

III. Confined to arthropods as pathogens or symbiotes. Not pathogenic for vertebrates.

Tribe III. *Wolbachieae*, p. 711

TRIBE I. **RICKETTSIEAE** PHILIP 1953, 486[AL]

Rick.ett′si.e.ae. M.l. fem. n. *Rickettsia* type genus of the tribe; *-eae* ending to designate a tribe; M.L. fem. pl. n. *Rickettsieae* the *Rickettsia* tribe.

Small, pleomorphic, usually intracellular organisms found in arthropods and pathogenic for man and certain other vertebrate hosts.

The tribe *Rickettsieae* includes three genera, *Rickettsia*, *Rochalimaea* and *Coxiella*. Some of their differential characteristics are listed in Table 9.2.

Genus I. **Rickettsia** *da Rocha-Lima 1916, 567*[AL] (*Nom. gen. cons. Opin. 19, Jud. Comm. 1958, 158*)

EMILIO WEISS AND JAMES W. MOULDER

Rick.ett′si.a. M.L. fem. n. *Rickettsia* named after Howard Taylor Ricketts, who first associated organisms of this description with spotted fever and typhus, and who died of typhus contracted in the course of his studies.

Short rods, 0.3–0.5 μm in diameter and 0.8–2.0 μm in length, sometimes longer when cell division is impaired. Often surrounded by protein microcapsular layer and slime layer. Gram-negative. Retain basic fuchsin when stained by the method of Giménez. Nonmotile; flagella do not occur. Aerobic. **Have not been cultivated in the absence of host cells.** Growth occurs in the cytoplasm, sometimes in the nucleus, of certain vertebrate and arthropod cells. Do not grow in phagocytic vacuoles. Generally unstable when separated from host components; stability enhanced by certain proteins, sucrose, and reagents that tend to maintain the integrity of outer membranes and ATP level. Best preserved by rapid freezing and storage below −50°C. Rapidly inactivated at 56°C. Derive energy from the metabolism of glutamate via the citric acid cycle, but do not utilize glucose. Transport and metabolize phosphorylated compounds, but do not synthesize or degrade nucleoside monophosphates. Natural cycle generally involves a vertebrate and an invertebrate host. **Etiological agents of human diseases, such as typhus, spotted fever, or scrub typhus.** The mol% G + C of the DNA is 29–33 (T_m, Bd).

Type species: *Rickettsia prowazekii* da Rocha-Lima 1916, 567.

Further Descriptive Information

A detailed description of the genus often requires the recognition of three groups: the **typhus group** (3 species); the **spotted fever group** (at least 8 species); and the **scrub typhus group** (single species, *R. tsutsugamushi*, with 3 major serovars). The typhus and the spotted fever groups are more closely related to each other than to *R. tsutsugamushi*.

Cell structure. Although rickettsiae are relatively small, they closely resemble other Gram-negative bacteria. In smears from yolk sacs, tissues, or cell cultures, rickettsiae are best visualized by the Giménez (1964) stain. By this procedure, rickettsiae stain bright red with basic fuchsin, while the background is decolorized and stains a pale greenish blue with the counterstain malachite green (Fig. 9.1, *A* and *B*). *R. tsutsugamushi* requires preliminary destaining with ferric nitrate and is counterstained with fast green. This species is also satisfactorily stained by Giemsa's stain preceded by Carnoy's fixation. All species thus far examined stain well with acridine orange buffered at low pH (Kronvall and Myhre, 1977).

Ultrathin sections viewed by electron microscopy reveal typical envelopes consisting of cell wall and cytoplasmic membrane and internal structures (Fig. 9.1, *E–G*) analogous to the ribosomes and DNA strands identified in other microorganisms. In *R. prowazekii*, in particular, the morphology of the rickettsial cell changes during the course of infection. The older cells are generally smaller, more electron dense. They occasionally fail to divide and become 3–4 μm long. The most striking feature of the older cells is the increase in the number of translucent vacuole-like structures, which in some cases appear to occupy as much as 25% of the cytoplasmic space (Wisseman and Waddell, 1975; Silverman et al., 1980). Cells of *R. rickettsii* are considerably more uniform in appearance (Silverman and Wisseman, 1979).

The outer layers of *R. tsutsugamushi* differ considerably from those of the other rickettsiae. The outer leaflet of the cell wall is considerably thicker than the inner leaflet, while the opposite is true of the other rickettsiae (Fig. 9.1, *F* and *G*) (Silverman and Wisseman, 1978). This property of *R. tsutsugamushi* is associated with much greater tenacity of adherence to host cell membranes, which renders separation from host components much more difficult. On the other hand, typhus and spotted fever rickettsiae are surrounded by substantial slime layers, presumably of carbohydrate nature, generally recognized as halos surrounding the rickettsiae in their intracellular location. They are readily lost during laboratory manipulations. They are best preserved by gently releasing the microorganisms in a medium containing high-titer specific antiserum, prior to fixation for electron microscopy (fig. 9.1*E*) (Silverman et al., 1978). *R. tsutsugamushi* does not appear to have a slime layer of comparable magnitude.

Cultivation. Rickettsiae are most commonly cultivated in the yolk sacs of developing chicken embryos (Cox, 1941). Although some of their differential characteristics are reflected in their growth in eggs (Table 9.3), their biological properties can be studied more conveniently on primary or established cell culture monolayers. Chicken embryo fibroblasts, mouse L cells, and golden hamster BHK-21 cells are employed most frequently, but a great variety of other cells have been used, including monocytes and polymorphonuclear leukocytes. The host cells are often subjected to ionizing radiation prior to infection to prevent multiplication without interfering with short term support of the growth of rickettsiae (Weiss and Dressler, 1958). There is considerable direct (Rikihisa and Ito, 1980) and indirect evidence that the entry of rickettsiae into professional and nonprofessional phagocytes is by phagocytosis or induced phagocytosis. However, to avoid digestion rickettsiae must quickly escape from the phagosome into the cytoplasm (Fig. 9.1*D*). Metabolic activity is required by the rickettsiae to interact with host cell membranes to induce phagocytosis or to escape from the phagosome. The hemolytic activity of the typhus rickettsiae and mouse toxicity demonstrated with most species are believed to reflect comparable interactions with host cell membranes (Weiss, 1973).

In irradiated chicken embryo cells, *R. prowazekii* at 34°C multiplies with a generation time of about 9 h, until a very high rickettsial density is reached (Fig. 9.1*C*). If the inoculum is derived from rickettsiae harvested past their logarithmic growth phase, exponential growth is preceded by a lag phase. The nucleus is not invaded. The rickettsiae are released by the disruption of packed cells (Wisseman and Waddell, 1975). Growth in other cell types is qualitatively comparable.

The infection cycle of *R. rickettsii* differs from that of *R. prowazekii* in two respects. Following a short exponential accumulation in the cytoplasm, the rickettsiae escape into the extracellular spaces and infect other host cells. A high intracytoplasmic cell density is usually not achieved. Occasionally a nucleus is invaded, and in this location multiplication proceeds to high density (Wisseman et al., 1976). These features of the growth of *R. rickettsii* were observed in many cells of mammalian, avian, and arthropod origin. The high rate of traffic from one cell to the other has been associated with high virulence for the chicken embryo and formation of large plaques on cell monolayers (Table 9.3).

R. tsutsugamushi achieves a high intracytoplasmic density, especially in the perinuclear region, and does not invade the nucleus, as is the case in the typhus group rickettsiae. It often bulges outward from the surface of the host cells and is released by a process of extrusion with some host cell cytoplasm attached to it. In the cytoplasm of a host cell it is sometimes seen surrounded by a membrane derived from a previous host cell (Rikihisa and Ito, 1980).

Plaques have been produced on chicken embryo cell monolayers by procedures similar, but somewhat more difficult, than those used in virology, since monolayers must be maintained for 5–17 days (Table 9.3) and the introduction of antibiotics must be carefully avoided (see Wike et al., (1972)).

Nutrition and metabolism. The nutritional requirements of the rickettsiae, as distinct from those of their host cells, are not known, except that rickettsiae will grow in heavily irradiated cells (Weiss and Dressler, 1958) and in the presence of low levels of cycloheximide. Under these conditions rickettsiae incorporate exogenous amino acids and adenine, but not thymidine (Weiss et al., 1972).

Much information has accumulated on the metabolic activities of rickettsiae separated from host constituents. Detailed experiments have been done only with a few species, but there is circumstantial evidence that there are no major differences in metabolic pathways within the genus. Optimum metabolic activity requires a high concentration of K^+ and physiological levels of Mg^{2+}. The activity is stabilized by the presence of a protein, such as bovine plasma albumin (BPA). The chief substrate is glutamate (Bovarnick and Snyder, 1949) which is utilized via a glutamate-oxaloacetate transaminase, glutamate dehydrogenase, and the enzymes of the tricarboxylic acid cycle (see Weiss (1973)). Glutamate metabolism is essential for the maintenance of a high adenylate energy charge (Williams and Weiss, 1978). Glutamine and pyruvate are also utilized, but to a lesser extent (Weiss, 1973). Glucose and glucose-6-phosphate are not utilized at all and rickettsiae do not appear to have any of the enzymes commonly associated with glucose metabolism (Coolbaugh et al., 1976).

As first shown by Bovarnick (see Weiss (1973)), rickettsiae interact with exogenous phosphorylated compounds. They have a carrier-mediated transport system for ADP and ATP (Winkler, 1976) and transport AMP without prior dephosphorylation (Williams, 1980). AMP is neither synthesized nor degraded, and this seems to be true of the other nucleoside monophosphates as well (Williams and Peterson, 1976). Thus, rickettsiae depend for their nutrition on the monophosphate compounds of the host, but they can also utilize their di- and triphosphates. It was also shown by Bovarnick et al. (see Weiss, (1973)) that rickettsiae are capable of in vitro synthesis of very low levels of protein and lipid. For protein synthesis they appear to require a full set of amino acids, but it is not known if every one of the 20 natural amino acids is actually needed. Rickettsiae, unlike chlamydiae, do not form morphologically recognizable vegetative forms. However, the possibility that key enzymes might be induced during their interaction with host cells cannot be excluded. Axenic cultivation does not seem an insurmountable goal, but it has not yet been achieved.

The DNA of the typhus and spotted fever group rickettsiae has a molecular weight of $1.0–1.5 \times 10^9$. The DNA of *R. tsutsugamushi* has not yet been investigated.

Antigenic structure. Studies of the antigenic structure of rickettsiae have played an important role in their classification. The oldest, simplest, but not entirely reliable serological test is the Weil-Felix reaction, based on the reaction of sera from primary rickettsial infections with the somatic antigens of three strains of *Proteus*, OX 19, OX 2, and OX K (Table 9.4). The location of the cross-reacting antigens on the rickettsial cell is not known, but the antigens are most likely surface carbohydrates. They do not elicit an anamnestic antibody response in recrudescent typhus. The Weil-Felix test is no longer recommended if more specific tests can be performed.

Rickettsial complement fixation (CF), agglutination or microagglutination (MA), and erythrocyte-sensitizing substance (ESS) antigens, used in indirect hemagglutination tests, have been prepared with varying degrees of purity and following different procedures of inactivation and extraction. In general, soluble ether-extracted CF antigens and ESS, obtained by heating soluble CF antigens in alkali, are group specific, while washed cells used in CF or MA retain a moderate degree

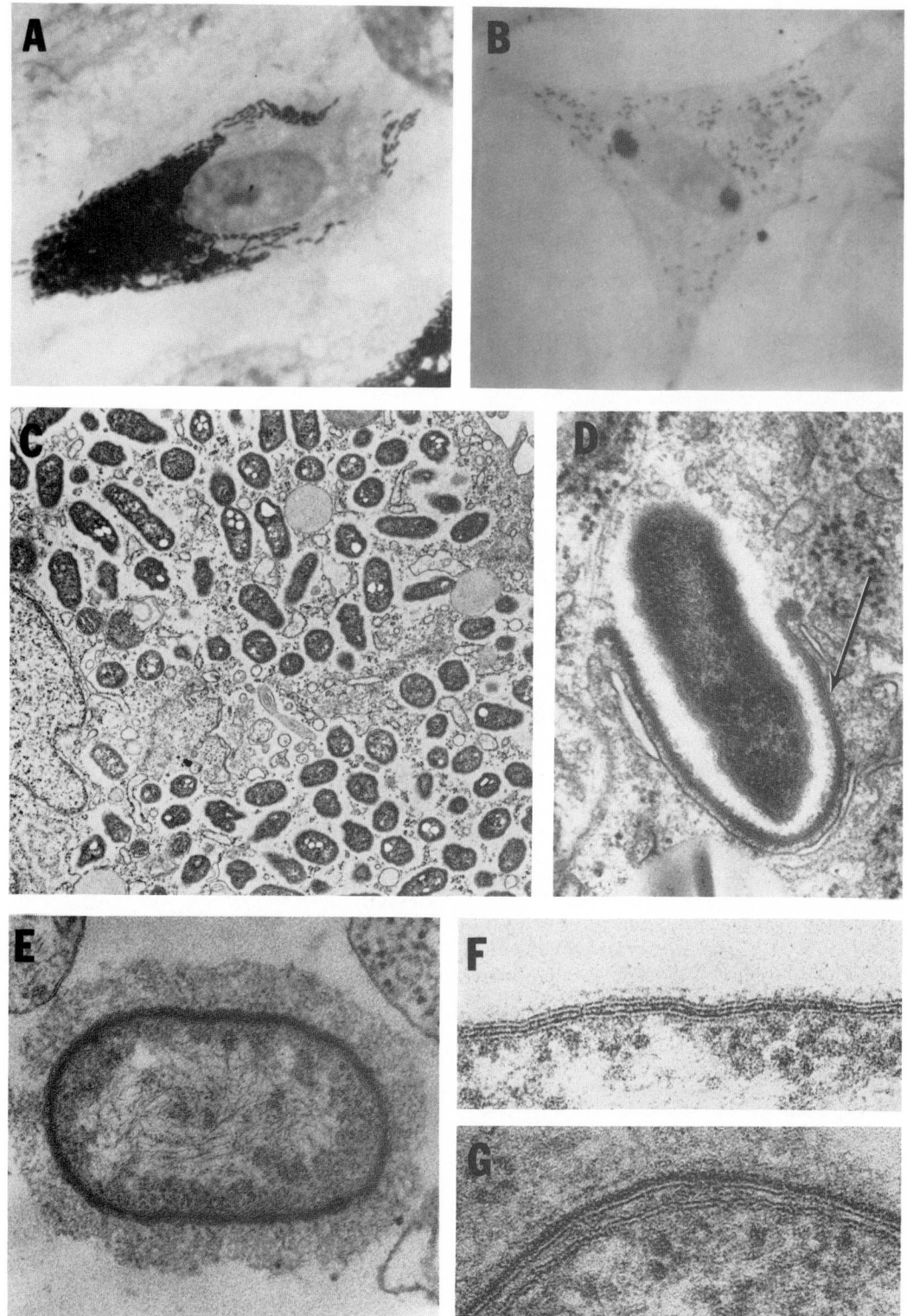

Figure 9.1. Interaction with host cells and fine structure of rickettsiae. (*A* and *B*) Giménez stained preparations. (C–G) Transmission electron micrographs of ultrathin sections stained with uranyl acetate and lead citrate. *A*, Human F-1000 fibroblast infected with the Breinl strain of *Rickettsia prowazekii.* A large mass of cytoplasmic rickettsiae comprises the entire left portion of the cell with some perinuclear organisms evident. No nuclear involvement is apparent (× 1200). *B*, Secondary chicken embryo fibroblast infected with the Sheila Smith strain of *Rickettsia rickettsii.* Note the sparse, diffusely distributed cytoplasmic organisms and the two distinct compact masses of the organisms within the nuclear region (× 1200). *C*, *R. prowazekii*-infected secondary chicken embryo fibroblast late in infection, showing large numbers of free cytoplasmic rickettsiae without vacuolar membrane, most of which contain vacuole-like structures characteristic of organisms in the stationary phase of growth (× 13,600). *D*, *R. prowazekii* (Breinl strain) exiting from a disrupted phagosome (*arrow*) into the cytoplasm of a human macrophage (× 46,000). *E*, *R. prowazekii* (Breinl strain) released from an infected host cell and treated with specific human immune serum to demonstrate the slime layer on the surface of the organism (× 67,000). *F*, Cell envelope of *R. prowazekii* including microcapsular layer. The envelope of *R. rickettsii* is morphologically similar (× 196,000). *G*, Cell envelope of *Rickettsia tsutsugamushi.* Note the thickened outer leaflet of the envelope compared with that of *R. prowazekii* (above) (× 171,000). (From the collection of Charles L. Wisseman, Jr., and David J. Silverman. Illustrated in greater detail in the following references: *A*, Wisseman and Waddell, 1975; *B*, Wisseman et al., 1976; *C*, Silverman et al., 1980; *D*, Meyer and Wisseman, to be published; *E*, Silverman et al., 1978; *F* and *G*, Silverman and Wisseman, 1978.)

Table 9.3.

Differentiation of the major groups and the species of the genus **Rickettsia**

Characteristics	Typhus group			Spotted fever group								Scrub typhus group
	1. *R. prowazekii*	2. *R. typhi*	3. *R. canada*	4. *R. rickettsii*	5. *R. sibirica*	6. *R. conorii*	7. *R. parkeri*	8. *R. australis*	9. *R. akari*	10. *R. montana*	11. *R. rhipicephali*	12. *R. tsutsugamushi*
Geographic distribution:												
Worldwide	+	+	−	−	−	−	−	−	+	−	−	−
Only Western hemisphere	−	−	+	+	−	−	+	−	−	+	+	−
Only Eastern hemisphere	−	−	−	−	+	+	−	+	−	−	−	+
Arthropod host:												
Louse	+	−	−	−	−	−	−	−	−	−	−	−
Flea	−	+	−	−	−	−	−	−	−	−	−	−
Tick	−	−	+	+	+	+	+	+	−	+	+	−
Mite	−	−	−	−	−	−	−	−	+	−	−	+
Intracellular location:												
Cytoplasm	+	+	+	+	+	+	+	+	+	+	+	+
Nucleus	−	−	+	+	+	+	+	+	+	+	+	−
Cultivation in chicken embyros:												
Optimum temperature:												
35°C	+	+	+	−	−	−	−	−	−	−	NA[a]	+
32–34°C	−	−	−	+	+	+	+	+	+	+	NA[a]	−
Peak titer occurs												
Just prior to death	+	+	+	−	−	−	−	−	−	−	NA[a]	+
24–72 h after death	−	−	−	+	+	+	+	+	+	+	NA[a]	−
Plaque formation:												
Days required:												
5–8	−	−	−	+	+	+	+	+	+	+	NA[b]	−
8–10	+	+	+	−	−	−	−	−	−	−	NA[b]	−
11–17	−	−	−	−	−	−	−	−	−	−	NA[b]	+
Size:												
1 mm	+	+	+	−	−	−	−	−	−	−	NA[b]	+
2–3 mm	−	−	−	+	+	+	+	+	+	+	NA[b]	−
Hemolytic activity[c]	+	+	+	−	−	−	−	−	−	−	−	−
Susceptibility to infection:												
Guinea pig[d]	+	++	−	+++	+++	++	++	++	++	−	−	+
Mouse[e]	−	+		−	−	−	−	++	++	−	−	++[f]
Mol% G + C of DNA:												
29–30	+	+	+	−	−	−				−	−	
32–33	−	−	−	+	+	+				+	+	

[a] NA = not applicable to *R. rhipicephali*: growth in chicken embryos is unsatisfactory.

[b] NA = not applicable to *R. rhipicephali*: plaques are small.

[c] For sheep or rabbit erythrocytes.

[d] Symbols: −, no symptoms except with very large doses; +, fever only; ++, fever and scrotal swelling; +++, fever and scrotal necrosis.

[e] Symbols: −, limited infection or virulence except with very large doses; +, infection with small doses; ++, extensive growth, animal of choice for inoculation.

[f] Great variation in virulence occurs among the rickettsial strains; also, great variation in susceptibility occurs among inbred mouse strains.

of specificity (Ormsbee, 1980; Philip et al., 1977). Comparable antigens of *R. tsutsugamushi* have been prepared only with considerable difficulty. With this organism, immunofluorescence (IF) has been the test of choice (Bozeman and Elisberg, 1963). Highly satisfactory for the antigenic analysis of the spotted fever group is the microimmunofluorescence test (micro-IF) done with sera elicited in mice. These animals respond to the species specific antigens more strongly than other animals (Philip et al., 1978). Results obtained by micro-IF are in agreement with the results of other tests used for species differentiation. The enzyme-linked immunosorbent assay (ELISA) has been applied to both the typhus and scrub typhus groups (Halle et al., 1977; Dasch et al., 1979). These tests require highly purified antigens obtainable,

for example, by Renografin density gradient centrifugation (Weiss et al., 1975) and disruption by a French pressure cell. Results with *R. tsutsugamushi* were entirely comparable to those obtained by IF, indicating that the surface antigens detected by IF reflect the composition of the disrupted whole cell preparations.

It has recently been shown (Dasch, 1981) that the species specific antigens of the typhus group are heat-labile surface antigens that are readily extracted from the cells in hypotonic solutions. There is good evidence that these antigens elicit protective immunity (Dasch and Bourgeois, 1981; Bourgeois and Dasch, 1981).

Drug and antibiotic susceptibility. Rickettsiae are unaffected by the sulfonamides; rather, most of them are inhibited by *p*-aminobenzoic

Table 9.4.

Weil-Felix reactions in rickettsial infection[a]

Group	Proteus antigens		
	OX 19	OX 2	OX K
Typhus	++++	+	−
Spotted fever[b]	++++	+	−
Spotted fever[c]	+	++++	
Scrub typhus	−	−	++++

[a] Symbols: −, no reaction; +, slight reaction; ++++, strong reaction.
[b] Rocky mountain spotted fever.
[c] Other tickborne infections.

acid (pAB), which in other microorganisms acts as an antagonist of sulfonamide inhibition. pAB was used therapeutically for a brief period before the advent of the broad-range antibiotics. Its inhibitory effect has been attributed to interference with the utilization of p-hydroxy-benzoic acid (pHB). pAB resistant isolates of *R. prowazekii*, obtained through serial passage in eggs in the presence of increasing concentrations of pAB, display an increased susceptibility to salicylic and acetylsalicylic acids. Inhibition by these compounds is competitively reversed by pAB, while the further addition of pHB reestablishes inhibition (Weiss, 1973).

Penicillin, streptomycin, and erythromycin are of no therapeutic value. However, even moderate levels of penicillin and streptomycin inhibit the growth of rickettsiae in cell culture, except *R. tsutsugamushi*. *R. prowazekii* is highly susceptible to erythromycin in vitro, but strains completely resistant to this antibiotic are quickly established (Weiss and Dressler, 1960).

Most of the broad-range antibiotics are efficacious against rickettsial infection. Chloramphenicol acts as a metabolic inhibitor, preventing the multiplication of *R. tsutsugamushi* but not its survival in cell cultures (Hopps et al., 1959). The tetracycline class of compounds has some in vitro rickettsiacidal effect, in addition to metabolic inhibition, and is more effective than chloramphenicol for therapy: a course of several days of treatment with tetracycline is generally prescribed. A single dose of doxycycline is also effective and is recommended for patients who are not under strict surveillance (Brown et al., 1978). Antibiotics are generally believed to clear infection only in concert with the development of the immune response.

Pathogenicity. The genus *Rickettsia* is notorious for its virulence for man. Before the advent of the broad-range antibiotics, three diseases—epidemic typhus, Rocky Mountain spotted fever and scrub typhus, caused by *R. prowazekii*, *R. rickettsii* and *R. tsutsugamushi*, respectively—had a very high case fatality rate. Epidemic typhus has on many occasions changed the course of history (Zinsser, 1935). Rocky Mountain spotted fever and epidemic typhus have exacted a heavy toll on the early investigators of their etiologies. Scrub typhus interfered with military operations in Southeastern Asia during World War II. Other rickettsiae, such as *R. typhi* and several species of the spotted fever group, although not nearly as lethal, have been and some still are responsible for significant human disease. The evidence that *R. canada* is associated with disease of man is circumstantial, and some of the spotted fever rickettsiae are probably not pathogenic for man.

Although each disease has some unique feature, a basic mechanism of pathogenesis appears to apply to all pathogenic rickettsiae. A toxin separable from the rickettsial cell has not been isolated. However, viable rickettsiae of several species display toxin-like action. This is best demonstrated by injecting at least 10^6 rickettsiae intravenously into mice. Depending on the concentration and species, the animals die within 1–8 h. The effect is described as damage to the endothelial cells accompanied by increased capillary permeability, flow of plasma into the tissues, hemoconcentration, and eventual collapse of the blood circulation. The factors required for the retention of toxin-like activity are the same as those that enhance hemolytic activity (by the typhus rickettsiae), penetration into cells, and metabolic activity (see Weiss, 1960). A comparable phenomenon is the immediate cytotoxicity for L

cells demonstrable with high multiplicities of rickettsiae (Winkler and Miller, 1981). Thus, the toxin-like activity of rickettsiae appears to reflect the mechanism by which rickettsiae gain access into the cytoplasm of host cells.

Although toxic deaths produced by massive doses of rickettsiae, prior to extensive multiplication, have not been reproduced in other experimental animals, the disease in man and animals is characterized by the same kind of vascular damage. The lesions occur in small blood vessels and result in petechial hemorrhages in the skin (thus the designations exanthematic typhus, Fleckfieber, spotted fever), in the brain, in the lung, and in other organs. The role of lipopolysaccharides in rickettsial pathogenesis is not known.

Ecology. Survival of rickettsiae is dependent on an ecological cycle that involves a vertebrate and an arthropod host. Depending on the species, either one or the other type of host is the primary one and the importance of the other host varies considerably. In some cases the vertebrate host does little more than furnish a blood meal to the arthropod. With one notable exception, man is an incidental host playing no role in the natural survival of the rickettsiae.

On one end of the spectrum is *R. prowazekii*, which has man as its primary host. Individuals who recover from the epidemic typhus often retain small numbers of organisms, presumably in their lymph nodes, which may occasionally give rise to a mild form of typhus, called Brill-Zinsser disease. During primary typhus infection, as well as during Brill-Zinsser disease, patients have sufficiently high rickettsiemia to infect human body lice, if they happen to be present. Lice play a key role as vectors but succumb to the infection and do not contribute directly to the interepidemic survival of the rickettsia. The effects of modern antibiotic therapy and widespread use of insecticides on the ecology of *R. prowazekii* have not yet been evaluated. It is generally believed, however, that this rickettsia could ravage humankind once again in the case of a major world catastrophe. It has recently been shown that *R. prowazekii* has a natural cycle in the United States that involves the southern flying squirrel and its ectoparasites (Bozeman et al., 1975; 1981) and has occasionally resulted in human infection (McDade et al., 1980, Duma et al., 1981).

R. typhi is another example of a rickettsia that has a vertebrate host, the urban rat, as its primary host. The rat louse and flea are the chief transmitters of the rickettsia from rat to rat and to other vertebrates including man. The flea survives rickettsial infection, possibly because, in contrast to the louse, it is able to renew its gut epithelial layer (Ito et al., 1975), which is denuded by the discharge of heavily infected cells into the feces.

A tick is the primary host of most of the species of the spotted fever group. Since transovarian passage has been demonstrated on numerous occasions, the role of the vertebrate host in the survival of the rickettsia has been regarded as secondary. This view was strengthened by recent evidence that casts some doubt on the role of the tick's host, the cottontail rabbit, as an important vertebrate reservoir host of the rickettsia, responsible for the geographic distribution of spotted fever (Burgdorfer et al., 1980). The infection of ticks with nonpathogenic rickettsiae, such as *R. montana*, appears to be the principal factor limiting infection with *R. rickettsii* (Burgdorfer et al., 1981). On the other hand, it was shown by Burgdorfer and Brinton (1975) that continuous transovarian passage of *R. rickettsii* in the tick adversely affects it. Thus, the number of infected ticks that transmit rickettsiae transovarially can be expected to steadily decline unless there are compensating factors. It can be surmised, therefore, that an ecological equilibrium is maintained by a slow rate of infection of ticks from vertebrate hosts.

At the other end of the spectrum is *R. tsutsugamushi* which is transmitted by mites, parasites of rural rats. Since mites feed on their host only once, the role of the rodents in the maintenance of the natural cycle of rickettsial infection appears even more remote than in the case of the spotted fever rickettsiae. Although transovarian passage has been repeatedly seen in naturally infected mites, it has not been demonstrated in mites infected in the laboratory by feeding on infected rats. Thus, it is believed that aquisition of rickettsiae by mites from vertebrates is a relatively infrequent occurrence, although, possibly, ecologically important (Traub et al., 1975).

Isolation Procedures

Isolation of rickettsiae from survey specimens or from clinical material requires strict adherence to safety precautions. A controlled access facility and an approved safety cabinet in which negative air pressure is maintained are recommended.

The hemolymph test (Burgdorfer, 1970) provides a rapid and simple technique for detecting rickettsiae in some of the larger engorged ticks. A small drop of hemolymph is obtained by amputating the distal portion of one or more legs and touching the wound to a slide. The slide is then stained by the Giménez method or fluorescent antibody. The tick generally survives the amputation and can be saved for isolation attempts. Unengorged ticks should be incubated for 24–48 h at 37°C to allow multiplication of rickettsiae within hemocytes. Some of the smaller ticks and mites must be dissected and smears prepared from their organs.

Rickettsiae are most commonly isolated in laboratory animals. The guinea pig is the animal of choice for rickettsiae of the typhus and spotted fever group. Burgdorfer et al. (1974) found the meadow vole (*Microtus pennsylvanicus*) particularly susceptible and valuable for the isolation of spotted fever rickettsiae. The mouse is the animal of choice for *R. australis*, *R. akari* and *R. tsutsugamushi*, although a number of *R. tsutsugamushi* strains were isolated in guinea pigs. Rickettsial species vary considerably in virulence for the guinea pig and mouse (Table 9.3). Spleens and other organs are collected from guinea pigs during the 2nd and 3rd day of fever or from moribund mice and are passed into eggs. When a species of low virulence is expected, such as *R. montana*, the spleen is passed on the 10th–12th day after injection, even in the absence of fever.

Chicken embryos are generally used for the production of seed of rickettsiae isolated in laboratory animals. They can be used for primary isolation if the inoculum is free from adventitious agents (blood or ticks whose exterior surfaces have been decontaminated by immersion in Merthiolate and by repeated washings). Chicken embryos must be obtained from flocks maintained on a rigorous antibiotic-free diet. Yolk sac inoculation during the 5th–7th d of embryo development is the only satisfactory route for rickettsiae (Cox, 1941). The optimal conditions for cultivation and harvest vary with the group (Table 9.3).

With the possible exception of *R. tsutsugamushi*, yields of rickettsiae are seldom as bountiful from cell cultures as from eggs. Provided the level of contamination in the original inoculum is low and the contaminants are slow-growing, cell cultures offer the advantage of plaque isolation and quantification of viable rickettsiae. Primary monolayers from 10-d-old-chicken embryos are used most frequently. For rickettsiae that require long periods of incubation (Table 9.3) minimal exposure of cells to trypsin and introduction in the medium of homologous serum (from chickens not fed antibiotics) is recommended (Woodman et al., 1977).

For further detail see Ormsbee (1980) and Weiss (1981).

Maintenance Procedures

Many seeds of rickettsiae have been maintained for decades at −70°C as crude yolk sac suspensions. When diluted or separated from host constituents, viability declines at rates that depend on the degree of purification and on the choice of suspending fluid. The diluent most frequently used is SPG, which consists of 0.22 M sucrose, 0.01 M potassium phosphate, pH 7.0, and 0.005 M potassium glutamate (Bovarnick et al., 1950). Stability is further enhanced by the addition of Mg^{2+} and a protein, such as brain-heart infusion broth (BHI), BPA or Renografin. Storage in liquid nitrogen is recommended for specimens of unusual value. *R. tsutsugamushi* is the most unstable of the rickettsial species.

Rickettsiae lose viability when they are repeatedly frozen and thawed, maintained at +4°C for a few days, or kept at room temperature for several hours. However, they may be quite stable in desiccated arthropods or louse feces and have, on occasion, been a source of unexpected infection (see Horsfall and Tamm (1965)).

Differentiation of the genus **Rickettsia** from other genera

Characteristics useful for the differentiation of the genus *Rickettsia* from the other genera of the family *Rickettsiaceae* are listed in Table 9.2.

Taxonomic Comments

Historically, the designation "rickettsia" has been used for species of this genus and also, indiscriminately, for other small rods that were not cultivated and were not otherwise identified. Most frequently, but not always, these organisms were seen in association with arthropods. The last two editions of the *Manual* reflect the considerable progress that has been made in establishing a more precise definition of the genus *Rickettsia* and in eliminating species that do not fit the definition. This process continues with this edition. *R. sennetsu* was removed from consideration as a *species incertae sedis* in this section and has been reclassified in the genus *Ehrlichia*.

The term "rickettsia" is still widely used for the three genera *Rickettsia*, *Rochalimaea* and *Coxiella*. Although distinguished by fundamental phenotypic differences (Table 9.2), *Rickettsia* and *Rochalimaea* are genotypically related, as indicated by 25–33% hybridization between the DNAs of *Rochalimaea quintana* and *Rickettsia prowazekii* or *R. typhi* (Myers and Wisseman, 1980). An evolutionary relationship between the two genera can be explained on the basis of a common niche in man and its ectoparasites. There is no genotypic relationship and only superficial phenotypic similarity between *Rickettsia* and *Coxiella*. The inclusion of these two genera in the same taxon is a matter of history and convenience. Many of the methods of laboratory investigation and some facets of epidemiological survey of the two genera have been the same. *Coxiella* is a monospecific genus for which the next of kin have not yet been found.

The subdivision of the genus *Rickettsia* into three groups is illustrated in Table 9.3. Within the typhus group, the rather tenuous differentiation between *R. prowazekii* and *R. typhi* has been strengthened by the demonstration of differences in DNA base sequences (Myers and Wisseman, 1980) and protein migration patterns (Dasch et al., 1978). *R. canada* remains a species which shares properties with the typhus and spotted fever rickettsiae and could be placed in either group (Myers and Wisseman, 1981). It is retained in the typhus group. Among the spotted fever rickettsiae, the biological significance of *R. montana* has been further recognized (Feng et al., 1980). For some of the other isolates two new names have been proposed: "*R. slovaca*" (Úrvölgyi and Brezina, 1978) and *R. rhipicephali* (Burgdorfer et al., 1978). Other strains have been isolated that are not serologically identical to named species. Further investigation is required to determine whether "*R. slovaca*" and the unclassified strains are serological variants of named species or deserve to be classified as separate species. Some of the same uncertainty has not been entirely removed with regard to other species of the spotted fever group. Investigations on the biological properties of the third group—scrub typhus—consisting of the single species *R. tsutsugamushi*, have not proceeded as rapidly as those conducted with the other two groups. Therefore, changes in classification, possibly desirable in the future, are not recommended at this time.

Acknowledgments

We are indebted to D. J. Silverman and C. L. Wisseman, Jr., Department of Microbiology, School of Medicine, University of Maryland, Baltimore, for providing the illustrations.

Further Reading

Horsfall, F.L., Jr. and I. Tamm (Editors). 1965. *Viral and Rickettsial Infections of Man*, 4th Ed., J. B. Lippincott, Philadelphia, pp. 1059–1143.

Ormsbee, R.A. 1969. Rickettsiae (as organisms). Annu. Rev. Microbiol. *23:* 275–292.

Weiss, E. 1973. Growth and physiology of rickettsiae. Bacteriol. Rev. *27:* 259–283.

Weiss, E. 1981. The *Rickettsiaceae*: the human pathogens. In Starr, Stolp, Trüper, Balows and Schlegel (Editors), *The Prokaryotes: A Handbook on Habitats,* *Isolation and Identification of Bacteria.* Springer-Verlag, New York, pp. 2137–2160.

Weiss, E. 1982. The biology of the rickettsiae. Annu. Rev. Microbiol. *36:* 345–370.

Zdrodovskii, P.F. and H. M. Golinevich. 1960. *The Rickettsial Diseases* (English translation). Pergamon Press, New York.

Differentiation of the major groups and the species of the genus **Rickettsia**

The differential characteristics of the three groups of *Rickettsia* and some of the species characteristics are presented in Table 9.3. Although most species display some obvious differential characteristics, final identification requires serological tests with species-specific antisera.

List of the species of the genus **Rickettsia**

1. **Rickettsia prowazekii** da Rocha-Lima 1916, 567.[AL] *Nom. cons.* Opin. 19, Jud. Comm. 1958, 158.

pro.wa.ze′ki.i. M.L. gen. n. *prowazekii* of Prowazek; named after Stanislav von Prowazek, an early investigator of the etiology of typhus who died of typhus contracted in the course of his studies.

Most widely studied species of the genus. The description of the genus and, in particular, of the typhus group within the genus is based to a large extent on studies of this species. Although cells are generally small, variation in size and morphology are most pronounced in this species. Unusually long cells (4 μm, single or in short chains) and cells with prominent vacuoles appear with moderate frequency in stationary cultures.

Highly infectious for chicken embryos, which die within 4–13 days after inoculation, depending on the size of the inoculum. Optimum yields of rickettsiae are obtained by inoculating the embryos with small inocula (~10 viable rickettsiae per egg) and harvesting the yolk sacs 10–14 days later, just before the embryos die.

Highly infectious also for monolayers of chick embryo fibroblasts, mouse L cells, and other cells. Plaques are usually small (0.5–1.5 mm in diameter) and turbid (see Table 9.3). The virulent Breinl strain, but not the avirulent E strain, multiplies to a moderate extent in human macrophage cultures (Gambrill and Wisseman, 1973).

R. prowazekii is the etiologic agent of epidemic typhus fever acquired through contact with lice. It is also the etiologic agent of recrudescent typhus (called Brill-Zinsser disease), and occasionally of sporadic typhus in individuals who have been in contact with flying squirrels (McDade et al., 1980; Duma et al., 1981). The most effective arthropod vector of epidemic typhus is the human body louse, *Pediculus humanus,* possibly because it takes frequent large blood meals and because it tends to desert febrile hosts to seek new ones. The head louse is equally susceptible to infection but has not been implicated in typhus fever transmission, possibly because it imbibes only very small amounts of blood (Murray and Torrey, 1975). The rickettsiae grow profusely in the cells of the gut epithelium of the louse, even when the ingested human blood contains high levels of antibodies (Wisseman et al., 1975). Heavily infected epithelial cells are released into the lumen, are discharged with the feces of the louse, and are the source of human infection when the louse feces are driven into the skin by scratching. Lice invariably succumb to infection, generally within 1–2 weeks, rarely surviving longer than 3 weeks. When louse infestation is very heavy or when large volumes of rickettsial suspensions are processed in the laboratory, infection by aerosols may occur.

The guinea pig is highly susceptible to infection, but develops a mild disease, usually manifested only by fever lasting approximately a week. The cotton rat, *Sigmodon hispidus,* is also highly susceptible to infection, but signs of disease and death are produced only by doses in excess of 3 × 10⁵ viable cells. Inapparent infection elicits solid immunity to the homologous species and to *R. typhi.* True infection is not usually demonstrated in the mouse, but acute toxic death is produced by the intravenous injection of at least 10⁶ viable cells.

Although the existence of an extrahuman reservoir of *R. prowazekii* has been suspected for the past 25 years (Reiss-Gutfreund, 1956, 1966), it was clearly demonstrated only in 1975 (Bozeman et al., 1975) in southern flying squirrels (*Glaucomys volans*) captured in Florida and Virginia. It is transmitted by the louse of the flying squirrel (*Neohaematopinus sciuropteri*). The squirrel flea (*Orchopeas howardi*) and fleas of other mammals are highly susceptible and might be the vectors involved in human infection (Bozeman et al., 1981).

Of the laboratory-induced variants of *R. prowazekii,* the most notable is strain E, isolated from a typhus patient in Madrid in 1941 and passed in quick succession in eggs 255 times. This strain has limited virulence for the guinea pig and low virulence for man. It has been used as a living vaccine (Fox, 1956). Other laboratory-induced variants have been obtained from strain E. They include strains unaffected by erythromycin and strains of increased resistance to *p*AB and chloramphenicol (Weiss, 1960; Weiss and Dressler, 1962).

The mol% G + C of the DNA is 28.5–29.7 with a mean of 29.2 (T_m), or 29.7–30.3 with a mean of 30.0 (Bd) (Tyeryar et al., 1973; Myers and Wisseman, 1980). The genome size of the strains examined is 106–114 × 10⁷, with a mean of 110 × 10⁷. Strains isolated from various geographic locations are remarkably similar to each other in biological properties. Differences in % DNA/DNA hybridization among strains isolated in Poland, Spain and Burundi from human sources, and from the United States from flying squirrels, were negligible (range = 92–100; mean = 97) (Myers and Wisseman, 1980). Minor difference in protein migration patterns separating strains derived from Eastern Europe, Spain or Africa, or from American flying squirrels, have been shown by isoelectric focusing in polyacrylamide gels. Another minor difference between the attenuated E strain and the virulent strains in protein migration patterns has been noted (Dasch et al., 1978).

Type strain: ATCC VR-142 (Breinl strain) (Maxcy, 1929).

2. **Rickettsia typhi** (Wolbach and Todd 1920) Philip 1943, 304[AL] (*Dermacentroxenus typhi* Wolbach and Todd 1920, 158). (*Rickettsia mooseri* Monteiro 1931, 97.)

ty′phi. Gr. n. *typhus* cloud, hence stupor arising from fever; M.L. n. *typhus* fever, typhus; M.L. gen. n. *typhi* of typhus.

Although the name *R. typhi* is the only valid name for the species, appearing on the Approved Lists of Bacterial Names (Skerman et al., 1980), some investigators prefer to call this species *R. mooseri* in honor of Herman Mooser, who clearly differentiated the species from *R. prowazekii* on the basis of its virulence for the guinea pig (Mooser, 1928). *R. typhi* is more virulent than *R. prowazekii* for guinea pigs and also for mice (Table 9.3) and, unlike *R. prowazekii,* can be passed indefinitely in the rat and may persist for months in rat brain.

Rats (*Rattus norvegicus* and *R. rattus,* especially those located in urban areas) and other rodents are the primary reservoirs. *R. typhi* has a worldwide distribution and has been reported from all countries where investigators have competently searched for it. The rat louse, *Polyplax spinulosus,* and the rat flea, *Xenopsylla cheopis,* are the chief transmitters of the rickettsia from rat to rat. The human flea, *Pulex irritans,* and the human body louse are highly susceptible to infection and may play roles in transmission in populations with high ectoparasitic infestation. The human disease, murine or endemic typhus, also known by names that reflect the geographic location, is a mild form of typhus. "Tabardillo" in Mexico possibly is murine typhus transmitted by the louse. It is reported with variable frequency (in the United States, 5400 cases in 1944, less than 100 in 1958) and, undoubtedly, because of the availability of antibiotics, many cases are not properly recognized (Traub et al., 1978).

Some of the antigens of *R. typhi*, demonstrable in serologial tests, are very similar if not identical to those of *R. prowazekii*, while others are species specific. The common antigens can be obtained by various methods, usually by treating the cells with ether and extracting the soluble aqueous fraction. The specific antigens, reacting weakly with heterologous antisera, are obtained from the insoluble fraction of viable, formalin- or ether-treated cells, washed repeatedly. It was recently shown, however, that the major portion of the specific antigen (10–15% of the total cellular protein) is released as a soluble fraction, when the cells are suspended in hypotonic solution lacking Mg^{2+} and are incubated at 45°C for 20 min (Dasch, 1981). The two species can also be distinguished immunologically by cross-challenge of vaccinated guinea pigs or by mouse toxin neutralization tests.

There have been no extensive attempts to isolate mutant strains of *R. typhi* in the laboratory.

Differences between *R. typhi* and *R. prowazekii* in cell morphology and in the mechanism of interaction with eucaryotic cells undoubtedly exist, but objective criteria of differentiation based on such characteristics have not been developed. Neither have differences in metabolic activities been described, and the two species have been used interchangeably for metabolic investigations.

The mol% G + C of the DNA of *R. typhi*, and also the genome size, are identical to those of *R. prowazekii*. Although no differences in the degree of DNA/DNA hybridization have been found between strains of *R. typhi*, the degree of hybridization between *R. typhi* and *R. prowazekii* is 70–79% (Myers and Wisseman, 1980). Similarly, no differences in electrophoretic protein migration patterns have been noted between strains of *R. typhi*, but consistent differences in the migration patterns of malate dehydrogenase and of several unidentified proteins have been found between *R. typhi* and *R. prowazekii* (Dasch and Weiss, 1977; Dasch et al., 1978).

Type strain: ATCC VR-144 (Wilmington strain) (Maxcy, 1929).

3. **Rickettsia canada** McKiel, Bell and Lackman 1967, 509.[AL]
ca′na.da. M.L. n. *canada* Canada, the country where the organism was first isolated.

There is still tenuous justification for maintaining this species in the typhus group, although two important characteristics militate against such a classification: an ecological niche in the tick and growth in the nucleus. Furthermore, *R. canada* can be grown in laboratory-reared ticks (Burgdorfer and Brinton, 1970), but grows sparingly in the louse (Weyer and Reiss-Gutfreund, 1973). However, unlike some of the species of the spotted fever group, *R. canada* is not highly cytopathic and is cultivated in chicken embryos by the procedure used for the typhus group (Table 9.3). Plaques on monolayers are small as are those of the typhus group (Woodman et al., 1977a). Virulence for the guinea pig and mouse is quite low: at least 10^6 viable cells are required to induce fever in the guinea pig and 10^5 viable cells for serological conversion in mice (Ormsbee et al., 1978). There is serological evidence that human infection has occurred (Bozeman et al., 1970).

Hemolytic activity measured with sheep erythrocytes and the rate of CO_2 production from glutamate are approximately the same as those obtained with *R. prowazekii* and *R. typhi* (Woodman et al., 1977a).

R. canada reacts strongly with guinea pig and rabbit sera prepared against the typhus group rickettsiae, but weakly with sera against spotted fever group rickettsiae (McKiel et al., 1967). Specific antigen can readily be demonstrated with mouse antisera (Philip et al., 1978).

Two strains were isolated from pools of engorged rabbit ticks (*Haemaphysalis leporispalustris*) collected near Richmond, Ontario, from an indicator rabbit and from a wild snowshoe hare (McKiel et al., 1967). Only one of these two strains is still available. Another strain was isolated recently in California from a *H. leporispalustris* tick feeding on a California jackrabbit (*Lepus californicus*) (Lane et al., 1981).

Although the mol% G + C of the DNA of *R. canada* is identical to the values for *R. prowazekii* and *R. typhi*, the genome size is distinctly higher (149×10^7 daltons). Thus, the extent of DNA/DNA hybridiza-

tions between *R. canada* and the other two species of the typhus group varies somewhat with the size of the genome (38% and 46% hybridization with *R. prowazekii* and *R. typhi*, respectively). A similar degree of DNA/DNA hybridization occurs between *R. canada* and *R. rickettsii* as between *R. canada* and *R. typhi* (Myers and Wisseman, 1981). When compared to *R. prowazekii* and *R. typhi* on the basis of migration patterns in polyacrylamide gel electrophoresis of solubilized proteins, *R. canada* displays a number of differences, although a basic similarity can still be recognized (Dasch et al., 1978).

Type strain: ATCC VR-610 (strain 2678; McKiel et al., 1967).

4. **Rickettsia rickettsii** (Wolbach 1919) Brumpt 1922, 757.[AL] (*Dermacentroxenus rickettsii* Wolbach 1919, 87.)
rick.ett′si.i. M.L. gen. *rickettsii* of Ricketts; named after H. T. Ricketts for his classic studies of the etiology of Rocky Mountain spotted fever.

Most widely studied species of the spotted fever group. The cells are slightly smaller and more uniform in size than those of *R. prowazekii*. Because of the high virulence of *R. rickettsii* for chicken embryos, the embryos die before extensive growth has taken place and the procedure for optimal harvests from yolk sacs differs considerably from the one employed for the typhus group (Table 9.3). It was defined as follows (Stoenner et al., 1962). Embryos, 4.5 days old, are inoculated with a sufficient number of viable cells to kill most embryos within 4.5 days. Eggs are incubated at 33.5°C (lowest temperature compatible with survival of most embryos) and maintained in incubator at 32°C for 2 days after death of the embryos. Even under best conditions yields of *R. rickettsii* cells from yolk sac are smaller than those of *R. prowazekii* or *R. typhi* but are somewhat higher than yields obtainable from cell cultures. Multiplication occurs primarily in the cytoplasm, but intranuclear growth is sufficiently prominent to have stimulated early investigators to use it as a criterion for the classification of the spotted fever group. Because of higher cytotoxicity, plaques on monolayers occur earlier and are larger than in the case of the typhus group, but sheep or rabbit erythrocytes are not hemolyzed (Table 9.3).

Strains of this species vary considerably in virulence for the guinea pig and for man. Virulence for these two hosts appears to vary independently.

The guinea pig is highly susceptible to infection. The more virulent strains induce fever, scrotal necrosis, and infection is often fatal. In man, *R. rickettsii* is the etiological agent of Rocky Mountain spotted fever, the most severe disease of the spotted fever group. The disease is characterized by high fever and damage to the small blood vessels, resulting in skin rash, intravascular thrombosis with consequent lesions in various organs and necrosis of the extremities. Before the era of broad-range antibiotics and the dramatic change in the geographic distribution of the disease, the case fatality rates were as high as 90% in the Bitterroot Valley of Montana and as low as 5% in the Snake River Valley of Idaho and 25% in Long Island. Since the disease responds well to tetracyclines, present mortality rates are relatively low, provided the disease is recognized and treated promptly. Dogs are susceptible to natural infection and in some cases develop clinical signs (Lissman and Benach, 1980). The mouse is quite resistant to infection and doses in excess of 10^6 viable cells are required to produce a significant antibody response.

Although the metabolism is not different from that of other rickettsiae, biochemical investigations have focused on loss and restoration of viability, and it was shown that metabolic activity is best maintained under microaerophilic conditions or in the presence of reduced glutathione or protein. Reversible changes in rickettsial activity occur in the tick: prolonged refrigeration reduces the virulence of the rickettsiae for the guinea pig, but virulence is restored when the ticks have a blood meal or are incubated at 37°C for 24 h.

R. rickettsii has group- and species-specific antigens, demonstrable by methods comparable to those used for the typhus group. Species specific antigens can best be demonstrated using mouse antisera in micro-IF tests (Philip et al., 1978).

In a series of brilliant experiments conducted in 1906 and 1907,

Ricketts (1911) clearly established the basic features of the ecology of the agent of Rocky Mountain spotted fever in the tick. *R. rickettsii* is confined to the Western Hemisphere (Table 9.3), but at present is encountered much more frequently in the Eastern United States, and in particular in the Piedmont plateau of North Carolina and Virginia, than in the Rocky Mountains. In the western United States the most common human vector is the wood tick, *Dermacentor andersoni*. In the eastern United States, the dog tick, *Dermacentor variabilis*, is the chief vector. Numerous other tick species were found naturally infected in the United States, but of these only the lone star tick, *Amblyomma americanum*, has been implicated in human infection (Burgdorfer, 1975). The brown dog tick, *Rhipicephalus sanguineus*, and *Amblyomma cajannense* are among the ticks that have been most commonly implicated in human infection in Mexico and South America.

Although there is considerable serological evidence that *R. rickettsii* is widespread among wild vertebrates, demonstration by recovery of the microorganisms has been difficult, because sufficient numbers of rickettsiae are present in the vertebrate host only during brief periods. Natural infection among vertebrates was first demonstrated in the 1930s in Brazil in the house and wild dog (*Canis brasiliensis*), the opossum (*Didelphis marsupialis*), the wild rabbit (*Sylvilagus minensis*), and the Brazilian cavy (*Cavia aperea*) (Moreira and de Magalhaes, 1937). The first recovery in the United States was reported in 1954 (Gould and Miesse) from a meadow vole (*Microtus pennsylvanicus*) trapped near Alexandria, Virginia. Subsequent recoveries in the eastern United States included the meadow vole, opossum, cotton rat, cottontail rabbit (*S. floridanus*), whitefooted mouse (*Peromyscus* sp.) and pine vole (*Pitymis pinetorum*) (Bozeman et al., 1967). In the western United States, *R. rickettsii* was isolated from the chipmunk (*Eutamias amoenus*), snowshoe hare (*Lepus americanus*) and from the golden mantled ground squirrel (*Citellus lateralis tescorum*) (Burgdorfer et al., 1962). Serological evidence of infection of the domestic dog was found more frequently in areas where Rocky Mountain spotted fever is endemic, than in nonendemic areas.

Transmission to man occurs through the bite of an infected tick that remains attached to the skin for a number of hours. Although man is only an incidental host, his infection reflects a changing ecology. From 1910 to 1930 most of the cases, about 100–600 per year, occurred within the area of distribution of *D. andersoni* in the Rocky Mountain region. Since 1930, there has been a shift toward the eastern and southeastern parts of the United States and to the transmission by *D. variabilis*. From 1948 to 1959 the number of cases decreased from an annual rate of 500 to 200, but there has been an increase since 1960 to an annual rate greater than 1000 cases in 1977, but the number has remained approximately constant since then. A strain indistinguishable from *R. rickettsii* has recently been isolated from patients in Costa Rica (Fuentes, 1979).

The mol% G + C of the DNA based on three strains is 32.0–33.2 with a mean of 32.6 (T_m, Bd) (Tyeryar et al., 1973; Myers and Wisseman, 1981). The genome size based on one strain is 130×10^7 daltons. Preliminary DNA/DNA hybridization studies indicate 25–50% relatedness to *R. prowazekii* (highest %), *R. canada* and *R. typhi* (lowest %) (Myers and Wisseman, 1981). Limited studies of the migration patterns of solubilized proteins by polyacrylamide gel electrophoresis has shown no differences between strains of *R. rickettsii* (Anacker et al., 1980); however, differences as well as similarities have been demonstrated in comparisons between species within the spotted fever group (Obijeski et al., 1974; Pedersen and Walters, 1978).

Type strain: ATCC VR-149 [strain Sheila Smith, isolated from a patient in Montana (Bell and Pickens, 1953)]. Reference strain: strain R, isolated from *Dermacentor andersoni*.

5. Rickettsia sibirica Zdrodovskii 1949, 20.[AL]

si.bi'ri.ca. M.L. fem. adj. *sibirica* pertaining to Siberia.

Closely resembles *R. rickettsii*, although most strains are less virulent for animals and man. Cinematographic observations by Kokorin et al. (1978) revealed intense mobility of the rickettsiae within their host cells, and some movement from cell to cell, but limited cytotoxicity.

The species is cultivated by procedures identical to those used for *R. rickettsii*.

Virulence for the guinea pig is quite variable and in some cases only fever of short duration is produced. An occasional rickettsial strain kills mice and hamsters (Bázliková and Brezina, 1978) and inbred strains of mice vary in susceptibility to this rickettsia (Kekcheeva et al., 1978). The disease in man resembles moderately severe or mild Rocky Mountain spotted fever. It is called Siberian or North Asian tick typhus.

This species is differentiated from the other strains of the spotted fever group by various serological tests: the micro-IF test with sera prepared in mice is the simplest and is highly specific (Philip et al., 1978). Toxin neutralization tests in mice demonstrate some cross reactions with *R. rickettsii* (Lackman et al., 1965).

Its habitat consists of foci extending from the Pacific maritime regions of the U.S.S.R. through a wide area of southern and northern Siberia to the Armenian Republic. The foci are usually associated with steppe landscapes with low rainfall close to foothills and mountain ranges, and they may extend to the dry slopes of the mountains. Nine tick species have been found naturally infected, and transovarian passage was demonstrated in *Dermacentor nuttalli*, *D. marginatus*, *D. pictus*, *D. silvarum*, *Haemophysalis punctata*, *H. concinna* and *Rhipicephalus sanguineus*. These ticks feed on numerous small wild rodents and domestic animals. At least 18 kinds of mammals have been found infected, among them, Siberian squirrels or susliks (genus *Citellus*), chipmunks (*Eutamias*), hamsters (*Cricetus*), lemmings (*Lagurus*), hares (*Lepus*), domestic and field mice, and voles (Hoogstraal, 1967). Attempts to isolate the rickettsia from domestic animals have been unsuccessful. Man is infected through the bite of a tick.

Strains identical to *R. sibirica* have been isolated from various mammals and *D. marginatus* ticks in Czechoslovakia (Yablonskaya, 1978). Other *D. marginatus* isolates are serologically distinguishable and were named "*R. slovaca*" (Úrvölgyi and Brezina, 1978). Whether these strains are serological variants of *R. sibirica* or are sufficiently different to warrant the establishment of a new species is not known at this time.

The mol% G + C of the DNA is 32.5 (T_m) (Schramek, 1974). DNA/DNA hybridization experiments have not been performed. The electrophoretic migration patterns in polyacrylamide gel of the solubilized proteins are similar but not identical to those of *R. rickettsii* (Pedersen and Walters, 1978).

Type strain: ATCC VR-151 [strain 246, isolated from *Dermacentor nuttalli* in the U.S.S.R. about 1949 (Bell and Stoenner, 1960)].

6. Rickettsia conorii Brumpt 1932, 1199.[AL]

co.no'ri.i. M.L. gen. n. *conorii* of Conor; named after A. Conor who, in collaboration with A. Bruch, provided the first description of fièvre boutonneuse.

Resembles *R. rickettsii*, but is antigenically distinct and less virulent for animals and man. Growth in cell cultures has the basic features of other rickettsiae of the spotted fever group, including early release from host cells (Oaks and Osterman, 1979), and it is cultivated by procedures identical to those used for *R. rickettsii*.

This species can be differentiatied from related species of the spotted fever group by appropriate serological and cross-immunity tests, but *R. conorii* strains isolated in various parts of the world appear to be antigenically identical (Bozeman et al., 1960; Bell and Stoenner, 1960).

The human disease varies in severity but is seldom fatal. It is called fièvre boutonneuse, Marseilles fever, Kenya tick typhus, Indian tick typhus, or other names that designate the locality of occurrence. It is normally transmitted by the bite of the tick, but may also be acquired through the skin or eyes when the ticks are crushed. In the 1930s in certain localities of Central Africa virtually every newcomer became infected but at present the disease is of limited public health importance (Hoogstraal, 1967). However, an outbreak has recently been reported in Italy (Scaffidi, 1981).

R. conorii is the most ubiquitous rickettsia of the spotted fever group.

It has been recognized in most of the regions bordering on the Mediterranean Sea and Black Sea, in Kenya and other parts of Central Africa, South Africa, and certain parts of India. Strains of low virulence which are not antigenically identical and may or may not be closely related have been isolated in Switzerland (Burgdorfer et al., 1979) and elsewhere. The brown dog tick, *Rhipicephalus sanguineus* is the prevailing vector, but other ticks parasitizing dogs, cattle, and smaller animals have also been implicated, *Dermacentor reticulatus, D. marginatus,* other species of the genera *Rhipicephalus, Ixodes, Haemaphysalis* and *Amblyomma.* In addition to dogs, several species of urban and wild rats and mice have been found infected. The involvement of rabbits in maintaining an infected tick population was suggested by the drop in human disease following the myxomatosis epizootic that reduced the rabbit population in Europe (Hoogstraal, 1967, 1981).

The mol% G + C of the DNA is 33 (T_m, Bd) (Tyeryar et al., 1973). The electrophoretic profiles of the soluble proteins is similar but not identical to those of *R. rickettsii* and *R. sibirica* (Pedersen and Walters, 1978; Anacker et al., 1980).

Type strain: NIAID Malish 7 (South African patient strain, isolated by J. H. Gear in 1946). This strain and the Moroccan strain (ATCC VR-141) are used most frequently in laboratory experiments.

7. **Rickettsia parkeri** Lackman, Bell, Stoenner and Pickens 1965, 137.[AL]

par′ke.ri. M.L. gen. n. *parkeri* of Parker; named after Ralph R. Parker, a founder of the Rocky Mountain Laboratory.

This species is cultivated in eggs by the same procedure used for the other species of the spotted fever group. It is identified by appropriate serum neutralization and cross vaccination tests, as described for other species of the spotted fever group. It produces a nonfatal disease in the guinea pig characterized by fever and reddening of the scrotum; the most abundant growth takes place in the testicular tissue (Parker et al., 1939). Human infection has not been reported.

The species has been isolated from *Amblyomma maculatum* ticks collected from domestic animals in Texas, Georgia and Mississippi (Lackman et al., 1949).

On the basis of some cross-reaction in toxin neutralization tests in mice, this species is regarded as closely related to *R. conorii* (Lackman et al., 1965). The electrophoretic migration pattern in polyacrylamide gel of the solubilized proteins (Pedersen and Walters, 1978) and micro-IF tests (Philip et al., 1978) are compatible with such a classification. The mol% G + C of the DNA is not known, and DNA/DNA hybridization experiments have not been performed.

Type strain: NIAID maculatum 20 [isolated in 1948 from *A. maculatum* ticks collected from sheep in Mississippi (Bell and Pickens, 1953)].

8. **Rickettsia australis** Philip 1950, 786.[AL]

aus.tra′lis. L. fem. adj. *australis* southern.

Very similar to the other species of the spotted fever group that are not highly virulent for the guinea pig and produce a relatively mild disease in man. The only notable difference is the high virulence for the newborn mouse, which has been successfully used for isolation (Campbell and Domrow, 1974), and low toxicity for the adult mouse (as distinct from infectivity) (Bell and Pickens, 1953). Growth characteristics are not appreciably different from those of related species. Identification, as with other species, is based on serological tests and/or incomplete cross-protection with other species. The disease in man is called Queensland tick typhus.

R. australis was first isolated in 1944–1945 from the blood of two military patients on field exercises in belts of dense forest interspersed in grassy savannah in North Queensland, Australia (Andrew et al., 1946). Cases have also been recognized along the southeastern coast of Queensland. The scrub tick, *Ixodes holocyclus,* has been implicated in these infections. Two strains were isolated from this tick, one from unfed ticks collected by drawing a cloth over the herbage, and one from a tick removed from a dog. A strain was also isolated from *I. tasmani* collected from a wild rat (Campbell and Domrow, 1974). There is serological evidence that some marsupials are also infected.

The mol% G + C of the DNA is not known. DNA/DNA hybridization experiments have not been performed. Protein migration patterns are not appreciably different from those of related species (Pedersen and Walters, 1978).

Type strain: NIAID Phillips 32 (isolated from a patient by Andrew et al., 1946).

9. **Rickettsia akari** Huebner, Jellison and Pomerantz 1946, 1682.[AL]

a.ka′ri. Gr. neut. n. *akari* a mite.

With the possible exception of the nonpathogenic species described below, *R. akari* is regarded as the species of the spotted fever group most distantly related to *R. rickettsii.* It grows somewhat more profusely in the yolk sac of chicken embryos than *R. rickettsii* (Shepard et al., 1976) and is somewhat more cytotoxic when grown in cell cultures (Weiss et al., 1972, Kokorin et al., 1978), but attempts to demonstrate a toxic factor for mice have not been successful (Bell and Pickens, 1953). The mouse, however, is highly susceptible to infection and is the animal of choice for isolation. Differences in virulence between strains of this species and differences in susceptibility between inbred strains of mice have been demonstrated (Anderson and Osterman, 1980). The virulence for the guinea pig is about the same as that of *R. conorii.* The disease in man, designated rickettsialpox, is characterized by a generalized rash, fever, malaise and lymphadenitis which are resolved in about a week.

This species is easily distinguished from the other members of the spotted fever group by one-way toxin neutralization tests (Bell and Stoenner, 1960) and IF (Philip et al., 1978), although cross-reacting antigens do exist.

The main vector is the mite, *Allodermanyssus sanguineus,* which transmits the rickettsia transovarially. The nymph and adult stages of the mite feed on the house mouse, *Mus musculus,* but may attack other animals and man. The rickettsia has been isolated also from rats and from a wild Korean rodent, *Microtus fortis pelliceus* (Jackson et al., 1957). The disease in man was first observed in 1946 in New York City (Huebner et al., 1946a). It has been reported from urban areas along the Atlantic coast of the United States, in the Crimean and southern Ukranian regions of the U.S.S.R., in Korea, and there is circumstantial evidence that it occurs in Africa. In the mid-1940s, about 180 cases were reported annually in the United States, but only sporadic cases have been recognized in recent years (Wong et al., 1979).

The mol% G + C of the DNA is 32.3 (Bd) and 33.2 (T_m) (Tyeryar et al., 1973.). DNA/DNA hybridization experiments have not been performed, but differences in electrophoretic migration patterns of the solubilized proteins between *R. akari* and the other species of the spotted fever group are easily visualized (Obijeski et al., 1974; Pedersen and Walters, 1978).

Type strain: ATCC VR-148 [Kaplan; MK; isolated from the blood of a patient in New York City (Heubner et al., 1946)]. Strain 7 (ATCC VR-612) isolated from a mouse by Fuller et al. (1951) and a strain isolated from a mite in the U.S.S.R. (received from Zhdanov in 1956) are also frequently used.

10. **Rickettsia montana** (*ex* Lackman, Bell, Stoenner and Pickens 1965) nom. rev.

mon.ta′na. M.L. n. *montana* Montana, the state where the organism was first isolated.

Resembles other members of the spotted fever group in growth characteristics in the chicken embryo and antigenic composition, but is avirulent for the guinea pig and the mouse. Although antibodies directed against this species have been detected in dogs, there is no evidence that *R. montana* is associated with spotted fever signs or symptoms in dogs or man. It is distinguished from other rickettsiae by limited cross-reactions in one-way toxin neutralization (Bell et al., 1963) and micro-IF tests (Philip et al., 1978).

R. montana was first isolated in Eastern Montana, repeatedly from *D. variabilis,* occasionally from *D. andersoni.* It has been also isolated from rodents (genera *Microtus*, *Peromiscus*) and from *D. variabilis* isolated in various parts of the U.S.A. (Lackman, et al., 1965; Philip et al., 1978). It has been isolated with notable frequency from *D. variabilis* ticks in Cape Cod, Massachusetts (Feng et al., 1980) and also in Clermont County, Ohio (Linnemann et al., 1980)

The genetic features of this species are not known.

Type strain: ATCC VR-611 [Tick strain (Bell et al., 1963)].

11. **Rickettsia rhipicephali** (*ex* Burgdorfer, Brinton, Krynski and Philip 1978) nom. rev.

rhi.pi.ce'pha.li. M.L. gen. n. *rhipicephali* of rhipicephalus; named after its natural tick host *Rhipicephalus sanguineus*.

This species is regarded as a member of the spotted fever group on the basis of similarities in antigenic composition, morphology and protein migration patterns (Burgdorfer et al., 1975; Hayes and Burgdorfer, 1979; Anacker et al., 1980). In contrast to the other rickettsiae of this group, however, it is cultivated with difficulty in the chicken embryo as well as the guinea pig. It is readily isolated in male meadow voles (*Microtus pennsylvanicus*) where it produces massive infection in the tissues of the tunica vaginalis. This species can also be cultivated on monolayers of chick embryo fibroblasts, Vero and mouse L cells. Although growth is profuse, damage to the cells is limited and plaques are small and turbid. There is no evidence that *R. rhipicephali* is pathogenic for the dog or for man, but the possibility cannot be excluded that this species, as well as *R. montana*, may provide partial protection of dogs against *R. rickettsii* (Burgdorfer et al., 1975).

R. rhipicephali has been detected in about 19% of the brown dog ticks (*R. sanguineus*) removed from dogs in central and northern Mississippi and also from ticks collected in Texas and North Carolina (Burgdorfer et al, 1975; 1978). It has also been isolated with high frequency from *Dermacentor andersoni* ticks collected in western Montana (Philip and Casper, 1981).

The mol% G + C of the DNA is 32.2 (T_m) (Anacker et al., 1980).

Type strain: none designated. Reference strain: 3-7-Q6, isolated in 1973 in Mississippi by Burgdorfer et al. (1975).

12. **Rickettsia tsutsugamushi** (Hayashi 1920) Ogata 1931, 252.[AL] (*Theileria tsutsugamushi* Hayashi 1920, 63; *Rickettsia orientalis* Nagayo, Tamiya, Mitamura and Sato 1930, 317.)

tsu.tsu.ga.mu.shi. M.L. n. *tsutsugamushi* popular name of the disease caused by this species, generally interpreted to mean mite disease.

As the sole member of the scrub typhus group within the genus *Rickettsia*, some of its biological characteristics are given in the section describing the genus. This rickettsia is somewhat smaller than the other species, averaging 1.2 μm in length, seldom exceeding 1.5 μm. It is stained by a modification of the Giménez procedure used for the other rickettsiae or by Giemsa's stain. The outer leaflet of the cell wall adheres tenaciously to host membranes and host components are not readily removed by the methods used for the other rickettsiae. It is cultivated well in the yolk sac of chicken embryos, provided the inoculum is relatively large (10^4–10^6 viable cells per egg) and the rickettsiae are harvested before the death of the embryos. It is also cultivated well in cell cultures and produces small plaques on cell monolayers following 11–17 days of incubation. It grows in the cytoplasm of the host cell, achieving high density in the perinuclear region and acquiring a host-membrane coat as it emerges from the cell surface (Ewing et al., 1978), but does not penetrate into the nucleus. CO_2 enrichment is not necessary for intracellular growth, in contrast to other rickettsiae (Kopmans-Gargantiel and Wisseman, 1981), but otherwise the metabolism of *R. tsutsugamushi* appears to be similar to that of other rickettsiae (Weiss et al., 1973; Dasch and Weiss, 1978).

R. tsutsugamushi strains vary considerably in antigenic composition and in some cases in virulence and other biological properties. Three main antigenic types are generally recognized: Gilliam, Karp and Kato. There is considerable cross reactions among the three types, but a soluble common antigen useful in CF tests is not obtained as readily with this species as with the other rickettsiae. In general, patients and experimental animals surviving infection are immune to all strains for a few months, but only homologous immunity persists for longer periods. Differences are even more marked in cross neutralization and cross-vaccination tests (Bennett et al., 1949; Eisenberg and Osterman, 1979). Most useful for antigenic type identification are the micro-IF test with whole cells (Bozeman and Elisberg, 1963) and the ELISA test done with disrupted cell extracts (Dasch et al., 1979). Higher titers are usually obtained in the homologous than in the heterologous reaction. Analysis of the antigens by polyacrylamide gel electrophoresis indicates that there are six major antigens located on the cell envelope and two of these from Karp and Kato do not react with heterologous sera (Eisemann and Osterman, 1981).

Virulent strains of *R. tsutsugamushi* injected intraperitoneally into mice cause peritonitis, splenomegaly and death in 10–24 days, but strains vary greatly in virulence. The Karp strain, for example, is more virulent than the Gilliam strain for most outbred mice. Certain inbred strains are highly resistant to the Gilliam strain and this resistance was shown to be controlled by a single, autosomal, dominant gene and not to involve susceptibility to the Karp strain (Groves et al., 1980). The disease in man, which also varies considerably in severity, is characterized by a cutaneous lesion, the eschar, at the site of the bite of the infected larval mite, and in lesions of the small blood vessels of various organs, as in other rickettsial diseases. The infection is not lethal when properly treated, but before the availability of the broad-range antibiotics, mortality varied in different localities from 1 to 40%.

R. tsutsugamushi is encountered in an area of the Orient that extends from India and Pakistan in the West to Japan, the northern portions of Australia, and the intervening islands in the Pacific Ocean in the east and including southeastern Siberia, Korea, Southeast Asia, Southern China, the Philippines and Indonesia. The rickettsia is usually found in circumscribed foci or "ecological islands," which have the proper vegetation and proper concentration of mites and their wild rat hosts. The habitats are usually characterized by the presence of changing ecological conditions, wrought by man or nature, and expressed by transitional types of vegetation (Traub and Wisseman, 1974).

The mite most commonly associated with scrub typhus is *Leptotrombidium deliense*, but several other trombiculid mites, including *L. fletcheri*, *L. akamushi*, *L. arenicola*, and *L. scutellare*, were shown to be naturally infected and to transmit the rickettsia transovarially. The six-legged larva, or chigger, shortly after its emergence from the egg, remains in the soil or travels up a few centimeters on debris or dead vegetation until it can burrow into the skin of any animal it happens to contact. Following a meal of tissue juices, it returns to the soil to resume a free-living existence. The vertebrates most commonly infected are rodents of the genus *Rattus*, although isolations from temperate-zone rodents, including *Apodemus* and *Microtus*, have been reported. The wide dissemination of *R. tsutsugamushi* on islands that are separated from each other and from the mainland by large bodies of water can best be explained by assuming that migrant birds play a role in the transport of the chiggers (Traub and Wisseman, 1974).

The genetic features of the species are not known.

Type strain: ATCC VR-150 [Karp strain (Derrick and Brown, 1949)]. Reference strains: ATCC VR-312 [Gilliam strain (Bennett et al., 1949)]; ATCC VR-609 [Kato strain (Shishido et al., 1958)].

Genus II. **Rochalimaea** *(Macchiavello 1947) Krieg 1961, 162*[AL]

EMILIO WEISS AND JAMES W. MOULDER

Ro.cha.li′mae.a. M.L. fem. n. *Rochalimaea* named after Henrique da Rocha-Lima, one of the early investigators of the etiology of rickettsial diseases.

Short rods, ~0.3–0.5 by 1.0–1.7 μm, closely resembling those of the genus *Rickettsia* in morphology and staining properties. Gram-negative.

Aerobic. **Can be cultivated in host cell-free media** of moderate complexity, such as blood agar or broth enriched with amino acids, yeast extract, and fetal bovine serum or hematin. Succinate, pyruvate, glutamine, or glutamate, but not glucose, serve as sources of energy. **Grow profusely on the surface of eucaryotic cells**, such as mouse L cells, or cells lining the intestines of the invertebrate host. **Natural cycle involves a vertebrate and invertebrate host.** The mol% G + C of the DNA is 38.5–39 (T_m, Bd).

Type species: *Rochalimaea quintana* (Schmincke 1917) Krieg 1961, 163.

Further Descriptive Information

By electron microscopy, the cells possess a trilaminar cell wall and plasma membrane, typical of Gram-negative bacteria. Flagella, pili or organs of adherence to cell surfaces have not been described. Although the organisms can be cultivated in axenic media, they have been grown to a moderate extent in the yolk sac of chicken embryos, in human HEP cells, and profusely on irradiated mouse L cells. When associated with eucaryotic cells, the location is epicellular (Fig. 9.2). The few intracellular forms appear to be degenerating (Ito and Vinson, 1965; Merrell et al., 1978; Vinson and Fuller, 1961; Weiss et al., 1978).

In 1961 Vinson and Fuller demonstrated that *Rochalimaea quintana* can be grown on blood agar. Prior to that time the organisms were often maintained in lice which were fed on human volunteers, who often became infected (Strong, 1918; Mooser et al., 1948). Subsequent work by Vinson (1966) established that increased CO_2 pressure, aerobic conditions, and a factor in red blood cells are essential, while serum is stimulatory. Colonies are small, round, and translucent. They appear after incubation for 12–14 days at 37°C on primary isolation from trench fever patients, but the period is greatly shortened after repeated passages on blood agar.

Myers et al. (1969; 1972) demonstrated that the erythrocyte requirement could be met by hemoglobin or hemin and that serum stimulation could be duplicated by a "detoxifying agent" such as starch or charcoal. Mason (1970) showed that fetal calf serum, but not calf serum, could replace the red blood lysate requirement. More recently, Weiss (1981) cultivated both species in a liquid medium consisting of amino acids, fetal bovine serum, yeast extract, and succinate. *R. quintana* but not

R. vinsonii required sodium bicarbonate as a source of CO_2. The generation time in this and other media is 4–7 h.

Rochalimaea, like the genus *Rickettsia*, appears to be devoid of glycolytic enzymes. Both species of *Rochalimaea* catabolize succinate, pyruvate and glutamine (Huang, 1967; Weiss et al., 1978).

Clinical isolates of *R. quintana* have thus far appeared to be identical to each other. Cells of the only available strain of *R. vinsonii* autoagglutinate readily and precipitate to the bottom of suspensions. This strain may be regarded as a " rough" variant. It is not known whether this is a species characteristic or a variation that has emerged during 44 yolk sac passages.

Serological cross-reactions between *R. quintana* and *R. vinsonii* are extensive. Common as well as specific antigens have been demonstrated. Some serological cross-reactions with the typhus and scrub typhus groups of rickettsieae have been reported (Hollingdale et al., 1978), but these findings require further confirmation.

A limited study of antibiotic susceptibility indicates that *R. quintana* is highly susceptible to tetracycline, moderately susceptible to penicillin, and relatively resistant to streptomycin (Vinson and Fuller, 1961).

R. quintana is the etiological agent of trench fever, an illness which made its first appearance in 1915 and afflicted at least 1,000,000 military personnel during World War I. The disease was characterized by sudden onset of chill and fever, headache, muscular pain, and generalized weakness which typically lasted five days but often recurred and incapacitated the patients for 6–8 weeks, although it was not fatal. The disease was considerably milder in more recent experiments with human volunteers (Varela et al., 1969; Vinson et al., 1969). Laboratory animals are not susceptible to infection, but a subclinical bacteremia has been produced in the rhesus monkey (Mooser and Weyer, 1953). *R. vinsonii* infects some of the common laboratory animals without signs of illness, and there is no record of human infection.

R. quintana has an ecological niche which appears to be identical to that of *Rickettsia prowazekii*. Man is the primary host and the human body louse, *Pediculus humanus*, the principal vector. The life of the louse is not shortened by the infection, but the microorganisms are not transmitted to the progeny. The disease in man becomes apparent only

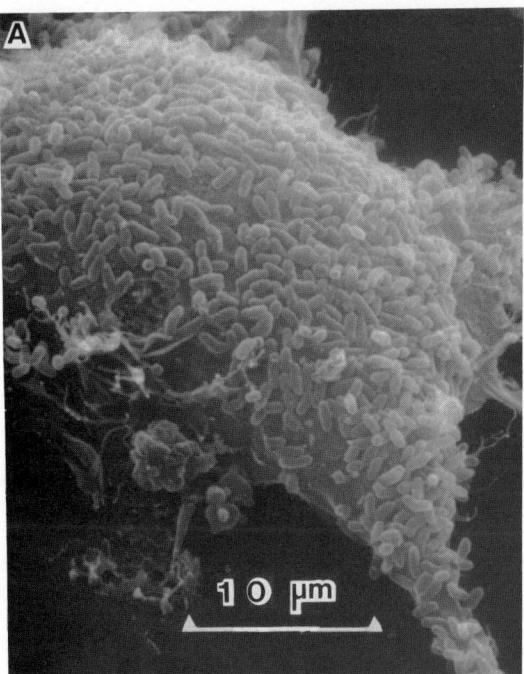
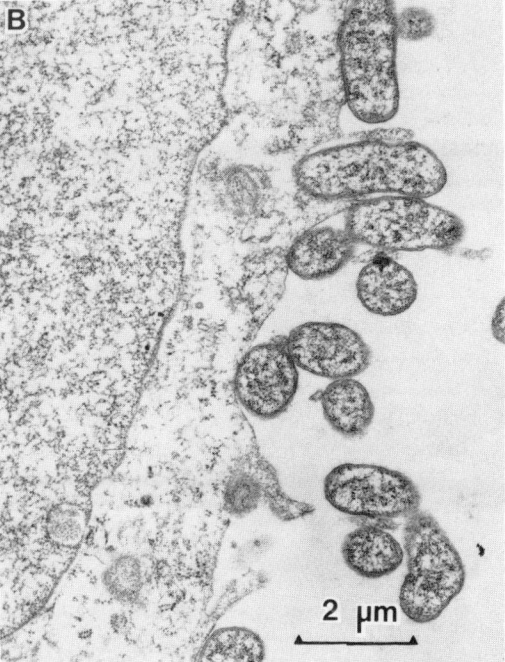

Figure 9.2. *A*, A scanning electron micrograph of irradiated mouse L cells 96 h after infection with the Fuller strain of *Rochalimaea quintana*. *B*, Transmission electron micrograph of portions of an irradiated L cell 48 h after infection with *Rochalimaea vinsonii*. (Reproduced with permission from B. R. Merrell et al., *Journal of Bacteriology, 135:* 633–640, 1978, © American Society for Microbiology, Washington, D.C.)

under circumstances of high louse infestation and, possibly severe stress. The disease disappeared as a clinical entity after World War I and reappeared on a more limited scale in World War II. A worldwide distribution of *R. quintana* is indicated by its repeated recovery from Mexico and evidence of its presence in other locations (Vinson, 1966).

R. vinsonii, on the other hand, apparently has a niche restricted to the vole (*Microtus pennsylvanicus*) that inhabits a small island in the St. Lawrence River. The island, Grosse Isle, Quebec, had served as a quarantine station in the 19th century. In 1847, at the height of the immigration from Ireland, thousands of immigrants died of typhus and were buried there. Baker (1946) examined the local fauna in search of a vestige of the 19th century epidemic, isolated a microorganism, but recognized that his isolate was not a typhus rickettsia. However, surveys of vole populations in other geographic locations for rickettsiae and related organisms have provided no indication that *R. vinsonii* is present elsewhere (see Weiss et al. (1978)). If *R. quintana* and *R. vinsonii* are indeed evolutionarily related, as suggested by genotypic and phenotypic similarities, two explanations can be offered for their different ecologies: (i) Both infections are of human origin, the vole infection on Grosse Isle originating from typhus patients buried on that island in the 19th century, who were also infected with the ancestor of *R. vinsonii*. (ii) Although *R. quintana* is well established in man, it originated in a vole. At the start of the World War I epidemic, trench fever was believed to have derived from the enormous number of voles present in the trenches (Rutherford, 1916). The vole involved was probably *Arvicola terrestris*, a rodent closely allied to *Microtus* (William L. Jellison, personal communication). Thus, the original niche of both species might have been a vole.

Isolation Procedures

The recommended procedure for the isolation of *Rochalimaea* is to place the suspected specimen, patients' blood or spleen suspension of voles, on blood agar (consisting of blood agar base, 4% fetal bovine serum, and 6% lysed sheep erythrocytes). The agar plates are incubated at 35–37°C in an atmosphere of 10% CO_2 and air for as long as 2 weeks. If small, translucent colonies appear after several days, the microorganisms are checked for typical morphological and staining properties and identified with fluorescent antibodies.

Maintenance Procedures

Seed has been produced (Weiss, 1981) in a liquid medium consisting of Hanks' balanced salt solution containing 25 mM HEPES buffer (*N*-2-hydroxyethylpiperazine-*N'*-2-ethanesulfonic acid) at pH 7.2, 0.8% casamino acids (Difco), 0.1% yeast extract, 10% fetal bovine serum, 2 mM succinate, and 4.5 mM sodium bicarbonate. The seed is best maintained by storage at −70°C.

Table 9.5.
Differential characteristics of the species of the genus **Rochalimaea**

Characteristics	1. *R. quintana*	2. *R. vinsonii*
Hosts:		
Humans and the human body louse (*Pediculus humanus*)	+	−
Voles (*Microtus pennsylvanicus*)	−	+
Cultivation:		
In the yolk sac of chicken embryos, satisfactory growth	−	+
Axenic, CO_2 required	+	−
Metabolism:		
Energy source such as pyruvate, succinate or glutamine required for transport of glutamate or ornithine	+	−
Ornithine decarboxylase:		
Constitutive	+[a]	−
Inducible	−	+[a]

[a] Weak activity.

Taxonomic Comments

An evolutionary relationship to the genus *Rickettsia* is indicated by DNA/DNA hybridization between *R. quintana* and the typhus group in the range of 25–33% (Myers and Wisseman, 1980). The degree of hybridization between *R. quintana* and *R. vinsonii* is 31–42% (Myers et al., 1979). The finding that the vole agent (*R. vinsonii*) and *R. quintana* are related raises the possibility that there are other microorganisms in nature that can be classified in this genus. One example may be *Wolbachia melophagi* (see the genus *Wolbachia*) which is seen as a strictly epicellular parasite in the lumen of the intestines of its sheep ked host, *Melophagus ovinus*. However, *W. melophagi* cannot be reclassified at this time without considerable additional investigation.

Further Reading

See the Further Reading for the article on the genus *Rickettsia*. Also see the following:

Strong, R.P. (Editor). 1918. *Trench fever.* Report of Commission, Medical Research Committee, American Red Cross. Oxford University Press, Oxford, England.

Differentiation of the species of the genus **Rochalimaea**

Characteristics useful for the differentiation of the two species of *Rochalimaea* are listed in Table 9.5.

List of the species of the genus **Rochalimaea**

1. **Rochalimaea quintana** (Schmincke 1917) Krieg 1961, 163[AL] (*Rickettsia quintana* Schmincke 1917, 961).

quin.ta'na. M.L. fem. adj. *quintana* fifth, referring to 5-day fever, one of the vernacular names of the fever caused by the species.

The morphological and cultural characteristics are as described for the genus.

Increased CO_2 pressure is required for growth.

A source of energy, such as pyruvate, succinate or glutamine is required for the transport of certain substrates, such as ornithine (Weiss et al., 1982) or glutamate (Weiss and Dasch, 1982).

Growth in the yolk sac of chicken embryos or laboratory animals is unsatisfactory.

Etiological agent of trench fever.

The mol% G + C of the DNA is ~39 (Tyeryar et al., 1973). The genome size is ~100 × 10⁷ daltons (Myers and Wisseman, 1980).

Type strain: ATCC VR-358 [Fuller strain (Vinson and Fuller, 1961).]

2. **Rochalimaea vinsonii** Weiss and Dasch 1982, 313,[VP]*. (Canadian vole agent Baker 1946, 37.)

vin.so'ni.i. M.L. gen. n. *vinsonii* of Vinson; named after J. William Vinson, who in collaboration with Henry S. Fuller demonstrated that *Rochalimaea* can be cultivated on blood agar.

The morphological and cultural characteristics are as described for the genus.

The organisms autoagglutinate in a suspending fluid or medium as readily as do "rough" strains of other species.

---

* *VP* denotes that this name has been validly published in the official publication, International Journal of Systematic Bacteriology.

CO_2 is not required for growth.

Glutamate is utilized without an added energy source (Weiss and Dasch, 1982). Weak inducible ornithine decarboxylase activity has been demonstrated (Weiss et al., 1982).

Has been grown in the yolk sac of chicken embryos and was recovered from experimental infections of laboratory animals, although no obvious illness was produced (Baker, 1946).

The mol% G + C of the DNA is ~39 (Weiss et al., 1978). The genome size is ~133 × 10[7] daltons (Myers et al., 1979).

Type strain: ATCC VR-152 [Baker strain (Baker, 1946)].

Genus III. **Coxiella** *(Philip 1943) Philip 1948, 58[AL]*

EMILIO WEISS AND JAMES W. MOULDER

(Subgenus *Coxiella* Philip 1943, 306)

Co.xi.el′la. M.L. fem. dim. ending -*ella*; M.L. fem. n. *Coxiella* named after Herold R. Cox, who in collaboration with G. E. Davis first isolated this organism in the United States shortly after its discovery in Australia and who introduced the technique of yolk sac inoculation of the chick embryo which greatly facilitated the study of this and other genera.

Short rods, usually 0.2–0.4 μm by 0.4–1.0 μm, resembling organisms of the genus *Rickettsia* in staining properties, dependence on host cells for growth, and close natural association with arthropod and vertebrate hosts. **Grows preferentially in the vacuoles of the host cell** (rather than in the cytoplasm or the nucleus as do the species of *Rickettsia*). Unlike *Chlamydia*, *Coxiella* does not prevent the formation of phagolysosomes. **Grows well in the yolk sac of chicken embryos**, where it undergoes a cycle of development which includes formation of an endospore-like body. **Has high resistance to chemical agents and elevated temperatures** that generally kill *Rickettsia* species. Although axenic growth has not been obtained, outside the host *Coxiella* metabolizes glutamate, glucose, and other substrates, provided the pH is low. This property may reflect an adaptation to the environment of the phagolysosome. **The distribution in ticks and various mammals is worldwide** and infection is particularly prevalent in cattle, sheep and goats. **Etiological agent of the aerosol-borne human disease Q fever.** The mol% G + C of the DNA is ~43 (chemical analysis; Smith and Stoker (1951)).

Type species: *Coxiella burnetii* (Derrick 1939) Philip 1948, 58.

Further Descriptive Information

Coxiella organisms are generally regarded as Gram-negative rods, smaller than those of the genus *Rickettsia*. They have no flagella or capsule and are not always retained by ordinary bacteriological filters (Cox, 1939). They stain well by the method of Giménez (1964) and may appear Gram-positive when ethanol-iodine is used as the mordant in the Gram stain (Giménez, 1965). Some cells are acid-fast by the Kinyoun carbol-fuchsin method used for tubercle bacilli (McCaul and Williams, 1981).

Electron microscopy of cells separated from host cell components shows great variation in morphology (Fig. 9.3*A*), and two distinct cell variants can be separated by density gradient centrifugation (Fig. 9.3*B*). Both types are infectious and convert to a mixture of cell types when cultured separately (Wiebe et al., 1972). Although the large (LCV) and small (SCV) cell variants have comparable outer membranes, they differ considerably in internal structure. The LCV has large periplasmic spaces and low electron density (Fig. 9.3*C*), whereas the SCV has complex internal membranous inclusions originating from the cytoplasmic membrane (Fig. 9.3*D*) and has high electron density. Both variants are capable of cell division (Fig. 9.3*A*). The SCV originates from the LCV as an electron dense "cap" in the periplasmic space (Fig. 9.3*E*) which progressively develops into an SCV (Fig. 9.3*F*) and is released upon lysis of the LCV. Forms reflecting transition from SCV to LCV are also seen. SCV stains with some of the spore stains, but dipicolinic acid has not been demonstrated. Thus, the morphogenesis of *C. burnetii* is comparable, although not identical, to a cycle of sporogenic and vegetative cell differentiation (Fig. 9.3*A*) (McCaul and Williams, 1981).

In addition to endospore-like differentiation, *C. burnetii*, unlike species of *Rickettsia*, undergoes phase variation, somewhat similar to the smooth-rough variation seen in other bacteria (Stoker and Fiset, 1956; Fiset and Ormsbee, 1968; Brezina, 1978). Naturally occurring strains are in phase I. Upon repeated passage in the yolk sac of chicken embryos, the rickettsiae gradually convert to phase II. Complete conversion may require as many as 100 egg passages, but most of the characteristics are acquired by the 10th passage. Except for the "pure" phase II strains, conversion back to phase I occurs rapidly, often during a single passage in laboratory animals. Phase I cells are hydrophilic, are stable in suspension, and are not ingested by phagocytic cells in the absence of antibodies. Phase II cells are hydrophobic, autoagglutinable, have greater affinity for acridine orange and hematoxylin, and are readily phagocytized. Phase II adhere more tenaciously to host cell components than phase I, but both phases can be obtained free of host cell contamination by procedures involving Renografin gradient centrifugation (Williams et al., 1981). The major chemical difference between the two phases stems from the presence of glucuronic acid in the surface antigen in phase I but not in phase II cells (Jerrells et al., 1974).

C. burnetii has a high degree of resistance to physical and chemical agents, comparable to that of sporogenic bacteria. This resistance might be attributable to the endospore-like forms and is responsible in part for the high hazard of infection. At 4°C viability is retained for 1 or more years in dried fomites such as tick feces or wool, as well as in sterile skim milk or unchlorinated water. Meats remain infected for at least 1 month. Complete inactivation is not always accomplished by exposure to 63°C for 30 min or 85–90°C for a few seconds. The viability of yolk sac suspensions is not entirely destroyed by 1% formalin or phenol for 24 h. *C. burnetii* is rapidly inactivated by diethyl ether, but not by ethanol (Ransom and Huebner, 1951; Babudieri, 1959).

The guinea pig is the animal of choice for isolation and the chicken embryo, inoculated via the yolk sac, is most valuable for the preparation of seed and antigen (see below). *C. burnetii* is also quite infectious for monolayers prepared from chicken embryos. Plaques, approximately 1 mm in diameter, are produced following incubation for 16 days. As in the case of *Rickettsia* and *Chlamydia*, infection is increased by centrifugation of the inoculum onto the cell monolayers (Ormsbee et al., 1978). *C. burnetii* is not highly infectious for mouse L-cells, and phase I cells grow less well than phase II (Kordova et al., 1970). However, persistent infections have been established in these cells as well as in green monkey kidney cells (Vero) (Burton et al., 1978). Phagocytosis seems to be the main mechanism of entry of *C. burnetii* into host cells. In contrast to *Rickettsia*, *C. burnetii* does not escape from the phagosome and often virtually the entire cytoplasm is converted into a large vacuole (Weiss, 1973). In contrast to *Chlamydia*, phagolysosome formation is not prevented. In the endodermal cell of the yolk sac, the normal lysis of the limiting membrane of the phagolysosome permits the release of *C. burnetii* into the cytoplasm (Khavkin et al., 1981). Although in some cases excessive levels of lysosomal activity may be detrimental to *C. burnetii* (Burton et al., 1971, 1978), the normal cycle of development occurs in the presence of lysosomal hydrolases (Ariel et al., 1973).

Investigators of the metabolic activities of cells separated from host components have been confronted with a paradox. While the list of enzymatic activities demonstrable with cell-free extracts has steadily grown and has included key cell functions (Baca, 1978; Donahue and Thompson, 1981), experiments with intact cells have been relatively

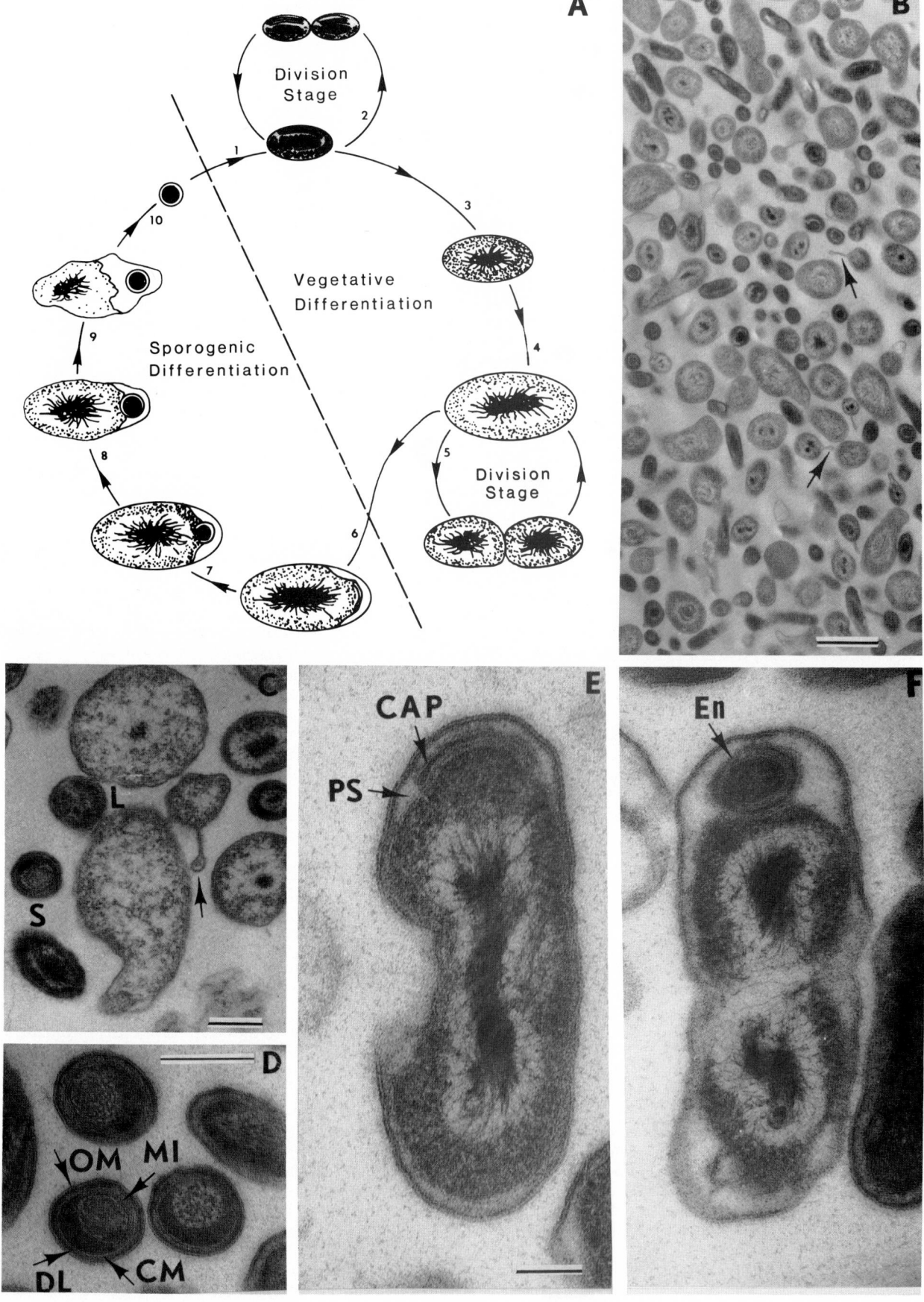

disappointing. Ormsbee and Peacock (1964) have shown that *C. burnetii* produces low levels of CO_2 from pyruvate, glutamate, and some of the intermediates of the citric acid cycle. Recently, Hackstadt and Williams (1981, 1981a) have shown that these activities are greatly enhanced when the pH is lowered to 4.5. Under these conditions glucose is also utilized but optimum activity requires the presence of other substrates, such as glutamate. This finding may remove an important block towards axenic cultivation, though the existence of a cycle of development may represent another major difficulty.

Despite the worldwide distribution of *C. burnetii* in invertebrate and vertebrate populations, naturally occurring variation in biological properties is small. Differences are reflected primarily in virulence for the guinea pig and total yield from yolk sacs. An attenuated series of strains (M-44) was developed in the laboratory by repeated serial passage in chicken embryos (Genig, 1968; Robinson and Hasty, 1974).

Although phase II cells are laboratory induced, phase II antigen is important in diagnostic procedures. When injected into animals phase I cells are much better immunogens than phase II cells (Ormsbee et al., 1964). Surprisingly, the antibodies first to appear in these animals or in human infection are directed against phase II and only later are phase I antibodies produced. Phase II cells induce only a slow homologous response. Fiset and Ormsbee (1968) suggested that the phase I carbohydrate acts primarily as an adjuvant for the production of phase II antibodies and only secondarily as an antigen that stimulates antibodies against itself. Phase I antigen is obtained from phase I cells by extraction with trichloroacetic acid. Phase II antigen is also obtained most conveniently from phase I cells by treating the antigen with potassium periodate. This procedure is believed to destroy the phase I carbohydrate and expose the phase II antigen (Schramek et al., 1972; Ormsbee, 1980).

Although a number of animals are susceptible, it is not certain that infection results in overt disease, except in the guinea pig and in man. Sporadic cases or outbreaks of the human disease, Q fever, occur primarily among individuals involved in the transport of infected domestic animals and in the meat and dairy industries. The infection is acquired by the aerosol route and the disease is characterized by fever, severe headache, and interstitial pneumonia. The disease is seldom fatal, but may extend to other organs. A rare but highly fatal complication is Q fever endocarditis. The tetracycline compounds are the antibiotics of choice for treatment. *C. burnetii* is also inhibited by rifampin (Spicer et al., 1981), chloramphenicol, but not by *p*-aminobenzoic acid, and only to a very limited extent by penicillin, streptomycin or erythromycin.

The distribution of the microorganism is worldwide (Babudieri, 1959). The most commonly infected arthropod is the tick. *Amblyomma, Dermacentor, Haemaphysalis, Myalomma, Ixodes, Ornithodorus* and *Rhipicephalus* are among the genera most frequently mentioned as carriers of *C. burnetii*. Although many wild mammals and birds are undoubtedly infected, isolations from these animals have not been numerous. Several isolations were made from the Australian bandicoot, *Isoodon torosus*, which was believed to play a role in the epidemiology of Q fever (Derrick, 1953). *C. burnetii* is isolated most frequently from domestic animals, in particular cattle, sheep, and goats. The microorganisms grow particularly well in the placenta of these animals, where they may reach a titer of 10^9 viable organisms/g. Highly infectious dust from tick feces deposited on the hides of animals and from dried placenta following parturition are the chief sources of infection. There are, however, other vehicles of infection, including milk.

Isolation Procedures

Isolation of *C. burnetii* from survey or clinical specimens, as well as any other laboratory work with this microorganism, requires even stricter adherence to safety precautions than in the case of *Rickettsia*. This is also true of certain other procedures, such as experiments on pregnant ewes that have not been tested for freedom from *C. burnetii*. An experimental phase I vaccine is available (Ormsbee et al., 1964), but it must be used with caution, since it may induce the formation of sterile abscesses in sensitized individuals. The hazard of infection is in large measure due to the previously described high degree of resistance of *C. burnetii* to physical and chemical agents.

Procedures for isolation are entirely similar to those used for *Rickettsia*. The guinea pig is the animal of choice for isolation. The chief clinical manifestation is fever that starts 5–12 days after inoculation, lasts 2–8 days, and terminates in death only with relatively large inocula of the more virulent strains. The spleen obtained on the second or third day of fever is used for passage into chicken embryos. Growth in the yolk sac is relatively slow. Maximum yields require relatively large inocula or a second and sometimes a third passage.

Maintenance Procedures

C. burnetii is most conveniently maintained at −70°C. Phase II cells are obtained following repeated serial egg passage. Phase I are also most conveniently grown in yolk sac following a previous passage in the guinea pig.

Differentiation of **Coxiella** from other closely related genera

Characteristics useful for the differentiation of *Coxiella* from the other genera of the tribe *Rickettsieae* are listed in Table 9.2.

Taxonomic Comments

The phylogenetic relationships of *Coxiella* to other members of the tribe *Rickettsieae* are not known. Nucleic acid studies, such as DNA/DNA hybridization, DNA/rRNA hybridization, or rRNA oligonucleotide cataloging, would be helpful in this regard.

Microorganisms that might be genotypically or phenotypically re-

lated to *C. burnetii* have not been identified, but the recent elucidation of what appears to be a sporogenic cycle in *C. burnetii* (McCaul and Williams, 1981) and the discovery of a low pH optimum for exogenous metabolism (Hackstadt and Williams, 1981) may stimulate the search.

Acknowledgments

We are greatly indebted to Mrs. Donna M. Boyle for her expert preparation of the manuscripts for this portion of the chapter on *Rickettsiales*.

Figure 9.3. Developmental cycle which suggests vegetative and sporogenic differentiation in *Coxiella burnetii*. *A*, Schematic representation of developmental cycle within the phagolysosome of eucaryotic cells. *B–F*, Thin sections of *C. burnetii* suspension following Renografin gradient separation from yolk sac components. The cells were fixed with a primary fixative and stained with potassium permanganate. *B*, Illustration of the pleomorphic nature of the organisms. There are about equal numbers of large cell variants (LCV) (corresponding to No. 3, 4, or 5 in *A*) and small cell variants (SCV) (No. 10, 1 or 2 in *A*). Some forms are intermediate, some LCV appear to be deteriorating, and some are losing their outer membranes (*arrows*). Bar, 1.0 μm. *C*, Comparison of LCV (*L*) and SCV (*C*) at higher magnification. Outer membrane bleb formation can be seen in a deteriorating LCV (*arrow*). Bar, 0.3 μm. *D*, Detail of SCV. *OM*, outer membrane; *DL*, dense layer; *CM*, cytoplasmic membrane; *MI*, membranous intrusions. Bar, 0.2 μm. *E*, Initial stage of endospore formation (No. 6 in *A*). Note enlarged periplasmic space (*PS*) and cap (*CAP*) formation. Bar, 0.1 μm. *F*, Formation of an endospore (*En*) in a LCV (No. 8 in *A*) concurrently undergoing cell division (No. 5 in *A*). Same magnification as in *E*. (Reproduced with permission from T. F. McCaul and J. C. Williams: *Journal of Bacteriology, 147:* 1063–1076, 1981, © American Society for Microbiology, Washington, D.C.) (The plate was kindly prepared by T. F. McCaul and J. C. Williams.)

Further Reading

See the Further Reading section for the genus *Rickettsia*. In addition, see the following references:

Babudieri, B. 1959. Q fever: a zoonosis. Adv. Vet. Sci. *5:* 81–182.
McCaul, T.F. and J.C. Williams. 1981. Developmental cycle of *Coxiella burnetii*: structure and morphogenesis of vegetative and sporogenic differentiations. J. Bacteriol. *147:* 1063–1076.
Paretsky, D. 1968. Biochemistry of rickettsiae and their infected hosts, with special reference to *Coxiella burnetii*. Zentralbl. Bakteriol. Parasitenk. Infektionskr. Hyg. Abt. Orig. *206:* 284–291.

List of the species of the genus **Coxiella**

1. **Coxiella burnetii** (Derrick 1939) Philip 1948, 58[AL] (*Rickettsia burneti (sic)* Derrick 1939, 14).

bur.ne'ti.i.M.L. gen. n. of Burnet; named after Frank MacFarlane Burnet, who first studied the properties of this organism.

The characteristics are as described for the genus.

The mol% G + C of the DNA is ~43.

Type strain: ATCC VR-615 [strain Nine Mile Phase I (Davis and Cox, 1938). Reference strains: ATCC VR-145 [strain Henzerling (Robbins et al., 1946); essentially in Phase II]; ATCC VR-616 [strain Nine Mile Q (Davis and Cox, 1938); essentially Phase II].

TRIBE II. **EHRLICHIEAE** PHILIP 1957, 948[AL]

MIODRAG RISTIC AND DAVID L. HUXSOLL

(Equivalent to family *Ehrlichiaceae* Moshkovski 1945, 18.)

Ehr.lich'ie.ae. M.L. fem. n. *Ehrlichia* type genus of the tribe; *-eae* ending to denote a tribe; M.L. fem. pl. n. *Ehrlicheae* the *Ehrlichia* tribe.

Rickettsial organisms pathogenic for certain mammals, including man. The predominant host cells are reticulendothelial cells, including circulating leukocytes, but not erythrocytes. The organisms grow in the cytoplasm but not in the nucleus; they usually appear as compact inclusions containing a number of individual organisms. There may be more than one inclusion per cell. The berry-like appearance of the inclusions has led to the use of the term "morula." Single organisms may occasionally be observed scattered throughout the cytoplasm. Gram-negative. Nonmotile. No Weil-Felix antibodies are elicited. Certain species are adapted to existence in ticks or, in one case, trematodes; vectors are not known for some species.

The differential characteristics of the genera of the tribe *Ehrlichieae* are presented in Table 9.6.

Table 9.6.
Differential characteristics of the genera of the tribe **Ehrlichieae**

Characteristics	I. Genus *Ehrlichia*	II. Genus *Cowdria*	III. Genus *Neorickettsia*
Mammalian hosts	Dogs, cattle, sheep, goats, horses and man	Sheep, goats and cattle	*Canidae*
Tissues parasitized in mammals	Circulating leukocytes	Vascular endothelial cells	Reticular cells of lymphoid tissue
Invertebrate vectors	Ticks: *Rhipicephalus* spp., *Ixodes* spp., *Hyalomma* spp.	Ticks: *Amblyomma* spp.	Trematodes: *Nanophyetus salmincola*
Maintenance in vectors	Transovarial and/or transstadial only	Transstadial only	Transovarial and through all stages
Distribution	Some species worldwide; others mainly in Great Britain, the Netherlands and Finland, or limited to western Japan	Africa	Pacific coast of Northern California, Oregon and Washington

Genus IV. **Ehrlichia** Moshkovski 1945, 18[AL]

MIODRAG RISTIC AND DAVID L. HUXSOLL

Ehr.lich'i.a. M.L. fem. n. *Ehrlichia* named after Paul Ehrlich, a German bacteriologist.

Small, often pleomorphic, coccoid to ellipsoidal organisms occurring intracytoplasmically, either singly or in compact inclusions (morulae) **in circulating leukocytes** of susceptible mammalian hosts. **Vectors, when known, are ticks.** Organisms may grow in tick vector. Nonmotile. Not cultivable in cell-free media or chicken embryos. Some species cultivable in blood monocyte cultures. The etiological agents of diseases of **dogs, cattle, sheep, goats, horses, and man.** The mol% G + C of the DNA is unknown.

Type species: *Ehrlichia canis* (Donatien and Lestoquard 1935) Moshkovski 1945, 18.

Further Descriptive Information

Ehrlichiae stain bluish-purple when stained by Romanowsky methods. They occur in membrane-bounded vacuoles in the cytoplasm of leukocytes, forming inclusions that contain variable numbers of organisms (Simpson, 1972, 1974; Hildebrandt, 1973) (see Fig. 9.4). The

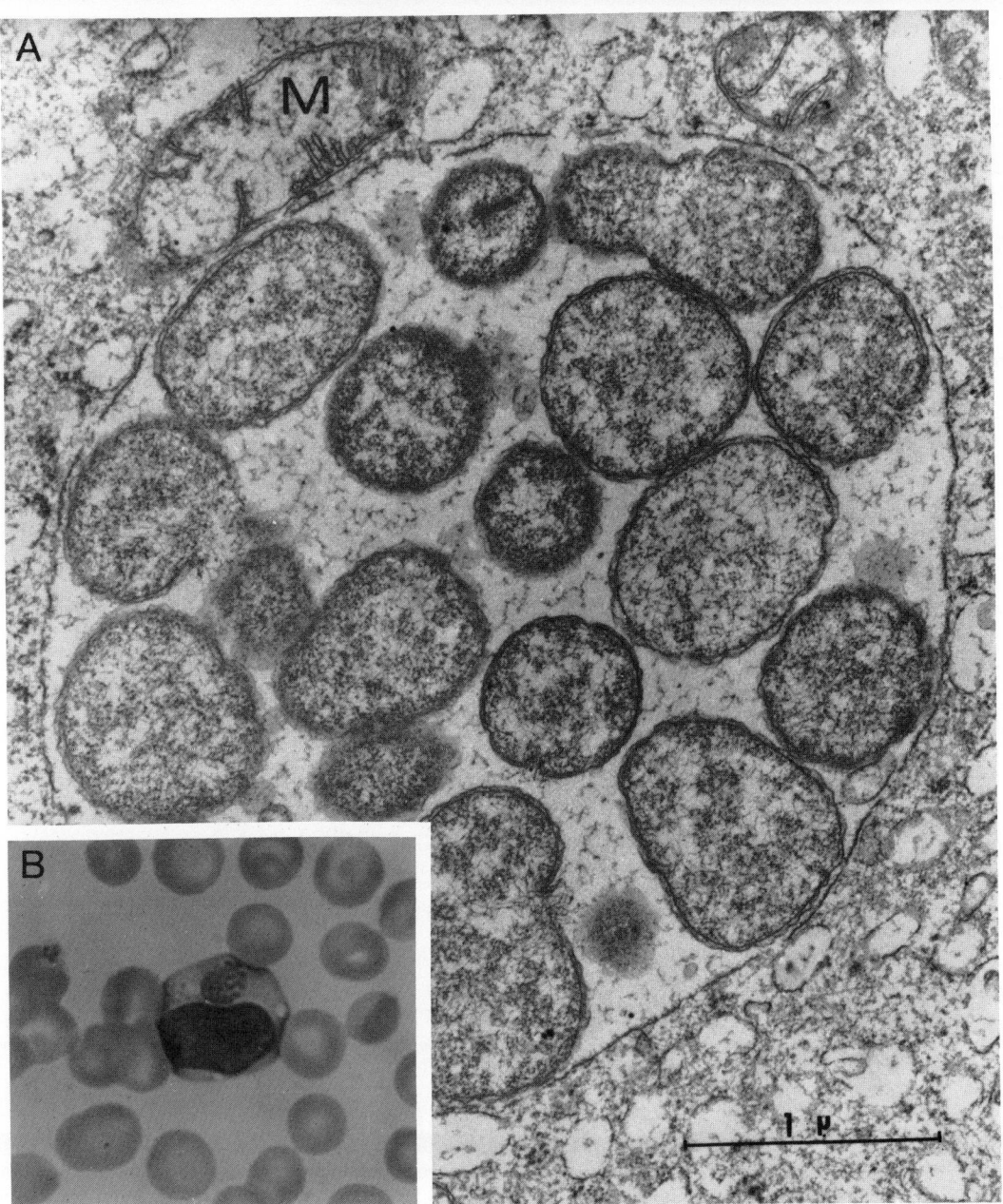

Figure 9.4. *Ehrlichia canis. A,* Ultrathin section of an infected canine blood monocyte with an intracytoplasmic inclusion body. Numerous elementary bodies with distinct plasma membranes and rippled outer cell walls are depicted. *M* = mitochondria (×70,000). (Reproduced with permission from P. K. Hildebrandt et al., *Infection and Immunity, 7:* 265–271, 1973, © American Society for Microbiology, Washington, D.C. *B,* An intramonocytic inclusion body stained by the Giemsa method (×800). (Reproduced with permission from Dr S. A. Ewing, Oklahoma State University, Stillwater.)

individual organisms are approximately 0.5 μm in diameter, and the inclusions (morulae) range in size up to 4.0 μm in diameter.

E. canis has been successfully grown in canine monocyte cultures (Nyindo et al., 1971; Stephenson and Osterman, 1977; Hemelt et al., 1980). *E. sennetsu* can be propagated in various cell cultures. No *Ehrlichia* species has been cultivated on cell-free media.

E. canis is the causative agent of canine ehrlichiosis, a febrile disease of dogs characterized by serous nasal and ocular discharges, anorexia, depression, loss of weight, elevation of the erythrocyte sedimentation rate, and pancytopenia. In many instances it is a mild disease and may go unrecognized. A severe, hemorrhagic disease, sometimes referred to as "tropical canine pancytopenia," is associated with aplastic anemia which occurs in some dogs 60 days or more following infection with *E.*

canis. Certain breeds of dogs (e.g., the German Shepherd) are more susceptible to the severe hemorrhagic form of the disease (Huxsoll, 1976). Canine ehrlichiosis has a worldwide distribution.

E. phagocytophila is the cause of natural infections of sheep and cattle and has been reported chiefly in Great Britian, the Netherlands and Finland. The organism has also been isolated from wild deer in Great Britain. Goats are susceptible. Infection with *E. phagocytophila* is sometimes referred to as "tick fever."

E. equi is the cause of equine ehrlichiosis, a febrile disease characterized by edema, lymphadenopathy, and thrombocytopenia.

E. sennetsu, originally called *Rickettsia sennetsu,* is the causative agent of human "sennetsu rickettsiosis." This disease has also been referred to as "infectious mononucleosis," "glandular fever," "Hyuga fever," and "Kagami fever." The clinical aspects may vary from mild headache, slight back pain, and a low-degree fever to a severe form of the disease characterized by persistent high fever, anorexia, lethargy, lymphadenopathy, and prominent hematologic abnormalities. The disease is limited to Western Japan.

Tetracyclines are effective in the treatment of canine ehrlichiosis and sennetsu rickettsiosis.

Isolation Procedures

In the case of *E. sennetsu*, isolation of the organism is accomplished by the inoculation of white mice with blood or diseased tissues from patients.

Maintenance Procedures

E. canis has been successfully cultured in canine monocyte cultures. Adaptation to small laboratory animals and chicken embyros has not yet been accomplished. *E. phagocytophila* can be passed serially in sheep and cattle, and strains have been adapted to growth in intact guinea pigs and splenectomized albino mice. The organisms have remained infectious in citrated blood for 10 days at room temperature, 14 days at 4–8°C, and 18 months at −79°C in infectious blood treated with glycerol and dimethylsulfoxide. *E. phagocytophila* has not been cultured in chick embyros. *E. equi* has induced experimental infections in the donkey, sheep, goat, dog, cat and monkey, but not cattle (Gribble, 1969; Lewis et al., 1975). It has not been grown in vitro. *E. sennetsu* has been propagated in vitro in African green monkey kidney cells, human amnion tissue-derived cells and in human and canine blood monocyte cultures. Dogs develop subclinical infections and the organisms can be found in monocyte cell cultures derived from the infected dogs for a period of several months.

Differentiation of the genus **Ehrlichia** from other closely related genera

See Table 9.6 for characteristics useful in differentiating *Ehrlichia* from other genera of the tribe *Ehrlichieae*.

Taxonomic Comments

E. sennetsu is listed on the Approved Lists of Bacterial Names as *Rickettsia sennetsu*; however, unlike the members of the family *Rickettsiaceae*, growth takes place in membrane-lined vacuoles in the cytoplasm of mononuclear cells (Anderson et al., 1965). These morphological attributes are characteristic of the genus *Ehrlichia*. Moreover, *E. sennetsu* shares antigens with *E. canis* (Ristic et al., 1981). Consequently, the organism has been reclassified as an *Ehrlichia* species in this edition of the *Manual*.

Further Reading

Farrell, R.K. 1966. Canine rickettsiosis. In Kirk (Editor), *Current Veterinary Therapy 1966–1967*, W. B. Saunders Co., Philadelphia, pp. 285–288.
Gribble, D.H. 1969. Equine ehrlichiosis. J. Amer. Vet. Med. Assoc. *155:* 462–469.
Hemelt, I.E., G.E. Lewis, Jr., D.L. Huxsoll and E.H. Stephenson. 1980. Serial propagation of *Ehrlichia canis* in primary canine peripheral blood monocyte cultures. Cornell Vet. *70:* 38–72.
Hildebrandt, P.K., J.D. Conroy, A.E. McKee, M.B.A. Nyindo and D.L. Huxsoll. 1973. Ultrastructure of *Ehrlichia canis*. Infect. Immun. *7:* 265–271.
Hoilien, C. 1980. Cultural, morphologic and immunologic properties of *Rickettsia sennetsu* propagated in primary human blood monocyte culture. M.S. Thesis, University of Illinois, Urbana.
Huxsoll, D.L. 1976. Canine ehrlichiosis (tropical canine pancytopenia): a review. Vet. Parasitol. *2:* 49–60.

Differentiation of the species of the genus **Ehrlichia**

Characteristics useful for distinguishing the species of *Ehrlichia* are listed in Table 9.7.

List of the species of the genus **Ehrlichia**

1. **Ehrlichia canis** (Donatien and Lestoquard 1935) Moshkovski 1945, 18.[AL] (*Rickettsia canis* Donatien and Lestoquard 1935, 419.)
ca′nis. M.L. gen. n. *canis* of the dog.

The characteristics are as given for the genus and as listed in Table 9.7.

Causative agent of canine ehrlichiosis (see Further Descriptive Information). Manifestations of the disease may be immunopathologic in nature (Nyindo et al., 1980). Rickettsiemia has been reported to last up to 5 years without accompanying clinical manifestations.

Using infected, cultured monocytes as the source of *E. canis* antigen, Ristic et al. (1972) developed an indirect fluorescent antibody test (IFA) for detection and titration of antibody to *E. canis*. In the IFA test, *E. canis* did not cross-react with a number of other rickettsial agents and common canine pathogens (Ristic et al., 1972). More recent serological studies have shown that *E. canis* cross-reacts with *E. equi* and *E. sennetsu*. The IFA test has proven useful in confirming the diagnosis of canine ehrlichiosis and in identification of clinically infected dogs.

Transstadial transmission in the tick vector *Rhipicephalus sanguineus* has been demonstrated.

The tetracyclines are effective and widely used in treatment of the disease. Dogs cleared of the infection are susceptible to reinfection and may develop disease in spite of demonstrable antibody. Low doses of tetracycline (3 mg/lb body weight/day) are effective prophylactically.

Table 9.7.
Differential characteristics of the species of the genus **Ehrlichia**

Characteristics	1. *E. canis*	2. *E. phagocytophila*	3. *E. equi*	4. *E sennetsu*
Leukocytes infected	Monocytes, lymphocytes, rarely neutrophils	Neutrophils, eosinophils, basophils, monocytes	Granulocytes	Mononuclear cells
Natural hosts	Dog (canine ehrlichiosis)	Sheep and cattle; wild deer; bison	Horses (equine ehrlichiosis) and possibly dogs	Man (sennetsu rickettsiosis)
Distribution	Worldwide	Mainly Great Britain, the Netherlands, Finland, and Austria		Western Japan and possibly Malaysia
Vector (if known)	*Rhipicephalus sanguineus*	*Ixodes ricinus* in Europe		

The mol% G + C of the DNA is unknown.

Type strain: no culture isolated.

2. Ehrlichia phagocytophila (Foggie 1951) Philip 1962, 42[AL]
(*Rickettsia phagocytophila* Foggie 1951, 4).

pha.go.cy.to′phi.la. Gr. inf. *phagein* to eat, devour; Gr. n. *kytos* a
vessel, enclosure; Gr. inf. *philein* to love; M.L. adj. *phagocytophila* fond
of devouring cells (in microbiology, attractive to phagocytes).

The characteristics are as described for the genus and as listed in
Table 9.7.

Electron microscopy of thin sections of infected sheep and calf
leukocyte concentrates reveals that the organism and inclusions are
similar to those of *E. canis.*

A relationship between Scottish and Finnish strains has been demonstrated by immunofluorescence (Tuomi, 1966).

Can be passed serially in sheep and cattle. Cross-immunity between
bovine and ovine strains is usually incomplete. From 6 to 50% of
granulocytes may show the parasites during the disease.

Transmission in the tick vector is transstadial, not transovarial.

The mol% G + C of the DNA is unknown.

Type strain: no culture isolated.

3. Ehrlichia equi Lewis, Huxsoll, Ristic and Johnson 1975, 85.[VP]

e′qui. L. gen. n. *equi* of the horse.

The characteristics are as described for the genus and as listed in
Table 9.7. See also Figure 9.5.

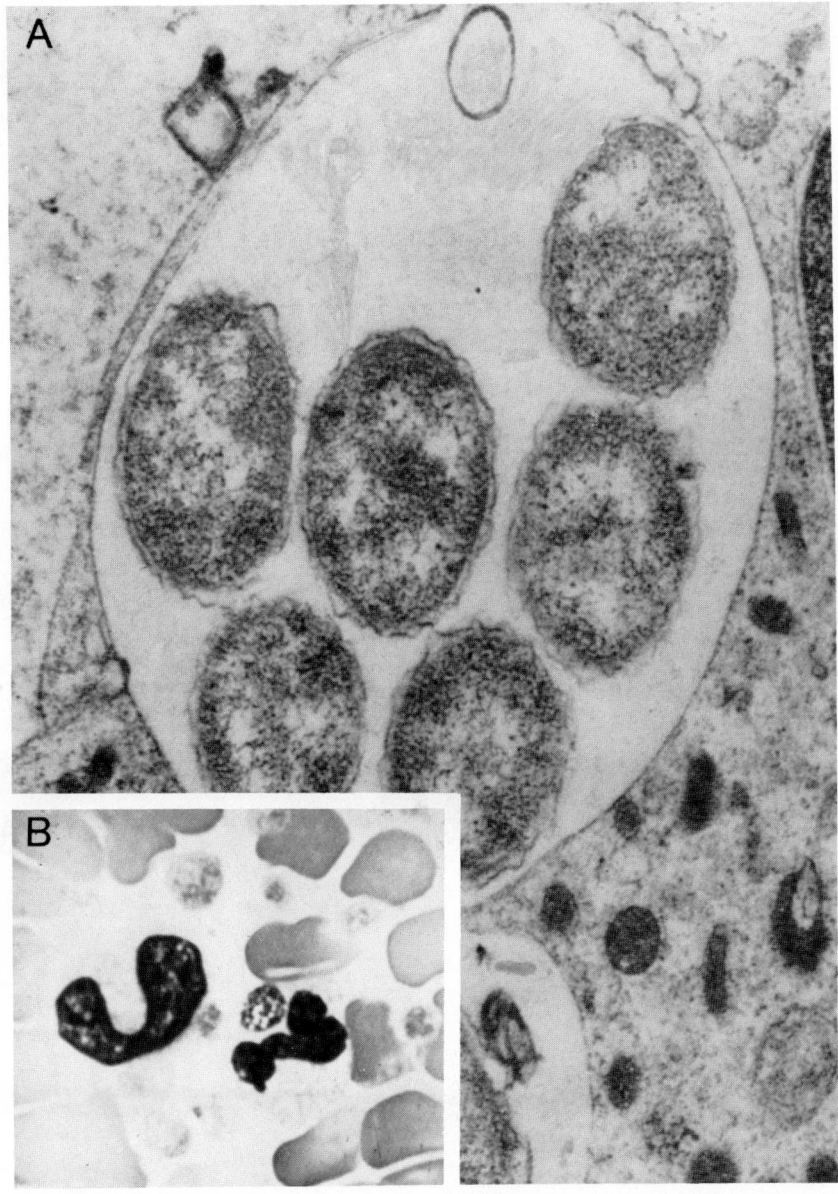

Figure 9.5. *Ehrlichia equi. A,* Ultrathin section of an infected equine blood granulocyte with an intracyto-
plasmic inclusion body. Several single organisms bound by a rippled cell wall and plasma membrane are evident
(×68,000). (Reproduced with permission from D. M. Sells et al., *Infection and Immunity, 13:* 273–280, 1976 ©
American Society for Microbiology, Washington, D. C.). *B,* An intragranulocytic inclusion body stained by the
Giemsa method (×1,200). (Reproduced with permission from Ms. Cynthia Holland, College of Veterinary
Medicine, University of Illinois, Urbana.)

The causative agent of equine ehrlichiosis (see Further Descriptive Information). Sterile immunity results from infection. Diagnosis of the disease in the horse can made on the basis of clinical and hematological observations. An indirect fluorescent antibody test, employing as antigen the granulocytes collected from acutely infected horses, can be used to confirm the diagnosis (Nyindo et al., 1978).

The mol% G + C of the DNA is unknown.

Type strain: no culture isolated.

Further Comments. A strain of *Ehrlichia* which infects granulocytes has been isolated from dogs (Ewing, 1971). The relationship of this organism to *E. equi,* which experimentally has been shown to infect dog granulocytes, has not been established.

4. **Ehrlichia sennetsu** (Misao and Kobayashi 1956) comb. nov.

sen′ne′tsu. M.L. n. *sennetsu* from Japanese, meaning glandular fever.

The characteristics are as described for the genus and as listed in Table 9.7. The morphology is depicted in Figure 9.6.

The causative agent of human sennetsu rickettsiosis (see Further Descriptive Information).

The organism shares antigens with *E. canis* (Ristic et al., 1981). Recent serological studies have indicated that *E. sennetsu* or an anti-

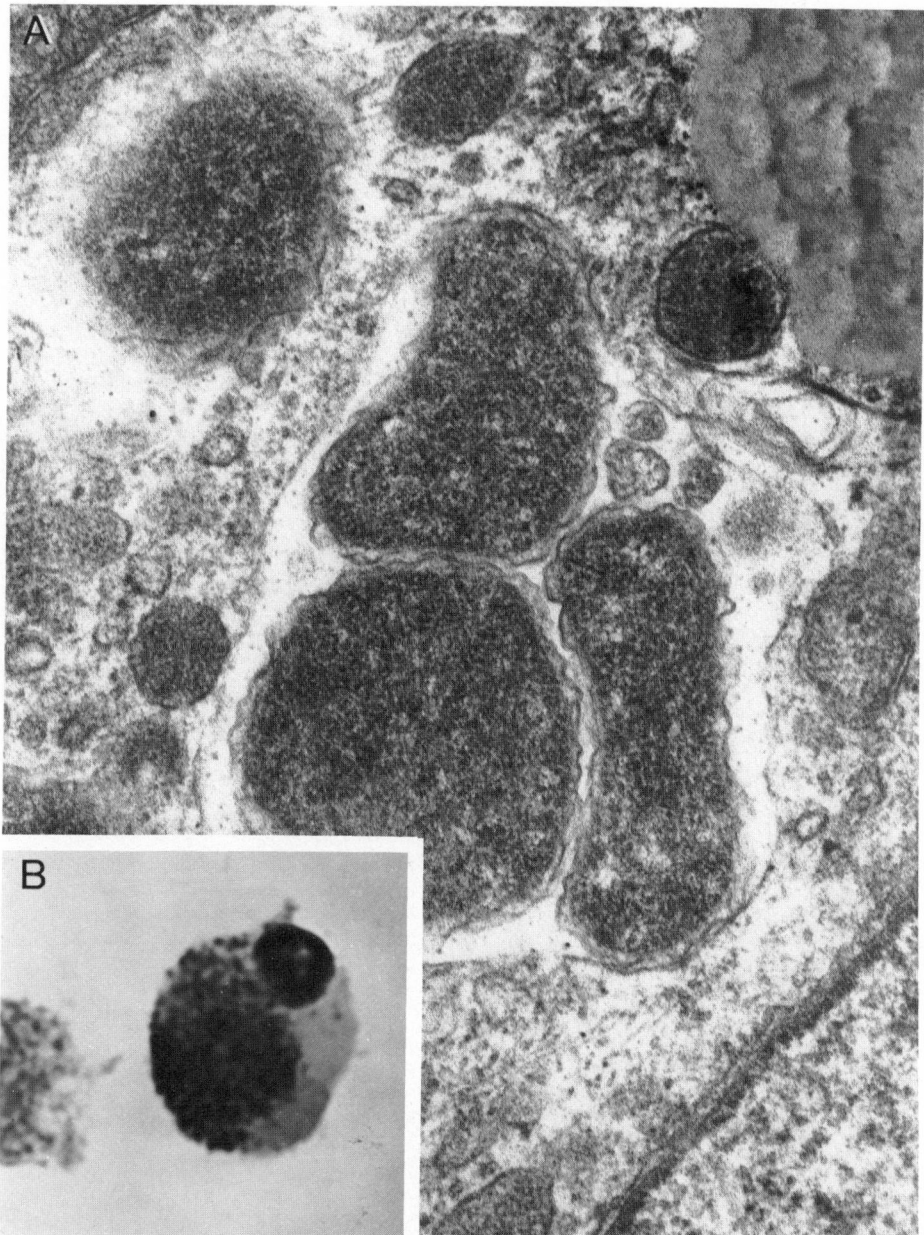

Figure 9.6. *Ehrlichia sennetsu. A,* Ultrathin section of an infected, cultured, canine blood monocyte with an intracytoplasmic inclusion body. Three elementary bodies are contained within the membrane-lined vacuole. Each elementary body is surrounded by a distinct plasma membrane and rippled outer cell wall (×75,000). *B,* Numerous individual organisms in the cytoplasm of cultured human blood monocytes stained by the Giemsa method. (Reproduced with permission from Mr. Adeyinka Cole, Ms. Cynthia Holland, and Ms. Catherine Hoilien, College of Veterinary Medicine, University of Illinois, Urbana.)

genically related organism may occur in Malaysia (C. Hoilien, M.S. Thesis, University of Illinois, Urbana, 1980).

Antigen derived from human (Hoilien et al., 1982) and canine blood monocyte cell cultures has been used in an indirect fluorescent antibody test for detection and titration of antibodies to *E. sennetsu* (C. J. Holland, M.S. Thesis, University of Illinois, Urbana, 1980).

Like other rickettsial infections, sennetsu rickettsiosis is effectively treated with tetracyclines.

The mol% G + C of the DNA is unknown.

Type strain: no culture isolated.

Species Incertae Sedis

The taxonomic relationship of the following organisms to the genus *Ehrlichia* is not known.

a. "*Ehrlichia bovis*" (Donatien and Lestoquard 1936) Moshkovski 1945, 18. (*Rickettsia bovis* Donatien and Lestoquard 1936, 1061.)

Originally isolated from circulating monocytes of Moroccan cattle to which *Hyalomma* ticks were transferred from cattle imported from Iran. Rioche (1967) has indicated a morphologic relationship of this organism to other members of the genus. However, information on immunologic and vector relationships is still inadequate.

b. "*Ehrlichia ovina*" (Lestoquard and Donatien 1936) Moshkovski 1945, 18. (*Rickettsia ovina* Lestoquard and Donatien 1936, 108.)

Observed originally in the peripheral monocytes of Moroccan sheep injected with ticks (*Rhipicephalus bursa*) from other sheep. Occurrence of the organisms in morula-like colonies resembles "*E. bovis*" but tick transmission by bite and adequate immunological relationships are unreported.

c. "*Ehrlichia kurlovi*" Moshkovski 1937, 382.

Observed in circulating monocytes of guinea pigs. The pathogenicity and rickettsial nature of so-called Kurlov's bodies has been disputed since their description by Moshkovski and later Zhdanov (1953), who compared their structure to *Cowdria ruminantium*. Bozeman et al. (1968) isolated rickettsia-like agents from guinea pigs. The nature and characteristics of these agents are described under the genus *Legionella*. These, however, should not be confused with the non-organismal "intracytoplasmic inclusion called the Kurloff body" discussed by Berendsen and Telford (1966) in light and electron micrographic studies which included supravital preparations.

d. "*Cytoecetes microti*" Tyzzer 1938, 254.

Observed in "granular leucocytes and rarely the lymphocytes" of a vole in Massachusetts. No subsequent observations have been reported. Although originally differentiated from *Aegyptianella pullorum* Carpano, microscopic similarity to *Ehrlichia* is obvious.

e. "*Cytoecetes ovis* var. *decani*" Raghavarhari and Reddy 1959, 69.

An organism observed in the blood of sheep and goats in India was reported to cause "tick-borne fever" but otherwise insufficiently characterized for systemic assignment. Since a typical "*C. ovis*" has not been described, "var. *decani*" will remain illegitimate for any future use.

f. "*Rickettsia delpyi*" Rousselot 1948, 110. (*Donatienella delpyi* Rousselot 1948, 112.)

This is an organism about which there has been no further information since its original isolation from the leukocytes of a splenectomized Iranian gerbille. However, the relationships are closer to "*Cytoecetes*" or *Ehrlichia* than to *Rickettsia sensu stricto*.

g. "*Rickettsia belgaumi*" Manjrekar 1954, 219.

Organisms found in monocytes which cause "rickettsiosis in sheep and goats in the State of Bombay, India." Differentiated from tick-borne fever and "*R. ovina*" by infection of guinea pigs and white rats. Although originally described in a mimeographed thesis, the name was effectively published in an abstract.

h. "*Ehrlichia* platys" French and Harvey 1982 (submitted to *American Journal of Veterinary Research*).

pla'tys. Gr. adj. *platys* flat, the word from which platelet is derived.

A platelet-specific parasite of dogs. The organisms stain blue with Wright-Giemsa stain. They range from 0.4 to 1.2 μm in diameter. They are round, oval, or bean-shaped and are surrounded by a double membrane. Platelets have been observed to contain from one to three single membrane-lined vacuoles with one to eight organisms per vacuole (Harvey et al., 1978). The organisms reproduce by binary fission. Although similar to *E. canis* ultrastructually, serologic studies have not demonstrated cross-reactivity between these agents (French and Harvey, unpublished results).

Following intravenous injections, parasitemias and subsequent thrombocytopenias reoccur at 1–2-week intervals. Thrombocytopenias are severe, but transient. Dogs infected with *E. platys* are not ill and evidence of hemorrhage is usually absent (Harvey et al., 1978).

The natural mode of transmission is likely to be by ticks. This agent has not been cultured in vitro. Serologic test results indicate that the organism occurs in many states of the U.S.A. (French and Harvey, unpublished results).

Genus V. **Cowdria** *Moshkovski 1947, 62*[AL]

MIODRAG RISTIC AND DAVID L. HUXSOLL

Cow'dri.a. M.L. fem. n. *Cowdria* named after E. V. Cowdry, who first described the organism in heartwater diseases of sheep, goats and cattle.

Small pleomorphic, coccoid or ellipsoidal, occasionally rod-shaped, organisms **occurring intracytoplasmically** but not intranuclearly, **and characteristically localized in clusters inside vacuoles in the cytoplasm of vascular endothelial cells of ruminants.** Not passed transovarially in **tick vectors.** Gram-negative. Nonmotile. Have not been cultivated in cell-free media. The etiological agent of **heartwater,** a septicemic disease of domestic ruminants in Africa. The mol% G + C of the DNA is not known.

Type species: *Cowdria ruminantium* (Cowdry 1925) Moshkovski 1947, 62.

Further Descriptive Information

The organisms differ morphologically from typical typhus-like rickettsiae, and usually show coccoid and ellipsoidal forms and occasionally short bacillary forms. Irregular pleomorphic forms also occur, sometimes in densely packed masses. In the vascular endothelial cells of animals, the cocci measure 0.2–0.5 μm in diameter; in tick tissues they measure 0.2–0.3 μm. Bacillary forms are 0.2–0.3 by 0.4–0.5 μm, and pairs are 0.2 by 0.8 μm. The organisms stain dark blue with Giemsa's stain; they can also be stained with methylene blue and other basic aniline dyes.

The organisms persist in recovered cattle, sheep and goats for periods up to 60 days. Immunity may last 3 months to 5 years after termination of the carrier state. During the period of sterile immunity, *C. ruminantium*—when reintroduced by infected ticks (*Amblyomma* spp.)—results in temporary subclinical infection. A significant characteristic of the agent is its transstadial, but not transovarial, transmission in the tick vectors. At least 5 species of wild ruminants have been shown to be susceptible. Microscopic evidence suggests that the organisms persist up to 90 days as inapparent infections in laboratory mice and rats, but cannot be serially passed in them.

Diagnosis of the disease is made by microscopic detection of the organism in the capillaries of the brain cortex obtained by biopsy or subinoculation of blood from infected animals intravenously into a susceptible ruminant. A capillary flocculation test using brain extract of infected animals as antigen enables serologic detection of recently recovered animals (A. A. Ilemobade, Doctoral Thesis, Ahmadu Bello University, Zaria, 1976).

Maintenance Procedures

Two methods are useful for the preservation of infectious organisms: (i) preservation of whole blood collected from animals during acute infection mixed with 10% dimethyl sulfoxide (DMSO) and stored in liquid nitrogen, and (ii) tissue homogenates from infected ticks mixed with 10% DMSO and stored in liquid nitrogen.

Differentiation of the genus **Cowdria** from other closely related genera

Characteristics useful for differentiation of *Cowdria* from other members of the tribe *Ehrlichieae* are presented in Table 9.6.

Taxonomic Comments

The phylogenetic relationship of the genus *Cowdria* to other members of the family *Rickettsiaceae* is not known. Nucleic acid studies are difficult because the organisms have not been cultivated in cell-free media.

Further Reading

Gilenberg, G. 1981. Heartwater disease. In Ristic and McIntyre (Editors), *Disease of Cattle in the Tropics*, Martinus Nijhoff Publishers, The Hague, pp. 345–360.

List of the species of the genus **Cowdria**

1. **Cowdria ruminantium** (Cowdry 1925) Moshkovski 1947, 62.[AL] [*Rickettsia ruminantium* Cowdry 1925, 231].

ru.mi.nan'ti.um. M.L. gen. pl. n. *ruminantium* of *Ruminantia*, formerly a common name for cud-chewing animals.

The characteristics are as described for the species.
The mol% G + C of the DNA is unknown.
Type strain: no culture isolated.

Genus VI. **Neorickettsia** Philip, Hadlow and Hughes 1953, 257[AL]

Miodrag Ristic and David L. Huxsoll

Ne.o.rick.ett'si.a. Gr. pref. *neo-* new; M.L. fem. n. *Rickettsia* type genus of the family *Rickettsiaceae*; M.L. fem. n. *Neorickettsia* the new *Rickettsia*.

Small, coccoid, often pleomorphic, intracytoplasmic organisms which occur primarily in reticular cells of lymphoid tissues of *Canidae*. Also seen in certain tissues of mature **fluke vectors**; all other fluke stages, eggs, rediaecercariae and metacercariae have been proven infectious by injection into susceptible vertebrate hosts, which confirms that the infectious cycle includes transovarial transmission in the vector. Gram-negative. Nonmotile. Not cultivable in cell-free media or in chicken embryos. Sensitive to tetracycline antibiotics. Has been reported from the Pacific Coast of Northern California, Oregon, and Washington. The mol% G + C of the DNA is not known.

Type species: *Neorickettsia helminthoeca* Philip, Hadlow and Hughes 1953, 257.

Further Descriptive Information

N. helminthoeca is found in the cytoplasm of reticuloendothelial cells of the lymphoid tissues of infected *Canidae*. In Giemsa-stained preparations the organisms are seen in single or multiple colonies, or as single organisms scattered throughout the cytoplasm. Coccoid forms 0.3–0.4 μm in diameter are most common, but pleomorphic forms include ellipsoids, short rods, and even clubs and rings.

The organisms are not detected microscopically in circulating lymphocytes, although blood is infectious by inoculation of susceptible hosts. The organism has been propagated in canine monocytes in culture (Brown et al., 1972). Infected cultured canine monocytes have been used as an antigen in an indirect fluorescent antibody test for the detection and titration of antibodies. *N. helminthoeca* does not persist after recovery of the canine host from clinical disease. Infection cannot be reactivated by splenectomy.

The natural cycle of this pathogen was shown (Nybert et al., 1967) to include passage through trematode eggs, other trematode stages in snails, salmonid fish and the dog. The dog appears to be an accidental host. Up to 90% mortality occurs in untreated dogs. The disease in dogs is known as "salmon poisoning" disease. Much remains to be learned about the epizootiology of *N. helminthoeca*.

Another fluke-borne disease of dogs occurring in the same region as salmon poisoning disease and transmitted by the same trematodes is referred as Elokomin fluke fever (Farrell, 1966). Animals recovered from Elokomin fluke fever are not immune to salmon poisoning disease.

Maintenance Procedures

The organism is preserved by storage in liquid nitrogen of infected primary blood monocyte cultures and/or by preparing a 20% suspension of homogenized infected lymph nodes and spleen tissues in Snyder I solution containing bovine serum albumin and storing at −80°C.

Differentiation of the genus **Neorickettsia** from other closely related genera

Differential characteristics are listed in Table 9.6. The organism is distinguished from both tick-borne *Ehrlichia* and *Cowdria* in its adaptation to the complicated life cycle of the trematode vector and its geographic isolation on the Pacific coast of the United States.

Taxonomic Comments

The phylogenetic relationship of *Neorickettsia* to other members of the family *Rickettsiaceae* is not known. However, as indicated above, the adaptation of the organism to flukes and the geographic isolation suggest that the genus is probably phylogenetically distinct from the other members of the tribe *Ehrlichieae*.

Further Reading

Brown, J.L., D.L. Huxsoll, M. Ristic and P.K. Hildebrandt. 1972. In vitro cultivation of *Neorickettsia helminthoeca*, the causative agent of salmon poisoning disease. Amer. J. Vet. Res. *33*: 1695–1700.

Nyberg, P.A., S.E. Knapp and R.E. Milleman. 1967. "Salmon poisoning: disease. IV. Transmission of the disease to dogs by *Nanophyetus salmincola* eggs. J. Parasitol. *53*: 694–699.

List of the species of the genus Neorickettsia

1. **Neorickettsia helminthoeca** Philip, Hadlow and Hughes 1953, 257.[AL]

hel.minth'oe.ca. Gr. n. *helmins, helminthis* worm; Gr. n. *oikos* house; M.L. fem. adj. *helminthoeca* worm-dwelling.

The characteristics are as described for the genus. The mol% G + C of the DNA is unknown. *Type strain:* no culture isolated.

TRIBE III. **WOLBACHIEAE** PHILIP 1956, 266[AL]

EMILIO WEISS, GREGORY A. DASCH AND KWANG-POO CHANG

Wol.ba'chi.e.ae. M.L. fem. n. *Wolbachia* type genus of the tribe; *-eae* ending to denote a tribe; M.L. fem. pl. n. *Wolbachieae* the *Wolbachia* tribe.

A miscellaneous group of organisms associated with invertebrates, mostly arthropods, but not usually invading vertebrates. Most of these organisms have not been cultivated outside their hosts, and only in rare instances have they been grown on bacteriologic media.

Four genera have been described, but characterization is usually not as adequate as for the preceding genera of the order *Rickettsiales*. There is no evidence of an evolutionary relationship among the genera of this tribe, or of any of these genera to the organisms described in Section 11 of this *Manual*. However, the possibility of some evolutionary relationships cannot be excluded.

Key to the genera of the tribe **Wolbachieae***

I. Associated with a number of arthropods. Not obviously pathogenic or beneficial to their hosts, although in some instances they may interfere with reproduction. Do not reside in specialized cells or organs.

Genus VII. *Wolbachia*

II. Pathogenic for insect larvae and for other invertebrate hosts.

Genus VIII. *Rickettsiella*

Genus VII. **Wolbachia** Hertig 1936, 472[AL]

EMILIO WEISS, GREGORY DASCH AND KWANG-POO CHANG

Wol.ba'chi.a. M.L. fem. n. *Wolbachia* named after S. Burt Wolbach, who described the rickettsial agent of Rocky Mountain spotted fever and, in collaboration with Marshall Hertig, studied the rickettsia-like microorganisms of insects.

Heterogeneous group of small rods and coccoid forms which morphologically resemble either *Rickettsia* or *Chlamydia*. **Gram-negative.** Stain well with Giemsa and some with Giménez stain, but poorly with most bacterial stains. Cultural requirements quite variable. **Associated with arthropods.** May induce reproductive incompatibility in their hosts, but otherwise not pathogenic for invertebrates or vertebrates. Extracellular or intracellular, but seldom developing inside mycetomes.

Type species: *Wolbachia pipientis* Hertig 1936, 472.

Further Descriptive Information

Although it is useful to examine the broad characteristics of the genus *Wolbachia*, as is done for the other genera, it is well to take into account that the degree of phenotypic similarity between species is small and no information is available on genetic relationships.

The morphological features are given in Table 9.8. *W. pipientis* stains most reliably by prolonged application of Giemsa's stain, diluted in buffer (pH 7.2–7.4). *W. persica* also stains best by the Giemsa method following Carnoy's fixation. *W. melophagi* closely resembles *Rochalimaea* in morphology, size and staining properties.

The fine structure of *W. pipientis* is typical of Gram-negative bacteria, but a peptidoglycan layer has not been observed and the cells display some plasticity (Yen and Barr, 1974; Wright and Wang, 1980; Wright and Barr, 1980). Bacteriophage-like particles have been seen in some instances (Wright et al., 1978). The outer surface of *W. persica* is not clearly separated into cell wall and cytoplasmic membrane and the cell displays considerable plasticity and instability in suspending media of low ionic strength. The presence of at least a remnant of cell wall in *W. persica* is indicated, however, by the marked effect that penicillin has on its morphology (Suitor and Weiss, 1961; Suitor, 1964; Burgdorfer et al., 1973).

W. pipientis multiplies by binary fission in the vacuoles of host cells and is normally surrounded by a membrane of host origin. *W. persica* also is located within host cell vacuoles, and the organisms form aggregates that are typically surrounded by a membrane of host origin. In contrast to these two species, *W. melophagi* occurs extracellularly in the lumen of the gut of its insect host.

Only one species, *W. melophagi*, has been cultivated on a non-living medium—a blood-glucose-bouillon agar—by Nöller (1917), who obtained colonies 0.4–0.6 mm in diameter after incubation for 35–40 days, by Hertig and Wolbach (1924), who detected minute colonies after 3–5 days, and by Kligler and Aschner (1931), who cultivated organisms

---

* *Editorial Note.* The genus *Blattabacterium* Hollande and Favre 1931, 754 (symbionts of cockroaches) and the genus *Symbiotes* Philip 1956, 267 (symbionts of bedbugs) were included in the Tribe *Wolbachieae* in the eighth edition of the *Manual*. However, it is not at all certain that they belong to this tribe. In the present edition of the *Manual*, therefore, the description of *Blattabacterium* is given in the article on *Endosymbionts of Insects* in Section 11. With regard to *Symbiotes*, recent information suggests that the definition of the genus as one containing pleomorphic organisms may not be correct; accordingly, no formal description of this genus has been included in this edition of the *Manual*. However, see a summary of the available information concerning symbionts of bedbugs (Family Cimicidae) in the article on *Endosymbionts of Insects* in Section 11.

derived from the sheep ked as well as from diptera infecting the goat, horse and dog. It has also been grown in the yolk sac of chicken embryos by Steinhaus (1946) and, more recently, by Henneberg and Wolff (1963). *W. persica* also can be cultivated in the yolk sac of chicken embryos, where it grows profusely and kills the embryos in 5–10 days depending on the concentration of the inoculum. *W. persica* also grows well in several types of mammalian or insect cell cultures. In contrast to *W. melophagi* and *W. persica*, *W. pipientis* has never been successfully cultivated outside its natural host or in cell cultures derived from the host.

The natural host range differs among the three species of *Wolbachia* (see Table 9.8). *W. pipientis* was originally seen in the gonads of the mosquito, *Culex pipiens* (Hertig and Wolbach, 1924; Hertig, 1936). It has been described in all members of the *C. pipiens* group examined and in a large number of the *Aedes scutellaris* group (Beckett et al., 1978; Wright and Wang, 1980), and also has been detected in the almond moth, *Ephestia cautella*, a cosmopolitan pest of dry fruits, grains, nuts, and other stored vegetable products (Kellen et al., 1981). *W. melophagi* occurs in virtually every sheep ked, *Melophagus ovinus*, and occupies an extracellular position in the lumen of the gut of the host. It is seen as a closely packed row of small rods lining the intestinal epithelium. *W. persica* was first isolated from the tick *Argas persicus* (which was later reclassified as *A. arboreus*), feeding on the buff-backed heron, *Bubulcus ibis* in Egypt. A related organism was isolated from *Dermacentor andersoni,* and *W. persica*-like organisms were seen in *Rhipicephalus sanguineus* and several other species of ticks collected in the United States (Hayes and Burgdorfer, 1979, 1981), *Ornithodorus moubata* originating in Tanzania (Reinhardt et al., 1972) and from several other genera of ticks in other parts of the world.

Concerning pathogenicity, *W. pipientis* does not cause obvious damage to its host cells; however, interest in this organism is derived from its role in incompatibility between strains (or subspecies) of the host. The wolbachiae have been detected in eggs, sperm cells, ovaries and associated epithelium of the insects, but seldom in other cells, and matings of insects from different geographical areas sometimes produce eggs that do not hatch, since females cannot be fertilized with the infected sperm of certain strains. The insects can be rendered aposymbiotic (nearly or completely freed of microorganisms) by the addition of 17–50 μg/ml of tetracycline to the diet of the larvae or by rearing the larvae at 32–33°C for 5–7 days. Incompatibility is eliminated by rendering the males aposymbiotic. This phenomenon has been demonstrated in both groups of mosquito and in the almond moth, and microbial reproductive incompatibility has been suggested as a possible approach to pest control (Yen and Barr, 1974; Wright and Wang, 1980; Kellen et al., 1981). It should be noted that in other insects aposymbiosis has the opposite effect, i.e., it results in infertility (see the article on Endosymbionts of Insects, in Section 11.

With regard to pathogenicity of *W. melophagi,* there is no evidence that the organism injures its host or infects sheep heavily infested with

sheep keds. Except for a single report of infection of guinea pigs (Henneberg and Wolff, 1963), the common laboratory animals are said to be insusceptible.

W. persica, also, is not pathogenic for its tick host in most cases; however, the organism, which normally resides in the Malpighian tubules and ovaries of its host, has been reported to invade other tissues. Intracoelomic injection of a heavy suspension results in a lethal infection (Burgdorfer et al., 1973), but reinfection of aposymbiotic ticks by feeding on heavily infected chicken embryos does not appear to be harmful (Suitor, 1964a). The *Argas* and *Dermacentor* isolates are pathogenic for vertebrates only when injected in very high concentrations.

Little is known of the antigenic properties of wolbachiae, except that the sera of guinea pigs and hamsters surviving infection with the *Dermacentor* isolate of *W. persica* cross-react in fluorescent antibody tests with the *Argas* isolate (Burgdorfer et al., 1973).

Isolation Procedures

Suitor and Weiss (1961) and Suitor (1964a) repeatedly isolated *W. persica* by the following procedure. Three or four ticks are immersed in a 0.1% solution of Merthiolate for 5 min (1 h by Burgdorfer et al., 1973) and subsequently washed 4 times in a buffered salt solution to eliminate surface bacteria. The Malpighian tubules and other organs are removed aseptically under a dissecting microscope and triturated with a glass rod in a tube. Sufficient salt solution is added to inoculate 4–10 chicken embryos, via the yolk sac, by the procedure commonly used for rickettsiae. Embryos dying within the first 3 days after inoculation are discarded, those dying later, usually on the 6th day, are examined individually for the presence of coccoid bodies and freedom from other bacteria. The coccoid bodies are most clearly visualized by Giemsa staining following Carnoy's fixation. For the production of seed and other tests the isolates are passed in eggs.

For the isolation of *W. melophagi,* Kligler and Aschner (1931) immersed the flies in 5% tincture of iodine for 5–10 sec, quickly washed them with 95% ethanol and rinsed them repeatedly with saline. The intestines were dissected out aseptically and placed on various types of blood agar media. Somewhat similar sterilization and dissection procedures were used for the inoculation of chicken embryos (Steinhaus, 1946; Henneberg and Wolff, 1963).

There is no record of isolation of *W. pipientis.* It might be worthwhile to attempt to cultivate mosquito cells containing numerous *W. pipientis.* If antibiotics are needed, the technique employed for the production of aposymbiosis can be applied to identify the antibiotics to which *W. pipientis* is resistant.

Maintenance Procedures

The viability of suspensions of *W. persica* in sucrose, phosphate buffer, glutamate solution (SPG, Bovarnick et al., 1950) was retained at least a decade at −70°C. No systematic effort has been made to retain the other two species.

Differentiation of the genus **Wolbachia** from other closely related genera

See the key to the tribe *Wolbachieae* for characteristics useful in differentiating *Wolbachia* from the other genus of the tribe.

Taxonomic Comments

Although strains of *W. pipientis* have been detected in the gonads of a number of insects other than the original *Culex pipiens,* they have not been successfully cultivated outside the host or in cell cultures derived from the host, and identification by serological methods has not been reported. Therefore the degree of relatedness of these strains to one another remains unknown. For purposes of description in this edition of the *Manual,* they have been regarded as strains of the same species, *W. pipientis.* See previous editions of the *Manual* for the other specific names that have been proposed.

W. melophagi appears to exhibit a number of similarities to the genus *Rochalimaea,* including morphology, cell size, staining properties, extracellular location in the lumen of the gut of its insect host, and

cultivation on nonliving media. The relationship of this species to *Rochalimaea* deserves to be investigated.

Occasionally tick symbiotes do not resemble *W. persica* morphologically (Řeháček et al., 1976); moreover, symbiotes that do resemble *W. persica* are seen in arthropods other than ticks. The proper classification of such symbiotes remains unknown and deserves investigation.

Further Reading

Beckett, E.B., B. Boothroyd and W.W. MacDonald. 1978. A light and electron microscope study of rickettsia-like organisms in the ovaries of mosquitoes of the *Aedes scutellaris* group. Ann. Trop. Med. Parasitol. *72:* 277–283.

Burgdorfer, W., L.P. Brinton and L.E. Hughes. 1973. Isolation and characterization of symbiotes from the Rocky Mountain wood tick, *Dermacentor andersoni.* J. Invert. Pathol. *22:* 424–434.

Kellen, W.R., D.F. Hoffman and R.A. Kwock. 1981. *Wolbachia* sp. (*Rickettsiales: Rickettsiaceae*) a symbiont of the almond moth, *Ephestia cautella*: ultrastructure and influence on host fertility. J. Invert. Pathol. *37:* 273–283.

Table 9.8.
Differential characteristics of the species of the genus **Wolbachia**

Characteristics	1. *W. pipientis*	2. *W. melophagi*	3. *W. persica*
Morphology	Coccoid forms <1 μm in diameter, or rod forms 0.3–1.5 μm	Coccoid forms, 0.4–0.6 μm, or short rods, 0.3–1.0 μm	Coccoid, ~1 μm in diameter, often elongated (up to 3.4 μm in length) especially when undergoing cell division
Cultivation			
Nonliving media	−	+[a]	−
Mammalian or insect cell cultures	−	−	+
Yolk sac of chicken embryos	−	+	+
Natural hosts	Mosquitoes: *Culex pipiens, Aedes scutellaris* Almond moth: *Ephestia cautella*	Sheep keds: *Melophagus ovinus* Diptera infesting horses, goats, and pigs	Ticks: *Argas* spp., *Dermacentor andersoni, Rhipicephalus sanguineus, Ornithodorus moubata,* and other ticks
Growth in host:	In vacuoles of host cells	Extracellular in lumen of gut	In vacuoles of host cells

[a] On a blood-glucose-bouillon agar.

Steinhaus, E.A. 1946. *Insect Microbiology.* Comstock Publishing Co., Ithaca, New York.

Suitor, E.C., Jr., and E. Weiss. 1961. Isolation of a rickettsialike microorganism (*Wolbachia persica* n. sp.) from *Argas persicus* (Oken). J. Infect. Dis. *108:* 95–106.

Wright, J.D., and A.R. Barr. 1980. The ultrastructure and symbiotic relationships of *Wolbachia* of mosquitoes of the *Aedes scutellaris* group. J. Ultrastruct. Res. *72:* 52–64.

Yen, J.H. and A.R. Barr. 1974. Incompatibility in *Culex pipiens. In* Pal and Whitten (Editors), *The Use of Genetics in Insect Control,* Elsevier, Amsterdam, pp. 97–118.

Differentiation of the species of the genus **Wolbachia**

Characteristics useful for differentiation of the species are listed in Table 9.8.

List of the species of the genus **Wolbachia**

1. **Wolbachia pipientis** Hertig 1936, 472[AL]
pi.pi.en′tis. M.L. n. *pipiens* specific epithet of the host mosquito, *Culex pipiens;* M.L. gen. n. *pipientis* of *pipiens.*

The characteristics are as described for the genus and as listed in Table 9.8.

The mol% G + C of the DNA is unknown.

Type strain: no culture isolated.

2. **Wolbachia melophagi** (Nöller 1917) Philip 1956, 267.[AL] (*Rickettsia melophagi* Nöller 1917, 70.)
me.lo.pha′gi. M.L. gen. n. *melophagi* of *Melophagus;* named after the genus of its natural host, *Melophagus ovinus,* a wingless fly commonly called sheep ked (sometimes incorrectly sheep tick).

The characteristics are as described for the genus and as listed in Table 9.8.

This species was of great interest to some of the investigators of the pathogenic rickettsiae during the second and third decades of this century, but interest has dwindled in recent times, possibly because sheep ked infestation has sharply declined.

The mol% G + C of the DNA is unknown.

Type strain: no culture isolated.

3. **Wolbachia persica** Suitor and Weiss 1961, 105.[AL]
per′si.ca. L. fem. adj. *persica* from the specific epithet of the reputed host tick, *Argas persicus.*

The characteristics are as described for the genus and as listed in Table 9.8.

W. persica actively metabolizes glucose (in contrast to *Rickettsia* species), as well as serine and glutamine, and other substrates, including glutamate (Weiss et al., 1962, 1964).

Erythromycin, chloramphenicol, tetracycline and *p*-aminobenzoic acid inhibit growth in chick embryos.

The mol% G + C of the DNA is approximately 30 (T_m, Bd) (Kingsbury and Weiss, 1968).

Type strain: ATCC VR-331 [first *Argas persicus* (*arboreus*) isolate (Suitor and Weiss, 1961)].

Genus VIII. **Rickettsiella** Philip 1956, 267[AL]

EMILIO WEISS, GREGORY A. DASCH AND KWANG-POO CHANG

Rick.ett.si.el′la. M.L. dim. ending *-ella;* M.L. fem. n. *Rickettsia* genus of parasitic bacteria; M.L. fem. n. *Rickettsiella* a small *Rickettsia.*

The infectious forms are **Gram-negative rod- or disk-shaped organisms,** usually smaller than those of the genus *Rickettsia,* but developing intracellularly into larger particles that multiply and reform the smaller forms in a cycle that resembles that of *Chlamydia.* Sometimes **produce or induce the formation of large crystalline bodies.** Growth takes place in cell vacuoles of the fat body, hepatopancreas, and other organs of invertebrate hosts. In some instances they have been cultivated in invertebrates other than the host of origin and rarely and only for a few passages in vertebrate and invertebrate cell cultures. Have not been grown in cell-free media. **Pathogenic for their larval hosts and young and mature stages of other invertebrate hosts,** but of little virulence for vertebrates. Natural hosts include insects, crustaceans, and arachnids.

Type species: *Rickettsiella popilliae* (Dutky and Gooden 1952) Philip 1956, 267.

Further Descriptive Information

Most of the knowledge of this genus is based on light and electron microscopic observations. Studies of host specificity, attempts at cultivation in cell or axenic media, or antigenic analyses have not been

very extensive. Table 9.9 lists the species which to the best of our limited knowledge appear to belong to this genus.

All strains that have been studied in some detail have been shown to undergo a cycle of development. The infectious form is a small, dense particle which gains entrance into a vacuole of its host cell by phagocytosis. In its intracellular environment it enlarges into intermediate and large forms of much lesser electron density which are capable of multiplication. As the host-cell vacuole becomes crowded with multiplying bacteria, the large forms condense to reform the small dense particles which eventually escape from the vacuole to start another cycle. There are striking differences between the cycles of *R. popilliae* and *R. chironomi*. The cycle of *R. grylli* resembles that of *R. popilliae*.

R. popilliae cells, although displaying some plasticity, retain typical

Table 9.9.

Ecology of the species of the genus **Rickettsiella** *and nomenclature proposed*

Nomenclature Used in Present Edition of the *Manual*	Hosts (Class, order, common name, and genus and species)			Location	Subjective Synonyms	References
1. *R. popilliae*	Insecta;	Coleoptera;	Japanese beetle (*Popillia japonica*)	U.S.A.		Dutky and Gooden, 1952; Philip, 1956
			Other scarabeid and carabid beetles	U.S.A.; U.K.	Unnamed	Sutter and Kirk, 1968; Carter and Luff, 1977
			Cockchafer (*Melolontha melolontha*)[a]	Germany	"*Rickettsiella (Rickettsia) melolonthae*"	Krieg, 1955; Philip, 1956
			Mealworm (*Tenebrio molitor*)	Germany	"*Rickettsiella tenebrionis*"	Krieg, 1965
			Cetonid beetle (*Cetonia* sp.)	Madagascar	"*Rickettsiella cetonidarum*"	Meynadier and Monsarrant, 1969
		Diptera;	Crane fly (*Tipula paludosa*)	Germany	"*Rickettsiella tipulae*"	Müller-Kogler, 1958
		Dictyoptera;	Cockroach (*Blatta orientalis*)	Germany	"*Rickettsiella blattae*"	Huger, 1964
2. *R. grylli*	Insecta;	Orthoptera;	Cricket (*Gryllus bimaculatus*)[a]	France	*Rickettsiella grylli*	Vago and Martoja, 1963
			Desert locust (*Schistocerca gregaria*)	Jordan	"*Rickettsiella schistocercae*"	Vago and Meynadier, 1965
	Crustacea;	Isopoda;	Isopode (*Armadillidium vulgare*)[a]	France	"*Rickettsiella armadillidii*"	Vago et al., 1970
		Amphipoda;	Amphipod (*Crangonyx floridanus*)	U.S.A.	Unnamed	Federici et al., 1974
3. *R. chironomi*	Insecta;	Diptera;	Midge (*Chironomus tentans*)[a]	Germany	*Rickettsiella chironomi*	Weiser, 1963
	Arachnida;	Aranea;	Spider (*Argyrodes gibbosus*)	Spain (Canary Islands)	Unnamed	Meynadier et al., 1974
			Spider (*Pisaura mirabilis*)	France	Unnamed	Morel, 1977
		Scorpionida;	Scorpion (*Buthus occitanus*)	France	"*Porochlamydia buthi*"	Mortel, 1976
Organisms for which insufficient data exist to support assignment to a species	Insecta;	Coleoptera;	Stethorus beetle (*Stethorus punctum*)	Morocco	*Rickettsiella stethorae*	Hall and Badgley, 1957
		Lepidoptera;	Saturnid moth (*Samia cynthia*)	U.K.	Unnamed	Entwhistle et al., 1968
			Navel orangeworm (*Paramyelois transitella*)	U.S.A.	Unnamed	Kellen et al., 1972
	Arachnida;	Acarina;	Mite (*Phytoseiulus persimilis*)	U.S.S.R.	"*Rickettsiella phytoseiuli*"	Sůtáková and Rüttgen, 1978
	Crustacea;	Decapoda;	Crab (*Carcinus mediterraneus*)	France (Mediterranean sea)	Unnamed	Bonami and Pappalardo, 1980

[a] And related species.

outer membranes and a rod-shape appearance throughout most of the cycle (Devauchelle et al., 1972). Some large forms become giant round cells from which bipyramidal crystalline bodies (0.8 by 1.8 by 3.8 μm in size) arise. The roles of the bacteria and of the host cell in the production of the crystalline bodies are not entirely clear. Except for the absence of tyrosine, these bodies have approximately the same amino acid composition as the albuminoid spheres, found in greatest number in late larval and pupal stages in normal insects. It has been postulated that the crystalline bodies derive from the albuminoid reserve as the result of a disturbance of host cell metabolism (Huger, 1959; Krieg, 1959). Other evidence suggests a close association with the bacterial cycle (Huger, 1962; Devauchelle et al., 1972), but also host dependence. A *Cetonia* strain (Table 9.9) produces crystalline bodies in Coleoptera, but not in Orthoptera or Lepidoptera (Meynadier and Monsarrat, 1969) and a *Melolontha* strain does not produce crystalline bodies in mammalian cell cultures (Pourquier et al., 1963).

In *R. chironomi* the infectious particle is a disk-shaped elementary body 0.06 by 0.6 μm in size. This enlarges into a spherical initial body 1 μm in diameter which divides by binary fission. The initial bodies are eventually reduced to intermediate forms which subsequently condense to reform the elementary bodies. Occasionally the initial bodies continue to grow to particles 1.5 by 2.0 μm in diameter which divide equally or unequally (Federici, 1980), or which may undergo multiple divisions resulting in as many as 30 elementary bodies (Götz, 1972). However, unlike *R. popilliae*, the giant cells in *R. chironomi* do not evolve into typical crystalline bodies (Weiser and Žižka, 1968).

Virulence for invertebrate hosts, other than the host of original isolation, has been studied in a few cases. Most of the strains of *R. popilliae* have been grown in *Melolontha*. The *Cetonia* strain of *R. popilliae* and *R. grylli* have been grown in Orthoptera, Coleoptera, and Lepidoptera (Meynadier and Monsarrat, 1965). *R. popilliae* (*Melolontha* strain) is virulent for the mouse only when massive doses are inoculated intraperitoneally (Krieg, 1955a). Successful infection by the intranasal route has also been demonstrated (Giroud et al., 1958). *R. grylli* also multiplies in the mouse inoculated by the intraperitoneal route or by inhalation, but the mouse overcomes the infection and the microorganisms disappear in about two weeks (Croizier and Meynadier, 1972).

Limited growth of *R. popilliae* has been obtained in chicken embryo entodermal cell cultures and in McCoy cells incubated at 28 or 32°C. Typical organisms were first detected after 1 week and their numbers increased during the following 2 weeks. During serial passage infectivity was reduced progressively and no growth was demonstrated after four or five passages (Suitor, 1964b). *R. popilliae* has also been passed three times in other mammalian cell lines (Pourquier et al., 1963). *R. grylli* was grown in cricket cardiac cell cultures (Meynadier et al., 1967).

The chief interest in rickettsiellae stems from their effect on laboratory insectaries and other animal collections, as well as experimentation as control agents for agricultural pests. Rickettsiellae are not as virulent and persistent as some of the viral and mycotic pathogens of agricultural pest insects, but are maintained in the soil for years. Infection of offspring is effected through contamination of soil, rather than by transovarian passage (Hurpin and Robert, 1972, 1976, 1977). The wide distribution of *Rickettsiella* geographically and in arthropod taxa suggests an early appearance in the course of evolution.

Isolation Procedures

Successful propagation of *Rickettsiella* has been achieved only by the inoculation of healthy larvae of the same or compatible species with infected blood or other infectious material. Japanese beetles or cockchafers inoculated with *R. popilliae* are incubated under appropriate conditions of temperature, food and moisture for about 35 days. The organisms are obtained from the surviving larvae by grinding following surface sterilization. The bacteria can be separated from tissue components by cycles of adsorption and dilution from Celite and filtration through diatomaceous filters of medium porosity (Dutky, 1959) or by comparable more recently developed methods.

Isolation in cell culture may be useful for the study of antigenic cross-reactions by fluorescent antibody techniques or for other comparative investigations. However, unlimited growth in cell culture has not been demonstrated.

Maintenance Procedures

Provided the *R. popilliae* suspensions are not contaminated with other bacteria, viability at 4°C is retained for 1 year, but not 3 years. Viability is retained for least 3 years and presumably much longer at −70°C.

Differentiation of the genus **Rickettsiella** from other closely related genera

Characteristics useful for the differentiation of *Rickettsiella* from the other genus of the tribe *Wolbachieae* are given in the key to the tribe.

Taxonomic Comments

Any attempt at classification of this group of organisms must take into account that much of the information that is generally used for bacterial classification is not available. Under the circumstances underclassification is a lesser sin than overclassification.

Many of the investigators feel that these organisms more closely resemble *Chlamydia* than *Rickettsia* and should be transferred to the order *Chlamydiales*. Points of similarity with *Chlamydia* are as follows: (i) they multiply in the vacuoles of their host cells; (ii) they undergo a developmental cycle that involves the formation of a small dense particle whose function is to infect, and a much larger particle of lower density whose function is to multiply. Points of difference with *Chlamydia* are the rod- or disk-shape of the infectious particles and the formation of large crystalline bodies. Resemblance to the *Rickettsiales* include a close association with arthropods and the intravacuolar location during multiplication, which is true of *Ehrlichia* and other *Rickettsiales* genera. Thus, the question of the proper classification of the *Rickettsiella* in the order *Rickettsiales* or *Chlamydiales* can be answered either way. However, there is sufficient similarity among the organisms discussed in this section to maintain them in the same major taxon. Dividing them into two widely separated taxonomic groups would only tend to reduce much needed communication among investigators.

A highly conservative classification of the species of the genus *Rickettsiella* is presented in Table 9.10. Six specific names previously proposed (Table 9.9) are placed in synonymy with the first described species, *R. popilliae*, since no major differences were encountered among

Table 9.10.
Differential characteristics of the species of the genus **Rickettsiella**[a]

Characteristics	1. *R. popilliae*	2. *R. grylli*	3. *R. chironomi*
Morphology of infectious particles:			
Rod	+	+	−
Disk	−	−	+
Crystalline body production	+	+	−
Serological reaction with anti-*R. popilliae* serum	+	−	−
Natural hosts:			
Insecta	+	+	+
Coleoptera	+	−	−
Diptera	+	−	+
Orthoptera	−	+	−
Crustacea	−	+	−
Arachnida	−	−	+

[a] Symbols: see standard definitions.

them. The *Popillia*, *Melolontha*, *Tipula*, and *Cetonia* pathogens have common antigens (Krieg, 1958; Croizier and Meynadier, 1971). The pathogens of *Tenebrio* and *Blatta* have not been extensively studied.

The separation of *R. popilliae* from *R. grylli* and the placement of "*R. schistocercae*" and "*R. armadillidii*" in synonymy with *R. grylli* lies on tenuous ground. It is based primarily on the serological cross-reactions between the cricket and isopode pathogens and lack of reaction with *R. popilliae* antisera (Croizier and Meynadier, 1971). The developmental cycles of the cricket and isopode pathogens appear to be quite similar (Louis et al., 1977a; Louis et al., 1977b), but it is not clear they differ from that of *R. popilliae* (Devauchelle et al., 1972). The hosts from which *R. grylli* have been isolated—Orthoptera and Crustacea— may not constitute a valid criterion for classification, but may serve as a guide for the provisional placement of the less well-studied pathogens of the desert locust and amphipod (Table 9.9).

The third species, *R. chironomi*, can be clearly separated from the other two by its unusual cycle involving the formation of flat disks instead of rods. There is a remarkable similarity between the cycle of the midge pathogen and the cycles of the arachnid pathogens (Morel, 1977; Federici, 1980), although differences have been described (Louis et al., 1979). Placing the midge and scorpion pathogens in separate genera (Table 9.9) is not justified at this time.

The stethorus beetle pathogen, for which the name *R. stethorae* has been assigned (Table 9.9), is obviously different in morphology from the above listed species of *Rickettsiella*, but information on obligate intracellular parasitism is inadequate for proper classification. The Leptidoptera pathogens listed in Table 9.9 appear to be typical rickettsiellae, but additional information is required for appropriate placement in a species. The same is true of the mite pathogen listed in Table 9.9, for which a specific name has been previously proposed, and the crab pathogen.

It should be noted that of the three species recognized in the present edition of the *Manual*, *R. popilliae*, *R. grylli*, and *R. chironomi*, only *R. popilliae* was included on the Approved Lists of Bacterial Names in 1980. Since the names *R. grylli* and *R. chironomi* presently have no standing in nomenclature, we propose that the names be revived. It should also be noted that the name *R. stethorae* was included on the Approved Lists of Bacterial Names in 1980; however, as indicated above, we do not believe that sufficient evidence is available at present to warrant considering the organism as a distinct species.

Further Reading

Devauchelle, G., G. Meynadier and C. Vago. 1972. Etude ultrastructurale du cycl de multiplication de *Rickettsiella melolonthae* (Krieg), Philip, dans les hémocytes de son hôte. J. Ultrastruct. Res. *38:* 134–148.
Dutky, S.R. 1959. Insect microbiology. Adv. Appl. Microbiol. *1:* 175–200.
Federici, B.A. 1980. Reproduction and morphogenesis of *Rickettsiella chironomi*, an unusual intracellular parasite of midge larvae. J. Bacteriol. *143:* 995–1002.
Götz, P. 1972. *Rickettsiella chironomi*: an unusual bacterial pathogen which reproduces by multiple cell division. J. Invert. Pathol. *20:* 22–30.
Louis, C., G. Morel, G. Nicolas and G. Kuhl. 1979. Étude compareé des caractères ultrastructuraux de rickettsies d'arthropodes, révélés par cryodécapage et cytochimie. J. Ultrastruct. Res. *66:* 243–253.
Morel, G. 1977. Étude d'une *Rickettsiella* (Rickettsie) se développant chez un arachnide, l'araigneé *Pisaura mirabilis*. Ann. Microbiol. (Institut Pasteur) *128A:* 49–59.
Suitor, E.C., Jr. 1964. Propagation of *Rickettsiella popilliae* (Dutky and Gooden) Philip and *Rickettsiella melolonthae* (Krieg) Philip in cell cultures. J. Insect. Pathol. *6:* 31–40.

Differentiation of the species of the genus **Rickettsiella**

Characteristics useful for the differentiation of *Rickettsiella* species are listed in Table 9.10.

List of the species of the genus **Rickettsiella**

1. **Rickettsiella popilliae** (Dutky and Gooden 1952) Philip 1956, 267.[AL] (*Coxiella popilliae* Dutky and Gooden 1952, 749.) This species includes the subjective synonyms listed in Table 9.9.

pop.il′li.ae. M.L. gen. n. *popilliae* of *Popillia* generic name of the Japanese beetle, one of its hosts.

The form usually obtained by separation procedures from infected host tissues is a rod 0.2 by 0.6 μm, or slightly larger; the rods may be oval, curved or kidney-shaped, with rounded edges. Satisfactorily stained by the Giemsa or Macchiavello methods (1947) and, presumably, by the Giménez (1964) method. The developmental cycle, the formation of crystalline bodies, and attempts at cultivation outside the host have been discussed in Further Descriptive Information.

R. popilliae is the etiological agent of blue disease of insect larvae, a name reflecting the discoloration of the infected larvae. Other names reflecting either the locality where the disease was discovered or the host have also been used. Infection is most commonly encountered among larvae, but is not confined to this stage. The infection starts in the fat body, but ultimately spreads to the blood and the other organs. Experimentally the larvae have been infected by injection, by feeding, or by holding them in soil inoculated with suspensions of the organism. By injection, <6 organisms are sufficient to infect a Japanese beetle larva with its natural pathogen. The time required for appearance of symptoms is dependent on dosage and on the temperature of incubation of the larvae (optimum ∼27°C). With an injection of 10^4 organisms per larva, the symptoms appear after 19 days and death occurs in 19 to 26 days after the first appearance of the symptoms. Larvae can be protected by injection of streptomycin (20 μg/larva) or sulfadiazine (200 μg/larva) but not by penicillin or chlortetracycline (Dutky, 1959).

There is considerable variation in the time of appearance of symptoms and death of the larvae infected by the various strains of *R. popilliae*. To what extent this variation reflects strain virulence or host susceptibility is not known.

The organisms can be inactivated at 60°C for 10 min.

The mol% G + C of the DNA is unknown.

Type strain: no culture isolated.

2. **Rickettsiella grylli** (*ex* Vago and Martoja 1963) nom. rev. This species includes the subjective synonyms listed in Table 9.9.

gryl′li. M.L. gen. n. *grylli* of *Gryllus* the generic name of the cricket, one of its hosts.

Resembles *R. popilliae* in morphology of its infectious particle and subsequent developmental stages and crystalline body production. It is differentiated from *R. popilliae* by serological tests and, to a certain extent, by the natural hosts infected, although it does not have a high degree of host specificity (Table 9.10).

The disease in the naturally infected hosts affects both larvae and adults and develops very slowly. In the cricket and locust it is characterized by swelling of the abdomen and, at later stages, by turgidity of the intersegmentary membranes and by ventral inclination of the head. The time of death following first appearance of symptoms is quite variable. In the isopod death is preceded by loss of weight and a whitish coloration of the intersegmentary membranes, due to the accumulation of iridescent fluid in the body cavities. Diseased amphipods are also recognized by their opaque pale green iridescence. Most of the patently diseased amphipods die within six weeks.

The mol% G + C of the DNA is unknown.

Type strain: no culture isolated.

3. **Rickettsiella chironomi** (*ex* Weiser 1963) nom. rev. This species includes the subjective synonyms listed in Table 9.9.

chi.ro.no′mi. M.L. gen. n. *chironomi* of *Chironomus* the generic name of the midge, one of its hosts.

The infectious stage has an unusual disk-shape morphology, 0.06 by 0.6 μm, as indicated in the discussion of the developmental cycle under

Further Descriptive Information. The similarity between the microorganism seen in the midge and those seen in spiders and a scorpion is most surprising. It can be surmised that there is greater diversity among the strains of this species than among the strains of the other two species of *Rickettsiella*; however, on the basis of present information, further speciation is not advisable.

In the midge the disease affects primarily the last two instars and the pupal stages. The infection starts in the fat body, progresses to lobes around the gut, which eventually rupture and fill the whole body with white fluid. Experimental infection results in death in 7–14 days (Weiser and Žižka, 1968). In the arachnids the hepatopancreas and intestinal diverticula are the organs primarily affected.

The mol% G + C of the DNA is unknown.

Type strain: no culture isolated.

FAMILY II. **BARTONELLACEAE** GIESZCZYKIEWICZ 1939, 25[AL]

MIODRAG RISTIC AND JULIUS P. KREIER

Bar.to.nel.la′ce.ae. M.L. fem. n. *Bartonella* type genus of the family; *-aceae* ending to denote family; M.L. fem. pl. n. *Bartonellaceae* the *Bartonella* family.

Parasites of erythrocytes of man and other vertebrates. Rod-shaped, coccus, ring- or disk-shaped bacteria, often beaded or filamentous and less than 3 μm in their greatest diameter. Erythrocytic forms stain lightly with many aniline dyes but distinctly with Giemsa's stain after methanol fixation. Gram-negative. Not acid-fast with acid-alcohol. One genus has unipolar flagella. Cultivable in vitro on nonliving media. Arthropod transmission has been established. Cause bartonellosis in man and grahamellosis in other vertebrates.

Key to the genera of the family **Bartonellaceae**

I. Occur in or on erythrocytes and within fixed tissue cells of vertebrates. Flagella are present on cultured organisms. Found in man and in *Phlebotomus* spp.

Genus I. *Bartonella*

II. Occur within erythrocytes. Not known to multiply in fixed tissue cells of vertebrates. Not found in man.

Genus II. *Grahamella*

Genus I. **Bartonella** Strong, Tyzzer and Sellards 1915, 808[AL]

MIODRAG RISTIC AND JULIUS P. KREIER

(*Bartonia* Strong, Tyzzer, Brues, Sellards and Gastiaburú 1913, 1715.)

Bar.to.nel′la. M.L. dim. ending *-ella*; M.L. fem. dim. n. *Bartonella* named after Dr. A. L. Barton, who described these organisms in 1909.

In stained blood films the organisms appear as rounded or ellipsoidal forms or as slender, straight, curved or bent rods, occurring singly or in groups. They characteristically occur in **chains of several segmenting organisms,** sometimes swollen at one or both ends and frequently beaded. **In the tissues they are situated within the cytoplasm of endothelial cells** as isolated elements or are grouped in rounded masses. Gram-negative. Not acid-fast. Stain poorly or not at all with many aniline dyes, but satisfactorily with Romanowsky's or Giemsa's stain. **In cultures the cells possess unipolar flagella.** Aerobic. May be cultivated on cell-free media. Growth occurs at 28 and 37°C, with greater longevity at 28°C. The organisms occur spontaneously in **man and in arthropod vectors (*Phlebotomus* spp.); found only in the Andes region of South America. Etiological agent of human bartonellosis.** The mol% G + C of the DNA is unknown.

Type species: *Bartonella bacilliformis* (Strong, Tyzzer, Brues, Sellards and Gastiaburú 1913, 1715) Strong, Tyzzer and Sellards 1915, 808.

Further Descriptive Information

B. bacilliformis consists of small, polymorphic forms. The maximum morphological range is seen in the blood of man, where the organisms appear as red-violet rod or coccus forms situated on or in the red cells when stained with Giemsa's stain. Bacilliform bodies are the most typical, measuring 0.25–0.5 by 1.0–3.0 μm. The cells are often curved and may show polar enlargement and granules at one or both ends. Rounded organisms measure ~0.75 μm in diameter, and a ringlike variety is sometimes abundant. By light microscopy or by "stripping" in the pseudoreplica technique for electron microscopy (Peters and Wigand, 1955), the organisms appear to be situated on the surface of erythrocytes; however, they have also been reported to occur within erythrocytes in thin sections observed by electron microscopy (Cuadra and Takano, 1969).

In semisolid media, a mixture of rods and granules appears. The organisms may occur singly or in large and small, irregular, dense collections measuring up to 25 μm or more in length. Punctiform, spindle-shaped and ellipsoidal forms occur which vary in size from 0.2 to 0.5 by 0.3 to 3.0 μm.

The organisms have a cell wall, the formation of which can be inhibited by penicillin.

In cultures the cells possess a tuft of 1–10 unipolar flagella (Peters and Wigand, 1955). Flagella have not been demonstrated in tissues.

B. bacilliformis may be cultivated in semisolid agar containing fresh rabbit serum and rabbit hemoglobin or containing the blood of man, horse or rabbit, with and without the addition of fresh tissue and carbohydrates. It may also be cultivated in other culture media containing blood, serum or plasma, in Huntoon's hormone agar, in semisolid gelatin media, and in blood-glucose-cystine agar. It can also be grown in certain tissue cultures and in the chorioallantoic fluid and yolk sac of the chicken embryo.

Gelatin is not liquefied. H_2S is not detected with lead acetate.

No acid or gas production occurs from amygdalin, arabinose, dextrin, dulcitol, fructose, galactose, glucose, inulin, lactose, maltose, mannitol, mannose, raffinose, rhamnose, salicin, sucrose or xylose.

Antigenically distinct strains of *B. bacilliformis* have not been identified. Immune sera fix complement in the presence of the organisms. When various isolates have been employed, no significant titer differences have been found in quantitative tests. Immune rabbit sera do not agglutinate *Proteus* strains OX19, OX2 or OXK at titers above 1:20. Agglutination of suspensions of *B. bacilliformis* by sera from convalescent patients has been reported.

Human bartonellosis may be manifested as a progressive anemia (Oroya fever) or as a cutaneous eruption (Verruga peruana); the latter usually follows the former. The transmission of Oroya fever is dependent on the ecology of the *Phlebotomus* fly vector and is therefore confined to an elevation of 2500–8000 feet above sea level in a band less than 100 miles wide and ~1000 miles long on the western slopes of the Andes mountains in Peru, Ecuador and Colombia. Human infection is acquired when the female flies take the organism contained in the blood meal during the nighttime.

Experimental Oroya fever has not been successfully produced in animals, except rarely in an atypical form in monkeys. Experimental Verruga peruana has been produced in man and in a number of species of monkeys.

B. bacilliformis is resistant in vivo to neosalvarsan and in general to other arsenical compounds. It is sensitive to penicillin, streptomycin, chloramphenicol and oxytetracycline. When grown with penicillin, the organism produces L forms (Sharp, 1968).

Isolation Procedures

In human cases of Oroya fever the organisms can be isolated from blood and from endothelial cells of lymph nodes, spleen and liver. In cases of Verruga peruana they are found in the blood and in the eruptive lesions. The organisms can also be isolated from the sand fly vector (*Phlebotomus* spp.)

In semisolid media growth first appears just below the surface of the medium in ~10 days.

Maintenance Procedures

During serial transfers, the greatest longevity is achieved at a temperature of 28°C. Cultures have also remained viable for 5 years when stored at −70°C.

Differentiation of the genus **Bartonella** from closely related genera

See the key to the genera of the family *Bartonellaceae* for characteristics useful in differentiating *Bartonella* from the genus *Grahamella*.

Taxonomic Comments

The phylogenetic relationships of *Bartonella* to *Grahamella* or other members of the order *Rickettsiales* are not known. DNA/DNA or DNA/rRNA hybridization experiments could yield valuable information in this regard.

Members of *Bartonella* are not identical with rods observed on the red blood cells of anemic patients in Thailand. Such rods have not been proven to be microorganisms.

Further Reading

Cuadra, M. and J. Takano. 1969. The relationship of *Bartonella bacilliformis* to the red blood cell as revealed by electron microscopy. Blood *33:* 708–716.
Kreier, J.P. and M. Ristic. 1981. The biology of hemotropic bacteria. Annu. Rev. Microbiol. *35:* 325–338.
Peter, D. and R. Wigand. 1955. Bartonellaceae. Bacteriol. Rev. *19:* 150–155.
Strong, R.P. and A.W. Sellards. 1915. Oroya fever. Second Report. J. Amer. Med. Assoc. *64:* 806–808.
Wigand, R. 1958. *Morphologische Biologische und Serologische Eigenschaften der Bartonellen*. Georg Thieme, Stuttgart.

List of the species of the genus **Bartonella**

1. **Bartonella bacilliformis** (Strong, Tyzzer, Brues, Sellards and Gastiaburú 1913) Strong, Tyzzer and Sellards 1915, 808.[AL] (*Bartonia bacilliformis* Strong, Tyzzer, Brues, Sellards and Gastiaburú 1913, 1715.)

ba.cil.li.for′mis. L. dim. n. *bacillus* a small staff, rodlet; L. noun *forma* shape, form; M.L. adj. *bacilliformis* rod-shaped.

The characteristics are as described for the genus.

The mol % G + C of the DNA is not known.

Type strain: no culture available.

Genus II. **Grahamella** (*ex* Brumpt 1911) nom. rev.

MIODRAG RISTIC AND JULIUS P. KREIER

Gra.ha.mel′la. M.L. dim. ending -*ella*; M.L. fem. dim. n. *Grahamella* named after Dr. G. S. Graham Smith, who discovered these organisms in the blood of old world voles in 1905.

Long or short rod-shaped organisms, sometimes curved, lying within the red blood cells of various vertebrates. Not observed in man. May multiply in fixed tissue cells of arthropod vectors; not known to multiply in fixed tissue cells of vertebrates. Morphologically resemble members of the genus *Bartonella* but **do not have flagella.** Gram-negative. Not acid-fast. Aerobic. Can be cultivated on nonliving media; growth is favored by addition of hemoglobin. Have a wide geographical distribution. The mol% G + C of the DNA is unknown.

Type species: *Grahamella talpae* (*ex* Brumpt 1911) nom. rev.

Further Descriptive Information

In blood smears stained by the Giemsa method the organisms stain bluish-purple and are seen lying within the erythrocytes. They are occasionally free in the plasma. Grahamellae have been cultivated on nonliving media and their growth is favored by supplementation with hemoglobin. In cultures, rods and coccoid forms with indistinct contours commonly appear together in dense masses. Infective cultures were first grown by Tyzzer (1942); Wu Lien-Teh and Jettmar may have cultivated the organisms in 1930 but their cultures failed to infect.

Grahamellae are not affected by arsenicals.

Isolation Procedures

The organism can be isolated in the semisolid *Leptospira* medium of Noguchi, and also on beef infusion-peptone agar base media to which 5–20% defibrinated blood has been added.

Maintenance Procedures

The organism can be maintained infectious for various periods of time by preservation of cultures at 4°C or by carrier infection in rodents (e.g., mice).

For long term preservation, storage of the organism in liquid nitrogen is preferred.

Differentiation of the genus **Grahamella** from other closely related genera

See the key to genera of the family *Bartonellaceae* for characteristics differentiating *Grahamella* from the genus *Bartonella*.

Taxonomic Comments

The phylogenetic relationships of *Grahamella* to *Bartonella* or other genera of the order *Rickettsiales* are not known. Nucleic acid hybridization experiments would be helpful in elucidating these relationships.

Grahamellae have been observed in a variety of animal hosts but it has not been demonstrated that they are different species. The distinction between the two presently recognized species, based on host, might be an artificial one.

The name of the genus and the names of the species included therein were not included on the Approved Lists of Bacterial Names in 1980 and are being revived herein.

Further Reading

Anderson, J.F., L.A. Magnarelli and J. Kurz. 1979. Intraerythrocytic parasites in rodent populations of Connecticut: *Babesia* and *Grahamella* species. J. Parasitol. *65:* 599–604.

Tyzzer, E.E. 1942. A comparison study of grahamellae, haemobartonellae and eperythrozoa in small mammals. Proc. Amer. Phil. Soc. *85:* 359–398.

Weinman, D. and J.P. Kreier. 1977. *Bartonella* and *Grahamella. In* J.P. Kreier (Editor), *Parasitic Protozoa.* Academic Press, New York, pp. 197–233.

Differentiation of the species of the genus **Grahamella**

The distinction presently made between *G. talpae* and *G. peromysci* is that the former are found in moles of the genus *Talpa*, whereas the natural host for the latter species is the deer mouse (*Peromyscus leucopus*).

List of the species of the genus **Grahamella**

1. **Grahamella talpae** (*ex* Brumpt 1911) nom. rev.

tal′pae. M.L. fem. n. *Talpa* a genus of moles; M.L. gen. n. *talpae* of *Talpa.*

The morphological characteristics are as described for the genus. In blood smears, most of the infected corpuscles contain between 6 and 20 organisms, but relatively few erythrocytes are infected (rarely more than 1%; Graham Smith (1905)).

Infective for moles of the genus *Talpa* and found in these animals.

The mol% G + C of the DNA is unknown.

Type strain: no culture available.

2. **Grahamella peromysci** (*ex* Tyzzer 1942) nom. rev.

pe.ro.mys′ci. M.L. masc. n. *Peromyscus* a genus of mice; M.L. gen. n. *peromysci* of *Peromyscus.*

Not morphologically distinct from *G. talpae.* Grows on nonliving media containing blood, at temperatures varying from 20 to 28°C under aerobic conditions. Colonies rarely exceed 1.5 mm in diameter and are composed of rods as long as 1.5 μm and which vary in thickness from 0.25 to 0.75 μm; coccoid forms also occur, 0.25–1.0 μm in diameter, occurring together in compact clumps. Older cultures may contain chains of rods and globoid bodies, the latter up to 12 μm in diameter. Organisms in cultures stain poorly with alkaline methylene blue solution (Loeffler's) but well with Giemsa's stain. Blood is not hemolyzed.

The natural host, the deer mouse (*Peromyscus leucopus novaboracensis*), may be infected by blood or cultures. The white Swiss mouse appears resistant. Monkeys (*Macaca mulatta*) are not infected by cultures.

The mol% G + C of the DNA is unknown.

Type strain: no culture available.

Species Incertae Sedis

The following organism requires restudy; it may be a *Grahamella:* "*Haemobartonella tyzzeri*" (Weinman and Pinkerton 1938) Groot 1942, 279. (*Bartonella tyzzeri* Weinman and Pinkerton 1938, 217.)

tyz′zer.i. M.L. gen. n. *tyzzeri* of Tyzzer; named after E. E. Tyzzer.

This organism has been grown in vitro and therefore might more properly be placed in the genus *Grahamella* (Weinman, 1957).

FAMILY III. **ANAPLASMATACEAE** PHILIP 1957, 980[AL]

Miodrag Ristic and Julius P. Kreier

A.na.plas.ma.ta′ce.ae. M.L. neut. n. *Anaplasma* type genus of the family; *-aceae* ending to denote a family; M.L. fem. pl. n. *Anaplasmataceae* the *Anaplasma* family.

Obligately parasitic organisms found within or on erythrocytes or free in the plasma of various wild and domestic vertebrates. No multiplication occurs in other tissues.

In blood smears treated with Giemsa's stain, appear as rod-shaped, spherical, coccoid- or ring-shaped bodies staining reddish violet and measuring 0.2–0.4 μm in diameter. May occur in short chains or in groups in blood plasma, on erythrocytes, and within erythrocytes. Morphologically, the organisms resemble rickettsiae. Multiply by binary fission. Gram-negative. Not acid-fast. Have not been cultivated in cell-free media. Organisms transmitted by arthropods. Blood or blood-containing tissue homogenates can cause infection by any parenteral route.

Infection may or may not cause disease but usually results in long term persistence of the agent, with concomitant resistance to clinically demonstrable reinfection. Anemia is the most prominent feature of the clinical disease. Members of the family occur throughout the world.

Growth is inhibited by tetracycline compounds, but not by penicillin or streptomycin.

Key to the genera of the family **Anaplasmataceae**

I. Organisms form inclusions in erythrocytes.
    A. Inclusions in erythrocytes are round, 0.3–1.0 μm in diameter. Several organisms may be found in each inclusion. In some species there are comet-like projections extending from the inclusion body into the cytoplasm. Infect ruminants only.

                Genus I.   *Anaplasma*

    B. Inclusions in erythrocytes are 0.3–4.0 μm in diameter. Infect birds.

                Genus II. *Aegyptianella*

II. Parasites within, or outside, erythrocytes.
    A. Parasites in and on erythrocytes are firmly attached. Stain intensely by Romanowsky methods. Ring structures are rare or absent.

                Genus III. *Haemobartonella*

    B. Parasites on erythrocytes are loosely attached and also occur in the plasma. Ring forms common in preparations stained by Romanowsky methods.

                Genus IV. *Eperythrozoon*

Genus I. **Anaplasma** *Theiler 1910, 7*[AL]

MIODRAG RISTIC AND JULIUS P. KREIER

(Includes *Paranaplasma* Kreier and Ristic 1963, 701.)

A.na.plas'ma. Gr. Prefix *an* without; Gr. n. *plasma* anything formed or molded; M.L. neut. n. *Anaplasma* a thing without form.

In blood smears stained by Romanowsky methods, the organisms appear in the erythrocytes as **dense, homogeneous, bluish-purple, round inclusions 0.3–1.0 μm in diameter.** Each inclusion contains from 1 to 8 subunits or initial bodies, which are the actual parasitic bacteria, each being 0.3–0.4 μm in diameter. The inclusion bodies of some *Anaplasma* species have appendages. Spores or resistant stages are not formed. Apparently aerobic. **Obligate parasites of vertebrates. Transmitted by arthropod vectors. The host range is limited to ruminants.** The mol% G + C of the DNA of one species has been reported as 51.

Type species: *Anaplasma marginale* Theiler 1910, 7.

Further Descriptive Information

The organism (initial body) enters the erythrocyte by causing an invagination of the cytoplasmic membrane and subsequent formation of a vacuole. In the vacuole the initial body multiplies by binary fission and forms an inclusion. Electron microscopy reveals that the inclusions are separated from the cytoplasm of the erythrocyte by a limiting membrane (Fig. 9.7). The formation of the inclusion body, which is

most frequently encountered during the acute and convalescent phases of infection, represents only a phase in the developmental cycle of the initial body (Ristic, 1981). By electron microscopy, each initial body appears as a dense aggregate of fine granular material embedded in an electron-lucid plasma and all enclosed in a double membrane (Ristic and Watrach, 1961).

Inclusion bodies and initial bodies are morphologically indistinguishable among species of *Anaplasma* in Giemsa-stained blood films studied by light microscopy. By techniques using wet preparations of blood or fluorescent antibody, and also preparations for electron microscopy, the inclusion bodies of some *Anaplasma* species may be shown to have appendages. The appendages may resemble tails, loops or rings or may connect two inclusion bodies in a "dumbbell" form. Fluorescent antibody and cross-immunity studies show organisms having such appendages to be in part antigenically distinct from those lacking them. Anaplasmas having appendages on their inclusions are infectious for cattle but not deer and they may or may not grow in sheep (Kreier and Ristic, 1963).

Anaplasma marginale has been shown to reproduce in vitro in eryth-

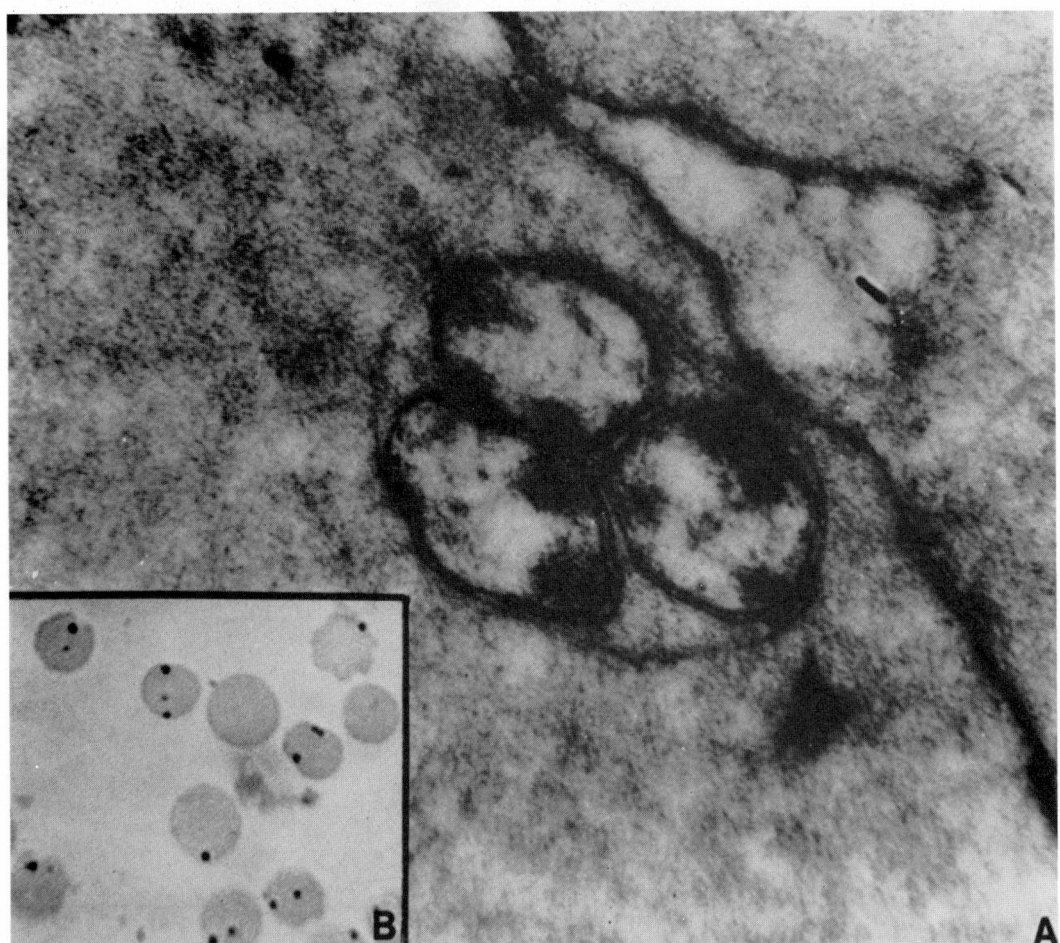

Figure 9.7. *Anaplasma marginale. A,* Ultrathin section of an infected erythrocyte. An inclusion body containing three initial bodies is depicted (×60,000). *B,* Intraerythrocytic inclusion bodies stained by Giemsa's method (×959).

rocyte cultures for short periods of time (Kessler and Ristic, 1979; Kessler et al., 1979).

The organisms are apparently aerobic (Pilcher et al., 1961) and produce catalase (Wallace and Dimopoullos, 1965). They do not produce pigments. They can incorporate amino acids from plasma in vitro (Mason and Ristic, 1966). The ATP and glutathione concentrations in erythrocytes remain essentially unchanged regardless of the intensity of infection and only a small quantity of methemoglobin occurs in parasitized erythrocytes. Histochemical analysis of the parasites reveals the presence of DNA, RNA, protein and organic iron (Moulton and Christensen, 1955).

Anaplasma species are transmitted by arthropod vectors. Ticks are probably biological vectors, and other biting arthropods are mechanical vectors. The host range is limited to ruminants. Common laboratory animals are all refractory, e.g., rabbits, guinea pigs, rats, mice, ferrets, dogs, cats and chickens. The subject has been reviewed by Ristic (1968, 1977, 1980).

A. marginale is the most pathogenic of the *Anaplasma* species and is the causative agent of severe bovine anaplasmosis. The disease is widespread where cattle are raised and where the appropriate arthropod vectors exist. *A. caudatum* occurs frequently in mixed infection with *A. marginale*. *A. centrale* causes anaplasmosis of sheep and goats in various regions of the world.

Maintenance Procedures

A. marginale in bovine blood can be preserved for many months by freezing infected blood to which glycerol has been added.

Differentiation of the genus **Anaplasma** from other closely related genera

See the key for the family *Anaplasmataceae* for the features that distinguish *Anaplasma* from the other genera of the family.

Taxonomic Comments

Species differentiation in this genus is based on host range and on the location and characteristics of the inclusions in the erythrocytes. Morphologically, the initial bodies of the species are indistinguishable by conventional and electron microscopy. Some antigens are common to all members of the genus; others are unique to one species (Kreier and Ristic, 1963; Schindler et al., 1966; Kuttler, 1966, 1967). Antigenic analysis, if developed, could be used for species differentiation.

Further Reading

Ristic, M. and J.P. Kreier. 1979. Hemotrophic bacteria. N. Eng. J. Med. *301:* 937–939.
Ristic, M. 1981. Anaplasmosis. *In* Ristic and McIntyre (Editors), *Diseases of Cattle in the Tropics.* Martinus Nijhoff Publ., Boston, pp. 327–344.

Differentiation of the species of the genus **Anaplasma**

Characteristics useful for the differentiation of the species of *Anaplasma* are listed in Table 9.11.

List of the species of the genus **Anaplasma**

1. **Anaplasma marginale** Theiler 1910, 7.[AL]

mar.gi.na′le. L. n. *margo, marginis* edge, margin; M.L. neut. adj. *marginale* marginal, referring to location of the organism within erythrocytes.

The characteristics are as described for the genus and as listed in Table 9.11.

Infectious for cattle, zebu, water buffalo (*Babalus babalis*), bison (*Bison bison*), African antelopes, gnu (*Connochaetes gnou*), blesbuck (*Damaliscus pygargus albifrons*) and duiker (*Sylvicapra grimmia*), American deer (southern black-tailed, Rocky Mountain mule deer, Virginia white-tailed deer), elk and camel (*Camelus bactrianus*). Sheep and goats usually develop a subclinical infection. The African buffalo (*Syncerus caffer*) is refractory (Ristic, 1981).

Infectivity of *A. marginale* can be destroyed by heating the organism at 60°C for at least 50 min, by exposure to sonic oscillation at 35°C for at least 90 min (Bedell and Dimopoullos, 1965) or by exposure to x-ray doses of 100,000 roentgens or higher (Wallace and Dimopoullos, 1965).

Tetracycline compounds and dithiosemicarbazones inhibit multiplication and ameliorate the disease course (Ristic, 1981; Barrett et al. 1965). Penicillin, streptomycin, sulfonamides and arsenicals are inactive.

Protective immunity is generally associated with persistence of subclinical infections. Following elimination of the agent by chemotherapy a sterile immunity may exist for a period of time.

The parasite is spread by blood-sucking arthropods. A biological cycle may exist in ticks.

The mol% G + C of the DNA has been reported to be 51 (Senitzer et al., 1972).

Type strain: no culture isolated.

Further comments. In cattle infected with *A. marginale*, antibodies are found in IgG and IgM classes of immunoglobulins during the acute and convalescing phases of the infection (Murphy et al. 1966). Erythrocytes from cows acutely infected with *A. marginale* yield two distinct antigens, a nonsedimentable one (NS) and a sedimentable one (S). The NS antigen is soluble and lipoproteinaceous and is serologically active

Table 9.11.
Differential characteristics of the species of the genus **Anaplasma**

Species	Host	Location in Erythrocyte	Appendages on Inclusions	Disease
1. *A. marginale*	Cow, deer	Predominantly marginal	No	Severe anaplasmosis
2. *A. centrale*	Cow	Predominantly central	No	Generally mild anaplasmosis
3. *A. caudatum*	Cow	Predominantly marginal	Yes	Mild to severe anaplasmosis
4. *A. ovis*	Sheep, deer, goats	Predominantly marginal	No	Mild to severe disease

in the gel-precipitation system (Ristic et al. 1963; Ristic and Mann, 1963). The particulate "S" antigen is a suspension of initial bodies (subunits of the marginal bodies) (Ristic, 1962).

Friedhoff and Ristic (1966), using the fluorescent antibody technique, found that *A. marginale* organisms occur in the gut contents and in the Malpighian tubules of engorged tick nymphs (*D. andersoni*). Transstadial transmission of *A. marginale* was demonstrated by feeding the newly molted adult ticks on susceptible cattle and by inoculation of gut homogenates collected from adult ticks at postattachment day 6 (Kocan et al, 1980).

2. Anaplasma centrale (*ex* Theiler 1911) nom. rev.

cen.tra′le. L. neut. adj. *centrale* central, referring to the location of the organism within erythrocytes.

The characteristics are as described for the genus and as listed in Table 9.11.

The disease produced in cattle is generally considered mild; however, severe disease may occur.

In addition to common generic antigens, this species possesses specific antigens (Schindler et al., 1966).

The mol% G + C of the DNA is unknown.

Type strain: no culture isolated.

3. Anaplasma caudatum (Kreier and Ristic 1963) comb. nov.

(*Paranaplasma caudatum* Kreier and Ristic 1963, 701.)

cau′da.tum. L. n. *cauda* tail; L. neut. adj. *caudatum* tailed, with a tail.

The characteristics are as described for the genus and as listed in Table 9.11. The inclusion bodies have appendages, usually in the form of a tapering tail, a loop, disk, or a ring; the appendages are visible only through use of special techniques. In electron microscopic studies the tail which is associated with the inclusion appears comet-shaped and may or may not be physically attached (Simpson et al., 1965). The inclusion body is typical of *Anaplasma*. The tail portion of the inclusion body is of parasite origin, since it may be stained with fluorescent labeled antibody from cattle which have recovered from infection with the parasite (Kreier and Ristic, 1963).

Infective for cattle. Will not grow in deer or sheep. All isolations have been from mixed infections with *A. marginale*.

The mol% G + C of the DNA is unknown.

Type strain: no culture isolated.

4. Anaplasma ovis Lestoquard 1924, 784.[AL]

o′vis. L. gen. n. *ovis* of the sheep.

The characteristics are as described for the genus and as listed in Table 9.11. Similar to *A. marginale* except for the hosts infected. *A. ovis* produces disease in sheep and goats and may produce mild or inapparent infections in deer and cattle.

A. ovis shares some antigens with *A. marginale*.

The mol% G + C of the DNA is unknown.

Type strain: no culture isolated.

Genus II. Aegyptianella Carpano 1929, 12[AL]

RAINER GOTHE AND JULIUS P. KREIER

Ae.gyp′ti.a.nel′la. Dim. ending -*ella*; M.L. fem. dim. n. *Aegyptianella* named after Egypt where the organism was described in 1929.

In blood smears stained by Romanowsky methods, the organisms appear in erythrocytes as **purple intracytoplasmic inclusions 0.3–4.0 μm in diameter.** Each inclusion contains up to 26 initial bodies, which are the actual parasitic bacteria, each measuring up to 0.8 μm in diameter. **Obligate parasites of domestic and wild birds. Transmitted by arthropod vectors.** The mol% G + C of the DNA is not known.

Type species: *Aegyptianella pullorum* Carpano 1929, 12.

Further Descriptive Information

In blood smears stained with Giemsa, inclusions appear in the host's erythrocytes in a variety of forms: compact, round and oval, ring- or horseshoe-shaped, polygonal or polymorphic, violet-reddish in color, with a diameter of 0.3–4.0 μm. In larger inclusion (marginal bodies), clearly defined smaller round organisms (initial bodies) measuring up to 0.8 μm in diameter can be distinguished. The initial bodies reproduce by binary fission.

The parasites may also be found free in the plasma and in phagocytic cells. Parasites in phagocytic cells are probably ingested (Gothe, 1967b, 1971).

In erythrocytes, the inclusions are separated from the cytoplasm by a single membrane. The internal structure of the initial bodies consists of dense aggregates of fine granular material embedded in an electron-lucid substance (Gothe, 1967a, 1971) (Fig. 9.8); the internal morphology is strikingly similar to that of *Anaplasma*. Scanned *Aegyptianella* preparations have produced static evidence of an endocytosis followed by an erythrocytic vesiculation as the possible mode of entrance of initial bodies into erythrocytes. The exit of initial bodies from parasitized erythrocytes appears to be the invasive mechanism in reverse order, an exocytosis. Generally, however, the affected erythrocytes are injured by the parasites, resulting in host cell lysis and release of the parasites into the plasma (Gothe and Burkhardt, 1979).

Organismal RNA can easily be demonstrated histochemically, but DNA-specific staining is possible only after treatment of the smears with ribonuclease.

Multiplication of *A. pullorum* has not been observed in cell-free media or in tissue cultures. Attempts at continuous propagation of the organism in chicken embryos have not been successful (Gothe, 1971).

Only broad-spectrum antibiotics of the tetracycline series, dithiosemicarbazones and pleuromutilins have an aegyptianellicidal efficacy with a significant chemotherapeutic influence on the course of the infection in chickens (Gothe, 1971; Gothe and Mieth, 1979).

Chickens are naturally infected with *A. pullorum* by the ticks. *Argas (Persicargas) walkerae, Argas (Persicargas) sanchezi* and *Argas (Persicargas) radiatus* may also act as biological vectors. Experimental infection can be achieved by subcutaneous, intramuscular, intravenous and intraperitoneal inoculation or by scarification with infected blood. Natural infections have also been described in geese, ducks, quails and in the ostrich. Wild birds which have been experimentally infected are: *Turtur erythrophrys* and *Balearica pavonina* (Curasson and Andrjesky, 1929), *Turtur senegalensis, Milvus aegyptiacus* and *Vidua principalis* (Curasson, 1938).

Infection is transstadial in ticks (Gothe, 1967c, 1971; Hadani and Dinur, 1968). Transovarial transmission has also been observed (Hadani and Dinur, 1968; Gothe, 1971).

Maintenance Procedures

The infectivity of *A. pullorum* in chicken blood can be preserved up to nearly 7 years by storage in liquid nitrogen (Raether and Seidenath, 1977). Cryopreservation does not affect the ability of the parasites to propagate in the vector tick *Argas (Persicargas) walkerae* (Gothe and Hartmann, 1979).

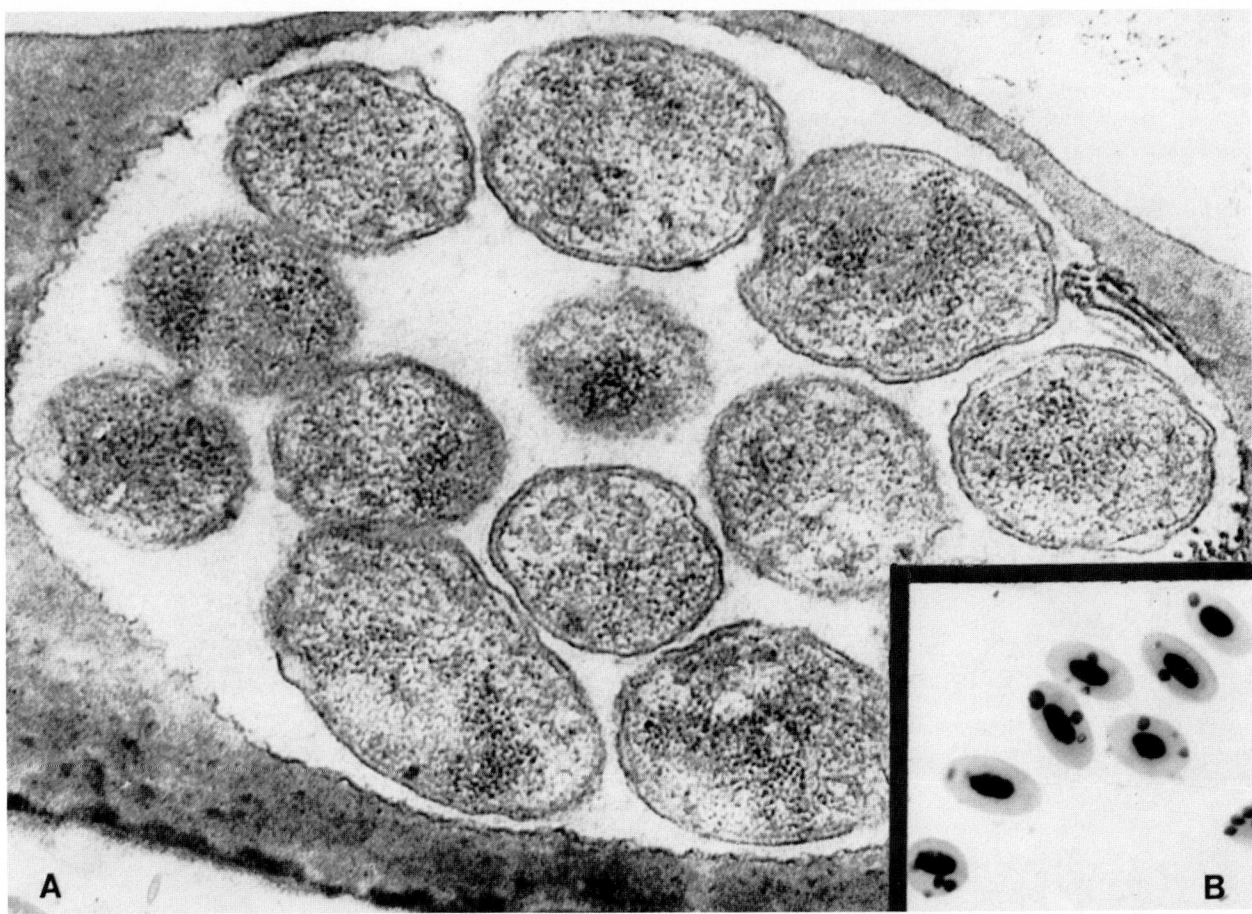

Figure 9.8. *Aegyptianella pullorum.* A, ultrathin section of an infected erythrocyte, showing an inclusion body with 12 initial bodies (×100,000). B, intraerythrocytic inclusion bodies stained by the Giemsa method. (Reproduced with permission from R. Gothe, *Zeitschrift fur Parasitenkunde, 29:* 119–129, 1967, Springer-Verlag, Stut Berlin.)

Differentiation of the genus Aegyptianella *from other closely related genera*

Characteristics useful for differentiating *Aegyptianella* from the other members of the family *Anaplasmataceae* are provided in the key to this family.

Taxonomic Comments

At present only a single species, *A. pullorum*, is recognized in the genus *Aegyptianella*. Other species have been described but neither their true identify nor their relationship to *A. pullorum* has been established (see *species incertae sedis*).

Other parasites having some resemblance to *Aegyptianella* have been reported in a number of species of domestic and wild birds, but their relation to *A. pullorum* is not known (Gothe, 1978).

Further Reading

Gothe, R. 1978. New aspects of the epizootiology of aegyptianellosis in poultry. *In* Wilde (Editor), *Tick-borne Diseases and Their Vectors,* Centre for Tropical Veterinary Medicine, University of Edinburgh, Lewis Reprints Ltd., Tonbridge, pp 201–204.

Gothe, R. and E. Burkhardt. 1979. The erythrocytic entry- and exit-mechanisms of *Aegyptianella pullorum* Carpano, 1928. Z. Parasitenkd. *60:* 221–227.

Gothe, R. and J.P. Kreier. 1977. *Aegyptianella, Eperythrozoon,* and *Haemobartonella. In* J.P. Kreier (Editor), *Parasitic Protozoa,* Vol. IV, Academic Press, New York, pp. 251–294.

List of the species of the genus Aegyptianella

1. **Aegyptianella pullorum** Carpano 1929, 12.[AL]
pul.lo′rum. L. gen. pl. n. *pullorum* of young fowls.
The characteristics are as described for the genus.
The mol% G + C of the DNA is unknown.
Type strain: no culture isolated.

Species Incertae Sedis

The following species have been described but neither their true identity nor their relationship to *A. pullorum* has been established.

a. "*Tunetella emydis*" Brumpt and Lavier, 1935, 548.

b. "*Sogdianella moshkovskii*" Shchurenkova 1938, 936. ["*Babesia moshkovskii*" (Shchurenkova 1938) Laird and Lari 1957, 794.]

c. "*Aegyptianella carpani*" Batelli 1947, 212.

Genus III. **Haemobartonella** Tyzzer and Weinman 1939, 143[AL]

JULIUS P. KREIER AND MIODRAG RISTIC

Hae.mo.bar.to.nel′la. Gr. n. *haema* blood; M.L. fem. n. *Bartonella* a genus of the family *Bartonellaceae*; M.L. dim. ending *-ella*; M.L. dim. fem. n. *Haemobartonella* blood (-inhabiting) bartonella.

In blood smears stained by Romanowsky methods, the organisms appear **coccus or rod-shaped and are located on or within the erythrocytes.** They occur singly, in pairs or in groups in shallow or deep indentations on the erythrocyte surface, sometimes in vacuoles within the erythrocytes, rarely in the plasma. **Ring forms are rare or absent.** Obligate parasites of many vertebrate species. One species is known to be transmitted by arthropods. The mol% G + C of the DNA is unknown.

Type species: *Haemobartonella muris* (Mayer 1921) Tyzzer and Weinman 1939, 143.

Further Descriptive Information

The organisms stain well with Romanowsky-type stains but poorly with many other aniline dyes. The cells possess a single or double limiting membrane and do not have a membrane-bounded nucleus (Tanaka et al., 1965). They are Gram-negative and non-acid-fast.

They have not been cultivated outside the host.

The experimental host range of any species is restricted; a species occurring naturally in a given host may be infective for closely related host species but not for the whole range of animals susceptible to other species of *Haemobartonella*. Morphology of the organism may vary in different hosts. Growth is inhibited by arsenicals and tetracyclines but not by penicillin or streptomycin.

Most species are pathogenic but clinical disease, characterized by anemia, is usually not apparent unless the animal is splenectomized, except in the case of *H. felis* in cats. *H. muris* is known to be transmitted by arthropods (the rat louse); transmission by such vectors is suspected but not yet established for the other species. In the case of *H. felis*, the organism can be transmitted by ingestion of blood (Flint et al., 1958).

Maintenance Procedures

The organism can be preserved indefinitely in 10% dimethyl sulfoxide when stored in liquid nitrogen.

Differentiation of the genus **Haemobartonella** from other closely related genera

See the key to the family *Anaplasmataceae* for characteristics that distinguish *Haemobartonella* from other members of the family.

Taxonomic Comments

Bacterial-type organisms which use the blood as their primary site of development are rarely observed in humans. In man, only *Bartonella bacilliformis*, the causative agent of Oroya fever and Verruga peruana, has been adequately described. In animals other than man, by contrast, blood-inhabiting bacteria of the genera *Haemobartonella*, *Eperythrozoon*, *Grahamella*, *Aegyptianella*, and *Anaplasma* are fairly common. Recently, there have been several reports of hemotropic procaryotic microorganisms infecting man in the United States (Kallick et al., 1972; Archer et al., 1979; Gretillate and Konarzemski, 1978). The

taxonomic position of these organisms is still uncertain. The possible relationship between these agents and the more familiar hemotropic bacteria has been discussed by Ristic and Kreier (1979).

Further Reading

Gothe, R. and J.P. Kreier. 1977. *Aegyptianella, Eperythrozoon, and Haemobartonella. In* J.P. Kreier (Editor), *Parasitic Protozoa*, Vol. IV, Academic Press, New York, pp. 251–294.

Kreier, J.P. and M. Ristic. 1968. Haemobartonellosis, eperythrozoonosis, grahamellosis and ehrlichiosis. *In* Weinman and Ristic (Editors), *Infectious Blood Diseases of Man and Animals*, Academic Press, New York, pp. 397–472.

Ristic, M. and J.P. Kreier. 1979. Hemotropic bacteria. N. Eng. J. Med. *301:* 937–939.

Differentiation of the species of the genus **Haemobartonella**

The three species listed in this chapter are differentiated on the basis of host range. *H. muris* infects rats, mice and hamsters; *H. felis* infects domestic cats; *H. canis* infects dogs. As indicated previously, a species

occurring naturally in a given host may be infective for closely related host species but not for the whole range of animals susceptible to other *Haemobartonella* species.

List of the species of the genus **Haemobartonella**

1. **Haemobartonella muris** (Mayer 1921) Tyzzer and Weinman 1939, 143.[AL] (*Bartonella muris* Mayer 1921, 151.)

mu′ris. L. n. *mus* the mouse; L. gen. n. *muris* of the mouse.

Causes infections in albino rats, albino mice, some wild mice, and hamsters. In Romanowsky-stained blood films the organisms appear as slender rods with rounded ends. They frequently show granules or swellings at one or both ends and may appear as dumbbell, coccus or diplococcus forms (Fig. 9.9A). They occur singly, in pairs, or in chains of 3 or 4 and, when abundant, in parallel groupings. The rods measure 0.1 by 0.3–0.7 μm; cocci are 0.1–0.2 μm in diameter. Electron micrographs have shown the rods to be composed of cocci 0.3–0.5 μm in diameter. The organisms are nonmotile and have no flagella (Wigand and Peters, 1952).

By the Giemsa staining method, the organisms appear bluish-purple. They stain faintly with methyl green, pyronin or fuchsin. They can be

stained with methylene blue. Enzyme treatment and histochemical staining indicate the presence of both DNA and RNA, the former in greater quantity (Peters and Wigand, 1955).

Nonfilterable with Seitz or Berkefeld filters.

There are no convincing reports of cultivation in vitro.

Susceptible to organic arsenical compounds, chlortetracycline and oxytetracycline; these compounds will destroy the organisms in the host in both latent and clinical infections.

Serum from *H. muris*-infected rats does not yield a positive Weil-Felix reaction (Wigand, 1958). Weak complement binding in the presence of infected rat serum has been obtained using *Anaplasma marginale* and *Eperythrozoon coccoides* antigens.

Following exposure, either patent infection or a carrier state may be produced; the determining factors are not all known, although splenectomized or young animals seem more susceptible to patent infection.

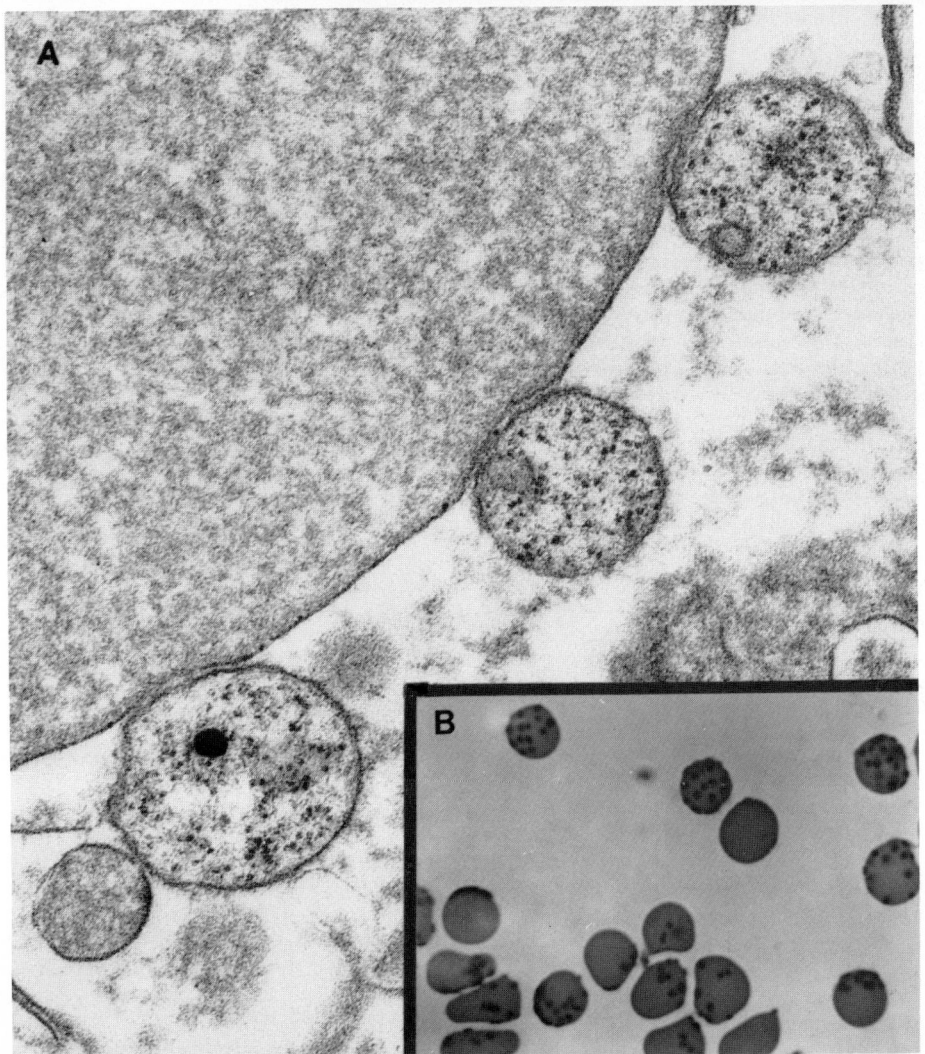

Figure 9.9. *Haemobartonella muris. A,* Ultrathin section showing the parasite on rat erythrocyte. There is some indentation of the erythrocyte membrane at the point of attachment (×63,920). (Reproduced with permission from H. Tanaka et al., *Journal of Bacteriology, 90:* 1735–1749, 1965, American Society for Microbiology, Washington, D.C.) *B, Haemobartonella felis* on erythrocytes of an infected cat (polychrome stain, ×950). (Reproduced with permission from J. C. Flint and D. K. McKelvie, *Proceedings of the 92nd Annual Meeting of the American Veterinary Medical Association,* pp. 240–242, 1955.)

Latent infections may become patent infections following splenectomy or other processes which disturb the function of the reticuloendothelial and immunological systems of the host.

H. muris is found worldwide in the blood of susceptible species. The rat louse (*Polypax spinulosa*) is an important vector.

The mol% G + C of the DNA is unknown.

Type strain: no culture isolated.

2. **Haemobartonella felis** (*ex* Flint and McKelvie 1956) nom. rev. (*Eperythrozoon felis* Clark 1942, 16.)

fe′lis. L. gen. n. *felis* of the cat.

Parasitic for the domestic cat. The organisms occur as short rods or cocci, the latter sometimes in chains (Fig. 9.9*B*). The cocci are 0.1–0.8 μm in diameter; the rods are 0.2–0.5 by 0.9–1.5 μm. They do not have flagella. Each organism is surrounded by two membranes, which may show erosion at the point of contact (Small and Ristic, 1967). Attachment of two or more erythrocytes to a common organism brings about erythrocyte sequestration in blood capillaries (Simpson et al., 1978).

The organisms stain deep purple with Giemsa's stain. Fluorescent antibody or acridine orange stains may reveal the organism when it cannot be demonstrated with Giemsa's stain. When stained with acridine orange the organisms fluoresce bright orange with possibly an undertone of yellow-green, indicating a high RNA content and presence of DNA (Small and Ristic, 1967).

The species has not been cultivated in vitro.

H. felis is unique among species of this genus in its clinical importance. The organism produces patent parasitemia and severe, sometimes fatal, anemia in intact animals under field conditions. Splenectomy has relatively little effect on the course of the infection (Splitter et al., 1956). The disease has been produced in susceptible cats by inoculations of pooled blood from clinically normal cats (Splitter et al., 1956), an indication that the infection can produce a carrier state as well as patent disease.

The infection can be transmitted by the intraperitoneal, intravenous or oral routes (Flint et al. 1959); intrauterine infection may occur (Harbutt, 1963). It may be spread by biting during cat fights. Transmission by arthropod vectors has not yet been established. The organism is not infective for rats, mice, swine, cattle, sheep or dogs (Splitter et al., 1956).

Chloramphenicol, tetracycline, oxytetracycline and neoarsphenamine are effective in suppressing the infection (Flint and McKelvie, 1956).

The mol% G + C of the DNA is unknown.

Type strain: no culture isolated.

Further comments. "*Eperythrozoon felis*" was described by Clark (1942) as a predominantly ring- or oval-shaped organism, 0.5–1.0 μm in diameter, and staining a pale violet by the Giemsa strain with nonstaining centers. However, *H. felis* stains a uniform deep purple. The simultaneous appearance of organisms of both types has been reported (Splitter et al., 1956). Clark's description was based on a single blood film from a dead cat. The parasites have never been available for further study.

3. **Haemobartonella canis** (*ex* Tyzzer and Weinman 1939) nom. rev. (*Bartonella canis* Kikuth 1928, 1730.)

can' is. L. gen n. *canis* of the dog.

The characteristics are as described for the genus. The organism is a parasite usually seen on the erythrocytes of dogs following splenectomy. It does not produce disease.

The organism has been reported widely, but the classification is confused. Antigenic studies and additional investigation of its morphology are needed. The reported infectivity of *H. canis* for cats (Lumb, 1961) suggests a close relationship to *H. felis*.

The mol% G + C of the DNA is unknown.

Type strain: no culture isolated.

Species Incertae Sedis

The following organism has been grown in vitro and might more properly be placed in the genus *Grahamella*.

"*Haemobartonella tyzzeri*" (Weinman and Pinkerton 1938) Groot 1942, 279. (*Bartonella tyzzeri* Weinman and Pinkerton 1938, 217.)

A number of other species have been described, most of them named after the host (see Kreier and Ristic (1968)). The classification of many of these is in doubt.

Genus IV. **Eperythrozoon** Schilling 1928, 1854.[AL]

JULIUS P. KREIER AND MIODRAG RISTIC

Ep.e.ryth′ro. zo′on. Gr. Prefix *epi-* on; Gr. adj. *erythrus* red; Gr. n. *zoum* or *zoon* living thing, animal; M.L. neut. n. *Eperythrozoon* (presumably intended to mean) animals on red (blood cells).

In blood smears stained by Giemsa's and Romanowsky-type stains, the organisms appear as **bluish or pinkish violet rings or cocci, 0.4–1.5 μm in diameter.** They occur **on erythrocytes and free in plasma** with about equal frequency. Obligate parasites of various vertebrate species including some **rodents, ruminants, and pigs.** Some species have been shown to be **transmitted by arthropods.** The mol% G + C of the DNA is unknown.

Type species: *Eperythrozoon coccoides* Schilling 1928, 1854.

Further Descriptive Information

The organisms characteristically are round with numerous annular or disk-shaped elements, rarely rods. In stained preparations there is no differentiation of nucleus and cytoplasm. Swarmlike clusters of ring-shaped eperythrozoa may occur on the surfaces of erythrocytes. Rod-shaped forms may occur partly or entirely circling an erythrocyte. In fresh preparations, observed by darkfield or phase-contrast microscopy, cocci but no ring forms are seen (Wigand, 1958; Kreier and Ristic, 1963). By electron microscopy the organisms appear pleomorphic; each is surrounded by a single limiting membrane, with no cell wall, nucleus or other organelles (Tanaka et al., 1965). A high affinity of *E. coccoides* for pyronin dye indicates the presence of RNA, and the action of deoxyribonuclease on the organisms suggests the presence of DNA as well (Wigand and Peters, 1954).

E. coccoides has been reported to pass through collodion filters having an average pore size of 0.36 μm (Niven et al., 1952), and *E. parvum* passes through gradocol membranes having pore sizes of 0.57 and 0.41 μm, but not 0.36 μm (Seamer, 1959). *E. suis* has been reported to pass

through "N" Berkefeld and 8- and 14-lb Mandler filters (Splitter, 1952), and *E. parvum* will pass through 8-, 12- and 14-lb Mandler filters. Although this suggests that some of the cells of *Eperythrozoon* species may have a smaller size than indicated by microscopy, another likely explanation is that the cells, being nonrigid, may be drawn through pores which would retain rigid organisms of the same volume.

Eperythrozoon species have not been cultivated in cell-free media. Experimental transmission by inoculation of hosts with blood can be readily accomplished; blood from animals infected with *E. coccoides* has been reported to be infective by the oral route. In addition to propagation in host animals, *E. coccoides* may be propagated in embryonated hen's eggs (Seamer, 1959).

One species, *E. suis*, is the causative agent of "icteroanemia" of swine, an economically important disease which produces anemia, weakness, stunted growth, and sometimes death in intact pigs under field conditions.

Some species have been shown to be transmitted by arthropod vectors.

Growth of eperythrozoa is inhibited by arsenicals and tetracyclines but not by penicillin or streptomycin.

Maintenance Procedures

The organism can be preserved in infected erythrocytes admixed with 10% glycerol or dimethyl sulfoxide and stored at −70°C. For preservation beyond 3 months, the preparations must be stored in liquid nitrogen.

Differentiation of the genus **Eperythrozoon** from other closely related genera

Characteristics useful for differentiating *Eperythrozoon* from other members of the family *Anaplasmataceae* are given in the key to the family.

Differentiation of *Eperythrozoon* from *Haemobartonella* is in many cases difficult and possibly arbitrary. Differentiation is based on the fact that haemobartonellae rarely occur as ring forms, whereas eperythrozoa commonly do; moreover, eperythrozoa occur with about equal

frequency on the erythrocytes and free in the plasma, whereas haemobartonellae rarely occur free in the plasma.

Taxonomic Comments

A number of species of *Eperythrozoon* have been described, most of them on the basis of observations of eperythrozoon-like bodies in stained blood films. The nature and classification of many of these is in doubt. Their names, and references to the original descriptions, are listed by Kreier and Ristic (1968).

Further Reading

Gothe, R. and J.P. Kreier. 1977. *Aegyptianella, Eperythrozoon,* and *Haemobartonella.* In J.P. Kreier (Editor), *Parasitic Protozoa,* Vol. IV, Academic Press, New York, pp. 251–294.

Kreier, J.P. and M. Ristic. 1968. Haemobartonellosis, eperythrozoonosis, grahamellosis and erhlichiosis. In Weinman and Ristic (Editors), *Infectious Blood Diseases of Man and Animals.* Academic Press, New York, pp. 387–472.
Splitter, E.J. 1950. *Eperythrozoon suis,* the etiologic agent of icteroanemia—an anaplasmosis-like disease in swine. Am. J. Vet. Res. *11:* 324–329.
Tanaka, H., W.T. Hall, J.B. Scheffield and O.H. Moore. 1965. Fine structure of *Haemobartonella muris* as compared with *Eperythrozoon coccoides* and *Mycoplasma pulmonis.* J. Bacteriol. *90:* 1735–1749.

Differentiation of the species of the genus **Eperythrozoon**

Characteristics useful for distinguishing the various species of *Eperythrozoon* are presented in Table 9.12.

List of the species of the genus **Eperythrozoon**

1. **Eperythrozoon coccoides** Schilling 1928, 1854.[AL]

coc.coi'des. Gr. n. *coccus* a berry; M.L. n. *coccus* a coccus; Gr. n. *eidus* shape; M.L. adj. *coccoides* coccus-shaped.

The characteristics are as described for the genus and as listed in Table 9.12.

Electron microscope studies of thin sections show only slight erosion of the erythrocyte surface at the point of contact with the parasite. In fresh blood strained by a vital acridine orange technique, the organisms are light yellow, coccoid and nonmotile. Reproduction is probably by binary fission.

Infection is usually maximal in young animals or in splenectomized adults. The organism has been reported to cause fatal mouse hepatitis when associated with another etiological agent (a virus): otherwise, moderate to no anemic changes are reported. Virus titers are increased 100-fold in combined infections (Niven et al., 1952).

The immunological state is of the infection immunity (premunition) type. *E. coccoides* antigen has been reported nonreactive with serum from humans having bartonellosis, with serum from cattle infected with *Anaplasma marginale,* and with serum from rats infected with *Haemobartonella muris.* Serum from mice infected with *E. coccoides* reacts with *A. marginale* and with *H. muris* antigens (Wigand, 1958).

Neoarsphenamine is a very effective therapeutic agent. Chlortetracycline and oxytetracycline are active, but sulfonamides, sulfones and penicillin show little or no activity.

The mouse louse (*Polyplax serrata*) is a natural vector. *E. coccoides* has been reported from Europe, North and South America, and Africa.

The mol% G + C of the DNA is unknown.

Type strain: no culture isolated.

2. **Eperythrozoon ovis** Neitz, Alexander and du Toit 1934, 267.[AL]

o'vis. L. n. *ovis* a sheep.

The characteristics are as described for the genus and as listed in Table 9.12.

Stains pale purple with Giemsa's stain. If rod forms occur they are most commonly attached to the margin of the erythrocyte and may partly or completely surround it (Fig. 9.10*B*). By phase-contrast microscopy, *E. ovis* appears as spheres, with neither rings nor rods apparent; no motility occurs (Kreier and Ristic, 1963). In thin sections observed with the electron microscope, the organisms are seen as round or oval structures 0.3–0.4 μm in diameter and are partially embedded in the erythrocyte. A peripheral dense region 20–30 nm thick is probably a surrounding membrane. No nucleus, endoplasmic reticulum, mitochondria or other organelles are seen (Kreier and Ristic, 1963). The suggested mode of multiplication is binary fission.

E. ovis can provoke mild symptoms in normal animals or receptive species without splenectomy. Mortality from infection is rare, but anemia and failure to gain weight occur in young lambs (Ilemabode and Blotkamp, 1977).

E. ovis has common antigens with *E. wenyonii* and possibly with *Anaplasma marginale* (Kreier and Ristic, 1963).

Transmission by horsefly bites and other arthropods may occur in nature. *E. ovis* has been reported from Africa, Australia, several European countries, and is common throughout the United States.

The mol% G + C of the DNA is unknown.

Type strain: no culture isolated.

3. **Eperythrozoon suis** Splitter 1950, 513.[AL]

su'is. L. gen. n. *suis* of the pig.

Table 9.12
Differential characteristics of the species of the genus **Eperythrozoon**[a]

Characteristics	1. *E. coccoides*	2. *E. ovis*	3. *E. suis*	4. *E. parvum*	5. *E. wenyonii*
Appearance in stained blood films	Rings, cocci, disks, rods, but mainly rings of regular outline with clear centers	Delicate rings or disks; irregularly shaped forms may be ovoid-, comma-, rod-, dumbbell- or tennis racket-shaped	Rods, rings, cocci	Disk, coccus forms, rings	Disks or rings; rarely rods or oval forms
Size of organisms in their greatest dimension (μm)	0.4–0.5	0.5–1.0	Most 0.8–1.0, some up to 2.5	0.5–0.8	0.3–1.5
Natural hosts	Albino and wild mice, albino rats, rabbits, hamsters	Domestic sheep, goats, deer, and a species of antelope	Domestic pigs	Domestic pigs	Domestic cattle
Appearance in fresh blood stained by acridine orange	Light yellow	Bright orange			Bright orange
Pathogenic for pigs (icteroanemia)	−	−	+	−	−

[a] Symbols: see standard definitions.

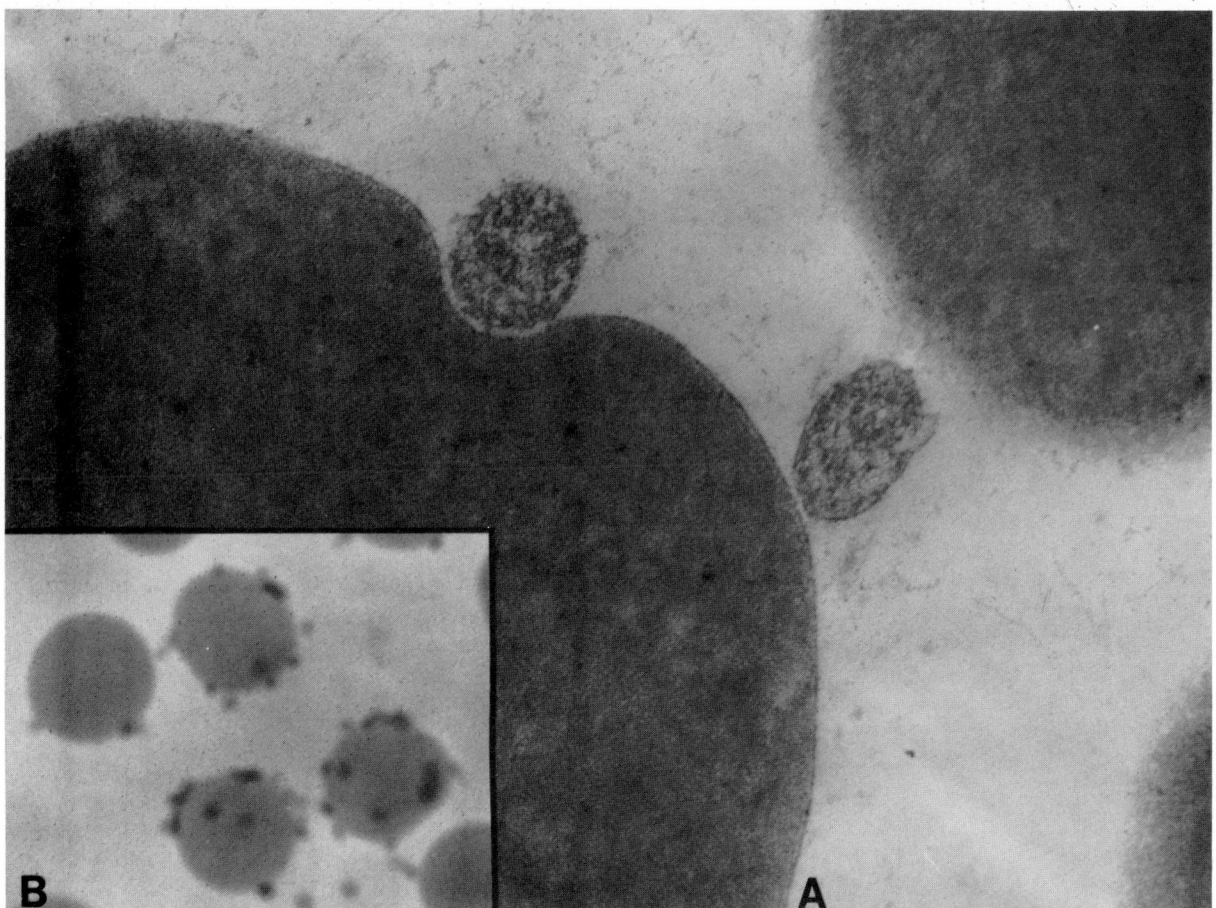

Figure 9.10. *Eperythrozoon wenyonii. A,* Ultrathin section showing the parasite on bovine erythrocytes. Note erosion of the erythrocyte at the point of attachment (×37,000). (Reproduced by permission from Adeyinka Cole, College of Veterinary Medicine, University of Illinois, Urbana.) *B, Eperythrozoon ovis* on infected sheep blood (Giemsa strain, ×2400). (Reproduced with permission from J.P. Kreier and M. Ristic, *American Journal of Veterinary Research, 24:* 488–500, 1963, American Veterinary Medical Association.)

The characteristics are as described for the genus and as listed in Table 9.12. The species is morphologically the largest in the genus.

The causative agent of icteroanemia in swine. Splenectomy causes relapse in carrier pigs, and an unusually high parasitemia occurs in pigs splenectomized before infection (Splitter, 1950). Infected animals produce a complement-fixing antibody (Splitter, 1958), possibly directed against erythrocyte stromal antigens. An indirect fluorescent antibody test was developed for diagnosis of infections caused by *E. suis* (Smith and Rahn, 1975). Neoarsphenamine controls the infection. The organism is common throughout the United States. There is a single report of *H. suis* from the Belgian Congo (Jansen, 1952).

It should be noted that *E. suis* can be differentiated from *E. parvum,* a nonpathogenic parasite of the pig, on the basis of size and morphology.

The mol% G + C of the DNA is unknown.

Type strain: no culture isolated.

Further comments. The parasite described in Taiwan as "*Anaplasma taiwanensis*" (Sugimoto, 1935) may be *E. suis.*

4. **Eperythrozoon parvum** Splitter 1950, 513.[AL]

par'vum. L. neut. adj. *parvum* small.

The characteristics are as described for the genus and as listed in Table 9.12.

Nonpathogenic parasite of domestic pigs.

Does not occur in cells or tissues other than blood.

Splenectomy causes relapse of carrier pigs.

Susceptible to arsenicals and tetracyclines but not to penicillin, streptomycin or sulfonamides.

May be readily transmitted by all parenteral routes and rarely by massive oral inoculations.

The pig louse (*Haematopinus suis*) may transmit the disease (Jansen, 1952).

Infection has been reported from the United States, Europe and Africa.

The mol% G + C of the DNA is unknown.

Type strain: no culture isolated.

5. **Eperythrozoon wenyonii** Adler and Ellenbogen 1934, 220.[AL]

wen.yo'ni.i. M.L. gen. n. *wenyonii* of Wenyon; named after Dr. C.M. Wenyon, an investigator of these organisms.

The characteristics are as described for the genus and as listed in Table 9.12. See also Figure 9.10.

Morphologically similar to *E. coccoides.* Stains pinkish purple with Giemsa's stain. May be attached to erythrocytes or platelets, or may occur free in the plasma (Kreier and Ristic, 1968). A number of ring forms may be attached to a single erythrocyte. Rod or coccus forms may completely encircle an erythrocyte. When stained with acridine orange, the organisms appear bright orange, primarily as cocci 0.5 μm in diameter and sometimes clustered like bunches of grapes. The rod-shaped structures seen in blood films stained by polychrome methods appear as chains of cocci following acridine orange staining. Electron microscope observations of dried, metal-shadowed blood preparations show flattened or collapsed spheres or short rods, indicating lack of a rigid cell wall. Masses in the cytoplasm, probably nucleoids, have been observed (Kreier and Ristic, 1963).

The organisms produce parasitemia and a mild anemia, but no other clinical signs occur. Latent infection is made manifest by splenectomy. Sheep, goats and deer (*Odocoileus virginiana*) are not susceptible.

E. wenyonii shares antigens with *E. ovis*, *Anaplasma marginale* and *A. ovis.*

Infection with *E. wenyonii* can be produced by parenteral injection of blood from a latently or patently infected animal. The natural means of transmission has not been determined. The distribution of the organism appears to be worldwide.

ORDER II. **CHLAMYDIALES** STORZ AND PAGE 1971, 334[AL]

JAMES W. MOULDER

Chla.my.di.a'les. M.L. n. *Chlamydia* type genus of the order; *-ales* ending to denote an order; M.L. fem. pl. n. *Chlamydiales* the *Chlamydia* order.

Coccoid microorganisms whose obligately intracellular mode of multiplication within cytoplasmic vacuoles is characterized by change of small, rigid-walled infectious forms (elementary bodies) into larger, flexible-walled noninfectious forms (reticulate bodies) that divide by fission. The developmental cycle is complete when daughter cells reorganize into elementary bodies which survive extracellularly to infect new host cells. Metabolically limited, Gram-negative parasites of eucaryotes in which they may cause various diseases.

FAMILY I. **CHLAMYDIACEAE** RAKE 1957, 957[AL]

JAMES W. MOULDER

Chla.my.di.a'ce.ae. M.L. fem. n. *Chlamydia* type genus of the family; *-aceae* ending to denote family; M.L. fem. pl. n. *Chlamydiaceae* the *Chlamydia* family.

At present this family comprises but a single genus, *Chlamydia*, whose description follows below.

Genus I. **Chlamydia** Jones, Rake and Stearns 1945, 55[AL]

JAMES W. MOULDER, THOMAS P. HATCH, CHO-CHOU KUO, JULIUS SCHACHTER AND JOHANNES STORZ

Chla.my'di.a. Gr. fem. n. *chlamys, chlamydis* a cloak; M.L. fem. dim. n. *Chlamydia* a cloak.

Nonmotile coccoid organisms, 0.2–1.5 μm, multiplying only within membrane-bounded vacuoles in the cytoplasm of host cells by means of a unique developmental cycle characterized by change of small elementary bodies into larger reticulate bodies that divide by fission. The cycle is complete when reticulate bodies reorganize and condense into a new generation of elementary bodies which survive extracellularly to infect other host cells by an unusual phagocytic mechanism that does not involve fusion of chlamydia-containing phagosomes with lysosomes. Transition between elementary bodies and reticulate bodies is gradual, and intermediate forms exist. **Elementary bodies** are 0.2–0.4 μm in diameter. They contain electron-dense nuclear material and few ribosomes, are surrounded by rigid trilaminar walls, and are infectious. **Reticulate bodies** are 0.6–1.5 μm in diameter. As compared to elementary bodies, reticulate bodies have less dense, fibrillar nuclear material, more ribosomes, and thinner and more flexible trilaminar walls. Infectivity of reticulate bodies for new host cells has not been demonstrated. Gram-negative. Cell walls are similar in structure and composition to walls of other Gram-negative bacteria, but muramic acid is either absent or is present only in traces. There is a genus-specific, wall-associated, lipopolysaccharide antigen that contains a 2-keto-3-deoxyoctanoic acid-like substance. Walls of elementary and reticulate bodies exhibit regular hexagonal arrays of subunits on the inner surface and single patches of hexagonally ordered hemispheric projections on their outer ones. **Chlamydiae cause a variety of diseases in humans, other mammals, and birds. Multiplication of chlamydiae outside of host cells has not been achieved.** They

may be propagated in laboratory animals, yolk sacs of chicken embryos, or in cell cultures. Chlamydiae rely on their hosts for high-energy compounds and low molecular weight synthetic intermediates which they use to synthesize their own DNA, RNA and protein, as well as smaller molecules specific for chlamydiae and not made by host cells. The mol% G + C of the DNA is 41–44 (T_m). The genome size is among the smallest of all procaryotes, 4–6 × 10^8 daltons.

Type species: *Chlamydia trachomatis* (Busacca 1935) Rake 1957, 958.

Two species are presently recognized, *C. psittaci* and *C. trachomatis*, the latter being further subdivided into three biovars. In the sections immediately following, if the chlamydial property being described is not common to all members of the genus, the species or biovar to which the description applies will be stated.

*Further Descriptive Information**

Cell morphology and developmental cycle. Chlamydiae are highly pleomorphic. Their morphology, composition, and activities are so highly dependent on the stage in the chlamydial developmental cycle at which they are observed that chlamydial properties can be described only within the context of the developmental cycle. The properties of the two principal cell types are summarized in Table 9.13. Elementary bodies are nonmultiplying cells specialized for extracellular transit and entry into new host cells, whereas reticulate bodies are non-infectious cells specialized for intracellular multiplication. Preparations of a single cell type essentially free of the other are obtained by taking advantage

---

* The voluminous chlamydial literature is exhaustively reviewed in the books and articles listed under Further Reading. Only very recent original publications have been individually cited.

Table 9.13.
Properties of chlamydial elementary bodies and reticulate bodies[a]

Characteristics	Elementary body	Reticulate body
General:		
Diameter, μm	0.2–0.4	0.5–1.5
Density, g/cm^3	1.21	1.18
Time of appearance in developmental cycle	Late	Early
Infectivity for new host cells	+	−
Intracellular multiplication	−	+
Intravenous lethality for mice	+	−
Immediate toxicity for cells in culture	+	−
Cell wall:		
Susceptibility to		
Mechanical stress	−	+
Osmotic stress	−	+
Lysis by trypsin	−	+
Trilaminar structure	+	+
Patches of hemispheric projections on outer surface	+	+
Hexagonal arrays of subunits on inner surface	+	+
Muramic acid in stoichiometric amounts	−	−
Synthesis inhibited by penicillin	+	−
Genus-specific antigen	+	+
Hemagglutinin	+	−
40,000-dalton outer membrane protein	+	+
Nucleic acids:		
DNA	Compact	Disperse
RNA/DNA ratio	1	3–4
Ribosomes	Scanty	Abundant
Metabolism:		
Net generation of ATP	−	−
ATP/ADP transport system	−	+
ATP-dependent liberation of CO_2 from glucose 6-P, pyruvate, glutamate, and aspartate	+[b]	+[b]
ATP-dependent host-free synthesis of protein	−	+

[a] Symbols: see standard definitions.
[b] These experiments were done mixtures of the two chlamydial cell types, so it is not certain whether only one or both types were responsible for the observed metabolic activity.

of (i) the predominance of reticulate bodies early, and elementary bodies late, in the developmental cycle and (ii) the slight difference in density between the two cell types.

The developmental cycle may be divided into three phases: (i) entry of elementary bodies into host cells and their reorganization into reticulate bodies (Fig. 9.11, A and B), (ii) multiplication of reticulate bodies (Fig. 9.11C), and (iii) conversion of a large fraction of the reticulate body population into a new generation of elementary bodies which are released from the host cells (Fig. 9.11D). The cycle is far from synchronous, and each of the first two phases overlaps into the next succeeding one.

The developmental cycle is started when extracellular elementary bodies attach to host cells. Evidence suggests the presence of specific receptors on the surface of both host cells and chlamydiae, but they have not been identified. The host cells then ingest the attached elementary bodies by a phagocytic mechanism that is unusual in that it is not inhibited by cytochalasin B and thus probably does not require the participation of microfilaments. The rate of attachment and inges-

tion varies widely, depending on the chlamydial strain, the host cell, and the multiplicity of infection. In both natural hosts and in cell cultures, chlamydiae are taken into cells that are not usually active phagocytes. Ingestion requires expenditure of energy by the host but not by the parasite. It is a temperature-dependent process. The ingested elementary bodies enter host cells inside of phagosomes, which are cytoplasmic vacuoles surrounded by membranes derived from the plasma membranes of host cells. Some unidentified property of chlamydiae prevents at all times in the developmental cycle the sequence of events that usually follows ingestion of microorganisms: fusion of lysosomes with phagosomes, release of lysosomal hydrolases into the phagosome, and destruction and digestion of the phagocytized organism. Within the phagosomes, the elementary bodies almost immediately begin to rearrange into reticulate bodies. They lose their infectivity, their dense nucleoids disperse into more evenly distributed fibrillar DNA, their ribosomes increase in number, and their walls become thinner, more flexible, and more fragile. Completion of the reorganization is signaled by division of reticulate bodies, which is first seen 8–10 h after infection.

Multiplication of chlamydiae generally occurs by binary fission of reticulate bodies without apparent septation. Nuclear segregation is accomplished by separation of two zones of low electron density filled with a fine fibrillar material. From approximately 10–20 h after infection, a large fraction of the reticulate bodies are in some stage of division. The proportion of dividing forms drops off thereafter, but some division occurs throughout the developmental cycle. Multiplication of reticulate bodies takes place within an expanding membrane-bound vacuole, or inclusion, that is an extension in time and space of the phagocytic vacuole in which the parent elementary body was brought into the host cell. Although chlamydiae spend their entire intracellular life within the confines of an inclusion membrane, next to nothing is known about the structure, behavior, or biosynthetic origin of these structures. By 12–15 h after infection, chlamydial inclusions are large enough to be seen by ordinary light microscopy of stained cells or by phase contrast microscopy of live ones. Near the end of the developmental cycle, an inclusion may contain hundreds of chlamydial cells and fill nearly the entire cytoplasm of the host cell.

At 20–25 h after infection, host cell-associated infectivity begins to rise, and elementary bodies are again seen in electron micrographs. The conversion of reticulate bodies into elementary bodies appears to retrace the path followed in the opposite conversion. Mature elementary bodies do not divide, but dividing dense-centered transition forms have been seen. Reticulate bodies continue to divide and to reorganize into elementary bodies until the host cell can no longer support chlamydial multiplication. Therefore, there is no clear-cut termination to the developmental cycle, although the end is usually considered to come 48–72 h after infection, depending mainly on the chlamydial strain. The mechanism whereby chlamydiae are released from host cells is poorly understood. Infected cells may simply burst from the sheer size of their inclusions. Terminal release of lysosomal enzymes may speed dissolution of moribund cells. The yield of infectious chlamydiae per host cell varies from 10 to nearly 1000, depending on many factors, such as the chlamydial strain, the host cell, the cultural conditions, and the way in which infectivity is measured.

Cell walls. Chlamydial cell walls exhibit monolayers of hexagonally arrayed subunits lining their inner surfaces (Fig. 9.12A) and spatially related patches of hemispheric projections on their outer surfaces (Fig. 9.12B). It was originally thought that these structures occur only on the walls of elementary bodies, but it now appears that they are also found on reticulate bodies. Outer membrane complexes similar to those of free-living Gram-negative bacteria have been obtained from chlamydial cells (Fig. 9.12C) (Caldwell et al., 1981; Hatch et al., 1981). One-half of the protein content of these complexes is in the form of a molecule with a molecular weight of approximately 40,000 and having species-specific and type-specific immunoreactivity.

Nutrition and growth. The nutritional requirements of chlamydiae cannot be defined because they have not been grown outside of host cells. Requirements for growth of chlamydiae above and beyond the

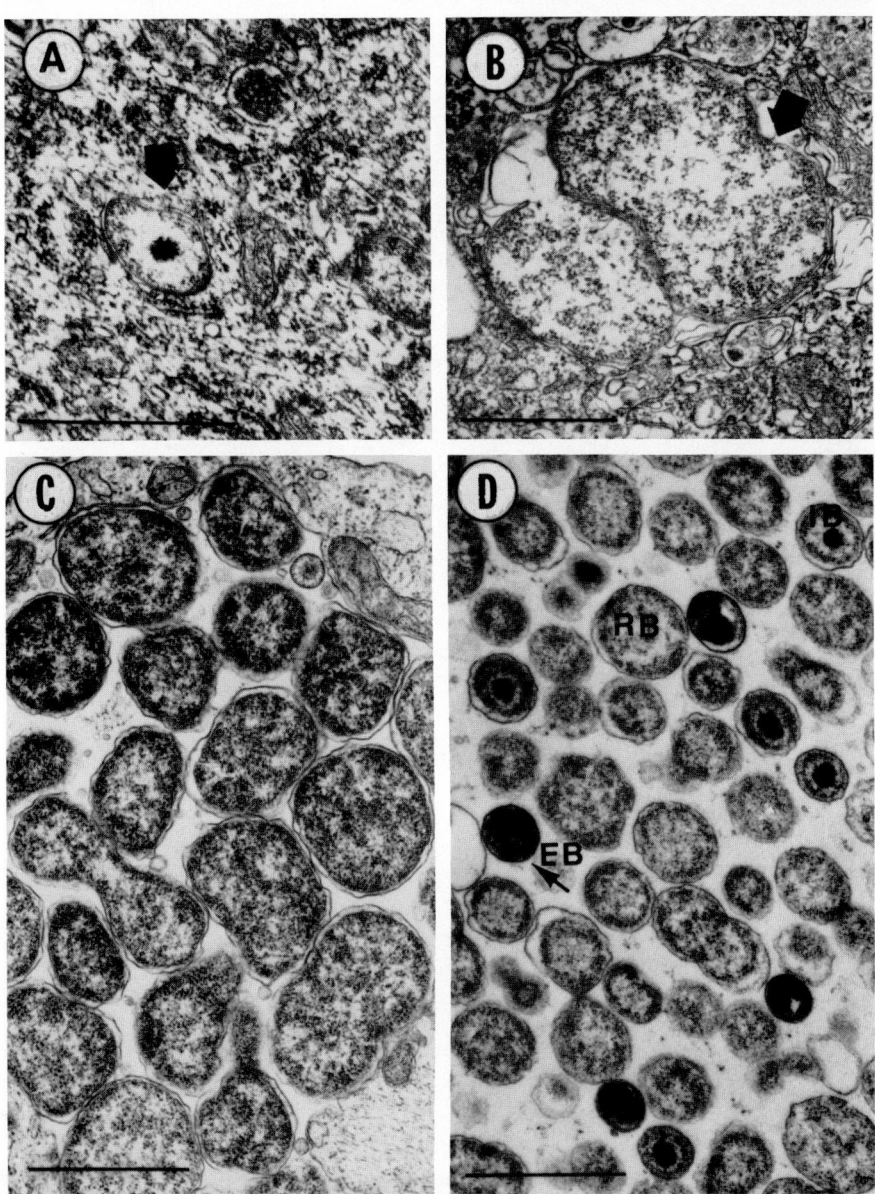

Figure 9.11. The chlamydial developmental cycle. Transmission electron micrographs of thin sections of mouse fibroblasts (L cells) infected with *Chlamydia psittaci* (strain meningopneumonitis). Bars, 1.0 μm. *A*, At 2.5 h after infection. The *arrow* points to an elementary body that has just begun to differentiate into a reticulate body. *B*, at 12 h after infection. The *arrow* points to a dividing reticulate body. *C*, At 20 h after infection. The chlamydial population consists almost entirely of dividing reticulate bodies. *D*, At 30 h after infection. Some reticulate bodies are still dividing, but others have begun to reorganize into elementary bodies. *EB*, elementary body; *RB*, reticulate body; IB, intermediate body—a chlamydial cell intermediate in appearance between an elementary body and a reticulate body. (Reproduced with permission from I. I. E. Tribby et al., *Journal of Infectious Diseases, 127:* 155–163, 1973, © University of Chicago Press, Chicago.)

growth needs of their hosts have not been demonstrated. Chlamydiae compete with host cell macromolecule-synthesizing machinery for biosynthetic intermediates in the metabolic pools of their hosts. They have free access to some constituents of these pools, such as the nucleoside triphosphates, but only limited access to others, such as the pyrimidine deoxynucleosides.

Metabolism. Investigations on the metabolism of host-free chlamydiae must be interpreted with two complications in mind. (i) Metabolic activities associated with "host-free" preparations of chlamydiae must be proven conclusively not to be due to contamination with host enzymes. (ii) Failure to detect an activity may be the result of examining the wrong cell type. Most metabolic studies have been carried out with chlamydial populations in which elementary bodies predominated, although it has long been apparent that the reticulate body is the metabolically active cell type.

When separated from host cells, chlamydiae have few metabolic activities. They do not catabolize glucose, but they produce carbon dioxide from glucose-6-phosphate, glutamate, pyruvate and aspartate in the presence of ATP, NADP and other cofactors. In contrast to rickettsiae, chlamydiae do not metabolize glutamate beyond succinate, and glutamate does not serve as an energy source. Cytochromes, flavins,

and reactions that result in a net gain of ATP have not been demonstrated. Because chlamydiae appear to be unable to synthesize their own high-energy compounds, they have been described as "energy parasites." Host-free reticulate bodies have an ATP/ADP transport system, and, in the presence of exogenous ATP, they synthesize chlamydial protein (T. P. Hatch, unpublished results). When supplied with host-generated precursors, intracellular chlamydiae synthesize their own DNA, RNA and protein. Glycogen is produced and deposited within cytoplasmic inclusions of *C. trachomatis* but not in amounts detected by iodine staining in inclusions of *C. psittaci*. Folic acid and its derivatives are synthesized by most strains of *C. trachomatis* but not by most strains of *C. psittaci*. Chlamydiae also synthesize lipids such as phosphatidyl glycerol and branched-chain fatty acids that are characteristic of procaryotes and are not found in their host cells.

Genetics. The results of studies on DNA base composition and DNA hybridization are discussed under Taxonomic Comments. With regard to genome size, chlamydia have relatively small genomes compared to most procaryotes. The isolated DNA of a *C. trachomatis* strain had a molecular weight of 660×10^6 (10×10^5 nucleotide pairs) as measured by zonal centrifugation in sucrose gradients. Measurement of the kinetics of DNA reassociation gave a genome size of 360×10^6 daltons

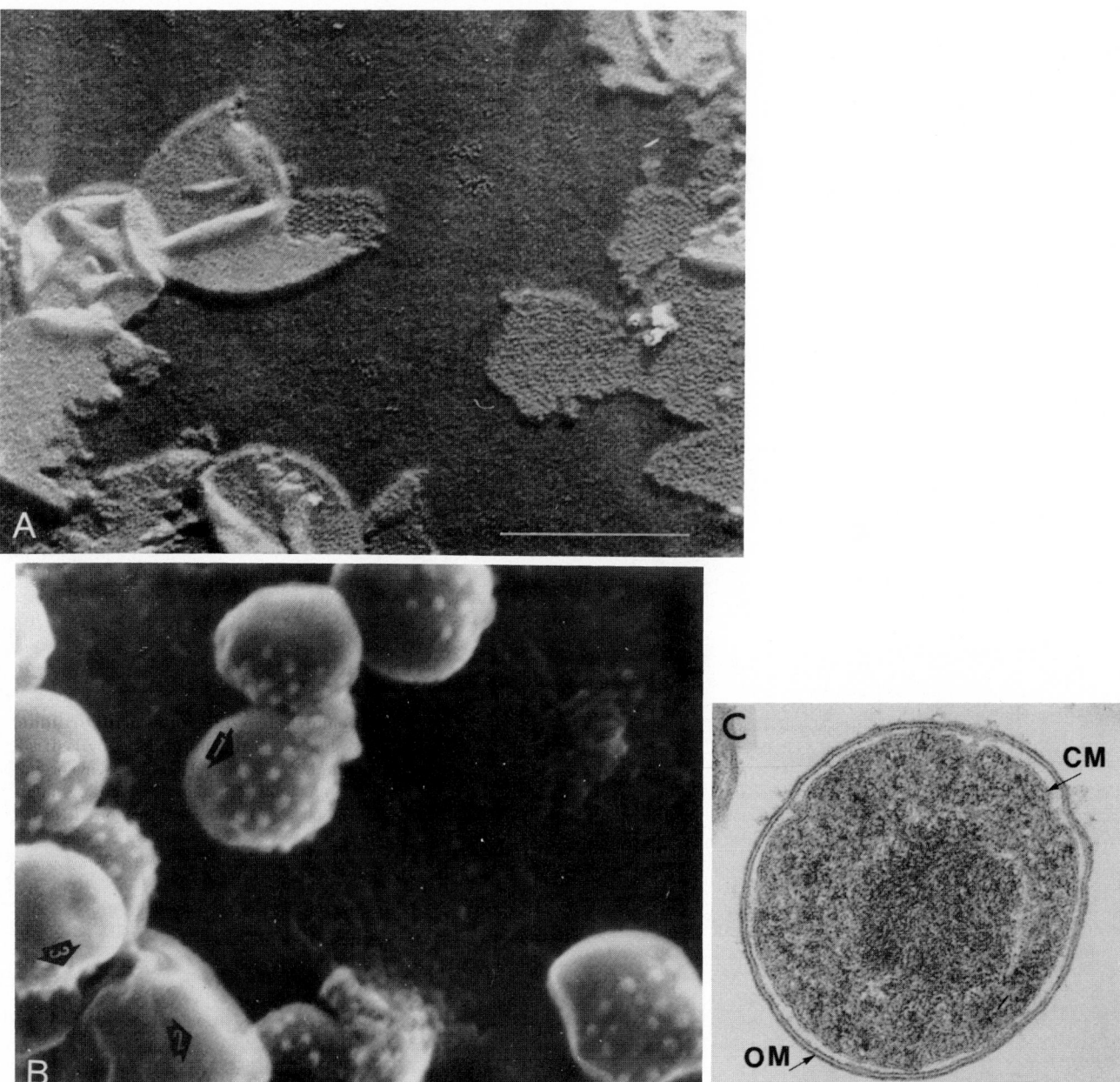

Figure 9.12. Chlamydial cell walls. *A*, Transmission electron micrograph of the cell walls of *Chlamydia psittaci* (strain meningopneumonitis) shadow-cast with platinum-palladium alloy. Note the hexagonally packed 100 Å units on the inner cell wall surfaces only. Bar, 1.0 μm. (Reproduced with permission from A. Matsumoto and G. P. Manire, *Journal of Bacteriology 104:* 1332–1337, 1970, © American Society for Microbiology, Washington, D.C.) *B*, Scanning electron micrograph of elementary bodies of *C. psittaci* (strain 6BC). *Arrows* point to (1) an elementary body with a patch of regularly arrayed hexagonal projections, (2) a reticulate body without such a patch, and (3) a hemispheric projection in profile, projecting from the surface of the elementary body. Bar, 1.0 μm. (Reproduced with permission from W. W. Gregory et al., *Journal of Bacteriology, 138:* 241–244, 1979, © American Society for Microbiology, Washington, D.C. *C*, Transmission electron micrograph of a thin section of an elementary body of *Chlamydia trachomatis* (biovar LGV, serovar L:2) showing the outer membrane (*OM*) and the cytoplasmic membrane (*CM*). There is no evidence of a peptidoglycan layer between the two membranes. Bar, 0.5 μm. (Reproduced with permission from H. D. Caldwell et al., *Infection and Immunity, 31:* 1161–1176, 1981, © American Society for Microbiology, Washington, D.C.)

(6.0×10^5 nucleotide pairs) for a *C. psittaci* strain and 500×10^6 daltons (8.5×10^5 nucleotide pairs) for a strain of *C. trachomatis*. Both *Chlamydia* species have a 4.4-megadalton plasmid. Cleavage of the plasmids with restriction nucleases showed that they are different in the two species. Mutants resistant to a number of antimicrobials have been obtained, but other genetic markers are not available.

Antigenic structure. All members of the genus share a common heat-stable antigen demonstrable by complement fixation that is associated with the cell wall and is present throughout the developmental cycle. It is a lipopolysaccharide-protein complex with an acidic polysaccharide as the antigenic determinant. The immunodominant group is a 2-keto-3-deoxyoctanoic acid similar to but not identical with that of *Salmonella* lipopolysaccharide. All chlamydiae appear to produce a hemagglutinin for certain rodent and fowl erythrocytes that is related to the group antigen in specificity and structure. It is released extracellularly during the developmental cycle and may be liberated from cell walls of elementary bodies (but not reticulate bodies) by sonic disintegration. Antigens of generic specificity are responsible for the Frei test, which demonstrates delayed hypersensitivity to chlamydial infection with a skin-test antigen that has been boiled or treated with phenol or formalin.

Species-specific chlamydial antigen may be demonstrated by complement fixation after removing genus-specific antibody by absorption of antisera with boiled antigen. Species-specific antigens have been liberated from sonicated cell walls and assayed by indirect hemagglutination or double immunodiffusion. *C. trachomatis* and *C. psittaci* have at least 15–18 different species-specific antigens. Most of them are heat labile.

The trachoma and lymphogranuloma venereum (LGV) biovars of *C. trachomatis* may be separated into serovars by a microimmunofluorescence test. Unknown strains are typed by testing antiserum prepared by intravenous immunization of mice against prototype antigens in an indirect immunofluorescent antibody test. Although the type-specific antigens form a continuous spectrum of cross-reactivity across both the trachoma and LGV biovars, the highest titers are obtained with homologous antigens and the pattern of heterologous cross-reaction is definitive for each serovar (see Fig. 9.14). Serovars belonging to the trachoma biovar are designated by the letters A through K plus Ba. These serovars tend to cluster according to geographical and clinical (site of infection) distribution. Thus, ocular isolates from trachoma-endemic areas are mainly A through C and ocular and genital isolates from nontrachoma areas are chiefly D through G. LGV serovars are designated L:1, L:2, and L:3. One type-specific antigen has been purified and found to be a protein of molecular weight 30,000. *C. psittaci* and the mouse biotype of *C. trachomatis* have not been serotyped systematically.

Antibiotic sensitivity. The growth of chlamydiae in the yolk sac cells of chicken embryos and in cells in culture is inhibited by the tetracyclines, erythromycin and rifampin. Chlamydial multiplication is not blocked by the aminoglycosides, bacitracin, vancomycin or ristocetin. Most *C. trachomatis* strains are sensitive to sulfonamides, whereas most strains of *C. psittaci* are not. The penicillins interrupt the developmental cycle by preventing the maturation of reticulate bodies into elementary bodies. However, if these antibiotics are removed, the cycle proceeds normally. D-Cycloserine acts similarly to penicillin, but *C. trachomatis* is much more susceptible to growth inhibition by this antimicrobial than is *C. psittaci*. The mode of action of the inhibitors of peptidoglycan synthesis on organisms without muramic acid is not known. Chlamydial infections are usually treated with tetracycline, sulfonamides, erythromycin and rifampin. Chlamydial strains resistant to sulfonamides, penicillin, chlortetracycline and rifampin have been obtained by in vitro passage in the presence of these drugs.

Pathogenesis. Chlamydiae are among the most widely distributed of all parasites of animals, and they produce protean clinical symptoms. Humans are the natural hosts for *C. trachomatis*, with the exception of the mouse biovar which occurs as a latent respiratory infection of mice. In developing countries, some serovars of the trachoma biovar of *C. trachomatis* cause millions of cases of trachoma, a chronic, potentially blinding keratoconjunctivitis. In industrialized countries, other serovars of *C. trachomatis* (trachoma biovar) are among the most common sexually transmitted agents of disease. They may cause relatively mild infections (cervicitis, urethritis, conjunctivitis) or more serious invasive ones (epididymitis, salpingitis). Infected mothers may transmit *C. trachomatis* (trachoma biovar) to their babies during the birth process with the production of conjunctivitis and pneumonia. The LGV biovar of *C. trachomatis* causes the sexually transmitted disease lymphogranuloma venereum which is marked by invasion of lymph nodes in the genito-anal region.

C. psittaci is a parasite of animals other than man. Practically all species of birds, most domestic mammals, and many wild mammals are natural hosts. *C. psittaci* may induce a wide range of clinical manifes-

tations, including respiratory and intestinal infections, conjunctivitis, polyarthritis, fetal death and abortion, as well as genital disease. Arthropods may harbor *C. psittaci*, but their role as agents of transmission has not been proven. Mammalian strains of *C. psittaci* do not readily infect humans, but some avian strains are highly infectious and produce the human disease psittacosis.

All chlamydial infections tend to persist in chronic or clinically inapparent forms. Persistent infection of cell cultures with both *C. trachomatis* (Lee and Moulder, 1981) and *C. psittaci* (Moulder et al., 1980) has been achieved.

The nature and effectiveness of the immune response of hosts to chlamydial infection is poorly defined. Vaccines have a short-lived and modest effect in preventing the onset of trachoma and in reducing the intensity of established disease. Vaccines against psittacosis are ineffective. Little is known about immunity to genital infections with *C. trachomatis* in humans. In infections of mammals with *C. psittaci*, different investigations have yielded conflicting results on the value of vaccinations.

Intravenous inoculation of mice with very large numbers of elementary bodies kills them in a few hours, before the injected chlamydiae have started to multiply. In cell lines such as L or HeLa cells, the damage incurred by host cells is a function of the multiplicity of infection. At high multiplicities (100–1000 infectious units per host cell), the cells are killed so rapidly that chlamydial multiplication does not occur. At moderate multiplicities of infection (10–100), host-cell multiplication is strongly inhibited, but infectious chlamydiae are produced, and at very low multiplicities (< 1), infected host cells support chlamydial multiplication and are still able to divide. Daughter cells may or may not be infected. Mouse macrophages are more susceptible to damage by chlamydiae than are established cell lines. The relation of injury to cells in culture to the pathogenic manifestations of chlamydial infections in natural hosts is unknown.

Enrichment and Isolation Procedures

All known strains of *Chlamydia* grow in the yolk sacs of chicken embryos. Most strains of *C. psittaci* and the LGV and mouse biovars of *C. trachomatis* multiply well in a variety of cell cultures. However, strains of the trachoma biovars of *C. trachomatis* readily enter and multiply in cells in culture only with the aid of infection-promoting procedures such as centrifugation of the inoculum onto host cell monolayers, pretreatment of host cells with polycations, and inhibition of the synthesis of host macromolecules. Mice are efficiently infected with many *C. psittaci* strains of avian origin and less efficiently with most mammalian strains of that species. They may also be readily infected with the LGV biovar of *C. trachomatis* and less so with the trachoma biovar. The hazard of laboratory infection with avian strains of *C. psittaci* is so great that attempts to isolate *C. psittaci* from birds and humans should be made only by experienced workers in specially equipped laboratories.

Collection and processing of specimens. Selection of an appropriate specimen for isolation of *Chlamydia* depends on the disease, the host, and the chlamydial agent. Because chlamydiae are susceptible to a number of antimicrobial agents, the choice of drugs for inhibiting the growth of other microorganisms must be selective. Frequently used isolation mixtures contain gentamicin (or streptomycin), vancomycin, and amphotericin B. Clinical specimens are usually suspended in sucrose-phosphate buffer or cell culture growth medium containing a suitable set of antimicrobial drugs, and intracellular chlamydiae are released by shaking with glass beads. Specimens may be kept at 4°C if they are to be processed within 24 h. If they must be held longer, they should be frozen (−60°C or lower). Strains of *C. psittaci* are more stable on storage than are strains of *C. trachomatis*. Strains of the trachoma biovar of *C. trachomatis* are the least stable of all.

Isolation in cell culture. The cell lines most commonly used for isolation of chlamydiae are McCoy cells (a heteroploid mouse line) and HeLa 229 cells (derived from a human cervical carcinoma). When isolations are made in McCoy cells, the monolayers are treated with cycloheximide, iododeoxyuridine, or cytochalasin B to inhibit host macromolecular synthesis either before or immediately after they are infected by centrifuging the inoculum onto the monolayers at 1000–3000 × g for 1 h at 20–35°C. The centrifugation is essential for the isolation of *C. trachomatis* strains of the trachoma biovar and optional for other chlamydiae. The monolayers are incubated for 2–3 days at 35–37°C and examined for the presence of chlamydial inclusions by staining with iodine (*C. trachomatis* only), Giemsa, or fluorescent antibody. If the results are negative or inconclusive, additional passages may be made. When HeLa 229 cells are used, they are treated before infection with diethylaminoethyl-dextran. The inoculum is centrifuged onto the monolayers, and inhibitors of host macromolecular synthesis may or may not be added.

Isolation in chicken embryos. Fertile hen eggs from a laying flock maintained on antibiotic-free feed are incubated 6–7 days at 38–39°C and candled for viability. Eggs with living embryos are then inoculated into their yolk sacs and reincubated at 35–37°C for another 14 days. On the 3rd day of incubation, eggs containing dead embryos are discarded. The eggs are candled daily thereafter, and all dead embryos are examined for the presence of chlamydiae by staining impression smears of the yolk sac with Gimenez or modified Macchiavello. Negative primary passages may be repassed blind. Criteria for positive isolation are the presence of elementary bodies and serially transmissable embryo mortality in the absence of contaminating bacteria or viruses. Serological confirmation of isolation may also be obtained. For isolation of human strains of *C. trachomatis*, the yolk sac is less sensitive than cell culture. For isolation of some mammalian strains of *C. psittaci*, it is currently the best procedure.

Isolation in mice. Mice of a strain known to be susceptible to chlamydiae and from a colony free of latent chlamydial infection must be used. Most avian strains of *C. psittaci* grow in mouse brain after intracranial inoculation and in liver and spleen after intraperitoneal injection. Mice are inferior to chicken embryos for isolation of most strains of *C. psittaci* from mammals. The LGV biovar of *C. trachomatis* grows in the brains of mice, but the trachoma biovar readily infects only primates. Gross pathological signs of infection are frequently absent, and isolation must be confirmed by identification of chlamydiae in impression smears of infected organs.

Maintenance Procedures

Chlamydiae are maintained by appropriate modification of the isolation procedures just described. Maintenance in cell culture is usually most economical of time and materials. Chlamydiae are best preserved by freezing at −60°C or lower.

Chlamydiae can be propagated to a high titer by serial passage in chicken embryo yolk sac cultures or cell monolayer cultures in a large bottle. For example, it is possible to achieve 100% infection of HeLa 229 cells by serial passage without the assistance of centrifugation or inhibitors of host macromolecular synthesis, even with the trachoma biovar of *C. trachomatis*.

Procedures for Testing for Special Characters

For titration of chlamydial infectivity, one can take advantage of the ability of most strains of *C. psittaci* and the LGV and mouse biovars of *C. trachomatis* to form plaques on monolayers of L cells (a heteroploid mouse line). Infectivity may, therefore, be expressed in terms of plaque-forming units. Strains of the trachoma biovar of *C. trachomatis*, however, do not form plaques with any known cell type. Their infectivity is usually measured by infecting a monolayer by centrifugation, incubating for 72 h, and staining with iodine or Giemsa as already described. Then the number of chlamydial inclusions in each of a number of microscopic fields is counted, and infectivity is expressed in terms of the inclusion count, the mean number of inclusions per microscopic field. Chlamydial infectivity in cell culture, chicken embryos, and mice may be expressed in terms of the 50% infectious dose or the 50% lethal dose. Infectivity for chicken embryos inoculated via the yolk sac may also be estimated from the nearly linear relation between the logarithm of the inoculum size and the survival time of the inoculated embryos.

Total chlamydial counts may be obtained by electron microscopic examination of grids of known geometry onto which chlamydial cells have been deposited.

Differentiation of the orders Chlamydiales *and* Rickettsiales

It is hard to compare the monogeneric order *Chlamydiales* with the polygeneric order *Rickettsiales*. The two orders were long united in the single order *Rickettsiales*, although an evolutionary relation between the two taxa was never claimed. The confusion of members of the two orders in the minds of many microbiologists is probably the reflection of the fact that all these microorganisms are, with one exception, Gram-negative procaryotes that are obligate intracellular parasites of eucaryotic hosts. Table 9.14 lists three differential characteristics of the two orders.

Some authors have casually used the epithet "neorickettsiae" as roughly synonymous with chlamydiae of mammalian origin. However, no evolutionary relation between chlamydiae and the organism known as *Neorickettsia helminthoeca* has ever been suggested.

Taxonomic Problems

The history of chlamydial taxonomy may be divided into two periods. In the first, relationships were slowly established among a group of infectious agents isolated from a variety of hosts and responsible for diverse diseases. This period culminated in the gathering together of all chlamydiae into a single genus (*Chlamydia*), the setting up of a separate order for that genus (*Chlamydiales*), and the popularization of a useful trivial epithet ("chlamydia," s. n.; "chlamydiae," pl. n.; "chlamydial," adj.). The second period, in which chlamydial taxonomists are still laboring, is marked by recognition, on the one hand, that the two species of *Chlamydia* are filled with strains very heterogeneous with respect to characters not used to define the species, and on the other, that presently available taxonomic characters are inadequate for further subdivision of the genus. Another obstacle to progress in chlamyd-

ial taxonomy is that these microorganisms are still so difficult to maintain in laboratory culture that only a very small fraction of the available isolates have ever been examined from a taxonomic point of view.

Thermal denaturation measurements give mol% G + C values of 41–44 for various chlamydial strains, with *C. psittaci* strains at the lower end of an almost continuous spectrum and *C. trachomatis* (trachoma biovar) strains at the upper. Some of the chlamydial DNA preparations may have been contaminated with host DNA. The degree of DNA relatedness between the two species of *Chlamydia* is 10% by the thermostability test. The intraspecies homology among strains of *Chlamydia* is near 100%, except for the mouse biovar of *C. trachomatis*, which shows only 30–60% relatedness to other strains of that species.

The species *C. trachomatis*, with the exception of a handful of closely

Table 9.14.
Differentiation of the Orders **Chlamydiales** *and* **Rickettsiales**[a]

Characteristics	Chlamydiales	Rickettsiales
Developmental cycle with alternation of morphologically recognizable specialized cell types	+	−
Cell walls with stoichiometric amounts of muramic acid	−	+
Oxidation of glutamate with net generation of ATP	−	+

[a] Symbols: see standard definitions.

related isolates from rodents, contains only strains of human origin. However, the human isolates of *C. trachomatis*, despite their close genetic relatedness and extensive serological cross-reactions, fall into two distinct groups on the basis of cultural characteristics and pathogenic potentials. This dichotomy has created an awkward nomenclatural difficulty which we have tried to remedy by proposing the establishment of three biovars within this species. The biovars may form the basis for some future creation of new species of *Chlamydia*, but subdivision of *C. trachomatis* into additional species is not recommended at this time. Selecting appropriate epithets for two of the biovars, "mouse" and "lymphogranuloma venereum," was easy, and the names chosen are self-explanatory. However, finding an acceptable name for the third was more difficult. This biovar could not be named after its host because the lymphogranuloma venereum biovar is also strictly human in its natural habitat. Neither could it be without hesitation named for a disease because no single disease is representative of the pathogenic potentials of the biovar as a whole. However, in the absence of a more acceptable alternative, the name "trachoma" was finally chosen for the third biovar on the justification that trachoma was the first disease for which the chlamydial agents of this biotype were proven responsible.

Initial attempts to differentiate *C. psittaci* isolates from birds and mammals have revealed significant pathogenic, antigenic, and cultural differences. However, much more information will be needed before additional species can be proposed.

The many striking and detailed phenotypic resemblances in structure and behavior among all chlamydial strains, no matter what their derivation, suggests a common evolutionary origin for members of the genus *Chlamydia*. The differences between *C. psittaci* and *C. trachomatis* and the differences among the biovars of the latter species may be accounted for in terms of long-term adaptation to different intracellular habitats and to different ways of getting from one host cell to another. However, in light of these strong phenotypic similarities, it is surprising

that the demonstrated degree of DNA relatedness between the two species of *Chlamydia* is so low (~10%).

Further Reading

Becker, Y. 1978. The chlamydiae: molecular biology of procaryotic obligate parasites of eucaryotes. Microbiol. Rev. *42:* 274–306.

Gordon, F.B. (Editor). 1962. *The Biology of the Trachoma Agent.* New York Academy of Sciences, New York.

Grayston, J.T. and S.P. Wang, 1975. New knowledge of the chlamydiae and the diseases they causes. J. Infect. Dis. *132:* 87–105.

Hobson, D. and K.K. Holmes (Editors). 1977. *Nongonococcal Urethritis and Related Infections.* American Society for Microbiology, Washington, D.C.

Jawetz, E. and P. Thygeson. 1965. Chapter 48. Trachoma and inclusion conjunctivitis agents. In Horsfall and Tamm (Editors), *Viral and Rickettsial Diseases of Man,* 4th Ed., J. B. Lippincott, Philadelphia, pp. 1042–1058.

Mårdh, P.-A., K.K. Holmes, J.D. Oriel, P. Piot, and J. Schachter (Editors). 1982. *Chlamydial Infections,* Elsevier Biomedical Press, Amsterdam.

Meyer, K.F. 1965. Chapter 47. Psittacosis-lymphogranuloma venereum agents. In Horsfall and Tamm (Editors), *Viral and Rickettsial Diseases of Man,* 4th Ed., J.B. Lippincott, Philadelphia, pp. 1006–1041.

Moulder, J.W. 1962. *The Biochemistry of Intracellular Parasitism.* University of Chicago Press, Chicago.

Moulder, J.W. 1964. *The Psittacosis Group as Bacteria.* John Wiley & Sons, New York.

Moulder, J.W. 1966. The relation of the psittacosis group (chlamydiae) to bacteria and viruses. Annu. Rev. Microbiol. *20:* 107–130.

Nichols, R.L. (Editor). 1971. *Trachoma and Related Disorders Caused by Chlamydial Agents.* Excerpta Medica, Amsterdam.

Page, L.A. 1981. Obligately intracellular bacteria: the genus *Chlamydia.* In Starr, Stolp, Trüper, Bulows and Schlegal (Editors). *The Procaryotes,* Springer-Verlag, New York, pp. 2210–2222.

Schachter, J. 1980. Chlamydiae (psittacosis-lymphogranuloma venereum-trachoma group). Chapter 29. *In* Lennette, Balows, Hausler and Truant (Editors). *Manual of Clinical Microbiology,* 3rd Ed., American Society for Microbiology, Washington, D.C., pp. 357–364.

Schachter, J. and H.D. Caldwell. 1980. Chlamydiae. Annu. Rev. Microbiol. *34:* 285–310.

Schachter, J. and C.R. Dawson. 1978. *Human Chlamydial Infections.* PSG Publishing Company, Littleton, Mass.

Storz, J. 1971. *Chlamydia and Chlamydia-Induced Diseases.* Charles C Thomas, Springfield, Ill.

Differentiation of the species and biovars of the genus **Chlamydia**

Characteristics useful for differentiation of the two species of *Chlamydia* are listed in Table 9.15. Differentiation of the three biovars of *C. trachomatis* is indicated in Table 9.16. Other characteristics of the species and biovars of *Chlamydia* are listed in Table 9.17.

List of the species of the genus **Chlamydia**

1. **Chlamydia trachomatis** (Busacca 1935) Rake 1957, 958[AL] (*Rickettsia trachomae (sic)* Busacca 1935, 567).

tra.cho'ma.tis. Gr. n. *trachoma* roughness; M.L. n. *trachoma* the disease trachoma; M.L. gen. n. *trachomatis* of trachoma.

The characteristics are as described for the genus and as listed in Tables 9.15–9.18. The inclusion morphology is depicted in Figure 9.13*A*.

All *C. trachomatis* isolates have come from humans except for a handful of apparently identical isolates from mice; these latter strains comprise biovar mouse. Human isolates of *C. trachomatis* are classified into two biovars, trachoma and lymphogranuloma venerum, on the basis of their pathogenic proclivities in humans and laboratory animals and their growth characteristics (Table 9.16). The close relation of the trachoma and lymphogranuloma venereum biovars is shown by the complete DNA homology between all strains so far tested and by the extensive cross-reactions among many of their type-specific antigens (Fig. 9.14). Whereas strains of the trachoma biovar parasitize mucous membranes almost exclusively, lymphogranuloma venereum strains are more invasive and cause systemic disease. This difference in behavior of the two human biovars in their natural hosts is paralleled by their behavior in laboratory animals and cultured cells which are more readily infected with LGV biovar. Further information concerning the three biovars of *C. trachomatis* is given below:

(i) *Biovar mouse.* These strains, usually referred to as those of mouse pneumonitis, latently infect mice from which they may be recovered by serial passage of apparently healthy lung tissue. Iodine-positive

staining may be hard to demonstrate. The mouse biovar strains cross-neutralize extensively, but they do not share type-specific antigens with the humans strain of *C. trachomatis.*

A single strain of the mouse biovar showed 30 to 60% DNA homology with a single trachoma biovar strain, depending on the conditions of reassociation of single strands of DNA, and the duplexes formed were thermolabile.

Table 9.15.

Differentiation of the Species of the Genus **Chlamydia**[a]

Characteristics	1. *C. trachomatis*	2. *C. psittaci*
Natural hosts	Humans, mice	Birds, mammals other than humans
Inclusion morphology[b]:		
Oval, vacuolar	+	−
Variable shape, dense	−	+
Glycogen in inclusions[c]	+	−
Folate biosynthesis[d]	+	−

[a] See Page (1966). For symbols see standard definitions.

[b] As seen in Giemsa-stained host cells. See Figure 9.13.

[c] As evidenced by positive iodine-staining of inclusions.

[d] As demonstrated by inhibition of chlamydial growth in chick embryo yolk sac by addition of 1 mg of sulfadiazine per embryo.

Table 9.16.
Differentiation of the three biovars of **Chlamydia trachomatis**[a]

Characteristics	Biovar		
	Trachoma	Lympho-granuloma venereum	Mouse
Natural host:			
Humans	+	+	−
Mice	−	−	+
Preferred site of infection in natural hosts:			
Squamocolumnar epithelium cells	+	−	−
Lymph nodes	−	+	−
Lungs	−	−	+
Behavior in laboratory animals:			
Intracerebral lethality for mice	−	+	−
Follicular conjunctivitis in primates	+	−	−
Behavior in cell culture:			
Plaques in L cells	−	+	+
Infection markedly enhanced by:			
Centrifugation onto cell sheet	+	−	−
Treatment of host cells with diethyl-aminoethyl-dextran	+	−	−
No. of serovars	12	3	NA
% DNA homology with trachoma	100	100	30–60

[a] Symbols: see standard definitions; also NA = data not available.

The mouse biovar is not an important pathogen of wild or laboratory mice, but as the only non-human example of *C. trachomatis*, it may prove vital to the unraveling of the evolutionary history of the genus.

The mol% G + C of the DNA of one mouse strain was 42.2 (T_m); for a second strain it was 43.6.

Reference strain: ATCC VR-123 (strain Nigg II).

(ii) *Biovar trachoma.* This biovar, an exclusively human parasite with no animal reservoirs, is primarily a pathogen of the squamocolumnar cells of mucous membranes. Transmission from person to person occurs in two distinct patterns. In areas where blinding trachoma occurs, the organism is spread from child to child or among family members having close contact. Serovars A, B, Ba and C are associated with this form of endemic blinding trachoma. In other communities, the agent is spread by sexual contact and causes diseases in the genital tract or in newborns exposed during passage through an infected birth canal. Serovars D through K are most commonly associated with this transmission pattern. Table 9.18 lists the diseases caused by the trachoma biovar of *C. trachomatis.*

The mean mol% G + C of the DNA of 5 strains was 44.0 (T_m).

Reference strain: ATCC VR-576 (strain PK-2, serovar C).

(iii) *Biovar lymphogranuloma venereum (LGV).* Strains of this biovar are sexually transmitted pathogens that cause systemic infections. There is no animal reservoir. The disease primarily involves lymphatic tissue, although protean clinical manifestations may occur. The typical presentation is that of inguinal lymphadenopathy in the male in which the disease predominates in a ratio >10:1.

Three LGV strains had a mean mol% G + C value of 41.9 (T_m).

Reference strain: ATCC VR-902 (strain 434, serovar L2).

Type strain of the species: ATCC VR-571 (strain HAR-13, serovar A, biovar trachoma).

2. **Chlamydia psittaci** (Lillie 1930) Page 1968, 60[AL] (*Rickettsia psittaci* Lillie 1930, 778).

psit′ta.ci. Gr. n. *psittacus* a parrot; M.L. gen. n. *psittaci* of a parrot.

C. psittaci has a much broader host range than *C. trachomatis.* Strains of this species have been isolated from a very large number of avian and mammalian species. The degree of host specificity is usually low, and a single strain may produce more than one disease entity in a single host species. There is much evidence of heterogeneity among *C. psittaci* isolates. Strains from cattle and sheep fall into two main antigenic groups, serovars 1 and 2. Antibodies against either of these serovars do not neutralize the infectivity of *C. psittaci* from turkeys, parrots, and pigeons, and vice versa. The avian strains can, in turn be subdivided into antigenic groups on the basis of plaque neutralization tests.

People become infected with *C. psittaci* by exposure to infective discharges from birds. The human disease, called psittacosis (and sometimes ornithosis), occurs in two forms, one a pneumonic form and the other a toxemic form without a respiratory component. Person to person transmission is rare. Human infections with mammalian strains have been reported but are rare.

Avian and mammalian diseases caused by *C. psittaci* have been studied mainly in parrots, poultry and the common farm animals. A number of these diseases are economically important. Young animals are especially prone to infection with *C. psittaci.* Clinically inapparent infections with prolonged shedding of infectious chlamydiae are com-

Table 9.17.
Other characteristics of the species and biovars of the genus **Chlamydia**[a]

Characteristics	1. C. trachomatis			2. C. psittaci
	Biovar trachoma	Biovar lympho-granuloma venereum	Biovar mouse	
Major surface protein (molecular weight = ~40,000) present on cell surface	+	+	NA	+
Metabolism:				
CO$_2$ from glucose-6-phosphate, glutamate, aspartate, or pyruvate	+	+	+	+
Glutamate metabolized beyond succinate	−	−	−	−
Net synthesis of ATP	−	−	−	−
Genetics:				
Mol% G + C of DNA	44.0	41.9	42.9	41.3
DNA, molecular weight (10^6)	NA	660	NA	360
% DNA homology (with trachoma biovar as reference)	100	100	30–60[b]	10
Presence of 4 × 10^6 dalton plasmid	+	+	NA	+
Antigens:				
Genus-specific antigen demonstrated by complement fixation	+	+	+	+
Type-specific antigens demonstrated by microimmunofluorescence[c]	A-K	L$_1$, L$_2$, L$_3$	NA	NA
Natural infections: natural hosts	Humans	Humans	Mice	Birds, lower mammals
Preferred site of infection:				
Squamocolumnar epithelial cells	+	−	−	−
Lymph nodes	−	+	−	−
Lungs	−	−	+	−
Multiple sites	−	−	−	+[d]
Diseases produced:				
Many kinds	+[e]	−	−	+[f]
Lymphogranuloma venereum	−	+	−	−
Latent respiratory disease in mice	−	−	+	−
Laboratory infections:				
Animals:				
Mouse lethality intracerebrally	−	+	−	+[f]
Follicular conjunctivitis in lower primates	+	−	−	−
Cell culture:				
Enhancement of infectivity by:				
Centrifugation of inoculum onto monolayer	+	−	−	−[g]
Treatment of host cells with DEAE-dextran	+	−	−	Weak
Plaques on L cells	−	+	+	+

[a] Symbols: see standard definitions; also NA = data not available
[b] Unstable duplexes.
[c] See Figure 9.14.
[d] See Table 9.19.
[e] See Table 9.18.
[f] See Table 9.19.
[g] Infectivity of some strains is enhanced by centrifugation.

mon. Table 9.19 lists the characteristics of a number of these diseases.

Four strains of *C. psittaci* had a mean mol% G + C value of 41.3 (T_m).

Type strain: ATCC VR-125 (strain 6BC). *Note:* this strain, although designated as the type strain on the Approved Lists of Bacterial Names, is not typical of the species, in that its growth in chicken embryo yolk sac is inhibited by sulfadiazine. Reference strain ATCC VR-351 (strain Texas Turkey) does exhibit typical characters and is recommended for purposes of comparison with new isolates.

Table 9.18.

Human diseases caused by **Chlamydia trachomatis**

Biovar	Disease or syndrome
Trachoma	Trachoma
	Inclusion conjunctivitis in newborns, children, and adults
	Otitis media
	Pneumonia in infants and normal (?) and immunosuppressed adults
	Nongonococcal urethritis in men
	Epididymitis
	Urethral syndrome in women
	Mucopurulent cervicitis
	Salpingitis
	Proctitis
	Perihepatitis and peritonitis (Fitz-Hugh-Curtis syndrome) in young women (?)
	Bartholinitis (?)
	Endocarditis (?)
Lymphogranuloma venereum	Lymphogranuloma venereum
Mouse	None

Table 9.19.

Diseases caused by **Chlamydia psittaci** *in nonhuman hosts*

Site of infection	Hosts	Characteristics of disease
Intestinal and respiratory tracts	Parrots, parakeets, pigeons, turkeys, geese, other birds[a]	Broadly referred to as avian chlamydiosis. Clinically inapparent intestinal infection with prolonged fecal shedding. Respiratory infection leading to pneumonia, air saculitis, pericarditis; splenomegaly, conjunctivitis and diarrhea
Intestinal tract	Lambs, calves, piglets	Diarrhea in newborns
Genital tract	Sheep and cattle	Bulls and rams have seminal vesculitis and orchitis with detrimental effect on semen quality. Chlamydiae in semen infect endometrial cells of female. Fertilization not affected but embryos die early
Placenta and fetus	Sheep, goats, cattle	Contracted from mothers with chlamydemia. Fetal death and abortion. Birth of stillborn or infected weak offspring
Respiratory tract	Mice, kittens, lambs, calves, piglets, foals	Pneumonia accompanied by tracheitis and sometimes sinusitis and rhinitis
Eye	Guinea pigs, lambs, calves, piglets, koala bears, cats	Conjunctivitis or keratoconjunctivitis
Synovial tissue	Lambs, calves, foals, piglets	Polyarthritis and polyserositis after initial intestinal infections. Chlamydiae spread systemically and multiply in cells of synovial tissue. Sporadic cases of encephalomyelitis in cattle

[a] One hundred thirty-one species of birds belonging to twelve orders.

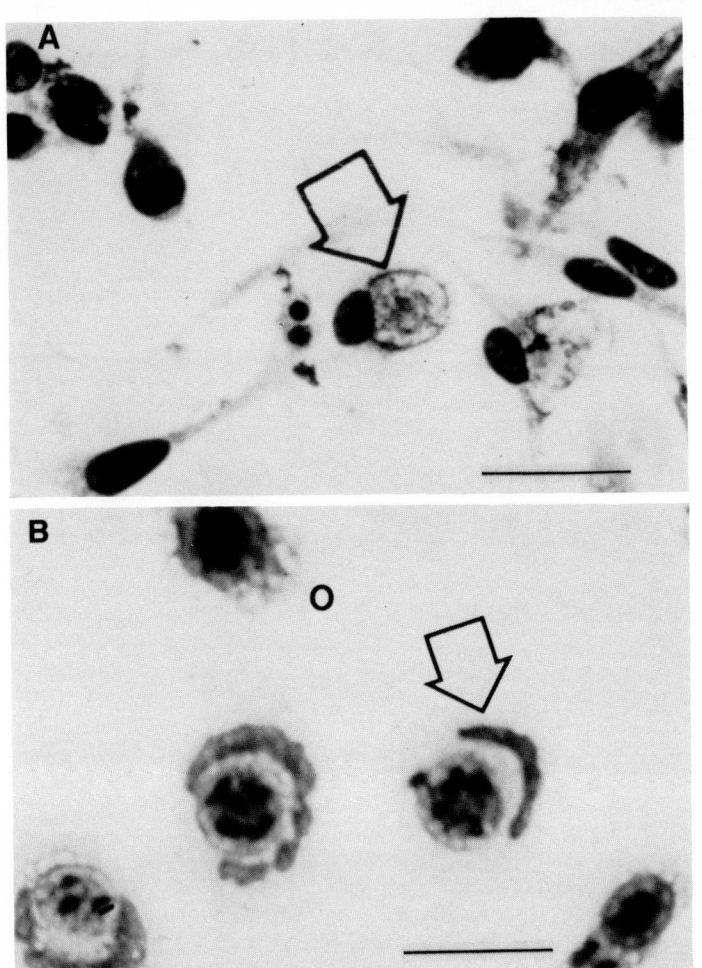

Figure 9.13. Inclusion morphology of *Chlamydia trachomatis* and *Chlamydia psittaci*. Bars, 20 μm. Infected host cell monolayers were fixed in absolute alcohol and stained with Giemsa. *Arrows* point to typical inclusions. *A, C. trachomatis* (biovar trachoma, strain G17) in McCoy cells 72 h after infection. *B, C. psittaci* (strain 6BC) in L cells 24 h after infection.

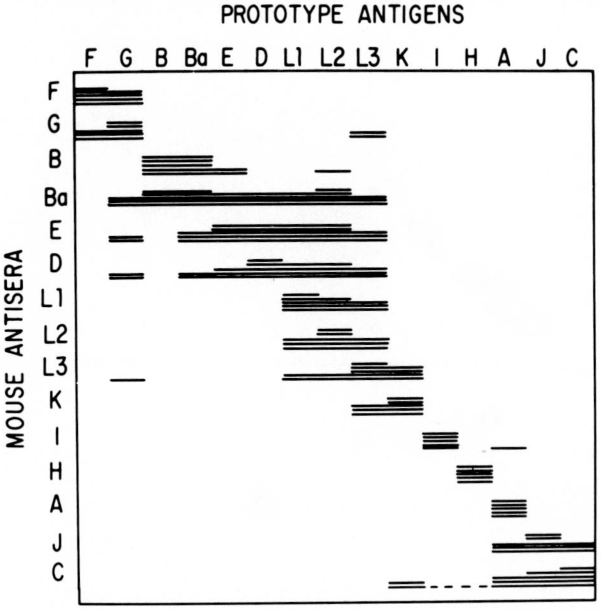

Figure 9.14. Microimmunofluorescence patterns of *Chlamydia trachomatis* antisera prepared in mice. A difference of one line expresses a 2-fold difference from the homologous titer which is shown by 5 lines. (Reproduced with permission from S. P. Wang et al., *Infection and Immunity 7*: 356–360, 1973, © American Society for Microbiology, Washington, D.C.)

Important Notes for Users of this Edition

1. Always read both generic and species descriptions because characters listed in the generic description are not usually listed in the species descriptions.

2. Unless otherwise indicated in footnotes to tables, the meanings of symbols are as follows:

+ 90% or more of strains are positive

− 90% or more of strains are negative

d 11–89% of strains are positive

v strain instability (*not* equivalent to "d")

D different reactions in different taxa (species of a genus or genera of a family)

3. All other symbols are defined in footnotes to tables.

SECTION 10

The Mycoplasmas

DIVISION **TENERICUTES** DIV. NOV. (*q.v.* p. 36)

CLASS I. **MOLLICUTES** EDWARD AND FREUNDT 1967, 267[AL]*

SHMUEL RAZIN AND E. A. FREUNDT

Mol.li.cu'tes or Mol.li'cu.tes. L. adj. *mollis* soft, pliable; L. fem. n. *cutis* skin; M.L. fem. pl. n. *Mollicutes* class with pliable cell boundary.

Very small procaryotes totally **devoid of cell walls.** Bounded by a plasma membrane only. Incapable of synthesis of peptidoglycan and its precursors. Consequently resistant to penicillin and its analogues, and sensitive to lysis by osmotic shock, detergents, alcohols and specific antibody plus complement. **Pleomorphic,** varying in shape from spherical or pear-shaped structures (0.3–0.8 μm in diameter) to **branched or helical filaments.** Genome replication precedes, but is not necessarily synchronized with, cell division. Thus, budding forms and chains of beads may be observed, as well as classical binary fission. **Usually nonmotile,** but some species show **gliding motility** on liquid-covered surfaces. Other species that occur as helical filaments show **rotary, flexional and translational motility.** No resting stages known. **Gram-negative.**

The species recognized thus far can be grown on artificial cell-free media of diverse complexity. Most species **require sterols** and fatty acids for growth. However, certain strains may grow poorly in artificial media and may be more readily isolated by cell-culture procedures. Most species are **facultatively anaerobic,** but some are **obligate anaerobes** that are killed by exposure to minute quantities of oxygen. One species tentatively associated with *Mollicutes* is thermoacidophilic, capable of growth at pH 1.0–2.0 and at 55–60°C. Colonies on solid media are minute, usually much smaller than 1 mm in diameter. There is a tendency for the organisms to penetrate and grow inside the medium. Under suitable conditions, almost all species **form colonies that have a characteristic "fried egg" appearance.**

Apart from the free-living thermoacidophiles all mollicutes are **parasites, commensals or saprophytes,** and many are **pathogens of man, animals, plants and insects.** Genome size is about 5×10^8 or 1×10^9 daltons, among the smallest recorded in procaryotes. The mol% G + C of the DNA is low, ranging from ~23% to ~46% (Bd, T_m).

Further Comments

Table 10.1 summarizes the present classification of the *Mollicutes* into families and genera and provides the major distinguishing characteristics of these taxa. The trivial names acholeplasma, ureaplasma, anaeroplasma, spiroplasma and thermoplasma are commonly used when reference is made to members of the corresponding genus. However, the trivial term mycoplasma has been used to denote any species included in the class *Mollicutes.* If the proposal to introduce mollicute(s) as a trivial name for all members of the class (Subcommittee meeting,

1980) gains acceptance, the trivial name mycoplasma can be retained for members of the genus *Mycoplasma* only.

Two genera within the class have not yet been assigned a definite taxonomic position, namely, *Anaeroplasma* and *Thermoplasma.* Sterol is required for the growth of the type strains of *A. bactoclasticum* and *A. abactoclasticum,* and they can therefore be tentatively assigned to the family *Mycoplasmataceae.* However, definite classification of *Anaeroplasma* at the family level was complicated by the proposal of Robinson et al. (1975) that sterol-nonrequiring anaerobic strains should be classified together with the sterol-requiring strains in the same genus, despite the fact that sterol requirement is a major property distinguishing the *Mycoplasmataceae* from *Acholeplasmataceae* (see Table 10.1). To solve this problem Robinson (this volume) has suggested that the sterol-nonrequiring strains be excluded from the genus *Anaeroplasma* and their taxonomic status be left open.

The inclusion of *Thermoplasma* strains in the *Mollicutes* is problematic. On the one hand their total lack of a cell wall and peptidoglycan supports their inclusion in *Mollicutes.* However, they differ from all other members of the class in many important properties including: nonparasitic mode of life; adaptation for natural existence under extreme environmental conditions (Darland et al., 1970); much simpler nutritional requirements (Smith et al., 1975); presence of quinones and cytochromes (Holländer et al., 1977); flagellar motility (Black et al., 1979); and DNA of a relatively high G + C content, associated with a histone-like protein (Christiansen et al., 1975; Searcy and Doyle, 1975; Searcy and Stein, 1980). These properties and the finding of peculiar diglycerol tetraether lipids in *Thermoplasma* (Langworthy, 1977) as well as their 16S rRNA nucleotide sequence characterization (Woese et al., 1980) indicate that *Thermoplasma* is widely separated from all other members of the *Mollicutes* and supports their inclusion in the archaeobacteria (Fox et al., 1980).

The characteristic morphology, dimensions, osmotic fragility, colony shape and filterability through 450-nm pore diameter membrane filters distinguish well the mycoplasmas from all other procaryotes. However, it may not be easy to distinguish them from the wall-defective or wall-less L-phase variants of bacteria which share with mycoplasmas the "fried egg" colony shape, total resistance to penicillin and lysozyme, osmotic fragility and in some cases also the total lack of a cell wall (Clasener, 1972; Butler and Blakey, 1975). Nevertheless, there are several fundamental differences separating the L-phase variants from

---

* *AL* denotes the inclusion of this name on the Approved Lists of Bacterial Names (1980).

mycoplasmas. These include the marked heterogeneity of cell size and the presence of the so-called "large bodies" in the L-phase cultures, the absence of filaments resembling those of mycoplasmas, and the presence of precursors of cell wall polymers and penicillin-binding proteins in the L-phase membranes (Reusch and Panos, 1976; Kroll et al., 1980; Martin et al., 1980) but not in those of mycoplasmas. L-phase variants are usually laboratory artifacts, which can revert to the bacterial form once the inducing substance (e.g. penicillin, lysozyme, antibody) is removed from the growth medium. Since high concentrations of penicillin are usually included in mycoplasma media, the possibility that a new isolate is an L-phase variant should be ruled out by subculturing it at least five consecutive times in media that do not contain antibiotics or other substances known to induce bacterial L-phase variants. This examination should preferably be made as soon as possible after isolation of the organism and not later than the first three to five passages on a medium containing antibacterial agents to prevent the transition of the isolate into a stable L-phase which has lost its ability to revert (Subcommittee, 1979).

Although *Mycoplasma* and *Ureaplasma* species have the smallest genomes recorded for procaryotes, their genetics is still largely unknown. Some reports indicating the presence of plasmids in some *Mycoplasma*, *Acholeplasma* and *Spiroplasma* species are available (reviewed in Razin, 1978; see also Ranhand et al., 1980). Viruses infecting species belonging to the above genera are known, and some have also been isolated and characterized (Cole, 1979; Maniloff et al., 1979; Howard et al., 1980; Liss and Cole, 1981).

Further Reading

Barile, M.F., S. Razin, J.G. Tully and R.F. Whitcomb (Editors). 1979. *The Mycoplasmas* (three volumes), Academic Press, New York.
Freundt, E.A, H. Ernø and R.M. Lemcke. 1980. Identification of mycoplasmas. In Norris and Gibbons (Editors), *Methods in Microbiology*, Vol. 13, Academic Press, New York, pp. 377–434.
International Committee on Systematic Bacteriology, Subcommittee on the Taxonomy of *Mollicutes*. 1979. Proposal of minimal standards for descriptions of new species of the class *Mollicutes*. Int. J. Syst. Bacteriol. *29:* 172–180.
Razin, S. 1978. The mycoplasmas. Microbiol. Rev. *42:* 414–470.
Tully, J.G. and S. Razin. 1977. The Mollicutes (mycoplasmas). In Laskin and Lechevalier (Editors), *CRC Handbook of Microbiology*, 2nd Ed., CRC Press, Boca Raton, Fl., pp. 405–459.

Table 10.1.
Taxonomy and properties of organisms included in the class **Mollicutes**[a]

Classification	Current Number of Recognized Species	Genome Size (10^8 daltons)	Mol% G + C of DNA	Cholesterol Requirement	Location of NADH Oxidase		Distinctive Properties	Habitat
					Cyt	Mem		
Order I. *Mycoplasmatales*								
Family I. *Mycoplasmataceae*								
Genus I. *Mycoplasma*	69	5	23–41	+	+	−		Animals
Genus II. *Ureaplasma*	2	5	27–30	+			Urease-positive	Animals
Family II. *Acholeplasmataceae*								
Genus I. *Acholeplasma*	8	10	27–36	−	−	+		Animals and possibly on plants
Family III. *Spiroplasmataceae*								
Genus I. *Spiroplasma*	3	10	25–31	+	+	−	Helical filaments	Arthropods and plants
Genera of uncertain taxonomic position:								
Genus *Anaeroplasma*	2		29–34	+			Anaerobic; some digest bacteria	Rumens of cattle and sheep
Genus *Thermoplasma*	1	10	46	−	−	+	Thermophilic (optimum temperature, 59°C) and acidophilic (optimum pH, 1.0–2.0)	Burning coal refuse piles

[a] Adapted from Razin (1978). Abbreviations: Cyt, cytoplasm; Mem, plasma membrane; NADH, reduced nicotinamide adenine dinucleotide. Symbols: see standard definitions.

ORDER I. **MYCOPLASMATALES** FREUNDT 1955, 71[AL]

SHMUEL RAZIN AND E. A. FREUNDT

My.co.plas.ma.ta'les. M.L. neut. n. *Mycoplasma* type genus of the order; *-ales* ending to denote an order; M.L. fem. n. *Mycoplasmatales* the *Mycoplasma* order.

Only one order, *Mycoplasmatales*, is accepted in the class *Mollicutes*; the description of the order is therefore the same as for the class. Three families are accepted in the order *Mycoplasmatales*: *Mycoplasmataceae*, *Acholeplasmataceae* and *Spiroplasmataceae*. Members of *Acholeplasmataceae* are distinguished by their lack of a sterol requirement, while members of the *Spiroplasmataceae* are distinguished by their characteristic helical morphology and motility (see Table 10.1).

Procedures for Testing Sterol Requirements

Quantitative growth responses to cholesterol are assessed in serum-free media that contain supplements of fatty acids, albumin, and increasing concentrations (1–20 μg/ml) of solubilized cholesterol (Razin and Tully, 1970; Edward, 1971). Growth of members of the families *Mycoplasmataceae* and *Spiroplasmataceae* (as measured by increase in total cellular protein or by colony-forming units enumerated on solid medium) is negligible or minimal in the absence of added cholesterol, but is considerably enhanced by increasing cholesterol concentrations. Growth response may, however, decrease at higher cholesterol concentrations (20 μg/ml). *Acholeplasma* species grow readily in serum-free media, but the addition of fatty acids is sometimes required (Razin and Tully, 1970). It should be emphasized that sterol growth requirements cannot be determined simply by a few passages of the organisms on serum-free agar or broth media because cholesterol carried over in the inoculum may permit growth of mycoplasmas with low sterol requirements such as *M. mycoides* subsp. *capri*, or *M. capricolum* (Rottem et al., 1973, 1976). Also, components of some serum-free basal media may contain sufficient amounts of sterols, in which case limited growth of sterol-requiring mycoplasmas may be observed. In critical tests it is therefore recommended to include organisms with known sterol requirements. The requirement for cholesterol can be satisifed by related sterols, such as stigmasterol and cholestanol, that share with cholesterol a planar ring structure, a free hydroxyl group at the 3β position, and a hydrocarbon side chain (Razin, 1981). Nevertheless, several sterols lacking one of these structural features have enabled *M. capricolum* to grow, but to a lesser extent than cholesterol (Odriozola et al., 1978; Dahl et al., 1980).

Testing the susceptibility of the mycoplasmas to digitonin (Freundt et al., 1973) provides an indirect way to establish the cholesterol requirement. The sterol-requiring mycoplasmas incorporate much more cholesterol into their membranes than the sterol-nonrequiring acholeplasmas (Razin et al., 1980) and are, therefore, much more susceptible to lysis by digitonin, which specifically forms a complex with cholesterol in the membrane causing disorganization of the lipid bilayer (Rottem and Razin, 1972). The digitonin test may be carried out on a solid medium inoculated with 0.01 ml of a liquid culture containing approximately 10^6 colony-forming units per milliliter. The inoculum is allowed to run down the surface of a slightly tilted plate and allowed to dry. Dried filter paper disks previously impregnated with 0.02 ml of a 1.5% (w/v) ethanolic solution of digitonin are placed in the center of the inoculated area. Members of the families *Mycoplasmataceae* and *Spiroplasmataceae* exhibit zones of inhibition of 3–12 mm wide (as measured from the rim of the disk to the edge of the area of growth), whereas no inhibition or slight inhibition (no more than 1 mm) is obtained with acholeplasmas. It is essential that the test medium contains 10–20% serum, even if the test is performed on a strain that does not require serum for growth.

FAMILY I. **MYCOPLASMATACEAE** FREUNDT 1955, 71[AL]

SHMUEL RAZIN AND E. A. FREUNDT

My.co.plas.ma.ta′ce.ae. M.L. neut. n. *Mycoplasma* type genus of the family; *-aceae* ending to denote a family; M.L. fem. pl. n. *Mycoplasmataceae* the *Mycoplasma* family.

Sterol is required for growth. Other characteristics are as described for the class and order.

Two genera are accepted in this family: *Mycoplasma* and *Ureaplasma*. Members of *Ureaplasma* are distinguished by their ability to hydrolyze urea (see Table 10.1).

Procedure for Testing Urea Hydrolysis

The demonstration of urea hydrolysis is a minimum requirement for assigning a new isolate to the genus *Ureaplasma* (Shepard et al., 1974). Since urease activity results in the hydrolysis of urea to CO_2 and ammonia, the simplest way to detect this activity is by observing

alkalinization of the culture medium supplemented with 1% urea instead of glucose or L-arginine (which are not attacked by ureaplasmas). However, this simple test may suffer from the deficiency that arginine-positive mycoplasmas, which do not possess urease activity, may cause alkalinization of the growth medium by hydrolysis of the significant quantities of L-arginine present in conventional mycoplasma media. More specific tests for urease activity are therefore recommended. These include flooding of the colonies with a urea-manganous chloride solution. Urease-positive colonies stain dark brown (Shepard and Howard, 1970). A sensitive, specific and quantitative test is based on breakdown and disappearance of radioactive urea added to the culture medium (Masover et al., 1976).

Genus I. **Mycoplasma** *Nowak 1929, 1349 Nom. cons. Jud. Comm. Opin. 22, 1958, 166*[AL]

E. A. FREUNDT AND SHMUEL RAZIN

My.co.plas′ma. Gr. masc. n. *myces* a fungus; Gr. neut. n. *plasma* something formed or molded, a form; M.L. neut. n. *Mycoplasma* fungus form.

Pleomorphic, varying in shape from spherical, slightly ovoid or pear shaped (0.3–0.8 μm in diameter) to **slender branched filaments** of uniform diameter, ranging in length from a few to 150 μm. Cells **lack a cell wall** and are bounded by a plasma membrane only. Gramnegative. **Usually nonmotile, but gliding motility has been described in some species. Facultatively anaerobic,** possessing a truncated flavin-terminated electron transport chain **devoid of quinones and cytochromes.** Colonies are very small (usually less than 1 mm in diameter). **The typical colony,** under adequate growth conditions, **has a "fried egg" appearance.** Catalase-negative. **Chemoorganotrophic,** using either sugars or arginine as the major energy source. **Require cholesterol** or related sterols for growth. **Parasites and pathogens** of a wide range of mammalian and avian hosts. The mol% G + C of the DNA ranges from 23–40 (T_m and Bd), and the genome size of the species examined is ~5 × 10^8 daltons.

Type species: *Mycoplasma mycoides* (Borrel, Dujardin-Beaumetz, Jeantet and Jouan 1910) Freundt 1955, 73.

Further Descriptive Information

Cell shape of mycoplasmas appears to depend upon the nutritional qualities and the osmotic pressure of the growth medium, as well as the growth phase of the culture. Filamentous growth is usually associated with young logarithmic cultures growing under optimum conditions. However, the filamentous phase is transitory and the filaments transform into chains of cocci (Figs. 10.1 and 10.2) which later break apart (Freundt, 1958; Razin and Cosenza, 1966; Bredt et al., 1973).

Many of the minute and plastic cells in mycoplasma cultures can be squeezed through membrane filters of 450 nm pore diameter. In fact, filterability through such a filter constitutes one of the properties used to define a new isolate as a mycoplasma (Subcommittee, 1979). Earlier information that led to the erroneous concept of extremely small viable cells (i.e. ~ 0.1 μm in diameter) in mycoplasma cultures was based upon sizing data obtained by filtration or by electron microscopy. The limitations of these techniques with regard to sizing plastic organisms

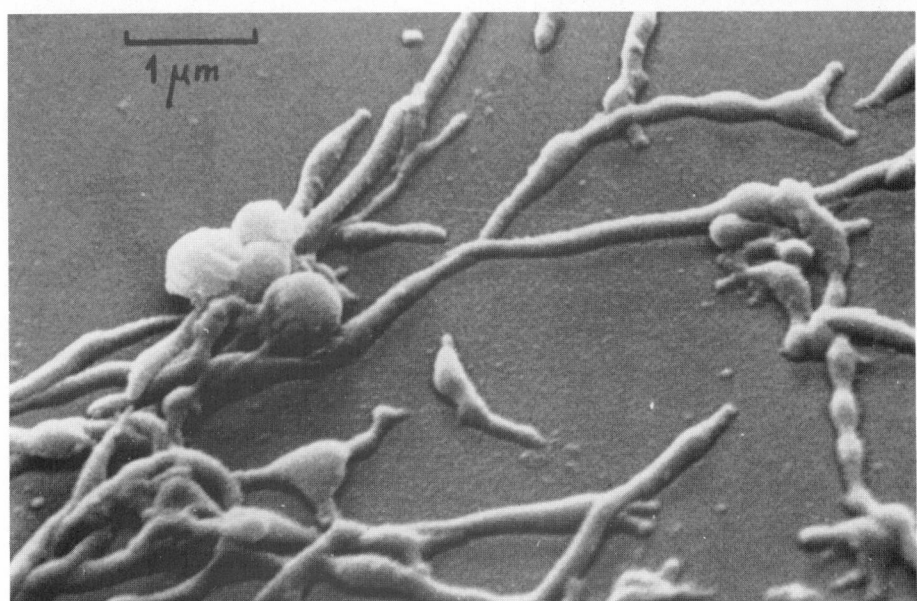

Figure 10.1 Scanning electron micrograph of a 6-day-old culture of *Mycoplasma pneumoniae* growing on the surface of a plastic Petri dish. The characteristic morphological elements of mycoplasma cultures can be seen, including branched filaments, chains of cocci, and elongated cells with terminal blebs. (Reproduced with permission from G. Biberfeld and P. Biberfeld, *Journal of Bacteriology 102*: 855–861, 1970, © American Society for Microbiology.)

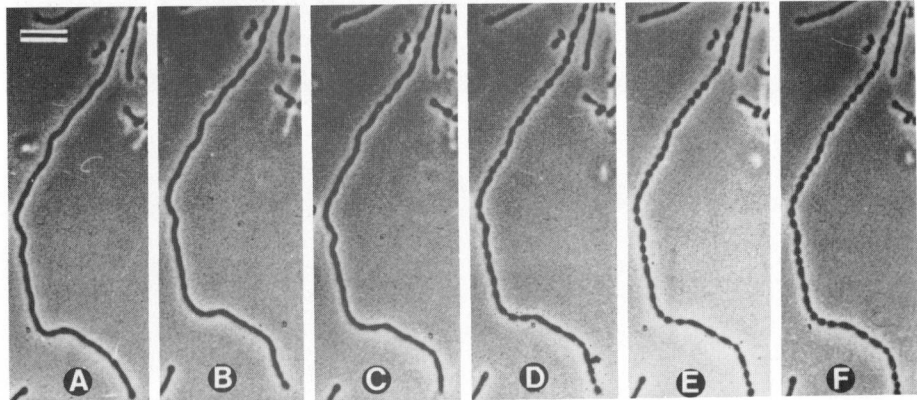

Figure 10.2 Transformation of a mycoplasma filament into a chain of cocci. Cinematographic pictures of a culture of *Mycoplasma hominis* growing in a cover slip chamber under the phase-contrast microscope. Time intervals of pictures from *A*, in minutes are: *B*, 2.8; *C*, 3.3; *D*, 4.8; *E*, 5.3; *F*, 5.5. *Bar*, 10 μm. (Reproduced with permission from W. Bredt et al., *Journal of Bacteriology 113*: 1223–1227, 1973, © American Society for Microbiology.)

has since been noted and discussed (Razin, 1969; Lemcke, 1971). The more recent experimental data as well as theoretical considerations have led to the conclusion that the smallest mycoplasma cell capable of reproduction is about 0.3 μm in diameter (Razin, 1978).

Phase-contrast or dark-field microscopy of young log-phase broth cultures is the recommended procedure for microscopic examination of mycoplasmas, as it introduces minimal distortions in the shape of the plastic cells. Moreover, it enables observation of the gliding motility that characterizes several *Mycoplasma* species. Examination of methanol-fixed organisms stained with Giemsa solution is preferable to the Gram stain. For electron microscopy special attention should be paid to the osmolarity of the fixatives (Lemcke, 1972) and of the buffers (Cole et al., 1973), as these may drastically alter the actual size and shape of the plastic organisms. Ideally the morphology and ultrastructure of mycoplasmas should be based upon close correlation of the appearance of the organisms under phase-contrast or dark-field microscopy with their appearance in the electron microscope (Boatman, 1979).

Demonstration of a single membrane having a trilaminar shape in properly fixed and sectioned cells is an essential requirement for defining a new isolate as a mycoplasma (Subcommittee, 1979). The lack of a cell wall and the rigid peptidoglycan polymer explains the resistance of the organisms to lysis by lysozyme and to growth inhibition by penicillin (Plackett, 1959; Razin and Argaman, 1963) as well as their susceptibility to lysis by osmotic shock and various agents causing the lysis of bacterial protoplasts (Razin and Argaman, 1963).

The mycoplasma membrane is a typical procaryotic plasma membrane built of amphipathic lipids (phospholipids, glycolipids, lipoglycans, sterols) and proteins. The ease of its isolation in a pure form and the ability to introduce controlled alterations in its lipid composition have made this membrane a most useful tool in membrane studies (Razin and Rottem, 1976; Razin, 1981). In many *Mycoplasma* species the cell surface is covered by a capsular material or nap, some of which can be stained with ruthenium red, a general stain for polyanions used to demonstrate the polysaccharide glycocalyxes of eucaryotic cells. In *M. mycoides* subsp. *mycoides* the slime layer is made of galactan (Plackett and Buttery, 1964; Gourlay and Thrower, 1968).

Thin sections of mycoplasmas reveal that the cells are built essentially of three organelles only, the cell membrane, the ribosomes, and the characteristic procaryotic genome. There is no evidence for any intracellular membranous structures, such as mesosomes (Fig. 10.3). In several species, however, specialized tip structures have been observed (Fig. 10.4). These tip structures appear to play a role in the attachment of the mycoplasmas to host cells (Zucker-Franklin et al., 1966; Wilson and Collier, 1976; Razin et al., 1980). The tip structures appear also to be involved in the gliding motility of mycoplasmas as the leading direction of movement is always tip first (Bredt, 1979). Brief treatment of *M. pneumoniae* with Triton X-100 indicates that the rod-like core of the terminal tip of this organism (Fig. 10.4) consists of a bundle of fibrils, which may extend throughout the cell, forming a "cytoskeleton" not reported so far in any other procaryote (Meng and Pfister, 1980). The finding of a "cytoskeleton" in mycoplasmas may be

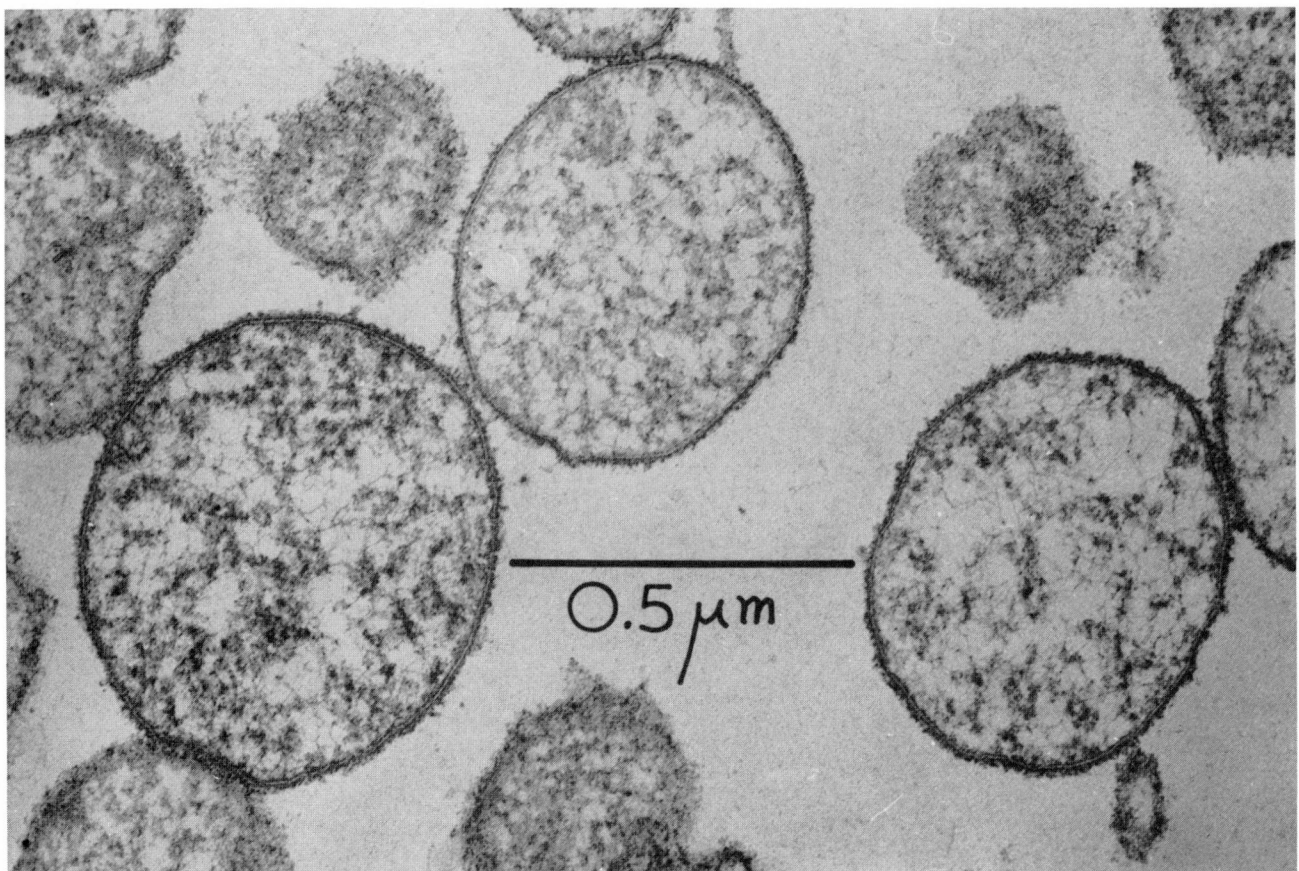

Figure 10.3. Electron micrograph of thin-sectioned mycoplasma cells. The cells are bounded by a single membrane showing in section the characteristic trilaminar structure. The cytoplasm contains thin threads representing the sectioned genome, and dark granules representing ribosomes. Such sections show no evidence for the occurrence of a cell wall or mesosomes. (Reproduced with permission from R. M. Cole.)

relevant to understanding of the contractile process and motility (Bredt et al., 1973).

Growth of mycoplamas in liquid media usually produces barely visible opacity. Absorbance of cultures at 640 nm rarely exceeds 0.3 and is usually much lower. Colonies of almost all *Mycoplasma* species show the typical "fried egg" shape, consisting of an opaque, granular central zone embedded in the agar and a flat translucent peripheral zone on its surface (Fig. 10.5). It should be stressed that colonial appearances are very dependent on growth conditions, age of culture, agar concentration, etc. Thus, on nutritionally poor media, or with inadequate pH or atmospheric conditions, or when the surface of the medium is too dry, the initial central "downgrowth" may occur without formation of the peripheral surface growth (Razin and Oliver, 1961). It should also be stated that a very small minority of *Mycoplasma* species have not been shown to form "fried egg" colonies under any cultural conditions tested so far.

The mode of reproduction of mycoplasmas has been a matter of dispute. However, study of the molecular biology of these organisms showed clearly that replication of their genome, which must precede cell division, follows the same pattern as with other procaryotes dividing by binary fission (Morowitz and Wallace, 1973). For binary fission to occur, however, cytoplasmic division must be fully synchronized with genome replication, which is not always the case with mycoplasmas. In mycoplasmas cytoplasmic division may lag behind genome replication, resulting in the formation of multinucleate filaments (Peterson et al., 1973; Bredt et al., 1973). The subsequent division of the cytoplasm by constriction of the membrane at sites in between the genomes leads to the formation of characteristic chains of beads which later fragment to give single cells. Budding, frequently seen in mycoplasma cultures

(Anderson and Barile, 1965; Razin and Cosenza, 1966), may actually be a form of binary fission in which the cytoplasm is not equally divided between the daughter cells. The mode of reproduction of mycoplasmas is presented schematically in Figure 10.6.

The mycoplasmas have limited biosynthetic abilities, probably reflecting their small genome and parasitic mode of life. Consequently they require complex media for growth, consisting of beef heart infusion, peptone, yeast extract and serum (Edward, 1947; Hayflick, 1965; Freundt et al., 1980). The serum provides, among other nutrients, fatty acids and cholesterol for membrane synthesis in an assimilable, nontoxic form. Defined media have been developed for only two mycoplasmas, *M. mycoides* and *Acholeplasma laidlawii* (Rodwell, 1979). Both organisms require a complex assortment of amino acids, nucleic acid precursors, lipids, vitamins and inorganic ions, and glucose as an energy source. Certain *M. hyorhinis* strains grow poorly or not at all in the conventional artificial media due to their sensitivity to toxic factors found mostly in yeast extract (Del Giudice et al., 1980). These "noncultivable" strains may be more readily isolated and identified by cell culture procedures (Hopps et al., 1973).

Most *Mycoplasma* species are facultatively anaerobic. Since cultures from primary tissue specimens frequently grow out only under anaerobic conditions, an atmosphere of 95% N_2 + 5% CO_2 is preferred for primary isolation. The initial pH of the growth medium should be adjusted to ~8.0 for the fermentative mycoplasmas, and to 6.0–6.5 for the nonfermentative arginine-utilizing mycoplasmas. The temperature range for growth varies according to species and fatty acid and cholesterol composition of the growth medium from ~20°C to ~40°C. The optimum temperature for most animal mycoplasmas is 36–37°C.

Glucose, or other metabolizable carbohydrates, can serve as an energy

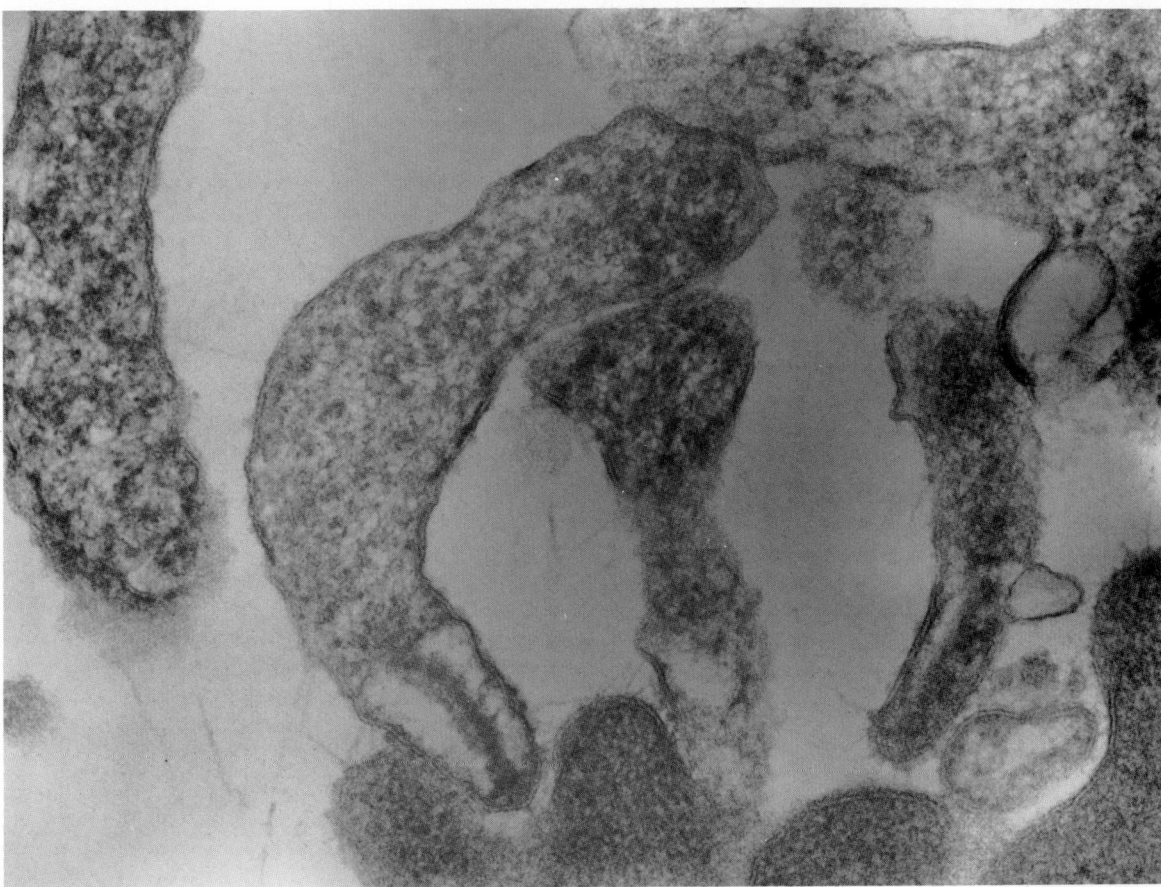

Figure 10.4. Electron micrograph of sectioned *Mycoplasma pneumoniae* cells adhering to the surface of an epithelial cell. The peculiar ultrastructure of the specialized tip of the mycoplasma filament and its proximity to the eucaryotic cell can be seen (× 136,000). (Reproduced with permission from M. H. Wilson and A. M. Collier, *Journal of Bacteriology 125*: 332–339, 1976, © American Society for Microbiology.)

source for the fermentative mycoplasmas possessing the Embden-Meyerhof glycolytic pathway. The major end products are lactic acid and, to a smaller extent, pyruvic acid, acetic acid and acetyl methyl carbinol (Pollack, 1979). Arginine degradation by the arginine dihydrolase pathway has been proposed as the major source for ATP in nonfermentative mycoplasmas (Schimke et al., 1966). This pathway may not, however, be the sole energy-yielding pathway in nonfermentative mycoplasmas, since ATP can also be derived from acetyl-coenzyme A through reactions catalyzed by phosphate acetyltransferase and acetate kinase (Kahane et al., 1978).

Antigenic diversity exists within certain *Mycoplasma* species, especially *M. hominis* (Razin, 1968; Hollingdale and Lemcke, 1970; Lin and Kass, 1974), *M. pulmonis* (Forshaw and Fallon, 1972) and *M. arginini* (Thirkill and Kenny, 1975). Such antigenic heterogeneity has been noted with these species in a variety of serological tests and appears to be correlated with differences between membrane antigens or membrane proteins of individual strains (Alexander and Kenny, 1980).

Mycoplasmas are usually resistant to benzyl penicillin and other β-lactam antibiotics whose primary target is the biosynthesis of peptidoglycan. The mycoplasmas are usually sensitive to antibiotics that specifically inhibit protein synthesis in procaryotes, such as tetracyclines and chloramphenicol. Susceptibility to other antibiotics is variable, particularly to erythromycin and certain other macrolides, a property which may be useful for differentiation of species (Taylor-Robinson, 1967; Arai et al., 1967; Niitu et al., 1974; Mårdh, 1975; Williams, 1978; Brunner and Weidner, 1981). Most mycoplasmas tolerate 1:2000 to 1:4000 thallium acetate, which is incorporated, therefore, as a selective agent in mycoplasma media.

Mycoplasma species are parasites of the mucous membranes and joints. Mycoplasma infections have been most frequently associated with diseases of the respiratory and urogenital tracts, where the parasites firmly adhere to and colonize the epithelial lining. The intimate association between the adhering mycoplasmas and their host cells provides an environment in which local concentrations of toxic by-products excreted by the organisms (i.e. H_2O_2, NH_3) can accumulate and cause tissue damage (Razin, 1978; Razin et al., 1981). Since no cell wall separates the plasma membrane of the parasite from that of its host, exchange of antigens may occur between the two membranes, an event which may trigger immunological responses of serious consequences to the host (Wise et al., 1978). The intimate association between mycoplasmas and host cell membranes is reflected also by the "capping" of mycoplasmas adhering to lymphocytes, followed by shedding of membrane vesicles of presumed host origin (Stanbridge and Weiss, 1978). This phenomenon is apparently related to the well-known induction of blast transformation by mycoplasmas (Naot, 1982).

Enrichment and Isolation Procedures

For a comprehensive description of the methods used in mycoplasma isolation, cultivation, and characterization, see Razin and Tully (1983) and Tully and Razin (1983).

Swabs, tissues or fluid from infected organs should be put immediately into liquid and on solid mycoplasma medium, or in a suitable transport medium. Conventional mycoplasma broth containing 500–1000 units of penicillin per milliliter is generally an effective transport medium. Tissues should be minced coarsely and added to mycoplasma broth medium. At least two 10-fold dilutions of the initial suspension

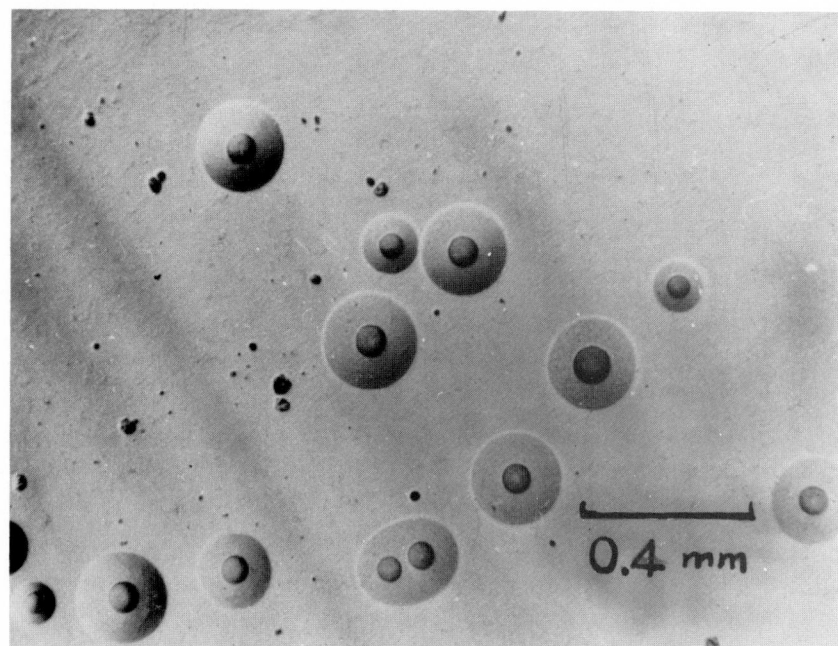

Figure 10.5. "Fried egg" colonies of mycoplasmas growing for 4 days on a solid medium. (Reproduced with permission from S. Razin and O. Oliver, *Journal of General Microbiology 24*: 225–237, 1961, © Society for General Microbiology.)

should be made in broth to reduce concentrations of inhibitory substances (lysolecithin, enzymes, antibodies, etc.) present in tissues (Tully and Rask-Nielsen, 1967; Kaklamanis et al., 1969). The most common medium for *Mycoplasma* isolation is the modification of Edward medium (1947) described by Hayflick (1965). This medium is supplemented with ~20% horse serum and penicillin (1000 IU/ml) and thallium acetate (1:2000) to inhibit growth of contaminating bacteria.* For slow growing and more fastidious mycoplasmas, such as *M. pneumoniae*, a diphasic SP-4 medium† (Tully et al., 1979) has recently been recommended. One milliliter of SP-4 agar prepared with 0.8% Noble agar is added to the bottom of a sterile vial. The agar is allowed to solidify and 2 ml of SP-4 broth is overlayed onto the agar. The inoculated diphasic broth cultures are incubated at 37°C and observed every 2 or 3 days for change in the color of the pH indicator. Samples (0.2 ml) of broth supernatant are added to agar plates, incubated in a 95% N_2 + 5% CO_2 atmosphere, and examined for the appearance of colonies.

To obtain pure cultures the organism should be cloned. This is accomplished by gentle filtration of a broth culture through a membrane filter (220 nm or 450 nm pore diameter) and culturing the filtrate on a solid medium. An isolated colony is picked up from a plate on which few colonies develop and is transferred to fresh broth. The cloning procedure is performed at least three times, in the hope that the colonies formed after the third cloning derive from single organisms and represent a pure clone (Subcommittee, 1979).

Maintenance Procedures

Mycoplasma broth cultures can be kept alive for long periods (months or even years) at −70°C. Damage to cells during freezing and thawing can be reduced by the addition of dimethyl sulfoxide or glycerol to the culture (Raccach et al., 1975). Mycoplasma cultures can also be successfully lyophilized, and under proper storage conditions (vacuum in ampoules and storage temperature between 4 and −20°C) viability can be kept for years. Techniques for mycoplasma lyophilization have been described in detail and evaluated (Working group document, 1974a).

Procedures for Testing Special Characters

The Subcommittee on the Taxonomy of *Mollicutes* has published a proposal of minimal standards for description of new species of the class *Mollicutes* (Subcommittee, 1979). Accordingly, there are obligatory and optional tests to be used in identification of mycoplasmas.

A. Obligatory Tests

DNA base composition. The guanine + cytosine content of the mycoplasmal DNA can be determined by thermal denaturation or buoyant density methods, or by high-performance liquid chromatography (Razin and Razin, 1980).

Fermentation of glucose. Breakdown of glucose or other carbohydrates is indicated by acid production and change in the color of a pH indicator incorporated into the carbohydrate-containing medium (Al-

---

* Hayflick medium (as modified by FAO/WHO Collaborating Centre for Animal Mycoplasmas, Institute of Medical Microbiology, University of Aarhus). *Liquid medium:* Dissolve 2.85 g of dehydrated heart infusion broth (Difco) in 90 ml of distilled water and sterilize at 121°C for 20 min. Aseptically add: horse serum (sterilized by filtration), 20.0 ml; fresh yeast extract solution (25%, w/v), 10.0 ml; calf thymus DNA solution (0.2%, w/v (Sigma Chemical Co., St. Louis, Missouri), 1.2 ml; thallium acetate solution (1%, w/v), 1.0 ml; and benzylpenicillin solution (20,000 IU/ml), 0.25 ml. Adjust the pH to 6.8 or 7.8 (depending on the particular strain). *Solid medium:* This is prepared by adding ~0.8% Noble agar (Difco) or other washed or purified agars (agarose). *Fresh yeast extract* may be prepared by the following procedure (Hayflick, 1965): suspend 250 g of brewers' yeast in 1000 ml of distilled water and gently boil the suspension for 15 min. Allow the sediment to settle, or centrifuge at 250 × g for 15 min. Filter the supernatant fluid through Whatman No. 1 filter paper. Adjust the pH to 8.0 with 1 N NaOH. Dispense 10-ml portions, sterilize at 121°C for 15 min, and store at −20°C. Quality control tests of the yeast extract are advisable.
† SP-4 medium: dehydrated Mycoplasma broth base (BBL), 3.5 g; Tryptone (Difco), 10.0 g; peptone (Difco), 5.3 g; glucose, 5.0 g; deionized water, 615 ml; CMRL 1066 medium (10× with glutamine; Grand Island Biological Co. (Gibco), New York, New York, Cat. No. 154), 50 ml; fresh yeast extract solution (25%; Microbiological Associates, Bethesda, Maryland), 35 ml; Yeastolate solution (2%, sterile; Difco), 100 ml; fetal bovine serum (heated at 56°C for 1 h; Flow Laboratories, Rockville, Maryland), 170 ml; penicillin G solution (100,000 U/ml), 10 ml; and phenol red solution (0.1%), 20 ml; pH 7.0–7.4.

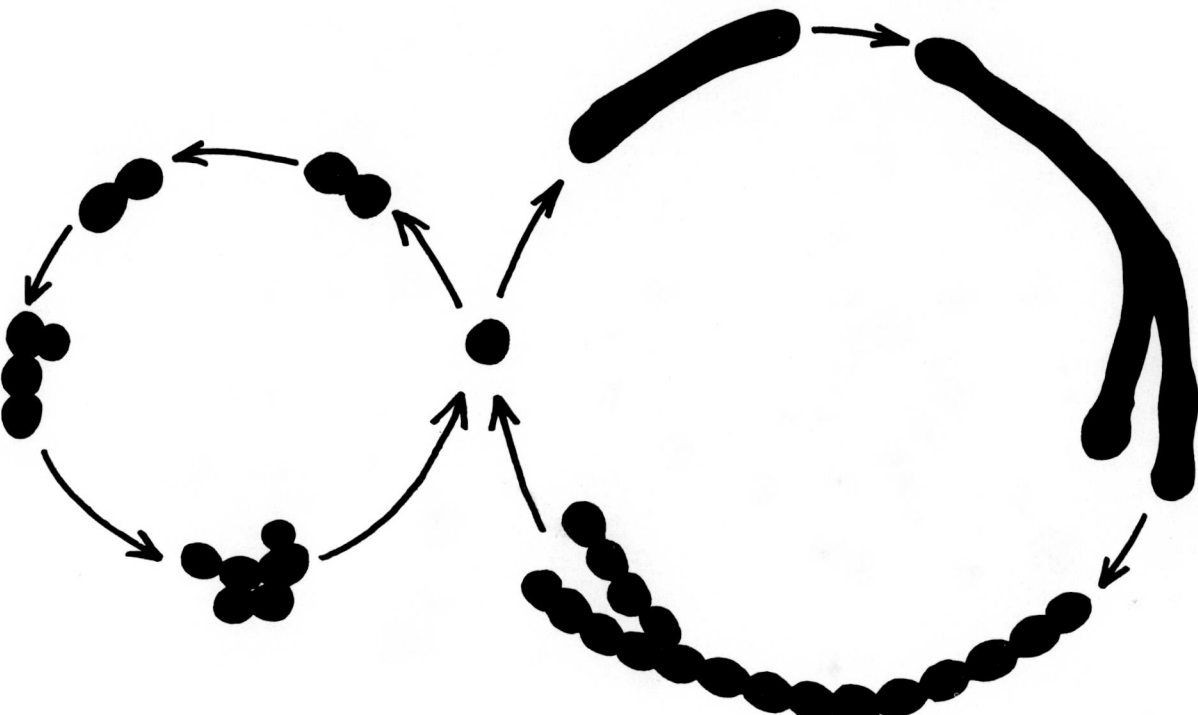

Figure 10.6. Schematic presentation of the mode of mycoplasma replication. Cells may either divide by the regular binary fission or elongate first to multinucleate filaments which subsequently break up into coccoid bodies. (Reproduced with permission from S. Razin, in B.K. Ghosh (Editor), *Organization of Prokaryotic Cell Membranes*, Vol. 1, pp. 165–250, 1981, © The Chemical Rubber Co., CRC Press, Inc., Boca Raton, Fla.)

uotto et al., 1970). Some difficulties with certain nonfermentative mycoplasmas have been observed in this procedure since a slight decrease in pH can also occur in control cultures grown in the absence of glucose. More sensitive and specific tests have therefore been devised including: glucose disappearance as determined by the glucose-oxidase reaction (Edward and Moore, 1975), determination of hexokinase activity, or detection of acid fermentation products from radioactive glucose (Cirillo and Razin, 1973). A detailed description and evaluation of these techniques is available (Working group document, 1974b).

Hydrolysis of arginine. The growth medium is supplemented with 0.2% (w/v) of L-arginine instead of glucose. The initial pH of the medium is adjusted to 7.0–7.3. A distinct alkaline shift (0.5 pH units or more) indicates a positive reaction. The problems encountered with determination of glucose metabolism according to pH changes may also be encountered in tests for arginine hydrolysis, particularly with strains which utilize both glucose and arginine. In this case the acid produced from glucose utilization may mask the alkali produced by arginine hydrolysis. Determination of the breakdown product citrulline may in this case serve as a specific and sensitive test for arginine hydrolysis. For details of the technique see Weickmann and Fahrney, 1977.

Hydrolysis of urea. This test, described in detail in the section on *Mycoplasmataceae*, is negative for all *Mycoplasma* species.

Production of pigmented carotenoids. This test is described in detail in the section on *Acholeplasmataceae* and is negative with all *Mycoplasma* species.

Serological tests. According to the recommendations of the Subcommittee on the Taxonomy of *Mollicutes* (Subcommittee, 1979) any candidate for a new species should be compared serologically with reference strains of other named species. If it can be assigned to an existing genus, it should as a minimum be compared with all species of this genus, at least by the two following serological methods: the growth inhibition test and the immunofluorescence test. The growth inhibition test (Clyde, 1964; Working group document, 1976), based on inhibition of growth on agar around disks saturated with specific antiserum, is of the most general use in identification of *Mycoplasma* species, as it is the most specific serological test. However, it requires highly potent sera, which are not always available. The second recommended method is based on identification of mycoplasma colonies on agar by direct or indirect fluorescent antibody tests (Del Giudice et al., 1967; Rosendal and Black, 1972; Tully, 1965). This is, perhaps, the most rapid test for diagnosis. It has proved also to be very specific, and is the only test capable of distinguishing between a mixture of colonies of different serotypes on the same plate—a most important feature in checking the purity of a clone or in testing clinical material, which very frequently contains more than one species or serotype.

B. Optional Tests

Genome size. Although special facilities are required for determination of genome size (Bak et al., 1969; Morowitz, 1969), the Subcommittee on the Taxonomy of *Mollicutes* strongly encourages the performance of this test for all organisms where generic or family affinity is uncertain. The genome size of all *Mycoplasma* species examined so far is about 5×10^8 daltons.

DNA homology. Determination of genetic relatedness by nucleic acid homology studies (McGee et al., 1967; Somerson et al., 1967; Christiansen et al., 1979; Aulakh et al., 1979; Junca et al., 1980) may be necessary to separate closely related species, and determine relatedness at the subspecies level.

Electrophoretic patterns of cell proteins. Since the synthesis of cell proteins is genetically directed, their electrophoretic patterns are likely to reflect the genetic identity or nonidentity of microorganisms. Electrophoresis of solubilized cell proteins can be carried out in polyacrylamide cylindrical (Razin and Rottem, 1967) or slab gels (Daniels and Meddins, 1973; Working Group document, 1975a; Mouches et al., 1979). Better definition is obtained by two-dimensional electrophoresis in which cell proteins are separated in one dimension by isoelectric focusing according to charge, and in the second dimension, in the presence of sodium dodecyl sulfate, according to size. The protein maps obtained are most useful in identification of mycoplasmas and determination of relatedness at the species and subspecies level (Rodwell and Rodwell, 1978; Mouches et al., 1979).

Fermentation of carbohydrates other than glucose. This test may be of taxonomic value. The substrates include: mannose, mannitol, lactose, sucrose, salicin, fructose, galactose, xylose, sorbitol, glycerol, and cellobiose (Barber and Fabricant, 1971).

Phosphatase activity. This property is determined according to the ability of mycoplasmas to hydrolyze phenolphthalein diphosphate incorporated into the growth medium (Aluotto et al., 1970; Black, 1973), or by the ability of washed cell suspensions to hydrolyze *p*-nitrophenylphosphate (Makki, 1971).

Film and spot reaction. This test is indicative of the lipolytic activity of the organisms, a property that has some diagnostic value. During growth of certain mycoplasmas on media containing horse serum, egg yolk (Edward, 1954; Fabricant and Freundt, 1967) or Tween 80 and $CaCl_2$ (Razin and Rottem, 1963) a characteristic wrinkled, pearly film appears on the medium surface together with tiny black spots beneath and around the colonies. The film contains cholesterol and phospholipids, whereas the spots consist of calcium and magnesium salts of fatty acids liberated by the mycoplasma lipases.

Proteolytic activity. This property is usually determined by the ability of the organisms to digest gelatin, coagulated serum or casein (Aluotto et al., 1970).

Tetrazolium reduction. The test is based on the ability of many mycoplasmas to reduce 2,3,5-triphenyltetrazolium chloride. More strains are capable of reducing tetrazolium under anaerobic than under aerobic conditions (Aluotto et al., 1970).

Hemolysis. This property can be tested by covering colonies with a thin layer of sheep or guinea pig blood agar (Aluotto et al., 1970) or by inoculating concentrated suspensions of organisms onto blood agar (Cole et al., 1968). Weak to strong α and β hemolysis was shown almost throughout to result from the production of peroxide by the organisms (Cole et al., 1968; Lind, 1970; Sobeslavsky and Chanock, 1968; Cohen and Somerson, 1969). Since hemolytic activity is shared by most mycoplasmas, and since its degree and type seem to depend on minor differences in the technique used, demonstration of the property *per se* is of secondary importance.

Hemadsorption. This can be tested microscopically by determining the adsorption of erythrocytes to mycoplasma colonies (Sobeslavsky et al., 1968; Manchee and Taylor-Robinson, 1968, 1969). This property may be relevant to pathogenicity as it measures the ability of mycoplasmas to adhere to eucaryotic cells. *M. pneumoniae* strains that lost this property lost their virulence to hamsters (Lipman et al., 1969).

Optional serological tests. The metabolism inhibition test is much more sensitive than the growth inhibition test, but may be more difficult to read. It is based upon determination of the amount of growth inhibition by antibody as measured by inhibition of certain metabolic activities of the mycoplasmas, specifically glucose fermentation, arginine hydrolysis or tetrazolium reduction (Purcell et al., 1966; Senterfit and Jensen, 1966; Taylor-Robinson et al., 1966). The procedure is described in detail and evaluated (Working group document, 1975b). The indirect agglutination test (Krogsgaard-Jensen, 1971; Lind, 1968; Morton, 1966) may be used as an alternative optional test. Complement-fixation tests with whole-cell antigens or cell extracts (Kenny and Grayston, 1965; Taylor-Robinson et al., 1963) and immunodiffusion tests with extracts of mycoplasma cells (Lemcke, 1965; Taylor-Robinson et al., 1963) have been utilized in early work. More recently it has been found that the crossed immunoelectrophoresis test (Alexander and Kenny, 1980) has the advantage of resolving antigens not measurable by other tests, and better reveals antigenic relatedness among strains.

Differentiation of the genus **Mycoplasma** from other genera

See Table 10.1 for characteristics useful in differentiating *Mycoplasma* from related genera.

Taxonomic Comments

The genus *Mycoplasma* is currently subdivided into 64 different species (see Table 10.2). Ideally, mycoplasma species may be regarded, in the wording of the Subcommittee on the Taxonomy of *Mollicutes* (1979), as "clusters of morphologically similar isolates whose genomes exhibit a high degree of relatedness." In practice, characterization and differentiation of species of the genus *Mycoplasma* is based on a variety of morphological, cultural, nutritional, biochemical, serological and genetic properties. The morphological markers that can be utilized in species characterization include the length of filaments under defined growth conditions, possible possession of an external capsular substance, and the occurrence in some species of highly structured organelles. Gliding motility distinguishes a few *Mycoplasma* species. Colonial morphology is sufficiently characteristic for certain species to be of some help in classification. Well-defined nutritional requirements, such as the dependence of *M. synoviae* on β-nicotinamide dinucleotide as a growth factor, may occasionally add to the species description. The biochemical and other physiological properties that are most useful to characterize species of *Mycoplasma* have already been described in the preceding section and appear also in Table 10.2.

Final classification of members of genus *Mycoplasma* depends on determination of the serological relatedness, for which purpose a number of tests of varying specificity and sensitivity are available. The growth inhibition, direct or indirect immunofluorescence, and/or the metabolism inhibition tests are the methods most commonly used for species differentiation. Double immunodiffusion in agar and the growth-precipitation test (Krogsgaard-Jensen, 1972) may reveal the presence of common antigens in species that are otherwise serologically distinct. Thus, interspecific cross-reactivity has been demonstrated within groups of glycolytic *Mycoplasma* species as well as among *Mycoplasma* species that hydrolyze arginine, but only very rarely between members of each of these two biochemically distinct groups (Taylor-Robinson et al., 1963; Lemcke, 1965; Fox et al., 1969; Kenny, 1973; Ernø and Salih, 1980; Jordan et al., 1981). Similar observations were made by comparison of glycolytic and arginine-utilizing *Mycoplasma* species by two-dimensional immunoelectrophoresis (Thirkill and Kenny, 1974).

Determination of the guanine plus cytosine content of the DNA by T_m and buoyant density is considered by the Subcommittee (1979) a necessary requirement for species characterization and has in fact been carried out for the vast majority of *Mycoplasma* species (Table 10.2). Determination of genetic relatedness by nucleic acid homology studies has been performed only to a limited extent (Reich et al., 1966; McGee et al., 1967; Somerson et al., 1967; Peterson and Pollock, 1969; Askaa and Ernø, 1976; Sugino et al., 1980). The relatedness reported in these studies between different *Mycoplasma* species ranges from less than 10 (in most cases) to 40%. Analysis of genotypic variations within individual *Mycoplasma* species by means of nucleic acid hybridization techniques has been made only in a few instances (Somerson et al., 1966; Askaa et al., 1978).

In the genus *Mycoplasma*, species names frequently have ecological significance; many *Mycoplasma* species are restricted to or are predominantly associated with a particular host, but some, as for example *M. arginini*, are found in very diverse hosts.

The conceptual basis for establishment of a subspecies has been formulated by the Subcommittee (1979) in the words: "The rank of subspecies in *Mollicutes* should be reserved for important strains which differ consistently in a number of major properties but which nevertheless prove to be too closely related by serological, or especially by nucleic acid hybridization, tests to warrant species rank." However, precise standards of serological and nucleic acid hybridization relatedness for subspecies differentiation have not yet been established. At present, only one species of the genus *Mycoplasma*, viz. *M. mycoides*, is formally subdivided into subspecies.

Differentiation of the species of the genus **Mycoplasma**

Table 10.2 lists characteristics useful for differentiating the 64 presently recognized species of *Mycoplasma*.

List of the species of the genus **Mycoplasma**

1. **Mycoplasma agalactiae** (Wroblewski 1931) Freundt 1955, 73.[AL] (*Anulomyces agalaxiae* (sic) Wroblewski 1931, 111.)

a.ga.lac'ti.ae. Gr. n. *agalactia* want of milk, agalactia; M.L. gen. n. *agalactiae* of agalactia.

Coccoid and short to moderately long, branched filamentous cells. Unlike the more transient filamentous forms seen in cultures of *M. mycoides* subsp. *mycoides*, the filaments of *M. agalactiae* usually persist for at least 8 days.

Distinctive physiological characteristics of the species are presented in Table 10.2. Agar colonies adsorb red blood cells from guinea pigs and cattle, but not HeLa cells or human spermatozoa.

Shares antigens(s) with *M. bovis* as revealed by double immunodiffusion in agar (DID) and growth precipitation and by a weak one-way cross-reaction in the metabolism inhibition (MI) test, but the two species are completely distinct by growth inhibition (GI) and immunofluorescence (IMF) tests. Serologically distinct also by two or more of the said tests from all other *Mycoplasma* and *Acholeplasma* species of ovine-caprine and bovine sources, and from a number of other species tested.

The etiological agent of contagious agalactia of sheep and goats, a disease that is particularly widespread in Mediterranean countries but has been reported recently also from other regions of Southern Europe and North Africa, from the Soviet Union, the Middle East, Asia and South America. The mastitis may be associated with a short transient mycoplasmemia and is frequently complicated by a painful nonsuppurative, edematous arthritis. Also associated with an outbreak of pleurisy and pneumonia and with granular vulvovaginitis, both of which conditions can be reproduced experimentally.

The mol% G + C of the DNA is reported in the range of 30.5 (T_m) to 34.2 (Bd). The DNA/DNA homology between *M. agalactiae* (strain PG2) and *M. bovis* (strain Donetta) is about 40%.

Type strain: NCTC 10123 (strain PG2, Edward and Freundt, 1973).

2. **Mycoplasma alkalescens** (sic) Leach 1973, 149.[AL] (Bovine serogroup 8, Leach 1967).

al.ca.les'cens. M.L. v. *alcalesco* make alkaline; M.L. part. adj. *alcalescens* alkaline making, referring to the reaction produced in arginine-containing media.

Coccoid to coccobacillary cells.

Distinctive physiological characteristics of the species are presented in Table 10.2.

Sharing of antigens with a number of other arginine-metabolizing species (*M. arginini*, *M. gateae*, *M. gallinarum*, *M. hominis* and *M. spumans*) has been demonstrated by DID and/or growth precipitation (GP) tests. Serologically distinct from these and all other previously established *Mycoplasma* and *Acholeplasma* species by GI and MI tests.

Pathogenic to cattle. Isolated from synovial fluid, lymph nodes, liver and umbilical lesions of calves with severe febrile arthritis. Intraarticular inoculation of cultures in calves produced fibrinopurulent arthritis associated with fever; intravenous inoculation induced only fever. Also isolated from milk samples from cows with clinical mastitis and occasionally from the milk of recovered animals; experimental production of mastitis has apparently not been attempted. Isolations from the prepuce of bulls and from commercial bovine serum have also been reported.

The mol% G + C of the DNA is 25.9 (T_m).

Type strain: NCTC 10135 (ATCC 29103; strain D12 (PG51) of Leach, 1973).

3. **Mycoplasma alvi** Gourlay, Wyld and Leach 1977, 95.[AL]

al'vi. L. n. *alvis* bowel, womb, stomach; L. gen. n. *alvi* of the bowel.

Coccoid to coccobacillary elements and very short branched filaments. Elongated, flask- or club-shaped cells with well-defined blunt protuberances can be seen by electron microscopy of negatively stained specimens prepared from broth cultures. A dense central core extending into a polar terminal structure closely resembling those of *M. pneumoniae* and *M. gallisepticum* is visible in thin-sectioned cells stained with uranyl acetate or ruthenium red.

Facultative anaerobe, although growth on solid medium is better anaerobically in H_2 or N_2 with 1% CO_2. Growth occurs within the range of 25 to ~40°C, with optimum at ~37°C. Optimum pH is between 6.0 and 7.0. Growth is slow, colonies attaining a diameter of up to 50 μm after 1 week, and up to 290 μm after 2 weeks of incubation under optimal conditions. Grows well in SP-4 broth (see under Enrichment and Isolation Procedures), whereas classic horse serum media are very inhibitory (J. G. Tully, personal communication, 1981).

Distinctive physiological characteristics of the species are presented in Table 10.2.

The type strain is serologically distinct from all previously named *Mycoplasma* and *Acholeplasma* species, as determined by MI tests supplemented in some cases by GI and IMF tests. Serologically distinct also by GI, MI and IMF tests from *Anaeroplasma bactoclasticum* and *Anaeroplasma abactoclasticum*, except for a low order cross-reaction with strain 161 of the latter species.

Susceptible to digitonin (0.3 μg), but unlike most other sterol-dependent mycoplasmas resistant to sodium polyanetholesulfonate (1.0 μg), as determined by a disk growth inhibition method on solid medium.

Pathogenicity is not known. Inoculation of cultures of the organism into the mammary glands of cows resulted in only moderate milk changes and excretion of the mycoplasmas for 4–5 days.

Found in the lower alimentary tract, feces, bladder and vagina of cows. Apparently a common inhabitant of the bovine alimentary tract from the abomasum to the rectum. Also found in the intestinal tract of voles.

The mol% G + C of the DNA is 26.4 (Bd).

Type strain: NCTC 10157 (ATCC 29626; strain Ilsley of Gourlay, Wyld and Leach, 1977).

Further Comments

Strains which closely resemble *M. alvi* in their cultural and physiological characteristics have been isolated from the intestinal tract of wild field mice and shrews. They are serologically distinct from other murine *Mycoplasma* species but have apparently not been serologically compared with *M. alvi*.

4. **Mycoplasma anatis** Roberts 1964, 471.[AL]

a.na'tis. L. n. *anas* a duck; L. gen. n. *anatis* of a duck.

Morphology poorly defined.

Growth occurs at 30 and 37°C, but not at 25 or 44°C.

Distinctive physiological characteristics of the species are presented in Table 10.2. Acid is produced also from fructose, maltose, dextrin, starch and (inconsistently) from sucrose, but not from lactose, galactose, mannitol, sorbitol, cellobiose, salicin or xylose.

Sharing of antigens with *M. gallinaceum* and *M. gallopavonis* has been demonstrated by DID. Serologically distinct by GI and IMF from these and other avian and a number of nonavian *Mycoplasma* species tested.

The type strain is susceptible to digitonin (0.3 μg), but unlike most other sterol-dependent mycoplasmas resistant to sodium polyanetholesulfonate (1.0 μg), as determined by a disk growth inhibition method on solid medium.

Pathogenicity is uncertain. First isolated together with influenza A virus from the infraorbital sinuses of a duck with sinusitis. Attempts to reproduce the disease experimentally in ducks with the mycoplasma alone or with the virus given 14 days later were largely unsuccessful.

Table 10.2.

Differential characteristics of the species of the genus **Mycoplasma**[a]

Characteristic	Glucose Catabolism	Mannose Catabolism	Arginine Hydrolysis	Phosphatase	Film and Spots	Tetrazolium Reduction (Ae/An)[b]	Gelatin Hydrolysis	Coagulated Serum Digestion	Casein Digestion	Hemadsorption[c]	Mol% G + C of DNA
1. *M. agalactiae*	−	−	−	+	d	+/+	−	−		+	30.5–34.2
2. *M. alkalescens*	−		+	+	−	−/−		−		−	25.9
3. *M. alvi*	+		+		−	−/+					26.4
4. *M. anatis*	+	d	−	+	+	−/+				−	
5. *M. arginini*	−		+	−		−/+		−		−	27.6–28.6
6. *M. arthritidis*	−	−	+	+	−	−/−	+	−	−	−	30.0–32.6
7. *M. bovigenitalium*	−	−	−	+	+	−/+	−	−	−	d	28.1–30.4
8. *M. bovirhinis*	+	−	−	d	−	+/+	−	d	+	d	24.5–27.3
9. *M. bovis*	−	−	−	+	d	+/+	−	−		x	27.8–32.9
10. *M. bovoculi*	+	d	d	+		+/+	−	−		−	29.0
11. *M. buccale*	−	−	+	+	−	−/+	−	−	−	−	25.0–26.4
12. *M. californicum*	−		−	+	−	−/[d]					31.9
13. *M. canadense*	−	−	+	W	−	−/+				−	29.0
14. *M. canis*	+	−	−	−	−	−/+	d	−	−	+	28.4–29.1
15. *M. capricolum*	+	+	d	+	−	+/+		+		−	24.1–25.5
16. *M. caviae*	+	v	+	+		−/[d]				−	
17. *M. citelli*	+	+	−	+	+	W/W	−			−	
18. *M. columbinasale*	−		+	+	+	−/−				−[e]	32.0
19. *M. columbinum*	−	−	+	−	+	−/+	−		−	−[e]	
20. *M. columborale*	+	−	−	−	−	−/+	−		−	−[e]	
21. *M. conjunctivae*	+	+	−	−	−	W/+				−	
22. *M. cynos*	+	+	−	+	+	W/+				+	
23. *M. dispar*	+	+	−	−	−	+/+				−	28.5–29.3
24. *M. edwardii*	+	−	−	−	+	−/+	−	−		d	29.2
25. *M. equigenitalium*	+	+	−	+	+	−/−				−	
26. *M. equirhinis*	−		+	+		−/−				−	
27. *M. fastidiosum*	W	−	−	+		−/−				+	
28. *M. faucium*	−	−	+	−		−/−				+[f]	
29. *M. feliminutum*	x	x	−	−	d	−/+				−	
30. *M. felis*	+	−	−	+	+	−/+	−	−	−	−	25.2
31. *M. fermentans*	+	−	+	d	+	−/+	−	−	−	−	27.5–28.7
32. *M. flocculare*	x	x	−	−	W	−/W	−			−	
33. *M. gallinaceum*	+	−	−	−	−	−/−				−[e]	28.0
34. *M. gallinarum*	−	−	+	−	+	+/+	−	−	−	−	26.5–28.0
35. *M. gallisepticum*	+	+	−	−	−	+/+	−	−	−	+	31.8–35.7
36. *M. gallopavonis*	+	−	−	−		d/d				−	27.0
37. *M. gateae*	−	−	+	−	−	−/W	−	−	−		28.5
38. *M. hominis*	−	−	+	−	−	−/−	−	−	−	−	27.3–33.7
39. *M. hyopneumoniae*	x	x	−	−	W	−/W	−			−	
40. *M. hyorhinis*	+	−	−	+	−	+/+	−	−	−	−	27.3–27.8
41. *M. hyosynoviae*	−	−	+	−	+	−/−		−		−	
42. *M. iners*	−	−	+	−	+	−/−				−	29.1–29.6
43. *M. iowae*	+	+	−	−	−	+/+				+	25.0
44. *M. lipophilum*	−	−	+	−	+					−	
45. *M. maculosum*	−	−	+	+	+	−/+	−	−	−	−	26.7–29.6
46. *M. meleagridis*	−	−	+	+	−	−/+				d	27.0–28.6
47. *M. moatsii*	+	+				−/−					
48. *M. molare*	+	+	−	−	+	+/+					
49a. *M. mycoides* subsp. *mycoides*	+	+	−	−	−	+/+	+[g]	+ or W[h]	+ or W[h]	−	26.1–27.1
49b. *M. mycoides* subsp. *capri*	+	+	−	−	−	+/+	+	+	+	−	24.0–26.0
50. *M. neurolyticum*	+	+	−	−	−	−/+	−	−	−	−	22.8–26.2
51. *M. opalescens*	−	−	+	+	+	−/−				−	
52. *M. orale*	−	−	+	−	−	−/−	−	−	−	+[f]	24.0–28.2
53. *M. ovipneumoniae*	+	−	−	−	−	W/+	−	−	−		
54. *M. pneumoniae*	+	+	−	−	−	+/+	−	−	−	+	38.6–40.8
55. *M. primatum*	−	−	+	+	−	−/−	−	−	−	−	28.6

Table 10.2—*continued*

Characteristic	Glucose Catabolism	Mannose Catabolism	Arginine Hydrolysis	Phosphatase	Film and Spots	Tetrazolium Reduction (Ae/An)[b]	Gelatin Hydrolysis	Coagulated Serum Digestion	Casein Digestion	Hemadsorption[c]	Mol% G + C of DNA
56. *M. pullorum*	+	–	–	–	–	–/–				–[e]	29.0
57. *M. pulmonis*	+	+	–	–	+	–/+	–	–	–	d	27.5–29.2
58. *M. putrefaciens*	+	+	–	+	+	W/+	–		–	+	28.9
59. *M. salivarium*	–	–	+	–	+	–/W	–	–	–	–	27.3–31.4
60. *M. spumans*	–	–	+	+	–	–/–	–	–	–	+	28.4–29.1
61. *M. sualvi*	+	+	–			–/+					23.7
62. *M. subdolum*	–	–	+	v	d	–/–	–	–	–	–	
63. *M. synoviae*	+	–	–	+		–/W				d	34.2
64. *M. verecundum*	–	–	–	+	+	–/[d]					27.0–29.2
Other species:[i]											
M. cricetuli	+	+	–	+	+	+/+		–		–	
M. genitalium	+	–	–	–		W/+		–		+	32.4
M. lipofaciens	+	+	–	+		–/+		–			24.5
M. muris	–	+	–	+		–/+		–		+	24.9
M. mustelae	+	–	+	+		–/+		–			28.2

[a] Symbols: see standard definitions; also x, not definitely settled; W, weak reaction.

[b] Ae, aerobically; An, anaerobically.

[c] Hemadsorption: unless otherwise indicated, the test is performed with red blood cells (RBC) from different animal species (often with RBC from the animal species from which the mycoplasma originated, plus guinea pig RBC; sometimes with a variety of RBC).

[d] Not tested anaerobically.

[e] Only chicken RBC tested.

[f] Adsorbs chicken, but not human, monkey, rat or guinea pig RBC.

[g] Hydrolysis of gelatin not specified for large colony (LC) and small colony (SC) strains.

[h] LC strains digest coagulated serum and casein more vigorously than SC strains.

[i] See Note Added in Proof at end of article.

Inoculation in another study of cultures of the type strain into the thoracic air sac of 1- to 2-day-old ducklings resulted in air sacculitis and some retardation in growth.

Isolated from the respiratory tract and particularly the cloaca of ducks, including Peking ducks and, on one occasion, a teal (*Anas crecca*); also found in a scaup (*Aythya marila*).

The G + C content of the DNA has not been determined.

Type strain: ATCC 25524 (NCTC 10156; strain 1340 of Roberts, 1964).

5. **Mycoplasma arginini** Barile, Del Giudice, Carski, Gibbs and Morris 1968, 490.[AL]

ar.gin.i'ni. Eng. n. *arginine* an amino acid; M.L. gen. n. *arginini* of arginine, referring to its hydrolysis.

Morphology is poorly defined.

Distinctive physiological characteristics of the species are presented in Table 10.2.

Two major surface antigens have been detected and characterized in *M. arginini* by a combination of quantitative immunoelectrophoresis and SDS-polyacrylamide gel electrophoresis. Three of four strains examined possessed immunologically strain-specific antigens. Another surface antigen was shared by all four strains (Alexander and Kenny, 1980). Sharing of antigens by this and other arginine-metabolizing *Mycoplasma* species has been demonstrated by DID and by two-dimensional immunoelectrophoresis, whereas no cross-reactions were observed between *M. arginini* and glucose-metabolizing species (Thirkill and Kenny, 1974). Serologically distinct by GI and IMF tests from previously named *Mycoplasma* species.

Pathogenicity is not known.

Mammalian parasite with an apparently wide host range. Found with a high frequency in the respiratory tract and less frequently the conjunctivae and genital tract of sheep, goats and cattle. Occasional isolations reported from pigs, horses, chamois, dogs, domesticated cats and certain wild large cats. Frequently isolated from commercial bovine serum.

The mol% G + C of the DNA is 27.6 (T_m) to 28.6 (Bd).

Type strain: ATCC 23838 (NCTC 10129; strain G230 of Barile et al., 1968; Edward and Freundt, 1973).

6. **Mycoplasma arthritidis** (Sabin 1941) Freundt 1955, 73.[AL] (*Murimyces arthritidis* Sabin 1941, 57.)

ar.thri'ti.dis. Gr. n. *arthritis* gout, arthritis; M.L. gen. n. *arthritidis* of arthritis.

Filaments vary from short (2–5 μm) to moderately long (10–30 μm).

Differences in colonial morphology and rate of growth between virulent and avirulent strains have been reported.

Distinctive physiological characteristics of the species are presented in Table 10.2.

Although a fairly homogeneous group antigenically, variant strains differing in some ways serologically have been described (Lemcke, 1961, 1964). Sharing of an antigen with *M. columbinasale* has been demonstrated by DID. Serologically distinct by at least GI, MI and IMF.

A relatively common pathogen of laboratory and wild rats causing purulent polyarthritis, sometimes occurring as localized outbreaks in stocks of laboratory rats. Repeatedly isolated also from submandibular abscesses, middle ear infections, ocular lesions, the nasal mucosa in purulent rhinitis, lung lesions (often together with *M. pulmonis*), paraovarian abscesses, as well as from the oropharynx of apparently healthy rats. Subcutaneous inoculation of the organisms suspended in agar emulsions produces localized abscesses and septicemia in rats. Localized arthritis may result from inoculation into the footpads, while widespread infection characterized by suppurative polyarthritis, rhinitis, conjunctivitis and urethritis can be produced by footpad or intravenous inoculation of particularly virulent strains. Flaccid paralysis related to inflammation of the interspinal articulations occurs in some animals. Inoculation into the kidneys of rats can induce pyelonephritis associated with production of local antibody. Although *M. arthritidis* does not infect mice under natural conditions chronic arthritis can be produced in these animals by intravenous inoculation. Arthritis has also been induced in rabbits by inoculation into knee joints. The species has been found to induce interferon production and to suppress antibody formation and lymphocyte transformation in vitro and to some extent in vivo. These and other effects may be manifested when infected animals are used for experimental purposes. (Also see Cassel and Hill, 1979, and Cassel et al., 1979). Has also been isolated on some occasions from rhesus monkeys, other monkeys and from bush babies, and reportedly from the genital tract and joints of humans. The significance of these last mentioned isolations remains to be determined.

The mol% G + C of the DNA is 30.0 (T_m) to 32.6 (Bd). Genome size is 4.4×10^8 daltons. No genetic relatedness is found between this species and *M. fermentans* and *M. gallisepticum* by nucleic acid hybridization (McGee et al., 1967).

Type strain: ATCC 19611 (NCTC 10162; strain PG6 (Preston); Edward and Freundt, 1973).

Further Comments

The Campo (PG27) strain and a small group of related strains previously classified as *M. hominis* type 2 were later reclassified on the basis of serological observations (Lemcke, 1964) as *M. arthritidis* (Edward and Freundt, 1965). The justification of this reclassification was subsequently confirmed by nucleic acid homology studies, by enzyme analysis and by examination of the electrophoretic patterns of the cell proteins of the strains involved.

7. **Mycoplasma bovigenitalium** Freundt 1955, 73.[AL]

bo.vi.ge.ni.ta′li.um. L. n. *bos, bovis* the ox; M.L. pl. n. *genitalia* the genitals; M.L. pl. gen. n. *bovigenitalium* of bovine genitalia.

Coccoid to short filamentous cells, ~2–5 µm in length.

Growth on primary isolation and in early subcultures is markedly enhanced by DNA (0.002%, w/v).

Distinctive physiological characteristics of the species are presented in Table 10.2. Agar colonies adsorb, in addition to various species of red blood cells, bovine, but not human, spermatozoa.

Sharing of an antigen with *M. columbinasale* has been demonstrated by DID. Serologically distinct by GI, MI, GP and DID tests from other *Mycoplasma* species of bovine habitat (including *M. bovirhinis*) and from a number of other *Mycoplasma* species tested by one or more of these methods and/or by IMF.

Some strains are pathogenic to cattle. Occasionally associated with outbreaks of mastitis. The disease, which can be reproduced experimentally, is characterized in the acute stage by extensive accumulations of eosinophilic leucocytes in the interstitia and alveoli of the mammary glands. Has been isolated in some cases from the semen of bulls with chronic seminal vesiculitis. Eosinophilic acute vesiculitis and epididymitis has been produced experimentally with strains recovered from field cases. Arthritic lesions characterized by esoinophilic granulomas of the joint capsules were produced on intravenous inoculation of calves. Occasionally isolated from aborted bovine and equine fetuses. The possible implication of the organism in infertility of cows and bulls is uncertain, as it is found very frequently in the vagina, prepuce and semen of both normal and infertile individuals. A statistically significant correlation between the presence of *M. bovigenitalium* in bull semen and spermatozoan motility has been reported in one study (Jurmanova and Sterbova, 1977). Strains that are identical with, or related to, *M. bovigenitalium* have been isolated from the respiratory and genital tracts of dogs.

The mol% G + C of the DNA is 28.1 (Bd) to 30.4 (T_m).

Type strain: ATCC 19852 (NCTC 10122; strain PG11 (B2) of Edward, 1955).

8. **Mycoplasma bovirhinis** Leach 1967, 313.[AL]

bo.vi.rhi′nis. L. n. *bos, bovis* the ox; Gr. n. *rhis, rhinis* nose; M.L. gen. n. *bovirhinis* of the nose of the ox.

Morphology is poorly defined.

Growth at 37°C, but not at 22°C.

Distinctive physiological characteristics of the species are presented in Table 10.2. Acid produced from glucose, but not from mannose, galactose, cellobiose, xylose or sorbitol. Agglutinates guinea pig but not bovine red blood cells, although neither are adsorbed by agar colonies.

Serologically distinct by DID, GP, GI and MI tests from other *Mycoplasma* species of bovine origin, and from a number of other species tested by at least GI and IMF.

Pathogenicity is uncertain. Although frequently isolated from the lungs and other tissues of calves with pneumonia there is no conclusive experimental evidence regarding a possible pathogenic role in respiratory disease of cattle. Isolated occasionally from mastitic milk; inoculation of the bovine mammary glands with cultures of the organism has resulted in mild mastitis in some cases. No cytopathic effect produced in bovine fetal tracheal organ cultures.

A most common inhabitant of the upper respiratory tract of cattle, but found only infrequently in the bovine urogenital tract.

The mol% of the DNA is reported as 24.5 (Bd) to 27.3 (T_m). A virus, MVBrl, capable of producing plaques on lawns of the species of origin has been isolated from a wild strain of *M. bovirhinis*; morphologically, it is composed of a polyhedral head (~72 nm across) and a tail (~71 nm long), ending in a base plate (Howard et al., 1980).

Type strain: ATCC 27748 (NCTC 10118; strain PG43 (5M331) of Leach, 1967; Edward and Freundt, 1973).

9. **Mycoplasma bovis** (Hale, Helmboldt, Plastridge and Stula 1962) Askaa and Ernø 1976, 325.[AL] (*Mycoplasma agalactiae* subsp. *bovis* Hale, Helmboldt, Plastridge and Stula 1962, 591; *Mycoplasma bovimastitidis* Jain, Jasper and Dellinger 1967, 409; Bovine serotype 5, Leach 1967, 312.)

bo′vis. L. n. *bos* the ox; L. gen. n. *bovis* of the ox.

Coccoid and very short filamentous cells.

Growth optimum about 37°C; no growth at 28°C.

Distinctive physiological characteristics of the species are presented in Table 10.2. Observations regarding hemadsorption are conflicting, the type strain being reported as positive for guinea pig and bovine red blood cells by some authors (Manchee and Taylor-Robinson, 1968) and negative by others (Ernø and Stipkovits, 1973).

The serological reactivity is associated, at least in part, with the lipid cell fraction (Andrew and Carter, 1977). Sharing of antigen(s) with *M. agalactiae* is revealed by DID and growth precipitation, and a low order one-way cross-reaction is demonstrable by MI, but the two species are completely distinct by GI and IMF tests. Serologically distinct also by two or more of the said tests from all other *Mycoplasma* and *Acholeplasma* species of bovine origin and a number of nonbovine species tested.

Next to *M. mycoides* subsp. *mycoides* probably the most pathogenic of the bovine *Mycoplasma* species. Associated with outbreaks of severe bovine mastitis in several parts of the world and undoubtedly the cause of at least some of these outbreaks. The disease can be reproduced experimentally by inoculation of cultures into the mammary gland. It has been isolated also from the joints of cattle with arthritis, and arthritis can be induced following inoculation directly into the joints, intravenously, into the lungs, or into the udder. Lesions have been produced in the internal genitals following intrauterine instillation of cultures, and inoculation into the amniotic cavity has been found to induce abortion. However, a possible causal relationship between *M. bovis* and infertility and abortion in cattle under natural conditions remains to be determined. Neither has the significance of relatively frequent isolations of the organism from the respiratory tract of cattle with respiratory disease been fully resolved as yet. There is some evidence of pathogenicity for goats. A mild cytopathogenic effect has been demonstrated in bovine cell cultures.

The mol% G + C of the DNA is 27.8–32.9 (T_m). The DNA/DNA homology between the type strain of this species and the type strain of *M. agalactiae* is about 40%.

Type strain: ATCC 25523 (NCTC 10131; strain Donetta (PG45) of Hale, Helmboldt, Plastridge and Stula, 1962).

Further Comments

Mycoplasma bovimastitidis Jain et al. 1967 is a later synonym of *Mycoplasma agalactiae* subsp. *bovis* Hale et al. 1962, the elevation of which to species rank, under the name of *Mycoplasma bovis* Askaa and Ernø 1976, was proposed on the basis of serology and nucleic acid homology studies.

10. **Mycoplasma bovoculi** Langford and Leach 1973, 1443.[AL] (*Mycoplasma oculi* Leach 1973, 137.)

bov.o′cu.li. L. n. *bos*, ox; L. *oculus* the eye; M.L. gen. n. *bovoculi* of the bovine eye.

Coccoid to coccobacillary; filamentous forms have not been observed.

Temperature range for growth is 30–40.5°C, with optimum at 36–37.5°C.

Distinctive physiological characteristics of the species are presented in Table 10.2. Acid is produced also from maltose, but not from lactose, sucrose, dulcitol, mannitol, salicin or xylose.

Serologically distinct by GI, MI and/or IMF from other *Mycoplasma* species tested, including previously established species of bovine habitat.

Associated with outbreaks of bovine conjunctivitis and keratoconjunctivitis in Canada, the United States, several European countries, and the Ivory Coast, with an isolation rate ranging from 15 to nearly 100%. Experimental inoculation of the organism into the conjunctival sac of healthy calves has resulted in some studies in the production of mild conjunctivitis with serous lacrimation (Langford and Leach, 1973; Rosenbusch and Knudtson, 1980). *M. bovoculi* has been shown in another study (Friis and Pedersen, 1979) to have an enhancing effect on the production of experimental keratitis in colostrum-deprived calves by *Moraxella bovis*.

Found only occasionally in the eyes of healthy cattle. Has also been isolated from commercial bovine serum.

The mol% of the DNA is 29.0 (T_m).

Type strain: NCTC 10141 (ATCC 29104; strain M165/69 of Langford and Leach, 1973).

Further Comments

M. bovoculi was previously referred to, without any accompanying description, as *Mycoplasma oculi* Leach 1973, 135. Following the almost simultaneous description of a new *Acholeplasma* species, under the name of *A. oculusi* Al-Aubaidi, Dardiri, Muscoplatt and McCauley 1973, 126 (*oculusi* orthographic error for *oculi* Al-Aubaidi 1975, 126) the name *Mycoplasma oculi* was abandoned to avoid any possible future confusion between these two *Mycoplasmatales* (Langford and Leach, 1973).

11. **Mycoplasma buccale** Freundt, Taylor-Robinson, Purcell, Chanock and Black 1974, 252.[AL]

buc.ca′le. L. adj. *buccalis* buccal, pertaining to the cheek.

Coccoid, coccobacillary and very short filamentous cells.

Growth appears to be stimulated by DNA (0.02%, w/v).

Distinctive physiological characteristics of the species are presented in Table 10.2.

Sharing of antigens with other arginine metabolizing human *Mycoplasma* species has been demonstrated by DID. Serologically distinct by GI, MI, IMF, indirect hemagglutination (IHA) and complement fixation (CF) tests from other *Mycoplasma* species of human source, including *M. orale* and *M. faucium*, as well as from a number of other *Mycoplasma* and *Acholeplasma* species tested. However, a marked cross-reaction has been demonstrated in the CF tests between *M. buccale* and *M. orale*.

Pathogenicity is not known.

An apparently infrequent parasitic inhabitant of the human oropharynx, comprising approximately 1–2% of the *Mycoplasma* species found in the oropharynx and being present in only 2% of the human population. It is the predominant species, on the other hand, of the mycoplasmal flora of the oropharynx of several nonhuman primates.

The mol% G + C of the DNA has been reported as 25 (Bd) to 26.4 (T_m). The relative relatedness to other human *Mycoplasma* species, including *M. orale*, as determined by nucleic acid homology tests is ≦10% (Reich et al., 1966; also see Stanbridge and Reff, 1979).

Type strain: ATCC 23636 (NCTC 10136; strain CH20247 of Taylor-Robinson, Fox and Chanock, 1965).

12. **Mycoplasma californicum** Jasper, Ernø, Dellinger and Christiansen 1981, 344.[VP*]

ca.li.for′ni.cum. M.L. adj. *californicum* pertaining to California.

Coccoid to small filamentous cells.

Colonies are conical in shape with distinct small centers; the surface appears to be smooth but with a slight granularity around the center. The maximum colony size after 72 h of growth is usually about 0.3 mm.

Distinctive physiological characteristics of the species are presented in Table 10.2.

Serologically distinct by the indirect IMF test from all other *Mycoplasma* species and serogroups. Antiserum to the type strain has been found to inhibit the growth of *M. fermentans*, *M. maculosum*, *M. meleagridis* and *M. synoviae*, while the type strain was slightly inhibited by antiserum to *M. sualvi*.

Identified as a frequent cause of acute bovine mastitis in California. The disease is characterized by variable swelling of the udder, slightly yellow, watery milk with sandy sediment, flakes or clots or a seropurulent secretion, and either a marked reduction in milk flow or agalactia. Morbidity may be prolonged; the organisms tend to disappear with clinical recovery and may not always be recovered upon culture during the clinical stages.

The mol% G + C of the DNA is 31.9 (Bd).

Type strain: ATCC 33461 (strain ST-6 Jasper et al., 1981).

13. **Mycoplasma canadense** Langford, Ruhnke and Onoviran 1976, 218.[AL]

can.a.den′se. M.L. neut. adj. *canadense* pertaining to Canada.

Coccoid to coccobacillary cells.

Growth occurs in conventional Hayflick-type medium (see under Enrichment and Isolation Procedures) modified to contain 0.024% (w/v) of thymus deoxyribonucleic acid (Sigma, No. D-1501). The temperature range for growth is 35–42°C, with optimum at 37.5–40°C.

Distinctive physiological characteristics of the species are presented in Table 10.2.

The type strain of the species shares antigens with strains of a number of other arginine-hydrolyzing species, as demonstrated by growth precipitation and counterimmune electrophoresis, but is distinct by these methods from the glucose-metabolizing *Mycoplasma* species of bovine habitat. Serologically distinct by GI and MI tests from these and all other *Mycoplasma* species tested.

Causes mastitis and possibly other diseases in cattle. Isolated from milk samples and the udders, and occasionally from arthritic joints, during outbreaks of bovine mastitis in Canada, California and England. Clinical mastitis can be produced by inoculation of cultures of the organism into the bovine mammary gland. Also found in the vagina, semen and upper respiratory tract of cattle.

The mol% G + C of the DNA is 29.0 (T_m).

Type strain: NCTC 10152 (ATCC 29418; strain 275C of Langford, Ruhnke and Onoviran, 1976).

---

* *VP* denotes that this name has been validly published in the official publication *International Journal of Systematic Bacteriology*.

14. **Mycoplasma canis** Edward 1955, 90.[AL]

ca'nis. L. n. *canis* a dog; L. gen. n. *canis* of a dog.

Filaments short (2–5 μm) with only occasional branching and early disintegration into coccal elements.

Two variants of colonies recovered from the same source have been described: "smooth" colonies with nongranular appearance and round edges and "rough" colonies with granular appearance and irregular or crenated edges; each form maintained its characteristic appearance during repeated subculturing.

Distinctive physiological characteristics of the species are presented in Table 10.2. Acid is produced also from maltose, glycogen, dextrin and starch, but not from mannose, sucrose or galactose.

Antigens are shared with other glucose-metabolizing canine *Mycoplasma* species and *M. felis*, as shown by DID and (in the case of *M. felis*) also by the CF test. The type strain is serologically distinct by GI, MI and IMF tests from a number of other *Mycoplasma* species tested, including those of canine and feline sources.

Pathogenicity is not known. Experimental inoculation by various routes with a strain isolated from the pericardium of a dog failed to produce any lesions in dogs and small rodents.

Found as a common parasitic inhabitant of the mucous membranes of the upper respiratory tract, conjunctivae and genitals of dogs. Occasionally isolated from nonhuman primates and in one instance from the throats of humans who had been in close contact with their pet dog.

The mol% G + C of the DNA is 28.4 (T_m) to 29.1 (Bd).

Type strain: ATCC 19525 (NCTC 10146; strain PG14 (C55) of Edward, 1955).

15. **Mycoplasma capricolum** Tully, Barile, Edward, Theodore and Ernø 1974, 116.[AL]

ca.pri.co'lum. L. n. *caper, capri* the male goat; L. v. *incolere* to dwell; L. part. adj. *capricolum* dwelling in a male goat.

Coccobacillary and very short filamentous cells. Intracytoplasmic structures, similar to the ρ (rho) forms found in *M. mycoides*, have been demonstrated in freshly isolated strains of the species.

Colonies on solid medium may attain a diameter of several millimeters and are thus similar in size to those of *M. mycoides* subsp. *capri* and LC strains of *M. mycoides* subsp. *mycoides*. Strong turbidity is produced in broth cultures after 24 h.

This species may grow with low amounts of cholesterol and the sterol requirements can be partially met by the alkyl-substituted precursors of cholesterol, lanosterol, 4,4-dimethylcholestanol, β-methylcholestanol, 4α-methylcholestanol and cholestanol, their efficacy as growth factors and their effect on the physical state of the mycoplasma membrane progressing in the said order. However, whereas cholesterol allows the cells to grow on media containing a wide variety of fatty acid supplements, lanosterol supports growth only with certain fatty acid combinations. Also, low levels of cholesterol unable to support the growth of *M. capricolum* have been found to produce a synergistic effect on growth when combined with lanosterol. None of the growth supporting sterol derivatives tested are metabolically modified (Odriozola et al., 1978; Dahl et al., 1980a, 1980b).

Distinctive physiological characteristics of the species are presented in Table 10.2. Acid produced from glucose and mannose, but not from esculin. Cultures grown in a conventional mycoplasma medium enriched with FeSO$_4$ to a final concentration of 4 μg/ml incorporate considerable quantities of iron into the cell membrane (Bauminger et al., 1980).

Sharing of antigens with the two subspecies of *M. mycoides* has been demonstrated by the growth precipitation test. Serologically distinct from this and all other established *Mycoplasma* and *Acholeplasma* species by GI and IMF tests.

Pathogenic for goats under natural and experimental conditions and experimentally for sheep and pigs. First isolated from the joints and spleen of goats during an outbreak in California of a severe disease characterized by sudden onset of symptoms with septic fever, diminu-

tion or cessation of lactation, painful swelling of the leg joints and conjunctivitis. Often rapidly fatal for kids. The pathology in experimentally infected goats is similar to that seen in natural cases with fibrinopurulent polyarthritis as the outstanding lesion. The organism is also found in the nasopharynx of healthy goats and sheep and has been isolated on one occasion from the uterine mucosa of a slaughtered sheep.

The mol% G + C of the DNA is 24.1 (T_m) to 25.5 (Bd).

Type strain: ATCC 27343 (NCTC 10154; strain California Kid of Tully, Barile, Edward, Theodore and Ernø, 1974).

16. **Mycoplasma caviae** Hill 1971a, 112.[AL]

ca'vi.ae. M.L. n. *cavia* the guinea pig; M.L. gen. n. *caviae* of a guinea pig.

Morphology has not been defined.

Distinctive physiological characteristics of the species are presented in Table 10.2.

Sharing of antigen(s) with *M. neurolyticum* has been demonstrated by DID. Serologically distinct from this and a number of other *Mycoplasma* species tested by MI and/or GI and IMF tests.

Pathogenicity is unknown. Experimental inoculation of guinea pigs, rabbits, rats and normal and athymic mice by numerous routes has failed to produce any lesions.

Usually present in the nasopharynx and genital tract of guinea pigs. Found in many different sites in several other animal species, including rats.

The mol% G + C content of the DNA has not been determined.

Type strain: ATCC 27108 (NCTC 10126; strain G122 of Hill 1971b; Edward and Freundt, 1973).

17. **Mycoplasma citelli** Rose, Tully and Langford 1978, 571.[AL]

ci'tel.li. M.L. n. *Citellus* a genus of ground squirrel; M.L. gen. n. *citelli* of *Citellus*.

Pleomorphic, round or coccobacillary forms, with few filaments apparent in cultures of young or old cells.

Optimum temperature for growth is 37°C; no growth at 25°C.

Distinctive physiological characteristics of the species are presented in Table 10.2. Acid is produced also from fructose and sucrose.

Serologically distinct by GI and IMF tests from previously described *Mycoplasma*, *Acholeplasma* and *Spiroplasma* species.

Pathogenicity is unknown. Isolated on two occasions from ground squirrels (*Citellus richardsonii richardsonii*).

The mol% G + C content of the DNA has not been determined.

Type strain: ATCC 29760 (strain RG-2C of Rose, Tully and Langford, 1978).

18. **Mycoplasma columbinasale** Jordan, Ernø, Cottew, Hinz and Stipkovits 1982, 114.[VP] (Avian serotype L, Yoder and Hofstad 1964.)

co.lum.bi.na.sa'le. L. n. *columbus*, a pigeon; L. gen. n. *columbi*, of a pigeon; L. neut. adj. *nasale*, pertaining to the nose; M.L. neut. adj. *columbinasale*, pertaining to the nose of a pigeon.

Description of the species is based exclusively on the type strain.

Coccoid to coccobacillary cells.

Distinctive physiological characteristics of the species are presented in Table 10.2.

Antigens are shared with *M. gallinarum*, *M. pullorum*, *M. gallinaceum*, *M. iners*, *M. gallopavonis*, *M. meleagridis* and *M. iowae*, and with several *Mycoplasma* species of mammalian source, as demonstrated by DID tests. Serologically distinct, however, by GI, MI and IMF from all avian, and by GI from all nonavian *Mycoplasma* species.

The mol% G +C of the DNA is 32 (Bd).

Type strain: ATCC 33549 (NCTC 10184; strain 694 of Yoder and Hofstad, 1964; see Jordan et al., 1982).

19. **Mycoplasma columbinum** Shimizu, Ernø and Nagatomo 1978, 545.[AL]

co.lum.bi'num. L. neut. adj. *columbinum* pertaining to a pigeon.

Mainly coccoid cells.

Distinctive physiological characteristics of the species are presented in Table 10.2.

Serologically distinct from previously established *Mycoplasma* species and serogroups as determined by GI and IMF, supplemented in some cases with the IMF test.

Pathogenicity is not known.

Found with a high incidence in the trachea and oropharynx of pigeons (*Columba livia* var. *domestica*).

The mol% G + C content of the DNA has not been determined.

Type strain: ATCC 29257 (NCTC 10178; strain MMP1 of Shimizu et al., 1978).

20. **Mycoplasma columborale** Shimizu, Ernø and Nagatomo 1978, 545.[AL]

co.lumb.o.ra′le. L. n. *columba* pigeon; L. n. *os*, *oris* the mouth, M.L. neut. adj. *orale* of the mouth; M.L. neut. adj. *columborale* of the pigeon mouth.

Mainly coccoid cells.

Distinctive physiological characteristics of the species are presented in Table 10.2. Acid is produced also from maltose, glycogen, starch and (by some strains) sucrose, but not from mannose, lactose, mannitol, sorbitol, salicin, arabinose or xylose.

Pathogenicity is not known.

Found in the trachea and oropharynx of pigeons (*Columba livia* var. *domestica*).

The mol% G + C content of the DNA has not been determined.

Type strain: ATCC 29258 (NCTC 10179; strain MMP4 of Shimizu et al., 1978).

21. **Mycoplasma conjunctivae** Barile, Del Giudice and Tully 1972, 74.[AL]

con.junc.tiv′ae. M.L. n. *conjunctiva* the membrane joining the eyeball to the lids; M.L. gen. n. *conjunctivae* of conjunctiva.

Coccoid to coccobacillary; clusters of 2–10 spherical cells joined together by short filaments may be seen by phase-contrast microscopy of broth cultures.

Colonies grown on solid medium may have elevated centers and a greenish, brownish or olive color. Growth on medium containing pig serum may be surrounded by a zone of milky precipitate within the agar that is apparently different from the film and spots formed on horse serum agar medium.

Distinctive physiological characteristics of the species are presented in Table 10.2.

The type strain is serologically distinct by GI, MI and IMF tests from other *Mycoplasma* species of ovine-caprine source and from a number of other *Mycoplasma* species tested.

Pathogenic to sheep and goats causing conjunctivitis and keratoconjunctivitis ("pink eye"). Associated with outbreaks of the disease in United States, Canada and Australia. Also isolated from inflamed eyes of chamois in Switzerland. Mild conjunctivitis to severe keratoconjunctivitis can be induced experimentally by inoculation of cultures subconjunctivally or by instillation into the conjunctival cul-de-sac of goats and sheep. Normal habitat has not been determined.

The mol% G + C content of the DNA has not been determined.

Type strain: ATCC 25834 (NCTC 10147; strain HRC581 of Barile, Del Giudice and Tully, 1972).

Further comments

The strain (67R) which Surman (1968) named *M. conjunctivae* var. *ovis* but did not describe has been identified as *M. arginini* (Barile et al., 1972).

22. **Mycoplasma cynos** Rosendal 1973, 53.[AL]

cy′nos. Gr. n. *cyon* a dog; Gr. gen. n. *cynos* of a dog.

Coccoid to coccobacillary cells.

Distinctive physiological characteristics of the species are presented in Table 10.2.

Antigens are shared with *M. pullorum*, *M. gallinaceum*, *M. gallopavonis*, *M. columbinasale* and *M. synoviae* as demonstrated by DID. Serologically distinct by GI, MI and IMF from other previously named canine and glucose-metabolizing noncanine *Mycoplasma* species.

Apparently pathogenic to dogs. The type strain (H831) was isolated from the lungs of a dog with pneumonia; during an outbreak of distemper four additional strains have later been recovered, although in association with one or two other canine *Mycoplasma* species, from pneumonic lungs. Purulent bronchitis and bronchiolitis with severe destruction of epithelial cilia and lung lesions, characterized, inter alia, by peribronchial and perivascular lymphocytic-plasmacytic hyperplasia, were produced by endobronchial inoculation of puppies with cloned cultures of *M. cynos*, but not with the other *Mycoplasma* species isolated.

A relatively infrequent inhabitant of the upper respiratory tract, conjunctivae and genital tracts of male and female dogs.

The mol% G + C content of the DNA has not been determined.

Type strain: ATCC 27544 (NCTC 10142; strain H831 of Rosendal, 1973).

23. **Mycoplasma dispar** Gourlay and Leach 1970, 121.[AL]

dis′par. L. adj. *dispar* dissimilar, different.

Coccoid to very short filamentous cells. An extracellular capsule can be visualized by electron microscopy following staining with ruthenium red.

Colonies on solid medium have a granular, lacy or reticulated appearance with no or a poorly defined central area on primary isolation and in early subcultures; after repeated subculturing they achieve the typical "fried egg" appearance.

Obviously more exacting than most other species. Growth is obtained in GS* and FF† broth and agar media containing ampicillin (0.05 mg/ml) in place of benzylpenicillin. Good growth is also provided by the SP-4 medium (see under Enrichment and Isolation Procedures).

Distinctive physiological characteristics of the species are presented in Table 10.2. Acid produced from glucose and mannose, but apparently not from galactose, sorbitol, cellobiose or xylose.

Forms a serologically homogeneous group as shown by GI and IHA tests. A common antigen in *M. dispar* and *M. bovoculi* has been demonstrated by DID. Moreover, comparison of fluorocarbon-extracted antigens by DID and counterimmunodiffusion techniques has revealed a close reciprocal relationship among *M. dispar*, *M. hyopneumoniae*

---

* GS broth (adapted from Gourlay and Leach, 1970): combine Hanks' buffered salt solution (Burroughs Wellcome), 40 ml; lactalbumin hydrolysate solution (Nutritional Biochemicals Corp., Cleveland, Ohio, 5%, w/v), 10 ml; and Hartley digest broth, 20 ml. Sterilize at 121°C for 20 min and aseptically add the following: fetal calf serum (heated at 56°C for 30 min), 20 ml; glucose solution (50%, w/v), 2 ml; calf thymus DNA solution (Sigma, Type 1, Cat. No. 1501; 0.2%, w/v), 1 ml; thallium acetate solution (5%, w/v), 0.5 ml; phenol red solution (1%, w/v), 0.2 ml; and sufficient ampicillin to give a final concentration of 0.05 mg/ml. Adjust the pH to 7.8. For a solid medium add 0.65% agarose.

† FF broth (Friis, 1971; modified by Rose, Tully and Wittler, 1979): combine Hanks' balanced salt solution (10X; Flow Laboratories, Rockville, Maryland), 30 ml; deionized water, 720 ml; brain-heart infusion (Difco), 5 g; mycoplasma broth base (BBL), 7.5 g; lactalbumin hydrolysate (Nutritional Biochemicals), 1.25 g; yeast extract (Difco), 0.5 g; and phenol red solution (0.1%), 13.7 ml. Adjust the pH to 7.8 and sterilize at 121°C for 20 min. Add aseptically the following sterile supplements: fresh yeast extract solution (25%; Microbiological Associates), 36.5 ml; glucose solution (50%), 2.5 ml; thallium acetate solution (1:50), 5.5 ml; horse serum (Flow Laboratories), 100 ml; and porcine serum (Gibco; heat-inactivated at 56°C for 30 min), 100 ml. For a solid medium, add 0.6% purified agar (Oxoid).

It is recommended to omit, as far as possible, penicillins from any of the media. If the addition of penicillin is needed, e.g. for the purpose of primary isolation from contaminated material, ampicillin (0.05 mg/ml) is preferable to benzylpenicillin.

and *M. ovipneumoniae*, and a lesser degree of crossreaction between *M. dispar*, *M. bovoculi* and *M. hyorhinis* (Ball and Todd, 1978). Otherwise serologically distinct from these species and distinct also by GI, MI, growth precipitation and DID from all other *Mycoplasma* and *Acholeplasma* species of bovine origin, and from a number of other *Mycoplasma* species tested.

Inhibited by benzylpenicillin (200 IU/ml), but apparently to a lesser degree by ampicillin (not being inhibited by 0.05 mg/ml).

Associated with pneumonia of calves in Great Britain, Denmark, the United States, Australia and Japan. Subclinical pneumonia characterized by interstitial alveolitis around the bronchioli has been produced by endobronchial inoculation of the organism in conventionally reared and gnotobiotic calves. Six out of seven strains tested induced clinical mastitis on inoculation into the bovine mammary gland. Inoculation of bovine fetal tracheal organ cultures resulted in greatly reduced or abolished ciliary activity within 24–48 h, together with progressive sloughing of epithelial cells.

Found as the most common microbial species in the respiratory tract of calves in at least some geographical areas.

The mol% G + C of the DNA is 28.5–29.3 (T_m).

Type strain: ATCC 27140; NCTC 10125 (strain 462/2 of Gourlay and Leach, 1970).

24. Mycoplasma edwardii Tully, Barile, Del Giudice, Carski, Armstrong and Razin 1970, 349.[AL]

ed.ward'i.i. M.L.. gen. n. *edwardii* of Edward; named after D.G. ff. Edward, who first isolated this organism.

Coccobacillary and short filamentous cells.

Distinctive physiological characteristics of the species are presented in Table 10.2. Acid is produced also from maltose, glycogen, dextrin and starch, but not from mannose, sucrose or galactose.

Sharing of antigens has been demonstrated by DID with the glucose fermenting *M. gallinaceum*, *M. gallopavonis* and *M. pullorum*. The type strain is serologically distinct by IMF and/or GI tests from previously established species of *Mycoplasma* and *Acholeplasma*, including those of canine habitat.

Pathogenicity not known although occasionally isolated from pneumonic lungs of dogs.

Frequently found in the upper respiratory and urogenital tracts of male and female dogs. Has also been isolated from shrews.

The mol% G + C of the DNA is 29.2 (T_m).

Type strain: ATCC 23462 (NCTC 10132; strain PG24 (C21) of Edward and Fitzgerald, 1951).

25. Mycoplasma equigenitalium Kirchhoff 1978, 500.[AL]

e.qui.ge.ni.ta'li.um. L. n. *equus, equi* the horse; M.L. pl. n. *genitalia* the genitals; M.L. pl. gen. n. *equigenitalium* of equine genitalia.

Pleomorphic cells with coccoid elements dominating.

Growth at 37°C, but not at 22°C.

Distinctive physiological characteristics of the species are presented in Table 10.2. Production of acid is reported also from fructose, lactose, maltose, sucrose, galactose, arabinose, sorbitol, rhamnose, xylose, trehalose, dulcitol, inositol and salicin, although the results observed with some substrates were not always reproducible.

The type strain and several other strains tested formed a serologically homogeneous group by GI, MI and IMF tests. Serologically distinct from previously established *Mycoplasma* species, as demonstrated by GI, MI and IMF tests in the case of glucose-metabolizing species and by IMF (sometimes supplemented with the MI method) in the case of the remaining species.

Pathogenicity is not known.

Frequently found in the urogenital tract and semen, and occasionally the upper respiratory tract of horses; on one occasion isolated from an aborted foal.

The mol% G + C content of the DNA has not been determined.

Type strain: ATCC 29869 (NCTC 10176; strain T37 of Kirchhoff, 1978).

26. Mycoplasma equirhinis Allam and Lemcke 1975, 405.[AL]

e.qui.rhi'nis. L. n. *equus, equi* a horse; Gr. n. *rhis, rhinis* nose; M.L. gen. n. *equirhinis* of the nose of a horse.

Coccoid and coccobacillary cells.

Distinctive physiological characteristics of the species are presented in Table 10.2.

Sharing of antigen with *M. columbinasale* has been demonstrated by DID. A low order cross-reaction has been demonstrated by MI and CF tests between strain M432/72 and *M. hyosynoviae*, but the two organisms are distinct by GI and in the electrophoretic patterns of their cell proteins. The type strain is serologically distinct by GI and MI from all other arginine-hydrolyzing *Mycoplasma* species tested.

Pathogenicity is not known, although several strains have been isolated from horses with respiratory disease. Horses are readily colonized with the organism following experimental inoculation into the respiratory tract, but do not develop clinical disease.

Found as a common parasitic inhabitant of the equine upper respiratory tract. Isolated once from the nasopharynx of a cow.

The mol% G + C content of the DNA has not been determined.

Type strain: ATCC 29420 (NCTC 10148; strain M432/72 of Allam and Lemcke, 1975).

27. Mycoplasma fastidiosum Lemcke and Poland 1980, 161.[VP]

fas.tid.i.o'sum. Lat. adj. *fastidiosus* fastidious, referring to the nutritionally fastidious nature of the organism on primary isolation.

Cells are highly filamentous. Filaments are twisted at intervals along their length, as demonstrated by electron microscopy of negatively stained specimens and thin sections prepared from broth cultures, but regular helical forms like those of *Spiroplasma* species are not produced. In thin sections, an electron-dense, structureless layer is visible outside the cell membrane, together with diagonal striations that may be due to some fibrillar substance wound helically around the cell (Fig. 10.7). Motility has not been detected.

On primary isolation, colonies develop more quickly and are slightly larger when incubated in 5% CO_2 in N_2 rather than in air. They fail to develop in a strictly anaerobic atmosphere of 5% CO_2 in H_2. Optimum growth conditions are provided by HuS medium* and SP-4 broth (see under Enrichment and Isolation Procedures). In addition to cholesterol, an as yet unidentified nutritional factor supplied by animal serum or PPLO serum fraction (Difco) is required for growth.

Distinctive physiological characteristics of the species are presented in Table 10.2

Serologically distinct by GI and MI tests from other established *Mycoplasma* species and from *Spiroplasma citri*.

Tests in broth indicate that all strains examined are susceptible to erythromycin at concentrations as low as 0.1 μg/ml.

Pathogenicity is unknown. Found with a low isolation rate in the nasopharynx of horses.

The mol% G + C content of the DNA has not been determined.

Type strain: NCTC 10180 (strain 4822 of Lemcke and Poland, 1980).

28. Mycoplasma faucium Freundt, Taylor-Robinson, Purcell, Chanock and Black 1974, 252.[AL]

fau.ci'um. L. n. *fauces* the throat; L. gen. pl. n. *faucium* of throats.

Coccoid, coccobacillary and very short filamentous cells.

Colonies on solid medium develop somewhat more superficially and are more loosely attached to the agar surface than the colonies of most other mycoplasmas.

Very fastidious. Fresh yeast extract is required for growth; DNA and

---

* HuS broth (Lemcke and Poland, 1980): aseptically combine PPLO broth (Difco), 70 ml; aqueous extract of dried yeast, (25%) 10 ml; and heat-inactivated human serum (sterilized by filtration), 20 ml. Add sodium deoxyribonucleate, ampicillin, thallium acetate, and phenol red to give final concentrations of 0.002%, 300 μg/ml, 0.0005%, and 0.002%, respectively. Adjust the pH to 7.8. For a solid medium add 1% Noble agar (Difco).

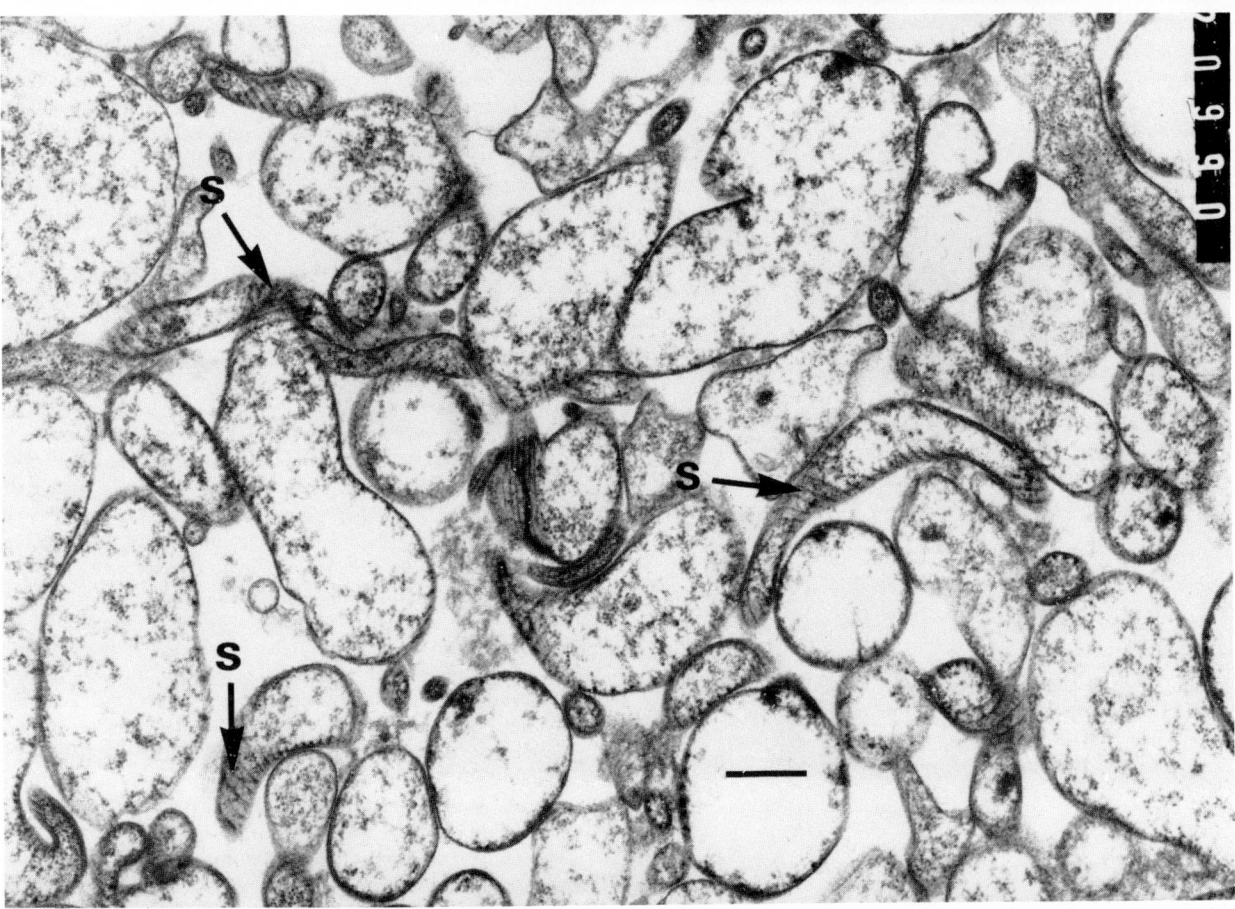

Figure 10.7. Electron micrograph of a 3-day-old broth culture of *Mycoplasma fastidiosum*; thin-section preparation stained with ruthenium red. Polymorphism, twisting of filaments, and diagonal striations (*S*) of organisms are visible. *Bar* represents 400 nm. (Reproduced with permission from R. M. Lemcke and J. Poland, *International Journal of Systematic Bacteriology 30*: 151–162, 1980, © American Society for Microbiology.)

L-cysteine stimulate growth significantly. These growth requirements are satisfied by the BACY broth and agar media.*

Distinctive physiological characteristics of the species are presented in Table 10.2.

Sharing of antigens by this species and other arginine-metabolizing *Mycoplasma* species, including *M. orale* and *M. buccale*, is demonstrable by DID. Serologically distinct by GI, MI, IHA and CF tests from these and other *Mycoplasma* species of human source, and by two or more of these methods from a number of other *Mycoplasma* species tested.

Pathogenicity is not known.

Apparently a rare member of the normal flora of the human oropharynx, comprising less than 2% of the total number of mycoplasmas recovered. Commonly found in the oropharynx of several species of nonhuman primates.

The mol% G + C content of the DNA has not been determined.

Type strain: ATCC 25293 (NCTC 10174; strain DC-333 of Fox, Purcell and Chanock, 1969).

29. **Mycoplasma feliminutum** Heyward, Sabry and Dowdle 1969, 621.[AL]

fe.li.mi.nu'tum. L. n. *felis* a cat; L. neut. part. adj. *minutum* small; M.L. neut. adj. *feliminutum* apparently intended to designate a small colony organism isolated from cats.

Morphology is poorly defined.

Colonies are relatively small, with an average diameter of 140 μm on conventional Hayflick-type medium, and develop slowly compared to other mycoplasmas of feline origin. They are irregular in shape and lack a well-defined central spot on primary isolation. The typical "fried egg" appearance develops on repeated subculturing.

Grows well in SP-4 medium (see under Enrichment and Isolation Procedures).

Distinctive physiological characteristics of the species are presented in Table 10.2. Reports regarding carbohydrate catabolism are conflicting, the observation of production of acid from glucose and mannose (J.G. Tully and M.F. Barile, personal communication, 1981) remaining as yet unconfirmed by others.

Serologically distinct by GI and/or IMF tests from other *Mycoplasma* species of feline and canine origin and from a number of other *Mycoplasma* and *Acholeplasma* species tested.

Pathogenicity is not known.

Only very few isolates of the species are known: recovered once from the oropharynx of a cat, on a few occasions from the upper respiratory tract, the lungs and the genital tract of dogs, and once from the respiratory tract of a horse.

The mol% G + C content of the DNA has not been determined.

Type strain: ATCC 25749 (NCTC 10159; strain Ben of Heyward, Sabry and Dowdle 1969).

---

* BACY broth: dissolve 2.25 g of heart infusion broth (Difco) in 90 ml of distilled water and sterilize at 121°C for 20 min. Add aseptically horse serum (unheated, sterilized by filtration), 20 ml; yeast extract solution (25%; prepared according to Taylor-Robinson et al., 1963), 10 ml; deoxyribonucleic acid solution (Sigma, 0.2%, w/v), 1.3 ml; L-cysteine-HCl solution (10%, w/v), 1.24 ml; L-arginine solution (30%, w/v), 1.0 ml; phenol red solution (0.06%, w/v), 5.0 ml; thallium acetate solution (1.0%, w/v), 1.0 ml; and benzylpenicillin solution (20,000 IU/ml), 0.25 ml. Adjust the pH to 7.3. For a solid medium add ~0.8% Noble agar (Difco) or other washed or purified agar.

30. **Mycoplasma felis** Cole, Golightly and Ward 1967, 1456.[AL]
(*Mycoplasma equipharyngis* Kirchhoff 1974, 208.)

fe'lis. L. n. *felis* a cat, L. gen.n. *felis* of a cat.

Coccobacillary to short filamentous cells.

Distinctive physiological characteristics of the species are presented in Table 10.2.

Antigens are shared with *M. canis* and *M. gallopavonis*, as shown by DID, but the type strain is serologically distinct from these and a number of other *Mycoplasma* species by GI and/or IMF tests.

Associated with conjunctivitis in cats, being recovered significantly more frequently from the conjunctivae of diseased cats than in convalescent or healthy animals. Conjunctivitis can be induced experimentally by conjunctival or intranasal instillation of cultures of *M. felis*. However, the available evidence that the organism may be pathogenic in its own right is considered inconclusive.

A relatively common parasitic inhabitant of the upper respiratory and lower genital tracts of cats. Strains related to *M. felis* have frequently been isolated from the tonsils and several other regions of the respiratory tract of healthy horses, as well as from horses with respiratory disease.

The mol% G + C of the DNA is 25.2 (T_m).

Type strain: ATCC 23391 (NCTC 10160; strain CO of Cole, Golightly and Ward, 1967; Edward and Freundt, 1973).

Further Comments

Mycoplasma equipharyngis Kirchhoff 1974 is a later synonym of *M. felis*. Allam and Lemcke (1975) consider that the physiological and serological similarity of a representative strain of *M. equipharyngis* and four British strains to *M. felis* prohibits the recognition of the equine strains as a new species.

31. **Mycoplasma fermentans** Edward 1955, 90.[AL]

fer.men'tans. L. part.adj. *fermentans* fermenting.

Short (1.5 μm) to moderately long (10–40 μm) filaments.

While on primary isolation and in early subcultures growth is definitely enhanced by anaerobic conditions (5% CO_2 in N_2), it is equally good aerobically on repeated subculturing.

Distinctive physiological characteristics of the species are presented in Table 10.2. Acid produced also from fructose, maltose, dextrin, starch and glycogen, but not from mannose or galactose.

Sharing of antigen(s), as demonstrated by DID, occurs with a number of glycolytic or arginine-metabolizing species, including *M. hominis*, *M. salivarium*, *M. orale*, *M. primatum*, *M. arthritidis*, *M. canis* and *M. felis*. Otherwise serologically distinct.

Pathogenicity to humans is uncertain. The first isolations were made from patients with balanitis and vulvovaginitis. Isolated on one occasion in pure culture (strain S38) from the uterine tube of a patient with subacute salpingitis. Inoculation of this and another strain directly into the uterine tubes of grivet monkeys (*Cercopithecus aethiops*), or into the uterine cavity followed by curettage of the endometrium, produced a self-limiting acute salpingitis and parametritis (Møller et al., 1980). The location and characteristics of the inflammatory lesions were very similar to those seen in acute pelvic disease induced experimentally in grivet monkeys with *M. hominis*. The very frequent association of *M. fermentans* with rheumatoid arthritis reported by Williams et al. (1970) remains largely unconfirmed, although the organism has been isolated also by others from the synovia of patients with rheumatoid and nonrheumatoid arthritis. Intraperitoneal inoculation of large quantities of *M. fermentans* cells or washed membranes is lethal to mice and

produces symptoms resembling the endotoxic effect of Gram-negative bacteria. A cytotoxic effect is exerted on cultured mouse thymocytes.

A rare inhabitant of the urogenital tract of humans, reports on its isolation ranging from less than 1–5% of the genital mycoplasmal flora. Recovered only occasionally from simian tissues.

The mol% G + C of the DNA is 27.5 (Bd) to 28.7 (T_m). Genome size is 4.8×10^8 daltons.

Type strain: ATCC 19989 (NCTC 10117; strain PG18 (G), Edward 1955).

32. **Mycoplasma flocculare** Meyling and Friis 1972, 289.[AL]

floc'cu.la.re. M.L. dim. n. *flocculus* a small floc or tuft of wool; apparently treated as the neuter *flocculare* of an adj. *floccularis* like a small floc of wool.

A comprehensive characterization of this species has been presented by Rose et al. (1979).

Cells are coccoid or coccobacillary to short filamentous.

Colonies are very small, up to 0.5–1.0 mm in diameter after 4–7 days incubation on solid media. They are slightly convex, devoid of a central nipple and with a coarsely granular surface. A "fried egg" appearance is unusual. Colonies are not readily produced on solid media on primary isolation. Aggregates of cells may be produced during growth in broth, appearing as small floccular elements upon gentle shaking of the culture.

Facultative anaerobe, although growth is best under aerobic conditions with supplements of 5% CO_2. Very fastidious; growth occurs in A26* and FF (see Footnote † on p. 755) broth and agar media at 25–37°C, but not at 45°C.

Distinctive physiological characteristics of the species are presented in Table 10.2. Reports regarding glucose fermentation are conflicting, the available evidence for and against utilization of this carbohydrate being essentially as described for *M. hyopneumoniae*.

The polyacrylamide gel electrophoresis (PAGE) pattern of the cell proteins of the type strain of the species shows some similarities to that of the type strain of *M. hyopneumoniae*, but there are some distinctive bands. In the growth precipitation test, only weak cross-reactions have been observed with *M. hyopneumoniae* (Friis, 1977). Serologically distinct by GI and IMF tests from this and other established *Mycoplasma* species.

Observations on the inhibitory activity of benzylpenicillin and ampicillin are in accord with those described for *M. hyopneumoniae*.

Experimental infection of piglets by exposure to aerosols results in the development of small pneumonic lesions with peribronchial and perivascular accumulations of mononuclear cells, slight mononuclear and polymorphonuclear cell accumulations and damage to the surface epithelium of the nasal cavity. The brain, pleural and pericardial cavities are infected on rare occasions. The organism has been isolated from the nasal cavity and pneumonic lungs of swine in Denmark and United Kingdom. Also found in the conjunctivae of swine.

The mol% G + C content of the DNA has not been determined.

Type strain: ATCC 27399 (NCTC 10143; strain Ms 42 of Meyling and Friis, 1972; Edward and Freundt, 1973).

33. **Mycoplasma gallinaceum** Jordan, Ernø, Cottew, Hinz and Stipkovits 1982, 114.[VP] (Avian serotype D, Kleckner 1960.)

gal.li.na'ce.um. L. n. *gallus* a domestic fowl; L. neut. adj. *gallinaceum*, pertaining to a domestic fowl.

Description of the species is based exclusively on the type strain.

Coccoid to coccobacillary cells.

Distinctive physiological characteristics of the species are presented in Table 10.2.

---

* A26 broth medium (Rose et al., 1979): combine Hanks' balanced salt solution [10X, without $NaHCO_3$ or phenol red (Gibco)], 4 ml; Hartley digest broth (Oxoid), 30 ml; lactalbumin hydrolysate solution (Nutritional Biochemicals), 5% (w/v) in Dulbecco phosphate-buffered salt solution (Gibco), 10 ml; fresh yeast extract solution (Microbiological Associates), 2 ml; porcine serum (Gibco; acid-treated by the method of Switzer, 1972), 20 ml; phenol red solution (0.25% aqueous), 1 ml; and deionized water, 36 ml. Adjust the pH to 7.4 and sterilize by filtration. For a solid medium, add 0.7% purified agar (Oxoid) to the deionized water component, sterilize the agar base at 121°C for 15 min, then add the filter-sterilized broth components to the melted agar.

Antigens are shared with *M. gallopavonis, M. meleagridis, M. iowae, M. columbinasale, M. synoviae, M. anatis, M. cynos, M. felis* and *M. edwardii,* as demonstrated by DID tests. Serologically distinct by GI, MI and IMF tests from all avian, and by GI from all other *Mycoplasma* species.

The type strain was isolated from the trachea of a chicken. Little is known of the pathogenicity.

The mol% G + C of the DNA is 28 (Bd).

Type strain: ATCC 33550 (NCTC 10183; strain DD of J. Fabricant; see Jordan, 1979, and Jordan et al., 1982).

34. Mycoplasma gallinarum Freundt 1955, 73.[AL]

gal.li.na′rum. L. n. *gallina* a hen; L. gen. pl. n. *gallinarum* of hens.

Coccoid to coccobacillary cells.

Distinctive physiological characteristics of the species are presented in Table 10.2.

Sharing of antigens with *M. iners, M. columbinasale* and *M. meleagridis* has been demonstrated by DID. Serologically distinct by GI, MI and IMF from these and a number of other *Mycoplasma* species tested.

Susceptible to digitonin (0.3 µg), but unlike most other sterol-dependent mycoplasmas resistant to sodium polyanetholesulfonate (1.0 µg), as determined by a disk growth inhibition method on solid medium.

Although generally accepted as apathogenic for chickens and turkeys, recent experimental observations suggest that some strains may be mildly pathogenic, producing low mortality in chick embryos and a cytopathogenic effect in cell and tracheal organ cultures. A synergistic effect with certain viruses has been reported.

A relatively common parasitic inhabitant of the respiratory tract of chickens and turkeys of all ages. The natural host range of *M. gallinarum* among domesticated and wild birds appears to be fairly wide, and the widest of all avian *Mycoplasma* species (Shimizu et al., 1979).

The mol% G + C of the DNA is 26.5–28.0 (T_m and Bd). DNA/DNA homology between these species and *M. gallisepticum* and *M. fermentans* is ≦ 10% (McGee et al., 1967).

Type strain: ATCC 19708 (NCTC 10120; strain PG16 (Fowl) of Edward, 1955).

35. Mycoplasma gallisepticum Edward and Kanarek 1960, 699.[AL]

gal.li.sep′ti.cum. L. n. *gallina* a hen; L. adj. *septicus,* from Gr. adj *septicos* putrefactive, septic; M.L. adj. *gallisepticum* hen-poisoning (infecting).

Cells are coccoid, ovoid and elongated pear shaped. They possess a highly structured polar body protruding from one or both ends and known as the "bleb." The bleb is hemispherical, about 800 × 1250 Å. In thin sections, two dense areas that are intensely stained with uranyl salts can be seen. The larger of these zones is apparently elliptical and located just beneath the plasma membrane, parallel to the convex surface of the bleb. It is connected with a lower circular flat plate by a series of fine, densely stained threads. Between the bleb and the rest of the cell is the infrableb region which is about 2000 Å in diameter. The infrableb region and the periphery of the bleb are rich in basic proteins and are the sites of enzymatic activities. The bleb also seems to be the site of attachment to cells of the animal host. Cells of freshly isolated organisms exhibit very slow gliding motility.

Distinctive physiological characteristics of the species are presented in Table 10.2. Acid is produced also from maltose, dextrin, glycogen, starch and (by some strains) from fructose, but not from galactose, lactose, sucrose, mannitol, dulcitol, sorbitol, salicin, cellobiose or xylose. Most strains agglutinate red blood cells (RBC) from a variety of animal species including chicken, turkey, guinea pig and man. Agar colonies adsorb RBC and tracheal epithelial cells from monkey, rat, guinea pig and chicken, as well as spermatozoa from man and bull, and HeLa cells. Adsorption is specifically inhibited by antiserum. Sialic acid moieties have been found to be responsible for most, if not all, of the attachment of *M. gallisepticum* to human erythrocytes, indicating that the specific receptor on the RBC membrane is glycophorin, a glycoprotein which carries over 90% of the sialic acid residues of the erythrocytes. However, studies of the influence of ionic strength and pH on

attachment and the failure of neuraminidase to detach the mycoplasmas bound to erythrocytes suggest that additional bonds (probably of the London-van der Waals type) are formed between the mycoplasmas and the host cell membrane after the initial specific attachment has occurred (Banai et al., 1978). The claim by Sethi and Müller (1972) that *M. gallisepticum* possesses neuraminidase activity has not been confirmed by others (Glasgow and Hill, 1980; M. Banai, I. Kahane and S. Razin, unpublished results).

Antigens are shared with *M. iners, M. pullorum* and *M. synoviae,* as demonstrated by DID. Serologically distinct from these and other avian, as well as a number of mammalian *Mycoplasma* species tested, by at least GI, MI and IMF tests.

Unlike most other mycoplasmas usually highly sensitive to erythromycin; the minimum inhibitory concentration ranges from 0.02–2.0 µg/ml. Cell division is inhibited by cytochalasin B, a fungal metabolite known otherwise to have an inhibiting effect on a variety of eucaryotic, but not procaryotic, cell processes (Ghosh et al., 1978).

A common pathogen of poultry causing tracheitis and air sacculitis in chickens and air sacculitis and sinusitis in turkeys. Mild conjunctivitis and less frequently arthritis and synovitis, particularly of the hocks, may occur in both chickens and turkeys. The arthritis is characterized by, *inter alia,* hypertrophy of the synovial membrane with round cell infiltration, excess joint fluid, and edematous swelling of the periarticular tissues. Lesions of the brain associated with polyarteritis of the cerebral and meningeal arteries and arterioles, together with occlusion of cerebral capillaries due to acute swelling of the endothelial cells (Manuelidis and Thomas, 1973), have been reported in turkeys. The neurological disease and characteristic pathology can be reproduced experimentally by intravenous inoculation into turkeys of high doses of washed organisms of the S6 strain, but not by cell-free filtrates of the culture medium. The pathogenesis of the vascular lesions is uncertain. Infection of the oviduct, usually without development of disease of the reproductive organs, is a frequent cause of egg transmission of the mycoplasma. Inoculation of chick embryo tracheal organ cultures results in abolition of cilial activity and degeneration of ciliated epithelium. Associated also under natural conditions with respiratory disease of other gallinaceous birds, including quails and peafowls. Air sacculitis, but no clinical disease, has been produced experimentally in day-old ducklings. Recovered occasionally from mammalian tissues.

Reports on the mol% G + C of the DNA range from 31.8 (T_m) to 35.7 (Bd). Genome size is 4.9 × 10⁸ daltons. DNA/DNA homology between this species and *M. arthritidis, M. fermentans, M. gallinarum* and *M. pneumoniae* is less than 10%; no genetic relatedness is found with *Acholeplasma laidlawii* (McGee et al., 1967).

Type strain: ATCC 19610 (NCTC 10115; strain PG31 (X95) of Edward and Kanarek, 1960; Edward and Freundt, 1973).

36. Mycoplasma gallopavonis Jordan, Ernø, Cottew, Hinz and Stipkovits 1982, 114.[VP] (Avian serotype F, Kleckner 1960.)

gal.lo.pa.vo′nis. L. n. *gallopavo,* a turkey; L. gen. n. *gallopavonis,* of a turkey.

Description of this species is based exclusively on the type strain.

Coccoid to coccobacillary cells.

Distinctive physiological characteristics of the species are presented in Table 10.2. The type strain has been reported to agglutinate erythrocytes from a variety of animal species to high titers, although colonies of this strain do not adsorb any red blood cells tested.

Antigens are shared with *M. anatis, M. cynos, M. felis, M. edwardii* and *M. pneumoniae* as shown by DID. Serologically distinct from all previously established *Mycoplasma* species by the GI test, and from other avian *Mycoplasma* species also by MI and IMF.

The type strain was isolated from air sac lesions in adult turkeys. The organism causes air sacculitis in turkeys, but not chickens, following inoculation into the air sac, but its role as a primary pathogen is doubtful.

The mol% G + C of the DNA is 27 (Bd).

Type strain: ATCC 33551 (NCTC 10186; strain WR1 of Roberts, 1963; see Jordan et al., 1982).

37. **Mycoplasma gateae** Cole, Golightly and Ward 1967, 1456.^{AL}

ga'te.ae. Probably from Spanish *gato*, the cat.

Morphology is poorly defined.

Colonies are vacuolated and lack a well-defined central spot on primary isolation and in early subcultures. On repeated subculturing they achieve the "fried egg" appearance that is typical of mycoplasmas.

Distinctive physiological properties of the species are presented in Table 10.2.

Common antigens have been demonstrated by DID in this and certain other arginine-metabolizing *Mycoplasma* species. Serologically distinct from these and a number of other *Mycoplasma* species tested by, *inter alia*, GI and IMF.

Sensitive to digitonin (0.3 μg) but, unlike most other sterol-dependent mycoplasmas, resistant to sodium polyanetholesulfonate (1.0 μg), as determined by a disk growth inhibition test on solid medium.

Pathogenicity is not known.

Found as a common inhabitant of the upper respiratory tract, the conjunctivae and the genital mucosae of clinically healthy and sick cats. Occasionally isolated from the canine respiratory tract and vagina, and more frequently from the urogenital tract of male dogs. A few isolations are reported from the bovine male and female genital tracts.

The mol% G + C of the DNA is 28.5 (T_m).

Type strain: ATCC 23392 (NCTC 10161; strain CS of Cole, Golightly and Ward, 1967; Edward and Freundt, 1973).

Further Comments

M. gateae has been divided, on the basis of plate immunofluorescence studies, into two serogroups (Rosendal, 1974b). However, *M. gateae* serogroup 2 was later reclassified by GI and IMF tests as *Mycoplasma arginini* (Rosendal, 1979).

38. **Mycoplasma hominis** (Freundt 1953) Edward 1955, 90.^{AL}

ho'mi.nis. L. n. *homo* man; L. gen. n. *hominis* of man.

Coccoid, coccobacillary and filamentous cells; filaments are usually very short but occasionally attain a length of up to 30 μm (Fig. 10.2). Living cells are able to change their shape rapidly and reversibly, suggesting the presence of contractile structures.

Distinctive physiological characteristics of the species are presented in Table 10.2. Colonies on solid medium adsorb HeLa and chick embryo tissue culture cells, but not erythrocytes or tracheal epithelial cells from monkey, rat, guinea pig and chicken and not spermatozoa from man or bull. Adsorption can be prevented by specific antisera. The receptors on HeLa and chick cells are not destroyed by neuraminidase; presently available evidence suggests that protein receptors in the membrane of *M. hominis* may be involved in the adsorption mechanism (Kihara et al., 1981).

Three membrane antigens have been partially characterized: two of them were identified as proteins on the basis of their lability and susceptibility to proteolytic enzymes. A third component was stable to 100°C and relatively stable to both pronase and trypsin (Hollingdale and Lemcke, 1972; see also Razin et al., 1972). The proteinaceous nature of the predominant membrane antigens received confirmation by Lin (1980). The results of agglutination, GI and MI tests, together with comparison of the PAGE patterns of cell proteins of a number of strains classified within this species, suggest the existence of a fairly extensive antigenic heterogeneity of the species. Extension and further support of these observations were obtained by Lin and Kass (1974), and Lin et al. (1975) by means of a complement-dependent mycoplasmacidal test and an agglutination test carried out during growth of the organisms, combined with conventional growth inhibition. Lin (1980) was able by these methods to detect 7–14 different strain-specific antigens, but no common surface antigens, in each of 7 *M. hominis* strains that were tentatively allocated to as many different serovars. Antigens are shared, as demonstrated by DID, with a number of other arginine-utilizing *Mycoplasma* species, including *M. salivarium, M. orale, M. buccale, M. faucium, M. arthritidis* and *M. primatum*, and with the arginine- plus glucose-metabolizing *M. fermentans*. Otherwise serologically distinct.

Potentially pathogenic as suggested by its recovery from inflamed uterine tubes, tuboovarian and pelvic abscesses, and from the blood of patients with postpartum fever or septic arthritis. Particularly strong evidence for the etiological role of this species in inflammatory pelvic disease was provided by its isolation in pure culture, in 8% of cases of acute salpingitis, from samples collected from the uterine tubes by laparoscopy (Mårdh and Weström, 1970a). Serological studies further confirmed an etiologic relationship (Lemcke and Csonka, 1962; Mårdh and Weström, 1970b). A self-limiting acute salpingitis and parametritis has been induced in grivet monkeys (*Cercopithecus aethiops*) by inoculation of *M. hominis* directly into the uterine tubes during laparotomy, into the cervical epithelium, or into the uterine cavity followed by curettage of the endometrium. The gross pathology was characterized by pronounced edematous swelling and hyperemia of the tubes and parametria. Microscopically, cellular infiltrations of lymphocytes were found in the acute phase in the subserosa and muscularis of the tubes and in the parametria. Granulation tissue and fat necrosis appeared at a later stage in the parametria (Møller et al., 1978; Møller and Freundt, 1979). There is convincing evidence, moreover, that *M. hominis* is the cause of some cases of acute pyelonephritis (Thomsen, 1978a; 1978b; Ernø and Thomsen, 1980). Any possible etiological implication of the organism in a number of other disease conditions of the male and female urogenital tracts, including nongonococcal urethritis, vaginitis, prostatitis, epididymitis, fetal wastage and infertility, remains to be determined. Has been identified as an apparently rare cause of neonatal meningitis.

A very common inhabitant of the mucosae of the lower urogenital tract of humans; more rarely encountered in the oropharynx. Also found in the genital tract and the oropharynx of at least four nonhuman primate groups.

Reports of the mol% G + C of the DNA vary from 27.3 (Bd) to 33.7 (T_m). The heterogeneity within the species has been revealed also by DNA/rRNA hybridization experiments between different strains, yielding relatedness values in the range of 39–100%. The genetic relatedness between *M. hominis* and other *Mycoplasma* species, including *M. arthritidis, M. salivarium, M. orale, M. buccale, M. pneumoniae* and *M. gallisepticum*, was found by the same technique to be ≦10% (Reich et al., 1966; Somerson et al., 1966; also see Stanbridge and Reff, 1979). The genome size is 4.5 × 10^8 daltons. Plasmid DNA with a molecular weight of about 18 × 10^6 daltons has been demonstrated in one strain (Zouzias et al., 1973). Acquired resistance to tetracycline was ascribed in another case to the presence of plasmids (Hjelm et al., 1980).

Type strain: ATCC 23114 (NCTC 10111; strain PG21 (H50) of Edward, 1955; Edward and Freundt 1973).

Further Comments

Originally two serovars, designated 1 and 2, were recognized for this species. The Campo (PG27) and other strains of serovar 2 were later identified, on the basis of serology, nucleic acid homology, enzyme analysis and the electrophoretic patterns of cell proteins, as *M. arthritidis*. In consequence, *M. hominis* serovar 2 was withdrawn (Edward and Freundt, 1965).

39. **Mycoplasma hyopneumoniae** Mare and Switzer 1965, 841.^{AL} (*Mycoplasma suipneumoniae* Goodwin, Pomeroy and Whittlestone, 1965, 1249).

hy.o.pneu.mo'ni.ae. Gr. n. *pneumonia* pneumonia; M.L. gen. n. *hyopneumoniae* of hog pneumonia.

A comprehensive characterization of this species has been presented by Rose, Tully and Wittler (1979).

Cells are coccoid or coccobacillary to short filamentous.

Colonies are very small, up to about 0.5 mm in diameter after 7–10 days incubation on solid medium. They are usually convex with a granular surface. No central portion burrowing into the medium has been detected, but in some cultures the older colonies develop a slight central depression. The typical "fried egg" appearance of most mycoplasma colonies is unusual in this species. Colonies are not produced, or produced only very inconsistently, on agar medium on primary isolation.

Facultative anaerobe, although growth is best under aerobic conditions with supplements of about 5% CO_2, as obtained in a candle jar. Very fastidious; growth occurs in A26 and FF (see footnotes on pp. 758 and 755) broth and agar media at 25–37°C, but not at 45°C.

Distinctive physiological characteristics of the species are presented in Table 10.2. Reports regarding utilization of glucose are conflicting. A fall in pH occurs during incubation in the complex media required for growth of the organism, but a fall of similar magnitude may be seen in media without glucose. A decline in the glucose content of the test medium has been reported (Jensen et al., 1978), and high levels of hexokinase activity have been demonstrated in soluble cell fractions of several strains of the species (Jensen et al., 1978; Rose et al., 1979). However, the organism does not appear to contain the highly efficient sugar transport system found in a few other *Mycoplasma* species tested, i.e. the phosphoenolpyruvate-dependent phosphotransferase system (Cirillo and Razin, 1973; Rose et al., 1979).

Serologically distinct by GI and IMF tests from other established *Mycoplasma* species. Sharing of antigens with *M. dispar*, *M. ovipneumoniae*, *M. bovoculi* and *M. hyorhinis* has been demonstrated by DID and counterimmunodiffusion techniques (Ball and Todd, 1978).

Unlike most other mycoplasmas, this species is inhibited by benzylpenicillin (penicillin G) in concentrations of 10–100 IU/ml. The inhibitory effect is mycoplasmastatic rather than mycoplasmacidal. Reports of susceptibility tests with ampicillin are somewhat conflicting. Some strains are susceptible to erythromycin at levels of 5–25 μg/ml.

The most important primary cause of enzootic pneumonia of swine. The natural disease can be reproduced experimentally. In pneumonic lungs the organisms are found in greatest numbers adjacent to the cilia and plasma membrane of epithelial cells lining the bronchioles and bronchi. Infection of porcine tracheal ring organ cultures in combination with monolayer cultures of fetal lung fibroblasts produces progressive epithelial ciliostasis, loss of parallel arrangements and gentle curvature of cilia, cilial exfoliation, and some sloughing of epithelial cells.

The organism has been detected only in the porcine respiratory tract and associated structures, including the nasal cavity and tonsils.

The mol% G + C content of the DNA has not been determined.

Type strain: ATCC 25934 (NCTC 10110; strain J of Goodwin, Pomeroy and Whittlestone 1965; see Rose, Tully and Wittler, 1979).

40. **Mycoplasma hyorhinis** Switzer 1955, 544.[AL]

hy.o.rhi′nis. Gr. n. *hys*, *hyos* a swine, Gr. n. *rhis*, *rhinis* nose; M.L. gen. n. *hyorhinis* of a hog's nose.

Coccoid and coccobacillary to very short filamentous cells with occasional branching.

Some strains are apparently very exacting. There is solid evidence to suggest the in vivo occurrence of porcine field strains which resist attempts at cultivation. Up to 60% of all strains contaminating cell cultures are noncultivable on conventional mycoplasma media. Such strains can be grown on agar medium containing viable BHK-21 cells or filtered freeze-thawed cell lysate. In addition, fresh yeast extract or yeast autolysate (250 μg/ml), and to a slight extent other medium components, have been found to inhibit the growth of strains hitherto regarded as noncultivable (Del Giudice et al., 1980).

Distinctive physiological characteristics of the species are presented in Table 10.2.

Although comparison of the PAGE patterns of cell proteins has failed to reveal significant differences between a number of *M. hyorhinis* strains, the same and other strains tested by GI, MI and latex agglutination (Gois et al., 1974), and to some extent also by IMF (Friis, 1976), show a rather considerable degree of serological heterogeneity. One or more antigens are shared with a number of other glucose metabolizing species including *M. felis*, *M. pulmonis*, *M. neurolyticum* and *M. edwardii*, as demonstrated by DID. Serologically distinct by, *inter alia*, GI, MI, IMF and CF tests from these and a number of other *Mycoplasma* species tested.

M. hyorhinis is a common secondary invader in preexisting pneumonia of pigs primarily caused by *M. hyopneumoniae*. There is some indication that *M. hyorhinis* may on its own be the cause of some cases of pneumonia occurring under natural conditions. Associated with at least some cases of porcine arthritis and polyserositis (pericarditis, pleuritis and peritonitis), having been isolated in one study from arthritic joints of slaughtered pigs. Experimental intranasal inoculation of cultures of a particular virulent strain (S218) in gnotobiotic piglets induced clinical disease with more or less extensive pneumonia, arthritis and/or polyserositis in a high percentage of the animals. Arthritis and serositis may also develop following intraperitoneal inoculation. The *M. hyorhinis* arthritis is characterized by mononuclear cell infiltration, serofibrinous effusion, intra- and periarticular proliferative lesions and development of pannus and bone and cartilage erosions. Irregular mortality and pericardial and peritoneal lesions may be produced in chick embryos. Many strains induce a well defined cytopathic effect in a great variety of cell cultures.

A very common inhabitant of the nasal cavity of both healthy and diseased pigs, occurring in some herds with a frequency of 50–60%. One of the most frequent mycoplasmal contaminants of cell cultures.

The mol% G + C of the DNA is 27.3–27.8 (T_m). Whereas a number of different strains of the species were indistinguishable by DNA/rRNA hybridization, no genetic relatedness was found between *M. hyorhinis* and *M. pulmonis*, *M. hominis*, *M. gallisepticum* or *A. laidlawii* (Somerson et al., 1966; Stanbridge and Reff, 1979).

Type strain: ATCC 17981 (NCTC 10130; strain BTS-7 of Switzer, 1955; Edward and Freundt, 1973).

Further Comments

Additional studies are required in substantiation of claims that serological variants of *M. hyorhinis* may deserve subspecies rank.

M. hyorhinis cultivar α Del Giudice, Gardella and Hopps 1980, 78, has been proposed as an infraspecific designation of formerly noncultivable strains in recognition of cultural properties that differ from the species description.

41. **Mycoplasma hyosynoviae** Ross and Karmon 1970, 710.[AL]

hyo.syn′ov.i.ae. Gr. n. *hys*, *hyos* a swine; M.L. n. *synovia* fluid in the joints; M.L. gen. n. *hyosynoviae* of joint fluid of swine.

Coccoid to short filamentous cells.

A granular deposit and a waxy surface pellicle are produced during growth in broth.

Growth is stimulated by swine gastric mucin (Mucin Bacteriological, Difco), 0.5%.

Distinctive physiological characteristics of the species are presented in Table 10.2.

Several differences have been demonstrated in the PAGE patterns of cell proteins of different strains, including strain S16, although a majority of bands are shared by the organisms (Wreghitt et al., 1974). The patterns are clearly distinct from those of a number of other *Mycoplasma* and *Acholeplasma* species tested, including *A. granularum* (Ross and Karmon, 1970; Wreghitt et al., 1974). Also, some intraspecific heterogeneity is demonstrable by the MI test, whereas it is rather homogeneous by IMF. Serologically distinct by GI, MI and IMF from a number of other *Mycoplasma* and *Acholeplasma* species tested, including *A. granularum*.

A cause of uncomplicated nonsuppurative arthritis of young pigs characterized by sudden onset of lameness and gross lesions consisting of increased serofibrinous to serosanguineous synovial fluid and swelling and hyperemia of synovium in large diarthrodial joints. *M. hyosynoviae* can be isolated from the synovial fluid, lymph nodes and mucous membrane secretions during the acute stage of the disease. It persists indefinitely in the tonsils and pharynx, and less frequently in nasal secretions of convalescent and adult pigs. The disease can be induced experimentally by intravenous and intranasal inoculation, although not always readily so. There is evidence that some strains of the species are invasive and more likely to produce arthritis than others. Outbreaks of the natural disease are relatively frequent in the United States, but apparently rare and sporadic in other parts of the world. Yet, the organism is often found in the nasopharynx of pigs outside the United States.

The mol% G + C content of the DNA has not been determined.

Type strain: ATCC 25591 (NCTC 10162; strain S16 of Ross and Karmon, 1970; Edward and Freundt, 1973).

Further Comments

A number of strains isolated from arthritic joints of swine that were tentatively identified as *Acholeplasma granularum* have been reclassified as *M. hyosynoviae* (Ross and Karmon, 1980).

42. **Mycoplasma iners** Edward and Kanarek 1960, 699.[AL]

in'ers. L. adj. *iners* inactive, inert.

Morphology poorly defined.

Distinctive physiological characteristics of the species are presented in Table 10.2.

Antigens are shared with *M. columbinasale*, *M. gallisepticum*, *M. gallinarum* and *M. meleagridis*, as demonstrated by DID. Serologically distinct by, *inter alia*, GI, MI and IMF from all other avian and nonavian *Mycoplasma* species tested.

Susceptible to digitonin (0.3 μg/ml); the type strain is resistant, but several other strains tested are moderately susceptible to sodium polyanetholesulfonate (1.0 μg/ml), as determined by a disk growth inhibition method on solid medium.

Pathogenicity is not known.

A parasitic inhabitant of the respiratory tract of chickens and turkeys. Occasionally isolated from mammalian tissues.

The mol% G + C of the DNA is 29.1 (T_m) to 29.6 (Bd).

Type strain: ATCC 19705 (NCTC 10165; strain PG30 (M) of Edward and Kanarek, 1960; Edward and Freundt, 1973).

43. **Mycoplasma iowae** Jordan, Ernø, Cottew, Hinz and Stipkovits 1982, 114.[VP] (Avian serotype I, Yoder and Hofstad 1964.)

i.o.wae'. M.L. gen. n. *iowae*, of Iowa.

Coccoid to coccobacillary cells.

Distinctive physiological characteristics of the species are presented in Table 10.2.

The former serotype I included strains I, J, K, N, Q and R of Dierks et al. (1967). Although there are quantitative differences in antigenic components, these strains are all antigenically sufficiently related to be included in one species (Jordan, 1979). Sharing of antigen with *M. gallinaceum*, *M. columbinasale* and *A. laidlawii* has been demonstrated by DID. Serologically distinct by GI, MI and IMF tests from these and other avian *Mycoplasma* species, and by GI from all other *Mycoplasma* species.

The type strain was isolated from air sac lesions of pipped turkey embryos. The organism kills turkeys and chicken embryos and causes mild air sac lesions in turkeys and exudative lesions in the hock and footpad of chicks following inoculation at the respective sites. Arthritis, tendosynovitis and partial or complete rupture of the digital flexor tendons of chickens after infection with *M. iowae* has been reported (Bradbury and McCarthy, 1981).

The mol% G + C of the DNA is 25 (Bd).

Type strain: ATCC 33552 (NCTC 10185; strain 695 of Yoder and Hofstad 1964; see Jordan, 1979, and Jordan et al., 1982).

44. **Mycoplasma lipophilum** Del Giudice, Purcell, Carski and Chanock 1974, 152.[AL]

li.po'phi.lum.Gr. n. *lipos* animal fat; Gr. adj. *philos* loving; M.L. adj. *lipophilus* fat-loving.

Mainly coccoid cells.

Colonies growing near a paper disk impregnated with Oil Red-O, Sudan Black or Sudan IV take up the stain and appear red or black, respectively. The strong affinity for lipid stains is unique for the species. In addition to a heavy production of film and spots, older colonies develop numerous internal particles of crystalline appearance. Neither the film nor the particles show any affinity for lipid stains. Microco-

lonies, both free and adhering to the glass surface, are formed in liquid medium, and a film similar to that produced on agar medium develops on the surface of the broth. In addition to the coccoid cells seen by microscopy of broth cultures, spherical structures (5–10 μm in diameter) and rough rods (1–2 μm in diameter and 20 μm in length), which do not stain with specific fluorescein-conjugated antibody, are observed.

Very fastidious even after prolonged serial passage. DNA (0.02%, w/v) from calf thymus has an enhancing effect in media containing yeast extract, but DNA enhancement varies with the lots of yeast. Growth is obtained on BACY broth and agar media (see footnote* on p. 757).

Distinctive physiological characteristics are presented in Table 10.2.

Serologically distinct from a considerable number of other *Mycoplasma* and *Acholeplasma* species tested, as demonstrated by GI and MI, supplemented in several cases with MI and CF tests.

Pathogenicity is unknown.

Normal habitat is uncertain. So far, only five strains have been recognized, two of them recovered from the human respiratory tract and three from the throat of rhesus monkeys.

The mol% G + C content of the DNA has not been determined.

Type strain: ATCC 27104 (NCTC 10173; strain MaBy of Del Giudice et al., 1974).

45. **Mycoplasma maculosum** Edward 1955, 90.[AL]

ma.cu.lo'sum. L. adj. *maculosus* spotted.

Filaments short (2–5 μm) with occasional branching.

Distinctive physiological characteristics of the species are presented in Table 10.2.

Antigens are shared with the arginine metabolizing species *M. spumans* and *M. gateae* as demonstrated by DID. The type strain is serologically distinct from these and other *Mycoplasma* species tested by GI, MI and IMF tests.

Susceptible to digitonin (0.3 μg), but some strains are (unlike most other sterol-dependent mycoplasmas) relatively resistant to sodium polyanetholesulfonate (1.0 μg) as determined by a disk growth inhibition method on solid medium.

Pathogenicity is not known, although isolated rather frequently from pneumonic lungs of dogs.

A common parasitic inhabitant of the upper respiratory tract of dogs; found only occasionally in the conjunctivae and urogenital tract.

The mol% G + C of the DNA is 26.7 (T_m) to 29.6 (Bd).

Type strain: ATCC 19327 (NCTC 10168; strain PG15 (C27) of Edward, 1955).

46. **Mycoplasma meleagridis** Yamamoto, Bigland and Ortmayer 1965, 47.[AL]

me.le.a'gri.dis. L. n. *meleagris* a turkey; L. gen. n. *meleagridis* of a turkey.

Coccoid to coccobacillary cells. An extracellular amorphous layer has been demonstrated by electron microscopy of sectioned cells stained with ruthenium red.

Growth is markedly stimulated by biotin at concentrations of 20–200 μg/ml, but not or only slightly so by pantothenate or folate (Bigland and Warenycia, 1978).

Distinctive physiological characteristics of the species are presented in Table 10.2.

Antigens are shared with *M. gallinaceum*, *M. gallinarum*, *M. iners* and *M. pullorum*, as demonstrated by DID. Serologically distinct by GI, MI and IMF from these and a number of other *Mycoplasma* species tested.

A primary pathogen for turkeys, in which it produces air sacculitis and skeletal abnormalities, including perosis, and occasionally synovitis of the hock joint. Infection of the oviduct is frequent and may originate from the air sacs or develop as an ascending infection from a focus in the cloacal region. The infection of the reproductive organs is apparently not associated with disease, but often results in egg transmission of the mycoplasma. *M. meleagridis* is not pathogenic for chickens.

Readily isolated from the upper respiratory tract and the cloaca of turkeys. Has not been found in other avian species.

The mol% G + C of the DNA is 27.0 (T_m) to 28.6 (Bd). Genome size is about 4.2×10^8 daltons.

Type strain: ATCC 25294 (NCTC 10153; strain 17529 of Yamamoto et al., 1965; Edward and Freundt, 1973).

47. Mycoplasma moatsii Madden, Moats, London, Matthew and Sever 1974, 464.[AL]

moat'si.i. M.L. gen. n. *moatsii* of Moats; named after K.E. Moats, whose primary interest has been in the mycoplasmas of nonhuman primates.

Morphology poorly defined.

Distinctive physiological characteristics of the species are presented in Table 10.2.

Serologically distinct by MI and IMF tests from previously named *Mycoplasma* species of human and nonhuman primate source, and from a number of other established *Mycoplasma* species with which it has been compared.

Pathogenicity is not known.

Found in the reproductive tracts and the throats of male and female grivet monkeys (*Cercopithecus aethiops*). Apparently specific for this host, as it has not been isolated from other primate species or from humans.

The mol% G + C content of the DNA has not been determined.

Type strain: ATCC 27625 (NCTC 10158; strain MK 405 of Madden, Moats, London, Matthew and Sever, 1974).

48. Mycoplasma molare Rosendal 1974a, 130.[AL]

mo.la're. L. adj. *molaris* millstone-like, referring to the heavy film reaction which resembles the pattern on the surface of a millstone.

Mainly coccoid cells.

Distinctive physiological characteristics of the species are presented in Table 10.2. A lipoid film of characteristic appearance (Fig. 10.8) develops on the surface and along the circumference of colonies grown on the egg-yolk medium of Fabricant and Freundt (1967).

Serologically distinct by GI, MI and IMF tests from previously established canine and glucose-metabolizing noncanine *Mycoplasma* species.

Pathogenicity is not known.

Found on a few occasions in the pharynx of healthy dogs with mild respiratory disease, and once from the genital tract of a bitch.

The mol% G + C content of the DNA has not been determined.

Type strain: ATCC 27746 (NCTC 10144; strain H542 of Rosendal, 1974a).

49. Mycoplasma mycoides (Borrel, Dujardin-Beaumetz, Jeantet and Jouan, 1910), Freundt 1955, 73.[AL] (*Asterococcus mycoides* Borrel, Dujardin-Beaumetz, Jeantet and Jouan 1910, 179).

my.co.i'des. Gr. n. *myces* a fungus; Gr. n. *eidos* shape; M.L. adj. *mycoides* fungus-like.

This species is the type species of the genus *Mycoplasma*.

Most strains have a potential for producing repeatedly branching filaments of greatly varying length. Unique structures, so-called ρ (rho) forms (Fig. 10.9), are found in all recently isolated strains, and in some stock strains, when grown in medium of the following composition: dehydrated PPLO broth (Difco), 2.1%, Albimi yeast autolysate (Pfizer), 0.6%; calf thymus DNA, 0.002%; ox serum, 10% (v/v); glucose, 0.72%; sodium phosphate, 0.9% (pH 7.8); and penicillin, 100 U/ml. The ρ-form organelles are characterized by an intracytoplasmic axial fiber, about 40–120 nm in diameter, which extends throughout the length of the cell. The fiber is composed of parallel aligned fibrils with a diameter of approximately 3 nm and terminates at one or both ends in a plate-like structure located just beneath, and apparently attached to, the plasma membrane of the cell. The function of the organelle is unknown (Peterson et al., 1973; Rodwell et al., 1975). Cell motility has not been demonstrated.

Sharing of antigens with *M. capricolum* has been demonstrated by growth precipitation and CF tests. The species is otherwise serologically distinct from other established *Mycoplasma* and *Acholeplasma* species tested.

Further Comments

Two subspecies are recognized; they are serologically distinct by agglutination, GI, MI and IMF tests; extensive to complete cross-reactivity is demonstrated by CF and IHA tests. Common antigens are also demonstrable by DID and by the growth precipitation test. The PAGE patterns of the cell proteins of the type strains of the two subspecies resemble each other, but are not identical. Hybridization experiments by the filter paper method between DNAs from the type strains have revealed homology values of 75 and 98%, respectively, corresponding to a relatedness value of 0.70.

49a. Mycoplasma mycoides subsp. mycoides (Borrel, Dujardin-Beaumetz, Jeantet and Jouan 1910) Freundt 1955, 73.[AL]

Most strains have a potential for producing very long, repeatedly branching filaments up to 150 μm in length, average 40–50 μm. A well-defined capsular layer, apparently made of the polysaccharide galactan, is demonstrated by electron microscopy of specimens stained with ruthenium red.

According to the size of the colonies, two groups of strains are recognized for the subspecies: small colony (SC) strains which includes the type of the subspecies, strain PG1, and large colony (LC) forming strains represented by strain Y-goat (Cottew and Yeats, 1978). LC strains resemble *M. mycoides* subsp. *capri* in producing fast-growing large colonies, with a diameter of several millimeters, and greater turbidity than SC strains in liquid medium. With most strains of the subspecies, inoculation of small numbers of organisms in broth medium results in the formation of "comets" and "threads," macroscopically visible structures consisting of whitish islets of growth connected by threads of filamentous growth. On prolonged incubation the culture becomes uniformly turbid.

A completely defined medium allowing the definition of minimal nutritional requirements has been described for the subspecies, strain Y (Rodwell 1969, 1979).

Distinctive physiological characteristics of the subspecies are presented in Table 10.2. Acid is produced also from fructose, maltose, dextrin, glycogen and starch, but not from galactose, cellobiose or xylose. LC strains differ from SC strains in producing acid from sorbitol, digesting casein and liquefying heat-coagulated serum more vigorously, and in surviving longer at 45°C. Agglutinates guinea pig, but not bovine red blood cells, although neither are adsorbed by agar colonies.

Antigenicity depends to a major extent on the capsular substance galactan, a galactose polymer with β-galactosyl configuration (Buttery and Plackett, 1960). SC and LC strains are serologically very closely related to each other by conventional methods. Analysis and comparison of the cell proteins by two-dimensional gel electrophoresis have shown that LC strains are more closely related to *M. mycoides* subsp. *capri* (strain PG3) than to SC strains (Rodwell and Rodwell, 1978). However, another study (Archer, 1979) based on immunoprecipitation of the cell proteins followed by two-dimensional electrophoresis revealed that many of the dominant protein antigens of PG1 (SC strain) and Y-goat (LC strain) are shared and that some of the proteins are also shared by PG3. In terms of mouse-protective antigens, the SC and LC strains are related but not identical (Smith et al., 1980; Smith and Oliphant, 1981a, b).

M. mycoides subsp. *mycoides* is the etiological agent of contagious bovine pleuropneumonia (CBPP), most if not all of the isolates associated with this disease being of the SC type. The disease that is still endemic in cattle and water buffaloes in many countries in Africa and some parts of India is characterized by fibrinous pneumonia and pleurisy. Calves under about 6 weeks old may develop septic arthritis with few or no lung lesions. Several SC strains of the subspecies have been isolated from goats; their pathogenicity for goats is unclear, but some of them are pathogenic for cattle. The LC strains are associated with a variety of infections of goats, such as arthritis, mastitis, peritonitis, abscesses and septicemia, and reportedly with contagious caprine pleuropneumonia (CCPP) in some endemic areas (Cottew, 1979); they have been recovered only occasionally from cattle. Contagious pleuro-

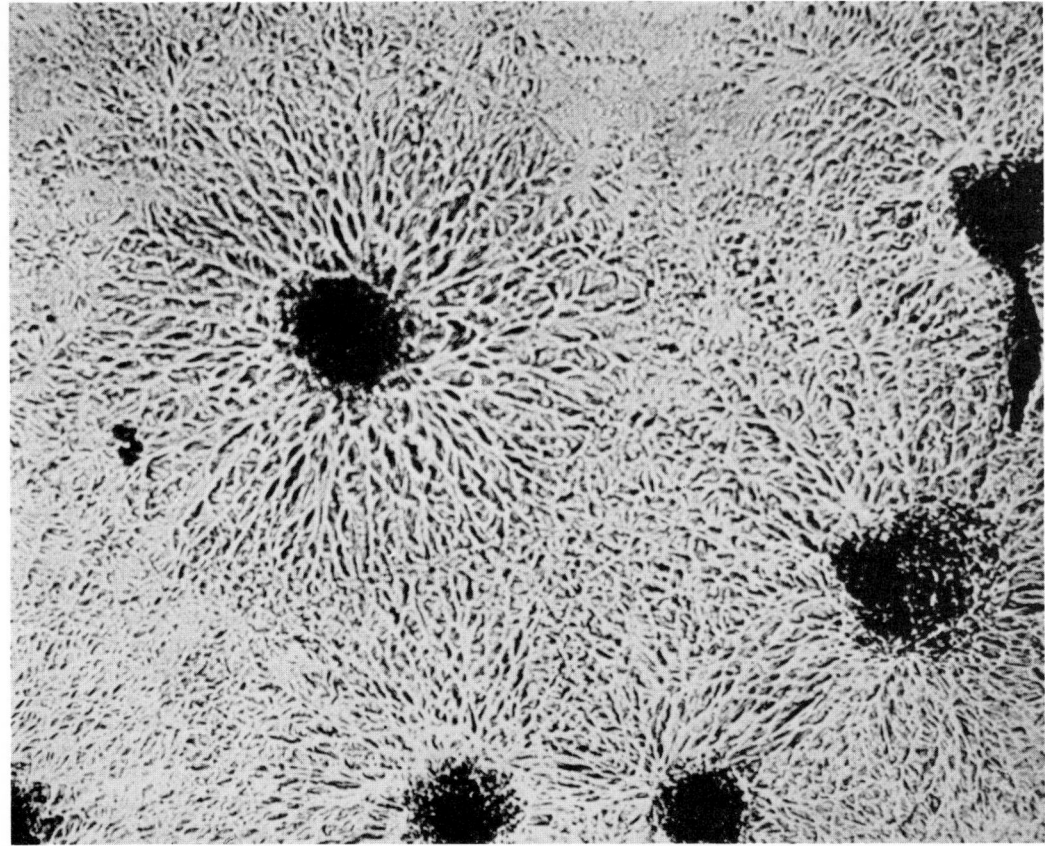

Figure 10.8. *Mycoplasma molare.* Four-day-old colonies grown on egg-yolk medium showing the heavy film reaction endowing the colonies with the millstone-like pattern that has given name to this species (× 40). (Reproduced with permission from S. Rosendal, *International Journal of Systematic Bacteriology 24*: 125–130, 1974, © American Society for Microbiology.)

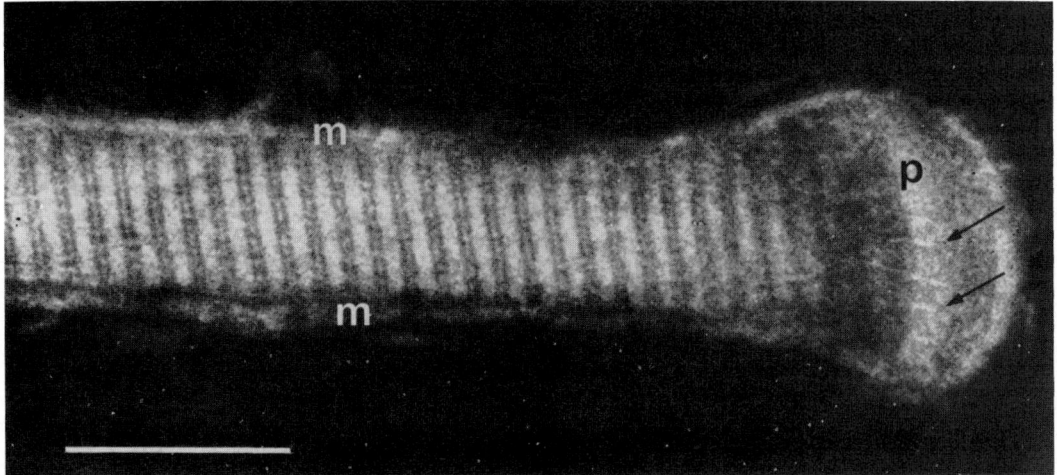

Figure 10.9. Electron micrograph of negatively stained specimen of *Mycoplasma mycoides* subsp. *mycoides*, strain Y. Terminal part of ρ-form cell showing the presence of an intracytoplasmic axial fiber associated with a terminal structure. The fiber almost completely fills the volume bounded by the plasma membrane (*m*) and presents a pattern of periodic transverse light and dark major bands, the dark band being divided by a light minor band. The terminal broad, electron-translucent plate (*p*) is penetrated by several longitudinal fibrils (*arrows*). *Bar* represents 100 nm. (Reproduced with permission from J. F. Peterson, A. W. Rodwell and S. Rodwell. *Journal of Bacteriology 115*: 411–425, 1973, © American Society for Microbiology.)

pneumonia is not recognized as a natural disease in sheep, but fibrinous pleuropneumonia or peritonitis can be induced experimentally with LC strains. An LC strain isolated from a goat with polyarthritis proved considerably more virulent for goats and sheep on intravenous inoculation than for calves (Rosendal, 1981). Small laboratory animals are only slightly susceptible to either LC or SC strains. However, SC strains differ from almost all LC strains in their ability to produce mycoplasmemia readily in mice on intraperitoneal inoculation (Smith et al., 1980; Smith and Oliphant, 1981a, b).

The galactan has some marked pathophysiological effects. Given intravenously to cattle in nonpyrogenic doses before or after subcutaneous infection with a M. mycoides subsp. mycoides strain of moderate virulence, it results in prolonged mycoplasmemia and the development of lesions in joints and kidneys. A single intravenous inoculation (100 μg/kg) in healthy noninfected calves produces, within a few seconds, severe respiratory and cardiovascular symptoms associated with pulmonary edema, distention of distal airway and capillary thrombosis.

The mol% G + C of the DNA is 26.1–27.1 (T_m and Bd).

Type strain: NCTC 10114 (strain PG1, Edward and Freundt 1973). This is an SC strain. The *reference strain* of the LC group is Y-goat (Cottew and Yeats, 1978).

49b. **Mycoplasma mycoides** subsp. **capri** (Edward 1953) Freundt 1955, 73.[AL] (*Asterococcus mycoides* subsp. *capri* Edward 1953, 873).

ca'pri. L. gen. n. *capri* of a (male) goat.

Filaments of moderate length, 10–30 μm; filamentation is stimulated by addition of oleic acid (about 50 μg/ml) to the growth medium.

Colonies on serum agar are large, with a diameter of 1.5–2.5 mm after 3 days, and may reach a maximum of 4 mm. Produces strong opalescence in a fluid medium.

Less exacting in its nutritional requirements than most other *Mycoplasma* species, growing to a slight extent in media without serum due to its requirement for minimal amounts of cholesterol for growth.

Distinctive physiological characteristics of the subspecies are presented in Table 10.2. Has an extended scheme of carbohydrate catabolism as for *M. mycoides* subsp. *mycoides*. Resembles the LC strains of *M. mycoides* subsp. *mycoides* in producing acid from sorbitol and in its strong proteolytic activities. Agglutinates guinea pig, but not bovine red blood cells, although neither are adsorbed by agar colonies.

Glucan is a major antigenic component.

This subspecies has been regarded for many years as the etiological agent of classical contagious caprine pleuropneumonia (CCPP). However, there is now convincing evidence (McMartin et al., 1980) that the disease is caused by an as yet unclassified mycoplasm represented by strain F38 (see below). Organisms isolated occasionally from outbreaks of highly fatal edema and cellulitis in goats may either belong to *M. mycoides* subsp. *capri* or represent LC strains of *M. mycoides* subsp. *mycoides*. Apparently nonpathogenic to cattle. Inhibition of the activity of epithelial cilia has been demonstrated by infection of chicken embryo tracheal organ cultures.

The mol% G + C of the DNA is ~24–26 (chemical analysis, T_m and Bd).

Type strain: NCTC 10137 (strain PG3, Edward and Freundt, 1973).

Organisms Related to **Mycoplasma mycoides.** The bovine serogroup 7 (reference strain PG50) of Leach 1967 and bovine serogroup L (reference strain B144P) of Al-Aubaidi and Fabricant 1971 crossreact by GI, IMF and DID and show a one-way cross-reaction in the MI test. Both groups are related to *M. mycoides*, the growth of strains PG50 and B144P being inhibited by PG3 antiserum or globulin fraction. Also, sharing of antigens by strains PG50, B144P, PG1 and PG3 has been demonstrated by the growth precipitation tests. DNA/DNA homology values between PG50 and B144P are 94% and 88%; between PG1 and PG50/B144P within the range of 79–93%, and between PG3 and PG50/B144P at the level of 66–89% (Askaa et al., 1978).

Strain F38 associated with CCPP in Kenya (MacOwan and Minette, 1976) is serologically closely related to strain PG50 and cross-reacts in the growth precipitation test with all members of the *M. mycoides*

complex. Similar strains have been isolated from outbreaks of CCPP in Sudan and certain other African countries. The organism is fastidious, requiring a growth medium based on goat meat and liver digest, with the addition of 30–50% (v/v) goat serum for primary isolation (MacOwan and Minette, 1976).

50. **Mycoplasma neurolyticum** (Sabin 1941) Freundt 1955, 73.[AL] (*Musculomyces neurolyticus* Sabin 1941, 57).

neu.ro.ly'ti.cum. Gr. n. *neuron* nerve; Gr. adj. *lyticos* able to loosen, (dissolve); M.L. neut. adj. *neurolyticum* nerve-destroying.

Filaments vary in length from very short (2.5 μm) to extremely long structures (up to 160 μm reported).

Apparently less exacting than most other *Mycoplasma* species as optimum growth is reported to be supported by as little as 1% whole serum or 2% agamma horse serum.

Distinctive physiological characteristics of the species are presented in Table 10.2. Acid is produced also from maltose, dextrin, glycogen and starch, but not from fructose.

Lipoglycans have been found in the membranes of both toxigenic and nontoxigenic strains (see below), but the neutral sugar content of toxigenic strains is greater. Sharing of one or more antigens with other glucose-metabolizing species, including *M. edwardii*, *M. hyorhinis* and *M. caviae*, has been demonstrated by DID. Serologically distinct by GI, MI and IMF from these and a number of other *Mycoplasma* and *Acholeplasma* species tested.

The type strain of the species has been reported to be inhibited by moderate concentrations of penicillin G (40 IU/ml and above) although it can be adapted to growth at high levels of penicillin on stepwise transfer to increasing concentrations. The effect of penicillin seems to be bacteriostatic rather than bactericidal.

A true exotoxin with neurotoxic effect on mice is produced by some but not all strains during the early growth phase. The toxin appears to be a protein with a molecular weight in excess of 200,000. It is thermolabile, being destroyed at 50°C in 10–30 min or at 45°C in 15–90 min. It is also inactivated by 0.025 mg/ml of trypsin within 10 min, or by binding in vitro to a sedimentable component of brain tissue, possibly a ganglioside. The toxin is antigenic and specifically neutralized in vitro by rabbit antiserum. The toxigenicity is often retained even after more than 100 subcultures although it may be lost on repeated subculturing. Cell-free filtrates of growing cultures are effective only on intravenous, but not on intraperitoneal, subcutaneous or intracerebral inoculation of young mice, producing characteristic symptoms of rolling disease and neuropathological lesions as well as pulmonary hemorrhage. In contrast, the neurotoxicity of suspensions of washed living mycoplasmas is the same when given intravenously and intraperitoneally, and substantially greater following intracerebral inoculation. Less characteristic neurological symptoms are elicited in young rats, while hamsters, guinea pigs and chickens are insusceptible to the toxin. The mode of action of the toxin has not been fully elucidated and appears to be rather complex. (Tully, 1964; Thomas et al., 1966; Thomas and Bitensky, 1966).

Rolling disease does not occur in mice that are naturally infected with *M. neurolyticum*, and the only disease associated with the organism under natural conditions is conjunctivitis. However, although mice are readily infected by conjunctival or nasopharyngeal inoculation of freshly isolated strains, experimental infection does not result in the development of conjunctivitis. Lesions can be produced by direct inoculation of *M. neurolyticum* into the cornea.

A common inhabitant of healthy and diseased mice, apparently primarily localized in the mucous membranes of the upper respiratory tract. The organisms are apparently transmitted to young mice shortly after birth; these then become carriers of latent mycoplasmas. Frequently isolated in association with experimental infection of mice with a variety of other agents. For example, the first isolations reported were made from the brain of mice that had developed "rolling disease" during the course of intracerebral passage of *Toxoplasma gondii* (Sabin, 1938) and lymphocytic choriomeningitis and yellow fever viruses (Findlay et al., 1938). Also repeatedly recovered from pneumonic lesions of mouse lungs after nasal instillation under anesthesia of various materials, and from the blood and tissues of mice under stress with malignant disease.

Reports on the mol% G + C of the DNA range from 22.8 (Bd) to 26.2 (T_m).

Type strain: ATCC 19988 (NCTC 10166; "type A" strain of Sabin, 1938; Edward and Freundt, 1973; see also Tully and Ruchman, 1964).

51. **Mycoplasma opalescens** Rosendal 1975, 469.[AL]

o.pa.le'scens. L. part. adj. *opalescens* opalescent, referring to the opalascing film produced on solid medium.

Morphology is not defined.

Distinctive physiological characteristics of the species are presented in Table 10.2.

Sharing of an antigen with *M. columbinasale* has been demonstrated by DID. Serologically distinct by DID, GI and IMF tests from other canine *Mycoplasma* species, and by GI and IMF from previously established arginine-metabolizing species of *Mycoplasma*.

Pathogenicity is not known.

Isolated only on a few occasions from the throat and urinary tract of dogs.

The mol% G + C content of the DNA has not been determined.

Type strain: ATCC 27921 (NCTC 10149; strain MH5408 of Rosendal, 1975).

52. **Mycoplasma orale** Taylor-Robinson, Canchola, Fox and Chanock 1964, 141.[AL]

o.ra'le. L. adj. *oralis* oral, pertaining to the mouth.

Coccoid and filamentous cells up to 8–10 μm long. Living cells of the organism are able to change their shape rapidly and reversibly, suggesting the presence of a contractile substance forming a sort of cytoskeleton (Bredt, 1979).

Yeast is required for the growth of at least fresh isolates.

Distinctive physiological characteristics of the species are presented in Table 10.2. Colonies on solid medium adsorb chicken red blood cells, but not red blood cells from man, monkey, rat or guinea pig, and not tracheal epithelial cells from monkey, rat or chicken. Receptor sites on chicken erythrocytes are not destroyed by neuraminidase. Observations by Kihara et al. (1981) suggest that protein receptors in the membrane of *M. orale* may be involved in the adsorption mechanism.

Antigens are shared by the type strain and strains of other arginine-utilizing human *Mycoplasma* species as demonstrated by DID. Serologically distinct by, *inter alia*, GI, MI and IMF tests from other *Mycoplasma* species of human origin and from a number of other *Mycoplasma* and *Acholeplasma* species tested.

Pathogenicity is not known.

A common parasitic inhabitant of the human oropharynx. Isolated only occasionally from nonhuman primates.

The mol% G + C of the DNA is within the range of 24.0–28.2 (T_m and Bd). Genome size is 4.7×10^8 daltons. DNA/rRNA hybridization studies revealed no relatedness between this species and *M. salivarium* (Reich et al., 1966; Stanbridge and Reff, 1979).

Type strain: ATCC 23714 (NCTC 10112; strain CH 19299 of Taylor-Robinson et al., 1964; Edward and Freundt, 1973).

53. **Mycoplasma ovipneumoniae** Carmichael, St. George, Sullivan and Horsfall 1972, 677.[AL]

o.vi.pneu.mo'ni.ae. Lat. fem. n. *ovis* a sheep; Gr. n. *pneumonia* pneumonia; M.L. gen. n. *ovipneumoniae* of sheep pneumonia.

Morphology is poorly defined.

Colonies on solid media usually lack the central downgrowth that contributes to the "fried egg" appearance, unless grown on media with a relatively low agar concentration, i.e. about 0.7%. The colonies are further characterized by a lacy or vacuolated appearance.

Growth occurs in conventional mycoplasma media, but the isolation rate is improved by use of A26 broth and agar media (see footnote* on p. 758).

Distinctive physiological characteristics of the species are presented in Table 10.2.

Some intraspecific heterogeneity has been demonstrated by subjecting the results of cross-reactions obtained between a number of strains of the species by GI and MI to a relatedness analysis by the method of Gois et al. (1974). However, the PAGE patterns of cell proteins of the same strains were very similar to each other. Serologically distinct by a variety of different methods, including GI and MI, from several other *Mycoplasma* species tested (Carmichael et al., 1972; Jones et al., 1976; Leach et al., 1976). However, sharing of antigen(s) with *M. bovoculi*, *M. dispar*, *M. hyopneumoniae*, and *M. hyorhinis* has been demonstrated by DID and counterimmunodiffusion techniques (Ball and Todd, 1978).

All strains tested so far have been found to be resistant to 5 μg of erythromycin in sensitivity tests on solid medium.

Associated with chronic proliferative interstitial pneumonia and pulmonary adenomatosis of sheep, being frequently isolated from the lungs, trachea, nose, and occasionally the conjunctivae of diseased animals. Has also been recovered from pneumonic lesions and pleural exudate of goats. Experimental infection of lambs has been produced with cultures of the organism given intravenously or in the form of aerosols. The resulting pneumonic lesions resembled those of the natural disease.

Also found in the respiratory tract of healthy sheep, although apparently less frequently than in diseased animals.

The mol% G + C content of the DNA has not been determined.

Type strain: NCTC 10151 (ATCC 29419; strain Y98 of Carmichael et al., 1972).

54. **Mycoplasma pneumoniae** Somerson, Taylor-Robinson and Chanock 1963, 122.[AL]

pneu.mo'ni.ae. Gr. n. *pneumonia* pneumonia, M.L. gen. n. *pneumoniae* of pneumonia.

Coccoid to short branched filamentous cells. In broth cultures grown in bottles or Petri dishes young filamentous cells attach to the glass or plastic surface, the tapered terminal portion of the organisms being oriented perpendicular to the surface. Sections of such organisms fixed and embedded in situ have revealed an electron-dense rod in the long axes of the tip, ending with a plate-like structure and surrounded by a mantle of electron-lucent cytoplasm limited by an outer membrane. Attachment of the mycoplasma cells to respiratory epithelium by means of the terminal structure is demonstrable also in tracheal organ cultures infected with *M. pneumoniae* (Fig. 10.4). Moreover, the terminal structure probably plays a role in the gliding motility exhibited by the organism. An actin-like protein similar to the contractile proteins found in muscles has been identified in extracts of *M. pneumoniae* cells (Neimark, 1977). Electron microscopy of Triton X-100 treated cells revealed a type of cytoskeleton composed of filaments resembling the actin seen in higher organisms and reacting in immunofluorescence and immunodiffusion tests with antiserum to rabbit muscle actin (Meng and Pfister, 1980).

Freshly isolated colonies on solid medium usually lack the light peripheral zone, appearing as circular dome-shaped, granular structures measuring 50–100 μm in diameter. Grows more slowly than most other mycoplasmas, the colonies being demonstrable 5–10 days or more after inoculation. On adaptation to artificial media colonies with the typical "fried egg" appearance predominate and growth is more rapid.

Fresh yeast extract is required for growth on primary isolation and in early subcultures. The replacement of crude horse serum by agamma horse serum in the growth medium has been reported to support growth more consistently. The SP-4 medium (see under Enrichment and Isolation Procedures) originally developed for cultivation of a spiroplasma has been shown to enhance the isolation rate of *M. pneumoniae* significantly.

The temperature range is ~30–39°C with optimum at ~36–38°C.

Distinctive physiological characteristics of the species are presented in Table 10.2. Agglutinates human erythrocytes to low titers. Colonies on solid medium adsorb red blood cells and tracheal epithelial cells from monkey, rat, guinea pig and chicken, as well as HeLa cells and cells from calf kidney and chick embryo tissue cultures and spermatozoa from man and bull. Adsorption occurs most readily at 37°C, less rapidly and less extensively at 22°C; it is specifically inhibited by antiserum. The attachment of *M. pneumoniae* erythrocytes and epithelial cells is mediated by sialic acid receptors that are destroyed by neuraminidase. The binding site on the *M. pneumoniae* cells may be lipid or lipoprotein in nature, since glycerophospholipid haptens can block hemadsorption-inhibiting antibodies. Isolated mycoplasma membranes apparently attach to tracheal epithelium in a manner distinct from the receptor site mediation common to intact cells (Gabridge et al., 1977). Observations by electron microscopy show that *M. pneumoniae* posseses the ability to attach to guinea pig red blood cells by sites other than the terminal tip (Brunner et al., 1979). The strong affinity of the species for respiratory epithelium is believed to play a role in its virulence in the way, *inter alia*, that the firm binding of the organisms to the epithelial cell sheet may provide a situation in which local concentrations of the hydrogen peroxide produced by the mycoplasma can build up and cause cell damage before being destroyed by peroxidase present in extracellular fluids. Studies with avirulent strains or strains of decreased virulence, e.g. temperature-sensitive mutants (see below), revealed that attachment was less pronounced, or absent, in these organisms, as compared to virulent wild strains.

Major membrane antigens are found in the lipid fraction and have been partly identified as glyceroglycolipid haptens, although some of the serologically reactive components do not contain carbohydrate. Forms a serologically homogeneous group. Whereas no precipitation lines are demonstrable by DID between this and other *Mycoplasma* species of human source, a sharing of antigens has been demonstrated by this method with *M. columbinasale* and *M. gallopavonis*. Otherwise serologically quite distinct, as far as known, from other *Mycoplasma* species. The cold-hemagglutinins induced by *M. pneumoniae* in patients infected with the organism are directed against the antigen-I determinant of erythrocytes (Feizi and Taylor-Robinson, 1967). They are probably the result of antigenic changes in the red blood cells caused by the mycoplasma, rather than being due to the possible presence of cross-reacting antigens between the organism and the blood cells (see Barile, 1979).

Unlike most other mycoplasmas, *M. pneumoniae* is highly sensitive to erythromycin, and minimum inhibitory concentrations range from 0.001–1.0 μg/ml. Resistance to erythromycin in concentrations of up to >100 μg/ml may be acquired both in vitro and in vivo (Niitu et al., 1974; Stopler et al., 1980); resistance may further develop in vitro to chloramphenicol and streptomycin, but has not been observed with tetracycline (Stopler et al., 1980).

The etiological agent of cold-hemagglutinin associated primary atypical pneumonia of man, characterized *inter alia*, by interstitial pneumonitis, desquamative bronchitis and bronchiolitis, and peribronchial infiltrations of mononuclear cells. Infection, which is very rarely fatal, may result in all degrees of respiratory involvement from inapparent infection to pneumonia, the incidence of clinically apparent lower respiratory tract involvement varying in different reports from 3–10% and up to about 80% of infections. Bullous myringitis, first observed in experimentally infected volunteers, occurs occasionally in naturally acquired infections. Involvement of the central and peripheral nervous system, as well as a number of other nonrespiratory complications, such as pancreatitis, myocarditis, pericarditis, hemolytic anemia and exanthema, have been reported with increasing frequency during recent years. Pneumonic lesions can be produced experimentally in cotton rats, hamsters, young guinea pigs and gnotobiotic mice. Rhesus monkeys and marmosets are relatively resistant to experimental infection. Produces distinct cytopathology, including loss of organized ciliary activity, in the respiratory epithelium of hamster trachea in organ culture or in ciliated respiratory epithelial monolayers (Gabridge et al., 1978), effects that distinguish this species from other mycoplasmas of human provenance. Cytotoxic effects have also been demonstrated with isolated membrane preparations (Gabridge et al., 1974) and ex-

tracts of *M. pneumoniae* (Chandler and Barile, 1980). Man is the only natural host of *M. pneumoniae*.

The mol% G + C of the DNA is unusually high compared to those of other mycoplasmas, viz. 38.6 (T_m) to 40.8 (Bd). There is no cross-reaction between this and other species of human source by nucleic acid homology tests. Genome size is 4.8×10^8 daltons. Temperature-sensitive (*ts*) mutants, induced by exposure of the organism to N-methyl-N'-nitro-nitrosoguanidine and exhibiting a 100-fold or greater decrease in efficiency of colony formation at 38°C relative to 32°C, have been utilized in an attempt to develop an attenuated vaccine against *M. pneumoniae* infection (Steinberg et al., 1969). Mutants that were incapable of hemadsorption have also been isolated by means of chemical mutagenesis, using nitrosoguanidine (Hansen et al., 1979).

Type strain: ATCC 15531 (NCTC 10119; strain FH of Somerson et al., 1963; Edward and Freundt, 1973).

55. **Mycoplasma primatum** Del Giudice, Carski, Barile, Lemcke and Tully 1971, 442.[AL]

pri.mat'um. M.L. n. *primas, primatis* chief, from which *primates*, the highest order of mammals; M.L. pl. gen. n. *primatum* of primates.

Spherical and coccobacillary to short filamentous cells.

Distinctive physiological characteristics of the species are presented in Table 10.2.

Serologically distinct by either GI, IMF or CF tests from previously established *Mycoplasma* and *Acholeplasma* species.

Found as a common inhabitant of the oral cavity and the urogenital tract of at least four distinct primate groups (baboons, African green monkeys, rhesus and squirrel monkeys). Two isolations have been reported from humans, one from an inflammatory skin lesion of the umbilicus (strain Navel) and another one from the genital tract.

Pathogenicity is not known.

The mol% G + C of the DNA is 28.6 (T_m).

Type strain: ATCC 25948 (NCTC 10163; strain HRC292 of Del Guidice et al., 1971.)

56. **Mycoplasma pullorum** Jordan, Ernø, Cottew, Hinz and Stipkovits 1982, 114.[VP] (Avian serotype C, Adler et al., 1958.)

pul.lo'rum. L. n. *pullus* a young animal, especially chicken; L. gen. pl. n. *pullorum* of chickens.

Description of the species is based exclusively on the type strain.

Coccoid to coccobacillary cells.

Distinctive physiological characteristics of the species are presented in Table 10.2.

Sharing of antigens with *M. gallinaceum, M. iowae, M. columbinasale, M. cynos, M. felis* and *M. edwardii* has been demonstrated by DID. Serologically distinct by GI, MI and IMF tests from all avian, and by GI from all nonavian *Mycoplasma* species.

The type strain was isolated from the trachea of a chicken. Little is known of the pathogenicity.

The mol% G + C of the DNA is 29 (Bd).

Type strain: ATCC 33553 (NCTC 10187; strain CKK of J. Fabricant, see Jordan, 1979, and Jordan et al., 1982).

57. **Mycoplasma pulmonis** (Sabin 1941) Freundt 1955, 73.[AL] (*Murimyces pulmonis* Sabin 1941, 57.)

pul.mo'nis. L. n. *pulmo* the lung; L. gen. n. *pulmonis* of the lung.

Two forms of motile cells have been observed: (a) a round cell with a protruding flexible stalk, often slightly thickened at the distal end, and (b) an elongated cell with a tapering leading end. The length of the cells vary from 1.5 to more than 5 μm. The terminal structures probably serve as organelles of attachment during the gliding motility exhibited by the cells. Constant rotation of coccoid cells is also described. Motility is usually lost on repeated subcultivation of the organism. An extracellular capsular matrix can be demonstrated by staining with ruthenium red. Surface projections similar to those of myxoviruses have also been observed by electron microscopy.

Colonies on solid medium have a coarsely granulated and vacuolated appearance; they have little tendency to grow into the agar, and the central spot is consistently less well defined than in most other *Myco-plasma* species.

Distinctive physiological characteristics of the species are presented in Table 10.2. Acid is produced also from maltose, dextrin, glycogen and starch, but not from fructose, while conflicting results are reported for galactose. Colonies on solid medium adsorb red blood cells, but not tracheal epithelial cells from monkey, rat, guinea pig and chicken; they also adsorb spermatozoa from man and bull. The red blood cell receptor sites are not destroyed by neuraminidase.

Although a large majority of murine strains of this species appear to constitute a fairly homogeneous group serologically, minor differences were demonstrable, in gel diffusion tests, between the C strain of Sabin (1941) and other murine strains. Also, the demonstration of divergencies in the antigenic composition of certain strains of nonmurine source (Deeb and Kenny, 1967) suggests the existence of serological variants of this species. Sharing of antigen(s) with other glucose-metabolizing species, including *M. neurolyticum*, *M. canis*, *M. felis* and *M. hyorhinis*, has been demonstrated by DID. Otherwise serologically distinct from these and a number of other *Mycoplasma* species tested.

Highly pathogenic to mice and rats. A primary cause of chronic respiratory disease (murine respiratory mycoplasmosis), a very common slowly progressing disease that is characterized in its subacute and chronic stages by rhinitis, otitis media, conjunctivitis, laryngo-tracheitis, and bronchopneumonia leading to bronchiectasis. Experimental infection of mice by intranasal inoculation is dose-dependent and results in either minimal transitory lesions, acute fatal disease characterized by exudation of fluid and large numbers of neutrophils in alveolar spaces, or chronic bronchopneumonia. The experimental disease in rats is not dose-dependent and is characterized by slowly developing chronic lung disease. In rats, the natural disease is accompanied in up to 30% of female animals by oophoritis and salpingitis that can be reproduced experimentally in pathogen-free rats by intravenous inoculation. A small percentage of male rats with epididymitis and urethritis harbour *M. pulmonis* in the vas deferens. Natural genital tract disease in mice has not been reported, but lesions similar to those seen in rats have been produced experimentally. Acute purulent arthritis, which may lead to chronic arthritis dominated by mononuclear cell infiltration and hyperplasia of synovial tissue, may be produced in both rats and mice following intravenous inoculation. Intracerebral inoculation of *M. pulmonis* into newborn rats induces hydrocephalus. Acute to chronic arthritis can be produced in rabbits by inoculation into the knee joints (Cassel and Hill, 1979; Cassel et al., 1979).

A very common inhabitant of the respiratory tract of laboratory and wild mice and rats. The infection is usually acquired by the offspring from their mothers during the early weeks of life. Also found in the nasopharynx, conjunctivae and genital tract of guinea pigs, Chinese and Syrian hamsters and very occasionally in rabbits. A few strains have been recovered from the nasopharynx of horses with acute febrile respiratory disease.

The mol% G + C of the DNA is 27.5–29.2 (Bd). Less than 10% nucleic acid homology has been found between this species and *M. hominis* and negligible homology with *A. laidlawii* (Peterson and Pollock, 1969).

Type strain: ATCC 19612 (NCTC 10139; strain Ash (PG34), Edward and Freundt, 1973).

Further Comments

The murine strain C, originally assigned to a separate species, *Musculomyces histotropicus* Sabin 1941, 58, thought for several years to be lost, was rediscovered in 1963 (Tully and Ruchman, 1964) and renamed *Mycoplasma histotropicus* (sic) Tully 1965, 184. Subsequent studies of the electrophoretic pattern of the cell proteins (Razin, 1969) and of the serology (Lemcke et al., 1969) of Sabin's type C strain clearly identified it as *Mycoplasma pulmonis*.

Leach and Butler (1966) showed *Mycoplasma mergenhagen* Grace et al., 1965, 1369 to be a synonym of *M. pulmonis*.

58. **Mycoplasma putrefaciens** Tully, Barile, Edward, Theodore and Ernø 1974, 116.[AL]

pu.tre.fa'ci.ens. L. v. *putrefacere* make rotten; L. part. adj. *putrefaciens* making rotten or putrefying, connoting the production of a putrid odor in broth and agar cultures.

Predominantly coccoid cells with only a few short filaments.

Unique among mycoplasmas in producing a strong odor of putrefaction during growth in broth and, to a lesser degree, on solid medium. Uniform heavy turbidity produced in broth cultures after 24 h.

Cholesterol requirement tests indicate a narrow concentration range (at the level of ~1 μg/ml) for optimally enhanced growth, larger amounts (5–20 μg/ml) appearing to be inhibitory.

Distinctive physiological properties of the species are presented in Table 10.2. Acid produced from glucose and mannose, but not from esculin.

The type strain is serologically distinct from previously established *Mycoplasma* and *Acholeplasma* species.

Associated with outbreaks of caprine mastitis in California and France; the disease is characterized by a heavy leucocytosis in the milk followed by an abrupt drop in milk production, but without any other clinical signs. As few as 50 organisms inoculated into the udder of lactating goats can induce purulent mastititis. The type strain of the species was supposedly isolated from the arthritic joint of a goat. Also found in healthy goats.

The mol% G + C of the DNA is 28.9 (T_m).

Type strain: ATCC 15718 (NCTC 10155; strain KS1 of Tully, Barile, Edward, Theodore and Ernø, 1974).

59. **Mycoplasma salivarium** Edward 1955, 90.[AL]

sa.li.va'ri.um. L. adj. *salivarius* salivary, slimy; intended to mean of saliva.

Coccoid to coccobacillary cells.

Distinctive physiological characteristics of the species are presented in Table 10.2. Confluent colonies on solid medium adsorb HeLa cells and chick embryo tissue culture cells, but not red blood cells from a variety of animal species, tracheal epithelial cells from monkey, rat or chicken and not spermatozoa from man or bull. The receptor sites on HeLa and chick embryo cells are not destroyed by neuraminidase.

Common antigens have been demonstrated by DID in this and certain other arginine-utilizing *Mycoplasma* species of human and nonhuman origin. Serologically distinct by several other methods from these and a number of other *Mycoplasma* species tested.

Pathogenicity is not known. Isolation of the organism with a significantly higher incidence from the gingival sulci of individuals with periodontal disease (87%) than in persons with healthy periodontium (32%) has stimulated interest in its possible role in periodontal pathology (Engel and Kenny, 1970; Forest, 1979). Also, it is the predominant mycoplasmal species in dental plaques.

A common parasitic inhabitant of the oropharynx of humans and several nonhuman primates. Occasionally isolated from the respiratory tract of horses.

The mol% G + C of the DNA is 27.3 (Bd) to 31.4 (T_m). The genome size is 4.7×10^8 daltons.

Type strain: ATCC 23064 (NCTC 10113; strain PG20 (H110) of Edward, 1955).

60. **Mycoplasma spumans** Edward 1955, 90.[AL]

spu'mans. L. part. adj. *spumans* foaming.

Coccoid to short filamentous cells, about 2–4 μm.

Colonies on horse serum agar are characterized, upon primary isolation and in early subcultures, by a coarsely reticulated and vacuolated appearance, the central spot tending to be hidden by the coarse markings. Typical fried egg appearance of the colonies develops on repeated subculturing.

Distinctive physiological characteristics of the species are presented in Table 10.2.

Antigens are shared with the arginine-metabolizing species *M. co-*

lumbinasale, M. maculosum and *M. gateae*, as shown by DID in agar. The type strain is serologically distinct from these and other *Mycoplasma* species tested by GI, MI and IMF.

Pathogenicity is uncertain. Associated with granulomatous colitis of boxer dogs, being isolated from the colon mucosa and lymph nodes of diseased dogs. Inoculation of *M. spumans* into the colon of puppies via a rectal catheter failed to produce any lesions (Rosendal et al., 1980). Isolated fairly frequently from pneumonic lungs of dogs.

Found as a common parasitic inhabitant of the upper respiratory tract, and infrequently in the conjunctivae and urogenital tract of dogs.

The mol% G + C of the DNA is 28.4 (T_m) to 29.1 (Bd).

Type strain: ATCC 19526 (NCTC 10169; strain PG13 (C48) of Edward, 1955).

Further Comments

M. spumans has been divided, on the basis of plate immunofluorescence studies, into two serogroups, I and II, represented by strains PG13 and H764, respectively (Rosendal, 1974b).

61. Mycoplasma sualvi Gourlay, Wyld and Leach 1978, 292.[AL]

su.al'vi. L. n. *sus* the hog; L. n. *alvus* bowel, womb, stomach; M. L. gen. n. *sualvi* of the bowel of the hog.

Mainly coccobacillary cells. Elongated, flask- and club-shaped cells, resembling those of *M. alvi*, are seen by electron microscopy of negatively stained culture deposits. Terminal structures, located at one end of elongated cells, are visible in thin sections of preparations stained with uranyl acetate. These structures are less well defined than those of *M. alvi*, and the cells lack the dense central core that is typical of this latter species.

Cultural characteristics are similar to those described for *M. alvi*. Optimum conditions are provided by an anaerobic atmosphere with 10% added CO_2 and a temperature of about 37°C; growth is less rapid at 30°C and does not occur at 25°C. Grows very heavily in SP-4 broth medium (see under Enrichment and Isolation Procedures) (J. G. Tully, personal communication, 1981).

Distinctive physiological characteristics of the species are presented in Table 10.2.

The serologically identical clones A and B of the Mayfield strain are serologically distinct by MI and/or GI and IMF tests from all previously named *Mycoplasma* species, including *M. alvi*, and *Acholeplasma* species. Serologically distinct also from *Anaeroplasma bactoclasticum* and *Anaeroplasma abactoclasticum*.

Susceptible to digitonin (0.3 μg), but unlike most other sterol-dependent mycoplasmas resistant to sodium polyanetholesulfonate (1.0 μg), as determined by a disk growth inhibition method on solid medium.

Pathogenicity is not known.

Found in the rectum, colon, small intestines and vagina of pigs.

The mol% G + C of the DNA is 23.7 (Bd).

Type strain: NCTC 10170 (strain Mayfield clone B of Gourlay, Wyld and Leach, 1978).

62. Mycoplasma subdolum Lemcke and Kirchhoff 1979, 49.[AL]

sub'do.lum. L. adj. *subdolus* somewhat deceptive; alludes to the deceptive color change produced by the organism in urea-containing broth that led to the original erroneous description of the strains as urea-hydrolyzing.

Coccoid and coccobacilliary cells. Small amounts of amorphous material are visible by electron microscopy outside of the membrane of thin-sectioned cells stained with ruthenium red, but there is no true capsule.

Distinctive physiological characteristics of the species are presented in Table 10.2.

The type strain and strains compared to it by PAGE of cell proteins form a homogeneous group by this method and also serologically. Serologically distinct by at least two tests (GI, MI and/or IMF) from all other established *Mycoplasma* species and three species of *Acholeplasma* found in horses.

Pathogenicity is not known. Isolated from the genital tract of both normal and infertile mares and stallions and from the lungs of aborted fetuses. Occasionally found in the nasopharynx of healthy horses.

The mol% G + C content of the DNA has not been determined.

Type strain: ATCC 29870 (NCTC 10175; strain TB of Lemcke and Kirchhoff, 1979).

63. Mycoplasma synoviae Olson, Kerr and Campbell 1964, 209.[AL]

syn.ov'i.ae. M.L. n. *synovia* the joint fluid, M.L. gen. n. *synoviae* of joint fluid.

Coccoid cells. An extracellular amorphous layer is seen by electron microscopy of sectioned cells stained with ruthenium red.

Nicotinamide adenine dinucleotide (diphosphopyridine nucleotide; coenzyme I) in the reduced form has been defined as an essential nutritional requirement, the addition of 0.01% of this coenzyme to 10% swine serum medium ensuring satisfactory growth. Growth is further stimulated by the addition of 0.01% cysteine. These requirements are satisfied by the Frey broth and agar media* and by the SP-4 medium (see under Enrichment and Isolation Procedures).

Distinctive physiological characteristics of the species are presented in Table 10.2. Acid is produced also from maltose, but not from lactose or sucrose. Some strains, including the type strain, agglutinate chicken erythrocytes. Specific hemagglutinating activity has been found in essentially cell-free supernatants of centrifuged broth cultures of the organism; the activity is retained for more than 18 months when stored at −20°C or freeze-dried and kept at +4°C (Cullen and Snell, 1976; Snell and Cullen, 1978).

Sharing of antigens with *M. columbinasale*, *M. gallisepticum*, *M. gallopavonis* and *M. pullorum* has been demonstrated by DID. Serologically distinct from these and all other established *Mycoplasma* species by GI, MI and IMF tests (Jordan et al., 1982).

Pathogenic to chickens, turkeys and guinea fowls causing disease in a wide variety of organs and tissues, including the joints, bursae and tendon sheaths ("infectious synovitis"), and the respiratory tract. Arthritic lesions are characterized by turbid to caseous exudate, infiltration with mononuclear cells and plasma cells, hyperplasia of the synovial lining, and sometimes erosion of articular cartilage. Lesions in the respiratory tract are similar to those produced by *M. gallisepticum*. The tissue tropism of different strains and the disease pattern appears to differ in different geographical areas. Thus, whereas infectious synovitis caused by *M. synoviae* appears to be relatively common in the United States, there is no correlation between *M. synoviae* infection and the occurrence of synovitis of poultry in the Netherlands (Goren, 1979; personal communication, 1980). Inoculation of chick embryo tracheal organ cultures with the mycoplasma results in reduction of ciliary activity, but apparently to a less degree than seen with *M. gallisepticum*.

Reported to occur rather frequently in the respiratory tract of chickens and turkeys without clinical disease.

The mol% G + C of the DNA is 34.2 (Bd).

Type strain: ATCC 25204 (NCTC 10124; strain WVU 1853 of Olson et al., 1967; Edward and Freundt, 1973).

---

* Frey broth: dissolve 2.25 g of dehydrated PPLO broth w/o CV (Difco) in 90 ml of distilled water. Sterilize at 121°C for 20 min. Aseptically add Eagle's essential vitamins (100X; Vitamins for Basal Medium Eagle (modified), Flow Laboratories, Cat. No. 16-004-49), 0.025 ml; glucose solution (50%, w/v), 2.0 ml; and swine serum (inactivated at 56°C for 30 min), 12.0 ml. Mix together β-nicotinamide adenine dinucleotide solution (reduced form; Sigma, Grade III, 1% (w/v), 1.0 ml, and cysteine-HCl solution (1%, w/v), 1.0 ml; after 10 min add the mixture to the other ingredients. Add phenol red solution (0.06%), 5 ml, and penicillin solution (20,000 IU/ml), 0.25 ml. A solid medium can be prepared by adding 1.4 g of purified agar (Oxoid, Code L28) to the broth base before autoclaving.

64. **Mycoplasma verecundum** Gourlay, Leach and Howard 1974, 483.[AL]

ve.re.cun'dum. L. adj. *verecundus* shy, unobtrusive, free from extravagance, alluding to the lack of obvious biochemical characteristics of the species.

Coccoid and short filamentous cells.

Growth occurs within the unusually large pH range of 4.7–7.8, with optimum at 6.0. Growth at 30°C is virtually as good as at 37°C, and occurs down to 20°C.

Distinctive physiological characteristics of the species are presented in Table 10.2.

Antigenically distinct by GI, MI and IMF tests from other established *Mycoplasma* species tested.

Pathogenicity is uncertain. Isolated from the eyes of two affected calves during an outbreak of severe conjunctivitis. Attempts to produce conjunctivitis in gnotobiotic calves with cultures of the organism were largely unsuccessful. Also found, although with a low isolation rate, in preputial wash samples from apparently healthy bulls.

The mol% G + C of the DNA is 27 (T_m) to 29.2 (Bd).

Type strain: ATCC 27862 (NCTC 10145; strain 107 of Gourlay, Leach and Howard, 1974).

Note Added in Proof

After this manuscript was completed, five new *Mycoplasma* species were validly published. Brief descriptions are presented below and other characteristics are listed in Table 10.2.

Mycoplasma cricetuli Hill 1983, 117.[VP]

cri.ce.tu'li. M.L. n. *Cricetulus* generic name of the Chinese hamster, *Cricetulus griseus*; M.L. gen. n. *cricetuli* of *Cricetulus*.

Growth occurs in conventional mycoplasma media. Acid is produced from glucose and certain other sugars. Agglutinates sheep erythrocytes but not human or guinea pig erythrocytes. Recovered from the conjunctivae and nasopharynx of Chinese hamsters. Pathogenicity not known. The mol% G + C of the DNA has not been determined. *Type strain:* NCTC 10190 (strain CH of Hill, 1983).

Mycoplasma genitalium Tully, Taylor-Robinson, Rose, Cole and Bové 1983, 395.[VP]

ge.ni.ta'li.um. L. n. *genitalis* genitals; M.L. adj. *genitalium* pertaining to the genitals.

Cells predominantly flask shaped, possessing a terminal structure apparently associated with attachment to host cells and inert surfaces. A prominent outer fibrillar layer, 8–12 nm thick, extends distally from the tip for ~40–60% of the cell length. Growth occurs in SP-4 medium (see under Enrichment and Isolation Procedures) or in conventional mycoplasma media with 20% fetal bovine serum. Susceptible to thallium acetate in concentrations (1:2000–1:4000) tolerated by most other mycoplasmas. Acid is produced from glucose. Foci of vacuolization and cell lysis are produced in Vero monkey kidney monolayers. Pathogen-

icity not known. Isolated from human urogenital tract. The mol% G + C of the DNA is 32.4 ± 2.0 (Bd.). *Type strain:* ATCC 33530 (strain G-37 of Tulley et al., 1983).

Mycoplasma lipofaciens Bradbury, Forrest and Williams 1983, 334.[VP]

li.po.fa'ci.ens. Gr. n. *lipos* lipid; L. v. *facere* to make; M.L. part. adj. *lipofaciens* lipid-making, intended to refer to the production of a lipid film on solid media.

Growth occurs in conventional mycoplasma media. Glucose and arginine are catabolized. Isolated from the infraorbital sinus of a chicken. Chicken and turkey embryo mortality has been produced experimentally. The mol% G + C of the DNA is 24.5 (Bd.). *Type strain:* ATCC 35015 (BNCTC 10191; strain R171 of Bradbury et al., 1983).

Mycoplasma muris McGarrity, Rose, Kwiatkowsky, Dion, Phillips and Tully 1983, 355.[VP]

mu'ris. L. n. *mus* mouse; L. gen. n. *muris* of a mouse.

SP-4 medium (see under Enrichment and Isolation Procedures) is required for primary isolation but the organism can be adapted to grow in conventional media with 20% fetal bovine serum. Colonies are small (50–100 μm in diameter) and granular with a few "fried egg" types. Apparently a strict anaerobe. Hydrolyzes arginine and exhibits uridine phosphorylase activity. Adsorbs guinea pig erythrocytes. Isolated so far only from the vagina of mouse strain RIII. Pathogenicity is unknown. The mol% G + C of the DNA is 24.9 ± 1.0 (chemical analysis). *Type strain:* ATCC 33757.

Mycoplasma mustelae Salih, Friis, Arseculeratne, Freundt and Christiansen 1983, 478.[VP]

mu.ste'lae. M.L. n. *Mustela* the generic name of the mink *Mustela vison*; M.L. gen. n. *mustelae* of *Mustela*.

Growth occurs in conventional mycoplasma media. Acid is produced from glucose. Isolated from the trachea and lungs of minks. Pathogenicity is not known. The mol% G + C of the DNA is 28.2 ± 0.3 (Bd.). *Type strain:* ATCC 35214 (NCTC 10193; strain MX9 of Salih et al., 1983).

References

Bradbury, J.M., M. Forrest and A. Williams. 1983. *Mycoplasma lipofaciens*, a new species of avian origin. Int. J. Syst. Bacteriol. *33:* 329–335.

Hill, A.C. 1983. *Mycoplasma cricetuli*, a new species from the conjunctivas of Chinese hamsters. Int. J. Syst. Bacteriol. *33:* 113–117.

Mcgarrity, G.J., D.L. Rose, V. Kwiatkowski, A.S. Dion, D.M. Phillips and J.G. Tully. 1983. *Mycoplasma muris*, a new species from laboratory mice. Int. J. Syst. Bacteriol. *33:* 350–355.

Salih, M.M., N.F. Friis, S.N. Arseculeratne, E.A. Freundt and C. Christiansen. 1983. *Mycoplasma mustelae*, a new species from mink. Int. J. Syst. Bacteriol *33:* 476–479.

Tully, J.G., D. Taylor-Robinson, D.L. Rose, R.M. Cole and J.M. Bové. 1983. *Mycoplasma genitalium*, a new species from the human urogenital tract. Int. J. Syst. Bacteriol. *33:* 387–396.

Genus II. **Ureaplasma** Shepard, Lunceford, Ford, Purcell, Taylor-Robinson, Razin and Black 1974, 167[AL]

DAVID TAYLOR-ROBINSON AND R. NIGEL GOURLAY

U.re.a.plas'ma. M.L. fem. n. *urea*; Gr. neut. n. *plasma* something formed, or a form; M.L. neut. n. *Ureaplasma* urea-form, i.e. intended to denote a form of mycoplasma that requires or utilizes urea.

Cells from 18- to 24-h-old cultures are round or coccobacillary, ~330 nm in diameter. A variety of **pleomorphic forms** may be seen depending on the strain, age of the culture, and the method of examination. Gram-negative. Nonmotile. Microaerophilic. Optimum temperature, 37°C; poor growth occurs at 22°C and no growth at 42°C. **Optimum pH, ~6.0. Colonies generally are small, ~15 to ~60 μm in diameter,** and may not have zones of surface growth (and therefore may lack the "fried egg" appearance of colonies of most other members of the *Mollicutes*). Growth is retarded by thallous acetate (0.05%), 5-iodo-2'-deoxyuridine (125 μg/ml), hydroxyurea (500 μg/ml),

acetohydroxamic acid (1 mM), the tetracyclines, erythromycin, streptomycin, chloramphenicol, gentamicin and kanamycin, but not by the penicillins. **All strains hydrolyze urea with the production of ammonia.** Arginine and the usual carbohydrates are not metabolized. **Occur predominantly in the mouth, respiratory tract and urogenital tract of humans and various animal species.** The mol% G + C of the DNA of ureaplasmas of human origin is 26.9–28.0 and of bovine origin is 28.7–30.2.

Type species: *Ureaplasma urealyticum* Shepard, Lunceford, Ford, Purcell, Taylor-Robinson, Razin and Black 1974, 167.

Further Descriptive Information

Cells of ureaplasmas of human and animal origin in young broth cultures (18- to 24-h-old) are morphologically similar to those of other mycoplasmas. By most staining methods, they appear as dense, round to ovoid elements ~330 nm in diameter, with a range of 100–850 nm (Taylor-Robinson et al., 1968; Rottem et al., 1971; Black et al., 1972a; Black, 1973a; Whitescarver and Furness, 1975). The cells are seen singly and in various combinations; by phase-contrast microscopy, they have been seen to occur singly or in pairs (Razin et al., 1977). Pleomorphic forms, such as branching filaments, occur in older cultures but are seen less frequently than in cultures of other members of the *Mycoplasmataceae*. Motility is not observed and organelles usually associated with motility are not detected by electron microscopy. In electron micrographs of embedded, ultrathin sections, and in negatively stained preparations, the cells are usually round with a diameter ranging from 120–1000 nm (Williams, 1967; Taylor-Robinson et al., 1968; Rottem et al., 1971; Black et al., 1972a; Whitescarver and Furness, 1975). Filamentous forms up to 2000 nm in length and 50–300 nm wide have been seen by some workers (Rottem et al., 1971; Black et al., 1972a) but not by others (Whitescarver and Furness, 1975). *U. urealyticum* is bounded by a single membrane (7.5–10 nm thick) with no distinguishable cell wall. However, an extramembranous layer, 20–30 nm thick, has been observed by ruthenium-red staining (Robertson and Smook, 1976) and structures morphologically similar to short pili have been seen radiating from the membrane surface (Williams, 1967; Black et al., 1972a; Whitescarver and Furness, 1975).

Ureaplasmas possess a variety of other properties (Table 10.3), many of which are common to other mycoplasmas. Of these features, the production of small colonies (15 to ~60 μm) on agar media, the optimum pH for growth (6.0 ± 0.5), the relatively small number of organisms attained on growth in broth medium (usually not more than 10^7) and, above all, the possession of ureases (Delisle, 1977), distinguish ureaplasmas from other mycoplasmas.

Ureaplasmas of human and animal origin may be grown in various media containing serum and yeast extract, such as those described by Shepard and Lunceford (1970),* Taylor-Robinson et al. (1971),† and Howard et al. (1978a).§ It should be emphasized, however, that the quality of the medium components is at least as important as the exact medium formulation (Taylor-Robinson and Furr, 1981) and those containing the lowest levels of heavy metals are recommended.

The various mammalian and avian hosts from which ureaplasmas have been isolated are indicated in Table 10.4. Some other animal species have been examined without success (Taylor-Robinson and Furr, 1973; Koshimizu et al., 1981).

At least 14 serovars have been delineated among human strains and at least 11 serovars among bovine strains. The human serovars are distinct from the bovine serovars and from strains isolated from other animal species (see Table 10.5 and also under Taxonomic Comments).

Ureaplasmas cause urethritis in man, pneumonia in cattle, and urogenital disease in cattle and some other animal species (Taylor-Robinson, 1979).

Enrichment and Isolation Procedures

Inoculation of a specimen into liquid medium and subsequent subculture to liquid and agar media provides the most sensitive method for the isolation of ureaplasmas (Taylor-Robinson et al., 1969; Braun et al., 1970; Shepard and Lunceford, 1970). Colonies sometimes fail to develop or are not seen when a specimen is plated directly on agar media, whereas the organisms are detected in liquid media. This is achieved by taking advantage of the urease activity of ureaplasmas (Purcell et al., 1966; Shepard, 1966; Shepard and Lunceford, 1967; Ford and MacDonald, 1967; Masover et al., 1976; Vinther, 1976). The clinical material (urine, urine deposit, expressed swab, homogenized tissue, etc.) is diluted in serial 10-fold steps, to a dilution of at least 10^{-3}, in a medium supplemented with phenol red and urea contained in screw-capped vials (Taylor-Robinson and Purcell, 1966; Taylor-Robinson and Furr, 1981). The vials are incubated at 37°C with the caps tight. Because ureaplasmas possess ureases which hydrolyze urea to ammonia and so cause the pH of the medium to rise, a change in color from yellow to pink signals the presence of a ureaplasma. Of course, other pH indicators, apart from phenol red, may be used (Robertson and Stemke, 1979). On agar, colonies develop best in an atmosphere of 5–15% carbon dioxide in nitrogen (Shepard et al., 1974), although 100% carbon dioxide is claimed to be conducive to the development of large colonies of some serovars (Razin et al., 1977). Ureaplasmas were termed "T-strains" or "T-mycoplasmas" originally (Shepard, 1956) because of the very small colonies (T for tiny) they produced (15–30 μm in diameter). With improved media colonies of 60 μm or even greater may now be seen. Because these colonies may lack surface peripheral growth and, hence, may not have a "fried-egg" appearance, they may not be recognized easily. As an aid to detecting them, 0.05 M HEPES buffer (N-2-hydroxyethylpiperazine-N'-2-ethanesulfonic acid) and MnSO₄, a sensitive indicator of ammonia, may be added at a final concentration of 0.015% (w/v) to the agar medium.¶ On this medium, ureaplasmas form dark brown colonies (Shepard and Lunceford, 1976).

Some workers (Braun et al., 1970) have incorporated lincomycin in media to isolate *U. urealyticum* more easily and distinguish it from *Mycoplasma hominis* and *M. fermentans*. This approach is based on the resistance of *U. urealyticum* to lincomycin and the sensitivity of *M. hominis* and *M. fermentans* (Shepard et al., 1974). However, since mycoplasmas may become resistant to antibiotics, the procedure is not useful for taxonomic purposes.

---

* U9 medium of Shepard and Lunceford (1970) consists of (g/100 ml of deionized water): Tryptic digest broth powder (BBL or Difco), 0.75; NaCl, 0.5; and KH₂PO₄, 0.02. This broth solution is adjusted to pH 5.5 with 2 N HCl and sterilized at 121°C for 15 min. The following sterile components are added aseptically to 95 ml of the sterile broth: unheated horse serum, 5.0 ml; urea solution (10%, w/v), 0.5 ml; phenol red solution (1.0%, w/v), 0.1 ml; and potassium penicillin G solution (100,000 U/ml), 1.0 ml. L-Cysteine·HCl solution (1.0%, w/v), 1.0 ml, is added also to medium referred to as U9B (Shepard and Lunceford, 1976). The final pH should be 6.0 ± 0.2.
† The medium of Taylor-Robinson et al. (1971) consists of sterile beef heart infusion broth (Difco PPLO broth) to which the following components are added from sterile stock solutions (per 70 ml of broth): yeast extract solution (see below), 10 ml; unheated horse serum (No. 6, Wellcome), 20 ml; urea solution (10%, w/v), 1.0 ml; phenol red solution (0.1%, w/v), 2.0 ml; thallous acetate solution (2.5%, w/v), 1.0 ml; and benzyl penicillin solution (Glaxo, 100,000 U/ml), 1.0 ml. The final pH is adjusted to 6.0–6.5 with 0.1 N HCl. The *yeast extract solution* is prepared by suspending 250 g of dried yeast (Distillers Co. Ltd.) in 1 liter of distilled water. After steaming at 100°C for 30 min and centrifuging at 600 × g for 60 min, the supernatant is removed and sterilized by Seitz filtration. The extract is stored at −20°C.
§ U4 medium of Howard et al. (1978a) consists of the following sterile components: Hanks balanced salt solution (10× concentrate, Wellcome), 4.0 ml; Hartley's digest broth, 20 ml; fetal calf serum (Flow Laboratories), 15 ml; yeast extract solution (see below), 10 ml; phenol red solution (1.0%, w/v), 0.2 ml; urea solution (20%, w/v), 0.25 ml; thallous acetate solution (5%, w/v), 0.5 ml; MgSO₄ solution (250 μg/ml), 1.0 ml; benzyl penicillin (Glaxo, 200,000 U/ml), 0.5 ml; and sufficient glass-distilled water to bring the final volume to 100 ml. The final pH is adjusted to 6.0–6.2 with 1 N HCl. The *yeast extract solution* is prepared as described in footnote † except that it is boiled for 2 min (not steamed for 30 min) and is sterilized by autoclaving at 10 psi (115°C) for 20 min.
¶ The preparation of differential agar medium A7 is described in detail by Shepard and Lunceford (1976). Briefly, it consists of the liquid medium U9 described previously (see footnote *) supplemented with 0.015% MnSO₄·H₂O and 1.0% agar (Gibco No. M00010). (The A7 differential agar is now additionally supplemented with putrescine dihydrochloride (to 0.01 M); see Shepard and Combs, 1979.) Taylor-Robinson et al. (1971) used Oxoid Ionagar No. 2, but now use Gibco purified agar (No. M00010). Howard et al. (1978a) solidified their medium with 0.8% agarose (Miles Laboratories) and also included 0.05 M HEPES buffer, 0.01% putrescine, and 0.009% L-cysteine.

Table 10.3.

Properties of ureaplasmas[a]

Property	Result or Reaction
Cells are spherical to ovoid; filamentous forms are rare	+
Cell diameter:	
Range, nm	100–850
Average	330
Colony diameter, μm	15–~60
Grow best in 5–15% CO_2 in air, N_2 or H_2	+
Grow poorly under aerobic conditions	+
Optimum pH[b]	6.0 ± 0.5
Usually not more than 10^7 viable cells per ml produced in broth cultures	+
Mean generation time at 37°C (10 different strains),[c] min	50–105
Genome size, daltons[d]	$4.1–4.8 \times 10^8$
Mol% G + C of DNA (Bd)[d,e]	26.9–30.2
Optimum temperature, °C	37
Cholesterol required for growth[f]	+
Sensitive to digitonin[f]	+
Enzyme activity:	
Urease[g]	+
Arginine deiminase[h]	−
Aminopeptidase[i]	+
Esterase[j]	+
α-Glycerophosphate dehydrogenase[j]	+
L-Histidine ammonia-lyase[k]	+
Malate dehydrogenase[j]	+
Lactate dehydrogenase[j]	−
Adenosine triphosphatase (ATPase)[l]	+
Ribonuclease (RNase)[l]	+
Deoxyribonuclease (DNase)[l]	+
Phosphatase[m]	+
Catalase[n]	−
Proteolytic activity[o]	+
Fermentation of carbohydrates	−[p]
Hemolysis of erythrocytes[q]	+
Hemadsorption of erythrocytes[r,s]	+[t]
Sensitive to thallous acetate[n]	+
Sensitive to erythromycin and tetracyclines[n]	+[u]
Sensitive to lincomycin[n]	−
Tetrazolium reduction (aerobic and anaerobic)[s]	−

[a] Symbols: see standard definitions.
[b] Shepard and Lunceford (1965).
[c] Furness (1975).
[d] Black et al. (1972b).
[e] Howard et al. (1974).
[f] Rottem et al. (1971).
[g] Shepard and Lunceford (1967).
[h] Woodson et al. (1965).
[i] Vinther and Black (1974).
[j] Delisle (1977).
[k] Ajelli et al.(1977).
[l] Romano and La Licata (1978).
[m] Black (1973b).
[n] Shepard et al. (1974).
[o] Watanabe et al. (1973).
[p] Hexokinase-negative.
[q] Manchee and Taylor-Robinson (1970).
[r] Manchee and Taylor-Robinson (1969).
[s] Black (1973a).
[t] Only human serovar 3 and squirrel monkey ureaplasmas are positive.
[u] About 10% of human strains are resistant to tetracyclines (Evans and Taylor-Robinson, 1978).

Table 10.4.

Isolation of ureaplasmas from various mammalian and avian hosts[a]

Host	Isolation Reported From:				Isolation First Reported by:
	Respiratory Tract		Genital Tract		
	Upper	Lower	Upper	Lower	
Human[b]	+	−	+	+	Shepard (1954)
Bovine[b]	+	+	+	+	Taylor-Robinson et al. (1967)
Canine	+	+	−	+	Taylor-Robinson et al. (1971)
Simian:					
Squirrel monkey	+	−	−	−	Taylor-Robinson et al. (1971)
Talapoin, patas,					
cynomolgus monkeys	−	−	−	+	Kundsin et al. (1975)
Chimpanzee	+	−	−	+	Brown et al. (1976)
Marmoset	+	−	−	+	Furr et al. (1976)
Green monkey	+	−	−	+	Ogata et al. (1981)
Feline	+	−	−	+	Tan and Markham (1971)
Caprine	−	−	−	+	Gourlay et al. (1973)
Ovine	−	−	+	+	Livingston and Gauer (1975)
Avian:					
Chicken	−	+	−	−	Stipkovits and Rashwan (1976)
Jungle fowl	+	−	−	−	Koshimizu and Magaribuchi (1977)
Turkey	−	−	−	+	Stipkovits et al. (1978a)
Porcine[c]	−	−	−	+	Stipkovits et al. (1978b)
Mink	−	+	−	−	Friis et al. (1980)

[a] Symbols: +, positive; −, negative or not tested.

[b] Human and bovine strains comprise the species *Ureaplasma urealyticum* and *U. diversum*, respectively. Strains from other hosts have not yet been given a species designation.

[c] One isolate only.

Table 10.5.

Comparison and relationship of ureaplasmas from various animal species

Host	Serological Comparison[a] with:		Mol% G + C of DNA	Polyacrylamide Gel Protein Analysis
	U. urealyticum	*U. diversum*		
Human	NA[b]	Distinct[c,d,e,f]	26.9–28.0[g,h]	Two groups (A and B)[i,j]
Bovine	Distinct[c,d,e]	NA	28.7–30.2[h]	Three groups distinct from human groups[i,j]
Canine	Distinct[e,k]	Distinct[e,f,k]	27.2–27.8[h]	NT[l]
Feline	Distinct[m]	Distinct[m]	27.9[h]	NT
Simian[n]:				
Squirrel monkey	Distinct[e,o]	Distinct[e,f,o,p]	28.3[h]	One strain tested; distinct from human and bovine groups[j]
Talapoin monkey	?Distinct[o]	NT	NT	Two strains tested; closely related to human group A[j]
Patas monkey	NT	NT	NT	NT
Cynomolgus monkey	?Distinct[o]	NT	NT	NT
Green monkey	?Distinct[o,q]	Distinct[o,p]	NT	NT
Chimpanzee	Distinct[o]	Distinct[o,p]	NT	One strain tested; closely related to human group A[j]
Marmoset	Distinct[o,r]	Distinct[o,p]	29.6[h]	Two strains tested; distinct from human, bovine and squirrel monkey ureaplasmas[j]
Caprine	Distinct[s]	Distinct[s]	31.4–31.6[h]	NT
Ovine	Distinct[s]	Distinct[s]	30.6–31.4[h]	NT
Avian:				
Chicken	Distinct[t]	Distinct[p,t]	NT	Single strain[u] closely related to human group A[j]
Jungle fowl	Distinct[t]	Distinct[p,t]	NT	NT
Turkey	Distinct[v]	NT	NT	NT
Porcine	NT	NT	NT	NT
Mink	Distinct[w,x]	NT	NT	NT

[a] Metabolism inhibition, growth inhibition, and immunofluorescence tests used singly or in combination.

[b] NA, not applicable; first 8 serovars used in tests against ureaplasmas from other animal species.

[c] Howard et al. (1978a).

[d] Ogata et al. (1979).

[e] Taylor-Robinson and Furr (1973).

[f] Howard and Gourlay (1973).

[g] Black et al. (1972b).

[h] Howard et al. (1978b).

[i] Howard et al. (1981).

[j] Mouches et al. (1981).

[k] Kotani and Ogata (1979).

[l] NT, not tested.

[m] Kotani et al. (1980a).

[n] Only one or two strains from each monkey tested.

[o] Ogata et al. (1981).

[p] Not tested against all bovine serovars.

[q] Møller et al. (1981).

[r] Taylor-Robinson (unpublished data).

[s] Kotani et al. (1980b).

[t] Koshimizu and Magaribuchi (1978).

[u] Stipkovits and Rashwan (1976).

[v] Stipkovits et al. (1978a).

[w] Tested against serovar 8 only.

[x] Friis et al. (1980).

Maintenance Procedures

The rate of multiplication of ureaplasmas is reduced at temperatures below 37°C (e.g. 30°C) and incubation at a low temperature is sometimes helpful in serial subcultivation (Ford and MacDonald, 1967). Broth cultures remain viable for longer than 2 weeks at 4°C and for up to 5 days at room temperature (Shepard et al., 1974), particularly if they contain a large number of organisms (10^5 or more) and the pH of the medium is low (less than pH 7.0). Thus, organisms freshly inoculated into medium of low pH may be transported successfully for 5 days or more without freezing. Young broth cultures retain viability on storage at −20°C for 3 months although this is less reliable than storage at −70°C (Shepard and Masover, 1979), at which temperature the organisms may survive for as long as 10 years. In liquid nitrogen, they may be kept indefinitely. Ureaplasmas may be preserved also by lyophilization of young broth cultures, without further additives, and such dried cultures have been found viable after at least 6 years of storage at −20 to −60°C (Shepard et al., 1974) and after at least 2 years at 37°C (Addey et al., 1970).

Procedures for Testing Special Characters

Although ureaplasmas produce small colonies on agar and have a low pH optimum, these features are less characteristic than the posses-sion of ureases, which is unique among members of the *Mycoplasmatales*. A strong suspicion that the organisms possess ureases is a color change from yellow to red in medium containing phenol red and 0.1% urea after incubation at 37°C for 24–48 h. However, this procedure is less specific than flooding colonies with a urea-$MnCl_2$ solution which results in urease-positive colonies staining dark brown (Shepard and Howard, 1970). This, in turn, is less specific than the sensitive quantitative test based on the breakdown and disappearance of $^{14}CO_2$ from the ^{14}C-urea added to the culture medium. The technique described by Masover et al. (1976) required the use of Erlenmeyer flasks to trap $^{14}CO_2$, but was later modified (Swanberg et al., 1978) to enable the test to be undertaken in test tubes. The procedure is as follows: ^{14}C-urea is added to an 18- to 24-h-old ureaplasma broth culture in stoppered glass tubes to produce a final concentration of about 1.0 μCi/ml. The same amount of ^{14}C-urea is added also to broth medium which has been incubated at 37°C for 18–24 h without ureaplasmas (control). The medium containing the organisms, and the uninoculated control, are incubated at 37°C for 24–48 h, the enzyme reaction is stopped by adding 3 N H_2SO_4 (0.2 ml/1.0 ml broth), and the broths are kept at room temperature for a further 2 h and agitated occasionally to facilitate removal of CO_2. Samples are then mixed with scintillation fluid, the radioactivity is counted, and the amount of urea which has been hydrolyzed is calculated.

Differentiation of the genus **Ureaplasma** from other genera

As indicated previously, the unique ability to hydrolyze urea differentiates the genus *Ureaplasma* from the genus *Mycoplasma* within the family *Mycoplasmataceae*. There is evidence also for the possession of an antigenic component common to human, bovine and several other ureaplasmas of animal origin, which has not been detected in several *Mycoplasma* and *Acholeplasma* species (Nagatomo et al., 1980). Further, ureaplasmas require sterols for growth and thus are different from organisms of the genus *Acholeplasma* in the family *Acholeplasmataceae*, while the smaller genome size (5×10^8 daltons) of ureaplasmas and their lack of helical shape differentiate them from the genus *Spiroplasma* within the family *Spiroplasmataceae* (genome size 1×10^9 daltons). The finding of helical, twisted, rope-like elements in one human ureaplasma strain (K510) by Klainer and Pollack (1973), while inviting further exploration, is insufficient to modify the statement concerning the nonhelical shape of ureaplasmas. Organisms within the genus *Ureaplasma* are not strict anaerobes, in contrast to those of the genus *Anaeroplasma*.

Taxonomic Comments

Ureaplasmas of human origin. The possession of ureases and hence the ability to hydrolyze urea sets ureaplasmas apart from all other members of the *Mycoplasmatales*. In view of this, a new genus was established (Shepard et al., 1974) within the family *Mycoplasmataceae* in which to classify ureaplasmas isolated from man and other animals. At this time, the existence of a number of distinct human serovars and the possible existence of more suggested that it would be unwise to consider each serovar as a separate species; therefore, a single human species was formulated (*U. urealyticum*) containing eight serovars. These serovars were defined and given Roman numerals by Black (1973c) who used specific rabbit antisera to cloned organisms in metabolism inhibition, growth inhibition, indirect hemagglutination and indirect immunofluorescence tests. Other serovars were thought to exist (Lin et al., 1972), although it was not clear how many might have been different from the eight serovars described by Black. Subsequently, serovars other than the original eight have been recognized (Robertson and Stemke, 1979; Lin and Kass, 1980) and, currently, there is evidence for the existence of at least 14 (Lin and Kass, 1980; Robertson and Stemke, 1982). This makes the continued use of Roman numerals instead of Arabic numerals unnecessarily cumbersome.

The mol% G + C of the DNA of the original eight serovars of *U. urealyticum* was found to range from 26.9–28.0 (Black et al., 1972b). This narrow range suggests genetic homogeneity among these urea-plasma strains. However, on the basis of one-dimensional polyacrylamide gel electrophoresis studies (Howard et al., 1981; Mouches et al., 1981) and two-dimensional mapping (Mouches et al., 1981), it is clear that the human strains so far tested comprise two groups, A and B. Group A contains serovars 2, 4, 5, 7 and 8 (of Black, 1973c), serovar 9 (of Robertson and Stemke, 1979) and strains K2 and U24 (of Lin and Kass, 1980); group B contains serovars 1, 3 and 6 (of Black, 1973c) and strain U26 (of Lin and Kass, 1980). This subdivision is consistent with the results of recent hybridization experiments with DNA from the original eight serovars of *U. urealyticum* (Christiansen et al., 1981) and with the results of analyzing the DNA of the first nine serovars by restriction endonucleases (Razin et al., 1983). Whether the ureaplasmas of human origin should be separated into two species is a moot point.

Ureaplasmas of bovine origin and their comparison with **U. urealyticum**. Bovine ureaplasmas are serologically heterogeneous also, and it has been suggested (Howard et al., 1975; Ogata et al., 1979), on the basis of tests with antisera raised in rabbits, that they exist as three clusters of serologically similar, but not identical, strains. Furthermore, strains that could be regarded as serovars, representing the whole range of antigens synthesized by bovine ureaplasmas, have been proposed (Howard et al., 1975; Howard et al., 1978a; Ogata et al., 1979). A large number of bovine ureaplasma strains were identified by an immunofluorescence test with antisera raised in gnotobiotic calves to three strains which represented the three clusters (Howard and Gourlay, 1981). This suggests that all bovine ureaplasmas could be placed in three groups. Bovine strains are serologically distinct from the human ureaplasma serovars (Table 10.5). Furthermore, the range of values for the G + C content of 10 bovine ureaplasmas (28.7–30.2 mol %) (Howard et al., 1974; Howard et al., 1978b) does not overlap with the values obtained for the original eight serovars of *U. urealyticum* (26.9–28.0 mol %). In addition, the results of one-dimensional (Howard et al., 1981; Mouches et al., 1981) and two-dimensional (Mouches et al., 1981) polyacrylamide gel analyses show that the bovine strains can be placed in three groups consistent with the clusterings detected by serological examination. The fact that the protein profiles and maps of the bovine ureaplasma strains are different from those of the human strains, the G + C content of the bovine and human ureaplasma strains is different, and that there are differences in the specificity of human and bovine ureaplasmas for animals (Howard et al., 1973; Furr et al., 1978) indicates that it would be reasonable to formulate a species name for the bovine ureaplasmas. The name *U. diversum* has been proposed (Howard and Gourlay, 1982).

Ureaplasmas of other animal species. Canine ureaplasmas have been classified into four serogroups (Kotani and Ogata, 1979). Feline ureaplasmas also show heterogeneity (Kotani et al., 1980a). Simian ureaplasmas comprise four serological groups: a) squirrel monkey, b) crab-eating, cynomolgus, green and talapoin monkeys, c) chimpanzee and d) marmoset (Ogata et al., 1981). Caprine and ovine ureaplasmas appear to form a single serological group, apart from one ovine strain (Kotani et al., 1980b). Avian ureaplasmas isolated by Koshimizu and Magaribuchi (1978) fall into one serogroup and are different from the avian isolate of Stipkovits and Rashwan (1976). As mentioned previously, an antigenic component has been found common to human, bovine, canine, feline, caprine and avian ureaplasmas but not to some *Mycoplasma* and *Acholeplasma* species (Nagatomo et al., 1980). Despite this, the ureaplasmas from the different animal species appear to be serologically distinct from *U. urealyticum* and the bovine ureaplasmas, except for the one avian isolate of Stipkovits and Rashwan (1976) which cross reacts with the human serovar 8 and some strains isolated from cynomolgus, green and talapoin monkeys which cross-react to various degrees with antisera to the human serovars (Ogata et al., 1981; Møller et al., 1981) (Table 10.5). The question arises, therefore, of whether ureaplasmas derived from other animal species might be regarded as sufficiently distinct from the human and bovine ureaplasmas to also justify separation into distinct species. Apart from the serological diversity, the data presented in Table 10.5, particularly

those concerning the G + C content and polyacrylamide gel analysis of proteins, are derived from examination of too few strains and are, as yet, insufficient to indicate that ureaplasmas from any other host warrant species designation. This is emphasized by the protein patterns of ureaplasmas from talapoin monkeys, a chimpanzee and a chicken which are similar to the human ureaplasmas of group A, the most closely related being the chimpanzee ureaplasma. In the latter case, there is certainly no reason to contemplate a new species formulation. Indeed, if the source of the ureaplasma isolated from the chimpanzee had not been known, it would have been regarded as of human origin. It would seem that ureaplasma species designation should reflect the inherent characteristics of the ureaplasma and not necessarily the animal species from which it originates.

Further Reading

Mouches, C., D. Taylor-Robinson, L. Stipkovits and J.M. Bové. 1981. Comparison of human and animal ureaplasmas by one- and two-dimensional protein analysis on polyacrylamide slab gel. Ann. Microbiol. (Inst. Pasteur) *132B:* 171–196.

Shepard, M.C. and G.K. Masover. 1979. Special features of ureaplasmas. In Barile and Razin (Editors), *The Mycoplasmas,* Vol. 1, Academic Press, New York, pp. 451–494.

Taylor-Robinson, D. 1979. Pathogenicity of ureaplasmas for animals and man. Zentralbl. Bakteriol. Parasitenkd. Infektionskr. Hyg., Abt. I Orig. A *245:* 150–163.

List of the species of the genus **Ureaplasma**

1. **Ureaplasma urealyticum** Shepard, Lunceford, Ford, Purcell, Taylor-Robinson, Razin and Black 1974, 167.[AL]

u.re.a.ly'ti.cum. M.L. fem. n. *urea* urea; Gr. adj. *lyticus* dissolving; M.L. neut. adj. *urealyticum* urea-dissolving.

The cellular morphology and colonial characteristics are as described for the genus. Other biological and biochemical features are presented in Table 10.3.

At least 14 serologically different serovars exist. The protein patterns established for 12 of these by polyacrylamide gel electrophoresis indicate that they may be placed in two groups (Table 10.5). These serovars are distinct from the three groups of *U. diversum* and from the (as yet unnamed) ureaplasmas isolated from several other animal species (Table 10.5).

Found in the human genitourinary tract, oropharynx and anal canal.

The mol% G + C of the DNA is 26.9–28.0.

Type strain: ATCC 27618 (T-strain 960 (CX8) of M. C. Shepard; serovar 8; cloned 8 times).

2. **Ureaplasma diversum** Howard and Gourlay 1982, 446.[VP]

di.ver'sum. L. neut. adj. *diversum* different, distinct, heterogenous.

The cellular morphology and colonial characteristics are as described for the genus. Other biological and biochemical features are presented in Table 10.3.

Isolates are serologically heterogeneous, but serological tests and protein patterns indicate that they may be placed in three groups of similar but not identical strains (Table 10.5). These groups are distinct from *U. urealyticum* serovars and groups (Table 10.5).

Found in the bovine respiratory tract, genitourinary tract and eyes.

The mol% G + C of the DNA is 28.7–30.2.

Type strain: NCTC 10182 (A417 of Gourlay and Thomas; group A; cloned 6 times).

FAMILY II. **ACHOLEPLASMATACEAE** EDWARD AND FREUNDT 1970, 1[AL]

JOSEPH G. TULLY

(*Saprophytaceae* (*sic*) Sabin 1941, 59; *Sapromycetaceae* Sabin 1941, 334)

A.cho.le.plas.ma.ta'ce.ae. M.L. neut. n. *Acholeplasma* type genus of the family; *-aceae* ending to denote a family; M.L. fem. pl. n. *Acholeplasmataceae* the *Acholeplasma* family.

Sterol not required for growth. Reduced nicotinamide adenine dinucleotide (NADH₂) oxidase activity located in the cell membrane. Possess lactic dehydrogenases specifically activated by fructose-1,6-di-phosphate. Capable of fatty acid biosynthesis from acetate. Other characters as for the class and order.

Type genus: *Acholeplasma* Edward and Freundt 1970, 1.

Genus I. **Acholeplasma** *Edward and Freundt 1970, 1*[AL]

JOSEPH G. TULLY

A.cho.le.plas'ma. Gr. pref. *a* not; Gr. *chole* combining form denoting relationship to the bile; Gr. neut. n. *plasma* something formed or molded, a form; M.L. neut. n. *Acholeplasma* name intended to indicate that cholesterol, a constituent of bile, is not required.

Cells spherical, with a minimum diameter of ∼300 nm, **and filamentous,** usually ∼2–5 μm in length. Modes of reproduction are as for the genus *Mycoplasma*. **Cells are bounded by a plasma membrane only.** Generally more susceptible to lysis by osmotic shock at 37°C than other mycoplasmas. Gram-negative. Nonmotile. Facultative anaerobes, having a fermentative type of metabolism. **Colonies**

on solid media containing animal serum usually show a "fried egg" appearance and may reach 2–3 mm in diameter. Temperature range for growth, 20–40°C. Chemoorganotrophic. Carbohydrates serve as the fermentable substrates. Carbohydrate transport occurs through an active carrier-mediated process different from the phosphoenolpyruvate-dependent phosphotransferase system (PEP-PTS) found in some *Mycoplasma* species. **Arginine and urea are not hydrolyzed.** Phosphatase activity is weak or negative. Reduced nicotinamide adenine dinucleotide (NADH₂) activity is located in the plasma membrane. Possess lactic dehydrogenases specifically activated by fructose-1,6-diphosphate. **Serum or cholesterol not required for growth.** Agar colonies do not show hemadsorption of red blood cells from a variety of hosts. All species are resistant, or only very slightly sensitive, to 1.5% digitonin. Absolutely resistant to penicillin, not being inhibited by penicillin G, ampicillin, cloxacillin or methacillin in concentrations of at least 4000 μg/ml. Apparently parasites of a wide range of vertebrate hosts. Pathogenicity has not been clearly established. May also be a part of plant and insect flora. The mol% G + C of the DNA is ~26–35.7 (T_m, Bd). The genome size is ~1.0×10^9 daltons. Other characters are as for the genus *Mycoplasma*.

Type species: *Acholeplasma laidlawii* (Sabin 1941) Edward and Freundt 1970, 1.

Further Descriptive Information

Cells of acholeplasmas typically appear as pleomorphic coccoid, coccobacillary, or short filamentous forms when grown in mycoplasma broth containing 20% horse serum* or 1% bovine serum fraction.† Viable spherical cells probably have a minimum diameter of 300 nm (Lemcke, 1971). Filaments usually are 2–5 μm in length, but some longer filaments and branching filaments occur in some strains. Filaments often show beading with eventual development of coccoid forms from these filaments. Cellular morphology may also depend upon the ratio of unsaturated to saturated fatty acids in the medium. Cells of *A. laidlawii* grown on serum-containing medium supplemented with oleate are filamentous, while cells grown in presence of palmitate are spherical (Razin, et al., 1966; McElhaney and Tourtellotte, 1969).

Ultrastructural studies by transmission electron microscopy have confirmed the occurrence of short filamentous forms and coccoid cells. Individual cells are bounded by a 7- to 8-nm membrane and the cytoplasm contents consist of nuclear material, ribosomes, and intracellular granules (Maniloff, 1970). Shape and ultrastructure varies with the osmolar concentration of the fixative, so adjustment of preparative materials to the osmolarity of the culture medium is necessary for proper examination (Lemcke, 1972). Morphologic examination by scanning electron microscopy reveals a collection of pleomorphic forms, including irregularly shaped cells, spherical cells, chains of beads, and long filaments (Polak-Vogelzang et al., 1979; Gallagher and Rhoades, 1979). Multiplication of acholeplasmas resembles that described for the class *Mollicutes* and appears to occur primarily by binary fission (Maniloff, 1970; Virkola, 1972).

Most acholeplasmas exhibit heavy turbidity when grown aerobically in mycoplasma broth containing 5–20% horse serum, or when grown in 1% bovine serum fraction broth at 37°C. Variable turbidity occurs when some mycoplasmas are cultured at 37°C in serum-free broth.§ Most strains also grow at room temperature (25°C) in all of the above

media, but may require 7 days to reach turbidity observed in 24 h at 37°C. Strains of *A. morum* appear to be inhibited in medium containing 20% horse serum, but the organism will grow well in broth prepared with 5–10% horse serum or in 1% serum fraction broth. Strains of some acholeplasmas (*A. morum, A. modicum, A. axanthum*) may not grow well in serum-free medium unless carbohydrate (glucose) and some fatty acids (Tween 80 and palmitic acid) are included in the medium.

Colonies of acholeplasmas grown on solid media containing serum, serum fraction, or albumin-Tween 80 supplements reflect cultural differences noted with liquid medium. Most acholeplasmas grown on serum or serum fraction medium for 24–72 h at 37°C show classical "fried egg" colonies varying from 100–200 μm in diameter. Colony growth may increase over time so that after 5–7 days of incubation, some colonies may measure 2–3 mm in diameter (Le Normand et al., 1971; Meloni et al., 1980). Colonies of *A. axanthum* strains grown on serum fraction or serum-free media usually show only central zones of growth into the agar (Tully and Razin, 1969). *A. morum* grown on serum fraction or serum-free agar media exhibit rough irregular colonies 60–150 μm in diameter, and with a mulberry-like appearance (Rose et al., 1980).

Nutritional requirements have only been established for *A. laidlawii*, based upon a fully defined medium for the organism (Tourtellotte et al., 1964). *A. laidlawii* requires a large number of amino acids, glucose or some fermentable sugar, fatty acids, and some nucleic acid precursors, but does not require sterol (Rodwell and Mitchell, 1979).

All strains of *Acholeplasma* species form acid from glucose. Most acholeplasmas do not catabolize mannose, although a few *A. laidlawii* strains have been observed to ferment this carbohydrate. Hydrolysis of esculin by a β-D-glucosidase occurs with *A. axanthum, A. oculi,* and with most *A. laidlawii* strains (Williams and Wittler, 1971; Stipkovits et al., 1973; Bradbury, 1977). *A. equifetale* strains may or may not hydrolyze esculin. Arbutin hydrolysis is a useful diagnostic test for differentiation of *A. axanthum* and *A. oculi* from other acholeplasmas (Ernø and Stipkovits, 1973; Tully, 1979). Strong color reactions (4+) occur in the test system with all *A. axanthum* strains, while *A. oculi* and *A. morum* strains produce less intense (1+ to 2+) reactions and a few *A. laidlawii* strains may produce a light tint only (± reaction) (Tully, 1979).

Carotenoid pigments, principally neurosporene, occur in *A. laidlawii*, *A. granularum* and *A. oculi* (Rothblatt and Smith, 1961; Razin and Cleverdon, 1965; Tully and Razin, 1968; Tully, 1973). A rapid test, based upon absorption at 438 nm by lipid extracts of cell pellets prepared from small volumes (500 ml) of culture fluid (Razin and Cleverdon, 1965; Tully and Razin, 1968), is not sensitive enough to detect carotenoids in *A. axanthum* or *A. modicum* strains. Carotenoids have been identified in each of the latter acholeplasmas when cell pellets from 60–100 liters of culture are extracted (Mayberry et al., 1974; Smith and Langworthy, 1979). Other established *Acholeplasma* species are negative in the test for pigmented carotenoids (see Table 10.7). The film and spot reaction, which occurs in a number of *Mycoplasma* and two *Acholeplasma* species (see Table 10.7) is thought to be related to the production of crystallized calcium soaps of fatty acids on the surface of agar plates (Edward, 1954; Fabricant and Freundt, 1967). Fatty acids are liberated from the large amounts (20%) of horse serum

---

* Mycoplasma medium with 20% horse serum (per liter): Mycoplasma broth base (BBL), 15 g; water, 670 ml. The base medium is sterilized by autoclaving. Sterile supplements include: fresh yeast extract solution (25%) (Flow Laboratories, Rockville, Md.), 100 ml; horse serum, 200 ml; arginine solution (42%), 5 ml; glucose solution (50%), 10 ml; penicillin G (100,000 units/ml), 5 ml; phenol red solution (0.1%), 20 ml. Final pH 7.8. For solid medium add 0.8% Noble agar (Difco) and omit phenol red solution.

† Serum fraction medium (per liter): Mycoplasma broth base (BBL), 21 g; H₂O, 885 ml. The base medium is sterilized by autoclaving. Sterile supplements include: fresh yeast extract (25%), 100 ml; PPLO bovine serum fraction (Difco), 10 ml; penicillin (100,000 units/ml), 5 ml. Final pH, 7.8. For solid medium add 0.8% Noble agar (Difco) and omit phenol red solution.

§ Serum-free broth medium (per liter): Mycoplasma broth base, 21 g; water, 810 ml. The base medium is sterilized by autoclaving. Sterile supplements include: fresh yeast extract (25%), 100 ml; penicillin (100,000 units/ml), 5 ml; glucose solution (50%), 10 ml; Tween 80 (10% solution), 1 ml; palmitic acid (10 mg/ml), 1 ml; albumin (10% solution), 50 ml; phenol red solution (0.10%), 20 ml. Final pH 7.8. For solid medium add 0.8% Noble agar and omit phenol red.

in the agar medium by the lipolytic activity of the organism. Incorporation of 10% egg yolk emulsion into the agar medium enhances the film and spot reaction (Fabricant and Freundt, 1967; Thorns and Boughton, 1978).

Three morphologically different viruses have been described in strains of *A. laidlawii*. Group 1 viruses are nonlytic, naked bullet-shaped particles consisting of circular, single-stranded DNA with a molecular weight of 1.5×10^6. Group 2 viruses are nonlytic, spherical, enveloped particles consisting of circular double-stranded DNA with a molecular weight of 7.8×10^6. Group 3 viruses are lytic, polyhedral particles with short tails, consisting of linear double-stranded DNA with a molecular weight of 25.8×10^6 (Maniloff et al., 1979; Cole, 1979). A new virus, serologically distinct from those recovered from *A. laidlawii*, has recently been isolated from a strain (PG-49) of *A. modicum* (Congdon et al., 1979).

Antisera to filter-cloned whole cell antigens are utilized in several serological techniques to assess the antigenic structure of acholeplasmas. No one serological procedure combines specificity, sensitivity, and ease of performance. The three techniques most useful include: growth inhibition (GI) (Clyde, 1964; Tully, 1973), plate immunofluorescence (FA) (DelGiudice et al., 1967; Tully, 1973), and metabolism inhibition (MI) (Taylor-Robinson et al., 1966).

The acholeplasmas are sensitive to the following antibiotics (minimum inhibitory concentration range in $\mu g/ml$): tetracycline, 0.5–25.0; erythromycin, 0.03–1.0; lincomycin, 0.25–1.0; tylosin tartarate, 0.1–12.5; kanamycin, 20–200 (Ogata et al., 1971; Kato et al., 1972; Lewis and Poland, 1978).

No clear evidence exists at this time that acholeplasmas play a pathogenic role in any animal, insect or plant disease. The widespread distribution of acholeplasmas in both healthy and diseased animal tissues, in addition to antibody against acholeplasmas in most animal sera, complicates experimental pathogenicity studies. Strains of *A. axanthum* and *A. laidlawii* are pathogenic to chicken embryos. Inoculation of acholeplasmas into leafhoppers, including those known to be vectors of plant mycoplasma diseases, shows multiplication and prolonged persistence of organisms in host tissues (Whitcomb et al., 1973; Whitcomb and Williamson, 1975).

Mycoplasmas now designated as acholeplasmas were first isolated from sewage, compost and soil, and they were considered to be true saprophytes, although little evidence was supplied that the organism could persist in these environments. Subsequent isolation of the organisms from the bovine genital tract (Edward, 1950) stimulated others to examine animal tissues. Acholeplasmas have been found in almost every type of vertebrate examined, and in a variety of tissues and organs (Tully, 1979). The frequent occurrence of acholeplasmas in animal tissues and sera has created a special problem where these organisms are now the most frequent contaminant of tissue cells in culture. The size and filterability of acholeplasmas has complicated their removal from animal serum used in growth and nutrition of cell cultures. The widespread distribution of acholeplasmas in animal tissues and secretions raises questions about their existence as saprophytes, since recovery from soil and sewage could be related to their presence in animal wastes (Tully, 1973). However, recovery of acholeplasmas from plant material is being reported with increasing frequency. While no clear role for acholeplasmas in plant diseases has been established, there is evidence accumulating that some acholeplasmas may be part of the normal plant flora. These may include not only established species, such as *A. axanthum* or *A. oculi* (Eden-Green and Tully, 1979), but some acholeplasmas not related to known species (McCoy et al., 1981).

Enrichment and Isolation Procedures

Animal tissues or plant materials are divided into two portions and each gently minced directly into tubes containing either mycoplasma broth with 20% horse serum or serum fraction broth. Extra supplements of penicillin (final concentration 1,000 units/ml) or thallium acetate (final concentration 1:2,000) may be added to the broth for-

mulation to suppress bacterial growth. The tissues are allowed to soak for 1 h and at least two 10-fold dilutions of each primary culture are made in the same type broth. Fluid samples (blood, secretions) should be added at about 1:10 dilution to the same type of primary broth cultures and similar 10-fold dilutions made. All tubes are incubated aerobically at 37°C for 14 days and examined periodically for turbidity. Tubes showing turbidity are plated to agar medium, prepared from the same basic broth medium formulation, and the plates incubated at 37°C in an atmosphere of 95% N_2, 5% CO_2. Tubes without obvious turbidity should be plated at the end of the 14-day incubation period. Colonies appearing on the medium in 2–5 days may be identified directly with specific fluorescein-conjugated antisera to *Acholeplasma* species in the plate immunofluorescence test. Colonies on other agar plates should be transferred to broth and the cultures purified by filtration-cloning techniques. Identification is confirmed by additional biochemical and serological tests.

Maintenance Procedures

Stock cultures can be maintained in either mycoplasma broth medium containing 5–20% horse serum or in the serum fraction broth formulation at 25–30°C, with weekly transfer. Maintenance is best in broth medium devoid of glucose, since excess acid production reduces viability. Agar colonies can also be maintained for 1–2 weeks at room temperature (25°C) if plates are sealed to prevent drying.

For optimum preservation, acholeplasmas should be lyophilized directly in the culture medium when the broth cultures reach mid-logarithmic phase (usually 1–2 days at 37°C). Lyophilized cultures should be sealed under vacuum and stored at 4°C. Stock cultures in broth medium may also be maintained indefinitely when frozen at −70°C.

Procedures for Testing Special Characteristics

Sterol test. The growth response of acholeplasmas to cholesterol is most accurately assessed by measuring growth in a number of serum-free medium preparations (100 ml each) to which various concentrations of solubilized cholesterol are added (Razin and Tully, 1970; Edward, 1971). Acholeplasmas usually show no significant growth response with increasing cholesterol levels, whether the response is measured by total protein yields of pellets recovered from liquid cultures or by numbers of colony-forming units on solid medium. Tween 80 and palmitic acid should be included in the base medium for these tests, since a number of acholeplasmas (*A. axanthum*, *A. morum*) require additional essential fatty acids in the fatty acid-poor base medium before adequate cellular growth occurs. Sterol requirements cannot be determined adequately by passage of the organism once or twice on a serum-free agar or broth. Small amounts of sterols are often passed with the initial inoculum or some sterol may occur in other media components employed.

Arbutin hydrolysis. Arbutin is added to mycoplasma agar medium prepared with 20% horse serum. Twelve milliliters of 10% solution are sterilized by filtration and added to a liter of sterile agar medium before plates are poured (Ernø and Stipkovits, 1973). The test is performed by inoculating arbutin agar plates with undiluted and diluted (1:1000) stock cultures. After the plates dry, disks soaked with 0.02 ml of 5% solution of ferric citrate are placed in the center of the plate. The plates are incubated at 37°C and examined daily for 14 days for the appearance of brown or black coloration. *A. axanthum* strains produce a strong coloration (4+) after 3–4 days incubation, while *A. oculi* (2+) and *A. morum* (1+) produce less intense coloration after 3–7 days incubation. A few *A. laidlawii* strains show a mild tint in the medium after 7–14 days.

Carotenoid pigment test. The organisms are grown in serum fraction broth containing 2% bovine serum fraction and 0.06 M sodium acetate. Organisms are harvested by centrifugation after 24–48 h incubation at 37°C and washed twice in 0.25 M sodium chloride. The pellet of washed cells is extracted with 20 volumes of boiling ethanol for 10 min. The extracted cells are removed by centrifugation and the absorption spectrum of the supernatant fluid is determined at a wave length of 438 nm. The amount of carotenoid pigment in cells is expressed as absorbancy at 438 nm × 1000/mg of cell protein. Values of 4–37 are obtained with acholeplasmas producing large amounts of carotenoid pigments (Tully and Razin, 1968).

Plate immunofluorescence test. Antisera prepared against purified type strains of *Acholeplasma* species are conjugated with fluorescein isothiocyanate (DelGiudice et al., 1967). Recommended procedures for preparation of potent and specific antisera to the acholeplasmas have been described earlier (Tully, 1979). Conjugated antisera for the plate immunofluorescence test should be diluted to a concentration that gives strong fluorescence with homologous strains and little or no reaction with heterologous strains. Usually, dilutions of 1:10–1:30 are satisfactory for acholeplasma conjugates. Colonies should be well spaced on the agar plates and tests performed only on young (1- to 2-d-old) colonies. The plates are soaked for 20–30 m with phosphate-buffered saline and then drained dry. About 1 ml of the diluted conjugate is added to each plate and cultures incubated at room temperature for 20 m. The conjugate is drained off, the plates allowed to dry in an inverted position for 20–30 m, and examined in the fluorescence microscope using incident illumination with ultraviolet light.

Differentiation of the genus **Acholeplasma** from other genera

The characteristics of *Acholeplasma* species that distinguish them from other genera with members having similar morphological or physiological features are given in the chapter on the class *Mollicutes*.

Taxonomic Comments

Acholeplasmas may possess a number of other biological characteristics that distinguish them from other organisms included in the class *Mollicutes*. These include polyterpenol synthesis (Smith, 1979), positional distribution of fatty acids (Rottem and Markowitz, 1979), and presence of superoxide dismutase (Kirby et al., 1980; Lynch and Cole, 1980; O'Brien et al., 1981). However, most of these features (with exception of superoxide dismutase) have not been established for even a majority of *Acholeplasma* species, so it is perhaps premature to include these as characteristics of the family *Acholeplasmataceae*. The presence of superoxide dismutase in 5 *Acholeplasma* species and the absence of the enzyme in 17 *Mycoplasma* species suggests it may have some taxonomic relevance (O'Brien et al., 1981).

The finding that a fructose 1,6-diphosphate (FDP)-activated nicotinamide adenine dinucleotide (NAD)-dependent L(+)-lactate dehydrogenase (LDH) occurs in *A. laidlawii* has suggested some important evolutionary relationships of acholeplasmas to streptococci (Neimark, 1979). This highly specific activation is unusual and previously was known to occur only within the *Lactobacillaceae*, specifically in streptococci, where it has been found in every streptococcal group examined. Properties of purified FDP-activated LDH from *A. laidlawii* were shown to resemble those of the LDHs from streptococci (Neimark and Tung, 1973). Neimark (1973) subsequently demonstrated FDP-activated LDHs in *A. granularum*, *A. axanthum*, *A. modicum*, and *A. oculi*. Confirmation of the suspected evolutionary relationship between acholeplasmas and streptococci was supported by further observations that antisera to specific *Streptococcus* aldolases showed an immunological cross-reaction to antigens of *A. laidlawii* and a number of other *Acholeplasma* species (Neimark, 1979). The position of the acholeplasma aldolases on a phylogenetic map constructed from various streptococcal aldolases was taken to suggest that the acholeplasmas appear to have diverged from streptococci and appear not to be ancestral to lactic acid bacteria. It seems clear at this point that the acholeplasmas could not have evolved from streptococci simply by loss of a cell wall, since there are a number of significant genetic differences (genome size, base ratio, etc.) between these groups of organisms.

More recent studies on the phylogenetic relationships of the acholeplasmas suggest a close link to the genus *Clostridium*. Comparative analysis of 16S ribosomal RNA oligonucleotide sequences have led some to conclude that acholeplasmas and mycoplasmas did not share a common ancestor and that the acholeplasmas cluster closely to two species of *Clostridium* (Woese, Maniloff, and Zablen, 1980).

Acholeplasma species have been compared recently by DNA/DNA hybridization techniques and the results confirm the species separation made on the basis of biochemical and serological markers (Aulakh et al., 1979; Stephens et al., 1983). This technique can also be used effectively to assess genetic distinctions of candidate *Acholeplasma* species (Stephens et al., 1981).

Further Reading

Tully, J. G. 1973. Biological and serological characteristics of the acholeplasmas. Ann. N. Y. Acad. Sci. *225:* 74–93.

Tully, J. G. 1979. Special features of the Acholeplasmas. In Barile and Razin (Editors), *The Mycoplasmas*, Vol. 1, Cell Biology, Academic Press, New York, pp. 431–449.

Tully, J. G. and S. Razin. 1977. The Mollicutes: Acholeplasmas, Spiroplasmas, Thermoplasmas, and Anaeroplasmas. In Laskin and Lechevalier (Editors), *Handbook of Microbiology*, 2nd Ed., Vol. I, Bacteria, CRC Press, Cleveland, pp. 445–459.

Differentiation of the species of the genus **Acholeplasma**

The differential characteristics of *Acholeplasma* species are listed in Table 10.6.

Table 10.6.
Characteristics differentiating **Acholeplasma** *species[a]*

Characteristics	1. *A. laidlawii*	2. *A. granularum*	3. *A. axanthum*	4. *A. modicum*	5. *A. oculi*	6. *A. equifetale*	7. *A. hippikon*	8. *A. morum*
Fermentation of mannose	[−]	−	−	−	−	d	+	−
Esculin hydrolysis	[+]	−	+	−	+	d	−	+
Arbutin hydrolysis	[±]	−	4+	−	2+	−	−	1+
Carotenoid pigments produced[b]	+	+	−	−	+	−	−	−
Film and spots	−[c]	−	−	−	−	+	+	−
Mol% G + C of DNA	31–36	30–32	31	29	26–27			34

[a] Symbols: +, all strains positive; [+], most strains positive; [±], most strains give a weak reaction; d, differs among strains; [−], most strains negative; −, all strains negative; blank space, not determined.

[b] Presence of carotenoids based upon light absorption at 438 nm and when the cell yield from 100–500 ml of culture medium is tested.

[c] Positive film and spot reactions occur when the organisms are grown in serum fraction medium supplemented with Tween 80 and CaCl₂ (Razin and Rottem, 1973).

List of the species of the genus **Acholeplasma**

1. **Acholeplasma laidlawii** (Sabin 1941*) Edward and Freundt 1970, 1.[AL] (*Sapromyces laidlawii* Sabin 1941, 59.)

laid.law'i.i. M.L. gen. n. *laidlawii* of Laidlaw; named for P. Laidlaw, one of the microbiologists who first isolated this species.

Filaments usually relatively short, 2–5 μm, although much longer branched filaments may develop in media with a proper ratio of saturated to unsaturated fatty acids. Coccoid forms may predominate in certain cultures (tissue cell cultures). Cells are very sensitive to osmotic lysis.

Agar colonies are large (200 μm to 3 mm in diameter) and exhibit well developed central zones and peripheral growth on horse serum agar. On serum-free agar the colonies are smaller and may show only the central zone of growth into the agar. Relatively strong turbidity is produced during growth in fluid media containing 5–20% horse serum.

Other characteristics are listed in Tables 10.6 and 10.7.

Agar colonies produce zones of β-hemolysis of sheep and guinea pig red blood cells by the overlay technique. The hemolysin has been identified as hydrogen peroxide.

Minimal nutritional requirements: potassium, magnesium and phosphate ions; glucose; 13 amino acids; nucleic acid precursors; nicotinic acid, riboflavin, folinic acid, pyridoxine, pyridoxal and thiamine; long chain fatty acids.

Although facultatively anaerobic, freshly isolated strains grow best aerobically.

Temperature range, ~20–41°C with optimum at 37°C, even for strains recovered from plant or nonanimal sources.

Serologically distinct from most other *Acholeplasma* species. Partially related to *A. granularum* strains in some serological tests, including complement fixation, double immunodiffusion, growth inhibition and metabolism inhibition tests.

Isolated from sewage, manure, humus, soil, and from almost all animals surveyed, including avian, bovine, caprine, equine, feline, murine, ovine, porcine and primate (including man) hosts. Frequently recovered from the oral cavity, respiratory and genital tract secretions, eye, lymph nodes, semen and serum. Pathogenicity has not been well established, although some strains are lethal for chicken embryos.

The mol% G + C of the DNA is 31.7–35.7 (T_m, Bd). Genome size: 1.0×10^9 daltons. The genetic relatedness among strains of the species

Table 10.7.

Other characteristics of the species of the genus **Acholeplasma**[a]

Characteristics	1. A. laidlawii	2. A. granularum	3. A. axanthum	4. A. modicum	5. A. oculi	6. A. equifetale	7. A. hippikon	8. A. morum
Acid production from:								
Cellobiose	+		+		+			
Dextrin	+		+					
Dulcitol	−		−					
Fructose			−		+	+	+	
Galactose			+		+	+	+	
Glucose	+	+	+	+	+	+	+	+
Glycerol	−		+		−			
Glycogen	+		+					
Lactose	−		−					
Maltose	+		+			+	+	
Mannitol	−		−		−			
Salicin	−		+		−			
Sorbitol	−		−		−			
Starch	+		+					
Sucrose	−		−		−	+	+	
Xylose	−		−		+			
Gelatin liquefaction	d	−				−		−
Digestion of:								
Casein	−	−				−		−
Coagulated blood serum	−	−	−	−		−		−
Adsorption of red blood cells:								
Guinea pig	−	−	−	−		−		−
Human			−					
Ox				−				
Bovine						−		−
Ovine						−		−
Canine						−		−
Rabbit						−		−
Chicken						−		−
Equine						−		−
Reduction of tetrazolium:								
Aerobically	Weak	Weak	+	+	+	−[b]	−[b]	
Anaerobically	+	+	+	+	+	−[b]	−[b]	+
Phosphatase activity	−	−	−	−				−

[a] For symbols see standard definitions.

[b] Variable reduction of tetrazolium occurs in liquid media and no reduction occurs on solid media.

---

* Incorrectly cited as "Freundt 1955" on the Approved Lists of Bacterial Names.

is not well established, although two strains (A and B) show ~70% relatedness by various techniques. These strains exhibit ~20% relatedness to *Acholeplasma granularum* (strain BTS-39).

Type strain: ATCC 23206 (NCTC 10116; strain PG-8 of Laidlaw and Elford, 1936; see Edward and Freundt, 1970).

2. **Acholeplasma granularum** (Switzer 1964) Edward and Freundt 1970, 2.[AL] (*Mycoplasma granularum* Switzer 1964, 504.)

gra.nu.la′rum. L. neut. n. *granulum* a small grain, a granule; L. gen. pl. n. *granularum* of small grains, made up of granules, granular.

Pleomorphic, with short filaments and coccoid cells.

Large colonies up to 1 mm in diameter with clearly marked centers are formed on horse serum agar; colonies on serum-free medium are smaller and may lack the peripheral growth around their centers.

Agar colonies produce a zone of α-hemolysis by the overlay technique, using sheep red blood cells.

Other characteristics are listed in Tables 10.6 and 10.7.

Minimum nutritional requirements have not been defined.

Temperature range, 22–37°C.

Serologically distinct from most other *Acholeplasma* species by plate immunofluorescence tests. Partially related to *A. laidlawii* strains in some serological tests, including complement fixation, double immunodiffusion, growth inhibition and metabolism inhibition tests.

Isolated frequently from the nasal cavity of swine. Also recovered from the lung and feces of swine, and from the conjunctivae and nasopharynx of horses. Pathogenicity has not been defined, although aerosol challenge of specific pathogen-free pigs did not induce clinical or histologic evidence of disease.

The mol% G + C of the DNA is 30.5–32.4 (T_m, Bd). DNA/DNA hybridization studies show 20–22% relatedness to *A. laidlawii* but no homology to other acholoplasmas.

Type strain: ATCC 19168 (NCTC 10128; strain BTS-39 of Switzer, 1964).

3. **Acholeplasma axanthum** Tully and Razin 1970, 751.[AL]

a.xan′thum. Gr. pref. *a* not, without; Gr. adj. *xanthus* yellow; M.L. neut. adj. *axanthum* without yellow (pigment).

Predominantly coccobacillary and coccoid with a few short myceloid elements usually 2–5 μm in length. The cells are very sensitive to osmotic lysis.

Large colonies with clearly marked centers are formed on horse serum agar; colonies on serum-free media are smaller and usually lack the peripheral growth around their center. In general, growth in media devoid of serum or serum fraction is much poorer than for other acholeplasmas.

Can be shown to synthesize carotenoid pigments only when large volumes (100 liters) of culture are tested. Produces sphingolipids.

Agar colonies produce zones of β-hemolysis by the overlay technique.

Other characteristics are listed in Tables 10.6 and 10.7.

Minimum nutritional requirements are poorly defined. A marked stimulation of growth by Tween 80 suggests a requirement for fatty acids.

Temperature range, 22–37°C.

Serologically distinct from other *Acholeplasma* species by complement fixation, growth inhibition and immunofluorescence tests.

Originally recovered from murine leukemia tissue culture cell lines, but subsequent isolation from bovine serum suggests cell culture contamination was of bovine origin. Also isolated from other bovine tissues (nasal cavity, lymph node and kidney), porcines (lung and peribronchial lymph nodes), equine oral cavity, and goose embryos. Also isolated from some plant tissues, including normal coconut palm and palms infected with "lethal yellowing" disease.

Intranasal challenge of specific pathogen-free piglets induces mild clinical symptoms, gross lesions and histologic changes in lungs. Pathogenic for goose and chicken embryos.

The mol% G + C of the DNA is 31 (Bd). No DNA/DNA hybridization occurs between this species and other acholeplasmas.

Type strain: ATCC 25176 (NCTC 10138; strain S-743 of Tully and Razin, 1970).

4. **Acholeplasma modicum** Leach 1973, 147.[AL]

mo.di′cum. L. neut. adj. *modicum* moderate, referring to moderate growth.

Pleomorphic, spherical, ring-shaped and coccobacillary forms.

Agar colonies are distinctly smaller than those of most other acholeplasmas. Very small colonies without peripheral zones of growth are noted on serum-free agar. Very light turbidity occurs in serum-free broth. Does not produce film and spots.

Can be shown to synthesize carotenoid pigments only when large volumes of culture (50 liters) are tested.

Agar colonies produce zones of α- or β-hemolysis by the overlay technique, using ox, sheep and guinea pig red blood cells. Guinea pig cells are agglutinated.

Other characteristics are listed in Tables 10.6 and 10.7.

Minimum nutritional requirements have not yet been defined.

The temperature range and optimum have not been exactly defined, but growth does occur at both 22 and 37°C.

Serologically distinct from other species of *Acholeplasma* by growth inhibition and plate immunofluorescence tests.

Isolated from the blood, bronchial lymph nodes, thoracic fluids, lungs, and semen of cattle. Also recovered from porcine nasal secretions. Pathogenicity remains to be determined.

The mol% G + C of the DNA is 29.3 (T_m). No DNA/DNA hybridization occurs between this species and other acholeplasmas.

Type strain: ATCC 29102 (NCTC 10134; strain Squire (PG-49) of Leach, 1973).

5. **Acholeplasma oculi** Al-Aubaidi, Dardiri, Muscoplatt and McCauley 1973, 117.[AL] (*Acholeplasma oculusi* (sic) Al-Aubaidi, Dardiri, Muscoplatt and McCauley 1973, 117; orthographic error corrected by Al-Aubaidi 1975, 221.)

o.cu′li. L. n. *oculus* the eye; L. gen. *oculi* of the eye.

Pleomorphic, spherical, ring-shaped and coccobacillary forms.

Medium-size colonies (200–700 μm in diameter) with clearly marked centers are formed on horse serum agar; colonies on serum-free media are smaller and may lack the peripheral growth around the center.

Other characteristics are listed in Tables 10.6 and 10.7.

Agar colonies produce zones of hemolysis by the overlay technique, using sheep red blood cells.

The minimum nutritional requirements have not been determined.

The temperature range and optimum have not been exactly defined, but growth does occur at both 25 and 37°C.

Resistant to 100 μg/ml of kanamycin; sensitive to 10 μg/ml of erythromycin. Sensitive to 1% bile salts.

Differs antigenically from other *Acholeplasma* species by metabolism inhibition and plate immunofluorescence tests.

Isolated from the conjunctivae of goats with keratoconjunctivitis, porcines (nasal secretions), equines (nasopharynx, lung, spinal fluid, joint, and semen), and external genitalia of guinea pigs. Intravenous inoculation of conventional goats produces signs of pneumonia and death within 6 days. Conjunctival inoculation of conventional goats produces mild conjunctivitis.

The mol% G + C of the DNA is 27 (T_m).

Type strain: ATCC 27350 (strain 19-L of Al-Aubaidi et al., 1973).

6. **Acholeplasma equifetale** Kirchhoff* 1974, 207.[AL]

eq.ui′fe.ta′le. L. n. *equus* horse; L. adj. *fetalis* fetal; M.L. adj. *equifetale* pertaining to the horse fetus.

Pleomorphic, predominantly coccoid.

Agar colonies are small (100–300 μm in diameter) but exhibit marked central zone of growth and peripheral growth on horse serum agar. On serum-free agar colony growth is similar but the colony size is smaller (60–100 μm in diameter). Film and spot reaction produced on horse serum agar plates containing 10% egg yolk emulsion after 7 days, but not produced on swine serum agar with egg yolk emulsion.

---

* Incorrectly cited as "Kirchoff" on the Approved Lists of Bacterial Names.

Other characteristics are listed in Tables 10.6 and 10.7.

Potassium tellurite is reduced when colonies are sparse.

Agar colonies produce zones of β-hemolysis by the overlay technique with equine, bovine, ovine, canine, rabbit, guinea pig and chicken red blood cells.

The minimum nutritional requirements have not been established. Growth occurs at 22 and 37°C.

Serologically distinct from other *Acholeplasma* species by growth inhibition, metabolism inhibition and plate immunofluorescence tests.

Isolated from the lung and liver of aborted horse fetuses. Also recovered from the respiratory tract of apparently normal horses and the respiratory tract and cloacae of broiler chickens. Pathogenicity has not been defined.

The mol% G + C of the DNA has not been determined. No DNA/DNA hybridization between this species and other acholeplasmas.

Type strain: ATCC 29724 (strain C112 of Kirchhoff, 1978).

7. **Acholeplasma hippikon** Kirchhoff (see previous footnote) 1974, 210.[AL]

hip.pi'kon. Gr. adj. *hippikon* pertaining to the horse.

Pleomorphic, with predominantly coccoid elements.

Agar colonies are small to medium (200–700 μm in diameter) and exhibit marked central zones and peripheral growth on horse serum agar. Smaller colonies are observed on serum-free agar. Light turbidity occurs in serum-free broth. Film and spot reaction produced on horse serum agar plates containing 10% egg yolk emulsion after 7 days, but not on swine serum agar with egg yolk emulsion.

Potassium tellurite is reduced when the colonies are sparse.

Other characteristics are listed in Tables 10.6 and 10.7.

Agar colonies produce zone of β-hemolysis by the overlay technique with equine, bovine, ovine, canine, rabbit, guinea pig and chicken red blood cells.

The minimum nutritional requirements have not been established. Growth occurs at 22 and 37°C.

Serologically distinct from other *Acholeplasma* species by growth inhibition, metabolism inhibition, and plate immunofluorescence tests.

Isolated from the lung of aborted horse fetuses. Pathogenicity has not been defined.

The mol% G + C of the DNA has not been determined. No DNA/DNA hybridization between this species and other acholeplasmas.

Type strain: ATCC 29725 (strain C1 of Kirchhoff, 1978).

8. **Acholeplasma morum** Rose, Tully and DelGiudice 1980, 653.[VP]

mo'rum. L. n. *morum* mulberry; denoting the mulberry-like appearance of agar colonies of the organism.

Pleomorphic, predominantly coccoid or coccobacillary forms, and with some beaded filaments.

Agar colony growth is depressed or completely inhibited on 20% horse serum agar. Medium size colonies (100–150 μm diameter) are formed on bovine serum fraction agar, but they have only a central zone of growth. Small colonies (60–100 μm) are formed on serum-free agar and also lack peripheral growth around the central zone. Poor growth occurs in broth cultures containing 20% horse serum. Good growth occurs in broth containing 5–10% horse serum or 1% bovine serum fraction. Growth in serum-free broth requires fatty acid supplements (palmitic acid, Tween 80).

Other characteristics are listed in Tables 10.6 and 10.7.

The minimum nutritional requirements have not been defined.

Although facultatively anaerobic, the organism grows best aerobically.

Temperature range, 23–37°C.

Serologically distinct from other acholeplasmas by growth inhibition and immunofluorescence tests.

Originally recovered from bovine serum and from calf kidney cell cultures containing fetal calf serum. Pathogenicity has not been defined, although calf kidney cell cultures containing the organism showed cytopathogenic effects.

The mol% G + C of the DNA is 34.0 (T_m). No DNA/DNA hybridization occurs between this species and other acholeplasmas.

Type strain: ATCC 33211 (strain 72-043 of Rose et al., 1980).

FAMILY III. **SPIROPLASMATACEAE** SKRIPAL 1983, 408[VP]

ROBERT F. WHITCOMB AND JOSEPH G. TULLY

Spi.ro.plas.ma.ta'ce.ae. M.L. neut. n. *Spiroplasma* type genus of the family; -aceae ending to denote a family; M.L. fem. pl. n. *Spiroplasmataceae* the *Spiroplasma* family.

Cells helical during logarithmic growth, with rotatory, flexional and translational motility. Genome size, 1×10^9 daltons. Sterols required for growth. Possess a phosphoenolpyruvate phosphotransferase system for glucose. Reduced nicotinamide adenine dinucleotide (NADH) oxidase activity is located only in the cytoplasm. Unable to synthesize fatty acids from acetate. Other characteristics are as described for the class and order.

Type genus: *Spiroplasma* Saglio, L'hospital, Laflèche, Dupont, Bové, Tully and Freundt 1973, 191.

Genus I. **Spiroplasma** Saglio, L'hospital, Laflèche, Dupont, Bové, Tully and Freundt 1973, 191[AL]

ROBERT F. WHITCOMB AND JOSEPH G. TULLY

Spi.ro.plas'ma. Gr. n. *spira* a coil, spiral; Gr. n. *plasma* something formed or molded, a form; M.L. neut. n. *Spiroplasma* spiral form.

Cells pleomorphic, varying in size and shape from **helical** and branched nonhelical filaments to spherical or ovoid. The helical forms, usually 100–200 nm in diameter and 3–5 μm in length, typically occur during the logarithmic phase of growth and in some species persist during the stationary phase. Spherical cells ~300 nm in diameter and nonhelical filaments are frequently seen in the stationary phase or in all growth phases in suboptimal growth media. **Helical filaments are motile, with flexional and twitching movements, and often show an apparent rotatory motility. Flagella, periplasmic fibrils, or other organelles of locomotion are not present, but intracellular fibrils have been demonstrated.** Cells divide by binary fission. Facultatively anaerobic. Temperature range for growth, 20–37°C. **Colonies are frequently diffuse,** reflecting the motility of the cells during active growth. The colony type is strongly dependent on the agar concentration. Colony sizes vary from 0.1–4.0 mm. Colonies formed on solid media by nonmotile variants or by cultures growing on inadequate media, typically attain diameters of 200 μm or less and exhibit a typical umbonate appearance. Colonies of motile, fast-growing spiroplasmas are diffuse, often with satellite colonies developing from foci adjacent to the initial site of colony development. Light to heavy turbidity is produced in liquid cultures. Chemoorganotrophic. Acid is produced from glucose. Variable fermentation of other carbohydrates occurs. Most strains hydrolyze arginine. There is no hydrolysis of urea, arbutin or esculin. Phosphatase-positive. No liquefaction of coagulated horse serum occurs. Adsorption of guinea pig red blood cells is variable. **Cholesterol** (or possibly other sterols) **is required for growth.** Some species require an optimum osmolality of the culture medium for primary growth, usually in the range of 300–800 mOsm. Media containing mycoplasma broth base, serum, and other supplements are required for primary growth, but on adaptation growth may occur in less complex media. Defined media are available for some strains. Resistant to 10,000 U/ml penicillin. Sensitive to erythromycin and tetracycline. Sensitive to 1.5% digitonin in the disk test. **Isolated from ticks, the hemolymph and guts of insects, from vascular plant fluids and insects that feed on the fluids, and from the surfaces of flowers and other plant parts.** The type species is pathogenic for citrus (grapefruit and orange), producing "stubborn" disease; experimental or natural infections are also established in corn, horseradish, periwinkle, radish, broad bean, etc. Some species are pathogenic under experimental conditions for insects, chicken embryos, and for a variety of suckling rodents (rats, mice, hamsters and rabbits). The mol% G + C of the DNA is 25–31 (T_m, Bd).

Type species: *Spiroplasma citri* Saglio, L'hospital, Laflèche, Dupont, Bové, Tully and Freundt 1973, 191.

Further Descriptive Information

In the logarithmic phase in liquid media, most spiroplasma cells are helically coiled filaments 90–250 nm in diameter and of variable length (Cole et al., 1973). The cells increase in length and divide by constriction; such division often occurs when the parental helix contains as few as four turns (Garnier et al., 1981). The cells contain ribosomes ~17 nm in diameter. One end of the typical helical filament often appears blunt, and the other end tapered (Williamson and Whitcomb, 1974; Garnier et al., 1981). In the stationary or death phase, the cells are usually distorted and often are nonhelical or spherical. The bounding membrane of the cells is 6.7–8.7 nm in width; inner and outer layers are often observed abutting it. An outer delicate "nap," 5–7 nm in width, consists of short, apparently periodic projections that can be seen by negative staining (Cole et al., 1973). An outer layer has also been seen in freeze-etched cells (Razin et al., 1973). An internal layer ~6.7 nm wide has been seen in sectioned cells (Cole et al., 1973). Regions in the membranes of aging spiroplasma cells with a barred periodicity of units ~4 nm in width and with a center-to-center distance of ~5 nm are frequently noted in negatively stained preparations (Cole et al., 1973).

Localization of enzymatic activities in spiroplasma cells has proved to be of taxonomic significance. For example, NADH oxidase activity is located in the cytoplasm (Kahane et al., 1977; Mudd et al., 1977) in contrast to its membrane location in acholeplasmas.

Helical spiroplasma cells exhibit flexing, twitching, and apparent rotation about the longitudinal axis (Cole et al., 1973; Davis and Worley, 1973). Spiroplasmas show little translational movement in media of low viscosity, but translational movement (Davis, 1978) occurs in semisolid agar media, in liquid media at high viscosity, or in hemo-

lymph from infected insects. Spiroplasmas have been shown to undergo chemotactic movement toward such substances as carbohydrates and amino acids (Daniels et al., 1980). A nonhelical variant (ASP-1) of *S. citri* that occurred naturally in diseased sweet orange was nonmotile (Townsend et al., 1977, 1980b). Although the variant twitched erratically, it showed none of the other types of motility observed with other spiroplasmas. The Y32 spiroplasma isolated from *Ixodes* ticks grew in predominantly nonhelical form throughout all growth phases in the best available media (Tully et al., 1981).

Spiroplasma cells may exhibit microfibrils 3.6 nm in width, with repeat intervals of 9 nm along their lengths (Williamson, 1974). These membrane-associated fibrils, which have been purified (Townsend et al., 1980a), may be involved in spiroplasma motility. The possible presence of actin-like proteins in spiroplasmas has been studied. Antisera prepared against SDS-denatured invertebrate actin coupled to horseradish peroxidase specifically stained cells of *S. citri* (Williamson et al., 1979a). Also, a protein with a molecular weight similar to actin has recently been isolated and reacts with IgG directed against rabbit actin (C. Mouches, A. Menara, B. Geny, D. Charlemagne and J. M. Bové, Rev. Inf. Dis., vol. 4 (Suppl.) S 277, 1982).

Recent development of a classification scheme for spiroplasmas (Junca et al., 1980; Whitcomb et al., 1982b) has permitted delineation of intrageneric groups and subgroups (Table 10.8). In the scheme, "groups" have been defined as clusters of similar organisms, all of which possess negligible DNA/DNA homology with representatives of other groups, but moderate to high levels (20–100%) with each other. Such groups are therefore, in effect, unnamed species. This level of molecular genetic differentiation has been correlated with substantial differences in two-dimensional PAGE (polyacrylamide gel electrophoresis) patterns, and in serology. "Subgroups" have been defined as clusters of spiroplasma strains showing intermediate levels of intragroup DNA/DNA homology (10–80%) and possessing corollary relationships in PAGE patterns and growth inhibition serology. "Serovars" have been defined as spiroplasmas varying substantially in metabolic inhibition and deformation serology, but that may be insufficiently differentiated from members of existing groups or subgroups to warrant separation.

Spiroplasmas reach titers of 10^8–10^{10} cells/ml (enumerated microscopically) in media containing 10–20% horse serum or fetal bovine serum. Growth rates of strains within groups tend to be similar. Fast-growing species (e.g. *S. floricola*, which has a doubling time of 1.8 h) achieve high titers with heavy turbidity even in simple media; slower growing species, such as *S. mirum*, may achieve titers of 10^8–10^9 organisms/ml, but have growth rates less than 1/10 that of *S. floricola*, and require more complex media (Tully et al., 1982).

S. citri can be cultivated on a relatively simple medium that utilizes sorbitol to adjust osmolality (Saglio et al., 1971). A modification of this medium (BSR) was recently reported (Bové and Saillard, 1979) in which the horse serum content was lowered to 10%, and fresh yeast extract and DNA were omitted.* Other simple media are also effective (Fudl Allah et al., 1971, 1972).

Certain even simpler variations, such as C-3G† (Liao and Chen, 1977) and DSM4§ (Davis et al., 1981) media, are suitable for maintenance or large-batch cultivation of fast-growing spiroplasmas. However, cultivation of more fastidious spiroplasmas is best achieved in M1A medium¶ (Jones et al., 1977) if they derive from plant or insect

---

* BSR medium contains: beef heart infusion (Difco), 25 g; glucose, 1.0 g; fructose 1.0 g; sucrose, 10.0 g; sorbitol, 70 g; phenol red, 20 mg; and deionized water, 900 ml. The base medium is sterilized at 121°C for 15–20 min. Horse serum (100 ml) is added as a sterile supplement to the cooled medium. The final pH should be ~7.6.
† C-3G medium contains: mycoplasma broth base (Difco), 15 g; sucrose, 120 g; agamma horse serum, 200 ml; and water, 720 ml.
§ DSM4 medium for *S. floricola* contains: mycoplasma broth base (Difco), 15 g; sucrose, 80 g; phenol red, 25 mg; horse serum (inactivated at 56°C for 30 min), 100 ml; and water, 820 ml.
¶ The base for M1A medium contains: beef heart infusion, 2.0 g; Bacto-tryptone (Difco), 3.3 g; Bacto-peptone (Difco), 6.0 g; sucrose, 3.3 g; glucose, 1.3 g; fructose, 0.3 g; sorbitol, 23.3 g; Yeastolate (Difco), 1.0 g; phenol red, 20 mg; and deionized water, 260 ml. The base medium is sterilized at 121°C for 15–20 min. The final pH should be ~7.6. The following sterile supplements are added to the cooled base medium: Schneider's *Drosophila* medium (Gibco), 533 ml; fetal bovine serum (heat-inactivated at 56°C for 1 h), 167 ml; fresh yeast extract solution (25%), 33 ml; and penicillin solution (250,000 U/ml), 8.3 ml.

Table 10.8.

Hosts, pathogenicity, and relatedness of the species, serogroups and subgroups of the genus **Spiroplasma**[a]

	Species or Serogroup[b]	Subgroup	Principal Strains	G + C (mol %)	DNA/DNA Homology[c]	Principal Host	Disease Incited	References
I	*Spiroplasma citri*	I-1	R8A2 [27556][d] C189 [27665] Israel	25–27	100%	Dicots, Leaf-hoppers	Citrus stubborn	Saglio et al., 1973 Markham et al., 1974
		I-2	AS 576 [29416] BC-3 [33219]	25–27	65%	Honey bee[e]	Honey bee spiro-plasmosis	T. B. Clark, 1977
		I-3	I-747 [29051] E275 [29320] B655 [33289]	25–27	53%	Maize, Leaf-hoppers	Corn stunt	Granados, 1969 Chen and Liao, 1975 Williamson and Whit-comb, 1975
		I-4	277F [29761]	25–27	21%	Rabbit tick	None known	Pickens et al., 1968 Brinton and Burgdorfer, 1976 Stalheim et al., 1978
		I-5[f]	LB-12 [33649]	nd	nd	Green leaf bug	None known	Lei et al., 1979
		I-6[f]	M55 [33502]	nd	nd	Flowers	None known	Whitcomb et al., 1982c
		I-7[f]	N525 [33287]	nd	nd	Plant sur-faces	None known	Eden-Green et al., 1981, 1983
II	SROs		WSRO NSRO ESRO	25–27	nd	*Drosophila*	Sex-ratio trait	Poulson and Sakaguchi, 1961 Williamson and Poulson, 1979
III	*Spiroplasma floricola*		23-6 [29989] OBMG [33221] BNR1 [33220] SLH	25–27	0%	Isolated from flow-ers and *Melo-lontha* beetles	"Lethargy disease" of *Melolontha*	Davis et al., 1981 T. B. Clark, 1978 Giannotti et al., 1981
IV	*Spiroplasma apis*[g]		B31 [33834] SR 3 [33095] PPS1 [33450]	29–31	0%	Isolated from flow-ers and honey bee	Honey bee disease	Mouches et al., 1982, 1983 Davis, 1978 McCoy et al., 1979
V	*Spiroplasma mirum*		SMCA [29335] GT-48 [29334] TP-2 [33503]	29–31	0%	Rabbit tick	Suckling mouse cataract syn-drome	H F. Clark, 1964 Tully et al., 1977
(VI)[h]			Y32[33835]	nd	nd	*Ixodes* ticks	None known	Tully et al., 1981
(VII)			MQ-1 [33825]	nd	nd	*Monobia* wasp	None known	T. B. Clark, 1982
(VIII)			EA-1 [33826]	nd	nd	Syrphid fly	None known	T. B. Clark, 1982
(IX)			CN-5 [33827]	nd	nd	*Cotinus* beetle	None known	T. B. Clark et al, 1982
(X)			AES-1 [35112]	nd	nd	*Aedes* mosquito	None known	T. A. Chen, unpublished data
(XI)			MQ-4	nd	nd	*Monobia* wasp	None known	T. B. Clark, unpublished data

[a] Modified from Junca et al. (1980), Whitcomb et al. (1982b) and Williamson and Whitcomb (1983) with additional data from other studies.

[b] Serogroups represent species or putative species and are assigned on the basis of failure to cross-react in growth inhibition, metabolic inhibition, and deformation tests.

[c] Reference DNA from *S. citri* strain R8A2.

[d] Accession numbers from American Type Culture Collection in brackets.

[e] Latin binomials for insect species: honey bee, *Apis mellifera*; rabbit tick, *Haemaphysalis leporispalustris*; green leaf bug, *Trigonotylus ruficornis*; wasp, *Monobia quadridens*; syrphid fly, *Eristalis arbustorum*.

[f] Subgroups I-5, I-6 and I-7 are proposed as designations for three previously unclassified serovars (J. M. Bové, C. Mouches, P. Carle-Junca, J. R. Degorce-Dumas, J. G. Tully and R. F. Whitcomb, personal communication).

[g] Serovars of *S. apis* are serologically heterogeneous (Tully et al., 1980).

[h] Groups in parentheses assigned on basis of serological evidence only, without molecular genetic data.

habitats, or in SP-4 medium‖ (Tully et al., 1977) if they derive from tick habitats. SM-1 medium** (Clark, 1982) has also been successfully employed for many spiroplasmas.

Spiroplasmas from ticks and certain strains from insects or flowers multiply at temperatures ranging from 20–37°C, with broad optima. Spiroplasmas from plants and leafhoppers tend to have optima at 30–32°C and have narrower ranges for growth.

All spiroplasmas studied have been shown to produce acid from glucose, although the rate may vary. Some strains of *S. citri* (i.e. members of subgroup I-4) and all strains of *S. mirum* ferment glucose slowly. Only one isolate, the PPS1 strain of *S. apis*, has been reported to lack the ability to hydrolyze arginine (McCoy et al., 1982). Arginine hydrolysis by some spiroplasmas may be observed only if glucose is also present in the medium. In these instances, there is an initial acidic shift followed by an eventual rise to alkaline pH values (Townsend, 1976). In other instances, arginine hydrolysis may occur rapidly, without the requirement for addition of glucose; this may be especially true if the cultures have been adapted to arginine or other amino acids prior to the test. In general, care should be taken in performing utilization tests to assure that the organisms have been given adequate opportunity to adapt for the utilization being studied. Fermentation of other carbohydrates has been difficult to study in a comparative way, because defined media have not been available that support growth of a wide variety of strains, and the complex constituents (e.g. yeast derivatives) that are required for growth of more fastidious strains contain fermentable substances that are utilized in control media (Tully et al., 1982).

S. floricola and some strains of *S. apis* have been cultivated in chemically defined media (Chang and Chen, 1982). These media contain inorganic salts, keto acids, nucleosides and nucleotides, carbohydrates, amino acids, vitamins, lipids, NAD phosphate, flavin adenine dinucleotide (FAD), glutathione, coenzymes, acetate, glucuronate, bovine serum albumin, and HEPES (N-2-hydroxyethylpiperazine-N'-2-ethanesulfonic acid) buffer. *S. floricola* achieves growth yields in excess of 10^9 cells/ml in 2 days in the defined medium, whereas more fastidious spiroplasmas have not been grown in this medium.

General characters of the spiroplasma genome have been examined with special reference to taxonomy. The genome size of five strains (including the type strain) has been determined to be $\sim 10^9$ daltons (Saglio et al., 1973; Lee and Davis, 1980). The G + C content of the DNA has been studied and found to be bimodally distributed (Bové and Saillard, 1979; Christiansen et al., 1979; Lee and Davis, 1980). The type species (*S. citri*: subgroup I-1), group II, and *S. floricola* (group III) have a content of 25–27 mol% G + C, as determined by both T_m and Bd. *Spiroplasma mirum* (group V) and strains of *S. apis* (Group IV) have a content of ~ 29–31 mol% G + C.

Cells of many spiroplasma species contain rod-shaped viruses (SVC1 = SpV1) that are associated with nonlytic infections. The dimensions of the SpV1 virus, which is frequently attached to the outer envelope of spiroplasma cells, are 10–15 × 230–280 nm. A second virus, reminiscent of a type B tailed bacterial virus that is possibly associated with lytic infection, occurs in a small number of *S. citri* strains (Cole et al., 1973). The virus (SVC2 = SpV2) has heads that are 48–51 nm between flat sides and 25–28 nm between vertices. The tails are 6–8 × 78–83 nm (Cole et al., 1974; Cole, 1979). A third virus (SVC3 = SpV3), whose virions are polyhedrons with short tails, has been found in all spiroplasma strains examined (Cole et al., 1974; Cole, 1979). The SpV1 and SpV3 viruses produce lytic cross-infections in other spiroplasma strains (Liss and Cole, 1981, 1982) that are detected by plaquing on lawns of *S. citri* strain 750. A fourth virus (SpV4) with isometric virions 25 nm in diameter was discovered in a single isolate of subgroup I-2 (B.

Table 10.9.
Antibiotic sensitivities of spiroplasmas[a]

Agent	Minimum Inhibitory Concentration
	μg/ml
Tetracycline, oxytetracycline, erythromycin	0.1–0.2
Tylosin	0.2
Carbomycin	0.4
Tobramycin	1.0
Lincomycin	1.6
Oleandomycin, filipin	3.1
Chloramphenicol	3.0–12.5
Kanamycin	5.0–25.0
Neomycin	10.0
Chlortetracycline	10.0–12.5
Nalidixic acid	25.0
Novobiocin	40.0
Streptomycin	50.0
Rifampicin, actinomycin D	100.0
Kasugamycin, hygromycin, spectinomycin	>200.0
Cycloheximide	>300.0
Sulfanilamide, penicillin	>500.0

[a] Sensitivities given for citrus stubborn spiroplasma (subgroup I-1) and honey bee spiroplasma BC-3 (subgroup I-2). Data from Saglio et al. (1973), Bowyer and Calavan (1974), and Liao and Chen (1981a).

Ricard, M. Garnier and J. M. Bové, Rev. Inf. Dis., vol. 4 (Suppl.) S 275, 1982). The state of carriage of all spiroplasma viruses is essentially unknown. Covalently closed circular DNA molecules of various sizes found in 10 spiroplasma strains including *S. citri* and subgroups I-2 and I-3 have been termed cryptic plasmids (Ranhand et al., 1980).

Although no group antigen has been conclusively demonstrated to distinguish all spiroplasmas from other mycoplasma genera, serum prepared against fibril protein (Townsend and Archer, 1983) has a clear potential to do so. Growth inhibition (GI) tests (Working Group, 1976) have been used to separate species (or groups), whereas metabolic inhibition (MI) (Working Group, 1975) and deformation (DF) tests (Williamson et al., 1978) have been used to identify subgroups (Williamson et al., 1979b; Davis et al., 1979). Serological heterogeneity between related strains, such as the various serovars of *S. apis* (Group IV) (Table 10.8), is shown best by MI procedures (Tully et al., 1980).

Antibiotic sensitivities are indicated in Table 10.9. Strains have been isolated that are permanently resistant to kanamycin, neomycin, gentamicin, erythromycin and several tetracycline antibiotics (Liao and Chen, 1981a).

Spiroplasmas have often been found to be associated with arthropods. Several spiroplasmas occur in a biological cycle that involves plant phloem and homopterous insects (Saglio and Whitcomb, 1979). In the course of passage through the insect, these organisms pass through, accumulate, or multiply in gut epithelial cells and salivary cells. Spiroplasma cells may also accumulate in the insect neurolemma. Large accumulations of spiroplasma cells frequently occur in the hemolymph, where they undoubtedly multiply. Koch's postulates have been fulfilled for the etiologic roles of *S. citri* in "stubborn" disease of citrus (Markham et al., 1974) and corn stunt spiroplasma (Chen and Liao, 1975; Williamson and Whitcomb, 1975). *S. citri* causes symptoms in a wide range of crops and wild plants (Calavan and Oldfield, 1979). Plants infected with *S. citri* are often stunted, chlorotic and malformed. The pathogenesis of *S. citri* in plants suggests toxin involvement (Daniels, 1979a,b), but such toxins are difficult to assay and have not been

---

‖ The base for SP-4 medium contains: mycoplasma broth base (BBL), 3.5 g; Bacto-tryptone (Difco), 10.0 g; Bacto-peptone (Difco), 5.3 g; glucose, 5.0 g; phenol red, 20 mg; and deionized water, 615 ml. The base medium is sterilized at 121°C for 15–20 min. The final pH should be 7.6. The following sterile supplements are added to the base medium: CMRL 1066 tissue culture supplement (10×), 50 ml; fresh yeast extract solution (25%), 35 ml; Yeastolate solution (2%; Difco), 100 ml; and fetal bovine serum (heat-inactivated at 56°C for 1 h), 170 ml.
** SM-1 medium contains (per liter): CaCl₂·2H₂O, 250 mg; KCl, 250 mg; MgCl₂·6H₂O, 125 mg; NaCl, 8.75 g; NaHCO₃, 150 mg; NaH₂PO₄·H₂O, 250 mg; glucose, 5.0 g; lactalbumin hydrolysate, 8.125 g; and Yeastolate (Difco), 6.25 g. Fetal bovine serum (20%, v/v) is added as a sterile supplement.

purified. Plant spiroplasmas may also be pathogenic for their vectors (Whitcomb and Williamson, 1979).

Two serologically distinct spiroplasmas have been isolated from the rabbit tick. *S. mirum* has been isolated from rabbit ticks from Georgia (Clark, 1964) and Maryland (Stiller et al., 1981), and the 277F spiroplasma (subgroup I-4) has apparently been isolated once from that host. However, in light of the apparent relatedness of this organism to subgroups I-6 and I-7 from plant surfaces (Eden-Green and Waters, 1981; Whitcomb et al., 1982c; Eden-Green et al., 1983), it seems desirable to confirm its natural habitat. *S. mirum* is experimentally pathogenic for a variety of suckling animals, causing cataract and other ocular symptoms and neural pathology (Clark and Rorke, 1979; Tully, 1982). A spiroplasma was recently isolated from *Ixodes* ticks (Tully et al., 1981), but no data on possible vertebrate reservoirs is available. The existence of tick spiroplasmas able to multiply at 37°C suggests a possible role of vertebrates in the natural maintenance of these organisms.

Flower and other plant surfaces represent another major spiroplasma habitat (Clark, 1978; Davis, 1978; McCoy et al., 1979; Vignault et al., 1980). Members of at least three spiroplasma groups, including the honey bee subgroup (I-2), *S. floricola* (Group III), and *S. apis* (Group IV), have been isolated from flower habitats. Strains from two of these groups of spiroplasmas have also been isolated from insects. Isolates serologically similar to *S. floricola* cause a lethargy disease in the beetle *Melolontha* (Giannotti et al., 1981; Bové, 1981), and the subgroup I-2 spiroplasmas (strains AS/576, BC-3, etc.) cause disease in honey bees (Clark, 1977). Strains of *S. apis* are also found in honey bees (Vignault et al., 1979; Tully et al., 1980), and can be pathogenic (Mouches et al., 1982, 1983). It is not known whether any of the so-called "flower spiroplasmas" can exist as true epiphytes. Recent isolations of spiroplasmas from various insects (Clark, 1982) suggest that it is likely that many or most of these flower isolates are deposited passively by arthropods that visit flowers.

Enrichment and Isolation Procedures

The success of spiroplasma isolations is influenced strongly by the titer of the inoculum. Spiroplasma isolations from infected plants are best obtained from sap expressed from vascular bundles of hosts showing early disease symptoms. Plant sap often contains spiroplasmastatic substances (Liao et al., 1979) whose presence in primary cultures may necessitate blind passage or serial dilution. Simple media that consist of horse serum, PPLO broth and sucrose (Liao and Chen, 1977) may be adequate (but possibly suboptimal) for primary isolation and continuous growth of *S. citri*, the corn stunt spiroplasma, and most spiroplasmas from flower habitats. On the other hand, several constituents of the M1A medium (including α-ketoglutaric acid, phosphate, and amino acids) influence the success of primary isolation of the corn stunt spiroplasma from insect hemolymph (Jones et al., 1977). Similarly, the rich CMRL 1066 component of SP-4 medium is necessary for primary isolation of *S. mirum* from fluids of the embryonated egg (Tully et al., 1982). Only the SP-4 medium is adequate for isolation of the Y32 *Ixodes* spiroplasma (Group VI) (Tully et al., 1981).

Maintenance Procedures

Most spiroplasmas can be adapted to a wide variety of medium formulations. Upon transfer to new media, spiroplasmas commonly undergo several changes in growth rates. Initial reduction in growth rate is probably related to a combination of differences in nutrients, pH, osmolality, etc. Often isolates may grow at only slightly reduced rates during the first 1 to 5 passages in a new medium. However, if the new medium has major deficiencies, the growth rate will decrease precipitously after 5 to 10 passages. Continuous careful passaging may result in recovery to levels similar to the initial growth rate. For such adaptations, best results are achieved by starting with a 1:1 ratio of old and new media, and gradually withdrawing the old formulation. All culture work with spiroplasmas should take into account their high stability on surfaces (Stanek et al., 1981; G. Stanek, G. Laber and A. Hirschl, Rev. Inf. Dis., vol. 4 (Suppl.) S 263, 1982).

Spiroplasmas are routinely preserved by lyophilization (Working Group, FAO/WHO, 1974a).

Procedures for Testing for Special Characters

Most test procedures with spiroplasmas are similar to those used for the genus *Mycoplasma*. Species separations are based to a large extent on growth inhibition tests (Whitcomb et al., 1982a). Infraspecific distinctions are best made by the metabolism inhibition and deformation tests (Williamson et al., 1979b); in these tests spiroplasmas that show some common proteins (Mouches et al., 1979) or significant levels of DNA/DNA homology (Bové and Saillard, 1979) may show significant serological differences. Species, of course, do not cross-react in these specific tests. The metabolism inhibition test (Working Group, FAO/WHO, 1975b) is performed as with *Mycoplasma*, although guinea pig complement has not been added to the system by all workers. The deformation test (Williamson et al., 1978, 1979b), however, is unique to spiroplasmas.

Deformation test. Spiroplasmas for the deformation test are taken from mid- to late-logarithmic phase cultures. All dilutions of serum or antigen are prepared in a medium that has been filtered (using a 300 nm pore size) that has the same formulation as that used for cultivation of the organisms. Media containing unaltered horse serum should be avoided, but agamma or heated horse serum is satisfactory. The M1A or SP-4 media have proved to be especially suitable. The titer of the antigen is adjusted to ~40–60 organisms per microscope field (× 1,250). Antisera are heated at 56°C for 30 min and passed through membrane filters (450 nm pore size) before use. Dilutions of sera are made in plastic tubes or in microtiter plates. The antigen and serum dilutions are mixed in a 1:1 ratio, and after incubation for 30 min at room temperature the results are read under dark-field microscopy. The deformation endpoint is the final antiserum dilution (taking into account the dilution introduced by mixing the reactants) in which approximately one-half of the helical organisms are deformed. Deformation is defined as entire or partial loss of helicity. At the end-point, cells are often seen in which an unaffected part of the helical filament exhibits flexing motility despite the presence of a bleb on another part of the cell. The DF titer is the reciprocal of this final antiserum dilution.

Differentiation of the genus **Spiroplasma** from other closely related taxa

Spiroplasmas can be clearly differentiated from all other microorganisms by their sterol requirement, their genome size, their unique properties of helicity and motility, and the complete absence of periplasmic fibrils, cell walls, or cell wall precursors (Saglio et al., 1973; Bové and Saillard, 1979). However, spiroplasmas may be nonhelical under some environmental conditions, or when cultures are in the stationary phase of growth. Morphological study of the organisms in the logarithmic phase of growth has usually revealed characteristic helical forms. Although the occurrence of spiroplasmas composed primarily of nonhelical forms (*Ixodes* spiroplasma: Group (VI)) suggests the theoretical possibility that an organism derived phylogenetically from the spiroplasmas could totally lack helicity or motility, such a "spiroplasma" has not yet been discovered.

Taxonomic Comments

The term "spiroplasma" was first coined as a trivial term to describe helical organisms associated with corn stunt disease that could not be cultivated by existing technology (Davis and Worley, 1973). When similar organisms associated with citrus stubborn disease were cultivated (Saglio et al., 1971; Fudl-Allah et al., 1972) and characterized (Saglio et al., 1973), the trivial term was adopted as the generic name. Spiroplasmas were first visualized by dark-field microscopy much earlier (Poulson and Sakaguchi, 1961), but were mistaken for spirochetes. The organism that eventually designated *S. mirum* (Tully et al., 1982) was isolated in embryonated chicken eggs soon thereafter (Clark, 1964), but as a result of its filterability was mistaken for a virus. The

277F spiroplasma was cultivated in 1968, but was mistaken for a spirochete (Pickens et al., 1968).

The examination of mollicute phylogeny by rRNA cataloging techniques (Fox et al., 1980; Woese et al., 1980) suggests that, despite differences in genome sizes, *Mycoplasma* and *Spiroplasma* species are phylogenetically related. With the possible exception of the Y32 spiroplasma (Group (VI)) from *Ixodes* (Tully et al., 1981), cultivable spiroplasmas share several basic genomic, morphological and nutritional features that would suggest that they represent a single genus. The taxonomic position of noncultivable spiroplasmas associated with *Drosophila* or other insects is unclear (Williamson and Poulson, 1979; Clark, 1982).

Recent isolations of spiroplasmas from flower surfaces (e.g., Clark, 1978; Davis, 1978; McCoy et al., 1979; Vignault et al., 1980) and arthropods (Clark, 1982) have added many new strains, including some that have been well studied. A classification of spiroplasma strains has been proposed (Junca et al., 1980) that is based on serological reactions of the organisms in growth inhibition, deformation, and metabolic inhibition tests, on characteristics of their genomes, and on analysis of their PAGE protein patterns. The original classification was recently updated (Whitcomb et al., 1982b; Williamson and Whitcomb, 1983). "Groups" in this classification (Table 10.8) represent, essentially, unnamed species, since the inclusion in that category implies serological distinctiveness in growth inhibition as well as metabolism inhibition and deformation tests, lack of significant DNA/DNA homology with members of other groups, and distinct PAGE patterns. Species descriptions (Saglio et al., 1973; Davis et al., 1981; Tully et al., 1982) have been in accord with recommendations of minimum standards proposed by the Subcommittee on the Taxonomy of *Mollicutes* (1979). All species, groups and subgroups are represented by type strains on deposit at the NCTC or ATCC. Progress in isolation of new spiroplasmas from insects (Clark, 1982) has been very rapid, and it is likely that many new groups will be discovered.

In practice, DNA/DNA hybridization experiments with spiroplasmas have proved difficult to standardize. Methods using nitrocellulose filters (Christiansen et al., 1979; Rahimian and Gumpf, 1980; Liao and Chen, 1981b), hydroxyapatite columns (Bové and Saillard, 1979), and enzymatic cleavage (Lee and Davis, 1980) have been published. Esti-mates of hydridization between *S. citri* and the spiroplasma causing the corn stunt disease (subgroup I-3) have varied between 30 and 70%, depending on the method employed (Bové and Saillard, 1979; Christiansen et al., 1979; Lee and Davis, 1980; Rahimian and Gumpf, 1980; Liao and Chen, 1981b). Recent DNA/DNA hybridization studies with hydroxyapatite columns have shown that labeled DNA from subgroup I-1 organisms hybridizes at 63.4% against honey bee spiroplasma (subgroup I-2), 49% against corn stunt spiroplasma E275 (subgroup I-3), and 21.2% against the 277F spiroplasma (subgroup I-4) (Bové et al., 1982). These results, which are in good agreement with various serological comparisons, indicate strongly that group I may consist of aggregates of strains with widely varying degrees of relatedness to each other.

Acknowledgments

We thank R. M. Cole for critical review of this manuscript.

Further Reading

Cole, R. M. 1977. Spiroplasma viruses. In Maramorosch (Editor), *The Atlas of Insect and Plant Viruses Including Mycoplasma Viruses and Viroids*, Academic Press, New York, pp. 451–464.
Bové, J.M. 1980. Les spiroplasmes: nouveax mycoplasmes pathogènes des végétaux et des animaux. In C. R. Journées Francaises sur les Maladies des Plantes, Assoc. Coord. Tech. Agric. Paris., pp. 379–427.
Daniels, M. and P.G. Markham (Editors). 1981. *Plant and Insect Mycoplasma Techniques*. Croom Helm Ltd., London.
Davis, R.E. 1979. *Spiroplasma*: newly recognized arthropod-borne pathogens. In Maramorosch and Harris (Editors), *Leafhopper Vectors and Plant Disease Agents*. Academic Press, New York, pp. 451–484.
Razin, S. and J.G. Tully (Editors). 1983. *Methods in Mycoplasmology*, Vol. 1, Academic Press, New York.
Tully, J.G. and R.F. Whitcomb. 1981. The genus *Spiroplasma*. In Starr, Stolp, Trüper, Balows and Schlegel (Editors), *The Prokaryotes: A Handbook on Habitats, Isolation and Identification of Bacteria*, Springer-Verlag, Berlin, Heidelberg, New York, pp. 2271–2282.
Tully, J.G., and S. Razin (Editors). 1983. *Methods in Mycoplasmology*, Vol. 2, Academic Press, New York.
Whitcomb, R.F. 1980. The genus *Spiroplasma*. Annu. Rev. Microbiol. *34*: 677–709.
Whitcomb, R. F. and J. G. Tully (Editors). 1979. *The Mycoplasmas*, Vol. 3, Plant and Insect Mycoplasmas, Academic Press, New York.
Whitcomb, R.F. 1981. The biology of spiroplasmas. Annu. Rev. Entomol. *26*: 397–425.

Differentiation of the species, groups and subgroups of the genus Spiroplasma

Differentiation is based on serological reactions (growth inhibition, metabolic inhibition, and deformation tests) according to the scheme indicated in Table 10.8.

List of the species of the genus Spiroplasma

1. **Spiroplasma citri** Saglio, L'hospital, Laflèche, Dupont, Bové, Tully and Freundt 1973, 191.[AL]

cit'ri. M.L. n. *Citrus* generic name; M.L. gen. n. *citri* of *Citrus*, to denote the plant host.

The morphology is as described for the genus. Helical filaments are usually 100–200 nm in diameter and 2–4 μm in length.

Colonies on solid media containing 20% horse serum and 0.8% Noble agar (Difco) are umbonate, 60–150 μm in diameter. Moderate turbidity is produced in liquid cultures.

Physiological properties are listed in Table 10.10.

The mininum nutritional requirements are poorly defined.

Serologically distinct from other *Spiroplasma* species by growth inhibition, metabolism inhibition, plate immunofluorescence, and enzyme-like immunosorbent tests.

Isolated from leaves, seeds and fruits of citrus plants (orange and grapefruit) infected with "stubborn" disease, and from other naturally infected plants (e.g. periwinkle, horseradish or brassicaceous weeds) or insects (leafhoppers).

Pathogenic for citrus plants and a variety of plant hosts (aster, periwinkle, broad bean) following transmission by experimentally inoculated insects (leafhoppers).

The mol% G + C of the DNA is 25–27 (T_m, Bd). DNA hybridization tests between *S. citri* and other established species in the genus show no significant genetic relatedness (<1%).

Type strain: ATCC 27556 (strain Morocco (R8A2) of Saglio et al., 1973).

2. **Spiroplasma mirum** Tully, Whitcomb, Rose and Bové 1982, 99[VP]

mir'um. L. neut. adj. *mirum* extraordinary.

The morphology is as described for the genus. Helical filaments measure 100–200 nm in diameter and 3–8 μm in length.

Colonies on solid media containing fetal bovine serum and 0.8–2.25% Noble agar (Difco) are diffuse and without central zones of growth into the agar. Solid media prepared with 2.25% agar, and in which fetal bovine serum has been replaced with bovine serum fraction, yield colonies with central zones of growth into the agar and no peripheral growth on the surface of the medium. Moderate turbidity is produced during growth in liquid media.

Physiological properties are listed in Table 10.10.

The minimum nutritional requirements are poorly defined.

Serologically distinct from other *Spiroplasma* species by growth inhibition, metabolism inhibition, plate immunofluorescence, and enzyme-linked immunosorbent tests.

Table 10.10.
Physiological characteristics of the species of the genus **Spiroplasma**[a]

Characteristics	1. S. citri	2. S. mirum	3. S. floricola	4. S. apis
Acid production from:				
Glucose	+	+[b]	+	+
Mannose	d	d	+	
Fructose	+	d	+	
Sucrose, trehalose, glycerol		d		
Lactose, mannitol, xylose, cellobiose, galactose, salicin		−		
Arginine hydrolysis	+	+	+	+
Tetrazolium, anaerobic reduction	+	+		+
Arbutin hydrolysis		−		
Phosphatase activity	+	+	−	−
Film and spot reaction	d	+		+
Adsorption of guinea pig red blood cells	d	−		−
Liquefaction of coagulated serum	−	−		
Optimum temperature, °C	32	30–37	34	32

[a] Symbols: see standard definitions.
[b] Fermentation is very slow in broth without amino acid supplement.

Isolated from rabbit ticks (*Haemaphysalis leporispalustris*) collected in Georgia and Maryland. Produces experimental ocular and nervous system disease and death in intracerebrally inoculated suckling animals (rats, mice, hamsters and rabbits). Pathogenic for chicken embryos via yolk sac inoculation. Experimentally pathogenic for the wax moth (*Galleria mellonella*).

The mol% G + C of the DNA is 29–31 (T_m). DNA from *S. mirum* does not hybridize significantly with DNA from *S. citri*.

Type strain: ATCC 29335 (strain SMCA of Tully et al., 1982).

3. **Spiroplasma floricola** Davis, Lee and Worley 1981, 456.[VP]

flor.i′co.la. L. n. *flora* flower; L. substantive ending -*cola* dweller; M.L. n. *floricola* flower-dweller.

The morphology is as described for the genus. Helical cells are 150–200 nm in diameter and 2–5 μm in length.

Colonies on solid media have granular central regions surrounded by "satellite" colonies that probably form after migration of cells from the central focus.

Physiological properties are listed in Table 10.10.

The minimum nutritional requirements are poorly defined.

Serologically distinct from other *Spiroplasma* species by growth inhibition, metabolism inhibition, and deformation tests.

Isolated from flowers of tulip tree and magnolia in Maryland.

Experimentally pathogenic for insects and embryonated chicken eggs.

The mol% G + C of the DNA is ~25 (T_m). The genome size is ~10^9 daltons. DNA of *S. floricola* does not hybridize significantly with that of other *Spiroplasma* species.

Type strain: ATCC 29989 (strain 23-6 of Davis et al., 1981).

4. **Spiroplasma apis** (effective publication: Mouches, Bové, Tully, Rose, McCoy, Carle-Junca, Garnier, and Saillard 1983, 383).

a′ pis. M.L. n. *apis*, named after the generic name of the honey bee, *Apis mellifera*, the insect host for this species.

The morphology is as described for the genus. Helical filaments are usually 100–150 nm in diameter and 3–10 μm in length.

Colonies on solid medium containing 20% fetal bovine serum and 0.8% Noble agar (Difco) are usually diffuse, rarely exhibiting central zones of growth into the agar. Colonies on solid medium with 2.25% Noble agar and 1–5% bovine serum fraction are smaller but exhibit central zones of growth into the agar and some surface peripheral growth around the central zones. Marked turbidity is produced during growth in most spiroplasma media (BSR, M1A, SP-4).

Physiological properties are listed in Table 10.10

The minimum nutritional requirements are poorly defined.

Serologically distinct from other *Spiroplasma* species by growth inhibition, deformation, metabolism inhibition, plate immunofluorescence, and enzyme-linked immunosorbent tests.

Isolated from honey bees (*Apis mellifera*) and from flower surfaces in widely separated geographic regions (France, Corsica, Morocco, United States). Etiologic agent of an infection (May disease) of honey bees in Southwestern France. Various strains of the organism exhibit experimental pathogenicity for young honey bees in feeding experiments.

The mol% G + C of the DNA is 29–31 (T_m, Bd).

Type strain: ATCC 33834 (strain B31 of Mouches et al., 1983).

OTHER GENERA

Genus **Anaeroplasma** Robinson, Allison and Hartman 1975, 173[AL]

ISADORE M. ROBINSON

An.a.e.ro.plas′ma. Gr. prefix *an* without; Gr. masc. n. *aer* air; Gr. neut. n. *plasma* a form; M.L. neut. n. *Anaeroplasma* intended to denote "anaerobic mycoplasma."

Cells of young (16- to 18-h-old) cultures are **coccoid**, 0.5–2.0 μm in diameter. Older cells have a variety of pleomorphic forms. **Gram-negative. Nonmotile. Obligately anaerobic**; the inhibitory effect of oxygen on growth is not alleviated during repeated subcultures. Optimum temperature, 37°C; no growth at 26 and 47°C. Optimum pH, 6.5–7.0. Surface colonies have a dense center with a translucent periphery, or "fried egg" appearance. Subsurface colonies are golden, irregular and often multilobed. Strains vary in their ability to ferment various carbohydrates. The products of carbohydrate fermentation include acids (generally acetic, formic, propionic, lactic and succinic), ethanol, and gases (primarily CO_2, but some strains also produce H_2). Bacteriolytic and nonbacteriolytic strains of anaerobic mycoplasmas are described. **Occur in the bovine and ovine rumen**. The mol% G + C of the DNA is 29–34 (T_m, Bd).

Type species: *Anaeroplasma abactoclasticum* Robinson, Allison and Hartman 1975, 173.

Further Descriptive Information

Cells of *Anaeroplasma* from 16- to 18-h-old cultures examined by phase contrast microscopy appear as single cells, clumps, dumbbell forms and clusters of 2–10 coccoid forms joined by short filaments. Motility is not observed, and organelles usually associated with motility are not detected in electron micrographs. In electron micrographs of negatively stained preparations, pleomorphic forms are observed; these include filamentous cells, budding cells, and cells with bleb-like structures. Electron micrographs of thin sections show cells bounded by a plasma membrane (7.5–10 nm thick), but no distinguishable cell wall (Robinson and Hungate, 1973; Robinson et al., 1975).

Anaeroplasma is obligately anaerobic. The roll tube anaerobic culture technique (Hungate, 1966) with prereduced medium maintained in a system for exclusion of oxygen is used to culture the organisms (Robinson and Allison, 1975; Robinson et al., 1975; Robinson and Hungate,

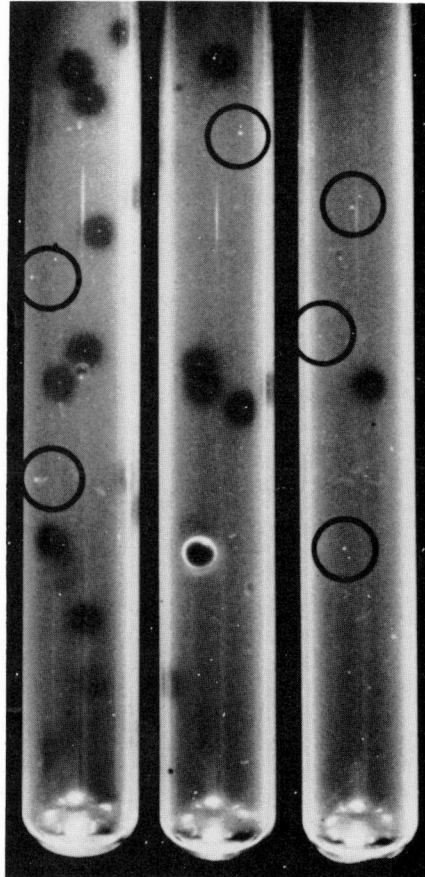

Figure 10.10. Appearance of colonies of *Anaeroplasma bactoclasticum* (*clear zones*) and *A. abactoclasticum* (*circles*) in PIM when the medium is supplemented with 0.5% (wet weight) of *E. coli* cells.

Table 10.11
Composition of media for anaerobic mycoplasmas

Component	Percentage in Medium		
	PIM	CRFB	D
Clarified rumen fluid[a] (v/v)	40.0	40.0	
Glucose (w/v)	0.05	0.2	
Cellobiose (w/v)	0.05	0.2	
Starch (w/v)	0.05	0.2	0.2
Mineral solution[b] (v/v)	3.75	3.75	3.75
Trypticase (w/v)	0.2	0.2	0.2[c]
Yeast extract (w/v)	0.1	0.1	0.1[d]
Volatile fatty acid solution[e] (v/v)			0.31
Resazurin (w/v)	0.0001	0.0001	0.0001
Lipopolyaccharide[f] (w/v)			0.025[g]
Cholesterol solution[h] (v/v)			2.0
E. coli cells (w/v)	0.5		
Na$_2$CO$_3$ (w/v)	0.4	0.4	0.4
Cysteine-HCl (w/v)	0.05	0.05	0.05
Agar (w/v)	1.5		
Benzylpenicillic acid[i] (w/v)	0.0006		

[a] Rumen contents are strained through several layers of cheesecloth, autoclaved, and clarified by centrifugation.
[b] Mineral solution (g/liter): K$_2$HPO$_4$, 6.0; KH$_2$PO$_4$, 6.0; NaCl, 12.0; (NH$_4$)$_2$SO$_4$, 12.0; CaCl$_2$, 1.2; and MgSO$_4 \cdot$7H$_2$O, 2.5.
[c] In some experiments, trypticase was replaced by amino acids (0.02% of each of the following): L-alanine, L-arginine, L-aspartic acid, L-asparagine, L-glutamic acid, L-glutamine, glycine, L-histidine, L-isoleucine, L-leucine, L-lysine, L-methionine, L-phenylalanine, L-proline, L-serine, L-threonine, L-tryptophan, L-tyrosine and L-valine.
[d] In some experiments, yeast extract was replaced by the following vitamins (mg/100 ml): thiamine hydrochloride, 0.2; calcium pantothenate, 0.2; niacinamide, 0.2; riboflavin, 0.2; pyridoxal, 0.2; p-aminobenzoic acid, 0.01; biotin, 0.005; folic acid, 0.005; thioctic acid, 0.005; and B$_{12}$, 0.002.
[e] Volatile fatty acid solution (g/100 ml): acetic acid, 56.2; propionic acid, 21.1; n-butyric acid, 12.4; isobutyric acid, 2.5; n-valeric acid, 3.0; isovaleric acid, 3.0; and DL-α-methylbutyric acid, 3.0.
[f] Lipopolysaccharide (Boivin), Difco *E. coli* 0111:B4.
[g] In some experiments, LPS was replaced by phosphatidyl choline (soybean); final concentration in medium, 0.05%.
[h] Cholesterol: 20 mg in 1 ml of ethanol, made to 20 ml with water.
[i] Final concentration, 1000 U/ml.

1973). Anaerobic mycoplasmas in a sewage sludge digester were cultured in an anaerobic cabinet (Rose and Pirt, 1981).

Anaeroplasma may be grown in a clarified rumen fluid broth (CRFB) or in a rumen fluid-free medium (Medium D) in which growth factors supplied in rumen fluid are replaced by lipopolysaccharide (Boivin) and cholesterol (Robinson et al., 1975). The compositions of these media are given in Table 10.11. Growth also occurs in a completely defined medium in which the trypticase, yeast extract and lipopolysaccharide of Medium D are replaced by amino acids, vitamins and phosphatidyl choline esterified with unsaturated fatty acids (Robinson, 1979).

Bacteriolytic and nonbacteriolytic anaeroplasmas occur. Those which are bacteriolytic form colonies which, when grown on agar media containing a suspension of *Escherichia coli* cells, are surrounded by a clear zone due to lysis of the suspended cells by a diffusible enzyme(s) (Fig. 10.10). On media lacking suspended cells, bacteriolytic and nonbacteriolytic strains of *Anaeroplasma* cannot be distinguished from each other on the basis of colonial or cellular morphology.

Growth of *Anaeroplasma* is inhibited by thallous acetate (0.2%), bacitracin (1000 μg/ml), streptomycin (200 μg/ml) and D-cycloserine (500 μg/ml), but not by benzylpenicillic acid (1000 U/ml).

The serological grouping of anaerobic mycoplasmas (see Table 10.13) is compatible with group separations based upon cultural, biochemical and biophysical properties (Robinson and Rhoades, 1977).

Enrichment and Isolation Procedures

Primary isolation medium (PIM) is used to grow and detect anaerobic mycoplasmas (Robinson and Hungate, 1973; Robinson et al., 1975). The composition of this medium is given in Table 10.11. Pure cultures are established by picking individual colonies from PIM roll tubes with a Pasteur pipette and subculturing into CRFB medium.

Maintenance Procedures

Cultures are viable after storage for as long as 5 years at −40°C in CRFB medium. They may also be preserved by lyophilization.

Procedures for Testing for Special Characters

For testing for bacteriolysis, *E. coli* cells are grown in trypticase soy broth, harvested by centrifugation, washed once with a mineral dilution solution (Bryant and Burkey, 1953), and stored as a frozen pellet. For use, the pellet is thawed and the cells are added (0.5%, w/v) to PIM before the medium is sterilized. Zones of clearing around *Anaeroplasma* colonies cultured on this medium indicate bacteriolysis (Fig. 10.10).

Differentiation of the genus **Anaeroplasma** from other genera

Table 10.12 indicates the characteristics of *Anaeroplasma* that differentiate this genus from aerobic mycoplasmas of the genera *Mycoplasma* and *Acholeplasma*.

Taxonomic Comments

The anaerobic nature of *Anaeroplasma* is a unique, stable characteristic among mycoplasmas and is the basis for establishing the genus

(Robinson et al., 1975). Moreover, the bacteriolytic capability possessed by some of the anaeroplasmas has not been reported for other mycoplasmas. Plasmalogens (alk-1-enyl-glyceryl ethers), which are found in various anaerobic bacteria but not aerobic bacteria, are major components of polar lipids from anaeroplasmas (Langworthy et al., 1975); this further supports the contention that anaeroplasmas are distinct from aerobic mycoplasmas.

The type strain of *A. bactoclasticum* (ATCC 27112, strain JR) was transferred from *Acholeplasma bactoclasticum* to the genus *Anaeroplasma* when it was discovered that cholesterol was required for growth (Robinson and Allison, 1975).

The taxonomic position of the unassigned group of anaerobic mycoplasmas (Table 10.13; also see under Other Organisms) has not been decided (Subcommittee on the Taxonomy of *Mycoplasmatales*, 1977). These mycoplasmas share some biologic properties with species of the genus *Anaeroplasma*, but differ in the following respects: (a) sterol is not required for growth; (b) they are serologically unrelated to the established species of *Anaeroplasma*; and (c) the mol% G + C content of the DNA of representative strains is higher than that of *Anaeroplasma* (Robinson and Allison, 1975). The existence of these unclassified mycoplasmas as a significant part of the anaerobic mycoplasma population is supported by the finding that 17 of 28 isolates from cattle and sheep (Robinson and Rhoades, unpublished results) were grouped serologically with the nonsterol-requiring strains (Table 10.13). Further work is needed to determine the taxonomic placement of these anaerobic organisms that resemble *Anaeroplasma*.

Table 10.12.

*Differential characteristics of the genus **Anaeroplasma** and other genera of mycoplasmas*[a]

Characteristics	Anaero-plasma	Myco-plasma	Achole-plasma
Obligately anaerobic	+	−	−
Polar lipids contain plasmalogens	+	−	−
Bacteriolytic activities	+	−	−
Sterol required for growth	+	+	−
Phospholipids required for growth	+	−	−

[a] Symbols: see standard definitions.

Table 10.13

Differential characteristics of anaerobic mycoplasmas[a]

Species	Serovar	No. of Strains	Bacteriolytic	Requirement for Sterol	Mol% G + C of DNA
Anaeroplasma bactoclasticum	1	1	+	+	33.7 (1)
	2	4	+	+	32.5 (1)
Anaeroplasma abactoclasticum	*	2	−	+	29.3 –
					29.5 (2)
Unassigned	**	3	−	−	40.3 –
					40.5 (2)

[a] Numbers in parentheses are the number of strains tested for mol% G + C; for symbols see standard definitions; also *, serologically distinct from serovar 1 and 2; **, serologically distinct from serovar 1 and 2 and *A. abactoclasticum* (Robinson and Rhoades, 1977).

Differentiation and characteristics of the species of **Anaeroplasma** and other similar anaerobic mycoplasmas

Table 10.13 presents differential characteristics. Table 10.14 indicates other characteristics of the two named species.

List of the species of the genus **Anaeroplasma**

1. **Anaeroplasma abactoclasticum** Robinson, Allison and Hartman 1975, 173.[AL]

a.bac′to.clas′ti.cum. Gr. prefix *a* without; *bact* part of the stem of the Gr. dim. n. *bacterium* a small rod; Gr. adj. *clasticus* breaking; M.L. adj. *abactoclasticum* intended to denote "not bacteriolytic."

The cellular morphology and colonial characteristics are as described for the genus. Physiological characteristics are given in Tables 10.13 and 10.14.

Growth is inhibited by digitonin (20 μg/ml), but not by penicillin (1000 U/ml).

Occur in the bovine and ovine rumen.

The mol% G + C of the DNA of the type strain is 29.3 (Bd).

Type strain: ATCC 27879.

2. **Anaeroplasma bactoclasticum** (Robinson and Hungate 1973) Robinson and Allison 1975, 182.[AL] (*Acholeplasma bactoclasticum* Robinson and Hungate 1973, 171.)

bac′to.clas′ti.cum. *bact* part of the stem of the Gr. dim. n. *bacterium* a small rod; Gr. adj. *clasticus* breaking; M.L. adj. *bactoclasticum* bacteria-breaking.

The cellular morphology and colonial characteristics are as described

Table 10.14.

*Other characteristics of **Anaeroplasma abactoclasticum** and **Anaeroplasma bactoclasticum***[a]

Characteristics	1. A. abacto-clasticum	2. A. bacto-clasticum
Arginine hydrolyzed	−	−
Urease	−	−
Hemolysis	−	−
Digest casein	−	+
Growth in presence of digitonin (20 μg/ml)	−	−
Acid production from substrates:		
Glucose, maltose, starch	+	+
Cellobiose	+	−
Arabinose, galactose, lactose, ribose, sucrose, xylose	−	+
Products of fermentation:		
Acetate, formate, lactate, ethanol, carbon dioxide	+	+
Succinate	+	−
Hydrogen, propionate	−	+

[a] For symbols see standard definitions.

for the genus. Physiological characteristics are given in Tables 10.13 and 10.14.

Optimum temperature, between 30 and 47°C.

Growth is inhibited by digitonin (20 μg/ml), but not by penicillin (5000 U/ml).

Occur in the bovine and ovine rumen.

The mol% G + C of the DNA is 32.5 to 33.7 (T_m, Bd).

Type strain: ATCC 27112.

Other organisms

Anaerobic mycoplasma strains having characteristics of the genus *Anaeroplasma* have been isolated from the rumen of cattle and sheep (Robinson et al., 1975). They are similar to *A. abactoclasticum* in that they lack extracellular bacteriolytic and proteolytic enzymes. However, unlike either of the two established species of *Anaeroplasma*, they do not require sterol for growth and their growth is not inhibited by digitonin (200 μg/ml). Moreover, no serologic cross-reaction has been observed between these sterol-independent anaerobic mycoplasmas and strains of *Anaeroplasma*. The mol% G + C contents of the two strains so far analyzed are 40.3 and 40.5, respectively, in contrast to the values of 29–33 for *Anaeroplasma*. Strain 161 has been deposited in the ATCC under the number 161 and is representative of the nonsterol-requiring strains. The taxonomic status of these organisms has not yet been determined (Subcommittee on the Taxonomy of *Mycoplasmatales*, 1977).

Two species of obligately anaerobic mycoplasmas were the major components of a methanogenic glucose-limited enrichment culture from a sewage sludge digester (Rose and Pirt, 1981). In pure culture, one of these organisms, tentatively named *Anaeroplasma* sp. strain London, fermented glucose primarily to butyric acid, hydrogen, and carbon dioxide; the other mycoplasma produced methane from hydrogen and carbon dioxide and was named "*Methanoplasma elizabethii*." Both species of mycoplasma were classified in the family *Mycoplasmataceae* on the basis of colonial and cellular morphology, ability to pass through a 0.45-μM membrane filter, and resistance to penicillin G.

Genus **Thermoplasma** Darland, Brock, Samsonoff and Conti 1970, 1418[AL]

THOMAS A. LANGWORTHY AND PAUL F. SMITH

Ther.mo.plas'ma. Gr. n. *thermus* heat; Gr. neut. n. *plasma* something formed or molded, a form; M.L. neut. n. *Thermoplasma* heat (loving) mycoplasma.

Pleomorphic, varying in shape from spherical (0.1–0.3 μm) to filamentous structures. Cells lack a cell wall and are surrounded by a cytoplasmic membrane only, ~7 nm thick and exhibiting a trilaminar shape in section. The membrane contains ether lipids based on 40-carbon, isopranoidbranched **diglycerol tetraethers**. Resting stages not known. **Gram-negative.** Generally nonmotile. Strict aerobe. **Obligate thermoacidophile.** Optimum growth at 55–59°C and pH 1–2. **Cells undergo lysis near neutrality.** On agar at pH 2, colonies attain a diameter of about 0.3 mm, are dark brown in color, flat, coarsely granular and some exhibit a typical "fried egg" apearance with a translucent peripheral zone. Biochemical and nutritional characteristics relatively poorly defined. Do not require cholesterol. Apparently chemoorganotrophic but have an **absolute requirement for yeast extract for growth and reproduction. Occur free living in self-heating coal refuse piles.** The mol% G + of the DNA is 46 (Bd, T_m).

Type species: *Thermoplasma acidophilum* Darland, Brock, Samsonoff and Conti 1970, 1418.

Further Descriptive Information

When grown in a liquid medium* consisting of Allan's basal salts solution (Allan, 1959) adjusted to pH 2 and supplemented with 0.1% yeast extract and 1.0% glucose, *T. acidophilum* exhibits a typical mycoplasmal morphology by light and phase microscopy. Cells appear as pleomorphic spheres varying in size from 0.1–0.3 μm to occasional large cells up to 5 μm in diameter. Filamentous structures exhibiting budding characteristics are also common, particularly in young cultures.

A cell wall is absent, evidenced by electron micrographs of thinsectioned cells (Darland et al., 1970; Belly et al., 1973; Langworthy, 1979).

The surrounding cytoplasmic membrane averages about 7 nm in thickness. Membrane lipids lack fatty acid ester residues. The lipids are ether-linked C_{40} biphytanyl diglycerol tetraethers (Langworthy, 1977). Similar ether lipids occur in *Sulfolobus*, *Halobacteriaceae* and *Methanobacteriaceae* (Kates, 1978; Langworthy, 1979; Tornebene and Langworthy, 1979). No internal membranes or organelles are present.

Cells are generally nonmotile, although flagella-like structures and swimming motility have been observed in two isolates, including the type strain (Black et al., 1979).

Growth on an agar surface is unreliable and difficult to achieve due to drying out at high temperature and hydrolysis of the agar in the presence of acid. These difficulties can sometimes be overcome by combining double-strength liquid medium and agar after cooling to 45–50°C followed by incubation in a humidified atmosphere (see under Thermoplasma Medium). When growth can be initiated, colonies are small (about 0.3 mm diameter) and some show a "fried egg" appearance (Belly et al., 1973). Colonies are typically flat, coarsely granular and dark brown in color. Scanning electron microscopy shows individual cells to have an imbricate surface texture characteristic of cells which lack a cell wall (Mayberry-Carson et al., 1974).

The optimum temperature for growth is 59°C. The maximum temperature for growth is 62°C and the minimum about 40°C. Growth is slow to slight at the extremes.

The optimum pH for growth is 2. Growth occurs between the pH limits of 1–4, but growth is very slow at the extremes.

Hydrogen ions are specifically required for maintenance of cellular stability. Cells undergo lysis at neutral pH (Belly and Brock, 1972; Smith et al., 1973). Other monovalent cations, divalent cations, or osmotic stabilizers do not substitute for the hydrogen ion requirement. The phenomenon is analogous to the sodium ion requirement by certain

---

* Thermoplasma medium (g/liter deionized water): KH_2PO_4, 3.0; $MgSO_4$, 0.5; $CaCl_2 \cdot 2H_2O$, 0.25; $(NH_4)_2SO_4$, 0.2; yeast extract (Difco), 1.0. Adjust to pH 2 with 10 N H_2SO_4. After autoclaving, add 10 g of separately sterilized glucose (25 ml of a 40% glucose solution), to give a final concentration of 1.0%. For agar medium, mix equal volumes of separately sterilized double-strength liquid medium and 5.6% Ionagar No. 2 (Consolidated Laboratories) after cooling to 45°C to give a final agar concentration of 2.8%. Incubate in a sealed and humidified atmosphere.

Halobacteriaceae for the maintenance of cellular integrity. The intracellular hydrogen ion concentration of *T. acidophilum*, however, is not in equilibrium with the external environment, but the internal pH is near neutrality (Hsung and Haug, 1975; Searcy, 1976).

T. acidophilum requires oxygen. It appears to possess cytochromes and menaquinone-7 suggesting the presence of a complete respiratory chain (Belly et al., 1973; Holländer, 1978). Growth is stimulated by slight aeration, but excessive aeration inhibits growth (Smith et al., 1973). Because the amount of dissolved oxygen at 59°C is low, *T. acidophilum* can be considered microaerophilic.

Nutritionally, *T. acidophilum* has an absolute requirement for yeast extract for growth and thus far no other compounds have been found to substitute (Belly et al., 1973; Smith, Langworthy and Smith, 1975). At yeast extract concentrations below 0.025% no growth occurs. At concentrations higher than 0.25% growth is inhibited. Between the limiting concentrations, growth is proportional to the concentration of yeast extract used. Growth rates and yields also vary, depending upon the manufacturer and the lot of yeast extract employed. The component(s) supplied by yeast extract for growth appears to be a basic oligopeptide(s) (Smith et al., 1975). No growth occurs on elemental sulfur or ferrous iron.

Cell yields are influenced by the inoculum size. Total cell yields decrease with inoculum sizes of less than 5% (v/v). Under optimal conditions, *T. acidophilum* has a generation time of about 5 h (Belly et al., 1973; Smith et al., 1973). Cell numbers increase with optical density (540 nm) to the stationary phase reaching about 1×10^9 cells/ml, at which point there is a drastic loss in viability, although no great reduction in optical density.

Sucrose, glucose, galactose, mannose and fructose, when added in 0.1% concentration to the basal medium containing the growth-limiting concentration of yeast extract (0.025%), appear to stimulate growth of *T. acidophilum* (Belly et al., 1973).

No lysis occurs when cells are suspended in distilled water, when heated to 100°C for 30 m, or when treated with EDTA, primary alcohols, digitonin, lysozyme, trypsin or pronase. Cells are rapidly lysed by sodium lauryl sulfate and more slowly by cetyl trimethylammonium bromide (Belly and Brock, 1972; Smith et al., 1973).

T. acidophilum is resistant to the cell wall inhibitors vancomycin and ristocetin (in concentrations of at least 5 mg/ml and 1 mg/ml, respectively). Cells are inhibited by novobiocin at a concentration of 0.1 mg/ml. Sensitivity to penicillin has not been determined because of the acid lability of this antibiotic (Darland et al., 1970; Belly et al., 1973; Brock, 1978).

Molecular characteristics which further distinguish *T. acidophilum* include: a membrane-associated linear lipoglycan containing 24 mannose residues and 1 glucose residue (Smith, 1980); a mannosyl membrane glycoprotein (Yang and Haug, 1979); small genome size of about 8.4×10^8 to 1×10^9 daltons (Searcy and Doyle, 1975; Christiansen et al., 1975); a histone-like protein associated with the DNA (DeLang et

al., 1981); a 7-subunit DNA-dependent RNA polymerase that is resistant to rifampicin, streptolydigine and α-amanitine (Sturm et al., 1980); unusual modification pattern in tRNAs (Gupta and Woese, 1980); unusual nucleotide sequences in the 16S rRNA (Woese et al., 1980); a 5S rRNA secondary structure which does not conform to the usual models employed for either procaryotic or eucaryotic 5S rRNAs (Luehrsen et al., 1981); and a cytoplasmic membrane which may exist as a lipid monolayer rather than a lipid bilayer and which may account for the characteristic cross-fracture rather than tangential fracture of freeze-etched cells (Langworthy, 1979).

T. acidophilum occurs in self-heating coal refuse piles in southern Indiana and western Pennsylvania. It is found in regions of the piles where temperatures range from 32–80°C and the pH ranges from 1.17–5.21. So far this is the only known natural habitat. It has not been detected in acidic geothermal regions which harbor the thermoacidophiles *Sulfolobus* and *Bacillus acidocaldarius* (Belly et al., 1973).

Enrichment and Isolation Procedures

Thermoplasma was originally isolated from a coal refuse pile at the Friar Tuck mine in southwestern Indiana by inoculating 20 ml of Thermoplasma medium with 1.0 g of coal refuse (Darland et al., 1970). The isolation medium has been modified to include the acid-stable antibiotic vancomycin at a concentration of 1,225 μg/ml to inhibit the growth of rod-shaped bacteria such as *B. acidocaldarius* (Belly et al., 1973). A procedure was also reported for the isolation of *Thermoplasma* from liquid samples of coal refuse material by filtration through a membrane filter (0.45-μm pore size) followed by passage of the filtrate through a second filter (0.22-μm pore size), with subsequent incubation in culture medium (Belly et al., 1973). Isolation samples are incubated at 55°C for 4–6 weeks, or until the development of visible turbidity. The presence of *Thermoplasma* is confirmed by microscopic examination. Cultures are purified by dilution in liquid medium since reproducible growth of colonies on agar has not been obtained. The above procedures are selective for the isolation of *Thermoplasma* with the possible exception of *Sulfolobus* which can be distinguished by its lobed shape and physiological characteristics.

Maintenance Procedures

The most reliable procedure is to maintain actively growing cultures by continuous passage after 2 or 3 d of incubation. A 10–20% inoculum should be used and multiple culture tubes incubated, since growth is sometimes spurious. Glassware should be free of any trace of detergent or soap residue which will kill the cells. Cells remain viable at room temperature for 10–15 d but die upon refrigeration. Sometimes cells can be recovered from the frozen state, but sufficient time (1–2 weeks) is required for development of visible turbidity. Cells are killed by lyophilization, and neutralization of cultures prior to preservation is precluded by cell lysis.

Differentiation of the genus **Thermoplasma** from other genera

Thermoplasma is distinguished from other genera of mycoplasmas by its stability, by its requirement for hot acid, and by its molecular features. Table 10.15 provides the primary characteristics of *Thermoplasma* that distinguish it from other genera morphologically or physiologically similar thermoacidophilic bacteria.

Taxonomic Comments

Thermoplasma is by definition a mycoplasma by virtue of being a self-replication procaryote which lacks a cell wall. Biochemical characteristics, however, also indicate that *Thermoplasma* is similar to the extremely halophilic bacteria, the methanogenic bacteria, and the thermoacidophile *Sulfolobus*. Although lacking a cell wall, *Thermoplasma* shares in common the presence of isopranoid (phytanyl-based) ether lipids, unusual but similar RNA polymerases, and certain nucleotide

sequences in the 16S rRNA, 5S rRNA and tRNA. It has been proposed that *Thermoplasma*, *Sulfolobus*, methanogens and extreme halophiles comprise a diverse but distinct phylogenetic group (Archaebacteria) which has evolved differently from either procaryotic or eucaryotic cells (Fox et al., 1980). The precise extent to which *Thermoplasma* is related to the physiologically distinct halophiles and methanogens requires further studies. The lack of DNA/DNA sequence homology (<0.25%) indicates absence of a close genetic relationship between *Thermoplasma* and *Sulfolobus*; i.e. *Thermoplasma* is not merely a stable L form derived from *Sulfolobus* (Christiansen et al., 1981). Thus, on biochemical grounds and on the basis of RNA studies, *Thermoplasma* is broadly related to *Sulfolobus*, halophiles and methanogens, but is not closely related to these or any other known organisms.

Table 10.15.
Differential characteristics of the genus **Thermoplasma** *and other genera of thermoacidophilic bacteria*[a]

Characteristics	Thermo-plasma	Sulfolobus	Bacillus acido-caldarius
Shape:			
Pleomorphic spheres	+	−	−
Lobed spheres	−	+	−
Rods, filaments	−	−	+
Cell wall present	−	+[b]	+
Ether lipids present	+	+	−
Endospores formed	−	−	+
Nutrition:			
Requires yeast extract	+	−	−
Facultative autotroph	−	+	−
Lysis at neutral pH	+	−	−
Mol% G + C of DNA	46	40	60–64

[a] Symbols: see standard definitions.
[b] *Sulfolobus* has a cell wall that lacks peptidoglycan.

Further Reading

Belly, R.T., B.B. Bohlool and T.D. Brock. 1973. The genus *Thermoplasma*. Ann. N.Y. Acad. Sci. *225:* 94–107.
Brock, T.D. 1978. *Thermophilic Microorganisms and Life at High Temperatures,* Springer-Verlag, New York.
Fox, G.E., E. Stackebrandt, T.B. Hespell, J. Gibson, J. Maniloff, T.A. Dyer, R.S. Wolfe, W.E. Balch, R.R. Tanner, L.J. Magrum, L.B. Zablen, R. Blakemore, R. Gupta, L. Bonen, B.R. Lewis, D.A. Stahl, K.R. Luehrsen, K.N. Chen and C.R. Woese. 1980. The phylogeny of prokaryotes. Science *209:* 457–463.
Langworthy, T.A. 1979. Special features of thermoplasmas. In Barile and Razin (Editors), *The Mycoplasmas* I: Cell Biology, Academic Press, New York, pp. 495–513.
Langworthy, T.A. 1979. Membrane structure of thermoacidophilic bacteria. In Shilo (Editor), *Strategies of Microbial Life in Extreme Environments*, Dahlem Konferenzen, Verlag-Chemie, Weinheim, Berlin, pp. 417–432.

Differentiaton and characteristics of species of **Thermoplasma**

Species diversity among *Thermoplasma* isolates has not been established. *Thermoplasma* isolates do exhibit serological diversity and can be differentiated into five antigenic groups by immunofluorescence and immunodiffusion analysis (Belly et al., 1973; Bohlool, and Brock, 1974).

Of the variety of original isolates obtained by Darland et al. (1970) and Belly et al. (1973) the following appear to be extant: 122–1B2, 122–1B3, 3–24, and 124–1.

List of the species of the genus **Thermoplasma**

1. **Thermoplasma acidophilum** Darland, Brock, Samsonoff and Conti 1970, 1418.[AL] (*Thermoplasma acdiophila* (sic) Darland et al. 1970, 1418.)

a.ci.do′phi.lum. M.L. n. *acidum* an acid; Gr. adj. *philus* loving; M.L. neut. adj. *acidophilum* acid-loving.

The characteristics are as described for the genus and as listed in Table 10.15.

Occur free living in self-heating coal refuse piles.

The mol% G + C of the DNA is 46 (T_m, Bd).

Type strain: ATCC 25905 = AMRC-C 165 (isolate 122-1B2 of Darland et al., 1970). *Reference strains:* ATCC 27658 (isolate 122-1B3), ATCC 27657 (isolate 3-24), ATCC 27656 (isolate 124-1).

Mycoplasma-Like Organisms of Plants and Invertebrates

RANDOLPH E. McCOY

Wall-less, polymorphic organisms 0.2–0.8 μm in diameter, morphologically resembling members of the *Mollicutes*. Often **filamentous and branched. Regularly observed within sieve tube elements of plants affected by yellows diseases, and in the salivary glands of insect vectors of these diseases.** Mycoplasma-like organisms (MLO) exist in nature in a dual host system characterized by alternate passage between plant and invertebrate hosts. In vitro culture has not been demonstrated for any nonhelical MLO as of 1983. Without confirmation of identity through cultural techniques, these agents are relegated to the uncertain aggregation of organisms termed MLO. **Tetracycline sensitivity** provides additional evidence supporting inclusion in the *Mollicutes*.

Further Descriptive Information

The MLO have a diameter of less than 1 μm and are polymorphic. In ultrathin sections (Fig. 10.11) they appear circular to oblong to filamentous in electron micrographs of plant and insect tissues (Doi et al., 1967; Hirumi and Maramorosch, 1969). Depending on orientation of the microscope section, dumbbell-shaped forms may be observed as well as small dense round forms of ∼0.1 μm diameter. Occasional forms have been seen apparently containing membrane-bound inclusions. Much confusion has developed concerning the morphology of these organisms as determined from the study of ultrathin sections. Subsequent studies using semi-thick (0.3 μm) sections (Thomas, 1979), serial

sections (Florance and Cameron, 1978; Chen and Hiruki, 1979; Waters and Hunt, 1980), and scanning electron microscopy (Haggis and Sinha, 1979) have done much to clarify the gross cellular morphology of the MLO. The MLO range from spherical to filamentous, often with extensive branching reminiscent of *Mycoplasma mycoides*. Small dense forms formerly considered to be "elementary bodies" in thin section were shown to be constriction points in filamentous forms of overall greater length. The dumbbell-shaped forms thought to be "dividing" were actually branch points of filamentous MLO. Forms thought to have internal vescicles were shown to be involuted and oriented such that the plane of the section cut through the cell membrane twice.

The MLO lack a cell wall and are only bounded by a plasma membrane; they contain typical procaryotic ribosomes and fibrillar DNA-like strands. No additional internal structures have been verified. The organisms are seen to degenerate, with loss of cellular content, in plants treated with tetracycline antibiotics (Sinha and Peterson, 1972). Penicillin has no apparent effect on MLO (Ishiie et al., 1967; Davis and Whitcomb, 1969). Several recent reviews on MLO are available (McCoy, 1979a; Sinha, 1979a; Tsai, 1979).

The MLO in plants have been observed with certainty only within sieve elements (McCoy, 1979a). Reports of MLO in mesophyll or in parenchyma or companion cells are unsubstantiated or in need of verification. Sieve elements are specialized living cells functioning in transport of photosynthate from leaves to growing points. They are

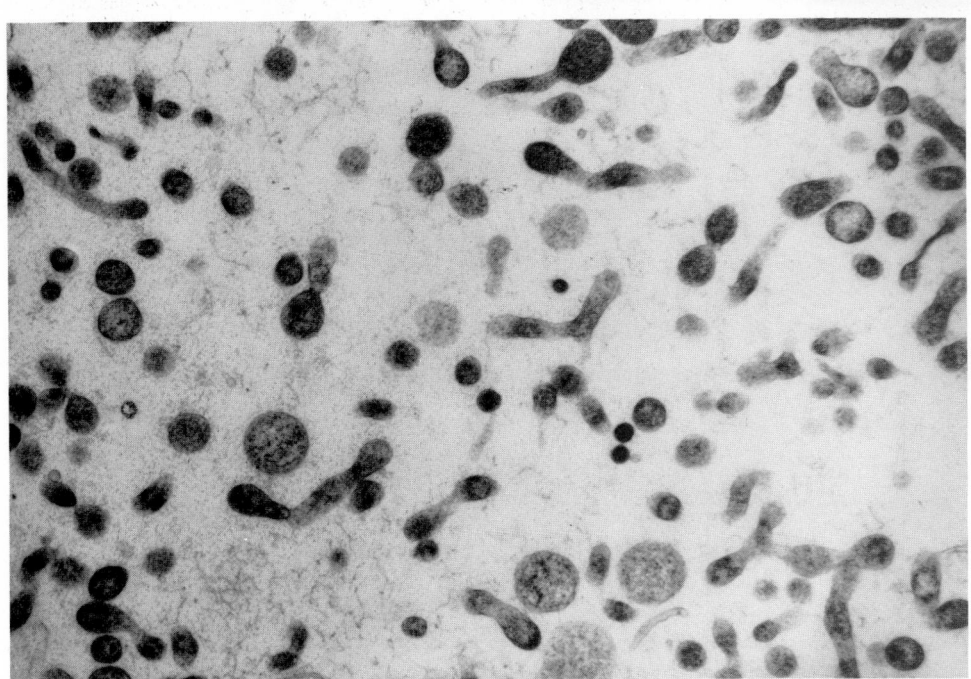

Figure 10.11. Polymorphic mycoplasma-like organisms within a sieve tube element of a witches'-broom diseased Madagascar periwinkle. Electron micrograph of an ultrathin section of leaf petiole. Magnification ×23,000. (Reproduced with permission from R. C. Norris.)

unique in that they contain from 12–30% sucrose, and are under high hydrostatic (turgor) pressure, up to 15 atm or more (Evert, 1977). The end plates of these cells are perforate and allow passage of photosynthate from sieve cell to sieve cell. The sieve pores have an average diameter of ∼2 μm and are sufficiently large to allow ready passage of spherical or filamentous MLO from cell to cell (McCoy, 1979b).

The chemical environment of the sieve tube is highly complex, containing, along with sucrose, minerals, free amino acids, proteins, and ATP (Evert, 1977). This rich milieu, with its high osmotic and hydrostatic pressures, serves to support extensive multiplication of MLO in vivo.

The plant yellows agents are also found and appear to multiply in the internal tissues and organs of their insect hosts. Insect hemolymph is similar in many respects to plant phloem sap (Saglio and Whitcomb, 1979), both containing a high percentage of organic nutrients. The leafhopper vector becomes infected after ingestion of phloem sap from infected plants. After an incubation period of one to several weeks the agent reaches a high titer in the salivary glands and the insect becomes capable of infecting the phloem of healthy plants on which it feeds (Sinha and Chiykowski, 1967).

Enrichment and Isolation Procedures

Mechanical separation and purification of MLO from plant host tissues has been investigated by Sinha (1974) using differential centrifugation and adsorption techniques. In absence of the ability to culture these organisms, such methods may be necessary to gather sufficient MLO for characterization studies (Sinha, 1979b; Sinha and Madhosingh, 1980).

Extraction of MLO from plant or, preferably, insect tissue is possible by grinding in various media. Viability may be assayed by infectivity tests in which aliquots of media containing extracted MLO are injected into vector insects which are subsequently fed on healthy indicator plants (Black, 1941). A logical approach to culture is through determination of the duration of infectivity of extracted MLO in various media and medium components (Caudwell, 1977). In this manner,

media may be formulated which can yield maximal viability of extracted MLO.

Another approach to isolation and multiplication of MLO is through experimental infection of unusual invertebrate hosts. Williamson and Whitcomb (1975) first isolated the corn stunt spiroplasma from fruit flies injected with infectious extracts of diseased plants. McCoy et al. (1981) have shown larvae of the greater wax moth to support multiplication of large numbers of spiroplasmas. Such approaches may be of use for the nonhelical MLO as well.

Maintenance Procedures

Storage of MLO has been accomplished for up to 2 years in vector insects frozen at −70°C (Chiykowski, 1977a), and in extracts of infected insects lyophilized in a $MgCl_2$-glycine medium at pH 7.2 and osmotically adjusted to 800 milliosmoles/kg with sucrose (A. J. Smith and R. E. McCoy, unpublished observations). Reconstitution of the infectious agent is accomplished by grinding the frozen insects or rehydrating the freeze-dried material in the extraction medium above, followed by injection into suitable vector insects.

Without these procedures, MLO must be maintained in diseased plants kept in a greenhouse. Transmission from plant to plant is accomplished by vector insects, grafting, or by vegetative reproduction. While greenhouse maintenance of MLO strains is the most commonly used procedure, it can result in strain attenuation over a period of time (Chiykowski, 1977b).

Taxonomic Comments

In the absence of culture, species identification of MLO has not been possible. While mechanical extraction techniques may eventually allow development of antisera suitable for differentiation of strains, such techniques have not been developed to a practicable level. Presently, distinction among MLO strains is based on differences in host preference, symptomatology, and vector specificity. Such techniques are insufficient to resolve the large number of MLO reported to be associated with the plant yellows diseases.

Important Notes for Users of this Edition

1. Always read both generic and species descriptions because characters listed in the generic description are not usually listed in the species descriptions.

2. Unless otherwise indicated in footnotes to tables, the meanings of symbols are as follows:

+ 90% or more of strains are positive

− 90% or more of strains are negative

d 11–89% of strains are positive

v strain instability (*not* equivalent to "d")

D different reactions in different taxa (species of a genus or genera of a family)

3. All other symbols are defined in footnotes to tables.

SECTION 11

Endosymbionts

A. ENDOSYMBIONTS OF PROTOZOA

John R. Preer, Jr. and Louise B. Preer

Intracellular symbiosis is common among protozoans. For a review see Preer and Preer (1977). Endosymbionts consist not only of bacteria, but of fungi, algae and even other protozoa. All appear well adapted to their intracellular habitat where they find a favorable environment. Of the bacteria, only a few have been cultured; apparently their requirements for growth are not easily duplicated in vitro. Infectivity through the medium varies, but usually is rare. In many, perhaps even in most cases, infectivity does not occur under normal conditions and transmission is by cellular heredity alone.

The effects of the symbionts on their hosts are generally not known. Some symbionts, such as omicron in *Euplotes* and X-bacteria of *Amoeba proteus* are necessary for survival of their hosts, but the mechanisms have not been discovered. Others, such as lambda in *Paramecium* and bipolar bodies in *Crithidia*, have been shown to produce important

metabolites utilized by the hosts. Some symbionts confer on their hosts the ability to produce toxins capable of killing sensitive strains of protozoa. If the toxins are liberated into the medium, the toxin producers are called "killers" and their victims "sensitives." If the toxins act only during cell-to-cell contact at conjugation, the toxin producers are called "mate killers." It has been found in many cases that not only do the symbionts make their hosts capable of producing the toxins, but they also confer upon them specific resistance to the toxins they produce. Toxin-producing symbionts may provide their hosts with a competitive selective advantage over sensitive nonsymbiont bearers (Landis, 1981). Although most symbionts are not harmful to their hosts, some are. Even in normally nonharmful relations an altered environment can upset the symbiotic balance and host or endosymbiont can be harmed. For example, when paramecia containing certain kappa

Table 11.1.

Some endosymbionts of protozoa[a]

Name (common name)	Host (typical strain)	Site	Killing
Holospora elegans	*Paramecium caudatum* (C101)	Mi	NK
Holospora undulata (omega)	*P. caudatum* (C204)	Mi	NK
Holospora obtusa (iota)	*P. caudatum* (C103)	Ma	NK
Holospora caryophila (alpha)	*P. aurelia*, sp.2(562)	Ma	NK
Caedibacter taeniospiralis (kappa)	*P. aurelia*, sp.4(51)	Cy	K
Caedibacter varicaedens (kappa)	*P. aurelia*, sp.2(7)	Cy	K
Caedibacter pseudomutans (kappa)	*P. aurelia*, sp.4(51m1)	Cy	K
Caedibacter paraconjugatus (mu)	*P. aurelia*, sp.2(570)	Cy	MK
Caedibacter species	*P. caudatum*	Ma	K
Pseudocaedibacter conjugatus (mu)	*P. aurelia*, sp.1,8(540)	Cy	MK
Pseudocaedibacter minutus (gamma)	*P. aurelia*, sp.8(214)	Cy	K
Pseudocaedibacter falsus (nu, pi)	*P. aurelia*, sp.2,4,5(1010)	Cy	NK
Lyticum flagellatum (lambda)	*P. aurelia*, sp.4,8(299)	Cy	K
Lyticum sinuosum (sigma)	*P. aurelia*, sp.2 (114)	Cy	K
Tectibacter vulgaris (delta)	*P. aurelia*, sp.1,2,4,6,8(225)	Cy	NK
(ms-2)	*P. caudatum* (C201)	Ma	NK
(Omicron)	*Euplotes* (7 species)	Cy	NK
(Epsilon)	*P. multimicronucleatum*	Cy	NK
(Epsilon)	*Euplotes minuta*	Cy	K
(Eta)	*E. crassus*	Cy	K
(Unnamed)	*E. crassus, E. patella*	Cy	MK
(Unnamed)	*Spirostomum*	Cy	NK
(Xenosomes)	*Pauronema*	Cy	NK
(Bipolar bodies)	*Crithidia oncopelti*	Cy	NK
(Diplosomes)	*Blastocrithidia culicis*	Cy	NK
(X-bacteria)	*Amoeba proteus*	Cy	NK
(Unnamed)	*Pelomyxa palustris*	Cy	NK

[a] Symbols: Mi, micronucleus; NK, nonkiller; Ma, macronucleus; Cy, cytoplasm; K, killer; and MK, mate killer.

symbionts are placed into various bacteria-free media, the symbionts increase in numbers more rapidly than their hosts and kill them, dying along with their hosts. Conversely when protozoa are brought in from the wild and grown under laboratory conditions they often lose their symbionts. Thus, forms vary from the noninfective, obligate, intracellular symbionts that are transmitted by heredity and benefit the host, to true parasites that are transmitted by infection and harm the host. However, the latter are clearly less frequent.

Studies on the kinetics of renaturation of the DNA of several of the endosymbionts reveal that the genome is small, comparable in size to that of the mycoplasmas. Two other interesting observations also relate to the genome of the endosymbionts. The first is that the DNA appears to be dispersed in most. Only in a few forms such as omicron in *Euplotes* is a clear nucleoid visible. The second is that the genome appears to be present in a rather high number of copies (5–20) per cell. This observation has been made for lambda, pi and mu of *Paramecium* (Soldo and Godoy, 1973a, 1974), omicron of *Euplotes* (Schmidt and Heckmann, 1980), and the bacterial endosymbionts of *Blastocrithidia* (Tuan and Chang, 1975). It is not clear at this time whether the endosymbionts are really exceptional in these respects or not.

Since the endosymbionts generally cannot be cultured, their taxon-omy is based primarily on the identity of the host and on cytological observations. In fact, intracellular symbionts were observed by the earliest microscopists in the 1800s, and cataloguing began early. Rod-shaped structures observed in the protozoa were at first thought to be an integral part of the cell concerned with sexuality and were not recognized as infective agents. Earlier work on endosymbionts in the Protozoa is reviewed by Kirby (1941), Ball (1969), and in *Paramecium* by Wichterman (1953). More recently the electron microscope has aided characterization, and biochemical data such as the buoyant density of the DNA, degree of homology in DNA hybridization studies and data on the plasmids and bacteriophages found within the endosymbionts are becoming available. Accordingly some have now been given binomial names. We have found it desirable to consider the endosymbionts of protozoa in two ways in this article. First we treat the symbionts according to the taxonomy of the host, listing those with binomial names and briefly discussing the unclassified forms (see Table 11.1). In some cases, where only a single endosymbiont has been described in a given protozoan species and nothing unique about the endosymbiont is known, we give only a reference. The second part of the article consists of a formal consideration of the 5 genera and 14 species that are now recognized as valid.

1. ENDOSYMBIONTS OF CILIATES

The ciliates bearing endosymbionts are generally cultivated in bacterized media of various kinds. Growth of ciliates in bacteria-free media often leads to loss of the endosymbionts. The endosymbionts reach their highest population levels when the ciliates are given sufficient food to multiply rather slowly (less than one doubling per day for many). Some ciliates with the symbionts they bear can be kept frozen in liquid nitrogen. They are maintained in this way at the American Type Culture Center in Rockville, Md.

Endosymbionts of ciliates show a high degree of host specificity, so that host specificity is an important taxonomic character. Consequently, the allocation of symbiotic forms to the same species or genus of bacteria because they appear very similar is perilous when these symbionts are found in hosts of different species. We have considered two such cases, both involving endosymbionts of *Paramecium caudatum* and *P. biaurelia*.

A description of alpha, found in *P. biaurelia* and initially designated *Cytophaga caryophila* (Preer et al., 1974) is very similar to that of *Holospora undulata*, found in *P. caudatum*. Although alpha is very infectious in the stock of *P. biaurelia* in which it is found, only a few other stocks in the *P. aurelia* complex are susceptible to infection by alpha, and they are all *P. biaurelia*. Evidence also exists that a specific gene in the host is necessary for the maintenance of alpha. Moreover, *Holospora elegans* specifically infects *P. caudatum*, and fails to infect many species of the *P. aurelia* complex (Görtz and Dieckmann, 1980). Nevertheless the features that these unusual forms have in common, we believe, warrant the inclusion of alpha in the genus *Holospora*.

The second case is the kappa-like form, parasitic in the macronucleus of *P. caudatum* (Estève, 1978). Electron micrographs of this bacterium appear very similar to those of the kappas of stock 7 and other stocks of *P. biaurelia*, designated *Caedibacter varicaedens* and stock 51m1 of *P. tetraurelia*, designated *Caedibacter pseudomutans*, even to the spherical phages inside the refractile body of the endosymbiont. Furthermore, strains of *P. caudatum* bearing the symbiont cause killing typical of kappa-bearers of the *P. aurelia* complex. Does the gene that supports kappa, or prevents its destruction in the *P. aurelia* complex have its counterpart in *P. caudatum*? We assume so, and suggest that the kappa-like symbiont in *P. caudatum* is a species of *Caedibacter*.

a. Paramecium caudatum

Intracellular symbionts are frequently found in *P. caudatum* (often mistakenly referred to by earlier workers as *P. aurelia*). Their study began with the discovery of slender threads in the macronucleus by Müller (1856). Many others subsequently described symbionts in this species, including Bütschli (1876), who also found them in the macro-nucleus, Hafkine (1890) in the macronucleus and micronucleus, and Petschenko (1911) in the cytoplasm. Petschenko noted that the cytoplasmic symbiont he named *Drepanospira mülleri* was much like *Holospora*, the nuclear symbionts discovered earlier. For lack of recent information, *Drepanospira mülleri* is not considered further.

List of the endosymbionts of Paramecium caudatum

Five endosymbionts of *P. caudatum* are listed here. All are found either in the micronucleus or the macronucleus. The first three, which have been assigned binomial names, are described later under the formal descriptions of valid species.

1. *Holospora undulata.*
2. *Holospora obtusa.*
3. *Holospora elegans.*
4. *Caedibacter* species. A parasitic form found in the macronucleus, described by Estève (1978) appears to belong to this genus and is treated as an undescribed species. The killer trait is manifested when *P. caudatum* bearing this symbiont is mixed with symbiont-free sensitives. The sensitive paramecia spin to the right and eventually die. The symbiont has two forms. One is similar to kappas lacking refractile bodies; it has a typical bacterial form, 1.5–3.0 μm × 0.3–0.4 μm. The other is of greater diameter (0.8 μm) and is similar to kappas with refractile bodies. It contains a laminated refractile structure, 0.5 μm in diameter, within which spherical viruses are found. Estève made a study of two populations of *P. caudatum* cultured under starvation conditions, one a symbiont bearer and the other symbiont free. He

concluded that the relationship of the bacterium to *P. caudatum* is parasitic, for the infected population died within 25 days, while the uninfected population was still viable at this time.

5. ms-2. This organism is found in stock C201 from Ringköbing

Fjord, Denmark studied by Görtz (1980). It is a small rod, 0.6–2.0 µm (rarely 3.0) long. Distinct reproductive and infectious forms have not been observed. It is sensitive to penicillin. ms-2 is found in *P. caudatum* in the macronucleus, but not coexistent with other forms.

b. *The* **Paramecium aurelia** *complex*

In the 1940s T. M. Sonneborn began a series of genetic studies on killer and sensitive paramecia of the *P. aurelia* complex. His experiments led him to the conclusion that some strains of paramecia contain an invisible element of cytoplasmic heredity called kappa. It was several years later that cytological investigations initiated by the present authors revealed that kappa is an obligate bacterial endosymbiont. At about this time, the discovery of DNA in mitochondria and chloroplasts led to renewed speculations about their possible prokaryotic evolutionary origin (a notion proposed originally in the 1800s). Furthermore, it was noted that many viruses have the capacity to exist in their hosts in nonvirulent forms, in some cases even integrated into the chromosome of the host. Thus, the studies begun with kappa in *Paramecium* led finally to the understanding that a clear distinction between elements of heredity and elements of endosymbiosis is not always easy, either technically or conceptually.

The *Paramecium aurelia* complex consists of 14 sibling species. Species 1 is *P. primaurelia*; species 2 is *P. biaurelia*; species 3 is *P. triaurelia*, etc. (See Sonneborn, 1975, for characteristics of the species.) Many strains of the *P. aurelia* complex have been found to carry bacterial endosymbionts—up to 50% in collections from ponds and streams in many areas (Fig. 11.1). All species have not been studied with the same thoroughness, nevertheless it is remarkable that endosymbionts have been found only in species 1, 2, 4, 5, 6 and 8. They are especially common in species 2 (*P. biaurelia*).

The formal classification of the endosymbionts (Preer et al., 1974) was based initially on morphology, killing activity and DNA base ratios. Subsequently, Quackenbush (1977, 1978) carried out DNA/DNA hybridizations between many of the strains. The results of these experiments have led to the revised classification given here. (See list below). Quackenbush showed that the four species of *Caedibacter* and the three species of *Pseudocaedibacter* in the revised classification each consists of a homogeneous group of strains, judged by hybridization. Hybridization between species is generally very low (less than 25%). In a few cases the degree of hybridization among strains within a species is perhaps lower than might be desired (down to 40% in a few cases). However, it has seemed wise not to produce separate taxonomic names without other more easily determined criteria. One relationship not reflected in the classification is Quackenbush's finding that *Pseudocaedibacter falsus* (nu and pi) are related by 40–60% homology with *Pseudocaedibacter conjugatus* (mu).

More recently, restriction endonuclease maps of different strains of the plasmids of *Caedibacter taeniospiralis* have become available (Quackenbush, 1981). The different strains show remarkable similarities and a few differences as well. The studies are relevant because many of the important characteristics of the strains (presence and structure of refractile bodies and nature of the killer phenotype) are probably determined by the genomes of the plasmids and phages rather than the chromosomes of the endosymbionts. In this connection it is interesting to note that Quackenbush (1978) has shown that although the phage in strain 562 of *C. varicaedens* shows 73% homology with the phage of strain 51m1 found in *C. pseudomutans*, the bacterial chromosomal DNAs show only 15% homology. He suggests that the

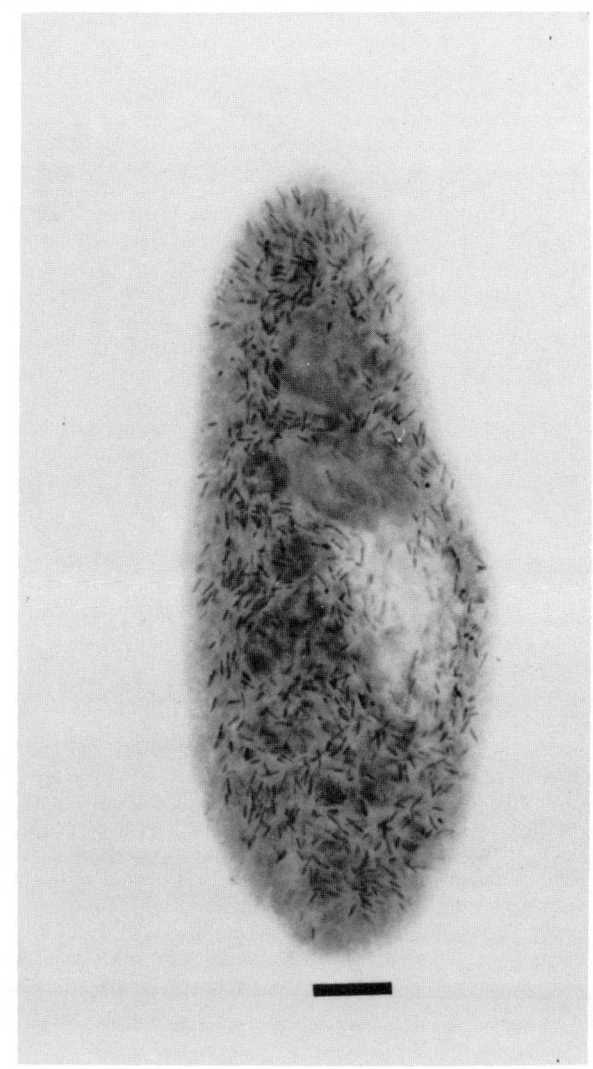

Figure 11.1. *Paramecium tetraurelia* stock 239 bearing endosymbiont *Lyticum flagellatum*. The numerous black rods throughout the cytoplasm are the endosymbionts. Osmium-lacto-orcein preparation, whole mount, dark phase-contrast (*bar*, 20 µm). (Reproduced with permission from J. R. Preer, Jr., L. B. Preer and A. Jurand, Bacteriological Reviews *38:* 113–163, 1974, © American Society for Microbiology.)

genus *Caedibacter* (the kappas) may consist of a diverse group of bacteria that contain extrachromosomal elements (phages and plasmids) with common genetic elements.

List of endosymbionts of the **Paramecium aurelia** *complex.*

See section on formal description of valid species below.

1. *Caedibacter taeniospiralis.*
2. *Caedibacter varicaedens.*
3. *Caedibacter pseudomutans.*
4. *Caedibacter paraconjugatus.*
5. *Pseudocaedibacter conjugatus.*
6. *Pseudocaedibacter minutus.*
7. *Pseudocaedibacter falsus.*
8. *Lyticum flagellatum.*
9. *Lyticum sinuosum.*
10. *Tectibacter vulgaris.*
11. *Holospora caryophila.*

c. **Paramecium multimicronucleatum**

Jenkins (1970) described epsilon, a short rod, nearly coccoid, 0.35–0.7 μm in length, located only within bulbous distensions of the outer membrane of the micronucleus and macronucleus of *P. multimicronu-* *cleatum*. It resembles Gram-negative bacteria in structure. Clear regions in the protoplasm may be nuclear areas, and membranous whorls, mesosome-like structures.

d. **Euplotes**

The many species of the genus *Euplotes* comprise both fresh-water and marine forms. Endosymbiotic bacteria are very common in *Eu-* *plotes*. None have been given binomial names, but several are well characterized.

List of endosymbionts of **Euplotes**

1. Omicron is the best studied form. See Heckmann (1980) and Schmidt and Heckmann (1980) for summaries. Omicron, found in the cytoplasm of *Euplotes aediculatus*, is a slightly curved rod with a diameter of approximately 0.3 μm and a length of 2.5–7.5 μm (Fig. 11.2). It is Gram-negative. Unlike most of the endosymbiotic bacteria, its DNA, instead of being dispersed throughout the cell, occurs in nucleoids. The mol% G + C is 45 (Bd). In addition to the two typical membranes of the Gram-negative bacteria, most cells are surrounded by another membrane, apparently originating from the host. It produces no killing activity. Heckmann has shown that omicron is essential for the life of its host, for *Euplotes* treated with penicillin, which destroys omicron, eventually die. Fauré-Fremiet (1952) made the same observation and reached the same conclusion. Decisive proof for this interpretation, however, was provided by Heckmann. He showed that penicillin-treated cells, doomed to die because of elimination of their endosymbionts, can be rescued by reinfecting with omicron, which the ciliates are able to take up from homogenates of omicron bearers.

Heckmann finds omicron or omicron-like endosymbionts in a group of closely related species of freshwater *Euplotes* (*E. aediculatus*, *E. eurystomus*, *E. patella*, *E. plumipes*, *E. woodruffi*, *E. diadaleos* and *E. octocarinatus*), but not in other unrelated species.

2. Several endosymbionts produce the killer phenotype. At least two strains of killers with associated bacteria in the cytoplasm have been described. One, designated epsilon, was found in *Euplotes minuta* and studied by Heckmann et al. (1967). Another, designated eta, was reported in the marine ciliate *Euplotes crassus* by Rosati et al. (1976).

3. Endosymbionts have been found in the cytoplasm of mate killers and are presumed to be the basis of the trait in *Euplotes patella* (Katashima, 1965) and in *Euplotes crassus* (Rosati, et al., 1976).

4. Several endosymbionts with no known effects on their hosts have also been reported in *Euplotes crassus*. While most are found in the cytoplasm (Rosati, et al., 1976), one is found in the macronucleus (Rosati and Verni, 1975).

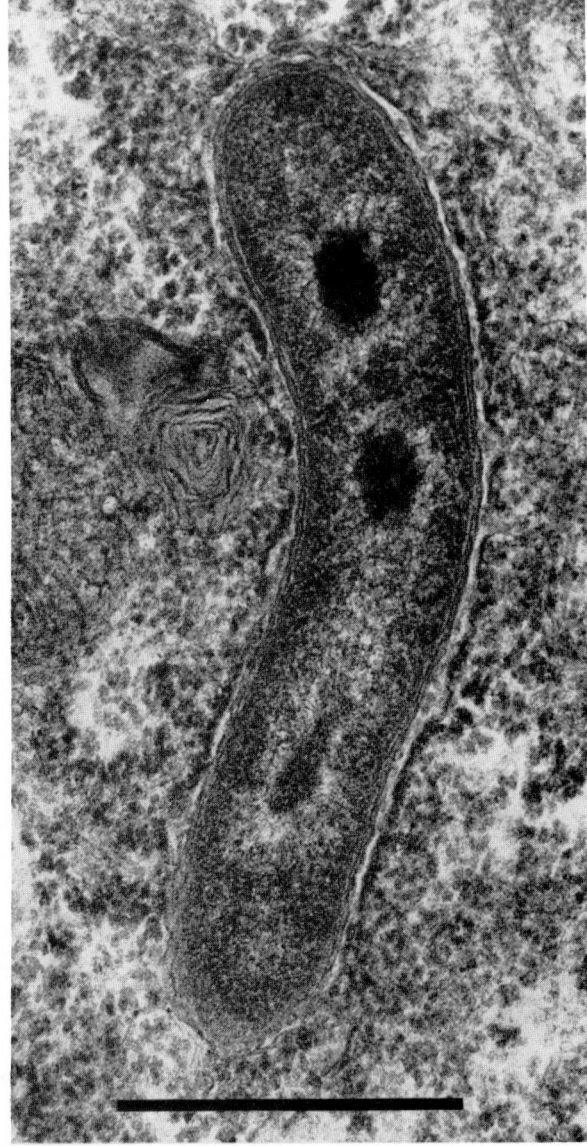

Figure 11.2. Omicron, endosymbiont of *Euplotes aediculatus*. Longitudinal section showing nucleoids (*bar*, 0.5 μm). (Reproduced with permission from K. Heckmann, Journal of Protozoology *22:* 97–104, 1975, © Society of Protozoologists.)

e. **Spirostomum**

Harrison et al. (1976a, b) have studied a Gram-variable bacterium in the macronucleus of *Spirostomum ambiguum*. It was originally discovered by Inaba (1960). It is pleomorphic with a diameter of 0.3–2.5 μm and a length of 0.6–3.5 μm. It has no known effect on its host. It was present in all of the strains of *Spirostomum ambiguum* examined, even though a search for symbiont-free strains was made.

A morphologically similar organism formed colonies when lysates of *Spirostomum* were passed through a filter and plated on blood agar.

Similar organisms could not be obtained from the fluid in which the *Spirostomum* had been cultured. The identity of the cultured bacteria with the endosymbionts in the macronucleus was confirmed by fluorescent antibody tests.

Further attempts to classify the organism revealed that it could be cultured on several standard bacteriological media. A number of its properties were determined (see Harrison et al., 1976b). Several lines of evidence suggest that the organism is flexible and motile, moving without flagella by amoeboid movement. No name has been assigned to the bacterium and its taxonomic position is unknown.

f. *Other ciliates*

Berezina (1975) has studied an endosymbiont in the cytoplasm of *Blepharisma japonicum*. The endosymbionts enlarge when the ciliates are exposed to visible light and develop numerous phage-like structures.

Soldo (1974) has described infective structures which he calls xenosomes in the marine ciliate *Paraurorema acutum*. It appears to be a small bacterium. The ciliate containing xenosomes grows well in axenic medium.

Reports of bacterial endosymbionts in many other genera of ciliates have been made, but in most cases the strains no longer exist and very little information is available. The interested reader is referred to Kirby (1941), Wichterman (1953) and Ball (1969). More recent cases include forms in *Paramecium bursaria* (Golikova, 1978) and in *Urostyla* (Ruthmann and Noll-Altmann, 1980).

2. ENDOSYMBIONTS OF FLAGELLATES

The flagellate *Crithidia oncopelti*, an inhabitant of the gut of a hemipteran, the milk-weed bug, contains bacterial endosymbionts known as bipolar bodies in its cytoplasm, generally two per host cell. Similar endosymbionts, called diplosomes are found in the related flagellate *Blastocrithidia culicis* (Fig. 11.3). Both flagellates can be cultured readily on bacteria-free medium, and both can be stored in liquid nitrogen. *Crithidia deanei*, found in another hemipteran, contains similar forms (Mundim et al., 1974). *Trypanosoma cobitis*, a parasite of fishes also contains bacterial endosymbionts (Lewis and Ball, 1981): Unlike the other flagellates, *T. cobitis* has a secondary host (a leech), and the endosymbionts are present in all stages of the life cycle of *T. cobitis*. Although the nature of these forms was a matter of controversy for many years, a consideration of their properties now leaves little doubt about their bacterial origin. See Tuan and Chang (1975), Chang (1974, 1976) and Chang and Dave (1980) for discussion and references.

Bipolar bodies and diplosomes appear as small rods in the cytoplasm of their hosts. The cell wall is much reduced or absent. The mol% G + C is 34–36 (Bd). It has been possible to eliminate bipolar bodies and diplosomes from the flagellates and study the effects of the endosymbionts on their hosts. The endosymbiont-free strains require additional factors for growth, hemin and nicotinamide (Mundim and Roitman, 1977). It appears that the endosymbionts contribute these substances to their hosts.

Bacterial endosymbionts appear in great abundance in several kinds of flagellates that live within the guts of termites and certain wood roaches. Unfortunately the protozoans are not readily cultivated free of their hosts. Consequently pure lines have not been established in the laboratory and most work has been restricted to microscopic observations on mixed wild strains. The many early observations (Kirby, 1941) were made with the light microscope, but in more recent years the electron microscope has been used. See the review by Ball (1969). Although some forms are endosymbionts, others are exosymbionts, attaching to the cell surface in great abundance. Specialized attachment structures are in some cases a part of the bacterium; in others they are part of the host (Bloodgood and Fitzharris, 1976). It is not known whether any of the endosymbionts or exosymbionts constitute part of the large population of free bacteria within the insect gut. Cleveland and Grimstone (1964) made the remarkable observation that in the flagellate *Mixotricha paradoxa* from the termite *Mastotermes darwiniensis*, spirochetes adhering to the cell surface appear to contribute to the motility of the flagellates in a coordinated way.

3. ENDOSYMBIONTS OF AMOEBAS

Bacterial endosymbionts have been reported in *Amoeba proteus*, *A. discoides*, *Acanthamoeba castellani*, *Pelomyxa palustris* and *P. illinoisiensis*. See Ball (1969) for a reveiw of the early literature. The best characterized are in *A. proteus* and *P. palustris*.

a. **Amoeba proteus**

Endosymbiosis in *A. proteus* has been reviewed by Jeon (1980). X-bacteria are present in the cytoplasm of strain xD of *Amoeba proteus*. They are rods 0.5 × 2.0 μm, and are enclosed in vacuoles, singly or in groups in the host. The endosymbionts are found in large numbers, 42,000/amoeba. X-bacteria appeared spontaneously in the laboratory in symbiont-free strain D of *A. proteus*. At first they acted like a parasite, but with the passage of time they became essential to this strain (Lorch and Jeon, 1980a). Experiments have shown that X-bacteria can establish symbiosis with noninfected amoebas of the original or different strains, and within 200 cell generations newly infected amoebas may become dependent upon X-bacteria. X-bacteria are Gram-negative, possess tough cell walls that are resistant to 2N NaOH, are sensitive to antibacterial agents such as chloroamphenicol and trimethoprim, and die at temperatures above 26.5°C. Nuclear compatibility studies involving nuclear transplants show that after X-bacteria have grown in newly infected amoebas for 4 weeks (10–15 generations of amoeba), nuclei of the amoebas can no longer survive in cytoplasm of the original endosymbiont-free strain D (Lorch and Jeon, 1980b). D nuclei, on the other hand, persist in infected amoebas. Infectivity of isolated X-bacteria is retained for more than 72 h at 20°C, 7 days at 4°C, and many months at −20°C. X-bacteria carry two plasmids, pHJ11 and pHJ12 (Han and Jeon, 1980) that have been isolated and characterized by gel electrophoresis. Their molecular weights are 39×10^6 and 11×10^6, respectively.

A different kind of endosymbiont in *A. proteus* called DNA-bodies (Rabinovitch and Plaut, 1962) is described as spherical with a diameter of 0.5 μm. It occurs singly in vacuoles in the cytoplasm, and is specific for the strain in which it is found. It can be further differentiated from X-bacteria by its solubility in 2N NaOH, incorporation of exogenous thymine, and the fact that it sustains reversible damage when treated with chloramphenicol (Jeon, 1980).

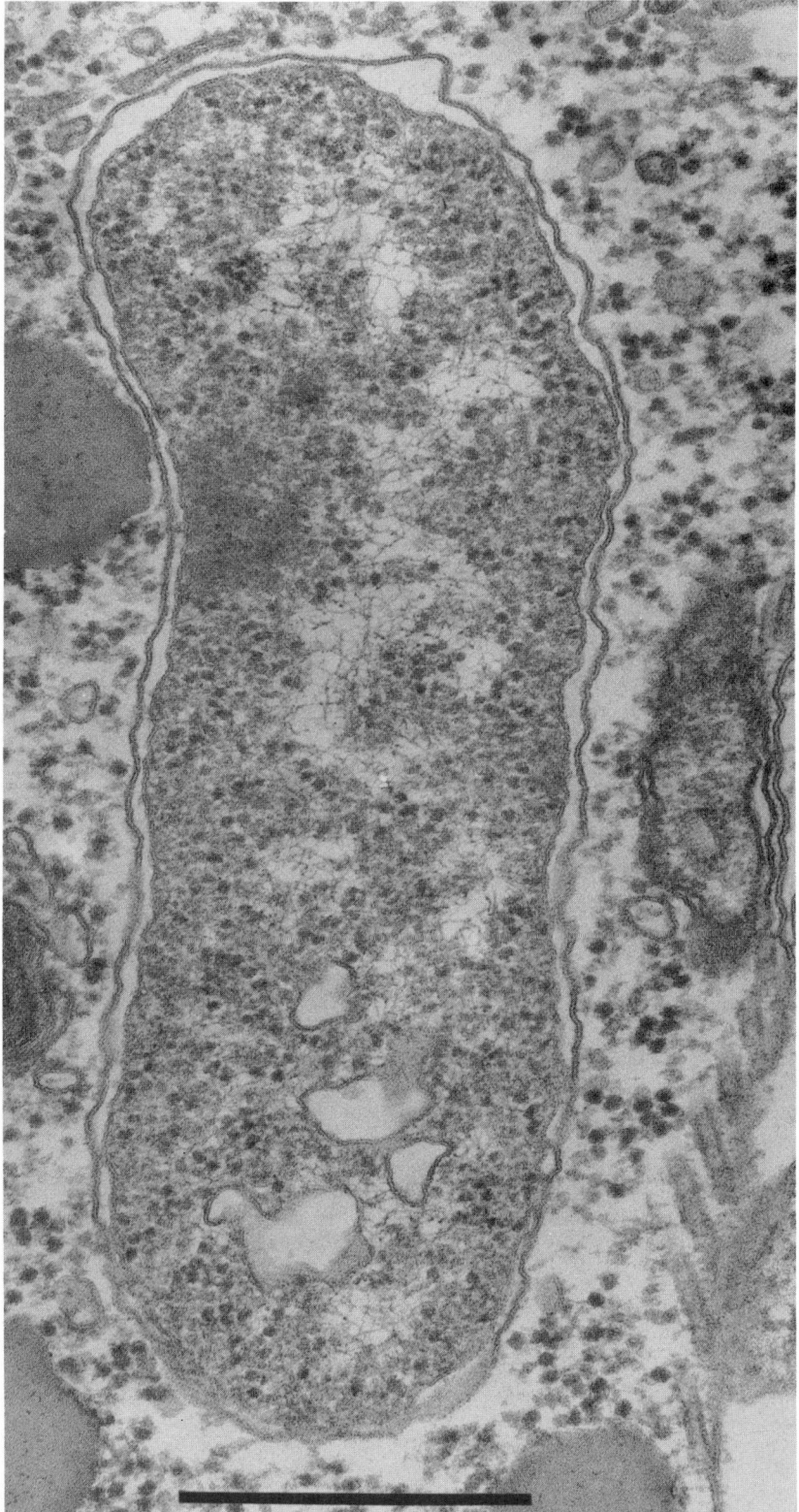

Figure 11.3. Diplosome of *Blastocrithidia culicis.* Longitudinal section (*bar*, 0.5 μm). (Reproduced with permission from K.-P. Chang, Journal of Protozoology *21:* 699–707, 1974, © Society of Protozoologists.)

b. Pelomyxa palustris

Two types of endosymbionts have been noted as constant constituents in strains of the giant amoeba *P. palustris* from North America and Europe (Chapman-Andresen, 1971). For descriptive detail beyond that abstracted below, and for references, see Whatley (1976). One type is a small, osmiophilic bacterium distributed throughout the cytoplasm.

The more distinctive type is large, coccoid or rod shaped, measuring 0.3 μm × 3 μm (Fig. 11.4). It is found in great numbers surrounding the nucleus and, to a much lesser extent, the glycogen bodies of the amoeba. The end of the bacterium is shaped like a truncated dome or pyramid and a deep axial cleft, invaginating through the bacterial

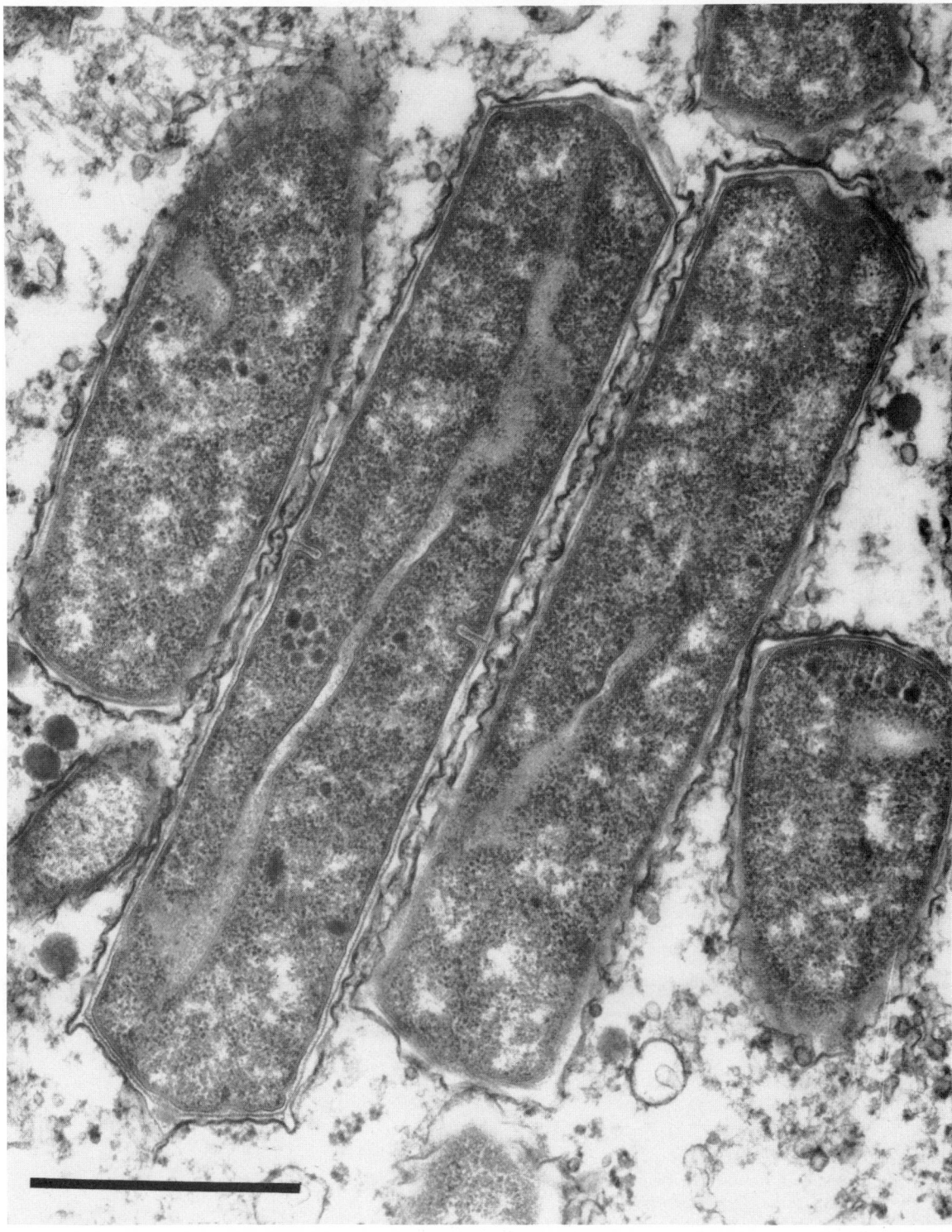

Figure 11.4.   Large bacterium with distinctive axial cleft, an endosymbiont in *Pelomyxa palustris*. This bacterium may be infected with a virus. Longitudinal section (*bar*, 1 μm). (Reproduced with permission from J. Whatley, New Phytologist *76:* 111–120, 1976, © Blackwell Scientific Publications Ltd.)

plasma membrane and cell wall extends the length of the cell. The bacterium divides by invagination of the plasma membrane and cell wall to form a thick cross-wall. Surrounding each bacterium is a close-fitting vacuolar membrane. Adjacent vacuoles are joined to each other and to the nuclear membrane by tubules. The symbiont appears to be infected with a virus. A structural element in the axial cleft of the endosymbiont is similar to an element in the cell wall of the only bacterium it resembles at all, *Methanospirillum hungatii*, and to an element in the cristae of mitochondria. The interesting suggestion has been made that these endosymbionts may function as mitochondria; mitochondria are not present in *P. palustris* (Daniels et al., 1966, John and Whatley, 1977).

FORMAL DESCRIPTIONS OF THE GENERA AND SPECIES OF THE ENDOSYMBIONTS OF PROTOZOA

Genus I. Holospora (ex Hafkine 1890) Gromov and Ossipov 1981, 351[VP]*

Ho.los'po.ra. Gr. n. *holos* whole; Gr. n. *spora* a spore; M.L. fem. n. *Holospora* whole spore.

Symbiont is present specifically in the **micronucleus or macronucleus of *Paramecium***. Exists in two forms: the **reproductive form is a short, fusiform rod** 1.0–3.0 μm long and 0.5–1.0 μm wide. It undergoes binary fission and gives rise to the **long infective form** measuring 5.0–20.0 μm with rounded or tapered ends that can infect *Paramecium* and become established in the nucleus. The infective form (three of four species) is **differentiated into a refractile portion** of electron-dense, fine granular material with a **less electron dense pale tip**, and a posterior part that contains typical bacterial protoplasm with ribosomes and mesosome-like structures. **The latter part** appears dark with dark phase-contrast microscopy and **stains with DNA-specific dyes**. Gram-negative, nonmotile, obligate symbiont. No toxic effect of *Holospora*-bearing paramecia on paramecia lacking the symbiont. Paramecia can outgrow *Holospora*, but **reinfection occurs readily**; mass cultures show up to 100% infection.

Type species: *Holospora undulata* (ex Hafkine 1890) Gromov and Ossipov 1981, 351

Further Descriptive Information

Symbionts in *Paramecium* that appear to belong to the genus *Holospora* (Hafkine, 1890) have been noted frequently (Wichterman, 1953).

The three species of *Holospora* described by Hafkine are presumed to be the same as those being studied by recent workers. Ossipov and coworkers (Ossipov and Ivakhnyuk, 1972; Ossipov et al., 1973, 1975; Ossipov and Podlipaev, 1977; Podlipaev and Ossipov, 1979; Gromov and Ossipov, 1981) in the U.S.S.R. have studied *H. undulata* and *H. obtusa*. Görtz (1980) and Görtz and Dieckmann (1980) in Germany have described *Holospora elegans* as well as the other two species.

The progress of the infective form of *H. obtusa* and *H. undulata* to the nucleus is described by Ossipov and Podlipaev (1977). The infective form of the bacterium enters the food vacuole of the host, where changes in refractility of the bacterium occur. Transport of the endosymbiont from the food vacuole through the cytoplasm into the micronucleus or macronucleus is effected by the cell membranes of the host. Görtz (1980) reports that after the holosporas enter the nucleus the long form becomes constricted at several points and divides to produce short forms. The short form undergoes binary fission to produce eventually the long form that may remain in the nucleus and repeat the cycle or be released at cell division to infect other paramecia. He found that the short form of three species of *Holospora* and endosymbiont ms-2 are sensitive to penicillin; the long forms are not.

Differentiation of the genus Holospora from other genera

Table 11.2 provides the necessary information.

Taxonomic Comments

Hafkine described *Holospora* as a spore former and reported that it reproduced to some extent by budding. The infectious form of the symbiont that Hafkine, and more recently Gromov and Ossipov

(1981), believed to be a spore may not be a spore in the true sense. Görtz and Dieckmann (1980) consider it a specialization to permit the organism to infect; what Hafkine considered budding, they regard as a misinterpretation. A fourth species, alpha, previously designated *Cytophaga caryophila* (Preer et al., 1974), a macronuclear symbiont of *P. biaurelia*, is now included in the genus *Holospora*.

Table 11.2.
Differentiation of the genera of the endosymbionts of ciliates[a]

Characteristics	I. *Holospora*	II. *Caedibacter*	III. *Pseudocaedibacter*	IV. *Lyticum*	V. *Tectibacter*
Found in nuclei	+	D[b]	−	−	−
Contain R bodies	−	+	−	−	−
Hosts are killers or mate killers	−	+	D[c]	+	−
Contain a thick cell wall	−	−	−	−	+
Flagellated	−	−	−	+	+
Infective	+	−	−	−	−

[a] Symbols: see standard definitions.
[b] One strain of *Caedibacter* is found in the macronucleus; the others are cytoplasmic.
[c] Some hosts are killers, some mate killers, some nonkillers.

Differentiation and characteristics of the species of Holospora

Table 11.3 presents the characteristics differentiating the species of this genus.

List of the species of the genus Holospora

1. **Holospora undulata** (ex Hafkine 1890) Gromov and Ossipov 1981, 351.[VP]
un.du.la'ta. L. fem. adj. *undulata* undulated.
Short, spindle-shaped reproductive form up to 3.0 μm long that

divides transversely, developing into **spiral infective form**, approximately 16 μm long with tapered ends. Found in *P. caudatum* **in the micronucleus.**

Type: the original description and illustration of Hafkine (1890).

---

* VP, denotes that this name has been validly published in the official publication, International Journal of Systematic Bacteriology.

Table 11.3.

Characteristics differentiating species of the genus **Holospora**[a]

Characteristics	1. *H. undulata*	2. *H. obtusa*	3. *H. elegans*	4. *H. caryophila*
Common name	Omega	Iota		Alpha
Length of infective form:				
5–6 μm	–	–	–	+
7–20 μm	+	+	+	–
Shape of infective form:				
Spiral, tapered ends	+	–.	–	+
Rod, rounded ends	–	+	–	–
Thin rods, tapered ends	–	–	+	–
Habitat:				
Micronucleus	+	–	+	–
Macronucleus	–	+	–	+

[a] For symbols see standard definitions.

Clone M1-48 of *P. caudatum* containing *H. undulata* in its micronuclei has been deposited in the collection of the Laboratory of Invertebrate Zoology, Biological Research Institute, Leningrad University (Gromov and Ossipov, 1981).

2. **Holospora obtusa** (ex Hafkine 1890) Gromov and Ossipov 1981, 351.[VP]

ob.tu'sa. L. fem. part. adj. *obtusa* blunt.

Short fusiform reproductive rod about 3.0 μm long that undergoes binary fission and grows to **infective form at least 18 μm long.**

Ends of rod rounded (Fig. 11.5). Found in *P. caudatum* **in the macronucleus.**

Type: the original description and illustration of Hafkine (1890). Clone M-115 of *P. caudatum* containing *H. obtusa* in its macronuclei has been deposited in the culture collection of the Laboratory of Invertebrate Zoology, Biological Research Institute, Leningrad University (Gromov and Ossipov, 1981).

3. **Holospora elegans** (ex Hafkine 1890) Preer and Preer 1982, 140.[VP]

e'le.gans. L. adj. *elegans* choice, elegant.

Thin reproductive rod 0.6–0.9 μm in width with an average length of 2.5 μm. It divides transversely to give rise to the **long, slender infective form** 1.4–1.8 μm wide and **10–20 μm long** with **tapered ends.** Found in *P. caudatum* **in the micronucleus.**

Type strain: In ATCC strain 50008 (stock C 101 of *P. caudatum*, Görtz and Dieckmann, 1980).

4. **Holospora caryophila** (Preer, Preer and Jurand 1974) Preer and Preer 1982, 141.[VP] (*Cytophaga caryophila* Preer, Preer and Jurand 1974, 156.)

ca.ry.o'phi.la. Gr. noun *caryum* nut, kernel, nucleus; Gr. adj. *philus* loving, M.L. fem. adj. *caryophila* nucleus loving.

Reproductive rod 0.3–0.5 × 1.0–3.0 μm. Spiral infective **form 5.0–6.0 μm long** with tapered ends. Is highly infective to a few stocks of *Paramecium biaurelia*; found **in the macronucleus** (Figs. 11.6 and 11.7). Known previously as alpha.

Type strain: In ATCC strain 30694 (stock 562 of *P. tetraurelia*; see Preer et al., 1974).

Genus II. **Caedibacter** (*ex* Preer, Preer and Jurand 1974) *Preer and Preer 1982, 140*[VP] (*Caedobacter* (*sic*) Preer, Preer and Jurand 1974, 157)

Cae'di.bac.ter. L. n. *caedes* act of killing; M.L. masc. n. *bacter* the masc. equivalent of the Gr. neut. n. *bactrum* a rod; M.L. masc. noun *Caedibacter* the bacterium which kills.

Straight rods or coccobacilli 0.4–1.0 μm in diameter and 1.0–4.0 μm in length. Up to 50% (usually less than 10%) of the cells in any given population **contain single (rarely double) refractile inclusion bodies (R bodies).** The R body is a proteinaceous ribbon, approximately 10 μm long, 0.5 μm wide and 13 nm thick, which is tightly rolled up within the cell (Figs. 11.8–11.10). Cells containing R bodies are usually larger than cells that do not contain R bodies and contain many spherical phage-like structures or covalently closed circular DNA plasmids. **All species are toxic** to certain sensitive strains of paramecia. Gram-negative. Nonmotile. Occurs in *Paramecium biaurelia* and *P. tetraurelia*. The mol% G+ C of the DNA is 40–44 (Bd).

Type species: *Caedibacter taeniospiralis* (*ex* Preer, Preer and Jurand 1974) Preer and Preer 1982, 140.

Further Descriptive Information

The distinguishing characteristic of *Caedibacter* is the presence of refractile (R) bodies. The coiled R body, a distinctive structure, is seen with bright phase-contrast microscopy either as doughnut-shaped or as a pair of parallel rods, depending upon its orientation (Fig. 11.11). Although the structure is readily resolved in bright phase-contrast, it is often obscure in dark phase-contrast, because dark phase optics are usually of such high contrast that phase reversal occurs in most aqueous media. Refractile bodies themselves appear to result from the induction of phage-like or plasmid-like extrachromosomal DNAs. The induction

appears to be lethal: R body-containing cells do not have the capacity to reproduce. It is interesting that R bodies have been observed in free-living bacteria belonging to the genus *Pseudomonas* (Lalucat et al., 1979). R body-containing *Pseudomonas* have been shown to be toxic to paramecia (Lalucat et al., 1980). The toxicity of *C. taeniospiralis*, *C. varicaedens* and *C. pseudomutans*, commonly known as kappa, is produced by the ingestion of R body-containing kappas by sensitive strains of paramecia. R bodies can be induced to unroll (Fig. 11.12) and in some cases, reroll; they have been shown to unroll in the food vacuoles of sensitive paramecia. In some species of *Caedibacter* the R body unrolls from the inside (Fig. 11.13); in others, from the outside. Rupture of the membrane of the food vacuole occurs, and the contents of the food vacuole, including the unrolling R bodies, pass into the cytoplasm of the paramecium. The toxins themselves have never been obtained in soluble form and their nature is unknown. Ingestion of the fourth species, *C. paraconjugatus*, does not produce toxic effects on sensitives. Instead it is a mate-killer, and sensitive paramecia die only after contact with paramecia-bearing *C. paraconjugatus* during conjugation. The endosymbiont discovered by Estève (1978) in the macronucleus of a strain of *P. caudatum* appears to belong to this genus. It was found to be deleterious to its host and produces spin killing. Kappas have been studied very extensively and the reader is referred to two reviews (Preer et al., 1974; Soldo 1974) as well as more recent papers dealing with their taxonomy (Quackenbush 1977, 1978, 1981).

Differentiation of the genus **Caedibacter** *from other genera*

Table 11.2 provides the necessary information.

Differentiation and characteristics of species of **Caedibacter**

Table 11.4 presents the characteristics differentiating the four species of the genus.

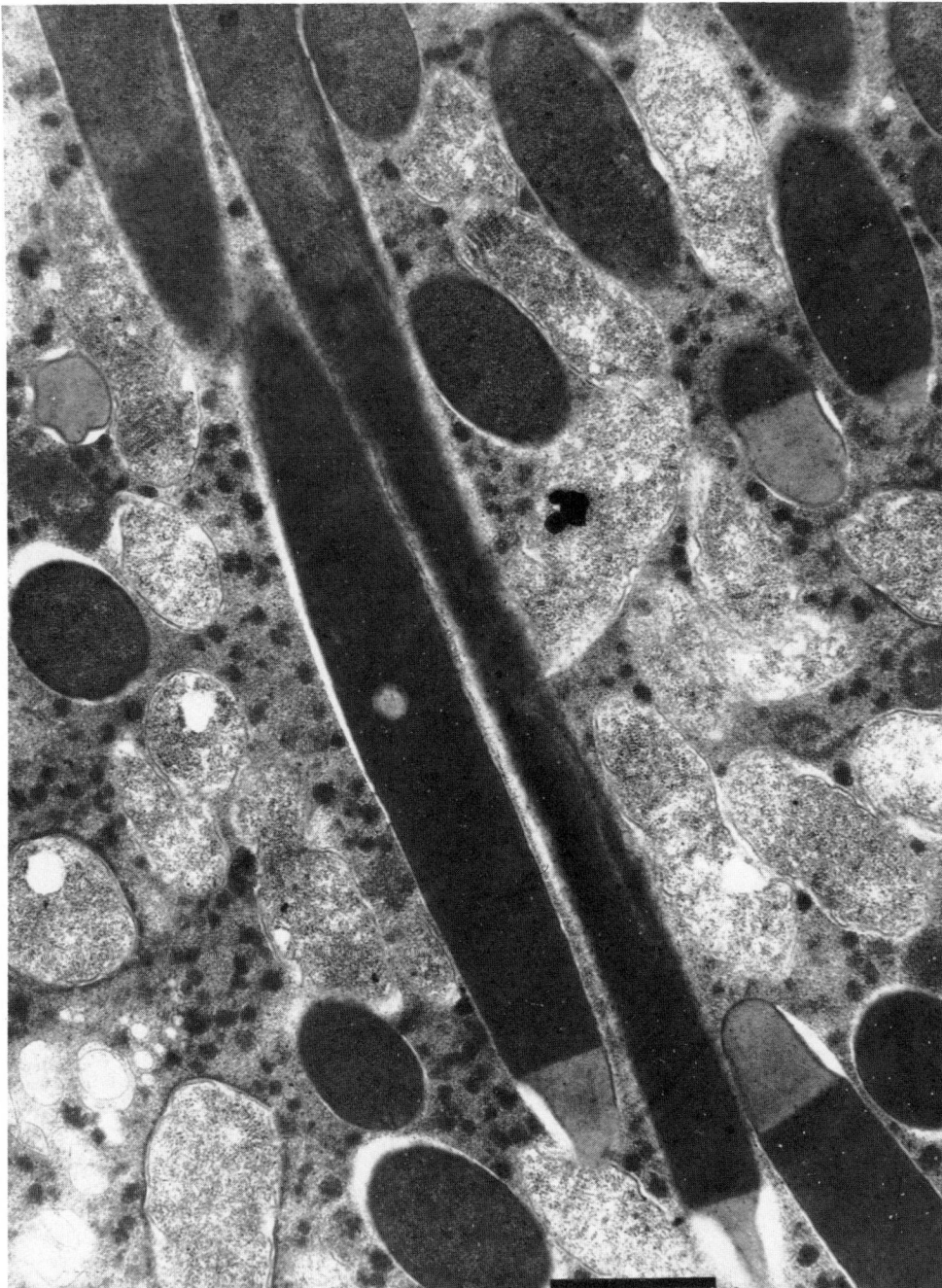

Figure 11.5. *Holospora obtusa*, macronuclear symbiont of *Paramecium caudatum*. Infectious form stains darkly; the noninfectious form is light (*bar*, 1 μm). (Reproduced with permission from H.-D. Görtz, *in* W. Schwemmler and H. E. A. Schenk (Editors), *Endocytobiology: endosymbiosis and cell biology*, Vol. 1, pp. 381–392, 1980, © Walter deGruyter and Co.)

List of the species of the genus **Caedibacter**

1. **Caedibacter taeniospiralis** (*ex* Preer, Preer and Jurand 1974) Preer and Preer 1982, 140.[VP] (*Caedobacter* (*sic*) *taeniospiralis* Preer, Preer and Jurand 1974, 157.)

taen.i.o.spi.ral′is. L. n. *taenia* ribbon; L. adj. *spiralis* coiled; M.L. masc. adj. *taeniospiralis* coiled ribbon.

Rods 0.4–0.7 μm in diameter and 1.0–2.5 μm long. Ingestion of R body-containing cells by sensitive paramecia usually results in the **development of an aboral blister or hump** preceding the death of the paramecium. **R bodies unroll from the inside. Found in** *P. tetraurelia* **exclusively**. Contain plasmids (Dilts, 1977).

The mol% G + C of the DNA is 41 (Bd).

Type strain: In ATCC strain 30632 (stock 51 of *P. tetraurelia*; see Preer et al., 1974).

2. **Caedibacter varicaedens** Quackenbush 1982, 266.[VP] (Effective publication: Quackenbush 1978, 186.)

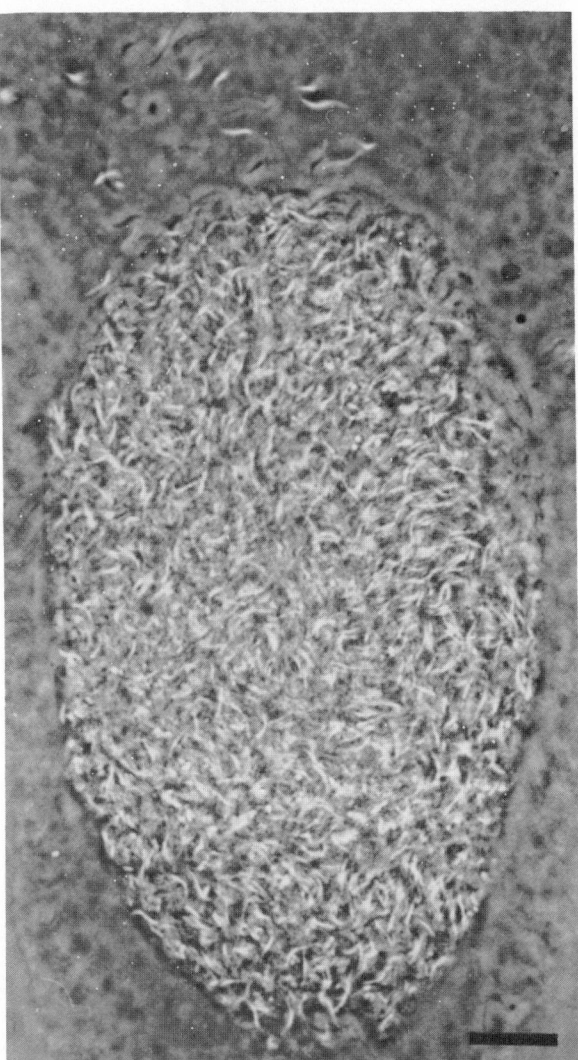

Figure 11.6. Macronucleus of *Paramecium biaurelia* stock 562. The spiral endosymbiont filling the macronucleus is *Holospora caryophila.* Osmium-lacto-orcein preparation, whole mount, bright phase-contrast (*bar*, 10 μm). (Reproduced with permission from L. B. Preer, Journal of Protozoology *16:* 570–578, 1969, © Society of Protozoologists.)

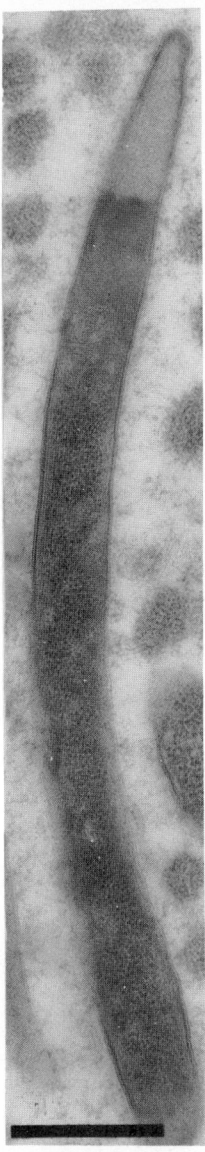

Figure 11.7. *Holospora caryophila* of *Paramecium biaurelia* stock 562, spiral form. Longitudinal section (*bar*, 0.5 μm). (Reproduced with permission from L. B. Preer, Journal of Protozoology *16:* 570–578, 1969, © Society of Protozoologists.)

Type strain: In ATCC strain 30633 (51m1 of *P. tetraurelia*; see Preer et al., 1974).

4. **Caedibacter paraconjugatus** Quackenbush 1982, 266.[VP] (Effective publication: Quackenbush 1978, 186.)

par.a.con.ju.ga′tus. Gr. prep. *para* alike; L. part. adj. *conjugatus* conjugated, also the specific epithet of a mate killer (*Pseudocaedibacter conjugatus*); M.L. masc. part. adj. *paraconjugatus* similar to mate killers.

Small rods. Less than 1% of the cells contain R bodies, which are smaller than those found in the other species of *Caedibacter*. Ingestion of cells by sensitive strains of paramecia does not produce any observable toxic effects. Cell-to-cell contact between host and sensitive paramecia is required for toxic effects (**mate killing**) to be observed in the sensitive paramecia. Found in *P. biaurelia*. Contain phage-like structures.

Type strain: In ATCC strain 30638 (570 of *P. biaurelia*; see Preer et al., 1974).

var.i.cae′dens. L. adj. *varis* different; L. v. *caedo* to kill; M.L. part. adj. *varicaedens* killing in different ways.

Rods 0.4–1.9 μm in diameter and 2.0–4.0 μm long. Different strains cause in sensitives either **vacuolization or paralysis or rapid reverse rotation** while swimming (spin-killing). R bodies unroll from the outside. One of the commonest killers of *P. biaurelia*. Most strains contain spherical phagelike structures (Fig. 11.9).

The mol% G + C of the DNA is 40–41 (Bd).

Type strain: In ATCC strain 30637 (stock 7 of *P. biaurelia*; see Preer et al., 1974).

3. **Caedibacter pseudomutans** Quackenbush 1982, 266.[VP] (Effective publication: Quackenbush 1978, 186.)

pseu.do.mu′tans. Gr. adj. *pseudo* false; L. part. adj. *mutans* changing; M.L. part. adj. *pseudomutans* false changing, referring to the fact that it was once thought to be a mutant of *C. taeniospiralis* (Dippel, 1950).

Cigar-shaped rods approximately 0.5 μm in diameter and 1.5 μm long. Found in *P. tetraurelia*.

The mol% G + C of the DNA is 44 (Bd).

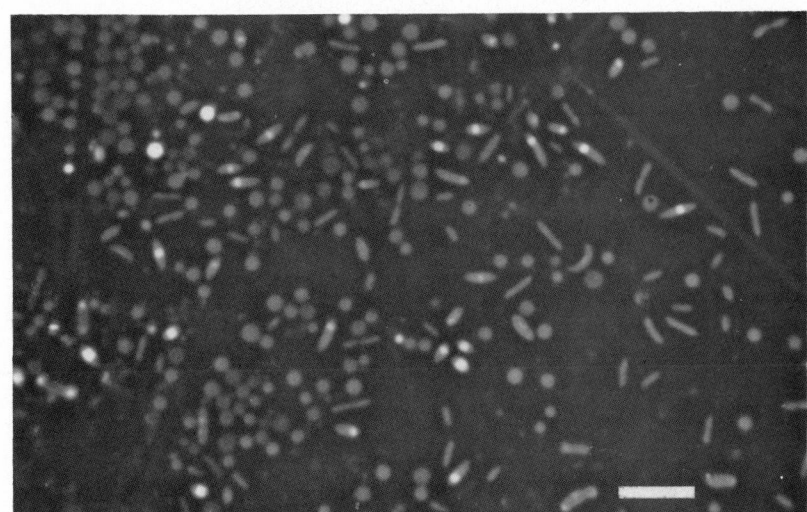

Figure 11.8. Fresh squash of *Paramecium biaurelia* stock 7, washed free of bacteria, viewed by bright phase-contrast. The spherical structures are mitochondria. Note *Caedibacter varicaedens* present in two forms: small rods and larger spindle-shaped cells containing refractile (R) bodies characteristic of the genus (*bar*, 5 μm).

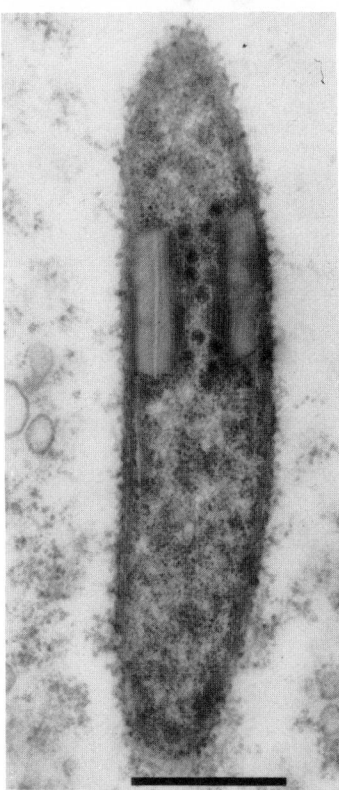

Figure 11.9. *Caedibacter varicaedens*, endosymbiont of *Paramecium biaurelia* stock 7. Note spherical phages inside the coiled R body. Longitudinal section (*bar*, 0.5 μm). (Reproduced with permission from J. R. Preer, Jr. and A. Jurand, The relation between virus-like particles and R-bodies of *Paramecium aurelia*, Genetical Research *12:* 331–340, 1968, © Cambridge University Press.)

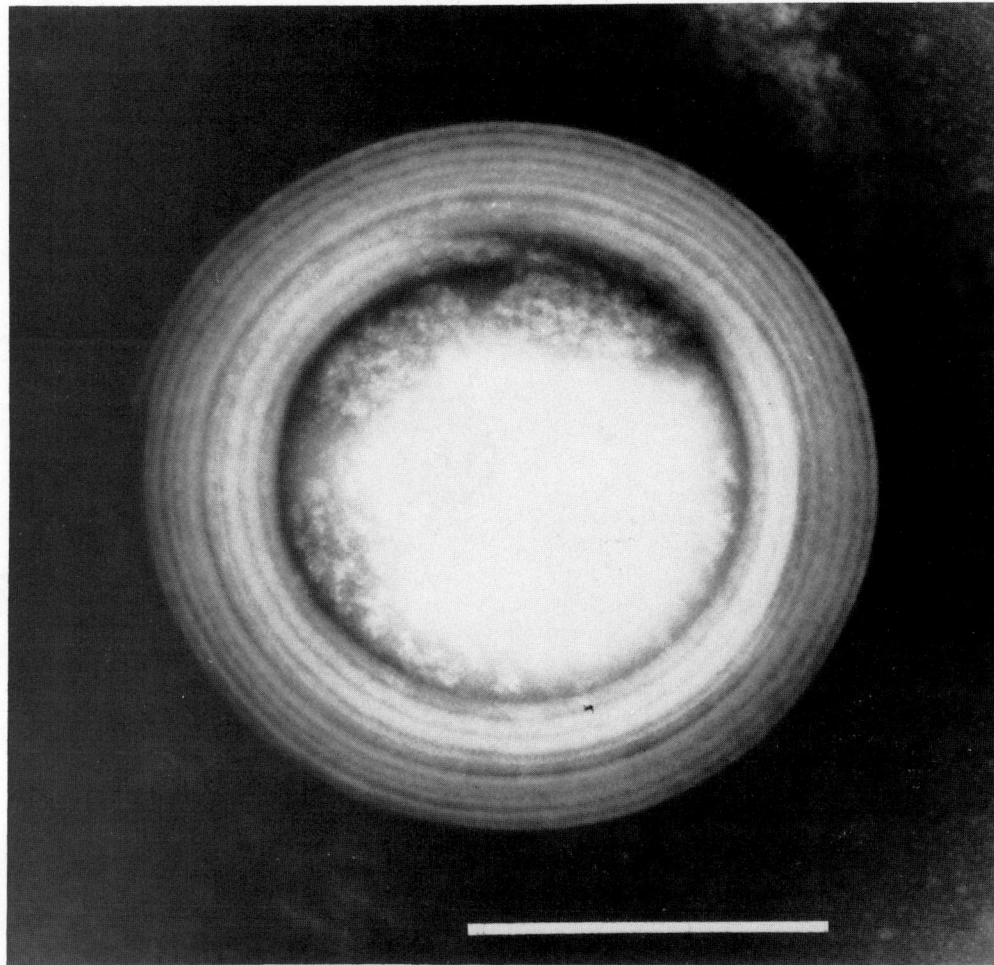

Figure 11.10. Intact R body from *Caedibacter varicaedens* of *Paramecium biaurelia* stock 511. Phosphotungstic acid (*bar*, 2.5 μm). (Reproduced with permission from J. R. Preer, Jr., L. B. Preer and A. Jurand, *Bacteriological Reviews 38:* 113–163, 1974, © American Society for Microbiology.)

Genus III. **Pseudocaedibacter** *Quackenbush 1982, 267^VP* (Effective publication: Quackenbush 1978, 186)

Pseu.do.cae′di.bac.ter. Gr. adj. *pseudo* false; M.L. n. *Caedibacter* genus of endosymbionts that includes organisms commonly known as kappa; M.L. masc. n. *Pseudocaedibacter* false kappa particles.

Rods 0.25–0.7 μm in diameter and 0.5–4.0 μm long. Do not produce R body-containing cells. May or may not confer a killer trait upon their host paramecia. Gram-negative. Nonmotile. Occurs in *Paramecium primaurelia*, *P. biaurelia*, *P. tetraurelia*, *P. pentaurelia* and *P. octaurelia*. The mol% G + C of the DNA is 35–39 (Bd).

Type species: *Pseudocaedibacter conjugatus* Quackenbush 1982, 267.^VP

Further Descriptive Information

Pseudocaedibacter is much like *Caedibacter* except that **no R bodies** are found. It includes the typical **mate killing** forms (*P. conjugatus*, called mu), the **small killing forms** of *Paramecium octaurelia* (*P. minuta*, gamma), and the nondescript forms with **no killing action** (*P. falsus*, nu).

Differentiation of the genus **Pseudocaedibacter** *from other genera*

See Table 11.2.

Differentiation and characteristics of species of **Pseudocaedibacter**

Table 11.5 presents the characteristics differentiating the three species of the genus.

List of the species of the genus **Pseudocaedibacter**

1. **Pseudocaedibacter conjugatus** (Preer, Preer and Jurand 1974) Quackenbush, 267.^VP (Effective publication: Quackenbush 1978, 187.) (*Caedobacter* (*sic*) *conjugatus* Preer, Preer and Jurand 1974, 157.)

con.ju.ga′tus. L. masc. part. adj. *conjugatus* conjugated.

Rods 0.3–0.5 μm in diameter and 1.0–4.0 μm long. They are called mu and (except for *Caedibacter paraconjugatus*) are the only symbiont

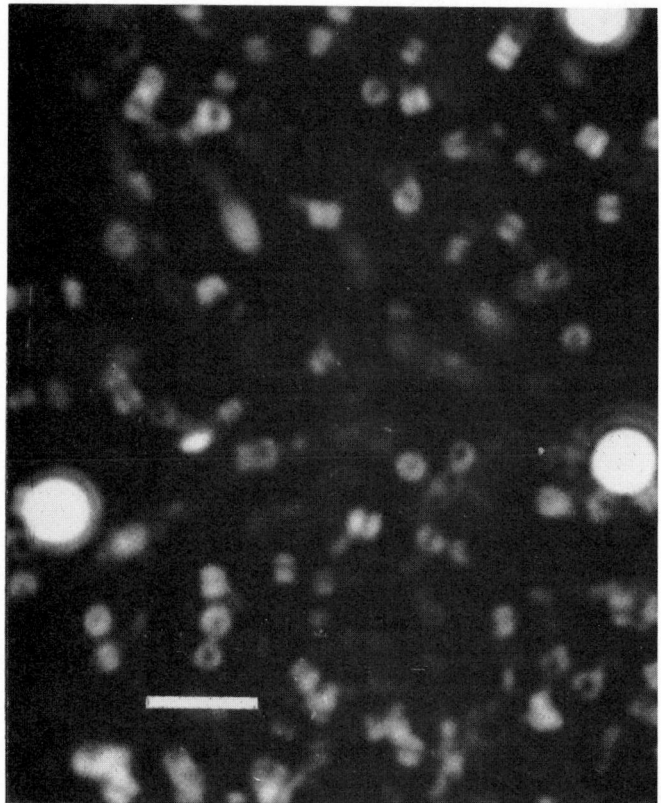

Figure 11.11. Isolated R bodies from *Caedibacter varicaedens* of *Paramecium biaurelia* stock 7, viewed with bright phase-contrast. R bodies appear doughnut-shaped when viewed on end and as two parallel rods when viewed from the side. The large bright spheres are latex particles (*bar*, 2 μm).

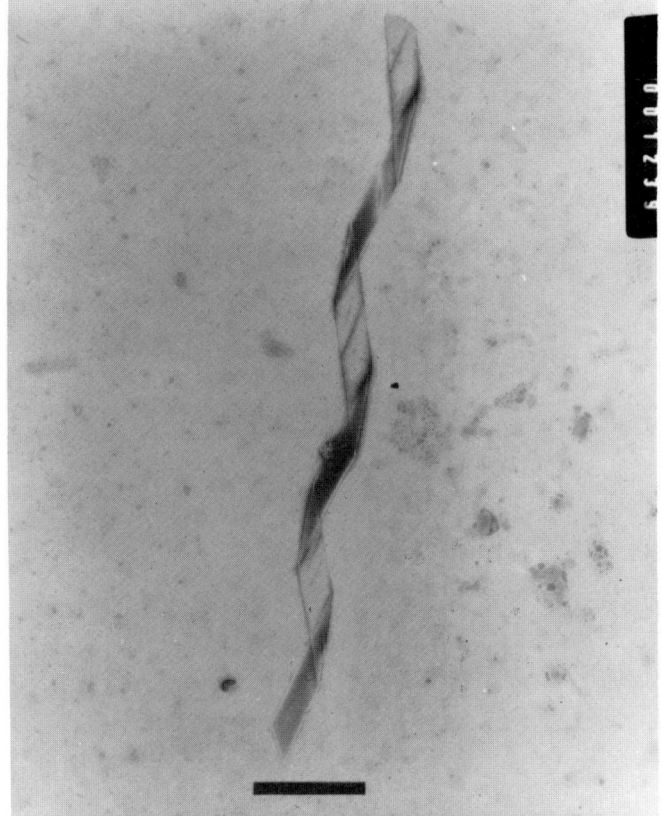

Figure 11.12. Unrolled R body isolated from *Caedibacter varicaedens* of *Paramecium biaurelia* stock 1039. Phosphotungstic acid (*bar*, 1 μm). (Reproduced with permission from J. R. Preer, Jr., L. B. Preer and A. Jurand, Bacteriological Reviews *38:* 113–163, 1974, © American Society for Microbiology.)

responsible for the **mate-killer** phenotype in the *Paramecium aurelia* complex (Fig. 11.14). They produce a toxin capable of killing sensitive strains of *Paramecium* only after cell-to-cell contact between killers and sensitives at conjugation. Cultivation free of the cytoplasm of *Paramecium* has been reported to occur on a very complex medium (Williams, 1971). Found in the cytoplasm of *Paramecium primaurelia* and *Paramecium octaurelia*.

The mol% G + C of the DNA is 35–37 (Bd).

Type strain: In ATCC strain 30796 (stock 540 of *Paramecium primaurelia*; see Preer et al., 1974).

2. **Pseudocaedibacter minutus** (Preer, Preer and Jurand 1974) Quackenbush 1982, 267.[VP] (Effective publication: Quackenbush 1978, 187.) (*Caedobacter* (*sic*) *minutus* Preer, Preer and Jurand 1974, 157.) mi.nu′tus. L. masc. adj. *minutus* small.

Rods often double, 0.25–0.35 μm in diameter and 0.5–1.0 μm long (singles). This very small cell is unique among the endosymbionts of *Paramecium* in being **surrounded by an extra set of membranes**, apparently continuous with the endoplasmic reticulum of its host (Fig. 11.15). Although they never rise to high concentrations in the cytoplasm, the paramecia which bear them are nevertheless very strong

killers. Found only in the cytoplasm of *Paramecium octaurelia*. Originally called gamma.

The mol% G + C of the DNA is 38 (Bd).

Type strain: In ATCC strain 30699 (stock 214 of *Paramecium octaurelia*; see Preer et al., 1974).

3. **Pseudocaedibacter falsus** (Preer, Preer and Jurand 1974) Quackenbush 1982, 267.[VP] (Effective publication: Quackenbush 1978, 187.) (*Caedobacter* (*sic*) *falsus* Preer, Preer and Jurand 1974, 157.) fal′sus. L. masc. adj. *falsus* false.

Rods 0.4–0.7 μm in diameter and 1.0–1.5 μm long. **No toxic actions known**, although the forms (called nu) found in *Paramecium pentaurelia* are said to increase the resistance of their hosts to the toxin produced by *Lyticum flagellatum* (Holtzman, 1959). The strains found in *Paramecium tetraurelia* were once regarded as mutants of *Caedibacter taeniospiralis* and called pi. Found also in the cytoplasm of *Paramecium biaurelia*.

The mol% G + C of the DNA is 36 (Bd).

Type strain: In ATCC strain 30640 (stock 1010 of *P. biaurelia*; see Preer et al., 1974).

Genus IV. **Lyticum** (*ex* Preer, Preer and Jurand 1974) *Preer and Preer 1982, 141*[VP] (*Lyticum* Preer, Preer and Jurand 1974, 157)

Ly′ti.cum. L. adj. *lyticus* dissolving; M.L. neut. n. *Lyticum* dissolver.

Large rods 0.6–0.8 μm in diameter, straight, curved or spiral. Length of single forms 3.0–5.0 μm. **Numerous peritrichous flagella.** Al-

though cultivation free of *Paramecium* has been reported, it has not been confirmed. Produce labile toxins which kill sensitive strains of

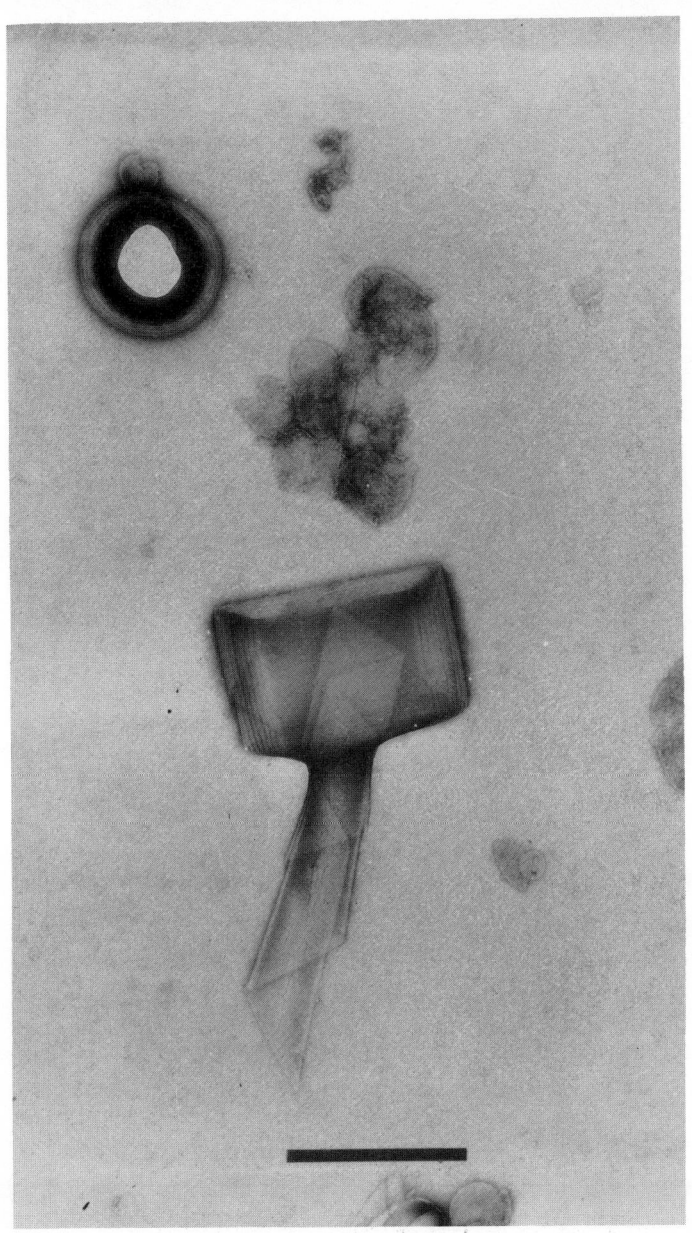

Figure 11.13. R bodies isolated from *Caedibacter taeniospiralis* of *Paramecium tetraurelia* stock 51. Note intact doughnut-shaped R body and, below, an R body unrolling from the inside. Phosphotungstic acid (*bar*, 0.5 μm). (Reproduced with permission from L. B. Preer, A. Jurand, J. R. Preer, Jr., and B. M. Rudman, Journal of Cell Science, *11:* 581–600, 1972, © Company of Biologists Ltd.)

Table 11.4.
Characteristics differentiating species of the genus **Caedibacter**[a]

Characteristics	1. C. taenio-spiralis	2. C. vari-caedens	3. C. pseudo-mutans	4. C. para-conjugatus
R bodies unroll from inside	+	−	−	−
Hump killers	+	−	−	−
Mate killers	−	−	−	+
R bodies very small	−	−	−	+
Paramecium aurelia complex species number	4	2	4	2
Mol% G + C of DNA	41	40–41[b]	44[b]	−

[a] For symbols see standard definitions.
[b] Quackenbush (1978) reports that the DNA of *C. varicaedens* and that of *C. pseudomutans* show only very weak cross-hybridization.

Table 11.5.
Characteristics differentiating species of the genus
Pseudocaedibacter[a]

Characteristics	1. *P. conjugatus*	2. *P. minutus*	3. *P. falsus*
Mate killing	+	−	−
Toxin liberated into medium	−	+	−
No toxins produced	−	−	+
Paramecium aurelia complex species number	1, 8	8	1, 4, 5

[a] For symbols see standard definitions.

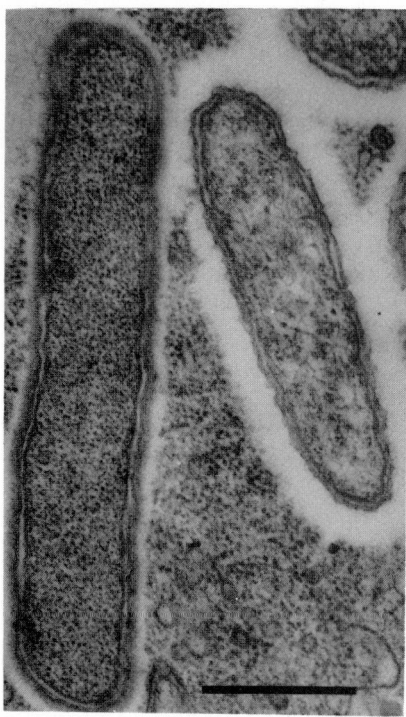

Figure 11.14. Two endosymbionts present in a single cell of *Paramecium octaurelia* stock 131. *Left, Tectibacter vulgaris* (note electron-dense material surrounding outer membrane); *right, Pseudocaedibacter conjugatus*. Longitudinal section (*bar*, 0.5 μm). (Reproduced with permission from J. R. Preer, Jr., L. B. Preer and A. Jurand, Bacteriological Reviews *38*: 113–163, 1974, © American Society for Microbiology.)

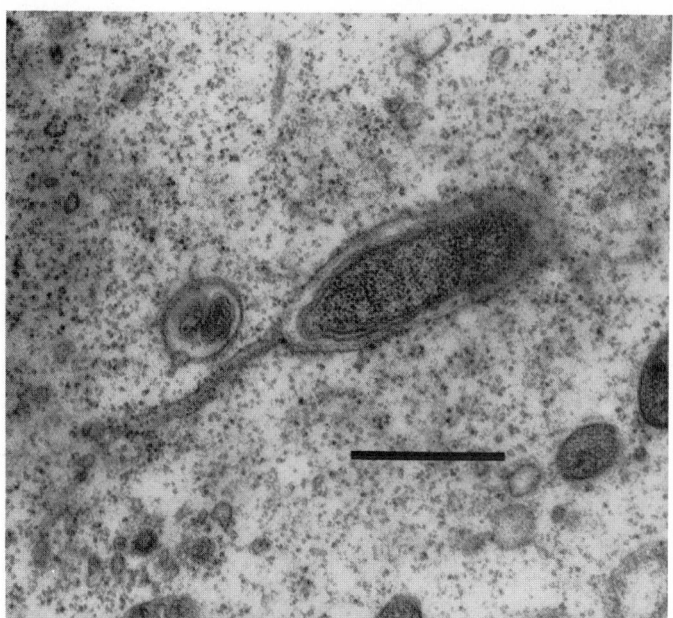

Figure 11.15. *Pseudocaedibacter minutus*, endosymbiont of *Paramecium octaurelia* stock 565. Note additional outer membrane. Longitudinal section (*bar*, 0.5 μm). Reproduced with permission from A. Jurand.)

paramecia very quickly by **lysis**. Gram-negative. Nonmotile or almost so, in spite of numerous, well developed flagella. Occurs in *Paramecium biaurelia, P. tetraurelia, P. octaurelia*.

The mol% G + C is 27 and 45–49 (Bd).

Type species: *Lyticum flagellatum* (*ex* Preer, Preer and Jurand 1974) Preer and Preer 1982, 140.

Further Descriptive Information

These very large endosymbionts with their conspicuous flagella and rapidly acting toxins (less than 30 min at room temperature) make them unique. *Paramecium triaurelia, P. pentaurelia* and *P. nonaurelia* are particularly sensitive to the toxins. The rather different reports of mol% G + C given by Behme and by Soldo (see Preer et al., 1974 for a discussion) are unresolved.

Differentiation of the genus **Lyticum** *from other genera*

See Table 11.2.

Differentiation and characteristics of species of **Lyticum**

Lyticum flagellatum is a straight rod found in *Paramecium tetraurelia* and *P. octaurelia*, while *L. sinuosum* is a strikingly curved rod found in *Paramecium biaurelia*.

List of the species of the genus **Lyticum**

1. **Lyticum flagellatum** (*ex* Preer, Preer and Jurand 1974) Preer and Preer 1982, 140.[VP] (*Lyticum flagellatum* Preer, Preer and Jurand 1974, 157.)

fla.gel.la′tum. L. neut. part. adj. *flagellatum* flagellated.

Straight rods 0.6–0.8 μm in diameter and 2.0–4.0 μm long (Fig. 11.16). Originally called lambda. Found within the cytoplasm of *Paramecium tetraurelia* and *P. octaurelia*. Stock 299 of *P. octaurelia* containing lambda does not require folic acid, whereas symbiont-free lines of 299 do (Soldo and Godoy, 1973b).

The mol% G + C of the DNA is 27 in one strain (Soldo) and 49 (Bd) in another (Bloomington).

Type strain: In ATCC strain 30700 (Stock 299 of *P. octaurelia*; see Preer et al., 1974).

2. **Lyticum sinuosum** (*ex* Preer, Preer and Jurand 1974) Preer and Preer 1982, 140.[VP] (*Lyticum sinuosum* Preer, Preer and Jurand 1974, 158.)

sin′u.o.sum. L. neut. adj. *sinuosum* winding, sinuous.

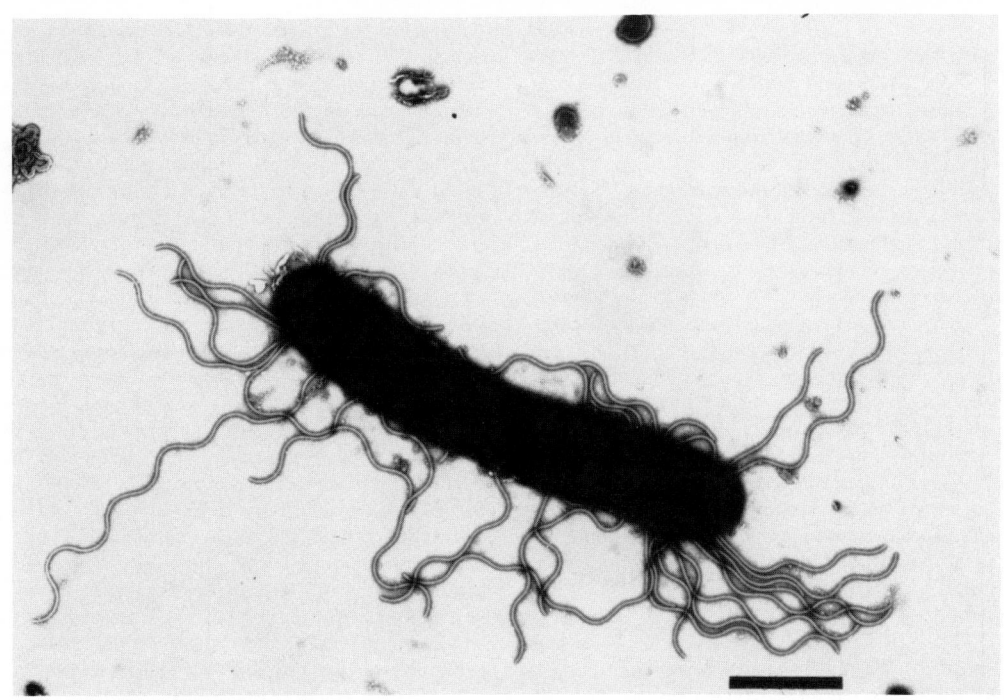

Figure 11.16. *Lyticum flagellatum* isolated from *Paramecium octaurelia* stock 327. Phosphotungstic acid (*bar*, 1 μm). (Reproduced with permission from J. R. Preer, Jr., L. B. Preer and A. Jurand, Bacteriological Reviews *38:* 113–163, 1974, © American Society for Microbiology.)

Curved or spiral rods 0.7–0.9 μm in diameter and 2.0–10.0 μm long, sometimes forming chains of 2–3 cells. Originally called sigma. Found within the cytoplasm of *Paramecium biaurelia*.

The mol% G + C of the DNA is 45 (Bd).

Type strain: In ATCC strain 30696 (stock 114 of *P. biaurelia*; see Preer et al., 1974).

Genus V. **Tectibacter** (*ex* Preer, Preer and Jurand 1974) *Preer and Preer 1982, 140[VP]* (*Tectobacter* (*sic*) Preer, Preer and Jurand 1974, 158)

Tec.ti.bac′ter. L. masc. n. *tectum* covering; M.L. masc. n. *bacter* the masculine equivalent of the Gr. neut. n. *bactrum* a rod. M.L. masc. n. *Tectibacter* the bacterium with a covering.

Straight rods 0.4–0.7 μm in diameter and 1.0–2.0 μm long. Distinguished by **outer covering around its cell wall** visible in sections with the electron microscope. Sparsely peritrichous. No known strains are toxic to protozoa. Gram-negative. Often observed to be motile. Occurs widely among strains of *Paramecium primaurelia*, *P. biaurelia*, *P. tetraurelia*, *P. sexaurelia*, and *P. octaurelia*, often with other symbionts (Fig. 11.14).

Type species: *Tectibacter vulgaris* (*ex* Preer, Preer and Jurand 1974) Preer and Preer 1982, 140.[VP]

Differentiation of the genus **Tectibacter** from other genera

Table 11.2 provides the necessary information.

List of the species of the genus **Tectibacter**

1. **Tectibacter vulgaris** (*ex* Preer, Preer and Jurand 1974) Preer and Preer 1982, 140.[VP] (*Tectobacter* (*sic*) *vulgaris* Preer, Preer and Jurand 1974, 158.)
vul.gar′is. L. masc. adj. *vulgaris* common.

The chracteristics are as described for the genus.

Type strain: in ATCC strain 30697 (stock 225 of *P. sexaurelia*; see Preer et al., 1974).

B. ENDOSYMBIONTS OF INSECTS

GREGORY A. DASCH, EMILIO WEISS, AND KWANG-POO CHANG

" . . . cooperative living of animals with microorganisms, regulated as it is to the last detail, must play a significant role in the economy of the host."

From Buchner, 1965.

Although most microorganisms associated with the external surfaces of insects are adventitious, a wide variety of procaryotes, including pathogens of vertebrates (see order *Rickettsiales*) and of plants (see mycoplasmas), are found in recurrent associations with insects. The relationship of these organisms to their insect hosts varies greatly. Although some procaryotes (*Rickettsia prowazekii, Rickettsiella, Bacillus thuringiensis*) are lethal for their hosts and a few (see *Wolbachia pipientis, Spiroplasma*) interfere with their hosts' reproduction, many (other *Rickettsiales*, mycoplasmas, and rickettsia-like plant pathogens) appear to have no effect on their hosts. This section is limited to those procaryotes that are regarded as true insect endosymbionts,* because their association with their hosts is so intimate that it is thought to be beneficial to both insect and symbiont. Such symbionts are found in at least 10% of the insect species. Similar associations also occur commonly in ticks and mites (*Acarina*). Within arthropods a wide range of types of endosymbioses may occur. These include the intriguing complex of endosymbionts of the flagellated protozoans of the termite (see the article in Section 1 of this *Manual* on Hindgut Spirochetes of Termites and *Cryptocercus punctulatus*), the various intracellular and extracellular bacteria, yeast and fungi associated with the insect gut, and the diverse "obligate" intracellular procaryotes which have become integrated to varying degrees into the developmental cycle of their hosts. Some of the latter are so bizarre that they may resemble eucaryotic organelles more closely than procaryotes. This resemblance has fueled speculations on procaryotic evolutionary origins of mitochondria and chloroplasts as proposed in the serial endosymbiosis hypothesis.

Our knowledge of most insect endosymbionts is tempered by the failure to cultivate them axenically, in cell culture, or in other insects. Most claims of in vitro cultivation have not been adequately confirmed. Our information is based primarily on four lines of investigation. (a) The insects have been examined to determine which organs and cells contain the endosymbionts and how the distribution and frequency changes with the stage of insect development. These observations lead to an understanding of the degree of adaptation of the endosymbionts to their host and mechanisms of hereditary transmission. (b) Staining and light and electron microscopy have been the principal methods of characterization of the microorganisms and of their relationships to the structures of their host cells. Fluorescent antibody staining has been used to a limited extent. (c) The insects have been rendered aposymbiotic (free from endosymbionts) by treatment with antibiotics, lysozyme, elevated temperatures, or other methods to determine the role of symbionts in their host's physiology and in connection with attempts to cultivate the microorganisms and satisfy Koch's postulates. (d) More recently, the endosymbionts have been separated from host components and subjected to studies of DNA base ratios, of enzymatic activities, and chemical composition. These studies have not been extensive enough to provide evidence of evolutionary relationships of endosymbionts to each other or to other bacteria, although such relationships can not be excluded. Classification attempts have of necessity been quite limited.

Elaborate insect structures and behavioral mechanisms often insure inheritance of the endosymbionts. Although they may be found extracellularly or epicellularly during brief periods of the insect's life cycle, they usually reside within the membranes of the phagosomes or phagolysosomes or traverse the membrane and lie free in the cytoplasm or nucleus. Most are harbored in specialized cells (mycetocytes) or organs (mycetomes). Little is known of the factors that contribute to the stability of the endosymbiont populations. Growth may be restricted by host-regulated nutrition, or excess endosymbionts in the mycetomes may be destroyed by the formation of autophagic vacuoles. Endosymbionts that escape the mycetomes may be eliminated by lysozymes or phagocytic hemocytes. Hereditary mechanisms are rather straightforward in insects which have extracellular symbionts in gut diverticula or intracellular endosymbionts that are readily shed into the gut lumen from the lining epithelium. In a process akin to infection, newly eclosed larvae recolonize their guts by ingesting packets or films of the symbionts which are deposited at oviposition on the egg by the parent.

Deposition may occur simply by defecation or by means of highly complex accessory sexual organs. Although heredity in insects with intracellular symbionts may also parallel maternal transmission of the mitochondria in a strict germ cell cycle, more often they involve complex transfers of symbionts between extraovarial mycetomes and the oocytes or between trophic and germ line cells in the ovary. Similarly, complex sequestration and transfer mechanisms, including extracellular "infective" phases of the symbiont life cycle, may occur during both embryonic development and metamorphosis. These mechanisms insure the correct localization of the endosymbionts in specific tissues and cells. Since some insects may harbor as many as six types of endosymbionts simultaneously, hereditary mechanisms may be exceedingly complex.

The endosymbionts have the same range of staining and morphological characteristics that are encountered in other procaryotes. The Gram stain has been reported to be positive, variable, and negative and occasionally has given rise to controversy, possibly due to improperly destained host membranes surrounding the bacterium. The fine structure of the cell wall ranges from that typical of Gram-negative or Grampositive bacteria to the single unit membrane of mycoplasmas. In some cases, murein and diaminopimelic acid have been demonstrated chemically and spheroplasts have been produced by treatment with lysozyme or antibiotics. Morphologically they have been described as small cocci, small or long rods, helical, pleomorphic, or bizarre forms not readily recognized as procaryotic. Sometimes the morphology changes with the stage of development, especially during embryogenesis and metamorphosis. Flagella, pili, and internal structures consistent with the presence of ribosomes, DNA fibrils, and various kinds of cytoplasmic inclusions have also been described. Division is often by binary fission, but a variety of other mechanisms have been described.

Since endosymbionts are found most frequently in insects that feed on restricted diets, it has been postulated that they provide to their hosts essential nutrients (vitamins, amino acids, sterol) or regulatory compounds for their synthetic, developmental and degradative functions. The direct test of this postulation by examination of endosymbionts cultivated in vitro has not been possible in most cases. Although an attractive idea, attempts at cultivation in cell lines established from infected mycetomes, embryonic, or ovarial tissue have not been more successful than axenic cultivation. The enzymatic capabilities of endosymbionts separated from their hosts have been studied only to a very limited extent. Consequently, the elucidation of the physiologic roles of individual endosymbionts has depended on the study of aposymbiotic insects. Altered insect physiology and added nutritional requirements for reproduction and normal growth are generally attributed to the loss of endosymbionts. However, some of these results must be treated with caution since many of the procedures which render the insects aposymbiotic are harsh and may directly damage the insect.

In the few instances studied, the mol% $G + C$ of the DNA ranges from 24–70.

Further Reading

Brooks, M.A. 1963. The microorganisms of healthy insects. In Steinhaus (Editor), *Insect Pathology: an Advanced Treatise*. Vol. I, Academic Press, London, pp. 215–250.

Buchner, P. 1965. *Endosymbiosis of Animals with Plant Microorganisms*. John Wiley & Sons, New York.

Bulla, L.A., Jr. and T.C. Cheng (Editors). 1975. Pathobiology of invertebrate vectors of disease. Ann. N.Y. Acad. Sci. *266:* 1–540.

Dadd, R.H. 1977. Quantitative requirements and utilization of nutrients: insects. In Rechcigl (Editor), *CRC Handbook Series in Nutrition and Food*. Section D. Nutritional Requirements, Vol. I, CRC Press, Cleveland, pp. 305–346.

Fredrick, J.F. (Editor). 1981. Origins and evolution of eukaryotic intracellular organelles. Ann. N.Y. Acad. Sci. *361:* 1–512.

Hopkins, D.L. 1977. Plant diseases caused by leafhopper-borne, rickettsia-like bacteria. Annu. Rev. Phytopathol. *15:* 277–294.

Houk, E.J. and G.W. Griffiths. 1980. Intracellular symbiotes of the Homoptera. Annu. Rev. Entomol. *25:* 161–187.

---

* Spelled "endosymbiotes" by some authors.

Koch, A. 1960. Intracellular symbiosis in insects. Annu. Rev. Microbiol. *14:* 121–140.

Koch, A. 1967. Insects and their endosymbionts. In Henry (Editor), *Symbiosis,* Vol. II, Academic Press, New York, pp. 1–106.

Lanham, U.N. 1968. The Blochmann bodies: hereditary intracellular symbionts of insects. Biol. Rev. (Cambridge Philos. Soc.). *43:* 269–286.

Margulis, L. 1981. *Symbiosis in cell evolution.* W.H. Freeman and Company, San Francisco.

Nienhaus, F. and R.A. Sikora. 1979. Mycoplasmas, spiroplasmas, and rickettsia-like organisms as plant pathogens. Annu. Rev. Phytopathol. *17:* 37–58.

Richards, A.G. and M.A. Brooks. 1958. Internal symbioses in insects. Ann. Rev. Entomol. *3:* 37–56.

Roberts, D.W. and M.A. Strand (Editors). 1977. Pathogens of medically important arthropods. Bull W.H.O. *55 (Suppl. 1):* 1–419.

Schwemmler, W. 1979. *Mechanismen der Zellevolution. Grundriss einer modernen Zelltheorie.* Walter de Gruyter, Berlin.

Schwemmler, W. and H.E.A. Schenk (Editors). 1980. *Endocytobiology. Endosymbiosis and Cell Biology. A Synthesis of Recent Research.* Walter de Gruyter, Berlin.

Smith, D.C. 1979. From extracellular to intracellular: the establishment of a symbiosis. Proc. R. Soc. London *B204:* 115–130.

Steinhaus, E.A. 1946. *Insect Microbiology.* Comstock Publishing Company, Ithaca, New York.

General Taxonomic Comments

Despite the diversity among insect endosymbionts only two monospecific genera are listed in the eighth edition of *Bergey's Manual. Blattabacterium cuenoti* clearly denotes an endosymbiont commonly seen in the cockroach. *Symbiotes lectularius,* on the other hand, refers to a "pleomorphic" endosymbiont of the bedbug (*Cimex*). Recent information suggests that the "pleomorphic" endosymbiont actually consists of more than one species and that the designation *S. lectularius* can not be properly used without further definition.

In view of the limited information on the microbiological properties of endosymbionts, further classification is not recommended at this time. The most useful descriptive treatment is according to the taxonomy and nutritional type of the host insect. This approach has the merit that it permits both direct discussion of divergent evolutionary relationships between symbionts present in closely related taxa and of convergent solutions provided by the endosymbionts to identical nutritional problems posed in the parallel evolution of unrelated insects.

BLOOD-SUCKING INSECTS

Information on the distribution of symbionts in hematophagous insects has largely accrued from incidental observations by bacteriologists and medical entomologists who were primarily interested in insects as vectors of disease (Table 11.6). Although extensive compilations of viral, parasitic, fungal and bacterial pathogens in hematophagous insects have been prepared from such observations (Steinhaus, 1946; Bulla and Cheng, 1975; Roberts and Strand, 1977), information on the precise nature of the continuum of vertebrate or insect pathogens, adventitious insect associates, and intimate mutualistic endosymbionts is inadequate. For example (Table 11.6), many hematophagous insects simultaneously harbor pathogens (*Rickettsia, Rochalimaea,* or *Wolbachia*), rickettsia-like companion forms, and several endosymbionts. Each may be pleomorphic, present in several tissue locations, and transmissible by transovarial passage. The rickettsia-like organisms are usually distinguished from endosymbionts solely on the basis of their smaller size, irregular distribution in host tissues and often among individuals of a species, and lack of obvious morphologic adaptations by the host to ensure their hereditary transmission, rather than from any information on pathogenicity. Many pathogens and most of the symbionts have not been cultivated, so that any classification, even when based on ultrastructural examination, is quite provisory and possibly incorrect. Indeed taxonomic relationships between bacteria present even in closely related species or genera of insects are often obscure.

In spite of these considerations, the nutritional hypothesis of Buchner (1965), that true mutualistic endosymbionts are present only in those hematophagous insects that feed exclusively on vertebrate blood throughout their life cycle, appears to have validity. Vertebrate blood is usually believed to contain inadequate levels of B vitamins and other growth factors required for normal insect development and reproduction. Therefore, these factors must be provided to hematophagous insects either by exogenous microorganisms or alternative dietary sources, or by resident endosymbionts. The hypothesis is well exemplified by the numerous symbiont-free hematophagous Diptera which have larval stages that are aquatic or feed on decaying vegetation. *Dasyhelea* (Ceratopogonidae) appears an exception to this symbiont-free pattern, but most sap-sucking insects harbor endosymbionts (see below) as this diet is also nutritionally inadequate. Other dipterans that feed only on blood and have viviparous larvae (*Glossina,* Pupipara) either harbor endosymbionts in midgut mycetomes or have other bacteria that appear to have similar functions. Members of the Streblidae have only intracellular bacteria that are randomly associated with various tissues much as the typical companion forms of other Pupipara (Hippoboscidae, Nycteribiidae) and *Glossina.* In *Eucampsipoda* these companion forms are found in the milk gland, the organ of symbiont transmission in other Pupipara, but they do not occupy mycetocytes as in more advanced symbioses. Buchner (1965) provides additional detail about these interesting variations in bacterial associations of the Pupipara. It is thus worthwhile to speculate that loosely associated companion forms may be analogous to those procaryotes that may have been the progenitors of both the various Rickettsiales and the endosymbionts of hematophagous insects.

Genus Glossina (tsetse flies)

Glossina species harbor large endosymbionts, 1.0–1.8 μm in width by 3–9 μm in length. They fill the cytoplasm of hypertrophied epithelial cells of the midgut and form parallel packets perpendicular to the gut lumen (Stuhlmann, 1907; Roubaud, 1919; Wigglesworth, 1929). They are not surrounded by a host cell membrane. The cell envelope, 25–30 nm thick, and internal structures are typical of Gram-negative bacteria. The symbionts are sometimes surrounded by an electron dense coat 10 nm thick (Reinhardt et al., 1972). No significant differences in endosymbiont structure and location have been noted in the approximately 10 species that have been examined, including the more notorious carriers of pathogenic trypanosomes, *G. morsitans* and *G. palpalis.*

The mechanism of hereditary transmission of the endosymbionts is not clearly understood. Endosymbionts are not generally seen in the embryo. The larvae of *Glossina,* as well as those of the Pupipara, develop in an intrauterine location, where they feed on the nutritive fluids secreted by special glands, commonly called "milk glands." Large extracellular bacteria commonly seen in the lumen of the milk glands are believed to colonize the larval midgut mycetomes as has been observed in the Pupipara (Zacharias, 1928). However, in *Glossina* they are not identical to the endosymbionts of the midgut in cell wall thickness and internal structure, and in addition, possess numerous fimbriae having a size 2 × 5–7 nm (Ma and Denlinger, 1974). Fimbriae may facilitate attachment and penetration into new cells and reflect an infective stage in a cycle of endosymbiont development, but this has not been proven. Not known, also, is the mechanism by which the endosymbionts infect the milk glands.

Oxygen consumption by isolated endosymbionts is enhanced most effectively by a combination of pyruvate and succinate or pyruvate and malate (Wink, 1979). The endosymbiont respiration, in contrast to that of the host cells, is insensitive to cyanide, but is inhibited by

Table 11.6.

Distribution and tissue location in blood sucking insects of symbionts and other bacteria that are not necessarily endosymbiotic (Rickettsiales and rickettsia-like organisms)

Order Family	Representative Genus	Common Name	Larval Food Source	Symbiont Type	Symbiont Tissue Location	Rickettsiales and Rickettsia-like Forms Type	Rickettsiales and Rickettsia-like Forms Tissue Location (Species)
Diptera							
Muscidae	*Glossina*	Tsetse fly	Viviparous	Intracellular Extracellular	Midgut Milk gland	Intracellular	Ovary, midgut
	Stomoxys	Stable fly	Decaying vegetation	None		None	
	Haemotobia	Horn fly	Cow dung	None		None	
Hippoboscidae	*Melophagus*	Sheep ked	Viviparous	Intracellular	Midgut	Extracellular	(*Wolbachia melophagi*)
				Extracellular	Milk gland	Intracellular	Widespread
Nycteriidae	*Nycteribia*	Bat fly	Viviparous	Intracellular	Abdominal mycetocytes	Intracellular	Widespread
				Extracellular	Milk gland		
	Eucampsipoda	Bat fly	Viviparous	Extracellular	Milk gland	None	
Streblidae	*Nycteribosca*	Bat fly	Viviparous	None		Intracellular	Widespread
Culicidae	*Culex*	Mosquito	Aquatic	None		Intracellular	(*Wolbachia pipientis*)
Tabanidae	*Tabanus*	Horse fly	Aquatic	None		Rare, intracellular	Pericardial cells
	Chrysops	Deer fly	Aquatic	None		None	
Psychodidae	*Phlebotomus*	Sand fly	Decaying vegetation	None		Primarily epicellular	(*Bartonella bacilliformis*)
	Lutzomyia	Sand fly	Decaying vegetation	None		Rare, intracellular	Ovary
Simulidae	*Simulium*	Black fly	Aquatic	None		None	
Ceratopogonidae	*Dasyhelea*	Punky	Tree sap	Intracellular	Abdominal mycetome	None	
	Culicoides	Punky	Aquatic	None		Extracellular	Hemolymph
Hemiptera							
Cimicidae	*Cimex*	Bedbug		Intracellular	Abdominal mycetome	Intracellular	Ovary
				Extracellular	Hemolymph		
Reduviidae	*Triatoma*	Kissing bug	Blood	Extracellular	Gut lumen	Intracellular	Widespread
Anoplura	*Pediculus*	Human louse	Blood	Intracellular	Abdominal mycetome	Intracellular	(*Rickettsia prowazekii*)
						Extracellular	(*Rochalimaea quintana*)
Mallophaga	*Columbicula*	Pigeon louse	Hair, feathers	Intracellular	Abdominal mycetocytes	Extracellular	Midgut
Siphonaptera	*Xenopsylla*	Rat flea	Organic debris, blood	None		Intracellular	(*Rickettsia typhi*)

rotenone and antimycin A. The claim that the endosymbionts can be cultivated in a serum-broth-yeast extract medium (Southwood et al., 1975) has not been confirmed (Wink, 1979); however, the bacteria have been maintained viable for over 85 days and have shown some elongation and cell division in a mycoplasma medium supplemented with pyruvate, succinate, and combination of nucleotides (Wink, 1979). ATP stimulates motility. Although endosymbionts display some chemotaxis for vertebrate and tsetse fly cells cultivated in vitro, they do not multiply in them. The relatively small number of bacteria that are incorporated are lysed and digested (Wink, 1979).

Aposymbiosis was first noted in *Glossina* feeding on rabbits treated with the coccidiostatic agent sulfaquinoxaline (Jordan and Trewern, 1973, 1976; Pell and Southern, 1976). It has also been produced with tetracycline compounds and other antimicrobial agents introduced either into the rabbits or into the blood used for membrane feeding of the flies (Nogge, 1976; Schlein, 1977), or by the oral administration of lysozyme (Nogge, 1976). With the exception of tetracycline compo-

nents, none of these compounds reduces the longevity of the adults. Of these studies, the use of rabbits immunized against endosymbionts separated from host antigen components by affinity chromatography is perhaps the most novel approach. Rabbits so immunized, in contrast to rabbits immunized with fly antigens, contain antibodies that are not toxic for the adult fly but specifically eliminate the endosymbionts, as demonstrated by fluorescent antibody staining (Nogge, 1978, 1980).

The above-described procedures eliminate primarily the midgut endosymbionts and this effect is associated with loss of fecundity. Prolonged treatment results in degeneration of the ovaries. The biochemical basis for these changes is thought to be the loss of vitamins provided by the endosymbionts. This belief is strengthened by the demonstration, by microbiological assays, that the endosymbionts are capable of producing B vitamins, and by the fact that loss of the endosymbionts by the flies after lysozyme treatment can be compensated by supplementation of the blood diet with the vitamins (Nogge, 1976, 1978).

In addition to the large endosymbionts, variable numbers of small bacteria (often designated as rickettsia-like) have been described in tissues of the ovaries, in midgut epithelium not containing the large endosymbionts, in the surrounding musculature, and in cells associated with the fat body (Reinhardt et al., 1972; Pinnock and Hess, 1974; Huebner and Davey, 1974; Pell and Southern, 1975, 1976). These microorganisms measure 0.5 × 2.0 μm, have a typical Gram-negative type cell wall, and are surrounded by a host cell membrane. Since they are seen in the embryo, they are believed to be transmitted transovarially. Although they have been seen in all species of *Glossina* examined, their distribution and numbers vary. Greater numbers have been found in the ovaries of *G. morsitans* than in *G. austeni* and this difference has been associated with higher retention of fecundity by *G. morsitans* following sulphaquinoxaline treatment (Pell and Southern, 1976).

Family Cimicidae (bedbugs)

Arkwright et al. (1921) were the first to describe highly pleomorphic intracellular bacteria present in smears of the midgut, Malpighian tubules, ovaries, testes, organ of Berlese, and in the embryos of five different stocks of the common human bedbug, *Cimex lectularius* L., and in *Cimex hirundinis*. They believed the various lanceolate, rickettsial, bacillary, and thread-like forms to be variants of the same nonpathogenic microorganism, which they named *Rickettsia lectularius*. Later investigators discovered pleomorphic symbionts located in small paired abdominal mycetomes in *Cimex lectularius* L., *C. rotundatus* Signoret, and in species of other genera of Cimicidae, *Oeciacus, Ornithocoris*, and *Leptocimex*. The primitive cimicid, *Primicimex cavernius* Barber, however, lacks mycetomes, but has symbionts in its midgut epithelium (Buchner, 1965). The relationship between microorganisms found within and without the mycetome of *C. lectularius* remains in doubt (Buchner, 1965; Louis et al., 1973; Chang, 1975). At least three ultrastructurally distinct symbionts have been described in *Cimex*: (a) The mycetomes and ovaries harbor the "typical" pleomorphic *Cimex* endosymbiont, about 1 μm in diameter, which has diffuse DNA fibrils and ribosomes and a Gram-negative type cell wall 23 nm thick. The endosymbiont occurs within a host cell membrane. The division mechanism is somewhat unusual in that the plasma membrane invaginates prior to the formation of the division furrow. These forms are present in all species of *Cimex* (Chang and Musgrave, 1973; Chang, 1975; Louis et al., 1973). (b) A more uniform rickettsia-like organism 0.3 × 1.0 μm is also seen in the mycetome and ovary of *C. lectularius*, but not in all strains. The Gram-negative type cell wall of this microorganism is surrounded by an electron dense microcapsule layer reminiscent of those found in typhus rickettsia. The organism is not enclosed by host cell membranes. It often penetrates the mycetome nuclei, and it divides by ordinary binary fission. (c) A "guest" symbiont is also occasionally found in the mycetome, but it is seen with greatest frequency in the ovaries, in other tissues, and even free in the hemolymph. It is a motile, Gram-negative rod with peritrichous flagella and is enclosed by host cell membranes. Although it is always found in *C. lectularius*, it is not seen in *C. hemipterus*, a tropical bedbug closely related to the temperate *C. lectularius* (Chang, 1975; Louis et al., 1973). Since Arkwright et al. (1921) obviously saw all three microorganisms, but believed them to be stages of the same pleomorphic species, the term *Symbiotes lectularius* must be redefined or discarded.

Attempts to culture the endosymbionts of *Cimex* either on media or on tissue culture cells have been unsuccessful (Buchner, 1965). De Meillon and his co-workers (1946, 1947) used rats fed on diets deficient in riboflavin, thiamine, or folic acid to assess the nutritional requirements of bedbugs. Bedbug longevity was unaffected by the deficient diets, but their fecundity declined with the thiamine and folic acid deficiency, not, however, with riboflavin deficiency. These authors suggested that the endosymbionts provided their hosts with sufficient riboflavin, but insufficient thiamine and folic acid for both longevity and fecundity. Chang (1974) later demonstrated that bedbugs rendered aposymbiotic by rearing at 36°C also lost their fecundity, even though their longevity was unaffected. Endosymbionts appeared to be highly sensitive to increased temperature. Their loss at 36°C did not appear to be due to increased host autophagic activity, which controls the endosymbiotic population at the normal ambient temperature of the bugs.

Family Reduviidae (assassin bugs)

Most species of Reduviidae are predaceous on other insects and do not have symbionts (Goodchild, 1955). In contrast, all the strictly blood-sucking species in the subfamily Triatominae harbor Gram-positive bacterial symbionts in the lumen of the anterior midgut (Nyirady, 1973). The triatomids, also known as "kissing bugs" because of their tendency to bite about the mouth, are important medically in South America as carriers of *Trypanosoma cruzi* (the etiologic agent of Chagas' disease), *Yersinia pestis* and *Leptospira*. Although the midgut location of the triatomid symbionts differs greatly from the complex posterior midgut caeca common for the Gram-negative symbionts of the plant sap sucking Hemiptera (see below), both groups are similar in the way they inherit the symbionts; newly eclosed larvae ingest fecal packets of the symbionts which are deposited directly on the eggs or in the immediate environment. Surprisingly, although the triatomid symbioses are among the most intensively studied microbiologically, it is still unclear whether the lumen symbionts may also occur intracellularly in the midgut epithelium, either in vacuoles or free in the cytoplasm (Buchner, 1965).

Most triatomids do not have a consistent association with a single microorganism. Gumpert and Schwartz (1962) have summarized the information on the isolates obtained by early investigators, primarily on *Triatoma infestans*. Both pure and mixed cultures of *Pseudomonas, Alcaligenes faecalis, Streptomyces, Staphylococcus albus, Streptococcus faecalis, Corynebacterium pseudodiphtheriticum*, and "*Nocardia rhodnii*"-like bacteria have been obtained in different laboratories from both feces and midgut samples. A somewhat wider but comparable range of predominantly Gram-positive microorganisms has been obtained from other *Triatoma* species (Varela and Aparicio, 1951; Marchette and Hatie, 1965; Cavanagh and Marsden, 1969). It is unclear whether the wide range of symbiont species found in laboratory-maintained Triatominae is representative of wild insects, since overzealous maintenance of "clean" stock cultures may inadvertently prevent the transmission of the normal symbiont flora and lead to the selective emergence of other bacteria (Brecher and Wigglesworth, 1944). Gumpert and Schwartz (1962) have also detected intracellular rickettsia-like symbionts (thin threads, 0.2–0.4 × 2–10 μm) in the posterior midgut, or small cocci (0.2–0.4 × 0.3–0.8 μm) in the proventriculus, but these symbionts could not be detected in eggs.

Triatoma species, freed of bacteria by surface sterilization of eggs, usually cannot develop beyond the fourth larval instar when fed on guinea pig blood (Gumpert and Schwartz, 1962). Symbiont-free *T. vitticeps* and *T. megista* that have been reinfected with a variety of Gram-positive species (*Nocardia, Mycobacterium, Corynebacterium*) or even of *Pseudomonas* species can develop normally, even when the bacteria have not been isolated from the insects (Gumpert, 1962). Finally, normal development may also occur in symbiont-free *Triatoma* following a single injection of pantothenic acid into the stomach, although the precursors of pantothenate, or folic acid, nicotinamide, pyridoxin, thiamine, vitamin B12, and biotin are ineffective. Indeed, culture filtrates of "*Nocardia rhodnii*" and other species which permit normal development contain concentrations of pantothenate exceeding those required for normal development, at least 100-fold higher than the normal levels in vertebrate blood (Gumpert and Schwartz, 1963). In conclusion, the symbiosis of *Triatoma* species appears rather primitive in that a variety of Gram-positive species and *Pseudomonas* serve equally well in providing essential pantothenic acid to the host insect.

Rhodnius prolixus appears to have a more intimate and complex relationship with its symbiont than Triatominae. Although the actinomycete "*Nocardia rhodnii*" was commonly reported to be in monoxenic culture in the midgut of *Rhodnius* (Brecher and Wigglesworth, 1944; Baines, 1956; Lake and Friend, 1967; Hill et al., 1976), *Streptococcus faecalis, Mycobacterium fortuitum, Corynebacterium* and *Staphylococcus* have also been cultivated from both midgut and feces (Gumpert and Schwartz, 1962; Cavanagh and Marsden, 1969). Although pure cultures of 10^9 "*N. rhodnii*"/gut are obtained within 5–7 days after a blood meal in instar V larvae, other bacteria are also commonly cultured both immediately after ecdysis, when titers of "*N. rhodnii*" are as low as 10^3/gut, and 10 or more days after a blood meal, when "*N. rhodnii*" titers begin to decline in late stationary phase (Dasch, 1975). "*N. rhodnii*" undergoes similar changes of concentration in instar IV (Hill et al., 1976). "*N. rhodnii*" might have an "antibiotic" effect toward other potential symbionts (Goodchild, 1955), but this effect could be due to the bactericidal activity of the host since a variety of bacterial species are unable to colonize the gut of *Rhodnius* even when it is free of "*N. rhodnii*" (Gumpert, 1962).

On solid media, "*N. rhodnii*" forms microcolonies consisting of an extensively branched mycelium which, after 24 h, fragments to successively shorter branched bacillary components (Erikson, 1935). Coccoid to bacillary forms 0.5–0.7×0.7–1.5 μm (some up to 5–10 μm long), predominate in the *Rhodnius* gut. No ultrastructure studies on "*N. rhodnii*" have been reported. On nutrient agar at 30°C, colonies 1–2 mm in diameter first appear at 48 h. These colonies gradually become pink and then brown after a week. Short aerial hyphae are present and microcysts may be formed in older cultures. Although growth is abundant on all three defined liquid media of Kwapinski and Horsman (1973), the presence of both a hard pellicle and sediment distinguishes "*N. rhodnii*" from the six previously defined cultural groups of nocardiae (Dasch, 1975). Lake and Friend (1967) and Gumpert and Schwartz (1962) have characterized the growth characteristics of "*N. rhodnii*" on a variety of other media. Most strains do not produce acid or gas from hexoses or 6-carbon alcohol sugars except mannitol and sorbitol (Bewig and Schwartz, 1956; Gumpert and Schwartz, 1962; Dasch, 1975). Malate, succinate, adipate, propionate, butyrate, and tartrate can be used as sole carbon sources. Strains using glucose can be selected since they grow more rapidly than the usual isolates (Hill et al., 1976). In *Rhodnius*, lactic acid generated by continued glycolysis by the ingested erythrocytes might be able to support replication of "*N. rhodnii*." Strains vary in their ability to liquify gelatin and to grow at 10°C or 45°C.

The role of "*N. rhodnii*" in providing B vitamins to its host has been examined by a variety of techniques in much greater detail in *Rhodnius* than with *Triatoma* species. These studies illustrate the real difficulties inherent in dissecting the basis of metabolic interactions in symbioses and suggest that the view of a simple key role of pantothenic acid in *Triatoma* symbioses is probably an oversimplification. Aposymbiotic *Rhodnius*, obtained by disinfection of eggs, are arrested in development by late instar IV (Brecher and Wigglesworth, 1944). Full development of aposymbiotic nymphs occurs when they are fed on mice infected with any of a variety of vitamin mixtures, each lacking a single B vitamin (Baines, 1956). Only folic acid and thiamine are required for development of membrane-fed aposymbiotic nymphs (Harington, 1960b); however, use of an improved membrane blood-feeding technique has shown that although full development of aposymbiotic nymphs does eventually occur on blood supplemented with different vitamins, their rate of development is generally slower than in controls (Lake and Friend, 1968). Consequently, although supplemental pantothenic acid is clearly necessary for *Rhodnius*, as in *Triatoma*, when fed on guinea pigs (Gumpert and Schwartz, 1963), it is not sufficient for normal development with rabbit blood. In vitro, "*N. rhodnii*" synthesizes significant amounts of folic acid, thiamine and pantothenic acid (Harington, 1960a; Gumpert and Schwartz, 1963). Germ-free CFW mice, but not rabbits, can support aposymbiotic *Rhodnius* development that differs little from normal development with symbionts (Nyirady, 1973; Auden, 1974). This may be due to their rich vitamin diet or possibly to differences in species concentrations of growth factors. The most novel approach that has been used is the selection of mutants of "*N. rhodnii*" which are unable to grow without added thiamine, nicotinamide, folic acid, pyridoxine, biotin, riboflavin, or *p*-aminobenzoic acid (Hill et al., 1976). Since infection of aposymbiotic *Rhodnius* with each of these mutants promotes normal development, it appears either that "*N. rhodnii*" mutants still produce enough other factors that can

spare the specific vitamin deficiency in *Rhodnius* or that it is capable of alternative nonvitamin-dependent metabolic transformations that are essential to the host. However, this metabolic capacity is certainly not unique to "*N. rhodnii*" since 9 of 62 species of bacteria, particularly other *Nocardia* and *Mycobacterium* species, have been found to be capable of replacing "*N. rhodnii*" in promoting full development of *Rhodnius* (Gumpert, 1962).

"*N. rhodnii*" is sensitive to chlortetracycline, tetracycline, chloramphenicol, and streptomycin, and somewhat less so to penicillin (Harington, 1960a).

The mol% G + C of the DNA from cultured "*N. rhodnii*" and the DNA obtained directly by extraction of *Rhodnius* midguts is 69.6% (T_m and Bd) with a minor satellite (plasmid?) of 35% (Dasch, 1975). This value conforms to the DNA base composition of other nocardiae (62–72 mol% G + C).

Orders Anoplura and Mallophaga (sucking and chewing lice)

Lice are small external parasites of birds and mammals. The Anoplura consists of about 250 species whose mouth parts are adapted for piercing the cuticle of their host and sucking blood. Most sucking lice that have been examined harbor intracellular bacterial symbionts in tissue locations suggesting a continuum of host adaptations (Table 11.7). The simplest location consists of individual mycetocytes inserted into the midgut epithelium (*Haematopinus*), or loosely aggregated in a narrow ring inserted between the midgut epithelium and basal membrane (*Pedicinus*). More complex arrangements include both the single abdominal mycetome of *Polyplax*, which consists solely of mycetocytes, and the complex, enveloped syncytial chambers of other species. In *Pediculus* the single syncytial chamber is tightly associated with the midgut as the "stomach disc" of early microscopists, while in *Haematomyzus* paired mycetomes lie freely in the abdomen. Most of the 1200 chewing lice in Mallophaga feed on hair, feathers, and only incidentally on blood while chewing on cuticle. Only the Philopteridae (at least 11 species) are known to harbor symbionts (Table 11.7). Despite their nonhematophagous nature, their symbionts greatly resemble those of Anoplura in their mode of transmission, intracellular location, and the close taxonomic relationship of the two orders. Indeed *Haematomyzus* has characteristics that are intermediate between the two orders and has been included in either order by different authorities. The symbionts of Philopteridae are all harbored in mycetocytes dispersed in the abdomen, but these may be localized somewhat differently in each species. In *Coloceras* the arrangement is reminiscent of the fat body mycetocytes of cockroaches.

Besides intracellular symbionts, a variety of rickettsia-like agents have been described in Mallophaga. These agents differ in cell morphology and tissue location from the well known rickettsia of *Pediculus*, *Rickettsia prowazekii*, and are perhaps more closely related to the trench fever agent, *Rochalimaea quintana*, or to *Wolbachia melophagi*. Such bacteria occur frequently in the intestinal lumen of the goat louse, *Trichodectes caprae*, the horse louse, *Trichodectes pilosus*, the pigeon louse, *Columbicola columbae*, and a *Trinoton* species (Menoponidae) (Steinhaus, 1946). The most regular and spectacular association of this type is found in the chicken louse, *Menopon biseriatum* (*Eomenacanthus stramineus*), where the chitinous lining of the crop is heavily infected with small rickettsia-like rods (0.2–0.3×0.3–1.0 μm) (Ries, 1931a). *M. pallidum*, which feeds less often on blood than *M. biseriatum*, also harbors small numbers of an apparently unrelated diplococcus (0.3–0.7 μm in diameter) in its gut lumen. Little is known about these agents.

The intracellular symbioses of Anoplura and Mallophaga resemble each other closely in the remarkably complex processes of hereditary transmission. In almost all cases the symbionts are transmitted to the progeny from special organs, called ovarial ampullae, which lie between the ovary and oviduct. The structure of these ampullae and the mechanism by which they are filled with symbionts varies with the species, in part because of the variation in the location of the mycetomes. Complex movements and translocations of the symbionts also occur

Table 11.7.

Distribution and tissue location of symbioses in lice

(Order) Suborder Family	Louse Species	Vertebrate Host	Tissue Location of Symbiont
(Anoplura)			
Echinophthiriidae	*Echinophthirius* sp.	Marine mammals	Not known if present
Haematomyzidae	*Haematomyzus elephantis*	Elephant	Paired, enveloped abdominal mycetomes
Haematopinidae	*Haematopinus suis*	Hog	Mycetocytes inserted in gut epithelium
	H. eurysternus	Ox	Same as above
	H. macrocephalus	Horse	Same as above
	Polyplax spinulosus	Rat	Single, unenveloped abdominal mycetome
	Linognathus tenuirostris	Cattle	Same as above
	L. piliferus	Dog	Same as above
	Haemodipsus ventricosus	Rabbit	None found
	Hoplopleura acanthopus	Field mouse	None found
Phthiriidae	*Phthirius pubis*	Man	Unpaired, enveloped mycetome attached to gut
Pediculidae	*Pediculus humanus*	Man	Same as above
	Pedicinus rhesi	Catarrhine ape	Ring of mycetocytes beneath gut epithelium
(Mallophaga)			
Amblycera			
Gyropidae	*Gyropus*	Rodents	None found
Boopidae	*Boopia*	Kangaroos	Not known if present
Trimenoponidae	*Trimenopon*	Rodents	Not known if present
Menoponidae	*Menopon gallinae*	Chicken	None found
Laemobothriidae	*Laemobothrion*	Water birds	Not known if present
Ricinidae	*Ricinus*	Passerine birds	None found
Ischnocera			
Trichodectidae	*Trichodectes canis*	Dog	None found
Philopteridae	*Columbicola columbae*	Pigeon	Individual mycetocytes dispersed in fat body
	Sturnidoecus sturni	Starling	Same as above
	Turdinirmus merulensis	Blackbird	Same as above

during embryonic development (Buchner, 1965). The best studied louse symbiosis is that of *Pediculus humanus* (Puchta, 1956; Ries, 1931b). The symbionts of freshly emerged larvae are giant bacteria (2 μm × 2–30 μm) which are harbored in large vacuoles of the syncytial layer of the enveloped stomach disk. The small forms arise by distal fragmentation of the larger forms. The symbionts increase in number and length throughout larval development, reaching a length of over 100 μm in third instar male larvae and adults. In adult males the symbionts then undergo progressive degeneration, becoming increasingly misshapen, inclusion filled, and pleomorphic. In females the symbionts are released from the stomach disk in thick clumps during the molts to adults, finally disappearing from the mycetome. The large symbionts increase in number preparatory to migration and following entry into the vacuoles of epithelial cells in the ovarial ampullae. In the oocyte, the large tubelike symbionts quickly decrease in size and eventually rodlets 2–4 μm long predominate. The egg symbionts are initially sequestered in provisional mycetocytes by filaments radiating from an aster, and subsequently transfer to the vacuoles of the definitive mycetome. Rapid increases in symbiont size again occur during late embryonic development in the final mycetome. In summary, the *Pediculus* symbionts undergo three changes in tissue location, three cycles of proliferation, and well regulated changes in size or degeneration during their life cycle.

The symbiosis of *Pediculus* is atypical of lice in some respects since the symbionts are harbored in large vacuoles and are destroyed in male adults. In most cases the symbionts of lice fill the cytoplasm of the

mycetocytes and are retained in the mycetocytes even in adult males (Buchner, 1965). Furthermore, in most Anoplura and Mallophaga, the ovarial ampullae are infected by migration of the mycetocytes themselves rather than by clumps of symbionts. However, in *Haematopinus*, symbiont transmission is even more complex than in *Pediculus* since five distinct cellular sites and two extracellular stages occur during the symbiont life cycle. The morphology of the pleomorphic symbionts in different lice also varies significantly, even at comparable stages in their complex life cycles. The cytoplasm of mycetomal symbionts of *Haematopinus* appears honeycombed rather than homogeneous as in *Pediculus* symbionts. The infectious forms of *Linognathus* are filamentous and contain a prominent chromatic granule while those of *Haematomyzus* are short fat tubes with a very prominent, strongly staining, ring-shaped structure in the middle of the bacterium. Similarly, the mycetocyte symbionts of Mallophaga are sparse, weakly staining sausage-like forms in *Columbicola*, oval-to-round forms in *Anaticola*, very thin filaments in *Sturnidoecus*, and short densely staining rodlets in *Coloceras*. The ultrastructure of the symbionts has not been described.

The few attempts to culture the symbionts of Anoplura appear to have been unsuccessful. Bewig and Schwartz (1956) and Puchta (1955) found that the symbionts of *Haematopinus suis* and *Pediculus humanus*, respectively, are very sensitive to nonisotonic culture media. Although Puchta (1955) and Kotter (1955) both cultured a pleomorphic diphtheroid organism from *Pediculus*, the low frequency of its isolation suggests that it is not the symbiont.

Direct evidence for an essential physiological role for the symbionts in *Pediculus* was obtained in the pioneering work of Aschner (1932, 1934) and Aschner and Ries (1933). The stomach disk can be readily removed surgically from third instar females before transfer of the symbionts to the ovarial ampullae occurred. Although males are relatively unaffected by this treatment, the aposymbiotic females survive for only a short time, have fewer eggs than symbiotic females, and their eggs do not develop. Females whose symbionts have already transferred to the ampullae are unaffected by removal of the sterile mycetome. Similarly, removal of the mycetome during the first or second larval instar leads to death late in the following instar for both sexes. Centrifugation of eggs to displace the stomach disk before it has received its normal complement of symbionts leads to destruction of the sym-

bionts. These sterile larvae do not survive the first larval instar. Aposymbiotic larvae obtained by centrifugation can complete normal development if they are fed through membranes with blood supplemented with yeast extract, or with defibrinated, hemolyzed blood supplemented with growth factors (Puchta, 1954, 1955). Although deletion of thiamine, riboflavin, and pteroylglutamic acid has no effect, deletion of pyridoxal phosphate results in death by the third instar, and deletion of nicotinic acid, pantothenic acid or biotin permits little development. These findings suggest that vitamins, present only in low levels in vertebrate blood, are essential for *Pediculus* development and that they are normally supplied to *Pediculus* by their symbionts.

PLANT SAP-SUCKING INSECTS

Among insects, with the exception of the beetles (Coleoptera), endosymbionts are found most frequently in plant sap sucking members of the closely related orders Hemiptera and Homoptera, a diverse group of species with piercing-sucking mouth parts formed into a beak. The Homoptera consists of at least 32,000 species which are all terrestrial and plant sap sucking, and which uniformly harbor intracellular sym-

bionts that are inherited transovarially (Table 11.8). The suborder Coleorrhyncha includes the rare, primitive Peloridiids which resemble predaceous Hemiptera, although their endosymbiosis is typically Homopteran. The suborder Auchenorrhyncha contains the larger, more generalized Homoptera. In the superfamilies Cicadoidea (cicadas, tree hoppers, spittle bugs, and leafhoppers) and in the planthoppers of the

Table 11.8.
Some of the best studied Homoptera harboring endosymbionts

(Suborder) Superfamily	Species	Common Name (plant host)	Reference
(Coleorrhyncha) Peloridiidae	*Hemiodoecus fidelis*	Peloridiid (moss)	Buchner, 1965
(Auchenorrhyncha) Cicadoidea	*Euscelis plebejus*	Leafhopper (clover)	Houk and Griffiths, 1980
	Graphocephala coccinea	Leafhopper (food plants)	Kaiser, 1980
	Helochara communis	Leafhopper (sedge)	Houk and Griffiths, 1980
	Nephotettix cincticeps	Leafhopper (green rice)	Mitsuhashi and Kono, 1975
Fulgoroidea	*Fulgora europea*	Planthopper (grass)	Buchner, 1965
	Cixius nervosus	Planthopper (grass)	Buchner, 1965
	Laodelphax striatellus	Planthopper (rice)	Noda, 1977
(Sternorrhyncha) Aleyrodoidea	*Trialeurodes vaporariorum*	Whitefly (vegetables, greenhouse)	Buchner, 1965
Psylloidea	*Psylla pyricola*	Psyllid or jumping louse (pear)	Houk and Griffiths, 1980
Coccoidea	*Pseudococcus citri*	Mealybug (citrus plants)	Houk and Griffiths, 1980; Tremblay and Tripodi, 1980
	Icerya purchasi	Giant coccid (citrus plants)	Buchner, 1965
Aphidoidea	*Acyrthosiphon pisum*	Aphid (pea)	Houk and Griffiths, 1980
	Brevicoryne brassicae	Aphid (cabbage)	Houk and Griffiths, 1980
	Lachnus tropicalis	Aphid (chestnut)	Ishikawa 1977, 1978
	Macrosiphum rosae	Aphid (rose)	Houk and Griffiths, 1980
	Myzus persicae	Aphid (green peach)	Houk and Griffiths, 1980
	Neomyzus circumflexus	Aphid (crescent-marked lily)	Houk and Griffiths, 1980
	Phodoron humuli	Aphid (hop)	Amiressami, 1980

superfamily Fulgoroidea, the beaks arise posteriorly from the head. The suborder Sternorryncha consists of smaller, highly specialized species whose beak lies between the forelegs. Both groups include numerous species of great economic importance because of their direct damage to crops and because they transmit a variety of plant diseases. Homopterans suck large volumes of plant sap in order to otain their required vitamins and micronutrients and excrete the well known sugary honeydews. Recent work on the intracellular symbionts of the Homoptera, particularly those of aphids and leafhoppers, has been reviewed extensively by Houk and Griffiths (1980) and the description given below is based largely on this review. Additional details on host adaptations to the symbiosis and the evolution of multiple symbioses in the Homoptera are provided by Buchner (1965, 1969) and Müller (1969, 1972).

The Hemiptera, better known as the true bugs, are the most diverse group of insects among those with incomplete metamorphosis. The order includes 30,000 species of relatively large insects which differ from Homoptera primarily in the structure of the beak and in having half-membranous forewings. However, compared to the Homoptera, the Hemiptera are less uniform with respect to morphological adaptations, transmission mechanisms, and distribution of symbionts (Table 11.9). Symbiotic relationships in the Hemiptera correlate reasonably well with diet. None of the predaceous aquatic Hemiptera harbor endosymbionts, including superfamilies in both suborders of Hemiptera: all of suborder Cryptocerata (water bugs, backswimmers, water boatmen) and superfamilies Gerroidea (water striders) and Aradoidea (water treaders, velvet water bugs) of suborder Gymnocerata. In the remainder of Gymnocerata, bugs predaceous on other arthropods also lack symbionts even though some blood sucking (bedbugs, Cimicoidea; reduviids, Reduvioidea) and plant sucking (most Scutelleroidea) bugs in the same superfamilies do have symbionts. These include the Anthocoridae (flower or pirate bugs, Cimicoidea), Nabidae (damsel bugs, Reduvioidea), Asopinae (predaceous stink bugs of family Pentatomidae, Scutelleroidea), and the Aradidae (fungus bugs, Aradoidea). Furthermore, even the leaf and plant bugs of the large family Miridae (Cimicoidea), which only supplement their usual phytophagous diet with small invertebrates, generally lack symbionts. Although symbionts are uniformly present among the plant sucking Scutelleroidea, some Lygaeoidea and Coreoidea apparently do not have them and symbionts are always absent in the grass bugs in the Corizidae (Coreoidea) and the superfamily Tingoidea. However, it is quite possible that endosymbiosis in the plant sucking Hemiptera has been underestimated because most workers have examined primarily the gut for obvious caeca, and therefore may have overlooked the less common intracellular symbionts present either in the gut epithelium or in abdominal mycetomes.

Bournier (1960) and Louis and Bournier (1971) have described a *Pediculus*-like symbiosis in several species of the small order of thrips, Thysanoptera. Species which suck the sap of fungi have symbionts, while other predaceous or plant-eating forms lack them. The symbionts of *Gaudothrips buffai* occur intracellularly in an abdominal mycetome, and are inherited by transovarial passage. Ultrastructurally, the thin thread-like symbionts have a Gram-negative type cell wall and are enclosed by a host membrane.

Homoptera

Morphology and Structure

Virtually all Homoptera have at least one symbiont, designated "a" (primary), but many insects have more than one and as many as six symbionts, [variously called "t" (auxiliary) and "H" (for Hefen (yeast) symbionts) and one or more "companion" symbionts. In some instances "g" (globular) symbionts (Change and Musgrave, 1972) and "s" (small) symbionts (Kaiser, 1980) have also been described. All these bacteria appear as typical Gram-negative organisms, except that the a and t symbionts are highly pleomorphic and undergo a cycle of development of vegetative and infective forms. The vegetative forms are distinguished from the infective forms by their lower electron density and

greater irregularity of shape, especially the t symbionts which are multilobated (Korner, 1976). The a symbionts are the largest (1.5 × 7.0 μm in *Helochara communis* and 4 by 10 μm in *Euscelis plebejus*). The t symbionts are somewhat smaller (1.5 × 4.0 μm and 3.0 × 9.0 μm in the two species, respectively). The companion symbionts are usually typical rods, 0.3 × 1.0–0.8 × 4.0 μm. By freeze-fracture ultrastructural techniques, Louis and Nicolas (1976) found that the a symbionts of *Euscelis* exhibit variant morphologies, possibly related to changes occurring between the vegetative and infective forms. In both thin sections and freeze-fracture preparations, the membranes of t symbionts also differ greatly from those of the a symbionts (Louis et al., 1976). The t symbionts have unique parallel fibrils associated with undulations of the plasma membrane and an unusually small number of particles on the cytoplasmic face of the plasma membrane.

Most symbionts have three membranes, a cytoplasmic membrane, a cell wall, and a third membrane, which is undoubtedly of host origin. In some instances it has been shown that the third membrane continues with the endoplasmic reticulum of the host cell; in other instances it is lost during the process of infection of another host cell and is reformed in the new cell. In the cytoplasm of the g symbiont of *Helochara communis*, parallel tubular formations are seen, which possibly are mesosome-like unfoldings of the cytoplasmic membrane (Chang and Musgrave, 1972). These are relatively uncommon structures, but they have been encountered in other endosymbionts (see the next article in this *Manual* on Endosymbionts of Fungi and Invertebrates other than Arthropods, particularly the part on Helminthes).

Ribosomes are readily recognized morphologically in a number of endosymbionts. When endosymbiont-rich mycetomes of the chestnut aphid, *Lachnus tropicalis*, are maintained in vitro, curiously only the production of procaryotic rRNA can be demonstrated, to the exclusion of eucaryotic rRNA. This synthetic activity is inhibited by low levels of rifampicin (Ishikawa, 1978).

Several of the earlier workers failed to detect DNA in some of the symbionts and cast doubt on their microbial nature or regarded them as degenerate bacteria. There is now good evidence that all symbionts have DNA, although there is some question about its molecular weight. Schwemmler et al. (1975) isolated the DNA from a combination of a and t symbionts derived from *Euscelis plebejus* and, on the basis of sedimentation analysis and electron microscopy, found the predominant sizes to be 2.2 and 2.6 × 10^7 daltons; such sizes are intermediate between those of mitochondria and plastids. These results are suprising and need to be confirmed by techniques that are not affected by the shearing of the DNA, such as reassociation kinetics. Houk et al. (1980) examined the DNA of the primary symbionts of the pea aphid by bouyant density centrifugation. The peak value and a shoulder indicate a mol% G + C value of 31 with a subpopulation extending to 41. These studies need to be extended to other Homopteran symbionts.

The primary symbionts of the pea aphid contain phosphatidyl ethanolamine, phosphatidyl choline, cholesterol, and glycerides. The fatty acids are predominantly C10, C12, C13, and C14. The host insect has the same classes of lipids, but different fatty acids (Houk, 1974a).

Paracrystalline inclusions, some shown to be digestible by Pronase or chymotrypsin, have been detected in the cytoplasm of a and t symbionts of *Euscelis plebejus* and in the t and g symbionts of *Helochara communis* (Houk and Griffiths, 1980). Similar inclusions have been seen in the genus *Rickettsiella* (see The Rickettsias and Chlamydias, Tribe *Wolbachiae*, in this *Manual*). Some of the microscopic observations are consistent with the possibility that the symbionts have polyphosphate crystals.

Transmission Cycles in the Host

Leafhoppers. The extensive investigations on the transmission cycles of the a and t symbionts have been summarized by Schwemmler (1980). In the cicada, *Euscelis incisus*, the adult female inserts a symbiont ball consisting of the infectious stages of both symbionts between the outer coat and cells of its eggs. The a symbionts are soon taken up by blastodermal host cells, a_1 mycetocytes, which disintegrate and the

Table 11.9.

Distribution and characteristics of symbioses in the plant sap sucking Hemiptera

Superfamily Family (Subfamily)	Common Name	Representative Genera	Symbiont Frequency No. present/ total species	Tissue Location of Symbionts	Mechanism of Hereditary Transmission of Symbiont
Scutelleroidea					
Scutelleridae	Shield-backed bugs	*Eurygaster, Homaemus*	12/12	Lumen of multiple midgut caeca in four rows	Egg smearing from gut
Popopidae	Terrestrial turtle bugs	*Podops, Graphosoma*	4/4	As above	As above
Pentatomidae (Pentatominae)	Stink bugs	*Murgantia, Aelia, Euschistus*	44/44	As above	As above
(Acanthosominae)		*Acanthosoma, Elasmucha*	9/9	Lumen of multiple blind midgut caeca in two rows	Special abdominal egg smearing organs
Corimelaenidae	Negro bugs	*Thyreocoris*	5/5	Lumen of multiple midgut caeca in two rows	Egg smearing from gut
Cydnidae	Burrower bugs	*Cydnus, Aethus*	8/8	As above	As above
		Brachypelta		As above, adult tract closed	Larval anal trophallaxis
		Chilocoris		Intracellular, midgut epithelial mycetocytes	Transovarial
Plataspidae	Plataspids	*Coptosoma*	5/5	Lumen of multiple midgut caeca in two rows; adult tract interrupted	Symbiont capsule
Cimicoidea					
Miridae	Leaf bugs	*Stenotus*	1/46	Intracellular, midgut epithelium	?
Coreoidea					
Coreidae	Leaf footed bugs	*Anasa, Mesocerus*	16/20	Lumen of multiple midgut caeca in two rows	Eggs smeared from gut
Coriscidae	Broad-headed bugs	*Alydus*	3/3	As above	As above
Corizidae	Grass bugs	*Corizus*	0/10	None present	None present
Tingoidea					
Tingidae	Lace bugs	*Piesma*	0/7	None present	None present
Piesmidae	Ash-gray leaf bugs	*Phymata*	0/3	None present	None present
Lygaeoidea					
Berytidae	Stilt bugs	*Jalysus, Berytus*	4/4	Lumen of 6 fused, finger-shaped midgut caeca in 2 rows	Eggs smeared from gut
Pyrrhocoridae (Pyrrhocorinae)	Stainers		6/6		
		Pyrrhocoris, Dysdercus		Lumen of gut or in 2–7 short, pouched midgut caeca in 2 rows	As above
(Largiinae)		*Largus, Euryophthalmus*		Lumen of multiple midgut caeca in two rows	As above
Lygaeidae (Lygaeinae)	Seed sucking bugs	*Nysius, Arocatus*	8/31	Intracellular, abdominal mycetome	Transovarial
		Arocatus		Lumen, brush border of Malpighian tubules	Transovarial
		Tropidothorax		Lumen, intracellular in gut and Malpighian tubule epithelium	Eggs smeared from gut
(Aphaninae)		*Stilbocoris*	21/23	Lumen of midgut caeca	Larval symbiont packet
		Aphanus		Lumen and intracellular in fused irregular finger-like gut caeca	As above
		Gastrodes		Lumen, intracellular in sparse ramified tubelike midgut caeca	As above
(Blissinae)		*Blissus*	3/3	Lumen of 5–7 individual fingerlike midgut caeca in two rows	As above

Table 11.9—*continued*

Superfamily Family (Subfamily)	Common Name	Representative Genera	Symbiont Frequency No. present/ total species	Tissue Location of Symbionts	Mechanism of Hereditary Transmission of Symbiont
		Ischnodemus		Intracellular, paired mycetome	Transovarial
(Cyminae)		*Ischnorrhynchus*	2/3	Intracellular, unpaired mycetome	Transovarial
(Heterogasterinae)		*Heterogaster*	2/3	Lumen of multiple midgut caeca in two rows	Eggs smeared from gut
(Geocorinae)		*Geocoris*	1/3	Intracellular, fat body	?
(Oxycareninae)		*Oxycarenus*	0/1	None present	None present

symbionts then enter binucleated mesodermal gonad bud cells, a_2 mycetocytes. The t symbionts enter separate mesodermal cells with large single nuclei, t mycetocytes. Within their respective cells, the a and t symbionts differentiate into replicating vegetative forms. The mycetocytes are then organized into a common organ, the mycetome. Soon thereafter, the mycetome divides into two lateral mycetomes, each consisting of a t mycetome surrounded by an a mycetome. They are located at each side of the abdomen during the remaining course of embryonic development. The cells of the mycetome are quite crowded with symbionts, but are not prevented from occasional cell division. In the adult female, the mycetomes develop into a considerably larger size than in the male. Ovary cells migrate into the a mycetome and form infectious cell mounds, in which the a symbionts convert to the infectious form. The t symbionts do the same in migratory mycetocytes close to the ovary. Both infectious forms enter the hemolymph of the sexually mature female and from there enter special ovarial cells, the wedge cells, which are involved in the infection of the eggs and the initiation of a new cycle.

The cycles of the a and t symbionts appear to be quite similar in other leafhoppers (Chang and Musgrave, 1972; Mitsuhashi and Kono, 1975). Other symbionts, often called companion or accessory because of their close association with the a and t symbionts, have less defined cycles. They are easily distinguished from the a and t symbionts by their smaller size and are sometimes called "rickettsia-like." They have been seen in mycetomes or free in the cytoplasm or nuclei of various types of cells. There is no evidence that they are pathogenic for their hosts. They are not transmitted to the next generation in infectious mounds, but just as individual cells invading ovarial tissue.

In addition to providing for the symbionts' growth and transmission, the host must regulate their numbers. It is not clear how this is done. The rate of bacterial multiplication in mycetomes is low and this may be accomplished, although there is no direct evidence for it, by host-induced enzyme repression. Buchner (1965) noted that the number of dividing symbionts increased prior to embryogenesis and transovarial infection. Other mechanisms involve phagocytosis and digestion of the symbionts that escape the mycetomes or intracellular destruction in phagolysosomes in specialized cells, such as wedge cells, when infection of the eggs is completed. Subtle biochemical control appears to be more prevalent in the case of the primary symbionts, destruction of excess bacteria in the instance of the accessory symbionts.

Aphids. The method of transmission of the symbionts of the cabbage aphid, *Brevicoryne brassicae,* and of other aphids was studied by electron microscopy by Hinde (1971a). In viviparae, the symbionts infect the embryos after the establishment of the blastoderm, entering them through a transient pore in the follicular epithelium. The symbionts develop in mycetocytes and those that are released into the hemolymph are digested by phagocytic hemocytes. However, even in the mycetocytes there are degenerating symbionts that are destroyed by action of the lysosomes. A similar phenomenon has been described in the pea aphid (Griffiths and Beck, 1973). The primary symbionts develop in large mycetocytes making up the bulk of the mycetome. The smaller

secondary symbionts are restricted to the syncytial sheath enclosing the primary mycetocytes. Both types of symbionts are subject to destruction by lysosomal action. In the primary mycetocytes, small numbers of symbionts are reduced to compact residual bodies. In the secondary mycetocytes, destruction can be more extensive.

Coccids. The cycle in *Pseudococcus citri,* which is representative of other coccids, has been summarized by Tremblay and Caltagirone (1973) and Tremblay and Tripodi (1980). The females have a voluminous yellow-colored mycetome, situated under the gut, as large as one-third of the total body length. The organ has a complex structure which includes an outer layer, tracheae, pigment, glycogen granules, and mitochondria. They also have large mycetocytes which enclose mucous spherular masses containing the symbionts. Some of the mucous spherules leave the mycetome and penetrate into the "ovariole neck" between the nurse cells and the oocyte and enter the oocyte. The symbionts are eventually engulfed in embryonic giant cells, formed from the fusion of polar bodies and cells from the blastoderm. The polar bodies are haploid daughter cells of the primary and secondary oocyte produced during egg maturation. Prior to participation in the mycetocyte formation they fuse and divide and become polyploid. The giant cells, produced in this highly involved sequence, are incorporated into the mycetome and the cycle is repeated.

Evolution of Endosymbiosis

Homoptera offer an excellent opportunity to study evolution of endosymbiosis. According to Buchner (1965), the presence of endosymbionts has been investigated in 405 species of cicadas and leafhoppers, of a total of approximately 30,000. A majority (55%) have two endosymbionts and a large number (30%) have three. The geological time of appearance of a symbiont can be judged by the number of insect species that have it, aberration in morphology (indicating adaptation to a specialized environment), and sophistication in the mechanism of transmission. By these criteria a symbiont is by far the most ancient and best entrenched. It is seen as the only symbiont in *Hemiodoecus,* which is considered to be a relict from the Carboniferous age. It is not replaced by other symbionts, with the rare exception of H symbionts (yeasts) which amply provide for the nutritional deficiency of the diet of the insects. The t and other auxiliary symbionts appear to be more recent acquisitions, each involving more restricted taxonomic groups of insects. The other symbionts are regarded as more recent still, because of their smaller numbers and casual manner in which they are transmitted to the next generation.

In the aphids, the presence of two symbiont types is the most common occurrence, with few species able to accommodate a third one. In the coccids, despite the conspicuous mycetome, the symbionts vary greatly in morphology in the different taxonomic groups and are believed to have been acquired more recently than in leafhoppers or aphids. Attempts to trace the evolutionary history of Homoptera on the basis of their symbionts have given results that are remarkably similar to those obtained by conventional methods.

Symbiont Function

The contribution of the symbionts to the nutrition and physiology of the host has been studied essentially in three ways: the insects have been rendered aposymbiotic, the in vitro metabolic activities of the symbionts have been determined, and the symbiont population has been examined in the insecticide-resistant insects.

Leafhoppers. The most obvious effect of aposymbiosis is seen in the embryonic development of *Euscelis*. The eggs are rendered aposymbiotic either by ligation of the symbiont ball immediately after oviposition or by interruption of infection of the eggs by application of tetracycline or lysozyme to the females. From these eggs "cephalothorax embryos" develop which lack an abdomen. In the adult insects levels of antibiotics or lysozyme that reduce the number of symbionts result in lower fecundity and growth, delay in development, and reduced longevity. Total, or nearly total, aposymbiosis results in the death of the insects within a week. The longevity of the insects can be increased to 2 weeks by supplementing the diet with additional ions, vitamins and amino acids. The death of the aposymbiotic insects is due not only to insufficient diet, but also to the accumulation of uric acid crystals. Thus, the symbionts seem to play a role both in the anabolism and catabolism of the insects (Schwemmler et al., 1973; Schwemmler, 1980). A role of the symbionts in the regulation of the host's adenylate energy charge as well as pH and osmolarity has also been suggested (Schwemmler and Hermann, 1979, 1980; Schwemmler, 1980).

Aphids. The introduction of antibiotics in the diet of the pea aphid affects its reproduction and survival, but the nature of this effect must be interpreted with caution. Streptomycin increases food consumption, but penicillin and chloramphenicol reduce it and, when the concentration of penicillin is 1%, the aphids quickly fall off the feeding membrane. Furthermore, 0.1% chlortetracycline damages the insect's mitochondria as well as destroying the symbionts. Thus, only the effects of 0.01% of most antibiotics or 0.002% chlortetracycline can reliably be attributed to their action on the symbionts. At these concentrations, penicillin and streptomycin have little effect on growth, but reduce fecundity and survival, while with chlortetracycline the larvae fail to develop. Chloramphenicol and neomycin reduce growth and survival, and the aphids do not reach the adult stage. The effects of streptomycin and penicillin are not entirely irreversible, since aphids maintained for 7 days on 0.1% of such antibiotics and then restored to normal diets may gain weight, reach maturity and reproduce, although progeny survival is low (Griffiths and Beck, 1974; Srivastava and Auclair, 1976).

Cells of the cabbage aphid and their symbionts have been maintained in vitro in Grace's insect medium for 2–3 weeks, and in one instance for as long as 16 weeks (Hinde, 1971b). Although some of the symbionts occur free in the medium, their persistence is assumed to be dependent on the presence of the aphid cells. There is some evidence that the symbionts can multiply. Pea aphid symbionts have been separated from host components (Houk and McLean, 1974) and such isolated symbionts, maintained in Grace's medium (as modified by Tokumitsu and Maramorosch), are suitable for the study of metabolic activities (Houk, 1974b).

The evidence for and against the hypothesis that the symbionts contribute specific amino acids and vitamins to the nutrition of aphids has been summarized by Houk and Griffiths (1980). Most of the recent investigations have been concerned with lipid metabolism. The insects as a group are not known to synthesize sterols de novo, but the aphid *Neomyzus circumflexus* has been maintained for 11 generations on a synthetic sterile medium which lacks sterols or any other lipids (Ehrhardt, 1968). ^{14}C-Acetate is incorporated by aphids with symbionts, but not by aposymbiotic aphids (Ehrhardt, 1968). The implication that the symbionts contribute lipids to the aphids is strengthened by the finding that purified suspensions of the primary symbionts of the pea aphid synthesize cholesterol from acetate and mevalonate and that free fatty acids, mono- and diglycerides, phosphatidyl choline, and phosphatidyl ethanolamine are also synthesized (Houk et al., 1976). With low concentrations of symbionts, lipid synthesis is stimulated by the addition of ATP and reduced nicotinamide adenine dinucleotide phosphate (NADPH). The transport of high energy phosphorylated compounds by intact cells and the synthesis of cholesterol are properties not often encountered among bacteria.

The possibility that much of the cholesterol synthesized is transferred to the host has been investigated ultrastructurally by treating aphids with digitonin, a compound that complexes with sterols and that can be used to identify cholesterol in electron micrographs (Griffiths and Beck, 1977a, b). By this procedure the membranes of both primary and secondary symbionts have high concentrations of cholesterol. When ^{3}H-mevalonate (a cholesterol precursor) is used instead of digitonin and the sections subjected to autoradiography, the same symbiont membranes become labeled. When the time of incubation of the aphids in labeled medium is increased, the number of silver grains on the membranes remains constant, but it increases in the surrounding tissues. This suggests that the cholesterol can be transferred to the aphid cells. The mechanism of transfer of this highly insoluble compound is unknown but possibly is mediated by vesicles that are present between the cell wall and the third membrane (Griffiths and Beck, 1975).

Understanding the mechanism of insecticide resistance is one of the outstanding problems in agricultural entomology. An attempt has been made to determine if the symbionts play a role in such resistance. No great differences are found between the mycetomes of normal and parathion-resistant peach aphids by light or electron microscopy, but the symbionts in the resistant aphids tend to be larger and the proportion of the very large forms (6 μm in diameter) is greater, although the total number of symbionts is reduced somewhat (Amiressami and Petzold, 1976, 1977). Some changes are also seen in the mycetocytes, i.e. greater vacuolation and larger nuclei. The symbionts of the hop aphid (*Phodoron humuli*), which is resistant to Demeton-*S*-methyl, appear to have undergone changes comparable to those seen in the peach aphid (Amiressami, 1980). The significance of these changes is not clear. They have not been observed in another study of peach aphids resistant to Demeton-*S*-methyl (Ball and Bailey, 1978).

Coccids. Differentiation of the host and symbiont metabolism in *Pseudococcus* can be achieved by injecting ^{3}H-leucine with or without chloramphenicol or cycloheximide and examining tissue sections by autoradiography (Louis and Kuhl, 1972). Chloramphenicol greatly reduces the number of silver grains in the location of the symbionts, whereas cycloheximide has a similar effect on the insect cells. Starvation of *Pseudococcus* for 7–30 days induces marked changes in the morphology of the symbionts (Tremblay and Tripodi, 1980). In the starved insects, the symbionts are less electron dense, more pleomorphic, and the mucous masses containing them are reduced in size. This might indicate either that the symbionts have a reduced metabolism or that they serve as a food reserve in starved insects.

Comment

The work that has been done thus far on Homoptera illustrates the high degree of interdependence of insects and their endosymbionts. It has also given us a glimpse of the mechanisms by which the two partners in this symbiosis regulate each other's activities. Much remains to be done. It would be unwise at this point to assume that interactions are similar in the three major groups (leafhoppers, aphids, and coccids) or that they are essentially the same within each group. With present technology it would be entirely feasible to determine the degrees of DNA relatedness of a number of symbionts to each other and to differentiate their metabolic capabilities. With this information we would greatly advance the understanding of the physiology of this very important group of agricultural pests.

Hemiptera

Symbionts in Midgut Caeca

The endosymbionts of Hemiptera are most often extracellular and are located in the lumen of midgut caeca which, in different taxonomic groups, vary considerably in number and arrangement (Table 11.9). Information on these symbionts is based on the lengthy surveys by Rosenkranz (1939) and Schneider (1940). Relatively little has been added since then. The most complex midgut caeca occur in the Scutelleroidea in which they are arranged in either two or four very regular rows of multiple (as many as 1400), short, laterally flattened crypts. The bacteria-filled lumen of the caeca is usually connected to the gut canal by a very narrow channel, but becomes occluded during larval development of the Acanthosominae. A similar complex double row of caeca is typical in the Coreoidea, and is occasionally found in the Lygaeoidea (Largiinae, Heterogasterinae). Midgut caeca are present in a variety of forms in the other Lygaeoidea. In the Pyrrhocorinae, female *Dysdercus*, but not males, harbor the symbionts in two rows of six or seven small, pouch-like caeca. The caeca are even more rudimentary in *Pyrrhocoris* and the symbionts are not found in the gut caeca, but only in the canal. In Aphanines like *Gastrodes* and *Tropistethus*, the caeca may appear like large ramified Malpighian tubules or as six or seven shorter, more regular, fingerlike outpocketings as in the Blissinae. In the Berytids and other Aphanines, these fingerlike forms are more numerous and less regular in their arrangements and may become partially fused laterally into palmate, leaf-like structures. The caeca of Aphaninae are clearly exceptional in harboring symbionts both in the lumen and intracellularly in the epithelium.

Symbionts harbored in the midgut caeca of Hemiptera are generally inherited by simple superficial contamination of the egg shell. Hemipteran larvae generally exhibit a marked resting behavior upon hatching: they remain associated with their egg shells, constantly probing them with their beaks. Symbionts ingested during this period become associated with the alimentary tract and proliferate in the lumen of the sterile midgut caeca. In many females, crypt development is more complex than in males. In some Pentatomidae the terminal crypts become highly enlarged and specialized in characteristic species-specific patterns for the production of the fecal symbionts. Indeed, the adult tracts of *Brachypelta* (Schorr, 1957) and *Coptosoma* (Schneider, 1940) are interrupted completely anteriorly, thus isolating the crypts as a mycetome-like transmission organ. Several interesting variations on symbiont transmission occur. In the Acanthosominae, where the caecal lumen is sealed from the gut canal, ventral abdominal smearing organs become infected with symbionts in late larval development. In *Brachypelta*, a Cydnid which burrows in sand, the eggs are not smeared, the larvae associate with the parent and receive symbionts directly in the anal fluid—a possible adaptation to their dry habitat. Similarly, in *Stilbocoris* (Lygaeidae), the viviparous larvae are provided with a drop of symbionts immediately upon emergence (Carayon, 1963). Finally, perhaps the most remarkable behavioral and anatomical arrangement for transmission occurs in the Plataspid, *Coptosoma*, in which each egg is provided a symbiont-filled fluid capsule. Upon hatching, the larva immediately punctures the capsule with its proboscis and ingests the symbionts (Schneider, 1940; Müller, 1956).

Most symbionts of plant sap-sucking Hemiptera stain Gram-negative and many are highly pleomorphic in their hosts. Regular changes occur in symbiont morphology within a given host, both in response to the season in temperate zone insects, and to stages in the hereditary transmission cycle (Rosenkranz, 1939; Schneider, 1940). In general, transmission forms in the female are short rodlets, 0.7 μm × 1–2 μm long, whereas they are considerably longer in mid- to late larval stages and in adult males, the maximum size depending on the host species (Buchner, 1965). In many Pentatomidae, the maximum length is usually 5–10 μm, but tubes 1.2 μm wide and 5 to 50 μm long occur in *Peribalus limbolarius*, and may even appear as highly twisted and convoluted giants 1–3 μm wide and 30–100 μm long in *Murgantia histrionica* (Glasgow, 1914).

The external carriage and transmission of the symbionts of the midgut caeca have simplified attempts at their culture and permitted simple nondestructive methods for interrupting their transmission and for obtaining sterile aposymbiotic larvae, much along the lines of such studies on the blood-sucking Hemiptera. Table 11.10 summarizes es-

sential features of those Hemipteran symbionts which have been cultivated and characterized. Symbionts from several Hemiptera have been grown easily on nutrient agar supplemented with glucose (Glasgow, 1914; Steinhaus et al., 1956; Huber-Schneider, 1957). The isolated cultures are similar to the caecal symbionts in morphological and tinctorial properties, and are routinely isolated in nearly pure culture and in large numbers only from the midgut caeca. Antiserum raised against the cultures agglutinates symbionts obtained from the midgut caeca. Although the symbionts of *Mesocerus* lack flagella and fail to produce acid or gas from carbohydrates, they greatly resemble the various *Pseudomonas* species obtained by Steinhaus et al. (1956) both in their small size (0.8–1.0 × 1.0–2.5 μm) and other properties (Huber-Schneider, 1957). The symbionts of *Chelinidea* have been isolated from a variety of specimens either as noninterconvertible mucoid or nonmucoid, occasionally mixed, strains of otherwise similar properties. The cultured symbiont of *Narnia pallidicornis* (Coreidae) resembles the nonmucoid variety of *Chelinidea* symbiont. In *Anasa tristis*, two varieties (types I and II), that differ greatly in properties have also been obtained regularly in primary cultures. Type I metabolizes a large number of substrates and is remarkably similar to the symbiont of *Euryophthalmus*, while type II has much more limited metabolic activity. The cultivation of the *Euschistus* symbiont needs confirmation because it has only been obtained in 14 of 59 attempts, particularly since the symbionts of other Pentatomidae in which the larger and more irregular forms predominate, have resisted culture. Finally, although most Coreidae appear to have Gram-negative symbionts, the symbiont of *Cletus signatus* is a Gram-positive spore-former, tentatively identified as "*Bacillus cereus* var. *signatus*" (Singh, 1974). This microorganism is auxotrophic for all essential vitamins and can degrade the insecticides DDT, parathion, and carbaryl (Singh and Pant, 1974).

The nature of the contribution of the symbionts to the nutrition of their hosts has been the subject of some controversy. The following observations support the idea that the symbionts are essential for their hosts. Although sterile larvae of the Plataspid *Coptosoma* and the Pentatomid *Eurydema* have been reared to adults (Müller, 1956; Bonnemaison, 1946), the *Coptosoma* larvae grow slowly and show a relatively high mortality compared to normal controls, and their growth can be enhanced by feeding them on seedlings rich in growth factors (Müller, 1956). Sterile larvae of *Mesocerus marginatus* (Coreidae) and *Brachypelta aterrima* (Cydnidae) fail to survive beyond the second instar (Huber-Schneider, 1957; Schorr, 1957). Furthermore, the *Mesocerus* symbiont contains significant amounts of all B vitamins except nicotinic acid, the only vitamin present in reasonable levels in the sieve tube sap of the red oak (*Quercus rubra*), a host for *Mesocerus* (Buchner, 1965). As counterarguments to an essential role of the symbionts, Goodchild (1966) has pointed out that multiple species of symbiont may occur in some hosts, that symbionts may be absent in some males, and that caeca are certainly not obligate structures for harboring symbionts, as is the case in *Rhodnius*. He has suggested that the caeca actually evolved primarily as an organ for efficient water excretion and micronutrient retention, rather than to harbor the symbionts. Nonetheless, for many species it is hard to avoid the view that both roles of the caeca are important. In the Pentatominae, the larval alimentary tract is usually interrupted anteriorly to the caeca for a time; symbionts undergoing apparent destruction have been seen in the anterior ampulla. This might provide a mechanism for recirculation to the host of micronutrients trapped in the symbionts. This is even further developed in Plataspidae, some Coreidae and Lygaeidae (Blissinae), and Dinidorinae where the gut is completely isolated posteriorly, and is converted into a pseudomycetome. The isolated symbionts in these species may play a significant role there in trapping micronutrients, releasing vitamins, and possibly even fixing nitrogen (Goodchild, 1966).

Symbionts in Midgut Epithelium, Abdominal Mycetomes and Elsewhere

Other Hemipteran symbionts have been described in both noncaecal gut associations and in abdominal mycetomes (Table 11.9). In the

Table 11.10.

Characteristics of cultivated symbionts of Hemiptera[a]

Property	"Pseudomonas excibis" n. sp. (Chelinidea vittiger, Coreidae)	Pseudomonas sp. (Anasa tristis, Coreidae) Type I	Type II	Pseudomonas sp. (Euschistus conspersus, Pentatomidae)	"Pseudomonas nactus", n. sp. (Euryophthalmus cinctus californicus, Pyrrhocoridae)	"Achromobacter" sp. (Mesocerus marginatus, Coreidae)
Gram-negative small rods	+	+	+	+	+	+
Motile (polar flagella)	+	+	−	+	+	−
Gelatin liquefaction	−	+	Slow	+	+	
Nitrate reduction	+	+	−	+	+	+
Indole	−	−	−	−	−	
Voges-Proskauer	−	+	−	+	+	−
Citrate	+	+	−	+	+	
Starch hydrolysis	−	−		−	−	−
Acid produced from:						
Glucose	Weak	+[b]	+	+[b]	+[b]	−
Mannose	−	+[b]		+[b]	+	
Arabinose	Weak	+[b]	−	−	+[b]	−
Xylose	−	+[b]	−	−	+[b]	−
Rhamnose	−	−	−	−	+	−
Sucrose	−	+[b]	−	+[b]	+[b]	−
Maltose	−	+[b]	−	+	+[b]	−
Lactose	−	−	−	−	−	−
Trehalose	−	+[b]	−	+	+[b]	−
Inulin	−	−	−	−	−	−
Raffinose	−	+[b]	−	−	+[b]	−
Mannitol	−	+[b]	−	+	+[b]	−
Growth at 6°C	+	+		−	+	+

[a] Adapted from a table from Steinhaus et al. (1956) and additional data from Huber-Schneider (1957). For symbols see standard definitions.

[b] Gas is also produced.

Cydnid, *Chilocoris somalicus*, Carayon (1974) has described rod-shaped bacterial symbionts that thickly fill the cytoplasm of unusual syncytial mycetocytes that are irregularly inserted in the midgut epithelium. The symbionts are also found in the trophic zone of the ovarioles from which they pass directly into the cytoplasm of the oocytes at the anterior pole. Carayon (1974) has also described two different forms of unusual symbiosis in three species of *Arocatus* (Lygaeinae) while a fourth, *A. rusticus*, had no symbionts. In *A. roeselii* and *A. longiceps* the slender, flexible symbionts 8 μm long are harbored intracellularly in abdominal mycetomes. Individual mycetocytes transfer their symbionts to an infection zone of the trophic chamber from which the oocytes are infected. In contrast the symbionts of *A. continctus*, which are 1–2 μm wide and 4–10 μm long, are found exclusively in the distal end of the Malpighian tubules in the lumen and brush border of the epithelium. Although the precise manner of their transmission is unknown, it appears to resemble the transovarial transmission from Malpighian tubules found in the Apionidae (Coleoptera). A somewhat parallel symbiosis occurs in the Lygaeine *Tropidothorax* (Pierantoni, 1954). However, in this insect the symbionts are found both in the lumen and epithelium of the Malpighian tubules and midgut, and transmission of the symbionts is by egg smearing. A singular instance of rickettsia-like microorganisms occurs in the midgut epithelium of

the Mirid, *Stenotus binotatus*, but nothing is known of the life cycle or hereditary transmission (Chang and Musgrave, 1970). Ultrastructurally, these symbionts are generally rodlets 0.3–0.5 μm wide × 0.5–2.5 μm long which exhibit some pleomorphism. The rods have a Gram-negative type cell wall, are not enveloped by a host cell membrane, and are sometimes found in indentations of mitochondrial membranes.

Schneider (1940) has given the most complete descriptions of Lygaeid symbionts in abdominal mycetomes. The morphology and development of the mycetomes of *Ischnodemus sabuleti* (Blissinae) strongly resembles those of the bedbugs (Cimicoidea) and many Homoptera. The pleomorphic symbionts thickly fill the cytoplasm of the large paired syncytial mycetomes. Some are large, round, poorly staining spheroplast-like forms while the majority are 4–7 μm long, thin, lightly granulated, well staining forms which are usually present in slightly curved pairs. Both long chains and coccoidal forms (0.5 μm in diameter) are also present. Transmission occurs by early infection of the gonads; the symbionts pass through the nutrient cords to the oocytes from the infected syncytial trophic chamber. Thin symbionts 3–11 μm long occur in abdominal mycetomes of several species of *Nysius* (Lygaeinae) and *Ischnorrhynchus* (Cyminae) and are inherited transovarially much as described above for *Arocatus* (Schneider, 1940). Variable numbers of small coccoidal forms, believed to be accessory symbionts, are also present in the ovaries and mycetomes of *Nysius* and *Ischnodemus* (Schneider, 1940). *Geocoris grylloides* (Geocorinae) harbors intracellular bacteria in its fat body but they have not been studied further (Schneider, 1940).

CELLULOSE AND STORED GRAIN FEEDERS

Among the insects, the preeminent group to exploit cellulose-rich plants and plant products are the Coleoptera. The beetles comprise the largest order of insects, including 60% of the known species, over 600,000 classified in at least 128 families. Although many species are predaceous and a few even parasitic, the majority have strong mandibles suitable for crushing seeds or gnawing wood, or, as in the weevils (Curculionidae)—the largest family with over 50,000 species—the mouth parts are modified into a snout for drilling into hard plant substrata. Consequently, it not surprising that the widest range and most irregular distribution of symbiotic associations among the insects are found in the beetles (Table 11.11). Wood wasps and tipulids are included because they have great similarities in their habitats and symbioses to some Coleoptera. Such symbioses are rather unusual for Hymenoptera and Diptera, respectively. Symbiont-harboring Coleoptera are of great economic importance because of their widespread destruction of valuable crops, timber, stored foods, and even finished wood products.

Despite the economic importance of beetles, from a microbiological standpoint only sketchy details are available for most of the symbioses (Buchner, 1965). The most informative studies have been those of the

Table 11.11.
Symbiotic associations in insects feeding on diets rich in cellulose[a]

Family	Representative Genus	Common Name	Frequency in Family	Symbiont			Tissue location				Hereditary transmission	
				Yeast	Fungus	Bacterium	Gut caeca		Ecto-dermal Pouch	Abdominal Mycetocyte	External	Transo-varial
							Lumen	Epithelium				
Scolytidae	*Xyloterus*	Ambrosia beetle	+	+	+				+		+	
	Xyleborus	Ambrosia beetle	+	+	+				+		+	
	Xyleborus	Ambrosia beetle	−			+	+[b]					+
	Ips	Bark beetle	+		+				+		+	
	Dendroctonus	Bark beetle	+		+				+		+	
	Scolytus	Bark beetle	+		+	+	+[c]	+[c]	+		+	+
Platypodidae	*Platypus*	Ambrosia beetle	+		+				+		+	
Lymexylidae	*Hylecoetus*	Ship timber beetle	+		+				+		+	
Siricidae	*Sirex*	Wood wasp	+		+				+		+	
Xiphydriidae	*Xiphydria*	Wood wasp	?		+				+		+	
Scarabaeidae	*Potosia*	Scarab beetle	±			+	+	+			+	
Lucanidae	*Sinodendron*	Stag beetle	+			+	+	+			+	
Tipulidae	*Tipula*	Crane fly	+			+	+	+			+	
Anobiidae	*Anobium*	Furniture beetle	+	+			+	+			+	
	Sitodrepa	Drugstore beetle	+	+			+	+			+	
Cerambycidae	*Leptura*	Long-horned wood-boring beetle	+	+			+	+			+	
Buprestidae	*Trachys*	Metallic wood-boring beetle	+			+	+	?			+	
Lagriidae	*Lagria*	Bark beetle	+			+				+	+	
Chrysomelidae	*Cassida*	Leaf beetle	+			+	+[c]	+[c]			+	
Cantharidae	*Dasytes*	Soldier beetle	?			+	+	+			+	
Curculionidae	*Sitophilus*	Snout beetle	+			+	+[c]	+[c]		+	+	+
Silvanidae	*Oryzaephilus*	Saw-toothed grain beetle	−			+				+		+
Bostrichidae	*Rhizopertha*	Twig-boring beetle	+			+				+		+
Lyctidae	*Lyctus*	Powder post beetle	+			+				+		+
Throsidae	*Throscus*	Throscid beetle	?			+				+		?
Nosodendridae	*Nosodendron*	Wounded tree beetle	?			+				+		+

[a] Symbols for frequency: +, common; ±, variable; −, rare; ?, few observations. Symbols for rest of table: +, indicates observed type of symbiosis; ?, not certain.

[b] Caecum absent.

[c] Some species with modified Malpighian tubules.

biochemistry and taxonomy of the yeast symbionts of Anobiidae and Cerambycidae (Bismanis, 1976; Jurzitza, 1974; Koch, 1967) and the fungi of bark and ambrosia beetles (Scolytoidea), ship timber beetles (Lymexylidae), and wood wasps (Siricidae, Xiphydriidae) (Francke-Grosmann, 1967). Such studies have established essential food and nutritional roles for the symbionts. Relatively little is known, however, about the roles of bacterial symbionts in these insects even though they are clearly present in *Xyleborus* and *Scolytus* (Scolytidae). Among the Buprestidae, Chrysomelidae, Cantharidae, and some Curculionidae, bacterial symbionts are harbored intracellularly in epithelial mycetocytes of gut caeca. They are passed extracellularly via the gut lumen to various egg-smearing mechanisms which are associated with the hindgut or ovipositor, much as in the well studied fungal symbioses of the Anobiidae and Cerambycidae. *Lagria* (Lagriidae) has an interesting variation in which the symbionts are enclosed in the lumen of abdominal mycetomes, even though hereditary transmission occurs by egg contamination. Some species of Scolytidae (*Coccotrypes*), Chrysomelidae (*Bromius, Donacia*), and Curculionidae (Apioninae) harbor symbionts both intracellularly and in the lumen of unusual modified Malpighian tubules; these symbionts are inherited either transovarially (Apioninae, *Coccotrypes*) or by external passage through the gut (Chrysomelidae). Finally, the most complex symbioses, including many in the Curculionidae, involve intracellular symbionts in abdominal mycetomes. Hereditary transmission involves transovarial passage, often of "infectious" transmission forms, and complex embryonic sequestration mechanisms. The Lyctidae and Throscidae harbor two and three bacterial symbionts, respectively. Nothing is known of taxonomic relationships among these Coleopteran bacterial symbionts, nor whether they can be cultured free of their insect hosts. The nutritional role of the bacterial symbionts, at least among the few cases examined, in contrast to yeasts, appears more auxiliary than essential: aposymbiotic insects are usually fecund and develop normally but are less able to use a wide variety of diets. Further details on these complex symbiotic relationships are given by Buchner (1965).

Genus Sitophilus (grain weevils, family Curculionidae)

Several of the various destructive pests of stored grains in the genus *Sitophilus* are morphologically similar. This has resulted in a synonymy that is often confusing. *Sitophilus* is identical to the *Calandra* of older European usage. *S. granarius* is known as the large granary weevil and is found in two variants: a large, dark and highly fecund form, typical of temperate climate, which is relatively intolerant of heat and harbors midgut endosymbionts; a smaller, lighter form (var. *africanus*) which is found in the tropics and does not harbor endosymbionts. The rice weevil, *S. oryza* L., actually consists of two morphologically similar and often misidentified species, *S. oryzae* L., the small rice weevil, and *S. zeamais* Mots., the large rice weevil. *S. oryzae* L., is identical to *S. sasakii* Tak. Different strains of this species vary significantly in the number of rod-shaped endosymbionts that they harbor. Finally, *S. zeamais*, the larger rice weevil or corn weevil, harbors distinctive pleomorphic spiral endosymbionts. The interrelationship of these insects with their endosymbionts has been studied extensively.

All *Sitophilus* species examined have similar morphological adaptations for their endosymbionts. The large larval mycetome, which partially surrounds the foregut, dissociates into individual mycetocytes during pupal metamorphosis. The mycetocytes in turn penetrate the gut musculature and occupy crypts in the large caeca of the anterior forgut. The caecal endosymbionts infect adjacent areas of gut epithelium and are shed into the gut lumen, possibly to participate in digestion (Mansour, 1930). According to Schneider (1956) they disappear by 20 days post eclosion, but Musgrave et al. (1964), in their strains, did not observe significant loss of the caecal endosymbionts during adult development. However, Musgrave and Grinyer (1968) later reported that in *S. zeamais* significant destruction of the endosymbionts occurred during the larval and adult stages, concomitant with the formation of large membrane masses from the disintegrating bacteria. Similar degeneration may occur in the ovarial symbiotes of *S. oryzae* (Nardon, 1971).

The controlled destruction of symbionts may provide significant metabolic benefit to the host. However, the presence of midgut endosymbionts in the adult is not essential for their hereditary transmission, since the ovaries are infected early in embryogenesis (Mansour, 1930; Scheinert, 1933). The endosymbionts replicate actively during oogenesis and are passed along with other nutrients from the mycetocytes in the trophic chambers to the maturing oocytes. As shown by reciprocal crosses between the symbiotic and naturally asymbiotic strains of *S. granarius*, the endosymbionts are transmitted exclusively by the female (Musgrave and Miller, 1956).

Although there is little variation in the organs and cells of *Sitophilus* that harbor endosymbionts, the endosymbionts themselves in the various strains and species differ substantially (Table 11.12). Only the large endosymbionts of *S. granarius* have been characterized extensively by light and electron microscopy. They are described as cells undergoing pronounced cyclic changes in morphology from small coccoidal resting forms to predominantly large vegetative forms. "Degenerative" forms, such as spheroplasts, coiled involution forms and branched rods also occur, possibly as the result of host disruption (Bhatnagar and Musgrave, 1970) or unphysiologic conditions (Schneider, 1956).

The rod forms are Gram-negative, nonacid fast, have no flagella, and are encapsulated. Their thin cell walls stain differentially with tannic acid and crystal violet, or Congo red and hematoxylin. "Chromatin" particles visible with Giemsa following acid hydrolysis or modified Feulgen stains correspond to the pronounced nucleoid areas evident by electron microscopy, and appear to divide in parallel with binary fission (Musgrave and Singh, 1965; Singh and Musgrave, 1966; Grinyer and Musgrave, 1966). In well fixed mycetocytes, symbionts are surrounded by an electron lucent area, possibly the polysaccharide capsule, but not by host cell membranes. The bacterial cell wall is sometimes absent (Grinyer and Musgrave, 1966).

Despite the pleomorphism of the endosymbionts, DNA preparations from the midgut endosymbionts of two strains of *S. granarius* have shown a normal buoyant density distribution and a mol% G + C of 49.8 (Dasch, 1975). This value is unlike that of the fruiting (67–71%) or nonfruiting (31–42%) Myxobacterales, to which the endosymbionts have been compared because of morphological similarities (Bhatnagar and Musgrave, 1970).

An endosymbiont of shorter length is also present in low numbers in two strains of *S. granarius*. Its tissue distribution is more generalized than in the case of the large endosymbiont (Musgrave et al., 1964; Grinyer and Musgrave, 1964) but its fine structure has not been well defined.

Table 11.12.
*Differential characteristics of endosymbionts of **Sitophilus** weevils[a]*

Characteristics	Host		
	S. granarius	S. oryzae	S. zeamais
Morphology:			
Pleomorphic	+	−	+
Uniform	−	+	−
Rod shaped:	+	+	+
6 μm in length	+	+	+
15 μm in length	+	−	+
Spiral forms	−	−	+
Number present:			
Usually high	d	−	+
Usually moderate	−	d	−
Aposymbiosis at:			
31°C	+	−	−
33°C	+	+	+
Mol% G + C of DNA			
49.8	+	−	+
54.5–55.0	−	+	+

[a] For symbols see standard definitions.

The endosymbionts of *S. oryzae* and *S. zeamais* differ from those of *S. granarius* in cell size and shape. In *S. oryzae*, the endosymbionts occasionally attain lengths of 15 μm, similar to *S. granarius*; however, only rarely are they 25–30 μm long, and the majority are shorter and less pleomorphic than those of *S. granarius* (Mansour, 1930, 1935; Musgrave and Homan, 1962; Dang-Gabrani, 1971; Nardon, 1971). The outer membranes are typical of Gram-negative bacteria, but occasionally membrane vesicles are shed from the cell wall. They are surrounded by a polysaccharide or glycoprotein capsule. Endosymbionts of *S. zeamais* are markedly pleomorphic, and spiral and C-shaped forms are common. As is the case in *S. oryzae*, membrane masses are produced from large numbers of degenerating forms (Musgrave and Grinyer, 1968; Musgrave and Homan, 1962).

The mol% G + C of the DNA of the *S. oryzae* endosymbionts is 55.0 (Bd), which is significantly different from that of *S. granarius* endosymbionts. On the other hand, the DNA of the endosymbionts of *S. zeamais* bands at two densities, corresponding to mol% G + C of 49.8 and 54.5. This suggests the presence of two discrete populations of endosymbionts (Dasch, 1975). DNA/DNA hybridization relationships among *Sitophilus* endosymbionts have not been established.

Attempts at cultivation of *Sitophilus* endosymbionts in a variety of media have not been successful (Musgrave and McDermott, 1961). The positive cultures that have been obtained are probably derived from the natural exogenous microflora of the weevils. They include bacteria of the *Bacillus cereus* group, "*Micrococcus freudenreichii*" Guillebeau, and *Corynebacterium* species (Crawford et al., 1960). Dang-Gabrani (1971) claimed to have cultured the endosymbiont of *S. oryzae* and found it to resemble *Bacillus circulans*. Although it was repeatedly isolated from ovary, eggs, and mycetomes (Pant and Dang, 1972), this sporeformer is clearly different from the Gram-negative endosymbionts. Similarly, Morris (1979a) found a high rate of internal carriage in *S. zeamais* of a *Bacillus* and two *Pseudomonas* species. Although not present in the external microflora, these bacteria did not resemble the endosymbionts and *Bacillus* was usually isolated only from adults and *Pseudomonas* only from larvae or pupae.

The contribution of the endosymbionts to the metabolism of their hosts can be studied by comparing the nutritional requirements of weevils that harbor endosymbionts to tropical strains of *S. granarius* that naturally lack endosymbionts (asymbiotic) (Mansour, 1935; Musgrave and Miller, 1956; Lum and Baker, 1973) and to aposymbiotic weevils. Aposymbiosis is achieved by rearing the weevils at 31°C or higher temperatures (Schneider, 1956; Musgrave et al., 1963) (Table 11.12) or by selection of weevils with increased resistance to methyl bromide (Musgrave et al., 1965). The C-shaped symbionts of *S. zeamais* are more susceptible to elimination at 33°C than the spiral forms (Morris, 1979b). Although chlortetracycline and penicillin, but not streptomycin or bacitracin, are quite effective in eliminating endosymbionts, they are often toxic (Baker and Lum, 1973). Asymbiotic and aposymbiotic weevils are similar in appearance and nutritional requirements. Weevils without endosymbionts are generally smaller, less intensely colored, and less fecund than their symbiotic counterpart.

When maintained on adequate diets, symbiotic and asymbiotic *S. granarius* and symbiotic *S. oryzae* and *S. zeamais* differ little in total lipid, phospholipid, or fatty acid profiles (Yadava and Musgrave, 1972a and b; Yadava et al., 1972). Heat-induced aposymbiotic *S. granarius* can be maintained for 7 generations on a rich diet of wheat flour and yeast, although the weevils are smaller, develop more slowly, and are less fecund than the parent strain (Nardon, 1973). Differences are particularly apparent when the diet is less than adequate. Aposymbiotic *S. granarius* strains are less able to utilize a wide range of cereals for normal development and survival than are symbiotic strains (Schneider, 1956; Musgrave et al., 1963). The symbionts may provide an essential growth factor, which is present in sorghum, but not in other grains (Schneider, 1956). Symbiotic *S. oryzae* strains are able to utilize a wider range of dietary sterols and a number of artificial diets more effectively than the asymbiotic *S. granarius* (Baker, 1974a, b). When reared on a defined artificial diet, *S. oryzae* strains, despite the presence of endosymbionts, require a number of B vitamins. Although endosym-

bionts do contribute to the nutrition of the host, it is not certain that they contribute B vitamins; it is possible that they exert a sparing effect on the needs for riboflavin, pantothenic acid, and other compounds contained in Brewer's yeast and wheat germs (Baker, 1975; Brown and Chippendale, 1975).

S. oryzae has β-N-acetylglucosaminidase activity (Nardon et al., 1978; Wicker and Nardon, 1980; Wicker, 1980). Both symbiotic and aposymbiotic weevils have this enzyme, but the levels are higher in aposymbiotic insects and in insects not reared on wheat or sorghum. The enzyme has bacteriolytic properties and may be implicated in the control of the endosymbionts. The mechanism by which this happens is not known.

In conclusion, endosymbionts undoubtedly contribute to the nutrition of the host, but are not essential. Possibly because of their heat sensitivity, they are absent in some of the tropical strains. The natural diet of the weevils is not deficient in a readily identifiable factor. The contribution of the endosymbionts, still to be defined, appears to be of an auxiliary nature.

Oryzaephilus surinamensis L. (saw-toothed grain beetle, family Silvanidae)

Among the Cucujidae (flat bark beetles) and their near relatives, the Silvanidae (flat grain beetles), only the silvanid genus *Oryzaephilus* is known to have bacterial symbionts (Buchner, 1965). The Cucujidae are predators of mites and small insects which they find under bark. Some Silvanidae live under bark but most, including *Oryzaephilus surinamensis*, are troublesome pests of stored grains. *Oryzaephilus* also feeds readily on seeds, nuts, dried fruits and even yeast, sugar and tobacco.

In *Oryzaephilus* the bacterial endosymbionts are harbored in two pairs of abdominal mycetomes of unique structure. In the larva, the mycetome consists of 12–15 syncytial chambers which are enclosed in a large enveloping syncytial tissue. Both tissues contain single large 64- or 128-ploid nuclei as well as numerous diploid nuclei. During larval stages the endosymbionts are present as very large, curved, vacuolated, tubular forms 15–30 μm in length. During pupation, the mycetomes begin to increase in size. At this time most of the symbionts also begin to elongate to 60–70 μm, although others become branched, more vacuolate pleomorphic forms which later degenerate (Buchner, 1965). Despite their size, the endosymbionts have the appearance of bacteria rather than fungi, but their Gram-staining properties have not been investigated. Huger (1956) found that the mycetomes and endosymbionts of males begin progressive degeneration after eclosion. By 15 weeks only small mycetomes with pleomorphic degenerated symbionts can be found. In contrast, the female mycetomes continue to increase in size and the endosymbionts rapidly divide into small infectious forms 3 to 6 μm long. Following the rupture of the greatly enlarged mycetome, small packets of these infectious forms pass through the body cavity to the ovary. The symbionts then pass between the follicular epithelium into the posterior cytoplasm of the oocytes. During embryonic development the symbionts are subjected to a complex sequestration in provisional mycetocytes before they infect the sterile rudiments of the larval mycetomes (Buchner, 1965). The ultrastructure of these symbionts has not been studied. Although Pant et al. (1957) claimed that the symbionts can be cultured on a lactose nutrient broth, these cultures have not been characterized.

Despite the complexity of the symbiont life cycle and morphological adaptations of *Oryzaephilus* to ensure their hereditary transmission, the endosymbionts do not appear to be essential for the survival of their host. Degeneration of the mycetomal symbionts can be accomplished by holding the insects at 4°C for extended periods, a procedure rarely able to induce aposymbiosis, and by dietary treatments with chlortetracycline or oxytetracycline; however, only a few of the F1 generation are completely sterile (Huger, 1956). In contrast, aposymbiosis can be induced by use of elevated temperatures (Koch, 1936). At 36°C hereditary transmission of the symbionts is effectively prevented, since the small infectious forms are thermolabile. Interestingly, the large nontransmitted mycetomal forms are not as heat sensitive, since they merely become elongated. This phenomenon may have its counterpart in nature: Huger (1956) found about 45% of a population of *Oryzaephilus* from a very hot granary were free of endosymbionts. Centrifugation of eggs fails to disrupt symbiont sequestration, but long wave ultraviolet irradiation of the eggs causes significant destruction of the symbionts with little effect on egg viability and subsequent pupation (Kolya and Pant, 1962). No differences occur for at least 20–25 generations in the responses of symbiotic and heat-induced aposymbiotic *Oryzaephilus* to either an extremely unbalanced potato starch diet or to starvation diets (Koch, 1936). Although evidence has been reported that suggests that the symbionts contribute significant nicotinic acid, pyridoxin, and choline (Fraenkel and Blewett, 1943; Pant and Fraenkel, 1954), later studies have not substantiated these claims (Pant and Dang, 1972). Indeed, numerous careful dietary studies have shown that *Oryzaephilus* does not significantly differ in its dietary requirements from other Coleoptera which lack symbionts (Dadd, 1977).

Genus Xyleborus (ambrosia beetles, family Scolytidae)

Xyleborus ferrugineus, the ambrosia beetle, has long been known for its interesting cultivation of a complex of fungi, yeasts, and bacteria on which it feeds. The bacterial component of this complex, a *Staphylococcus* species, plays a role in the activation of this insect's unusual arrhenotokous parthenogenesis (Peleg and Norris, 1972a, b; 1973). In *Xyleborus* the oocytes are activated to begin parthenogenetic development while still in the follicle. Males arise from these eggs, while the activated oocytes that are subsequently fertilized become female.

Staphylococci are normally present in the lumen of the gut during the larval, pupal and adult stages of *Xyleborus* and are passed transovarially to the embryo. During this passage the bacteria are intimately associated with the nuclei of maturing oocytes. An additional rod-shaped symbiont of unknown significance has also been observed in the oocytes (Norris and Chu, 1980).

When pupae are surface-sterilized to remove the ectosymbiotic ambrosia complex and raised on an artificial diet supplemented with either chlortetracycline, streptomycin, or penicillin, or without antibiotic, the antibiotic treatment (particularly chlortetracycline and streptomycin) results in a significant reduction of oviposition and maturation of oviposited eggs to pupae and adults. Antibiotic-treated *Xyleborus* no longer have staphylococci in their ovaries nor do their oocytes undergo the normal parthenogenetic activation even though yolk deposition had occurred (Norris, 1972). Subsequent removal of the antibiotics

from the diet leads to rapid recolonization of the ovary by the few remaining symbionts, to activation of the oocytes, and to production of viable progeny (Peleg and Norris, 1972a, b; 1973).

Rhizopertha dominica (lesser grain borer, family Bostrichidae)

The branch and twig-boring beetles of the family Bostrichidae include about 600 species in 50 genera, the majority of tropical or subtropical habitat. Both larvae and imagoes generally infest either living or dead wood, sometimes even finished furniture. A few species, including the pernicious *Rhizopertha dominica* Fabricius, are cosmopolitan storage pests of cereals, tubers and nuts. Although distinctive bacterial symbionts are found in unusual paired abdominal mycetomes in each of the five genera of Bostrichidae examined, the symbioses differ somewhat in morphological details (Buchner, 1965). In *Rhizopertha*, the larval symbionts consists of loose rosettes of small coccoid elements held together by delicate threads. They fill the interior of the large oval syncytial mycetomes which are each surrounded by a thin cell layer. In the female imagoes, the symbionts undergo a transformation to larger, denser-staining, irregular coccoidal transmission elements. The transformation initially occurs in sections of the mycetomes but eventually encompasses the entire organ. The infection forms traverse the mycetome epithelium into the body cavity, pass between the follicle cells, and penetrate the oolemma to the outer surface of the mature egg. Following embryonic sequestration and reinfection of the rudiments of the larval mycetome, the symbionts reform the typical rosette stage. Rather than transforming to infectious forms, the symbionts of males undergo progressive total degeneration in pupae and adults of *Rhizopertha* and *Synoxylon*, partial degeneration in *Apate*, but no change in *Scobicia*.

Symbionts can be eliminated completely from *Rhizopertha* mycetomes by rearing the insects at 38°C and 85–95% relative humidity—conditions which this tropical beetle can easily withstand (Huger, 1956). Successive filial generations are gradually repopulated by the symbionts, which apparently arise from remnants of more heat-stable transmission forms. Although low temperatures are also deleterious to the symbionts, the beetles tolerate such temperatures poorly. Centrifugation of the eggs to displace the symbionts is also unsuccessful. Oxytetracycline and chlortetracycline are as effective as heat in destroying mycetome symbionts, but resistant forms can repopulate the mycetomes in the offspring (Huger, 1956).

INSECTS FEEDING ON COMPLEX DIETS

It is not difficult to envision the nature of the selective forces that promoted the widespread establishment of mutualistic symbionts in the blood-sucking and plant-sucking insects, and in the specialized feeders which have been described in the preceding sections. In contrast, most of the remaining insect groups appear to have adequate diets and endosymbiosis is relatively uncommon. For example, endosymbionts have not been found among any of the detritus scavengers in the primitive orders Protura (telsontails), Thysanura (bristletails, silverfish), and Collembola (springtails) of the Subclass Apterygota, or in the more advanced orders Dermaptera (earwigs), Embioptera (webspinners), Psocoptera (booklice and barklice), Zoraptera (termite-like zorapterans), or Mecoptera (scorpionflies).

Among the detritus scavengers, the omnivorous cockroaches (Suborder Blattaria, Dictyoptera) are a major exception. The presence of comparable bacterial endosymbionts in the fat body of all cockroaches suggests that they were acquired from a common ancestor whose diet was more restricted. Indeed, both the wood-eating primitive cockroaches, *Cryptocercus* and *Parcoblatta*, and the primitive Australian termite, *Mastotermes darwiniensis* Froggat, possess two sets of symbionts: the usual cockroach fat body bacterial symbionts and the

complex hindgut symbiotic flagellated protozoans typical of many termites. Since termites (Isoptera) and Blattaria probably evolved from a common ancestor by the early Carboniferous period, the symbiotic relationships in both groups are probably at least 300 million years old. With their adaptation to diets more generalized than wood, most cockroaches have lost their flagellates and associated capacity for utilization of wood, while retaining their highly integrated bacterial symbionts. Conversely, while retaining their essential flagellates, most advanced wood-feeding termites have lost their bacterial symbionts. Furthermore, the most specialized termites, which maintain fungus gardens as their food source, have even lost their flagellates. The forces promoting the original establishment of the cockroach endosymbiosis, of course, are uncertain and symbionts may not be strictly necessary for cockroach survival today. It is clear, however, that they can provide a distinct advantage to their hosts in competition with aposymbiotic arthropods in the more efficient utilization of marginal diets.

As in the case of most detritus-eating insects, none of the insects with aquatic nymphs that have either phytophagous, scavenging, or predaceous habits are known to harbor symbionts. These include the orders Ephemeroptera (mayflies), Odonata (dragonflies, damselflies),

Plecoptera (stoneflies), Neuroptera (dobsonflies, alderflies) and Trichoptera (caddisflies). Symbiosis is also unusual among the endoparasitic Strepsiptera (stylopids) and Hymenoptera (ichneumons, braconids, chalcids), or the predaceous members of the orders Thysanoptera (thrips), Neuroptera (antlions, snake flies), Mecoptera (bittacids), and Hymenoptera (vespid and spider wasps). The exceptions include the interesting symbioses of the green lacewings, *Chrysopa carnea* Stephens (Chrysopidae, Neuroptera), which have extracellular yeasts and those of Formicine ants (Formicidae, Hymenoptera) which have intracellular bacteria. Although lacewing larvae are predaceous, the adults feed primarily on homopterous honeydews and require the yeast for high fecundity and fertility (Hagen and Tassan, 1972). Similarly, although ants are generally omnivorous and often predaceous, many Formicine ants resemble the Chrysopidae in their exploitation of homopterous honeydews as a mainstay of their diet. Indeed, the acquisition of symbionts may have played a significant role in permitting the remarkably competitive Formicine ants to extend into new areas and exploit limited diets. Most of the presently existing *Formica* species which are primarily predaceous, omnivorous, or slavemakers of ants with symbionts, have lost the characteristic *Formica* endosymbionts.

Finally, a number of other insects which also feed on apparently adequate diets have extracellular bacteria which appear to play critical roles in the utilization of these diets. The bacterial wilt pathogens of corn, *Erwinia stewartii*, and cucurbits (melons, cucumbers, squash), *Erwinia tracheiphila*, are transmitted, respectively, by the corn flea beetle, *Chaetocnema pulicaria* Melsheimer, and the cucumber beetles, *Acalymma vittata* Fabr. and *Diabrotica undecimpunctata howardi* Baker (Steinhaus, 1946). These pathogens overwinter in the alimentary tracts of the adult beetles and are transmitted to the plants by fecal contamination of the wounds produced on the leaves by chewing. The pathogens multiply in the vascular system of these plants and cause wilting, yellowing, reduced produce yield and even death. The larvae of these Chrysomelid beetles feed also on the roots and underground stems of the host plants, but a role for the pathogens in larval nutrition has not been described.

The evidence for essential participation of plant pathogens in insect food utilization is much stronger for many Anthomyiidae (genus *Hylemya*) and Tephritidae (Diptera). In these flies the pathogens are harbored extracellularly in all life stages and are transmitted by fecal contamination of the eggs or habitat or by specialized egg smearing mechanisms. Infected larvae inoculate the pathogens by repeatedly cutting the plant tissues, thus permitting the rapid decomposition by the pathogens. Sterile larvae from surface disinfected eggs are unable to grow on their usual food sources. However, unlike many flies, which feed on bacteria that are growing on decaying organic material, or *Drosophila* which use fruit yeasts, the *Hylemya* and tephritid larvae feed on the decomposed vegetable tissue rather than the bacteria themselves. Finally, the adult tephritids resemble the Chrysopidae and perhaps the formicid ants, in feeding on homopterous honeydews, and in requiring supplementation of that inadequate diet by their microflora for full egg development and fecundity (Hagen and Tassan, 1972).

The descriptions below are limited to the major endosymbiotic relationships.

Subfamily Formicinae (ants)

Among the first clear descriptions of endosymbiosis in insects were those of Blochmann (1884, 1886) who found very large intracellular bacteria in the ovaries of *Camponotus ligniperda* and *Formica fusca*. Buchner and his associates (1965) and, more recently, Jungen (1968) and Dasch (1975) have added considerably to our understanding of endosymbiosis in ants. Endosymbionts have been described only in the subfamily Formicinae (family Formicidae, order Hymenoptera), but this subfamily includes the well known large carpenter ants in the tribe Camponotini, the diverse field and wood ants in the complex genus *Formica*, and the more distantly related genus *Plagiolepis*.

All the 20 species in eight genera of Camponotini examined, of diverse geography, habitat, and morphology, have numerous large Gram-negative rods in the midgut epithelium and ovaries of workers and queens. Kolb (1959) demonstrated that the morphology of the symbionts of *Camponotus ligniperda* depends on the stage of ant embryological development or metamorphosis. Among the morphologic types there are characteristic spheroplast-like forms. In the adult midgut the endosymbionts are straight rods 1 μm wide × 5–15 μm long that form paracrystalline arrays. Each rod has a very large subapical intracellular granule. The peculiar twig-dwelling species, *Colobopsis mississippiensis*, has atypical midgut endosymbionts since their morphology resembles those of the *Formica* mycetomal endosymbionts described below (Dasch, 1975).

In *Formica* and *Plagiolepis* the endosymbionts are not in the midgut, but rather in symmetrical unicellular layers of mycetocytes on either side of the midgut epithelium (Buchner, 1965). Although they are Gram-negative and have apical granules and spheroplast forms like those of Camponotini, endosymbionts are significantly shorter (3–4 μm) and distinctly crescent-shaped. The presence of endosymbionts in *Formica* is erratic: some species always have them, others never and, in some species, their presence varies with the colony and even with the individual. Some of this variability may be due to the competitive status of the colony, since endosymbionts appear to be destroyed during periods of starvation (Jungen, 1968; Dasch, 1975). Finally, some species, which have never been found to harbor symbionts, have unusual embryonic cells identical to those specialized cells that harbor symbionts in closely related ants. These embryonic cell relicts suggest that these ants, also, once harbored symbionts (Buchner, 1965).

Despite basic differences in the symbiosis of Camponotini and *Formica*, their endosymbionts may have had a common origin since (a) their distribution is restricted to closely related ants; (b) their morphology is similar; (c) in both groups the endosymbionts are transmitted to the ovaries by massive invasion and proliferation; and (d) at least one genus, *Colobopsis*, has intermediate properties. Species similar to *Formica* and *Camponotus* have been discovered in Baltic amber dated to be 35–40 million years old, so it would appear that the Formicine endosymbiosis, if it is indeed derived from a common ancestor, predates that age.

The ultrastructure of the endosymbionts of *Formica glacialis* and *Camponotus pennsylvanicus* has been examined by Dasch (1975). Both types of endosymbionts have Gram-negative type cell walls and are not enclosed by host vacuolar membranes. In *C. pennsylvanicus* the apical granules as well as the numerous smaller granules dispersed throughout the cytoplasm consist of volutin. This conclusion is based on metachromatic staining, electron density, and chemical identification of polyphosphate. The cytoplasm contains dispersed DNA fibrils, ribosomes, tubular mesosomes, but no nucleoids. Cell wall septa are particularly obvious in division forms of the *Camponotus* endosymbionts. Somewhat greater variation in morphology of the endosymbionts is seen in the large autophagic cells in the upper ovarioles of pupae during invasion of the oocytes. The cells contain typical rods, spheroplasts with a central volutin granule, and stages intermediate between the two forms, as well as numerous electron dense, multilamellar bodies believed to be lysosomal rest bodies. The number of spheroplasts in the midgut of workers can be increased up to 90% and the number of rod forms correspondingly reduced by feeding the ants chlortetracycline or more effectively by injecting lyzozyme.

The DNA base composition of endosymbionts of six species in three genera of Camponotini and two species of *Formica* has been determined by Dasch (1975). Partially purified symbiont preparations, separated mainly from host cell nuclei and mitochondria, can be obtained by differential centrifugation of homogenates of tissues containing numerous endosymbionts. DNA is then purified from both the endosymbionts and from symbiont-free host tissues and characterized by Bd and T_m. When the ants are small or their number limited, the symbionts and host tissues are treated with lysozyme, detergent, and Pronase, and the extracts subjected to Bd analysis only. The two procedures yield results that are entirely comparable. The DNA of Camponotini endosymbionts varies from 29.7–31.6 mol% G + C, whereas the host

DNA varies from 34.2–38.6%. In contrast, the mol% G + C of the DNA of *Formica* endosymbionts is 40.8, whereas the host DNA, as well as the DNA of symbiont-free *Formica* species, has multiple satellites besides the main band of DNA which has a value of 36.1.

The difference of about 10% in DNA base ratio between the Camponotini and *Formica* endosymbionts does not in itself preclude a distant common origin and does not contradict the morphological observations.

Lamparter (1967) and Steiger (1968) have described a small *Wolbachia*-like endosymbiont in the wood ant, *Formica lugubris*. It is a small-Gram negative rod, 0.8 μm in diameter, which is enveloped by a host-derived membrane, and is unrelated to the large endosymbiont since it is found in a variety of tissues including the central nervous system, muscles, and fat body. It is transmitted transovarially.

Suborder Blattaria (cockroaches).

The following genus of endosymbionts has been formally named and had been listed under the *The Rickettsias and Chlamydias* in the order *Rickettsiales*, family *Rickettsiaceae*, tribe *Wolbachieae*, in the eighth edition of *Bergey's Manual*. It seems more appropriate, however, to describe this genus here in the context of other endosymbionts.

Genus **Blattabacterium** Hollande and Favre 1931, 754^AL

Blat.ta.bac.te′ri.um. L. fem. n. *Blatta* generic name cockroach; M.L. neut. n. *Blattabacterium* a bacterium found in cockroaches.

Plump, slightly curved and straight rods, about 1 μm in diameter and 1.6–9 μm length, usually with rounded ends. Length varies somewhat with cockroach species. Within given cockroach species, longer forms occur intracellularly in specialized cells of abdominal fat body called **mycetocytes,** shorter forms in gonads and embryos. **Binary fission** pairs common in mycetocytes. **Enveloped by host cell membranes.** Gram-variable to Gram-positive staining but has a Gram-negative type cell wall structure. **Not motile or flagellated.** Possesses enzymes of the tricarboxylic acid cycle and respiratory cytochromes. Cultivation in vitro or with tissue culture cells has not been demonstrated. Believed to contribute nutritional factors to the host. Maintenance in host insect inhibited by treatment with penicillin, streptomycin, chloramphenicol, chlortetracycline, oxytetracycline, sulfathiazole, egg white lysozyme, incubation at 37°C, and lindane, but not aerosporin. Present only as **intracellular symbionts in all species of cockroaches** (Blattaria, Dictyoptera). The mol% G + C of the DNA is 26–28 (Bd).

Type species: *Blattabacterium cuenoti* (Mercier 1906) Hollande and Favre 1931, 754.

Further Descriptive Information

Cell structure and life cycle. The light microscopic and ultrastructural appearances of cockroach endosymbionts vary somewhat among different cockroach species, in different tissues of a single species, or in response to the recurrent hormonal changes of the host's incomplete metamorphosis. Primary differences among species of cockroaches are those of symbiont length. In *Blatta orientalis* they are 2.5–5.3 μm long; in *Cryptocercus punctulatus*, 2.5–8.1 μm; in *Heterogamia*, 5.3–9.0 μm; in *Blatta aethiopica*, 1.6 μm; and in *Blatta germanica*, 3.0 μm (Buchner, 1965). The longer forms, which are usually slightly curved, occur in the mycetocytes. The shorter forms occur in the ovaries and during embryonic development. Very small dense forms 0.3 × 1.0 μm are occasionally found mixed with the usual larger endosymbionts in the fat body mycetocytes of *Byrsotria*, *Blaberus*, *Nauphoeta*, and *Blattella*, but these have not been further characterized (Milburn, 1966; Meyer and Frank, 1960). The large endosymbionts may stain Gram-positive or Gram-variable and appear speckled or barred; the differences are possibly due to their physiologic state. Unlike most Gram-positive bacteria, dilute sodium dodecyl sulfate abolishes the Gram-positive staining property of the symbionts (Malke and Bartsch, 1966). Cytoplasmic vacuoles and cross-walls are evident in freshly isolated symbionts. The Feulgen reaction of the symbionts is positive but diffuse (Rizki, 1954). The symbionts are not acid fast, are not encapsulated, have no flagella,

but stain well with Giemsa and Delafield or Heidenhain's hematoxylin. They do not stain with Sudan dyes, with the Baker acid hematin test for phospholipids, with periodic acid Schiff reagent for polysaccharides, nor are they fluorescent or birefringent (Richards and Brooks, 1958). Poorly fixed specimens or symbionts isolated in hypotonic solutions may appear swollen and yeastlike. Nucleoid areas with associated metachromatic granules that divide simultaneously with transverse fission have been observed (Buchner, 1965). Ultrastructurally, these appear to correspond to the electron-dense bodies and complex mesosome-like structures that are often associated with division septa (Daniel and Brooks, 1972; Bush and Chapman, 1961; Gromov and Mamkaeva, 1980; Brooks, 1970; Anderson, 1964).

Other than size, differences are negligible in the ultrastructure of the symbionts found in the larval fat body mycetocytes of the genera *Periplaneta*, *Cryptocercus*, *Blatta*, *Byrsotria*, *Blaberus*, and *Nauphoeta*, these genera being representative of five very different families of Blattaria (Brooks, 1970). This is in contrast to the location of the mycetocytes themselves, which may be isolated (*Cryptocercus*, and in the termite, *Mastotermes*) or loosely scattered in the fat body (*Pycnoscelus*, *Blatta aethiopica*), or banded in a single (*Blatta orientalis*, *Rhicnoda*), double (*Nauphoeta*), or triple row (*Blattella*), or cluster (*Ectobia*), each surrounded by a monolayer of fat body cells (Koch, 1967; Buchner, 1965). Compared to the host cytoplasm, the symbiont cytoplasm is electron dense, although it is otherwise unremarkable in its content of ribosomes, loose DNA fibrils, and mesosome-like membrane invaginations. Each fat body mycetocyte symbiont is enclosed by at least one tightly adhering host cell-derived membrane. The symbiont cell wall is usually thin (5–10 nm) for a Gram-positive bacterium and, indeed, has an ultrastructure characteristic of Gram-negative bacteria (Anderson, 1964; Brooks, 1970; Daniel and Brooks, 1972). The additional host cell membrane might be responsible for the Gram-positive staining properties. The encapsulating host membranes are lost from the symbionts only during the extracellular phases of their transfer from the fat body mycetocytes to the oocyte cytoplasm to insure their hereditary transmission (Anderson, 1964; Daniel, 1973). At this time the symbionts are completely enveloped by the highly microvillous surface of the oocyte oolemma until egg maturation is complete. A similar symbiont transfer occurs in the rudimentary ovaries of males, but some tissue tropism is involved since the testes are not infected (Brooks and Kurtti, 1972). During embryonic development complex extracellular and transmembrane movements of the symbionts occur between the time of their initial penetration of the oolemma and ultimate sequestration in the larval mycetocytes (Buchner, 1965).

By the freeze-fracture technique, alterations occur in the disposition of membrane particles on the convex surface of the plasma membrane of the symbionts present at the oocyte-follicle interface, following changes in the levels of host juvenile hormone (Liu, 1973, 1974). An apparent hormonally related transformation of the usual rod-shaped symbionts into large rounded forms filled with concentric lamellar membranes has also been described during the later instar ecdyses and oocyte development of false ovoviviparous cockroaches (Milburn, 1966).

Muramic acid and glucosamine have been identified in the cell wall of the symbionts of *Periplaneta americana*, indicating the occurrence of peptidoglycan (Daniel and Brooks, 1967). This is consistent with the finding of naturally occurring symbiont spheroplasts (Milburn, 1966) and formation of spheroplasts by injection of lysozyme into the host (Daniel and Brooks, 1972). However, the precise mechanism of lysozyme spheroplast-forming ability is not clear, since denatured lysozyme is equally effective.

Cultivation. Numerous reports of successful cultivation of the symbionts of cockroaches have appeared as well as studies on the biochemistry of these cultures (Brooks, 1970). Gier (1947) and Brooks and Richards (1966) have carefully reviewed the problems inherent in preventing contamination with the diverse normal cockroach flora and in fulfilling Koch's postulates for the cultivation of symbionts. For none of the microorganisms so far cultivated is there a convincing relationship to the symbionts, as measured by definitive criteria such as common antigens, DNA base composition, or common distinctive metabolic properties or enzymes. Limited extracellular culture of symbionts in the presence of embryo cell cultures has been reported (Landureau, 1966) but not intracellular growth of symbionts in either vertebrate or invertebrate tissue culture systems. Both of these promising approaches may be frustrated by the ubiquitous production of lysozyme-like enzymes (chitinases) by many insect cell cultures (Bernier et al., 1974; Brooks, 1975). These enzymes may adversely affect the symbiont cell wall. Similarly, the release of such enzymes from damaged cells and mycetocytes during the isolation of symbionts might account for their apparent failure to grow on bacteriologic media. To circumvent some of these problems, Kurtti and Brooks (1976) have developed a procedure for obtaining individual cockroach mycetocytes, a step preparatory to attempting to culture them in vitro. However, since mycetocytes normally divide only in bursts during the latter half of each host instar (Brooks and Richards, 1955), continuous mycetocyte replication may require precise hormonal conditions.

Nutrition and metabolism. Definitive characterization of the physiological properties of the endosymbionts of cockroaches and their precise role in the metabolism of the host, despite considerable effort, remain quite elusive (Buchner, 1965; Brooks, 1970; Brooks and Kringen, 1972). Numerous methods have been utilized to free cockroaches of their symbionts. Perhaps the most satisfactory approach has been to interfere with the maternal hereditary transmission of the symbionts. Symbionts are not transmitted to the oocytes when the parent generation is raised on diets containing chlortetracycline, rancid linoleic acid, high concentrations of urea, or diets which are deficient in manganese or zinc or high in calcium (which antagonizes the utilization of the manganese and zinc) (Brooks, 1963). Although these diets significantly reduce cockroach fecundity, aposymbiotic nymphs may be obtained which survive and reproduce on diets containing yeast or liver. Some symbionts are still present since they gradually repopulate the cockroaches after several generations, even when the diets contain chlortetracycline to which the symbionts have become resistant (Brooks, 1963). Aposymbiotic cockroaches are smaller, lighter colored, slower in maturing and less fecund than those with symbionts. Brooks and Kringen (1972) excluded B vitamins, sterol, nucleic acid, choline, carnitine, or either sulfur-containing or aromatic amino acids as the growth factor in yeast. A peptide or growth factor which is also contained in β-lactoglobulin, protease peptone (Difco) or hydrolyzed lactalbumin appears to be the essential factor. In addition, a second factor present in water-soluble yeast extract is required for full egg viability with aposymbiotic cockroaches. A decline in the levels of cobalamine, but not riboflavin or pantothenic acid, occurs following induction of aposymbiosis by lysozyme injection (Malke and Schwartz, 1966). Cockroach vitellogenin and an additional protein are lost coincident with the loss of symbionts following pencillin injection (Garthe and Elliott, 1971). The loss of yolk protein in the female is accompanied by a striking loss of egg size and viability. However, these changes might be secondary toxic side effects of the repeated antibiotic injection, much as has been noted with repeated lysozyme treatment (Wharton and Lola, 1969).

Because claims for the culture of the cockroach symbionts have not been substantiated, reports of the presence of guanase, uricase, urease, xanthine oxidase, malic and lactic dehydrogenases, and synthesis of ascorbic acid based on such cultures are questionable. However, direct histochemical studies on symbionts present in the mycetocytes and respirometric studies on isolated symbionts indicate that the symbionts do reduce tetrazolium dyes, succinate, and cytochrome *c* while they consume oxygen. These results suggest that they have a complete set of respiratory cytochromes and an aerobic metabolism. The symbionts also transaminate alanine to pyruvate and glutamate to aspartate (Brooks, 1970). In many other studies fractions enriched in symbionts have been compared with similarly prepared fractions from cockroaches which had been freed of their symbionts. Although the greater glycolytic and tricarboxylic acid cycle activities of such enriched symbiont fractions have been attributed to the symbionts (Laudani et al., 1974), the symbionts might merely eliminate an inhibitor which depresses such host activities. Lysozyme-induced aposymbiosis results in up to 20-fold elevation of uric acid levels in the fat bodies. (Malke and Schwartz, 1966). Since only the tissues containing symbionts exhibit significant respiration in the presence of added uric acid, the symbionts might play an important role in the uric acid catabolism of the cockroach.

Symbiotic insects, even when deprived of intestinal bacteria, but not aposymbiotic cockroaches (derived from chlortetracycline-treated adults), are able to synthesize labeled cysteine, glutathione, methionine, and taurine from ^{35}S-sulfate and to transfer labeled sulfur from cysteine to methionine (Block and Henry, 1961). Fur-

thermore, only symbiotic insects can label the essential amino acids tyrosine, phenylalanine, isoleucine, valine, threonine, and arginine to a significant level from [U-14C]glucose; the nonessential amino acids are labeled as well (Henry, 1962). The lighter melanization of aposymbiotic cockroaches may in part be due to their inability to synthesize tyrosine. Similarly, significant aromatic ring cleavage of phenylalanine occurs in symbiotic but not in aposymbiotic cockroaches, although this is not a major metabolic pathway even in the symbiotic cockroaches (Murdock et al., 1970).

The mol% G + C of the DNA (Bd) has been determined for symbionts and their hosts from nine species of cockroaches in six different subfamilies. Symbiont DNA base compositions vary from 25.8–28.3 mol% G + C except for 21.8% obtained in a single determination on the symbionts of *Supella longipalpa*. Although no phylogenetic trends are obvious among the various species, the DNA compositions of the symbionts of two *Parcoblatta* species and two *Blaberus* species are identical (27.3 and 26.5%, respectively). In contrast, the cockroach DNAs differ greatly in the number (one to three), amount, and base compositions of the satellite DNA peaks, whereas the main band of DNA only varies from 33.5–36.3 (Dasch, 1975).

Antibiotic sensitivity. No significant differences have been reported in the antibiotic sensitivities of symbionts of different cockroach species, subjected to either prolonged feeding or repeated injections. Penicillin, oxytetracycline, chlortetracycline, streptomycin, sigmamycin, penetracin, chloramphenicol, and sulfathiazole, but not aerosporin, are effective (Brooks, 1970). Although significantly reduced, the symbiont populations are rarely completely eliminated. Aposymbiosis is most readily obtained by interruption of the hereditary transmission of the symbionts to the ovary. Resistant symbiont strains have arisen with continued antibiotic treatment.

Pathogenicity. The symbionts are not known to be infectious or pathogenic for insects, other invertebrates, or vertebrates. Indeed, reinfection of aposymbiotic cockroaches with isolated symbionts, to fulfill Koch's postulates, has not been accomplished (Brooks and Richards, 1966).

Taxonomic Comments

In the eighth edition of *Bergey's Manual*, this genus was placed in the Tribe *Wolbachieae*, Family *Rickettsiaceae*, Order *Rickettsiales*. The information on this genus that has since accrued does not contribute to reclassification. However, it is reasonable to examine this genus in the context of other endosymbionts, as is done in this subsection.

With regard to classification within the genus, the basic similarity in the DNA base composition, ultrastructure, mechanism of transmission to subsequent insect generations, and antibiotic sensitivities of the symbionts from different cockroaches, suggests that there is no strong basis for further speciation. Additional studies of the apparently discrepant *Supella longipalpa* symbiont (mol% G + C is 21.8) or the distant symbiont of the termite, *Mastotermes darwiniensis*, particularly by DNA/DNA hybridization procedures with the symbionts of the other species, may add to the understanding of the 300-million-year evolution of this genus and form the basis for further speciation.

Further Reading

Brooks, M.A. 1970. Comments on the classification of intracellular symbiotes of cockroaches and a description of the species. J. Invert. Pathol. *16:* 249–258.

List of the species of the genus **Blattabacterium**

1. **Blattabacterium cuenoti** (Mercier 1906) Hollande and Favre 1931, 754.[AL] (*Bacillus cuenoti* Mercier 1906, 684).

cu.en.ot'i. L. gen. n. *cuenoti* of Cuenot; named after L. Cuenot, who studied intracellular inclusions in orthopteran insects.

Description as for genus. Originally studied in *Blatta orientalis*, the oriental cockroach.

The mol% G + C of the DNA is 27.4 (Bd).

Type strain: none isolated.

Genus Hylemya (vegetable maggots)

The larvae of the diverse flies in the family Anthomyiidae use a wide variety of foods including dung, living insects, and decaying animal and plant matter. The existence of bacterial symbionts is only known among the most economically important members of the family, particulary the root and stem pests in the genus *Hylemya*, *H. brassicae* Bouche (cabbage maggot), *H. cilicrura* Rondani (seed-corn maggot), *H. trichodactyla* Rondani (seed-potato maggot), *H. antiqua* Meig. (onion maggot), and the celery leaf miners, *Scaptomyza graminum* Fall. and *Elachiptera costata* Leow. These flies have essentially similar symbiotic relationships with the important plant pathogen, *Erwinia carotovora* (Leach, 1930; Bonde, 1939a), for which they serve as important vectors of disease: potato blackleg, heartrot of celery, and the bacterial soft rots of Cruciferae (carrot, cabbage, kohlrabi, turnip, mustard, radish) and onions (Leach, 1927, 1930; Bonde, 1939b). Indeed, each *Hylemya* species may attack and transmit the pathogen to more than one plant species. Besides *Erwinia carotovora*, which may or may not be present, the internal flora of *H. cilicrura* contains a complex of saprophytic nonpathogenic species, notably *Pseudomonas fluorescens* and "*P. non-liquefaciens*" (Leach, 1931). Johnson (1930) routinely isolated fluorescent bacteria from *H. brassicae*, including *P. fluorescens* and an unidentified species, as well as *Xanthomonas campestris* and various cellulose-decomposing bacteria. Although the transmission and frequency of *E. carotovora* carriage is well known, scant attention has been paid to the role of the more regular saprophytic members and little is known of their importance to the host. In many studies, no differentiation has been made between pathogen and saprophytic species.

The continuous association of bacteria with *H. cilicrura* and *H. brassicae* has been shown in a series of studies by Leach (1926, 1931, 1933) and Johnson (1930) and is presumed to be similar in the other hosts. Eggs are deposited in the soil near the tuber or cabbage stem. The eggshells have a sticky surface fluid which is heavily contaminated with the intestinal flora of the ovipositing female. As the larvae hatch they become contaminated with these bacteria externally and internally, and these are, in turn, transmitted to the plant host. It is uncertain how successfully larvae can initiate plant decomposition in healthy hosts with this endogenous flora. Possibly, oviposition occurs primarily near plants previously damaged by exogenous soil flora that serve as chemotactic attractants to the larvae (Bonde, 1939b). However, it is certain that larval growth does not occur readily in the absence of pathogens and saprophytes. Indeed, most plant injuries are limited by the rapid formation of dense cork walls. The larvae serve to disseminate the pathogens and cause plant decomposition by repeated laceration and inoculation of the plant tissue. Survival of the pathogens over winter appears to be aided by carriage in the lumen of the midintestine, in the castout linings of the larval gut, and in the shed larval cuticle of the sclerotized puparium. Selective growth and retention of limited bacterial species occurs in both pupal and adult stages (Leach, 1931, 1933). Bacterial carriage is strictly extracellular in the intestinal lumen in all stages or superficially on the larval and adult cuticles.

Sterile *H. cilicrura* larvae from surface sterilized eggs cannot grow on sterile beef extract agar or potato plugs, whereas the development of larvae contaminated with the symbiont complex proceeds normally on either medium (Leach, 1926). Viable bacteria, per se, killed bacteria, or bacterial filtrates are not required for larval development; rather, these bacteria release a growth factor by the decomposition of potatoes (Huff, 1928). A similar growth factor occurs in actively growing bean or pea seedlings. *E. carotovora* alone is unable to replace the symbiotic complex in producing this factor on potato plugs (Bonde, 1939b). Sterile onion maggots, *H. antiqua*, have been grown on a defined medium containing 19 amino acids and 11 vitamins, nucleosides, cholesterol, glucose, and agar. However, a variety of bacteria (*E. coli*, *Bacillus*) effectively accelerate larval development on this diet or diets deficient in some of the growth factors (Friend et al., 1979).

Family Tephritidae (fruitflies, gallflies, peacock flies).

Tephritids (Trypetidae of older European usage) comprise a diverse group of flies in three subfamilies of disparate morphology and habitats. The Dacinae include agricultural pests of major economic importance. The larvae feed on the flesh of tropical and subtropical fruits and some vegetables (genus *Dacus*). The Tephritinae commonly parasitize the

flower head of composites, devouring the immature seeds and receptacles, and sometimes forming galls. Finally, the Trypetinae include a wide array of distantly related species which include leaf miners of composites and umbellifers, stem borers, some of which are gall formers, and the very important pests of fleshy fruits in the genera *Rhagoletis* (cherries, apples, prunes), *Ceratitis* (citrus fruits), and *Anastrepha* (grapes, peaches, guavas).

Over 40 species of Tephritidae, including eight of the palearctic tribes, clearly have symbiotic relationships with microorganisms at some stage in their life cycle (Stammer, 1929; Hellmuth, 1956). The symbiosis of Dacinae has been studied most extensively. The most elaborate, morphologically advanced Tephritid symbiosis occurs in the olive fruit fly, *Dacus oleae* Gmelin (Petri, 1904–1910, see Buchner, 1965). In the larvae of this species the symbionts are harbored extracellularly in the lumen of four spacious round caeca located at the anterior midgut. In imagoes, the *Dacus* symbionts are found in the lumen of the cephalic organ, a deep saclike evagination of the esophagus, and in 20 finger-like evaginations of the hindgut. The hindgut evaginations are juxtaposed to the vagina, so that each egg is effectively smeared with symbionts as it is oviposited. The micropyle of the egg contains large spaces in which the transmitted symbionts multiply and from which they subsequently infect the larval gut rudiments as soon as they form in embryogenesis. During pupation, although most gut symbionts are expelled, the few remaining infect the cephalic organ and subsequently the rectal organs. *D. dorsalis* Hendel (oriental fruit fly) and *D. cucurbitae* (melon fly), lack such complex symbiotic organs, although they do harbor symbionts and transmit them by fecal contamination during oviposition (Bateman, 1972).

Among the Tephritinae, morphological adaptations for maintaining symbionts are common and in the tribes Tephritini and Schistopterini even approach those of *Dacus oleae*. In *Tephritis* species, the larval caeca resemble those of *Dacus* and, although no cephalic organ has been described in the adult, a differentiated midgut crypt zone is inhabited by symbionts. In other genera, the adult crypt zone may surpass that of the larvae in complexity (*Sphenella*) or be poorly defined (*Trypanea*, *Paroxyna*, *Euarestella*). Among the remaining Tephritini (tribes Terellini, Xyphosiini, and Ditrichini) as well as the Trypetidae, both larval and adult symbiotic organs are usually lacking, although a cephalic organ has been described for *Rhagoletis pomonella* Walsh. The symbiotic bacteria merely lie freely in the intestines, and although few species have even simple rectal folds and modified micropyles, fecal transmission of the symbionts is quite effective.

Although a variety of yeasts, fungi, and Gram-positive bacteria have been isolated from Tephritidae, the majority are considered random contaminants and not the symbiont species. In *Dacus oleae*, "*Agrobacterium luteum*" may be commonly acquired from the surface of olive trees and increase in numbers in the latter larval stages. However, no provision is made for its hereditary transmission. *Pseudomonas syringae* pathovar *savastanoi* has generally been regarded as the symbiont of *D. oleae* (Hellmuth, 1956). This agent causes olive knot, a gall disease, even in the absence of *D. oleae* (Hagen, 1966).

Some doubt about *P. syringae* pv. *savastanoi*'s role as the symbiont of *D. oleae* was cast by Yamvrias et al. (1970) who could not isolate it from adult flies in Greece. Many of their bacterial isolates were not even *Pseudomonas*. Further, Poinar et al. (1975) described an unusual structure for the symbionts of the adult cephalic organ. The organ lumen was bounded by a cuticle layer 0.6–1.3 μm thick which separated the symbionts from the epithelial wall. The symbionts were of uniform rod morphology, 0.6–0.9 μm by 3–7 μm, with few division forms and no spores. They were covered by numerous unusual filamentous (10–29 nm diameter by up to 2.6 μm, or catenulate appendages, which differ from fimbriae, flagella or prosthecae) as described for other bacteria.

The symbionts of *D. oleae* may have an essential role for their hosts (Hagen, 1966). Adults fed diets containing streptomycin lay sterile eggs from which larvae can be obtained. The larvae are unable to mature on olives even though they develop normally on an artificial diet which contained protein hydrolysate, but not on a diet which contains intact protein. Normal aposymbiotic larval development does ensue in olives stored at 4°C (Fytizas and Tzanakakis, 1966), further suggesting that the symbionts are essential in breaking down the olive tissues for larval nutrition, possibly by proteolysis. The presence of protein hydrolysates in foods of adult Tephritids (Hagen and Tassan, 1972) is also critical for full fecundity of laboratory reared insects, since their symbionts are destroyed by the acidic artificial larval media that are used (Hagen, 1966). Treatment of adults with tetracyclines or viomycin is completely effective, polymyxin and neomycin are partially effective, and kanamycin, chloramphenicol, erythromycin, and nystatin are ineffective in preventing larval development (Fytizas, 1970). The time-dependent changes in effects of polymyxin and neomycin have been ascribed to altered composition of the symbiont population (Fytizas, 1970). Oxytetracycline or sulfanilamide are also effective in eliminating the symbionts of *D. cucurbitae* Coq. (Chinnarajan et al., 1972). Although metamorphosis of treated insects occurs, all developmental stages have reduced size and nutritional status.

Among the other Tephritidae, only the symbiont of *Rhagoletis pomonella* Walsh, the apple maggot, has been studied extensively. As in the *Hylemya* symbiosis, the symbiont "*Pseudomonas melophthora*" (Allen and Riker, 1932), causes an appple rot that accompanies the larval infestation. The symbionts are carried both externally and in the alimentary tract of all stages. Larvae derived from surface disinfected eggs neither cause apple rot nor develop to maturity. A strain and species-specific antiserum that was raised against a "*P. melophthora*" isolate, reacted with the symbionts of only 40% of adults and 10% of the larvae, although additional positive reactions could be obtained against symbionts cultured from the organs (Baerwald and Boush, 1968). It is not known whether this low reactivity was due to the specificity of the serum, antigenic differences that may occur in the symbionts in vivo and in vitro, or to the presence of unidentified bacterial symbionts in *Rhagoletis*. However, significant numbers of "*P.*

melophthora" were found in the hemocoele and between muscle bundles. "*P. melophthora*" can synthesize cysteine and methionine which are not present in soluble apple extracts and this may indicate a role for the symbiont in the amino acid nutrition of *Rhagoletis* (Miyazaki et al., 1968). The symbiont significantly degrades organophosphate insecticides and attacks chlorinated hydrocarbons and carbamates to a limited extent (Boush and Matsumura, 1967). An actual role in insecticide resistance of *Rhagoletis* has not been demonstrated.

Although external decomposition of plant tissues by Tephritid symbionts is common and well studied with the symbiont of *Rhagoletis pomonella*, Hellmuth (1956) has claimed that all the Tephritinae and Trypetinae (including other *Rhagoletis* species) that she examined, harbored the same bacterium, "*Pseudomonas mutabilis*" n. sp., which she could not isolate from the host plant tissues. Antisera raised against symbionts isolated directly from *Tephritis conura* agglutinated cultures of "*P. mutabilis*" from this and two other species of Tephritids.

Acknowledgments

The preparation of this chapter was in part supported by the Naval Medical Research and Development Command, Research Unit No. MR00001.001.1271. The opinions and assertions contained herein are the private ones of the writers and are not to be construed as official or reflecting the views of the Navy Department or the Naval Service at large. We are greatly indebted to Donna M. Boyle for typing much of the first draft of this chapter and its several revisions. The senior author (G.A.D.) gratefully acknowledges the guidance of Dr. Gerard R. Wyatt in his symbiont studies mentioned in this chapter, the training support of a National Science Foundation Graduate Fellowship, and research support from grants to Dr. Wyatt from the Whitehall Foundation, the United States Public Health Service, and the National Research Council of Canada.

C. ENDOSYMBIONTS OF FUNGI AND INVERTEBRATES OTHER THAN ARTHROPODS

KWANG-POO CHANG, GREGORY A. DASCH AND EMILIO WEISS

This section deals with fungal and invertebrate endosymbionts which have not been classified in the order *Rickettsiales*. Their discovery has often been a by-product of studies of their hosts, many of which are of agricultural, medical, and economic importance. In some cases it has been the result of exciting new discoveries in biology. Recognition of endosymbionts as procaryotic microorganisms and interpretation of their relations to the hosts are based largely on morphological examinations by light and electron microscopy and on biochemical reactions not usually associated with animal or plant tissues. Many of these endosymbionts, like those found in insects and protozoa, are intracellular and have not been cultivated axenically. Interrelationships of the endosymbionts with their host are sometimes clearly symbiotic, but more often they are obscure. Mutualism is proposed for some of these associations, but in several cases the procaryotes are pathogenic. Descriptions of the endosymbionts provided below are based on work published in the last decade. Earlier literature is summarized from monographs whenever appropriate.

Further Reading

Berger, B., G. Thorington and L. Margulis. 1979. Two aeromonads: growth of symbionts from *Hydra viridis*. Curr. Microbiol. *3:* 5–10.
Buchner, P. 1965. *Endosymbiosis of animals with plant microorganisms.* John Wiley, New York.

Cole, R.M., C.S. Richards, and T.J. Ponkin. 1977. Novel bacterium infecting an African snail. J. Bacteriol. *132:* 950–966.
Endo, B.Y. 1979. The ultrastructure and distribution of an intracellular bacterium-like microorganism in tissue of larvae of the soybean cyst nematode, *Heterodera glycines*. J. Ultrastruct. Res. *67:* 1–14.
Jones, M.L. and associates. 1981. (Series of papers on *Riftia pachyptila*). Science *213:* 333–346.
Kozek, W.J. and H.F. Marroquin. 1977. Intracytoplasmic bacteria in *Onchocerca volvulus*. Amer. J. Trop. Med. Hyg. *26:* 663–678.
McLaren, D.A., M. J. Worms, B.R. Laurence and M.G. Simpson. 1975. Microorganisms in filarial larvae (Nematoda). Trans. Roy. Soc. Trop. Med. Hyg. *69:* 509–514.
Otto, S.V., J.C. Harshbarger and S.C. Chang. 1977. Status of selected unicellular eucaryote pathogens, and prevalence and histopathology of inclusions containing obligate procaryote parasites, in commercial bivalve mollusks from Maryland estuaries. Haliotis *8:* 285–295.
Sayre, R.M. and W.P. Wergin. 1977. Bacterial parasite of a plant nematode: morphology and ultrastructure. J. Bacteriol. *129:* 1091–1101.
Schwemmler, W. and H.E.A. Schenk (Editors). 1980. *Endocytobiology. Endosymbiosis and Cell Biology. A Synthesis of Recent Research.* Walter de Gruyter, Berlin, New York.
Scott, D.A. and A.J. Musgrave. 1971. Aspects of the fine structure of symbiotes and related host tissues in nephridia of *Allolobophora caliginosa typica* (Annelida; Lumbricidae). J. Invert. Pathol. *18:* 51–60.
Trytek, R.E., and W.V. Allen. 1980. Synthesis of essential amino acids by bacterial symbionts in the gills of the shipworm *Bankia setacea* (Tryon). Comp. Biochem. Physiol. *67A:* 419–427.

FUNGI

The mycorrhizal fungus, *Endogone* sp., an obligate parasite of plant roots, harbors endosymbionts in spores and hyphae (Mosse, 1970).

These microorganisms resemble actinomycetes and occur free in the cytoplasm, where they undergo binary fission. They are not pathogenic.

In *Humicola* sp., each fungal cell of certain strains has about a dozen mycoplasma-like entities, 0.3 μm in diameter (Lepidi et al., 1975). Their occurrence appears to alter colony morphology when the fungus is grown on malt agar, and also sensitivity to ultraviolet (UV) light. Occasionally, colony sectors are formed which are recognized by their white color and abundant aerial mycelium. UV treatment increases the frequency of appearance of these sectors. The number of mycoplasma-like particles is significantly lower in these sectors.

The plasmodia of the true slime mold, *Didymium squamulosus*, harbor rickettsia-like bodies of two different sizes, 0.4 × 1.0 and 1.0 × 3.0 μm, respectively. The organisms are first located between mito-chondrial membranes, but later are seen free in the cytoplasm. They damage the mitochondria and eventually the plasmodia degenerate, but the fungus regenerates from the sexual sporangia. The bacteria have unusual tubular inclusions and some appear to be degenerating. They are sensitive to chloramphenicol but not to penicillin or tetracycline. In aposymbiotic fungi the mitochondria become numerous again (Duval, 1966; 1970; 1972).

There is no indication of a symbiotic relationship between the fungi and their intracellular bacteria and it is not known how often fungi harbor these bacteria.

SPONGES

Symbiotic associations of various microorganisms with marine sponges of different geographic origin have been described in 13 of the 17 orders in the Classes *Calcarea* and *Desmospongiae* (Wilkinson, 1980). Most common are the unicellular cyanobacteria, e.g. *Aphanocapsa feldmanni* and *A. raspaigellae*. These organisms are, respectively, spherical to oval with dimensions of 1–3 × 2–5 μm, and spherical forms, 6–12 μm in diameter. They are Gram-negative, but are characterized by extensive thickening of the cell wall. They are located intracellularly in cells called bacteriocytes in some species, e.g. *Siphonochalin tabernacula*, or in the intercellular matrix of other hosts, e.g. *Theonella swinhoei*, where they may also occur in specialized amoebocytes, termed cyanocytes. Multicellular cyanobacteria have also been encountered.

Cyanobacteria occur only in tissue exposed to illumination and are thought to protect their hosts from light and enhance their ability to extend into photic zones. Species of sponges which have cyanobacteria grow equally well in both light and shaded conditions, while those without cyanobacteria grow more profusely in the shade. Cyanobacteria also provide their hosts with nutrients via photosynthetic and N_2-fixing activities (Wilkinson, 1980; Wilkinson and Fay, 1979). Keratose sponges, *Verongia*, apparently harbor large numbers of extracellular Gram-negative bacteria in their tissues, which may constitute as much as 33% of the volume of the living material. They are maternally inherited (Vacelet, 1975). Presumably, these bacteria benefit their hosts, too, by supplying them with essential nutrients.

COELENTERATES

The symbiotic association of *Hydra* with algae of the genus *Chlorella* is well established and need not be discussed here. However, in addition to chlorellae, in three strains of the green hydra, *Hydra viridis*, Gram-negative rods, 1.2 × 5 μm, with single flagella have been observed coexisting with algal symbionts in gastrodermal cells. They also occur in the ovary without associated chlorellae (Margulis et al., 1978). Two species of bacteria, believed to be the bacterial endosymbionts, have been isolated from ruptured bacteria-containing vesicles flushed out of the enteron of surface-sterilized hydras. They have been identified as

Aeromonas punctata and *A. hydrophila* (Berger et al., 1979). Both the algal and bacterial endosymbionts are deposited by the female on the surface of the fertilized eggs. Newly hatched hydras acquire the algae and bacteria by ingestion (Thorington and Margulis, 1980). Hydras lacking bacteria have been obtained by treatment with streptomycin and polymyxin B. In treated hydras uptake of exogenous phosphate is curtailed by 55% (Wilkerson, 1980). It is postulated that both chlorellae and aeromonads are involved in the acquisition, storage, and provision of phosphate to the hydras when exogenous sources are limited.

HELMINTHES

Trematodes. Bacteria are seldom seen in sections of freshly collected ectoparasitic monogenean trematodes; they are found, however, in the caeca of flukes that have been maintained for several days in culture media without antibiotics. They are also found in some of the endoparasitic digenean trematodes. These bacteria are regarded as adventitious rather than endosymbiotic. In addition, intracellular mycoplasma-like organisms have been occasionally seen in *Diclidophora merlangi*, a fluke that parasitizes the gills of the whiting (a marine food fish). The organisms are oval or spherical, 0.5 μm in diameter, and are devoid of a cell wall. They are located in gland cells of the anterior alimentary canal in freshly caught flukes, as well as in flukes that have been maintained in the laboratory. They are not affected by penicillin or streptomycin and do not seem to be pathogenic for their hosts (Morris and Halton, 1975). Intracellular bacteria have also been seen in *Euzetrema knoepffleri*, a monogenean fluke of a urodelean amphibian. These organisms are elongated, 0.1 × 1.5 μm, and have a cell wall. They are seen in all stages of development of the host and, in some cases, they cause cell damage (Fournier et al., 1975). Interest in the

bacteria of trematodes was in part stimulated by the discovery that a fluke is involved in the cycle of an important pathogen of the dog (see *Neorickettsia helminthoeca*).

Filarias. Because of their importance as pathogens of man and domestic animals, filarias have been subjected to extensive investigations. Intracellular microorganisms have been recognized in *Dirofilaria immitis* (the dog heartworm), *Brugia pahangi* and *B. malayi* (Malayan filarial worms of cats), and *Onchocerca volvulus* (a human pathogen causing eye disease and blindness in tropical countries) (McLaren et al., 1975; Vincent et al., 1975; Kozek and Marroquin, 1977). They have also been seen but not recognized as procaryotes in other investigations on *D. immitis* (Harada et al., 1970; Kozek, 1971; Lee, 1975). They have not been seen in *Loa loa* and in a number of other filarias (McLaren et al., 1975).

The bacteria in *D. immitis* are 0.3–0.5 μm in diameter and up to 4.5 μm in length. They have a typical Gram-negative morphology. They appear to be aligned in the long axis of the hypodermal tissue and are individually enclosed in membrane-bounded vacuoles. They are seen in all developmental stages of the filaria, including fertilized ova, but not in adults (McLaren et al., 1975). The bacteria in *Brugia* are somewhat smaller, 2 μm in length, and less numerous, but otherwise identical in morphology and location in hypodermal tissues (Vincent et al., 1975; McLaren et al., 1975). The bacteria in *O. volvulus* are also located in hypodermal tissues in the very broad lateral cords; however, the morphology appears to be different. The predominant form, which multiplies by binary fission, is bacillary, 0.7 × 1.5 μm. The smallest form is spheroid, up to 0.3 μm in diameter, and the third form is intermediate between the two. All three forms have two trilaminar membranes, but are often surrounded by other membranes, which are most likely of host origin. The existence of a cycle akin to that of *Chlamydia* has been postulated. In addition to hypodermal tissue, they have been seen in the germinal tissues of the female and presumably are transmitted transovarially. They are not associated with obvious pathology (Kozek and Marroquin, 1977).

Plant nematodes. Endoparasites have been seen in the root-knot nematode, *Meilondogyne incognita*, which parasitizes the pepper, and in the cyst nematodes of the potato, *Globodera rostochiensis*, of the pea, *Heterodera goettingiana*, and of the soybean, *H. glycines* (Sayre and Wergin, 1977; Shepherd et al., 1973; Endo, 1979; Walsh et al., 1979).

The parasite of the root-knot nematode appears to be a member of the order *Actinomycetales*. The infective stage begins with attachment of an endospore to the surface of the nematode. A germ tube then penetrates the cuticle, and mycelial colonies form in the pseudocoelom. The mycelium eventually forms sporangia and spores which are released into the soil to start another cycle (Sayre and Wergin, 1977). The intracellular bacteria of the cyst nematodes from Bolivia and England (Shepherd et al., 1973) and from the U.S.A. (Endo, 1979) appear to be quite similar. They are small rods, 0.3–0.5 μm in width by 1–5 μm in length, with typical Gram-negative outer cell membranes. They are not usually surrounded by membranes of host origin. The most notable feature of the organisms is that they have rod or tubular structures 14 to 17 nm in diameter that extend from the plasma membrane into the cytoplasm, or even right across the width of the cell. The microorganisms are located in most of the larval tissues and are particularly abundant in the reproductive system. Although they produce little if any cytopathic effect, treatment of the nematodes with penicillin, to which these microorganisms are susceptible, increases the multiplication rate of the nematode (Walsh et al., 1979). Thus, these microorganisms may be a factor in the suppression of field populations of nematodes (Endo, 1979).

ANNELIDS

Endosymbionts of earthworms and blood-sucking leeches have been reviewed by Buchner (1965). Bacteria are present regularly in excretory organs (lumen and/or epithelium of nephridial ampullae or urinary bladder) of the earthworms examined, i.e. in all 30 species of lumbricids and in 7 of 8 of the related glossoscolecid species. The organisms are short or slender rods, 0.3–0.7 × 1.5–5.0 μm. They often exhibit bipolar staining properties and are covered with mucoidal substances. They are transmitted transovarially via the cocoon fluid. Ultrastructural studies of *Allolobophora caliginosa typica* have revealed extracellular Gram-negative bacteria which appear to be embedded in a ground substance covering the inner surface of the nephridial ampullae. However, few bacteria are seen free in the lumen (Scott and Musgrave, 1971). In other work quoted by Buchner (1965), samples of ampullar fluid of *Lumbriculus* sp., *Lumbricus* sp., and *Eisenia* sp., cultivated on ascitic fluid or bovine serum agar, have given rise to a nonmotile, nonspore-forming, Gram-negative bacterium which has lipolytic, proteolytic and nitrate-reducing activities. Disposal of waste products for earthworms is thought to be the function of these endosymbionts largely because of their existence in excretory organs.

Filamentous intracellular bacteria have been described in evaginations of the esophagus of the rhynchobdellid leech of the turtle, *Placob-*

della catenigera. After a blood meal, the microorganisms are seen in large numbers in the gut and posterior intestines in association with digested blood cells. Transmission to progeny is believed to occur through infection of the cocoon fluid. Similar observations have been made in the fish leeches *Piscicola geometrica*, *P. punctata*, and *Cystibranchus respirans*, and in the electric ray, *Branchellion torpedinis*, except that symbionts in these hosts appeared to be located in an extracellular position (Buchner, 1965).

From the gut of the medicinal leech, *Hirudo medicinalis*, a microorganism has been isolated which grows luxuriantly on usual culture media and which has been named "*Pseudomonas hirudinis*." It is hemolytic and it displays antibiotic activity. It is believed to aid the host in the digestion of blood and to protect it from invasion by other bacteria. Leeches rendered aposymbiotic by treatment with chloramphenicol do not digest blood. *H. medicinalis* and *H. sanguisuga* (the horse leech) also have endosymbionts in the urinary tract; a short rod has been cultivated on bacteriologic media and named "*Corynebacterium nephridii*." On the basis of experiments with aposymbiotic hosts, this bacterium is believed to participate in the breakdown of protein fragments (Buchner, 1965).

MARINE WORMS AND MOLLUSKS

Echiurida. An unusual filamentous bacterium, 0.03–0.05 × 10–18 μm, has been described in *Borellia viridis*. It is seen in constant association with its host in the peripheral mucus layer and internal intercellular spaces (Bosch, 1976).

Pogonophora. A major recent biological discovery is the dense animal population clustered around the newly explored hydrothermal vents near the Galapagos Islands at depths of approximately 2500 m. The most conspicuous animal is the large, redplumed, vestimentiferan tube worm, *Riftia pachyptila* Jones, which is 38 mm in diameter and up to 1.5 m long. A striking feature of the phylum Pogonophora is the lack of mouth, gut, and anus. Nutrition depends on the uptake of dissolved organic and inorganic material through the epidermis and utilization of this material in the trophosome. Besides the gonads and related

structures, the trophosome is the chief internal organ. It consists mainly of a large number of closely packed bacteria, which measure approximately 3–5 μm in diameter and have a typical Gram-negative cell wall and lipopolysaccharide. They number approximately 3.7×10^9 cells/g of wet weight of trophosome and are believed to be responsible for most of the enzymatic activities of the trophosome. They derive their energy from H_2S, which is present in the vents in concentrations as high as 0.16 mM. The extended vascular system within the trophosome, the high oxygen affinity of the hemoglobin, and the relative insensitivity of the blood oxygen-carrying capacity to changes in temperature and CO_2 concentration, insures a supply of O_2 and CO_2 to the trophosome bacteria. Two diagnostic enzymes of the Calvin-Benson cycle of CO_2 fixation have been demonstrated in the trophosome. The data are consistent with a mode of nutrition of the worm which is autotrophic and entirely dependent on their endosymbiotic bacteria (Cavanaugh et al., 1981).

Recent investigations indicate that similar endosymbiosis exists in the small pogonophora of the more common habitats: *Siboglinum atlanticum* and *Oligobrachia gracilis* from the Bay of Biscay, and *S. ekmani* and *S. fiordicum* from the fjords near Bergen, and a new species of *Siboglinum* from the Skagerrak. The $^{13}C/^{14}C$ ratios of the animals are the lowest of any contemporary marine organic material, which is strong indication of isotope fractionation during chemoautotrophic biosynthesis. The bacteria are typical Gram-negative rods, smaller than those in *Riftia*, 0.16–0.30 × 1–3 μm in four species, and more rounded, 1.4 × 3.2 μm in the new *Siboglinum* species. They are more common in female than male pogonophores. Their specific metabolic activity is somewhat lower than in the giant *Riftia*, possibly because in the smaller hosts dilution by tissue components is greater (Southward et al., 1981).

Snails. Bacteria-like organisms have been reported in snails and the earlier information is summarized by Buchner (1965). In *Cyclostoma elegans*, endosymbionts have been described as nonmotile curved rods (2–5 μm in length), occasionally filamentous, Gram-negative bacteria. They reside in purinocytes (cells rich in uric acids and the purines xanthine, hypoxanthine, and adenine) of storage kidney or concrement glands. Similar endosymbionts occur in other species of the same genus in different geographic areas. In annulariid snails, e.g. *Tudora putre* and *Chondropoma subreticulatum*, bacteria are present in the same anatomical location, but have a coccal morphology. Endosymbiotic bacteria are thought to utilize nitrogen waste products of their hosts for protein biosynthesis. From *Cyclostoma fluorescens* a uricase-positive, Gram-negative bacterium has been isolated, but its identity as an endosymbiont is uncertain.

Some of the bacteria associated with snails are clearly pathogenic and one, "*Bacillus pinottii*," has been considered as a possible biological control agent (reviewed by Cole et al., 1977). Dean et al. (1970) identified "*Aeromonas liquefaciens*" as the etiological agent of a disease in the giant African snail. In a more recent study, Cole et al. (1977) have described a bacterium infecting the African planorbid snail *Bulinus jousseaumei*; this bacterium could not be cultivated, but seems to have unique morphological features. It produces nodules in the superficial tissue, but does not interfere with the longevity or reproduction of the infected snail. It readily infects snails of the same or different species, but is not transmitted transovarially. It forms intracellular packets of Gram-negative rods, 0.3–0.4 × 3–6 μm, without peripheral organelles, or occurs as somewhat smaller rods with numerous flagella-like structures arranged in an unusual manner. The presence of intermediate forms suggests that a single bacterial species is involved which undergoes morphological changes. The flagellar form is characterized by a "head" from which emerge long, thick, rigid, helically constituted organelles with a core and an outer component that is not an extension of the bacterial envelope. The function of the "cephalotrichous" arrangement of the filaments and the nature of the interaction of the bacterium with its host remain unknown.

Shipworms. Until recent times, seafarers dreaded the marine wood-boring bivalves commonly called shipworms or Toredo. These animals actively metabolize cellulose and depend for their nutrition on wood, diatoms, and nanoplankton. In a recent study on the shipworm *Bankia setacea*, Trytek and Allen (1980) obtained good evidence that some of the nutritional requirements are derived from endosymbionts. The bacteria are located in mycetocytes in glandular structures of the gills. They are small, have typical Gram-negative outer membranes, and contain diaminopimelic acid. When isolated gills are incubated with ^{14}C-glucose, most of the essential amino acids (as defined in animal nutrition) of the free and protein-bound pools become labeled. In contrast, with mantle tissue of *B. setacea*, devoid of mycetocytes, only nonessential amino acids are labeled. In an earlier study, Carpenter and Culliney (1975) had found that four species of shipworms fix nitrogen. The rate of N_2 fixation was particularly high in *Teredora malleolus* from the Sargasso Sea. Since N_2 fixation is generally associated with procaryotes, this finding, too, supports the view that endosymbionts are an important source of nutrition to shipworms.

Edible bivalves. Marine mollusks of commercial value have recently been found to harbor intracellular procaryotes. The microorganisms have been seen in the cells of the digestive tubule of both the hard clam (*Mercenaria mercenaria*), soft clams (*Mya arenaria*), and in gut goblet cells of the American oyster (*Crassostrea virginica*) collected in the Chesapeake and Chincoteague Bays. They appear as amorphous, basophilic, finely granular intracytoplasmic inclusions and they are described ultrastructurally as rickettsia-like, chlamydia-like, or mycoplasma-like organisms (Harshbarger et al., 1976; Otto et al., 1977). They have also been seen in *M. mercenaria* collected in the Great South Bay, New York. The microorganism in this host appears to undergo a cycle of development somewhat similar to what has been described in *Chlamydia*, and it fluoresces weakly with fluorescent antibodies to known chlamydial agents. It does not stain satisfactorily with the Giménez stain (Meyers, 1979).

Rickettsia-like and chlamydia-like organisms have been seen by electron microscopy in a number of other bivalves collected in the U.S.A. and Europe: in mussel (*Brachidontes recurvus*), scallops (*Argopectin irridiens* and *Plecopecten megallanicus*), other species of clams (*Macoma balthica*, *Rangia cuneata*, *Spisula solidissima*, *Tellina tenuis* and *Artica islandica*) and oysters (*C. gigas* and *C. angulata*) (Buchanan, 1978; 1979; Comps et al., 1973; 1977; 1979a,b). The occurrence of these microorganisms has not been correlated with water temperature or with proximity to population centers. They are not believed to be pathogenic for their hosts or to be involved in tumor formation, which is very common in bivalves. They have not been associated with human disease.

Bibliography

Abbot, J.D. and R. Shannon. 1958. A method for typing *Shigella sonnei*, using colicine production as a marker. J. Clin. Pathol. *11:* 71-77.

Abd-al-Malek, Y. and Y.Z. Ishac. 1966. Longevity of *Azotobacter*. Plant Soil. *24:* 325-327.

Abdelal, A.T. and H.G. Schlegel. 1974. Purification and regulatory properties of fructose 1,6-diphosphatase from *Hydrogenomonas eutropha*. J. Bacteriol. *120:* 304-310.

Abdel-Ghaffar, A.S. and H.L. Jensen. 1966. The rhizobia of *Lupinus densiflorus* Benth., with some remarks on the classification of root nodule bacteria. Arch. Mikrobiol. *54:* 393-405.

Abd-el-Malek, Y. and S.H. Rizk. 1958. Counting of sulphate-reducing bacteria in mixed bacterial populations. Nature (London) *182:* 538.

Abel, R. 1893. Bakteriologische Studien über Ozaena simplex. Zentralbl. Bakteriol. Parasitenk. Infektionskr. Hyg. Abt. I. Orig. *13:* 161-173.

Abram, D., J. Castro e Melo and D. Chou. 1974. Penetration of *Bdellovibrio bacteriovorus* into host cells. J. Bacteriol. *118:* 663-680.

Abram, D., and B. K. Davis. 1970. Structural properties and features of parasitic *Bdellovibrio bactériovorus*. J. Bacteriol. *104:* 948-965.

Abramson, I.J. and R.M. Smibert. 1971. Bactericidal activity of antimicrobial agents for treponemes. Brit. J. Vener. Dis. *47:* 413-418.

Abramson, I.J. and R.M. Smibert. 1971. Inhibition of growth of treponemes by antimicrobial agents. Brit. J. Vener. Dis. *47:* 407-412.

Abshire, R.L., G.L. Lombard and V.R. Dowell, Jr. 1977. Fluorescent-antibody studies on selected strains of *Bacteroides fragilis* subsp. *fragilis*. J. Clin. Microbiol. *6:* 425-432.

Abuchowski, A., D. Kafkewitz and F.F. Davis. 1979. A rapid purification procedure for L-asparaginase from *Vibrio succinogenes*. Prep. Biochem. *9:* 205-211.

Ackerman, H.W. and F. Poty. 1969. Relationship between coliphages T₂ and T₃ and phage PST of *Pasteurella pseudotuberculosis*. Rev. Canad. Biol. *28:* 201-204.

Adams, G.A. and S.M. Martin. 1964. Extracellular polysaccharides of *Serratia marcescens*. Can. J. Biochem. *42:* 1403-1413.

Adams, G.A., C. Quadling, M. Yaguchi and T.G. Tornabene. 1970. The chemical composition of cell wall lipopolysaccharides from *Moraxella duplex* and *Micrococcus calco-aceticus*. Can. J. Microbiol. *16:* 1-8.

Adams, G.A. and R. Young. 1965. Capsular polysaccharides of *Serratia marcescens*. Can. J. Biochem. *43:* 1499-1512.

Addey, J.P., D. Taylor-Robinson and M. Dimic. 1970. Viability of mycoplasmas after storage in frozen or lyophilised states. J. Med. Microbiol. *3:* 137-145.

Aderhold, R. and W. Ruhland. 1905. Ueber ein durch Bakterien hervorgerufenes Kirschensterben. Zentralbl. Bakteriol. Parasitenk. Infektionskr. Hyg. Abt. II, *15:* 376-377.

Adler, H.E., J. Fabricant, R. Yamamoto and J. Berg. 1958. Isolation and identification of pleuropneumonia-like organisms of avian origin. Am. J. Vet. Res. *19:* 440-447.

Adler, H.E., M. Shifrine and H. Ortmayer. 1961. *Mycoplasma inocuum* sp. n., a saprophyte from chickens. J. Bacteriol. *82:* 239-240.

Adler, S. and V. Ellenbogen. 1934. A note on two new blood parasites of cattle: *Eperythrozoon* and *Bartonella*. J. Comp. Pathol. *47:* 220-221.

Ahrens, R. 1968. Taxonomische Untersuchungen an sternbildenden *Agrobacterium*-Arten aus der westlichen Ostsee. Kiel. Meeresforsch. *24:* 147-173.

Ahrens, R. and G. Rheinheimer. 1967. Über einige sternbildenden Bakterien aus der Ostsee. Kiel. Meeresforsch. *23:* 127-136.

Ahvonen, P., E. Jansson and K. Aho. 1969. Marked cross-agglutination between *Brucella* and a sub-type of *Yersinia enterocolitica*. Acta Pathol. Microbiol. Scand. *75:* 291-295.

Ainsworth, G.C. and P.H.A. Sneath (Editors). 1962. Microbial classification: Appendix I. Symp. Soc. Gen. Microbiol. *12:* 456-463.

Ajelli, F., N. Romano and M.F. Massenti. 1977. L-histidine ammonia-lyase from a T-strain mycoplasma (*Ureaplasma urealyticum*). Boll. Ist. Sierotes Milan *56:* 343-350.

Ajello, G.W. and A.W. Hoadley. 1976. Fluorescent pseudomonads capable of growth at 41°C but distinct from *Pseudomonas aeruginosa*. J. Clin. Microbiol. *4:* 443-449.

Al-Aubaidi, J.M. 1975. Orthographic error in the name *Acholeplasma oculusi*. Int. J. Syst. Bacteriol. *25:* 221.

Al-Aubaidi, J.M., A.H. Dardiri, C.C. Muscoplatt and E.H. McCauley. 1973. Identification and characterization of *Acholeplasma oculusi* spec. nov. from the eyes of goats with keratoconjunctivitis. Cornell Vet. *63:* 117-129.

Al-Aubaidi, J.M. and J. Fabricant. 1971. Characterization and classification of bovine mycoplasma. Cornell Vet. *61:* 490-518.

Albanese, E. and D. Kafkewitz. 1978. Effect of medium composition on the growth and asparaginase production of *Vibrio succinogenes*. Appl. Environ. Microbiol. *36:* 25-30.

Albrecht, H. and A. Ghon. 1901. Ueber die Aetiologie und pathologische Anatomie der *Meningitis cerebrospinalis* epidemica. Wien. Klin. Wochenschr. *14:* 984-996.

Albrecht, H. and A. Ghon. 1903. Zür Frage der morphologischen und biologischen Charakterisierung des *Meningococcus intracellularis*. Zentralbl. Bakteriol. Parasitenkd. Infektionskr. Hyg. Abt. I Orig. *33:* 496-510.

Albritton, W.L.J., L. Brunton, L. Slaney and I.W. Maclean. 1982. Plasmid-mediated sulfonamide resistance in *Haemophilus ducreyi*. Antimicrob. Agents Chemother. *21:* 159-165.

Alcorn, S.M. 1961. Some hosts of *Erwinia carnegieana*. Plant Dis. Rep. *45:* 587-590.

Aldová, E., K. Láznicková, E. Stepánková and J. Lietava. 1968. Isolation of nonagglutinable vibrios from an enteritis outbreak in Czechoslovakia. J. Infect. Dis. *118:* 25-31.

Aldred, P., A.C. Hill and C. Young. 1974. The isolation of *Streptobacillus moniliformis* from cervical abscesses of guinea pigs. Lab. Anim. *8:* 275-277.

Aleksandrushkina, N.I. and L.A. Egorova. 1978. Nucleotide composition of the DNA of thermophilic bacteria of the genus *Thermus*. Microbiology (U.S.S.R.) *47:* 203-205. (English translation, Plenum Publishing Corp., New York).

Alexander, A.D. 1980a. *Leptospira*. *In* Lennette, Balows, Hausler and Truant (Editors), Manual of Clinical Microbiology, 3rd Ed., American Society for Microbiology, Washington, D.C., pp. 376-382.

Alexander, A.D. 1980b. Serological diagnosis of leptospirosis. *In* Rose and Friedman (Editors), Manual of Clinical Immunology, 2nd Ed., American Society for Microbiology, Washington, D.C. pp. 542-546.

Alexander, A.D., E.F. Lessel, L.B. Evans, E. Franck and S.S. Green. 1972. Preservation of leptospiras by liquid-nitrogen refrigeration. Int. J. Syst. Bacteriol. *22:* 165-169.

Alexander, A.G. and G.E. Kenny. 1980. Characterization of the strain-specific and common surface antigens of *Mycoplasma arginini*. Infect. Immun. *29:* 442-451.

Alexander, H.E. 1958. The *Hemophilus* group. *In* Dubos (Editor), Bacterial and Mycotic Infections of Man, 3rd Ed., Lippincott Co., Philadelphia and Montreal, p. 474.

Alexander, H.E. and G. Leidy. 1951. Induction of heritable new type in type-specific strains of *H. influenzae*. Proc. Soc. Exp. Biol. Med. *78:* 625-626.

Alexander, H.E., G. Leidy and E. Hahn. 1954. Studies on the nature of *Hemophilus influenzae* cells susceptible to heritable changes by desoxyribonucleic acids. J. Exp. Med. *99:* 505-533.

Alexander, T.J.L., K. Thornton and G. Boon. 1980. Medicated early weaning to obtain pigs free from pathogens endemic in the herd of origin. Vet. Rec. *106:* 114-119.

Al-Jumaili, I.J. 1975. Bacteriocine typing of *Proteus*. J. Clin. Pathol. *28:* 784-787.

Al-Jumaili, I.J. and G.A. Fenwick. 1978. Bacteriocine typing of *Providencia* isolates. Zentralbl. Bakteriol. Parasitenk. Infektionskr. Hyg. Abt. I Orig. A *240:* 202-207.

Al-Khoja, M.S. and J.H. Darrell. 1979. The skin as the source of *Acinetobacter* and *Moraxella* species occurring in blood cultures. J. Clin. Pathol. *32:* 497-499.

Allam, N.M. and R.M. Lemcke. 1975. Mycoplasmas isolated from the respiratory tract of horses. J. Hyg. *74:* 385-408.

Allan, M.B. 1959. Studies with *Cyanidium caldarium* an anomalously pigmented chlorophyte. Arch. Mikrobiol. *32:* 270-277.

Allen, O.N. and A.J. Holding. 1974. Genus II. *Agrobacterium* Conn 1942, 359. *In* Buchanan and Gibbons (Editors), Bergey's Manual of Determinative Bacteriology, 8th Ed. The Williams and Wilkins Co., Baltimore, pp. 264-267.

Allen, R.D. and P. Baumann. 1971. Structure and arrangement of flagella in species of the genus *Beneckea* and *Photobacterium fischeri*. J. Bacteriol. *107:*

295-302.

Allen, T.C. and A.J. Riker. 1932. A rot of apple fruit caused by *Phytomonas melophthora*, n. sp., following invasion by the apple maggot. Phytopathology *22:* 557-571.

Allison, M.J., I.M. Robinson and A.L. Baetz. 1979. Synthesis of α-ketoglutarate by reductive carboxylation of succinate in *Veillonella*, *Selenomonas*, and *Bacteroides* species. J. Bacteriol. *140:* 980-986.

Allison, M.J., I.M. Robinson, J.A. Bucklin and G.D. Booth. 1979. Comparison of populations of the pig cecum and colon based upon enumeration with specific energy sources. Appl. Environ. Microbiol. *37:* 1142-1151.

Allsopp, A. 1969. Phylogenetic relationships of the Procaryota and the origin of the eucaryotic cell. New Phytol. *68:* 591-612.

Alonso, J.M., A. Joseph-Francois, D. Mazigh, H. Bercovier and H.H. Mollaret. 1978. Résistance à la peste de souris experimentalement infectées par *Yersinia enterocolitica*. Ann. Microbiol. (Inst. Pasteur) *129B:* 203-207.

Althauser, M., W. A. Samsonoff, C. Anderson and S. F. Conti. 1972. Isolation and preliminary characterization of bacteriophages for *Bdellovibrio bacteriovorus*. J. Virol. *10:* 516-523.

Altmann, G. and B. Bogokovsky. 1971. In-vitro sensitivity of *Flavobacterium meningosepticum* to antimicrobial agents. J. Med. Microbiol. *4:* 296-299.

Altmeir, W.A., III and E.M. Ayoub. 1977. Erythromycin prophylaxis in pertussis. Pediatrics *59:* 623-625.

Alton, G.G., L.M. Jones and D.E. Pietz. 1975. Laboratory Techniques in Brucellosis, 2nd Ed., World Health Organization, Geneva.

Altson, R.A. 1936. Studies on *Azotobacter* in Malayan soils. J. Agric. Sci. (Cambridge) *26:* 268-280.

Aluotto, B.B., R.G. Wittler, C.O. Williams and J.E. Faber. 1970. Standardized bacteriologic techniques for the characterization of *Mycoplasma* species. Int. J. Syst. Bacteriol. *20:* 35-58.

Amako, K., K. Yasumaka and K. Takeya. 1970. Relationship between rhapidosomes and pyocin in *Pseudomonas fluorescens*. J. Gen. Microbiol. *62:* 107-112.

Amaral, J.F., C. Teixeira and E.D. Pinheiro. 1956. O bactério causador da mancha aureolada do cafeeiro. Arg. Inst. Biol. (São Paulo) *23:* 151-155.

Ambler, R.P. 1973. Bacterial cytochrome *c* and molecular evolution. Syst. Zool. *22:* 554-565.

Ambler, R.P. 1974. The evolutionary stability of cytochrome c-551 in *Pseudomonas aeruginosa* and *Pseudomonas fluorescens* biotype C. Biochem. J. *137:* 3-14.

Ambler, R.P. 1976. Amino acid sequences of prokaryotic cytochromes *c*. *In* Fasman (Editor), Handbook of Biochemistry and Molecular Biology, Proteins, Vol. 3, 3rd Ed., CRC Press, Cleveland, Ohio, pp. 292-307.

Ambler, R.P. 1981. The structure and classification of cytochromes *c*. *In* Robinson and Kaplan (Editors), From Cyclotrons to Cytochromes. Academic Press, New York, pp. 263-280.

Ambler, R.P., M. Daniel, J. Hermoso, T.E. Meyer, T.G. Bartsch and M.D. Kamen. 1979. Cytochrome c_2 sequence variation among the recognized species of purple nonsulphur photosynthetic bacteria. Nature (London) *278:* 659-660.

Ambler, R.P., T.E. Meyer and M.D. Kamen. 1979. Anomalies in amino acid sequences of small cytochromes *c* and cytochromes c^1 from two species of purple photosynthetic bacteria. Nature (London) *278:* 661-662.

Ameyama, M. 1975. *Gluconobacter oxydans* subsp. *sphaericus*, new subspecies isolated from grapes. Int. J. Syst. Bacteriol. *25:* 365-370.

Ameyama, M. and K. Kondo. 1966. Carbohydrate metabolism by the acetic acid bacteria. Part V. On the vitamin requirements for the growth. Agr. Biol. Chem. *30:* 203-211.

Amin, P.M. and S.V. Ganapati. 1967. Occurrence of *Zoogloea* colonies and protozoans at different stages of sewage purification. Appl. Microbiol. *15:* 17-21.

Amiressami, M. 1980. Investigation of the light microscopical and ultrastructure of the Demeton-S-Methyl resistance aphids under consideration of the mycetome symbionts of the *Phorodon humuli* Schrank. *In* Schwemmler and Schenk (Editors), Endocytobiology. Endosymbiosis and Cell Biology. A Synthesis of Recent Research. W. de Gruyter, Berlin, New York, pp. 425-443.

Amiressami, M. and H. Petzold. 1976. Licht- und elektronenmikroskopische Untersuchungen über das Verhalten der Mycetomsymbionten bei insektizidresistenten und normal-sensiblen Pfirsichblattläusen *Myzus persicae* Sulz. Z. Angew. Zool. *63:* 273-289.

Amiressami, M. and H. Petzold. 1977. Symbioseforschung und Insektizidresistenz. Ein licht- und elektronenmikroskopischer Beiträg zür Klärung den Insektizid-resistenz von Aphiden unter Berücksichtigung der Mycetom-Symbionten bei *Myzus persicae* Sulz. Z. Angew. Entomol. *82:* 252-259.

Anacker, R.L., T.F. McCaul, W. Burgdorfer and R.K. Gerloff. 1980. Properties of selected rickettsiae of the spotted fever group. Infect. Immun. *27:* 468-474.

Anacker, R.L. and E.J. Ordal. 1959. Studies on the myxobacterium *Chondrococcus columnaris*. I. Serological typing. J. Bacteriol. *78:* 25-32.

Anagnostopoulos, G.D. and H.S. Sidhu. 1978. Helical growth of *Bacillus stearothermophilus* in low water activity media. Microbios Lett. *5:* 115-121.

Anand, V.K. and G.T. Heberlein. 1977. Crown gall tumorigenesis in potato tuber tissue. Amer. J. Bot. *64:* 153-158.

Anderson, A.R. and L.W. Moore. 1979. Host specificity in the genus *Agrobacterium*. Phytopathology *69:* 320-323.

Anderson, D.R. and M.F. Barile. 1965. Ultrastructure of *Mycoplasma hominis*. J. Bacteriol. *90:* 180-192.

Anderson, D.R., H.E. Hopps, M.F. Barile and B.C. Bernheim. 1965. Comparison of the ultrastructure of several rickettsiae, ornithosis virus and mycoplasma in tissue culture. J. Bacteriol. *90:* 1387-1404.

Anderson, E. 1964. Oocyte differentiation and vitellogenesis in the roach *Periplaneta americana*. J. Cell Biol. *20:* 131-155.

Anderson, E.S. 1964. The phage typing of *Salmonellae* other than *S. typhi*. *In* Van Oye (Editor) The world problem of salmonellosis. Junk, The Hague, pp. 84-110.

Anderson, E.S. 1975. The problem and implications of chloramphenicol resistance in the typhoid bacillus. J. Hyg. Camb. *74:* 289-299.

Anderson, E.S., G.O. Humphreys and G.A. Willshaw. 1975. The molecular relatedness of R factors in enterobacteria of human and animal origin. J. Gen. Microbiol. *91:* 376-382.

Anderson, G. and E.A. North. 1943. The relation of pertussis endotoxin to pertussis immunity in the mouse. Aust. J. Exp. Biol. Med. Sci. *21:* 1-8.

Anderson, G.R. 1966. Identification of *Beijerinckia* from Pacific Northwest soils. J. Bacteriol. *91:* 2105-2106.

Anderson, G.W., Jr. and J.V. Osterman. 1980. Host defenses in experimental rickettsialpox: resistance of C3H mouse sublines. Acta Virol. *24:* 294-296.

Anderson, H. 1954. The reddening of salted hides and fish. Appl. Microbiol. *21:* 64-69.

Anderson, J.D. and H. Smith. 1965. The metabolism of erythritol by *Brucella abortus*. J. Gen. Microbiol. *38:* 109-124.

Anderson, J.F., L.A. Magnarelli and J. Kurz. 1979. Intraerythrocytic parasites in rodent populations of Connecticut: *Babesia* and *Grahamella* species. J. Parasitol. *65:* 599-604.

Anderson, J.I.W. and D.A. Conroy. 1970. *Vibrio* disease in marine fishes. *In* Snieszko (Editor) A symposium on diseases of fishes and shell fishes. Special Publication No. 5, American Fisheries Society, Washington, pp. 266-272.

Anderson, K. 1944. The cultivation from granuloma inguinale of microorganisms having characteristics of Donovan bodies in yolk sac of chick embryos. Science *97:* 560.

Anderson, K., W. de Monbreun and E. Goodpasture. 1945. An etiologic consideration of *Donovania granulomatis* cultivated from granuloma inguinale (three cases) in embryonic egg yolk. J. Exp. Med. *81:* 25-40.

Anderson, R.S. and E.J. Ordal. 1972. Deoxyribonucleic acid relationships among marine vibrios. J. Bacteriol. *109:* 696-706.

Andreesen, M. and H.G. Schlegel. 1974. A new coryneform bacterium: *Corynebacterium autotrophicum* strain 7 c. II. Isolation of a slime-free mutant. Arch. Microbiol. *100:* 351-361.

Andreev, L.V., Y.A. Trotsenko and V.F. Galchenko. 1977. Fatty acid composition of methylotrophic bacteria. *In* Skryabin, Ivanov, Kondratjeva, Zavarzin, Trotsenko and Nesterov (Editors), Microbial growth on C_1-compounds. Abstracts Second Internat. Symp. Microbial Growth on C_1-compounds, USSR Academy of Sciences, Pushchino, USSR, pp. 12-15.

Andrew, A.T. and P.B. Carter. 1977. Serological reactivity of chemical fractions of *Mycoplasma bovis*. Can. J. Microbiol. *23:* 852-853.

Andrew, R., J.M. Bonnin and S. Williams. 1946. Tick typhus in North Queensland. Med. J. Aust. *2:* 253-258.

Andrewes, A.G., S. Hertzberg, S. Liaaen-Jensen and M.P. Starr. 1973. The *Xanthomonas* "carotenoids" - non-carotenoid, brominated, aryl-polyene esters. Acta Chem. Scand. *27:* 2383-2395.

Andrewes, A.G., C.L. Jenkins, M.P. Starr, M.P. Shepherd and H. Hope. 1976. Structure of xanthomonadin I, a novel dibrominated aryl polyene pigment produced by the bacterium *Xanthomonas juglandis*. Tetrahedr. Lett. *45:* 4023.

Andron, L.A., II and H.T. Eigelsbach. 1975. Biochemical and immunological properties of ribonucleic acid-rich extracts from *Francisella tularensis*. Infect. Immun. *12:* 137-142.

Antheunisse, J. 1972. Preservation of Microorganisms. Antonie van Leeuwenhoek J. Microbiol. Serol. *38:* 617-622.

Antheunisse, J. 1973. Viability of lyophilized microorganisms after storage. Antonie van Leeuwenhoek J. Microbiol. Serol. *39:* 243-248.

Aoki, T., T. Arai and S. Egusa. 1975. R factors detected from *Vibrio anguillarum* and marine *Vibrio*. *In* Mitsuhashi and Hashimoto (Editors) Microbial Drug Resistance. University Park Press, Baltimore, pp. 223-228.

Aoki, T. and S. Egusa. 1971. Detection of resistance factors in fish pathogen *Aeromonas liquefaciens*. J. Gen. Microbiol. *65:* 343-349.

Aoki, T., S. Egusa, C. Yada and T. Watanabe. 1972. Studies of drug resistance and R factors in bacteria from pond-cultured salmonids. I. Amago (*Oncorhynchus rhodurus macrostomus*) and Yamame (*Oncorhynchus masou ishikawae*). Jpn. J. Microbiol. *16:* 233-238.

Aoki, Y., S.-T. Hsu and D. Chun. 1967. Distribution of *Vibrio parahaemolyticus* in the sea and harbors in Southeast Asia and Central Pacific. Endem. Dis. Bull. Nagasaki Univ. *8:* 191-202.

Appel, M. and D. Bemis. 1977. Canine respiratory disease complex. *In* Kirk (Editor), Current Veterinary Therapy. VI. Small Animal Practice. W.B. Saunders, Philadelphia. pp. 1287-1292.

Appelbaum, P.C., J. Stavitz, M.S. Bentz and L.C. von Kuster. 1980. Four methods for identification of gram-negative nonfermenting rods: organisms more commonly encountered in clinical specimens. J. Clin. Microbiol. *12:* 271-278.

Aragao, H. and G. Vianna. 1913. Pesquizas sobre o *Granuloma venereo*. (Untersuchungen ueber das *Granuloma venereum*). Mem. Inst. Oswaldo Cruz. *5:* 211-238.

Aragno, M. 1975. Mise en évidence d'hydrogénobactéries corynéformes auxohétérotrophes pour la biotine dans l'eau d'un lac eutrophe. Ann. Microbiol. (Inst. Pasteur) 126A: 539-542.

Aragno, M. and H.G. Schlegel. 1977. *Alcaligenes ruhlandii* (Packer and Vishniac) comb. nov., a peritrichous hydrogen bacterium previously assigned to *Pseudomonas*. Int. J. Syst. Bacteriol. 27: 279-281.

Aragno, M. and H. G. Schlegel. 1978. *Aquaspirillum autotrophicum*, a new species of hydrogen-oxidizing, facultatively autotrophic bacteria. Int. J. Syst. Bacteriol. 28: 112-116.

Aragno, M. and H.G. Schlegel. 1981. The hydrogen-oxidizing bacteria. *In* Starr, Stolp, Trüper, Balows and Schlegel (Editors), The Prokaryotes, A Handbook on Habitats, Isolation and Identification of Bacteria, Springer-Verlag, Berlin, pp. 865-893.

Aragno, M., A. Walther-Mauruschat, F. Mayer and H.G. Schlegel. 1977. Micromorphology of gram-negative hydrogen bacteria. 1. Cell morphology and flagellation. Arch. Microbiol. 114: 93-100.

Arai, T., N. Ikejima, T. Itoh, S. Sakai, T. Shimada and R. Sakazaki. 1980. A survey of *Plesiomonas shigelloides* from aquatic environment, domestic animals, pets and humans. J. Hyg. 84: 203-211.

Arai, S., K.Y. Yuri, A. Kudo, M. Kikuchi, K. Kumagai and N. Ishida. 1967. Effect of antibiotics on the growth of various strains of *Mycoplasma*. J. Antibiot. Ser. A 20: 246-253.

Aranki, A. and R. Freter. 1972. Use of anaerobic glove boxes for the cultivation of strictly anaerobic bacteria. Amer. J. Clin. Nutr. 25: 1329-1334.

Archer, D.B. 1979. Immunoprecipitation of triton X-100-solubilized *Mycoplasma mycoides* proteins. J. Gen. Microbiol. 115: 111-116.

Archer, G.L., P.H. Coleman, R.M. Cole, R.J. Duma and C.L. Johnston. 1979. Human infection from an identified erythrocyte-associated bacterium. N. Engl. J. Med. 301: 897-900.

Argaman, M., T.-Y. Liu and J.B. Robbins. 1974. Polyribitol phosphate: an antigen of four gram-positive bacteria cross-reactive with the capsular polysaccharide of *Haemophilus influenzae*, type b. J. Immunol. 112: 649-655.

Ariel, B.M., T.N. Khavkin and N.I. Amosenkova. 1973. Interaction between *Coxiella burnetii* and the cells in experimental Q-rickettsiosis. Pathol. Microbiol. 39: 412-423.

Ark, P.A. 1939. Bacterial leaf spot of maple. Phytopathology 29: 968-970.

Ark, P.A. 1940. Bacterial stalk rot of field corn caused by *Phytomonas lapsa* n. sp. Phytopathology 30: 1.

Ark, P.A. and J.T. Barrett. 1946. A new bacterial leaf-spot of greenhouse-grown gardenias. Phytopathology 36: 865-868.

Ark, P.A. and M.W. Gardner. 1936. Bacterial leaf spot of *Primula*. Phytopathology 26: 1050-1055.

Ark, P.A. and H.E. Thomas. 1946. Bacterial leaf spot and bud rot of orchids caused by *Phytomonas cattleyae*. Phytopathology 36: 695-698.

Ark, P.A. and C.M. Tompkins. 1946. Bacterial leaf blight of bird's nest fern. Phytopathology 36: 758-761.

Arkwright, J.A., E.E. Atkin and A. Bacot. 1921. An hereditary *Rickettsia*-like parasite of the bed bug (*Cimex lectularius*). Parasitology 13: 27-36.

Arnaud, G. 1920. Une maladie bactérienne du lierre (*Hedera helix* L.). C. R. Hebd. Séances Acad. Sci. Ser. D 171: 121-122.

Arseculeratne, S.N. 1962. Actinobacillosis in joints of rabbits. J. Comp. Pathol. 72: 33-39.

Arthur, L.O., L.A. Bulla, Jr., G. St. Julian and L.K. Nakamura. 1973. Carbohydrate metabolism in *Agrobacterium tumefaciens*. J. Bacteriol. 116: 304-313.

Arthur, L.O., L.K. Nakamura, G. St. Julian and L.A. Bulla, Jr. 1975. Carbohydrate catabolism of selected strains in the genus *Agrobacterium*. Appl. Microbiol. 30: 731-737.

Asai, T. 1935. Taxonomic studies on acetic acid bacteria and allied oxidative bacteria isolated from fruits. A new classification of the oxidative bacteria. J. Agr. Chem. Soc. Jpn. 11: 499-513, 610-620, 674-708.

Asai, T. 1968. Acetic acid bacteria. Classification and biochemical activities. University of Tokyo Press, Tokyo; University Park Press, Baltimore.

Asai, T., K. Aida, Z. Sugisaki and N. Yakeishi. 1955. On α-ketoglutaric acid fermentation. J. Gen. Appl. Microbiol. (Japan) 1: 308-346.

Asai, T., H. Iizuka and K. Komagata. 1962. The flagellation of genus *Kluyvera*. J. Gen. Appl. Microbiol. (Japan) 8: 187-191.

Asai, T., H. Iizuka and K. Komagata. 1964. The flagellation and taxonomy of the genera *Gluconobacter* and *Acetobacter* with reference to the existence of intermediate strains. J. Gen. Appl. Microbiol. 10: 95-126.

Asai, T., S. Okumura and T. Tsunoda. 1956. On a new genus, *Kluyvera*. Proc. Japan Acad. 32: 488-493.

Asai, T., S. Okumura and T. Tsunoda. 1957. On the classification of the α-ketoglutaric acid accumulating bacteria in aerobic fermentation. J. Gen. Appl. Microbiol. (Japan) 3: 13-33.

Asai, T. and K. Shoda. 1958. The taxonomy of *Acetobacter* and allied oxidative bacteria. J. Gen. Appl. Microbiol. 4: 289-311.

Aschner, M. 1932. Experimentelle Untersuchungen über die Symbiose der Kleiderlaus. Naturwissenschaften 20: 501-505.

Aschner, M. 1934. Studies on the symbiosis of the body louse. 1. Elimination of the symbionts by centrifugalisation of the eggs. Parasitology 26: 309-314.

Aschner, M. and E. Ries. 1933. Das Verhalten der Kleiderlaus bei Ausschaltung ihrer Symbionten. Eine experimentelle Symbiosestudie. Z. Morphol. Oekol. Tiere 26: 529-590.

Ashby, S.F. 1929. Gumming disease of sugar cane. Trop. Agr. Trinidad 6: 135-138.

Askaa, G. and H. Ernø. 1976. Elevation of *Mycoplasma agalactiae* subsp. *bovis* to species rank: *Mycoplasma bovis* (Hale et al.) comb. nov. Int. J. Syst. Bacteriol. 26: 323-325.

Askaa, G., H. Ernø and M.O. Ojo. 1978. Bovine mycoplasmas: classification of groups related to *Mycoplasma mycoides*. Acta Vet. Scand. 19: 166-178.

Asonganyi, T.M. and P.M. Meadow. 1980. Biosynthesis of the core part of the lipopolysaccharide of *Pseudomonas aeruginosa*. J. Gen. Microbiol. 117: 1-7.

Auden, D.T. 1974. Studies on the development of *Rhodnius prolixus* and the effects of its symbiote *Nocardia rhodnii*. J. Med. Entomol. 11: 68-71.

Audureau, A. 1940. Étude du genre *Moraxella*. Ann. Inst. Pasteur (Paris) 64: 6-166.

Aulakh, G.S., J.G. Tully and M.F. Barile. 1979. Differentiation among some acholeplasmas by nucleic acid homology. Curr. Microbiol. 2: 91-94.

Auling, G., M. Dittbrenner, M. Maarzahl, T. Nokhal and M. Reh. 1980. Deoxyribonucleic acid relationships among hydrogen-oxidizing strains of the genera *Pseudomonas*, *Alcaligenes*, and *Paracoccus*. Int. J. Syst. Bacteriol. 30: 123-128.

Auling, G., F. Mayer and H.G. Schlegel. 1977. Isolation and partial characterization of normal and defective bacteriophages of gram-negative hydrogen bacteria. Arch. Microbiol. 115: 237-247.

Auling, G., M. Reh, C.M. Lee and H.G. Schlegel. 1978. *Pseudomonas pseudoflava*, a new species of hydrogen-oxidizing bacteria: its differentiation from *Pseudomonas flava* and other yellow-pigmented, Gram-negative, hydrogen oxidizing species. Int. J. Syst. Bacteriol. 28: 82-95.

Austen, R.A. and N.W. Dunn. 1980. Regulation of the plasmid-specified naphthalene catabolic pathway of *Pseudomonas putida*. J. Gen. Microbiol. 117: 521-528.

Austen, R.A. and T.J. Trust. 1980. Detection of plasmids in the related group of the genus *Campylobacter*. FEMS Microbiol. Lett. 8: 201-204.

Austin, B. and M. Goodfellow. 1979. *Pseudomonas mesophilica*, a new species of pink bacteria isolated from leaf surfaces. Int. J. Syst. Bacteriol. 29: 373-378.

Austin, B., M. Goodfellow and C.H. Dickinson. 1978. Numerical taxonomy of phylloplane bacteria isolated from *Lolium perenne*. J. Gen. Microbiol. 104: 139-155.

Austin, B., C.J. Rodgers, J.M. Forns and R.R. Colwell. 1981. *Alcaligenes faecalis* subsp. *homari* subsp. nov., a new group of bacteria from moribund lobsters. Int. J. Syst. Bacteriol. 31: 72-76.

Austin, F.E., J.T. Barbieri, R.E. Corin, K.E. Grigas and C.D. Cox. 1981. Distribution of superoxide dismutase, catalase, and peroxidase activities among *Treponema pallidum* and other spirochetes. Infect. Immun. 33: 372-379.

Ayers, T.T., C.L. Lefebvre and H.W. Johnson. 1939. Bacterial wilt of lespedeza. U.S. Dept. Agr. Tech. Bull. 704: 1-22.

Ayres, P.A. and G.I. Barrow. 1978. The distribution of *Vibrio parahaemolyticus* in British coastal waters: report of a collaborative study 1975-76. J. Hyg. Camb. 80: 281-294.

Azad, H.R. and C.I. Kado. 1980. Numerical and DNA:DNA reassociation analyses of *Erwinia rubrifaciens* and other members of the *Enterobacteriaceae*. J. Gen. Microbiol. 120: 117-129.

Azuma, I., T. Taniyama, Y. Yamamura, Y. Yanagihara, S. Hattori, S. Yasuda and I. Mifuchi. 1975. Chemical studies on the cell walls of *Leptospira biflexa* strain Urawa and *Treponema pallidum* strain Reiter. Jpn. J. Microbiol. 19: 45-51.

Baars, J.K. 1930. Over sulfaatreduktie door bacteriën. Dissertation: W.D. Meinema, N.V. Delft, Holland, pp. 1-164.

Babb, J.L. and C.S. Cummins. 1978. Encapsulation of *Bacteroides* species. Infect. Immun. 19: 1088-1091.

Babb, J.L. and C.S. Cummins. 1981. Relationships between serological groups and deoxyribonucleic acid homology groups in *Bacteroides fragilis* and related species. J. Clin. Microbiol. 13: 369-379.

Baboolal, R. 1969. Cell wall analysis of oral filamentous bacteria. J. Gen. Microbiol. 58: 217-256.

Babudieri, B. 1959. Q fever: a zoonosis. Adv. Vet. Sci. 5: 81-182.

Babudieri, B. 1961. Studio serologico del gruppo Semaranga - Patoc de *Leptospira biflexa*. *In* Atti dell' XI Congresso Societa Italiana di Microbiologia, Cagliari - Sassari, October 1961, Naples, pp. 9-12.

Babudieri, B. 1973. Experimental infections by spirilla. *In* Eichler (Editor), Handbuch der Experimentellen Pharmakologie, New Series, Vol. 17 IIB, Springer-Verlag, New York, Berlin, pp. 43-49.

Baca, O.G. 1978. Comparison of ribosomes from *Coxiella burnetii* and *Escherichia coli* by gel electrophoresis, protein synthesis, and immunological techniques. J. Bacteriol. 136: 429-432.

Bächi, B. and L. Ettlinger. 1974. Cytochrome difference spectra of acetic acid bacteria. Int. J. Syst. Bacteriol. 24: 215-220.

Bachman, B.J. and K.B. Cow. 1980. Linkage map of *Eschericha coli* K12. Microbiol. Rev. 44: 1-56.

Back, W., N. Weiss and H. Seidel. 1979. Isolation and systematic classification of Gram negative bacteria which are harmful to beer. II. Gram negative anaerobic rods (English Translation) Brauwissenschaft 32: 233-238.

Bacon, M.F., W.G. Overend, P.H. Lloyd and A.R. Peacocke. 1967. The isolation, composition and physiochemical properties of deoxyribonucleic acid from *Bordetella pertussis*. Arch. Biochem. Biophys. 118: 352-361.

Bader, R.E. 1954. Über die Herstellung eines agglutinierenden Serums gegen due Rundform von *Shigella sonnei* mit einem Stamm der Gattung *Pseudomonas*. Z. Hyg. Infektionskr. 140: 450-456.

Badger, S.J., T. Butler, C.K. Kim and K.H. Johnston. 1979. Experimental *Eikenella corrodens* endocarditis in rabbits. Infect. Immun. 23: 751-757.

Badger, S.J. and A.C.R. Tanner. 1981. Serological studies of *Bacteroides gracilis, Campylobacter concisus, Wolinella recta,* and *Eikenella corrodens,* all from humans with periodontal disease. Int. J. Syst. Bacteriol. *31:* 446-451.

Bae, B.H.C., S.B. Sureka and J.A. Ajamy. 1981. Enteric Group 15 *(Enterobacteriaceae)* associated with pneumonia. J. Clin. Microbiol. *14:* 596-597.

Baechler, C.A. and R.S. Berk. 1972. Ultrastructural observations of *Pseudomonas aeruginosa:* rhapidosomes. Microstructures 3: 24-28, 34.

Baerwald, R.J. and G.M. Boush. 1968. Demonstration of the bacterial symbiote *Pseudomonas melophthora* in the apple maggot, *Rhagoletis pomonella,* by fluorescent-antibody technique. J. Invertebr. Pathol. *11:* 251-259.

Bag, J. 1973. Studies on a phosphoenolpyruvate hexose phosphotransferase system in *Vibrio cholerae.* Indian J. Biochem. Biophys. *10:* 257-260.

Bag, J. 1974. Diauxic growth of *Vibrio cholerae:* Effect of glucose on the transport and phosphoenolpyruvate dependent phosphorylation of galactose and fructose. Indian J. Biochem. Biophys. *11:* 148-151.

Bagley, D.H., J.C. Alexander, V.J. Gill, R. Dolin and A.S. Ketcham. 1976. Late *Flavobacterium* species meningitis after craniofacial exenteration. Arch. Intern. Med. *136:* 229-231.

Bagley, S.T. and R.J. Seidler. 1978. Primary *Klebsiella* identification with MacConkey-inositol-carbenicillin agar. Appl. Environ. Microbiol. *36:* 536-538.

Bagley, S.T., R.J. Seidler and D.J. Brenner. 1982. *In* Validation of the publication of new names and new combinations previously effectively published outside the IJSB. List No. 8. Int. J. Syst. Bacteriol. *32:* 266-268.

Bahr, L. 1919. Paratyfus hos Honningbien samt nogle undersøgelser verdrørende Forekomsten af Bakterierhenhorende til Coli-tyfus gruppen. i. Honningbiens tarm. Scand. Vet. Tidskrift. *9:* 25-40, 45-60.

Bailie, W.E. 1966. Characterization of *Haemophilus somnus,* new species, a microorganism isolated from infectious thromboembolic meningoencephalitis of cattle. Ph.D. Dissertation, Kansas State University, Manhattan, Kansas.

Bailie, W.E., H.D. Anthony and K.D. Weide. 1976. Infectious thromboembolic meningoencephalitis (sleeper syndrome) in feedlot cattle. J. Am. Vet. Med. Assoc. *148:* 162-166.

Bailie, W.E., E.H. Coles and K.D. Weide. 1973. Deoxyribonucleic acid characterization of a microorganism isolated from infectious thromboembolic meningoencephalitis of cattle. Int. J. Syst. Bacteriol. *23:* 231-237.

Baillie, A., W. Hodgkiss and J.R. Norris. 1962. Flagellation of *Azotobacter* spp. as demonstrated by electron microscopy. J. Appl. Bacteriol. *25:* 116-119.

Baine, W.B., J.K. Rasheed, J.C. Feeley, G.W. Gorman and L.E. Casida, Jr. 1978. Effect of supplemental L-tyrosine on pigment production in cultures of the Legionnaires' disease bacterium. Curr. Microbiol. *1:* 93-94.

Baines, S. 1956. The role of the symbiotic bacteria in the nutrition of *Rhodnius prolixus* (Hemiptera). J. Exp. Biol. *33:* 533-541.

Baird-Parker, A.C. 1957. Isolation of *Leptotrichia buccalis* and *Fusobacterium* species from oral material. Nature (London) *180:* 1056-1057.

Baird-Parker, A.C. 1960. The classification of fusobacteria from the human mouth. J. Gen. Microbiol. *22:* 458-469.

Bak, A.L., C. Christiansen and A. Stenderup. 1970. Bacterial genome sizes determined by DNA renaturation studies. J. Gen. Microbiol. *64:* 377-380.

Baker, D.A. and R.W.A. Park. 1975. Changes in morphology and cell wall structure that occur during growth of *Vibrio* sp. NCTC 4716 in batch culture. J. Gen. Microbiol. *86:* 12-28.

Baker, J.A. 1946. A rickettsial infection in Canadian voles. J. Exp. Med. *84:* 37-51.

Baker, J.E. 1974a. Differential net food utilization by larvae of *Sitophilus oryzae* and *Sitophilus granarius.* J. Insect Physiol. *20:* 1937-1942.

Baker, J.E. 1974b. Differential sterol utilization by larvae of *Sitophilus oryzae* and *Sitophilus granarius.* Ann. Entomol. Soc. Am. *67:* 591-594.

Baker, J.E. 1975. Vitamin requirements of larvae of *Sitophilus oryzae.* J. Insect Physiol. *21:* 1337-1342.

Baker, J.E. and P.T.M. Lum. 1973. Development of aposymbiosis in larvae of *Sitophilus oryzae* (Coleoptera: Curculionidae) by dietary treatment with antibiotics. J. Stored Prod. Res. *9:* 241-245.

Baker, P.J. and J.B. Wilson. 1965. Chemical composition and biological properties of endotoxin of *Brucella abortus.* J. Bacteriol. *90:* 895-902.

Bakos, K. 1955. Studien über *Haemophilus suis,* mit besonderer Berücksichtigung der serologischen Differenzierung seiner Stämme. Uppsala, Appelberg's Boktrykkeri AB.

Bakos, K., A. Nilsson and E. Thal. 1952. Untersuchungen über *Haemophilus suis.* Nord. Vet. Med. *4:* 241-255.

Balakrishnan, C.V. and N. Langerman. 1977. The isolation of a bacterial glycoprotein with luciferase activity. Arch. Biochem Biophys. *181:* 680-682.

Balch, W.E., G.E. Fox, L.J. Magrum, C.R. Woese and R.S. Wolfe. 1979. Methanogens: Reevaluation of a unique biological group. Microbiol. Rev. *43:* 260-296.

Baldani, V. L. D. and J. Döbereiner. 1980. Host plant specificity in the infection of cereals with *Azospirillum* spp. Soil Biol. Biochem. *12:* 433-440.

Baldwin, I.L. and E.B. Fred. 1929. Nomenclature of the root-nodule bacteria of the Leguminosae. J. Bacteriol. *17:* 141-150.

Baldwin, T.O., M.M. Ziegler and D.A. Powers. 1979. Covalent structure of subunits of bacterial luciferase: NH₂-terminal sequence demonstration of subunit homology. Proc. Natl. Acad. Sci. U.S.A. *76:* 4887-4889.

Balke, E., A. Weber and B. Fronk. 1977. Untersuchungen des Aminosäurestoffwechsels mit der Dünnschichtchromatographie zür Differenzierung von Bru-

cellen. Zentralbl. Bakteriol. Parasitenkd. Infektionskr. Hyg. Abt. I Orig. A *237:* 523-529.

Balkwill, D. L., D. Maratea and R. P. Blakemore. 1980. Ultrastructure of a magnetotactic spirillum. J. Bacteriol. *141:* 1399-1408.

Ball, B.V. and L. Bailey. 1978. The symbiotes of *Myzus pepsicae* (Sulz.) in strains resistant and susceptible to Demeton-S-Methyl. Pestic. Sci. Soc. Chem. Ind. *9:* 522-524.

Ball, G.H. 1969. Organisms living on and in Protozoa. Res. Protozool. *3:* 567-718.

Ball, H.J. and T. Todd. 1978. Comparison of antigens of pneumonia-associated *Mycoplasma* species by gel diffusion. Infect. Immun. *21:* 954-958.

Ballard, R.W., M. Doudoroff, R.Y. Stanier and M. Mandel. 1968. Taxonomy of the aerobic pseudomonads: *Pseudomonas diminuta* and *P. vesiculare.* J. Gen. Microbiol. *53:* 349-361.

Ballard, R.W., N.J. Palleroni, M. Doudoroff, R.Y. Stanier and M. Mandel. 1970. Taxonomy of the aerobic pseudomonads: *Pseudomonas cepacia, P. marginata, P. alliicola,* and *P. caryophylli.* J. Gen. Microbiol. *60:* 199-214.

Ballester, M., J.M. Ballester and J.P. Belaich. 1977. Isolation and characterization of a high molecular weight antibiotic produced by a marine bacterium. Microb. Ecol. *3:* 289-303.

Balsdon, M.J., L. Pead, G.E. Taylor and R. Maskell. 1980. *Corynebacterium vaginale* and vaginitis: A controlled trial of treatment. Lancet *i:* 501-504.

Baltazard, M. 1964. La conservation de la Peste en foyer invétérré. Med. Hyg. *22:* 172-174.

Bamforth, C.W. and J.R. Quayle. 1978. Aerobic and anaerobic growth of *Paracoccus denitrificans* on methanol. Arch. Microbiol. *119:* 91-97.

Banai, M., I. Kahane, S. Razin and W. Bredt. 1978. Adherence of *Mycoplasma gallisepticum* to human erythrocytes. Infect. Immun. *21:* 365-372.

Banerjee, A.K. 1966. Physiologische Untersuchungen an *Micrococcus denitrificans* Beijerinck und auxotropher Mutanten. Isolierung auxotropher Mutanten und Spaltung des cystathionins. Arch. Mikrobiol. *53:* 107-131.

Banerjee, A.K. and H.G. Schlegel. 1966. Zür Rolle des Hefeextraktes während chemolithotrophen Wachstums von *Micrococcus denitrificans.* Arch. Mikrobiol. *53:* 132-153.

Bang, S.S., P. Baumann and K.H. Nealson. 1978a. Phenotypic characterization of *Photobacterium logei* (sp. nov.), a species related to *P. fischeri.* Curr. Microbiol. *1:* 285-288.

Bang, S.S., L. Baumann, M.J. Woolkalis and P. Baumann. 1981. Evolutionary relationships in *Vibrio* and *Photobacterium* as determined by immunological studies of superoxide dismutase. Arch. Microbiol. *130:* 111-120.

Bang, S.S., M.J. Woolkalis and P. Baumann. 1978b. Electrophoretic mobilities of superoxide dismutases from species of *Photobacterium, Beneckea, Vibrio,* and selected terrestrial enterobacteria. Curr. Microbiol. *1:* 371-376.

Banning, F. 1902. Zür Kenntnis der Oxalsäurebildung durch Bakterien. Zentralbl. Bakteriol. Parasitenkd. Infektionskr. Hyg. Abt. II 8: 395-398.

Barber, C. and E. Eylan. 1976. Immunochemical relations of *Yersinia enterocolitica* with *Yersinia pestis* and their connection with other *Enterobacteriaceae.* Microbios Lett. *3:* 25-29.

Barber, T.L. and J. Fabricant. 1971. Identification of *Mycoplasmatales:* characterization procedures. Appl. Microbiol. *21:* 600-605.

Barbieri, J.T. and C.D. Cox. 1979. Pyruvate oxidation by *Treponema pallidum.* Infect. Immun. *25:* 157-163.

Barbieri, J.T. and C.D. Cox. 1981. Influence of oxygen on respiration and glucose catabolism by *Treponema pallidum.* Infect. Immun. *31:* 992-997.

Barile, M.F. 1979. Mycoplasma-tissue cell interactions. *In* Tully and Whitcomb (Editors), The Mycoplasmas, Vol. II, Academic Press, New York, pp. 425-474.

Barile, M.F., R.A. Del Giudice, T.R. Carski, C.J. Gibbs and J.A. Morris. 1968. Isolation and characterization of *Mycoplasma arginini:* spec. nov. Proc. Soc. Exp. Biol. Med. *129:* 489-494.

Barile, M.F., R.A. Del Giudice and J.G. Tully. 1972. Isolation and characterization of *Mycoplasma conjunctivae* sp. n. from sheep and goats with keratoconjunctivitis. Infect. Immun. *5:* 70-76.

Barker, B.T.P. 1948. Some recent studies on the nature and incidence of cider sickness. Annu. Rep. Agric. Hort. Res. Stn. Long Ashton Bristol, pp. 174-181.

Barker, B.T.P. and V.F. Hillier. 1912. Cider sickness. J. Agric. Sci. 5: 67-85.

Barksdale, L. 1970. *Corynebacterium diphtheriae* and its relatives. Bacteriol. Rev. *34:* 378-422.

Barksdale, L. and S.B. Arden. 1974. Persisting bacteriophage infections, lysogeny, and phage conversions. Annu Rev. Microbiol. *28:* 265-299.

Barnes, D.M. and D.K. Sorensen. 1975. Salmonellosis. *In* H.W. Dunne and A.D. Leman (Editors), Diseases of Swine, 4th edition, The Iowa State University Press, Ames, pp. 554-564.

Barnes, E.M. and G.C. Burton. 1970. The effect of hibernation on the caecal flora of the thirteen-lined ground squirrel *(Citellus tridecemlineatus).* J. Appl. Bacteriol. *33:* 505-514.

Barnes, E.M. and C.S. Impey. 1968. Anaerobic gram negative nonsporing bacteria from the caeca of poultry. J. Appl. Bacteriol. *31:* 530-541.

Barnes, L.A. and W.B. Cherry. 1946. A group of paracolon organisms having apparent pathogenicity. Amer. J. Publ. Health 36: 481-483.

Baron, E.S. and A.K. Saz. 1978. Genetic transformation of piliation and virulence into *Neisseria gonorrhoeae* T4. J. Bacteriol. *13:* 972-998.

Barooah, P.P. and A. Sen. 1959. Studies on *Beijerinckia* from some acid soil in India. Ind. J. Agric. Sci. *29:* 36-51.

Baross, J.A., J. Liston and R.Y. Morita. 1978a. Incidence of *Vibrio parahaemo-*

lyticus bacteriophages and other *Vibrio bacteriophages* in marine samples. Appl. Environ. Microbiol. 36: 492-499.

Baross, J.A., J. Liston and R.Y. Morita. 1978b. Ecological relationship between *Vibrio parahaemolyticus* and digesting vibrios as evidenced by bacteriophage susceptibility patterns. Appl. Environ. Microbiol. *36:* 500-505.

Baross, J.A., P.A. Tester and R.Y. Morita. 1978c. Incidence, microscopy, and etiology of exoskeleton lesions in the tanner crab, *Chionoecetes tanneri.* J. Fish. Res. Board Can. *35:* 1141-1149.

Barrett, P.A., E. Beveridge, P.L. Bradley, C.G.D. Brown, S.R.M. Bushby, M.L. Clarke, R.A. Neal, R. Smith and J.K.H. Wilde. 1965. Biological activities of some α-dithiosemicarbozones. Nature (London) *206:* 1340-1341.

Bartlett, K.H., T.J. Trust and H. Lior. 1977. Small pet aquarium frogs as a source of *Salmonella.* Appl. Environ. Microbiol. *33:* 1026-1029.

Barua, D. and W. Burrows (Editors). 1974. Cholera. W.B. Saunders Co., Philadelphia-London-Toronto.

Bascomb, S., S.P. Lapage, M.A. Curtis and W.R. Willcox. 1971. Numerical classification of the tribe *Klebsielleae.* J. Gen. Microbiol. *66:* 279-295.

Bascombe, S. and R.M. Jackson. 1965. *Rhizobium* culture collection. Rothamsted Exp. Stn. Rep. *1964:* 86-87.

Baseman, J.B. and C.D. Cox. 1969a. Intermediate energy metabolism of *Leptospira.* J. Bacteriol. *97:* 992-1000.

Baseman, J.B. and C.D. Cox. 1969b. Terminal electron transport in *Leptospira.* J. Bacteriol. *97:* 1001-1004.

Bashford, D.J., T.J. Donovan, A.L. Furniss and J.V. Lee. 1979. *Vibrio cholerae* in Kent. Lancet *1:* 436-437.

Basnyat, S.R. and Y.S. Kulkarni. 1979. New bacterial leafspot of *Centella asiatica* L. Urban. Biovigyanam *5:* 179-180.

Bass, J.W., E.L. Klenk, J.B. Kotheimer, C.C. Linnemann and M.H.D. Smith. 1969. Antimicrobial treatment of pertussis. J. Pediatr. *75:* 768-781.

Bassot, J.M. 1975. Les organes lumineux à bactéries symbiotiques de quelques téléostéens léiognathides. Arch. Zool. Exp. Gen., Notes Rev. *116:* 359-373.

Basu, S. and S. Mukerjee. 1968. Bacteriophage typing of *Vibrio eltor.* Experientia *24:* 299-300.

Bateman, M.A. 1972. The ecology of fruit flies. Annu. Rev. Entomol. *17:* 493-518.

Bates, L.B. and J.H. St. John. 1922. Suggestion of *Spirochaeta neotropicalis* as name for spirochaete of relapsing fever found in Panama. J. Amer. Med. Assoc. *79:* 575-576.

Battelli, C. 1947. Si di un. piroplasma della *Naia nigrocollis. (Aegyptianella carpani* n. sp.) Riv. Parassit. *8:* 205-212.

Baturo, A.P. and V.P. Raginskaya. 1978. Antigenic schema for the hafniae. Int. J. Syst. Bacteriol. *28:* 126-127.

Batzing, B.L. and G.W. Claus. 1973. Fine structural changes of *Acetobacter suboxydans* during growth in a defined medium. J. Bacteriol. *113:* 1455-1461.

Bauer, A.W., W.M.M. Kirby, J.C. Sherris and M. Turck. 1966. Antibiotic susceptibility testing by a standardized single disk method. Amer. J. Clin. Pathol. *45:* 493-496.

Baughn, R.E. and B.A. Freeman. 1966. Antigenic structure of *Brucella suis* spheroplasts. J. Bacteriol. *92:* 1298-1303.

Baum, D.H. and L.A. Joens. 1979. Serotypes of beta-hemolytic *Treponema hyodysenteriae.* Infect. Immun. *25:* 792-796. 133-138.

Baumann, L., S.S. Bang and P. Baumann. 1980. Study of relationship among species of *Vibrio, Photobacterium,* and terrestrial enterobacteria by an immunological comparison of glutamine synthetase and superoxide dismutase. Curr. Microbiol. *4:* 133-138.

Baumann, L. and P. Baumann. 1973. Regulation of aspartokinase activity in the genus *Beneckea* and marine, luminous bacteria. Arch. Microbiol. *90:* 171-188.

Baumann, L. and P. Baumann. 1973. Enzymes of glucose catabolism in cell-free extracts of non-fermentative marine eubacteria. Can. J. Microbiol. *19:* 302-304.

Baumann, L. and P. Baumann. 1974. Regulation of aspartokinase activity in non-fermentative, marine eubacteria. Arch. Microbiol. *95:* 1-18.

Baumann, L. and P. Baumann. 1975a. Catabolism of D-fructose and D-ribose by *Pseudomonas doudoroffii.* II. Properties of 1-phosphofructokinase and 6-phosphofructokinase. Arch. Microbiol. *105:* 241-248.

Baumann, L. and P. Baumann. 1976. Study of the relationship among marine and terrestrial enterobacteria by means of in vitro DNA/ribosomal RNA hybridization. Microbios. Lett. *3:* 11-20.

Baumann, L. and P. Baumann. 1978. Studies of relationship among terrestrial *Pseudomonas, Alcaligenes,* and enterobacteria by an immunological comparison of glutamine synthetase. Arch. Microbiol. *119:* 25-30.

Baumann, L. and P. Baumann. 1980. Immunological relationships of glutamine synthetases from marine and terrestrial enterobacteria. Curr. Microbiol. *3:* 191-196.

Baumann, L., P. Baumann, M. Mandel and R.D. Allen. 1972. Taxonomy of aerobic marine eubacteria. J. Bacteriol. *110:* 402-429.

Baumann, P. 1968. Isolation of *Acinetobacter* from soil and water. J. Bacteriol. *96:* 39-42.

Baumann, P., S.S. Bang and L. Baumann. 1978. Phenotypic characterization of *Beneckea anguillara* biotypes I and II. Curr. Microbiol. *1:* 85-88.

Baumann, P. and L. Baumann. 1975b. Catabolism of D-fructose and D-ribose by *Pseudomonas doudoroffii.* I. Physiological studies and mutant analysis. Arch. Microbiol. *105:* 225-240.

Baumann, P. and L. Baumann. 1977. Biology of the marine enterobacteria: genera *Beneckea* and *Photobacterium.* Annu. Rev. Microbiol. *31:* 39-61.

Baumann, P. and L. Baumann. 1981. The marine Gram-negative eubacteria. *In*

Starr, Stolp, Trüper, Balows and Schlegel (Editors), The Prokaryotes, a handbook on habitats, isolation and identification of bacteria. Springer-Verlag, New York, pp. 1352-1394.

Baumann, P., L. Baumann, S.S. Bang and M.J. Woolkalis. 1980b. Reevaluation of the taxonomy of *Vibrio, Beneckea,* and *Photobacterium:* abolition of the genus *Beneckea.* Curr. Microbiol. *4:* 127-132.

Baumann, P., L. Baumann, S.S. Bang and M.J. Woolkalis. 1981a. *In* List No. 6, Validation of the publication of new names and combinations previously effectively published outside the IJSB. Int. J. Syst. Bacteriol. *31:* 215-218.

Baumann, P., L. Baumann, S.S. Bang and M.J. Woolkalis. 1982. *In* List No. 8, Validation of the publication of new names and combinations previously effectively published outside the IJSB. Int. J. Syst. Bacteriol. *32:* 266-268.

Baumann, P., L. Baumann and M. Mandel. 1971a. Taxonomy of marine bacteria: the genus *Beneckea.* J. Bacteriol. *107:* 268-294.

Baumann, P., L. Baumann, M. Mandel and R.D. Allen. 1971b. Taxonomy of marine bacteria: *Beneckea nigrapulchrituda* sp. n. J. Bacteriol. *108:* 1380-1383.

Baumann, P., L. Baumann and J.L. Reichelt. 1973. Taxonomy of marine bacteria: *Beneckea parahaemolytica* and *Beneckea alginolytica.* J. Bacteriol. *113:* 1144-1155.

Baumann, P., M. Doudoroff and R.Y. Stanier. 1968a. Study of the *Moraxella* group. I. Genus *Moraxella* and the *Neisseria catarrhalis* group. J. Bacteriol. *95:* 58-73.

Baumann, P., M. Doudoroff and R.Y. Stanier. 1968b. A study of the *Moraxella* group II. Oxidative-negative species (Genus *Acinetobacter*). J. Bacteriol. *95:* 1520-1541.

Baumgarten, J., M. Reh and H.G. Schlegel. 1974. Taxonomic studies on some Gram-positive coryneform hydrogen bacteria. Arch. Microbiol. *100:* 207-217.

Bauminger, E.R., S.G. Cohen, F.L. de Kanter, A. Levy, S. Offer, M. Kessel and S. Rottem. 1980. Iron storage in *Mycoplasma capricolum.* J. Bacteriol. *141:* 378-381.

Bauwens, M. and J. De Ley. 1981. Improvements in the taxonomy of *Flavobacterium* by DNA:rRNA hybridizations. *In* Reichenbach and Weeks (Editors), The Flavobacterium-Cytophaga Group, Verlag Chemie, Weinheim, pp. 27-31.

Baxter, R.M. and N.E. Gibbons. 1956. Effects of sodium and potassium chloride on certain enzymes of *Micrococcus halodenitrificans* and *Pseudomonas salinaria.* Can. J. Microbiol. *2:* 599-606.

Bayley, S.T. and R.A. Morton. 1978. Recent developments in the molecular biology of extremely halophilic bacteria. Crit. Rev. Microbiol. *6:* 151-205.

Bayliss, M.E., S.H. Wilkes and J.M. Prescott. 1980. *Aeromonas* neutral protease: specificity toward extended substrates. Arch. Biochem. Biophys. *204:* 214-219.

Baynes, I.D. and G.C. Simmons. 1960. Ovine epididymitis caused by *Actinobacillus seminis* n. sp. Aust. Vet. J. *36:* 454-459.

Bázliková, M. and R. Brezina. 1978. Some biological properties of rickettsiae isolated in Armenian SSR. *In* Kazár, Ormsbee, and Tarasevich (Editors), Rickettsiae and Rickettsial Diseases, VEDA, Bratislava, pp. 155-159.

Beardsley, R.E. 1955. Phage production by crown-gall bacteria and the formation of plant tumors. Am. Natur. *89:* 175-176.

Beardsley, R.E. 1962. Glycine resistance in *Agrobacterium tumefaciens.* J. Bacteriol. *83:* 6-13.

Beattie, K.L. and J.K. Setlow. 1970. Transformation between *Haemophilus influenzae* and *Haemophilus parainfluenzae.* J. Bacteriol. *104:* 390-400.

Beck, E., J. Wieczorek and W. Reinecke. 1980. Purification and properties of hamamelosekinase. Eur. J. Biochem. *107:* 485-489.

Becker, A.H. 1962. Infections due to *Proteus mirabilis* in newborn nursery. Am. J. Dis. Child. *104:* 355-359.

Beckett, E.B., B. Boothroyd and W.W. MacDonald. 1978. A light and electron microscope study of rickettsia-like organisms in the ovaries of mosquitoes of the *Aedes scutellaris* group. Ann. Trop. Med. Parasitol. *72:* 277-283.

Becking, J.H. 1959. Nitrogen-fixing bacteria of the genus *Beijerinckia* in South African soils. Plant Soil *11:* 193-206.

Becking, J.H. 1961a. Studies on nitrogen-fixing bacteria of the genus *Beijerinckia.* I. Geographical and ecological distribution in soils. Plant Soil *14:* 49-81.

Becking, J.H. 1961b. Studies on nitrogen-fixing bacteria of the genus *Beijerinckia.* II. Mineral nutrition and resistance to high levels of certain elements in relation to soil type. Plant Soil *14:* 297-322.

Becking, J.H. 1962. Species differences in molybdenum and vanadium requirements and combined nitrogen utilization by *Azotobacteraceae.* Plant Soil *16:* 171-201.

Becking, J. H. 1963. Fixation of molecular nitrogen by an aerobic *Vibrio* or *Spirillum.* Antonie van Leeuwenhoek J. Microbiol. Serol. *29:* 326.

Becking, J.H. 1971. Biological nitrogen fixation and its economic significance. *In* Nitrogen-15 in soil-plant studies, Symposium Sofia, December 1979, Int. Atomic Energy Agency, Vienna, IAEA-PL-341/*14:* 189-222.

Becking, J.H. 1974a. Family II. *Azotobacteraceae* Pribram 1933. Genus III. *Beijerinckia* Derx 1950. Genus IV. *Derxia* Jensen, Petersen, De and Bhattacharya 1960. *In* Buchanan and Gibbons (Editors), Bergey's Manual of Determinative Bacteriology, 8th Ed., The Williams and Wilkins Co., Baltimore, pp. 253, 256-260, 260-261.

Becking, J.H. 1974b. Nitrogen-fixing bacteria of the genus *Beijerinckia.* Soil Sci. *18:* 196-212.

Becking, J.H. 1978. *Beijerinckia* in irrigated rice soils. *In* Environmental role of nitrogen-fixing blue-green algae and asymbiotic bacteria. Ecol. Bull. (Stockholm) 26: 116-129.

Becking, J.H. 1981. The family Azotobacteraceae. In Starr, Stolp, Trüper, Balows and Schlegel (Editors), The Prokaryotes: a handbook on habitats, isolation, and identification of bacteria, Springer-Verlag, Berlin, pp. 795-817.

Beckman, W. and T.G. Lessie. 1980. Response of Pseudomonas cepacia to beta-lactam antibiotics; utilization of penicillin G as the carbon source. J. Bacteriol. 140: 1126-1128.

Bedell, D.M. and G.T. Dimopoullus. 1965. Biologic properties and characteristics of Anaplasma marginale: effects of temperature on infectivity of whole blood preparations. Am. J. Vet. Res. 23: 618-625.

Beer, S.V. and A.K. Vidaver. 1978. Bacteriocins produced by Erwinia herbicola inhibit Erwinia amylovora. In W. Laux (Editor) Abstr. Papers, 3rd Inter. Congr. Plant Pathol., München Aug. 16-22. Deutsche Phytomedizinische Gesselschaft, Göttingen, F. R. Germany, p. 75.

Beerens, H., L. Fievez and P. Wattre. 1971. Observations concernant 7 souches appartenant aux espèces Sphaerophorus necrophorus, Sphaerophorus funduliformis, Sphaerophorus pseudonecrophorus. Ann. Inst. Pasteur (Paris) 121: 37-41.

Beerens, H. and M. Tahon-Castel. 1965. Infections humaines à bactéries anaérobies non toxigènes. Presses Academiques Européennes, Brussels.

Behki, R.M. 1967. Metabolism of amino acids in Agrobacterium tumefaciens. III. Uptake of L-proline. Canad. J. Biochem. 45: 1819-1830.

Behki, R.M. and R.M. Hochster. 1967. Metabolism of amino acids in Agrobacterium tumefaciens. II. Uptake of L-valine by growing cells. Can. J. Biochem. 45: 165-170.

Behling, U.H., P. Phan and A. Nowotny. 1979. Biological activity of the slime and endotoxin of the periodontopathic organism Eikenella corrodens. Infect. Immunol. 26: 580-584.

Behrman, E.J. 1962. Tryptophan metabolism in Pseudomonas. Nature 196: 150-152.

Beijerinck, M.W. 1888. Cultur des Bacillus radicola aus den Knöllchen. Bot. Ztg. 46: 740-750.

Beijerinck, M.W. 1889. Le Photobacterium luminosum. Bactérie luminosum de la Mer Nord. Arch. Neer. Sci. 23: 401-427.

Beijerinck, M.W. 1895. Über Spirillum desulfuricans als Ursache von Sulfatreduktion. Zentralbl. Bakteriol. Parasitenkd. Infektionskr. Hyg. Abt. I Orig. 1: 1-9; 49-59; 104-114.

Beijerinck, M.W. 1898. Ueber die Arten der Essigbakterien. Zentralbl. Bakteriol. Parasitenkd. Infektionskr. Hyg. Abt II 4: 209-216.

Beijerinck, M.W. 1900. On different forms of hereditary variation in microbes. Proc. Acad. Sci. Amst. 3: 352-365.

Beijerinck, M.W. 1901. Ueber oligonitrophile Mikroben. Zentralbl. Bakteriol. Parasitenkd. Infektionskr. Hyg. Abt II 7: 561-582.

Beijerinck, M.W. 1903. Sur des microbes oligonitrophiles. Arch. Neerl. Sci. Ser. 2 8: 190-217.

Beijerinck, M.W. 1911. Pigments as products of oxidation by bacterial action. Proc. K. Ned. Akad. Wet. 13: 1066-1077.

Beijerinck, M.W. 1916. Formation of pyruvic acid from malic acid by microbes. Verslag gewone Vergad. Akad. Amst. 18: 1198-2000.

Beijerinck, M. 1922. Azotobacter chroococcum als indikator van de vruchtbarrheid van den grond. K. Ned. Akad. Wet. Versl. Gewone Vergad. Afd. Natuurkd. 30: 431-438.

Beijerinck, M.W. 1925. Über ein Spirillum, welches freien Stickstoff binden kann? Zentralbl. Bakteriol. Parasitenkd. Infektionskr. Hyg. Abt, 2, 63: 353-359.

Beijerinck, M.W. and A. van Delden. 1902. Über die Assimilation des freien Stickstoffs durch Bakterien. Zentralbl. Bakteriol. Parasitenk. Infektionskr. Hyg. Abt. II, 9: 3-43.

Bein, S.J. 1954. A study of certain chromogenic bacteria isolated from "Red Tide" water with a description of a new species. Bull. Mar. Sci. Gulf Caribb. 4: 110-119.

Belaich, J.P. 1963. Thermogénèse et croissance de Pseudomonas lindneri en glucose limitant. C.R. Soc. Biol. 157: 316-322.

Belaich, J.P. and J.C. Senez. 1965. Influence of aeration and pantothenate on growth yields of Zymomonas mobilis. J. Bacteriol. 89: 1195-1200.

Beljanski, M. 1955. Isolement des mutants d'Escherichia coli streptomycino-résistants dépourvus d'enzymes respiratoires. Action de l'hémine sur la formation de ces enzymes chez le mutant H₇. C.R. Hebd. Acad. Sci. Paris 240: 374-377.

Bell, D.P. and P.C. Elmes. 1969. Effects of certain organisms associated with chronic respiratory disease on SPF and conventional rats. J. Med. Microbiol. 2: 511-519.

Bell, E.J., G.M. Kohls, H.G. Stoenner and D.B. Lackman. 1963. Nonpathogenic rickettsias related to the spotted fever group isolated from ticks, Dermacentor variabilis and Dermacentor andersoni from Eastern Montana. J. Immunol. 90: 770-781.

Bell, E.J. and E.G. Pickens. 1953. A toxic substance associated with the rickettsias of the spotted fever group. J. Immunol. 70: 461-472.

Bell, E.J. and H.G. Stoenner. 1960. Immunologic relationships among the spotted fever group of rickettsias determined by toxin neutralization tests in mice with convalescent animal serums. J. Immunol. 84: 171-182.

Bell, S.M. and D. Plowman. 1980. Mechanisms of ampicillin resistance in Haemophilus influenzae from respiratory tract. Lancet (1) 8163: 279-280.

Belly, R.T., B.B. Bohlool and T.D. Brock. 1973. The genus Thermoplasma. Ann. N.Y. Acad. Sci. 225: 94-107.

Belly, R.T. and T.D. Brock. 1972. Cellular stability of a thermophilic, acidophilic mycoplasma. J. Gen. Microbiol. 73: 465-469.

Belly, R.T. and G.W. Claus. 1972. Effect of amino acids on the growth of Acetobacter suboxydans. Arch. Mikrobiol. 83: 237-245.

Belser, W.L. and M.I. Bunting. 1956. Studies on a mechanism providing for genetic transfer in Serratia marcescens. J. Bacteriol. 72: 582-592.

Belyavin, G. 1951. Cultural and serological phases of Proteus vulgaris. J. Gen. Microbiol. 5: 197-207.

Bemis, D.A., H.A. Greisen and M.J.G. Appel. 1977a. Pathogenesis of canine bordetellosis. J. Infect. Dis. 135: 753-762.

Bemis, D.A., H.A. Greisen and M.J.G. Appel. 1977b. Bacteriological variation among Bordetella bronchiseptica isolates from dogs and other species. J. Clin. Microbiol. 5: 471-480.

Ben-Gurion, R. and I. Hertman. 1958. Bacteriocin-like material produced by Pasteurella pestis. J. Gen. Microbiol. 19: 289-297.

Ben-Gurion, R. and A. Shafferman. 1981. Essential virulence determinants of different Yersinia species are carried on a common plasmid. Plasmid 5: 183-187.

Bennett, B.L., J.E. Smadel and R.L. Gauld. 1949. Studies on scrub typhus (tsutsugamushi disease) IV. Heterogeneity of strains of R. tsutsugamushi as demonstrated by cross-neutralization tests. J. Immunol. 62: 453-461.

Benson, S. and J. Shapiro. 1978. TOL is a broad-host-range plasmid. J. Bacteriol. 135: 278-280.

Bercovier, H., J.M. Alonso, Z. Bentaiba, J. Brault and H.H. Mollaret. 1979. Contribution to the definition and taxonomy of Yersinia enterocolitica. Contr. Microbiol. Immunol. 5: 12-22.

Bercovier, H., J. Brault, N. Barre, M. Treignier, J.M. Alonso and H.H. Mollaret. 1978. Biochemical, serological and phage typing characteristics of 459 Yersinia strains isolated from a terrestrial ecosystem. Curr. Microbiol. 1: 353-357.

Bercovier, H., D.J. Brenner, J. Ursing, A.G. Steigerwalt, G.R. Fanning, J.M. Alonso, G.P. Carter and H.H. Mollaret. 1980a. Characterization of Yersinia enterocolitica sensu stricto. Curr. Microbiol. 4: 201-206.

Bercovier, H., H.H. Mollaret, J.M. Alonso, J. Brault, G.R. Fanning, A.G. Steigerwalt and D.J. Brenner. 1980b. Intra- and interspecies relatedness of Yersinia pestis by deoxyribonucleic acid hybridization and its relationship to Yersinia pseudotuberculosis. Curr. Microbiol. 4: 225-230.

Bercovier, H., P. Perreaux, F. Escande, J. Brault, M. Kiredjian and H.H. Mollaret. 1981. Characterization of Pasteurella aerogenes isolated in France. In Kilian, Frederiksen and Biberstein (Editors), Haemophilus, Pasteurella and Actinobacillus. Academic Press, London and New York, pp. 175-183.

Bercovier, H., J. Ursing, D.J. Brenner, A.G. Steigerwalt, G.R. Fanning, G.P. Carter and H.H. Mollaret. 1980. Yersinia kristensenii: a new species of Enterobacteriaceae composed of sucrose-negative strains (formerly called atypical Yersinia enterocolitica or Yersinia enterocolitica-like). Curr. Microbiol. 4: 219-224.

Bercovier, H., J. Ursing, D.J. Brenner, A.G. Steigerwalt, G.R. Fanning, G.P. Carter and H.H. Mollaret. 1981. In Validation of the publication of new names and new combinations previously effectively published outside the IJSB. List No. 6. Int. J. Syst. Bacteriol. 31: 215-218.

Berdal, B.P. and E. Søderlund. 1977. Cultivation and isolation of Francisella tularensis on selective chocolate agar, as used routinely for the isolation of gonococci. Acta Pathol. Microbiol. Scand. Sect. B: Microbiol. 85: 108-109.

Berendsen, P.B. and I.R. Teford. 1966. A light and microscopic study of Kurloff bodies in the blood and spleen of the guinea pig. Anat. Rec. 156: 104-118.

Berezina, I.G. 1975. An electron microscope study of endosymbionts in Blepharisma japonicum. Acta Protozool. 13: 365-369.

Berg, R. L., J.W. Jutila and B. D. Firehammer. 1971. A revised classification of Vibrio fetus. Am. J. Vet. Res. 32: 11-22.

Berg, R. H., M. E. Tyler, N. J. Novick, V. Vasil and I. K. Vasil. 1980. Biology of Azospirillum-sugarcane association: enhancement of nitrogenase activity. Appl. Environ. Microbiol. 39: 642-649.

Berg, R. H., V. Vasil and I. K. Vasil. 1979. The biology of Azospirillum-sugarcane association. II. Ultrastructure. Protoplasma 101: 143-163.

Bergan, T. 1978. Phage typing of Pseudomonas aeruginosa. In Bergan and Norris (Editors) Methods in Microbiology Vol. 10, Academic Press, London, pp. 169-199.

Bergan, T. 1979. Bacteriophage typing of Shigella. In Bergan and Norris (Editors), Methods in Microbiology, Vol. 13, Academic Press, New York, pp. 177-286.

Bergan, T. 1981. Human- and animal-pathogenic members of the genus Pseudomonas. In Starr, Stolp, Trüper, Balows and Schlegel (Editors) The Prokaryotes, a handbook on habitats, isolation and identification of bacteria, Springer-Verlag, Berlin, pp. 666-700.

Bergeman, A.M. 1909. Die rote Beulenkrankheit des Aals. Ber. Bayer Biol. Vers. Sta. 2: 10-54.

Berger, U. 1960. Neisseria animalis nov. spec. Z. Hyg. Infektionskr. 147: 158-161.

Berger, U. 1961. Untersuchungen ueber die Pigmentbildung durch Neisseria. Z. Hyg. 147: 461-469.

Berger, U. 1962. Über das Vorkommen von Neisserien bei einigen Tieren. Z. Hyg. Infektionskr. 148: 445-457.

Berger, U. 1963. Die anspruchslosen Neisserien. Ergebn. Mikrobiol. Immun. Exp. Ther. 36: 97-167.

Berger, U. 1970. Untersuchungen yur Reduktionvon Nitrat und Nitrit durch Neisseria gonorrhoeae und Neisseria meningitidis. Z. Med. Mikrobiol. Immunol. *156:* 86-89.

Berger, U. 1971. *Neisseria mucosa* var. *heidelbergensis.* Z. Med. Mikrobiol. Immunol. *156:* 154-158.

Berger, U., I. Aboulkchair and W. Rottman. 1974. Sepsis und meningitis durch *Neisseria mucosa* var. *heidelbergensis.* Infection *2:* 108-110.

Berger, U. and H. Brunhoeber. 1961. *Neisseria flava* (Bergey et al. 1923). Art oder Varietät? Z. Hyg. Infektionskr. *148:* 39-44.

Berger, V. and B.W. Catlin. 1975. Biochemical differentiation between *N. sicca* and *N. perflava.* Zentralbl. Bakteriol. Parasitenkd. Infektionskr. Hyg. Abt. I. Orig. Reike A *232:* 129-130.

Berger, U. and E. Falsen. 1976. Über die Artenverteilung von *Moraxella* und *Moraxella* - ahnlichen Keimen im Nasopharynx gesunder Erwachsener. Med. Mikrobiol. Immunol. *162:* 239-249.

Berger, U. and R. Issi. 1971. Resistenz gegen Acetazolamid als taxonomisches Kriterium bei *Neisseria.* Arch. Hyg. *154:* 540-544.

Berger, U. and M. Miersch. 1970. Zuna normalen Vorkommen von *Neisseria mucosa* (Véron et al. 1959) Z. Med. Mikrobiol. Immunol. *155:* 186-191.

Berger, U. and H.D. Piotrowski. 1974. Die biochemische Diagnose von *Neisseria elongata* (Bøvre und Holten, 1970). Med. Microbiol. Immunol. *159:* 309-316.

Berger, U. and K. Schlez. 1970. Untersuchungen auf Meningokokkentrager in Kindergarten und Kinderheimen. Z. Kinderheilk. *108:* 54-60.

Berger, B., G. Thorington and L. Margulis. 1979. Two aeromonads: growth of symbionts from *Hydra viridis.* Curr. Microbiol. *3:* 5-10.

Berger, U. and B. Wulf. 1961. Untersuchungen an saprophytischen Neisserien. Z. Hyg. Infektionskr. *147:* 257-268.

Bergersen, F.J. 1980. Methods for evaluating biological nitrogen fixation. John Wiley and Sons, Chichester.

Bergey, D.H. and R.S. Breed. 1948. Genus III. *Flavobacterium.* Bergey et al. *In* Breed, Murray and Hitchens (Editors), Bergey's Manual of Determinative Bacteriology, 6th ed. The Williams and Wilkins Co., Baltimore, pp. 427-442.

Bergey, D.H., R.S. Breed, B.W. Hammer, F.M. Huntoon, E.G.D. Murray and F.C. Harrison. 1934. Bergey's Manual of Determinative Bacteriology, 4th Ed. The Williams and Wilkins Co., Baltimore, pp. 1-664.

Bergey, D.H., R.S. Breed, E.G.D. Murray and A.P. Hitchens (Editors). 1939. Bergey's Manual of Determinative Bacteriology, 5th Ed. The Williams and Wilkins Co., Baltimore.

Bergey, D.H., F.C. Harrison, R.S. Breed, B.W. Hammer and F.M. Huntoon. 1923. Bergey's Manual of Determinative Bacteriology, 1st ed. The Williams and Wilkins Co., Baltimore. pp. 1-442.

Bergey, D.H., F.C. Harrison, R.S. Breed, B.W. Hammer and F.M. Huntoon. 1925. Bergey's Manual of Determinative Bacteriology, 2nd ed. The Williams and Wilkins Co., Baltimore. pp. 1-462.

Bergey, D.H., F.C. Harrison, R.S. Breed, B.W. Hammer and F.M. Huntoon. 1930. Bergey's Manual of Determinative Bacteriology, 3rd Ed., The Williams and Wilkins Co., Baltimore, pp. 1-589.

Bergonzini, C. 1881. Sopra un nuovo bactério colorato. Annuar. Soc. Nat. Modena, Ser. 2, *14:* 149-158.

Beringer, J.E. and D.A. Hopwood. 1976. Chromosomal recombination and mapping in *Rhizobium leguminosarum.* Nature (London) *264:* 291-293.

Berkeley, C. 1933. The oxidase and dehydrogenase systems of the crystalline style of *Mollusca.* Biochem. J. *27:* 1357-1365.

Berkeley, C. 1959. Some observations on *Cristispira* in the crystalline style of *Saxidomus giganteus* Deshayes and in that of some other *Lamellibranchiata.* Can. J. Zool. *37:* 53-58.

Berkeley, C. 1962. Toxicity of plankton to *Cristispira* inhabiting the crystalline style of a mollusk. Science *135:* 664-665.

Berkowitz, D.M. and W.S. Lee. 1973. A selective medium for isolation and identification of *Serratia marcescens.* Abstr. Annu. Mtg. Amer. Soc. Microbiol., p. 105.

Berlier, Y. M. and P. A. Lespinat. 1980. Mass-spectrometric kinetic studies of the nitrogenase and hydrogenase activities in *in vivo* cultures of *Azospirillum brasilense.* Arch Microbiol. *125:* 67-72.

Bernaerts, M.J. and J. De Ley. 1960a. Microbial formation and preparation of 3-ketoglycosides from disaccharides. J. Gen. Microbiol. *22:* 129-136.

Bernaerts, M.J. and J. De Ley. 1960b. The structure of 3-keto-glycosides formed from disaccharides by certain bacteria. J. Gen. Microbiol. *22:* 137-146.

Bernaerts, M.J. and J. De Ley. 1963. A biochemical test for crown gall bacteria. Nature (London) *197:* 406-407.

Bernard, F.R. 1970. Occurrence of the spirochaete genus *Cristispira* in western Canadian marine bivalves. Veliger *13:* 33-36.

Bernard, U., I. Probst and H.G. Schlegel. 1974. The cytochromes of some hydrogen bacteria. Arch. Microbiol. *95:* 29-37.

Bernardi, G. 1969a. Chromatography of nucleic acids on hydroxyapatite. I. Chromatography of native DNA. Biochim. Biophys. Acta *174:* 423-434.

Bernardi, G. 1969b. Chromatography of nucleic acids on hydroxyapatite. II. Chromatography of denatured DNA. Biochim. Biophys. Acta *174:* 435-448.

Berndt, H., D.J. Lowe and Yates, M.G. 1978. The nitrogen-fixing system of *Corynebacterium autotrophicum.* Purification and properties of the nitrogenase components and two ferredoxins. Eur. J. Biochem. *86:* 133-142.

Berndt, H., K.P. Ostwal, J. Lalucat, C. Schumann, F. Mayer and H.G. Schlegel. 1976. Identification and physiological characterization of the nitrogen fixing bacterium *Corynebacterium autotrophicum* GZ29. Arch. Microbiol. *108:* 17-26.

Berndt, H. and D. Wölfle. 1979. Hydrogenase: its role as electron generating

enzyme in the nitrogen fixing bacterium *Xanthobacter autotrophicus. In* Schlegel and Schneider (Editors), Hydrogenases: Their Catalytic Activity, Structure and Function, Erich Goltze KG, Göttingen, pp. 327-351.

Bernheim, F., M.L.C. Bernheim and M.D. Webster. 1935. Oxidation of certain amino acids by "resting" *Bacillus proteus.* J. Biol. Chem. *110:* 165-172.

Berniac, M. 1974. Une maladie bactérienne de *Xanthosoma sagittifolium* (L.) Schott. Ann. Phytopathol. *6:* 197-202.

Bernier, I., J.-C. Landureau, P. Grellet and P. Jolles. 1974. Characterization of chitinases from haemolymph and cell cultures of cockroach (*Periplaneta americana*). Comp. Biochem. Physiol. B Comp. Biochem. *47:* 41-44.

Berridge, E.M. 1924. The influence of hydrogen-ion concentration on the growth of certain bacterial plant parasites and saprophytes. Ann. Appl. Biol. *11:* 73-85.

Berthet, J.A. and G. Bondar. 1915. Molestia bacteriana da mandioca. Bol. Agr. São Paulo *16:* 513-524.

Bertschinger, H.U. and P. Seifert. 1978. Isolation of a *Pasteurella haemolytica*-like organism from porcine necrotic pleuropneumonia. 5th I.P.V.S. World Congress on Hyology and Hyiatrics, Zagreb, M19.

Bettelheim, K.A., J.F. Gordon and J. Taylor. 1968. The detection of a strain of *Chromobacterium lividum* in the tissues of certain leaf-nodulated plants by the immunofluorescence technique. J. Gen. Microbiol. *54:* 177-184.

Betz, J.L., P.R. Brown, M.J. Smyth and P.H. Clarke. 1974. Evolution in action. Nature *247:* 261-264.

Betz, J.L. and P.H. Clarke. 1972. Selective evolution of phenylacetamide-utilizing strains of *Pseudomonas aeruginosa.* J. Gen. Microbiol. *73:* 161-174.

Beuchat, L.R. 1977. Evaluation of enrichment broths for enumerating *Vibrio parahaemolyticus* in chilled and frozen crab meat. J. Food Protect. *40:* 592-595.

Bevan, L.G.W. 1930. Blood culture in undulant fever. Brit. Med. J. *2:* 267.

Beveridge, T. J. and R. G. E. Murray. 1976. Dependence of the superficial layers of *Spirillum putridiconchylium* on Ca^{2+} or Sr^{2+}. Can. J. Microbiol. *22:* 1233-1244.

Beveridge, W.I.B. 1938. Foot-rot in sheep: a preliminary note on the probable causal agent. J. Counc. Sci. Indust. Res. Aust. *11:* 1-3.

Beveridge, W.I.B. 1941. Foot-rot in sheep: a transmissible disease due to infection with *Fusiformis nodosus* (n. sp.). Studies on its causes, epidemiology and control. Counc. Sci. Indust. Res. Aust. Bull. *140:* 1-56.

Bewig, F. and W. Schwartz. 1956. Untersuchungen über die Symbiose von Tieren mit Pilzen und Bakterien. VII. Über die Physiologie der Symbiose bei einigen blutsaugenden Insekten. Arch. Mikrobiol. *24:* 174-208.

Bey, R.F. and R.C. Johnson. 1978. Protein-free and low-protein media for the cultivation of *Leptospira.* Infect. Immun. *19:* 562-569.

Bhaskaran, K. 1958. Genetic recombination in *Vibrio cholerae.* J. Gen. Microbiol. *19:* 71-75.

Bhaskaran, K. 1959. Observations of the nature of genetic recombination in *Vibrio cholerae.* Indian J. Med. Res. *47:* 253-260.

Bhat, J.V. and K. Rijsinghani. 1955. Studies on *Acetobacter.* I. Isolation and characterization of the species. Proc. Indian Acad. Sci. *41:* 209-219.

Bhat, P., R.M. Myers and J.P. George. 1967. *Edwardsiella tarda* in a study of juvenile diarrhoea. J. Hyg. Camb. *65:* 293-298.

Bhat, P., S. Shanthakumari and D. Rajan. 1974. The characterization and significance of *Plesiomonas shigelloides* and *Aeromonas hydrophila* isolated from an epidemic of diarrhoea. Indian J. Med. Res. *62:* 1051-1060.

Bhatnagar, R.D.S. and A.J. Musgrave. 1970. Cytochemistry, morphogenesis, and tentative identification of mycetomal microorganisms of *Sitophilus granarius* L. (Coleoptera). Can. J. Microbiol. *16:* 1357-1362.

Bhattacharyya, F.K. 1975. *Vibrio cholerae* flagellar antigen: a sero-diagnostic test, functional implications of H-reactivity and taxonomic importance of cross-reactions within the *Vibrio* genus. Med. Microbiol. Immunol. *162:* 29-41.

Bhonghbhibat, N., S. Elberg and T.H. Chen. 1970. Characterisation of *Brucella* skin test antigens. J. Infect. Dis. *122:* 70-82.

Bibb, W.F., R.J. Sorg, B.M. Thomason, M. Hicklin, A.G. Steigerwalt, D.J. Brenner and M.R. Wulf. 1981. Recognition of a second serogroup of *Legionella longbeachae.* J. Clin. Microbiol. *14:* 674-677.

Biberfeld, G. and P. Biberfeld. 1970. Ultrastructural features of *Mycoplasma pneumoniae.* J. Bacteriol. *102:* 855-861.

Biberstein, E.L. 1978. Biotyping and serotyping of *Pasteurella haemolytica. In* Bergan and Norris (Editors), Methods in Microbiology, Vol. 10, Academic Press, New York, pp. 253-269.

Biberstein, E.L. 1981. *Haemophilus-Pasteurella-Actinobacillus:* Their significance in veterinary medicine. *In* Kilian, Frederiksen and Biberstein (Editors), *Haemophilus, Pasteurella* and *Actinobacillus.* Academic Press, London, pp. 57-76.

Biberstein, E.L. and C.K. Francis. 1968. Nucleic acid homologies between the A and T types of *Pasteurella haemolytica.* J. Med. Microbiol. *1:* 105-108.

Biberstein, E.L., M. Gills and H. Knight. 1960. Serological types of *Pasteurella hemolytica.* Cornell Vet. *50:* 283-300.

Biberstein, E.L., A. Gunnarsson and B. Hurvell. 1977. Cultural and biochemical criteria for the identification of *Haemophilus* cultures from swine. Am. J. Vet. Res. *38:* 7-11.

Biberstein, E.L., M.E. Meyer and P.C. Kennedy. 1958. Colonial variation of *Pasteurella haemolytica* isolated from sheep. J. Bacteriol. *76:* 445-452.

Biberstein, E.L., P.D. Mini and M.G. Gills. 1963. Action of *Haemophilus* cultures on α-aminolevulinic acid. J. Bacteriol. *86:* 814-819.

Biberstein, E.L. and P.D. Spencer. 1962. Oxidative metabolism of *Haemophilus* species grown at different levels of hemin supplementation. J. Bacteriol. *84:* 916-920.

Biberstein, E.L. and D.C. White. 1969. A proposal for the establishment of two new *Haemophilus* species. J. Med. Microbiol. *2:* 75-78.

Bidan, P. 1959. La maladie du framboisé dans les cidres. Ind. Aliment. Agric. *76:* 31-33.

Biebl, H. and N. Pfennig. 1977. Growth of sulfate-reducing bacteria with sulfur as electron acceptor. Arch. Microbiol. *112:* 115-117.

Bieger, C.D. and I.P. Crawford. 1978. Genes of tryptophan biosynthesis in the marine luminous bacterium *Beneckea harveyi*. Abst. Annu. Meet. Am. Soc. Micro. 1978, K204, p. 160.

Biggins, D.R. and J.R. Postgate. 1969. Nitrogen fixation by cultures and cell-free extracts of *Mycobacterium flavum* 301. J. Gen. Microbiol. *56:* 181-193.

Biggins, D.R. and J.R. Postgate. 1971. Nitrogen fixation by extracts of *Mycobacterium flavum* 301. Use of natural electron donors and oxygen-sensitivity of cell-free preparations. Eur. J. Biochem. *19:* 408-415.

Bigland, C.H. and M.W. Warenycia. 1978. Effects of biotin, folic acid, and pantothenic acid on the growth of *Mycoplasma meleagridis*, a turkey pathogen. Poultry Sci. *57:* 611-618.

Bignell, D.E. and J.M. Anderson. 1980. Determination of pH and oxygen status in the guts of lower and higher termites. J. Insect Physiol. *26:* 183-188.

Billing, E. 1963. The value of phage sensitivity tests for the identification of phytopathogenic *Pseudomonas* species. J. Appl. Bacteriol. *26:* 193-210.

Billing, E. 1970a. Further studies on the phage sensitivity and the determination of phytopathogenic *Pseudomonas* species. J. Appl. Bacteriol. *33:* 478-491.

Billing, E. 1970b. *Pseudomonas viridiflava* (Burkholder 1930; Clara 1934). J. Appl. Bacteriol. *33:* 492-500.

Billing, E. and L.A.E. Baker. 1963. Characteristics of *Erwinia*-like organisms found in plant material. J. Appl. Bacteriol. *26:* 58-65.

Bilos, Z.J., A. Kucharchuk and W. Metzger. 1978. *Eikenella corrodens* in human bites. Clinical Orthopaedics *134:* 320-324.

Bird, R.G. and P.C.C. Garnham. 1967. *Aegyptianella pullorum* Carpano 1928—fine structure and taxonomy. J. Protozool. *14:* 42 (Suppl.).

Bird, R.G. and P.C.C. Garnham. 1969. *Aegyptianella pullorum* Carpano 1928—fine structure and taxonomy. Parasitology 59: 745-752.

Bird, C.W., C.M. Lynch, F.J. Pirr, W.W. Reid, C.J.W. Brooks and B.C. Middleditch. 1971. Steroids and squalene in *Methylococcus capsulatus* grown on methane. Nature (London) *230:* 473-474.

Birkeland, N.K. and T. Hofstad. 1982. Chemical composition, ultrastructure and some serological properties of lipopolysaccharides from *Leptotrichia buccalis*. Acta Pathol. Microbiol. Immunol. Scand. Sect. B *90:* 329-334.

Bisgaard, M. 1975. Characterization of atypical *Actinobacillus lignieresii* isolated from ducks with salpingitis and peritonitis. Nord. Veterinaermed. *27:* 378-383.

Bisgaard, M., R. Sakazaki and T. Shimada. 1978. Prevalence of non-cholera vibrios in cavum nasi and pharynx of ducks. Acta Pathol. Microbiol. Scand. Sect. B *86:* 261-266.

Bishop, P.E., J.K. Gordon, V.K. Shah and W.J. Brill. 1977. Transformation of nitrogen fixation genes in *Azotobacter*. In A. Hollaender (Editor), Genetic Engineering for Nitrogen Fixation. Plenum Publishing Co., N.Y. pp. 67-76.

Bismanis, J.E. 1976. Endosymbionts of *Sitodrepa panicea*. Can. J. Microbiol. *22:* 1415-1424.

Bisset, K.A. and C.M.F. Hale-McCaughey. 1967. The electron microscopy of ultra-(thin) sections of the swollen elements in a strain of *Streptobacillus*. G. Microbiol. *15:* 137-139.

Biswas, G.D., T. Sox, E. Blackman and P.F. Sparling. 1977. Factors affecting genetic transformation of *Neisseria gonorrhoeae*. J. Bacteriol. *129:* 983-992.

Bizio, B. 1823. Lettera di Bartolomeo Bizio al chiarissimo canonico Angelo Bellani sopra il fenomeno della polenta porporina. Biblioteca Italiana o sia Giornale di Letteratura Scienze e Arti (Anno VIII), *30:* 275-295.

Black, F.T. 1973a. Biological and physical properties of human T-mycoplasmas. Ann. N.Y. Acad. Sci. *225:* 131-143.

Black, F.T. 1973b. Phosphatase activity in T-mycoplasmas. Int. J. Syst. Bacteriol. *23:* 65-66.

Black, F.T. 1973c. Modifications of the growth inhibition test and its application to human T-mycoplasmas. Appl. Microbiol. *25:* 528-533.

Black, F.T., A. Birch-Andersen and E.A. Freundt. 1972a. Morphology and ultrastructure of human T mycoplasmas. J. Bacteriol. *111:* 254-259.

Black, F.T., C. Christiansen and G. Askaa. 1972b. Genome size and base composition of deoxyribonucleic acid from eight human T-mycoplasmas. Int. J. Syst. Bacteriol. *22:* 241-242.

Black, F.T., E.A. Freundt, O. Vinther and C. Christiansen. 1979. Flagellation and swimming motility of *Thermoplasma acidophilum*. J. Bacteriol. *137:* 456-460.

Black, L.M. 1941. Further evidence for multiplication of the aster-yellows virus in the aster leafhopper. Phytopathology *31:* 120-135.

Blackburn, T.H. 1968. Protease production by *Bacteroides amylophilus* strain H18. J. Gen. Microbiol. *53:* 27-36.

Blackburn, T.H. and P.N. Hobson. 1962. Further studies on the isolation of proteolytic bacteria from the sheep rumen. J. Gen. Microbiol. *29:* 69-81.

Blackkolb, F. and H.G. Schlegel. 1968. Regulation der Glucose-6-phosphat-Dehydrogenase aus *Hydrogenomonas* H 16 durch ATP und NADH₂. Arch. Mikrobiol. *63:* 177-196.

Blackwood, A.-C., G. Guimberteau and E. Peynaud. 1969. Sur les bactéries acetiques isolées de raisins. C. R. Hebd. Séances Acad. Sci. Ser. D *269:* 802-804.

Bladen, H.A., M.P. Bryant and R.N. Doetsch. 1961. A study of bacterial species from the rumen which produce ammonia from protein hydrolysate. Appl. Microbiol. *9:* 175-180.

Blair, E.B. 11970. Media, test procedures and chemical reagents. In Bodily, Updike and Mason (Editors), Diagnostic Procedures for Bacterial, Mycotic and Parasitic Infections, 5th Ed. American Public Health Association, New York, pp. 816-817.

Blake, P.A., D.T. Allegra, J.D. Snyder, T.J. Barrett, L. McFarland, C.T. Caraway, J.C. Feeley, J.P. Craig, J.V. Lee, N.D. Puhr and R.A. Feldman. 1980a. Cholera - a possible endemic focus in the United States. N. Engl. J. Med. *302:* 305-309.

Blake, P.A., M.H. Mersen, R.E. Weaver, D.G. Hollis and P.C. Heublein. 1979. Disease caused by a marine *Vibrio*. Clinical characteristics and epidemiology. N. Engl. J. Med. *300:* 1-5.

Blake, P.A., R.E. Weaver and D.G. Hollis. 1980b. Diseases of humans (other than cholera) caused by vibrios. Annu. Rev. Microbiol. *34:* 341-367.

Blakemore, R.P. and E. Canale-Parola. 1973. Morphological and ecological characteristics of *Spirochaeta plicatilis*. Arch. Mikrobiol. *89:* 273-289.

Blakemore, R.P. and E. Canale-Parola. 1976. Arginine catabolism by *Treponema denticola*. J. Bacteriol. *128:* 616-622.

Blakemore, R. P., D. Maratea and R. S. Wolfe. 1979. Isolation and pure cullture of a freshwater magnetic spirillum in chemically defined medium. J. Bacteriol. *140:* 720-729.

Blanchard, R. 1906. Spirilles, spirochetes et autres microorganismes a corps spirale. Sem. Med. *26:* 1-5.

Blaser, M. J., C. W. Moss and R. E. Weaver. 1980. Cellular fatty acid composition of *Campylobacter fetus* J. Clin. Microbiol. *11:* 448-451.

Bloch, M. 1918. Beiträg zür Untersuchungen über die *Zoogloea ramigera* (Itzigsohn) auf Grund von Reinkulturen. Zentralbl. Bakteriol. Parasitenkd. Infektionskr. Hyg. Abt. II *48:* 44-62.

Blochmann, F. 1884. Ueber eine Metamorphose der Ovarialeiern und über den Beginn der Blastoderm-bildung bei den Ameisen. Verh. Naturhist.-med. Ver. Heidelb. *3:* 243-247.

Blochmann, F. 1886. Ueber die Reifung der Eier bei Ameisen und Wespen. Naturhist.-med. Ver. Heidelb. Festschr. Ruperto-Carola, pp. 141-172.

Block, R.J. and S.M. Henry. 1961. Metabolism of the sulphur amino-acids and of sulphate in *Blattella germanica*. Nature (Lond.) *191:* 392-393.

Bloodgood, R.A. and T.P. Fitzharris. 1976. Specific associations of prokaryotes with symbiotic flagellate protozoa from the hindgut of the termite *Reticulitermes* and the wood-eating roach *Cryptocercus*. Cytobios 17: 103-122.

Bloodgood, R.A. and T.P. Fitzharris. 1976. Specific associations of prokaryotes with symbiotic flagellate Protozoa from the hindgut of the termite *Reticulotermes* and the wood-eating roach *Cryptocercus*. Cytobios *17:* 103-122.

Bloodgood, R.A., K.R. Miller, T.P. Fitzharris and J.R. McIntosh. 1974. The ultrastructure of *Pyrsonympha* and its associated microorganisms. J. Morphol. *143:* 77-106.

Board, R.G. 1965. The properties and classification of the predominant bacteria in rotten eggs. J. Appl. Bacteriol. *28:* 437-453.

Boatman, E.S. 1979. Morphology and ultrastructure of the Mycoplasmatales. In Barile and Razin (Editors), The Mycoplasmas, Vol. I, Academic Press, New York, pp. 63-102.

Bobo, R.A. and R.G. Eagon. 1968. Lipids of cell walls of *Pseudomonas aeruginosa* and *Brucella abortus*. Can. J. Microbiol. *14:* 503-513.

Bockemühl, J. 1972. Die Lysosensibilität von Stämmen der *Salmonella* Subgenera I-IV gegenüber dem Phagen 0-1. Ihre mögliche Bedeutung für die Klassifikation des Genus *Salmonella*. Med. Microbiol. Immunol. *158:* 44-53.

Boe, J. and T. Thjotta. 1944. The position of *Fusobacterium* and *Leptotrichia* in the bacteriological system. Acta Pathol. Microbiol. Scand. *21:* 441-450.

Boháček, J. and O. Mraz. 1967. Basengehalt der Desoxyribonukleinsäure bei den Arten *Pasteurella haemolytica*, *Actinobacillus lignieresii* und *Actinobacillus equuli*. Zentralbl. Bakteriol. Parasitenkd. Infektionskr. Hyg. Abt. 1, Orig. *202:* 468-478.

Bohlool, B.B. and T.D. Brock. 1974. Immunodiffusion analysis of membranes of *Thermoplasma acidophilum*. Infect. Immun. *10:* 280-281.

Bohn, G.W. and J.C. Maloit. 1946. Bacterial spot of native golden currant (*Ribes aureum*). J. Agr. Res. *73:* 281-290.

Boisvert, H., R. Chatelain and M.-J. Bassot. 1967. Étude d'un *Photobacterium* isolé de l'organe lumineux de poissons *Leiognathidae*. Ann. Inst. Pasteur Paris *112:* 520-524.

Bokkenheuser, V. D., N. J. Richardson, J. H. Bryner, D. J. Rouy, A. R. Shutte, H. J. Koornhof, I. Freiman and E. Hartman. 1979. Detection of enteric campylobacteriosis in children. J. Clin. Microbiol. *9:* 227-232.

Boling, M.E., D.P. Allison and J.K. Setlow. 1973. Bacteriophage of *Haemophilus influenzae*. III. Morphology, DNA homology, and immunity properties of HPlcl, S2, and the defective bacteriophage from strain Rd. J. Virol. *11:* 585-591.

Bollag, J.M. and S. Russel. 1976. Aerobic versus anaerobic metabolism of halogenated anilines by a *Paracoccus* sp. Microb. Ecol. *3:* 65-73.

Bollgren, I., G. Källenius, C.E. Nord and J. Winberg. 1979. Periurethral anaerobic microflora of healthy girls. J. Clin. Microbiol. *10:* 419-424.

Bomhoff, G., P.M. Klapwijk, H.C.M. Kester, R.A. Schilperoort, J.P. Hernalsteens and J. Schell. 1976. Octopine and nopaline synthesis breakdown genetically controlled by a plasmid of *Agrobacterium tumefaciens*. Molec. Gen. Genet. *145:* 177-181.

Bonami, J.R. and R. Pappalardo. 1980. Rickettsial infection in marine crustacea. Experientia *36:* 180-181.

Bonde, R. 1939a. Comparative studies of the bacteria associated with blackleg and seed-piece decay. Phytopathology *29:* 831-851.

Bonde, R. 1939b. The role of insects in the dissemination of potato blackleg and seed-piece decay. J. Agric. Res. *59:* 889-917.

Bonnemaison, L. 1946. Remarques sur la symbiose chez les Pentatomidae (Hem.). Bull. Soc. Entomol. Fr. *51:* 40-42.

Boone, D.R. and M.P. Bryant. 1980. Propionate-degrading bacterium, *Syntrophobacter wolinii* sp. nov. gen. nov., from methanogenic ecosystems. Appl. Environ. Microbiol. *40:* 626-632.

Booth, S.J., J.L. Johnson and T.D. Wilkins. 1977. Bacteriocin production by strains of *Bacteroides* isolated from human feces and the role of these strains in the bacterial ecology of the colon. Antimicrob. Agents Chemother. *11:* 718-724.

Bopp, M. 1965. Die Hemmung von *Agrobacterium tumefaciens* durch D-Aminosäuren. Z. Naturforschg. *20:* 899-905.

Bopp, C.A., J.W. Sumner, G.K. Morris and J.G. Wells. 1981. Isolation of *Legionella* from environmental water samples by low pH treatment and use of selective medium. J. Clin. Microbiol. *13:* 714-719.

Bordet, J. and O. Gengou. 1906. Le microbe de la coqueluche. Ann. Inst. Pasteur (Paris) *20:* 731-741.

Borg-Petersen, C. 1971. A thermo-labile antigen in the leptospira strain Ictero No. 1. Trop. Geogr. Med. *23:* 282-285.

Borg-Petersen, C. 1974. Thermo-labile agglutinogens in leptospires of the Pomona serogroup. Folia Fac. Med. Univ. Comenianae Bratisl. *12:* (supplementum): 111-121.

Borman, E.K., C.A. Stuart and K. Wheeler. 1944. Taxonomy of the family *Enterobacteriaceae*. J. Bacteriol. *48:* 351-367.

Borowski, J. and M. Zaremba. 1973. Some problems connected with *Yersinia pseudotuberculosis* resistance to antibiotics. Contr. Microbiol. Immunol. *2:* 196-202.

Borrall, R. and J.M. Larkin. 1978. *Flectobacillus marinus* (Raj) comb. nov., a marine bacterium previously assigned to *Microcyclus*. Int. J. Syst. Bacteriol. *28:* 341-343.

Borrel, A., E. Dujardin-Beaumetz, Jeantet and C. Jouan. 1910. Le microbe de la péripneumonie. Ann. Inst. Pasteur (Paris) *24:* 168-179.

Bosanquet, W.C. 1911. Brief notes on the structure and development of *Spirochaeta anodontae* Keysselitz. Quart. J. Microsc. Sci. *56:* 387-394.

Bosch, C. 1976. Sur un nouveau type de symbiose chez la Bonellie *(Bonellia viridis*, Echiurien). C.R. Acad. Sci. Paris, Ser. D *282:* 2179-2182.

Boskamp, E. 1922. Ueber Bau, Lebensweise und systematische Stellung von *Selenomonas palpitans* (Simons). Zentralbl. Bakteriol. Parasitenkd. Infektionskr. Hyg. Abt. I Orig. *88:* 58-73.

Bothe, H. and M.G. Yates. 1976. The electron transport to nitrogenase in *Mycobacterium flavum*. Arch. Microbiol. *107:* 25-31.

Bottone, E.J. 1977. *Yersinia enterocolitica*: a panoramic view of a charismatic microorganism. Crit. Rev. Microbiol. *5:* 211-241.

Bottone, E.J., J. Kittick and S.S. Schneierson. 1973. Isolation of Bacillus HB-1 from human clinical sources. Amer. J. Clin. Pathol. *59:* 560-566.

Bottone, E.J., K.K. Sandh and M.A. Pisano. 1979. *Yersinia intermedia*: Temperature dependent bacteriocin production. J. Clin. Microbiol. *10:* 433-436.

Boucher, C. and L. Sequeira. 1978. Evidence for the cotransfer of genetic markers in *Pseudomonas solanacearum*. Can. J. Microbiol. *24:* 69-72.

Bouley, G. 1966. Étude d'une souche d'*Actinobacillus suis* (van Dorssen et Jaartsveld) isolée en Normandie. Recl. Med. Vet. Ec. Alfort *142:* 25-29.

Bourgault, A.-M. and J.E. Rosenblatt. 1979. Characterization of anaerobic Gram-negative bacilli by using rapid slide tests for beta lactamase production. J. Clin. Microbiol. *9:* 654-656.

Bourgeois, A.L. and G.A. Dasch. 1981. The species-specific surface protein antigen of *Rickettsia typhi*: immunogenicity and protective efficacy in guinea pigs. *In* Burgdorfer and Anacker (Editors), Rickettsiae and Rickettsial Diseases, Academic Press, New York, pp. 71-80.

Bournier, A. 1960. Sur l'existence et l'evolution d'un mycétome au cours de l'embryologenèse de *Caudothrips buffai* Karny. Proc. XI Int. Congr. Entomol. *1:* 353-354.

Bousfield, E.G., G.G.H. Wright and T.K. Walker. 1947. Oxidation of glycerol by *Acetobacter* species. J. Inst. Brew. *53:* 258-262.

Boush, G.M. and F. Matsumura. 1967. Insecticidal degradation by *Pseudomonas melophthora*, the bacterial symbiote of the apple maggot. J. Econ. Entomol. *60:* 918-920.

Bovarnick, M.R., J.C. Miller and J.C. Snyder. 1950. The influence of certain salts, amino acids, sugars and proteins on the stability of rickettsiae. J. Bacteriol. *59:* 509-522.

Bovarnick, M.R. and J.C. Snyder. 1949. Respiration of typhus rickettsiae. J. Exp. Med. *89:* 561-565.

Bové, J. 1981. Mycoplasma infections in plants. Israel J. Clin. Sci. *17:* 572-585.

Bové, J.M. and C. Saillard. 1979. Cell biology of spiroplasmas. *In* Whitcomb and Tully (Editors), The Mycoplasmas, Vol. 3, Academic Press, New York, pp. 83-153.

Bové, J.M., C. Saillard, P. Junca, J.R. DeGorce-Dumas, B. Ricard, A. Nhami, R.F. Whitcomb, D. Williamson and J.G. Tully. 1982. Guanine-plus-cytosine

content, hybridization percentages, and EcoRI restriction enzyme profiles of spiroplasmal DNA. Rev. Inf. Dis., *4* (Suppl.): S129–S136.

Bovell, C. 1967. The effect of sodium nitrite on the growth of *Micrococcus denitrificans*. Arch. Mikrobiol. *59:* 13-19.

Bøvre, K. 1963. Affinities between *Moraxella* spp. and a strain of *Neisseria catarrhalis* as expressed by transformation. Acta Pathol. Microbiol. Scand. *58:* 528.

Bøvre, K. 1964. Studies on transformation in *Moraxella* and organisms assumed to be related to *Moraxella*. 2. Quantitative transformation reactions between *Moraxella nonliquefaciens* strains, with streptomycin resistance marked DNA. Acta Pathol. Microbiol. Scand. *62:* 239-248.

Bøvre, K. 1965a. Studies on transformation in *Moraxella* and organisms assumed to be related to *Moraxella*. 3. Quantitative streptomycin resistance transformation between *Moraxella bovis* and *Moraxella nonliquefaciens* strains. Acta Pathol. Microbiol. Scand. *63:* 42-50.

Bøvre, K. 1965b. Studies on transformation in *Moraxella* and organisms assumed to be related to *Moraxella*. 4. Streptomycin resistance transformation between asaccharolytic Neisseria strains. Acta Pathol. Microbiol. Scand. *64:* 229-242.

Bøvre, K. 1965c. Studies on transformation in *Moraxella* and organisms assumed to be related to *Moraxella*. 5. Streptomycin resistance transformation between serum-liquefying, nonhaemolytic moraxellae, *Moraxella bovis* and *Moraxella nonliquefaciens*. Acta Pathol. Microbiol. Scand. *65:* 435-449.

Bøvre, K. 1965d. Studies on transformation in *Moraxella* and organisms assumed to be related to *Moraxella*. 6. A distinct group of *Moraxella nonliquefaciens*-like organisms (the 19116/51 group). Acta Pathol. Microbiol. Scand. *65:* 641-652.

Bøvre, K. 1967. Transformation and DNA base composition in taxonomy with special reference to recent studies in *Moraxella* and *Neisseria*. Acta Pathol. Microbiol. Scand. *69:* 123-144.

Bøvre, K. 1970. Pulse-RNA-DNA hybridization between rod-shaped and coccal species of the *Moraxella-Neisseria* groups. Acta Pathol. Microbiol. Scand. Sect. B *78:* 565-574.

Bøvre, K. 1979. Proposal to divide the genus *Moraxella* Lwoff 1939 emend. Henriksen and Bøvre 1968 into two subgenera - subgenus *Moraxella* (Lwoff 1939) Bøvre 1979 and subgenus *Branhamella* (Catlin 1970) Bøvre 1979. Int. J. Syst. Bacteriol. *29:* 403-406.

Bøvre, K. 1980. Progress in classification and identification of *Neisseriaceae* based on genetic affinity. *In* Goodfellow and Board (Editors), Microbial Classification and Identification, Academic Press, London - New York, pp. 55-72.

Bøvre, K., T. Bergan and L.O. Frøhlm. 1970. Electron microscopical and serological characteristics associated with colony type in *Moraxella nonliquefaciens*. Acta Pathol. Microbiol. Scand. *78B:* 765-779.

Bøvre, K., M. Fiandt and W. Szybalski. 1969. DNA base composition of *Neisseria*, *Moraxella*, and *Acinetobacter*, as determined by measurement of buoyant density. Can. J. Microbiol. *15:* 335-338.

Bøvre, K. and L.O. Frøhlm. 1970. Correlation between the fimbriated state and competence of genetic transformation in *Moraxella nonliquefaciens* strains. Acta Pathol. Microbiol. Scand. *78B:* 526-528.

Bøvre, K. and L.O. Frøhlm. 1971. Competence of genetic transformation correlated with the occurrence of fimbriae in three bacterial species. Nature New Biol. *234:* 151-152.

Bøvre, K. and L.O. Frøhlm. 1972a. Variation of colony morphology reflecting fimbriation in *Moraxella bovis* and two reference strains of *M. nonliquefaciens*. Acta Pathol. Microbiol. Scand. *80B:* 629-640.

Bøvre, K. and L.O. Frøhlm. 1972b. Competence in genetic transformation related to colony type and fimbriation in three species of *Moraxella*. Acta Pathol. Microbiol. Scand. *80B:* 649-659.

Bøvre, K., L.O. Frøhlm, S.D. Henriksen and E. Holten. 1977. Relationship of *Neisseria elongata* subsp. *glycolytica* to other members of the family *Neisseriaceae*. Acta Pathol. Microbiol. Scand. *85B:* 18-26.

Bøvre, K., J.E. Fuglesang, N. Hagen, E. Jantzen and L.O. Frøhlm. 1976. *Moraxella atlantae* sp. nov. and its distinction from *Moraxella phenylpyruvica*. Int. J. Syst. Bacteriol. *26:* 511-521.

Bøvre, K., J.E. Fuglesang and S.D. Henriksen. 1972. *Neisseria elongata*. Presentation of new isolates. Acta Pathol. Microbiol. Scand. *80B:* 919-922.

Bøvre, K., J.E. Fuglesang, S.D. Henriksen, S.P. Lapage, H. Lautrop and J.J.S. Snell. 1974. Studies on a collection of Gram-negative bacterial strains showing resemblance to moraxellae: examination by conventional bacteriological methods. Int. J. Syst. Bacteriol. *24:* 438-446.

Bøvre, K. and N. Hagen. 1981. The family *Neisseriaceae*: rod-shaped species of the genera *Moraxella*, *Acinetobacter*, *Kingella*, and *Neisseria*, and the *Branhamella* group of cocci. *In* Starr, Stolp, Trüper, Balows and Schlegel (Editors), The Prokaryotes: A handbook on habitats, isolation and and identification of bacteria, Springer-Verlag, Berlin, pp. 1506-1529.

Bøvre, K., N. Hagen, B.P. Berdal and E. Jantzen. 1977b. Oxidase positive rods from cases of suspected gonorrhoea. A comparison of conventional, gas chromatographic and genetic methods of identification. Acta Pathol. Microbiol. Scand. *85B:* 27-37.

Bøvre, K. and S.D. Henriksen. 1967a. A new *Moraxella* species, *Moraxella osloensis*, and a revised description of *Moraxella nonliquefaciens*. Int. J. Syst. Bacteriol. *17:* 127-135.

Bøvre, K. and S.D. Henriksen. 1967b. A revised description of *Moraxella poly-*

morpha Flamm 1957, with a proposal of a new name, *Moraxella phenylpyrouvica* for this species. Int. J. Syst. Bacteriol. *17:* 343-360.

Bøvre, K. and S.D. Henriksen. 1976. Minimal standards for description of new taxa within the genera *Moraxella* and *Acinetobacter:* Proposal by the subcommittee on *Moraxella* and allied bacteria. Int. J. Syst. Bacteriol. *26:* 92-96.

Bøvre, K., S.D. Henriksen and V. Jonsson. 1974. Correction of specific epithet *kingii* in the combinations *Moraxella kingii* Henriksen and Bøvre 1968 and *Pseudomonas kingii* Jonsson 1970 to *kingae.* Int. J. Syst. Bacteriol. *24:* 307.

Bøvre, K. and E. Holten. 1970. *Neisseria elongata* sp. nov., a rod-shaped member of the genus *Neisseria.* Re-evaluation of cell shape as a criterion in classification. J. Gen. Microbiol. *60:* 67-75.

Bowden, G.H. and J.M. Hardie. 1971. Anaerobic organisms from the human mouth. *In* Shapton and Board (Editors), Isolation of Anaerobes. Academic Press, London and New York, pp. 177-205.

Bowdre, J. H., N. R. Krieg, P. S. Hoffman and R. M. Smibert. 1976. Stimulatory effect of dihydroxyphenyl compounds on the aerotolerance of *Spirillum volutans* and *Campylobacter fetus* subspecies *jejuni.* Appl. Environ. Microbiol. *31:* 127-133.

Bowien, B., A.M. Cook and H.G. Schlegel. 1974. Evidence for the in vivo regulation of glucose-6-phosphate dehydrogenase activity in *Hydrogenomonas eutropha* H 16 from measurements of the intracellular concentrations of metabolic intermediates. Arch. Microbiol. *97:* 273-281.

Bowien, B. and F. Mayer. 1978. Further studies on the quaternary structure of D-ribulose-1,5-biphosphate carboxylase from *Alcaligenes eutrophus.* Eur. J. Biochem. *88:* 97-107.

Bowien, B., F. Mayer, G.A. Codd and H.G. Schlegel. 1976. Purification, some properties and quaternary structure of the D-ribulose 1,5-diphosphate carboxylase of *Alcaligenes eutrophus.* Arch. Microbiol. *110:* 157-166.

Bowien, B., F. Mayer, E. Spiess, A. Pähler, U. Englisch and W. Saenger. 1980. On the structure of crystalline ribulosebiphosphate carboxylase from *Alcaligenes eutrophus.* Eur. J. Biochem. *106:* 405-410.

Bowien, B. and H.G. Schlegel. 1981. Physiology and biochemistry of aerobic, hydrogen-oxidizing bacteria. Annu. Rev. Microbiol. *35:* 405-452.

Bowser, P.R., R. Rosemark and C.R. Reiner. 1980. A preliminary report of vibriosis in cultured American lobsters, *Homarus americanus.* J. Invertebr. Pathol. *37:* 80-85.

Bowser, D.V., R.W. Wheat, J.W. Foster and D. Leong. 1976. Occurrence of quinovosamine in lipopolysaccharides of *Brucella* species. Infect. Immun. *9:* 772-774.

Bowyer, J.W. and E.C. Calavan. 1974. Antibiotic sensitivity in vitro of the mycoplasmalike organism associated with citrus stubborn disease. Phytopathology *64:* 346-349.

Boyce, J.M.H., J. Frazer and K. Zinnemann. 1969. The growth requirements of *Haemophilus aphrophilus.* J. Med. Microbiol. *2:* 55-62.

Boyce, K.J. and A.W. Edgar. 1966. Production of freeze-dried *Brucella abortus* Strain 19 vaccine using cells produced by continuous culture. J. Appl. Bacteriol. *29:* 401-408.

Boyd, R.J., A.C. Hildebrandt and O.N. Allen. 1970a. Electron microscopy of phages for *Agrobacterium tumefaciens.* Arch. Mikrobiol. *73:* 47-54.

Boyd, R.J., A.C. Hildebrandt and O.N. Allen. 1970b. Specificity patterns of *Agrobacterium tumefaciens* phages. Arch. Mikrobiol. *73:* 324-330.

Boyer, C.I., Jr., D.W. Bruner and J.A. Brown. 1958. A streptobacillus, the cause of tendon-sheath infection in turkeys. Avian Dis. *2:* 418-427.

Boyer, G. and F. Lambert. 1893. Sur deux nouvelles maladies du mûrier. C. R. Hebd. Séances Acad. Sci. *117:* 342-343.

Boyle, A.M. 1949. Further studies of the bacterial necrosis of the giant cactus. Phytopathology *39:* 1029-1052.

Bozeman, F.M. and B.L. Elisberg. 1963. Serological diagnosis of scrub typhus by indirect immunofluorescence. Proc. Soc. Exp. Biol. Med. *112:* 568-573.

Bozeman, F.M., B.L. Elisberg, J.W. Humphries, K. Runcik and D.B. Palmer, Jr. 1970. Serologic evidence of *Rickettsia canada* infection of man. J. Infect. Dis. *121:* 367-371.

Bozeman, F.M., J.W. Humphries and J.M. Campbell. 1968. A new group of rickettsia-like agents recovered from guinea pigs. Acta Virol. (Praha) *12:* 87-93.

Bozeman, F.M., J.W. Humphries, J.M. Campbell and P.L. O'Hara. 1960. Laboratory studies of the spotted fever group of rickettsiae. *In* Wisseman (Editor), Symposium on the spotted fever group of rickettsiae, Med. Sci. Publ. 7, Walter Reed Army Inst. Res., Washington, pp. 7-11.

Bozeman, F.M., S.A. Masiello, M.S. Williams and B.L. Elisberg. 1975. Epidemic typhus rickettsiae isolated from flying squirrels. Nature (London) *255:* 545-547.

Bozeman, F.M., A. Shirai, J.W. Humphries and H.S. Fuller. 1967. Ecology of Rocky Mountain spotted fever. II. Natural infection of wild mammals and birds in Virginia and Maryland. Am. J. Trop. Med. Hyg. *16:* 48-59.

Bozeman, F.M., D.E. Sonenshine, M.S. Williams, D.P. Chadwick, D.M. Lauer and B.L. Elisberg. 1981. Experimental infection of ectoparasitic arthropods with *Rickettsia prowazekii* (GvF-16 strain) and transmission to flying squirrels. Am. J. Trop. Med. Hyg. *30:* 253-263.

Bradbury, J.M. 1977. Rapid biochemical tests for characterization of the *Mycoplasmatales.* J. Clin. Microbiol. *5:* 531-534.

Bradbury, J.M. and J. McCarthy. 1981. Rupture of the digital flexor tendons of chickens after infection with *Mycoplasma iowae.* Vet. Rec. *109:* 428–429.

Brade, H. and H. Brunner. 1979. Serological cross-reactions between *Acinetobacter calcoaceticus* and chlamydiae. J. Clin. Microbiol. *10:* 819-822.

Bradley, D.E. 1967. Ultrastructure of bacteriophages and bacteriocins. Bacteriol. Rev. *31:* 230-314.

Bradley, D.E. 1972a. Evidence for the retraction of *Pseudomonas aeruginosa* RNA phage pili. Biochem. Biophys. Res. Comm. *47:* 142-149.

Bradley, D.E. 1972b. Stimulation of pilus formation in *Pseudomonas aeruginosa* by RNA bactériophage adsorption. Biochem. Biophys. Res. Comm. *47:* 1080-1087.

Bradley, D.E. 1974. The adsorption of *Pseudomonas aeruginosa* pilus-dependent bacteriophages to a host mutant with non-retractile pili. Virology *58:* 149-163.

Bradley, D.E. 1980a. Mobilization of chromosomal determinants for the polar pili of *Pseudomonas aeruginosa* PAO by FP plasmids. Can. J. Microbiol. *26:* 155-160.

Bradley, D.E. 1980b. A function of *Pseudomonas aeruginosa* polar pili: twitching motility. Can. J. Microbiol. *26:* 146-154.

Bradley, D.E. and T.L. Pitt. 1975. An immunological study of the pili of *Pseudomonas aeruginosa.* J. Hyg. (London) *74:* 419-430.

Bradshaw, M.W., R. Schneerson, J.C. Parke and J.B. Robbins. 1971. Bacterial antigens cross-reactive with the capsular polysaccharide of *Haemophilus influenzae* type b. Lancet *1:* 1095-1096.

Brancaccio, M. and G.G. Legendre. 1979. *Megasphaera elsdenii* endocarditis. J. Clin. Microbiol. *10:* 72-74.

Brandis, H. 1978. Vibriocin typing. *In* Bergan and Norris (Editors), Methods in Microbiology, Vol. 12. Academic Press Inc., London-New York-San Francisco, pp. 117-126.

Branefors, P. 1964. Transformation of streptomycin-resistance in *Bordetella pertussis.* Acta Pathol. Microbiol. Scand. *62:* 249-254.

Branefors-Helander, P. 1972. Serological studies of *Haemophilus influenzae.* II. Variation in the capsular antigen of type e strains. Int. Arch. Allergy Appl. Immunol. *43:* 908-920.

Branefors-Helander, P. 1977. The structure of the capsular antigen from *Haemophilus influenzae* type A. Carbohydr. Res. *56:* 117-121.

Branefors-Helander, P. 1979. Cross reactivity of the O antigens among *Haemophilus influenzae* type b strains. Int. Arch. Allergy Appl. Immunol. *52:* 150-154.

Branefors-Helander, P., B. Clausen, L. Kenne and B. Lindberg. 1979. Structural studies of the capsular antigen of *Haemophilus influenzae* type c. Carbohydr. Res. *76:* 197-202.

Branefors-Helander, P., L. Keene and B. Lindqvist. 1980. Structural studies of the capsular antigen from *Haemophilus influenzae* type f. Carbohydr. Res. *79:* 308-312.

Branham, S. 1930. A new meningococcus-like organism *(Neisseria flavescens* n. sp.) from epidemic meningitis. U.S. Public Health Serv. Rep. *45:* 845-846.

Braude, A.I. 1951. Studies in the pathology and pathogenesis of experimental brucellosis. I. A comparison of the pathogenicity of *Brucella abortus, Brucella melitensis,* and *Brucella suis* for guinea pigs. J. Infect. Dis. *89:* 76-86.

Braude, A.I. and J. Siemienski. 1960. Role of bacterial urease in experimental pyelonephritis. J. Bacteriol. *80:* 171-179.

Braun, A.C. 1978. Plant tumors. Biochem. Biophys. Acta *516:* 167-191.

Braun, K. 1928. Bericht über das autreten von Schädlingen und Krankheiten im Obstbau im Regierungsbezirk Stade wärend der Monate Juni, Juli, August 1927. Rev. Appl. Mycol. *7:* 177.

Braun, P., J.O. Klein, Y.H. Lee and E.H. Kass. 1970. Methodologic investigations and prevalence of genital mycoplasmas in pregnancy. J. Infect. Dis. *121:* 391-400.

Braunstein, H., M. Tomasulo, S. Scott and M.P. Chadwick. 1980. A biotype of *Enterobacteriaceae* intermediate between *Citrobacter* and *Enterobacter.* Amer. J. Clin. Pathol. *73:* 114-116.

Brecher, G. and V.B. Wigglesworth. 1944. The transmission of *Actinomyces rhodnii* Erikson in *Rhodnius prolixus* Stål (Hemiptera) and its influence on the growth of the host. Parasitology *35:* 220-224.

Bredt, W. 1979. Motility. *In* Barile and Razin (Editors). The Mycoplasmas, Vol. I, Academic Press, New York, pp. 141-155.

Bredt, W., H.H. Heunert, K.H. Höfling and B. Milthaler. 1973. Microcinematographic studies of *Mycoplasma hominis* cells. J. Bacteriol. *113:* 1223-1227.

Breed, R.S. 1939. Genus II. *Malleomyces* Pribram. *In* Bergey, Breed, Murray and Hitchens (Editors), Bergey's Manual of Determinative Bacteriology, 5th Ed., The Williams and Wilkins Co., Baltimore, pp. 298-300.

Breed, R.S. 1948. Genus I. *Pseudomonas* Migula. *In* Breed, Murray and Hitchens (Editors) Bergey's Manual of Determinative Bacteriology, 6th Ed., The Williams and Wilkins Co., Baltimore, pp. 82-150.

Breed, R.S. 1957. Genus II. *Aerobacter* Beijerinck, 1900. *In* Breed, Murray and Smith (Editors), Bergey's Manual of Determinative Bacteriology, 7th Ed., The Williams and Wilkins Co., Baltimore, pp. 341-344.

Breed, R.S. 1957. The genus *Pasteurella. In* Breed, Murray and Smith (Editors), Bergey's Manual of Determinative Bacteriology, 7th Ed., The Williams and Wilkins Co., Baltimore, pp. 395-402.

Breed, R.S. and E.F. Lessel. 1954. The classification of luminescent bacteria. Antonie van Leeuwenhoek J. Microbiol. Serol. *20:* 58-64.

Breed, R.S. and E.G.D. Murray. 1957. Family IV. *Enterobacteriaceae* Rahn, 1937. *In* Breed, R.S., E.G.D. Murray and N.R. Smith (Editors), Bergey's Manual of Determinative Bacteriology, 7th ed. The Williams and Wilkins Co., Baltimore, pp. 332-335.

Breed, R.S. and E.G.D. Murray. 1957. Genus VII. *Serratia* Bizio, 1823, emend. Breed and Breed, 1927. *In* Breed, R.S., E.G.D. Murray and N.R. Smith (Editors), Bergey's Manual of Determinative Bacteriology, 7th ed. The Williams and Wilkins Co., Baltimore, pp. 359-364.

Breed, R.S., E.G.D. Murray and A.P. Hitchens (Editors). 1948. Bergey's Manual of Determinative Bacteriology, 6th ed. Williams and Wilkins Co., Baltimore.

Breinl, A. 1906. On the specific nature of the spirochaete of the African tick fever. Lancet *1:* 1690-1691.

Brendle, J.J.., M. Rogul and A.D. Alexander. 1974. Deoxyribonucleic acid hybridization among leptospiral serotypes. Int. J. Syst. Bacteriol. *24:* 205-214.

Brennan, P.C., T.E. Fritz and R.J. Flynn. 1965. *Pasteurella pneumotropica*: cultural and biochemical characteristics and its association with disease in laboratory animals. Lab. Anim. Care *15:* 302-312.

Brenner, D.J. 1978. Characterization and clinical identification of *Enterobacteriaceae* by DNA hybridization. Prog. Clin. Pathol. *7:* 71-117.

Brenner, D.J. 1981. Introduction to the family *Enterobacteriaceae*. *In* M.P. Starr, H. Stolp, H.G. Trüper, A. Balows and H.G. Schlegel (Editors), The Prokaryotes, A handbook on habitats, isolation, and identification of bacteria, Springer-Verlag, New York, pp. 1105-1127.

Brenner, D.J., H. Bercovier, J. Ursing, J.M. Alonso, A.G. Steigerwalt, G.R. Fanning, G.P. Carter and H.H. Mollaret. 1980a. *Yersinia intermedia*: a new species of *Enterobacteriaceae* composed of rhamnose-positive strains (formerly called atypical *Yersinia enterocolitica* or *Yersinia enterocolitica*-like). Curr. Microbiol. *4:* 207-212.

Brenner, D.J., H. Bercovier, J. Ursing, J.M. Alonso, A.G. Steigerwalt, G.R. Fanning, B.P. Carter and H.H. Mollaret. 1981. *In* Validation of the publication of new names and new combinations previously effectively published outside the IJSB. List No. 6. Int. J. Syst. Bacteriol. *31:* 215-218.

Brenner, D.J., G.R. Fanning, G.V. Miklos and A.G. Steigerwalt. 1973. Polynucleotide sequence relatedness among *Shigella* species. Int. J. Syst. Bacteriol. *23:* 1-7.

Brenner, D.J., G.R. Fanning, A.V. Rake and K.E. Johnson. 1969. Batch procedure for thermal elution of DNA from hydroxyapatite. Anal. Biochem. *28:* 447-459.

Brenner, D.J., G.R. Fanning, F.J. Skerman and S. Falkow. 1972. Polynucleotide sequence divergence among strains of *Escherichia coli* and closely related organisms. J. Bacteriol. *109:* 953-965.

Brenner, D.J., G.R. Fanning and A.G. Steigerwalt. 1972. Deoxyribonucleic acid relatedness among species of *Erwinia* and between *Erwinia* species and other enterobacteria. J. Bacteriol. *110:* 12-17.

Brenner, D.J., G.R. Fanning and A.G. Steigerwalt. 1974. Deoxyribonucleic acid relatedness among erwiniae and other *Enterobacteriaceae*: the gall, wilt and dry necrosis organisms (genus *Erwinia* Winslow et al., *sensu stricto*). Int. J. Syst. Bacteriol. *24:* 197-204.

Brenner, D.J., G.R. Fanning and A.G. Steigerwalt. 1974. Polynucleotide sequence relatedness in *Edwardsiella tarda*. Int. J. Syst. Bacteriol. *24:* 186-190.

Brenner, D.J., G.R. Fanning, A.G. Steigerwalt, I. Ørskov and F. Ørskov. 1972. Polynucleotide sequence relatedness among three groups of pathogenic *Escherichia coli* strains. Infect. Immun. *6:* 308-315.

Brenner, D.J., G.R. Fanning, A.G. Steigerwalt, M.A. Sodd and B.P. Doctor. 1977. Conservation of transfer RNA and 5S RNA cistrons in *Enterobacteriaceae*. J. Bacteriol. *129:* 1435-1439.

Brenner, D.J. and J.J. Farmer, III. 1981. *In* Farmer, J.J., III, The genus *Citrobacter*, *In* M.P. Starr, H. Stolp, H.G. Trüper, A. Balows and H.G. Schlegel (Editors), The Prokaryotes, A handbook on habitats, isolation, and identification of bacteria, Springer-Verlag, pp. 1140-1147.

Brenner, D.J. and J.J. Farmer, III. 1982. *In* List No. 8, Validation of the publication of new names and new combinations previously published outside of the IJSB. Int. J. Syst. Bacteriol. *32:* 266-268.

Brenner, D.J., J.J. Farmer, III, G.R. Fanning, A.G. Steigerwalt, P. Klykken, H.G. Wathen, F.W. Hickman and W.H. Ewing. 1978. Deoxyribonucleic acid relatedness in species of *Proteus* and *Providencia*. Int. J. Syst. Bacteriol. *28:* 269-282.

Brenner, D.J., J.J. Farmer III, F.W. Hickman, M.A. Asbury and A.G. Steigerwalt. 1977. Taxonomic and nomenclature changes in *Enterobacteriaceae*. HEW Publication No. (CDC) 79-8356. Center for Disease Control, Atlanta.

Brenner, D.J., C. Richard, A.G. Steigerwalt, M.A. Asbury and M. Mandel. 1980. *Enterobacter gergoviae* sp. nov.: a new species of *Enterobacteriaceae* found in clinical specimens and the environment. Int. J. Syst. Bacteriol. *30:* 1-6.

Brenner, D.J., A.G. Steigerwalt, D.P. Falcão, R.E. Weaver and G.R. Fanning. 1976. Characterization of *Yersinia enterocolitica* and *Yersinia pseudotuberculosis* by deoxyribonucleic acid hybridization and by biochemical reactions. Int. J. Syst. Bacteriol. *26:* 180-194.

Brenner, D.J., A.G. Steigerwalt and G.R. Fanning. 1972. Differentiation of *Enterobacter aerogenes* from klebsiellae by deoxyribonucleic acid reassociation. Int. J. Syst. Bacteriol. *22:* 193-200.

Brenner, D.J., A.G. Steigerwalt, G.W. Gorman, R.E. Weaver, J.C. Feeley, L.C. Cordes, H.W. Wilkinson, C. Patton, B.M. Thomason and K.R. Lewallen Sasseville. 1980. *In* Validation of the publication of new names and new combinations previously effectively published outside the IJSB. List No. 5. Int. J. Syst. Bacteriol. *30:* 676-677.

Brenner, D.J., A.G. Steigerwalt, G.W. Gorman, R.E. Weaver, J.C. Feeley, L.G. Cordes, H.W. Wilkinson, C. Patton, B.M. Thomason and K.R. Lewallen Sasseville. 1980. *Legionella bozemanii* sp. nov. and *Legionella dumoffii* sp. nov.: classification of two additional species of *Legionella* associated with human pneumonia. Curr. Microbiol. *4:* 111-116.

Brenner, D.J., A.G. Steigerwalt and J.E. McDade. 1979. Classification of the Legionnaires' disease bacterium: *Legionella pneumophila*, genus novum, species nova of the family *Legionellaceae*, familia nova. Ann. Intern. Med. *90:* 656-658.

Brenner, D.J., A.G. Steigerwalt, G.V. Miklos and G.R. Fanning. 1973. Deoxyribonucleic acid relatedness among erwiniae and other *Enterobacteriaceae*: the soft-rot organisms (genus *Pectobacterium* Waldee). Int. J. Syst. Bacteriol. *23:* 205-216.

Brenner, D.J., A.G. Steigerwalt, S. Pohl, H. Behrens, W. Mannheim and R.E. Weaver. 1981. Lack of relatedness of *Legionella pneumophila* to *Cytophagaceae*, "*Pasteurellaceae*" and *Kingella*. Int. J. Syst. Bacteriol. *31:* 89-90.

Brenner, D.J., A.G. Steigerwalt, R.E. Weaver, J.E. McDade, J.C. Feeley and M. Mandel. 1978. Classification of the Legionnaires' disease bacterium: an interim report. Curr. Microbiol. *1:* 71-75.

Brenner, D.J., J. Ursing, H. Bercovier, A.G. Steigerwalt, G.R. Fanning, J.M. Alonso and H.H. Mollaret. 1980b. Deoxyribonucleic acid relatedness in *Yersinia enterocolitica* and *Yersinia enterocolitica*-like organisms. Curr. Microbiol. *4:* 195-200.

Breuil, C. and D.J. Kushner. 1975. Lipase and esterase formation by psychrophilic and mesophilic *Acinetobacter* species. Can. J. Microbiol. *21:* 423-433.

Breuil, C., T.J. Novitsky and D.J. Kushner. 1975. Characteristics of a facultatively psychrophilic *Acinetobacter* species isolated from river sediment. Can. J. Microbiol. *21:* 2103-2108.

Brezina, R. 1978. Phase variation phenomenon in *Coxiella burnetii*. *In* Kazár, Ormsbee and Tarasevich (Editors), Rickettsiae and Rickettsial Diseases, VEDA, Bratislava, pp. 221-235.

Breznak, J.A. 1973. Biology of nonpathogenic, host-associated spirochetes. Crit. Rev. Microbiol. *2:* 457-489.

Breznak, J.A. and E. Canale-Parola. 1969. *Spirochaeta aurantia*, a pigmented, facultatively anaerobic spirochete. J. Bacteriol. *97:* 386-395.

Breznak, J.A. and E. Canale-Parola. 1972a. Metabolism of *Spirochaeta aurantia*. I. Anaerobic energy-yielding pathways. Arch. Mikrobiol. *83:* 261-277.

Breznak, J.A. and E. Canale-Parola. 1972b. Metabolism of *Spirochaeta aurantia*. II. Aerobic oxidation of carbohydrates. Arch. Mikrobiol. *83:* 278-292.

Breznak, J.A. and E. Canale-Parola. 1975. Morphology and physiology of *Spirochaeta aurantia* strains isolated from aquatic habitats. Arch. Microbiol. *105:* 1-12.

Breznak, J.A. and H.S. Pankratz. 1977. In situ morphology of the gut microbiota of wood-eating termites *Reticulitermes flavipes* (Kollar) and *Coptotermes formosanus* Shiraki. Appl. Environ. Microbiol. *33:* 406-426.

Bridré, J. and A. Donatien. 1923. Le microbe de l'agalaxie contagieuse et sa culture in vitro. C.R. Acad. Sci. Paris *177:* 841-843.

Brill, W.J. 1980. Biochemical genetics of nitrogen fixation. Microbiol. Rev. *44:* 449-467.

Brimacombe, R., G. Staffer and H.G. Wittmann. 1978. Ribosome structure. Annu. Rev. Biochem. *47:* 217-249.

Brindle, C.S. and S.T. Cowan. 1951. Flagellation and taxonomy of Whitmore's bacillus. J. Pathol. Bacteriol. *63:* 571-575.

Brinton, C.C. 1965. The structure, function, synthesis and genetic control of bacterial pili and a molecular model for DNA and RNA transport in Gram-negative bacteria. Trans. N.Y. Acad. Sci. Ser. II *27:* 1003-1054.

Brinton, L.P. and W. Burgdorfer. 1976. Cellular and subcellular organization of the 277F agent: a spiroplasma from the rabbit tick, *Haemaphysalis leporis-palustris*. Int. J. Syst. Bacteriol. *26:* 554-560.

Brion, A. and H. Kayser. 1902. Ueber eine Erkrankung mit dem Befund eines typhus-ähn lichen Bakteriums (Paratyphus). Muenchen Med. Wochenschr. *49:* 611-615.

Brisbane, P.G. and A. Kerr. 1981. Selective media for agrobacteria. Abstract n° 28 of the Conference on crown gall: a plant cancer caused by *Agrobacterium tumefaciens* - Sydney, Australia.

Brisou, J. 1955. Microbiologie du milieu marin. E. Flammarion and Co., Paris.

Brisou, J. 1961. Études de quelques *Pseudomonas* chomogènes isolés à Diego-suarez. Bull. Soc. Pathol. Exot. *54:* 746-755.

Brisou, J. and A.R. Prévot. 1954. Étude de systématique bactérienne. X. Révision des espèces réunies dans le genre *Achromobacter*. Ann. Inst. Pasteur (Paris) *86:* 722-728.

Brisou, B., C. Richard and A. Lénriot. 1972. Intérêt taxonomique de la recherche de la β-xylosidase chez les *Enterobacteriaceae*. Ann. Inst. Pasteur (Paris) *123:* 341-347.

Brisou, J., C. Tysset and A. Jacob. 1960. Étude d'un germe de la famille des *Pseudomonadaceae* (Tribu des *Chromobactereae*) *Empedobacter aquatile* isolé d'un produit frais de charcuterie. Arch. Inst. Pasteur Alg. *38:* 353-360.

Brisou, J., C. Tysset and B. Vacher. 1959. Étude de trois souches microbiennes famille des *Pseudomonadaceae* dont la synergie provoque une maladie de caractère septicémique chez les poissons blancs de la Dordogne, du Lot et de leurs affluents. Ann. Inst. Pasteur, Paris *96:* 689-696.

Brock, T.D. 1978. Thermophilic microorganisms and life at high temperatures. Springer-Verlag, New York.

Brock, T.D. and K.L. Boylen. 1973. Presence of thermophilic bacteria in laundry and domestic hot-water heaters. Appl. Microbiol. *25:* 72-76.

Brock, T.D. and M.R. Edwards. 1970. Fine structure of *Thermus aquaticus*, an extreme thermophile. J. Bacteriol. *104:* 509-517.

Brock, T.D. and H. Freeze. 1969. *Thermus aquaticus* gen. n. and sp. n., a non-sporulating extreme thermophile. J. Bacteriol. *98:* 289-297.

Brock, T.D. and A.H. Rose. 1969. Psychrophiles and thermophiles. *In* Norris and Ribbons (Editors), Methods in Microbiology, Vol. 3B, Academic Press, New York, pp. 161-168.

Broda, P. 1979. Plasmids. W.H. Freeman Co., London - San Francisco.

Brokopp, C.D. and J.J. Farmer. 1979. Typing methods for *Pseudomonas aeruginosa*. *In* Doggett (Editor) *Pseudomonas aeruginosa*. Clinical manifestations of infection and current therapy. Academic Press, New York, pp. 89-133.

Brooks, B.W., R.G.E. Murray, J.L. Johnson, E. Stackebrandt, C.R. Woese and G.E. Fox. 1980. Red-pigmented micrococci: a basis for taxonomy. Int. J. Syst. Bacteriol. *30:* 627-646.

Brooks, G.F., J.M. O'Donoghue and J.P. Rissing. 1974. *Eikenella corrodens*, a recently recognized pathogen: infections in medical-surgical patients and in association with methylphenylate abuse. Medicine (Baltimore) *53:* 325-342.

Brooks, M.A. 1963. The microorganisms of healthy insects. *In* Steinhaus (Editor), Insect Pathology: an Advanced Treatise. Vol. I, Academic Press, London, New York, pp. 215-250.

Brooks, M.A. 1963. Symbiosis and aposymbiosis in arthropods. Symp. Soc. Gen. Microbiol. *13:* 200-231.

Brooks, M.A. 1970. Comments on the classificationn of intracellular symbiotes of cockroaches and a description of the species. J. Invertebr. Pathol. *16:* 249-258.

Brooks, M.A. 1975. Symbiosis and attenuation. Ann. N.Y. Acad. Sci. *266:* 166-172.

Brooks, M.A. and W.B. Kringen. 1972. Polypeptides and proteins as growth factors for aposymbiotic *Blattella germanica* (L.). *In* Rodriguez (Editor), Insect and Mite Nutrition. North Holland, Amsterdam, pp. 353-364.

Brooks, M.A. and T.J. Kurtti. 1972. Male rudimentary ovaries: a case of cellular symbiosis in *Blattella germanica* (L.). (Dictyopter: Blattellidae). Int. J. Insect Morphol. Embryol. *1:* 169-179.

Brooks, M.A. and A.G. Richards. 1955. Intracellular symbiosis in cockroaches. I. Production of aposymbiotic cockroaches. Biol. Bull. 109: 22-39.

Brooks, M.A. and K. Richards. 1966. On the in vitro culture of intracellular symbiotes of cockroaches. J. Invertebr. Pathol. *8:* 150-157.

Broome, C.V., W.B. Cherry, W.C. Winn, Jr. and B.R. MacPherson. 1979. Rapid diagnosis of Legionnaires' disease by direct immunofluorescent staining. Ann. Intern. Med. *90:* 1-4.

Brotherson, T., T. Lees and R.D. Feigin. 1976. Susceptibility of *Haemophilus influenzae* type b to cefatrizine, ampicillin, and chloramphenicol. Antimicrob. Agents Chemother. *10:* 322-324.

Brown, A.D. 1963. The peripheral structure of Gram-negative bacteria. IV. The cation-sensitive dissolution of the cell membrane of the halophilic bacterium, *Halobacterium halobium*. Biochim. Biophys. Acta 75: 435-435.

Brown, A.D. 1976. Microbial water stress. Bacteriol. Rev. *40:* 832-846.

Brown, A.D. and K.Y. Cho. 1970. The walls of the extremely halophilic cocci: Gram-positive bacteria lacking muramic acid. J. Gen. Microbiol. *62:* 267-270.

Brown, A.D. and C.D. Shorey. 1963. The cell envelopes of two extremely halophilic bacteria. J. Cell Biol. *18:* 681-689.

Brown, A.J. 1886. On an acetic ferment which forms cellulose. J. Chem. Soc. (London) 49: 432-439.

Brown, A.T. and C. Wagner. 1970. Regulation of enzymes involved in the conversion of tryptophan to nicotinamide adenine dinucleotide in a colorless strain of *Xanthomonas pruni*. J. Bacteriol. *101:* 456-463.

Brown, C.L., M.J. Hill and P. Richards. 1971. Bacterial ureases in uremic men. Lancet *2:* 406-408.

Brown, D.W. and W.E.C. Moore. 1960. Distribution of *Butyrivibrio fibrisolvens* in nature. J. Dairy Sci. *43:* 1570-1574.

Brown, G.M., C.R. Ranger and D.J. Kelley. 1971. Selective media for the isolation of *Brucella ovis*. Cornell Vet. *61:* 265-280.

Brown, G.W., J.P. Saunders, S. Singh, D.L. Huxsoll and A. Shirai. 1978. Single dose doxycycline therapy for scrub typhus. Trans. Royal Soc. Trop. Med. Hyg. 72: 412-416.

Brown, H.J. and N.E. Gibbons. 1955. The effect of magnesium, potassium, and iron on the growth and morphology of red halophilic bacteria. Can. J. Microbiol. *62:* 267-270.

Brown, J.E., P.R. Brown and P.H. Clarke. 1969. Butyramide-utilizing mutants of *Pseudomonas aeruginosa* 8602 which produce an amidase with altered substrate specificity. J. Gen. Microbiol. *57:* 273-285.

Brown, J.E. and P.H. Clarke. 1972. Amino acid substitution in an amidase produced by an acetanilide-utilizing mutant of *Pseudomonas aeruginosa*. J. Gen. Microbiol. *70:* 287-298.

Brown, J.J. and G.M. Chippendale. 1975. Survival of the adult maize weevil, *Sitophilus zeamais*: role of nutrients, larval reserves and symbionts. Comp. Biochem. Physiol. A Comp. Physiol. *50:* 83-90.

Brown, J.L., D.L. Huxsoll, M. Ristic and P.K. Hildebrandt. 1972. In vitro cultivation of *Neorickettsia helminthoeca*, the causative agent of salmon poisoning disease. Am. J. Vet. Res. *33:* 1695-1700.

Brown, L.N., R.C. Dillman and R.E. Dierks. 1970. The *Haemophilus somnus* complex. U.S. Animal Hlth. Assoc. Proc. *74:* 94-108.

Brown, L.R. and R.J. Strawinski. 1958. Intermediates in the oxidation of methane. Bacteriol. Proc. *58:* 96-132.

Brown, M.R.W. and J. Melling. 1969. Role of divalent cations in the action of polymyxin B and EDTA on *Pseudomonas aeruginosa*. J. Gen. Microbiol. *59:* 263-274.

Brown, N.A. 1918. Some bacterial diseases of lettuce. J. Agr. Res. *13:* 367-388.

Brown, N.A. 1923. Bacterial leafspot of geranium in the eastern United States. J. Agr. Res *23:* 361-372.

Brown, N.A. and C.O. Jamieson. 1913. A bacterium causing a disease of sugarbeet and nasturtium leaves. J. Agr. Res. *1:* 189-210.

Brown, R.M., J.H.M. Willison and C.L. Richardson. 1976. Cellulose biosynthesis in *Acetobacter xylinum*: visualization of the site of synthesis and direct measurement of the in vivo process. Proc. Natl. Acad. Sci. U.S.A. *73:* 4565-4569.

Brown, T.McP. and J.C. Nunemaker. 1942. Rat-bite fever. A review of the American cases with reevaluation of etiology; report of cases. Bull. Johns Hopkins Hosp. *70:* 201-236.

Brown, W.J., N.F. Jacobs, E.S. Arum and R.J. Arko. 1976. T-strain mycoplasma in the chimpanzee. Lab. Anim. Sci. *26:* 81-83.

Brubaker, R.R. 1972. The genus *Yersinia*: Biochemistry and genetics of virulence. Curr. Top. Microbiol. Immunol. *57:* 111-158.

Brubaker, R.R. and M.J. Surgalla. 1962. Pesticins II. Production of pesticin I and II. J. Bacteriol. *84:* 539-545.

Bruce, D. 1893. Sur une nouvelle forme de fièvre rencontrée sur les bords de la Mediterranée. Ann. Inst. Pasteur (Paris) *7:* 289-304.

Brumfitt, W. 1959. Some growth requirements of *Haemophilus influenzae* and *Haemophilus pertussis*. J. Pathol. Bacteriol. *77:* 95-100.

Brumpt, E. 1910. Précis de Parasitologie. 1st Ed., Masson and Co., Paris.

Brumpt, E. 1911. Note suz le parasite des hematies de la taupe: *Grahamella talpae* n. g. n. sp. Bull. Soc. Pathol. Exot. *4:* 514-517.

Brumpt, E. 1921. Les parasites des invertébrés hematophages. *In* Lavier, Thèse, Paris, p. 207.

Brumpt, E. 1922. Les spirochetoses. *In* Roger, Widal and Teissier (Editors), Nouveau Traité de Medicin, Fasc. IV. Masson, Paris. pp. 491-531.

Brumpt, E. 1922. Précis de parasitologie. 3rd Ed., Masson and Co., Paris.

Brumpt, E. 1930. Rechutes parasitaires intenses, dues a la splenectomie, au cours d'infections latentes a *Aegyptianella*, chez la poule. C.R. Acad. Sci. (Paris) *191:* 1028-1030.

Brumpt, E. 1932. Longévité de virus de la fièvre bout neuse *(Rickettsia conori*, n.sp.) chez la tique, *Rhipicephalus sanguineus*. C.R. Séances Soc. Biol. Filiales *110:* 1199-1209.

Brumpt, E. 1933. Étude du Spirochaeta turicatae, n. sp. agent de la fièvre récurrente sporadique des Etats-Unis transmis par *Ornithodoros turicata*. C. R. Soc. Biol. (Paris) *113:* 1369-1372.

Brumpt, E. 1938. Rickettsia intracellulaire stomacle (*Rickettsia culices* n.sp.) of *Culex fatigans*. Ann. Parasitol. Hum. Comp. *16:* 153-158.

Brumpt, E. 1939. Un nouveau treponeme parasite de l'homme: *Treponema carateum*, agent des carates ou "mal del Pinto". C. R. Soc. Biol. (Paris) *130:* 942-945.

Brumpt, E. and G. Lavier. 1935. Sur un piroplasmide nouviau, parasite de tortue *Tunetella emydis* N. G., N. Sp. Ann. Parasit. Hum. Comp. *13:* 544-550.

Bruner, D.W., P.R. Edwards and A.S. Hopson. 1949. The Ballerup group of paracolon bacteria. J. Infect. Dis. *85:* 290-294.

Bruner, D.W. and J. Fabricant. 1954. A strain of *Moraxella anatipestifer (Pfeifferella anatipestifer)* isolated from ducks. Cornell Vet. *44:* 461-464.

Bruner, D.W. and J.H. Gillespie. 1973. Hagan's Infectious Diseases of Domestic Animals, 6th edition, Cornell University Press, Ithaca, 1385 pp.

Brunner, H., H. Krauss, H. Schaar and H.-G. Schiefer. 1979. Electron microscopic studies on the attachment of *Mycoplasma pneumoniae* to guinea pig erythrocytes. Infect. Immun. *24:* 906-911.

Brunner, H. and W. Weidner. 1981. Chemotherapy of human mycoplasma disease. Isr. J. Med. Sci. *17:* 656-660.

Brunton, J.L., I. MacLean, A.R. Ronald and W.L. Albritton. 1979. Plasmid mediated ampicillin resistance in *H. ducreyi*. Antimicrob. Agents Chemother. *15:* 294-299.

Bryan, M.K. 1926. Bacterial leaf-spot on hubbard squash. Science (Washington) *63:* 165.

Bryan, M.K. 1932. Color variations in bacterial plant pathogens. Phytopathology *22:* 787-788.

Bryan, M.K. 1933. Bacterial speck of tomatoes. Phytopathology *23:* 897-904.

Bryan, L.E. 1978. Transferable chloramphenicol and ampicillin resistance in a strain of *Haemophilus influenzae*. Antimicrob. Agents Chemother. *14:* 154-156.

Bryan, L.E. 1979. Resistance to antimicrobial agents: the general nature of the problem and the basis of resistance. *In* Doggett (Editor) *Pseudomonas aeruginosa*. Clinical manifestations of infection and current therapy. Academic Press, New York, pp. 219-270.

Bryan, L.E., R. Haraphongse and H.M. van Den Elzen. 1976. Gentamicin resistance in clinical isolates of *Pseudomonas aeruginosa* associated with diminished gentamicin accumulation and no detectable enzymatic modifications. J. Antibiot. *29:* 743-453.

Bryan, M.K. and F.P. McWhorter. 1930. Bacterial blight of poppy caused by *Bacterium papavericola* sp. nov. J. Agr. Res. *40:* 1-9.

Bryan, L.E., S.D. Semaka, H.M. van Den Elzen, J.E. Kinnear and R.L.S. Whitehouse. 1973. Characteristics of R931 and other *Pseudomonas aeruginosa* R factors. Antimicrob. Agents Chemother. *3:* 625-637.

Bryan, L.E., M.S. Shahrabadi and H.M. van Den Elzen. 1974. Gentamicin resistance in *Pseudomonas aeruginosa*: R-factor-mediated resistance. Antimicrob. Agents Chemother. *6:* 191-199.

Bryan, L.E., H.M. van Den Elzen and M.S. Shahrabadi. 1975. The relationship of aminoglycoside permeability to streptomycin and gentamicin susceptibility of *Pseudomonas aeruginosa*. *In* Mitsuhashi and Hashimoto (Editors), Microbial Drug Resistance, University of Tokyo Press, Tokyo, pp. 475-490.

Bryan, L.E., H.M. van Den Elzen and J.T. Tseng. 1972. Transferable drug resistance in *Pseudomonas aeruginosa*. Antimicrob. Agents Chemother. *1:* 22-29.

Bryant, M.P. 1956. The characteristics of strains of *Selenomonas* isolated from bovine rumen contents. J. Bacteriol. *72:* 162-167.

Bryant, M.P. 1972. Commentary on the Hungate technique for culture of anaerobic bacteria. Am. J. Clin. Nutr. *25:* 1324-1328.

Bryant, M.P. 1973. Nutritional requirements of the predominant rumen cellulolytic bacteria. Fed. Proc. *32:* 1809-1813.

Bryant, M.P. 1974a. Genus *Butyrivibrio*. *In* Buchanan and Gibbons (Editors), Bergey's Manual of Determinative Bacteriology, 8th Ed. The Williams and Wilkins Co., Baltimore, pp. 420-421.

Bryant, M.P. 1974b. Genus *Succinivibrio*. *In* Buchanan and Gibbons (Editors), Bergey's Manual of Determinative Bacteriology, 8th Ed. The Williams and Wilkins Co., Baltimore, p. 422.

Bryant, M.P. 1974c. *Succinimonas*. *In* Buchanan and Gibbons (Editors), Bergey's Manual of Determinative Bacteriology, 8th Ed. The Williams and Wilkins Co., Baltimore, p. 422.

Bryant, M.P., B.F. Barrentine, J.F. Sykes, I.M. Robinson, C.B. Shawver and L.W. Williams. 1960. Predominant bacteria in the rumen of cattle on bloat provoking ladino clover pasture. J. Dairy Sci. *43:* 1435-1444.

Bryant, M.P. and L.A. Burkey. 1953. Cultural methods and some characteristics of some of the numerous groups of bacteria in the bovine rumen. J. Dairy Sci. *36:* 205-217.

Bryant, M.P. and R.N. Doetsch. 1954. A study of actively cellulolytic rod-shaped bacteria of the bovine rumen. J. Dairy Sci. *37:* 1176-1183.

Bryant, M.P. and R.N. Doetsch. 1955. Factors necessary for the growth of *Bacteroides succinogenes* in the volatile acid fraction of rumen fluid. J. Dairy Sci. *38:* 340-350.

Bryant, M.P. and I.M. Robinson. 1962. Some nutritional characteristics of predominant culturable ruminal bacteria. J. Bacteriol. *84:* 605-614.

Bryant, M.P. and I.M. Robinson. 1963. Apparent incorporation of ammonia and amino acid carbon during growth of selected species of ruminal bacteria. J. Dairy Sci. *46:* 150-154.

Bryant, M.P., I.M. Robinson, C. Bouma and H. Chu. 1958. *Bacteroides ruminicola* n. sp. and the new genus and species *Succinimonas amylolytica*. Species of succinic acid-producing anaerobic bacteria of the bovine rumen. J. Bacteriol. *76:* 15-23.

Bryant, M.P., I.M. Robinson and H. Chu. 1959. Observations on the nutrition of *Bacteroides succinogenes* - ruminal cellulolytic bacterium. J. Dairy Sci. *42:* 1831-1847.

Bryant, M.P., I.M. Robinson and I.L. Lindahl. 1961. A note on the flora and fauna in the rumen of steers fed a feedlot bloat-provoking ration and the effect of penicillin. Appl. Microbiol. *9:* 511-515.

Bryant, M.P. and N. Small. 1956. The anaerobic monotrichous butyric acid-producing curved rod-shaped bacteria of the rumen. J. Bacteriol. *72:* 16-21.

Bryant, M.P. and N. Small. 1956. Characteristics of two new genera of anaerobic curved rods isolated from the rumen of cattle. J. Bacteriol. *72:* 22-26.

Bryant, M.P., N. Small, C. Bouma and H. Chu. 1958. *Bacteroides ruminicola* sp. nov. and *Succinimonas amylolytica* gen. nov., species of succinic acid-producing anaerobic bacteria of the bovine rumen. J. Bacteriol. *76:* 1-23.

Bryn, K., E. Jantzen and K. Bøvre. 1977. Occurrence and patterns of waxes in *Neisseriaceae*. J. Gen. Microbiol. *102:* 33-43.

Bryner, J.H., A.H. Frank and P.A. O'Berry. 1962. Dissociation studies of *Vibrio* from the bovine genital tract. Am. J. Vet. Res. *23:* 32-41.

Bryner, J. H., P. A. O'Berry and A. H. Frank. 1964. Vibrio infection of the digestive organs of cattle. Am. J. Vet. Res. *25:* 1048-1050.

Bryner, J. H., A. E. Ritchie, G. D. Booth and J. W. Foley. 1973. Lytic activity of vibrio phages on strains of *Vibrio fetus* isolated from man and animals. Appl. Microbiol. *26:* 404-409.

Bryner, J.H., A.E. Ritchie, J.W. Foley and D.T. Berman. 1970. Isolation and characterization of a bacteriophage for *Vibrio fetus*. J. Virol. *6:* 94-99.

Brysk, M.M., W.A. Corpe and L.V. Hanks. 1969. β-cyanoalanine formation by *Chromobacterium violaceum*. J. Bacteriol. *97:* 322-327.

Brzin, B. 1968. Tellurite reduction in *Yersinia*. Experientia *24:* 405.

Buchanan, J.S. 1978. Cytological studies on a new species of rickettsia found in association with a phage in the digestive gland of the marine bivalve mollusc, *Tellina tenuis* (da Costa). J. Fish Dis. *1:* 27-43.

Buchanan, J.S. 1979. Ultrastructural studies of a rickettsia-like organism (with phage) from the digestive gland of the marine bivalve, *Tellina tenuis* (da Costa). Haliotis *8:* 309-316.

Buchanan, R.E. 1917. Studies on the nomenclature and classification of the bacteria. III. The families of the *Eubacteriales*. J. Bacteriol. *2:* 347-350.

Buchanan, R.E. 1918. Studies in the nomenclature and classification of the bacteria. V. Subgroups and genera of the *Bacteriaceae*. J. Bacteriol. *3:* 27-61.

Buchanan, R.E. 1925. General systematic bactériology. The Williams and Wilkins Co., Baltimore.

Buchanan, R.E. 1926. What names should be used for the organisms producing nodules on the roots of leguminous plants? Proc. Iowa Acad. Sci. *33:* 81-90.

Buchanan, R.E. and N.E. Gibbons. 1974. Bergey's manual of determinative bacteriology, 8th Ed., The Williams and Wilkins Co., Baltimore.

Buchanan, R.E., J.G. Holt and E.F. Lessel, Jr. (Editors). 1966. Index Bergeyana. Williams and Wilkins Co., Baltimore.

Buchanan, T.M. 1975. Antigenic heterogeneity of gonococcal pili. J. Exp. Med. *141:* 1470-1475.

Buchanan, T.M. 1977. Surface antigens: pili. *In* Roberts (Editor), The Gonococcus, John Wiley and Sons, New York, pp. 255-272.

Buchanan, T.M. and W.A. Pearce. 1979. Pathogenic aspects of outer membrane components of Gram negative bacteria. *In* Inouye (Editor), Bacterial Outer Membranes. Biogenesis and Function. John Wiley and Sons, New York, pp. 475-514.

Buchner, P. 1965. Endosymbiosis of animals with plant microorganisms. John Wiley and Sons, New York.

Buchner, P. 1969. Endosymbiosestudien an Schildläusen. VIII. Symbiosen der Palaeococcoidea. 3. Z. Morphol. Tiere *64:* 201-308.

Buck, J.D., S.P. Meyers and E. Leifson. 1963. *Pseudomonas (Flavobacterium) piscicida* Bein comb. nov. J. Bacteriol. *86:* 1125-1126.

Buckmire, F. L. A. and R. G. E. Murray. 1970. Studies on the cell wall of *Spirillum serpens*. I. Isolation and partial purification of the outermost cell wall layer. Can. J. Microbiol. *16:* 1011-1022.

Buddenhagen, I.W. 1965. The relation of plant-pathogenic bacteria to the soil. *In* Baker and Snyder (Editors), Ecology of Soil-borne Pathogens, Prelude to Biological Control. University of California Press, Berkeley, pp. 269-284.

Buddenhagen, I.W. and T.A. Elsasser. 1962. An insect spread bacterial wilt epiphytotic on Bluggoe banana. Nature *194:* 164.

Buddenhagen, I.W. and A. Kelman. 1964. Biological and physiological aspects of bacterial wilt caused by *Pseudomonas solanacearum*. Annu. Rev. Phytopathol. *2:* 203-230.

Buddle, M.B. 1956. Studies on *Brucella ovis* (n. sp.), a cause of genital disease of sheep in New Zealand and Australia. J. Hyg. Camb. *54:* 351-364.

Buissière, J., G. Brault and L. Le Minor. 1981. Intérêt taxonomique de la recherche de l'utilisation du 2-cétogluconate par les *Enterobacteriaceae*. Ann. Microbiol. Inst. Pasteur (Paris) *132A:* 191-195.

Bulkacz, J., M.G. Newman, S.S. Socransky, E. Newbrun and D.F. Scott. 1979. Phospholipase A activity of micro-organisms from dental plaque. Microbios Letters *10:* 79-88.

Bulla, L.A., Jr. and T.C. Cheng (Editors). 1975. Pathobiology of invertebrate vectors of disease. Ann. N.Y. Acad. Sci. *266:* 1-540.

Bunt, J.S. and Y.T. Tchan. 1955. Estimation of protozoan population in soil by direct microscopy. Proc. Linn. Soc. N.S.W. *80:* 148-153.

Burgdorfer, W. 1970. Hemolymph test. Am. J. Trop. Med. Hyg. *19:* 1010-1014.

Burgdorfer, W. 1975. A review of Rocky Mountain spotted fever (tick-borne typhus), its agent, and its tick vectors in the United States. J. Med. Ent. *12:* 269-278.

Burgdorfer, W., A. Aechlimann, O. Peter, S.F. Hayes and R.N. Philip. 1979. *Ixodes ricinus*: vector of a hitherto undescribed spotted fever group agent in Switzerland. Acta Trop. *36:* 357-367.

Burgdorfer, W. and L.P. Brinton. 1970. Intranuclear growth of *Rickettsia canada*, a member of the typhus group. Infect. Immun. *30:* 112-114.

Burgdorfer, W. and L.P. Brinton. 1975. Mechanisms of transovarial infection of spotted fever rickettsiae in ticks. Ann. N.Y. Acad. Sci. *266:* 61-72.

Burgdorfer, W., L.P. Brinton and L.E. Hughes. 1973. Isolation and characterization of symbiotes from the Rocky Mountain wood tick, *Dermacentor andersoni*. J. Invert. Pathol. *22:* 424-434.

Burgdorfer, W., L.P. Brinton, W.L. Krynski and R.N. Philip. 1978. *Rickettsia rhipicephali*, a new spotted fever group rickettsia from the brown dog tick, *Rhipicephalus sanguineus*. *In* Kazár, Ormsbee and Tarasevich (Editors), Rickettsiae and Rickettsial Diseases, VEDA, Bratislava, pp. 307-316.

Burgdorfer, W., J.C. Cooney, A.J. Mavros, W.L. Jellison and C. Maser. 1980. The role of cotton tail rabbits (*Sylvilagus* spp.) in the ecology of *Rickettsia rickettsii* in the United States. Am. J. Trop. Med. Hyg. *29:* 686-690.

Burgdorfer, W., J.C. Cooney and L.A. Thomas. 1974. Zoonotic potential (Rocky Mountain spotted fever and tularemia) in the Tennessee Valley region. II. Prevalence of *Rickettsia rickettsii* and *Francisella tularensis* in mammals and ticks from Land Between the Lakes. Am. J. Trop. Med. Hyg. *23:* 109-117.

Burgdorfer, W., S.F. Hayes and A.J. Mavros. 1981. Nonpathogenic rickettsiae in *Dermacentor andersoni*: a limiting factor for the distribution of *Rickettsia rickettsii*. *In* Burgdorfer and Anacker (Editors), Rickettsiae and Rickettsial Diseases, Academic Press, New York, pp. 585-594.

Burgdorfer, W., V.F. Newhouse, E.G. Pickens and D.B, Lackman. 1962. Ecology of Rocky Mountain spotted fever in Western Montana. I. Isolation of *Rickettsia rickettsii* from wild mammals. Am. J. Hyg. *76:* 293-301.

Burgdorfer, W., D.J. Sexton, R.K. Gerloff, R.L. Anacker, R.N. Philip and L.A. Thomas. 1975. *Rhipicephalus sanguineus*: Vector of a new spotted fever group rickettsia in the United States. Infect. Immun. *12:* 205-210.

Burgess, N.R.H., S.N. McDermott and I. Whiting. 1973. Aerobic bacteria occurring in the hind-gut of the cockroach, *Blatta orientalis*. J. Hyg. *71:* 1-7.

Burgwitz, G.K. 1935. Phytopathogenic bacteria (Russ.). Akad. Nauk. USSR, Leningrad, pp. 1-252.

Burke, J.P., D. Ingall, J.O. Klein, H.M. Gezon and M. Finland. 1971. *Proteus mirabilis* infections in a hospital nursery traced to a human carrier. N. Eng. J. Med. *184:* 115-121.

Burkholder, W.H. 1926. A new bacterial disease of the bean. Phytopathology *16:* 915-928.

Burkholder, W.H. 1930. The bacterial diseases of the bean. Cornell Agr. Expt.

Sta. Mem. *127:* 1-88.

Burkholder, W.H. 1941. The black rot of *Barbarea vulgaris*. Phytopathology *31:* 347-348.

Burkholder, W.H. 1942. Three bacterial plant pathogens: *Phytomonas caryophylli* sp. n. *Phytomonas alliicola* sp. n., and *Phytomonas manihotis* (Arthaud-Berthet et Bondar) Viegas. Phytopathology *32:* 141-149.

Burkholder, W.H. 1944. *Xanthomonas vignicola* sp. nov. pathogenic on cowpeas and beans. Phytopathology *34:* 430-432.

Burkholder, W.H. 1948. Genus I. *Erwinia* Winslow et al. *In* Breed, Murray and Hitchens (Editors), Bergey's Manual of Determinative Bacteriology, 6th ed. The Williams and Wilkins Co., Baltimore, pp. 463-478.

Burkholder, W.H. 1948. Genus I. *Pseudomonas* Migula. *In* Breed, Murray and Hitchens (Editors), Bergey's Manual of Determinative Bacteriology, 6th Ed., The Williams and Wilkins Co., Baltimore, pp. 82-150.

Burkholder, W.H. 1950. Sour skin, a bacterial rot of onion bulbs. Phytopathology *40:* 115-117.

Burkholder, W.H. 1957. Genus VI. *Erwinia* Winslow et al. 1917. *In* Breed, Murray and Smith (Editors), Bergey's Manual of Determinative Bacteriology, 7th ed. The Williams and Wilkins Co., Baltimore, pp. 349-359.

Burkholder, W.H. 1960. Some observations on *Erwinia tracheiphila*, the causal agent of the cucurbit wilt. Phytopathology *50:* 179-180.

Burkholder, W.H. 1960. A bacterial brown rot of parsnip roots. Phytopathology *50:* 280-282.

Burkholder, W.H. and C.E.F. Guterman. 1935. Bacterial leaf spot of carnations. Phytopathology *25:* 114-120.

Burkholder, W.H., L.A. McFadden and A.W. Dimock. 1953. A bacterial blight of chrysanthemum. Phytopathology *43:* 522-526.

Burnham, J.C., T. Hashimoto and S.F. Conti. 1968. Electron microscopic observations on the penetration of *Bdellovibrio bacteriovorus* into Gram-negative bacterial hosts. J. Bacteriol. *96:* 1366-1381.

Burnham, J.C., T. Hashimoto and S.F. Conti. 1970. Ultrastructure and cell division of a facultatively parasitic strain of *Bdellovibrio bacteriovorus*. J. Bacteriol. *101:* 997-1004.

Burnham, J.C. and J. Robinson. 1974. Genus *Bdellovibrio* Stolp and Starr. *In* Buchanan and Gibbons (Editors), Bergey's Manual of Determinative Bacteriology, 8th Ed., The Williams and Wilkins Co., Baltimore, pp. 212-214.

Burrill, T.J. 1882. The Bacteria: an account of their nature and effects, together with a systematic description of the species. Illinois Indust. Univ. 11th Rep., pp. 93-157.

Burrill, T.J. 1883. New species of *Micrococcus*. Amer. Naturalist *17:* 319-320.

Burrows, T.W. 1963. Virulence of *Pasteurella pestis* and immunity to plague. Ergebn. Mikrobiol. *37:* 59-113.

Burrows, T.W. and W.A. Gillett. 1966. The nutritional requirement of some *Pasteurella* species. J. Gen. Microbiol. *45:* 333-345.

Burt, S., S. Meldrum, D.R. Woods and D.T. Jones. 1978. Colonial variation, capsule formation, and bacteriophage resistance in *Bacteroides thetaiotaomicron*. Appl. Environ. Microbiol. *35:* 439-443.

Burton, P.R., N. Kordova and D. Paretsky. 1971. Electron microscopic studies of the rickettsia *Coxiella burnetii*: Entry, lysosomal response, and fate of rickettsial DNA in L-cells. Can. J. Microbiol. *17:* 143-150.

Burton, P.R., J. Stueckemann, R.M. Welsh and D. Paretsky. 1978. Some ultrastructural effects of persistent infections by the rickettsia *Coxiella burnetii* in mouse L cells and green monkey kidney (Vero) cells. Infect. Immun. *21:* 556-566.

Busacca, A. 1935. Un germe caractères de rickettsies *(Rickettsia trachome)* dans tissus trachomateux. Arch. Ophthalmol. *52:* 567-572.

Bush, G.L. and G.B. Chapman. 1961. Electron microscopy of symbiotic bacteria in developing oocytes of American cockroach, *Periplaneta americana*. J. Bacteriol. *81:* 267-276.

Bushby, S.R. 1973. *Haemophilus influenzae* apparently resistant to trimethoprim. Brit. Med. J. *3:* 50-51.

Büsing, K.H., W. Döll and K. Freytag. 1953. Die Bakterienflora der medizinische Blutegel. Arch. Mikrobiol. *19:* 52-86.

Butler, H.M. and J.L. Blakey. 1975. A review of bacteria in L-phase and their possible clinical significance. Med. J. Aust. *2:* 463-467.

Butler, L.O. 1962. A defined medium for Haemophilus influenzae and *Haemophilus parainfluenzae*. J. Gen. Microbiol. *27:* 51-60.

Bütschli, O. 1876. Studien über die dersten Entwicklungsvorgänge der Eizelle, die Zelltheilung und die Conjugation der Infusorien. Abh. Senckenb. Naturforsch. Ges. *10:* 213-464.

Butterfield, C.T. 1935. Studies of sewage purification. II. A zoogloea-forming organism found in activated sludge. Pub. Health Rep. *50:* 671-684.

Buttery, S.H. and P. Plackett. 1960. A specific polysaccharide from *Mycoplasma mycoides*. J. Gen. Microbiol. *23:* 357-368.

Buttiaux, R., R. Osteux, R. Fresnoy and J. Moriamez. 1954. Les propriétés biochemiques caractèristiques du genre *Proteus*. Inclusion souhaitable des *Providencia* dans celui-ci. Ann. Inst. Pasteur (Paris) *87:* 375-386.

Buttiaux, R. and C. Voisin. 1958/1959. Coexistence de cils polaires et péritriches chez un bacille halophile. Influence de la composition du milieu sur cette association. Ann. Inst. Pasteur (Lille) *10:* 151-158.

Butzler, J. P. and M. B. Skirrow. 1979. Campylobacter enteritis. Clinics in Gastroenterology 8: 737-765.

Buxton, A.E., R.L. Anderson, D. Werdegar and E. Atlas. 1978. Nosocomial respiratory tract infection and colonization with *Acinetobacter calcoaceticus*. Amer. J. Med. *65:* 507-513.

Buxton, J. B. 1929. A note on *Vibrio foetus ovis* in the ram. First Report of Director, Inst. Anim. Pathol., Univ. of Cambridge, 1929-1930, pp. 47-51.

Byng, G.S., R.J. Whitaker, R.L. Gherna and R.A. Jensen. 1980. Variable enzymological patterning in tyrosine biosynthesis as a means of determining natural relatedness among the *Pseudomonadaceae*. J. Bacteriol. *144:* 247-257.

Cadmus, M.C., C.A. Knutson, A.A. Lagoda, J.E. Pittsley and K.A. Burton. 1978. Synthetic media for production and quality of xanthan gum in 20 liter fermentors. Biotechnol. Bioeng. *20:* 1003-1014.

Calavan, E.C. and G.N. Oldfield. 1979. Symptomatology of spiroplasmal plant diseases. *In* Whitcomb and Tully (Editors), The Mycoplasmas, Vol. 3, Academic Press, New York, pp. 37-64.

Calderone, J.G. and M.J. Pickett. 1965. Characterization of brucella phages. J. Gen. Microbiol. *39:* 1-10.

Caldwell, D.R. and M.P. Bryant. 1966. Medium without rumen fluid for nonselective enumeration and isolation of rumen bacteria. Appl. Microbiol. *14:* 794-801.

Caldwell, D.R., M. Keeney and P.J. van Voest. 1969. Effects of carbon dioxide on growth and maltose fermentation by *Bacteroides amylophilus*. J. Bacteriol. *98:* 668-676.

Caldwell, D.R., D.C. White, M.P. Bryant and R.N. Doetsch. 1965. Specificity of the heme requirement for growth of *Bacteroides ruminicola*. J. Bacteriol. *90:* 1645-1654.

Caldwell, H.D., J. Kromhout and J. Schachter. 1981. Purification and partial characterization of the major outer membrane protein of *Chlamydia trachomatis*. Infect. Immun. *31:* 1151-1176.

Caldwell, M.E. and D.C. Ryerson. 1939. Salmonellosis in certain reptiles. J. Infect. Dis. *65:* 242-245.

Callahan, W.S. and K. Koroma. 1979. Use of the Juni-Janik genetic homology technique in the identification of *Yersinia enterocolitica*. J. Amer. Med. Technol. *41:* 229-231.

Callies, E. and W. Mannheim. 1978. Classification of the *Flavobacterium-Cytophaga* complex on the basis of respiratory quinones and fumarate respiration. Int. J. Syst. Bacteriol. *28:* 14-19.

Callies, E. and W. Mannheim. 1980. Deoxyribonucleic acid relatedness of some menaquinone-producing *Flavobacterium* and *Cytophaga* strains. Antonie van Leeuwenhoek J. Microbiol. Serol. *46:* 41-49.

Camin, J.H. 1948. Mite transmission of hemorrhagic septicemia in snakes. J. Parasitol. *34:* 345-354.

Campbell, A. 1981. Evolutionary significance of accessory DNA elements in bacteria. Annu. Rev. Microbiol. *35:* 55-83.

Campbell, L.L. 1957. Genus *Beneckea* Campbell. *In* Breed, Murray and Smith (Editors), Bergey's Manual of Determinative Bacteriology, 7th Ed. The Williams and Wilkins Co., Baltimore, pp. 328-332.

Campbell, L.L., M.A. Kasprzycki and J.R. Postgate. 1966. *Desulfovibrio africanus* sp. n., a new dissimilatory sulfate-reducing bacterium. J. Bacteriol. *92:* 1122-1127.

Campbell, L.L., Jr. and O.B. Williams. 1951. A study of chitin-decomposing micro-organisms of marine origin. J. Gen. Microbiol. *5:* 894-905.

Campbell, L.M. and I.L. Roth. 1975. Methyl violet: a selective agent for differentiation of *Klebsiella pneumoniae* from *Enterobacter aerogenes* and other Gram-negative organisms. Appl. Microbiol. *30:* 258-261.

Campbell, L.M., I.L. Roth and R.D. Klein. 1976. Evaluation of double agar in the isolation of *Klebsiella pneumoniae* from river water. Appl. Environ. Microbiol. *31:* 213-215.

Campbell, R.W. and R. Domrow. 1974. Rickettsioses in Australia: Isolation of *Rickettsia tsutsugamushi* and *R. australis* from naturally infected arthropods. Roy. Soc. Trop. Med. Hyg. 68: 397-402.

Campêlo, A.B. and J. Döbereiner. 1970. Ocorrêntia de *Derxia* sp. em solos de alguns Estados Brasileiro. Pesqui. Agropecuária Bras. *5:* 327-332.

Canale-Parola, E. 1973. Isolation, growth and maintenance of anaerobic free-living spirochetes, p. 61-73. *In* J.R. Norris and D.W. Ribbons (Editors), Methods in Microbiology, vol. 8. Academic Press Inc., New York.

Canale-Parola, E. 1977. Physiology and evolution of spirochetes. Bacteriol. Rev. *41:* 181-204.

Canale-Parola, E. 1978. Motility and chemotaxis of spirochetes. Annu. Rev. Microbiol. *32:* 69-99.

Canale-Parola, E. 1980. Revival of the names *Spirochaeta litoralis*, *Spirochaeta zuelzerae*, and *Spirochaeta aurantia*. Int. J. Syst. Bacteriol. *30:* 594.

Canale-Parola, E. 1981. Proposal that *Spirochaeta stenostrepta* Zuelzer replace *Spirochaeta plicatilis* Ehrenberg as the type species of the genus *Spirochaeta* Ehrenberg. Int. J. Syst. Bacteriol. *31:* 105-106.

Canale-Parola, E., S.C. Holt and Z. Udris. 1967. Isolation of free-living, anaerobic spirochetes. Arch. Mikrobiol. *59:* 41-48.

Canale-Parola, E., S. L. Rosenthal and D. G. Kupfer. 1966. Morphological and physiological characteristics of *Spirillum gracile* sp. n. Antonie van Leeuwenhoek J. Microbiol. *32:* 113-124.

Canale-Parola, E., Z. Udris and M. Mandel. 1968. The classification of free-living spirochetes. Arch. Mikrobiol. *63:* 385-397.

Cantu, C. 1911. Le bacillus *Proteus* sa distribution dans la nature. Ann. Inst. Pasteur (Paris) *25:* 852-864.

Caraway, B. H. and N. R. Krieg. 1974. Aerotaxis in *Spirillum volutans*. Can. J.

Microbiol. 20: 1367-1377.

Carayon, J. 1963. La transmission héréditaire des bactéries symbiotiques chez les Lygaeidae vivipares (Heteroptera). Proc. XVI Int. Congr. Zool. 1: 145.

Carayon, J. 1974. Formes nouvelles d'endosymbiose chez les Hémiptères. C.R. Hebd. Séances Acad. Sci. 278: 11495-1498.

Carleton, O., N.W. Charon, P. Allender and S. O'Brien. 1979. Helix handedness of Leptospira interrogans as determined by scanning electron microscopy. J. Bacteriol. 137: 1413-1416.

Carlile, M.J., J.F. Collins and B.E.B. Moseley (Editors). 1981. Molecular and Cellular Aspects of Microbial Evolution, Cambridge University Press, Cambridge.

Carlsson, J., J. Wrethén and G. Beckman. 1977. Superoxide dismutase in Bacteroides fragilis and related Bacteroides species. J. Clin. Microbiol. 6: 280-284.

Carmichael, L.E. and D.W. Bruner. 1968. Characteristics of a newly-recognized species of Brucella responsible for infectious canine abortion. Cornell Vet. 58: 579-592.

Carmichael, L.E., R. Flores-Castro and S. Zoha. 1980. Brucellosis caused by Brucella canis (Br. canis): An update of infection in animals and man. World Health Organization Brucellosis Document: WHO/BRUC/80.361.

Carmichael, L.E., T.D. St. George, N.D. Sullivan and N. Horsfall. 1972. Isolation, propagation, and characterization studies of an ovine Mycoplasma responsible for proliferative interstitial pneumonia. Cornell Vet. 62: 654-679.

Carpano, M. 1929. Su di un piroplasma osservato nei polli egitto (Aegyptianella pullorum). Bull. Minist. Agric. Egypt 86: 1-12.

Carpenter, E.J. and J.L. Culliney. 1975. Nitrogen fixation in marine shipworms. Science 187: 551-552.

Carpenter, K.P. 1961. The relationship of the Enterobacterium A12 (Sachs) to Shigella boydii 14. J. Gen. Microbiol. 26: 535-542.

Carpenter, K.P. 1963. Report of the Subcommittee on Taxonomy of the Enterobacteriaceae. Int. Bull. Bacteriol. Nomencl. Taxon. 13: 69-93.

Carpenter, K.P. 1964. The Proteus-Providence group. In S.D. Dyke (Editor), Recent Advances in Clinical Pathology, Series IV, Little, Brown and Co., Boston, pp. 13-24.

Carr, J.G. 1958. Acetobacter estunense nov. spec. An addition to Frateur's ten basic species. Antonie van Leeuwenhoek J. Microbiol. Serol. 24: 157-160.

Carr, J.G. 1968. Methods for identifying acetic acid bacteria. In Gibbs and Shapton (Editors), Identification Methods for Microbiologists, Academic Press, London.

Carr, J.G. 1974. Genus Zymomonas Kluyver and van Niel. In Buchanan and Gibbons (Editors), Bergey's Manual of Determinative Bacteriology, 8th Ed. The Williams and Wilkins Co., Baltimore, pp. 352-353.

Carr, J.G. and S.M. Passmore. 1971. Discovery of the "cider sickness" bacterium Zymomonas anaerobia in apple pulp. J. Inst. Brew. London 77: 462-466.

Carric, L. and R.S. Berk. 1971. Membranous inclusions of Pseudomonas aeruginosa. J. Bacteriol. 106: 250-256.

Carruthers, M.M. and H.M. Sommers. 1973. Eikenella corrodens osteomyelitis. Ann. Intern. Med. 79: 900.

Carter, G.R. 1955. Studies on Pasteurella multocida. I. A hemagglutination test for the identification of serological types. Am. J. Vet. Res. 16: 481-484.

Carter, G.R. 1956. A serological study of Pasteurella haemolytica. Can. J. Microbiol. 2: 483-488.

Carter, G.R. 1957. Studies on Pasteurella multocida. II. Identification of antigenic and colonial characteristics. Am. J. Vet. Res. 18: 210-213.

Carter, G.R. 1958. Some characteristics of type A strains of Pasteurella multocida. Brit. Vet. J. 114: 356-357.

Carter, G.R. 1961. A new serological type of Pasteurella multocida from central Africa. Vet. Rec. 73: 1052.

Carter, G.R. 1967. Pasteurellosis: Pasteurella multocida and Pasteurella haemolytica. Adv. Vet. Sci. 11: 321-379.

Carter, G.R. 1976. A proposal for five biotypes of Pasteurella multocida. 19th Annu. Proc. Amer. Assoc. Vet. Lab. Diagnosticians, pp. 189-196.

Carter, G.R. 1981. Pasteurelloses. In Balows and Hausler (Editors), Diagnostic Procedures in Bacterial, Mycotic and Parasitic Infections, 6th Ed., Amer. Pub. Health Assoc., Washington, D.C., pp. 551-563.

Carter, G.R. and E. Annau. 1953. Isolation of capsular polysaccharides from colonial variants of Pasteurella multocida. Am. J. Vet. Res. 14: 475-478.

Carter, G.R. and R.V.S. Bain. 1960. Pasteurellosis (Pasteurella multocida). A review stressing recent developments. Vet. Rev. Annot. 6: 105-128.

Carter, G.R. and M.M. Chengappa. 1981a. Recommendations for a standard system of designating serotypes of Pasteurella multocida. 24th Annu. Proc. Amer. Vet. Lab. Diagnosticians, pp. 37-42.

Carter, G.R. and M.M. Chengappa. 1981b. Identification of types B and E of Pasteurella multocida by counterimmunoelectrophoresis. Vet. Rec. 108: 145-146.

Carter, H. V. 1888. Note on the occurrence of a minute blood-spirillum in an Indian rat. Sci. Mem. Offrs. Army India 3: 45-48.

Carter, J.C. 1945. Wetwood of elms. Ill. Nat. Hist. Surv. Bull. 23: 407-448.

Carter, J.B. and M.L. Luff. 1977. Rickettsia-like organisms infecting Harpalus rufipes (Coleoptera: Carabidae). J. Invert. Pathol. 30: 99-101.

Carty, C.E. and C.D. Litchfield. 1978. Characterization of a new marine sedimentary bacterium as Flavobacterium oceanosedimentum sp. nov. Int. J. Syst. Bacteriol. 28: 561-566.

Cary, S.G. and D.H. Hunter. 1967. Isolation of bacteriophages active against

Neisseria meningitidis. J. Virol. 1: 538-542.

Case, A.C. 1965. Conditions controlling Flavobacterium proteus in brewery fermentations. J. Inst. Brew. 71: 250-256.

Cassel, G.H. and A. Hill. 1979. Murine and other small-animal mycoplasmas. In Tully and Whitcomb (Editors), The Mycoplasmas, Vol. II, Academic Press, New York, pp. 235-273.

Cassel, G.H., J.R. Lindsey, H.J. Baker and J.K. Davis. 1979. Mycoplasmal and rickettsial diseases. In Baker, Lindsey and Weisbroth (Editors), The Laboratory Rat, Vol. I, Academic Press, New York, pp. 243-269.

Castañeda, M.R. 1947. A practical method for routine blood cultures in brucellosis. Proc. Soc. Exp. Biol. Med. 64: 114-115.

Castaño, J.J., H.D. Thurston and L.V. Crowder. 1964. Transmissión de gomosis en los pastos Micay e Imperial. Agr. Trop. 20: 379-387.

Castellani, A. 1905. On the presence of spirochetes in some cases of parangi (yaws, Framboesia tropica) Preliminary note. J. Ceylon Brit. Med. Assoc. 2: 54.

Castellani, A. and A.J. Chalmers. 1910. Manual of tropical medicine, 1st ed. Baillière, Tindall and Cox, London.

Castellani, A. and A.J. Chalmers. 1913. Manual of tropical medicine, 2nd ed. Baillière, Tindall and Cox, London.

Castellani, A. and A.J. Chalmers. 1919. Manual of tropical medicine, 3rd ed. Williams Wood and Co., New York.

Castenholz, R.W. 1969. Thermophilic blue-green algae and the thermal environment. Bacteriol. Rev. 33: 476-504.

Cate, J.C. 1972. Isolation of Serratia marcescens from stools with an antibiotic plate. In Hejzlar, Semonsky, and Masák (Editors) Advances in antimicrobial and antineoplasic chemotherapy. Progress in research and clinical application. Proceeding of the 7th International Congress of Chemotherapy, Vol. 1/2. Urban and Schwarzenberg, Munich, pp. 763-764.

Catlin, B.W. 1960. Transformation reactions within and between species of Neisseria. Bacteriol. Proc. p. 74 (G 76).

Catlin, B.W. 1961. Affinities among Neisseria as revealed by studies of DNAs and DNAses. Bacteriol. Proc. p. 90 (G 73).

Catlin, B.W. 1964. Reciprocal genetic transformation between Neisseria catarrhalis and Moraxella nonliquefaciens. J. Gen. Microbiol. 37: 369-379.

Catlin, B.W. 1970. Transfer of the organism named Neisseria catarrhalis to Branhamella gen. nov. Int. J. Syst. Bacteriol. 20: 155-159.

Catlin, B.W. 1971. Report (1966-1970) of the Subcommittee on the taxonomy of the Neisseriaceae to the International Committee on Nomenclature of Bacteria. Int. J. Syst. Bacteriol. 21: 154-155.

Catlin, B.W. 1973. Nutritional profiles of Neisseria gonorrhoeae, Neisseria meningitidis, Neisseria lactamica in chemically defined media and use of growth requirements for gonococcal typing. J. Infect. Dis. 128: 178-194.

Catlin, B.W. 1977. Nutritional requirements and auxotyping. In Roberts (Editor), The Gonococcus, John Wiley and Sons, New York, pp. 92-109.

Catlin, B.W. 1978. Characterization and auxotyping of Neisseria gonorrhoeae. In Bergan and Norris (Editors), Methods in Microbiology, Vol. 10, Academic Press Inc., New York, pp. 345-380.

Catlin, B.W., J.W. Bendler III and S.H. Goodgal. 1972. The type b capsulation locus of Haemophilus influenzae: map location and size. J. Gen. Microbiol. 70: 411-422.

Catlin, B.W. and L.S. Cunningham. 1964. Genetic transformation of Neisseria catarrhalis by deoxyribonucleate preparations having different average base compositions. J. Gen. Bacteriol. 37: 341-352.

Catlin, B.W. and L.S. Cunningham. 1964. Transforming activities and base composition of deoxyribonucleates from strains of Moraxella and Mima. J. Gen. Microbiol. 37: 353-367.

Catlin, B.W. and V.R. Tartagni. 1969. Delayed multiplication of newly capsulated transformants of Haemophilus influenzae detected by immunofluorescence. J. Gen. Microbiol. 56: 387-401.

Cato, E.P. and E.M. Barnes. 1976. Designation of the neotype strain of Bacteroides hypermegas Harrison and Hansen. Int. J. Syst. Bacteriol. 26: 494-497.

Cato, E.P., D.E. Hash, L.V. Holdeman and W.E.C. Moore. 1982. Electrophoretic study of Clostridium species. J. Clin. Microbiol. 15: 688-702.

Cato, E.P., L.V. Holdeman and W.E.C. Moore. 1979. Proposal of neotype strains for seven non-saccharolytic Bacteroides species. Int. J. Syst. Bacteriol. 29: 427-434.

Cato, E.P. and J.L. Johnson. 1976. Reinstatement of species rank for Bacteroides fragilis, B. ovatus, B. distasonis, B. thetaiotaomicron, and B. vulgarus: designation of neotype strains for Bacteroides fragilis (Veillon and Zuber) Castellani and Chalmers and Bacteroides thetaiotaomicron (Distaso) Castellani and Chalmers. Int. J. Syst. Bacteriol. 26: 230-237.

Cato, E.P., R.W. Kelley, W.E.C. Moore and L.V. Holdeman. 1982. Bacteroides zoogloeformans (Weinberg, Nativelle and Prévot 1937) corrig., comb. nov.: emended description. Int. J. Syst. Bacteriol. 32: 271-274.

Cato, E.P., W.E.C. Moore and M.P. Bryant. 1978. Designation of neotype strains for Bacteroides amylophilus Hamlin and Hungate 1956 and Bacteroides succinogens Hungate 1950. Int. J. Syst. Bacteriol. 28: 491-495.

Cato, E.P. and C.W. Salmon. 1976. Transfer of Bacteroides clostridiiformis subsp. clostridiiformis (Burri and Ankersmit) Holdeman and Moore and Bacteroides clostridiiformis subsp. girans (Prévot) Holdeman and Moore to the genus Clostridium as Clostridium clostridiiforme (Burri and Ankersmit) comb. nov.: emendation of description and designation of neotype strain. Int. J. Syst.

Bacteriol. 26: 205-211.

Cato, E.P., C.W. Salmon and W.E.C. Moore. 1974. *Fusobacterium prausnitzii* (Hauduroy et al.) Moore and Holdeman: emended description and designation of neotype strain. Int. J. Syst. Bacteriol. 24: 225-229.

Catsaras, M., J. Antoniewski and R. Buttiaux. 1965. Sur la production de colonies a centre orange par *Proteus rettgeri* et *Providencia* sur la gelose au desoxycholate-citrate-lactose. Ann. Inst. Pasteur (Lille) 16: 99-101.

Catsaras, M. and R. Buttiaux. 1965. Les *Aeromonas* dans les matières fécales humaines. Ann. Inst. Pasteur (Lille) 16: 85-88.

Caudwell, A. 1977. Aspects statistiques des épreuves d'infectivité chez les Jaunisses (Yellows) des plantes et chez les viroses transmises selon le mode persistant. Intéret de la Feve (Vicia faba) comme plante-test pour les Jaunisses. Ann. Phytopathol. 9: 141-159.

Cavanagh, P. and P.D. Marsden. 1969. Bacteria isolated from the gut of some Reduviid bugs. Trans. R. Soc. Trop. Med. Hyg. 63: 415-416.

Cavanagh, P., C.A. Morris and N.J. Mitchell. 1975. Chloramphenicol resistance in *Haemophilus* species. Lancet 1: 696.

Cavanaugh, C.M., S.L. Gardiner, M.L. Jones, H.W. Jannasch and J.B. Waterbury. 1981. Prokaryotic cells in the hydrothermal vent tubeworm, *Riftia pachyptila* Jones: possible chemoautotrophic symbionts. Science 213: 340-342.

Cavara, F. 1905. Bacteriosi del fico. Atti Accad. Gioenia Sci. Matur. Catania 18: mem. 14: 1-17.

Center for Disease Control. 1977. National nosocomial infections study report, annual summary 1974. Center for Disease Control Publication, Atlanta.

Center for Disease Control. 1977. Morbidity and Mortality Weekly Report, Center for Disease Control, Atlanta 26: 93.

Centifanto, Y.M. and W.S. Silver. 1964. Leaf nodule symbiosis. I. Endophyte of *Psychotria bacteriophila*. J. Bacteriol. 88: 776-781.

Certes, A. 1882. Notes sur les parasites et les commensaux de l'huitre. Bull. Soc. Zool. France 7: 347-353.

Certes, A. 1889. Note sur les microorganismes de la panse des ruminants. Bull. Soc. Zool. Fr. 14: 70-73.

Chaby, R., G. Ayme, M. Caroff, R. Donikian, N. Haeffner-Cavaillon, A. Le Dur, M. Moreau, M.-C. Mynard, M. Roumiantzeff and L. Szabo. 1979. Structural features and separation of some of the biological activities of the *Bordetella pertussis* endotoxin by chemical fractionation. *In* Manclark and Hill (Editors), International Symposium on Pertussis. U.S. Government Printing Office, Washington, D.C., pp. 185-190.

Chakrabarty, A.M. 1981. Microorganisms having multiple compatible degradative energy-generating plasmids and preparation thereof. U.S. Patent 4,259,444 (March 31).

Chakrabarty, A.M., C.F. Gunsalus and I.C. Gunsalus. 1968. Transduction and the clustering of genes in fluorescent pseudomonads. Proc. Natl. Acad. Sci. U.S.A. 60: 168-175.

Chakrabarty, A.N., S. Adhya, J. Basu and S.G. Dastidar. 1970. Bacteriocin typing of *Vibrio cholerae*. Infect. Immun. 1: 293-299.

Chakrabarty, A.N., S.G. Dastidar and S. Adhya. 1971. Bacteriocine typing of *Vibrio cholerae* - supplementary typing scheme. Proceedings of the Symposium on Immunity and Immunoprophylaxis in Cholera (March 1970, Calcutta). Technical Report Series No. 9 Indian Council of Medical Research, New Delhi, pp. 190-199.

Chamberlain, R.E. 1965. Evaluation of live tularemia vaccine prepared in a chemically defined medium. Appl. Microbiol. 13: 232-235.

Chamoiseau, G. 1967. Note sur le pouvoir pathogène d'*Edwardsiella tarda*. Un cas de septicémie mortelle du pigeon. Rev. Elev. Med. Vet. Pays Trop. (Paris) 20: 493-495.

Champion, A.B., E.L. Barrett, N.J. Palleroni, K.L. Soderberg, R. Kunisawa, R. Contopoulou, A.C. Wilson and M. Doudoroff. 1980. Evolution in *Pseudomonas fluorescens*. J. Gen. Microbiol. 120: 485-511.

Chan, K.Y., L. Baumann, M.M. Garza and P. Baumann. 1978. Two new species of *Alteromonas: Alteromonas espejiana* and *Alteromonas undina*. Int. J. Syst. Bacteriol. 28: 217-222.

Chan, Y. K., L. M. Nelson and R. Knowles. 1980. Hydrogen metabolism of *Azospirillum brasilense* in nitrogen-free medium. Can. J. Microbiol. 26: 1126-1131.

Chandler, D.K.F. and M.F. Barile. 1980. Ciliostatic, hemagglutinating, and proteolytic activities in a cell extract of *Mycoplasma pneumoniae*. Infect. Immun. 29: 1111-1116.

Chandler, F.W., J.A. Blackmon, M.D. Hicklin, R.M. Cole and C.S. Callaway. 1979. Ultrastructure of the agent of Legionnaires' disease in the human lung. Am. J. Clin. Pathol. 71: 43-50.

Chandler, F.W., M.D. Hicklin and J.A. Blackmon. 1977. Demonstration of the agent of Legionnaires' disease in tissue. N. Engl. J. Med. 297: 1218-1220.

Chandler, F.W., I.L. Roth, C.S. Callaway, J.L. Bump, B.M. Thomason and R.E. Weaver. 1980. Flagella on Legionnaires' disease bacteria. Ultrastructural observations. Ann. Intern. Med. 93: 711-714.

Chandler, F.W., B.M. Thomason and G.A. Hébert. 1980. Flagella on Legionnaires' disease bacteria in the human lung. Ann. Intern. Med. 93: 715-716.

Chang, C.J. and T.A. Chen. 1982. *Spiroplasma*: cultivation in chemically defined medium. Science 215: 1121-1122.

Chang, K.-P. 1974. Ultrastructure of symbiotic bacteria in normal and antibiotic treated *Blastocrithidia culicis* and *Crithidia oncopelti*. J. Protozool. 21: 699-707.

Chang, K.-P. 1974. Effects of elevated temperature on the mycetome and sym-

biotes of the bed bug *Cimex lectularius* (Heteroptera). J. Invertebr. Pathol. 23: 333-340.

Chang, K.-P. 1975. Haematophagous insect and haemoflagellate as host for prokaryotic endosymbionts. Symp. Soc. Exp. Biol. 29: 407-428.

Chang, K.-P. 1976. Symbiote-free hemoflagellates, *Blastocrithidia culicis* and *Crithidia oncopelti*. J. Protozool. 23: 241-244.

Chang, K.-P. and C. Dave. 1980. Modulation of polyamine level and biosynthetic enzymes by bacterial endosymbiotes in trypanosomatid Protozoa. *In* Schwemmler and Schenk (Editors), Endosymbiosis and Cell Biology, Vol. 1, de Gruyter, Berlin, pp. 350-359.

Chang, K.-P. and A.J. Musgrave. 1970. Ultrastructure of rickettsia-like microorganisms in the midgut of a plant bug, *Stenotus binotatus* Jak. (Heteroptera: Miridae). Can. J. Microbiol. 16: 621-632.

Chang, K.-P. and A.J. Musgrave. 1972. Multiple symbiosis in a leafhopper, *Helochara communis* Fitch: envelopes, nucleoids and inclusions of the symbiotes. J. Cell. Sci. 11: 275-293.

Chang, K.-P. and A.J. Musgrave. 1973. Morphology, histochemistry, and ultrastructure of mycetome and its rickettsial symbiotes in *Cimex lectularius* L. Can. J. Microbiol. 19: 1075-1081.

Chang, W. and J. E. Ogg. 1970. Transduction in *Vibrio fetus*. Am. J. Vet. Res. 31: 919-924.

Chang, W. and J. E. Ogg. 1971. Transduction and mutation to glycine tolerance in *Vibrio fetus*. Am. J. Vet. Res. 32: 649-653.

Chang, W.H. and G.R. Carter. 1976. Multiple drug resistance in *Pasteurella multocida* and *Pasteurella haemolytica* from cattle and swine. J. Am. Vet. Med. Assoc. 169: 710-712.

Chang, Y.-F. and E. Adams. 1971. Induction of separate catabolic pathways for L-lysine and D-lysine in *Pseudomonas putida*. Biochem. Biophys. Res. Comm. 45: 570-577.

Chang, Y.F. and D.S. Feingold. 1970. D-Glucaric acid and galactaric acid catabolism by *Agrobacterium tumefaciens*. J. Bacteriol. 102: 85-96.

Chapman, G., J. Hillier and F.H. Johnson. 1951. Observations on the bacteriophagy of *Erwinia carotovora*. J. Bacteriol. 61: 261-268.

Chapman, J.A., R.G.E. Murray and M.R.J. Salton. 1963. The surface anatomy of *Lampropedia hyalina*. Proc. Roy. Soc. B. 158: 498-513.

Chapman, P.J. and R.G. Duggleby. 1967. Dicarboxylic acid catabolism by bacteria. Biochem. J. 103: 7C-9C.

Chapman-Andresen, C. 1971. Biology of the large Amoebae. Annu. Rev. Microbiol. 25: 27-48.

Charney, A.N., R.E. Gots, S.B. Formal and R.A. Gianella. 1976. Activation of intestinal mucosal adenylate cyclase by *Shigella dysenteriae* 1 enterotoxin. Gastroenterology 70: 1085-1090.

Charudattan, R., R.E. Stall and D.L. Batchelor. 1973. Serotypes of *Xanthomonas vesicatoria* unrelated to its pathotypes. Phytopathology 63: 1260-1265.

Chatelain, R., H. Bercovier, A. Guiyole, A. Richard and H.H. Mollaret. 1979. Intérêt du composé vibriostatique 0/129 pour différencier les genres *Pasteurella* et *Actinobacillus* de la famille *Enterobacteriaceae*. Ann. Microbiol. (Paris) 130A: 449-454.

Chatterjee, B.D. and K.N. Neogy. 1972. Studies on *Aeromonas* and *Plesiomonas* species isolated from cases of choleric diarrhoea. Indian J. Med. Res. 60: 520-524.

Chatterjee, A.K. and M.P. Starr. 1980. Genetics of *Erwinia* species. Annu. Rev. Microbiol. 34: 645-676.

Chatty, H.B. and T.L. Gavan. l968. *Edwardsiella tarda*--identification and clinical significance. Report of two cases. Cleveland Clinic Quarterly 35: 223-228.

Chen, C.H., R.B. Dienst and R.B. Greenblatt. 1949. Skin reaction of patients to *Donovania granulomatis*. Amer. J. Syph. Gonorrhea Vener. Dis. 33: 60-64.

Chen, M. and M.J. Wolin. 1981. Influence of heme and vitamin B_{12} on growth and fermentations of *Bacteroides* species. J. Bacteriol. 145: 466-471.

Chen, M.H. and C. Hiruki. 1977. Effects of dark treatment on the ultrastructure of the aster yellows agent in situ. Phytopathology 67: 321-324.

Chen, T.A. and C.H. Liao. 1975. Corn stunt spiroplasma: isolation, cultivation, and proof of pathogenicity. Science 188: 1015-1017.

Cheng, K.-J. and J.W. Costerton. 1977. Ultrastructure of *Butyrivibrio fibrisolvens*: a gram-positive bacterium? J. Bacteriol. 129: 1506-1512.

Cheng, K.-J., D. Dinsdale and C.S. Stewart. 1979. Maceration of clover and grass leaves by *Lachnospira multiparus*. Appl. Environ. Microbiol. 38: 723-729.

Cheng, K.-J., G.A. Jones, F.J. Simpson and M.P. Bryant. 1959. Isolation and identification of rumen bacteria capable of anaerobic rutin degradation. Can. J. Microbiol. 15: 1365-1371.

Chern, C.-K., A. Ando, I. Kusaka and S. Fukui. 1976a. A succinate dehydrogenase-deficient mutant of *Agrobacterium tumefaciens*. Agr. Biol. Chem. 40: 144-149.

Chern, C.-K., I. Kusaka and S. Fukui. 1976b. Significance of pyruvate carboxylase in sugar metabolism of *Agrobacterium tumefaciens*. Agr. Biol. Chem. 40: 136-143.

Cherry, W.B. and W.H. Ewing. 1966. Phillip Rarick Edwards, 1901-1966. J. Bacteriol. 92: 531-535.

Cherry, W.B., P.L. Lentz and L.A. Barnes. 1946. Implication of *Proteus mirabilis* in an outbreak of gastroenteritis. Am. J. Public Health 36: 484-488.

Cherry, W.B. and R.M. McKinney. 1979. Detection of Legionnaires' disease bacteria in clinical specimens by direct immunofluorescence. *In* Jones and

Hébert (Editors), "Legionnaires'" the disease, the bacterium and methodology. Center for Disease Control, Atlanta, pp. 92-103.

Cherry, W.B., B. Pittman, P.P. Harris, G.A. Hébert, B.M. Thomason, L. Thacker and R.E. Weaver. 1978. Detection of Legionnaires' disease bacteria by direct immunofluorescent staining. J. Clin. Microbiol. 8: 329-338.

Chester, B. and L.H. Cooper. 1979. *Achromobacter* species (CDC group Vd): morphological and biochemical characterization. J. Microbiol. 9: 425-436.

Chester, B. and G. Stotzky. 1976. Temperature dependent cultural and biochemical characteristics of rhamnose positive *Yersinia enterocolitica*. J. Clin. Microbiol. 3: 119-127.

Chester, F.D. 1897. Report of the mycologist: bacteriological work. Del. Agr. Exp. Sta. Bull. 9: 38-145.

Chester, F.D. 1901. A manual of determinative bacteriology. The Macmillan Co., New York, pp. 1-401.

Chester, F.D. 1939. Genus IV. *Erwinia* Winslow et al. *In* Bergey, Breed, Murray and Hitchens (Editors), Bergey's Manual of Determinative Bacteriology, 5th ed. The Williams and Wilkins Co., Baltimore, pp. 404-420.

Chester, I. R. and R. G. E. Murray. 1975. Analysis of the cell wall and lipopolysaccharide of *Spirillum serpens*. J. Bacteriol. 124: 1168-1176.

Chester, I. R. and R. G. E. Murray. 1978. Protein-lipid-lipopolysaccharide association in the superficial layer of *Spirillum serpens* cell walls. J. Bacteriol. 133: 932-941.

Child, J. J. and W. G. W. Kurz. 1978. Inducing effect of plant cells on nitrogenase activity by *Spirillum* and *Rhizobium* in vitro. Can. J. Microbiol. 24: 143-148.

Chilton, M.-D., A.L. Montoya, D.J. Merlo, M.H. Drummond, R. Nutter, M.P. Gordon and E.W. Nester. 1978. Restriction endonuclease mapping of a plasmid that confers oncogenicity upon *Agrobacterium tumefaciens* strain B6-806. Plasmid 1: 254-269.

Chilton, M.-D., R.K. Saiki, N. Yadav, M.P. Gordon and F. Quetier. 1980. T-DNA from *Agrobacterium* Ti plasmid is in the nuclear DNA fraction of crown gall tumor cells. Proc. Natl. Acad. Sci. USA 77: 4060-4064.

Chinnarajan, A.M., S. Jayaraj and K. Narayanan. 1972. Destruction of endosymbionts with oxytetracycline and sulphanilamide in the gourd fruitfly, *Dacus cucurbitae* Coq. (Trypetidae, Diptera). Hind. Antibiot. Bull. 15: 16-22.

Chiykowski, L.N. 1977a. Cryopreservation of aster yellows agent in whole leafhoppers. Can. J. Microbiol. 23: 1038-1040.

Chiykowski, L.N. 1977b. Reduction in the transmissibility of a greenhouse-maintained isolate of aster yellows agent. Can. J. Bot. 55: 1783-1786.

Chladeck, D.W. and G.R. Ruth. 1976. Isolation of *Actinobacillus lignieresi* from an epidural abscess in a horse with progressive paralysis. J. Am. Vet. Med. Assoc. 168: 64-66.

Chladeck, D.W. 1975. Bovine abortion associated with *Haemophilus somnus*. Am. J. Vet. Res. 36: 1041.

Chludzinski, A.M., D.S. Slater and D. Nasser. 1972. Feedback regulation of 3-deoxy-D-*arabino*-heptulosonate-7-phosphate synthetase from a marine bacterium, *Vibrio* MB22. J. Bacteriol. 109: 1162-1169.

Cho, J.J., A.C. Hayward and K.G. Rohrbach. 1980. Nutritional requirements and biochemical activities of pineapple pink disease bacterial strains from Hawaii. Antonie van Leeuwenhoek J. Microbiol. Serol. 46: 191-204.

Chorpenning, F.W., D.H. Schmidt, H.B. Stamper and P.R. Dugan. 1978. Antigenic relationships among floc-forming *Pseudomonadaceae*. Ohio J. Sci. 78: 29-33.

Chow, A.W., D. Bednorz and L.B. Guze. 1977. Susceptibility of obligate anaerobes to metronidazole: an extended study of 1,054 clinical isolates. *In* S.M. Finegold (Editor), Metronidazole. Excerpta Medica, Princeton, N.J., pp. 286-292.

Chow, A.W., V. Patten and L.B. Guze. 1975. Rapid screening of *Veillonella* by ultraviolet fluorescence. J. Clin. Microbiol. 2: 546-548.

Christensen, P. 1977. Synonymy of *Flavobacterium pectinovorum* Dorey with *Cytophaga johnsonae* Stanier. Int. J. Syst. Bacteriol. 27: 122-132.

Christensen, P. 1980. Description and taxonomic status of *Cytophaga heparina* (Payza and Korn) comb. nov. (basionym: *Flavobacterium heparinum* Payza and Korn 1956). Int. J. Syst. Bacteriol. 30: 473-475.

Christian, J.H.B. and J.A. Waltho. 1962. Solute concentrations within cells of halophilic and non-halophilic bacteria. Biochim. Biophys. Acta 65: 506-508.

Christiansen, A.H. 1964. Studies on the antigenic structure of *T. pallidum*. Acta Pathol. Microbiol. Scand. 60: 123-130.

Christiansen, C., G. Askaa, E.A. Freundt and R.F. Whitcomb. 1979. Nucleic acid hybridization experiments with *Spiroplasma citri* and the corn stunt and suckling mouse cataract spiroplasmas. Curr. Microbiol. 2: 323-326.

Christiansen, C., F.T. Black and E.A. Freundt. 1981. Hybridization experiments with DNA from *Ureaplasma urealyticum*, serovars I to VIII. Int. J. Syst. Bacteriol. 31: 259-262.

Christiansen, C., E.A. Freundt and F.T. Black. 1975. Genome size and deoxyribonucleic acid base composition of *Thermoplasma acidophilum*. Int. J. Syst. Bacteriol. 25: 99-101.

Christiansen, C., E.A. Freundt and O. Vinther. 1981. Lack of deoxyribonucleic acid homology between *Thermoplasma acidophilum* and *Sulfolobus acidocaldarius*. Int. J. Syst. Bacteriol. 31: 346-347.

Christiansen, M. 1917. En ejendommelig pyaemisk Lidelse hos Faar. Maanedsskr. Dyrlaeg 29: 449-458.

Christison, M.H., J. Mackenzie and T.J. Mackie. 1938. *Bacillus salmonicida*

bacteriophage: with particular reference to its occurrence in water and the question of its application in controlling *B. salmonicida* infection. Fisheries Scotland, Salmon Fish No. 5, H.M. Stationery Office, Edinburgh.

Christofferson, F.A. and H.E. Ottosen. 1961. Recent staining methods. Skand. Vettidskr. 31: 599-607.

Christopher, W.N. and C.W. Edgerton. 1930. Bacterial stripe diseases of sugarcane in Louisiana. J. Agr. Res. 41: 259-267.

Chumakova, R.I., B.F. Vanyushin, N.A. Kokurina, T.I. Vorobleva and S.E. Medvedeva. 1973. Composition of DNA and taxonomy of the luminescent bacteria. Microbiology 41: 539-545.

Chung, K.-T., G.M. Anderson and G.E. Fulk. 1975. Formation of indoleacetic acid by intestinal anaerobes. J. Bacteriol. 124: 573-575.

Cinco, M., M. Tamaro and L. Cociancich. 1975. Taxonomical, cultural and metabolic characteristics of halophilic leptospirae. Zentralbl. Bakteriol. Parasitenkd. Infektionskr. Hyg., Abt. 1 Orig., Reihe A 233: 400-405.

Cioglia, L. 1950. Antigeni communi a brucelle e salmonelle. G. Batt. Immunol. 42: 81-90.

Cirillo, V.P. and S. Razin. 1973. Distribution of a phosphoenolpyruvate-dependent sugar phosphotransferase system in mycoplasmas. J. Bacteriol. 113: 212-217.

Citarella, R.V. and R.R. Colwell. 1970. Polyphasic taxonomy of the genus *Vibrio*: polynucleotide sequence relationships among selected *Vibrio* species. J. Bacteriol. 104: 434-442.

Claflin, J.L. and C.L. Larson. 1972. Infection-immunity in tularemia: specificity of cellular immunity. Infect. Immun. 5: 311-318.

Clara, F.M. 1930. A new bacterial leaf disease of tobacco in the Philippines. Phytopathol. 20: 691-706.

Clara, F.M. 1932. A new bacterial disease of pears. Science 75: 111.

Clara, F.M. 1934. A comparative study of the green-fluorescent bacterial plant pathogens. Cornell Agr. Exptl. Sta. Mem. 159: 1-36.

Clark, A.G. 1969. A selective medium for the isolation of *Agrobacterium* species. J. Appl. Bacteriol. 32: 348-351.

Clark, H F. 1964. Suckling mouse cataract agent. J. Infect. Dis. 114: 476-487.

Clark, H F. and L.B. Rorke. 1979. Spiroplasmas of tick origin and their pathogenicity. *In* Whitcomb and Tully (Editors), The Mycoplasmas, Vol. 3, Academic Press, New York, pp. 155-174.

Clark, P.H. and S.T. Cowan. 1952. Biochemical methods for bacteriology. J. Gen. Microbiol. 6: 187-197.

Clark, R. 1942. *Eperythrozoon felis* (sp. nov.) in a cat. J. Afr. Vet. Med. Assoc. 13: 15-16.

Clark, T.B. 1977. *Spiroplasma* sp., a new pathogen in honey bees. J. Invertebr. Pathol. 29: 112-113.

Clark, T.B. 1978. Honey bee spiroplasmosis, a new problem for beekeepers. Am. Bee J. 118: 18-19, 23.

Clark, T.B. 1982. Spiroplasmas: diversity of arthropod reservoirs and host-parasite relationships. Science 217: 57-59.

Clark, T.B., R.F. Whitcomb and J.G. Tully. 1982. Spiroplasmas from coleopterous insects: new ecological dimensions. Microbial Ecol. 8: 401-409.

Clark, W.A. and A.G. Steigerwalt. 1977. Deoxyribonucleic acid reassociation experiments with a halophilic, lactose-fermenting *Vibrio* isolated from blood cultures. Int. J. Syst. Bacteriol. 27: 194-199.

Clarke, P.H. 1976. Mutant isolation. *In* MacDonald (Editor) Second International Symposium on the Genetics of Industrial Microorganisms. Academic Press, London, pp. 15-28.

Clarke, P.H. and L.N. Ornston. 1975. Metabolic pathways and regulation (I and II) *In* Clarke and Richmond (Editors), Genetics and Biochemistry of *Pseudomonas*, J. Wiley and Sons, London, pp. 191-340.

Clarke, R.T.J. 1979. Niche in pasture-fed ruminants for the large rumen bacteria *Oscillospira*, *Lampropedia*, and Quinn's and Eadie's ovals. Appl. Environ. Microbiol. 37: 654-657.

Clark-Walker, G.D. 1969. Association of microcyst formation in *Spirillum itersonii* with the spontaneous induction of a defective bacteriophage. J. Bacteriol. 97: 885-892.

Clark-Walker, G.D. and J. Lascelles. 1970. Cytochrome c550 from *Spirillum itersonii*: purification and some properties. Arch. Biochem. Biophys. 136: 153-159.

Clark-Walker, G.D. and S.B. Primrose. 1971. Isolation and characterization of a bacteriophage Si 1 for *Spirillum itersonii*. J. Gen. Virol. 11: 139-145.

Clark-Walker, G.D., B. Rittenberg and J. Lascelles. 1967. Cytochrome synthesis and its regulation in *Spirillum itersonii*. J. Bacteriol. 94: 1648-1655.

Clasener, H. 1972. Pathogenicity of the L-phase of bacteria. Ann. Rev. Microbiol. 26: 55-84.

Claus, D. 1967. Taxonomy of some highly pleomorphic bacteria. Spisy Prirodoved. Fak. Univ. J. E. Purkyne Brno. 40: 254-257.

Claus, D., J.E. Bergendahl and M. Mandel. 1968. DNA base composition of *Microcyclus* species and organisms of similar morphology. Arch. Mikrobiol. 63: 26-28.

Claus, G.W., B.L. Batzing, C.A. Baker and E.M. Goebel. 1975. Intracytoplasmic membrane formation and increased oxidation of glycerol during growth of *Gluconobacter oxydans* J. Bacteriol. 123: 1169-1183.

Claxton, P.D. and R.E. Everett. 1966. Recovery of an organism resembling *Histophilus ovis* from a ram. Aust. Vet. J. 42: 457-458.

Claydon, T.J. and B.W. Hammer. 1939. A skunk-like odor of bacterial origin in butter. J. Bacteriol. 37: 252-258.

Cleveland, L.R. and A.V. Grimstone. 1964. The fine structure of the flagellate *Mixotrichia paradoxa* and its associated micro-organisms. Proc. R. Soc. Lond. B. *159:* 668-686.

Clewell, D.B. 1981. Plasmids, drug resistance and gene transfer in the genus *Streptococcus*. Microbiol. Rev. *45:* 409-436.

Cline, T.W. 1978. Isolation and characterization of luminescence system mutants in bacteria. *In* DeLuca (Editor), Methods in Enzymology, Vol. 57. Academic Press, Inc., London-New York-San Francisco, pp. 166-171.

Clyde, W.A., Jr. 1964. *Mycoplasma* species identification based upon growth inhibition by specific antisera. J. Immunol. *92:* 958-965.

Cobb, N.A. 1893. Plant diseases and their remedy. Agr. Gaz. N. S. W. *4:* 777-798.

Cobet, A.B., C. Wirsen and G.E. Jones. 1970. The effect of nickel on a marine bacterium, *Arthrobacter marinus* sp. nov. J. Gen. Microbiol. *62:* 159-169.

Cocchi, P. and A. Ulivelli. 1968. Meningitis caused by *Neisseria catarrhalis*. Acta Paediatr. Scand. *57:* 451-453.

Cocks, G.T. and A.C. Wilson. 1972. Enzyme evolution in *Enterobacteriaceae*. J. Bacteriol. *110:* 793-802.

Codd, G.A., B. Bowien and H.G. Schlegel. 1976. Glycollate production and excretion by *Alcaligenes eutrophus*. Arch. Microbiol. *110:* 167-171.

Coder, D. M. and M. P. Starr. 1978. Antagonistic association of the chlorellavorus bacterium ("*Bdellovibrio*" *chlorellavorus*) with *Chlorella vulgaris*. Curr. Microbiol. *1:* 59-64.

Cody, R.M. 1978. Preservation and storage of pathogenic *Neisseria*. Health Lab. Sci. *15:* 206-209.

Coerper, F.M. 1919. Bacterial blight of soybean. J. Agr. Res. *18:* 179-194.

Coetzee, J.N. 1963. Lysogeny in *Proteus rettgeri* and the host-range of *P. rettgeri* and *P. hauseri* bacteriophages. J. Gen. Microbiol. *31:* 219-229,

Coetzee, J.N. 1967. Bacteriocinogeny in strains of Providence and *Proteus morganii*. Nature (London) *213:* 614-616.

Coetzee, J.N. 1972. Genetics of the *Proteus* group. Ann. Rev. Microbiol. *26:* 23-54.

Coetzee, J.N. and H.C. De Klerk. 1964. Effect of temperature on flagellation, motility and swarming of *Proteus*. Nature (London) *202:* 211-212.

Coetzee, J.N. and T.G. Sacks. 1960. Morphological variants of *Proteus hauseri*. J. Gen. Microbiol. *23:* 209-216.

Coffey, J.D., Jr., A.D. Martin and H.N. Booth. 1967. *Neisseria catarrhalis* in exudate otitis media. Arch. Otolaryngol. *86:* 403-406.

Cohen, G. and N.L. Somerson. 1969. Glucose-dependent secretion and destruction of hydrogen peroxide by *Mycoplasma pneumoniae*. J. Bacteriol. *98:* 547-551.

Cohen, G.N., R.Y. Stanier and G. LeBras. 1969. Regulation of the biosynthesis of amino acids of the aspartate family in coliform bacteria and pseudomonads. J. Bacteriol. *99:* 791-801.

Cohen, G.N., R.Y. Stanier and G. LeBras. 1969. Regulation of the biosynthesis of amino acids of the aspartate family in coliform bacteria and the pseudomonads. J. Bacteriol. *99:* 791-801.

Cohen, G.N., R.Y. Stanier and G. LeBras. 1969. Regulation of the biosynthesis of amino acids of the aspartate family in coliform bacteria and the pseudomonads. J. Bacteriol. *99:* 791-801.

Cohen, R.L., R.G. Wittler and J.E. Faber. 1968. Modified biochemical tests for characterization of L-phase variants of bacteria. Appl. Microbiol. *16:* 1655-1662.

Cohn, F. 1872. Untersuchungen über Bakterien. Beitr. Biol. Pflanz. 1875 *1*(Heft 2): 127-224.

Cohn, F. 1878. Letter of J. Penn which describes *Micrococcus phosphoreum*. Versameling van stukken betreffende het geneeskundig staats toerzitch, 126-130.

Colby, J., H. Dalton and R. Whittenbury. 1979. Biological and biochemical aspects of microbial growth on C_1 compounds. Annu. Rev. Microbiol. *33:* 481-517.

Cole, B.C., L. Golightly and J.R. Ward. 1967. Characterization of mycoplasma strains from cats. J. Bacteriol. *94:* 1451-1558.

Cole, B.C., J.R. Ward and C.H. Martin. 1968. Hemolysin and peroxide activity of *Mycoplasma* species. J. Bacteriol. *95:* 2022-2030.

Cole, J. A. 1972. Base composition of deoxyribonucleic acid from *Spirillum volutans*, *Spirillum serpens*, and *Spirillum itersonii*. J. Gen. Microbiol. *72:* 411-413.

Cole, J.A. and S.C. Rittenberg. 1971. A comparison of respiratory processes in *Spirillum volutans*, *Spirillum itersonii*, and *Spirillum serpens*. J. Gen. Microbiol. *69:* 375-383.

Cole, R.M. 1979. Mycoplasma and spiroplasma viruses: ultrastructure. *In* Barile and Razin (Editors), The Mycoplasmas, Vol. I, Academic Press, New York, pp. 385-410.

Cole, R.M., C.S. Richards and T.J. Ponkin. 1977. Novel bacterium infecting an African snail. J. Bacteriol. *132:* 950-966.

Cole, R.M., J.G. Tully and T.J. Popkin. 1974. Virus-like particles in *Spiroplasma citri*. Colloq. Inst. Natl. Santé Rech. Méd. *33:* 125-132.

Cole, R.M., J.G. Tully, T.J. Popkin and J.M. Bove. 1973. Morphology, ultrastructure and bacteriophage infection of the helical mycoplasma-like organism (*Spiroplasma citri* gen. nov., sp. nov.) cultured from "Stubborn" disease of citrus. J. Bacteriol. *115:* 367-386.

Colebrook, L. 1920. The mycelial and other microorganisms associated with human actinomycosis. Br. J. Exp. Pathol. *1:* 197-212.

Coles, J.D.W.A. 1931. A rickettsia-like organism in the conjunctiva of sheep. Report Vet. Serv. Anim. Ind. Union S. Afr. 17th, 175-186.

Coles, J.D.W.A. 1936. A rickettsia-like organism in the conjunctival epithelium of cattle. J.S. Afr. Vet. Med. Assoc. *7:* 221-225.

Coles, J.D.W.A. 1940. Conjunctivitis of the domestic fowl and an associated rickettsia-like organism in the conjunctival epithelium. Onderstepoort J. Vet. Sci. *14:* 469-478.

Coles, J.D.W.A. 1953. Classification of rickettsia pathogenic to vertebrates. Ann. N.Y. Acad. Sci. *56:* 457-483.

Collier, R.J. and J.J. Mekalanos. 1980. ADP-Ribosylating exotoxins. *In* Bisswanger and Schmincke-Ott (Editors), Multifunctional Proteins, Wiley and Sons, New York, pp. 261-291.

Collier, W.A. 1921. *Cristispira helgolandica* nov. spec. und ihre Fortpflanzung. Zentralbl. Bakteriol. Parasitenkd. Infektionskr. Hyg. Abt. I. Orig. *86:* 132-134.

Collin, B. 1913. Sur en ensemble de protistes parasites des bactraciens (note préliminaire). Arch. Zool. Exp. Gen. Notes Rev. *51:* 59-76.

Collins, M.D., M. Goodfellow and D.E. Minnikin. 1982. A survey of the structures of mycolic acids in *Corynebacterium* and related taxa. J. Gen. Microbiol. *128:* 129-149.

Collins, M.D. and D. Jones. 1981. Distribution of isoprenoid quinone structural types in bacteria and their taxonomic implications. Microbiol. Rev. *45:* 316-354.

Collins, M.D., H.N.M. Ross, B.J. Tindall and W.D. Grant. 1981. Distribution of isoprenoid quinones in halophilic bacteria. J. Appl. Bacteriol. *50:* 559-565.

Colloc, M.L., O. Masure, Y. Perramant, B. Lejeune and C. Chastel. 1980. Actualités des infections à *Eikenella corrodens*. Medicine et Maladies Infectieuses. *10:* 387-390.

Colvin, J.R. 1977. A new look at cellulose biosynthesis in relation to structure and industrial use. Tappi *60:* 59-62.

Colwell, R.R. 1970. Polyphasic taxonomy of the genus *Vibrio*: numerical taxonomy of *Vibrio cholerae*, *Vibrio parahaemolyticus*, and related *Vibrio* species. J. Bacteriol. *104:* 410-433.

Colwell, R.R. 1973. Genetic and phenetic classification of bacteria. Adv. Appl. Microbiol. *16:* 137-175.

Colwell, R.R. (Editor). 1976. The Role of Culture Collections in the Era of Molecular Biology. American Society for Microbiology, Washington, D.C.

Colwell, R.R., J. Kaper, R. Seidler, M.J. Voll, L.A. McNicol, S. Garges, H. Lockman, D. Maneval, E. Remmers, S.W. Joseph, H. Bradford, N. Roberts, I. Huq and A. Huq. 1980. Isolation of 01 and non-01 *Vibrio cholerae* from estuaries and brackish water environments. Proc. 15th Joint Conf. Cholera US - Japan Coop. Med. Sci. Prog., pp. 44-56.

Colwell, R.R., C.D. Litchfield, R.H. Vreeland, L.A. Kiefer and N.E. Gibbons. 1979. Taxonomic studies of red halophilic bacteria. Int. J. Syst. Bacteriol. *29:* 379-399.

Colwell, R.R. and R.Y. Morita. 1964. Reisolation and emendation of description of *Vibrio marinus* (Russell) Ford. J. Bacteriol. *88:* 831-837.

Colwell, R.R. and A.K. Sparks. 1967. Properties of *Pseudomonas enalia*, a marine bacterium pathogenic for the invertebrate *Crassostrea gigas* (Thunberg). Appl. Microbiol. *15:* 980-986.

Comai, L. and T. Kosuge. 1980. Involvement of plasmid deoxyribonucleic acid in indole-acetic acid synthesis in *Pseudomonas savastanoi*. J. Bacteriol. *143:* 950-957.

Committee on the Serological Typing of *Vibrio parahaemolyticus*. 1970. New serotypes of *Vibrio parahaemolyticus*. Jpn. J. Microbiol. *14:* 249-250.

Comps, M., J.R. Bonami and C. Vago. 1977. Mise en évidence d'une infection rickettsienne chez les huitres. C.R. Acad. Sci., Ser. D *285:* 427-429.

Comps, M. and J.-P. Deltreil. 1979. Un microorganisme de type rickettsien chez l'huitre portugaise, *Crassostrea angulata* Lmk. C.R. Acad. Sci. Ser. D *289:* 169-171.

Comps, M., H. Quiros-Ramos and D. Razet. 1973. Sur une tumeur du manteau de *Crassostrea gigas* (Thunberg). Rev. Trav. Inst. Pech. Marit. *37:* 383-386.

Comps, M., G. Tige, J.-L. Duthoit and H. Grizel. 1979. Microorganismes de type rickettsien chez les huitres, *Crassostrea gigas* Th. et *Ostrea edulis* L. Haliotis *8:* 317-321.

Congdon, A.L., E.S. Boatman and G.E. Kenny. 1979. Mycoplasmatales virus, MV-M1: discovery in *Acholeplasma modicum* and preliminary characterization. Curr. Microbiol. *3:* 111-115.

Conn, H.J. 1939. Genus III. *Alcaligenes* Castellani and Chalmers. *In* Bergey, Breed, Murray and Hitchens (Editors), Bergey's Manual of Determinative Bacteriology, 5th Ed. The Williams and Wilkins Co., Baltimore, pp. 95-102.

Conn, H.J. 1942. Validity of the genus *Alcaligenes*. J. Bacteriol. *44:* 353-360.

Cook, G.T. 1948. Urease and other biochemical reactions of the *Proteus* group. J. Pathol. Bacteriol. *60:* 171-181.

Cook, G.T. 1950. A plate test for nitrate reduction. J. Clin. Pathol. *3:* 359-362.

Cook, I., R.W. Campbell and G. Barrow. 1966. Brucellosis in North Queensland rodents. Aust. Vet. J. *42:* 5-8.

Cooksey, K.E. and C. Rainbow. 1962. Metabolic patterns in acetic acid bacteria. J. Gen. Microbiol. *27:* 135-142.

Coolbaugh, J.C., J.J. Progar and E. Weiss. 1976. Enzymatic activities of cell-free extracts of *Rickettsia typhi*. Infect. Immun. *14:* 298-305.

Cooper, B.S. 1977. Differences in morphology in *Bacteroides nodosus* attributable to culture media. N.Z. Vet. J. *25:* 16-20.

Cooper, J.E. 1982. Acid production, acid tolerance and growth rate of *Lotus* rhizobia in laboratory media. Soil Biol. Biochem. *14:* 127–131.

Cooper, K.E., J. Davies and Jean Wiseman. 1941. An investigation of an outbreak of food poisoning associated with organisms of the *Proteus* group. J. Pathol. Bacteriol. *52:* 91-98.

Cooper, R. and G. Brown. 1968. *Plesiomonas shigelloides* in South Australia. J. Clin. Pathol. *21:* 715-718.

Cooper, R.A. and H.L. Kornberg. 1964. The utilization of itaconate by *Pseudomonas* species. Biochem. J. *91:* 82-91.

Coote, J.G. and H. Hassal. 1973. The degradation of L-histidine, imidazolyl-L-lactate and imidazolyl-propionate by *Pseudomonas testosteroni*. Biochem. J. *132:* 409-422.

Corbel, M.J. 1975. The serological relationship between *Brucella* spp. *Yersinia enterocolitica* serotype IX and *Salmonella* serotypes of Kauffmann-White group N. J. Hyg. Camb. *75:* 151-171.

Corbel, M.J. 1977a. Production of a phage variant lytic for non-smooth *Brucella* strains. Ann. Sclavo. *19:* 99-108.

Corbel, M.J. 1977b. Isolation and partial characterization of a phage receptor from *Brucella neotomae* 5K33. Ann. Sclavo. *19:* 131-142.

Corbel, M.J. 1979. Isolation and properties of a phage lytic for non-smooth *Brucella* organisms. J. Biol. Stand. *7:* 349-360.

Corbel, M.J. 1982. International Committee on Nomenclature of Bacteria Subcommittee on the Taxonomy of *Brucella*. Minutes of meeting, 4 and 5 September, 1978. Int. J. Syst. Bacteriol. *32:* 260-261.

Corbel, M.J., C.D. Bracewell, E.L. Thomas and K.P.W. Gill. 1979. Techniques in the identification and classification of *Brucella* species. *In* Skinner and Lovelock (Editors), Identification Methods for Microbiologists, 2nd Ed., Soc. for Appl. Bacteriol. Technical Series No. 14, Academic Press, London, pp. 71-122.

Corbel, M.J. and G.A. Cullen. 1970. Differentiation of the serological response to *Yersinia enterocolitica* serotype IX and *Brucella abortus* in cattle. J. Hyg. Camb. *68:* 519-530.

Corbel, M.J., K.P.W. Gill and E.L. Thomas. 1978. Methods for the identification of *Brucella*. Ministry of Agriculture, Fisheries and Food: Pinner, Middlesex.

Corbel, M.J. and J.A. Morris. 1974. Studies on a smooth phage-resistant variant of *Brucella abortus*. I. Immunological properties. Br. J. Exp. Pathol. *55:* 78-87.

Corbel, M.J. and J.A. Morris. 1975. Studies on a smooth phage-resistant variant of *Brucella abortus*. 2. Mechanism of phage resistance. Br. J. Exp. Pathol. *56:* 1-7.

Corbel, M.J., A.C. Scott and H.M. Ross. 1980. Properties of a cell-wall-defective variant of *Brucella abortus* of bovine origin. J. Hyg. Camb. *85:* 103-113.

Corbel, M.J. and E.L. Thomas. 1976. Properties of some new *Brucella* phage isolates: evidence for lysogeny within the genus. Develop. Biol. Stand. *31:* 38-45.

Corbel, M.J. and E.L. Thomas. 1980. The brucella-phages: their properties, characterization and applications. Ministry of Agriculture, Fisheries and Food; Pinner, Middlesex.

Corboz, L. and J. Nicolet. 1975. Infektionen mit sogenannten *Haemophilus somnus* beim Rind: Isolierung und Characterisierung von Stämmen aus Respirations- und Geschlechtsorganen. Schweiz. Arch. Tierheilk. *117:* 493-502.

Corboz, L. and P. Wild. 1981. Epidemiologie der *Haemophilus somnus* Infektion beim Rind: Vergleich von Stämmen in der Polyacrilamid-Elektrophorese. Schweiz. Arch. Tierheilk *123:* 79-88.

Cordes, L.G. and D.W. Fraser. 1980. Legionellosis: Legionnaires' disease; Pontiac fever. Med. Clin. North America *64:* 395-416.

Cordes, L.G., A.M. Wiesenthal and G.W. Gorman. 1981. Isolation of *Legionella pneumophila* from hospital shower heads. Ann. Intern. Med. *94:* 195-197.

Cordes, L.G., H.W. Wilkinson, G.W. Gorman, B.J. Fikes and D.W. Fraser. 1979. Atypical *Legionella*-like organisms: fastidious water-associated bacteria pathogenic for man. Lancet *ii:* 927-930.

Corey, R.R. and M.P. Starr. 1957a. Colony types of *Xanthomonas phaseoli*. J. Bacteriol. *74:* 137-140.

Corey, R.R. and M.P. Starr. 1957b. Genetic transformation of colony type in *Xanthomonas phaseoli*. J. Bacteriol. *74:* 141-145.

Corey, R.R. and M.P. Starr. 1956c. Genetic transformation of streptomycin resistance in *Xanthomonas phaseoli*. J. Bacteriol. *74:* 146-150.

Corin, R.E., E. Boggs and C.D. Cox. 1978. Enzymatic degradation of H_2O_2 by *Leptospira*. Infect. Immun. *22:* 672-675.

Cornelis, G. and E.P. Abraham. 1975. β-Lactamases from *Yersinia enterocolitica*. J. Gen. Microbiol. *87:* 273-284.

Cornelis, G., P. Bennett and J. Grinsted. 1976. Properties of pGC1, a lac plasmid originating in *Yersinia enterocolitica* 842. J. Bacteriol. *127:* 1058-1062.

Cornelis, G. and Colson, C. 1975. Restriction of DNA in *Yersinia enterocolitica* detected by recipient ability for a derepressed R Factor from *Escherichia coli*. J. Gen. Microbiol. *87:* 285-291.

Cornelis, G., G. Wauters and E.G. Bruynogh. 1973. Résistance transférables chez des souches sauvages de *Yersinia enterocolitica*. Ann. Microbiol. (Inst. Pasteur) *124A:* 299-309.

Corpe, W.A. 1951. A study of the wide spread distribution of *Chromobacterium* species in soil by a simple technique. J. Bacteriol. *62:* 515-517.

Corpe, W.A. 1963. Extracellular accumulation of pyrroles in bacterial cultures. Appl. Microbiol. *11:* 145-150.

Corpe, W.A. 1964. Factors influencing growth and polysaccharide formation by

strains of *Chromobacterium violaceum*. J. Bacteriol. *88:* 1433-1441.

Corstvet, R.E., R.J. Panciera, H.B. Rinker, B.L. Starks and C. Howard. 1973. Survey of tracheas of feedlot cattle for *Haemophilus somnus* and other selected bacteria. J. Am. Vet. Med. Assoc. *163:* 870-873.

Cosbie, A.J.C., J. Tosic and T.K. Walker. 1942. *Acetobacter turbidans*, a new species of acetic acid-producing bacterium. J. Inst. Brew. *48:* 82-86.

Cosenza, B.J. and J.D. Buck. 1966. Simple device for enumeration and isolation of luminescent bacterial colonies. Appl. Microbiol. *14:* 692.

Cosenza, B.J. and J.D. Podgwaite. 1966. A new species of *Proteus* isolated from larvae of the gypsy moth *Porthetria dispar* (L). Antonie van Leeuwenhoek. J. Microbiol. Serol. *32:* 187-191.

Costerton, J.W., J.M. Ingram and K.J. Cheng. 1974. Structure and function of the cell envelope of Gram-negative bacteria. Bacteriol. Rev. *38:* 87-110.

Cottew, G.S. 1979. Caprine-ovine mycoplasmas. *In* Tully and Whitcomb (Editors), The Mycoplasmas, Vol. II, Academic Press, New York, pp. 103-132.

Cottew, G.S. and J. Francis. 1954. The isolation of *Shigella equuli* and *Salmonella newport* from normal horses. Aust. Vet. J. *30:* 301-304.

Cottew, G.S. and F.R. Yeats. 1978. Subdivision of *Mycoplasma mycoides* subsp. *mycoides* from cattle and goats into two types. Aust. Vet. J. *54:* 293-296.

Coty, V.F. 1967. Atmospheric nitrogen fixation by hydrocarbon-oxidizing bacteria. Biotechnol. Bioengin. *9:* 25-32.

Coulton, J. W. and R. G. E. Murray. 1978. Cell envelope associations of *Aquaspirillum serpens* flagella. J. Bacteriol. *136:* 1037-1049.

Counotte, G.H.M. 1981. Regulation of lactate metabolism in the rumen. Ph.D. Thesis, University of Utrecht.

Counts, G.W., L. Seeley and H.N. Beaty. 1971. Identification of an epidemic strain of *Neisseria meningitidis* by bacteriocin typing. J. Infect. Dis. *124:* 26-32.

Cowan, S.T. 1968. A Dictionary of Microbial Taxonomic Usage, Oliver and Boyd, Edinburgh.

Cowan, S.T. 1970. Heretical taxonomy for bacteriologists. J. Gen. Microbiol. *61:* 145-154.

Cowan, S.T. 1974. Family I. *Enterobacteriaceae* Rahn 1937, 281, *Nom. gen. cons.* Opin. 15, Jud. Comm. 1958, 73. *In* Buchanan, R.E. and N.E. Gibbons, (Editors), Bergey's Manual of Determinative Bacteriology, 8th ed. The Williams and Wilkins Co., Baltimore, pp. 290-293.

Cowan, S.T. 1974. Cowan and Steel's manual for the identification of medical bacteria, 2nd Ed., Cambridge University Press, London.

Cowan, S.T. 1978. *In* Hill (Editor), A dictionary of microbial taxonomy, Cambridge University Press, Cambridge, United Kingdom.

Cowan, S.T. and K.J. Steel. 1965. Manual for the identification of medical bacteria. Cambridge Univ. Press, London.

Cowan, S.T., K.J. Steel, C. Shaw and J.P. Duguid. 1960. A classification of the *Klebsiella* group. J. Gen. Microbiol. *23:* 601-612.

Cowdry, E.V. 1925. Studies on the etiology of heartwater. I. Observation of a rickettsia, *Rickettsia ruminantium* (n. sp.) in the tissues of infected animals. J. Exp. Med. *42:* 231-252.

Cox, C.D. and M.K. Barber. 1974. Oxygen uptake by *Treponema pallidum*. Infect. Immun. *10:* 123-127.

Cox, C.D. and R. Graham. 1979. Isolation of an iron-binding compound from *Pseudomonas aeruginosa*. J. Bacteriol. *137:* 357-364.

Cox, C.D. and A.D. Larson. 1957. Colonial growth of leptospirae. J. Bacteriol. *73:* 587-589.

Cox, C.D., K.L. Rinehart, M.L. Moore and J.C. Cook. 1981. Pyochelin: novel structure of an iron-chelating growth promoter for *Pseudomonas aeruginosa*. Proc. Natl. Acad. Sci. U.S.A. *78:* 4256-4260.

Cox, H.R. 1939. Studies of a filter-passing infectious agent isolated from ticks. V. Further attempts to cultivate in cell-free media. Suggested classification. Pub. Health Rep. *54:* 1822-1827.

Cox, H.R. 1941. Cultivation of rickettsia of Rocky Mountain spotted fever, typhus and Q-fever groups in the embryonic tissues of developing chicks. Science (Washington) *94:* 399-403.

Cox, R.B. and J.R. Quayle. 1975. The autotrophic growth of *Micrococcus denitrificans* on methanol. Biochem. J. *150:* 569-571.

Coxon, D.T., A.M.C. Davies, G.R. Fenwick, R. Self, J.L. Firmin, D. Lipkin and N.F. Janes. 1980. Agropine, a new amino acid derivative from crown gall tumours. Tetrahedron Lett. *21:* 495-498.

Coykendall, A.L. and F.S. Kaczmarek. 1980. DNA homologies among *Eikenella corrodens* strains. J. Periodontal Res. *15:* 615-620.

Coykendall, A.L., F.S. Kaczmarek and J. Slots. 1980. Genetic heterogeneity in *Bacteroides asaccharolyticus* (Holdeman and Moore 1970) Finegold and Barnes 1977 (Approved Lists, 1980) and proposal of *Bacteroides gingivalis* sp. nov. and *Bacteroides macacae* (Slots and Genco) comb. nov. Int. J. Syst. Bacteriol. *30:* 559-564.

Crabtree, K., W. Boyle, E. McCoy and G.A. Rohlich. 1966. A mechanism of floc formation by *Zoogloea ramigera*. J. Water Pollut. Control Fed. *38:* 1968-1980.

Crabtree, K. and E. McCoy. 1967. *Zoogloea ramigera* Itzigsohn, identification and description. Request for an opinion as to the status of the generic name *Zoogloea*. Int. J. Syst. Bacteriol. *17:* 1-10.

Cradock-Watson, J.E. 1965. The production of bacteriocines by *Proteus* species. Zentrabl. Bakteriol. Parasitenk. Infektionskr. Hyg. Abt. 1 Orig. *196:* 385-388.

Craig, A.S., R.M. Greenwood and K.I. Williamson. 1973. Ultrastructural inclusions of rhizobial bacteroids of *Lotus* nodules and their taxonomic significance. Arch. Mikrobiol. *89:* 23-32.

Craigie, J. and C.H. Yen. 1938. The demonstration of types of *B. typhosus* by means of preparation of type II Vi phage. Can. J. Publ. Health *29:* 448-463; 484-496.

Crawford, I.P. 1975. Gene rearrangements in the evolution of the tryptophan synthetic pathway. Bacteriol. Rev. *39:* 87-120.

Crawford, I.P. and K. Nealson. 1976. The tryptophan genes and enzymes of a marine luminous bacterium. Fed. Proc., Fed. Am. Soc. Exp. Biol. *35:* 546.

Crawford, I.P., B.P. Nichols and C. Yanofsky. 1980. Nucleotide sequence of the trpB gene in *Escherichia coli* and *Salmonella typhimurium*. J. Mol. Biol. *142:* 489-502.

Crawford, R.E., L.A. McDermott and A.J. Musgrave. 1960. Microbial isolations from the granary weevil *Sitophilus granarius* (L.). (Coleoptera: Curculionidae). Can. Entomol. *92:* 577-581.

Crisel, R.M., R.S. Baker and D.E. Dorman. 1975. Capsular polymer of *Haemophilus influenzae* type b. I. Structural characterization of the capsular polymer of strain Egan. J. Biol. Chem. *250:* 4926-4930.

Criswell, B.S., J.H. Marston, W.A. Stenback, S.H. Black and H.L. Gardner. 1971. *Haemophilus vaginalis* 594, a Gram-negative organism? Can. J. Microbiol. *17:* 865-869.

Criswell, B.S., W.A. Stenback, S.H. Black and H.L. Gardner. 1972. Fine structure of *Haemophilus vaginalis*. J. Bacteriol. *109:* 930-932.

Croizier, G. and G. Meynadier. 1971. Recherche d'antigènes de groupe chez des rickettsies de la tribu des *Wolbachia* par la technique d'agglutination des corps élémentaires. Entomophaga *16:* 11-17.

Croizier, G. and G. Meynadier. 1972. Etude en immunofluorescence de l'infection expérimentale de la souris par *Rickettsiella grylli*. Ann. Rech. Vétér. *3:* 373-380.

Crosa, J.H. 1980. A plasmid associated with virulence in the marine fish pathogen *Vibrio anguillarum* specifies an iron-sequestering system. Nature (London) *284:* 566-568.

Crosa, J.H., D.J. Brenner, W.H. Ewing and S. Falkow. 1973. Molecular relationships among the *Salmonelleae*. J. Bacteriol. *115:* 307-315.

Crosa, J.H., D.J. Brenner and S. Falkow. 1973. Use of a single-strand specific nuclease for analysis of bacterial and plasmid deoxyribonucleic acid homo- and heteroduplexes. J. Bacteriol. *115:* 904-911.

Crosa, J.H., A.G. Steigerwalt, G.R. Fanning and D.J. Brenner. 1974. Polynucleotide sequence divergence in the genus *Citrobacter*. J. Gen. Microbiol. *83:* 271-282.

Crosse, J.E. and C.M.E. Garrett. 1963. Studies on the bacteriophagy of *Pseudomonas morsprunorum*, *Pseudomonas syringae* and related organisms. J. Appl. Bacteriol. *26:* 159-177.

Crosse, J.E. and R.N. Goodman. 1973. A selective medium for a definitive colony characteristic of *Erwinia amylovora*. Phytopathology *63:* 1425-1426.

Crow, V.L., B.D.W. Jarvis and R.M. Greenwood. 1981. Deoxyribonucleic acid homology among acid-producing strains of *Rhizobium*. Int. J. Syst. Bacteriol. *31:* 152-172.

Crues, J.V., B.E. Murray and R.C. Moellering. 1979. *In vitro* activity of three tetracycline antibiotics against *Acinetobacter calcoaceticus* subsp. *anitratus*. Antimicrob. Agents Chemother. *16:* 690-692.

Cruickshank, J.C. 1935. A study of the so-called *Bacterium typhi flavum*. J. Hyg. *35:* 354.

Cruickshank, R. 1965. Medical Microbiology, 11th Ed., E.S. Livingstone, Edinburgh.

Cuadra, M. and J. Takano. 1969. The relationship of *Bartonella bacilliformis* to the red blood cell as revealed by electron microscopy. Blood *33:* 708-716.

Cullen, G.A. and G.C. Snell. 1976. A cell-free haemagglutinating antigen of *Mycoplasma synoviae* and its use in haemagglutination inhibition tests. J. Biol. Standard. *4:* 203-207.

Cullum, J. and H. Saedler. 1981. DNA rearrangements and evolution. *In* Carlile, Collins and Moseley (Editors), Molecular and Cellular Aspects of Microbial Evolution, Symposium No. 32 of the Society for General Microbiology, Cambridge University Press, London - New York, pp. 131-150.

Cummins, C.S. 1962. Immunochemical specificity and the location of antigens in the bacterial cell. *In* Ainsworth and Sneath (Editors), Microbial Classification, 12th Symposium of the Society for General Microbiology, Cambridge University Press, United Kingdom.

Cummins, C.S. and H. Harris. 1956. The chemical composition of the cell wall in some Gram-positive bacteria and its possible value as a taxonomic character. J. Gen. Microbiol. *14:* 583-600.

Cuppels, D. and A. Kelman. 1974. Evaluation of selective media for isolation of soft-rot bacteria from soil and plant tissue. Phytopathology *64:* 468-475.

Curasson, G. 1938. Notes sur la piroplasmose aviaire en E.O.F. Bull. Serv. Zootech. Epiz. A.O.F. *1:* 33-35.

Curasson, G. and P. Adnrjesky. 1929. Sur les "corps de Balfour" du sang de la poule. Bull. Soc. Pathol. Exot. *22:* 316-317.

Currier, T.C. and E.W. Nester. 1976. Evidence for diverse types of large plasmids in tumor-inducing strains of *Agrobacterium*. J. Bacteriol. *126:* 157-165.

Curtiss, R. 1969. Bacterial conjugation. Annu. Rev. Microbiol. *23:* 69-136.

Cutlip, R.C., W.C. Amtower and M.R. Zinober. 1972. Septic embolic actinobacillosis of swine: a case report and laboratory reproduction of the disease. Am. J. Vet. Res. *33:* 1621-1626.

Cwyk, W.M. and E. Canale-Parola. 1979. *Treponema succinifaciens* sp. nov., an anaerobic spirochete from the swine intestine. Arch. Microbiol. *122:* 231-239.

Cwyk, W.M. and E. Canale-Parola. 1981. *In* Validation of the publication of new names and new combinations previously effectively published outside the IJSB. List No. 7. Int. J. Syst. Bacteriol. *31:* 382-383.

Czukás, Z. 1976. Reisolation and characterization of *Haemophilus influenzae-murium*. Acta Microbiol. Acad. Sci. Hung. *23:* 89-96.

Czygan, F.C. and W. Heumann. 1967. Die Zusammensetzung und Biogenese der Carotinoide in *Pseudomonas echinoides* und einige Mutanten. Arch. Mikrobiol. *57:* 123-134.

Dadd, R.H. 1977. Quantitative requirements and utilization of nutrients: insects. *In* Rechcigl (Editor), CRC Handbook Series in Nutrition and Food. Section D. Nutritional Requirements, Vol. I, CRC Press, Cleveland, pp. 305-346.

Dadds, M.J.S. 1971. The detection of *Zymomonas anaerobia*. *In* Shapton and Board (Editors), Isolation of anaerobes. The Society for Applied Bacteriology, Technical Series No. 5, Academic Press, Inc., New York, pp. 219-222.

Dadds, M.J.S., P.A. Martin and J.G. Carr. 1973. The doubtful status of the species *Zymomonas anaerobia* and *Z. mobilis*. J. Appl. Bacteriol. *36:* 531-539.

Dagley, S. and P.W. Trudgill. 1965. The metabolism of galactarate, D-glucarate and various pentoses by species of *Pseudomonas*. Biochem. J. *95:* 48-58.

Dahl, C.E., J.S. Dahl and K. Bloch. 1980a. Effect of alkyl-substituted precursors of cholesterol on artificial and natural membranes and on the viability of *Mycoplasma capricolum*. Biochemistry *19:* 1462-1467.

Dahl, J.S., C.E. Dahl and K. Bloch. 1980b. Sterols in membranes: growth characteristics and membrane properties of *Mycoplasma capricolum* cultured on cholesterol and lanosterol. Biochemistry *19:* 1467-1472.

Dahlberg, J.E. and R.M. Franklin. 1970. Structure and synthesis of a lipid-containing bacteriophage. IV. Electron microscopic studies of PM2-infected *Pseudomonas* BAL-31. Virology *42:* 1073-1086.

Dahlen, G., H. Nygren and H.-A. Hansson. 1978. Immunoelectron microscopic localization of lipopolysaccharides in the cell wall of *Bacteroides oralis* and *Fusobacterium nucleatum*. Infect. Immun. *19:* 265-271.

Dailey, H. A., Jr. 1976. Membrane-bound respiratory chain of *Spirillum itersonii*. J. Bacteriol. *127:* 1286-1291.

Dailey, H. A., Jr. and J. Lascelles. 1974. Ferrochelatase activity in wild-type and mutant strains of *Spirillum itersonii*. Arch. Biochem. Biophys. *160:* 523-529.

Daily, O.P., R.M. Debell and S.W. Joseph. 1978. Superoxide dismutase and catalase levels in halophilic vibrios. J. Bacteriol. *134:* 375-380.

Dainty, R.H., D.J. Etherington, B.G. Shaw, J. Barlow and G.T. Banks. 1978. Studies on the production of extracellular proteinases by a non-pigmented strain of *Chromobacterium lividum* isolated from abbatoir effluent. J. Appl. Bacteriol. *45:* 111-124.

Dakin, W.P.H., D.J. Howell, R.G.A. Sutton, M.F. O'Keefe and P. Thomas. 1974. Gastroenteritis due to non-agglutinable (non-cholera) vibrios. Med. J. Aust. *2:* 487-490.

Dalton, H. 1980. Chemoautotrophic nitrogen fixation. *In* Stewart and Gallon (Editors), Nitrogen Fixation, Annu. Proc. Phytochem. Soc. Eur., Academic Press, London, pp. 177-195.

D'Amato, R.F., L.A. Eriquez, K.M. Tomfahrde and E. Singerman. 1978. Rapid identification of *Neisseria gonorrhoeae* and *Neisseria meningitidis* by using enzymatic profiles. J. Clin. Microbiol. *7:* 77-81.

Damon, S.R. 1926. A note on the spirochaetes of termites. J. Bacteriol. *11:* 31-36.

Dangeard, P.A. 1926. Recherches sur les tubercles radicaux des Légumineuses. Botaniste (Paris) *16:* 1-275.

Dang-Gabrani, K. 1971. On the functions of intracellular symbiotes of *Sitophilus oryzae* Linn. Experientia (Basel) *27:* 107.

Dang Van, A., G. Bieth and D.H. Bouanchaud. 1975. Résistance plasmidique à la tetracycline chez *Haemophilus influenzae*. C.R. Acad. Sci. Paris, Ser. D *280:* 1321-1323.

Dang Van, A., F. Goldstein, J.F. Acar and D.H. Bouanchaud. 1975. A transferable kanamycin resistance plasmid isolated from *Haemophilus influenzae*. Ann. Microbiol. (Inst. Pasteur) *126A:* 397-399.

Daniel, R.M., A.W. Limmer, K.W. Steele and I.M. Smith. 1982. Anaerobic growth, nitrate reduction and denitrification in 46 *Rhizobium* strains. J. Gen. Microbiol. *128:* 1811-1815.

Daniel, R.S. 1973. Inheritance of intracellular bacteroids in *Periplaneta americana*. Proc. Electron Microsc. Soc. Am. *31:* 510-511.

Daniel, R.S. and M.A. Brooks. 1967. Chromatographic evidence for murein from the bacteroid symbiotes of *Periplaneta americana*. (L). Experientia (Basel) *23:* 499-502.

Daniel, R.S. and M.A. Brooks. 1972. Intracellular bacteroides: electron microscopy of *Periplaneta americana* injected with lysozyme. Exp. Parasitol. *31:* 232-246.

Daniels, E.W., E.P. Breyer and R.R. Kudo. 1966. *Pelomyxa palustris*, Greeff. II. Its ultrastructure. Z. Zellforsch. *73:* 367-383.

Daniels, M.J. 1979a. The pathogenicity of mycoplasmas for plants. Zentralbl. Bakteriol. Parasitenkd. Infektionskr. Hyg. Abt. 1 Orig. Reihe A *245:* 184-199.

Daniels, M.J. 1979b. Mechanisms of spiroplasma pathogenicity. *In* Whitcomb and Tully (Editors), The Mycoplasmas, Vol. 3, Academic Press, New York, pp. 209-227.

Daniels, M.J., J.M. Longland and J. Gilbart. 1980. Aspects of motility and chemotaxis in spiroplasmas. J. Gen. Microbiol. *118:* 429-436.

Daniels, M.J. and B.M. Meddins. 1973. Polyacrylamide gel electrophoresis of mycoplasma proteins in sodium dodecyl sulfate. J. Gen. Microbiol. *76:* 239-242.

Danielsson, D. and J. Maeland. 1978. Serotyping and antigenic studies of *Neisseria gonorrhoeae. In* Bergan and Norris (Editors), Methods in Microbiology, Vol. 10, Academic Press, New York, pp. 315-344.

Danielsson, D. and S. Normark (Editors). 1980. Genetics and immunobiology of pathogenic *Neisseria.* Proceedings of an EMBO Workshop, Hemaven, Sweden.

Danielsson, D. and E. Sandstrom. 1980. Serology of *Neisseria gonorrhoeae* demonstration by co-agglutination and immunoelectrophoresis of antigenic differences associated with colour/opacity colony variants. Acta Pathol. Microbiol. Scand. Sect. B *88:* 39-46.

Danner, D.B., R.A. Deich, K.L. Sisco and H.O. Smith. 1980. An eleven-base-pair sequence determines the specificity of DNA uptake in *Haemophilus* transformation. Gene *11:* 311-318.

D'Aoust, J.Y. and N.N. Gerber. 1974. Isolation and purification of prodigiosin from *Vibrio psychroerythrus.* J. Bacteriol. *118:* 756-757.

D'Aoust, J.Y. and D.J. Kushner. 1971. Structural changes during lysis of a psychrophilic marine bacterium. J. Bacteriol. *108:* 916-927.

D'Aoust, J.Y. and D.J. Kushner. 1972. *Vibrio psychroerythrus* sp. n.: classification of the psychrophilic marine bacterium, NRC 1004. J. Bacteriol. *111:* 340-342.

Darland, G., T.D. Brock, W. Samsonoff and S.F. Conti. 1970. A thermophilic, acidophilic mycoplasma isolated from a coal refuse pile. Science *170:* 1416-1418.

da Rocha-Lima, H. 1916. Zür Aetiologie des Fleck-fiebers. Berlin Klin. Wochenschr. *53:* 567-569.

Das, P.K., M. Basu and G.C. Chatterjee. 1979. Lipid profile of the strains of *Agrobacterium tumefaciens* in relation to agrocin resistance. J. Gen. Appl. Microbiol. *25:* 1-9.

Das, J. and S.N. Chatterjee. 1966. Electron microscopic studies on some ultrastructural aspects of *Vibrio cholerae.* Indian J. Med. Res. *54:* 330-338.

Das, J. and S.N. Chatterjee. 1969. Morphological changes of *Vibrio cholerae* organisms in glucose saline. J. Gen. Microbiol. *54:* 445-450.

Dasch, G.A. 1975. Morphological and molecular studies on intracellular bacterial symbiotes of insects. Ph.D. Thesis, Yale University, New Haven, Connecticut.

Dasch, G.A. 1981. Isolation of species-specific protein antigens of *Rickettsia typhi* and *Rickettsia prowazekii* for immunodiagnosis and immunoprophylaxis. J. Clin. Microbiol. *14:* 333-341.

Dasch, G.A. and A.L. Bourgeois. 1981. Antigens of the typhus group of rickettsiae: importance of the species-specific surface protein antigens in eliciting immunity. *In* Burgdorfer and Anacker (Editors), Rickettsiae and Rickettsial Diseases, Academic Press, New York, pp. 61-70.

Dasch, G.A., S. Halle and A.L. Bourgeois. 1979. Sensitive microplate enzyme-linked immunosorbent assay for detection of antibodies against the scrub typhus rickettsia, *Rickettsia tsutsugamushi.* J. Clin. Microbiol. *9:* 38-48.

Dasch, G.A., J.R. Samms and E. Weiss. 1978. Biochemical characteristics of typhus group rickettsiae with special attention to the *Rickettsia prowazekii* strains isolated from flying squirrels. Infect. Immun. *19:* 676-685.

Dasch, G.A. and E. Weiss. 1977. Characterization of the Madrid E strain of *Rickettsia prowazekii* purified by Renografin density gradient centrifugation. Infect. Immun. *15:* 280-286.

Dasch, G.A. and E. Weiss. 1978. Factors affecting the viability of *Rickettsia tsutsugamushi* purified from yolk sacs and L cells. *In* Kazár, Ormsbee and Tarasevich (Editors), Rickettsiae and Rickettsial Diseases, VEDA, Bratislava, pp. 115-127.

Datta, N. and R.W. Hedges. 1973. R factors of compatibility group A. J. Gen. Microbiol. *74:* 335-336.

Davey, J.F., R. Whittenbury and J.F. Wilkinson. 1972. The distribution in the methylobacteria of some key enzymes concerned with intermediary metabolism. Arch. Mikrobiol. *87:* 359-366.

Davies, D.A.L. 1958. The smooth and rough antigens of *Pasteurella pseudotuberculosis.* J. Gen. Microbiol. *18:* 118-128.

Davies, G., C.N. Hébert and A.D. Casey. 1973. Preservation of *Brucella abortus* (Strain 544) in liquid nitrogen and its virulence when subsequently used as a challenge. J. Biol. Stand. *1:* 165-170.

Davies, S.L. and R. Whittenbury. 1970. Fine structure of methane- and other hydrocarbon-utilizing bacteria. J. Gen. Microbiol. *61:* 227-232.

Davis, B.R. and W.H. Ewing. 1963. Serologic relations that may lead to erroneous diagnosis of *Escherichia coli* infections by means of fluorescent antibody technics. Amer. J. Clin. Pathol. *39:* 198-202.

Davis, B.R., G.R. Fanning, J.M. Madden, A.G. Steigerwalt, H.B. Bradford, Jr., H.L. Smith and D.J. Brenner. 1981. Characterization of biochemically atypical *Vibrio cholerae* strains and designation of a new pathogenic species, *Vibrio mimicus.* J. Clin. Microbiol. *14:* 631-639.

Davis, B.R., G.R. Fanning, J.M. Madden, A.G. Steigerwalt, H.B. Bradford, Jr., H.L. Smith, Jr. and D.J. Brenner. 1982. *In* Validation of the publication of new names and new combinations previously effectively published outside the IJSB. List no. 8. Int. J. Syst. Bacteriol. *32:* 266-268.

Davis, C.M. and C. Collins. 1969. Granuloma inguinale: ultrastructural study of *Calymmatobacterium granulomatis.* J. Invest. Dermatol. *53:* 315-321.

Davis, C.P., D. Cleven, J. Brown and E. Balish. 1976. *Anaerobiospirillum,* a new genus of spiral-shaped bacteria. Int. J. Syst. Bacteriol. *26:* 498-504.

Davis, D.H. 1969. *In* Davis, D.H., M. Doudoroff, R.Y. Stanier and M. Mandel. 1969. Proposal to reject the genus *Hydrogenomonas.* Taxonomic implications. Int. J. Syst. Bacteriol. *19:* 375-390.

Davis, D.H., M. Doudoroff, R.Y. Stanier and M. Mandel. 1969. Proposal to reject the genus *Hydrogenomonas:* taxonomic implications. Int. J. Syst. Bacteriol. *19:* 375-390.

Davis, D. H., R. Y. Stanier, M. Doudoroff and M. Mandel. 1969. Proposal to reject the genus *Hydrogenomonas,* taxonomic implications. Arch. Mikrobiol. *70:* 1-13.

Davis, D.H., R.Y. Stanier, M. Doudoroff and M. Mandel. 1970. Taxonomic studies on some Gram negative polarly flagellated "hydrogen bacteria" and related species. Arch. Mikrobiol. *70:* 1-13.

Davis, D.J., M. Pitmann and J.J. Griffitts. 1950. Hemagglutination by the Koch-Weeks bacillus (*Hemophilus aegyptius*). J. Bacteriol. *59:* 427-431.

Davis, G.E. 1942. Species unity or plurality of the relapsing fever spirochetes. Amer. Assoc. Advan. Soc. Pub. No. 18, pp. 41-47.

Davis, G.E. 1948. The spirochetes. Annu. Rev. Microbiol. *2:* 305-334.

Davis, G.E. 1952. Observations on the biology of the Argasid tick, *Ornithodoros brasiliensis.* Aragao. 1923 with the recovery of a spirochete, *Borrelia brasiliensis* n. sp. J. Parasitol. *38:* 473-476.

Davis, G.E. 1956. A relapsing fever spirochete, *Borrelia mazzottii* (sp. nov.) from *Ornithodorus talaje* from Mexico. Amer. J. Hyg. *63:* 13-17.

Davis, G.E. 1957. Order IX. *Spirochaetales* Buchanan 1918. *In* Breed, Murray and Smith (Editors), Bergey's Manual of Determinative Bacteriology, 7th Ed. The Williams and Wilkins Co., Baltimore, pp. 892-907.

Davis, G.E. and H.R. Cox. 1938. A filter-passing infectious agent isolated from ticks. I. Isolation from *Dermacentor andersoni,* reactions in animals, and filtration experiments. Pub. Health Rep. *53:* 2259-2267.

Davis, G.H.G. and A.C. Baird-Parker. 1959. Cell-wall composition of *Leptotrichia* spp. Nature (London) *183:* 1206-1207.

Davis, G.H.G. and R.W.A. Park. 1962. A taxonomic study of certain bacteria currently classified as *Vibrio* species. J. Gen. Microbiol. *27:* 101-119.

Davis, J.B., V.F. Coty and J.P. Stanley. 1964. Atmospheric nitrogen fixation by methane-oxidizing bacteria. J. Bacteriol. *88:* 468-472.

Davis, J.M., M.M. Peel and J.A. Gillians. 1979. Colonization of an amputation site by *Flavobacterium odoratum* after gentamicin therapy. Med. J. Aust. *2:* 703-704.

Davis, R.E. 1978. Spiroplasma associated with flowers of the tulip tree (*Liriodendron tulipifera* L.) Can. J. Microbiol. *24:* 954-959.

Davis, R.E. 1979. Spiroplasmas: newly recognized arthropod-borne pathogens. *In* Maramorosch and Harris (Editors), Leafhopper vectors and plant disease agents, Academic Press, New York, pp. 451-484.

Davis, R.E., I.M. Lee and L.K. Basciano. 1979. Spiroplasmas: serological grouping of strains associated with plants and insects. Can. J. Microbiol. *25:* 861-866.

Davis, R.E., I.M. Lee and J.F. Worley. 1981. *Spiroplasma floricola,* a new species isolated from surfaces of flowers of the tulip tree, *Liriodendron tulipifera* L. Int. J. Syst. Bacteriol. *31:* 456-464.

Davis, R.E. and R.F. Whitcomb. 1969. Spectrum of antibiotic sensitivity of aster yellows disease in insects and plants. Phytopathology *59:* 1556.

Davis, R.E. and J.F. Worley. 1973. Spiroplasma: motile helical microorganism associated with corn stunt disease. Phytopathology *63:* 403-408.

Davis, W.A., II, J.H. Chretien, V.F. Garagusi and M.A. Goldstein. 1978. Snake-to-human transmission of *Aeromonas shigelloides* resulting in gastroenteritis. South Med. J. *71:* 474-476.

Davis, W.A., J.G. Kane and V.G. Garagusi. 1978. Human *Aeromonas* infections: a review of the literature and a case report of endocarditis. Medicine *57:* 267-277.

Davýdov, N.N. 1961. Properties of *Brucella* isolated from reindeer. (In Russian.) Trudý Vsyesoyuz. Inst. Eksp. Vet. *27:* 24-31.

Dawes, E.A., M. Midgley and M. Ishaq. 1970. The endogenous metabolism of anaerobic bacteria. Final technical report (Dec. 1970) for Contract no. DAJA37-67-C-0567, European Research Office, U.S. Army.

Dawes, E.A., D.W. Ribbons and D.A. Rees. 1966. Sucrose utilization by *Zymomonas mobilis:* formation of a levan. Biochem. J. *98:* 804.

Day, N.P., S.M. Scotland and B. Rowe. 1981. Comparison of an HEp-2 tissue culture test with the Sèrèny test for detection of enteroinvasiveness in *Shigella* spp. and *Escherichia coli.* J. Clin. Microbiol. *13:* 596-597.

Day, W.R. 1924. The watermark disease of the cricket-bat willow. Oxford For. Mem. *3:* 1-30.

Dayhoff, M.O. 1976. Atlas of protein sequence and structure, Vol. 5, Suppl. 2, National Biomedical Research Foundation, Washington, D.C.

Deacon, W.E. 1952. Antigenic study of certain slow lactose fermenting *Aerobacter cloacae* cultures. Proc. Soc. Exp. Biol. Med. *81:* 165-170.

Deacon, W.E., W.L. Peacock, E.M. Freeman and A. Harris. 1959. Identification of *Neisseria gonorrhoeae* by means of fluorescent antibodies. Proc. Soc. Exp. Biol. Med. *101:* 322-325.

Dean, H.F. and A.T. Morgan. 1983. Integration of R91-5::Tn*501* into the *Pseudomonas putida* chromosome and genetic circularity of the chromosomal map. J. Bacteriol. *153:* 485-497.

Dean, W.W., A.R. Mead and W.T. Northey. 1970. *Aeromonas liquefaciens* in the giant African snail, *Achatina fulica.* J. Invert. Pathol. *16:* 346-351.

de Araujo, W.C., E. Varah and S.E. Mergenhagen. 1963. Immunochemical analysis of human oral strains of *Fusobacterium* and *Leptotrichia*. J. Bacteriol. *86:* 837-844.

Debette, J. and R. Blondeau. 1980. Présence de *Pseudomonas maltophilia* dans la rhizosphère de quelques plantes cultivées. Can. J. Microbiol. *26:* 460-463.

De Boer, S.H., R.J. Copeman and H. Vruggink. 1979. Serogroups of *Erwinia carotovora* potato strains determined with diffusible somatic antigens. Phytopathology *69:* 316-319.

de Boer, W.E., C. Golten and W.A. Scheffers. 1975a. Effects of some chemical factors on flagellation and swarming of *Vibrio alginolyticus*. Antonie van Leeuwenhoek J. Microbiol. Serol. *41:* 385-403.

de Boer, W.E., C. Golten and W.A. Scheffers. 1975b. Effects of some physical factors on flagellation and swarming of *Vibrio alginolyticus*. Nether. J. Sea Res. *9:* 197-213.

De Bont, J.A.M. and M.W.M. Leijten. 1976. Nitrogen fixation by hydrogen-utilizing bacteria. Arch. Microbiol. *107:* 235-240.

De Bord, G.G. 1942. Descriptions of *Mimeae* trib. nov. with three genera and three species and two new species of *Neisseria* from conjunctivitis and vaginitis. Iowa State Coll. J. Sci. *16:* 471-480.

de Buen, S. 1926. Note préliminaire sur l'épidémiologia de la fièvre récurrente espagnole. Ann. Parasitol. Hum. Comp. *4:* 185-192.

De Cleene, M. 1979. Crown gall: economic importance and control. Zentralbl. Bakteriol. Parasitenk. Infektionskr. Hyg. II Abt. *134:* 551-554.

De Cleene, M. and J. De Ley. 1976. The host range of crown gall. Bot. Rev. *42:* 389-466.

De Cleene, M. and J. De Ley. 1981. The host range of infectious hairy-root. Bot. Rev. *47:* 147-194.

Dedonder, R.A. and W.Z. Hassid. 1964. The enzymatic synthesis of a β-1, 2-0-linked glucan by an extract of *Rhizobium japonicum*. Biochim. Biophys. Acta *90:* 239-248.

Deeb, B.J. and G.E. Kenny. 1967. Characterization of *Mycoplasma* variants isolated from rabbits. J. Bacteriol. *93:* 1416-1424.

Dees, S.B., D.G. Hollis, R.E. Weaver and C.W. Moss. 1981. Cellular fatty acids of *Brucella canis* and *Brucella suis*. J. Clin. Microbiol. *14:* 111-112.

deGraaf, J.L., L. Elwell and S. Falkow. 1976. Molecular nature of two beta-lactamase-specifying plasmids isolated from *Haemophilus influenzae* type b. J. Bacteriol. *126:* 439-446.

De Greve, H., H. Decraemer, J. Seurinck, M. Van Montagu and J. Schell. 1981. The functional organization of the octopine *Agrobacterium tumefaciens* plasmid pTiB6S3. Plasmid *6:* 235-248.

Degryse, E., N. Glansdorff and A. Pierard. 1978. A comparative analysis of extreme thermophilic bacteria belonging to the genus *Thermus*. Arch. Microbiol. *117:* 189-196.

Dehority, B.A. 1966. Characterization of several bovine rumen bacteria isolated with a xylan medium. J. Bacteriol. *91:* 1724-1729.

Dehority, B.A. 1969. Pectin-fermenting bacteria isolated from the bovine rumen. J. Bacteriol. *99:* 189-196.

Deibel, R.H. and J.B. Evans. 1960. Modified benzidine test for the detection of cytochrome-containing respiratory systems in microorganisms. J. Bacteriol. *79:* 356 360.

Deich, R.A. and H.O. Smith. 1980. Mechanism of homospecific DNA uptake in *Haemophilus influenzae* transformation. Mol. Gen. Genet. *177:* 369-374.

Deinema, M.H. and L.P.T.M. Zevenhuizen. 1971. Formation of cellulose fibrils by Gram-negative bacteria and their role in bacterial flocculation. Arch. Mikrobiol. *78:* 42-57.

de Jongh, P. 1938. On the symbiosis of *Ardisia crispa* DC. Verh. Kon. Ned. Akad. Wet. II *37:* 1-74.

De Kam, M. 1976. *Erwinia salicis*: its metabolism and variability *in vitro*, and a method to demonstrate the pathogen in the host. Antonie van Leeuwenhoek J. Microbiol. Serol. *42:* 421-428.

Dekker, R.F.H. and G.P. Candy. 1979. The β-mannanases elaborated by the phytopathogen *Xanthomonas campestris*. Arch. Microbiol. *122:* 297-299.

Delafield, F.P., M. Doudoroff, N.J. Palleroni, C.J. Lusty and R. Contopoulos. 1965. Decomposition of poly-β-hydroxybutyrate by pseudomonads. J. Bacteriol. *90:* 1455-1466.

DeLang, R.J., G.R. Green and D.G. Searcy. 1981. A histone-like protein (HTa) from *Thermoplasma acidophilum*. I. Purification and properties. J. Biol. Chem. *256:* 900-904.

Delaporte, B. 1964a. Étude comparée de grands spirilles format des spores: *Sporospirillum (Spirillum) praeclarum* (Collin) n. g. , *Sporospirillum gyrini* n. sp. et *Sporospirillum bisporum* n. sp. Ann. Inst. Pasteur Paris *107:* 246-262.

Delaporte, B. 1964b. Étude descriptive de bactéries de très grandes dimensions. Ann. Inst. Pasteur Paris *107:* 845-862.

Delaporte, B. and P. Daste. 1956. Une bactérie du sol capable de décomposer la fraction fixe de certaines oléorésines *Flavobacterium resinovorum* n. sp. C.R. Acad. Sci. *242:* 831-834.

Delaporte, B., M. Raynaud and P. Daste. 1965. Une bactérie du sol capable d'utiliser, comme source de carbone, la fraction fixe de certain oléorésines, *Pseudomonas resinovorans* n. sp. C. R. Hebd. Séances Acad. Sci. Paris *252:* 1073-1075.

Del Bene. V.E. and W.E. Farrar. 1973. Cephalosporinase activity in *Bacteroides* microb. Agents Chemother. *3:* 369-372.

Comparative carbohydrate metabolism and a proposal for a

phylogenetic relationship of the acetic acid bacteria. J. Gen. Microbiol. *24:* 31-50.

De Ley, J. 1968. DNA base composition and hybridization in the taxonomy of the phytopathogenic bacteria. Annu. Rev. Phytopathol. *6:* 63-90.

De Ley, J. 1970. Re-examination of the association between melting point, buoyant density, and chemical base composition of deoxyribonucleic acid. J. Bacteriol. *101:* 738-754.

De Ley, J. 1972. *Agrobacterium*: intrageneric relationships and evolution. *In* Proc. 3rd Int. Conf. Plant Path. Bact., Wageningen, 1971, pp. 251-259. Wageningen, The Netherlands: Pudoc.

De Ley, J. 1974. Phylogeny of procaryotes. Taxon. *23:* 291-300.

De Ley, J. 1978. Modern molecular methods in bacterial taxonomy: evaluation, application, prospects. Proc. 4th Int. Conf. Plant Path. Bact. Angers 1978, 347-357.

De Ley, J., M. Bernaerts, A. Rassel and J. Guilmot. 1966. Approach to an improved taxonomy of the genus *Agrobacterium*. J. Gen. Microbiol. *43:* 7-17.

De Ley, J., H. Cattoir and A. Reynaerts. 1970. The quantitative measurement of DNA hybridization from renaturation rates. Eur. J. Biochem. *12:* 133-142.

De Ley, J. and J. Frateur. 1970. The status of the generic name *Gluconobacter*. Int. J. Syst. Bacteriol. *20:* 83-95.

De Ley, J. and J. Frateur. 1974. *Gluconobacter. In* Buchanan and Gibbons (Editors), Bergey's Manual of Determinative Bacteriology, 8th Ed., The Williams and Wilkins Co., Baltimore, pp. 251-253.

De Ley, J. and J. Frateur. 1974. *Acetobacter. In* Buchanan and Gibbons (Editors), Bergey's Manual of Determinative Bacteriology, 8th Ed., The Williams and Wilkins Co., Baltimore.

De Ley, J., M. Gillis, C.F. Pootjes, K. Kersters, R. Tytgat and M. Van Braekel. 1972. Relationship among temperate *Agrobacterium* phage genomes and coat proteins. J. Gen. Virol. *16:* 199-214.

De Ley, J. and K. Kersters. 1964. Oxidation of aliphatic glycols by acetic acid bacteria. Bacteriol. Rev. *28:* 164-180.

De Ley, J., K. Kersters, J. Khan-Matsubara and J.M. Shewan. 1970. Comparative D-gluconate metabolism and DNA base composition in *Achromobacter* and *Alcaligenes*. Antonie van Leeuwenhoek J. Microbiol. Serol. *36:* 193-207.

De Ley, J. and I.W. Park. 1966. Molecular biological taxonomy of some free-living nitrogen-fixing bacteria. Antonie van Leeuwenhoek J. Microbiol. Serol. *32:* 6-16.

De Ley, J. and A. Rassel. 1965. DNA base composition, flagellation and taxonomy of the genus *Rhizobium*. J. Gen. Microbiol. *41:* 85-91.

De Ley, J., P. Segers and M. Gillis. 1978. Intra- and intergeneric similarities of *Chromobacterium* and *Janthinobacterium* ribosomal ribonucleic acid cistrons. Int. J. Syst. Bacteriol. *28:* 154-168.

De Ley, J. and J. Swings. 1976. Phenotypic description, numerical analysis and a proposal for an improved taxonomy and nomenclature of the genus *Zymomonas* Kluyver and van Niel 1936. Int. J. Syst. Bacteriol. *26:* 146-157.

De Ley, J., R. Tijtgat, J. De Smedt and M. Michiels. 1973. Thermal stability of DNA:DNA hybrids within the genus *Agrobacterium*. J. Gen. Microbiol. *78:* 241-252.

Del Giudice, R.A., T.R. Carski, M.F. Barile, R.M. Lemcke and J.G. Tully. 1971. Proposal for classifying human strain navel and related simian mycoplasmas as *Mycoplasma primatum* sp. n. J. Bacteriol. *108:* 439-445.

Del Giudice, R.A., R.S. Gardella and H.E. Hopps. 1980. Cultivation of formerly noncultivable strains of *Mycoplasma hyorhinis*. Curr. Microbiol. *4:* 75-80.

Del Giudice, R.A., R.H. Purcell, T.R. Carski and R.M. Chanock. 1974. *Mycoplasma lipophilum* sp. nov. Int. J. Syst. Bacteriol. *24:* 147-153.

Del Giudice, R.A., N.F. Robillard and T.R. Carski. 1967. Immunofluorescence identification of *Mycoplasma* on agar by use of incident illumination. J. Bacteriol. *93:* 1205-1209.

Delisle, G.J. 1977. Multiple forms of urease in cytoplasmic fractions of *Ureaplasma urealyticum*. J. Bacteriol. *130:* 1390-1392.

De Louvois, J. 1969. Serotyping and the Dienes reaction on *Proteus mirabilis* from hospital infections. J. Clin. Pathol. *22:* 263-268.

Demarco de Hormaeche, R. and M.J. Thornley. 1978. Demonstration by light and electron microscopy of capsules on gonococci recently grown in vivo. J. Gen. Microbiol. *106:* 81-91.

DeMeillon, B. and L. Goldberg. 1946. Nutritional studies on blood-sucking arthropods. Nature (Lond.) *158:* 269-270.

DeMeillon, B. and L. Goldberg. 1947. Preliminary studies on the nutritional requirements of the bedbug (*Cimex lectularius* L.) and the tick *Ornithodorus moubata* Murray. J. Exp. Biol. *24:* 41-63.

DeMeillon, B., J.M. Thorp and F. Hardy. 1947. The relationship between ectoparasite and host. 1. The development of Cimex lectularius and *Ornithodorus moubata* on riboflavin deficient rats. S. Afr. J. Med. Sci. *12:* 111-116.

deMello, F. 1921. Protozoaires parasites du *Pachelebra moesta* Reeve. C. R. Soc. Biol. (Paris) *84:* 241-242.

DeMello, F.J. and M.S. Leonard. 1979. *Eikenella corrodens*, a new pathogen. Oral Surg. *45:* 401-404.

De Meuron, P.A. and R. Peduzzi. 1979. Caractérisation de souches du genre *Aeromonas* isolées chez des poissons d'eau douce et quelques reptiles. Zentralbl. Vet. Med. *26:* 153-167.

DeMoss, R.D. 1967. Violacein. *In* D. Gottlieb and P. Shaw (Editors), Mechanisms of Action and Biosynthesis of Antibiotics, Vol. 2. Springer-Verlag, New York, pp. 77-81.

d'Empaire, M. 1969. Les facteurs de croissance des *Edwardsiella tarda*. Ann. Inst. Past. (Paris) *116:* 63-68.

Dénarie, J., G. Truchet and B. Bergeron. 1976. Effects of some mutations on symbiotic properties of *Rhizobium. In* Nutman (Editor), Symbiotic Nitrogen Fixation in Plants. International Biological Programme 7. Cambridge University Press, London. pp. 47-61.

Den Dooren de Jong, L.E. 1926. Bijdrage tot de kennis van het mineralisatie-proces, Nijgh and van Ditmar Uitgevers-Mij, Rotterdam, pp. 1-200.

Deneer, H.G., L. Slaney, I.W. Maclean and W.L. Albritton. 1982. Mobilization of nonconjugative antibiotic resistance plasmids in *Haemophilus ducreyi*. J. Bacteriol. *149:* 726-732.

Denhardt, D.T. 1966. A membrane-filter technique for the detection of complementary DNA. Biochem. Biophys. Res. Comm. *23:* 641-646.

De Paula Gomes, A. 1959. Observações sôbre a utilizaçao de *Zymomonas mobilis* (Lindner) Kluyver and van Niel, 1936 *(Termobacterium mobile,* Lindner, 1928: *Pseudomonas lindneri Kluyver and Hoppenbrouwers, 1931),* na terapêutica humana. Rev. Inst. Antibiot. Univ. Recife *2:* 77-81.

De Petris, S. 1967. Ultrastructure of the cell wall of *Escherichia coli* and the chemical nature of its constituent layers. Ultrastruct. Res. *19:* 45-83.

De Petris, S., G. Karlsbad and R.W.I. Kessel. 1964. Ultra structure of S and R variants of *Brucella abortus* grown on a lifeless medium. J. Gen. Microbiol. *35:* 373-382.

Depicker, A., M. De Wilde, G. De Vos, R. De Vos, M. Van Montagu and J. Schell. 1980. Molecular cloning of overlapping segments of the nopaline Ti-plasmid pTiC58 as a means to restriction endonuclease mapping. Plasmid *3:* 193-211.

Depicker, A., M. Van Montagu and J. Schell. 1978. Homologous DNA sequences in different Ti-plasmids are essential for oncogenicity. Nature *275:* 150-153.

de Plessis, S.J. 1940. Bacterial blight of vines (Vlamiekte) in South Africa caused by *Erwinia vitivora* (Bacc.) Du P. S. Afr. Dep. Agric. Tech. Serv. Sci. Bull. *214:* 1-105.

De Polli, H., B. B. Bohlool and J. Döbereiner. 1980. Serological differentiation of *Azospirillum* species belonging to different host-plant specificity groups. Arch. Microbiol. *126:* 217-222.

Derby, H.A. and B.W. Hammer. 1931. Bacteriology of butter. IV. Bacteriological studies of surface taint butter. Iowa Agr. Exp. Sta., Res. Bull. *145:* 387-416.

Derijcke, J., L. Devriese, J. Hoorens, P. de Rose and F. Castryck. 1978. *Haemophilus pleuropneumoniae* infecties bij het varken. Vlaams Diergeneesk. Tijdschr. *47:* 405-417.

de'Rossi, G. 1927. Microbiologia agraria e technica. Unione Tipographico-Editrice Torinese, Torino.

Derrick, E.H. 1939. *Rickettsia burneti:* The cause of "Q" fever. Med. J. Aust. *1:* 14.

Derrick, E.H. 1953. The epidemiology of Q fever: A review. Med. J. Aust. *1:* 245-253.

Derrick, E.H. and H.E. Brown. 1949. Isolation of the Karp strain of *R. tsutsugamushi*. Lancet *2:* 150-151.

Derx, H.G. 1950a. *Beijerinckia,* a new genus of nitrogen-fixing bacteria occurring in tropical soils. Proc. Kon. Ned. Akad. Wet., Ser. C *53:* 140-147.

Derx, H.G. 1950b. Further researches on *Beijerinckia*. Ann. Bogor. *1:* 1-12.

Derx, H.G. 1951. L'accumulation spécifique de l'*Azotobacter agile* Beijerinck et de l'*Azotobacter vinelandii* Lipman. Proc. Sect. Sci. K. Ned. Akad. Wet. C*54:* 624-635.

Derx, H.G. 1951a. *Azotobacter insigne* spec. nov. fixateur d'azote à flagellation polaire. Proc. Sect. Sci. K. Ned. Akad. Wet C*54:* 342-350.

Desai, S.G., A.B. Gandi, M.K. Patel and W.V. Kotasthane. 1966. A new bacterial leaf-spot and blight of *Azadirachta indica*. A. Juss. Indian Phytopathol. *19:* 322-323.

Desai, M.V. and H.M. Shah. 1959. A new bacterial leaf spot of *Crotalaria juncea* L. Curr. Sci. *28:* 377-378.

Desai, M.V. and H.M. Shah. 1960. Bacterial leaf spot disease of *Desmodium rotundifolium* DC. Curr. Sci. *29:* 65-66.

Desai, M.V., M.J. Thirumalachar and M.K. Patel. 1965. Bacterial blight disease of *Eleusine coracana* Gaertn. Indian Phytopathol. *18:* 384-386.

Desmarchelier, P.M. and J.L. Reichelt. 1981. Phenotypic characterization of clinical and environmental isolates of *Vibrio cholerae* from Australia. Curr. Microbiol. *5:* 123-127.

De Smedt, J., M. Bauwens, R. Tijtgat and J. De Ley. 1980. Intra- and intergeneric similarities of ribosomal ribonucleic acid cistrons of free-living, nitrogen-fixing bacteria. Int. J. Syst. Bacteriol. *30:* 106-122.

De Smedt, J. and J. De Ley. 1977. Intra- and inter-generic similarities of *Agrobacterium* ribosomal ribonucleic acid cistrons. Int. J. Syst. Bacteriol. *27:* 222-240.

De Souza, C. and L.A.G. De Souza. 1973. Colpitis and vulvovaginitis treatment using *Zymomonas mobilis* var. *recifensis*. Rev. Inst. Antibiot. Univ. Recife *13:* 85-87.

de Toni, J.B. and V. Trevisan. 1889. Schizomycetaceae Naeg., *In* Saccardo (Editor), Sylloge fungorum omnium hujusque cognitorum, Vol. 8, pp. 923-1087.

Detrick-Hooks, B. and E.R. Kennedy. 1974. Immunological cross reactions among strains of *Hydrogenomonas, Pseudomonas* and *Alcaligenes*. Antonie van Leeuwenhoek J. Microbiol. Serol. *40:* 577-584.

Devauchelle, G., G. Meynadier and C. Vago. 1972. Étude ultrastructurale du cycle de multiplication de *Rickettsiella melolonthae* (Krieg), Philip, dans les hém-ocytes de son hôte. J. Ultrastruc. Res. *38:* 134-148.

DeVay, J.E. and W.C. Schnathorst. 1963. Single-cell isolation and preservation of bacterial cells. Nature *199:* 775-777.

DeVoe, I.W. and J.E. Gilchrist. 1975. Pili on meningococci from primary culture nasopharyngeal carriers and cerebrospinal fluid of patients with acute disease. J. Exp. Med. *141:* 297-305.

De Vos, G., M. De Beuckeleer, M. Van Montagu and J. Schell. 1981. Restriction endonuclease mapping of the octopine tumor inducing plasmid pTiAch5 of *Agrobacterium tumefaciens*. Plasmid *6:* 249-253.

De Vos, P. 1980. Intrageneric and intergeneric similarities of ribosomal RNA cistrons of the genus *Pseudomonas* and the implications for taxonomy. Antonie van Leeuwenhoek J. Microbiol. Serol. *46:* 96.

De Vos, P. 1981. Intra- and intergeneric similarities of ribosomal ribonucleic acid cistrons in and with the genus *Pseudomonas*. Med. K. Acad. Wetensch. *43:* 23-60.

de Vries, W., W.M.C. van Wijck-Kapteijn and S.K.H. Oosterhuis. 1974. The presence and function of cytochromes in *Selenomonas ruminantium, Anaerovibrio lipolytica* and *Veillonella alcalescens*. J. Gen. Microbiol. *81:* 69-78.

Dhanvantari, B.N. 1977. A taxonomic study of *Pseudomonas papulans* Rose 1917. N. Z. J. Agr. Res. *20:* 577-561.

Dhawan, V.K., V.R. Rajashekaraiah, W.I. Metzger, T.W. Rice and C.A. Kallick. 1980. Spontaneous bacterial peritonitis due to a group IIk-2 strain. J. Clin. Microbiol. *11:* 492-495.

Dias, F.F. and J.V. Bhat. 1964. Microbial ecology of activated sludge. Appl. Microbiol. *12:* 412-417.

Diaz, R. and N. Bosseray. 1973. Identification d'un composé antigénique spécifique de la phase rugueuse (R) des *Brucella*. Ann. Rech. Vet. *4:* 283-292.

Diaz, R. and I. Dorronsoro. 1971. Contribución al diagnóstico serológico de brucelosis y yersiniosis. I. Utilidad de la reaccion de precipitacion en gel. Rev. Clin. Esp. *121:* 367-372.

Diaz, R., P. Garatea, L.M. Jones and I. Moriyon. 1979. Radial immunodiffusion test with a *Brucella* polysaccharide antigen for differentiating infected from vaccinated cattle. J. Clin. Microbiol. *10:* 37-41.

Diaz, R., L.M. Jones, D. Leong and J.B. Wilson. 1968. Surface antigens of smooth brucellae. J. Bacteriol. *96:* 893-901.

Diaz, R., L.M. Jones and J.B. Wilson. 1967. Antigenic relationship of *Brucella ovis* and *Brucella melitensis*. J. Bacteriol. *93:* 1262-1268.

Diaz, R., L.M. Jones and J.B. Wilson. 1968. Antigenic relationship of the Gram-negative organism causing canine abortion to smooth and rough brucellae. J. Bacteriol. *95:* 618-624.

Dickerson, R.E. 1980. Cytochrome *c* and the evolution of energy metabolism. Sci. Amer. *242:* 137-153.

Dickerson, R.E., R. Timkovich and R.J. Almassy. 1976. The cytochrome fold and the evolution of bacterial energy metabolism. J. Molec. Biol. *100:* 473-491.

Dickey, R.S. 1979. *Erwinia chrysanthemi:* a comparative study of phenotypic properties of strains from several hosts and other *Erwinia* species. Phytopathology *69:* 324-329.

Dickey, R.S. 1981. *Erwinia chrysanthemi:* reaction of eight plant species to strains from several hosts and to strains of other *Erwinia* species. Phytopathology *71:* 23-29.

Dickey, R.S. and J.I. Victoria. 1980. Taxonomy and emended description of strains of *Erwinia* isolated from *Musa paradisiaca* Linnaeus. Int. J. Syst. Bacteriol. *30:* 129-134.

Diedrich, D.L. and E.H. Cota-Robles. 1974. Heterogeneity in lipid composition of the outer membrane and cytoplasmic membrane of *Pseudomonas* BAL-31. J. Bacteriol. *119:* 1006-1018.

Diedrich, D. L., C. F. Denny, T. Hashimoto and S. F. Conti. 1970. Facultatively parasitic strain of *Bdellovibrio bacteriovorus*. J. Bacteriol. *101:* 989-996.

Dienes, L. 1946. Reproductive processes in *Proteus* cultures. Proc. Soc. Exp. Biol. Med. *63:* 265-270.

Dienst, R.B., R.B. Greenblatt and C.H. Chen. 1948. Laboratory diagnosis of Granuloma Inguinale and studies on the cultivation of the Donovan Body. Amer. J. Syph. Gonorrhea Vener. Dis. *32:* 301-306.

Dienst, R.B., R.B. Greenblatt and E.S. Sanderson. 1938. Cultural studies on the "Donovan bodies" of granuloma inguinale. J. Inf. Dis. *62:* 112-114.

Dierks, R.E., J.A. Newman and B.S. Pomeroy. 1967. Characterization of avian *Mycoplasma*. Ann. N.Y. Acad. Sci. *143:* 170-189.

Diguid, J.P. 1964. Functional anatomy of *E. coli* with special reference to enteropathogenic *E. coli*. Rev. Latinoam. Microbiol. *7:* 1-16.

Dilts, J.A. 1977. Chromosomal and extrachromosomal deoxyribonucleic acid from four bacterial endosymbionts derived from stock 51 of *Paramecium tetraurelia*. J. Bacteriol. *129:* 888-894.

Dimitroff, V.T. 1926. Spirochaetes in Baltimore market oysters. J. Bacteriol. *12:* 135-177.

Dimock, W.W., P.R. Edwards and D.W. Bruner. 1947. Infections of fetuses and foals. Ky. Agr. Exp. Sta. Bull. *509:* 1-40.

Dippell, R.V. 1950. Mutation of the killer cytoplasmic factor in *Paramecium aurelia*. Heredity *4:* 165-187.

Distaso, A. 1912. Contribution à l'étude sur l'intoxication intestinale. Zentralbl. Bakteriol. Parasitenkd. Infektionskr. Hyg., I Abt. Orig. *62:* 433-468.

Dixon, M., D.M. Jackson and I.M. Richards. 1979. The effect of a respiratory tract infection on histamine-induced changes in lung mechanics and irritant receptor discharge in dogs. Am. Rev. Respir. Dis. *120:* 843-848.

Djambazov, B., Y. Hamon and Y. Péron. 1971. Propriétés générales de quelques souches appartenant au genre *Alcaligenes*. Intérêt taxonomique de cette étude. C. R. Hebd. Acad. Sci. Paris 272: 339-342.

Dobell, C.C. 1910. On some parasitic protozoa from Ceylon. Spolia Zeylanica 7: 65-87.

Dobell, C.C. 1911. On *Cristispira veneris* nov. spec. and the affinities and classification of spirochaets. Quart. J. Microsc. Sci. 56: 507-542.

Dobell, C. 1912. Researches on the spirochaets and related organisms. Arch. Protistenk. 26: 117-240.

Döbereiner, J. 1961. Nitrogen-fixing bacteria of the genus *Beijerinckia* Derx in the rhizosphere of sugar cane. Plant and Soil 15: 211-216.

Döbereiner, J. 1966. *Azotobacter paspali* sp. n. uma bacteria fixadora de nitrogenio na rizosfera de *Paspalum*. Pesqui. Agropecu. Bras. 1: 357-365.

Döbereiner, J. 1970. Further research on *Azotobacter paspali* and its variety specific occurrence in the rhizosphere of *Paspalum notatum* Flügge. Zentralbl. Bakteriol. Parasitenkd. Infektionskr. Hyg. Abt. II 124: 224-230.

Döbereiner, J. and R. Alvahydo. 1959. Sôbre a influência da cana-de-acúcar na ocorrência de '*Beijerinckia*' no solo. II. Influência das diversas partes do vegetal. Rev. Bras. Biol. 19: 401-412.

Döbereiner, J. and V. L. D. Baldani. 1979. Selective infection of maize roots by streptomycin-resistant *Azospirillum lipoferum* and other bacteria. Can. J. Microbiol. 25: 1264-1269.

Döbereiner, J. and J.M. Day. 1975. Nitrogen fixation in the rhizosphere of tropical grasses. *In* Stewart (Editor), Nitrogen Fixation by Free-living Microorganisms. Int. Biol. Programme 6., Cambridge University Press, Cambridge, U.K., pp. 39-56.

Döbereiner, J. and J.M. Day. 1976. Associative symbioses in tropical grasses: characterization of microorganisms and dinitrogen fixing sites. *In* Newton and Nymans (Editors), Symposium on Nitrogen Fixation, Washington State University Press, Pullman, Washington, pp. 518-538.

Döbereiner, J. and H. De Polli. 1980. Diazotrophic rhizocoenoses. *In* Stewart and Gallon (Editors). Nitrogen Fixation, Academic Press, London, pp. 301-333.

Döbereiner, J., I. E. Marriel and M. Nery. 1976. Ecological distribution of *Spirillum lipoferum* Beijerinck. Can. J. Microbiol. 22: 1464-1473.

Döbereiner, J. and A.P. Ruschel. 1958. Uma nova espécie de *Beijerinckia*. Rev. Biol. 1: 261-272.

Dobrogosz, W.J., J.W. Ezzell, W.E. Kloos and C.R. Manclark. 1979. Physiology of *Bordetella pertussis*. *In* Manclark and Hill (Editors), International Symposium on Pertussis. U.S. Government Printing Office, Washington, D.C., pp. 86-93.

Dodson, R.F., G.S. Fritz, W.R. Hubler, A.H. Rudolph, J.M. Knox and L. W-F. Chu. 1973. Donovanosis: a morphologic study. J. Invest. Dermatol. 62: 611-614.

Doern, G.V. and F.L.A. Buckmire. 1976. Ultrastructural characterization of capsulated *Haemophilus influenzae* type b and two spontaneous nontypable mutants. J. Bacteriol. 127: 523-535.

Doern, G.V. and S.A. Morse. 1980. *Branhamella (Neisseria) catarrhalis*: criteria for laboratory identification. J. Clin. Microbiol. 11: 193-195.

Doetsch, R.N., B.H. Howard, S.O. Mann and A.E. Oxford. 1957. Physiological factors in the production of an iodophilic polysaccharide from pentose by a sheep rumen bacterium. J. Gen. Microbiol. 16: 156-168.

Doggett, R.G. 1969. Incidence of mucoid *Pseudomonas aeruginosa* from clinical sources. Appl. Microbiol. 18: 936-937.

Doggett, R.G. (Editor) 1979. *Pseudomonas aeruginosa*. Clinical manifestations of infection and current therapy. Academic Press, New York, pp. 1-504.

Doi, Y., M. Teranaka, K. Yora and H. Asuyama. 1967. Mycoplasma or PLT group-like microorganisms found in the phloem elements of plants infected with mulberry dwarf, potato witches' broom, aster yellows, or paulownia witches' broom. Ann. Phytopathol. Soc. Jap. 33: 259-266.

Doidge, E.M. 1920. A tomato canker. J. Dep. Agr. S. Afr. 1: 718-721.

Dolby, J.M. and C.J. Bronne-Shanbury. 1975. The use of spheroplasts-derived strains to differentiate between *Bordetella pertussis* heat-labile agglutinogens and protective antigen in mice. J. Biol. Stand. 3: 89-100.

Dolman, C.E., D.E. Kerr, H. Chang and A.R. Shearer. 1951. Two cases of rat-bite fever due to *Streptobacillus moniliformis*. Can. J. Public Health 42: 228-241.

Dommerques, Y. 1963. Distribution des *Azotobacter* et des *Beijerinckia* dans les principaux types de sol de l'ouest Africain. Ann. Inst. Pasteur (Paris) 105: 179-187.

Don, R.H. and J.M. Pemberton. 1981. Properties of six pesticide degradation plasmids isolated from *Alcaligenes paradoxus* and *Alcaligenes eutrophus*. J. Bacteriol. 145: 681-686.

Donahue, J.P. and H.A. Thompson. 1981. Efficacy of translation in the rickettsia *Coxiella burnetii*. J. Bacteriol. 146: 808-812.

Donatien, A. and G. Gayot. 1942. Conjonctivite rickettsienne du porc. Bull. Soc. Pathol. Exot. 35: 325.

Donatien, A. and F. Lestoquard. 1935. Existence en Algérie d'une Rickettsia du chien. Bull. Soc. Pathol. Exot. 28: 418-419.

Donatien, A. and F. Lestoquard. 1936. *Rickettsia bovis* Nouvelle espèce pathogène pour le boeuf. Bull. Soc. Pathol. Exot. 29: 1057-1061.

Donoghue, N.A. and P.W. Trudgill. 1975. The metabolism of cyclohexanol by *Acinetobacter* NCIB 9871. Eur. J. Biochem. 60: 1-7.

Donovan, C. 1905. Ulcerating granuloma of the pudenda. Indian Med. Gazette 40: 414.

Donovan, C. 1909. Kala-azar in Madras, especially with regard to its connection with the dog and the bugs (*Conorrhinus*). Lancet 177: 1495-1496.

Doolittle, R.E. 1981. Similar amino acid sequences: chance or common ancestry? Science (Washington) 214: 149-159.

Dorff, G.F., L.J. Jackson and M.W. Rytel. 1974. Infections with *Eikenella corrodens*, a newly recognized human pathogen. Ann. Intern. Med. 80: 305-309.

Dorofe'ev, K.A. 1947. Classification of the causative agent of tularemia. Symp. Res. Works Inst. Epidemiol. Mikrobiol. (Chita) (Russ.) 1: 170-180.

Doudoroff, M. 1940. The oxidative assimilation of sugars and related substances by *Pseudomonas saccharophila* with a contribution to the problem of the direct respiration of di- and polysaccharides. Enzymologia 9: 59-72.

Doudoroff, M. 1942. Studies on the luminous bacteria. II. Some observations on the anaerobic metabolism of facultatively anaerobic species. J. Bacteriol. 44: 461-467.

Doudoroff, M. 1974. Genus *Paracoccus*. *In* Buchanan and Gibbons (Editors), Bergey's Manual of Determinative Bacteriology, 8th Ed., Williams and Wilkins Co., Baltimore, pp. 438-440.

Doudoroff, M. and N.J. Palleroni. 1974. Genus *Pseudomonas*. *In* Buchanan and Gibbons (Editors), Bergey's Manual of Determinative Bacteriology, 8th Ed., The Williams and Wilkins Co., Baltimore, pp. 217-243.

Douglas, J.T. and S.S. Elberg. 1976. Isolation of *Brucella melitensis* phage of broad biotype and species specificity. Infect. Immun. 14: 306-308.

Douglas, J.T. and S.S. Elberg. 1978. Properties of the Berkeley phage lytic for *Brucella melitensis* and other species. Ann. Sclavo 20: 681-691.

Dowell, V.R., Jr. and T.M. Hawkins. 1974. Laboratory methods in anaerobic bacteriology. CDC Laboratory Manual. CDC Publications no. 77-8272. Centers for Disease Control, Atlanta, Ga.

Downs, C.M., L.L. Coriell, S.S. Chapman and A. Klauber. 1947. The cultivation of *Bacterium tularense* in embryonated eggs. J. Bacteriol. 53: 89-100.

Dowson, W.J. 1939. On the systematic position and generic names of the Gram negative bacterial plant pathogens. Zentrabl. Bakteriol. Parasitenk. Infektionskr. Hyg. Abt. II 100: 177-193.

Dowson, W.J. 1943. On the generic names *Pseudomonas*, *Xanthomonas* and *Bacterium* for certain bacterial plant pathogens. Trans. Brit. Mycol. Soc. 26: 1-14.

Dowson, W.J. 1957. Plant diseases due to bacteria. 2nd ed., University Press, Cambridge.

Doyle, L. P. 1948. The etiology of swine dysentery. Am. J. Vet. Res. 9: 50-51.

Dranovskaya, E.A. and V.M. Kushnarev. 1968. Cytochromes of *Brucella*. (In Russian.) Zh. Mikrobiol. Epidemiol. Immunobiol. 12: 3-5.

Dranovskaya, E.A. and P.A. Vershilova. 1977. The endotoxin of *Brucellae* Ann. Sclavo 19: 109-116.

Drasar, B.S. and M.J. Hill. 1974. Human Intestinal Flora. Academic Press, New York.

Dresel, E.G. and O. Stickl. 1928. Über reversible Mutationformen der Typusbazillen beim Menschen. Deut. Med. Wochenschr. 54: 517-519.

Dreyfus, B.L. and Y.R. Dommergues. 1980. Non-inhibition de la fixation d'azote atmosphérique par l'azote combiné chez une légumineuse à nodules caulinaires. C. R. Acad. Sci. Paris, Ser. D, 291: 767-770.

Dreyfus, B.L. and Y.R. Dommergues. 1981. Nodulation of *Acacia* species by fast- and slow-growing tropical strains of *Rhizobium*. Appl. Environ. Microbiol. 41: 97-99.

Drozhevkina, M.S. and Y.I. Arutyunov. 1979. Phage typing of *Vibrio cholerae* using a new collection of phages. J. Hyg. Epidem. Microbiol. Immunol. 23: 340-347.

Drummond, M. 1979. Crown gall disease. Nature 281: 343-347.

Drzeniek, R., W. Scharmann and E. Balke. 1972. Neuraminidase and N-acetyl-neuraminate pyruvate-lyase of *Pasteurella multocida*. J. Gen. Microbiol. 72: 357-368.

Dschunkowsky, E. 1913. Das Rukfallfieber in Persien. Deut. Med. Wochenschr. 39: 419-420.

Dschunkowsky, E. 1937. Balfoursche Granula als echte Geflugelparasiten, ihre Natur und Stellung in der Systematik: *Aegyptianella pullorum* Carpano, *Balfouria* n. genus, *Balfouria anserina* n. sp. und *Balfouria gallinarum* n. sp. Zentralbl. Bakteriol. Parasitenkd. Infektionskr. Hyg. Abt. I Orig. 140: 131-136.

Dubnau, D., I. Smith, P. Morell and J. Marmur. 1965. Gene conservation in *Bacillus* species. I. The location of genes concerned with the synthesis of ribosomal components and soluble RNA. Proc. Nat. Acad. Sci. U.S.A. 54: 491-498.

Duboscq, O. and P. Grassé. 1927. Flagelles et schizophytes de *Calotermes (Glyptotermes) irridipennis* Frogg. Arch. Zool. Expt. Gen. 66: 451-496.

DuBow, M.S. and T. Blumenthal. 1975. Host Factor for coliphage QβRNA replication is present in *Pseudomonas putida*. Molec. Gen. Genet. 141: 113-119.

DuBow, M.S. and T. Ryan. 1977. Host Factor for coliphage QβRNA replication as an aid in elucidating phylogenetic relationships: the genus *Pseudomonas*. J. Gen. Microbiol. 102: 263-268.

Dubray, G. 1972. Étude ultrastructurale des bactéries de colonies lisses (S) et rugueuses (R) du genre *Brucella*. Ann. Inst. Pasteur (Paris) 123: 171-193.

Dubray, G. 1976. Localization cellulaire des polyosides des bactéries des genres *Brucella* et *Escherichia* en phase lisse (S) ou rugueuse (R). Ann. Microbiol. (Paris) 1276: 133-140.

Dubray, G. and M. Plommet. 1976. Structure et constituants des *Brucella*.

Characterization des fractions et propriétés biologiques. Develop. Biol. Stand. *31:* 68-91.

Dudley, J.P., E.J.C. Goldstein, W.L. George, B.V. Bock, B.D. Kirby and S.M. Finegold. 1978. Sinus infection due to *Eikenella corrodens*. Arch. Otolaryngol. *104:* 462-463.

Dudley, M.V. and E.B. Shotts. 1979. Medium for the isolation of *Yersinia enterocolitica*. J. Clin. Microbiol. *10:* 180-183.

Dudman, W.F. 1976. The extracellular polysaccharides of *Rhizobium japonicum*: compositional studies. Carbohydr. Res. *46:* 97-110.

Dudman, W.F. 1978. Structural studies of the extracellular polysaccharides of *Rhizobium japonicum* strain 71A, CC708 and CB1795. Carbohydr. Res. *66:* 9-23.

Dudman, W.F. and A.J. Jones. 1981. The extracellular glucans of *Rhizobium japonicum* strain 3I1b71a. Carbohydr. Res. *84:* 358-364.

Dugan, P.R. and D.G. Lundgren. 1960. Isolation of the floc-forming organism *Zoogloea ramigera* and its culture in complex and synthetic media. Appl. Microbiol. *8:* 357-361.

Duggar, B.M. 1909. Fungous diseases of plants. Ginn and Co., Boston.

Duguid, J.P. 1959. Fimbriae and adhesive properties in *Klebsiella* strains. J. Gen. Microbiol. *21:* 271-286.

Dujardin, F. 1841. Histoire naturelle des Zoophytes. Infusoires, comprenant la physiologie et la classification de ces animaux. De Roret, Paris.

Dulaney, A.D., K. Guto and H. Packer. 1948. *Donovania granulomatis*: cultivation, antigenic preparation, and immunological tests. J. Immunol. *59:* 335-340.

Dulaney, A.D. and H. Packer. 1947. Complement-fixation studies with pus antigen in granuloma inguinale. Proc. Soc. Exp. Biol. Med. *65:* 254-256.

Duma, R.J., D.E. Sonenshine, F.M. Bozeman, J.M. Veazey, Jr., B.L. Elisberg, D.P. Chadwick, N.I. Stocks, T.M. McGill, G.B. Miller and J.N. MacCormack. 1981. Epidemic typhus in the United States associated with flying squirrels. J. Amer. Med. Assoc. *245:* 2318-2323.

Dumoff, M. 1979. Direct in vitro isolation of the Legionnaires' disease bacterium in two fatal cases. Cultural and staining characteristics. Ann. Intern. Med. *90:* 694-696.

Du Moulin, G.C. 1979. Airway colonization by *Flavobacterium* in an intensive care unit. J. Clin. Microbiol. *10:* 155-160.

Duncan, J.R., F.K. Ramsey and W.P. Switzer. 1966. Pathology of experimental *Bordetella bronchiseptica* infection in swine: Pneumonia. Am. J. Vet. Res. *27:* 467-472.

Duncan, M.J. 1979. L-arabinose metabolism in *Rhizobium*. J. Gen. Microbiol. *113:* 177-179.

Dundas, I.D. 1977. Physiology of *Halobacteriaceae*. In Rose and Tempést (Editors), Advances in Microbial Physiology, Vol. 15. Academic Press, London, pp. 85-120.

Dundas, I.D. and H. Larsen. 1962. The physiological role of the carotenoid pigments of *Halobacterium salinarium*. Arch. Mikrobiol. *44:* 233-239.

Dundas, I.D., V.R. Srinivasan and H.O. Halvorson. 1963. A chemically defined medium for *Halobacterium salinarium* strain 1. Can. J. Microbiol. *9:* 619-624.

Dunican, L.K. and F.C. Cannon. 1971. The genetic control of symbiotic properties in *Rhizobium* -- evidence for plasmid control. Plant Soil, 1971 (Special Volume) 73-79.

Dunican, L.K., F. O'Gara and A.B. Tierney. 1976. Plasmid control of effectiveness in *Rhizobium*: transfer of nitrogen-fixing genes on a plasmid from *Rhizobium trifolii* to *Klebsiella aerogenes*. In Nutman (Editor), Symbiotic Nitrogen Fixation in Plants. Cambridge University Press, pp. 77-90.

Dunkelberg, W.E., Jr. and I. McVeigh. 1969. Growth requirements of *Haemophilus vaginalis*. Antonie van Leeuwenhoek J. Microbiol. Serol. *35:* 129-145.

Dunkelberg, W.E., Jr., R. Skaggs and D.S. Kellogg, Jr. 1970. Method for isolation and identification of *Corynebacterium vaginale* (*Haemophilus vaginalis*). Appl. Microbiol. *19:* 47-52.

du Plessis, J.L., C.M. Cameron and E. Langen. 1967. Focal necrotic pneumonia and rumeno-enteritis in Afrikaner calves. J. S. Afr. Vet. Med. Assoc. *38:* 121-128.

Dupouey, P. 1963. Étude immunologique de six especies de treponemes anaerobies d'origine genitale: *Treponema phagedenes*, *refringens*, *calligyra*, *minutum*, Reiter, et *pallidum*. Ann. Inst. Pasteur (Paris) *105:* 725-736; 949-970.

Dupuy, P. 1957. Les Acetobacters du vin. Identification de quelques souches. Ann. Technol. *2:* 217-233.

Durbin, R.D., T.F. Uchytil, J.A. Steele and R. de L.D. Ribeiro. 1978. Tabtoxine-beta-lactam from *Pseudomonas tabaci*. Phytochemistry *14:* 147.

Durgapal, J.C. 1977. Albinism in *Xanthomonas sesami*. Curr. Sci. *46:* 274.

Durlakowa, I., Z. Lachowicz and S. Slopek. 1967. Biochemical properties of *Klebsiella* bacilli. Arch. Immunol. Ther. Exp. *15:* 490-496.

Dutky, S.R. 1959. Insect microbiology. Adv. Appl. Microbiol. *1:* 175-200.

Dutky, S.R. and E.L. Gooden. 1952. *Coxiella popilliae*, n. sp., a rickettsia causing blue disease of Japanese beetle larvae. J. Bacteriol. *63:* 743-750.

Dutton, A.A.C. and M. Ralston. 1957. Urinary tract infection in a male urological ward with special reference to the mode of infection. Lancet *1:* 115-119.

Duval, J.C. 1966. Mise en évidence de bactéries intramitochondriales chez *Didymium squamulosum* Fr. (Didymiaceae). C.R. Acad. Sci., Ser. D *262:* 1441-1443.

Duval, J.C. 1970. Bactéries d'allure rickettsienne parasites d'un Myxomycete. I. Cycle de l'infection et obtention d'une souche saine par action du chloramphenicol. J. Microscop. *9:* 185-200.

Duval, J.C. 1972. Bactéries d'allure rickettsienne parasites d'un Myxomycete. II. Mise en évidence d'elements tubulaires dans des parasites intramitochon-

driaux. J. Microscop. *13:* 31-46.

Dworkin, M. and J.W. Foster. 1956. Studies on *Pseudomonas methanica* (Söhngen) nov. comb. J. Bacteriol. *72:* 646-659.

Dye, D.W. 1960. Pectolytic activity in *Xanthomonas*. N. Z. J. Sci. *3:* 61-69.

Dye, D.W. 1962. The inadequacy of the usual determinative tests for the identification of *Xanthomonas* spp. N. Z. J. Sci. *5:* 393-416.

Dye, D.W. 1963. The taxonomic position of *Xanthomonas uredovorus* Pon et al., 1954. N. Z. J. Sci. *6:* 146-149.

Dye, D.W. 1963. The taxonomic position of *Xanthomonas stewartii* (Erw. Smith, 1914) Dowson 1939. N. Z. J. Sci. *6:* 495-506.

Dye, D.W. 1963. A bacterial disease of pukatea (*Laurelia novae-zelandiae* A. Cunn.) caused by *Xanthomonas laureliae* n. sp. N. Z. J. Sci. *6:* 179-185.

Dye, D.W. 1964. The taxonomic position of *Xanthomonas trifolii* (Huss, 1907) James 1955. N. Z. J. Sci. *7:* 261-269.

Dye, D.W. 1966a. A comparative study of some atypical "xanthomonads". N. Z. J. Sci. *9:* 843-854.

Dye, D.W. 1966b. Cultural and biochemical reactions of additional *Xanthomonas* spp. N. Z. J. Sci. *9:* 913-919.

Dye, D.W. 1968. A taxonomic study of the genus *Erwinia*. I. The "amylovora" group. N. Z. J. Sci. *11:* 590-607.

Dye, D.W. 1969. A taxonomic study of the genus *Erwinia*. II. The "carotovora" group. N. Z. J. Sci. *12:* 81-97.

Dye, D.W. 1969. A taxonomic study of the genus *Erwinia*. III. The "herbicola" group. N. Z. J. Sci. *12:* 223-236.

Dye, D. 1969. A taxonomic study of the genus *Erwinia*. IV. "Atypical erwinias." N.Z.J. Sci. *12:* 833-839.

Dye, D.W. 1978. Genus IX *Xanthomonas* Dowson 1939. In J.M. Young, D.W. Dye, J.F. Bradbury, C.G. Panagopoulos and C.F. Robbs. A proposed nomenclature and classification for plant pathogenic bacteria. N. Z. J. Agr. Res. *21:* 153-177.

Dye, D.W. 1980. *Xanthomonas*. In N.W. Schaad (Editor), Laboratory guide for identification of plant pathogenic bacteria. American Phytopathological Society, St. Paul, Minn. pp. 45-49.

Dye, D.W. 1981. A numerical taxonomic study of the genus *Erwinia*. N. Z. J. Agric. Res. *24:* 223-229.

Dye, D.W., J.F. Bradbury, R.S. Dickey, M.S. Goto, C.N. Hale, A.C. Hayward, A. Kelman, R.A. Lelliott, P.N. Patel, D.C. Sands, M.N. Schroth, D.R.W. Watson and J.M. Young. 1975. Proposals for a reappraisal of the status of names of plant-pathogenic *Pseudomonas* species. Int. J. Syst. Bacteriol. *25:* 252-257.

Dye, D.W., J.F. Bradbury, M. Goto, A.C. Hayward, R.A. Lelliott and M.N. Schroth. 1980. International standards for naming pathovars of phytopathogenic bacteria and a list of pathovar names and pathotype strains. Rev. Plant Pathol. *59:* 153-168.

Dye, D.W., M.P. Starr and H. Stolp. 1964. Taxonomic clarification of *Xanthomonas vesicatoria* based upon host specificity, bacteriophage sensitivity and cultural characteristics. Phytopathol. Z. *51:* 394-407.

Eadie, J.M. 1962. The development of rumen microbial populations in lambs and calves under various conditions of management. J. Gen. Microbiol. *29:* 563-578.

Eagon, R.G. 1962. *Pseudomonas natriegens*, a marine bacterium with a generation time of less than 10 minutes. J. Bacteriol. *83:* 736-737.

Eagon, R.G. and C.H. Wang. 1962. Dissimilation of glucose and gluconic acid by *Pseudomonas natriegens*. J. Bacteriol. *83:* 879-886.

Eaton, M.D., G. Meiklejohn and W. van Herick. 1944. Studies on the etiology of primary atypical pneumonia. A filterable agent transmissible to cotton rats, hamsters and chick embryos. J. Exp. Med. *79:* 649-668.

Eberhardt, U. 1966. Über das Wasserstoff aktivierende System von *Hydrogenomonas* H16. I. Verteilung der Hydrogenase-Aktivität auf zwei Zellfraktionen. Arch. Mikrobiol. *53:* 288-302.

Eberhardt, U. 1969. On chemolithotrophy and hydrogenase of a Gram-positive Knallgas bacterium. Arch. Mikrobiol. *66:* 91-104.

Eberhardt, U. 1971. The cell wall as the site of carotenoid in the "Knallgas" bacterium 12/60/x. Arch. Mikrobiol. *80:* 32-37.

Echandi, E. and J.W. Moyer. 1979. Production, properties, and morphology of bacteriocins from *Erwinia chrysanthemi*. Phytopathology *69:* 1204-1207.

Eddy, B.P. 1960. Cephalotrichous, fermentative, Gram-negative bacteria: the genus *Aeromonas*. J. Appl. Bacteriol. *23:* 216-249.

Eddy, B.P. 1962. Further studies on *Aeromonas*. I. Additional strains and supplementary biochemical tests. J. Appl. Bacteriol. *25:* 137-146.

Eddy, B.P. and K.P. Carpenter. 1964. Further studies on *Aeromonas*. II. Taxonomy of *Aeromonas* and C-27 strains. J. Appl. Bacteriol. *27:* 96-109.

Edelstein, P.H. and S.M. Finegold. 1979. Use of a semiselective medium to culture *Legionella pneumophila* from contaminated lung specimens. J. Clin. Microbiol. *10:* 141-143.

Edelstein, P.H., R.D. Meyer and S.M. Finegold. 1979. Isolation of *Legionella pneumophila* from blood. Lancet *i:* 750-751.

Eden-Green, S.J., D.B. Archer, J.G. Tully and H. Waters. 1983. Further studies on a spiroplasma isolated from coconut palms in Jamaica. Ann. Appl. Biol. *102:* 127-134.

Eden-Green, S. and J.G. Tully. 1979. Isolation of *Acholeplasma* spp. from coconut palms affected by lethal yellowing disease in Jamaica. Curr. Microbiol. *2:* 311-316.

Eden-Green, S.J. and H. Waters. 1981. Isolation and preliminary characterization of a spiroplasma from coconut palms in Jamaica. J. Gen. Microbiol. *124:* 263-270.

Edenharder, R. 1976. The significance of the bacterial steroid degradation for the etiology of large bowel cancer. VI. Degradation of deoxycholic acid by saccharolytic *Bacteroides*-species. Zentralbl. Bakteriol. Parasitenkd. Infektionskr. Hyg., I. Abt. Orig. B *162:* 519-527.

Edenharder, R., S. Stubenrauch and J. Slemrova. 1976. The significance of the bacterial steroid degradation for the etiology of large bowel cancer. V. Transformation of chenodeoxycholic acid by saccharolytic *Bacteroides*-species. Zentralbl. Bakteriol. Parasitenkd. Infektionskr. Hyg., I Abt. Orig. B *162:* 506-518.

Ederer, G.M. and M.L. Schurr. 1971. Optimal bacitracin concentration for selective isolation medium for haemophilus. Am. J. Med. Technol. *37:* 304-305.

Edmondson, A.S. and E.M. Cooke. 1979. The development and assessment of a bacteriocin typing method for *Klebsiella*. J. Hyg. London *82:* 207-223.

Edmunds, P.N. 1962. The biochemical, serological and haemagglutinating reactions of "*Haemophilus vaginalis*". J. Pathol. Bacteriol. *83:* 411-422.

Edward, D.G. ff. 1947. A selective medium for pleuropneumonia-like organisms. J. Gen. Microbiol. *1:* 238-243.

Edward, D.G. ff. 1950a. An investigation of pleuropneumonia-like organisms isolated from the bovine genital tract. J. Gen. Microbiol. *4:* 4-15.

Edward, D.G. ff. 1950b. An investigation of the biological properties of organisms of the pleuropneumonia group, with suggestions regarding the identification of strains. J. Gen. Microbiol. *4:* 311-329.

Edward, D.G. ff. 1953. Organisms of the pleuropneumonia group causing disease in goats. Vet. Rec. *65:* 873-874.

Edward, D.G. ff. 1954. The pleuropneumonia group of organisms: a review, together with some new observations. J. Gen. Microbiol. *10:* 27-64.

Edward, D.G. ff. 1955. A suggested classification and nomenclature for organisms of the pleuropneumonia group. Int. Bull. Bacteriol. Nomencl. Taxon. *5:* 85-93.

Edward, D.G. ff. 1963. Organisms of the pleuropneumonia group causing disease in goats. Vet. Rec. *65:* 873-874.

Edward, D.G. ff. 1971. Determination of sterol requirement for *Mycoplasmatales*. J. Gen. Microbiol. *69:* 205-210.

Edward, D.G. ff. and W.A. Fitzgerald. 1951. The isolation of organism of the pleuropneumonia group from dogs. J. Gen. Microbiol. *5:* 566-575.

Edward, D.G. ff. and E.A. Freundt. 1965. A note on the taxonomic status of strains like "Campo" hitherto classified as *Mycoplasma hominis*, type 2. J. Gen. Microbiol. *41:* 263-265.

Edward, D.G. ff. and E.A. Freundt. 1969. Proposal for classifying organisms related to *Mycoplasma laidlawii* in a Family *Sapromycetaceae*, Genus *Sapromyces*, within the *Mycoplasmatales*. J. Gen. Microbiol. *57:* 391-395.

Edward, D.G. ff. and E.A. Freundt. 1970. Amended nomenclature for strains related to *Mycoplasma laidlawii*. J. Gen. Microbiol. *62:* 1-2.

Edward, D.G. ff. and E.A. Freundt. 1973. Type strains of species of the order *Mycoplasmatales*, including designation of neotypes for *Mycoplasma mycoides* subsp. *mycoides,* *Mycoplasma agalactiae* subsp. *agalactiae*, and *Mycoplasma arthritidis*. Int. J. Syst. Bacteriol. *23:* 55-61.

Edward, D.G. ff. and A.D. Kanarek. 1960. Organisms of the pleuropneumonia group of avian origin: their classification into species. Ann. N.Y. Acad. Sci. *79:* 696-702.

Edward, D.G. ff. and W.B. Moore. 1975. A method for determining the utilization of glucose by mycoplasmas. J. Med. Microbiol. *8:* 451-454.

Edwards, C. E. and R. Kraus. 1960. *Spirillum serpens* meningitis: report of case. N. Engl. J. Med. *262:* 458-460.

Edwards, P.R. 1931. Studies on *Shigella equirulis* (*Bacterium viscosum equi*). Ky. Agr. Exp. Sta. Bull. *320:* 289-330.

Edwards, P.R. and W.H. Ewing. 1972. Identification of *Enterobacteriaceae*, 3rd edition, Burgess Publishing Co., Minneapolis, 362 pp.

Edwards, P.R., M.A. Fife and W.H. Ewing. 1965. Antigenic schema for the genus *Arizona*. Monograph Communicable Disease Center, Atlanta, Georgia.

Edwards, P.R., M.A. Fife and C.H. Ramsey. 1959. Studies on the Arizona group of *Enterobacteriaceae*. Bacteriol. Rev. *23:* 155-174.

Edwards, P.R., M.G. West and D.W. Bruner. 1947. *Arizona* group of paracolon bacteria. Ky. Agr. Exp. Sta. Bull. *499:* 3-32.

Edwards, P.R., M.G. West and D.W. Bruner. 1948. Antigenic studies of a paracolon bacteria (Bethesda group). J. Bacteriol. *55:* 711-719.

Egan, W., F. Tsui and R. Schneerson. 1980a. Structural studies of the *Haemophilus influenzae* type f capsular polysaccharide. Carbohydr. Res. *79:* 271-277.

Egan, W., F.-P. Tsui, P.A. Climenson and R. Schneerson. 1980b. Structural and immunological studies of the *Haemophilus influenzae* type c capsular polysaccharide. Carbohydr. Res. *80:* 305-316.

Eggerth, A.H. and B.H. Gagnon. 1933. The bacteroides of human feces. J. Bacteriol. *25:* 389-413.

Eggleton, P., S.R. Eldsen and N. Gough. 1943. The estimation of creatine and diacetyl. Biochem. J. *37:* 526-529.

Egli, T., M. Goto and D. Schmidt. 1975. Bacterial wilt, a new forage grass disease. Phytopathol. Z. *82:* 111-121.

Ehrenberg, C. G. 1832. Beiträge zür Kenntnis der Organization der Infusorien und ihrer geographischen Verbreitung, besonders in Sibirien. Abh. Konig. Akad. Wiss. Berlin. 1830, pp. 1-88.

Ehrenberg, C.G. 1835. Dritter Beiträg zür Erkenntniss grosser Organisation in der Richtung des Kleinsten Raumes. Abh. Preuss. Akad. Wiss. Phys. Kl. Berlin aus den Jahre 1833-1835, pp. 143-336.

Ehrenberg, C.G. 1840. Charakteristik von 274 neuen Arten von Infusorien. Ber. Bekannt Verhandl. Königl. Preuss Akad. Wiss. Berlin *1840:* 197-219.

Ehrhardt, P. 1968. Nachweis einer durch symbiontische Mikroorganismen bewirkten Sterinsynthese in künstlich ernährten Aphiden (Homoptera, Rhynchota, Insecta). Experientia (Basel) *24:* 82-83.

Eicholz, W. 1902. Erdbeerbacillus (*Bacterium fragi*) Zentralbl. Bakteriol. Parasitenkd. Infektionskr. Hyg. Abt. II, *9:* 425-428.

Eidels, L., P.L. Edelmann and J. Preiss. 1970. Biosynthesis of bacterial glycogen. VIII. Activation and inhibition of the adenosine diphosphoglucose pyrophosphorylases of *Rhodopseudomonas capsulata* and of *Agrobacterium tumefaciens*. Arch. Biochem. Biophys. *140:* 60-74.

Eigelsbach, H.T. 1974. *Francisella tularensis. In* Lennette, Spaulding and Truant (Editors), Manual of Clinical Microbiology, 2nd Ed., American Society for Microbiology, Washington, pp. 316-319.

Eigelsbach, H.T., W. Braun and R.D. Herring. 1951. Studies on the variation of *Bacterium tularense*. J. Bacteriol. *61:* 557-569.

Eigelsbach, H.T. and C.M. Downs. 1961. Prophylactic effectiveness of live and killed tularemia vaccines. I. Production of vaccine and evaluation in the white mouse and guinea pig. J. Immunol. *87:* 415-425.

Eigelsbach, H.T. and R.D. Herring. 1952. Studies on immunogenic properties of *Bacterium tularense* variants. J. Infect. Dis. *91:* 86-91.

Eigelsbach, H.T. and V.G. McGann. 1981. The genus *Francisella. In* Starr, Stolp, Trüper, Balows and Schlegel (Editors), The Prokaryotes: a handbook on habitats, isolation and identification of bacteria. Springer-Verlag, Berlin, pp. 620-633.

Eiken, M. 1958. Studies on an anaerobic rod-shaped Gram-negative microorganism: *Bacteroides corrodens* N. sp. Acta Pathol. Microbiol. Scand. *43:* 404-416.

Eisemann, C.S. and J.V. Osterman. 1981. Antigens of scrub typhus rickettsiae: Separation by polyacrylamide gel electrophoresis and identification by enzyme-linked immunosorbent assay. Infect. Immun. *32:* 525-533.

Eisenberg, G.H.G., Jr. and J.V. Osterman. 1979. Gamma-irradiated scrub typhus immunogens: Broad-spectrum immunity with combinations of rickettsial strains. Infect. Immun. *26:* 131-136.

Eisenberg, J. 1891. Bacteriologische Diagnostik Hiflstabellen zum Gebrauche beim Praktischen Arbeiten. 3 Aufl. Leopold Voss, Hamburg.

Eisenberg, R.C., S.J. Butters, S.C. Quay and S.B. Friedman. 1974. Glucose uptake and phosphorylation in *Pseudomonas fluorescens*. J. Bacteriol. *120:* 147-153.

Eisenstein, B.I., T. Sox, G. Biswas, E. Blackman and P.F. Sparling. 1977. Conjugal transfer of the gonococcal penicillinase plasmid. Science *195:* 998-1000.

Eksztejn, J. and M. Varon. 1977. Elongation and cell division in *Bdellovibrio bacteriovorus*. Arch. Microbiol. *114:* 175-181.

Elander, R.P., J.A. Mabe, R.H. Hamill and M. Gorman. 1968. Metabolism of tryptophans by *Pseudomonas aureofaciens*. VI. Production of pyrrolnitrin by selected *Pseudomonas* species. Appl. Microbiol. *16:* 753-758.

Elazari-Volcani, B. 1939. On *Pseudomonas indigofera* (Voges) Migula and its pigment. Arch. Mikrobiol. *10:* 343-358.

Elazari-Volcani, B. 1940. Studies on the microflora of the Dead Sea. Doctoral thesis. Hebrew University, Jerusalem, pp. 1-116 and i-xiii.

Elazari-Volcani, B. 1957. Genus XII. *Halobacterium. In* Breed, Murray and Smith (Editors), Bergey's Manual of Determinative Bacteriology, 7th Ed., The Williams and Wilkins Co., Baltimore, pp. 207-212.

El Banoby, F.E. and K. Rudolph. 1979. Induction of water-soaking in plant leaves by extracellular polysaccharides from phytopathogenic pseudomonads and xanthomonads. Physiol. Plant Pathol. *15:* 341-349.

Eldering, G., C. Hornbeck and J. Baker. 1957. Serological study of *Bordetella pertussis* and related species. J. Bacteriol. *74:* 133-136.

Eldering, G. and P. Kendrick. 1938. *Bacillus parapertussis:* A species resembling both *Bacillus pertussis* and *Bacillus bronchisepticus* but identical with neither. J. Bacteriol. *35:* 561-572.

Elhâg, K.M. and S. Tabaqchali. 1978. The distribution of *Bacteroides fragilis* serotypes amongst clinical strains. J. Hyg. Camb. *81:* 89-97.

El-Helaly, A.F., M.K. Abo-El-Dahab, M.A. Goorani and M.R.M. Gabr. 1979. Production of cellulase (Cₓ) by different species of *Erwinia*. Zentralbl. Bakteriol. Parasitenkd. Infektionskr. Hyg. Abt. II. *134:* 187-192.

Elkan, G.H. and I. Kwik. 1968. Nitrogen, energy and vitamin nutrition of *Rhizobium japonicum*. J. Appl. Bacteriol. *31:* 399-404.

Elkan, G.H. and R.A. Usanis. 1971. Theoretical deoxyribonucleic acid homology between strains of *Rhizobium japonicum*. Int. J. Syst. Bacteriol. *21:* 295-298.

Ellinghausen, H.C. and W.G. McCullough. 1965. Nutrition of *Leptospira pomona* and growth of 13 other serotypes: fractionation of oleic albumin complex and a medium of bovine and polysorbate 80. Amer. J. Vet. Res. *26:* 45-51.

Ellingworth, S., J.W. McLeod and J. Gordon. 1929. Further observations on the oxidation by bacteria of compounds of the paraphenylene diamine series. J. Pathol. Bacteriol. *32:* 173-183.

Elliott, C. 1920. Halo-blight of oats. J. Agr. Res. *19:* 139-172.

Elliott, C. 1923. A bacterial stripe disease of proso millet. J. Agr. Res. *26:* 151-160.

Elliott, C. 1927. Bacterial stripe blight of oats. J. Agr. Res. *35:* 811-824.

Elliott, C. 1930. Manual of Bacterial Plant Pathogens. The Williams and Wilkins Co., Baltimore, pp. 1-349.

Elliott, C. 1930. Bacterial streak disease of sorghums. J. Agr. Res. *40:* 963-976.

Ellis, J.G., A. Kerr, J. Tempé and A. Petit. 1979a. Arginine catabolism: a new function of both octopine and nopaline Ti-plasmids of *Agrobacterium*. Molec. Gen. Genet. *173:* 263-269.

Ellis, J.G., A. Kerr, M. Van Montagu and J. Schell. 1979b. *Agrobacterium*: genetic studies on agrocin 84 production and the biological control of crown gall. Physiol. Plant Pathol. *15:* 311-319.

Ellner, P.D., P.A. Granato and C.B. May. 1973. Recovery and identification of anaerobes: a system suitable for the routine clinical laboratory. Appl. Microbiol. *26:* 904-913.

Ellner, P.D. and L.R. McCarthy. 1973. *Aeromonas shigelloides* bacteremia: a case report. Amer. J. Clin. Pathol. *59:* 216-218.

Elrod, R.P. 1946. The serological relationship between *Erwinia tracheiphila* and species of *Shigella*. J. Bacteriol. *52:* 405-410.

Elrod, R.P. and A.C. Braun. 1947a. Serological studies of the genus *Xanthomonas* I. Cross agglutination relationships. J. Bacteriol. *53:* 509-518.

Elrod, R.P. and A.C. Braun. 1947b. Serological studies of the genus *Xanthomonas* II. *Xanthomonas translucens* group. J. Bacteriol. *53:* 519-524.

Elsden, S.R. and D. Lewis. 1953. The production of fatty acids by a Gram-negative coccus. Biochem. J. *55:* 183-189.

Elsden, S.R., B.E. Volcani, F.M.C. Gilchrist and D. Lewis. 1956. Properties of a fatty acid forming organism isolated from the rumen of sheep. J. Bacteriol. *72:* 681-689.

Elwell, L.P., J.D. DeGraaff, D. Seibert and S. Falkow. 1975. Plasmid-linked ampicillin resistance in *Haemophilus influenzae* type b. Infect. Immun. *12:* 404-410.

Elwell, L.P., J.R. Saunders, M.H. Richmond and S. Falkow. 1977. Relationships among some R plasmids found in *Haemophilus influenzae*. J. Bacteriol. *131:* 356-362.

Emerson, H. and C. Norris. 1905. "Red-leg", an infectious disease of frogs. J. Exp. Med. *7:* 32-58.

Emerson, B.B., A.L. Smith, A.L. Harding and D.H. Smith. 1975. *Haemophilus influenzae* type b susceptibility to 17 antibiotics. J. Pediatr. *86:* 617-620.

Emery, R.S., C.K. Smith and L.F. To. 1957. Utilization of inorganic sulfate by rumen microorganisms. II. The ability of single strains of rumen bacteria to utilize inorganic sulfate. Appl. Microbiol. *5:* 363-366.

Emmerich, R. and C. Weibel. 1894. Ueber eine durch Bakterien erzugte Seuche unter den Forellen. Arch. Hyg. *21:* 1-21.

Emslie-Smith, A.H. 1961. *Hafnia alvei* strains possessing alpha antigen of Stamp and Stone. J. Pathol. Bacteriol. *81:* 534-536.

Endo, B.Y. 1979. The ultrastructure and distribution of an intracellular bacterium-like microorganism in tissue of larvae of the soybean cyst nematode, *Heterodera glycines*. J. Ultrastruct. Res. *67:* 1-14.

Endoh, M., T. Takezawa and Y. Nakase. 1980. Adenylate cyclase activity of *Bordetella* organisms. Its production in liquid medium. Microbiol. Immunol. *24:* 95-104.

Engel, L.D. and G.E. Kenny. 1970. *Mycoplasma salivarium* in human gingival sulci. J. Periodont. Res. *5:* 1-9.

England, A.C. III, R.M. McKinney, P. Skaliy and G.W. Gorman. 1980. A fifth serogroup of *Legionella pneumophila*. Ann. Intern. Med. *93:* 58-59.

Enigk, K. 1942. Eine Rickettsieninfektion beim Bison (A rickettsial infection in bison). Berl. Münch. Tierarztl. Wschr. Jan. 23rd, pp. 25-27.

Enomoto, K. and M.D. Gill. 1980. Cholera toxin activation of adenylate cyclase. J. Biol. Chem. *255:* 1252-1258.

Ensminger, P.W. 1953. Pigment production by *Haemophilus parapertussis*. J. Bacteriol. *63:* 509-510.

Entner, N. and M. Doudoroff. 1952. Glucose and gluconic acid oxidation of *Pseudomonas saccharophila*. J. Biol. Chem. *196:* 853-862.

Entwistle, P.F., J.S. Robertson and B.E. Juniper. 1968. The ultrastructure of a rickettsia pathogenic to a saturnid moth. J. Gen. Microbiol. *54:* 97-104.

Ercolani, G.L. and M. Caldarola. 1972. *Pseudomonas ciccaronei* sp. n. agente de una maculatura fogliare del carrubo in Puglia. Phytopathol. Mediterr. *11:* 71-73.

Ercolani, G.L., D.J. Hagedorn, A. Kelman and R.E. Rand. 1974. Epiphytic survival of *Pseudomonas syringae* on hairy vetch in relation to epidemiology of bacterial brown spot of bean in Wisconsin. Phytopathology *64:* 1330-1339.

Erikson, D. 1935. The pathogenic aerobic organisms of the Actinomyces group. Spec. Rep. Ser. Med. Res. Coun. London, No. 203, 61 pp.

Ernø, H. and M.M. Salih. 1980. The growth precipitation test as a diagnostic method for differentiation of *Mycoplasma* and *Acholeplasma* species. Acta Vet. Scand. *21:* 469-481.

Ernø, H. and L. Stipkovits. 1973. Bovine mycoplasmas: cultural and biochemical studies. Acta Vet. Scand. *14:* 450-463.

Ernø, H. and A.C. Thomsen. 1980. Immunoglobulin classes of urinary and serum antibodies in mycoplasmal pyelonephritis. Acta Pathol. Microbiol. Scand. Sect. C *88:* 237-240.

Escalante-Semerena, J. C., R. P. Blakemore and R. S. Wolfe. 1980. Nitrate dissimilation under microaerophilic conditions by a magnetic spirillum. Appl. Environ. Microbiol. *40:* 429-430.

Escherich, T. 1885. Die Darmbakterien des Neugeborenen und Säuglingen. Fortschr. Med. *3:* 515-528 and 547-554.

Eskew, D. L., D. D. Focht and I. P. Ting. 1977. Nitrogen fixation, denitrification and pleomorphic growth in a highly pigmented *Spirillum lipoferum*. Appl. Environ. Microbiol. *34:* 583-585.

Espejo, R.T. and E.S. Canelo. 1968a. Properties and characterization of the host bacterium of bacteriophage PM2. J. Bacteriol. *95:* 1887-1891.

Espejo, R.T. and E.S. Canelo. 1968b. Properties of bacteriophage PM2: a lipid-containing bacterial virus. Virology *34:* 738-747.

Esteve, J.-C. 1978. Une population de type "killer" chez *Paramecium caudatum* (Ehrenberg). Protistologica *14:* 201-207.

Eutick, M.L., P. Veivers, R.W. O'Brien and M. Slaytor. 1978. Dependence of the higher termite, *Nasutitermes exitiosus* and the lower termite, *Coptotermes lacteus* on their gut flora. J. Insect Physiol. *24:* 363-368.

Evaldson, G., G. Carlström, A. Lagrelius, A.-S. Malmborg and C.E. Nord. 1980. Microbiological findings in pregnant women with premature rupture of the membranes. Med. Microbiol. Immunol. *168:* 283-297.

Evans, A.C. 1918. Further studies on *Bacterium abortus* and related bacteria. J. Infect. Dis. *22:* 580-593.

Evans, N.M., D.D. Smith and A.J. Wicken. 1974. Haemin and nicotinamide adenine dinucleotide requirements of *Haemophilus influenzae* and *parainfluenzae*. J. Med. Microbiol. *7:* 359-365.

Evans, R.T. and D. Taylor-Robinson. 1978. The incidence of tetracycline-resistant strains of *Ureaplasma urealyticum*. J. Antimicrob. Chemother. *4:* 57-63.

Eveland, W.C. and J.E. Faber. 1953. Antigenic studies on a group of paracolon bacteria (32011 group). J. Infect Dis. *93:* 226-236.

Evert, R.F. 1977. Phloem structure and physiology. Annu. Rev. Pl. Physiol. *28:* 199-222.

Every, D. and T.M. Skerman. 1980. Ultrastructure of the *Bacteroides nodosus* cell envelope layers and surface. J. Bacteriol. *141:* 845-857.

Ewing, E.P., A. Takeuchi, A. Shirai and J.V. Osterman. 1978. Experimental infection of mouse peritoneal mesothelium with scrub typhus rickettsiae: an ultrastructural study. Infect. Immun. *19:* 1068-1075.

Ewing, W.H. 1949. Shigella nomenclature. J. Bacteriol. *57:* 633-638.

Ewing, W.H. 1960. Biochemical method for group differentiation. Center for Disease Control Publication, Atlanta, Georgia.

Ewing, W.H. 1962. The Tribe *Proteeae*: Its nomenclature and taxonomy. Int. Bull. Bacteriol. Nomencl. Taxon. *12:* 93-102.

Ewing, W.H. 1966. *Enterobacteriaceae*: taxonomy and nomenclature. Monograph Communicable Disease Center, Atlanta, Georgia.

Ewing, W.H. 1967. Revised definitions for the family *Enterobacteriaceae*, its tribes and genera. Monograph Communicable Disease Center, Atlanta, Georgia.

Ewing, W.H. 1969. Excerpts from: an evaluation of the *Salmonella* problem. Center for Disease Control Publication, Atlanta.

Ewing, W.H. and B.R. Davis. 1972. Biochemical characteristics of *Citrobacter diversus* (Burkey) Werkman and Gillen and designation of the neotype strain. Int. J. Syst. Bacteriol. *22:* 12-18.

Ewing, W.H., B.R. Davis, M.A. Fife and E.F. Lessel. 1973. Biochemical characterization of *Serratia liquefaciens* (Grimes and Hennerty) Bascomb et al. (formerly *Enterobacter liquefaciens)* and *Serratia rubidaea* (Stapp) comb. nov. and designation of type and neotype strains. Int. J. Syst. Bacteriol. *23:* 217-225.

Ewing, W.H., J.J. Farmer, III and D.J. Brenner. 1980. Proposal of *Enterobacteriaceae* nom. rev. to replace *Enterobacteriaceae* Rahn 1937, Nom. fam. cons. Opin. 15, Jud. Comm. 1958, which lost standing in nomenclature on January 1, 1980. Int. J. Syst. Bacteriol. *30:* 664-665.

Ewing, W.H. and M.A. Fife. 1968. *Enterobacter hafniae* (the Hafnia group). Int. J. Syst. Bacteriol. *18:* 263-271.

Ewing, W.H. and M.A. Fife. 1971. *Enterobacter agglomerans* the Herbicola-Lathyri bacteria. Center for Disease Control Publication, Atlanta, Ga.

Ewing, W.H. and M.A. Fife. 1972. *Enterobacter agglomerans* (Beijerinck) comb. nov. (the Herbicola-Lathyri bacteria). Int. J. Syst. Bacteriol. *22:* 4-11.

Ewing, W.H., R. Hugh and J.C. Johnson. 1961. Studies on the *Aeromonas* group. U.S. Department of Health, Education and Welfare, Communicable Disease Center, Atlanta, Georgia.

Ewing, W.H. and W.J. Martin. 1967. The biochemical reactions of the genus *Escherichia*. Monograph, National Communicable Disease Center, Atlanta, Georgia.

Ewing, W.H. and W.J. Martin. 1974. *Enterobacteriaceae*. *In* Lennette, Spaulding and Truant (Editors), Manual of Clinical Microbiology, 2nd Ed., Am. Soc. Microbiology, Washington, D.C., pp. 189-221.

Ewing, W.H., A.C. McWhorter, M.R. Escobar and A.H. Lubin. 1965. *Edwardsiella*, a new genus of *Enterobacteriaceae* based on a new species, *E. tarda*. Int. Bull. Bacteriol. Nomen. Taxon. *15:* 33-38.

Ewing, W.H., R.W. Reavis and B.R. Davis. 1958. Provisional *Shigella* serotypes. Can. J. Microbiol. *4:* 89-107.

Ewing, S.A., W.R. Robertson, R.G. Buckner and C.S. Hayat. 1971. A new strain of *Ehrlichia canis*. J. Am. Vet. Med. Assoc. *159:* 1771-1774.

Ewing, W.H., A.J. Ross, D.J. Brenner and G.R. Fanning. 1978. *Yersinia ruckeri* sp. nov., the redmouth (RM) bacterium. Int. J. Syst. Bacteriol. *28:* 37-44.

Ewing, W.H., K.E. Tanner and D.A. Dennard. 1954. The Providence group: An intermediate group of enteric bacteria. J. Infect. Dis. *94:* 134-140.

Eykyn, S. and I. Phillips. 1978. Carbon dioxide-dependent *Escherichia coli*. Br. Med. J. *1:* 576.

Eymers, J.G. and K.L. van Schouwenburg. 1937. On the luminescence of bacteria. Enzymologia *3:* 235-241.

Eyre, J.W. 1900. A clinical and bacteriological study of diplobacillary conjunctivitis. J. Pathol. Bacteriol. *6:* 1-13.

Fabricant, J. and E.A. Freundt. 1967. Importance of extension and standardization of laboratory tests for the identification and classification of mycoplasma. Ann. N.Y. Acad. Sci. *14:* 50-58.

Faibich, M.M. 1959. The problem of increasing the antigenicity of live tularemia vaccine. J. Microbiol. Epidemiol. Immunobiol. *30:* 23-27.

Faibich, M.M. and T.S. Tamarkina. 1946. Dry living tularemia vaccine of NIIEG of the Red Army. Report I. (Russ.) Zh. Mikrobiol. Epidemiol. Immunobiol. *17:* 59-63.

Faine, S. 1960. Catalase activity in pathogenic *Leptospira.* J. Gen. Microbiol. *22:* 1-9.

Falkler, W.A., Jr. and C.E. Hawley. 1976. Antigens of *Leptotrichia buccalis.* I. Their serologic reaction with human sera. J. Periodontal Res. *10:* 211-215.

Falkler, W.A., Jr. and C.E. Hawley. 1977. Hemagglutinating activity of *Fusobacterium nucleatum.* Infect. Immun. *15:* 230-238.

Falkler, W.A., Jr., C.E. Hawley and J.R. Mongiello. 1978. *Leptotrichia buccalis* hemagglutinin in cell binding and salivary inhibition studies. J. Periodontal Res. *13:* 425-432.

Falkow, S. 1975. Infectious multiple drug resistance. Pion, London.

Falkow, S., I.R. Ryman and O. Washington. 1962. Deoxyribonucleic acid base composition of *Proteus* and Providence organisms. J. Bacteriol. *83:* 1318-1321.

Faltus, I.I. and Y.G. Kishko. 1980. Certain biological and physicochemical properties of the virulent and temperate phages of *Erwinia carotovora.* (Russian). Mikrobiol. Zh. *42:* 226-231.

Fang, C.T., H.C. Ren, T.Y. Chen, Y.K. Chu, H.C. Faan and S.C. Wu. 1957. A comparison of the rice bacterial leaf blight organism with the bacterial leaf streak organism of rice and *Leersia hexandra* Swartz. Acta Phytopathol. Sinica *3:* 99-124.

Fanning, G.R., B.R. Davis, J.M. Madden, H.B. Bradford, Jr., A.G. Steigerwalt and D.J. Brenner. 1981. *Vibrio mimicus:* a newly recognized, cholera-like organism. Abstr. Annu. Meet. Soc. Microbiol. p. 50.

Fanning, G.R., B.R. Davis, A.G. Steigerwalt, I.K. Wachsmuth, F.W. Hickman, J.J. Farmer III and D.J. Brenner. 1980. Biochemical and genetic parameters of *Vibrio cholerae.* Abstr. Annu. Meet. Amer. Soc. Microbiol., p. 47.

Fanning, G.R., J.J. Farmer III, J.N. Parker, G.P. Huntley-Carter and D.J. Brenner. 1979. *Kluyvera:* A new genus in *Enterobacteriaceae.* Abstr. Ann. Meet. Amer. Soc. Microbiol., p. 100, I 30.

Fantham, H.B. 1908. *Spirochaeta (Trypanosoma) balbianii* (Certes) and *Spirochaeta anodontae* (Keysselitz): their movements, structure, and affinities. Quart. J. Microsc. Sci. *52:* 1-73.

Fantham, H.B. 1911. Some researches on the life-cycle of spirochaetes. Ann. Trop. Med. Parasitol. *5:* 479-496.

Faparusi, S.I. 1973. Origin of initial microflora of palm wine from oil palm trees (*Elaeis guineensis*). J. Appl. Bacteriol. *36:* 559-565.

Faparusi, S.I. 1974. Microorganisms from oil palm tree (*Elaeis guineensis*) tap holes. J. Food Sci. *39:* 755-757.

Farabaugh, P.J., V. Schmeissner, M. Hofer and J.H. Miller. 1978. Genetic studies of the lac repressor. VII. On the molecular nature of spontaneous hotspots in the lacI gene of *Escherichia coli.* J. Mol. Biol. *126:* 847-863.

Fargie, B. and B.W. Holloway. 1965. Absence of clustering of functionally related genes in *Pseudomonas aeruginosa.* Genet. Res. *6:* 284-299.

Farlow, W.G. 1880. On the nature of the peculiar reddening of salted codfish during the summer season. Report of the Commissioner for 1878, U.S. Commission of Fish and Fisheries, pp. 969-974.

Farlow, W.G. 1886. Vegetable parasites of codfish. Bull. U.S. Fish Commission *6:* 1-4.

Farmer, J.J. III. 1974. Lysotypie de *Serratia marcescens.* Arch. Roum. Pathol. Exp. Microbiol. *34:* 189.

Farmer, J.J. III. 1980. Revival of the name *Vibrio vulnificus.* Int. J. Syst. Bacteriol. *30:* 656.

Farmer, J.J. III. 1981. The genus *Citrobacter, In* M.P. Starr, H. Stolp, H.G. Trüper, A. Balows and H.G. Schlegel (Editors), The Prokaryotes, A handbook on habitats, isolation and identification of bacteria, Springer-Verlag, New York, pp. 1140-1147.

Farmer, J.J. III, M.A. Asbury, F.W. Hickman, D.J. Brenner and the Enterobacteriaceae study group. 1980. *Enterobacter sakazakii:* a new species of *Enterobacteriaceae* isolated from clinical specimens. Int. J. Syst. Bacteriol. *30:* 569-584.

Farmer, J.J. III and D.J. Brenner. 1977. Concept of a bacterial species. *In* Hoadley and Dutka (Editors), Bacterial Indicators/health hazards associated with water. Amer. Soc. Testing and Materials, Philadelphia, pp. 37-47.

Farmer, J.J. III, D.J. Brenner and W.A. Clark. 1976. Proposal to conserve the specific epithet tarda over the specific epithet anguillimortiferum in the name of the organism presently known as *Edwardsiella tarda:* Request for an opinion. Int. J. Syst. Bacteriol. *26:* 293-294.

Farmer, J.J. III, D.J. Brenner and W.H. Ewing. 1980. Opposition to the recent proposals which would reject the family name *Enterobacteriaceae* and reject *Escherichia* as its type genus. Int. J. Syst. Bacteriol. *30:* 660-673.

Farmer, J.J. III, G.R. Fanning, G.P. Huntley-Carter, B. Holmes, F.W. Hickman, C. Richard and D.J. Brenner. 1981a. *Kluyvera:* A new (redefined) genus in the family *Enterobacteriaceae:* Identification of *Kluyvera ascorbata* sp. nov. and *Kluyvera cryocrescens* sp. nov. in clinical specimens. J. Clin. Microbiol. *13:* 919-933.

Farmer, J.J. III, G.R. Fanning, G.P. Huntley-Carter, B. Holmes, F.W. Hickman, C. Richard and D.J. Brenner. 1981b. *In* List No. 7, Validation of the publication of new names and new combinations previously effectively pub-

lished outside the IJSB. Int. J. Syst. Bacteriol. *31:* 382-383.

Farmer, J.J. III, P.A.D. Grimont, F. Grimont and M.A. Asbury. 1980b. *Cedecea:* A new genus in *Enterobacteriaceae.* Abstr. Ann. Meet. Am. Soc. Microbiol., p. 295, Abstract C 123.

Farmer, J.J. III, F. Silva and D.R. Williams. 1973. Isolation of *Serratia marcescens* on deoxyribonuclease-toluidine blue-cephalothin agar. Appl. Microbiol. *25:* 151-152.

Faro, S., C. Walker and R.L. Pierson. 1980. Amnionitis with intact amniotic membranes involving *Streptobacillus moniliformis.* Obstet. Gynecol. *55:* 9S-11S.

Farrah, S.R. and R.F. Unz. 1975. Fluorescent-antibody study of natural fingerlike zoogloeae. Appl. Microbiol. *30:* 132-139.

Farrah, S.R. and R.F. Unz. 1976. Isolation of exocellular polymer from *Zoogloea* strains MP6 and 106 and from activated sludge. Appl. Environ. Microbiol. *32:* 33-37.

Farrell, I.D. 1974. The development of a new selective medium for the isolation of *Brucella abortus* from contaminated sources. Res. Vet. Sci. *16:* 280-286.

Farrell, R.K. 1966. Canine rickettsiosis. *In* Kirk (Editor), Current Veterinary Therapy 1966-1967, W.B. Saunders Co., Philadelphia, pp. 285-288.

Farrington, D.O. and R.D. Jorgenson. 1976. Prevalence of *Bordetella bronchiseptica* in certain wild mammals and birds in central Iowa. J. Wildl. Dis. *12:* 523-525.

Faur, Y.C., M.H. Weisburd, M.E. Wilson and P.S. May. 1973. A new medium for the isolation of pathogenic *Neisseria* (NYC medium). I. Formulation and comparisons with standard media. Health Lab. Sci. *10:* 44-54.

Fauré-Fremiet, E. 1952. Symbiontes bactériens des ciliés du genre *Euplotes.* C.R. Acad. Sci. *235:* 402-403.

Fautz, E., L. Grotjahn and H. Reichenbach. 1981. Hydroxy fatty acids as valuable chemosystematic markers in gliding bacteria and flavobacteria. *In* Reichenbach and Weeks (Editors), The *Flavobacterium-Cytophaga* Group (Proceedings of the International Symposium on Yellow-Pigmented Gram-negative Bacteria of the *Flavobacterium-Cytophaga* Group, Braunschweig, July 8 to 11, 1980). Verlag Chemie, Weinheim, pp. 127-133.

Favero, M.S., L.A. Carson, W.W. Bond and N.J. Petersen. 1971. *Pseudomonas aeruginosa:* growth in distilled water from hospitals. Science *173:* 836-838.

Federici, B.A. 1980. Reproduction and morphogenesis of *Rickettsiella chironomi,* an unusual intracellular procaryotic parasite of midge larvae. J. Bacteriol. *143:* 995-1002.

Federici, B.A., E.I. Hazard and D.W. Anthony. 1974. Rickettsia-like organism causing disease in a crangonid amphipod from Florida. Appl. Microbiol. *28:* 885-886.

Federov, M.V. and T.A. Kalininskaya. 1961. A new species of a nitrogen-fixing mycobacterium and its physiological peculiarities. Mikrobiologiya *30:* 9-14.

Feeley, J.C. 1969. Somatic O antigen relationship of *Brucella* and *Vibrio cholerae.* J. Bacteriol. *99:* 645-649.

Feeley, J.C., R.J. Gibson, G.W. Gorman, N.C. Langford, J.K. Rasheed, D.C. Mackel and W.B. Baine. 1979. CYE agar: a primary isolation medium for *Legionella pneumophila.* J. Clin. Microbiol. *10:* 436-441.

Feeley, J.C. and G.W. Gorman. 1980. Legionellae. *In* Lennette, Balows, Hausler, and Truant (Editors), Manual of Clinical Microbiology, 3rd Ed., American Society for Microbiology, Washington, D.C., pp. 318-324.

Feeley, J.C., G.W. Gorman and R.J. Gibson. 1979. Primary isolation media and methods. *In* Jones and Hébert (Editors), "Legionnaires'" the disease, the bacterium and methodology. Center for Disease Control, Atlanta, pp. 78-84.

Feeley, J.C., G.W. Gorman, R.E. Weaver, D.C. Mackel and H.W. Smith. 1978. Primary isolation media for the Legionnaires' disease bacterium. J. Clin. Microbiol. *8:* 320-325.

Fein, J.E. and R.A. MacLeod. 1975. Characterization of neutral amino acid transport in a marine pseudomonad. J. Bacteriol. *124:* 1177-1190.

Feist, C.F. and G.D. Hegeman. 1969. Phenol and benzoate metabolism by *Pseudomonas putida:* regulation of tangential pathways. J. Bacteriol. *100:* 869-877.

Feizi, T. and D. Taylor-Robinson. 1967. Cold agglutinin anti-I and *Mycoplasma pneumoniae.* Immunology *13:* 405-409.

Felix, A. and B.R. Callow. 1943. Typing of paratyphoid B bacilli by means of Vi bacteriophage. Brit. Med. J. *2:* 127.

Fellman, J.H. and R.C. Mills. 1960. Succinoxidase system of *Pasteurella tularensis.* J. Bacteriol. *79:* 800-806.

Felter, R. A., R. R. Colwell and G. B. Chapman. 1969. Morphology and round body formation in *Vibrio marinus.* J. Bacteriol. *99:* 326-335.

Felter, R.A, S.F. Kennedy, R.R. Colwell and G.B. Chapman. 1970. Intracytoplasmic membrane structures in *Vibrio marinus.* J. Bacteriol. *102:* 552-560.

Feng, W.C., E.S. Murray, W. Burgdorfer, J.M. Spielman, G. Rosengerg, K. Dang, C. Smith, C. Spickert and J.L. Waner. 1980. Spotted fever group rickettsiae in *Dermacentor variabilis* from Cape Cod, Massachusetts. Am. J. Trop. Med. Hyg. *29:* 691-694.

Fensom, A.H., W.M. Kurowski and S.J. Pirt. 1974. The use of ferricyanide for the production of 3-ketosugars by non-growing suspensions of *Agrobacterium tumefaciens.* J. Appl. Chem. Biotechnol. *24:* 457-467.

Ferber, D.M. and R.R. Brubaker. 1981. Plasmids in *Yersinia pestis.* Infect. Immun. *31:* 839-841.

Ferguson, G., D.R. Pollard, J.M. Robertson, G.O.P. Doherty, N.B. Haynes, D.W. Mathieson, W.B. Whalley and T.H. Simpson. 1980. The bacterial pigment from *Pseudomonas lemonnieri.* Part I. Structure of a degradation product, 3 *n*-octanamido pyridine-2,5,6-trione, by X-ray crystallography. J. Chem. Soc. Perkin Trans. I, *8:* 1782-1787.

Ferguson, W.E. and V.W. Nuttall. 1964. The preservation of *Erwinia tracheiphila* by freeze-drying. Can. J. Botany *42:* 333-335.

Ferguson, W.W. and N.D. Henderson. 1947. Description of strain C27: a motile organism with the major antigen of *Shigella sonnei* phase I. J. Bacteriol. *54:* 179-181.

Fernández-Borrero, O. and S. López-Duque. 1970. Pudricion acuosa del suedo tallo del plátana (*Musa paradisiaca*) causada por *Erwinia paradisiaca*, n. sp. Cenicafe *21:* 3-44.

Ferragut, C., D. Izard, F. Gavini, B. Lefebvre and H. Leclerc. 1981. *Buttiauxella*, a new genus of the family *Enterobacteriaceae*. Zentralbl. Bakteriol. Parasitenkd. Infektionskr. Hyg. Abt. 1 Orig. C *2:* 33-44.

Ferragut, C. and H. Leclerc. 1978. Study of motile and negative acetoin *Klebsiella pneumoniae* strains. Antonie van Leeuwenhoek J. Microbiol. Serol. *44:* 407-424.

Ferraris, T. 1926. Trattato di Patologia e Terapia Vegetale. 3rd Ed., Hoepli, Milan.

Ferry, N.S. 1911. Etiology of canine distemper. J. Infect. Dis. *8:* 399-420.

Ferry, N.S. 1912. Further studies on the *Bacillus bronchicanis*, the cause of canine distemper. Am. Vet. Rev. *41:* 77-79.

Ferry, N.S. 1912. *Bacillus bronchisepticus (bronchicanis)*: the cause of distemper in dogs and a similar disease in other animals. Vet. J. *68:* 376-391.

Fewson, C.A. and D.J.D. Nicholas. 1961. Respiratory enzymes in *Micrococcus denitrificans*. Biochim. Biophys. Acta *48:* 208-210.

Field, M. 1979. Modes of action of enterotoxins from *Vibrio cholerae* and *Escherichia coli*. Rev. Infect. Dis. *1:* 918-925.

Fieldsteel, A.H., F.A. Becker and J.G. Stout. 1977. Prolonged survival of virulent *Treponema pallidum* (Nichols strain) in cell-free and tissue culture systems. Infect. Immun. *18:* 173-182.

Fieldsteel, A.H., D.L. Cox and R.A. Moeckli. 1981. Cultivation of virulent *Treponema pallidum* in tissue culture. Infect. Immun. *32:* 908-915.

Fieldsteel, A.H., J.G. Stout and F.A. Becker. 1979. Comparative behavior of virulent strains of *Treponema pallidum* and *Treponema pertenue* in gradient cultures of various mammalian cells. Infect. Immun. *24:* 337-345.

Fildes, P. 1924/25. The growth requirements of haemolytic influenzae bacilli and the bearing of these upon the classification of related organisms. Brit. J. Exp. Pathol. *5:* 69-74.

Filion, R., S. Cloutier, E.R. Vrancken and G. Bernier. 1967. Infection respiratoire du dindonneau causée par un microbe apparente au *Bordetella bronchiseptica*. Can. J. Comp. Med. Vet. Sci. *31:* 129-134.

Finan, T.M., J.M. Wood and D.C. Jordan. 1981. Succinate transport in *Rhizobium leguminosarum*. J. Bacteriol. *148:* 193-202.

Findlay, G.M., E. Klieneberger, F.O. MacCallum and R.D. Mackenzie. 1938. Rolling disease. New syndrome in mice associated with a pleuro-pneumonia-like organism. Lancet *2:* 1511-1513.

Finegold, S.M. 1977. Anaerobic bacteria in human disease. Academic Press, New York.

Finegold, S.M. and E.M. Barnes. 1977. Proposal that the saccharolytic and asaccharolytic strains at present classified in the species as *Bacteroides melaninogenicus* and *Bacteroides asaccharolyticus*. Int. J. Syst. Bacteriol. *27:* 388-391.

Finegold, S.M., W.J. Martin and E.G. Scott. 1978. Bailey and Scott's diagnostic microbiology, 5th Ed., C.V. Mosby Co., St. Louis.

Finkelstein, R.A. and S. Mukerjee. 1963. Hemagglutination: a rapid method for differentiating *Vibrio cholerae* and *El Tor* vibrios. Proc. Soc. Exp. Biol. Med. *112:* 355-359.

Finn, V.G. 1978. The characteristics and clinical importance of bacteria of the genus *Citrobacter* isolated from patients with acute intestinal infections in the territory of Volgograd. J. Hyg. Epid. Microbiol. Immunol. *22:* 338-343.

Finstein, M.S. 1967. Growth and flocculation in a *Zoogloea* culture. Appl. Microbiol. *15:* 962-963.

Firehammer, B. D. 1965. The isolation of vibrios from ovine feces. Cornell Vet. *55:* 482-495.

Fiset, P. and R.A. Ormsbee. 1968. The antibody response to antigens of *Coxiella burnetii*. Zentralbl. Bakteriol. Parasitenkd. Infektionskr. Hyg. I Orig. *206:* 321-329.

Fitzgerald, J.M. 1977. Classification of luminous bacteria from the light organ of the Australian Pinecone Fish, *Cleidopus gloriamaris*. Arch. Microbiol. *112:* 153-156.

Fitzgerald, T.J., P. Cleveland, R.C. Johnson, J.N. Miller and J.A. Sykes. 1977. Scanning electron microscopy of *Treponema pallidum* (Nichols strain) attached to cultured cells. J. Bacteriol. *130:* 1333-1344.

Fitzgerald, T.J. and R.C. Johnson. 1979a. Surface mucopolysaccharides of *Treponema pallidum*. Infect. Immun. *24:* 244-251.

Fitzgerald, T.J. and R.C. Johnson. 1979. Mucopolysaccharidase of *Treponema pallidum*. Infect. Immun. *24:* 261-268.

Fitzgerald, T.J., R.C. Johnson and D.M. Ritzi. 1979b. Relationship of *Treponema pallidum* to acidic mucopolysaccharides. Infect. Immun. *24:* 252-260.

Fitzgerald, T.J., R.C. Johnson, J.A. Sykes and J.N. Miller. 1977. Interaction of *Treponema pallidum* (Nichols strain) with cultured mammalian cells: Effects of oxygen, reducing agents, serum supplements and different cell types. Infect. Immun. *15:* 444-452.

Flamm, H. 1957. Eine weitere neue Species des Genus *Moraxella M. polymorpha*

sp. n. Zentralbl. Bakteriol. Parasitenk. Infektionskr. Hyg. Abt. I Orig. *168:* 261-267.

Flamm, H. 1966. *Moraxella saccharolytica* (sp. n.) aus dem Liquor eines Kindes mit meningitis. Zentralbl. Bakteriol. Parasitenk. Infektionskr. Hyg. Abt. I Orig. *166:* 498-502.

Flesher, A.R. and R. Insel. 1978. Characteristics of lipopolysaccharide of *Haemophilus influenzae*. J. Infect. Dis. *138:* 719-730.

Flesher, A.R., I. Susumu, B.J. Mansheim and D.L. Kasper. 1979. The cell envelope of the Legionnaires' disease bacterium. Ann. Intern. Med. *90:* 628-630.

Fletcher, R. D. 1965. Activity and morphology of *Vibrio coli* phage. Am. J. Vet. Res. *26:* 361-364.

Fletcher, W. 1928. Recent work on leptospirosis, tsutsugamushi disease, and tropical typhus in the Federated Malay States. Trans. Roy. Soc. Trop. Med. Hyg. *21:* 265-288.

Fliermans, C.B., W.B. Cherry, L.H. Orrison, S.J. Smith, D.L. Tison and D.H. Pope. 1981. Ecological distribution of *Legionella pneumophila*. Appl. Environ. Microbiol. *41:* 9-16.

Flint, J.C. and D.K. McKelvie. 1955. Feline infectious anemia -- diagnosis and treatment. Proc. 92nd Ann. Mtg. Vet. Assoc., pp. 240-242.

Flint, J.C., M.H. Roepke and R. Jensen. 1958. Feline infectious anemia. I. Clinical aspects. Am. J. Vet. Res. *19:* 164-168.

Flint, J.C., M.H. Roepke and R. Jensen. 1959. Feline infectious anemia. II. Experimental cases. Am. J. Vet. Res. *20:* 33-40.

Florance, E.R. and H.R. Cameron. 1978. Three-dimensional structure and morphology of mycoplasmalike bodies associated with the albino disease of *Prunus avium*. Phytopathology *68:* 75-80.

Florent, A. 1959. Les deux vibrioses genitales de la bete bovine: La vibriose venerienne, due a *Vibrio foetus venerealis*, et la vibriose d'origine intestinale due a *V. foetus intestinales*. Proc. 10th Int. Vet. Cong. Madrid *2:* 953-957.

Florenzano, G., W. Balloni and R. Materassi. 1968. Nitrogen-fixing bacteria of the genus *Beijerinckia* in Venezuelan soils. *In* Transactions 9th Int. Congr. Soil Sci., Adelaide, Australia, *2:* 125-128.

Flügge, C. 1886. Die Microorganismen. F. C. W. Vogel, Leipzig.

Flynn, J. and M.J. McEntegart. 1972. Bacteriocins from Neisseria gonorrhoeae and their possible role in epidemiological studies. J. Clin. Pathol. *25:* 60-61.

Foggie, A. 1949. Studies on tick-borne fever in sheep. J. Gen. Microbiol. *3:* v-vi.

Foggie, A. 1951. Studies on the infectious agent of tick-borne fever in sheep. J. Pathol. Bacteriol. *63:* 1-15.

Foggie, A. 1962. Studies on tick pyaemia and tick-borne fever. Symp. Zool. Soc. London *6:* 51-58.

Foley, H. and L. Panot. 1937. Sur la rickettsia du trachome. C.R. Soc. Biol. Paris *124:* 230-232.

Follett, E.A.C. and J. Gordon. 1963. An electron microscope study of *Vibrio* flagella. J. Gen. Microbiol. *32:* 235-239.

Ford, D.K. and J. MacDonald. 1967. Influence of urea on the growth of T-strain mycoplasmas. J. Bacteriol. *93:* 1509-1512.

Ford, W.W. 1927. Text-book of bacteriology. Saunders, Philadelphia.

Forest, N. 1979. Caractérisation de *Mycoplasma salivarium* dans les parodontopathies. J. Biol. Buccale *7:* 321-330.

Forget, P. and F. Pichinoty. 1965. Le cycle tricarboxylique chez une bactérie denitrifiante obligatoire. Ann. Inst. Pasteur (Paris) *108:* 364-377.

Forrest, W.W. 1967. Energies of activation and uncoupled growth in *Streptococcus faecalis* and *Zymomonas mobilis*. J. Bacteriol. *94:* 1459-1463.

Forsberg, C.W. 1978. Effects of heavy metals and other trace elements on the fermentative activity of the rumen microflora and growth of functionally important rumen bacteria. Can. J. Microbiol. *24:* 298-306.

Forsberg, C.W., J.W. Costerton and R.A. MacLeod. 1970. Separation and localization of cell wall layers of a gram-negative bacterium. J. Bacteriol. *104:* 1338-1353.

Forshaw, K.A. and R.J. Fallon. 1972. Serological heterogeneity of *Mycoplasma pulmonis*. J. Gen. Microbiol. *72:* 501-510.

Förster, H.J., K. Biemann, W.G. Haigh, N.H. Tattrie and J.R. Colvin. 1973. The structure of novel C_{35} pentacyclic terpenes from *Acetobacter xylinum*. Biochem. J. *135:* 133-143.

Foshay, L. and W. Hesselbrock. 1945. Some observations on the filtrability of *Bacterium tularense*. J. Bacteriol. *49:* 233-236.

Foster, J.W. and R.H. Davis. 1966. A methane-dependent coccus, with notes on classification and nomenclature of obligate, methane-utilizing bacteria. J. Bacteriol. *91:* 1924-1931.

Fothergill, J.C. and J.R. Guest. 1977. Catabolism of L-lysine by *Pseudomonas aeruginosa*. J. Gen. Microbiol. *99:* 139-155.

Fournier, A., C. Combes and C. Vago. 1975. Mise en évidence de bactéries endocellulaires pathogèns chez le monogene, *Euzetrema knoepffleri*. C.R. Acad. Sci., Ser. D *281:* 1895-1896.

Fox, E.N. and K. Higuchi. 1958. Synthesis of the fraction 1 antigenic protein by *Pasteurella pestis*. J. Bacteriol. *75:* 209-216.

Fox, G.E., E. Stackebrandt, R.B. Hespell, J. Gibson, J. Maniloff, T.A. Dyer, R.S. Wolfe, W.E. Balch, R.S. Tanner, L.J. Magrum, L.B. Zablen, R. Blakemore, R. Gupta, L. Bonen, B.J. Lewis, D.A. Stahl, K.R. Luehrsen, K.N. Chen and C.R. Woese. 1980. The phylogeny of prokaryotes. Science *209:* 457-463.

Fox, H., R.H. Purcell and R.M. Chanock. 1969. Characterization of a newly identified mycoplasma (*Mycoplasma orale* type 3) from the human oropharynx. J. Bacteriol. *98:* 36-43.

Fox, J.P. 1956. Immunization against epidemic typhus. Am. J. Trop. Med. Hyg.

5: 464-479.

Fraenkel, G. and M. Blewett. 1943. Intracellular symbionts of insects as a source of vitamins. Nature (Lond.) 152: 506-507.

France, D.R. and N.P. Markham. 1968. Epidemiological aspects of Proteus infections with particular reference to phage typing. J. Clin. Pathol. 21: 97-102.

Franche, C., E. Canelo, D. Gauthier and C. Elmerich. 1981. Mobilization of the chromosome of Azospirillum brasilense by plasmid R68-45. FEMS Microbiol. Lett. 10: 199-202.

Franche, C. and C. Elmerich. 1981. Physiological properties and plasmid content of several strains of Azospirillum brasilense and A. lipoferum. Ann. Microbiol. (Inst. Pasteur) 132: 3-18.

Francis, E. and A.C. Evans. 1926. Agglutination, cross-agglutination and agglutinin absorption in tularaemia. Public Health Rep. 41: 1273-1295.

Francke-Grosmann, H. 1967. Ectosymbiosis in wood-inhabiting insects. In Henry (Editor), Symbiosis, Vol. II, Academic Press, New York, pp. 141-205.

Frank, B. 1879. Ueber die Parasiten in den Wurzelanschwillungen der Papilionaceen. Ber. Deut. Bot. Ges. 37: 376-387; 394-399.

Frank, B. 1889. Ueber die Pilzsymbiose der Leguminosen. Ber. Deut. Bot. Ges. 7: 332-346.

Frank, G.H. and G.E. Wersman. 1978. Rapid plate agglutination procedure for typing Pasteurella haemolytica. J. Clin. Microbiol. 7: 142-145.

Franke, I. 1968. Beobachtungen über eine seltene, gesmacklich kaum festelbare Infection eines Orangenfruchtsaftgetränkes. Erfrischungsgetränk. 21: 1099-1104.

Frankland, G.C. and P.F. Frankland. 1889. Ueber einige typische Mikroorganismen im Wasser und im Boden. Z. Hyg. 6: 373-400.

Franklin, R.M., A. Datta, J.E. Dahlberg and S.N. Braunstein. 1971. The cell membranes of a marine pseudomonad, Pseudomonas BAL-31; physical, chemical and biochemical properties. Biochim. Biophys. Acta 233: 521-537.

Franklin, R.M., R. Hinnen, R. Schäfer and N. Tsukagoshi. 1976. Structure and assembly of lipid-containing viruses, with special reference to bacteriophage PM2 as one type of model system. Philos. Trans. R. Soc. London Ser. B 276: 63-80.

Frantz, J.C. and R.E. McCallum. 1979. Growth yields and fermentation balance of Bacteroides fragilis cultured in glucose-enriched medium. J. Bacteriol. 137: 1263-1270.

Frantzen, E. 1950. Biochemical and serological studies on alkalescens and dispar strains. Acta Pathol. Microbiol. Scand. 27: 236-248.

Franze de Fernández, M.T., W.S. Hayward and J.T. August. 1972. Bacterial proteins required for replication of phage Qβ ribonucleic acid. Purification and properties of Host Factor I, a ribonucleic acid-binding protein. J. Biol. Chem. 247: 824-831.

Frasch, C.E. 1977. Role of protein serotype antigens in protection against disease due to Neisseria meningitidis. J. Infect. Dis. 136: (Suppl.) 84-90.

Frasch, C.E. 1979. Noncapsular surface antigens of Neisseria meningitidis. In Weinstein and Fields (Editors), Seminars in Infectious Diseases, Vol. 2, Stratton Intercontinental Medical Book Corp., New York, pp. 304-337.

Frasch, C.E. 1980. Role of lipopolysaccharide in wheat germ agglutinin-mediated agglutination of Neisseria meningitidis and Neisseria gonorrhoeae. J. Clin. Microbiol. 12: 498-501.

Fraser, D.W., D.C. Deubner, D.L. Hill and D.K. Gilliam. 1979. Nonpneumonic, short-incubation-period legionellosis (Pontiac fever) in men who cleaned a steam turbine condenser. Science 205: 690-691.

Fraser, D.W., T.F. Tsai, W. Orenstein, W.E. Parkin, H.J. Beecham, R.G. Sharrer, J. Harris, G.F. Mallinson, S.M. Martin, J.E. McDade, C.C. Shepard, P.S. Brachman and the Field Investigation Team. 1977. Legionnaires' disease: description of an epidemic. N. Engl. J. Med. 297: 1189-1197.

Fraser, D.W., I.K. Wachsmuth, C. Bopp, J.C. Feeley and T.F. Tsai. 1978. Antibiotic treatment of guinea pigs infected with agent of Legionnaires' disease. Lancet i: 175-177.

Frateur, J. 1950. Essai sur la systématique des Acetobacters. La Cellule 53: 287-392.

Frateur, J. and P. Simonart. 1952. Étude de la flore bactérienne d'un acétificateur de vinaigre d'alcool. IX Congresso Internazionale Industrie Agrarie Roma.

Frazer, J., K. Zinnemann and J.M.H. Boyce. 1975. The agglutination reactions of Haemophilus paraphrophilus and H. paraphrohaemolyticus, and some observations on the agglutination of H. aphrophilus and H. haemoglobinophilus (H. canis). J. Med. Microbiol. 8: 89-96.

Frazier, W.C. 1926. A method for the detection of changes in gelatin due to bacteria. J. Infect. Dis. 39: 302-309.

Fred, E.B., I.L. Baldwin and E. McCoy. 1932. Root nodule bacteria and leguminous plants. University of Wisconsin Studies in Science, Number 5. University of Wisconsin Press, Madison.

Fredericq, P. 1948. Actions antibiotiques réciproques chez les Enterobacteriaceae. Rev. Belge Pathol. 19: suppl. 4; 1-107.

Frederiksen, W. 1964. A study of some Yersinia pseudotuberculosis-like bacteria (Bacterium enterocoliticum and Pasteurella X). Proc. XIV Scand. Cong. Pathol. Microbiol., pp. 103-104. Universitetsforlaget, Oslo.

Frederiksen, W. 1970. Citrobacter koseri (n. sp.), a new species within the genus Citrobacter, with a comment on the taxonomic position of Citrobacter intermedium (Werkman and Gillen). Publ. Fac. Sci. Univ. J. E. Purkyne, Brno 47: 89-94.

Frederiksen, W. 1973. Pasteurella taxonomy and nomenclature. In Winblad

(Editor), Contributions to microbiology and immunology, Vol. 2. Yersinia, Pasteurella and Francisella. Karger, Basel, pp. 170-176.

Frederiksen, W. 1981. Gas-producing species within Pasteurella and Actinobacillus. In Kilian, Frederiksen and Biberstein (Editors), Haemophilus, Pasteurella and Actinobacillus. Academic Press, London and New York, pp. 185-196.

Frederiksen, W. and M. Kilian. 1981. Haemophilus, Pasteurella and Actinobacillus: Their significance in human medicine. In Kilian, Frederiksen and Biberstein (Editors), Haemophilus, Pasteurella and Actinobacillus. Academic Press, London, pp. 39-55.

Fredrick, J.F. (Editor). 1981. Origins and evolution of eukaryotic intracellular organelles. Ann. N.Y. Acad. Sci. 361: 1-512.

Freeman, B.A., J.R. McGhee and R.E. Baughn. 1970. Some physical, chemical and taxonomic features of the soluble antigens of the Brucellae. J. Infect. Dis. 121: 522-527.

Freundt, E.A. 1953. The occurrence of Micromyces (pleuropneumonia-like organisms) in the female genito-urinary tract. Acta Pathol. Microbiol. Scand. 32: 468-480.

Freundt, E.A. 1954. Morphological and biochemical investigations of human pleuropneumonia-like organisms (Micromyces). Acta Pathol. Microbiol. Scand. 34: 127-144.

Freundt, E.A. 1955. The classification of the pleuropneumonia group of organisms (Borrelomycetales). Int. Bull. Bacteriol. Nomencl. Taxon. 5: 67-78.

Freundt, E.A. 1956a. Streptobacillus moniliformis infection in mice. Acta Pathol. Microbiol. Scand. 38: 231-245.

Freundt, E.A. 1956b. Experimental investigations into pathogenicity of L-phase variant of Streptobacillus moniliformis. Acta Pathol. Microbiol. Scand. 38: 246-258.

Freundt, E.A. 1958. The Mycoplasmataceae. Munksgaard, Copenhagen.

Freundt, E.A., B.E. Andrews, H. Erno, M. Kunze and F.T. Black. 1973. The sensitivity of Mycoplasmatales to sodium-polyanethol-sulfonate and digitonin. Zentralbl. Bakteriol. Parasitenkd. Infektionskr. Hyg. I Abt. Orig. A 225: 104-112.

Freundt, E.A., H. Ernø and R.M. Lemcke. 1980. Identification of mycoplasmas. In Norris and Gibbons (Editors), Methods in Microbiology, Vol. 13, Academic Press, New York, pp. 377-434.

Freundt, E.A., D. Taylor-Robinson, R.H. Purcell, R.M. Chanock and F.T. Black. 1974. Proposal of Mycoplasma buccale nom. nov. and Mycoplasma faucium nom. nov. for Mycoplasma orale "types" 2 and 3, respectively. Int. J. Syst. Bacteriol. 24: 252-255.

Friedberger, E. 1902/03. Über ein neues zür Gruppe des Influenzabacillus gehöriges häemoglobinophiles Bakterium (Bacillus haemoglobinophilus-canis). Zentralbl. Bakteriol. Parasitenkd. Infektionskr. Hyg. Abt. I Orig. 33: 401-406.

Friedhoff, K.T. and M. Ristic. 1966. Anaplasmosis XIX. A preliminary study of Anaplasma marginale in Dermacentor andersoni (Stiles) by fluorescent antibody technique. Am. J. Vet. Res. 27: 643-646.

Friedman, B.A. and P.R. Dugan. 1968. Identification of Zoogloea species and the relationship to zoogloeal matrix and floc formation. J. Bacteriol. 95: 1903-1909.

Friedman, B.A., P.R. Dugan, R.M. Pfister and C.C. Remsen. 1968. Fine structure and composition of the zoogloeal matrix surrounding Zoogloea ramigera. J. Bacteriol. 96: 2144-2153.

Friedman, B.A., P.R. Dugan, R.M. Pfister and C.C. Remsen. 1969. Structure of exocellular polymers and their relationship to bacterial flocculation. J. Bacteriol. 98: 1328-1334.

Friedrich, B., C. Hogrefe and H.G. Schlegel. 1981. Naturally occurring genetic transfer of the hydrogen-oxidizing ability between strains of Alcaligenes eutrophus. J. Bacteriol. 147: 198-205.

Friedrich, C.G., B. Bowien and B. Friedrich. 1979. Formate and oxalate metabolism in Alcaligenes eutrophus. J. Gen. Microbiol. 115: 185-192.

Friedrich, C.G., B. Friedrich and H.G. Schlegel. 1976. Aromatic amino acid biosynthesis in Alcaligenes eutrophus H 16. III. Properties and regulation of anthranilate synthase. Arch. Microbiol. 107: 125-131.

Friedrich, C.G. and G. Mitrenga. 1981. Oxidation of thiosulfate by Paracoccus denitrificans and other hydrogen bacteria. FEMS Microbiol. Lett. 10: 209-212.

Friend, W.G., E.H. Salkeld and I.L. Stevenson. 1959. Nutrition of onion maggots, larvae of Hylemya antiqua (Meig.), with reference to other members of the genus Hylemya. Ann. N.Y. Acad. Sci. 77: 384-393.

Friis, N.F. 1971. Mycoplasmas cultivated from the respiratory tract of Danish pigs. Acta Vet. Scand. 12: 69-79.

Friis, N.F. 1976. A serological variant of Mycoplasma hyorhinis recovered from the conjunctivae of swine. Acta Vet. Scand. 17: 343-353.

Friis, N.F. 1977. Mycoplasma suipneumoniae and Mycoplasma flocculare in the growth precipitation test. Acta Vet. Scand. 18: 168-175.

Friis, N.F. and K.B. Pedersen. 1979. Isolation of Mycoplasma bovoculi from cases of infectious bovine keratoconjunctivitis. Acta Vet. Scand. 20: 51-59.

Friis, N.F., K.B. Pedersen and B. Bloch. 1980. Ureaplasma isolated from the respiratory tract of mink. Acta Vet. Scand. 21: 134-136.

Fritsche, D. 1975. Investigations on the structure of the sphingolipids of the genus Bacteroides. Zentralbl. Bakteriol. Parasitenkd. Infektionskr. Hyg., I. Abt. Orig. A 233: 64-71.

Fritzsch, W. and W. Abadjieff. 1966. Beiträg zür Differenzierung innerhalt des

Spezies *Brucella abortus* mittels Antibiotika. Zentralbl. Bakteriol. Parasitenkd. Infektionskr. Hyg. Abt. I Orig. A *201:* 493-505.

Frøholm, L.O. and K. Bøvre. 1972. Fimbriation associated with the spreading-corroding colony type of *Moraxella kingii.* Acta Pathol. Microbiol. Scand. Sect. B *80:* 641-648.

Frøholm, L.O., K. Jyssum and K. Bøvre. 1973. Electron microscopical and cultural features of *Neisseria meningitidis* competence variants. Acta Pathol. Microbiol. Scand. Sect. B *81:* 525-537.

Frosch, P. 1923. Zür Morphologie des Lungenseucheerregers. II. Mitteilung. Arch. Wiss. Prakt. Tierheilk. *49:* 273-282.

Frosch, P. and W. Kolle. 1896. Die Mikrokokken. *In* Flügge (Editor), Die Mikroorganismen, 3 Aufl., 2 Teil, Verlag von Vogel, Leipzig, pp. 154-155.

Fry, J. C. and D. G. Staples. 1976. Distribution of *Bdellovibrio bacteriovorus* in sewage works, river waters, and sediments. Appl. Environ. Microbiol. *31:* 469-474.

Fudl-Allah, A.E.A., E.C. Calavan and E.C.K. Igwegbe. 1972. Culture of a mycoplasmalike organism associated with stubborn disease of citrus. Phytopathology *62:* 729-731.

Fuentes, L.G. 1979. Primer caso de fiebre de las Montañas Rocosas en Costa Rica, America Central. Rev. Lat-amer. Microbiol. *21:* 167-172.

Fuerst, J.A. and A.C. Hayward. 1969a. Surface appendages similar to fimbriae (pili) on *Pseudomonas* species. J. Gen. Microbiol. *58:* 227-237.

Fuerst, J.A. and A.C. Hayward. 1969b. The sheathed flagellum of *Pseudomonas stizolobii.* J. Gen. Microbiol. *58:* 239-245.

Fuerst, J.A. and A.C. Hayward. 1980. The effect of temperature on the formation of sheathed flagella by *Pseudomonas stizolobii.* J. Gen. Microbiol. *117:* 111-117.

Fujino, T., Y. Okuno, D. Nakada, A. Aoyama, K. Fukai, T. Mukai and T. Ueho. 1951. On the bacteriological examination of shirasu food poisoning (in Japanese). J. Jpn. Assoc. Infect. Dis. *25:* 11-12.

Fujino, T., G. Sakaguchi, R. Sakazaki and Y. Takeda (Editors). 1974. International symposium on *Vibrio parahaemolyticus.* Saikon Publishing Co., Tokyo.

Fukui, S., R.M. Hochster, R. Durbin, E.E. Grebner and D.S. Feingold. 1963. The conversion of sucrose to α-D-ribohexopyranosyl-3-ulose-β-D-fructofuranoside by cultures of *Agrobacterium tumefaciens.* 1963. Bull. Res. Counc. Israel *11A4:* 314-320.

Fulbright, D.W. and J.V. Leary. 1978. Linkage analysis of *Pseudomonas glycinea.* J. Bacteriol. *136:* 497-500.

Fuller, H.S., E.S. Murray, J.C. Ayres, J.C. Snyder and L. Potash. 1951. Studies of rickettsialpox. I. Recovery of the causative agent from house mice in Boston, Massachusetts. Am. J. Hyg. *54:* 82-100.

Fulton, M. 1943. The identity of *Bacterium columbensis* Castellani. J. Bacteriol. *46:* 79-82.

Fulton, H. R., P. O. Sikorowski and B. R. Morment. 1974. A survey of North Mississippi mosquitoes for pathogenic microorganisms. Mosq. News *34:* 86-90.

Funnell, G.R. and Y.T. Tchan. 1977. Nitrogenase activity during the life cycle of *Azotobacter.* Proc. Linn. Soc. N.S.W. *102:* 35-42.

Furness, G. 1968. Analysis of the growth cycle of *Mycoplasma orale* by synchronized division and by ultraviolet radiation. J. Infect. Dis. *118:* 436-442.

Furness, G. 1970. The growth and morphology of mycoplasmas replicating in synchrony. J. Infect. Dis. *122:* 146-158.

Furness, G. 1975. T-mycoplasmas: growth patterns and physical characteristics of some human strains. J. Infect Dis. *132:* 592-596.

Furness, G., F.J. Pipes and M.J. McMurtrey. 1968. Analysis of the life-cycle of *Mycoplasma pneumoniae* by synchronized division and by ultraviolet and X-irradiations. J. Infect. Dis. *118:* 7-13.

Furniss, A.L., J.V. Lee and T.J. Donovan. 1978. The vibrios. Public Health Laboratory Service Monograph Series, Her Majesty's Stationery Office, London.

Furr, P.M., C.M. Hetherington and D. Taylor-Robinson. 1978. Studies of the specificity of ureaplasmas for marmosets. J. Med. Microbiol. *11:* 537-540.

Furr, P.M., D. Taylor-Robinson and C.M. Hetherington. 1976. The occurrence of ureaplasmas in marmosets. Lab. Anim. *10:* 393-398.

Furth, A. 1975. Purification and properties of a constitutive beta-lactamase from *Pseudomonas aeruginosa* strain Dalgleish. Biochim. Biophys. Acta *377:* 431-443.

Fyfe, J.A.M. and J.R.W. Govan. 1980. Alginate synthesis in mucoid *Pseudomonas aeruginosa:* a chromosomal locus involved in control. J. Gen. Microbiol. *119:* 443-450.

Fytizas, E. 1970. Action de quelques antibiotiques sur les adultes de *Dacus oleae* et leur descendance. Z. Angew. Entomol. *65:* 453-458.

Fytizas, E. and M.E. Tzanakakis. 1966. Some effects of streptomycin, when added to the adult food, on the adults of *Dacus oleae* (Diptera: Tephritidae) and their progeny. Ann. Entomol. Soc. Am. *59:* 269-273.

Gabridge, M.G., Y.D. Barden-Stahl, R.B. Polisky and J.A. Engelhardt. 1977. Differences in the attachment of *Mycoplasma pneumoniae* cells and membranes to tracheal epithelium. Infect. Immun. *16:* 766-772.

Gabridge, M.G., H. Gunderson, S.L. Schaeffer and Y.D. Barden-Stahl. 1978. Ciliated respiratory epithelial monolayers: new model for *Mycoplasma pneumoniae* infection. Infect. Immun. *21:* 333-336.

Gabridge, M.G., C.K. Johnson and A.M. Cameron. 1974. Cytotoxicity of *Mycoplasma pneumoniae* membranes. Infect. Immun. *10:* 1127-1134.

Gadberry, J.L., K. Clemmons and K. Drumm. 1980. Evaluation of methods to detect oxidase activity in the genus *Pasteurella.* J. Clin. Microbiol. *12:* 220-225.

Galarneault, T.P. and E. Leifson. 1956. Taxonomy of *Lophomonas* n. gen. Can. J. Microbiol. *2:* 102-110.

Galarneault, T.P. and E. Leifson. 1964. *Pseudomonas vesiculare* (Büsing et al.) nov. comb. Int. Bull. Bacteriol. Nomencl. Taxon. *14:* 165-168.

Galau, G.A., R.J. Britten and E.H. Davidson. 1977. Studies on nucleic acid reassociation kinetics: rate of hybridization on excess RNA with DNA, compared to the rate of DNA renaturation. Proc. Nat. Acad. Sci. USA *74:* 1020-1023.

Gallagher, J.E. and K.R. Rhoades. 1979. Simplified preparation of mycoplasmas, an acholeplasma, and a spiroplasma for scanning electron microscopy. J. Bacteriol. *137:* 972-976.

Gallavan, M. and W.E. Goodpasture. 1937. Infection of chick embryos with *H. pertussis* reproducing pulmonary lesions of whooping cough. Am. J. Pathol. *13:* 927-938.

Gamaléia, M.N. 1888. *Vibrio metschnikovi* (n. sp.) et ses rapports avec le microbe du choléra asiatique. Ann. Inst. Pasteur (Paris) *2:* 482-488.

Gaman, G.A., W.C. Cates, C.F.T. Snelling, B. Lank and A.R. Ronald. 1976. Emergence of gentamicin- and carbenicillin-resistant *Pseudomonas aeruginosa* in a hospital environment. Antimicrob. Agents Chemother. *9:* 474-480.

Gamble, T.N., M.R. Betlach and J.M. Tiedje. 1977. Numerically dominant denitrifying bacteria from world soils. Appl. Environ. Microbiol. *33:* 926-939.

Gambrill, M.R. and C.L. Wisseman, Jr. 1973. Mechanisms of immunity in typhus infections. II. Multiplication of typhus richettsiae in human macrophage cell cultures in the nonimmune system: Influence of virulence of rickettsial strains and of chloramphenicol. Infect. Immun. *8:* 519-527.

Gan, K.H. and S.K. Tija. 1963. A new method for the differentiation of *Vibrio comma* and *Vibrio eltor.* Amer. J. Hyg. *77:* 184-186.

Ganaway, J.R., A.M. Allen and C.W. McPherson. 1965. Prevention of acute *Bordetella bronchiseptica* pneumonia in a guinea pig colony. Lab. Anim. Care *15:* 156-162.

Gangulee, P.C., G.P. Sen and G.L. Sharma. 1966. Serological diagnosis of glanders by hemagglutination test. Indian Vet. J. *43:* 386-391.

Garcia-Delgado, G.A., P.B. Little and D.A. Barnum. 1976. A comparison of various *Haemophilus somnus* strains. Canad. J. Comp. Med. *41:* 380-388.

Gardner, A.D. and K.V. Venkatraman. 1935. The antigens of the cholera group of vibrios. J. Hyg. Camb. *35:* 262-282.

Gardner, H.L. and C.D. Dukes. 1955. *Haemophilus vaginalis* vaginitis. A newly defined specific infection previously classified "nonspecific" vaginitis. Am. J. Obstet. Gynecol. *69:* 962-976.

Gardner, J.M. and C.I. Kado. 1972. Comparative base sequence homologies of the deoxyribonucleic acids of *Erwinia* species and other *Enterobacteriaceae.* Int. J. Syst. Bacteriol. *22:* 201-209.

Gardner, J.M. and C.I. Kado. 1976. Polygalacturonic acid *trans*-eliminase in the osmotic shock fluid of *Erwinia rubrifaciens:* characterization of the purified enzyme and its effect on plant cells. J. Bacteriol. *127:* 451-460.

Gardner, M.W. and J.B. Kendrick. 1923. Bacterial spot of cowpea. Science *57:* 275.

Gargani, G. 1977. Attempts at a system of numerical taxonomy for the genus *Brucella.* Ann. Sclavo *19:* 61-66.

Garibaldi, J.A. 1967. Media for the enhancement of fluorescent pigment production by *Pseudomonas* species. J. Bacteriol. *94:* 1296-1299.

Garibaldi, J.A. 1971. Influence of temperature on the iron metabolism of a fluorescent pseudomonad. J. Bacteriol. *105:* 1036-1038.

Garibaldi, A. and D.F. Bateman. 1971. Pectic enzymes produced by *Erwinia chrysanthemi* and their effects on plant tissues. Physiol. Plant Pathol. *1:* 25-40.

Garnham, P.C.C. 1947. A new blood spirochaete in the grivet monkey, *Cercopithecus aethiops.* East Afr. Med. J. *24:* 47-51.

Garnier, M., M. Clerc and J.M. Bové. 1981. Growth and division of spiroplasmas: morphology of *Spiroplasma citri* during growth in liquid medium. J. Bacteriol. *147:* 642-652.

Garrard, W. T. 1971. Selective release of proteins from *Spirillum itersonii* by tris(hydroxymethyl)aminomethane and ethylenediaminetetraacetate. J. Bacteriol. *105:* 93-100.

Garrard, W. T. 1972. Synthesis, assembly, and localization of periplasmic cytochrome c. J. Biol. Chem. *247:* 5935-5943.

Garrett, C.M.E., C.G. Panagopoulos and J.E. Crosse. 1966. Comparison of plant pathogenic pseudomonads from fruit trees. J. Appl. Bacteriol. *29:* 342-356.

Garrod, L.P. and P.M. Waterworth. 1969. Effect of medium composition on the apparent sensitivity of *Pseudomonas aeruginosa* to gentamicin. J. Clin. Pathol. *22:* 534-538.

Garthe, W.A. and M.W. Elliott. 1971. Role of intracellular symbionts in the fatbody of cockroaches: influence on hemolymph proteins. Experientia (Basel) *27:* 593.

Gaspar, A.J., H.B. Tresselt and M.K. Ward. 1961. New solid medium for enhanced growth of *Pasteurella tularensis.* J. Bacteriol. *82:* 564-569.

Gaumann, E. 1923. Ueber zwei Bananenkrankheiten in Niederlandisch Indien. Z. Pflanzenkr. Pflanzenpathol. Pflanzenschutz. 33: 1-17.

Gauthier, D. K., G. D. Clark-Walker, W. T. Garrard, Jr. and J. Lascelles. 1970. Nitrate reductase and soluble cytochrome c in *Spirillum itersonii.* J. Bacteriol. *102:* 797-803.

Gauthier, M.J. 1976a. *Alteromonas rubra* sp. nov., a new marine antibiotic-producing bacterium. Int. J. Syst. Bacteriol. 26: 459-466.

Gauthier, M.J. 1976b. Morphological, physiological, and biochemical characteristics of some violet-pigmented bacteria isolated from seawater. Can. J. Microbiol. 22: 138-149.

Gauthier, M.J. 1976c. Modification of bacterial respiration by a macromolecular polyanionic antibiotic produced by a marine Alteromonas. Antimicrob. Agents and Chemother. 9: 361-366.

Gauthier, M.J. 1977. Alteromonas citrea, a new Gram-negative, yellow-pigmented species from seawater. Int. J. Syst. Bacteriol. 27: 349-354.

Gauthier, M.J. 1982. Validation of the name Alteromonas luteoviolacea. Int. J. Syst. Bacteriol. 32: 82-86.

Gauthier, M.J. and V.A. Breittmayer. 1979. A new antibiotic-producing bacterium from seawater: Alteromonas aurantia sp. nov. Int. J. Syst. Bacteriol. 29: 366-372.

Gauthier, M.J. and G.N. Flatau. 1976. Antibacterial activity of marine violet-pigmented Alteromonas with special reference to the production of brominated compounds. Can. J. Microbiol. 22: 1612-1619.

Gauthier, M.J., J.M. Shewan, D.M. Gibson and J.V. Lee. 1975. Taxonomic position and seasonal variations in marine neritic environment of some Gram-negative antibiotic-producing bacteria. J. Gen. Microbiol. 87: 211-218.

Gavini, F., C. Ferragut, D. Izard, P.A. Trinel, H. Leclerc, Lefebvre and D.A.A. Mossel. 1979. Serratia fonticola, a new species from water. Int. J. Syst. Bacteriol. 29: 92-101.

Gavini, F., C. Ferragut, B. Lefebre and H. Leclerc. 1976. Étude taxonomique d'enterobactéries appartenant ou apparentées au genre Enterobacter. Ann. Microbiol. (Inst. Past.) 127B: 317-335.

Gavini, F., H. Leclerc, B. Lefèbvre, C. Ferragut and D. Izard. 1977. Étude taxonomique d'entérobactéries appartenant ou apparentées au genre Klebsiella. Ann. Microbiol. (Inst. Pasteur) 128B: 45-49.

Gay, F.W., M.E. Maguire and A. Baskerville. 1972. Etiology of chronic pneumonia in rats and a study of the experimental disease in mice. Infect. Immun. 6: 83-91.

Gee, D.L., P. Baumann and L. Baumann. 1975. Enzymes of D-fructose catabolism in species of Beneckea and Photobacterium. Arch. Microbiol. 103: 205-207.

Gehring, F. 1962. Untersuchungen über den Infektionsverlauf einer durch Pectobacterium parthenii (Starr) Hellmers var. dianthicola Hellmers verursachten Nelkenbakteriose sowie über enzymatische Eigenschaften dieses Bakteriums im Vergleich mit Pseudomonas caryophylli (Burkholder) Starr et Burkholder und einigen typischen Nassfäuleerregern. Phytopathol. Zeitschr. 43: 383-407.

Geilinger, H. 1921. Experimentalle Beiträge zür Mikrobiologie der Getreidamehl. I. Ueber koliartige Mehlbakterien. Mitt. Lebensm. Hyg., Bern 12: 49-81; 105-119; 231-262.

Gemski, P., J.R. Lazere and T. Casey. 1980a. Plasmid associated with pathogenicity and calcium dependency of Yersinia enterocolitica. Infect. Immun. 27: 682-685.

Gemski, P., J.R. Lazere, T. Casey and I.A. Wohlhieter. 1980b. Presence of a virulence-associated plasmid in Yersinia pseudotuberculosis. Infect. Immun. 28: 1044-1047.

Genig, V.A. 1968. A live vaccine 1/M-44 against Q-fever for oral use. J. Hyg. Epidemiol. Microbiol. Immunol. 12: 265-273.

George, H. A., P. S. Hoffman, R. M. Smibert and N. R. Krieg. 1978. Improved media for growth and aerotolerance of Campylobacter fetus. J. Clin. Microbiol. 8: 36-41.

George, J.R., L. Pine, M.W. Reeves and W.K. Harrell. 1980. Amino acid requirements of Legionella pneumophila. J. Clin. Microbiol. 11: 286-291.

George, W.L., B.D. Kirby, V.L. Sutter, D.M. Citron and S.M. Finegold. 1981. Gram-negative anaerobic bacilli: Their role in infections and patterns of susceptibility to antimicrobial agents. II. Little-known Fusobacterium species and miscellaneous genera. Rev. Inf. Dis. 3: 599-626.

Geraci, J.E., P.E. Hermans and J.A. Washington II. 1974. Eikenella corrodens endocarditis. Mayo Clin. Proc. 49: 950-953.

Gerber, N.N. and M.J. Gauthier. 1979. New prodigiosin-like pigment from Alteromonas rubra. Appl. Environ. Microbiol. 37: 1176-1179.

Gerhardt, P. and C.G. Heden. 1961. Concentrated culture of gonococci in clear liquid medium. Proc. Soc. Exp. Biol. Med. 105: 49-51.

Ghai, S.K., M. Hisamatsu, A. Amemura and T. Harada. 1981. Production and chemical composition of extracellular polysaccharides of Rhizobium. J. Gen. Microbiol. 122: 33-40.

Gharagozlou, I.D. 1968. Aspect infrastructural de Diplocalyx calotermitidis nov. gen. nov. sp., spirochaetale de l'intestin de Calotermes flavicollis. C. R. Acad. Sci. (Paris) 266: 494-496.

Ghidini, G.M. and I. Archeti. 1939. Studi sulle termite: 2 - Le spirochete presenti in Reticulitermes lucifugus Rossi. Riv. Biol. Coloniale 2: 125-140.

Ghoniem, N., C. Amstberg and W. Bisping. 1973. Vergleichende Darstellung Desbiochemischen Reaktionsspektrums von Pasteurella multocida-Stämmen von Hund und Schwein. Zentralbl. Veterinaermed. Reihe B 20: 310-317.

Ghosh, A., J. Maniloff and D.A. Gerling. 1978. Inhibition of mycoplasma cell division by cytochalasin B. Cell 13: 57-64.

Giammanco, G., J. Buissière, M. Toucas, G. Brault and L. Le Minor. 1980. Intérêt taxonomique de la recherche de la γ-glutamyltransférase chez les Enterobacteriaceae. Ann. Microbiol. Inst. Pasteur (Paris) 131A: 181-187.

Giannotti, J., C. Vago, D. Giannotti and C. Legoff. 1981. Étude comparée in vivo et in vitro des diverses formes del l'agent mollicute de la léthargie de

coléoptères. C.R. Acad. Sci. Paris Ser. D 292: 1043-1049.

Giard, R. and N.A. Vedros. 1981. Bi-phasic resin system for growing concentrated cultures of microorganisms and its application to the cultivation of Neisseria gonorrhoeae. Appl. Environ. Microbiol. 41: 846-847.

Gibbins, A.M. and K.F. Gregory. 1972. Relatedness among Rhizobium and Agrobacterium species determined by three methods of nucleic acid hybridization. J. Bacteriol. 111: 129-141.

Gibbins, L.N. 1978. Erwinia herbicola: a review and perspective. In Station de Pathologie Végétale et Phytobactériolgie (Editor), Proc. IVth Int. Conf. on Plant Pathogenic Bacteria, Gilbert-Clarey, Tours, France. pp. 403-431.

Gibbons, N.E. 1969. Isolation, growth and requirements of halophilic bacteria. In Norris and Ribbons (Editors), Methods in Microbiology, Vol. 3B, Academic Press, London, pp. 169-183.

Gibbons, N.E. 1974. Family V. Halobacteriaceae fam. nov. In Buchanan and Gibbons (Editors), Bergey's Manual of Determinative Bacteriology, 8th Ed., The Williams and Wilkins Co., Baltimore, pp. 269-273.

Gibbons, N.E. 1974. Reference collections of bacteria - the need and requirements for type and neotype strains. In Buchanan and Gibbons (Editors), Bergey's Manual of Determinative Bacteriology, 8th Ed., The Williams and Wilkins Co., Baltimore, pp. 14-17.

Gibbons, N.E. and R.G.E. Murray. 1978. Proposals concerning the higher taxa of bacteria. Int. J. Syst. Bacteriol. 28: 1-6.

Gibbons, N.E., K.B. Pattee and J.G. Holt (Editors). 1981. Supplement to Index Bergeyana. Williams and Wilkins Co., Baltimore.

Gibbons, R.J., S.S. Socransky, S. Sawyer, B. Kapsimalis and J.B. MacDonald. 1963. The microbiota of the gingival crevice area of man. II. The predominant cultivable organisms. Arch. Oral Biol. 8: 281-289.

Gibson, J., E. Stackebrandt, L. B. Zablen, R. Gupta and C. R. Woese. 1979. A phylogenetic analysis of the purple photosynthetic bacteria. Curr. Microbiol. 3: 59-64.

Gier, H.T. 1947. Intracellular bacteroids in the cockroach (Periplaneta americana Linn). J. Bacteriol. 53: 173-189.

Giesberger, G. 1936. Beiträge zür Kenntnis der Gattung Spirillum. Ehrb. Inaug. Dissertation Utrecht, pp. 1-136.

Gieszczykiewicz, M. 1939. Zagadniene systematihki w bakteriologii -- Zür Frage der Bakterien-Systematic. Bull. Acad. Polon. Sci., Ser. Sci. Biol. 1: 9-27.

Gilardi, G.L. 1971. Characterization of Pseudomonas species isolated from clinical specimens. Appl. Microbiol. 21: 414-419.

Gilardi, G.L. 1971a. Antimicrobial susceptibility as a diagnostic aid in the identification of nonfermenting Gram-negative bacteria. Appl. Microbiol. 22: 821-823.

Gilardi, G.L. 1971b. Characterization of nonfermentative nonfastidious Gram-negative bacteria encountered in medical bacteriology. J. Appl. Bacteriol. 34: 623-644.

Gilardi, G.L. 1973. Nonfermentative Gram-negative bacteria encountered in clinical specimens. Antonie van Leeuwenhoek J. Microbiol. Serol. 39: 229-242.

Gilardi, G.L. 1978. Identification of miscellaneous glucose nonfermenting Gram-negative bacteria. In Gilardi (Editor), Glucose Nonfermenting Gram-negative Bacteria in Clinical Microbiology. CRC Press Inc., West Palm Beach, pp. 45-65.

Gilardi, G.L., E. Bottone and M. Birnbaum. 1970. Unusual fermentative, gram-negative bacilli isolated from clinical specimens. I. Characterization of Erwinia strains of the "lathyri-herbicola group". Appl. Microbiol. 20: 151-155.

Gilder, H. and S. Granick. 1947. Studies on the Haemophilus group of organisms. Quantitative aspects of growth on various porphin compounds. J. Gen. Physiol. 31: 103-117.

Gilenberg, G. 1981. Heartwater disease. In Ristic and McIntyre (Editors), Disease of Cattle in the Tropics, Martinus Nijhoff Publishers, The Hague, pp. 345-360.

Gill, J.W. and K.W. King. 1958. Nutritional characteristics of a butyrivibrio. J. Bacteriol. 75: 666-673.

Gilleland, H.E. and R.G.E. Murray. 1976. Ultrastructure study of polymyxin-resistant isolates of Pseudomonas aeruginosa. J. Bacteriol. 125: 267-281.

Gilleland, H.E., J.D. Stinnett, I.L. Roth and R.G. Eagon. 1973. Freeze-etch study of Pseudomonas aeruginosa: localization within the cell wall of an ethylenediaminetetraacetate-extractable component. J. Bacteriol. 113: 417-432.

Gillespie, D. and S. Spiegelman. 1965. A quantitative assay for DNA-RNA hybrids with DNA immobilized on a membrane filter. J. Mol. Biol. 12: 829-842.

Gilliland, R.B. and J.P. Lacey. 1964. Lethal action by an Acetobacter on yeasts. Nature 202: 727-728.

Gilliland, R.B. and J.P. Lacey. 1966. On Acetobacter lethal to yeasts in bottled beer. J. Inst. Brew. 72: 291-303.

Gillis, M. and J. De Ley. 1980. Intra- and intergeneric similarities of the ribosomal ribonucleic acid cistrons of Acetobacter and Gluconobacter. Int. J. Syst. Bacteriol. 30: 7-27.

Gilman, R.H., M. Madasamy, E. Gan, M. Mariappan, C.E. Davis and K.A. Kyser. 1971. Edwardsiella tarda in jungle diarrhoea and a possible association with Entamoeba histolytica. Southeast Asian Jour. Trop. Med. Publ. Health 2: 186-189.

Gilmour, M.N., A.H. Howell, Jr. and B.G. Bibby. 1961. The classification of organisms termed Leptotrichia (Leptotrix) buccalis. I. Review of the literature and proposed separation into Leptotrichia buccalis Trevisan, 1879 and Bacterionema gen. nov. B. matruchotii (Mendel, 1919) comb. nov. Bacteriol. Rev. 25: 131-141.

Giménez, D.F. 1964. Staining rickettsiae in yolk-sac cultures. Stain Technol. *39:* 135-140.

Giménez, D.F. 1965. Gram staining of *Coxiella burnetii.* J. Bacteriol. *90:* 834-835.

Gingrich, W. and F. Schlenk. 1944. Codehydrogenase I and other pyridinium compounds as V factor for *Haemophilus influenzae* and *Haemophilus parainfluenzae.* J. Bacteriol. *47:* 535-550.

Ginoza, H.S. and T.A. Matney. 1963. Transmission of a resistance transfer factor from *Escherichia coli* to two species of *Pasteurella.* J. Bacteriol. *85:* 1177-1178.

Ginther, C.L. 1978. Genetic analysis of *Acinetobacter calcoaceticus* proline auxotrophs. J. Bacteriol. *133:* 439-441.

Girard, A.E. 1971. A comparative study of the fatty acids of some micrococci. Can. J. Microbiol. *17:* 1503-1508.

Girard, G. 1953. Méthodes permettant de différencier *P. pestis* de *P. pseudotuberculosis.* Bull. Wld. Hlth. Org. *9:* 645-653.

Girard, G. and J. Gallut. 1957. *Pasteurella novicida* sp. nov. ne constitue-t-elle pas une variété de *Pasteurella tularensis?* Ann. Inst. Pasteur (Paris) *92:* 544-547.

Giroud, P., N. Dumas and B. Hurpin. 1958. Essais d'adaptation a la souris blanche de la rickettsie agent de la maladie bleue de *Melolontha melolontha* L.: voie pulmonaire et voie buccale. C.R. Acad. Science, Series D *247:* 2499-2501.

Glasgow, H. 1914. The gastric caeca and the caecal bacteria of the Heteroptera. Biol. Bull. *26:* 101-170.

Glasgow, L.R. and R.L. Hill. 1980. Interaction of *Mycoplasma gallisepticum* with sialyl glycoproteins. Infect. Immun. *30:* 353-361.

Glauert, A.M., D. Kerridge and R.W. Horne. 1963. The fine structure and mode of attachment of the sheathed flagellum of *Vibrio metschnikovii.* J. Cell. Biol. *18:* 327-336.

Glew, R.H., R.C. Moellering and L.J. Kunz. 1977. Infections with *Acinetobacter calcoaceticus (Herellea vaginicola)*: Clinical and laboratory studies. Medicine *56:* 79-97.

Glick, T.H., M.B. Gregg, B. Berman, G. Mallison, W.W. Rhodes, Jr. and I. Kassanoff. 1978. Pontiac fever. An epidemic of unknown etiology in a health department: I. clinical and epidemiologic aspects. Am. J. Epidemiol. *107:* 149-160.

Glock, R.D., D.L. Harris and J.P. Kluge. 1974. Localization of spirochetes with the structural characteristics of *Treponema hyodysenteriae* in the lesions of swine dysentery. Infect. Immun. *9:* 167-178.

Glünder, G., K.-H. Hinz, H. Lüders and B. Stiburek. 1979. Zür therapeutischen Wirksamkeit von Tetracyclin·HCl und Sulfaquinoxalin/Trimethoprim bei der Puten-Bordetellose. Zentrabl. Veterinaermed. Reihe B *26:* 591-602.

Gochnauer, M.B. and D.J. Kushner. 1969. Growth and nutrition of extremely halophilic bacteria. Can. J. Microbiol. *15:* 1157-1165.

Gochnauer, M.B. and D.J. Kushner. 1971. Potassium binding, growth and survival of an extremely halophilic bacterium. Can. J. Microbiol. *17:* 17-23.

Godfrey, C.A. 1972. The carotenoid pigment and deoxyribonucleic base ratio of a *Rhizobium* which nodulated *Lotononis bainesii* Baker. J. Gen. Microbiol. *72:* 399-402.

Goedert, M. 1973. *Agrobacterium tumefaciens* (Smith et Town) Conn: Détermination des concentrations minimales inhibitrices de différents antibiotiqes et sulfamides. Ann. Microbiol. (Inst. Pasteur) *124A:* 237-241.

Gogotov, J.N. and H.G. Schlegel. 1974. N₂-fixation by chemoautotrophic hydrogen bacteria. Arch. Microbiol. *97:* 359-362.

Gois, M., F. Kuksa, J. Franz and D. Taylor-Robinson. 1974. The antigenic differentiation of seven strains of *Mycoplasma hyorhinis* by growth-inhibition, metabolism inhibition, latex-agglutination, and polyacrylamide-gel-electrophoresis tests. J. Med. Microbiol. *7:* 105-115.

Golberg, R.L. and J.A. Washington II. 1976. Comparison of isolation of *Haemophilus vaginalis (Corynebacterium vaginale)* from peptone-starch-dextrose agar and Columbia colistin-nalidixic acid agar. J. Clin. Microbiol. *4:* 245-247.

Goldberg, H.S., E.M. Barnes and A.B. Charles. 1964. Unusual *Bacteroides*-like organism. J. Bacteriol. *87:* 737-742.

Goldberg, J. 1959. Studies on granuloma inguinale. IV. Growth requirements of *Donovania granulomatis* and its relationship to the natural habitat of the organism. Brit. J. Vener. Dis. *35:* 266-268.

Goldberg, J. 1962. Studies on granuloma inguinale. V. Isolation of a bacterium resembling *Donovania granulomatis* from the feces of a patient with Granuloma inguinale. Brit. J. Vener. Dis. *38:* 99-102.

Goldstein, E.J.C., E.O. Agyare and R. Silletti. 1981. Comparative growth of *Eikenella corrodens* on fifteen media in three atmospheres of incubation. J. Clin. Microbiol. *13:* 951-953.

Goldstein, E.J.C., D.M. Citron, B. Wield, U. Blachman, V.L. Sutter, T.A. Miller and S.M. Finegold. 1978. Bacteriology of Human and Animal Bite Wounds. J. Clin. Microbiol. *8:* 667-672.

Goldstein, E.J.C., V.L. Sutter and S.M. Finegold. 1978. The susceptibility of *Eikenella corrodens* to 10 cephalosporins. Antimicrob. Agents Chemother. *14:* 404.

Goldstein, F.W., A. Boisivon, P. Leclerc and J.F. Acar. 1977. Sensibilité d'Hémophilus sp. aux antibiotiques. Transfert de résistance a *Escherichia coli.* Pathol. Biol. (Paris) *25:* 323-332.

Golikova, M.N. 1978. Intranuclear parasitic bacteria in micro and macronuclei of the ciliate *Paramecium bursaria.* Tsitologiya *20:* 576-580.

Golten, C. and W.A. Scheffers. 1975. Marine vibrios isolated from water along the Dutch Coast. Nether. J. Sea Res. *9:* 351-364.

Gomes, L. de S. 1944. Sobre una nova especie do genero *Eberthella* Buchanan, isolada de fezes patologicas de crianca. Rev. Inst. Lutz *4:* 183-195.

Gonçalves de Lima, O., J.M. De Araújo, I.E. Schumacher and E. Cavalcanti Da Silva. 1970. Estudos de microorganismos antagonistas presentes nas bebidas fermentadas usadas pelo povo do Recife. I. Sôbre uma variedade de *Zymomonas mobilis* (Lindner) (1928) Kluyver e van Niel (1936): *Zymomonas mobilis* var. *recifensis* (Gonçalves de Lima, Araújo, Schumacher and Cavalcanti) (1970), isolada de bebida popular denominada "caldo-de-cana picado". Rev. Inst. Antibtio. Univ. Recife *10:* 3-15.

Gonçalves de Lima, O., I.E. Schumacher and J.M. De Araújo. 1968. Novas observaçoes sôbre e açao antagonista de *Zymomonas mobilis* (Lindner) (1928), Kluyver e van Niel (1936). Rev. Inst. Antibiot. Univ. Recife *8:* 19-48.

Gonçalves de Lima, O., I.E. Schumacher and J.M. De Araújo. 1972. New observations about the antagonistic effects of *Zymomonas mobilis* var. *recifensis.* Rev. Inst. Antibiot. Univ. Recife *12:* 57-69.

Gonder, R. 1908. Spirochäten aus dem Darmtraktus von *Pinna: Spirochaete pinnae* nov. spec. und *Spirochaete hartmanni* nov. spec. Zentralbl. Bakteriol. Parasitenk. Infektionskr. Abt. I. Orig. *47:* 491-494.

Gonzalez, C., C. Buttierrez and C. Ramirez. 1978. *Halobacterium vallismortis* sp. nov. An amylolytic and carbohydrate-metabolizing, extremely halophilic bacterium. Can. J. Microbiol. *24:* 710-715.

González, C.F. and A.K. Vidaver. 1977. Syringomycin and holcus spot of maize: plasmid associated properties. Proc. Amer. Phytopathol. Soc. (4), 107.

Goodchild, A.J.P. 1955. The bacteria associated with *Triatoma infestans* and some other species of Reduviidae. Parasitology *45:* 441-448.

Goodchild, A.J.P. 1966. Evolution of the alimentary canal in the Hemiptera. Biol. Rev. Camb. Philos. Soc. *41:* 97-140.

Goodfellow, M. and R.G. Board (Editors). 1980. Microbiological classification and identification. Society for Applied Bacteriology Symposium Series No. 8. Academic Press, London.

Goodman, A.D. 1977. *Eikenella corrodens* isolated in oral infections of dental origin. Oral Surg. *44:* 128-134.

Goodnow, R.A. 1980. Biology of *Bordetella bronchiseptica.* Microbiol. Rev. *44:* 722-738.

Goodwin, R.F.W., A.P. Pomeroy and P. Whittlestone. 11965. Production of enzootic pneumonia in pigs with a mycoplasma. Vet. Rec. *77:* 1247-1249.

Gordon, J.K. and W.J. Brill. 1972. Mutants that produce nitrogenase in the presence of ammonia. Proc. Nat. Acad. Sci. U.S.A. *69:* 3501-3503.

Gordon, M., D.M. Donaldson and G.G. Wright. 1964. Immunization of mice with irradiated *Pasteurella tularensis.* J. Infect. Dis. *114:* 435-440.

Gordon, M.P. 1981. Tumor formation in plants. *In* Marcus (Editor), Proteins and Nucleic Acids, Vol. 6, Academic Press, New York, pp. 531-570.

Gordon, R.E. 1967. The taxonomy of soil bacteria. *In* Gray and Parkinson (Editors), The Ecology of Soil Bacteria. An International Symposium. University of Toronto Press, Toronto, pp. 293-321.

Gordon, R.E. and J.M. Mihm. 1957. A comparative study of some strains received as nocardiae. J. Bacteriol. *73:* 15-27.

Goren, E. 1979. *Mycoplasma synoviae* infectie bij pluimvee. Doctoral Thesis, Publicatie van de Stichting Gezondheidsdienst voor Pluimvee, Doorn.

Gorin, P.A.J., J.F.T. Spencer and D.W.S. Westlake. 1961. The structure and resistance to methylation of 1-2-β glucans from species of *Agrobacterium.* Can. J. Chem. *39:* 1067-1073.

Gorman, G.W., V.L. Yu, A. Brown, J.A. Hall, L.K. Corcoran, W.T. Martin, W.F. Bibb, G.K. Morris, M.H. Magnussen and D.W. Fraser. 1980. Isolation of Pittsburgh pneumonia agent from nebulizers used in respiratory therapy. Ann. Intern. Med. *93:* 572-573.

Görtz, H.-D. 1980. Nucleus-specific symbionts in *Paramecium caudatum. In* Schwemmler and Schenk (Editors), Endocytobiology: endosymbiosis and cell biology, Vol. 1, de Gruyter, Berlin, pp. 381-392.

Görtz, H.-D. and J. Dieckmann. 1980. Life cycle and infectivity of *Holospora elegans* Haffkine a micronucleus-specific symbiont of *Paramecium caudatum* (Ehrenberg). Protistologica *16:* 591-603.

Goslings, W.R.O. and E.P. Snijders. 1936. Undersuchungen über das Schleroma respiratorium (Sklerom). IV. Mitteilung. Die antigène Struktur der Skleromstämme im Vergleich mit den anderen Kapselbakterien. Zentralbl. Bakteriol. Parasitenk. Infektionskr. Hyg. Abt. I. Orig. *136:* 1-24.

Gosselé, F., J. Swings and J. De Ley. 1980. Growth factor requirements of *Gluconobacter.* Zentralbl. Bakteriol. Parasitenkd. Infektionskr. Hyg. Abt. I Orig. Ser. C *1:* 348-350.

Gosselé, F., J. Swings, K. Kersters and J. De Ley. 1983. Numerical analysis of phenotypic features and protein gel electropherograms of *Gluconobacter* Asai 1935 emend. mut. char. Asai, Iizuka, and Komagata 1964. Int. J. Syst. Bacteriol. *33:* 65–81.

Gossling, J. 1966. The bacteria isolated from lesions of embolic meningoencephalitis of cattle. Illinois Vet. *9:* 14-18.

Gossling, J. and W.E.C. Moore. 1975. *Gemmiger formicilis,* n. gen., n. sp., an anaerobic budding bacterium from intestines. Int. J. Syst. Bacteriol. *25:* 202-207.

Gothe, R. 1967a. Ein Beiträg zür systematischen Stellung von *Aegyptianella pullorum* Carpano 1928. Z. Parasitenkd. *29:* 119-129.

Gothe, R. 1967b. Untersuchungen über die Entwicklung und den Infektionsverlauf von *Aegyptianella pullorum* Carpano 1928, im Huhn. Z. Parasitenkd. *29:* 149-158.

Gothe, R. 1967c. Zür Entwicklung von *Aegyptianella pullorum* Carpano 1928, in der Lederzecke *Argas (Persicargas) persicus* (Oken, 1818) und Übertragung. Z. Parasitenkd. *29:* 103-118.

Gothe, R. 1971. Wirt-Parasit-Verhaltnis von *Aegyptianella pullorum* Carpano 1928, im biologischen Übertrager *Argas (Persicargas) persicus* (Oken, 1818) und im Wirbeltierwirt *Gallus gallus domesticus* L. Fortschr. Veterinaermed. Heft 16.

Gothe, R. 1978. New aspects of the epizootiology of aegyptianellosis in poultry. *In* Wilde (Editor), Tick-borne Diseases and their Vectors, Centre for Tropical Veterinary Medicine, University of Edinburgh, Lewis Reprints Ltd., Tonbridge, pp. 201-204.

Gothe, R. and E. Burkhardt. 1979. The erythrocytic entry- and exit-mechanism of *Aegyptianella pullorum* Carpano 1928. Z. Parasitenkd. *60:* 221-227.

Gothe, R. and S. Hartmann. 1979. The viability of cryopreserved *Aegyptianella pullorum* Carpano 1928 in the vector *Argas (Persicargas) walkerae* Kaiser and Hoogstraal 1969. Z. Parasitenkd. *58:* 189-190.

Gothe, R. and H. Mieth. 1979. Zür Wirksamkeit von Pleuromutilinen bei *Aegyptianella pullorum*-Infektionen der Kuken. Tropenmed. Parasitol. *30:* 323-327.

Goto, E., T. Kodama and Y. Minoda. 1977. Isolation and culture conditions of thermophilic hydrogen bacteria. Agr. Biol. Chem. (Japan) *41:* 685-690.

Goto, E., T. Kodama and Y. Minoda. 1978. Growth and taxonomy of thermophilic hydrogen bacteria. Agr. Biol. Chem. (Japan) *42:* 1305-1308.

Goto, M. 1976. *Erwinia mallotivora* sp. nov., the causal organism of bacterial leaf spot of *Mallotus japonicus* Muell. Arg. Int. J. Syst. Bacteriol. *26:* 467-473.

Goto, M. and T. Makino. 1977. Emendation of *Pseudomonas cissicola*, the causal organism of bacterial leaf spot of *Cayratia japonica* (Thunb.) Gagn. and designation of the neotype strain. Ann. Phytopathol. Soc. Japan *43:* 40-45.

Goto, M. and N. Okabe. 1958. Bacterial plant diseases in Japan IX. 1. Bacterial stem rot of pea. 2. Halo blight of bean. 3. Bacterial spot of physalis plant. Rept. Fac. Agr. Shizuoka Univ. *8:* 33-49.

Goto, M. and M.P. Starr. 1971. A comparative study of *Pseudomonas andropogonis, P. stizolobii* and *P. alboprecipitans*. Ann. Phytopathol. Soc. Japan *37:* 233-241.

Goto, M., T. Takahasi and T. Okajima. 1980. A comparative study of *Erwinia milletiae* and *Erwinia herbicola*. Ann. Phytopathol. Soc. Japan *46:* 185-192.

Goto, M., A. Toyoshima and S. Tanaka. 1978. Studies on saprophytic survival of *Xanthomonas citri* (Hasse) Dowson. 3. Inoculum density of the bacterium surviving in the saprophytic form. Ann. Phytopathol. Soc. Japan *44:* 197-201.

Gottschalk, G. 1965. Die Verwertung organischer Substrate durch *Hydrogenomonas* in Gegenwart von molekularem Wasserstoff. Biochem. Z. *341:* 260-270.

Gottschalk, G., U. Eberhardt and H.G. Schlegel. 1964. Verwertung von Fructose durch *Hydrogenomonas* H 16 (I.). Arch. Mikrobiol. *48:* 95-108.

Götz, P. 1972. *Rickettsiella chironomi*: an unusual bacterial pathogen which reproduces by multiple cell division. J. Invert. Pathol. *20:* 22-30.

Gould, D.J. and M.L. Miesse. 1954. Recovery of a rickettsia of the spotted fever group from *Microtus pennsylvanicus* from Virginia. Proc. Soc. Exp. Biol. Med. *85:* 558-561.

Gourlay, R.N., J. Brownlie and C.J. Howard. 1973. Isolation of T-mycoplasmas from goats, and the production of subclinical mastitis in goats by the intramammary inoculation of human T-mycoplasmas. J. Gen. Microbiol. *76:* 251-254.

Gourlay, R.N. and R.H. Leach. 1970. A new *Mycoplasma* species isolated from pneumonic lungs of calves (*Mycoplasma dispar* sp. nov.). J. Med. Microbiol. *3:* 111-123.

Gourlay, R.N., R.H. Leach and C.J. Howard. 1974. *Mycoplasma verecundum*, a new species isolated from bovine eyes. J. Gen. Microbiol. *81:* 475-484.

Gourlay, R.N. and L.H. Thomas. 1969. The isolation of large colony and T-strain mycoplasmas from cases of bovine kerato-conjunctivitis. Vet. Res. *84:* 416-417.

Gourlay, R.N. and K.J. Thrower. 1968. Morphology of *Mycoplasma mycoides* thread-phase growth. J. Gen. Microbiol. *54:* 155-159.

Gourlay, R.N., S.G. Wyld and R.H. Leach. 1977. *Mycoplasma alvi*, a new species from bovine intestinal and urogenital tracts. Int. J. Syst. Bacteriol. *27:* 86-96.

Gourlay, R.N., S.G. Wyld and R.H. Leach. 1978. *Mycoplasma sualvi*, a new species from the intestinal and urogenital tracts of pigs. Int. J. Syst. Bacteriol. *28:* 289-292.

Govan, J.R.W. 1978. Pyocin typing in *Pseudomonas aeruginosa*. *In* Bergan and Norris (Editors), Methods in Microbiology, Vol. 10, Academic Press, London, pp. 61-91.

Govan, J.R.W. and J.A.M. Fyfe. 1978. Mucoid *Pseudomonas aeruginosa* and cystic fibrosis: resistance of the mucoid form to carbenicillin, flucoxacillin and tobramycin and the isolation of mucoid variants *in vitro*. J. Antimicr. Chemother. *4:* 233-240.

Grace, J.T., J.S. Horoszewicz, T.B. Stim, E.A. Mirand and C. James. 1965. Mycoplasmas (PPLO) and human leukemia and lymphoma. Cancer *18:* 1369-1376.

Gradin, J.L. and J.A. Schmitz. 1977. Selective medium for isolation of *Bacteroides nodosus*. J. Clin. Microbiol. *6:* 298-302.

Graevenitz, A. and A.H. Mensch. 1968. The genus *Aeromonas* in human bacteriology. Report of 30 cases and review of the literature. N. Engl. J. Med. *278:* 245-249.

Graham, D.C. 1964. Taxonomy of the soft rot coliform bacteria. Annu. Rev. Phytopathol. *2:* 13-42.

Graham, D.C. 1972. Identification of soft rot coliform bacteria. *In* H. P. Maas Geesteranus (Editor), Proc. 3rd Int. Conf. on Plant Pathogenic Bacteria, Centre for Agricultural Publishing and Documentation (Pudoc), Wageningen, pp. 273-279.

Graham, D.C. and W.J. Dowson. 1960. The coliform bacteria associated with potato blackleg and other soft rots. I. Their pathogenicity in relation to temperature. Ann. Appl. Biol. *48:* 51-57.

Graham, D.C. and W. Hodgkiss. 1967. Identity of Gram negative, yellow pigmented, fermentative bacteria isolated from plants and animals. J. Appl. Bacteriol. *30:* 175-189.

Graham, P.H. 1963. Antigenic affinities of the root-nodule bacteria of legumes. Antonie van Leeuwenhoek J. Microbiol. Serol. *29:* 281-291.

Graham, P.H. 1964a. The application of computer techniques to the taxonomy of the root nodule bacteria of legumes. J. Gen. Microbiol. *35:* 511-517.

Graham, P.H. 1964b. Studies on the utilization of carbohydrates and Krebs cycle intermediates by rhizobia, using an agar plate method. Antonie van Leeuwenhoek J. Microbiol. Serol. *30:* 68-72.

Graham, P.H. 1971. Serological studies with *Agrobacterium radiobacter, A. tumefaciens*, and *Rhizobium* strains. Arch. Mikrobiol. *78:* 70-75.

Graham, P.H. and C.A. Parker. 1964. Diagnostic features in the characterization of the root-nodule bacteria of legumes. Plant Soil *20:* 383-396.

Graham-Smith, G.S. 1905. A new form of parasite found in the red blood corpuscles of moles. J. Hyg. *5:* 453-459.

Granados, R.R. 1969. Electron microscopy of plants and insect vectors infected with corn stunt disease agent. Contrib. Boyce Thompson Inst. *24:* 1773-187.

Grant, R.B., J.L. Penner, J.N. Hennessy and B.J. Jackowski. 1981. Transferable urease activity in *Providencia stuartii*. J. Clin. Microbiol. *13:* 561-565.

Grassberger, R. 1897. Beiträge zür Bakteriologie der Influenza. Z. Hyg. Infektionskr. *25:* 453-475.

Grassi, B. and A. Sandias. 1896-1897. The constitution and development of the society of termites: observations on their habits; with appendices on the parasitic protozoa of Termitidae, and on the Embiidae. Quart. J. Microsc. Sci. *39:* 245-322, *40:* 1-75.

Gray, M.W. and W.F. Doolittle. 1982. Has the endosymbiont hypothesis been proven? Microbiol. Rev. *46:* 1-42.

Gray, P.H.H. 1928. The formation of indigotin from indole by soil bacteria. Proc. Roy. Soc., B, *102:* 263-280.

Gray, P.H.H. and H.G. Thornton. 1928. Soil bacteria that decompose certain aromatic compounds. Zentralbl. Bakteriol. Parasitenk. Infektionskr. Hyg. Abt. II, *73:* 74-96.

Grebe, H.H. and K.-H. Hinz. 1975. Vorkommen von Bakterien der Gattung *Haemophilus* bei verschiedenen Vogelarten. Zentralbl. Veterinaermed. Riehe. B *22:* 749-757.

Green, M.J., D.M. Anderson, D.M. Norris and S.L. Gubbins. 1979. Antimicrobial resistance in *Haemophilus* species. N.Z. Med. J. *90:* 29.

Greenberg, E.P. and E. Canale-Parola. 1975. Carotenoid pigments of facultatively anaerobic spirochetes. J. Bacteriol. *123:* 1006-1012.

Greenberg, E.P. and E. Canale-Parola. 1976. *Spirochaeta halophila* sp. n., a facultative anaerobe from a high-salinity pond. Arch. Microbiol. *110:* 185-194.

Greenberg, E.P. and E. Canale-Parola. 1977a. Chemotaxis in *Spirochaeta aurantia*. J. Bacteriol. *130:* 485-494.

Greenberg, E.P. and E. Canale-Parola. 1977b. Relationship between cell coiling and motility of spirochetes in viscous environments. J. Bacteriol. *131:* 960-969.

Greenberg, E.P. and E. Canale-Parola. 1977c. Motility of flagellated bacteria in viscous environments. J. Bacteriol. *132:* 356-358.

Greenberg, E.P. and E. Canale-Parola. 1977d. *Spirochaeta halophila*, new species. *In* Announcement of the valid publication of new names and new combinations previously effectively published outside the IJSB. List No. 1. Int. J. Syst. Bacteriol. *27:* 306.

Greenfield, S. and G.W. Claus. 1972. Nonfunctional tricarboxylic acid cycle and the mechanism of glutamate biosynthesis in *Acetobacter suboxydans*. J. Bacteriol. *112:* 1295-1301.

Greenwood, J.R. and M.J. Pickett. 1979. Salient features of *Haemophilus vaginalis*. J. Clin. Microbiol. *9:* 200-204.

Greenwood, J.R. and M.J. Pickett. 1980. Transfer of *Haemophilus vaginalis* Gardner and Dukes to a new genus, *Gardnerella*: G. *vaginalis* (Gardner and Dukes) comb. nov. Int. J. Syst. Bacteriol. *30:* 170-178.

Greenwood, J.R., M.J. Pickett, W.J. Martin and E.G. Mack. 1977. *Haemophilus vaginalis* (*Corynebacterium vaginale*): method for isolation and rapid biochemical identification. Health Lab. Sci. *14:* 102-106.

Greer, P.W., F.W. Chandler and M.D. Hicklin. 1980. Rapid demonstration of *Legionella pneumophila* in unembedded tissue: an adaptation of the Giménez stain. Am. J. Clin. Pathol. *73:* 788-790.

Gregory, E.M. and C.H. Dapper. 1980. Chemical and physical differentiation of superoxide dismutases in anaerobes. J. Bacteriol. *144:* 967-974.

Gregory, E.M., W.E.C. Moore and L.V. Holdeman. 1978. Superoxide dismutase in anaerobes: survey. Appl. Environ. Microbiol. *35:* 988-991.

Grehn, M. 1976. Über die Toxizität einiger Spezies sog. nicht fermentativer gramnegativer Bakterien. Zentralbl. Bakteriol. Parasitenkd. Infektionskr. Hyg. Abt. I Orig. A *235:* 84-89.

Gress, F.M., R.L. Myerowitz, A.W. Pasculle, C.R. Rinaldo, Jr. and J.N. Dowling. 1980. The ultrastructural morphologic features of Pittsburgh pneumonia agent. Am. J. Pathol. *101:* 63-69.

Gresshoff, P.M., M.L. Skotnicki and B.G. Rolfe. 1979. Crown gall teratoma formation is plasmid and plant controlled. J. Bacteriol. *137:* 1020-1021.

Gretillat, S. and B. Konarzewski. 1939. Presense d'un procarote du genre *Haemobartosiphon* Tyzzer et Weinman dans le sang de Nigeriens de la region de Niamey. Bull. Soc. Pathol. Exot. *71:* 412-416.

Gribble, D.H. 1969. Equine ehrlichiosis. J. Am. Vet. Med. Assoc. 155: 462-469.

Griffin, P.J., S.F. Snieszko and S.B. Friddle. 1953. A more comprehensive description of *Bacterium salmonicida*. Trans. Amer. Fish. Soc. *82:* 129-138.

Griffith, D.P., Musher, D.M. and J.W. Campbell. 1973. Inhibition of bacterial urease. Invest. Urol. *11:* 234-238.

Griffiths, G.W. and S.D. Beck. 1973. Intracellular symbiotes of the pea aphid, *Acyrthosiphon pisum*. J. Insect. Physiol. *19:* 75-84.

Griffiths, G.W. and S.D. Beck. 1974. Effects of antibiotics on intracellular symbiotes in the pea aphid, *Acyrthosiphon pisum*. Cell. Tissue Res. *148:* 287-300.

Griffiths, G.W. and S.D. Beck. 1975. Ultrastructure of pea aphid mycetocytes: evidence for symbiote secretion. Cell Tissue Res. *159:* 351-367.

Griffiths, G.W. and S.D. Beck. 1977a. In vivo sterol biosynthesis by pea aphid symbiotes as determined by digitonin and electron microscopic autoradiography. Cell Tissue Res. *176:* 179-190.

Griffiths, G.W. and S.D. Beck. 1977b. Effect of dietary cholesterol on the pattern of osmium deposition in the symbiote-containing cells of the pea aphid. Cell Tissue Res. *176:* 191-203.

Grimes, M. and A.J. Hennerty. 1931. A study of bacteria belonging to the subgenus *Aerobacter*. Sci. Proc. R. Dublin Soc. *20:* 89-97.

Grimont, P.A.D. 1977. Le genre *Serratia*. Taxonomie et approche ecologique. Thesis, University of Bordeaux I, 377 pp.

Grimont, P.A.D. and H.L.C. Dulong de Rosnay. 1972. Numerical study of 60 strains of *Serratia*. J. Gen. Microbiol. *72:* 259-268.

Grimont, P.A.D. and F. Grimont. 1978a. The genus *Serratia*. Annu. Rev. Microbiol. *32:* 221-248.

Grimont, P.A.D. and F. Grimont. 1978b. Biotyping of *Serratia marcescens* and its use in epidemiological studies. J. Clin. Microbiol. *8:* 73-83.

Grimont, P.A.D. and F. Grimont. 1981. The genus *Serratia*. *In* Starr, Stolp, Trüper, Balows and Schlegel (Editors), The Prokaryotes. A handbook on habitats, isolation, and identification of bacteria. Springer-Verlag, New York, pp. 1187-1203.

Grimont, P.A.D., F. Grimont and H.L.C. Dulong de Rosnay. 1977a. Characterization of *Serratia marcescens*, *S. liquefaciens*, *S. plymuthica*, and *S. marinorubra* by electrophoresis of their proteinases. J. Gen. Microbiol. *99:* 301-310.

Grimont, P.A.D., F. Grimont, H.L.C. Dulong de Rosnay and P.H.A. Sneath. 1977b. Taxonomy of the genus *Serratia*. J. Gen. Microbiol. *98:* 39-66.

Grimont, P.A.D., F. Grimont, J.J. Farmer III and M.A. Asbury. 1981a. *Cedecea davisae* gen. nov., sp. nov. and *Cedecea lapagei* sp. nov., new *Enterobacteriaceae* from clinical specimens. Int. J. Syst. Bacteriol. *31:* 317-326.

Grimont, P.A.D., F. Grimont and K. Irino. 1982b. Biochemical characterization of *Serratia liquefaciens sensu stricto*, *Serratia proteamaculans*, and *Serratia grimesii* sp. nov. Curr. Microbiol. *7:* 69-74.

Grimont, P.A.D., F. Grimont, S. Le Minor, B. Davis and F. Pigache. 1979a. Compatible results obtained from biotyping and serotyping in *Serratia marcescens*. J. Clin. Microbiol. *10:* 425-432.

Grimont, P.A.D., F. Grimont, C. Richard, B.R. Davis, A.G. Steigerwalt and D.J. Brenner. 1978a. Deoxyribonucleic acid relatedness between *Serratia plymuthica* and other *Serratia* species with a description of *Serratia odorifera* sp. nov. (holotype: ICPB 3995). Int. J. Syst. Bacteriol. *28:* 453-463.

Grimont, P.A.D., F. Grimont, C. Richard and R. Sakazaki. 1980. *Edwardsiella hoshinae*, a new species of *Enterobacteriaceae*. Curr. Microbiol. *4:* 347-351.

Grimont, P.A.D., F. Grimont, C. Richard and R. Sakazaki. 1981. *In* List No. 6, Validation of the publication of new names and new combinations previously effectively published outside the IJSB. Int. J. Syst. Bacteriol. *31:* 215-218.

Grimont, P.A.D., F. Grimont and M.P. Starr. 1978b. *Serratia proteamaculans* (Paine and Stanfield) comb. nov., a senior subjective synonym of *Serratia liquefaciens* (Grimes and Hennerty) Bascomb et al. Int. J. Syst. Bacteriol. *28:* 503-510.

Grimont, P.A.D., F. Grimont and M.P. Starr. 1979b. *Serratia ficaria* sp. nov., a bacterial species associated with Smyrna figs and the fig wasp *Blastophaga psenes*. Curr. Microbiol. *2:* 277-282.

Grimont, P.A.D., F. Grimont and M.P. Starr. 1981b. Comment on the request to the Judicial Commission to conserve the specific epithet *liquefaciens* over the specific epithet *proteamaculans* in the name of the organisms currently known as *Serratia liquefaciens*. Int. J. Syst. Bacteriol. *31:* 211-212.

Grimont, P.A.D., F. Grimont and M.P. Starr. 1981c. *In* Validation of the publication of new names and new combinations previously effectively published outside the IJSB. List No. 6. Int. J. Syst. Bacteriol. *31:* 215-218.

Grimont, P.A.D., K. Irino and F. Grimont. 1982a. The *Serratia liquefaciens-S. proteamaculans-S. grimesii* complex: DNA relatedness. Curr. Microbiol. *7:* 63-68.

Grimont, P.A.D., M.Y. Popoff, F. Grimont, C. Coynault and M. Lemelin. 1982. Reproducibility and correlation study of three deoxyribonucleic acid hybridization procedures. Curr. Microbiol. *4:* 325-330.

Grimstone, A.V. 1963. A note on the fine structure of a spirochaete. Quart. J. Microsc. Sci. *104:* 145-153.

Grinyer, I. and A.J. Musgrave. 1964. Microorganisms and mitochondria in the Malpighian tubules of *Sitophilus* (Coleoptera). Can. J. Microbiol. *10:* 805-806.

Grinyer, I. and A.J. Musgrave. 1966. Ultrastructure and peripheral membranes of the mycetomal microorganisms of *Sitophilus granarius* (L.). (Coleoptera). J. Cell Sci. *1:* 181-186.

Grob, P.R. 1981. Prophylactic erythromycin for whooping-cough contacts. Lancet *1:* 772.

Gromet, Z., M. Schramm and S. Hestrin. 1957. Synthesis of cellulose by *Acetobacter xylinum* 4. Enzyme systems present in a crude extract of glucose-grown cells. Biochem. J. *67:* 679-689.

Gromov, B.V. 1963. A new bacterium of the genus *Microcyclus*. Dokl. Akad. Nauk. SSSR *152:* 733-734.

Gromov, G. V. and K. A. Mamkaeva. 1972. Electron microscope examination of *Bdellovibrio chlorellavorus* parasitism on cells of the green alga *Chlorella vulgaris*. Tsitologiya 14: 256-260.

Gromov, B. V. and K. A. Mamkaeva. 1980. Proposal of a new genus *Vampirovibrio* for *chorellavorus* bacteria previously assigned to *Bdellovibrio*. Mikrobiologia *49:* 165-167.

Gromov, B.V. and K.A. Mamkaeva. 1980. *Blattabacterium* in the fatbody of the Maritime Territory relic roach, *Cryptocercus relictus*. Mikrobiologiya *49:* 1005-1007.

Gromov, B.V. and D.V. Ossipov. 1981. *Holospora* (ex Hafkine 1890) nom. rev., a genus of bacteria inhabiting the nuclei of paramecia. Int. J. Syst. Bacteriol. *31:* 348-352.

Groot, H. 1942. *Haemobartonella tyzzeri* in Columbia. Proc. Soc. Exp. Biol. Med. *51:* 279.

Gross, D.C. and J.E. De Vay. 1977. Production and purification of syringomycin, a phytotoxin produced by *Pseudomonas syringae*. Physiol. Plant Pathol. *11:* 13-28.

Gross, J. 1910. *Cristispira* nov. gen. Ein Beiträg zür Spirochätenfrage. Mitt. Zool. Sta. Neapel 20: 41-93.

Gross, J. 1912. Zür Nomenklatur der *Spirochaeta pallida*. Schaud. u. Hoffm. Arch. Protistenk 24: 109-118.

Gross, J. 1912. Ueber systematik struktur und fortpflanzung der *Spironemaceae*. Zentralbl. Bakteriol. Parasitenk. Infectionskr. Hyg. Abt. I Orig. *65:* 83-98.

Gross, R.J., T. Cheasty and B. Rowe. 1977. Isolation of bacteriophages specific for the K1 polysaccharide antigen of *Escherichia coli*. J. Clin. Microbiol. *6:* 548-550.

Gross, R.J. and B. Rowe. 1974. The serology of *Citrobacter koseri*, *Levinea malonatica* and *Levinea amalonatica*. J. Med. Microbiol. *7:* 155-162.

Gross, R.J. and B. Rowe. 1975. *Citrobacter koseri*. I. An extended antigenic scheme for *Citrobacter koseri* (syn. *C. diversus*, *Levinea malonatica*). J. Hyg. *75:* 121-127.

Gross, R.J., B. Rowe, T. Cheasty and L.V. Thomas. 1981. Increase in drug resistance among *Shigella dysenteriae*, *Sh. flexneri* and *Sh. boydii*. Brit. Med. J. *283:* 575.

Gross, R.J., B. Rowe and J.A. Easton. 1973. Neonatal meningitis caused by *Citrobacter koseri*. J. Clin. Pathol. 26: 138-139.

Gross, R.J., B. Rowe, I. Sechter, D. Cahan and G. Altman. 1981. Antigenic scheme for *Citrobacter koseri* (syn. *C. diversus*, *Levinea malonatica*); three new antigens recognized in strains from Israel. J. Hyg. *86:* 111-115.

Grossberg, S.E., R.G. Petersdorf, J.A. Curtin and I.L. Bennett. 1962. Factors influencing the species and antimicrobial resistance of urinary pathogens. Am. J. Med. *32:* 44-55.

Grossman, J.P. and J.R. Postgate. 1953. Cultivation of sulphate-reducing bacteria. Nature (London) *171:* 600-674.

Groves, M.G., D.L. Rosenstreich, B.A. Taylor and J.V. Osterman. 1980. Host defenses in experimental scrub typhus: mapping the gene that controls natural resistance in mice. J. Immunol. *125:* 1395-1399.

Gruber, T. 1905. Ein weiterer Beiträg zür Aromabildung speziell zür Bildung des Erdbeergeruches in der Gruppe "*Pseudomonas*". *Pseudomonas fragariae* II. Zentralbl. Bakteriol. Parasitenk. Infektionskr. Hyg. Abt. II, *14:* 122-123.

Guelin, A. M., I. E. Michustina, L. V. Andreev, M. A. Bobyk and V. A. Lambina. 1978. Some problems of the ecology and taxonomy of marine microvibrios. Biol. Bull. Acad. Sci. USSR. *5:* 336-340.

Guelin, A., I. E. Michustina, S. A. Goulevskaya, N. V. Petchnikov and L. A. Ledoya. 1977. Étude sur les microvibrions marins de Roscoff (*Microvibrio marinus roscoffensis*). C. R. Acad. Sci. (Paris) *284:* 2171-2174.

Guerinot, M.L. and D.G. Patriquin. 1981. N_2-fixing vibrios isolated from the gastrointestinal tract of sea urchins. Can. J. Microbiol. *27:* 311-317.

Guerinot, M.L., P.A. West, J.V. Lee and R.R. Colwell. 1982. *Vibrio diazotrophicus* sp. nov., a marine nitrogen-fixing bacterium. Int. J. Syst. Bacteriol. *32:* 350-357.

Guerrant, G.O., M.S. Lambert and C.W. Moss. 1979. Identification of diaminopimelic acid in the Legionnaires' bacterium. J. Clin. Microbiol. *10:* 815-818.

Guerrant, R.L., M.D. Dickens, R.P. Wenzel and A.Z. Kapikian. 1976. Toxigenic bacterial diarrhea: Nursery outbreak involving multiple bacterial strains. J. Pediat. 89: 885-891.

Guerry, P. and R.R. Colwell. 1977. Isolation of cryptic plasmid deoxyribonucleic acid from Kanagawa-positive strains of *Vibrio parahaemolyticus*. Infect. Immun. *16:* 328-334.

Guignard, L. and C. Sauvageau. 1894. Sur un nouveau microbe chromogène, le

Bacillus chlororaphis. C. R. Soc. Biol. Paris. Sér. 10, *1:* 841-843.

Guillebeau, A. 1890. Studien über Milchfehler un Enterentzundungen bei Rindern und Ziegen. I. Über Ursachen der Enterentzundung. Landwirt. Jahrb. Schweiz. *4:* 27-44.

Guillermet, F.N. 1981. *Fusobacterium fusiforme* et *nucleatum.* Rev. Inst. Pasteur (Lyon) *14:* 87-93.

Guinée, P.A.M., W.H. Jansen, H.M.E. Maas, L. Le Minor and R. Beaud. 1981. An unusual H antigen (z_{66}) in strains of *S. typhi.* Ann. Microbiol. (Inst. Pasteur) *132A:* 331-334.

Guinée, P.A.M. and J.J. Valkenburg. 1968. Diagnostic value of a *Hafnia* specific bacteriophage. J. Bacteriol. *96:* 564.

Guittonneau, G., G. Mocquot and J. Tavernier. 1939. La cause microbiologique de la maladie des cidres dit framboisés: production d'éthanol par actions conjuguées de levures alcooliques et de bactéries acétiques. C.R. Acad. Sci. *209:* 809-811.

Gumbrell, R.C. and J.M.B. Smith. 1974. Deoxyribonucleic acid base composition of ovine actinobacilli. J. Gen. Microbiol. *84:* 399-402.

Gumpert, J. 1962. Untersuchungen über die Symbiose von Tieren mit Pilzen und Bakterien. X. Die Symbiose der Triatominen. 2. Infektion symbiontenfreier Triatominen mit symbiontischen und saprophytischen Mikroorganismen und gemeinsame Eigenschaften der symbiontischen Stämme. Z. Allg. Mikrobiol. *2:* 290-302.

Gumpert, J. and W. Schwartz. 1962. Untersuchungen über die Symbiose von Tieren mit Pilzen und Bakterien. X. Die Symbiose der Triatominen. 1. Aufzucht symbiontenhaltiger und symbiontenfreier Triatominen und Eigenschaften der bei Triatominen vorkommenden Mikroorganismen. Z. Allg. Mikrobiol. *2:* 209-225.

Gumpert, J. and W. Schwartz. 1963. Untersuchungen über die Symbiose von Tieren mit Pilzen und Bakterien. X. Die Symbiose der Triatominen. 3. Pantothensaurelieferung als Funktion der Symbionten. Z. Allg. Mikrobiol. *3:* 1-14.

Gunnarsson, A., E.L. Biberstein and B. Hurvell. 1977. Serologic studies on porcine strains of *Haemophilus parahaemolyticus (pleuropneumoniae):* agglutination reactions. Am. J. Vet. Res. *38:* 1111-1114.

Gunnarsson, A., B. Hurvell and E.L. Biberstein. 1978. Serologic studies of porcine strains of *Haemophilus parahaemolyticus (pleuropneumoniae):* antigenic specificity and relationship between serotypes. Am. J. Vet. Res. *39:* 1286-1292.

Gunnison, J.B., A. Larson and A.S. Lazarus. 1951. Rapid differentiation between *Pasteurella pestis* and *Pasteurella pseudotuberculosis* by action of bacteriophage. J. Infect. Dis. *88:* 251-255.

Gunsalus, I.C., C.A. Tyson, R.L. Tsai and J.D. Lipscomb. 1971. P-450$_{cam}$ hydroxylase: substrate-effector and electron-transport reactions. Chem. Biol. Interact. *4:* 75-78.

Guo, M.M.S. and P.V. Liu. 1965. Serological specificities of ureases of *Proteus* species. J. Gen. Microbiol. *38:* 417-422.

Gupta, R. and C.R. Woese. 1980. Unusual modification patterns in the transfer ribonucleic acids of Archaebacteria. Curr. Microbiol. *4:* 245-249.

Gustafson, R.L. and A.V. Kroeger. 1962. Antigenic characteristics of *Leptotrichia buccalis (Fusobacterium fusiforme).* J. Bacteriol. *84:* 1313-1320.

Gustafson, R.L., A.V. Kroeger, J.L. Gustafson and E.M.K. Vaichulis. 1966. The biological activity of *Leptotrichia buccalis* endotoxin. Arch. Oral Biol. *11:* 1149-1162.

Guthrie, J.W. 1968. The serological relationship of races of *Pseudomonas phaseolicola.* Phytopathology *58:* 716-717.

Gutierrez, J., R.E. Davis, I.H. Lindahl and E.J. Warwick. 1959. Bacterial changes in the rumen during the onset of feed-lot bloat of cattle and characteristics of *Peptostreptococcus elsdenii* n. sp. Appl. Microbiol. *7:* 16-22.

Guyon, P., M.-D. Chilton, A. Petit and J. Tempé. 1980. Agropine in "null-type" crown gall tumors: Evidence for generality of the opine concept. Proc. Natl. Acad. Sci. USA *77:* 2693-2697.

Gwynn, C.M. and R.H. George. 1973. Neonatal citrobacter meningitis. Arch. Dis. Childh. *48:* 455-458.

Gyles, C., M. So and S. Falkow. 1974. The enterotoxin plasmids of *Escherichia coli.* J. Infect. Dis. *130:* 40-49.

Haas, D. and B.W. Holloway. 1976. R factor variants with enhanced sex factor activity in *Pseudomonas aeruginosa.* Mol. Gen. Genet. *144:* 243-251.

Haas, D. and B.W. Holloway. 1978. Chromosome mobilization by the R plasmid R68.45: a tool in *Pseudomonas* genetics. Mol. Gen. Genet. *158:* 229-237.

Habs, I. 1957. Untersuchungen über die O-antigene von *Pseudomonas aeruginosa.* Z. Hyg. Infektionskr. *144:* 218-228.

Habs, H. and R.H.W. Schubert. 1962. Über die biochemischen Merkmale und die taxonomische Stellung von *Pseudomonas shigelloides* (Bader). Zentralbl. Bakteriol. Parasitenkd. Infektionskr. Hyg. Abt. I Orig. *186:* 316-327.

Hacking, M.A. and L. Sileo. 1977. Isolation of a hemolytic *Actinobacillus* from waterfowl. J. Wildl. Dis. *13:* 69-73.

Hackman, A.S. and T.D. Wilkins. 1975. In vivo protection of *Fusobacterium necrophorum* from penicillin by *Bacteroides fragilis.* Antimicrob. Agents Chemother. *7:* 698-703.

Hackstadt, T. and J.C. Williams. 1981. Biochemical stratagem for obligate parasitism of eukaryotic cells by *Coxiella burnetii.* Proc. Natl. Acad. Sci. U.S.A. *78:* 3240-3244.

Hackstadt, T. and J.C. Williams. 1981a. Incorporation of macromolecular precursors by *Coxiella burnetii* in an axenic medium. *In* Burgdorfer and Anacker

(Editors), Rickettsiae and Rickettsial Diseases, Academic Press, New York, pp. 431-440.

Hadani, A. and Y. Dinur. 1968. Studies on the transmission of *Aegyptianella pullorum* by the tick *Argas persicus.* J. Protozool. *15,* Suppl., p. 45, No. 186.

Hadi, A.W. and C. Russell. 1969. Fusiforms in gingival material. Quantitative estimations from normal individuals and cases of periodontal disease. Brit. Dent. J. *126:* 82-84.

Hadley, P., M.W. Elkins and D.N. Caldwell. 1918. The colon-typhoid intermediates as causative agents of disease in birds. 1. The paratyphoid bacteria. Bull. R. I. Agr. Exp. Sta. *174:* 1-216.

Hafkine, M.W. 1890. Maladies infectieuses des Paramécies. Ann. Inst. Pasteur *4:* 148-162.

Hagborg, W.A.F. 1942. Classification revision in *Xanthomonas translucens.* Can. J. Res. *20:* 312-326.

Hagen, K.S. 1966. Dependence of the olive fly, *Dacus oleae,* larvae on symbiosis with *Pseudomonas savastanoi* for the utilization of olive. Nature (Lond.) *209:* 423-424.

Hagen, K.S. and R.L. Tassan. 1972. Exploring nutritional roles of extracellular symbiotes on the reproduction of honeydew feeding Chrysopids and Tephritids. *In* Rodriguez (Editor), Insect and Mite Nutrition. North-Holland, Amsterdam, pp. 323-351.

Haggis, G.H. and R.C. Sinha. 1978. Scanning electron microscopy of mycoplasmalike organisms after freeze fracture of plant tissues affected with clover phyllody and aster yellows. Phytopathology *68:* 677-680.

Haigh, W.G., H.J. Förster, K. Biemann, N.L. Tattrie and J.R. Colvin. 1973. Induction of orientation of bacterial cellulose microfibrils by a novel terpenoid from *Acetobacter xylinum.* Biochem. J. *135:* 145-149.

Haikara, A., L. Penttila, T.-M. Enari and K. Lounatmaa. 1981. Microbiological, biochemical and electron microscopic characterization of a *Pectinatus* strain. Appl. Environ. Microbiol. *41:* 511-517.

Hajna, A.A. 1955. A new enrichment broth medium for Gram-negative organisms of the intestinal group. Public Health Lab. *13:* 83-89.

Hale, H.H., C.F. Helmboldt, W.N. Plastridge and E.F. Stula. 1962. Bovine mastitis caused by a *Mycoplasma* species. Cornell Vet. *52:* 582-591.

Haley, R., S.P. Davis and J.M. Hyde. 1967. Environmental stress and *Aeromonas liquefaciens* in American and threadfin shad mortalities. Prog. Fish. Cult. *29:* 193-194.

Hall, I.M. and M.E. Badgley. 1957. A rickettsial disease of larvae of species of *Stethorus* caused by *Rickettsiella stethorae,* n. sp. J. Bacteriol. *74:* 452-455.

Hall, J.A. and R.K.S. Wood. 1970. Plant cells killed by soft rot parasites. Nature (London) *227:* 1266-1267.

Hall, M.M., C.A. Mueske, D.M. Ilstrup and J.A. Washington II. 1979. Evaluation of a biphasic medium for blood cultures. J. Clin. Microbiol. *10:* 673-676.

Hall, W.J. 1965. Fowl typhoid. *In* H.E. Biester and L.H. Schwarte (Editors), Diseases of Poultry, 5th edition, The Iowa State University Press, Ames, pp. 329-358.

Hall, W.J., K.L. Heddleston, D.H. Legenhausen and R.W. Hughes. 1955. Studies on pasteurellosis. I. A new species of *Pasteurella* encountered in chronic fowl cholera. Am. J. Vet. Res. *16:* 598-603.

Hallé, J. 1898. Recherches sur la bactériologie du canal génital de la femme (état normal et pathologique). Thesis, Paris.

Halle, S., G.A. Dasch and E. Weiss. 1977. Sensitive enzyme-linked immunosorbent assay for detection of antibodies against typhus rickettsiae, *Rickettsia prowazekii* and *Rickettsia typhi.* J. Clin. Microbiol. *6:* 101-110.

Hallock, F.A. 1960. The life cycle of *Vibrio alternans* (sp. nov.) Trans. Am. Microsc. Soc. *79:* 404-411.

Halsey, N.A., M.A. Welling and R.M. Lehman. 1980. Nosocomial pertussis: A failure of erythromycin treatment and phophylaxis. Am. J. Dis. Child. *134:* 521-522.

Hambleton, P., C.G.T. Evans, A.M. Hood and R.E. Strange. 1974. Vaccine potencies of the live vaccine strain of *Francisella tularensis* and isolated bacterial components. Br. J. Exp. Pathol. *55:* 363-373.

Hamburger, M. and H.C. Knowles. 1953. *Streptobacillus moniliformis* infection complicated by acute bacterial endocarditis. Arch. Intern. Med. *92:* 216-220.

Hamilton, R.D. and K.E. Austin. 1967. Physiological and cultural characteristics of *Chromobacterium marinum* sp. n. Antonie van Leeuwenhoek J. Microbiol. Serol. *33:* 257-264.

Hamilton, R.D. and S.A. Zahler. 1957. A study of *Leptotrichia buccalis.* J. Bacteriol. *73:* 386-393.

Hamilton, R.L. and W.J. Brown. 1972. Bacteriophage typing of clinically isolated *Serratia marcescens.* Appl. Microbiol. *24:* 899-906.

Hamlin, L.J. and R.E. Hungate. 1956. Culture and physiology of a starch-digesting bacterium (*Bacteroides amylophilus* n. sp.) from the bovine rumen. J. Bacteriol. *72:* 548-554.

Hammann, R., H. von Nicolai and H. Werner. 1981. Neuraminidases of *Bacteroidaceae.* Zentralbl. Bakteriol. Parasitenkd. Infektionskr. Hyg., I Abt. Orig. A *248:* 526-531.

Hammann, R. and H. Werner. 1981. Presence of diaminopimelic acid in propionate-negative *Bacteroides* species and in some butyric acid-producing strains. J. Med. Microbiol. *14:* 205-212.

Hammerschlag, F. 1979. Determination of feasibility of selecting geranium cells for resistance to a toxin produced by *Xanthomonas pelargonii.* Abs. Phytopathology *69:* 1030.

Hammond, G.W., C.J. Lian, J.C. Wilt, W. Albritton and A.R. Ronald. 1978.

Determination of the hemin requirement of *Haemophilus ducreyi*: evaluation of the porphyrin test and media used in the satellite growth test. J. Clin. Microbiol. *7:* 243-246.

Hammond, G.W., C.J. Lian, J.C. Wilt and A.R. Ronald. 1978. Antimicrobial susceptibility of *Haemophilus ducreyi*. Antimicrob. Agents Chemother. *13:* 608-612.

Hamon, Y., A. Kayser, L. Le Minor and J. Maresz. 1969. Les bactériocines d'*Edwardsiella tarda*. Intérêt taxonomique de l'étude de ces antibiotiques. C. R. Acad. Sci. Paris *268:* 2517-2520.

Hamon, Y., L. Le Minor and Y. Péron. 1970. Les bactériocines d'*Enterobacter liquefaciens*. Intérêt taxonomique de leur étude. C. R. Acad. Sci. Serie D. *270:* 886-889.

Hamon, Y. and Y. Péron. 1961. Étude de la propriété bactériocinogène dans le genre *Serratia*. Ann. Inst. Pasteur (Paris) *100:* 818.

Hamon, Y. and Y. Péron. 1966. La propriété bactériocinogène dans la tribu des *Salmonelleae*. Ann. Inst. Pasteur (Paris) *110:* 389-402.

Hamon, Y. and Y. Péron. 1979. Bacteriocines et (ou) phages létaux de *Serratia marcescens, S. liquefaciens*, et *S. marinorubra*. Ann. Microbiol. *130A:* 403.

Han, J. and K.W. Jeon. 1980. Isolation and partial characterization of two plasmid deoxyribonucleic acids from endosymbiotic bacteria of *Amoeba proteus*. J. Bacteriol. *141:* 1466-1469.

Hancock, R.E.W. and H. Nikaido. 1978. Outer membranes of Gram-negative bacteria. XIX. Isolation from *Pseudomonas aeruginosa* PAO1 and use in reconstitution and definition of the permeability barrier. J. Bacteriol. *136:* 381-390.

Hansen, A.J., A. Ingebritsen and O.B. Weeks. 1963. Flagellation of *Flavobacterium piscicida*. J. Bacteriol. *86:* 602-603.

Hansen, A.J., O.B. Weeks and R.R. Colwell. 1965. Taxonomy of *Pseudomonas piscicida* (Bein) Buck, Meyers and Leifson. J. Bacteriol. *89:* 752-761.

Hansen, E.C. 1879. Bidrag til kundskab om hvilke organismer der kunne forekomme og leve i øl og ølurt. Medd. Carlsberg Lab. *1:* 185-234.

Hansen, E.C. 1894. Undersøgelser over Eddikesyrerbaktenner. (Andenafhandeling). Medd. Carlsberg Lab. *4:* 326-327.

Hansen, E.C. 1911. *In* Klöcker (Editor), Gesammelte theoretische Abhandlungen über Gärungsorganismen. Gustav Fischer Verlag. Jena.

Hansen, E.J., R.M. Wilson and J.B. Baseman. 1979. Isolation of mutants of *Mycoplasma pneumoniae* defective in hemadsorption. Infect. Immun. *23:* 903-906.

Hansen, S.L. and B.S. Stewart. 1978. Slide catalase. A reliable test for differentiation and presumptive identification of certain clinically significant anaerobes. Am. J. Clin. Pathol. *69:* 36-40.

Hanson, L.E., D.N. Tripathy, L.B. Evans and A.D. Alexander. 1974. An unusual *Leptospira*, serotype *illini* (a new serotype). Int. J. Syst. Bacteriol. *24:* 355-357.

Hanson, R.S. 1980. Ecology and diversity of methylotrophic organisms. Adv. Appl. Microbiol. *26:* 3-39.

Hanus, F.J., R.J. Maier and H.J. Evans. 1979. Autotrophic growth of hydrogen uptake-positive strains of *Rhizobium japonicum* in an atmosphere supplied with hydrogen gas. Proc. Nat. Acad. Sci. U.S.A. *76:* 1788-1792.

Hara, T., A. Aumayr and S. Ueda. 1981. Genetic transformation of *Pseudomonas aeruginosa* with extracellular DNA. J. Gen. Appl. Microbiol. *27:* 109–114.

Harada, R., T. Maeda, A. Nakashima, M. Sadakata, M. Ando, K. Yonomine, Y. Otsuji and H. Sato. 1970. Electronmicroscopical studies on the mechanism of oogenesis and fertilization in *Dirofilaria immitis*. *In* Sasa (Editor), Recent Advances in Researches on Filariasis and Schistosomiasis in Japan, University Park Press, Baltimore, pp. 99-121.

Harbutt, P.R. 1963. A clinical appraisal of feline infectious anemia and its transmission under natural conditions. Aust. Vet. J. *39:* 401-404.

Hardy, K. 1981. Bacterial plasmids. *In* Cole and Knowles (Editors), Aspects of Microbiology Series No. 4, Thomas Nelson and Sons, Ltd., Walton-on-Thames, United Kingdom.

Hardy, P.C., G.M. Ederer and J.M. Matsen. 1970. Contamination of commercially packed urinary catheter kits with pseudomonad EO-1. N. Engl. J. Med. *282:* 33-35.

Harhoff, N. 1949. Studies on bacteria of the Ballerup group. Acta Path. Microbiol. Scand. *26:* 167-174.

Harington, J.S. 1960a. Studies on *Rhodnius prolixus*: growth and development of normal and sterile bugs, and the symbiotic relationship. Parasitology *50:* 279-286.

Harington, J.S. 1960b. Synthesis of thiamine and folic acid by *Nocardia rhodnii*, the micro-symbiont of *Rhodnius prolixus*. Nature (Lond.) *188:* 1027-1028.

Harrington, A.A. and R.E. Kallio. 1960. Oxidation of methanol and formaldehyde by *Pseudomonas methanica*. Can. J. Microbiol. *6:* 1-7.

Harrington, R., Jr., D.R. Bond and G.M. Brown. 1977. Smooth-phage-resistant *Brucella abortus* from bovine tissue. J. Clin. Microbiol. *5:* 663-664.

Harris, D.L., R.D. Glock, C.R. Christensen and J.M. Kinyon. 1972. Swine dysentery 1. Inoculation of pigs with *Treponema hyodysenteriae* (new species) and reproduction of the disease. Vet. Med. Small Anim. Clin. *67:* 61-69.

Harris, D.L., R.F. Ross and W.P. Switzer. 1969. Incidence of certain microorganisms in nasal cavities of swine in Iowa. Am. J. Vet. Res. *30:* 1621-1624.

Harris, J.I. and H.M. Steinman. 1977. Amino acid sequence homologies among superoxide dismutases. *In* Michelson, McCord and Fridovich (Editors), Superoxide and Superoxide Dismutases, Academic Press, New York, pp. 225-230.

Harris, M.A. and C.A. Reddy. 1977. Hydrogenase activity and the H_2-fumarate electron transport system in *Bacteroides fragilis*. J. Bacteriol. *131:* 922-928.

Harris, N.M. 1901. *Bacillus mortiferus* (nov. spec.). J. Exp. Med. *6:* 519-547.

Harrison, A. and L.N. Gibbins. 1975. The isolation and characterization of a temperate phage, Y46/(E2), from *Erwinia herbicola* Y46. Can. J. Microbiol. *21:* 937-944.

Harrison, A.P., Jr. and P.A. Hansen. 1963. *Bacteroides hypermegas* nov. spec. Antonie van Leeuwenhoek J. Microbiol. Serol. *29:* 22-28.

Harrison, D.N., C.H. Dorsey and C.A. Brown. 1976a. Studies on a macronuclear endosymbiont of *Spirostomum ambiguum*. II. Ultrastructural comparison of the in situ and the cultivated endosymbiont. Trans. Amer. Micros. Soc. *95:* 565-568.

Harrison, D.N., C.H. Dorsey and H.E. Finley. 1976b. Studies on a macronuclear endosymbiont of *Spirostomum ambiguum*. I. Isolation of the microorganism from the macronucleus. Trans. Amer. Micros. Soc. *95:* 560-564.

Harrison, F.C. 1929. The discoloration of halibut. Can. J. Res. *1:* 214-239.

Harrison, F.C. and M.E. Kennedy. 1922. The red discolouration of cured codfish. Trans. Roy. Soc. Canada, Sect. V, *16:* 101-152.

Harshbarger, J.C., S.C. Chang and S.V. Otto. 1977. Chlamydiae (with phages), mycoplasmas, and rickettsiae in Chesapeake Bay bivalves. Science *196:* 666-668.

Hartingsveldt, J., M.G. Marinus and A.H. Stouthamer. 1971. Mutants of *Pseudomonas aeruginosa* blocked in nitrate or nitrite dissimilation. Genet. *67:* 469-472.

Hartman, P.E., Z. Hartman, R.C. Stahl and B.N. Ames. 1971. Classification and mapping of spontaneous and induced mutations in the histidine operon of *Salmonella*. Adv. Genet. *16:* 1-34.

Hartman, R., H.-D. Sickinger and D. Oesterhelt. 1980. Anaerobic growth of halobacteria. Proc. Natl. Acad. Sci. U.S.A. *77:* 3821-3825.

Harvey, S. and M.J. Pickett. 1980. Comparison of adansonian analysis and deoxyribonucleic acid hybridization results in the taxonomy of *Yersinia enterocolitica*. Int. J. Syst. Bacteriol. *30:* 86-102.

Harvey, S. M. 1980. Hippurate hydrolysis by *Campylobacter fetus*. J. Clin. Microbiol. *11:* 435-437.

Harvey, S. M. and J. Lascelles. 1980. Respiratory systems and cytochromes in *Campylobacter fetus* subsp. *intestinalis*. J. Bacteriol. *144:* 917-922.

Harwood, C.S. 1978. *Beneckea gazogenes* sp. nov., a red, facultatively anaerobic, marine bacterium. Curr. Microbiol. *1:* 233-238.

Harwood, C.R. 1980. Plasmids. *In* Goodfellow and Board (Editors), Microbiological Classification and Identification, Academic Press, London - New York, pp. 27-53.

Harwood, C.S., S.S. Bang, S. Baumann and K.H. Nealson. 1980. *Photobacterium logei*. sp. nov., nom. rev., *Beneckea nereida* sp. nov., nom. rev. and *Beneckea gazogenes* sp. nov., nom. rev. Int. J. Syst. Bacteriol. *30:* 655.

Harwood, C.S. and E. Canale-Parola. 1981. Amino acid catabolism as a source of maintenance energy in a marine spirochete. Abst. Annu. Mtg. Amer. Soc. Microbiol., p. 101.

Hase, S. and E.T. Rietschel. 1976. Isolation and analysis of the lipid A backbone. Lipid A structure of lipopolysaccharides from various bacterial groups. Eur. J. Biochem. *63:* 101-107.

Hashimoto, T., D.L. Diedrich and S.F. Conti. 1970. Isolation of a bacteriophage for *Bdellovibrio bacteriovorus*. J. Virol. *5:* 97-98.

Hassal, H. 1966. The adaptive degradation of imidazolelactic acid by *Pseudomonas acidovorans*. Biochem. J. *101:* 22P.

Hassal, H. and F. Rabie. 1966. The bacterial metabolism of imidazolepropionate. Biochim. Biophys. Acta *115:* 521-523.

Hasse, C.H. 1915. *Pseudomonas citri*, the cause of citrus canker. J. Agr. Res. *4:* 97-100.

Hastings, J.W., T.O. Baldwin and M.Z. Nicoli. 1978. Bacterial luciferase: assay, purification and properties. *In* DeLuca (Editor), Methods in Enzymology, Vol. 57, Academic Press Inc., London-New York-San Francisco, pp. 135-152.

Hastings, J.W. and K.H. Nealson. 1977. Bacterial bioluminescence. Annu. Rev. Microbiol. *31:* 549-595.

Hastings, J.W. and K.H. Nealson. 1981. The symbiotic luminous bacteria. *In* Starr, Stolp, Trüper, Balows and Schlegel (Editors), The Prokaryotes, a handbook on habitats, isolation and identification of bacteria, Springer-Verlag, New York, pp. 1332-1345.

Hatch, T.P., D.W. Vance, Jr. and E. Al Hossainy. 1981. Identification of a major envelope protein in *Chlamydia* spp. J. Bacteriol. *146:* 426-429.

Hatten, B.A. 1973. Growth characteristics of microorganisms occurring in penicillin-treated *Brucella abortus* cultures. Proc. Soc. Exp. Biol. Med. *142:* 909-914.

Hatten, B.A. and R.D. Brodeur. 1978. Soluble antigens of virulent and attenuated biotypes of *Brucella abortus*. Infect. Immun. *22:* 956-962.

Hauduroy, P., G. Ehringer, G. Guilott, J. Magrou, A.R. Prévot, Rosset and A. Urbain. 1953. Dictionnaire des bactéries pathogènes, 2nd Ed., Masson and Co., Paris.

Hauduroy, P., G. Ehringer, A. Urbain, G. Guillot and J. Magrou. 1937. Dictionnaire des bactéries pathogènes. Masson and Co., Paris.

Hauge, J.G. 1960. Kinetics and specificity of glucose dehydrogenase from *Bacterium anitratum*. Biochim. Biophys. Acta *45:* 263-269.

Hauge, J.G., T.E. King and V.H. Cheldelin. 1955. Oxidation of dihydroxyacetone via the pentose cycle in *Acetobacter suboxydans*. J. Biol. Chem. *214:* 11-26.

Haupt, H. 1932. Der gegenwärtige Stand der Systematik und Benennung der

Bakterien und ihre Anwendung in der medizinischen Bakteriologie. Ergeb. Hyg. Bakteriol. *13:* 641-685.

Haupt, H. 1934. Zür Frage der Verwandtschaft des *Actinobacillus Lignièresii* Brumpt 1910, des *Bacillus equuli* van Straaten 1918 und des *Bacillus mallei* Flügge 1886. Arch. Wiss. Prakt. Tierheilk *67:* 513-524.

Haupt, H. 1935. Zür Systematik der Bakterien. Die für Mensch und Tier pthogenen gram-negativen alkali-bildenden Stäbchen-bakterien *(Aerobactereae* Pribram 1929 em.) Ergeb. Hyg. Bakteriol. *17:* 175-230.

Hauser, G. 1885. Über Faulnisbakterien und deren Beziehungen zür Septicamie. Ein Beiträg zür Morphologie der Spaltpilze, Vogel, Leipzig.

Hawke, J.P. 1979. A bacterium associated with disease of pond cultured channel catfish, *Ictalurus punctatus.* J. Fisheries Res. Board Canada *36:* 1508-1512.

Hawke, J.P., A.G. McWhorter, A.G. Steigerwalt and D.J. Brenner. 1981. *Edwardsiella ictaluri* sp. nov., the causative agent of enteric septicemia of catfish. Int. J. Syst. Bacteriol. *31:* 396-400.

Hawley, C.E. and W.A. Falkler, Jr. 1976. Antigens of *Leptotrichia buccalis.* II. Their reaction with complement fixing IgM in human sera. J. Periodontal Res. *10:* 216-223.

Hayano, K. and S. Fukui. 1967. Purification and properties of 3-keto-sucrose-forming enzyme from the cells of *Agrobacterium tumefaciens.* J. Biol. Chem. *242:* 3665-3672.

Hayano, K. and S. Fukui. 1970. α-3-Ketoglucosidase of *Agrobacterium tumefaciens.* J. Bacteriol. *101:* 692-697.

Hayano, K., Y. Tsubouchi and S. Fukui. 1973. 3-Ketoglucose reductase of *Agrobacterium tumefaciens.* J. Bacteriol. *113:* 652-657.

Hayashi, N. 1920. Etiology of tsutsugamushi disease. J. Parasitol. *7:* 53-68.

Hayashi, K., T. Kodaira, K. Baba and K. Kikuchi. 1966. Adansonian taxonomy and relationship of microorganisms based on the concept of similar value and center species. III. Studies on the tribe *Neisserieae* in the family *Coccaceae* and isolation and certification of species in a new genus *Halococcus* induced theoretically. Jpn. J. Bacteriol. *21:* 633-639.

Hayes, N.S., K.E. Muse, A.M. Collier and J.B. Baseman. 1977. Parasitism by virulent *Treponema pallidum* of host cell surfaces. Infect. Immun. *17:* 174-186.

Hayes, P.R. 1977. A taxonomic study of flavobacteria and related Gram negative yellow pigmented rods. J. Appl. Bacteriol. *43:* 345-367.

Hayes, S.F. and W. Burgdorfer. 1979. Ultrastructure of *Rickettsia rhipicephali,* a new member of the spotted fever group rickettsiae in tissues of the host vector *Rhipicephalus sanguineus.* J. Bacteriol. *137:* 605-613.

Hayes, S.F. and W. Burgdorfer. 1981. Ultrastructural comparisons of *Wolbachia*-like symbiotes in ticks (Acari: Ixodidae). *In* Burgdorfer and Anacker (Editors), Rickettsiae and Rickettsial Diseases. Academic Press, New York, pp. 281-289.

Hayflick, L. 1965. Tissue cultures and mycoplasmas. Texas Rep. Biol. Med. *23:* (Suppl. 1) 285-303.

Haynes, W.C. and W.H. Burkholder. 1957. Genus I *Pseudomonas* Migula 1894. *In* Breed, Murray and Smith (Editors) Bergey's Manual of Determinative Bacteriology, 7th Ed., The Williams and Wilkins Co., Baltimore, pp. 89-152.

Hayward, A.C. 1962. Studies on bacterial pathogens of sugar cane. II. Differentiation, taxonomy and nomenclature of the bacteria causing red stripe and mottled stripe diseases. Mauritius Sugar Ind. Res. Inst. Occas. Pap. No. 13, 13-27.

Hayward, A.C. 1964. Characteristics of *Pseudomonas solanacearum.* J. Appl. Bacteriol. *27:* 265-277.

Hayward, A.C. 1972. A bacterial disease of clover in Hawaii. Plant Dis. Rep. *56:* 446-450.

Hayward, A.C. 1974. Latent infection by bacteria. Ann. Rev. Phytopathol. *12:* 87-97.

Hayward, A.C. 1977. Occurrence of glycoside hydrolases in plant pathogenic and related bacteria. J. Appl. Bacteriol. *43:* 407-411.

Hayward, H.R. and T.C. Stadtman. 1959. Anaerobic degradation of choline by an anaerobic, cytochrome-producing bacterium, *Vibrio cholinicus,* n. sp. J. Bacteriol. *78:* 557-561.

Hazen, T.C., C.B. Fliermans, R.P. Hirsch and G.W. Esch. 1978. Prevalence and distribution of *Aeromonas hydrophila* in the United States. Appl. Environ. Microbiol. *36:* 731-738.

Hazeu, W., W.H. Batenburg-van der Vegte and C. de Bruyn. 1980b. *In* Validation of the publication of new names and new combinations previously published outside the IJSB. List No. 5. Int. J. Syst. Bacteriol. *30:* 676-677.

Hazeu, W., W.H. Batenburg-van de Vegte and C. de Bruyn. 1980a. Some characteristics of *Methylococcus mobilis* sp. nov. Arch. Microbiol. *124:* 211-220.

Hazeu, W. and P.J. Steenis. 1970. Isolation and characterization of two vibrio-shaped methane-oxidizing bacteria. Antonie van Leeuwenhoek J. Microbiol. Serol. *36:* 67-72.

Heberlein, G.T., J. De Ley and R. Tijtgat. 1967. Deoxyribonucleic acid homology and taxonomy of *Agrobacterium, Rhizobium* and *Chromobacterium.* J. Bacteriol. *94:* 116-124.

Hébert, G.A. 1980. Room temperature storage of *Legionella* cultures. J. Clin. Microbiol. *12:* 807-809.

Hébert, G.A. 1981. Hippurate hydrolysis by *Legionella pneumophila.* J. Clin. Microbiol. *13:* 240-242.

Hébert, G.A., C.W. Moss, L.K. McDougal, F.M. Bozeman, R.M. McKinney and D.J. Brenner. 1980a. The rickettsia-like organisms TATLOCK (1943) and HEBA (1959): bacteria phenotypically similar to but generally distinct from *Legionella pneumophila* and the WIGA bacterium. Ann. Intern. Med. *92:* 45-52.

Hébert, G.A., A.G. Steigerwalt and D.J. Brenner. 1980. *Legionella micdadei* species nova: classification of a third species of *Legionella* associated with human pneumonia. Curr. Microbiol. *3:* 255-258.

Hébert, G.A., A.G. Steigerwalt and D.J. Brenner. 1980. *In* Validation of the publication of new names and new combinations previously effectively published outside the IJSB. List No. 5. Int. J. Syst. Bacteriol. *30:* 676-677.

Hébert, G.A., B.M. Thomason, P.P. Harris, M.D. Hicklin and R.M. McKinney. 1980b. "Pittsburgh pneumonia agent": a bacterium phenotypically similar to *Legionella pneumophila* and identical to the TATLOCK bacterium. Ann. Intern. Med. *92:* 53-54.

Hechtman, P. and C.R. Scriver. 1970. Neutral amino acid transport in *Pseudomonas fluorescens.* J. Bacteriol. *104:* 857-863.

Heckly, R. 1961. Preservation of bacteria by lyophilization. Adv. Appl. Microbiol. *3:* 1-76.

Heckmann, K. 1975. Omikron, ein essentieller Endosymbiont von *Euplotes aediculatus.* J. Protozool. *22:* 97-104.

Heckmann, K. 1980. Omikron, an essential endosymbiont of *Euplotes aediculatus. In* Schwemmler and Schenk (Editors), Endocytobiology: endosymbiosis and cell biology, Vol. 1, de Gruyter, Berlin, pp. 393-400.

Heckmann, K., J.R. Preer, Jr. and W.H. Straetling. 1967. Cytoplasmic particles in the killers of *Euplotes minuta* and their relationship to the killer substance. J. Protozool. *14:* 360-363.

Heddleston, K.L. 1975. Pasteurellosis. *In* Hitchner, Domermuth, Purchase and Williams (Editors), Isolation and Identification of Avian Pathogens, Arnold Printing Corp., Ithaca, New York, pp. 38-50.

Heddleston, K.L., J.E. Gallagher and P.A. Rebers. 1972. Fowl cholera: gel diffusion precipitin test for serotyping *Pasteurella multocida* from avian species. Avian Dis. *16:* 925-936.

Hedén, C. and T. Illéni (Editors). 1975. New approaches to the identification of microorganisms. John Wiley & Sons, New York.

Hedges, F. 1922. Bacterial pustule of soy bean. Science (Washington) *56:* 111-112.

Hedges, R.W. 1980. R factors of *Serratia. In* Von Graevenitz and Rubin (Editors) The genus *Serratia.* CRC Press, Boca Raton, Florida, pp. 139-153.

Hedges, R.W. and A.E. Jacob. 1977. *In vivo* translocation of genes of *Pseudomonas aeruginosa* onto a promiscuously transmissible plasmid. FEMS Microbiol. Lett. *2:* 15-19.

Hedges, R.W., A.E. Jacob and J.T. Smith. 1974. Properties of an R factor for *Bordetella bronchiseptica.* J. Gen. Microbiol. *84:* 199-204.

Heefner, D.L. and G.W. Claus. 1976. Change in quantity of lipids and cell size during intracytoplasmic membrane formation in *Gluconobacter suboxydans.* J. Bacteriol. *125:* 1163-1171.

Heefner, D.L. and G.W. Claus. 1978. Lipid and fatty acid composition of *Gluconobacter oxydans* before and after intracytoplasmic membrane formation. J. Bacteriol. *134:* 38-47.

Hegazi, N.A. and V. Jensen. 1973. Study of *Azotobacter* bacteriophage in Egyptian soil. Soil Biol. Biochem. *5:* 231-243.

Heidelberger, M., A. Das and E. Juni. 1969. Immunochemistry of the capsular polysaccharide of an *Acinetobacter.* Proc. Nat. Acad. Sci. U.S.A. *63:* 47-50.

Heidelberger, M. and W. Nimmich. 1976. Immunochemical relationships between bacteria belonging to two separate families: pneumococci and *Klebsiella.* Immunochemistry *13:* 67-80.

Heilman, F.R. 1941. A study of *Asterococcus muris (Streptobacillus moniliformis)* I. Morphologic aspects and nomenclature. J. Infect. Dis. *69:* 32-44.

Heinrich, S. and G. Pulverer. 1959a. Zür Aetiologie und Mikrobiologie der Aktinomykose. II. Definition und praktische Diagnostik des *Actinobacillus actinomycetem-comitans.* Zentralbl. Bakteriol. Parasitenkd. Infektionskr. Hyg. Abt. 1, Orig. *174:* 123-135.

Heinrich, S. and G. Pulverer. 1959b. Zür Ätiologie und Mikrobiologie der aktinomykose. III. Die pathogène Bedeutung des *Actinobacillus actinomycetem-comitans* unter den "Begleitbakterien" des *Actinomyces israeli.* Zentralbl. Bakteriol. Parasitenkd. Infektionskr. Hyg. Abt. 1, Orig. *176:* 91-101.

Heinrichsen, J. 1972. Bacterial surface translocation: a survey and a classification. Bacteriol. Rev. *36:* 478-503.

Heisch, R.B. 1953. On a spirochaete isolated from *Ornithodoros graingeri.* Parasitology *43:* 133-135.

Heisch, R.B., E.R.N. Cooke, A.E.C. Harvey and F. DeSouza. 1963. The isolation of *Brucella suis* from rodents in Kenya. East Afr. Med. J. *40:* 132-133.

Hellmann, G. 1913. Über die im Excretionsorgan der Ascidien der Gattung *Caesira (Molgula)* vorkommenden Spirochäten: *Spirochaeta caesirae septentrionalis* n. sp. und *Spirochaeta caesirae retortiformis* n. sp. Arch. Protistenk. *29:* 22-38.

Hellmers, E. 1955. Bacterial leaf spot of African marigold *(Tagetes erecta)* caused by *Pseudomonas tagetis* sp. n. Acta Agr. Scandin. *5:* 185-200.

Hellmers, E. 1958. Four wilt diseases of perpetual-flowering carnations in Denmark. Dan. Bot. Ark. *18:* 1-200.

Hellmers, E. 1959. *Pectobacterium carotovorum* var. *atrosepticum* (van Hall) Dowson, the correct name of the potato blackleg pathogen: a historical and critical review. Eur. Potato J. *2:* 251-271.

Hellmers, E. and W.J. Dowson. 1953. Further investigations of potato blackleg. Acta Agr. Scand. *3:* 103-112.

Hellmuth, H. 1956. Untersuchungen zür Bakteriensymbiose der Trypetiden (Diptera). Z. Morphol. Oekol. Tiere *44:* 483-517.

Hemelt, I.E., G.E. Lewis, Jr., D.L. Huxsoll and E.H. Stephenson. 1980. Serial propagation of *Ehrlichia canis* in primary canine peripheral blood monocyte cultures. Cornell Vet. *70:* 38-42.

Henderson, C. 1968. A study of the lipase of *Anaerovibrio lipolytica*: a rumen bacterium. Ph.D. Thesis, University of Aberdeen.

Henderson, C. 1970. The lipases produced by *Anaerovibrio lipolytica* in continuous culture. Biochem. J. *119:* 5-6.

Henderson, C. 1971. A study of lipase produced by *Anaerovibrio lipolytica*: a rumen bacterium. J. Gen. Microbiol. *65:* 81-89.

Henderson, C. 1973a. The effects of fatty acids on pure cultures of rumen bacteria. J. Agric. Sci. *81:* 107-112.

Henderson, C. 1973b. An improved method for enumerating and isolating lipolytic rumen bacteria. J. Appl. Bacteriol. *36:* 187-188.

Henderson, C. 1980. The influence of extracellular hydrogen on the metabolism of *Bacteroides ruminicola*, *Anaerovibrio lipolytica* and *Selenomonas ruminantium*. J. Gen. Microbiol. *119:* 485-491.

Henderson, C. and W. Hodgkiss. 1973. An electron-microscopic study of *Anaerovibrio lipolytica* (strain 5S) and its lipolytic enzyme. J. Gen. Microbiol. *76:* 389-393.

Hendley, J.O., K.R.A. Powell, R. Rodewald, H.H. Holzgrefe and R. Lyles. 1977. Demonstration of a capsule on *Neisseria gonorrhoeae*. N. Eng. J. Med. *296:* 608-611.

Hendrickson, A.A., I.L. Baldwn and A.J. Riker. 1934. Studies on certain physiological characters of *Phytomonas tumefaciens*, *Phytomonas rhizogenes* and *Bacillus radiobacter*. Part II. J. Bacteriol. *28:* 597-618.

Hendrickson, J.M. and K.F. Hilbert. 1932. A new and serious septicemic disease of young ducks with a description of the causative organism, *Pfeifferella anatipestifer*. Cornell Vet. *22:* 239-252.

Hendrie, M.S., W. Hodgkiss and J.M. Shewan. 1970. The identification, taxonomy and classification of luminous bacteria. J. Gen. Microbiol. 64: 151-169.

Hendrie, M.S., A.J. Holding and J.M. Shewan. 1974. Emended descriptions of the genus *Alcaligenes* and of *Alcaligenes faecalis* and proposal that the generic name *Achromobacter* be rejected; status of the named species of *Alcaligenes* and *Achromobacter*. Int. J. Syst. Bacteriol. *24:* 534-550.

Henneberg, W. 1897. Beiträge zür Kenntnis der Essigbakterien. Zentralbl. Bakteriol. Parasitenkd. Infektionskr. Hyg. Abt. II *3:* 223-231.

Henneberg, W. 1898. Weitere Untersuchungen ueber Essigbakteriën. Zentralbl. Bakteriol. Parasitenkd. Infektionskr. Hyg. Abt. II *4:* 14-20, 67-73, 138-147.

Henneberg, W. 1906. Zür Kenntnis der Schnellessig und Weinessigbakteriën. Deut. Essigindustrie *10:* 89-93, 98-99, 106-108, 113-116, 121-124, 129-132, 137-140, 146-148.

Henneberg, G. and I. Wolff. 1963. Experimente mit *Rickettsia melophagi*. Zentralbl. Bakteriol. Parasitenkd. Infektionskr. Hyg. Abt. I Orig. *188:* 487-493.

Henrichsen, J. 1972. Bacterial surface translocation: a survey and a classification. Bacteriol. Rev. *36:* 478-503.

Henrichsen, J. 1975. The occurrence of twitching motility among Gram-negative bacteria. Acta Pathol. Microbiol. Scand. Sect. B 83: 171-178.

Henrichsen, J. and J. Blom. 1975. Examination of fimbriation of some Gram-negative rods with and without twitching and gliding motility. Acta Pathol. Microbiol. Scand. Sect. B 83: 161-170.

Henrichsen, J., L.O. Frøhlm and K. Bøvre. 1972. Studies on bacterial surface translocation 2. Correlation of twitching motility and fimbriation in colony variants of *Moraxella nonliquefaciens*, *M. bovis* and *M. kingii*. Acta Pathol. Microbiol. Scand. Sect. B *80:* 445-452.

Henriksen, S.D. 1948. Studies on Gram-negative anaerobes II. Gram-negative anaerobic rods with spreading colonies. Acta Pathol. Microbiol. Scand. *25:* 368.

Henriksen, S.D. 1950. A comparison of the phenylpyruvic acid reaction and the urease test in the differentiation of *Proteus* from other enteric organisms. J. Bacteriol. *60:* 225-231.

Henriksen, S.D. 1952. *Moraxella*: Classification and taxonomy. J. Gen. Microbiol. *6:* 318-328.

Henriksen, S.D. 1969. Corroding bacteria from the respiratory tract. I. *Moraxella kingii*. Acta Pathol. Microbiol. Scand. *75:* 85-90.

Henriksen, S.D. 1969a. Corroding bacteria from the respiratory tract; 2. *Bacteroides corrodens*. Acta Pathol. Microbiol. Scand. *75:* 91-96.

Henriksen, S.D. 1969b. Designation of the type strain of *Bacteroides corrodens* Eiken 1958. Int. J. Syst. Bacteriol. *19:* 165-166.

Henriksen, S.D. 1973. *Moraxella*, *Acinetobacter*, and the *Mimeae*. Bacteriol. Rev. *37:* 522-561.

Henriksen, S.D. and K. Bøvre. 1968. *Moraxella kingii* spec. nov., a haemolytic, saccharolytic species of the genus *Moraxella*. J. Gen. Microbiol. *51:* 377-385.

Henriksen, S.D. and K. Bøvre. 1968. The taxonomy of the genera *Moraxella* and *Neisseria*. J. Gen. Microbiol. *51:* 387-392.

Henriksen, S.D. and K. Bøvre. 1976. Transfer of *Moraxella kingae* Henriksen and Bøvre to the genus *Kingella* gen. nov. in the family *Neisseriaceae*. Int. J. Syst. Bacteriol. *26:* 447-450.

Henriksen, S.D. and L.O. Frøhlm. 1975. A fimbriated strain of *Pasteurella multocida* with spreading and corroding colonies. Acta Pathol. Microbiol. Scand. Sect. B 83: 129-132.

Henriksen, S.D. and J. Henrichsen. 1975. Twitching motility and possession of fimbrae in spreading *Streptococcus sanguis* isolates from the human throat.

Acta Pathol. Microbiol. Scand. Sect. B *83:* 133-140.

Henriksen, S.D. and E. Holten. 1976. *Neisseria elongata* subsp. *glycolytica* subsp. nov. Int. J. Syst. Bacteriol. *26:* 478-481.

Henriksen, S.D. and K. Jyssum. 1960. A new variety of *Pasteurella haemolytica* from the human respiratory tract. Acta Pathol. Microbiol. Scand. *50:* 443.

Henriksen, S.D. and K. Jyssum. 1961. A study of some *Pasteurella* strains from the human respiratory tract. Acta Pathol. Microbiol. Scand. *51:* 354-368.

Henry, B.S. 1933. Dissociation in the genus *Brucella*. J. Infect. Dis. *52:* 374-402.

Henry, R.A. and R.C. Johnson. 1978. Distribution of the genus *Leptospira* in soil and water. Appl. Environ. Microbiol. *35:* 492-499.

Henry, S.M. 1962. The significance of microorganisms in the nutrition of insects. Trans. N.Y. Acad. Sci. *24:* 676-683.

Herbert, R.B. and F.G. Holliman. 1964. Aeruginosin B. A naturally occurring phenazinesulfonic acid. Proc. Chem. Soc. *1964:* 19.

Herbst, E.J. and E.E. Snell. 1949. The nutritional requirements of *Haemophilus parainfluenzae*. J. Bacteriol. *58:* 379-386.

Herman, N.J. and E. Juni. 1974. Isolation and characterization of a generalized transducing bacteriophage for *Acinetobacter*. J. Virol. *13:* 46-52.

Hernandez, F.J., B.D. Kirby, T.M. Stanley and P.H. Edelstein. 1980. Legionnaires' disease: postmortem findings of 20 cases. Amer. J. Clin. Pathol. *73:* 488-495.

Herring, P.J. and J.G. Morin. 1978. Bioluminescence in fishes. *In* Herring (Editor), Bioluminescence in Action, Academic Press Inc., London-New York-San Francisco, pp. 273-329.

Herriott, R.M., E.Y. Meyer, M. Vogt and M. Modan. 1970. Defined medium for growth of *Haemophilus influenzae*. J. Bacteriol. *101:* 513-516.

Hertig, M. 1936. The rickettsia, *Wolbachia pipientis* (gen. et sp. n.) and associated inclusions of the mosquito, *Culex pipiens*. Parasitol. *28:* 453-486.

Hertig, M. and S.B. Wolbach. 1924. Studies on rickettsia-like microorganisms in insects. J. Med. Res. *44:* 329-374.

Hertman, I. 1964. Bacteriophage common to *Pasteurella pestis* and *Escherichia coli*. J. Bacteriol. *88:* 1002-1005.

Hertzberg, S., G. Borch and S. Liaaen-Jensen. 1976. Bacterial carotenoids. L Absolute configuration of zeaxanthin dirhamnoside. Arch. Microbiol. *110:* 95-99.

Hespell, R. B. 1976. Glycolytic and tricarboxylic acid cycle enzyme activities during intraperiplasmic growth of *Bdellovibrio bacteriovorus* on *Escherichia coli*. J. Bacteriol. *128:* 677-680.

Hespell, R.B. 1977. *Serpens flexibilis* gen. nov., sp. nov., an unusually flexible lactate-oxidizing bacterium. Int. J. Syst. Bacteriol. *27:* 371-381.

Hespell, R.B. and E. Canale-Parola. 1970a. Carbohydrate metabolism in *Spirochaeta stenostrepta*. J. Bacteriol. *103:* 216-226.

Hespell, R.B. and E. Canale-Parola. 1970b. *Spirochaeta litoralis* sp. n., a strictly anaerobic marine spirochete. Arch. Mikrobiol. *74:* 1-18.

Hespell, R.B. and E. Canale-Parola. 1971. Amino acid and glucose fermentation by *Treponema denticola*. Arch. Mikrobiol. *78:* 234-251.

Hespell, R.B. and E. Canale-Parola. 1973. Glucose and pyruvate metabolism of *Spirochaeta litoralis*, an aerobic marine spirochete. J. Bacteriol. *116:* 931-937.

Hespell, R. B. and M. Mertens. 1977. Effects of nucleic acid compounds on viability and cell composition of *Bdellovibrio bacteriovorus* during starvation. Arch. Microbiol. *116:* 151-159.

Hespell, R. B., G. F. Miozzari and S. C. Rittenberg. 1975. Ribonucleic acid destruction and synthesis during intraperiplasmic growth of *Bdellovibrio bacteriovorus*. J. Bacteriol. *123:* 481-491.

Hespell, R. B., R. A. Rosson, M. F. Thomashow and S. C. Rittenberg. 1973. Respiration of *Bdellovibrio bacteriovorus* strain 109J and its energy substrates for intraperiplasmic growth. J.Bacteriol. *113:* 1280-1288.

Hespell, R. B., M. F. Thomashow and S. C. Rittenberg. 1974. Changes in cell composition and viability of *Bdellovibrio bacteriovorus* during starvation. Arch. Microbiol. *97:* 313-327.

Hess, A., R. Holländer and W. Mannheim. 1979. Lipoquinones of some spore-forming rods, lactic acid bacteria and actinomycetes. J. Gen. Microbiol. *115:* 246-252.

Hesselbrock, W. and L. Foshay. 1945. The morphology of *Bacterium tularense*. J. Bacteriol. *49:* 209-231.

Heumann, W. 1962. Die Metodik der Kreuzung sternbildender Bakterien. Biol. Zentralbl. *81:* 341-354.

Heuschmann-Brunner, G. 1965. Ein Beiträg zür Erregerfrage der infektiosen Bauchwassersucht des Karpfens in: Der Fisch in Wissenschaft und Praxis. Festschrift, herausgegeben anlässlich des 50-jahrigen Bestehens der teichwirtschaftlichen Abteilung Wielenbach der Bayerischen Biologischen Versuchsanstalt (Demoll-Hofer-Institut), München, pp. 41-49.

Hevesi, M. and S.F. Mashaal. 1975. Contributions to the mechanism of infection of *Erwinia uredovora*, a parasite of rust fungi. Acta Phytopathol. Acad. Sci. Hung. *10:* 275-280.

Hewetson, L., H.M. Dunn and N.W. Dunn. 1978. Evidence for a transmissible catabolic plasmid in *Pseudomonas putida* encoding the degradation of p-cresol *via* the protocatechuate *ortho* pathway. Genet. Res. *32:* 249-255.

Hewett, M.J., A.J. Wicken, K.W. Knox and M.E. Sharpe. 1976. Isolation of lipoteichoic acids from *Butyrivibrio fibrisolvens*. J. Gen. Microbiol. *94:* 126-130.

Hewlett, E.L., L.H. Underhill, S.A. Vargo, J. Wolff and C.R. Manclark. 1979.

Bordetella pertussis adenylate cyclase: Regulation of activity and its loss in degraded strains. *In* Manclark and Hill (Editors), International Symposium on Pertussis, U.S. Government Printing Office, Washington, D.C., pp. 81-85.

Heyward, J.T., M.Z. Sabry and W.R. Dowdle. 1969. Characterization of *Mycoplasma* species of feline origin. Am. J. Vet. Res. *30:* 615-622.

Hickman, D. D. and A. W. Frenkel. 1965. Observations on the structure of *Rhodospirillum molischianum.* J. Cell Biol. *25:* 261-278.

Hickman, D. D. and A. W. Frenkel. 1965. Observations on the structure of *Rhodospirillum rubrum.* J. Cell Biol. *25:* 279-291.

Hickman, F.W. and J.J. Farmer III. 1976. Differentiation of *Proteus mirabilis* by bacteriophage typing and the Dienes reaction. J. Clin. Microbiol. *3:* 350-353.

Hickman, F.W. and J.J. Farmer, III. 1978. *Salmonella typhi.*: Identification, antibiograms, serology, and bacteriophage typing. Amer. J. Med. Technol. *44:* 1149-1159.

Hickman, F.W., J.J. Farmer III, A.G. Steigerwalt and D.J. Brenner. 1980. Unusual groups of *Morganella* ("*Proteus*") *morganii* isolated from clinical specimens: Lysine-positive and ornithine-negative biogroups. J. Clin. Microbiol. *12:* 88-94.

Hickman, J. and G. Ashwell. 1966. Isolation of bacterial lipopolysaccharide from *Xanthomonas campestris* containing 3-acetamido-3,6-dideoxy-D-galactose and D-rhamnose. J. Biol. Chem. *241:* 1424-1428.

Higashi, S. 1967. Transfer of clover infectivity of *Rhizobium trifolii* to *Rhizobium phaseoli* as mediated by an episomic factor. J. Gen. Appl. Microbiol. *13:* 391-403.

Higuchi, K. and J.L. Smith. 1961. Studies on the nutrition and physiology of *Pasteurella pestis*: VI. A differential plating medium for the estimation of the mutation rate to avirulence. J. Bacteriol. *81:* 605-608.

Hildebrand, D.C. 1971. Pectate and pectin gels for differentiation of *Pseudomonas* sp. and other bacterial plant pathogens. Phytopathology *61:* 1430-1436.

Hildebrand, D.C., N.J. Palleroni and M. Doudoroff. 1973. Synonymy of *Pseudomonas gladioli* Severini 1913 and *Pseudomonas marginata* (McCulloch 1921) Stapp 1928. Int. J. Syst. Bacteriol. *23:* 433-437.

Hildebrand, D.C. and M.N. Schroth. 1964. Beta-glucosidase activity in phytopathogenic bacteria. Appl. Microbiol. *12:* 487-491.

Hildebrand, D.C. and M.N. Schroth. 1967. A new species of *Erwinia* causing the drippy nut disease of live oaks. Phytopathology *57:* 250-253.

Hildebrand, E.M. 1940. Cane gall of brambles caused by *Phytomonas rubi* n. sp. J. Agr. Res. *61:* 685-696.

Hildebrandt, P.K., J.D. Conroy, A.E. McKee, M.B.A. Nyindo and D.L. Huxsoll. 1973. Ultrastructure of *Ehrlichia canis.* Infect. Immun. *7:* 265-271.

Hilger, F. 1965. Études sur la systematique du genre *Beijerinckia* Derx. Ann. Inst. Pasteur (Paris) *109:* 406-423.

Hill, A. 1971a. *Mycoplasma caviae,* a new species. J. Gen. Microbiol. *65:* 109-113.

Hill, A. 1971b. Incidence of mycoplasma infection in guinea pigs. Nature (London) *232:* 560.

Hill, A. 1974. Experimental and natural infection of the conjunctiva of rats. Lab. Anim. *8:* 305-310.

Hill, F. and H.G. Schlegel. 1969. Die α-Isopropylmalat-Synthetase bei *Hydrogenomonas* H 16. Arch. Mikrobiol. *68:* 1-17.

Hill, G.B. and S. Osterhout. 1974. In vitro and in vivo effects of hyperbaric oxygen on nonsporeforming anaerobic bacteria. *In* W.G. Trapp et al. (Editors), Fifth International Hyperbaric Congress Proceedings, Simon Fraser Univ., Burnaby 2, B.C., Canada, pp. 562-568.

Hill, L.R. 1966. An index to desoxyribonucleic acid base compositions of bacterial species. J. Gen. Microbiol. *44:* 419-437.

Hill, L.R., J.J.S. Snell and S.P. Lapage. 1970. Identification and characteristics of *Bacteroides corrodens.* J. Med. Microbiol. *3:* 483-491.

Hill, M.J. and B.S. Drasar. 1968. Degradation of bile salts by human intestinal bacteria. Gut *9:* 22-27.

Hill, P., J.A. Campbell and I.A. Petrie. 1976. *Rhodnius prolixus* and its symbiotic actinomycete: a microbiological, physiological and behavioural study. Proc. R. Soc. Lond. B Biol. Sci. *194:* 501-525.

Hill, S. 1971. Influence of oxygen concentration on the colony type of *Derxia gummosa* grown on nitrogen-free media. J. Gen. Microbiol. *67:* 77-83.

Hill, S. 1975. Acetylene reduction by *Klebsiella pneumoniae* in air related to colony dimorphism on low fixed nitrogen. J. Gen. Microbiol. *91:* 207-209.

Hill, S. 1976. Influence of atmospheric oxygen concentration on acetylene reduction and efficiency of nitrogen fixation in intact *Klebsiella pneumoniae.* J. Gen. Microbiol. *93:* 335-345.

Hill, S. and J.R. Postgate. 1969. Failure of putative nitrogen-fixing bacteria to fix nitrogen. J. Gen. Microbiol. *58:* 277-285.

Hilpert, R., J. Winter, W. Hammes and O. Kandler. 1981. The sensitivity of archaebacteria to antibiotics. Zentralbl. Bakteriol. Hyg. Abt. I Orig. C *2:* 11-20.

Hinchliffe, E. and A. Vivian. 1980. Naturally occurring plasmids in *Acinetobacter calcoaceticus*: a P class R factor of restricted host range. J. Gen. Microbiol. *116:* 75-80.

Hinde, R. 1971a. The control of the mycetome symbiotes of the aphids *Brevicoryne brassicae, Myzus persicae* and *Macrosiphum rosae.* J. Insect Physiol. *17:* 1791-1800.

Hinde, R. 1971b. Maintenance of aphid cells and the intracellular symbiotes of aphids in vitro. J. Invertebr. Pathol. *17:* 333-338.

Hines, W.D., B.A. Freeman and G.A. Pearson. 1964. Production and characteri-

zation of *Brucella* spheroplasts. J. Bacteriol. *87:* 438-445.

Hingorani, M.K. and N.J. Singh. 1959. *Xanthomonas punicae* sp. nov. on *Punica granatum* L. Indian J. Agr. Sci. *29:* 45-48.

Hinz, K.-H. 1973. Beiträg zür Differenzierung von *Haemophilus*-Stämmen aus Huhnern. 1. Mitteilung: Kulturelle und biochemische Untersuchungen. Avian Pathol. *2:* 211-229.

Hinz, K.-H. 1980. Heat-stable antigenic determinants of *Haemophilus paragallinarum.* Zentralbl. Veterinämed. *27:* 668-676.

Hinz, K.-H., G. Glünder and H. Lüders. 1978. Acute respiratory disease in turkey poults caused by *Bordetella bronchiseptica*-like bacteria. Vet. Rec. *103:* 262-263.

Hinz, K.-H., G. Glünder, B. Stiburek and H. Lüders. 1979. Experimentelle Untersuchungen zür Bordetellose der Pute. Zentrabl. Veterinaermed. Reihe B *26:* 202-213.

Hinz, K.-H. and C. Kunjara. 1977. *Haemophilus avium,* a new species from chickens. Int. J. Syst. Bacteriol. *27:* 324-329.

Hinz, K.-H. and H.E. Müller. 1977. Neuraminidase und N-Acylneuraminat-Pyruvat-Lyase bei *Haemophilus paragallinarum* und *Haemophilus paravium* n. sp. Zentralbl. Bakteriol. Parasitenkd. Infektionskr. Hyg. I Abt. Orig. A *237:* 72-79.

Hippe, H. 1967. Abbau und Wiederverwertung von Poly-β-hydroxybuttersäure durch *Hydrogenomonas* H 16. Arch. Mikrobiol. *56:* 248-277.

Hirota, Y., H. Suzuki, Y. Nishimura and S. Yasuda. 1977. On the process of cellular division in *Escherichia coli*: A mutant of *E. coli* lacking a mureinlipoprotein. Proc. Natl. Acad. Sci. U.S.A. *74:* 1417-1420.

Hirsch, P. 1963. CO_2-Fixierung durch Knallgasbakterien. II. Chromatographischer Nachweis der frühzeitigen Fixierungsprodukte. Arch. Mikrobiol. *46:* 53-78.

Hirsch, P. 1977. Ecology and morphogenesis of *Thiopedia* spp. in ponds, lakes, and laboratory cultures. *In* Codd and Stewart (Editors), Proceedings of the Second International Symposium on Photosynthetic Prokaryotes. Dundee, Scotland, pp. 13-15.

Hirsch, P. 1981. The genus *Brachyarcus. In* Starr, Stolp, Trüper, Balows and Schlegel (Editors), The Prokaryotes: a handbook of habitats, isolation and identification of bacteria, Springer-Verlag, New York, pp. 649-651.

Hirsch, P. 1981. The genus *Pelosigma. In* Starr, Stolp, Trüper, Balows and Schlegel (Editors), The Prokaryotes: a handbook of habitats, isolation and identification of bacteria, Springer-Verlag, New York, pp. 645-648.

Hirsch, P., G. Georgiev and H.G. Schlegel. 1963. CO_2-Fixierung durch Knallgasbakterien. III. Autotrophe und organotrophe CO_2-Fixierung. Arch. Mikrobiol. *46:* 79-95.

Hirsch, P. and H.G. Schlegel. 1963. CO_2-Fixierung durch Knallgasbakterien. I. Einbau und Fraktionierung. Arch. Mikrobiol. *46:* 44-52.

Hirschfeld, L. 1919. A new germ of paratyphoid. Lancet *196:* 296-297.

Hirsh, D.C., L.D. Martin and M.C. Libal. 1981. Plasmid-mediated antimicrobial resistance in *Haemophilus pleuropneumoniae.* Am. J. Vet. Res. *43:* 269-272.

Hirumi, H. and K. Maramorosch. 1969. Mycoplasma-like bodies in the salivary glands of insect vectors carrying the aster yellows agent. J. Virology *3:* 82-84.

Hisamatsu, M., J. Abe, A. Amemura and T. Harada. 1980. Structural elucidation on succinoglycan and related polysaccharides from *Agrobacterium* and *Rhizobium* by fragmentation with two special β-D-glycanases and methylation analysis. Agr. Biol. Chem. *44:* 1049-1055.

Hisatsune, K., A. Kiuye, S. Kondo and K. Takeya. 1980a. Chemical composition of O-antigenic lipopolysaccharides isolated from *Vibrio parahaemolyticus* (Part II). Proc. 15th Joint Conf. Cholera US-Japan Coop. Med. Sci. Prog., pp. 166-184.

Hisatsune, K., S. Kondo, T. Iguchi and K. Takeya. 1980b. Comparative study on sugar composition of O-antigenic lipopolysaccharides isolated from *Vibrio cholerae,* group F vibrios, *Aeromonas, V. metschnikovii* and *V. proteus.* Proc. 15th Joint Conf. Cholera US-Japan Coop. Med. Sci. Prog., pp. 148-165.

Hitchens, A.P. 1957. Genus *Dialister* Bergey et al. 1923. *In* Breed, Murray and Smith (Editors), Bergey's Manual of Determinative Bacteriology, 7th Ed. The Williams and Wilkins Co., Baltimore, pp. 440-441.

Hitzig, W. M. and A. Liebesman. 1944. Subacute endocarditis associated with infection by a spirillum. Arch. Intern. Med. *73:* 415-424.

Hjelm, E., G. Jonsell, T. Linglöf, P.-A. Mårdh, B. Møller and G. Sedin. 1980. Meningitis in a newborn infant caused by *Mycoplasma hominis.* Acta Paediatr. Scand. *69:* 415-418.

Ho, Y. K. and J. Lascelles. 1971. δ-Aminolevulinic acid dehydratase of *Spirillum itersonii* and the regulation of tetrapyrrole synthesis. Arch. Biochem. Biophys. *144:* 734-740.

Hobson, P.N. 1965. Continuous culture of some anaerobic and facultatively anaerobic rumen bacteria. J. Gen. Microbiol. *38:* 167-180.

Hobson, P.N. 1969. Rumen bacteria. *In* Norris and Ribbons (Editors), Methods in Microbiology, Vol. 3B. Academic Press, London and New York, pp. 133-149.

Hobson, P.N. and S.O. Mann. 1961. The isolation of glycerol-fermenting and lipolytic bacteria from the rumen of the sheep. J. Gen. Microbiol. *25:* 227-240.

Hobson, P.N., S.O. Mann and W. Smith. 1962. Serological tests of a relationship between rumen selenomonads in vitro and in vivo. J. Gen. Microbiol. *29:* 265-270.

Hobson, P.N. and R. Summers. 1966. Effect of growth rate on the lipase activity of a rumen bacterium. Nature 209: 736-737.

Hobson, P.N. and R. Summers. 1967. The continuous culture of anaerobic bacteria. J. Gen. Microbiol. 47: 53-65.

Hochster, R.M. and N.B. Madsen. 1959. The breakdown of adenosine phosphates in extracts of Xanthomonas phaseoli. Can. J. Biochem. Physiol. 37: 639-649.

Hochster, R.M. and C.G. Nozzolillo. 1960. Respiratory carriers and the nature of the reduced diphosphopyridine nucleotide oxidase system in Xanthomonas phaseoli. Can. J. Biochem. Physiol. 38: 79-93.

Hodgkiss, W. and J.M. Shewan. 1968. Problems and modern principles in the taxonomy of marine bacteria. Adv. Microbiol. Sea 1: 127-166.

Hof, T. 1935. An investigation of the micro-organisms commonly present in salted beans. Rec. Trav. Bot. Néerl. 32: 151-173.

Hofer, A.W. 1941. A characterization of Bacterium radiobacter (Beijerinck and van Delden) Löhnis. J. Bacteriol. 41: 193-224.

Hofer, A.W. 1944. Flagellation of Azotobacter. J. Bacteriol. 48: 697-701.

Hoffman, H. 1957. Genus Fusobacterium Knorr. In Breed, Murray and Smith (Editors), Bergey's Manual of Determinative Bacteriology, 7th Ed. The Williams and Wilkins Co., Baltimore, pp. 436-440.

Hoffman, H.P., S.C. Geftic, H. Heymann and F.W. Adair. 1973. Mesosomes in Pseudomonas aeruginosa. J. Bacteriol. 114: 434-438.

Hoffman, P. S., H. A. George, N. R. Krieg and R. M. Smibert. 1979a. Studies of the microaerophilic nature of Campylobacter fetus subsp. jejuni. II. Role of exogenous superoxide anions and hydrogen peroxide. Can. J. Microbiol. 25: 8-16.

Hoffman, P. S., N. R. Krieg and R. M. Smibert. 1979b. Studies of the microaerophilic nature of Campylobacter fetus subsp. jejuni. I. Physiological aspects of enhanced aerotolerance. Can. J. Microbiol. 25: 1-7.

Hofherr, L., H. Votava and D.J. Blazevic. 1978. Comparison of three methods for identifying nonfermenting Gram-negative rods. Can. J. Microbiol. 24: 1140-1144.

Hofstad, T. 1970. Leptotrichia buccalis. A Gram-negative bacterium. Int. J. Syst. Bacteriol. 20: 175-177.

Hofstad, T. 1974. The distribution of heptose and 2-keto-3-deoxy-octonate in Bacteroidaceae. J. Gen. Microbiol. 85: 314-320.

Hofstad, T. 1979. Serological responses to antigens of Bacteroidaceae. Microbiol. Rev. 43: 103-115.

Hofstad, T. 1981. Leptotrichia buccalis. In Starr, Stolp, Trüper, Balows and Schlegel (Editors), The Prokaryotes. A handbook on habitats, isolation, and identification of bacteria, Springer-Verlag, Berlin-Heidelberg-New York, pp. 1475-1478.

Hofstad, T. and P. Aasjord. Eubacterium plautii (Séguin 1928) comb. nov. Int. J. Syst. Bacteriol. 32: 346-349.

Hofstad, T. and E. Jantzen. 1982. Fatty acids Leptotrichia buccalis: taxonomic implications. J. Gen. Microbiol. 128: 151-153.

Hofstad, T. and K.A. Selvig. 1969. Ultrastructure of Leptotrichia buccalis. J. Gen. Microbiol. 56: 23-26.

Hofstad, T. and N. Skaug. 1980. Fatty acids and neutral sugars present in lipopolysaccharides isolated from Fusobacterium species. Acta Pathol. Microbiol. Scand. Sect. B 88: 115-120.

Høiby, N. 1975. Cross-reactions between Pseudomonas aeruginosa and thirty-six other bacterial species. Scand. J. Immunol. 4: Suppl. 2, 187-196.

Høiby, N. 1979. Immunity- Humoral response. In Doggett (Editor), Pseudomonas aeruginosa. Clinical manifestations of infection and current therapy. Academic Press, New York, pp. 157-189.

Hoilien, C. 1982. Rickettsia sennetsu in human blood monocyte culture. Similarities to the growth cycle of Ehrlichia canis. Infect. Immun. 35: 314-319.

Hoke, C. and N.A. Vedros. 1982a. Taxonomy of the Neisseriae. Deoxyribonucleic acid (DNA) base composition, enterspecific transformation, and DNA hybridization. Int. J. Syst. Bacteriol. 32: 57-66.

Hoke, C. and N.A. Vedros. 1982b. Taxonomy of the Neisseriae. Fatty acid analysis, aminopeptidase activity, and pigment extraction. Int. J. Syst. Bacteriol. 32: 51-56.

Hoke, C. and N.A. Vedros. 1982c. Characterization of "atypical" aerobic Gram-negative cocci isolated from humans. J. Clin. Microbiol. 15: 906-914.

Holbrook, W.P., B.I. Duerden and A.G. Deacon. 1977. The classification of Bacteroides melaninogenicus and related species. J. Appl. Bacteriol. 42: 259-273.

Holbrook, W.P. and C. McMillan. 1977. The hydrolysis of dextran by gram negative non-sporing anaerobic bacilli. J. Appl. Bacteriol. 43: 369-374.

Holdeman, L.V., E.P. Cato and W.E.C. Moore. 1977. Anaerobe laboratory manual, 4th Ed. Virginia Polytechnic Institute and State University, Blacksburg, Virginia.

Holdeman, L.V., I.J. Good and W.E.C. Moore. 1976. Human fecal flora: variation in bacterial composition within individuals and a possible effect of emotional stress. Appl. Environ. Microbiol. 31: 359-375.

Holdeman, L.V. and J.L. Johnson. 1977. Bacteroides disiens sp. nov. and Bacteroides bivius sp. nov. from human clinical infections. Int. J. Syst. Bacteriol. 27: 337-345.

Holdeman, L.V. and J.L. Johnson. 1982. Description of Bacteroides loescheii sp. nov. and emendation of the descriptions of Bacteroides melaninogenicus (Oliver and Wherry) Roy and Kelly, 1939, and Bacteroides denticola Shah and Collins 1981. Int. J. Syst. Bacteriol. 32: 399-409.

Holdeman, L.V. and W.E.C. Moore. 1970. Bacteroides. In E.P. Cato, C.S. Cummins, L.V. Holdeman, J.L. Johnson, W.E.C. Moore, R.M. Smibert and L.DS.

Smith (Editors), Outline of Clinical Methods in Anaerobic Bacteriology, 2nd rev. Virginia Polytechnic Institute, Anaerobe Laboratory, Blacksburg, Virginia.

Holdeman, L.V. and W.E.C. Moore. 1974a. Genus Bacteroides Castellani and Chalmers 1919, 959. In R.E. Buchanan and N.E. Gibbons (Editors), Bergey's Manual of Determinative Bacteriology, 8th Ed. The Williams and Wilkins Co., Baltimore, pp. 385-404.

Holdeman, L.V. and W.E.C. Moore. 1974b. New genus, Coprococcus, twelve new species, and emended descriptions of four previously described species of bacteria from human feces. Int. J. Syst. Bacteriol. 24: 260-277.

Holdeman, L.V. and W.E.C. Moore (Editors). 1975. Anaerobe Laboratory Manual, 3rd Ed. Virginia Polytechnic Institute and State University, Blacksburg, Virginia.

Holdeman, L.V., W.E.C. Moore, P.J. Churn and J.L. Johnson. 1982. Bacteroides oris and Bacteroides buccae, new species from human periodontitis and other human infections. Int. J. Syst. Bacteriol. 32: 125-131.

Holding, A.J. and J.G. Collee. 1971. Routine biochemical tests. In Norris and Ribbons (Editors), Methods in Microbiology, 6A, Academic Press Inc., New York, pp. 1-33.

Holding, A.J. and J.M. Shewan. 1974. Genus Alcaligenes Castellani and Chalmers 1919, 936. In Buchanan and Gibbons (Editors), Bergey's Manual of Determinative Bacteriology, 8th Ed., The Williams and Wilkins Co., Baltimore, pp. 273-275.

Holland, D.F. 1920. V. Generic index of the commoner forms of bacteria. In C.-E.A. Winslow, J. Broadhurst, R.E. Buchanan, C. Krumwiede, Jr., L.A. Rogers and G.H. Smith. The families and genera of the bacteria. J. Bacteriol. 5: 191-229.

Holland, J.W., L.R. Stauffer and W.A. Altemeier. 1979. Fluorescent antibody test kit for rapid detection and identification of members of the Bacteroides fragilis and Bacteroides melaninogenicus groups in clinical specimens. J. Clin. Microbiol. 10: 121-127.

Hollande, A.C. 1922. Les spirochètes des termites; processus de division: formation du schizoplaste. Arch. Zool. Expt. Gén. Notes Rev. 61: 23-28.

Hollande, A.C. and R. Favre. 1931. La structure cytologique de Blattabacterium cuenoti (Mercier) N. G., symbiote du tissue adipeux des Blattides. C.R. Séances Soc. Biol. Fil. 107: 752-754.

Hollande, A.C. and I. Gharagozlou. 1967. Morphologie infrastructurale de Pillotina calotermitidis nov. gen., nov. sp., spirochaetale de l'intestin de Calotermes praecox. C. R. Acad. Sci. (Paris) 265: 1309-1312.

Holländer, R. 1976a. Energy metabolism of some representatives of the Haemophilus group. Antonie van Leeuwenhoek J. Microbiol. Serol. 42: 429-444.

Holländer, R. 1976b. Correlation of the function of demethylmenaquinone in bacterial electron transport with its redox potential. FEBS Lett. 72: 98-100.

Holländer, R. 1978. The cytochromes of Thermoplasma acidophilum. J. Gen. Microbiol. 108: 165-167.

Holländer, R. 1980. Charakterisierung von Erwinia-Stämmen insbesondere der Herbicola-Gruppe durch Chinone der Atmungskette und Enzyme des Fumarat-Stoffwechsels. Zentralbl. Bakteriol. Parasitenkd. Infektionskr. Hyg. Abt. I Orig. C 1: 243-256.

Holländer, R., A. Hess-Reihse and W. Mannheim. 1981. Respiratory quinones in Haemophilus, Pasteurella and Actinobacillus: Pattern, functions and taxonomic evaluation. In Kilian, Frederiksen and Biberstein (Editors), Haemophilus, Pasteurella and Actinobacillus. Academic Press, London, pp. 83-97.

Holländer, R. and W. Mannheim. 1975. Characterization of hemophilic and related bacteria by their respiratory quinones and cytochromes. Int. J. Syst. Bacteriol. 25: 102-107.

Holländer, R. and S. Pohl. 1980. Deoxyribonucleic acid base composition of bacteria. Zentralbl. Bakteriol. Parasitenkd. Infektionskr. Hyg. Abt. I Orig. A 246: 236-275.

Holländer, R., G. Wolf and W. Mannheim. 1977. Lipoquinones of some bacteria and mycoplasmas, with considerations on their functional significance. Antonie Van Leeuwenhoek J. Microbiol. Serol. 43: 177-185.

Holliman, F.G. 1957. Pigments of a red strain of Pseudomonas aeruginosa. Chem. Ind. 28: 1668.

Hollingdale, M.R. and R.M. Lemcke. 1970. Antigenic differences within the species Mycoplasma hominis. J. Hyg. (Camb.) 68: 469-477.

Hollingdale, M.R. and R.M. Lemcke. 1972. Membrane antigens of Mycoplasma hominis. J. Hyg. (Camb.) 70: 85-98.

Hollingdale, M.R., J.W. Vinson and J.E. Herrmann. 1980. Immunochemical and biological properties of the outer membrane-associated lipopolysaccharide and protein of Rochalimaea quintana. J. Infect. Dis. 141: 672-679.

Hollis, A.B., W.E. Kloss and G.H. Elkan. 1981. DNA:DNA hybridization studies of Rhizobium japonicum and related Rhizobiaceae. J. Gen. Microbiol. 123: 215-222.

Hollis, D.G., F.W. Hickman and G.R. Fanning. 1981a. In D.G. Hollis, F.W. Hickman, G.R. Fanning, J.J. Farmer III, R.E. Weaver and D.J. Brenner. Tatumella ptyseos gen. nov., sp. nov., a member of the family Enterobacteriaceae found in clinical specimens. J. Clin. Microbiol. 14: 79-88.

Hollis, D.G., F.W. Hickman and G.R. Fanning. 1982. In List No. 8, validation of the publication of new names and new combinations previously effectively published outside the IJSB. Int. J. Syst. Bacteriol. 32: 266-268.

Hollis, D.G., F.W. Hickman, G.R. Fanning, D.J. Brenner and R.E. Weaver. 1980. EF-9: A newly described group of Enterobacteriaceae. Abstr. Ann. Meet. Am. Soc. Microbiol. p. 295, Abstract C 122.

Hollis, D.G., F.W. Hickman, G.R. Fanning, J.J. Farmer III, R.E. Weaver and D.J. Brenner. 1981b. *Tatumella ptyseos* gen. nov., sp. nov., a membeer of the family *Enterobacteriaceae* found in clinical specimens. J. Clin. Microbiol. *14:* 79-88.

Hollis, D.G., R.E. Weaver, C.N. Baker and C. Thornsberry. 1976. Halophilic *Vibrio* species isolated from blood cultures. J. Clin. Microbiol. *3:* 425-431.

Hollis, D.G., G.L. Wiggins and R.E. Weaver. 1969. *Neisseria lactamica* sp. n., a lactose-fermenting species resembling *Neisseria meningitidis.* Appl. Microbiol. *17:* 71-77.

Hollis, D.G., G.L. Wiggins and R.E. Weaver. 1972. An unclassified Gram-negative rod isolated from the pharynx on Thayer-Martin medium (selective agar). Appl. Microbiol. *24:* 772-777.

Holloway, B.W. 1960. Grouping *Pseudomonas aeruginosa* by lysogenicity and pyocinogenicity. J. Pathol. Bacteriol. *80:* 448-450.

Holloway, B.W. 1969. Genetics of *Pseudomonas.* Bacteriol. Rev. *33:* 419-443.

Holloway, B.W. 1974. *Pseudomonas. In* King (Editor), Handbook of Genetics. Vol. I: Bacteria, bacteriophages and fungi, Plenum, New York, pp. 59-68.

Holloway, B.W. 1975. Genetic organization of *Pseudomonas. In* Clarke and Richmond (Editors), Genetics and Biochemistry of *Pseudomonas,* John Wiley and Sons, London, pp. 133-161.

Holloway, B.W. 1978. Isolation and characterization of an R' plasmid in *Pseudomonas aeruginosa.* J. Bacteriol. *133:* 323-330.

Holloway, B.W. 1979a. Role of formal genetics in medical microbiology. *In* Doggett (Editor), *Pseudomonas aeruginosa.* Clinical manifestations of infection and current therapy. Academic Press, New York, pp. 9-39.

Holloway, B.W. 1979b. Plasmids that mobilize bacterial chromosome. Plasmid *2:* 1-19.

Holloway, B.W. and V. Krishnapillai. 1975. Bacteriophages and bacteriocins. *In* Clarke and Richmond (Editors), Genetics and Biochemistry of *Pseudomonas,* John Wiley and Sons, London, pp. 99-132.

Holloway, B.W., V. Krishnapillai and A.F. Morgan. 1979. Chromosomal genetics of *Pseudomonas.* Microbiol. Rev. *43:* 73-102.

Holloway, B.W., V. Krishnapillai and V. Stanisich. 1971. *Pseudomonas* genetics. Annu. Rev. Genet. *5:* 425-446.

Holm, P. 1950. Studies on the etiology of human actinomycosis. I. The "other microbes" and their importance. Acta Pathol. Microbiol. Scand. *27:* 736-751.

Holm, P. 1954. The influence of carbon dioxide on the growth of *Actinobacillus actinomycetemcomitans (Bacterium actinomycetem comitans* (Klinger 1912)). Acta Pathol. Microbiol. Scand. *34:* 235-248.

Holm, S.E., A. Tärnvik and G. Sandström. 1980. Antigenic composition of a vaccine strain of *Francisella tularensis.* Int. Arch. Allergy Appl. Immunol. *61:* 144-144.

Holmes, B. 1980. Proposal to conserve the specific epithet *liquefaciens* over the specific epithet *proteamaculans* in the name of the organism currently known as *Serratia liquefaciens* (Grimes and Hennerty 1931) Bascomb et al. 1971. Request for an opinion. Int. J. Syst. Bacteriol. *30:* 220-222.

Holmes, B. and M.S. Ahmed. 1981. Group EF-4: a pasteurella-like organism. *In* Kilian, Frederiksen and Biberstein (Editors), *Haemophilus, Pasteurella, and Actinobacillus.* Proceedings of an International Symposium held at Statens Seruminstitut, Copenhagen, August 20-22, 1980. Academic Press, London and New York, pp. 161-174.

Holmes, B., A. King, I. Phillips and S.P. Lapage. 1974. Sensitivity of *Citrobacter freundii* and *Citrobacter koseri* to cephalosporins and penicillins. J. Clin. Pathol. *27:* 729-733.

Holmes, B., S.P. Lapage and H. Malnick. 1975. Strains of *Pseudomonas putrefaciens* from clinical material. J. Clin. Pathol. *28:* 149-155.

Holmes, B. and R.J. Owen. 1979. Proposal that *Flavobacterium breve* be substituted as the type species of the genus in place of *Flavobacterium aquatile* and emended description of the genus *Flavobacterium:* status of the named species of *Flavobacterium.* Request for an Opinion. Int. J. Syst. Bacteriol. *29:* 416-426.

Holmes, B. and R.J. Owen. 1981. Emendation of the genus *Flavobacterium* and the status of the genus. Developments after the 8th edition of Bergey's Manual. *In* Reichenbach and Weeks (Editors), The *Flavobacterium-Cytophaga* Group (Proceedings of the International Symposium on Yellow-Pigmented Gram-Negative Bacteria of the *Flavobacterium-Cytophaga* Group, Braunschweig, July 8 to 11, 1980). Verlag Chemie, Weinheim. pp. 17-26.

Holmes, B. and R.J. Owen. 1982. *Flavobacterium breve* sp. nov., nom. rev. Int. J. Syst. Bacteriol. *32:* 233-234.

Holmes, B., R.J. Owen, A. Evans, H. Malnick and W.R. Willcox. 1977. *Pseudomonas paucimobilis,* a new species isolated from human clinical specimens, the hospital environment, and other sources. Int. J. Syst. Bacteriol. *27:* 133-146.

Holmes, B., R.J. Owen and D.G. Hollis. 1982. *Flavobacterium spiritivorum,* a new species isolated from human clinical specimens. Int. J. Syst. Bacteriol. *32:* 157-165.

Holmes, B., R.J. Owen and R.E. Weaver. 1981. *Flavobacterium multivorum,* a new species isolated from human clinical specimens and previously known as Group IIk, biotype 2. Int. J. Syst. Bacteriol. *31:* 21-34.

Holmes, B. and P. Roberts. 1981. The classification, identification and nomenclature of agrobacteria. Incorporating revised descriptions for each of *Agrobacterium tumefaciens* (Smith and Townsend) Conn 1942, *Agrobacterium rhizogenes* (Riker *et al.*) Conn 1942, and *Agrobacterium rubi* (Hildebrand) Starr and Weiss 1943. J. Appl. Bacteriol. *50:* 443-467.

Holmes, B., J.J.S. Snell and S.P. Lapage. 1977a. Strains of *Achromobacter xylosoxidans* from clinical material. J. Clin. Pathol. *30:* 595-601.

Holmes, B., J.J.S. Snell and S.P. Lapage. 1977b. Revised description, from clinical isolates, of *Flavobacterium odoratum* Stutzer and Kwaschnina 1929, and designation of the neotype strain. Int. J. Syst. Bacteriol. *27:* 330-336.

Holmes, B., J.J.S. Snell and S.P. Lapage. 1978. Revised description, from clinical strains, of *Flavobacterium breve* (Lustig) Bergey et al. 1923 and proposal of the neotype strain. Int. J. Syst. Bacteriol. *28:* 201-208.

Holmes, B., J.J.S. Snell and S.P. Lapage. 1979. *Flavobacterium odoratum:* a species resistant to a wide range of antimicrobial agents. J. Clin. Pathol. *32:* 73-77.

Holmes, R.K., M.G. Bramucci and E.M. Twiddy. 1979. Genetics of toxinogenesis in *Vibrio cholerae* and *Escherichia coli.* Contr. Microbiol. Immunol. *6:* 165-177.

Holsters, M., B. Silva, F. Van Vliet, C. Genetello, M. De Block, P. Dhaese, A. Depicker, D. Inzé, G. Engler, R. Villaroel, M. Van Montagu and J. Schell. 1980. The functional organization of the nopaline A. *tumefaciens* plasmid pTiC58. Plasmid *3:* 212-230.

Holt, L.B. 1962. The growth factor requirements of *Haemophilus influenzae.* J. Gen. Microbiol. *27:* 317-322.

Holt, S.C. 1978. Anatomy and chemistry of spirochetes. Microbiol. Rev. *42:* 114-160.

Holt, S.C. and E. Canale-Parola. 1968. Fine structure of *Spirochaeta stenostrepta,* a free-living, anaerobic spirochete. J. Bacteriol. *96:* 822-835.

Holt, S.C., A.C.R. Tanner and S.S. Socransky. 1980. Morphology and ultrastructure of oral strains of *Actinobacillus actinomycetemcomitans* and *Haemophilus aphrophilus.* Infect. Immun. *30:* 588-600.

Holten, E. 1974. Immunological comparison of NADP-dependent glutamate dehydrogenase and malate dehydrogenase in genus *Neisseria.* Acta Pathol. Microbiol. Scand. *82B:* 849-859.

Holtzman, H.E. 1959. A kappa-like particle in a non-killer stock of *Paramecium aurelia,* syngen 5. J. Protozool. *6:* (Suppl.) 26.

Holwerda, J. 1971. Symposium on pertussis immunization in honor of Dr. Pearl L. Kendrick in her eightieth year: Current diagnostic procedures in whooping cough. Health Lab. Sci. *8:* 206-209.

Holzworth, G. and E.B. Prestridge. 1977. Multistranded helix in xanthan polysaccharide. Science *197:* 757-759.

Homma, J.Y. and N. Suzuki. 1966. The protein moiety of the endotoxin of *Pseudomonas aeruginosa.* Ann. N.Y. Acad. Sci. *133:* 508-526.

Hon-nami, K. and T. Oshima. 1977. Purification and some properties of cytochrome *c*-552 from an extreme thermophile, *Thermus thermophilus* HB8. J. Biochem. *82:* 769-776.

Hood, A.M. 1977. Virulence factors of *Francisella tularensis.* J. Hyg. *79:* 47-60.

Hoogstraal, H. 1967. Ticks in relation to human diseases caused by rickettsia species. Annu. Rev. Entomol. *12:* 377-420.

Hoogstraal, H. 1981. Changing patterns of tickborne diseases in modern society. Annu. Rev. Entomol. *26:* 75-99.

Hoover, D.L. 1979. Tularemia. *In* Conn (Editor), Current Therapy, W.B. Saunders Co., Philadelphia, pp. 75-76.

Hopkins, D.L. 1977. Plant diseases caused by leafhopper-borne, rickettsia-like bacteria. Annu. Rev. Phytopathol. *15:* 277-294.

Hopkins, J.C.F. and W.J. Dowson. 1949. A bacterial leaf and flower disease of *Zinnia* in Southern Rhodesia. Trans. Brit. Mycol. Soc. *32:* 252-254.

Hopps, H.E., E.B. Jackson, J.X. Danauskas and J.E. Smadel. 1959. Study on the growth of rickettsiae IV. Effect of chloramphenicol and several metabolic inhibitors on the multiplication of *Rickettsia tsutsugamushi* in tissue culture cells. J. Immunol. *82:* 172-181.

Hopps, H.E., B.C. Meyer, M.F. Barile and R.A. Del Giudice. 1973. Problems concerning "noncultivable" mycoplasma contaminants in tissue cultures. Ann. N.Y. Acad. Sci. *225:* 265-276.

Hori, H. and S. Osawa. 1979. Evolutionary change in 5S RNA secondary structure and a phylogenetic tree of 54 5S RNA species. Proc. Natl. Acad. Sci. U.S.A. *76:* 381-385.

Hori, S. 1911. A bacterial leaf-disease of tropical orchids. Zentrabl. Bakteriol. Parasitenk. Infektionskr. Hyg. Abt. II *31:* 85-92.

Hori, S. 1915. An important disease of tea plants caused by a bacterium. J. Plant Protect., Tokyo *2:* 1-7.

Horisberger, M. 1977. Structure of the peptidoglycans of *Moraxella glucidolytica* and *Moraxella lwoffi* grown on hydrocarbons. Arch. Microbiol. *112:* 297-302.

Hormaeche, E. and P. R. Edwards. 1960. A proposed genus *Enterobacter.* Int. Bull. Bacteriol. Nomen. Taxon. *10:* 71-74.

Horner, H.T., Jr. and N.R. Lersten. 1972. Nomenclature of bacteria in leaf nodules of the families *Myrsinaceae* and *Rubiaceae.* Int. J. Syst. Bacteriol. *22:* 117-122.

Hornick, R.B., A.T. Dawkins, H.T. Eigelsbach and J.J. Tulis. 1967. Oral tularemia vaccine in man. *In* Hobby (Editor), Antimicrobial Agents and Chemotherapy - 1966, American Society for Microbiology, Ann Arbor, Michigan, pp. 11-14.

Hornick, R.B. and H.T. Eigelsbach. 1966. Aerogenic immunization of man with live tularemia vaccine. Bacteriol. Rev. *30:* 532-538.

Horowitz, A. T., M. Kessel and M. Shilo. 1974. Growth cycle of predacious bdellovibrios in host-free extract systems and some properties of the host extract. J. Bacteriol. *117:* 270-282.

Horsfall, F.L., Jr. and I. Tamm (Editors). 1965. Viral and rickettsial infections of man. 4th Ed., J.B. Lippincott Co., Philadelphia and Montreal, pp. 1059-

1143.

Hoshina, T. 1962. On a new bacterium, *Paracolobactrum anguillimortiferum* n. sp. Bull. Japanese Society Scien. Fisheries *28:* 162-164.

Houk, E.J. 1974a. Lipids of the primary intracellular symbiote of the pea aphid, *Acyrthosiphon pisum.* J. Insect Physiol. *20:* 471-478.

Houk, E.J. 1974b. Maintenance of the primary symbiote of the pea aphid *Acyrthosiphon pisum* in liquid media. J. Invertebr. Pathol. *24:* 24-28.

Houk, E.J. and G.W. Griffiths. 1980. Intracellular symbiotes of the Homoptera. Annu. Rev. Entomol. *25:* 161-187.

Houk, E.J., G.W. Griffiths and S.D. Beck. 1976. Lipid metabolism in the symbiotes of the pea aphid, *Acyrthosiphon pisum.* Comp. Biochem. Physiol. B Comp. Biochem. *54:* 427-431.

Houk, E.J. and D.L. McLean. 1974. Isolation of the primary intracellular symbiote of the pea aphid, *Acyrthosiphon pisum.* J. Invertebr. Pathol. *23:* 237-241.

Houk, E.J., D.L. McLean and R.S. Criddle. 1980. Pea aphid primary symbiote deoxyribonucleic acid. J. Invertebr. Pathol. *35:* 105-106.

Hovig, B. and E.H. Aandahl. 1969. A selective method for the isolation of *Haemophilus* in material from the respiratory tract. Acta Pathol. Microbiol. Scand. *77:* 676-684.

Hovind-Hougen, K. 1976. Determination by means of electron microscopy of morphological criteria of value for classification of some spirochetes, in particular treponemes. Acta Pathol. Microbiol. Scand. Sect. B, Supp. No. 255, 1-41.

Hovind-Hougen, K. 1979. *Leptospiraceae*, a new family to include *Leptospira* Noguchi 1917 and *Leptonema* gen. nov. Int. J. Syst. Bacteriol. *29:* 245-251.

Hovind-Hougen, K. 1983. *In* Validation of the publication of new names and new combinations previously effectively published outside the IJSB. List No. 10. Int. J. Bacteriol. *33:* 348–440.

Howard, C.J., J. Brownlie, R.N. Gourlay and J. Collins. 1975. Presence of a dialysable fraction in normal bovine whey capable of killing several species of bovine mycoplasmas. J. Hyg. *74:* 261-270.

Howard, C.J. and R.N. Gourlay. 1973. Serological comparison of bovine T-mycoplasmas. J. Gen. Microbiol. *79:* 129-134.

Howard, C.J. and R.N. Gourlay. 1981. Identification of ureaplasmas from cattle using antisera prepared in gnotobiotic calves. J. Gen. Microbiol. *126:* 365-369.

Howard, C.J. and R.N. Gourlay. 1982. Proposal for a second species within the genus *Ureasplasma, Ureaplasma diversum* sp. nov. Int. J. Syst. Bacteriol. *32:* 446–452.

Howard, C.J., R.N. Gourlay and J. Brownlie. 1973. The virulence of T-mycoplasmas, isolated from various animal species, assayed by intramammary inoculation of cattle. J. Hyg. *71:* 163-170.

Howard, C.J., R.N. Gourlay and J. Collins. 1975. Serological comparison between twenty-five bovine *Ureaplasma* (T-mycoplasma) strains by immunofluorescence. Int. J. Syst. Bacteriol. *25:* 155-159.

Howard, C.J., R.N. Gourlay and J. Collins. 1978a. Serological studies with bovine ureaplasmas (T-mycoplasmas). Int. J. Syst. Bacteriol. *28:* 473-477.

Howard, C.J., R.N. Gourlay, D.J. Garwes, D.H. Pocock and J. Collins. 1974. Base composition of deoxyribonucleic acid from bovine T-mycoplasmas. Int. J. Syst. Bacteriol. *24:* 373-374.

Howard, C.J., R.N. Gourlay and S.G. Wyld. 1980. Isolation of a virus, MVBrl, from *Mycoplasma bovirhinis.* FEMS Lett. *7:* 163-165.

Howard, C.J., D.H. Pocock and R.N. Gourlay. 1978b. Base composition of deoxyribonucleic acid from ureaplasmas isolated from various animal species. Int. J. Syst. Bacteriol. *28:* 599-601.

Howard, C.J., D.H. Pocock and R.N. Gourlay. 1981. Comparison, by polyacrylamide gel electrophoresis, of the polypeptides from ureaplasmas isolated from cattle and man. Int. J. Syst. Bacteriol. *31:* 128-130.

Hoyer, B.H., B.J. McCarthy and E.T. Bolton. 1964. A molecular approach in the systematics of higher organisms. Science (Washington) *144:* 959-967.

Hoyer, B.H. and N.B. McCullough. 1968a. Polynucleotide homologies of *Brucella* nucleic acids. J. Bacteriol. *95:* 444-448.

Hoyer, B.H. and N.B. McCullough. 1968b. Homologies of deoxyribonucleic acids from *Brucella ovis*, canine abortion organisms and other *Brucella* species. J. Bacteriol. *96:* 1783-1790.

Hranitzky, K.W., A. Mulholland, A.D. Larson, E.R. Eubanks and L.T. Hart. 1980. Characterization of a flagellar sheath protein of Vibrio cholerae. Infect. Immun. *27:* 597-603.

Hromatka, O. and U. Leutner. 1963. Untersuchung der Bakterienflora submerser Essiggärungen. Die Branntweinwirtschaft. *103:* 6, 174.

Hsung, J.C. and A. Haug. 1975. Intracellular pH of *Thermoplasma acidophila.* Biochim. Biophys. Acta *389:* 477-482.

Huang, K.-Y. 1967. Metabolic activity of the trench fever rickettsia, *Rickettsia quintana.* J. Bacteriol. *93:* 853-859.

Huang, T.-C., F.-H. Lin and T.-T. Kuo. 1975. Properties of membrane-bound adenosine triphosphatase from *Xanthomonas oryzae* Bot. Bull. Acad. Sinica *16:* 36-44.

Hubbert, W.T. and M.N. Rosen. 1970a. I. *Pasteurella* infection due to animal bite. Am. J. Publ. Health *60:* 1103-1108.

Hubbert, W.T. and M.N. Rosen. 1970b. II. *Pasteurella multocida* in man unrelated to animal bite. Am. J. Publ. Health *60:* 1109-1117.

Huber-Schneider, L. 1957. Morphologische und physiologische Untersuchungen an der Wanze *Mesocerus marginatus* L. und ihren Symbionten (Heteroptera). Z. Morphol. Oekol. Tiere *46:* 433-480.

Huddleson, I.F. 1929. The differentiation of the species of the genus *Brucella.* Bull. Mich. Agric. Exp. Sta. *100:* 1-6.

Huddleson, I.F. 1940. The presence of a capsule on *Brucella* cells. J. Am. Vet. Med. Assoc. *96:* 708-709.

Huddleson, I.F. 1943. Brucellosis in Man and Animals. Revised Edition. The Commonwealth Fund: New York.

Huddleson, I.F. 1954. Effect of killed *Brucella* cells and an extract from sonically disintegrated *Brucella* cells in culture mediums on the growth of *Brucella abortus.* Quart. Bull. Mich. St. Univ. Agric. Exp. Sta. *37:* 14-22.

Huddleson, I.F. 1957. Genus III *Brucella* Meyer and Shaw 1920. *In* Breed, Murray and Smith (Editors), Bergey's Manual of Determinative Bacteriology, 7th Ed., The Williams and Wilkins Co., Baltimore, pp. 404-406.

Hudson, H.P., A.A. Lindberg and B.A.D. Stocker. 1978. Lipopolysaccharide core defects in *Salmonella typhimurium* mutants which are resistant to Felix 0 phage but retain smooth character. J. Gen. Microbiol. *109:* 97-112.

Hudson, J.R., G.S. Cottew and H.E. Adler. 1967. Diseases of goats caused by *Mycoplasma.* A review of the subject with some new findings. Ann. N.Y. Acad. Sci. *143:* 287-297.

Huebner, E. and K.G. Davey. 1974. Bacteroids in the ovaries of a tsetse fly. Nature (Lond.) *249:* 260-261.

Huebner, R.J., W.L. Jellison and C. Pomerantz. 1946. Rickettsialpox, newly recognized rickettsial disease. IV. Isolation of a rickettsia apparently identical with the causative agent of rickettsialpox from *Allodermanyssus sanguineus*, a rodent mite. Pub. Health Rep. *61:* 1677-1682.

Huebner, R.J., P. Stamps and C. Armstrong. 1946. Rickettsialpox--a newly recognized rickettsial disease. I. Isolation of the etiological agent. Pub. Health Rep. *61:* 1605-1614.

Huet, M. and H. Thouvenot. 1964. Étude d'un bacteriophage actif sur une bactérie anaérobie: *Sphaerophorus varius.* Ann. Inst. Pasteur *106:* 867.

Huff, C.G. 1928. Nutritional studies on the seed-corn maggot, *Hylemyia cilicrura* Rondani. J. Agric. Res. *36:* 625-630.

Huger, A. 1956. Experimentelle Untersuchungen über die kunstliche Symbionten-elimination bei Vorratsschädlingen: *Rhizopertha dominica* F. (Bostrychidae) und *Oryzaephilus surinamensis* L. (Cucujidae). Z. Morphol. Oekol. Tiere *44:* 626-701.

Huger, A. 1959. Histological observations on the development of crystalline inclusions of the rickettsial disease of *Tipula paludosa* Meigen. J. Insect Pathol. *1:* 60-66.

Huger, A. 1962. Zür Genese der Begleitkristalle bei *Rickettsiella*-Infektionen von Insekten. Naturwissenschaften *49:* 358-360.

Huger, A. 1964. Eine Rickettsiose der Orientalischen Schabe, *Blatta orientalis* L., verursacht durch *Rickettsiella blattae* nov. spec. Naturwissenschaften *51:* 22.

Hugh, R. 1965. A comparison of *Pseudomonas testosteroni* and *Comamonas terrigena.* Int. Bull. Bacteriol. Nomencl. Taxon. *15:* 125-132.

Hugh, R. 1981. *Pseudomonas maltophilia* sp. nov., nom. rev. Int. J. Syst. Bacteriol. *31:* 195.

Hugh, R. and G.L. Gilardi. 1980. *Pseudomonas. In* Lenette, Balows, Hausler and Truant (Editors), Manual of Clinical Microbiology 3rd Ed., American Society for Microbiology, Washington, D.C., pp. 289-317.

Hugh, R. and P. Ikari. 1964. The proposed neotype strain of *Pseudomonas alcaligenes* Monias (1928). Int. Bull. Bacteriol. Nomencl. Taxon. *14:* 103-107.

Hugh, R. and E. Leifson. 1953. The taxonomic significance of fermentative versus oxidative metabolism of carbohydrates by various Gram-negative bacteria. J. Bacteriol. *66:* 24-26.

Hugh, R. and E. Ryschenkow. 1960. An alcaligenes-like *Pseudomonas* species. Bacteriol. Proc., p. 78.

Hughes, D.E. and G.W. Pugh. 1970. Isolation and description of a *Moraxella* from horses with conjunctivitis. Amer. J. Vet. Res. *31:* 457-462.

Hughes, M.L. 1893. The natural history of certain fevers occurring in the Mediterranean. Mediterranean Nat. *2:* 299-300; 325-327; 332-334.

Humm, H.J. 1946. Marine agar-digesting bacteria of the South Atlantic coast. Bull. Duke Univ. Mar. Sta. *3:* 45-75.

Humphrey, B., J.M. Vincent and V. Skerdleta. 1973. Group antigens in slow-growing *Rhizobium.* Arch. Mikrobiol. *89:* 79-82.

Hungate, R.E. 1950. The anaerobic mesophilic cellulolytic bacteria. Bacteriol. Rev. *14:* 1-49.

Hungate, R.E. 1966. The rumen and its microbes. Academic Press, New York.

Hungate, R.E. 1969. A roll tube method for cultivation of strict anaerobes. *In* Norris and Ribbons (Editors), Methods in Microbiology, Vol. 3B, Academic Press, London and New York, pp. 117-132.

Hunt, J.C. and P.V. Phibbs, Jr. 1981. Failure of *Pseudomonas aeruginosa* to form membrane-associated glucose dehydrogenase activity during anaerobic growth with nitrate. Biochim. Biophys. Res. Comm. *102:* 1393-1399.

Hunter, M.I.S., T.L. Olawoye and D.A. Saynor. 1981. The effect of temperature on the growth and lipid composition of the extremely halophilic coccus, *Sarcina marina.* Antonie van Leeuwenhoek J. Microbiol. Serol. *47:* 25-40.

Huq, M.I., A.K.M.J. Alam, D.J. Brenner and G.K. Morris. 1980a. Isolation of *Vibrio*-like group, EF-6, from patients with diarrhea. J. Clin. Microbiol. *11:* 621-624.

Huq, M.I., A.R.M.A. Alim, L.N. Mutanda, M.D. Yunus and M.U. Khan. 1980b. Multiply antibiotic-resistant O-group 1 *Vibrio cholerae.* Bangladesh Morb. Mortal Wkly. Rep. *29:* 109-110.

Hurlbert, R.E. and W.B. Jakoby. 1965. Tartaric acid metabolism. J. Biol. Chem. *240:* 2772-2777.

Hurpin, B. and P.H. Robert. 1972. Comparison of the activity of certain pathogens of the cockchafer *Melolontha melolontha* in plots of natural meadowland. J. Invert. Pathol. *19:* 291-298.

Hurpin, B. and P.H. Robert. 1976. Conservation dans le sol de trois germes pathogènes pour les larvaes de *Melolontha melolontha (Col: Scarabaeidae).* Entomophaga 21: 73-80.

Hurpin, B. and P.H. Robert. 1977. Effets en population naturelle de *Melolontha melolontha (Col: Scarabaeidae)* d'une introduction de *Rickettsiella melolontahe* et de *Entomopoxvirus melolonthae.* Entomophaga 22: 85-92.

Hurvell, B. and A.A. Lindberg. 1973. Immunochemical studies on the cross-reactions between *Brucella* species and *Yersinia enterocolitica* type 9. Contr. Microbiol. Immunol. *2:* 159-168.

Hurvell, B. and A.A. Lindberg. 1973. Serological cross-reactions between different *Brucella* species and *Yersinia enterocolitica.* Immunochemical studies on phenol-water-extracted lipopolysaccharides from *Brucella abortus* and *Yersinia enterocolitica* type IX. Acta Pathol. Microbiol. Scand. B*81:* 113-119.

Hutchinson, P.B. 1949. A bacterial disease of *Dysoxylum spectabile* caused by the pathogen *Pseudomonas dysoxyli* n. sp. N. Z. J. Sci. Technol. *B30:* 274-286.

Huxsoll, D.L. 1976. Canine ehrlichiosis (tropical canine pancytopenia): a review. Vet. Parasitol. *2:* 49-60.

Hvid-Hansen, N. 1951. Sulfate-reducing and hydrocarbon-producing bacteria in ground-water. Acta Pathol. Microbiol. Scand. *29:* 266-289.

Hwang, M. N. and G. M. Ederer. 1975. Rapid hippurate hydrolysis method for presumptive identification of Group B Streptococci. J. Clin. Microbiol. *1:* 114-115.

Hylemon, P. B., N. R. Krieg and P. V. Phibbs, Jr. 1974. Transport and catabolism of D-fructose by *Spirillum itersonii.* J. Bacteriol. *117:* 144-150.

Hylemon, P.B. and E.J. Stellwag. 1976. Bile acid biotransformation rates of selected Gram-positive and Gram-negative intestinal anaerobic bacteria. Biochem. Biophys. Res. Commun. *69:* 1088-1094.

Hylemon, P. B., J. S. Wells, Jr., J. H. Bowdre, T. O. MacAdoo and N. R. Krieg. 1973a. Designation of *Spirillum volutans* Ehrenberg 1832 as type species of the genus *Spirillum* Ehrenberg 1832 and designation of the neotype strain of *S. volutans.* Request for an opinion. Int. J. Syst. Bacteriol. *23:* 20-27.

Hylemon, P. B., J. S. Wells, Jr., N. R. Krieg and H. W. Jannasch. 1973b. The genus *Spirillum:* a taxonomic study. Int. J. Syst. Bacteriol. *23:* 340-380.

Hynes, M. 1942. The isolation of intestinal pathogens by selective media. J. Pathol. Bacteriol. *54:* 193-207.

Igarashi, Y., T. Kodama and Y. Minoda. 1980. Identification and physiological characterization of a new amylolytic hydrogen bacterium, *Pseudomonas hydrogenovora.* Agr. Biol. Chem. (Japan) 44: 1277-1281.

Igra-Siegman, Y., H. Chmel and C. Cobbs. 1980. Clinical and laboratory characteristics of *Achromobacter xylosoxidans* infection. J. Clin. Microbiol. *11:* 141-145.

Iida, T. and Y. Ajiki. 1975. The effect of 2,4-dinitrophenol on the growth of *Bordetella pertussis* on chick tracheal organ culture. Jpn. J. Microbiol. *19:* 381-386.

Iida, T., N. Kusano, A. Yamamoto and H. Shiga. 1962. Studies on experimental infection with *Bordetella pertussis.* Bacteriological and pathological studies on the mode of infection in mouse brain. Jpn. J. Exp. Med. *32:* 471-492.

Iino, T. and J. Lederberg. 1964. Genetics of *Salmonella. In* Van Oye (Editor) The world problem of salmonellosis. Junk, The Hague, pp. 110-142.

Iizuka, H. and K. Komagata. 1963. New species of *Pseudomonas* belonged to fluorescent group (Studies on the microorganisms of cereal grains. Part V). J. Agr. Chem. Soc. Japan 37: 137-141.

Iizuka, H. and K. Komagata. 1964a. Microbiological studies on petroleum and natural gas. I. Determination of hydrocarbon-utilizing bacteria. J. Gen. Microbiol. *10:* 207-221.

Iizuka, H. and K. Komagata. 1964b. Microbiological studies on petroleum and natural gas. II. Determination of pseudomonads isolated from oil-brines and related materials. J. Gen. Appl. Microbiol. *10:* 223-231.

Ikemoto, S., K. Suzuki, T. Kaneko and K. Komagata. 1980. Characterization of strains of *Pseudomonas maltophilia* which do not require methionine. Int. J. Syst. Bacteriol. *30:* 437-447.

Ilemobade, A.A. and C. Blotkamp. 1978. *Eperythrozoon ovis* as a possible cause of anemia in Nigerian sheep. Vet. Rec. *101:* 153-154.

Imada, A., K. Kintaka and K. Haibara. 1980a. Antibiotic SB-72310. U.S. Patent 4,225,586 (Sept. 30, 1980).

Imada, A., K. Kitano and M. Asai. 1980b. Antibiotic G-6302. U.S. Patent 4,229,436 (October 21, 1980).

Imada, A., K. Kitano, K. Kintaka, M. Muroi and M. Asai. 1981. Sulfazecin and isosulfazecin, novel beta-lactam antibiotics of bacterial origin. Nature 289: 590-591.

Imanaka, H., M. Kousaka, G. Tamura and K. Arima. 1965. Studies on pyrrolnitrin, a new antibiotic. II. Taxonomic studies on pyrrolnitrin-producing strain. J. Antibiot. *18:* 205-206.

Inaba, F. 1960. The fine structure of the nuclei of *Spirostomum ambiguum* seen by the electron microscope. Biol. J. Nara Women's Univ. *10:* 26-29.

Inoue, K. and K. Komagata. 1976. Taxonomic study on obligately psychrophilic bacteria isolated from Antarctica. J. Gen. Appl. Microbiol. *22:* 165-176.

International Committee of Bacteriological Nomenclature. 1958. International Code of Nomenclature of Bacteria and Viruses: Bacteriological Code. The Iowa State University Press, Ames.

International Committee on Systematic Bacteriology Subcommittee on the Taxonomy of *Mollicutes.* 1979. Proposal of minimal standards for descriptions of new species of the class *Mollicutes.* Int. J. Syst. Bacteriol. *29:* 172-180.

International Salmonella Subcommittee. 1934. The genus *Salmonella* Lignières 1900. J. Hyg. *34:* 333-350.

Irgens, R.L. 1977. *Meniscus,* a new genus of aerotolerant, gas-vacuolated bacteria. Int. J. Syst. Bacteriol. *27:* 38-43.

Irons, L.I. and A.P. MacLennan. 1979. Substrate specificity and the purification by affinity combination methods of the two *Bordetella pertussis* haemagglutinins. *In* Manclark and Hill (Editors), International Symposium on Pertussis, U.S. Government Printing Office, Washington, D.C., pp. 338-349.

Isaac, L. and G. C. Ware. 1974. The flexibility of bacterial cell walls. J. Appl. Bacteriol. *37:* 335-339.

Isayama, Y., R. Azuma, S. Tanaka and T. Suto. 1977. The pathogenicity and antigenicity of *Brucella canis* QE13 for experimental animals. Ann. Sclavo *19:* 89-98.

Iseki, S.K. and K. Kashiwagi. 1955. Induction of somatic 1 antigen by bacteriophage in *Salmonella* B group. Proc. Jpn. Acad. *31:* 558-563.

Iseki, S.K. and K. Kashiwagi. 1957. Lysogenic conversions and transduction of genetic characters by temperate phage iota in *Salmonella.* Proc. Jpn. Acad. *33:* 481-485.

Iseki, S.K. and T. Sakai. 1953. Artificial transformation of O antigens in *Salmonella* E group. Proc. Jpn. Acad. *29:* 121-126; 127-131.

Isenberg, H.D. and J. Sampson-Scherer. 1977. Clinical laboratory evaluation of a system approach to the recognition of nonfermentative or oxidase-producing Gram-negative rod-shaped bacteria. J. Clin. Microbiol. *5:* 336-340.

Ishiie, T., Y. Doi, K. Yora and H. Asuyama. 1967. Suppressive effects of antibiotics of tetracycline group on symptom development of mulberry dwarf disease. Ann. Phytopathol. Soc. Jap. *33:* 267-275.

Ishikawa, H. 1977. RNA synthesis in aphids, *Lachnus tropicalis.* Biochem. Biophys. Res. Commun. *78:* 1418-1423.

Ishikawa, H. 1978. Intracellular symbiont as a major source of the ribosomal RNAs in the aphid mycetocytes. Biochem. Biophys. Res. Commun. *81:* 993-999.

Ishiyama, S. 1922. Studies of bacterial leaf blight of rice. Rept. Imperial Agr. Sta. Konosu *45:* 233-261.

Ito, H. and H. Iizuka. 1971. Taxonomic studies on radio-resistant *Pseudomonas.* Part XII. Studies on the microorganisms of cereal grains. Agr. Biol. Chem. (Japan) *35:* 1566-1571.

Ito, S. and J.W. Vinson. 1965. Fine structure of *Rickettsia quintana* cultivated in vitro and in the louse. J. Bacteriol. *89:* 481-495.

Ito, S., J.W. Vinson and T.J. Mcguire, Jr. 1975. Murine typhus rickettsiae in the oriental rat flea. Ann. N.Y. Acad. Sci. *266:* 35-60.

Itoh, Y., K. Izaki and H. Takahashi. 1978. Purification and characterization of a bacteriocin from *Erwinia carotovora.* J. Gen. Appl. Microbiol. *24:* 27-39.

Itzigsohn, H. 1868. Entwicklungsvorgange von *Zoogloea, Oscillaria, Synedra, Staurastrum, Spirotaenia* und *Chroolepus,* p. 30-31. S. B. Ges. Naturf. Fr. Berlin, 19 Nov. 1967.

Iuchi, S., Y. Kubota and S. Tanaka. 1975. Mutants defective in binding activity for cyclic adenosine-3',5'-monophosphate in *Vibrio parahaemolyticus.* J. Bacteriol. *124:* 567-569.

Iveson, J.B. 1971. Strontium chloride B and E.E. enrichment broth media for the isolation of *Edwardsiella, Salmonella* and *Arizona* species from tiger snakes. J. Hyg. Camb. *69:* 323-330.

Iveson, J.B. 1973. Enrichment procedures for the isolation of *Salmonella, Arizona, Edwardsiella* and *Shigella* from faeces. J. Hyg. Camb. *71:* 349-361.

Iwasa, S., S. Asakawa, S. Ishida and M. Kurakawa. 1966. The study on lymphocytosis-promoting factor produced by *Bordetella pertussis* (in Japanese). Proc. of the 21st general meeting of Kanto Region of Japanese Bacteriological Society. Jpn. J. Bacteriol. *22:* 233-234.

Izard, D., C. Ferragut, F. Gavini, K. Kersters, J. De Ley and H. Leclerc. 1981. *Klebsiella terrigena,* a new species from soil and water. Int. J. Syst. Bacteriol. *31:* 116-127.

Izard, D., F. Gavini and H. Leclerc. 1980a. Polynucleotide sequence relatedness and genome size among *Enterobacter intermedium* and the species *Enterobacter cloacae* and *Klebsiella pneumoniae.* Zentralbl. Bakteriol. Parasitenkd. Infektionskr. Hyg. Abt. I, Orig. C *1:* 51-60.

Izard, D., F. Gavini and H. Leclerc. 1980. *In* Validation of the publication of new names and new combinations previously effectively published in the IJSB. Int. J. Syst. Bacteriol. *30:* 601.

Izard, D., F. Gavini, P.A. Trinel, F. Krubwa, A. and H. Leclerc. 1980. Contribution of DNA-DNA hybridization to the transfer of *Enterobacter aerogenes* to the genus *Klebsiella* as *K. mobilis.* Zentralbl. Bakteriol. Parasitenkd. Infektionskr. Hyg. Abt. I. Orig. C *1:* 257-263.

Izard, D., F. Gavini, P.A. Trinel and H. Leclerc. 1979. *Rahnella aquatilis,* nouveau membre de la famille des *Enterobacteriaceae.* Ann. Microbiol. (Inst. Pasteur) *130A:* 163-177.

Izard, D., F. Gavini, P.A. Trinel and H. Leclerc. 1981. Deoxyribonucleic acid relatedness between *Enterobacter cloacae* and *Enterobacter amnigenus* sp. nov. Int. J. Syst. Bacteriol. *31:* 35-42.

Izard, D., F. Gavini, P.A. Trinel and H. Leclerc. 1981. In List No. 7, Validation of the publication of new names and new combinations previously effectively published outside the IJSB. Int. J. Syst. Bacteriol. *31:* 382-383.

Izdebska-Szymona, K., E. Monczak and B. Lemczak. 1971. Preliminary scheme of phage typing of *Proteus mirabilis* strains. Exp. Med. Microbiol. *23:* 18-22. (Originally published in Polish.)

Jackins, H.C. and H.A. Barker. 1951. Fermentative processes of the fusiform bacteria. J. Bacteriol. *61:* 101-114.

Jackson, E.B., T.T. Crocker and J.E. Smadel. 1952. Studies on two rickettsialike agents probably isolated from guinea pigs. Bacteriol. Proc., p. 119.

Jackson, E.B., J.X. Danauskas, M.C. Coale and J.E. Smadel. 1957. Recovery of *Rickettsia akari* from the Korean vole *Microtus fortis pelliceus*. Am. J. Hyg. *66:* 301-308.

Jackson, F.L. and Y.E. Goodman. 1972. Transfer of the facultatively anaerobic organism *Bacteroides corrodens* Eiken to a new genus, *Eikenella*. Int. J. Syst. Bacteriol. *22:* 73-77.

Jackson, F.L. and Y.E. Goodman. 1978. *Bacteroides ureolyticus,* a new species to accommodate strains previously identified as "*Bacteroides corrodens,* anaerobic." Int. J. Syst. Bacteriol. *28:* 197-200.

Jackson, F.L., Y.E. Goodman, F.R. Bel, P.C. Wong and R.L.S. Whitehouse. 1971. Taxonomic status of facultative and strictly anaerobic "corroding bacilli" that have been classified as *Bacteroides corrodens.* J. Med. Microbiol. *4:* 171-184.

Jackson, F.L., Y.E. Goodman and D.E. Rhodes. 1971. A new type of anaerobic Gram-negative 'Corroding Bacillus'. Bacteriol. Proc., p. 108, M-262.

Jackson, F.L., R.L.S. Whitehouse, Y. Goodman and P.C. Wong. 1970. Comparison of certain agar-pitting organisms designated *Bacteroides corrodens.* Bacteriol. Proc. *75:* (M3).

Jackson, T.J., R.F. Ramaley and W.G. Meinschein. 1973. *Thermomicrobium,* a new genus of extremely thermophilic bacteria. Int. J. Syst. Bacteriol. *23:* 28-36.

Jacob, A.E., J.M. Cresswell and R.W. Hedges. 1977. Molecular characterization of the P group plasmid R68 and variants with enhanced chromosome mobilizing ability. FEMS Microbiol. Lett. *1:* 71-74.

Jacobs, D.R. 1980. A descriptive microbiological study of periapical infection. Thesis, Harvard University, Boston, Massachusetts.

Jacobsthal, E. 1920. Untersuchungen über eine syphilisahnliche spontanerkrankung des kaninchens (*Paralues-cuniculi*). Derm. Wochenschr. *71:* 569-571.

Jacoby, G.A. 1974a. Properties of R plasmids determining gentamicin resistance by acetylation in *Pseudomonas aeruginosa.* Antimicrob. Agents Chemother. *6:* 239-252.

Jacoby, G.A. 1974b. Properties of an R plasmid in *Pseudomonas aeruginosa* producing amikicin (BB-K8), butirosin, kanamycin, tobramycin, and sisomicin resistance. Antimicrob. Agents Chemother. *6:* 807-810.

Jacoby, G.A. 1977. Classification of plasmids in *Pseudomonas aeruginosa. In* Schlessinger (Editor), "Microbiology-1977", American Society for Microbiology, Washington, D.C., pp. 119-126.

Jacoby, G.A. 1979. Plasmids of *Pseudomonas aeruginosa. In* Doggett (Editor), *Pseudomonas aeruginosa.* Clinical manifestations of infection and current therapy. Academic Press, New York, pp. 271-309.

Jacoby, G.A. and M. Matthew. 1979. The distribution of beta-lactamase genes on plasmids found in *Pseudomonas.* Plasmid *2:* 41-47.

Jacoby, G.A. and J.A. Shapiro. 1977. Plasmids studied in *Pseudomonas aeruginosa* and other pseudomonads. *In* Bukhari, Shapiro and Adhya (Editors), DNA insertion elements, plasmids and episomes. Cold Spring Harbor Laboratory, Cold Spring Harbor, New York, pp. 639-656.

Jagger, I.C. 1921. Bacterial leaf spot disease of celery. J. Agr. Res. *21:* 185-188.

Jahn, T.L. and M.D. Landman. 1965. Locomotion of spirochetes. Trans. Am. Microsc. Soc. *84:* 395-406.

Jain, K., K. Radsak and W. Mannheim. 1974. Differentiation of the oxytocum group from *Klebsiella* by deoxyribonucleic acid-deoxyribonucleic acid hybridization. Int. J. Syst. Bacteriol. *24:* 402-407.

Jain, K.C. and W.B. Whalley. 1980. The bacterial pigment from *Pseudomonas lemonnieri.* Part 2. The synthesis of 3 *n*-octanamidopyridine-2,5,6-trione: the structure and synthesis of lemonnierin. J. Chem. Soc. Perkins Trans. I, *8:* 1788-1794.

Jain, N.C., D.E. Jasper and J.D. Dellinger. 1967. Cultural characters and serological relationships of some mycoplasmas isolated from bovine sources. J. Gen. Microbiol. *49:* 401-410.

James, J. and J. Swanson. 1978. Studies on gonococcus infection. XIII. Occurrence of color/opacity colonial variants in clinical cultures. Infect. Immun. *19:* 332-340.

James, J.F. and J. Swanson. 1977. The capsule of the gonococcus. J. Exp. Med. *145:* 1082-1086.

James, N. 1955. Yellow chromogenic bacteria on wheat. II. Determinative studies. Can. J. Microbiol. *1:* 479-485.

Jamieson, A.F., R.L. Bieleski and R.E. Mitchell. 1981. Plasmids and phaseolotoxin production in *Pseudomonas syringae* pv. *phaseolicola.* J. Gen. Microbiol. *122:* 161-165.

Jandl, G. and K. Linke. 1976. Bericht über zwei Fälle von akuter Gastroenteritis durch *Plesiomonas shigelloides.* Zentralbl. Bakteriol. Parasitenkd. Infektionskr. Hyg. Abt. I Orig. A *236:* 136-140.

Janke, A. 1916. Studien über die Essigsäurebakterien-flora von Lagerbieren des Wiener Handels. Zentralbl. Bakteriol. Infektionskr. Hyg. Abt. II *45:* 1-48.

Janke, A. 1950. *Acetobacter lafarianum* nov. nom. Arch. Mikrobiol. *15:* 116-118.

Janke, A. 1957. Zür Systematik der Essigbakterien. Zentralbl. Bakteriol. Parasitenkd. Infektionskr. Hyg. Abt. II *110:* 728-739.

Janke, A. 1960. Die Essigsäuregärung. Handbuch der Pflanzenphysiologie, pp. 670-746.

Jann, K. and B. Jann. 1977. Bacterial polysaccharide antigens. *In* Sutherland (Editor), Surface Carbohydrate of the Prokaryotic Cell. Academic Press, London, pp. 247-287.

Jannasch, H. W. 1963. Studies on the ecology of a marine spirillum in the chemostat. *In* Oppenheimer (Editor), 1st International Symposium on Marine Microbiology. C. C. Thomas, Springfield, Ill., pp. 558-566.

Jannasch, H. W. 1965. Die Isolierung heterotropher aquatischer Spirillen. *In* Schlegel (Editor), Anreicherungskultur und Mutantenauslese, Gustav Fischer Verlag, Stuttgart, pp. 198-203.

Jannasch, H. W. 1967. Enrichments of aquatic spirilla in continuous culture. Arch. Mikrobiol. *59:* 165-173.

Jansen, B.C. 1952. The occurrence of *Eperythrozoon parvum* Splitter in South African swine. Onderstenpoort J. Vet. Res. *24:* 5-6.

Janssen, W.A. and H.J. Surgalla. 1968. Morphology, physiology and serology of a *Pasteurella* species pathogenic for white perch. J. Bacteriol. *96:* 1606-1610.

Jansson, E., H. Kenne, B. Lindberg, H. Ljunggren, J. Lonngren, V. Ruden and S. Svensson. 1977. Demonstration of an octasaccharide repeating unit in the extracellular polysaccharide of *Rhizobium meliloti* by sequential degradation. J. Amer. Chem. Soc. *99:* 3812-3815.

Jantzen, E., B.P. Berdal and T. Omland. 1979. Cellular fatty acid composition of *Francisella tularensis.* J. Clin. Microbiol. *10:* 928-930.

Jantzen, E., B.P. Berdal and T. Omland. 1981. Cellular and fatty acid taxonomy of *Haemophilus, Pasteurella* and *Actinobacillus. In* Kilian, Frederiksen and Biberstein (Editors), *Haemophilus, Pasteurella* and *Actinobacillus,* Academic Press, London and New York, pp. 197-203.

Jantzen, E., K. Bryn, T. Bergan and K. Bøvre. 1974a. Gas chromatography of bacterial whole cell methanolysates. IV. A procedure for fractionation and identification of fatty acids and monosaccharides of cellular structures. Acta Pathol. Microbiol. Scand. Sect. B *82:* 753-766.

Jantzen, E., K. Bryn, T. Bergan and K. Bøvre. 1974b. Gas chromatography of bacterial whole cell methanolysates. V. Fatty acid composition of *Neisseriae* and *Moraxellae.* Acta Pathol. Microbiol. Scand. Sect. B *82:* 767-779.

Jantzen, E., K. Bryn, T. Bergan and K. Bøvre. 1975. Gas chromatography of bacterial whole cell methanolysates. VII. Fatty acid composition of *Acinetobacter* in relation to the taxonomy of *Neisseriaceae.* Acta Pathol. Microbiol. Scand. *83B:* 569-580.

Jantzen, E., K. Bryn and K. Bøvre. 1976. Cellular monosaccharide patterns of *Neisseriaceae.* Acta Pathol. Microbiol. Scand. *84B:* 177-188.

Jantzen, E. and T. Hofstad. 1981. Fatty acids of *Fusobacterium* species: taxonomic implications. J. Gen. Microbiol. *123:* 163-171.

Jantzen, E. and J. Lassen. 1980. Characterization of *Yersinia* species by analysis of whole cell fatty acids. Int. J. Syst. Bacteriol. *30:* 421-428.

Jarvis, B.D.W., A.G. Dick and R.M. Greenwood. 1980. Deoxyribonucleic acid homology among strains of *Rhizobium trifolii* and related species. Int. J. Syst. Bacteriol. *30:* 42-52.

Jarvis, B.D.W., C.E. Pankhurst and J.J. Patel. 1982. *Rhizobium loti,* a new species of legume root nodule bacteria. Int. J. Syst. Bacteriol. *32:* 378-380.

Jasper, D.E., H. Erno, J.D. Dellinger and C. Christiansen. 1981. *Mycoplasma californicum,* a new species from cows. Int. J. Syst. Bacteriol. *31:* 339-345.

Jawetz, E. 1950. A pneumotropic pasteurella of laboratory animals. I. Bacteriological and serological characteristics of the organism. J. Infect. Dis. *86:* 172-183.

Jeannes, A. 1974. Applications of extracellular microbial polysaccharide-polyelectrolytes: Review of literature, including patents. J. Polym. Sci. Polym. Symp. No. *45:* 209-227.

Jeffrey, C. 1977. Biological Nomenclature, 2nd Ed., Arnold, London.

Jeffries, L., M.A. Cawthorne, M. Harris, B. Cook and A.T. Diplock. 1969. Menaquinone determination in the taxonomy of *Micrococcaceae.* J. Gen. Microbiol. *54:* 365-380.

Jellison, W.L. 1974. Tularemia in North America 1930-1974. University of Montana Foundation, Missoula, Montana.

Jenkins, C.L., A.G. Andrewes, T.J. McQuade and M.P. Starr. 1979. The pigment of *Pseudomonas paucimobilis* is a carotenoid (nostoxanthin), rather than a brominated aryl-polyene (xanthomonadin). Curr. Microbiol. *3:* 1-4.

Jenkins, R.A. 1970. The fine structure of nuclear envelope associated endosymbiont of *Paramecium.* J. Gen. Microbiol. *61:* 355-359.

Jennings, H.J., A.K. Bhattacharjee, L. Kenne, C.P. Kenny and G. Calver. 1980. The R-type lipopolysaccharides of *Neisseria meningitidis.* Can. J. Biochem. *58:* 128-136.

Jennison, H.M. 1923. Potato blackleg with special reference to the etiological agent. Ann. Rep. Mo. Bot. Gard. *10:* 1-72.

Jensen, H.L. 1955. *Azotobacter macrocytogenes* n. sp. a nitrogen-fixing bacterium resistant to acid reactions. Acta Agric. Scand. *5:* 280-294.

Jensen, H.L., E.J. Petersen, P.K. De and R. Bhattacharya. 1960. A new nitrogen-fixing bacterium: *Derxia gummosa* nov. gen. nov. spec. Arch. Mikrobiol. *36:* 182-195.

Jensen, J. and E. Thofern. 1953. Chlorhämin als Bakterienwuchsstoff I. Z. Naturforsch. *8b:* 599-603.

Jensen, M.J., P. Baumann, M. Mandel and J.V. Lee. 1980. Characterization of facultatively anaerobic marine bacteria belonging to group F of Lee, Donovan, and Furniss. Curr. Microbiol. *3:* 373-376.

Jensen, M.J., B.M. Tebo, P. Baumann, M. Mandel and K.H. Nealson. 1980. Characterization of *Alteromonas hanedai* (sp. nov.), a nonfermentative luminous species of marine origin. Curr. Microbiol. *3:* 311-315.

Jensen, M.J., B.M. Tebo, P. Baumann, M. Mandel and K.H. Nealson. 1981. *In* List No. 7, Validation of the publication of new names and combinations previously effectively published outside the IJSB. Int. J. Syst. Bacteriol. *31:* 382-383.

Jensen, P.T., C. Wolstrup and N.F. Friis. 1978. Utilisation of glucose by *Mycoplasma suipneumoniae* and *Mycoplasma flocculare*. Acta Vet. Scand. *19:* 179-183.

Jensen, R. 1974. Diseases of Sheep. Lea and Febiger, Philadelphia, 389 pp.

Jensen, R.A., D.S. Nasser and E.W. Nester. 1967. Comparative control of a branchpoint enzyme in microorganisms. J. Bacteriol. *94:* 1582-1593.

Jensen, R.A. and J.L. Rebello. 1970. Comparative allostery of microbial enzymes at metabolic branch-points: evolutionary implications. *In* Corum (Editor), Devel. Indust. Microbiol. Vol. 11, Plenum Press, New York, pp. 105-121.

Jensen, R.A. and S.L. Stenmark. 1970. Comparative allostery of 3-deoxy-D-*arabino*-heptulosonate-7-phosphate synthetase as a molecular basis for classification. J. Bacteriol. *101:* 763-769.

Jensen, V. 1955. The *Azotobacter*-flora of some Danish watercourses. Bot. Tidsskr. *52:* 143-157.

Jensen, V. and E. Holm. 1975. Associative growth of nitrogen-fixing bacteria with other micro-organisms. *In* Stewart (Editor), Nitrogen Fixation by Free-living Micro-organisms, Cambridge University Press, Cambridge, pp. 101-119.

Jensen, V. and E.J. Petersen. 1954. Studies on the occurrence of *Azotobacter* in Danish forest soils. Åsskr. K. Vet.-Landbøhojsk, 94-101.

Jensen, V. and E.J. Petersen. 1955. Taxonomic studies on *Azotobacter chroococcum* Beijerinck and *Azotobacter beijerinckii* Lipman. Åsskr. K. Vet.-Landbøhojsk, 107-126.

Jensen, W.I. and R.M. Duncan. 1980. *Bordetella bronchiseptica* associated with pulmonary disease in mountain vole (*Microtus montanus*). J. Wildl. Dis. *16:* 11-14.

Jensen, W.I., C.R. Owen and W.J. Jellison. 1969. *Yersinia philomiragia* sp. n., a new member of the *Pasteurella* group of bacteria, naturally pathogenic for the muskrat (*Ondatra zibethica*). J. Bacteriol. *100:* 1237-1241.

Jeon, K.W. 1980. Symbiosis of bacteria with *Amoeba*. *In* Cook, Pappas and Rudolph (Editors), Cellular Interactions in Symbiotic and Parasitic Relationships, Ohio State University Press, Columbus, pp. 245-262.

Jermoljewa, S. 1926. *Vibrio phosphorescens* beim klinishen Bilde der Cholera und sein Zusammenhang mit anderen Vibrionen. Zentralbl. Bakteriol. Parasitenkd. Infektionskr. Hyg. Abt. I Orig. *100:* 170-177.

Jerrels, J.R., D.J. Hinricks and L.P. Mallavia. 1974. Cell envelope analysis of *Coxiella burnetii* phase I and phase II. Can. J. Microbiol. *20:* 1465-1470.

Jessen, J. 1934. Studien über gramnegative Kokken. Zentralbl. Bakteriol. Parasitenkd. Infektionskr. Hyg. Abt. I Orig. *133:* 73-88.

Jessen, O. 1965. *Pseudomonas aeruginosa* and other green fluorescent pseudomonads. A taxonomic study. Munksgaard, Copenhagen, pp. 1-244.

Jindal, J.K., P.N. Patel and R. Singh. 1972. Bacterial leaf spot disease on *Amorphophallus campanulatus*. Indian Phytopathol. *25:* 374-377.

Joens, L.A., R.D. Glock and J.M. Kinyon. 1980. Differentiation of *Treponema hyodysenteriae* from *T. innocens* by enteropathogenicity testing in the CF1 mouse. Vet. Record. *107:* 527-529.

John, A., H.R. Isaacson and M.P. Bryant. 1974. Isolation and characteristics of a ureolytic strain of *Selenomonas ruminantium*. J. Dairy Sci. *57:* 1003-1014.

John, P. and F.R. Whatley. 1975a. *Paracoccus denitrificans* and the evolutionary origin of the mitochondrion. Nature *254:* 495-498.

John, P. and F.R. Whatley. 1975b. *Paracoccus denitrificans*: a present-day bacterium resembling the hypothetical free-living ancestor of the mitochondrion. Symp. Soc. Exp. Biol. *29:* 39-40.

John, P. and F.R. Whatley. 1977a. The bioenergetics of *Paracoccus denitrificans*. Biochim. Biophys. Acta *463:* 129-153.

John, P. and F.R. Whatley. 1977b. *Paracoccus denitrificans* Davis (*Micrococcus denitrificans* Beijerinck) as a mitochondrion. Adv. Bot. Res. *4:* 51-115.

Johnsen, J. 1977. Utilization of benzylpenicillin as carbon, nitrogen and energy source by a *Pseudomonas fluorescens* strain. Arch. Microbiol. *115:* 271-275.

Johnson, B.F. and R.Y. Stanier. 1971. Regulation of the β-ketoadipate pathway in *Alcaligenes eutrophus*. J. Bacteriol. *107:* 476-485.

Johnson, D.A., U.H. Behling, C.H. Lain, M. Listgarten, S. Socransky and A. Nowotny. 1978. Role of bacterial products in periodontitis: immune response in gnotobiotic rats monoinfected with *Eikenella corrodens*. Infect. Immun. *19:* 246-253.

Johnson, D.E. 1930. The relation of the cabbage maggot and other insects to the spread and development of soft rot of Cruciferae. Phytopathology *20:* 857-872.

Johnson, F.H. and I.V. Shunk. 1936. An interesting new species of luminous bacteria. J. Bacteriol. *31:* 585-592.

Johnson, F.H., N. Zworykin and G. Warren. 1943. A study of luminous bacterial cells and cytolysates with the electron microscope. J. Bacteriol. *46:* 167-185.

Johnson, G.V., H.J. Evans and T.M. Ching. 1966. Enzymes of the glyoxylate cycle in rhizobia and nodules of legumes. Plant Physiol. *41:* 1330-1336.

Johnson, J. 1923. A bacterial leafspot of tobacco. J. Agr. Res. *23:* 481-494.

Johnson, J.L. 1973. Use of nucleic acid homologies in the taxonomy of anaerobic bacteria. Int. J. Syst. Bacteriol. *23:* 308-315.

Johnson, J.L. 1978. Taxonomy of the *Bacteroides*. I. Deoxyribonucleic acid homologies among *Bacteroides fragilis* and other saccharolytic *Bacteroides* species. Int. J. Syst. Bacteriol. *28:* 245-256.

Johnson, J.L. 1980. Classification of anaerobic bacteria. *In* Proceedings of International Symposium on Anaerobes (Tokyo, Japan, June 22, 1980). Nippon Merck-Banyu Co., Ltd., Tokyo, pp. 19-29.

Johnson, J.L. 1981. Genetic characterization. *In* Gerhardt et al. (Editors), Manual of Methods for General Bacteriology, American Society for Microbiology, Washington, D.C., pp. 450-472.

Johnson, J.L., R.S. Anderson and E.J. Ordal. 1970. Nucleic acid homologies among oxidase-negative *Moraxella* species. J. Bacteriol. *101:* 568-573.

Johnson, J.L. and D.A. Ault. 1978. Taxonomy of the *Bacteroides*. II. Correlation of phenotypic characteristics with deoxyribonucleic acid homology groupings for *Bacteroides fragilis* and other saccharolytic *Bacteroides* species. Int. J. Syst. Bacteriol. *28:* 257-268.

Johnson, J.L. and L.V. Holdeman. 1983. *Bacteroides intermedius* comb. nov., and descriptions of *B. corporis* sp. nov. and *B. levii* sp. nov. Int. J. Syst. Bacteriol. *33:* 15-25.

Johnson, J.L. and E.J. Ordal. 1968. Deoxyribonucleic acid homology in bacterial taxonomy: effect of incubation temperature on reaction specificity. J. Bacteriol. *95:* 893-900.

Johnson, K.G., I.J. McDonald and M.B. Percy. 1976. Studies on the cellular and free lipopolysaccharides from *Branhamella catarrhalis*. Can. J. Microbiol. *22:* 460-466.

Johnson, K.G., M.B. Perry and I.J. McDonald. 1976. Studies of the cellular and free lipopolysaccharides from *Neisseria sicca* and *N. subflava*. Can. J. Microbiol. *22:* 189-196.

Johnson, R., R.R. Colwell, R. Sakazaki and K. Tamura. 1975. Numerical taxonomy study of the family Enterobacteriaceae. Int. J. Syst. Bacteriol. *25:* 12-37.

Johnson, R. and P.H.A. Sneath. 1973. Taxonomy of *Bordetella* and related organisms of the families Achromobacteraceae, Brucellaceae and Neisseriaceae. Int. J. Syst. Bacteriol. *22:* 381-404.

Johnson, R.C. (Editor) 1976. The biology of parasitic spirochetes. Academic Press, N.Y.

Johnson, R.C. 1977. The spirochetes. Annu. Rev. Microbiol. *31:* 89-106.

Johnson, R.C. 1981. Aerobic Spirochetes: The Genus *Leptospira*. *In* Starr, Stolp, Trüper, Balows and Schlegel (Editors), The Prokaryotes, a handbook on habitats, isolation and identification of bacteria. Springer-Verlag, New York, pp. 582-591.

Johnson, R.C. and L.M. Eggebraten. 1971. Fatty acid requirements of the Kazan 5 and Reiter strains of *Treponema pallidum*. Infect. Immun. *3:* 723-726.

Johnson, R.C. and V.G. Harris. 1967. Differentiation of pathogenic and saprophytic leptospires. I. Growth at low temperature. J. Bacteriol. *94:* 27-31.

Johnson, R.C. and V.G. Harris. 1968. Purine analogue sensitivity and lipase activity of leptospires. Appl. Microbiol. *16:* 1584-1590.

Johnson, R.C. and P. Rogers. 1964a. 5-Fluorouracil as a selective agent for the growth of leptospirae. J. Bacteriol. *88:* 422-426.

Johnson, R.C. and P. Rogers. 1964b. Differentiation of pathogenic and saprophytic leptospires with 8-azaguanine. J. Bacteriol. *88:* 1618-1623.

Johnson, R.C., J. Walby, R.A. Henry and N.E. Auran. 1973. Cultivation of parasitic leptospires: effect of pyruvate. Appl. Microbiol. *26:* 118-119.

Johnson, R.M. 1968. Characteristics of a marine *Vibrio*-bacteriophage system. J. Ariz. Acad. Sci. *5:* 28-33.

Johnson, S.M. and G.A. Pankey. 1976. *Eikenella corrodens* osteomyelitis, arthritis and cellulitis of the hand. South Med. J. *69:* 535-539.

Johnson, S.R. and W.R. Romig. 1979. Transposon-facilitated recombinations in *Vibrio cholerae*. Mol. Gen. Genet. *170:* 93-101.

Johnson, S.R., R. Sublett and W.R. Romig. 1980. Transposon-facilitated recombination in *Vibrio cholerae*: genetic mapping of el tor and classical biotypes. Proc. 15th Joint Conf. Cholera US-Japan Coop. Med. Sci. Prog., pp. 401-414.

Johnston, A.W.B., J.L. Beynon, A.V. Buchanan-Wallaston, S.M. Setchell, P.R. Hirsch and J.E. Beringer. 1978. High frequency transfer of nodulating ability between strains and species of *Rhizobium*. Nature (London) *276:* 634-636.

Johnston, A.W.B., M.J. Bibb and J.E. Bevinger. 1978. Tryptophan genes in *Rhizobium*—their organization and their transfer to other bacterial genera. Mol. Gen. Genet. *165:* 323-330.

Johnstone, D.B. 1974. Genus *Azotobacter* Beijerinck. *In* Buchanan and Gibbons (Editors), Bergey's Manual of Determinative Bacteriology, 8th Ed., The Williams and Wilkins Co., Baltimore, pp. 254-255.

Joklik, W.K., H.P. Willett and D.B. Amos (Editors). 1980. Zinsser Microbiology, 17th ed. Appleton-Century-Crofts, New York.

Jones, A.L., R.F. Whitcomb, D.L. Williamson and M.E. Coan. 1977. Comparative growth and primary isolation of spiroplasmas in media based on insect tissue culture formulations. Phytopathology *67:* 738-746.

Jones, C.W. 1980. Cytochrome patterns in classification and identification including their relevance to the oxidase test. *In* Goodfellow and Board (Editors), Microbiological Classification and Identification, Academic Press, London - New York, pp. 127-138.

Jones, D. 1978. Composition and differentiation of the genus *Streptococcus*. *In* Skinner and Quesnel (Editors), Streptococci, Academic Press, London - New York, pp. 1-49.

Jones, D. and P.H.A. Sneath. 1970. Genetic transfer and bacterial taxonomy.

Bacteriol. Rev. *34:* 40-81.

Jones, D.M. 1962. A pasteurella-like organism from the human respiratory tract. J. Pathol. Bacteriol. *83:* 143-151.

Jones, D.M. and P.M. O'Connor. 1962. *Pasteurella haemolytica* var. *ureae* from human sputum. J. Clin. Pathol. *15:* 247-248.

Jones, F.S., M. Orcutt and R.B. Little. 1931. Vibrios *(Vibrio jejuni* n. sp.) associated with intestinal disorders of cows and calves. J. Exp. Med. *53:* 853-864.

Jones, G.E., A. Foggie, D.L. Mould and S. Livitt. 1976. The comparison and characterization of glycolytic mycoplasmas isolated from the respiratory tract of sheep. J. Med. Microbiol. *9:* 39-52.

Jones, G.L. and G.A. Hébert. 1979. Laboratory safety: precautions and practices. *In* Jones and Hébert (Editors), "Legionnaires'" the disease, the bacterium and methodology. Center for Disease Control, Atlanta, pp. 134-136.

Jones, H., G. Rake and B. Stearns. 1945. Studies on lymphogranuloma venereum. III. The action of the sulfonamides on the agent of lymphogranuloma venereum. J. Infect. Dis. *76:* 55-69.

Jones, J.L. and D.A. Romig. 1979. *Eikenella corrodens*: a pathogen in head and neck infections. Oral Surg. *47:* 501-505.

Jones, L.M. 1967. Report to the International Committee on Nomenclature of Bacteria by the Sub-committee on Taxonomy of *Brucellae.* Minutes of Meeting, 22-23 July 1966. Int. J. Syst. Bacteriol. *17:* 371-375.

Jones, L.M. and D.T. Berman. 1951. The pathogenicity of mucoid variants of brucellae for guinea pigs. J. Infect. Dis. *89:* 214-223.

Jones, L.M. and W.J.B. Morgan. 1958. A preliminary report on a selective medium for the culture of *Brucella*, including fastidious types. Bull. W.H.O. *19:* 200-203.

Jones, L.M. and W. Wundt. 1971. International Committee on Nomenclature of Bacteria Sub-committee on the Taxonomy of *Brucella*. Minutes of Meeting, 7 August 1970. Int. J. Syst. Bacteriol. *21:* 126-128.

Jones, L.R. 1901. *Bacillus carotovorus* n. sp., die Ursache einer weichen Faulnis der Mohre. Zentrabl. Bakteriol. Parasitenk. Infektionskr. Hyg. Abt. II *7:* 12-21.

Jones, L.R., A.G. Johnson and C.S. Reddy. 1917. Bacterial blight of barley. J. Agr. Res. *11:* 625-644.

Jones, L.R., M.M. Williamson, F.A. Wolf and L. McCulloch. 1923. Bacterial leafspot of clovers. J. Agr. Res. *25:* 471-490.

Jones, R.N., J. Slepack and J. Bigelow. 1976. Ampicillin resistant *Haemophilus paraphrophilus* epiglottitis. J. Clin. Microbiol. *4:* 405-407.

Jonnson, V. 1970. Proposal of a new species *Pseudomonas kingii.* Int. J. Syst. Bacteriol. *20:* 255-257.

Jordan, A.M. and M.A. Trewern. 1973. Sub-lethal effect of sulphaquinoxaline on the tsetse fly, *Glossina austeni* Newst. Nature (Lond.) *245:* 462.

Jordan, A.M. and M.A. Trewern. 1976. Sulphaquinoxaline in host diet as the cause of reproductive abnormalities in the tsetse fly *(Glossina* spp.). Entomol. Exp. Appl. *19:* 115-129.

Jordan, D.C. 1982. Transfer of *Rhizobium japonicum* Buchanan 1980 to *Bradyrhizobium* gen. nov., a genus of slow-growing, root-nodule bacteria from leguminous plants. Int. J. Syst. Bacteriol. *32:* 136-139.

Jordan, D.C. and O.N. Allen. 1974. Family III. *Rhizobiaceae* Conn 1938, *In* Buchanan and Gibbons (Editors), Bergey's Manual of Determinative Bacteriology, 8th Ed. The Williams and Wilkins Co., Baltimore, pp. 261-264.

Jordan, D.C. and P.J. McNicol. 1977. Identification of *Beijerinckia* in the High Arctic (Devon Island, Northwest Territories). Appl. Environ. Microbiol. *35:* 204-205.

Jordan, F.T.W. 1979. Avian mycoplasmas. *In* Tully and Whitcomb (Editors), The *Mycoplasmas*, Vol. II, Academic Press, New York, pp. 1-48.

Jordan, F.T.W., H. Erno, G.S. Cottew, K.H. Hinz and L. Stipkovits. 1982. The characterization and taxonomic description of five mycoplasma serotypes of avian origin and their elevation to species rank and further evaluation of the taxonomic status of *Mycoplasma synoviae*. Int. J. Syst. Bacteriol. *32:* 108-115.

Jordan, G.W. and W.K. Hadley. 1969. Human infections with *Edwardsiella tarda*. Ann. Intern. Med. *70:* 283-288.

Joseph, R. and E. Canale-Parola. 1972. Axial fibrils of anaerobic spirochetes: ultrastructure and chemical characteristics. Arch. Mikrobiol. *81:* 146-168.

Joseph, R., S.C. Holt and E. Canale-Parola. 1970. Ultrastructure and chemical composition of the cell wall of *Spirochaeta stenostrepta*. Bacteriol. Proc. p. 57.

Joseph, R., S.C. Holt and E. Canale-Parola. 1973. Peptidoglycan of free-living anaerobic spirochetes. J. Bacteriol. *115:* 426-435.

Joubert, J.J., D.C. Hildebrand and M.N. Schroth. 1970. Nonutilization of beta-glucosides for growth by fluorescent pseudomonads. Phytopathology *60:* 502-505.

Joubert, J.J. and S.J. Truter. 1972. A variety of *Xanthomonas campestris* pathogenic to *Zantedeschia aethiopica*. Netherlands J. Plant Pathol. *78:* 212-217.

Joyce, G.H. and P.R. Dugan. 1970. The role of floc-forming bacteria in BOD removal from waste water. Develop. Ind. Microbiol. *11:* 377-386.

Judd, W. 1979. The secretions and fine structure of bivalve crystalline style sacs. Ophelia *18:* 205-233.

Judicial Commission. 1954. Opinion 11. Nomenclature of species in the bacterial genus *Shigella*. Int. Bull. Bacteriol. Nomencl. Taxon. *4:* 148-149.

Judicial Commission. 1958. Conservation of the family name *Enterobacteriaceae*, of the name of the type genus, and designation of the type species. Int. Bull.

Bacteriol. Nomen. Taxon. *8:* 73-74.

Judicial Commission. 1958. Opinion 21. Conservation of the generic name *Selenomonas* von Prowazek. Int. Bull. Bacteriol. Nomencl. Taxon. *8:* 163-165.

Judicial Commission. 1958. Opinion 22. Status of the generic name *Asterococcus* and conservation of the generic name *Mycoplasma*. Int. Bull. Bacteriol. Nomencl. Taxon. *8:* 166-168.

Judicial Commission. 1963. Opinion 26. Designation of neotype strains (cultures) of type species of the bacterial genera *Salmonella, Shigella, Arizona, Escherichia, Citrobacter,* and *Proteus* of the family *Enterobacteriaceae*. Int. Bull. Bacteriol. Nomencl. Taxon. *13:* 35-36.

Judicial Commission. 1970. Opinion 33. Conservation of the generic name *Agrobacterium* Conn 1942. Int. J. Syst. Bacteriol. *20:* 10.

Judicial Commission. 1971. Opinion 42. Conservation of the specific epithet "*phenylpyruvica*" in the name *Moraxella phenylpyruvica* Bøvre and Henriksen. Int. J. Syst. Bacteriol. *21:* 107.

Judicial Commission. 1979. Minutes of the meeting, 3 September 1978, Munich, West Germany. Int. J. Syst. Bacteriol. *29:* 267-269.

Judicial Commission. 1981. Present standing of the family name *Enterobacteriaceae* Rahn 1937. Int. J. Syst. Bacteriol. *31:* 104.

Julianelle, L.A. 1926. A biological classification of *Encapsulatus pneumoniae* (Friedländer's bacillus). J. Exp. Med. *44:* 113-128.

Junca, P., C. Saillard, J. Tully, O. Garcia-Jurado, J.-R. Degorce-Dumas, C. Mouches, J.-C. Vignault, R. Vogel, R. McCoy, R. Whitcomb, D. Williamson, J. Latrille and J.-M. Bove. 1980. Caractérisation de spiroplasmes isolés d'insectes et de fleurs de France continentale, de Corse et du Maroc. Proposition pour une classification des spiroplasmes. C.R. Acad. Sci. Paris, Ser. D *290:* 1209-1211.

Jungano, M. 1909. Sur la flore anaérobie du rat. C.R. Soc. Biol. Paris *66:* 112-114; 122-124.

Jungen, H. 1968. Endosymbionten bei Ameisen. Insect. Soc. *15:* 227-232.

Jungmann, P. and M.H. Kuczynski. 1918. Zür aetiologie und Pathogenese des Wolhynischer Fiebers und Fleckfiebers. Zeitschr. Klin. Med. *85:* 251-272.

Juni, E. 1972. Interspecies transformation of *Acinetobacter:* Genetic evidence for a ubiquitous genus. J. Bacteriol. *112:* 917-931.

Juni, E. 1974. Simple genetic transformation assay for rapid diagnosis of *Moraxella osloensis*. Appl. Microbiol. *27:* 16-24.

Juni, E. 1977. Genetic transformation assays for identification of strains of *Moraxella urethralis*. J. Clin. Microbiol. *5:* 227-235.

Juni, E. 1978. Genetics and Physiology of *Acinetobacter*. Annu. Rev. Microbiol. *32:* 349-371.

Juni, E. and G.A. Heym. 1977. Simple method for distinguishing gonococcal colony types. J. Clin. Microbiol. *6:* 511-517.

Juni, E. and G.A. Heym. 1980. Transformation assay for identification of psychrotrophic achromobacters. Appl. Environ. Microbiol. *40:* 1106-1114.

Juni, E. and A. Janik. 1969. Transformation of *Acinetobacter calco-aceticus (Bacterium anitratum)*. J. Bacteriol. *98:* 281-288.

Jurmanova, K. and J. Sterbova. 1977. Correlation between spermatozoan motility and mycoplasma findings in bull semen. Vet. Rec. *100:* 157-158.

Jurzitza, G. 1974. Über die Lieferung von Sterinen durch die hefeartigen Endosymbionten von *Lasioderma serricorne* F. (Coleoptera, Anobiidae) und die ökologische Bedeutung dieser Leistung für den Wirt. Oecologia (Berl.) *16:* 163-172.

Jüttner, R.R., R.M. Lafferty and H.J. Knackmuss. 1975. A simple method for the determination of poly-β-hydroxybutyric acid in microbial biomass. Eur. J. Appl. Microbiol. *1:* 233-237.

Jyssum, K. 1959. Assimilation of nitrogen in meningococci grown with the ammonium ion as sole nitrogen source. Acta Pathol. Microbiol. Scand. *46:* 320-332.

Jyssum, S. 1971. Utilization of thymine, thymidine and TMP by *Neisseria meningitidis*. 2. Lack of enzymes for specific incorporation of exogenous thymine, thymidine and TMP into DNA. Acta Pathol. Microbiol. Scand. *79B:* 778-788.

Jyssum, S. 1974. Search for thymidine phosphorylase, nucleoside deoxyribosyltransferase and thymidine kinase in genus *Neisseria*. Acta Pathol. Microbiol. Scand. *82B:* 53-56.

Jyssum, S. and K. Bøvre. 1974. Search for thymidine phosphorylase, nucleoside deoxyribosyltransferase and thymidine kinase in *Moraxella, Acinetobacter*, and allied bacteria. Acta Pathol. Microbiol. Scand. *82B:* 57-66.

Jyssum, K. and P.E. Joner. 1965. Growth of *Bacterium anitratum* (B5W) with nitrate or nitrite as nitrogen source. Acta Path. Microbiol. Scand. *64:* 381-386.

Jyssum, K. and S. Jyssum. 1968. Isolation of variants with increased mutability from *Neisseria meningitidis*. Acta Pathol. Microbiol. Scand. *74:* 93-100.

Kaczmarek, F.S. and A.L. Coykendall. 1980. Production of phenylacetic acid by strains of *Bacteroides asaccharolyticus* and *Bacteroides gingivalis* (sp. nov.). J. Clin. Microbiol. *12:* 288-290.

Kado, C.I. and M.G. Heskett. 1970. Selective media for isolation of *Agrobacterium, Corynebacterium, Erwinia, Pseudomonas*, and *Xanthomonas*. Phytopathology *60:* 969-976.

Kado, C.I. and S.-T. Liu. 1981. Rapid procedure for detection and isolation of large and small plasmids. J. Bacteriol. *145:* 1365-1373.

Kadota, H. 1951. Studies on the biochemical activities of marine bacteria. I. On the agar-decomposing bacteria in the sea. Memoirs Coll. Agric., Kyoto Univ. *59:* 54-67.

Kafkewitz, D. 1975. Improved growth media for *Vibrio succinogenes*. Appl. Microbiol. 29: 121-122.

Kafkewitz, D. and D. Goodman. 1974. L-Asparaginase production by the rumen anaerobe *Vibrio succinogenes*. Appl. Microbiol. 27: 206-209.

Kahane, I., S. Greenstein and S. Razin. 1977. Carbohydrate content and enzymic activities in the membrane of *Spiroplasma citri*. J. Gen. Microbiol. 101: 173-176.

Kahane, I., S. Razin and A. Muhlrad. 1978. Possible role of acetate kinase in ATP generation in *Mycoplasma hominis* and *Acholeplasma laidlawii*. FEMS Lett. 3: 143-145.

Kahlon, R.S. and S.R. Vyas. 1971. Isolation and identification of acetic acid bacteria from different ecosystems. Proc. Indian Acad. Sci. Sect. B. 74: 293-300.

Kaiser, B. 1980. Licht- und elektronenmikroskopische Untersuchung der Symbionten von *Graphocephala coccinea* Forstier (Homoptera: Jassidae). Int. J. Insect Morphol. Embryol. 9: 79-88.

Kaklamanis, E., L. Thomas, L. Stavropoulos, K. Borman and C. Boshwitz. 1969. Mycoplasmacidal action of normal tissue extracts. Nature 221: 860-862.

Kalina, G.P., A.G. Somova, L.S. Podosinnikova and T.I. Grafova. 1980. *Allomonas*, a new group of microorganisms of the family *Vibrionaceae*. Communication I. Methods of study and preliminary results of differentiating *Allomonas* from *Aeromonas* and *Vibrio* (Russian). Microbiol. Epidemiol. Immunobiol. (Eng. Transl.) 1: 40-46.

Kallick, C.A., K.T. Reddi and W.L. Landau. 1972. Systemic lupus erythematosus associated with *Haemobartonella*-like organisms. Nat. New Biol. 236: 145-146.

Kallings, L.O. 1967. Sensitivity of various *Salmonella* strains to Felix 01 phage. Acta Pathol. Microbiol. Scand. 70: 446-454.

Kamal, A.M. and Y. Zinnaka. 1971. Outbreak of gastroenteritis by nonagglutinable (NAG) vibrios in the republic of Sudan. J. Egypt Public Health Assoc. 46: 125-174.

Kamio, Y. and H. Takahashi. 1980. Isolation and characterization of outer and inner membranes of *Selenomonas ruminantium*: Lipid compositions. J. Bacteriol. 141: 888-898.

Kamme, C. 1969. Susceptibility in vitro of *Haemophilus influenzae* to penicillin G., penicillin V. and ampicillin. Incubation of strains from acute otitis media in air and in CO₂. atmosphere. Acta Pathol. Microbiol. Scand. 75: 611-621.

Kanazawa, Y. and K. Ikemura. 1979. Isolation of *Yersinia enterocolitica* and *Yersinia pseudotuberculosis* from human specimens and their drug resistance in the Niigata district in Japan. Contr. Microbiol. Immunol. 5: 106-114.

Kandler, O. 1981. Archaebakterien und Phylogenie der Organismen. Naturwissenschaften 68: 183-192.

Kandler, O. and K.-H. Schleifer. 1980. Taxonomy I: Systematics of bacteria. *In* Ellenberg, Esser, Kubitzki, Schnepf and Ziegler (Editors), Progress in Botany, Fortschritte der Botanik, Vol. 42. Springer-Verlag, Berlin and Heidelberg, pp. 234-252.

Kandler, O., K.H. Schleifer, E. Niebler, M. Nakel, H. Zahradnik and M. Ried. 1970. Murein types in micrococci and similar organisms. Publ. Fac. Sci. Univ. J. E. Purkyné, Brno, K47: 143-156.

Kanegasaki, S. and H. Takahashi. 1967. Function of growth factors for rumen microorganisms. I. Nutritional characteristics of *Selenomonas ruminantium*. J. Bacteriol. 93: 456-463.

Kaneko, T. and R.R. Colwell. 1978. The annual cycle of *Vibrio parahaemolyticus* in Chesapeake Bay. Microb. Ecol. 4: 135-155.

Kaneuchi, C. and T. Mitsuoka. 1978. *Bacteroides microfusus*, a new species from the intestines of calves, chickens, and Japanese quails. Int. J. Syst. Bacteriol. 28: 478-481.

Kaneuchi, C., T. Miyazato, T. Shinjo and T. Matsuoka. 1979. Taxonomic study of helically coiled, sporeforming anaerobes from the intestines of humans and other animals: *Clostridium cocleatum* sp. nov. and *Clostridium spiroforme* sp. nov. Int. J. Syst. Bacteriol. 29: 1-12.

Kaneuchi, C., K. Watanabe, A. Terada, Y. Benno and T. Mitsuoka. 1976. Taxonomic study of *Bacteroides clostridiiformis* subsp. *clostridiiformis* (Burri and Ankersmit) Holdeman and Moore and of related organisms: proposal of *Clostridium clostridiiformis* (Burri and Ankersmit) comb. nov. and *Clostridium symbiosum* (Stevens) comb. nov. Int. J. Syst. Bacteriol. 26: 195-204.

Kaper, J., H. Lockman, R.R. Colwell and S.W. Joseph. 1979. Ecology, serology, and enterotoxin production of *Vibrio cholerae* in Chesapeake Bay. Appl. Environ. Microbiol. 37: 91-103.

Kaplan, J.M., G.H. McCracken and J.D. Nelson. 1973. Infections in children caused by the HB group of bacteria. J. Pediatr. 82: 398-403.

Kaplan, R. 1980. *Campylobacter*. *In* Lennette, Balows, Hausler and Truant (Editors). Manual of Clinical Microbiology. Third Ed. American Society for Microbiology, Washington, D. C., pp. 235-241.

Kaplan, R.W. and M. Brendel. 1969. Formation of prototrophs in mixtures of two auxotropic mutants of *Serratia marcescens* HY by a transducing bacteriophage produced by some auxotrophs. Molec. Gen. Genetics 104: 27-39.

Kapperud, G. 1977. *Yersinia enterocolitica* and *Yersinia*-like microbes isolated from mammals and water in Norway and Denmark. Acta Pathol. Microbiol. Scand. Sect. B 85: 129-135.

Kapperud, G. 1980. Studies on the pathogenicity of *Yersinia enterocolitica* and *Y. enterocolitica*-like bacteria. 1. Enterotoxin production at 22°C and 37°C by environmental and human isolates from Scandinavia. Acta Pathol. Microbiol. Scand. Sect. B 88: 287-291.

Karami, Y., K. Hovind-Hougen, A. Birch-Andersen and M. Asmar. 1979. *Borrelia persica* and *B. baltazardi* sp. nov.; experimental pathogenicity for some animals and comparison of the ultra structure. Ann. Microbiol. (Inst. Pasteur) 130B: 157-168.

Karami, Y., K. Hovind-Hougen, A. Birch-Andersen and M. Asmar. 1983. *In* Validation of the publication of new names and new combinations previously effectively published outside the IJSB. List No. 10. Int. J. Syst. Bacteriol. 33: 438-440.

Karlsson, K.A. 1964. Serologishe Studien von *Aeromonas salmonicida*. Zentralbl. Bakteriol. Parasitenkd. Infektionskr. Hyg. Abt. I Orig. 194: 73-80.

Karmali, M. A., A. K. Allen and P. C. Fleming. 1981. Differentiation of catalase-positive campylobacters with special reference to morphology. Int. J. Syst. Bacteriol. 31: 64-71.

Karmali, M. A., S. DeGrandis and P. C. Fleming. 1981. Antimicrobial susceptibility of *Campylobacter jejuni* with special reference to resistance patterns of Canadian isolates. Antimicrob. Agents Chemother. 19: 593-597.

Kasai, G.J. 1961. A study of *Leptotrichia buccalis*. I. Morphology and preliminary observations. J. Dental Res. 40: 800-811.

Kasai, G.J. 1965. A study of *Leptotrichia buccalis*. II. Biochemical and physiological observations. J. Dental Res. 44: 1015-1022.

Kaska, M. 1976. The toxicity of extracellular protease of the bacterium *Serratia marcescens* for larvae of greater wax moth *Galleria mellonella*. J. Invertebr. Pathol. 27: 271.

Kaspar, D.L. 1976a. The polysaccharide capsule of *Bacteroides fragilis* subsp. *fragilis*: immunochemical and morphologic definition. J. Infect. Dis. 133: 79-87.

Kaspar, D.L. 1976b. Chemical and biological characterization of the lipopolysaccharide of *Bacteroides fragilis* subspecies *fragilis*. J. Infect. Dis. 134: 59-66.

Kaspar, D.L., M.E. Hayes, B.G. Reinap, F.O. Craft, A.B. Onderdonk and B.F. Polk. 1977. Isolation and identification of encapsulated strains of *Bacteroides fragilis*. J. Infect. Dis. 136: 75-81.

Katashima, R. 1965. Mate-killing in *Euplotes patella*, syngen 1. Annot. Zool. Jpn. 38: 207-215.

Kater, J.C., S.C. Marshall and W.J. Hartley. 1962. A specific suppurative synovitis and pyaemia in lambs. N.Z. Vet. J. 10: 143-144.

Kates, M. 1978. The phytanyl ether-linked polar lipids and isoprenoid lipids of extremely halophilic bacteria. Progr. Chem. Fats Lipids 15: 301-342.

Kates, M., S.N. Sehgal and N.E. Gibbons. 1961. The lipid composition of *Micrococcus halodenitrificans* as influenced by salt concentration. Can. J. Microbiol. 7: 427-435.

Kato, H., T. Murakami, S. Takase and K. Ono. 1972. Sensitivities in vitro to antibiotics of mycoplasma isolated from canine sources. Jpn. J. Vet. Sci. 34: 197-206.

Kato, K. and H. Tsubahara. 1962. Infectious coryza of chickens. II. Identification of isolates. Bull. Nat. Inst. An. Health 45: 21-26.

Kato, K., T. Umemoto, H. Fukuhara, H. Sagawa and S. Kotani. 1981. Variation of dibasic amino acid in the cell wall peptidoglycan of bacteria of genus *Fusobacterium*. FEMS Microbiol. Lett. 10: 81-85.

Kato, K., T. Umemoto, H. Sagawa and S. Kotani. 1979. Lanthionine as an essential constituent of cell wall peptidoglycan of *Fusobacterium nucleatum*. Curr. Microbiol. 3: 147-151.

Katznelson, H. 1955. The metabolism of phytopathogenic bacteria. I. Comparative studies on the metabolism of representative species. J. Bacteriol. 70: 469-475.

Katznelson, H. 1958. Metabolism of phytopathogenic bacteria. II. Metabolism of carbohydrates by cell-free extracts. J. Bacteriol. 75: 540-543.

Katznelson, H. and A.G. Lochhead. 1952. Growth factor requirements of halophilic bacteria. J. Bacteriol. 64: 97-103.

Katznelson, R. and S. Ulitzur. 1977. Control of luciferase synthesis in a newly isolated strain of *Photobacterium leiognathi*. Arch. Microbiol. 115: 347-351.

Katznelson, H. and A.C. Zagallo. 1957. Metabolism of rhizobia in relation to effectiveness. Can. J. Microbiol. 3: 879-884.

Kauffman, C.A., A.G. Bergman and C.S. Hertz. 1979. Antimicrobial resistance of *Haemophilus* species in patients with chronic bronchitis. Am. Rev. Resp. Dis. 120: 1382-1385.

Kauffmann, F. 1935. Weitere erfahrungen mit den Kombinierten Anreichurungsverfahren für Salmonellabazillus. Z. Hyg. Infektionskr. 117: 26-32.

Kauffmann, F. 1941a. Die Bakteriologie der Salmonella-Gruppe. E. Munksgaard, Copenhagen.

Kauffmann, F. 1941b. Über mehrere neue *Salmonella* types. Acta Pathol. Microbiol. Scand. 18: 351-366.

Kauffmann, F. 1947. The serology of the coli group. J. Immunol. 57: 71-100.

Kauffmann, F. 1949. On the serology of the *Klebsiella* group. Acta Pathol. Microbiol. Scand. 26: 381-406.

Kauffmann, F. 1951. *Enterobacteriaceae*. E. Munksgaard, Copenhagen.

Kauffmann, F. 1954. *Enterobacteriaceae*. 2nd Ed. E. Munksgaard, Copenhagen.

Kauffmann, F. 1956. Zür biochemischen und serologischen Gruppen- und Typen-Einteilung der *Enterobacteriaceae*. Zentralbl. Bakteriol. Parasitenkd. Infektionskr. Hyg. Abt. I Orig. 165: 344-354.

Kauffmann, F. 1960. Two biochemical subdivisions of the genus *Salmonella*. Acta Pathol. Microbiol. Scand. 49: 393-396.

Kauffmann, F. 1961. The species-definition in the *Enterobacteriaceae*. Int. Bull. Bacteriol. Nomencl. Taxon. 11: 5-6.

Kauffmann, F. 1963a. Zür differential diagnose der *Salmonella* sub-genera I, II und III. Acta Pathol. Microbiol. Scand. *58:* 109-113.

Kauffmann, F. 1963b. On the species definition. Int. Bull. Bacteriol. Nomencl. Taxon. *13:* 181-186.

Kauffmann, F. 1964. Vereinfachtes Antigen-Schema der *Salmonella* sub-genera II, III. Acta Pathol. Microbiol. Scand. *62:* 68-72.

Kauffmann, F. 1965. Die Diagnose von *Arizona*-Kulturen nach dem originalen Kauffmann-White Schema. Pathol. Microbiol. *28:* 575-580.

Kauffmann, F. 1966a. The bacteriology of *Enterobacteriaceae*. E. Munksgaard, Copenhagen.

Kauffmann, F. 1966b. Das *Salmonella* Sub-genus IV. Ann. Immunol. Hung. *9:* 77-80.

Kauffmann, F. 1966. The Bacteriology of *Enterobacteriaceae*. Williams and Wilkins Co., Baltimore.

Kauffmann, F. 1971. On the classification and nomenclature of the genus *Salmonella*. Acta Pathol. Microbiol. Scand. *79:* 421-422.

Kauffmann, F. 1973. Zür Klassification und Nomenklatur der *Salmonella* species. Zentralbl. Bakteriol. Parasitenkd. Infektionskr. Hyg. I Abt. Orig., *A223:* 508-512.

Kauffmann, F. 1978. Das Fundament. E. Munksgaard, Copenhagen.

Kauffmann, F. and P.R. Edwards. 1952. Classification and nomenclature of *Enterobacteriaceae*. Int. Bull. Bacteriol. Nomencl. Taxon. *2:* 2-8.

Kauffmann, F. and E. Møller. 1940. A new type of *Salmonella (S. ballerup)* with Vi antigen. J. Hyg. *40:* 246-251.

Kauffmann, F. and R. Rohde. 1962. Eine Vereinfachung der serologischen *Arizona*, Diagnose. Acta Pathol. Microbiol. Scand. *54:* 473-478.

Kauffmann, J. and P. Toussaint. 1951a. Un nouveau germe fixateur de l'azote atmosphérique: *Azotobacter lacticogenes*. C.R. Acad. Sci. (Paris) *223:* 710-711.

Kauffmann, J. and P. Toussaint. 1951b. Un nouveau germe fixateur de l'azote atmosphérique: *Azotobacter lacticogenes*. Rev. Gen. Bot. *58:* 553-561.

Kaulfers, P.M., R. Laufs and G. Lahn. 1978. Molecular properties of transmissible R factors of *Haemophilus influenzae* determining tetracycline resistance. J. Gen. Microbiol. *105:* 243-252.

Kawai, Y. and E. Yabuuchi. 1975. *Pseudomonas pertucinogena* sp. n., an organism previously misidentified as *Bordetella pertussis*. Int. J. Syst. Bacteriol. *25:* 317-323.

Kawamura, Y. 1926. A coryne-bacillus as a cause of abscess in the feet of hens. J. Jap. Soc. Vet. Sci. *5:* 22.

Kawamura, E. 1934. Bacterial leaf spot of sunflower. Ann. Phytopathol. Soc. Japan *4:* 25-28.

Kawasaki, H., K. Takasu and S. Omata. 1978. Metabolism of glucose in extremely halophilic bacteria. J. Agr. Chem. Soc. Japan *52:* 433-440.

Kay, W.W. and A.F. Gronlund. 1969. Amino acid transport in *Pseudomonas aeruginosa*. J. Bacteriol. *97:* 273-281.

Kayser, H. 1902. Das Wachstum der zurischen *Bacterium typhi* und *coli* stehenden Spaltpilze auf dem v. Drigalski-Conradi'schen Agarboden. Zentrabl. Bakteriol. Parasitenk. Infektionskr. Hyg. Abt. I Orig. *31:* 426-429.

Keane, P.J., A. Kerr and P.B. New. 1970. Crown gall of stone fruit. II. Identification and nomenclature of *Agrobacterium* isolates. Aust. J. Biol. Sci. *23:* 585-595.

Keddie, R.M. and I.J. Bousfield. 1980. Cell wall composition in the classification and identification of coryneform bacteria. *In* Goodfellow and Board (Editors), Microbiological Classification and Identification, Academic Press, London - New York, pp. 167-188.

Keeble, J.R. and T. Cross. 1977. An improved medium for the enumeration of *Chromobacterium* in soil and water. J. Appl. Bacteriol. *43:* 325-327.

Keel, J.A., W.R. Finnerty and J.C. Feeley. 1979. Fine structure of the Legionnaires' disease bacterium. Ann. Int. Med. *90:* 652-655.

Keele, B.B., P.B. Hamilton and G.H. Elkan. 1969. Glucose catabolism in *Rhizobium japonicum*. J. Bacteriol. *97:* 1184-1191.

Keele, B.B., Jr., P.B. Hamilton and G.H. Elkan. 1970. Gluconate catabolism in *Rhizobium japonicum*. J. Bacteriol. *101:* 698-704.

Keeler, R. F., A. E. Ritchie, J. H. Bryner and J. Elmore. 1966. The preparation and characterization of cell walls and the preparation of flagella of *Vibrio fetus*. J. Gen. Microbiol. *43:* 439-462.

Keilich, G., J. Roppel and H. Mayer. 1976. Characterization of a diaminohexose (2,3-diamino-2,3-dideoxy-D-glucose) from *Rhodopseudomonas viridis* lipopolysaccharides by circular dichroism. Carbohydr. Res. *51:* 129-134.

Kekcheeva, N.G., I.N. Kokorin and E.D. Miskarova. 1978. Study of rickettsial infection in inbred mice. *In* Kazar, Ormsbee and Tarasevich (Editors), Rickettsiae and Rickettsial Diseases, VEDA, Bratislava, pp. 189-196.

Kellen, W.R., D.F. Hoffmann and R.A. Kwock. 1981. *Wolbachia* sp. (*Rickettsiales: Rickettsiaceae*) a symbiont of the almond moth, *Ephestia cautella*: Ultrastructure and influence on host fertility. J. Invert. Pathol. *37:* 273-283.

Kellen, W.R., J.E. Lindegren and D.F. Hoffmann. 1972. Developmental stages and structure of a *Rickettsiella* in the navel orangeworm, *Paramyelois transitella* (Lepidoptera: Phycitidae). J. Invert. Pathol. *20:* 193-199.

Kellerman, G.D., J. Foster and F.F. Badakhsh. 1970. Comparison of chemical components of cell walls of *Brucella abortus* strains of low and high virulence. Infect. Immun. *2:* 237-243.

Kellogg, D.S., J.L. Peacock, Jr., W.L. Deacon, L. Brown and C.I. Pirkle. 1963. *Neisseria gonorrhoeae*. I. Virulence genetically linked to clonal variation. J. Bacteriol. *85:* 1274-1279.

Kelly, C.D. 1957. Genus *Bacteroides* Castellani and Chalmers 1919. *In* Breed, Murray and Smith (Editors), Bergey's Manual of Determinative Bacteriology, 7th Ed. The Williams and Wilkins Co., Baltimore, pp. 424-436.

Kelly, R.T. 1971. Cultivation of *Borrelia hermsi*. Science *173:* 443-444.

Kelly, R.T. 1976. Cultivation and physiology of relapsing fever borreliae. *In* R.C. Johnson (Editor), The Biology of Parasitic Spirochetes. Academic Press, New York, pp. 87-94.

Kelman, A. 1953. The bacterial wilt caused by *Pseudomonas solanacearum*. North Carol. Agric. Expt. Sta., Tech. Bull. No. 99, pp. 1-194.

Kelman, A. 1954. The relationship of pathogenicity of *Pseudomonas solanacearum* to colony appearance on a tetrazolium medium. Phytopathology *44:* 693-695.

Kelman, A. and R.S. Dickey. 1980. Soft rot of 'carotovora' group. *In* Schaad (Editor), Laboratory guide for identification of plant pathogenic bacteria, American Phytopathological Society, St. Paul, pp. 31-35.

Kelterborn, E. 1967. Salmonella-species. Hirzel, Leipzig.

Kemenes, F. and B. Markói. 1959. *Aktinobacillus Lignièresi* okozta tályog kutya szájüregében. Magy Allatorv Lapja *14:* 31-32.

Kendrick, J.B. 1934. Bacterial blight of carrot. J. Agr. Res. *49:* 493-510.

Kendrick, J.B. and K.F. Baker. 1942. Bacterial blight of garden stocks and its control by hot water seed treatment. Calif. Agr. Exp. Sta. Bull. *665:* 1-23.

Kendrick, P.L. and G. Eldering. 1969. Microbiology of whooping cough. *In* Ocklitz (Editor), Der Keuchusten, Vol. 8. VI. Infectionskrankheiten und ihre Erreger. Veb Gustav Fischer, Jena, pp. 259-282.

Kendrick, P.L., G. Eldering, M.K. Dixon and J. Misner. 1947. Mouse protection tests in the study of pertussis vaccine. Am. J. Publ. Health *37:* 803-810.

Kennedy, B.W. and T.H. King. 1962. Angular leaf spot of strawberry caused by *Xanthomonas fragariae*. sp. nov. Phytopathology *52:* 873-875.

Kennedy, L.D. 1976. Isolation of 3-0-methyl-D-ribose from *Rhizobium* polysaccharide. Carbohydr. Res. *52:* 259-261.

Kennedy, L.D. and R.W. Bailey. 1976. Monomethyl sugars in extracellular polysaccharides from slow-growing rhizobia. Carbohydr. Res. *49:* 451-454.

Kennedy, P.C., E.L. Biberstein, J.A. Howarth, L.M. Frazier and D.L. Dungworth. 1960. Infectious meningoencephalitis in cattle, caused by a *Haemophilus*-like organism. Am. J. Vet. Res. *21:* 403-409.

Kennedy, P.C., L.M. Frazier, G.H. Theilen and E.L. Biberstein. 1958. A septicemic disease of lambs caused by *Hemophilus agni* (new species). Am. J. Vet. Res. *19:* 645-654.

Kennedy, S.F., R.R. Colwell and G.B. Chapman. 1970. Ultrastructure of a marine psychrophilic *Vibrio*. Can. J. Microbiol. *16:* 1027-1031.

Kenny, G.E. 1973. Serological heterogeneity of the Mycoplasmatales. J. Infect. Dis. *127:* (Suppl.): S2-S5.

Kenny, G.E. and J.T. Grayston. 1965. Eaton pleuropneumonia-like organism (*Mycoplasma pneumoniae*) complement-fixing antigen: extraction with organic solvents. J. Immunol. *95:* 19-25.

Kerr, A. 1969. Transfer of virulence between isolates of *Agrobacterium*. Nature, Lond. *223:* 1175-1176.

Kerr, A. 1971. Acquisition of virulence by non-pathogenic isolates of *Agrobacterium radiobacter*. Physiol. Plant Pathol. *1:* 241-246.

Kerr, A. 1980. Biological control of crown gall through production of agrocin 84. Plant Disease *64:* 25-30.

Kerr, A. and K. Htay. 1974. Biological control of crown gall through bacteriocin production. Physiol. Plant Pathol. *4:* 37-44.

Kerr, A. and C.G. Panagopoulos. 1977. Biotypes of *Agrobacterium radiobacter* var. *tumefaciens* and their biological control. Phytopath. Z. *90:* 172-179.

Kerr, A., J.M. Young and C.G. Panagopoulos. 1978. Genus II. *Agrobacterium* Conn 1942. *In* Young, J.M., D.W. Dye, J.F. Bradbury, C.G. Panagoulos and C.F. Robbs. A proposed nomenclature for plant pathogenic bacteria. N.Z. J. Agr. Res. *21:* 153-177.

Kersters, K. 1978. Taxonomy of *Alcaligenes* and *Achromobacter* by polyacrylamide gel electrophoresis of their soluble proteins. Antonie van Leeuwenhoek J. Microbiol. Serol. *44:* 116-117.

Kersters, K. and J. De Ley. 1963. The oxidation of glycols by acetic acid bacteria. Biochim. Biophys. Acta *71:* 311-331.

Kersters, K. and J. De Ley. 1968. The occurrence of the Entner-Doudoroff pathway in bacteria. Antonie Van Leeuwenhoek J. Serol. Microbiol. *34:* 393-408.

Kersters, K. and J. De Ley. 1975. Identification and grouping of bacteria by numerical analysis of their electrophoretic protein patterns. J. Gen. Microbiol. *87:* 333-342.

Kersters, K. and J. De Ley. 1980. Classification and identification of bacteria by electrophoresis of their proteins. *In* Goodfellow and Board (Editors), Microbiological Classification and Identification. The Society for Applied Bacteriology Symposium Series No. 8, Academic Press, New York, pp. 273-297.

Kersters, K., J. De Ley, P.H.A. Sneath and M. Sackin. 1973. Numerical taxonomic analysis of *Agrobacterium*. J. Gen. Microbiol. *78:* 227-239.

Kerwar, S.S., V.H. Cheldelin and L.W. Parks. 1964. Valine-leucine metabolism in *Acetobacter suboxydans* and the inhibition of growth by valine. J. Bacteriol. *88:* 179-186.

Kessel, M. and F. Klink. 1980. Archaebacterial elongation factor is ADP-ribosylated by diphtheria toxin. Nature (London) *287:* 250-251.

Kessell, M. and M. Shilo. 1976. Relationship of *Bdellovibrio* elongation and fission to host cell size. J. Bacteriol. *128:* 477-480.

Kessler, R.H. and M. Ristic. 1979. In vitro cultivation of *Anaplasma marginale*: invasion and development in noninfected erythrocytes. Am. J. Vet. Res. *40:*

1774-1776.

Kessler, R.H., M. Ristic, D.M. Sells and C.A. Carson. 1979. In vitro cultivation of *Anaplasma marginale*: growth pattern and morphologic appearance. Am. J. Vet. Res. *40:* 1767-1773.

Keynan, A., K. Nealson, H. Sideroupoulos and J.W. Hastings. 1974. Marine transducing bacteriophage attacking a luminous bacterium. J. Virol. *14:* 333-340.

Keyser, H.H., B.B. Bohlool, T.S. Hu and D.F. Weber. 1982. Fast-growing rhizobia isolated from root nodules of soybean. Science *215:* 1631–1632.

Keysselitz, G. 1906. *Spirochaeta anodontae* nov. spec. Arb. Gesundh Amt. Berl. *23:* 566-569.

Khairat, O. 1940. Endocarditis due to a new species of *Haemophilus*. J. Pathol. Bacteriol. *50:* 497-505.

Khairat, O. 1967. *Bacteroides corrodens* isolated from bacteremias. J. Pathol. Bacteriol. *94:* 29-40.

Khan, A.W., T.M. Trottier, G.B. Patel and S.M. Martin. 1979. Nutrient requirement for the degradation of cellulose to methane by a mixed population of anaerobes. J. Gen. Microbiol. *112:* 365-372.

Khavkin, T., V. Sukhinin and N. Amosenkova. 1981. Host-parasite interaction and development of infraforms in chicken embryos infected with *Coxiella burnetii* via the yolk sac. Infect. Immun. *32:* 1281-1291.

Kihara, K., K. Yamada, T. Sasaki, M. Shintani and H. Okumura. 1981. Attachment of *Mycoplasma hominis* and *M. orale* to human diploid lung fibroblasts. Microbiol. Immunol. *25:* 745-749.

Kikuth, W. 1928. Über Einen neuen Anämeerreger, *Bartonella canis* nov. spec. Klin. Wochenschr. *7:* 1729-1730.

Kilian, M. 1974. A rapid method for the differentiation of *Haemophilus* strains. The porphyrin test. Acta Pathol. Microbiol. Scand. Sect. B *82:* 835-842.

Kilian, M. 1976. A taxonomic study of the genus *Haemophilus* with the proposal of a new species. J. Gen. Microbiol. *93:* 9-62.

Kilian, M. 1978. Rapid identification of *Actinomycetaceae* and related bacteria. J. Clin. Microbiol. *8:* 127-133.

Kilian, M. and W. Frederiksen. 1981. Ecology of *Haemophilus*, *Pasteurella* and *Actinobacillus*. In Kilian, Frederiksen and Biberstein (Editors), *Haemophilus*, *Pasteurella* and *Actinobacillus*. Academic Press, London, pp. 11-38.

Kilian, M., J. Mestecky, R. Kulhavy, M. Tomana and W.T. Butler. 1980. IgA1 proteases from *Haemophilus influenzae*, *Streptococcus pneumoniae*, *Neisseria meningitidis*, and *Streptococcus sanguis*: comparative immunochemical studies. J. Immunol. *124:* 2596-2600.

Kilian, M., J. Mestecky and R.E. Schrohenloher. 1979. Pathogenic species of the genus *Haemophilus* and *Streptococcus pneumoniae* produce immunoglobulin A1 protease. Infect. Immun. *26:* 143.

Kilian, M., C.-H. Mordhorst, C.R. Dawon and H. Lautrop. 1976. The taxonomy of haemophili from conjunctivae. Acta Pathol. Microbiol. Scand. Sect. B *84:* 132-138.

Kilian, M., J. Nicolet and E.L. Biberstein. 1978. Biochemical and serological characterization of *Haemophilus pleuropneumoniae* (Matthews and Pattison 1961) Shope 1964 and proposal of a neotype strain. Int. J. Syst. Bacteriol. *28:* 20-26.

Kilian, M. and C.R. Schiøtt. 1975. Haemophili and related bacteria in the human oral cavity. Arch. Oral Biol. *20:* 791-796.

Kilian, M. and J. Theilade. 1975. Cell wall ultrastructure of strains of *Haemophilus ducreyi* and *Haemophilus piscium*. Int. J. Syst. Bacteriol. *25:* 351-356.

Kilian, M. and J. Theilade. 1978. Amended description of *Haemophilus segnis* Kilian 1977. Int. J. Syst. Bacteriol. *28:* 411-415.

Kim, B.H. 1976. Studies on *Actinobacillus equuli*. Ph.D. Thesis. University of Edinburgh.

Kim, B.H., J.E. Phillips and J.G. Atherton. 1976. *Actinobacillus suis* in the horse. Vet. Rec. *98:* 239.

Kimura, T. 1969. A new subspecies of *Aeromonas salmonicida* as an etiological agent of furunculosis on "Sakuramasu" (*Oncorhynchus masou*) and Pink Salmon (*O. gorbuscha*) rearing for maturity. Part 1. On the morphological and physiological properties. Part 2. On the serological properties. Fish Pathol. (Tokyo) *3:* 34-44; 45-52.

Kinaidy, H.K. 1973. Two new infectious blood diseases of cattle in Austria. Wien Tieraerztl. Monatsschr. *60:* 164-366.

King, A., B. Holmes, I. Phillips and S.P. Lapage. 1979. A taxonomic study of clinical isolates of *Pseudomonas pickettii*, '*P. thomasii*' and 'Group IVd' bacteria. J. Gen. Microbiol. *114:* 137-147.

King, B.M. and D.L. Adler. 1964. A previously undescribed group of *Enterobacteriaceae*. Amer. J. Clin. Pathol. *41:* 230-232.

King, E.O. 1959. Studies on a group of previously unclassified bacteria associated with meningitis in infants. Am. J. Clin. Pathol. *31:* 241-247.

King, E.O. 1964. The identification of unusual pathogenic Gram-negative bacteria. Communicable Disease Center, Atlanta, Georgia.

King, E.O. 1964. Round Table. Current trends in diagnostic microbiology. The identification of unusual pathogenic Gram-negative bacteria. Dept. of Health, Education and Welfare, Public Health Service, Center for Disease Control, Atlanta, Georgia.

King, E.O. 1972. The identification of unusual pathogenic Gram negative bacteria (Revised by R.E. Weaver, H.W. Tatum and D.G. Hollis). Training materials. Center for Disease Control, Atlanta, Georgia.

King, E.O. and H.W. Tatum. 1962. *Actinobacillus actinomycetemcomitans* and *Haemophilus aphrophilus*. J. Infect. Dis. *111:* 85-94.

King, E.O., W.K. Ward and D.E. Raney. 1954. Two simple media for the demonstration of pyocyanin and fluorescein. J. Lab. Clin. Med. *44:* 301-307.

King, S. and W.I. Metzger. 1968. A new plating medium for the isolation of enteric pathogens. Appl. Microbiol. *16:* 577-578.

King, T.E., E.H. Kawasaki and V.H. Cheldelin. 1956. Tricarboxylic acid cycle activity in *Acetobacter pasteurianum*. J. Bacteriol. *72:* 418-421.

Kingsbury, D.T. 1966. Bacteriocin production by strains of *Neisseria meningitidis*. J. Bacteriol. *91:* 1696-1699.

Kingsbury, D.T. 1967. Deoxyribonucleic acid homologies among species of the genus *Neisseria*. J. Bacteriol. *94:* 870-874.

Kingsbury, D.T., G.R. Fanning, K.E. Johnson and D.J. Brenner. 1969. Thermal stability of interspecies *Neisseria* DNA duplexes. J. Gen. Microbiol. *55:* 201-208.

Kingsbury, D.T. and E. Weiss. 1968. Deoxyribonucleic acid homology between species of the genus *Chlamydia*. J. Bacteriol. *96:* 1421-1423.

Kingsley, V.V. and J.F.M. Hoeniger. 1973. Growth, structure, and classification of *Selenomonas*. Bacteriol. Rev. *37:* 479-521.

Kinyon, J.M. and D.L. Harris. 1974. Growth of *Treponema hyodysenteriae* in liquid medium. Vet. Record. *95:* 219-220.

Kinyon, J.M. and D.L. Harris. 1979. *Treponema innocens*, a new species of intestinal bacteria and emended description of the type strain of *Treponema hyodysenteriae* Harris et al. Int. J. Syst. Bacteriol. *29:* 102-109.

Kippax, P.W. 1957. A study of *Proteus* infections in a male urological ward. J. Clin. Pathol. *10:* 211-213.

Kirakosyan, A.V. and Zh.S. Melkonyan. 1964. New *Azotobacter agile* varieties from the soils of ARMSSR. (R) Dokl. Akad. Nauk. Armyan. S.S.R. *17:* 33-42.

Kirby, B.D., W.L. George, V.L. Sutter, D.M. Citron and S.M. Finegold. 1980. Gram-negative anaerobic bacilli: their role in infection and patterns of susceptibility to antimicrobial agents. I. Little-known *Bacteroides* species. Rev. Infect. Dis. *2:* 914-951.

Kirby, B.D., K.M. Snyder, R.D. Meyer and S.M. Finegold. 1978. Legionnaires' disease: clinical features of 24 cases. Ann. Intern. Med. *89:* 297-309.

Kirby, H., Jr. 1941. Organisms living on and in Protozoa. In Calkins and Summers (Editors), Protozoa in Biological Research, Columbia University Press, New York, pp. 1009-1113.

Kirby, T., J. Blum, I. Kahane and I. Fridovich. 1980. Distinguishing between Mn-containing and Fe-containing superoxide dismutases in crude extracts of cells. Arch. Biochem. Biophys. *201:* 551-555.

Kirchhoff, H. 1974. Neue Spezies der Fam. *Acholeplasmataceae* und der Fam. *Mykoplasmataceae* bei Pferden. Zbl. Vet. Med. B *21:* 207-210.

Kirchhoff, H. 1978. *Acholeplasma equifetale* and *Acholeplasma hippikon*, two new species from aborted horse fetuses. Int. J. Syst. Bacteriol. *28:* 76-81.

Kirchhoff, H. 1978. *Mycoplasma equigenitalium*, a new species from the cervix region of mares. Int. J. Syst. Bacteriol. *28:* 496-502.

Kirchner, O. 1896. Die Wurzelknöllchen der Sojabohne. Beitr. Biol. Pflanz. *7:* 213-224.

Kiredjian, M., M. Popoff, C. Coynault, M. Lefèvre and M. Lemelin. 1981. Taxonomie du genre *Alcaligenes*. Ann. Microbiol. (Inst. Pasteur) *132B:* 337–374.

Kistner, A. 1954. Conditions determining the oxidation of carbon monoxide and of hydrogen by *Hydrogenomonas carboxydovorans*. Proc. Acad. Sci. Amst. Ser. C, *57:* 186-195.

Kitos, P.A. 1956. Terminal oxidation pathways in *Acetobacter suboxydans*. Thesis, Oregon State College.

Kitos, P.A., C.H. Wang, B.A. Mohler, T.E. King and V.H. Cheldelin. 1958. Glucose and gluconate dissimilation in *Acetobacter oxydans*. J. Biol. Chem. *233:* 1295-1298.

Kjems, E. 1955. Studies on five bacterial strains of the genus *Pseudomonas*. Acta Pathol. Microbiol. Scand. *36-37:* 531-536.

Klainer, A.S. and J.D. Pollack. 1973. Scanning electron microscopy techniques in the study of the surface of mycoplasmas. Ann. N.Y. Acad. Sci. *225:* 236.

Klaviter, E.C. and R.C. Johnson. 1979. Isolation of the outer envelope. Chemical components and ultrastructure of *Borellia hermsi* grown in vitro. Acta Trop. *36:* 123-131.

Klebahn, H. 1919. Die Schadlinge des Klippfisches. Ein Beiträg zür Kenntnis der salzliebenden Organismen. Mitt. Inst. Allg. Bot. Hamburg Vol. 4., Otto Meissners Verlag, Hamburg, pp. 11-69.

Kleckner, A.L. 1960. Serotypes of avian pleuropneumonia-like organisms. Am. J. Vet. Res. *21:* 274-280.

Kleczkowska, J., P.S. Nutman, F.A. Skinner and J.M. Vincent. 1968. The identification and classification of *Rhizobium*. In Gibbs and Shapton (Editors), Identification Methods for Microbiologists, Part B. Academic Press, New York, pp. 51-65.

Kleeberger, Von Alfons. 1979. Untersuchungen zür Taxonomie von Enterobakterien und Pseudomonaden aus Hachfleisch. Arch. Lebensmittelhyg. *30:* 130-137.

Klein, D. and G.H. Luginbuhl. 1979. Simplified media for growth of *Haemophilus influenzae* from clinical and normal flora sources. J. Gen. Microbiol. *113:* 409-411.

Klein, D. A. and L. E. Casida. 1967. Occurrence and enumeration of *Bdellovibrio bacteriovorus* in soil capable of parasitizing *Escherichia coli* and indigenous soil bacteria. Can. J. Microbiol. *13:* 1235-1241.

Klein, E. 1889. Ueber epidemische Krankheit der Hühner, verursacht durch einen

Bacillus - *Bacillus gallinarum*. Zentrabl. Bakteriol. Parasitenk. Infektionskr. Hyg. Abt. I Orig. *5:* 689-693.

Klein, J.R. 1940. The oxidation of 1(-) aspartic and 1(+) glutamic acids by *Haemophilus parainfluenzae*. Note on the preparation of pyridine nucleotides from baker's yeast by the method of Warburg and Christian. J. Biol. Chem. *134:* 43-57.

Klein, R.M. and I.L. Tenebaum. 1955. A quantitative bioassay for crown-gall tumor formation. Am. J. Bot. *42:* 709-712.

Kleinig, H., R. Schmitt, W. Meister, G. Englert and H. Thommen. 1979. New C₃₀-carotenoic acid glucosyl esters from *Pseudomonas rhodos*. Z. Naturforsch. Sect. C: Biosciences *34:* 181.

Kleinmeier, H. and E. Schafer. 1956. Beiträg zür Pathogenitätsfrage der Bethesda-Ballerup Stämme. Zentralb. Bakt. Hyg. Parasitenk. Infektionskr. Abt. I. Orig. *165:* 97-107.

Klieneberger, E. 1938. Pleuropneumonia-like organisms of diverse provenance: some results of an enquiry into methods of differentiation. J. Hyg. *38:* 458-475.

Klieneberger, E. 1942. Some new observations bearing on the nature of the pleuropneumonia-like organism known as L₁ associated with *Streptobacillus moniliformis*. J. Hyg. *42:* 485-497.

Klieneberger, E. and D.B. Steabben. 1937. On a pleuropneumonia-like organism in lung lesions of rats, with notes on the clinical and pathological features of the underlying condition. J. Hyg. *37:* 143-152.

Klieneberger-Nobel, E. 1948. Capsules and mucoid envelopes of bacteria. J. Hyg. *46:* 345-348.

Kligler, I.J. and M. Aschner. 1931. Cultivation of rickettsia-like microorganisms from certain blood-sucking pupipara. J. Bacteriol. *22:* 103-117.

Klinger, R. 1912. Untersuchungen über menschliche Aktinomykose. Zentralbl. Bakteriol. Parasitenkd. Infektionskr. Hyg. Abt. 1, Orig. *62:* 191-200.

Klipstein, F.A., R.F. Engert and H.B. Short. 1977. Relative enterotoxigenicity of coliform bacteria. J. Infect. Dis. *136:* 205-215.

Klipstein, F.A., L.V. Holdeman, J.J. Corcino and W.E.C. Moore. 1973. Enterotoxigenic intestinal bacteria in tropical sprue. Ann. Int. Med. *79:* 632-636.

Kloepper, J.W., J. Leong, M. Teintze and M.N. Schroth. 1980. Enhanced plant growth by siderophores produced by plant growth-promoting rhizobacteria. Nature *286:* 885-886.

Klontz, G.W., W.T. Yasutake and A.J. Ross. 1966. Bacterial disease of the *Salmonidae* in the Western United States. Pathogenesis of furunculosis in rainbow trout. Am. J. Vet. Res. *27:* 1455-1460.

Kloos, W.E., W.J. Dobrogosz, J.W. Ezzell, B.R. Kimbro and C.R. Manclark. 1979. DNA-DNA hybridization, plasmids, and genetic exchange in *Bordetella*. *In* Manclark and Hill (Editors), International Symposium on Pertussis, U.S. Government Printing Office, Washington, D.C., pp. 70-80.

Klucas, R. 1972. Nitrogen fixation by *Klebsiella* grown in the presence of oxygen. Can. J. Microbiol. *18:* 1845-1850.

Kluyver, A.J. 1956. *Pseudomonas aureofaciens* nov. spec. and its pigments. J. Bacteriol. *72:* 406-411.

Kluyver, A.J. 1956. Life's flexibility; microbial adaptation. *In* Kluyver and Van Niel (Editors), The Microbe's Contribution to Biology. Harvard University Press, Cambridge, Massachusetts, pp. 93-129.

Kluyver, A.J. 1957. *Zymomonas* Kluyver and van Viel 1936, *In* Breed, Murray and Smith (Editors), Bergey's Manual of Determinative Bacteriology, 7th Ed. The Williams and Wilkins Co., Baltimore, pp. 199-200.

Kluyver, A.J. and J.H. Becking. 1955. Some observations on the nitrogen-fixing bacteria of the genus *Beijerinckia* Derx. Ann. Acad. Sci. Fennicae A II *60:* 367-380.

Kluyver, A.J. and W.J. Hoppenbrouwers. 1931. Ein merkwürdiges Gärungsbakterium: Lindner's *Termobacterium mobile*. Arch. Mikrobiol. *2:* 245-260.

Kluyver, A.J. and A. Manten. 1942. Some observations on the metabolism of bacteria oxidizing molecular hydrogen. Antonie van Leeuwenhoek J. Microbiol. Serol. *8:* 71-85.

Kluyver, A.J. and M.T. van den Bout. 1936. Notiz über *Azotobacter agilis* Beijerinck. Arch. Microbiol. *7:* 261-263.

Kluyver, A.J. and C.B. van Niel. 1936. Prospects for a natural system of classification of bacteria. Zentrabl. Bakteriol. Parasitenkd. Infektionskr. Hyg. Abt. II *94:* 369-403.

Kmety, E. 1967. Factoreanalyze von Leptospiren der icterohaemorrhagiae und einiger verwandter Serogruppen. Edition of Scientific Committee for General and Special Biology of the Slovak Acad. Sci., Vol. 13, no. 3, Bratislava.

Kmety, E. 1972. A Vi antigen in leptospirae. Zentralbl. Bakteriol. Parasitenkd. Infektionskr. Hyg., Abt. Orig. A *221:* 343-351.

Knapp, W. 1968. Serologische kreuzreaktionen zwischen *Pasteurella pseudotuberculosis* (syn. *Yersinia pseudotuberculosis), Escherichia coli* and *Enterobacter cloacae*. Prog. Immunobiol. Standard *2:* 179-186.

Knapp, W. and G. Lebek. 1967. Übertragung der infektiösen resistenz auf Pasteurellen. Pathol. Microbiol. *30:* 103-121.

Knipp, L.H. and J.R. Sokatch. 1969. The chemical composition of the cell envelope of *Streptobacillus moniliformis*. Can. J. Microbiol. *15:* 665-669.

Knorr, M. 1922. Über die fusospirilläre Symbiose, die Gattung *Fusobacterium* (K.B. Lehmann) und *Spirillum sputigenum*. Zugleich ein Beiträg zür Bakteriologie der Mundhohle. II. Mitteilung. Die Gattung *Fusobacterium*. Zentralbl. Bakteriol. Parasitenkd. Infektionskr. Hyg., I Abt. Orig. *89:* 4-22.

Knösel, D. 1961. Eine an Kohl blattfleckenerzeugende Varietas von *Xanthomonas campestris* (Pammel) Dowson. Z. Pflanzenkr. Pflanzenpathol. Pflanzen-

schutz *68:* 1-6.

Knösel, D. 1962. Prüfung von Bakterien auf Fähigkeit zür Sternbildung. Zentralbl. Bakteriol. Parasitenkd. Infektionskr. Hyg. II Abt. *116:* 79-100.

Knox, K.W. and R.V.S. Bain. 1960. The antigens of *Pasteurella multocida* type 1. I. Capsular polysaccharides. Immunol. *3:* 352-362.

Knox, K.W. and R.B. Parker. 1973. Isolation of a phenol-soluble endotoxin from *Leptotrichia buccalis*. Arch. Oral Biol. *18:* 85-93.

Knudson, G.B. and P. Mikesell. 1980. A plasmid in *Legionella pneumophila*. Infect. Immunol. *29:* 1092-1095.

Ko, C.Y., J.L. Johnson, L.B. Barnett, H.M. McNair and J.R. Vercellotti. 1977. A sensitive estimation of the percentage of guanine plus cytosine in deoxyribonucleic acid by high performance liquid chromatography. Anal. Biochem. *80:* 183-192.

Kobayashi, T., S. Enomoto, R. Sakazaki and S. Kuwahara. 1963. A new selective medium for pathogenic vibrios, TCBS agar (modified Nakanishi's agar). Jpn. J. Bacteriol. *18:* 387-391.

Kocan, K.M., K.D. Teel and J. Hair. 1980. Demonstration of *Anaplasma marginale* Theiler in ticks by tick transmission, animal inoculation and fluorescent antibody studies. Am. J. Vet. Res. *41:* 183-186.

Koch, A. 1936. Symbiosestudien. II. Experimentelle Untersuchungen an *Oryzaephilus surinamensis* L. Z. Morphol. Oekol. Tiere *32:* 137-180.

Koch, A. 1960. Intracellular symbiosis in insects. Annu. Rev. Microbiol. *14:* 121-140.

Koch, A. 1967. Insects and their endosymbionts. *In* Henry (Editor), Symbiosis, Vol. II, Academic Press, New York, pp. 1-106.

Koch, A.L. 1981. Evolution of antibiotic resistance gene function. Microbiol. Rev. *45:* 335-378.

Koch, R. 1877. Untersuchungen über Bacterien. VI. Verfahren zür Untersuchung zum Conserviren und Photographiren der Bacterien. Beitr. Biol. Pflanz. *2:* 399-440.

Kocur, M. and J. Boháček. 1972. DNA base composition of extremely halophilic cocci. Arch. Mikrobiol. *82:* 280-282.

Kocur, M. and W. Hodgkiss. 1973. Taxonomic status of the genus *Halococcus* Schoop. Int. J. Syst. Bacteriol. *23:* 151-156.

Kocur, M., T. Martinec and K. Mazanec. 1968. Fine structure of *Micrococcus denitrificans* and *M. halodenitrificans* in relation to their taxonomy. Antonie van Leeuwenhoek J. Microbiol. Serol. *34:* 19-26.

Kocur, M., B. Smid and T. Martinec. 1972. The fine structure of extreme halophilic cocci. Microbios *5:* 101-107.

Kodama, T., Y. Igarashi and Y. Minoda. 1975. Isolation and culture conditions of a bacterium grown on hydrogen and carbon dioxide. Agr. Biol. Chem. *39:* 77-82.

Koekman, B.P., G. Ooms, P.M. Klapwijk and R.A. Schilperoort. 1979. Genetic map of an octopine Ti-plasmid. Plasmid *2:* 347-357.

Kohlert, R. 1968. Untersuchungen zür Ätiologie der Eileiterentzündung beim Huhn. Monatsh Veterinaermed. *23:* 392-395.

Kohn, J. 1953. Preliminary report of new gelatin liquefaction method. J. Clin. Pathol. *6:* 249.

Koka, P. and J. Lee. 1979. Separation and structure of the prosthetic group of the blue fluorescence protein from the bioluminescent bacterium *Photobacterium phosphoreum*. Proc. Natl. Acad. Sci. U.S.A. *76:* 3068-3072.

Kokorin, I.N., E.D. Miskarova, O.S. Gudima, E.A. Kabanova and Truong dinh Kiet. 1978. Intracellular development of rickettsiae. *In* Kazar, Ormsbee and Tarasevich (Editors), Rickettsiae and Rickettsial Diseases, VEDA, Bratislava, pp. 197-203.

Kolb, G. 1959. Untersuchungen über die kernverhaltnisse und morphologischen Eigenschaften symbiontischer Mikroorganismen bei verschiedenen Insekten. Z. Morphol. Oekol. Tiere *48:* 1-71.

Koliaditskaia, L.S., K.V. Kuchina and A.A. Schmurygina. 1959. Tularaemia phage. J. Microbiol. Epidemiol. Immunobiol. *30:* 14-18.

Kolya, A.K. and N.C. Pant. 1962. Effect of ultraviolet irradiation and centrifugation on the mycetomes and symbiotes of *Oryzaephilus surinamensis*. Indian J. Entomol. *24:* 191-198.

Kondo, W., M. Sato and H. Ozawa. 1976. Haemagglutinating activity of *Leptotrichia buccalis* cells and their adherence to saliva-coated enamel powder. Arch. Oral Biol. *21:* 363-369.

Kondo, W., M. Sato and N. Sato. 1978. Properties of the human salivary aggregating factor for *Leptotrichia buccalis* cells. Arch. Oral Biol. *23:* 453-458.

Kondorosi, A., G.B. Kiss, T. Forrai, E. Vincze and Z. Banfalvi. 1977. Circular linkage map of the *Rhizobium meliloti* chromosome. Nature (London) *268:* 525-527.

König, C., I. Sammler, E. Wilde and H.G. Schlegel. 1969. Konstitutive Glucose-6-phosphat-Dehydrogenase bei Glucose verwertenden Mutanten von einem kryptischen Wildstamm. Arch. Mikrobiol. *67:* 51-57.

Koning, H.C. 1938. Bacterial canker of the poplar. Chron. Bot. *4:* 11-12.

Kono, M., M. Sasatsu and K. Makino. 1980. R-plasmid conferring resistance to ampicillin, carbenicillin and erythromycin in a *Flavobacterium odoratum* strain. Microbios Lett. *14:* 55-58.

Konopka, A.E., J.C. Lara and J.T. Staley. 1977. Isolation and characterization of gas vesicles from *Microcyclus aquaticus*. Arch. Microbiol. *112:* 133-140.

Konopka, A.E., R.L. Moore and J.T. Staley. 1976. Taxonomy of *Microcyclus* and other nonmotile ring-forming bacteria. Int. J. Syst. Bacteriol. *26:* 505-510.

Kontaxis, D.G. and A.C. Hayward. 1978. The pathogen and symptomatology of pink disease of pineapple fruit in the Philippines. Plant Dis. Rep. *62:* 446-450.

Kopmans-Gargantiel, A.I. and C.L. Wisseman, Jr. 1981. Differential requirements for enriched atmospheric carbon dioxide content for intracellular growth in cell culture among selected members of the genus *Rickettsia.* Infect. Immun. *31:* 1277-1280.

Kopp, R., J. Muller and R. Lemme. 1966. Inhibition of *Proteus* by sodium tetradecyl sulphate, β-phenethyl alcohol and p-nitrophenylglycerol. Appl. Microbiol. *14:* 872-878.

Kordová, N., P.R. Burton, C.M. Downs, D. Paretsky and E. Kováčová. 1970. The interaction of *Coxiella burnetii* phase I and phase II in Earle's cells. Can. J. Microbiol. *16:* 125-133.

Korfhagen, T.R., L. Sutton and G.A. Jacoby. 1978. Classification and physical properties of *Pseudomonas* plasmids. *In* Schlessinger (Editor) "Microbiology - 1978", American Society for Microbiology, Washington, D.C., pp. 221-224.

Kornberg, H.L., J.F. Collins and D. Bigley. 1960. The influence of growth substrates on metabolic pathways in *Micrococcus denitrificans.* Biochim. Biophys. Acta *39:* 9-24.

Kornberg, H.L. and N.B. Madsen. 1958. The metabolism of C_2 compounds in microorganisms. 3. Synthesis of malate from acetate via the glyoxylate cycle. Biochem. J. *68:* 549-557.

Korner, H.K. 1976. On the host-symbiont-cycle of a leafhopper *(Euscelis plebejus)* endosymbiosis. Experientia (Basel) *32:* 463-464.

Kornman, K.S. and S.C. Holt. 1981. Physiological and ultrastructural characterization of a new *Bacteroides* species (*Bacteroides capillus*) isolated from severe localized periodontitis. J. Periodontal Res. *16:* 542-555.

Kornman, K.S. and S.C. Holt. 1982. *In* Validation of the publication of new names and new combinations previously effectively published outside the IJSB. List No. 8. Int. J. Syst. Bacteriol. *32:* 266-268.

Korol, A.G. and J. Parnas. 1967. Ein neue Serobiotyp von *Brucella - Brucella murium* (Korol). Zschr. Ges. Hyg. *13:* 791-800.

Korth, H., I. Ørskov and P. Pulverer. 1969. Farbstoffbildende Klebsiella -Stämme. Zentralbl. Bakteriol. Parasitenk. Infektionskr. Hyg. Abt. I, Orig. A *211:* 105-107.

Koshimizu, K. and T. Magaribuchi. 1977. Isolation of ureaplasma (T-mycoplasma) from the chicken and jungle-fowl. Jpn. J. Vet. Sci. *39:* 195-199.

Koshimizu, K. and T. Magaribuchi. 1978. Biological and serological characterization of ureaplasmas isolated from domestic fowls and red jungle fowls. Jpn. J. Vet. Sci. *40:* 719-727.

Koshimizu, K., T. Magaribuchi, M. Ito, H. Kotani and M. Ogata. 1981. Distribution of ureaplasmas in various laboratory animals. Vet. Rec. *108:* 309-310.

Kotani, H., R. Harasawa, K. Yamamoto and M. Ogata. 1980a. Serological studies with feline ureaplasmas. Microbiol. Immun. *24:* 83-86.

Kotani, H., H. Nagatomo and M. Ogata. 1980b. Isolation and serological comparison of ureaplasmas from goats and sheep. Jpn. J. Vet. Sci. *42:* 31-40.

Kotani, H. and M. Ogata. 1979. Isolation and serological grouping of ureaplasmas from dogs. Jpn. J. Vet. Sci. *41:* 639-646.

Kotasthane, W.V., A.C. Padhya and M.K. Patel. 1965. Utilization of amino acids as sole source of carbon and nitrogen by some xanthomonads. Indian Phytopathol. *18:* 154-159.

Kotte, W. 1930. Eine bakterielle Blattfäule der Winter-Endivie *(Cichorium endivia* L.). Phytopathol. Z. pp. 605-613.

Kottel, R.H. and H.D. Raj. 1973. Pathways of carbohydrate metabolism in *Microcyclus* species. J. Bacteriol. *113:* 341-349.

Kotter, L. 1955. Bakteriologische und mikrochemische Untersuchungen an der Magenscheibe von *Pediculus vestimenti* Burm. Arch. Mikrobiol. *23:* 38-66.

Kouno, K. and A. Ozaki. 1975. Distribution and identification of methanol-utilizing bacteria. *In* Organizing Committee (Editors), Microbial Growth on C_1-Compounds, Society of Fermentation Technology, Osaka, Japan, pp. 11-21.

Kovács, N. 1956. Identification of *Pseudomonas pyocyanea* by the oxidase reaction. Nature (London) *178:* 703.

Koval, S.F. and P.M. Meadow. 1977. The isolation and characterization of lipopolysaccharide-defective mutants of *Pseudomonas aeruginosa* PAC1. J. Gen. Microbiol. *98:* 387-398.

Kovochevich, R. and W.A. Wood. 1955. Carbohydrate metabolism by *Pseudomonas fluorescens.* J. Biol. Chem. *213:* 745-756, 757-767.

Kowal, J. 1961. *Spirillum* fever: report of a case and review of the literature. N. Engl. J. Med. *264:* 123-128.

Kozek, W.J. 1971. Ultrastructure of the microfilaria of *Dirofilaria immitis.* J. Parasitol. *57:* 1052-1067.

Kozek, W.J. and H.F. Marroquin. 1977. Intracytoplasmic bacteria in *Onchocerca volvulus.* Amer. J. Trop. Med. Hyg. *26:* 663-678.

Kozulis, J.A. and R.H. Parsons. 1958. *Acetobacter alcoholophilus* n. sp. - a new species isolated from storage beer. J. Inst. Brew. *64:* 47-50.

Kraepelin, G. and D. Passern. 1980. Gallertlager einer besonderen Mikroorganismengesellschaft an verbautem Grubenholz. Zeit. Allgem. Mikrobiol. *20:* 303-314.

Kraepilin, G. and J.-U. Gravenstein. 1980. Experimentella Induktion von "rotund bodies" bei *Thermus aquaticus.* Z. Allg. Mikrobiol. *20:* 33-45.

Kramer, T. T. and J. M. Westergaard. 1977. Antigenicity of *Bdellovibrio.* Appl. Environ. Microbiol. *33:* 967-970.

Krasil'nikov, N.A. 1949. Opredelitel' bakterii i aktinomitsetov. Izdatel'stvo Akademii Nauk SSSR, Moskva-Leningrad, pp. 1-830.

Krasil'nikov, N.A. 1949. Guide to the bacteria and actinomycetes. Akad. Nauk SSSR, Moscow, pp. 1-830.

Krasil'nikov, N.A. 1949. Diagnostik der Bakterien und Actinomyceten. German translation by R. Wittwer and R. Dickscheit. Veb. Gustav Fischer Verlag Jena, 1959. English translation of section on Actinomycetes by Chas. Pfizer and Company, Inc. (J.B. Routien, Editor), 1959.

Kraut, M.S., J.R. Attenberry, S.M. Finegold and V.L. Sutter. 1972. Detection of *Haemophilus aphrophilus* in the human oral flora with a selective medium. J. Infect. Dis. *126:* 189–192.

Krehan, M. 1930. Beiträge zür Physiologie und Systematik der Essigbakteriën. Arch. Mikrobiol. *1:* 493-536.

Kreier, J.P., N. Dominguez, H.E. Krampitz, R. Gothe and M. Ristic. 1981. The hemotrophic bacteria: the families *Bartonellaceae* and *Anaplasmataceae. In* Starr, Stolp, Trüper, Balows and Schlegel (Editors), The Prokaryotes: a handbook on habitats, isolation and identification of bacteria, Springer-Verlag, Berlin - Heidelberg - New York, pp. 2189-2209.

Kreier, J.P. and M. Ristic. 1963. Anaplasmosis. X. Morphologic characteristics of the parasites present in the blood of calves infected with the Oregon strain of *Anaplasma marginale.* Am. J. Vet. Res. *24:* 676-687.

Kreier, J.P. and M. Ristic. 1963. Anaplasmosis. XI. Immunoserologic characteristics of the parasites present in the blood of calves infected with the Oregon strain of *Anaplasma marginale.* Am. J. Vet. Res. *24:* 688-696.

Kreier, J.P. and M. Ristic. 1963. Anaplasmosis. XII. The growth and survival in deer and sheep of the parasites present in the blood of calves infected with the Oregon strain of *Anaplasma marginale.* Am. J. Vet. Res. *24:* 697-702.

Kreier, J.P. and M. Ristic. 1968. Haemobartonellosis, eperythrozoonosis, grahamellosis and ehrlichiosis. Chapter 22. *In* Weinman and Ristic (Editors), Infectious Blood Diseases of Man and Animals, Academic Press, New York, pp. 387-472.

Kreier, J.P. and M. Ristic. 1972. Definition and taxonomy of *Anaplasma* species with emphasis on morphologic and immunologic features. Z. Tropenmed. Parasitol. *23:* 88-89.

Kreier, J.P. and M. Ristic. 1973. Organisms of the family *Anaplasmataceae* in the forthcoming 8th edition of Bergey's Manual. *In* Jones (Editor), Proc. 6th National Anaplasmosis Conference, Heritage Press, Stillwater, Oklahoma, pp. 24-28.

Kreutzer, D.L., C.S. Buller and D.C. Robertson. 1979. Chemical characterization and biological properties of lipopolysaccharides isolated from smooth and rough strains of *Brucella abortus.* Infect. Immun. *23:* 811-818.

Kreutzer, D.L. and D.C. Robertson. 1979. Surface macromolecules and virulence in intracellular parasitism: comparison of cell envelope components of smooth and rough strains of *Brucella abortus.* Infect. Immun. *23:* 819-828.

Kreutzer, D.L., J.W. Scheffel, L.R. Draper and D.C. Robertson. 1977. Mitogenic activity of cell wall components from smooth and rough strains of *Brucella abortus.* Infect. Immun. *15:* 842-845.

Krichevsky, M.I. and L.M. Norton. 1974. Storage and manipulation of data by computers for determinative bacteriology. Int. J. Syst. Bacteriol. *24:* 525-531.

Krieg, A. 1955a. Untersuchungen zür Wirbeltier-Pathogenität und zum serologischen Nachweis der *Rickettsia melolonthae* im Arthropod-Wirt. Naturwissenschaften *42:* 609-610.

Krieg, A. 1955. Licht- und elektronenmikroskopische Untersuchungen zür Pathologie der "Lorscher Erkrankung" von Engerlingen und zür Zytologie der *Rickettsia melolonthae* nov. spec. Z. Naturforsch. *10b:* 34-37.

Krieg, A. 1958. Vergleichende taxonomische, morphologische und serologische Untersuchungen an insektenpathogènen Rickettsien. Z. Naturforsch. *13b:* 555-557.

Krieg, A. 1959. On the problem of crystals associated with *Rickettsiella* infections. J. Insect Pathol. *1:* 95.

Krieg, A. 1961. Grundlagen der Insektenpathologie; Viren-, Rickettsien- und Bakterien- Infektionen. Steinkopff, Darmstadt.

Krieg, A. 1965. Über eine neue Rickettsie aus Coleopteren, *Rickettsiella tenebrionis* nov. spec. Naturwissenschaften *52:* 144-145.

Krieg, N. R. 1976. Biology of the chemoheterotrophic spirilla. Bacteriol. Rev. *40:* 55-115.

Krieg, N. R. 1980. The genera *Spirillum, Aquaspirillum,* and *Oceanospirillum. In* Starr, Stolp, Trüper, Balows and Schlegel (Editors), The Prokaryotes, A handbook on habitats, isolation, and identification of bacteria. Springer-Verlag, New York, pp. 2601-2644.

Krieg, N.R. and P. Gerhardt. 1981. Solid culture. *In* Gerhardt, Murray, Costilow, Nester, Wood, Krieg and Phillips (Editors), Manual of Methods for General Bacteriology. Amer. Soc. Microbiol., Washington, D.C.

Krishnapillai, V. and L.S. Baron. 1964. Alterations in the mouse virulence of *Salmonella typhimurium* by genetic recombination. J. Bacteriol. *87:* 593-605.

Kröger, A., E. Dorrer and E. Winkler. 1980. The orientation of the substrate sites of formate dehydrogenase and fumarate reductase in the membrane of *Vibrio succinogenes.* Biochim. Biophys. Acta *589:* 118-136.

Kröger, A., E. Winkler, A. Innerhofer, H. Hackenberg and H. Schagger. 1979. The formate dehydrogenase involved in electron transport from formate to fumarate in *Vibrio succinogenes.* Eur. J. Biochem. *94:* 465-475.

Krogsgaard-Jensen, A. 1971. Indirect hemagglutination with *Mycoplasma* antigens: effects of pH on antigen sensitization of tanned fresh and formalinized sheep erythrocytes. Appl. Microbiol. *22:* 756-759.

Krogsgaard-Jensen, A. 1972. Growth precipitation as a serodiagnostic method. Appl. Microbiol. *23:* 553-558.

Kroll, H.-P., J. Gmeiner and H.H. Martin. 1980. Membranes of the protoplast L-form of *Proteus mirabilis*. Arch. Microbiol. *127:* 223-229.

Kronvall, G. and E. Myhre. 1977. Differential staining of bacteria in clinical specimens using acridine orange buffered at low pH. Acta Pathol. Microbiol. Scand. Sect. B *85:* 249-254.

Kropinski, A. M. 1975. A chemically defined medium for *Aquaspirillum aquaticum* ATCC 11330. Can. J. Microbiol. *21:* 1886-1889.

Krumwiede, E. and A.G. Kuttner. 1938. A growth-inhibiting substance for the influenza group of organisms in the blood of various animal species. J. Exp. Med. *67:* 429-441.

Kruse, W. 1896. *Bacillus ulceris cancrosi* (Bacillus des weichen Schankers) *In* Flügge (Editor), Die Mikroorganismen, unter spezieller Berücksichtigung der Ätiologie infektiöser Krankheiten 3. Aufl. *2:* 456-458.

Krych, V.A., J.L. Johnson and A.A. Yousten. 1980. Deoxyribonucleic acid homologies among strains of *Bacillus sphaericus*. Int. J. Syst. Bacteriol. *30:* 476-484.

Krywolap, G.N. and L.R. Page. 1977. Oral *Fusobacterium, Leptotrichia* and *Bacterionema:* II. Pathogenicity: a review of the literature. J. Baltimore Coll. Dent. Surg. *32:* 26-32.

Kuberski, T., J.M. Papadimitriou and P. Phillips. 1980. Ultrastructure of *Calymmatobacterium granulomatis* in lesions of granuloma inguinale. J. Infect. Dis. *142:* 744-749.

Kubomura, K. 1969. Fructose medium for the cultivation of *Cristispira* sp., a flagellate living in the crystalline style of bivalves. Sci. Rep. Saitama Univ. Ser. B. *5:* 1-5.

Kubota, Y., S. Iuchi, A. Fujisawa and S. Tanaka. 1979. Separation of four components of the phosphoenolpyruvate: glucose phosphotransferase system in *Vibrio parahaemolyticus*. Microbiol. Immunol. *23:* 131-146.

Kuenen, J. G. and S. C. Rittenberg. 1975. Incorporation of long- chain fatty acids of the substrate organism by *Bdellovibrio bacteriovorus* during intraperiplasmic growth. J. Bacteriol. *121:* 1145-1157.

Kuhn, D.A. 1974. Genus II. *Cristispira* Gross 1910, 44. *In* Buchanan and Gibbons (Editors), Bergey's Manual of Determinative Bacteriology 8th Ed., Williams and Wilkins Co., Baltimore, pp. 171-174.

Kuhn, D.A. and M.P. Starr. 1965. Clonal morphogenesis of *Lampropedia hyalina*. Arch. Microbiol. *52:* 360-375.

Kuhn, R., M.P. Starr, D.A. Kuhn, H. Bauer and H.-J. Knackmuss. 1965. Indigoidine and other pigments related to 3,3'-bipyridyl. Arch. Mikrobiol. *51:* 71-84.

Kulka, D., J.M. Preston and T.K. Walker. 1949. Giant colonies of *Acetobacter* species as an aid to identification. J. Inst. Brew. *55:* 141-146.

Kulka, D., J. Singh, R.M. Nattrass, A.N. Hall and T.K. Walker. 1958. Studies of vinegar bacteria. J. Sci. Food Agr. *9:* 487-492.

Kulka, D. and T.K. Walker. 1946. Capsule formation by *Acetobacter* species. J. Inst. Brew. *52:* 129-131.

Kulkarni, Y.S., M.K. Patel and S.G. Abhyankar. 1950. A new bacterial leaf spot and stem canker of pigeon pea. Curr. Sci. *19:* 384.

Kulkarni, Y.S., M.K. Patel and G.W. Dhande. 1951. *Xanthomonas cassiae* a new bacterial disease of *Cassia tora* L. Curr. Sci. *20:* 47.

Kulp, W.L. and D.G. Borden. 1942. Further studies on *Proteus hydrophilus*, the etiological agent in "red-leg" disease of frogs. J. Bacteriol. *44:* 673-685.

Kumar, R., A. K. Banerjee, J. H. Bowdre, L. J. McElroy and N. R. Krieg. 1974. Isolation, characterization, and taxonomy of *Aquaspirillum bengal* sp. nov. Int. J. Syst. Bacteriol. *24:* 453-458.

Kume, K., A. Sawata and Y. Nakase. 1980. Immunologic relationship between Page's and Sawata's serotype strains of *Haemophilus paragallinarum*. Am. J. Vet. Res. *41:* 757-760.

Kundsin, R.B., T. Rowell, M.C. Shepard, A. Parreno and C.D. Lunceford. 1975. T-strain mycoplasmas and reproductive failure in monkeys. Lab. Anim. Sci. *25:* 221-224.

Kunin, C.M. 1963. Separation, characterization and biological significance of a common antigen in *Enterobacteriaceae*. J. Exp. Med. *118:* 565-586.

Kunin, C.M., M.V. Beard and N.E. Halmaggi. 1962. Evidence for a common hapten associated with endotoxin fractions of *Escherichia coli* and other *Enterobacteriaceae*. Proc. Soc. Exp. Biol. *111:* 160-166.

Kunitsa, G.M., M.A. Aikimbaev, M.K. Tleugobylov, I.V. Rodionova and I.S. Meshcheryakova. 1972. Biochemical characteristics of *Francisella tularensis* strains of the Central Asian geographic race. (Russ.). Zh. Mikrobiol. Epidemiol. Immunobiol. *49:* 124-127.

Kunsman, J.E. 1973. Characterization of the lipids of six strains of *Bacteroides ruminicola*. J. Bacteriol. *113:* 1121-1126.

Kunsman, J.E. and D.R. Caldwell. 1974. Comparison of the spingolipid content of rumen *Bacteroides* species. Appl. Microbiol. *28:* 1088-1089.

Kunz, H.H. 1974. *Leptotrichia buccalis* - Kultur, Biochemie und Antibiotica. Zentralbl. Bakteriol. Parasitenkd. Infektionskr. Hyg., I Abt. Orig. A *228:* 110-115.

Kurita, T. and H. Tabei. 1967. On the pathogenic bacterium of bacterial grain rot of rice. Ann. Phytopathol. Soc. Japan *33:* 111.

Kurowski, W.M., A.H. Fensom and S.J. Pirt. 1975. Factors influencing the formation and stability of D-glucoside 3-dehydrogenase activity in cultures of *Agrobacterium tumefaciens*. J. Gen. Microbiol. *90:* 191-202.

Kurowski, W.M. and S.J. Pirt. 1971. The iron requirement of *Agrobacterium tumefaciens* for growth and 3-ketosucrose production. The removal of iron from solutions by Seitz filters. J. Gen. Microbiol. *68:* 65-69.

Kurtti, T.J. and M.A. Brooks. 1976. Preparation of mycetocytes for culture in vitro. J. Invertebr. Pathol. *27:* 209-214.

Kurtz, W.G.W. and T.A. La Rue. 1975. Nitrogenase activity in rhizobia in absence of host plant. Nature (London) *256:* 407-409.

Kurtzman, C.P., M.J. Smiley, C.J. Johnson, L.B. Wickerham and G.B. Fuson. 1980. Two new and closely related heterothallic species, *Pichia amylophila* and *Pichia mississippiensis:* characterization by hybridization and deoxyribonucleic acid reassociation. Int. J. Syst. Bacteriol. *30:* 208-216.

Kusek, J.W. and L.G. Herman. 1980. Typing of *Proteus mirabilis* by bacteriocin production and sensitivity as a possible epidemiological marker. J. Clin. Microbiol. *12:* 112-120.

Kushner, D.J. 1978. Life in high salt and solute concentrations: halophilic bacteria. *In* Kushner (Editor), Microbial Life in Extreme Environments, Academic Press, London, pp. 317-368.

Kushner, D.J., S.T. Bayley, J. Boring, M. Kates and N.E. Gibbons. 1964. Morphological and chemical properties of cell envelopes of the extreme halophile, *Halobacterium cutirubrum*. Can. J. Microbiol. *10:* 483-497.

Kushwaha, S.C., M.B. Gochnauer, D.J. Kushner and M. Kates. 1974. Pigments and isoprenoid compounds in extremely and moderately halophilic bacteria. Can. J. Microbiol. *20:* 241-245.

Kuttler, K.L. 1966. Clinical and hematologic comparison of *Anaplasma marginale* and *Anaplasma centrale* infections in cattle. Am. J. Vet. Res. *27:* 941-946.

Kuttler, K.L. 1967. Serological relationship of *Anaplasma marginale* and *Anaplasma centrale* as measured by the complement fixation and capillary tube agglutination tests. Res. Vet. Sci. *8:* 207-211.

Kuttler, K.L. and L.G. Adams. 1977. Influence of dexamethasone on the recrudescence of *Anaplasma marginale* in splenectomized calves. Am. J. Vet. Res. *38:* 1327-1330.

Kützing, F.T. 1834. Algarium aquae dulcis germanicarum. Halis Saxonum, Decas *12:* 1-2.

Kützing, F.T. 1849. Species Algarum. Lipsiae. pp. 1-922.

Kwapinski, J.B.G. and G. Horsman. 1973. Cultural characterization and differentiation of Nocardiae. Can. J. Microbiol. *19:* 895-899.

Laanbroek, H. J. 1978. Ecology and physiology of L-aspartate and L-glutamate fermenting bacteria. Ph.D. Thesis, University of Groningen, The Netherlands, pp. 1-98.

Laanbroek, H. J., W. Kingma and H. Veldkamp. 1977. Isolation of an aspartate fermenting, free-living *Campylobacter* species. FEMS Microbiol. Lett. *1:* 99-102.

Laanbroek, H.J., L.J. Stal and H. Veldkamap. 1978. Utilization of hydrogen and formate by *Campylobacter* spec. under aerobic and anaerobic conditions. Arch. Microbiol. *119:* 99-102.

Labaw, L.W. and V.M. Mosley. 1955. Periodic structure of the flagella of *Brucella bronchiseptica*. Biochim. Biophys. Acta *17:* 322-324.

Labbé, M., N. Delamare, F. Pepersack, F. Crokaert and E. Yourassowsky. 1980. Detection of *Bacteroides fragilis* and *Bacteroides melaninogenicus* by direct immunofluorescence. J. Clin. Pathol. *33:* 1189-1192.

Labbé, M., W. Hansen, E. Schoutens and E. Yourassowsky. 1977. Isolation of *Bacteroides corrodens* and *Eikenella corrodens* from human clinical specimens. Comparative study of incidence and methods of identification. Infection *5:* 159-162.

LaCave, C., J. Asselineau, A. Serre and J. Roux. 1969. Comparison de la composition chimique d'une fraction lipopolysaccharidique et d'une fraction polysaccharidique isolées de *Brucella melitensis*. Eur. J. Biochem. *9:* 189-198.

Lackman, D.B., E.J. Bell, H.G. Stoenner and E.G. Pickens. 1965. The Rocky Mountain spotted fever group of rickettsias. Health Lab. Sci. *2:* 135-141.

Lackman, D.B., R.R. Parker and R.K. Gerloff. 1949. Serological characteristics of a pathogenic rickettsia occurring in *Amblyomma maculatum*. Pub. Health Rep. *64:* 1342-1349.

Lacy, B.W. 1951. Antigenic modulation of *Haemophilus pertussis*. J. Gen. Microbiol. *5:* xxi.

Lacy, B. W. 1960. Antigenic modulation of *Bordetella pertussis*. J. Hyg. *58:* 57-93.

Lacy, G.H. and J.V. Leary. 1979. Genetic systems in phytopathogenic bacteria. Annu. Rev. Phytopathol. *17:* 181-202.

Lai, C.-H., M.A. Listgarten, A.C.R. Tanner and S.S. Socransky. 1981. Ultrastructures of *Bacteroides gracilis, Campylobacter concisus, Wolinella recta,* and *Eikenella corrodens,* all from human periodontal disease. Int. J. Syst. Bacteriol. *31:* 465-475.

Lai, M., N.J. Panopoulos and S. Shaffer. 1977a. Transmission of R Plasmids among *Xanthomonas* spp. and other plant pathogenic bacteria. Phytopathology *67:* 1044-1050.

Lai, M., S. Shaffer and N.J. Panopoulos. 1977b. Stability of plasmid-borne antibiotic resistance in *Xanthomonas vesicatoria* in infected tomato leaves. Phytopathology *67:* 1527-1530.

Laidlaw, P.P. and W.J. Elford. 1936. A new group of filterable organisms. Proc. Roy. Soc. (London) Ser. B *120:* 292-303.

Laird, M. and F.A. Lari. 1957. The avian blood parasite *Babesia moshkovskii* (Schurenkova, 1938), with a record from *Corvus splejdens* Vieillot in Pakistan. Can. J. Zool. *35:* 783-795.

Laird, W.J. and D.C. Cavanaugh. 1980. Correlation of autoagglutination and virulence of Yersiniae. J. Clin. Microbiol. *11:* 430-432.

Lake, P. and W.G. Friend. 1967. A monoxenic relationship, *Nocardia rhodnii* Erikson in the gut of *Rhodnius prolixus* Stahl. (Hemiptera: Reduviidae).

Proc. Entomol. Soc. Ont. *98:* 53-57.

Lake, P. and W.G. Friend. 1968. The use of artificial diets to determine some of the effects of *Nocardia rhodnii* on the development of *Rhodnius prolixus.* J. Insect. Physiol. *14:* 543-562.

Lakso, J.U. and M.P. Starr. 1970. Comparative injuriousness to plants of *Erwinia* spp. and other enterobacteria from plants and animals. J. Appl. Bacteriol. *33:* 692-707.

Lalucat, J., A. Alvarez, R. Pares and H.G. Schlegel. 1980. R bodies in *Pseudomonas. In* Schwemmler and Schenk (Editors), Endosymbiosis and Cell Biology, Vol. 1, de Gruyter, Berlin, pp. 409-416.

Lalucat, J., O. Meyer, F. Mayer, R. Pares and H.G. Schlegel. 1979. R-bodies in newly isolated free-living hydrogen-oxidizing bacteria. Arch. Microbiol. *121:* 9-15.

La Macchia, E.H. and M.J. Pelczar, Jr. 1966. Analysis of deoxyribonucleic acid of *Neisseria caviae* and other *Neisseria.* J. Bacteriol. *91:* 514-516.

Lambe, D.W., Jr. 1974. Determination of *Bacteroides melaninogenicus* serogroups by fluorescent antibody staining. Appl. Microbiol. *28:* 561-567.

Lambe, D.W., Jr. 1979. Characterization of a polyvalent conjugate of *Bacteroides fragilis* by fluorescent antibody staining. Am. J. Clin. Pathol. *71:* 97-101.

Lambe, D.W., Jr. 1980. Serology of *Bacteroidaceae. In* Lambe, Genco and Mayberry-Carson (Editors), Anaerobic Bacteria: Selected Topics, Plenum Press, New York, pp. 141-153.

Lambe, D.W., Jr. and R.C. Jerris. 1976. Description of a polyvalent conjugate and a new serogroup of *Bacteroides melaninogenicus* by fluorescent antibody staining. J. Clin. Microbiol. *3:* 506-512.

Lambe, D.W., Jr., A.M. McPhedran, J. Mertz and P. Stewart. 1973. *Streptobacillus moniliformis* isolated from a case of Haverhill fever: biochemical characterization and inhibitory effect of sodium polyanethiol sulfonate. Amer. J. Clin. Pathol. *60:* 854-860.

Lambe, D.W., Jr. and D.M. Moroz. 1976. Serogrouping of *Bacteroides fragilis* subsp. *fragilis* by the agglutination test. J. Clin. Microbiol. *3:* 586-592.

Lamfrom, H., A. Sarabhai and J. Abelson. 1978. Cloning of *Beneckea* genes in *Escherichia coli.* J. Bacteriol. *133:* 354-363.

Lämmler, G. and R. Gothe. 1967. Zür Chemotherapie der *Aegyptianella pullorum-* Infektion des Huhnes. Tropenmed. Parasitol. *18:* 479-488.

Lamparter, H.E. 1967. Intrazelluläre symbiontische Bakterien im Zentralnervensystem der Ameise. Z. Zellforsch. *81:* 1-11.

Lancefield, R.C. 1933. A serological differentiation of human and other groups of hemolytic streptococci. J. Exp. Med. *57:* 571-595.

Lancefield, R.C. 1934. A serological differentiation of specific types of bovine hemolytic streptococci (Group B). J. Exp. Med. *59:* 441-458.

Landay, M.E., G.G. Wright, J.D. Pulliam and M.J. Finegold. 1968. Toxicity of *Pasteurella tularensis* killed by ionizing radiation. J. Bacteriol. *96:* 804-810.

Landis, W.G. 1981. The ecology, role of the killer trait, and interactions of five species of the *Paramecium aurelia* complex inhabiting the littoral zone. Can. J. Zool. *59:* 1734-1743.

Landureau, J.C. 1966. Des cultures de cellules embryonnaires de Blattes permettent d'obtenir la multiplication in vitro des bactéries symbiotiques. C.R. Hebd. Séances Acad. Sci. Ser. D Sci. Nat. *262:* 1484-1487.

Lane, R.S., R.N. Philip and E.A. Casper. 1981. Ecology of tick-borne agents in California. II. Further observations on rickettsiae. *In* Burgdorfer and Anacker (Editors), Rickettsiae and Rickettsial Diseases, Academic Press, New York, pp. 575-584.

Lange, E. and D. Knösel. 1970. Zür Bedeutung pektolytischer, cellulolytischer und proteolytischer Enzyme für die Virulenz phytopathogèner Bakterien. Phytopathol. Z. *69:* 315-329.

Langford, E.V. and R.H. Leach. 1973. Characterization of a mycoplasma isolated from bovine keratoconjunctivitis: *M. bovoculi* sp. nov. Can. J. Microbiol. *19:* 1435-1444.

Langford, E.V., H.L. Ruhnke and O. Onoviran. 1976. *Mycoplasma canadense,* a new bovine species. Int. J. Syst. Bacteriol. *26:* 212-219.

Langworth, B.F. 1977. *Fusobacterium necrophorum:* its characteristics and role as an animal pathogen. Bacteriol. Rev. *41:* 373-390.

Langworthy, T.A. 1977. Long-chain diglycerol tetraethers from *Thermoplasma acidophilum.* Biochim. Biophys. Acta *487:* 37-50.

Langworthy, T.A. 1978. Special Features of Thermoplasmas. *In* Barile and Razin (Editors), The Mycoplasmas I: Cell Biology, Academic Press, Inc., New York, pp. 495-513.

Langworthy, T.A. 1979. Membrane structure of thermoacidophilic bacteria. *In* Shilo (Editor), Strategies of Microbial Life in Extreme Environments, Berlin: Dahlem Konferenzen, Verlag-Chemie, Weinheim, pp. 417-432.

Langworthy, T.A., W.R. Mayberry, P.F. Smith and I.M. Robinson. 1975. Plasmalogen composition of *Anaeroplasma.* J. Bacteriol. *122:* 788-797.

Lanham, U.N. 1968. The Blochmann bodies: hereditary intracellular symbionts of insects. Biol. Rev. Camb. Philos. Soc. *43:* 269-286.

Lankford, C.E. and W. Burrows. 1965. Oblique light microscopy as an aid to rapid detection of enteric pathogens. Proc. Cholera Res. Symp. (Jan. 24-29, 1965 Honolulu). Public Health Serv. Publ. No. 1328. U.S. Gov. Printing Office, Washington, D.C., pp. 45-50.

Lányi, B. 1956. Serological typing of *Proteus* strains from infantile enteritis and other sources. Acta Microbiol. Acad. Sci. Hung. *3:* 417-428.

Lányi, B. 1957. Serological typing of *Proteus* strains. Sensitivity of serotypes to antibiotics. Acta Microbiol. Acad. Sci. Hung. *4:* 447-457.

Lányi, B. 1970. Serological properties of *Pseudomonas aeruginosa.* II. Type-specific thermolabile (flagellar) antigens. Acta Microbiol. Acad. Sci. Hung. *17:* 35-48.

Lányi, B. and T. Bergan. 1978. Serological characterization of *Pseudomonas aeruginosa. In* Bergan and Norris (Editors), Methods in Microbiology, Vol. 10. Academic Press, London, pp. 93-168.

Lányi, J.K. 1974. Salt-dependent properties of proteins from extremely halophilic bacteria. Bacteriol. Rev. *38:* 272-290.

Lányi, J. 1978. Light energy conversion in *Halobacterium halobium.* Microbiol. Rev. *42:* 682-706.

Lányi, J. 1980. Light-driven primary sodium ion transport in *Halobacterium halobium* membranes. J. Supramolec. Struct. *13:* 83-92.

Lanzi, M. 1876. I Batteri parassiti di funghi. Nuovo G. Bot. Ital. 8: 256-261.

Lapage, S.P. 1971. Culture collections of bacteria. Biol. J. Linnean Soc. *3:* 197-210.

Lapage, S.P. 1974. Species incertae sedis. *Haemophilus vaginalis* Gardner and Dukes 1955, 963. *In* Buchanan and Gibbons (Editors), Bergey's Manual of Determinative Bacteriology, 8th Ed., The Williams and Wilkins Co., Baltimore, pp. 368-370.

Lapage, S.P. 1974. Genus *Cardiobacterium* Slotnick and Daugherty 1964, 271. *In* Buchanan and Gibbons (Editors), Bergey's Manual of Determinative Bacteriology, 8th Ed. The Williams and Wilkins Co., Baltimore, pp. 377-378.

Lapage, S.P. 1975. Report of the World Federation for Culture Collections. Int. J. Syst. Bacteriol. *25:* 90-94.

Lapage, S.P., S. Bascomb, W.R. Willcox and M.A. Curtis. 1973. Identification of bacteria by computer. I. General aspects and perspectives. J. Gen. Microbiol. *77:* 273-290.

Lapage, S.P., L.R. Hill and J.D. Reeve. 1968. *Pseudomonas stutzeri* in pathological material. J. Med. Microbiol. *1:* 195-202.

Lapage, S.P., J.E. Shelton, T.G. Mitchell and A.R. MacKenzie. 1970. Culture collections and the preservation of bacteria. *In* Norris and Ribbons (Editors), Methods in Microbiology, Vol. 3A, Academic Press, London, pp. 135-228.

Lapage, S.P., P.H.A. Sneath, E.F. Lessel, V.B.D. Skerman, H.P.R. Seeliger and W.A. Clark (Editors). 1975. International Code of Nomenclature of Bacteria. 1976 revision. American Society for Microbiology, Washington, D.C.

Larkin, J.M. and R. Borrall. 1978. *Spirosomaceae,* a new family to contain the genera *Spirosoma* Migula 1894, *Flectobacillus* Larkin et al, 1977, and *Runella* Larkin and Williams 1978. Int. J. Syst. Bacteriol. *28:* 595-596.

Larkin, J.M. and R. Borrall. 1979. Proposal of ATCC 25396 as the neotype strain of *Microcyclus aquaticus* Ørskov 1928. Int. J. Syst. Bacteriol. *29:* 414-415.

Larkin, J.M. and P.M. Williams. 1978. *Runella slithyformis* gen. nov., sp. nov., a curved, nonflexible, pink bacterium. Int. J. Syst. Bacteriol. *28:* 32-36.

Larkin, J.M., P.M. Williams and R. Taylor. 1977. Taxonomy of the genus *Microcyclus* Ørskov 1928: Reintroduction and emendation of the genus *Spirosoma* Migula 11894 and proposal of a new genus, *Flectobacillus.* Int. J. Syst. Bacteriol. *27:* 147-156.

Larsen, H. 1962. Halophilism. *In* Gunsalus and Stanier (Editors), The Bacteria, Vol. IV, Academic Press, New York, pp. 297-342.

Larsen, H. 1967. Biochemical aspects of extreme halophilism. *In* Rose and Wilkinson (Editors), Advances in Microbial Physiology, Vol. 1, Academic Press, London, pp. 97-132.

Larsen, H. 1980. Ecology of hypersaline environments. *In* Nissenbaum (Editor), Hypersaline Brines and Evaporitic Environments. Elsevier, Amsterdam, pp. 23-39.

Larsen, H. 1981. The family *Halobacteriaceae. In* Starr, Stolp, Trüper, Balows and Schlegel (Editors), The Prokaryotes, A handbook on habitats, isolation, and identification of bacteria. Springer-Verlag, Berlin, pp. 985-994.

Larson, C.L., C.B. Philip, W.C. Wicht and L.E. Hughes. 1951. Precipitin reactions with soluble antigens from suspensions of *Pasteurella pestis* or from tissues of animals dead of plague. J. Immunol. *67:* 289-298.

Larson, C.L., W. Wicht and W.L. Jellison. 1955. A new organism resembling P. *tularensis* isolated from water. Public Health Rep. *70:* 253-258.

Larson, W.P. and J.P. Sedgwick. 1913. The complement fixation reaction of the blood of children and infants, using the *Bacillus abortus* antigen. Proc. Seventh Annu. Conf. Am. Assoc. Med. Milk Commission, pp. 199-210.

Larsson, P. and S. Olling. 1977. O antigen distribution and sensitivity to the bactericidal effect of normal human serum of *Proteus* strains from clinical specimens. Med. Microbiol. Immunol. *163:* 77-82.

LaScolea, L.J., Jr. and F.E. Young. 1974. Development of defined minimal medium for the growth of *Neisseria gonorrhoeae.* Appl. Microbiol. *28:* 70-76.

Lasseur, P. 1913. Contribution à l'étude de *Bacillus lemonnieri,* nov. spec. C. R. Soc. Biol. Paris *74:* 47-48.

Latham, M.J., B.E. Brooker, G.L. Pettipher and P.J. Harris. 1978. Adhesion of *Bacteroides succinogenes* in pure culture and in the presence of *Ruminococcus flavefaciens* to cell walls in leaves of perennial ryegrass (*Lolium perenne*). Appl. Environ. Microbiol. *35:* 1166-1173.

Lattimer, G.L., C. McCrone and J. Galgon. 1978. Diagnosis of Legionnaires' disease from transtracheal aspirate by direct fluorescent-antibody staining and isolation of the bacterium. N. Engl. J. Med. *299:* 1172-1173.

Laudani, U., G.P. Frizzi, C. Roggi and A. Montani. 1974. The function of the endosymbiotic bacteria of Blattoidea. Experientia (Basel) *30:* 882-883.

Laudien, L. 1923. Kotuntersuchung bei Pferden auf die Anwesenheit des *Bakt. pyosepticum equi* und von paratyphusbazillen. Inaug. Diss. Hannover, Germany.

Laufs, R. and P.-M. Kaulfers. 1977. Molecular characterization of a plasmid specifying ampicillin resistance and its relationship to other R factors from *Haemophilus influenzae.* J. Gen. Microbiol. *103:* 277-286.

Laufs, R., P.-M. Kaulfers and G. Jahn. 1978. Infektiöse Antibiotikaresistenz bei *Haemophilus influenzae.* Dtsch. Med. Wochenschr. *103:* 658-662.

Laufs, R., P.-M. Kaulfers, G. Jahn and U. Teschner. 1979. Molecular characterization of a small *Haemophilus influenzae* plasmid specifying β-lactamase and its relationship to R factors from *Neisseria gonorrhoeae.* J. Gen. Microbiol. *111:* 223-231.

Laughon, B. E. and N. R. Krieg. 1974. Sugar catabolism in *Aquaspirillum gracile.* J. Bacteriol. *119:* 691-697.

Laughon, B.E., S.A. Syed and W.J. Loesche. 1982. Rapid identification of *Bacteroides gingivalis.* J Clin. Microbiol. *15:* 345-346.

Lauterborn, R. 1913. Zür Kenntnis einiger sapropelischer Schizomyceten. Allg. Bot. Z. *19:* 97-100.

Lauterborn, R. 1915. Die sapropelische Lebewelt. Ein Beispiel zür Biologie des Faulschlammes natürlicher Gewässer. Verh. Naturkundl. Medizin. Verein. Heidelberg *13:* 395-481.

Lauterborn, R. 1916. Die sapropelische Lebewelt. Ein Beiträg zür Biologie der Faulschlammes natürlicher Gewasser. Verh. Naturh. Mediz. Ver. Heidelberg *13:* 395-481.

Lautrop, H. 1956a. A modified Kohn's test for the demonstration of bacterial gelatin liquefaction. Acta Pathol. Microbiol. Scand. *39:* 357-369.

Lautrop, H. 1956b. Gelatin-liquefying *Klebsiella* strains (*Bacterium oxytocum* Flügge). Acta Pathol. Microbiol. Scand. *39:* 375-384.

Lautrop, H. 1956. Gliding motility in bacteria as a taxonomic criterion. Publ. Fac. Sci. Univ. J. E. Purkyne, Ser. K, *35:* 322-327.

Lautrop, H. 1960. Laboratory diagnosis of whooping-cough or *Bordetella* infections. Bull. W.H.O. *23:* 15-31.

Lautrop, H. 1961. *Aeromonas hydrophila* isolated from human faeces and its possible pathological significance. Acta Pathol. Microbiol. Scand. *51:* 299-301.

Lautrop, H. 1967. *Agrobacterium* spp. isolated from clinical specimens. Acta Path. Microbiol. Scand., Suppl. *187:* 63-64.

Lautrop, H. 1974. Genus X. *Proteus* Hauser 1885, 12. *In* Buchanan, R.E. and N.E. Gibbons (Editors), Bergey's Manual of Determinative Bacteriology, 8th edition, The Williams and Wilkins Co., Baltimore, pp. 327-330.

Lautrop, H. 1974. Genus III. *Moraxella* Lwoff 1939, 173. *In* Buchanan and Gibbons (Editors), Bergey's Manual of Determinative Bacteriology, 8th Ed., The Williams and Wilkins Co., Baltimore, pp. 433-438.

Lautrop, H., K. Bøvre and W. Frederiksen. 1970. A *Moraxella*-like microorganism isolated from the genitourinary tract of man. Acta Pathol. Microbiol. Scand. *78B:* 255-256.

Lautrop, H. and O. Jessen. 1964. On the distinction between polar monotrichous and lophotrichous flagellation in green fluorescent pseudomonads. Acta Pathol. Microbiol. Scand. *60:* 588-598.

Lautrop, H., with an annex by B.W. Lacy. 1960. Laboratory diagnosis of whooping-cough or *Bordetella* infections. Bull. W.H.O. *23:* 15-35.

Lautrop, H., I. Ørskov and K. Gaarslev. 1971. Hydrogen sulphide producing variants of *Escherichia coli.* Acta Pathol. Microbiol. Scand. B *79:* 641-650.

Laveran, A. 1903. Sur la spirillose des bovides. C. R. Acad. Sci. Paris *136:* 939-941.

Law, I.J. 1979. Resistance of *Rhizobium* specific for *Lotononis bainesii* to ultraviolet radiation. Soil Biol. Biochem. *11:* 87-88.

Lawson, G. H. K., J. L. Leaver, G. W. Pettigrew and A. C. Rowland. 1981. Some features of *Campylobacter sputorum* subsp. *mucosali* subsp. nov., nom. rev. and their taxonomic significance. Int. J. Syst. Bacteriol. *31:* 385-391.

Lawson, G. H. K. and A. C. Rowland. 1974. Intestinal adenomatosis in the pig: a bacteriological study. Res. Vet. Sci. *17:* 331.

Lawson, G. H. K., A. C. Rowland and L. Roberts. 1977. The surface antigens of *Campylobacter sputorum* subspecies *mucosalis.* Res. Vet. Sci. *23:* 378-382.

Lawson, G. H. K., A. C. Rowland and P. Wooding. 1975. The characterization of *Campylobacter sputorum* subspecies *mucosalis* isolated from pigs. Res. Vet. Sci. *18:* 121-126.

Lawson, G. McL. 1940. Modified technique for staining capsules of *Haemophilus pertussis.* J. Lab. Clin. Med. *25:* 435-438.

Lawton, W.D., G.M. Fukiji and M.J. Surgalla. 1960. Studies on the antigens of *Pasteurella pestis* and *Pasteurella pseudotuberculosis.* J. Immunol. *84:* 475-479.

Lawton, W.D., B.C. Morris and T.W. Burrows. 1968a. Gene transfer by conjugation in *P. pseudotuberculosis* and *P. pestis.* Progr. Immunobiol. Standard *9:* 285-292.

Lawton, W.D., B.C. Morris and T.W. Burrows. 1968b. Gene transfer in strains of *Pasteurella pseudotuberculosis.* J. Gen. Microbiol. *52:* 25-34.

Lawton, W.D. and H.B. Stull. 1971. Chromosome mapping of *Pasteurella pseudotuberculosis* by interrupted mating. J. Bacteriol. *105:* 855-863.

Layne, P., A.S. Hu, A. Balows and B.R. Davis. 1971. Extrachromosomal nature of hydrogen sulfide production in *Escherichia coli.* J. Bacteriol. *106:* 1029-1030.

Leach, J.G. 1926. The relation of the seed-corn maggot (*Phorbia fusciceps* Zett.) to the spread and development of potato blackleg in Minnesota. Phytopathology *16:* 149-176.

Leach, J.G. 1927. The relation of insects and weather to the development of heart rot of celery. Phytopathology *17:* 663-667.

Leach, J.G. 1930. The identity of the potato blackleg pathogene. Phytopathology *20:* 743-751.

Leach, J.G. 1931. Further studies on the seed-corn maggot and bacteria with special reference to potato blackleg. Phytopathology *21:* 387-406.

Leach, J.G. 1933. The method of survival of bacteria in the puparia of the seed-corn maggot (*Hylemyia cilicrura* Rond.). Z. Angew. Entomol. *20:* 150-161.

Leach, J.G. 1964. Observations on cucumber beetles as vectors of cucurbit wilt. Phytopathology *54:* 606-607.

Leach, J.G., V.G. Lilly, H.A. Wilson and M.R. Purvis, Jr. 1957. Bacterial polysaccharides: The nature and function of the exudate produced by *Xanthomonas phaseoli.* Phytopathology *47:* 113-120.

Leach, R.H. 1967. Comparative studies of mycoplasma of bovine origin. Ann. N.Y. Acad. Sci. *143:* 305-316.

Leach, R.H. 1973. Further studies on classification of bovine strains of *Mycoplasmatales,* with proposals for new species, *Acholeplasma modicum* and *Mycoplasma alkalescens.* J. Gen. Microbiol. *75:* 2135-153.

Leach, R.H. 1973. Further studies on classification of bovine strains of *Mycoplasmatales,* with proposals for new species, *Acholeplasma modicum* and *Mycoplasma alkalescens.* J. Gen. Microbiol. *75:* 135-153.

Leach, R.H. and M. Butler. 1966. Comparison of mycoplasmas associated with human tumors, leukemia and tissue cultures. J. Bacteriol. *91:* 934-940.

Leach, R.H., G.S. Cottew, B.E. Andrews and D.G. Powell. 1976. Atypical mycoplasmas from sheep in Great Britain and Australia identified as *Mycoplasma ovipneumoniae.* Vet. Res. *98:* 377-379.

Leadbetter, E.R. 1974. Family *Methylomonadaceae. In* Buchanan and Gibbons (Editors), Bergey's Manual of Determinative Bacteriology, 8th Ed., The Williams and Wilkins Company, Baltimore, pp. 267-269.

Leadbetter, E.R., S.C. Holt and S.S. Socransky. 1979. *Capnocytophaga:* new genus of Gram-negative gliding bacteria. I. General characteristics, taxonomic considerations and significance. Arch. Microbiol. *122:* 9-16.

Leben, C. 1974. Survival of plant pathogenic bacteria. Ohio Agr. Res. Development Center Special Circular 100.

Leblanc, D. 1981. Caractérisation sérologique des *Aeromonas* mobiles. Thèse de M. Sc., Université de Montréal, Montréal, Canada.

Lechevalier, M.P. 1977. Lipids in bacterial taxonomy - a taxonomist's view. Crit. Rev. Microbiol. *5:* 109-210.

Leclerc, H. 1962. Étude biochemique d'*Enterobacteriaceae* pigmentées. Ann. Inst. Pasteur (Paris) *102:* 726-741.

Leclerc, H. and R. Buttiaux. 1962. Fréquence des *Aeromonas* dans le eaux d'alimentation. Ann. Inst. Pasteur (Paris) *103:* 97-100.

Lee, C.-C. 1975. *Dirofilaria immitis:* Ultrastructural aspects of oocyte development and zygote formation. Exp. Parasitol. *37:* 449-468.

Lee, C.K. and J.W. Moulder. 1981. Persistent infection of mouse fibroblasts (McCoy cells) with a trachoma strain of *Chlamydia trachomatis.* Infect. Immun. *32:* 822-829.

Lee, H.A. 1917. A new bacterial citrus disease. J. Agr. Res. *9:* 1-8.

Lee, H.A., H.A. Purdy, C.C. Barnum and J.P. Martin. 1925. A comparison of red-stripe of sugar cane and other grasses. *In* Red-stripe Disease Studies, Bull. Expt. Sta. Hawaiian Sugar Planters' Assoc., pp. 1-99.

Lee, I.M. and R.E. Davis. 1980. DNA homology among diverse spiroplasma strains representing several serological groups. Can. J. Microbiol. *26:* 1356-1363.

Lee, J., E.D. Small, Y.-M. Liu and S. Sinha. 1979a. High molecular weight blue fluorescence protein from the bioluminescent bacterium *Photobacterium fischeri.* Biochem. Biophys. Res. Comm. *86:* 1241-1247.

Lee, J.V., T.J. Donovan and A.L. Furniss. 1978. Characterization, taxonomy, and emended description of *Vibrio metschnikovii.* Int. J. Syst. Bacteriol. *28:* 99-111.

Lee, J.V. and A.L. Furniss. 1981. The phage-typing of *V. cholerae* serovar 01. *In* Holmgren, Holme, Merson and Mollby (Editors), Acute Enteric Infections of Children. Elsevier/North Holland, Amsterdam.

Lee, J.V., D.M. Gibson and J.M. Shewan. 1977. A numerical taxonomic study of some *Pseudomonas*-like marine bacteria. J. Gen. Microbiol. *98:* 439-451.

Lee, J.V., D.M. Gibson and J.M. Shewan. 1981. *Alteromonas putrefaciens* sp. nov. *In* Validation of the publication of new names and new combinations previously published outside the IJSB. List No. 6. Int. J. Syst. Bacteriol. *31:* 215-218.

Lee, J.V., M.S. Hendrie and J.M. Shewan. 1979b. Identification of *Aeromonas, Vibrio* and related organisms. *In* Skinner and Lovelock (Editors), Identification Methods for Microbiologists. Soc. Appl. Bacteriol. Tech. Ser. No. 14. Academic Press, London, pp. 151-166.

Lee, J.V., P. Shread, A.L. Furniss and T. Bryant. 1981a. Taxonomy and description of *Vibrio fluvialis* sp. nov. (synonym group F vibrios, group EF6). J. Appl. Bacteriol. *50:* 73-94.

Lee, J.V., P. Shread, A.L. Furniss and T. Bryant. 1981b. *In* List No. 6, Validation of the publication of new names and combinations previously effectively published outside the IJSB. Int. J. Syst. Bacteriol. *31:* 215-218.

Lee, K.J., D.E. Tribe and P.L. Rogers. 1979. Ethanol production by *Zymomonas mobilis* in continuous culture at high glucose concentrations. Biotechnol. Lett. *1:* 421-426.

Lee, M. and A.C. Chandler. 1941. A study of the nature, growth and control of bacteria in cutting compounds. J. Bacteriol. *41:* 373-386.

Lee, S.Y., M.S. Mabee and N.O. Jangaard. 1978. *Pectinatus,* a new genus of the family *Bacteroidaceae.* Int. J. Syst. Bacteriol. *28:* 582-594.

Lee, S.Y., M.S. Mabee, N.O. Jangaard and E.K. Horiuchi. 1980. *Pectinatus*, a new genus of bacteria capable of growth in hopped beer. J. Inst. Brew. *86:* 28-30.

Lee, S.Y., S.E. Moore and M.S. Mabee. 1981. Selective-differential medium for isolation and differentiation of *Pectinatus* from other brewery microorganisms. Appl. Environ. Microbiol. *41:* 386-387.

Lee, W.H. 1977. Two plating media modified with tween 80 for isolating *Yersinia enterocolitica*. Appl. Environ. Microbiol. *33:* 215-216.

Lee, W.H., M.E. Harris, D. McClain, R.E. Smith and R.W. Johnston. 1980. Two modified selenite media for the recovery of *Yersinia enterocolitica* from meats. Appl. Environ. Microbiol. *39:* 205-209.

Leedle, J.A.Z. and R.B. Hespell. 1980. Differential carbohydrate media and anaerobic replica plating techniques in delineating carbohydrate-utilizing subgroups in rumen bacterial populations. Appl. Environ. Microbiol. *39:* 709-719.

Legakis, N.J., J.T. Papavassiliou and M.E. Xilinas. 1976. Inositol as a selective substrate for the growth of *Klebsiellae* and *Serratiae*. Zentralbl. Bakteriol. Infektionskr. Parasitenkd. Hyg. Abt. I, Orig. A *235:* 453-458.

Leger, A. 1917. Spirochaete de la musaraigne (*Crocidura stampfli* Tentink.) Bull. Soc. Pathol. Exot. *10:* 280-281.

Le Goffic, F.L., N. Moreau, S. Siegrist, F.W. Goldstein and J. Acar. 1977. La résistance plasmidique de *Haemophilus* sp. aux antibiotiques aminoglycosidiques: isolément et étude d'une nouvelle phosphotransférase. Ann. Microbiol. (Paris) *128A:* 383-391.

Lehmann, K.B. and R. Neuman. 1896. Atlas und Grundriss der Bakteriologie und Lehrbuch der speciellen bakteriologischen Diagnostik. 1st Ed., J.F. Lehmann, München.

Lehmann, K.B. and R. Neumann. 1899. Lehmann's Medizin, Handatlanten. X. Atlas und Grundriss der Bakteriologie und Lehrbuch der speziellen Bakteriologischen Diagnostik. 3. Aufl.

Lehmann, K.B. and R. Neumann. 1907. Atlas und Grundriss der Bakteriologie und Lehrbuch der speciellen Bakteriologischen Diagnostik. 4. Aufl. Teil 2. Lehmann, München.

Lehmann, K.B. and R. Neumann. 1912. Lehmann's Medizin, Handatlanten, X. Atlas und Grundriss der Bakteriologischen Diagnostik. 5. Aufl.

Lehmann, K.B. and R. Neumann. 1927. Bakteriologie insbesondere Bakteriologische Diagnostik. II. Allgemeine und spezielle Bakteriologie. 7 Aufl. J.F. Lehmann, München.

Lehmann, V. 1971. Phospholipase activity of *Acinetobacter calcoaceticus*. Acta Path. Microbiol. Scand. Sec. B *79:* 372-376.

Lei, J.D., H.J. Su and T.A. Chen. 1979. Spiroplasmas isolated from green leafbug, *Trigonotylus ruficornis* Geoffroy. *In* Proc. R.O.C. U.S. Coop. Sci. Semin. Mycoplasma Dis. Plants, NSC Symp. Ser. I, Nat'l. Sci. Council, Taipei, pp. 89-97.

Leidigh, B.J. and M.L. Wheelis. 1973. Genetic control of the histidine dissimilatory pathway in *Pseudomonas putida*. Mol. Gen. Genet. *120:* 201-210.

Leidy, G., E. Hahn and H.E. Alexander. 1956. On the specificity of the desoxyribonucleic acid which induces streptomycin resistance in *Hemophilus*. J. Exp. Med. *104:* 305-320.

Leidy, G., E. Hahn and H.E. Alexander. 1959. Interspecific transformation in *Hemophilus*: A possible index of relationship between *H. influenzae* and *H. aegyptius*. Proc. Soc. Exp. Biol. Med. *102:* 86-88.

Leidy, G., I. Jaffee and H.E. Alexander. 1965. Further evidence of a high degree of genetic homology between *H. influenzae* and *H. aegyptius*. Proc. Soc. Exp. Biol. Med. *118:* 671-679.

Leidy, J. 1877. On intestinal parasites of *Termes flavipes*. Proc. Acad. Nat. Sci. (Phila.) *29:* 146-149.

Leidy, J. 1881. The parasites of the termites. J. Acad. Nat. Sci. (Phila.) *8:* 425-447.

Leifson, E. 1936. New Selenite typhoid and paratyphoid (*Salmonella*) bacilli. Amer. J. Hyg. *24:* 423-432.

Leifson, E. 1954. The flagellation and taxonomy of species of *Acetobacter*. Antonie van Leeuwenhoek J. Microbiol. Serol. *20:* 102-110.

Leifson, E. 1956. Morphological and physiological characters of the genus *Chromobacterium*. J. Bacteriol. *71:* 393-400.

Leifson, E. 1960. Atlas of Bacterial Flagellation. Academic Press, New York and London. pp. 1-171.

Leifson, E. 1962. The bacterial flora of distilled and stored water. III. New species of the genera *Corynebacterium, Flavobacterium, Spirillum* and *Pseudomonas*. Int. Bull. Bacteriol. Nomencl. Taxon. *12:* 161-170.

Leifson, E. 1962a. *Pseudomonas spinosa* n. sp. Int. Bull. Bacteriol. Nomencl. Taxon. *12:* 88-92.

Leifson, E. 1970. Motile marine bacteria. IV. Ionic relationship of marine and terrestrial bacteria. Zentralbl. Bakteriol. Parasitenkd. Infektionskr. Hyg., II Abt. *125:* 170-206.

Leifson, E. and R. Hugh. 1954. A new type of polar monotrichous flagellation. J. Gen. Microbiol. *10:* 68-70.

Leifson, E. and R. Hugh. 1954. *Alcaligenes denitrificans* n. sp. J. Gen. Microbiol. *11:* 512-513.

Leisinger, T. 1965. Untersuchungen zu Systematik und Stoffwechsel der Essigsäurebakteriën. Thesis, Rudolstadt (Thür.).

Leisinger, T. 1965. Untersuchungen zu Systematik und Stoffwechsel der Essigsäurebakteriën. Zentralbl. Bakteriol. Parasitenkd. Infektionskr. Hyg. Abt II *119:* 329-376.

Leisinger, T., D. Hass and M.V. Hegarty. 1972. Indospicine as an arginine antagonist in *Escherichia coli* and *Pseudomonas aeruginosa*. Biochim. Biophys. Acta *262:* 214-219.

Leisinger, T. and R. Margraff. 1979. Secondary metabolites of the fluorescent pseudomonads. Microbiol. Rev. *43:* 422-442.

Leisinger, T., A. Wiemken and L. Ettlinger. 1966. Ueber cellulosefreie Mutanten von *Acetobacter xylinum*. Arch. Mikrobiol. *54:* 21-36.

Lelliott, R.A. 1956. Slow wilt of carnations caused by a species of *Erwinia*. Plant Pathol. *5:* 19-23.

Lelliott, R.A. 1965. The preservation of plant pathogenic bacteria. J. Appl. Bacteriol. *28:* 181-193.

Lelliott, R.A. 1968. The diagnosis of fireblight (*Erwinia amylovora*) and some diseases caused by *Pseudomonas syringae*. Rep. Eur. Medit. Plant Prot. Orgn. Coaf. Fireblight, 1967, *45:* 27-34.

Lelliott, R.A. 1974. Genus XII. *Erwinia* Winslow, Broadhurst, Buchanan, Krumweide, Rogers and Smith 1920, 209. *In* R.E. Buchanan and N.E. Gibbons (Editors), Bergey's Manual of Determinative Bacteriology, 8th edition, The Williams and Wilkins Co., Baltimore, pp. 332-339.

Lelliott, R.A., E. Billing and A.C. Hayward. 1966. A determinative scheme for the fluorescent plant pathogenic pseudomonads. J. Appl. Bacteriol. *29:* 470-489.

Lelliott, R.A. and M.M. Wallace. 1955. A bacterial disease of Shirley poppies in Tanganyika. Trans. Brit. Mycol. Soc. *38:* 88-91.

Lemcke, R.M. 1961. Association of PPLO infection and antibody response in rats and mice. J. Hyg. *59:* 401-412.

Lemcke, R.M. 1964. The serological differentiation of *Mycoplasma* strains (pleuropneumonia-like organisms) from various sources. J. Hyg. *62:* 199-219.

Lemcke, R.M. 1965. A serological comparison of various species of mycoplasma by an agar gel double-diffusion technique. J. Gen. Microbiol. *38:* 91-100.

Lemcke, R.M. 1971. Sizing small organisms. Nature (London) *229:* 492 493.

Lemcke, R.M. 1972. Osmolar concentration and fixation of mycoplasmas. J. Bacteriol. *110:* 1154-1162.

Lemcke, R.M., J. Bew, M.R. Burrows and R.J. Lysons. 1979. The growth of *Treponema hyodysenteriae* and other porcine intestinal spirochaetes in a liquid medium. Res. Vet. Sci. *26:* 315-319.

Lemcke, R.M. and M.R. Burrows. 1980. Sterol requirement for the growth of *Treponema hyodysenteriae*. J. Gen. Microbiol. *116:* 539-543.

Lemcke, R. and G.W. Csonka. 1962. Antibodies against pleuropneumonia-like organisms in patients with salpingitis. Br. J. Vener. Dis. *38:* 212-217.

Lemcke, R.M., K.A. Forshaw and R.J. Fallon. 1969. The serological identity of Sabin's murine type C *Mycoplasma* and *Mycoplasma pulmonis*. J. Gen. Microbiol. *58:* 95-98.

Lemcke, R.M. and H. Kirchhoff. 1979. *Mycoplasma subdolum*, a new species isolated from horses. Int. J. Syst. Bacteriol. *29:* 42-50.

Lemcke, R.M. and J. Poland. 1980. *Mycoplasma fastidiosum*: a new species from horses. Int. J. Syst. Bacteriol. *30:* 151-162.

Le Minor, L. 1965. Conversions antigéniques chez les *Salmonella*. VI. Acquisitions des facteurs 6,14 par les sérotypes du groupe K (0:18) sous l'effet de la lysogénisation. Ann. Inst. Pasteur (Paris) *108:* 805-811.

Le Minor, L. 1968. Conversions antigéniques chez les *Salmonella*. Ann. Inst. Pasteur (Paris) *109:* 505-515.

Le Minor, L. 1968. Lysogénie et classification des *Salmonella*. Int. J. Syst. Bacteriol. *18:* 197-201.

Le Minor, L., J. Buissière and G. Brault. 1979. Intérêt de la recherche de la fermentation du galacturonate pour différencier les *Salmonella* des sousgenres I et III monophasiques des autres *Salmonella* des sous-genres II, III diphasiques, IV, et de *Citrobacter* et *Hafnia alvei*. Ann. Microbiol. Inst. Pasteur (Paris) *130B:* 305-312.

Le Minor, L., A.M. Chalon and M. Véron. 1972. Recherches sur la présence de l'antigène commun des *Enterobacteriaceae* (antigène Kunin) chez les *Yersinia, Levinea, Aeromonas* et *Vibrio*. Ann. Inst. Pasteur *123:* 761-774.

Le Minor, L., G. Chamoiseau, E. Barbe, Ch. Charié-Marsaines and L. Egrou. 1969. Dix nouveaux sérotypes de Salmonella isolés au Tchad. Ann. Inst. Pasteur (Paris) *116:* 775-780.

Le Minor, L., M. Chippaux, F. Pichinoty, C. Coynault and M. Piéchaud. 1970. Méthodes simples permettant de rechercher la tétrathionate-reductase en cultures liquides ou sur colonies isolées. Ann. Inst. Pasteur (Paris) *119:* 733-737.

Le Minor, L. and C. Coynault. 1976. Déterminisme plasmidique du caractère atypique lactose positif de *Enterobacter hafniae* et de *Proteus morganii*. Ann. Microbiol. (Inst. Past.) *127A:* 213-221.

Le Minor, L., C. Coynault and N. Guiso. 1977. Discordance entre la positivite du test a l'O.N.P.G. et la présence d'une β-galactosidase chez les *Enterobacteriaceae* et autres bacilles gram negatif à métabolisme fermentatif. Ann. Microbiol. (Inst. Pasteur) *128B:* 35-43.

Le Minor, L., C. Coynault and G. Pessoa. 1974. Déterminisme plasmidique du caractère atypique "lactose positif" de souches de *S. typhimurium* et *S. oranienburg* isolées au Brésil lors d'épidémies de 1971 à 1973. Ann. Microbiol. (Inst. Pasteur) *125A:* 261-285.

Le Minor, L., C. Coynault, R. Rohde, B. Rowe and S. Aleksic. 1973. Localisation plasmidique du déterminant génétique du caractère atypique "saccharose +" des *Salmonella*. Ann. Microbiol. (Inst. Pasteur) *124B:* 295-306.

Le Minor, L., C. Coynault and M. Schwartz. 1974. Déterminisme plasmidique du caractère "lactose positif" de *Serratia liquefaciens*. Ann. Microbiol. *125B:*

357-366.

Le Minor, L., S. Le Minor, A.M. Chalon and C. Coynault. 1972b. Conversions antigéniques chez les *Salmonella*. XIV. Étude sur le groupe X (O=47). Ann. Inst. Pasteur *122:* 19-30

Le Minor, L., M. Piéchaud, F. Pichinoty and C. Coynault. 1969. Étude par transduction sur les nitrate-tétrathionate et thiosulfate réductases de *S. typhimurium*. Ann. Inst. Pasteur (Paris) *117:* 637-644.

Le Minor, L. and R. Rohde. 1974. Genus IV. *Salmonella* Lignières 1900, 389. *In* R.E. Buchanan and N.E. Gibbons (Editors), Bergey's Manual of Determinative Bacteriology, 8th edition, The Williams and Wilkins Co., Baltimore, pp. 298-319.

Le Minor, L., R. Rohde and J. Taylor. 1970. Nomenclature des *Salmonella*. Ann. Inst. Pasteur (Paris) *119:* 206-210.

Le Minor, S. and F. Pigache. 1977. Étude antigénique de souches de *Serratia marcescens* isolées en France. I. Antigènes H: individualisation de six nouveaux facteurs H. Ann. Microbiol. *128B:* 207-214.

Le Minor, S. and F. Pigache. 1978. Étude antigénique de souches de *Serratia marcescens* isolées en France. II. Caractérisation des antigènes 0 et individualisation de 5 nouveaux facteurs, frequencne des sérotypes et désignation des nouveaux facteurs H. Ann. Microbiol. *129B:* 407-423.

Le Minor, S. and F. Sauvageot-Pigache. 1981. Nouveaux facteurs antigéniques H (H21-H25) et 0 (021) de *Serratia marcescens*. Subdivision des facteurs 05, 010, 016. Ann. Microbiol. *132A:* 239-252.

Lemmers, M., M. De Beuckeleer, M. Holsters, P. Zambryski, A. Depicker, J.P. Hernalsteens, M. Van Montagu and J. Schell. 1980. Internal organization, boundaries and integration of Ti-plasmid DNA in nopaline crown gall tumours. J. Mol. Biol. *144:* 355-378.

LeNormand, M., J.-P. Gourret and P.-L. Maillet. 1971. Ultrastructure et developpement des colonies d'une souche de *Mycoplasma laidlawii* en milieu solide. C.R. Acad. Sci. (Paris) *273:* 2016-2019.

Leopold, S. 1953. Heretofore undescribed organism isolated from the genitourinary system. U.S. Armed Forces Med. J. *4:* 263-266.

Lepidi, A.A., C. Filippi, M. Giovannetti and M.P. Nuti. 1975. Sensitivity to U.V. treatment and nuclear size of mycoplasma-like organism infected *Humicola* sp. Experientia (Basel) *31:* 1155-1157.

Lepo, J.E., F.J. Hanus and H.J. Evans. 1980. Chemoautotrophic growth of hydrogen-uptake positive strains of *Rhizobium japonicum*. J. Bacteriol. *141:* 664-670.

Leschine, S.B. and E. Canale-Parola. 1980. Rifampin as a selective agent for isolation of oral spirochetes. J. Clin. Microbiol. *12:* 792-795.

Leschine, S.B. and E. Canale-Parola. 1980. Ornithine dissimilation by *Treponema denticola*. Curr. Microbiol. *3:* 305-3110.

Lesher, R.J. and W.H. Jones. 1978. Urease production from clinical isolates of beta-hemolytic *Escherichia coli*. J. Clin. Microbiol. *8:* 344-345.

Leslie, P.H. and A.D. Gardner. 1931. The phases of *Hemophilus pertussis*. J. Hyg. *31:* 423-434.

Lessel, E.F. 1971. Status of the name *Proteus morganii* and designation of the neotype strain. Int. J. Syst. Bacteriol. *21:* 55-57.

Lessel, E.F. 1971. Minutes of the meeting. International committee on nomenclature of bacteria. Subcommittee on the taxonomy of *Moraxella* and allied bacteria. Int. J. Syst. Bacteriol. *21:* 213-214.

Lessel, E.F. and R.S. Breed. 1954. *Selenomonas* Boskamp, 1922. Genus that includes species showing an unusual type of flagellation. Bacteriol. Rev. *18:* 165-169.

Lessie, T.G. and H.R. Whiteley. 1969. Properties of threonine deaminase from a bacterium able to use threonine as sole source of carbon. J. Bacteriol. *100:* 878-889.

Lestoquard, F. 1924. Seuscieme note sue les piraplasmoses du mouton en Algerie. L'anaplasmose. *Anaplasma ovis* nov. sp. Bull. Soc. Pathol. Exot. *17:* 784-787.

Lestoquard, F. and A. Donatien. 1936. Sur une nouvelle *Rickettsia* du mouton. Bull. Soc. Pathol. Exot. *29:* 105-108.

Lev, M. 1958. Apparent requirement for vitamin K of rumen strains of *Fusiformis nigrescens*. Nature (London) *181:* 203.

Lev, M. 1959. The growth promoting activity of compounds of the vitamin K group and analogues for a rumen strain of *Fusiformis nigrescens*. J. Gen. Microbiol. *20:* 697-703.

Lev, M. 1968. Vitamin K deficiency in *Fusiformis nigrescens*. I. Influence on whole cells and cell envelope characteristics. J. Bacteriol. *95:* 2317-2324.

Lev, M. 1977. Casamino acids enhance growth of *Bacteroides melaninogenicus*. J. Bacteriol. *129:* 562-564.

Lev, M. 1980. Glutamine-stimulated amino acid and peptide incorporation in *Bacteroides melaninogenicus*. J. Bacteriol. *143:* 753-760.

Lev, M., K.C. Keudell and A.F. Milford. 1971. Succinate as a growth factor for *Bacteroides melaninogenicus*. J. Bacteriol. *108:* 175-178.

Lev, M. and A.F. Milford. 1972. Effect of vitamin K depletion and restoration on sphingolipid metabolism in *Bacteroides melaninogenicus*. J. Lipid Res. *13:* 364-370.

Lev, M. and A.F. Milford. 1973. The 3-keto dihydrosphingosine synthetase of *Bacteroides melaninogenicus*: induction by vitamin K. Arch. Biochem. Biophys. *157:* 500-508.

Lev, M. and A.F. Milford. 1977. Energy-dependent incorporation of sphingolipid precursors and fatty acids in *Bacteroides melaninogenicus*. J. Bacteriol. *130:* 445-454.

Lev, M. and A.F. Milford. 1978. Role of nucleosides, 5-phosphoribosyl-1-pyro-

phosphate and ribose-1-phosphate in the biosynthesis of phosphosphingolipids in *Bacteroides melaninogenicus*. Arch. Biochem. Biophys. *185:* 82-87.

Levaditi, C., S. Nicolau and P. Poincloux. 1925. Sur le rôle étiologique de *Streptobacillus moniliformis (nov. spec.) dans l'erythème polymorph aigu septicémique*. C.R. Hebd. Séances Acad. Sci. (Paris) *180:* 1188-1190.

Levaditi, C., R.F. Selbie and R. Schoen. 1932. Le rheumatisme infectieux spontané de la souris provoqué par le *Streptobacillus moniliformis*. Ann. Inst. Pasteur (Paris) *48:* 308-343.

Levaditi, J.C., F. Roger and P. Destombes. 1964. Tentative de classification des *Chlamydiaceae* (Rake 1955) tenate compte de leurs affinités tissulaires et de leur épidemiologie. Ann. Inst. Pasteur (Paris) *107:* 656-662.

Levey, J.S. and S. Levey. 1948. Chemotherapy of joint involvement in mice produced by *Streptobacillus moniliformis*. Proc. Soc. Exp. Biol. Med. *68:* 314-317.

Levin, R.E. 1972. Correlation of DNA base composition and metabolism of *Pseudomonas putrefaciens* isolates from food, human clinical specimens and other sources. Antonie van Leeuwenhoek J. Microbiol. Serol. *38:* 121-127.

Levin, R. E. and R. H. Vaughn. 1968. Spontaneous spheroplast formation by *Desulfovibrio aestuarii*. Can. J. Microbiol. *14:* 1271-1276.

Levine, H.B. and R.L. Maurer. 1958. Immunization with an induced avirulent autotrophic mutant of *Pseudomonas pseudomallei*. J. Immunol. *81:* 433-438.

Levine, M. 1920. Dysentery and allied bacilli. J. Infect. Dis. *27:* 31-39.

Levine, M. and D.Q. Anderson. 1932. Two new species of bacteria causing mustiness in eggs. J. Bacteriol. *23:* 337-347.

Levine, M.G. and R.E. Hoyt. 1945. *Proteus* speciation. J. Bacteriol. *49:* 523.

Levine, M.G., M.J. Pickett and M. Mandel. 1980. Taxonomy of nonfermentative bacilli: the IIk-2 group. Curr. Microbiol. *4:* 41-44.

Levy, H.L. and D. Ingall. 1967. Meningitis in neonates due to *Proteus mirabilis*. Am. J. Dis. Child. *114:* 320-324.

Lewallen, K.R., R.M. McKinney, D.J. Brenner, C.W. Moss, D.H. Dail, B.M. Thomason and R.A. Bright. 1979. A newly identified bacterium phenotypically resembling, but genetically distinct from, *Legionella pneumophila*: an isolate in a case of pneumonia. Ann. Intern. Med. *91:* 831-834.

Lewis, G.E.J., D.L. Huxsoll, M. Ristic and A.J. Johnson. 1975. Experimentally induced infections of dogs, cats, and nonhuman primates with *Ehrlichia equi*, etiologic agent of equine ehrlichiosis. Am. J. Vet. Res. *36:* 85-88.

Lewis, I.M. 1930. Growth of plant pathogenic bacteria in synthetic culture media with special reference to *Phytomonas malvaceara*. Phytopathology *20:* 723-731.

Lewis, J. and J. Poland. 1978. Sensitivity of mycoplasmas of the respiratory tract of pigs and horses to erythromycin and its use in selective media. Res. Vet. Sci. *24:* 121-123.

Lewis, J.W. and S.J. Ball. 1981. Micro-organisms in *Trypanosoma cobitis*. Int. J. Parasitol. *11:* 121-125.

Lewis, V.J., W.L. Thacker, C.C. Shepard and J.E. McDade. 1978. In vivo susceptibility of Legionnaires' disease bacterium to ten antimicrobial agents. Antimicrob. Agents Chemother. *13:* 419-422.

Liao, C.H., C.J. Chang and T.A. Chen. 1979. Spiroplasmostatic action of plant tissue extracts. *In* Proc. R.O.C. U.S. Coop. Sci. Semin. Mycoplasma Dis. Plants, NSC Symp. Ser. I, Nat'l. Sci. Council, Taipei, pp. 99-103.

Liao, C.H. and T.A. Chen. 1977. Culture of corn stunt spiroplasma in a simple medium. Phytopathology *67:* 802-807.

Liao, C.H. and T.A. Chen. 1981a. In vitro susceptibility and resistance of two spiroplasmas to antibiotics. Phytopathology *71:* 442-445.

Liao, C.H. and T.A. Chen. 1981b. Deoxyribonucleic acid hybridization between *Spiroplasma citri* and the corn stunt spiroplasma. Curr. Microbiol. *5:* 83-86.

Librach, I.M. 1968. *Proteus* meningitis. Develop. Med. Child. Neurol. *10:* 392-394.

Lickfield, K.G., H. Achterrath, F. Hentrich, L. Kolehmainen-Seveus and A. Persson. 1972. Die Feinstrukturen von *Pseudomonas aeruginosa* in ihrer Deutung durch die Gefrierätztechnik, Ultramikrotomie und Kryo-Ultramikrotomie. J. Ultrastruct. Res. *38:* 27-45.

Lieske, R. 1921. Morphologie und Biologie der Strahlenpilze (Actinomyceten). Borntraeger Bros., Leipzig.

Lieske, R. 1928. Untersuchungen über die Krebskrankheit bei Pflanzen, Tieren und Menschen. Zentralbl. Bakteriol. Parasitenk. Infektionskr. Hyg. Abt. I Orig. *108:* 118-146.

Lignières, J. 1900. Maladies du porc. Bull. Soc. Cent. Med. Vétérin. *18:* 389-431.

Lignières, J. 1914. L-anaplasmose bovine en Argentine. Contribution à l'étude de cette maladie. Zentralbl. Bakteriol. Parasitenkd. Infektionskr. Hyg. Abt. I Orig. *74:* 133-162.

Lignières, J. and G. Spitz. 1902. L'actinobacillose. Bull. Mém. Soc. Centr. Méd. Vét. *20:* 487-535; 546-565.

Lignières, J. and J. Spitz. 1902. Contribución al estudio de las afecciones conocidas bajo el nombre de actinomicosis: Actinobacilosis. Bol. Agr. Ganad. Buenos Aires *2:* 169-230.

Lillie, R.D. 1930. Psittacosis-rickettsia-like inclusions in man and in experimental animals. Pub. Health. Rep. *45:* 773-778.

Lim, S.T., K. Andersen, R. Tait and R.C. Valentine. 1980. Genetic engineering in agriculture: hydrogen uptake (hup) genes. Trends Biochem. Sci. *5:* 167-170.

Lin, B.-C., H.-J. Day, S.-J. Chen and M.-C. Chien. 1979. Isolation and characterization of plasmids in *Xanthomonas manihotis*. Bot. Bull. Acad. Sinica *20:* 157-171.

Lin, J.-S. 1980. An antigenic analysis for membranes of *Mycoplasma hominis* by cross-absorption. J. Gen. Microbiol. *116:* 187-193.

Lin, J.-S., S. Alpert and K.M. Radnay. 1975. Combined type-specific antisera in the identification of *Mycoplasma hominis*. J. Infect. Dis. *131:* 727-730.

Lin, J.-S. and E.H. Kass. 1974. Serological reactions of *Mycoplasma hominis*: differences among mycoplasmacidal, metabolic inhibition, and growth agglutination tests. Infect. Immun. *10:* 535-540.

Lin, J.-S.L. and E.H. Kass. 1980. Fourteen serotypes of *Ureaplasma urealyticum* (T-mycoplasmas) demonstrated by the complement-dependent mycoplasmacidal test. Infection 8: 152-155.

Lin, J.-S.L., M.I. Kendrick and E.H. Kass. 1972. Serologic typing of human genital T-mycoplasmas by a complement-dependent mycoplasmacidal test. J. Infect. Dis. *126:* 658-663.

Lind, K. 1968. An indirect haemagglutination test for serum antibodies against *Mycoplasma pneumoniae* using formalinized, tanned sheep erythrocytes. Acta Pathol. Microbiol. Scand. *73:* 459-472.

Lind, K. 1970. A simple test for peroxide secretion by Mycoplasma. Acta Pathol. Microbiol. Scand. Sect. B *78:* 256-257.

Lindberg, A.A. and T. Holme. 1969. Influence of O side chain on the attachment of the Felix 0-1 bacteriophage of *Salmonella* bacteria. J. Bacteriol. *99:* 513-519.

Lindner, J.G.E.M., J.H. Marcelis, N.M. de Vos and J.A.A. Hoogkamp-Korstanje. 1979. Intracellular polysaccharide of *Bacteroides fragilis*. J. Gen. Microbiol. *111:* 93-99.

Lindner, P. 1895. Mikroscopische Betriebskontrolle in den Gärungsgewerben. Aufl. I.P. Parey. Berlin I-IX: 1-278.

Lindner, P. 1928a. Gärungsstudien über pulque in Mexiko. Ber. Westpreuss. Bot. Zool. Ver. *50:* 253-255.

Lindner, P. 1928b. Atlas der mikroskopischen Grundlagen der Gärungskunde, Tafel 68, 3rd Ed., Berlin.

Lindner, P. 1929. Allgemeine Betrachtungen über Gärung und Fäulnis und die Anwendung von Gärungsmikroben in der Milchwirtschaft. Süddeutsche Molk. *50:* 889-891.

Lindner, P. 1931. *Termobacterium mobile*, ein mexikanisches Bakterium als neues Einsäuerungsbakterium für Rübenschnitzel. Z. Ver. Dsch. Zuckerind. *81:* 25-36.

Lindqvist, K. 1960. A *Neisseria* species associated with infectious keratoconjunctivitis of sheep - *Neisseria ovis* nov. spec. J. Infect. Dis. *106:* 162-165.

Linehan, B., C.C. Scheifinger and M.J. Wolin. 1978. Nutritional requirements of *Selenomonas ruminantium* for growth on lactate, glycerol or glucose. Appl. Environ. Microbiol. *35:* 317-322.

Lingens, F., P. Vollprecht and V. Gildemeister. 1966. Zür Biosynthese der Nicotinsaure in *Xanthomonas* und *Pseudomonas*- Arten, *Mycobacterium phlei* und Rotaigen. Biochem. Z. *344:* 462-477.

Linn, D. M. and N. R. Krieg. 1978. Occurrence of two organisms in cultures of the type strain of *Spirillum lunatum*: proposal for rejection of the name *Spirillum lunatum* and characterization of *Oceanospirillum maris* subsp. *williamsae* and an unclassified vibrioid bacterium. Int. J. Syst. Bacteriol. *28:* 132-138.

Linnemann, C.C., Jr., J.W. Bass and M.H.D. Smith. 1968. The carrier state in pertussis. Am. J. Epidemiol. *88:* 422-427.

Linnemann, C.C., Jr., A.E. Schaeffer, W. Burgdorfer, L. Hutchinson and R.N. Philip. 1980. Rocky Mountain spotted fever in Clermont County, Ohio. II. Distribution of population and infected ticks in an endemic area. Am. J. Epidemiol. *111:* 31-36.

Linton, J.D., A.T. Bull and D.E.F. Harrison. 1977. Determination of the apparent K_m for oxygen of *Beneckea natriegens* using the respirograph technique. Arch. Microbiol. *114:* 111-113.

Linton, J.D. and J. Vokes. 1978. Growth of the methane utilizing bacterium *Methylococcus* NCIB 11083 in mineral salts medium with methanol as the sole source of carbon. FEMS Lett. *4:* 125-128.

Lior, H., D. L. Woodward, J. A. Edgar and L. J. LaRoche. 1981. Serotyping by slide agglutination of *Campylobacter jejuni* and epidemiology. Lancet 2(8255): 1103-1104.

Lipman, J.G. 1903. Experiments on the transformation and fixation of nitrogen by bacteria. Rep. N.J. St. Agric. Exp. Stat. *24:* 217-285.

Lipman, J.G. 1904. Soil bacteriological studies. Further contributions to the physiology and morphology of the members of Azotobacter group. Rep. N.J. St. Agric. Exp. Stat. *25:* 237-289.

Lipman, R.P., W.A. Clyde, Jr. and F.W. Denny. 1969. Characteristics of virulent, attenuated, and avirulent *Mycoplasma pneumoniae* strains. J. Bacteriol. *100:* 1037-1043.

Lippincott, J.A., R. Beiderbeck and B.B. Lippincott. 1973. Utilization of octopine and nopaline by *Agrobacterium*. J. Bacteriol. *116:* 378-383.

Lippincott, J.A. and G.T. Heberlein. 1965. The quantitative determination of the infectivity of *Agrobacterium tumefaciens*. Amer. J. Bot. *52:* 856-863.

Lippincott, J.A. and B.B. Lippincott. 1969. Tumor initiating ability and nutrition in the genus *Agrobacterium*. J. Gen. Microbiol. *59:* 57-75.

Lippincott, J.A. and B.B. Lippincott. 1975. The genus *Agrobacterium* and plant tumorigenesis. Annu. Rev. Microbiol. *29:* 377-405.

Liss, A. and R.M. Cole. 1981. Spiroplasmavirus group 1: isolation, growth, and properties. Curr. Microbiol. *5:* 357-362.

Liss, A. and R.M. Cole. 1982. Spiroplasmaviruses: group 1 characteristics. Rev. Infect. Dis. *4*(Suppl.): S115-S119.

Lissman, B.A. and J.L. Benach. 1980. Rocky Mountain spotted fever in dogs. J. Am. Vet. Med. Assoc. *176:* 994-995.

Listgarten, M.A., D. Johnson, A. Nowotny, A.C.R. Tanner and S.S. Socransky. 1978. Histopathology of periodontal disease in gnotobiotic rats monoinfected with *Eikenella corrodens*. J. Periodontol. Res. *13:* 134-148.

Listgarten, M.A. and C.-H. Lai. 1975. Unusual cell wall ultrastructure of *Leptotrichia buccalis*. J. Bacteriol. *123:* 747-749.

Listgarten, M.A. and C.-H. Lai. 1979. Comparative ultrastructure of *Bacteroides melaninogenicus* subspecies. J. Periodontal Res. *14:* 332-340.

Liston, J., W. Weibe and R.R. Colwell. 1963. Quantitative approach to the study of bacterial species. J. Bacteriol. *85:* 1061-1070.

Litkenhous, C. and P.V. Liu. 1967. Bacteriocin produced by *Bordetella pertussis*. J. Bacteriol. *93:* 1484-1488.

Little, P.A. and B.M. Lyon. 1943. Demonstration of serological types within the nonhemolytic pasteurellae. Am. J. Vet. Res. *4:* 110-112.

Little, T.W.A., D.G. Pritchard and J.E. Shreeve. 1980. Isolation of *Haemophilus* species from the oropharynx of British sheep. Res. Vet. Sci. *29:* 41-44.

Litwack, K. and K. Borchardt. 1973. Osteomyelitis of the tibia caused by a corrodens-like organism (HB-1). Clin. Med. *80:* 21.

Liu, P.V. 1979. Toxins of *Pseudomonas aeruginosa*. *In* Doggett (Editor). *Pseudomonas aeruginosa*. Clinical Manifestations of Infection and Current Therapy. Academic Press, New York, pp. 63-88.

Liu, P.V. and H. Hsieh. 1973. Exotoxins of *Pseudomonas aeruginosa*. III. Characteristics of antitoxin A. J. Infect. Dis. *128:* 520-526.

Liu, T.P. 1973. The influence of juvenile hormone on the plasma membrane of symbiotic bacteria. Protoplasma *79:* 409-412.

Liu, T.P. 1974. The effect of corpora allata on the plasma membrane of the symbiotic bacteria of the oocyte surface of *Periplaneta americana* L. Gen. Comp. Endocrinol. *23:* 118-123.

Livermore, B.P., R.F. Bey and R.C. Johnson. 1978. Lipid metabolism of *Borrelia hermsi*. Infect. Immun. *20:* 215-220.

Livermore, B.P. and R.C. Johnson. 1974. Lipids of the *Spirochaetales*: comparison of the lipids of several members of the genera *Spirochaeta*, *Treponema*, and *Leptospira*. J. Bacteriol. *120:* 1268-1273.

Livingston, C.W. and B.B. Gauer. 1975. Isolation of T-strain mycoplasma from sheep and goats in Texas. Am. J. Vet. Res. *36:* 313-314.

Lochhead, A.G. 1934. Bacteriological studies on the red discoloration of salted hides. Can. J. Res. *10:* 275-286.

Lochhead, A.G. 1943. Note on the taxonomic position of the red chromogenic halophilic bacteria. J. Bacteriol. *45:* 574-575.

Lockhart, W.R. and J. Liston. 1970. Methods for numerical taxonomy. American Society for Microbiology, Washington, D.C.

Lockwood, L.B., B. Tabenkin and G.E. Ward. 1941. The production of gluconic acid and 2-keto gluconic acid from glucose by species of *Pseudomonas* and *Phytomonas*. J. Bacteriol. *42:* 51-61.

Lode, E.T. and M.J. Coon. 1971. Enzymatic ω-oxidation. V. Forms of *Pseudomonas oleovorans* rubredoxin containing one or two iron atoms: structure and function in ω-oxidation. J. Biol. Chem. *246:* 791-802.

Loeb, M.R., A.L. Zachary and D.H. Smith. 1981. Isolation and partial characterization of outer and inner membranes from encapsulated *Haemophilus influenzae* type b. J. Bacteriol. *145:* 596-604.

Loesche, W.J. 1969. Oxygen sensitivity of various anaerobic bacteria. Appl. Microbiol. *18:* 723-727.

Loesche, W.J. and R.J. Gibbons. 1965. A practical scheme for identification of the most numerous oral Gram negative anaerobic rods. Arch. Oral Biol. *10:* 723-725.

Loesche, W. J., R. J. Gibbons and S. S. Socransky. 1965. Biochemical characteristics of *Vibrio sputorum* and relationship to *Vibrio bubulus* and *Vibrio fetus*. J. Bacteriol. *89:* 1109-1116.

Loesche, W.J., S.S. Socransky and R.J. Gibbons. 1964. *Bacteroides oralis*, proposed new species isolated from the oral cavity of man. J. Bacteriol. *88:* 1329-1337.

Loginova, N.V., B.B. Namsaraev and Y.A. Trotsenko. 1978. Autotrophic metabolism of methanol in *Microcyclus aquaticus*. Microbiology (English translation) *47:* 134-135.

Löhnis, F. 1905. Beiträge zür Kenntnis der Stickstoffbacterien. Zentralbl. Bakteriol. Parasitenk. Infektionskr. Hyg. Abt. II *14:* 87-101; 582-604.

Löhnis, F. 1911. Landwirtschaftlich-bakteriologishes Praktikum. Gebrüder Borntraeger, Berlin, pp. 1-156.

Löhnis, F. and J. Hanzawa. 1914. Die Stellung von *Azotobacter* im System. Zentralbl. Bakteriol. Parasitenkd. Infektionskr. Hyg. Abt. II *42:* 1-8.

Loitsianskaya, M.S. and B. Lebentrau. 1964. Einige Fragen der Morphologie von *Acetobacter aceti*. I. Ueber die Involutionsformen von *Acetobacter aceti*. Z. Allg. Mikrobiol. *4:* 13-21.

Loitsianskaya, M.S., G.V. Pavlenko and A.I. Ivchenko. 1979. Studies on the taxonomy of acetic acid bacteria. Mikrobiologiya *48:* 545-551.

Loitsianskaya, M.S., G.V. Pavlenko and G.A. Zolatareva. 1977. Type of flagellation in *Acetobacter* and *Gluconobacter*. Vestn. Leningr. Gos. Univ. *3:* 99-105.

London, J. and K. Kline. 1973. Aldolases of lactic acid bacteria: a case history in the use of an enzyme as an evolutionary marker. Bacteriol. Rev. *37:* 453-478.

Long, H.F. and B.W. Hammer. 1941. Distribution of *Pseudomonas putrefaciens*. J. Bacteriol. *41:* 100-101.

Longley, E.O. 1940. Contagious pleuropneumonia of goats. Int. J. Vet. Sci. Delhi *10:* 127-197.

Longley, E.O. 1951. Contagious caprine pleuropneumonia. A study of the disease in Nigeria. Colon. Res. Pub. (London) 7: 23.

Loper, J.E. and C.I. Kado. 1979. Host range conferred by the virulence-specifying plasmid of Agrobacterium tumefaciens. J. Bacteriol. 130: 591-596.

Lopez, M. 1978. Characteristics of French isolates of Agrobacterium. Proc. 4th Int. Conf. Plant Path. Bact., Angers 1978, 233-237.

Lopez, R. and J.H. Becking. 1968. Polysaccharide production by Beijerinckia and Azotobacter. Microbiol. Esp. 21: 53-75.

Lorch, I.J. and K.W. Jeon. 1980a. Rapid induction of cellular stain specificity by newly acquired cytoplasmic components in amoebas. Science 211: 949-951.

Lorch, I.J. and K.W. Jeon. 1980b. Resuscitation of amoebae deprived of essential symbiotes: micrurgical studies. J. Protozool. 27: 423-426.

Louis, C. and A. Bournier. 1971. Ultrastructure des symbiotes de Thysanopteres. J. Microsc. (Paris) 11: 74-75.

Louis, C., G. Croizier and G. Meynadier. 1977a. Trame cristalline des inclusions proteiques chez une Rickettsiella. Biol. Cell. 29: 77-80.

Louis, C. and G. Kuhl. 1972. Synthese in situ de protéines chez les symbiotes et chez les mycetocytes de Pseudococcus obscurus (Coccidae-Pseudococcinae) en absence ou en présence d'inhibiteurs spécifiques. C.R. Hebd. Séances Acad. Sci. 274: 715-718.

Louis, C., M. Laporte, J. Carayon and C. Vago. 1973. Mobilité, ciliature et caractères ultrastructuraux des micro-organismes symbiotiques endo et exocellulaires de Cimex lectularius L. (Hemiptera Cimicidae). C.R. Hebd. Séances Acad. Sci. Ser. D Sci. Nat. 277: 607-611.

Louis, C., G. Morel, G. Nicholas and G. Kuhl. 1979. Étude comparée des caractères ultrastructuraux de rickettsies d'arthropodes, révélés par cryodécapage et cytochimie. J. Untrastruc. Res. 66: 243-253.

Louis, C. and G. Nicolas. 1976. Ultrastructure of the endocellular procaryotes of arthropods as revealed by freeze-etching. I. A study of "a"-type endosymbionts of the leafhopper Euscelis plebejus Fall. (Homoptera, Jassidae). J. Microsc. Biol. Cell. 26: 121-126.

Louis, C., G. Nicolas and M. Pouphile. 1976. Ultrastructure of the endocellular procaryotes of arthropods as revealed by freeze-etching. II. "t"-type endosymbionts of the leafhopper Euscelis plebejus Fall. (Homoptera, Jassidae). J. Microsc. Biol. Cell. 27: 53-58.

Louis, C., A. Yousfi, C. Vago and G. Nicolas. 1977b. Étude par cytochimie et cryodécapage de l'ultrastructure d'une Rickettsiella de crustacé. Ann. Microbiol. 128B: 177-205.

Loutit, J.S. 1971. Investigations of the mating system of Pseudomonas aeruginosa strain I. VI. Mercury resistance associated with the sex factor (FP). Genet. Res. 16: 179-184.

Lo Verde, P.T., C. Amento and G.I. Higashi. 1980. Parasite interaction of Salmonella typhimurium and Shistosoma. J. Infect. Dis. 141: 177-185.

Lovrekovich, L. and Z. Klement. 1965. Serological and bacteriophage sensitivity studies on Xanthomonas vesicatoria strains isolated from tomato and pepper. Phytopathol. Z. 52: 222-228.

Lovrekovich, L., Z. Klement and W.J. Dowson. 1963. Serological investigation of Pseudomonas syringae and Pseudomonas morsprunorum strains. Phytopathol. Z. 47: 19-24.

Lowbury, E.J.L., H.A. Lilly, A. Kidson, G.A.J. Ayliffe and R.J. Jones. 1969. Sensitivity of Pseudomonas aeruginosa to antibiotics: emergence of strains highly resistant to carbenicillin. Lancet 1969: 448-452.

Lucas, L.T. and R.G. Grogan. 1969a. Serological variation and identification of Pseudomonas lachrymans and other phytopathogenic Pseudomonas nomenspecies. Phytopathology 59: 1908-1912.

Lucas, L.T. and R.G. Grogan. 1969b. Some properties of specific antigens of Pseudomonas lachrymans and other Pseudomonas nomenspecies. Phytopathology 59: 1913-1917.

Ludden, P. W., Y. Okon and R. H. Burris. 1978. The nitrogenase system of Spirillum lipoferum. Biochem. J. 173: 1001-1003.

Lüderitz, O., A.M. Staub and O. Westphal. 1966. Immunochemistry of O and R antigens of Salmonella and related Enterobacteriaceae. Bacteriol. Rev. 30: 192-255.

Lüderitz, O., O. Westphal, A.M. Staub and H. Nikaido. 1971. Isolation and chemical and immunological characterization of bacterial lipopolysaccharides. In Weinbaum, Kadis and Ajl (Editors), Microbial Toxins, Vol. IV, Academic Press, New York, pp. 145-233.

Ludwig, F. 1898. Review of Hoyer (1898). Bijdrage tot de kennis van de azijnzuurbakterien. Zentralbl. Bakteriol. Parasitenkd. Infektionskr. Hyg. Abt. II 4: 867-875.

Luehrsen, K., G.E. Fox, M.W. Kilpatrick, R.T. Walker, H. Domdey, G. Krupp and H.J. Gross. 1981. The nucleotide sequence of the 5S rRNA from the archaebacterium Thermoplasma acidophilum. Nucleic Acids Res. 9: 965-970.

Luisetti, J., J.-P. Prunier and L. Gardan. 1972. Un milieu pour la mise en évidence de la production d'un pigment fluorescent par Pseudomonas mors-prunorum f. sp. persicae. Ann. Phytopathol. 4: 295-296.

Lum, P.T.M. and J.E. Baker. 1973. Development of mycetomes in larvae of Sitophilus granarius and S. oryzae. Ann. Entomol. Soc. Am. 66: 1261-1263.

Lumb, W.V. 1961. Canine haemobartonellosis and its feline counterpart. Cal. Vet. 5: 24-25.

Lund, B.M. 1969. Properties of some pectolytic, yellow pigmented, Gram negative bacteria isolated from fresh cauliflowers. J. Appl. Bacteriol. 32: 60-67.

Lund, M.E., J.M. Matsen and D.J. Blazevic. 1974. Biochemical and antibiotic susceptibility studies of H₂S-negative Citrobacter. Appl. Microbiol. 28: 22-25.

Lustig, A. 1890. Diagnostica dei batteri delle acque con una guida alle ricerche batteriologiche e microscopiche. Rosenberg and Sellier, Torino.

Lusty, C.J. and M. Doudoroff. 1966. Poly-beta-hydroxybutyrate depolymerases of Pseudomonas lemoignei. Proc. Nat. Acad. Sci. U.S.A. 56: 960-965.

Lutwick, L.I. 1976. Pancreatic abscess with Haemophilus influenzae and Eikenella corrodens. J. Am. Med. Assoc. 236: 2091-2092.

Lutz, A., O. Grootten and T. Wurch. 1956. Étude des characteres culturaux et biochimiques de bacilles du type Hemophilus hemolyticus vaginalis. Rev. Immunol. 20: 132-138.

Lwoff, A. 1939. Revision et démembrement des Hemophilae le genre Moraxella nov. gen. Ann. Inst. Pasteur 62: 168-176.

Lwoff, A. 1944. L' évolution physiologique. Étude des pertes de fonctions chez les microorganismes. Actualites Scientifiques et Industrielles, Hermann Cie, Paris.

Lwoff, A. and M. Lwoff. 1937. Rôle physiologique de l'hémine pour Haemophilus influenzae Pfeiffer. Ann. Inst. Pasteur 59: 129-136.

Lynch, M.J., A.E. Wopat and M.L. O'Connor. 1980. Characterization of two new facultative methanotrophs. Appl. Environ. Microbiol. 40: 400-407.

Lynch, R.E. and B.C. Cole. 1980. Mycoplasma pneumoniae: a prokaryote which consumes oxygen and generates superoxide but which lacks superoxide dismutase. Biochem. Biophys. Res. Commun. 96: 98-105.

Lysenko, O. 1974. Bacterial exoenzymes toxic for insects: Proteinase and lecithinase. J. Hyg. Epidemiol. Microbiol. Immunol. 18: 347-352.

Lysenko, O. 1976. Chitinase of Serratia marcescens and its toxicity for insects. J. Invertebr. Pathol. 27: 385-386.

Lysko, P.G. and C.D. Cox. 1977. Terminal electron transport in Treponema pallidum. Infect. Immun. 16: 885-890.

Lysko, P.G. and C.D. Cox. 1978. Respiration and oxidative phosphorylation in Treponema pallidum. Infect. Immun. 21: 462-473.

Ma, W.-C. and D.L. Denlinger. 1974. Secretory discharge and microflora of milk gland in tsetse flies. Nature (Lond.) 247: 301-303.

Macchiavello, A. 1937. Estudios sobre tifus exantematico. III. Un nuevo metodo para tenir Rickettsia. Rev. Chilena Hig. Med. Prev. 1: 101-106.

Macchiavello, A. 1947. Notes on the taxonomy of the rickettsias and the classification of the rickettsioses. Prim. Reunion Interamer. del Tifo, Mexico, pp. 405-426.

MacDonald, J.B. 1953. The Motile Non-Sporulating Anaerobic Rods of the Oral Cavity. Ph.D. Thesis, U. of Toronto, pp. 1-95.

MacDonald, J.B. and E.M. Madlener. 1957. Studies on the isolation of Spirillum sputigenum. Can. J. Microbiol. 3: 679-686.

MacDonald, J.B., E.M. Madlener and S.S. Socransky. 1959. Observations on Spirillum sputigenum and its relationship to Selenomonas species with special reference to flagellation. J. Bacteriol. 77: 559-565.

Macé, E. 1889. Traité Pratique de Bactériologie, 1st Ed., Ballière, Paris, pp. 1-711.

Macé, E. 1913. Traité Practique de Bactériologie, 6th ed. Baillière, Paris. pp. 1-918.

Machado, W.C. and J. Döbereiner. 1969. Estudos complementares sobre a fisiologia de Azotobacter paspali e sua dependencia da planta (Paspalum notatum). Pesqui. Agropecu. Brasil. 4: 53-58.

Macierevicz, M. 1966. Propozycja nowej grupy (Rodzaju) paleczek Enterobacteriaceae. Med. Dosw. Mikrobiol. 18: 333-339.

MacInnes, J.I., T.J. Trust and J.H. Crosa. 1979. Deoxyribonucleic acid relationships among members of the genus Aeromonas. Can. J. Microbiol. 25: 579-586.

MacKay, R.M., L.B. Zablen, C.R. Woese and W.F. Doolittle. 1979. Homologies in processing and sequence between the 23S ribosomal ribonucleic acids of Paracoccus denitrificans and Rhodopseudomonas sphaeroides. Arch. Microbiol. 123: 165-172.

Mackie, T.J., C.E. Van Rooyen and E. Gilroy. 1933. Epizootic disease occurring in breeding stock of mice: bacteriological and experimental observations. Brit. J. Exp. Pathol. 14: 132-136.

Maclagan, R.M. and D.C. Old. 1980. Haemagglutins and fimbriae in different serotypes and biotypes of Yersinia enterocolitica. J. Appl. Bacteriol. 49: 353-360.

MacLaren, D.M. 1968. The significance of urease in Proteus pyelonephritis: A bacteriological study. J. Pathol. Bacteriol. 96: 45-56.

MacLeod, R.A. 1968. On the role of inorganic ions in the physiology of marine bacteria. Adv. Microbiol. Sea 1: 95-126.

MacLeod, R.A. and A. Hori. 1960. Nutrition and metabolism of marine bacteria. VIII. Tricarboxylic acid cycle enzymes in a marine bacterium and their response to inorganic salts. J. Bacteriol. 80: 464-471.

MacLeod, R.A., A. Hori and S.M. Fox. 1960a. Nutrition and metabolism of marine bacteria. IX. Ion requirements for obtaining and stabilizing isocitric dehydrogenase from a marine bacterium. Can. J. Biochem. Physiol. 38: 693-701.

MacLeod, R.A., A. Hori and S.M. Fox. 1960b. Nutrition and metabolism of marine bacteria. X. The glyoxylate cycle in a marine bacterium. Can. J. Microbiol. 6: 639-644.

MacOwan, K.J. and J.E. Minette. 1976. A mycoplasma from acute contagious caprine pleuropneumonia in Kenya. Trop. Anim. Health Prod. 8: 91-95.

MacPhee, D.G., V. Krishnapillai, R.J. Roantree and B.A.D. Stocker. 1975. Mutations in Salmonella typhimurium conferring resistance to Felix O phage without loss of smooth character. J. Gen. Microbiol. 87: 1-10.

Macrae, A.D., P.W. Greaves and P. Platts. 1979. Isolation of *Legionella pneumophila* from blood culture. Brit. Med. J. *2:* 1189-1190.

Macrae, R.M. and J.F. Wilkinson. 1958. Poly-β-hydroxybutyrate metabolism in washed suspensions of *Bacillus cereus* and *Bacillus megaterium*. J. Gen. Microbiol. *19:* 210-222.

Macy, J.M., L.G. Ljungdahl and G. Gottschalk. 1978. Pathway of succinate and propionate formation in *Bacteroides fragilis*. J. Bacteriol. *134:* 84-91.

Macy, J.M., I. Probst and G. Gottschalk. 1975. Evidence for cytochrome involvement in fumarate reduction and adenosine 5'-triphosphate synthesis by *Bacteroides fragilis* grown in the presence of hemin. J. Bacteriol. *123:* 436-442.

Madden, D.L., K.E. Moats, W.T. London, E.B. Matthew and J.L. Sever. 1974. *Mycoplasma moatsii*, a new species isolated from recently imported grivet monkeys. (*Cercopithecus aethiops*). Int. J. Syst. Bacteriol. *24:* 459-464.

Madsen, N.B. and R.M. Hochster. 1959. The tricarboxylic acid and glyoxylate cycles in *Xanthomonas phaseoli*. (XP8). Can. J. Microbiol. *5:* 1-8.

Maeland, J.A. and A. Digranes. 1975. Common enterobacterial antigen in *Yersinia enterocolitica*. Acta Path. Microbiol. Scand. Sect. B *83:* 382-386.

Magalhães, L. M. S., C. A. Neyra and J. Döbereiner. 1978. Nitrate and nitrite reductase negative mutants of N₂-fixing *Azospirillum* spp. Arch. Microbiol. *117:* 247-252.

Mager, J., A. Traub and N. Grossowicz. 1954. Cultivation of *Pasteurella tularensis* in chemically defined media: effect of buffers and spermine. Nature (London) *174:* 747-748.

Maggi, L. 1866. Essai d'une classification protistologique des ferments vivants. J. Micrographie *10:* 80-85, 173-178, 327-333.

Magheru, A. 1923. Recherches expérimentales sur le Bacille de Morgan. C. R. Hebd. Séances Soc. Biol. *89:* 643-645.

Magrou, J. 1937. *In* Hauduroy, P., G. Ehringer, A. Urbain, G. Guillot and J. Magrou (Editors). Dictionnaire des bactéries pathogènes pour l'homme les animaux et les plantes. Masson and Co., Paris, pp. 195-220; 326-437.

Magrum, L.J., K.R. Luehrsen and C.R. Woese. 1978. Are extreme halophiles actually "bacteria"? J. Mol. Evol. *11:* 1-8.

Maia, A., F.W. Goldstein, J.F. Acar and F. Roland. 91980. Isolation of *Eikenella corrodens* from human infections. J. Infect. *2:* 347-353.

Maier, R.J. 1981. *Rhizobium japonicum* mutant strains unable to grow chemoautotrophically with H₂. J. Bacteriol. *145:* 533-540.

Maier, R.J., P.E. Bishop and W.J. Brill. 1978. Transfer from *Rhizobium japonicum* to *Azotobacter vinelandii* of genes required for nodulation. J. Bacteriol. *134:* 1199-1201.

Maino, A.L., M.N. Schroth and N.J. Palleroni. 1974. Degradation of xylan by bacterial plant pathogens. Phytopathology *64:* 881-885.

Mair, N.S. and E. Fox. 1973. An antigenic relationship between *Yersinia pseudotuberculosis* type 6 and *Escherichia coli* O-group 55. Contr. Microbiol. Immunol. *2:* 180-183.

Mair, N.S., C.J. Randall, G.W. Thomas, J.F. Harbourne, C.T. McCrea and K.P. Cowl. 1974. *Actinobacillus suis* infection in pigs: a report of four outbreaks and two sporadic cases. J. Comp. Pathol. *84:* 113-119.

Mäkelä, P.H., V.V. Valtonen and M. Valtonen. 1973. Role of O-antigens (lipopolysaccharide) factors in the virulence of *Salmonella*. J. Infect. Dis. *128 Suppl.:* 81-85.

Makki, M.A. 1971. Phosphatase activity of some *Mycoplasma* species. J. Hyg. Epidemiol. Microbiol. Immunol. *15:* 417-423.

Makula, R.A. 1978. Phospholipid composition of methane utilizing bacteria. J. Bacteriol. *134:* 771-777.

Makulu, A., F. Gatti and J. Vandepitte. 1973. *Edwardsiella tarda* infections in Zaire. Ann. Soc. Belge Med. Trop. (Bruxelles) *53:* 165-172.

Malashenko, Y.R., V.A. Romanovskaya, V.N. Bogachenko and A.D. Shved. 1975. Thermophilic and thermotolerant methane-assimilating bacteria. Mikrobiologiya *44:* 844-850.

Malashenko, Y.R., V.A. Romanovskaya and E.I. Kvasnikov. 1972. Taxonomy of bacteria utilizing gaseous hydrocarbons. Mikrobiologiya *41:* 871-879.

Male, C. 1979. Immunoglobulin A1 protease production by *Haemophilus influenzae* and *Streptococcus pneumoniae*. Infect. Immun. *26:* 254-261.

Málek, I. and Kazdová-Kožišková. 1946. *Pseudomonas* n. sp., a new microbe discovered from diagnostic material. Sb. Lek. *47:* 189-194.

Málek, I., M. Radochová and O. Lysenko. 1963. Taxonomy of the species *Pseudomonas odorans*. J. Gen. Microbiol. *33:* 349-355.

Malik, R.J. 1975. Preservation of Knallgas bacteria. *In* Dellweg (Editor), Fifth International Fermentation Symposium, Berlin, Westkreuz Druckerei und Verlag Berlin-Bonn, p. 180.

Malik, K.A. and D. Claus. 1979. *Xanthobacter flavus*, a new species of nitrogen-fixing hydrogen bacteria. Int. J. Syst. Bacteriol. *29:* 283-287.

Malik, K. A. and H.G. Schlegel. 1980. Enrichment and isolation of new nitrogen-fixing hydrogen bacteria. FEMS Microbiol. Lett. *8:* 101-104.

Malik, K.A. and H.G. Schlegel. 1981. Chemolithotrophic growth of bacteria able to grow under N₂-fixing conditions. FEMS Microbiol. Lett. *11:* 63-67.

Malinowski, F. 1966. A primary isolation medium for the differentiation of genus *Proteus* from other non-lactose and lactose fermenters. Can. J. Med. Technol. *28:* 118-121.

Maliszewski, C.R. and S.J. Badger. 1980. Group and type antigens of *Eikenella corrodens*. Abstracts of the Annual Meeting of the American Society for Microbiology, p. 94.

Malke, H. and G. Bartsch. 1966. Elektronenoptische Untersuchung zür intracel-

lulären Bakteriensymbiose von *Nauphoeta cinerea* (Olivier) (Blattariae). Z. Allg. Mikrobiol. *6:* 163-176.

Malke, H. and W. Schwartz. 1966. Untersuchungen über die Symbiose von Tieren mit Pilzen und Bakterien. XI. Die Rolle des Wirtslysozyms in der Blattidensymbiose. Arch. Mikrobiol. *53:* 17-32.

Malkoff, K. 1906. Weitere Untersuchungen über die Bakterienkrankheit auf *Sesamum orientale*. Zentralbl. Bakteriol. Parasitenk. Infektionskr. Hyg. Abt. II *16:* 664-666.

Mallman, W.L. 1930. The interagglutinability of members of the *Brucella* and *Pasteurella* genera. J. Am. Vet. Med. Assoc. *77:* 636-638.

Malone, B.H., M. Schreiber, N.J. Schneider and L.V. Holdeman. 1975. Obligately anaerobic strains of *Corynebacterium vaginale (Haemophilus vaginalis)*. J. Clin. Microbiol. *2:* 272-275.

Manasse, R.J. and W.A. Corpe. 1967. Chemical composition of cell envelopes from *Agrobacterium tumefaciens*. Can. J. Microbiol. *13:* 1591-1603.

Manasse, R.J., R.C. Staples, R.R. Granados and E.G. Barnes. 1972. Morphological, biological, and physical properties of *Agrobacterium tumefaciens* bacteriophages. Virology *47:* 375-384.

Manchee, R.J. and D. Taylor-Robinson. 1968. Haemadsorption and haemagglutination by mycoplasmas. J. Gen. Microbiol. *50:* 465-478.

Manchee, R.J. and D. Taylor-Robinson. 1969. Enhanced growth of T-strain mycoplasmas with N-2-hydroxyethylpiperazine-N'-2-ethanesulfonic acid buffer. J. Bacteriol. *100:* 78-85.

Manchee, R.J. and D. Taylor-Robinson. 1969. Studies on the nature of receptors involved in attachment of tissue culture cells to mycoplasmas. Br. J. Exp. Pathol. *50:* 66-75.

Manchee, R.J. and D. Taylor-Robinson. 1970. Lysis and protection of erythrocytes by T-mycoplasmas. J. Med. Microbiol. *3:* 539-546.

Mandel, M. 1966. Deoxyribonucleic acid base composition in the genus *Pseudomonas*. J. Gen. Microbiol. *43:* 273-292.

Mandel, M., L. Igambi, J. Bergendahl, M.L. Dodson and E. Scheltgen. 1970. Correlation of melting temperature and cesium chloride buoyant density of bacterial deoxyribonucleic acid. J. Bacteriol. *101:* 333-338.

Mandel, M., C.L. Schildkraut and J. Marmur. 1968. Use of CsCl density gradient analysis for determining the guanine plus cytosine content of DNA. Methods Enzymol. *12B:* 184-195.

Mandel, M., O.B. Weeks and R.R. Colwell. 1965. Deoxyribonucleic acid base composition of *Pseudomonas piscicida*. J. Bacteriol. *90:* 1492-1493.

Mandrell, R.E. and W.D. Zollinger. 1977. Lipopolysaccharide serotyping of *Neisseria meningitidis* by hemagglutination inhibition. Infect. Immun. *16:* 471-475.

Maniloff, J. 1970. Ultrastructure of *Mycoplasma laidlawii* during culture development. J. Bacteriol. *102:* 561-572.

Maniloff, J., J. Das, R.M. Putzrath and J.A. Nowak. 1979. Mycoplasma and spiroplasma viruses: molecular biology. *In* Barile and Razin (Editors), The Mycoplasmas, Vol. I, Academic Press, New York, pp. 411-430.

Manjrekar, S.L. 1954. Rickettsia of domesticated animals. Indian J. Vet. Sci. Anim. Husb. *24:* 217-222.

Mannheim, W. 1981. Taxonomic implications of DNA relatedness and quinone patterns in *Actinobacillus, Haemophilus*, and *Pasteurella*. *In* Kilian, Frederiksen and Biberstein (Editors), *Haemophilus, Pasteurella* and *Actinobacillus*, Academic Press, London, pp. 265-280.

Mannheim, W., S. Pohl and R. Holländer. 1980. Zür Systematik von *Actinobacillus, Haemophilus* und *Pasteurella*: Basenzusammensetzung der DNS, Atmungschinone und kulturell-biochemische Eigenschaften repräsentativer Sammlungsstämme. Zentralbl. Bakteriol. Parasitenkd. Infektionskr. Hyg. Abt. I Orig. A *246:* 512-540.

Mannheim, W., W. Stieler, G. Wolf and R. Zabel. 1978. Taxonomic significance of respiratory quinones and fumarate respiration in *Actinobacillus* and *Pasteurella*. Int. J. Syst. Bacteriol. *28:* 7-13.

Manniello, J.M., H. Heymann and F.N. Adair. 1979. Isolation of atypical lipopolysaccharides from purified walls of *Pseudomonas cepacia*. J. Gen. Microbiol. *112:* 397-400.

Manns, T.F. 1909. The blade blight of oats. A bacterial disease. Bull. Ohio Agr. Expt. Sta. *210:* 91-167.

Mansheim, B.J. and S.E. Coleman. 1980. Immunochemical differences between oral and nonoral strains of *Bacteroides asaccharolyticus*. Infect. Immun. *27:* 589-596.

Mansheim, B.J. and D.L. Kaspar. 1977. Purification and immunochemical characterization of the outer membrane complex of *Bacteroides melaninogenicus* subsp. *asaccharolyticus*. J. Infect. Dis. *135:* 787-799.

Mansour, K. 1930. Preliminary studies on the bacterial cell-mass (accessory cell-mass) of *Calandra oryzae* (Linn.): the rice weevil. Q. J. Microsc. Sci. *73:* 421-436.

Mansour, K. 1935. On the micro-organism-free and the infected *Calandra granaria* (Lin.). Bull. Soc. Roy. Entomol. Egypte *19:* 290-307.

Manuelidis, E.E. and L. Thomas. 1973. Occlusion of brain capillaries by endothelial swelling in mycoplasma infections. Proc. Natl. Acad. Sci. U.S.A. *70:* 706-709.

Maraite, H. and J. Weyns. 1979. Distinctive physiological, biochemical and pathogenic characteristics of *Xanthomonas manihotis* and *X. cassavae*. *In* Maraite and Meyer, J.A. (Editors) Diseases of Tropical Food Crops. Universite Catholique de Louvain, Louvain-la-Neuve, Belgium. pp. 103-117.

Maratea, D. and R. P. Blakemore. 1981. *Aquaspirillum magnetotacticum* sp.

nov., a magnetic spirillum. Int. J. Syst. Bacteriol. *31:* 452-455.

Marbach, A., M. Varon and M. Shilo. 1976. Properties of marine bdellovibrios. Microb. Ecol. *2:* 284-295.

Marchette, N.J. and C. Hatie. 1965. Microbial isolates from the digestive tract of *Triatoma protracta* (Uhler) (Reduviidae). J. Invertebr. Pathol. *7:* 45-48.

Marchette, N.J. and P.S. Nicholes. 1961. Virulence and citrulline ureidase activity of *Pasteurella tularensis.* J. Bacteriol. *82:* 26-32.

Marcus, B.B., S.B. Samuels, B. Pittman and W.B. Cherry. 1969. A serologic study of *Herellea vaginicola* and its identification by immunofluorescent staining. Am. J. Clin. Pathol. *52:* 309-319.

Marcus, H.R. and C.M. Phelps. 1977. *Eikenella corrodens* subacute bacterial endocarditis: mixed infection in amphetamine user. N.Y. State J. Med. *77:* 2259-2261.

Marcus, L.C. 1971. Infectious diseases of reptiles. J. Am. Vet. Med. Assoc. *159:* 1629-1631.

Marcus, P.I. and P. Talalay. 1956. Induction and purification of alpha- and beta-hydroxysteroid dehydrogenases. J. Biol. Chem. *218:* 661-674.

Mårdh, P.-A. 1975. Human respiratory tract infections with mycoplasmas and their in vitro susceptibility to tetracyclines and some other antibiotics. Chemotherapy *21:* 47-57.

Mårdh, P.-A. and L. Weström. 1970a. Tubal and cervical cultures in acute salpingitis with special reference to *Mycoplasma hominis* and T-strain mycoplasmas. Br. J. Vener. Dis. *46:* 179-186.

Mårdh, P.-A. and L. Weström. 1970b. Antibodies to *Mycoplasma hominis* in patients with genital infections and in healthy controls. Br. J. Vener. Dis. *46:* 390-397.

Mare, C.J. and W.P. Switzer. 1965. *Mycoplasma hyopneumoniae,* a causative agent of virus pig pneumonia. Vet. Med. *60:* 841-845.

Maré, I.J. and J.N. Coetzee. 1964. Antibiotics of *Alcaligenes faecalis.* Nature (London) *203:* 430-431.

Maré, I.J., H.C. de Klerk and O.W. Prozesky. 1966. The morphology of *Alcaligenes faecalis* bacteriophages. J. Gen. Microbiol. *44:* 23-26.

Margherita, S.S. and R.E. Hungate. 1963. Serological analysis of *Butyrivibrio* from the bovine rumen. J. Bacteriol. *86:* 855-860.

Margulis, L. 1981. Symbiosis in cell evolution. W.H. Freeman and Company, San Francisco, California.

Margulis, L., D. Chase and L.P. To. 1979. Possible evolutionary significance of spirochaetes. Proc. R. Soc. Lond. B *204:* 189-198.

Margulis, L., G. Thorington, B. Berger and J. Stolz. 1978. Endosymbiotic bacteria associated with the intracellular green algae of *Hydra viridis.* Curr. Microbiol. *1:* 227-232.

Margulis, L., L. To and D. Chase. 1978. Microtubules in prokaryotes. Science *200:* 1118-1124.

Marini, F. and C. Spalla. 1964. Un nuovo fattore di crescita per un batterio marino (*Flavobacterium tirrenicum* n. sp.) presente nella farina di pesce e prodotto da microrganismi. G. Microbiol. *12:* 35-44.

Markham, P.G., R. Townsend, M. Bar-Joseph, M.J. Daniels, A. Plaskitt and B.M. Meddins. 1974. Spiroplasmas are the causal agents of citrus little-leaf disease. Ann. Appl. Biol. *78:* 49-57.

Markowits, A., H.P. Klein and E.H. Fischer. 1956. Purification, crystallization and properties of the alpha-amylase of *Pseudomonas saccharophila.* Biochim. Biophys. Acta *19:* 267-273.

Markowitz, S.M. 1980. Isolation of an ampicillin-resistant, non-betalactamase-producing strain of *Haemophilus influenzae.* Antimicrob. Agents Chemother. *17:* 80-83.

Marmur, J. 1961. A procedure for the isolation of DNA from microorganisms. J. Mol. Biol. *3:* 208-218.

Marmur, J. and P. Doty. 1961. Thermal renaturation of deoxyribonucleic acids. J. Mol. Biol. *3:* 585-594.

Marmur, J. and P. Doty. 1962. Determination of the base composition of deoxyribonucleic acid from its thermal denaturation temperature. J. Mol. Biol. *5:* 109-118.

Marsden, H.B. and W.A. Hyde. 1971. Isolation of *Bacteroides corrodens* from infections in children. J. Clin. Pathol. *24:* 117-119.

Marshall, V.P. and J.R. Sokatch. 1972. Regulation of valine catabolism in *Pseudomonas putida.* J. Bacteriol. *110:* 1073-1081.

Marshall, R.B., E.B. Walton and A.J. Robinson. 1981. Identification of leptospira serovars by restriction-endonuclease analysis. J. Med. Microbiol. *14:* 163-166.

Martin, G. and F. Jacob. 1962. Transfert de l'épisome sexuel d'*Escherichia coli* à *Pasteurella pestis.* C. R. Acad. Sci. *254:* 3589-3590.

Martin, H.H., W. Schilf and H.-G. Schiefer. 1980. Differentiation of *Mycoplasmatales* from bacterial protoplast L-forms by assay for penicillin-binding proteins. Arch. Microbiol. *127:* 297-299.

Martin, J.E., J.H. Armstrong and P.B. Smith. 1974. New system for cultivation of *Neisseria gonorrhoeae.* Appl. Microbiol. *27:* 802-805.

Martin, J.P., J. Fleck, M. Mock and J.M. Ghuysen. 1973. The wall peptidoglycans of *Neisseria perflava, Moraxella glucidolytica, Pseudomonas alcaligens* and *Proteus vulgaris* strain P18. Eur. J. Biochem. *38:* 301-306.

Martin, S.M. (Editor). 1963. Culture Collections: Perspectives and Problems. Proceedings of the Specialists' Conference on Culture Collections, Ottawa, 1962. University of Toronto Press, Toronto.

Martin, S.M. and V.B.D. Skerman. 1972. World Directory of Collections of Cultures of Microorganisms. Wiley-Interscience, New York.

Martinec, T. and M. Kocur. 1963. Taxonomicka studie rodu *Erwinia.* Folia Biol. (Praha) *4:* 1-163.

Martinec, T. and M. Kocur. 1964. A taxonomic study of *Erwinia amylovora* (Burrill 1882) Winslow et al. 91920. Int. Bull. Bacteriol. Nomen. Taxon. *14:* 5-14.

Martínez, J. and P.H. Clarke. 1975. R factor mediated gene transfer in *Pseudomonas putida.* Proc. Soc. Gen. Microbiol. *3:* 51-52.

Martinez, R. J. 1963. On the nature of the granules of the genus *Spirillum.* Arch. Microbiol. *44:* 334-343.

Martinenz-De Drets, G. and A. Arias. 1970. Metabolism of some polyols by *Rhizobium meliloti.* J. Bacteriol. *103:* 97-103.

Martinez-De Drets, G. and A. Arias. 1972. Enzymatic basis for differentiation of *Rhizobium* into fast- and slow-growing groups. J. Bacteriol. *109:* 467-470.

Martzinovski, E.-J. 1911. De l'etiologie de la peripneumonie. Ann. Inst. Pasteur (Paris) *25:* 914-917.

Maruashvili, G.M. 1945. On the tick borne relapsing fever. Med. Parazitol. Parazit. Bolez. *14:* 24-27.

Mashimo, P.A., R.J. Genco and S.A. Ellison. 1976. Antibodies reactive with *Leptotrichia buccalis* in human serum from infancy to adulthood. Arch. Oral Biol. *21:* 277-283.

Maskell, J.P. 1981. The pathogenicity of *Bacteroides fragilis* and related species by intracutaneous infection in the guinea pig. J. Med. Microbiol. *14:* 131-140.

Mason, R.A. 1970. Propagation and growth cycle of *Rickettsia quintana* in a new liquid medium. J. Bacteriol. *103:* 184-190.

Mason, R.A. and M. Ristic. 1966. In vitro incorporation of glycine by bovine erythrocytes infected with *Anaplasma marginale.* J. Infect. Dis. *116:* 335-342.

Masover, G.K., J.E. Sawyer and L. Hayflick. 1976. Urea hydrolyzing activity of a T-strain mycoplasma: *Ureaplasma urealyticum.* J. Bacteriol. *125:* 581-587.

Materassi, R., G. Florenzano, W. Balloni and F. Flavilli. 1966. Su una nuova specie di *Beijerinckia (Beijerinckia venezuelae* nov. sp.) isolata da terreni venezuelani. Ann. Microbiol. Enzimol. *16:* 201-215.

Mathey, W. J. 1956. A diphtheroid stomatitis of chickens apparently due to a spirillum, *Spirillum pulli,* species nova. Am. J. Vet. Res. *17:* 742-746.

Mathey, W. J. and A. C. Rissberger. 1964. A turkey sinus vibrio (*Vibrio maleagridis* n. sp.) compared with the avian hepatitis vibrio (*Vibrio hepaticus,* n. sp.). Poultry Sci. *43:* 1339.

Mathias, R.G., A.R. Ronald, M.J. Garwith, D.W. McCullough, H.G. Stiver, J. Berger, C.Y. Cates, L.M. Fox and B.A. Lank. 1976. Clinical evaluation of amikacin in treatment of infections due to Gram-negative aerobic bacilli. J. Infect. Dis. *134:* Suppl., S394-S401.

Matin, A. and S. C. Rittenberg. 1972. Kinetics of deoxyribonucleic acid destruction and synthesis during growth of *Bdellovibrio bacteriovorus* strain 109D on *Pseudomonas putida* and *Escherichia coli.* J. Bacteriol. *111:* 664-673.

Matsumoto, H. 1963. Studies on the *Hafnia* isolated from normal human. Japan J. Microbiol. *7:* 105-114.

Matsumoto, H. 1964. Additional new antigens of Hafnia group. Jpn. J. Microbiol. *8:* 139-141.

Matsumoto, H. and T. Tazaki. 1970. Genetic recombination in *Klebsiella pneumoniae.* Jpn. J. Microbiol. *14:* 129-141.

Matsumoto, H. and T. Tazaki. 1973. FP5 factor, an undescribed sex factor of *Pseudomonas aeruginosa.* Jpn. J. Microbiol. *17:* 409-417.

Matsumoto, H., T. Tazaki and S. Hosogaya. 1973. A generalized transducing phage of *Serratia marcescens.* Jpn. J. Microbiol. *17:* 473-479.

Matsumoto, K., S. Iuchi, A. Fujisawa and S. Tanaka. 1973. Enrichment of mutants lacking the phosphoenolpyruvate-dependent phosphotransferase system of *Vibrio parahaemolyticus* by screening with methyl-α-D-glucoside. J. Bacteriol. *119:* 632-634.

Matthews, P.R.J. and I.H. Pattison. 1961. The identification of a *Hemophilus*-like organism associated with pneumonia and pleurisy in the pig. J. Comp. Pathol. *71:* 44-52.

Matthysse, A.G., K.V. Holmes and R.H.G. Gurlitz. 1981. Elaboration of cellulose fibrils by *Agrobacterium tumefaciens* during attachment to carrot cells. J. Bacteriol. *145:* 583-595.

Matzuschita, T. 1902. Bacteriologische Diagnostik. Gustav Fisher, Jena.

Maxam, A.M. and W. Gilbert. 1977. A new method for sequencing DNA. Proc. Nat. Acad. Sci. USA *74:* 560-564.

Maxcy, K.F. 1929. Endemic typhus fever of the Southeastern United States: Reaction of the guinea pig. Publ. Health Rep. *44:* 589-600.

May, J.R. 1973. *Haemophilus influenzae* apparently resistant to trimethoprim. Br. Med. J. *3:* 407-408.

May, J.R. and J. Davies. 1972. Resistance of *Haemophilus influenzae* to trimethoprim. Br. Med. J. *3:* 376-377.

Mayberry, W.R. 1980a. Hydroxy fatty acids in *Bacteroides* species: D-(-)-3-hydroxy-15-methylhexadecanoate and its homologs. J. Bacteriol. *143:* 582-587.

Mayberry, W.R. 1980b. Cellular distribution and linkage of D-(-)-3-hydroxy fatty acids in *Bacteroides* species. J. Bacteriol. *144:* 200-204.

Mayberry, W.R., D.W. Lambe, Jr. and K.P. Ferguson. 1982. Identification of *Bacteroides* species by cellular fatty acid profiles. Int. J. Syst. Bacteriol. *32:* 21-27.

Mayberry, W.R., P.F. Smith and T.A. Langworthy. 1974. Heptose-containing pentaglycosyl diglyceride among the lipids of *Acholeplasma modicum.* J. Bacteriol. *118:* 898-904.

Mayberry-Carson, K.J., I.L. Roth, J.L. Harris and P.F. Smith. 1974. Scanning electron microscopy of *Thermoplasma acidophilum.* J. Bacteriol. *120:* 1472-1475.

Mayer, M. 1921. Über einige bakterienähnliche Parasiten der Erythrozyten bei Menschen und Tieren. Arch. Schiffs-Trop. Hyg. *25:* 150-152.

Mayo, J.B. and L.R. McCarthy. 1977. Antimicrobial susceptibility of *Haemophilus parainfluenzae.* Antimicrob. Agents Chemother. *11:* 844-847.

Mayrand, D. 1979. Identification of clinical isolates of selected species of *Bacteroides*: production of phenylacetic acid. Can. J. Microbiol. *25:* 927-928.

Mayrand, D., B.C. McBride, T. Edwards and S. Jensen. 1980. Characterization of *Bacteroides asaccharolyticus* and *B. melaninogenicus* oral isolates. Can. J. Microbiol. *26:* 1178-1183.

Mays, T.D., L.V. Holdeman, W.E.C. Moore, M. Rogosa and J.L. Johnson. 1982. Taxonomy of the genus *Veillonella* Prévot. Int. J. Syst. Bacteriol. *32:* 28-36.

Mazzotti, L. 1949. Sobre una nueva espiroqueta de la fiebre recurrente, encontrada en Mexico. Rev. Inst. Salubr. Inst. Inst. Enferm. Trop. Mex. *10:* 277-281.

McAllister, H.A. and G.R. Carter. 1974. An aerogenic *Pasteurella*-like organism recovered from swine. Am. J. Vet. Res. *35:* 917-922.

McCall, J.O. and R.K. Sizemore. 1979. Description of a bacteriocinogenic plasmid in *Beneckea harveyi.* Appl. Environ. Microbiol. *38:* 974-979.

McCarthy, B.J. and E.T. Bolton. 1963. An approach to the measurement of genetic relatedness among organisms. Proc. Nat. Acad. Sci. USA *50:* 156-164.

McCarthy, D.H. 1975. *Aeromonas proteolytica*--a halophilic aeromonad? Can. J. Microbiol. *21:* 902-904.

McCarthy, D.H. 1975. The bacteriology and taxonomy of *Aeromonas liquefaciens.* Technical Report Series, Fish Diseases Laboratory, Ministry of Agriculture, Weymouth, Dorset.

McCarthy, L.R., P.A. Mickelsen and E.G. Smith. 1979. Antibiotic susceptibility of *Haemophilus vaginalis (Corynebacterium vaginale)* to 21 antibiotics. Antimicrob. Agents Chemother. *16:* 186-189.

McCarty, C., M.L. Snyder and R.B. Parker. 1965. The indigenous oral flora of man. I. The newborn to the 1-year-old infant. Arch. Oral Biol. *10:* 61-70.

McCaul, T.F. and J.C. Williams. 1981. Developmental cycle of *Coxiella burnetii*: Structure and morphogenesis of vegetative and sporogenic differentiations. J. Bacteriol. *147:* 1063-1076.

McClung, C. R. and D. G. Patriquin. 1980. Isolation of a nitrogen-fixing *Campylobacter* species from the roots of *Spartina alterniflora* Loisel. Can. J. Microbiol. *26:* 881-886.

McComb, J.A., J. Elliot and M.J. Dilworth. 1975. Acetylene reduction by *Rhizobium* in pure culture. Nature (London) *256:* 409-410.

McCormack, W.M., W.E. DeWitt, P.E. Bailey, G.K. Morris, P. Soeharjono and E.J. Gangarosa. 1974. Evaluation of Thiosulphate-Citrate-Bile salts-Sucrose agar, a selective medium for the isolation of *Vibrio cholerae* and other pathogenic vibrios. J. Infect. Dis. *129:* 497-500.

McCoy, E. C., D. Doyle, K. Bruda, L. B. Corbeil and A. J. Winter. 1975. Superficial antigens of *Campylobacter* (Vibrio) *fetus*: characterization of an antiphagocytic component. Infect. Immun. *11:* 517-525.

McCoy, G.W. and C.W. Chapin. 1912. Further observations on a plague-like disease of rodents with a preliminary note on the causative agent: *Bacterium tularense.* J. Infect. Dis. *10:* 61-72.

McCoy, R.E. 1979a. Mycoplasmas and yellows diseases. *In* Whitcomb and Tully (Editors), The Mycoplasmas, Vol. 3, Academic Press, New York, pp. 229-263.

McCoy, R.E. 1979b. Passage of mycoplasmalike organisms through sieve pores? Nat. Sci. Counc. (Taiwan). Symp. Ser. No. 2. pp. 45-48.

McCoy, R.E., H.G. Basham and R.E. Davis. 1982. Power puff spiroplasma: a new epiphytic mycoplasma. Microbial. Ecol. *8:* 169–180.

McCoy, R.E., M.J. Davis and R.V. Dowell. 1981. In vivo multiplication of spiroplasmas in larvae of the greater wax moth. Phytopathology *71:* 408-411.

McCoy, R.E., D.S. Williams and D.L. Thomas. 1979. Isolation of mycoplasmas from flowers. *In* Proc. R.O.C. U.S. Coop. Sci. Semin. Mycoplasma Dis. Plants, NSC Symp. Ser. I, Natl. Sci. Council, Taipei, pp. 75-81.

McCraw, B.M. 1952. Furunculosis of fish. U.S. Department of the Interior, Fish and Wildlife Service, Special Scientific Report: Fisheries No. 84.

McCrumb, F. 1961. Aerosol infection of man with *Pasteurella tularensis.* Bacteriol. Rev. *25:* 262-267.

McCulloch, L. 1911. A spot disease of cauliflower. Bull. U.S. Depart. Agr. Bur. Plant Ind. No. 225: 1-15.

Mc Culloch, L. 1918. A morphological and cultural note on the organism causing Stewart's disease of sweet corn. Phytopathology *8:* 440-441.

McCulloch, L. 1920. Basal glumerot of wheat. J. Agr. Res. *18:* 543-552.

McCulloch, L. 1921. A bacterial disease of gladiolus. Science *54:* 115-116.

McCulloch, L. 1924. A bacterial blight of gladioli. J. Agr. Res. *27:* 225-230.

McCulloch, L. 1929. A bacterial leaf spot of horse-radish caused by *Bacterium campestre* var. *armoraciae* n. var. J. Agr. Res. *38:* 269-287.

McCulloch, L. 1937. An iris leaf disease caused by *Bacterium tardicrescens* n. sp. Phytopathology *27:* 135.

McCullough, N.B. and L.A. Dick. 1943. Growth of *Brucella* in a simple chemically defined medium. Proc. Soc. Exp. Biol. Med. *52:* 310-311.

McDade, J.E. 1979. Primary isolation using guinea pigs and embryonated eggs. *In* Jones and Hébert (Editors), "Legionnaires' " the disease, the bacterium and methodology. Center for Disease Control, Atlanta, pp. 70-76.

McDade, J.E., D.J. Brenner and F.M. Bozeman. 1979. Legionnaires' disease bacterium isolated in 1947. Ann. Intern. Med. *90:* 659-661.

McDade, J.E., C.C. Shepard, D.W. Fraser, T.R. (sic) Tsai, M.A. Redus, W.R. Dowdle and the Laboratory Investigation Team. 1977. Legionnaires' disease: isolation of a bacterium and demonstration of its role in other respiratory disease. N. Engl. J. Med. *297:* 1197-1203.

McDade, J.E., C.C. Shepard, M.A. Redus, V.F. Newhouse and J.D. Smith. 1980. Evidence of *Rickettsia prowazekii* infections in the United States. Am. J. Trop. Med. Hyg. *29:* 277-284.

McDermott, E. N. 1928. Rat-bite fever: a study of the experimental disease, with a critical review of the literature. Q. J. Med. *21:* 433-458.

McDermott, W., M.M. Leask and M. Benoiti. 1945. *Streptobacillus moniliformis* as cause of subacute bacterial endocarditis: case treated with penicillin. Ann. Intern. Med. *23:* 414.

McDonald, I.J. and K.G. Johnson. 1975. Nutritional requirements of some non-pathogenic *Neisseria* grown in simple synthetic media. Can. J. Microbiol. *21:* 1198-1204.

McElhaney, R.M. and M.E. Tourtellotte. 1969. Mycoplasma membrane lipids: Variations in fatty acid composition. Science *164:* 433-434.

McElroy, L. J. and N. R. Krieg. 1972. A serological method for the identification of *Spirilla.* Can. J. Microbiol. *18:* 57-64.

McFadden, B.A. and Howes, W.V. 1961. *Pseudomonas indigofera.* J. Bacteriol. *81:* 858-862.

McGee, Z.A., R.R. Dourmashkin and J.G. Gross. 1977. Relationship of pili to colonial morphology among pathogenic and nonpathogenic species of *Neisseria.* Infect. Immun. *15:* 594-600.

McGhee, J.R. and B.A. Freeman. 1970a. Osmotically sensitive *Brucella* in infected, normal and immune macrophages. Infect. Immun. *1:* 146-150.

McGhee, J.R. and B.A. Freeman. 1970b. Separation of soluble *Brucella* antigens by gel-filtration chromatography. Infect. Immun. *2:* 48-53.

McInerney, M.J., M.P. Bryant, R.B. Hespell and J.W. Costerton. 1981. *Syntrophomonas wolfei* gen. nov. sp. nov., an anaerobic, syntrophic, fatty acid-oxidizing bacterium. Appl. Environ. Microbiol. *41:* 1029-1039.

McInerney, M.J., M.P. Bryant, R.B. Hespell and J.W. Costerton. 1982. *In* Validation of the publication of new names and new combinations previously effectively published outside the IJSB. List No. 8. Int. J. Syst. Bacteriol. *32:* 266-268.

McIntosh, A.F. 1962. A serological examination of some acetic acid bacteria. Antonie van Leeuwenhoek J. Microbiol. Serol. *28:* 49-62.

McIntyre, O.R. and J.C. Feeley. 1965. Characteristics of non-cholera vibrios isolated from cases of human diarrhoea. Bull. W.H.O. *32:* 627-632.

McKiel, J.A., E.J. Bell and D.B. Lackman. 1967. *Rickettsia canada*: A new member of the typhus group of rickettsiae isolated from *Haemophysalis leporispalustris* ticks in Canada. Can. J. Microbiol. *13:* 503-510.

McKinney, R.M., R.K. Porschen, P.H. Edelstein, M.L. Bisset, P.P. Harris, S.P. Bondell, A.G. Steigerwalt, R.E. Weaver, M.E. Ein, D.S. Lindquist, R.S. Kops and D.J. Brenner. 1982. *In* Validation of the publication of new names and new combinations previously effectively published outside the IJSB. List No. 8. Int. J. Syst. Bacteriol. *32:* 266-268.

McKinney, R.M., R.K. Porschen, P.H. Edelstein, M.L. Bissett, P.P. Harris, S.P. Bondell, A.G. Steigerwalt, R.E. Weaver, M.E. Ein, D.S. Lindquist, R.S. Kops and D.J. Brenner. 1981. *Legionella longbeachae* sp. nov., another etiologic agent of human pneumonia. Ann. Intern. Med. *94:* 739-743.

McKinney, R.M., H.W. Wilkinson, H.M. Sommers, B.J. Fikes, K.R. Sasseville, M.M. Yungbluth and J.S. Wolf. 1980. *Legionella pneumophila* serogroup six: isolation from cases of legionellosis, identification by immunofluorescence staining, and immunologic response to infection. J. Clin. Microbiol. *12:* 395-401.

McLaren, D.A., M.J. Worms, B.R. Laurence and M.G. Simpson. 1975. Microorganisms in filarial larvae (Nematoda). Trans. Roy. Soc. Trop. Med. Hyg. *69:* 509-514.

McMahon, P.C. 1973. Mapping the chromosome of *Yersinia pseudotuberculosis* by interrupted mating. J. Gen. Microbiol. *77:* 61-69.

McMartin, D.A., K.J. MacOwan and L.L. Swift. 1980. A century of classical caprine pleuropneumonia: from original description to aetiology. Brit. Vet. J. *136:* 507–515.

McMeekin, T.A. 1977. Ultraviolet light sensitivity as an aid for the identification of gram-negative, yellow-pigmented rods. J. Gen. Microbiol. *103:* 149-151.

McMeekin, T.A., J.T. Patterson and J.G. Murray. 1971. An initial approach to the taxonomy of some Gram-negative yellow pigmented rods. J. Appl. Bacteriol. *34:* 699-716.

McMeekin, T.A. and J.M. Shewan. 1978. Taxonomic strategies for *Flavobacterium* and related genera. J. Appl. Bacteriol. *45:* 321-332.

McMillan, S.A. 1972. Bacteriuria of elderly women in hospitals. Occurrence and drug resistance. Lancet *2:* 452-455.

McNicol, L.A., J.B. Kaper, H.A. Lockman, E.F. Remmers, W.M. Spira, M.J. Voll and R.R. Colwell. 1980. R-Factor carriage in a group F vibrio isolated from Bangladesh. Antimicrob. Agents Chemother. *17:* 512-515.

McWethy, S.J. and P.A. Hartman. 1977. Purification and some properties of an extracellular alpha-amylase from *Bacteroides amylophilus.* J. Bacteriol. *129:* 1537-1544.

Mead, A.R. 1969. *Aeromonas liquefaciens* in the leukodermia syndrome of *Achatina fulica.* Malacologia *9:* 43.

Meade, H. and E. Singer. 1977. Genetic mapping of *Rhizobium meliloti*. Proc. Natl. Acad. Sci. U.S.A. *74:* 2076-2078.

Meadow, P.M. 1975. Wall and membrane structures in the genus *Pseudomonas*. *In* Clarke and Richmond (Editors), Genetics and Biochemistry of *Pseudomonas*. John Wiley and Sons, London, pp. 67-98.

Meadow, P.M. and P.L. Wells. 1978. Receptor sites for R-type pyocins and bacteriophage E79 in the core part of the lipopolysaccharide of *Pseudomonas aeruginosa* PAC1. J. Gen. Microbiol. *108:* 339-342.

Mee, B.J. and B.T.O. Lee. 1967. An analysis of histidine requiring mutants in *Pseudomonas aeruginosa*. Genetics *55:* 709-722.

Megraud, F., J.L. Traissac and J. Latrille. 1981. Abscès à *Eikenella corrodens*: à propos d'un cas grave. Medicine et Maladies Infectieuses *11:* 39-43.

Meiklejohn, J. 1954. Notes on nitrogen-fixing bacteria from East African soils. Proc. 5th Int. Congr. Soil Sci. *3:* 123-125.

Mekalanos, J.J. and J.R. Murphy. 1980. Regulation of cholera toxin production in *Vibrio cholerae*: genetic analysis of phenotypic instability in hypertoxinogenic mutants. J. Bacteriol. *141:* 570-576.

Mekalanos, J.J., R.D. Sublett and W.R. Romig. 1979. Genetic mapping of toxin regulatory mutations in *Vibrio cholerae*. J. Bacteriol. *139:* 859-865.

Melly, M.A., Z.A. McGee, R.G. Horn, F. Morris and A.D. Glick. 1979. An electron microscopic India ink technique for demonstrating capsules on microorganisms: Studies with *Streptococcus pneumoniae*, *Staphylococcus aureus*, and *Neisseria gonorrhoeae*. J. Infect. Dis. *140:* 605-609.

Meloni, G.A., G. Bertoloni, F. Busolo and L. Conventi. 1980. Colony morphology, ultrastructure and morphogenesis in *Mycoplasma hominis*, *Acholeplasma laidlawii* and *Ureaplasma urealyticum*. J. Gen. Microbiol. *116:* 435-443.

Mendelson, N.H. 1976. Helical growth of *Bacillus subtilis*: a new model of cell growth. Proc. Natl. Acad. Sci. USA. *73:* 1740-1744.

Meng, K.E. and R.M. Pfister. 1980. Intracellular structures of *Mycoplasma pneumoniae* revealed after membrane removal. J. Bacteriol. *144:* 390-399.

Mercier, L. 1906. Les corps bactéroides de la Blatte *(Periplaneta orientalis)*: *Bacillus Cuenoti* (n. sp. L. Mercier). (Note préliminaire). C.R. Séances Soc. Biol. Fil. *61:* 682-684.

Mergeay, M. and J. Gerits. 1978. F′ plasmid transfer from Escherichia coli to *Pseudomonas fluorescens*. J. Bacteriol. *135:* 18-28.

Mergenhagen, S.E., W.C. de Araujo and E. Varah. 1965. Antibody to *Leptotrichia buccalis* in human sera. Arch. Oral Biol. *10:* 29-33.

Merkel, G.J., D.R. Durham and J.J. Perry. 1980. The atypical cell wall composition of *Thermomicrobium roseum*. Can. J. Microbiol. *26:* 556-559.

Merkel, G.J., S.S. Stapleton and J.J. Perry. 1978. Isolation and peptidoglycan of Gram-negative hydrocarbon utilizing bacteria. J. Gen. Microbiol. *109:* 141-148.

Merkel, G.J., W.C. Underwood and J.J. Perry. 1978. Isolation of thermophilic bacteria capable of growth solely on long-chain hydrocarbons. FEMS Microbiol. Lett. *3:* 81-83.

Merkel, J.R., E.D. Traganza, B.B. Mukherjee, T.B. Griffin and J.M. Prescott. 1964. Proteolytic activity and general characteristics of a marine bacterium, *Aeromonas proteolytica* sp. n. J. Bacteriol. *87:* 1227-1233.

Merkel, J.R. and E. Tranganza. 1958. Possible symbiotic role of proteolytic and cellulytic bacteria found in the digestive system of a marine isopod. Bacteriol. Proc. pp. 53-54.

Merlo, D.J. and E.W. Nester. 1977. Plasmids in avirulent strains of *Agrobacterium*. J. Bacteriol. *129:* 76-80.

Merrell, B.R., E. Weiss and G.A. Dasch. 1978. Morphological and cell association characteristics of *Rochalimaea quintana*: Comparison of the vole and Fuller strains. J. Bacteriol. *135:* 633-640.

Mescher, M.F. and J.L. Strominger. 1976a. Purification and characterization of a prokaryotic glycoprotein from the cell envelope of *Halobacterium salinarium*. J. Biol. Chem. *251:* 2005-2014.

Mescher, M.F. and J.L. Strominger. 1976b. Structural (shape-maintaining) role of the cell surface glycoprotein of *Halobacterium salinarium*. Proc. Natl. Acad. Sci. U.S.A. *73:* 2687-2691.

Mesnil, F. 1930. Ohne Titelangabe. Bull. Inst. Pasteur *28:* 125-126.

Mesnil, F. and M. Caullery. 1916. Sur un organisme spirochétoide (*Cristispira polydorae* n. sp.) de l'intestin d'une annélide polychéte. C. R. Soc. Biol. (Paris) *79:* 1118-1121.

Metzner, P. 1920. Die Bewegung und Reizbeantwortung der bipolar gegeisselten *Spirillen*. Jahr. Wiss. Bot. *59:* 325-412.

Meyer, D.J. and C.W. Jones. 1973. Distribution of cytochromes in bacteria: relationship to general physiology. Int. J. Syst. Bacteriol. *23:* 459-467.

Meyer, F.P. and G.L. Bullock. 1973. *Edwardsiella tarda*, a new pathogen of channel catfish (*Ictalurus punctatus*). Appl. Microbiol. *25:* 155-156.

Meyer, G.F. and W. Frank. 1960. Elektronenmikroskopische Studien über symbiontische Einrichtungen bei Insekten. Proc. Int. Congr. Electronenmicrosc. *2:* 539-542.

Meyer, J.M. 1977. Pigment fluorescent et métabolisme du fer chez *Pseudomonas fluorescens*. Thèse d'Etat, Strasbourg.

Meyer, J.M. and M.A. Abdallah. 1978. The fluorescent pigment of *Pseudomonas fluorescens*: biosynthesis, purification and physico-chemical properties. J. Gen. Microbiol. *107:* 319-328.

Meyer, J.M. and M.A. Abdallah. 1980. The siderochromes of non-fluorescent pseudomonads: production of nocardamine by *Pseudomonas stutzeri*. J. Gen. Microbiol. *118:* 125-129.

Meyer, J.M. and J.M. Hornsperger. 1978. Role of pyoverdine$_{Pf}$, the iron-binding fluorescent pigment of *Pseudomonas fluorescens*, in iron transport. J. Gen.

Microbiol. *107:* 329-331.

Meyer, K.F. 1931. Public Health control of infectious abortion in certified milk. Amer. J. Public Health *21:* 503-514.

Meyer, K.F. 1953. Psittacosis group. Ann. N.Y. Acad. Sci. *56:* 545-556.

Meyer, K.F. and E.B. Shaw. 1920. A comparison of the morphological, cultural and biochemical characteristics of *B. abortus* and *B. melitensis*. Studies on genus *Brucella* nov. gen. J. Infect. Dis. *27:* 173-184.

Meyer, M.C. and S.C. Pueppke. 1980. Differentiation of *Rhizobium japonicum* strain derivatives by antibiotic sensitivity patterns, lectin binding, and utilization of biochemicals. Can. J. Microbiol. *26:* 607-612.

Meyer, M.E. 1961. Metabolic characterization of the genus *Brucella*. III. Oxidative metabolism of strains that show anomalous characteristics by conventional determinative methods. J. Bacteriol. *82:* 401-410.

Meyer, M.E. 1962. Metabolic and bacteriophage identification of *Brucella* strains described as *Brucella melitensis* from cattle. Bull. W.H.O. *26:* 829-831.

Meyer, M.E. 1969. *Brucella* organisms isolated from dogs: comparison of characteristics of members of the genus *Brucella*. Amer. J. Vet. Res. *30:* 1751-1756.

Meyer, M.E. 1976a. Evolution and taxonomy in the genus *Brucella*: concepts of the contemporary species. Am. J. Vet. Res. *37:* 199-202.

Meyer, M.E. 1976b. Evolution and taxonomy in the genus *Brucella*: steroid hormone induction of filterable forms with altered characteristics after reversion. Am. J. Vet. Res. *37:* 207-210.

Meyer, M.E. 1976c. Evolution and taxonomy in the genus *Brucella*; progesterone induction of filterable forms of *Brucella abortus* type 2 with revertant characteristics essentially indistinguishable in vitro from those of *Brucella ovis*. Am. J. Vet. Res. *37:* 211-214.

Meyer, M.E. and H.S. Cameron. 1957. Species metabolic patterns in morphologically similar Gram negative pathogens. J. Bacteriol. *73:* 158-161.

Meyer, M.E. and H.S. Cameron. 1961a. Metabolic characterization of the genus *Brucella*. I. Statistical evaluation of the oxidative rates by which type I of each species can be identified. J. Bacteriol. *82:* 387-395.

Meyer, M.E. and H.S. Cameron. 1961b. Metabolic characterization of the *Brucella*. II. Oxidative metabolic patterns of the described species. J. Bacteriol. *82:* 396-400.

Meyer, M.E. and W.J.B. Morgan. 1962. Metabolic characterization of *Brucella* strains that show conflicting identity by biochemical and serological methods. Bull. W.H.O. *26:* 823-827.

Meyer, O., J. Lalucat and H.G. Schlegel. 1980. *Pseudomonas carboxydohydrogena* (Sanjieva and Zavarzin) comb. nov., a monotrichous, non budding, strictly aerobic, carbon monoxide-utilizing hydrogen bacterium previously assigned to *Seliberia*. Int. J. Syst. Bacteriol. *30:* 189-195.

Meyer, O. and H.G. Schlegel. 1978. Reisolation of the carbon monoxide utilizing hydrogen bacterium *Pseudomonas carboxydovorans* (Kistner) comb. nov. Arch. Microbiol. *118:* 35-43.

Meyer, P.E. and E.F. Hunter. 1967. Antigenic relationships of 14 treponemes demonstrated by immunofluorescence. J. Bacteriol. *93:* 784-789.

Meyers, M.B., G. Cherry, B.B. Bornside and G.H. Bornside. 1969. Ultraviolet red fluorescence of *Bacteroides melaninogenicus*. Appl. Microbiol. *17:* 760-762.

Meyers, S.P., M.H. Baslow, S.J. Bein and C.E. Marks. 1959. Studies of *Flavobacterium piscicida* Bein. I. Growth, toxicity, and ecological considerations. J. Bacteriol. *78:* 225-230.

Meyers, T.R. 1979. Preliminary studies on a chlamydial agent in the digestive diverticular epithelium of hard clams, *Mercenaria mercenaria* (L.) from Great South Bay, New York. J. Fish Dis. *2:* 179-189.

Meyling, A. and N.F. Friis. 1972. Serological identification of a new porcine *Mycoplasma* species, *M. flocculare*. Acta Vet. Scand. *13:* 287-289.

Meynadier, G., A. Lopez and J.-L. Duthoit. 1974. Mise en évidence de Rickettsiales chez une araignée (*Argyrodes gibbosus* Lucas), Ananeae, Theridiidae. C.R. Acad. Science, Series D *278:* 2365-2367.

Meynadier, G. and P. Monsarrat. 1969. Une rickettsiose chez une cétoine de Madagascar. Entomophaga *14:* 401-406.

Meynadier, G., J.M. Quiot and C. Vago. 1967. Infection "in vitro" de cellules cardiaques d'invertébrés. Second Int. Coll. Invert. Tissue Culture, Fondazione Baselli, Milano, pp. 218-226.

Meynell, E.W. 1961. A phage, $\rho\chi$, which attacks motile bacteria. J. Gen. Microbiol. *25:* 253-290.

Mez, C. 1898. Mikroskopische Wasseranalyse, Anleitung zür Untersuchung des Wassers mit besonderer Berücksichtigung von Trink- und Abwasser. J. Springer, Berlin, pp. 1-69.

Mhalu, F.S., P.W. Mmari and J. Ijumba. 1979. Rapid emergence of eltor *Vibrio cholerae* resistant to antimicrobial agents during first six months of fourth cholera epidemic in Tanzania. Lancet *1:* 345-347.

Miao, R. and A.H. Fieldsteel. 1978. Genetics of *Treponema*: Relationship between *Treponema pallidum* and five cultivable treponemes. J. Bacteriol. *133:* 101-107.

Miao, R. and A.H. Fieldsteel. 1980. Genetic relationship between *Treponema pallidum* and *Treponema pertenue*, two non-cultivable human pathogens. J. Bacteriol. *141:* 427-429.

Miao, R.M., A.H. Fieldsteel and D.L. Harris. 1978. Genetics of *Treponema*: characterization of *Treponema hyodysenteriae* and its relationship to *Treponema pallidum*. Infect. Immun. *22:* 736-739.

Mickelsen, P.A., L.R. McCarthy and M.E. Mangum. 1977. New differential medium for the isolation of *Corynebacterium vaginale*. J. Clin. Microbiol. *5:* 488-489.

Midgley, J., S.P. LaPage, B.A.G. Jenkins, G.I. Barrow, M.E. Roberts, and A.G.

Buck. 1970. *Cardiobacterium hominis* endocarditis. J. Med. Microbiol. *3:* 91-98.

Midtvedt, T. and A. Norman. 1967. Bile acid transformation by microbial strains belonging to genera found in intestinal contents. Acta Pathol. Microbiol. Scand. *71:* 629-638.

Miehe, H. 1911. Javanische Studien. V. Die Bakterienknoten en den Blattändern der *Ardisia crispa* A. DC. Abh. Kgl. Sächs Ges. Wiss. Math.-Phys. (Leipzig) *32:* 339-431.

Miehe, H. 1914. Weitere Untersuchungen über die Bakteriensymbiose bei *Ardisia crispa.* I. Die Mikroorganismen. Jahrb. Wiss. Bot. *53:* 1-54.

Migula, W. 1894. Über ein neues System der Bakterien. Arb. Bakteriol. Inst. Karlsruhe. *1:* 235-238.

Migula, W. 1895. *Bacteriaceae* (Stabchenbactérien) *In* Engler and Prantl (Editors), Pflanzenfamilien, W. Engelmann, Leipzig, Teil I, Abt. 1a, pp. 20-30.

Migula, W. 1900. System der Bakterien, Vol. 2. Gustav Fischer, Jena.

Mikesell, P., J.W. Ezzell and G.B. Knudson. 1981. Plasmid isolation in *Legionella pneumophila* and *Legionella*-like organisms. Infect. Immun. *31:* 1270-1272.

Milburn, N.S. 1966. Fine structure of the pleomorphic bacteroids in the mycetocytes and ovaries of several genera of cockroaches. J. Insect Physiol. *12:* 1245-1254.

Miles, D.G., H.D. Anthony and S.M. Dennis. 1972. *Haemophilus somnifer* infection in sheep. Am. J. Vet. Res. *33:* 431-435.

Miles, D.O., J.K. Dyer and J.C. Wong. 1976. Influence of amino acids on the growth of *Bacteroides melaninogenicus*. J. Bacteriol. *127:* 899-903.

Miles, E.M. and A.A. Miles. 1951. The identity of *Proteus hydrophilus* Bergey et al. and *Proteus melanovogenes* Miles and Halvan, and their relation to the genus *Aeromonas* Kluyver and Van Niel. J. Gen. Microbiol. *5:* 298-306.

Millar, W.N. 1973. Heterotrophic bacterial population in acid coal mine water: *Flavobacterium acidurans* sp. n. Int. J. Syst. Bacteriol. *23:* 142-150.

Millard, W.A. 1924. Crown rot of rhubarb. Yorks Council Agr. Ed. Bull. *134:* 1-28.

Miller, D.L. and V.W. Rodwell. 1971. Metabolism of basic amino acids in *Pseudomonas putida*. J. Biol. Chem. *246:* 2758-2764.

Miller, J.D.A., J.E. Hughes, G.E. Saunders and L.L. Campbell. 1968. Physiological and biochemical characteristics of some strains of sulfate-reducing bacteria. J. Gen. Microbiol. *52:* 173-178.

Miller, J.D.A. and A.M. Saleh. 1964. A sulphate-reducing bacterium containing cytochrome c_3 but lacking desulfoviridin. J. Gen. Microbiol. *37:* 419-423.

Miller, P.W., W.B. Bollen, J.E. Simmons, H.N. Gross and H.P. Barss. 1940. The pathogen of filbert nut bacteriosis compared with *Phytomonas juglandis*, the cause of walnut blight. Phytopathology *30:* 713-733.

Miller, R.J., H.W. Renshaw and J.A. Evans. 1975. *Haemophilus somnus* complex: antigenicity and specificity of fractions of *H. somnus.* Am. J. Vet. Res. *36:* 1123-1128.

Miller, R.V., J.M. Pemberton and A.J. Clark. 1977. Prophage F 116: evidence for extrachromosomal location in *Pseudomonas aeruginosa* strain PAO. J. Virol. *22:* 844-847.

Miller, T.D. and M.N. Schroth. 1972. Monitoring the epiphytic population of *Erwinia amylovora* on pear with a selective medium. Phytopathology *62:* 1175-1182.

Miller, T.L. 1978. The pathway of formation of acetate and succinate from pyruvate by *Bacteroides succinogenes*. Arch. Microbiol. *117:* 145-152.

Millis, N.F. 1951. Some bacterial fermentations of cider. Ph.D. thesis, University of Bristol, Bristol, U.K.

Millis, N.F. 1956. A study of cider-sickness bacillus - a new variety of *Zymomonas anaerobia*. J. Gen. Microbiol. *15:* 521-528.

Mills, R.C., H. Berthelsen, E. Donaldson and P.L. Wilhelm. 1949. Nutritional requirements of *Pasteurella tularense*. Bacteriol. Proc., 37.

Milstead, J.E. 1980. Pathophysiological influences of the *Heterorhabditis bacteriophora* complex on fifth-instar larvae of the red humped caterpillar, *Schizura concinna*: changes in feeding rate, larval weight, and frass production. J. Invertebr. Pathol. *35:* 260-264.

Minnikin, D.E. and M. Goodfellow. 1980. Lipid composition in the classification and identification of acid fast bacteria. *In* Goodfellow and Board (Editors), Microbiological Classification and Identification, Academic Press, London - New York, pp. 189-256.

Misaghi, I. and R.G. Grogan. 1969. Nutritional and biochemical comparisons of plant-pathogenic and saprophytic fluorescent pseudomonads. Phytopathology *59:* 1436-1450.

Misao, R. and Y. Kobayashi. 1956. Infectious mononucleosis (glandular fever). J. Jpn. Assoc. Infect. Dis. *30:* 453-465.

Misenheimer, T.J., R.F. Anderson, A.A. Lagoda and D.D. Tyler. 1965. Production of 2-ketogluconic acid by *Serratia marcescens*. Appl. Microbiol. *13:* 393-396.

Mishra, A. K., P. Roy and S. Bhattacharya. 1979. Deoxyribonucleic acid-mediated transformation of *Spirillum lipoferum*. J. Bacteriol. *137:* 1425-1427.

Mitchell, R.E. 1976. Isolation and structure of a chlorosis-inducing toxin of *Pseudomonas phaseolicola*. Phytochemistry *15:* 1941-1947.

Mitchell, R.G. and S.K.R. Clarke. 1965. An *Alcaligenes* species with distinctive properties isolated from human sources. J. Gen. Microbiol. *40:* 343-348.

Mitchell, T.G., M.S. Hendrie and J.M. Shewan. 1969. The taxonomy, differentiation and identification of *Cytophaga* species. J. Appl. Bacteriol. *32:* 40-50.

Mitruka, B.J. 1976. Methods of detection and identification of bacteria. CRC Press, Cleveland, Ohio.

Mitsuhashi, J. and Y. Kono. 1975. Intracellular microorganisms in the green rice leafhopper, *Nephotettix cincticeps* Uhler (Hemiptera: Deltocephalidae). Appl. Entomol. Zool. *10:* 1-9.

Mitsuoka, T. 1980. The World of Intestinal Bacteria - The Isolation and Identification of Anaerobic Bacteria; A Color Atlas of Anaerobic Bacteria (English title given). Sobunsha (Sobun Press), Tokyo.

Mitsuoka, T., A. Terada, K. Watanabe and K. Uchida. 1974. *Bacteroides multiacidus*, a new species from the feces of humans and pigs. Int. J. Syst. Bacteriol. *24:* 35-41.

Mittal, K.R. and I.R. Tizard. 1981. Serological cross-reactions between *Brucella abortus* and *Yersinia enterocolitica* serotype 09. Vet. Bull. *51:* 501-505.

Miura, H., M. Horiguchi and T. Matsumoto. 1980. Nutritional interdependence among rumen bacteria, *Bacteroides amylophilus*, *Megasphaera elsdenii*, and *Ruminococcus albus*. Appl. Environ. Microbiol. *40:* 294-300.

Miwatani, T., S. Shinoda and T. Fujino. 1970. Purification of monotrichous flagella of *Vibrio parahaemolyticus* Biken J. *13:* 149-155.

Miwatani, T. and Y. Takeda. 1976. *Vibrio parahaemolyticus* a causative bacterium of food poisoning. Saikon Publishing Co., Tokyo.

Miyagawa, E., R. Azuma and T. Suto. 1978. Distribution of sphingolipids in *Bacteroides* species. J. Gen. Appl. Microbiol. *24:* 341-348.

Miyagawa, E., R. Azuma and T. Suto. 1979a. Cellular fatty acid composition in Gram-negative obligately anaerobic rods. J. Gen. Microbiol. *25:* 41-51.

Miyagawa, E., R. Azuma and T. Suto. 1981. Peptidoglycan composition of Gram-negative obligately anaerobic rods. J. Gen. Appl. Microbiol. *27:* 199-208.

Miyagawa, E., R. Azuma, T. Suto and I. Yano. 1979b. Occurrence of free ceramides in *Bacteroides fragilis* NCTC 9343. J. Biochem. *86:* 311-320.

Miyagawa, E. and T. Suto. 1980. Cellular fatty acid composition in *Bacteroides oralis* and *Bacteroides ruminicola*. J. Gen. Appl. Microbiol. *26:* 331-343.

Miyamoto, S. and K. Kuroda. 1975. Lethal effect of fresh sea water on *Vibrio parahaemolyticus* and isolation of *Bdellovibrio* parasitic against the organism. Jpn. J. Microbiol. *19:* 303- 317.

Miyamoto, Y., T. Kato, Y. Obara, S. Akiyama, K. Takizawa and S. Yamai. 1969. In vitro hemolytic characteristic of *Vibrio parahaemolyticus*: Its close correlation with human pathogenicity. J. Bacteriol. *100:* 1147-1149.

Miyamoto, Y., K. Nakamura and K. Takizawa. 1961. Pathogenic halophiles. Proposals of a new genus "*Oceanomonas*" and of the amended species names. Jpn. J. Microbiol. *5:* 477-486.

Miyazaki, S., G.M. Boush and R.J. Baerwald. 1968. Amino acid synthesis by *Pseudomonas melophthora*, bacterial symbiote of *Rhagoletis pomonella* (Diptera). J. Insect Physiol. *14:* 513-518.

Miyazawa, Y. and C.A. Thomas. 1965. Composition of short segments of DNA molecules. J. Mol. Biol. *11:* 223-237.

Mizuhara, M. and M. Yamanaka. 1961. The cytochrome of *Pasteurella tularensis*. Nature (London) *190:* 1024-1025.

Möbius, K. 1883. *Trypanosoma balbianii* Certes im Krystallstiel schleswigholsteinischer Austern. Zool. Anz. *6:* 148.

Moffett, M.L. and R.R. Colwell. 1968. Adansonian analysis of the *Rhizobiaceae*. J. Gen. Microbiol. *51:* 245-266.

Mohamed, Y.W., P.D. Moorhead and E.H. Bohl. 1969. Natural *Streptobacillus moniliformis* infection of turkeys and attempts to infect turkeys, sheep and pigs. Avian Dis. *13:* 379-385.

Moillo, A.M. 1973. Isolation of a transducing phage forming plaques on *Pseudomonas maltophilia* and *Pseudomonas aeruginosa*. Genet. Res. *21:* 287-289.

Molin, G. and A. Ternstrom. 1982. Numerical taxonomy of psychrotropic pseudomonads. J. Gen. Microbiol. *128:* 1249-1264.

Mollaret, H.H. 1965. Sur la nomenclature et la taxinomie du bacille de Malassez et Vignal. Int. Bull. Bacteriol. Nomencl. Taxon. *15:* 97-106.

Mollaret, H.H., H. Bercovier and J.M. Alonso. 1979. Summary of the data received at the W.H.O. reference center for *Yersinia enterocolitica*. Contr. Microbiol. Immunol. *5:* 174-184.

Mollaret, W.H., Y. Karimi, M. Eftekhari and M. Baltazard. 1963. La peste de Fouissement. Bull. Soc. Path. Exot. *56:* 1186-1193.

Møller, B.R., F.T. Black and E.A. Freundt. 1981. Attempts to produce gynaecological disease in grivet monkeys with *Ureaplasma urealyticum*. J. Med. Microbiol. *14:* 475-478.

Møller, B.R. and E.A. Freundt. 1979. Experimental infection of the genital tract of female grivet monkeys by *Mycoplasma hominis*: effects of different routes of infection. Infect. Immun. *26:* 1123-1128.

Møller, B.R., E.A. Freundt, F.T. Black and P. Frederiksen. 1978. Experimental infection of the genital tract of female grivet monkeys by *Mycoplasma hominis*. Infect. Immun. *20:* 248-257.

Møller, B.R., E.A. Freundt, F.T. Black and F. Melsen. 1980. Experimental infection of the upper genital tract of female grivet monkeys with *Mycoplasma fermentans*. J. Med. Microbiol. *13:* 145-149.

Møller, V. 1954. Distribution of amino acid decarboxylase in *Enterobacteriaceae*. Acta Pathol. Microbiol. Scand. *35:* 259-277.

Møller, V. 1955. Simplified test for some amino acid decarboxylases and arginine dihydrolase system. Acta Pathol. Microbiol. Scand. *36:* 158-172.

Monias, B.L. 1928. Classification of *Bacterium alcaligenes, pyocyaneum* and *fluorescens*. J. Infect. Dis. *43:* 330-334.

Moniz, L. and M.K. Patel. 1958. Three new bacterial diseases of plants from Bombay State. Curr. Sci. *27:* 494-495.

Moniz, L., J.E. Sabley and W.D. More. 1964. A new bacterial canker of *Carissa congesta* in Maharashtra. Indian Phytopathol. *17:* 256.

Monsur, K.A. 1963. Bacteriological diagnosis of cholera under field conditions.

Bull. W.H.O. 28: 387-389.

Monteiro, J.L. 1931. Estudos sobre o typho exanthematico de São Paulo. Mem. Inst. Butantan São Paulo 6: 3-135.

Monteverde, J.J. 1944. A new type of *Salmonella* genus. Nature 154: 676.

Montgomerie, J.Z. 1979. Epidemiology of *Klebsiella* and hospital-associated infections. Rev. Infect. Dis. 1: 736-753.

Montoya, A.L., M.-D. Chilton, M.P. Gordon, D. Sciaky and E.W. Nester. 1977. Octopine and nopaline metabolism in *Agrobacterium tumefaciens* and crown gall tumor cells: role of plasmid genes. J. Bacteriol. 129: 101-107.

Montoya, A.L., L.W. Moore, M.P. Gordon and E.W. Nester. 1978. Multiple genes coding for octopine-degrading enzymes in *Agrobacterium*. J. Bacteriol. 136: 909-915.

Moon, H.W., D.M. Barnes and J.M. Higbee. 1969. Septic embolic actinobacillosis. A report of 2 cases in New World monkeys. Pathol. Vet. 6: 481-486.

Moore, L.W., A. Anderson and C.I. Kado. 1980. *Agrobacterium. In* Schaad (Editor), Laboratory Guide for Identification of Plant Pathogenic Bacteria. American Phytopathological Society, St. Paul, Minnesota, pp. 17-25.

Moore, L.W. and R.V. Carlson. 1975. Liquid nitrogen storage of phytopathogenic bacteria. Phytopathology 65: 246-250.

Moore, L.W. and G. Warren. 1979. *Agrobacterium radiobacter* strain 84 and biological control of crown gall. Annu. Rev. Phytopathol. 17: 163-179.

Moore, L.W., G. Warren and G. Strobel. 1979. Involvement of a plasmid in the hairy root disease of plants caused by *Agrobacterium rhizogenes*. Plasmid 2: 617-626.

Moore, R.L. and B.J. McCarthy. 1969. Characterization of the deoxyribonucleic acid of various strains of halophilic bacteria. J. Bacteriol. 99: 248-254.

Moore, W.E.C., E.P. Cato, I.J. Good and L.V. Holdeman. 1981. The effect of diet on the human fecal flora. *In* Banbury Report 7: Gastrointestinal Cancer: Endogenous Factors, pp. 11-24.

Moore, W.E.C., D.E. Hash, L.V. Holdeman and E.P. Cato. 1980. Polyacrylamide slab gel electrophoresis of soluble proteins for studies of bacterial floras. Appl. Environ. Microbiol. 39: 900-907.

Moore, W.E.C. and L.V. Holdeman. 1969. Anaerobic Gram-negative non-spore-forming rods. *In* Cato, Cummins, Holdeman, Johnson, Moore, Smibert and Smith (Editors), Outline of Clinical Methods in Anaerobic Bacteriology, 1st rev. Virginia Polytechnic Institute Anaerobe Laboratory, Blacksburg, Virginia.

Moore, W.E.C. and L.V. Holdeman. 1970. *Fusobacterium. In* Cato, Cummins, Holdeman, Johnson, Moore, Smibert and Smith (Editors), Outline of Clinical Methods in Anaerobic Bacteriology, 2nd rev. Virginia Polytechnic Institute Anaerobe Laboratory, Blacksburg, Virginia.

Moore, W.E.C. and L.V. Holdeman. 1972. *Fusobacterium. In* Holdeman and Moore (Editors), Anaerobe Laboratory Manual. Virginia Polytechnic Institute Anaerobe Laboratory, Blacksburg, Virginia.

Moore, W.E.C. and L.V. Holdeman. 1973. New names and combinations in the genera *Bacteroides* Castellani and Chalmers, *Fusobacterium* Knorr, *Eubacterium* Prévot, *Propionibacterium* Orla-Jensen, and *Lactobacillus* Beijerinck. Int. J. Syst. Bacteriol. 23: 69-74, 1974; 24: 311.

Moore, W.E.C. and L.V. Holdeman. 1974a. Genus *Fusobacterium* Knorr 1922, 4. *In* Buchanan and Gibbons (Editors), Bergey's Manual of Determinative Bacteriology, 8th Ed. The Williams and Wilkins Co., Baltimore, pp. 404-416.

Moore, W.E.C. and L.V. Holdeman. 1974b. Human fecal flora: the normal flora of 20 Japanese-Hawaiians. Appl. Microbiol. 27: 961-979.

Moore, W.E.C., J.L. Johnson and L.V. Holdeman. 1976. Emendation of *Bacteroides* and *Butyrivibrio* and descriptions of *Desulfomonas* gen. nov. and ten new species in the genera *Desulfomonas, Butyrivibrio, Eubacterium, Clostridium*, and *Ruminococcus*. Int. J. Syst. Bacteriol. 26: 238-252.

Mooser, H. 1928. Experiments relating to the pathology and the etiology of Mexican typhus (tabardillo). J. Infect. Dis. 43: 241-272.

Mooser, H., A. Leemann, S.H. Chao and H.U. Gubler. 1948. Beobachtungen an Funftagefieber. Schweiz. Z. Allg. Pathol. Bakteriol. 11: 513-522.

Mooser, H. and F. Weyer. 1953. Experimental infection of *Macacus rhesus* with *Rickettsia quintana* (trench fever). Proc. Soc. Exper. Biol. Med. 83: 699-701.

Moran, A.B. and D.W. Bruner. 1949. Further studies on the Bethesda group of paracolon bacteria. J. Bacteriol. 58: 695-700.

Morcos, Z. 1935. Preliminary studies in fowl spirochaetosis in Egypt. Vet. J. 91: 161-171.

Moreira, J.A., O. de Magalhaes. 1937. Typho exanthematico de Minas Geraes. Brasil-Med. 51: 583-584.

Moreira-Jacob, M. 1968. New group of virulent bacteriophages showing differential affinity for brucella species. Nature (London) 219: 752-753.

Morel, G. 1976. Studies on *Porochlamydia buthi* g. n., sp. n., an intracellular pathogen of the scorpion *Buthus occitanus*. J. Invert. Pathol. 28: 167-175.

Morel, G. 1977. Étude d'une *Rickettsiella* (Rickettsie) se développant chez un arachnide, l'araignée *Pisaura mirabilis*. Ann. Microbiol. 128A: 49-59.

Morello, J.A. and M. Bonhoff. 1980. *Neisseria* and *Branhamella. In* Lennette, Balows, Hausler and Truant (Editors), Manual of Clinical Microbiology, 3rd Ed., American Society for Microbiology, Washington, D.C., pp. 111-130.

Moreno, E., M.W. Pitt, L.M. Jones, G.G. Schurig and D.T. Berman. 1979. Purification and characterization of smooth and rough lipopolysaccharides from *Brucella abortus*. J. Bacteriol. 138: 361-369.

Moreno, E., S.L. Speth, L.M. Jones and D.T. Berman. 1981. Immunochemical characterization of *Brucella* lipopolysaccharides and polysaccharides. Infect.

Immun. 31: 214-222.

Moreno-López, M. 1952. El genero *Bordetella*. Microbiol. Esp. 5: 117-181.

Morgan, A.T. 1982. Isolation and characterization of *Pseudomonas aeruginosa* R' plasmids constructed by interspecific mating. J. Bacteriol. 149: 654-661.

Morgan, H. de R. 1906. Upon the bacteriology of the summer diarrhoea of infants. Brit. Med. J. 1: 908-912.

Morgan, W.J.B. and S.G.M. Gower. 1966. Techniques in the identification and classification of *Brucella. In* Gibbs and Skinner (Editors), Identification Methods of Microbiologists, Academic Press, London, pp. 35-50.

Morgenstein, A.A., D.M. Citron, B. Orisek and S.M. Finegold. 1980. Serious infection with *Leptotrichia buccalis*. Report of a case and review of the literature. Am. J. Med. 69: 782-785.

Morihara, K. 1964. Production of elastase and proteinase by *Pseudomonas aeruginosa*. J. Bacteriol. 88: 745-757.

Morihara, K., H. Tsuzuki, T. Oka, H. Inoue and M. Ebata. 1965. *Pseudomonas aeruginosa* elastase. Isolation, crystallization, and preliminary characterization. J. Biol. Chem. 240: 3295-3304.

Morinaga, Y., S. Yamanaha, S. Otsuka and Y. Hirose. 1976. Characteristics of a newly isolated methane-oxidizing bacterium, *Methylomonas flagellata* nov. sp. Agric. Biol. Chem. 40: 1539-1545.

Morita, R. Y. 1975. Psychrophilic bacteria. Bacteriol. Rev. 39: 144-167.

Morowitz, H.J. 1969. The genome of mycoplasmas. *In* Hayflick (Editor), The Mycoplasmatales and the L Phase of Bacteria, Appleton-Century-Crofts, New York, pp. 405-412.

Morowitz, H.J. and D.C. Wallace. 1973. Genome size and life cycle of the mycoplasma. Ann. N.Y. Acad. Sci. 225: 62-73.

Morris, E.R., D.A. Rees, G. Young, M.D. Walkinshaw and A. Darke. 1977. Order-disorder transition for a bacterial polysaccharide in solution. A role for polysaccharide conformation in recognition between *Xanthomonas* pathogen and its plant host. J. Mol. Biol. 110: 1-16.

Morris, G.K., C.M. Patton, J.C. Feeley, S.E. Johnson, G. Gorman, W.T. Martin, P. Skaliy, G.F. Mallison, B.D. Politi and D.C. Mackel. 1979. Isolation of Legionnaires' disease bacterium from environmental samples. Ann. Intern. Med. 90: 664-666.

Morris, G.K., A. Steigerwalt, J.C. Feeley, E.S. Wong, W.T. Martin, C.M. Patton and D.J. Brenner. 1980. *Legionella gormanii* sp. nov. J. Clin. Microbiol. 12: 718-721.

Morris, G.P. and D.W. Halton. 1975. The occurrence of bacteria and mycoplasma-like organisms in a monogenean parasite, *Diclidophora merlangi*. Int. J. Parasitol. 5: 495-498.

Morris, G.W. 1979a. Microbial isolations from *Sitophilus zeamais* (Coleoptera: Curculionidae). Proc. Entomol. Soc. Ont. 110: 93-96.

Morris, G.W. 1979b. An attempt to produce aposymbiotic *Sitophilus zeamais* (Coleoptera: Curculionidae) by rearing at 33°C. Proc. Entomol. Soc. Ont. 110: 107-108.

Morris, J.A. 1973. The use of polyacrylamide gel electrophoresis in taxonomy of *Brucella*. J. Gen. Microbiol. 76: 231-237.

Morris, J.A. and M.J. Corbel. 1973. Properties of a new phage lytic for *Brucella suis*. J. Gen. Virol. 21: 539-544.

Morris, M.B. and J.B. Roberts. 1959. A group of pseudomonads able to synthesize poly-beta-hydroxybutyric acid. Nature 183: 1538-1539.

Morse, E.V., M. Ristic, G.W. Robertstad and D.W. Schneider. 1953. Cross-agglutination reactions among *Brucella, Vibrio* and other microorganisms. Am. J. Vet. Res. 14: 324-327.

Morse, J.H. and S.I. Morse. 1970. Studies on the ultrastructure of *Bordetella pertussis*. I. Morphology, origin, and biological activity of structures present in the extracellular fluid of liquid cultures of *Bordetella pertussis*. J. Exp. Med. 131: 1342-1357.

Morse, S.A. 1980. Sexually transmitted diseases. *In* Lennette, Balows, Hausler and Truant (Editors) Manual of Clinical Microbiology, 3rd Ed., American Society for Microbiology, Washington, D.C.

Morse, S.A., A.F. Cacciapuoti and P.G. Lyski. 1979. Physiology of *Neisseria gonorrhoeae. In* Rose and Morris (Editors), Advances in Microbial Physiology. 20: 251-320.

Morse, S.I. 1976. Biologically active components and properties of *Bordetella pertussis. In* Perlman (Editor), Advances in Applied Microbiology, Vol. 20, Academic Press, New York, pp. 9-26.

Morse, S.I. and J.H. Morse. 1976. Isolation and properties of the leukocytosis- and lymphocytosis-promoting factor of *Bordetella pertussis*. J. Exp. Med. 143: 1483-1502.

Morton, H.E. 1966. *Mycoplasma*-latex agglutination reaction. J. Bacteriol. 92: 1196-1205.

Moshkovski, C. 1937. Sur l'existence chez le cobaye, d'une rickettsiose chronique determinee par *Ehrlichia (Rickettsia) kurlovi* subg. nov. sp. nov. C.R. Soc. Biol. (Paris) 126: 379-382.

Moshkovski, S.D. 1945. Cytotropic inducers of infection and the classification of the *Rickettsiae* with *Chlamydozoa* (in Russian, English summary). Adv. Mod. Biol. (Moscow) 19: 1-44.

Moshkovski, S.D. 1947. Comments by readers. Science (Washington) 106: 62.

Mosing, H. 1936. Une nouvelle infection a *Rickettsia: Rickettsia weigli* nov. sp. Arch. Inst. Pasteur Tunis 25: 373-387.

Moss, C.W. and S.B. Dees. 1978. Cellular fatty acids of *Flavobacterium meningosepticum* and *Flavobacterium* species group IIb. J. Clin. Microbiol. 8: 772-774.

Moss, C.W. and S.B. Dees. 1979. Cellular fatty acid composition of WIGA, a rickettsia-like agent similar to the Legionnaires' disease bacterium. J. Clin. Microbiol. *10:* 390-391.

Moss, C.W. and W.E. Dunkelberg, Jr. 1969. Volatile and cellular fatty acids of *Haemophilus vaginalis.* J. Bacteriol. *100:* 544-546.

Moss, C.W., S.B. Samuels, J. Liddle and R.M. McKinney. 1973. Occurrence of branched-chain hydroxy fatty acids in *Pseudomonas maltophilia.* J. Bacteriol. *114:* 1018-1024.

Moss, C.W., R.E. Weaver, S.B. Dees and W.B. Cherry. 1977. Cellular fatty acid composition of isolates from Legionnaires' disease. J. Clin. Microbiol. *6:* 140-143.

Moss, F.J. and Y.T. Tchan. 1958. Studies of nitrogen-fixing bacteria. VII. Cytochromes of *Azotobacteraceae.* Proc. Linn. Soc. N.S.W. *83:* 161-164.

Moss, J. and M. Vaughn. 1979. Activation of adenylate cyclase by choleragen. Annu. Rev. Biochem. *48:* 581-600.

Moss, M.O. and C. Ryall. 1981. The genus *Chromobacterium.* In Starr, Stolp, Trüper, Balows and Schlegel (Editors), The Prokaryotes, A handbook on habitats, isolation, and identification of bacteria. Springer-Verlag, Berlin, pp. 1355-1364.

Moss, M.O., C. Ryall and N.A. Logan. 1978. The classification and characterization of chromobacteria from a lowland river. J. Gen. Microbiol. *105:* 11-21.

Moss, M.O., C. Ryall and N.A. Logan. 1981. *In* Validation of the publication of new names and new combinations previously effectively published in the IJSB. List No. 6. Int. J. Syst. Bacteriol. *31:* 215-218.

Mosse, B. 1970. Honey-coloured, sessile *Endogone* spores: II. Changes in fine structure during spore development. Arch. Microbiol. *74:* 129-145.

Mossie, K.G., D.T. Jones, F.T. Robb and D.R. Woods. 1979. Characterization and mode of action of a bacteriocin produced by a *Bacteroides fragilis* strain. Antimicrob. Agents Chemother. *16:* 724-730.

Mouches, C., J.M. Bové, J. Albisetti, T.B. Clark and J.G. Tully. 1982. A spiroplasma of serogroup IV causes a May-disease-like disorder of honeybees in Southwestern France. Microbial Ecol. *8:* 387–399.

Mouches, C., J.M. Bové, J.G. Tully, D.L. Rose, R.E. McCoy, P. Carle-Junca, M. Garnier and C. Saillard. 1983. *Spiroplasma apis,* a new species from the honey bee (*Apis mellifera*). Ann. Microbiol. (Inst. Pasteur) *134A:* 383–397.

Mouches, C., A. Menara, B. Geny, D. Charlemagne and J.M. Bové. 1982. Synthesis of *Spiroplasma citri* protein specifically recognized by rabbit immunoglobulin to rabbit actin. Rev. Infect. Dis. *4:* 5277.

Mouches, C., D. Taylor-Robinson, L. Stipkovits and J.M. Bové. 1981. Comparison of human and animal ureaplasmas by one and two-dimensional protein analysis on polyacrylamide slab gel. Ann. Microbiol. (Inst. Pasteur) *132B:* 171-196.

Mouches, C., J.C. Vignault, J.G. Tully, R.F. Whitcomb and J.M. Bové. 1979. Characterization of spiroplasmas by one- and two-dimensional protein analysis on polyacrylamide slab gels. Curr. Microbiol. *2:* 69-74.

Moulder, J.W., N.J. Levy and L.P. Schulman. 1980. Persistent infection of mouse fibroblasts (L cells) with *Chlamydia psittaci.* Evidence for a cryptic chlamydial form. Infect. Immun. *30:* 874-883.

Moulton, J.E. and J.F. Christensen. 1955. The histochemical nature of *Anaplasma marginale.* Am. J. Vet. Res. *16:* 377-380.

Mount, M.S., D.F. Bateman and H.G. Basham. 1970. Induction of electrolyte loss, tissue maceration, and cellular death of potato tissue by an endopolygalacturonate trans-eliminase. Phytopathology *60:* 924-931.

Mouton, C., P. Hammond, J. Slots and R. Genco. 1980. Evaluation of Fluoretec-M for detection of oral strains of *Bacteroides asaccharolyticus* and *Bacteroides melaninogenicus.* J. Clin. Microbiol. *11:* 682-686.

Mráz, O. 1963. Schizomycetes. *In* Mráz, Tesarcik and Varejka (Editors), Nomina und Synonyma der Pathogenen und Saprophytaren Mikroben, Isoliert aus den Wirtschaftlich oder Epidemiologischen Bedeutenden Wirbeltieren und Lebensmitteln Tierischer Herkunft. VEB Gustav Fischer Verlag, Jena, pp. 53-334.

Mráz, O. 1968. Reevaluation of original strains *Actinobacillus suis* and haemolytic strains *Actinobacillus lignieresii* ATCC isolated from organs of diseased pigs. Acta Univ. Agric. Fac. Vet. *37:* 277-290.

Mráz, O. 1969a. Vergleichende Studie der Arten *Actinobacillus lignieresii* und *Pasteurella haemolytica.* I. *Actinobacillus lignieresii* Brumpt, 1910; emend. Zentralbl. Bakteriol. Parasitenkd. Infektionskr. Hyg. Abt. 1, Orig. *209:* 212-232.

Mráz, O. 1969b. Vergleichende Studie der Arten *Actinobacillus lignieresii* und *Pasteurella haemolytica.* III. *Actinobacillus haemolyticus* (Newsom und Cross, 1932) comb. nov. Zentralbl. Bakteriol. Parasitenkd. Infektionskr. Hyg. Abt. 1, Orig. *209:* 349-364.

Mráz, O. 1975. Differentiation possibilities between Pasteurellae and Actinobacilli. Acta Vet. Brno *44:* 105-113.

Mráz, O. 1977. Antigenní vztahy mezi pasteurelami a aktinobacily. Vet. Med. (Prague) *22:* 121-132.

Mráz, O., P. Jelen and J. Bohacek. 1977. On species characteristics of *Pasteurella gallinarum.* Acta Vet. (Brno) *46:* 135-147.

Mráz, O., P. Vladik and J. Boháček. 1976. Actinobacilli in domestic fowl. Zentralbl. Bakteriol. Parasitenkd. Infektionskr. Hyg. Abt. I, Orig. A *236:* 294-307.

Mudd, J.B., M. Ittig, B. Roy, J. Latrille and J.M. Bove. 1977. Composition and enzyme activities of *Spiroplasma citri* membranes. J. Bacteriol. *129:* 1250-1256.

Mukerjee, S. 1978. Principles and practice of typing *Vibrio cholerae.* In Bergan

and Norris (Editors), Methods in Microbiology, Vol. 12, Academic Press Inc., London-New York-San Francisco, pp. 51-115.

Mukerjee, S., D.K. Guha and U.K. Guha Roy. 1957. Studies on typing of cholera by bacteriophage. Part 1. Phage-typing of *Vibrio cholerae* from Calcutta epidemics. Ann. Biochem. Exp. Med. *17:* 161-176.

Mukoo, H. 1955. On the bacterial blacknode of barley and wheat and its causal bacteria. *In* Jubilee Publication in Commemoration of the Sixtieth Birthdays of Prof. Yoshihiko Tochinai and Prof. Teikichi Fukushi. Sapporo, Japan, pp. 153-157.

Mulks, M.H., S.J. Kornfeld, B. Frangione and A.G. Plaut. 1982. Relationship between the specificity of IgA proteases and serotypes of *Haemophilus influenzae.* J. Infect. Dis. *146:* 266–174.

Mullakhanbhai, M.F. and H. Larsen. 1975. *Halobacterium volcanii* spec. nov., a Dead Sea halobacterium with a moderate salt requirement. Arch. Microbiol. *104:* 207-214.

Müller, G. 1977. Non-agglutinable cholera vibrios (NAG) in sewage, riverwater, and seawater. Zentralbl. Bakteriol. Parasitenkd. Infektionskr. Hyg. Abt. I Orig. B *165:* 487-497.

Müller, H.E. 1972. The aerobic fecal flora of reptiles with special reference to the enterobacteria of snakes. Zentralbl. Bakteriol. Parasitenk. Infektionskr. Hyg. Abt. I. Orig. *222:* 487-495.

Müller, H.E. and K.-H. Hinz. 1977. Über das Vorkommen von Neuraminidase und N-Acetyl-neuraminat-Pyruvat-Lyase bei human-pathogenen *Haemophilus*-Arten. Zentralbl. Bakteriol. Parasitenkd. Infektionskr. Hyg. Abt. I, Orig. A *239:* 231-239.

Müller, H.E. and H. Werner. 1970. In vitro-untersuchungen über das vorkommen von neuraminidase bei *Bacteroides*-arten. Pathol. Microbiol. *36:* 135-152.

Müller, H.J. 1956. Experimentelle Studien an der Symbiose von *Coptosoma scutellatum* Geoffr. (Hem. Heteropt.) Z. Morphol. Oekol. Tiere *44:* 459-482.

Müller, J. 1856. Beobachtungen an Infusorien. Monatsber. Dtsch. Akad. Wiss. Berlin *1856:* 389-393.

Müller, J. 1969. Untersuchungen über die intrazellulare Symbiose einiger Aetalionidae, Eurymelidae und Cicadellidae (Homoptera-Auchenorrhyncha). Zool. Jahrb. Abt. Syst. Oekol. Geogr. Tiere *96:* 558-608.

Müller, J. 1972. Die intrazellulare Symbiose der Zikaden mit Mikroorganismen. Biol. Rundsch. *10:* 46-57.

Muller, L. 1923. Un nouveau milieu d'enrichissement pour la recherche des bacilles typhiques et paratyphiques. C. R. Soc. Biol. *89:* 434-437.

Müller, O.F. 1773. Vermium Terrestrium et Fluviatilium, seu Animalium Infusoriorum, Helminthicorum et Testaceorum, non Marionorum, Succincta Historia *1 (1):* 1-135.

Müller, O. F. 1786. Animalcula Infusoria Fluviatilia et Marina, Quae Detexit, Systematice Descripsit et Ad Vivum Delineari Curavit, pp. 1-367.

Müller-Kogler, E. 1958. Eine Rickettsiose von *Tipula paludosa* Meig. durch *Rickettsiella tipulae* nov. spec. Naturwissenschaften *45:* 248-250.

Mulongoy, K. and G.H. Elkan. 1977. Glucose catabolism in two derivatives of a *Rhizobium japonicum* strain differing in nitrogen-fixing efficiency. J. Bacteriol. *131:* 179-187.

Mundim, M.H. and I. Roitman. 1977. Extra nutritional requirements of artificially produced aposymbiotic *Crithidia deanei.* J. Protozool. *24:* 329-331.

Mundim, M.H., I. Roitman, M.A. Hermans and E.W. Kitajima. 1974. Simple nutrition of *Crithidia deanei,* a reduviid trypanosomatid with an endosymbiont. J. Protozool. *21:* 518-521.

Munk, F. and H. da Rocha-Lima. 1917. Klinik und Aetiologie des sogennanten "wolhinischen Fiebers" (Werner-Hissche Krankheit). II. Ergebnis der aetiologischen Untersuchungen und deren Beziehungen zür Fleckfieberforschung. Münch. Med. Wochschr. *64:* 1422-1426.

Munoz, J.J., H. Arai and R.L. Cole. 1981. Mouse-protecting and histamine-sensitizing activities of pertussigen and fimbrial hemagglutinin from *Bordetella pertussis.* Infect. Immun. *32:* 243-250.

Muraschi, T.F., M. Friend and D. Bolles. 1965. *Erwinia*-like microorganisms isolated from animal and human hosts. Ann. Microbiol. *13:* 128-131.

Murase, N., A. Juraku, J. Tien, F. Shimizu and H. Ashizawa. 1952. Epizootiological observations on glanders of horses. II. Studies on the relation between immunological reactions and pathological observations. Jpn. J. Vet. Sci. *14:* 187-188.

Murata, N. and M.P. Starr. 1974. Intrageneric clustering and divergence of *Erwinia* strains from plants and man in the light of desoxyribonucleic and segmental homology. Can. J. Microbiol. *20:* 1545-1565.

Murdock, L.L., T.L. Hopkins and R.A. Wirtz. 1970. Phenylalanine metabolism in cockroaches, *Periplaneta americana:* intracellular symbionts and aromatic ring cleavage. Comp. Biochem. Physiol. *34:* 143-146.

Murphy, I.A., J.W. Osebold and O. Aalund. 1966. Kinetic of the antibody response to *Anaplasma marginale* injection. J. Infect. Dis. *116:* 99-111.

Murray, B.E. and R.C. Moellering. 1979. Aminoglycoside-modifying enzymes among clinical isolates of *Acinetobacter calcoaceticus* subs. *anitratus* (*Herellea vaginicola*): Explanation for high-level aminoglycoside resistance. Antimicrob. Agents Chemother. *15:* 190-199.

Murray, E.G.D. 1939. Family *Neisseriaceae.* In Bergey, Breed, Murray and Hitchens (Editors), Bergey's Manual of Determinative Bacteriology, 5th Ed., The Williams and Wilkins Co., Baltimore, pp. 278-288.

Murray, E.G.D. 1939. Family *Parvobacteriaceae.* In Bergey, Breed, Murray and Hitchens (Editors), Bergey's Manual of Determinative Bacteriology, 5th Ed., The Williams and Wilkins Co., Baltimore, p. 309.

Murray, E.G.D. 1948. Genus II. *Moraxella* Lwoff. In Breed, Murray, and Hitchens

(Editors), Bergey's Manual of Determinative Bacteriology, 6th Ed., The Williams and Wilkins Co., Baltimore, pp. 590-592.

Murray, E.S. and S.B. Torrey. 1975. Virulence of *Rickettsia prowazeki* for head lice. Ann. N.Y. Acad. Sci. *266:* 25-34.

Murray, R.G.E. 1962. Fine structure and taxonomy of bacteria. *In* Ainsworth and Sneath (Editors), Microbial Classification. Cambridge University Press, Cambridge.

Murray, R.G.E. 1963. Role of superficial structures in the characteristic morphology of *Lampropedia hyalina.* Can. J. Microbiol. *9:* 593-600.

Murray, R.G.E. 1968. Microbial structure as an aid to microbial classification and taxonomy. Spisy (Faculte des Sciences de l'Universite J.E. Purkyne, Brno) *43:* 249-252.

Murray, R.G.E. 1974. A place for bacteria in the living world. *In* Buchanan and Gibbons (Editors), Manual of Determinative Bacteriology, 8th Ed., The Williams and Wilkins Co., Baltimore, pp. 4-9.

Murray, R. G. E. and A. Birch-Andersen. 1963. Specialized structure in the region of the flagella tuft in *Spirillum serpens.* Can. J. Microbiol. *9:* 393-401.

Murray, R. G. E., P. Steed and H. E. Elson. 1965. The location of the mucopeptide in sections of the cell wall of *Escherichia coli* and other Gram-negative bacteria. Can. J. Microbiol. *11:* 547-560.

Musgrave, A.J., G.C. Ashton and R. Homan. 1963. Quantitative and qualitative effects of temperature and type of grain on populations of *Sitophilus* (Coleoptera: Curculionidae) and on their mycetomal microorganisms. Can. J. Zool. *41:* 1245-1261.

Musgrave, A.J. and I. Grinyer. 1968. Membranes associated with the disintegration of mycetomal micro-organisms in *Sitophilus zeamais* (Mots.) (Coleoptera). J. Cell Sci. *3:* 65-70.

Musgrave, A.J. and R. Homan. 1962. *Sitophilus sasakii* (Tak.) (Coleoptera: Curculionidae) in Canada: anatomy and mycetomal symbiotes as valid taxonomic characters. Can. Entomol. *94:* 1196-1197.

Musgrave, A.J., R. Homan and I. Grinyer. 1964. Mycetomal and other microorganisms in young and aging *Sitophilus* (Coleoptera: Curculionidae). Can. J. Microbiol. *10:* 806-808.

Musgrave, A.J. and L.A. McDermott. 1961. Some media used in an attempt to isolate and culture the mycetomal microorganisms of *Sitophilus* weevils. Can. J. Microbiol. *7:* 842-843.

Musgrave, A.J. and J.J. Miller. 1956. Some micro-organisms associated with the weevils *Sitophilus granarius* (L.) and *Sitophilus oryza* (L.), (Coleoptera). II. Population differences of mycetomal and micro-organisms in different strains of *S. granarius.* Can. Entomol. *88:* 97-100.

Musgrave, A.J., H.A.U. Monro and E. Upitis. 1965. Apparent elimination of symbiotes in successive generations of *Sitophilus* (Coleoptera) fumigated with methyl bromide. J. Invertebr. Pathol. *7:* 506-511.

Musgrave, A.J. and S.B. Singh. 1965. Histochemical evidence of nuclear equivalents in mycetomal microorganisms in *Sitophilus granarius* (Linnaeus). J. Invertebr. Pathol. *7:* 269-270.

Musher, D.M., D.P. Griffith, D. Yawn and R.D. Rossen. 1975. Role of urease in pyelonephritis resulting from urinary tract infection with Proteus. J. Infect. Dis. *131:* 177-181.

Muyembe, T., J. Vandepitte and J. Desmyter. 1973. Natural colistin resistance in *Edwardsiella tarda.* Antimicrob. Agents Chemother. *4:* 521-524.

Myers, W.F., L.D. Cutler and C.L. Wisseman, Jr. 1969. Role of erythrocytes and serum in the nutrition of *Rickettsia quintana.* J. Bacteriol. *97:* 663-666.

Myers, W.F., J.V. Osterman and C.L. Wisseman, Jr. 1972. Nutritional studies of *Rickettsia quintana*: Nature of the hematin requirement. J. Bacteriol. *109:* 89-95.

Myers, W.F. and C.L. Wisseman, Jr. 1980. Genetic relatedness among the typhus group of rickettsiae. Int. J. Syst. Bacteriol. *30:* 143-150.

Myers, W.F. and C.L. Wisseman, Jr. 1981. The taxonomic relationship of *Rickettsia canada* to the typhus and spotted fever groups of the genus *Rickettsia.* *In* Burgdorfer and Anacker (Editors), Rickettsiae and Rickettsial Diseases, Academic Press, New York, pp. 313-325.

Myers, W.F., C.L. Wisseman, Jr., P. Fiset, E.V. Oaks and J.F. Smith. 1979. Taxonomic relationship of vole agent to *Rochalimaea quintana.* Infect. Immun. *26:* 976-983.

Mylroie, J.R., D.A. Friello and A.M. Chakrabarty. 1978. Transformation of *Pseudomonas putida.* Biochem. Biophys. Res. Comm. *82:* 281-288.

Mylroie, J.R., D.A. Friello, T.V. Siemens and A.M. Chakrabarty. 1977. Mapping of *Pseudomonas putida* chromosome genes with recombinant sex-factor plasmid. Mol. Gen. Genet. *157:* 231-237.

Năcescu, N. and C. Ciufecu. 1978. Serotypes of NAG vibrios isolated from clinical and environmental sources. Zentralbl. Bakteriol. Parasitenkd. Infektionskr. Hyg. Abt. I Orig. A *240:* 334-338.

Naemura, L.G., S.T. Bagley, R.J. Seidler, J.B. Kaper and R.R. Colwell. 1979. Numerical taxonomy of *Klebsiella pneumoniae* strains isolated from clinical and nonclinical sources. Curr. Microbiol. *2:* 175-180.

Nagarkoti, M.S., A.K. Banerjee and J. Swarup. 91973. *Xanthomonas convolvuli* spec. nov. causing leaf spot of *Convolvulus arvensis* in India. Indian J. Mycol. Plant Pathol. *3:* 105.

Nagatomo, H., H. Kotani, M. Ogata and T. Shimizu. 1980. Serological studies of bovine ureaplasmas by agar-gel precipitation test. Jpn. J. Vet. Sci. *42:* 9-17.

Nagayo, M., T. Tamiya, T. Mitamura and K. Sato. 1930. On the virus of tsutsugamushi disease and its demonstration by a new method. Jpn. J. Exp. Med. *8:* 309-318.

Nagesh, K.G., K.P. Poulose and G.M. Rao. 1977. Liver abscess, *Eikenella corro-*

dens and streptococci. J. Kans. Med. Soc. *78:* 340-342.

Nagle, S.C., Jr., R.E. Anderson and N.D. Gary. 1960. Chemically defined medium for the growth of *Pasteurella tularensis.* J. Bacteriol. *79:* 566-571.

Nakada, N. and K. Takimoto. 1923. Bacterial blight of hibiscus. Ann. Phytopathol. Soc. Japan *1:* 13-19.

Nakamura, L.K. and D.D. Tyler. 1977. Induction of D-aldohexoside:cytochrome *c* oxidoreductase in *Agrobacterium tumefaciens.* J. Bacteriol. *129:* 830-835.

Nakanishi, H., Y. Iida, K. Maeshima, T. Teramoto, Y. Hosaka and M. Ozaki. 1966. Isolation and properties of bacteriophages of *Vibrio parahaemolyticus.* Biken J. *9:* 149-157.

Nakanishi, I., K. Kimura, T. Suzuki, M. Ishikawa, I. Banno, T. Sakane and T. Harada. 1976. Demonstration of curdlan-type polysaccharide and some other β-1,3-glucan in microorganisms with aniline blue. J. Gen. Appl. Microbiol. *22:* 1-11.

Nakanishi, H., L. Leistner and H. Hechelmann. 1969. Das Vorkommen von enteropathogenen, gramnegativen Stäbchen in Patientenstühlen. Fleischwirtschaft *49:* 1501.

Nakano, K. 1919. Soybean leaf spot. J. Plant Prot. *6:* 217-221.

Nakase, Y. 1957. Studies on *Hemophilus bronchisepticus.* IV. Serological relation of *H. bronchisepticus* from guinea pig, dog and human. Kitasoto Arch. Exp. Med. *30:* 85-94.

Nakayama, T. and J. De Ley. 1965. Localisation and distribution of alcohol-cytochrome 553 reductase in acetic acid bacteria. Antonie van Leeuwenhoek J. Microbiol. Serol. *31:* 205-219.

Nakazawa, H., H. Enei, S. Okumura and H. Yamada. 1972. Synthesis of L-tryptophane from pyruvate and indole. Agric. Biol. Chem. *36:* (*Supplement 13*): 2523-2528.

Nakazawa, M., Y. Kagemori and R. Azuma. 1979. Serological variants of *Actinobacillus lignieresii* in slaughtered cattle. Jpn. J. Vet. Sci. *41:* 89-90.

Nakimovskaya, M.I. 1948. *Pseudomonas aurantiaca* nov. sp. Mikrobiologiya *17:* 58-65.

Namioka, S. and R. Sakazaki. 1958. Étude sur les *Rettgerella.* Ann. Inst. Pasteur (Paris) *94:* 485-499.

Namioka, S. and K. Sakazaki. 1959. New K antigen (C antigen) possessed by *Proteus* and *Rettgerella* cultures. J. Bacteriol. *78:* 301-306.

Namsaraev, B.B. 1973. Growth of *Microcyclus* on methanol. Microbiology (English translation) *42:* 986-987.

Namsaraev, B.B. and A.N. Nozhevnikova. 1978. Autotrophic growth of *Microcyclus aquaticus* in an atmosphere of hydrogen. Microbiology (English translation) *47:* 315-318.

Nandadasa, H.G., M. Andreesen and H.G. Schlegel. 1974. The utilization of 2-ketogluconate by *Hydrogenomonas eutropha* H 16. Arch. Microbiol. *99:* 15-23.

Naot, Y. 1982. In vitro studies on mitogenic activity of mycoplasmas towards lymphocytes. Rev. Infect. Dis. *Suppl. 4:* S205-S209.

Napoli, C., R. Sanders, R. Carlson and P. Albersheim. 1980. Host-symbiont interactions: recognizing *Rhizobium.* *In* Newton and Orme-Johnson (Editors), Nitrogen Fixation, Vol. II, University Park Press, Baltimore, pp. 189-203.

Nara, T., R. Okachi and M. Misawa. 1971. Enzymatic synthesis of D(-)-α-aminobenzylpenicillin by *Kluyvera citrophila.* J. Antibiot. (Tokyo) *24:* 321-323.

Nardon, P. 1971. Contribution a l'étude des symbiotes ovariens de *Sitophilus sasakii*: localisation, histochimie et ultrastructure chez la femelle adulte. C.R. Hebd. Séances Acad. Sci. Ser. D Sci. Nat. *272:* 2975-2978.

Nardon, P. 1973. Obtention d'une souche asymbiotique chez le charançon *Sitophilus sasakii* Tak.: differentes methodes et comparaison avec la souche symbiotique d'origine. C.R. Hebd. Séances Acad. Sci. Ser. D Sci. Nat. *277:* 981-984.

Nardon, P., C. Wicker, A.-M. Grenier and P. Laviolette. 1978. Contrôle de la proliferation des symbiotes par l'hôte: étude préliminaire de l'exo-β-N-acetyl glucosaminidase chez le curculionide *Sitophilus oryzae* L. C.R. Hebd. Séances Acad. Sci. Ser. D Sci. Nat. *287:* 1157-1160.

Nayudu, M.V. 1972. *Pseudomonas viticola* sp. nov. incitant of a new bacterial disease of grape vine. Phytopathol. Z. *73:* 183-186.

Nazly, N., I.S. Carter and C.J. Knowles. 1980. Adenine nucleotide pools during starvation of *Beneckea natriegens.* J. Gen. Microbiol. *116:* 295-303.

Neal, D.J. and S.G. Wilkinson. 1979. Lipopolysaccharides from *Pseudomonas maltophilia*: structural studies of the side-chain polysaccharide from strain NCTC 10257. Carbohydr. Res. *69:* 191-201.

Nealson, K.H. 1978. Isolation, identification, and manipulation of luminous bacteria. *In* DeLuca (Editor), Methods in Enzymology, Vol. 57. Academic Press Inc., London-New York-San Francisco, pp. 153-166.

Nealson, K.H. and J.W. Hastings. 1979. Bacterial bioluminescence: its control and ecological significance. Microbiol. Rev. *43:* 496-518.

Neil, D.H., M.M. Garcia and K.A. McKay. 1970. A stable L-form of *Haemophilus pleuropneumoniae.* Can. J. Comp. Med. *34:* 50-58.

Neill, S. D., W. A. Ellis and J. J. O'Brien. 1979. Designation of aerotolerant *Campylobacter*-like organisms from porcine and bovine abortions to the genus *Campylobacter.* Res. Vet. Sci. *27:* 180-186.

Neilson, A.H. and L. Sparell. 1976. Acetylene reduction (nitrogen fixation) by *Enterobacteriaceae* isolated from paper mill process waters. Appl. Microbiol. *32:* 197-205.

Neimark, H. 1973. Molecular evolutionary studies on mycoplasmas and acholeplasmas. Ann. N.Y. Acad. Sci. *225:* 14-21.

Neimark, H.C. 1977. Extraction of an actin-like protein from the prokaryote *Mycoplasma pneumoniae*. Proc. Natl. Acad. Sci. U.S.A. *74:* 4041-4045.

Neimark, H. 1979. Phylogenetic relationships between mycoplasmas and other prokaryotes. *In* Barile and Razin (Editors), The Mycoplasmas, Vol. I, Cell Biology, Academic Press, New York, pp. 43-61.

Neimark, H. and M.C. Tung. 1973. Properties of a fructose-1,6-diphosphate-activated lactate dehydrogenase from *Acholeplasma laidlawii* Type A. J. Bacteriol. *114:* 1025-1033.

Neish, A.C., A.C. Blackwood, F.M. Robertson and G.A. Ledingham. 1948. Production and properties of 2,3-butanediol. XXV. Dissimilation of glucose by bacteria of the genus *Serratia*. Can. J. Research *26:* 335-342.

Neitz, W.O., R.A. Alexander and P.J. du Toit. 1934. *Eperythrozoon ovis* (sp. nov.) infection in sheep. Onderstepoort J. Vet. Sci. *3:* 263-274.

Nelson, A.D., L.E. Barber, J. Tjepkema, S.A. Russell, R. Powelson and H.J. Evans. 1976. Nitrogen fixation associated with grasses in Oregon. Can. J. Microbiol. *22:* 523-530.

Nelson, E.L. and M.J. Pickett. 1951. The recovery of L-forms of *Brucella* and their relation to *Brucella* phage. J. Infect. Dis. *89:* 226-232.

Nelson, J.B. 1940. Infectious catarrh of the albino rat. J. Exp. Med. *72:* 645-654.

Nelson, L. M. and R. Knowles. 1978. Effect of oxygen and nitrate on nitrogen fixation and denitrification by *Azospirillum brasilense* grown in continuous culture. Can. J. Microbiol. *24:* 1395-1403.

Nelson, T.C. 1918. On the origin, nature, and function of the crystalline style of lamellibranchs. J. Morphol. *31:* 53-111.

Nemeč, B. 1932, Prag 1933. Über Bakteriensymbiose bei *Ardisia crispa*. Mem. Soc. Roy. Sci. Bohéme *19:* 23.

Nestorescu, N., M. Popovici, L. Szégli, A. Negut, M. Negut and E. Barbulescu. 1964. Untersuchungen einiger hackpathogèner atypischer *Citrobacter*-stamme. Zentralb. Bakteriol. Parasitol. Infektionskr. Hyg. Abt. I. Orig. *194:* 443-450.

Neter, E. and D.M. Dryja. 1981. Use of penicillin and nafcillin discs as an aid to identification of *Pasteurella multocida*. J. Med. *12:* 285-287.

Neu, H.C., C.E. Cherubin, E.D. Longo and W. Winter. 1975. Antimicrobial resistance of *Shigella* isolated in New York City in 1973. Antimicrob. Agents Chemother. *7:* 833-835.

Neveu-Lemaire, M. 1921. Précis de la Parasitologie Humaine, 5th Ed., J. Lemaire, Paris.

New, P.B. and A. Kerr. 1971. A selective medium for *Agrobacterium radiobacter* biotype 2. J. Appl. Bacteriol. *34:* 233-236.

New, P.B. and A. Kerr. 1972. Biological control of crown gall: field measurements and glass-house experiments. J. Appl. Bacteriol. 35: 279-287.

New, P.B. and Y.T. Tchan. 1982. *Azomonas macrocytogenes* (ex. Baillie, Hodgkiss, and Norris 1962, 118) nom. rev. Int. J. Syst. Bacteriol. *32:* 381.

Newsom, I.E. and F. Cross. 1932. Some bipolar organisms found in pneumonia of sheep. J. Am. Vet. Med. Assoc. *80:* 711-719.

Newton, J.W., A.G. Marr and J.B. Wilson. 1954. Fixation of $C^{14}O_2$ into nucleic acid constituents by *Brucella abortus*. J. Bacteriol. *67:* 233-236.

Neyra, C. A., J. Döbereiner, R. LaLande and R. Knowles. 1977. Denitrification by N_2 fixing *Spirillum lipoferum*. Can. J. Microbiol. *23:* 300-305.

Neyra, C. A. and P. Van Berkum. 1977. Nitrate reduction and nitrogenase activity in *Spirillum lipoferum*. Can. J. Microbiol. *23:* 306-310.

Nguyen-Van-Ai, Nguyen-Duc-Hanh, Le-Tien-Van, Nguyen-Van-Le and Nguyen-Thi Lan-Huong. 1975. Contribution a l'étude des *Edwardsiella tarda* isolés au Viet-Nam. Bull. Soc. Pathol. Exot. *68:* 355-359.

Nicholes, P.S. 1946. On the antigenic structure of *Bacterium tularense*. Ph.D. Thesis, University of Cincinnati, Cincinnati, Ohio.

Nicholls, K.M., J.V. Lee and T.J. Donovan. 1976. An evaluation of commercial Thiosulphate Citrate Bile salt Sucrose agar (TCBS). J. Appl. Bacteriol. *41:* 265-269.

Nichols, J.C. and J.B. Baseman. 1975. Carbon sources utilized by virulent *Treponema pallidum*. Infect. Immun. *12:* 1044-1050.

Nicol, C.S. and D.G. ff. Edward. 1953. Role of organisms of the pleuropneumonia group in human genital infections. Brit. J. Vener. Dis. *29:* 141-150.

Nicolet, J. 1971. Sur l'hémophilose du porc. III. Differentiation sérologique de *Haemophilus parahaemolyticus*. Zentralbl. Bakteriol. Parasitenkd. Infektionskr. Hyg. Abt. I, Orig. *216:* 487-495.

Nicolet, J. and M. Krawinkler. 1981. Polyacrylamide gel electrophoresis, a possible taxonomical tool for *Haemophilus*. *In* Kilian, Frederiksen and Biberstein (Editors), *Haemophilus, Pasteurella* and *Actinobacillus*. Academic Press, London, pp. 205-212.

Nicolet, J., P. Paroz and M. Krawinkler. 1980. Polyacrylamide gel electrophoresis of whole-cell proteins of porcine strains of *Haemophilus*. Int. J. Syst. Bacteriol. *30:* 69-76.

Nicolle, P., H.H. Mollaret and J. Brault. 1973. Recherches sur la lysogènie, la lysosensibilité, la lysotypie et la sérologie de *Yersinia enterocolitica*. Contr. Microbiol. Immunol. *2:* 54-58.

Niederhauser, J.S. 1943. A bacterial leaf spot and blight of the Russian dandelion. Phytopathology *33:* 959-961.

Niederman, R.A. and M.J. Wolin. 1972. Requirement of succinate for the growth of *Vibrio succinogenes*. J. Bacteriol. *109:* 546-549.

Niekus, H. G. D., W. De Vries and A. H. Stouthamer. 1977. The effect of different dissolved oxygen tensions on growth and enzyme activities of *Campylobacter sputorum* subspecies *bubulus*. J. Gen. Microbiol. *103:* 215-222.

Niekus, H. G. D., E. van Doorn, W. DeVries and A. H. Stouthamer. 1980a. Aerobic growth of *Campylobacter sputorum* subspecies *bubulus* with formate.

J. Gen. Microbiol. *118:* 419-428.

Niekus, H. G. D., E. van Doorn and A. H. Stouthamer. 1980b. Oxygen consumption by *Campylobacter sputorum* subspecies *bubulus* with formate as substrate. Arch. Microbiol. *127:* 137-143.

Nielsen, R. 1979. *Haemophilus parahaemolyticus* serotypes. Serological response. Nord. Vet.-Med. *31:* 401-406.

Nielsen, R. 1982. *Haemophilus pleuropneumoniae* infections in pigs. Carl Fr. Mortensen A/S, Copenhagen, p. 108.

Nienhaus, F. and R.A. Sikora. 1979. Mycoplasmas, spiroplasmas, and rickettsia-like organisms as plant pathogens. Annu. Rev. Phytopathol. *17:* 37-58.

Nigg, C. 1963. Serological studies on subclinical melioidosis. J. Immunol. *91:* 18-28.

Niitu, Y., H. Kubota, S. Hasegawa, S. Kumatsu, M. Horikawa and T. Suetake. 1974. Susceptibility of *Mycoplasma pneumoniae* to antibiotics in vitro. Jpn. J. Microbiol. *18:* 149-155.

Nikitin, D.I. 1971. A new soil microorganism - *Renobacter vacuolatum*. Gen. et Sp. n. Dokl. Akad. Nauk SSSR Ser. Biol. *2:* 296-301.

Niklewski, B. 1910. Über die wasserstoffoxydation durch Mikroorganismen, Jahrb. Wiss. Bot. *48:* 113-142.

Nilehn, B. 1969. Studies on *Yersinia enterocolitica* with special reference to bacterial diagnosis and occurrence in human enteric disease. Acta Pathol. Microbiol. Scand. Suppl. *206:* 1-48.

Nimmich, W. 1968. Zür Isolierung und qualitativen Bausteinanalyse der K-Antigène von Klebsiellen. Z. Med. Mikrobiol. Immunol. *154:* 117-131.

Nimmich, W. 1971. Über die spezifischen Polysaccharide (K-Antigène) der *Klebsiella* Typen K73-K80. Acta Biol. Med. Germ. *26:* 397-403.

Ninane, G., J. Joly and M. Kraytman. 1978. Bronchopulmonary infection due to *Branhamella catarrhalis*: 11 cases assessed by transtracheal puncture. Brit. Med. J. *1:* 276-278.

Nitzan, D., J.F. Sperry and T.D. Wilkins. 1978. Fibrinolytic activity of oral anaerobic bacteria. Arch. Oral Biol. *23:* 465-470.

Niven, D.F., P.A. Collins and C.J. Knowles. 1975. The respiratory system of *Chromobacterium violaceum* grown under conditions of high and low cyanide evolution. J. Gen. Microbiol. *90:* 271-285.

Niven, J.S.F., G.W.A. Dick, A.W. Gledhill and C.H. Andrewes. 1952. Further light on mouse hepatitis. Lancet *263:* 1061.

Nocard, E. and E. Roux. 1898. Le microbe de la péripneumonie. Ann. Inst. Pasteur (Paris) *12:* 240-262.

Noda, H. 1977. Histological and histochemical observation of intracellular yeast-like symbiotes in the fat body of the smaller brown planthopper, *Laodelphax striatellus* (Homoptera: Delphacidae). Appl. Entomol. Zool. *12:* 134-141.

Noel, K.D. and W.J. Brill. 1980. Diversity and dynamics of indigenous *Rhizobium japonicum* populations. Appl. Environ. Microbiol. *40:* 931-938.

Nogge, G. 1976. Sterility in tsetse flies (*Glossina morsitans* Westwood) caused by loss of symbionts. Experientia (Basel) *32:* 995.

Nogge, G. 1978. Aposymbiotic tsetse flies, *Glossina morsitans morsitans* obtained by feeding on rabbits immunized specifically with symbionts. J. Insect Physiol. *24:* 299-304.

Nogge, G. 1980. Elimination of symbiots of tsetse flies (*Glossina m. morsitans* Westwood) by help of specific antibodies. *In* Schwemmler and Schenk (Editors), Endocytobiology. Endosymbiosis and Cell Biology. A Synthesis of Recent Research. W. deGruyter, Berlin, New York, pp. 445-452.

Noguchi, H. 1912. Cultural studies on mouth spirochaetae (*Treponema microdentium* and *macrodentium*). J. Exp. Med. *15:* 81-89.

Noguchi, H. 1912. Pure cultivation of *Spirochaeta phagedenis* (new species), a spiral organism found in phagedenic lesions on human external genitalia. J. Exp. Med. *16:* 216-268.

Noguchi, H. 1912. *Treponema mucosum* (new species), a mucin-producing spirochaete from pyorrhea alveolaris, grown in pure culture. J. Exp. Med. *16:* 194-198.

Noguchi, H. 1913. Cultivation of *Treponema calligyrum* (n. sp.) from condylomata of man. J. Exp. Med. *17:* 89-98.

Noguchi, H. 1917. *Spirochaeta icterohaemorrhagiae* in American wild rats and its relation to the Japanese and European strains. J. Exp. Med. *25:* 755-763.

Noguchi, H. 1918. The spirochetal flora of the normal male genitalia. J. Exp. Med. *27:* 667-678.

Noguchi, H. 1918. Morphological characteristics and nomenclature of *Leptospira* (*Spirochaeta*) *icterohaemorrhagiae* (Inada and Ido). J. Exp. Med. *27:* 575-592.

Noguchi, H. 1921. *Cristispira* in North American shellfish. A note on a spirillum found in oysters. J. Exp. Med. *34:* 295-315.

Noguchi, H. 1923. Laboratory diagnosis of syphilis. Hoeber, New York.

Noguchi, H. 1928. The Spirochetes. *In* Jordan and Falk (Editors), The Newer Knowledge of Bacteriology and Immunology, University of Chicago Press, Chicago, pp. 452-497.

Noguchi, T.T., R. Nachum and C.A. Lawrence. 1963. Acute purulent meningitis caused by chromogenic *Neisseria*. A case report and literature review. Med. Art. Sci. *17:* 11-18.

Nokhal, T.H. and F. Mayer. 1979. Structural analysis of four strains of *Paracoccus denitrificans*. Antonie van Leeuwenhoek J. Microbiol. Serol. *45:* 185-197.

Noller, W. 1917. Blut- und Insektenflagellaten Zuchtung auf Platten. Arch. Schiffs- u. Tropen-Hyg. *21:* 53-94.

Normore, W.M. 1973. Guanine-plus-cytosine (GC) composition of the DNA of bacteria, fungi, algae and protozoa. *In* Laskin and Lechavalier (Editors), CRC Handbook of Microbiology, Vol II: Microbial Composition, Chemical Rubber Co. Press, Cleveland, Ohio, pp. 585-740.

Norris, D.M. 1972. Dependence of fertility and progeny development of *Xyleborus ferrugineus* upon chemicals from its symbiotes. *In* Rodriguez (Editor), Insect and Mite Nutrition. North-Holland, Amsterdam, pp. 229-310.

Norris, D.M. and H. Chu. 1980. Symbiote-dependent arrhenotokous parthenogenesis in the eukaryotic *Xyleborus*. *In* Schwemmler and Schenk (Editors), Endocytobiology. Endosymbiosis and Cell Biology. A Synthesis of Recent Research. W. deGruyter, Berlin, New York, pp. 453-460.

Norris, D.O. 1956. Legumes and the legume symbiosis. Emp. J. Exp. Agric. *24:* 247-270.

Norris, D.O. 1958. A red strain of *Rhizobium* from *Lotononis bainesii* Baker. Aust. J. Agric. Res. *9:* 629-632.

Norris, D.O. 1963. A porcelain bead method for storing *Rhizobium*. J. Exp. Agric. *31:* 255-258.

Norris, D.O. 1965. Acid production by *Rhizobium*. A unifying concept. Plant Soil *22:* 143-166.

Norris, J.R. and H.L. Jensen. 1958. Calcium requirement of *Azotobacter*. Arch. Mikrobiol. *31:* 198-205.

Norris, J.R. and W.H. Kingham. 1968. The classification of *Azotobacter*. *In* Festskrift til Hans Laurits Jensen. Demvig, Denmark, Gadgaard Nielsens Bogtrykkeri, pp. 95-105.

Norris, S.J., J.N. Miller, J.A. Sykes and T.J. Fitzgerald. 1978. Influence of oxygen tension, sulfhydryl compounds and serum on the motility and virulence of *Treponema pallidum* (Nichols strain) in a cell-free system. Infect. Immun. *22:* 689-697.

Novick, R.P. 1969. Extrachromosomal inheritance in bacteria. Bacteriol. Rev. *33:* 210-235.

Novotny, P. and J.E. Brookes. 1975. The use of *Bordetella pertussis* preserved in liquid nitrogen as a challenge suspension in the Kendrick mouse protection test. J. Biol. Stand. *3:* 11-29.

Novotny, P. and K. Cownley. 1979. Effect of growth conditions on the composition and stability of the outer membrane of *Bordetella pertussis*. *In* Manclark and Hill (Editors), International Symposium on Pertussis, U.S. Government Printing Office, Washington, D.C., pp. 99-123.

Novotny, P. and W.H. Turner. 1975. Immunological heterogeneity of pili of *Neisseria gonorrhoeae*. J. Gen. Microbiol. *89:* 87-92.

Novy, F.G. and R.E. Knapp. 1906. Studies on *Spirillum obermeiri* and related organisms. J. Int. Dis. *3:* 291-393.

Nowak, J. 1929. Morphologie, nature et cycle évolutif du microbe de la péripneumonie des bovidés. Ann. Inst. Pasteur (Paris) *43:* 1330-1352.

Nozhevnikova, A.N. and G.A. Zavarzin. 1974. On the taxonomy of CO-oxidizing Gram-negative bacteria. Izv. Akad. Nauk SSSR, Ser. Biolog. *3:* 436-440.

Nur, I., Y. Okon and Y. Henis. 1980. Comparative studies of nitrogen-fixing bacteria associated with grasses in Israel with *Azospirillum brasilense*. Can. J. Microbiol. *26:* 714-718.

Nur, I., Y. L. Steinitz, Y. Okon and Y. Henis. 1981. Carotenoid composition and function in nitrogen-fixing bacteria of the genus *Azospirillum*. J. Gen. Microbiol. *122:* 27-32.

Nuti, M.P., A.A. Lepidi, R.K. Prakash, R.A. Schilperoot and F.C. Cannon. 1979. Evidence for nitrogen fixation (nif) genes on indigenous *Rhizobium* plasmids. Nature (London) *282:* 533-535.

Nutter, J.E. 1971. Antigens of *Pasteurella tularensis*: preparative procedures. Appl. Microbiol. *22:* 44-48.

Nyberg, P.A., S.E. Knapp and R.E. Milleman. 1967. "Salmon poisoning" disease. IV. Transmission of the disease to dogs by *Nanophyetus salmincola* eggs. J. Parasitol. *53:* 694-699.

Nygaard, A.P. and B.D. Hall. 1963. A method for detection of RNA-DNA complexes. Biochem. Biophys. Res. Comm. *12:* 98-104.

Nyindo, M.B.A., D.L. Huxsoll, I. Ristic, I. Kakoma, J.L. Brown, C.A. Carson and E.H. Stephenson. 1980. Cell-mediated and humoral immune responses of German Shepherd dogs and beagles to experimental infection with *Ehrlichia canis*. Am. J. Vet. Res. *41:* 250-254.

Nyindo, M.B.A., M. Ristic, D.L. Huxsoll and A.R. Smith. 1971. Tropical canine pancytopenia: in vitro cultivation of the causative agent - *Ehrlichia canis*. Am. J. Vet. Res. *32:* 1651-1658.

Nyindo, M.B.A., M. Ristic, G.E. Lewis, Jr., D.L. Huxsoll and E.H. Stephenson. 1978. Immune response of ponies to experimental infections with *Ehrlichia equi*. Am. J. Vet. Res. *39:* 71-76.

Nyirady, S.A. 1973. The germfree culture of three species of Triatominae: *Triatoma protracta* (Uhler), *Triatoma rubida* (Uhler), and *Rhodnius prolixus* Stål. J. Med. Entomol. *10:* 417-448.

Oaks, S.C., Jr. and J.V. Osterman. 1979. The influence of temperature and pH on the growth of *Rickettsia conorii* in irradiated mammalian cells. Acta Virol. *23:* 67-72.

Oberhofer, T.R. and A.E. Back. 1979. Biotypes of *Haemophilus* encountered in clinical laboratories. J. Clin. Microbiol. *10:* 168-174.

Oberhofer, T.R., J.W. Rowen and G.F. Cunningham. 1977. Characterization and identification of gram-negative, nonfermentative bacteria. J. Clin. Microbiol. *5:* 208-220.

Obijeski, J.F., E.L. Palmer and T. Tzianabos. 1974. Proteins of purified rickettsiae. Microbios *11:* 61-76.

O'Brien, A.D., M.R. Thompson, P. Gemski, B.P. Doctor and S.B. Formal. 1977. Biological properties of *Shigella flexneri* 2a toxin and its serological relationship to *Shigella dysenteriae* 1 toxin. Infect. Immun. *15:* 796-798.

O'Brien, C.H. and R.K. Sizemore. 1979. Distribution of the luminous bacterium *Beneckea harveyi* in a semitropical estuarine environment. Appl. Environ. Microbiol. *38:* 928-933.

O'Brien, S.J., J.M. Simonson, M.W. Grabowski and M.F. Barile. 1981. Analysis of multiple isoenzyme expression among twenty-two species of *Mycoplasma* and *Acholeplasma*. J. Bacteriol. *146:* 222-232.

Ochiai, K., T. Yamanaka, K. Kimura and O. Sawada. 1959. Studies on inheritance of drug resistance between *Shigella* strains and *E. coli* strains. (In Japanese.) Nihon Ija Shimpo *1861:* 34-46.

Odriozola, J.M., E. Waitzkin, T.L. Smith and K. Bloch. 1978. Sterol requirement of *Mycoplasma capricolum*. Proc. Natl. Acad. Sci. U.S.A. *75:* 4107-4109.

Oeding, P. and H. Pederson. 1950. *Streptothrix muris ratti (Streptobacillus moniliformis)* isolated from a brain abscess. Acta Pathol. Microbiol. Scand. *27:* 436-442.

Oesterhelt, D. and W. Stoeckenius. 1974. Isolation of the cell membrane of *Halobacterium halobium* and its fractionation into red and purple membrane. *In* Fleischer and Packer (Editors) Methods in Enzymology, Vol. 31, Biomembranes Part A, Academic Press, New York, pp. 667-678.

O'Farrell, P. 1975. High resolution two-dimensional electrophoresis of proteins. J. Biol. Chem. *250:* 4007-4021.

Ogata, M., H. Atobe, H. Kushida and K. Yamamoto. 1971. In vitro sensitivity of mycoplasmas isolated from various animals and sewage to antibiotics and nitrofurans. J. Antibiot. *24:* 443-451.

Ogata, M., H. Kotani, K. Koshimizu and T. Magaribuchi. 1981. Isolation and serological characterization of ureaplasmas from nonhuman primates. Jpn. J. Vet. Sci. *43:* 521-529.

Ogata, M., H. Kotani and K. Yamamoto. 1979. Serological comparison of bovine ureaplasmas. Jpn. J. Vet. Sci. *41:* 629-637.

Ogata, N. 1931. Aetiologie der Tsutsugamuchi-Krankheit: *Rickettsia tsutsugamushi*. Zentralbl. Bakteriol. Parasitenkd. Infektionskr. Hyg. Abt. I Orig. *122:* 249-253.

Ogawa, H., A. Nakamura, R. Nakaya, K. Mise, S. Honjo, M. Takasaka, T. Fujiwara and K. Imaizumi. 1967. Virulence and epithelial cell invasiveness of dysentery bacilli. Jpn. J. Med. Sci. Biol. *20:* 315-328.

Ogg, J. E. 1962. Studies on the coccoid form of ovine *Vibrio fetus*. I. Cultural and serologic investigations. Am. J. Vet. Res. *23:* 354-358.

Ogg, J.E., S.Y. Lee and B.J. Ogg. 1979. A modified tube method for the cultivation and enumeration of anaerobic bacteria. Can. J. Microbiol. *25:* 987-990.

Ogg, J.E., T.L. Timme and M.M. Alemohammad. 1981. General transduction in *Vibrio cholerae*. Infect. Immun. *31:* 737-741.

Ogimoto, K. 1972. Über *Selenomonas* aus dem Caecum von Ratten. Zentralbl. Bakteriol. Parasitenkd. Hyg. Abt. 1 Orig. *221:* 467-473.

O'Grady, F., M.J. Lewis and N.J. Pearson. 1976. Global surveillance of antibiotic sensitivity of *Vibrio cholerae*. Bull. W.H.O. *54:* 181-185.

Ohad, I., D. Danon and S. Hestrin. 1962. Synthesis of cellulose by *Acetobacter xylinum*. V. Ultrastructure of polymer. J. Cell. Biol. *12:* 31-46.

Ohara, S., T. Sato and M. Homma. 1974. Serological studies on *Francisella tularensis, Francisella novicida, Yersinia philomiragia*, and *Brucella abortus*. Int. J. Syst. Bacteriol. *24:* 191-196.

Ohi, K., N. Takada, S. Komemushi, M. Okazaki and Y. Miura. 1979. A new species of hydrogen-utilizing bacterium. J. Gen. Appl. Microbiol. *25:* 53-58.

Ohuchi, A. and T. Tominaga. 1973. Pectolytic enzymes secreted by soft rot and saprophytic pseudomonads. Ann. Phytopathol. Soc. Japan *39:* 417-424.

Ohuchi, A. and T. Tominaga. 1975. Histochemical changes of cell walls during the macerating action by pectolytic enzyme, endo-PTE, of soft rot pseudomonad. Bull. Natl. Inst. Agric. Sci. Ser. C, No. 29: 45-63.

Okabe, N. 1933. Bacterial diseases of plants occurring in Formosa. II. Bacterial leaf spot of tomato. J. Soc. Trop. Agric., Taiwan *5:* 26-36.

Okabe, N. 1935. Bacterial diseases of plants occurring in Taiwan (Formosa). V. A bacterial disease of chicory. J. Soc. Trop. Agric., Formosa *7:* 57-66.

Okabe, N. and M. Goto. 1952. Studies on *Bacterium solanacearum* with special reference to the kinds of strains and their classification and with special reference to the pathogenicity of strains. Shizuoka Univ. Fac. Agr. Rep. *2:* 64-114.

Okabe, N. and M. Goto. 1961. Studies on *Pseudomonas solanacearum*. XI. Pathotypes in Japan. Shizuoka Univ. Fac. Agr. Rep. *11:* 25-42.

Okabe, N. and M. Goto. 1963. Bacteriophages of plant pathogens. Annu. Rev. Phytopathol. *1:* 397-418.

Okamoto, K., T. Inoue, H. Ichikawa, Y. Kawamoto and A. Miyama. 1981. Partial purification and characterization of heat-stable enterotoxin produced by *Yersinia enterocolitica*. Infect. Immunol. *31:* 554-559.

Okon, Y., S. L. Albrecht and R. H. Burris. 1976. Carbon and ammonia metabolism of *Spirillum lipoferum*. J. Bacteriol. *128:* 592-597.

Okon, Y., S. Cakmakci, I. Nur and I. Chet. 1980. Aerotaxis and chemotaxis of *Azospirillum brasilense*: a note. Microb. Ecol. *6:* 277-280.

Olander, H.J. 1963. A septicaemic disease of swine and its causative agent *Haemophilus parahaemolyticus*. Ph.D. Thesis, University of California, Davis.

O'Leary, W.M., C. Panos and G.E. Helz. 1956. Studies on the nutrition of *Bacterium salmonicida*. J. Bacteriol. *72:* 673-676.

Olijve, W. and J.J. Kok. 1979a. An analysis of the growth of *Gluconobacter oxydans* in chemostat cultures. Arch. Mikrobiol. *121:* 291-297.

Olijve, W. and J.J. Kok. 1979b. Analysis of growth of *Gluconobacter oxydans* in glucose containing media. Arch. Mikrobiol. *121:* 283-290.

Olitsky, P.K. and F.L. Gates. 1921. Experimental studies of the nasopharyngeal secretions from influenza patients. J. Exp. Med. *33:* 713-729.

Oliver, W.W. and W.B. Wherry. 1921. Notes on some bacterial parasites of the human mucous membranes. J. Infect. Dis. *28:* 341-344.

Olsen, R.H. and P. Shipley. 1973. Host range and properties of the *Pseudomonas aeruginosa* R factor R 1822. J. Bacteriol. *113:* 772-780.

Olson, L.D. and E.L. McCune. 1968. Histopathology of chronic otitis media in the rat. Lab. Anim. Care *18:* 478-485.

Olson, N.O., K.M. Kerr and A. Campbell. 1964. Control of infectious synovitis. 13. The antigen study of three strains. Avian Dis. *8:* 209-214.

Olsson, B., C.-E. Nord and T. Wadstrom. 1976. Formation of beta-lactamase in *Bacteroides fragilis*: cell bound and extracellular activity. Antimicrob. Agents Chemother. *9:* 727-735.

Olsufiev, N.G., O.S. Emelyanova and T.N. Dunayeva. 1959. Comparative study of strains of *B. tularense*. J. Hyg. Epidemiol. Microbiol. Immunol. (Praha) *3:* 138-149.

Omata, K.R. and M.N. Disraely. 1956. A selective medium for oral fusobacteria. J. Bacteriol. *72:* 677-680.

Omelianski, V.L. 1923. Aroma-producing microörganisms. J. Bacteriol. *8:* 393-419.

Omland, T. 1964. Serological Studies of *Haemophilus influenzae* and related species. VIII. Examination of ultrasonically prepared *Haemophilus* antigens by means of immunoelectrophoresis. Acta Pathol. Microbiol. Scand. *62:* 83-106.

Onderdonk, A.B., D.K. Kaspar, R.L. Cisneros and J.G. Bartlett. 1977. The capsular polysaccharide of *Bacteroides fragilis* as a virulence factor: comparison of the pathogenic potential of encapsulated and unencapsulated strains. J. Infect. Dis. *136:* 82-89.

Onderdonk, A.B., N.E. Moon, D.L. Kaspar and J.G. Bartlett. 1978. Adherence of *Bacteroides fragilis* in vivo. Infect. Immun. *19:* 1083-1087.

Onderka, D.K. and M. Kierstead. 1979. *Actinobacillus septicemia* in a black swan (*Cygnus atratus*). J. Wildl. Dis. *15:* 363-366.

Onishi, H., M.E. McCance and N.E. Gibbons. 1965. A synthetic medium for extremely halophilic bacteria. Can. J. Microbiol. *11:* 365-373.

Onogawa, T., T. Terayama, H. Zen-yoji, Y. Amano and K. Suzuki. 1976. Distribution of *Edwardsiella tarda* and hydrogen sulfide-producing *Escherichia coli* in healthy persons. Jpn. Assoc. Infec. Dis. (Tokyo) *50:* 10-17.

Ooyama, J. 1971. Simultaneous fixation of CO_2 and N_2 in the presence of H_2 and O_2 by a bacterium. Rep. Ferment. Res. Inst. *39:* 41-44.

Ooyama, J. 1976. Bacterial culture by CO_2 or CO and N_2 as sole source of C and N. *In* Schlegel, Gottschalk and Pfennig (Editors), Microbial Production and Utilization of Gases, E. Goltze K.G., Göttingen, pp. 237-245.

Opitz, R. and H.G. Schlegel. 1978. Allosteric inhibition by phosphoenolpyruvate of glucose-6-phosphate dehydrogenase from bacteria and its taxonomic importance. Biochem. Syst. Ecol. *6:* 149-155.

Oppenheimer, C. H. and H. W. Jannasch. 1962. Some bacterial populations in turbid and clear sea water near Port Aransas, Texas. Publ. Inst. Mar. Sci. Univ. Tex. *8:* 56-60. *8:* 56-60.

Orian, G. 1962. A disease of *Paspalum dilatatum* Poir in Mauritius caused by a species of bacterium closely resembling *Xanthomonas albilineans* (Ashby) Dowson. Rev. Agr. Sucr. Île Maurice *41:* 7-24.

Orla-Jensen, S. 1909. Die Hauptlinien des natürlichen Bakterien-systems. Zentralbl. Bakteriol. Parasitenkd. Infektionskr. Hyg. Abt. II *22:* 97-98, 305-346.

Orla-Jensen, S. 1919. The lactic acid bacteria. Høst, Copenhagen.

Orla-Jensen, S. 1921. The main lines of the natural bacterial system. J. Bacteriol. *6:* 263-273.

Ormsbee, R.A. 1969. Rickettsiae (as organisms). Annu. Rev. Microbiol. *23:* 275-292.

Ormsbee, R.A. 1980. Chapter 92. Rickettsiae. *In* Lennette, Balows, Hausler and Truant (Editors), Manual of Clinical Microbiology, 3rd Ed., American Society for Microbiology, Washington, pp. 922-933.

Ormsbee, R.A., E.J. Bell, D.B. Lackman and G. Tallent. 1964. The influence of phase on the protective potency of Q fever vaccine. J. Immunol. *92:* 404-412.

Ormsbee, R.A. and C.L. Larson. 1955. Studies on *Bacterium tularense* antigens. II. Chemical and physical characteristics of protective antigen preparations. J. Immunol. *74:* 359-370.

Ormsbee, R.A. and M.G. Peacock. 1964. Metabolic activity in *Coxiella burnetii*. J. Bacteriol. *88:* 1205-1210.

Ormsbee, R., M. Peacock, R. Gerloff, G. Tallent and D. Wike. 1978. Limits of rickettsial infectivity. Infect. Immun. *19:* 239-245.

Ornstein, M. 1921. Zür Bakteriologie des Schmitzbacillus. Z. Hyg. Infektionskr. *91:* 152-178.

Ornston, L.N. 1971. Regulation of catabolic pathways in *Pseudomonas*. Bacteriol. Rev. *35:* 87-116.

Orrison, L.H., W.B. Cherry and D. Milan. 1981. Isolation of *Legionella* from cooling tower water by filtration. Appl. Environ. Microbiol. *41:* 1202-1205.

Ørskov, F. and I. Ørskov. 1975. *Escherichia coli* O:H serotypes isolated from human blood. Acta Pathol. Microbiol. Scand. B *83:* 595-600.

Ørskov, I. 1957. Biochemical types in the *Klebsiella* group. Acta Pathol. Microbiol. Scand. *40:* 155-162.

Ørskov, I. 1974. The genus *Klebsiella*. *In* Buchanan and Gibbons (Editors), Bergey's Manual of Determinative Bacteriology, 8th Ed., The Williams and Wilkins Co., Baltimore, pp. 321-324.

Ørskov, I., A. Ferencz and F. Ørskov. 1980a. Tamm-Horsfall protein or uromucoid is the normal urinary slime that traps type 1 fimbriated *Escherichia coli*. Lancet *I:* 887.

Ørskov, I. and F. Ørskov. 1977. Special O:K:H serotypes among enterotoxigenic *E. coli* strains from diarrhoea in adults and children. Occurrence of the CF (colonization factor) antigen and of haemagglutinating abilities. Med. Microbiol. Immunol. *163:* 99-110.

Ørskov, I. and F. Ørskov. 1980. Significance of surface antigens in relation to enterotoxigenicity of *E. coli*. *In* Ouchterlony and Holmgren (Editors) Cholera and related diarrheas, pp. 134-141. Karger, Basel.

Ørskov, I., F. Ørskov and A. Birch-Andersen. 1980b. Comparison of *Escherichia coli* fimbrial antigen F7 with type 1 fimbriae. Infect. Immun. *27:* 657-666.

Ørskov, I., F. Ørskov, A. Birch-Andersen, M. Kanamori and C. Svanborg-Edén. 1982. O, K, H and fimbrial antigens in *Escherichia coli* serotypes associated with pyelonephritis and cystitis. Scand. J. Infect. Dis. Suppl. *33:* 18-25.

Ørskov, I., F. Ørskov, B. Jann and K. Jann. 1977. Serology, chemistry and genetics of O and K antigens of *Escherichia coli*. Bacteriol. Rev. *41:* 667-710.

Ørskov, I., F. Ørskov, W.J. Sojka and J.M. Leach. 1961. Simultaneous occurrence of *E. coli* B and L antigens in strains from diseased swine. Acta Pathol. Microbiol. Scand. *53:* 404-422.

Ørskov, J. 1928. Beschreibung eines neuen Mikroben, *Microcyclus aquaticus*, mit eigentumlicher Morphologie. Zentralbl. Bakteriol. Parasitenk. Infektionskr. Hyg. Abt. I Orig. *107:* 180-184.

Ørskov, J. 1953. *Microcyclus*. Riassunti d. communicazioni. VI. Cong. Int. Microbiol. Roma *1:* 24-25.

Osbaldiston, G.W. and R.D. Walker. 1972. Enteric actinobacillosis in calves. Cornell Vet. *62:* 364-371.

Osborne, F.H. and H.L. Ehrlich. 1976. Oxidation of arsenite by a soil isolate of *Alcaligenes*. J. Appl. Bacteriol. *41:* 295-305.

Oshima, M. and T. Yamakawa. 1972. Isolation and partial characterization of a novel glycolipid from an extremely thermophilic bacterium. Biochem. Biophys. Res. Commun. *49:* 185-191.

Oshima, T. 1975. Thermine: a new polyamine from an extreme thermophile. Biochem. Biophys. Res. Commun. *63:* 1093-1098.

Oshima, T. and K. Imahori. 1974. Description of *Thermus thermophilus* (Yoshida and Oshima) comb. nov., a non-sporulating thermophilic bacterium from a Japanese thermal spa. Int. J. Syst. Bacteriol. *24:* 102-112.

Ossipov, D.V., B.V. Gromov and K.A. Mamkaeva. 1973. Electron microscope examination of omega-particles (bacterial symbionts of the micronucleus) and nucleolar apparatus of *Paramecium caudatum* clone Mi-48. Tsitologia *15:* 97-103.

Ossipov, D.V. and I.S. Ivakhnyuk. 1972. Omega-particles--micronuclear symbiotic bacteria of *Paramecium caudatum* clone Mi-48. Tsitologia *14:* 1414-1419.

Ossipov, D.V. and S.A. Podlipaev. 1977. Electron microscope examination of early stages of infection of *Paramecium caudatum* by bacterial symbionts of the macronucleus (iota-bacteria). Acta Protozool. *16:* 289-308.

Ossipov, D.V., I.I. Skoblo and M.S. Rautian. 1975. Iota-particles, macronuclear symbiotic bacteria of ciliate *Paramecium caudatum* clone M-115. Acta Protozool. *14:* 263-280.

O'Sullivan, J. 1974a. The effect of some single amino acids on the growth of *Acetobacter aceti*. Pathol. Microbiol. *41:* 179-180.

O'Sullivan, J. 1974b. Growth inhibition of *Acetobacter aceti* by L-threonine and L-homoserine: the primary regulation of the biosynthesis of amino acids of the aspartate family. J. Gen. Microbiol. *85:* 153-159.

Otis, V.S. and J.L. Behler. 1973. The occurrence of salmonellae and *Edwardsiella* in the turtles of the New York Zoological Park. J. Wildlife Dis. *9:* 4-6.

Otta, J.D. 1977. Occurrence and characteristics of *Pseudomonas syringae* on winter wheat. Phytopathology *67:* 22-26.

Otta, J.D. and H. English. 1971. Serology and pathology of *Pseudomonas syringae*. Phytopathology *61:* 443-452.

Otto, L.A. and U. Blachman. 1979. Nonfermentative bacilli: evaluation of three systems for identification. J. Clin. Microbiol. *10:* 147-154.

Otto, L.A. and M.J. Pickett. 1976. Rapid method for identification of Gram-negative, nonfermentative bacilli. J. Clin. Microbiol. *3:* 566-575.

Otto, S.V., J.C. Harshbarger and S.C. Chang. 1977. Status of selected unicellular eucaryote pathogens, and prevalence and histopathology of inclusions containing obligate procaryote parasites, in commercial bivalve mollusks from Maryland estuaries. Haliotis *8:* 285-295.

Ottow, J.C.G. and W. Zolg. 1974. Improved procedure and colorimetric test for the detection of *ortho*- and *meta*-cleavage of protocatechuate by *Pseudomonas* isolates. Can. J. Microbiol. *20:* 1059-1061.

Ouelette, C.A., R.H. Burris and P.W. Wilson. 1969. Deoxyribonucleic acid base composition of species of *Klebsiella*, *Azotobacter* and *Bacillus*. Antony van Leeuwenhoek J. Microbiol. Serol. *35:* 275-286.

Overbeck, J. 1974. Microbiology and Biochemistry. Mitt. Internat. Verein. Limnol. *20:* 198-288.

Overturf, E.D., J. Wilkins and R. Ressler. 1974. Emergence of resistance of *Providencia stuartii* to multiple antibiotics: Speciation and biochemical characterization of *Providencia*. J. Infect. Dis. *129:* 353-357.

Owen, C.R. 1974. *Francisella*. *In* Buchanan and Gibbons (Editors), Bergey's Manual of Determinative Bacteriology, 8th Ed., The Williams and Wilkins Co., Baltimore, pp. 283-285.

Owen, C.R., E.O. Buker, W.L. Jellison, D.B. Lackman and J.F. Bell. 1964. Comparative studies of *Francisella tularensis* and *Francisella novicida* J. Bacteriol. *87:* 676-683.

Owen, R.J. and B. Holmes. 1978. Heterogeneity in the characteristics of deoxyribonucleic acid from *Flavobacterium odoratum*. FEMS Microbiol. Lett. *4:* 41-46.

Owen, R.J. and B. Holmes. 1980. Differentiation between strains of *Flavobacterium breve* and allied bacteria by comparisons of deoxyribonucleic acids. Curr. Microbiol. *4:* 7-11.

Owen, R.J. and B. Holmes. 1981. Identification and classification of *Flavobacterium* species from clinical sources. *In* Reichenbach and Weeks (Editors), The *Flavobacterium-Cytophaga* Group (Proceedings of the International Symposium on Yellow-Pigmented Gram-negative Bacteria of the *Flavobacterium-Cytophaga* Group, Braunschweig, July 8 to 11, 1980). Verlag Chemie, Weinheim, pp. 39-50.

Owen, R.J. and S.P. Lapage. 1974. A comparison of strains of King's group IIb of *Flavobacterium* with *Flavobacterium meningosepticum*. Antonie van Leeuwenhoek. J. Microbiol. Serol. *40:* 255-264.

Owen, R.J., R.M. Legros and S.P. Lapage. 1978. Base composition, size and sequence similarities of genome deoxyribonucleic acids from clinical isolates of *Pseudomonas putrefaciens*. J. Gen. Microbiol. *104:* 127-138.

Owen, R.J. and J.J.S. Snell. 1973. Comparison of group IIf with *Flavobacterium* and *Moraxella*. Antonie van Leeuwenhoek. J. Microbiol. Serol. *39:* 473-480.

Owen, R.J. and J.J.S. Snell. 1976. Deoxyribonucleic acid reassociation in the classification of flavobacteria. J. Gen. Microbiol. *93:* 89-102.

Owen, R.W., R.F. Bilton and M.E. Tenneson. 1977. The degradation of cholic acid and deoxycholic acid by *Bacteroides* species under strict anaerobic conditions. Biochem. Soc. Trans. *5:* 1711-1713.

Owens, D.R., S.L. Nelson and J.B. Addison. 1974. Isolation of *Edwardsiella tarda* from swine. Appl. Microbiol. *27:* 703-705.

Owens, J.D. and R.M. Keddie. 1969. The nitrogen nutrition of soil and herbage coryneform bacteria. J. Appl. Bacteriol. *32:* 338-347.

Oyaizu, H. and K. Komagata. 1981. Chemotaxonomic and phenotypic characterization of the strains of species in the *Flavobacterium-Cytophaga* complex. J. Gen. Appl. Microbiol. *27:* 57-107.

Ozaki, M., S. Mizushima and M. Nomura. 1969. Identification and functional characterization of the protein controlled by the streptomycin-resistant locus in *E. coli*. Nature (London) *222:* 333-339.

Pace, B. and L.L. Campbell. 1971. Homology of ribosomal ribonucleic acid of *Desulfovibrio* species with *Desulfovibrio vulgaris*. J. Bacteriol. *106:* 717-719.

Pacini, F. 1854. Osservazione microscopiche e deduzioni patologiche sul Cholera Asiatico. Gaz. Med. Ital. Toscana Firenze *6:* 405-412.

Packer, L. and W. Vishniac. 1955. Chemosynthetic fixation of carbon dioxide and characteristics of hydrogenase in resting cell suspensions of *Hydrogenomonas ruhlandii* nov. spec. J. Bacteriol. *70:* 216-223.

Padgett, P.J., W.H. Cover and N.R. Krieg. 1982. The microaerophile *Spirillum volutans*: cultivation on complex liquid and solid media. Appl. Environ. Microbiol. *43:* 469-477.

Padgett, P.J., M.W. Friedman and N.R. Krieg. 1983. Straight mutants of *Spirillum volutans* can swim. J. Bacteriol. *153:* 1543-1544.

Padhya, A.C. and M.K. Patel. 1962. A new bacterial leaf-spot on *Alangium lamarckii* Thw. Curr. Sci. *31:* 196-197.

Padhya, A.C. and M.K. Patel. 1963a. A new bacterial leaf-spot on *Ionidum heterophyllum* Went. Indian Phytopathol. *16:* 98-99.

Padhya, H.C. and M.K. Patel. 1963b. A new bacterial leaf spot on *Corchorus acutangulus* Lam. Curr. Sci. *32:* 326.

Padhya, A.C. and M.K. Patel. 1964. Bacterial leaf spot on *Triumfetta pilosa* Roth. Curr. Sci. *33:* 342.

Padhya, A.C., M.K. Patel and W.V. Kotasthane. 1965a. A new bacterial leaf spot disease of *Bauhinia racemosa* Lamk. Curr. Sci. *34:* 224-225.

Padhya, A.C., M.K. Patel and W.V. Kotasthane. 1965b. A new bacterial leaf-spot on *Vitis trifolia*. Curr. Sci. *34:* 462-463.

Pagan, J.D., J.J. Child, W.R. Scowcroft and A.H. Gibson. 1975. Nitrogen fixation by *Rhizobium* cultured in a defined medium. Nature (London) *256:* 406-407.

Page, L.A. 1961. Experimental ulcerative stomatitis in King snakes. Cornell Vet. *51:* 258-266.

Page, L.A. 1962. Acetylmethylcarbinol production and the classification of aeromonads associated with ulcerative diseases of ectothermic vertebrates. J. Bacteriol. *84:* 772-777.

Page, L.A. 1962. *Haemophilus* in chickens. I. Characteristics of 12 *Haemophilus* isolates recovered from diseased chickens. Am. J. Vet. Res. *23:* 85-95.

Page, L.A. 1966. Revision of the family Chlamydiaceae Rake (Rickettsiales): unification of the psittacosis-lymphogranuloma venereum-trachoma group of organisms in the genus *Chlamydia* Jones, Rake, and Stearns 1945. Int. J. Syst. Bacteriol. *16:* 223-252.

Page, L.A. 1968. Proposal for the recognition of two species in the genus *Chlamydia* Jones, Rake, and Stearns 1945. Int. J. Syst. Bacteriol. *18:* 51-66.

Page, L.R. and G.N. Krywolap. 1976. Determination of the deoxyribonucleic acid composition and deoxyribonucleic acid-deoxyribonucleic acid hybridization of *Fusobacterium fusiforme*, *Fusobacterium polymorphum*, and *Leptotrichia buccalis*: taxonomic considerations. Int. J. Syst. Bacteriol. *26:* 301-304.

Page, L.R. and G.N. Krywolap. 1977. Oral *Fusobacterium*, *Leptotrichia* and *Bacterionema*: I. Historical survey and taxonomic considerations. J. Baltimore Coll. Dent. Surg. *32:* 12-24.

Page, M.I. and E.O. King. 1966. Infection due to *Actinobacillus actinomycetemcomitans* and *Haemophilus aphrophilus* N. Engl. J. Med. *275:* 181-188.

Page, W.J. 1978. Transformation of *Azotobacter vinelandii* strains unable to fix nitrogen with *Rhizobium* spp. DNA. Can. J. Microbiol. *24:* 209-214.

Pagel, J.E. and P.L. Seyfried. 1976. Numerical taxonomy of aquatic *Acinetobacter* isolates. J. Gen. Microbiol. *95:* 220-232.

Pai, C.H. and V. Mors. 1978. Production of enterotoxin by *Yersinia enterocolitica*.

Infect. Immunol. *19:* 909-911.

Paine, S.G. 1919. Studies on bacteriosis. II. A brown blotch disease of cultivated mushrooms. Ann. Appl. Biol. *5:* 206-219.

Paine, S.G. and H. Stanfield. 1919. Studies in bacteriosis. III. A bacterial leaf-spot disease of *Protea cynaroides* exhibiting a host reaction of possibly bacteriolytic nature. Ann. Appl. Biol. *6:* 27-39.

Pal, S.C., G.V.S. Murty, C.G. Pandit, D.K. Murty and J.B. Shrivastav. 1967. A comparative study of enrichment media in the bacteriological diagnosis of cholera. Indian J. Med. Res. *55:* 318-324.

Palleroni, N.J. 1975. General properties and taxonomy of the genus *Pseudomonas*. *In* Clarke and Richmond (Editors), Genetics and Biochemistry of *Pseudomonas*, John Wiley and Sons, London, pp. 1-36.

Palleroni, N.J. 1977. *Pseudomonas*. *In* Laskin and Lechevalier (Editors), CRC Handbook of Microbiology, 2nd Ed., Vol. 1, CRC Press, Inc., Cleveland, Ohio, pp. 247-258.

Palleroni, N.J. 1978. The *Pseudomonas* group. Patterns of Progress, Meadowfield Press Ltd., Shildon Co., Durham, pp. 1-80.

Palleroni, N.J. 1980. Isolation and properties of a new hydrogen bacterium related to *Pseudomonas saccharophila*. J. Gen. Microbiol. *117:* 155-161.

Palleroni, N.J. 1981. Introduction to the family Pseudomonadaceae. *In* Starr, Stolp, Trüper, Balows and Schlegel (Editors), The Prokaryotes, A handbook on habitats, isolation and identification of bacteria. Springer-Verlag, Berlin, pp. 655-665.

Palleroni, N.J., R.W. Ballard, E. Ralston and M. Doudoroff. 1972. Deoxyribonucleic acid homologies among some *Pseudomonas* species. J. Bacteriol. *110:* 1-11.

Palleroni, N.J. and M. Doudoroff. 1956. Mannose isomerase of *Pseudomonas saccharophila*. J. Biol. Chem. *218:* 535-548.

Palleroni, N.J. and M. Doudoroff. 1965. Identity of *Pseudomonas saccharophila*. J. Bacteriol. *89:* 264.

Palleroni, N.J. and M. Doudoroff. 1971. Phenotypic characterization and deoxyribonucleic acid homologies of *Pseudomonas solanacearum*. J. Bacteriol. *107:* 690-696.

Palleroni, N.J. and M. Doudoroff. 1972. Some properties and taxonomic subdivisions of the genus *Pseudomonas*. Annu. Rev. Phytopathol. *10:* 73-100.

Palleroni, N.J., M. Doudoroff, R.Y. Stanier, R.E. Solanes and M. Mandel. 1970. Taxonomy of the aerobic pseudomonads: the properties of the *Pseudomonas stutzeri* group. J. Gen. Microbiol. *60:* 215-231.

Palleroni, N.J. and B. Holmes. 1981. *Pseudomonas cepacia* sp. nov., nom. rev. Int. J. Syst. Bacteriol. *31:* 479-481.

Palleroni, N.J., R. Kunisawa, R. Contopoulou and M. Doudoroff. 1973. Nucleic acid homologies in the genus *Pseudomonas*. Int. J. Syst. Bacteriol. *23:* 333-339.

Palleroni, N.J. and A.V. Palleroni. 1978. *Alcaligenes latus*, a new species of hydrogen-utilizing bacteria. Int. J. Syst. Bacteriol. *28:* 416-424.

Palleroni, N.J. and R.Y. Stanier. 1964. Regulatory mechanisms governing synthesis of the enzymes for tryptophan oxidation in *Pseudomonas fluorescens*. J. Gen. Microbiol. *35:* 319-334.

Pammel, L.H. 1895. Bacteriosis of rutabaga, (*Bacillus campestris* n. sp.). Iowa State Coll. Agr. Exp. Sta. Bull. *27:* 130-134.

Panagopoulos, C.G. 1969. The disease "Tsilik Marasi" of grapevine, its description and identification of the causal agent (*Xanthomonas ampelina* sp. nov.). Ann. Inst. Phytopathol. Benaki *9:* 59-81.

Panagopoulos, C.G. and P.G. Psallidas. 1973. Characteristics of Greek isolates of *Agrobacterium tumefaciens* (E.F. Smith and Townsend) Conn. J. Appl. Bacteriol. *36:* 233-240.

Panagopoulos, C.G., P.G. Psallidas and A.S. Alivizatos. 1978. Studies on biotype 3 of *Agrobacterium radiobacter* var. *tumefaciens*. Proc. 4th Int. Conf. Plant Path. Bact., Angers 1978, 221-228.

Panciera, R.J., R.R. Dahlgren and H.B. Rinker. 1968. Observations on septicemia of cattle caused by a *Hemophilus*-like organism. Pathol. Vet. *5:* 212-226.

Pande, P.G. and P.C. Sekariah. 1960. A preliminary note on the isolation of *Moraxella caprae* nov. sp. from an outbreak of infectious keratoconjunctivitis in goats. Curr. Sci. *29:* 276-277.

Pandit, V.M. and Y.S. Kulkarni. 1979. Bacterial leaf-spot of *Clitoria biflora* Dalz. Biovigyanam *5:* 9-20.

Pangborn, J. and M.P. Starr. 1966. Ultrastructure of *Lampropedia hyalina*. J. Bacteriol. *91:* 2025-2030.

Pankhurst, C.E. 1979. Some antigenic properties of cultured cell and bacteroid forms of fast- and slow-growing strains of *Lotus rhizobia*. Microbios *24:* 19-28.

Pant, N.C. and K. Dang. 1972. Physiology and elimination of intracellular symbiotes in some stored product beetles. *In* Rodriguez (Editor), Insect and Mite Nutrition. North-Holland, Amsterdam, pp. 311-322.

Pant, N.C. and G. Fraenkel. 1954. On the function of the intracellular symbionts of *Oryzaephilus surinamensis* L. (Cucujidae: Coleoptera). J. Zool. Soc. India *6:* 173-177.

Pant, N.M. and Y.S. Kulkarni. 1976a. Bacterial leaf-spot of *Desmodium laxiflorum* DC. Biovigyanam *2:* 97-98.

Pant, N.M. and Y.S. Kulkarni. 1976b. Bacterial leaf-spot of *Merremia gangetica* (L.) Cufod. Biovigyanam *2:* 207-208.

Pant, N.C., J.K. Nayar and P. Gupta. 1957. On the isolation and cultivation of intracellular symbiotes of *Oryzaephilus surinamensis* L. (Cucujidae: Coleoptera). Experientia (Basel) *13:* 241.

Paraskeva, C. 1979. Transfer of kanamycin resistance mediated by plasmid

R68.45 in *Paracoccus denitrificans*. J. Bacteriol. *139:* 1062-1064.

Paraskeva, C. and F.R. Whatley. 1980. Cysteine biosynthesis in *Paracoccus denitrificans*. J. Gen. Microbiol. *118:* 79-84.

Parfentjev, I.A. and M.A. Goodine. 1948. Histamine shock in mice sensitized with *Hemophilus pertussis* vaccine. J. Pharmacol. Exp. Ther. *92:* 411-413.

Park, W.E. and A.W. Williams. 1917. Pathogenic microorganisms, 6th ed. Lea and Febiger, New York.

Parker, C., D. Gauthier, A. Tate, K. Richardson and W.R. Romig. 1979. Expanded linkage map of *Vibrio cholerae*. Genetics *91:* 191-214.

Parker, C.D. 1979. The genetics and physiology of *Bordetella pertussis*. *In* Manclark and Hill (Editors), International Symposium on Pertussis. U.S. Government Printing Office, Washington, D.C., pp. 65-69.

Parker, F., Jr. and N.P. Hudson. 1926. The etiology of Haverhill fever (Erythema arthriticum epidemicum). Amer. J. Pathol. *2:* 357-379.

Parker, R.R., G.M. Kohls, G.W. Cox and G.E. Davis. 1939. Observations on an infectious agent from *Amblyomma maculatum*. Pub. Health Rep. *54:* 1482-1484.

Parsons, A.B. and P.R. Dugan. 1971. Production of extracellular polysaccharide matrix by *Zoogloea ramigera*. Appl. Microbiol. *21:* 657-661.

Pasculle, A.W., J.C. Feeley, R.J. Gibson, L.G. Cordes, R.L. Myerowitz, C.M. Patton, G.W. Gorman, C.L. Carmack, J.W. Ezzel and J.N. Dowling. 1980. *In* Validation of the publication of new names and new combinations previously effectively published outside the IJSB. List No. 5. Int. J. Syst. Bacteriol. *30:* 676-677.

Pasculle, A.W., J.C. Feeley, R.J. Gibson, L.G. Cordes, R.L. Myerowitz, C.M. Patton, G.W. Gorman, C.L. Carmack, J.W. Ezzell and J.N. Dowling. 1980. Pittsburgh pneumonia agent: direct isolation from human lung tissue. J. Infect. Dis. *141:* 727-732.

Pasculle, A.W., R.L. Myerowitz and C.R. Rinaldo, Jr. 1979. New bacterial agent of pneumonia isolated from renal-transplant recipients. Lancet *ii:* 58-61.

Pask-Hughes, R. and R.A.D. Williams. 1975. Extremely thermophilic Gramnegative bacteria from hot tap water. J. Gen. Microbiol. *88:* 321-328.

Pask-Hughes, R.A. and R.A.D. Williams. 1977. Yellow-pigmented strains of *Thermus* spp. from Icelandic hot springs. J. Gen. Microbiol. *102:* 375-383.

Pask-Hughes, R.A. and R.A.D. Williams. 1978. Cell envelope components of strains belonging to the genus *Thermus*. J. Gen. Microbiol. *107:* 65-72.

Passmore, S.M. 1973. The acetic acid bacteria: ecology, taxonomy and morphology. Ph.D. Thesis. University of Bristol.

Passmore, S.M. and J.G. Carr. 1975. The ecology of the acetic acid bacteria with particular reference to cider manufacture. J. Appl. Bacteriol. *38:* 151-158.

Paster, B.J. and E. Canale-Parola. 1980. Involvement of periplasmic fibrils in motility of spirochetes. J. Bacteriol. *141:* 359-364.

Pasteur, L. 1864. Mémoire sur la fermentation acétique. Ann. Sci. Ec. norm. sup. Paris *1:* 113-158.

Patel, A.M., J.M. Chauhan, W.V. Kotasthane and M.V. Desai. 1969. A new bacterial disease of *Biophytum sensitivum*. Curr. Sci. *38:* 274-275.

Patel, A.M. and W.V. Kotasthane. 1969a. Bacterial leaf-spot disease of *Corchorus fascicularis* caused by *Xanthomonas nakatae* var. *fascicularis*. Curr. Sci. *38:* 596-597.

Patel, A.M. and W.V. Kotasthane. 1969b. Bacterial blight of *Leea edgeworthii* incited by *Xanthomonas leeanum* nov. sp. Curr. Sci. *38:* 519-520.

Patel, G.B. and C. Breuil. 1981. Isolation and characterization of *Bacteroides polypragmatus* sp. nov., an isolate which produces carbon dioxide, hydrogen and acetic acid during growth on various organic substrates. *In* Moo-Young and Robinson (Editors), Advances in Biotechnology, Vol. II. Pergamon Press, Toronto.

Patel, G.B. and C. Breuil. 1981. Accumulation of an iodophilic polysaccharide during growth of *Acetivibrio cellulolyticus* on cellobiose. Arch. Microbiol. *129:* 265-267.

Patel, G.B. and C. Breuil. 1982. *In* Validation of the publication of new names and new combinations previously effectively published outside the IJSB. List No. 8. Int. J. Syst. Bacteriol. *32:* 266-268.

Patel, G.B., A.W. Khan, B.J. Agnew and J.R. Colvin. 1980. Isolation and characterization of an anaerobic, cellulolytic microorganism *Acetivibrio cellulolyticus* gen. nov., sp. nov. Int. J. Syst. Bacteriol. *30:* 179-185.

Patel, M.K. 1948. *Xanthomonas uppalli* sp. nov. pathogenic on *Ipomoea muricata*. Indian Phytopathol. *1:* 67-69.

Patel, M.K. 1949. *Xanthomonas desmodii*, a new bacterial leaf-spot of *Desmodium diffusum* DC. Curr. Sci. *18:* 213.

Patel, M.K., V.V. Bhatt and Y.S. Kulkarni. 1951. Three new bacterial diseases of plants from Bombay. Curr. Sci. *20:* 326-327.

Patel, M.K., S.G. Desai and A.J. Patel. 1968. A new bacterial leaf-spot on *Veronia cinerea* Less. Science and Culture *34:* 220-221.

Patel, M.K., G.W. Dhande and Y.S. Kulkarni. 1953. Bacterial leaf-spot of *Cyamopsis tetragonoloba* (L.) Taub. Curr. Sci. *22:* 183.

Patel, M.K. and Y.S. Kulkarni. 1949. Nitrogen utilization by *Xanthomonas malvacearum* (Sm.) Dowson. Indian Phytopathol. *2:* 62-64.

Patel, M.K. and Y.S. Kulkarni. 1951. Nomenclature of bacterial plant pathogens. Indian Phytopathol. *4:* 74-84.

Patel, M.K. and Y.S. Kulkarni. 1951. A new bacterial leaf spot on *Vitis woodrowii* Stapf. Curr. Sci. *20:* 132.

Patel, M.K., Y.S. Kulkarni and G.W. Dhande. 1950. *Xanthomonas badrii* sp. nov., on *Xanthium strumarium* L in India. Indian Phytopathol. *3:* 103-104.

Patel, M.K., Y.S. Kulkarni and G.W. Dhande. 1951. Three bacterial diseases of plants. Curr. Sci. *20:* 106.

Patel, M.K., Y.S. Kulkarni and G.W. Dhande. 1952a. Two new bacterial diseases of plants. Curr. Sci. *21:* 74-75.

Patel, M.K., Y.S. Kulkarni and G.W. Dhande. 1952b. Some new bacterial diseases of plants. Curr. Sci. *21:* 345-346.

Patel, M.K. and L. Moniz. 1948. *Xanthomonas desmodii-gangeticii*, sp. nov., Uppal, Patel and Moniz; a new bacterial leaf-spot of *Desmodium gangeticum* DC. Curr. Sci. *17:* 268.

Patel, M.K., L. Moniz and Y.S. Kulkarni. 1948. A new bacterial disease of *Mangifera indica* L. Curr. Sci. *17:* 189-190.

Patel, M.K., B.N. Wankar and Y.S. Kulkarni. 1952. Bacterial leaf-spot of *Amaranthus viridis* L. Curr. Sci. *21:* 346-347.

Patel, P.N. and J.K. Jindal. 1972. Bacterial leaf spot on *Pedalium murex* L. caused by a new albino species of *Xanthomonas*. Indian Phytopathol. *25:* 318-320.

Paterson, W.D., D. Douey and D. Desautels. 1980. Relationship between selected strains of typical and atypical *Aeromonas salmonicida, Aeromonas hydrophila*, and *Haemophilus piscium*. Can. J. Microbiol. *26:* 588-598.

Paterson, W.D., R.J. Douglas, I. Grinyer and L.A. Mac Dermott. 1969. Isolation and preliminary characterization of some *Aeromonas salmonicida* bacteriophages. J. Fish. Res. Bd. Canada *26:* 629-632.

Pathak, R.C. and M. Ristic. 1962. Detection of an antibody to *Actinobacillus lignieresi* in infected human beings and the antigenic characterization of isolates of human and bovine origin. Am. J. Vet. Res. *23:* 310-314.

Patil, S.S. 1974. Toxins produced by phytopathogenic bacteria. Annu. Rev. Phytopathol. *12:* 259-279.

Paton, A.M. 1959. An improved method for preparing pectate gels. Nature (London) *183:* 1812-1813.

Paton, A.M. 1959. Enhancement of pigment production by *Pseudomonas*. Nature *184:* 1254.

Patriquin, D. G. and J. Döbereiner. 1978. Light microscopy observations of tetrazolium-reducing bacteria in the endorhizosphere of maize and other grasses in Brazil. Can. J. Microbiol. *24:* 734-742.

Patt, T.E., G.C. Cole and R.S. Hanson. 1976. *Methylobacterium*, a new genus of facultatively methylotrophic bacteria. Int. J. Syst. Bacteriol. *26:* 226-229.

Pauley, E. H. and N. R. Krieg. 1974. Long-term preservation of *Spirillum volutans*. Int. J. Syst. Bacteriol. *24:* 292-293.

Paulin, J.P. and N.A. Nassan. 1978. Lysogenic strains and phage-typing in *Erwinia chrysanthemi*. *In* Station de Pathologie Végétale et Phytobactériolgie (Editor), Proc. IVth Conf. on Plant Pathogenic Bacteria, Gilbert-Clarey, Tours, France, pp. 539-545.

Pavarino, G.L. 1911. Malattie causate da bacteri nelle orchidee. Atti Accad. Lincei *20:* 233-237.

Pavlatou, M., E. Hassikou-Kaklamani and E. Zantioti. 1965. Lysotypie du genre *Proteus*. Ann. Inst. Pasteur (Paris) *108:* 402-407.

Payne, S.M. and R.A. Finkelstein. 1975. Pathogenesis and immunology of experimental gonococcal infection: Role of iron in virulence. Infect. Immun. *12:* 1313-1318.

Payne, S.M. and R.A. Finkelstein. 1977. Detection and differentiation of ironresponsive avirulent mutants in Congo red agar. Infect. Immun. *18:* 94-98.

Payne, S.M. and R.A. Finkelstein. 1978. The critical role of iron in host-bacterial interactions. J. Clin. Invest. *61:* 1428-1440.

Payne, W.J., R.G. Eagon and A.K. Williams. 1961. Some observations on the physiology of *Pseudomonas natriegens* nov. spec. Antonie van Leeuwenhoek J. Microbiol. Serol. *27:* 121-128.

Payza, A.N. and E.D. Korn. 1956. The degradation of heparin by bacterial enzymes. I. Adaptation and lyophilized cells. J. Biol. Chem. *223:* 853-858.

Pearce, W.A. and T.M. Buchanan. 1978. Attachment role of gonococcal pili. Optimum conditions and quantitation of adherence of isolated pili to human cells in vitro. J. Clin. Invest. *61:* 931-943.

Pease, P. 1979. Observations on L-forms of *Yersinia enterocolitica*. J. Med. Microbiol. *12:* 337-346.

Peattie, D.A. 1979. Direct chemical method for sequencing RNA. Proc. Nat. Acad. Sci. USA *76:* 1760-1764.

Pecher, T. and A. Böck. 1981. In vivo susceptibility of halophilic and methanogenic organisms to protein synthesis inhibitors. FEMS Microbiol. Lett. *10:* 295-297.

Pecknold, P.C. and R.G. Grogan. 1973. Deoxyribonucleic acid homology groups among phytopathogenic *Pseudomonas* species. Int. J. Syst. Bacteriol. *23:* 111-121.

Pedersen, C.E., Jr. and V.D. Walters. 1978. Comparative electrophoresis of spotted fever group rickettsial proteins. Life Sci. *22:* 583-587.

Pedersen, K.B. 1975. The serology of *Bordetella bronchiseptica* isolated from pigs compared with strains from other animal species. Acta Pathol. Microbiol. Scand. Sect. B *83:* 590-594.

Pedersen, K.B., L.O. Frøhlm and K. Bøvre. 1972. Fimbriation and colony type of *Moraxella bovis* in relation to conjunctival colonization and development of keratoconjunctivitis in cattle. Acta Pathol. Microbiol. Scand. *80B:* 911-918.

Pederson, C.S. 1957. Necrology - Otto Rahn. Bacteriological News *24(No. 2):* 21-22.

Pedrosa, F.O., J. Döbereiner and M.G. Yates. 1980. Hydrogen-dependent growth and autotrophic carbon dioxide fixation in *Derxia*. J. Gen. Microbiol. *119:*

547-551.

Pelczar, M.J. 1953. *Neisseria caviae* nov. spec. J. Bacteriol. *65:* 744.

Pelczar, M.J. and J.R. Porter. 1940. Pantothenic acid and nicotinic acid as essential growth substances for Morgan's bacillus *(Proteus morganii).* Proc. Soc. Exp. Biol. Med. *43:* 151-154.

Peleg, B. and D.M. Norris. 1972a. Bacterial symbiote activation of insect parthenogenetic reproduction. Nat. New Biol. *236:* 111-112.

Peleg, B. and D.M. Norris. 1972b. Symbiotic interrelationships between microbes and ambrosia beetles. VII. Bacterial symbionts associated with *Xyleborus ferrugineus.* J. Invertebr. Pathol. *20:* 59-65.

Peleg, B. and D.M. Norris. 1973. Oocyte activation in *Xyleborus ferrugineus* by bacterial symbionts. J. Insect Physiol. *19:* 137-145.

Pell, P.E. and D.I. Southern. 1975. Symbionts in the female tsetse fly *Glossina morsitans morsitans.* Experientia (Basel) *31:* 650-651.

Pell, P.E. and D.I. Southern. 1976. Effect of the coccidiostat, sulphaquinoxaline, on symbiosis in the tsetse fly, *Glossina* species. Microbios Lett. *2:* 203-211.

Peloux, Y., Y. Carcassonne et F. Ouilichini. 1977. A propos d'un cas d'abcès sous cutané provoqué par *Eikenella corrodens* chez un sujet splenectomisé. Pathol. Biol. (Paris) *25:* 245-246.

Penn, C.W. and L.K. Nagy. 1976. Isolation of a protective, non-toxic capsular antigen from *Pasteurella multocida,* types B and E. Res. Vet. Sci. *20:* 90-96.

Penner, J.L., P.C. Fleming, G.R. Whiteley and J.N. Hennessy. 1979a. O-serotyping *Providencia alcalifaciens.* J. Clin. Microbiol. *10:* 761-765.

Penner, J.L. and J.N. Hennessy. 1979. Application of O-serotyping in a study of *Providencia rettgeri (Proteus rettgeri)* isolated from human and non-human sources. J. Clin. Microbiol. *10:* 834-840.

Penner, J.L. and J.N. Hennessy. 1979. O antigen grouping of *Morganella morganii (Proteus morganii)* by slide agglutination. J. Clin. Microbiol. *10:* 8-13.

Penner, J. L. and J. N. Hennessy. 1980. Passive hemagglutination technique for serotyping *Campylobacter fetus* subsp. *jejuni* on the basis of soluble heat-stable antigens. J. Clin. Microbiol. *12:* 732-737.

Penner, J.L. and J.N. Hennessy. 1980. Separate O-grouping schemes for serotyping clinical isolates of *Proteus vulgaris* and *Proteus mirabilis.* J. Clin. Microbiol. *12:* 304-309.

Penner, J.L., N.A. Hinton, I.B.R. Duncan, J.N. Hennessy and G.R. Whiteley. 1979b. O-serotyping of *Providencia stuartii* isolates collected from twelve hospitals. J. Clin. Microbiol. *9:* 11-14.

Penner, J.L. and M.A. Preston. 1980. Differences among *Providencia* species in their in vitro susceptibilities to five antibiotics. Antimicrob. Agents Chemother. *18:* 868-871.

Pepper, E.H. 1967. Stewart's bacterial wilt of corn. Monograph No. 4. American Phytopathological Society, St. Paul, Minnesota, 36 pp.

Perch, B. 1948. On the serology of the Proteus group. Acta Pathol. Microbiol. Scand. *25:* 703-714.

Pereira, A.L.G. 1969. Uma nova doença bacteriana do maracujá *(Passiflora edulis* Sims) causada por *Xanthomonas passiflorae* n. sp. Arq. Inst. Biol. São Paulo *36:* 163-174.

Pereira, A.L.G., F.O. Paradella and A.G. Zagetto. 1971. Uma nova doença bacteriana da mandioquinha salsa *(Arracacia Xanthorrhiza)* causada por *Xanthomonas arracaciae* n. sp. Arq. Inst. Biol. São Paulo *38:* 99-108.

Perrin, W.S. 1906. Researches upon the life-history of *Trypanosoma balbianii* (Certes). Arch. Protistenk. *7:* 131-156.

Perry, L.B. 1973. Gliding motility in some nonspreading flexibacteria. J. Appl. Bacteriol. *36:* 227-232.

Persley, G.J. 1978. Epiphytic survival of *Xanthomonas manihotis* in relation to the disease cycle of cassava bacterial blight. Proc. 4th Internat. Conf. on Plant Pathol. Bact. *II:* 401-429.

Peschkov, J.J. and V. Feodorov. 1978. Comparative study on the ultrastructure of L-forms obtained from S and R variants of *Brucella suis* 1330. Zentralbl. Bakteriol. Parasitenkd. Infektionskr. Hyg. Abt. I Orig. A *240:* 94-105.

Peters, D. and R. Wigand. 1955. *Bartonellaceae.* Bacteriol. Rev. *19:* 150-155.

Petersen, E.J. 1959. Serological investigations on *Azotobacter* and *Beijerinckia.* Kgl. Vet.-Landbohjsk. Arsskr. 70-90.

Petersen, E.J. and E. Holmes. 1964. On nitrogen fixation in Danish deciduous forests. Roy. Vet. Agric. Coll. Copenhagen Yearbook, pp. 209-226.

Peterson, A.M. and M.E. Pollock. 1969. Deoxyribonucleic acid homology and relative genome size in *Mycoplasma.* J. Bacteriol. *99:* 639-644.

Peterson, J.A. 1970. Cytochrome content of two pseudomonads containing mixed-function oxidase systems. J. Bacteriol. *103:* 714-721.

Peterson, J.E., A.W. Rodwell and E.S. Rodwell. 1973. Occurrence and ultrastructure of a variant (rho) form of *Mycoplasma.* J. Bacteriol. *115:* 411-425.

Pethybridge, C.H. and P.A. Murphy. 1911. A bacterial disease of the potato plant in Ireland and the organism causing it. Proc. Roy. Ir. Acad. *29:* 1-37.

Petit, A., S. Delhaye, J. Tempé and G. Morel. 1970. Recherches sur les guanidines des tissus de crown gall. Mise en évidence d'une relation biochimique spécifique entre les souches d'*Agrobacterium* et les tumeurs qu'elles induisent. Physiol. Vég. *8:* 205-213.

Petit, A., Y. Dessaux and J. Tempé. 1978. The biological significance of opines. I. A study of opine catabolism by *Agrobacterium tumefaciens.* Proc. IVth Int. Conf. Plant Path. Bact., Angers 1978, 143-152.

Petrovskaya, V.G. and V.M. Bondarenko. 1977. Recommended corrections to the classification of *Shigella flexneri* on a genetic basis. Int. J. Syst. Bacteriol. *27:* 171-175.

Petruschky, J. 1896. *Bacillus faecalis alcaligenes* n. sp. Zentralbl. Bakteriol. Parasitenk. Infektionskr. Hyg. Abt. I *19:* 187-191.

Petschenko, B. 1911. *Drepanospira mülleri* n. g., n. sp. parasite des paraméciums; contribution à l'étude de la structure des bactériés. Arch. Protistenk. *22:* 248-298.

Petter, H.F.M. 1932. Over roode en andere bacteriën van gezouten visch. Doctoral thesis. Rijks-Universiteit te Utrecht, Utrecht, pp. 1-116.

Peynaud, E. and S. Domercq. 1961. Présence de bactéries lactiques sur les raisins mûrs. C. R. Hebd. Sean. Acad. Sci. Ser. D *252:* 3343-3344.

Pezzlo, M.A., P.J. Valter and M.J. Burns. 1979. Wound infection associated with *Vibrio alginolyticus.* Am. J. Clin. Pathol. *71:* 476-478.

Pfeifer, F., G. Weidinger and W. Goebel. 1981. Characterization of plasmids in halobacteria. J. Bacteriol. *145:* 369-374.

Pfeiffer, A. 1889. Ueber die bäcillare Pseudotuberkulose bei Nagethieren. Thieme, Leipzig.

Pfeiffer, R. 1896. Die Spirillen *In* Flügge (Editor), Die Mikroorganismen, 3rd Ed., 2 Thiel. F.C.W. Vogel, Leipzig, pp. 527-599.

Pfennig, N. 1974. *Rhodopseudomonas globiformis,* sp. n., a new species of the *Rhodospirillaceae.* Arch. Microbiol. *100:* 197-206.

Pfennig, N. 1978. *Rhodocyclus purpureus* gen. nov. and sp. nov. a ring-shaped, vitamin B_{12}-requiring member of the family *Rhodospirillaceae.* Int. J. Syst. Bacteriol. *28:* 283-288.

Pfennig, N. and H. Biebl. 1976. *Desulfuromonas acetoxidans* gen. nov. and sp. nov., a new anaerobic, sulfur-reducing, acetate-oxidizing bacterium. Arch. Microbiol. *110:* 3-12.

Pfennig, N. and H. Biebl. 1977. *In* List No. 1, Announcement of the valid publication of new names and new combinations previously effectively published outside the IJSB. Int. J. Syst. Bacteriol. *27:* 306.

Pfennig, N. and H.G. Trüper. 1974. The phototrophic bacteria. *In* Buchanan and Gibbons (Editors), Bergey's Manual of Determinative Bacteriology, 8th Ed., The Williams and Wilkins Co., Baltimore, pp. 24-60.

Pfennig, N. and F. Widdel. 1981. Ecology and physiology of some anaerobic bacteria from the microbial sulfur cycle. *In* Bothe and Trebst (Editors), Biology of inorganic nitrogen and sulfur, Springer Verlag, Berlin - Heidelberg, pp. 169-177.

Phelps, L.N. 1967. Isolation and characterization of bacteriophages for *Neisseria.* J. Gen. Virol. *1:* 529-532.

Philip, C.B. 1943. Nomenclature of the pathogenic rickettsiae. Am. J. Hyg. *37:* 301-309.

Philip, C.B. 1948. Comments on the name of the Q fever organism. Pub. Health Rep. *63:* 58.

Philip, C.B. 1950. Miscellaneous human rickettsioses. *In* R.L. Pullen, Communicable Diseases. Lea and Febiger Co., Philadelphia, pp. 781-788.

Philip, C.B. 1953. Nomenclature of the *Rickettsiaceae* pathogenic to vertebrates. Ann. N.Y. Acad. Sci. *56:* 484-494.

Philip, C.B. 1956. Comments on the classification of the order *Rickettsiales.* Can. J. Microbiol. *2:* 261-270.

Philip, C.B. 1962. Appendix G. Summary of tick-borne rickettsioses. Rep. 2nd Mtg. FAO/OIE Expert Panel Tick-borne Dis. Livestock, Cairo. United Nations, Rome 1962, pp. 41-43.

Philip, C.B., W.J. Hadlow and L.E. Hughes. 1953. *Neorickettsia helmintheca,* a new rickettsia-like disease agent in dogs in western United States transmitted by a helminth. Riass. Comun. VI Congr. Int. Microbiol., Roma *2:* 256-257.

Philip, R.N. and E.A. Casper. 1981. Serotypes of spotted fever group rickettsiae isolated from *Dermacentor andersoni* (Stiles) ticks in Western Montana. Amer. J. Trop. Med. Hyg. *30:* 230-238.

Philip, R.N., E.A. Casper, W. Burgdorfer, R.K. Gerloff, L.E. Hughes and E.J. Bell. 1978. Serologic typing of rickettsiae of the spotted fever group by microimmunofluorescence. J. Immunol. *121:* 1961-1968.

Philip, R.N., E.A. Casper, J.N. MacCormack, D.J. Sexton, L.A. Thomas, R.L. Anacker, W. Burgdorfer and S. Vick. 1977. A comparison of serologic methods for diagnosis of Rocky Mountain spotted fever. Am. J. Epidemiol. *105:* 56-67.

Philippon, A. 11968. Identification of *Brucella abortus:* metabolism et lysotypie. Ann. Inst. Pasteur *115:* 367-378.

Phillips, I. and S. Eykin. 1972. Contaminated drip fluids. British Med. J. i, 746.

Phillips, I., S. Eykyn, B.A. King, C. Jenkins, C.A. Warren and K.P. Shannon. 1977. The in-vitro antibacterial activity of nine aminoglycosides and spectinomycin on clinical isolates of common Gram-negative bacteria. J. Antimicrob. Chemother. *3:* 403-410.

Phillips, J.E. 1955. The experimental pathogenicity in mice of strains of *Proteus* of animal origin. J. Hyg. *53:* 212-216.

Phillips, J.E. 1955. In vitro studies of *Proteus* organisms of animal origin. J. Hyg. *53:* 26-31.

Phillips, J.E. 1961. The commensal role of *Actinobacillus lignieresi.* J. Pathol. Bacteriol. *82:* 205-208.

Phillips, J.E. 1964. Commensal actinobacilli from the bovine tongue. J. Pathol. Bacteriol. *87:* 442-444.

Phillips, J.E. 1967. Antigenic structure and serological typing of *Actinobacillus lignieresi.* J. Pathol. Bacteriol. *93:* 463-475.

Phillips, S.E. and M.L. Taylor. 1976. Oxidation of arsenite to arsenate by *Alcaligenes faecalis.* Appl. Environ. Microbiol. *32:* 392-399.

Phillips, W.E. and J.J. Perry. 1976. *Thermomicrobium fosteri* sp. nov., a hydrocarbon-utilizing obligate thermophile. Int. J. Syst. Bacteriol. *26:* 220-225.

Pichinoty, F. 1964. A propos des nitrate-reductases d'une bacteria denitrifiante. Biochim. Biophys. Acta *89:* 378-381.

Pichinoty, F. 1970. Les nitrate-reductases bactériennes. IV. Regulation de la biosynthese et de l'activite de l'enzyme B. Arch. Mikrobiol. *71:* 116-122.

Pichinoty, F. 1971. Les nitrate-reductases bactériennes. VIII. Étude préliminaire de l'enzyme de *Micrococcus halodenitrificans.* Arch. Mikrobiol. *76:* 83-90.

Pichinoty, F., E. Azoulay, P. Couchoud-Beaumont, L. Le Minor, C. Rigano, J. Bigliardi-Rouvier and M. Piéchaud. 1969. Recherche des nitrate-reductases bactériennes A et B: resultats. Ann. Inst. Pasteur *116:* 27-42.

Pichinoty, F. and R. Chatelain. 1973. Réduction du nitrate, du nitrite et de l'oxyde nitreux par *Alcaligenes denitrificans* et *Alcaligenes odorans.* Ann. Microbiol. (Inst. Pasteur) *124:* 445-449.

Pichinoty, F., M. Mandel and J.-L. Garcia. 1977. Étude de six souches de *Agrobacterium tumefaciens* et *A. radiobacter.* Ann. Microbiol. (Inst. Pasteur) *128A:* 303-310.

Pichinoty, F., M. Mandel and J.L. Garcia. 1977. Étude physiologique et taxonomique de *Paracoccus denitrificans.* Ann. Microbiol. (Inst. Pasteur) *128B:* 243-251.

Pichinoty, F., M. Mandel, B. Greenway and J.-L. Garcia. 1977. Isolation and properties of a denitrifying bacterium related to *Pseudomonas lemoignei.* Int. J. Syst. Bacteriol. *27:* 346-348.

Pichinoty, F. and M. Piéchaud. 1968. Recherche des nitrate-réductases bactériennes A et B: methodes. Ann. Inst. Pasteur *114:* 77-98.

Pichinoty, F., M. Véron, M. Mandel, M. Durand, C. Job and J.-L. Garcia. 1978. Étude physiologique et taxonomique du genre *Alcaligenes: A. denitrificans, A. odorans* et *A. faecalis.* Can. J. Microbiol. *24:* 743-753.

Pickens, E.G., R.K. Gerloff and W. Burgdorfer. 1968. Spirochete from the rabbit tick *Haemaphysalis leporispalustris* (Packard). J. Bacteriol. *95:* 291-299.

Pickett, J. and R. Kelly. 1974. Lipid catabolism of relapsing fever borreliae. Infect. Immun. *9:* 279-285.

Pickett, M.J. and J.R. Greenwood. 1980. A study of the Va-1 group of pseudomonads and its relationship to *Pseudomonas pickettii.* J. Gen. Microbiol. *120:* 439-446.

Pickett, M.J. and E.L. Nelson. 1950. *Brucella* bacteriophage. J. Hyg. Camb. *48:* 500-503.

Pickett, M.J. and E.L. Nelson. 1955. Speciation within the genus *Brucella.* IV. Fermentation of carbohydrates. J. Bacteriol. *69:* 333-336.

Pickett, M.J. and M.M. Pedersen. 1970. Salient features of nonsaccharolytic and weakly saccharolytic nonfermentative rods. Can. J. Microbiol. *16:* 401-409.

Pierantoni, U. 1954. Nuovi aspetti della convivenza fisiologica fra insetti e microrganismi. Boll. Zool. *21:* 447-452.

Pierce, N.B. 1901. Walnut bacteriosis. Bot. Gaz. *31:* 272-273.

Piéron, R. and Y. Mafart. 1977. Infection d'un kyste branchial à *Eikenella corrodens.* Sem. Hop. *53:* 1087-1091.

Piggott, A. and L. Hochholzer. 1970. Human melioidosis. A histopathologic study of acute and chronic melioidosis. Arch. Pathol. *90:* 101-111.

Pilacinski, W.P. and E.L. Schmidt. 1981. Plasmid transfer within and between serologically distinct strains of *Rhizobium japonicum,* using antibiotic resistant mutants and auxotrophs. J. Bacteriol. 145: 1025-1030.

Pilcher, K.S., W.G. Wu and O.H. Muth. 1961. Studies on the morphology and respiration of *Anaplasma marginale.* Am. J. Vet. Res. *22:* 298-307.

Pillich, J., Z. Hradečna and M. Kocur. 1964. An attempt at phage typing in the genus *Serratia.* J. Appl. Bacteriol. *27:* 65-68.

Pillot, J. 1965. Contribution a l'étude du genre *Treponema.* Structures anatomique et antigénique. Lons-le-Saunier, Paris.

Pillot, J. and A. Ryter. 1965. Structure des spirochètes. I. Étude des genres *Treponema, Borrelia* et *Leptospira* au microscope électronique. Ann. Inst. Pasteur *108:* 791-804.

Pine, L., I.R. George, M.W. Reeves and W.K. Harrell. 1979. Development of a chemically defined liquid medium for the growth of Legionnaires' disease bacterium. J. Clin. Microbiol. *9:* 615-626.

Pinkerton, H. 1936. Criteria for the accurate classification of the rickettsial diseases (rickettsioses) and of their etiological agents. Parasitol. *28:* 172-189.

Pinkwart, M., H. Bahl, M. Reimer, D. Wölfle and H. Berndt. 1979. Activity of the H₂-oxidizing hydrogenase in different N₂-fixing bacteria. FEMS Microbiol. Lett. *6:* 177-181.

Pinnock, D.E. and R.T. Hess. 1974. The occurrence of intracellular rickettsialike organisms in the tsetse flies, *Glossina morsitans, G. fuscipes, G. brevipalpis* and *G. pallidipes.* Acta Trop. *31:* 70-79.

Pintér, M. and M. Kántor. 1974. Comparison of *Alcaligenes faecalis* and *Alcaligenes odorans* var. *viridans* by carbon source utilization test. Acta Microbiol. Acad. Sci. Hung. *21:* 293-295.

Piot, P., E. Van Dyck, M. Goodfellow and S. Falkow. 1980. A taxonomic study of *Gardnerella vaginalis (Haemophilus vaginalis)* Gardner and Dukes 1955. J. Gen. Microbiol. *119:* 373-396.

Piot, P., E. van Dyck and S.R. Patty. 1977. Sensibilité d'*Haemophilus influenza* à 5 antibiotiques et détection rapide de sa résistance à l'ampicilline. Pathol. Biol. (Paris) *25:* 83-87.

Pitt, T.L. and D.E. Bradley. 1975. The antibody response to the flagella of *Pseudomonas aeruginosa.* J. Med. Microbiol. *8:* 97-106.

Pittman, K.A. and M.P. Bryant. 1964. Peptides and other nitrogen sources for growth of *Bacteroides ruminicola.* J. Bacteriol. *88:* 401-410.

Pittman, K.A., S. Lakshmanan and M.P. Bryant. 1967. Oligopeptide uptake by *Bacteroides ruminicola.* J. Bacteriol. *93:* 1499-1508.

Pittman, M. 1931. Variation and type specificity in the bacterial species *Haemophilus influenzae.* J. Exp. Med. *53:* 471-492.

Pittman, M. 1953. A classification of the hemolytic bacteria of the genus *Haemophilus: Haemophilus haemolyticus* Bergey et al. and *Haemophilus parahaemolyticus* nov. spec. J. Bacteriol. *65:* 750-751.

Pittman, M. 1970. *Bordetella pertussis* - bacterial host factors in the pathogenesis and prevention of whooping cough. *In* Mudd (Editor), Infectious Agents and Host Reactions, W.B. Saunders Co., Philadelphia, pp. 239-270.

Pittman, M. 1979. Pertussis toxin: the cause of the harmful effects and prolonged immunity of whooping cough. A hypothesis. Rev. Infect. Dis. *1:* 401-412.

Pittman, M. and D.J. Davis. 1950. Identification of the Koch-Weeks bacillus *(Hemophilus aegyptius).* J. Bacteriol. *59:* 413-426.

Pittman, M., B.L. Furman and A.C. Wardlaw. 1980. *Bordetella pertussis* respiratory tract infection in the mouse: pathophysiological responses. J. Infect. Dis. *142:* 56-66.

Place, E.H., L.E. Sutton and O. Willner. 1926. Erythema arthriticum epidemicum; preliminary report. Boston Med. Surg. J. *194:* 285-287.

Plackett, P. 1959. On the probable absence of "mucocomplex" from *Mycoplasma mycoides.* Biochim. Biophys. Acta *35:* 260-262.

Plackett, P. and S.H. Buttery. 1964. A galactofuranose disaccharide from the galactan of *Mycoplasma mycoides.* Biochem. J. *90:* 201-205.

Platt, H. and J.G. Atherton. 1976. *Klebsiella* and *Enterobacter* organisms isolated from horses. J. Hyg. (Camb.) *77:* 401-408.

Plesko, I. and Z. Hlavata. 1971. Cross-immunity studies with lipase negative strains of leptospires. Biologia (Bratislava) *26:* 689-693.

Ploss, M., J. Erber and F. Eschenbecher. 1979. Die Essigsäurebakteriën in der Brauerei. Eur. Brew. Conv. Proceedings of the 17th Congress. Berlin (West), pp. 521-531.

Podlipaev, S.A. and D.V. Ossipov. 1979. Early stages of infection of *Paramecium caudatum* micronuclei by symbiotic bacteria-omega-particles (electron microscope examination). Acta Protozool. *18:* 465-480.

Pohl, S. 1979. Reklassifizierung der Gattung *Actinobacillus* Brumpt 1910, *Haemophilus* Winslow et al. 1917 und *Pasteurella* Trevisan 1887 anhand phänotypischer und molekular Daten, insbesondere der DNS-Verwandtschaften bei DNS:DNS-Hybridisierung in vitro und Vorschlag einer neuen Familie, *Pasteurellaceae.* Inaug.-Diss. Phillips-Universität Marburg/Lahn. Mauersberger, Marburg.

Pohl, S. 1981. DNA relatedness among members of *Actinobacillus, Haemophilus* and *Pasteurella. In* Kilian, Frederiksen and Biberstein (Editors), *Haemophilus, Pasteurella* and *Actinobacillus,* Academic Press, pp. 246-253.

Pohl, S. 1981a. *In* Validation of the publication of new names and new combinations previously effectively published outside the IJSB. List No. 7. Int. J. Syst. Bacteriol. *31:* 382-383.

Poinar, G.O., Jr. 1966. The presence of *Achromobacter nematophilus* in the infective stage of a *Neoaplectana* sp. *(Steinernematidae: Nematoda).* Nematologica *12:* 105-108.

Poinar, G.O., Jr., R.T. Hess and J.A. Tsitsipis. 1975. Ultrastructure of the bacterial symbiotes in the pharyngeal diverticulum of *Dacus oleae* (Gmelin) (Trypetidae; Diptera). Acta Zool. (Stockh.) *56:* 77-84.

Poinar, G.O., Jr. and G.M. Thomas. 1965. A new bacterium, *Achromobacter nematophilus* sp. nov. *(Achromobacteriaceae: Eubacteriales)* associated with a nematode. Int. Bull. Bacteriol. Nomencl. Taxon. *15:* 249-252.

Poinar, G.O., Jr. and G.M. Thomas. 1966. Significance of *Achromobacter nematophilus* Poinar and Thomas *(Achromobacteraceae: Eubacteriales)* in the development of the nematode, DD-136 *(Neoaplectana* sp. *Steinernematidae).* Parasitology *56:* 385-390.

Poinar, G.O., Jr. and G.M. Thomas. 1967. The nature of *Achromobacter nematophilus* as an insect pathogen. J. Invertebr. Pathol. *9:* 510-514.

Poinar, G.O., Jr., G.M. Thomas and R. Hess. 1977. Characteristics of the specific bacterium associated with *Heterorhabditis bacteriophora (Heterorhabditidae; Rhabditida).* Nematologica *23:* 97-102.

Poinar, G.O., Jr., G.M. Thomas, G.V. Veremtschuk and D.E. Pinnock. 1971. Further characterization of *Achromobacter nematophilus* from American and Soviet populations of the nematode Neoaplectana carpocapsae Weiser. Int. J. Syst. Bacteriol. *21:* 78-82.

Polak-Vogelzang, A.A., R.A. Samson and G.T.N. DeLeeuw. 1979. Scanning electron microscopy of *Acholeplasma* colonies on agar. Can. J. Microbiol. *25:* 1373-1380.

Poland, J. and R. Lemcke. 1978. Mycoplasmas of the respiratory tract of horses and their significance in upper respiratory tract disease. Proceedings of the 4th International Conference on Equine Infectious Diseases, Lyon, 1978. J. Equine Med. Surg. Suppl. *1:* 438-446.

Pollack, J.D. 1979. Respiratory pathways and energy-yielding mechanisms. *In* Barile and Razin (Editors), The Mycoplasmas, Vol. I, Academic Press, New York, pp. 187-211.

Pollard, A. 1959. Le framboisé du cidre et la "cider sickness". Ind. Aliment. Agric., 537.

Pollitzer, R. 1954. Plague. W.H.O. Monograph Series No. 22. Geneva, World Health Organization.

Pollitzer, R. 1967. History and incidence of tularemia in the Soviet Union. Institute of Contemporary Russian Studies, Fordham University, New York.

Polster, M. and M. Svobodova. 1964. Production of reddish-brown pigment from dl-tryptophan by Enterobacteria of the *Proteus-Providencia* group. Experientia *20:* 637-638.

Pon, D.S., C.E. Townsend, G.E. Wessmann, C.G. Schmitt and C.H. Kingsolver. 1954. A *Xanthomonas* parasite on uredia of cereal rusts. Phytopathology *44:* 707-710.

Pootjes, C.F. 1964. Isolation of a bacteriophage for *Hydrogenomonas facilis.* J. Bacteriol. *87:* 1259.

Pootjes, C.F., R.B. Mayhew and B.D. Korant. 1966. Isolation and characterization of *Hydrogenomonas facilis* bacteriophages under heterotrophic growth conditions. J. Bacteriol. *92:* 1787-1791.

Popkhadze, N.Z. and T.G. Abashidze. 1957. Characteristics of the *Brucella* bacteriophage isolated at the Tbilisi NIIVS. (In Russian). Bakteriofagiya *5:* 321-325.

Popoff, M. 1969. Étude sur les *Aeromonas salmonicida.* I. Caractéres biochimiques et antigéniques. Rech. Vet. *3:* 49-57.

Popoff, M. 1971. Étude sur les *Aeromonas salmonicida.* II. Caractérisation des bacteriophages actifs sur les *Aeromonas salmonicida* et lysotypie. Ann. Rech. Vet. *2:* 33-45.

Popoff, M.Y., C. Coynault, M. Kiredjian and M. Lemelin. 1981. Polynucleotide sequence relatedness among motile *Aeromonas* species. Curr. Microbiol. *5:* 109-114.

Popoff, M. and Y. Davaine. 1971. Facteurs de résistance transférables chez *Aeromonas salmonicida.* Ann. Inst. Pasteur (Paris) *121:* 337-342.

Popoff, M. and C. Richard. 1975. O and H antigens of *Levinea malonatica.* Ann. Microbiol. *126B:* 17-23.

Popoff, M. and M. Véron. 1976. A taxonomic study of the *Aeromonas hydrophila - Aeromonas punctata* group. J. Gen. Microbiol. *94:* 11-22.

Popoff, M. and M. Véron. 1981. *Aeromonas sobria* sp. nov. *In* Validation of the publication of new names and new combinations previously effectively published outside the IJSB. List No. 6. Int. J. Syst. Bacteriol. *31:* 215.

Popoff, M. and J.F. Vieu. 1970. Bactériophages et lysotypie des *Aeromonas salmonicida.* C.R. Acad. Sci. Paris *270:* 2219-2222.

Popova, L.Y. and A.N. Shenderov. 1979a. Genetic research of *Photobacterium mandapamensis.* I. Obtaining and the description of the collection of auxotrophic and dark mutant strains (Russian). Genetika (Moscow) *15:* 56-61.

Popova, L.Y. and A.N. Shenderov. 1979b. Genetic investigation of *Photobacterium mandapamensis.* II. Classification of mutants with increased luminescence intensity for their sensitivity to exogenous aldehyde (Russian). Genetika (Moscow) *15:* 1555-1560.

Popovici, M., L. Szegli, C. Racovitza, E. Badulescu, D. Florescu, M. Negut, A. Negut, E. Thomas and S. Masek. 1967. Über die aetiologische Bedeutung und Haufigkeit der Citrobacter-gruppe bei Enteritiden. Zentralbl. Bakteriol. Parasitenkd. Infektionskr. Hyg. Abt. I Orig. *204:* 112-121.

Porschen, R.K. and P. Chan. 1977. Anaerobic vibrio-like organisms cultured from blood: *Desulfovibrio desulfuricans* and *Succinivibrio* species. J. Clin. Microbiol. *5:* 444-447.

Porschen, R.K. and S. Sonntag. 1974. Extracellular deoxyribonuclease production by anaerobic bacteria. Appl. Microbiol. *27:* 1031-1033.

Porschen, R.K. and E.H. Spaulding. 1974. Phosphatase activity of anaerobic organisms. Appl. Microbiol. *27:* 744-747.

Porschen, R.K. and D.R. Stalons. 1976. Evaluation of simplified dichotomous schemata for the identification of anaerobic bacteria from clinical material. J. Clin. Microbiol. *3:* 161-171.

Porter, I. A. and T. M. S. Reid. 1980. A milk-borne outbreak of *Campylobacter* infection. J. Hyg. Camb. *84:* 415-419.

Porter, J.R. 1976. The world view of culture collections. *In* Colwell (Editor), The Role of Culture Collections in the Era of Molecular Biology. American Society for Microbiology, Washington, D.C., pp. 62-72.

Portnoy, D.A., S.L. Moseley and S. Falkow. 1981. Characterization of plasmids and plasmid-associated determinants of *Yersinia enterocolitica* pathogenesis. Infect. Immunol. *31:* 775-782.

Portwood, L.M. 1946. Catalase activity of *Hemophilus pertussis.* J. Bacteriol. *51:* 265-266.

Postgate, J.R. 1963. A strain of *Desulfovibrio* able to use oxamate. Arch. Mikrobiol. *46:* 287-295.

Postgate, J.R. 1965. Enrichment and isolation of sulphate-reducing bacteria. Zentralbl. Bakteriol. Suppl. 1: Anreicherungskultur und Mutantenauslese, pp. 190-197.

Postgate, J.R. 1967. Report of the Subcommittee on sulfate-reducing bacteria (1962-1966) to the International Committee on Nomenclature of Bacteria. Int. J. Syst. Bacteriol. *17:* 111-112.

Postgate, J.R. 1974. Genus *Desulfovibrio. In* Buchanan and Gibbons (Editors), Bergey's Manual of Determinative Bacteriology, 8th Ed. The Williams and Wilkins Co., Baltimore, pp. 418-420.

Postgate, J. 1978. Nitrogen fixation. Edward Arnold Ltd., London.

Postgate, J.R. 1979. The sulphate-reducing bacteria. Cambridge University Press, Cambridge-London-New York-Melbourne.

Postgate, J.R. 1981. Microbiology of the free-living nitrogen-fixing bacteria, excluding Cyanobacteria. *In* Gibson and Newton (Editors), Current Perspectives in Nitrogen Fixation, Proc. 4th Int. Symp. on Nitrogen Fixation, Canberra, Australia, Dec. 1-5, 1980, pp. 217-227.

Postgate, J.R. and L.L. Campbell. 1963. Identification of Coleman's sulfate-reducing bacterium as a mesophilic relative of *Clostridium nigrificans.* J. Bacteriol. *86:* 274-279.

Postgate, J.R. and L.L. Campbell. 1966. Classification of *Desulfovibrio* species, the nonsporulating sulfate-reducing bacteria. Bacteriol. Rev. *30:* 732-738.

Potrikus, C.J. and J.A. Breznak. 1980a. Anaerobic degradation of uric acid by gut bacteria of termites. Appl. Environ. Microbiol. *40:* 125-132.

Potrikus, C.J. and J.A. Breznak. 1980b. Uric acid-degrading bacteria in guts of termites *Reticulitermes flavipes* (Kullar). Appl. Environ. Microbiol. *40:* 117-124.

Poulsen, V.A. 1879. Om nogle mikroskopiske Planteorganismer. Vidensk. Medd. Naturhist. Foren. Kjöbenhavn *1879-80:* pp. 231-254.

Poulson, D.F. and B. Sakaguchi. 1961. Nature of "sex ratio" agent in *Drosophila.* Science *133:* 1489-1490.

Pourquier, M., J. Mandin and C. Vago. 1963. Développement d'une rickettsie de coléoptère en culture de tissus de vertébrés. Ann. Epiphyties *14:* 193-197.

Powell, D.A. 1979. Structure, solution properties and biological interactions of some microbial extracellular polysaccharides. *In* Berkeley, Gooday and Ellwood (Editors), Microbial Polysaccharides and Polysaccharases, Academic Press, New York, pp. 117-160.

Powell, P.E., G.R. Cline, C.P.P. Reid and P.J. Szaniszlo. 1980. Occurrence of hydroxamate siderophore iron chelators in soils. Nature *287:* 833-834.

Preer, J.R., Jr. and A. Jurand. 1968. The relation between virus-like particles and R bodies of *Paramecium aurelia.* Genet. Res. *12:* 331-340.

Preer, J.R., Jr. and L.B. Preer. 1982. Revival of names of protozoan endosymbionts and proposal of *Holospora caryophila* nom. nov. Int. J. Syst. Bacteriol. *32:* 140-141.

Preer, J.R., Jr., L.B. Preer and A. Jurand. 1974. Kappa and other endosymbionts in *Paramecium aurelia.* Bacteriol. Rev. *38:* 113-163.

Preer, L.B., A. Jurand, J.R. Preer, Jr. and B.M. Rudman. 1972. The classes of kappa in *Paramecium aurelia.* J. Cell. Sci. *11:* 581-600.

Preer, L.B. and J.R. Preer, Jr. 1977. Inheritance of infectious elements. *In* Goldstein and Prescott (Editors), Cell Biology: A Comprehensive Treatise, Vol. 1, Academic Press, New York, pp. 319-373.

Prefontaine, G. and F.L. Jackson. 1972. Cellular fatty acid profiles as an aid to the classification of "corroding bacilli" and certain other bacteria. Int. J. Syst. Bacteriol. *22:* 210-217.

Prescott, S.C. and C.G. Dunn. 1959. The acetic acid bacteria and some of their biochemical activities. *In* Industrial Microbiology. McGraw-Hill, New York, Toronto, London, pp. 428-473.

Pretorius, W. A. 1963. A systematic study of genus *Spirillum* which occurs in oxidation ponds, with a description of a new species. J. Gen. Microbiol. *32:* 403-408.

Prévot, A.R. 1933. Études de systématique bactérienne. I. Lois générales. II. Cocci anaérobius. Ann. Sci. Natur. Zool. Biol. Anim. *15:* 23-260.

Prévot, A.R. 1938. Études de systématique bactérienne. Ann. Inst. Pasteur (Paris) *60:* 285-307.

Prévot, A. R. 1940. Manual de classification et de détermination des bactéries anaérobies. Masson and Co., Paris, pp. 1-223.

Prévot, A.R. 1948. Manuel de classification et de détermination des bactéries anaérobies, 2nd Ed. Masson and Co., Paris.

Prévot, A.R. 1961. Traité de Systématique Bactérienne, Vol. 2, Dunod, Paris, pp. 1-771.

Prévot, A.R. 1966. Manual for the classification and determination of the anaerobic bacteria. 1st Amer. Ed., transl. by V. Fredette. Lea and Febiger, Philadelphia.

Prévot, A.R., A. Turpin and P. Kaiser. 1967. Les bactéries anaérobies. Dunod, Paris.

Pribram, E. 1933. Klassifikation der Schizomyceten. F. Deuticke, Leipzig, pp. 1-143.

Price, K.W. and M.J. Pickett. 1981. Studies of clinical isolates of flavobacteria. *In* Reichenbach and Weeks (Editors), The *Flavobacterium-Cytophaga* Group (Proceedings of the International Symposium on Yellow-Pigmented Gram-negative Bacteria of the *Flavobacterium-Cytophaga* Group, Braunschweig, July 8 to 11, 1980). Verlag Chemie, Weinheim, pp. 63-77.

Pridham, T.G. 1974. Micro-organism Culture Collections: Acronyms and Abbreviations. ARS-NC-17. Agricultural Research Service, US Department of Agriculture, North Central Region, Peoria, Illinois.

Priest, F.G., H.J. Somerville, J.A. Cole and J.S. Hough. 1973. The taxonomic position of *Obesumbacterium proteus,* a common brewery contaminant. J. Gen. Microbiol. *75:* 295-307.

Prince, G.H. and J.E. Smith. 1966a. Antigenic studies on *Pasteurella multocida* using immunodiffusion techniques. I. Identification of soluble antigens of a bovine hemorrhagic septicemia strain. J. Comp. Pathol. *76:* 303-314.

Prince, G.H. and J.E. Smith. 1966b. Antigenic studies on *Pasteurella multocida* using immunodiffusion techniques. II. Relationship with other Gram-negative species. J. Comp. Pathol. *76:* 315-329.

Prince, G.H. and J.E. Smith. 1966c. Antigenic studies on *Pasteurella multocida* using immunodiffusion techniques. III. Relationships between strains of *Pasteurella multocida.* J. Comp. Pathol. *76:* 321-332.

Pringsheim, E.G. 1955. *Lampropedia hyalina* Schroeter 1886 and *Vanniella aggregata* n.g., n.sp., with remarks on natural and on organized colonies in bacteria. J. Gen. Microbiol. *13:* 285-291.

Pringsheim, E.G. 1966. *Lampropedia hyalina* Schroeter, eine apochlorotische

Merismopedia (Cyanophyceae). Kleine Mitteilungen über Flagellaten und Algen. XII. Arch. Mikrobiol. *55:* 200-208.

Pringsheim, E.G. 1967. Bakterien und Cyanophyceen. Oesterr. Bot. Z. *114:* 324-340.

Prins, R.A. 1971. Isolation, culture and fermentation characteristics of *Selenomonas ruminantium* var. *bryanti* var. n. from the rumen of sheep. J. Bacteriol. *105:* 820-825.

Prins, R.A., A. Lankhorst, P. van der Meer and C.J. van Nevel. 1975. Some characteristics of *Anaerovibrio lipolytica*, a rumen lipolytic organism. Antonie van Leeuwenhoek J. Microbiol. Serol. *41:* 1-11.

Prinsloo, H.E. 1966. Bacteriocins and phages produced by *Serratia marcescens*. J. Gen. Microbiol. *45:* 205-212.

Pritchard, M. A., D. Langley and S. C. Rittenberg. 1975. Effects of methotrexate on intraperiplasmic and axenic growth of *Bdellovibrio bacteriovorus*. J. Bacteriol. *121:* 1131-1136.

Pritchett, I.W. and E.G. Stillman. 1919. The occurrence of *Bacillus influenzae* in throats and saliva. J. Exp. Med. *29:* 259-266.

Privitera, G., F. Fayolle and M. Sebald. 1981. Resistance to tetracycline, erythromycin, and clindamycin in the *Bacteroides fragilis* group: Inducible versus constitutive tetracycline resistance. Antimicrob. Agents Chemother. *20:* 314-320.

Privitera, G., M. Sebald and F. Fayolle. 1979. Common regulatory mechanism of expression and conjugative ability of a tetracycline resistance plasmid in *Bacteroides fragilis*. Nature (London) *278:* 657-659.

Probst, J., M. Bruschi, N. Pfennig and J. LeGall. 1977. Cytochrome c-551.5 (c-7) from *Desulfuromonas acetoxidans*. Biochim. Biophys. Acta *460:* 58-64.

Progulske, A. and S.C. Holt. 1980. Transmission-scanning electron microscopic observations of selected *Eikenella corrodens* strains. J. Bacteriol. *143:* 1003-1018.

Prunier, J.-P., J. Luisetti and L. Gardan. 1970. Études sur les bactérioses des arbres fruitiers. II. Caractérisation d'un *Pseudomonas* non-fluorescent agent d'une bactériose nouvelle du pêcher. Ann. Phytopathol. *2:* 181-197.

Psallidas, P.G. and C.G. Panagopoulos. 1975. A new bacteriosis of almond caused by *Pseudomonas amygdali* sp. nov. Ann. Inst. Phytopathol. Benaki (N.S.) *11:* 94-108.

Pshenin, L.N. 1964. *Azotobacter miscellum* nov. sp. an inhabitant of the Black Sea. Microbiology (Engl. Transl.) *33:* 615-620.

Puchta, O. 1954. Experimentelle Untersuchungen über die Symbiose der Kleiderlaus *Pediculus vestimenti* Burm. Naturwissenschaften *41:* 71-72.

Puchta, O. 1955. Experimentelle Untersuchungen über die Bedeutung der Symbiose der Kleiderlaus *Pediculus vestimenti* Burm. Z. Parasitenkd. *17:* 1-40.

Puchta, O. 1956. Zuchtungsversuche an den Symbionten von *Pediculus vestimenti* Burm. nebst physiologischen und morphologischen Beobachtungen. Z. Morphol. Oekol. Tiere *44:* 416-441.

Puentes, R., M. Cerda, B. Orellana and T.M. Lopez. 1975. Sepsis a *Citrobacter* en el lactante. Rev. Chil. Pediat. *46:* 211-217.

Pueppke, S.G. and U.K. Benny. 1981. Induction of tumors on *Solanum tuberosum* L. by *Agrobacterium*: quantitative analysis, inhibition by carbohydrates, and virulence of selected strains. Physiol. Plant Pathol. *18:* 169-179.

Puget, K., F. Lavelle and A.M. Michelson. 1977. Superoxide dismutases from procaryote and eucaryote bioluminescent organisms. *In* Michelson, McCord and Fridovitch (Editors), Superoxide and Superoxide Dismutases. Academic Press Inc., London-New York-San Francisco, pp. 139-150.

Pulverer, G. and H.L. Ko. 1970. *Actinobacillus actinomycetem-comitans*: fermentative capabilities of 140 strains. Appl. Microbiol. *20:* 693-695.

Purcell, R.H., D. Taylor-Robinson, D. Wong and R.M. Chanock. 1966. Color test for the measurement of antibody to T-strain mycoplasmas. J. Bacteriol. *92:* 6-12.

Purcell, R.H., D. Taylor-Robinson, D.C. Wong and R.M. Chanock. 1966. A color test for the measurement of antibody to the non-acid-forming human mycoplasma species. Am. J. Epidemiol. *84:* 51-66.

Puttlitz, D.H. and H.W. Seeley, Jr. 1968. Physiology and nutrition of *Lampropedia hyalina*. J. Bacteriol. *96:* 931-938.

Quackenbush, R.L. 1977. Phylogenetic relationships of bacterial endosymbionts of *Paramecium aurelia*: polynucleotide sequence relationships of 51 kappa and its mutants. J. Bacteriol. *129:* 895-900.

Quackenbush, R.L. 1978. Genetic relationships among bacterial endosymbionts of *Paramecium aurelia*: deoxyribonucleotide sequence relationships among members of *Caedobacter*. J. Gen. Microbiol. *108:* 181-187.

Quackenbush, R.L. 1981. Plasmid DNA from six strains of *Caedobacter taeniospiralis*. Fed. Proc. *40:* 1752.

Quackenbush, R.L. 1982. *In* Validation of the publication of new names and new combinations previously effectively published outside the IJSB. List No. 9. Int. J. Syst. Bacteriol. *32:* 384-385.

Quan, S.F., W. Knapp, M.I. Goldenberg, B.W. Hudson, W.D. Lawton, T.H. Chen and L. Kartman. 1965. Isolation of a strain of *Pasteurella pseudotuberculosis* from Alaska identified as *Pasteurella pestis*; an immunofluorescent false positive. Am. J. Trop. Med. Hyg. *14:* 424-432.

Quayle, J.R. 1972. The metabolism of one-carbon compounds by micro-organisms. *In* Rose and Tempest (Editors), Advances in Microbial Physiology, Vol. 7, Academic Press, London, pp. 119-203.

Queener, S.F. and I.C. Gunsalus. 1970. Anthranilate synthetase enzyme system and complementation in *Pseudomonas* species. Proc. Natl. Acad. Sci. U.S.A. *67:* 1225-1232.

Quincke, G. 1967. Untersuchungen über die O-Antigène der Plesiomonaden. Arch. Hyg. (Berlin) *151:* 525-529.

Rabinovitch, M. and W. Plaut. 1962. Cytoplasmic DNA synthesis in *Amoeba proteus*. II. On the behavior and possible nature of the DNA-containing elements. J. Cell. Biol. *15:* 535-540.

Raccach, M., S. Rottem and S. Razin. 1975. Survival of frozen mycoplasmas. Appl. Microbiol. *30:* 167-171.

Radcliffe, C.W., D. Kafkewitz and A. Abuchowski. 1979. Asparaginase production by human clinical isolates of *Vibrio succinogenes*. Appl. Environ. Microbiol. *38:* 761-762.

Raether, W. and H. Seidenath. 1977. Survival of *Aegyptianella pullorum*, *Anaplasma marginale* and various parasitic protozoa following prolonged storage in liquid nitrogen. Z. Parasitenkd. *53:* 41-46.

Rafaeli-Eshkol, D. 1968. Studies on halotolerance in a moderately halophilic bacterium: Effect of growth conditions on salt resistance of the respiratory system. Biochem. J. *109:* 679-685.

Raghavarhari, K. and A.M.K. Reddy. 1959. *Cytocetes ovis* var. *decani* (n. sp.) as the cause of tick-borne fever in sheep in India. Indian J. Vet. Sci. *29:* 69-86.

Ragland, T.E., T. Kawasaki and J.M. Lowenstein. 1966. Comparative aspects of some bacterial dehydrogenases and transhydrogenases. J. Bacteriol. *91:* 236-244.

Rahaley, R.S. and W.E. White. 1977. *Histophilus ovis* infection in sheep in western Victoria. Aust. Vet. J. *53:* 124-127.

Rahimian, H. and D.J. Gumpf. 1980. Deoxyribonucleic acid relationship between *Spiroplasma citri* and the corn stunt spiroplasma. Int. J. Syst. Bacteriol. *30:* 605-608.

Rahn, O. 1937. New principles for the classification of bacteria. Zentralbl. Bakteriol. Parasitenkd. Infektionskr. Hyg. Abt. II *96:* 273-286.

Rainbow, C. 1971. Spoilage organisms in breweries. Process Biochemistry *6:* 15-18.

Raj, H.D. 1970. A new species-*Microcyclus flavus*. Int. J. Syst. Bacteriol. *20:* 61-81.

Raj, H.D. 1976. A new species: *Microcyclus marinus*. Int. J. Syst. Bacteriol. *26:* 528-544.

Raj, H.D. 1977. *Microcyclus* and related ring-forming bacteria. Crit. Rev. Microbiol. *5:* 243-269.

Rake, G. 1948. Family III. *Chlamydozaceae* Moshkovsky. *In* Breed, Murray and Hitchens (Editors), Bergey's Manual of Determinative Bacteriology, 6th Ed., The Williams and Wilkins Co., Baltimore, pp. 114-120.

Rake, G. 1948. The antigenic relationships of *Donovania granulomatis* (Anderson) and the significance of this organism in granuloma inguinale. Amer. J. Syph. Gonorrhea Vener. Dis. *32:* 150-158.

Rake, G. 1957. Family *Chlamydiaceae* fam. nov. *In* Breed, Murray and Smith (Editors), Bergey's Manual of Determinative Bacteriology, 7th Ed., The Williams and Wilkins Co., Baltimore, pp. 957-968.

Ralston, E., N.J. Palleroni and M. Doudoroff. 1973. *Pseudomonas pickettii*, a new species of clinical origin related to *Pseudomonas solanacearum*. Int. J. Syst. Bacteriol. *23:* 15-19.

Ralston-Barrett, E., N.J. Palleroni and M. Doudoroff. 1976. Phenotypic characterization and deoxyribonucleic acid homologies of the "*Pseudomonas alcaligenes*" group. Int. J. Syst. Bacteriol. *26:* 421-426.

Ramaley, R.F. and J. Hixson. 1970. Isolation of a non-pigmented, thermophilic bacterium similar to *Thermus aquaticus*. J. Bacteriol. *103:* 527-528.

Ramaley, R.F., F.R. Turner, L.E. Malick and R.B. Wilson. 1978. The morphology and surface structure of some extremely thermophilic bacteria found in slightly alkaline hot springs. *In* Friedman (Editor), Biochemistry of Thermophily. Academic Press, N.Y.

Rangaswami, G. and K.S.S. Easwaran. 1962. A bacterial leafspot disease of bhendi or okra. Andhra Agr. J. *9:* 1-2.

Ranhand, J.M., W.O. Mitchell, T.J. Popkin and R.M. Cole. 1980. Covalently closed circular deoxyribonucleic acids in spiroplasmas. J. Bacteriol. *143:* 1194-1199.

Ransom, S.E. and R.J. Huebner. 1951. Studies on the resistance of *Coxiella burneti* to physical and chemical agents. Am. J. Hyg. *53:* 110-119.

Rao, R.M.R. and J.L. Stokes. 1953. Nutrition of the acetic acid bacteria. J. Bacteriol. *65:* 405-412.

Rao, Y.P. and S.K. Mohan. 1970. A new bacterial leaf stripe disease of arecanut *(Areca catechu)* in Mysore State. Indian Phytopathol. *23:* 702-704.

Rarick, H.R., P.S. Riley and R. Martin. 1978. Carbon substrate utilization studies of some cultures of *Alcaligenes denitrificans*, *Alcaligenes faecalis*, and *Alcaligenes odorans* isolated from clinical specimens. J. Clin. Microbiol. *8:* 313-319.

Rauch, H.C. and M.J. Pickett. 1961. *Bordetella bronchiseptica* bacteriophage. Can. J. Microbiol. *7:* 125-133.

Rauss, K.F. 1936. The systematic position of Morgan's bacillus. J. Pathol. Bacteriol. *42:* 183-192.

Rauss, K.F., H. Puzova, L. Dubay, D. Velin, M. Doliak and S. Vörös. 1975. New serotypes of *Morganella morganii*. Acta Microbiol. Acad. Sci. Hung. *22:* 315-321.

Rauss, K.F. and S. Voros. 1959. The biochemical and serological properties of *Proteus morganii*. Acta Microbiol. Acad. Sci. Hung. *6:* 233-246.

Ravin, A.W. 1963. Experimental approaches to the study of bacterial phylogeny. Am. Natur. *97:* 307-318.

Ray, P.H., D.C. White and T.D. Brock. 1971a. Effect of temperature on the fatty acid composition of *Thermus aquaticus*. J. Bacteriol. *106:* 25-30.

Ray, P.H., D.C. White and T.D. Brock. 1971b. Effect of growth temperature on the lipid composition of *Thermus aquaticus*. J. Bacteriol. *108:* 227-235.

Raybould, T.J.G., J.E. Beesley and S. Chantler. 1981. Ultrastructural localization of characterized antigens of *Brucella abortus* and distribution among different biotypes. Infect. Immun. *32:* 318-322.

Raybould, T.J.G and S. Chantler. 1980. Serological differentiation between infected and vaccinated cattle by using purified soluble antigens from *Brucella abortus* in a hemagglutination system. Infect. Immun. *29:* 435-441.

Razin, A. and S. Razin. 1980. Methylated bases in mycoplasmal DNA. Nucleic Acid Res. *8:* 1383-1390.

Razin, S. 1968. *Mycoplasma* taxonomy studied by electrophoresis of cell proteins. J. Bacteriol. *96:* 687-694.

Razin, S. 1969. Structure and function in mycoplasma. Annu. Rev. Microbiol. *23:* 317-356.

Razin, S. 1978. The mycoplasmas. Microbiol. Rev. *42:* 414-470.

Razin, S. 1981. The mycoplasma membrane. *In* Ghosh (Editor), Organization of Prokaryotic Cell Membranes, Vol. I, CRC Press, Boca Raton, Florida. pp. 165-250.

Razin, S. and M. Argaman. 1963. Lysis of mycoplasma, bacterial protoplasts, spheroplasts and L-forms by various agents. J. Gen. Microbiol. *30:* 155-172.

Razin, S., M. Banai, H. Gamliel, A. Polliack, W. Bredt and I. Kahane. 1980. Scanning electron microscopy of mycoplasmas adhering to erythrocytes. Infect. Immun. *30:* 538-546.

Razin, S. and C. Boschwitz. 1968. The membrane of the *Streptobacillus moniliformis* L-phase. J. Gen. Microbiol. *54:* 21-32.

Razin, S. and R.C. Cleverdon. 1965. Carotenoids and cholesterol in membranes of *Mycoplasma laidlawii*. J. Gen. Microbiol. *41:* 409-415.

Razin, S. and B.J. Cosenza. 1966. Growth phases of *Mycoplasma* in liquid media observed with phase-contrast microscope. J. Bacteriol. *91:* 858-869.

Razin, S., B.J. Cosenza and M.E. Tourtellotte. 1966. Variations in mycoplasma morphology induced by long-chain fatty acids. J. Gen. Microbiol. *42:* 139-145.

Razin, S., R. Harasawa and M.F. Barile. 1983. Cleavage patterns of the mycoplasma chromosome, obtained by using restriction endonucleases, as indicators of genetic relatedness among strains. Int. J. Syst. Bacteriol. *33:* 201-206.

Razin, S., M. Hasin, Z. Ne'eman and S. Rottem. 1973. Isolation, chemical composition, and ultrastructural features of the cell membrane of the mycoplasma-like organism *Spiroplasma citri*. J. Bacteriol. *116:* 1421-1435.

Razin, S., I. Kahane, M. Banai and W. Bredt. 1981. Adhesion of mycoplasmas to eukaryotic cells. *In* Elliott, O'Conner and Whelan (Editors), Adhesion and Microorganism Pathogenicity. Ciba Foundation Symp. 80, Pitman Medical, London, pp. 98-113.

Razin, S., I. Kahane and J. Kovartovsky. 1972. Immunochemistry of mycoplasma membranes. *In* Elliott and Birch (Editors), Pathogenic Mycoplasmas, Ciba Foundation Symposium, Elsevier Exerpta Medica North-Holland, Amsterdam, pp. 93-122.

Razin, S., S. Kutner, H. Efrati and S. Rottem. 1980. Phospholipid and cholesterol uptake by mycoplasma cells. Biochim. Biophys. Acta *598:* 628-640.

Razin, S., G.K. Masover, M. Palant and L. Hayflick. 1977. Morphology of *Ureaplasma urealyticum* (T-mycoplasma) organisms and colonies. J. Bacteriol. *130:* 464-471.

Razin, S. and O. Oliver. 1961. Morphogenesis of mycoplasmal and bacterial L-form colonies. J. Gen. Microbiol. *24:* 225-237.

Razin, S. and S. Rottem. 1963. Fatty acid requirements of *Mycoplasma laidlawii*. J. Gen. Microbiol. *33:* 459-470.

Razin, S. and S. Rottem. 1967. Identification of *Mycoplasma* and other microorganisms by polyacrylamide gel electrophoresis of cell proteins. J. Bacteriol. *94:* 1807-1810.

Razin, S. and S. Rottem. 1976. Techniques for the manipulation of mycoplasma membranes. *In* Maddy (Editor), Biochemical Analysis of Membranes, Chapman and Hall, London, pp. 3-26.

Razin, S. and Z. Shafer. 1969. Incorporation of cholesterol by membranes of bacterial L-phase variants; with an appendix on the determination of the L-phase parentage by the electrophoretic patterns of cell protein. J. Gen. Microbiol. *58:* 327-339.

Razin, S. and J.G. Tully. 1970. Cholesterol requirement of mycoplasmas. J. Bacteriol. *102:* 306-310.

Razin, S. and J. G. Tully (Editors). 1983. Methods in Mycoplasmology, Vol. I. Mycoplasma characterization. Academic Press, New York.

Reanney, D. 1976. Extrachromosomal elements as possible agents of adaptation and development. Bacteriol. Rev. *40:* 552-590.

Reddy, C.A. and M.P. Bryant. 1977. Deoxyribonucleic acid base composition of certain species of the genus *Bacteroides*. Can. J. Microbiol. *23:* 1252-1256.

Reddy, C.S. and J. Godkin. 1923. A bacterial disease of brome-grass. Phytopathology *13:* 75-86.

Reddy, C.S., J. Godkin and A.G. Johnson. 1924. Bacterial blight of rye. J. Agr. Res. *28:* 1039-1040.

Redfearn, M.S. 1964. Toxic lysolipid isolation from *Pseudomonas pseudomallei*. Science *146:* 648-649.

Redfearn, M.S. and N.J. Palleroni. 1975. Glanders and melioidosis. *In* Hubbert, McCulloch and Schnurrenberger (Editors), Diseases Transmitted from Animals to Man, 6th Ed., C.C. Thomas, Springfield, Illinois, pp. 110-128.

Redfearn, M.S., N.J. Palleroni and R.Y. Stanier. 1966. A comparative study of *Pseudomonas pseudomallei* and *Bacillus mallei*. J. Gen. Microbiol. *43:* 293-313.

Redmond, D.L. and E. Kotcher. 1963. Cultural and serological studies on *Haemophilus vaginalis*. J. Gen. Microbiol. *33:* 77-87.

Redway, K.F. and Lapage, S.P. 1974. Effect of carbohydrates and related compounds on the long-term preservation of freeze-dried bacteria. Cryobiology *11:* 73-79.

Reeve, E.C.R. and J.A. Braithwaite. 1973. Lac⁺ plasmids are responsible for the strong lactose-positive phenotype found in many strains of *Klebsiella* species. Genet. Res. *22:* 329-333.

Regendanz, P. and W. Kikuth. 1928. Sur la *Bartonella muris ratti* (Meyer). C.R. Soc. Biol. (Paris) *98:* 1578-1579.

Reh, M. and H.G. Schlegel. 1969. Die Biosynthese von Isoleucin und Valin in *Hydrogenomonas* H 16. Arch. Mikrobiol. *67:* 110-127.

Řeháček, J., V. Pospisil and F. Ciampor. 1976. First record of bacillary rickettsia-like organisms in European tick *Dermacentor marginatus* (Sulzer). Folia Parasitologica (Praha) *23:* 301-307.

Reich, P.R., N.L. Somerson, C.J. Hybner, R.M. Chanock and S.M. Weissman. 1966. Genetic differentiation by nucleic acid homology. I. Relationships among *Mycoplasma* species of man. J. Bacteriol. *92:* 302-310.

Reichelt, J.L. and P. Baumann. 1973. Taxonomy of the marine, luminous bacteria. Arch. Mikrobiol. *94:* 283-330.

Reichelt, J.L. and P. Baumann. 1973. Change of the name *Alteromonas marinopraesens* (ZoBell and Upham) Baumann et al. to *Alteromonas haloplanktis* (ZoBell and Upham) comb. nov. and assignment of strain ATCC 23821 (*Pseudomonas enalia*) and strain c-A1 of De Voe and Oginsky to this species. Int. J. Syst. Bacteriol. *23:* 438-441.

Reichelt, J. L. and P. Baumann. 1974. Effect of sodium chloride on the growth of heterotrophic marine bacteria. Arch. Microbiol. *97:* 329-345.

Reichelt, J.L. and P. Baumann. 1975. *Photobacterium mandapamensis* Hendrie et al., a later subjective synonym of *Photobacterium leiognathi* Boisvert et al. Int. J. Syst. Bacteriol. *25:* 208-209.

Reichelt, J.L., P. Baumann and L. Baumann. 1976. Study of genetic relationships among marine species of the genera *Beneckea* and *Photobacterium* by means of in vitro DNA/DNA hybridization. Arch. Microbiol. *110:* 101-120.

Reichelt, J.L., P. Baumann and L. Baumann. 1979. *In* List No. 2, Validation of the publication of new names and combinations previously effectively published outside the IJSB. Int. J. Syst. Bacteriol *29:* 79-80.

Reichelt, J.L., K.H. Nealson and J.W. Hastings. 1977. The specificity of symbiosis: pony fish and luminescent bacteria. Arch. Microbiol. *112:* 157-161.

Reichenbach, H., W. Kohl and H. Achenbach. 1981. The flexirubin-type pigments, chemosystematically useful compounds. *In* Reichenbach and Weeks (Editors), The *Flavobacterium-Cytophaga* Group (Proceedings of the International Symposium on Yellow-Pigmented Gram-negative Bacteria of the *Flavobacterium-Cytophaga* Group, Braunschweig, July 8 to 11, 1980). Verlag Chemie, Weinheim, pp. 101-108.

Reid, D.H. 1938. Grease-spot of passion-fruit. N.Z. J. Sci. Technol. *A20:* 260-265.

Reiderer-Henderson, M.A. and P.W. Wilson. 1970. Nitrogen-fixation by sulphate-reducing bacteria. J. Gen. Microbiol. *61:* 27-32.

Reinhardt, C., A. Aeschlimann and H. Hecker. 1972. Distribution of rickettsia-like microorganisms in various organs of an *Ornithodorus moubata* laboratory strain (Ixodoidea, Argasidae) as revealed by electron microscopy. Z. Parasitenkd. *39:* 201-209.

Reinhardt, C., R. Steiger and H. Hecker. 1972. Ultrastructural study of the midgut mycetome-bacteroids of the tsetse flies *Glossina morsitans, G. fuscipes,* and *G. brevipalpis* (Diptera, Brachycera). Acta Trop. *29:* 280-288.

Reinhold, L. 1966. Untersuchungen an *Bacteroides corrodens* (Eiken, 1958). Zentralbl. Bakteriol. Parasitenkd. Infektionskr. Hyg. Abt. 1 *201:* 49-57.

Reiss-Gutfreund, R.J. 1956. Un nouveau réservoir de virus pour *Rickettsia prowazeki*: Les animaux domestiques et leur tiques. Bull. Soc. Pathol. Exotique *49:* 946-1023.

Reiss-Gutfreund, R.J. 1966. The isolation of *Rickettsia prowazeki* and *mooseri* from unusual sources. J. Trop. Med. Hyg. *15:* 943-949.

Reistad, R. 1970. On the composition and nature of the bulk protein of extremely halophilic bacteria. Arch. Mikrobiol. *71:* 353-360.

Reistad, R. 1975. Amino sugar and amino acid constituents of the cell wall of the extremely halophilic cocci. Arch. Mikrobiol. *102:* 71-73.

Reistad, R. 1978. Extremely halophilic bacteria. A study with emphasis on halococcal cell walls. Doctoral thesis. Universitetet i Oslo, Oslo, pp. 1-59.

Remsen, C. C., S. W. Watson, J. B. Waterbury and H. G. Trüper. 1968. Fine structure of *Ectothiorhodospira mobilis* Pelsh. J. Bacteriol. *95:* 2374-2392.

Rennie, R.P. and J.B.R. Duncan. 1974. Combined biochemical and serological typing of clinical isolates of Klebsiella. Appl. Microbiol. *28:* 534-539.

Renoux, G. 1952a. La classification des *Brucella*. Remarques à propos de l'identification de 2,598 souches. Ann. Pasteur (Paris) *82:* 289-298.

Renoux, G. 1952b. Une nouvelle "espèce" de *Brucella intermedia*. Ann. Inst. Pasteur (Paris) *83:* 814-815.

Renoux, G., M. Renoux and R. Tinelli. 1973. Phenol-water fractions from smooth *Brucella abortus* and *Brucella melitensis*: immuno-chemical analysis and biologic behaviour. J. Infect. Dis. *127:* 139-148.

Repaske, R. and R. Mayer. 1976. Dense autotrophic cultures of *Alcaligenes eutrophus*. Appl. Environ. Microbiol. *32:* 592-597.

Rest, R.F. and D.C. Robertson. 1974. Glucose transport in *Brucella abortus*. J.

Bacteriol. *118:* 250-259.

Rest, R.F. and D.C. Robertson. 1975. Characterization of the electron transport system in *Brucella abortus.* J. Bacteriol. *122:* 139-144.

Retailliau, H.F., A.W. Hightower, R.E. Dixon and J.R. Allen. 1979. *Acinetobacter calcoaceticus:* A nosocomial pathogen with an unusual seasonal pattern. J. Infect. Dis. *139:* 371-375.

Rettger, L.F. 1909. Further studies on fatal septicaemia in young chickens or "white diarrhoea". J. Med. Res. *21:* 115-123.

Reusch, V.W. and C. Panos. 1976. Defective synthesis of lipid intermediates for peptidoglycan formation in a stabilized L-form of *Streptococcus pyogenes.* J. Bacteriol. *126:* 300-311.

Reyn, A. 1974. Family I. *Neisseriaceae* Prévot 1933, 119. *In* Buchanan and Gibbons (Editors), Bergey's Manual of Determinative Bacteriology, 8th Ed. The Williams and Wilkins Co., Baltimore, p. 427.

Reyn, A. 1974. The genus *Neisseria. In* Buchanan and Gibbons (Editors), Bergey's Manual of Determinative Bacteriology, 8th Ed., The Williams and Wilkins Co., Baltimore, pp. 428-432.

Reyn, A., A. Birch-Andersen and S.P. Lapage. 1966. An electron microscope study of thin sections of *Haemophilus vaginalis* (Gardner and Dukes) and some possibly related species. Can. J. Microbiol. *12:* 1125-1136.

Reyn, A., A. Birch-Andersen and R.G.E. Murray. 1971. The fine structure of *Cardiobacterium hominis.* Acta Pathol. Scand. Sect. B Microbiol. Immunol. *79:* 51-60.

Reyn, A., A.E. Jephcott and H. Raun. 1971. Brief Report. *Neisseria gonorrhoeae.* Colony variation II. Acta Pathol. Microbiol. Scand. Sect. B *79:* 435-436.

Reynders, L. and K. Vlassak. 1979. Conversion of tryptophan to indoleacetic acid by *Azospirillum brasilense.* Soil Biol. Biochem. *11:* 547-548.

Rhodes, M.E. 1958. The cytology of *Pseudomonas* spp. as revealed by a silver-plating staining method. J. Gen. Microbiol. *18:* 639-648.

Ribeiro, C.D., P. Davis and D.M. Jones. 1976. *Citrobacter koseri* meningitis in a special care baby unit. J. Clin. Pathol. *29:* 1094-1096.

Ribi, E. and C. Shepard. 1955. Morphology of *Bacterium tularense* during its growth cycle in liquid medium as revealed by the electron microscope. Exp. Cell Res. *8:* 474-487.

Ricard, B., M. Garnier and J.M. Bove. 1982. Characterization of spiroplasmal virus 3 from spiroplasmas and discovery of a new spiroplasmal virus (SpV4). Rev. Infect. Dis. *4:* S275.

Richard, C. 1965. Mesure de l'activite uréasique des *Proteus* au moyen de la réaction phenol-hypochlorite de Bertholet. Ann. Inst. Pasteur (Paris). *109:* 516-524.

Richard, C. 1972. Méthode rapide pour l'étude des réactions de rouge de méthyle et Voges-Proskauer. Ann. Inst. Pasteur (Paris) *122:* 979-986.

Richard, C. 1975. A propos de "nouvelles" entérobactéries: *Edwardsiella tarda, Levinea malonatica* et *amalonatica* et *Enterobacter agglomerans.* Bull. Inst. Pasteur (Paris) *73:* 357-381.

Richard, C. 1977. La tétrathionate-réductase (TTR) chez les bacilles à gram négatif: intérêt diagnostique et épidémiologique. Bull. Inst. Pasteur (Paris) *75:* 369-382.

Richard, C. 1977. Présence chez *Enterobacter aerogenes* d'antigènes capsulaires apparentés à ceux de *Klebsiella:* intérêt de l'utilisation du métahydroxybenzoate dans le diagnostic différentiel *E. aerogenes - K. pneumoniae.* Ann. Microbiol. (Inst. Pasteur) *128A:* 289-295.

Richard, C. 1978. Isolation and diagnosis of *Erwinia herbicola (Enterobacter agglomerans)* in medical bacteriology (a study of 380 strains). Proc. 4th Int. Conf. Plant Pathol. Bacteriol. Angers (France): 451-452.

Richard, C. 1979. Enterobactéries inhabituelles. Bull. Inst. Pasteur *77:* 83-98.

Richard, C., B. Joly, J. Sirot, G.H. Stoleru and M. Popoff. 1976. Étude de souches de *Enterobacter* appartenant à un groupe particulier proche de *E. Aerogenes.* Ann. Inst. Pasteur (Paris) *127A:* 545-548.

Richard, C., H. Monteil and B. Laurent. 1979a. Individualisation des six nouveaux types antigéniques de *Flavobacterium meningosepticum.* Ann. Microbiol. (Inst. Pasteur) *130B:* 141-144.

Richard, C., H. Monteil, A. LeFaou and B. Laurent. 1979b. Étude de souches de *Flavobacterium* isolées à Strasbourg dans un service de réanimation médicale. Individualisation d'un nouveau sérotype (G) de *Flavobacterium meningosepticum.* Med. Mal. Infect. *9:* 124-128.

Richards, A.G. and M.A. Brooks. 1958. Internal symbioses in insects. Annu. Rev. Entomol. *3:* 37-56.

Richards, C.S. 1978. Spirochaetes in planorbid molluscs. Trans. Amer. Microsc. Soc. *97:* 191-198.

Richards, K.L. and S.D. Douglas. 1978. Pathophysiological effects of *Vibrio cholerae* and enterotoxigenic *Escherichia coli* and their exotoxins on eucaryotic cells. Microbiol. Rev. *42:* 592-613.

Richards, M. and D.A. Corbey. 1974. Isolation of *Zymomonas* from primed beer. J. Inst. Brewing *80:* 241-244.

Richardson, M.J. 1979. An annotated list of seed-borne diseases. Phytopathol. Pap. *23:* 1-320.

Richardson, W.P. and J.C. Sadoff. 1977. Production of a capsule by *Neisseria gonorrhoeae.* Infect. Immun. *15:* 663-664.

Ricketts, H.T. 1911. Contributions to medical science by Howard Taylor Ricketts. 1870-1910. Chicago: University of Chicago Press.

Ride, M. 1958. Sur l'étiologie du chancre suintant du peuplier. C. R. Hebd. Séances Acad. Sci. Paris *246:* 2795-2798.

Ride, M. and S. Ride. 1978. *Xanthomonas populi* (Ride) comb. nov. (syn. *Aplan-obacter populi* Ride), specificité, variabilité et absence de relations avec *Erwinia cancerogena* Ur. Eur. J. For. Path. *8:* 310-333.

Ride, M. and S. Ride. 1979. The causal agent of the bacterial canker of poplar (ex *Aplanobacter populi* Ride): *Xanthomonas populi* or *Xanthomonas campestris* pathovar *populi*?. Proc. 4th Internat. Conf. Plant Pathol. Bact., Angers *I:* 365-370.

Rieger-Hug, D. and S. Stirm. 1981. Comparative study of host capsule depolymerases associated with *Klebsiella* bacteriophages. Virology *113:* 363-378.

Riendeau, D. and E. Meighen. 1979. Evidence for a fatty acid reductase catalyzing the synthesis of aldehydes for the bacterial bioluminescent reaction. J. Biol. Chem. *254:* 7488-7490.

Ries, E. 1931a. Ueber ein regelmässiges Rickettsienvorkommen bei der Hühnerlaus. Zentralbl. Bakteriol. Parasitenkd. Infektionskr. Hyg. Abt. I Orig. *121:* 40-49.

Ries, E. 1931b. Die Symbiose der Läuse und Federlinge. Z. Morphol. Oekol. Tiere *20:* 233-367.

Rietschel, E.T., S. Hase, M.-T. King, J. Redmond and V. Lehmann. 1977. Chemical structure of lipid A. *In* Schlessinger (Editor) "Microbiology - 1977", American Society for Microbiology, Washington D.C., pp. 262-268.

Rietschel, E.T., O. Lüderitz and W.A. Volk. 1975. Nature, type of linkage, and absolute configuration of (hydroxy) fatty acids in lipopolysaccharides from *Xanthomonas sinensis* and related strains. J. Bacteriol. *122:* 1180-1188.

Riker, A.J., W.M. Banfield, W.H. Wright, G.W. Keitt and H.E. Sagen. 1930. Studies on infectious hairy root of nursery apple trees. J. Agr. Res. *41:* 507-540.

Riker, A.J., F.R. Jones and M.C. Davis. 1935. Bacterial leaf spot of alfalfa. J. Agr. Res. *51:* 177-182.

Rikihisa, Y. and S. Ito. 1980. Localization of electron-dense tracers during entry of *Rickettsia tsutsugamushi* into polymorphonuclear leukocytes. Infect. Immun. *30:* 231-243.

Riley, P.S., H.W. Tatum and R.E. Weaver. 1972. *Pseudomonas putrefaciens* isolates from clinical specimens. Appl. Microbiol. *24:* 798-800.

Riley, P.S., H.W. Tatum and P.E. Weaver. 1973. Identity of HB-1 of King and *Eikenella corrodens* (Eiken) Jackson and Goodman. Int. J. Syst. Bacteriol. *23:* 75-76.

Riley, P.S. and R.E. Weaver. 1974. Observation of nitrate reduction in some non-saccharolytic strains of *Acinetobacter.* Appl. Microbiol. *28:* 1071-1072.

Riley, P.S. and R.E. Weaver. 1977. Comparison of thirty-seven strains of Vd-3 bacteria with *Agrobacterium radiobacter:* morphological and physiological observations. J. Clin. Microbiol. *5:* 172-177.

Rimler, R.B., E.B. Shotts, J. Brown and R.B. Davis. 1976. The effect of atmospheric conditions on the growth of *Haemophilus gallinarum* in a defined medium. J. Gen. Microbiol. *92:* 405-409.

Rimler, R.B., E.B. Shotts, J. Brown and R.B. Davis. 1977. The effect of sodium chloride and NADH on the growth of 6 strains of *Haemophilus* species pathogenic to chickens. J. Gen. Microbiol. *98:* 349-354.

Rioche, M. 1967. Lesions microscopiques de la rickettsiose generale bovine a *Rickettsia (Ehrlichia) bovis* (Donatien et Lestoquard 1936). Rev. Elevage. Med. Vet. des Pays. Trop. (Paris) *20:* 415-427.

Riou, J.Y. 1977. Diagnostic bactériologiques des espèces des genres *Neisseria* et *Branhamella.* Ann. Biol. Clin. *35:* 73-87.

Rische, H., W. Beer, G. Seltmann, E. Thal and G. Horn. 1973. Die zusammensetzung der lipopolysaccharide von *Yersinia enterocolitica* und *Yersinia pseudotuberculosis* und die empfindlichkeit gegenüber bakteriophagen. Contr. Microbiol. Immunol. *2:* 23-26.

Ristic, M. 1960. Anaplasmosis. Adv. Vet. Sci. *6:* 112-192.

Ristic, M. 1968. Anaplasmosis. *In* Weinman and Ristic (Editors), Infectious Blood Diseases of Man and Animals, Vol. II, Academic Press, New York, pp. 474-542.

Ristic, M. 1977. Bovine anaplasmosis. *In* Kreier (Editors), Parasitic Protozoa, Vol. IV, Academic Press, New York, pp. 235-239.

Ristic, M. 1981. Anaplasmosis. *In* Ristic and McIntyre (Editors), Diseases of Cattle in the Tropics, Martinus Nijhoff Publishers, The Hague, pp. 327-344.

Ristic, M. and C.A. Carson. 1977. Methods of immunoprophylaxis against bovine anaplasmosis with emphasis on use of the attenuated *Anaplasma marginale* vaccine. *In* Miller, Pino and McKelvey (Editors), Immunity to Blood Parasites of Animals and Man, Advances in Experimental Medicine and Biology 93, Plenum Press, New York, pp. 151-188.

Ristic, M., M. Herzberg, D.A. Sanders and J.W. Williams. 1956. Actinobacillosis: I. An evaluation of cultural characteristics of selected strains of *Actinobacillus lignieresi.* Am. J. Vet. Res. *17:* 555-562.

Ristic, M., D.L. Huxsoll, N. Tachibana and G. Rapmund. 1981. Evidence of a serologic relationship between *Ehrlichia canis* and *Rickettsia sennetsu.* J. Trop. Med. Hyg. *30:* 1324-1328.

Ristic, M., D.L. Huxsoll, R.M. Weisiger, P.K. Hildebrandt and M.B.A. Nyindo. 1972. Serological diagnosis of tropical canine pancytopenia by indirect immunofluorescence. Infect. Immun. *6:* 226-231.

Ristic, M. and J.P. Kreier. 1974. *Anaplasmataceae* Philip 1957. *In* Buchanan and Gibbons (Editors), Bergey's Manual of Determinative Bacteriology, 8th Ed., The Williams and Wilkins Co., Baltimore, pp. 906-914.

Ristic, M. and J.P. Kreier. 1979. Hemotrophic bacteria. N. Engl. J. Med. *301:* 937-939.

Ristic, M. and D.K. Mann. 1963. Anaplasmosis. IX. Immunoserologic properties of soluble *Anaplasma* antigens. Am. J. Vet. Res. *24:* 478-482.

Ristic, M. and A. Watrach. 1961. Studies in anaplasmosis. II. Electron microscopy of *Anaplasma marginale* in deer. Am. J. Vet. Res. *22:* 109-116.

Ristroph, J.D., K.W. Hedlund and R.G. Allen. 1980. Liquid medium for growth of *Legionella pneumophila*. J. Clin. Microbiol. *11:* 19-21.

Ritchie, A. E., R. F. Keeler and J. H. Bryner. 1966. Anatomical features of *Vibrio fetus*: an electron microscope survey. J. Gen. Microbiol. *43:* 427-438.

Ritchie, D.F. and E.J. Klos. 1979. Some properties of *Erwinia amylovora* bacteriophages. Phytopathology *69:* 1078-1083.

Rittenberg, B. T. and S. C. Rittenberg. 1962. The growth of *Spirillum volutans* Ehrenberg in mixed and pure cultures. Arch. Mikrobiol. *42:* 138-153.

Rittenberg, S.C. 1969. The roles of exogenous organic matter in the physiology of chemolithotrophic bacteria. Adv. Microb. Physiol. *3:* 159-196.

Rittenberg, S.C. and R.B. Hespell. 1975. Energy efficiency of intraperiplasmic growth of *Bdellovibrio bacteriovorus*. J. Bacteriol. *121:* 1158-1165.

Rittenberg, S. C. and D. Langley. 1975. Utilization of nucleoside monophosphates *per se* for intraperiplasmic growth of *Bdellovibrio bacteriovorus*. J. Bacteriol. *121:* 1137-1144.

Rittenberg, S. C. and M. F. Thomashow. 1979. Intraperiplasmic growth-life in a cozy environment. *In* Schlessinger (Editor), Microbiology - 1979, American Society for Microbiology, Washington, D. C., pp. 80-86.

Rittenhouse, H.G., J.B. Rodda and B.A. McFadden. 1973. Immunologically cross-reacting proteins in cell walls of many bacteria. J. Bacteriol. *113:* 1400-1403.

Ritter, D.B. and R.K. Gerloff. 1966. Deoxyribonucleic acid hybridization among some species of the genus *Pasteurella* J. Bacteriol. *92:* 1838-1839.

Rivers, T.M. 1922. Influenza-like bacilli. Growth of influenza-like bacilli on media containing only an autoclave-labile substance as an accessory food factor. Johns Hopkins Hosp. Bull. *33:* 149-151.

Rizki, M.T.M. 1954. Desoxyribose nucleic acid in the symbiotic microorganisms of the cockroach, *Blattella germanica*. Science (Wash. D.C.) *120:* 35-36.

Robbins, J.B. 1978. Vaccines for the prevention of encapsulated bacterial diseases: current status, problems and prospects for the future. Immunochem. *15:* 839-854.

Robbins, F.C., R. Rustigian, M.J. Snyder and J.E. Smadel. 1946. Q fever in the Mediterranean area: report of its occurrence in Allied troops. III. The etiological agent. Am. J. Hyg. *44:* 51-63.

Robbs, C.F. 1956. Uma nova doenca bacteriana do mamoeiro. Rev. Soc. Brasil. Agron. *12:* 73-76.

Robbs, C.F., R. de L.D. Ribiero and O. Kimura. 1974. Sobre a posicao taxonomica de *Pseudomonas mangiferaeindicae* Patel et al. 1948, agente causal da "Mancha bacteriana das folhas da mangueira *(Mangifera indica* L.). Arq. Univ. Fed. Rural, Rio de Janeiro. *4:* 11-14.

Robert-Gero, M., M. Poiret and R.Y. Stanier. 1969. The function of the beta-keto-adipate pathway in *Pseudomonas acidovorans*. J. Gen. Microbiol. *57:* 207-214.

Roberts, D.E., A. Ingold, S.V. Want and J.R. May. 1974. Osmotically stable L forms of *Haemophilus influenzae* and their significance in testing sensitivity to penicillins. J. Clin. Pathol. *27:* 560-564.

Roberts, D.H. 1963. The isolation of a previously unreported serotype and some observations on the incidence of mycoplasmas in poultry. Vet. Rec. *75:* 665-667.

Roberts, D.H. 1964. The isolation of an influenza A virus and a *Mycoplasma* associated with duck sinusitis. Vet. Res. *76:* 470-473.

Roberts, D.S. 1956. A new pathogen from a ewe with mastitis. Aust. Vet. J. *32:* 330-332.

Roberts, D.S. and J.R. Egerton. 1969. The aetiology and pathogenesis of ovine foot rot. II. The pathogenic association of *Fusiformis nodosus* and *F. necrophorus*. J. Comp. Pathol. *79:* 217-227.

Roberts, D.W. and M.A. Strand (Editors). 1977. Pathogens of medically important arthropods. Bull. W.H.O. *55* (Suppl. 1): 1-419.

Roberts, G.P., W.T. Leps, L.E. Silver and W.J. Brill. 1980. Use of two-dimensional polyacrylamide gel electrophoresis to identify and classify *Rhizobium* strains. Appl. Environ. Microbiol. *39:* 414-422.

Roberts, M. and S. Falkow. 1978. Plasmid-mediated chromosomal gene transfer in *Neisseria gonorrhoeae*. J. Bacteriol. *134:* 66-70.

Roberts, P. 1974. *Erwinia rhapontici* (Millard) Burkholder associated with pink grain of wheat. J. Appl. Bacteriol. *37:* 353-358.

Roberts, P. and C.M. Scarlett. 1981. *Pseudomonas corrugata* sp. nov. *In* Validation of the publication of new names and new combinations previously effectively published outside the IJSB. List No. 6. Int. J. Syst. Bacteriol. *31:* 216.

Roberts, R.J. 1976. Restriction endonucleases. Crit. Rev. Biochem. *4:* 123-164.

Roberts, W.P., M.E. Tate and A. Kerr. 1977. Agrocin 84 is a 6-N-phosphoramidate of an adenine nucleotide analogue. Nature, London *265:* 379-381.

Robertson, B., P. Åman, A.G. Darvill, M. McNeil and P. Albersheim. 1981. Host-symbiont interactions V. The structure of the acidic extracellular polysaccharides secreted by *Rhizobium leguminosarum* and *Rhizobium trifolii*. Pl. Physiol. *67:* 389-400.

Robertson, A. 1924. Observation of the causal organism of rat-bite fever in man. Ann. Trop. Med. Parasitol. *18:* 157-175.

Robertson, D.C. and W.G. McCullough. 1968a. The glucose catabolism of the genus *Brucella*. I. Evaluation of pathways. Arch. Biochem. Biophys. *127:* 263-273.

Robertson, D.C. and W.G. McCullough. 1968b. The glucose catabolism of the genus *Brucella*. II. Cell-free studies with *B. abortus* (S-19). Arch. Biochem. Biophys. *127:* 445-456.

Robertson, J. and E. Smook. 1976. Cytochemical evidence of extramembranous carbohydrates on *Ureaplasma urealyticum* (T-strain mycoplasma). J. Bacteriol. *128:* 658-660.

Robertson, J.A. and G.W. Stemke. 1979. Modified metabolic inhibition test for serotyping strains of *Ureaplasma urealyticum* (T-strain mycoplasma). J. Clin. Microbiol. *9:* 673-676.

Robertson, J.A. and G.W. Stemke. 1983. An expanded serotyping scheme for strains of *Ureaplasma urealyticum* isolated from humans. Int. J. Syst. Bacteriol. (In press)

Robin, C. 1853. Histoire naturelle des végétaux parasites qui crossent sur l'homme et sur les animaux vivants. J.-B. Baillière, Paris.

Robinson, D.M. and S.E. Hasty. 1974. Production of a potent vaccine from the attenuated M-44 strain of *Coxiella burneti*. Appl. Microbiol. *27:* 777-783.

Robinson, I.M. 1979. Special features of *Anaeroplasma*. *In* Barile and Razin (Editors), The Mycoplasmas, Vol. I, Academic Press, Inc., New York, pp. 515-528.

Robinson, I.M. and M.J. Allison. 1975. Transfer of *Acholeplasma bactoclasticum* Robinson and Hungate to the genus *Anaeroplasma (Anaeroplasma bactoclasticum* Robinson and Hungate comb. nov.): emended description of the species. Int. J. Syst. Bacteriol. *25:* 182-186.

Robinson, I.M., M.J. Allison and J.A. Bucklin. 1981. Characterization of the cecal bacteria of normal pigs. Appl. Environ. Microbiol. *41:* 950-955.

Robinson, I.M., M.J. Allison and P.A. Hartman. 1975. *Anaeroplasma abactoclasticum* gen. nov., sp. nov.: an obligately anaerobic mycoplasma from the rumen. Int. J. Syst. Bacteriol. *25:* 173-181.

Robinson, I.M. and K.R. Rhoades. 1977. Serological relationships between strains of anaerobic mycoplasmas. Int. J. Syst. Bacteriol. *27:* 200-203.

Robinson, I.M. and A.E. Ritchie. 1981. Emendation of *Acetivibrio* and description of *Acetivibrio ethanolgignens,* a new species from the colons of pigs with dysentery. Int. J. Syst. Bacteriol. *31:* 333-338.

Robinson, J. and N.E. Gibbons. 1952. The effect of salts on the growth of *Micrococcus halodenitrificans* (n. sp.). Can. J. Bot. *30:* 147-154.

Robinson, J., N.E. Gibbons and F.S. Thatcher. 1952. A mechanism of halophilism in *Micrococcus halodenitrificans*. J. Bacteriol. *64:* 69-77.

Robinson, J.P. and R.E. Hungate. 1973. *Acholeplasma bactoclasticum* sp. n., an anaerobic mycoplasma from the bovine rumen. Int. J. Syst. Bacteriol. *23:* 171-181.

Robinson, J.V.A. and A.L. James. 1973. Some serological studies on *Bacteroides corrodens*. J. Gen. Microbiol. *78:* 193-197.

Robinson, J.V.A. and A.L. James. 1974. In vitro susceptibility of *Bacteroides corrodens* to ten chemotherapeutic agents. Antimicrob. Agents and Chemother. *6:* 545-548.

Robyt, J.F. and R.J. Ackerman. 1971. Isolation, purification, and characterization of a maltotetraose producing amylase from *Pseudomonas stutzeri*. Arch. Biochem. Biophys. *145:* 105-114.

Roché, C., H. Albertyn, N.O. van Gylswyk and A. Kistner. 1973. The growth response of cellulolytic acetate-utilizing and acetate-producing butyrivibrios to volatile fatty acids and other nutrients. J. Gen. Microbiol. *78:* 253-260.

Rode, L.M., B.R.S. Genthner and M.P. Bryant. 1981. Synthrophic association by cocultures of the methanol- and CO_2-H_2-utilizing species, *Eubacterium limosum,* and pectin-fermenting *Lachnospira multiparus* during growth in a pectin medium. Appl. Environ. Microbiol. *42:* 20-22.

Rodina, A. G. 1956. Aquatic spirilla fixing molecular nitrogen. Microbiology (U.S.S.R.) *25:* 144-149.

Rodionova, A.V. 1976. Study of the catalase activity in the causative agent of tularemia. (Russ.). Zh. Mikrobiol. Epidemiol. Immunobiol. *53:* 60-63.

Rodrigues, L.P., R. Schneerson and J.B. Robbins. 1971. Immunity to *Haemophilus influenzae* type b. I. The isolation and some physicochemical, serologic and biologic properties of the capsular polysaccharide of *Haemophilus influenzae* type b. J. Immunol. *107:* 1071-1080.

Rodriguez-Valera, F., F. Ruiz-Berraquero and A. Ramos-Cormenzana. 1980. Isolation of extremely halophilic bacteria able to grow in defined inorganic media with single carbon source. J. Gen. Microbiol. *119:* 535-538.

Rodwell, A.W. 1969. A defined medium for mycoplasma strain Y. J. Gen. Microbiol. *58:* 39-47.

Rodwell, A.W. 1979. Nutrition, growth, and reproduction. *In* Barile and Razin (Editors), The Mycoplasmas, Vol. I, Academic Press, New York, pp. 103-139.

Rodwell, A.W. and A. Mitchell. 1979. Nutrition, growth, and reproduction. *In* Barile and Razin (Editors), The Mycoplasmas, Vol. I., Cell Biology, Academic Press, New York, pp. 103-139.

Rodwell, A.W., J.E. Peterson and E.S. Rodwell. 1975. Striated fibers of the *rho* form of *Mycoplasma*: in vitro reassembly, composition and structure. J. Bacteriol. *122:* 1216-1229.

Rodwell, A.W. and E.S. Rodwell. 1978. Relationships between strains of *Mycoplasma mycoides* subsp. *mycoides* and *capri* studied by two-dimensional gel electrophoresis of cell proteins. J. Gen. Microbiol. *109:* 259-263.

Roessler, W.G., T.H. Sanders, J. Dulberg and C.R. Brewer. 1952. Anaerobic glycolysis by enzyme preparations of *Brucella suis*. J. Biol. Chem. *194:* 207-213.

Rogers, P.L., K.J. Lee, M.L. Skotnicki and D.E. Tribe. 1980. Ethanol fermentation by highly productive strains of *Zymomonas mobilis*. 6th IFS Conference London, Ontario. July 20-25, 1980.

Rogers, P.L., K.J. Lee and D.E. Tribe. 1979. Kinetics of alcohol production by

Zymomonas mobilis at high sugar concentrations. Biotechnol. Lett. *1:* 165-170.

Rogers, R.C., R.G.C.J. Cuffe, Y.M. Cossins, D.M. Murphy and A.T.C. Bourke. 1980. The Queensland cholera incident of 1977. 2. The epidemiological investigation. Bull. W.H.O. *58:* 665-669.

Rogol, M., I. Sechter, L. Grinberg and C.B. Gerichter. 1979. Pril-xylose-ampicillin agar, a new selective medium for the isolation of *Aeromonas hydrophila.* J. Med. Microbiol. *12:* 229-231.

Rogosa, M. 1964. The genus *Veillonella.* II. Nutritional studies. J. Bacteriol. *87:* 574-580.

Rogosa, M. 1964. The genus *Veillonella.* I. General cultural, ecological and biochemical considerations. J. Bacteriol. *87:* 162-170.

Rogosa, M. 1965. The genus *Veillonella.* IV. Serological groupings and genus and species emendations. J. Bacteriol. *90:* 704-709.

Rogosa, M. 1969. *Acidaminococcus fermentans* sp. nov., anaerobic Gram-negative diplococci using amino acids as the sole energy source for growth. J. Bacteriol. *98:* 756-766.

Rogosa, M. 1970. *Streptobacillus moniliformis* and *Spirillum minor. In* Blair, Lennette and Truant (Editors), Manual of Clinical Microbiology, 1st Ed., American Sciety for Microbiology, Bethesda, Maryland, pp. 226-231.

Rogosa, M. 1971. Transfer of *Peptostreptococcus elsdenii* Gutierrez et al. to a new genus, *Megasphaera M. elsdenii* (Gutierrez et al.) comb. nov.. Int. J. Syst. Bacteriol. *21:* 187-189.

Rogosa, M. 1971. Transfer of *Veillonella* Prévot and *Acidaminococcus* Rogosa from *Neisseriaceae* to *Veillonellaceae* fam. nov. and the inclusion of *Megasphaera* Rogosa in *Veillonellaceae.* Int. J. Syst. Bacteriol. *21:* 231-233.

Rogosa, M. 1974a. Gram-negative anaerobic cocci. *In* Buchanan and Gibbons (Editors), Bergey's Manual of Determinative Bacteriology, 8th Ed. The Williams and Wilkins Co., Baltimore, pp. 445-449.

Rogosa, M. 1974b. Genus *Veillonella. In* Buchanan and Gibbons (Editors), Bergey's Manual of Determinative Bacteriology, 8th Ed. The Williams and Wilkins Co., Baltimore, pp. 446-447.

Rogosa, M. 1980. *Streptobacillus moniliformis* and *Spirillum minor. In* Lennette, Balows, Hausler and Truant (Editors), Manual of Clinical Microbiology, 3rd ed., Amer. Soc. Microbiol., Washington, D. C., pp. 350-356.

Rogul, M., J.J. Brendle, D.K. Haapala and A.D. Alexander. 1970. Nucleic acid similarities among *Pseudomonas pseudomallei, Pseudomonas multivorans,* and *Actinobacillus mallei.* J. Bacteriol. *101:* 827-835.

Rohde, R. 1965. The identification, epidemiology and pathogenicity of the salmonellae of subgenus II. J. Appl. Bacteriol. *28:* 368-372.

Rohde, R. 1966. Neue serologische Befunde hinsichtlich der Subgenus Einteilung der Salmonellen. Zentrabl. Bakteriol. Parasitenk. Infektionskr. Hyg. Abt. I Orig. *202:* 484-503.

Rohde, R. 1967. Zür serologischen Differential-diagnose der *Salmonella* Subgenera I-IV. Zentrab. Bakteriol. Parasitenk. Infektionskr. Hyg. Abt. I Orig. *205:* 404-424.

Rohmer, M., P. Bouvier and G. Ourisson. 1979. Molecular evolution of biomembranes: structural equivalents and phylogenetic precursors of sterols. Proc. Natl. Acad. Sci. U.S.A. *76:* 847-851.

Rohrbach, K.G. and J.B. Pfeiffer. 1975. The field induction of bacterial pink disease. Phytopathology *65:* 803-805.

Rohrbach, K.G. and J.B. Pfeiffer. 1976. The interaction of four bacteria causing pink disease of pineapple with several pineapple cultivars. Phytopathology *66:* 369-399.

Romano, N. and R. La Licata. 1978. Cell fractions and enzymatic activities of *Ureaplasma urealyticum.* J. Bacteriol. *136:* 833-838.

Romanovskaya, V.A., Y.R. Malashenko and V.N. Bogachenko. 1978. Corrected diagnoses of genera and species of methane-assimilating bacteria. Mikrobiologiya *47:* 120-130.

Romanovskaya, V.A., Y.R. Malashenko and V.N. Bogachenko. 1981. *In* Validation of the publication of new names and new combinations previously published outside the IJSB. List. No. 7. Int. J. Syst. Bacteriol. *31:* 382-383.

Ronson, C.W. and S.B. Primrose. 1979. Carbohydrate metabolism in *Rhizobium trifolii.* Identification and symbiotic properties of mutants. J. Gen. Microbiol. *112:* 77-88.

Roppel, J., H. Mayer and J. Weckesser. 1975. Identification of a 2,3-diamino-2,3-dideoxyhexose in the lipid A component of lipopolysaccharides of *Rhodopseudomonas viridis* and *Rhodopseudomonas palustris.* Carbohydr. Res. *40:* 31-40.

Rosati, G. and F. Verni. 1975. Macronuclear symbionts in *Euplotes crassus (Ciliata Hypotrichida).* Boll. Zool. *42:* 231-232.

Rosati, G., F. Verni and P. Luporini. 1976. Cytoplasmic bacteria-like endosymbionts in *Euplotes crassus* (Dujardin) (*Ciliata Hypotrichida*). Monitore Zool. Ital. (N. S.) *10:* 449-460.

Rose, C.S. and S.J. Pirt. 1981. Conversion of glucose to fatty acids and methane: roles of two mycoplasmal agents. J. Bacteriol. *147:* 248-254.

Rose, D.H. 1917. Blister spot of apples and its relation to a disease of apple bark. Phytopathology *7:* 198-208.

Rose, D.L., J.G. Tully and R.A. Del Guidice. 1980. *Acholeplasma morum,* a new non-sterol-requiring species. Int. J. Syst. Bacteriol. *30:* 647-654.

Rose, D.L., J.G. Tully and E.V. Langford. 1978. *Mycoplasma citelli,* a new species from ground squirrels. Int. J. Syst. Bacteriol. *28:* 567-572.

Rose, D.L., J.G. Tully and R.G. Wittler. 1979. Taxonomy of some swine mycoplasmas: *Mycoplasma suipneumoniae* Goodwin et al. 1965, a later, objective synonym of *Mycoplasma hyopneumoniae,* Mare and Switzer 1965, and the status of *Mycoplasma flocculare* Mayling and Friis 1972. Int. J. Syst. Bacteriol. *29:* 83-91.

Rosen, H.R. 1922. The bacterial pathogen of corn stalk rot. Phytopathology *12:* 497-499.

Rosen, H.R. 1922. A bacterial disease of foxtail *(Chaetochloa lutescens).* Ann. Mo. Bot. Gard. *9:* 333-402.

Rosen, H.R. 1926. Bacterial stalk rot of corn. Phytopathology *16:* 241-267.

Rosenberg, H., A.H. Ennor and V.F. Morrison. 1956. The estimation of arginine. Biochem. J. *63:* 153-159.

Rosenberg, S.L. and G.D. Hegeman. 1969. Clustering of functionally related genes in *Pseudomonas aeruginosa.* J. Bacteriol. *99:* 353-355.

Rosenbusch, C.T. and I.A. Merchant. 1939. A study of the hemorrhagic septicemia *Pasteurellae.* J. Bacteriol. *37:* 69-89.

Rosenbusch, R.F. and W.U. Knudtson. 1980. Bovine mycoplasmal conjunctivitis: experimental reproduction and characterization of the disease. Cornell Vet. *70:* 307-320.

Rosendal, S. 1973. *Mycoplasma cynos,* a new canine *Mycoplasma* species. Int. J. Syst. Bacteriol. *23:* 49-54.

Rosendal, S. 1974. *Mycoplasma molare,* a new canine *Mycoplasma* species. Int. J. Syst. Bacteriol. *24:* 125-130.

Rosendal, S. 1974b. Canine mycoplasmas II: biochemical characterization and serological identification. Acta Pathol. Microbiol. Scand. Sect. B *82:* 25-32.

Rosendal, S. 1975. Canine mycoplasmas: serological studies of type and reference strains, with a proposal for the new species, *Mycoplasma opalescens.* Acta Pathol. Microbiol. Scand. Sect. B *83:* 463-470.

Rosendal, S. 1979. Canine and feline mycoplasmas. *In* Tully and Whitcomb (Editors), The Mycoplasmas, Vol. 2, Academic Press, New York-London, pp. 217-234.

Rosendal, S. 1981. Experimental infection of goats, sheep and calves with the large colony type of *Mycoplasma mycoides* subsp. *mycoides.* Vet. Pathol. *18:* 71-81.

Rosendal, S. and F.T. Black. 1972. Direct and indirect immunofluorescence of unfixed and fixed mycoplasma colonies. Acta Pathol. Microbiol. Scand. Sect. B *80:* 615-622.

Rosendal, S., J. Patterson and H.J. van Kruiningen. 1980. Studies on the association of mycoplasmas with colitis in dogs. Third Conference International Organization for Mycoplasmology, Custer, S.D., Abstracts, paper No. 11.

Rosenfeld, H. and P. Feigelson. 1969. Synergistic and product induction of the enzymes of tryptophan metabolism in *Pseudomonas aeruginosa.* J. Bacteriol. *97:* 697-704.

Rosenkranz, W. 1939. Die Symbiose der Pentatomiden (Hemiptera Heteroptera). Z. Morphol. Oekol. Tiere *36:* 279-309.

Rosenstein, I.J., J.M. Hamilton-Miller and W. Brumfitt. 1981. Role of urease in the formation of infection stones: comparison of ureases from different sources. Infect. Immun. *32:* 32-37.

Roslycky, E.B. 1967. Bacteriocin production in the rhizobia bacteria. Can. J. Microbiol. *13:* 431-432.

Roslycky, E.B., O.N. Allen and E. McCoy. 1963. Serological properties of phages of *Agrobacterium radiobacter.* Canad. J. Microbiol. *9:* 709-717.

Ross, A.J. 1962. Isolation of a pigment-producing strain of *Aeromonas liquefaciens* from silver salmon *(Oncorhynchus kisutch).* J. Bacteriol. *84:* 590-591.

Ross, A.J., R.R. Rucker and W.H. Ewing. 1966. Description of a bacterium associated with redmouth disease of rainbow trout *(Salmo gairdneri).* Can. J. Microbiol. *12:* 763-770.

Ross, H.M. and M.J. Corbel. 1980. Isolation of a cell-wall-defective strain of *Brucella abortus* from bovine tissue. Vet. Rec. *106:* 242.

Ross, R.F., J.E. Hall, A.P. Orning and S.E. Dale. 1972. Characterization of an *Actinobacillus* isolated from the sow vagina. Int. J. Syst. Bacteriol. *22:* 39-46.

Ross, R.F. and J.A. Karmon. 1970. Heterogeneity among strains of *Mycoplasma granularum* and identification of *Mycoplasma synoviae,* sp. n. J. Bacteriol. *103:* 707-713.

Rosso, J.P., P. Forget and F. Pichinoty. 1973. Les nitrate-reductases bactériennes solubilisation, purification et propriétés de l'enzyme a de *Micrococcus halodenitrificans.* Biochim. Biophys. Acta *321:* 443-455.

Rothblatt, G.H. and P.F. Smith. 1961. Nonsaponifiable lipids of representative pleuropneumoniae-like organisms. J. Bacteriol. *82:* 479-491.

Rottem, S., M.C. Hardegree, M.W. Grabowski, R. Fornwald and M.F. Barile. 1976. Interaction between tetanolysin and mycoplasma cell membrane. Biochim. Biophys. Acta *455:* 876-888.

Rottem, S. and O. Markowitz. 1979. Unusual positional distribution of fatty acids in phosphatidglycerol of sterol-requiring mycoplasmas. FEBS Letters *107:* 379-382.

Rottem, S., E.A. Pfendt and L. Hayflick. 1971. Sterol requirements of T-strain mycoplasmas. J. Bacteriol. *105:* 323-330.

Rottem, S. and S. Razin. 1972. Isolation of mycoplasma membranes by digitonin. J. Bacteriol. *110:* 699-705.

Rottem, S., J. Yashouv, Z. Ne'eman and S. Razin. 1973. Cholesterol in mycoplasma membranes. Composition, ultrastructure and biological properties of membranes from *Mycoplasma mycoides* var. *capri* cells adapted to grow with low cholesterol concentrations. Biochim. Biophys. Acta *323:* 495-508.

Rouboud, E. 1919. Les particularités de la nutrition et la vie symbiotique chez les mouches tsétsés. Ann. Inst. Pasteur (Paris) *33:* 489-537.

Roughgarden, J.W. 1965. Antimicrobial therapy of rat-bite fever; a review. Arch. Intern. Med. *116:* 39-54.

Rousselot, R. 1948. *Rickettsia (Donatienella) delpyi* n. sp. n. subgen. Bull. Soc. Pathol. Exot. *41:* 110-112.

Roux, J. and J. Sassine. 1971. Étude d'une souche fixée de sphéroplastes (formes L) de *Brucella melitensis.* Ann. Inst. Pasteur (Paris) *120:* 174-185.

Rowatt, E. 1955. Amino acid metabolism of the genus *Bordetella.* J. Gen. Microbiol. *13:* 552-560.

Rowatt, E. 1957. The growth of *Bordetella pertussis*: a review. J. Gen. Microbiol. *17:* 297-326.

Rowe, B. 1979. The role of *Escherichia coli* in gastroenteritis. Clin. Gastroenterol. *8:* 625-643.

Rowe, B., R.J. Gross and E. van Oye. 1975. An organism differing from *Shigella boydii* 13 only in its ability to produce gas from glucose. Int. J. Syst. Bacteriol. *25:* 301-303.

Roy, A.B. 1958. A new species of *Azotobacter* producing heavy slime and acid. Nature (London) *182:* 120-121.

Roy, A.B. and S. Sen. 1962. A new species of *Derxia.* Nature (London) *194:* 604-605.

Roy, T.E. and C.D. Kelly. 1939. Genus *Bacteroides* Castellani and Chalmers. *In* Bergey, Breed, Murray and Hitchens (Editors), Bergey's Manual of Determinative Bacteriology, 5th Ed. The Williams and Wilkins Co., Baltimore, pp. 556-558.

Royle, P.L., H. Matsumoto and B.W. Holloway. 1981. Genetic circularity of the *Pseudomonas aeruginosa* PAO chromosome. J. Bacteriol. *145:* 145-155.

Rozanova, E.P. and A.I. Khudyakova. 1974. A new nonspore-forming thermophilic sulfate-reducing organism *Desulfovibrio thermophilus* nov. sp. Mikrobiologiya *43:* 1069-1075.

Rozanova, E.P. and T.N. Nazina. 1976. A mesophilic, sulfate-reducing, rod-shaped nonspore-forming bacterium. Mikrobiologiya *45:* 825-830.

Rubenchick, L.I. 1959. A contribution to the systematics of bacteria of the *Azotobacteriaceae* family. Microbiologiya *28:* 328-335.

Rubenstein, J.E., M.F. Lieberman and N. Gadoth. 1976. Central nervous system infection with *Eikenella corrodens*: Report of two cases. Pediatrics *57:* 264-265.

Rubin, S.J., P.A. Granato and B.L. Wasilauskas. 1980. Glucose-nonfermenting Gram-negative bacteria. *In* Lennette (Editor) Manual of Clinical Microbiology 3rd Ed., American Society for Microbiology, Washington, pp. 263-287.

Ruby, E.G., E.P. Greenberg and J.W. Hastings. 1980. Planktonic marine luminous bacteria: species distribution in the water column. Appl. Environ. Microbiol. *39:* 302-306.

Ruby, E.G. and J.G. Morin. 1978. Specificity of symbiosis between deep-sea fishes and psychrotrophic luminous bacteria. Deep-Sea Res. *25:* 161-167.

Ruby, E.G. and J.G. Morin. 1979. Luminous enteric bacteria of marine fishes: a study of their distribution, densities, and dispersion. Appl. Environ. Microbiol. *38:* 406-411.

Ruby, E.G. and K.H. Nealson. 1976. Symbiotic association of *Photobacterium fischeri* with the marine luminous fish *Monocentris japonica*: a model of symbiosis based on bacterial studies. Biol. Bull. Woods Hole, Mass. *151:* 574-586.

Ruby, E.G. and K.H. Nealson. 1977. A luminous bacterium that emits yellow light. Science (Washington) *196:* 432-434.

Ruby, E.G. and K.H. Nealson. 1978. Seasonal changes in the species composition of luminous bacteria in nearshore seawater. Limnol. Oceanogr. *23:* 530-533.

Ruckelshausen, R. and R. Holländer. 1978. On the phenotypical characteristics of *Haemophilus* isolates from human respiratory tracts. Zentralbl. Bakteriol. Parasitenkd. Infektionskr. Hyg. Abt. I Orig. A *242:* 500-511.

Rucker, R.R. 1966. Redmouth disease in rainbow trout *(Salmo gairdneri).* Bull. Off. Int. Epiz. *65:* 825-830.

Rudek, W. and R.-U. Haque. 1976. Extracellular enzymes of the genus *Bacteroides.* J. Clin. Microbiol. *4:* 458-460.

Rüger, H.-J. and T.L. Tan. 1983. Separation of *Alcaligenes denitrificans* sp. nov., nom. rev. from *Alcaligenes faecalis* on the basis of DNA base composition, DNA homology, and nitrate reduction. Int. J. Syst. Bacteriol. *33:* 85-89.

Ruinen, J. 1956. Occurrence of *Beijerinckia* species in the "phyllosphere". Nature (London) *177:* 220-221.

Ruinen, J. 1961. The phyllosphere. I. An ecologically neglected milieu. Plant and Soil. *15:* 81-109.

Ruiter, M. and H.M.M. Wentholt. 1952. The occurrence of a pleuropneumonia-like organism in fusospirillary infections of the human genital mucosa. J. Invest. Dermatol. *18:* 313-325.

Russ, V.R. 1905. Über ein Influenzabacillenahnliches anaerobes. Stabchen. Zentralbl. Bakteriol. Parasitenkd. Infektionskr. Hyg., I Abt. Orig. *39:* 357-359.

Russel, E.G. and E.P. Straube. 1979. Streptobacillary pleuritis in a koala *(Phascolarctos cinereus).* J. Wildl. Dis. *15:* 391-394.

Russell, F.H. 1898. An epidemic, septicemic disease among frogs due to the *Bacillus hydrophilus fuscus.* J. Am. Med. Assoc. *30:* 1442-1449.

Russell, R.R.B., K.G. Johnson and I.J. McDonald. 1975. Envelope proteins in *Neisseria.* Can. J. Microbiol. *21:* 1519-1534.

Rustigian, R. and C.A. Stuart. 1943. Taxonomic relationships in the genus *Proteus.* Proc. Soc. Exp. Biol. Med. *53:* 241-243.

Rustigian, R. and C.A. Stuart. 1943. The biochemical and serological relationships of the organisms of the genus *Proteus.* J. Bacteriol. *45:* 198-199.

Rustigian, R. and C.A. Stuart. 1945. The biochemical and serological relationships

of the organisms of the genus *Proteus.* J. Bacteriol. *49:* 419-436.

Rutherford, W.J. 1916. Trench fever: The field vole a possible origin. Brit. Med. J. *2:* 386-387.

Ruthman, A. and G. Noll-Altmann. 1980. Effect of a bacterial symbiont on cell division in a ciliate. *In* Schwemmler and Schenk (Editors), Endosymbiosis and Cell Biology, Vol. 1, de Gruyter, Berlin, pp. 361-370.

Ryall, C. and M.O. Moss. 1975. Selective media for the enumeration of *Chromobacterium* spp. in soil and water. J. Appl. Bacteriol. *38:* 53-59.

Ryan, W.J. 1968. An X-factor requiring *Haemophilus* species. J. Gen. Microbiol. *52:* 275-286.

Ryter, A. and J. Pillot. 1965. Structure des spirochètes. II. Étude du genre *Cristispira* au microscope optique et au microscope électronique. Ann. Inst. Pasteur *109:* 552-562.

Ryzhkov, V.L. 1950. Study on systematics of viruses. Vop. Med. Virusol. *3:* 9-19.

Sabath, L.D., M. Jago and E.P. Abraham. 1965. Cephalosporinase and penicillinase activities of a beta-lactamase from *Pseudomonas pyocyanea.* Biochem. J. *96:* 739-752.

Sabath, L.D., L.L. Stumpf, S.J. Wallace and M. Finland. 1970. Susceptibility of *Diplococcus pneumoniae, Haemophilus influenzae,* and *Neisseria meningitidis* to 23 antibiotics. Antimicrob. Agents Chemother. *10:* 53-56.

Sabbaj, J., V.L. Sutter and S.M. Finegold. 1970. Urease and deaminase activities of fecal bacteria in hepatic coma. Antimicrob. Agents Chemother. *10:* 181-185.

Sabet, K.A. 1957. Studies in the bacterial diseases of Sudan crops I. Bacterial leaf spot of Jute *(Corchorus olitorius* L.). Ann. Appl. Biol. *45:* 516-520.

Sabet, K.A. 1959. Studies in the bacterial diseases of Sudan crops IV. Bacterial leaf spot and canker disease of mahogany *(Khaya senegalensis* (Desr.) A. Juss and *K. grandifoliola* C. DC). Ann. Appl. Biol. *47:* 658-665.

Sabet, K.A. and W.J. Dowson. 1960. Bacterial leaf spot of sesame, *(Sesamum orientale* L.). Phytopathol. Z. *37:* 252-258.

Sabet, K.A., F. Ishag and O. Khalil. 1969. Studies on the bacterial diseases of Sudan crops VII. New records. Ann. Appl. Biol. *63:* 357-369.

Sabin, A.B. 1938. Identification of the filtrable transmissible neurolytic agent isolated from toxoplasma-infected tissue as a new pleuropneumonia-like microbe. Science *88:* 575-576.

Sabin, A.B. 1941. The filtrable microorganisms of the pleuropneumonia group. Bacteriol. Rev. *5:* 1-66.

Sabin, A.B. 1941. The filtrable microorganisms of the pleuropneumonia group (appendix on classification and nomenclature). Bacteriol. Rev. *5:* 331-335.

Sackett, W.G. 1916. A bacterial stem blight of field and garden peas. Bull. Colo. Agr. Sta. No. *218:* 1-43.

Šafeřsteijn, D.L. 1973. Some proposals concerning the taxonomy of the genus *Brucella.* WHO Brucellosis Document WHO/Bruc./73.337.

Saglio, P., D. Laflèche, C. Bonissol and J.M. Bové. 1971. Isolement, culture et observation au microscope électronique des structures de type mycoplasme associées à la maladie du stubborn des agrumes et leur comparaison avec les structures observées dans le cas de la maladie du greening des agrumes. Physiol. Vég. *9:* 569-582.

Saglio, P., M. L'hospital, D. Laflèche, G. Dupont, J.M. Bové, J.G. Tully and E.A. Freundt. 1973. *Spiroplasma citri* gen. and sp. n: a mycoplasma-like organism associated with "stubborn" disease of citrus. Int. J. Syst. Bacteriol. *23:* 191-204.

Saglio, P.H.M. and R.F. Whitcomb. 1979. Diversity of wall-less prokaryotes in plant vascular tissue, fungi, and invertebrate animals. *In* Whitcomb and Tully (Editors), The Mycoplasmas, Vol. 3, Academic Press, New York, pp. 1-36.

Saiki, T., R. Kimura and K. Arima. 1972. Isolation and characterization of extremely thermophilic bacteria from hot springs. Agr. Biol. Chem. *36:* 2357-2366.

Saisawa, K. 1912. Über den Erreger und die Diagnose des Maltafiebers. Z. Hyg. Infektionskr. *70:* 177-203.

Sakaki, Y. and T. Oshima. 1975. Isolation and characterization of a bacteriophage infectious to an extreme thermophile, *Thermus thermophilus* HB8. J. Virol. *15:* 1449-1453.

Sakazaki, R. 1961. Studies on the Hafnia group of *Enterobacteriaceae.* Japan J. Med. Sci. Biol. *14:* 223-241.

Sakazaki, R. 1965. A proposed group of the family *Enterobacteriaceae,* the Asakusa group. Int. Bull. Bacteriol. Nomencl. Taxon. *15:* 45-47.

Sakazaki, R. 1965. *Vibrio parahaemolyticus.* A non-choleragenic enteropathogenic vibrio. Proc. Cholera Res. Symp. (Jan. 24-29, 1965, Honolulu). Public Health Serv. Publication No. 1328. U.S. Gov. Printing Office, Washington, D.C., pp. 30-34.

Sakazaki, R. 1967. Studies on the Asakusa group of *Enterobacteriaceae (Edwardsiella tarda)* Japan. J. Med. Sci. Biol. *20:* 205-212.

Sakazaki, R. 1968. Proposal of *Vibrio alginolyticus* for the biotype 2 of *Vibrio parahaemolyticus.* Jpn. J. Med. Sci. Biol. *21:* 359-362.

Sakazaki, R. 1974. Genus IX. *Serratia* Bizio 1823, 288. *In* Buchanan, R.E. and N.E. Gibbons (Editors), Bergey's Manual of Determinative Bacteriology, 8th edition, The Williams and Wilkins Co., Baltimore, p. 326.

Sakazaki, R. 1979. Vibrio infections. *In* Riemann and Bryan (Editors), Food-Borne Infections and Intoxications, 2nd Ed., Academic Press, New York-San Francisco-London, pp. 173-209.

Sakazaki, R., S. Iwanami and H. Fukumi. 1963. Studies on the enteropathogenic facultatively halophilic bacteria *Vibrio parahaemolyticus.* I. Morphological, cultural and biochemical properties and its taxonomical position. Jpn. J. Med. Sci. Biol. *16:* 161-188.

Sakazaki, R. and Y. Murata. 1962. The new group of the *Enterobacteriaceae*, the Asakusa group. Jpn. J. Bacteriol. *17:* 617-618.

Sakazaki, R. and S. Namioka. 1957. *Citrobacter* possessing O antigen of *Salmonella*. Bull. Nat. Inst. Anim. Health *32:* 1-6.

Sakazaki, R. and S. Namioka. 1957. Biochemical studies on Voges-Proskauer positive enteric bacteria. Japan. J. Exp. Med. *27:* 273-282.

Sakazaki, R. and S. Namioka. 1960. Serological studies on the *Cloaca (Aerobacter)* group of enteric bacteria. Jpn. J. Med. Sci. Biol. *13:* 1-12.

Sakazaki, R., S. Namioka, R. Nakaya and H. Fukumi. 1959. Studies on so-called Paracolon C27 (Ferguson). Jpn. J. Med. Sci. Biol. *12:* 355-363.

Sakazaki,R., S. Namioka, A. Osada and C. Yamada. 1960. A problem on the pathogenic role of *Citrobacter* of enteric bacteria. Japan J. Exp. Med. *30:* 13-21.

Sakazaki, R. and K. Tamura. 1975. Priority of the specific epithet *anguillimortiferum* over the specific epithet *tarda* in the name of the organism presently known as *Edwardsiella tarda*. Int. J. Syst. Bacteriol. *25:* 219-220.

Sakazaki, R. and K. Tamura. 1978. Comment on a proposal of Farmer et al., to conserve the specific epithet *tarda* over the specific epithet *anguillimortiferum* in the name of the organism known as *Edwardsiella tarda*. Int. J. Syst. Bacteriol. *28:* 130-131.

Sakazaki, R., K. Tamura, C.Z. Gomez and R. Sen. 1970. Serological studies on the cholera group of vibrios. Jpn. J. Med. Sci. Biol. *23:* 13-20.

Sakazaki, R., K. Tamura, R. Johnson and R.R. Colwell. 1976. Taxonomy of some recent described species in the family *Enterobacteriaceae*. Int. J. Syst. Bacteriol. *26:* 158-179.

Sakazaki, R., K. Tamura, T. Kato, Y. Obara, S. Yamai and K. Hobo. 1968. Studies of the enteropathogenic facultatively halophilic bacteria, *Vibrio parahaemolyticus*. III. Enteropathogenicity. Jpn. J. Med. Sci. Biol. *21:* 325-331.

Sakazaki, R., K. Tamura and M. Murase. 1971. Determination of the hemolytic activity of *Vibrio cholerae*. Jpn. J. Med. Sci. Biol. *24:* 83-91.

Sakazaki, R., K. Tamura, A. Nakamura, T. Kurata, A. Gohda and S. Takeuchi. 1974. Enteropathogenicity and enterotoxigenicity of human enteropathogenic *Escherichia coli*. Jpn. J. Med. Sci. Biol. *27:* 19-33.

Sakharoff, M.N. 1891. *Spirochaeta anserina* et la septicémie des oies. Ann. Inst. Pasteur (Paris) *5:* 564-566.

Salanitro, J.P., P.A. Muirhead and J.R. Goodman. 1976. Morphological and physiological characteristics of *Gemmiger formicilis* isolated from chicken ceca. Appl. Environ. Microbiol. *32:* 623-632.

Salcher, O. and F. Lingens. 1980. Metabolism of tryptophan by *Pseudomonas aureofaciens* and its relationship to pyrrolnitrin biosynthesis. J. Gen. Microbiol. *121:* 465-471.

Saleh, A.M. 1964. Differences in the resistance of sulphate-reducing bacteria to inhibitors. J. Gen. Microbiol. *37:* 113-121.

Salkinoja-Salonen, M. and R. Boeck. 1978. Characterization of lipopolysaccharides isolated from *Agrobacterium tumefaciens*. J. Gen. Microbiol. *105:* 119-125.

Salmon, D.E. 1885. On swine plague. Annu. Rep. U.S. Dept. Agric. Bureau of Animal Industry *2:* 212-246.

Saltet, R.H. 1900. Ueber Reduktion von Sulfaten in Brackwasser durch Bakterien. Zentralbl. Bakteriol. Parasitenkd. Infektionskr. Hyg. Abt. II *6:* 648-651; 695-703.

Salton, M.R.J. 1964. The bacterial cell wall. Elsevier Publishing Co., Amsterdam.

Salyers, A.A., J.R. Vercelloti, S.E.H. West and T.D. Wilkins. 1977a. Fermentation of mucin and plant polysaccharides by strains of *Bacteroides* from the human colon. Appl. Environ. Microbiol. *33:* 319-322.

Salyers, A.A., J. Wong and T.D. Wilkins. 1977b. Beta-lactamase activity in strains of *Bacteroides melaninogenicus* and *Bacteroides oralis*. Antimicrob. Agents Chemother. *11:* 142-146.

Sambon, L. 1907. *Spiroschaudinnia. In* Manson (Editor), Tropical diseases, 4th Ed, Cassell, London, p. 833.

Sampaio, M-J.A.M., E.M.R. da Silva, J. Döbereiner, M.G. Yates and F. O. Pedrosa. 1981. Autotrophy and methylotrophy in *Derxia gummosa, Azospirillum brasilense* and *A. lipoferum. In* Gibson and Newton (Editors), Current Perspectives in Nitrogen Fixation, Proc. 4th Int. Symp. on Nitrogen Fixation, Canberra, Australia, Dec. 1-5, 1980, p. 444.

Samson, R. and N. Nassan-Agha. 1978. Biovar and serovars among 129 strains of *Erwinia chrysanthemi. In* Station de Pathologie Végétale et Phytobactériologie (Editor), Proc. IVth Int. Conf. on Plant Pathogenic Bacteria, Gilbert-Clarey, Tours, France, pp. 547-553.

Samuels, S.B., C.W. Moss and R.E. Weaver. 1973. The fatty acids of *Pseudomonas multivorans (Pseudomonas cepacia)* and *Pseudomonas kingii*. J. Gen. Microbiol. *74:* 275-279.

Samuels, S.B., B. Pittman and W.B. Cherry. 1969. Practical physiological schema for the identification of *Herellea vaginicola* and its differentiation from similar organisms. Appl. Microbiol. *18:* 1015-1024.

Sand, F. 1976. *Gluconobacter*, boissons plates et emballages en matiére plastique. Bios. *7:* 7-14.

Sand, F.E.M.J. 1971. Zür Bakterien-Flora von Erfrischungsgetränken. Brauwelt. *111:* 252-264.

Sanderson, K.E. and P.E. Hartman. 1978. Linkage map of *S. typhimurium*, 5th Ed., Microbiol. Rev. *42:* 471-519.

Sandok, P.L, H.M. Jenkin, H.M. Mathews and M.S. Roberts. 1978. Unsustained multiplication of *Treponema pallidum* (Nichols virulent strain) in vitro in the presence of oxygen. Infect. Immun. *19:* 421-429.

Sands, D.C., F.H. Gleason and D.C. Hildebrand. 1967. Cytochromes of *Pseudomonas syringae*. J. Bacteriol. *94:* 1785-1786.

Sands, D.C., F.H. Gleason and D.C. Hildebrand. 1967. Cytochromes of *Pseudomonas syringae*. J. Bacteriol. *94:* 1785-1786.

Sands, D.C., L. Hankin and M. Zucker. 1972. A selective medium for pectolytic fluorescent pseudomonads. Phytopathology *62:* 998-1000.

Sands, D.C. and A.D. Rovira. 1970. Isolation of fluorescent pseudomonads with a selective medium. Appl. Microbiol. *20:* 513-514.

Sands, D.C. and A.D. Rovira. 1971. *Pseudomonas fluorescens* biotype G, the dominant fluorescent pseudomonad in South Australian soils and wheat rhizospheres. J. Appl. Bacteriol. *34:* 261-275.

Sands, D.C., M.N. Schroth and D.C. Hildebrand. 1970. Taxonomy of phytopathogenic pseudomonads. J. Bacteriol. *101:* 9-23.

Sands, D.C., M.N. Schroth and D.C. Hildebrand. 1980. *Pseudomonas. In* Schaad (Editor), Laboratory Guide for Identification of Plant Pathogenic Bacteria. American Phytopathological Society, St. Paul, Minn., pp. 36-44.

Sanger, F., G.G. Brownless and B.G. Barrell. 1965. A two-dimensional fractionation procedure for radioactive nucleotides. J. Mol. Biol. *13:* 373-398.

Sanger, F., S. Nicklen and A.R. Coulson. 1977. DNA sequencing with chain-terminating inhibitors. Proc. Nat. Acad. Sci. USA *74:* 5463-5467.

Sanjieva, E.U. and G.A. Zavarzin. 1971. Oxidation of carbon monoxide by *Seliberia carboxydohydrogena*. Dokl. Akad. Nauk SSSR *196:* 956-958.

Sano, Y. and M. Kageyama. 1977. Transformation of *Pseudomonas aeruginosa* by plasmid DNA. J. Gen. Appl. Microbiol. *23:* 183-186.

Sanyal, S., S. Singh and P.C. Sen. 1975. Enteropathogenicity of *Aeromonas hydrophila* and *Plesiomonas shigelloides*. J. Med. Microbiol. *8:* 195-198.

Sarff, L., G.H. McCracken Jr., M.S. Shiffer, M.P. Glode, J.B. Robbins, I. Ørskov and F. Ørskov. 1975. Epidemiology of *Escherichia coli* K1 in healthy and diseased newborns. Lancet *I:* 1099-1104.

Saslaw, S. and H.N. Carlisle. 1961. Studies with tularemia vaccine in volunteers. IV. *Brucella* agglutinins in vaccinated and nonvaccinated volunteers challenged with *Pasteurella tularensis*. Am. J. Med. Sci. *242:* 166-172.

Saslaw, S., H.T. Eigelsbach, J.A. Prior, H.E. Wilson and S. Carhart. 1961. Tularemia vaccine study. II. Respiratory challenge. Arch. Intern. Med. *107:* 702-714.

Sato, G., M. Asagi, C. Oka, N. Ishiguro and N. Terakado. 1978. Transmissible citrate-utilizing ability in *Escherichia coli* isolated from pigeons, pigs and cattle. Microbiol. Immunol. *22:* 357-360.

Sato, M. and K. Takahashi. 1972. Ecological studies on the bacterial blight of mulberry. I. The overwintering of the pathogen, *Pseudomonas mori* (Boyer and Lambert) Stevens. J. Seric. Sci. Jpn. *41:* 285-293.

Sato, Y., K. Izumiya, H. Sato, J.L. Cowell and C.R. Manclark. 1981. Role of antibody to leucocytosis-promoting factor hemagglutinin and to filamentous hemagglutinin in immunity to pertussis. Infect. Immun. *31:* 1223-1231.

Satoh, T., Y. Hoshino and H. Kitamura. 1976. *Rhodopseudomonas sphaeroides* forma sp. *denitrificans*, a denitrifying strain as a subspecies of *Rhodopseudomonas sphaeroides*. Arch. Microbiol. *108:* 265-269.

Savage, D.D., R.L. Kagan, N.A. Young and A.E. Horvath. 1977. *Cardiobacterium hominis* endocarditis: description of two patients and characterization of the organism. J. Clin. Microbiol. *5:* 75-80.

Săvulescu, T. 1947. Contribution à la classification des bactériacées phytopathogénes. Anal. Acad. Romane Ser. III *22*(4): 1-26.

Sawicki, L., H.M. Bruce and C.H. Andrewes. 1962. *Streptobacillus moniliformis* as a probable cause of arrested pregnancy and abortion in laboratory mice. Brit. J. Exp. Pathol. *43:* 194-197.

Sawula, R.V. and I.P. Crawford. 1972. Mapping of the tryptophan genes of *Acinetobacter calcoaceticus* by transformation. J. Bacteriol. *112:* 797-805.

Sawyer, M.H., P. Baumann and L. Baumann. 1977a. Pathways of D-fructose and D-glucose catabolism in marine species of *Alcaligenes, Pseudomonas marina* and *Alteromonas communis*. Arch. Microbiol. *112:* 169-172.

Sawyer, M.H., P. Baumann, L. Baumann, S.M. Berman, J.L. Canovas and R.H. Berman. 1977b. Pathways of D-fructose metabolism in species of *Pseudomonas*. Arch. Microbiol. *112:* 49-55.

Sawyer, S.J., J.B. Macdonald and R.J. Gibbons. 1962. Biochemical characteristics of *Bacteroides melaninogenicus*. A study of thirty-one strains. Arch. Oral Biol. *7:* 685-691.

Sayre, R.M. and W.P. Wergin. 1977. Bacterial parasite of a plant nematode: morphology and ultrastructure. J. Bacteriol. *129:* 1091-1101.

Scaffidi, V. 1981. Attuale espansione endemo-epidemica della febbre bottonosa in Italia. Minerva Med. *72:* 2063-2070.

Scarlett, M. 1916. Infections cornéennes a diplobacilles. Notes sur deux diplobacilles non encore décrit (*Bacillus duplex nonliquefaciens* et *Bacillus duplex josefi*). Ann. Ocul. *153:* 100-111.

Scarlett, C.M., J.T. Fletcher, P. Roberts and R.A. Lelliott. 1978. Tomato pith necrosis caused by *Pseudomonas corrugata* n. sp. Ann. Appl. Biol. *88:* 105-114.

Schaad, N.W. 1976. Immunological comparison and characterization of ribosomes of *Xanthomonas vesicatoria*. Phytopathology *66:* 770-776.

Schaad, N.W. 1978. Use of direct and indirect immunofluorescence tests for identification of *Xanthomonas campestris*. Phytopathology *68:* 249-252.

Schaad, N.W. 1979. Serological identification of plant pathogenic bacteria. Annu.

Rev. Phytopathol. *17:* 123-147.

Schaad, N.W. (Editor). 1980. Laboratory guide for identification of plant pathogenic bacteria. Bacteriological Committee of American Phytopathological Society, St. Paul, Minn.

Schaad, N.W., C.I. Kado and D.R. Sumner. 1975. Synonymy of *Pseudomonas avenae* Manns 1905 and *Pseudomonas alboprecipitans* Rosen 1922. Int. J. Syst. Bacteriol. *25:* 133-137.

Schaad, N.W., G. Sowell, R.W. Goth, R.R. Colwell and R.E. Webb. 1978. *Pseudomonas pseudoalcaligenes* subsp. *citrulli* subsp. nov. Int. J. Syst. Bacteriol. *28:* 117-125.

Schad, G.A., R. Knowles and E. Meerovitch. 1964. The occurrence of *Lampropedia* in the intestines of some reptiles and nematodes. Can. J. Microbiol. *10:* 801-804.

Schaeffer, P. 1958. Interspecific reactions in bacterial transformation. Gene *11:* 311-318.

Schank, S. C., R. L. Smith, G. C. Weiser, D. A. Zuberer, J. H. Bouton, K. H. Quesenberry, M. E. Tyler, J. R. Milam and R. C. Littell. 1979. Fluorescent antibody technique to identify *Azospirillum brasilense* associated with roots of grasses. Soil Biol. Biochem. *11:* 287-295.

Scharmann, W., R. Drzeniek and H. Blobel. 1970. Neuraminidase of *Pasteurella multocida.* Infect. Immun. *1:* 319-320.

Schatz, A. and C.R. Bovell. 1952. Growth and hydrogenase activity of a new bacterium, *Hydrogenomonas facilis.* J. Bacteriol. *63:* 87-98.

Schaub, I.G. and F.D. Hauber. 1948. A biochemical and serological study of a group of identical unidentifiable gram-negative bacilli from human sources. J. Bacteriol. *56:* 379-385.

Schaudinn, F. 1905. Korrespondenzen. Deut. Med. Wochenschr. *31:* 1728.

Schaudinn, F. and E. Hoffmann. 1905. Vorlaufiger bericht über das vorkommen for spirochaeten in syphilitischen krankheitsprodukten und bei papillomen. Arb. Gesundh Amt. Berlin *22:* 528-534.

Scheifinger, C.C., M.J. Latham and M.J. Wolin. 1975. Relationship of lactate dehydrogenase specificity and growth rate to lactate metabolism by *Selenomonas ruminantium.* Appl. Microbiol. *30:* 916-921.

Scheifinger, C.C., B. Linehan and M.J. Wolin. 1975. H$_2$ production by *Selenomonas ruminantium* in the absence and presence of methanogenic bacteria. Appl. Microbiol. *29:* 480-483.

Scheifinger, C.C. and M.J. Wolin. 1973. Propionate formation from cellulose and soluble sugars by combined cultures of *Bacteroides succinogenes* and *Selenomonas ruminantium.* Appl. Microbiol. *26:* 789-795.

Scheinert, W. 1933. Symbiose und Embryonalentwicklung bei Russelkafern. Z. Morphol. Oekol. Tiere *27:* 76-128.

Schell, J., M. Van Montagu, M. De Beuckeleer, M. De Block, A. Depicker, M. De Wilde, G. Engler, C. Genetello, J.P. Hernalsteens, M. Holsters, J. Seurinck, A. Silva, F. Van Vliet and R. Villaroel. 1979. Interactions and DNA transfer between *Agrobacterium tumefaciens,* the Ti-plasmid and the plant host. Proc. R. Soc. Lond. B *204:* 251-266.

Schell, R.F., J.L. Lefrock, J.K. Chan and O. Bagasra. 1980. LSH hamster model of syphilitic infection. Infect. Immun. *28:* 909-913.

Schellack, C. 1909. Studien zür Morphologie und Systematik der Spirochaeten aus Muscheln. Arb. Gesundh Amt. Berl. *30:* 379-428.

Scherff, R. H., J. E. DeVay and T. W. Carroll. 1966. Ultrastructure of host-parasite relationships involving reproduction of *Bdellovibrio bacteriovorus* in host bacteria. Phytopathology. *56:* 627-632.

Schiemann, D. 1979. Synthesis of a selective agar medium for *Yersinia enterocolitica.* Can. J. Microbiol. *25:* 1298-1304.

Schiewe, M.H., J.H. Crosa and E.J. Ordal. 1977. Deoxyribonucleic acid relationships among marine vibrios pathogenic to fish. Can. J. Microbiol. *23:* 954-958.

Schiff, J., L.S. Suter, R.D. Gourley and W.D. Sutliff. 1961. *Flavobacterium* infection as a cause of bacterial endocarditis. Ann. Intern. Med. *55:* 499-506.

Schildkraut, C.L., J. Marmur and P. Doty. 1962. Determination of the base composition of deoxyribonucleic acid from its buoyant density in CsCl. J. Mol. Biol. *4:* 430-443.

Schiller, N.L. and C.D. Cox. 1977. Catabolism of glucose and fatty acids by virulent *Treponema pallidum.* Infect. Immun. *16:* 60-68.

Schilling, V. 1928. *Eperythrozoon coccoides,* eine neue durch Splenektomie aktivierbare Dauerinfektion der weissen Maus. Klin. Wochenschr. *7:* 1853-1855.

Schimizu, T., T. Furuki, T. Waki and K. Ichikawa. 1978. Metabolic characteristics of denitrification by *Paracoccus denitrificans.* J. Ferment. Technol. *56:* 207-213.

Schimke, R.T., C.M. Berlin, E.W. Sweeney and W.R. Carroll. 1966. The generation of energy by the arginine dihydrolase pathway in *Mycoplasma hominis* 07. J. Biol. Chem. *241:* 2228-2236.

Schindler, J. 1964. Die Synthese von Poly-β-hydroxybuttersäure durch *Hydrogenomonas* H 16: Die au β-Hydroxybutyryl-Coenzym A führende Reaktionsschritte. Arch. Mikrobiol. *49:* 236-255.

Schindler, R., M. Ristic and R. Wokatsch. 1966. Vergleichende Untersuchungen mit *Anaplasma marginale* and *A. centrale.* Z. Tropenmed. Parasitol. *17:* 337-360.

Schink, B. and H.G. Schlegel. 1979. The membrane-bound hydrogenase of *Alcaligenes eutrophus.* I. Solubilization, purification, and biochemical properties. Biochim. Biophys. Acta *567:* 315-324.

Schink, B. and H.G. Schlegel. 1980. The membrane-bound hydrogenase of *Alcaligenes eutrophus:* II. Localization and immunological comparison with other

hydrogenase systems. Antonie van Leeuwenhoek J. Microbiol. Serol. *46:* 1-14.

Schiøtz, P.O., N. Høiby and J.B. Hertz. 1979. Cross-reaction between *Haemophilus influenzae* and nineteen other bacterial species. Acta Pathol. Microbiol. Scand. Sect. B, *87:* 337-344.

Schlegel, H.G. and G. Gottschalk. 1965. Verwertung von Glucose durch eine Mutante von *Hydrogenomonas* H 16. Biochem. Z. *341:* 249-259.

Schlegel, H.G., H. Kaltwasser and G. Gottschalk. 1961. Ein Submersverfahren zür Kultur wasserstoffoxydierender Bakterien: Wachstumsphysiologische Untersuchungen. Arch. Mikrobiol. *38:* 209.

Schlegel, H.G. and R.M. Lafferty. 1971. Novel energy and carbon sources. A. The production of biomass from hydrogen and carbon dioxide. Adv. Biochem. Eng. *1:* 143-168.

Schlegel, H.G. and D. Vollbrecht. 1980. Formation of dehydrogenases for lactate, ethanol and butanediol in the strictly aerobic bacterium *Alcaligenes eutrophus.* J. Gen. Microbiol. *117:* 475-481.

Schleifer, K.H. and R. Joseph. 1973. A directly cross-linked L-ornithine-containing peptidoglycan in cell walls of *Spirochaeta stenostrepta.* FEBS Lett. *36:* 83-86.

Schleifer, K.H. and O. Kandler. 1967. Zür chemischen zusammensetzung der zellwand der Streptokokken. I. Die aminosäuresequenz des mureins von *Str. thermophilus* und *Str. faecalis.* Arch. Mikrobiol. *57:* 335-364.

Schleifer, K.H. and O. Kandler. 1972. Peptidoglycan types of bacterial cell walls and their taxonomic implications. Bacteriol. Rev. *36:* 407-477.

Schleifstein, J. and M.B. Coleman. 1943. *Bacterium enterocoliticum.* N. Y. State Dep. Health Div. Lab. Res. Annu. Rep. 1943, 56.

Schlein, Y. 1977. Lethal effect of tetracycline on tsetse flies following damage to bacterioid symbionts. Experientia (Basel) *33:* 450-451.

Schlesinger, D. (Editor) 1975. Microbiology - 1975. American Society for Microbiology, Washington, D.C.

Schmid, E.E., T. Velaudapillai and G.R. Niles. 1954. Study of Paracolon organisms with the major antigen of *Shigella sonnei,* Form I. J. Bacteriol. *68:* 50-52.

Schmidt, H.J. and K. Heckmann. 1980. DNA of omicron. *In* Schwemmler and Schenk (Editors), Endosymbiosis and Cell Biology, Vol. 1, de Gruyter, Berlin, pp. 401-407.

Schmidt, J. 1901. Familie *Bacteriaceac. In* Schmidt and Weis Bakterienne. Naturhistorisk Grundlag for det bakteriologiske Studium. Morten Porsild, København 1899-1901, p. 266.

Schmidt, J. and F. Weis. 1902. Die Bakterien. Naturhistorische Grundlage für das bakteriologische Studium. Verlag von Gustav Fischer, Jena.

Schmidt, W.C. and C.D. Jeffries. 1974. Bacteriophage typing of *Proteus mirabilis, Proteus vulgaris* and *Proteus morganii.* Appl. Microbiol. *27:* 47-53.

Schminke, A. 1917. Histopathologischer Befund in Roseolen der Haut bei wolhynischem Fieber. Münch. Med. Wochschr. *64:* 961.

Schmitz, J.A. and J.L. Gradin. 1980. Serotypic and biochemical characterization of *Bacteroides nodosus* isolates from Oregon. Can. J. Comp. Med. *44:* 440-446.

Schnaitman, C.A. 1970. Comparison of the envelope protein compositions of several Gram-negative bacteria. J. Bacteriol. *104:* 1404-1405.

Schnathorst, W.C. 1966. Unaltered specificity in several xanthomonads after repeated passage through *Phaseolus vulgaris.* Phytopathology *56:* 58-60.

Schneider, G. 1940. Beiträge zür Kenntnis der symbiontischen Einrichtungen der Heteropteren. Z. Morphol. Oekol. Tiere *36:* 595-644.

Schneider, H. 1956. Morphologische und experimentelle Untersuchungen uber die Endosymbiose der Korn-und Reiskäfer (*Calandra granaria* L. und *Calandra oryzae* L.). Z. Morphol. Oekol. Tiere *44:* 555-625.

Schneider, K., R. Cammack, H.G. Schlegel and D.O. Hall. 1979. The iron-sulphur centres of soluble hydrogenase from *Alcaligenes eutrophus.* Biochim. Biophys. Acta *578:* 445-461.

Schneider, K., V. Rudolph and H.G. Schlegel. 1973. Description and physiological characterization of a coryneform hydrogen bacterium, strain 14g. Arch. Mikrobiol. *93:* 179-193.

Schneider, K. and H.G. Schlegel. 1976. Purification and properties of soluble hydrogenase from *Alcaligenes eutrophus* H 16. Biochim. Biophys. Acta *452:* 66-80.

Schneider, K. and H.G. Schlegel. 1977. Localization and stability of hydrogenases from aerobic hydrogen bacteria. Arch. Microbiol. *112:* 229-238.

Schneider, K. and H.G. Schlegel. 1978. Identification and quantitative determination of the flavin component of soluble hydrogenase from *Alcaligenes eutrophus.* Biochem. Biophys. Res. Commun. *84:* 564-571.

Schnepf, E., E. Hegewald and C.-J. Soeder. 1974. Electron microscopic observations on parasites of *Scenedesmus* mass culture. 4. Bacteria. Arch. Microbiol. *98:* 133-145.

Schocher, A.J., H. Kuhn, B. Schlindler, N.J. Palleroni, C.W. Despreaux, M. Boublik and P.A. Miller. 1979. *Acetobacter* bacteriophage A-1. Arch. Microbiol. *121:* 193-197.

Schoop, G. 1935a. Obligat halophile Mikroben. Zentralbl. Bakteriol. Parasitenk. Infektionskr. Hyg. Abt. I *134:* 14-26.

Schoop, G. 1935b. *Halococcus litoralis,* ein obligat halophiler Farbstoffbildner. Deut. Tierärztl. Wochenschr. *43:* 817-820.

Schorr, H. 1957. Zür Verhaltensbiologie und Symbiose von *Brachypelta aterrima* Först. (Cydnidae, Heteroptera). Z. Morphol. Oekol. Tiere *45:* 561-602.

Schramek, S. 1974. Deoxyribonucleic acid base composition of rickettsiae belonging to the Rocky Mountain spotted fever group isolated in Czechoslovakia. Acta Virol. *18:* 173-174.

Schramek, S., R. Brezina and J. Urvolgyi. 1972. A new method of preparing diagnostic Q fever antigen. Acta Virol. *16:* 487-482.

Schramm, M. and S. Hestrin. 1954. Synthesis of cellulose by *Acetobacter xylinum.* I. Micromethod for the determination of celluloses. Biochem. J. *56:* 163-166.

Schrautemeier, B. 1981. The role of ferredoxin in the nitrogen-fixing hydrogen bacterium *Xanthobacter autotrophicus.* FEMS Microbiol. Lett. *12:* 153-157.

Schröder, M. 1932. Die Assimilation des Luftstickstoff durch einige Bakterien. Zentralbl. Bakteriol. Parasitenkd. Infektionskr. Hyg. Abt. 2 *85:* 177-212.

Schroeter, J. 1872. Ueber einige durch Bacterien gebildete Pigmente. *In* F. Cohn (1875) Beiträge zür Biologie der Pflanzen, J.U. Kern's Verlag, Breslau, pp. 109-126.

Schroeter, J. 1885-1889. *In* F. Cohn, Kryptogamenflora von Schlesien. Bd. 3, Heft 4, Pilze J.U. Kern's Verlag, Breslau, pp. 1-814.

Schroter, G. 1975. The detection of twitching motility in *Eikenella corrodens.* Z. Med. Microbiol. Immunol. (Berl.) *161:* 41-46.

Schroter, G. and J. Stawru. 1970. Die bedeutung von Bacteroides corrodens. Eiken, 1958, in Rhamen der Tonsillenflora. Z. Med. Mikrobiol. Immunol. *155:* 241-247.

Schroth, M.N. and D.C. Hildebrand. 1980. *E. amylovora* or true erwiniae group. *In* Schaad (Editor), Laboratory guide for identification of plant pathogenic bacteria, American Phytopathological Society, St. Paul, pp. 26-30.

Schroth, M.N., D.C. Hildebrand and M.P. Starr. 1981. Phytopathogenic members of the genus *Pseudomonas. In* Starr, Stolp, Trüper, Balows and Schlegel (Editors), The Prokaryotes, A handbook on habitats, isolation and identification of the bacteria. Springer-Verlag, Berlin, pp. 701-718.

Schroth, M.N., J.P. Thompson and D.C. Hildebrand. 1965. Isolation of *Agrobacterium tumefaciens - A. radiobacter* group from soil. Phytopathology *55:* 645-647.

Schroth, M.N., A.R. Weinhold, A.H. McCain, D.C. Hildebrand and N. Ross. 1971. Biology and control of *Agrobacterium tumefaciens.* Hilgardia *40:* 537-552.

Schubert, R.H.W. 1960. Untersuchungen über die merkmale der gattung *Aeromonas.* Zentralbl. Bakteriol. Parasitenkd. Infektionskr. Hyg. Abt. I Orig. *180:* 310-327.

Schubert, R.H.W. 1963. Zür Morphologie der Geisseln von *Plesiomonas shigelloides.* Zentralbl. Bakteriol. Parasitenkd. Infektionskr. Hyg. Abt. I Orig. *191:* 376-382.

Schubert, R.H.W. 1964. Zür Taxonomie der Voges-Proskauer negativen "hydrophila-ähnlichen" Aeromonaden. Zentralbl. Bakteriol. Parasitenkd. Infektionskr. Hyg. Abt. I Orig. *193:* 482-490.

Schubert, R.H.W. 1967. The taxonomy and nomenclature of the genus *Aeromonas* Kluyver and Van Niel 1936. Part I. Suggestions on the taxonomy and nomenclature of the aerogenic *Aeromonas* species. Int. J. Syst. Bacteriol. *17:* 23-37.

Schubert, R.H.W. 1967. The taxonomy and nomenclature of the genus *Aeromonas* Kluyver and Van Niel 1936. Part II. Suggestions on the taxonomy and nomenclature of the anaerogenic *Aeromonas* species. Int. J. Syst. Bacteriol. *17:* 273-279.

Schubert, R.H. 1969. *Aeromonas hydrophila* subsp. *proteolytica* (Merkel et al. 1964) comb. nov. Zentralbl. Bakteriol. Parasitenkd. Infektionskr. Hyg. Abt. I Orig. *211:* 409-412.

Schubert, R.H.W. 1969. Infrasubspezifische Taxonomie von *Aeromonas hydrophila* (Chester 1901) Stanier 1943. Zentralbl. Bakteriol. Parasitenkd. Infektionskr. Hyg. Abt. I Orig. *211:* 406-409.

Schubert, R.H.W. 1977. Über den Nachweis von *Plesiomonas shigelloides* Habs and Schubert 1962 und ein Elektivmedium, den Inositol-Brillantgrün-Gallesalz-Agar. E. Rodenwaldt Arch. *4:* 97-103.

Schubert, R.H.W. 1981. Zür Ökologie von *Plesiomonas shigelloides.* Zentralbl. Bakteriol. Parasitenkd. Infektionskr. Hyg. Abt. I Orig. B *172:* 528-533.

Schubert, R.H.W. and G. Kexel. 1964. Der Ausfall der Butandioldehydrogenase-Reaktion bei einigen *Pseudomonadaceen* und *Vibrionen.* Zentralbl. Bakteriol. Parasitenkd. Infektionskr. Hyg. Abt. I Orig. *194:* 130-132.

Schubert, R.H.W., E. Schafer and W. Meiser. 1972. Vergleichende untersuchungen über die eliminierung von Poliomyelitis-Impfvirus und Aeromonads an einer halbtechnischen belebschlam-manlage des Grossen erftverbandes in Bergheim. Gas. Wasserf. *113:* 132-134.

Schultze, W.H. 1910. Über eine neue Methode zum Nachweis von Reduktionsund Oxydationswirkungen der Bakterien. Zentralbl. Bakteriol. Parasitenkd. Infektionskr. I Abt. Orig. *56:* 544-551.

Schulz, S., S. Pohl and W. Mannheim. 1977. Mischinfektion von Albinoratten durch *Pasteurella pneumotropica* und eine neue pneumotype *Pasteurella* species. Zentralbl. Vet. Med. *24:* 476-485.

Schurenkova, A. 1938. *Sogdianella moshkowskii* gen. nov.—a parasite belonging to the Piroplasmidea in a raptororial bird—*Gypäetus barbatus* L. Med. Parazit. (Mosk) *7:* 932-937.

Schuster, M.I. and D.P. Coyne. 1974. Survival mechanisms of phytopathogenic bacteria. Ann. Rev. Phytopathol. 12: 199-221.

Schwabacher, H., D.R. Lucas and C. Rimington. 1947. *Bacterium melaninogenicum,* a misnomer. J. Gen. Microbiol. *1:* 109-120.

Schwach, T. 1979. Case report. Clin. Microbiol. Newsletter *1 (Number 16):* 4-5.

Schwartz, R.M. and M.O. Dayhoff. 1978. Origins of prokaryotes, eukaryotes, mitochondria, and chloroplasts. Science (Washington) *199:* 395-403.

Schwartz, W. 1959. Bakterien- und Actinomyceten-Symbiosen. Hb. Pflphysiol. *11:* 546-576.

Schwemmler, W. 1979. Mechanismen der Zellevolution. Grundriss einer modernen Zelltheorie. W. deGruyter, Berlin, New York.

Schwemmler, W. 1980. Endocytobiosis: general principles. Biosystems *12:* 111-122.

Schwemmler, W., J.-L. Duthoit, G. Kuhl and C. Vago. 1973. Sprengung der Endosymbiose von *Euscelis plebejus* F. und Ernährung aposymbiontischer Tiere mit synthetischer Diät (Hemiptera, Cicadidae). Z. Morphol. Tiere *74:* 297-322.

Schwemmler, W. and M. Hermann. 1979. Oszillationen im Energiestoffwechsel von Wirt und Symbiont eines Zikadeneis. I. Analyse möglicher stoffwechselphysiologischer Korrelationen beider Systeme. Cytobios *25:* 45-62.

Schwemmler, W. and M. Hermann. 1980. Oszillationen im Energiestoffwechsel von Wirt und Symbiont eines Insekteneies. II. Analyse möglicher endogener Rhythmen beider Systeme. Cytobios *27:* 193-208.

Schwemmler, W., G. Hobom and M. Egel-Mitani. 1975. Isolation and characterization of leafhopper endosymbiont DNA. Cytobiologie *10:* 249-259.

Schwemmler, W. and H.E.A. Schenk (Editors). 1980. Endocytobiology. Endosymbiosis and cell biology. A synthesis of recent research. W. deGruyter, Berlin, New York.

Scott, C.C.L., R.A. Makula and W.R. Finnerty. 1976. Isolation and characterization of membranes from a hydrocarbon-oxidizing *Acinetobacter* sp. J. Bacteriol. *127:* 469-480.

Scott, D.A. and A.J. Musgrave. 1971. Aspects of the fine structure of symbiotes and related host tissues in nephridia of *Allolobophora caliginosa typica* (Annelida; Lumbricidae). J. Invert. Pathol. *18:* 51-60.

Scott, D. B., C. A. Scott and J. Döbereiner. 1979. Nitrogenase activity and nitrate respiration in *Azospirillum* spp. Arch. Microbiol. *121:* 141-145.

Scott, S.S. and D.C. Old. 1981. Mannose-resistant and eluting (MRE) haemagglutinins, fimbriae and surface structures in strains of *Haemophilus.* FEMS Microbiol. Lett. *10:* 235-240.

Scully, D. A. and N. C. Dondero. 1973. Estimation with several culture media of spirilla of 11 natural sources. Can. J. Microbiol. *19:* 983-989.

Seamer, J. 1959. The propagation and preservation of *Eperythrozoon coccoides.* J. Gen. Microbiol. *21:* 244-351.

Searcy, D.G. 1976. *Thermoplasma acidophilum:* Intracellular pH and potassium concentration. Biochim. Biophys. Acta *451:* 278-286.

Searcy, D.G. and E.K. Doyle. 1975. Characterization of *Thermoplasma acidophilum* deoxyribonucleic acid. Int. J. Syst. Bacteriol. *25:* 286-289.

Searcy, D.G. and D.B. Stein. 1980. Nucleoprotein subunit structure in an unusual prokaryotic organism: *Thermoplasma acidophilum.* Biochim. Biophys. Acta *609:* 180-196.

Sebald, M. 1962. Étude sur les bactéries anaérobies gram-négatives asporulées. Thèses de L'université Paris, Imprimerie Barnéoud S.A. Laval, France.

Sebald, M. and M. Véron. 1963. Teneur en bases de l'ADN et classification des vibrions. Ann. Inst. Pasteur (Paris) *105:* 897-910.

Sebek, O.K. 1965. Microbiological method for the determination of L-tryptophan. J. Bacteriol. *90:* 1026-1031.

Sedlák, J. 1957. Über die Eigenschaften und Bedeutung der Stämmen der *Citrobacter- (E. freundii*-Ballerup-Bethesda) Gruppe. Zentralbl. Bakteriol. Parasitenkd. Infektionskr. Hyg. Abt. Orig. *166:* 11-20.

Sedlák, J. 1974. The genus *Citrobacter. In* Buchanan and Gibbons (Editors), Bergey's Manual of Determinative Bacteriology, 8th Ed. The Williams and Wilkins Co., Baltimore, pp. 296-298.

Sedlák, J. and Slajsová. 1966. Antigenstruktur und Antigenbeziehungen der Gattung *Citrobacter.* Zentralbl. Bakteriol. Parasitenkd. Infektionskr. Hyg. Abt. I Orig. *200:* 369-374.

Sedlák, J. and M. Slajsová. 1966. On the antigenic relationships of certain *Citrobacter* and *Hafnia* cultures. J. Gen. Microbiol. *43:* 151-158.

Sedlák, J. and M. Slajsová. 1967. Taxonomie due "coliforme 1433". Ann. Inst. Pasteur *112:* 119-121.

Seeley, H.W., Jr. 1974. Genus *Lampropedia* Schroeter 1886, 151. *In* Buchanan and Gibbons (Editors) Bergey's Manual of Determinative Bacteriology, 8th Ed., The Williams and Wilkins Co., Baltimore, pp. 440-441.

Seghal, S.N. and N.E. Gibbons. 1960. Effect of some metal ions on the growth of *Halobacterium cutirubrum.* Can. J. Microbiol. *6:* 165-169.

Seidler, R.J., M.D. Knittel and C. Brown. 1975. Potential pathogens in the environment: Cultural reactions and nucleic acid studies of *Klebsiella pneumoniae* from clinical and environmental sources. Appl. Microbiol. *29:* 819-825.

Seidler, R. J., M. Mandel and J. N. Baptist. 1972. Molecular heterogeneity of the bdellovibrios: evidence of two new species. J. Bacteriol. *109:* 209-217.

Seidler, R. J. and M. P. Starr. 1968. Structure of the flagellum of *Bdellovibrio bacteriovorus.* J. Bacteriol. *95:* 1952-1955.

Seidler, R. J. and M. P. Starr. 1969. Isolation and characterization of host-independent bdellovibrios. J. Bacteriol. *100:* 769-785.

Sellers, W. 1964. Medium for differentiating the Gram-negative nonfermenting bacilli of medical interest. J. Bacteriol. *87:* 46-48.

Sellwood, J. and R.A. Lelliott. 1978. Internal browning of hyacinth caused by *Erwinia rhapontici.* Plant Pathol. *27:* 120-124.

Semancik, J.S., A.K. Vidaver and J.L. van Etten. 1973. Characterization of a segmented double-helical RNA from bacteriophage ϕ6. J. Molec. Biol. *78:* 617-625.

Sen, G.P., G. Singh and T.P. Joshi. 1968. Comparative efficacy of serological tests in the diagnosis of glanders. Indian Vet. J. 45: 286-292.

Senff, L.M., W.S. Wegener, G.F. Brooke, W.R. Finnerty and R.A. Makula. 1976. Phospholipid composition and phospholipase activity of Neisseria gonorrhoeae. J. Bacteriol. 127: 874-880.

Senior, B.W. 1977. The Dienes phenomenon: Identification of the determinants of compatibility. J. Gen. Microbiol. 102: 236-244.

Senior, B.W. 1977. Typing of Proteus strains by proticine production and sensitivity. J. Med. Microbiol. 10: 7-17.

Senitzer, D., G.T. Dimopoullos, B.R. Brinkley and M. Mandel. 1972. Deoxyribonucleic acid of Anaplasma marginale. J. Bacteriol. 109: 434-436.

Senterfit, L.B. and K.E. Jensen. 1966. Antimetabolic antibodies to Mycoplasma pneumoniae measured by tetrazolium reduction inhibition. Proc. Soc. Exp. Biol. Med. 122: 786-790.

Sereny, B. 1957. Experimental kerato-conjunctivitis shigellosa. Acta Microbiol. Acad. Sci. Hung. 4: 367-376.

Serizawa, N., K. Nakagawa, S. Kamimura, T. Miyadera and M. Arai. 1979. Stereo-specific synthesis of 3-trifluromethylcephalosporin derivative by microbial acylase. J. Antibiot. (Tokyo) 32: 1016-1018.

Serrano, F.B. 1928. Bacterial fruitlet brown-rot of pineapple in the Philippines. Philippine J. Sci. 36: 271-305.

Sertic, V. and N.A. Boulgakov. 1936. Bactériophages spécifiques pour des variétés bactériennes flagellées. C. R. Soc. Biol. Paris 123: 887-888.

Sethi, K.K. and H.E. Müller. 1972. Neuraminidase activity in Mycoplasma gallisepticum. Infect. Immun. 5: 260-262.

Setlow, J.K., D.C. Brown, M.E. Boling, A. Mattingly and M.P. Gordon. 1968. Repair of deoxyribonucleic acid in Haemophilus influenzae I. X-ray sensitivity of ultraviolet-irradiated bacteriophage and transforming deoxyribonucleic acid. J. Bacteriol. 95: 546-558.

Seubert, W. 1960. Degradation of isoprenoid compounds by microorganisms. I. Isolation and characterization of an isoprenoid-degrading bacterium, Pseudomonas citronellolis n. sp. J. Bacteriol. 79: 426-434.

Severin, V. 1978. Ein neues pathogènes Bakterium an Hanf—Xanthomonas campestris pathovar cannabis. Arch. Phytopathol. Pflanzenschutz 14: 7-15.

Severini, G. 1913. Una bactériosi dell' Ixia maculata e del Gladiolus coluilli. Ann. Bot. (Rome) 11: 413-424.

Sevin, A. and Buttiaux, R. 1939. The characters and systematic position of Morgan's bacillus. J. Pathol. Bacteriol. 49: 457-466.

Sgorbati, B. 1979. Preliminary quantification of immunological relationships among the transaldolases of the genus Bifidobacterium. Antonie van Leeuwenhoek J. Microbiol. Serol. 45: 557-564.

Sgorbati, B. and V. Scardovi. 1979. Immunological relationships among transaldolases in the genus Bifidobacterium. Antonie van Leeuwenhoek J. Microbiol. Serol. 49: 129-140.

Shah, H.N., R. Bonnett, B. Mateen and R.A.D. Williams. 1979. The porphyrin pigmentation of subspecies of Bacteroides melaninogenicus. Biochem. J. 180: 45-50.

Shah, H.N. and M.D. Collins. 1980. Fatty acid and isoprenoid quinone composition in the classification of Bacteroides melaninogenicus and related taxa. J. Appl. Bacteriol. 48: 75-87.

Shah, H.N. and M.D. Collins. 1981. Bacteroides buccalis, sp. nov., Bacteroides denticola, sp. nov., and Bacteroides pentosaceus, sp. nov., new species of the genus Bacteroides from the oral cavity. Zentralbl. Bakteriol. Mikrobiol. Hyg., I Abt. Orig. C2: 235-241.

Shah, H.N. and M.D. Collins. 1982. In Validation of the publication of new names and new combinations previously effectively published outside the IJSB. List No. 8. Int. J. Syst. Bacteriol. 32: 266-268.

Shah, H.N., R.A.D. Williams, G.H. Bowden and J.M. Hardie. 1976. Comparison of the biochemical properties of Bacteroides melaninogenicus from human dental plaque and other sites. J. Appl. Bacteriol. 41: 473-492.

Shamberger, R.J. 1960. Amino acid requirement of Acetobacter suboxydans. Thesis. Oregon State University, Corvallis.

Shane, B.S., L. Gouws and A. Kistner. 1969. Cellulolytic bacteria occurring in the rumen of sheep conditioned to low-protein teff hay. J. Gen. Microbiol. 55: 445-457.

Sharp, J.T. 1968. Isolation of L-forms of Bartonella bacilliformis. Proc. Soc. Exp. Biol. Med. 128: 1072-1075.

Sharpe, M.E., J.H. Brock and B.A. Phillips. 1975. Glycerol teichoic acid as an antigenic determinant in a Gram-negative bacterium Butyrivibrio fibrisolvens. J. Gen. Microbiol. 88: 355-363.

Shaw, C. and P.H. Clarke. 1955. Biochemical classification of Proteus and Providence cultures. J. Gen. Microbiol. 13: 155-161.

Shaw, D.H. and H.J. Hodder. 1978. Lipopolysaccharides of the motile aeromonads; core oligosaccharide analysis as an aid to taxonomic classification. Can. J. Microbiol. 24: 864-868.

Shaw, N. 1975. Bacterial glycolipids and glycophospholipids. Adv. Microb. Physiol. 12: 141-167.

Shaw, W.V., D.H. Bouanchaud and F.W. Goldstein. 1978. Mechanism of transferable resistance to chloramphenicol in Haemophilus parainfluenzae. Antimicrob. Agents Chemother. 13: 326-330.

Sheikholeslam, S., B.-C. Lin and C.I. Kado. 1979. Multiple-size plasmids in Agrobacterium radiobacter and A. tumefaciens. Phytopathology 69: 54-58.

Shenberg, E. 1967. Growth of pathogenic Leptospira in chemically defined media. J. Bacteriol. 93: 1598-1606.

Shepard, M.C. 1954. The recovery of pleuropneumonia-like organisms from Negro men with and without nongonococcal urethritis. Am. J. Syph. 38: 113-124.

Shepard, M.C. 1956. T-form colonies of pleuropneumonialike organisms. J. Bacteriol. 71: 362-369.

Shepard, M.C. 1966. Human mycoplasma infections. Health Lab. Sci. 3: 163-169.

Shepard, M.C. and R.S. Combs. 1979. Enhancement of Ureaplasma urealyticum growth on a differential agar medium (A7B) by a polyamine, putrescine. J. Clin. Microbiol. 10: 931-933.

Shepard, M.C. and D.R. Howard. 1970. Identification of "T" mycoplasmas in primary agar cultures by means of a direct test for urease. Ann. N.Y. Acad. Sci. 174: 809-819.

Shepard, M.C. and C.D. Lunceford. 1965. Effect of pH on human mycoplasma strains. J. Bacteriol. 89: 265-270.

Shepard, M.C. and C.D. Lunceford. 1967. Occurrence of urease in T strains of Mycoplasma. J. Bacteriol. 93: 1513-1520.

Shepard, M.C. and C.D. Lunceford. 1970. Urease color test medium U-9 for the detection and identification of "T" mycoplasmas in clinical material. Appl. Microbiol. 20: 539-543.

Shepard, M.C. and C.D. Lunceford. 1976. Differential agar medium (A7) for identification of Ureaplasma urealyticum (human T mycoplasmas) in primary cultures of clinical material. J. Clin. Microbiol. 3: 613-625.

Shepard, M.C., C.D. Lunceford, D.K. Ford, R.H. Purcell, D. Taylor-Robinson, S. Razin and F.T. Black. 1973. Ureaplasma urealyticum gen. nov., sp. nov.: proposed nomenclature for the human T (T-strain) mycoplasmas. Int. J. Syst. Bacteriol. 24: 160-171.

Shepard, M.C. and G.K. Masover. 1979. Special features of ureaplasmas. In Barile and Razin (Editors), The Mycoplasmas, Vol. 1, Academic Press, New York, pp. 451-494.

Shepard, C.C., M.A. Redus, T. Tzianabos and D.T. Warfield. 1976. Recent experience with the complement fixation test in the laboratory diagnosis of rickettsial diseases in the United States. J. Clin. Microbiol. 4: 277-283.

Shepherd, A.M., S.A. Clark and A. Kempton. 1973. An intracellular microorganism associated with tissues of Heterodera spp. Nematologia 19: 31-34.

Sherris, J.C., J.G. Shoesmith, M.T. Parker and D. Breckon. 1959. Tests for the rapid breakdown of arginine by bacteria: their use in the identification of pseudomonads. J. Gen. Microbiol. 21: 389-396.

Sherwin, R.P. and J. Wilkins. 1973. The ultrastructure of Hemophilus influenzae. In Sell (Editor), Hemophilus influenzae, Vanderbilt University Press, Nashville, Tennessee, pp. 143-151.

Sheth, N.K. and V.P. Kurup. 1975. Evaluation of tyrosine medium for the identification of Enterobacteriaceae. J. Clin. Microbiol. 1: 483-485.

Shewan, J.M. and M. Véron. 1974. Genus Vibrio Pacini. In Buchanan and Gibbons (Editors), Bergey's Manual of Determinative Bacteriology, 8th Ed. The Williams and Wilkins Co., Baltimore, pp. 340-345.

Shibata, S., Y. Isayama and T. Shimizu. 1962. A possibility of variation in Brucella abortus from type II to type I. Natl. Inst. Anim. Health Q. (Tokyo) 2: 10-14.

Shieh, H.S. and B.H. Reddy. 1972. Arginine catabolism in the causative agent of furunculosis, Aeromonas salmonicida. Int. J. Biochem. 3: 705-708.

Shiga, K. 1898. Ueber den Dysenteriebacillus (Bacillus dysenteriae). Zentralbl. Bakteriol. Infektionskr. Hyg. Abt. I Orig. 24: 817-828.

Shigeta, S., Y. Yasunaga, K. Honzumi, H. Okamura, R. Kumata and S. Endo. 1978. Cerebral ventriculitis associated with Achromobacter xylosoxidans. J. Clin. Pathol. 31: 156-161.

Shigidi, M.A. and A.B. Hoerlein. 1970. Characterization of the Haemophilus-like organism of infectious thromboembolic meningoencephalitis of cattle. Am. J. Vet. Res. 31: 1017-1022.

Shilo, M. 1969. Morphological and physiological aspects of the interaction of Bdellovibrio with host bacteria. Curr. Top. Microbiol. Immunol. 50: 174-204.

Shilo, M. (Editor). 1979. Strategies of microbial life in extreme environments. Berlin: Dahlem Konferenzen, Verlag Chemie, Weinheim, New York.

Shilo, M. and R.Y. Stanier. 1957. The utilization of the tartaric acids by pseudomonads. J. Gen. Microbiol. 16: 482-490.

Shilo, M. and T. Yetinson. 1979. Physiological characteristics underlying the distribution patterns of luminous bacteria in the Mediterranean Sea and the Gulf of Elat. Appl. Environ. Microbiol. 38: 577-584.

Shimada, T. and R. Sakazaki. 1973. R antigen of Vibrio cholerae. Jpn. J. Med. Sci. Biol. 26: 155-160.

Shimada, T. and R. Sakazaki. 1977. Additional serovars and inter-O antigenic relationships of Vibrio cholerae. Jpn. J. Med. Sci. Biol. 30: 275-277.

Shimada, T. and Sakazaki, R. 1978. On the serology of Plesiomonas shigelloides. Jpn. J. Med. Sci. Biol. 31: 135-142.

Shimizu, T., H. Ernø and H. Nagatomo. 1978. Isolation and characterization of Mycoplasma columbinum and Mycoplasma columborale, two species from pigeons. Int. J. Syst. Bacteriol. 28: 538-546.

Shimizu, T., K. Numano and K. Uchida. 1979. Isolation and identification of mycoplasmas from various birds: an ecological study. Jpn. J. Vet. Sci. 41: 273-282.

Shimwell, J.L. 1936. A study of the common rod bacteria of brewers' yeast. J. Inst. Brew. 42: 119-127.

Shimwell, J.L. 1937. Study of a new type of beer disease bacterium (Achromobacter anaerobium sp. nov.) producing alcoholic fermentation of glucose. J. Inst. Brew. London 43: 507-509.

Shimwell, J.L. 1948. Brewing Bacteriology. IV. The acetic acid bacteria (Family *Acetobacteriaceae*: Genus *Acetobacter*). Wallerstein Lab. Commun. *11:* 27-39.

Shimwell, J.L. 1948. Brewing bacteriology V. Gram-negative wort, yeast and beer bacteria. Wallerstein Lab. Commun. *11:* 135-145.

Shimwell, J.L. 1950. *Saccharomonas*, a proposed new genus for bacteria producing a quantitative alcoholic fermentation of glucose. J. Inst. Brew. London *56:* 179-182.

Shimwell, J.L. 1958. Flagellation and taxonomy of *Acetobacter* and *Acetomonas*. Antonie van Leeuwenhoek J. Microbiol. Serol. *24:* 187-192.

Shimwell, J.L. 1959. A re-assessment of the genus *Acetobacter*. Antonie van Leeuwenhoek J. Microbiol. Serol. *25:* 49-67.

Shimwell, J.L. 1963. *Obesumbacterium* gen. nov. Brewers' J. *99:* 759-760.

Shimwell, J.L. 1964. *Obesumbacterium*, a new genus for the inclusion of '*Flavobacterium proteus*'. J. Inst. Brewing *70:* 247-248.

Shimwell, J.L. and J.G. Carr. 1959. The genus *Acetomonas*. Antonie van Leeuwenhoek. J. Microbiol. Serol. *25:* 353-368.

Shimwell, J.L. and M. Grimes. 1936. The distinguishing characters of *Flavobacterium proteum* (sp. nov.), the common rod bacterium of brewers' yeast. J. Inst. Brew. *42:* 348-350.

Shinde, P.A. and F.L. Lukezic. 1974. Isolation, pathogenicity and characterization of fluorescent pseudomonads associated with discolored alfalfa roots. Phytopathology *64:* 865-871.

Shinde, P.A. and F.L. Lukezic. 1974. Characterization and serological comparisons of bacteria of the genus *Erwinia* associated with discolored alfalfa roots. Phytopathology *64:* 871-876.

Shinhar, E., J. Silver, R. Yeivin and M. Shapiro. 1980. Polymicrobial endocarditis due to *Eikenella corrodens* and group B beta-haemolytic streptococci. Israel J. Med. Sci. *16:* 458-459.

Shinoda, S., T. Honda, Y. Takeda and T. Miwatani. 1974a. Antigenic difference between polar monotrichous and peritrichous flagella of *Vibrio parahaemolyticus*. J. Bacteriol. *120:* 923-928.

Shinoda, S., R. Kariyama, M. Ogawa, Y. Takeda and T. Miwatani. 1976. Flagellar antigens of various species of the genus *Vibrio* and related genera. Int. J. Syst. Bacteriol. *26:* 97-101.

Shinoda, S., T. Miwatani, T. Honda and Y. Takeda. 1974b. Antigenicity of flagella of *Vibrio parahaemolyticus*. In Fujino, Sakaguchi, Sakazaki and Takeda (Editors), International Symposium on *Vibrio parahaemolyticus*. Saikon Publishing Co., Tokyo, pp. 193-197.

Shinoda, S. and K. Okamoto. 1977. Formation and function of *Vibrio parahaemolyticus* lateral flagella. J. Bacteriol. *129:* 1266-1271.

Shirarata, K., T. Deguchi, T. Hayashi, I. Matsubara and T. Suzuki. 1970. The structures of fluopsins C and F. J. Antibiot. *23:* 546-550.

Shishido, A., M. Ohtawara, S. Tateno, S. Mizuno, M. Ogura and M. Kitaoka. 1958. The nature of immunity against scrub typhus in mice. I. The resistance of mice, surviving subcutaneous infection of scrub typhus rickettsia, to intraperitoneal reinfection of the same agent. Jpn. J. Med. Sci. Biol. *11:* 383-399.

Shmilovitz, M., O. Kretzer and E. Levy. 1974. The anerogenic serotype 147 as an etiologic agent of dysentery in Israel. Isr. J. Med. Sci. *10:* 1425-1429.

Shope, R.E. 1964. Porcine contagious pneumonia. I. Experimental transmission, etiology and pathology. J. Exp. Med. *119:* 357-368.

Short, J.A., C.M. Thorley and P.D. Walker. 1976. An electron microscope study of *Bacteroides nodosus*: ultrastructure of organisms from primary isolates and different colony types. J. Appl. Bacteriol. *40:* 311-315.

Shortland-Webb, W.R. 1968. *Proteus* and coliform meningo-encephalitis in neonates. J. Clin. Pathol. *21:* 422-431.

Shotts, E.B., J.L. Gaines, C. Martin and A.K. Prestwood. 1972. *Aeromonas* induced deaths among fish and reptile in an eutrophic inland lake. J. Am. Vet. Med. Assoc. *161:* 603-607.

Shotts, E.B. and R. Rimler. 1973. Medium for the isolation of *Aeromonas hydrophila*. Appl. Microbiol. *26:* 550-553.

Shotts, E.B. and S.F. Snieszko. 1976. Selected bacterial fish diseases. In L.A. Page (Editor), Wildlife Diseases, Plenum Publishing Co., New York, pp. 143-151.

Shrivastava, R., V.B. Sinha and B.S. Shrivastava. 1980. Events in the pathogenesis of experimental cholera: role of bacterial adherence and multiplication. J. Med. Microbiol. *13:* 1-9.

Sias, S.R., A.H. Stouthamer and J.L. Ingraham. 1980. The assimilatory and dissimilatory nitrate reductases of *Pseudomonas aeruginosa* are encoded by different genes. J. Gen. Microbiol. *118:* 229-234.

Sickles, G.M. and M. Shaw. 1934. A systematic study of microorganisms which decompose the specific carbohydrates of the pneumococcus. J. Bacteriol. *28:* 415-431.

Siddiqui, A. and I.D. Goldberg. 1975. Intergenic transformation of *Neisseria gonorrhoeae* and *Neisseria perflava* to streptomycin and nutritional independence. J. Bacteriol. *124:* 1359-1365.

Sieburth, J.M. 1979. Sea microbes. Oxford Univ. Press, New York.

Sierra, G. 1957. A simple method for the detection of lipolytic activity of microorganisms and some observations on the influence of the contact between cells and fatty substrates. Antonie van Leeuwenhoek J. Microbiol. Serol. *23:* 15-22.

Sierra, G. and N.E. Gibbons. 1962. Production of poly-β-hydroxybutyric acid granules in *Micrococcus halodenitrificans*. Can. J. Microbiol. *8:* 249-253.

Sierra, G. and N.E. Gibbons. 1962. Role and oxidation pathway of poly-β-hydroxybutyric acid in *Micrococcus halodenitrificans*. Can. J. Microbiol. *8:* 255-269.

Sierra, G. and N.E. Gibbons. 1963. Sodium requirement of poly-β-hydroxybutyric acid depolymerase of *Micrococcus halodenitrificans*. Can. J. Microbiol. *9:* 491-497.

Sijderius, R. 1946. Heterotrophe bacteriën, die thiosulfaat oxydeeren. Thesis, Univ. Amsterdam, pp. 1-146.

Sil, J. and F.K. Bhattacharyya. 1979. A rapid test for the identification of all serotypes of *Vibrio cholerae* (including "non-agglutinating" vibrios). J. Med. Microbiol. *19:* 63-70.

Silverman, D.J. and C.L. Wisseman, Jr. 1978. Comparative ultrastructural study on the cell envelopes of *Rickettsia prowazeki*, *Rickettsia rickettsii*, and *Rickettsia tsutsugamushi*. Infect. Immun. *21:* 1020-1023.

Silverman, D.J. and C.L. Wisseman, Jr. 1979. In vitro studies of rickettsia-host cell interactions: Ultrastructural changes induced by *Rickettsia rickettsii* infection of chicken embryo fibroblasts. Infect. Immun. *26:* 714-727.

Silverman, D.J., C.L. Wisseman, Jr. and A. Waddell. 1980. In vitro studies of rickettsia-host cell interactions: Ultrastructural study of *Rickettsia prowazekii*-infected chicken embryo fibroblasts. Infect. Immun. *29:* 778-790.

Silverman, D.J., C.L. Wisseman, Jr., A.D. Waddell and M. Jones. 1978. External layers of *Rickettsia prowazekii* and *Rickettsia rickettsii*: Occurrence of a slime layer. Infect. Immun. *22:* 233-246.

Silvestri, L., M. Turri, L.R. Hill and E. Gilardi. 1962. A quantitative approach to the systematics of actinomycetes based on overall similarity. Symp. Soc. Gen. Microbiol. *12:* 333-360.

Simberkoff, M.S. 1980. Experimental *Serratia marcescens* infection and defense mechanisms. In von Graevenitz and Rubin (Editors), The genus *Serratia*. CRC Press, Boca Raton, Florida, pp. 157-164.

Simon, P.C. 1975. A simple method for rapid identification of *Sphaerophorus necrophorus* isolates. Can. J. Comp. Med. *39:* 349-353.

Simon, R.D. 1978. *Halobacterium* strain 5 contains a plasmid which is correlated with the presence of gas vacuoles. Nature (London) *273:* 314-317.

Simons, H. 1920. Eine saprophytische Oscillarie im Darm des Meerschweinschens. Zentralbl. Bakteriol. Parasitenkd. Infektionskr. Hyg. Abt. II *50:* 356-368.

Simons, H. 1921. Ueber *Selenomonas palpitans*. Zentralbl. Mikrobiol. Parasitenkd. Infektionskr. Hyg. Abt. I Orig. *87:* 50.

Simons, H. 1922. Saprophytische Oscillatorien des Menschen und der Tiere. Zentralbl. Bakteriol. Parasitenkd. Infektionskr. Hyg. Abt. I Orig. *88:* 501-510.

Simpson, C.F. 1972. Structure of *Ehrlichia canis* in blood monocytes of a dog. Am. J. Vet. Res. *33:* 2451-2454.

Simpson, C.F. 1974. Relationship of *Ehrlichia canis*-infected mononuclear cells to blood vessels of lungs. Infect. Immun. *10:* 590-596.

Simpson, C.F., J.M. Gaskin and J.W. Harvey. 1978. Ultrastructure of erythrocytes parasitized by *Haemobartonella felis*. J. Parasitol. *64:* 504-511.

Simpson, C.F., J.M. Kling and F.C. Neal. 1965. The nature of bonds in parasitized bovine erythrocytes. J. Cell. Biol. *27:* 225-235.

Simpson, F. J. and J. Robinson. 1968. Some energy producing systems in *Bdellovibrio bacteriovorus*, strain 6-5-S. Can. J. Biochem. *46:* 865-873.

Sims, M.A. 1974. *Flavobacterium meningosepticum*: a probable cause of meningitis in a cat. Vet. Record *95:* 567-569.

Sims, W. 1970. Oral haemophili. J. Med. Microbiol. *3:* 615-625.

Sims, W. 1972. Pathogenicity of human oral strains of haemophili: enzymes anaerobiosis and effects on mice and rabbits. Arch. Oral. Biol. *17:* 745-750.

Sinclair, M.I. and A.F. Morgan. 1978. Transformation of *Pseudomonas aeruginosa* strain PAO with bacteriophage and plasmid DNA. Aust. J. Biol. Sci. *31:* 679-688.

Sinden, S.L. and R.D. Durbin. 1970. A comparison of the chlorosis-inducing toxin from *Pseudomonas coronafaciens* with wildfire toxin from *Pseudomonas tabaci*. Phytopathology *60:* 360-364.

Sindermann, C.J. 1970. Principal diseases of marine fish and shellfish. Academic Press, New York-London.

Singer, J. and J. Bar-Chay. 1954. Biochemical investigation of Providence strains and their relationship to the *Proteus* group. J. Hyg. *52:* 1-8.

Singer, J. and B.E. Volcani. 1955. An improved ferric chloride test for differentiating Proteus-Providence group from other *Enterobacteriaceae*. J. Bacteriol. *69:* 303-306.

Singh, G. 1974. Endosymbiotic microorganisms in *Cletus signatus* Walker (Coreidae: Heteroptera). Experientia (Basel) *30:* 1406-1407.

Singh, G. and N.C. Pant. 1974. Degradation of insecticides by the cultured symbiotes of *Cletus signatus* Walker (Coreidae: Heteroptera). Curr. Sci. (Bangalore) *43:* 624-625.

Singh, S.B. and A.J. Musgrave. 1966. Some studies on the chromatin and cell wall of the mycetomal microorganisms of *Sitophilus granarius* (L.) (Coleoptera). J. Cell Sci. *1:* 175-180.

Sinha, R.C. 1974. Purification of mycoplasmalike organisms from China aster plants affected with clover phyllody. Phytopathology *64:* 1156-1158.

Sinha, R.C. 1979a. Chemotherapy of mycoplasmal plant diseases. In Whitcomb and Tully (Editors), The Mycoplasmas, Vol. 3, Academic Press, New York, pp. 309-335.

Sinha, R.C. 1979b. Lipid composition of mycoplasmalike organisms purified from clover phyllody and aster yellows diseased plants. Phytopathol. Z. *96:* 132-139.

Sinha, R.C. and L.N. Chiykowski. 1967. Initial and subsequent sites of aster yellows virus infection in a leafhopper vector. Virology 33: 702-708.

Sinha, R.C. and C. Madhosingh. 1980. Proteins of mycoplasma-like organisms purified from clover phyllody and aster yellows-affected plants. Phytopathol. Z. 99: 294-300.

Sinha, R.C. and E.A. Peterson. 1972. Uptake and persistence of oxytetracycline in aster plants and vector leafhoppers in relation to inhibition of clover phyllody agent. Phytopathology 62: 377-383.

Sirot, D., J. Sirot, M. Chanal and M. Cluzel. 1980. Étude bactériostatique comparée de 7 céphalosporines: pour un choix rationnel au niveau de l'antibiogramme en pratique hospitalière. Méd. Mal. Infect. 10: 146-155.

Sivendra, R. and S.H. Tan. 1977. Pathogenicity of nonpigmented cultures of Chromobacterium violaceum. J. Clin. Microbiol. 5: 514-516.

Skerman, T.M. 1975. Determination of some in vitro growth requirements of Bacteroides nodosus. J. Gen. Microbiol. 87: 107-119.

Skerman, T.M., S.K. Erasmuson and D. Every. 1981. Differentiation of Bacteroides nodosus biotypes and colony variants in relation to their virulence and immuno-protective properties in sheep. Infect. Immun. 32: 788-795.

Skerman, V.B.D. 1967. A guide for the identification of the genera of bacteria, 2nd Ed., The Williams and Wilkins Co., Baltimore.

Skerman, V.B.D. 1969. Abstracts of microbiological methods. Wiley-Interscience, New York.

Skerman, V.B.D. 1974. A key for the determination of the generic position of organisms listed in the Manual. In Buchanan and Gibbons (Editors), Bergey's Manual of Determinative Bacteriology, 8th Ed., Williams & Wilkins, Baltimore, pp. 1098-1146.

Skerman, V.B.D., V. McGowan and P.H.A. Sneath. 1980. Approved lists of bacterial names. Int. J. Syst. Bacteriol. 30: 225-420.

Skinner, F.A. 1977. An evaluation of the nile blue test for differentiating rhizobia from agrobacteria. J. Appl. Bacteriol. 43: 91-98.

Skinner, F.A. and D.W. Lovelock. 1979. Identification methods for microbiologists. Society for Applied Bacteriology Technical Series No. 14. Academic Press, New York.

Skirrow, M.B. 1969. The Dienes (mutual inhibition) test in the investigation of Proteus infections. J. Med. Microbiol. 2: 471-477.

Skirrow, M. B. and J. Benjamin. 1980. '1001' campylobacters: Cultural characteristics of intestinal campylobacters from man and animals. J. Hyg. Camb. 85: 427-442.

Skirrow, M. B. and J. Benjamin. 1981. Differentiation of enteropathogenic campylobacter. J. Clin. Path. 33: 1122.

Sklarow, S.S., R.R. Colwell, G.B. Chapman and S.F. Zane. 1973. Characteristics of a Vibrio parahaemolyticus bacteriophage isolated from Atlantic coast sediment. Can. J. Microbiol. 19: 1519-1520.

Skotnicki, Mary L. and B.G. Rolfe. 1977. Differential stimulation and inhibition of growth of Rhizobium trifolii strain T1 and other Rhizobium species by various carbon sources. Microbios 20: 15-28.

Skotnicki, M.L., D.E. Tribe and P.L. Rogers. 1980. R-plasmid transfer in Zymomonas mobilis. Appl. Environ. Microbiol. 40: 7-12.

Skripal, I.G. 1974. On improvement of taxonomy of the class Mollicutes and establishment in the order Mycoplasmatales of the new family Spiroplasmataceae Fam. Nova. Microbiol. Zh. Kiev 36: 462-467.

Skripal, I.G. 1983. Revival of the name Spiroplasmataceae fam. nov., nom. rev., omitted from the 1980 Approved Lists of Bacterial Names. Int. J. Syst. Bacteriol. 33: 408.

Skuja, H. 1964. Grundzüge der Algenflora und Algenvegetation der Fjeldgegenden um Abisko in Schwedisch-Lappland. Nova Acta Reg. Soc. Sci. Upsal. Ser. IV 18: 1-465.

Slabas, A.R. and L.R. Whatley. 1977. Metabolic regulation of pyruvate kinase isolated from autotrophically and heterotrophically grown Paracoccus denitrificans. Arch. Microbiol. 115: 67-71.

Slack, G.L. and G.H. Bowden. 1965. Preliminary studies of experimental dental plaque in vivo. Adv. Fluor. Res. Dent. Caries Prev. 3: 193-215.

Slanetz, L.W. and L.F. Rettger. 1933. A systematic study of the fusiform bacteria. J. Bacteriol. 26: 599-617.

Slee, A.M. and J.M. Tanzer. 1978. Selective medium for isolation of Eikenella corrodens from periodontal lesions. J. Clin. Microbiol. 8: 459-462.

Sleesman, J.P. and C. Leben. 1978. Preserving phytopathogenic bacteria at -70°C or with silica gel. Plant Dis. Rep. 62: 910-913.

Slepecky, R.A. and J.H. Law. 1960. A rapid spectrophotometric assay of alpha, beta unsaturated acids and beta-hydroxy acids. Anal. Chem. 32: 1697-1699.

Sleytr, U. and M. Kocur. 1973. Structure of Micrococcus denitrificans and M. halodenitrificans revealed by freeze-etching. J. Appl. Bacteriol. 36: 19-22.

Slopek, S. 1978. Phage typing of Klebsiella. In Bergan and Norris (Editors), Methods in Microbiology, Vol. II. Academic Press, London, pp. 193-222.

Slopek, S. and I. Durlakowa. 1967. Studies on the taxonomy of Klebsiella bacilli. Arch. Immunol. Ther. Exp. 15: 481-487.

Slopek, S. and J. Maresz-Babczyszyn. 1967. A working scheme for typing Klebsiella bacilli by means of pneumocins. Arch. Immunol. Ther. Exp. 15: 525-529.

Slopek, S., A. Przonko-Hessek, A. Milch and S. Deak. 1967. A working scheme for bacteriophage typing of Klebsiella bacilli. Arch. Immunol. Ther. Exp. 15: 589-599.

Slotnick, I.J. 1968. Cardiobacterium hominis in genitourinary specimens. J. Bacteriol. 95: 1175.

Slotnick, I.J. and M. Dougherty. 1964. Further characterization of an unclassified group of bacteria causing endocarditis in man: Cardiobacterium hominis gen. et sp. n. Antonie van Leeuwenhoek J. Microbiol. Serol. 30: 261-272.

Slotnick, I.J. and M. Dougherty. 1965. Unusual toxicity of riboflavin and flavin mononucleotide for Cardiobacterium hominis. Antonie van Leeuwenhoek J. Microbiol. Serol. 31: 355-360.

Slotnick, I.J., J.A. Mertz and M. Dougherty. 1964. Fluorescent antibody detection of human occurrence of an unclassified bacterial group causing endocarditis. J. Infect. Dis. 114: 503-505.

Slots, J. 1981. Enzymatic characterization of some oral and nonoral Gram-negative bacteria with the API ZYM System. J. Clin. Microbiol. 14: 288-294.

Slots, J. and R.J. Genco. 1979. Direct hemagglutination technique for differentiating Bacteroides asaccharolyticus oral strains from nonoral strains. J. Clin. Microbiol. 10: 371-373.

Slots, J. and R.J. Genco. 1980. Bacteroides melaninogenicus subsp. macacae, a new subspecies from monkey periodontopathic indigenous microflora. Int. J. Syst. Bacteriol. 30: 82-85.

Slyter, L.L., D.L. Kern and J.M. Weaver. 1976. Effect of pH on ruminal lactic acid utilization and accumulation in vitro. J. Anim. Sci. 43: 333-334 (Abstract).

Small, E. and M. Ristic. 1967. Morphological features of Haemobartonella felis. J. Am. Vet. Med. Assoc. 28: 845-851.

Smalley, D.L., P.A. Jaquess and J.P. Layne. 1980. Selenium enriched medium for Legionella pneumophila. J. Clin. Microbiol. 12: 32-34.

Smibert, R. M. 1965. Vibrio fetus var. intestinalis isolated from fecal and intestinal contents of clinically normal sheep: Biochemical and cultural characteristics of microaerophilic vibrios isolated from the intestinal contents of sheep. Am. J. Vet. Res. 26: 315-319.

Smibert, R. M. 1969. Vibrio fetus var. intestinalis isolated from the intestinal content of birds. Am. J. Vet. Res. 30: 1437-1442.

Smibert, R. M. 1970. Cell wall composition in the classification of Vibrio fetus. Int. J. Syst. Bacteriol. 20: 407-412.

Smibert, R.M. 1973. The Spirochaetales, a review. Crit. Rev. Microbiol. 2: 491-552.

Smibert, R. 1974. Campylobacter. In Buchanan and Gibbons (Editors). Bergey's Manual of Determinative Bacteriology. 8th ed. The Williams and Wilkins Co., Baltimore, pp. 207-212.

Smibert, R. 1974. Genus Treponema. In Buchanan and Gibbons (Editors), Bergey's Manual of Determinative Bacteriology, 8th ed., The Williams and Wilkins Co., Baltimore, pp. 175-184.

Smibert, R.M. 1976. Classification of non-pathogenic treponemes. Borrelia and Spirochaeta. In R.C. Johnson (Editor), The biology of parasitic spirochetes, Academic Press, New York, pp. 121-131.

Smibert, R. M. 1978. The genus Campylobacter. Annu. Rev. Microbiol. 32: 673-709.

Smibert, R.M. 1981. The Genus Campylobacter. In Starr, Stolp, Trüper, Balows, and Schleger (Editors). The Prokaryotes, A handbook on habitats, isolation and identification of bacteria. Springer-Verlag, New York, pp. 609-617.

Smibert, R.M. 1981. The Genus Treponema. In Starr, Stolp, Trüper, Balows and Schlegel (Editors). The Prokaryotes, a handbook on habitats, isolation and identification of bacteria. Springer-Verlag, New York, pp. 564-577.

Smibert, R.M. and L.V. Holdeman. 1976. Clinical isolates of anaerobic Gram-negative rods with a formate-fumarate energy metabolism: Bacteroides corrodens, Vibrio succinogenes, and unidentified strains. J. Clin. Microbiol. 3: 432-437.

Smibert, R.M. and N.R. Krieg. 1981. General characterization. In Gerhardt, Murray, Costilow, Nester, Wood, Krieg and Phillips (Editors), Manual of methods for general bacteriology, American Society for Microbiology, Washington, D.C., pp. 409-443.

Smiles, J. and M.J. Dobson. 1956. Direct ultra-violet and ultra-violet negative phase contrast micrography of bacteria from the stomachs of sheep. J. Roy. Microscop. Soc. Series III 75: 244-253.

Smit, J.A. and J.N. Coetzee. 1967. Serological specificities of phenylalanine deaminases of the Proteus-Providence group. Nature 214: 1238-1239.

Smith, A.R. and R. Rahn. 1975. An indirect hemagglutination test for the diagnosis of Eperythrozoon suis infection in swine. Am. J. Vet. Res. 36: 1319-1321.

Smith, C.O. 1913. Black pit of lemon. Phytopathology 3: 277-281.

Smith, D.C. 1979. From extracellular to intracellular: the establishment of a symbiosis. Proc. R. Soc. Lond. B Biol. Sci. 204: 115-130.

Smith, E.F. 1895. Bacillus tracheiphilus sp. nov. die Ursache des Verwelkens verschiedener Curcurbitaceen. Zentrabl. Bakteriol. Parasitenk Infektionskr. Hyg. Abt. II 1: 364-373.

Smith, E.F. 1896. A bacterial disease of the tomato, egg plant and Irish potato (Bacillus solanacearum n. sp.) U.S. Dept. Div. Veg. Phys. Pathol. Bull. 12: 1-28.

Smith, E.F. 1897. Pseudomonas campestris (Pammel). The cause of a brown rot in cruciferous plants. Zentrabl. Bakteriol. Infektionskr. Hyg. Abt. II 3: 284-291; 408-415; 478-485.

Smith, E.F. 1897. Description of Bacillus phaseoli n. sp. Bot. Gaz. 24: 192.

Smith, E.F. 1898. Notes on Stewart's sweet corn germ, Pseudomonas stewarti n. sp. Proc. Amer. Assoc. Adv. Sci. 47: 422-426.

Smith, E.F. 1901. The cultural characters of Pseudomonas hyacinthi, Ps. campestris, Ps. phaseoli and Ps. stewarti—four one-flagellate yellow bacteria parasitic on plants. U.S. Dept. Agr. Div. Veg. Phys. Pathol. Bull. *28:* 1-153.

Smith, E.F. 1903. Observations on a hitherto unreported bacterial disease, the cause of which enters the plant through ordinary stomata. Science (Washington) *17:* 456-457.

Smith, E.F. 1904. Bacterial leaf spot diseases. Science *19:* 417-418.

Smith, E.F. 1908. Recent studies of the olive-tubercle organism. Bull. U.S. Dept. Agr., Bur. Plant Ind. No. *131:* 25-43.

Smith, E.F. 1911. Bacteria in relation to plant diseases. Carnegie Inst. Wash. Publ. *2:* 1-368.

Smith, E.F. 1914. Bacteria in relation to plant diseases. Carnegie Institute, Washington, Publ. 3, pp. 1-309.

Smith, E.F. and M.K. Bryan. 1915. Angular leaf-spot of cucumbers. J. Agr. Res. *5:* 465-476.

Smith, E.F., L.R. Jones and C.S. Reddy. 1919. The black chaff of wheat. Science (Washington) *50:* 48.

Smith, E.F. and C.O. Townsend. 1907. A plant-tumor of bacterial origin. Science (New York) *25:* 671-673.

Smith, F. 1948. Genus I. *Salmonella* Lignières. Appendix II. *In* Breed, Murray and Hitchens (Editors), Bergey's Manual of Determinative Bacteriology, 6th ed. The Williams and Wilkins Co., Baltimore, p. 533.

Smith, F.B. 1938. An investigation of a taint in rib bones of bacon. The determination of halophilic vibrios. Proc. Roy. Soc. Queensland *49:* 29-52.

Smith, G.R. 1959. Isolation of two types of *Pasteurella haemolytica* from sheep. Nature (London) *183:* 1132-1133.

Smith, G.R. 1961. The characteristics of two types of *Pasteurella haemolytica* associated with different pathological conditions in sheep. J. Pathol. Bacteriol. *81:* 431-440.

Smith, G.R. and J.C. Oliphant. 1981a. The ability of *Mycoplasma mycoides* subsp. *mycoides* and closely related strains from goats and sheep to immunize mice against subspecies *capri*. J. Hyg. (Camb.) *87:* 321-329.

Smith, G.R. and J.C. Oliphant. 1981b. Observations on the antigenic differences between the so-called SC and LC strains of *Mycoplasma mycoides* subsp. *mycoides*. J. Hyg. (Camb.) *87:* 437-442.

Smith, G.R., J.M. Hooker and R.A. Milligan. 1980. Further studies on caprine and ovine mycoplasmas related to *Mycoplasma mycoides* subsp. *mycoides*. J. Hyg. (Camb.) *85:* 247-256.

Smith, H.E. and H.J. Arnott. 1974. Epi- and endobiotic bacteria associated with *Pyrsonympha vertens*, a symbiotic protozoon of the termite *Reticulitermes flavipes*. Trans. Amer. Microsc. Soc. *93:* 180-194.

Smith, H.E., H.E. Buhse and S.J. Stamler. 1975. Possible formation and development of spirochaete attachment sites found on the surface of symbiotic polymastigote flagellates of the termite *Reticulitermes flavipes*. BioSystems *7:* 374-379.

Smith, H.E., S.J. Stamler and H.E. Buhse. 1975. A scanning electron microscope survey of the surface features of polymastigote flagellates from *Reticulitermes flavipes*. Trans. Amer. Microsc. Soc. *94:* 401-410.

Smith, H.L., Jr. 1979. Serotyping of non-cholera vibrios. J. Clin. Microbiol. *10:* 85-90.

Smith, H.W. and S. Halls. 1968. The transmissible nature of the genetic factor in *Escherichia coli* that controls enterotoxin production. J. Gen. Microbiol. *52:* 319-334.

Smith, I.M. and A.J. Baskerville. 1979. A selective medium facilitating the isolation and recognition of *Bordetella bronchiseptica* in pigs. Rev. Vet. Sci. *27:* 187-192.

Smith, I.W. 1963. The classification of *Bacterium salmonicida*. J. Gen. Microbiol. *33:* 263-274.

Smith, J.E. 1958. Studies on *Pasteurella septica*. II. Some cultural and biochemical properties of strains from different host species. J. Comp. Pathol. Ther. *68:* 315-323.

Smith, J.E. 1974. The genus *Pasteurella*. *In* Buchanan and Gibbons (Editors), Bergey's Manual of Determinative Bacteriology, 8th Ed., The Williams and Wilkins Company, Baltimore, pp. 370-373.

Smith, J.E. and M.G.P. Stoker. 1951. The nucleic acids of *Rickettsia burneti*. Brit. J. Exp. Pathol. *32:* 433-441.

Smith, J.E. and E. Thal. 1965. A taxonomic study of the genus *Pasteurella* using a numerical technique. Acta Pathol. Microbiol. Scand. *64:* 213-223.

Smith, J.L. and D.R. Persetsky. 1967. The current status of *Treponema cuniculi*. Review of the literature. Brit. J. Vener. Dis. *43:* 117-127.

Smith, L. 1961. Cytochrome systems in aerobic electron transport. *In* Gunsalus and Stanier (Editors), The Bacteria: a treatise on structure and function, Vol. II. Metabolism. Academic Press, New York, London.

Smith, P.F. 1979. Membrane lipids and lipopolysaccharides. *In* Barile and Razin (Editors), The Mycoplasmas, Vol. I, Cell Biology, Academic Press, New York, pp. 231-257.

Smith, P.F. 1980. Sequence and glycoside bond arrangement of sugars in lipopolysaccharide from *Thermoplasma acidophilum*. Biochim. Biophys. Acta *619:* 367-373.

Smith, P.F. and T.A. Langworthy. 1979. Existence of carotenoids in *Acholeplasma axanthum*. J. Bacteriol. *137:* 185-188.

Smith, P.F., T.A. Langworthy, W.R. Mayberry and A.E. Hougland. 1973. Characterization of the membranes of *Thermoplasma acidophilum*. J. Bacteriol. *116:* 1019-1028.

Smith, P.F., T.A. Langworthy and M.R. Smith. 1975. Polypeptide nature of growth requirement in yeast extract for *Thermoplasma acidophilum*. J. Bac-

teriol. *124:* 884-892.

Smith, R.D. and A.A. Salyers. 1981. Incorporation of leucine into phospholipids of *Bacteroides thetaiotaomicron*. J. Bacteriol. *145:* 8-13.

Smith, R.F. 1975. New medium for isolation of *Corynebacterium vaginale* from genital specimens. Health Lab. Sci. *12:* 219-224.

Smith, T. 1894. The hog-cholera group of bacteria. U.S. Bur. Anim. Ind. Bull. *6:* 6-40.

Smith, T. 1921. The etiological relation of *Bacillus actinoides* to bronchopneumonia. J. Exp. Med. *33:* 441-469.

Smith, T. and M. S. Taylor. 1919. Some morphological and biochemical characteristics of the spirilla (*Vibrio fetus*, n. sp.) associated with disease of the fetal membranes in cattle. J. Exp. Med. *30:* 299-311.

Smithies, W.R., N.E. Gibbons and S.T. Bayley. 1955. The chemical composition of the cell and cell wall of some halophilic bacteria. Can. J. Microbiol. *1:* 605-613.

Sneath, P.H.A. 1956. Cultural and biochemical characteristics of the genus *Chromobacterium*. J. Gen. Microbiol. *15:* 70-98.

Sneath, P.H.A. 1960. A study of the bacterial genus *Chromobacterium*. Iowa St. J. Sci. *34:* 243-500.

Sneath, P.H.A. 1972. Computer taxonomy. *In* Norris and Ribbons, Methods in Microbiology, Vol. 7A, Academic Press, London - New York, pp. 29-98.

Sneath, P.H.A. 1974. Phylogeny of microorganisms. Symp. Soc. Gen. Microbiol. *24:* 1-39.

Sneath, P.H.A. 1977. A method for testing the distinctness of clusters: a test for the disjunction of two clusters in euclidean space as measured by their overlap. J. Int. Assoc. Math. Geol. *9:* 123-143.

Sneath, P.H.A. 1977. The maintenance of large numbers of strains of microorganisms, and the implications for culture collections. FEMS Microbiology Letters *1:* 333-334.

Sneath, P.H.A. 1978. Classification of microorganisms. *In* Norris and Richmond (Editors), Essays in Microbiology, John Wiley, Chichester, United Kingdom, pp. 9/1-9/31.

Sneath, P.H.A. 1978. Identification of microorganisms. *In* Norris and Richmond (Editors), Essays in microbiology, John Wiley, Chichester, England, pp. 10/1-10/32.

Sneath, P.H.A. 1979. Identification methods applied to *Chromobacterium*. *In* Skinner and Lovelock (Editors), Identification Methods for Microbiologists, 2nd Ed., Academic Press, London and New York, pp. 167-175.

Sneath, P.H.A. 1979a. BASIC program for a significance test for clusters in UPGMA dendrograms obtained from square euclidean distances. Comput. Geosci. *5:* 127-137.

Sneath, P.H.A. 1979b. BASIC program for a significance test for two clusters in euclidean space as measured by their overlap. Comput. Geosci. *5:* 143-155.

Sneath, P.H.A. and V.G. Collins. 1974. A study in test reproducibility between laboratories: report of a *Pseudomonas* working party. Antonie van Leeuwenhoek J. Microbiol. Serol. *40:* 481-527.

Sneath, P.H.A. and R. Johnson. 1973. Numerical taxonomy of *Haemophilus* and related bacteria. Int. J. Syst. Bacteriol. *23:* 405-418.

Sneath, P.H.A. and V.B.D. Skerman. 1966. A list of type and reference strains of bacteria. Int. J. Syst. Bacteriol. *16:* 1-133.

Sneath, P.H.A. and R.R. Sokal. 1973. Numerical taxonomy. The principles and practice of numerical classification. W.H. Freeman, San Francisco.

Sneath, P.H.A., M. Stevens and M.J. Sackin. 1982. Numerical taxonomy of *Pseudomonas* based on published records of substrate utilization. Antonie van Leeuwenhoek J. Microbiol. Serol. *47:* 423-448.

Snell, G.C. and G.A. Cullen. 1978. The storage qualities of a cell-free haemagglutinating antigen of *Mycoplasma synoviae* for use in haemagglutination tests. J. Biol. Standard. *6:* 283-288.

Snell, J.J.S., L.R. Hill and S.P. Lapage. 1972. Identification and characterization of *Moraxella phenylpyruvica*. J. Clin. Pathol. *25:* 959-965.

Snell, J.J.S. and S.P. LaPage. 1976. Transfer of some saccharolytic *Moraxella* species to *Kingella* Henriksen and Bøvre 1976, with descriptions of *Kingella indologenes* sp. nov. and *Kingella denitrificans* sp. nov. Int. J. Syst. Bacteriol. *26:* 451-458.

Snyder, T.L., R.A. Penfield, F.B. Engley, Jr. and J.C. Creasy. 1946. Cultivation of *Bacterium tularense* in peptone media. Proc. Soc. Exp. Biol. Med. *63:* 26-30.

Sobeslavsky, O. and R.M. Chanock. 1968. Peroxide formation by mycoplasmas which infect man. Proc. Soc. Exp. Biol. Med. *129:* 531-535.

Sobeslavsky, O., B. Prescott and R.M. Chanock. 1968. Adsorption of *Mycoplasma pneumoniae* to neuraminic acid receptors of various cells and possible role in virulence. J. Bacteriol. *96:* 695-705.

Socolofsky, M.D. and O. Wyss. 1961. Cysts of *Azotobacter*. J. Bacteriol. *81:* 946-954.

Socransky, S.S. 1977. Microbiology of periodontal disease—present status and future considerations. J. Periodontol. *48:* 497-504.

Socransky, S.S., M. Listgarten, C. Hubersak, J. Cotmore and A. Clark. 1969. Morphological and biochemical differentiation of three types of small oral spirochetes. J. Bacteriol. *98:* 878-882.

Söder, G. 1980. Vergleichende wachstumphysiologische und taxonomische Untersuchungen on Stämmen Kohlenmonoxid oxidierender Bakterien. Thesis, Gottingen University.

Sofiev, M.S. 1941. *Spirochaeta latyschewi* n. sp. of relapsing fever type. Med. Parasitol. Moscow *10:* 337-373.

Söhngen, N.L. 1906. Über Bakterien welche Methan als Kohlenstoffnahrung und Energiequelle gebrauchen. Zentralbl. Bakteriol. Parasitenk. Infektionskr. Hyg. Abt. II *15:* 513-517.

Soldo, A.T. 1974. Intracellular particles in *Paramecium aurelia. In* van Wagtendonk (Editor), *Paramecium:* A Current Survey, Elsevier, Amsterdam, pp. 375-442.

Soldo, A.T. and G.A. Godoy. 1973a. The molecular complexity of *Paramecium* symbiont lambda DNA: evidence for the presence of a multicopy genome. J. Mol. Biol. *73:* 93-108.

Soldo, A.T. and G.A. Godoy. 1973b. Observations on the production of folic acid by symbiont lambda particles of *Paramecium aurelia* stock 299. J. Protozool. *20:* 502.

Soldo, A.T. and G.A. Godoy. 1974. The molecular complexity of mu and pi symbiont DNA of *Paramecium aurelia*. Nucleic Acids Res. *1:* 387-396.

Soliman, G.S.H. and H.G. Trüper. 1981. *Halobacterium pharaonis* sp. nov., a new extremely haloalkaliphilic archaebacterium with low magnesium requirement. Zentralbl. Bakteriol. Abt. I Orig. C, *3:* 318-329.

Soliman, G.S.H. and H.G. Trüper. 1983. *In* Validation of the publication of new names and new combinations previously effectively published outside the IJSB. List No. 10. Int. J. Syst. Bacteriol. *33:* 438-440.

Soltesz, L.V., C. Schalen and P.A. Mårdh. 1980. An effective selective medium for *Yersinia enterocolitica* containing sodium oxalate. Acta Pathol. Microbiol. Scand. Sect. B *88:* 11-16.

Somerson, N.L., P.R. Reich, R.M. Chanock and S.M. Weissman. 1967. Genetic differentiation by nucleic acid homology. III. Relationships among mycoplasma, L-forms, and bacteria. Ann. N.Y. Acad. Sci. *143:* 9-20.

Somerson, N.L., P.R. Reich, B.E. Walls, R.M. Chanock and S.M. Weissman. 1966. Genetic differentiation by nucleic acid homology. II. Genotypic variations within two *Mycoplasma* species. J. Bacteriol. *92:* 311-317.

Somerson, N.L., D. Taylor-Robinson and R.M. Chanock. 1963. Hemolysin production as an aid in the identification of Eaton agent *(Mycoplasma pneumoniae)*. Am. J. Hyg. *77:* 122-128.

Sommer, E.C., W.S. Silver and L.C. Vining. 1961. Studies of pigmentation by *Pseudomonas indigofera*. Can. J. Microbiol. *5:* 577-585.

Sompolinsky, D., J.B. Hertz, N. Høiby, K. Jensen, B. Mansa and Z. Samra. 1980a. An antigen in common to a wide range of bacteria. I. The isolation of a 'common antigen' from *Pseudomonas aeruginosa*. Acta Pathol. Microbiol. Scand., Sect. B, *88:* 143-149.

Sompolinsky, D., J.B. Hertz, N. Høiby, K. Jensen, B. Mansa, V.B. Pedersen and Z. Samra. 1980b. An antigen common to a wide range of bacteria. 2. A biochemical study of a 'common antigen' from *Pseudomonas aeruginosa*. Acta Pathol. Microbiol. Scand. Sect. B, *88:* 253-260.

Sonea, S. 1971. A tentative unifying view of bacteria. Rev. Can. Biol. *30:* 239-244.

Sonea, S. and M. Panisset. 1976. Pour une nouvelle bactériologie. Rev. Can. Biol. *35:* 103-167.

Sonea, S. and M. Panisset. 1980. Introduction a la nouvelle bactériologie. Les Presses de l'Université de Montréal et Masson, Montréal and Paris.

Songer, J.G., J.M. Kinyon and D.L. Harris. 1976. Selective medium for isolation of *Treponema hyodysenteriae*. J. Clin. Microbiol. *4:* 57-60.

Sonneborn, T.M. 1975. The *Paramecium aurelia* complex of fourteen sibling species. Trans. Am. Microsc. Soc. *94:* 155-178.

Sonnenshein, C. 1927. Die Mucosus-Form des Pyocyaneus-Bakteriums, *Bacterium pyocyaneum mucosum*. Zentralbl. Bakteriol. Parasitenk. Infektionskr. Hyg. Abt. I, Orig. *104:* 365-373.

Sorrell, W.B. and L.V. White. 1953. Acute bacterial endocarditis caused by a variant of the genus *Herellea*. Am. J. Clin. Pathol. *23:* 134-138.

Sottile, M.I., J.N. Baldwin and R.E. Weaver. 1973. Deoxyribonucleic acid hybridization studies on *Flavobacterium meningosepticum*. Appl. Microbiol. *26:* 535-539.

Sourek, J. 1968. On some findings concerning Dienes' phenomenon in swarming *Proteus* strains. Zentralbl. Bakteriol. Parasitenkd. Infektionskr. Hyg. Abt. I Orig. *208:* 419-427.

Sourek, J. and E. Aldová. 1976. Serotyping of strains belonging to the *Citrobacter-Levinea* group isolated from diagnostic material. Zentralbl. Bakteriol. Parasitenkd. Infektionskr. Hyg. Abt. I Orig. A *234:* 480-490.

Southern, P.M., Jr. 1975. Bacteremia due to *Succinivibrio dextrinosolvens*. Am. J. Clin. Pathol. *64:* 540-543.

Southward, A.J., E.C. Southward, P.R. Dando, G.H. Rau, H. Felbeck and H. Flugel. 1981. Bacterial symbionts and low $^{13}C/^{12}C$ ratios in tissues of *Pogonophora* indicate unusual nutrition and metabolism. Nature *293:* 614-620.

Southwood, T.R.E., S. Khalaf and R.E. Sinden. 1975. The micro-organisms of tsetse flies. Acta Trop. *32:* 259-266.

Sowden, L.C. and J.R. Colvin. 1978. Morphology, microstructure and development of colonies of *Acetobacter xylinum*. Can. J. Microbiol. *24:* 772-779.

Sox, T.E., W. Mohammed and P.F. Sparling. 1979. Transformation derived *Neisseria gonorrhoeae* plasmids with altered structure and function. J. Bacteriol. *138:* 510-518.

Spark, R.P., M.L. Fried, C. Perry and C. Watkins. 1979. *Vibrio alginolyticus* wound infection: case report and review. Ann. Clin. Lab. Sci. *9:* 133-138.

Sparling, P.F. 1966. Genetic transformation of *Neisseria gonorrhoeae* to streptomycin resistance. J. Bacteriol. *92:* 1364-1371.

Spaulding, E.H. and L.F. Rettger. 1937. *Fusobacterium* genus I. Biochemical and serological classification. J. Bacteriol. *34:* 535-548.

Spaulding, E.H. and L.F. Rettger. 1937. The *Fusobacterium* genus. I. Biochemical and serological classification. J. Bacteriol. *34:* 535-548.

Spelhaug, D. R., M. J. R. Gilchrist and J. A. Washington, II. 1981. Bactericidal activity of antibiotics against *Campylobacter fetus* subspecies *intestinalis*. J. Infect. Dis. *143:* 500.

Spence, K.D., J.L. Fryer and K.S. Pilcher. 1965. Active and passive immunization of certain salmonid fishes against *Aeromonas salmonicida*. Can. J. Microbiol. *11:* 397-407.

Spencer, R. 1960. Indigenous marine bacteriophages. J. Bacteriol. *79:* 614.

Spencer, R. 1963. Bacterial viruses in the sea. *In* Oppenheimer (Editor), Symposium on Marine Microbiology, Thomas, Springfield, Illinois, pp. 350-365.

Sperry, J.F. and T.D. Wilkins. 1977. Presence of cytochrome *c* in *Desulfomonas pigra*. J. Bacteriol. *129:* 554-555.

Spicer, A.J., M.G. Peacock and J.C. Williams. 1981. Effectiveness of several antibiotics in suppressing chick embryo lethality during experimental infections by *Coxiella burnetti*, *Rickettsia typhi* and *R. rickettsii*. *In* Burgdorfer and Anacker (Editors), Rickettsiae and Rickettsial Diseases, Academic Press, New York, pp. 375-383.

Spiff, E.D. and C.T. Odu. 1973. Acetylene reduction by *Beijerinckia* under various partial pressures of oxygen and acetylene. J. Gen. Microbiol. *78:* 207-209.

Spira, W.M. and R.R. Daniel. 1980. Biotype clusters formed on the basis of virulence characteristics in non-O group 1 *Vibrio cholerae*. Proc. 15th Joint Conf. Cholera US-Japan Coop. Med. Sci. Prog., pp. 440-458.

Spira, W.M., R.R. Daniel, Q.S. Ahmed, A. Huq, A. Yusuf and D.A. Sack. 1979. Clinical features and pathogenicity of O group I nonagglutinating *Vibrio cholerae* and other vibrios isolated from cases of diarrhea in Dacca, Bangladesh. Proc. 14th Joint Conf. Cholera U.S.-Japan Coop. Med. Sci. Prog., pp. 137-153.

Splitter, E.J. 1950. *Eperythrozoon suis*, the etiologic agent of icteroanemia—an anaplasmosis-like disease in swine. Am. J. Vet. Res. *11:* 324-329.

Splitter, E.J. 1952. Eperythrozoonosis in swine—filtration studies. Am. J. Vet. Res. *13:* 290-297.

Splitter, E.J. 1958. The complement fixation test in diagnosis of eperythrozoonosis in swine. J. Am. Vet. Med. Assoc. *132:* 47-49.

Splitter, E.J., E.R. Castro and W.L. Kanawyer. 1956. Feline infectious anemia. Vet. Med. *51:* 17-22.

Srinivasan, M.C. and M.K. Patel. 1956. Three undescribed species of *Xanthomonas*. Curr. Sci. *25:* 366-367.

Srinivasan, M.C. and M.K. Patel. 1957. Two new phytopathogenic bacteria on verbenaceous hosts. Curr. Sci. *26:* 90-91.

Srinivasan, M.C., M.K. Patel and M.J. Thirumalachar. 1961a. A new bacterial blight disease of *Argemone mexicana*. Proc. Nat. Inst. Sci. India Part B, Biol. Sci. *27:* 104-107.

Srinivasan, M.C., M.K. Patel and M.J. Thirumalachar. 1961b. A bacterial blight disease of coriander. Proc. Indian Acad. Sci. Sect. B *53:* 298-301.

Srinivasan, M.C., M.K. Patel and M.J. Thirumalachar. 1962. Two bacterial leafspot diseases on *Physalis minima* and studies on their relationship to *Xanthomonas vesicatoria* (Doidge) Dowson. Proc. Indian Acad. Sci. Sect. B *56:* 93-96.

Srivastava, P.N. and J.L. Auclair. 1976. Effects of antibiotics on feeding and development of the pea aphid, *Acyrthosiphon pisum* (Harris) (Homoptera: Aphidae). Can. J. Zool. *54:* 1025-1029.

Stableforth, A.W. 1959. Diseases due to bacteria. *In* Stableforth and Galloway (Editors), Infectious Diseases of Animals, Vol. 1, Butterworths, London, p. 60.

Stableforth, A.W. and L.M. Jones. 1963. Report of the sub-committee on Taxonomy of the genus *Brucella*. Speciation in the genus *Brucella*. Int. Bull. Bacteriol. Nomencl. Taxon. *13:* 145-158.

Stackebrandt, E. and C.R. Woese. 1979. A phylogenetic dissection of the family *Micrococcaceae*. Curr. Microbiol. *2:* 317-322.

Stackebrandt, E. and C.R. Woese. 1981. The evolution of prokaryotes. *In* Carlile, Collins and Moseley (Editors), Molecular and Cellular Aspects of Microbial Evolution, Cambridge University Press, Cambridge, pp. 1-31.

Stainer, D.W. and M.J. Scholte. 1970. A simple chemically defined medium for the production of Phase I *Bordetella pertussis*. J. Gen. Microbiol. *63:* 211-220.

Staley, J.T. 1968. *Prosthecomicrobium* and *Ancalomicrobium*: new prosthecate freshwater bacteria. J. Bacteriol. *95:* 1921-1942.

Staley, J.T. 1974. Genus *Microcyclus*. *In* Buchanan and Gibbons (Editors), Bergey's Manual of Determinative Bacteriology, 8th ed. The Williams and Wilkins Co., Baltimore, p. 214-215.

Staley, T.E. and R.R. Colwell. 1973a. Polynucleotide sequence relationships among Japanese and American strains of *Vibrio parahaemolyticus*. J. Bacteriol. *114:* 916-927.

Staley, T.E. and R.R. Colwell. 1973b. Deoxyribonucleic acid reassociation among members of the genus *Vibrio*. Int. J. Syst. Bacteriol. *23:* 316-332.

Stalheim, O.H.V., A.E. Ritchie and R.F. Whitcomb. 1978. Cultivation, serology, ultrastructure, and virus-like particles of spiroplasma 277F. Curr. Microbiol. *1:* 365-370.

Stalon, V., F. Ramos, A. Pierard and J.-M. Wiame. 1972. Regulation of the catabolic ornithine carbamoyltransferase of *Pseudomonas fluorescens*. A comparison with the anabolic transferase and with a mutationally modified catabolic transferase. Eur. J. Biochem. *29:* 25-35.

Stamm, W.E., J.J. Colella, R.L. Anderson and R.E. Dixon. 1975. Indwelling arterial catheters as a source of nosocomial bacteremia. New Engl. J. Med.

292: 1099-1102.

Stämmer, H.J. 1929. Die Bakteriensymbiose der Trypetiden (Diptera). Z. Morphol. Oekol. Tiere *15:* 481-523.

Stamp, J.T., A.D. McEwen, I.A.A. Watt and D.I. Nisbet. 1950. Enzootic abortion in ewes. Transmission of the disease. Vet. Rec. *62:* 251-254.

Stamp, L. and D.M. Stone. 1944. An agglutinogen common to certain strains of lactose and non-lactose fermenting coliform bacilli. J. Hyg. *43:* 266-272.

Stanbridge, E.J. and M.E. Reff. 1979. The molecular biology of mycoplasmas. *In* Barile and Razin (Editors), The Mycoplasmas, Vol. I, New York, pp. 157-185.

Stanbridge, E.J. and R.L. Weiss. 1978. Mycoplasma capping on lymphocytes. Nature (London) *276:* 583-587.

Standring, E.T. 1942. *In* Lightle, Standring and Brown. A bacterial necrosis of the giant cactus. Phytopathology *32:* 303-313.

Stanek, G., A. Hirschl and G. Laber. 1981. Sensitivity of various spiroplasma strains against ethanol, formalin, glutaraldehyde, and phenol. Zentralbl. Bakteriol. Parasitenkd. Infektionskr. Hyg. Abt. Orig. B *174:* 348-354.

Stanek, G., G. Laber and A. Hirschl. 1982. Survival of spiroplasma on different carriers and resistance against three disinfectants. Rev. Infect. Dis. *4:* S263.

Stanghellini, M.E., D.C. Sands, W.C. Kronland and M.M. Mendonca. 1977. Serological and physiological differentiation among isolates of *Erwinia carotovora* from potato and sugarbeet. Phytopathology *67:* 1178-1182.

Stanier, R.Y. 1943. A note on the taxonomy of *Proteus hydrophilus.* J. Bacteriol. *46:* 213-214.

Stanier, R. Y. 1961. La place des bactéries dans le monde vivant. Ann. Inst. Pasteur (Paris) *101:* 297-303.

Stanier, R.Y. 1970. Some aspects of the biology of cells and their possible evolutionary significance. *In* Charles and Knight (Editors), Organization and Control in Procaryotic and Eucaryotic Cells, Cambridge University Press, Cambridge.

Stanier, R. Y., N. J. Palleroni and M. Doudoroff. 1966. The aerobic pseudomonads: a taxonomic study. J. Gen. Microbiol. *43:* 159-271.

Stanier, R.Y. and C.B. van Niel. 1962. The concept of a bacterium. Arch. Mikrobiol. *42:* 17-35.

Stanier, R.Y., D. Wachter, D. Gasser and A.C. Wilson. 1970. Comparative immunological studies of two *Pseudomonas* enzymes. J. Bacteriol. *102:* 351-362.

Stanley, J. and Dunican, L.K. 1979. Intergenetic mobilization of *Rhizobium nif* genes to *Agrobacterium* and *Klebsiella.* Mol. Gen. Genet. *174:* 211-220.

Stanton, A.T. and W. Fletcher. 1921. Melioidosis, a new disease of the tropics. Trans. 4th Congr. Far-East Assoc. Trop. Med. *2:* 196-198.

Stanton, T.B. and E. Canale-Parola. 1979. Enumeration and selective isolation of rumen spirochetes. Appl. Environ. Microbiol. *38:* 965-973.

Stanton, T.B. and E. Canale-Parola. 1980. *Treponema bryanttii* sp. nov., a rumen spirochete that interacts with cellulolytic bacteria. Arch. Microbiol. *127:* 145-156.

Stanton, T.B. and E. Canale-Parola. 1981. *In* Validation of the publication of new names and new combinations previously effectively published outside the IJSB. List No. 5. Int. J. Syst. Bacteriol. *30:* 676-677.

Staples, D. G. and J. C. Fry. 1973. Factors which influence the enumeration of *Bdellovibrio bacteriovorus* in sewage and river water. J. Appl. Bacteriol. *36:* 1-11.

Stapp, C. 1928. Schizomycetes (Spaltpilze oder Bakterien). *In* Sorauer (Editor) Handbuch der Pflanzenkrankheiten, 5th ed., Vol. 2, Paul Parey, Berlin, pp. 1-295.

Stapp, C. 1935. Contemporary understanding of bacterial plant diseases and their causal organisms. Bot. Rev. *1:* 405-418.

Stapp, C. 1940. *Bacterium rubidaeum* nov. spec. Zentralbl. Bakteriol. Parasitenkd. Infektionskr. Hyg. Abt. II *102:* 252-260.

Stapp, C. 1940. *Azotomonas insolita* ein neuer aerober stickstoffbindender Mikroorganismus. Zentralbl. Bakteriol. Parasitenkd. Infektionskr. Hyg. Abt. II *102:* 142-150.

Stapp, C. and D. Knösel. 1956. Fortgeführte Untersuchungen über den Entwicklungscyclus und die Karyologie sternbildender Bakterien. Zentralbl. Bakteriol. Parasitenkd. Infektionskr. Hyg. Abt. II *109:* 416-428.

Starkey, R.L. and P.K. De. 1939. A new species of *Azotobacter.* Soil Sci. *47:* 329-343.

Starr, M.P. 1946. The nutrition of phytopathogenic bacteria I. Minimal nutritive requirements of the genus *Xanthomonas.* J. Bacteriol. *51:* 131-143.

Starr, M.P. 1946. The nutrition of phytopathogenic bacteria. II. The genus *Agrobacterium.* J. Bacteriol. *52:* 187-194.

Starr, M.P. 1947. The causal agent of bacterial root and stem disease of guavule. Phytopathology *37:* 291-300.

Starr, M.P. 1958. The blue pigment of *Corynebacterium insidiosum.* Arch. Mikrobiol. *30:* 325-334.

Starr, M.P. 1981. The Genus *Lampropedia. In* Starr, Stolp, Trüper, Balows and Schlegel (Editors), The Prokaryotes: a handbook of habitats, isolation and identification of bacteria. Springer-Verlag, Berlin, pp. 1530-1536.

Starr, M. P. and N. L. Baigent. 1966. Parasitic interaction of *Bdellovibrio bacteriovorus* with other bacteria. J. Bacteriol. *91:* 2006-2017.

Starr, M.P. and W.H. Burkholder. 1942. Lipolytic activity of phytopathogenic bacteria determined by means of spirit blue agar and its taxonomic significance. Phytopathology *32:* 598-604.

Starr, M.P. and A.K. Chatterjee. 1972. The genus *Erwinia:* enterobacteria path-

ogenic to plants and animals. Ann. Rev. Microbiol. *26:* 389-426.

Starr, M.P., A.K. Chatterjee, P.B. Starr and G.E. Buchanan. 1977. Enzymatic degradation of polygalacturonic acid by *Yersinia* and *Klebsiella* species in relation to clinical laboratory procedures. J. Clin. Microbiol. *6:* 379-386.

Starr, M.P. and D. Folsom. 1951. Bacterial fireblight of raspberry. Phytopathology *41:* 915-919.

Starr, M.P. and C. Garces. 1950. El agente causante de la gomosis bacterial del pasto imperial en Colombia. Rev. Fac. Nac. Agron., Medellin, Colombia *11:* 73-83.

Starr, M.P., P.A.D. Grimont, F. Grimont and P.B. Starr. 1976. Caprylate-thallous agar medium for selectively isolating *Serratia* and its utility in the clinical laboratory. J. Clin. Microbiol. *4:* 270-276.

Starr, M.P., C.L. Jenkins, L.B. Bussey and A.G. Andrewes. 1977. Chemotaxonomic significance of the xanthomonadins, novel brominated aryl polyene pigments produced by bacteria of the genus *Xanthomonas.* Arch. Microbiol. *113:* 1-9.

Starr, M.P., H.G. Knackmuss and G. Cosens. 1967. The intracellular blue pigment of *Pseudomonas lemonnieri.* Arch. Mikrobiol. *59:* 287-294.

Starr, M.P. and M. Mandel. 1969. DNA base composition and taxonomy of phytopathogenic and other enterobacteria. J. Gen. Microbiol. *56:* 113-123.

Starr, M.P. and J.M. Schmidt. 1981. Prokaryote diversity. *In* Starr, Stolp, Trüper, Balows and Schlegel (Editors), The Prokaryotes, a handbook of habitats, isolation and identification of bacteria, Springer-Verlag, Berlin - Heidelberg - New York.

Starr, M. P. and R. J. Seidler. 1971. The bdellovibrios. Annu. Rev. Microbiol. *25:* 649-678.

Starr, M.P. and W.L. Stephens. 1964. Pigmentation and taxonomy of the genus *Xanthomonas.* J. Bacteriol. *87:* 293-302.

Starr, M.P. and J.E. Weiss. 1943. Growth of phytopathogenic bacteria in a synthetic asparagin medium. Phytopathology *33:* 314-318.

Stauffer, L.R., E.O. Hill, J.W. Holland and W.A. Altemeier. 1975. Indirect fluorescent antibody procedure for the rapid detection and identification of *Bacteroides* and *Fusobacterium* in clinical specimens. J. Clin. Microbiol. *2:* 337-341.

Steadman, J.R., C.R. Maier, H.F. Schwartz and E.D. Kerr. 1975. Pollution of surface irrigation waters by plant pathogenic organisms. Water Resour. Bull. *11:* 796-804.

Steber, J. and K.H. Schleifer. 1975. *Halococcus morrhuae*: a sulfated heteropolysaccharide as the structural component of the bacterial cell wall. Arch. Microbiol. *105:* 173-177.

Steel, K.J. and S.T. Cowan. 1964. Le rattachement de *Bacterium anitratum, Moraxella lwoffii, Bacillus mallei* et *Haemophilus parapertussis* au genre *Acinetobacter* Brisou et Prévot. Ann. Inst. Pasteur (Paris) *106:* 479-483.

Steel, K.J. and J. Midgeley. 1962. Decarboxylase and other reactions of some Gram-negative rods. J. Gen. Microbiol. *29:* 171-178.

Steensland, H. and H. Larsen. 1971. The fine structure of the extremely halophilic cocci. Kgl. Norske Vidensk. Selsk. Skrifter No. 8. Universitetsforlaget, Oslo, pp. 1-5.

Steiger, U. 1968. Intrazelluläre Mikroorganismen bei der Waldameise im Ei-, Larven- und Puppenstadium. Acta Trop. *25:* 263-266.

Steigerwalt, A.G., G.R. Fanning, M.A. Fife-Asbury and D.J. Brenner. 1976. DNA relatedness among species of *Enterobacter* and *Serratia.* Can. J. Microbiol. *22:* 121-137.

Steinberg, P., R.L. Horswood and R.M. Chanock. 1969. Temperature-sensitive mutants of *Mycoplasma pneumoniae.* I. In vitro biologic properties. J. Infect. Dis. *120:* 217-224.

Steiner, S., S. F. Conti and R. L. Lester. 1973. Characterization of the major phospholipids of *Bdellovibrio bacteriovorus*; occurrence of phosphonosphingolipids in strain UKi2. J. Bacteriol. *116:* 1199-1211.

Steinhardt, W.L. and R.M. Herriott. 1968. Genetic integration in the heterospecific transformation of *Haemophilus influenzae* cells by *Haemophilus parainfluenzae* DNA. J. Bacteriol. *96:* 1725-1731.

Steinhaus, E.A. 1941. A study of the bacteria associated with thirty species of insects. J. Bacteriol. *42:* 757-790.

Steinhaus, E.A. 1946. Insect microbiology. Comstock Pub. Co., Ithaca, New York.

Steinhaus, E.A., M.M. Batey and C.L. Boerke. 1956. Bacterial symbiotes from the caeca of certain Heteroptera. Hilgardia *24:* 495-518.

Stemshorn, B. and K. Nielsen. 1977. The bovine immune response to *Brucella abortus.* 1. A water soluble antigen precipitated by sera of some naturally infected cattle. Can. J. Comp. Med. *41:* 152-159.

Stenzel, W. 1978. Problems of *Escherichieae* systematics and the classification of atypical dysentery bacilli. Int. J. Syst. Bacteriol. *28:* 597-598.

Stenzel, W., H. Bürger and W. Mannheim. 1972. Zür Systematik und Differentialdiagnostik der *Klebsiella*-Gruppe mit besonderer Berucksichtigung der sogenannten Oxytocum-Typen. Zentralbl. Bakteriol. Parasitenkd. Infektionskr. Hyg. Abt. I Orig. A *219:* 193-203.

Stepan, D.E. and R.C. Johnson. 1981. Helical conformation of *Treponema pallidum* (Nichols strain), *Treponema paraluis-cuniculi, Treponema denticola, Borrelia turicatae* and unidentified oral spirochetes. Infect. Immun. *32:* 937-940.

Stephens, E.B., G.S. Aulakh, R.E. McCoy, D.L. Rose, J.G. Tully and M.F. Barile. 1981. Lack of genetic relatedness among animal and plant acholeplasmas by nucleic acid hybridization. Curr. Microbiol. *5:* 367-370.

Stephens, E.B., G. Aulakh, D.L. Rose, J.G. Tully and M.F. Barile. 1983. Intraspecies genetic relatedness among strains of *Acholeplasma laidlawii* and of *Acholeplasma axanthum* by nucleic acid hybridization. J. Gen. Microbiol. *129:* 1929-1934.

Stephens, J.W.W. and S.R. Christophers. 1904. The practical study of malaria and other blood parasites. Williams and Northgate, London.

Stephens, W.L. and M.P. Starr. 1963. Localization of carotenoid pigment in the cytoplasmic membrane of *X. juglandis.* J. Bacteriol. *86:* 1070-1074.

Stephenson, E.H. and J.V. Osterman. 1977. Canine peritoneal macrophages: cultivation and infection with *Ehrlichia canis.* Am. J. Vet. Res. *38:* 1815-1819.

Stevens, C.L., P. Blumbergs, F.A. Daniher, R.W. Wheat, A. Kujomoto and E.L. Rollins. 1963. The identification and synthesis of the 4-amino sugar from *Chromobacterium violaceum.* J. Am. Chem. Soc. *85:* 3061.

Stevens, F.L. 1925. Plant Disease Fungi. Macmillan Co., New York, pp. 1-469.

Stevenson, R.M.W. 1979. Amino acid uptake systems in *Bacteroides ruminicola.* Can. J. Microbiol. *25:* 1161-1168.

Stewart, C.S., C. Paniagua, D. Dinsdale, K.-J. Cheng and S.H. Garrow. 1981. Selective isolation and characteristics of *Bacteroides succinogenes* from the rumen of a cow. Appl. Environ. Microbiol. *41:* 504-510.

Stewart, D.J. 1978. The role of various antigenic factors of *Bacteroides nodosus* in eliciting protection against foot rot in vaccinated sheep. Res. Vet. Sci. *24:* 14-19.

Stewart, D.J. and J.R. Egerton. 1979. Studies on the ultrastructural morphology of *Bacteroides nodosus.* Res. Vet. Sci. *26:* 227-235.

Stewart, J.E., R.E. Kallio, D.P. Stevenson, A.C. Jones and D.O. Schissler. 1959. Bacterial hydrocarbon oxidation I. Oxidation of n-hexadecane by a gram-negative coccus. J. Bacteriol. *78:* 441-448.

Stewart, M., T. J. Beveridge and R. G. E. Murray. 1980. Structure of the regular surface layer of *Spirillum putridiconchylium.* J. Mol. Biol. *137:* 1-8.

Stewart, W.W. 1971. Isolation and proof of structure of wildfire toxin. Nature. *229:* 174-178.

Stiles, C.W. and C.A. Pfender. 1905. The generic name *Spironema* Vuillemin 1905 (not Meek, 1964, Mollusk) - *Microspironema* Stiles and Pfender 1905 of the parasite of syphilis. Amer. Med. *10:* 936.

Stiller, D., R.F. Whitcomb, M.E. Coan and J.G. Tully. 1981. Direct isolation in cell-free medium of a spiroplasma from *Haemaphysalis leporispalustris* (Acari:Ixodidae) in Maryland. Curr. Microbiol. *5:* 339-342.

Stillman, E.G. and J.M. Bourn. 1920. Biological study of the hemophilic bacilli. J. Exp. Med. *32:* 665-682.

Stimson, A.M. 1907. Note on an organism found in yellow-fever tissue. Public Health Rep., Wash. *22:* 541.

Stipkovits, L. and A. Rashwan. 1976. Isolation of ureaplasmas from chickens. Abstracts of 1st Conf. Int. Org. Mycoplasmology. Proc. Soc. Gen. Microbiol. *3:* 158.

Stipkovits, L., A. Rashwan, J. Takács and K. Lapis. 1978a. Detection of ureaplasmas in turkey semen. Abstracts of 2nd Conf. Int. Org. Mycoplasmology. Zentralbl. Bakteriol. Parasitenkd. Infektionskr. Hyg. I Abt. Orig. A *241:* 257.

Stipkovits, L., A. Rashwan, J. Takács and K. Lapis. 1978b. Occurrence of ureaplasmas in swine semen. Zentralbl. Veterinaermed. *B25:* 605-608.

Stipkovits, L., L. Varga and D. Schimmel. 1973. Isolation of *Acholeplasma axanthum* from swine. Acta Vet. Acad. Sci. Hung. *23:* 361-368.

Stocker, B.A.D. 1958. Lysogenic conversion by the A phages of *Salmonella typhimurium.* J. Gen. Microbiol. *18:* IX.

Stocker, B.A.D. and P.A. Mäkelä. 1971. Genetic aspect of biosynthesis and structure of *Salmonella* lipopolysaccharide. *In* Weinbaum, Kadis and Ajl (Editors), Microbial toxins, Academic Press, New York, 369-438.

Stocker, B.A.D. and P.H. Mäkelä. 1978. Genetics of the (Gram negative) bacterial surface. Proc. Roy. Soc. Lond. B. *202:* 5-30.

Stoeckenius, W. and W.H. Kunau. 1968. Further characterization of particulate fractions from lysed cell envelopes of *Halobacterium halobium* and isolation of gas vacuole membranes. J. Cell Biol. *38:* 337-357.

Stoenner, H.G. and D.B. Lackman. 1957. A new species of *Brucella* isolated from the desert wood rat, *Neotoma lepida* Thomas. Amer. J. Vet. Res. *18:* 942-951.

Stoenner, H.G., D.B. Lackman and E.J. Bell. 1962. Factors affecting the growth of rickettsias of the spotted fever group in fertile hens' eggs. J. Infect. Dis. *110:* 121-128.

Stoker, M.G.P. and P. Fiset. 1956. Phase variation of the Nine Mile and other strains of *Rickettsia burnetii.* Can. J. Microbiol. *2:* 310-321.

Stokes, J.F., I.R. Gray and E.J. Stokes. 1951. *Actinomyces muris* endocarditis treated with chloramphenicol. Brit. Heart J. *13:* 247-251.

Stoleru, G.H., L. Le Minor and A.M. Lhéritier. 1976. Polynucleotide sequence divergence among strains of *Salmonella* subgenus IV and closely related organisms. Ann. Microbiol. (Inst. Pasteur), *127A:* 477-486.

Stolp, H. 1973. The bdellovibrios: bacterial parasites of bacteria. Annu. Rev. Phytopath. *11:* 53-76.

Stolp, H. and D. Gadkari. 1981. Non-pathogenic members of the genus *Pseudomonas. In* Starr, Stolp, Trüper, Balows and Schlegel (Editors) The Prokaryotes, a handbook on the habitats, isolation and identification of the bacteria, Vol. I, Springer-Verlag, Berlin, pp. 719-741.

Stolp, H. and M. P. Starr. 1963. *Bdellovibrio bacteriovorus* gen. et sp. n., a predatory, ectoparasitic, and bacteriolytic microorganism. Antonie van Leeuwenhoek J. Microbiol. Serol. *29:* 217-248.

Stone, R.L., C.G. Gulbertson and H.M. Powell. 1956. Studies of a bacteriophage active against a chromogenic *Neisseria.* J. Bacteriol. *71:* 516-520.

Stonier, T., J. McSharry and T. Speitel. 1967. *A. tumefaciens* Conn. IV. Bacteriophage PB2 and its inhibitory effect on tumor induction. J. Virol. *1:* 268-273.

Stopler, T., C.P. Gerichter and D. Branski. 1980. Antibiotic-resistant mutants of *Mycoplasma pneumoniae.* Isr. J. Med. Sci. *16:* 169-173.

Story, P. 1954. *Proteus* infections in hospital. J. Pathol. Bacteriol. *68:* 55-62.

Storz, J. and L.A. Page. 1971. Taxonomy of the chlamydiae: reasons for classifying organisms of the genus *Chlamydia*, family *Chlamydiaceae*, in a separate order *Chlamydiales* ord. nov. Int. J. Syst. Bacteriol. *21:* 332-334.

Stout, J.D. 1960. Biological studies of some tussock-grassland soils. XV. Bacteria of two cultivated soils. N. Z. J. Agr. Res. *3:* 214-223.

Stouthamer, A.H. 1959. Oxidative possibilities in the catalase positive *Acetobacter* species. Antonie van Leeuwenhoek J. Microbiol. Serol. *25:* 241-264.

Stouthamer, A.H. 1960. Koolhydratenstofwisseling van de azijnzuurbakteriën. Thesis. Utrecht.

Stouthamer, A.H. 1980. Bioenergetic studies on *Paracoccus denitrificans.* Trends Biochem. Sci. *5:* 164-166.

Stouthamer, A.H., J.H. Van Boom and A.J. Bastiaanse. 1963. Metabolism of C_2 compounds in *Acetobacter aceti.* Antonie van Leeuwenhoek J. Microbiol. Serol. *29:* 393-406.

Stovall, I. and M. Cole. 1978. Organic acid metabolism by isolated *Rhizobium japonicum* bacteroids. Plant Physiol. *61:* 787-790.

Straley, S. C. and S. F. Conti. 1974. Chemotaxis in *Bdellovibrio bacteriovorus.* J. Bacteriol. *120:* 549-551.

Straley, S. C. and S. F. Conti. 1977. Chemotaxis by *Bdellovibrio bacteriovorus* toward prey. J. Bacteriol. *132:* 628-640.

Strandberg Pederson, N., N.H. Axelsen, B.B. Jorgensen and C. Sand Petersen. 1980. Antibodies in secondary syphilis against five of forty Reiter treponeme antigens. Scand. J. Immunol. *11:* 629-633.

Strandberg Pederson, N., N.H. Axelsen and C. Sand Petersen. 1981. Antigenic analysis of *Treponema pallidum:* Cross-reactions between individual antigens of *T. pallidum* and *T. reiter.* Scand. J. Immunol. *13:* 143-150.

Strandskov, F.B., H.W. Baker and J.B. Bockelmann. 1953. A study of the Gram-negative bacterial rod infection of brewery yeast and brewery fermentations. Wallerstein Laboratories Communications. *16:* 261-270.

Strandskov, F.B. and J.B. Bockelmann. 1955. Nutritional requirements of brewing microorganisms. I. The nutritional requirements of *Flavobacterium proteus.* Wallerstein Laboratories Communications *18:* 275-282.

Strangeways, W.I. 1933. Rats as carriers of *Streptobacillus moniliformis.* J. Pathol. Bacteriol. *37:* 45-51.

Strauss, J.M., A.D. Alexander, G. Rapmund, E. Gan and A.E. Dorsey. 1969. Melioidosis in Malaysia. III. *Pseudomonas pseudomallei* antibodies in the human population. Am. J. Trop. Med. *18:* 703-707.

Strength, W. J., B. Isani, D. M. Linn, F. D. Williams, G. E. Vandermolen, B. E. Laughon and N. R. Krieg. 1976. Isolation and characterization of *Aquaspirillum fasciculus* sp. nov., a rod-shaped, nitrogen-fixing bacterium having unusual flagella. Int. J. Syst. Bacteriol. *26:* 253-268.

Strobel, G.A. 1977. Bacterial phytotoxins. Annu. Rev. Microbiol. *31:* 205-224.

Strong, R.P. (Editor). 1918. Trench fever. Report of commission. Medical research committee American Red Cross. Oxford University Press, Oxford.

Strong, R.P., E.E. Tyzzer, C.T. Brues, A.W. Sellards and J.A. Gastiaburú. 1913. Verruga peruviana, Oroya fever and uta. J. Am. Med. Assoc. *61:* 1713-1716.

Strong, R.P., E.E. Tyzzer and A.W. Sellards. 1915. Oroya fever, second report. J. Am. Med. Assoc. *64:* 806-808.

Strzelcowa, A. 1968. The use of the technique of van Schreven for the taxonomy of *Rhizobium* strains. Acta Microbiol. Polon. *17:* 263-268.

Stuart, C.A. and R. Rustigian. 1943. Further studies on one type of paracolon organisms. Amer. J. Publ. Health *33:* 1323-1325.

Stuart, C.A., K.M. Wheeler, R. Rustigian and A. Zimmerman. 1943. Biochemical and antigenic relationships of the paracolon bacteria. J. Bacteriol. *45:* 101-109.

Stuart, R.D. 1946. The preparation and use of a simple culture medium for leptospirae. J. Pathol. Bacteriol. *58:* 343-349.

Stuhlmann, F. 1907. Beiträge zür Kenntnis der Tsetsefliege *(Glossina fusca* und *Gl. tachinoides).* Arbeit. K. Gesundhamt. *26:* 301-383.

Stumpf, P.K. and D.E. Green. 1944. L-amino acid oxidase of *Proteus vulgaris.* J. Biol. Chem. *153:* 387-399.

Sturges, W.S. and A.G. Heideman. 1924. Studies of halophilic organisms II. The flora of meat-curing solutions. Abstr. Bacteriol. *8:* 14-15.

Sturm, S., U. Schönefeld, W. Zillig, D. Janekovic and K.O. Stetter. 1980. Structure and function of the DNA dependent RNA polymerase of the Archaebacterium *Thermoplasma acidophilum.* Zentralbl. Bakteriol. Parasitenkd. Infektionskr. Hyg. Abt. Orig. C1, 12-25.

Stutzer, M.J. 1923. Zür Frage über die Fäulnisbakterien im Darm. Zentralbl. Bakteriol. Parasitenkd. Infektionskr. Hyg. Abt. I Orig. 87-90.

Stutzer, M. and A. Kwaschnina. 1929. *In* Aussaaten aus den Fäzes des Menschen gelbe Kolonien bildende Bakterien (Gattung *Flavobacterium* u.a.) Zentralbl. Bakteriol. Parasitenkd. Infektionskr. Hyg. Abt. I Orig. *113:* 219-225.

Stuy, J.H. 1976. Restriction enzymes do not play a significant role in *Haemophilus* homospecific or heterospecific transformation. J. Bacteriol. *128:* 212-220.

Stuy, J.H. 1978. On the nature of nontypable *Haemophilus influenzae.* Antonie van Leeuwenhoek J. Microbiol. Serol. *44:* 367-376.

Stuy, J.H. 1979. Plasmid transfer in *Haemophilus influenzae.* J. Bacteriol. *139:* 520-529.

Subcommittee on the *Enterobacteriaceae* of the International Committee on

Bacteriological Nomenclature. 1958. Recommended biochemical method for group differentiation within the *Enterobacteriaceae*. Int. Bull. Bacteriol. Nomen. Taxon. *8:* 25-70.

Subcommittee on the Taxonomy of *Mollicutes*. 1979. Proposal of minimal standards for descriptions of new species of the class *Mollicutes*. Int. J. Syst. Bacteriol. *29:* 172-180.

Subcommittee on the Taxonomy of *Mollicutes*. 1981. Minutes of Meeting on September 2, 1980 (in preparation).

Subcommittee on the Taxonomy of *Mycoplasmatales*. 1977. Int. J. Syst. Bacteriol. *27:* 392-394.

Sugihara, P.T., V.L. Sutter, H.R. Attebery, K.S. Bricknell and S.M. Finegold. 1974. Isolation of *Acidaminococcus fermentans* and *Megasphaera elsdenii* from normal human feces. Appl. Microbiol. *27:* 274-275.

Sugimoto, M. 1935. Anaplasmosis-like disease in Formosan swine. J. Soc. Trop. Agr. (Taiwan) *7:* 240-244.

Sugino, W.M., R.C. Wek and D.T. Kingsbury. 1980. Partial nucleotide sequence similarity within species of *Mycoplasma* and *Acholeplasma*. J. Gen. Microbiol. *121:* 333-338.

Suitor, E.C., Jr. 1964. Studies on the cell envelope of *Wolbachia persica*. J. Infect. Dis. *114:* 125-134.

Suitor, E.C., Jr. 1964a. The relationship of *Wolbachia persica* Suitor and Weiss to its host. J. Insect Pathol. *6:* 111-124.

Suitor, E.C., Jr. 1964b. Propagation of *Rickettsiella popilliae* (Dutky and Gooden) Philip and *Rickettsiella melolonthae* (Krieg) Philip in cell cultures. J. Insect Pathol. *6:* 31-40.

Suitor, E.C., Jr. and E. Weiss. 1961. Isolation of a rickettsialike microorganism (*Wolbachia persica* n. sp.) from *Argas persicus* (Oken). J. Infect. Dis. *108:* 95-106.

Süle, S. 1978. Biotypes of *Agrobacterium tumefaciens* in Hungary. J. Appl. Bacteriol. *44:* 207-213.

Sulzer, C.R. and W.L. Jones. 1976. Leptospirosis. *In* Methods in laboratory diagnosis (revised edition). Publication No.(CDC) 74-8275. Department of Health, Education and Welfare, Washington, D.C.

Summers, A.O., G.A. Jacoby, M.N. Swartz, G. McHugh and L. Sutton. 1978. Metal cations and oxyanion resistances in plasmids of Gram-negative bacteria. *In* Schlessinger (Editor), "Microbiology - 1978", American Society for Microbiology, Washington D.C., pp. 128-131.

Sumner, D.R. and N.W. Schaad. 1977. Epidemiology and control of bacterial leaf blight of corn. Phytopathology *67:* 1113-1118.

Sundqvist, G. 1976. Bacteriological studies of necrotic dental pulps. Ph.D. Thesis, Umea University Odontological Dissertations, No. 7. Umea, Sweden.

Suomalainen, H., E.J.A. Keränen and J. Kangasperko. 1965. Production of spirit vinegar by the quick process with a pure culture of *Acetobacter rancens* Beijerinck. J. Inst. Brew. *71:* 41-45.

Surgalla, M.J., A.W. Andrews and D.M. Cavanaugh. 1968. Studies on virulence factors of *Pasteurella pestis* and *Pasteurella pseudotuberculosis*. Symp. Ser. Immunol. Stand. *9:* 293-302.

Surman, P.G. 1968. Cytology of "pink-eye" of sheep, including a reference to trachoma of man, by employing acridine orange and iodine stains, and isolation of *Mycoplasma* agents from infected sheep eyes. Aust. J. Biol. Sci. *21:* 447-467.

Šuťaková, G. and F. Rüttgen. 1978. *Rickettsiella phytoseiuli* and virus-like particles in *Phytoseiulus persimilis* (Gamasoidea:Phytoseiidae) mites. Acta Virol. *22:* 333-336.

Sutherland, I.W. 1966. The production of azurin and similar proteins. Arch. Mikrobiol. *54:* 350-357.

Sutherland, I.W. 1977. Bacterial exopolysaccharides: their nature and production. *In* I.W. Sutherland (Editor), Surface Carbohydrate of the Procaryotic Cell. Academic Press, New York, pp. 27-96.

Sutherland, I.W. 1979. Microbial exopolysaccharides: control of synthesis and acylation. *In* Berkeley, Gooday and Ellwood (Editors), Microbial Polysaccharides and Polysaccharases, Academic Press, New York, pp. 1-34.

Sutherland, I.W. and D.C. Ellwood. 1979. Microbial exopolysaccharides—industrial polymers of current and future potential. Symp. Soc. Gen. Microbiol. *29:* 107-150.

Šutič, D. and W.J. Dowson. 1959. An investigation of a serious disease of hemp (*Cannabis sativa* L.) in Jugoslavia. Phytopathol. Z. *34:* 307-314.

Šutič, D. and Ž. Tešic. 1958. Judna nova bakterioza bresta izazivač *Pseudomonas ulmi* n. sp. Zăstita Bilja *445:* 13-25.

Suto, T. 1954. An acid fast *Azotobacter* in a volcanic ash soil. Sci. Rep. Res. Inst. Tohoku Univ. *6:* 25-31.

Suto, T. 1957. Some properties of an acid tolerant *Azotobacter, Azotobacter indicum*. Tohoku J. Agric. Res. *7:* 369-382.

Sutter, G.R. and V.M. Kirk. 1968. Rickettsialike particles in fat-body cells of carabid beetles. J. Invert. Pathol. *10:* 445-449.

Sutter, V.L., D.M. Citron and S.M. Finegold. 1980. Wadsworth anaerobic bacteriology manual, 3rd Ed. C.V. Mosby Co., St. Louis, Mo.

Sutter, V.L. and Foecking, F.J. 1962. Biochemical characteristics of lactose-fermenting *Proteus rettgeri* from clinical specimens. J. Bacteriol. *83:* 933-935.

Sutter, V.L., V.L. Vargo and S.M. Finegold. 1975. Wadsworth anaerobic bacteriology manual, 2nd Ed. Anaerobe Bacteriology Laboratory, Wadsworth Hospital Center, Veterans Administration, and Department of Medicine, UCLA School of Medicine, Los Angeles, Ca.

Sutton, A., R. Schneerson, S. Kendall-Morris and J.B. Robbins. 1982. Differential complement resistance mediates virulence of *Haemophilus influenzae* type b. Infect. Immun. *35:* 95-104.

Sutton, R.G.A., M.F. O'Keeffe, M.A. Bundock, J. Jeboult and M.P. Tester. 1972. Isolation of a new *Moraxella* from a corneal abscess. J. Med. Bacteriol. *5:* 148-150.

Suzuki, A. and M. Goto. 1971. Isolation and characterization of pteridines from *Pseudomonas ovalis*. Bull. Chem. Soc. (Japan) *44:* 1869-1872.

Suzuki, A. and M. Goto. 1972. The structure of a new pteridine compound produced by *Pseudomonas ovalis*. Bull. Chem. Soc. Japan *45:* 2198-2199.

Svedhem, A. and B. Kaijser. 1981. Isolation of *Campylobacter jejuni* from domestic animals and pets: probable origin of human infection. J. Infect. Dis. *3:* 37-40.

Svoboda, K.H., S. Pohl and W. Mannheim. 1981. Untersuchungen zür Phylogenie von *Pasteurella multocida*: DNS-Basensequenz-Verwandtschaften zwischer Vertreten der Serogruppen A bis E, sowie Abtrennung der Biovarietät 6 (sog. dog-type-Stämme). Zentralbl. Bakteriol. Parasitenkd. Infektionskr. Hyg. Abt. I Orig. A *248:* 494-501.

Swan, M.A. 1982. Trailing flagella rotate faster than leading flagella in unipolar cells of *Spirillum volutans*. J. Bacteriol. *150:* 377–380.

Swanberg, S.L., G.K. Masover and L. Hayflick. 1978. Some characteristics of *Ureaplasma urealyticum*. Urease activity in a simple buffer: effect of metal ions and sulphydryl inhibitors. J. Gen. Microbiol. *108:* 221-225.

Swaney, L.M. and S.S. Breese. 1980. Ultrastructure of *Haemophilus equigenitalis*, causative agent of contagious equine metritis. Am. J. Vet. Res. *41:* 127-132.

Swaney, L.M. and S.P. Sahu. 1978. CEM: bacteriological methods. Vet. Rec. *102:* 43.

Swann, A.I., C.L. Garby, P.R. Schnurrenberger and R.R. Brown. 1981. Safety aspects in preparing suspensions of field strains of *Brucella abortus* for serological identification. Vet. Rec. *109:* 254-255.

Swanson, J. 1973. Studies on gonococcus infection. IV. Pili: their role in attachment of gonococci to tissue culture cells. J. Exp. Med. 137: 571-589.

Swanson, J. 1978. Studies on gonococcus infection. XII. Colony color and opacity variations of gonococci. Infect. Immun. *19:* 320-331.

Swellengrebel, N.H. 1907. Sur la cytologie comparée des spirochètes et des spirilles. Ann. Inst. Pasteur (Paris) 21: 448-466; 562-586.

Swindlehurst, C.A., H.N. Shah, C.W. Parr and R.A.D. Williams. 1977. Sodium dodecyl sulphate-polyacrylamide gel electrophoresis of polypeptides from *Bacteroides melaninogenicus*. J. Appl. Bacteriol. *43:* 319-324.

Swingle, D.B. 1925. Center rot of "French endive" or wilt of chicory (*Chichorium intybus* L.). Phytopathology *15:* 730.

Swings, J. and J. De Ley. 1975. Genome deoxyribonucleic acid of the genus *Zymomonas* Kluyver and van Niel 1936: base composition, size, and similarities. Int. J. Syst. Bacteriol. *25:* 324-328.

Swings, J. and J. De Ley. 1977. The biology of *Zymomonas*. Bacteriol. Rev. *41:* 1-46.

Swings, J. and J. De Ley. 1981. The genera *Gluconobacter* and *Acetobacter*. *In* Starr, Stolp, Trüper, Balows and Schlegel (Editors), The Prokaryotes, a handbook on habitats, isolation and identification of bacteria. Springer-Verlag, Berlin, pp. 771-778.

Swings, J., M. Gillis, K. Kersters, P. De Vos, F. Gosselé and J. De Ley. 1980. *Frateuria*, a new genus for "*Acetobacter aurantius*". Int. J. Syst. Bacteriol. *30:* 547-556.

Swings, J., K. Kersters and J. De Ley. 1976. Numerical analysis of electrophoretic protein patterns of *Zymomonas* strains. J. Gen. Microbiol. *93:* 266-271.

Swings, J., K. Kersters and J. De Ley. 1977. Taxonomic position of additional *Zymomonas mobilis* strains. Int. J. Syst. Bacteriol. *27:* 271-273.

Swings, J. and W. Van Pee. 1977. Infra-red spectroscopy of *Zymomonas* cells. J. Gen. Appl. Microbiol. *23:* 297-301.

Switzer, W.P. 1955. Studies on infectious atrophic rhinitis. IV. Characterization of a pleuropneumonia-like organism isolated from the nasal cavities of swine. Am. J. Vet. Res. *16:* 540-544.

Switzer, W.P. 1963. Elimination of *Bordetella bronchiseptica* from the nasal cavity of swine by sulphonamide therapy. Vet. Med. *58:* 571-574.

Switzer, W.P. 1964. Mycoplasmosis. *In* Dunne (Editor), Diseases of Swine, 2nd Ed., Iowa State Univ. Press, Ames, Iowa, pp. 498-507.

Switzer, W.P. 1972. Mycoplasmal pneumonia of swine. J. Am. Vet. Assoc. *160:* 651-653.

Switzer, W.P., C.J. Maré and E.D. Hubbard. 1966. Incidence of *Bordetella bronchiseptica* in wildlife and man in Iowa. Am. J. Vet. Res. *27:* 1134-1136.

Sykes, R.B. and M. Matthew. 1976. The beta-lactamases of of Gram-negative bacteria and their role in resistance to beta-lactam antibiotics. J. Antimicrob. Chemother. *2:* 115-157.

Sykes, R.B. and M.H. Richmond. 1970. Intergeneric transfer of a beta-lactamase gene between *Ps. aeruginosa* and *E. coli*. Nature *226:* 952-954.

Szeness, L., L. Sey and A. Szeness. 1979. Bacteriological studies of the intestinal content of aquatic birds, fishes, and frogs with special reference to the presence of non-cholerae vibrios (NCV). Zentralbl. Bakteriol. Parasitenkd. Infektionskr. Hyg. Abt. I Orig. A *45:* 89-95.

Szita, J., A. Svidró, H. Smith, É. Czirók and K. Solt. 1979. Incidence of non-cholera vibrios in Hungary. Acta Microbiol. Acad. Sci. Hung. *26:* 71-83.

Tahara, Y., M. Kameda, Y. Yamada and K. Kondo. 1976a. An ornithine containing lipid isolated from *Gluconobacter cerinus*. Biochem. Biophys. Acta *450:* 225-230.

Tahara, Y., M. Kameda, Y. Yamada and K. Kondo. 1976b. A new lipid; the ornithine and taurine-containing "cerilipin". Agric. Biol. Chem. *40:* 243-244.

Tahara, Y., Y. Yamada and K. Kondo. 1976c. Phospholipid compositiion of *Gluconobacter cerinus*. Agric. Biol. Chem. *40:* 2355-2360.

Takahashi, I. and N.E. Gibbons. 1959. Effect of salt concentration on the morphology and chemical composition of *Micrococcus halodenitrificans*. Can. J. Microbiol. *5:* 25-35.

Takahashi, J., Y. Ichikawa, H. Sagae, I. Komura, H. Kanou and K. Yamada. 1980. Isolation and identification of n-butane-assimilating bacterium. Agr. Biol. Chem. *44:* 1835-1840.

Takahashi, T. 1906-1908. Studies on diseases of Saké. Bull. Coll. Agr. Tokyo Imp. Univ. *7:* 531-563.

Takeda, K., S. Motomatsu, Y. Hachiya, S. Fukuoka and Y. Takahara. 1974. Characterization and culture conditions for a methane-oxidizing bacterium. J. Ferm. Technol. *52:* 793-798.

Takimoto, S. 1920. On the bacterial leaf-spot of *Antirrhinum majus* L. Bot. Magaz., Tokyo *34:* 253-257.

Takimoto, S. 1927. Bacterial black spot of burdock. J. Plant Prot. *14:* 519-523.

Takimoto, S. 1931. Bacterial bud rot of loquat. J. Plant. Prot., Tokyo *18:* 349-355.

Takimoto, S. 1933. The Bacterial disease of New Zealand flax. J. Plant Prot. *20:* 774-778.

Takimoto, S. 1934. Leaf spot of begonia. J. Plant Prot. *21:* 258-262.

Takimoto, S. 1939. Bacterial leaf spot of *Cissus japonica* Willd. Ann. Phytopathol. Soc. Jpn. *9:* 41-43.

Tall, B.D. and R.K. Nauman. 1981. Scanning electron microscopy of *Cristispira* species in Chesapeake Bay oysters. Appl. Environ. Microbiol. *42:* 336-343.

Talmadge, M.B. and R.M. Herriott. 1960. A chemically defined medium for growth, transformation, and isolation of nutritional mutants of *Haemophilus influenzae*. Biophys. Biochem. Res. Comm. *2:* 203-206.

Tambarlane, W.V. and E.V. Soto. 1975. *Citrobacter diversus* meningitis: A case report. Pediat. *55:* 739-741.

Tamura, K., S. Shimada and L.M. Prescott. 1971. Vibrio agar: a new plating medium for isolation of *Vibrio cholerae*. Jap. J. Med. Sci. Biol. *24:* 125-127.

Tamura, M., K. Nogiomori, M. Yajima, K. Ase and M. Ui. 1983. A role of the B-oligomer moiety of islet activating protein, pertussis toxin, in the development of the biological effects of intact cells. J. Biol. Chem. *258:* 6756-6761.

Tan, R.J.S. and J.G. Markham. 1971. Feline T-strain mycoplasmas. Jpn. J. Exp. Med. *41:* 247-248.

Tanaka, H., W.T. Hall, J.B. Scheffield and O.H. Moore. 1965. Fine structure of *Haemobartonella muris* as compared with *Eperythrozoon coccoides* and *Mycoplasma pulmonis*. J. Bacteriol. *90:* 1735-1749.

Tanaka, S., T. Suto, Y. Isayama, R. Azuma and H. Hatakeyama. 1977. Chemotaxonomical studies on fatty acids of *Brucella* species. Ann. Sclavo *19:* 67-82.

Tanner, A.C.R., S. Badger, C.-H. Lai, M.A. Listgarten, R.A. Visconti and S.S. Socransky. 1981. *Wolinella* gen. nov., *Wolinella succinogenes (Vibrio succinogenes* Wolin et al.) comb. nov., and description of *Bacteroides gracilis* sp. nov., *Wolinella recta* sp. nov., *Campylobacter concisus* sp. nov., and *Eikenella corrodens* from humans with periodontal disease. Int. J. Syst. Bacteriol. *31:* 432-445.

Tanner, A.C.R., C. Haffer, G.T. Bratthall, R.A. Visconti and S.S. Socransky. 1979. A study of the bacteria associated with advancing periodontitis in man. J. Clin. Periodontol. *6:* 278-307.

Tanner, A.C.R., R.A. Visconti, L.V. Holdeman, G. Sundquist and S.S. Socransky. 1982. Similarity of *Wolinella recta* strains isolated from periodontal pockets and root canals. J. Endodontics *8:* 294-300.

Taran, I.F., N.A Pogorelov, G.G. Kulikova, A.Z. Kutsemakina, M.M. Rudnev, N.M. Nelyapin, V.A. Rudneva and A.E. Suvarova. 1966. Studies on brucella cultures isolated from mouselike rodents and their ectoparasites. (In Russian.) Zh. Mikrobiol. Epidemiol. Immunobiol. *43:* 70-74.

Targowski, S. and H. Targowski. 1979. Characterization of a *Haemophilus paracuniculus* isolated from gastrointestinal tracts of rabbits with mucoid enteritis. J. Clin. Microbiol. *9:* 33-37.

Tarrand, J. J., N. R. Krieg and J. Döbereiner. 1978. A taxonomic study of the *Spirillum lipoferum* group, with descriptions of a new genus, *Azospirillum* gen. nov. and two species, *Azospirillum lipoferum* (Beijerinck) comb. nov. and *Azospirillum brasilense* sp. nov. Can. J. Microbiol. *24:* 967-980.

Tarrand, J. J., N. R. Krieg and J. Döbereiner. 1979. *In* List No. 2, Validation of the publication of new names and new combinations previously published outside the IJSB. Int. J. Syst. Bacteriol. *29:* 79-80.

Tartakowsky, J. 1910. Piroplasmose bei Fledermäusen *(Vespertilio noctula)* und ihre Vermittler. Arb. IXth Int. Tierärztl. Kongr. in Haag. *4:* 242-244.

Tatlock, H. 1944. A rickettsia-like organism recovered from guinea pigs. Proc. Soc. Exp. Biol. Med. *57:* 95-99.

Tatum, H.W., W.H. Ewing and R.E. Weaver. 1974. Miscellaneous Gram-negative bacteria. *In* Lennette, Spaulding and Truant (Editors), Manual of Clinical Microbiology, 2nd Ed., American Society for Microbiology, Washington, D.C., pp. 270-294.

Taylor, A.G., T.G. Harrison, M.W. Dighero and C.M.P. Bradstreet. 1979. False positive reactions in the indirect fluorescent antibody test for Legionnaires' disease eliminated by use of formolised yolk-sac antigen. Ann. Intern. Med. *90:* 686-689.

Taylor, C.E.D., R.O. Rosenthal, D.E.J. Brown, S.P. Lapage, L.R. Hill and R.M. Legros. 1978. The causative organism of contagious equine metritis 1977. Proposal for a new species to be known as *Haemophilus equigenitalis*. Equine Vet. J. *10:* 136-144.

Taylor, D. E., S. A. DeGrandis, M. A. Karmali and P. C. Fleming. 1981. Transmissiblle plasmids from *Campylobacter jejuni*. Antimicrob. Agents Chemother. *19:* 831-835.

Taylor, J., H.J. Bensted, J.S.K. Boyd, K.P. Carpenter, W.J. Dowson, R. Lovell, E.W. Taylor, H.G. Thornton, G.S. Wilson and C. Shaw. 1952. Classification of the *Bacteriaceae*. Int. Bull. Bacteriol. Nomencl. Taxon. *2:* 137-140.

Taylor, J.D. 1972. Specificity of bacteriophages and antiserum for *Pseudomonas pisi*. N. Z. J. Agr. Res. *15:* 421-431.

Taylor, R.M., M. Lisbonne and G. Roman. 1932. Recherches sur l'identification des *Brucella* isolées en France par l'action bactériostatic des matières colorantes et la production d'hydrogène sulfuré (Huddleson). Ann. Inst. Pasteur (Paris) *49:* 284-302.

Taylor, S. 1977. Evidence for the presence of ribulose 1,5-biphosphate carboxylase and phosphoribulokinase in *Methylococcus capsulatus* (Bath.) FEMS Microbiol. Lett. *2:* 305-307.

Taylor, V. I., P. Baumann, J. L. Reichelt and R. D. Allen. 1974. Isolation, enumeration, and host range of marine bdellovibrios. Arch. Microbiol. *98:* 101-114.

Taylor, W.H. and E. Juni. 1961. Pathways for biosynthesis of a bacterial capsular polysaccharide II. Carbohydrate metabolism and terminal oxidation mechanisms of a capsule-producing coccus. J. Bacteriol. *81:* 694-703.

Taylor, W.I. and D. Achanzar. 1972. Catalase test as an aid to the identification of *Enterobacteriaceae*. Appl. Microbiol. *24:* 58-61.

Taylor-Robinson, D. 1967. Mycoplasmas of various hosts and their antibiotic sensitivity. Postgraduate Med. J. *43:* (Suppl. 43): 100-104.

Taylor-Robinson, D. 1979. Pathogenicity of ureaplasmas for animals and man. Zentralbl. Bakteriol. Parasitenkd. Infektionskr. Hyg. I Abt. Orig. *A245:* 150-163.

Taylor-Robinson, D., J.P. Addey and C.S. Goodwin. 1969. Comparison of techniques for the isolation of T-strain mycoplasmas. Nature (London) *222:* 274-275.

Taylor-Robinson, D., J. Canchola, H. Fox and R.M. Chanock. 1964. A newly identified oral mycoplasma (*M. orale*) and its relationship to other human mycoplasmas. Am. J. Hyg. *80:* 135-148.

Taylor-Robinson, D., H. Fox and R.M. Chanock. 1965. Characterization of a newly identified mycoplasma from the human oropharynx. Am. J. Epidemiol. *81:* 180-191.

Taylor-Robinson, D. and P.M. Furr. 1973. The distribution of T-mycoplasmas within and among various animal species. Ann. N.Y. Acad. Sci. *225:* 108-117.

Taylor-Robinson, D. and P.M. Furr. 1981. Recovery and identification of human genital tract mycoplasmas. Isr. J. Med. Sci. *17:* 648-653.

Taylor-Robinson, D., D.A. Haig and M.H. Williams. 1967. Bovine T-strain mycoplasma. Ann. N.Y. Acad. Sci. *143:* 517-518.

Taylor-Robinson, D., C. Martin-Bourgon, T. Watanabe and J.P. Addey. 1971. Isolation of T-mycoplasmas from dogs and squirrel monkeys: biological and serological comparison with those isolated from man and cattle. J. Gen. Microbiol. *68:* 97-107.

Taylor-Robinson, D. and R.H. Purcell. 1966. Mycoplasmas of the human urogenital tract and oropharynx and their possible role in disease: a review with some recent observations. Proc. Roy. Soc. Med. *59:* 1112-1116.

Taylor-Robinson, D., R.H. Purcell, D.C. Wong and R.M. Chanock. 1966. A colour test for the measurement of antibody to certain mycoplasma species based upon the inhibition of acid production. J. Hyg. Camb. *64:* 91-104.

Taylor-Robinson, D., N.L. Somerson, H.C. Turner and R.M. Chanock. 1963. Serological relationships among human mycoplasmas as shown by complement-fixation and gel diffusion. J. Bacteriol. *85:* 1261-1273.

Taylor-Robinson, D., M.H. Williams and D.A. Haig. 1968. The isolation and comparative biological and physical characteristics of T-mycoplasmas of cattle. J. Gen. Microbiol. *54:* 33-46.

Tchan, Y.T. 1952. Studies of N-fixing bacteria. I. A note on the estimation of *Azotobacter* in the soil. Proc. Linn. Soc. N.S.W. *77:* 89-91.

Tchan, Y.T. 1952. Studies of nitrogen-fixing bacteria. II. The presence of aerobic non-symbiotic nitrogen-fixing bacteria in soil of the Sydney district. Proc. Linn. Soc. N.S.W. *77:* 92-97.

Tchan, Y.T. 1953. Studies of N-fixing bacteria. IV. Taxonomy of genus *Azotobacter* (Beijerinck, 1901). Proc. Linn. Soc. N.S.W. *78:* 85-89.

Tchan, Y.T. 1953. Studies of nitrogen-fixing bacteria. V. Presence of *Beijerinckia* in Northern Australia and geographic distribution of non-symbiotic nitrogenfixing microorganisms. Proc. Linn. Soc. N.S.W. *78:* 172-178.

Tchan, Y.T. 1957. Studies of nitrogen-fixing bacteria. VI. A new species of nitrogen-fixing bacteria. Proc. Linn. Soc. N.S.W. *82:* 314-316.

Tchan, Y.T. 1968. Importance of systematics of *Azotobacteriaceae* in the study of its ecology. Trans. 9th Int. Congr. Soil Sci. Adelaide *2:* 115-124.

Tchan, Y.T., A. Birch-Anderson and H.L. Jensen. 1962. The ultrastructure of vegetative cells and cysts of *Azotobacter chroococcum*. Arch. Mikrobiol. *43:* 50-66.

Tchan, Y.T. and R.R. de Ville. 1970. Application de l'immunofluorescence à l'étude des *Azotobacter* du sol. Ann. Inst. Pasteur (Paris) *118:* 665-673.

Tchan, Y.T. and H.L. Jensen. 1963. Studies of N-fixing bacteria. VIII. Influence of N-content of the media on the N-fixation capacity and colony variation of *Derxia gummosa* Jensen et al (1960). Proc. Linn. Soc. N.S.W. *88:* 379-385.

Tchan, Y.T., Z. Wyszomirska-Dreher, P.B. New and J.-C. Zhou. 1983. Taxonomy of the *Azotobacteraceae* determined by using immunoelectrophoresis. Int. J. Syst. Bacteriol. *33:* 147-156.

Tchan, Y.T., Z. Wyszomirska-Dreher and J.M. Vincent. 1980. Preliminary study of taxonomy of *Azotobacter* and *Azomonas* using rocket line immunoelectrophoresis. Curr. Microbiol. *4:* 265-270.

Tebo, B.M., D.S. Linthicum and K.H. Nealson. 1979. Luminous bacteria and light emitting fish: ultrastructure of the symbiosis. BioSystems *11:* 269-280.

Terada, A., K. Uchida and T. Mitsuoka. 1976. Die Bacteroidaceenflora in den faeces von schweinen. Zentralbl. Bakteriol. Parasitenkd. Infektionskr. Hyg., I Abt. Orig. *234:* 362-370.

Terakado, N. and S. Mitsuhashi. 1974. Properties of R factors from *Bordetella bronchiseptica*. Antimicrob. Agents Chemother. *6:* 836-840.

Terasaki, Y. 1958. Studies on *Cristispira* in the crystalline style of a fresh water snail, *Semisulcospira libertina* Gould. I. The morphological characters and living condition within the style. Bull. Suzugamine Women's Coll. (Hiroshima) 5 Ser. 7-19.

Terasaki, Y. 1961. on *Spirillum putridiconchylium* nov. sp. Bot. Mag. *74:* 79-85.

Terasaki, Y. 1961. On two new species of *Spirillum*. Bot Mag. *74:* 220-227.

Terasaki, Y. 1963. On the isolation of *Spirillum*. Bull. Suzugamine Women's Coll., Nat. Sci. *10:* 1-10.

Terasaki, Y. 1970. Über die Anhäufung von in Süsswasser und Meerwasser vorkommenden Spirillum. Bull. Suzugamine Women's Coll., Nat. Sci. *15:* 1-7.

Terasaki, Y. 1972. Studies on the genus *Spirillum* Ehrenberg. I. Morphological, physiological, and biochemical characteristics of water spirilla. Bull. Suzugamine Women's Coll., Nat. Sci. *16:* 1-146.

Terasaki, Y. 1973. Studies on the genus *Spirillum* Ehrenberg. II. Comments on type and reference strains of *Spirillum* and descriptions of new species and new subspecies. Bull. Suzugamine Women's Coll., Nat. Sci. *17:* 1-71.

Terasaki, Y. 1975. Freeze-dried cultures of water spirilla made on experimental basis. Bull. Suzugamine Women's Coll., Nat. Sci. *19:* 1-10.

Terasaki, Y. 1979. Transfer of five species and two subspecies of *Spirillum* to other genera (*Aquaspirillum* and *Oceanospirillum*), with emended descriptions of the species and subspecies. Int. J. Syst. Bacteriol. *29:* 130-144.

Terasaki, Y. 1980. Enrichment and isolation of aerobic chemoheterotrophic spirilla from mud and sand samples. J. Gen. Appl. Microbiol. *26:* 395-402.

Terzaghi, B.E. 1980a. Ultra-violet sensitivity and mutagenesis of *Azotobacter*. J. Gen. Microbiol. *118:* 271-273.

Terzaghi, B.E. 1980b. A method for the isolation of *Azotobacter* mutants derepressed for *nif*. J. Gen. Microbiol. *118:* 275-278.

Tezuka, Y. 1973. A zoogloea bacterium with gelatinous mucopolysaccharide matrix. J. Water Pollut. Control Fed. *45:* 531-536.

Thacker, L., R.M. McKinney, C.W. Moss, H.M. Sommers, M.L. Spivack and T.F. O'Brien. 1981. Thermophilic sporeforming bacilli that mimic fastidious growth characteristics and colonial morphology of legionellae. J. Clin. Microbiol. *13:* 794-797.

Thal, E. 1973. Observation on immunity in *Yersinia pseudotuberculosis*. Contr. Microbiol. Immunol. *2:* 190-195.

Thal, E. and W. Knapp. 1971. A revised antigenic scheme of *Yersinia pseudotuberculosis*. Progr. Immunobiol. Standard *15:* 219-222.

Thanbichler, A. and E. Beck. 1974. Catabolism of hamamelose. The anaerobic dissimilation of D-hamamelose by *Kluyvera citrophila* 627. Eur. J. Biochem. *50:* 191-196.

Thayer, D.W. 1978. Carboxymethylcellulase produced by facultative bacteria from the hind-gut of the termite *Reticulitermes hesperus*. J. Gen. Microbiol. *106:* 13-18.

Theiler, A. 1910. *Anaplasma marginale* (gen. and spec. nov.). The marginal points in the blood of cattle suffering from specific disease. Transvaal S. Afr. Rep. Vet. Bacteriol. Dept. Agr. *1908-9:* 7-64.

Theiler, A. 1911. Further investigations into anaplasmosis of South African cattle. 1st Rep. Dir. Vet. Res. August *1911:* 7-46.

Thibault, P. and L. Le Minor. 1957. Méthodes simples de recherche de la lysinedecarboxylase et de la tryptophane-desaminase a l'aide des milieux pour differenciation rapide des *Enterobacteriacees*. Ann. Inst. Pasteur (Paris) *92:* 551-554.

Thiele, O.W. and W. Kehr. 1969. Die "freien" Lipide aus *Brucella abortus* Bang. Eur. J. Biochem. *9:* 167-175.

Thirkill, C.E. and G.E. Kenny. 1975. Serological comparison of five arginineutilizing *Mycoplasma* species by two-dimensional immunoelectrophoresis. Infect. Immun. *10:* 624-632.

Thjøtta, T. 1920. On the bacillus of Morgan No. 1 - a meta colon-bacillus. J. Bacteriol. *5:* 67-77.

Thjøtta, T. and O.T. Avery. 1921. Studies on bacterial nutrition. II. Growth accessory substances in the cultivation of hemophilic bacilli. III. Plant tissue, as a source of growth accessory substances in the cultivation of *Bacillus influenzae*. J. Exp. Med. *34:* 97-114.

Thjøtta, T., O. Hartmann and J. Bøe. 1939. A study of the *Leptotrichia* Trevisan. History, morphology, biological and serological characteristics. Skr. Nor. Vidensk.-Akad. Oslo I. Mat.-Naturvidensk. Kl. *5:* 1-199.

Thjøtta, T. and S. Sydnes. 1951. *Actinobacillus actinomycetem comitans* as the sole infecting agent in a human being. Acta Pathol. Microbiol. Scand. *28:* 27-35.

Thomas, D.L. 1979. Mycoplasmalike bodies associated with lethal declines of palms in Florida. Phytopathology *69:* 928-934.

Thomas, G.M. and G.O. Poinar, Jr. 1979. *Xenorhabdus* gen. nov., a genus of entomopathogenic nematophilic bacteria of the family *Enterobacteriaceae*. Int. J. Syst. Bacteriol. *29:* 352-360.

Thomas, L., F. Aleu, M.W. Bitensky, M. Davidson and B. Gesner. 1966. Studies of PPLO infection. II. The neurotoxin of *Mycoplasma neurolyticum*. J. Exp. Med. *124:* 1067-1082.

Thomas, L. and M.W. Bitensky. 1966. Studies of PPLO infection. IV. The neurotoxicity of intact mycoplasmas, and their production of toxin in vivo and in vitro. J. Exp. Med. *124:* 1089-1098.

Thomas, M.E.M. and H.E. Tillett. 1973. Dysentery in general practice: a study of cases and their contacts in Enfield and an epidemiological comparison with salmonellosis. J. Hyg. *71:* 373-389.

Thomashow, M.F., R. Nutter, A.L. Montoya, M.P. Gordon and E.W. Nester. 1980a. Integration and organization of Ti plasmid sequences in crown gall tumors. Cell *19:* 729-739.

Thomashow, M.F., C.G. Panagopoulos, M.P. Gordon and E.W. Nester. 1980b. Host range of *Agrobacterium tumefaciens* is determined by the Ti-plasmid. Nature (London) *283:* 794-796.

Thomashow, M.F. and S.C. Rittenberg. 1978a. Penicillin-induced formation of osmotically stable spheroplasts in non-growing *Bdellovibrio bacteriovorus*. J. Bacteriol. *133:* 1484-1491.

Thomashow, M.F. and S.C. Rittenberg. 1978b. Intraperiplasmic growth of *Bdellovibrio bacteriovorus* 109J: Solubilization of *Escherichia coli* peptidoglycan. J. Bacteriol. *135:* 998-1007.

Thomashow, M.F. and S.C. Rittenberg. 1978c. Intraperiplasmic growth of *Bdellovibrio bacteriovorus* 109J: N-deacylation of *Escherichia coli* peptidoglycan amino sugars. J. Bacteriol. *135:* 1008-1014.

Thomashow, M.F. and S.C. Rittenberg. 1978d. Intraperiplasmic growth of *Bdellovibrio bacteriovorus* 109J: Attachment of long chain fatty acids to *Escherichia coli* peptidoglycan. J. Bacteriol. *135:* 1015-1023.

Thomason, B.M., F.W. Chandler and D.G. Hollis. 1979. Flagella on Legionnaires' disease bacteria: an interim report. Ann. Intern. Med. *91:* 224-226.

Thompson, J., J.W. Costerton and R.A. MacLeod. 1970. K⁺-Dependent deplasmolysis of a marine pseudomonad plasmolyzed in a hypotonic solution. J. Bacteriol. *102:* 843-854.

Thompson, J. and R.A. MacLeod. 1973. Na⁺ and K⁺ gradients and and α-aminoisobutyric acid transport in a marine pseudomonad. J. Biol. Chem. *248:* 7106-7111.

Thompson, J. and R.A. MacLeod. 1974. Potassium transport and the relationship between intracellular potassium concentration and amino acid uptake by cells of a marine pseudomonad. J. Bacteriol. *120:* 598-603.

Thompson, J.P. 1968. The occurrence of nitrogen-fixing bacteria of the genus *Beijerinckia* in Australia outside the tropical zone. *In* Transactions 9th Int. Congr. Soil Sci., Adelaide, Australia, *2:* 129-139.

Thompson, J.P. and V.B.D. Skerman. 1979. Azotobacteraceae: The taxonomy and ecology of the aerobic nitrogen-fixing bacteria. Academic Press, London.

Thompson, J.P. and V.B.D. Skerman. 1981. *Azorhizophilus paspali* comb. nov. *In* Validation of the publication of new names and new combinations previously effectively published outside the IJSB. List No. 6. Int. J. Syst. Bacteriol. *311:* 215-218.

Thompson, J.P. and V.B.D. Skerman. 1981. *Azotobacter armeniacus* sp. nov. *In* Validation of the publication of new names and new combinations previously effectively published outside the IJSB. List No. 6. Int. J. Syst. Bacteriol. *31:* 215-218.

Thompson, J.P. and V.B.D. Skerman. 1981. *Azomonotrichon macrocytogenes* comb. nov. *In* Validation of the publication of new names and new combinations previously effectively published outside the IJSB. List No. 6. Int. J. Syst. Bacteriol. *31:* 215-218.

Thompson, J.P. and V.B.D. Skerman. 1981. *Beijerinckia indica* subspecies *lacticogenes. Beijerinckia derxia* subspecies *venezuelae*. *In* Validation of the publication of new names and new combinations previously effectively published outside the IJSB. List No. 6. Int. J. Syst. Bacteriol. *31:* 215-218.

Thompson, L. and F.A. Willius. 1932. *Actinobacillus* bacteraemia. J. Am. Med. Assoc. *99:* 298-301.

Thompson, R.D. 1852. Ueber die Natur und die chemische Wirkungen der Essigmutter. Liebigs Ann. Chem. *83:* 89-93.

Thomsen, A. 1929. Smitsom kasthingsenzooti (Bang-Infektion) blandt soer i Midtjylland. Mskr. Dyrläg. *41:* 386.

Thomsen, A.C. 1978a. Occurrence of mycoplasmas in urinary tracts of patients with acute pyelonephritis. J. Clin. Microbiol. *8:* 84-88.

Thomsen, A.C. 1978b. Mycoplasmas in human pyelonephritis: demonstration of antibodies in serum and urine. J. Clin. Microbiol. *8:* 197-202.

Thomson, K.S., T.A. McMeekin and C.J. Thomas. 1981. Electron microscopic observations of *Flavobacterium aquatile* NCIB 8694 (= ATCC 11947) and *Flavobacterium meningosepticum* NCTC 10016 (= ATCC 13253). Int. J. Syst. Bacteriol. *31:* 226-231.

Thomson, S.V., D.C. Hildebrand and M.N. Schroth. 1981. Identification and nutritional differentiation of the *Erwinia* sugar beet pathogen from members of *Erwinia carotovora* and *Erwinia chrysanthemi*. Phytopathology *71:* 1037-1042.

Thorington, G. and L. Margulis. 1980. Transmission of the algal and bacterial symbionts of green hydra through the host sexual cycle. *In* Schwemmler and Schenk (Editors), Endocytobiology. Endosymbiosis and Cell Biology. A

Synthesis of Recent Research. Walter de Gruyter, Berlin, pp. 175-224.

Thornberry, H.H. and H.W. Anderson. 1931a. A bacterial disease of barberry caused by *Phytomonas berberidis* n. sp. J. Agr. Res. *43:* 29-36.

Thornberry, H.H. and H.W. Anderson. 1931b. Bacterial leaf spot of viburnum. Phytopathology *21:* 907-912.

Thornberry, H.H. and H.W. Anderson. 1937. Some bacterial diseases of plants in Illinois. Phytopathology *27:* 946-949.

Thorne, G.M. and W.E. Farrar, Jr. 1975. Transfer of ampicillin resistance between strains of *Haemophilus influenzae* type b. J. Infect. Dis. *132:* 276-281.

Thornley, M.J. 1960. The differentiation of *Pseudomonas* from other Gram-negative bacteria on the basis of arginine metabolism. J. Appl. Bacteriol. *23:* 37-52.

Thornley, M.J. 1975. Cell envelopes with regularly arranged surface subunits in *Acinetobacter* and related bacteria. Crit. Rev. Microbiol. *4:* 65-100.

Thorns, C.J. and E. Boughton. 1978. Studies on film production and its specific inhibition with special reference to *Mycoplasma bovis (M. agalactiae* var. *bovis).* Zentralbl. Vet. Med. B *25:* 657-667.

Thornsberry, C., C.N. Baker and L.A. Kirven. 1978. In vitro activity of antimicrobial agents on the Legionnaires' disease bacterium. Antimicrob. Agents Chemother. *13:* 78-80.

Thornsberry, C. and L.A. Kirven. 1978. β-lactamase of the Legionnaires' bacterium. Curr. Microbiol. *1:* 51-54.

Thorpe, T.C. and R.D. Miller. 1980. Negative enrichment procedure for isolation of *Legionella pneumophila* from seeded cooling tower water. Appl. Environ. Microbiol. *40:* 849-851.

Thouvenot, H. and A. Florent. 1954. Étude d'un anaerobie du sperme du Taureau et du vagin de la vache *Vibrio bubulus* Florent 1953. Ann. Inst. Pasteur (Paris) *86:* 237-240.

Threlfall, E.J., R. Rowe and I. Huq. 1980. Plasmid-encoded multiple antibiotic resistance in *Vibrio cholerae* eltor from Bangladesh. Lancet *1:* 1247-1248.

Thurner, K. and M. Busse. 1978. Numerisch taxonomische untersuchungen an Enterobakterien aus Oberflachenwasser. Zentralbl. Bakteriol. Parasitenkd. Infektionskr. Hyg. Abt. I *167:* 262-271.

Tien, T. M., H. G. Diem, M. H. Gaskins and D. H. Hubbell. 1981. Polygalacturonic acid transeliminase production by *Azospirillum* species. Can. J. Microbiol. *27:* 426-431.

Tien, T. M., M. H. Gaskins and D. H. Hubbell. 1979. Plant growth substances produced by *Azospirillum brasilense* and their effect on the growth of pearl millet. Appl. Environ. Microbiol. *37:* 1016-1024.

Tierno, P.M. and P. Steinberg. 1975. Isolation and characterization of lactose-positive strain of *Proteus morganii.* J. Clin. Microbiol. *1:* 108-109.

Tilak, K. V. B. R., M. Lakshmi-Kumari and C. S. Nautiyal. 1979. Survival of *Azospirillum brasilense* in different carriers. Curr. Sci. *48:* 412-413.

Till, D.H. and F.P. Palmer. 1960. A review of actinobacillosis with a study of the causal organism. Vet. Rec. *72:* 527-534.

Tindall, B.J., A.A. Mills and W.D. Grant. 1980. An alkalophilic red halophilic bacterium with a low magnesium requirement from a Kenyan soda lake. J. Gen. Microbiol. *116:* 257-260.

Tisdale, W.B. and M.M. Williamson. 1923. Bacterial leaf spot of lima bean. J. Agr. Res. *25:* 141-154.

Tissier, H. 1900. Recherches sur la flore intestinale des nourrissons. Thèses, Paris.

Tissier, H. 1905. Répartition des microbes dans l'intestin du nourrisson. Ann. Inst. Pasteur *19:* 109-123.

Tissier, H. 1908. Recherches sur la flore intestinale normale des enfants agés d'un an à cinq ans. Ann. Inst. Pasteur (Paris) *22:* 189-208.

Tiwari, A.D., M.P. Bryant and R.S. Wolfe. 1969. Simple method for isolation of *Selenomonas ruminantium* and some nutritional characteristics of the species. J. Dairy Sci. *52:* 2054-2056.

't Mannetje, L. 1967. A re-examination of the taxonomy of the genus *Rhizobium* and related genera using numerical analysis. Antonie van Leeuwenhoek J. Microbiol. Serol. *33:* 477-491.

To, L.P., L. Margulis, D. Chase and W.L. Nutting. 1980. The symbiotic microbial community of the Sonoran Desert termite: *Pterotermes occidentis.* BioSystems *13:* 109-137.

To, L., L. Margulis and A.T.W. Cheung. 1978. Pillotinas and Hollandinas: distribution and behaviour of large spirochaetes symbiotic in termites. Microbios *22:* 103-133.

Toala, P., Y.H. Lee, C. Wilcox and M. Finland. 1970. Susceptibility of *Enterobacter aerogenes* and *Enterobacter cloacae* to 19 antimicrobial agents in vitro. Amer. J. Med. Sci. *260:* 41-55.

Tobin, J.O'H., R.A. Swann and C.L.R. Bartlett. 1981. Isolation of *Legionella pneumophila* from water systems: methods and preliminary results. Brit. Med. J. *282:* 515-517.

Todorov, T. and P. Koleva-Todorova. 1971. Brucellacins and their formation. (In Bulgarian.) Second Congress in Microbiology, Sofia. p. 219.

Tomlinson, G.A. and L.I. Hochstein. 1976. *Halobacterium saccharovorum* sp. nov., a carbohydrate-metabolizing, extremely halophilic bacterium. Can. J. Microbiol. *22:* 587-591.

Tomlinson, G.A., T.K. Koch and L.I. Hochstein. 1974. The metabolism of carbohydrates by extremely halophilic bacteria: glucose metabolism via a modified Entner-Doudoroff pathway. Can. J. Microbiol. *20:* 1085-1091.

Topley, W.W.C. and G.S. Wilson. 1929. The principles of bacteriology and immunity, 1st Ed. Edward Arnold and Co., London, pp. 1-587.

Tornabene, T.G. 1973. Lipid composition of selected strains of *Yersinia pestis* and *Yersinia pseudotuberculosis.* Biochim. Biophys. Acta *306:* 173-185.

Tornebene, T.G. and T.A. Langworthy. 1979. Diphytanyl and dibiphytanyl glycerol ether lipids of methanogenic archaebacteria. Science *203:* 51-53.

Torrella, F., R. Guerrero and R. J. Seidler. 1978. Further taxonomic characterization of the genus *Bdellovibrio.* Can. J. Microbiol. *24:* 1387-1394.

Torsvik, T. and I. Dundas. 1974. Bacteriophage of *Halobacterium salinarium.* Nature (London) *248:* 680-681.

Tošić, J. and T.K. Walker. 1944. *Acetobacter* infection. I. *Acetobacter mobile* (sp. nov.). J. Inst. Brew. *50:* 296-300.

Tošić, J. and T.K. Walker. 1950. *Acetobacter acidum - mucosum* Tosic and Walker, n. sp., an organism forming a starch-like polysaccharide. J. Gen. Microbiol. *4:* 192-197.

Tourtellotte, M.E., H.J. Morowitz and P. Kasimer. 1964. Defined medium for *Mycoplasma laidlawii.* J. Bacteriol. *88:* 11-15.

Towner, K.J. and A. Vivian. 1976a. RP4-mediated conjugation in *Acinetobacter calcoaceticus.* J. Gen. Microbiol. *93:* 355-360.

Towner, K.J. and A. Vivian. 1976b. RP4 fertility variants in *Acinetobacter calcoaceticus.* Genet. Res. *28:* 301-306.

Townsend, R. 1976. Arginine metabolism by *Spiroplasma citri.* J. Gen. Microbiol. *94:* 417-420.

Townsend, R. and D.B. Archer. 1983. A fibril protein antigen specific to *Spiroplasma.* J. Gen. Microbiol. *129:* 199-206.

Townsend, R., D.B. Archer and K.A. Plaskitt. 1980a. Purification and preliminary characterization of spiroplasma fibrils. J. Bacteriol. *142:* 694-700.

Townsend, R., J. Burgess and K.A. Plaskitt. 1980b. Morphology and ultrastructure of helical and non-helical strains of *Spiroplasma citri.* J. Bacteriol. *142:* 973-981.

Townsend, R., P.G. Markham, K.A. Plaskitt and M.J. Daniels. 1977. Isolation and characterization of a non-helical strain of *Spiroplasma citri.* J. Gen. Microbiol. *100:* 15-21.

Tramont, E.C. 1976. Specificity of inhibition of epithelial cell adhesion of *Neisseria gonorrhoeae.* Infect. Immun. *14:* 593-595.

Traub, A., J. Mager and N. Grossowicz. 1955. Studies on the nutrition of *Pasteurella tularensis.* J. Bacteriol. *70:* 60-69.

Traub, R. and C.L. Wisseman, Jr. 1974. The ecology of chigger-borne rickettsiosis (scrub typhus). J. Med. Entomol. *11:* 237-303.

Traub, R., C.L. Wisseman, Jr. and A. Farhang-Azad. 1978. The ecology of murine typhus - a critical review. Trop. Dis. Bull. *75:* 237-317.

Traub, R., C.L. Wisseman, Jr., M.R. Jones and J.J. O'Keefe. 1975. The acquisition of *Rickettsia tsutsugamuchi* by chiggers (trombiculid mites) during the feeding process. Ann. N.Y. Acad. Sci. *266:* 91-114.

Traub, W.H. 1972. Studies on group A bacteriocins of *Serratia marcescens:* preliminary characterization of two subgroups of bacteriocins. Zentralbl. Bakteriol. Parasitenkd. Infektionskr. Hyg. Abt. 1: Orig. Reihe A. *222:* 232-244.

Traub, W.H. 1980. Bacteriocin and phage typing of *Serratia. In* Von Graevenitz and Rubin (Editors), The genus *Serratia.* CRC Press, Boca Raton, Florida, pp. 79-100.

Traub, W.H. and P.I. Fukushima. 1979. Serotyping of *Serratia marcescens:* Current status of seven recently described flagellar (H) antigens. J. Clin. Microbiol. *10:* 56-63.

Tremblay, E. and L.E. Caltagirone. 1973. Fate of polar bodies in insects. Annu. Rev. Entomol. *18:* 421-444.

Tremblay, E. and G. Tripodi. 1980. Ultrastructural data on pseudococcid endosymbionts (Homoptera, Coccoidea). *In* Schwemmler and Schenk (Editors), Endocytobiology. Endosymbiosis and Cell Biology. A Synthesis of Recent Research. W deGruyter, Berlin, New York, pp. 419-423.

Trevisan, V. 1879. Prime linee d'introduzione allo studio dei Batterj italiani. Rend. Reale Ist. Lombardo Sci., Ser. II, 12: 133-151.

Trevisan, V. 1884. A proposito del bacillo del cholera. Koch o Pacini? Intorno al modo di algire del Bacille nel corpo umano. Gaz. Med. Ital. Milano (Ser. 8) *6:* 373-376.

Trevisan, V. 1885. Caratteri di alcuni nuovi generi di Batteriacee. Atti Accad. Fis-Med-Stat. Milano (Ser.4) *3:* 92-107.

Trevisan, V. 1885. Il fungo del cholera asiatico. Questioni risolte. Atti. Accad. Fis.-Med.-Stat. Milano (Ser. 4) *3:* 78-91.

Trevisan, V. 1887. Sul micrococco della rabbia e sulla possibilità di riconoscere durante il periode d'incubazione, dall'esame del sangue della persona moricata, se ha contratta l'infezione rabbica. Rend. Ist. Lombardo (Ser. 2) *20:* 88-105.

Trevisan, V. 1889. I generi e le specie delle Batteriacee. Zanaboni and Gabuzzi, Milan, pp. 1-35.

Tribondeau, L. and M. Fichet. 1916. Note sur les dysenteries des Dardanelles. Ann. Inst. Pasteur (Paris) *30:* 357-362.

Trinick, M.J. 1973. Symbiosis between *Rhizobium* and the non-legume *Trema aspera.* Nature (London) *244:* 459-460.

Trinick, M.J. 1976. *Rhizobium* symbiosis with a non-legume. *In* Newton and Nyman (Editors), Proceedings of the First International Symposium on Nitrogen Fixation. Washington State University Press, Pullman. pp. 507-517.

Trinick, M.J. 1980. Relationships among the fast-growing rhizobia of *Lablab purpureus, Leucaena leucocephala, Mimosa* spp., *Acacia farnesiana* and *Ses-*

bania grandiflora and their affinities with other rhizobial groups. J. Appl. Bacteriol. *49:* 39-53.

Truman, R. 1974. Die-back of *Eucalyptus citriodora* caused by *Xanthomonas eucalypti* sp. n. Phytopathology *64:* 143-144.

Trüper, H.G. and J.F. Imhoff. 1981. The genus *Ectothiorhodospira*. *In* Starr, Stolp, Trüper, Balows and Schlegel (Editors), The Prokaryotes, A handbook on habitats, isolation and identification of bacteria. Springer-Verlag, Berlin, pp. 274-278.

Trytek, R.E. and W.V. Allen. 1980. Synthesis of essential amino acids by bacterial symbionts in the gills of the shipworm *Bankia setacea* (Tryon) Comp. Biochem. Physiol. *67A:* 419-427.

Tsai, J.H. 1979. Vector transmission of mycoplasma agents of plant disease. *In* Whitcomb and Tully (Editors), The Mycoplasmas, Vol. 3, Academic Press, New York, pp. 265-307.

Tsui, F.-P., R. Schneerson and W. Egan. 1981a. Structural studies of the *Haemophilus influenzae* type e capsular polysaccharide. Carbohydr. Res. *88:* 85-92.

Tsui, F.-P., R. Schneerson, R.A. Boykins, A.B. Karpas and W. Egan. 1981b. Structural and immunological studies of the *Haemophilus influenzae* type d capsular polysaccharide. Carbohydr. Res. *97:* 293-306.

Tsukagoshi, N., M.H. Petersen and R.M. Franklin. 1975. Structure and synthesis of a lipid-containing bacteriophage. XVIII. Modification of the lipid composition of bacteriophage PM2. Virology *66:* 206-216.

Tuan, R.S. and K.-P. Chang. 1975. Isolation of intracellular symbiotes by immune lysis of flagellate Protozoa and characterization of their DNA. J. Cell. Biol. *65:* 309-323.

Tucker, D.N, I.J. Slotnick, E.O. King, B. Tynes, J. Nicholson and L. Crevasse. 1962. Endocarditis caused by a *Pasteurella*-like organism. N. Engl. J. Med. *267:* 913-916.

Tudor, J.J. 1980. Chemical analysis of the outer cyst wall and inclusion material of *Bdellovibrio* bdellocysts. Curr. Microbiol. *4:* 251-256.

Tudor, J.J. and S.F. Conti. 1977a. Characterization of bdellocysts of *Bdellovibrio* sp. J. Bacteriol. *131:* 314-322.

Tudor, J.J. and S.F. Conti. 1977b. Ultrastructural changes during encystment and germination of *Bdellovibrio*. J. Bacteriol. *131:* 323-330.

Tulasne, R. and J. Brisou. 1955. Les pleuropneumoniales. Taxonomie des pleuropneumonia-like organisms et des formes L. Ann. Inst. Pasteur (Paris) *88:* 237-239.

Tully, J.G. 1964. Production and biological characteristics of an extracellular neurotoxin from *Mycoplasma neurolyticum*. J. Bacteriol. *88:* 381–388.

Tully, J.G. 1965. Biochemical, morphological, and serological characterization of mycoplasma of murine origin. J. Infect. Dis. *115:* 171-185.

Tully, J.G. 1973. Biological and serological characteristics of acholeplasma. Ann. N.Y. Acad. Sci. *225:* 74-93.

Tully, J.G. 1979. Special features of the acholeplasmas. *In* Barile and Razin (Editors), The Mycoplasmas, Vol. I, Cell Biology, Academic Press, New York, pp. 431-449.

Tully, J.G. 1982. Interactions of spiroplasmas with plant, insect, and arthropod hosts. Rev. Infect. Dis. *4*(Suppl.): S193-199.

Tully, J.G., M.F. Barile, R.A. Del Giudice, T.R. Carski, D. Armstrong and S. Razin. 1970. Proposal for classifying strain PG-24 and related canine mycoplasmas as *Mycoplasma edwardii* sp. n. J. Bacteriol. *101:* 346-349.

Tully, J.G., M.F. Barile, D.G. ff. Edward, T.S. Theodore and H. Ernø. 1974. Characterization of some caprine mycoplasmas, with proposals for new species, *Mycoplasma capricolum* and *Mycoplasma putrefaciens*. J. Gen. Microbiol. *85:* 102-120.

Tully, J.G., R.A. Del Giudice and M.F. Barile. 1972. Synonymy of *Mycoplasma arginini* and *Mycoplasma leonis*. Int. J. Syst. Bacteriol. *22:* 47-49.

Tully, J.G. and R. Rask-Nielsen. 1967. Mycoplasma in leukemic and nonleukemic mice. Ann. N.Y. Acad. Sci. *143:* 345-352.

Tully, J.G. and S. Razin. 1968. Physiological and serological comparisons among strains of *Mycoplasma granularum* and *Mycoplasma laidlawii*. J. Bacteriol. *95:* 1504-1512.

Tully, J.G. and S. Razin. 1969. Characteristics of a new sterol-nonrequiring *Mycoplasma*. J. Bacteriol. *98:* 970-978.

Tully, J.G. and S. Razin. 1970. *Acholeplasma axanthum*, sp. n.: a new sterol-nonrequiring member of the *Mycoplasmatales*. J. Bacteriol. *103:* 751-754.

Tully, J.G. and S. Razin (Editors). 1983. Methods in Mycoplasmology, Vol. II. Diagnostic Mycoplasmology. Academic Press, New York.

Tully, J.G., D.L. Rose, O. Garcia-Jurado, J.C. Vignault, C. Saillard, J.M. Bové, R.E. McCoy and D.L. Williamson. 1980. Serological analysis of a new group of spiroplasmas. Curr. Microbiol. *3:* 369-372.

Tully, J.G., D.L. Rose, R.F. Whitcomb and R.P. Wenzel. 1979. Enhanced isolation of *Mycoplasma pneumoniae* from throat washings with a newly modified culture medium. J. Infect. Dis. *139:* 478-482.

Tully, J.G., D.L. Rose, C.E. Yunker, J. Cory, R.F. Whitcomb and D.L. Williamson. 1981. Helical mycoplasmas (spiroplasmas) from *Ixodes* ticks. Science *212:* 1043-1045.

Tully, J.G. and I. Ruchman. 1964. Recovery, identification and neurotoxicity of Sabin's type A and C mouse mycoplasma (PPLO) from lyophilized cultures. Proc. Soc. Exp. Biol. Med. *115:* 554-558.

Tully, J.G., R.F. Whitcomb, H F. Clark and D.L. Williamson. 1977. Pathogenic mycoplasmas: Cultivation and vertebrate pathogenicity of a new spiroplasma. Science *195:* 892-894.

Tully, J.G., R.F. Whitcomb, D.L. Rose and J.M. Bové. 1982. *Spiroplasma mirum*, a new species from the rabbit tick *Haemaphysalis leporispalustris*. Int. J. Syst. Bacteriol. *32:* 92-100.

Tunail, N. and H.G. Schlegel. 1972. Phosphoenolpyruvate, a new inhibitor of glucose-6-phosphate dehydrogenase. Biochem. Biophys. Res. Commun. *49:* 1554-1560.

Tunail, N. and H.G. Schlegel. 1974. A new coryneform hydrogen bacterium: *Corynebacterium autotrophicum* strain 7C. I. Characterization of the wild type strain. Arch. Microbiol. *100:* 341-350.

Tunevall, G. 1953. Studies on *Haemophilus influenzae* antigen studied by gel precipitation method. Acta Pathol. Microbiol. Scand. *32:* 193-197.

Tunnicliff, E.A. 1941. A study of *Actinobacillus lignieresi* from sheep affected with actinobacillosis. J. Infect. Dis. *69:* 52-58.

Tunnicliff, R. 1916. Streptothrix in bronchopneumonia of rats, similar to that in rat-bite fever. J. Infect. Dis. *19:* 767-771.

Tunnicliff, R. and L. Jackson. 1925. *Bacillus gonidiaformans* (n. sp.) An hitherto undescribed organism. J. Infect. Dis. *36:* 430-438.

Tuomi, J. 1966. Taxonomic position of pathogenic, tick-borne rickettsia-like organisms (in Finnish). Suomen Elainlaakarilehti 72: 415-422.

Tuomi, J. 1966. Studies on epidemiology of bovine tick-borne fever in Finland and a clinical description of field cases. Ann. Med. Exp. Biol. Fenn. *44:* Suppl. 6, 1-62.

Turk, D.C. and J.R. May. 1967. *Haemophilus influenzae*. Its clinical importance. The English Universities Press Ltd., London.

Turner, A.W. 1935. A study of the morphology and life cycles of the organism of *Pleuromoniae contagiosa boum* (*Borrelomyces peripneumoniae* nov. gen.) by observation in the living state under dark-ground illumination. J. Pathol. Bacteriol. *41:* 1-32.

Turner, A.W. 1954. Bacterial oxidation of arsenite. I. Description of bacteria isolated from arsenical cattle-dipping fluids. Aust. J. Biol. Sci. 7: 452-478.

Turner, A.W. 1960. Letter. Int. Bull. Bacteriol. Nomencl. Taxon. *10:* 255-256.

Turova, T.P. and G.F. Levanova. 1980. Genome characteristics of a new group of microorganisms belonging to the family *Vibrionaceae* (Russian). J. Microbiol. Epidemiol. Immunobiol. (Eng. Transl.) *3:* 27-29.

Turtura, G.C., F. Casaliccio and B. Biavati. 1973. Isolamento e identificazione di acetobatteri. Annali di Microbiologia ed Enzimologia 23: 157-164.

Tuyau, J.E. and W. Sims. 1975. Occurrence of haemophili in dental plaque and their association with neuraminidase activity. J. Dent. Res. *54:* 737-739.

Twarog, R. and L.E. Blouse. 1968. Isolation and characterization of transducing bacteriophage BP1 for a *Bacterium anitratum* (*Achromobacter* sp.). J. Virol. 2: 716-722.

Tweedy, J.M., R.W.A. Park and W. Hodgkiss. 1968. Evidence for the presence of fimbriae (pili) on *Vibrio* species. J. Gen. Microbiol. *51:* 235-244.

Tyeryar, F.J., Jr. and W.D. Lawton. 1969. Transformation in *Pasteurella novicida*. J. Bacteriol. *100:* 1112-1113.

Tyeryar, F.J., Jr. and W.D. Lawton. 1970. Factors affecting transformation of *Pasteurella novicida*. J. Bacteriol. *104:* 1312-1317.

Tyeryar, F.J., Jr., E. Weiss, D.B. Millar, F.M. Bozeman and R.A. Ormsbee. 1973. DNA base composition of rickettsiae. Science (Washington) *180:* 415-417.

Tyler, M.E., J.R. Milam, R.S. Smith, S.C. Schank and D.A. Zuberer. 1979. Isolation of *Azospirillum* from diverse geographic regions. Can. J. Microbiol. 25: 693-697.

Tyler, D.D. and L.K. Nakamura. 1971. Conditions for production of 3-ketomaltose from *Agrobacterium tumefaciens*. Appl. Microbiol. *21:* 175-180.

Tyzzer, E.E. 1938. *Cytoectes microti* N.G., (n. sp.) a parasite developing in granulocytes and infective for small rodents. Parasitology *30:* 242-257.

Tyzzer, E.E. 1942. A comparison study of *Grahamellae*, *Haemobartonellae* and *Eperythrozoa* in small mammals. Proc. Am. Phil. Soc. *85:* 359-398.

Tyzzer, E.E. and D. Weinman. 1939. *Haemobartonella* n. g. (*Bartonella olim* pro parte) *H. microti* n. sp. of the field vole, *Microtus pennsylvanicus*. Am. J. Hyg. *30:* 141-157.

Uchida, T., L. Bonen, H.W. Schaup, B.J. Lewis, L. Zablen and C.R. Woese. 1974. The use of ribonuclease U_2 in RNA sequence determination: some corrections in the catalog of oligomers produced by ribonuclease T1 digestion of *Escherichia coli* 16s ribosomal RNA. J. Mol. Evol. *3:* 63-77.

Ucke, A. 1898. Ein Beiträg zür Kenntnis der Anäeroben. Zentralbl. Bakteriol. Parasitenkd. Infektionskr. Hyg. Abt. I Orig. *23:* 996-1001.

Udaka, S. 1966. Pathway-specific pattern of control of arginine biosynthesis in bacteria. J. Bacteriol. *91:* 617-621.

Ueda, K., S. Ishikawa, T. Itami and T. Asai. 1952. Studies on the mesophilic cellulose-decomposing bacteria. Part 5-2. Taxonomical study on genus *Pseudomonas*. J. Agr. Chem. Soc. Japan 26: 35-41.

Uehara, K. and K. Arai. 1980. Canker of tea, a new disease and its causal bacterium *Xanthomonas campestris* pv. *theaecola* Uehara et Arai pv. nov. Bull. Fac. Agr. Kagoshima Univ. *30:* 17-21.

Ui, M., T. Katada and M. Yajima. 1979. Islet-activating protein in *Bordetella pertussis*: purification and mechanism of action. *In* Manclark and Hill (Editors), International Symposium on Pertussis. U.S. Government Printing Office, Washington, D.C., pp. 166-173.

Ulitzer, S. 1974. Induction of swarming in *Vibrio parahaemolyticus*. Arch. Microbiol. *101:* 357-363.

Ulitzer, S. 1975a. The mechanism of swarming of *Vibrio alginolyticus*. Arch. Microbiol. *104:* 67-71.

Ulitzer, S. 1975b. Effect of temperature, salts, pH and other factors on the development of peritrichous flagella in *Vibrio alginolyticus*. Arch. Microbiol. *104:* 285-288.

Ulitzer, S. and J.W. Hastings. 1978. Myristic acid stimulation of bacterial bioluminescence in "aldehyde" mutants. Proc. Natl. Acad. Sci. U.S.A. *75:* 266-269.

Ulitzer, S. and J.W. Hastings. 1980. Reversible inhibition of bacterial bioluminescence by long-chain fatty acids. Curr. Microbiol. *3:* 295-300.

Ulitzer, S. and M. Kessel. 1973. Giant flagellar bundles of *Vibrio alginolyticus* (NCMB 1803). Arch. Mikrobiol. *94:* 331-339.

Ullman, J.S. and B.J. McCarthy. 1973. The relationship between mismatched base pairs and the thermal stability of DNA duplexes. II. Effects of deamination of cytosine. Biochim. Biophys. Acta 294: 416-424.

Ullmann, U. 1979. Methods in *Campylobacter*. *In* Bergan and Norris (Editors). Methods in Microbiology. Vol 13. Academic Press, New York.

Ulloa, M. and T. Herrera. 1972. Descripcion de dos especies nuevas de bacterias aisladas del pozol: *Agrobacterium azotophilum* y *Achromobacter pozolis*. Rev. Latinoam. Microbiol. *14:* 15-24.

Umali-Garcia, M., D.H. Hubbell, M.H. Gaskins and F.B. Dazzo. 1980. Association of *Azospirillum* with grass roots. Appl. Environ. Microbiol. *39:* 219-226.

Umbreit, W.W., R.H. Burriss and J.F. Stauffer. 1972. Manometric techniques, 5th ed. Burgess Publishing Co., Minneapolis.

Unden, G., H. Hackenberg and A. Kröger. 1980. Isolation and functional aspects of the fumarate reductase involved in the phosphorylative electron transport of *Vibrio succinogenes*. Biochim. Biophys. Acta *591:* 275-288.

Unemoto, T. and M. Hayashi. 1979. NADH:quinone oxidoreductase as a site of Na$^+$-dependent activation in the respiratory chain of marine *Vibrio alginolyticus*. J. Biochem. *85:* 1461-1467.

Unemoto, T., M. Hayashi, Y. Kozuka and M. Hayashi. 1974. Localizations and salt modifications of phosphohydrolases in slightly halophilic *Vibrio alginolyticus*. *In* Colwell and Morita (Editors), Effect of the Ocean Environment on Microbial Activities, University Park Press, Baltimore-London-Tokyo, pp. 46-79.

Unger, L., A.K.M.M. Rahman and R.D. DeMoss. 1961. Anaerobic dissimilation of glucose by *Vibrio comma*. Can. J. Microbiol. *7:* 844-847.

Unz, R.F. 1971. Neotype strain of *Zoogloea ramigera* Itzigsohn. Request for an opinion. Int. J. Syst. Bacteriol. *21:* 91-99.

Unz, R.F. and N.C. Dondero. 1967a. The predominant bacteria in natural zoogloeal colonies. I. Isolation and identification. Can. J. Microbiol. *13:* 1671-1682.

Unz, R.F. and N.C. Dondero. 1967b. The predominant bacteria in natural zoogloeal colonies. II. Physiology and nutrition. Can. J. Microbiol. *13:* 1683-1691.

Unz, R.F. and S.R. Farrah. 1972. Use of aromatic compounds for growth and isolation of *Zoogloea*. Appl. Microbiol. *23:* 524-530.

Unz, R.F. and S.R. Farrah. 1976a. Observations on the formation of wastewater zoogloeae. Water Res. *10:* 665-671.

Unz, R.F. and S.R. Farrah. 1976b. Exopolymer production and flocculation by *Zoogloea* MP6. Appl. Environ. Microbiol. *31:* 623-626.

Urošević, B. 1966. Canker of poplar caused by *Erwinia cancerogena* n. sp. (Czech.). Lesn. Čas. *12:* 493-505.

Ursing, J. 1981. Deoxyribonucleic acid hybridization studies of gas producing pasteurellae. *In* Kilian, Frederiksen and Biberstein (Editors), *Haemophilus, Pasteurella* and *Actinobacillus*, Academic Press, London and New York, pp. 255-263.

Ursing, J., D.J. Brenner, H. Bercovier, G.R. Fanning, A.G. Steigerwalt, J. Brault and H.H. Mollaret. 1980a. *Yersinia frederiksenii*: a new species of *Enterobacteriaceae* composed of rhamnose positive strains (formerly called atypical *Yersinia enterocolitica* or *Yersinia enterocolitica*-like). Curr. Microbiol. *4:* 213-217.

Ursing, J., D.J. Brenner, H. Bercovier, G.R. Fanning, A.G. Steigerwalt, J. Brault and H.H. Mollaret. 1981. *In* Validation of the publication of new names and new combinations previously effectively published outside the IJSB. List No. 6. Int. J. Syst. Bacteriol. *31:* 215-218.

Ursing, J., A.G. Steigerwalt and D.J. Brenner. 1980b. Lack of genetic relatedness between *Yersinia philomiragia* (The "Philomiragia" bacterium) and *Yersinia* species. Curr. Microbiol. *4:* 231-233.

Úrvölgyi, J. and R. Brezina. 1978. *Rickettsia slovaca*: A new member of spotted fever group rickettsiae. *In* Kazar, Ormsbee and Tarasevich (Editors), Rickettsiae and Rickettsial Diseases, VEDA, Bratislava, pp. 299-305.

Vacelet, J. 1975. Étude en microscopie electronique de l'association entre bactéries et spongiaires du genre *Vérongia* (Dictyoceratida). J. Microscop. Biol. Cell. *23:* 272-288.

Vago, C. and R. Martoja. 1963. Une rickettsiose chez les Gryllidae (Orthoptera). C.R. Acad. Science Ser. D *256:* 1945-1947.

Vago, C. and G. Meynadier. 1965. Une rickettsiose chez le criquet pelerin (*Schistocerca gregaria* Forsk.). Entomophaga *10:* 307-310.

Vago, C., G. Meynadier, P. Juchault, J.-J. Legrand, A. Amargier and J.-J. Duthoit. 1970. Une maladie rickettsienne chez les crustaces isopodes. C.R. Acad. Science Ser. D *271:* 2061-2063.

Vakimoff, W.L. and W.S. Belawine. 1927. L'anaplasmose des bovides en Russie (USSR). Zentralbl. Bakteriol. Parasitenkd. Infektionskr. Hyg. Abt. I Orig. *103:* 419-421.

Valentine, F.C.O. and T.M. Rivers. 1927. Further observations concerning growth requirements of hemophilic bacilli. J. Exp. Med. *45:* 993-1002.

Valleé, A. 1959. Isolement de *Bacterium viscosum equi* chez deux lapins domestiques. Recl. Med. Vet. Ec. Alfort *135:* 821-822.

Valleé, A., J. Durieux, M. Durieux and B. Virat. 1960. Étude d'une pyodermite particulièrement rebelle chez le chien isolément d'*Actinobacillus equuli* associé à un staphylocoque. Bull. Acad. Vet. Fr. *33:* 153-156.

Valleé, A. and J.-A. Gaillard. 1953. Infection pyogène contagieuse de la souris determinée par *Bacillus actinomycetem comitans*. Ann. Inst. Pasteur (Paris) *84:* 647-649.

Valleé, A., M. Piéchaud, P. Destombes and L. Second. 1959. Enzootie d'adénites suppurées chez le rat blanc provoquée par un actinobacille. Ann. Inst. Pasteur (Paris) *97:* 346-352.

Valleé, A., P. Thibault and L. Second. 1963. Contribution à l'étude d'*A. lignieresii* et d'*A. equuli*. Ann. Inst. Pasteur (Paris) *104:* 108-114.

Valleé, A., R. Tinelli, J.-C. Guillon, A. le Priol and T. Cuong. 1974. Étude d'un actinobacillus isolé chez un cheval. Recl. Med. Vet. Ec. Alfort *150:* 695-700.

Van Assche, P.F. 1978. Differentiation of *Bacteroides fragilis* species by gas chromatographic detection of phenylacetic acid. J. Clin. Microbiol. *8:* 614-615.

Van Assche, P.F. and A.T. Wilssens. 1977. *Fusobacterium perfoetens* (Tissier) Moore and Holdeman 1973: description and proposed neotype strain. Int. J. Syst. Bacteriol. *27:* 1-5.

Van Beeumen, J. and J. De Ley. 1968. Hexopyranoside:cytochrome c oxidoreductase from *Agrobacterium tumefaciens*. Eur. J Biochem. *6:* 331-343.

Van Beeumen, J., P. Tempst, P. Stevens, D. Bral, J. Van Damme and J. De Ley. 1980. Cytochromes *c* of two different sequence classes in *Agrobacterium tumefaciens*. *In* Peeters (Editor), Protides of the Biological Fluids 28. Pergamon Press, Oxford, pp. 69-74.

van Bijsterveld, O.P. 1970. A new *Moraxella* strain isolated from angular conjunctivitis. Appl. Microbiol. *20:* 405-408.

Vančura, V., Y. Abd-el-Malek and M.N. Zayed. 1965. *Azotobacter* and *Beijerinckia* in the soils and rhizosphere of plants in Egypt. Folia Microbiol. *10:* 224-228.

van Damme, P.A., A.G. Johannes, H.C. Cox and W. Berends. 1960. On toxoflavin, the yellow poison of *Pseudomonas cocovenenans*. Recl. Trav. Chim. Pays-Bas Belg. *79:* 255-267.

van Delden, A. 1903. Beiträge zür Kenntnis der Sulfatredultion durch Bakterien. Zentralbl. Bakteriol. Parasitenkd. Infektionskr. Hyg. Abt. II *11:* 81-94.

Vanden Abeele, P., C. Van Keer, J. Swings, F. Gosselé and J. De Ley. 1980. Browning and rotting of apples caused by acetic acid bacteria. Meded. Fac. Landbouwwet. Gent. *45:* 391-397.

Vandepitte, J., J. Colaert, J. and C. Lamotte-Legrand and F. Perrin. 1953. Les ostéites à *Salmonella* chez les sicklanémiques: à propos de 5 observations. Ann. Soc. Belge Méd. Trop. *33:* 511-522.

Vandepitte, J., A. Makulu and F. Gatti. 1974. *Plesiomonas shigelloides*. Survey and possible association with diarrhoea in Zaire. Ann. Soc. Belge Med. Trop. *54:* 503-513.

Vandepitte, J., L. van Damme, Y. Fofana and J. Desmyter. 1980. *Edwardsiella tarda* et *Plesiomonas shigelloides*. Leur rôle comme agents de diarrhées et leur épidemiologie. Bull. Soc. Pathol. Exot. *73:* 139-149.

Van der Schaaf, A. and M. Rosa. 1940. Brucellosis oncho-cerciasis in verband met een chronisch gewrichtslijden bij runderen. Ned.-Ind. Blad. Diergeneesk. *52:* 1-20.

Vanderzant, C., R. Nickelson and J.C. Parker. 1971. Isolation of *Vibrio parahaemolyticus* from gulf coast shrimp. J. Milk Food Technol. *33:* 161-162.

van Dorssen, C.A. and F.H.J. Jaartsveld. 1962. *Actinobacillus suis* (novo species), een bij het varken voorkomende bactérie. Tijdschr. Diergeneeskd. *87:* 450-458.

Van Drimmelen, G.C. 1953. *Brucella melitensis* isolated from karakul sheep in south-west Africa. S. Afr. J. Sci. *49:* 299-302.

Van Ert, M. and J.T. Staley. 1971. Gas vacuolated strains of *Microcyclus aquaticus*. J. Bacteriol. *108:* 236-240.

Van Eys, J. 1960. Pyridine ribosidase from *Xanthomonas pruni*. J. Bacteriol. *80:* 386-393.

Van Golde, L.M.G., J. Akkermans-Kruyswijk, W. Franklin-Klein, A. Lankhorst and R.A. Prins. 1975. Accumulation of phosphatidylserine in strictly anaerobic lactate fermenting bacteria. FEBS Letters *53:* 57-60.

Van Gylswyk, N.O. 1980. *Fusobacterium polysaccharolyticum* sp. nov., a Gram-negative rod from the rumen that produces butyrate and ferments cellulose and starch. J. Gen. Microbiol. *116:* 157-163.

Van Gylswyk, N.O., E.J. Morris and H.J. Els. 1980. Sporulation and cell wall structure of *Clostridium polysaccharolyticum* comb. nov. (formerly *Fusobacterium polysaccharolyticum*). J. Gen. Microbiol. *121:* 491-493.

van Hall, C.J.J. 1902. Bijdragen tot de kennis der Bakterieele Plantenziekten. Inaug. Diss., Amsterdam.

Van Heyningen, S. 1977. Cholera toxin. Biol. Rev. *52:* 509-549.

van Hove, C. 1976. Bacterial leaf symbiosis and nitrogen fixation. *In* Nutman (Editor), Symbiotic Nitrogen Fixation in Plants, Cambridge University Press, Cambridge, U.K., pp. 551-560.

van Huyssteen, J.J. 1967. Gas chromatographic separation of digester gases using porous polymers. Water Res. *1:* 237-242.

Van Keer, C., K. Kersters and J. De Ley. 1976. L-Sorbose metabolism in *Agrobacterium tumefaciens*. Antonie van Leeuwenhoek J. Microbiol. Serol. *42:* 13-24.

Van Keer, C., P. Vanden Abeele, J. Swings, F. Gosselé and J. De Ley. 1981. Acetic acid bacteria as causal agents of browning and rot of apples and pears.

Zentralbl. Bakteriol. Parasitenkd. Infektionskr. Hyg. Abt. I Orig. C 2: 197-204.

van Klingeren, B., J.D.A. van Embden and M. Dessons-Kroon. 1977. Plasmid-mediated chloramphenicol resistance in *Haemophilus influenzae*. Antimicrob. Agents Chemother. 11: 383-387.

Van Larebeke, N., G. Engler, M. Holsters, S. Van Den Elsacker, I. Zaenen, R.A. Schilperoort and J. Schell. 1974. Large plasmid in *Agrobacterium tumefaciens* essential for crown gall-inducing ability. Nature (London) 252: 169-170.

Van Larebeke, N., C. Genetello, J. Schell, R.A. Schilperoort, A.K. Hermans, J.P. Hernalsteens and M. Van Montagu. 1975. Acquisition of tumour-inducing ability by non-oncogenic agrobacteria as a result of plasmid transfer. Nature (London) 255: 742-743.

Van Loghem, J.J. 1944. The classification of plague bacillus. Antonie van Leeuwenhoek J. Serol. Microbiol. 10: 15-16.

Van Montagu, M., M. Holsters, P.Z. O'Farrell, J.P. Hernalsteens, A. Depicker, M. De Beuckeleer, G. Engler, M. Lemmers, L. Willmitzer and J. Schell. 1980. The interaction of *Agrobacterium* Ti-plasmid DNA and plant cells. Proc. R. Soc. Lond. B 210: 351-365.

Van Montagu, M. and J. Schell. 1979. The plasmids of *Agrobacterium tumefaciens*. *In* Timmis and Pühler (Editors), Plasmids of Medical, Environmental and Commercial Importance. Elsevier/North-Holland Biomedical Press, Amsterdam, pp. 71-95.

van Niel, C.B. and M. Allen. 1952. A note on *Pseudomonas stutzeri*. J. Bacteriol. 64: 413-422.

van Oye, E. 1964. The world problem of salmonellosis. Junk, The Hague.

van Oye, E., M. Thevelin and C. Richard. 1975. Antigenic relationships between *Levinea amalonatica* and *Shigella dysenteriae* and *boydii*. Ann. Microbiol. 126A: 187-192.

Van Palenstein Helderman, W.H. and I. Rosman. 1976. Hydrogen-dependent organisms from the human gingival crevice resembling *Vibrio succinogenes*. Antonie van Leeuwenhoek J. Serol. Microbiol. 42: 107-118.

Van Pee, W. and J. Stragier. 1979. Evaluation of some cold enrichment and isolation media for the recovery of *Yersinia enterocolitica*. Antonie van Leeuwenhoek J. Serol. Microbiol. 45: 465-477.

Van Pee, W. and J. Swings. 1971. Chemical and microbiological studies on congolese palm wines (*Elaeis guineensis*). E. Afr. Agr. For. J. 36: 311-314.

Van Pee, W., M. Vanlaar and J. Swings. 1974. The nutrition of *Zymomonas*. Acad. R. Sci. Outre-Mer (Brussels) Bull. Séances 2: 206-211.

Van Rooyen, C.E. 1936. Biology, pathogenesis and classification of *Streptobacillus moniliformis*. J. Pathol. Bacteriol. 43: 455-472.

Van Steenbergen, T.J.M. 1981. Classification and virulence of black-pigmented *Bacteroides* strains. Vrije University, Amsterdam.

Van Steenbergen, T.J.M., J.J. de Soet and J. de Graaff. 1979. DNA base composition of various strains of *Bacteroides melaninogenicus*. FEMS Microbiol. Letters 5: 127-130.

Van Steenbergen, T.J.M., C.A. Vlaanderen and J. de Graaff. 1981. Confirmation of *Bacteroides gingivalis* as a species distinct from *Bacteroides asaccharolyticus*. Int. J. Syst. Bacteriol. 31: 236-241.

van Straaten, H. 1918. Bacteriologische bevindingen bij eenige gevallen van pyosepticaemie (Lähme) der veulens. Verslag van den Werksaambeden der Rijksseruminrichting voor 1916-1917, Rotterdam. pp. 71-76.

van Tonder, E.M. 1979. *Actinobacillus seminis* infection in sheep in the Republic of South Africa. III. Growth and cultural characteristics of *A. seminis*. Onderstepoort J. Vet. Res. 46: 141-148.

Van Verseveld, H.W. and A.H. Stouthamer. 1978. Growth yields and the efficiency of oxidative phosphorylation during autotrophic growth of *Paracoccus denitrificans* on methanol and formate. Arch. Microbiol. 118: 27-34.

van Vuuren, H.J.J. 1978. Identification and physiology of *Enterobacteriaceae* isolated from South African lager beer breweries. Ph.D. thesis, Rijksuniversiteit Gent, Belgium.

van Vuuren, H.J.J., K. Kersters, J. De Ley and D.F. Toerien. 1981. The identification of *Enterobacteriaceae* from breweries: Combined use and comparison of API 20E system, gel electrophoresis of proteins and gas chromotography of volatile metabolites. J. Appl. Bacteriol. 51: 51-65.

Vardanis, A. and R.M. Hochster. 1961. On the mechanism of glucose metabolism in the plant tumor-inducing organism *Agrobacterium tumefaciens*. Canad. J. Biochem. Physiol. 39: 1165-1182.

Varel, V.H. and M.P. Bryant. 1974. Nutritional features of *Bacteroides fragilis* subsp. *fragilis*. Appl. Microbiol. 28: 251-257.

Varela, G. and A. Aparicio. 1951. Intestinal bacteria found in *Triatoma* and *Ornithodoros*. Am. J. Trop. Med. Hyg. 31: 381-382.

Varela, G., J.W. Vinson and C. Molina-Pasquel. 1969. Trench fever. II. Propagation of *Rickettsia quintana* on cell-free medium from the blood of two patients. Am. J. Trop. Med. Hyg. 18: 708-712.

Vargo, V., M. Korzeniowski and E.H. Spaulding. 1974. Tryptic soy bile-kanamycin test for the identification of *Bacteroides fragilis*. Appl. Microbiol. 27: 480-483.

Varney, P.L. 1927. The serological classification of fusiform bacilli. J. Bacteriol. 13: 275-314.

Varon, M. and R. Levisohn. 1972. Three-membered parasitic system: a bacteriophage, *Bdellovibrio bacteriovorus*, and *Escherichia coli*. J. Virol. 9: 519-525.

Varon, M. and M. Shilo. 1980. Ecology of aquatic bdellovibrio. *In* Droop and Jannasch (Editors), Advances in Aquatic Microbiology, Vol. 2, Academic Press, New York, pp. 1-48.

Varon, M. and B. P. Zeigler. 1978. Bacterial predator-prey interaction at low prey density. Appl. Environ. Microbiol. 36: 11-17.

Vasil, M.L., R.K. Holmes and R.A. Finkelstein. 1975. Conjugal transfer of a chromosomal gene determining production of enterotoxin in *Vibrio cholerae*. Science 187: 849-850.

Vasstrand, E.N. 1981. Lysozyme digestion and chemical characterization of the peptidoglycan of *Fusobacterium nucleatum* Fev 1. Infect. Immun. 33: 75-82.

Vasstrand, E.N., H.B. Jensen, T. Miron and T. Hofstad. 1982. Composition of peptidoglycans in *Bacteroidaceae*: determination and distribution of lanthionine. Infect. Immun. 36: 114-122.

Vastine, D.W., C.R. Dawson, I. Hoshiwara, C. Yoneda, T. Daghfous and M. Messadi. 1974. Comparison of media for the isolation of *Haemophilus* species from cases of seasonal conjunctivitis associated with severe endemic trachoma. Appl. Microbiol. 28: 688-691.

Vaughn, R.H. 1942. The acetic acid bacteria. Wallerstein Lab. Comm. 5: 5-26.

Vaughn, R.H. and M. Levine. 1942. Differentiation of the "intermediate" coli-like bacteria. J. Bacteriol. 44: 487-505.

Vedros, N.A. 1978. Serology of the meningococcus. *In* Bergan and Norris (Editors), Methods in Microbiology, Vol. 10, Academic Press, Inc., New York, pp. 293-314.

Vedros, N.A. 1981. The Genus *Neisseria*. *In* Starr, Stolp, Trüper, Balows and Schlegel (Editors), The Prokaryotes: a handbook of habitats, isolation, and identification of bacteria. Springer-Verlag, Berlin, pp. 1497-1505.

Vedros, N.A., D.G. Johnston and P.I. Warren. 1973. *Neisseria* species isolated from dolphins. J. Wild. Dis. 9: 241-244.

Vedros, N.A., N. Ng and G. Culver. 1968. A new serological Group (E) of *Neisseria meningitidis*. J. Bacteriol. 95: 1300-1304.

Vedros, N.A., J. Quinlivan and R. Cranford. 1982. Bacterial and fungal flora of wild northern für seals (*Callorhinus ursinus*). J. Wildl. Dis. 18: 447-456.

Veillon, A. and A. Zuber. 1898. Recherches sur quelques microbes strictement anaérobies et leur role en pathologie. Arch. Med. Exp. 10: 517-545.

Veivers, P.C., R.W. O'Brien and M. Slaytor. 1980. The redox state of the gut of termites. J. Insect Physiol. 26: 75-77.

Vela, G.R. and O. Wyss. 1964. Improved stain for visualization of *Azotobacter* encystment. J. Bacteriol. 87: 467-477.

Veldkamp, H. 1960. Isolation and characteristics of *Treponema zuelzerae* nov. spec., an anaerobic, free-living spirochete. Antonie van Leeuwenhoek J. Microbiol. Serol. 26: 103-125.

Venezia, R.A. and R.G. Robertson. 1975. Bactericidal substance produced by *Haemophilus influenzae* b. Can. J. Microbiol. 21: 1587-1594.

Verder, E. and J. Evans. 1961. A proposed antigenic schema for the identification of strains of *Pseudomonas aeruginosa*. J. Infect. Dis. 109: 183-193.

Verger, J.M. and M. Grayon. 1977. Oxidative metabolic profiles of *Brucella* species. Ann. Sclavo 19: 45-60.

Verger, J.M., M. Grayon, M.P. Doutre and F. Sagna. 1979. *Brucella abortus* d'origine bovine au Sénégal: identification et typage. Rev. Elev. Med. Vet. Pays. Trop. 32: 25-32.

Verhoeven, W., A.L. Koster and M.C.A. van Nievelt. 1954. Studies on true dissimilatory nitrate reduction. III. *Micrococcus denitrificans* Beijerinck, a bacterium capable of using molecular hydrogen in denitrification. Antonie van Leeuwenhoek J. Microbiol. Serol. 20: 273-284.

Verkley, A.J., P.H.J.Th. Ververgaert, R.A. Prins and L.M.G. van Golde. 1975. Lipid-phase transitions of the strictly anaerobic bacteria *Veillonella parvula* and *Anaerovibrio lipolytica*. J. Bacteriol. 124: 1522-1528.

Véron, M.M. 1965. La position taxonomique des *Vibrio* et de certaines bactéries comparables. C.R. Acad. Sci. Paris 261: 5243-5246.

Véron, M. 1966. Taxonomie numérique des vibrions et de certaines bactéries comparables. II. Corrélation entre les similitudes phénétiques et al composition en bases de l'AND. Ann. Inst. Pasteur Paris 111: 671-709.

Véron, M. 1975. Nutrition et taxonomie des *Enterobacteriaceae* et bactéries voisines. I. Méthode d'étude des auxanogrammes. Ann. Microbiol. Inst. Pasteur (Paris) 126A: 267-274.

Véron, M. and P. Berche. 1976. Virulence et antigènes de *Pseudomonas aeruginosa*. Bull. Inst. Pasteur 74: 295-337.

Véron, M. and R. Chatelain. 1973. Taxonomic study of the genus *Campylobacter* Sebald and Véron and designation of the neotype strain for the type species, *Campylobacter fetus* (Smith and Taylor) Sebald and Véron. Int. J. Syst. Bacteriol. 23: 122-134.

Véron, M. and L. Le Minor. 1975a. Nutrition et taxonomie des *Enterobacteriaceae* et bactéries voisines. II. Résultats d'ensemble et classification. Ann. Microbiol. Inst. Pasteur (Paris) 126B: 111-124.

Véron, M. and L. Le Minor. 1975b. Nutrition et taxonomie des *Enterobacteriaceae*. III. Caracteres nutritionnels et différenciation des groupes taxonomiques. Ann. Microbiol. Inst. Pasteur (Paris) 126B: 125-147.

Véron, M., P. Thibault and L. Second. 1961. *Neisseria mucosa* (*Diplococcus mucosus* Lingelsheim). II. Étude antigénique et classification. Ann. Inst. Pasteur (Paris) 100: 166-179.

Vervliet, G., M. Holsters, H. Teuchy, M. Van Montagu and J. Schell. 1975. Characterization of different plaque-forming and defective temperate phages in *Agrobacterium* strains. J. Gen. Virol. 26: 33-48.

Vicente, M. and J.L. Cánovas. 1973. Glucolysis in *Pseudomonas putida*: physiological role of alternative routes from the analysis of defective mutants. J. Bacteriol. 116: 908-914.

Vickerstaff, J.M. and B.C. Cole. 1969. Characterization of *Haemophilus vaginalis*,

Corynebacterium cervicis, and related bacteria. Can. J. Microbiol. *15:* 587-594.

Victoria, J.I. and O. Barros. 1969. Etiologia de una nueva enfermdad bacterial del plátano (*Musa paradisiaca* L.) en Colombia. Inst. Colomb. Agropecu. Revista ICA *4:* 173-190.

Vidaver, A.K. 1976. Prospects for control of phytopathogenic bacteria by bacteriophages and bacteriocins. Annu. Rev. Phytopathol. *14:* 451-465.

Vidaver, A.K. and S. Buckner. 1978. Typing of fluorescent phytopathogenic pseudomonads by bacteriocin production. Can. J. Microbiol. *24:* 14-18.

Vidaver, A.K., R.K. Koski and J.L. van Etten. 1973. Bacteriophage φ6: a lipid-containing virus of *Pseudomonas phaseolicola.* J. Virol. *11:* 799-805.

Vidaver, A.K., M.L. Mathys, M.E. Thomas and M.L. Schuster. 1972. Bacteriocins of the phytopathogens *Pseudomonas syringae, P. glycinea* and *P. phaseolicola.* Can. J. Microbiol. *18:* 705-713.

Vieu, J-F. 1963. Distribution de la lysogenie parmi les *Proteus* et les *Providencia.* C. R. Acad. Sci. Paris *256:* 4317-4319.

Vieu, J-F. and M. Capponi. 1965. Lysotypie des *Proteus* OX19, OXK, OX2 and OXL. Ann. Inst. Pasteur (Paris) *108:* 103-106.

Vieu, J.F., O. Croissant and C. Dauguet. 1965. Structure des bactériophages responsables des phénomènes de conversion chez les *Salmonella.* Ann. Inst. Pasteur (Paris) *109:* 160-166.

Vignault, J.C., J.M. Bové, C. Saillard, R. Vogel, A. Farro, L. Venegas, W. Stemmer, S. Aoki, R. McCoy, A.S. Al-Beldawi, M. Larue, O. Tuzcu, M. Ozsam, A. Nhami, M. Abassi, J. Bonfils, G. Moutous, A. Fos, F. Poutiers and G. Viennot-Bourgin. 1980. Mise en culture de spiroplasmes à partir de matériel végétal et d'insectes provenant de pays circum-méditerraneens et du Proche-Orient. C.R. Acad. Sci. Paris Ser. D *290:* 775-778.

Vincent, J.M. 1970. A manual for the practical study of root nodule bacteria. International Biological Programme Handbook No. 12., Blackwell Scientific Pub. Oxford and Edinburgh.

Vincent, J.M. 1974. Root nodule symbioses with *Rhizobium. In* Quispel (Editor), The Biology of Nitrogen Fixation. North-Holland Pub. Co. Amsterdam, pp. 265-341.

Vincent, J.M. 1977. *Rhizobium* - General Microbiology. *In* Hardy and Silver (Editors), A Treatise in Dinitrogen Fixation, Section III. J. Wiley and Sons, New York, pp. 277-366.

Vincent, J.M. and B.A. Humphrey. 1970. Taxonomically significant group antigens in *Rhizobium.* J. Gen. Microbiol. *63:* 379-382.

Vincent, J.M., P.S. Nutman and F.A. Skinner. 1979. The identification and classification of *Rhizobium. In* Skinner and Lovelock (Editors), Identification Methods for Microbiologists. Soc. Appl. Bacteriol. Tech. Ser. 14, 2nd Ed., Academic Press, New York, London.

Vincent, A.L., J.K. Portaro and L.R. Ash. 1975. A comparison of the body wall ultrastructure of *Brugia pahangi* with that of *Brugia malayi.* J. Parasitol. *61:* 567-570.

Vinson, J.W. 1966. In vitro cultivation of the rickettsial agent of trench fever. Bull. WHO *35:* 155-164.

Vinson, J.W. and H.S. Fuller. 1961. Studies on trench fever I. Propagation of rickettsia-like microorganisms from a patient's blood. J. Path. Microbiol. *24:* 152-166.

Vinson, J.W., G. Varela and C. Molina-Pasquel. 1969. Trench fever III. Induction of clinical disease in volunteers inoculated with *R. quintana* propagated on blood agar. Am. J. Trop. Med. Hyg. *18:* 713-722.

Vinther, O. 1976. Localization of urease activity in *Ureaplasma urealyticum* cells. Acta Pathol. Microbiol. Scand., Sect. B *84:* 217-224.

Vinther, O. and F.T. Black. 1974. Aminopeptidase activity of *Ureaplasma urealyticum.* Acta Pathol. Microbiol. Scand., Sect. B *82:* 917-918.

Virkola, P. 1972. The growth and morphology of *Acholeplasma (Mycoplasma) laidlawii* in different media. Acta Pathol. Microbiol. Scand. B. *80:* 388-396.

Virtanen, A.I. and B. Bärlund. 1926. Die Oxydation des glycerins zu Dioxyaceton durch Bakterien. Biochem. Z. *169:* 169-177.

Viscontini, M. and M. Frater-Schröder. 1968. Isolierung von 6-Hydroxy-methylpterin aus kulturen von *Pseudomonas roseus-fluorescens* J.C. Marchal 1937. Helv. Chim. Acta *51:* 1554-1557.

Vishniac, W. and P.A. Trudinger. 1962. Symposium on autotrophy. V. Carbon dioxide fixation and substrate oxidation in chemosynthetic sulfur and hydrogen bacteria. Bacteriol. Rev. *26:* 168-175.

Visser't Hooft, F. 1925. Biochemische Onderzoekingen over het geslacht *Acetobacter.* Diss. Techn. Univ., Meinema, Delft, pp. 1-129.

Voges, O. 1893. Ueber einige im Wasser vorkommende Pigmentbakterien. Zentralbl. Bakteriol. Parasitenk. Infektionskr. Hyg. Abt. I, *14:* 301-314.

Vogt, M. 1965. Wachstumsphysiologische Untersuchungen an *Micrococcus denitrificans* Beij. Arch. Mikrobiol. *50:* 256-281.

Volk, W.A. 1966. Cell wall lipopolysaccharides from *Xanthomonas* spp. J. Bacteriol. *91:* 39-42.

Volk, W.A. 1968a. Isolation of D-galacturonic acid 1-phosphate from hydrolysates of cell wall lipopolysaccharide extracted from *Xanthomonas campestris.* J. Bacteriol. *95:* 782-786.

Volk, W.A. 1968b. Quantitative assay of polysaccharide components obtained from cell wall lipopolysaccharides of *Xanthomonas* species. J. Bacteriol. *95:* 980-982.

Volk, W.A., N.L. Salmonsky and D. Hunt. 1972. *Xanthomonas sinensis* cell wall lipopolysaccharides 1. Isolation of 4,7-anhydro- and 4,8-anhydro-3-deoxy octulosonic acid following hydrolysis. J. Biol. Chem. *247:* 3881-3887.

Volpon, A.G., H. De-Polli and J. Döbereiner. 1981. Physiology of nitrogen fixation in *Azospirillum lipoferum* Br 17 (ATCC 29709). Arch. Microbiol. *128:* 371-375.

von Faber, F.C. 1912. Das erbliche Zusammenleben von Bakterien und tropischen Pflanzen. Jarhb. Wiss. Bot. *51:* 285-375.

von Graevenitz, A. 1977. The role of opportunistic bacteria in human disease. Annu. Rev. Microbiol. *31:* 447-471.

von Graevenitz, A. 1978. Clinical role of infrequently encountered nonfermenters. *In* Gilardi (Editor), Glucose Nonfermenting Gram-negative Bacteria in Clinical Microbiology. CRC Press Inc., West Palm Beach, pp. 119-153.

von Graevenitz, A. 1981. Clinical significance and antimicrobial susceptibility of flavobacteria. *In* Reichenbach and Weeks (Editors), The *Flavobacterium-Cytophaga* Group (Proceedings of the International Symposium on Yellow-Pigmented Gram-Negative Bacteria of the *Flavobacterium-Cytophaga* Group, Braunschweig, July 8 to 11, 1980). Verlag Chemie, Weinheim, pp. 153-164.

von Graevenitz, A. and M. Grehn. 1977. Susceptibility studies on *Flavobacterium* II-b. FEMS Microbiol. Lett. *2:* 289-292.

von Graevenitz, A. and S.J. Rubin (Editors). 1980. The genus *Serratia.* CRC Press, Boca Raton, Florida.

von Graevenitz, A. and H. Spector. 1969. Observations on indol-positive *Proteus.* Yale J. Biol. Med. *41:* 434-445.

von Graevenitz, A. and L. Zinterhofer. 1970. The detection of *Aeromonas hydrophila* in stool specimens. Health Lab. Sci. *7:* 124-127.

von Lingelsheim, W. 1906. Die bakteriologischen Arbeiten der Kgl. Hygienischen Station zu Beuthen O.-Schl. wahrend der Genickstarreepedemie in Oberschlesien in Winter 1904/05. Klin. Jahrb. *15:* 373-489.

von Lingelsheim, W. 1908. Beiträge zür Atiologie der epidemischen Genickstarre nach Ergebnissen der letzten Jahre. Z. Hyg. Infektionskr. *59:* 457-476.

von Prowazek, S. 1910. Parasitische Protozoen aus Japan, gesammelt von Herrn Dr. Mine in Fukuoka, Arch. Schiffs-Trop. Hyg. *14:* 297-302.

von Prowazek, S. 1913. Zür Parasitologie von Westafrika. Zentralbl. Bakteriol. Parasitenkd. Infektionskr. Hyg. Abt. I Orig. *70:* 32-36.

Von Riesen, V.L. 1976. Pectinolytic, indole-positive strains of *Klebsiella pneumoniae.* Int. J. Syst. Bacteriol. *26:* 143-145.

Von Roekel, H.V. 1965. Pullorum disease. *In,* H.E. Biester and L.H. Schwarte (Editors), Diseases of Poultry, 5th edition, the Iowa State University Press, Ames, pp. 220-259.

Vosti, K.L., A.S. Monto, J.J. Older and L.A. Rantz. 1964. The serologic specificity of crude and purified antigen extracts of *Escherichia coli* in hemagglutination reaction with rabbit and human antisera. J. Immunol. *93:* 199-204.

Vreeland, R.H., C.D. Litchfield, E.L. Martin and E. Elliot. 1980. *Halomonas elongata,* a new genus and species of extremely salt-tolerant bacteria. Int. J. Syst. Bacteriol. *30:* 485-495.

Vreeland, R.H. and E.L. Martin. 1980. Growth characteristics, effects of temperature, and ion specificity of the halotolerant bacterium *Halomonas elongata.* Can. J. Microbiol. *26:* 746-752.

Vuillemin, P. 1905. Sur la denomination de l'agent presume de la syphilis. C. R. Acad. Sci. Paris *140:* 1567-1568.

Wachsmuth, I.K., B.R. Davis and S.D. Allen. 1979. Urealytic *Escherichia coli* of human origin: serologic, epidemiologic, and genetic analysis. J. Clin. Microbiol. *10:* 897-902.

Wadström, T., A. Aust-Kettis, D. Habte, J. Holmgren, G. Meeuwisse, R. Mollby and O. Soderlind. 1976. Enterotoxin-producing bacteria and parasites in stools of Ethiopian children with diarrhoeal disease. Arch. Dis. Childh. *51:* 865-870.

Wadström, T., A. Ljungh and B. Wretlind. 1976. Enterotoxin, haemolysin and cytotoxic protein in *Aeromonas hydrophila* from human infections. Acta Pathol. Microbiol. Scand. *84:* 112-114.

Wagenbreth, D. 1961. Ein Beiträg zür Systematischen Einordnung der Knollchenbakterien durch Bestimmung des relativen Basengehaltes ihrer Desoxyribonucleinsäuren. Flora *151:* 219-230.

Wagner, C. and A.T. Brown. 1970. Regulation of tryptophan pyrrholase activity in *Xanthomonas pruni.* J. Bacteriol. *104:* 90-97.

Wahren, A. and R.J. Gibbons. 1970. Amino acid fermentation by *Bacteroides melaninogenicus.* Antonie van Leeuwenhoek J. Microbiol. Serol. *36:* 149-159.

Wais, A.C., M. Kon, R.E. MacDonald and B.D. Stollar. 1975. Salt-dependent bacteriophage infecting *Halobacterium cutirubrum* and *H. halobium.* Nature (London) *256:* 314-315.

Wakabayashi, H. and S. Egusa. 1972. Characteristics of a *Pseudomonas* sp. from an epizootic of pond-cultured eels (*Anguillula japonica*). Bull. Jpn. Soc. Scient. Fisheries *38:* 577-587.

Wakabayashi, H. and S. Egusa. 1973. *Edwardsiella tarda (Paracolobactrum anguillimortiferum)* associated with pond-cultured eel disease. Bull. Japan. Soc. Scientific Fisheries *39:* 931-936.

Wakimoto, S. and T.W. Mew. 1979. Predicting the outbreak of bacterial blight of rice by the bacteriophage method. Philippine Phytopathol. *15:* 81-85.

Wakker, J.H. 1883. Vorläufige Mittheilungen über Hyacinthenkrankheiten Bot. Centrabl. *14:* 315-317.

Waldee, E.L. 1945. Comparative studies of some peritrichous phytopathogenic bacteria. Iowa State Coll. J. Sci. *19:* 435-484.

Waldhalm, D.G., R.F. Hall, W.A. Meinershagen, C.S. Card and F.W. Fran 1974. *Haemophilus somnus* infection in the cow as a possible contribut factor to weak calf syndrome: isolation and animal inoculation studies. J. Vet. Res. *35:* 1401-1403.

Walker, C.B., D. Ratliff, D. Muller, R. Mandel and S.S. Socransky. 1979. Medium for selective isolation of *Fusobacterium nucleatum* from human periodontal pockets. J. Clin. Microbiol. *10:* 844-849.

Walker, C.B. and T.D. Wilkins. 1976. Use of semisolid agar for initiation of pure *Bacteroides fragilis* infection in mice. Infect. Immun. *14:* 721-725.

Walker, C.N. and P.W. Smith. 11980. Ampicillin resistance in *Haemophilus parainfluenzae*. Am. J. Clin. Pathol. *74:* 229-232.

Wallace, A.L. and A. Harris. 1967. Reiter treponeme. A review of the literature. Bull. World Health Organ. *36:* Suppl. 2.

Wallace, J.J. and R.G. Petersdorf. 1971. Urinary tract infections. Postgraduate Medicine *50:* 138-144.

Wallace, W.R. and G.T. Dimopoullus. 1965. Biologic properties and characteristics of *Anaplasma marginale*. Effect of radiation on infectivity of partially purified marginal body preparation. J. Am. Vet. Res. *26:* 1356-1358.

Wallen, L.L. and E.N. Davis. 1972. Biopolymers of activated sludge. Environ. Sci. Technol. *6:* 161-164.

Wallin, J.R. and C.S. Reddy. 1945. A bacterial streak disease of *Phleum pratense* L. Phytopathology *35:* 937-939.

Wallnöfer, P. and R.L. Baldwin. 1967. Pathway of propionate formation in *Bacteroides ruminicola*. J. Bacteriol. *93:* 504-505.

Walsby, A.E. 1978. The gas vesicles of aquatic prokaryotes. *In* Stanier, Rogers and Ward (Editors), Relation between Structure and Function in the Prokaryotic Cell. 28th Symp. Soc. Gen. Microbiol., Cambridge University Press, Cambridge, pp. 327-358.

Walsh, J.A., D.L. Lee and A.M. Shepherd. 1979. Intracellular microorganisms parasitizing cyst nematodes and their pathogenic effect on the potato cyst nematode, *Globodera rostochiensis*. Parasitology *79:* xlvii-xlviii.

Walter, M.R. (Editor). 1977. Life in the Precambrian. Precambrian Res. *5* (2): 105-219.

Walther-Mauruschat, A., M. Aragno, F. Mayer and H.G. Schlegel. 1977. Micromorphology of Gram-negative hydrogen bacteria. II. Cell envelope, membranes and cytoplasmic inclusions. Arch. Microbiol. *114:* 101-110.

Wanick, M.C. and E. Cavalcanti Da Silva. 1971. Novas observações sobre e emprego de *Zymomonas mobilis* var. *recifensis* em infecções por *Neisseria gonorrhoeae*, *Candida albicans* e *Trichomonas vaginalis*. Rev. Inst. Antibiot. Univ. Recife *11:* 69-71.

Wanick, M.C., J.M. De Araújo, E. Cavalcanti Da Silva and I.E. Schumacher. 1970. Cura de vaginites de etiologia variada pelo emprêgo de cultura de *Zymomonas mobilis* (Lindner) (1928) Kluyver e van Niel (1936). Rev. Inst. Antibiot. Univ. Recife *10:* 47-50.

Ward, J.I., T.F. Tsai, G.A. Filice and D.W. Fraser. 1978. Prevalence of ampicillin- and chloramphenicol-resistant strains of *Haemophilus influenzae* causing meningitis and bacteremia: national survey of hospital laboratories. J. Infect. Dis. *138:* 421-424.

Wardlaw, A.C. and R. Parton. 1979. Changes in envelope proteins and correlation with biological activities of *B. pertussis*. *In* Manclark and Hill (Editors), International Symposium on Pertussis. U.S. Government Printing Office, Washington, D.C., pp. 94-98.

Warming, E. 1875. Om nogle ved Danmarks kyster levende bactérier. Vidensk. Medd. Dan. Naturhist. Foren. København, pp. 306-420.

Warren, S.H. and W.M. Scott. 1930. A new serological type of *Salmonella*. J. Hyg. *9:* 415-417.

Warren, W.J. and R.D. Miller. 1979. Growth of Legionnaires' disease bacterium (*Legionella pneumophila*) in a chemically defined medium. J. Clin. Microbiol. *10:* 50-55.

Warskow, A.L. and E. Juni. 1972. Nutritional requirements of *Acinetobacter* strains isolated from soil, water, and sewage. J. Bacteriol. *112:* 1014-1016.

Washington, J.A. II and M.D. Maker. 1975. Unclassified, lactose-fermenting, urease-producing member of the family *Enterobacteriaceae* resembling *Escherichia coli*. J. Clin. Microbiol. *2:* 70-71.

Washington, J.A. II and J.A. Timm. 1976. Unclassified, citrate-positive member of the family *Enterobacteriaceae* resembling *Escherichia coli*. J. Clin. Microbiol. *4:* 165-167.

Watanabe, H., Y. Kamita, T. Nakamura, A. Takimoto and T. Yamanaka. 1979. The terminal oxidase of *Photobacterium phosphoreum*. A novel cytochrome. Biochim. Biophys. Acta *547:* 70-78.

Watanabe, H., N. Mimura, A. Takimoto and T. Nakamura. 1975. Luminescence and respiratory activities of *Photobacterium phosphoreum*. Competition for cellular reducing power. J. Biochem. *77:* 1147-1155.

Watanabe, H., K. Yoshida, M. Takahashi, G. Tomita and T. Nakamura. 1976. Reaction mechanism of bacterial luciferase from *Photobacterium phosphoreum*. *In* Singer (Editor), Flavins and Flavoproteins, Elsevier, Amsterdam, pp. 62-76.

Watanabe, N. 1959. On four new halophilic species of *Spirillum*. Bot. Mag. (Tokyo) *72:* 77-86.

Watanabe, T., K. Mishima and T. Horikawa. 1973. Proteolytic activities of human mycoplasmas. Jpn. J. Microbiol. *17:* 151-153.

...ers, H. and P. Hunt. 1980. The in vivo three-dimensional form of a plant ...ycoplasmalike organism by the analysis of serial ultrathin sections. J. Gen. ...robiol. *116:* 111-131.

... T.C. Currier, M.P. Gordon, M.-D. Chilton and E.W. Nester. 1975. ... required for virulence of *Agrobacterium tumefaciens*. J. Bacteriol. ...4.

... B.W. Holloway. 1978a. Chromosome mapping in *Pseudomonas* ... acteriol. *133:* 1113-1125.

Watson, J.M. and B.W. Holloway. 1978b. Linkage map of *Pseudomonas aeruginosa* PAT. J. Bacteriol. *136:* 507-521.

Wauters, G. 1973. Improved methods for the isolation and the recognition of *Yersinia enterocolitica*. Contr. Microbiol. Immunol. *2:* 68-70.

Wauters, G., L. Le Minor, A. Chalon and J. Lassen. 1972. Supplement au schema antigénique de *Yersinia enterocolitica*. Ann. Inst. Pasteur *122:* 951-956.

Weaver, R.E. and J.C. Feeley. 1979. Cultural and biochemical characterization of the Legionnaires' disease bacterium. *In* Jones and Hébert (Editors), "Legionnaires' " the disease, the bacterium and methodology. Center for Disease Control, Atlanta, pp. 20-25.

Weber, F.H. and E.P. Greenberg. 1981. Rifampin as a selective agent for the enumeration and isolation of spirochetes from salt marsh habitats. Curr. Microbiol. *5:* 303-306.

Webster, J.A. and R. Hugh. 1979. *Flavobacterium aquatile* and *Flavobacterium meningosepticum*: glucose nonfermenters with similar flagellar morphologies. Int. J. Syst. Bacteriol. *29:* 333-338.

Weckesser, J., G. Drews, J. Roppel, H. Mayer and I. Fromme. 1974. The lipopolysaccharides (O-antigens) of *Rhodopseudomonas viridis*. Arch. Microbiol. *101:* 233-245.

Weeks, O.B. 1955. *Flavobacterium aquatile* (Frankland and Frankland) Bergey et al., type species of the genus *Flavobacterium*. J. Bacteriol. *69:* 649-658.

Weeks, O.B. 1974. Genus *Flavobacterium* Bergey et al., 1923. *In* Buchanan and Gibbons (Editors), Bergey's Manual of Determinative Bacteriology, 8th Ed. The Williams and Wilkins Co., Baltimore, pp. 357-364.

Weeks, O.B. 1981. Preliminary studies of the pigments of *Flavobacterium breve* NCTC 11099 and *Flavobacterium odoratum* NCTC 11036. *In* Reichenbach and Weeks (Editors), The *Flavobacterium-Cytophaga* Group (Proceedings of the International Symposium on Yellow-Pigmented Gram-negative Bacteria of the *Flavobacterium-Cytophaga* Group, Braunschweig, July 8 to 11, 1980). Verlag Chemie, Weinheim, pp. 109-114.

Weeks, O.B., S.M. Beck, M.D. Thomas and H.D. Isenberg. 1962. Pigment of *Flavobacterium piscicida*. J. Bacteriol. *84:* 1118.

Weeks, O.B. and R.S. Breed. 1957. Genus III. *Flavobacterium* Bergey et al., 1923. *In* Breed, Murray and Smith (Editors), Bergey's Manual of Determinative Bacteriology, 7th Ed. The Williams and Wilkins Co., Baltimore, pp. 309-322.

Weickmann, J.L. and D.E. Fahrney. 1977. Arginine deiminase from *Mycoplasma arthritidis*. J. Biol. Chem. *252:* 2615-2620.

Weidinger, G., G. Klotz and W. Goebel. 1979. A large plasmid from *Halobacterium halobium* carrying genetic information for gas vacuole formation. Plasmid *2:* 377-386.

Weimberg, R. 1962. Studies with a constitutive dehydrogenase in *Pseudomonas fragi*. Biochim. Biophys. Acta *67:* 349-358.

Weinberg, M., R. Nativelle and A.R. Prévot. 1937. Les microbes anaérobies. Masson and Co., Paris.

Weinman, D. 1957. *Bartonellaceae*. *In* Breed, Murray and Smith (Editors), Bergey's Manual of Determinative Bacteriology, 7th Ed., The Williams and Wilkins Co., Baltimore, p. 906.

Weinman, D. and A.H. Pinkerton. 1938. A *Bartonella* of the guinea pig, *Bartonella tyzzeri* sp. nov. Ann. Trop. Med. Parasitol. *33:* 215-224.

Weinrich, A.E. and V.E. Del Bene. 1976. Beta-lactamase activity in anaerobic bacteria. Antimicrob. Agents Chemother. *10:* 106-111.

Weiser, J. 1963. Diseases of insects of medical importance in Europe. Bull. WHO *28:* 121-127.

Weiser, J. and Z. Žižka. 1968. Electron-microscope studies of *Rickettsiella chironomi* in the midge *Camptochironomus tentans*. J. Invert. Pathol. *12:* 222-230.

Weiss, E. 1960. Some aspects of variation in rickettsial virulence. Ann. N.Y. Acad. Sci. *88:* 1287-1297.

Weiss, E. 1973. Growth and physiology of rickettsiae. Bacteriol. Rev. *37:* 259-283.

Weiss, E. 1981. The family *Rickettsiaceae*: Human pathogens. *In* Starr, Stolp, Trüper, Balows and Schlegel (Editors), The Prokaryotes: a handbook on habitats, isolation and identification of bacteria, Springer-Verlag, Berlin - Heidelberg - New York, pp. 2137-2160.

Weiss, E. 1981. Biochemistry and metabolism of rickettsiae: Current trends. *In* Burgdorfer and Anacker (Editors), Rickettsiae and Rickettsial Diseases, Academic Press, New York, pp. 387-400.

Weiss, E. 1982. The biology of the rickettsiae. Annu. Rev. Microbiol. *36:* 345-370.

Weiss, E., J.C. Coolbaugh and J.C. Williams. 1975. Separation of viable *Rickettsia typhi* from yolk sac and L cell host components by Renografin density gradient centrifugation. Appl. Microbiol. *30:* 456-463.

Weiss, E. and G.A. Dasch. 1982. Differential characteristics of strains of *Rochalimaea*: *Rochalimaea vinsonii* sp. nov., the Canadian vole agent. Int. J. Syst. Bacteriol. *32:* 305-314.

Weiss, E., G.A. Dasch, D.R. Woodman and J.C. Williams. 1978. Vole agent identified as a strain of the trench fever rickettsia, *Rochalimaea quintana*. Infect. Immun. *19:* 1013-1020.

Weiss, E. and H.R. Dressler. 1958. Growth of *Rickettsia prowazeki* in irradiated monolayer cultures of chick embryo entodermal cells. J. Bacteriol. *75:* 544-552.

Weiss, E. and H.R. Dressler. 1960. Selection of an erythromycin-resistant strain of *Rickettsia prowazekii*. Am. J. Hyg. *71:* 292-298.

Weiss, E. and H.R. Dressler. 1962. Increased resistance to chloramphenicol in *Rickettsia prowazekii* with a note on failure to demonstrate genetic interaction among strains. J. Bacteriol. *83:* 409-414.

Weiss, E., H.K. Mamay and G.A. Dasch. 1982. Ornithine metabolism in the genus *Rochalimaea*. J. Bacteriol. *150:* 245-250.

Weiss, E., W.F. Myers, E.C. Suitor, Jr. and E.M. Neptune, Jr. 1962. Respiration of a rickettsialike microorganism, *Wolbachia persica*. J. Infect. Dis. *110:* 155-164.

Weiss, E., E.M. Neptune, Jr. and J.A. Davies. 1964. Lipid metabolism of the rickettsialike microorganism *Wolbachia persica*. III. Comparison with other metabolic activities. J. Infect. Dis. *114:* 50-54.

Weiss, E., L.W. Newman, R. Grays and A.E. Green. 1972. Metabolism of *Rickettsia typhi* and *Rickettsia akari* in irradiated L cells. Infect. Immun. *6:* 50-57.

Weitzman, P.D.J. 1980. Citrate synthase and succinate thiokinase in classification and identification. *In* Goodfellow and Board (Editors) Microbiological Classification and Identification. Academic Press, London, New York, pp. 107-125.

Weitzman, P.D.J. and D. Jones. 1968. Regulation of citrate synthase and microbial taxonomy. Nature *219:* 270-272.

Weldin, J.C. 1927. The colon-typhoid group of bacteria and related forms. Relationships and classification. Iowa State J. Sci. *1:* 121-197.

Wells, J.S., Jr. and N.R. Krieg. 1965. Cultivation of *Spirillum volutans* in a bacteria-free environment. J. Bacteriol. *90:* 817-818.

Wenyon, C.M. 1926. Protozoology, Vol. 1. Baillière, Tindall and Cox, London, pp. 1-778.

Wenyon, C.M. 1926. Spirochaetes. *In* Wenyon (Editor), Protozoology, Vol. 2, William Wood and Co., New York, pp. 1233-1288.

Werber, M.M. and M. Mevarech. 1978. Induction of a dissimilatory reduction pathway of nitrate in *Halobacterium* of the Dead Sea. Arch. Biochem. Biophys. *186:* 60-65.

Werkman, C.H. and G.F. Gillen. 1932. Bacteria producing trimethylene glycol. J. Bacteriol. *23:* 167-182.

Werner, H. 1970a. Das kulturell-biochemische verhalten und die antibiotikaempfindlichkeit des *Bacteroides putredinis* (Weinberg et al. 1937) Kelly 1957. Zentralbl. Bakteriol. Parasitenkd. Infektionskr. Hyg. I Abt. Orig. A *215:* 327-332.

Werner, H. 1970b. Glutaminsäuredecarboxylaseaktivitat bei *Bacteroides* Arten. Zentralbl. Bakteriol. Parasitenkd. Infektionskr. Hyg., I Abt. Orig. A *215:* 320-326.

Werner, H. 1973. *Megasphaera elsdenii* -- a normal inhabitant of the human intestines? Zentralbl. Bakteriol. Parasitenkd. Infektionskr. Hyg. Abt. I Orig. A *223:* 343-347.

Werner, H. 1974. Demonstration of lysine decarboxylase activity in the obligately anaerobic bacterium *Sphaerophorus varius*. Zentralbl. Bakteriol. Parasitenkd. Infektionskr. Hyg., I Abt. Orig. A *226:* 364-368.

Werner, H., G. Pulverer and C. Reichertz. 1971. Biochemical properties and antibiotic susceptibility of *Bacteroides melaninogenicus*. Med. Microbiol. Immunol. *157:* 3-9.

Werner, H., G. Rintelen and H. Kunštek-Santos. 1975. A new butyric acid-producing *Bacteroides* species: *B. splanchnicus* n. sp. Zentralbl. Bakteriol. Parasitenkd. Infektionskr. Hyg., I. Abt. Orig. A *231:* 133-144.

West, M.G. and P.R. Edwards. 1954. The Bethesda-Ballerup group of paracolon bacteria. Public Health Monograph No. 22. U.S.D.H.E.W., Atlanta.

West, P.A., R.M. Daniel, C.J. Knowles and J.V. Lee. 1978. Tetramethyl-*p*-phenylenediamine (TMPD) oxidase activity and cytochrome distribution in the genus *Vibrio*. FEMS Microbiol. Lett. *4:* 339-342.

West, P.A., C.J. Knowles and J.V. Lee. 1980. Ecology of *Vibrio* species including *Vibrio cholerae*, in waters of Kent, United Kingdom. Soc. Gen. Microbiol. Quart. 7: 80.

Weston, J.A. and C.J. Knowles. 1974. The respiratory system of the marine bacterium *Beneckea natriegens*. I. Cytochrome composition. Biochim. Biophys. Acta *333:* 228-236.

Westphal, O. 1974. Bacterial endotoxins. Int. Arch. Allergy Appl. Immunol. *49:* 1-43.

Wetmore, P.W., J.F. Thiel, Y.F. Herman and J.R. Harr. 1963. Comparison of selected *Actinobacillus* species with a hemolytic variety of *Actinobacillus* from irradiated swine. J. Infect. Dis. *113:* 186-194.

Wetmur, J.G. 1976. Hybridization and renaturation kinetics of nucleic acids. Annu. Rev. Biophys. Bioeng. *5:* 337-361.

Wetmur, J.G. and N. Davidson. 1968. Kinetics of renaturation of DNA. J. Mol. Biol. *31:* 349-370.

Wetzler, T.F. 1970. Animal diseases transmissible to man, pseudotuberculosis *In* Bodily, H.L., E.L. Updyke and J.O. Mason (Editors), Diagnostic Procedures for Bacterial, Mycotic and Parasitic Infections, 5th ed., Amer. Pub. Health Assoc. Inc., New York, p. 449.

Weyer, F. and R.J. Reiss-Gutfreund. 1973. Verhalten von *Rickettsia montana* und *R. canada* in Kleiderlausen. Acta Tropica *30:* 177-192.

Whang, H.Y., M.E. Heller and E. Neter. 1972. Production by *Aeromonas* of common enterobacterial antigen and its possible taxonomic significance. J. Bacteriol. *110:* 161-164.

Whang, H.Y. and E. Neter. 1964. Immunological studies of a heterogenic enterobacterial antigen (Kunin). J. Bacteriol. *84:* 1245-1250.

Wharton, D.R.A. and J.E. Lola. 1969. Lysozyme action on the cockroach, *Periplaneta americana*, and its intracellular symbionts. J. Insect Physiol. *15:* 1647-1658.

Whatley, J.M. 1976. Bacteria and nuclei in *Pelomyxa palustris*: comments on the theory of serial endosymbiosis. New Phytol. *76:* 111-120.

Whatley, M.H., J.S. Bodwin, B.B. Lippincott and J.A. Lippincott. 1976. Role for

Agrobacterium cell envelope lipopolysaccharide in infection site attachment. Infect. Immunol. *13:* 1080-1083.

Wheelis, M.L. 1975. The genetics of dissimilatory pathways in *Pseudomonas*. Annu. Rev. Microbiol. *29:* 505-524.

Wheelis, M.L. and R.Y. Stanier. 1970. The genetic control of dissimilatory pathways in *Pseudomonas putida*. Genetics 66: 245-266.

Wherry, W.B. and W.W. Oliver. 1916. *Leptotrichia innominata* (Miller). J. Infect. Dis. *19:* 299-303.

Whitaker, R.J., G.S. Byng, R.L. Gherna and R.A. Jensen. 1981a. Comparative allostery of 3-deoxy-D-arabino-heptulosonate 7-phosphate synthetase as an indicator of taxonomic relatedness in pseudomonad genera. J. Bacteriol. *145:* 752-759.

Whitaker, R.J., G.S. Byng, R.L. Gherna and R.A. Jensen. 1981b. Diverse enzymological patterns of phenylalanine biosynthesis in pseudomonad bacteria are conserved in parallel with DNA/DNA homology groupings. J. Bacteriol. *147:* 526-534.

Whitby, G.E. and R.G.E. Murray. 1980. Defined medium for *Aquaspirillum serpens* VHL effective in batch and continuous culture. Appl. Environ. Microbiol. *39:* 20-24.

Whitcomb, R.F. 1980. The genus *Spiroplasma*. Annu. Rev. Microbiol. *34:* 677-709.

Whitcomb, R.F. and J.G. Tully (Editors). 1979. The Mycoplasmas, Vol. 3. Academic Press, New York.

Whitcomb, R.F., J.G. Tully, P. McCawley and D.L. Rose. 1982a. Application of the growth inhibition test to *Spiroplasma* taxonomy. Int. J. Syst. Bacteriol. *32:* 387-394.

Whitcomb, R.F., J.G. Tully, J.M. Bové and P. Saglio. 1973. Spiroplasmas and acholeplasmas: multiplication in insects. Science *182:* 1251-1253.

Whitcomb, R.F., J.G. Tully, J.M. Bové and T.B. Clark. 1982b. Revised serological classification of spiroplasmas. New provisional groups and recommendations for serotyping of isolates. Curr. Microbiol. *7:* 291-296.

Whitcomb, R.F., J.G. Tully, D.L. Rose, E.B. Stephens, A. Smith, R.E. McCoy and M.F. Barile. 1982c. Wall-less prokaryotes from fall flowers in central United States and Maryland. Curr. Microbiol. *7:* 285-290.

Whitcomb, R.F. and D.L. Williamson. 1975. Helical wall-free prokaryotes in insects: multiplication and pathogenicity. Ann. N.Y. Acad. Sci. *266:* 260-275.

Whitcomb, R.F. and D.L. Williamson. 1979. Pathogenicity of mycoplasmas for arthropods. Zentralbl. Bakteriol. Parasitenkd. Infektionskr. Hyg. Abt. 1 Orig. A *245:* 200-221.

White, A.H. 1940. A bacterial discoloration of print butter. Sci. Agr. *20:* 638-645.

White, B. 1926. Further studies of the *Salmonella* Group. Med. Res. Comm. Spec. Rep. *103:* 3-160.

White, D.C. 1963. Respiratory systems in the hemin-requiring *Haemophilus* species. J. Bacteriol. *85:* 84-96.

White, D.C. 1966. The obligatory involvement of the electron transport system in the catabolic metabolism of *Haemophilus parainfluenzae*. Antonie van Leeuwenhoek J. Microbiol. Serol. *32:* 139-158.

White, D.C., M.P. Bryant and D.R. Caldwell. 1962. Cytochrome linked fermentation in *Bacteroides ruminicola*. J. Bacteriol. *84:* 822-828.

White, D.C. and S. Granick. 1963. Hemin biosynthesis in *Haemophilus*. J. Bacteriol. *85:* 842-850.

White, D.C., G. Leidy, J.D. Jamieson and R.E. Shope. 1965. Porcine contagious pleuropneumonia III. Interrelationship of *Hemophilus pleuropneumonia* to other species of *Hemophilus*: nutritional, metabolic, transformation, and electron microscopy studies. J. Exp. Med. *120:* 1-12.

White, D.C. and P.R. Sinclair. 1971. Branched electron transport systems in bacteria. Adv. Microb. Physiol. *5:* 173-211.

White, F.F. and E.W. Nester. 1980a. Hairy root: plasmid encodes virulence traits in *Agrobacterium rhizogenes*. J. Bacteriol. *141:* 1134-1141.

White, F.F. and E.W. Nester. 1980b. Relationship of plasmids responsible for hairy root and crown gall tumorigenicity. J. Bacteriol. *144:* 710-720.

White, F.H., C.F. Simpson and L.E. Williams, Jr. 1973. Isolation of *Edwardsiella tarda* from aquatic animal species and surface waters in Florida. J. Wildlife Dis. *9:* 204-208.

White, G.A. and C.H. Wang. 1964a. The dissimilation of glucose and gluconate by *Acetobacter xylinum*. 1. The origin and the fate of triose phosphate. Biochem. J. *90:* 408-423.

White, G.A. and C.H. Wang. 1964b. The dissimilation of glucose and gluconate by *Acetobacter xylinum*. 2. Pathway evaluation. Biochem. J. *90:* 424-433.

White, H.E. 1930. Bacterial spot of radish and turnip. Phytopathology 20: 653-662.

White, H.L. 1936. Diseases of early vegetables. Rep. Exp. Sta. Cheshunt *1935:* 42-43.

White, J.N. and M.P. Starr. 1971. Glucose fermentation end products of *Erwinia* spp. and other enterobacteria. J. Appl. Bacteriol. *34:* 459-475.

White, L.O. 1972. The taxonomy of the crown-gall organism *Agrobacterium tumefaciens* and its relationship to rhizobia and other agrobacteria. J. Gen. Microbiol. *72:* 565-574.

White, P.G. and J.B. Wilson. 1951. Differentiation of smooth and non-smooth colonies of *Brucellae*. J. Bacteriol. *61:* 239-240.

Whitescarver, J. and G. Furness. 1975. T-mycoplasmas: a study of the morphology, ultrastructure and mode of division of some human strains. J. Med. Microbiol. *8:* 349-355.

Whitfield, C., I.W. Sutherland and R.E. Cripps. 1981. Surface polysaccharides in mutants of *Xanthomonas campestris*. J. Gen. Microbiol. *124:* 385-392.

Whitmore, A. 1913. An account of a glanders-like disease occurring in Rangoon. J. Hyg. *13:* 1-34.

Whittaker, P.A. 1971. Terminal respiration in *Moraxella lwoffi* (NCIB8250). Microbios *4:* 65-70.

Whittaker, R.H. and L. Margulis. 1978. Protist classification and the kingdoms of organisms. BioSystems *10:* 3-18.

Whittenbury, R., J. Colby, H. Dalton and H.C. Reed. 1976. Biology and ecology of methane oxidisers. *In* Schlegel, Gottschalk and Pfennig (Editors), Symposium on microbial production and utilisation of gases (H_2, CH_4, CO), Göttingen, Akademie der Wissenschaften, Göttingen, pp. 281-292.

Whittenbury, R., H. Dalton, M. Eccleston and H.L. Reid. 1975. The different types of methane oxidizing bacteria and some of their more unusual properties. *In* Microbial Growth on C_1-compounds, Society of Fermentation Technology, Japan, pp. 1-9.

Whittenbury, R., K.C. Phillips and J.F. Wilkinson. 1970. Enrichment, isolation and some properties of methane-utilizing bacteria. J. Gen. Microbiol. *61:* 205-218.

Wichterman, R. 1953. The Biology of *Paramecium*, Blakiston, New York, pp. 399-405.

Wicker, C. 1980. Influence of sex, developmental time and food on 2-N-acetyl glucosaminidase activity in the rice weevil *Sitophilus oryzae* L. Experientia (Basel) *36:* 1059-1060.

Wicker, C. and P. Nardon. 1980. Role des symbiotes et du genotype dans la regulation de l'activite de la 2-N-acetylglucosaminidase chez le coleoptere curculionide *Sitophilus oryzae* L. Bull. Soc. Zool. Fr. *105:* 191-198.

Widdel, F. 1980. Anaerober Abbau von Fettsäuren und Benzoesäure durch neu Isolierte Arten Sulfat-reduzierender Bakterien. Dissertation. Georg-August-Universität zu Göttingen. Lindhorst/Schaumburg-Lippe, Göttingen.

Widdel, F. 1981. *In* Validation of the publication of new names and new combinations previously published outside the IJSB. List No. 7. Int. J. Syst. Bacteriol. *31:* 382-383.

Wiebe, M.E., P.R. Burton and D.M. Shankel. 1972. Isolation and characterization of two cell types of *Coxiella burnetii* phase I. J. Bacteriol. *110:* 368-377.

Wiegel, J. 1981. Distinction between the Gram-reaction and the Gram-type of bacteria. Int. J. Syst. Bacteriol. *31:* 88.

Wiegel, J. and F. Mayer. 1978. Isolation of lipopolysaccharides and the effect of polymyxin B on the outer membrane of *Corynebacterium autotrophicum.* Arch. Microbiol. *118:* 67-69.

Wiegel, J. and H.G. Schlegel. 1976. Enrichment and isolation of nitrogen-fixing hydrogen bacteria. Arch. Microbiol. *107:* 139-142.

Wiegel, J. and H.G. Schlegel. 1977a. α-Isopropylmalate synthase from *Alcaligenes eutrophus* H 16. I. Purification and general properties. Arch. Microbiol. *112:* 239-246.

Wiegel, J. and H.G. Schlegel. 1977b. α-Isopropylmalate synthase from *Alcaligenes eutrophus* H 16. III. Endproduct inhibition and its relief by valine and isoleucine. Arch. Microbiol. *114:* 203-210.

Wiegel, J., D. Wilke, J. Baumgarten, R. Opitz and H.G. Schlegel. 1978. Transfer of the nitrogen-fixing hydrogen bacterium *Corynebacterium autotrophicum* Baumgarten et al. to *Xanthobacter* gen. nov. Int. J. Syst. Bacteriol. *28:* 573-581.

Wiehe, P.O. and W.J. Dowson. 1953. A bacterial disease of cassava *(Manihot utilissima)* in Nyasaland. Emp. J. Exp. Agr. *21:* 141-143.

Wieland, F., W. Dompert, G. Bernhardt and M. Sumper. 1980. Halobacterial glycoprotein saccharides contain covalently linked sulphate. FEBS Lett. *120:* 110-114.

Wigand, R. 1958. Morphologische Biologische und Serologische Eigenschaften der Bartonellen. Georg Thieme, Stuttgart.

Wigand, R. and D. Peters. 1952. Neuere Untersuchungen über *Bartonella muris* Mayer. II. Mittellung. Z. Tropenmed. Parasitol. *3:* 437-452.

Wigand, R. and D. Peters. 1954. Abbauversuche an *Haemobartonella muris* und *Eperythrozoon coccoides.* Z. Tropenmed. Parasitol. *5:* 482-492.

Wigglesworth, V.B. 1929. Digestion in the tsetse fly: a study of structure and function. Parasitology *21:* 288-321.

Wike, D.A., G. Tallent, M.G. Peacock and R.A. Ormsbee. 1972. Studies of the rickettsial plaque assay technique. Infect. Immun. *5:* 715-722.

Wilke, D. 1980. Conjugational gene transfer in *Xanthobacter autotrophicus* GZ29. J. Gen. Microbiol. *117:* 431-436.

Wilke, D. and H.G. Schlegel. 1979. A defective generalized transducing bacteriophage in *Xanthobacter autotrophicus* GZ29. J. Gen. Microbiol. *115:* 403-410.

Wilkerson, F.P. 1980. Symbionts involved in phosphate uptake by green hydra. *In* Schwemmler and Schenk (Editors), Endocytobiology. Endosymbiosis and Cell Biology. A Synthesis of Recent Research. Walter de Gruyter, Berlin, pp. 269-277.

Wilkie, P.J., D.W. Dye and D.R.W. Watson. 1973. Further hosts of *Pseudomonas viridiflava.* N.Z. J. Agr. Res. *16:* 315-323.

Wilkins, T.D., S.L. Chalgren, F. Jimenez-Ulate, C.R. Drake, Jr. and J.L. Johnson. 1976. Inhibition of *Bacteroides fragilis* on blood agar plates and reversal of inhibition by added hemin. J. Clin. Microbiol. *3:* 359-363.

Wilkins, T.D. and T. Thiel. 1973. A modified broth-disk method for testing the antibiotic susceptibility of anaerobic bacteria. Antimicrob. Agents Chemother. *3:* 350-356.

Wilkins, T.D., D.L. Wagner, B.J. Veltri, Jr. and E.M. Gregory. 1978. Factors affecting production of catalase by *Bacteroides.* J. Clin. Microbiol. *8:* 553-557.

Wilkins, T.D. and C.B. Walker. 1976. Fermentation of glucose-1-phosphate: a screening test for fermentative *Bacteroides* species. Appl. Environ. Microbiol. *31:* 320-321.

Wilkinson, B.J., M.R. Morman and D.C. White. 1972. Phospholipid composition and metabolism of *Micrococcus denitrificans.* J. Bacteriol. *112:* 1288-1294.

Wilkinson, C.R. 1980. Cyanobacteria symbiotic in marine sponges. *In* Schwemmler and Schenk (Editors), Endocytobiology. Endosymbiosis and Cell Biology. A Synthesis of Recent Research. Walter de Gruyter, Berlin, pp. 553-563.

Wilkinson, C.R. and P. Fay. 1979. Nitrogen fixation in coral reef sponges with symbiotic cyanobacteria. Nature *279:* 527-529.

Wilkinson, H.W., D.D. Cruce, B.J. Fikes, L.P. Yealy and C.E. Farshy. 1979. Indirect immunofluorescence test for Legionnaires' disease. *In* Jones and Hébert (Editors), "Legionnaires'" the disease, the bacterium and methodology. Center for Disease Control, Atlanta, pp. 111-116.

Wilkinson, H.W. and B.J. Fikes. 1980. Slide agglutination test for serogrouping *Legionella pneumophila* and atypical *Legionella*-like organisms. J. Clin. Microbiol. *11:* 99-101.

Wilkinson, H.W., B.J. Fikes and D.D. Cruce. 1979. Indirect immunofluorescence test for serodiagnosis of Legionnaires' disease: evidence for serogroup diversity of Legionnaires' disease bacterial antigens and for multiple specificity of human antibodies. J. Clin. Microbiol. *9:* 379-383.

Wilkinson, S.G. 1968. Studies on the cell walls of *Pseudomonas* spp. resistant to ethylenediaminetetra-acetic acid. J. Gen. Microbiol. *54:* 195-213.

Wilkinson, S.G. 1970. Cell walls of *Pseudomonas* species sensitive to ethylenediaminetetraacetic acid. J. Bacteriol. *104:* 1035-1044.

Wilkinson, S.G. and L. Galbraith. 1979. Polar lipids of *Pseudomonas vesicularis.* Presence of a heptosyldiacylglycerol. Biochim. Biophys. Acta *575:* 244-254.

Wilkinson, S.G., L. Galbraith and G.A. Lightfoot. 1973. Cell walls, lipids, and lipopolysaccharides of *Pseudomonas* species. Eur. J. Biochem. *33:* 158-174.

Wilkinson, S.G. and D.P. Taylor. 1978. Occurrence of 2,3-diamino-2,3-dideoxy-D-glucose in lipid A from lipopolysaccharide of *Pseudomonas diminuta.* J. Gen. Microbiol. *109:* 367-370.

Willcox, W.R., S.P. Lapage and B. Holmes. 1980. A review of numerical methods in bacterial identification. Antonie van Leeuwenhoek J. Microbiol. Serol. *46:* 233-239.

Willetts, N.S., C. Crowther and B.W. Holloway. 1981. The insertion sequence IS21 of R68.45 and the molecular basis for mobilization of the bacterial chromosome. Plasmid *6:* 30-52.

William, F. and A. Mahadevan. 1980. Degradation of aromatic compounds by *Xanthomonas* species. Z. Pflanzenkr. Pflanzenschutz *87:* 738-744.

Williams, C.O. and R.G. Wittler. 1971. Hydrolysis of aesculin and phosphatase production by members of the Order *Mycoplasmatales* which do not require sterol. Int. J. Syst. Bacteriol. *21:* 73-77.

Williams, C.O., R.G. Wittler and C. Burris. 1969. Deoxyribonucleic acid base compositions of selected mycoplasmas and L-phase variants. J. Bacteriol. *99:* 341-343.

Williams, F.D. 1973. Abolition of swarming of *Proteus* by p-nitrophenyl glycerin: Application to blood agar media. Appl. Microbiol. *25:* 751-754.

Williams, F.D. and R.H. Schwarzhoff. 1978. Nature of the swarming phenomenon in *Proteus.* Annu. Rev. Microbiol. *32:* 101-122.

Williams, J. 1971. The growth in vitro of killer particles from *Paramecium aurelia* and the axenic culture of this protozoan. J. Gen. Microbiol. *68:* 253-262.

Williams, J.C. 1980. Adenine nucleotide degradation by the obligate intracellular bacterium *Rickettsia typhi.* Infect. Immun. *28:* 74-81.

Williams, J.C., M.G. Peacock and T.F. McCaul. 1981. Immunological and biological characterization of *Coxiella burnetii*, phases I and II, separated from host components. Infect. Immun. *32:* 840-851.

Williams, J.C. and J.C. Peterson. 1976. Enzymatic activities leading to pyrimidine nucleotide biosynthesis from cell-free extracts of *Rickettsia typhi.* Infect. Immun. *14:* 439-448.

Williams, J.C. and E. Weiss. 1978. Energy metabolism of *Rickettsia typhi*: Pools of adenine nucleotides and energy charge in the presence and absence of glutamate. J. Bacteriol. *134:* 884-892.

Williams, J.E. 1965. Paratyphoid and *Arizona* infections. *In* H.E. Biester and L.H. Schwarte (Editors), Diseases of Poultry, 5th edition, Iowa State University Press, Ames, pp. 260-328.

Williams, M. A. 1959. Some problems in the identification and classification of *Spirillum.* Int. Bull. Bacteriol. Nomencl. Taxon. *9:* 35-55.

Williams, M. A. 1960. Flagellation in six species of *Spirillum* — a correction. Int. Bull. Bacteriol. Nomencl. Taxon. *10:* 193-196.

Williams, M.A. and S.C. Rittenberg. 1956. Microcyst formation and germination in *Spirillum lunatum.* J. Gen. Microbiol. *15:* 205-209.

Williams, M.A. and S.C. Rittenberg. 1957. A taxonomic study of the genus *Spirillum* Ehrenberg. Int. Bull. Bacteriol. Nomencl. Taxon. *7:* 49-111.

Williams, M.H. 1967. Electron microscopy of T-strains. Ann. N.Y. Acad. Sci. *143:* 397-400.

Williams, M.H., J. Brostoff and I.M. Roitt. 1970. Possible role of *Mycoplasma fermentans* in pathogenesis of rheumatoid arthritis. Lancet *2:* 277-280.

Williams, P.J. le B. and C. Rainbow. 1964. Enzymes of the tricarboxylic acid cycle in acetic acid bacteria. J. Gen. Microbiol. *35:* 237-247.

Williams, P.P. 1978. In vitro susceptibility of *Mycoplasma hyopneumoniae* and *Mycoplasma hyorhinis* to fifty-one antimicrobial agents. Antimicrob. Agents Chemother. *14:* 210-213.

Williams, R.A.D. 1975. Caldoactive and thermophilic bacteria and their thermostable proteins. Sci. Progr. (Oxford) *62:* 373-393.

Williams, R.A.D., G.H. Bowden, J.M. Hardie and H. Shah. 1975. Biochemical properties of *Bacteroides melaninogenicus* subspecies. Int. J. Syst. Bacteriol. *25:* 298-300.

Williams, R.P. and S.M.H. Qadri. 1980. The pigment of *Serratia. In* von Graevenitz and Rubin (Editors), The genus *Serratia,* CRC Press, Boca Raton, Florida, pp. 31-75.

Williamson, D.H. and J.F. Wilkinson. 1958. The isolation and estimation of poly-β-hydroxybutyrate inclusions in *Bacillus* species. J. Gen. Microbiol. *19:* 198-209.

Williamson, D.L. 1974. Unusual fibrils from the spirochete-like sex ratio organism. J. Bacteriol. *117:* 904-906.

Williamson, D.L., D.I. Blaustein, R.J.C. Levin and M.J. Elfvin. 1979a. Antiactin-peroxidase staining of the helical wall-free prokaryote *Spiroplasma citri.* Curr. Microbiol. *2:* 143-145.

Williamson, D.L. and D.F. Poulson. 1979. Sex ratio organisms (spiroplasmas) of *Drosophila. In* Whitcomb and Tully (Editors), The Mycoplasmas, Vol. 3, Academic Press, New York, pp. 175-208.

Williamson, D.L., J.G. Tully and R.F. Whitcomb. 1979b. Serological relationships of spiroplasmas as shown by combined deformation and metabolism inhibition tests. Int. J. Syst. Bacteriol. *29:* 345-351.

Williamson, D.L. and R.F. Whitcomb. 1974. Helical wall-free prokaryotes in *Drosophila,* leafhoppers, and plants. Colloq. Inst. Natl. Sante Rech. Med. *33:* 283-290.

Williamson, D.L. and R.F. Whitcomb. 1975. Plant mycoplasmas: a cultivable spiroplasma causes corn stunt disease. Science *188:* 1018-1020.

Williamson, D.L. and R.F. Whitcomb. 1983. Special serological tests for *Spiroplasma* identification. *In* Tully and Razin (Editors), Methods in Mycoplasmology, Vol. 2, Academic Press, New York, pp. 249-259.

Williamson, D.L., R.F. Whitcomb and J.G. Tully. 1978. The *Spiroplasma* deformation test, a new serological method. Curr. Microbiol. *1:* 203-207.

Williamson, G.M. and K. Zinnemann. 1951. The occurrence of two distinct capsular antigens in *Haemophilus influenzae* type e strains. J. Pathol. Bacteriol. *63:* 695-698.

Willliamson, G.M. and K. Zinnemann. 1954. The degradation of *Haemophilus influenzae* type e capsular antigens. J. Pathol. Bacteriol. *68:* 453-457.

Williamson, I.J.F. 1928. Furunculosis of the salmonidae. Fishery Board of Scotland, Salmon Fisheries, *5:* 1-17.

Willison, J.C. and P. John. 1979. Mutants of *Paracoccus denitrificans* deficient in c-type cytochromes. J. Gen. Microbiol. *115:* 443-450.

Willmitzer, L., M. De Beuckeleer, M. Lemmers, M. Van Montagu and J. Schell. 1980. DNA from Ti plasmid present in nucleus and absent from plastids of crown gall plant cells. Nature *287:* 359-361.

Wilson, A.C., S.S. Carlson and T.J. White. 1977. Biochemical evolution. Annu. Rev. Biochem. *46:* 573-639.

Wilson, E.E., M.P. Starr and J.A. Berger. 1957. Bark canker, a bacterial disease of the Persian walnut tree. Phytopathology *47:* 669-673.

Wilson, E.E., F.M. Zeitoun and D.L. Fredrickson. 1967. Bacterial phloem canker, a new disease of Persian walnut trees. Phytopathology *57:* 618-621.

Wilson, G.S. 1933. The classification of the *Brucella* group: a systematic study. J. Hyg. Camb. *33:* 516-541.

Wilson, G.S. 1934. The reputed antigenic relationship between organisms of the *Brucella* group on the one hand, and of the *Pasteurella, Pfeifferella* and *Proteus* groups on the other. J. Hyg. Camb. *34:* 361-371.

Wilson, G.S. and A.A. Miles. 1932. The serological differentiation of smooth strains of the *Brucella* group. Br. J. Exp. Pathol. *13:* 1-13.

Wilson, G.S. and A.A. Miles. 1946. Topley and Wilson's Principles of bacteriology and immunity, 3rd Ed., Vol. 1, Williams and Wilkins Co., Baltimore.

Wilson, G.S. and A.A. Miles. 1955. Topley and Wilson's Principles of Bacteriology and Immunity, 4th Ed. The Williams and Wilkins Co., Baltimore.

Wilson, G.S. and A.A. Miles. 1975. Topley and Wilson's Principles of Bacteriology, Virology and Immunity. 6th ed. E. Arnold Ltd., London, pp. 887-900.

Wilson, M.H. and A.M. Collier. 1976. Ultrastructural study of *Mycoplasma pneumoniae* in organ culture. J. Bacteriol. *125:* 332-339.

Wilson, P.W. 1940. The Biochemistry of Symbiotic Nitrogen Fixation. University of Wisconsin Press, Madison, Wisconsin.

Wilson, R.G. and L.M. Henderson. 1963. Tryptophan-niacin relationship in *Xanthomonas pruni.* J. Bacteriol. *85:* 221-229.

Winet, H. and S.R. Keller. 1976. Spirillum swimming: theory and observations of propulsion by the flagellar bundle. J. Exp. Biol. *65:* 577-602.

Wink, M. 1979. The endosymbionts of *Glossina morsitans* and *Glossina palpalis*: cultivation and experiments and some physiological properties. Acta Trop. *36:* 215-222.

Winkler, H.H. 1976. Rickettsial permeability - an ADP-ATP transport system. J. Biol. Chem. *251:* 389-396.

Winkler, H.H. and E.T. Miller. 1981. Immediate cytotoxicity and phospholipase A: the role of phospholipase A in the interaction of *R. prowazeki* and L-cells. *In* Burgdorfer and Anacker (Editors), Rickettsiae and Rickettsial Diseases, Academic Press, New York, pp. 327-333.

Winogradsky, S. 1932. Études sur la microbiologie du sol. 5ᵉ mémoire. Analyse microbiologique du sol. Principes d'une nouvelle méthode. Ann. Inst. Pasteur *48:* 89-134.

Winogradsky, S. 1938. Sur la morphologie et l'écologie des *Azotobacter.* Ann. Inst.

Pasteur (Paris) *60:* 351-400.

Winslow, C.-E.A., J. Broadhurst, R.E. Buchanan, C. Krumwiede, Jr., L.A. Rogers and G.H. Smith. 1917. The families and genera of the bacteria. Preliminary report of the Committee of the Society of American Bacteriologists on characterization and classification of bacterial types. J. Bacteriol. *2:* 506-566.

Winslow, C.-E.A., J. Broadhurst, R.E. Buchanan, C. Krumwiede, Jr., L.A. Rogers and G.H. Smith. 1920. The families and genera of the bacteria. Final report of the Committee of the Society of American Bacteriologists on characterization and classification of bacterial types. J. Bacteriol. *5:* 191-229.

Winslow, C.-E.A., I.J. Kliger and W. Rothberg. 1919. Studies on the classification of the colon-typhoid group of bacteria with special reference to their fermentative reactions. J. Bacteriol. *4:* 429-503.

Winslow, C.-E.A. and A. Winslow. 1908. The systematic relationships of the Coccaceae. John Wiley and Sons, New York, pp. 1-300.

Winter, A.J., E.C. McCoy, C.S. Fullmer, K. Burda and P.J. Bier. 1978. Microcapsule of *Campylobacter fetus*: Chemical and physical characterization. Infect. Immun. *22:* 963-971.

Winter, G. 1884. Die Pilze Deutschlands, Oesterreichs und der Schweiz. I. Abt. Schizomyceten, Saccharomyceten und Basidiomyceten. Rabenhorst's Kryptogamen-Flora *1:* 1-924.

Wise, K.S., G.H. Cassell and R.T. Acton. 1978. Selective association of murine T lymphoblastoid cell surface alloantigens with *Mycoplasma hyorhinis.* Proc. Nat. Acad. Sci. USA *75:* 4479-4483.

Wisseman, C.L., Jr., J.L. Boese, A.D. Waddell and D.J. Silverman. 1975. Modification of antityphus antibodies on passage through the gut of the human body louse with discussion of some epidemiologic and evolutionary implications. Ann. N.Y. Acad. Sci. *266:* 6-24.

Wisseman, C.L., Jr., E.A. Edlinger, A.D. Waddell and M.R. Jones. 1976. Infection cycle of *Rickettsia rickettsii* in chicken embryo and L-929 cells in culture. Infect. Immun. *14:* 1052-1064.

Wisseman, C.L., Jr. and A.D. Waddell. 1975. In vitro studies on rickettsia-host cell interactions: Intracellular growth cycle of virulent and attenuated *Rickettsia prowazeki* in chicken embryo cells in slide chamber cultures. Infect. Immun. *11:* 1391-1401.

Wistreich, G.A. and R.F. Baker. 1971. The presence of fimbriae (pili) in three species of *Neisseria.* J. Gen. Microbiol. *65:* 167-173.

Wittenberger, C.L. and R. Repaske. 1958. Studies on the electron transport system in *Hydrogenomonas eutropha.* Bacteriol. Proc., p. 106.

Wittenberger, C.L. and R. Repaske. 1961. Studies on hydrogen oxidation in cell-free extracts of *Hydrogenomonas eutropha.* Biochim. Biophys. Acta *47:* 542-552.

Wittler, R.G. and S.G. Cary. 1974. Genus *Streptobacillus* Levaditi, Nicolau and Poincloux. 1925, 1188. *In* Buchanan and Gibbons (Editors), Bergey's Manual of Determinative Bacteriology, 8th Ed. The Williams and Wilkins Co., Baltimore, pp. 378-381.

Wober, W., O.W. Thiele and D. Urbaschek. 1964. Die "Freien Lipide" aus *Brucella abortus* Bang. I. Untersuchung der Phosphatide. Biochem. Biophys. Acta *84:* 376-390.

Woese, C.R. 1981. Archaebacteria. Sci. Am. *244:* 98-122.

Woese, C.R., P. Blanz, R.B. Hespell and C.M. Hahn. 1982. Phylogenetic relationships among various helical bacteria. Curr. Microbiol. *7:* 119-124.

Woese, C.R. and G.E. Fox. 1977. The concept of cellular evolution. J. Mol. Evol. *10:* 1-6.

Woese, C.R., G.E. Fox, L. Zablen, T. Uchida, L. Bonen, K. Pechman, B.J. Lewis and D. Stahl. 1975. Conservation of primary structure in 16S ribosomal RNA. Nature (London) *254:* 83-86.

Woese, C.R., L.J. Magrum and G.E. Fox. 1978. Archaebacteria. J. Mol. Evol. *11:* 245-252.

Woese, C.R., J. Maniloff and L.B. Zablen. 1980. Phylogenetic analysis of the mycoplasmas. Proc. Nat. Acad. Sci. (USA) *77:* 494-498.

Woese, C.R., C.D. Pribula, G.E. Fox and L.B. Zablen. 1975. The nucleotide sequence of 5S ribosomal RNA from a *Photobacterium.* J. Mol. Evol. *5:* 35-46.

Wohlegemuth, K., R.L. Pierce and C.A. Kirkbride. 1972. Bovine abortion associated with *Aeromonas hydrophila.* J. Am. Vet. Med. Assoc. *160:* 1001-1002.

Wolbach, S.B. 1919. Studies on Rocky Mountain spotted fever. J. Med. Res. *41:* 1-197.

Wolbach, S.B. and C.A.L. Binger. 1914. Notes on a filterable spirochaete from fresh water, *Spirochaeta biflexa* (new species). J. Med. Res. *30:* 23-26.

Wolbach, S.B. and J.L. Todd. 1920. Note sur l'étiologie et l'anatomie pathologique du typhus exanthématique au Mexique. Ann. Inst. Pasteur (Paris) *34:* 153-158.

Wolf, F.A. 1920. A bacterial leaf spot of velvet bean. Phytopathology *10:* 73-80.

Wolf, F.A. and A.C. Foster. 1917. Bacterial leaf spot of tobacco. Science *46:* 361-362.

Wolfe, R. S. and N. Pfennig. 1977. Reduction of sulfur by spirillum 5175 and syntrophism with *Chlorobium.* Appl. Environ. Microbiol. *33:* 427-433.

Wolff, M. 1907. *Pedioplana haeckeli* n.g., n.sp. und *Planosarcina schaudinni* n.sp. zwei neue bewegliche Coccaceen. Zentralbl. Bakteriol. Parasitenkd. Infektionskr. Hyg. Abt. II *18:* 9-26.

Wolin, H.L. 1963. Defined medium for *Haemophilus influenzae* type b. J. Bacteriol. *85:* 253-254.

Wolin, M.J., E.A. Wolin and N.J. Jacobs. 1961. Cytochrome-producing anaerobic vibrio, *Vibrio succinogenes*, sp. n. J. Bacteriol. *81:* 911-917.

Wong, B., C. Singer, D. Armstrong and S.J. Millian. 1979. Rickettsialpox - case report and epidemiologic review. J. Am. Med. Assoc. *242:* 1998-1999.

Wong, D.H. and C.H. Chow. 1937. Group agglutinins of *Brucella abortus* and *Vibrio cholerae*. Chinese Med. J. *52:* 591-594.

Wong, J.C., J.K. Dyer and J.L. Tribble. 1977. Fermentation of L-aspartate by a saccharolytic strain of *Bacteroides melaninogenicus*. Appl. Environ. Microbiol. *33:* 69-73.

Wong, P.P., N.E. Stenberg and L. Edgar. 1980. Characterization of a bacterium of the genus *Azospirillum* from cellulolytic nitrogen-fixing mixed cultures. Can. J. Microbiol. *26:* 291-296.

Wong, P.T.S. 1969. Studies on the mechanism of Na$^+$ dependent transport in marine bacteria. Ph.D. thesis. McGill University, Montreal, 218 pp.

Wong, T.P., R.K. Shockley and K.H. Johnston. 1980. WSJM, a simple chemically defined medium for growth of *Neisseria gonorrhoeae*. J. Clin. Microbiol. *11:* 363-369.

Wong, W.C. and T.F. Preece. 1979. Identification of *Pseudomonas tolaasi*: the white line in agar and mushroom tissue block rapid pitting tests. J. Appl. Bacteriol. *47:* 401-407.

Woo, D.D.L., S.C. Holt and E.R. Leadbetter. 1979. Ultrastructure of *Bacteroides* species: *Bacteroides asaccharolyticus*, *Bacteroides fragilis*, *Bacteroides melaninogenicus* subspecies *melaninogenicus*, and *B. melaninogenicus* subspecies *intermedius*. J. Infect. Dis. *139:* 534-546.

Wood, J., R.C. Johnson and K. Palin. 1981. Surface colonies of *Leptospira interrogans*. J. Clin. Microbiol. *13:* 102-105.

Woodcock, H.M. and G. Lapage. 1914. On a remarkable type of protisten parasite. Quart. J. Microscop. Sci. *59:* 431-457.

Woodman, D.R., R. Grays and E. Weiss. 1977. Improved chicken embryo cell culture plaque assay for scrub typhus rickettsiae. J. Clin. Microbiol. *6:* 639-641.

Woodman, D.R., E. Weiss, G.A. Dasch and F.M. Bozeman. 1977a. Biological properties of *Rickettsia prowazekii* strains isolated from flying squirrels. Infect. Immun. *16:* 853-860.

Woodson, B.A., K.S. McCarty and M.C. Shepard. 1965. Arginine metabolism in mycoplasma and infected strain L929 fibroblasts. Arch. Biochem. Biophys. *109:* 364-371.

Woodward, B.W., M. Carter and R.J. Seidler. 1979. Most nonclinical *Klebsiella* strains are not *K. pneumoniae sensu stricto*. Curr. Microbiol. *2:* 181-185.

Woodward, T.E. 1979. Tularemia. *In* Beeson and McDermott (Editors), Textbook of Medicine, 15th Ed., W.B. Saunders Co., Philadelphia, pp. 465-468.

Working Group of the FAO/WHO Programme on Comparative Mycoplasmatology. 1974a. Preservation of mycoplasmas by lyophilization. World Health Organization Working Document, VPH/MIC/74.1.

Working Group of the FAO/WHO Programme on Comparative Mycoplasmology. 1974b. The determination of the metabolism of glucose. World Health Organization Working Document. VPH/MIC/74.2.

Working Group of the FAO/WHO Programme on Comparative Mycoplasmology 1975a. Identification of mycoplasmas by electrophoretic analysis of cell proteins. World Health Organization Working Document, VPH/MIC/75.3.

Working Group of the FAO/WHO Programme on Comparative Mycoplasmology. 1975b. The metabolism inhibition test. World Health Organization Working Document, VPH/MIC/75.6.

Working Group of the FAO/WHO Programme on Comparative Mycoplasmology. 1976. The growth inhibition test. World Health Organization Working Document, VPH/MIC/76.7.

Wormald, H. 1930. Bacterial diseases of stone fruit trees in Britain. II. Bacterial shoot wilt of plum trees. Ann. Appl. Biol. *17:* 725-744.

Wormald, H. 1931. Bacterial diseases of stone fruit trees in Britain. III. The symptoms of bacterial canker in plum trees. J. Pomol. Hortic. Sci. *9:* 239-256.

Wozny, M.A., M.P. Bryant, L.V. Holdeman and W.E.C. Moore. 1977. Urease assay and urease-producing species of anaerobes in the bovine rumen and human feces. Appl. Environ. Microbiol. *33:* 1097-1104.

Wreghitt, T.G., G.D. Windsor and M. Butler. 1974. Flat gel polyacrylamide electrophoresis of porcine mycoplasmas. Appl. Microbiol. *28:* 530-533.

Wright, J.D. and A.R. Barr. 1980. The ultrastructure and symbiotic relationships of *Wolbachia* of mosquitoes of the *Aedes scutellaris* group. J. Ultrastruct. Res. *72:* 52-64.

Wright, J.D., F.S. Sjostrand, J.K. Portaro and A.R. Barr. 1978. The ultrastructure of the rickettsia-like microorganism *Wolbachia pipientis* and associated virus-like bodies in the mosquito *Culex pipiens*. J. Ultrastruct. Res. *63:* 79-85.

Wright, J.D. and B.-T. Wang. 1980. Observations on *Wolbachiae* in mosquitoes. J. Invert. Pathol. *35:* 200-208.

Wright, J.H. 1895. Report on the results of an examination of the water supply of Philadelphia. Mem. Nat. Acad. Sci. *7:* 422-484.

Wroblewski, W. 1931. Morphologie et cycle évolutif des microbes de la péripneumonie des bovides et de l'agalaxie contagieuse des chèvres et des moutons. Ann. Inst. Pasteur (Paris) *47:* 94-115.

Wundt, W. 1959. Zür Frage der Antigen-gemeinschaften zwischen Brucellen und Bakterien anderer Gattungen. Z. Hyg. Infektionskr. *145:* 556-563.

Wundt, W. 1963. Stoffwechseluntersuchungen als experimentelle Grundlage zür Enteilung des Genus *Brucella*. Zentralbl. Bakteriol. Parasitenkd. Infektionskr. Hyg. Abt. I Orig. A *189:* 389-404.

Wundt, W. and W.J.B. Morgan. 1975. International Committee on Systematic Bacteriology Sub-Committee on Taxonomy of *Brucella*. Minutes of Meeting 3 September 1974. Int. J. Syst. Bacteriol. *25:* 235-236.

Wüst, J. and T.D. Wilkins. 1978. Effect of clavulanic acid on anaerobic bacteria resistant to beta-lactam antibiotics. Antimicrob. Agents Chemother. *13:* 130-133.

Wuthe, H.H. 1972. Ein Beiträg zum Vorkommen der *Plesiomonas shigelloides* beim Menschen. Zentralbl. Bakteriol. Parasitenkd. Infektionskr. Hyg. Abt. I Orig. A *220:* 546-549.

Wyss, O. and M.B. Wyss. 1950. Mutants of *Azotobacter* that do not fix nitrogen. J. Bacteriol. *59:* 287-291.

Xilinas, M.E., J.T. Papavassiliou and N.J. Legakis. 1975. Selective medium for growth of *Proteus*. J. Clin. Microbiol. *2:* 459-460.

Yablonskaya, V.A. 1978. Identification of rickettsiae isolated from the brain of rodents trapped in the endemic areas in Czechoslovakia. *In* Kazár, Ormsbee and Tarasevich (Editors), Rickettsiae and Rickettsial Diseases, VEDA, Bratislava, pp. 281-291.

Yabuuchi, E., T. Miwatani, Y. Takeda and M. Arita. 1974. Flagellar morphology of *Vibrio parahaemolyticus* (Fujino et al.) Sakazaki, Iwanami and Fukumi 1963. Jpn. J. Microbiol. *18:* 295-305.

Yabuuchi, E. and A. Ohyama. 1971. *Achromobacter xylosoxidans* sp. n. from human ear discharge. Jpn. J. Microbiol. *15:* 477-481.

Yabuuchi, E., E. Tanimura, A. Ohyama, I. Yano and A. Yamamoto. 1979. *Flavobacterium devorans* ATCC 10829: A strain of *Pseudomonas paucimobilis*. J. Gen. Appl. Microbiol. *25:* 95-107.

Yabuuchi, E. and I. Yano. 1981. *Achromobacter* gen. nov. and *Achromobacter xylosoxidans* (ex Yabuuchi and Ohyama 1971) nom. rev. Int. J. Syst. Bacteriol. *31:* 477-478.

Yabuuchi, E., I. Yano, S. Goto, E. Tanimura, T. Ito and A. Ohyama. 1974. Description of *Achromobacter xylosoxidans* Yabuuchi and Ohyama 1971. Int. J. Syst. Bacteriol. *24:* 470-477.

Yabuuchi, E., I. Yano, T. Kaneko and A. Ohyama. 1981. Classification of group IIk-2 and related bacteria. *In* Reichenbach and Weeks (Editors), The *Flavobacterium-Cytophaga* Group (Proceedings of the International Symposium on Yellow-Pigmented Gram-Negative Bacteria of the *Flavobacterium-Cytophaga* Group, Braunschweig, July 8 to 11, 1980). Verlag Chemie, Weinheim, pp. 79-90.

Yadava, R.P.S. and A.J. Musgrave. 1972a. Mycetomal micro-organisms and total lipid and phospholipid in granary weevils, *Sitophilus granarius* L. (Coleoptera). Comp. Biochem. Physiol. B Comp. Biochem. *41:* 425-431.

Yadava, R.P.S. and A.J. Musgrave. 1972b. Phospholipid patterns of two symbiote-harbouring weevils, the rice weevil, *Sitophilus oryzae* L., and the corn weevil, *Sitophilus zeamais* (Mots.) (Coleoptera: Curculionidae). Comp. Biochem. Physiol. B Comp. Biochem. *42:* 197-200.

Yadava, R.P.S., J.B.M. Rattray and A.J. Musgrave. 1972. Fatty acid profiles of two microbiologically different strains of granary weevil, *Sitophilus granarius* L. (Coleoptera). Comp. Biochem. Physiol. B Comp. Biochem. *43:* 383-391.

Yakrus, M. and N.W. Schaad. 1979. Serological relationships among strains of *Erwinia chrysanthemi*. Phytopathology *69:* 517-522.

Yale, M.W. 1939. Genus *Escherichia* Castellani and Chalmers. *In* Breed, Murray and Hitchens (Editors), Bergey's Manual of Determinative Bacteriology. 5th Ed. The Williams and Wilkins Co., Baltimore, pp. 389-396.

Yale, M.W. 1939. The genus *Proteus* Hauser. *In* Bergey, Breed, Murray and Hitchens (Editors), Bergey's Manual of Determinative Bacteriology, 5th Ed., The Williams and Wilkins Co., Baltimore, pp. 430-436.

Yamada, Y., K. Aida and T. Uemura. 1969. Enzymatic studies on the oxidation of sugar and sugar alcohol. V. Ubiquinone of acetic acid bacteria and its relation to classification of genera *Gluconobacter* and *Acetobacter*, especially of the so-called intermediate strains. J. Gen. Appl. Microbiol. *15:* 181-196.

Yamada, Y., G. Inouye, Y. Tahara and K. Kondo. 1976. The menaquinone system in the classification of coryneform and nocardioform bacteria and related organisms. J. Gen. Appl. Microbiol. *22:* 203-214.

Yamada, Y., Y. Okada and K. Kondo. 1976. Isolation and characterization of "polarly flagellated intermediate strains" in acetic acid bacteria. J. Gen. Appl. Microbiol. *22:* 237-245.

Yamada, Y., N. Seki, T. Kitahara, M. Takahashi and M. Matsui. 1970. The structure and synthesis of aeruginoic acid (2-o-hydroxyphenylthiazole-4-carboxylic acid). Agr. Biol. Chem. *34:* 780-783.

Yamamoto, R., C.H. Bigland and H.B. Ortmayer. 1965. Characteristics of *Mycoplasma meleagridis* sp. n. isolated from turkeys. J. Bacteriol. *90:* 47-49.

Yamamoto, T. 1967. Presence of rhapidosomes in various species of bacteria and their morphological characteristics. J. Bacteriol. *94:* 1746-1756.

Yamanaka, T. 1964. Identity of *Pseudomonas* cytochrome oxidase with *Pseudomonas* nitrite reductase. Nature *204:* 253-255.

Yamasato, K., M. Akagawa, N. Oishi and H. Kuraishi. 1982. Carbon substrate assimilation profiles and other taxonomic features of *Alcaligenes faecalis*, *Alcaligenes ruhlandii* and *Achromobacter xylosoxidans*. J. Gen. Appl. Microbiol. *28:* 195-213.

Yamvrias, C., C.G. Panagopoulos and P.G. Psallidas. 1970. Preliminary study of the internal bacterial flora of the olive fruit fly *Dacus oleae* (Gmelin). Ann. Inst. Phytopathol. Benaki *9:* 201-206.

Yancey, R.J. and R.A. Finkelstein. 1981a. Assimilation of iron by pathogenic *Neisseria* spp. Infect. Immun. *32:* 592-599.

Yancey, R.J. and R.A. Finkelstein. 1981b. Siderophore production by pathogenic *Neisseria* spp. Infect. Immun. *32:* 600-608.

Yancey, R.J., D.L. Willis and L.J. Berry. 1978. Role of motility in experimental cholera in adult rabbits. Infect. Immun. *22:* 387-392.

Yang, L.L. and A. Haug. 1979. Purification and partial characterization of a procaryotic glycoprotein from the plasma membrane of *Thermoplasma acidophilum*. Biochim. Biophys. Acta *556:* 265-277.

Yang, S.-E., F.H. Lin and T.-T. Kuo. 1975. The utilization of exogenously supplied nucleotide by *Xanthomonas oryzae*. Bot. Bull. Acad. Sinica *16:* 61-65.

Yano, T., A.F. Pestana de Castro, J.A. Lauritis and T. Namekata. 1979. Serological differentiation of bacteria belonging to the *Xanthomonas campestris* group by indirect haemagglutination test. Ann. Phytopathol. Soc. Japan. *45:* 1-8.

Yanofsky, C. 1954. The absence of a tryptophan-niacin relationship in *Escherichia coli* and *Bacillus subtilis*. J. Bacteriol. *68:* 577-584.

Yaphe, W. 1957. The use of agarase from *Pseudomonas atlantica* in the identification of agar in marine algae (*Rhodophyceae*). Can. J. Microbiol. *3:* 987-993.

Yasaki, Y. 1927. Bacteriological studies on bioluminescence. 1. Cause of luminescence in the fresh water shrimp, *Xiphocaridina compressa* (De Haan). J. Infect. Dis. *40:* 404-407.

Yasunaga, N. 1965. Studies on *Vibrio parahaemolyticus*. 4. On the distribution of *Vibrio parahaemolyticus* in fish in pelagic ocean to the south of the Hawaiian Archipelago, and fish and sea mud in Honolulu. Endemic Dis. Bull. Nagasaki Univ. *7:* 272-282.

Yates, M.G., J. O'Donnell, D.J. Lowe and H. Bothe. 1978. Ferredoxins from nitrogen-fixing bacteria. Physical and chemical characterisation of two ferredoxins from *Mycobacterium flavum* 301. Eur. J. Biochem. *85:* 291-299.

Yen, J.H. and A.R. Barr. 1974. Incompatability in *Culex pipiens*. In Pal and Whitten (Editors), The Use of Genetics in Insect Control. Elsevier, Amsterdam, pp. 97-118.

Yetinson, T. and M. Shilo. 1979. Seasonal and geographic distribution of luminous bacteria in the Eastern Mediterranean Sea and the Gulf of Elat. Appl. Environ. Microbiol. *37:* 1230-1238.

Yirgou, D. 1964. *Xanthomonas guizotiae* sp. nov. on *Guizotia abyssinica*. Phytopathology *54:* 1490-1491.

Yirgou, D. and J.F. Bradbury. 1968. Bacterial wilt of enset (*Ensete ventricosum*) incited by *Xanthomonas musacearum*. Phytopathology *58:* 111-112.

Yoder, H.W. and M.S. Hofstad. 1962. A previously unreported serotype of avian mycoplasma. Avian Dis. *6:* 147-160.

Yoder, H.W. and M.S. Hofstad. 1964. Characterization of avian mycoplasma. Avian Dis. *8:* 481-512.

Yordy, J.R. and T.L. Weaver. 1977. *Methylobacillus*: a new genus of obligately methylotrophic bacteria. Int. J. Syst. Bacteriol. *27:* 247-255.

York, W.S., M. McNeil, A.G. Darvill and P. Albersheim. 1980. Beta-2-linked glucans secreted by fast-growing species of *Rhizobium*. J. Bacteriol. *142:* 243-248.

York, W.S., M. McNeil, A.G. Darvill and P. Albersheim. 1980. Beta-2-linked glucans secreted by fast-growing species of *Rhizobium*. J. Bacteriol. *142:* 243-248.

Yoshida, K. and T. Nakamura. 1973. Studies on luciferase from *Photobacterium phosphoreum*. IV. Preparation and properties of stripped luciferase. J. Biochem. *74:* 915-922.

Yoshii, H. and S. Takimoto. 1928. Bacterial leafspot of castor bean and its pathogen. J. Plant Prot. *15:* 12-18.

Yoshinari, T. 1980. N$_2$O reduction by *Vibrio succinogenes*. Appl. Environ. Microbiol. *39:* 81-84.

Young, C. and A. Hill. 1974. Conjunctivitis in a colony of rats. Lab. Anim. *8:* 301-304.

Young, H.O., F. Chao, C. Turnbill and D.E. Philpott. 1972. Ultrastructure of *Pseudomonas saccharophila* at early and late log phase of growth. J. Bacteriol. *109:* 862-868.

Young, J.M. 1970. Drippy gill: a bacterial disease of cultivated mushrooms caused by *Pseudomonas agarici* n. sp. N. Z. J. Agr. Res. *13:* 977-990.

Young, J.M. 1978. Survival of bacteria on *Prunus* leaves. Proc. 4th Internat. Conf. Plant Path. Bact. *II:* 779-786.

Young, J.M., D.W. Dye, J.F. Bradbury, C.G. Panagopoulos and C.F. Robbs. 1978. A proposed nomenclature and classification for plant pathogenic bacteria. N. Z. J. Agric. Res. *21:* 153-177.

Young, J.M., D.W. Dye and J.P. Wilkie. 1978. Genus VII. *Pseudomonas* Migula 1894. In Young, J.M., D.W. Dye, J.F. Bradbury, C.G. Panagopoulos and C.F. Robbs. A proposed nomenclature and classification for plant pathogenic bacteria. N. Z. J. Agr. Res. *21:* 153-177.

Young, V.M., D.M. Kenton, B.J. Hobbs and M.R. Moody. 1971. *Levinea*, a new genus of the family *Enterobacteriaceae*. Int. J. Syst. Bacteriol. *21:* 58-63.

Zaar, K. 1979. Visualisation of pores (export sites) correlated with cellulose production in the envelope of Gram-negative bacterium *Acetobacter xylinum*. J. Cell. Biol. *80:* 773-777.

Zacharias, A. 1928. Untersuchungen über die intrazellulare Symbiose bei den Pupiparen. Z. Morphol. Oekol. Tiere *10:* 676-737.

Zachary, A. 1974. Isolation of bacteriophages of the marine bacterium *Beneckea natriegens* from coastal salt marshes. Appl. Microbiol. *27:* 980-982.

Zaenen, I., N. Van Larebeke, H. Teuchy, M. Van Montagu and J. Schell. 1974. Supercoiled circular DNA in crown gall inducing *Agrobacterium* strains. J. Mol. Biol. *86:* 109-127.

Zagallo, A.C. and C.H. Wang. 1967. Comparative glucose catabolism of *Xanthomonas* species. J. Bacteriol. *93:* 970-975.

Zahorchak, R.J., W.T. Charnetzky, R.V. Little and R.R. Brubaker. 1979. Consequences of Ca^{2+} deficiency on macromolecular synthesis and adenylate energy charge in *Yersinia pestis*. J. Bacteriol. *139:* 792-799.

Zajc-Satler, J., A.Z. Dragas and M. Kumelj. 1972. Morphological and biochemical studies of 6 strains of *Plesiomonas shigelloides* isolated from clinical sources. Zentralbl. Bakteriol. Parasitenkd. Infektionskr. Hyg. Abt. I Orig. A *219:* 514-521.

Zakharova, M.S. 1979. Theoretical outlines on the preparation of a noncellular pertussis vaccine. In Manclark and Hill (Editors), International Symposium on Pertussis. U.S. Government Printing Office, Washington, D.C., pp. 320-326.

Zambon, J.J., H.S. Reynolds and J. Slots. 1981. Black-pigmenting *Bacteroides* spp. in the human oral cavity. Infect. Immun. *32:* 198-203.

Zambryski, P., M. Holsters, K. Kruger, A. Depicker, J. Schell, M. Van Montagu and H.M. Goodman. 1980. Tumor DNA structure in plant cells transformed by *A. tumefaciens*. Science *209:* 1385-1391.

Zaremba, M. and E. Aldova. 1979. Sensitivity to chemotherapeutics of *Yersinia enterocolitica* strains. Arch. Immunol. Therap. Exp. *27:* 847-852.

Zarett, A.J. and R.N. Doetsch. 1949. A new selective medium for quantitative determination of members of the genus *Proteus* in milk. J. Bacteriol. *57:* 266.

Zavarzin, G.A. and A.N. Nozhevnikova. 1977. Aerobic carboxydobacteria. Microb. Ecol. *3:* 305-326.

Zdrodovskii, P.F. 1949. Systematics and comparative characterization of endemic rickettsioses. Zhur. Mikrobiol. Epidemiol. *10:* 19-28.

Zdrodovskii, P.F. and H.M. Golinevich. 1960. The rickettsial diseases (English translation), Pergamon Press, New York, Oxford, London, Paris.

Zeikus, J.G. 1979. Thermophilic bacteria: ecology, physiology and technology. Enz. Microb. Technol. *1:* 243-252.

Zeitoun, F.M. and E.E. Wilson. 1969. The relation of bacteriophage to the walnut-tree pathogens, *Erwinia nigrifluens* and *Erwinia rubrifaciens*. Phytopathology *59:* 756-761.

Zevenhuizen, L.P.T.M. 1971. Chemical composition of exopolysaccharides of *Rhizobium* and *Agrobacterium*. J. Gen. Microbiol. *68:* 239-243.

Zhdanov, V.M. 1953. Opredelitel' Virusov Cheloveka i Zhivotnykh. Isdatel' stro Akademii Meditsinkikh Nauk U.S.S.R., Moscow.

Zhdanov, V. and R.S. Korenblit. 1950. Systematics and nomenclature of viruses. Zhur. Mikrobiol. Epidemiol. Immunobiol. *9:* 40-44.

Zherebilo, O.E. and R.I. Gvozdyak. 1976. Decarboxylation of amino acids by bacteria of the genus *Erwinia* under aerobic conditions. Rv. Mikrobiol. Zh. (Kiev) *38:* 3-8.

Ziegler, M.M. and T.O. Baldwin. 1981. Biochemistry of bacterial bioluminescence. Curr. Top. Bioenerg. *12:* 131-170.

Zimmerer, R.P., R.H. Hamilton and C. Pootjes. 1966. Isolation and morphology of temperate *Agrobacterium tumefaciens* bacteriophage. J. Bacteriol. *92:* 746-750.

Zimmermann, A. 1902. Über Bakterienknoten in den Blättern einiger Rubiaceen. Jahrb. Wiss. Bot. *37:* 1-11.

Zimmermann, O.E.R. 1890. Die Bakterien unserer Trink- und Nutzwässer, insbesondere des Wassers der Chemnitzer Wasserleitung. Elfter Bericht. Naturwiss. Ges. Chemnitz., pp. 53-154.

Zimmermann, T. 1964. Untersuchungen über die Actinobazillose des Schweines. I Mitteilung: Isolierung und Characterisierung der Erreger. Dtsch. Tieraerztl. Wochenschr. *71:* 457-461.

Zimmermann, W. 1979. Penetration through the Gram-negative cell wall: a co-determinant of the efficacy of beta-lactam antibiotics. Int. J. Clin. Pharmacol. Biopharm. *17:* 131-134.

Zimmermann, W. 1980. Penetration of beta-lactam antibiotics into their target enzymes in *Pseudomonas aeruginosa*: comparison of a highly sensitive mutant with its parent strain. Antimicrob. Agents Chemother. *18:* 94-100.

Zinder, N.D. 1957. Lysogenic conversion in *S. typhimurium*. Science (Washington) *126:* 1237.

Zink, D.L., J.C. Feeley, J.G. Wells, C. Vanderzant, J.C. Vickery, W.D. Roof and G.A. O'Donovan. 1980. Plasmid-mediated tissue invasiveness in *Yersinia enterocolitica*. Nature (London) *283:* 224-226.

Zinnemann, K. and E.L. Biberstein. 1974. Genus *Haemophilus* Winslow, Broadhurst, Buchanan, Krumwiede, Rogers and Smith 1917, 561. In Buchanan and Gibbons (Editors), Bergey's Manual of Determinative Bacteriology, 8th Ed., Williams and Wilkins Co., Baltimore, pp. 364-368.

Zinnemann, K., K.B. Rogers, J. Frazer and J.M.H. Boyce. 1968. A new V-dependent *Haemophilus* species preferring increased CO$_2$ tension for growth and named *Haemophilus paraphrophilus*, nov. sp. J. Pathol. Bacteriol. *96:* 413-419.

Zinnemann, K., K.B. Rogers, J. Frazer and S.K. Devaraj. 1971. A haemolytic V-dependent CO$_2$-preferring *Haemophilus* species *Haemophilus paraphrohaemolyticus* nov. spec. J. Med. Microbiol. *4:* 139-143.

Zinnemann, K. and G.C. Turner. 1963. The taxonomic position of "*Haemophilus vaginalis*" (*Corynebacterium vaginale*). J. Pathol. Bacteriol. *85:* 213-219.

Zinner, S.H., A.K. Daly and W.M. McCormack. 1973. Isolation of *Eikenella corrodens* in a general hospital. Appl. Microbiol. *25:* 705-708.

Zinsser, H. 1935. Rats, lice, and history. Little, Brown and Co., Boston.

ZoBell, C.E. 1941. Studies on marine bacteria. I. The cultural requirements of heterotrophic aerobes. J. Mar. Res. *4:* 42-75.

ZoBell, C.E. 1948. Genus II. *Desulfovibrio* Kluyver and van Niel. In Breed, Murray and Hitchens (Editors), Bergey's Manual of Determinative Bacteriology, 6th

Ed. The Williams and Wilkins Co., Baltimore, pp. 207-209.

ZoBell, C.E. 1957. Genus II. *Desulfovibrio* Kluyver and van Niel, 1936. *In* Breed, Murray and Smith (Editors), Bergey's Manual of Determinative Bacteriology, 7th Ed. The Williams and Wilkins Co., Baltimore, pp. 248-249.

ZoBell, C.E. and K.F. Meyer. 1932. Metabolism studies on the *Brucella* group. VI. Nitrates and nitrite reduction. J. Infect. Dis. *51:* 99-108.

ZoBell, C.E. and H.C. Upham. 1944. A list of marine bacteria including descriptions of sixty new species. Bull. Scripps Inst. Oceanogr. Univ. Calif. *5:* 239-292.

Zoha, S.J. and L.E. Carmichael. 1981. Properties of *Brucella canis* su face (sic) antigens associated with colonial mucoidiness. Cornell Vet. *71:* 428-438.

Zolg, W. and J.C.G. Ottow. 1975. *Pseudomonas glathei* sp. nov., a new nitrogen-scavenging rod isolated from acid lateritic relicts in Germany. Z. Allg. Mikrobiol. *15:* 287-299.

Zollinger, W.D. and R.E. Mandrell. 1978. Outer-membrane protein and lipopolysaccharide serotyping of *Neisseria meningitidis* by inhibition of a solid-phase radioimmunoassay. Infect. Immun. *18:* 424-433.

Zoon, K.C. and J.J. Scocca. 1975. Constitution of the cell envelope of *Haemophilus influenzae* in relation to competence for genetic transformation. J. Bacteriol. *123:* 666-677.

Zopf, W. 1883. Die Spaltpilze. Edward Trewendt. Breslau.

Zopf, W. 1885. Die Spaltpilze, 3rd Ed., Edward Trewendt, Breslau, pp. 1-127.

Zousias, D., A.J. Mazaitis, M. Simberkoff and M. Rush. 1973. Extrachromosomal DNA of *Mycoplasma hominis*. Biochim. Biophys. Acta *312:* 484-491.

Zuber, H. (Editor). 1976. Enzymes and proteins from thermophilic microorganisms. Birkhauser Verlag, Basel.

Zucker-Franklin, D., M. Davidson and L. Thomas. 1966. The interaction of mycoplasmas with mammalian cells. I. Hela Cells, neutrophils, and eosinophils. J. Exp. Med. *124:* 521-531.

Zuckerkandl, E. and L. Pauling. 1965. Molecules as documents of evolutionary history. J. Theoret. Biol. *8:* 357-366.

Zuelzer, M. 1912. Über *Spirochaeta plicatilis* Ehrbg. und deren Verwandtschaftsbeziehungen. Arch. Protistenk. *24:* 1-59.

Zuelzer, M. 1925. Die spirochäten. *In* von Prowazek (Editor), Handbuch der Pathogenen Protozoen, 3rd ed., Barth, Leipzig, pp. 1627-1798.

Zumpt, F. and D. Organ. 1961. Strains of spirochaetes isolated from *Ornithodoros zumpti* Heisch and Guggisberg and from wild rats in the Cape province. A preliminary note. S. Afr. J. Lab. Clin. Med. *7:* 31-35.

Index of Scientific Names of Bacteria

Key to the fonts and symbols used in this index:

Nomenclature
    Lower case, Roman: Genera, species, and subspecies of bacteria. Every bacterial name mentioned in the *Manual* is listed in the Index. Specific epithets are listed individually and also under the genus.*

    CAPITALS, ROMAN: Names of taxa higher than genus (tribes, families, orders, classes, divisions, kingdoms).

Pagination
    Roman: Pages on which taxa are mentioned.

    Boldface: Indicates page on which the description of a taxon is given.†

---

* Infrasubspecific names, such as serovars, biovars, and pathovars, are not listed in the Index.
† A description may not necessarily be given in the *Manual* for a taxon that is considered as *incertae sedis* or that is listed in an addendum or note added in proof; however, the page on which the complete citation of such a taxon is given is indicated in boldface type.

Index of Scientific Names of Bacteria